FUNK & WAGNALLS

Standard
REGISTERED UNITED STATES PATENT OFFICE

DICTIONARY

OF THE ENGLISH LANGUAGE

International Edition

VOLUME TWO

FUNK & WAGNALLS COMPANY

NEW YORK

FUNK & WAGNALLS STANDARD DICTIONARY
International Edition

© 1958, 1959, 1960, AND 1961, by Funk & Wagnalls Company

"STANDARD" is our trademark registered in U. S. Patent Office

Library of Congress Catalog Card Number 58-11577

Copyright under the Articles of the Copyright Convention
of the Pan-American Republics and the United States.

Printed in the United States of America
61-B

TABLE OF CONTENTS

VOLUME ONE

	PAGE
EDITORIAL ADVISORY BOARD	iii
EDITORS AND CONSULTANTS	iv
INTRODUCTION—THE LEXICON OF THE LANGUAGE	v
USAGE LEVELS AND LABELS	vi
SEMANTIC CHANGE	vii
SYNONYMS, ANTONYMS, GEOGRAPHY, BIOGRAPHY, PRONUNCIATION	x
ETYMOLOGIES	xi
PERSPECTIVE ON THE ENGLISH LANGUAGE	xii
EXPLANATORY NOTES	xvi
TABLE OF ENGLISH SPELLINGS—SPELLING: PLURALS AND PARTICIPLES	xix
PRONUNCIATION KEY—ABBREVIATIONS USED IN THIS WORK	xx
DICTIONARY OF THE ENGLISH LANGUAGE—A THROUGH LOBAR	1

VOLUME TWO

DICTIONARY OF THE ENGLISH LANGUAGE—LOBATE THROUGH Z	747
THE ALFRED BERNHARD NOBEL PRIZES	1466
ABBREVIATIONS COMMONLY IN USE	1469
WORLD POPULATIONS	
UNITED STATES AND CANADA	1479
OTHER COUNTRIES	1493

EXPLANATORY NOTES
See pronunciation key on page xx.

Car·bo·run·dum (kär'bə·run'dəm) *n.* An abrasive of silicon carbide: a trade name.

-cidal *combining form* Killing; able to kill: *homicidal*. [< L *caedere* kill]

Col·o·ra·do (kol'ə·rä'dō, -rad'ō) A western State of the United States; 103,967 square miles; capital, Denver; entered the Union Aug. 1, 1876: nicknamed *Centennial State*: abbr. *Colo.*

com- *prefix* With; together: *combine, compare*. Also: *co-* before *gn, h*, and vowels; *col-* before *l*, as in *collide*; *con-* before *c, d, f, g, j, n, q, s, t, v*, as in *concur, confluence, connect, conspire*; *cor-* before *r*, as in *correspond*. [< L *com-* < *cum* with]

craal (kräl) See KRAAL.

crab[1] (krab) *n.* 1 Any of various species of ten-footed crustaceans of the suborder *Brachyura* in the order *Decapoda*, characterized by a small abdomen folded under the body, a flattened carapace, and short antennae. They can walk in any direction without turning, but usually move sideways. 2 The hermit crab. 3 The horseshoe crab. 4 A crab louse, *Phthirus pubis*. 5 *Aeron.* The lateral slant in an airplane needed to maintain a flight line in a cross-wind. 6 A form of windlass. 7 *pl.* The lowest throw of a pair of dice. — **to catch a crab** In rowing, to sink an oar blade too deeply; also, to miss the water entirely or skim the surface in making a stroke, and thus fall backward. — *v.* **crabbed, crab·bing** *v.i.* 1 To take or fish for crabs. 2 *U.S. Colloq.* To back out: to *crab* out of an agreement. 3 *Naut.* To drift sideways, as a ship. — *v.t.* 4 *Aeron.* To head (an airplane) across a contrary wind so as to compensate for drift. [OE *crabba*. Akin to CRAB[3].]

crab[2] (krab) *n.* 1 A crab apple. 2 A crab-apple tree. 3 An ill-tempered, surly, or querulous person. — *v.* **crabbed, crab·bing** *v.t.* 1 *Colloq.* To disparage; belittle; complain about. 2 *Colloq.* To ruin or spoil: He *crabbed* the entire act. 3 *Obs.* To make surly or sour; irritate. 4 *Brit. Dial.* To cudgel or beat, as with a crabstick. — *v.i.* 5 To be ill-tempered. [? < Scand. Cf. dial. Sw. *scrabba* wild apple.]

crab[3] (krab) *v.i.* **crabbed, crab·bing** To seize each other fiercely, as hawks when fighting; claw. [< MDu. *crabben* scratch. Akin to CRAB[1].]

Crab A constellation and sign of the Zodiac; Cancer.

crab angle The angle between the direction of movement of an airplane, rocket, or guided missile and the direction in which the nose points, resembling . . .

crab apple A kind of small, sour apple: also called *crab*.

deep-seat·ed (dēp'sē'tid) *adj.* So far in as to be ineradicable or almost ineradicable: said of emotions, diseases, etc.

Deep South The southernmost parts of Alabama, Georgia, Louisiana, and Mississippi, conventionally regarded as typifying Southern culture and traditions.

deer (dir) *n. pl.* **deer** 1 A ruminant (family *Cervidae*) having deciduous antlers, usually in the male only, as the moose, elk, and reindeer. Popularly, *deer* is used mainly of the smaller species. ◆ Collateral adjective: *cervine*. See FALLOW DEER, VENISON. 2 A deerlike animal. 3 Formerly, any quadruped; a wild animal. ◆ Homophone: *dear*. [OE *dēor* beast]

deer·fly (dir'flī') *n. pl.* **·flies** A bloodsucking fly (genus *Chrysops*), similar to a horsefly but smaller and with banded wings. For illustration see INSECTS (injurious).

di·eth·y·lene glycol (dī·eth'ə·lēn) *Chem.* An organic compound, $O(CH_2CH_2OH)_2$, used as an anti-freeze mixture and as an agent in many chemical processes for the . . . .

di·eth·yl·stil·bes·trol (dī·eth'əl·stil·bes'trōl) *n. Biochem.* Stilbestrol.

di·e·ti·tian (dī'ə·tish'ən) *n.* One skilled in the principles of dietetics and in their practical application in health and disease. Also **di·e·tet·ist** (dī'ə·tet'ist), **di·e·ti'cian**.

Dieu vous garde (dyœ' vōō' gàrd') *French* God protect you.

Diez (dēts), **Friedrich Christian**, 1794–1876, German philologist.

dif·fer (dif'ər) *v.i.* 1 To be unlike in quality, degree, form, etc.: often with *from*. 2 To disagree; dissent: often with *with*. 3 To quarrel: sometimes with *over* or *about*. [< OF *differer* < L *differre* < *dis-* apart + *ferre* carry. Doublet of DEFER[1].]

dif·fer·ence (dif'ər·əns, dif'rəns) *n.* 1 The state or quality of being other or unlike, or that in which two things are unlike; . . . . 8 *Her.* A device in blazons to distinguish persons bearing the same arms. — *v.t.* **·enced, ·enc·ing** 1 To make or mark as different; distinguish; discriminate. 2 *Her.* To add a mark of difference to.
— *Synonyms* (noun): contrariety, contrast, disagreement, discrepancy, discrimination, disparity, dissimilarity, dissimilitude, distinction, divergence, diversity, inconsistency, inequality, unlikeness, variation. A *difference* is in the things compared; . . . . *Diversity* involves more than two objects; *variation* is a *difference* in the condition or action of the same object at different times. *Antonyms*: agreement, consonance, harmony, identity, likeness, resemblance, sameness, similarity, uniformity, unity.

dif·fer·ent (dif'ər·ənt, dif'rənt) *adj.* 1 Not the same; distinct; other: A *different* clerk is there now. 2 Marked by a difference; completely or partly unlike; dissimilar. 3 Unusual. See synonyms under CONTRARY. [< F *différent* < L *differens, -entis*, ppr. of *differre*. See DIFFER.] — **dif'fer·ent·ly** *adv.* — **dif'fer·ent·ness** *n.* ◆ **different from, than, to** In American usage, *from* is established as the idiomatic preposition to follow *different*; when, however, a clause follows the connective, *than* is gaining increasing acceptance: a result *different* than (= *from that which* or *from what*) had been expected. This last is established British usage, which also accepts *to* . . . .

-dom *suffix of nouns* 1 State or condition of being: *freedom*. 2 Rank of; domain of: *kingdom*. 3 The totality of those having a certain rank, state or condition: *Christendom*. [OE *-dōm* < *dōm* state]

EXPLANATORY NOTES

Syllabication Division of words into syllables — as an indication of the points at which a word may be broken at the end of a line — is indicated by a centered dot (·) in the main bold–faced entries, as **ad·jec·ti·val.** In the secondary entries (run–on derivatives and variant forms) the centered dot is eliminated wherever the primary and secondary syllable stresses are marked, as in **ad′jec·ti′val·ly.** In hyphened compounds the hyphen takes the place of a centered dot. Phrasal entries of two or more words are not syllabified when each element is entered elsewhere, as in **caballine fountain.**

Pronunciations The pronunciation is shown in parentheses immediately following the bold–faced entry, as **di·chot·o·my** (dī·kot′ə·mē). When more than one pronunciation is recorded, the first given is usually the most widely used wherever it has been possible to determine extent of usage; often, however, usage may be almost equally divided. The order of the pronunciations is not intended to be an indication of preference; all pronunciations shown are valid for educated American speech.

The syllabication of the pronunciations follows, in general, the syllabic breaks heard in speech, rather than the conventional division of the bold–faced entry, as **bod·ing** (bō′ding), **grat·er** (grā′tər), **ju·di·cial** (jo͞o·dish′əl), **an·es·the·tize** (ə·nes′thə·tīz).

When a variant pronunciation differs merely in part from the first pronunciation recorded, only the differing syllable or syllables are shown, provided that there is no possibility of misinterpretation, as **eq·ua·bil·i·ty** (ek′wə·bil′ə·tē, ē′kwə-). Phrasal entries (those which consist of two or more words) are not pronounced if the individual elements are separately entered in proper alphabetic place.

Parts of speech These are shown in italics following the pronunciation for main entries, and are abbreviated as follows: *n.* (noun), *v.* (verb), *pron.* (pronoun), *adj.* (adjective), *adv.* (adverb), *prep.* (preposition), *conj.* (conjunction), *interj.* (interjection). When more than one part of speech is entered under a main entry, the additional designations are run in and preceded by a bold–faced dash, as **cor·ner** (kôr′nər) *n.* . . . **—** *v.t.* . . . **—** *v.i.* . . . **—** *adj.* . . .

Verbs used transitively are identified as *v.t.*, those intransitively as *v.i.*; those used both transitively and intransitively in all senses are designated *v.t.* & *v.i.*

Inflected forms These include the past tense, past participle, and present participle of verbs, the plural of nouns, and the comparative and superlative of adjectives and adverbs. The inflected forms are entered wherever there is some irregularity in spelling or form. They are shown in boldface type, with syllabication, immediately after the part of speech designation. Only the syllable affected is shown, provided there is no ambiguity possible, as **com·pute** (kəm·pyo͞ot′) *v.t.* **·put·ed, ·put·ing.** An inflected form that requires pronunciation or is alphabetically distant from the main entry may also be separately entered and pronounced in its proper vocabulary place.

Principal parts of verbs The order in which the principal parts are shown is past tense, past participle, and present participle, as **come** (kum) *v.* **came, come, com·ing.** Where the past tense and past participle are identical, only two forms are entered, as **bake** (bāk) *v.* **baked, bak·ing.** When alternative forms are given, the first form indicated is usually the one preferred, as **grov·el** (gruv′əl, grov′-) *v.i.* **grov·eled** or **grov·elled, grov·el·ing** or **grov·el·ling.** Variant forms not in the standard vocabulary are shown in parentheses and labeled, as **drink** (dringk) *v.* **drank** (*Obs.* **drunk**), **drunk** (*Obs.* **drunk·en**), **drink·ing.** Principal parts entirely regular in formation — those that add *-ed* and *-ing* directly to the infinitive without spelling modification — are not shown.

Plural of nouns Irregular forms are here preceded by the designation *pl.*, as **a·lum·nus** (ə·lum′nəs) *n.* *pl.* **·ni** (-nī); **co·dex** (kō′deks) *n.* *pl.* **co·di·ces** (kō′də·sēz, kod′ə-); **deer** (dir) *n. pl.* **deer.** When alternative plurals are given, the first shown is the preferred form, as **buf·fa·lo** (buf′ə·lō) *n.* *pl.* **·loes** or **·los; chrys·a·lis** (kris′ə·lis) *n.* *pl.* **chrys·a·lis·es** or **chry·sal·i·des** (kri·sal′ə·dēz). Words that have different plural forms for specific senses are shown as follows:

an·ten·na (an·ten′ə) *n.* *pl.* **an·ten·nae** (an·ten′ē) *for def.* **1, an·ten·nas** *for def.* **2. 1** *Entomol.* One of the paired, lateral, movable, jointed appendages on the head of an insect or other arthropod. **2** *Telecom.* A system of wires upheld in a vertical or horizontal position by a mast or tower, for transmitting or receiving electromagnetic waves in wireless telegraphy, telephony, and radio. . . . .

Comparison of adjectives and adverbs The comparatives and superlatives of adjectives and adverbs are shown immediately after the part of speech when there is some spelling modification or a complete change of form, as **mer·ry** (mer′ē) *adj.* **·ri·er, ·ri·est; bad**[1] (bad) *adj.* **worse, worst; well**[2] (wel) *adv.* **bet·ter, best.**

Definition numbers In entries for words having several senses, the order in which the definitions appear is, wherever possible, that of frequency of use, rather than semantic evolution. Each such definition is distinguished by a bold–faced number, the numbering starting anew after each part–of–speech designation when it is followed by more than one sense. Closely related meanings, especially those within a specific field or area of study, are defined under the same number and set apart by small bold–faced letters.

bol·ster (bōl′stər) *n.* **1** A long, narrow pillow as wide as a bed. **2** A pad used as a support or . . . . **3** Anything shaped like . . . . **4** *Archit.* **a** The lateral part of the volute of an Ionic capital. **b** A crosspiece of an arch centering . . . . **5** *Mech.* A steel block . . . . **—** *v.t.* **1** To support with a pillow. **2** To prop up . . . . **3** To furnish with padding. . . .

Restrictive labels Entries or particular senses of words and terms having restricted application are variously labeled according to: (1) usage level, as *Slang, Colloq.* (colloquial), *Dial.* (dialectal), *Poetic*, etc.; (2) localization, as *U.S.* (United States), *Brit.* (British), *Austral.* (Australian), *Scot.* (Scottish English), etc.; (3) field or subject, as *Astron.* (astronomy), *Geom.* (geometry), *Mining* (mining), *Naut.* (nautical), *Surg.* (surgery), etc.; (4) language of origin, as *Afrikaans, French, German, Latin,* etc.

These labels serve as a guide in the ready identification of special aspects of a word or term as a whole or of one or more of its parts. The usage labels qualify a word in terms of its relationship to standard English; the localized area designations identify the geographical region of the English–speaking world in which a word has originated or where it has particular application; the subject labels indicate that a word or definition has a specialized use in some field of work or study; and the foreign language labels reflect the fact that a word or phrase, although used in English speech and writing, has not yet undergone the process of Anglicization of pronunciation, meaning, or usage, and is still felt to be foreign (these foreign terms are usually italicized in writing).

Restrictive labels that apply to only one sense of a word are entered after the definition number, as:

beat (bēt) *v.* . . . . **9** *Music* To mark or measure with or as with a baton: to *beat* time. . . . . **13** *Colloq.* To baffle; perplex: It *beats* me. **14** *Slang* To defraud; swindle. . . . . **20** *Physics* To alternate in intensity so as to pulsate . . . . **24** *Naut.* To work against contrary winds or currents by tacking.

Labels entered immediately after the part of speech designation apply to all the senses for that part of speech; those shown directly after the pronunciation and before the first part–of–speech designation refer to the entire entry, as:

hal·i·dom (hal′ə·dəm) *n.* *Archaic* **1** Holiness. **2** A holy relic. **3** A holy place; sanctuary. . . .
grouch (grouch) *U.S. Colloq.* *v.i.* To grumble; be surly or discontented. **—** *n.* **1** A discontented, grumbling person. **2** A grumbling, sulky mood. . . . .

A complete list of the label abbreviations used in this dictionary will be found on page xx.

Variant forms In the case of words having more than one approved spelling (as *esthetic, aesthetic; center, centre*), the main entry is made under what is considered to be the form in more general use in the United States. The alternate form is cross–referred when it is not within close range of the alphabetic position of the main entry, and is also shown in italic type at the end of the main entry or after all the definitions for a particular part of speech to which it applies. Variant forms that do not require cross–reference are shown in the same position in the main entry but in boldface type with syllabication, stress marks, and, when necessary, pronunciation. A variant that applies to but one of several senses of a word is attached with a colon to the definition to which it pertains.

bach·e·lor (bach′ə·lər, bach′lər) *n.* . . . . **3** A young knight serving under another's banner: also **bach′e·lor–at–arms′.** . . . . **5** A young male fur seal kept from the breeding grounds by the older males: also called *holluschick.* . . . .

Forms that have some restricted usage are labeled accordingly, as **hon·or** . . . Also *Brit.* **hon′our.**

Phrasal entries Numerous phrases are entered and defined in the vocabulary section of this book in alphabetic place according to the first word of the phrase, as **bird of paradise, earth inductor, free verse, right of search.** In many other instances, however, it has been more expedient to enter and define such phrases under the main element. Thus, *Old English* and *Middle English* are run in as subordinate entries under **English,** preceded by a heavy dash; *alternating current, direct current,* and *eddy current* are entered and defined under **current.** All such entries are cross–referred in alphabetic place.

In some entries, particularly those for plants and animals, varying combinations of the word being defined are given within the main entry in boldface type and not entered in alphabetic place, as **leop·ard** · · · **2** Any similar cat, such as the **American leopard** or jaguar, the **hunting leopard** or cheetah, the **snow leopard** or ounce.

Encyclopedic entries A similar device has been employed where it has been advisable to bring together in one place as much logically related information as possible. Certain groups of terms are fully treated in an alphabetic boldface listing under the primary word common to each group. This treatment points up significant relationships between terms, facilitates comparison and selection of the term desired, and allows for the presentation of a range of information ordinarily characteristic of encyclopedias. For example, the entry for **time** contains a listing, with full definitions, of the various classifications from *astronomical time* through *zone time.* Similar listings have been included under the following entries: *angle, calendar, court, cross, current (ocean currents), diamond, fraction, glass, law, number, school, spaniel,* and *terrier.* All terms so entered are separately cross–referred.

Idiomatic phrases Often a main–entry word, when in conjunction with various prepositions, adverbs, adjectives, etc., will form a phrase distinct in sense from the meaning of the combined elements. Such idiomatic phrases are shown in smaller boldface type within the entry for the principal word in the phrase; they are preceded by a heavy dash, and follow all the definitions for the particular part of speech involved. Thus, under the verb **carry** will be found subordinate entries for the phrases *carry arms, carry away, carry off, carry on,* etc.; under the entry for the noun **hand** such phrases as *at first hand, by hand, to have one's hands full,*

add,āce,câre,pälm; end,ēven; it,īce; odd,ōpen,ôrder; to͝ok,po͞ol; up,bûrn; ə = a in *above,* e in *sicken,* i in *clarity,* o in *melon,* u in *focus*; yo͞o = u in *fuse*; oi,oil; ou,pout; ch,check; g,go; ng,ring; th,thin; th,this; zh,vision. Foreign sounds à,œ,ü,kh,ṅ; and ◆: see page xx. < from; + plus; ? possibly.

to lend a hand, etc., are set apart and defined in detail.

Collateral adjectives Because of the grafting of Norman French and late Renaissance Latin idioms on early English we find a good many English nouns which have adjectives closely connected with them in sense but not in form, such as *brachial* with *arm, cervical* with *neck, lacustrine* with *lake, hibernal* with *winter, diurnal* with *day, reticular* with *net,* and the like. Such adjectives which, through a collateral line of meaning, have come to express certain special adjectival senses, are, of course, listed in their regular alphabetic place, but, as a convenience for those who do not know or cannot recall them, a large number of such functionally related adjectives have been entered with their associated nouns, attached to the particular meaning of the noun to which they apply, in the form ◆ Collateral adjective: *brachial.*

Homophones Words identical in sound but different in spelling and meaning, such as the groups *altar / alter, filter / philter, hail / hale, principal / principle, right / rite / write,* and several hundred others, often lead to confusion in the writing of English. A large number of these have been listed just before the etymology of every relevant entry, in the form ◆ Homophone: *altar.* No entries have been made for such groups as *horse / hoarse, burrow / burro* which, because of variant pronunciations, are homophonic for some speakers but not for others.

Etymologies The etymologies are shown in brackets after the definitions and before the run-on derivatives. The following examples show the manner of entry and the use of cross-references: (1) **spe·cial** . . . [<OF *especial* <L *specialis* < *species* kind, species]; the etymology is to be read: derived from (<) the Old French (OF) word *especial* from the Latin (L) word *specialis* which in turn is from Latin *species* meaning "kind, species"; (2) **arroyo** . . . [<Sp.]; here the reading is to be: derived from (<) a Spanish (Sp.) word of the same form and meaning; (3) **has·sle** . . . [? <HAGGLE + TUSSLE]; this etymology is to be read: perhaps (?) derived from (<) a blending of "haggle" and (+) "tussle," the small capital letters indicating that the etymologies of these words will be found under their main entries; (4) **de·cep·tion** . . . [<L *deceptio, -onis* < *decipere.* See DECEIVE.]; (5) **bul·wark** . . . [<MHG *bolwerc.* Akin to BOULEVARD.]; (6) **a·dult** . . . [<L *adultus,* pp. of *adolescere* grow up < *ad-* + *alescere* grow. Related to ADOLESCENT.]; (7) **jour·nal** . . . [<OF <L *diurnalis.* Doublet of DIURNAL.]

Cross-references In (4), "See" directs attention to the etymology under "deceive" for further information; in (5), "Akin to" points to the fact that "boulevard" and "bulwark" are cognate words; in (6), "Related to" indicates that these words derive from different recorded stems of the same word; in (7), "Doublet of" marks the fact that "journal" and "diurnal" are ultimately derived from the same Latin word, but have come into English by different paths, in this case, the former through Old French and the latter directly from Latin.

For the complete list of the abbreviations which are used in the etymologies in this work, see page xx.

Run-on entries Words that are actually or ostensibly derived from other words by the addition or replacement of a suffix, and whose sense can be inferred from the meaning of the main word, are run on, in smaller boldface type, at the end of the appropriate main entries. The run-on entries are preceded by a heavy dash and followed by a part-of-speech designation. They are syllabified and stressed, and, when necessary, a full or partial pronunciation is indicated, as **in·sip·id** (in·sip′id) *adj.* . . . **— in·si·pid·i·ty** (in′si·pid′ə·tē), **in·sip′id·ness** *n.* **— in·sip′id·ly** *adv.*

Usage notes Points of grammar and idiom, when an integral part of definition, are included, following a colon, after the particular sense of a word to which they apply, as **anx·ious** . . . 3 Intent; eagerly desirous; solicitous: with *for* or the infinitive with *to*: *anxious* for success; *anxious* to succeed . . .

More extensive notes consisting of supplementary information on grammar, accepted usage, the relative status of variant forms, etc., are entered at the end of the relevant entries and prefaced with the symbol ◆. (See **anyone, Asiatic, can, have.**)

Synonyms and antonyms Extended discussions of the differentiation in shades of meaning within a group of related words, or in some cases simple lists of synonyms, are given at the end of relevant entries in paragraphed form after the run-on derivatives. Lists of antonyms are often added as well to point up further distinctions in meaning.

Alphabetization All entries in this dictionary (general vocabulary, affixes, geographical and biographical entries, foreign words and phrases, etc.) are in one alphabetic list, with the exception of an extended list of abbreviations given, for the sake of ready reference, in one section at the back of the book. Thus, the entry for **ampere** (the electrical unit) immediately precedes that for **André Marie Ampère,** for whom it was named; the entry for **Bridge of Sighs** precedes **Bridgeport** and follows **bridgehead;** the prefix **pro-** is entered immediately after the word **pro** (a professional) and before the word **proa** (a sailing vessel).

Hyphens The hyphen used in the spelling of hyphemes (hyphened words) in this dictionary is printed with extra length to distinguish it as a spelling characteristic, as **cap-a-pie, battle-scarred.** This lengthened hyphen is also used in the entries for prefixes, suffixes, and combining forms, as **un-, -less, hydro-.** The standard hyphen is utilized at the ends of lines to indicate syllabic breaks in words ordinarily written solid, as *com-prehensive,* while the lengthened hyphen is retained for the end-of line breaks in hyphemes.

Homographs These words, which are identical in spelling but differ in meaning and origin (and often in pronunciation), are separately entered and differentiated by the following superior figure, as **bushel**[1] (measure), **bushel**[2] (mend); **pink**[1] (color), **pink**[2] (stab), **pink**[3] (sailing vessel), **pink**[4] (fade). The numbering also serves to simplify cross-reference to such words.

Cross-references Cross-references are directions to see another entry for additional information. They direct the reader from a variant spelling, inflected form, subentry, etc., to a main entry. The entry to be sought is generally indicated in small capital letters, as **car·a·cul** . . . See KARAKUL; **aes·thete** . . . , **aes·thet·ic** . . . See ESTHETE, etc.; **cor·po·ra** . . . Plural of CORPUS; **Old English** See under ENGLISH.

Sometimes a cross-reference is made to a homograph entry or to a particular definition or part of speech of the main entry, as

> **flied** (flīd) Past tense and past participle of FLY[1] (def. 7).
> **taps** (taps) See TAP[2] (*n.* def. 3).

Some entries are defined by citing another form, as

> **se·pi·o·lite** . . . *n.* Meerschaum.

This is a type of cross-reference. Complete information will be found under the word or term used in the definition.

See the note on **Etymologies,** for the system of cross-referring there used.

Prefixes, suffixes, and combining forms These are entered and defined in regular alphabetic order. The prefix is followed by the lengthened hyphen, the suffix preceded by it, as **anti-** *prefix*; **-ical** *suffix.*

Similarly, the combining form is followed or preceded by the lengthened hyphen depending on its position in combination, as **proto-** *combining form*; **-cide** *combining form.*

Word lists The meaning of many combinations of words (hyphemes, solidemes, and two-word phrases) is easily deduced by combining the senses of their component parts. Such self-explaining compounds have been entered in list form under the first element — that is, under prefixes (**bi-, co-, non-, re-,** etc.) combining forms (**auto-, counter-, mid-,** etc.), and words (**corn, heart, man, peace,** etc.). These lists serve to indicate the preferred form of a compound — whether written solid, with a hyphen, or as a two-word phrase. The listings are not intended to be all-inclusive; most of the prefixes and combining forms so entered combine freely in English in the formation of new compounds based on existing forms.

Chemical formulas As an integral part of the definitions of the large number of chemical substances listed in this dictionary, the reader will find the formulas which, in the chemist's shorthand, indicate what constituents enter into a compound, and in what proportions. These formulas, though usually of the simple empirical type, help to prevent confusion in the identification of hundreds of the substances used in medicine, industry, and the arts, and are particularly useful in recognizing distinctions between the two broad fields of inorganic and organic chemicals.

Taxonomic classification Essential technical information regarding the many plants and animals described under their common names is provided by the listing, usually in italic type enclosed in parentheses, of one or more of the principal taxonomic categories — phylum or division, class, order, family, genus, and species — by which they are correctly identified in botanical and zoological usage. This information, checked against the latest and most reliable sources, is especially useful in showing relationships between seemingly very different types and in discriminating within and between deceptively similar groups, such as toads and frogs, spruce and hemlock, and moths and butterflies.

Trade names Of the thousands of words used to identify trademarked or proprietary articles, drugs, processes, and services, a generous number have been entered and defined in this book because of their wide public acceptance. In every such case the word is entered as though it were a proper name: that is, with an initial capital letter, and the added notation, "a trade name" or — chiefly for the pharmaceutical products — "a proprietary name (or brand) . . ." This technique is employed to alert the reader to the commercial status of a word and, by implication, to caution him against employing it in a generic sense that might involve him in legal difficulties with those claiming a prescriptive or legal right to its use. This treatment of proprietary names is in no sense to be interpreted as establishing a formal status within the meaning of any of the various statutes involving the protection and use of trade names, registered or otherwise.

Illustrations, maps, and tables The illustrations in this book — all of them carefully prepared line drawings — have been selected with the emphasis on their explanatory value rather than their pictorial or decorative effect. They often include informative captions, and are intended to supplement the definitions.

A certain number of geographic entries are accompanied by small, precise spot maps which show at a glance just how a given place, region, lake, island, or other feature is related to its immediate surroundings — as under *Congo* (river system), *English Channel, Holy Roman Empire, Suez.*

For a selected group of entries supplementary information has been provided in the form of charts and tables, as for foreign alphabets (see **alphabet**), the geological time scale (see **geology**), chemical elements (see **periodic table**), the major wars of history (see **war**), constellations, clouds, the endocrine glands, etc.

Abbreviations The abbreviations used in the body of this book (in labeling within entries, in etymologies, etc.) will be found listed on page xx. Where abbreviations of main-entry words are entered, they are shown at the end of the entry, as Abbr. *B.A., A.B.* (Bachelor of Arts); Abbr. *AWOL, awol, A.W.O.L., a.w.o.l.* (absent without leave); Abbr. *Dan.* (Danish). Biblical references are to the King James Bible and indicate book, chapter, and verse, in that order, as *Matt.* v 3–12. An extended list of standard abbreviations will be found at the end of the dictionary, following the vocabulary section.

TABLE OF ENGLISH SPELLINGS—SPELLING: PLURALS AND PARTICIPLES

TABLE OF ENGLISH SPELLINGS

FOLLOWING is a list of words exemplifying the possible spellings for the sounds of English. The sounds represented by these spellings are shown in the pronunciation symbols used in this dictionary, followed by their equivalents in the International Phonetic Alphabet.

a	æ	cat, plaid, calf, laugh
ā	eɪ,e	mate, bait, gaol, gauge, pay, steak, skein, weigh, prey
â(r)	ɛ,ɛr	dare, fair, prayer, where, bear, their
ä	ɑ	dart, ah, calf, laugh, sergeant, heart
b	b	boy, rubber
ch	tʃ	chip, batch, righteous, bastion, structure
d	d	day, ladder, called
e	ɛ	many, aesthete, said, says, bet, steady, heifer, leopard, friend, foetid
ē	i	Caesar, quay, scene, meat, see, seize, people, key, ravine, grief, phoebe
f	f	fake, coffin, cough, half, phase
g	g	gate, beggar, ghoul, guard, vague
h	h	hot, whom
hw	hw,ʍ	whale
i	ɪ	pretty, been, tin, sieve, women, busy, guilt, lynch
ī	aɪ	aisle, aye, sleight, eye, dime, pie, sigh, guile, buy, try, lye
j	dʒ	edge, soldier, modulate, rage, exaggerate, joy
k	k	can, accost, saccharine, chord, tack, acquit, king, talk, liquor
l	l	let, gall
m	m	drachm, phlegm, palm, make, limb, grammar, condemn
n	n	gnome, know, mnemonic, note, banner, pneumatic
ng	ŋ	sink, ring, meringue
o	ɑ,ɒ	watch, pot
ō	oʊ,o	beau, yeoman, sew, over, soap, roe, oh, brooch, soul, though, grow
ô	ɔ	ball, balk, fault, dawn, cord, broad, ought
oi	ɔɪ	poison, toy
ou	aʊ	out, bough, cow
ōō	u	rheum, drew, move, canoe, mood, group, through, fluke, sue, fruit
ŏŏ	ʊ	wolf, foot, could, pull
p	p	map, happen
r	r	rose, rhubarb, marry, diarrhea, wriggle
s	s	cite, dice, psyche, saw, scene, schism, mass
sh	ʃ	ocean, chivalry, vicious, pshaw, sure, schist, prescience, nauseous, shall, pension, tissue, fission, potion
t	t	walked, thought, phthisic, ptarmigan, tone, Thomas, butter
th	θ	thick
th	ð	this, bathe
u	ʌ	some, does, blood, young, sun
yōō	ju,ɪu	beauty, eulogy, queue, pew, ewe, adieu, view, fuse, cue, youth, yule
û(r)	ɜr,ɝ	yearn, fern, err, girl, worm, journal, burn, guerdon, myrtle
v	v	of, Stephen, vise, flivver
w	w	choir, quilt, will
y	j	onion, hallelujah, yet
z	z	was, discern, scissors, xylophone, zoo, muzzle
zh	ʒ	rouge, pleasure, incision, seizure, glazier
ə	ə	above, fountain, darken, clarity, parliament, cannon, porpoise, vicious, locus
ər	ər,ɚ	mortar, brother, elixir, donor, glamour, augur, nature, zephyr

SPELLING: Plurals and Participles

BASICALLY, plurals in English are formed by the addition of -s or -es (depending on the preceding sound) to the complete word; past participles are formed by the addition of -ed, and present participles by adding -ing. There are, however, many exceptions. In this book, all such exceptions (the "irregular" inflected forms) are indicated within the entry, in boldface immediately following the part-of-speech label.

fly (flī) *n. pl.* **flies** . . .
sheep (shēp) *n. pl.* **sheep** . . .
cal·ci·fy (kal′sə·fī) *v.t. & v.i.* **·fied, ·fy·ing** . . .
go (gō) *v.i.* **went, gone, go·ing** . . .

Some rules for the spelling of these forms (with the exception of nouns which form their plurals by some internal change and the so-called strong verbs) are listed below:

PLURALS

1. Nouns ending in *y* preceded by a consonant change *y* to *i* and add *-es*.

baby babies   story stories

2. Nouns ending in *y* preceded by a vowel add *-s* without change.

chimney chimneys   valley valleys

Note, however, that *money* may have either form in the plural — *moneys, monies*.

3. Some nouns ending in *f* or *fe* change this to *v* and add *-es*.

knife knives   shelf shelves
BUT: roof roofs   safe safes

Note: Some words may have alternate plural forms.

scarf   scarfs or scarves

4. Most words ending in *o* form a plural in *-os*.

cameo cameos   folio folios   halo halos

A few words ending in *o* (echo, hero, Negro, etc.) form the plural only in *-oes* (echoes, heroes, Negroes), but many others in this category have alternative plurals in both forms.

buffalo   buffalos or buffaloes
mosquito   mosquitos or mosquitoes
volcano   volcanoes or volcanos

PAST AND PRESENT PARTICIPLES

1. The final consonant is doubled for monosyllables or words accented on the final syllable when they end in a *single* consonant preceded by a *single* vowel.

control, controlled, controlling
hop, hopped, hopping
occur, occurred, occurring
quit, quitted, quitting (*Note: a* u *following* q *is not to be counted as an additional vowel.*)
BUT: help, helped, helping (*two consonants*)
seed, seeded, seeding (*two vowels*)

Some words *not* accented on the final syllable have a variant participial form with a doubled consonant; the single consonant form is preferred in the United States.

travel, traveled or travelled, traveling or travelling
worship, worshiped or worshipped, worshiping or worshipping

2. Words ending in silent or mute *e* drop the *e* before *-ed* and *-ing*, unless it is needed to avoid confusion with another word.

change, changed, changing   love, loved, loving
singe, singed, singeing   dye, dyed, dyeing

3. Verbs ending in *ie* usually change this to *y* before adding *-ing*.

die, died, dying   lie, lied, lying

4. Verbs ending in *c* add a *k* before *-ed* and *-ing*.

mimic, mimicked, mimicking
picnic, picnicked, picnicking

PRONUNCIATION KEY

The primary stress mark (′) is placed after the syllable bearing the heavier stress or accent; the secondary stress mark (′) follows a syllable having a somewhat lighter stress, as in **com·men·da·tion** (kom′ən·dā′shən).

a	add, map	f	fit, half	n	nice, tin	p	pit, stop	u	up, done	ə	the schwa, an un-stressed vowel representing the sound of
ā	ace, rate	g	go, log	ng	ring, song	r	run, poor	û(r)	urn, term		
â(r)	care, air	h	hope, hate	o	odd, hot	s	see, pass	yōō	use, few		
ä	palm, father	i	it, give	ō	open, so	sh	sure, rush	v	vain, eve		*a* in *above*
b	bat, rub	ī	ice, write	ô	order, jaw	t	talk, sit	w	win, away		*e* in *sicken*
ch	check, catch	j	joy, ledge	oi	oil, boy	th	thin, both	y	yet, yearn		*i* in *clarity*
d	dog, rod	k	cool, take	ou	out, now	th	this, bathe	z	zest, muse		*o* in *melon*
e	end, pet	l	look, rule	ōō	pool, food			zh	vision, pleasure		*u* in *focus*
ē	even, tree	m	move, seem	oo	took, full						

FOREIGN SOUNDS

à as in French *ami, patte*. This is a vowel midway in quality between (a) and (ä).

œ as in French *peu*, German *schön*. Round the lips for (ō) and pronounce (ā).

ü as in French *vue*, German *grün*. Round the lips for (ōō) and pronounce (ē).

kh as in German *ach*, Scottish *loch*. Pronounce a strongly aspirated (h) with the tongue in position for (k) as in *cool* or *keep*.

ṅ This symbol indicates that the preceding vowel is nasal. The nasal vowels in French are œṅ (*brun*), aṅ (*main*), äṅ (*chambre*), ôṅ (*dont*).

′ This symbol indicates that a preceding (l) or (r) is voiceless, as in French *débâcle* (dā·bä′kl′) or *fiacre* (fyà′kr′), or that a preceding (y) is pronounced consonantly in a separate syllable followed by a slight schwa sound, as in French *fille* (fē′y′).

Note on the accentuation of foreign words: Many languages do not employ stress in the manner of English; only an approximation can be given of the actual situation in such languages. As it is not possible to reproduce the tones of Chinese in a work of this kind, Chinese names have been here recorded with primary stress on each syllable and may be so pronounced. Japanese and Korean have been shown without stress and may be pronounced with a level accent throughout. French words are shown conventionally with a primary stress on the last syllable; however, this stress tends to be evenly divided among the syllables (except for those that are completely unstressed), with slightly more force and higher pitch on the last syllable.

ABBREVIATIONS USED IN THIS WORK

A.D.	year of our Lord	F, Fr.	French	Mech.	Mechanics	Philos.	Philosophy
adj.	adjective	fem.	feminine	Med.	Medicine, Medieval	Phonet.	Phonetics
adv.	adverb	freq.	frequentative			Phot.	Photography
Aeron.	Aeronautics	G, Ger.	German	Med. Gk.	Medieval Greek (600–1500)	Physiol.	Physiology
AF	Anglo-French	Gal.	Galatians			pl.	plural
Agric.	Agriculture	Gen.	Genesis	Med. L	Medieval Latin (600–1500)	pp.	past participle
Alg.	Algebra	Geog.	Geography			ppr.	present participle
alter.	alteration	Geol.	Geology	Metall.	Metallurgy	prep.	preposition
Am. Ind.	American Indian	Geom.	Geometry	Meteorol.	Meteorology	prob.	probably
Anat.	Anatomy	Gk.	Greek (Homer—A.D. 200)	MF	Middle French (1400–1600)	pron.	pronoun
Anthropol.	Anthropology					Prov.	Proverbs
appar.	apparently	Gmc.	Germanic	MHG	Middle High German (1100–1450)	Ps., Psa.	Psalms
Archeol.	Archeology	Govt.	Government			Psychoanal.	Psychoanalysis
Archit.	Architecture	Gram.	Grammar			Psychol.	Psychology
assoc.	association	Hab.	Habakkuk	Mic.	Micah	pt.	preterit
Astron.	Astronomy	Hag.	Haggai	Mil.	Military	ref.	reference
aug.	augmentative	Heb.	Hebrews	MLG	Middle Low German (1100–1450)	Rev.	Revelation
Austral.	Australian	Her.	Heraldry			Rom.	Romans
Bacteriol.	Bacteriology	HG	High German			Sam., Saml.	Samuel
B.C.	Before Christ	Hind.	Hindustani	n.	noun	Scand.	Scandinavian
Biochem.	Biochemistry	Hos.	Hosea	Nah.	Nahum	Scot.	Scottish
Biol.	Biology	Icel.	Icelandic	Naut.	Nautical	SE	Southeast
Bot.	Botany	Illit.	Illiterate	Nav.	Naval	sing.	singular
Brit.	British	imit.	imitative	N. Am. Ind.	North American Indian	Skt.	Sanskrit
c.	century	infl.	influence, influenced			Sociol.	Sociology
Can.	Canadian			NE	Northeast	S. of Sol.	Song of Solomon
cf.	compare	intens.	intensive	Neh.	Nehemiah	S. Am. Ind.	South American Indian
Chem.	Chemistry	interj.	interjection	neut.	neuter		
Chron.	Chronicles	Is., Isa.	Isaiah	NL	New Latin (after 1500)	Sp.	Spanish
Col.	Colossians	Ital.	Italian			Stat.	Statistics
Colloq.	Colloquial	Jas.	James	Norw.	Norwegian	superl.	superlative
compar.	comparative	Jer.	Jeremiah	Num., Numb.	Numbers	Surg.	Surgery
conj.	conjunction	Jon.	Jonah	NW	Northwest	Sw.	Swedish
Cor.	Corinthians	Josh.	Joshua	O	Old	SW	Southwest
Dan.	Daniel, Danish	Judg.	Judges	Ob., Obad.	Obadiah	Technol.	Technology
def.	definition	L, Lat.	Latin (Classical, 80 B.C.–A.D. 200)	Obs.	Obsolete	Telecom.	Telecommunication
Dent.	Dentistry			OE	Old English (before 1150)		
Deut.	Deuteronomy					Theol.	Theology
Dial.	Dialect, Dialectal	Lam.	Lamentations	OF	Old French (before 1400)	Thess.	Thessalonians
dim.	diminutive	Lev., Levit.	Leviticus			Tim.	Timothy
Du.	Dutch	LG	Low German	OHG	Old High German (before 1100)	Tit.	Titus
E	English	LGk.	Late Greek (200–600)			trans.	translation
Eccl.	Ecclesiastical					Trig.	Trigonometry
Eccles.	Ecclesiastes	Ling.	Linguistics	ON	Old Norse (before 1500)	ult.	ultimate, ultimately
Ecclus.	Ecclesiasticus	lit.	literally				
Ecol.	Ecology	LL	Late Latin (200–600)	orig.	original, originally	U.S.	United States
Econ.	Economics					v.	verb
Electr.	Electricity	M	Middle	Ornithol.	Ornithology	var.	variant
Engin.	Engineering	Mal.	Malachi	OS	Old Saxon (before 1100)	v.i.	verb intransitive
Entomol.	Entomology	masc.	masculine			v.t.	verb transitive
Eph.	Ephesians	Math.	Mathematics	Paleontol.	Paleontology	WGmc.	West Germanic
esp.	especially	Matt.	Matthew	Pet.	Peter	Zech.	Zechariah
Esth.	Esther	MDu.	Middle Dutch	Pg.	Portuguese	Zeph.	Zephaniah
Ex., Exod.	Exodus	ME	Middle English (1150–1500)	Phil.	Philippians	Zool.	Zoology
Ezek.	Ezekiel			Philem.	Philemon		

◆ Usage note; Homophone; Collateral adjective   < from   + plus   ? possibly

lo·bate (lō′bāt) *adj.* Composed of lobes; lobelike. Also **lo′bat·ed.** [<NL *lobatus*] — **lo′bate·ly** *adv.* — **lo·ba′tion** *n.*

lob·by (lob′ē) *n. pl.* **·bies** 1 A hall, vestibule, or corridor on the main floor of a large public building, as a theater or hotel; a lounge; foyer. 2 The part of an assembly-room of a legislative or deliberative body not appropriated to the official use of members, and to which outsiders have free entry. 3 *U. S.* The persons or groups of persons who accost or solicit legislators in order to influence the action of a legislative body in the interest of a special group, industry, etc.: so called because supposed to frequent lobbies. 4 A cold-storage chamber for the temporary storage of ice. — *v.* **·bied, ·by·ing** *v.i.* To attempt to influence a legislator or legislators in favor of one's own interests. — *v.t.* To attempt to obtain passage or defeat of (a bill, etc.) by such means. [<Med. L *lobia*. Doublet of LODGE, *n.,* LOGE, LOGGIA.]

lob·by·ism (lob′ē-iz′əm) *n. U. S.* The practice of lobbying. — **lob′by·ist** *n.*

lobe (lōb) *n.* 1 A protuberance, especially globular, as of an organ or part. 2 The soft lower extension of the external ear. 3 A principal division of a molar tooth. [<F <Gk. *lobos*]

lo·bec·to·my (lō-bek′tə-mē) *n. Surg.* The operation of removing a lobe, as of the brain, lung, etc. [<LOBE + -ECTOMY]

lobed (lōbd) *adj.* 1 Lobate; having lobes. 2 *Bot.* Having divisions that extend not more than half-way from the margin to the center and rounded lobes or sinuses: said of leaves, petals, etc.

lo·be·li·a (lō-bē′lē-ə, -bēl′yə) *n.* Any of a large genus (*Lobelia*) of herbaceous plants with showy flowers either axillary or in bracted racemes. [<NL, after Matthias de *Lobel*, 1538–1616, Flemish botanist]

lo·be·line (lō′bə-lēn) *n. Chem.* A white, crystalline alkaloid, $C_{22}H_{27}O_2N$, from the seeds of Indian tobacco. Its hydrochloride is used as a respiratory stimulant. [<LOBEL(IA) + -INE²]

Lo·ben·gu·la (lō′beng-gyōō′lə, lō-beng′gyə-lə), 1833–94, king of the Matabele.

lob·lol·ly (lob′lol-ē) *n. pl.* **·lies** *U. S.* 1 *Colloq.* A mudhole; oozy mire. 2 A loblolly bay or pine. [<dial. E *lob* bubble + *lolly* broth]

loblolly bay A tree (*Gordonia lasianthus*) with smooth, shining, lanceolate leaves and showy white flowers, growing in southern U. S. coastal swamps.

loblolly boy *Obs.* A ship's surgeon's assistant and dispenser. [obs. nautical slang *loblolly* medicine + *boy*]

loblolly pine See under PINE.

lo·bo (lō′bō) *n. pl.* **·bos** The timber wolf (*Canis nubilus*), of the western United States. [<Sp., wolf]

lo·bot·o·my (lō-bot′ə-mē) *n. Surg.* The operation of cutting into or across a lobe, especially of the brain. [<Gk. *lobos* a lobe + -TOMY]

lob·scouse (lob′skous) *n.* A dish consisting of salt meat, vegetables, and biscuit. Also **lob·scourse** (-skôrs, -skōrs). [Origin unknown]

lob·ster (lob′stər) *n.* 1 A marine crustacean (genus *Homarus*) much used as food, having a large first pair of ambulatory legs, which form the claws, and compound eyes carried on flexible stalks. 2 One of various other long-tailed crustaceans, as a spiny lobster or crayfish. [OE *loppestre* lobster, grasshopper <L *locusta* lobster, locust; infl. in form by *loppe* spider. Doublet of LOCUST.]

lobster pot A trap consisting of a cage with netting at the ends for catching lobsters.

lobster shift A work shift during the latter part of the night; a graveyard shift.

lob·ule (lob′yōōl) *n.* A small lobe, or lobe made small by separation from a larger lobe. [<NL *lobulus*, dim. of *lobus* a lobe] — **lob′u·lar, lob′u·late** (-lit, -lāt) *adj.*

lob·worm (lob′wûrm′) See LUGWORM.

lo·cal (lō′kəl) *adj.* 1 Pertaining to a prescribed place or a limited portion of space. 2 Restricted to or characteristic of a particular place. 3 Pertaining to place in general. 4 *Med.* Relating to or affecting a specific part or organ of the body: said of a disease or injury, or of the remedies used. 5 Relating to a locus. — *n.* 1 A subway or suburban train that stops at all the stations. 2 A local branch or unit of a trade union or fraternal organization. 3 An item of local interest in a newspaper. [<F <L *localis* <*locus* place] — **lo′cal·ly** *adv.*

local anesthesia Anesthesia restricted in action to a specific part of the body.

local color In literature and art, the presentation of the characteristic manners, speech, dress, scenery, etc., of a certain period or region so as to achieve a sense of realism.

lo·cale (lō-kal′, -käl′) *n.* Locality; specifically, a spot considered with reference to surrounding features or circumstances. [<F]

local government 1 Independent government in local affairs by the small political entity of a limited region. 2 The governing head or body of such a locality.

lo·cal·ism (lō′kəl-iz′əm) *n.* 1 A manner of acting or speaking particular to a place; local custom or idiom. 2 A word, a meaning of a word, a pronunciation, etc., peculiar to a locality, rather than in general usage. 3 Provincialism. 4 The state or condition of being local; influence exerted by a particular place.

lo·cal·i·ty (lō-kal′ə-tē) *n. pl.* **·ties** 1 A definite region in any part of space; geographical position. 2 Restriction to a particular place. See synonyms under NEIGHBORHOOD, PLACE. [<F *localité* <LL *localitas, -tatis*]

lo·cal·ize (lō′kəl-īz) *v.t.* **·ized, ·iz·ing** 1 To make local; limit or assign to a specific area or locality. 2 To determine the place of origin of. Also *Brit.* **lo′cal·ise.** — **lo′cal·i·za′tion** *n.*

local option The privilege granted to a political division, as a county or town, of determining whether something, especially the sale of intoxicants, shall be permitted within its limits.

local time See under TIME.

Lo·car·no (lō-kär′nō) A town on Lago Maggiore in SE central Switzerland; scene of an international conference on political problems following World War I, 1925.

lo·cate (lō′kāt, lō-kāt′) *v.* **·cat·ed, ·cat·ing** *v.t.* 1 To discover the position or source of; find. 2 To assign place or locality to: to *locate* a scene in a valley. 3 *U. S.* To establish in a place; situate: My office is *located* in Portland. 4 *U. S.* To survey, set, or designate the site or boundaries of, as a mining claim. — *v.i.* 5 *Colloq.* To establish oneself or take up residence; settle. See synonyms under SET. [<L *locatus*, pp. of *locare* <*locus* place]

lo·ca·tion (lō-kā′shən) *n.* 1 The act of locating, or the state of being located. 2 The exact position in space; place. 3 A plot of ground defined by boundaries; a mining claim. 4 *Law* A renting or letting for hire; also, the establishment or fixing of the boundaries of a tract of land. 5 A site selected for staging a scene in a motion picture. 6 Any one of five minor civil divisions in New Hampshire. See synonyms under PLACE. [<L *locatio, -onis*]

loc·a·tive (lok′ə-tiv) *adj.* 1 *Gram.* In certain inflected languages, as Latin, Greek, and Sanskrit, designating the case of the noun denoting place where or at which. 2 *Anat. & Zool.* Indicating relative position in a series. — *n. Gram.* 1 The locative case. 2 A word in this case. [<L *locatus*, pp. of *locare* locate; on analogy with L *vocativus* vocative]

lo·ca·tor (lō′kā·tər) *n.* 1 One who or that which locates. 2 One who locates land under the land laws, or is entitled to locate it. 3 *Law* One who lets a thing or services for hire. [<L]

locator card 1 A card bearing all essential facts about a person, especially an enlisted man or officer. 2 A similar card describing an article in a military depot.

loch (lokh, lok) *n. Scot.* 1 A lake. 2 A bay, or arm of the sea.

loch·an (lokh′ən) *n. Scot.* A little loch; pond.

lo·chi·a (lō′kē-ə, lok′ē-ə) *n. pl. Med.* The discharges from the vaginal passages after childbirth, continuing from two to three weeks. [<NL <Gk. *lochia*, neut. pl. of *lochios* pertaining to childbirth <*lochos* childbirth] — **lo′chi·al** *adj.*

Loch·in·var (lok′in-vär, lok′-) The young hero of a ballad in Scott's *Marmion* who abducts his sweetheart just before she is to be wedded to another.

lo·ci (lō′sī) Plural of LOCUS.

lock¹ (lok) *n.* 1 A device to fasten an object; specifically, one for so securing a door, drawer, or the like, as to prevent its being opened, except by a special key or combination. 2 A spring mechanism for exploding the charge of a firearm. 3 A section of a canal, etc., enclosed by gates at either end, within which the water level may be varied to raise or lower vessels from one level to another. 4 An intermingling or fastening together; hence, a hold, hug, or grapple in wrestling. 5 A lockup. 6 One of various mechanical devices for fixing something so that it may remain in place. 7 An airlock. 8 A device to prevent a carriage wheel from turning, as in descending a hill. 9 The oblique position of the fore axle with relation to the hind axle of a vehicle when turning or swerving. — **combination lock** A lock which can be opened only by combining its dial-operated tumblers in a certain sequence. — **cylinder lock** A lock fitted with a cylinder having a control element which can be actuated only by a key whose particularly shaped surface engages with the tumblers and sets them to the exact position

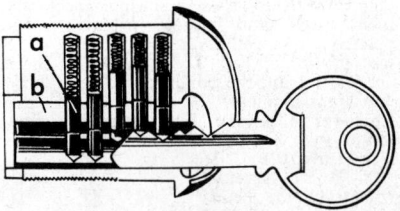

CYLINDER LOCK
Insertion of the proper key raises all the tumblers (a) to the exact position required to release the key-plug (b), thus permitting the key to turn (here partly inserted).

for unlocking. — *v.t.* 1 To secure or fasten by means of a lock. 2 To shut, confine, or exclude by means of a lock: with *in, up,* or *out.* 3 To join or unite securely; interlock; link: to *lock* arms. 4 To make immovable, as by jamming or by a lock. 5 To provide (a canal, etc.) with a lock or locks. 6 To move (a ship) through a waterway by means of locks. — *v.i.* 7 To become locked or fastened. 8 To become joined or linked. 9 To become jammed immovably. 10 To proceed by means of locks: said of ships. 11 To turn (wheels) under a carriage body. [OE *loc* fastening, enclosure]

— Synonyms (noun): bar, bolt, catch, clasp, fastening, hasp, hook, latch. A *bar* is a piece of wood or metal, usually of considerable size, by which an opening is obstructed, a door held fast, etc. A *bar* may be movable or permanent; a *bolt* is a movable rod or pin of metal sliding in a socket, and adapted for securing a door or window. A *lock* is an arrangement by which an enclosed *bolt* is shot forward or backward by a key, or other device; the *bolt* is the essential part of the *lock*. A *latch* or *catch* is an accessible *fastening* designed to be easily movable, and simply to secure against accidental opening of the door, cover, etc. A *hasp* is a metallic strap that fits over a staple, calculated to be secured by a padlock; a simple *hook* that fits into a staple is also called a *hasp*. A *clasp* is a *fastening* that can be sprung into place, to draw and hold the parts of some object firmly together, as the *clasp* of a book.

lock² (lok) *n.* 1 A tuft of hair; ringlet; tress. 2 *pl.* A head of hair. 3 A small quantity of hay, wool, etc. [OE *locc*]

lock·age (lok′ij) *n.* 1 Material going to form the lock of a canal. 2 The difference in level between the locks of a canal. 3 The toll levied for passing through a lock.

Locke (lok), **David Ross,** 1833–88, U. S. political satirist; pseudonym *Petroleum V. Nasby.* — **John,** 1632–1704, English philosopher and essayist.

lock·er (lok′ər) *n.* 1 One who or that which

locks. 2 A closet or receptacle fastened with a lock.
lock·et (lok′it) *n.* A small case, suspended on a necklace or chain, often holding a portrait. [<OF *locquet,* dim. of *loc* latch <Gmc.]
lock·fast (lok′fast′, -fäst′) *adj.* Securely held by some locked contrivance.
Lock·hart (lok′ərt, -härt), **John Gibson,** 1794–1854, Scottish writer and biographer.
lock·jaw (lok′jô′) *n. Pathol.* A spasmodic contraction of the muscles of the lower jaw; trismus. See TETANUS.
lock nut *Mech.* 1 An auxiliary nut used to prevent the loosening of another. 2 A nut that automatically locks when screwed tight.
lock-out (lok′out′) *n.* The closing of a factory or other place of business by employers to bring employees on strike to terms. Also **lock′out′**. Compare STRIKE.
lock·ram (lok′rəm) *n.* A coarse, cheap linen. [after *Locronan,* a village in Brittany]
locks·man (loks′mən) *n. pl.* **·men** (-mən) A warden; turnkey.
lock·smith (lok′smith′) *n.* A maker or repairer of locks. **—lock′smith′er·y, lock′smith′ing** *n.*
lock step A marching step in which each marcher follows as closely as possible the one in front of him.
lock stitch A stitch made by two interlocking threads, as on some sewing machines.
lock, stock, and barrel Altogether; completely; in its entirety.
lock·up (lok′up′) *n.* 1 A prison. 2 The act of locking up; the condition of being locked up; imprisonment.
Lock·yer (lok′yər), **Sir Joseph Norman,** 1836–1920, English astronomer.
loco (lō′kō) *n.* 1 Any one of several plants of the bean family (genera *Astragalus* and *Oxytropis*), often poisonous to livestock, and found in the western and SW United States: also called *crazyweed.* 2 Loco disease. —*adj. Slang* Crazy; insane. —*v.t.* 1 To poison with locoweed. 2 *U.S. Slang* To render insane. [<Sp., insane]

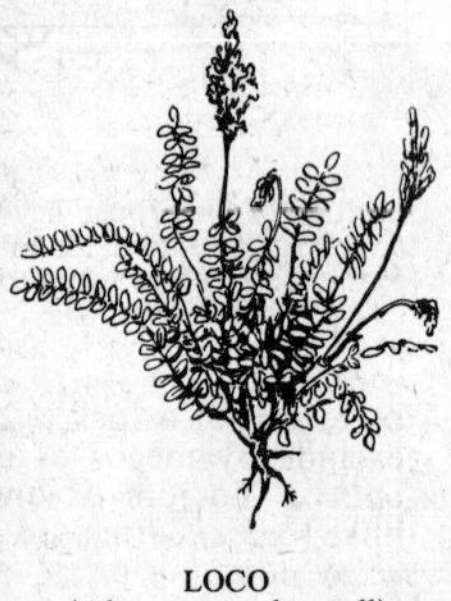
LOCO
(About 12 inches tall)

lo·co ci·ta·to (lō′kō sī·tā′tō) *Latin* In the place cited. Abbr. *loc. cit.*
loco disease An ailment attacking horses or other animals that have eaten the loco. It affects the brain, causing slowness of gait, loss of flesh, defective vision, delirium, and eventually, death: generally curable by careful and prolonged dieting.
Lo·co·fo·co (lō′kō·fō′kō) *n. pl.* **·cos** 1 The extreme section of the Democratic party of 1835, known as the Equal Rights party. 2 Any adherent of that party. [after *Locofoco,* trade name for a friction match; from the use of these matches to light the candles at one of their early meetings]
lo·co·mo·bile (lō′kə·mō′bil) *adj.* Self-propelling.
lo·co·mo·tion (lō′kə·mō′shən) *n.* The act or power of moving from one place to another. [<L *loco* from a place + *motio, -onis* movement]
lo·co·mo·tive (lō′kə·mō′tiv) *adj.* 1 Pertaining to locomotion. 2 Moving from one place to another. 3 Possessed of the power of moving. —*n.* A self-propelling electric, diesel, or steam engine on wheels, especially one for use on a railway.
lo·co·mo·tor (lō′kə·mō′tər) *adj.* Of or pertaining to locomotion. —*n.* One who or that which has the power of locomotion.
locomotor ataxia *Pathol.* A disease of the spinal cord, characterized by unsteadiness and inability to coordinate locomotor and other voluntary movements; tabes dorsalis.
lo·co·weed (lō′kō·wēd′) *n.* Loco (def. 1).
Lo·cri (lō′krī, lok′rī) *n. pl.* The people of Locris.
Lo·cris (lō′kris, lok′ris) An ancient state of eastern and central Greece, divided into **Eastern Locris,** NW of Boeotia, and **Western Locris,** north of the Gulf of Corinth. **—Lo′cri·an** *adj.*
loc·u·lus (lok′yə·ləs) *n. pl.* **·li** (-lī) A small cavity or chamber; a cell. [<L, dim. of *locus* place] **—loc′u·lar, loc′u·late** (-lāt, -lit), **loc′u·lat′ed** *adj.*
lo·cum te·nens (lō′kəm tē′nenz) *Chiefly Brit.* A temporary representative or substitute. [<L, holding the place <*locus* place + *tenens,* ppr. of *tenere* hold]
lo·cus (lō′kəs) *n. pl.* **·ci** (-sī) 1 A place; locality; area. 2 *Math.* A surface or curve regarded as traced by a line or point moving under specified conditions; any figure made up wholly of points or lines that satisfy given conditions. 3 A figure formed by the foci of a series of pencils of light. 4 A passage in a writing. 5 *pl.* A series of passages, as from the Scriptures, classified for reading or study; any book containing such passages. 6 *Genetics* The linear position of a gene on a chromosome. [<L. Doublet of LIEU.]
lo·cus clas·si·cus (lō′kəs klas′ə·kəs) *Latin* An illustrative or authoritative passage from a standard work.
lo·cus si·gil·li (lō′kəs si·jil′ī) *Latin* The place of the seal; abbreviated in legal documents, *L. S.*
lo·cust[1] (lō′kəst) *n.* 1 Any of a family *(Locustidae)* of widely distributed orthopterous insects resembling grasshoppers but having short antennae, especially those of migratory habits (*Locusta, Pachytylus, Melanoplus,* and related genera), which are destructive of grain and vegetation in many parts of the world. 2 A cicada or harvest fly. [<OF *locuste* <L *locusta.* Doublet of LOBSTER.]
lo·cust[2] (lō′kəst) *n.* 1 A North American tree *(Robinia pseudoacacia)* of the bean family, with a rough bark, odd-pinnate leaves, and loose, slender racemes of fragrant, white flowers; also, its wood. 2 Any of several other trees with similar pods, as the carob tree. 3 The honey locust. [<NL *locusta;* orig. applied to the carob pod from its fancied resemblance to the insect]
lo·cus·ta (lō·kus′tə) *n. Bot.* A spikelet in grasses. [<NL. See LOCUST[2].]
lo·cu·tion (lō·kyōō′shən) *n.* 1 A mode of speech. 2 An idiom; phrase. [<L *locutio, -onis* a speaking <*loqui* speak]
loc·u·to·ry (lok′yə·tôr′ē, -tō′rē) *n. pl.* **·ries** A place for conversation; specifically, a reception room in a monastery or convent. [<Med. L *locutorium* <L *locutor* speaker <*loqui* talk]
lode (lōd) *n.* 1 A somewhat continuous, unstratified, metal-bearing vein. 2 A tabular deposit of valuable mineral between definite boundaries of associated rock. 3 A reach of water, as in a canal. ◆ Homophone: *load.* [OE *lād* way, journey. Doublet of LOAD.]
lode·star (lōd′stär′) *n.* A guiding star; the polestar: also spelled *loadstar.*
lode·stone (lōd′stōn′) *n.* A variety of magnetite that shows polarity and acts like a magnet when freely suspended: also spelled *loadstone.*
Lo·de·wijk (lō′də·vīk) Dutch form of LOUIS. Also *Ital.* **Lo·do·vi·co** (lō′dō·vē′kō).
lodge (loj) *v.* **lodged, lodg·ing** *v.t.* 1 To furnish with temporary living quarters; house. 2 To rent a room or rooms to; take as a paying guest. 3 To serve as a shelter or dwelling for. 4 To deposit for safekeeping or storage. 5 To place or implant, as by throwing, thrusting, etc.: I *lodged* an arrow in the tree. 6 To place (a complaint, information, etc.) before proper authority. 7 To confer or invest (power, etc.). 8 To beat down (crops): said of rain, storms, etc. —*v.i.* 9 To take temporary shelter or quarters; pass the night. 10 To live in a rented room or rooms. 11 To become fixed in some place or position: The bullet *lodged* in his leg. See synonyms under ABIDE, ACCOMMODATE. —*n.* 1 A small house affording temporary accommodations; a hut. 2 A small dwelling appurtenant to a manor house, park, or the like. 3 The lair of a wild animal, especially of a group of beavers; also, collectively, the beavers themselves. 4 A local subdivision of a secret society, or its meeting place. 5 *U. S.* Among the American Indians, a small hut or tepee of skins, bark, and poles; also, its inhabitants. [<OF *logier* <*loge* <Med. L *lobia, laubia* porch, gallery <OHG *louba* <*loub* foliage; *n.,* doublet of LOBBY, LOGE, LOGGIA]
Lodge (loj), **Henry Cabot,** 1850–1924, U. S. politician and historian. **—Sir Oliver (Joseph),** 1851–1940, English physicist, author, and investigator of psychic phenomena. **—Thomas,** 1558?–1625, English dramatist and novelist.
lodge pole A pole used in building an American Indian lodge.
lodge-pole pine (loj′pōl′) A tall, slender tree *(Pinus contorta latifolia)* of the western United States.
lodg·er (loj′ər) *n.* Something or someone that lodges; especially, one who occupies a rented room or rooms in the house of another.
lodg·ing (loj′ing) *n.* 1 A place of temporary abode. 2 *pl.* A room or rooms hired as a place of residence in the house of another. 3 Accommodation, as a room: to include board and *lodging.*
lodging house A house other than a hotel where lodgings are let.
lodg·ment (loj′mənt) *n.* 1 The act of lodging or the state of being lodged. 2 A foothold gained in some place. 3 Lodgings; accommodation; a lodging house. Also **lodge′ment.**
Lo·di (lō′dē) A city in northern Italy; scene of Napoleon's defeat of the Austrians, 1796.
Łódź (wōōj) A city in central Poland. *Russian* **Lodz** (lôts′y′).
lo'e (lōō) *v.* & *n. Scot.* Love. Also spelled *loo.*
Loeb (lōb, *Ger.* lœp), **Jacques,** 1859–1924, U. S. physiologist born in Germany. **— James,** 1867–1933, U. S. philanthropist and banker, born in Germany; endowed the publication of the Loeb Classical Library.
loess (lō′is, lœs) *n. Geol.* A pale, yellowish clay or loam forming deposits along river valleys, etc., in Asia, Europe, and North America. [<G <*lösen* pour, dissolve]
Loe·wi (lœ′vē), **Otto,** born 1873, U. S. pharmacologist born in Germany.
Lo·fo·ten Islands (lō·fō′tən) A group of islands off the NW coast of Norway; 550 square miles.
loft (lôft, loft) *n.* 1 A low story directly under a roof. 2 A large storeroom. 3 An elevated floor or gallery within a large building, as a church or barn. 4 An incline on the face of a golf club tending to cause elevation in the trajectory of the ball; also, a stroke which lifts the ball high in the air. 5 A place for keeping pigeons; hence, a flock of pigeons. —*v.t.* 1 To provide with a loft. 2 In golf: **a** To give loft to (a club). **b** To strike (a ball) so that it rises or travels in a high arc. —*v.i.* 3 In golf, to strike the ball so that it rises in a high arc. [OE <ON, upper room, air, sky. Akin to LIFT.]
loft·er (lôf′tər, lof′-) *n.* In golf, an iron club used for lofting the ball. Also **lofting iron.**
loft·y (lôf′tē, lof′-) *adj.* **loft·i·er, loft·i·est** Elevated, as in position, character, language, or quality; exalted; haughty; stately. See synonyms under EMINENT, GRAND, HIGH, SUBLIME. **—loft′i·ly** *adj.* **—loft′i·ness** *n.*
log (lôg, log) *n.* 1 A bulky piece of timber cut down and cleared of branches. 2 Figuratively, a dull, stupid person. 3 *Naut.* A device for showing the speed of a vessel, consisting of a triangular board, the **log chip** or **ship,** weighted on one edge and attached to a line, the **log line,** that runs out from a reel, the **log reel,** on shipboard. 4 A record of the daily progress of a vessel and of the events of a voyage. 5 Any record of performance, as of an engine, oil well, aircraft, etc. —*v.* **logged, log·ging** *v.t.* 1 To cut (trees) into logs. 2 To cut down the trees of (a region) for timber. 3 *Naut.* **a** To enter in a logbook. **b** To travel (a specified distance) as shown by a log. **c** To travel at (a specified speed). —*v.i.* 4 To cut down trees and transport logs for sawing into lumber. [ME *logge,* prob. <Scand. Cf. ON *lāg,* Dan. *laag* felled tree.]
log- Var. of LOGO-.
Lo·gan (lō′gən), **Mount** A peak in SW Yukon Territory, Canada, the second highest in North America; 19,850 feet.
lo·ga·ni·a·ceous (lō·gā′nē·ā′shəs) *adj.* Designating or pertaining to a family *(Loganiaceae)* of poisonous herbs, shrubs, and trees, with opposite, entire, stipulate leaves and a cymose inflorescence of regular, perfect, four- or five-parted flowers. [<NL, after James *Logan,* 1674–1751, Irish botanist]
lo·gan·ber·ry (lō′gən·ber′ē) *n. pl.* **·ries** 1 A hybrid plant *(Rubus loganobaccus)* obtained by crossing the red raspberry with the blackberry: cultivated for its edible fruit. 2 The

logaoedic / **loiter**

fruit itself. [after J. H. *Logan* of California, the originator]
log·a·oe·dic (lôg′ə·ē′dik, log′-) *adj.* Prose-poetic; partaking of the nature of prose and poetry: applied to a meter composed of cyclic dactyls and trochees. — *n.* A logaoedic verse. [<LL *logaoedicus* <Gk. *logaoidikos* <*logos* speech + *aoidē* song]
log·a·pha·si·a (lôg′ə·fā′zhē·ə, -zhə, log′-) *n.* *Psychiatry* Inability to communicate ideas in speech. [<Gk. *logos* speech + APHASIA]
log·a·rithm (lôg′ə·rith′əm, log′-) *n.* *Math.* 1 The exponent of the power to which a fixed number, called the base, must be raised in order to produce a given number. For example, in decimal logarithms the base is 10 and the logarithm of 100 is 2 because 10 raised to the second power is 100; the logarithm of 1000 is 3 because 10 raised to the third power is 1000, and so on. 2 In a former and broader sense, one of any series of numbers whose members correspond, each to each, with the natural numbers, but are in arithmetical progression when the latter are in geometrical, so that, if the products of two sets of numbers are equal, the sums of the corresponding logarithms are also equal. [<NL <Gk. *logos* word, ratio + *arithmos* number] — **log′a·rith′mic** or **·mi·cal** *adj.* — **log′a·rith′mi·cal·ly** *adv.*
logarithmic curve *Math.* A curve traced by a point with ordinates increasing arithmetically and abscissas increasing geometrically. It is asymptotic to the negative axis of the dependent variable and passes through the coordinate (1,0) or (0,1) depending on which coordinate is assumed as the variable.
logarithmic spiral *Math.* The polar curve traced by a point moving so that the angle subtended between its radius vector and a tangent to the curve is equal to the modulus; a polar curve traced by a point with a polar angle proportional to the logarithm of the distance from the point to the pole: also called *equiangular spiral, logistic spiral.*

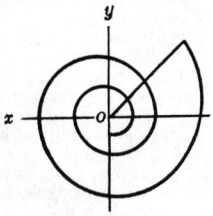

LOGARITHMIC SPIRAL
*o.*Origin.
x.,y. x–axis,*y*–axis.

log·book (lôg′book′, log′-) *n.* The book in which the official record of a ship or aircraft is entered; also, a book containing a similar record of performance of a small military unit. Also **log book.**
log cabin A small, rough house built of logs. Also **log house, log hut.**
log carriage The carriage in a sawing machine that moves the log back and forth before the saw.
loge (lōzh) *n.* A box in a theater; booth; stall. [<F <OF. Doublet of LOBBY, LODGE, *n.,* LOGGIA.]
log·gan (lôg′ən, log′-) *n.* A large boulder so balanced as to rock easily; a rocking stone. Also **lo·gan** (lō′gən), **loggan stone.** [<dial. E *rock,* move to and fro]
log·ger (lôg′ər, log′-) *n.* 1 A person engaged in logging; a lumberjack. 2 A machine used for loading logs on flat cars.
log·ger·head (lôg′ər·hed′, log′-) *n.* 1 A blockhead; dunce. 2 A large marine turtle (*genus Caretta*) of tropical Atlantic waters. 3 The loggerhead shrike. See under SHRIKE. 4 A post on the gunwale of a whaleboat around which the line is turned to retard the motion of a harpooned whale. — **at loggerheads** Engaged in a quarrel; unable to agree. [<dial. E *logger* log tied to a horse's leg to impede movement + HEAD]
log·gi·a (loj′ē·ə, lô′jə; *Ital.* lôd′jä) *n.* A covered gallery or portico having a colonnade on one or more sides, open to the air: compare PORCH, VERANDA. [<Ital. <OF *loge*. Doublet of LOBBY, LODGE, *n.,* LOGE.]
log·ging (lôg′ing, log′-) *n.* The business or occupation of felling timber and transporting logs to a mill or market.
log·i·a (log′ē·ə) *n.* *pl.* of **log·i·on** (log′ē·on) Collections of sayings attributed to a religious leader; especially, **Logia,** the maxims, doctrines, or truths ascribed to Jesus in the four Gospels; also, the agrapha, or collection of sayings ascribed to Jesus, but not found in the Bible. [<Gk., *pl.* of *logos* word]
log·ic (loj′ik) *n.* 1 The normative science which investigates the principles of valid reasoning and correct inference, either from the general to the particular, **deductive logic;** or from the particular to the general, **inductive logic.** Compare FORMAL LOGIC, SYMBOLIC LOGIC. 2 The basic principles of reasoning developed by and applicable to any field of knowledge: the *logic* of science. 3 Effective or convincing force: The *logic* of his argument is unassailable. 4 The system of thought or ideas governing conduct, belief, behavior, etc.: the *logic* of business enterprise. [<F *logique* <L *logica* <Gk. *logikē (technē)* logical (art) <*logikos* of speaking or reason <*logos* word, speech, thought]
log·i·cal (loj′i·kəl) *adj.* 1 Relating to or of the nature of logic. 2 Conformed to the laws of logic; consistent in point of reasoning: a *logical* conclusion. 3 Capable of or characterized by clear reasoning; versed in the principles of logic: a *logical* writer. — **log′i·cal·ly** *adv.* — **log·i·cal′i·ty** (-kal′ə·tē), **log′i·cal·ness** *n.*
-logical *combining form* Of or related to a (specified) science or study: *biological, geological, zoological.* [<-LOG(Y) + -ICAL]
logical positivism A movement in philosophy devoted to unifying the sciences, chiefly by creation of a unified terminology in which the statements of any science are expressible. Also called **logical empiricism.**
lo·gi·cian (lō·jish′ən) *n.* One versed in logic.
lo·gis·tic (lō·jis′tik) *adj.* 1 Of, pertaining to, or skilled in calculation. 2 Of or pertaining to proportion. 3 Sexagesimal. 4 Of or pertaining to logistics. Also **lo·gis′ti·cal.** — *n.* 1 The art of calculation; common practical arithmetic. 2 Sexagesimal arithmetic. [<Med. L *logisticus* <Gk. *logistikos* <*logizesthai* reckon <*logos* word, calculation]
lo·gis·tics (lō·jis′tiks) *n. pl.* (construed as singular) 1 The branch of military science that embraces the details of moving, evacuating, and supplying armies. 2 Logistic. [<F *logistique* <*logis* quarters, lodging <*loger* quarter <OF *logier.* See LODGE.]
logistic spiral Logarithmic spiral.
log jam A mass of logs that have become jammed in their course down a stream.
logo- *combining form* Word; speech: *logomachy.* Also, before vowels, *log-*. [<Gk. *logos* word, speech]
log·o·gram (lôg′ə·gram, log′-) *n.* 1 An abbreviation or other sign representing a word, as $ for dollar. 2 A form of versified word puzzle. — **log′o·gram·mat′ic** *adj.*
log·o·graph (lôg′ə·graf, -gräf, log′-) *n.* 1 A character, or combination of characters, used to represent a word; logogram. 2 A logotype. — **log′o·graph′ic** or **·i·cal** *adj.*
lo·gog·ra·phy (lō·gog′rə·fē) *n.* 1 *Printing* The use of logotypes. 2 The art of reporting speeches in longhand by several reporters, each taking down a few words in succession.
log·o·griph (lôg′ə·grif, log′-) *n.* 1 A word riddle, in which it is required to discover some word by a recombination of the letters or elements of various given words or by guessing and combining other words which (when correctly arranged) form the word to be guessed. 2 Any anagram or puzzle which involves an anagram. [<LOGO- + Gk. *griphos* riddle] — **log′o·griph′ic** *adj.*
log·o·ma·chy (lō·gom′ə·kē) *n. pl.* **·chies** 1 Strife over mere words; verbal contention. 2 Any of various games in which letters are arranged into words. [<Gk. *logomachia* <

LOGGIA

logos word + *machē* a battle] — **lo·gom′a·chist** *n.*
log·on (lôg′on, log′-) *n.* An elementary tone signal, of whatever frequency and intensity, distinguishable as such by the human ear: used especially in the quantitative study and analysis of auditory capacity. [<Gk. *logon,* accusative of *logos* word, utterance]
lo·gop·a·thy (lō·gop′ə·thē) *n.* *Pathol.* Any disorder affecting the speech. [<LOGO- + -PATHY]
Log·os (log′os, lôg′-) *n.* 1 In classical Greek and neo-Platonic philosophy, the cosmic reason giving order, purpose, and intelligibility to the world. 2 The creative Word of God, the second person of the Trinity, incarnate as Jesus Christ, identified with the cosmic reason. *John* i 1-14. [<Gk., word]
lo·go·tech·nics (lō′gō·tek′niks) *n.* The theory, art, and practice of forming new words, especially with reference to specific requirements in science, technology, medicine, etc. [<Gk. *logotechnēs* an artificer of words + -ICS; coined (1954) by R. W. Brown, U.S. geologist] — **lo′go·tech′ni·cal** *adj.*
log·o·thete (lôg′ə·thēt, log′-) *n.* In Byzantine history, any one of several officials, especially, the imperial auditor or accountant. [<Med. L *logotheta* <Gk. *logothetēs* <*logos* word, account + *tithenai* set]
log·o·type (lôg′ə·tīp, log′-) *n.* *Printing* A type bearing a syllable, a word, or words. Compare LIGATURE (def. 4).
log·o·typ·y (lôg′ə·tī′pē, log′-) *n.* The use of logotypes.
log-roll (lôg′rōl′, log′-) *v.t.* To obtain passage of (a bill) by log-rolling. — *v.i.* To engage in log-rolling. Also **log′roll.** [Back formation <LOG-ROLLING]
log-rolling (lôg′rō′ling, log′-) *n.* 1 Handling and removing of logs, as in clearing land. 2 *U.S.* A combining of politicians for mutual assistance on their separate projects. 3 Birling. — **log′-roll′er** *n.*
Lo·gro·ño (lō·grō′nyō) A city on the Ebro in northern Spain.
log scale A table showing the amount of lumber in board measure one inch thick contained in a round log of given length and diameter.
log·way (lôg′wā′, log′-) *n.* 1 An inclined chute or slide up which logs are moved from the water to the sawmill: also called *gangway.* 2 A corduroy road.
log·wood (lôg′wood′, log′-) *n.* 1 A Central American tree (*Haematoxylon campechianum*). 2 Its heavy, reddish wood: used as a dyestuff.
log·work (lôg′wûrk′, log′-) *n.* 1 The service of keeping a ship's log. 2 A structure of logs.
lo·gy (lō′gē) *adj.* **·gi·er, ·gi·est** *U.S. Colloq.* Dull; heavy; lethargic. [? <Du. *log* dull, heavy]
-logy *combining form* 1 The science or study of: *biology, conchology.* 2 A list or compilation of: *anthology, martyrology.* [<Gk. *-logia* <*logos* word, study <*legein* speak]
Lo·hen·grin (lō′ən·grin) In German medieval romances, son of Parsifal and a knight of the Grail: first mentioned in Wolfram von Eschenbach's *Parzival*; subject of an opera by Wagner.
loi·a·sis (loi′ə·sis) See LOASIS.
Loi·kaw (loi′kô′) The capital of Karenni State, Upper Burma.
loin (loin) *n.* 1 The part of the body between the lower rib and hip bone. ♦ Collateral adjective: *lumbar.* 2 *pl.* Figuratively as relating to man, physical or generative power. 3 The forepart of the hindquarters of beef, lamb, veal, etc., with the flank removed. — **to gird up one's loins** To prepare for action. [<OF *loigne, logne* <L *lumbus*]
loin·cloth (loin′klôth′, -kloth′) *n.* A piece or strip of cloth worn about the loins and hips; a waistcloth; breechcloth.
loir (lwär) *n.* A large European dormouse (*Glis glis*). [<F <L *glis, gliris*]
Loire (lwär) The longest river in France, flowing 620 miles NW from the Cévennes to the Bay of Biscay.
loi·ter (loi′tər) *v.i.* 1 To remain or pause idly or aimlessly; loaf. 2 To travel in a leisurely manner and with frequent pauses; linger on the way; dawdle. — *v.t.* 3 To pass (time) idly:

with *away.* [<Du. *leuteren* shake, totter, dawdle] — **loi′ter·er** *n.* — **loi′ter·ing** *adj. & n.*
Lo·ki (lō′kē) In Norse mythology, a god who created disorder and mischief: he appears occasionally as a helper but mostly as an enemy of the gods. [<ON]
loll (lol) *v.i.* 1 To lie or lean in a relaxed or languid manner; lounge. 2 To hang loosely; droop. — *v.t.* 3 To permit to droop or hang, as the tongue. — *n.* 1 The act of lolling. 2 An indolent person. [Cf. MDu. *lollen* sleep] — **loll′er** *n.*
Loll·and (lôl′än) See LAALAND.
Lol·lard (lol′ərd) *n.* The name applied to a follower of John Wyclif, an English religious reformer and precursor of Protestantism. Also **Lol′ler.** [<MDu. *lollaerd,* lit., grumbler, mumbler (of prayers) <*lollen* mumble, doze]
lol·li·pop (lol′ē·pop) *n.* A lump or piece of hard candy attached to the end of a stick. Also **lol′ly·pop.** [Prob. <dial. E *lolly* tongue + POP¹]
lol·lop (lol′əp) *v.i. Colloq.* 1 To lounge; sprawl. 2 To move with leaps or jumps. [<LOLL]
Lo·ma·mi (lō·mä′mē) A tributary of the Congo in SE and central Belgian Congo, flowing 900 miles north.
Lo·mas de Za·mo·ra (lō′mäs thä sä·mō′rä) A city in greater Buenos Aires, Argentina.
Lo·max (lō′maks), **Alan,** born 1915, U.S. folklorist. — **John Avery,** 1872?-1948, U.S. folklorist, father of the preceding.
Lombard (lom′bərd, -bärd, lum′-) *n.* 1 One of a Germanic tribe that established a kingdom, the modern Lombardy, in northern Italy, 568-774. 2 A native or inhabitant of Lombardy. [<OF <Ital. *Lombardo* <LL *Longobardus,* ? <OHG *lang* long + *bart* beard] — **Lombar′dic** *adj.*
Lom·bard (lom′bärd, -bərd, lum′-), **Peter,** 1100?-60, Italian theologian.
Lombard Street 1 A street in London where many banks and financial offices are located. 2 Figuratively, the world of London finance and financiers.
Lom·bar·dy (lom′bər·dē, lum′-) A region of northern Italy; 9,190 square miles; capital, Milan; formerly an independent kingdom. *Italian* **Lom·bar·di·a** (lôm·bär′dē·ä).
Lombardy poplar See under POPLAR.
Lom·blem (lom·blem′) One of the Lesser Sunda Islands east of Alor, the largest of the Solor group; 499 square miles: also *Kawula.*
Lom·bok (lom·bok′) One of the Lesser Sunda Islands east of Bali; 1,825 square miles; capital, Mataram.
Lombok Strait A channel connecting the Indian Ocean with the Java Sea between Bali and Lombok.
Lom·bro·si·an school (lom·brō′zē·ən) A school of criminologists who follow Cesare Lombroso's theory that the criminal is a definite, atavistic type of man.
Lom·bro·so (lôm·brō′sō), **Cesare,** 1836-1909, Italian criminologist.
Lo·mé (lô·mā′) A port on the Gulf of Guinea, capital of Togo.
lo·ment (lō′ment) *n. Bot.* An indehiscent legume with constrictions or transverse articulations between the seeds, as a peanut. Also **lo·men·tum** (lō·men′təm). [<L *lomentum* bean meal <*lotum,* pp. of *lavare* wash; because used as a cosmetic wash in antiquity] — **lo′men·ta′ceous** (-tā′shəs) *adj.*
Lo·mond (lō′mənd), **Loch** The largest lake of Scotland, in Dumbarton and Stirling counties; 23 miles long.
Lon·don (lun′dən), **Jack,** 1876-1916, U.S. author.
Lon·don (lun′dən) 1 A city and administrative county near the mouth of the Thames, England, capital of the United Kingdom and chief city of the British Empire. The **City of London** represents London within its ancient boundaries and is the business and financial center: often called *the City;* 1 square mile. The **County of London** comprises the cities of London and Westminster and 28 metropolitan boroughs; administered by the **London County Council;** 117 square miles. **Greater London,** embracing the City and Metropolitan Police Districts, comprises the County of London, parts of Middlesex, and parts of Surrey, Hertford, Essex, and Kent, an area roughly within a radius of 15 miles from Charing Cross; 693 square miles. *Ancient* **Lon·din·i·um** (lon·din′ē·əm). 2 A city on the Thames in SW Ontario, Canada.

Lon·don·der·ry (lun′dən·der′ē) A county of Ulster, Northern Ireland; 801 square miles; county town, Londonderry: also *Derry.*
Lon·dres (lon′dres) *n.* A cylindrically shaped cigar of medium or large size. [<F, London]
lone (lōn) *adj.* 1 Standing by itself; isolated. 2 Unaccompanied; solitary. 3 Single; unmarried; widowed. 4 Lonesome. 5 Lonely; unfrequented. ◆ Homophone: *loan.* [Aphetic var. of ALONE] — **lone′ness** *n.*
lone hand *n.* In some card games, a person playing without aid from a partner; also, the hand he plays.
lone·ly (lōn′lē) *adj.* **·li·er, ·li·est** 1 Deserted or unfrequented by human beings; sequestered: a *lonely* dell or gorge. 2 Solitary or addicted to solitude; living in seclusion. 3 Sad from lack of sympathy or friendship; lonesome; forlorn. See synonyms under SOLITARY. — **lone′li·ly** *adv.* — **lone′li·ness** *n.*
lone·some (lōn′səm) *adj.* 1 Depressed or sad because of loneliness. 2 Lonely or secluded; so sequestered as to cause uneasiness: a *lonesome* forest. See synonyms under SOLITARY. — **lone′some·ly** *adv.* — **lone′some·ness** *n.*
Lone Star State Nickname of TEXAS.
long¹ (lông, long) *adj.* 1 Having relatively great linear extension; not short: opposed to *short,* and distinguished from *broad* and *wide.* 2 Having relatively great extension in time; lasting. 3 Extended either in space or time to a specified degree: an hour *long,* a foot *long.* 4 Continued in a series to a great extent: a *long* list of grievances. 5 Delayed unexpectedly or unduly; dilatory. 6 Far-reaching; extending far in prospect or into the future; far away in time. 7 Holding for a rise, as stocks. 8 Denoting measure, weight, quantity, etc., in excess of a standard: a *long* five minutes. 9 *Phonet.* a Relatively more prolonged in sound: The sound (ē) is *longer* in *seed* than in *seat.* b Conventionally, indicating the sounds of *a, e, i, o, u* as they are pronounced in *mate, mene, nice, dote, fuse* (in diacritical systems, often written with a macron), as opposed to the "short" sounds of *bat, fed, pit, rot, cup:* in the early Old English period, each of these letters indicated a long or a short vowel sound of similar quality but different duration; this distinction no longer exists and the designation is now an arbitrary one. 10 In classical prosody, requiring relatively more time to pronounce: said of syllables containing a long vowel (*eta, omega,* etc.), a diphthong, or a vowel followed by two consonants or a double consonant. 11 In English prosody, accented. — *n.* 1 The whole extent of a thing; something that is characterized by length: used elliptically. 2 In medieval music, a note equal to four or sometimes six whole notes. 3 A long syllable. 4 *pl.* Those who have purchased securities or commodities and are holding them for an advance in price: opposed to *shorts.* — **the long and the short** The whole; the entire sum and substance. — *adv.* 1 To or at a great extent or period. 2 For a length of time. 3 Through the whole extent or duration. 4 At a point of duration far distant: *long* before or after. — **as** (or **so**) **long as** Under condition that; since. — **before long** Soon. — **for long** For a long time. — **so long** *Colloq.* Good-by. [OE *lang, long*]
Long, meaning "for a long time," may appear as a combining form in hyphemes:

long-accustomed	long-living
long-agitated	long-lost
long-awaited	long-neglected
long-borne	long-past
long-breathed	long-planned
long-buried	long-possessed
long-cherished	long-projected
long-contented	long-protracted
long-continued	long-resounding
long-delayed	long-settled
long-deserted	long-sought
long-desired	long-standing
long-enduring	long-suffered
long-established	long-suffering
long-expected	long-term
long-experienced	long-threatened
long-felt	long-time
long-forgotten	long-wandering
long-hidden	long-wedded
long-kept	long-wished
long-lasting	long-withheld

long² (lông, long) *v.i.* To have a strong or yearning desire; wish earnestly. [OE *langian* grow long]
long³ (lông, long) *v.i. Obs.* 1 To be fitting or proper. 2 To belong. [Aphetic form of OE *gelang* dependent on]
long⁴ (lông, long) *conj. U.S. Dial.* Because; on account with *of.* [Aphetic var. of ALONG]
Long (lông, long), **Crawford Williamson,** 1815-78, U.S. surgeon; pioneer in use of ether anesthesia in surgery. — **Huey,** 1893-1935, U.S. lawyer and politician.
lon·gan (long′gən) *n.* 1 A Chinese and East Indian tree (*Euphoria longan*) of the soapberry family. 2 Its small, edible fruit, resembling the litchi. [<NL *longanum* <Chinese *lungyen* dragon's-eye]
lon·ga·nim·i·ty (long′gə·nim′ə·tē) *n.* Disposition to endure long under offense; patience. [<LL *longanimitas* <*longus* long + *animus* mind]
long·boat (lông′bōt′, long′-) *n. Naut.* The largest boat carried on a sailing vessel.
long·bow (lông′bō′, long′-) *n.* A bow of great length drawn and discharged by hand: distinguished from the *crossbow.* — **to draw, use,** or **pull the longbow** To overstate; exaggerate.

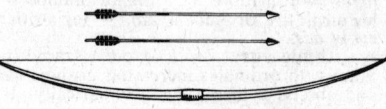

LONGBOW AND ARROWS

long·cloth (lông′klôth′, long′kloth′) *n.* A fine, soft, cotton cloth woven of loosely twisted yarns: used for children's garments.
long-dis·tance (lông′dis′təns, long′-) *adj.* 1 Connecting distant places: a *long-distance* telephone. 2 To or from a distant place; not local: a *long-distance* call.
long distance The telephone exchange or operator dealing with other than local or suburban connections.
long-drawn (lông′drôn′, long′-) *adj.* Protracted; prolonged; tedious. Also **long′-drawn′-out′** (-out′).
long-eared owl (lông′ird′, long′-) See under OWL.
lon·ge·ron (lon′jər·ən, *Fr.* lônzh·rôn′) *n. Aeron.* A main longitudinal member of the body of an airplane. [<F]
lon·gev·i·ty (lon·jev′ə·tē) *n.* 1 Great age, or length of life. 2 The tendency to live long. [<L *longaevitas, -tatis* <*longaevus* long-lived <*longus* long + *aevum* age]
longevity pay Additional pay for long service given to members of the U.S. armed forces. Also called *fogy pay.*
long face An expression of the face indicating exaggerated sadness. — **long-faced** (lông′fāst′, long′-) *adj.*
Long·fel·low (lông′fel·ō, long′-), **Henry Wadsworth,** 1807-82, U.S. poet.
Long·ford (lông′fərd, long′-) A county in NW Leinster province, Ireland; 403 square miles; county town, Longford.
long green *U.S. Slang* Money in the form of bills.
long·hair (lông′hâr′, long′-) *n. U.S. Slang* A person interested in classical rather than popular music. — *adj.* Of or pertaining to classical rather than popular music: *longhair* tastes.
long·hand (lông′hand′, long′-) *n.* Ordinary handwriting with the words spelled in full: distinguished from *shorthand.*
long·head (lông′hed′, long′-) *n.* A dolichocephalic person. — **long′head′ed** *adj.* — **long′head′ed·ly** *adv.* — **long′head′ed·ness** *n.*
long head *Colloq.* Shrewdness; foresight; common sense. — **long′-head′ed** *adj.*
long·horn (lông′hôrn′, long′-) *n.* 1 One of a breed of domestic cattle with long horns: also **Texas longhorn.** 2 *Entomol.* A longicorn beetle. 3 In the western United States, a seasoned inhabitant who knows the ways and cannot be tricked: opposed to *tenderfoot.*

TEXAS LONGHORN

long house Among North American Indians, especially the Iroquois, a council house or community dwelling sometimes 100 to 200 feet long.
longi- combining form Long: longipennate. [<L longus long]
lon·gi·corn (lon'ji·kôrn) n. adj. 1 Having long antennae. 2 Of or pertaining to a family of beetles (Cerambycidae), usually with long, filiform antennae the larvae are woodborers. [<LONGI- + L cornu horn]
lon·gi·lin·e·al (lon'ji·lin'ē·əl) adj. Anthropol. Designating a physical type characterized by length and relative slenderness of body; dolichomorphic: opposed to brevilineal.
long·ing (lông'ing, long'-) n. An eager, strong, or earnest craving. See synonyms under APPETITE, DESIRE. — **long'ing·ly** adv.
Lon·gi·nus (lon·jī'nəs), **Dionysius Cassius**, 213?-273, Greek Platonic philosopher.
lon·gi·pen·nate (lon'ji·pen'āt) adj. Ornithol. Having long wings or feathers, as certain birds. Also **lon'gi·pen'nine** (-pen'ēn, -īn). [<LONGI- + PENNATE]
lon·gi·ros·tral (lon'ji·ros'trəl) adj. Ornithol. Having a long bill, as the ibis and related birds. Also **lon'gi·ros'trate** (-trāt, -trit). [<LONGI- + ROSTRAL]
long·ish (lông'ish, long'-) adj. Rather long.
Long Island An island of SE New York just east of the mouth of the Hudson; 1,723 square miles; separated from Connecticut by **Long Island Sound**, a sheltered arm of the Atlantic 90 miles long.
lon·gi·tude (lon'ji·tōōd, -tyōōd) n. 1 Geog. Distance east or west on the earth's surface, measured by the angle which the meridian through a place makes with some standard meridian, as that of Greenwich or Paris. Longitude may be expressed either in time (**longitude in time**) or in degrees (**longitude in arc**). 2 Astron. The angular distance from the vernal equinox to the intersection with the ecliptic of the perpendicular from a heavenly body: usually termed **celestial longitude**, and distinguished as geocentric when the earth's center is assumed as the central point, and heliocentric when the sun's center is taken as the central point. 3 Math. The angle the radius vector makes with the initial meridian axis in the spherical coordinate system. [<F <L longitudo <longus long]
lon·gi·tu·di·nal (lon'ji·tōō'də·nəl, -tyōō'-) adj. 1 Pertaining to longitude or length. 2 Running lengthwise. 3 Biol. Of, pertaining to, or extending along the cephalocaudal axis of a body or part of a body. — **lon'gi·tu'di·nal·ly** adv.
long johns (jons) Slang Ankle-length, fitted underdrawers of knitted fabric.
long jump In sports, the broad jump.
long-leaf pine (lông'lēf', long'-) The southern yellow or Georgia pine (Pinus palustris), important as a source of turpentine.
long-lived (lông'līvd', -livd', long'-) adj. Having a long life. — **long'-lived'ness** n.
long measure Linear measure.
long moss Spanish moss.
long·neck (lông'nek', long'-) n. The pintail duck.
Lon·go·bar·di (long'gō·bär'dē) n. pl. The Lombards (def. 1). [<LL] — **Lon'go·bar'di·an, Lon'go·bard** adj.
long pig Human flesh, as prepared for a feast by cannibals: so called from Maori and Polynesian term.
long-range (lông'rānj', long'-) adj. 1 Designed to shoot or move over distances: a long-range projectile. 2 Taking account of, or extending over, a long span of future time: long-range plans.
long run An extended series of occurrences. — **in the long run** After the whole course of events; eventually.
long·shanks (lông'shangks', long'-) n. The black-necked stilt (Himantopus mexicanus).
long·shore (lông'shôr', -shōr', long'-) adj. Belonging, living, or working along a shore or waterside. [Aphetic var. of ALONGSHORE]
long·shore·man (lông'shôr'mən, -shōr'-, long'-) n. pl. **-men** (-mən) One who loads and unloads vessels; a stevedore. [<LONGSHORE + MAN]
long shot Colloq. 1 A bet made with little chance of winning and hence at great odds. 2 Something backed at great odds, as a horse. — **not by a long shot** Decidedly not; not at all.
long-sight·ed (lông'sī'tid, long'-) adj. 1 Seeing far or to a great distance; sagacious. 2 Far-sighted; hypermetropic. — **long'-sight'·ed·ness** n.
long·some (lông'səm, long'-) adj. Very long; hence, irksome; tedious. [OE langsum]
Longs Peak (lôngz, longz) A mountain in Rocky Mountain National Park, north central Colorado; 14,255 feet.
long·spur (lông'spûr', long'-) n. A fringilline bird (genera Calcarius and Rhynchophanes) with elongated hind plumes, found in the arctic regions and Great Plains of North America.
long-sta·ple (lông'stā'pəl, long'-) adj. Having a long fiber: said of cotton.
Long·street (lông'strēt, long'-), **James**, 1821-1904, American Confederate general.
long-suf·fer·ing (lông'suf'ər·ing, long'-) adj. Enduring injuries for a long time; patient; forbearing. — n. Patient and forbearing endurance of injuries or offense: also **long'-suf'fer·ance**.
long-tom (lông'tom', long'-) n. A cradle used by miners for washing gold-bearing dirt.
Long Tom 1 A long swivel gun mounted on the decks of old sailing vessels. 2 A long-range coastal gun. 3 Slang Any large cannon having a long range. [<long Tom Turk; 19th century naval slang]
long-wind·ed (lông'win'did, long'-) adj. Continuing for a long time in speaking or writing; hence, tedious; lacking conciseness. — **long'wind'ed·ly** adv. — **long'wind'ed·ness** n.
long·wise (lông'wīz', long'-) adv. Lengthwise. Also **long'ways'** (-wāz').
Long·xu·yen (loung'swē'un) A city in southern Vietnam.
loo¹ (lōō) n. 1 A game of cards resembling euchre, played by several persons with three or five cards apiece. 2 The deposit made in the pool in the game of loo. 3 The fact of failing to take a trick at loo. — v.t. To subject to a forfeit at loo. [Short for lanterloo <F lanturelu, name of the game; orig. a vaudeville refrain]
loo² (lōō) See LO'E.
loo·by (lōō'bē) n. pl. **-bies** Colloq. 1 A lubber; a lout. 2 The ruddy duck. [? Akin to LUBBER]
loof¹ (lōōf) See LUFF.
loof² (lōōf) n. Scot. The palm of the hand.
loo·fah (lōō'fə) n. 1 Any of a genus (Luffa) of Old World, tropical, cucurbitaceous herbs with the male flowers in racemes and the female solitary. 2 The ovate or oblong fruit of this herb, fibrous within, and often used to filter oil and grease from condensed steam, as well as for cleaning and scrubbing. Also **loo'fa**. Also called dishcloth gourd, vegetable sponge, luffa. [<Arabic lūfah]
look (lŏŏk) v.i. 1 To direct the eyes toward something in order to see. 2 To direct or turn one's attention or consideration. 3 To search; make examination or inquiry: to look through a desk. 4 To appear to be; seem: He looks trustworthy. 5 To have a specified direction or view; front: This house looks over the park. 6 To expect: with an infinitive: I didn't look to find you home. — v.t. 7 To turn or direct the eyes upon: He looked her up and down. 8 To express by looks: to look one's hatred. 9 To give the appearance of being (a specified age). 10 To influence by looks: to look someone into silence. — **to look after** To take care of. — **to look daggers (at)** To scowl; glare (at). — **to look down on** To regard condescendingly or contemptuously. — **to look for** 1 To search for. 2 To expect. — **to look forward to** To anticipate pleasurably. — **to look in (or in on)** To make a short visit to. — **to look into** To examine; make inquiry. — **to look like** 1 To have the appearance of being; resemble. 2 To indicate the probability of: It looks like rain. — **to look on** 1 To be a spectator. 2 To consider; regard. — **to look out** To be on the watch; take care. — **to look over** To examine; scrutinize. — **to look to** 1 To attend to. 2 To turn to, as for help, advice, etc. — **to look up** 1 To search for and find, as in a file, dictionary, etc. 2 Colloq. To discover the whereabouts of and make a visit to. 3 Colloq. To improve; become better. — **to look up to** To have respect for. — n. 1 The act of looking or seeing with voluntary attention: I will take a look at it. 2 pl. The appearance of the face or figure; cast of countenance. 3 Aspect; expression. 4 Appearance in general, either to the eye or understanding: I do not like the look of the thing. See synonyms under AIR, MANNER. [OE locian]
Synonyms (verb): behold, contemplate, descry, discern, gaze, glance, inspect, regard, scan, see, stare, survey, view, watch. To see is simply to become conscious of an object of vision; to look is to make a conscious and direct endeavor to see. To behold is to fix the sight and the mind with distinctness and consideration upon something that has come to be clearly before the eyes. We may look without seeing, as in darkness, and we may see without looking, as in the case of a flash of lightning. To gaze is to look intently, long, and steadily. To glance is to look casually or momentarily. To stare is to look with a fixed intensity. To scan is to look at minutely, to note every visible feature. To inspect is to go below the surface, study item by item. View and survey are comprehensive, survey expressing the greater exactness of measurement. Watch brings in the element of time; we watch for a movement or change.
look·er (lŏŏk'ər) n. 1 One who looks or watches. 2 U.S. Slang A handsome or good-looking person.
look·er-on (lŏŏk'ər·on') n. pl. **look·ers-on** A spectator; onlooker.
look-in (lŏŏk'in') n. Slang 1 A hasty, casual glance or visit. 2 U.S. Colloq. A chance: If he fights the champion, he hasn't a look-in.
look·ing-glass (lŏŏk'ing·glas', -gläs') n. A mirror.
look-out (lŏŏk'out') n. 1 The act of watching or looking out; especially, careful or alert watchfulness. 2 A post of observation; also, the person or the group on watch in or at such a post. 3 One engaged in the U.S. Forest Service to detect fires. 4 One assigned to watch for enemy aircraft, tanks, or troop movements. 5 Prospect; outlook; chances of good or bad to come: a good look-out ahead. 6 Colloq. Concern; care: It's your own look-out.
Look·out Mountain (lŏŏk'out') A ridge in SE Tennessee, near Chattanooga; scene of a Civil War battle, 1863.
look-see (lŏŏk'sē') n. Slang An inspection or survey: Have a look-see. [Orig. Pidgin English]
loom¹ (lōōm) v.i. 1 To appear or come into view indistinctly, as from below the horizon or through a mist, especially so as to seem large or ominous. 2 To appear to the mind as threatening or portentous: Great difficulties loom ahead. 3 To shine. — n. A gradual, vague appearance of something, as of a ship in the fog. [Origin uncertain. Cf. Sw. loma move slowly (toward).]
loom² (lōōm) n. 1 A machine in which yarn or thread is woven into a fabric, by the crossing of threads called chain or warp, running lengthwise, with others called weft, woof, or filling. See illustration on page following. 2 The art or technique of working with a loom; the occupation of a weaver; weaving. 3 Naut. a The shaft of an oar, as distinguished from the blade. b That part of an oar between the rowlock and the handle. [OE gelōma tool]
loom³ (lōōm) n. 1 A guillemot. 2 A loon (def. 1). [<ON lomr]
loom·ing (lōō'ming) n. A mirage that elevates and elongates a figure, especially when viewed across the water.
loon¹ (lōōn) n. 1 A diving, fish-eating waterfowl (genus Gavia), with short tail feathers and webbed feet. 2 A guillemot. — **common loon** The great northern diver (Gavia immer). [See LOOM³]

LOON (Length 31 to 36 inches)

loon² (lōōn) n. 1 A worthless, demented, or

stupid person; a lout; dolt. 2 A rogue; scamp. 3 *Obs.* A menial. Also *Obs.* & *Scot.* **loun, lown.** [Cf. Du. *loen* stupid fellow G *lümmel* lout]

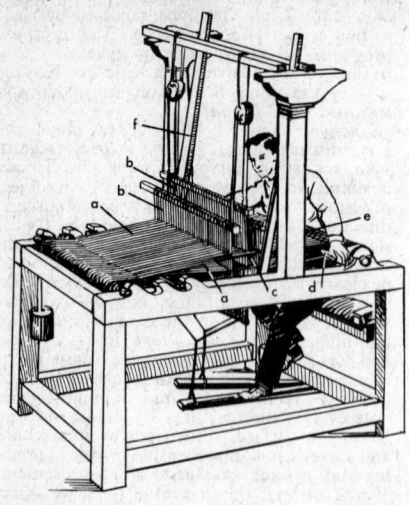

HAND LOOM
The two sets of warp threads (*a, a*), alternately raised and lowered by the heddle (*b, b*), form between them a tunnel of threads called the shed (*c*). Through this shed the shuttle (*d*) is passed, carrying across the weft thread (*e*), which is beaten against the finished fabric by the movable comblike frame or reed (*f*). When the heddle is shifted, the two sets of warp reverse position, binding the weft into the fabric and opening another shed.

loon·y (lōō′nē) *Slang. adj.* 1 Lunatic or demented. 2 Foolish; somewhat crazy; erratic; silly. — *n. pl.* **loon·ies** A demented or daft person. Also spelled **luny.** [<LUNATIC]
loop¹ (lōōp) *n.* 1 A fold or doubling, as of a string or rope, so as to form an eye or a bend through which something may be passed; a noose; bight; hence, a curve or bend of any kind. 2 A stitch used in crocheting and knitting. 3 *Electr.* A complete magnetic or electrical circuit. 4 A curve in a railroad, carried completely around to reach a different level; also, a branch from the main line, returning to it after making a detour. 5 One of the basic patterns by which fingerprints are classified and identified, consisting of one or more ridges whose terminal points are on or toward the same side of the design. 6 *Aeron.* A complete, vertical, circular turn made by an airplane in flight. 7 *Physics* The part of a vibrating string, column of air, or standing wave system which is between two nodes; an antinode. — **to loop the loop** To make a vertical, circular turn, as an airplane in flight. — *v.t.* 1 To form a loop or loops in or of. 2 To fasten, connect, or encircle by means of a loop or loops. 3 *Aeron.* To fly (an aircraft) in a loop or loops. — *v.i.* 1 To make a loop or loops. 5 To move by forming loops, as a measuring worm. [Cf. Irish *lub* loop]
loop² (lōōp) *n.* Iron of pasty consistency ready for rolling. [<F *loupe*]
loop³ (lōōp) *n.* Any small window or aperture; a loophole. [Cf. MDu. *lupen* lie in wait, peer]
Loop, The The principal business, financial, hotel, and theatrical district of downtown Chicago, near the lake front.
loop buttonhole A loop of crochet, cord, or fabric serving as a buttonhole.
loop·er (lōō′pər) *n.* 1 A bodkinlike instrument for making loops. 2 A measuring worm.
loop·hole (lōōp′hōl′) *n.* 1 A narrow opening through which small arms are fired. 2 A means of escape, or place of observation. — *v.t.* **·holed, ·hol·ing** To furnish with loopholes. [<LOOP² + HOLE]
loop stitch A stitch composed of a series of loops, the last one finished with a knot; also called *picot stitch.*
loop·y (lōō′pē) *adj.* 1 Having or full of loops. 2 *Slang* Crazy. 3 *Scot.* Shrewd; sly.
loose (lōōs) *adj.* **loos·er, loos·est** 1 Not fastened or confined; not bound or attached; unbound or untied; freed from normal bonds or restraint: *loose* tresses; to be *loose* from old habits. 2 Lax in power, character, qual-
ity, principle, or conduct; careless; slovenly; slack; relaxed; wanton; dissolute: *loose* bond; *loose* conduct. 3 Not precise or exact; vague; indefinite; rambling; unconnected: *loose* reasoning; a *loose* style. 4 Not close, compact, dense, tight, or crowded; lacking union of parts; slackly joined or tied; not compact in frame: a *loose* knot or bond; a *loose* array; a fabric of *loose* texture; a man of *loose* build. 5 Not tight; open: said of a cough when expectoration is without difficulty, or of an abnormal laxity of the bowels. 6 Designating a stable or stall in which the animals are kept untied. See synonyms under IMMORAL, VAGUE, VULGAR, WANTON. — **on the loose** At liberty; at large; unconfined. — *adv.* In a loose manner: to play fast and *loose.* — *v.* **loosed, loos·ing** *v.t.* 1 To set free, as from bondage, penalty, etc. 2 To untie or undo, as a knot or rope. 3 To make less tight; loosen; slacken. 4 *Naut.* To cast off; release, as a boat from its moorings. 5 To let fly; shoot or discharge, as arrows. — *v.i.* 6 To become loose. 7 To loose something. [<ON *lauss*] — **loose′ly** *adv.* — **loose′ness** *n.*
loose-joint·ed (lōōs′join′tid) *adj.* 1 Having joints not tightly articulated; hence, able to move with more than ordinary freedom. 2 Ungainly.
loose-leaf (lōōs′lēf′) *adj.* Designed for the easy insertion and removal of pages: a *loose-leaf* notebook.
loos·en (lōō′sən) *v.t.* 1 To untie or undo, as bonds or chains. 2 To set free; release. 3 To make less tight, firm, or compact: to *loosen* the stones of a wall. 4 To effect laxity in the action of (the bowels). 5 To relax the strictness of, as discipline. — *v.i.* 6 To become loose or looser. See synonyms under RELAX. — **loos′en·er** *n.*
loose sentence A sentence that is grammatically complete before its end: All compound sentences are *loose sentences.*
loose·strife (lōōs′strīf′) *n.* 1 Any one of various plants, mostly with four-cornered branches and regular or irregular flowers, as the **common loosestrife** (*Lysimachia vulgaris*) of the primrose family or the **purple loosestrife** (*Lythrum salicaria*) of the family Lythraceae. 2 Any plant of the loosestrife family. [<LOOSE + STRIFE; direct trans. of L *lysimachia* <Gk. *lysimachion* <*lyein* loose + *mache* battle]
loo·some (lōō′səm) *adj. Scot.* Lovable.
loot¹ (lōōt) *v.t.* 1 To plunder, as a conquered city; pillage. 2 To carry off as plunder. — *v.i.* 3 To engage in plundering. — *n.* 1 Booty taken from a sacked city by a victorious army; plunder. 2 Anything unlawfully taken, as by one in an official position. [<Hind. *lūt* <Skt. *lunt*] — **loot′er** *n.*
loot² (lōōt) *Scot.* Past tense of LET¹.
looves (lōōvz) Plural of LOOF.
lop¹ (lop) *v.t.* **lopped, lop·ping** 1 To cut or trim the branches, twigs, etc., from, as a tree. 2 To cut off, as branches or twigs. — *n.* A cutting from a tree; specifically, the trimmings or small twigs and branches not measured as timber; fagot. [Origin unknown] — **lop′per** *n.*
lop² (lop) *v.* **lopped, lop·ping** *v.i.* 1 To droop or hang down. 2 To hang or move about in an awkward or slouching manner. — *v.t.* 3 To permit to droop or hang down. — *adj.* Drooping. [Origin unknown]
Lo·pa·du·sa (lō′pə·dōō′sə, -dyōō′-) An ancient name for LAMPEDUSA.
lope (lōp) *v.t.* & *v.i.* **loped, lop·ing** To run or cause to run with a steady, swinging stride. — *n.* A slow, easy gallop. [<ON *hlaupa* leap, run] — **lop′er** *n.*
lop-eared (lop′ird′) *adj.* Having drooping or pendulous ears.
Lo·pe de Ve·ga (lō′pā thā vā′gä) See VEGA, LOPE DE.
lopho- combining form Crest; crested. Also, before vowels, **loph-**. [<Gk. *lophos* crest]
lo·pho·branch (lō′fō·brangk, lof′ō-) *adj.* Of or pertaining to a division of teleost fishes (*Lophobranchii*), especially an order with imperfect branchial arches and tuftlike gill elements, including pipefishes and sea horses. — *n.* One of the *Lophobranchii.* [<LOPHO- + Gk. *branchion* gill] — **lo′pho·bran′chi·ate** (-kē·it, -āt) *adj.* & *n.*
lo·phot·ri·chous (lō·fot′ri·kəs) *adj. Zool.* Having a tuft of cilia at one pole of the cell;
characteristic of some micro-organisms. Also **lo·pho·trich·ic** (lō′fō·trik′ik, lof′ō-). [<LOPHO- + Gk. *thrix, trichos* hair]
lop·per (lop′ər) *v.t.* & *v.i. Scot.* To curdle: also spelled *lapper.*
lop·py (lop′ē) *adj.* **·pi·er, ·pi·est** 1 Limp; pendulous; hanging down. 2 Choppy: said of water.
lop-sid·ed (lop′sī′did) *adj.* Heavy or hanging down on one side; lacking in symmetry. — **lop′-sid′ed·ly** *adv.* — **lop′-sid′ed·ness** *n.*
lo·qua·cious (lō·kwā′shəs) *adj.* Given to continual talking; chattering. See synonyms under GARRULOUS. [<L *loquax, loquacis* <*loqui* talk] — **lo·qua′cious·ly** *adv.* — **lo·qua′cious·ness, lo·quac·i·ty** (lō·kwas′ə·tē) *n.*
lo·quat (lō′kwot, -kwat) *n.* 1 A low-growing, pomaceous tree (*Eriobotrya japonica*), cultivated for its fruit, a small, yellow pome. 2 This fruit. [Cantonese alter. of Chinese *lu-chü* rush orange]
lo·qui·tur (lok′wi·tər) *Latin* He or she speaks.
lo·ran (lôr′an, lō′ran) *n.* A navigation system by which a ship or an aircraft may accurately determine its position by noting the time intervals between radio pulse signals transmitted from a network of ground stations and recorded on an oscilloscope. [<LO(NG) RA(NGE) N(AVIGATION)]
Lor·ca (lôr′kä) A city in Murcia, SE Spain.
Lor·ca (lôr′kä), **Federico García** See García Lorca, Federico.
lord (lôrd) *n.* 1 One possessing supreme power and authority; a ruler. 2 A title of respect, formerly given to any superior, as by a wife to a husband: still sometimes humorously so used. 3 In feudal law, the owner of a manor under grant from the crown; a landlord. 4 A title of honor or nobility in Great Britain, given generally to men noble by birth or ennobled by patent. This includes **lords spiritual** (archbishops and bishops), who are members of the House of Lords, and also **lords temporal**: marquises, earls, viscounts, and barons. The formal titles are as follows: *Baron X*, the *Marquis* of X, the *Earl* of X, *Viscount* X; informally all are addressed *Lord* X. The given name is mentioned only to distinguish holders of the same title at different periods. Where the names are homonymous, the locality of the peerage is stated, as *Viscount Grey of Fallodon*, who was spoken of as *Lord Grey*. The title is given by courtesy to the eldest sons of dukes, marquises, and earls, who each take by courtesy also an inferior title held by the father, frequently the second title, and to the younger sons of dukes and marquises, prefixed to their Christian name and surname; in these cases the Christian names must always be mentioned, coming after "Lord," to distinguish among brothers: *Lord* Robert Cecil. It is also a title of office, such as the *Lord* Lieutenant, the *Lord* Chancellor, *Lord* Privy Seal, *Lords* of the Treasury, Admiralty, Bedchamber. 5 In astrology, a controlling planet. — *v.t.* To invest with the title of lord. — **to lord it (over)** To act in a domineering or arrogant manner. [OE *hlāford, hlāfweard*, lit., bread keeper <*hlāf* loaf + *weard* keeper, ward] — **lord′less** *adj.* — **lord′like** *adj.*
Lord (lôrd) 1 God. 2 Jesus Christ. Also **Our Lord.**
Lord Advocate The principal public prosecutor in Scotland.
lord-and-la·dy (lôrd′ənd·lā′dē) *n. pl.* **lord-and-la·dies** The harlequin duck.
Lord Howe Island (hou) An Australian dependency NE of Sydney; 5 square miles.
lord·ing (lôr′ding) *n.* 1 A lordling. 2 *Obs.* Lord; master.
lord lieutenant 1 The chief executive officer of a county in England. 2 Until 1922, the English viceroy in Ireland.
lord·ling (lôrd′ling) *n.* A little lord; petty chieftain: generally used contemptuously.
lord·ly (lôrd′lē) *adj.* **·li·er, ·li·est** 1 Of, pertaining to, or like a lord; becoming a lord; lofty; noble; insisting on compliance: a *lordly* presence. 2 Characterized by undue loftiness; insolent: a *lordly* air or demeanor. See synonyms under IMPERIOUS. — *adv.* In a lordly manner. — **lord′li·ness** *n.*
Lord of Hosts Jehovah; God.
Lord of Misrule Formerly, a person chosen to preside over the revels, festivities, and

sports in English royal or aristocratic households.

lor·do·sis (lôr·dō′sis) *n. Pathol.* Inward curvature of a bone; specifically, curvature of the spine with the convexity forward. Also **lor·do′ma** (-mə). [<NL <Gk. *lordōsis* < *lordos* bent backward] — **lor·dot′ic** (-dot′ik) *adj.*

lords-and-la·dies (lôrdz′ənd·lā′dēz) *n.* **1** The cuckoo pint or wakerobin. **2** The jack-in-the-pulpit.

Lord's Day Sunday; the Sabbath. [Trans. of L *dies dominica*]

lord·ship (lôrd′ship) *n.* **1** The state or quality of a lord; hence, the title by which noblemen (excluding dukes), bishops, and judges in England are addressed or spoken of, preceded by *your* or *his.* **2** The jurisdiction of a lord; seigniory; domain; manor. **3** The dominion, power, or authority of a lord; hence, sovereignty in general; supremacy. [OE *hlafordscipe.* See LORD.]

Lord's Prayer The prayer taught by Jesus to his disciples. *Matt.* vi 9–13.

Lord's Supper *Eccl.* **1** The Eucharist; the Holy Communion. **2** The Last Supper.

Lord's Table The altar or table on which the elements of the Eucharist are laid during and after consecration.

lore[1] (lôr, lōr) *n.* **1** The body of traditional, popular, often anecdotal knowledge about a particular subject. **2** Learning or erudition. **3** *Archaic* Any special instruction; also, a lesson or the act of teaching. See synonyms under LEARNING, WISDOM. [OE *lar*]

lore[2] (lôr, lōr) *n.* **1** *Ornithol.* The side of a bird's head between the eye and the beak. **2** *Zool.* A corresponding part in fishes and reptiles. [<L *lorum* strap, thong]

Lo·re·lei (lôr′ə·lī, *Ger.* lō′rə·lī) In German folklore, a siren on a rock in the Rhine, who lured boatmen to shipwreck by her singing: also called *Lurlei.* [<G]

Lo·ren·gau (lō′rən·gou) A port on Manus island, chief city of the Admiralty Islands.

Lo·rentz (lō′rents), **Hendrik Antoon,** 1853–1928, Dutch physicist.

Lo·renz (lō·rents′) Danish and German form of LAWRENCE. Also **Lo·ren·zo** (lə·ren′zō; *Ital.* lō·ren′tsō, *Sp.* lō·ren′thō).

Lo·renz (lō′rents), **Adolf,** 1854–1946, Austrian orthopedic surgeon.

lor·gnette (lôr·nyet′) *n.* **1** A pair of eyeglasses with an ornamental handle into which they may be folded when not in use. **2** A long-handled opera glass. [<F < *lorgner* spy, peer <OF *lorgne* squinting]

lor·gnon (lôr·nyôṅ′) *n. French* **1** A monocle. **2** A lorgnette.

lo·ri·ca (lō·rī′kə) *n. pl.* **·cae** (-kē) **1** An ancient Roman cuirass or corselet. **2** *Zool.* A protective covering or shell, as in infusorians or rotifers. [<L < *lorum* thong]

lor·i·cate (lôr′ə·kāt, lor′-) *v.t.* **·cat·ed, ·cat·ing** To cover with a protective coating. — *adj.* Covered with a lorica or shell. [<L *loricatus,* pp. of *loricare* clothe in mail, harness < *lorica* corselet] — **lor′i·ca′tion** *n.*

Lo·rient (lô·ryäṅ′) A port of NW France on the Bay of Biscay.

lor·i·keet (lôr′ə·kēt, lor′-) *n.* Any of certain small Polynesian parrots resembling the lory. [<LORY + (PARA)KEET]

Lo·rin·da (lô·rin′də, lə-) See LAURA.

lo·ris (lôr′is, lō′ris) *n.* A small, arboreal, nocturnal, Asian lemur (genera *Loris* and *Nycticebus*), having the index finger small. *Loris gracilis* is the **slender loris** of southern India and Ceylon; *Nycticebus tardigradus* is the slow lemur or **East Indian loris.** [<F <Flemish *lorrias* lazy, the sloth]

lorn (lôrn) *adj.* **1** *Archaic* Without kindred or friends; forlorn. **2** *Obs.* Lost. [OE *loren,* pp. of *lēosan* lose. Akin to FORLORN.]

Lorne (lôrn), **Firth of** An arm of the Atlantic separating Mull island from the Scottish mainland.

Lor·rain (lô·raṅ′), **Claude** Pseudonym of *Claude Gelée,* 1600–82, French painter.

Lor·raine (lô·rān′, lō-, lə-; *Fr.* lô·ren′) A region and former province of eastern France; after the Franco-Prussian War part of Alsace-Lorraine; restored to France, 1919: German *Lothringen.* See ALSACE-LORRAINE.

lor·ry (lôr′ē, lor′ē) *n. pl.* **·ries** **1** A low, four-wheeled, platform wagon. **2** *Brit.* A similar motor vehicle for carrying heavy loads; a truck. [Prob. dial. E *lurry* pull, tug]

lo·ry (lō′rē, lôr′ē) *n. pl.* **·ries** Any of certain parrots of Australasia (genera *Lorius, Apnosmictus,* and others) with brilliant, scarlet plumage, long bill, and tongue ending in a form of brush. [<Malay *lūrī*]

Los An·ge·les (lôs an′jə·lēz, ang′gə·ləs, los) A port in southern California, the fourth largest city in the United States: also *Colloq. L.A.* A native of Los Angeles is known as an *Angeleno.*

Lo·schmidt number (lō′shmit) *Physics* **1** The number of molecules in 1 cubic centimeter of an ideal gas at 0° C. and normal atmospheric pressure, equal to 2.687 x 10^{19}. **2** The Avogadro number. [after Joseph *Loschmidt,* 1821–95, Austrian physicist]

lose (lo͞oz) *v.* **lost, los·ing** *v.t.* **1** To part with, as by accident or negligence, and be unable to find; mislay. **2** To fail to keep, control, or maintain: to *lose* one's footing; to *lose* one's mind. **3** To be deprived of; suffer the loss of, as by accident, death, removal, etc.: to *lose* a leg. **4** To fail to gain or win: to *lose* a prize; to *lose* a battle. **5** To fail to utilize or take advantage of; miss: to *lose* a chance. **6** To fail to see or hear; miss: I *lost* not a word of the speech. **7** To fail to keep in sight, memory, etc.: We *lost* him in the crowd. **8** To cease to have: to *lose* one's sense of duty. **9** To squander; waste, as time. **10** To wander from so as to be unable to find: to *lose* the path. **11** To outdistance or elude, as runners or pursuers. **12** To cause the loss of: His rashness *lost* him his opportunity. **13** To bring to destruction or death; ruin: usually in the passive: All hands were *lost.* — *v.i.* **14** To suffer loss. **15** To be defeated. — **to lose heart** To become discouraged. — **to lose oneself** **1** To lose one's way, as in a wood. **2** To disappear or hide: The thief *lost himself* in the crowd. **3** To become engrossed. **4** To become confused or bewildered. — **to lose one's heart (to)** To fall in love (with). — **to lose out** *U.S. Colloq.* To fail or be defeated. — **to lose sight of** **1** To fail to keep in sight. **2** To take no notice of; ignore or forget. [Fusion of OE *losian* be lost and *lēosan* lose] — **los′a·ble** *adj.*

lo·sel (lō′zəl, lo͞o′zəl, loz′əl) *adj.* Inclined to idleness and waste. — *n.* A worthless fellow. [OE *losen,* pp. of *losian* be lost]

los·er (lo͞o′zər) *n.* One who loses or fails to win in any transaction; a defeated contestant.

losh (losh) *interj. Scot.* An exclamation of surprise or deprecation. [Euphemistic alter. of LORD; cf. GOSH]

los·ing (lo͞o′zing) *n.* **1** The act or fact of letting go, missing, lacking, or being deprived. **2** *pl.* Money lost, especially in gambling. — *adj.* **1** That incurs loss: a *losing* business. **2** Not winning; defeated: the *losing* team.

loss (lôs, los) *n.* **1** The act or state of losing; failure to keep or win; privation: *loss* of fortune or friends. **2** *Usually pl.* That which is lost, or its amount: His *losses* were great; The *losses* of the army were severe. **3** The state of being lost, or of having suffered destruction: the *loss* of a ship at sea. **4** Useless application; futile expenditure; waste. **5** Injury or diminution of value within the limits provided in an insurance policy, or the sum payable on that account. **6** *Physics* That part of electrical or mechanical energy which is expended in overcoming friction, etc., and from which no productive work is obtained. — **at a loss** **1** At so low a price as to result in a loss. **2** In confusion or doubt; perplexed. — **dead loss** An irremediable loss; one without hope of salvage, insurance, or any mitigation. — **to bear a loss** **1** To sustain a loss bravely. **2** To make good a loss. [OE *los*]

Synonyms: damage, defeat, deprivation, destruction, detriment, disadvantage, failure, forfeiture, injury, misfortune, privation, waste. See INJURY.

lost (lôst, lost) *adj.* **1** Not to be found or recovered; parted with; missing; left, as by accident, in a forgotten place: *lost* goods or friends. **2** Not won, gained, used, or enjoyed; missed; wasted: *lost* opportunity. **3** Ruined physically, morally, or spiritually; damned; destroyed. **4** Having wandered from the way; also, abstracted: *lost* in thought; bewildered; perplexed. **5** No longer known or used: a *lost* art. — **to be lost upon** (or **on**) To have no effect upon (a person).

lost cause Any cause that cannot possibly succeed; specifically, in American history, the cause of the Confederate States.

lost motion Slackness in a mechanical connection resulting in an appreciable difference between the travel of the driving and the driven elements.

lost tribes Those members of the ten northern, or Israelitish, Hebrew tribes that were taken into Assyrian captivity (II *Kings* xvii 6), believed never to have returned.

lot (lot) *n.* **1** Anything, as a die or piece of paper, used in determining something by chance; also, the fact or process of deciding something in such manner. **2** The share that comes to one as the result of drawing lots. **3** The part in life that comes to one without his planning; chance; fate. **4** A collection or parcel of things separated from others: The auctioneer sold the goods in ten *lots.* **5** A parcel or quantity of land, as surveyed and apportioned for sale or other special purpose: a city *lot,* a wood *lot.* **6** *Colloq.* A kind of person: He is a bad *lot.* **7** *Often pl. Colloq.* A great quantity or amount; a number of things, collectively: a *lot* of money, *lots* of trouble. ◆ *Lot* or *lots* is construed as a singular if attributed to a singular word, plural if to a plural word: A *lot* of money was hidden, but, A *lot* of diamonds were hidden. **8** A proportion of taxes allotted to one. **9** A motion-picture studio and the space it uses. See synonyms under FLOCK. — *adv.* Very much: a *lot* worse. — *v.* **lot·ted, lot·ting** *v.t.* **1** To divide, as land, into lots. **2** To apportion by lots; allot. — *v.i.* **3** To cast lots. [OE *hlot*]

Lot (lot) A nephew of Abraham, who with his wife and daughters escaped the destruction of Sodom. His wife, disobeying a warning, looked back at the city and became a pillar of salt. *Gen.* xi 27, xix.

Lot (lôt) A river in south central France, flowing 300 miles west from the Cévennes to the Garonne: ancient *Oltis.*

lo·tah (lō′tə) *n. Anglo-Indian* A small, round pot, usually of brass or copper, used in India for drinking and ablution. Also **lo′ta.**

lote (lōt) *n.* The lotus. [<LOTUS]

loth (lōth) *adj.* Loath.

Lo·thair (lō·thâr′, *Fr.* lō·târ′) A masculine personal name. Also *Ger.* **Lo·thar** (lō·tär′), *Ital.* **Lo·tha·rio** (lō·tä′ryō). [<Gmc., famous warrior]

— **Lothair I,** 795–855, Holy Roman Emperor 840–55.

— **Lothair the Saxon,** 1060?–1137, Holy Roman Emperor 1133–37.

Lo·thar·i·o (lō·thâr′ē·ō) In Nicholas Rowe's *The Fair Penitent,* a young Genoese nobleman and gay rake; hence, a seducer.

Lo·thi·ans (lō′thē·ənz), **The** A division of SE Scotland including East Lothian, Midlothian, and West Lothian.

Loth·ring·en (lōt′ring·ən) The German name for LORRAINE.

Lo·ti (lō·tē′), **Pierre** Pseudonym of *Louis Marie Julien Viaud,* 1850–1923, French novelist.

lo·tion (lō′shən) *n.* **1** A medicated liquid for the skin, eyes, or any diseased or bruised part, for cleansing or healing. **2** *Archaic* A bathing or washing. [<L *lotio, -onis* washing < *lavare* wash]

Lo·toph·a·gi (lō·tof′ə·jī) *n. pl.* In the *Odyssey,* a people living on the northern coast of Africa who lived a life of indolence and forgetfulness induced by eating the fruit of the lotus: some of Odysseus' followers ate with them, and forgot their friends, homes, and native land. [<Gk. *lōtophagoi* < *lōtos* lotus + *phagein* eat]

Lot·ta (lot′ə), **Lot·tie** (lot′ē), **Lot·ty** Diminutives of CHARLOTTE.

lot·ter·y (lot′ər·ē) *n. pl.* **·ter·ies** A distribution of, or scheme for distributing, prizes as determined by chance or lot, especially where such chances are allotted by sale of

lotto

tickets; hence, any chance disposition of any matter. [<Ital. *lotteria* < *lotto* lottery, lot <F *lot* <Gmc.]
lot·to (lot′ō) *n.* A game of chance played with cards and disks: also called *keno.* Also **lo·to** (lō′tō). [<Ital. See LOTTERY.]
Lot·to (lôt′tō), **Lorenzo**, 1480?-1556, Venetian painter.
lo·tus (lō′təs) *n.* 1 One of various Old World plants of the waterlily family, noted for their large floating leaves and showy flowers; especially, the **white** and **blue lotus** of the Nile (respectively *Nymphaea lotus,* and *N. caerulea*); also, the **sacred lotus** of India (*Nelumbium nelumbo*) with fragrant pink or rose flowers. 2 The lotus tree. 3 Any of a genus (*Lotus*) of herbs or shrubs of the bean family. 4 *Archit.* A representation or conventionalization of the lotus flower, bud, or leaves. Also **lo′tos**. [<L <Gk. *lōtos*]

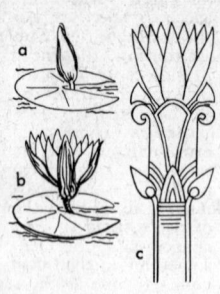

LOTUS
a. Bud and leaf.
b. Blossom and leaf.
c. Stylized lotus in Egyptian architecture and ornament.

lo·tus-eat·er (lō′təs·ē′tər) *n.* In the *Odyssey,* one who lives in irresponsible enjoyment from eating the fruit of the lotus tree; hence, a voluptuary. See LOTOPHAGI.
lotus tree The common jujube.
Loua·la·ba (lwä·lä·bä′) The French spelling of LUALABA.
loud (loud) *adj.* 1 Striking the auditory nerves with great force; having great volume or intensity of sound; noisy. 2 Making a great noise. 3 Pressing or urgent; clamorous: a *loud* demand. 4 Conspicuous or ostentatious without taste or refinement; vulgarly showy; flashy. — *adv.* With loudness; loudly. [OE *hlūd*] — **loud′ly** *adv.* — **loud′ness** *n.*
loud·en (loud′n) *v.t.* & *v.i.* To make or become louder.
loud-mouthed (loud′mouthd′, -moutht′) *adj.* Possessed of a loud voice; offensively clamorous or talkative.
loud·speak·er (loud′spē′kər) *n.* An electromagnetic device for amplifying the sounds transmitted by radio, public-address system, or the like.
lough (lokh) *n. Irish* 1 A loch; lake or pool. 2 A partially landlocked arm of the sea.
lou·is (lōō′ē) *n.* 1 A French gold coin worth twenty francs. 2 An old French gold coin worth about twenty-three francs, used through the reigns of Louis XIII–Louis XVI. Also *louis d'or.* [<F, after *Louis* XIII]
Lou·is (lōō′is, *Fr.* lōō·ē′) A masculine personal name. Also spelled *Lewis.* [<Gmc., famed warrior]
— **Louis I,** 778–840, Holy Roman Emperor 814–840; son and successor of Charlemagne: called "le Débonnaire."
— **Louis VIII,** 1187–1226, king of France 1223–26.
— **Louis IX,** 1214–70; king of France 1226–70; defeated and captured by the Saracens at Mansūra 1249; canonized 1297 as *Saint Louis.*
— **Louis XI,** 1423–83, king of France 1461–1483; made the power of the throne paramount.
— **Louis XIII,** 1601–43, king of France 1610–1643; son of Henry IV.
— **Louis XIV,** 1638–1715, king of France 1643–1715; son of the preceding: called "Le Grand Monarque," "le roi soleil."
— **Louis XV,** 1710–74, king of France 1715–1774; great-grandson of the preceding; ceded Canada to Great Britain.
— **Louis XVI,** 1754–93, king of France 1774–1792; grandson of Louis XV; dethroned by the revolution and guillotined.
— **Louis XVIII,** 1755–1824, brother of Louis XVI; king of France April, 1814–March, 1815, and June, 1815–24.
— **Louis Napoleon** See NAPOLEON III.
— **Louis Philippe,** 1773–1850, king of France 1830–48; abdicated: known as *The Citizen King.*

Lou·is (lōō′is), **Joe,** born 1914, U. S. heavyweight boxing champion 1937–49; real name *Joseph Louis Barrow.*
Lou·i·sa (lōō·ē′zə) A feminine personal name. Also **Lou·ise** (lōō·ēz′). [See LOUIS]
Lou·is·burg (lōō′is·bûrg) A town on the eastern coast of Cape Breton Island, Canada; captured from the French by British forces in 1745 and 1758. Also **Lou′is·bourg.**
lou·is d'or (lōō′ē dôr) A louis. [<F, gold louis]
Lou·ise (lōō·ēz′), **Lake** A lake in Banff National Park, SW Alberta, Canada.
Lou·i·si·ade Archipelago (lōō·ē′zē·äd′) A Papuan island group SE of New Guinea; 980 square miles.
Lou·i·si·an·a (lōō·ē′zē·an′ə) One of the United States, fronting south on the Gulf of Mexico; 48,523 square miles; capital, Baton Rouge; comprised of portions of West Florida and of the Louisiana Purchase of 1803, it entered the Union April 8, 1812; nickname, *Pelican State* or *Creole State*: abbr. *La.* — **Lou·i·si·an′i·an, Lou·i·si·an′an** *adj.* & *n.*
Louisiana Purchase The French colonial territory in west central North America, purchased from France by the United States in 1803 for $15,000,000: the land between the Mississippi River and the Rocky Mountains, the Gulf of Mexico and Canada.
Lou·is-Qua·torze (lōō′ē·kà·tôrz′) *adj.* Designating the style of architecture, interior decoration, and furniture which characterized the period of Louis XIV in France (1643–1715), marked by baroque architectural forms and ornate and grandiose decorative treatment employing animal and mythological forms richly carved. [<F]
Lou·is-Quinze (lōō′ē·kànz′) *adj.* Designating the style of architecture, decoration, and furniture characteristic of the period of Louis XV (1715–74), marked by the culmination of the rococo as expressed in flowing lines, rounded forms, and graceful shell, flower, and other ornaments. [<F]
Lou·is-Seize (lōō′ē·sez′) *adj.* Designating the style of architecture, decoration, and furniture which characterized the period of Louis XVI (1774–93): a reaction against the Louis-Quinze period, marked by simple, rectilinear forms, and using symmetrical Greco-Roman wreaths, birds, etc., as ornaments, or Greek egg-and-anchor or leaf motifs, achieving a plain internal decoration. [<F]
Lou·is-Treize (lōō′ē·trez′) *adj.* Designating the style of architecture, decoration, and furniture which characterized the period of Louis XIII (1610–43), marked by Renaissance forms, rich inlaid ornaments, geometric panels and deep moldings, and carving in the Flemish style. [<F]

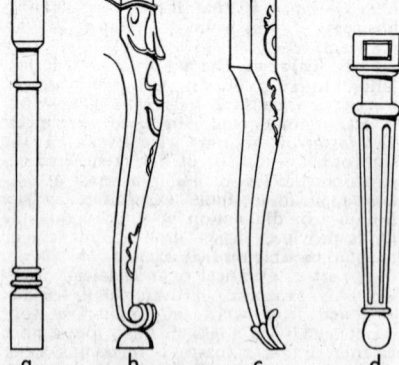

PERIOD STYLES ILLUSTRATED BY CHAIR LEGS
a. Louis XIII. b. Louis XIV. c. Louis XV.
d. Louis XVI.

Lou·is·ville (lōō′ē·vil) A city on the Ohio River in Kentucky.
loun (loun) *adj. Scot.* Lown.
lounge (lounj) *v.* **lounged, loung·ing** *v.i.* 1 To lie, lean, move, etc., in an idle or lazy manner. 2 To pass time indolently. — *v.t.* 3 To spend or pass indolently, as time. — *n.* 1 The act of lounging. 2 A room in a hotel, club, etc., for lounging. 3 A couch with little or no back; any sofa. [Origin unknown]

loung·er (loun′jər) *n.* One who lounges; an idler.
loup[1] (loup, lōōp) *v.t.* & *v.i.* **lap** or **loup·en, loup·ing** *Scot.* & *Brit. Dial.* To leap. [<ON *hlaupa*]
loup[2] (lōō) *n.* A light mask or half-mask made of silk: worn at masquerades. [<F]
loup-cer·vier (lōō·ser·vyā′) *n. French* The Canada lynx (*Lynx canadensis*). [<L *lupus cervarius* lynx < *lupus* wolf + *cervarius* stag-hunting < *cervus* stag]
loupe (lōōp) *n.* A small magnifying glass; a lens, especially one adapted for the use of jewelers and watchmakers. [<F]
loup-ga·rou (lōō′gà·rōō′) *n. pl.* **loup-ga·rous** (lōō′gà·rōō′) A werewolf. [<F *loup* wolf (<L *lupus*) + *garou* werewolf <Gmc.]
loup·ing ill (lou′ping, lō′-) *Scot.* An infectious virus disease of sheep; leaping evil. Also **loup ill.**

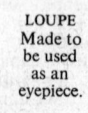

LOUPE
Made to be used as an eyepiece.

Loup River (lōōp) A river in east central Nebraska, flowing 300 miles east to the Platte River.
lour (lour), **lour·ing** (lour′ing), **lour·y** (lour′ē) See LOWER[1], etc.
Lourdes (lōōrd) A town in SW France; shrine and grotto of the Virgin (*Our Lady of Lourdes*).
Lou·ren·ço Mar·ques (lō·ren′sō mär′kəs, *Pg.* lō·rän′sōō mär′kish) A port on Delagoa Bay, capital of Mozambique.
louse (lous) *n. pl.* **lice** (līs) 1 A small, flat-bodied, wingless insect (order *Anoplura*) living as an external parasite on birds or mammals, especially the crab louse (*Phthirus pubis*) and the human body louse (genus *Pediculus*). 2 One of various other insects parasitic on other animals or plants, as the biting bird louse (order *Mallophaga*) and the plant louse. For illustrations see under INSECTS. (injurious). 3 *Slang* A contemptible person. — *v.t.* & *v.i. U.S. Slang* To ruin; bungle: with *up.* [OE *lūs*]
louse·wort (lous′wûrt′) *n.* Any one of a genus (*Pedicularis*) of scrophulariaceous woodland herbs; wood betony.
lous·y (lou′zē) *adj.* **lous·i·er, lous·i·est** 1 Infested with lice; pedicular. 2 *Slang* Dirty. 3 *Slang* Contemptible; foul; mean. 4 *Slang* Having plenty (of): usually with *with: lousy* with money.
lout[1] (lout) *n.* An awkward fellow; clown; boor. [? <ON *lutr* bent, stooped]
lout[2] (lout) *v.i. Obs.* To bow or curtsy; bend; stoop. [OE *lūtan* bow]
Louth (louth) A county of NE Leinster province, Ireland; 317 square miles; county town, Drogheda.
lou·ther (lōō′thər) *v.i. Scot.* To plod, as in mire or wet snow.
lout·ish (lou′tish) *adj.* Clumsy; awkward; boorish. — **lout′ish·ly** *adv.* — **lout′ish·ness** *n.*
Lou·vain (lōō·vān′, *Fr.* lōō·van′) A city in central Belgium. *Flemish* **Leu·ven** (lœ′vən, -və).
lou·ver (lōō′vər) *n.* 1 A window, as in a belfry tower, designed for ventilation and having slats (**louver boards**) sloped to keep out the rain: also **lou′ver-win′dow** (-win′dō). 2 A lantern-like cupola or turret on the roof of a medieval dwelling. 3 Any of several narrow openings, or the slatted piece having these, serving as outlets for heated air. [<OF *lover*]

LOUVER
a. Construction set in gable (b).

Lou·vre (lōō′vr′) An ancient royal palace in Paris, made a museum in the 18th century.
Lou·ÿs (lōō·ē′, *Fr.* lwē), **Pierre,** 1870–1925, French writer.
lov·a·ble (luv′ə·bəl) *adj.* Worthy of love; amiable; also, evoking love. See synonyms under AMIABLE, LOVELY. Also **love′a·ble.** — **lov′a·bil′i·ty, lov′a·ble·ness** *n.* — **lov′a·bly** *adv.*
lov·age (luv′ij) *n.* A European herb (*Levisticum officinale*) of the parsley family: used sometimes as a domestic remedy. [<OF *luvesche* <LL *levisticum,* alter. of L *ligusticum,* neut. of *ligusticus* Ligurian]
love (luv) *n.* 1 A strong, complex emotion or feeling causing one to appreciate, delight in,

and crave the presence or possession of another and to please or promote the welfare of the other; devoted affection or attachment. **2** Specifically, such feeling between husband and wife or lover and sweetheart. **3** One who is beloved; a sweetheart. **4** Sexual passion, or the gratification of it. **5** A very great interest or fondness: *love* of learning. **6** In tennis, a score of nothing. **— for the love of** For the sake of; in loving remembrance of: used in adjurations, solemn oaths, etc. **— for love or money** For any consideration; under any circumstances. **— in love** Experiencing love; enamored. **— no love lost between** No affection or liking between. **—** *v.* **loved, lov·ing** *v.t.* **1** To feel love or affection for. **2** To take pleasure or delight in; like very much. **3** To caress. **—** *v.i.* **4** To feel love, especially for one of the opposite sex; be in love. See synonyms under ADMIRE, LIKE. [OE *lufu*]
Synonyms (*noun*): affection, attachment, attraction, charity, devotion, esteem, feeling, fondness, friendship, liking, passion, regard, tenderness. *Affection* is kindly feeling, deep, tender, and constant, going out to some person or object, being less fervent and ardent than *love*. *Love* is the yearning or outgoing of soul toward something that is regarded as excellent, beautiful, or desirable; *love* may be briefly defined as strong and absorbing *affection* for and *attraction* toward a person or object. *Love* is more intense, absorbing, and tender than *friendship*, more intense, impulsive, and perhaps passionate than *affection*; we speak of fervent *love*, but of deep or tender *affection*, or of close, firm, strong *friendship*. *Love* is used specifically for personal *affection* between the sexes, and can never properly denote mere animal passion, which is expressed by such words as appetite, desire, lust. One may have *love* for animals, inanimate objects, or for abstract qualities that enlist the affections, as we speak of *love* for a horse or a dog, for mountains, woods, ocean, or of *love* of nature, and *love* of virtue. *Love* of articles of food is better expressed by *liking*, as *love*, in its full sense, denotes something spiritual and reciprocal, such as can have no place in connection with objects that minister merely to the senses. See ATTACHMENT, FRIENDSHIP. *Antonyms:* see synonyms for ANTIPATHY, ENMITY, HATRED.
Love (luv) A personification of love; Cupid or Eros.
love affair A romantic attachment between two people not married to each other.
love apple The tomato.
love·bird (luv′bûrd′) *n.* One of several small parrots (genera *Agapornis* and *Psitta*) often kept as cage birds: so called from the affection they appear to show for their mates.
love feast 1 A common devotional meal partaken of by early Christians, originally culminating in the Eucharist; agape. **2** A somewhat similar celebration observed in some modern churches, as the Methodist and Moravian. **3** A banquet held in common rejoicing over something.
love game In tennis, a game in which the winner has lost no point.
love-in-a-mist (luv′in·ə·mist′) *n.* A European ranunculaceous garden plant (*Nigella damascena*) with blue flowers.
love-in-i·dle·ness (luv′in·ī′dəl·nis) *n.* The pansy.
love·knot (luv′not′) *n.* A knot tied in pledge of love and constancy, or representation of it, as in jewelry.
Love·lace (luv′lās), **Richard,** 1618–58, English poet.
love·less (luv′lis) *adj.* Having no love; unloving; also, not lovable. **— love′less·ly** *adv.* **— love′less·ness** *n.*
love-lies-bleed·ing (luv′līz′blē′ding) *n.* Any of several species of amaranth.
love·lock (luv′lok′) *n.* A separate lock of hair worn curled and tied with ribbons; especially such a lock as worn formerly by courtiers.
love·lorn (luv′lôrn′) *adj.* Forsaken by or pining for a lover.
love·ly (luv′lē) *adj.* **·li·er, ·li·est 1** Possessing mental or physical qualities that inspire admiration and love; charming. **2** Beautiful.

3 Attractive; inviting. **4** *Colloq.* Delightful; pleasing: We had a *lovely* visit with them. **5** *Obs.* Affectionate; loving. [OE *luflic*] **— love′li·ness** *n.*
Synonyms: amiable, beautiful, charming, delectable, delightful, enchanting, lovable, pleasing, sweet, winning, winsome. See AMIABLE, BEAUTIFUL.
love·mak·ing (luv′mā′king) *n.* The act of making love; wooing; courtship.
love potion A magic draft or drink designed to arouse love toward a certain person in the one who drinks.
lov·er (luv′ər) *n.* **1** One who loves; a warm admirer; devoted friend. **2** One who is in love; specifically, a paramour: in the singular now used only of the man. **3** One who enjoys or is strongly attracted by some object or diversion. **— lov′er·ly** *adj.* & *adv.*
Lov·er (luv′ər), **Samuel,** 1797–1868, Irish novelist.
love seat A double chair or small sofa for two persons in Queen Anne and later styles.
love set In tennis, a set in which the winner has not lost a game.
love·sick (luv′sik′) *adj.* **1** Languishing with love. **2** Indicating or expressing such a condition: a *lovesick* serenade. **— love′sick′ness** *n.*
love·some (luv′səm) *adj. Obs.* **1** Inspiring love; lovable. **2** Expressing love; loving.
love·vine (luv′vīn′) *n.* The dodder.
lov·ing (luv′ing) *adj.* **1** Affectionate; devoted; kind: *loving* friends or brothers. **2** Indicative of love or kind feeling: *loving* looks and words. See synonyms under AMIABLE, CHARITABLE. **— lov′ing·ly** *adv.* **— lov′ing·ness** *n.*
loving cup 1 A wine cup, usually with two or more handles, meant to be passed from hand to hand around a circle of friends; a parting cup. **2** A trophy presented to the winner of an athletic or other kind of contest.
lov·ing-kind·ness (luv′ing·kīnd′nis) *n.* Kindness that comes from personal attachment; specifically, the loving care of God for his people.
low¹ (lō) *adj.* **1** Having relatively little upward extension; not high or tall. **2** Having relatively little elevation; raised only slightly above a recognized level: a *low* bridge; near the horizon: a *low* moon. **3** Situated below a recognized level; deep; depressed. **4** Having less than the normal or regular height, or less height than something taken as a standard; descending from the usual level. **5** Cut so as to expose the neck; décolleté. **6** Of sounds having depth of pitch; deep; also, having little volume or strength; soft. **7** Being less than the usual rate, amount, or reckoning; scant: *low* wages or interest. **8** Being little in degree, number, grade, station, quality, character, etc.; humble; moderate. **9** Being below the proper standard in refinement, moral character, principle, or condition; lacking pride, principle, force, dignity, or worth; vulgar; weak: *low* spirits; *low* standards of morality. **10** Dead; prostrate: He lies *low*. **11** *Phonet.* Pronounced with the tongue low and flat; open: said of vowels; opposed to *high*. **12** Little advanced in civilization or organic evolution. **13** Badly nourished; lacking in vigor; also, not nourishing; plain: a *low* diet. **14** *Mech.* Giving the least speed: *low* gear. **15** *Geog.* Pertaining to latitudes near the equator. **16** *Eccl.* **a** Pertaining to broad evangelical doctrine. **b** Denoting a service shorn of elaborate ritual: a *low* mass. **17** *Bot.* Growing close to the ground: a *low* herb. See synonyms under BASE, COMMON, HUMBLE, VULGAR. **—** *adv.* **1** In a low way; in or to a low position; not on high. **2** At a low price; cheap. **3** In a humble rank or degraded condition. **4** So as not to be loud in sound; softly; also, at a low pitch. **5** In such a path that the declination or the altitude is small; near the equator or the horizon. **— to lie low 1** To be dead or prostrate. **2** To be or remain in hiding. **3** To hold one's tongue till the proper moment; wait. **—** *n.* **1** A low area. **2** The first gear or speed of an automobile: Put it in *low*. [OE *lah* < ON *lagr*] **— low′ness** *n.*
low² (lō) *n.* The moo or bellow of cattle: also **low′ing.** **—** *v.i.* To bellow, as cattle; moo. **—** *v.t.* To utter by lowing. [OE *hlowan*]

low³ (lō) *n. Scot.* A glowing fire; blaze. Also **lowe.**
Low (lō), **David,** born 1891, English cartoonist.
Low Archipelago See TUAMOTU ARCHIPELAGO.
low area *Meteorol.* A region of low atmospheric pressure.
low-ar·e·a storm (lō′âr′ē·ə) A cyclone.
low-born (lō′bôrn′) *adj.* Of humble birth.
low·boy (lō′boi′) *n.* A table, often a dressing table, with drawers, similar to the lower part of a highboy.
low-bred (lō′bred′) *adj.* Vulgar; ill-bred.
low·brow (lō′brou′) *Colloq. n.* A person of uncultivated tastes; a non-intellectual: an uncomplimentary term. **—** *adj.* Of or suitable for such a person: also **low′browed′.** **— low′brow′ism** *n.*
Low-Church (lō′chûrch′) *adj.* Of or belonging to a group (**Low Church**) in the Anglican Church which stresses evangelical doctrine and is, in general, opposed to extreme ritualism.
Low-Church·man (lō′chûrch′mən) *n. pl.* **·men** (-mən) A member of the Low-Church group in the Anglican Church: an opponent of extreme ritualism.
low comedy Comedy in which both subject matter and presentation are in the broad style of farce or burlesque; comedy characterized by slapstick and lively physical action rather than by witty dialog.
Low Countries The region of NW Europe comprising the Netherlands, Belgium, and Luxemburg.
low-down¹ (lō′doun′) *n. Slang* Inside or secret information; the truth of a matter.
low-down² (lō′doun′) *adj. Colloq.* Degraded; mean.
low-down·er (lō′dou′nər) *n. Dial.* A poor white.
Low·ell (lō′əl) A distinguished Massachusetts family, descendants of **John,** 1743–1802, jurist, including **James Russell,** 1819–91, poet, essayist, editor, and diplomat; **Percival,** 1855–1916, astronomer; **Abbott Lawrence,** 1856–1943, educator; president of Harvard University 1909–33; brother of Percival; **Amy,** 1874–1925, poet and critic, sister of Percival.
Low·ell (lō′əl) A city on the Merrimac River in NE Massachusetts.
low·er¹ (lou′ər) *v.i.* **1** To look angry or sullen; scowl. **2** To appear dark and threatening, as the weather. **—** *n.* A scowl; a gloomy aspect. Also spelled *lour*. [Cf. G *lauern* lurk]
low·er² (lō′ər) Comparative of LOW. **—** *adj.* **1** Inferior in position, in value or rank; situated below something else: a *lower* berth. **2** *Geol.* Older; designating strata normally beneath the newer (and upper) rock formations. **—** *n.* That which is beneath something above; specifically, a lower berth. **—** *v.t.* **1** To bring to a lower position or level; let down, as a window. **2** To reduce in degree, quality, amount, etc.: to *lower* prices. **3** To bring down in estimation, rank, etc.; humble or degrade. **4** To change, as a sound, to a lower pitch or volume. **—** *v.i.* **5** To become lower; sink; decrease. See synonyms under ABASE, ABATE, DISPARAGE, WEAKEN.
Lower Austria An autonomous province of NE Austria; 7,097 square miles; capital, Vienna. German **Nie·der·ös·ter·reich** (nē′dər·œs′tə·rīkh).
Lower Burma See BURMA, UNION OF.
Lower California A peninsula of NW Mexico separating the Gulf of California from the Pacific; about 760 miles long; 55,634 square miles: divided into **Northern Territory;** 27,655 square miles; capital, Mexicali; and **Southern Territory;** 27,979 square miles; capital, La Paz. Spanish *Baja California*.
Lower Canada A former name for the province of QUEBEC.
low·er-case (lō′ər·kās′) *adj. Printing* Describing the small letters of a font of type which are kept in the low case (see CASE); small letters. **—** *v.t.* **-cased, -cas·ing** To set as or change to lower-case letters. [From their being kept in the lower two cases of type]
low·er-class·man (lō′ər·klas′mən, -kläs′-) *n. pl.* **men** (-mən) A freshman or a sophomore.
Lower Cretaceous *Geol.* The older part of the

Cretaceous period, or its representative rocks or fossils. See CRETACEOUS.

lower criticism The method of critical investigation which seeks to determine the original wording of a text: also called *textual criticism.* Compare HIGHER CRITICISM.

Lower Franconia An administrative division of NW Bavaria, Germany; 3,277 square miles; capital, Würzburg: formerly *Mainfranken. German* **Un·ter·fran·ken** (o͝on′tər·fräng′kən).

Lower House The popular division of a legislative body; in the United States, the House of Representatives, as opposed to the Upper House or Senate; in Great Britain, the House of Commons, as opposed to the House of Lords.

low·er·ing (lou′ər·ing) *adj.* **1** Overcast with clouds; gloomy; threatening. **2** Frowning or sullen. Also spelled *louring.* — **low′er·ing·ly** *adv.* — **low′er·ing·ness** *n.*

low·er·most (lō′ər·mōst′) *adj.* Lowest.

Lower Sax′ony A state of north central West Germany; 18,279 square miles; capital, Hanover. *German* **Nie·der·sach·sen** (nē′dər·zäk′sən).

Lower Silurian *Geol.* The former name for ORDOVICIAN.

lower world **1** The abode of the dead; hell; Hades. Also **lower regions.** **2** The earth.

low·er·y (lou′ər·ē) *adj.* Cloudy; threatening storm: also spelled *loury.*

Lowes (lōz), **John Livingston,** 1867–1945, U. S. scholar, critic, and educator.

lowest common multiple See under MULTIPLE.

Lowes·toft (lōs′tôft, -təft) *n.* A variety of porcelain originally made at Lowestoft, England.

Lowes·toft (lōs′tôft, -təft) A municipal borough and port in eastern Suffolk, England.

Low German See under GERMAN.

low·land (lō′lənd) *adj.* Pertaining to or characteristic of a low or level country. — *n.* (*also* lō′land′) *Usually pl.* Land lower than the adjacent country; level land. — **The Lowlands** The less elevated districts lying in the south and west of Scotland. — **Low′land′er** *n.*

Low·land (lō′lənd) *adj.* Pertaining to or belonging to the Lowlands. — *n.* (*also* lō′land′) The speech or dialect of the Scotch Lowlands.

Low Latin See under LATIN.

low·li·hood (lō′lē·ho͝od) *n. Obs.* The state of being lowly; modesty; humility. Also **low′li·head** (-hed).

low·ly (lō′lē) *adj.* **·li·er, ·li·est** **1** Situated or lying low, as land. **2** Having low rank or importance; unpretending; humble. See synonyms under HUMBLE. — *adv.* **1** In a manner appropriate to humble life; rudely; meanly. **2** In a meek or modest manner. — **low′li·ness** *n.*

low mass The ordinary form of mass without music.

low-mind·ed (lō′mīn′did) *adj.* Having low or mean thoughts, sentiments, or motives; base.

lown[1] (loun) *adj. Scot.* Sheltered; tranquil. — **lown′ly** *adv.*

lown[2] (lo͞on) See LOON[2].

low-necked (lō′nekt′) *adj.* Cut low in the neck; décolleté: said of a garment.

low-pitched (lō′picht′) *adj.* **1** Low in tone, key, or range of tone, as a voice. **2** Having little angular elevation: said of a roof.

low-pres·sure (lō′presh′ər) *adj.* **1** Requiring a low degree of pressure. **2** Having a condenser, as an engine. **3** *Meteorol.* Designating an atmospheric pressure below that which is normal at sea level.

low relief Bas-relief.

lowse (lōz) *adj. Scot.* Loose.

low-spir·it·ed (lō′spir′it·id) *adj.* Lacking spirit or animation; despondent; melancholy. — **low′-spir′it·ed·ly** *adv.* — **low′-spir′it·ed·ness** *n.*

Low Sunday The Sunday following Easter.

low-ten·sion (lō′ten′shən) *adj. Electr.* Having or characterized by a low electric potential, as a battery, magneto, or vacuum tube.

low-test (lō′test′) *adj.* Possessing a relatively low boiling point, as gasoline.

low tide The furthest recession of the tide at any point; also, the time of its occurrence. Also **low water.**

low-wa·ter (lō′wô′tər, -wot′ər) *adj.* Pertaining to low tide or its time or measure of recession.

lox (loks) *n.* Smoked salmon. [<Yiddish <G *lachs* salmon]

lox·o·drom·ic (lok′sə·drom′ik) *adj.* Pertaining to oblique sailing on the rhumb line. Also **lox′o·drom′i·cal.** [<Gk. *loxos* oblique + *dromos* a running]

lox·o·drom·ics (lok′sə·drom′iks) *n. pl.* (*construed as singular*) The art of oblique sailing. Also **lox·od·ro·my** (lok·sod′rə·mē).

loy·al (loi′əl) *adj.* **1** Constant and faithful in any relation implying trust or confidence; bearing true allegiance to constituted authority. **2** Professing or indicative of faithful devotion. **3** *Obs.* Legitimate; legal. See synonyms under FAITHFUL. [<OF *loial, leial* <L *legalis.* Doublet of LEGAL.] — **loy′al·ism** *n.* — **loy′al·ly** *adv.*

loy·al·ist (loi′əl·ist) *n.* One who adheres to and defends his sovereign or state.

Loy·al·ist (loi′əl·ist) *n.* **1** One who was loyal to the British crown during the American Revolution. **2** One who was loyal to the Union during the Civil War. **3** One who was loyal to the Republican Constitution during the Spanish Civil War.

loy·al·ty (loi′əl·tē) *n. pl.* **·ties** Devoted allegiance. See synonyms under ALLEGIANCE, FIDELITY. [<OF *loialte*]

Loyalty Islands A dependency of New Caledonia; total, 740 square miles. *French* **Îles Loy·au·té** (ēl lwȧ·yō·tā′).

Lo·yo·la (loi·ō′lə, *Sp.* lō·yō′lä), **Ignatius of,** 1491–1556, Spanish soldier and priest; founder, with Francisco Xavier, of the Society of Jesus (the Jesuits); canonized in 1622: original name *Iñigo de Oñez y Loyola.* — **Loy·o·lism** (loi′ə·liz′əm) *n.*

loz·enge (loz′inj) *n.* **1** *Math.* A rhombus with all sides equal, having two acute and two obtuse angles. **2** A small medicated or sweetened tablet, originally diamond-shaped. [<OF *losenge,*? <L *lapis, -idis* stone]

loz·en·ger (loz′in·jər) *n.* A lozenge.

LP (el′pē′) *adj.* Designating a phonograph record pressed with microgrooves and played at a speed of 33 1/3 revolutions per minute. — *n.* A long-playing phonograph record: a trade name. [< *l(ong)-p(laying)*]

Lu·a·la·ba (lo͞o′ä·lä′bä) The upper course of the Congo, flowing 400 miles north: French *Loualaba.*

Lu·an·da (lo͞o·än′də) A port, capital of Angola: formerly *Loanda, São Paulo de Loanda.*

Luang Pra·bang (lwäng prä·bäng′) A city of north central Laos; royal residence.

Luang·wa (lwäng′gwä) A river in eastern Northern Rhodesia, flowing 500 miles SW to the Zambesi: formerly *Loangwa.*

Lu·bang Islands (lo͞o·bäng′) A small Philippine island group off NE Mindoro; 88 square miles.

lub·ber (lub′ər) *n.* An awkward, ungainly fellow; specifically, a raw or green landsman on shipboard. [Origin uncertain. Cf. ON *lubba* short.] — **lub′ber·li·ness** *n.* — **lub′ber·ly** *adj.* & *adv.*

lubber line *Aeron.* A fixed line in an aircraft compass, gyro, direction finder, or similar instrument, parallel to the longitudinal axis of the aircraft and used as a base line.

lubber's hole *Naut.* A hole through the floor of a top, by which sailors can go aloft without climbing over the rim by the futtock shrouds.

Lub·bock (lub′ək), **Sir John,** 1834–1913, Lord Avebury, English statesman, scientist, financier, and writer.

Lü·beck (lo͞o′bek, *Ger.* lü′bek) A city in NW Germany.

Lu·blin (lo͞o′blin, *Polish* lyo͞o′blēn) A city in eastern Poland: Russian *Lyublin.*

lu·bric (lo͞o′brik) *adj. Obs.* Lubricous. Also **lu′bri·cal.** [<L *lubricus* slippery]

lu·bri·cant (lo͞o′brə·kənt) *adj.* Lubricating. — *n.* Anything that lubricates, as graphite, oil, or grease.

lu·bri·cate (lo͞o′brə·kāt) *v.t.* **·cat·ed, ·cat·ing** **1** To apply grease, oil, or other lubricant to so as to reduce friction and wear. **2** To make slippery or smooth. [<L *lubricatus,* pp. of *lubricare* make slippery < *lubricus* slippery] — **lu′bri·ca′tion** *n.* — **lu′bri·ca′tive** *adj.*

lu·bri·ca·tor (lo͞o′brə·kā′tər) *n.* **1** One who or that which lubricates. **2** A device, as an oil cup, by which a lubricant is fed or applied to a bearing surface.

lu·bric·i·ty (lo͞o·bris′ə·tē) *n.* **1** The state of being slippery; hence, shiftiness; instability; evanescence. **2** Lewdness. **3** Power for lubrication. [<F *lubricité* <L *lubricitas* < *lubricus* slippery]

lu·bri·cous (lo͞o′brə·kəs) *adj.* **1** Smooth and slippery. **2** Elusive; unstable. **3** Lascivious; lewd; wanton. Also **lu·bri·cious** (lo͞o·brish′əs). [<L *lubricus*]

lu·bri·fac·tion (lo͞o′brə·fak′shən) *n.* The act or process of lubricating or making slippery or smooth. [<L *lubricus* smooth + *factus,* pp. of *facere* make]

Luc (lük) French form of LUKE. Also *Ital.* **Lu·ca** (lo͞o′kä), **Lu·cas** (*Dan., Du., Ger., Sw.* lo͞o′käs; *Lat.* lo͞o′kəs), *Hungarian* **Lu·cáts** (lo͞o′·käch). [<L, light]

Lu·can (lo͞o′kən), 39–65, Roman poet: full name *Marcus Annaeus Lucanus.*

Lu·ca·ni·a (lo͞o·kā′nē·ə) The ancient name for BASILICATA.

lu·carne (lo͞o·kärn′) *n.* A dormer or garret window; also, a small window or light in a spire. [<F; ult. origin uncertain. Cf. OHG *lukka* opening.]

Luc·ca (lo͞ok′kä) A city in Tuscany, north central Italy.

luce (lo͞os) *n.* A fish, the pike, especially when fully grown. [<OF *lus* <L *lucius*]

Luce (lüs) French form of LUCIUS.

Luce (lo͞os), **Clare Boothe,** born 1903, U. S. dramatist and diplomat; wife of Henry R. — **Henry Robinson,** born 1898, U. S. editor and publisher.

lu·cent (lo͞o′sənt) *adj.* Showing radiance or brilliance; shining; luminous. [<L *lucens, -entis,* ppr. of *lucere* shine] — **lu′cen·cy** *n.* — **lu′cent·ly** *adv.*

lu·cer·nal (lo͞o·sûr′nəl) *adj.* Relating to a lamp or to any artificial light. [<L *lucerna* lamp < *lux* light]

lu·cerne (lo͞o·sûrn′) *n.* A tall, cloverlike herb of the pea family (*Medicago sativa*), used for forage: now commonly called *alfalfa, purple medic.* Also **lu·cern′.** [<F *luzerne*; ult. origin unknown]

Lu·cerne (lo͞o·sûrn′, *Fr.* lü·sern′) A canton of central Switzerland; 580 square miles; capital, Lucerne (German *Luzern*) on **Lake of Lucerne,** a lake in north central Switzerland; 44 square miles: also **Lake of the Four Forest Cantons** (German *Vierwaldstättersee*).

lu·ces (lo͞o′sēz) Plural of LUX.

Lu·cia (lo͞o′shə; *Ger.* lo͞o′tsē·ə, *Ital.* lo͞o·chē′ä, *Lat.* lo͞o′shē·ə, *Pg.* lo͞o′sē·ə, *Sp.* lo͞o·thē′ä) A feminine personal name: also *Lucy.* Also *Fr.* **Lu·cie** (lü·sē′, *Du.* lo͞o′sē·ā). [<L, light]

Lu·cian (lo͞o′shən), 125?–210?, Greek rhetorician and satirist. Also **Lu·ci·a·nus** (lo͞o′shē·ā′·nəs).

lu·cid (lo͞o′sid) *adj.* **1** Intellectually bright and clear; mentally sound; sane. **2** Easily understood; perspicuous; clear. **3** Giving forth light; shining. **4** Translucent; pellucid. **5** Smooth and shining. See synonyms under CLEAR, SANE, TRANSPARENT. [<L *lucidus* < *lucere* shine] — **lu·cid·i·ty** (lo͞o·sid′ə·tē), **lu′·cid·ness** *n.* — **lu′cid·ly** *adv.*

lu·ci·fer (lo͞o′sə·fər) *n.* A friction match. — *adj.* Emitting light. [after *Lucifer*]

Lu·ci·fer (lo͞o′sə·fər) **1** The planet Venus as the morning star. **2** Satan, especially as the leader of the revolt of the angels before his fall from heaven. [<L, light-bearer < *lux, lucis* light + *ferre* bear] — **Lu·cif·er·ous** (lo͞o·sif′ər·əs) *adj.*

lu·cif·er·ase (lo͞o·sif′ə·rās) *n. Biochem.* An enzyme present in fireflies and in certain other luminous organisms: it oxidizes luciferin to produce light. [<L *lucifer* light-bearing + -ASE]

lu·cif·er·in (lo͞o·sif′ər·in) *n. Biochem.* A water-soluble, heat-stable protein which produces heatless light, as in fireflies. [<L *lucifer* light-bearing + -IN]

lu·cif·er·ous (lo͞o·sif′ər·əs) *adj.* Emitting light; illuminating. [<L *lucifer* light-bearing + -OUS]

lu·ci·form (lo͞o′sə·fôrm) *adj.* Having the nature or appearance of light. [<L *lux, lucis* light + -FORM]

Lu·ci·na (lo͞o·sī′nə) In Roman mythology, the goddess presiding over childbirth. [<L *lucina,* fem. of *lucinus* bringing to the light < *lux, lucis* light]

Lu·cin·da (lo͞o·sin′də) See LUCY.

Lu·cite (lo͞o′sīt) *n.* A thermoplastic, transparent acrylic resin, easily machined into various shapes and having the property of transmitting light around curves and corners: a trade name.

Lu·cius (lo͞o′shəs; *Ger.* lo͞o′tsē·o͞os, *Fr.* lü·sē·üs′) A masculine personal name. Also *Ital.*

Lu·cio (lōō′chō, *Pg.* lōō′syōō; *Sp.* lōō′thyō). [<L, light]

lu·ci·vee (lōō′sə·vē) *n.* The loup-cervier. Also **lu′ci·fee** (-fē). [Alter. of LOUP-CERVIER]

luck (luk) *n.* 1 That which happens by chance; fortune or lot. 2 Happy chance; good fortune; success. [Prob. <MDu. *luk, geluk*]

luck·less (luk′lis) *adj.* Having no luck. — **luck′·less·ly** *adv.*

Luck·ner (luk′nər, *Ger.* lōōk′nər), **Count Felix von**, born 1881, German naval officer: known as the *Sea Devil*.

Luck·now (luk′nou) A city in central Uttar Pradesh, India; the capital of the State; British forces were besieged here during the Sepoy Rebellion, 1857.

luck·y[1] (luk′ē) *adj.* **luck·i·er, luck·i·est** 1 Favored by fortune; fortunate; successful. 2 Productive of luck; auspicious; favorable: said of events and things: a *lucky* penny or circumstance. See synonyms under AUSPICIOUS, FORTUNATE, HAPPY, WELL. — **luck′i·ly** *adv.* — **luck′i·ness** *n.*

luck·y[2] (luk′ē) *n. pl.* **luck·ies** *Scot.* A grandam or aged woman; goody; sometimes part of a person's name: *Lucky* Macpherson. Also **luck′ie.**

luck·y[3] (luk′ē) *adv. Scot.* More than sufficiently: *lucky* hot.

lu·cra·tive (lōō′krə·tiv) *adj.* Productive of wealth; highly profitable. See synonyms under PROFITABLE. [<L *lucrativus* < *lucratus*, pp. of *lucrari* gain < *lucrum* wealth] — **lu′cra·tive·ly** *adv.* — **lu′cra·tive·ness** *n.*

lu·cre (lōō′kər) *n.* Money, especially as the object of greed; gain. See synonyms under WEALTH. [<F <L *lucrum* gain]

Lu·cre·tia (lōō·krē′shə, *Ger.* lōō·krā′tsē·ə) A feminine personal name. Also *Fr.* **Lu·crèce** (lü·kres′), *Ital.* **Lu·cre·zi·a** (lōō·krā′tsē·ä), *Sp.* **Lu·cre·ci·a** (lōō·krā′thē·ä). [<L, wealth]

— **Lucretia** In Roman legend, the wife of Collatinus who killed herself after she had been violated by Sextus Tarquinius; subject of a poem by Shakespeare.

Lu·cre·tius (lōō·krē′shəs, -shē·əs), 96?–55 B.C. Roman poet and philosopher: full name *Titus Lucretius Carus.*

lu·cu·brate (lōō′kyōō·brāt) *v.i.* **·brat·ed, ·brat·ing** 1 To study or write laboriously, as at night. 2 To write in a learned manner. [<L *lucubratus*, pp. of *lucubrare* work by artificial light < *lux, lucis* light]

lu·cu·bra·tion (lōō′kyōō·brā′shən) *n.* 1 Close and earnest meditation or study. 2 The product of deep meditation or earnest study; a literary composition; often, a pedantic or over-elaborated work. [<L *lucubratio, -onis*] — **lu′cu·bra·tor** *n.* — **lu′cu·bra·to·ry** (-brə·tôr′ē, -tō′rē) *adj.*

lu·cu·lent (lōō′kyōō·lənt) *adj.* 1 Full of light; brilliant. 2 Clearly evident; lucid. [<L *luculentus*]

Lu·cul·lus (lōō·kul′əs), **Lucius Licinius**, 110?–57? B.C., Roman consul, proverbial for his wealth and luxury. — **Lu·cul′lan, Lu·cul·le·an** (lōō′kə·lē′ən), **Lu·cul′li·an** *adj.*

lu·cus a non lu·cen·do (lōō′kəs ā non lōō·sen′dō) *Latin* Literally, a grove because it is *not* light: in allusion to an ancient absurd derivation of *lucus*, grove, from *lucere*, be light. Hence, something absurdly reasoned out.

Lu·cy (lōō′sē) Variant of LUCIA.

Lucy Sto·ner (stō′nər) An advocate of women's rights; especially one advocating the retention of maiden names by married women. [after *Lucy Stone*, 1818–93, U.S. woman suffragist]

Lud·dite (lud′īt) *n.* One of a band of workmen who joined in riots (1811–16) for the destruction of machinery, under the belief that its introduction reduced wages and increased unemployment: called after Ned Lud, a feeble-witted mechanic who destroyed several stocking-frames: in English industrial history.

Lu·den·dorff (lōō′dən·dôrf), **Erich von**, 1865–1937, German general in World War I.

Lü·der·itz (lü′dər·its) A port in SW South-West Africa: formerly *Angra Pequeña.*

Lu·dhi·a·na (lōō′dē·ä′nə) A city in central Punjab, India.

lu·di·crous (lōō′də·krəs) *adj.* Calculated to excite laughter; incongruous; droll; ridiculous. See synonyms under ABSURD, HUMOROUS,
QUEER, RIDICULOUS. [<L *ludicrus* < *ludere* play] — **lu′di·crous·ly** *adv.* — **lu′di·crous·ness** *n.*

Lu·do·vi·ka (lōō·dō·vē′kä) German and Swedish form of LOUISE.

Lud·wig (*Ger.* lōōt′vikh, *Sw.* lōōd′vig) German and Swedish form of LOUIS. Also *Latin* **Lu·do·vi·cus** (lōō′dō·vē′kəs).

— **Ludwig II**, 1845–86, king of Bavaria, 1864–1886; patron of the arts; declared insane; committed suicide.

Lud·wigs·ha·fen (lōōt′viks·hä·fən) A city in the Rhineland Palatinate, West Germany. Also **Lud′wigs·ha′fen-am-Rhein** (-äm·rīn′).

lu·es (lōō′ēz) *n.* 1 Formerly, an infectious or pestilential disease. 2 Plague. 3 Syphilis. [<L, plague, discharge < *luere* flow] — **lu·et·ic** (lōō·et′ik) *adj.*

luff (luf) *n. Naut.* 1 The sailing of a ship close to the wind. 2 The rounded part of a vessel's bow. 3 The foremost edge of a fore-and-aft sail. — *v.i. Naut.* 1 To bring the head of a vessel nearer the wind; sail near the wind. 2 To bring the head of a vessel into the wind, with the sails shaking. Also spelled *loof.* [Origin uncertain]

luf·fa (luf′ə) See LOOFAH.

Luft·waf·fe (lōōft′väf′ə) *n. German* The German air force, as organized by the Nazi regime in 1939. [<G, lit., air weapon]

lug[1] (lug) *n.* 1 The lobe of the ear; the ear. 2 Hence, an earlike projection such as an ear or handle for carrying or supporting, or for insertion of a handle, ring, pole, etc.: the *lugs* of a kettle. 3 The loop at the side of the saddle of a harness which holds the shaft. [Origin uncertain. Cf. Sw. *lugg* forelock.]

lug[2] (lug) *n.* 1 The act or exertion of lugging; anything that is moved with slowness and difficulty. 2 A shallow box or container, 13 1/2 by 16 inches inside dimensions, for carrying fruit or vegetables. 3 *Naut.* A square sail bent to a yard and having no boom: also *lugsail.* 4 *Slang* An exaction, especially of money for political uses. — *v.t.* **lugged, lug·ging** 1 To carry or pull with effort; drag. 2 *Colloq.* To bring, as irrelevant topics, into a conversation, discussion, etc., to introduce unreasonably. [Prob. <Scand. Cf. Sw. *lugga* pull by the hair.]

lug[3] (lug) *n.* A lugworm. [Origin uncertain. Cf. LG *lug* slow, heavy.]

Lug (lōō) In Old Irish mythology, one of the Tuatha De Danann, the god of light and sun, patron of the arts; father of Cuchulain. Also transliterated **Lugh**. [<Irish *Lugh*, lit., swift, active]

Lug-aid (lōō′ē) In Old Irish mythology, a king of Munster who joined forces with Maeve, Queen of Connacht, in the Cattle Raid of Cuailgne; also, his son, who later slew Cuchulain.

Lu·gansk (lōō·gänsk′) An industrial city of eastern Ukrainian S.S.R.: formerly (1935–58) called *Voroshilovgrad.*

Lu·ger (lōō′gər) *n.* A German automatic pistol originally developed and adopted by the German army, widely used in Europe as a military sidearm: manufactured in various calibers, most commonly 7.65 mm. and 9 mm.: also called *Parabellum* in Europe. [after Georg *Luger*, 19th c. German engineer]

lug·gage (lug′ij) *n.* 1 Anything burdensome or heavy to carry. 2 Baggage. [<LUG[2]]

lugged (lugd) *adj. Scot.* Having ears or earlike appendages.

lug·ger (lug′ər) *n.* A one-, two-, or three-masted vessel having lugsails only. [<LUG(SAIL) + -ER[2]]

lug·gie (lug′ē, lōōg′ē, lōō′gē) *n. Scot.* A small wooden dish with lugs or ears.

Lu·go (lōō′gō) A city in Galicia, NW Spain.

lug·sail (lug′səl, -sāl′) *n.* Lug[2] (def. 3).

lu·gu·bri·ous (lōō·gōō′brē·əs, -gyōō′-) *adj.* 1 Exhibiting or producing sadness; doleful. 2 Exaggeratedly solemn or sad. See synonyms under
SAD. [<L *lugubris* < *lugere* mourn] — **lu·gu′bri·ous·ly** *adv.* — **lu·gu′bri·ous·ness** *n.*

lug·worm (lug′wûrm′) *n.* An annelid worm (genus *Arenicola*) with two rows of tufted gills on the back, living in the sand of seashores: much used for bait: also called *lobworm.*

Lu·i·gi (lōō·ē′jē) Italian form of LOUIS. Also *Sp.* **Lu·ís** (lōō·ēs′), *Pg.* **Luiz** (lōō·ēs′).

Lu·i·gia (lōō·ē′jä) Italian form of LOUISA.

Luim·neach (lim′nəkh) The Gaelic name for LIMERICK.

Lu·i·ni (lōō·ē′nē), **Bernardino**, 1470?–1530?, Italian painter.

Lu·i·sa (lōō·ē′sä) Italian and Spanish form of LOUISA. Also *Ger.* **Lu·i·se** (lōō·ē′zə), *Pg.* **Lu·i·za** (lōō·ē′zə).

Lu·it·pold (lōō′it·pōlt) Old German form of LEOPOLD.

Luke (lōōk) A masculine personal name. [<L, light]

— **Luke, Saint** Physician and companion of St. Paul; traditionally thought to be author of the third gospel and of *The Acts of the Apostles.*

luke·warm (lōōk′wôrm′) *adj.* 1 Moderately warm; tepid. 2 Hence, not ardent or hearty; indifferent. [Prob. OE *hlēow* warm. Cf. LG *luk* tepid.] — **luke′warm′ly** *adv.* — **luke′warm′ness** *n.*

Lü·le·bur·gaz (lü′lə·bōōr·gäz′, -gäs′) A town in central Turkey in Europe; scene of a battle of the First Balkan War, 1912. Also **Lu′le Bur·gas′.**

lull (lul) *v.t.* 1 To soothe or quiet; put to sleep. 2 To calm; allay, as suspicions. — *v.i.* 3 To become calm. — *n.* 1 An abatement of noise or violence; an interval of calm or quiet. 2 That which lulls or soothes. See synonyms under TRANQUILIZE. [Prob. <Sw. *lulla* sing to sleep; ult. imit.]

lull·a·by (lul′ə·bī) *n. pl.* **·bies** 1 A song to lull a child to sleep; a cradlesong; also, the music for such a song. 2 *Obs.* A goodnight or farewell. [<LULL]

Lul·ly (lü·lē′), **Jean Baptiste**, 1633?–87, French composer born in Italy. Also **Lul·li′.**

— **Lul·ly** (lul′ē), **Raymond**, 1235?–1315, Spanish ecclesiastic and philosopher.

lu·lu (lōō′lōō) *n. Slang* Anything of an exceptional or outstanding nature, as a difficult examination, a beautiful person, etc. [Prob. reduplication of *Lou*, the familiar form of LOUISE]

lum (lum) *n. Brit. Dial. & Scot.* A chimney.

lu·ma·chelle (lōō′mə·kel) *n.* A variety of limestone containing fragments of shells, fossils, etc., sometimes brilliantly iridescent. Also **lu′ma·chel, lu′ma·chel′la** (-kel′ə). [<Ital. *lumachella*, dim. of *lumaca* snail <L *limax, limacis*]

lum·ba·go (lum·bā′gō) *n. Pathol.* Rheumatic pain in the lumbar region of the back; backache. [<L <*lumbus* loin]

lum·bar (lum′bər, -bär) *adj.* Pertaining to or situated near the loins. — *n.* A lumbar vertebra or nerve. [<NL *lumbaris* <L *lumbus* loin]

lum·ber[1] (lum′bər) *n.* 1 *U.S. & Can.* Timber sawed into boards, planks, etc. 2 Disused articles laid aside. 3 Hence, rubbish. — *adj.* Made of, pertaining to, or dealing in, lumber. — *v.t.* 1 *U.S. & Can.* To cut down (timber); also, to cut down the timber of (an area). 2 To fill or obstruct with useless articles. 3 To heap in disorder. — *v.i.* 4 *U.S. & Can.* To cut down or saw timber for marketing. [Var. of *Lombard* in obs. sense of "moneylender, pawnshop"; hence, stored articles] — **lum′ber·er** *n.*

lum·ber[2] (lum′bər) *v.i.* 1 To move or proceed in a heavy or awkward manner. 2 To move with a rumbling noise. — *n.* A rumbling noise. [Origin uncertain. Cf. Sw. *lomra* resound, *loma* walk heavily.]

lum·bered (lum′bərd) *adj.* Having the timber cut off: said of land.

lum·ber·ing[1] (lum′bər·ing) *n.* The business of felling and shaping timber into logs, boards, etc.

lum·ber·ing[2] (lum′bər·ing) *adj.* Clumsily huge and heavy; moving heavily; rumbling. — **lum′ber·ing·ly** *adv.* — **lum′ber·ing·ness** *n.*

lum·ber·jack (lum′bər·jak) *n.* 1 A lumberman.

2 A boy's or man's short, straight jacket of heavy warm material: originally made in imitation of coats worn by lumbermen: also **lum′ber·jack′et** (-jak′it). [< LUMBER[1] + *jack man, boy*]
lum·ber·man (lum′bər·mən) *n. pl.* **·men** (-mən) A person engaged in the business of lumbering.
lum·ber-room (lum′bər·room′, -room′) *n.* A room for lumber or useless articles.
lum·ber·some (lum′bər·səm) *adj.* Cumbersome.
lum·ber·yard (lum′bər·yärd′) *n.* A yard for the storage or sale of lumber.
lum·bri·ca·lis (lum′brə·kā′lis) *n. pl.* **·les** (-lēz) *Anat.* 1 One of the four vermiform muscles in the palm of the hand. 2 One of four similar muscles in the sole of the foot. [< NL < *lumbricus* worm] — **lum′bri·cal** *adj.*
lum·bri·coid (lum′brə·koid) *adj.* 1 Of or pertaining to the common earthworm (genus *Lumbricus*). 2 Pertaining to or designating a roundworm (*Ascaris lumbricoides*) parasitic in the intestines of man. — *n.* The roundworm. [< L *lumbricus* earthworm + -OID]
lu·men (loo′mən) *n. pl.* **·mens** or **·mi·na** (-mə·nə) 1 A passageway or opening. 2 *Biol.* The inner cavity of a cell or tubular organ, as a gland. 3 *Physics* A unit for measuring luminous flux, equal to the flow through a steradian from a uniform point source of one international candle. [< L, light]
lum·head (lum′hed′) *n. Scot.* A chimney top.
Lu·mi·nal (loo′mə·nəl, -nôl) Proprietary name for a brand of phenobarbital.
lu·mi·nance (loo′mə·nəns) See BRIGHTNESS (def. 2).
lu·mi·nar·y (loo′mə·ner′ē) *n. pl.* **·nar·ies** 1 Any body that gives light; specifically, one of the heavenly bodies as a source of light. 2 One who enlightens men or makes clear any subject. 3 Any light or source of illumination. [< OF *luminaire* < LL *luminarium* candle, torch < *lumen* light]
lu·mi·nesce (loo′mə·nes′) *v.i.* **·nesced, ·nesc·ing** To become luminescent. [Back formation < LUMINESCENT]
lu·mi·nes·cence (loo′mə·nes′əns) *n.* A light-emission not directly attributable to the heat that produces incandescence, as *bioluminescence.*
lu·mi·nes·cent (loo′mə·nes′ənt) *adj.* Emitting or capable of emitting light apart from incandescence or high temperature. [< L *lumen* light + -ESCENT]
lu·mi·nif·er·ous (loo′mə·nif′ər·əs) *adj.* Producing or conveying light. [< L *lumen* light + -FEROUS]
lu·mi·nos·i·ty (loo′mə·nos′ə·tē) *n. pl.* **·ties** 1 The quality of being luminous. 2 Something luminous. 3 *Physics* The ratio of the luminous flux of an object to the corresponding radiant flux, expressed in lumens per watt.
lu·mi·nous (loo′mə·nəs) *adj.* 1 Giving or emitting light; shining. 2 Full of light; well lighted; bright. 3 Perspicuous; lucid. See synonyms under BRIGHT, VIVID. [< L *luminosus* < *lumen* light] — **lu′mi·nous·ly** *adv.* — **lu′mi·nous·ness** *n.*
luminous energy Light.
luminous flux *Physics* The time rate of the flow of visible light, expressed in lumens.
luminous intensity *Physics* The luminous flux emitted from a point source of light per spheradian.
lu·mis·ter·ol (loo·mis′tər·ôl) *n.* A sterol, $C_{28}H_{44}O$, resulting from the ultraviolet irradiation of ergosterol. [< *lumi-* (< L *lumen* light) + (ERGO)STEROL]
lum·mox (lum′əks) *n. U.S. Colloq.* A stupid, heavy, clumsy person. [Cf. G *lümmel* lout]
lump[1] (lump) *n.* 1 A shapeless mass of inert matter, especially a small mass. 2 A mass of things thrown together. 3 A protuberance. 4 A stupid person. 5 A heavy, ungainly person. See synonyms under MASS. — **in a** (or **the**) **lump** All together; with no distinction. — *v.t.* 1 To put together in one mass, group, etc. 2 To consider or treat collectively; to *lump* facts. 3 To make lumps in or on. — *v.i.* 4 To become lumpy; raise in lumps. 5 To move or fall heavily. [< ME, prob. < Scand. Cf. Dan. *lumpe* a lump.]
lump[2] (lump) *v.t. Colloq.* To put up with; endure (something disagreeable): You may like it or *lump* it. [Origin uncertain]
lump·er (lum′pər) *n.* One who lumps. 2 A stevedore.
lump·fish (lump′fish′) *n. pl.* **·fish** or **·fish·es** A fish of northern seas (*Cyclopterus lumpus*),

oval in shape and with the skin studded with three lateral rows of tubercles. Also **lump′suck′er** (-suk′ər). [So called from its bulk and clumsiness]
lump·ing (lum′ping) *adj.* Heavy; bulky.
lump·ish (lum′pish) *adj.* Like a lump; stupid. — **lump′ish·ly** *adv.* — **lump′ish·ness** *n.*
lump sum A full and single sum.
lump·y (lum′pē) *adj.* **lump·i·er, lump·i·est** 1 Full of lumps. 2 Lumpish or gross. 3 Running in confused, pounding waves that do not break: a *lumpy* sea. — **lump′i·ly** *adv.* — **lump′i·ness** *n.*
lumpy jaw Actinomycosis.
lu·na (loo′nə) *n.* In alchemy, silver. [< L, the moon]
Lu·na (loo′nə) In Roman mythology, the goddess of the moon and of months: identified with the Greek *Selene.* [< L]
lu·na·cy (loo′nə·sē) *n. pl.* **·cies** 1 An intermittent form of insanity: formerly supposed to depend on the changes of the moon. 2 In forensic psychiatry and law, mental unsoundness to the point of irresponsibility. 3 Wild foolishness; wanton and senseless conduct. See synonyms under INSANITY. [< LUNATIC]
Luna moth A large North American moth (*Tropaea luna*), having light-green wings with long tails and lunate spots. [So called from the crescent-shaped spots on its wings]

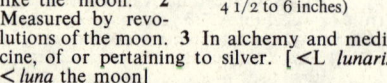
LUNA MOTH
(Wingspread from 4 1/2 to 6 inches)

lu·nar (loo′nər) *adj.* 1 Of or pertaining to the moon; crescented or orbed; like the moon. 2 Measured by revolutions of the moon. 3 In alchemy and medicine, of or pertaining to silver. [< L *lunaris* < *luna* the moon]
lunar caustic *Chem.* Silver nitrate formed into pencils and used for cauterizing. [< LUNA]
lunar day The period of the moon's rotation on its axis.
lu·nar·i·an (loo·nâr′ē·ən) *adj.* Pertaining to the moon. — *n.* 1 A supposed inhabitant of the moon. 2 An investigator of the moon. [< L *lunaris*]
lunar month See under MONTH.
lu·na·ry (loo′nər·ē) *n.* Connected with the moon; lunar. — *n.* A fern, the moonwort. [< L *lunaris*]
lunar year Twelve lunar months.
lu·nate (loo′nāt) *adj.* Crescent-shaped. Also **lu′nat·ed.** [< L *lunatus*]
lu·na·tic (loo′nə·tik) *adj.* 1 Affected with lunacy. 2 Characteristic of or resembling lunacy; crazy; insane. Also **lu·nat·i·cal** (loo·nat′i·kəl). See synonyms under INSANE. — *n.* An insane person. [< LL *lunaticus* < L *luna* moon]
lu·na·tion (loo·nā′shən) *n. Astron.* The interval between two returns of the new moon, averaging 29.53059 days. [< Med. L *lunatio, -onis* < L *luna* the moon]
lunch (lunch) *n.* 1 A light meal between other meals, as between breakfast and dinner. 2 Food provided for a lunch. — *v.i.* To eat lunch. — *v.t.* To furnish lunch for. [Short for LUNCHEON] — **lunch′er** *n.*
lunch·eon (lun′chən) *n.* 1 A bit of food taken between meals. 2 A lunch. [? Blend of dial. E *lunch* a lump of food + obs. *nuncheon* afternoon snack]
lunch·eon·ette (lun′chən·et′) *n.* A very light lunch; also, a place or counter where light lunches can be obtained.
lunch·room (lunch′room′, -room′) *n.* A restaurant serving light meals and refreshments.
Lun·dy's Lane (lun′dēz) A region near Niagara Falls, Ontario; scene of a battle (1814) in the War of 1812.
lune[1] (loon) *n.* 1 *Geom.* A figure bounded by two arcs of circles. 2 The moon. ◆ Homophone: *loon.* [< L *luna* the moon]
lune[2] (loon) *n.* A leash used in hawking. ◆ Homophone: *loon.* [< OF *loigne* leash < ML *longea* < L *longus* long]
Lü·ne·burg (lü′nə·boorkh) A city in Lower Saxony, West Germany.
Lü·nen (lü′nən) A city in North Rhine-Westphalia, West Germany.
lunes (loonz) *n. pl. Archaic* Outbreaks of lu-

nacy; mad freaks. [< Med. L *luna* moon, fit of lunacy]
lu·nette (loo·net′) *n.* 1 Something shaped like a half-moon. 2 *Mil.* A fieldwork or a detached work formed by two parallel flanks, two faces meeting in a salient angle, with an open gorge. 3 *Archit.* An arched opening in the side of a long vault formed by the intersection with it of a smaller vault, as for a window. 4 A ring on the tongue or trail plate of a vehicle, for attaching to a limber, motor truck, or other towing vehicle. Also **lu·net** (loo′nit), *n.* [< F, dim. of *lune* moon]
Lu·né·ville (lü·nā·vēl′) A town in NE France; the Treaty of Lunéville, effecting peace between Austria and France, was signed here in 1801.
lung (lung) *n. Anat.* 1 Either of the two sac-like organs of respiration in air-breathing vertebrates. ◆ Collateral adjective: *pulmonary.* 2 An analogous organ in invertebrates. [OE *lungen*]
lunge (lunj) *n.* 1 A sudden pass or thrust; specifically, a long thrust with a sword or a bayonet. 2 A sudden forward lurch; plunge. 3 A long rope used in training horses. — *v.* **lunged, lung·ing** — *v.i.* 1 To make a lunge or pass; thrust. 2 To move with a lunge. — *v.t.* 3 To thrust with or as with a lunge. 4 To cause (a horse) to circle at the end of a lunge. [Aphetic var. of obs. *allonge* < F *allonger* prolong < L *ad* + *longus* long]
lung·er (lung′ər) *n. Slang* A person having tuberculosis of the lungs.
lung·fish (lung′fish′) *n. pl.* **·fish** or **·fish·es** A dipnoan.
lun·gi (loon′gē) *n.* 1 In India, a loincloth. 2 The material from which it is made: also used for scarfs and turbans. Also **lun′gee.** [< Hind.]
Lung-ki (loong′kē′) A port of southern Fukien province, China: formerly *Changchow.*
lung·worm (lung′wûrm′) *n.* Any of several nematode worms parasitic in the lungs of animals, especially the common lungworm (genus *Dictyocaulus*) of horses, cattle, and sheep.
lung·wort (lung′wûrt′) *n.* 1 A European herb (genus *Pulmonaria*) of the borage family, having white blotches on the leaves. 2 The Virginia cowslip.
luni- *combining form* Of or pertaining to the moon; lunar. [< L *luna* moon]
lu·ni·so·lar (loo′ni·sō′lər) *adj.* Of or resulting from the combined action of the sun and moon.
lu·ni·ti·dal (loo′ni·tīd′l) *adj.* Relating to the tides as produced by the moon's attraction.
lunitidal interval The interval between the moon's passage of the meridian and lunar high tide.
lunk·head (lungk′hed′) *n. Slang* A slow-witted person. [Prob. alter. of LUMP + HEAD] — **lunk′head·ed** *adj.*
lunt (lunt, lônt) *n. Scot.* 1 A whiff of smoke. 2 A slow-burning match or torch.
Lunt (lunt), **Alfred,** born 1893, U.S. actor.
lu·nu·la (loo′nyə·lə) *n. pl.* **·lae** (-lē) 1 A crescentic structure or appearance. 2 The whitish area at the base of the nails: also **lu′nule** (-nyool). 3 A lune (arc). [< L, diminutive of *luna* moon]
lu·nu·lar (loo′nyə·lər) *adj.* Having the form of a small crescent. [< LUNULA]
lu·nu·late (loo′nyə·lāt, -lit) *adj.* 1 Having or approaching a crescent form. 2 Having crescentic markings. Also **lu′nu·lat·ed.**
lu·ny (loo′nē) See LOONY.
Lu·per·ca·li·a (loo′pər·kā′lē·ə, -kāl′yə) *n. pl.* A Roman festival celebrated on February 15, in honor of a rustic deity, **Lu·per·cus** (loo·pûr′kəs) or *Faunus,* at which after various ceremonies the priests ran half-clothed through the streets. Also **Lu′per·cal** (-kal). [< L, neut. pl. of *Lupercalis* of Lupercus < *Lupercus Faunus* < *lupus* wolf]
Lu·per·ci (loo·pûr′sī) *n. pl.* The priests of Lupercus or Faunus. [< L]
lu·pine[1] (loo′pin) *adj.* 1 Of or pertaining to a wolf; like a wolf; wolfish. 2 Pertaining to the group of canines that includes dogs and wolves. [< L *lupinus* < *lupus* wolf]
lu·pine[2] (loo′pin) *n.* A plant of the bean family (genus *Lupinus*) bearing terminal racemes of mostly blue or purple flowers, as the **white lupine** of the Old World (*L. albus*) whose seeds are edible. Also **lu′pin** (-pin). [< F *lupin*

lupulin 759 **Luxor**

[< L *lupinus* wolflike; reason for the name unknown]

lu·pu·lin (loō′pyə·lin) *n.* 1 A brittle crystalline compound forming the active principle of hops. 2 A yellow aromatic powder contained in the fruit of hops, and used medicinally as a sedative. [< NL *lupulus* the hop, dim. of L *lupus*]

lu·pus (loō′pəs) *n. Pathol.* A chronic tuberculous skin disease with warty nodules, generally about the nose. Also **lupus vul·ga·ris** (vul·gâr′is). [< L, wolf]

Lu·pus (loō′pəs) *n.* The Wolf, a constellation. See CONSTELLATION. [< L]

lurch[1] (lûrch) *v.i.* 1 To roll suddenly to one side, as a ship at sea. 2 To move unsteadily; stagger. — *n.* A sudden swaying or rolling to one side; hence; any sudden swinging movement. [Origin uncertain]

lurch[2] (lûrch) *n.* In cribbage, the state of a player who has made 30 holes or less while his opponent has won with 61; also a game thus won. — **to leave in the lurch** To leave in an embarrassing position. — *v.i.* In cribbage, piquet, and similar games, to win a double game. [< F *lourche*, name of a game < *lourche* deceived < Gmc. Cf. MDu. *lurz* left-handed, unlucky.]

lurch[3] (lûrch) *v.t.* & *v.i. Obs.* To swindle; cheat. [Var. of LURK]

lurch·er (lûr′chər) *n.* 1 One who lies in wait or lurks; a lurking thief; poacher. 2 A crossbred dog that hunts by scent and in silence.

lur·dan (lûr′dən) *Archaic adj.* Stupid; incapable. — *n.* A stupid person; blockhead. Also **lur′dane**. [< OF *lourdin* < *lourd* heavy]

lure (loor) *n.* 1 A device resembling a bird and sometimes baited with food: fastened to a falconer's wrist and used to recall the hawk. 2 In angling, an artificial bait; also, a decoy for animals. 3 Anything that invites by the prospect of advantage or pleasure. — *v.t.* **lured, lur·ing** 1 To attract or entice; allure. 2 To recall (a hawk) with a lure. See synonyms under ALLURE, DRAW. [< OF *leurre* < MHG *luoder* bait] — **lur′er** *n.*

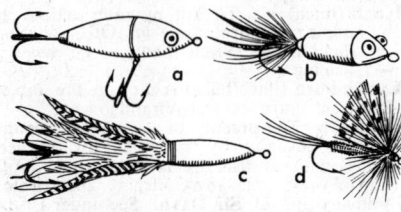

LURES
a. Wood or plastic lure.
b. Floating cork lure.
c. Feather lure.
d. Dry fly.

lu·rid (loor′id) *adj.* 1 Giving a ghastly or dull-red light; dismal. 2 Of a dingy, dirty-brown color. 3 Violent; terrible; sensational: a *lurid* crime. 4 Pale; wan; sallow; ghastly. [< L *luridus* sallow] — **lu′rid·ly** *adv.* — **lu′rid·ness** *n.*

lurk (lûrk) *v.i.* 1 To lie hidden, as in ambush. 2 To exist unnoticed or unsuspected. 3 To move secretly or furtively; slink. [Origin uncertain. Cf. Norw. *lurka* sneak off.] — **lurk′er** *n.* — **lurk′ing·ly** *adv.*

Lur·lei (loor′lī) See LORELEI.

Lu·sa·ka (loo·sä′kə) The capital of Northern Rhodesia, in the center of the territory.

Lu·sa·tia (loo·sā′shə) A region of SE Germany and SW Poland between the Elbe and the Oder. German **Lau·sitz** (lou′zits).

lus·cious (lush′əs) *adj.* 1 Very grateful to the sense of taste; rich, sweet, and delicious; pleasing to any sense or to the mind. 2 Sweet and rich to excess. See synonyms under DELICIOUS, SWEET. [? Blend of LUSH and DELICIOUS] — **lus′cious·ly** *adv.* — **lus′cious·ness** *n.*

lush[1] (lush) *adj.* 1 Full of juice or succulence; fresh and luxuriant; also, characterized by abundant growth. 2 Easily plowed and mellow, as ground. [? Var. of dial. E *lash* soft and watery < OF *lasche* < L *laxus* loose] — **lush′ness** *n.*

lush[2] (lush) *Slang n.* 1 Intoxicating, or strong alcoholic beverage. 2 A drunken man: also

lush′er. — *v.t.* & *v.i.* To drink (alcoholic liquor). [Origin unknown]

Lu·shun (lü′shoon′) The Chinese name for PORT ARTHUR.

lush·y (lush′ē) *adj. Slang* Drunk.

Lu·si·ta·ni·a (loō′sə·tā′nē·ə) An ancient name for PORTUGAL.

Lu·si·ta·ni·a (loō′sə·tā′nē·ə) *n.* A British passenger ship torpedoed and sunk by a German submarine off the coast of Ireland May 7, 1915, with loss of 1,134 lives including 114 Americans.

Lu·si·ta·ni·an (loō′sə·tā′nē·ən) *adj.* Of or pertaining to Lusitania. — *n.* A native or inhabitant of Lusitania.

lust (lust) *n.* 1 Vehement or longing affection or desire. 2 Inordinate desire for carnal pleasure. 3 *Obs.* Pleasure; inclination. See synonyms under APPETITE. — *v.i.* To have passionate or inordinate desire, especially sexual desire. [OE, pleasure]

lus·ter (lus′tər) *n.* 1 Natural or artificial brilliancy or sheen; refulgence; gloss. 2 Brilliancy of beauty, of character, or of achievements; splendor. 3 A glaze, varnish, or enamel applied to porcelain in a thin layer, and giving it a smooth, glistening surface: common in the phrase *metallic luster*. 4 A dress material having a highly finished surface. 5 A source or center of light; specifically, a branched candelabrum, chandelier, or the like. 6 *Mineral.* The quality of the surface of a mineral as regards the kind and intensity of the light it reflects. — *v.* **·tered** or **·tred, ·ter·ing** or **·tring** *v.t.* To give a luster or gloss to. — *v.i.* To be or become lustrous. Also **lus′tre**. [< F *lustre* < Ital. *lustro* < *lustrare* shine < L *lustrum* purification]

lus·ter·ware (lus′tər·wâr′) *n.* Pottery treated with metallic luster.

lust·ful (lust′fəl) *adj.* 1 Having carnal or sensual desire. 2 *Archaic* Vigorous; lusty. — **lust′ful·ly** *adv.* — **lust′ful·ness** *n.*

lust·i·hood (lus′tē·hood) *n. Archaic* Vigor of body; brawny strength. Also **lust′i·head** (-hed).

lus·tral (lus′trəl) *adj.* 1 Pertaining to or used in purification. 2 Pertaining to a lustrum. [< L *lustralis* < *lustrum* purification]

lus·trate (lus′trāt) *v.t.* **·trat·ed, ·trat·ing** To make pure by propitiatory offering or ceremony. [< L *lustratus*, pp. of *lustrare* purify by propitiatory offerings < *lustrum* purification]

lus·tra·tion (lus·trā′shən) *n.* The ancient rite of purification and expiation. [< L *lustratio*, *-onis*]

lus·trous (lus′trəs) *adj.* Having luster; shining; also, illustrious. See synonyms under BRIGHT. — **lus′trous·ly** *adv.* — **lus′trous·ness** *n.*

lus·trum (lus′trəm) *n.* 1 A lustration or purification; the solemn ceremony of purification of the entire Roman people made every five years. 2 A period of five years. [< L, ? < *luere* wash]

lust·y (lus′tē) *adj.* **lust·i·er, lust·i·est** 1 Full of vigor and health; able-bodied; robust. 2 *Obs.* Jolly; merry. 3 *Obs.* Agreeable; delightful. 4 *Obs.* Lustful. — **lust′i·ly** *adv.* — **lust′i·ness** *n.*

lu·sus na·tu·rae (loō′səs nə·tōōr′ē, -tyōōr′ē) A strange and abnormal natural production; a freak or sport of nature: also **lu′sus**. [< L, joke of nature]

Lut (loot) See DASHT-I-LUT.

lu·ta·nist (loō′tə·nist) *n.* One who plays the lute. Also **lu′te·nist**.

lute[1] (loot) *n.* A stringed musical instrument having a large, pear-shaped body like a mandolin, played by plucking the strings with the fingers. — *v.t.* & *v.i.* To play on the lute. [< OF *leüt* < Pg. *laut* < Arabic *al ′ud* the piece of wood]

lute[2] (loot) *n.* 1 A cementlike composition used to exclude air, as around pipe joints. 2 In

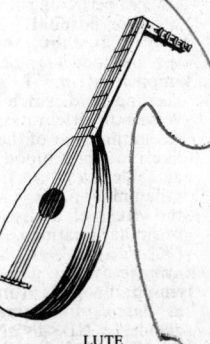

LUTE

brickmaking, a scraper having a cutting edge to smooth the surface of a drying yard. Also **lut′ing**. — *v.t.* **lut·ed, lut·ing** To seal with lute. [< OF *lut* < L *lutum* mud] — **lu·ta′tion** *n.*

lu·te·al (loō′tē·əl) *adj. Anat.* Pertaining to or similar to the cells of the corpus luteum. [< L *luteus* yellow]

lu·te·in (loō′tē·in, -tēn) *n. Biochem.* A yellow pigment, $C_{48}H_{56}O_2$, found in egg yolks, fat cells, and in the corpus luteum. [< L *luteum* egg yolk < *luteus* yellow] — **lu′te·in′ic** *adj.*

lu·te·o·lin (loō′tē·ə·lin) *n. Chem.* A yellow crystalline compound, $C_{15}H_{10}O_6$, obtained from dyer′s-broom, and used as a dyestuff. [< F *lutéoline* < NL (*Reseda*) *luteola* dyer's weed < L *luteus* yellowish < *luteus* yellow]

lu·te·ous (loō′tē·əs) *adj.* Of golden-yellowish color. [< L *luteus* < *lutum* weed used by dyers]

Lu·te·tia (loō·tē′shə) Ancient name for PARIS. Also **Lu·te′tia Pa·ri·si·o·rum** (pa·rē′sē·ō′rəm).

lu·te·tium (loō·tē′shəm) *n.* A metallic element (symbol Lu) of the lanthanide series, isolated from ytterbium. See ELEMENT. [from LUTETIA]

lu·te·um (loō′tē·əm) See CORPUS LUTEUM.

Lu·ther (loō′thər, Ger. loō′tər) *n.* A masculine personal name. Also *Latin* **Lu·ther·us** (loō′tər·əs). [< Gmc., famous warrior]

Lu·ther (loō′thər), **Martin**, 1483–1546, German monk, theologian, and leader of the Reformation; excommunicated in 1520 by Pope Leo X; composed many Lutheran hymns and translated the Bible into German.

Lu·ther·an (loō′thər·ən) *n.* A follower or disciple of Luther; a member of the Lutheran Church. — *adj.* 1 Pertaining to or believing in Martin Luther or his doctrines. 2 Pertaining to the Lutheran Church or its doctrines. — **Lu′ther·an·ism** *n.*

Lutheran Church A Protestant denomination founded in Germany by Martin Luther in the 16th century. Its chief doctrine is justification by faith alone. The Lutheran churches in America are grouped in three principal bodies: the **American Lutheran Church, Evangelical Lutheran Synodical Conference of North America**, and **United Lutheran Church in America**. See also MISSOURI SYNOD.

lu·thern (loō′thərn) *n.* A lucarne. [? Alter. of LUCARNE]

Lu·ton (loō′tən) A municipal borough of southern Bedford, England.

Lu·tu·a·mi·an (loō′tōō·ä′mē·ən) *n.* An independent linguistic stock of North American Indians, having two divisions, Klamath and Modoc, formerly dwelling in California and Oregon: now inhabiting reservations in Oregon and Oklahoma. Both dialects were most commonly known as *Klamath*.

Lüt·zen (lüt′sən) A town in the former state of Saxony-Anhalt, East Germany; scene of battles in the Thirty Years′ War (1632) and the Napoleonic Wars (1813).

Lu·wi·an (loō′wē·ən) *n.* An extinct language of Asia Minor of which few inscriptions remain: considered to be closely related to Hittite. — *adj.* Of this language. Also **Lu·vi·an** (loō′vē·ən).

lux (luks) *n. pl.* **lux·es** or **lu·ces** (loō′sēz) *Physics* The unit of illumination in the metric system, equivalent to the illumination on a surface of one square meter on which there is a uniformly distributed flux of one lumen. [< L, light]

lux·ate (luk′sāt) *v.t.* **·at·ed, ·at·ing** To put out of joint; dislocate. [< L *luxatus*, pp. of *luxare* dislocate < *luxus* dislocated] — **lux·a′tion** *n.*

luxe (looks, luks; *Fr.* lüks) *n.* Superfine quality; luxury: usually with *de*: edition *de luxe*. [< F < L *luxus* extravagance]

Lux·em·burg (luk′səm·bûrg, *Fr.* lük·sän·boōr′) 1 A constitutional grand duchy between Belgium, France, and Germany; 999 square miles; capital, Luxemburg. 2 A province of SE Belgium; 1,705 square miles; capital, Arlon. Also **Lux′em·bourg**.

Lux·em·burg (look′səm·boōrkh), **Rosa**, 1870–1919, German socialist leader.

lux·me·ter (luks′mē′tər) *n.* An instrument that measures illumination in terms of luxes.

lux mun·di (luks mun′dī) *Latin* Light of the world.

Lux·or (luk′sôr, loōk′-) A city on the Nile in Upper Egypt on the southern part of the site of ancient Thebes. *Arabic* **El Uk·sor** (al ŏŏk′sŏŏr).

lux·u·ri·ance (lug·zhŏŏr′ē·əns, luk·shŏŏr′-) *n.* Excessive growth; exuberance; lushness. Also **lux·u′ri·an·cy.**

lux·u·ri·ant (lug·zhŏŏr′ē·ənt, luk·shŏŏr′-) *adj.* **1** Exhibiting or characterized by luxuriance in growth; also, fertile, as soil. **2** Exuberant in fancy, invention, etc.; abundant, extravagant, or excessive in action or speech; ornate, florid, or rich in design. See synonyms under FERTILE. [<L *luxurians, -antis,* ppr. of *luxuriare.* See LUXURIATE.] **—lux·u′ri·ant·ly** *adv.*

lux·u·ri·ate (lug·zhŏŏr′ē·āt, luk·shŏŏr′-) *v.i.* **·at·ed, ·at·ing 1** To take great pleasure; indulge oneself fully. **2** To live sumptuously. **3** To grow profusely. [<L *luxuriatus,* pp. of *luxuriare* be fruitful, abound <*luxuria.* See LUXURY.] **—lux·u′ri·a′tion** *n.*

lux·u·ri·ous (lug·zhŏŏr′ē·əs, luk·shŏŏr′-) *adj.* **1** Pertaining or administering to luxury; voluptuous. **2** Supplied with luxuries. [<L *luxuriosus*] **—lux·u′ri·ous·ly** *adv.* **—lux·u′ri·ous·ness** *n.*

lux·u·ry (luk′shər·ē, *occasionally* lug′zhər·ē) *n. pl.* **·ries 1** A free indulgence in the pleasures that gratify the senses. **2** Anything that ministers to comfort or pleasure that is expensive or rare, but is not necessary to life, health, subsistence, etc.; a delicacy. [<OF *luxurie* <L *luxuria* <*luxus* extravagance]

Lu·zern (lōō·tsern′) The German name for LUCERNE.

Lu·zon (lōō·zon′, *Sp.* lōō·sōn′) The largest of the Philippines; 40,420 square miles; capital, Manila.

Luzon Strait The channel connecting the Pacific and the South China Sea: divided into the Bashi, Balintang, and Babuyan Channels.

Lvov (lvôf) A city in Ukrainian S.S.R.: German *Lemberg. Polish* **Lwów** (lvŏŏf).

-ly[1] *suffix of adjectives* Like; characteristic of; pertaining to: *manly, godly.* Compare -LIKE. [OE *-lic*]

-ly[2] *suffix of adverbs* **1** In a (specified) manner: used to form adverbs from adjectives, or (rarely) from nouns: *brightly, busily.* **2** Occurring every (specified interval): *hourly, yearly.* [OE *-lice* <*-lic* -LY[1]]

◆ In cases where an adjective already ends in *-ly,* the forms of the adjective and adverb are often identical: He spoke *kindly.* Occasionally, *-ly* is added to *-ly* (which is then changed to *-li*), as in *surlily,* but this is generally avoided as awkward. In the case of words derived from French adjectives in *-le,* the ending is dropped before adding *-ly,* as in *nobly, possibly.*

Lya·khov Islands (lyä′khôf) A southern group of the New Siberian Islands in Yakut Autonomous S.S.R.; total, 2,660 square miles.

Ly·all·pur (lī′əl·pŏŏr) A city in Punjab, West Pakistan.

ly·ard (lī′ərd) *adj. Scot.* Silver-gray or streaked with gray. Also **ly′art** (-ərt). [<OF *liart*]

Lyau·tey (lyō·tā′), **Louis Hubert,** 1854–1934, French marshal and colonial administrator.

Lyau·tey (lyō·tā′), **Port** See PORT-LYAUTEY.

ly·can·thrope (lī′kən·thrōp, lī·kan′thrōp) *n.* **1** A werewolf. **2** One afflicted with lycanthropy. [<Gk. *lykanthrōpos* <*lykos* wolf + *anthrōpos* a man]

ly·can·thro·py (lī·kan′thrə·pē) *n.* **1** The supposed power of turning a human being into a wolf or of becoming a wolf, by magic or witchcraft. **2** Belief in werewolves. **3** *Psychiatry* A mania in which the patient imagines himself to be a wolf or some other wild animal. [<Gk. *lykanthrōpia*] **—ly′can·throp′ic** (-throp′ik) *adj.*

Ly·ca·on (lī·kā′on) In Greek mythology, a king of Arcadia and father of Callisto, who, when Zeus sought his hospitality disguised as a poor man, tested the god's divinity by offering him human flesh as food: Zeus transformed him into a wolf. [<Gk. *Lykaōn* <*lykos* wolf]

Ly·ca·o·ni·a (lī′kā·ō′nē·ə, lik′ə-) An ancient district in the SE part of Phrygia, Asia Minor.

ly·cée (lē·sā′) *n. French* A public classical secondary school in France qualifying its students for a university; a lyceum. [<F <L *lyceum.* See LYCEUM.]

ly·ce·um (lī·sē′əm) *n. pl.* **·ce·ums** or **·ce·a** (-sē′ə) **1** An association for popular instruction by lectures, a library, debates, etc.; also, its building or hall. **2** An intermediate classical school. [<LYCEUM]

Ly·ce·um (lī·sē′əm) A grove near Athens in which Aristotle taught. [<L <Gk. *Lykeion* <*lykeios,* ? wolf-slaying, epithet of Apollo (whose temple was near this grove)]

ly·chee (lē′chē) See LITCHI.

lych·nis (lik′nis) *n.* Any plant of the genus *Lychnis* of erect ornamental herbs. *L. chalcedonica* is the scarlet lychnis, and *L. coronaria* the rose campion. [<L <Gk.]

Lyc·i·a (sis′ē·ə, lish′ə) An ancient country on the SW coast of Asia Minor; a Roman province.

Lyc·i·an (lish′ē·ən, lish′ən) *adj.* Belonging or relating to ancient Lycia. **—***n.* **1** A member of the people that inhabited Lycia. **2** Their language, believed to be related to Hittite.

Lyc·i·das (lis′ə·dəs) **1** In Virgil's *Eclogue* iii, a shepherd. **2** Subject of an elegy (1637) by Milton.

Ly·co·me·des (lī′kə·mē′dēz) In Greek legend, a king of Scyros; son of Apollo and guardian of Achilles.

ly·co·pod (lī′kə·pod) *n.* A plant of the club-moss family. [<Gk. *lykos* wolf + -POD; after the resemblance of the root to a claw]

ly·co·po·di·um (lī′kə·pō′dē·əm) *n.* **1** A plant (genus *Lycopodium*) of erect or creeping evergreen pteridophytes, the clubmosses. **2** An inflammable fine yellow powder, the spores of clubmosses. [<NL]

Ly·cur·gus (lī·kûr′gəs) A traditional Spartan lawgiver of the ninth century B.C.

Lyd·da (lid′ə) A town and site of Israel's chief airport in central Israel. *Arabic* **Ludd** (lŏŏd). *Hebrew* **Lud** (lŏŏd).

lyd·dite (lid′īt) *n.* A high explosive, chiefly of picric acid: used for torpedo and other shells which on explosion kill by shock or suffocation from its deadly fumes. [from *Lydd,* town in England, where first manufactured]

Lyd·gate (lid′gāt, -git), **John,** 1370?–1451?, English poet.

Lyd·i·a (lid′ē·ə; *Dan., Du.* lē′dē·ä, *Greek* lē·dē′ä) A feminine personal name. Also *Fr.* **Ly·die** (lē·dē′). [<Gk., Lydian girl]

Lyd·i·a (lid′ē·ə) An ancient country of western Asia Minor on the Aegean.

Lyd·i·an (lid′ē·ən) *adj.* **1** Belonging or relating to ancient Lydia in Asia Minor, famous for wealth, luxury and music; also, pertaining to its inhabitants. **2** By extension, effeminate; gentle: in reference to Lydian accomplishments and culture; also, sensuous or voluptuous. **—***n.* **1** One of the people of ancient Lydia. **2** Their language, probably related to Hittite. **3** One of the modes in early music.

lye (lī) *n.* **1** A solution leached from ashes, or derived from a substance containing alkali: used in making soap, preparing hominy, etc. **2** A lixivium. [OE *lēah*]

Ly·ell (lī′əl), **Sir Charles,** 1797–1875, English geologist.

ly·ing[1] (lī′ing) *n.* The practice of telling lies; untruthfulness. See synonyms under DECEPTION. **—***adj.* Addicted to, conveying, or constituting falsehood; mendacious; false. **—ly′ing·ly** *adv.*

ly·ing[2] (lī′ing) Present participle of LIE[1]. **—***adj.* Being in a horizontal position; prostrate.

ly·ing-in (lī′ing·in′) *n.* The confinement of women during childbirth; parturition. **—***adj.* Of or pertaining to childbirth: a *lying-in* home or hospital.

Lyl·y (lil′ē), **John,** 1554–1606, English dramatist and novelist. See EUPHUISM.

lymph (limf) *n.* **1** A transparent, colorless, alkaline fluid which circulates in the lymph vessels of vertebrates. It consists of a plasma resembling that of the blood and of corpuscles like the white blood corpuscles. **2** The coagulable exudation from the blood vessels in inflammation. **3** The virus, or a culture of the virus, of a disease, used in vaccination or similar treatment. **4** *Obs.* A spring; water. [<L *lympha,* var. of *limpa* water, ? <Gk. *nymphē* nymph, goddess of moisture]

lym·phad·e·ni·tis (lim·fad′ə·nī′tis, lim′fə·də-) *n. Pathol.* Inflammation of the lymphatic glands. [<NL <LYMPH(O)- + ADEN(O)- + -ITIS]

lym·phan·gi·al (lim·fan′jē·əl) *adj.* Of or pertaining to the lymphatic vessels; lymphatic. [<LYMPH(O)- + ANGI(O)- + -AL[1]]

lym·phat·ic (lim·fat′ik) *adj.* **1** Pertaining to, containing, or conveying lymph. **2** Caused by or affecting the lymphatic glands. **3** Having a phlegmatic temperament. **—***n.* A vessel that conveys lymph into the veins; an absorbent vessel. [<NL <L *lymphaticus* frantic, frenzied, trans. of Gk. *nympholēptos* caught by nymphs]

lym·pha·tism (lim′fə·tiz′əm) *n.* **1** Lymphatic temperament. **2** *Pathol.* An unhealthy condition due to an excess of lymphatic tissue. [<LYMPHATIC]

lym·pha·ti·tis (lim′fə·tī′tis) *n. Pathol.* Inflammation in any part of the lymphatic system. [<LYMPHAT(O)- + -ITIS]

lymphato- *combining form* Lymphatic. Also, before vowels, **lymphat-.**

lym·pha·tol·y·sis (lim′fə·tol′ə·sis) *n. Pathol.* Dissolution or breakdown of lymphatic tissue. [<LYMPHATO- + -LYSIS] **—lym·pha·to·lyt·ic** (lim′fə·tō·lit′ik) *adj.*

lymph cell A lymphocyte.

lymph gland *Anat.* One of the nodular bodies about the size of a pea, found in the course of the lymphatic vessels, composed of a reticulum containing lymphoid cells.

lympho- *combining form* Lymph; of or pertaining to lymph or the lymphatic system: *lymphocyte.* Also, before vowels, **lymph-.** [See LYMPH]

lym·pho·cyte (lim′fə·sīt) *n. Anat.* A variety of nucleated, colorless leucocyte in the lymphatic glands of all vertebrate animals, resembling a small white blood corpuscle.

lym·pho·cy·to·sis (lim′fə·sī·tō′sis) *n. Pathol.* An excess of lymphocytes in the blood. **—lym′pho·cy·tot′ic** (-tot′ik) *adj.*

lym·phoid (lim′foid) *adj.* Of, pertaining to, or resembling lymph or a lymphatic gland. [<LYMPH + -OID]

lymphoid tissue Tissue constituting the lymph glands; adenoid tissue.

lym·phor·rhe·a (lim′fə·rē′ə) *n. Pathol.* A flow of lymph from ruptured lymph vessels.

lyn (lin) See LIN.

lyn·ce·an (lin·sē′ən) *adj.* **1** Pertaining to or characteristic of the lynx. **2** Lynx-eyed; sharp-sighted. [<L *lynceus* <Gk. *lynkeios* <*lynx* lynx]

lynch (linch) *v.t.* To kill by mob action, as by hanging or burning. **—***adj.* Of or relating to lynching. [<LYNCH LAW] **—lynch′er** *n.* **—lynch′ing** *n.*

Lynch·burg (linch′bûrg) A city on the James River in south central Virginia.

lynch law The practice of administering punishment, usually by hanging, for alleged crimes without trial by law. [? after Capt. Wm. *Lynch,* 1742–1820, Virginia magistrate]

Lynd·say (lin′zē), **Sir David** See under LINDSAY.

Lynn (lin) **1** An industrial city in eastern Massachusetts. **2** See KING'S LYNN.

Lynn Canal A fiord in the mainland of SE Alaska near Alexander Archipelago.

Lynn Re·gis (rē′jis) See KING'S LYNN.

lynx (lingks) *n.* Any of several wildcats (genus *Lynx*) of Europe and North America, with short tails, tufted ears, and relatively long limbs; especially, the Canadian lynx (*L. canadensis*) and the North American bay lynx (*L. rufus*). [<L *lynx, lyncis* <Gk. *lynx*]

Lynx (lingks) A northern constellation. See CONSTELLATION.

lynx-eyed (lingks′īd′) *adj.* Having acute sight.

lyo- *combining form* A loosening; dissolution: *lyophilic.* Also, before vowels, **ly-.** [<Gk. *lyein* loosen]

Ly·on (lī′ən), **Mary,** 1797–1849, U. S. educator; founder of Mt. Holyoke College.

Ly·on·nais (lē·ô·nā′) A region and former province of east central France. Also **Ly·o·nais′.**

ly·on·naise (lī′ə·nāz′, *Fr.* lē·ô·nez′) *adj.* Made with finely sliced onions; especially, designating a method of preparing potatoes with fried onions. [<F, fem. of *lyonnais* (of Lyon]

Ly·on·nesse (lī′ə·nes′) In Arthurian legend, a region lying between Cornwall and the Scilly Islands, supposed to have sunk into the sea.

Ly·ons (lī′ənz, *Fr.* lē·ôṅ′) A city of east central France at the confluence of the Rhône and the Saône; the third largest city of France. Also **Ly·on′.** **—Ly·on·nais′** *adj.*

ly·o·phil·ic (lī′ə·fil′ik) *adj. Chem.* Pertaining to or designating a colloidal system either phase of which is more or less freely soluble

lyophobic / 761 / **macaroni**

in the other. [<LYO- + Gk. *philos* loving]
ly·o·pho·bic (lī'ə·fō'bik) *adj. Chem.* Resisting solution: said of colloidal systems neither phase of which will freely dissolve in the other. [<LYO- + Gk. *phobos* fear]
ly·o·sorp·tion (lī'ə·sôrp'shən) *n. Chem.* The adsorption of a solvent film upon the surface of suspended particles. [<LYO- + (AD)SORPTION]
ly·o·trop·ic (lī'ə·trop'ik, -trō'pik) *adj. Chem.* 1 Designating a series of ions, radicals, or salts arranged according to their influence on certain colloidal, physiological, or catalytic phenomena. 2 Having an affinity for entering into solution.
Ly·ra (lī'rə) An ancient constellation representing the lyre of Hermes or of Orpheus; the Harp. It contains the bright star Vega. See CONSTELLATION. [<L <Gk.]
ly·rate (lī'rāt) *adj. Bot.* Lyre-shaped, as a pinnatifid leaf having its upper lobes largest. Also **ly'rat·ed**. [<NL *lyratus*] — **ly'rate·ly** *adv.*
lyre (līr) *n.* An ancient harplike stringed instrument, having a hollow body and two horns bearing a crosspiece between which and the body were stretched the strings, generally seven. [<L *lyra* <Gk.]
Lyre (līr) The constellation Lyra.
lyre·bird (līr'bûrd') *n.* An Australian passerine bird (genera *Menura* and *Harriwhitea*) having the tail feathers of the males arranged in lyre shape.
lyr·ic (lir'ik) *adj.* 1 Originally, belonging to a lyre; hence, adapted for singing to a lyre. 2 Characterizing verse expressing the poet's personal emotions or sentiments; songlike: distinguished from *epic, dramatic.* 3 Pertaining to the writing of such verse; having written lyric poetry. 4 Musical; singing or to be sung. 5 *Music* Light, graceful, and having a flexible quality: opposed to *dramatic*: said of the voice in singing. Also **lyr'i·cal**. — *n.* 1 A lyric poem; a song. 2 Usually pl. The words of a song, especially as distinguished from the music. [<L *lyricus* <Gk. *lyrikos* <*lyra* a lyre] — **lyr'i·cal·ly** *adv.*

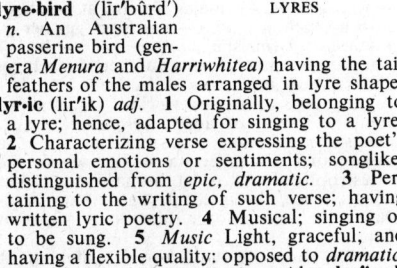
LYRES

lyr·i·cism (lir'ə·siz'əm) *n.* 1 Lyrical quality; a lyric composition. 2 Emotional expression, as of enthusiasm.
lyr·i·cist (lir'ə·sist) *n.* 1 One who writes the words of a song or the lyrics for a musical play. 2 A lyric poet.
lyric poetry Any form of poetry giving expression to the poet's personal emotions or sentiments. It includes the *sonnet*, the *elegy, ode, song, psalm, hymn,* etc.
ly·ri·form (lī'rə·fôrm) *adj.* Shaped like a lyre.
lyr·ism (lir'iz·əm) *n.* 1 Lyre-playing; hence, the singing of lyrics, or music in general. 2 Lyricism. [<F *lyrisme*]
lyr·ist (lir'ist) *n.* 1 One who plays the lyre. 2 A lyric poet. [<L *lyristes* <Gk. *lyristēs* <*lyrixein* play on a lyre]
Lys (lēs) A river in northern France and western Belgium, flowing 130 miles NE to the Scheldt. *Flemish* **Lei'e** (lī'ə).
Ly·san·der (lī·san'dər), died 395 B.C., Spartan general.
lyse (līs) *v.t. & v.i.* **lysed, lys·ing** To undergo or cause to undergo lysis. [<Gk. *lysis* loosing]
Ly·sen·ko (li·seng'kō, *Russian* lē'sin·kə), **Trofim,** born 1898, U.S.S.R. biologist and agronomist: also *Lisenko.*
Ly·sen·ko·ism (li·seng'kō·iz'əm) *n.* The theory regarding the nature and processes of inheritance which challenges the formal genetics of Mendel, Weismann, and Morgan, claiming especially that the genes may be permanently modified by somatic influences, thus permitting a selective inheritance of acquired characteristics. [after T. *Lysenko*]
lysi- *combining form* A loosening: *lysigenetic.* Also, before vowels, **lys-**. [<Gk. *lysis* a loosening]
Lys·i·as (lis'ē·əs), 458?–378? B.C., Athenian orator.
ly·si·ge·net·ic (lī'sə·jə·net'ik) *adj. Biol.* Produced or producing by the breakdown or absorption of intermediate or contiguous cells: said of certain intercellular spaces or of their mode of formation. Also **ly'si·gen'ic, ly·sig·e·nous** (lī·sij'ə·nəs).
Ly·sim·a·chus (lī·sim'ə·kəs), 361?–281 B.C., Macedonian general under Alexander the Great; king of Thrace.
ly·sim·e·ter (lī·sim'ə·tər) *n.* An instrument for determining the solubility of substances. [<LYSI- + METER]
ly·sin (lī'sin) *n.* A substance capable of destroying bacteria, blood corpuscles, etc. [<LYS(I)- + -IN]
ly·sine (lī'sēn, -sin) *n. Biochem.* An important amino acid, $C_6H_{14}O_2N_2$, produced on the hydrolysis of various proteins. Also **ly'sin**. [<LYS(I)- + -INE²]
Ly·sip·pus (li·sip'əs) Greek sculptor of the fourth century B.C. — **Ly·sip'pan** *adj.*
ly·sis (lī'sis) *n.* 1 *Pathol.* The gradual disappearing of a disease. Compare CRISIS. 2 *Biochem.* The process of disintegration or destruction of cells. [<NL <Gk., loosening <*lyein* loose, dissolve]
-lysis *combining form* A loosing, dissolving, etc.: *hydrolysis, paralysis.* [<Gk., loosening]
Ly·sol (lī'sôl, -sōl, -sol) *n.* Proprietary name for a saponified product of coal tar containing cresol: used as a disinfectant.
lys·sa (lis'ə) *n.* Hydrophobia; rabies. [<Gk., frenzy, madness]
lys·so·pho·bi·a (lis'ə·fō'bē·ə) *n. Psychiatry* A morbid fear of insanity. [<Gk. *lyssa* madness + -PHOBIA] — **lys'so·pho'bic** *adj.*
Lys·ter bag (lis'tər) A portable, waterproof bag for supplying disinfected drinking water to troops: also spelled *Lister bag.* [after W.J.L. *Lyster*, 1869–1947, U.S. Army surgeon]
Lys·tra (lis'trə) An ancient town of Lycaonia.
-lyte¹ *combining form* A substance decomposed by a (specified) process: *electrolyte.* [<Gk. *lytos* loosened, dissolved <*lyein* loosen]
-lyte² See -LITE.
-lytic *combining form* Loosing; dissolving: used in adjectives corresponding to nouns in *-lysis*: *hydrolytic paralytic*. [<Gk. *lytikos* loosing <*lysis* a loosening]
lyt·ta (lit'ə) *n. pl.* **lyt·tae** (lit'ē) *Anat.* A vermiform cartilage or fibrous band on the under surface of the tongue in carnivores, as the dog. [<L, a worm said to grow under a dog's tongue and cause madness <Gk., madness]
Lyt·ton (lit'ən), **Earl of,** 1831–91, Edward Robert Bulwer-Lytton, son of Lord Lytton; English poet and diplomat; pseudonym *Owen Meredith.* — **Lord,** 1803–73, Edward George Earle Lytton Bulwer-Lytton, first Baron Lytton, English novelist.
Lyu·blin (lyōō'blin) The Russian name for LUBLIN.

M

m, M (em) *n. pl.* **m's, M's** or **ms, Ms** or **ems** (emz) 1 The 13th letter of the English alphabet: from Phoenician *mem,* Greek *mu,* Roman *M.* 2 The sound of the letter *m,* usually a voiced, bilabial nasal. See ALPHABET. — *symbol* 1 The Roman numeral 1,000. See under NUMERAL. 2 *Printing* An em.
ma (mä, mô) *n.* Mama; mother. [Short for MAMA]
ma'am (mam, mäm, *unstressed* məm) *n.* 1 A respectful address to women, corresponding to *sir* in the case of men. 2 A dame; mistress: used in combination: *a schoolma'am.* 3 *Brit.* A term of respectful address used to the queen or to a royal princess. [Contraction of MADAM]
Maar·tens (mär'tənz), **Maarten** Pen name of Joost Marius Willem Van der Poorten-Schwartz, 1858–1915, Dutch novelist; wrote in English.
Maas (mäs) The Dutch name for the MEUSE.
Maas·tricht (mäs'trikht, mäs'trikt') See MAESTRICHT.
Mab (mab), **Queen** In English folklore, the queen of the fairies.
Ma·bel (mā'bəl) A feminine personal name. Also *Irish* **Mab,** *Fr.* **Ma·belle** (må·bel'), *Lat.* **Ma·bil·i·a** (mə·bil'ē·ə). [Short for AMABEL]
Mab·i·no·gi·on (mab'ə·nō'gē·ən) *n.* A collection of eleven medieval Welsh romances translated by Lady Charlotte Guest in 1838–1849. [<Welsh *mabinogion,* pl. of *mabinogi* bardic instructional material <*mabinog* a bard's apprentice]
Mac- *prefix* In Scottish and Irish names, son of: *MacDougal,* son of Dougal: abbr. *Mc, Mᶜ,* or *M'.* See also MC-. [<Scottish Gaelic and Irish, son]
ma·ca·bre (mə·kä'brə, -bər) *adj.* Suggestive of death; gruesome; frightful. Also **ma·ca'ber**. [<F <OF *(danse) macabré* (dance) of death, prob. alter. of *Macabé* <LL *Maccabaeus,* ? a character in a morality play]
ma·ca·co (mə·kä'kō) *n. pl.* **·cos** Any one of various lemurs, especially the black lemur of Madagascar (*Lemur macaco*). [<Pg. <Tupian *macaco, macaca* a monkey]
mac·ad·am (mə·kad'əm) *n.* 1 Broken stone for macadamizing a road. 2 A macadamized road. [after John L. *McAdam,* 1756–1836, British engineer]
mac·ad·am·ize (mə·kad'ə·mīz) *v.t.* **·ized, ·iz·ing** To pave with small consolidated broken stone on a soft or hard, but drained and convex, substratum. — **mac·ad'am·i·za'tion** *n.* — **mac·ad'am·iz'er** *n.*
Ma·cao (mə·kou', -kä'ō) A Portuguese overseas province at the mouth of the Canton River, comprising a peninsula of an island in the delta (often called **Macao Island**) and three small offshore islands, the **Tai·pa Islands** (tī'pä') and **Co·lo·a·ne** (kō·lō'ə·nē); total, 6 square miles; capital, Macao, on Macao Island. *Portuguese* **Ma·cáu** (mə·kou').
ma·caque (mə·käk') *n.* An Old World monkey (genus *Macaca*) with stout body, short tail, cheek pouches, and pronounced muzzle. [<F <Pg. *macaco.* See MACACO.]
mac·a·ro·ni (mak'ə·rō'nē) *n. pl.* **·nis** or **·nies** 1 An edible paste of wheat flour made into slender tubes. Compare SPAGHETTI, VERMICELLI. 2 A medley. 3 One of a body of Revolutionary soldiers from Maryland who wore a showy uniform. 4 An English dandy of the 18th century who affected the tastes and fashions prevalent in Continental society; a fop. Also **mac·ca·ro'ni**. [<

MACAQUE
(Body length: 2 to 3 feet, depending on species)

add, āce, câre, pälm; end, ēven; it, īce; odd, ōpen, ôrder; tōōk, pōōl; up, bûrn; ə = a in *above,* e in *sicken,* i in *clarity,* o in *melon,* u in *focus;* yōō = u in *fuse;* oi, oil; ou, pout; ch, check; g, go; ng, ring; th, thin; ŧħ, this; zh, vision. Foreign sounds å, œ, ü, kh, ṅ; and ◆: see page xx. < from; + plus; ? possibly.

Ital. *maccaroni, maccheroni,* pl. of *macherone* groats <LGk. *makaria* a broth with barley groats <Gk., blessedness < *makar* blessed]

mac·a·ron·ic (mak′ə·ron′ik) *adj.* **1** Consisting of a burlesque medley of real or coined words from various languages; hence, jumbled; mixed. **2** Pertaining to or like macaroni. Also **mac′a·ron′i·cal.** [<NL *macaronicus* <Ital. *maccheronico* <*maccheroni* MACARONI]

mac·a·roon (mak′ə·ro͞on′) *n.* A small cooky of ground almonds or almond paste, white of egg, and sugar; also, an imitation of this made with coconut. [<MF *macaron* <Ital. *maccarone,* var. of *macherone* MACARONI]

Mac·Ar·thur (mək·är′thər), **Douglas,** born 1880, U. S. general in World War II; commander in chief in the Pacific and in Japan.

Ma·cas·sar (mə·kas′ər) A port and the largest city of Celebes, on **Macassar Strait,** a channel between Borneo and Celebes: also *Makassar.*

Ma·cas·sar oil (mə·kas′ər) **1** A volatile oil distilled from the flowers of ylang-ylang, and originally obtained from Macassar. **2** A substitute made from perfumed castor oil, etc.: used as a hair dressing.

Ma·cau·lay (mə·kô′lē), **Thomas Babington,** first Baron Macaulay of Rothley, 1800–59, English historian, essayist, statesman.

ma·caw (mə·kô′) *n.* A large tropical American parrot (genus *Ara*), with a long tail, harsh voice, and brilliant plumage. [<Pg. *macao,* prob. <Tupian *maca(vuana)*]

Mac·beth (mək·beth′), died 1057, king of Scotland; hero of Shakespeare's tragedy of the same name.

Mac·ca·bees (mak′ə·bēz) A family of Jewish patriots who, beginning in the reign of Antiochus IV (175–164 B.C.), led a revolt against Syrian religious oppression, which resulted in the deliverance and freedom of Judea; also, four books written in Hebrew and translated into Greek, treating of Jewish oppression and persecution from 222 to 135 B.C. The first two books are regarded as canonical by Roman Catholics, but all as apocryphal by Protestants. [<LL *Machabaei,* pl. of *Machabaeus* <Gk. *Makkabaios,* prob. < Aramaic *maggābā* a hammer]

Mac·ca·be·us (mak′ə·bē′əs), **Judas,** died 160? B.C., Jewish patriot; leader against the Seleucids. **— Mac′ca·be′an** *adj.*

mac·ca·boy (mak′ə·boi) *n.* A finely ground, rose-scented, dark snuff. Also **mac′ca·baw** (-bô), **mac′co·boy.** [<F *macouba,* from *Macouba,* Martinique; because originally made from tobacco grown there]

mac·chia (mak′yə) See MAQUIS.

Mac·don·ald (mək·don′əld), **George,** 1824–1905, Scottish novelist and poet. **— James Ramsay,** 1866–1937, English labor leader and statesman; prime minister 1924, 1929–35.

Mac·Don·ough (mək·don′ō), **Thomas,** 1783–1825, U. S. naval officer in the War of 1812.

Mac·Dow·ell (mək·dou′əl), **Edward Alexander,** 1861–1908, U. S. composer.

Mac·Duff (mək·duf′) A character in Shakespeare's *Macbeth,* who kills Macbeth.

mace¹ (mās) *n.* **1** A club-shaped staff of office borne before officials or displayed on the table of a legislative body. **2** A medieval steel war club, often with spiked metal head, for use against armor. **3** A person who carries a mace, as in a ceremony. **4** A curriers' knobbed mallet for softening hides. **5** A flat-headed form of billiard cue, formerly used on pocket tables. [<OF *masse, mace.* Cf. LL *matteola* a mallet.]

mace² (mās) *n.* An aromatic spice made from the covering of nutmeg seed. [<OF *macis* <L *macir* <Gk. *maker* a spicy bark from India]

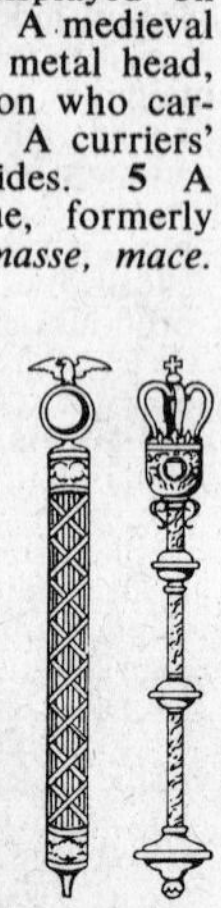

MACES

mace·bear·er (mās′bâr′ər) *n.* An official who carries a mace, as in a procession.

mac·é·doine (mas′i·dwän′, *Fr.* mȧ·sā·dwȧn′) *n.* A dish of mixed vegetables or fruits, served as a salad or dessert; also, any mixture; medley; olio. [<F, a type of parsley <OF *(perresel) macedoine* Macedonian (parsley)]

Mac·e·do·ni·a (mas′ə·dō′nē·ə) A region of SE Europe in the southern Balkan peninsula on the Aegean, divided among Bulgaria, Greece, and Yugoslavia; the ancient Greek kingdom of **Mac·e·don** (mas′ə·don) under Alexander the Great became a leading world power in the fourth century B.C. **— Mac′e·do′ni·an** *adj.* & *n.*

Ma·cei·ó (mä′sā·ô′) A port on the Atlantic, capital of Alagoas state, eastern Brazil.

mac·er (mā′sər) *n.* **1** A macebearer. **2** In Scotland, an officer who attends the courts and executes their orders. [<OF *maissier* < *masse, mace* mace]

mac·er·ate (mas′ə·rāt) *v.* **·at·ed, ·at·ing** *v.t.* **1 a** To reduce to a soft mass by soaking. **b** To separate the soft parts of by soaking; digest. **2** To make thin; emaciate. **—** *v.i.* **3** To undergo maceration. [<L *maceratus,* pp. of *macerare* make soft, knead] **— mac′er·at′er, mac′er·a′tor** *n.*

mac·er·a·tion (mas′ə·rā′shən) *n.* The process or action of steeping a solid material in liquid so as to soften it or break down its structure; digestion.

mach (mak, mäk) See MACH NUMBER.

Ma·cha·do y Mo·ra·les (mä·chä′thō ē mō·rä′·läs), **Gerardo,** 1871–1939, president of Cuba 1925–33.

mach effect (mak, mäk) *Physics* **1** Any effect produced by an object moving at transonic or supersonic speeds, as a shock wave. **2** The aggregate of such effects.

ma chère (mȧ shâr′) *French fem.* My dear.

ma·chet·e (mə·shet′ē, mə·shet′; *Sp.* mä·chā′tā) *n.* A heavy knife or cutlas used both as an implement and as a weapon by the natives of tropical America. [<Sp., dim. of *macho* an ax, a hammer <L *marculus,* dim. of *marcus* a hammer]

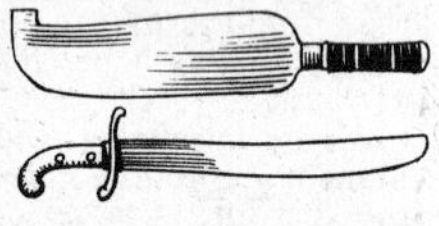

SPANISH MACHETES

Mach·i·a·vel·li (mäk′ē·ə·vel′ē, *Ital.* mä′kyä·vel′·lē), **Niccolò,** 1469–1527, Florentine statesman and writer on government.

Mach·i·a·vel·li·an (mak′ē·ə·vel′ē·ən) *adj.* Of or pertaining to the Florentine statesman and writer, Niccolò Machiavelli, or to the unscrupulous doctrines of political opportunism associated with his name. **—** *n.* A follower of Machiavelli. Also **Mach′i·a·vel′i·an. — Mach′i·a·vel′list** *adj.* & *n.*

Mach·i·a·vel·li·an·ism (mak′ē·ə·vel′ē·ən·iz′əm) *n.* The theory and practice of power politics elaborated from Machiavelli's *The Prince:* envisaging (1) seizure, maintenance, and extension of absolute power by the nicely graduated use of guile, fraud, force, and frightfulness respectively; (2) reliance on expediency and reasons of state as justifying any departures from morality needed to hold power; control being then maintained by the ruler of all avenues of communication, thus facilitating the deliberate molding of public opinion; (3) use of a common enemy as political cement in holding together allies needed in acquiring power, and the ruthless liquidation of these allies and all other rivals once power has been acquired; (4) the employment for surveillance and terrorist activities of subordinates who can be disowned and liquidated by the ruler, who thus escapes the blame for their atrocities. Also **Mach′i·a·vel′lism.**

ma·chic·o·late (mə·chik′ə·lāt) *v.t.* **·lat·ed, ·lat·ing** To furnish with machicolations. [< Med. L *machicolatus,* pp. of *machicolare* <OF *machicoler*]

ma·chic·o·la·tion (mə·chik′ə·lā′shən) *n.* **1** *Archit.* An opening between a wall and a parapet, to permit missiles or boiling liquids to be dropped upon an assailing enemy. **2** The act of showering missiles on an attacking party through such openings.

ma·chin·a·ble (mə·shē′nə·bəl) *adj. Metall.* Capable of being shaped, cut, or polished by high-speed machine tools, as certain metals and alloys. **— ma·chin′a·bil′i·ty** *n.*

mach·i·nate (mak′ə·nāt) *v.t.* & *v.i.* **·nat·ed, ·nat·ing** To plan or devise; scheme, especially with evil intent. [<L *machinatus,* pp. of *machinari* contrive < *machina* MACHINE]

mach·i·na·tion (mak′ə·nā′shən) *n.* The act of contriving a secret or hostile plan; also, such a plan; plot. See synonyms under ARTIFICE. **— mach′i·na′tor** *n.*

ma·chine (mə·shēn′) *n.* **1** Any combination of mechanisms for utilizing, modifying, applying, or transmitting energy, whether simple, as a lever and fulcrum, pulley, etc., or complex, as a Fourdrinier papermaking apparatus. **2** An automobile; also, any other mechanical vehicle, as a bicycle, airplane, etc. **3** In many trades or vocations, the construction principally used or typical of the trade. **4** One who acts in a mechanical manner; a robot. **5** An ancient theatrical contrivance, originating in Greece, for the creation of stage effects: also applied to such portions of the plot of a work of fiction as are introduced for the sake of effect. **6** The organization of the powers of any complex body: the *machine* of government; specifically, an organization within a political party, controlled by practical politicians, in which discipline and subordination are maintained chiefly by the use of patronage, as in the distribution of offices and contracts. **7** Formerly, a military engine. **—** *adj.* **1** Pertaining to, for, produced by, or producing a machine or machinery: *machine* knitting, *machine* shop, etc. **2** Typified by the application or predominance of machines: *machine* age. **3** Mechanical; lacking in originality or uniqueness. **—** *v.t.* **·chined, ·chin·ing** To shape, mill, make, etc., by machinery. [<F <L *machina* <Gk. *mēchanē* < *mēchos* a contrivance] **— ma·chin′al** *adj.*

machine element One of various simple mechanisms, as the lever, pulley, cam, gearwheel, etc., which, alone or in combination, facilitate conversion of energy to useful work. See illustration, page 763. Also **simple machine.**

ma·chine-gun (mə·shēn′gun′) *n.* An automatic gun that discharges small-arms ammunition in a rapid, continuous fire: also **machine gun. —** *v.t.* **-gunned, -gun·ning** To fire at or shoot with a machine-gun. **— ma·chine′-gun′ner** *n.*

ma·chin·er·y (mə·shē′nər·ē) *n. pl.* **·er·ies** **1** The parts of a machine or a number of machines and kindred appliances collectively. **2** Any combination of means working together; a complex system of appliances; the arrangements for effecting a specific end. **3** The supernatural or other means by which the catastrophe of a classical drama was brought about; hence, the incidents and events introduced to build up the plot in a work of fiction.

machine shop A workshop for making or repairing machines.

machine tool A power-driven tool, partly or wholly automatic in action, as a lathe, used for cutting and shaping the parts of a machine.

ma·chin·ist (mə·shē′nist) *n.* A maker or repairer of machines; one expert in their design or construction, or in using metalworking tools.

mach·me·ter (mak′mē′tər, mäk′-) *n. Aeron.* An instrument for determining the speed of airplanes, especially with reference to the mach number. [<MACH (NUMBER) + -METER]

mach number (mak, mäk) A number expressing the speed of an object moving through a fluid medium in relation to the speed of sound in the same medium: in aeronautics it is the ratio of the air speed to the speed of sound in the undisturbed medium, a mach number of 1 indicating a speed equal to the speed of sound. Also *mach.* [after Ernst *Mach,* 1838–1916, Austrian physicist]

ma·chree (mə·krē′) *n.* My heart; my love: a term of endearment. [<Irish < *mo* my + *croidhe* heart <OIrish *cride*]

Ma·chu·pic·chu (mä′cho͞o·pēk′cho͞o) A ruined pre-Incan city in south central Peru. Also **Machu Picchu.**

-machy *combining form* A fight between, or by means of: *logomachy.* [<Gk. *-machia* < *machē* a battle]

Ma·cie (mā′sē), **James Lewis** See SMITHSON, JAMES.

mac·in·tosh (mak′ən·tosh) See MACKINTOSH.

Mack·en·sen (mäk′ən·zən), **August von,** 1849–1945, German field marshal in World War I.

Mac·Ken·zie (mə·ken′zē) **1** A district in the Northwest Territories, Canada; 527,490 square miles. **2** A river in NW Canada, flowing generally north 2,514 miles to the Beaufort Sea.

Mac·Ken·zie (mə·ken′zē), **Sir Alexander,** 1755–1820, Scottish explorer in North America.

mack·er·el (mak′ər·əl) *n.* **1** An Atlantic food fish *(Scomber scombrus),* steel-blue above with blackish bars, and silvery beneath. **2** Some scombroid fish resembling it, as the

mackerel gull

Spanish mackerel (*Scomberomorus maculatus*). [<OF *makerel*; ult. origin unknown]
mackerel gull The common tern.
mackerel sky *Meteorol.* A cloud formation with numerous detached cloudlets resembling the markings on a mackerel's back.
Mack·i·nac (mak′ə·nô), **Straits of** A channel between Lake Michigan and Lake Huron; 4 miles wide at the narrowest point.
Mack·i·nac Island (mak′ə·nô) A resort city on the southern end of Mackinac Island in the Straits of Mackinac. Also **Mackinac**.
Mackinaw cloth A heavy-napped woolen fabric, or one having cotton or rayon mixed in the yarns: the two sides may be different colors, sometimes in plaids: used for lumbermen's jackets, sport clothes, windbreakers, etc.
Mackinaw coat A thick, short, double-breasted coat with a plaid pattern.
Mackinaw trout See NAMAYCUSH.
mack·in·tosh (mak′ən·tosh) *n.* 1 A waterproof overgarment or cloak. 2 Thin rubber-coated cloth. Also *macintosh*. [after Charles *Macintosh*, 1766–1843, Scottish chemist, inventor of the cloth]
mack·le (mak′əl) *n.* A spot or blemish; also, a blurred impression. — *v.t. & v.i.* **mack·led**, **mack·ling** To print with a blurred or double image; blur; blot. Also spelled *macule*. [<F *macule* <L *macula* a spot]
Mac·lar·en (mə·klar′ən), **Ian** Pseudonym of John Watson, 1850–1907, Scottish author.
Mac·Lau·rin (mə·klôr′in), **trisectrix of** See TRISECTRIX OF MACLAURIN.
mac·le (mak′əl) *n.* 1 A twin crystal, especially of a diamond. 2 Chiastolite. 3 A mackle. 4 Mascle. [<F <OF *mascle* < Med. L *mascula* a mesh of a net <L *macula* a spot]
Mac·Leish (mək·lēsh′), **Archibald**, born 1892, U.S. poet.
Mac·leod (mə·kloud′), **Fiona** Pseudonym of William Sharp.
— **John James Rickard**, 1876–1935, Scottish physiologist.
Ma·clu·ra (mə·kloor′ə) *n.* The Osage orange. [<NL, after Wm. *Maclure*, 1763–1840, U.S. geologist born in Scotland]
Mac-Ma·hon (måk·mȧ·ôṅ′), **Marie Edmé Patrice Maurice de**, 1808–93, Duke of Magenta, French marshal; president of France 1873–79.
Mac·Mil·lan (mək·mil′ən), **Donald Baxter**, born 1874, U.S. Arctic explorer.
Mac·mil·lan (mək·mil′ən), **Harold**, born 1894, British statesman; prime minister 1957–.
Mac·Mon·nies (mək·mon′ēz), **Frederick William**, 1863–1937, U.S. sculptor.
Ma·con (mā′kən) A city in central Georgia.
Mâ·con (må·kôṅ′) *n.* A wine produced in the neighborhood of Mâcon, a manufacturing city in east central France.
Mac·Pher·son (mək·fûr′sən), **James**, 1736–1796, Scottish writer. See OSSIAN.
Mac·quar·ie River (mə·kwôr′ē, -kwor′ē) A river in central New South Wales, Australia, flowing 590 miles NW to the Darling River.
mac·ra·mé (mak′rə·mā) *n.* A fringe, lace, or trimming of knotted thread or cord; knotted work. [Appar. <Turkish *maqramah* a towel <Arabic *miqramah* a veil]
Mac·rea·dy (mə·krē′dē), **William Charles**, 1793–1873, English tragedian.
mac·ren·ceph·a·ly (mak′ren·sef′ə·lē) *n. Pathol.* Hypertrophy of the brain. [<MACR(O)- + ENCEPHAL(O)- + -Y¹] — **mac′ren·ce·phal′ic** (-sə·fal′ik), **mac′ren·ceph′a·lous** *adj.*
macro– *combining form* Large or long in size or duration: *macrocephaly*, *macrobiosis*. Also, before vowels, **macr–**. [<Gk. *makros* large]
mac·ro·bi·o·sis (mak′rō·bī·ō′sis) *n.* Longevity. [<NL <Gk. *makros* long + *bios* life] — **mac′ro·bi·ot′ic** (-ot′ik) *adj.*
mac·ro·ceph·a·ly (mak′rō·sef′ə·lē) *n.* Excessive size of the head. [<MACRO- + Gk. *kephalē* head] — **mac′ro·ce·phal′ic** (-sə·fal′ik) *adj. & n.*, **mac′ro·ceph′a·lous** *adj.*
mac·ro·chem·is·try (mak′rō·kem′is·trē) *n.* The chemistry of large-scale operations or of reactions visible to the naked eye. — **mac′ro·chem′i·cal** (-i·kəl) *adj.*
mac·ro·cli·mate (mak′rō·klī′mit) *n. Meteorol.* The general climate prevailing over a large area considered as a unit. Compare MICROCLIMATE. — **mac′ro·cli·mat′ic** (-klī·mat′ik) *adj.*
mac·ro·cosm (mak′rə·koz′əm) *n.* 1 The great world; the universe: opposed to *microcosm*. 2 The whole of any sphere or department of nature or knowledge to which man is related. [<OF *macrocosme* <Med. L *macrocosmus* <Gk. *makros* long, great + *kosmos* world] — **mac′ro·cos′mic** *adj.*
mac·ro·cyst (mak′rə·sist) *n. Biol.* 1 An enlarged cyst. 2 *Bot.* A large reproductive cell in certain fungi.
mac·ro·cyte (mak′rə·sīt) *n.* An abnormally large erythrocyte. [<MACRO- + (ERYTHRO)-CYTE]
mac·ro·dome (mak′rə·dōm) See under DOME.
mac·ro·e·co·nom·ics (mak′rō·ē′kə·nom′iks, -ek′ə-) *n.* Economics studied in terms of large aggregates of data whose mutual relationships are interpreted with reference to the behavior of the system as a whole: developed by John Maynard Keynes. — **mac′ro·e·co·nom′ic** *adj.*
mac·ro·ga·mete (mak′rō·gə·mēt′, -gam′ēt) *n. Biol.* The female of two conjugating gametes: so called from its being the larger one. Compare MICROGAMETE. [<MACRO- + GAMETE]
ma·crog·a·my (mə·krog′ə·mē) *n. Biol.* Conjugation between gametes similar in structure to the original vegetative cells.
mac·ro·graph (mak′rə·graf, -gräf) *n.* A drawing or illustration of an object as seen with the unaided eye.
ma·crog·ra·phy (mə·krog′rə·fē) *n.* 1 Examination with the unaided eye. 2 Extremely large writing, often indicative of some nervous malady: contrasted with *micrography*.

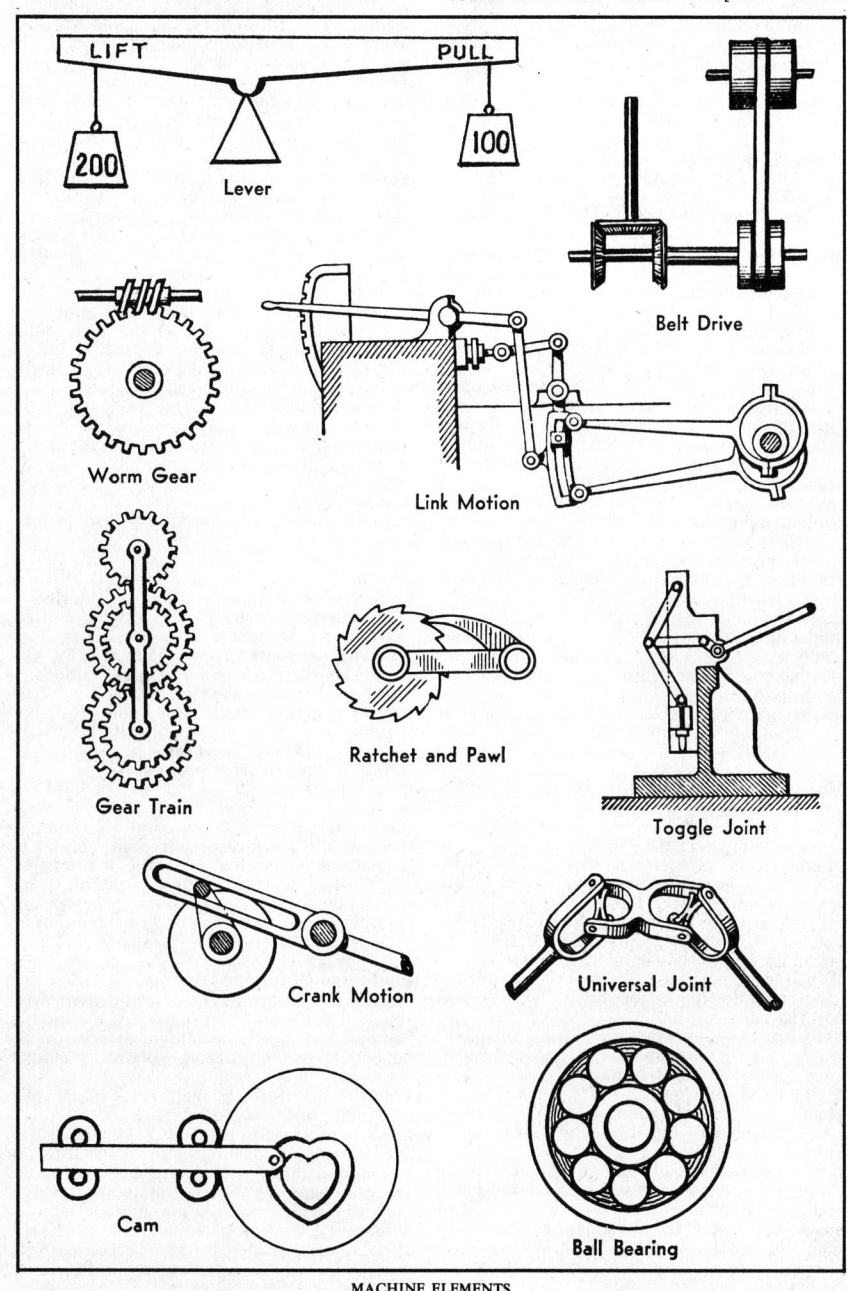

mack·i·naw (mak′ə·nô) A shortened form of MACKINAW BLANKET, MACKINAW BOAT, MACKINAW COAT. [<dial. F (Canadian) *Mackinac* <*Michilimackinac* Mackinac Island <Algonquian (Ojibwa) *mitchimakinak* a large turtle]
Mackinaw blanket A thick, heavy blanket formerly used by Indians and traders of the western United States.
Mackinaw boat A large, sharp-ended, flat-bottomed bateau, sometimes equipped with sails, and formerly used on the northern Great Lakes.

MACHINE ELEMENTS

add, āce, cāre, pälm; end, ēven; it, īce; odd, ōpen, ôrder; tŏŏk, pōōl; up, bûrn; ə = a in *above*, e in *sicken*, i in *clarity*, o in *melon*, u in *focus*; yōō = u in *fuse*; oi, oil; ou, pout; ch, check; g, go; ng, ring; th, thin; th, this; zh, vision. Foreign sounds à, œ, ü, kh, ń; and ♦: see page xx. < from; + plus; ? possibly.

ma·crom·e·ter (mə·krom′ə·tər) *n.* A range-finding instrument similar to a sextant, consisting of a telescope and two mirrors, for measuring distant objects.

mac·ro·me·te·or·ol·o·gy (mak′rō·mē′tē·ə·rol′ə·jē) *n.* The meteorology of large areas; world meteorology; also, the study of climatic phenomena over extended periods of time. Compare MICROMETEOROLOGY.

mac·ro·mor·phol·o·gy (mak′rō·môr·fol′ə·jē) *n.* The gross anatomy of plants.

ma·cron (mā′krən, -kron) *n.* A straight line (¯) over a vowel to show that it has a long sound, as, ā: opposed to *breve* (˘). [<Gk. *makron,* neut. of *makros* long]

mac·ro·nu·cle·us (mak′rō·nōō′klē·əs, -nyōō′-) *n. pl.* **·cle·i** (-klē·ī) *Biol.* The larger of two nuclei in certain protozoans, as *Paramecium.*

mac·ro·phys·ics (mak′rō·fiz′iks) *n.* The physics of masses, or of large bodies, without taking finer or molecular structure into account. **— mac′ro·phys′i·cal** *adj.*

mac·ro·po·di·an (mak′rə·pō′dē·ən) *n.* Any of a family *(Macropodidae)* of marsupials, especially those having six sharp upper incisors and two larger lower ones, enlarged saltatorial hind legs, and long tail: including wallabies and kangaroos. *— adj.* Of or pertaining to the *Macropodidae*: also **ma·crop·o·dine** (mə·krop′ə·din, -dīn). [<NL <Gk. *makropous, -podos* long-footed <*makros* long + *pous, podos* a foot]

ma·crop·si·a (mə·krop′sē·ə) *n. Pathol.* A condition of the eyes in which objects appear larger than they really are: often called *megalopsia.* Also **ma·cro′pi·a** (-krō′pē·ə). [<NL <Gk. *makros* large + *ōps, opos* eye]

mac·ro·scop·ic (mak′rə·skop′ik) *adj.* Visible to the naked eye: opposed to *microscopic.* Also **mac′ro·scop′i·cal.**

mac·ro·spe·cies (mak′rə·spē′shēz) *n. Biol.* A species of which the members exhibit a large range of variation. Compare MICROSPECIES.

mac·ro·spo·ran·gi·um (mak′rō·spə·ran′jē·əm) See MEGASPORANGIUM.

mac·ro·spore (mak′rə·spôr, -spōr) *n.* **1** *Bot.* A megaspore. **2** *Biol.* One of the larger of the two kinds of anisospores in ameboid protozoans.

mac·ro·struc·ture (mak′rə·struk′chər) *n.* The structure of metal as observed with the naked eye or under low magnification on a polished or etched surface. **— mac′ro·struc′tur·al** *adj.*

mac·ro·therm (mak′rə·thûrm) *n.* A tropical plant, or one requiring much heat and moisture for survival: also called *megatherm.* [<MACRO- + THERM]

ma·cru·ran (mə·krōōr′ən) *n.* One of a suborder *(Macrura)* of crustaceans with well-developed, elongated abdomens, including shrimps, prawns, lobsters, and crayfishes. [<NL <Gk. *makros* long + *oura* tail] **— ma·cru′ral, ma·cru′roid, ma·cru′rous** *adj.*

mac·ta·tion (mak·tā′shən) *n.* The killing of a sacrificial victim. [<L *mactatio, -onis* <*mactatus,* pp. of *mactare* kill]

mac·u·la (mak′yə·lə) *n. pl.* **·lae** (-lē) **1** A spot, as of color on the skin. **2** *Astron.* A dark spot on the sun's surface. **3** A fleck; blotch. [<L, a spot]

mac·u·late (mak′yə·lāt) *v.t.* **·lat·ed, ·lat·ing** To spot or blemish; stain. *— adj.* (-lit) Spotted; stained. [<L *maculatus,* pp. of *maculare* <*macula* a spot]

mac·u·la·tion (mak′yə·lā′shən) *n.* **1** The act of spotting, or a spotty condition. **2** The marking of a spotted animal or plant. **3** A soiling; defilement.

mac·ule (mak′yōōl) See MACKLE.

mad (mad) *adj.* **mad·der, mad·dest 1** Mentally deranged; insane. **2** Subject to an overpowering emotion; excited intensely or beyond self-control: *mad* with jealousy or grief. **3** *Colloq.* Angry; furious; enraged. **4** Rabid; having hydrophobia; hence, uncontrollable. **5** Proceeding from or indicating a disordered mind; rash: a *mad* project. **6** Tumultuous or uncontrollable in movement or action: said of things: a *mad* torrent. **7** Reckless; heedless. *— n. Slang* Anger; a fit of temper. **— to have a mad on** *Slang* To be angry. *— v.t. & v.i.* **mad·ded, mad·ding** *Rare* To madden. [Aphetic var. of OE *gemǣd, gemǣded,* pp. of *gemǣdan* make mad <*gemād* insane] **— mad′ly** *adv.* **— mad′ness** *n.*

Mad·a·gas·car (mad′ə·gas′kər) The world's fourth largest island, in the Indian Ocean off the SE coast of Africa; 227,602 square miles. *French* **Grande-Île** (grän·dēl′), **Grande-Terre** (gränd·târ′). See MALAGASY REPUBLIC. **— Mad′a·gas′can** *adj. & n.*

mad·am (mad′əm) *n. pl.* **mes·dames** (mā·däm′, *Fr.* mā·dȧm′) *for def.* **1**; **mad·ams** *for def.* **2**. **1** My lady; mistress: a title of courtesy originally addressed to a woman of rank or high position, but now used to any woman, as at the beginning of a letter. Compare SIR. See MA'AM. **2** The mistress of a brothel. [<OF *ma dame* <*ma* my (<L *meus*) + *dame* lady <L *domina.* Doublet of MADONNA, MADAME.]

mad·ame (mad′əm, *Fr.* mȧ·dȧm′) *n. pl.* **mes·dames** (mā·däm′, *Fr.* mā·dȧm′) The French title of courtesy for a married woman, equivalent to the English *Mrs.*: abbreviated *Mme.*; often used in English, especially in the plural. [<F <OF *ma dame.* Doublet of MADAM, MADONNA.]

Ma·da·ria·ga y Ro·jo (mä′thä·ryä′gä ē rō′hō), **Salvador de,** born 1886, Spanish writer and diplomat.

mad·cap (mad′kap′) *adj.* Wild; rattle-brained. *— n.* One who acts wildly or rashly.

mad·den (mad′n) *v.t. & v.i.* To make or become mad or insane; inflame with anger; enrage. **— mad′den·ing** *adj.* **— mad′den·ing·ly** *adv.*

mad·der[1] (mad′ər) Comparative of MAD.

mad·der[2] (mad′ər) *n.* **1** Any plant of the genus *Rubia* (family *Rubiaceae*); especially, an Old World shrubby, perennial, hairy herb *(R. tinctorum),* resembling the common bedstraw. **2** A brilliant red tinctorial extract from the root of the madder plant, used in dyeing and as a pigment in many lakes: the coloring principle, alizarin, is now made also synthetically. *— v.t.* To dye with madder. [OE *mæd-ere, mæddre*]

madder lake A pigment and dyestuff formerly made from the root of the madder plant, now largely replaced by alizarin.

mad·ding (mad′ing) *adj.* Being or growing mad; delirious; raging.

mad·dish (mad′ish) *adj.* Rather mad.

mad-dog skullcap (mad′dôg′, -dog′) A common American plant *(Scutellaria lateriflora)* having blue flowers: formerly believed to cure hydrophobia.

made (mād) Past tense and past participle of MAKE. *— adj.* **1** Fabricated; produced, especially artificially. **2** Assured of fortune. **3** Filled in: *made* land.

Ma·dé·casse (mȧ·dā·käs′) *adj. & n. French* Madagascan.

Ma·dei·ra (mə·dir′ə, *Pg.* mə·thā′rə) *n.* A fortified wine made in the Madeira islands.

Ma·dei·ra (mə·dir′ə, *Pg.* mə·thā′rə) **1** An archipelago west of Morocco, including Madeira, Porto Santo, and the Desertas and Selvagens islands, comprising an administrative district *(Funchal)* of Portugal; 305 square miles; capital, Funchal, on Madeira. **2** A river in NW Brazil, the most important tributary of the Amazon, flowing 2,100 miles NE. **— Ma·dei′ran** *adj.*

Madeira vine A vine *(Boussingaultia baselloides)* having a bulbous root, shiny leaves, and clusters of white flowers.

Mad·e·leine (mad′ə·lin, -lān, *Fr.* mȧ·dlen′) A feminine personal name. See MAGDALENE. Also **Mad′e·line** (-lin, -līn), **Mad·e·lon** (mad′ə·lon, *Fr.* mȧ·dlôn′).

mad·e·moi·selle (mad′ə·mə·zel′, *Fr.* mȧd·mwȧ·zel′) *n. pl.* **mad·e·moi·selles,** *Fr.* **mes·de·moi·selles** (mād·mwa·zel′) **1** The French title of courtesy for unmarried women, equivalent to the English *Miss*: abbreviated *Mlle.* **2** A French nurse or governess. [<F <*ma* my + *demoiselle.* See DEMOISELLE.]

Ma·de·ro (mä·thā′rō), **Francisco Indalecio,** 1873–1913, president of Mexico 1911–13.

made-up (mād′up′) *adj.* **1** Artificial; fictitious. **2** Complete; finished. **3** With make-up or cosmetics applied.

mad·house (mad′hous′) *n.* **1** A lunatic asylum. **2** A place of confusion, turmoil, and uproar; bedlam.

Ma·dhya Bha·rat (mu′dyə bä′rət) A former constituent state in west central India, formed in 1948 by merger of Gwalior and the states of Central India, most of which became merged into Madhya Pradesh, November 1, 1956; 46,710 square miles; capitals, Lashkar and Indore.

Ma·dhya Pra·desh (mu′dyə prə·dāsh′) A constituent State in central India on the northern Deccan Plateau; 171,201 square miles; capital, Bhopal: formerly *Central Provinces and Behar.*

Mad·i·son (mad′ə·sən) The capital of Wisconsin.

Mad·i·son (mad′ə·sən), **Dolly,** 1768–1849, *née* Payne, wife of James; celebrated as a hostess. **— James,** 1751–1836, president of the United States 1809–17.

Madison Avenue 1 A street running north and south through Manhattan borough of New York City, center of a commercial district containing many advertising agencies and offices of mass media. **2** The world of American advertising and mass media, its power, influence, policies, etc.

Madison's War The War of 1812.

mad·man (mad′man′, -mən) *n. pl.* **·men** (-men′, -mən) A lunatic; maniac.

Ma·doe·ra (mə·dōō′rä) The Dutch spelling of MADURA.

Ma·don·na (mə·don′ə) *n.* **1** My lady; madam: an Italian title of respect now replaced by *Signora.* **2** The Virgin Mary; also, a painting or statue of the Virgin Mary. [<Ital. <*ma* my (<L *meus*) + *donna* lady <L *domina.* Doublet of MADAM, MADAME.]

ma·dras (mə·dras′, -dräs′, mad′rəs) *n.* **1** A kind of cotton cloth, with thick strands at intervals throughout its weave, giving either a striped, corded, or checked effect. **2** A large, brightly colored kerchief, formerly worn as a headdress by Negroes. [from *Madras,* India, because originally made there]

Ma·dras (mə·dras′, -dräs′, mad′rəs) **1** A constituent state in southern India; 50,110 square miles. **2** Its capital, a port on the Bay of Bengal.

Madras hemp Sunn.

Madras States A former agency of princely states in southern India.

ma·dre (*Ital.* mä′drā, *Pg., Sp.* mä′thrā) Italian, Portuguese, and Spanish form of MOTHER.

Ma·dre de Dios (mä′thrā thā dyōs′) A river in the Amazon basin of Peru and Bolivia, flowing 700 miles NE to the Beni.

mad·re·po·rar·i·an (mad′rə·pô·râr′ē·ən) *n.* One of a suborder *(Madreporaria)* of anthozoans with a calcareous skeleton secreted by cells; one of the reef-forming stony corals of tropical seas. *— adj.* Of or pertaining to the *Madreporaria*: also **mad′re·po′ral** (-pôr′əl, -pō′rəl). [<NL <Ital. *madrepora* MADREPORE]

mad·re·pore (mad′rə·pôr, -pōr) *n.* **1** A branched reef coral (family *Acroporidae*); also, any perforate stone coral. **2** The animal which produces madrepore coral. **3** Limestone consisting of fossil madrepores. Also **mad′re·po′ra** (-pôr′ə, -pō′rə). [<F <Ital. *madrepora,* lit., mother stone <*madre* mother (<L *mater*) + *poro,* ? calcareous stone <L *porus* <Gk. *pōros*] **— mad′re·por′ic** (-pôr′ik, -por′ik) *adj.*

mad·re·po·rite (mad′rə·pô·rīt′) *n. Zool.* In starfishes and certain other echinoderms, the circular perforated plate by which the internal ambulacral system communicates with the outside. Also **madreporic plate.** [<MADREPOR(E) + -ITE[1]]

Ma·drid (mə·drid′, *Sp.* mä·thrēth′) The capital of Spain, in the central part.

mad·ri·gal (mad′rə·gəl) *n.* **1** A short lyric poem, usually dealing with a pastoral or amatory subject. **2** A musical setting for such a poem; an a capella part song characterized by elaborate rhythm and contrapuntal imitation. **3** Any part song. **4** Any song. [<Ital. *madrigale, mandrigale* <LL *matricale* original, chief <L *matrix* womb; infl. in form by Ital. *mandra* a flock <L, a herd <Gk., a stable, fold]

mad·ri·gal·ist (mad′rə·gəl·ist) *n.* One who writes, or performs in, madrigals.

ma·dri·lène (mad′rə·len′, *Fr.* mȧ·drē·len′) *n.* A consommé flavored with tomato and served hot or cold. [<F <Sp. *Madrileño* of Madrid]

ma·dro·ña (mə·drō′nyə) *n.* A large evergreen tree *(Arbutus menziesii)* of northern California with shining oval or oblong leaves, dense racemes of white flowers, and dry, yellow, edible berries, called **madroña apples.** Also **ma·dro′ño** (-nyō). [<Sp. *madroño*]

mad·stone (mad′stōn′) *n.* A stone formerly believed to cure hydrophobia.

Ma·du·ra (mä·dōō′rä) An Indonesian island in the Java Sea; 2,112 square miles including offshore islands: Dutch *Madoera.*

Madura foot

Madura foot Mycetoma.
Ma·du·rai (mə·jōō′rī, mə·dōōr′ī) A city in southern Madras State, India: also *Mathurai*. Formerly **Ma·du·ra** (mə·jōō′rə, mə·dōōr′ə).
Mad·u·rese (mə·jōō·rēz′, -rēs′) *adj.* Of or pertaining to Madurai, its people, or their language. — *n. pl.* **·rese** 1 A native or inhabitant of Madurai and eastern Java. 2 The Indonesian language of the Madurese.
Ma·du·ro (mə·dōōr′ō) *adj.* Matured; that is, of full strength and color: said of cigars. — *n.* Such a cigar. [<Sp. <L *maturus*]
mad·wom·an (mad′wŏŏm′ən) *n. pl.* **·wom·en** (-wim′in) An insane woman; lunatic.
mad·wort (mad′wûrt′) *n.* Any of various shrubs or herbs of the mustard family (genera *Alyssum* or *Lobularia*) with small alternate leaves and small yellow or white flowers in terminal racemes or clusters. [? Trans. of NL *Alyssum*, genus name <L <Gk. *alysson* < *a-* not + *lyssa* rabies]
mae (mā) *adj.* & *adv. Scot.* More.
Mae·an·der (mē·an′dər) See MEANDER. Also **Mæ·an′der.**
Mae·ce·nas (mi·sē′nəs), **Gaius Cilnius**, 73?–8 B.C., Roman statesman; friend and patron of Horace and Vergil; hence, any patron of the arts.
mael·strom (māl′strəm) *n.* Any irresistible movement or influence: from the **Maelstrom**, a whirlpool or current off the NW coast of Norway. [<Du. *maelstrom*, *maalstroom* < *malen* grind, whirl round + *stroom* a stream]
mae·nad (mē′nad) *n.* 1 A female votary or priestess of Dionysius; a bacchante. 2 Any woman beside herself with frenzy or excitement. Also spelled *menad*. [<L *Maenas*, *-adis* <Gk. *mainas* frenzied < *mainesthai* rave] — **mae·nad′ic** *adj.*
Mae Nam (ma′ näm′) See CHAO PHRAYA.
ma·es·to·so (mä′es·tō′sō) *adj.* & *adv. Music* With majesty; stately. [<Ital., majestic <L *majestas* greatness <*major*, compar. of *magnus* great]
Maes·tricht (mäs′trikht) A city in southern Netherlands, capital of Limburg province: also *Maastricht*.
ma·es·tro (mä·es′trō, mī′strō) *n.* A master in any art, especially in music. [<Ital., a master <L *magister*]
Mae·ter·linck (mā′tər·lingk), **Maurice**, 1862–1949, Belgian poet, dramatist, and essayist.
Mae West (mā) An inflatable vestlike life preserver used by aviators downed at sea in World War II. [after *Mae West*, born 1892, U. S. actress]
Maf·e·king (maf′ə·king) A town in NE Cape of Good Hope Province, Union of South Africa; capital of Bechuanaland Protectorate.
maf·fick (maf′ik) *v.i. Brit.* To celebrate uproariously. [Back formation <*mafficking* <MAFEKING, taken as a ppr.; from the extravagant celebration of the relief of the British garrison at Mafeking, May 17, 1900]
Ma·fi·a (mä′fē·ä, maf′ē·ə) *n.* A Sicilian secret society, characterized by hostility to and deliberate flouting of the law and its representatives; also, any similar organization of Italians and Sicilians believed to exist in other countries; Black Hand. Also **Maf′fi·a**. Compare CAMORRA. [<Ital. *maffia*; ult. origin uncertain]
ma foi (mȧ fwȧ′) *French* My faith; upon my word: an interjection.
mag (mag) *n. Brit. Slang* A halfpenny. [<dial. E *make* a halfpenny; infl. in form by *meg*, orig. a guinea]
Mag (mag) Diminutive of MARGARET.
Ma·gal·la·nes (mä′gä·yä′näs) See PUNTA ARENAS.
mag·a·zine (mag′ə·zēn′, mag′ə·zēn) *n.* 1 A house, room, or receptacle in which anything is stored; a depot; warehouse; specifically, a strong building for storing gunpowder and other military stores; also, a storeroom for gunpowder and shells aboard ship. 2 A receptacle in which the supply of reserve cartridges of a repeating rifle is placed; also, a case in which cartridges are carried. 3 A periodical publication containing sketches, stories, essays, etc. 4 A reservoir or supply chamber in a battery, camera, or the like. — *v.t.* **·zined**, **·zin·ing** To store up for future use. [<MF *magasin* <OF *magazin* <Arabic

765

makhāzin, pl. of *makhzan* a storehouse < *khazana* store up]
magazine gun A quick-firing small arm fitted with a case carrying a supply of reserve cartridges. See MAGAZINE (def. 2). Also **magazine pistol** or **rifle**.
mag·a·zin·ist (mag′ə·zē′nist) *n.* A contributor to or editor of a magazine. — **mag′a·zin′ism** *n.*
mag·da·len (mag′də·lin) *n.* A reformed prostitute. Also **mag·da·lene** (-lēn). [after *Mary Magdalene*]
Mag·da·le·na (mäg′dä·lā′nä) A river in Colombia, flowing 1,060 miles north to the Caribbean Sea.
Mag·da·lene (mag′də·lēn, *Ger.* mäg′dä·lā′nə) A feminine personal name. Also **Mag′da·len**, **Mag·da·le·na** (mag′də·lē′nə, *Du.*, *Pg.*, *Sp.*, *Sw.* mäg′dä·lā′nä). [<LL <Gk. *Magdalēnē*, lit., of Magdala <*Magdala* Magdala, a town on the Sea of Galilee]
— **the Magdalene** Mary Magdalene.
Mag·da·le·ni·an (mag′də·lē′nē·ən) *adj. Anthropol.* Describing an advanced culture stage of the upper Paleolithic in western Europe, immediately preceding the Mesolithic period: it is noted especially for its delicate carvings in bone and ivory and for the brilliant realism of its polychrome cave paintings. [<F *magdalenien*, from *La Madeleine* in west central France, where artifacts were found]
Mag·da·len Islands (mag′də·lin) An island group in the Gulf of Saint Lawrence, eastern Quebec province, Canada; total, 102 square miles. French **Îles Ma·de·leine** (ēl mȧ·dlen′).
Mag·de·burg (mag′də·bûrg, *Ger.* mäg′də·bōōrkh) A city of west central East Germany, in the former state of Saxony-Anhalt.
mage (māj) *n.* A magician. [<F <L *magus*. See MAGI.]
Ma·ge·lang (mä′gə·läng′) A town in central Java.
Ma·gel·lan (mə·jel′ən), **Fernando**, 1480–1521, Portuguese navigator in the Spanish service.
— **Ma·gel·lan·ic** (maj′ə·lan′ik) *adj.*
Magellan, Strait of The channel between the Atlantic and the Pacific separating the South American mainland from Tierra del Fuego.

Magellanic clouds *Astron.* Two aggregations of star clusters and nebulae near the south pole of the heavens, looking like detached fragments of the Milky Way. See under COALSACK.
Ma·gen·die (mȧ·zhän·dē′), **François**, 1783–1855, French physician and physiologist.
ma·gen·ta (mə·jen′tə) *n.* 1 A somewhat glaring red coal-tar dyestuff derived from aniline: also called *aniline red* and *fuchsin*. 2 The color given by the pigment, a strong purplish-rose or purplish-red. [from *Magenta*; so called because discovered just after the French victory (1859)]
Ma·gen·ta (mä·jen′tä) A town in Lombardy, northern Italy; scene of a French victory over Austrian forces, 1859.
Ma·gers·fon·tein (mä′gərs·fōn·tān′) A locality of NE Cape of Good Hope Province, Union of South Africa; scene of a Boer victory in the South African War.
Mag·gio·re (mäd·jō′rā), **Lago** A lake in northern Italy and Switzerland; 65 square miles.
mag·got (mag′ət) *n.* 1 The larva of a fly; a footless insect larva; a grub. 2 A whim; fancy. [Prob. alter. of ME *maddock*, *mathek* <ON *mathkr* a worm]
mag·got·y (mag′ət·ē) *adj.* 1 Infested with maggots; flyblown. 2 Whimsical. — **mag′got·i·ness** *n.*

Maglemosian

Ma·gi (mā′jī) *n. pl.* of **Ma·gus** (mā′gəs) 1 The priestly caste of the Medes and Persians. 2 Specifically, the three "wise men" who came "from the east" to Bethlehem to pay homage to the infant Jesus. *Matt.* ii 1. [<L, pl. of *magus* <Gk. *magos* <Persian *magu* a priest, a magician] — **Ma′gi·an** *adj.* &.
Ma·gi·an·ism (mā′jē·ən·iz′əm) *n.* The creed and cult of the Magi.
mag·ic (maj′ik) *n.* 1 Any supernatural art; sorcery; necromancy. 2 Sleight of hand. 3 Any agency that works with wonderful effect. See synonyms under SORCERY. — **black magic** Any of the branches of magic which invoke the aid of demons or spirits, as witchcraft or diabolism. — **like magic** As if by magic; instantly. — *adj.* 1 Of the nature of magic; used in magic; possessing supernatural powers; sorcerous. 2 Acting like magic. [<OF *magique* <LL *magica (ars)* magic (art) <Gk. *magikē (technē)* <*magikos* of the Magi]
mag·i·cal (maj′i·kəl) *adj.* Pertaining to or produced by or as by magic. ◆ The adjective *magic* is applied more commonly to the powers, influences, or practices, while *magical* is more frequently used of the effects of magic: *magic arts*, *a magic wand*, but *magical effect*, *a magical result*. — **mag′i·cal·ly** *adv.*
ma·gi·cian (mə·jish′ən) *n.* An expert in magic arts; sorcerer; wizard.
magic lantern A device for throwing magnified pictures upon a screen in a darkened room by means of a light placed behind a lens or lenses.
ma·gilp (mə·gilp′) *n.* A mixture used as a vehicle for oil colors, usually composed of a pale drying oil and a turpentine varnish, such as mastic: also spelled *megilp*. Also **ma·gilph′** (-gilf′). [? after *McGilp*, a surname]
Ma·gi·not line (mazh′ə·nō, *Fr.* mȧ·zhē·nō′) A system of French fortifications along the German frontier, built during 1925–1935. [after André *Maginot*, 1877–1932, French statesman]
mag·is·te·ri·al (maj′is·tir′ē·əl) *adj.* 1 Of or pertaining to a magistrate or magistracy; like or befitting a master; commanding; authoritative. 2 Hence, having an air of authority; dictatorial. 3 Domineering; pompous. 4 Pertaining to a chemist's or alchemist's magistery. See synonyms under DOGMATIC. [<Med. L *magisterialis* <LL *magisterius* <L *magister* a master] — **mag′is·te′ri·al·ly** *adv.* — **mag′is·te′ri·al·ness** *n.*
mag·is·te·ri·um (maj′is·tir′ē·əm) *n.* The authority of the church to teach dogmatically: a Roman Catholic usage. [<L, the office of a master <*magister* a master]
mag·is·ter·y (maj′is·tər·ē, -tər·ē) *n. pl.* **·ter·ies** 1 An authoritative statement or exposition; a magisterial decree. 2 A fundamental master principle of nature; also, a panacea. 3 In alchemy, the power to transmute metals or the product of transmutation. 4 A compound, as a precipitate, formed when two liquids are mixed, and differing in character from either: a term used by the older chemists and preserved in the phrase *magistery of bismuth*. [<Med. L *magisterium* the philosopher's stone <L. See MAGISTERIUM.]
mag·is·tra·cy (maj′is·trə·sē) *n. pl.* **·cies** 1 The office or dignity of a magistrate. 2 The district under a magistrate's jurisdiction. 3 Magistrates collectively.
mag·is·tral (maj′is·trəl) *adj.* 1 Like a magistrate; imperious or pedagogical; magisterial. 2 In pharmacy, specially compounded or prescribed; not kept in stock. 3 Having sovereign power as a medicine. 4 Chief; main: the *magistral* line. 5 *Mil.* The line from which the positions of the various members of a fortification are determined: also **magistral line**. [<F <L *magistralis* <*magister* a master]
mag·is·trate (maj′is·trāt, -trit) *n.* 1 One clothed with public civil authority; an executive or judicial officer. 2 Usually, when unqualified, a minor local justice. [<L *magistratus* magisterial office <L *magister* a master]
mag·is·tra·ture (maj′is·trə·chŏŏr′) *n.* 1 Magistracy; government. 2 The term of a magistrate's office. 3 Magistrates collectively. [<F <L *magistrat* <L *magistratus*. See MAGISTRATE.]
Ma·gle·mo·si·an (mä′glə·mō′zē·ən) *adj.* Of or

relating to the special phases or aspects of the Mesolithic forest culture of northern Europe as indicated by discoveries at Maglemose, Denmark.
mag·ma (mag′mə) *n. pl.* **·ma·ta** (-mə·tə) 1 Any soft, doughy mixture of organic and mineral materials. 2 *Geol.* The molten mass from which igneous rocks are formed. 3 The residue obtained after expressing the juice from fruits. [<L <Gk. <root of *massein* knead] — **mag·mat·ic** (mag·mat′ik) *adj.*
Mag·na Car·ta (mag′nə kär′tə) 1 The Great Charter of English liberties, delivered June 19, 1215, by King John, at Runnymede, on the demand of the English barons: first document of the English constitution. 2 Any fundamental constitution that secures personal liberty and civil rights. Also **Mag′na Char′ta**, the erroneous but common form. [<Med. L, lit., Great Charter]
mag·na cum lau·de (mag′nə kum lô′dē, mäg′·nä kŏŏm lou′də) See under CUM LAUDE.
mag·na est ve·ri·tas, et prae·va·let (mag′nə est ver′ə·tas et prē·vä′let) *Latin* Great is truth, and it shall prevail.
mag·na·flux test (mag′nə·fluks) A method for the detection of defects in metals and in engine parts by noting the arrangement of the particles of a magnetic powder scattered over the magnetized surface. [<MAGN(ETIC) + -*a*- + FLUX]
Magna Grae·ci·a (grē′shē·ə) See GRAECIA MAGNA. Also **Mag′na Græ′ci·a**.
mag·na·nim·i·ty (mag′nə·nim′ə·tē) *n. pl.* **·ties** 1 The quality of being magnanimous; greatness of soul; generosity in sentiment or conduct toward others; exaltation above mean or petty motives. 2 A magnanimous deed.
mag·nan·i·mous (mag·nan′ə·məs) *adj.* 1 Elevated in soul; scorning what is mean or base. 2 Dictated by magnanimity; unselfish: *magnanimous* candor. See synonyms under GENEROUS. [<L *magnanimus* <*magnus* great + *animus* mind, soul] — **mag·nan′i·mous·ly** *adv.* — **mag·nan′i·mous·ness** *n.*
mag·nate (mag′nāt) *n.* A person of rank or importance; a noble or grandee; one notable or powerful in any sphere: an industrial *magnate*. [<LL *magnas, -atis* <*magnus* great]
mag·ne·sia (mag·nē′zhə, -shə, -zē·ə) *n. Chem.* Magnesium oxide, MgO, a light, white, earthy powder, used in medicine as an antacid laxative, in glassmaking, etc. It can be made by burning magnesium or by igniting certain of the magnesium salts. — **milk of magnesia** A milk–white aqueous suspension of magnesium hydroxide, Mg(OH)$_2$: used as a laxative and antacid. [<Med. L <Gk. *Magnēsia (lithos)* (stone) of Magnesia <*Magnēsia* Magnesia (def. 2)] — **mag·ne′sian, mag·ne′sic** *adj.*
Mag·ne·sia (mag·nē′zhə, -shə) 1 An ancient city of Lydia, western Asia Minor, on the site of modern Manissa; scene of the defeat of Antiochus the Great by the two Scipios, 190 B.C. Also **Magnesia ad Si·py·lum** (ad sip′ə·lum). 2 In classical geography, the coastal district of Thessaly south of the Vale of Tempe, part of which forms a nome of modern Greece.
magnesia alba *Chem.* A light, white, hydrous magnesium carbonate. [<NL, lit., white magnesia]
mag·ne·site (mag′nə·sīt) *n.* A massive, granular carbonate of magnesium, MgCO$_3$. [<MAGNES(IA) (ALBA) + -ITE[1]]
mag·ne·si·um (mag·nē′zē·əm, -zhē-, -shē-) *n.* A light, malleable, ductile, silver–white, metallic element (symbol Mg), used to produce a brilliant light by its combustion, and also as an alloy metal. It occurs abundantly in combination, as in magnesite and dolomite. See ELEMENT. [<NL, magnesia]
magnesium light A brilliant and intense light obtained by the combustion of magnesium: used in signaling, photography, pyrotechnics, etc.
mag·net (mag′nit) *n.* 1 A lodestone. 2 Any mass of a material capable of attracting magnetic or magnetized bodies. Magnets are natural when, like lodestones, they are found already magnetized, or artificial when magnetism has been given to them by placing them in the field of another magnet or in that caused by an electric current. See ELECTROMAGNET. 3 Figuratively, a person or thing exercising a strong attraction. [<OF *magnete* <L *magnes, magnetis* <Gk. *Magnēs (lithos)* Magnesian (stone), i.e., a magnet <*Magnēsia* Magnesia (def. 2)]

mag·net·ic (mag·net′ik) *adj.* 1 Pertaining to a magnet or magnetism. 2 Able to attract or be attracted by a lodestone. 3 Capable of exerting or responding to magnetic force. 4 Pertaining to terrestrial magnetism. 5 Possessing personal attractiveness or magnetism. 6 Pertaining to mesmerism. 7 Employing electrical treatments: a *magnetic* physician. Also **mag·net′i·cal**. — *n.* A substance that has or may be given a magnetic field, as iron, steel, nickel, cobalt. — **mag·net′i·cal·ly** *adv.*
magnetic chart A chart indicating the variations in the earth's magnetic field for a given area.
magnetic dip The angle which a dipping needle or the lines of magnetic force at any place make with the horizon.
magnetic equator An irregular, unstable line on the earth's surface, encircling it nearly midway between the magnetic poles, where a free magnetic needle has no tendency to dip; the aclinic line. It coincides nearly with the geographical or terrestrial equator. See ACLINIC.
magnetic field That region in the neighborhood of a magnet or current–carrying body in which magnetic forces are observable.
magnetic flux The number of magnetic lines of force passing through a magnetic circuit.
magnetic hysteresis That property of a magnetic material by virtue of which the magnetic induction depends upon the previous conditions of magnetization.
magnetic induction The magnetic flux per unit of area in a magnetized body, expressed in gauss. Symbol, B.
magnetic latitude Latitude as measured from the magnetic equator.
magnetic lens *Physics* An assembly of coils and electromagnets so arranged as to produce a magnetic field which will constrain a stream of charged particles to follow a prescribed path.
magnetic meridian *Nav.* A grid line indicating any of the horizontal components of the earth's magnetic field and which passes through the magnetic poles.
magnetic mine An underwater mine containing a sensitive device which, in the presence of any large mass of magnetic material, as a ship, senses the change in the magnetic field and actuates an electrical circuit which detonates the mine.
magnetic moment A measure of the magnetizing force exerted by a magnetized body or electric current.
magnetic needle A freely movable, needle–shaped piece of magnetized material which tends to point to the north and south (magnetic) poles of the earth.
magnetic north That direction on the earth's surface toward which one end of the magnetic needle tends to point.
magnetic pole 1 Either of the poles of a magnet; more specifically, those points on the earth's surface where the lines of magnetic force are vertical, called the **North** and **South Magnetic Poles**. These slowly change position and do not coincide with the geographical poles. 2 A pole of a magnet.
magnetic pyrites See PYRRHOTITE.
mag·net·ics (mag·net′iks) *n. pl.* (construed as singular) The science of magnetism.
magnetic signature The specific magnetic susceptibility of a material, product, or structure: said especially of the change of field in the neighborhood of ships exposed to the danger of sensitive magnetic mines.
magnetic speaker A loudspeaker.
magnetic storm A sudden disturbance of the magnetic field surrounding the earth, occurring simultaneously over areas of the earth and apparently connected with sunspots.

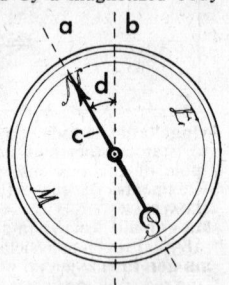

MAGNETIC NEEDLE
a. Magnetic north.
b. True north.
c. Magnetic needle.
d. Angle of variation.

magnetic tape *Electronics* A thin ribbon of paper or plastic, one side of which is coated with particles of iron oxide which form magnetic patterns corresponding to the electromagnetic impulses of a tape recorder.
mag·net·ism (mag′nə·tiz′əm) *n.* 1 The specific properties of a magnet, regarded as an effect of molecular interaction. 2 The science that treats of the laws and conditions of magnetic phenomena: also *magnetics*. 3 The amount of magnetic moment in a magnetized body. 4 The sympathetic personal quality that attracts or interests. 5 Mesmerism. Compare ANIMAL MAGNETISM.
mag·net·ite (mag′nə·tīt) *n.* A massive, granular, isometric, black iron oxide, Fe$_3$O$_4$; lodestone; an important ore of iron. [<MAGNET + -ITE[1]] — **mag′net·it′ic** (-tit′ik) *adj.*
mag·net·i·za·tion (mag′nə·tə·zā′shən) *n.* The amount of magnetism in or the magnetic moment of a material.
mag·net·ize (mag′nə·tīz) *v.t.* **·ized, ·iz·ing** 1 To communicate magnetic properties to. 2 To attract by strong personal influence; captivate. 3 *Obs.* To hypnotize. Also *Brit.,* **mag′net·ise**. — **mag′net·iz′a·bil′i·ty** *n.* — **mag′net·iz′a·ble** *adj.* — **mag′net·iz′er** *n.*
magnetizing force A vector quantity that measures the capacity of magnetized bodies or electric currents to produce magnetic induction. Symbol, H.
mag·ne·to (mag·nē′tō) *n. pl.* **·tos** Any magnetoelectric machine in which the rotation of a coil of wire between the poles of a permanent magnet induces a current of electricity in the coil; especially, a type of alternator, widely used as a means of igniting the explosive mixtures used in internal–combustion engines, as in automobiles. Also **mag·ne·to·dy·na·mo** (mag·nē′tō·dī′nə·mō), **mag·ne·to·gen·er·a·tor** (mag·nē′tō·jen′ə·rā′tər). [Short for *magnetoelectric machine*]
magneto– *combining form* Magnetic; magnetism: *magnetomotive*.
mag·ne·to·chem·is·try (mag·nē′tō·kem′is·trē) *n.* The science which treats of the interrelations between magnetic and chemical phenomena. — **mag·ne′to·chem′i·cal** *adj.*
mag·ne·to·e·lec·tric·i·ty (mag·nē′tō·i·lek·tris′·ə·tē) *n.* 1 Electricity generated by the inductive action of a magnet. 2 The science that treats of the principles and phenomena of such electricity. — **mag·ne′to·e·lec′tric, mag·ne′to·e·lec′tri·cal** *adj.*
mag·ne·to·graph (mag·nē′tə·graf, -gräf) *n.* A recording magnetometer.
mag·ne·to·hy·dro·dy·nam·ic (mag·nē′tō·hī′drō·dī·nam′ik) *adj. Physics* Of, pertaining to, or characterized by the interaction of electromagnetic, mechanical, thermal, and hydrodynamic forces, as by an electric arc in the generation of a plasma jet.
mag·ne·tol·y·sis (mag′nə·tol′ə·sis) *n.* Chemical action in a substance placed in a magnetic field: analogous to electrolysis. [<MAGNETO- + -LYSIS] — **mag·ne·to·lyt·ic** (mag·net′ə·lit′ik) *adj.*
mag·ne·tom·e·ter (mag′nə·tom′ə·tər) *n.* An instrument for measuring the intensity and direction of magnetic forces. — **mag′ne·tom′·e·try** *n.*
mag·ne·to·mo·tive (mag·nē′tō·mō′tiv, mag′·nə·tō-) *adj.* 1 Acting magnetically. 2 Characterizing a force producing magnetic flux: distinguished from *electromotive*.
mag·ne·ton (mag′nə·ton) *n. Physics* A unit of magnetic moment: The Bohr magneton has the value of 9.27×10^{-21} erg gauss^{-1}.
mag·ne·to·op·tics (mag·nē′tō·op′tiks) *n.* The study of the behavior of light waves in a magnetic field. — **mag·ne′to–op′tic, mag·ne′to·op′ti·cal** *adj.*
mag·ne·to·scope (mag·nē′tə·skōp, -net′ə-) *n.* An instrument designed to indicate the presence of magnetic force lines without measuring them.
mag·ne·to·stric·tion (mag·nē′tō·strik′shən) *n. Physics* The mechanical deformation produced in certain materials when subjected to the action of a magnetic field.
mag·ne·tron (mag′nə·tron) *n. Electronics* A vacuum tube in which the flow of electrons is subject to the control of an external magnetic field. [<MAGNET + (ELEC)TRON]
magni– *combining form* 1 Great; large: *magnirostrate*, large–beaked. 2 *Zool.* Long: *magnicaudate*, long–tailed. [<L *magnus* great]

mag·nif·ic (mag·nif'ik) *adj.* 1 Illustrious; magnificent; sumptuous. 2 Strikingly vast or dignified; imposing. 3 Of language, exalted; sublime; also, pompous; grandiloquent. Also **mag·nif'i·cal.** [<F *magnifique* <L *magnificus* <*magnus* great + *fic-*, stem of *facere* make] — **mag·nif'i·cal·ly** *adv.*

Mag·nif·i·cat (mag·nif'ə·kat) *n.* The hymn or canticle of the Virgin Mary, beginning with the word *Magnificat* in the Latin version. *Luke* i 46-55. [<L, it magnifies]

mag·ni·fi·ca·tion (mag'nə·fə·kā'shən) *n.* 1 The act, process, or degree of magnifying. 2 The state of being magnified. 3 Magnifying power. 4 The act of extolling or glorifying. [<L *magnificatio, -onis* <*magnificare* <*magnificus*. See MAGNIFIC.]

mag·nif·i·cence (mag·nif'ə·səns) *n.* 1 The state or quality of being magnificent; the exhibition of greatness of action, character, intellect, wealth, or power; brilliant or imposing appearance; display of splendor. 2 A title of courtesy in medieval Rome. [<OF <L *magnificentia* <*magnificus* noble. See MAGNIFIC.]

mag·nif·i·cent (mag·nif'ə·sənt) *adj.* 1 Grand or majestic in appearance, quality, character, or action; extremely fine or good; befitting the great, as in deeds, manners, or surroundings; great in effect, promise, or import: a *magnificent* prospect, pearl, plan, etc.; also, exalted; imposing: *magnificent* language. 2 Exhibiting magnificence; characterized by splendor: sometimes used as a title: Suleiman the *Magnificent*. See synonyms under GRAND, IMPERIAL, SUBLIME. [<OF <LL *magnificens,* var. of *magnificus*. See MAGNIFIC.] — **mag·nif'i·cent·ly** *adv.*

mag·nif·i·co (mag·nif'ə·kō) *n. pl.* **·coes** 1 A noble of the Venetian republic: an old title. 2 A lordly personage; one who affects state or splendor. [<Ital. <L *magnificus*. See MAGNIFICENCE.]

mag·ni·fy (mag'nə·fī) *v.t.* **·fied, ·fy·ing** 1 To increase the apparent size of, as by a microscope. 2 To increase the size of; enlarge. 3 To cause to seem greater or more important. 4 *Archaic* To extol; exalt. See synonyms under AGGRAVATE, INCREASE, PRAISE. [<OF *magnifier* <L *magnificare* <*magnificus*. See MAGNIFICENCE.] — **mag'ni·fi'a·ble** *adj.* — **mag'ni·fi'er** *n.*

mag·nil·o·quent (mag·nil'ə·kwənt) *adj.* Of bombastic, pompous style; vainglorious. [<L *magnus* great + *loquens, -entis,* ppr. of *loqui* speak] — **mag·nil'o·quence** *n.* — **mag·nil'o·quent·ly** *adv.*

Mag·ni·to·gorsk (mäg·nē'tô·gôrsk') A city on the Ural in Russian S.F.S.R.

mag·ni·tude (mag'nə·tōōd, -tyōōd) *n.* 1 Great size or extent; grandeur; importance. 2 *Math.* That which is conceived of as measurable. 3 The property of having size or extent. 4 *Astron.* A measure of the relative brightness of a star, ranging from one for the brightest to six for those just visible to the naked eye. The standard of reference is the polestar, with a brightness 2.512 times greater than the one next below it; magnitudes greater than one on the scale are expressed as minus quantities. Compare LIGHT RATIO. 5 Largeness in respect to relation or effect. [<L *magnitudo* <*magnus* large]
 — *Synonyms*: bigness, bulk, dimension, extent, greatness, hugeness, immensity, largeness, size, vastness. *Antonyms*: diminutiveness, littleness, pettiness, slightness, smallness.

mag·no·li·a (mag·nō'lē·ə, -nōl'yə) *n.* 1 Any of a genus (*Magnolia*) of trees or shrubs with large, fragrant, and often showy flowers. 2 The blossom of the evergreen magnolia (*M. grandiflora*), the State flower of Louisiana and of Mississippi. [<NL, genus name, after Pierre *Magnol*, 1638-1715, French botanist]

mag·no·li·a·ceous (mag·nō'lē·ā'shəs) *adj.* Of or pertaining to a family (*Magnoliaceae*) of polypetalous trees or shrubs, the magnolia family, often aromatic, with alternate, undivided, feather-veined leaves, and large, solitary, axillary or terminal flowers with calyx and corolla colored alike, in three or more rows of three each. [<NL *Magnoliaceae*, family name <*Magnolia* MAGNOLIA]

magnolia warbler The black-and-yellow warbler (*Dendroica magnolia*) of North America.

mag·num (mag'nəm) *n.* A wine bottle of twice the ordinary size, holding about two quarts; also, the quantity such a bottle will hold. [<L, neut. of *magnus* great]

magnum o·pus (ō'pəs) *Latin* The chief work of an artist; a masterpiece; literally, great work.

Mag·nus effect (mag'nəs) *Physics* The deflecting effect upon the normal path of a rotating cylinder or sphere caused by the transverse forces of wind or air currents circulating around it. [after H. G. *Magnus*, 1802-70, German physicist]

mag·nus hitch (mag'nəs) A knot used to fasten a rope to a spar, etc., similar to a clove-hitch but having one more turn. [Prob. <L *magnus* large]

Ma·gog (mā'gog) See GOG AND MAGOG.

ma·got (má·gō') *n.* The Barbary ape. [<F, ?<*Magog* MAGOG]

Mag·ot·ty-bay bean (mag'ə·tē·bā') A leguminous vine (*Chamaecrista fasciculata*) common on eastern American shores. [? from *Magothy* River, Maryland]

mag·pie (mag'pī) *n.* 1 A corvine bird (genus *Pica*), having a long, graduated tail. The common **European magpie** (*P. pica* or *caudata*) has iridescent black plumage with white scapulars, belly, sides, flanks, and inner web of flight feathers: often tamed and taught to speak words, and noted for its thievishness. The **American magpie** (*P. hudsonia*) is a variety of the European magpie. *P. nuttali* is the **yellow-billed magpie** of California. 2 An Australian crow shrike resembling a magpie. 3 A chatterbox; a garrulous gossip. [<*Mag*, diminutive of MARGARET, + PIE²]

Mag·say·say (mäg·sī'sī), **Ramón**, 1907-57, president of the Philippines 1953-57.

mag·uey (mag'wā, *Sp.* mä·gā'ē) *n.* 1 Any of various Mexican agave plants with fleshy leaves and edible cabbage-like heads, especially the century plant (*Agave americana*) and the pulque agave (*A. atrovirens*). 2 A fiber plant (genus *Furcraea*) related to the agave. 3 The fiber of these plants. [<Sp., prob. <Taino]

MAGUEY
(Leaves up to 6 feet long; 8 to 10 inches broad)

Ma·gus (mā'gəs) Singular of MAGI.

Mag·yar (mag'yär, *Hungarian* mud'yär) *n.* 1 One of a people who invaded and conquered Hungary at the end of the ninth century; a Hungarian. 2 The Finno-Ugric language of Hungary. — *adj.* Of or pertaining to the Magyars or their language.

Mag·yar·or·szåg (mud'yär·ôr'säg) The Hungarian name for HUNGARY.

Ma·ha·bha·ra·ta (mə·hä'bä'rə·tə) An ancient Hindu epic, written in Sanskrit and thought to date from 300 B.C., recounting the dynastic wars of two related families over a kingdom in northern India. This large work, probably the longest in the world, and the *Rāmayana* are the two great epics of ancient India. Also **Ma·ha'bha'ra·tam**. [<Skt. *Mahābhārata*, lit., the great story]

ma·hal·a (mə·hal'ə) *n.* In the western United States and Canada, an Indian squaw. [<N. Am. Ind. *muk'ela*]

Ma·hal·la el Ku·bra (mə·hal'ə el kōō'brə) A city in the Nile delta. Also **Ma·hal'la**.

Ma·han (mə·han'), **Alfred Thayer**, 1840-1914, U.S. admiral and naval historian.

Ma·ha·na·di (mə·hä'nə·dē) A river in eastern India, flowing 512 miles to the Bay of Bengal.

ma·ha·ra·ja (mä'hə·rä'jə, *Hind.* mə·hä'rjə) *n.* 1 A great Hindu prince; the title of some native rulers. 2 A prominent religious teacher of the Hindus. Also **ma'ha·ra'jah**. [<Hind. <Skt. *mahārāja* <*maha* great + *rāja* a king]

ma·ha·ra·ni (mä'hə·rä'nē, *Hind.* mə·hä'rä'nē) *n.* A Hindu princess; the wife of a maharaja. Also **ma'ha·ra'nee**. [<Hind. <*maha* great + *rānī* a queen]

ma·hat·ma (mə·hat'mə, -hät'-) *n.* In theosophy or esoteric Buddhism, an adept of the highest order; literally, great-souled one: a title of respect. [<Skt. *mahātman* <*maha* great + *ātman* soul] — **ma·hat'ma·ism** *n.*

Mah·di (mä'dē) *n.* The Moslem messiah, or one claiming the title; specifically, Mohammed Ahmed, 1843-85, who led a revolt in the Sudan, 1883. [<Arabic *mahdīy*, lit., he who is guided aright, pp. of *hadā* lead rightly] — **Mah'dism** *n.* — **Mah'dist** *adj.* & *n.*

Ma·hé (mä·hā') A free city and former French settlement in SW Madras, India, on the Arabian Sea; 23 square miles.

Mahé Island The principal island in the Seychelles group in the western Indian Ocean; 56 square miles.

Ma·hi·can (mə·hē'kən) *n.* One of a tribe of North American Indians of Algonquian linguistic stock formerly occupying the territory from the Hudson River to Lake Champlain; later, one of an Algonquian tribe between the Hudson River and Narragansett Bay, dialectally divided into the *Mohegans* of the Thames and lower Connecticut rivers, and the *Mohicans*, occupying both banks of the Hudson. [<Algonquian, lit., a wolf]

mah jong (mä'jong', jông') A game of Chinese origin, usually for four persons, played with 144 tiles marked in suits, dice, and counters. Also **mah jongg**. [<dial. Chinese <Chinese *ma ch'iao*, lit., a house sparrow; from the design on one of the tiles]

Mah·ler (mä'lər), **Gustav**, 1860-1911, Austrian composer and conductor.

mahl·stick (mäl'stik', môl'-) *n.* A staff with a ball at one end, used by painters to steady the hand while using the brush. Also spelled *maulstick*. [<Du. *maalstok* <*malen* paint + *stok* a stick]

Mah·moud (mä·mōōd', mä'mōōd) Persian form of MOHAMMED. Also **Mah·mud'**.
— **Mahmoud II**, 1785-1839, sultan of Turkey; defeated (1829) in Greek war for independence.

ma·hog·a·ny (mə·hog'ə·nē) *n.* 1 A large tropical American tree (genus *Swietenia*), with fine-grained, hard, reddish wood much used for cabinetwork. 2 The wood itself. 3 One of various trees yielding a similar wood, as African mahogany (*Khaya ivorensis*); also its wood. 4 Acajou. 5 Any of the various shades of brownish-red or reddish-brown of the finished wood. — *adj.* 1 Of or pertaining to, or consisting of mahogany. 2 Of a mahogany color. [<obs. Sp. *mahogani*, prob. <Arawakan]

Ma·hom·et (mə·hom'it) See MOHAMMED.

Ma·hom·e·tan (mə·hom'ə·tən), etc. See MOHAMMEDAN, etc.

Ma·hón (mə·hōn', *Sp.* mä·ōn') The chief city and port of Minorca, Balearic Islands: formerly *Port Mahon*.

Ma·hound (mə·hound', -hōōnd') *n.* 1 The prophet Mohammed. 2 *Scot.* Satan; an evil spirit. [<OF *Mahon, Mahum* <*Mahomet* Mohammed]

ma·hout (mə·hout') *n.* The keeper and driver of an elephant. [<Hind. *mahāut, mahāvat* <Skt. *mahāmātra*, lit., great in measure]

Mah·rat·i (mə·rat'ē) See MARATHI.

Mah·rat·ta (mə·rat'ə) *n.* One of a Hindu people of SW and central India. Also spelled *Maratha*. [<Hind. *Marhaṭa* <Marathi *Marathi* <Skt. *Mahārāṣṭra*, lit., great country]

Mäh·ren (mā'rən) The German name for MORAVIA.

Mäh·risch-Os·trau (mā'rish-ôs'trou) The German name for MORAVSKÁ OSTRAVA.

ma huang (mä hwäng') A Chinese species of joint fir. [<Chinese]

Mai·a (mā'yə, mī'ə) 1 In Greek mythology, the eldest of the Pleiades, mother by Zeus of Hermes. 2 In Roman mythology, a goddess of growth and spring: identified with the Greek *Maia* and the *Bona Dea*. Also **Maia Ma·jes·ta** (mə·jes'tə).

maid (mād) *n.* 1 Any unmarried woman; virgin; girl; lass. 2 A female servant: also **maidservant**. [Short for MAIDEN]

mai·dan (mī·dän') *n.* In Persia and India, a public plaza or parade ground; hence, an open space. [<Persian *maidān*]

Mai·da·nek (mī'də·nek) During World War II,

a Nazi concentration and extermination camp near Lublin, Poland.
maid·en (mād′n) *n.* 1 An unmarried woman, especially if young; a maid; virgin. 2 Something untried or unused, as a race horse that has never won an event. 3 A rude kind of beheading machine used in Scotland in the 16th and 17th centuries. — *adj.* 1 Pertaining to or suitable for a maiden. 2 Virgin; unmarried. 3 Initiatory; unused; untried. 4 Of or pertaining to the first use or experience: a *maiden* voyage. 5 Pure; sinless. [OE *mægden*, prob. dim. of *mægeth* a virgin] — **maid′en·li·ness** *n.* — **maid′en·ly** *adj. & adv.*
maid·en·hair (mād′n-hâr′) *n.* A very delicate and graceful fern (genus *Adiantum*) with an erect black stem, common in damp, rocky woods. Also **maidenhair fern**.
maidenhair tree The ginkgo.
maid·en·head (mād′n-hed′) *n.* 1 The hymen. 2 Maidenhood; virginity.
maid·en·hood (mād′n-ho͝od) *n.* The state of being a maiden; freshness; purity; virginity. Also **maid′hood**.
maiden name A woman's surname before marriage.
Maid Marian 1 The sweetheart and companion of Robin Hood. 2 A character in morris dances and other ancient sports: at first a May queen, later a grotesque buffoon, often impersonated by a boy.
maid of honor 1 An unmarried lady attendant upon an empress, queen, or princess: usually of noble birth. 2 The chief unmarried attendant of a bride at a wedding ceremony.
maid·ser·vant (mād′sûr′vənt) *n.* A female servant.
Maid·stone (mād′stən, -stōn) The county town of Kent, England.
Mai·du·gu·ri (mī-do͞o′go͞or-ē) A city in NE Nigeria, capital of Bornu province.
ma·ieu·tic (mā-yo͞o′tik) *adj.* Helping to bring forth ideas and truths: said of the Socratic method. Also **ma·ieu′ti·cal**. [<Gk. *maieutikos*, lit., obstetric < *maieuesthai* act as a midwife < *maia* a midwife]
ma·ieu·tics (mā-yo͞o′tiks) *n.* 1 The art of facilitating the bringing forth of ideas from the mind of a pupil by a series of pertinent questions; the Socratic method. 2 Midwifery; obstetrics.
mai·gre (mā′gər) *adj.* Not consisting of flesh or its juices: said of dishes used by Roman Catholics on days of abstinence; hence, of or appropriate for a fast. — *n.* A large marine food fish, the European *Sciaena aquila*. [<OF. See MEAGER.]
mai·hem (mā′hem) See MAYHEM.
Mai·kop (mī′kôp) A city in the NW Caucasus mountains, Russian S.F.S.R. Also **May′kop**.
mail[1] (māl) *n.* 1 The governmental system for handling letters, etc., by post. 2 Letters, magazines and other printed matter, parcels, etc., consigned and sent from place to place under a governmental post-office system. 3 The collection or delivery of postal matter at a specified time: My letter missed the *mail*. 4 Letters, papers, etc., received by, or for, a person: Your *mail* is on the table. 5 A conveyance, as a train, plane, etc., for carrying postal matter. — *adj.* Pertaining to or used in the process of conveying or handling mail. — *v.t.* To send by mail, as letters; deposit in a mailbox or at a post office; post. ◆ Homophone: *male*. [<OF *male* <OHG *malha* a wallet] — **mail′a·ble** *adj.*
mail[2] (māl) *n.* 1 Armor of chains, rings, or scales: often called **chain mail**. 2 Any strong covering or defense, as the shell of a turtle. — **coat of mail** A hauberk. — *v.t.* To cover with or as with mail. ◆ Homophone: *male*. [<OF *maille* <L *macula* spot, mesh of a net]
mail[3] (māl) *n. Scot. & Brit. Dial.* That which is paid; rent; wages. Also **maill**. ◆ Homophone: *male*. [OE *māl* <ON, speech, agreement; infl. in sense by ON *māle* contract, stipend]
mail·bag (māl′bag′) *n.* A bag or pack in which mail is transported.
mail·box (māl′boks′) *n.* 1 A box in which

CHAIN MAIL ARMOR

letters, etc., are posted for collection. 2 A box into which private mail is put when delivered. Also **mail box**.
mail car A railroad car for carrying mail.
mail·catch·er (māl′kach′ər) *n.* A mechanical device for transferring mailbags to or from a moving train.
mailed (māld) *adj.* 1 Covered or armed with mail. 2 *Zool.* Having a defensive armor, as scales. [<MAIL[2]]
mailed fist Menace of attack or violence; especially, a measure of aggressive war.
mail·er (mā′lər) *n.* 1 A mail boat. 2 An addressing machine: also **mailing machine**. 3 One who mails a letter.
mail·ing (mā′ling) *n. Scot.* A rented piece of ground; farm; homestead. Also **mail′en**.
Mail·lol (mä·yôl′), **Aristide**, 1861–1944, French sculptor.
mail·lot (mī·yō′) *n.* A tightly fitting garment which covers the torso, used by dancers, acrobats, and swimmers. [<F, dim. of *maille* knitted material, lit., mail[2]]
mail·man (māl′man′, -mən) *n. pl.* **·men** (-men′, -mən) A letter-carrier; postman.
mail order An order for goods, sent and filled through the agency of the mail. — **mail-or·der** (māl′ôr′dər) *adj.*
mail road A road along which mail was formerly transported.
mail route A regular route along which mail is collected and delivered.
mail station A stopping place on a mail route where passengers, drivers, and horses could procure food, rest, and shelter.
mail train 1 A railway train that carries mail. 2 A train of wagons that formerly carried mail in the western United States.
maim (mām) *v.t.* To deprive of the use of a bodily part; mutilate; disable. — *n. Rare* A crippling; mutilation; maiming. See MAYHEM. [<OF *mahaigner, mayner*; ult. origin uncertain] — **maim′er** *n.*
Mai·mon·i·des (mī-mon′ə-dēz), 1135–1204, Spanish rabbi, theologian, and philosopher: also called *Moses ben Maimon*.
main[1] (mān) *n.* 1 A chief conduit or conductor, as for conveying gas, water, etc. 2 The mainland. 3 *Poetic* The ocean. 4 Violent effort; strength: chiefly in the phrase *with might and main*. 5 The chief part; the most important point. — **in** (or **for**) **the main** For the most part; on the whole. — *adj.* 1 First or chief in size, rank, importance, strength, extent, etc.; principal; chief; leading: the *main* building, the *main* line of a railroad, with which branch lines connect. 2 Being concentrated or undivided; unqualified; full: by *main* force. 3 Designating any broad extent or expanse, as of land or sea. 4 *Naut.* Near or connected with the mainsail or mainmast. 5 *Brit. Dial.* Considerable; remarkable. 6 *Obs.* Vast; mighty; powerful. — *adv. Brit. Dial.* Very; greatly; extremely. ◆ Homophone: *mane*. [OE *mægen*] — **main′ly** *adv.*
main[2] (mān) *n.* 1 A match of several battles at cockfighting. 2 A hand or throw of dice; also, a number selected by the caster before he throws the dice in games of hazard and craps. ◆ Homophone: *mane*. [? <MAIN[1], as in *main chance*]
main[3] (mān) *Scot. v.i.* To moan. — *n.* A moan. ◆ Homophone: *mane*.
Main (mān, *Ger.* mīn) A river in western Germany, flowing 305 miles west to the Rhine.
main deck *Naut.* The chief deck of a vessel; on a warship, the topmost deck stretching from stem to stern; on a merchantman, the deck between and below the poop and forecastle decks.
main drag *U.S. Slang* The principal street or section of a city.
Maine (mān), **Sir Henry James**, 1822–88, English jurist.
Maine (mān) 1 A State of the NE United States bordering on Canada and the Atlantic; 32,562 square miles; capital, Augusta; entered the Union March 15, 1820; nickname, *Pine-Tree State*; abbr. *Me.* 2 An ancient province of western France.
Maine (mān) A U.S. battleship, blown up in Havana harbor, Feb. 15, 1898, with the loss of 260 lives: one of the events precipitating the Spanish-American War.
Main-fran·ken (mīn′fräng′kən) The former name of Lower Franconia, 1938–45.
main·land (mān′land′, -lənd) A principal body of land; a continent, as distinguished from an island.
Main·land (mān′lənd) 1 Pomona Island. 2 The largest of the Shetland Islands, Scotland; 406 square miles.
Main Line A traditionally wealthy and fashionable residential district outside of Philadelphia.
main·mast (mān′məst, -mast′, -mäst′) *n. Naut.* The principal mast of a vessel: ordinarily, the second mast from the bow.
main·or (mān′ər) *n. Law* 1 Formerly, a thing stolen found on a thief. 2 The act of theft. Also **main′our**. [<AF *mainoure, mainoevere*, OF *maneuvre*, lit., handwork. Doublet of MANEUVER, MANURE.]
mains (mānz) *n. Scot.* The principal or home farm. [See MANSE.]
main·sail (mān′səl, -sāl′) *n. Naut.* A sail bent to the main yard, or one carried on the mainmast: in a square-rigged vessel, called the **main course**.
main·sheet (mān′shēt′) *n. Naut.* The sheet by which the mainsail is trimmed and set.
main·spring (mān′spring′) *n.* 1 The principal spring of a mechanism, as of a watch. 2 The most efficient cause or motive.
main·stay (mān′stā′) *n. Naut.* 1 The rope from the mainmast head forward, used to steady the mast in that direction. 2 A chief support or dependence.
Main Street The principal business street of a small town: a symbol of its manners, customs, culture, typical thinking, etc.
main·tain (mān-tān′) *v.t.* 1 To carry on or continue; engage in, as a correspondence. 2 To keep unimpaired or in proper condition: to *maintain* roads; to *maintain* a reputation. 3 To supply with food or livelihood; support; pay for. 4 To uphold; claim to be true. 5 To assert or state; affirm. 6 To hold or defend, as against attack. See synonyms under AFFIRM, ALLEGE, ASSERT, JUSTIFY, KEEP, PRESERVE, RETAIN, SUPPORT. [<OF *maintenir* <L *manu tenere*, lit., hold in one's hand <*manu*, ablative of *manus* hand + *tenere* hold]
main·te·nance (mān′tə-nəns) *n.* 1 The act of maintaining. 2 Means of support. 3 *Law* The officious intermeddling in a suit, by assisting or maintaining either party, with money or otherwise. [<OF *maintenance* <*maintenir* MAINTAIN]
Main·te·non (mant·nôn′), **Marquise de**, 1635–1719, Françoise d'Aubigné; second wife of Louis XIV of France.
main·top (mān′top′) *n. Naut.* A platform at the head of the lower section of the mainmast.
main·top·gal·lant·mast (mān′tə-gal′ənt-məst, -mast′, -mäst′, -top-) *n. Naut.* On a square-rigged vessel, the mast next above the mainmast.
main·top·mast (mān′top′məst) *n. Naut.* 1 On a square-rigged vessel, the mast next above the maintopgallantmast. 2 On a fore-and-aft-rigged ship, the mast next above the mainmast.
main yard *Naut.* The lower yard on the mainmast.
Mainz (mīnts) A city on the Rhine, capital of Rhineland-Palatinate, West Germany: French *Mayence*.
ma·iol·i·ca (mä·yol′ē·kä) See MAJOLICA.
mai·o·sis (mī·ō′sis) See MEIOSIS (def. 1).
mair (mâr) *adj. & n. Scot.* More.
mai·son de san·té (mā·zôn′ də sän·tā′) *French* Sanitarium; private asylum.
maist (māst) *adj., n., & adv. Scot.* Most; almost.
mais·ter (mās′tər) *n. Scot.* Master.
Mai·sur (mī·so͞or′) See MYSORE.
Mait·land (māt′lənd), **Frederic William**, 1850–1906, English jurist and historian.
maî·tre (me′tr′) *n. French* Master.
maître d'hô·tel (dō·tel′) *French* 1 The proprietor or manager of a hotel. 2 A head-waiter; steward. 3 With a sauce composed of melted butter, parsley, and lemon juice or vinegar. 4 Served in the special manner of the house.
maize (māz) *n.* 1 A tall, stout food and forage plant (*Zea mays*). 2 Its grain; Indian corn. 3 A light, soft shade of yellow; any of the various yellow tints of ripe corn. ◆ Homophone: *maze*. [<Sp. *maíz* <Taino *mahiz*]
maize·bird (māz′bûrd′) *n.* A bird that feeds on Indian corn; specifically, the red-winged blackbird and the purple grackle.

ma·jes·tic (mə·jes′tik) *adj.* Having or exhibiting majesty; stately; royal, august. Also **ma·jes′ti·cal.** See synonyms under AWFUL, GRAND, IMPERIAL, KINGLY, SUBLIME. —**ma·jes′ti·cal·ly** *adv.* —**ma·jes′ti·cal·ness** *n.*

maj·es·ty (maj′is·tē) *n. pl.* **·ties** Exalted dignity; stateliness; grandeur; especially, **His, Her,** or **Your Majesty,** a title given to monarchs. [<OF *majesté* <L *majestas, -tatis* (related to *majus,* neut. compar. of *magnus* great)]

ma·jol·i·ca (mə·jol′i·kə, -yol′-) *n.* A kind of Italian pottery, glazed and decorated, usually in rich colors and Renaissance designs; any glazed Italian pottery; faience. Also spelled *maiolica.* [<Ital. *maiolica,* prob. < *Majolica,* early name of MAJORCA, where formerly made]

ma·jor (mā′jər) *adj.* **1** Greater in number, quantity, or extent. **2** Greater in dignity or importance; of primary consideration; principal; leading. **3** *Music* **a** Designating a chord or interval greater by a half-step than the preceding minor. **b** Containing a major third, sixth, and seventh. See INTERVAL. **4** *Logic* Designating the first premise of a syllogism, or the premise containing the first proposition. **5** *Law* Being of legal age. —*n.* **1** An officer in the U.S. Army, Air Force, or Marine Corps ranking next above a captain and next below a lieutenant colonel. **2** *Law* One who is of legal age. **3** *Music* A major key, chord, or interval. **4** *U.S.* The specialized course of study in a definite field which a college or university student follows to obtain his degree. **5** *U.S.* A student who follows a (specified) course of study: an English *major.* —*v.i.* To pursue a definite field of study: with *in*: to *major* in history. [<L, compar. of *magnus* great. Doublet of MAYOR.]

Ma·jor·ca (mə·jôr′kə) The largest of the Balearic Islands; 1,352 square miles; capital, Palma: Spanish *Mallorca.* —**Ma·jor′can** *adj.* & *n.*

ma·jor·do·mo (mā′jər·dō′mō) *n. pl.* **·mos** **1** The chief steward of a royal household. **2** A butler. **3** *SW U.S.* The overseer of a ranch. [<Sp. *mayordomo* <Med. L *major domus* chief of a house < *major* an elder (<L, greater) + *domus* a house]

major general See under GENERAL.

ma·jor·i·ty (mə·jôr′ə·tē, -jor′-) *n. pl.* **·ties** **1** More than half of a given number or group; the greater part. **2** The amount or number by which one group of things exceeds another group; excess. **3** The age at which the laws of a country permit a person to manage his or her own affairs: in most States of the United States, the age of 21 years. **4** The rank or commission of a major. **5** In U.S. politics, more than half of the people; more than half of the votes cast. **6** The number of votes cast for a candidate over and above the number cast for his nearest opponent; a plurality. **7** The party having the most power in a legislature. [<MF *majorité* <L *majoritas, -tatis* < *major.* See MAJOR.]

major key *Music* A key based on the tones of a major scale, the intervals being a half-step larger than minor intervals.

major league In baseball, either of the two main groups of professional teams in the United States. —**ma′jor-league′** *adj.*

major mode See under MODE.

major scale See under SCALE.

major suit In bridge, either of the sets of spades and hearts: so called from their higher point value.

major term See under TERM.

ma·jus·cule (mə·jus′kyōōl) *n.* A capital letter; especially, a large initial letter, as in old manuscripts. [<F <L *majuscula (littera),* fem. of *majusculus* somewhat larger, dim. of *major.* See MAJOR.] —**ma·jus′cu·lar** *adj.*

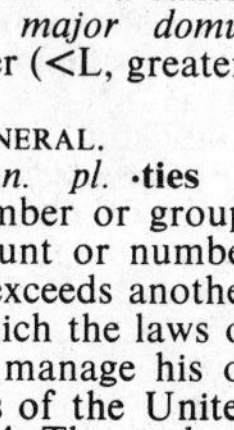

MAJUSCULE

mak (mak) *v.t.* & *v.i. Scot.* To make.

Ma·kas·sar (mə·kas′ər) See MACASSAR.

Ma·ka·te·a (mä′kä·tā′ə) An island dependency of Tahiti in the Society Islands.

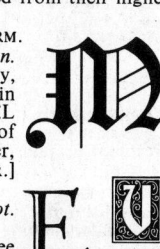

make[1] (māk) *v.* **made, mak·ing** *v.t.* **1** To bring about the existence of by the shaping or combining of materials; produce; build; construct; fashion. **2** To bring about; cause: Don't *make* trouble. **3** To bring to some state or condition; cause to be: The wind *made* him cold. **4** To appoint or assign; elect: They *made* him captain. **5** To form or create in the mind, as a plan. **6** To compose or create, as a poem or piece of music. **7** To understand or infer to be the meaning or significance; interpret; think: What do you *make* of it? **8** To put forward; advance: to *make* an offer. **9** To present, as for record; utter or express: to *make* a declaration. **10** To obtain for oneself; earn; accumulate: to *make* a fortune. **11** To amount to; add up to: Four quarts *make* a gallon. **12** To bring the total to: That *makes* five attempts. **13** To develop into; become: He *made* a good soldier. **14** To accomplish; effect or form: to *make* an agreement. **15** To estimate to be; reckon: He *made* the height twenty feet. **16** To induce or force; compel: He *made* me do it! **17** To draw up, enact, or frame, as laws, testaments, etc. **18** To prepare for use, as a bed. **19** To afford or provide: This brandy *makes* good drinking. **20** To be the essential element or determinant of: Stone walls do not a prison *make.* **21** To cause the success of: His speech *made* him politically. **22** To traverse; cover: to *make* fifty miles before noon. **23** To travel at the rate of: to *make* fifty miles per hour. **24** To arrive at; reach: to *make* Boston. **25** To board before departure: to *make* a train. **26** To earn so as to count on a score: to *make* a touchdown. **27** *Electr.* To complete (a circuit). **28** *Colloq.* To win a place on: to *make* the team. **29** In card games: **a** To declare as trump. **b** To capture a trick with (a card). **c** To shuffle: to *make* the deck. **30** In bridge, to win (a bid). **31** *U.S. Slang.* To seduce. —*v.i.* **32** To cause something to assume a specified condition: to *make* ready; to *make* fast. **33** To act or behave in a certain manner: to *make* merry. **34** To start: They *made* to go. **35** To go or extend in some direction: with *to* or *toward.* **36** To flow, as the tide; rise, as water. —**to make as if** (or **as though**) To pretend. —**to make away with** **1** To carry off; steal. **2** To get rid of; destroy. —**to make believe** To pretend. —**to make do** To get along with what is available, especially with an inferior substitute. —**to make for** **1** To go toward. **2** To attack; assail. **3** To have effect on; contribute to. —**to make heavy weather** *Naut.* To roll and pitch, as a ship in a storm. —**to make it** To succeed in doing something. —**to make off** To leave suddenly; run away. —**to make off with** To steal. —**to make out** **1** To see; discern. **2** To comprehend; understand. **3** To establish by evidence. **4** To fill out; draw up, as a bank draft. **5** To be successful; manage. —**to make over** **1** To put into new form; renovate. **2** To transfer title or possession of. —**to make up** **1** To compose; compound, as a prescription. **2** To be the parts of; comprise: These elements *make up* the structure. **3** To settle differences and become friendly again: to kiss and *make up.* **4** To devise; invent; fabricate: to *make up* an answer. **5** To supply what is lacking in. **6** To compensate for; atone for. **7** To settle; decide: to *make up* one's mind. **8** *Printing* To arrange, as lines, into columns or pages. **9** To put cosmetics on (the face). **10** In education, to repeat (an examination or course one has failed), or take (an examination) one has missed. —**to make up to** *Colloq.* To make a show of friendliness and affection toward; flatter. —*n.* **1** The manner in which parts or qualities are grouped to constitute a whole; constitution; structure; shape. **2** The operation or product of manufacture; brand: a new *make* of automobile. **3** The amount produced; yield. **4** The closing or completion of an electrical circuit. **5** A declaration (def. 3). —**on the make** *Slang* **1** Greedy for profit; interested only in making money. **2** Eager for amorous conquest. [OE *macian*]

Synonyms (verb): become, cause, compel, compose, constitute, constrain, construct, create, do, effect, establish, execute, fabricate, fashion, force, frame, get, manufacture, occasion, perform, reach, require, shape. *Make* is essentially causative; to the idea of cause all its various senses may be traced (compare synonyms for CAUSE *noun*). To *make* is to *cause* to exist, or to *cause* to exist in a certain form or in certain relations; it thus includes the idea of *create. Make* includes also the idea of *compose, constitute*; as, The parts *make* up the whole. Similarly, to *cause* a voluntary agent to do a certain act is to *make* him do it, or *compel* him to do it, *compel* fixing the attention more on the process, *make* on the accomplished fact. See COMPEL, GET, PRODUCE, REACH. *Antonyms*: see synonyms for ABOLISH, BREAK, DEMOLISH.

make[2] (māk) *n. Obs.* A mate. [OE *gemaca*]

make-and-break (māk′ənd·brāk′) *n. Electr.* A device for making and breaking an electrical circuit.

make-be·lieve (māk′bi·lēv′) *adj.* Pretended; unreal. —*n.* A mere pretense; sham. Also **make′-be·lief′** (-lēf′).

make-fast (māk′fast′, -fäst′) *n. Naut.* An iron ring or other object to which a boat is made fast.

make·less (māk′lis) *adj. Obs.* **1** Without mate. **2** Matchless. [<MAKE[2] + -LESS]

make·peace (māk′pēs′) *n.* A peacemaker.

mak·er (mā′kər) *n.* **1** One who makes. **2** A manufacturer. **3** *Law* One who signs a promissory note. **4** *Archaic* A poet.

Mak·er (mā′kər) *n.* God, as the creator of the universe.

-maker *combining form* One who or that which produces: *glassmaker.* [<MAKE]

make-read·y (māk′red′ē) *n. Printing* The operation of leveling up and adjusting a type form on a press so that all parts of it will give a clear, clean, and uniform impression when printed.

make·shift (māk′shift′) *adj.* Having the character of or being a temporary resource: also **make′shift′y.** —*n.* Something adopted as a temporary contrivance in any emergency.

make-up (māk′up′) *n.* **1** The arrangement or combination of the parts of which anything is composed; an aggregate of qualities. **2** *Printing* The arrangement of composed type in pages, columns, or forms, as in imposition. **3** The costumes, wigs, cosmetics, etc., used to assume a theatrical role; also, the art of applying or assuming them. **4** Lipstick, powder, rouge, etc., applied to a woman's face.

make·weight (māk′wāt′) *n.* **1** That which is thrown into a scale to increase weight. **2** Any person or thing made use of to fill up a deficiency.

Ma·ke·yev·ka (mä·kā′yif·kə) A city in the Donbas, eastern Ukrainian S.S.R.

Ma·khach·ka·la (mä·khäch′kä·lä′) A port on the Caspian Sea, capital of Dagestan Autonomous S.S.R.

Ma·kin (mā′kin, mug′in) See BUTARITARI.

mak·ing (mā′king) *n.* **1** The act of causing, fashioning, or constructing; workmanship. **2** That which contributes to improvement or success: He has the *making* of a fine character. **3** A quantity of anything made at one time; batch. **4** *Often pl.* The materials or ingredients required to make something. **5** Composition; structure; make.

-making *combining form* Act of causing or producing; creating or causing to be: *papermaking.* [<MAKE]

mal- *prefix* Bad; ill; evil; wrong; defective; imperfect; uneven: signifying also simple negation, and forming words directly from Latin and mediately through French: opposed to *bene-, eu-.* [<F *mal-* <L *male-* < *malus* bad]

Many words containing the prefix *mal-* are self-explaining, the prefix simply adding the meaning defective or evil:

maladaptation	**malconstruction**
maladjustment	**malexecution**
maladminister	**malinfluence**
malconformation	**malnutrition**

Mal·a·bar (mal′ə·bär) A district of SW India extending 450 miles NW from Cape Comorin to the Madras-Bombay border. Also **Malabar Coast.**

Ma·lac·ca (mə·lak′ə) A port on the Strait of Malacca and the capital of the **Settlement of**

Malacca, a civil division of the Federation of Malaya under British protection; 633 square miles. *Malay* **Ma·lak'a.**

Malacca, Strait of A channel between the Malay Peninsula and Sumatra connecting the Andaman Sea and the South China Sea.

Malacca cane A walking stick made from the wood of an Asian rattan palm *(Calamus rotang).* [from *Malacca*]

ma·la·ceous (mə·lā'shəs) *adj.* Designating a family *(Malaceae)* of trees, including the apple, pear, quince, etc. See POMACEOUS. [< NL, family name <L *malum* an apple <Gk. *mēlon*]

Mal·a·chi (mal'ə·kī) A masculine personal name. [<Hebrew *malākhī*, lit., my messenger] **—Malachi** A minor Hebrew prophet of the fifth century B.C.; also, the book containing his prophecies.

mal·a·chite (mal'ə·kīt) *n.* A green basic cupric carbonate, $CuCO_3 \cdot Cu(OH)_2$, found usually massive, rarely in crystals, and sometimes as an incrustation. It is one of the ores of copper. [<OF *melochite*, ult. <L *malache* a mallow <Gk. *malachē*; so called because resembling mallow leaves in color]

malachite green A pigment made of malachite, having an intense bluish-green color. Compare BICE GREEN.

malaco– *combining form* Soft; mucilaginous: *malacopterous.* Also, before vowels, **malac–.** [<Gk. *malakos* soft]

mal·a·coid (mal'ə·koid) *adj.* Having a soft texture. [<MALAC(O)- + -OID]

mal·a·col·o·gy (mal'ə·kol'ə·jē) *n.* The branch of zoology that treats of mollusks. [MALACO- + -LOGY]

mal·a·cop·ter·ous (mal'ə·kop'tər·əs) *adj.* Having soft fins, as certain fishes. [<MALACO- + -PTEROUS]

mal·a·cos·tra·can (mal'·ə·kos'trə·kən) *n.* Any of a division or subclass *(Malacostraca)* of crustaceans, embracing crabs, lobsters, crayfish, etc. *—adj.* Of or pertaining to the *Malacostraca*: also **mal'a·cos'tra·cous.** [<NL *Malacostraca*, subclass name <Gk. *malakostrakos* soft-shelled <*malakos* soft + *ostrakon* a shell]

mal·ad·dress (mal'ə·dres') *n.* Awkwardness or rudeness in speech or manner; lack of politeness or tact. [<F *maladresse* <*maladroit* MALADROIT]

mal·ad·just·ed (mal'ə·jus'tid) *adj.* 1 Imperfectly adjusted. 2 *Psychol.* Poorly adapted to one's environment through conflict between personal desires and external circumstances. **—mal'ad·just'ment** *n.*

mal·ad·min·is·ter (mal'əd·min'is·tər) *v.t.* To administer badly or dishonestly.

mal·ad·min·is·tra·tion (mal'əd·min'is·trā'shən) *n.* Bad management, as of public affairs.

mal·a·droit (mal'ə·droit') *adj.* Clumsy or blundering. See synonyms under AWKWARD. [<F <*mal-* MAL- + *adroit* clever] **—mal'a·droit'ly** *adv.* **—mal'a·droit'ness** *n.*

mal·a·dy (mal'ə·dē) *n. pl.* **·dies** 1 A disease, especially when chronic or deep-seated; sickness; illness. 2 Any disordered condition. See synonyms under DISEASE. [<OF *maladie* <LL *male habitus* <L *male* ill + *habitus*, pp. of *habere* have]

ma·la fi·de (mā'lə fī'dē) *Latin* In bad faith.

Mal·a·ga (mal'ə·gə) *n.* 1 A rich, sweet white wine made in Málaga, Spain. 2 A white, sweet grape of the muscat variety, grown in Spain and California.

Mál·a·ga (mal'ə·gə, *Sp.* mä'lä·gä) A port on the Mediterranean in Andalusia, southern Spain.

Mal·a·gas·y (mal'ə·gas'ē) *adj.* Of or pertaining to Madagascar, its inhabitants, or their language. *—n.* 1 A native of Madagascar. 2 The Indonesian language of Madagascar.

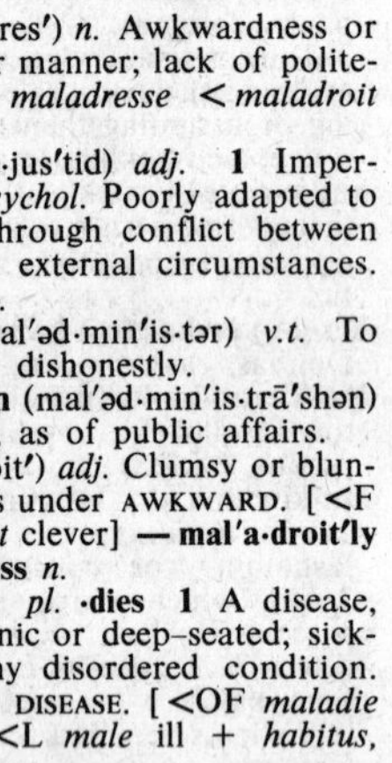

ESKIMO MALAMUTE (From 22 to 25 inches high at the shoulder)

Malagasy Republic An independent republic of the French Community comprising the island of Madagascar; 227,602 square miles; capital, Tananarive.

mal·aise (mal·āz', *Fr.* mà·lez') *n.* Uneasiness; indisposition. [<F <*mal* ill + *aise* EASE]

Ma·lai·ta (mə·lā'tə) One of the British Solomon Islands; 2,500 miles.

ma·la·mute (mä'lə·myōōt, mal'ə-) *n.* A large sled dog of Alaska, having a compact body, a thick, long coat, usually gray or black-and-white, a broad head, straight, big-boned forelegs, and well-cushioned feet. Also spelled *malemute, malemiut.* [Orig., name of an Innuit tribe, alter. of Eskimo (Malamute) *Mahlemut* <*Mahle,* the tribe's name + *mut* a village]

Ma·lan (mə·län'), **Daniel François,** born 1874, South African prime minister 1948–54.

mal·an·ders (mal'ən·dərz) *n. pl.* A scaly disease on the hock and at the bend of the knee of the foreleg of a horse: also spelled *mallenders.* Compare SALLENDERS. [<OF *malandre* a sore in a horse's knee <L *malandria*]

Ma·lang (mä·läng') A city in eastern Java.

mal·a·pert (mal'ə·pûrt) *adj.* Bold or forward; impudent; saucy. *—n.* A saucy person. [<OF <*mal-* MAL- + *apert*, *espert* clever, able <L *expertus.* See EXPERT.] **—mal'a·pert'ly** *adv.* **—mal'a·pert'ness** *n.*

Mal·a·prop (mal'ə·prop), **Mrs.** A character in Sheridan's *The Rivals*, who uses words inappropriately. [<MALAPROPOS]

mal·a·prop·ism (mal'ə·prop·iz'əm) *n.* The incorrect or inappropriate use of a word; a verbal blunder. [after Mrs. *Malaprop*] **—mal'a·pro'pi·an** (-prō'pē·ən) *adj.*

mal·ap·ro·pos (mal'ap·rə·pō') *adj.* Out of place; not appropriate. [<F *mal à propos* not to the point <*mal* ill + *à* to + *propos* purpose] **—mal'ap·ro·po'ism** *n.*

ma·lar (mā'lər) *adj. Anat.* Relating to or being in or near the cheek. *—n.* The cheek bone. [<NL *malaris* <L *mala* jaw, cheek]

Mäl·ar (mel'är) A lake in eastern Sweden; 440 square miles. *Swedish* **Mäl·ar·en** (mel'·är·ən).

ma·lar·i·a (mə·lâr'ē·ə) *n.* 1 *Pathol.* A disease caused by any of certain animal parasites (genus *Plasmodium*), which are introduced into the system by the bite of the infected anopheles mosquito and invade the red corpuscles of the blood, causing intermittent chills and fever. 2 Any foul or unwholesome air, as from decomposition; miasma; mephitis. [<Ital. *mal' aria, mala aria*, lit., bad air] **—ma·lar'i·al, ma·lar'i·an, ma·lar'i·ous** *adj.*

ma·lar·i·o·ther·a·py (mə·lâr'ē·ō·ther'ə·pē) *n. Med.* The treatment of cerebrospinal syphilis by inoculation with the parasite causing tertian malaria. [MALARI(A) + *-o-* + THERAPY]

ma·lar·ky (mə·lär'kē) *n. Slang* Insincere or senseless talk; bunk. Also **ma·lar'key.** [? <an Irish personal name]

mal·as·sim·i·la·tion (mal'ə·sim'ə·lā'shən) *n.* Imperfect or faulty assimilation.

mal·ate (mal'āt, mā'lāt) *n. Chem.* A salt or ester of malic acid. [<MAL(IC) + -ATE[3]]

mal·ax·ate (mal'ak·sāt) *v.t.* **·at·ed, ·at·ing** To knead to softness; soften. Also **mal'ax.** [<L *malaxatus*, pp. of *malaxare* make soft <Gk. *malassein* <*malakos* soft] **—mal'ax·a'tion** *n.*

Ma·lay (mā'lā, mə·lā') *n.* 1 A member of the dominant race in Malaysia; a Malayan. 2 The language spoken on the Malay Peninsula and widely used as a lingua franca throughout the East Indies, belonging to the Indonesian subfamily of Austronesian languages. 3 A variety of domestic fowl. *—adj.* Of or pertaining to the Malays; Malayan.

Ma·lay·a (mə·lā'ə), **Federation of** An independent federation of nine Malay states in the Commonwealth of Nations, including the former Federated Malay States and Unfederated Malay States, the two settlements of Malacca and Penang on the southern Malay Peninsula, and adjacent islands; 50,600 square miles; capital, Kuala Lumpur: formerly **Malayan Union.** Also **Malaya.**

Mal·a·ya·lam (mal'ə·yä'ləm) *n.* The language of the Malabar coast, India, related to Tamil, and belonging to the Dravidian family of languages.

Ma·lay·an (mə·lā'ən) *adj.* 1 Malay. 2 Indonesian. *—n.* 1 A Malay (def. 1). 2 An Indonesian. 3 The Indonesian subfamily of Austronesian languages.

Malay Archipelago An island group in the Indian and Pacific oceans SE of Asia, including Java, Borneo, Sumatra, Celebes, Timor, the Lesser Sunda Islands, and the Philippines; about 773,000 square miles: also the *East Indies, Malaysia.*

Ma·lay·o-Pol·y·ne·sian (mə·lā'ō·pol'ə·nē'zhən, -shən) *adj.* 1 Of or pertaining to the brown peoples of the Indian and Pacific oceans, including Malays and Polynesians. 2 Designating the languages of these peoples. *—n.* The Austronesian family of languages. [<MALAY + *-o-* + POLYNESIAN]

Malay Peninsula The southernmost peninsula of Asia, including the Federation of Malaya and part of Thailand.

Ma·lay·sia (mə·lā'zhə, -shə) See MALAY ARCHIPELAGO. **—Ma·lay'sian** *adj. & n.*

Mal·colm (mal'kəm) A masculine personal name. [<Celtic, servant of (St.) Columba]

mal·con·tent (mal'kən·tent) *adj.* Discontented, as with a government or economic system; dissatisfied; uneasy. *—n.* A person dissatisfied with the existing state of affairs; one rebellious against authority. [<OF <*mal-* MAL- + *content* CONTENT]

mal de dents (màl də dän') *French* Toothache.

mal de mer (màl də mâr') *French* Seasickness.

Mal·den Island (môl'dən) A British possession in the Line Islands; 35 square miles.

mal de tête (màl də tet') *French* Headache.

Mal·dive Islands (mal'dīv) A sultanate under British protection, comprising 12 islands SW of Ceylon; 115 square miles; capital, Malé Island.

mal du pays (màl dü pā·ē') *French* Homesickness.

male (māl) *adj.* 1 Pertaining to the sex that begets young; masculine. 2 Made up of men or boys. 3 *Bot.* Having stamens, but no pistil; also, adapted to fertilize, but not to produce fruit, as stamens. 4 Denoting some implement or object, as a gage or plug, which fits into a corresponding part known as *female.* 5 Indicating superiority of strength and quality of anything: distinguished from *female.* See synonyms under MASCULINE. *—n.* 1 An organism that produces sperm cells; a male person or animal. 2 *Bot.* A plant with only staminate flowers. ◆ Homophone: *mail.* [<OF *male, mascle* <L *masculus.* Doublet of MASCULINE.]

Ma·le·a (mə·lē'ə), **Cape** 1 The SE extremity of Lesbos Island. 2 The SE extremity of Peloponnesus, Greece, on the Aegean.

Male·branche (màl·bräṅsh'), **Nicolas de,** 1638–1715, French philosopher.

mal·e·dict (mal'ə·dikt) *adj. Obs.* Accursed. [<L *maledictus.* See MALEDICTION.]

mal·e·dic·tion (mal'ə·dik'shən) *n.* 1 An invocation of evil; a cursing: opposed to *benediction.* 2 Slander. 3 The state of being reviled. See synonyms under IMPRECATION, OATH. [<L *maledictio, -onis* <*maledictus*, pp. of *maledicere* <*male* ill + *dicere* speak] **—mal'e·dic'to·ry** *adj.*

mal·e·fac·tor (mal'ə·fak'tər) *n.* One who commits a crime. [<L <*malefactus*, pp. of *malefacere* <*male* ill + *facere* do] **—mal'e·fac'·tion** *n.* **—mal'e·fac'tress** *n. fem.*

male fern A fern *(Dryopteris filixmas)* used in medicine as a vermifuge.

ma·lef·ic (mə·lef'ik) *adj.* Occasioning evil or disaster. [<L *maleficus* <*malefacere.* See MALEFACTOR.]

ma·lef·i·cent (mə·lef'ə·sənt) *adj.* Causing or doing evil or mischief; harmful: opposed to *beneficent.* [<L *maleficus* MALEFIC]

male hormone Androgen.

ma·le·ic (mə·lē'ik) *adj. Chem.* Pertaining to or designating a white, crystalline, astringent acid, $C_4H_4O_4$, prepared by the catalytic oxidation of benzene: used as a dye for fabrics, etc. [<F *maléique* <*malique* MALIC]

Ma·lé Island (ma'lā) The capital of the Maldive Islands.

Ma·le·ku·la (mal'ə·kōō'lə) The second largest island of the New Hebrides group; 980 square miles.

ma·le·mute (mä'lə·myōōt, mal'ə-), **ma·le·miut** See MALAMUTE.

Mal·en·kov (mal'ən·kôv, mə·len'-; *Russian* mä'lyin·kôf), **Georgi,** born 1902, U.S.S.R. leader; premier 1953–55.

mal·en·ten·du (màl·äṅ·täṅ·dü') *n. French* Misunderstanding.

Mal·e·ven·tum (mal'ə·ven'təm) The ancient name for BENEVENTO.

ma·lev·o·lent (mə·lev'ə·lənt) *adj.* Having an evil disposition toward others; ill-disposed. See synonyms under MALICIOUS. [<OF <L *malevolens, -entis* <*male* ill + *volens, -entis*, ppr of *velle* wish, will] **—ma·lev'o·lence** *n.* **—ma·lev'o·lent·ly** *adv.*

mal·fea·sance (mal·fē′zəns) *n. Law* The performance of some act which is unlawful or wrongful or which one has specifically contracted not to perform: said usually of official misconduct. Compare MISFEASANCE, NONFEASANCE. [<AF *malfaisance* <OF *malfaisant* <*mal* ill + *faisant*, ppr. of *faire* do <L *facere*]

mal·fea·sant (mal·fē′zənt) *adj.* Guilty of malfeasance. — *n.* A person guilty of malfeasance.

mal·for·ma·tion (mal′fôr·mā′shən) *n.* Any irregularity, anomaly, or abnormal deformation in the structure of an organism. [<MAL- + FORMATION]

mal·formed (mal·fôrmd′) *adj.* Badly formed or made; deformed.

mal·func·tion (mal·fungk′shən) *n. Physiol.* Impairment or disturbance of any bodily function; dysfunction.

Mal·gache (mȧl·gȧsh′) *n. & adj.* French Madagascan.

mal·gré (mȧl·grā′) *prep.* French In spite of; notwithstanding. Also, *Obs., maugre.*

malgré lui (lwē) *French* In spite of himself or herself.

Mal·herbe (mȧ·lerb′), **François de**, 1555–1628, French poet.

Ma·li (mä′lē) A landlocked, independent republic in west Africa; 464, 873 miles; capital, Bamako; formerly French Sudan, an overseas territory. — **Ma′li** *adj. & n.*

mal·ic (mal′ik, mā′lik) *adj.* 1 Of, pertaining to, or obtained from apples. 2 *Chem.* Pertaining to or designating a deliquescent crystalline acid, $C_4H_6O_5$, with a pleasant taste: contained in the juice of many sour fruits and some plants, and also made synthetically. [<F *malique* <L *malum* an apple]

mal·ice (mal′is) *n.* 1 A disposition to injure another; evil intent; spite; ill will. 2 *Law* A wilfully formed design to do another an injury: also **malice aforethought**. See synonyms under ENMITY, HATRED. [<OF <L *malitia* <*malus* bad]

ma·li·cious (mə·lish′əs) *adj.* 1 Harboring malice, ill will, or enmity; spiteful. 2 Resulting from or prompted by malice. [<OF *malicios* <L *malitiosus* <*malitia* MALICE] — **ma·li′cious·ly** *adv.* — **ma·li′cious·ness** *n.*
— *Synonyms:* bitter, evil-disposed, evil-minded, hostile, ill-disposed, ill-natured, invidious, malevolent, malign, malignant, mischievous, rancorous, resentful, spiteful, venomous, virulent. The *malevolent* person wishes ill to another; the *malicious* person has the desire and intent to do evil, if possible, to another. The *malign* or *malignant* spirit has a deep, intense, and insatiable hostility, such as is indicated by *rancorous* or *venomous*, with or without active desire or intent to injure. *Spiteful* is a feeble word indicating the desire or intent to inflict petty, exasperating annoyance or injury. Compare ACRIMONY, BITTER, ENMITY, HATRED. *Antonyms:* amiable, amicable, beneficent, benevolent, benign, benignant, friendly, good-natured, kind, kind-hearted, kindly, sympathetic, tender, well-disposed.

ma·lign (mə·līn′) *v.t.* To speak slander of. See synonyms under ABUSE, ASPERSE, REVILE. — *adj.* 1 Having an evil disposition toward others; ill-disposed; malevolent: opposed to *benign.* 2 Tending to injure; pernicious. See synonyms under MALICIOUS. [<OF *malignier, maliner* plot, deceive <L *malignare* contrive maliciously <*malignus* evil-disposed <*malus* evil] — **ma·lign′ly** *adv.* — **ma·lign′er** *n.*

ma·lig·nant (mə·lig′nənt) *adj.* 1 Having or manifesting extreme malevolence or enmity. 2 Evil in nature, or tending to do great harm; also, malcontent. 3 *Pathol.* So aggravated as to threaten life: opposed to *benign*: a *malignant* tumor. 4 Boding ill; baleful; threatening. — *n.* A person of extreme enmity or evil intentions. [<L *malignans, -antis*, ppr. of *malignare.* See MALIGN.] — **ma·lig′nance, ma·lig′nan·cy** *n.* **ma·lig′nant·ly** *adv.*

ma·lig·ni·ty (mə·lig′nə·tē) *n. pl.* **·ties** 1 The state or quality of being malign; violent animosity. 2 Destructive tendency; virulence. 3 *Often pl.* An evil thing or event. See synonyms under ACRIMONY, ENMITY, HATRED.

ma·lines (mə·lēn′, *Fr.* mȧ·lēn′) *n.* 1 Lace made in Mechlin, Belgium: also called *Mechlin lace.* 2 A gauzelike veiling for trimming hats: also **ma·line′**. [<F, from *Malines* Mechlin]

Ma·lines (mȧ·lēn′) The French name for MECHLIN.

mal·in·ger (mə·ling′gər) *v.i.* To feign sickness or incapacity, especially so as to avoid work or duty. [<F *malingre* sickly <*mal* bad (<L *malus*) + OF *heingre* lean] — **ma·lin′ger·er** *n.*

Ma·li·now·ski (mä′li·nôf′skē), **Bronislaw Kasper**, 1884–1942, U.S. anthropologist born in Poland.

mal·i·son (mal′ə·zən, -sən) *n. Archaic* A malediction; curse. [<OF <L *maledictio* MALEDICTION]

mal·kin (mô′kin, môl′-, mal′-) *n. Obs.* 1 A kitchenmaid; a slattern. 2 A scarecrow representing a woman. 3 A cat. 4 *Scot.* A hare. Also spelled *maukin.* [Dim. of MATILDA, MAUD]

mall¹ (môl, mal) *n.* 1 A maul. 2 A war hammer. — *v.t.* To maul. [Var. of MAUL]

mall² (môl, mal, mel) *n.* 1 The game pall-mall, or a place in which it is played. 2 A level, shaded walk. [Short for PALL-MALL]

mal·lard (mal′ərd) *n.* 1 The common wild duck (*Anas platyrhynchos*), or, formerly, its drake. 2 Any wild duck. 3 *Obs.* The domesticated duck. [<OF *malart* <*masle* MALE]

Mal·lar·mé (mȧ·lȧr·mā′), **Stéphane**, 1842–98, French poet.

mal·le·a·ble (mal′ē·ə·bəl) *adj.* 1 Capable of being hammered or rolled out without breaking; ductile. 2 Hence, susceptible to the shaping power of surrounding influences; pliant. [<OF <L *malleare* MALLEATE] — **mal′le·a·bil′i·ty, mal′le·a·ble·ness** *n.* — **mal′le·a·bly** *adv.*

malleable iron 1 Cast iron that has been rendered tough and malleable by long-continued high heating and slow cooling. 2 Wrought iron; forged iron.

mal·le·ate (mal′ē·āt) *v.t.* **·at·ed, ·at·ing** To shape into a plate or leaf by beating; hammer. [<L *malleatus*, pp. of *malleare* <*malleus* a hammer] — **mal′le·a′tion** *n.*

mal·lee (mal′ē) *n. Austral.* 1 Any one of several scrubby species of eucalyptus of South Australia and Victoria; especially, *Eucalyptus dumosa* and *E. oleosa.* 2 Brushwood composed of such trees.

mal·le·in (mal′ē·in) *n. Biochem.* A poisonous, yellowish-white compound, obtained from the active metabolic products of the bacillus of glanders, used for the diagnosis of that disease. Also **mal′le·ine** (-in, -ēn). [<L *malle(us)* glanders + -IN]

mal·le·muck (mal′ə·muk) *n.* The southern albatross, fulmar, petrel, or other closely related bird. [<Du. *mallemok* <*mol* foolish + *mok* a gull]

mal·len·ders (mal′ən·dərz) See MALANDERS.

mal·le·o·lus (mə·lē′ə·ləs) *n. pl.* **·o·li** (-ə·lī) *Anat.* A hammer-shaped bony process on each side of the ankle. [<L, dim. of *malleus* a hammer] — **mal′le·o·lar** *adj.*

mal·let (mal′it) *n.* 1 A wooden hammer or light maul. 2 A light hammer, frequently of metal. 3 A long-handled wooden hammer used in the game of croquet. 4 A wooden-headed Malacca cane or stick used in the game of polo. [<OF *maillet*, dim. of *mail* a MAUL]

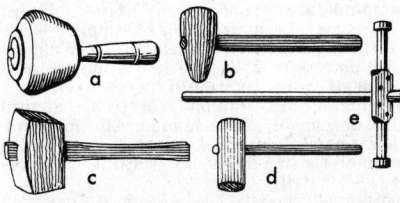

MALLETS
a. Mason's mallet. *b.* Bossing mallet.
c. Carpenter's mallet. *d.* Tinsmith's mallet.
e. Calking mallet.

mal·le·us (mal′ē·əs) *n. pl.* **·le·i** (-lē·ī) *Anat.* The club-shaped outermost ossicle of the middle ear, articulating with the incus. See illustration under EAR. [<L, a hammer]

Mal·lor·ca (mä·lyôr′kä) The Spanish spelling of MAJORCA.

mal·low (mal′ō) *n.* 1 Any plant of the genus *Malva.* The most common in the United States is the **running** or **dwarf mallow** (*M. rotundifolia*), a spreading herb with roundish leaves, small, pale-pink flowers, and flat, disklike fruit: the leaves are used in brewing a medicinal tea. 2 Any plant of the mallow family: Indian *mallow.* [OE *mealuwe* <L *malva.* Doublet of MAUVE.]

mallow rose See ROSEMALLOW.

malm (mäm) *n.* 1 A soft, friable, whitish limestone. 2 A whitish calcareous loam occurring in the southern counties of England; marl. [OE *mealm(stan)* sandstone or limestone]

Mal·mé·dy (mȧl·mā·dē′) A town on the Belgian-German frontier; awarded to Belgium under the Versailles Treaty.

Malmes·bur·y (mämz′bər·ē), **William of**, 1095?–1143?, English monk and historian.

Mal·mö (mal′mō, *Sw.* mäl′mœ) A port in SW Sweden.

malm·sey (mäm′zē) *n.* A rich sweet wine made in the Canary Islands, Madeira, Spain, and Greece. [<Med. L *malmasia* <Gk. *Monembasia* Monemvasia, Greece, a small Laconian coastal town formerly exporting wine]

mal·nu·tri·tion (mal′nōō·trish′ən, -nyōō-) *n.* Faulty or inadequate nutrition.

ma·lo (mä′lō) *n.* A loincloth or girdle worn by Hawaiian men, formerly made of tapa, now of brightly dyed cotton fabrics. [<Hawaiian, cloth]

mal·oc·clu·sion (mal′ə·klōō′zhən) *n. Dent.* Faulty closure of the upper and lower teeth. [<MAL- + OCCLUSION]

mal·o·dor (mal·ō′dər) *n.* An offensive odor.

mal·o·dor·ous (mal·ō′dər·əs) *adj.* Having a disagreeable smell, literally or figuratively; obnoxious. — **mal·o′dor·ous·ly** *adv.* — **mal·o′dor·ous·ness** *n.*

ma·lon·ic (mə·lon′ik, -lō′nik) *adj. Chem.* Of, pertaining to, or designating a white, crystalline acid, $CH_2(CO_2H)_2$, obtained chiefly by oxidizing malic acid. [<F *malonique* <*malique* MALIC]

mal·o·nyl·u·re·a (mal′ə·nil·yŏŏr′ē·ə) *n.* Barbituric acid. [<MALON(IC) + -YL + UREA]

Mal·o·ry (mal′ər·ē), **Sir Thomas**, died 1470?, English author and translator; compiled the *Morte d'Arthur*, a collection of stories about King Arthur translated from the French.

mal·pais (mal′pīs) *n. SW U.S.* Bad land; specifically, land having an under layer of basaltic lava. [<Sp. *mal, malo* bad (<L *malus*) + *país* a country <L *patria*]

Mal·pi·ghi (mäl·pē′gē), **Marcello**, 1628–94, Italian anatomist. — **Mal·pigh′i·an** (-pig′ē·ən) *adj.*

mal·pigh·i·a·ceous (mal·pig′ē·ā′shəs) *adj.* Of or pertaining to a family (*Malpighiaceae*) of trees, shrubs, or, more rarely, herbs, with hermaphrodite flowers, mostly native to the tropics. [<NL *malpighiaceae*, family name <*Malpighia*, genus name, after Marcello *Malpighi*]

Malpighian bodies *Anat.* A tuft of blood vessels at the commencement of the uriniferous tubules in the kidney. Also **Malpighian capsules** or **corpuscles.** [after Marcello *Malpighi*]

Malpighian layer *Anat.* The deeper, softer layer of the epidermis, comprising the active cells.

Malpighian tubes *Entomol.* The tubular portions of the excretory organ of an insect. Also **Malpighian vessels.**

Mal·pla·quet (mȧl·plȧ·ke′) A village in northern France; scene of Marlborough's victory over the French, 1709.

mal·po·si·tion (mal′pə·zish′ən) *n. Pathol.* A wrong or faulty position, as of the fetus. — **mal·posed** (-pōzd′) *adj.*

mal·prac·tice (mal·prak′tis) *n.* 1 Improper or illegal practice, as in medicine or surgery. 2 Improper or immoral conduct. — **mal·prac·ti·tion·er** (mal′prak·tish′ən·ər) *n.*

Mal·raux (mȧl·rō′), **André**, born 1901, French novelist and art critic.

malt (môlt) *n.* 1 Grain, usually barley, softened by water, artificially germinated and then dried: rich in carbohydrates and proteins and essential in brewing and as a nutrient. 2 *Slang* Malt liquor; beer or ale. — *v.t.* 1 To cause (grain) to germinate artificially, by moisture and heat, and become malt. 2 To treat with

Malta | 772 | **manacle**

malt, or extract of malt. —*v.i.* **3** To be changed into or become malt: said of grain. **4** To convert grain into malt. [OE *mealt.* Akin to MELT.]
Mal·ta (môl′tə) **1** A British colony comprising the Maltese Islands; Malta, Gozo, and Comino; 122 square miles; capital, Valletta. **2** The largest of these islands; 95 square miles: ancient *Melita.*
Malta fever Undulant fever.
mal·tase (môl′tās) *n. Biochem.* A digestive enzyme which hydrolyzes maltose into dextrose. [<MALT + -ASE]
malted milk 1 A powder made of dehydrated milk and malted cereals, soluble in milk or water. **2** The beverage made with this powder.
Mal·tese (môl·tēz′, -tēs′) *adj.* Of or pertaining to Malta, its inhabitants, or their language, or to the Knights of Malta. — *n. pl.* **·tese 1** A native of Malta; the people of Malta collectively. **2** The language of Malta, a dialectal Arabic with elements of Italian. **3** A Maltese cat or dog.
Maltese cat A domestic cat with long, silky, bluish-gray hair.
Maltese cross See under CROSS.
Maltese dog An ancient breed of toy spaniel, originating in Malta, with a long, silky, white coat.
mal·tha (mal′thə) *n.* **1** A thick mineral pitch, formed by drying petroleum. **2** A variety of ozocerite. **3** Any similar preparation used as a cement, stucco, or mortar. [<L <Gk., a mixture of wax and pitch]
malt house A building in which grains are malted and other preparations made for use in the brewer's trade.
Mal·thus (mal′thəs), **Thomas Robert**, 1766–1834, English political economist.
Mal·thu·si·an (mal·thōō′zē·ən, -zhən) *adj.* Of or pertaining to the theory of T. R. Malthus that population increases faster than the means of support and, unless checked by sexual restraint, is restricted only by famine, pestilence, war, etc. — *n.* A believer in the theories of Malthus. — **Mal·thu′si·an·ism** *n.*
malt·ose (môl′tōs) *n. Biochem.* A hard, white, dextrorotatory, crystalline sugar, $C_{12}H_{22}O_{11} \cdot H_2O$, formed by the action of amylase on starch. [<MALT + -OSE²]
mal·treat (mal·trēt′) *v.t.* To treat badly or unkindly; abuse. See synonyms under ABUSE. [<F *maltraiter* <*mal-* MAL- + *traiter* <OF *traitier* TREAT] — **mal·treat′ment** *n.*
malt·ster (môlt′stər) *n.* A maker of or dealer in malt.
malt·y (môl′tē) *adj.* **malt·i·er**, **malt·i·est** Of, pertaining to, containing, or resembling malt.
mal·va·ceous (mal·vā′shəs) *adj.* Pertaining or belonging to the mallow family (*Malvaceae*) of herbs, shrubs, or trees, with alternate palmately nerved leaves and regular flowers, including althea, cotton, okra, etc. [<LL *malvaceus* <*malva* a mallow]
mal·va·si·a (mal′və·sē′ə) *n.* Malmsey wine; also, the kind of grape from which malmsey wine is made. [<Ital. *Malvasia*, alter. of Gk. *Monembasia.* See MALMSEY.] — **mal′va·si′an** *adj.*
Mal·vern Hill (mal′vərn) A plateau near Richmond, Virginia; scene of a Confederate defeat in the Civil War, 1862.
Malvern Hills A range in western England on the Worcester–Hereford border; highest point, 1,307 feet.
mal·ver·sa·tion (mal′vər·sā′shən) *n.* Evil or corrupt conduct; misconduct, as in public office. [<MF <*malverser* <L *male versari* <*male* wrongly, ill + *versari* behave, passive freq. of *vertere* turn]
mal·voi·sie (mal′voi·zē, -vō-) *n.* Malmsey. [<OF *malvesia* <Ital. *malvasia* MALVASIA; refashioned after F *malvoisie*]
Mal·vo·li·o (mal·vō′lē·ō) In Shakespeare's *Twelfth Night*, the pompous majordomo of Olivia.
ma·ma (mä′mə, mə·mä′) *n.* Mother: a term of familiar address and endearment. Also **mam′ma**. [Repetition of infantile syllable *ma*]
mam·ba (mäm′bə) *n.* Any of certain long, venomous, arboreal snakes (genus *Dendraspis*) of southern Africa; especially, the common olive-green or black mamba (*D. angusticeps*). [<Zulu *in-amba*]
Mam·be·ra·mo (mäm′bə·rä′mō) A river in western New Guinea, flowing 500 miles NW to the Pacific.

mam·bo (mäm′bō) *n. pl.* **·bos** A form of popular music, derived from Cuban Negro styles, achieving its effects by syncopation of a four-beat rhythmic pattern, with accents on the second and fourth beats.
ma·melle (ma·mel′) *n.* In western United States, a rounded hillock. [F, breast]
mam·e·luke (mam′ə·lōōk) *n.* In Moslem countries, a male slave. [<F *mameluk* <Arabic *mamlūk* a slave, orig. pp. of *malaka* possess]
Mam·e·luke (mam′ə·lōōk) *n.* A member of a famous military caste, originally composed of slaves, which dominated Egypt from 1254 to 1811. Also **Mam′e·luke**.
ma·mey (mä·mā′, -mē′) *n.* A tropical American tree (*Mamea americana*) bearing edible, yellow fruits resembling the pomelo in size and shape: also **mamey de Santo Domingo**. **2** A fruit of this tree: also **mamey apple**. **3** The sapodilla or marmalade tree. Also **mam·mee** (mä·mā′, -mē′). [<Sp. <Taino]
mam·ma (mam′ə) *n. pl.* **·mae** (-ē) The milk-secreting organ of a mammal; a breast, udder, or bag. [<L, breast]
mam·mal (mam′əl) *n.* A vertebrate animal whose female suckles its young. [<MAMMALIA]
Mam·ma·li·a (ma·mā′lē·ə) *n. pl.* A class of vertebrates whose females have milk-secreting mammae to nourish their young, including man, all warm-blooded quadrupeds, seals, cetaceans, and whales. [<NL <LL *mammalis* of the breast <L *mamma* breast]
mam·ma·lif·er·ous (mam′ə·lif′ər·əs) *adj.* Containing remains of mammals, as geological strata. [<MAMMALI(A) + -FEROUS]
mam·mal·o·gy (ma·mal′ə·jē) *n.* The branch of zoology that treats of the *Mammalia*. [<MAMMAL + -LOGY]
mam·ma·ry (mam′ər·ē) *adj.* Of, pertaining to, or of the nature of a mamma or breast, or the mammae.
mammary gland The milk gland, which in a female forms the bulk of the breast or mamma. The mammary glands occur in both the male and female in all mammals, but are rudimentary in the male.
mam·mate (mam′āt) *adj.* Having mammae or breasts. Also **mam′me·at·ed** (-ē·ā′tid). [<L *mammatus* <*mamma* breast]
mam·ma·tus (ma·mā′təs) *n. Meteorol.* A cloud form characterized by pouchlike protuberances along the lower surface: noted especially in stratocumulus and cumulonimbus clouds. [<L. See MAMMATE.]
mam·mer (mam′ər) *v.i. Brit. Dial.* To stammer; hesitate; be confused. [Prob. imit.]
mam·mer·ing (mam′ər·ing) *n.* A state of doubt or perplexity.
mam·met (mam′it) See MAUMET.
mam·mie (mam′ē) See MAMMY.
mam·mif·er·ous (ma·mif′ər·əs) *adj.* Having mammae or breasts; mammalian. [<MAMM(A) + -(I)FEROUS]
mam·mil·la (ma·mil′ə) *n. pl.* **·lae** (-ē) *Anat.* A nipple or teat, or some nipplelike or teat-shaped structure or protuberance. [<L *mamilla, mammilla*, dim. of *mamma* breast]
mam·mil·lar·y (mam′ə·ler′ē) *adj.* **1** Of, pertaining to, or resembling a mammilla or a mamma. **2** Studded with or composed of breast-shaped or rounded protuberances: a *mammillary* mineral, a *mammillary* prairie.
mam·mil·late (mam′ə·lāt) *v.t.* **·lat·ed**, **·lat·ing** To shape like a breast or a nipple. — *adj.* **1** Having a mammilla, mammillae, or nipple-like processes. **2** Shaped like a nipple. Also **mam′mil·lat′ed**. [<MAMMILL(A) + -ATE²]
mam·mil·li·form (ma·mil′ə·fôrm) *adj.* Shaped like or resembling a mammilla. Also **mam·mil·loid** (mam′ə·loid). [MAMMILL(A) + -(I)FORM]
mam·mi·tis (ma·mī′tis) *n.* Mastitis. [<MAMM(A) + -ITIS]
mam·mock (mam′ək) *n. Archaic* A fragment; scrap. [< dial. E; ult. origin uncertain]
mam·mol·o·gy (ma·mol′ə·jē) *n.* Mammalogy. [<MAMM(ALIA) + -(O)LOGY]
mam·mon (mam′ən) *n.* **1** Riches; wealth. **2** Worldliness; avarice. *Matt.* vi 24; *Luke* xvi 9. [<LL <Gk. *mammōnas* <Aramaic *māmōnā* riches, prob. ult. <′*āman* trust]
mam′mon·ish *adj.*
Mam·mon (mam′ən) **1** The personification of riches, avarice, and worldly gain. **2** In Milton's *Paradise Lost*, one of the fallen angels.
mam·mon·ism (mam′ə·niz′əm) *n.* Devotion to the acquisition of wealth; worldliness. — **mam′mon·ist, mam′mon·ite** *n.*

mam·moth (mam′əth) *n.* **1** A large, once very abundant, extinct northern elephant (*Mammuthus primigenius*) of the Pleistocene, closely resembling the Indian elephant, with coarse outer hair, close woolly under hair, and enormous, usually curved tusks: its remains have been found in Alaska and Siberia. **2** The extinct imperial elephant (*M. imperator*) which ranged west of the Mississippi River. — *adj.* Huge; colossal. [<Russian *mammot, mamant*]

WOOLLY MAMMOTH
(About 9 feet high at the shoulder)

Mammoth Cave National Park A series of underground caverns in SW Kentucky, noted for their remarkable onyx formations; 79 square miles; established, 1936.
mam·my (mam′ē) *n. pl.* **·mies 1** Mother; mama. **2** A Negro nurse or foster mother of white children. Also spelled *mammie*. [Dim. of MAMA]
ma·mon·cil·lo (mä′mōn·sē′yō) *n.* A tropical American tree (*Melicocca bijuga*) with small, greenish-white, fragrant flowers and an edible fruit. [<Sp., dim. of *mamon*, orig., a suckling]
Ma·mo·ré (mä′mō·rā′) A river in Bolivia, flowing 1,200 miles north to the Madeira.
man (man) *n. pl.* **men** (men) **1** A member of the genus *Homo* (family *Hominidae*, class *Mammalia*, the most highly developed of the primates, differing from other animals in having erect posture, extraordinary development of the brain, and the power of articulate language: only existing species, *Homo sapiens*. Earlier forms include the following: CRO-MAGNON MAN, GRIMALDI MAN, HEIDELBERG MAN, MIDLAND MAN, MODJOKERTO MAN, NEANDERTHAL MAN, RHODESIAN MAN, SOLO MAN, WADJAK MAN. **2** The human race. **3** Any one, indefinitely. **4** An adult male of the human kind: distinguished from *woman*, *boy*, and *youth*. **5** The male part of the race collectively. **6** A male person who is manly; also, manhood. **7** An adult male servant, dependent, or vassal. **8** A piece, figure, disk, etc., used in playing certain games, as chess or checkers. **9** Sir; fellow: a familiar address, often expressing impatience or depreciation. **10** A ship or vessel: used in composition: a *man-of-war*; an *Indiaman*. **11** A husband; lover: *man* and wife; Her *man* is dead. — **to a man** Every one. — **to be one's own man** To be independent. — *v.t.* **manned**, **man·ning 1** To supply with men: to *man* a fort. **2** To take stations at, on, or in for work, defense, etc.: *Man* the pumps! **3** To accustom (a hawk) to the presence or handling of men. — **to man oneself** To prepare or brace oneself, as for an ordeal. [OE *mann*]
Man may appear as a combining form in hyphemes or in solidemes, meaning "of, by, for, or like a man":

man-abhorring	man-hating
man-baiting	man-high
man-bodied	man-hunt
man-born	man-hunting
man-catching	man-idolatry
man-changed	mankiller
man-created	mankilling
man-degrading	man-made
man-destroyer	man-ridden
man-destroying	man-shaped
man-devised	man-size
man-eating	man-stealing
man-enslaved	man-taught
man-fearing	mantrap
man-grown	man-worthy
man-hater	

Man (man), **Isle of** One of the British Isles in the Irish Sea between Northern Ireland and England; 220 square miles; capital Douglas: ancient *Mona*: a native of the island is known as a Manxman.
man-a·bout-town (man′ə·bout′toun′) *n.* A man who frequents night clubs, theaters, restaurants, bars, etc.; hence, a sophisticate.
man·a·cle (man′ə·kəl) *n.* **1** *Usually pl.* One of a connected pair of metallic instruments for confining or restraining the hands; a hand-

manage cuff. 2 Anything that constrains or fetters. — *v.t.* **·cled, ·cling** 1 To put manacles on. 2 To hamper; constrain. [< OF *manicle* < L *manicula*, dim. of *manus* hand]

man·age (man'ij) *v.* **·aged, ·ag·ing** *v.t.* 1 To direct or conduct the affairs or interests of: to *manage* a business. 2 To control the direction, operation, etc., of, as a machine. 3 To cause to do one's bidding, as by persuasion or flattery. 4 To bring about or contrive: to *manage* to do something. 5 To handle or wield, as a weapon or implement. 6 To train (a horse) in the exercises of a manège. — *v.i.* 7 To carry on or conduct business or affairs. 8 To contrive to get along: I'll *manage*. See synonyms under GOVERN, REGULATE. — *n.* *Obs.* 1 Management. 2 Behavior. 3 Manége; a riding school. [< Ital. *maneggiare* handle, train horses, ult. < L *manus* a hand]

man·age·a·ble (man'ij·ə·bəl) *adj.* Capable of being managed; tractable; docile. See synonyms under DOCILE. — **man'age·a·bil'i·ty, man'age·a·ble·ness** *n.* — **man'age·a·bly** *adv.*

man·age·ment (man'ij·mənt) *n.* 1 The act, art, or manner of managing, controlling, or conducting. 2 The skilful use of means to accomplish a purpose. 3 Managers or directors collectively. See synonyms under CARE, OVERSIGHT.

man·ag·er (man'ij·ər) *n.* 1 One who manages; especially, one who has the control of a business or a business establishment; a director. 2 An adroit schemer; intriguer. See synonyms under MASTER, SUPERINTENDENT. — **man'ag·er·ship** *n.*

man·ag·er·ess (man'ij·ər·is, -ij·ris; *Brit.* man'ij·ər·es') *n. Chiefly Brit.* A female manager.

man·a·ge·ri·al (man'ə·jir'ē·əl) *adj.* Of, pertaining to, or characteristic of a manager or management. — **man'a·ge'ri·al·ly** *adv.*

Ma·na·gua (mä·nä'gwä) The capital of Nicaragua, on Lake Managua, a lake in SW Nicaragua (390 square miles).

man·a·kin (man'ə·kin) *n.* 1 Any of numerous small tropical American birds (family *Pipridae*) of brilliant plumage. 2 A manikin. [Orig. var. of MANIKIN]

Ma·na·ma (mä·nä'mä) The capital of the Bahrein Islands, on Bahrein.

ma·ña·na (mä·nyä'nä) *n. & adv. Spanish* Tomorrow. — **land of mañana** Land of perpetual procrastination.

Man·a·sa·ro·war (man'ə·sə·rō'ər) A lake in SW Tibet; 200 square miles.

Ma·nas·sas (mə·nas'əs) A town in NE Virginia; scene of two battles of Bull Run in the Civil War, July 21, 1861, and Aug. 29–30, 1862.

Ma·nas·seh (mə·nas'ə) A masculine personal name. [< Hebrew, causing to forget]

— **Manasseh** A son of Joseph (*Gen.* xli 51); also, a tribe of Israel descended from him.

— **Manasseh** King of Judah, in the seventh century B.C.; restored idol-worship.

man-at-arms (man'ət·ärmz') *n.* *pl.* **men-at-arms** (men'-) A soldier; especially, a heavily armed soldier of medieval times.

man·a·tee (man'ə·tē') *n.* A sirenian (genus *Trichechus*) of the tropical Atlantic shores and rivers; a sea cow. [< Sp. *manati* < Cariban *manattoui*]

MANATEE
(From 8 to 12 feet in length over-all)

Ma·naus (mä·nous') A city on the Rio Negro in NW Brazil, capital of Amazonas. Formerly **Ma·ná·os** (mä·nä'ōs).

ma·nav·e·lins (mə·nav'ə·linz) *n. pl. Naut. Slang* Odds and ends; leftover scraps. Also **ma·nav'i·lins**. [Var. of nautical slang *manarvelings* < *manarvel* steal small stores]

Man·ches·ter (man'ches·tər, -chis-) 1 A county borough and city in Lancashire, England. A native of Manchester is known as a *Mancunian*. 2 A city in southern New Hampshire.

man·chet (man'chit) *n. Archaic* A small loaf of fine white bread. [ME *manchett*, ? dim. <OF (*painde*)*maine* < L *panis dominicus*, lit., lord's bread]

man-child (man'chīld') *n. pl.* **·chil·dren** (-chil'drən) A male child.

man·chi·neel (man'chi·nēl') *n.* A tropical American tree (*Hippomane mancinella*) having an acrid, milky, poisonous sap. [< F *mancenille* <Sp. *manzanilla*, dim. of *manzana* an apple < L (*mala*) *matiana* (apples) of Matius < *Matius*, a Roman culinary author]

Man·chu (man·chōō', man'chōō) *n.* 1 One of a Mongoloid people that conquered China in 1643 and established the dynasty overthrown in 1912. 2 The language of this people, belonging to the Manchu-Tungusic subfamily of Altaic languages. — *adj.* Of or pertaining to Manchuria, its people, or its language. Also **Man·choo'**. [<Manchu, lit., pure] — **Man·chu'ri·an** (-chōōr'ē·ən) *adj. & n.*

Man·chu·kuo (man'chōō·kwō', *Chinese* män'jō'kwō') A former empire in NE Asia, 1932–45, established under Japanese auspices; comprising Manchuria and the Chinese province of Jehol and part of Chahar in Inner Mongolia; about 503,013 square miles; capital, Hsinking. Also **Man'chou'kuo'**.

Man·chu·ri·a (man·chōōr'ē·ə) A major division of NE China; 585,000 square miles; capital, Mukden. See MANCHUKUO. — **Man·chu'ri·an** *adj. & n.*

Man·chu-Tun·gus·ic (man·chōō'tŏon·gōōz'ik) *n.* A subfamily of the Altaic languages, consisting of Manchu and Tungus.

man·cip·i·um (man·sip'ē·əm) *n.* In ancient Roman law, the legal status of a person conveyed by the paterfamilias to another by the ceremony of mancipation. If emancipated from the mancipium, the person became again subject to paternal power and so remained, unless sold three times, when the authority of the paterfamilias ceased. [< L < *manceps* a buyer < *manus* a hand + *capere* take]

man·ci·ple (man'sə·pəl) *n.* A steward, as of an English college. [<OF *manciple, mancipe* < L *mancipium* MANCIPIUM]

Man·cu·ni·an (man·kyōō'nē·ən) *n.* A native of Manchester, England. [<Med. L *Mancunium* Manchester]

man·cus (mang'kəs) *n.* An Anglo-Saxon monetary unit worth thirty pence. [OE]

-mancy *combining form* Divining, foretelling, or discovering by means of: *necromancy*. [<Gk. *manteia* power of divination]

Man·da·lay (man'də·lā, man'də·lā') A city on the Irrawaddy in central Burma.

man·da·mus (man·dā'məs) *n. Law* A writ originally (in England) of royal prerogative, now a writ of right, issued by courts of superior jurisdiction and directed to subordinate courts, corporations, or the like, commanding them to do something therein specified: modified by statute from the ordinary common law form in jurisdiction. — *v.t. Colloq.* To command by or serve with a mandamus. [<L, we command < *mandare*. See MANDATE.]

Man·dan (man'dan) *n.* 1 One of a tribe of North American Indians of Siouan stock, of the NW United States. 2 The Siouan language of this tribe.

man·da·rin (man'də·rin) *n.* 1 An official of the Chinese Empire, either civil or military: a title used by foreigners indiscriminately. The recognized official grades under the Empire were nine, each rank being distinguished by its official specific regalia, a conspicuous part of which is the **mandarin button**. 2 Any of certain small Chinese oranges, having a loose skin and very sweet pulp; a tangerine. 3 An orange or reddish-yellow dye. [<Pg. *mandarim* <Malay *mantrī* a minister of state <Hind. <Skt. *mantrin* a counselor <*mantra* counsel]

Man·da·rin (man'də·rin) *n.* 1 The Chinese language of north China, in the Peking dialect now the official language of the country. 2 Formerly, the court language of the Chinese Empire.

mandarin duck A crested duck (*Aix galericulata*), with variously colored plumage.

man·da·ta·ry (man'də·ter'ē) *n. pl.* **·ta·ries** One to whom a charge is given; an agent. See MANDATE (def. 2).

man·date (man'dāt, -dit) *n.* 1 An authoritative requirement, as of a sovereign; a command; order; charge. 2 A charge to a nation, authorizing the government, administration, and development of conquered territory, given by a congress or league of nations, to which the grantee is responsible; also, the territory given in charge. 3 A judicial command directed to an officer of the court to enforce an order of that court; a precept from an appellate court directing what a subordinate court shall do in an appealed case. 4 A rescript of the pope ordering that the person named shall have the first vacant benefice in the gift of the person addressed. 5 An instruction from an electorate to the legislative body, or its representative, to follow a certain course of action. See synonyms under LAW. — *v.t.* (-dāt) **·dat·ed, ·dat·ing** To assign (a colony or other territory) to a specific nation under a mandate. [<L *mandatum*, pp. neut. of *mandare* command <*manus* hand + *dare* give (to)]

man·da·tor (man·dā'tər) *n.* One who gives a mandate; a director. [<L <*mandatus*, pp. of *mandare*. See MANDATE.]

man·da·to·ry (man'də·tôr'ē, -tō'rē) *adj.* Expressive of positive command; obligatory. — *n. pl.* **·ries** 1 A mandatary. 2 A mandate.

Man·de·an (man·dē'ən) *n.* 1 A member of an ancient sect of Gnostics, Christians of St. John, still existing in Mesopotamia, who combine Judaism, Mohammedanism, and Christianity with the ancient Babylonian worship. 2 The Aramaic dialect used in the writings of the Mandeans. — *adj.* Of or pertaining to the Mandeans or their doctrines. Also **Mandae'an**. [<Mandean *mandayyā*, lit., having knowledge (< *mandā* knowledge), trans. of Gk. *gnostikoi* Gnostics] — **Man·de'ism** *n.*

man·del·ic (man·del'ik) *adj.* Pertaining to or designating a white crystalline acid, $C_8H_8O_3$, used in medicine as a urinary antiseptic. [<G *mandel* an almond]

Man·de·ville (man'də·vil), **Bernard**, 1670?–1733, English author born in Holland. — **Sir John**, 1300?–72, pseudonym of a reputed English traveler whose *Narrative of Travels* appeared in Latin and French, 1357–71, and in English in the 15th century.

man·di·ble (man'də·bəl) *n. Biol.* 1 The lower jaw bone, or its equivalent. 2 Either the upper or the lower part of the beak of a bird, or of the beak of a cephalopod. 3 Either one of the upper or outer pair of jaws in an insect. 4 The operculum of a polyzoan. [<LL *mandibula* <L *mandere* chew]

man·dib·u·lar (man·dib'yə·lər) *adj.* Of, pertaining to, or formed by a mandible: the *mandibular* arch of the fetal skull. Also **man·dib'u·lar'y** (-ler'ē). — *n.* The lower jaw, or mandible. [<LL *mandibul(a)* MANDIBLE + -AR]

man·dib·u·late (man·dib'yə·lit, -lāt) *adj.* Having mandibles adapted for biting and chewing: said of certain insects. — *n.* Any insect having chewing jaws.

Man·din·go (man·ding'gō) *n. pl.* **·gos** or **·goes** 1 One of a Negroid people of western Sudan, Africa. 2 Any of the Sudanic languages of the Mandingos. — **Man·din'gan** *adj. & n.*

man·do·la (man·dō'lə) *n.* A large mandolin. [<Ital. *mandola, mandora* <L *pandura*. See BANDORE.]

man·do·lin (man'də·lin, man'də·lin') *n.* A stringed musical instrument with an almond-shaped body and metal strings arranged like those of a violin. [<F *mandoline* <Ital. *mandolino*, dim. of *mandola* MANDOLA] — **man'do·lin'ist** *n.*

man·drake (man'drāk) *n.* 1 A short-stemmed Old World plant (*Mandragora officinarum*) of the nightshade family, with narcotic properties. 2 Its fleshy roots, sometimes having a fancied resemblance to the human form. 3 The May apple. Also **man·drag·o·ra** (man·drag'ə·rə). [Alter. of ME *mandrag(g)e*, <OE *mandragora* <LL <L *mandragoras* <Gk.; infl. in form by folk etymology <MAN + DRAKE[2] a dragon]

man·drel (man'drəl) *n. Mech.* 1 A shaft or spindle on which an object may be fixed for rotation. 2 A smooth, hard, cylindrical or conical core about which wire may be coiled or metal or glass forged. 3 A pattern or form against which metalwork is pressed in spinning. Also **man′dril**. [Prob. alter. of F *mandrin* a lathe]

man·drill (man′dril) *n.* A large, ferocious West African baboon (genus *Papio*), having large canine teeth and bony prominences on the cheeks striped with blue and scarlet. [<MAN + DRILL³]

MANDRILL
(About 30 inches at the shoulder; length 3 to 4 feet)

man·du·cate (man′-jōō·kāt) *v.t.* **·cat·ed**, **·cat·ing** *Rare* To chew; masticate. [<L *manducatus*, pp. of *manducare* chew]

Man·dy (man′dē) Diminutive of AMANDA.

mane (mān) *n.* The long hair growing on and about the neck of some animals, as the horse, lion, etc. ◆ Homophone: main. [OE *manu*] — **maned** *adj.* — **mane′less** *adj.*

man-eat·er (man′ē′tər) *n.* **1** A cannibal. **2** An animal that devours, likes to devour, or is supposed to devour human flesh; especially, a tiger or a lion. **3** A large shark with trenchant teeth; especially, *Carcharodon carcharias.*

ma·nège (ma·nezh′) *n.* **1** The art of training and riding horses. **2** A school of horsemanship; riding school; also, a school for the training of horses. **3** The style and movements of a trained horse. Also **ma·nege′**. [<F <Ital. *maneggio* <*maneggiare*. See MANAGE.]

ma·nes (mā′nēz) *n. pl.* **1** In ancient Roman religion, the spirits of the dead; especially, the spirits of dead ancestors. **2** Deified ancestral spirits collectively. [<L, ? orig. pl. of *manis* good]

Ma·nes (mā′nēz), 216?–276?, Persian prophet; founder of Manicheism: also called *Manicheus.*

Ma·net (ma·nā′, *Fr.* må·ne′), **Edouard,** 1832–1883, French painter.

Man·e·tho (man′ə·thō) An Egyptian priest and writer of the third century B.C.

ma·neu·ver (mə·nōō′vər, -nyōō′-) *n.* **1** A movement or change of position, as of troops or war vessels. **2** Any dexterous or artful proceeding. — *v.t.* **1** To put, as troops, through a maneuver or maneuvers. **2** To put, bring, make, etc., by a maneuver or maneuvers. **3** To manipulate; conduct adroitly. — *v.i.* **4** To perform a maneuver or maneuvers. **5** To use tricks or stratagems; manage adroitly. Also spelled *manoeuver, manoeuvre.* [<F *manoeuver* <OF *maneuvre* <LL *manopera* <*manoperare* <L *manu operari* work with the hand <*manus* hand + *operari* to work. See OPERATE. Doublet of MAINOR, MANURE.] — **ma·neu′ver·a·bil′i·ty** or **·vra·bil′i·ty** *n.* — **ma·neu′ver·a·ble** or **·vra·ble** *adj.*

ma·neu·vers (mə·nōō′vərz, -nyōō′-) *n. pl.* Large-scale military exercises involving the use of great numbers of troops under conditions simulating actual battle conditions; war games.

man·fash·ion (man′fash′ən) *adv.* **1** In a manly, confident, straightforward way. **2** Astride: to ride *manfashion.*

man Friday A person devoted or subservient to another, like Robinson Crusoe's servant of that name; a factotum.

man·ful (man′fəl) *adj.* Having a manly spirit; characterized by courage and perseverance; sturdy. See synonyms under MANLY, MASCULINE. — **man′ful·ly** *adv.* — **man′ful·ness** *n.*

mang (mang) *prep. Scot.* Among.

Man·ga·lore (mang′gə·lôr′, -lōr′) A port of western Madras, SW India.

Man·gan (mang′gən), **James Clarence,** 1803–1849, Irish poet.

man·ga·nate (mang′gə·nāt) *n. Chem.* A salt of manganic acid, such as those of sodium, potassium, and barium. [<MANGAN(IC) + -ATE³]

man·ga·nese (mang′gə·nēs, -nēz) *n. Chem.* A hard, brittle, metallic element (symbol Mn). In color, it is grayish-white tinged with red; it rusts like iron, but is not magnetic; it is widely distributed (in combination) in nature, and forms an important component of certain alloys, such as manganese steel. See ELEMENT. [<F *manganèse* <Ital. *manganese*, alter. of Med. L *magnesia* MAGNESIA]

manganese dioxide Pyrolusite.

manganese spar 1 Rhodonite. **2** Rhodochrosite.

manganese steel A very hard, ductile steel containing from 12 to 14 percent of manganese.

man·gan·ic (mang·gan′ik) *adj. Chem.* Of, pertaining to, containing, or obtained from manganese in its highest valence, as manganic acid, H_2MnO_4, known chiefly by its salts. [<MANGAN(ESE) + -IC]

man·ga·nite (mang′gə·nīt) *n.* **1** A dark, steel-gray to iron-black orthorhombic manganese hydroxide, MnO(OH). **2** *Chem.* Any salt obtained from certain hydroxides of manganese, and regarded as an acid.

man·ga·nous (mang′gə·nəs) *adj.* Of, pertaining to, or containing manganese in its lowest valence.

Man·ga·re·va (mäng′ä·rā′vä) **1** Largest of the Gambier Islands. **2** The Gambier Islands.

mange (mānj) *n.* An itching skin disease of dogs and other domestic animals, caused by burrowing parasitic mites. Also **the mange**. [<OF *manjue* an itch, eating <*manjuer, mangier.* See MANGER.]

man·gel·wur·zel (mang′gəl·wûr′zəl, -wûrt′səl) *n.* A large-rooted European beet (*Beta vulgaris*), fed to cattle. Also **man′gel**. [<G <*mangoldwurzel* <*mangold* a beet + *wurzel* a root]

man·ger (mān′jər) *n.* A trough or box, for feeding horses or cattle. [<OF *mangeoire, mangeure* <*mangier* eat <L *manducare* chew]

man·gle¹ (mang′gəl) *v.t.* **·gled**, **·gling 1** To disfigure or mutilate, as by cutting, bruising, or crushing; lacerate. **2** To mar or ruin; spoil: to *mangle* a word. See synonyms under REND. [<AF *mangler, mahangler,* appar. freq. of OF *mahaigher* MAIM]

man·gle² (mang′gəl) *n.* A machine for smoothing fabrics by pressing them between rollers. — *v.t.* **·gled**, **·gling** To smooth with a mangle. [<Du. *mangel* <MDu. *mange* <Ital. *mangano* <LL *manganum* <Gk. *manganon* a pulley, a war machine. Doublet of MANGONEL.]

man·gler (mang′glər) *n.* One who or that which mangles.

man·go (mang′gō) *n. pl.* **·goes** or **·gos 1** The edible, fleshy fruit of a tropical tree allied to the sumac. **2** The tree (*Mangifera indica,* family *Anacardiaceae*) producing the fruit. **3** A pickled green muskmelon. [<Pg. *manga* <Malay *manga* <Tamil *mān-kāy* <*mān* a mango tree + *kāy* a fruit]

man·go·nel (mang′gə·nel) *n.* A military engine formerly used for throwing stones and other missiles. [<OF, dim. of LL *manganum.* Doublet of MANGLE².]

man·go·steen (mang′gə·stēn) *n.* **1** The reddish-brown fruit of an East Indian tree, about the size of an apple, having a thick, fleshy rind, and a white, juicy pulp. **2** The tree (*Garcinia mangostana*) producing this fruit. [<Malay *mangustan*]

man·grove (mang′grōv, man′-) *n.* **1** A tropical tree (genus *Rhizophora,* especially *R. mangle*) which throws out many aerial roots from the lower branches and stem, forming dense thickets. **2** A shrub of the vervain family, as the **black mangrove** (*Avicennia marina*). [<Sp. *mangle* <Taino; infl. in form by GROVE]

mangrove cuckoo A cuckoo (*Coccyzus minor*) frequenting mangroves of the West Indies, Florida, etc.

man·gy (mān′jē) *adj.* **·gi·er**, **·gi·est 1** Affected with the mange. **2** Figuratively, poverty-stricken; squalid. — **man′gi·ly** *adv.* — **man′gi·ness** *n.*

man·han·dle (man′han′dəl) *v.t.* **·dled**, **·dling 1** To move by manpower without mechanical aids. **2** To handle with roughness, as in anger.

Man·hat·tan (mən·hat′ən, man-) *n.* **1** A cocktail made of whisky and vermouth, often with a dash of bitters and a cherry. **2** One of a tribe of Algonquian North American Indians, formerly inhabiting Manhattan Island.

Man·hat·tan Island (mən·hat′ən, man-) An island in SE New York at the mouth of the Hudson River; 22 square miles; comprising **Manhattan** borough of New York City.

man·hole (man′hōl′) *n.* An opening through which a man may enter a boiler, conduit, sewer, etc., for making repairs.

man·hood (man′hood) *n.* **1** Manly qualities collectively; manliness; courage. **2** The state of being of age; man's estate. **3** The state of being a man, or a human being as distinguished from other animals or beings. **4** Men collectively: the *manhood* of the nation.

man-hour (man′our′) *n.* A unit of measure equal to the amount of work one man can do in one hour.

ma·ni·a (mā′nē·ə, mān′yə) *n.* **1** Madness. **2** *Psychiatry* Exaggerated melancholia alternating periodically with an exaggerated sense of well-being, accompanied by excessive activity (both mental and physical): also called *manic-depressive psychosis.* **3** A strong, ungovernable desire; also, colloquially, a craze: a *mania* for rare books. See synonyms under FRENZY, INSANITY. [<L <Gk., madness < *mainesthai* rage. Akin to MANTIC.]

-mania *combining form* An exaggerated, persistent, or irrational craving for or infatuation with. [<Gk. *mania* madness] Below are some examples.

agromania	open country
bibliomania	books
cratomania	power
dipsomania	alcohol
kleptomania	stealing
megalomania	greatness, fame
mythomania	telling lies
nostomania	one's home
nymphomania	men
plutomania	wealth
pyromania	setting fires
toxicomania	poisons

ma·ni·ac (mā′nē·ak) *adj.* Having a mania; raving. See synonyms under INSANE. — *n.* A person wildly or violently insane; a madman. [<LL *maniacus* <L *mania* MANIA] — **ma·ni·a·cal** (mə·nī′ə·kəl) *adj.* — **ma·ni′a·cal·ly** *adv.*

-maniac *combining form* Forming adjectives (often used as nouns) from nouns in *-mania:* kleptomaniac.

man·ic (man′ik, mā′nik) *adj.* Pertaining to, like, or affected by mania. [<MAN(IA) + -IC]

man·ic-de·pres·sive (man′ik·di·pres′iv) *adj. Psychiatry* Denoting a mental disorder characterized by sudden fluctuations of depression and excitement. — *n.* One who suffers from this disorder.

Man·i·chee (man′ə·kē) *n.* A follower of the Persian Manes. [<LL *Manichaeus* <LGk. *Manichaios* Manes]

Man·i·che·ism (man′ə·kē′iz·əm) *n.* A dualistic religious philosophy developed by the Persian Manes and his followers in which goodness, typified as light, God, or the soul, is represented as in conflict with evil, typified by darkness, Satan, or the body: taught from the third to the seventh century. Also **Man′i·chae·ism.** — **Man′i·che′an**, **Man′i·chae′an** *adj.* & *n.*

man·i·cure (man′ə·kyoor) *n.* **1** The care or treatment of the hands and fingernails. **2** *Obs.* A manicurist. — *v.t.* & *v.i.* **·cured**, **·cur·ing** To take care of or treat (the hands and nails). [<F <L *manus* hand + *cura* care]

man·i·cur·ist (man′ə·kyoor′ist) *n.* One who cares for or treats the hands and fingernails.

man·i·fest (man′ə·fest) *v.t.* **1** An itemized statement of a vessel's cargo for the information of customs officers, stating the ports of lading and destination, and giving the names of consignees, passengers, etc. **2** Loosely, a waybill of lading. **3** Perishable goods or livestock traveling by fast freight. **4** *Obs.* A manifesto. — *v.t.* **1** To make plain to sight or understanding; reveal; display. **2** To prove; be evidence of. **3** To show the manifest of (a shipment). — *adj.* Plainly apparent to sight or understanding; evident; plain. [<L *manifestus* evident, lit., struck by the hand] — **man′i·fest′ly** *adv.*

Synonyms (adj.): apparent, bare, clear, conspicuous, distinct, evident, glaring, indubitable, obvious, open, overt, palpable, patent, plain, transparent, unmistakable, visible. See CLEAR, EVIDENT, NOTORIOUS, OVERT. *Antonyms:* concealed, covert, dark, hidden, impalpable, impenetrable, imperceptible, invisible, latent, obscure, occult, secret, undiscovered, unimagined, unknown, unseen.

man·i·fes·ta·tion (man′ə·fes·tā′shən) *n.* **1** The act of manifesting or making plain; the state or fact of being manifested; disclosure or display; a revelation. **2** Hence, a public or collective act by a government or party in order to emphasize its power, determination, or special views. 3 A revealing agency. 4 In spiritualism, the materialization of a spirit. See synonyms under MARK[1], SIGN. — **man'i·fes'tant** *n.*

man·i·fes·to (man'ə·fes'tō) *n. pl.* **·toes** A public, official, and authoritative declaration making announcement or explanation of intentions, motives, or principles of actions. [<Ital. <L *manifestus.* See MANIFEST.]

man·i·fold (man'ə·fōld) *v.t.* **1** To make more than one copy of at once, as with carbon paper on a typewriter. **2** To multiply. — *adj.* **1** Of great variety; numerous. **2** Manifested in many ways, or including many acts or elements; complex; being so in many ways or for many reasons. **3** Existing in great abundance. **4** Comprising or uniting several parts or channels of the same kind, as a pipe with several outlets. See synonyms under COMPLEX, MANY. — *n.* **1** A copy made by manifolding. **2** *Math.* A number of objects related under one system. **3** A tube with one or more inlets and two or more outlets; a T branch, as the pipe between carburetor and engine, in internal-combustion engines with more than one cylinder. [OE *manigfeald* varied, numerous] — **man'i·fold·ly** *adv.* — **man'i·fold·ness** *n.*

man·i·fold·er (man'ə·fōl'dər) *n.* **1** One who or that which manifolds. **2** A machine or apparatus for making manifold copies, as of a document.

Ma·ni·hi·ki (mä'nē·hē'kē) An island of the South Pacific administered by New Zealand with the Cook Islands: also *Humphrey Island.*

man·i·hot (man'ē·hot) *n.* Any of a large genus *(Manihot)* of tropical American, mainly Brazilian, herbs or woody plants of the spurge family. Brazilian or Ceará rubber is the product of *M. glaziovi.* [<NL <F <Tupian *mandioca* manioc root]

man·i·kin (man'ə·kin) *n.* **1** A model of the human body, showing its structure. **2** A dressmaker's assistant or lay figure wearing new costumes so as to display them for sale: also spelled *mannequin.* **3** A little man; dwarf. Also spelled *manakin, mannikin.* [<Du. *manneken,* dim. of *man* man]

ma·nil·a (mə·nil'ə) *n.* **1** A cheroot made in Manila. **2** The fiber of the abaca *(Musa textilis),* a tall, perennial herb of the same genus as the banana: also **Manila hemp, ma·nil'la.**

Ma·nil·a (mə·nil'ə, *Sp.* mä·nē'lä) A port on SW Luzon; former capital of the Philippines.

Manila Bay A landlocked inlet of the China Sea in SW Luzon, Philippines.

Manila paper A heavy, light-brown paper originally made of Manila hemp, now made of various fibers.

Manila rope Rope made of Manila hemp.

man·i·oc (man'ē·ok, mā'nē-) *n.* The bitter or sweet cassava. [<F <Tupian *mandioca* manioc root]

man·i·ple (man'ə·pəl) *n.* **1** *Eccl.* A band worn on the left arm as a vestment by the clergy of the Roman Catholic and sometimes of the Anglican Church. **2** A subdivision of the Roman legion containing 60 or 120 men. [<OF *maniple, manipule* <L *manipulus* a handful <*manus* hand + base of *plere* fill]

ma·nip·u·lar (mə·nip'yə·lər) *adj.* **1** Pertaining to manipulation or handling. **2** Pertaining to a maniple. [<L *manipularis* <*manipulus* a MANIPLE; infl. in sense by MANIPULATION]

ma·nip·u·late (mə·nip'yə·lāt) *v.t.* **·lat·ed, ·lat·ing** **1** To handle, operate, or use with or as with the hands, especially with skill. **2** To influence or control artfully or deceptively: to *manipulate* stocks. **3** To change or alter (figures, accounts, etc.), usually fraudulently. [Back formation <MANIPULATION <F, ult. <L *manipulus* a MANIPLE] — **ma·nip'u·la'tion** *n.* — **ma·nip'u·la'tive, ma·nip'u·la·to'ry** *adj.* — **ma·nip'u·la'tor** *n.*

Ma·ni·pur (mun'i·po͝or) A Union Territory of NE India, bordering on Burma; 8,628 square miles; capital, Imphal.

Man·i·pu·ri (mun'i·po͝or'ē) *n.* A native or inhabitant of Manipur.

ma·nis (mā'nis) *n.* A pangolin. [<NL, sing. <L *manes* MANES]

Ma·nis·sa (mä'nē·sä', mä'nē·sä) A city in eastern Turkey in Asia, on the site of ancient Magnesia. Also **Ma'ni·sa'.**

man·i·to (man'ə·tō) *n.* Among the Algonquian Indians, the unfathomable spirit or power behind life and the universe. See WAKANDA. Also **man·i·tou** (man'ə·to͞o), **man'i·tu.** — **Kitchi Manito** Great Spirit. [<Algonquian (Massachuset) *manitto* he is a god]

Man·i·to·ba (man'ə·tō'bə, -tō·bä') A province in central Canada; 246,512 square miles; capital, Winnipeg. — **Man'i·to'ban** *adj.* & *n.*

Manitoba, Lake A lake in SW Manitoba; 67 square miles.

Man·i·tou·lin Island (man'ə·to͞o'lin) A Canadian island in Lake Huron, south central Ontario; 1,600 square miles.

Ma·ni·za·les (mä'nē·sä'läs) A city in west central Colombia.

man·kind (man'kīnd', man'kīnd') *n.* **1** The whole human species. **2** (man'kīnd') Men collectively as distinguished from women. [<MAN + KIND[2]]

Synonyms: humanity, humankind, man, men. See HUMANITY.

man·like (man'līk') *adj.* Like a man; having the qualities proper to the human race, to the male sex, to manly character. See synonyms under MANLY, MASCULINE.

man·ly (man'lē) *adj.* **·li·er, ·li·est** Possessing the characteristics of a true man, as strength, frankness, and intrepidity. — *adv. Archaic* In a manner befitting a man. — **man'li·ly** *adv.* — **man'li·ness** *n.*

Synonyms: manful, manlike, mannish. *Manlike* may mean only having the outward appearance or semblance of a man, or it may be substantially equivalent to *manly. Manly* refers to all the qualities and traits worthy of a man; *manful* especially to the valor and prowess that become a man; we speak of a *manful* struggle; *manly* decision; we say *manly* gentleness, or tenderness; we could not say *manful* tenderness. *Mannish* is a depreciatory word referring to the mimicry or parade of some superficial qualities of manhood; as, a *mannish* woman. See MASCULINE.

Mann (man), **Horace,** 1796–1859, U. S. educator.

Mann (män), **Heinrich,** 1871–1950, German novelist active in the United States. — **Thomas,** 1875–1955, German novelist active in the United States; brother of the preceding.

man·na (man'ə) *n.* **1** The miraculously supplied food on which the Israelites subsisted in the wilderness. *Exodus* xvi 14–36. **2** Divine or spiritual nourishment. **3** A sweetish substance obtained from incisions in the stems of various trees or shrubs, especially the stems of the flowering ash *(Fraxinus ornus* and *F. rotundifolia)* of southern Europe. It is a mild laxative. [OE <LL <Gk. <Aramaic *mannā* <Hebrew *mān*]

Mann Act (man) A bill prohibiting the interstate transportation of women and girls for immoral purposes, enacted by Congress in June, 1910: also called *White-slave Act.* [after James Robert *Mann,* 1856–1922, U. S. congressman]

Man·nar (mə·när'), **Gulf of** An inlet of the Indian Ocean between southern Madras and Ceylon.

man·ne·quin (man'ə·kin) See MANIKIN (def. 2).

man·ner (man'ər) *n.* **1** The way of doing anything; method of procedure; mode. **2** The demeanor or bearing peculiar to one; personal carriage; mien; address. **3** *pl.* General modes of life or conduct; especially, social behavior. **4** *pl.* Polite, civil, or well-bred behavior. **5** Usual or ordinary practice; habit; custom; also, characteristic style in literature or art. **6** Sort or kind. **7** Character; guise. — **to the manner born** Familiar with something from birth. ◆ Homophone: *manor.* [<AF *manere,* OF *maniere,* ult. <L *manuarius* of the hand <*manus* hand]

Synonyms: appearance, aspect, carriage, demeanor, deportment, fashion, habit, look, mien, mode, practice, style, way. See ADDRESS, AIR, BEHAVIOR, CUSTOM, SYSTEM.

man·nered (man'ərd) *adj.* **1** Having (specified) manners: often used in combination: ill-*mannered.* **2** Affected.

Man·ner·heim (män'ər·hīm), **Baron Carl Gustaf Emil von,** 1867–1951, Finnish soldier and statesman.

man·ner·ism (man'ər·iz'əm) *n.* **1** Characteristic or marked adherence to an unusual or affected manner, style, or peculiarity. **2** A peculiarity of manner, as in behavior or speech.

man·ner·ist (man'ər·ist) *n.* A person addicted to mannerism; an imitator; specifically, an artist or writer whose work is marked by a persistent or extreme adherence to some style, manner, etc., as one of the school of painters of the 16th and 17th centuries who unduly emphasized and imitated the style of Michelangelo and other Italian painters.

man·ner·less (man'ər·lis) *adj.* Lacking manners; ill-mannered.

man·ner·ly (man'ər·lē) *adj.* Well-behaved; polite. — *adv.* With good manners; politely. — **man'ner·li·ness** *n.*

man·ners-bit (man'ərz·bit') *n.* A small amount left on a plate by a guest at table, for the sake of good manners, as indicating that the serving was abundant.

Mann·heim (män'hīm) A city on the Rhine in NW Baden-Württemberg, West Germany.

man·ni·kin (man'ə·kin) See MANIKIN.

Man·ning (man'ing), **Henry Edward,** 1808–92, English Roman Catholic cardinal and writer. — **William Thomas,** 1866–1949, U. S. Episcopalian divine; bishop of New York 1921–46.

man·nish (man'ish) *adj.* Resembling, characteristic of, or suitable to a man; aping manhood; masculine. See synonyms under MANLY, MASCULINE. [OE *menisc*] — **man'nish·ly** *adv.* — **man'nish·ness** *n.*

man·ni·tol (man'ə·tōl, -tol) *n. Biochem.* A slightly sweet crystalline alcohol, $C_6H_{14}O_6$, found widely distributed in nature, as in celery, sponges, sea grasses, and especially in the dried sap of the flowering ash. Also **man'nite** (-īt). [<MANNA (def. 3) + -ITE[3] + -OL[1]] — **man·nit·ic** (mə·nit'ik) *adj.*

man·nose (man'ōs) *n.* A hexose, $C_6H_{12}O_6$, obtained by the oxidation of mannitol. [<MANN(ITOL) + -OSE]

man·o·cry·om·e·ter (man'ō·krī·om'ə·tər) *n.* An instrument for determining the variations in the freezing or melting point of substances with changes in pressure. [<Gk. *manos* thin, rare + *kryos* an icy cold + -METER]

Ma·no·el (mä'nō·el') Portuguese form of EMMANUEL.

ma·noeu·ver (mə·no͞o'vər, -nyo͞o'-), **ma·noeu·vre** See MANEUVER.

Man of Destiny Napoleon I: so regarded by himself.

man of God **1** A saint, prophet, etc.; a holy man. **2** A clergyman.

Man of Sorrows A name supposed to allude to the Messiah (*Isa.* liii 3); hence, Jesus Christ.

man of straw One put forward as an irresponsible tool or as a fraudulent surety.

man-of-war (man'əv·wôr', man'ə-) *n. pl.* **men-of-war** (men'-) **1** A naval vessel armed for active hostilities. **2** The Portuguese man-of-war. Also **man'-o'-war'.**

man-of-war bird or **hawk** A frigate bird.

ma·nom·e·ter (mə·nom'ə·tər) *n.* An instrument for measuring pressure, as of gases and vapors; usually, a U-tube. [<Gk. *manos* thin, rare + -METER] — **man·o·met·ric** (man'ə·met'rik) or **·ri·cal** *adj.*

Ma·non Les·caut (mà·nôn' les·kō') Promiscuous heroine of Prévost's novel *Histoire du Chevalier Des Grieux et de Manon Lescaut.*

man·or (man'ər) *n.* **1** *Brit.* A nobleman's or gentleman's landed estate. **2** In old English law, a tract or district of land granted by the king to one as lord, with authority to exercise jurisdiction over it by a court-baron. **3** In Anglo-Saxon times, a thane's or lord's estate, composed of the land and of a part of the agricultural capital employed to till it, as well as the laborers, beasts, implements, etc., and having on it a community of serfs or villeins; later, an estate complying with the minimum requirements entitling the lord of the manor to hold a court-baron. **4** *U.S.* A tract of land originally granted as a manor and let by the proprietor to tenants in perpetuity or for a long term. See synonyms under HOUSE. ◆ Homophone: *manner.* [<OF *manoir* a

add,āce,câre,pälm; end,ēven; it,īce; odd,ōpen,ôrder; to͝ok,po͞ol; up,bûrn; ə = a in *above,* e in *sicken,* i in *clarity,* o in *melon,* u in *focus*; yo͞o = u in *fuse*; oi,oil; ou,pout; ch,check; g,go; ng,ring; th,thin; th,this; zh,vision. Foreign sounds à,œ,ü,kh,ṅ; and ◆: see page xx. < from; + plus; ? possibly.

manor–house 776 **Manutius**

dwelling, orig. a verb <L *manere* remain] — **ma·no·ri·al** (mə·nôr'ē·əl, -nō'rē-) *adj.*
man·or–house (man'ər·hous') *n.* The residence of the lord of the manor. Also **man'or–seat'** (-sēt').
man·pow·er (man'pou'ər) *n.* 1 Power supplied by the physical strength of a man or men. 2 The normal rate at which a man can work, generally equal to 1/10 horsepower. 3 The strength of all the men available for any service; specifically, the warpower of a nation in terms of the number of men available for military service, industry, agriculture, etc.
man·qué (män·kā') *adj. French* Defective; falling short of what is intended. — **man·quée'** *adj. fem.*
man·rope (man'rōp') *n. Naut.* A rope serving as a hand railing to a gangway, ladder, etc., on board ship.
man·sard (man'särd) *n.* 1 A roof with a double pitch on all sides: also **mansard roof.** 2 A room within such a roof; an attic. [<F *mansarde*, after François *Mansard*, 1598–1666, French architect who revived it]

MANSARD ROOF

manse (mans) *n.* 1 A clergyman's house, especially a Scottish Presbyterian minister's house; a parsonage. 2 A landholder's residence. [<Med. L *mansa, mansus*, orig. pp. of L *manere* remain, dwell]
man·ser·vant (man'sûr'vənt) *n.* An adult male servant.
Mans·field (manz'fēld, mans'-) A municipal borough of Nottinghamshire, England.
Mans·field (manz'fēld, mans'-), **Katherine** Pseudonym of Kathleen Beauchamp Murry, 1888–1923, English author. — **Richard,** 1854–1907, U. S. actor born in England.
Mans·field (manz'fēld, mans'-), **Mount** The highest peak of the Green Mountains, Vermont; 4,393 feet.
Man·ship (man'ship), **Paul,** born 1885, U. S. sculptor.
man·sion (man'shən) *n.* 1 A large or handsome dwelling; specifically, the house of the lord of a manor; a manor–house. 2 In astrology, one of the 12 divisions of the heavens; a house. 3 According to Oriental and medieval astronomers, one of the 28 divisions of the heavens occupied by the moon on successive days. 4 *Archaic* Any place of abode. 5 A small compartment, abode, or dwelling in a larger house or enclosure. 6 *pl. Brit.* An apartment house. See synonyms under HOUSE. [<OF <L *mansio, -onis* a dwelling <*mansus,* pp. of *manere* remain, dwell. Doublet of MENAGE.]
man–sized (man'sīzd') *adj. Colloq.* Of a size appropriate for a man; large.
man·slaugh·ter (man'slô'tər) *n.* The killing of man by man; especially, such killing without malice.
man·slay·er (man'slā'ər) *n.* One who commits homicide. — **man'slay'ing** *n.*
man–stop·ping (man'stop'ing) *adj.* Having force sufficient to stop a man's advance: said of a bullet. Compare DUMDUM BULLET.
man·sue·tude (man'swə·to͞od, -tyo͞od) *n.* Accustomed gentleness or mildness; tameness. [<L *mansuetudo, -inis* <*mansuetus,* pp. of *mansuesere* tame <*manus* hand + *suescere* accustom]
man·swear (man'swâr') *v.i.* **·swore, ·sworn, ·swear·ing** *Obs.* To swear falsely. [OE *mānswerian* <*mān* wickedness + *swerian* swear]
mant (mänt) *v.t. & v.i. Scot.* To stammer. — **mant'er** *n.*
man·ta (man'tə) *n.* 1 A coarse cotton cloth used by the lower classes of Spanish America for clothing; specifically, a woman's shawl or other article of clothing made of such material. 2 In the Rocky Mountains, the canvas covering of the load of a pack–animal. 3 A devilfish. 4 *Obs.* A mantelet. [<Sp., a blanket <LL *mantum* a cloak, back formation <L *mantellum.* See MANTLE.]
man·teau (man'tō, *Fr.* män·tō') *n. pl.* **·teaus** (-tōz) or **·teaux** (*Fr.* -tō') 1 A cloak or mantle worn by women; any mantle. 2 *Obs.* A woman's gown; a mantua. [<F <OF *mantel* MANTLE]

Man·te·gna (män·tā'nyä), **Andrea,** 1431–1506, Italian painter and engraver.
man·tel (man'təl) *n.* The facing about a fireplace, including the shelf above it; also, the shelf. ♦ Homophone: *mantle.* [Var. of MANTLE; infl. in meaning by F *manteau* a mantelpiece]

MANTEL

man·tel·et (man'təl·et, mant'lit) *n.* 1 A small mantle or short cloak. 2 *Mil.* A screen or shield, as in an embrasure, to protect the defenders; a movable roof to protect a besieging party; also, a shield or protection, made of metal, rope, or wood, placed at openings, portholes, etc., to protect the gunner from bullets or smoke. 3 In target–shooting, a bulletproof enclosure for observation. 4 A movable shelter used by hunters. Also spelled *mantlet.* [<OF, dim. of *mantel.* See MANTLE.]
Man·tell (man·tel'), **Robert Bruce,** 1854–1928, U. S. actor born in Scotland.
man·tel·let·ta (man'tə·let'ə) *n.* In the Roman Catholic Church, a sleeveless vestment reaching almost to the knees, worn by bishops and various church dignitaries. [<Ital., dim. of *mantello* <L *mantellum.* See MANTLE.]
man·tel·piece (man'təl·pēs') *n.* A mantel shelf.
man·tel·tree (man'təl·trē') *n.* A wooden mantel; also, the arch of a mantel.
man·tic (man'tik) *adj.* Relating to divination, soothsaying, or the supposed inspired condition of a soothsayer; prophetic: *mantic frenzy.* [<Gk. *mantikos* <*mantis* a prophet. Akin to MANIA.]
-mantic *combining form* Used to form adjectives corresponding to nouns ending in *-mancy: necromantic.* [<Gk. *mantikos* prophetic <*manteia* divination]
man·til·la (man·til'ə) *n.* 1 A woman's light scarf or head covering of lace, as worn in Spain, Mexico, Italy, etc. 2 Any short mantle. [<Sp., dim. of *manta* MANTA]
man·tis (man'tis) *n. pl.* **·tis·es** or **·tes** (-tēz) A carnivorous, orthopterous, long–bodied insect (family *Mantidae*) with large eyes and movable head, which assumes a position with its forelegs folded as if in prayer. Also called *praying mantis.* For illustration see INSECTS (beneficial). [<Gk., a prophet, also a kind of insect]

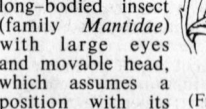
PRAYING MANTIS
(From 3 to 6 inches long)

man·tis·sa (man·tis'ə) *n. Math.* The decimal or fractional part of a logarithm: so named as being added to the integral part or characteristic. [<L, a makeweight, trifling addition, ? <Etruscan]
mantis shrimp A squill. Also **mantis crab.**
man·tle (man'təl) *n.* 1 A loose garment, usually without sleeves, worn over the other garments; a cloak. 2 Anything that clothes or envelops; hence, whatever covers or conceals: a *mantle* of darkness. 3 *Zool.* **a** The variously modified flap or folds of the membranous covering of a mollusk. It secretes the shell. **b** The back, scapulars, and folded wings of a bird, when distinguished by color, as in gulls. **c** The soft external body–wall in the tunic of ascidians. 4 The outer covering of a wall. 5 The outer masonry of a blast furnace. 6 A sheath of clay laid over a wax model, forming a mold when the wax is melted out. 7 A mantel. 8 A hood of network fabric, generally cylindrical, of the salts of certain rare refractory earths with high radiating power, as cerium oxide, intended to give light by incandescence, as in the flame of a Bunsen burner, or in the Welsbach burner. — *v.* **·tled, ·tling** *v.t.* 1 To cover with or as with a mantle; conceal. — *v.i.* 2 To overspread or cover the surface of something. 3 To be or become covered, overspread, or suffused. 4 To spread out one wing at a time over the corresponding outstretched leg: said of hawks. ♦ Homophone: *mantel.* [Fusion of OE *mentel* and

OF *mantel*, both <L *mantellum, mantelum* a cloak, cloth, towel]
mant·let (mant'lit) See MANTELET.
man·tu·a (man'cho͞o·ə, -to͞o·ə) *n.* A woman's loose cloak, worn about 1850. [Alter. of *manteau*; infl. in form by *Mantua*]
Man·tu·a (man'cho͞o·ə, -to͞o·ə) A city on the Mincio, Lombardy, Italy; birthplace of Vergil. *Italian* **Man·to·va** (män·tō·vä). — **Man'tu·an** *n. & adj.*
man·ty (man'tē) *n. Scot.* A mantle; gown.
Ma·nu·a Islands (mä·no͞o'ä) An island group in American Samoa; 22 square miles.
man·u·al (man'yo͞o·əl) *adj.* 1 Done, made, or used by the hand; of, relating to, or affecting the hand: *manual* employments. 2 *Law* Actually possessed; in one's own hands. 3 Resembling a manual; designed to be retained for reference: said of a book. — *n.* 1 A compact volume; handbook of instruction or directions. 2 A keyboard, as of an organ. 3 A systematic exercise in the handling of some military weapon. [<OF *manuel* <L *manualis* <L *manus* hand] — **man'u·al·ly** *adv.*
manual training In U. S. schools, training of pupils in carpentry, woodworking, and the like.
ma·nu·bri·um (mə·no͞o'brē·əm, -nyo͞o'-) *n. pl.* **·bri·a** (-brē·ə) 1 *Anat.* Some part or process like or likened to a handle; especially, the anterior part of the sternum in man and many mammals: also called *episternum.* 2 *Bot.* A cylindrical cell borne on the center of each of the eight shields that compose the globular antheridium of the stoneworts (*Characeae*). Each manubrium projects into the cavity of the antheridium and bears long, coiled, many–celled filaments. 3 *Mech.* A handle or haft of an organ stop. [<L, a handle, <*manus* hand] — **ma·nu'bri·al** *adj.*
Ma·nu·el (mä·no͞o·el') Portuguese and Spanish form of EMMANUEL.
man·u·fac·to·ry (man'yə·fak'tər·ē) *n. pl.* **·ries** A place or establishment where anything is manufactured; a factory. [<MANUFACTURE; on analogy with FACTORY]
man·u·fac·ture (man'yə·fak'chər) *v.t.* **·tured, ·tur·ing** 1 To make or fashion by hand or machinery, especially in large quantities. 2 To work into useful form, as wool or steel. 3 To create by artifice; invent falsely; concoct. 4 To produce in a mechanical way, as art, poetry, etc. See synonyms under MAKE[1], PRODUCE. — *n.* 1 The production of goods by hand or by industrial art or processes. 2 Anything made by industrial art or processes; manufactured articles collectively. 3 The making or contriving of anything. [<MF <Med. L *manufactura* <L *manus* hand + *factura* a making <*factus*, pp. of *facere* make] — **man'u·fac'tur·er** *n.* — **man'u·fac'tur·ing** *adj.*
man·u·mit (man'yə·mit') *v.t.* **·mit·ted, ·mit·ting** To free from bondage, as a slave; emancipate; liberate. [<L *manumittere* <*manu emittere,* lit., send forth from one's hand <*manus* hand + *emittere* <*ex* away from + *mittere* send] — **man'u·mis'sion** (-mish'ən) *n.*
ma·nure (mə·no͞or', -nyo͞or') *n.* Any substance, as dung, decaying animal or vegetable matter, or certain minerals, applied to fertilize soil. — *v.t.* **·nured, ·nur·ing** To apply manure or other fertilizer to, as soil. [<AF *maynoverer* work with the hands, OF *manouvrer* <LL *manoperare.* Doublet of MANEUVER, MAINOR.] — **ma·nur'er** *n.*
ma·nus (mā'nəs) *n. pl.* **ma·nus** 1 *Anat.* The hand, or the corresponding terminal part of a limb in vertebrates, as a forefoot, claw, hoof, or the like. 2 In ancient Roman law, authoritative control: said of persons rather than of things, as of the wife or children being *in manu* (literally, in the hand) of the husband or father. [<L, hand]
Ma·nus (mä'no͞os) Largest of the Admiralty Islands; 633 square miles.
man·u·script (man'yə·skript) *n.* 1 Matter written by hand with a pen or the like; a composition in handwriting or typewriting, as distinguished from a printed one. Abbr. *MS.,* plural *MSS.* 2 A roll or book written before the invention of printing. — *adj.* Written by hand. [<Med. L *manuscriptus* <L *manus* hand + *scriptus,* pp. of *scribere* write]
Ma·nu·ti·us (mə·no͞o'shē·əs, -shəs, -nyo͞o'-),

Aldus, 1450-1515, Italian printer; inventor of italic letters.

man·ward (man'wərd) *adv.* To or toward man. Also **man'wards.**

man·wise (man'wīz') *adv.* After the manner of a man or of men.

Manx (mangks) *adj.* Pertaining to the Isle of Man, its people, or their language. — *n.* 1 The people of the Isle of Man collectively: with *the.* 2 The Gaelic language of the Manx, virtually extinct. [<ON *manskr* of the Isle of Man, ult. <Celtic]

Manx cat A variety of domestic cat having a very short rudimentary tail.

Manx·man (mangks'mən) *n.* pl. **·men** (-mən) A native of the Isle of Man.

man·y (men'ē) *adj.* **more, most** Constituting a large number; numerous. — *n.* 1 A large number: *Many* of those present left early. 2 The masses; crowd; multitude: with *the.* — *pron.* A large number of persons or things. [OE *manig*]

◆ *Many* followed by *a, an,* or *another* indicates a great number thought of singly: *Many a* man has had to find this out for himself. The phrase *a great many* is idiomatic; it resembles a collective noun, but takes only a plural verb: *A great many* are involved in the plan.

Synonyms (adj.): divers, frequent, manifold, multifarious, multiplied, multitudinous, numerous, sundry, various. *Antonyms:* few, infrequent, rare, scarce, uncommon.

man·y·plies (men'i·plīz') *n. Zool.* The third stomach of a ruminant, whose lining membrane is raised into many closely set, longitudinal folds; omasum. [<MANY + *plies,* pl. of PLY¹, *n.*]

Man·za·na·res (män'thä·nä'räs) A river in central Spain, flowing 55 miles SE past Madrid to a tributary of the Tagus.

man·za·nil·la (man'zə·nil'ə, *Sp.* män'thä·nē'lyä) *n.* 1 A pale, dry sherry with low alcoholic content. 2 A bitter olive usually stuffed with pimentos. [<Sp., orig. name of several carduaceous plants, dim. of *manzana* an apple. See MANCHINEEL.]

Man·za·nil·lo (män'sä·nē'yō) A port in SE Cuba.

man·za·ni·ta (man'zə·nē'tə, *Sp.* män'sä·nē'tä) *n.* Any one of several shrubs or small trees (genus *Arctostaphylos*) of the western United States, including the bearberry. [<Sp., dim. of *manzana* an apple. See MANCHINEEL.]

Man·zi·kert (man'zi·kûrt) A village in eastern Turkey NW of Lake Van; scene of the decisive defeat of the Byzantine Empire by the Seljuk Turks, 1071.

Man·zo·ni (män·dzō'nē), **Alessandro,** 1785-1873, Italian novelist and poet.

Ma·o·ri (mä'ō·rē, mou'rē) *n.* 1 One of an aboriginal, light-brown people of New Zealand, chiefly Polynesian, mixed somewhat with Melanesian. 2 The Polynesian language of these people. — *adj.* Of or pertaining to the Maoris or their language.

Mao Tse-tung (mou' dzu'doong'), born 1893, Chinese Communist leader; chairman of the People's Republic of China 1949-59.

map (map) *n.* 1 A representation on a plane surface of any region, as of the earth's surface; a chart. 2 Figuratively, any exact delineation. 3 *Slang* The face. — **off the map** Out of existence; out of the running. — *v.t.* **mapped, map·ping** 1 To make a map of. 2 To plan in detail: often with *out.* [<OF *mappe-(monde)* <Med. L *mappa (mundi)* map (of the world) <L, cloth, napkin]

Map (map), **Walter,** 1140?-1209?, Welsh author: also called **Ma·pes** (mā'pēz).

Ma·phar·sen (mə·fär'sən) *n.* Proprietary name of a white, odorless, powdered arsenical compound used intravenously in the treatment of syphilis.

ma·ple (mā'pəl) *n.* 1 Any of a large genus (*Acer*) of deciduous trees of the north temperate zone, with opposite leaves and a fruit of two joined samaras. 2 Its wood, which is hard, light in color, and of close grain. 3 The amber-yellow color of the finished wood. 4 The flavor of the sap of the sugar maple. — **hard maple** 1 The sugar maple. 2 The black maple (*A. nigrum*). [OE *mapel(trēow), mapul(der)* a maple (tree)]

ma·qui (mä'kē) *n.* An ornamental evergreen shrub (*Aristotelia macqui*), used in Chile in making musical instruments. A medicinal wine is made from its purple acid berries. [<Sp. <Araucan]

ma·quis (mä·kē') *n.* A zone of shrubby, mostly evergreen plants in the Mediterranean region, transitional between steppe and forest growths: known as cover for game or bandits: also called **macchia.** [<F <Ital. *macchia* a thicket, orig. a spot <L *macula* a spot]

Ma·quis (mä·kē') *n.* pl. **·quis** The military branch of the French Underground, originally guerrilla bands, named for the maquis, chaparral-like brush of the Mediterranean coasts where they were first organized; by extension, the whole French Forces of the Interior, or French Underground, which developed after the fall of France in World War II.

mar (mär) *v.t.* **marred, mar·ring** 1 To do harm to; impair or ruin. 2 To injure so as to deface; disfigure. See synonyms under HURT. — *n.* A disfiguring mark; blemish; injury. [OE *merran, mierran* hinder, injure] — **mar'rer** *n.*

Mar (mär), **Earl of,** 1675-1732, John Erskine, Scottish politician and rebel, defeated by Argyle at Dunblane, 1715.

mar·a·bou (mar'ə·bōō) *n.* 1 A stork of the genus *Leptoptilos,* especially the African marabou (*L. crumeniferus*), whose soft, white, lower tail and wing feathers are used in millinery. 2 The adjutant bird. 3 A plume from the marabou. 4 A delicate white silk that can be dyed without being freed from gum. Also **mar'a·bout** (-bōōt). [<F *marabou, marabout.* See MARABOUT.]

MARABOU
(From 3 to 4 feet high at the shoulder)

Mar·a·bout (mar'ə·bōōt) *n.* A Moslem hermit or holy man of northern Africa, revered as a saint by the Berbers. [<F <Arabic *murābit* a hermit]

ma·ra·ca (mə·rä'kə, *Pg.* mä·rä'kä) *n.* A percussion instrument made of a gourd or gourd-shaped rattle with beans or beads inside it. [<Pg. *maracá* <Tupian]

Mar·a·cai·bo (mä'rä·kī'bō) A port and the second largest city of Venezuela, on the narrows between Lake Maracaibo (5,000 square miles) in NW Venezuela and the Gulf of Maracaibo (also *Gulf of Venezuela*), an inlet of the Caribbean in Venezuela and Colombia.

Mar·a·can·da (mar'ə·kan'də) Ancient name for SAMARKAND.

Mar·a·cay (mä'rä·kī') A city in northern Venezuela.

Ma·rah (mā'rə, mâr'ə) *n.* Bitter water: used in allusion to the meaning in *Exodus* xv 23. [<Hebrew *mārāh,* fem. of *mar* bitter]

mar·a·na·tha (mar'ə·nath'ə) See ANATHEMA MARANATHA.

Ma·ran·hão (mä'rə·nyouṅ') A state of NE Brazil on the Atlantic; 129,270 square miles; capital, São Luis de Maranhão.

Ma·ra·ñon (mä'rä·nyōn') One of the main sources of the Amazon in Peru, flowing 1,000 miles north.

ma·ran·ta (mə·ran'tə) *n.* Starch from arrowroot. [after Bartolommeo *Maranta,* died 1554, Italian physician and botanist]

ma·ras·ca (mə·ras'kə) *n.* A small, wild cherry (*Prunus cerasus marasca*) of the Dalmatian mountains. [<Ital., aphetic var. of *amarasca* <*amaro* bitter <L *amarus*]

mar·a·schi·no (mar'ə·skē'nō) *n.* A cordial distilled from the fermented juice of the marasca and flavored with the cracked pits. [<Ital. <*marasca* MARASCA]

maraschino cherries Cherries preserved in maraschino liqueur.

ma·ras·mus (mə·raz'məs) *n. Pathol.* A gradual and continuous wasting away of the body; emaciation, especially in infants and the aged. [<NL <Gk. *marasmos* <*marainein* waste] — **ma·ras'mic** *adj.*

Ma·rat (mà·rà'), **Jean Paul,** 1743-93, French revolutionary leader; killed by Charlotte Corday.

Ma·ra·tha (mə·rä'tə) See MAHRATTA.

Ma·ra·thi (mə·rä'tē) *n.* The Indic language of the Mahratta of India: also spelled *Mahrati.*

mar·a·thon (mar'ə·thon) *n.* 1 A footrace of 26 miles, 385 yards: a feature of the Olympic games: so called from a messenger's legendary run from Marathon to Athens to announce the Athenian victory over the Persians, 490 B.C. 2 Any endurance contest.

Mar·a·thon (mar'ə·thon) A plain in Attica, Greece, on the Aegean; scene of decisive victory of the Athenians over the Persians, 490 B.C.

ma·raud (mə·rôd') *v.i.* To rove in search of plunder; make raids for booty. — *v.t.* To invade for plunder; raid. — *n.* A foray. [<F *marauder* <*maraud* a rogue] — **ma·raud'er** *n.*

mar·a·ve·di (mar'ə·vā'dē) *n.* A former Spanish coin of little value. [<Sp. *maravedí* <Arabic *Murābitīn,* a Moorish dynasty of Spain, 1087-1147]

mar·ble (mär'bəl) *n.* 1 A compact, granular, partly crystallized limestone, occurring in many colors, valuable for building or ornamental purposes. ◆ Collateral adjective: *marmoreal.* 2 A sculptured or inscribed piece of this stone. 3 A small ball made of this stone, or of baked clay, glass, or porcelain. 4 *pl.* A boys' game played with such balls. 5 Marbling (def. 1). — *v.t.* **bled, ·bling** To color or vein in imitation of marble, as book edges. — *adj.* 1 Made of or like marble. 2 Without feeling; cold. See synonyms under PALE². [<OF *marble, marbre* <L *marmor* <Gk. *marmaros,* lit., sparkling-stone, orig. stone; infl. in sense by *marmairein* sparkle]

marble cake A cake made of light and dark batter mixed to give a marblelike appearance.

mar·bled (mär'bəld) *adj.* Veined, clouded, or variegated like marble.

mar·ble·ize (mär'bəl·īz) *v.t.* **·ized, ·iz·ing** *U.S.* To marble.

mar·ble·wood (mär'bəl·wood') *n.* 1 A large East Indian tree (*Diospyros marmorata*) of the ebony family, yielding a variegated wood. 2 Its wood.

mar·bling (mär'bling) *n.* 1 A marking, mottling, or coloring resembling that of marble. 2 The act or method of imitating marble.

mar·bly (mär'blē) *adj.* 1 Resembling or containing marble. 2 Still or rigid like marble.

Mar·burg (mär'bûrg, *Ger.* mär'bōōrkh) A city in west central Hesse, West Germany. Also **Mar'burg an der Lahn** (än der län').

marc (märk) *n.* 1 Solid refuse remaining from grapes or other fruit after pressing. 2 A brandy distilled from this. 3 Any insoluble residue after a substance has been treated with a solvent. [<F, prob. <*marcher* tread, press (grapes). See MARCH¹.]

Marc (märk), **Franz,** 1880-1916, German painter.

mar·ca·site (mär'kə·sīt) *n.* 1 A pale, bronze-yellow, orthorhombic iron disulfide, FeS_2; white iron pyrites. It is a dimorphous form of pyrites. 2 An ornament made of crystallized white pyrites or of highly polished steel. [<Med. L *marcasita,* prob. <Arabic *marqashīta* <Aramaic]

mar·cel (mär·sel') *v.t.* **·celled, ·cel·ling** To dress (the hair) in even, continuous waves by means of special irons. [after M. *Marcel,* 19th c. French hairdresser] — **mar·cel'ler** *n.*

marcel wave In hairdressing, a style of dressing the hair in rows of even, continuous waves.

Mar·cel·la (mär·sel'ə) A feminine personal name.

Mar·cel·lus (mär·sel'əs) A masculine personal

name; diminutive of MARCUS. Also *Fr.* **Mar·cel** (màr·sel′), *Ital.* **Mar·cel·lo** (mär·chel′lō). —**Marcellus, Marcus Claudius,** 268?–208 B.C., Roman general in the Second Punic War.

mar·ces·cent (mär·ses′ənt) *adj.* **1** Withering; withered. **2** *Bot.* Withering without falling off, as the corollas of heaths, etc. [<L *marcescens, -entis,* ppr. of *marcescere,* inceptive of *marcere* be faint, languid] —**mar·ces′cence** *n.*

march[1] (märch) *n.* **1** Movement together on foot and in time, as of soldiers; a stately, dignified walk. **2** A movement, as of soldiers, from one stopping place to another. **3** The distance thus passed over. **4** Onward progress: the *march* of events. **5** A piece of music suitable for regulating the movements of persons marching. —*v.i.* **1** To move with measured, regular steps, as a soldier; proceed in step, as troops. **2** To walk in a solemn or dignified manner. **3** To proceed steadily; advance. —*v.t.* **4** To cause to march. [<MF *marche* <*marcher* walk, orig. trample, ult. <LL *marcus* a hammer <L *marculus*] —**march′er** *n.*

march[2] (märch) *n.* **1** A region or district lying along a boundary line; frontier. **2** *pl.* The border regions of England and Wales, or of England and Scotland. **3** *Scot.* The boundary or boundary marks between lands or estates. See synonyms under BOUNDARY. [<OF *marche* <Gmc. Akin to MARK[1], MARGIN.]

March The third month of the year, containing 31 days. [<AF *marche,* OF *marz* <L *Martius (mensis)* (month) of Mars <*Mars* the god Mars]

March (märkh) A river in Czechoslovakia forming part of the boundary between Czechoslovakia and Austria and flowing 180 miles NW to the Danube: Czech *Morava.*

Marche (màrsh) A region and former province of central France.

Mär·chen (mer′khən) *n. pl.* **Mär·chen** *German* A story; especially, a fairy tale or folk tale.

march·er (mär′chər) *n.* **1** An officer who defended a march. **2** One who resides in a march. [<MARCH[2]]

Mar·ches (mär′chiz), **The** A region of central Italy; 3,741 square miles; capital, Ancona. *Italian* **Le Mar·che** (lā mär′kā).

mar·che·sa (mär·kā′zä) *n. pl.* **·che·se** (-kā′zā) *Italian* A marchioness.

mar·che·se (mär·kā′zā) *n. pl.* **·che·si** (-kā′zē) *Italian* A nobleman of the rank of a marquis.

Mar·ches·van (mär·khesh′vən) See HESHWAN.

March hare A hare in the breeding season: regarded as a symbol of madness from the supposed wildness of hares at this time.

mar·chion·ess (mär′shən·is) *n.* **1** The wife or widow of a marquis. **2** A woman having in her own right the rank corresponding to that of a marquis. [<Med. L *marchionissa,* fem. of *marchio, -onis* a captain of the marches <*marca* march[2] <Gmc.]

march·land (märch′land′) *n.* Land along the boundaries of adjacent countries; borderland. [<MARCH[2] + LAND]

march·pane (märch′pān) *n.* Marzipan. [<MF *marcepain* <Ital. *marzapane* MARZIPAN]

Mar·ci·a (mär′shē·ə, -shə) A feminine personal name.

Mar·co Po·lo (mär′kō pō′lō) See POLO, MARCO.

Mar·co·ni (mär·kō′nē) *adj.* Pertaining to or designating the system of wireless telegraphy, developed by Guglielmo Marconi.

Mar·co·ni (mär·kō′nē), **Guglielmo,** 1874–1937, Italian inventor of wireless telegraphy.

Marconi rig In yachting, a type of rig consisting of a triangular sail mounted on a tall mast and having no gaff and a relatively short boom: so called from the resemblance of the mast and its stays to a radio mast. Compare GAFF RIG.

mar·cot·tage (mär′kō·täzh′) *n.* A method for the vegetative propagation of plants in which a part of the stem or branch is packed with moss until roots have formed and the treated part is ready for independent growth. [<F <*marcotter* plant layers <*marcotte* a layer (def. 3) <L *mergus* a vine layer, a diver <*mergere* dip, bury]

Mar·cus (mär′kəs) See MARK. Also *Fr.* **Marc** (màrk), *Ital.* **Mar·co** (mär′kō). [<L, of MARS]

—**Marcus Antonius** See ANTONY, MARK.

—**Marcus Aurelius** See under AURELIUS.

Mar·cus Island (mär′kəs) A North Pacific island east of the Volcano Islands, administered by the United States; one square mile; held (1899–1945) by Japan as **Mi·na·mi To·ri Shi·ma** (mē·nä·mē tō·rē shē·mä).

Mar·cy (mär′sē), **Mount** The highest peak in New York and in the Adirondack Mountains; 5,344 feet.

Mar del Pla·ta (mär thel plä′tä) A city in SE Buenos Aires province, Argentina.

Mar·di gras (mär′dē grä′) Shrove Tuesday; last day before Lent: celebrated as a carnival in certain cities. [<F., lit., fat Tuesday]

Mar·duk (mär′dŏŏk) In Babylonian mythology, the chief deity, originally a local sun-god.

mare[1] (mâr) *n.* The female of the horse and other equine animals. [OE *mēre,* fem. of *mearh* a horse]

mare[2] (mâr) *n.* A hag or goblin supposed to produce nightmare; also, nightmare. [OE. Akin to MAR.]

ma·re clau·sum (mâr′ē klô′səm) *Latin* A closed sea; a sea subject to one nation: distinguished from an *open sea,* which is free to all.

Mare Island (mâr) An island in San Pablo Bay, northern California; site of a U. S. Navy yard.

ma·rem·ma (mə·rem′ə) *n. pl.* **·me** (-ē) **1** A fertile, marshy, but unhealthful piedmont region near the sea, as in Tuscany, Italy. **2** The miasmatic exhalations of such a region. [<Ital. <L *maritimus* MARITIME]

Ma·ren·go (mə·reng′gō) A village in Piedmont, NW Italy; scene of Napoleon's defeat of the Austrians, 1800.

ma·re nos·trum (mâr′ē nos′trəm) *Latin* Our sea: the Roman name for the Mediterranean.

mare's-nest (mârz′nest′) *n.* A seemingly important discovery that proves worthless, imaginary, or false.

mare's-tail (mârz′tāl′) *n.* **1** *Meteorol.* Long, fibrous, cirrus clouds, supposed to indicate rain. **2** A perennial aquatic herb *(Hippuris vulgaris)* with entire lineal leaves in whorls and minute flowers.

Mar·ga·ret (mär′gə·rət) **1** A feminine personal name. Also **Mar·ger·y** (mär′jər·ē), *Lat.* **Mar·ga·re·ta** (mär′gə·rē′tə), *Du.* **Mar·ga·re·tha** (mär′gä·rē′tä), *Ger.* **Mar·ga·re·the** (mär′gə·rā′tə), *Pg.* **Mar·ga·ri·da** (mär′gə·rē′thə), *Lat., Ital.* **Mar·ga·ri·ta** (mär′gä·rē′tä), *Ital.* **Mar·ghe·ri·ta** (mär′ge·rē′tä), *Fr.* **Mar·gue·rite** (màr·gə·rēt′). [<Gk., pearl] **2** In Goethe's *Faust,* Margarethe, the heroine: also *Gretchen.*

—**Margaret of Anjou,** 1430–82, wife of Henry VI of England; a leader of the Lancastrians in the Wars of the Roses.

—**Margaret of Navarre,** 1492–1549, queen of Navarre, 1544–49.

—**Margaret of Valois,** 1553–1615, wife of Henry of Navarre; divorced: also known as *Margaret of France.*

mar·gar·ic (mär·gar′ik, -gär′-) *adj.* Of, pertaining to, or resembling pearl; pearly. Also **mar·ga·rit·ic** (mär′gə·rit′ik). [<F *margarique* <Gk. *margaron* a pearl]

margaric acid A white, crystalline, fatty acid, $C_{17}H_{34}O_2$, obtained from the wax of lichens or made synthetically. [<F *margarique* MARGARIC; with ref. to the pearly luster of its crystals]

mar·ga·rine (mär′jə·rin, -rēn, -gə-) *n.* A blend of refined, edible vegetable oil or meat fat, or a combination of both, churned with cultured skim milk to the consistency of butter. Also called *oleomargarine.* [<the early mistaken belief that it contained a derivative of margaric acid]

mar·ga·rite (mär′gə·rīt) *n.* **1** A hydrated silicate of calcium and aluminum. **2** Minute spherical crystals arranged as a beadlike pattern in glassy igneous rocks. **3** *Obs.* A pearl. [<OF <L *margarita* a pearl. See MARGARET.]

Mar·gate (mär′git) A municipal borough, port, and resort in NE Kent, England.

mar·gay (mär′gā) *n.* One of various South and Central American striped and spotted wild cats; especially, the long-tailed *Felis tigrina.* [<F <*margaia* <Pg. *maracajá* <Tupian *mbaracaïa*]

marge (märj) *n. Obs.* A margin. [<MF <L *margo* a margin]

mar·gent (mär′jənt) *adj.* Marginal. —*n.* **1** A marginal note. **2** *Obs.* A margin. [Var. of MARGIN]

mar·gin (mär′jin) *n.* **1** A bounding line; border; verge; brink; edge. **2** An allowance, provision, or reservation for contingencies or changes, as of time or money. **3** Range or scope; provision for increase or progress. **4** The difference between the cost of an article and its selling price. **5** A sum of money or other security deposited with a broker to protect him against loss in buying and selling for his principal; also, the difference in value between the security and the loan. **6** The difference between price and cost of production; in the adjustment of the relations of capital and labor, the minimum profit which will enable an undertaking to continue active. **7** The part of a page left blank around the body of printed or written text. —*v.t.* **1** To furnish with a margin; form a margin or border to; border. **2** To enter, place, or specify on the margin of a page, as a note or comment. **3** In commerce, to deposit a margin upon; hold by giving an addition to or a deposit upon a margin. [<L *margo, -inis.* Akin to MARK[1], MARCH[2].]

Synonyms (noun): beach, border, boundary, brim, brink, confines, edge, limit, lip, marge, shore, skirt, verge. See BANK, BOUNDARY.

mar·gi·nal (mär′jə·nəl) *adj.* **1** Pertaining to or constituting a margin. **2** Written, printed, or placed on the margin. **3** *Psychol.* Relating to the fringe of consciousness. **4** *Econ.* Operating or furnishing goods at a rate barely meeting the costs of production. —**mar′gi·nal·ly** *adv.*

mar·gi·na·li·a (mär′jə·nā′lē·ə, -nāl′yə) *n. pl.* **1** Marginal notes. **2** *Zool.* Spicules forming a collar around the osculum of a sponge. [<NL, neut. pl. of *marginalis* marginal <L *margo, -inis* a margin]

marginal land *Econ.* Land capable of furnishing an economic return, but of such low fertility or productivity as to remain unused until the lack of more desirable land forces its development.

mar·gi·nate (mär′jə·nāt) *v.t.* **·nat·ed, ·nat·ing** To provide with a margin or margins. —*adj.* *Biol.* Having a margin, especially one of a distinct character, appearance, or color: also **mar′gi·nat′ed.** [<L *marginatus,* pp. of *marginare* <*margo, -inis* a margin] —**mar′gi·na′·tion** *n.*

mar·grave (mär′grāv) *n.* **1** A hereditary German title of nobility, corresponding to *marquis.* **2** Formerly, the lord or governor of a German mark, march, or border. **3** A hereditary title of certain princes of the Holy Roman Empire. [<MDu. *markgrave* <MHG *marcgrave* <*mark* a march[2] + *graf* a count]

mar·gra·vi·ate (mär·grā′vē·it) *n.* The territory of a margrave. Also **mar·gra·vate** (mär′grə·vāt). [<Med. L *margravius,* ? <MHG *marcgrave* a margrave]

mar·gra·vine (mär′grə·vēn) *n.* The wife or widow of a margrave. [<Du. *markgravin,* fem. of *markgraaf* <MDu. *markgrave* a margrave]

mar·gue·rite (mär′gə·rēt′) *n.* A flower of the composite family, as the common garden daisy and the oxeye daisy of the fields; also, several cultivated species of chrysanthemum, especially *Chrysanthemum frutescens.* [<F, a pearl, daisy, from a proper name. See MARGARET.]

Ma·ri·a (mə·rī′ə, -rē′ə; *Du., Ger., Ital., Sp., Sw.* mä·rē′ä; *Hungarian* mä′rē·ä) A feminine personal name. See also MARY. Also **Mar·i·an** (mar′ē·ən), **Ma·rie** (mə·rē′; *Dan.* mä·rē′e, *Fr.* mà·rē′), **Mar·i·on** (mar′ē·ən). [See MARY]

—**Maria de' Medici,** 1573–1642, wife of Henry IV of France. Also *Marie de Médicis.*

—**Maria Theresa,** 1717–80, wife of Francis I, Holy Roman Emperor.

—**Marie Antoinette,** 1755–93, wife of Louis XVI of France; guillotined.

—**Marie Louise,** 1791–1847, wife of Napoleon I; empress of the French.

ma·ri·age de con·ve·nance (mà·ryàzh′ də kôṅv·näṅs′) *French* A marriage of convenience; an advantageous marriage; not a love match.

Mar·i·an (mâr′ē·ən) *n.* **1** A worshiper or devotee of the Virgin Mary. **2** An adherent of Mary I, queen of England, or a defender of Mary Queen of Scots. —*adj.* **1** Of or pertaining to the Virgin Mary, or characterized by a special devotion to her. **2** Pertaining to Queen Mary of England, or to Mary Queen of Scots.

Mar·i·an (mar′ē·ən) A feminine personal name. See MARIA.

Ma·ri·a·na·o (mä′ryä·nä·ō′) A city in western Cuba.

Ma·ri·a·nas Islands (mä′rē·ä′näs) An archipelago in the western Pacific Ocean, including Guam, Saipan, Tinian, and Rota; comprising part of the United Nations Trust Territory of the Pacific Islands (excluding Guam); total, 450 square miles; Japanese mandate, 1919–1944; formerly *Ladrones*. Also **Ma′ri·a′na** or **Ma′ri·an′ne Islands** (-rē·än′ə).

Ma·ri·anne (mâr′ē·an′) 1 A feminine personal name. 2 The French Republic, as personified on coins, etc.

Ma·ri·a Ther·e·si·o·pel (mä·rē′ä ter·ā′zē·ō′pel) A German name for SUBOTICA.

Ma·ri Autonomous S.S.R. (mä′rē) An administrative division of central European Russian S.F.S.R.; 8,900 square miles; capital, Ioshkar-Ola.

Ma·rie (mə·rē′) See MARIA.

Ma·rie Ga·lante (mȧ·rē′ gȧ·länt′) An island dependency of Guadeloupe; 60 square miles. Also **Ma·rie′–Ga·lante′**.

Ma·ri·gna·no (mä′rē·nyä′nō) A former name for MELEGNANO.

mar·i·gold (mar′ə·gōld) *n.* 1 A plant of the composite family (genus *Tagetes*), with golden-yellow flowers; especially, the French marigold (*T. patula*), and the Aztec or African marigold (*T. erecta*). 2 The calendula. 3 The marsh marigold. [<MARY, prob. with ref. to the Virgin Mary + GOLD]

marigold yellow The bright orange-yellow color of various marigolds.

mar·i·hua·na (mar′ə·wä′nə, *Sp.* mä′rē·hwä′nä) *n.* The hemp plant (*Cannabis sativa*), whose dried leaves and flower tops yield a narcotic smoked in cigarettes. Also **ma′ri·jua′na.** [<Am. Sp. *marihuana, mariguana,* ? blend of N. Am. Ind. name and Sp. *Maria Juana* Mary Jane, a personal name]

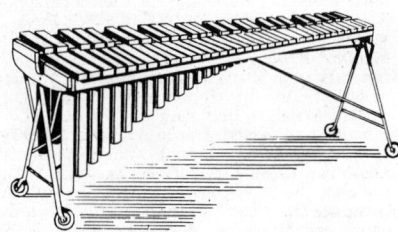

MODERN MARIMBA

ma·rim·ba (mə·rim′bə) *n.* A form of xylophone sometimes having calabash resonators. [<Bantu *marimba, malimba,* pl. of *limba,* a kind of musical instrument]

Ma·rin (mä′rin), **John,** 1870–1953, U.S. painter.

ma·ri·na (mə·rē′nə) *n. Naut.* A basin or safe anchorage for small vessels; especially, one at which provisions, supplies, etc., may be obtained. [<Ital., a seacoast <L *marinus* MARINE]

Ma·ri·na (mä·rē′nä) See ESPÍRITO SANTO.

mar·i·nade (mar′ə·nād′) *n.* 1 A brine pickle sometimes flavored with wine, spices, and herbs, in which meat or fish are placed before cooking, to improve their flavor. 2 Pickled meat or fish. [<F <Sp. *marinada* <*marinar* pickle in brine <*marino* marine <L *marinus* MARINE]

mar·i·nate (mar′ə·nāt) *v.t.* **·nat·ed, ·nat·ing** 1 To soak in oil and vinegar or brine preparatory to cooking; prepare with marinade. 2 To allow, as salad, to soak in French dressing before serving. [<MARIN(ADE) + -ATE¹]

Ma·rin·du·que (mä′rin·doo′kā) An island of the Philippines south of Luzon; 346 square miles.

ma·rine (mə·rēn′) *adj.* 1 Of or pertaining to the sea or matters connected with the sea; maritime. 2 Native to, existing in, or formed by the sea. 3 Intended for use at sea or in navigation; nautical; naval: *marine currents, marine law,* a *marine almanac.* 4 Employed on shipboard. See synonyms under NAUTICAL. — *n.* 1 A soldier trained for service at sea and on land; a member of the Marine Corps: also **Marine.** 2 Shipping, or shipping interests generally. See MERCHANT MARINE. 3 A picture or painting of the sea. [<OF *marin* <L *marinus* <*mare, maris* a sea]

Marine Corps A branch of the U.S. Navy made up of combat troops, air forces, etc., under their own officers: the oldest organized military or naval body in the United States, authorized 1775: officially, the *United States Marine Corps.*

mar·i·ner (mar′ə·nər) *n.* One who navigates or assists in navigating a ship; a sailor. [<AF <L *marinus* marine]

Mariner's Medal A medal awarded to any seaman who is wounded or suffers from exposure owing to enemy action while serving on a ship during a war.

Ma·ri·nism (mar′ə·niz′əm) *n.* An ornate and flamboyant literary style, of the type cultivated by the Italian poet **Giambattista Ma·ri·ni** (mä·rē′nē), 1569–1625.

Mar·i·ol·a·try (mâr′ē·ol′ə·trē) *n.* Worship of the Virgin Mary: an opprobrious term: also spelled *Maryolatry.* — **Mar·i·o·** (<MARY) + -LATRY] — **Mar′i·ol′a·trous** *adj.*

Mar·i·ol·o·gy (mâr′ē·ol′ə·jē) *n.* The whole body of religious belief and dogma relating to the Virgin Mary. Also spelled *Maryology.* [<*Mario-* (<MARY) + -LOGY]

Mar·i·on (mar′ē·ən) A masculine or feminine personal name. [<F, dim. of *Marie* MARY]

Mar·i·on (mar′ē·ən), **Francis,** 1732–95, American Revolutionary general: called "the Swamp Fox."

mar·i·o·nette (mar′ē·ə·net′) *n.* A puppet moved by strings. [<F *marionnette,* dim. of *Marion,* dim. of *Marie* MARY; prob. orig. a small image of the Virgin Mary]

Mar·i·po·sa (mar′ə·pō′zə) A county in central California, containing Yosemite Valley; 1,455 square miles.

Mariposa lily Any of a genus (*Calochortus*) of showy, colorful, liliaceous Mexican and Californian plants. Also **Mariposa tulip.**

mar·ish (mar′ish) *Obs. adj.* Marshy; boggy; growing in marshes, as plants. — *n.* A marsh; fen. [<OF *marais, mareis* <Med. L *mariscus* a marsh <Gmc.]

Mar·ist (mâr′ist) *adj.* 1 Of, pertaining to, or dedicated to the Virgin Mary. 2 Of or pertaining to the Marist Fathers, or to institutions founded by them. — *n.* A member of a Roman Catholic order devoted specifically to instruction and foreign missions: also **Marist Fathers.** [<(the VIRGIN) MAR(Y) + -IST]

Ma·ri·tain (mȧ·rē·tan′), **Jacques,** born 1882, French philosopher.

mar·i·tal (mar′ə·təl) *adj.* 1 Of or pertaining to marriage. 2 Of or pertaining to a husband. [<L *maritalis* <*maritus* a husband, orig. married] — **mar′i·tal·ly** *adv.*

mar·i·time (mar′ə·tīm) *adj.* 1 Situated on or near the sea. 2 Pertaining to the sea or matters connected with the sea; marine. 3 Characterized by pursuits, interests, or power at sea; nautical. See synonyms under NAUTICAL. [<F <L *maritimus* <*mare, maris* a sea]

Maritime Alps The part of the Alps between France and Italy, extending to the Mediterranean coast.

Maritime Provinces The provinces of New Brunswick, Nova Scotia, and Prince Edward Island on the Atlantic seaboard of eastern Canada.

Ma·ri·tza (mä·rē′tsä) A river in SE Europe, flowing 300 miles SE to the Aegean. Also **Ma·ri′tsa.** Turkish **Me·riç** (me·rēch′), Greek **Év·ros** (ev′ros).

Mar·itz·burg (mar′its·bûrg) A popular name for PIETERMARITZBURG.

Ma·ri·u·pol (mä·ryoo′pôl) The former name for ZHDANOV.

Mar·i·us (mâr′ē·əs), **Gaius,** 157–86 B.C., Roman general and consul.

Ma·ri·vaux (mȧ·rē·vō′), **Pierre Carlet de Chamblain de,** 1688–1763, French dramatist and novelist.

mar·jo·ram (mär′jər·əm) *n.* Any of several perennial herbs of the mint family (genus *Majorana*), with nearly entire leaves, dense oblong spikes of flowers, and colored bracts. *M. hortensis* is the **sweet marjoram,** used for seasoning in cookery. [<OF *majorane* <Med. L *majorana,* ? ult. <L *amaracus* <Gk. *amarakos*]

mark¹ (märk) *n.* 1 A visible trace, impression, or sign produced or left on any substance, as a line, scratch, dot, scar, spot, stain, or blemish; any physical peculiarity produced by drawing, indenting, stamping, or other process or agency. 2 A symbol or character, as a stamp, brand, or device, made on or attached to something to identify, distinguish, or call attention; a trademark. 3 A cross or other character made instead of a signature by one who cannot write. 4 A letter of the alphabet, number, or character by which excellence, defect, or quality is registered, as on a student's paper or record. 5 A symbol, written or printed: a *mark* of interrogation. 6 An object serving to guide, direct, or point out, as a boundary, a course, or a place in a book. 7 That which indicates the presence or existence of something; a characteristic; an evidence; a symptom. 8 That which is aimed at, or toward which effort is directed; something shot, fired, or thrown at, as a target; that which one strives to attain or achieve; a goal. 9 A proper bound or limit; standard. 10 Distinction; eminence: a person of *mark.* 11 A license to make reprisals. See LETTER OF MARQUE. 12 A person easily duped: an easy *mark.* 13 *Naut.* A strip of cloth or the like knotted or twisted into a lead line at intervals to indicate fathoms of depth. 14 An observing or noting; heed. 15 In medieval times, a piece of land held in common by a body of kindred freemen. 16 *Archaic* A boundary; limit. — **beside the mark** Pointless. — **bless the mark! save the mark!** Ejaculations of deprecation, irony, scorn, or humorous surprise: used originally of a good, then ironically of a bad, marksman. — **to make one's mark** To succeed. — **of mark** Famous; noteworthy; important. — **up to the mark** Up to standard; in good health or condition; etc. — *v.t.* 1 To make a mark or marks on. 2 To trace the boundaries of; limit. 3 To indicate by a mark or sign: X *marks* the spot. 4 To make or produce by writing, drawing, etc. 5 To be a characteristic of; typify. 6 To designate as if by marking; destine: He was *marked* for death. 7 To pay attention to; notice; remark. 8 To make known; manifest; show: to *mark* displeasure with a frown. 9 To apply a price, identification, etc., to. 10 To give marks or grades to; grade. — *v.i.* 11 To take notice; pay attention. 12 To keep score or count. 13 To make a mark or marks. See synonyms under INSCRIBE. Compare CIRCUMSCRIBE. — **to mark down** 1 To note down by writing or making marks. 2 To put a lower price on, as for a sale. — **to mark time** 1 To keep time by moving the feet but not advancing. 2 To pause in action or progress temporarily, as while awaiting developments. — **to mark up** 1 To make marks on; scar. 2 To increase the price of. [OE *mearc,* orig. boundary. Akin to MARGIN, MARCH².]

Synonyms (noun): badge, characteristic, fingerprint, footprint, impress, impression, indication, line, manifestation, print, sign, stamp, symbol, token, trace, track, vestige. See AIM, CHARACTERISTIC, LETTER, SIGN, TRACE.

mark² (märk) *n.* 1 The former monetary unit of Germany, a silver coin equivalent to 100 pfennige and valued at 23.8 cents: superseded in 1924 by the **reichs·mark** (rīkhs′märk′) and, after World War II, by the **deut·sche·mark** (doi′chə·märk′) in West Germany and the **ost·mark** (ôst′märk′) in East Germany. 2 A former silver coin of Scotland and England, worth 13s. 4d. 3 A former European unit of weight, equal to 256.27 grains. 4 A markka. [OE *marc* a unit of weight, prob. <LL *marca,* ? <Gmc. ? Akin to MARK¹.]

Mark (märk) A masculine personal name. Also Greek **Mar·kos** (mär′kos), Ger., Sw. **Mar·kus** (mär′koōs). [<L, of Mars]

— **Mark** The evangelist who wrote the second of the gospel narratives in the New Testament; also, the Gospel by him. Also **Saint Mark.**

— **Mark Antony** See under ANTONY.

marked (märkt) *adj.* Brought prominently to

marked man / **marram**

notice; distinguished; prominent. — **mark·ed·ly** (mär'kid-lē) *adv.* — **mark'ed·ness** *n.*
marked man One who is singled out by others, as for suspicion, vengeance, etc.
mark·er (mär'kər) *n.* 1 That which marks; specifically, a bookmark, a milestone, a gravestone, etc. 2 A scorekeeper. 3 A device for tracing the lines of a tennis court or other playing ground. 4 *Slang* A written promise to pay a specified amount, used as currency in gambling, speculation, etc.
mar·ket (mär'kit) *n.* 1 A place where merchandise is exposed for sale; specifically, an open space or a large building in a town or city, generally with stalls or designated positions occupied by different dealers, especially such a place for the sale of provisions: also **market place.** 2 A private store for the sale of provisions: a meat *market.* 3 The state of trade as determined by prices, supply, and demand; traffic: a brisk *market.* 4 A locality or country where anything can be bought or sold; a place where any commodity is in demand: the South American *markets.* 5 A gathering of people for selling and buying, especially of a particular commodity: the wheat *market.* 6 The value of a thing as determined by the price it will bring; value in general; worth. — *v.t.* 1 To take or send to market for sale; sell. — *v.i.* 2 To buy or sell in a market; sell or buy. 3 To buy food. [OE < AF < L *mercatus,* orig. pp. of *mercari* trade < *merx, mercis* merchandise. Doublet of MART.]
mar·ket·a·ble (mär'kit·ə·bəl) *adj.* 1 Suitable for sale; in demand. 2 Current in markets. 3 Of or pertaining to trading. — **mar'ket·a·bil'i·ty** *n.*
mar·ket·er (mär'kit·ər) *n.* One who buys or sells in a market.
market order An order to a broker to buy or sell at the current market price.
market price See under PRICE.
mar·ket·ripe (mär'kit·rīp') *adj.* Not quite ripe: said of slightly unripe fruits picked to reach the market in salable condition.
market value The price which may be expected for a given commodity, security, or service under the conditions of a given market: distinguished from *normal value,* which is the average of values over a long period.
Mark·ham (mär'kəm), **(Charles) Edwin,** 1852–1940, U.S. poet.
Markham, Mount An Antarctic peak on the west edge of Ross Shelf Ice; 15,100 feet.
mark·ing (mär'king) *n.* 1 A mark or an arrangement of marks; characteristic coloring. 2 The act of making a mark.
mark·ka (märk'kä) *n.* The monetary unit of Finland. [Finnish <Sw. *mark* <ON *mörk.* Akin to MARK².]
marks·man (märks'mən) *n. pl.* **·men** (-mən) 1 One skilled in hitting the mark, as with a rifle or other weapon. 2 In the U.S. Army, the lowest of three grades for skill in the use of small arms. 3 The soldier having this grade. Compare SHARPSHOOTER, EXPERT. — **marks'man·ship** *n.* **marks'wom'an** *n. fem.*
Mark Twain (märk twān) Pseudonym of Samuel Langhorne Clemens, 1835–1910, U.S. humorist and novelist.
mark·up (märk'up') *n.* 1 A raising of price. 2 The amount of price increase. 3 The sum added to cost, in computing selling price, to cover overhead and profit. 4 *Printing* The placement on copy of editorial directions for the printer or engraver.
marl¹ (märl) *v.t. Naut.* To wrap with marline, tying each turn with a hitch. [<Du. *marlen,* appar. freq. <MDu. *merren* tie]
marl² (märl) *n.* 1 An earthy deposit containing lime, clay, and sand, used as fertilizer. 2 A soft, earthy, crumbling deposit of varying composition. — *v.t.* To fertilize or spread with marl. [<OF *marle* <LL *margila,* dim. of L *marga,* ? <Celtic]
Marl·bor·ough (märl'bur'·ə, -bər·ə) A municipal borough in eastern Wiltshire, England.
Marl·bor·ough (märl'bur'ə, -bər·ə) **Duke of,** 1650–1722, John Churchill; English general who defeated the French at Blenheim, Aug. 13, 1704.
marled (märld) *adj. Scot.* Variegated; marbled; mottled.
mar·lin¹ (mär'lin) *n.* Any of various deep-sea game fishes of the genus *Makaira;* especially, *M. ampla* of the Atlantic and the black or striped marlins of the Pacific. [<MARLINE(SPIKE); so called because of its shape]
mar·lin² (mär'lin) *n. U.S. Dial.* A curlew. [Alter. of obs. *marling,* var. of MERLIN]
mar·line (mär'lin) *n. Naut.* A small rope of two strands loosely twisted together: used for winding ropes, cables, etc. [<Prob. fusion of Du. *marlijn* (<*marren* tie + *lijn* a line) and E *marling* a binding <Du. <*marlen,* freq. of *marren*]
mar·line·spike (mär'lin·spīk') *n. Naut.* A sharp-pointed iron pin used in splicing ropes. Also **mar'lin·spike', mar'ling·spike'** (-ling-).

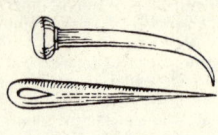
MARLINESPIKES

marl·ite (märl'īt) *n.* A variety of marl that differs from common marl by remaining solid on exposure to air. [<MARL² + -ITE¹] — **marl·it·ic** (märl·it'ik) *adj.*
Mar·lowe (mär'lō), **Christopher,** 1564–93, English dramatist. — **Julia** Stage name of *Sarah Frost Sothern,* 1866–1950, U.S. actress born in England; wife of E. H. Sothern.
marl·y (mär'lē) *adj.* Resembling or of the nature of marl; abounding in marl. Also **mar·la·ceous** (mär·lā'shəs).
mar·ma·lade (mär'mə·lād) *n.* A preserve made by boiling the pulp and part of the rind or skin of fruits, usually citrus fruits, with sugar to the consistency of jam. [<MF *marmelade* <Pg. *marmelada* <*marmelo* a quince <L *melimelum* <Gk. *melimēlon* <*meli* honey + *mēlon* an apple]
marmalade tree A tall, tropical American, sapotaceous evergreen tree (*Achras zapota*). It bears plumlike fruits used chiefly for preserving. Also called *sapodilla.*
Mar·ma·ra, Sea of A sea between Europe and Asia, leading by the Bosporus to the Black Sea, and by the Dardanelles to the Aegean: ancient *Propontis.* Also **Mar·mo·ra** (mär'mə·rə, mär·môr'ə, -mō'rə).
Mar·mi·on (mär'mē·ən), **Lord** The hero of *Marmion, A Tale of Flodden Field,* a romantic poem by Sir Walter Scott.
mar·mite (mär·mēt') *n.* A small, lidded ceramic or enameled cooking pot. [<F, a pot, kettle]
Mar·mo·la·da (mär'mō·lä'dä) The highest mountain in the Dolomites, northern Italy; 10,964 feet.
mar·mo·re·al (mär·môr'ē·əl, -mō'rē-) *adj.* Pertaining to, made of, or resembling marble. Also **mar·mo're·an** (-môr'-, -mō'rē-). [<L *marmoreus* <*marmor* MARBLE]
mar·mo·set (mär'mə·zet) *n.*
1 A small Central and South American monkey (family *Callithricidae*) with soft, woolly hair and a long hairy tail. 2 A related species, Goeldi's marmoset (*Callimico goeldii*). 3 The tamarin. 4 The ouistiti. [<OF *marmouset* a grotesque figure, prob. ult. <Gk. *mormō, -ous* a she-monster, bogey; with ref. to its appearance]

MARMOSET
(About the size of a large squirrel)

mar·mot (mär'mət) *n.* 1 A stout, short-tailed, burrowing rodent of mountain regions (genus *Marmota*). The species found in North America is known as a *woodchuck* or *ground hog.* 2 A related rodent, as the prairie dog. [<F *marmotte,* fusion of OF *marmotte* a monkey and Romansch *murmont* a marmot <L *mus, muris* a mouse + *mons, montis* a mountain]
Marne (märn) A river in NE France, flowing 325 miles NW to the Seine; scene of two decisive battles of World War I, 1914, 1918.
Ma·roc (mà·rôk') The French name for Morocco.
Ma·ro·ni (mə·rō'nē) A river in the Guianas, flowing 450 miles north to the Atlantic. Dutch **Ma·ro·wij·ne** (mä'rō·wī'nə).
ma·roon¹ (mə·roon') *v.t.* 1 To put ashore and abandon on a desolate island or coast. 2 To abandon; leave helpless. [<*n.* (def. 1)] — *n.* 1 One of a class of Negroes, chiefly fugitive slaves or their descendants, living wild in the mountains of some West Indies islands and of Guiana. 2 *Rare* A person left alone on an island; a marooned person. [<F *marron* a maroon (def. 1) <Sp. *cimarron* wild]
ma·roon² (mə·roon') *n.* 1 A dull, dark-red color. 2 A coal-tar dyestuff. — *adj.* Having a dull, dark-red color. [<F *marron* a chestnut, chestnut color <Ital. *marrone*]
Ma·ros (mô·rōsh') The Hungarian name for the MURES.
mar·plot (mär'plot') *n.* One who, by meddling, mars or frustrates a design or plan.
marque (märk) *n.* A license of reprisal upon an enemy, as at sea in wartime: chiefly in the phrase *letter of marque.* [<F, mark, imprint, stamp <Provençal *marca* seizure <*marcar* seize as a pledge <*marc* a pledge <Gmc. Akin to MARK¹.]
mar·quee (mär·kē') *n.* 1 A large field tent, especially one for an officer, or one used at lawn entertainments. 2 An awning or rooflike structure over the entrance to a hotel, theater, or other building. [<MARQUISE a canopy, mistaken as pl. <F]
Mar·que·san (mär·kā'sən) *adj.* Of the Marquesas Islands, the inhabitants, or their language. — *n.* 1 A native of the Marquesas Islands. 2 Their Polynesian language.
Mar·que·sas Islands (mär·kā'səs) A group of islands in French Oceania, in the South Pacific north of the Tuamotu group; 480 square miles. Also **Mar·que'zas Islands.** French **Îles Mar·quises** (ēl mär·kēz').
Marquesas Keys A group of Florida islets west of Key West.
mar·quess (mär'kwis) See MARQUIS.
mar·que·try (mär'kə·trē) *n.* Inlaid work of wood often interspersed with stones, ivory, etc., especially as used in the decoration of furniture. Also **mar'que·te·rie.** [<MF *marqueterie* <*marqueter* variegate, inlay <*marque* a mark¹, ult. <Gmc.]
Mar·quette (mär·ket'), **Jacques,** 1636–75, French Jesuit priest and missionary who explored Canada: called "Père Marquette."
mar·quis (mär'kwis, *Fr.* mär·kē') *n.* The title of a nobleman next in rank below a duke. Also, *Brit.,* **marquess.** [<OF *marchis, marquis* <Med. L *marchensis* a commander of the marches <*marca* a march² <Gmc.]
Mar·quis (mär'kwis) *n.* An important variety of wheat, first developed in Canada, widely grown in the United States.
Mar·quis (mär'kwis), **Don(ald) Robert Perry,** 1878–1937, U.S. journalist and humorist.
mar·quis·ate (mär'kwiz·it) *n.* The rank of a marquis.
mar·quise (mär·kēz') *n.* 1 The wife or widow of a French marquis. 2 An ornamental hood over a door; a marquee. 3 In gem-cutting, a pointed oval form, especially for diamonds. 4 A ring set with an oval cluster of gemstones. [<F, fem. of *marquis* a marquis]

MARQUISE

mar·qui·sette (mär'ki·zet', -kwi-) *n.* A lightweight, open-mesh fabric of cotton, silk, rayon, or nylon, or a combination of these, used for curtains and women's and children's garments. [<F, dim. of *marquise* a marquise]
Marquis of Queensberry Rules A widely observed boxing code devised in 1869 by John Sholto Douglas (1844–1900), eighth Marquis of Queensberry.
Mar·ra·kesh (mä·rä'kesh) 1 A city in SW Morocco: formerly *Morocco.* 2 A region of SW Morocco; 45,725 square miles. Also **Mar·ra'kech.**
mar·ram (mar'əm) *n.* 1 Beachgrass. 2 A dune overgrown and bound together by this grass. [<ON *maralmr* <*marr* a sea + *halmr* haulm]

marriage / 781 / **Martial**

mar·riage (mar′ij) n. 1 The act of marrying, or the state of being married; specifically, a compact entered into by a man and a woman, to live together as husband and wife; wedlock. ◆ Collateral adjectives: *hymeneal, marital.* 2 A wedding; a nuptial celebration: with *of* or *between.* 3 Figuratively, any close union. 4 In pinochle, the king and queen of any suit. [<OF *mariage,* ult. <L *maritus* a husband] Synonyms: matrimony, nuptials, union, wedding, wedlock. *Matrimony* denotes the state of those united in the *marriage* relation; *marriage* denotes primarily the act of so uniting, but is much used also for the state. *Wedlock,* a word of specific legal use, is the Saxon term for the state or relation denoted by *matrimony. Wedding* denotes the ceremony, with any attendant festivities, by which two persons are united as husband and wife, *nuptials* being the more formal and stately term to express the same idea. Antonyms: bachelorhood, celibacy, divorce, maidenhood, virginity, widowhood.
mar·riage·a·ble (mar′ij·ə·bəl) adj. Fitted by age, physical condition, etc., for marriage; nubile. — **mar′riage·a·bil′i·ty, mar′riage·a·ble·ness** n.
marriage lines Brit. A marriage certificate.
mar·ried (mar′ēd) adj. 1 Pertaining to marriage; connubial; conjugal: the *married* state. 2 Having a spouse; united by or as by matrimony; wedded: a *married* man.
mar·ron (mar′ən, Fr. ma·rôn′) n. A chestnut, especially when used as food and in confectionery. [<F. See MAROON[2].]
marron gla·cé (glà·sā′) French A candied chestnut.
mar·row[1] (mar′ō) n. 1 A soft vascular tissue found in the central cavities of bones. 2 The interior substance of anything; essence; pith. [OE *mearg, mearh*] — **mar′row·y** adj.
mar·row[2] (mar′ō) v.t. & v.i. Scot. & Brit. Dial. To associate; marry. [? <Scand. Cf. ON *margr* friendly, lit., many.]
mar·row·bone (mar′ō·bōn′) n. 1 A bone containing marrow. 2 pl. One's knees: used humorously. 3 pl. Crossbones, the piratical emblem.
mar·row·fat (mar′ō·fat′) n. A large, rich pea. Also **marrow pea.**
marrow squash A variety of squash having an ovoid, fine-grained body; vegetable marrow.
Mar·rue·cos (mär·rwä′kōs) The Spanish name for MOROCCO.
mar·ry[1] (mar′ē) v. ·ried, ·ry·ing v.t. 1 To join as husband and wife in marriage. 2 To take in marriage. 3 To give in marriage. 4 To unite closely. 5 *Naut.* To fasten end to end, as ropes, without increasing the diameter. — v.i. 6 To take a husband or wife. [<OF *marier* <L *maritare* <*maritus* a husband, married] — **mar′ri·er** n.
mar·ry[2] (mar′ē) interj. Archaic An exclamation of surprise or asseveration: a euphemistic variant of *Mary!* or *by Mary!* [Alter. of MARY; with ref. to the Virgin Mary]
Mar·ry·at (mar′ē·it, -at), **Frederick,** 1792–1848, English naval captain and novelist.
Mars (märz) n. *Obs.* In medieval alchemy, iron. [after *Mars,* the planet]
Mars (märz) n. 1 In Roman mythology, the god of war: identified with the Greek *Ares:* formerly called *Mavors.* 2 *Astron.* The fourth planet in order of distance from the sun, from which its mean distance is 141,500,000 miles, its least distance from the earth being about 35,000,000 miles. Mars has two satellites, a diameter of about 4,230 miles, a diurnal rotation of 24 hours, 37 minutes, 22.67 seconds, and a year of 686.9 days.
Mar·sa·la (mär·sä′lä) n. A light, sweet white wine originally made in Marsala.
Mar·sa·la (mär·sä′lä) A port of western Sicily.
Marse (märs) n. *U.S. Dial.* Master: formerly used by Negro slaves. Also **Mars.** [Alter. of MASTER]
Mar·seil·lais (mär′sə·lāz′, Fr. mar·sā·ye′) adj. Of or pertaining to Marseille, or to its inhabitants. — n. A native or inhabitant of Marseille. [<F <*Marseille* Marseille] — **Mar·seil·laise** (mär′sə·lāz′, Fr. mar·sā·yez′) & adj. fem.
Marseillaise The national anthem of the French Republic, written in 1792 by Rouget de Lisle. [<F, fem. of *Marseillais* MARSEILLAIS]
mar·seille (mär·sāl′) n. A thick cotton fabric having a raised weave, similar to piqué. Also **mar·seilles**(mär·sālz′). [from *Marseille,* France]
Mar·seille (mär·sā′y) A port on the Gulf of the Lion; the second largest city in France. Also **Mar·seilles** (mär·sā′, -sālz′).
marsh (märsh) n. A tract of low, wet land; swamp. ◆ Collateral adjective: *paludal.* [OE *mersc, merisc.* Akin to MORASS.] — **marsh′y** adj. — **marsh′i·ness** n.
Marsh (märsh), **Othniel Charles,** 1831–1899, U.S. paleontologist.
mar·shal (mär′shəl) n. 1 An officer authorized to regulate ceremonies, preserve order, etc. 2 An official of the United States courts; also, the head of the police force or fire department in some cities. 3 In some European countries, a military officer of high rank; a field marshal. 4 In medieval times, a groom or master of the horse; later, as an officer of the king, a judge in courts of chivalry, etc. — v.t. ·shaled or ·shalled, ·shal·ing or ·shal·ling 1 To arrange or dispose in order, as facts. 2 To array or draw up, as troops for battle. 3 To lead; usher. ◆ Homophone: *martial.* [<OF *mareschal* <Med. L *mariscalcus* <OHG *marahscalh,* lit., a horse-servant <*marah* a horse + *scalh* a servant] — **mar′shal·cy, mar′shal·ship** n. — **mar′shal·er, mar′shal·ler** n.
Mar·shall (mär′shəl), **Alfred,** 1842–1924, English economist. — **George Catlett,** 1880–1959, chief of staff of U.S. Army in World War II; secretary of state 1947–49. — **John,** 1755–1835, U.S. statesman; jurist; secretary of state 1800–01; chief justice of United States 1801–1835.
Mar·shall Islands (mär′shəl) An island group in the NW Pacific, including Bikini, Eniwetok, Kwajalein, Jaluit, and other islands, comprising part of the United Nations Trust Territory of the Pacific Islands; total, 65 square miles; capital, Jaluit; Japanese mandate, 1920–1944.
Marshall Plan A post–World War II recovery program of U.S. financial aid to certain European countries, initiated June, 1947.
Mar·shal·sea (mär′shəl·sē) *Brit.* 1 The court (abolished 1849) held by the marshal of the royal household, originally to administer justice to the sovereign's domestic servants. 2 A debtors' prison in Southwark, abolished in 1842. [Var. of *marshalcy* <AF *mareschalcie* <Med. L *marisalcia* <*mariscalcus* a marshal]
marsh bluebill A ring-necked duck.
marsh elder 1 Any of a genus (*Iva*) of American salt-marsh plants, especially the high-water shrub (*I. frutescens*) of the SE United States: also called *sumpweed.* 2 The guelder-rose.
marsh frog A pickerel frog.
marsh gas Methane.
marsh hare In Louisiana, the muskrat.
marsh harrier An Old World hawk (*Circus aeruginosus*) with a yellowish head and predominantly dark-brown plumage: it nests on swampy ground and feeds largely on frogs, snakes, and small waterfowl.
marsh hawk An American hawk (*Circus hudsonius*) inhabiting marshes where it preys on snakes, frogs, etc.
marsh hen 1 The northern clapper rail (*Rallus longirostris crepitans*), which inhabits saltwater marshes along the American Atlantic coast. 2 The American coot (*Fulica americana*).
marsh·mal·low (marsh′mal′ō, -mel′ō) n. 1 A plant of the mallow family (*Althaea officinalis*) growing in marshy places. 2 A sweetmeat formerly made from the root of this plant; now a confection made of starch, corn sirup, and gelatin, coated with powdered sugar.
marsh marigold A showy swamp plant (*Caltha palustris*) of the crowfoot family, having bright-yellow flowers, found in swamps and wet meadows: also called *cowslip.*
marsh poisoning Impaludism.
marsh rosemary 1 A seaside perennial herb (*Limonium carolinianum*), with a strongly astringent root: used medicinally. 2 The moorwort.
Mar·si·an (mär′sē·ən) adj. Of or pertaining to the Mar·si (mär′sī), a people of ancient Italy. — n. 1 One of the Marsi. 2 Their language, belonging to the Sabellian branch of the Italic languages.
Mar·si·van (mär′sē·vän′) See MERZIFON.
Mars-la-Tour (märs·là·tōōr′) A village in NE France; scene of a battle of the Franco-Prussian War, 1870.
Mar·ston (mär′stən), **John,** 1575?–1634, English dramatist and satirist.
Mar·ston Moor (mär′stən) An area in central Yorkshire, England, where Cromwell defeated the Royalists in a battle of the Civil War, July 2, 1644.
mar·su·pi·al (mär·sōō′pē·əl) n. Any member of an order (*Marsupialia*) of mammals, as the kangaroos, opossums, wombats, etc., whose females typically lack a placenta, carrying their young in a marsupium. — adj. 1 Having a marsupium. 2 Of or pertaining to the *Marsupialia,* or of the nature of a marsupium or pouch. [<NL *marsupialis* <L *marsupium.* See MARSUPIUM.] — **mar·su′pi·a′li·an** (-ā′lē·ən), **mar·su′pi·an** adj. & n.
mar·su·pi·um (mär·sōō′pē·əm) n. pl. ·pi·a (-pē·ə) A pouchlike invagination; specifically, a brood pouch or external receptacle for carrying young, as on the abdomen of marsupials. [<L, a pouch, purse <Gk. *marsypion,* dim. of *marsipos*]
Mars yellow A pigment consisting chiefly of hydrous oxide of iron and aluminum; iron yellow. Other pigments in the same group are **Mars orange, Mars red, Mars violet.**
Mar·sy·as (mär′sē·əs) In Greek mythology, a satyr and flute-player, who was defeated in a musical contest and flayed alive by Apollo.
mart[1] (märt) n. 1 A market. 2 *Archaic* A fair; also, traffic; trading. [<MDu. *markt, mart* <L *mercatus.* Doublet of MARKET.]
mart[2] (märt) n. *Scot.* An animal slaughtered at Martinmas.
Mar·ta (mär′tä) Italian and Spanish form of MARTHA.
Mar·ta·ban (mär′tə·bän′) A triangular inlet of the Andaman Sea in SE Burma.
mar·ta·gon (mär′tə·gon) See under LILY.
mar·tel·lo tower (mär·tel′ō) A circular tower of masonry, formerly erected on coasts for defense against invasion. Also **mar·tel′lo.** [from Cape *Mortella,* in Corsica, where such a tower was erected; infl. in form by Ital. *martello* a hammer]
mar·ten (mär′tən) n. 1 A weasel-like, fur-bearing carnivorous mammal (genus *Martes*) having arboreal habits, as the American pine marten (*M. americana*), and the large, sturdy fisher marten (*M. pennanti*). 2 The fur of the marten, often dyed and sold as an inferior sable. [<OF *martrine* of the marten <*martre* a marten <WGmc.]

MARTEN
(Species vary from 1 1/2 to 2 1/2 feet in body size)

mar·ten·site (mär′tən·zīt) n. *Metall.* A constituent of rapidly quenched steel, consisting of a solid solution of iron and up to two percent of carbon. [after Prof. A. *Martens,* German metallurgist + -ITE[1]] — **mar′ten·sit′ic** (-tən·zit′ik) adj.
Mar·tha (mär′thə, Dan., Du., Ger., Sw. mär′tä) A feminine personal name. Also Fr. **Marthe** (märt). [<Aramaic, a lady] — **Martha** Friend of Jesus, and sister of Lazarus and Mary, who served Jesus at Bethany (Luke x 40); hence, figuratively, a houseworker.
Martha's Vineyard An island off SE Massachusetts; 20 miles long.
mar·tial (mär′shəl) adj. 1 Pertaining to, connected with, or suggestive of war or military operations: opposed to *civil:* court *martial.* 2 Warlike; brave; characteristic of a warlike person. See synonyms under WARLIKE. ◆ Homophone: *marshal.* [<OF <L *martialis* pertaining to Mars <*Mars, Martis* Mars] — **mar′tial·ism** n. — **mar′tial·ist** n. — **mar′tial·ly** adv.
Mar·tial(mär′shəl) adj. 1 Pertaining to the god

Mars. 2 In astrology, under the evil influence of the planet Mars. 3 *Astron.* Martian.

Mar·tial (mär′shəl), 40?–100?, Latin epigrammatist: full name *Marcus Valerius Martialis.*

martial law Military jurisdiction exercised by a government temporarily governing the civil population of a locality through its military forces, without the authority of written law, as necessity may require.

Mar·tian (mär′shən) *adj.* 1 *Astron.* Pertaining to the planet Mars. 2 Martial (defs. 1 and 2). —*n.* One of the supposed inhabitants of Mars. [<L *Martius* <*Mars, Martis* Mars]

mar·tin (mär′tən) *n.* 1 Any of certain birds of the swallow family, having a tail less forked than the common swallows; specifically, the **purple martin** *(Progne subis),* the **sand martin** or bank swallow *(Riparia riparia),* etc. 2 Some bird likened to a true martin, as a kingbird or chimney swift. [<F, prob. <*Martin* MARTIN]

Mar·tin (mär′tən; *Fr.* mȧr·taṅ′, *Ger.* mär′tēn) A masculine personal name. Also **Mar·tyn** (mär′tən), *Du.* **Mar·tijn** (mär′tīn), *Pg.* **Mar·ti·nho** (mär·tē′nyo͞o), *Ital., Sp.* **Mar·ti·no** (mär·tē′nō), *Lat.* **Mar·ti·nus** (mär·tī′nəs). [<L *Martinus* warlike]

—**Martin, Saint,** 315?–399?, bishop of Tours. See MARTINMAS.

Mar·tin (mär′tən), **Homer Dodge,** 1836–97, U. S. painter.

Mar·tin du Gard (mȧr·taṅ′ dü gȧr′), **Roger,** born 1881, French novelist.

Mar·ti·neau (mär′tə·nō), **Harriet,** 1802–76, English writer. —**James,** 1805–1900, English theologian and writer; brother of the preceding.

mar·ti·net (mär′tə·net′) *n.* A strict disciplinarian, especially military or naval: often used in a derogatory sense. [after General *Martinet,* 17th c. French drillmaster]

Mar·ti·net (mȧr·tē·ne′), **André,** born 1908, French linguist.

mar·tin·gale (mär′tən·gāl) *n.* 1 A forked strap for holding down a horse's head by connecting the head gear with the bellyband. 2 *Naut.* A vertical spar under the bowsprit used in guying the stays. Also **mar′tin·gal** (-gal). [<F <Provençal *martengalo,* appar. fem. of *martengo* an inhabitant of Martigues, a town in Provence; ? with ref. to tight hose or breeches worn there]

mar·ti·ni (mär·tē′nē) *n.* A cocktail made of gin and dry vermouth, usually served with a green olive or a twist of lemon peel. [after *Martini* and Rossi, a company making vermouth]

Mar·ti·ni (mär·tē′nē), **Simone di,** 1283–1344, Sienese painter.

Mar·ti·nique (mär′ti·nēk′) An island in the Lesser Antilles, comprising an overseas department of France; 427 square miles; capital, Fort-de-France.

Mar·tin·mas (mär′tən·məs) *n.* The feast of St. Martin, November 11. [<Saint *Martin* + -MAS]

mart·let (märt′lit) *n.* 1 A martin. 2 *Her.* A martin or swallow without feet: used as a bearing, crest, or mark of cadency. [<F *martelet,* alter. of *martinet,* dim. of *martin* a martin; ? infl. in form by *roitelet* a wren]

mar·ton·ite (mär′tən·īt) *n.* Bromacetone.

mar·tyr (mär′tər) *n.* 1 One who submits to death rather than forswear his religion; specifically, one of the early Christians who suffered death for their religious principles. 2 One who dies or suffers for principles, or sacrifices all for a cause: a *martyr* to science. 3 One who suffers much or long, as from ill health or misfortune: a *martyr* to rheumatism. —*v.t.* 1 To put to death as a martyr. 2 To torture; persecute or torment. [OE <LL < Gk. *martyr,* Aeolic form of *martys, martyros* a witness]

mar·tyr·dom (mär′tər·dəm) *n.* 1 The condition or fate of a martyr. 2 Protracted or extreme suffering. [OE *martyrdōm*]

mar·tyr·ize (mär′tər·īz) *v.* **·ized, ·iz·ing** *v.t.* To make a martyr of. —*v.i.* To become a martyr. —**mar′tyr·i·za′tion** *n.* —**mar′tyr·iz′er** *n.*

mar·tyr·ol·o·gy (mär′tə·rol′ə·jē) *n. pl.* **·gies** A historical record of martyrs. [<Med. L *martyrologium* <LGk. *martyrologion* <*martyr* martyr + Gk. *logos* word, account] —**mar·tyr·o·log·ic** (mär′tər·ə·loj′ik) or **·i·cal** *adj.* —**mar′tyr·ol′o·gist** *n.*

mar·tyr·y (mär′tər·ē) *n.* 1 Martyrdom. 2 The spot where a martyr suffered or is buried, or a chapel erected there.

mar·vel (mär′vəl) *v.* **·veled** or **·velled, ·vel·ing** or **·vel·ling** *v.i.* To be filled with wonder, surprise, etc. —*v.t.* To wonder at or about: with a clause as object. —*n.* 1 That which excites wonder; a prodigy. 2 The emotion of wonder. 3 A miracle. See synonyms under PRODIGY. [<OF *merveillier* <*merveille* a wonder <L *mirabilia,* neut. pl. of *mirabilis* wonderful <*mirari* wonder at]

Mar·vell (mär′vəl), **Andrew,** 1621–78, English poet and satirist.

mar·vel-of-Pe·ru (mär′vəl·əv·pə·ro͞o′) *n.* The four-o'clock.

mar·vel·ous (mär′vəl·əs) *adj.* 1 Exciting astonishment; singular; wonderful. 2 Pertaining to the supernatural; miraculous. Also **mar′vel·lous.** See synonyms under EXTRAORDINARY. —**mar′vel·ous·ly, mar′vel·lous·ly** *adv.* —**mar′vel·ous·ness, mar′vel·lous·ness** *n.*

Mar·war (mär′wär) See JODHPUR.

Marx (märks), **Karl,** 1818–83, German socialist, revolutionary leader and writer on economics; author of *Das Kapital.* —**Marx′i·an** *adj.* & *n.*

Marx·ism (märk′siz·əm) *n.* The body of doctrine formulated by Karl Marx and Friedrich Engels in systematic form, including economic determinism, class conflicts leading inevitably to revolution in the transition from feudalism to capitalism and thence to communism under the dictatorship of the proletariat, and the predicted ultimate triumph of world communism as a result of destructive rivalries among the capitalist-imperialist powers.

Marx·ism-Len·in·ism (märk′siz·əm·len′in·iz′·əm) *n.* The philosophy of history and politics based upon the works of Marx and Lenin as primary sources: stressing Lenin's recension of Marxism in order to make it applicable to backward agrarian countries, his techniques for engineering consent through integrating Communist party organization on the soviet system with the operation of the state (democratic centralism), his fusion of theory, practice, and timing in interlacing domestic and international policy, and finally, his pragmatic use of reformist and piecemeal measures where he believed these necessary to assure the success of revolution. —**Marx′ist-Len′in·ist** *n.*

Marx·ist (märk′sist) *adj.* 1 Of or pertaining to Karl Marx or his theories; Marxian. 2 Like or developed from Marxian theory: *Marxist* socialism. —*n.* A Marxian; specifically, an adherent of Marxist socialism.

Mar·y (mâr′ē) A feminine personal name. Also *Polish* **Mar·ya** (mär′yä). See also MARIA. [<Hebrew, bitter]

—**Mary** The mother of Jesus: also *Virgin Mary.*

—**Mary** The sister of Lazarus and Martha.

—**Mary I,** 1516–58, queen of England, 1553–1558: known as *Mary Tudor* or *Bloody Mary.*

—**Mary II,** 1662–94, queen of England 1689–1694, ruling jointly with her husband, William III.

—**Mary Magdalene** A disciple from Magdala out of whom Jesus cast seven devils. *Luke* viii 2. She is often identified with the penitent sinner whom Jesus forgave. *Luke* vii 36–50.

—**Mary Queen of Scots,** 1542–87, queen of Scotland 1542–67; beheaded: also *Mary Stuart.*

Mar·y·land (mâr′i·lənd, mer′i-) An eastern State of the United States, on Chesapeake Bay; 12,327 square miles; capital, Annapolis; entered the Union April 28, 1788, one of the original thirteen States; nickname, *Cockade State* or *Old Line State*: abbr. *Md.*

Maryland yellowthroat See under YELLOWTHROAT.

Mar·y·le·bone (mâr′i·lə·bōn′, mâr′lə·bən) A metropolitan borough of NW London, England: also *St. Marylebone.*

Mar·y·ol·a·try (mâr′ē·ol′ə·trē), **Mar·y·ol·o·gy** (-ol′ə·jē) See MARIOLATRY, etc.

mar·zi·pan (mär′zə·pan) *n.* A confection of grated almonds, sugar, and white of eggs, usually made into a paste and molded into various shapes: formerly known as *marchpane.* [<G <Ital. *marzapane,* orig. a small box, a dry measure, a weight <Med. L *matapanus* a Venetian coin stamped with an enthroned Christ <Arabic *mauthabān* a seated king, a coin < *wathaba* sit]

-mas *combining form* Mass; a (specified) festival or its celebration: *Christmas.* [<MASS²]

Ma·sac·cio (mä·sät′chō), 1402–28, Florentine painter: real name *Tommaso Guidi.*

Ma·san (mä·sän) A port in southern Korea. Formerly **Ma·sam·po** (mä·säm·pō).

Ma·sa·ryk (mä′sä·rik), **Jan,** 1886–1948, Czechoslovak diplomat and politician. —**Tomás,** 1850–1937, philosopher; first president of Czechoslovakia 1918–35; father of the preceding.

Mas·ba·te (mäz·bä′tā) One of the Visayan Islands, Philippines; 1,262 square miles.

Mas·ca·gni (mäs·kä′nyē), **Pietro,** 1863–1945, Italian composer.

mas·ca·longe (mas′kə·lonj) See MUSKELLUNGE.

mas·car·a (mas·kar′ə) *n.* A cosmetic preparation used to darken the eyelashes, usually black, brown, or blue. [<Sp. *máscara* a mask <Arabic *maskharah* a buffoon]

Mas·ca·rene Islands (mas′kə·rēn) An island group, including Réunion and Mauritius, in the Indian Ocean.

mas·cle (mas′kəl) *n.* 1 A lozenge-shaped plate used in scale armor. 2 *Her.* A lozenge voided. Also spelled *macle.* [<OF. See MACLE.]

mas·cot (mas′kot, -kət) *n.* 1 A person, animal, or thing thought to bring good luck by its presence. 2 *Brit.* An ornament on the radiator of an automobile. [<F *mascotte* <Provençal *mascot,* dim. of *masco* a sorcerer, lit., a mask]

mas·cu·line (mas′kyə·lin) *adj.* 1 Having the distinguishing qualities of the male sex, or pertaining to males; specially suitable for men; manly: opposed to *feminine.* 2 *Gram.* Being of the male gender. 3 *Bot.* Male; staminate. —*n.* 1 A male person; that which is of the male sex. 2 *Gram.* The masculine gender or a word of this gender: opposed to *feminine* and *neuter.* [<OF *masculin* <L *masculinus* <*masculus* male <*mas.* Doublet of MALE.] —**mas′cu·line·ly** *adv.* —**mas′cu·lin′i·ty, mas′·cu·line·ness** *n.*

Synonyms (adj.): male, manful, manlike, manly, mannish, virile. *Male* is applied to the sex, *masculine* to the qualities, especially to the stronger, hardier, and more imperious qualities that distinguish the *male* sex; as applied to women, *masculine* has often the depreciatory sense of unwomanly, rude, or harsh; as, a *masculine* face or voice, or the like; still one may say in a commendatory way, she acted with *masculine* courage or decision. *Masculine* may apply to the distinctive qualities of the *male* sex at any age; *virile* applies to the distinctive qualities of mature manhood only, as opposed not only to *feminine* or *womanly* but to *childish,* and is thus an emphatic word for *sturdy, intrepid,* etc. See also under MANLY. *Antonyms*: see synonyms for FEMININE.

Mase·field (mās′fēld), **John,** born 1875, English poet, novelist, and dramatist; poet laureate 1930–.

ma·ser (mā′zər) *n.* A sound-amplifying device which uses a crystal of potassium-cobalt-cyanide to pick up radio waves emitted by remote celestial objects. [<*m(icrowave) a(mplification by) s(timulated) e(mission of) r(adiation)*]

mash (mash) *n.* 1 A soft, pulpy mass. 2 A mixture of meal, bran, etc., and water, fed warm to cattle. 3 Crushed or ground grain or malt, infused in hot water to produce wort. 4 In winemaking, the crushed grapes before fermentation begins. —*v.t.* 1 To crush or beat into a mash or pulp. 2 To convert into mash, as malt or grain, by infusing in hot water. [OE *māsc(wyrt)* mashwort, infused malt]

mash·er (mash′ər) *n.* 1 One who or that which mashes. 2 *Slang* A man who persistently annoys women unacquainted with him, as by attempting familiarities, etc.

Mash·er·brum (mush′ər·bro͞om) A peak in the Karakoram mountain system, Kashmir; 25,600 feet. Also **Mash′ar·brum.**

Mash·had (mâsh·hâd′) See MESHED.

mash·ie (mash′ē) *n.* An iron golf club with a deep, short blade and much loft: used in approaching. See illustration under GOLF CLUB. Also **mash′y.** [? Alter. of F *massue* a club <LL *mattiuca* (assumed), prob. <Celtic]

mashlin

mash·lin (mash′lin) *n. Scot.* Mixed grain.
Ma·sho·na·land (mə·shō′nə·land) A region of Southern Rhodesia; 80,344 square miles.
Mas·i·nis·sa (mas′ə·nis′ə), 238–148 B.C., king of Numidia; ally of Scipio against the Carthaginians.
mas·jid (mus′jid) *n. Arabic* A Moslem mosque.
mask[1] (mask, mäsk) *n.* 1 A cover or disguise, as for the features; a protective appliance for the face or head: a gas *mask.* 2 A subterfuge. 3 A cast of the face taken just after death. 4 A play or dramatic performance, in vogue in the 16th and 17th centuries, in which the actors were masked and represented allegorical or mythological subjects, originally in dumb show, but later in dialog, poetry, and song; also, a dramatic composition written for such a play: also spelled *masque.* 5 A masquerade: also spelled *masque.* 6 A masker. 7 An artistic covering for the face, used by Greek and Roman actors in comedy and tragedy. 8 *Archit.* A reproduction of a face or a face and neck, used as a gargoyle, a keystone of an arch, etc. 9 *Mil.* A screen of brush or the like for hiding a battery or any military operation from the enemy; camouflage. 10 *Zool.* Any formation about the head suggesting a mask; specifically, the enlarged lower lip of a larval dragonfly. 11 A fox's or dog's head. See synonyms under PRETENSE. — *v.t.* 1 To cover (the face, head, etc.) with a mask. 2 To hide or conceal with or as with a mask; disguise. — *v.i.* 3 To put on a mask; assume a disguise. [<F *masque* <Ital. *maschera, mascara* <Arabic *maskharah* a buffoon]
  Synonyms (verb): cloak, conceal, cover, disguise, dissemble, hide, masquerade, pretend, screen, shroud, veil. See HIDE. *Antonyms:* betray, communicate, declare, disclose, divulge, exhibit, expose, publish, reveal, show, tell.

MASKS
a. Greek tragedy.   *c.* Tibetan ceremonial.
b. Greek comedy.   *d.* Ancient Shinto.
      e. Domino.

mask[2] (mask, mäsk) *v.t. & v.i. Brit. Dial.* To infuse or be infused. [Var. of MASH]
masked (maskt, mäskt) *adj.* 1 Having the face covered with or as with a mask; disguised. 2 Personate. 3 *Zool.* Having markings resembling a mask, as various pupae of insects.
masked or **mask ball** A ball at which the dancers wear masks or dominos.
masked hunter The kissing bug.
mas·keg (mas′keg) See MUSKEG.
mask·er (mas′kər, mäs′-) *n.* One who wears a mask: also spelled *masquer.*
mask·ing (mas′king, mäs′-) *n. Scot.* A crushing into a mash.
masking tape An adhesive tape used to protect those parts of a surface not to be painted, sprayed, or otherwise treated.
mas·ki·nonge (mas′kə·nonj) See MUSKELLUNGE.
mas·lin (maz′lin) *n. Brit. Dial.* 1 Mixed grain, especially wheat and rye. 2 Bread made from such mixture. 3 A mixture; potpourri. [<OF *mesteillon* <LL *mistilio, -onis,* ult. <*mixtus,* pp. of *miscere* mix]
mas·o·chism (mas′ə·kiz′əm) *n. Psychiatry* A condition in which sexual gratification depends on being dominated, cruelly treated, beaten, etc. [after Leopold von Sacher-*Masoch,* 1835–95, Austrian novelist, who described this condition] — **mas′o·chist** *n.* — **mas′o·chis′tic** *adj.*
ma·son (mā′sən) *n.* One who lays brick and stone in building; also, a stonecutter. — *v.t.* To build of masonry. [<OF *masson, maçon* <Med. L *matio, macio, -onis,* prob. <Gmc.]
Ma·son (mā′sən) One of the order of Freemasons. — **Ma·son·ic** (mə·son′ik) *adj.*
Ma·son (mā′sən), **James Murray,** 1798–1871, Confederate statesman; with John Slidell, seized on board the British steamer *Trent,* 1861. — **Walt,** 1862–1939, U. S. humorist and poet.
mason bee A solitary bee (*Chalicodoma muraria*) of southern Europe which builds its nest of sand, clay, etc.
Ma·son–Dix·on line (mā′sən·dik′sən) The boundary between Pennsylvania and Maryland as surveyed by the Englishmen Charles Mason and Jeremiah Dixon in 1763–67: before the Civil War it was regarded as dividing Slave States from Free States, and is still used to distinguish the North from the South. Also **Mason and Dixon's line.**
Ma·son·ite (mā′sən·īt) *n.* A tough, dense, moisture-resistant fiberboard made from wood fibers exploded under high steam pressure: widely used as a building and construction material: a trade name.
Mason jar A glass jar having a tightly fitting screw top, used for canning and preserving: patented by John L. Mason of New York in 1857.
ma·son·ry (mā′sən·rē) *n. pl.* **·ries** 1 The art or work of building with brick or stone. 2 That which is built by masons or of materials which masons use.
Ma·son·ry (mā′sən·rē) *n.* Freemasonry.
Ma·so·ra (mə·sō′rə) *n.* 1 A collection of criticisms and marginal notes to the Old Testament, made by Jewish writers before the tenth century. 2 The tradition relied on by the Jews to preserve the Old Testament text from corruption. Also **Ma·so′rah.** [<Hebrew (modern) *māsōrah* <Hebrew *māsōreth* tradition, orig., bond (of the covenant)] — **Mas·o·ret·ic** (mas′ə·ret′ik) or **·i·cal,** *adj.* — **Mas′o·rete** (-rēt), **Mas′o·rite** (-rīt) *n.*
Mas·pe·ro (mäs′pə·rō, *Fr.* mäs·pə·rō′), **Gaston Camille Charles,** 1846–1916, French Egyptologist.
Mas·qat (mus′kat, mäs′kät) See MUSCAT.
masque (mask, mäsk), **mas·quer** (mas′kər, mäs′-) See MASK[1] (defs. 4 and 5), MASKER.
mas·quer·ade (mas′kə·rād′, mäs′-) *n.* 1 A social party composed of persons masked and costumed; also, the costumes and disguises worn on such an occasion. 2 A false show or disguise. 3 Formerly, a form of dramatic representation. — *v.i.* **·ad·ed, ·ad·ing** 1 To take part in a masquerade. 2 To wear a mask or disguise. 3 To disguise one's true character; assume a false appearance. See synonyms under MASK[1]. [Alter. of F *mascarade* <Sp. *mascarada* <*máscara* a mask. See MASCARA.] — **mas′quer·ad′er** *n.*
mass[1] (mas, mäs) *n.* 1 An assemblage of things that collectively make one quantity. 2 A body of concrete matter; a lump. 3 The principal part of anything. 4 Extent of volume; bulk; size. 5 *Physics* The measure or expression of the inertia of a body, as indicated by the acceleration imparted to it when acted upon by a given force: it is the quotient of the weight of a body divided by the acceleration due to gravity. 6 The homogeneous plastic combination of drugs prepared for making pills. — **the masses** The common people. See synonyms under MOB[1]. — *v.t. & v.i.* To form into a mass; assemble. [<OF *masse* <L *massa,* prob. <Gk. *maza* a barley cake]
  Synonyms (noun): aggregate, body, bulk, heap, lump, matter, substance, total, totality, whole. See AGGREGATE, HEAP, THRONG.
mass[2] (mas, mäs) *n. Eccl.* 1 The eucharistic liturgy in the Roman Catholic and some Anglican churches, regarded as a commemoration or repetition of Christ's sacrifice on Calvary. 2 A celebration of this. See HIGH MASS, LOW MASS. 3 A musical setting for the fixed portions of this liturgy, as the Credo, Sanctus, Kyrie eleison, etc. Also **Mass.** — **black mass** 1 A mass for the dead: so called because the celebrant wears black vestments. 2 A ceremony performed in so-called worship of Satan as a burlesque of the Christian mass. [OE *mæsse* <LL *messa* dismissal <L *missa,* pp. fem. of *mittere* send, dismiss <*ite, missa est* go, you are dismissed; said by the priest after the Eucharist is ended]
Mas·sa·chu·set (mas′ə·choo′sit) *n.* 1 One of a large and important tribe of North American Indians of Algonquian linguistic stock, formerly inhabiting the region around Massachusetts Bay. 2 The language of this tribe. Also **Mas′sa·chu′sett.** [<Algonquian (Massachuset) *Massa-adchu-es-et,* lit., at the big hill <*massa* big + *wadchu* hill + *es,* dim. suffix + *et* at the]
Mas·sa·chu·setts (mas′ə·chōō′sits) A NE State of the United States on the Atlantic; 8,266 square miles; capital, Boston; entered the Union Feb. 6, 1788, one of the thirteen original States; nickname, *Bay State:* abbr. *Mass.*
Massachusetts Bay A wide inlet of the Atlantic on the eastern coast of Massachusetts, extending from Cape Ann to Cape Cod.
mas·sa·cre (mas′ə·kər) *n.* The indiscriminate killing of human beings, as in savage warfare; slaughter. — *v.t.* **·cred** (-kərd), **·cring** To kill indiscriminately or in great numbers; slaughter. [<MF <OF *maçacre, macecle,* ? <*mache-col* butcher <*macher* smash + *col* neck <L *collum*] — **mas′sa·crer** *n.*
  Synonyms (noun): butchery, carnage, havoc, slaughter. A *massacre* is the indiscriminate killing in numbers of the unresisting or defenseless; *butchery* is the killing of men rudely and ruthlessly as cattle are killed in the shambles. *Havoc* may not be so complete as *massacre,* nor so coldly brutal as *butchery,* but is more widely spread and furious; it is destruction let loose, and may be applied to organizations, interests, etc., as well as to human life. *Carnage* refers to widely scattered or heaped up corpses of the slain; *slaughter* is similar in meaning, but refers more to the process, and *carnage* to the result.
mas·sage (mə·säzh′) *n.* A system of remedial treatment consisting of kneading, rubbing, and otherwise manipulating a part or the whole of the body with the hands. — *v.t.* **·saged, ·sag·ing** To treat by massage. [<F <*masser* massage <*masse* mass <L *massa,* ? <Gk. *massein* knead] — **mas·sag′er, mas·sag′ist** *n.* — **mas·sa·geuse** (mas′ə·zhœz′) *n. fem.*
mas·sa·sau·ga (mas′ə·sô′gə) *n.* The pigmy rattlesnake (*Sistrurus miliarius*) of the southern United States, seldom exceeding 20 inches in length, and living in dry, warm areas: also called *ground rattlesnake.* [<Ojibwa name of a river in Ontario, Canada]
Mas·sa·soit (mas′ə·soit), died 1661, American Indian chief of Massachusetts; friendly with the Pilgrims of Plymouth Colony.
Mas·sa·wa (mäs·sä′wä) The chief port of Eritrea, on the Red Sea; former capital of Eritrea. Formerly **Mas·so′wa.**
mass bell The Sanctus bell.
mass communication The simultaneous dissemination of a single item of information, advertising, propaganda, etc., to the largest possible audience by the use of mass media.
mass defect *Physics* The difference between the mass number of an isotope and the isotopic weight. Compare PACKING FRACTION.
mas·sé (ma·sā′) *n.* In billiards, a stroke with a cue held perpendicularly, causing the cue ball to return in a straight line or to describe a curve. Also **massé shot.** [<F <*masser* make a massé shot <*masse* billiard cue, lit., a mace]
Mas·sé·na (mà·sā·nà′), **André,** Prince d'Essling, 1758–1817, French marshal under Napoleon I.
Mas·se·net (mas′nā′, *Fr.* más·ne′), **Jules Émile Frédéric,** 1842–1912, French composer.
mas·se·ter (ma·sē′tər) *n. Anat.* A masticatory muscle connected with the lower jaw. [<NL <Gk. *masētēr* chewer <*masasthai* chew] — **mas·se·ter·ic** (mas′ə·ter′ik), **mas′se·ter′ine** (-ēn, -īn) *adj.*
mas·seur (ma·sûr′, *Fr.* mà·sœr′) *n. French* A man who practices or gives massage. — **mas··seuse** (ma·sōōz′, -sōōs′; *Fr.* mà·sœz′) *n. fem.*
mas·si·cot (mas′ə·kot) *n.* 1 Lead monoxide: a rare mineral associated with galena. 2 A yellowish paint pigment similar to litharge. [<F <Sp. *mexacote* <Arabic *shabb qubṭi* Coptic alum]
mas·sif (mas′if, *Fr.* mà·sēf′) *n. Geol.* 1 The dominant, central mass of a mountain ridge more or less defined by longitudinal or transverse valleys. 2 A diastrophic block of the earth's crust, isolated by boundary faults. [<F]

Mas·sif Cen·tral (må·sēf′ sän·tràl′) A plateau region of central France, covering one sixth of the country's surface.

Mas·si·mi·lia·no (mäs′sē·mē·lyä′nō) Italian form of MAXIMILIAN.

Mas·sine (mä·sēn′), **Léonide**, born 1896, Russian ballet dancer and choreographer.

Mas·sin·ger (mas′ən·jər), **Philip**, 1583–1640, English dramatist.

mas·sive (mas′iv) *adj.* 1 Constituting a large mass; ponderous; large: a *massive* forehead. 2 Belonging to the total mass of anything. 3 Being without definite form, as a mineral; amorphous. 4 *Geol.* Homogeneous: said of certain rock formations. 5 Figuratively, imposing in scope or degree; having considerable magnitude. See synonyms under LARGE. [<F *massif*] — **mas′sive·ly** *adv.* — **mas′sive·ness** *n.*

Massive Mount The second highest peak in the Rocky Mountains, in central Colorado; 14,418 feet. Also **Massive Mountain**.

mass media Newspapers, magazines, paperbound books, radio, television, and motion pictures, considered as means of reaching a very wide public audience.

mass meeting A large public gathering for the discussion or promotion of some topic or cause, usually political.

mass number *Physics* 1 The whole number nearest the mass of any isotope of an element: 3 and 4 are *mass numbers* of the isotopes of helium. 2 The number of protons and neutrons in an atom: distinguished from *atomic number*.

mas·so·ther·a·py (mas′ō·ther′ə·pē) *n.* Treatment by massage. [<Gk. *massein* knead + THERAPY] — **mas′so·ther′a·peu′tic** (-ther′ə·pyōō′tik) *adj.*

mass production The manufacture or production by machinery of goods or articles in great numbers or quantities.

mass ratio The ratio of the total mass of a rocket to its mass after the expenditure of fuel, calculated at approximately 2.72 to 1 for a rocket designed to travel at the exhaust velocity of its fuel.

mass spectrograph *Physics* An instrument for determining the relative masses of electrically charged particles by passing a stream of them through a magnetic field and noting the variable deflections from a straight path.

mass·y (mas′ē) *adj.* **mass·i·er, mass·i·est** Consisting of a mass or masses; big; bulky. — **mass′i·ness** *n.*

mast[1] (mast, mäst) *n.* 1 *Naut.* A pole or spar of round timber or tubular iron or steel, set upright in a sailing vessel to sustain the yards, sails, etc. 2 The upright pole of a derrick. 3 Any large, upright pole. — **before the mast** Occupying the position of, or serving as, a common sailor: from the fact that common sailors were quartered forward of the foremast. — *v.t.* To furnish with a mast or masts. [OE *mæst*] — **mast′less** *adj.*

mast[2] (mast, mäst) *n.* The fruit of the oak, beech, and other trees; acorns, etc. [OE *mæst*, fodder]

mast- Var. of MASTO-.

mas·ta·ba (mas′tə·bə) *n.* In ancient Egypt, an oblong building used as a mortuary chapel and place of offerings, with sloping sides and flat top, covering the mouth of a sepulchral pit. Also **mas′ta·bah**. [<Arabic *maṣṭabah* bench]

MASTABA
Of solid masonry, except for a small chapel and (unconnected) shaft to the mummy chamber.

mas·tax (mas′taks) *n. Zool.* The pharynx or gizzard of a rotifer. [<Gk., mouth, jaws < *masasthai* chew]

mas·tec·to·my (mas·tek′tə·mē) *n. Surg.* The operation of removing the breast. [<MAST- + -ECTOMY]

mas·ter (mas′tər, mäs′-) *n.* 1 A male person who has authority over others, as the principal of a school, an employer, the head of a household, the owner of a domestic animal, etc. 2 One who has control or disposal of something; an owner. 3 In the U.S. merchant marine, the captain of a vessel. 4 One who is familiar with all the details of an art, profession, science, or trade. 5 One who has charge of some special thing, place, or business. 6 A young gentleman. 7 An honorary title; specifically, **Master**, a scholastic title and rank between bachelor and doctor. 8 *Law* Any of various officers of the court who assist the judges by hearing evidence, reporting, etc.: a *master* in chancery. 9 *Scot.* The courtesy title of a viscount's (or baron's) eldest son. 10 One who has disciples or followers; a religious leader. 11 One who gains the victory; a victor. — **the** (or **our**) **Master** Christ. — *v.t.* 1 To overcome or subdue; bring under control; defeat. 2 To become expert in: to *master* Greek. 3 To control or govern as a master. See synonyms under CONQUER, GAIN, LEARN, SUBDUE. — *adj.* Having the mastery; chief; controlling. [Fusion of OE *magister* and OF *maistre*, both <L *magister*, orig. a double compar. of *magnus* great] — **mas′ter·dom** *n.* — **mas′ter·hood** *n.* — **mas′ter·less** *adj.*

Synonyms (noun): boss, captain, chief, commander, despot, director, employer, foreman, governor, head, leader, lord, manager, monarch, overseer, owner, prince, principal, proprietor, schoolmaster, sovereign, teacher. See CHIEF, SUPERINTENDENT, VICTOR. Antonyms: assistant, attendant, dependent, drudge, inferior, menial, retainer, serf, servant, servitor, slave, subaltern, subordinate, valet, waiter.

mas·ter-at-arms (mas′tər·ət·ärmz′, mäs′-) *n. pl.* **mas·ters-at-arms** A petty officer who maintains discipline and order on a naval vessel.

master builder 1 A contractor who employs men to build. 2 One who has charge of building operations; a foreman or architect.

mas·ter·ful (mas′tər·fəl, mäs′-) *adj.* 1 Having the power or characteristics of a master; domineering; arbitrary. 2 Showing mastery, as of an art, science, situation, etc. — **mas′ter·ful·ly** *adv.* — **mas′ter·ful·ness** *n.*

master hand 1 One skilled in his craft; an expert. 2 Great skill; expertness.

master key A key that will unlock two or more locks each of which has its own key that fits no other lock.

mas·ter·ly (mas′tər·lē, mäs′-) *adj.* Characteristic of a master; befitting a master. — *adv.* In a masterly manner; as befits a master. — **mas′ter·li·ness** *n.*

master mason A skilled mason.

Master Mason A Freemason of the third degree.

mas·ter·mind (mas′tər·mīnd′, mäs′-) *n.* A person of great intelligence and executive ability. — *v.t.* To plan and direct (a project) skilfully: to *mastermind* a plot.

Master of Arts 1 A degree given by a college or university to a person who has completed a prescribed course of graduate study in the humanities, social sciences, etc. 2 A person who has received this degree. Abbr. *M.A.*, *A.M.*

master of ceremonies A person presiding over an entertainment or dinner and introducing the performers or speakers: also, *emcee*.

Master of Science 1 A degree given by a college or university to a person who has completed a prescribed course of graduate study in science. 2 A person who has received this degree. Abbr. *M.S.*, *M.Sc.*

mas·ter·piece (mas′tər·pēs′, mäs′-) *n.* A work or piece of art or literature done with consummate skill or showing the hand of a master; a supreme accomplishment; chef-d'oeuvre. [after G *meisterstück*]

Mas·ters (mas′tərz, mäs′-), **Edgar Lee**, 1869–1950, U. S. poet.

mas·ter·ship (mas′tər·ship, mäs′-) *n.* 1 The state, personality, or character of a master; also, mastery. 2 Masterly skill; preeminence.

mas·ter·sing·er (mas′tər·sing′ər, mäs′-) See MEISTERSINGER.

mas·ter·stroke (mas′tər·strōk′, mäs′-) *n.* A masterly or decisive action or achievement.

mas·ter·work (mas′tər·wûrk′, mäs′-) *n.* A masterpiece.

mas·ter·work·man (mas′tər·wûrk′mən, mäs′-) *n. pl.* **-men** (-mən) A skilled workman, craftsman, or artist; also, a foreman or overseer over workmen.

mas·ter·wort (mas′tər·wûrt′, mäs′-) *n.* 1 Any of several herbs of the parsley family, especially those of the genus *Astrantia*. 2 A European plant (*Peucedanum ostruthium*), formerly used as a pot herb.

mas·ter·y (mas′tər·ē, mäs′-) *n.* 1 The condition of having power and control; dominion. 2 The knowledge or skill of a master. 3 Superiority in a contest; victory. See synonyms under VICTORY.

mast·head (mast′hed′, mäst′-) *n.* 1 *Naut.* **a** The top of a lower mast. **b** The head or top of a mast to which a flag is raised. **c** A sailor acting as look-out at the masthead. 2 That section of a newspaper or other periodical, published in each edition, stating the ownership, officers, and staff conducting it, the advertising, editorial, and publishing offices, etc. — *v.t.* 1 To raise to or display at the masthead, as a flag. 2 To send to a masthead for punishment.

mas·tic (mas′tik) *n.* 1 A small Mediterranean evergreen tree (*Pistacia lentiscus*) of the cashew family. 2 The aromatic resin obtained from this tree, used as a varnish and in medicine as a styptic. 3 A quick-drying cement. [<F <LL *mastichum* <Gk. *mastichē*]

mas·ti·cate (mas′tə·kāt) *v.t.* **·cat·ed, ·cat·ing** 1 To crush or grind (food) for swallowing; chew. 2 To reduce, as rubber, to a pulp by crushing or kneading. [<LL *masticatus*, pp. of *masticare* chew <Gk. *mastichaein* gnash the teeth <*mastax* jaw] — **mas′ti·ca′tion** *n.* — **mas′ti·ca′tor** *n.*

mas·ti·ca·to·ry (mas′ti·kə·tôr′ē, -tō′rē) *adj.* 1 Of, pertaining to, or used in mastication. 2 Adapted for chewing: the *masticatory* mouth of a bee. — *n. pl.* **·ries** A substance chewed to increase the secretion of saliva.

mas·tiff (mas′tif, mäs′-) *n.* One of an old English breed of large hunting dogs, with a thick-set, heavy body, straight forelegs, a broad skull, drooping ears, and pendulous lips. [<OF *mastin* <L *mansuetus* gentle < pp. of *mansuescere* tame < *manus* hand; infl. in form by OF *mestif* mongrel]

MASTIFF
(About 30 inches high at the shoulder)

mas·ti·goph·o·ran (mas′ti·gof′ər·ən) *n.* Any one of a class of protozoa (*Mastigophora*), many of which are parasitic in man and other animals, characterized by the presence of one or more whiplike organs of locomotion called flagellae. — *adj.* Of or pertaining to the Mastigophora. [<Gk. *mastix*, *mastigos* whip + *phoros* bearing < *pherein* bear]

mas·ti·tis (mas·tī′tis) *n.* 1 *Pathol.* Inflammation of the mammary gland. 2 Garget. [< MAST- + -ITIS]

masto- *combining form Med.* The breast or the mammary gland. Also, before vowels, *mast-*, as in *mastitis*. [<Gk. *mastos* the breast]

mas·to·don (mas′tə·don) *n.* A primitive elephantlike mammal (order *Proboscidea*), distinguished from elephants and mammoths chiefly by its molar teeth; especially, the extinct, shaggy-haired *Mammut americanus* once common in North America. [< NL <MAST- + Gk. *odous*, *odontos* tooth; from the nipple-shaped projections on its teeth]

MASTODON
(About 9 feet high at the shoulder)

mas·toid (mas′toid) *adj. Anat.* 1 Designating a process of the temporal bone behind the ear for the attachment of muscles. 2 Pertaining to or situated near this process. 3 Nipplelike; breastlike. — *n.* The mastoid process. [<Gk. *mastoeidēs* < *mastos* breast + *eidos* form]

mas·toid·ec·to·my (mas′toid·ek′tə·mē) *n. Surg.* Excision of mastoid cells or of the mastoid process. [<MASTOID + -ECTOMY]

mas·toid·i·tis (mas′toid·ī′tis) *n. Pathol.* Inflammation of the mastoid process. [<MASTOID + -ITIS]

mas·tur·bate (mas′tər·bāt) *v.i.* **·bat·ed, ·bat·ing** To perform masturbation. [<L *masturbatus*, pp. of *masturbari*; ult. origin uncertain]

mas·tur·ba·tion (mas′tər·bā′shən) *n.* Sexual

self-gratification; onanism: also called *self-abuse*. [<L *masturbatio, -onis*] — **mas′tur·ba′tor** *n*.

Ma·su·ri·a (mə·zŏŏr′ē·ə, -sŏŏr′-) A region of NE Poland on the U.S.S.R. border, containing the many **Masurian Lakes**: Polish *Mazury*. German **Ma·su·ren·land** (mä·zŏŏ′rən·länt).

ma·su·ri·um (mə·sŏŏr′ē·əm) *n*. A supposed metallic element whose place in the periodic table is now occupied by technetium. [from *Masuria*, where first found]

mat[1] (mat) *n*. 1 A flat article, woven or plaited, or made of some perforated or corrugated material, to be laid on a floor, and on which to wipe one's shoes or feet. 2 Any flat piece of lace, plaited straw, leather, etc., used as a floor covering, table protection, ornament, etc. 3 Any dense or twisted growth, as of hair or rushes. — *v.* **mat·ted, mat·ting** *v.t.* 1 To cover with or as with mats. 2 To knot or entangle into a mat. — *v.i.* 3 To become knotted or entangled together. [OE *matt(e)* <LL *matta*]

mat[2] (mat) *n*. 1 A lusterless, dull, or roughened surface; also, a tool for producing such a surface. 2 *Printing* A sheet of papier-mâché or wood fiber for recording the impression of type or cuts in stereotyping; a matrix. 3 A border of white cardboard, serving as the frame, or part of the frame, of a picture. — *v.t.* **mat·ted, mat·ting** To produce a dull surface on, as metal or glass. — *adj*. Presenting a lusterless surface. Also *matte*. [OF, defeated, overcome <Arabic *māt*]

Mat·a·be·le (mat′ə·bē′lē) *n. pl.* **·be·le** or **·be·les** (-bē′lēz) One of a Zulu people who were driven across the Vaal by the Boers in 1837 and occupied the region north of the Limpopo, later known as Matabeleland.

Mat·a·be·le·land (mat′ə·bē′lē·land′) A region and former SW division of Southern Rhodesia; 69,989 square miles.

mat·a·dor (mat′ə·dôr) *n*. 1 In bullfighting, the man who kills the bull with a thrust of a sword. 2 In various card games, one of the highest trumps. [<Sp., killer <*matar* slay]

Ma·ta·mo·ros (mä′tä·mō′rōs) A city near the mouth of the Rio Grande in NE Mexico.

Mat·a·nus·ka Valley (mat′ə·nōōs′kə) A region of southern Alaska, site of an experimental colonization project by the United States government in 1935.

Ma·tan·zas (mä·tän′säs) 1 A province in western Cuba; 3,260 square miles. 2 Its capital, a port.

Ma·ta·pan (mä′tä·pän′), **Cape** The southernmost point of the Greek mainland: also *Cape Tainaron*.

match[1] (mach) *n*. 1 One similar or equal in appearance, position, quality, or character; a suitable or fit associate; also, a possible mate. 2 A person or thing that is the equal of another in ability, strength, character, position, etc.; one able to cope with or oppose another; a peer. 3 A contest of skill, strength, etc., between persons or animals. 4 A counterpart; facsimile; also, either of two things harmonizing or corresponding. 5 A marriage or mating, or an agreement to marry or pair; a pairing or coupling. — *v.t.* 1 To be similar to or in accord with in quality, degree, etc.: His looks *match* his mood; The hat *matches* the coat. 2 To make or select as equals or as suitable for one another: to *match* pearls. 3 To place together as mates or companions; marry. 4 To cause to correspond; adapt: *Match* your efforts to your salary. 5 To compare so as to decide superiority; test: to *match* wits. 6 To set (equal opponents) in opposition: to *match* boxers. 7 To equal; oppose successfully: No one could *match* him. — *v.i.* 8 To be equal, similar, or corresponding; suit. 9 To be married; mate. [OE *gemæcca* companion] — **match′a·ble** *adj*. — **match′er** *n*.

match[2] (mach) *n*. 1 Any article manufactured for the purpose of starting or communicating a fire; specifically, a splinter of soft wood or a piece of waxed thread or cardboard tipped with a combustible composition that ignites by friction. 2 A fuse of cotton wicking prepared to burn quickly or slowly, and used for firing cannon. [<OF *mesche* wick, prob. <L *myxa* wick of a candle]

match·board (mach′bôrd′, -bōrd′) *n*. A board, specially cut with a groove along one edge and a tongue along the other, for close joining on floors, ceilings, etc.

match·book (mach′bŏŏk′) *n*. A small paper folder containing safety matches, with a strip of specially prepared surface at one end for striking them.

match·box (mach′boks′) *n*. A small box for containing matches.

matched (macht) *adj*. Having a tongue on one edge and a groove on the other: said of boards.

match·less (mach′lis) *adj*. That cannot be matched or equaled; peerless. — **match′less·ly** *adv*. — **match′less·ness** *n*.

match·lock (mach′lok′) *n*. 1 An old type of musket fired by placing a lighted match against the powder in the pan. 2 The gunlock on such a musket.

match·mak·er[1] (mach′mā′kər) *n*. 1 One who makes plans, or schemes, to bring about a marriage. 2 One who arranges matched games or contests. — **match′mak′ing** *adj. & n*.

match·mak·er[2] (mach′mā′kər) *n*. One who makes matches for lighting. — **match′mak′ing** *adj. & n*.

match·mark (mach′märk′) *n*. A distinguishing mark placed on separable parts of machinery as a guide for assembling. — *v.t.* To put a matchmark upon.

match play In golf, a form of competitive play in which the score is computed by totaling the number of holes won or lost by each side.

match player A golfer who competes in match play.

match point In tennis and similar games, the final point needed to win a match.

match·stick (mach′stik′) *n*. A piece of wood tipped with sulfur and used as a match.

match·wood (mach′wŏŏd′) *n*. 1 Wood suitable for making matches; also, splinters. 2 *Obs*. A combustible substance used as tinder; touchwood.

mate[1] (māt) *n*. 1 A companion or associate; comrade. 2 One that is paired or coupled with another, as in matrimony; also, the male or female of animals paired for propagation. 3 An equal in a contest; a match. 4 An officer of a merchant vessel, ranking next below the captain. 5 *Nav*. An assistant to a warrant officer; a petty officer: the boatswain's *mate*. See synonyms under ASSOCIATE. — *v.* **mat·ed, mat·ing** *v.t.* 1 To join as mates or a pair; match; marry. 2 To pair for breeding, as animals. 3 To associate; couple. — *v.i.* 4 To match; marry. 5 To pair. 6 To consort; associate. [<MLG <*gemate* <*ge-* together + *mat* meat, food. Prob. akin to MEAT.] — **mate′less** *adj*.

mate[2] (māt) *v.t.* **mat·ed, mat·ing** 1 In chess, to checkmate. 2 To defeat or confound. — *n*. A checkmate. — *interj*. Checkmate. [<CHECKMATE]

ma·té (mä′tā, mat′ā) *n*. 1 An infusion of the leaves of a Brazilian holly (*Ilex paraguariensis*), much used as a beverage in South America: also called *Paraguay tea, yerba*. 2 The Paraguay tea plant. [<Sp. <Quechua *mati* calabash (in which it was steeped)]

mat·e·lote (mat′ə·lōt) *n*. A stew of fish in wine and oil, with herb seasoning. Also **mat′e·lotte** (-lot). [<F <*matelot* sailor]

Ma·te·o (mä·tā′ō) Spanish form of MATTHEW.

ma·ter (mā′tər, mä′-) *n. Latin* Mother.

mater do·lo·ro·sa (dō′lə·rō′sə) *Latin* Sorrowful mother: in art or music, the Virgin Mary as the sorrowing mother.

ma·ter·fa·mil·i·as (mā′tər·fə·mil′ē·əs) *n. Latin* The mother of a family; a matron.

ma·te·ri·al (mə·tir′ē·əl) *adj*. 1 That of which anything is composed or may be constructed; matter considered as a component part of something. 2 Collected facts, impressions, ideas, or notes containing them, and sketches, etc., that may be used in completing a literary or an artistic production. 3 Matter regarded as the amorphous substratum of reality. 4 The tools, instruments, articles, etc., for doing something. See also MATERIEL. 5 A cloth or fabric. — *adj*. 1 Pertaining to matter; having a corporeal existence; physical. 2 Pertaining to matter in a corporeal relation; affecting the physical nature; also, pertaining to the body or the appetites; corporeal; sensuous; sensual. 3 Pertaining to the subject matter; essential; important. 4 Pertaining to matter as opposed to form. 5 Consisting of, relating to, or composed of matter regarded as the primary substance of the physical universe. 6 Replete with matter or good sense. See synonyms under IMPORTANT, PHYSICAL. [<LL *materialis* <L *materia* matter, stuff]

ma·te·ri·al·ism (mə·tir′ē·əl·iz′əm) *n*. 1 The doctrine that the facts of experience are all to be explained by reference to the reality, activities, and laws of physical or material substance. 2 Undue regard for material interests.

ma·te·ri·al·ist (mə·tir′ē·əl·ist) *n*. 1 A believer in the doctrine of materialism; also, a believer in the existence of matter. 2 One who takes interest exclusively, chiefly, or excessively in the material or bodily necessities and comforts of life. — **ma·te′ri·al·is′tic** *adj*. — **ma·te′ri·al·is′ti·cal·ly** *adv*.

ma·te·ri·al·i·ty (mə·tir′ē·al′ə·tē) *n. pl.* **·ties** 1 The quality or state of being material; physical as distinguished from psychical being. 2 Substance; matter; a material thing; a body. Also **ma·te′ri·al·ness**.

ma·te·ri·al·ize (mə·tir′ē·əl·īz′) *v.* **·ized, ·iz·ing** *v.t.* 1 To give material or actual form to; represent as material. 2 To cause (a spirit, etc.) to appear in visible form. 3 To make materialistic. — *v.i.* 4 To assume material or visible form; appear. 5 To take form or shape; be realized: Our plans never *materialized*. — **ma·te′ri·al·i·za′tion** *n*. — **ma·te′ri·al·iz′er** *n*.

ma·te·ri·al·ly (mə·tir′ē·əl·ē) *adv*. 1 In a material and important manner. 2 In essence or substance. 3 From a physical point of view; physically.

ma·te·ri·a med·i·ca (mə·tir′ē·ə med′i·kə) *Med*. 1 The branch of medical science that relates to medicinal substances, their nature, uses, effects, etc. 2 The substances employed as remedial agents. [<Med. L <*materia* matter + *medica*, fem. of *medicus* medical]

ma·te·ri·el (mə·tir′ē·el′) *n. Mil*. 1 All non-expendable ordnance, transport, and equipment of an army. 2 All material things of an army except personnel. Also *French* **ma·té·ri·el** (mȧ·tā·ryel′). [<F, material]

ma·te·ri·es mor·bi (mə·tir′i·ēz môr′bī) *Latin* The morbid substances, organisms, etc., which cause disease.

ma·ter·nal (mə·tûr′nəl) *adj*. 1 Pertaining to a mother; motherly. 2 Connected with or inherited from one's mother; coming through the relationship of a mother. [<F *maternel* <L *maternus* <*mater* a mother] — **ma·ter′nal·ly** *adv*.

ma·ter·nal·ize (mə·tûr′nəl·īz) *v.t.* **·ized, ·iz·ing** To make maternal.

ma·ter·ni·ty (mə·tûr′nə·tē) *n. pl.* **·ties** 1 The condition of being a mother. 2 The qualities of a mother; motherliness. [<F *maternité* <L *maternitas*]

maternity hospital A hospital for the care of women during childbirth and of newborn babies; lying-in hospital.

mate·ship (māt′ship) *n*. 1 The state or condition of being a mate or companion. 2 The position or authority of one holding the office of mate.

ma·tey (mā′tē) *Brit. Colloq. adj*. Friendly; chummy. — *n*. Friend; chum.

math[1] (math) *n*. A mowing, or that obtained by mowing: now rare except in *aftermath*. [OE *māth*]

math[2] (math) *n. Colloq*. Mathematics.

math·e·mat·i·cal (math′ə·mat′i·kəl) *adj*. 1 Pertaining to or of the nature of mathematics. 2 Rigidly exact or precise. Also **math′e·mat′ic**. [<L *mathematicus* <Gk. *mathēmatikos* disposed to learn, mathematical <*mathēma* learning <*manthanein* learn] — **math′e·mat′i·cal·ly** *adv*.

mathematical expectation *Stat*. The probability of the occurrence of a given event multiplied by the amount of money offered or wagered on its occurrence.

mathematical logic Symbolic logic.

math·e·ma·ti·cian (math′ə·mə·tish′ən) *n*. One versed in mathematics.

math·e·mat·ics (math′ə·mat′iks) *n.* The logical study of quantity, form, arrangement, and magnitude; especially, the methods for disclosing, by the use of rigorously defined concepts and self-consistent symbols, the properties and exact relations of quantities and magnitudes, whether in the abstract, **pure mathematics**, or in their practical connections, **applied mathematics**.

MATHEMATICAL SYMBOLS

$+$	Plus or positive; sign of addition
$-$	Minus or negative; sign of subtraction
$\pm$	Plus or minus
$\mp$	Minus or plus
$\times$ or $\cdot$	Multiplied by
$\div$ or $:$	Divided by
$=$ or $::$	Equals; is equal to; as
$\neq$ or $\not\equiv$	Does not equal
$\cong$	Approximately equal; congruent
$>$	Greater than
$<$	Less than
$\geq$	Greater than or equal to
$\leq$	Less than or equal to
$\sim$	Similar to; equivalent
$\therefore$	Therefore
$\because$	Since or because
$\equiv$	Identical; identically equal to
$\propto$	Directly proportional to; varies directly as
∞	Infinity
$\sqrt{}$	Square root
$\sqrt[n]{}$	nth root
e or ϵ	Base (2.718···) of natural system of logarithms
Σ	Summation of
$\int$	Integral of
$\int_b^a$	Integral between limits of a and b
$\doteq$ or $\rightarrow$	Approaches as a limit
$f(x), F(x), \phi(x)$	Function of x
Δ	Increment of, as Δy
d	Differential, as dy
$\frac{dy}{dx}$, or $f'(x)$	Derivative of $y = f(x)$ with respect to x
δ	Variation, as δy
π	pi; ratio of circumference of circle to diameter
$n!$ or $\lfloor n$	Factorial n or n factorial
$\angle, \angle s$	Angle, angles
$\perp, \perp s$	Perpendicular to, perpendiculars
$\parallel, \parallel s$	Parallel to, parallels
$\triangle, \triangle\!\triangle$	Triangle, triangles
$\llcorner, \llcorner\!\llcorner$	Right angle, right angles
▱, ▱▱	Parallelogram, parallelograms
□, □□	Rectangle, rectangles
□, □□	Square, squares
○ or ⊙	Circle, circumference
′	Minutes of arc; prime; minutes of time; feet
″	Seconds of arc; double prime; seconds of time; inches

Math·er (math′ər), **Cotton**, 1663?–1728?, American clergyman and writer. — **Increase**, 1639–1723, American clergyman and writer; father of Cotton.
Ma·thi·as (mä·tē′äs) German form of MATTHIAS. Also *Sp.* **Ma·tí·as** (mä·tē′äs).
Ma·thieu (má·tyœ′) French form of MATTHEW.
Ma·thu·ra (mu′tŏŏr·ə) A city in western Uttar Pradesh, India, sacred to Hindus as the birthplace of Krishna: formerly *Muttra*.
Ma·thu·rai (mə·tŏŏr′ī) See MADURAI.
ma·ti·co (mə·tē′kō) *n.* A tropical American shrub (*Piper angustifolium*) of the pepper family: its hairy leaves yield a volatile oil having stimulant and hemostatic properties. [<Sp., dim. of *Mateo* Matthew]
Ma·til·da (mə·til′də; *Ital.* mä·tēl′dä, *Sw.* mä·til′dä) A feminine personal name. Also **Ma·thil·da** (mə·til′də, *Du.* mä·til′dä), **Ma·thil·de** (*Dan.* mä·til′de, *Fr.* má·tēld′), **Ma·til·de** (*Ger.* mä·tēl′də, *Sp.* mä·tēl′dä). [<Gmc., mighty in battle]
— **Matilda**, 1102–67, daughter of Henry I of England; disputed the English throne with Stephen of Blois: known as *Maud*.
— **Matilda of Flanders**, died 1083, wife of William the Conqueror.
mat·in (mat′in) *n.* **1** *pl. Eccl.* The first of the canonical hours, usually said at midnight. **2** *pl.* In the Anglican Church, Morning Prayer. **3** Figuratively, any morning song, as of a bird. **4** *Obs.* Morning. — *adj.* Of or belonging to the morning: also **mat′in·al**. Also spelled *mattin*. [<OF *matin* early <L *matutinus (tempus)* (time) of the morning, appar. <*Matuta* goddess of morning]

mat·i·née (mat′ə·nā′) *n.* An entertainment or reception held in the daytime; specifically, a theatrical or cinematic performance given in the afternoon. Also **mat′i·nee′**. [<F <*matin* morning. See MATIN.]
mat·ing (mā′ting) *n.* The act of pairing or matching.
Ma·tisse (må·tēs′), **Henri**, 1869–1954, French painter.
mat-knife (mat′nīf′) *n. pl.* **-knives** (-nīvz′) A knife having the edge ground at an angle, used for cutting engravings, heavy paper, artist's matboard, etc. See illustration under KNIFE.
Ma·to Gros·so (mat′ə grō′sō, *Pg.* mä′tŏŏ grō′sō) A state of central and western Brazil, the second largest state in Brazil; 487,489 square miles; capital, Cuiabá; formerly *Matto Grosso*.
mat·rass (mat′rəs) *n.* **1** A long-necked, round-bodied glass vessel for distilling and digesting. See BOLTHEAD (def. 1). **2** A thin, hard glass tube used in blowpipe analysis. Also spelled *mattrass*. [<F *matras* bolt]
matri- combining form Mother: *matricide*. [<L *mater, matris* mother]
ma·tri·arch (mā′trē·ärk) *n.* A woman holding the position corresponding to that of a patriarch in her family or tribe. [<MATRI- + (PATRI)ARCH] — **ma′tri·ar′chal** *adj.* — **ma′tri·ar′chal·ism** *n.*
ma·tri·ar·chate (mā′trē·är′kit) *n.* Matriarchal government; a system of matriarchalism. [<MATRIARCH + -ATE²]
ma·tri·ar·chy (mā′trē·är′kē) *n. pl.* **-chies** A social organization having the mother as the head of the family, in which descent, kinship, and succession are reckoned through the mother, instead of the father; also, government by women. — **ma′tri·ar′chic** *adj.*
ma·tri·ces (mā′trə·sēz, mat′rə-) Plural of MATRIX.
mat·ri·cide (mat′rə·sīd) *n.* **1** The killing of one's mother. **2** One who kills his mother. [<L *matricidium* <*mater, matris* mother + *caedere* kill; def. 2 <L *matricida*] — **mat′ri·ci′dal** *adj.*
mat·ri·cli·nous (mat′rə·klī′nəs) *adj. Biol.* Showing hereditary characteristics inclined to the maternal side: opposed to *patriclinous*. Also **mat′ro·cli′nous**, **mat′ro·cli′nal**.
ma·tric·u·lant (mə·trik′yə·lənt) *n.* A candidate or applicant for matriculation.
ma·tric·u·late (mə·trik′yə·lāt) *v.t.* & *v.i.* **-lat·ed**, **-lat·ing** To enrol in a college or university as a candidate for a degree. — *n.* One who is so enrolled. [<Med. L *matriculatus*, pp. of *matriculare* enrol <*matricula* <dim. of *matrix* womb, origin, public roll <*mater* mother] — **ma·tric′u·la′tion** *n.* — **ma·tric′u·la′tor** *n.*
mat·ri·lin·e·al (mat′rə·lin′ē·əl) *adj.* Pertaining to or describing descent or derivation through the female line. Compare PATRILINEAL.
mat·ri·lo·cal (mat′rə·lō′kəl) *adj. Anthropol.* Denoting residence of a married couple in the wife's community, as in certain clan societies. Compare PATRILOCAL.
mat·ri·mo·ni·al (mat′rə·mō′nē·əl) *adj.* Pertaining to matrimony. — **mat′ri·mo′ni·al·ly** *adv.*
— Synonyms: bridal, conjugal, connubial, hymeneal, hymenean, nuptial, sponsal, spousal. — Antonyms: celibate, single, unespoused, unwedded.
mat·ri·mo·ny (mat′rə·mō′nē) *n. pl.* **-nies** **1** The union of a man and a woman in marriage; wedlock. **2** A card game played by any number of persons; also, a combination of king and queen in this and certain other games. See synonyms under MARRIAGE. [<L *matrimonium* <*mater, matris* mother]
matrimony vine The boxthorn.
ma·trix (mā′triks) *n. pl.* **ma·tri·ces** (mā′trə·sēz, mat′rə-) **1** That which contains and gives shape or form to anything. **2** The womb. **3** *Biol.* Intercellular substance; hence, the formative cells from which a structure grows. **4** A mold in which anything is cast or shaped, or that which encloses like a mold. **5** *Printing* A papier–mâché, plaster, or other impression of a form, from which a plate for printing may be made. **6** *Geol.* The impression or mold of the exterior of a crystal or other mineral left in the containing rock when the object is removed, or the mass in which a fossil, gemstone, mineral, etc., is embedded. **7** *Math.* A rectangular array of symbols or terms enclosed between parentheses or double vertical bars to facilitate the study of relationships. **8** The material used as a filler between the fragments of a shrapnel projectile. [<L, womb, breeding animal <*mater, matris* mother]
ma·tron (mā′trən) *n.* **1** A married woman; mother; also, a woman of established age and dignity. **2** A housekeeper, or a female superintendent, as of an institution. [<OF *matrone* <L *matrona* <*mater, matris* mother] — **ma′tron·al** *adj.* — **ma′tron·like′** *adj.*
ma·tron·age (mā′trən·ij) *n.* **1** The condition of being a matron. **2** Matronly attention or care. **3** Matrons collectively.
ma·tron·ize (mā′trən·īz) *v.t.* **-ized**, **-iz·ing** **1** To render matronlike. **2** To chaperon.
ma·tron·ly (mā′trən·lē) *adj.* Of or like a matron; elderly. — *adv.* In a matronly manner. — **ma′tron·li·ness** *n.*
matron of honor A married woman acting as chief attendant to a bride at her wedding. See MAID OF HONOR.
mat·ro·nym·ic (mat′rō·nim′ik) See METRONYMIC.
Ma·trûh (mə·trōō′) A Mediterranean port of western Egypt: also *Mersa Matrûh*.
Ma·tsu·ya·ma (mä·tsōō·yä·mä) A port on NW Shikoku island, Japan.
Matt (mat) Diminutive of MATTHEW.
matte¹ (mat) *n. Metall.* An impure metallic product containing sulfur, obtained in smelting sulfide ores, as of copper, lead, etc. [<F]
matte² (mat) See MAT².
mat·ted (mat′id) *adj.* **1** Covered with mats or matting. **2** Tangled like the fibers of a mat. — **mat′ted·ly** *adv.* — **mat′ted·ness** *n.*
mat·ter (mat′ər) *n.* **1** That which makes up the substance of anything, especially of material things. **2** The material of which a thing is composed. **3** *Physics* That aspect of reality conceived as existing prior to and independently of the mind and to have characteristics susceptible to precise measurement in terms of extension, force, mass, radiation, and energy. **4** That which constitutes the

essence or substance of a particular thing. **5** Something not exactly conceived or stated; an indefinite, often a comparatively small, amount. **6** A subject for discussion or feeling. **7** Something of moment and importance. **8** A condition of affairs, especially if unpleasant or unfortunate; case; difficulty; trouble: What's the *matter*? **9** The thought, or material of thought. **10** *Pus.* **11** *Philos.* The as yet undifferentiated substratum of those properties and changes of which the human senses take cognizance and which, by their differentiation and combination in an infinite variety of forms, constitute the separate existences and characteristic qualities of physical things. **12** *Printing* Type that is set or composed; also, material to be set up; copy. **13** Written or printed documents sent by mail. See synonyms under MASS[1], TOPIC. — *v.i.* **1** To be of concern or importance; signify. **2** To form or discharge pus; suppurate. See synonyms under INTEREST. [<F *matière* <L *materia* stuff]
Mat·ter·horn (mat'ər·hôrn, *Ger.* mä'ter·hôrn) A mountain in the Alps on the Swiss-Italian border; 14,780 feet: French *Mont Cervin*.
mat·ter-of-course (mat'ər·əv·kôrs', -kōrs') *adj.* Following or accepting as an expected conclusion or as a natural or logical result.
mat·ter-of-fact (mat'ər·əv·fakt') *adj.* Closely adhering to facts; not fanciful; unimaginative.
Mat·thew (math'yoo) A masculine personal name. Also *Lat.* **Mat·thae·us** (mə·thē'əs, *Dan.* mä·tä'əs), *Ger., Sw.* **Mat·thä·us** (mä·tä'ŏŏs), *Du.* **Mat·the·us** (mä·tä'əs), *Ital.* **Mat·te·o** (mät·tā'ō). [<Hebrew, gift of God]
— **Matthew** One of the Twelve Apostles, author of the first Gospel; also, the Gospel by him. Also **Saint Matthew**.
— **Matthew of Paris**, 1200-59?, English monk and chronicler.
Mat·thews (math'yōōz), **(James) Brander**, 1852-1929, U.S. scholar, author, and educator.
Mat·thi·as (mə·thī'əs, *Fr.* mȧ·tē·ȧs') A masculine personal name. Also *Du.* **Mat·thijs** (mä·tīs'). [See MATTHEW]
— **Matthias** The apostle chosen by lot to succeed Judas Iscariot. *Acts* i 23-26.
mat·tin (mat'in) See MATIN.
mat·ting (mat'ing) *n.* **1** A woven fabric of fiber, straw, etc., used as a floor covering, etc. **2** The act or process of making mats. **3** A dull, flat surface effect, as in gilding, etc.
mat·tock (mat'ək) *n.* A pickaxlike tool for digging and grubbing, having blades instead of points. [OE *mattuc*]
Mat·to Gros·so (mat'ə grō'sō, *Pg.* mä'tŏŏ grô'sŏŏ) A former spelling of MATO GROSSO.

MATTOCK

mat·toid (mat'oid) *n.* A person mentally unbalanced in one way or on one subject. [<Ital. *mattoide* <*matto* mad <L *mattus* intoxicated]
mat·trass (mat'rəs) See MATRASS.
mat·tress (mat'rəs) *n.* **1** A casing of ticking or other strong fabric filled with hair, cotton, or rubber, and used as a bed. **2** A mat woven of brush, poles, etc., used in protecting embankments, forming dikes, jetties, etc. [<OF *materas* <Ital. *materasso* <Arabic *maṭraḥ* place where something is thrown]
Mat·ty (mat'ē) Diminutive of MATILDA.
mat·u·rate (mach'ŏŏ·rāt, mat'yŏŏ-) *v.t.* ·**rat·ed**, ·**rat·ing** **1** To ripen or mature. **2** *Med.* To suppurate; form pus. [<L *maturatus*, pp. of *maturare* <*maturus* ripe, fully developed] — **mat'u·ra'tive** *adj.*
mat·u·ra·tion (mach'ŏŏ·rā'shən, mat'yŏŏ-) *n.* **1** *Med.* The formation of pus. **2** The process of ripening or coming to maturity; ripeness. **3** *Biol.* The final stages in the preparation of gametes for fertilization, during which reduction to one half in the number of chromosomes in the germ cells occurs; meiosis.
ma·ture (mə·chŏŏr', -tŏŏr', -tyŏŏr') *adj.* **1** Completely developed; perfectly ripe: *mature* grain; as applied to persons, fully developed in character and powers: a *mature* thinker. **2** Highly developed; approaching perfection. **3** Thoroughly elaborated or arranged; fully digested or considered; complete in detail: a *mature* scheme. **4** Due and payable; having reached its time limit: a *mature* bond. **5** *Geol.* Designating the maximum complexity and diversity of earth features, as achieved by the forces of erosion at full vigor; also, adjusted to local surroundings, as the course of a river. See synonyms under RIPE. — *v.* ·**tured**, ·**tur·ing** *v.t.* **1** To cause to ripen or come to maturity; bring to full development. **2** To perfect; complete. — *v.i.* **3** To come to maturity or full development; ripen. **4** To become due, as a note. [<L *maturus* of full age] — **ma·ture'ly** *adv.*
ma·tu·ri·ty (mə·chŏŏr'ə·tē, -tŏŏr'-, -tyŏŏr'-) *n.* **1** The state or condition of being mature: also **ma·ture'ness**. **2** Full development, as of the body. **3** The time at which a thing matures: a note payable at *maturity*. [<F *maturité* <L *maturitas*]
ma·tu·ti·nal (mə·tōō'tə·nəl, -tyōō'-) *adj.* Pertaining to morning; before noon; early. [<L *matutinalis* <*matutinus* early in the morning <*Matuta*, goddess of morning] — **ma·tu'ti·nal·ly** *adv.*
mat·zoon (mat·sōōn') See YOGURT.
mat·zoth (mät'sōth, -sōs) *n. pl.* of **mat·zo** (mät'sō) Wafers of unleavened bread. [<Hebrew *matstsōth*, pl. of *matstsāh* unleavened]
Mau·beuge (mō·bœzh') A city in northern France.
Maud (môd) A feminine personal name; also, diminutive of MAGDALENE and MATILDA. Also **Maude**.
maud·lin (môd'lin) *adj.* **1** Made foolish by liquor. **2** Foolishly and tearfully affectionate. [<*Maudlin* <OF *Maudelene*, *Madeleine* (Mary) Magdalen, who was often depicted with eyes swollen from weeping]
Maugham (môm), **(William) Somerset**, born 1874, English novelist and dramatist.
mau·gre (mô'gər) *prep.* Obsolete form of MALGRÉ. [<OF]
Mau·i (mou'ē) The second largest of the Hawaiian Islands; 728 square miles.
mau·kin (mô'kin) See MALKIN.
maul (môl) *n.* A heavy mallet for driving wedges, piles, etc. — *v.t.* **1** To beat and bruise; batter. **2** To handle roughly; manhandle; abuse. **3** *U.S.* To split by means of a maul and wedges, as logs or rails. Also spelled *mall*. [<OF *mail* <L *malleus* hammer. ? Akin to L *malleus* beat to pieces, crush.] — **maul'er** *n.*
Maul·main (môl·mān') See MOULMEIN.
maul·stick (môl'stik') See MAHLSTICK.

STEEL FENCE-POST MAUL

Mau Mau (mou'mou') *pl.* **Mau Mau** or **Mau Maus** A member of a secret organization of Kikuyu tribesmen, active from 1952 in terrorist activities directed against European colonists in Kenya, Africa. [< native name]
mau·met (mô'mit) *n.* **1** *Obs.* An idol; from the early belief that the Moslems worshiped images of Mohammed. **2** *Brit. Dial.* A scarecrow; a doll. Also spelled *mammet*. [Contraction of MAHOMET] — **mau'met·ry** *n.*
maun (môn) *v.i. Scot.* Must.
mau·na (mô'nə) *Scot.* Must not. Also **maun'na**.
Mau·na Ke·a (mou'nä kā'ä) An extinct volcano on Hawaii; highest mountain in the Hawaiian Islands; 13,825 feet.
Mau·na Lo·a (mou'nä lō'ä) An active volcano on Hawaii; 13,675 feet.
maund (mônd) *n.* A unit of weight used in India, varying from 24.7 to 82.28 lbs. avoirdupois. [<Hind. *mān*]
maun·der (môn'dər) *v.i.* **1** To talk in a wandering or incoherent manner; drivel. **2** To move dreamily or idly. [? Freq. of obs. *maund* beg; infl. in meaning by MEANDER] — **maun'der·er** *n.*
maun·dy (môn'dē) *n.* The religious ceremony of washing the feet of others, especially of inferiors: in commemoration of the washing of the disciples' feet by Christ; hence, **Maundy**, the service connected with such ceremony. [<OF *mandé* <L *mandatum* command; from the use of *mandatum* at the beginning of the ceremony. See MANDATE.]
Maundy Thursday The day before Good Friday, commemorating the Last Supper of Christ with his disciples, at which he washed their feet. [See MAUNDY]
Mau·pas·sant (mō·pȧ·sän'), **(Henri René Albert) Guy de**, 1850-93. French writer of novels and short stories.
Mau·re·ta·ni·a (môr'ə·tā'nē·ə) An ancient country of North Africa, including the northern part of modern Morocco and the western part of Algeria.
Mau·re·ta·ni·an (môr'ə·tā'nē·ən) *adj.* Pertaining to ancient Mauretania or its inhabitants. — *n.* **1** One of the ancient people of Mauretania. **2** The Libyan dialect spoken by these people.
Mau·riac (mō·ryȧk'), **François**, born 1885, French novelist.
Mau·rice (mə·rēs', môr'is, mor'is; *Fr.* mô·rēs') A masculine personal name. Also *Sp.* **Mau·ri·cio** (mou·rē'thyō), *Ital.* **Mau·ri·sio** (mou·rē'syō) or **Mau·ri·zio** (mou·rē'tsyō), **Mau·ri·ti·us** (*Du.* mou·rē'sĕ·əs, *Lat.* mou·rish'əs), *Du.* **Mau·rits** (mou'rits). [<L, Moorish]
— **Maurice**, 1521-53, Elector of Saxony; secured religious freedom in Germany.
— **Maurice of Nassau**, 1567-1625, Prince of Orange; Dutch general; son of William the Silent.
Mau·ri·ta·ni·a (môr'ə·tā'nē·ə), **Islamic Republic of** An independent republic of the French Community, on the coast of west Africa; 418,120 square miles; capital, Nouakchott; formerly a French overseas territory. — **Mau'ri·ta'ni·an** *adj.* & *n.*
Mau·ri·ti·us (mô·rish'ē·əs) An island in the Indian Ocean east of Madagascar; 720 square miles; comprising with its dependencies a British crown colony (804 square miles); capital, Port Louis. French **Mau·rice** (mō·rēs'). — **Mau·ri'ti·an** *adj.* & *n.*
Mau·rois (mō·rwȧ'), **André** Pseudonym of Émile Herzog, born 1885, French biographer and novelist.
Mau·ry (mô'rē), **Matthew Fontaine**, 1806-1873, U.S. naval officer and oceanographer.
Maur·ya (mour'yə) An Indian dynasty, 325?-184? B.C., founded by *Chandragupta I.*
Mau·ser (mou'zər) *n.* A magazine rifle having great range and high muzzle velocity; also, a type of automatic pistol: a trade name. [after P. P. *Mauser*, 1838-1914, German inventor]
mau·so·le·um (mô'sə·lē'əm) *n. pl.* ·**le·ums** or ·**le·a** (-lē'ə) A large, stately tomb. [<L <Gk. *Mausōleion*, tomb of King Mausolus of Caria, erected by Queen Artemisia at Halicarnassus about 350 B.C. See SEVEN WONDERS OF THE WORLD.] — **mau'so·le'an** *adj.*
maut (môt) *n. Scot.* Malt.
mau·vais goût (mō·ve' gōō') *French* Bad taste.
mauve (mōv) *n.* **1** A purple pigment and dye derived from mauvein. **2** Any of various purplish-rose shades. [<F, mallow <L *malva*. Doublet of MALLOW.]
mauv·e·in (mō'vin) *n. Chem.* A coal-tar violet dyestuff, $C_{27}H_{24}N_4$, obtained by oxidizing aniline: the first aniline dye of commerce. Also **mauve·ine** (mō'vin). [<MAUVE]
mav·er·ick (mav'ər·ik) *n.* **1** *U.S.* An unbranded or orphaned animal, particularly a calf, legitimately belonging to the first one to brand it. **2** *Colloq.* A person with no attachments or affiliations, especially political ones. [after Samuel A. *Maverick*, 1803-1870, Texas lawyer, who did not brand his cattle]
ma·vis (mā'vis) *n.* The European song thrush or throstle (genus *Turdus*). [<F *mauvis*]
Ma·vors (mā'vôrz) An early name for Mars, the Roman god of war.
ma·vour·neen (mə·vŏŏr'nēn, -vôr'-) *n.* My darling: an expression of affection. Also **ma·vour'nin**. [<Irish *mo muirnin*]
maw[1] (mô) *n.* **1** The craw of a bird. **2** The stomach. **3** The air bladder of a fish. **4** The gullet, jaws, or mouth of a voracious mammal or fish. [OE *maga* stomach]
maw[2] (mô) *v.t. Scot.* To mow, as hay.
mawk·ish (mô'kish) *adj.* **1** Productive of disgust or loathing; sickening or insipid. **2** Char-

Mawson acterized by false or feeble sentimentality; lacking in strength or vigor. See synonyms under FLAT. [< obs. *mawk* a maggot] — **mawk′ish·ly** *adv.* — **mawk′ish·ness** *n.*

Maw·son (mô′sən), **Sir Douglas**, born 1882, British explorer.

Max (maks) A masculine personal name; originally a diminutive of MAXIMILIAN. [See MAXIMILIAN]

max·il·la (mak·sil′ə) *n. pl.* **·lae** (-ē) **1** *Anat.* The upper jaw bone in vertebrates. **2** *Zool.* One of the pair or pairs of jaws behind the mandibles of an arthropod. [<L, dim. of *mala* jaw]

max·il·lar·y (mak′sə·ler′ē, mak·sil′ər·ē) *adj.* Of, pertaining to, or situated near the jaw or a maxilla: a *maxillary* artery. — *n. pl.* **·lar·ies** A maxilla or jaw bone.

max·il·li·ped (mak·sil′ə·ped) *n. Zool.* **1** One of the limbs of certain crustaceans, modified to serve as masticatory organs, and situated behind the maxillae. **2** One of a pair of poisonous claws situated near the mouth of a centipede. [<L *maxilla* jaw bone + -PED]

max·im (mak′sim) *n.* A brief statement of a practical principle or proposition; a proverbial saying. See synonyms under ADAGE, RULE. [<F *maxime* <L *maxima* (*sententia, propositio*) greatest (authority, premise), fem. of *maximus.* See MAXIMUM.]

Max·im (mak′sim), **Hiram Percy**, 1869–1936, U.S. inventor, son of Sir Hiram. — **Sir Hiram Stevens**, 1840–1916, U.S. inventor, civil, mechanical, and electrical engineer; became a British subject. — **Hudson**, 1853–1927, U.S. mechanical engineer and inventor; brother of Sir Hiram.

max·i·mal (mak′sə·məl) *adj.* Greatest; highest. See MAXIMUM.

Max·i·mal·ist (mak′sə·məl·ist) *n.* One of a former party or faction of extremist Russian revolutionists: distinguished from *Minimalist.* [<MAXIMAL + -IST]

Maxim gun A water-cooled machine-gun which utilizes the recoil of each shot to fire the next: invented by Sir Hiram S. Maxim.

Max·i·mil·ian (mak′sə·mil′yən, *Ger., Sw.* mäk′·si·mē′lē·än) A masculine personal name. Also *Sp.* **Max·i·mi·li·a·no** (mäk′sē·mē·lē·ä′nō), **Max·i·mi·li·a·nus** (*Du.* mäk′si·mē·lē·ä′nəs, *Lat.* mak′si·mil′ē·ā′nəs), *Pg.* **Max·i·mi·li·ão** (mäk′si·mē′lē·oun′), *Fr.* **Max·i·mi·lien** (mȧk·sē·mē·lyaṅ′). [<L, greatest Aemilius]

— **Maximilian I**, 1459–1519, Holy Roman Emperor 1493–1519.

— **Maximilian II**, 1527–76, Holy Roman Emperor 1564–76.

— **Maximilian (Ferdinand Joseph)**, 1832–67, archduke of Austria; emperor of Mexico 1864–67; executed.

max·im·ite (mak′sim·īt) *n.* A picric-acid high explosive used as a bursting charge for projectiles: invented by Hudson Maxim.

max·i·mize (mak′sə·mīz) *v.t.* **·mized**, **·miz·ing** To make as great as possible; increase to the maximum; intensify.

max·i·mum (mak′sə·məm) *n. pl.* **·ma** (-mə) **1** The greatest quantity, amount, degree, or magnitude that can be assigned. **2** *Math.* **a** A value of a varying quantity that is greater than any neighboring value. **b** The highest possible of all the values which a variable or a function can express; the point at which a varying quantity ceases to increase and begins to decrease. **3** *Astron.* The moment of greatest brightness in a variable star, or its magnitude at such time. — *adj.* **1** As large or great as possible. **2** Pertaining to a maximum: *maximum* weight. [<L *maximus,* superl. of *magnus* great]

ma·xi·xe (mə·shē′shä, mäk·sēks′) *n.* A Brazilian dance to the two-step. [<Brazilian Pg.]

Max Mül·ler (mäks mül′ər), **Friedrich** See MÜLLER.

max·well (maks′wel) *n.* The practical cgs unit of magnetic flux, equal to the flux through one square centimeter normal to a magnetic field with an induction of one gauss. [after James Clerk *Maxwell*]

Max·well (maks′wel), **James Clerk**, 1831–79, Scottish physicist.

Maxwell's demon *Physics* An imaginary being of molecular proportions, devised by James Clerk Maxwell to illustrate the kinetic theory of gases.

may[1] (mā) *v.* Present: *sing.* **may, may** (*Archaic* **may·est** or **mayst**), **may**, *pl.* **may**; past: **might** A defective verb now used only in the present and past tenses as an auxiliary followed by the infinitive without *to,* or elliptically with the infinitive understood, to express: **1** Permission or allowance: *May* I go? You *may*. **2** Desire or wish: *May* your tribe increase! **3** Contingency, especially in clauses of result, concession, purpose, etc.: He died that we might live. **4** Possibility: You *may* be right. **5** *Law* Obligation or duty: the equivalent of *must* or *shall.* **6** *Obs.* Ability; power: now usually *can.* ◆ See usage note under CAN. [OE *mæg*]

may[2] (mā) *n.* The English hawthorn (*Crataegus oxyacantha*), with white, rose, or crimson flowers. Also **may′bush′** (-boosh′). [<MAY, because it blooms in this month]

May (mā) **1** The fifth month of the year, containing 31 days. **2** The prime of life; youth. **3** May Day festivities. [<OF *mai* <L (*mensis*) *Maius* (month of) May, prob. <*Maia,* goddess of growth; so called because regarded as a month of growth]

May (mā) A feminine personal name: contraction of MARY.

May (mā), **Cape** The southernmost point in New Jersey between the Atlantic and Delaware Bay.

Ma·ya (mä′yə) **1** In Hindu religion, Devi, mother of the world; the personified active will of the Creator. **2** In Hindu philosophy, illusion, often personified as a maiden. [<Skt. *māyā* illusion]

Ma·ya (mä′yə) *n.* **1** One of a tribe of Central American Indians, the most important tribe of the Mayan linguistic stock, still comprising a large percentage of the population of Yucatán, northern Guatemala, and British Honduras. They were the first Indians to develop writing and an accurate astronomical calendar. **2** The language of the Mayas, in its historical and modern forms. — *adj.* Of or pertaining to the Mayas, their culture, or their language.

Ma·ya·güez (mä′yä·gwās′) A port in western Puerto Rico.

Ma·yan (mä′yən) *adj.* Of the Mayas, their culture, or their language. — *n.* **1** A Maya. **2** A family of Central American Indian languages, including Maya and Quiché.

May apple **1** The ovoid, oblong, yellowish fruit of a North American herb (*Podophyllum peltatum*). **2** The herb itself, of which the rhizomes yield podophyllin. Also called *mandrake root.*

Ma·ya·ri iron (mä·yä′rē) An iron made from Cuban ores, containing small percentages of chromium, vanadium, nickel, titanium, and certain other elements: used in high-grade machine castings. [from *Mayari,* a Cuban town]

MAY APPLE
a. Flower.
b. Fruit.

May basket A little basket of flowers left at the door of a friend as a May Day token.

may·be (mā′bē) *adv.* Perhaps; possibly. [<(*it*) *may be*]

May·bird (mā′bûrd) *n.* **1** In the southern United States, the bobolink. **2** In the eastern United States, the redbreast (def. 3).

May·bug (mā′bug′) *n.* **1** The cockchafer. **2** The June beetle. Also **May′bee′tle** (-bēt′l).

May·day (mā′dā′) *n.* The international radiotelephone call for immediate help sent out by an aircraft or ship in distress. [<F *m'aidez* help me]

May Day The first day of May, traditionally celebrated as a spring festival by crowning a May Queen, dancing around a Maypole, etc.: recently commemorated in many countries as a labor holiday.

Ma·yence (mȧ·yäṅs′) The French name for MAINZ.

Ma·yenne (mȧ·yen′), **Duc de**, 1554–1611, Charles of Lorraine; French general; defeated by Henry IV.

May·er (mī′ər), **Julius Robert von**, 1814–78, German physicist.

may·est (mā′ist) May: obsolescent or poetic second person singular, present tense of MAY: with *thou.* Also spelled *mayst.*

May·fair (mā′fâr) *n.* A fashionable residential district of the West End, London.

may·flow·er (mā′flou′ər) *n.* **1** The trailing arbutus, State flower of Massachusetts. **2** *Brit.* The hawthorn or may; also, the marsh marigold. **3** Any of various other plants which blossom in the spring.

Mayflower The ship on which the Pilgrims came to America in 1620.

may·fly (mā′flī′) *n. pl.* **·flies** **1** An ephemerid insect, which in the nymphal state inhabits water and is long-lived, and in the adult state merely propagates its kind and then dies. **2** An artificial fly in imitation of this insect. **3** *Brit.* A caddis fly.

may·hap (mā′hap) *adv. Archaic* It may chance or happen; very likely; peradventure; perhaps. Also **may′hap′pen** (-ən). [<(*it*) *may hap*(*pen*)]

may·hem (mā′hem) *n. Law* The offense of depriving a person by violence of any limb, member, or organ, or causing any mutilation of the body. Also spelled *maihem.* [<OF *mehaing, mahaigne.* Related to MAIM.]

May·ing (mā′ing) *n.* The celebration of May Day.

May·kop (mī·kôp′) See MAIKOP.

May·o (mā′ō) A maritime county of western Connacht, Ireland; 2,084 square miles; county town, Castlebar.

May·o (mā′ō), **Charles Horace**, 1865–1939, U.S. surgeon. — **William James**, 1861–1939, surgeon, brother of the preceding.

Ma·yon (mä·yôn′) An active volcano in SE Luzon, Philippines; 7,916 feet.

may·on·naise (mā′ə·nāz′, mī′-) *n.* A sauce or dressing made by beating together raw egg yolk, butter or olive oil, lemon juice or vinegar, and condiments; also, a cold dish of which it forms an element. [<F, ? <*Mahón,* Balearic port]

may·or (mā′ər, mâr) *n.* The chief magistrate of a city, borough, or municipal corporation. [<F *maire* <L *major* greater. Doublet of MAJOR.] — **may′or·al** *adj.*

may·or·al·ty (mā′ər·əl·tē, mâr′əl-) *n. pl.* **·ties** The office or term of a mayor. [<OF *mairalté*]

Ma·yotte (mȧ·yôt′) Easternmost of the Comoro Islands, NW of Madagascar; 140 square miles.

May·pole (mā′pōl′) *n.* A pole decorated with flowers and ribbons, around which dancing takes place on May Day.

may·pop (mā′pop) *n.* **1** The passionflower (*Passiflora incarnata*) of the southern United States. **2** The small yellow fruit of this plant. [Alter. of N. Am. Ind. *maracock.* Cf. Tupian *maracujá.*]

May Queen A young girl crowned with flowers in May Day festivities.

mayst (māst) See MAYEST.

May·thorn (mā′thôrn′) *n.* The hawthorn.

May·time (mā′tīm′) *n.* The month of May. Also **May′tide′** (-tīd′).

may·weed (mā′wēd′) *n.* A strong-scented, acrid weed (*Anthemis cotula*) of the composite family, bearing white-rayed flowers on a yellow disk; stinking camomile. [Alter. of *maidweed* < *maytheweed* <OE *magothe* + WEED]

May wine Any still, white wine flavored with steeped orange and pineapple slices and woodruff: named for the month of May, in which the woodruff flowers.

Maz·a·gan (maz′ə·gan′) A port in western Morocco.

maz·ard (maz′ərd) *n. Obs.* **1** A mazer. **2** The 'skull; the head. Also **maz′zard.** [Var. of MAZER]

Maz·a·rin (maz′ə·rin, *Fr.* mȧ·zȧ·raṅ′), **Jules**, 1602–61, French cardinal and statesman; prime minister under Louis XIV: real name *Giulio Mazzarini.*

Maz·a·rine (maz′ə·rēn) *adj.* Of or pertaining to Cardinal Mazarin, or named from him.

Mazarine blue A deep, rich blue: named after Cardinal Mazarin.

Maz·a·ru·ni (maz′ə·rōō′nē) A river in northern British Guiana, flowing 350 miles NE to the Essequibo.

Ma·zat·lán (mä′sä·tlän′) A port in NW Mexico, on the Gulf of California.

Maz·da·ism (maz′də·iz′əm) *n.* Zoroastrianism. Also **Maz′de·ism.** [<OPersian *Aura mazda* the principle of good]

maze (māz) *n.* **1** An intricate network of paths or passages; a labyrinth. **2** Embarrassment; uncertainty; perplexity. — *v.t. Archaic* To daze or stupefy; bewilder. ◆ Homophone: *maize.* [<AMAZE] — **maz′y** *adj.* — **maz′i·ly** *adv.* — **maz′i·ness** *n.*

Ma·zep·pa (mə·zep′ə), **Ivan Stephanovich**,

1644-1709, a Cossack chief; hero of a poem by Byron.
ma·zer (mā′zər) *n.* A bowl, goblet, or drinking cup made of hard wood. Also *mazard*. [<OF *masere* maple wood <Gmc.]
ma·zu·ma (mə-zōō′mə) *n. U.S. Slang* Money. [<Yiddish *m'zumon* "the ready necessary"]
ma·zur·ka (mə-zûr′kə, -zōōr′-) *n.* 1 A lively Polish round dance resembling the polka. 2 The music for such a dance. Also **ma·zour′ka**. [<Polish, woman from Mazovia, a province in Poland]
Ma·zu·ry (mä-zōō′rē) The Polish name for MASURIA.
ma·zut (mə-zōōt′) *n.* The residue from the distillation of Russian petroleum; used as a fuel oil. Also **ma·zout′**. [<Russian]
maz·zard cherry (maz′ərd) The fruit of the European wild sweet cherry (*Prunus avium*). [Earlier *mazer* maple wood; appar. from the hardness of the wood]
Maz·zi·ni (mät-tsē′nē), **Giuseppe**, 1805-72, Italian patriot and revolutionary; aided Garibaldi to unite Italy.
Mba·ba·ne (mbä-bä′nä) The capital of Swaziland protectorate.
M'bo·mu (mbō′mōō) See BOMU.
Mc- See also MAC-.
Mc·A·doo (mak′ə-dōō), **William Gibbs**, 1863-1941, U.S. lawyer and statesman; secretary of the treasury 1913-18.
Mc·Car·thy·ism (mə-kär′thē-iz′əm) *n.* 1 The practice of making public and sensational accusations of disloyalty or corruption, usually with little or no proof or with doubtful evidence. 2 The practice of conducting sensational inquisitorial investigations, ostensibly to expose subversion, especially pro-Communist activity. [after Joseph *McCarthy*, 1909-1957, U. S. Senator]
Mc·Clel·lan (mə-klel′ən), **George Brinton**, 1826-85, Union general in the Civil War.
Mc·Col·lum (mə-kol′əm), **Elmer Verner**, born 1879, American physiological chemist.
Mc·Cor·mack (mə-kôr′mik), **John**, 1884-1945, U.S. tenor born in Ireland.
Mc·Cor·mick (mə-kôr′mik), **Cyrus Hall**, 1809-1884, U.S. inventor of the reaping machine. — **Joseph Medill**, 1877-1925, U. S. newspaper publisher.
Mc·Coy (mə-koi′), **the (real)** *U.S. Slang* The authentic person or thing. [Appar. from an episode, existing in many versions, in which a celebrated American boxer, Kid McCoy, spectacularly established his identity]
Mc·Dou·gal (mək-dōō′gəl), **William**, 1871-1938, U. S. psychologist born in England.
Mc·Guf·fey (mə-guf′ē), **William Holmes**, 1800-1873, U. S. educator; author of a series of children's readers.
Mc·In·tire (mak′ən-tīr) *adj.* Pertaining to or naming the finely carved furniture and architectural embellishments in the neo-classic taste by Samuel McIntire, 1757-1811, woodcarver of Salem, Mass.
Mc·In·tosh (mak′ən-tosh) *n.* A red variety of early autumn apple. Also **McIntosh red**. [after John *McIntosh* of Ontario, who discovered it in the 18th century]
Mc·Kin·ley (mə-kin′lē), **Mount** A peak in central Alaska, highest in North America, 20,300 feet.
Mc·Kin·ley (mə-kin′lē), **William**, 1843-1901, president of the United States 1897-1901; assassinated.
Mc·Mas·ter (mək-mas′tər, -mäs′-), **John Bach**, 1852-1932, U. S. historian.
M-Day (em′dā′) *n. Mil.* Mobilization day; the day the Department of Defense orders mobilization for war.
Mdi·na (mə-dē′nə) The Maltese name for CITTÀ VECCHIA.
me (mē) *pron.* The objective case of *I*.
♦ **It's me, etc.** Anyone who answers the question "Who's there?" by saying "It's me" is using acceptable colloquial idiom. Here *It is I* would seem stilted, although at the formal level of writing it is expected: They have warned me that *it is I*, and not he, who will have to bear the brunt of the criticism. After a finite impersonal form of the verb *to be*, as *it is, it was*, etc., a personal pronoun should, according to prescriptive grammar, be in the same case as the subject: the nominative; accordingly, at the formal level, we find *It is he, It is we, It is they*, etc. Since only the personal and relative pronouns retain different inflected forms for the nominative and objective cases, the rule might appear to be invented to cover this one exceptional situation. The normal subject–verb–object order of the English sentence creates a strong expectation that what follows the verb will be in the objective case, and perhaps for this reason popular usage has long favored *It's me* over the formal *It is I*. British and American playwrights in recent years have also represented their characters, and not only the uneducated speakers among them, as saying *It's him, It's her, It's them*, etc. In spite of the exact parallel in construction with *It's me*, these expressions are not yet condoned to the same extent, even at the colloquial level.
me·a cul·pa (mē′ə kul′pə) *Latin* My fault; my blame.
mead[1] (mēd) *n.* A liquor of fermented honey and water to which malt, yeast, and spices are added; metheglin. See HYDROMEL. ♦ Homophone: *meed*. [OE *meodu*]
mead[2] (mēd) *n. Poetic* A meadow. ♦ Homophone: *meed*. [OE *mǣd*]
Mead (mēd), **Lake** A reservoir formed by Hoover Dam in Boulder Canyon of the Colorado River in Arizona and Nevada; 246 square miles; largest artificial lake in the world. See BLACK CANYON.
Mead (mēd), **Margaret**, born 1901, U. S. anthropologist.
Meade (mēd), **George Gordon**, 1815-72, Union general in the Civil War, born in Spain.
mead·ow (med′ō) *n.* A tract of low or level land, producing grass for hay. [OE *mǣdwe*, oblique case of *mǣd*] — **mead′ow·y** *adj.*
mead·ow·beau·ty (med′ō-byōō′tē) *n. pl.* ·ties Any of a genus (*Rhexia*) of low-growing herbs, usually with four-parted purple flowers; especially, the common species (*R. virginica*).
meadow bird The bobolink.
meadow hen The American bittern.
mead·ow·lark (med′ō-lärk′) *n.* An American songbird (genus *Sturnella*) related to the blackbird; especially, the southern meadowlark (*S. magna*), brownish or grayish above, marked with black and yellow beneath.
meadow lily The Canada lily.
meadow mouse The field mouse.
meadow rue Any species of the genus *Thalictrum* of the crowfoot family, having leaves like those of rue.
mead·ow·sweet (med′ō-swēt′) *n.* 1 A shrub of the rose family (genus *Spiraea*); especially, *S. salicifolia*, having alternate simple or pinnate leaves and white or rose-colored flowers: also **meadow queen**. 2 Any plant of a related genus (*Filipendula*).
meadow weed Wild oat (def. 1).
mea·ger (mē′gər) *adj.* 1 Deficient or destitute of quantity or quality; scanty; inadequate. 2 Deficient in or scantily supplied with fertility, strength, or richness. 3 Wanting in flesh; thin; emaciated. Also **mea′gre**. [<OF *maigre* <L *macer* lean] — **mea′ger·ly** *adv.* — **mea′ger·ness** *n.*
Synonyms: barren, emaciated, feeble, gaunt, jejune, lank, lean, poor, skinny, spare, starved, starveling, tame, thin. See GAUNT. Antonyms: bonny, bouncing, burly, chubby, corpulent, fat, fleshy, hearty, obese, plump, portly, stout.
meal[1] (mēl) *n.* 1 Comparatively coarsely ground grain. 2 Unbolted wheat flour; chop. 3 A powder produced by grinding; any powdery, meal-like material: sulfur *meal*. [OE *melu*]
meal[2] (mēl) *n.* 1 The portion of food taken at one time; a repast. 2 Its occasion or time. ♦ Collateral adjective: *prandial*. [OE *mǣl* measure, time, meal]
-meal *suffix* The quantity taken at one time or the unit of measurement: now obsolete except in *piecemeal*. [OE *-mǣlum*, oblique case of *mǣl* measure, time]
meal·ie (mē′lē) *n.* 1 An ear of maize. 2 *pl.* Maize; Indian corn. [<Afrikaans *milje* <Pg. *milho* millet <L *milium*]
meal·time (mēl′tīm′) *n.* The habitual time for eating a meal.
meal·worm (mēl′wûrm′) *n.* 1 The larva of a beetle (*Tenebrio molitor*) destructive to flour, meal, etc. 2 A meal-bred grub prepared as bait for ground fishing.
meal·y (mē′lē) *adj.* **meal·i·er, meal·i·est** 1 Having a resemblance to or the qualities or taste of meal; farinaceous. 2 Overspread or besprinkled with or as with meal; pale-colored; of anemic appearance. 3 Having the appearance of being covered with meal; farinose. 4 Mealy-mouthed. 5 Friable; floury: said of the endosperm of malt. 6 Flecked with white spots: said of cattle. 7 Smooth-flavored; not harsh to the taste: said of tea. — **meal′i·ness** *n.*
meal·y·bug (mē′lē-bug′) *n.* Any of a large cosmopolitan family (*Pseudococcidae*) of coccids whose soft, oval bodies are usually covered with a mealy wax secretion: they include many of the most destructive plant pests.
meal·y-mouthed (mē′lē-moutht′, -mouthd′) *adj.* Unwilling to express adverse opinions plainly; euphemistic; insincere.
mean[1] (mēn) *v.* **meant, mean·ing** *v.t.* 1 To have in mind as a purpose or intent: I *meant* to visit him. 2 To intend or design for some purpose, destination, etc.: Was that remark *meant* for me? 3 To intend to express or convey: That's not what I *meant*. 4 To have as the particular sense or significance; denote; portend: Those clouds *mean* snow. — *v.i.* 5 To have disposition or intention; be disposed: He *means* well. 6 To be of specified importance or influence: Her beauty *means* everything to her. See synonyms under IMPORT, PURPOSE. ♦ Homophone: *mien*. [OE *mǣnan* tell, wish, intend]
mean[2] (mēn) *adj.* 1 Low in grade, quality, or condition; of humble antecedents; lowly; also, indicative of or suited to low rank; poor; inferior; shabby. 2 *Colloq.* Disagreeable; unpleasant; unkind; wicked. 3 Ignoble in mind, character, and spirit; lacking magnanimity or honor; base; petty; also, miserly; stingy. 4 Worthy of no respect; slight or contemptible. 5 Of little value or efficiency. 6 Not fit for cultivation: said of land. 7 Vicious or unmanageable: said of horses. 8 *Colloq.* Irritable; ashamed; paltry; also, ill: to feel *mean*. See synonyms under BAD, BASE, COMMON, INSIGNIFICANT, LITTLE, SMALL, VULGAR. ♦ Homophone: *mien*. [OE (*ge*)*mǣne* common, ordinary] — **mean′ly** *adv.* — **mean′ness** *n.*
mean[3] (mēn) *n.* 1 The middle state between two extremes; hence, moderation; avoidance of excess; medium: the happy *mean*. 2 *Math.* A quantity having an intermediate value between two extremes, or between several quantities: the arithmetical *mean*. 3 *pl.* The medium through which anything is done; instrumentality: used often with singular construction: a *means* to an end. 4 *pl.* Money or property as a procuring medium; wealth. 5 A plan of procedure. 6 *Logic* The middle term in a syllogism. 7 *Obs.* An intermediary; mediator. See synonyms under AGENT. — **arithmetical mean** The quotient of the sum of two or more quantities divided by the number of quantities; the average. — **by all means** Without hesitation; certainly. — **by any means** 1 In any manner possible; somehow; at all. 2 By all means. — **by no manner of means** Most certainly not; not for any consideration; on no account whatever: also **by no means**. — **geometric (or geometrical) mean** The square root of the product of two given numbers. — **golden mean** A wise moderation; the avoidance of extremes. — *adj.* 1 Intermediate as to position between extremes. 2 Intermediate as to size, degree, or quality; medium; average. 3 Intermediate as to time; intervening. 4 Having an intermediate value between two extremes or among several values; average: the *mean* distance covered daily. ♦ Homophone: *mien*. [<OF *meien* <L *medianus* < *medius* middle. Doublet of MEDIAN, MESNE, MIZZEN.]
me·an·der (mē-an′dər) *v.i.* 1 To wind and turn in a course. 2 To wander aimlessly. — *n.* 1 A tortuous or winding course; hence, a maze; perplexity. 2 The Greek key or fret pattern. [<MEANDER] — **me·an′der·er** *n.* — **me·an′der·ing, me·an′drous** *adj.*
Me·an·der (mē-an′dər) The ancient name for

the MENDERES, proverbial for its windings: also *Maeander.*

mean·ing (mē′ning) *n.* **1** That which is intended; object; intention; aim. **2** That which is signified; significance; sense; acceptation; import; connotation. See synonyms under PURPOSE. — *adj.* **1** Having purpose or intention: usually in combination: well-*meaning.* **2** Significant; suggestive. — **mean′ing·ful** *adj.* — **mean′ing·less** *adj.* — **mean′ing·less·ly** *adv.* — **mean′ing·ly** *adv.* — **mean′ing·less·ness** *n.*

mean latitude Middle latitude.

mean sun *Astron.* A fictitious sun considered to be moving uniformly with respect to the equator: a concept used to facilitate the computation of time.

meant (ment) Past tense and past participle of MEAN[1].

mean·time (mēn′tīm′) *n.* Intervening time or occasion. — *adv.* In or during the intervening time.

mean time See under TIME. Also **mean solar time.**

mean·while (mēn′hwīl′) *n.* & *adv.* Meantime.

Mearns (mûrnz), **The** See KINCARDINE.

mea·sle (mē′zəl) *n.* **1** The larva *(Cysticercus)* of a certain tapeworm (genus *Taenia*) found in meat and generally producing measles in swine. **2** Any excrescence upon a plant or tree. [See MEASLES]

mea·sled (mē′zəld) *adj.* Affected with measles.

mea·sles (mē′zəlz) *n.* **1** An acute, highly contagious, generally self-immunizing virus disease affecting children and sometimes adults: it is characterized by chills, fever, severe coryza, and an extensive eruption of small red macules; rubeola. **2** Any of various eruptive diseases of a similar character: used with a qualifying adjective: French *measles.* **3** A disease affecting swine and cattle, caused by the presence of larval tapeworms in the flesh. — **French measles, German measles** Rubella. [ME *maseles,* pl. of *masel* a blister <LG. Cf. MDu. *masel* spot on the skin.]

mea·sly (mēz′lē) *adj.* **·sli·er, ·sli·est** **1** Affected with measles. **2** Containing tapeworm larvae: said of meat. **3** *Slang* Not fit to be touched; beneath contempt; also, mean; skimpy; stingy; scanty.

meas·ur·a·ble (mezh′ər·ə·bəl) *adj.* **1** Capable of being measured or of computation. **2** Limited; moderate: *measurable* severity. — **meas′·ur·a·bil′i·ty, meas′ur·a·ble·ness** *n.* — **meas′·ur·a·bly** *adv.*

meas·ure (mezh′ər) *n.* **1** The extent or dimensions of anything. **2** A standard of measurement. **3** Hence, any standard of criticism, comparison, judgment, or award. **4** A series of measure units; a system of measurements: dry *measure.* See WEIGHT. **5** An instrument or vessel of measurement. **6** The act of measuring; measurement. **7** A quantity measured, or regarded as measured. **8** Reasonable limits; moderation: beyond *measure*; also, degree; reasonable proportion: A *measure* of allowance should be made. **9** A certain proportion; relative extent. **10** A specific act or course; transaction; specifically, a legislative bill. **11** That which makes up a sum or total. **12** Any quantity regarded as a unit and standard of comparison with other quantities; a quantity of which some other given quantity forms an exact multiple. **13** *Music* **a** That division of time by which melody and rhythm are regulated; rate of movement; time. **b** The portion of music contained between two bar lines; bar. **14** In prosody, meter; a rhythmical period. **15** A slow and stately dance or dance movement. **16** *pl. Geol.* A series of related rock strata, having some common feature. — *v.* **·ured, ·ur·ing** *v.t.* **1** To take or ascertain the dimensions, quantity, capacity, etc., of, especially by means of a measure. **2** To set apart, mark off, etc., by measuring: often with *off* or *out.* **3** To estimate by comparison; judge; weigh: to *measure* a risk. **4** To serve as the measure of. **5** To bring into competition or comparison. **6** To traverse as if measuring; travel over. **7** To adjust; regulate: *Measure* your actions to your aspirations. — *v.i.* **8** To make or take measurements. **9** To yield a specified measurement: The table *measures* six by four feet. **10** To admit measurement. — **to measure one's length** To fall prostrate at full length. — **to measure out** To distribute or allot by measure. — **to measure swords** **1** To fight with swords. **2** To fight or contend as in a debate. — **to measure up to** To fulfil, as expectations. [<F *mesure* <L *mensura* < *metiri* measure] — **meas′ur·er** *n.*

meas·ured (mezh′ərd) *adj.* **1** Ascertained, adjusted, or proportioned by rule. **2** Uniform; slow and stately; rhythmical; deliberate. **3** In moderation; held in restraint; guarded. — **meas′ured·ly** *adv.* — **meas′ured·ness** *n.*

meas·ure·less (mezh′ər·lis) *adj.* Incapable of measurement; unlimited; immense. See synonyms under INFINITE.

meas·ure·ment (mezh′ər·mənt) *n.* **1** The process or result of measuring anything; mensuration. **2** The amount, capacity, or extent determined by measuring. **3** A system of measuring units.

measuring worm A geometrid.

meat (mēt) *n.* **1** The flesh of animals used as food: sometimes limited to the flesh of mammals, as opposed to fish or fowl; also, any animal killed for food: to hunt one's *meat.* **2** Anything eaten for nourishment; victuals; hence, that which nourishes. **3** The edible part of anything. **4** A meal; especially, dinner; the main meal. **5** The essence, gist, or pith: the *meat* of an essay. **6** *Slang* Anything one does with special ease or pleasure; forte. ◆ Homophones: *meet, mete.* [OE *mete*] — **meat′less** *adj.*

Meath (mēth, mēth) A county of NE Leinster province, Ireland; 903 square miles; county town, Trim.

me·a·tus (mē·ā′təs) *n. pl.* **·tus** or **·tus·es** *Anat.* A conspicuous passage or canal: the auditory *meatus*; nasal *meatus.* [<L, a passage < *meare* go]

meat·y (mē′tē) *adj.* **meat·i·er, meat·i·est** **1** Full of or resembling meat. **2** Having strength; nourishing. **3** Significant; pithy. — **meat′i·ness** *n.*

Meaux (mō) A city in north central France.

me·ca·te (mā·kä′tā) *n. SW U.S.* A rope made of maguey fiber, or sometimes of plaited horsehair: used for tying horses. [<Sp. < Nahuatl]

mec·ca (mek′ə) *n.* **1** A place visited by many people; any attraction. **2** The object of one's aspiration, yearning, or effort. [from *Mecca*]

Mec·ca (mek′ə) The capital of Hejaz and one of the capitals of Saudi Arabia; birthplace of Mohammed and a holy city of Islam to which Moslems make pilgrimages: also *Mekka.*

me·chan·ic (mə·kan′ik) *n.* **1** One engaged in mechanical employment; an artisan; handicraftsman. **2** *Obs.* A mean or lowly fellow. See synonyms under ARTIST. — *adj.* **1** Pertaining to mechanics; mechanical. **2** Materialistic; atomistic. **3** Pertaining to the artisan class. **4** Involving manual labor or skill. [<L *mechanicus* <Gk. *mēchanikos* < *mēchanē* a machine]

me·chan·i·cal (mə·kan′i·kəl) *adj.* **1** Pertaining to mechanics; in accordance with the laws of mechanics. **2** Produced by a machine. **3** Operated by mechanism. **4** Operating as if by a machine or machinery. **5** Doing or done involuntarily, by mere force of habit. **6** Automatic; not instinct with life; artificial. **7** Failing to show independence of thought; slavish. **8** Skilled in the use of tools and mechanisms. **9** Materialistic. — **me·chan′i·cal·ly** *adv.* — **me·chan′i·cal·ness** *n.*

mech·a·ni·cian (mek′ə·nish′ən) *n.* One who understands the science of mechanics; a designer of machinery.

me·chan·ics (mə·kan′iks) *n. pl.* **1** The branch of physics that treats of the phenomena caused by the action of forces on material bodies, including statics and kinetics: construed as singular. **2** The science of machinery and of its practical applications: construed as singular. **3** The mechanical or technical aspects: construed as plural.

Me·chan·ics·ville (mə·kan′iks·vil) A hamlet 7 miles NE of Richmond, Virginia; site of a battle (June 26, 1862) in the Civil War.

mech·a·nism (mek′ə·niz′əm) *n.* **1** The parts of a machine collectively; machinery in general. **2** A system which constitutes a working agency. **3** Technique; mechanical execution or action. **4** The theory that the forces that produce organic growth are the same physical and chemical agencies that operate in the inorganic world, differing from them only in degree: opposed to *vitalism.* **5** *Psychol.* The mental processes, conscious or unconscious, by which certain actions are effected. See synonyms under TOOL. — **mech′·a·nis′tic** *adj.* — **mech′a·nis′ti·cal·ly** *adv.*

mech·a·nist (mek′ə·nist) *n.* **1** A mechanician. **2** A believer in philosophical mechanism.

mech·a·ni·za·tion (mek′ə·nə·zā′shən, -nī-) *n.* **1** The act or process of applying machinery to the performance of specified operations. **2** The aggregate results of such application. **3** *Mil.* The maximum coordinated utilization of power-driven equipment, mechanized armament, and automatic weapons in the service of the combat personnel.

mech·a·nize (mek′ə·nīz) *v.t.* **·nized, ·niz·ing** **1** To make mechanical. **2** To convert (an industry, etc.) to machine production. **3** *Mil.* To equip with tanks, trucks, etc. — **mech′a·nized** *adj.*

mech·a·no·ther·a·py (mek′ə·nō·ther′ə·pē) *n. Med.* The treatment of disease by mechanical means, as massage. [< *mechano-* (<MECHANIC) + THERAPY]

Mech·lin (mek′lin) *n.* A lace with bobbin ground and designs outlined by thread or flat cord, made in Malines (Mechlin), Belgium: also called *malines.*

Mech·lin (mek′lin) A town of north central Belgium: French *Malines. Flemish* **Mech·e·len** (mekh′ə·lən).

Meck·len·burg (mek′lən·bûrg, *Ger.* mek′lən·bo͝orkh) A former state of northern Germany on the Baltic, comprising the former grand duchies of **Meck′len·burg-Schwe·rin′** (-shvā·rēn′), **Meck′len·burg-Stre′litz** (-shtrā′lits), and western Pomerania; 8,856 square miles; capital, Schwerin; divided, July 1952, into several districts of East Germany.

mec·on·ism (mek′ə·niz′əm, mē′kə-) *n. Pathol.* Addiction to opium; poisoning by opium. [<Gk. *mēkōn* a poppy]

me·cop·ter·ous (mə·kop′tər·əs) *adj.* Of, pertaining or belonging to an order *(Mecoptera)* of slender, predacious insects with biting mouth parts and long, narrow wings. [<NL <Gk. *mēkos* length + *pteron* wing]

med·al (med′l) *n.* A small piece of metal, bearing a device, usually commemorative of some event or deed of bravery, scientific research, or literary production, etc. — *v.t.* **·aled** or **·alled, ·al·ing** or **·al·ling** To confer a medal upon. ◆ Homophone: *meddle.* [<F *médaille* <Ital. *medaglia,* ult. <L *metallum.* Doublet of METAL.] — **me·dal·lic** (mə·dal′ik) *adj.*

Medal for Merit A medal awarded by the United States to civilians for exceptional services in times of peace or war.

med·al·ist (med′l·ist) *n.* **1** A collector, engraver, or designer of medals. **2** The recipient of a medal awarded for services or merit. **3** In golf, the winner at medal play. Also **med′·al·list.**

me·dal·lion (mə·dal′yən) *n.* **1** A large medal; also, a subject painted, engraved, etc., and set in a circular or oval frame or design: used as decorative elements in carpets, lace, etc. **2** Any one of several ancient Greek silver coins. [<F *médaillon* <Ital. *medaglione,* aug. of *medaglia* MEDAL] — **me·dal′lion·ist** *n.*

Medal of Honor The highest U. S. decoration, awarded to a member of the armed forces who risked his life in action beyond the call of duty. It is awarded in the name of Congress and usually presented personally by the president. Also called *Congressional Medal of Honor.*

medal play In golf, a form of competitive play in which the score is computed by counting only the total number of strokes played by each competitor in playing the designated number of holes.

Me·dan (me·dän′) A town in NE Sumatra.

med·dle (med′l) *v.i.* **·dled, ·dling** **1** To take part in or concern oneself with something without need or request: often with *in* or *with.* **2** *Obs.* To mix; mingle. See synonyms under INTERPOSE, MIX. ◆ Homophone: *medal.* [<OF *medler, mesdler,* var. of *mesler,* ult. <L *miscere* mix] — **med′dler** *n.* — **med′dling** *adj.* & *n.*

med·dle·some (med′l·səm) *adj.* Given to meddling; officiously inclined; interfering; intrusive. — **med′dle·some·ly** *adv.* — **med′dle·some·ness** *n.*

Synonyms: impertinent, intrusive, meddling, obtrusive, officious. The *meddlesome* person interferes unasked in the affairs of others; the *intrusive* person thrusts himself uninvited into their company or conversation; the *obtrusive* person thrusts himself or his opinions conceitedly and undesirably upon their notice; the *officious* person thrusts his services, unasked and undesired, upon others. Compare ACTIVE, INQUISITIVE, INTERPOSE. *Antonyms*: modest, reserved, retiring, shy, unassuming, unobtrusive.

Mede (mēd) *n.* One of an ancient Asiatic people who flourished in Media in the sixth century B.C.

Me·de·a (mə·dē′ə) In Greek legend, a sorceress of Colchis who aided Jason to obtain the Golden Fleece and, when deserted by him for Creusa, killed her rival and her own children and fled to Athens.

Me·del·lín (mā′thā·yēn′) A city in NW Colombia.

me·di·a (mē′dē·ə) *n. pl. of* **medium** 1 Means of disseminating information, entertainment, etc., such as books, newspapers, radio, television, motion pictures, and magazines. 2 In advertising, all means of communication that carry advertisements, including billboards, direct mail, catalogs, radio, etc.

Me·di·a (mē′dē·ə) An ancient country of SW Asia, corresponding to the NW Iranian plateau. — **Me′di·an** *adj.* & *n.*

me·di·a·cy (mē′dē·ə·sē) *n.* 1 The state or quality of being mediate. 2 Mediation.

me·di·ae·val (mē′dē·ē′vəl, med′ē-), **me·di·ae·val·ism**, etc. See MEDIEVAL, etc.

me·di·al (mē′dē·əl) *adj.* 1 Of or pertaining to the middle, in position or character or in calculation; mean. 2 Nearer the median plane of a body: opposed to *lateral.* 3 Designating a letter neither initial nor final. — *n. Phonet.* Any of a group of voiced stops (b, d, g), conceived as intermediate between the voiceless stops (p, t, k) and the rough or aspirate group (bh, dh, gh, ph, th, kh). [<LL *medialis* <L *medius* middle] — **me′di·al·ly** *adv.*

me·di·an (mē′dē·ən) *adj.* 1 Pertaining to the middle, or situated in the median plane. 2 *Stat.* Of, pertaining to, or designating that point in a series of values which divides the series into two groups, each containing the same number of entries: 8 is the *median* point of the series 2, 5, 8, 10, 13. [<L *medianus* <*medius* middle. Doublet of MEAN³, MESNE, MIZZEN.] — **me′di·an·ly** *adv.*

median plane That plane that divides a body longitudinally into symmetrical halves.

me·di·as·ti·num (mē′dē·as·tī′nəm) *n. pl.* **·na** (-nə) *Anat.* 1 A median partition or septum separating two cavities of the body. 2 The partition between the two pleural sacs of the chest, extending from the sternum to the thoracic vertebrae and downward to the diaphragm. [<NL <Med. L *mediastinus* medial <*medius*] — **me′di·as·ti′nal** *adj.*

me·di·ate (mē′dē·āt) *v.* **·at·ed, ·at·ing** *v.t.* 1 To settle or reconcile by mediation, as differences. 2 To bring about or effect by mediation. 3 To serve as the medium for effecting (a result) or conveying (an object, information, etc.). — *v.i.* 4 To act between disputing parties in order to bring about a settlement, compromise, etc. 5 To occur or be in an intermediate relation or position. See synonyms under INTERPOSE. — *adj.* (-it) 1 Acting as an intervening agency; indirect. 2 Occurring or effected as a result of indirect or median agency. 3 Intermediate. [<LL *mediatus*, pp. of *mediare* stand between, mediate] — **me′di·ate·ly** *adv.* — **me′di·ate·ness** *n.* — **me′di·a′tive** *adj.* — **me′di·a′tor** *n.* — **me′di·a·to′ri·al** (-tôr′ē·əl, -tō′rē·əl), **me′di·a·to′ry** *adj.*

me·di·a·tion (mē′dē·ā′shən) *n.* 1 The act of mediating; intercession; interposition. 2 A friendly intervention in the disputes of others, with their consent, for the purpose of adjusting differences. See CONCILIATION.

me·di·a·tize (mē′dē·ə·tīz′) *v.t.* **·tized, ·tiz·ing** 1 To reduce from a direct to a mediate relation: said of certain German states (and their princes) under the Holy Roman Empire, deprived of direct part in the government of the Empire by being annexed or subordinated to other states, while retaining a nominal sovereignty and local governmental powers. 2 To cause to take a mediate position. — **me′di·a·ti·za′tion** *n.*

med·ic¹ (med′ik) *n.* Any one of several plants of the bean family (genus *Medicago*), especially the lucerne or alfalfa. [<L *medicus* <Gk. *Mēdikē (poa)* Median (grass) <*Mēdos* a Mede]

med·ic² (med′ik) *n. Colloq.* 1 A doctor, physician, or intern. 2 A medical aide or corpsman.

med·i·ca·ble (med′ə·kə·bəl) *adj.* Capable of relief by medicine; curable.

med·i·cal (med′i·kəl) *adj.* 1 Pertaining to medicine or the practice of medicine. 2 Having curative properties. [<F *médicale* <LL *medicalis* <L *medicus* a physician] — **med′i·cal·ly** *adv.*

medical examiner 1 *Law* An official legally designated to examine the bodies of those dead as a result of violence or crime, to perform autopsies, and to establish the proximate cause and circumstances of the death. 2 A doctor authorized by a life insurance company to determine the physical fitness of a prospective insurant.

medical jurisprudence The branch of medicine and related sciences which deals with questions involving the applications of the civil and criminal law; forensic medicine. See under JURISPRUDENCE.

med·i·ca·ment (med′ə·kə·mənt, mə·dik′ə-) *n.* 1 Any substance for the alleviation of disease or wounds. 2 A healing agency. [<L *medicamentum* <*medicare.* See MEDICATE.] — **med′i·ca·men′tal** *adj.*

med·i·cate (med′ə·kāt) *v.t.* **·cat·ed, ·cat·ing** 1 To treat medicinally. 2 To tincture or impregnate with medicine. [<L *medicatus*, pp. of *medicare* heal <*medicus* a physician] — **med′i·ca′tion** *n.* — **med′i·ca′tive, med′i·ca·to′ry** (-kə·tôr′ē, -tō′rē) *adj.*

Med·i·ce·an (med′ə·sē′ən, -chē′-) *adj.* Of or pertaining to the Medici.

Med·i·ci (med′ə·chē, *Ital.* mā′dē·chē) A family of Florentine bankers who became rulers of Tuscany, patrons of the arts, literature, etc. Among its members were **Catherine de' Medici,** 1519–89, queen of Henry II of France, who planned the St. Bartholomew's Day Massacre; **Cosimo (Cosmo) de' Medici,** 1389–1464; **Cosimo (Cosmo) de' Medici,** 1519–74, first Grand Duke of Tuscany: called "Cosimo the Great"; **Lorenzo de' Medici,** 1448?–92, prince of Florence, statesman, educator, patron of art, literature, and printing: known as *Lorenzo the Magnificent*; **Maria de' Medici** (see under MARIA); **Giovanni de' Medici** (see LEO X).

me·dic·i·nal (mə·dis′ə·nəl) *adj.* Adapted to cure or mitigate disease. — **me·dic′i·nal·ly** *adv.*

med·i·cine (med′ə·sin) *n.* 1 A substance possessing, or reputed to possess, curative or remedial properties. 2 The healing art; the science of the preservation of health and of treating disease for the purpose of cure, specifically, as distinguished from surgery or obstetrics. 3 Among North American Indians, any agent or influence used to prevent or cure ills or to invoke supernatural protection or aid, varying from actual remedies (cinchona, etc.) to fetishes, prayers, and symbolic rites (**medicine dance, medicine song,** etc.); specifically, among the Algonquians, any mystery. See MANITO. 4 *Colloq.* Something distasteful or unpleasant: to give someone a dose of his own *medicine.* 5 *Obs.* Something used for other than healing purposes, as the philosopher's stone, elixirs, poisons, love philters, etc. — **to take one's medicine** To endure necessary hardship, discomfort, or punishment, or to do something unpleasant but required. [<L *medicina* <*medicus* physician <*mederi* heal]

medicine ball A large, heavy, leather-covered ball, thrown and caught for physical exercise.

medicine lodge A lodge in a North American Indian village, used for ritualistic, religious ceremonies.

medicine man Among North American Indians, one professing supernatural powers of healing and of invoking the spirits; a shaman; magician; wizard.

med·i·co (med′ə·kō) *n. pl.* **·cos** *Colloq.* A doctor or a medical student. [<Ital, a physician]

medico- *combining form* Pertaining to medical science and: *medico-legal*, medical and legal. Also, before vowels, **medic-.** [<L *medicus* a physician]

me·di·e·val (mē′dē·ē′vəl, med′ē-) *adj.* Belonging to or descriptive or characteristic of the Middle Ages: also spelled *mediaeval.* [<L *medius* middle + *aevum* age] — **me′di·e′val·ly** *adv.*

Medieval Greek See under GREEK.

me·di·e·val·ism (mē′dē·ē′vəl·iz′əm, med′ē-) *n.* The spirit or practices of the Middle Ages; the flavor or general tone of medieval life; devotion to the institutions, ideas, or traits of the Middle Ages; also, any custom, idea, etc., surviving from the Middle Ages. — **me′di·e′val·ist** *n.*

Medieval Latin See under LATIN.

Me·di·na (mə·dē′nə) A Moslem holy city in Hejaz, western Saudi Arabia; goal of Mohammed's Hegira and the place of his tomb.

medio- *combining form* Middle. Also, before vowels, **medi-.** [<L *medius* middle]

me·di·o·cre (mē′dē·ō′kər, mē′dē·ō′kər) *adj.* Of only middle quality; ordinary; commonplace. [<L *mediocris* <*medius* middle]

me·di·oc·ri·ty (mē′dē·ok′rə·tē) *n. pl.* **·ties** 1 Commonplace ability or condition. 2 A commonplace person.

med·i·tate (med′ə·tāt) *v.* **·tat·ed, ·tat·ing** *v.i.* To engage in continuous and contemplative thought; muse; cogitate. — *v.t.* To think about doing; plan; intend: to *meditate* mischief. See synonyms under CONSIDER, DELIBERATE, MUSE. [<L *meditatus*, pp. of *meditari* muse, ponder] — **med′i·tat′er, med′i·ta′tor** *n.* — **med′i·ta′tive** *adj.* — **med′i·ta′tive·ly** *adv.*

med·i·ta·tion (med′ə·tā′shən) *n.* The act of meditating; the turning or revolving of a subject in the mind; continuous thought; contemplation. See synonyms under REFLECTION, THOUGHT.

med·i·ter·ra·ne·an (med′ə·tə·rā′nē·ən) *adj.* Enclosed nearly or wholly by land, as a sea or other large body of water; landlocked. [<L *medius* middle + *terra* earth]

Med·i·ter·ra·ne·an (med′ə·tə·rā′nē·ən) *adj.* 1 Of or pertaining to the Mediterranean Sea. 2 Inhabiting the shores of the Mediterranean. — *n.* 1 The Mediterranean Sea. 2 One who lives in a Mediterranean country, or belongs to the Mediterranean race. — **Key of the Mediterranean** Gibraltar.

Mediterranean fever Undulant fever.

Mediterranean race A subdivision of the Caucasoid race, regarded as designating the peoples inhabiting the shores of the Mediterranean Sea, including the ancient Iberian, Ligurian, Pelasgian, and Hamitic stocks and their modern descendants.

Mediterranean Sea An arm of the Atlantic Ocean comprising a great inland sea between Europe, Asia Minor, and Africa; 965,000 square miles.

me·di·um (mē′dē·əm) *n. pl.* **·di·ums** *(always for def. 5)* or **me·di·a** (-dē·ə) 1 An intermediate quality, degree, or condition; mean. 2 A surrounding or enveloping element; condition of life; environment. 3 Any substance, as air, through or in which something may move or an effect be produced: Air is a *medium* of sound. 4 An intermediate means or agency; instrument: Radio is an advertising *medium.* 5 A person believed to be in communication with or controlled by the personality of someone deceased. 6 In painting, a liquid which gives fluency to the pigment. 7 A mathematical mean. 8 *Bacteriol.* A substance sterilized by heat in which bacteria, viruses, and other micro-organisms are developed: also **culture medium.** 9 A size of paper, usually 18 x 23 inches, between demy and royal. — **circulating medium** A money currency. — *adj.* 1 Intermediate in quantity, quality, position, size, or degree; middle. 2 Mediocre. [<L, orig. neut. sing. of *medius* middle]

medium bomber See under BOMBER.

me·di·um·is·tic (mē′dē·əm·is′tik) *adj.* Of or pertaining to spiritualist mediums or to their practices.

me·dji·di·e (me·jē′dē·e) *n.* 1 A modern Turkish silver coin equivalent to 19 piasters. 2 A Turkish gold coin, the lira. [<Turkish *mejīdī*]

med·lar (med′lər) *n.* 1 A small, European tree *(Mespilus germanica)* of the rose family. 2 Its edible fruit, hard and bitter when ripe, but agreeably acid when it begins to decay. [<OF *medler*, var. of *meslier* <*mesle* fruit of the medlar <L *mespila* <Gk. *mespilē*]

med·ley (med′lē) *n.* **1** A mingled and confused mass of ingredients, usually incongruous; a heterogeneous group; jumble. **2** A composition of different songs or parts of songs arranged to run as a continuous whole. **3** A cloth woven from yarn of mingled colors: properly including blue and black. — *adj.* **1** Mixed; confused. **2** Of mixed colors; motley. [<OF *medlee*, orig. fem. pp. of *medler*. See MEDDLE.]

Mé·doc (mā·dôk′) *n.* A red wine originally made in Médoc, France.

Mé·doc (mā·dôk′) A region of SW France.

me·dul·la (mə·dul′ə) *n. pl.* **·lae** (-ē) **1** *Anat.* **a** The inner portion of an organ or part, distinguished from the cortex. **b** The marrow of long bones. **c** The pith of a hair. **d** The spinal cord: also **medulla spi·na·lis** (spī·nā′lis). **2** The ganglion of the brain which, connecting with the spinal cord, controls breathing, swallowing, circulation, etc.: also **medulla ob·lon·ga·ta** (ob′lông·gā′tə). **3** *Bot.* The inner central columnar mass of parenchymatous tissue in the stems and roots of certain plants; also, in lichens, the middle layer of tissue composing the thallus, and in fungi proper, the central tissue within the rind of the fungus body. [<L <*medius* middle] — **med·ul·lar·y** (med′ə·ler′ē, mi·dul′ər·ē), **me·dul′lar** *adj.*

medullary rays 1 *Anat.* Extensions of the tubules of the kidney into the cortical substance. **3** *Bot.* The vertical bands or plates of cellular (parenchymatous) tissue, proceeding from the pith to the surface, and characteristic of the species of exogenous plants.

medullary sheath Myelin. See illustration under EXOGEN.

med·ul·lat·ed (med′ə·lā′tid, mi·dul′ā·tid) *adj.* Provided with a medullary sheath: said of nerve fibers.

me·du·sa (mə·dōō′sə, -zə, -dyōō′-) *n. pl.* **·sas** or **·sae** (-sē, -zē) A jellyfish. [<L] — **me·du′san** *adj.*

Me·du·sa (mə·dōō′sə, -zə, -dyōō′-) One of the Gorgons, killed by Perseus who gave her head to Athena.

me·du·soid (mə·dōō′soid, -zoid, -dyōō′-) *adj.* Resembling a medusa or jellyfish. — *n.* **1** A medusa-shaped gonophore of a hydrozoan. **2** Any medusa.

Med·way (med′wā) A river rising in SE Surrey and NE Sussex, England, and flowing 70 miles NE through western Kent to the Thames estuary at Sheerness.

meed (mēd) *n. Archaic* **1** A well-deserved reward; recompense. **2** A present, gift, or bribe. ◆ Homophone: *mead*. [OE *mēd*]

meek (mēk) *adj.* **1** Of gentle and long-suffering disposition. **2** Submissive; compliant; lacking spirit or backbone. **3** Humble; lowly. **4** *Obs.* Gentle; indulgent; kind; compassionate. See synonyms under HUMBLE, PACIFIC. — *adv.* Meekly. [<ON *miukr* gentle, soft] — **meek′ly** *adv.* — **meek′ness** *n.*

meer (mir) *n. Scot.* Mare.

meer·schaum (mir′shəm, -shôm, -shoum) *n.* **1** A soft, light, compact, heat-resisting magnesium silicate, $H_4Mg_2Si_3O_{10}$, used for carving into tobacco pipes, cigar holders, and the like; sepiolite. It is closely related to talc. **2** A pipe made from this material. [<G <*meer* sea + *schaum* foam]

Mee·rut (mē′rət) A city in NW Uttar Pradesh, India.

Meer van Delft (mār vän delft) See VERMEER, JAN.

meet[1] (mēt) *v.* **met, meet·ing** *v.t.* **1** To come upon; encounter. **2** To make the acquaintance of; be introduced to. **3** To be at the place of arrival of: We *met* him at the station. **4** To come into contact, conjunction, or intersection with; join: where the path *meets* the road. **5** To keep an appointment with. **6** To come into the view, hearing, etc., of: A ghastly sight *met* our eyes. **7** To experience; undergo: to *meet* bad weather. **8** To oppose in battle; fight with. **9** To face or counter: to *meet* a blow with a blow. **10** To deal with; refute or cope with: to *meet* an accusation. **11** To comply with; act or result in conformity with, as expectations or wishes. **12** To pay, as a bill. — *v.i.* **13** To come together, as from opposite or different directions. **14** To come together in contact, conjunction, or intersection; join. **15** To assemble. **16** To make acquaintance or be introduced. **17** To come together in conflict or opposition; contend. **18** To agree. — **to meet up with** *U.S. Colloq.* **1** To encounter. **2** To experience; undergo. — **to meet with 1** To come upon; encounter. **2** To deal or confer with. **3** To experience. **4** To be the subject or recipient of. — *n.* **1** An assembling together of huntsmen; also, the company or rendezvous. **2** An athletic contest: a track meet. ◆ Homophones: *meat, mete*. [OE *mētan*]

meet[2] (mēt) *adj.* Suitable, as to an occasion; adapted; fit. See synonyms under APPROPRIATE, BECOMING. — *adv. Obs.* Meetly; suitably. ◆ Homophones: *meat, mete*. [OE *gemǣte*] — **meet′ly** *adv.* — **meet′ness** *n.*

meet·ing (mē′ting) *n.* **1** A coming together. **2** An assembly of persons; specifically, a congregation of the Friends or Quakers; also, their meeting house. See synonyms under ASSEMBLY, COLLISION, COMPANY.

meeting house 1 A house used for public meetings of any kind, but especially for public worship. **2** A place of worship used by the Friends.

mega- *combining form* **1** Great; large; powerful: *megaphone*. **2** In the metric system, electricity, etc., a million, or a million times, as in the following:

megabar	mega-erg	megampere
megacycle	megafarad	megavolt
megadyne	megameter	megohm

Also, before vowels, **meg-**. [<Gk. *megas* large]

meg·a·ce·phal·ic (meg′ə·sə·fal′ik) *adj.* Large-headed; specifically, having a cranial capacity above the average, or more than 1500 cubic centimeters. Also **meg′a·ceph′a·lous** (-sef′ə·ləs).

Me·gae·ra (mə·jir′ə) In Greek mythology, one of the three Furies. Also **Me·gæ′ra**. [<L <Gk. *Megaira* < *magairein* grudge]

meg·a·ga·mete (meg′ə·gə·mēt′, -gam′ēt) *n.* A macrogamete.

meg·a·lith (meg′ə·lith) *n. Archeol.* A huge stone, especially one used in prehistoric monuments. — **meg′a·lith′ic** *adj.*

megalo- *combining form* Big; indicating excessive or abnormal size: *megalocephalic*. Also, before vowels, **megal-**. [<Gk. *megas, megalou* big]

meg·a·lo·car·di·a (meg′ə·lō·kär′dē·ə) *n. Pathol.* Morbid enlargement of the heart. [<MEGALO- + Gk. *kardia* heart]

meg·a·lo·ceph·a·ly (meg′ə·lō·sef′ə·lē) *n.* **1** Unusual largeness of the head. **2** *Pathol.* Progressive enlargement of the bones and soft tissues of the head. Also **meg·a·lo·ce·pha′li·a** (-sə·fā′lē·ə). [<MEGALO- + Gk. *kephalē* head] — **meg·a·lo·ce·phal′ic** (-sə·fal′ik), **meg′a·lo·ceph′a·lous** *adj.*

meg·a·lo·ma·ni·a (meg′ə·lō·mā′nē·ə, -mān′yə) *n.* **1** *Psychiatry* A mental disorder in which the subject thinks himself great or exalted; delusions of grandeur, power, etc. **2** A tendency to magnify and exaggerate. — **meg′a·lo·ma′ni·ac** *adj. & n.* — **meg′a·lo·ma·ni′a·cal** (-mə·nī′ə·kəl) *adj.*

meg·a·lop·si·a (meg′ə·lop′sē·ə) *n.* Macropsia.

Meg·a·lop·ter·a (meg′ə·lop′tər·ə) *n. pl.* An order of soft-bodied insects with large wings, long antennae, chewing mouth parts, and aquatic larvae; alderflies and hellgrammites or dobson flies. [<MEGALO- + Gk. *pteron* wing] — **meg′a·lop′ter·ous** *adj.*

meg·a·lo·saur (meg′ə·lō·sôr) *n. Paleontol.* A gigantic, terrestrial, carnivorous dinosaur (genus *Megalosaurus*) of the suborder *Therapoda*, which flourished in the Jurassic period. [<MEGALO- + Gk. *sauros* lizard] — **meg′a·lo·saur′i·an** *adj. & n.*

Meg·an·thro·pus (meg·an′thrə·pəs) *n. Zool.* An extinct giant hominoid primate of the Pleistocene, represented by massive fossil jawbones found in 1939 and 1941 in central Java. [<NL <Gk. *megas* great + *anthrōpos* man]

meg·a·phone (meg′ə·fōn) *n.* A funnel-shaped device for amplifying or directing sound. — *v.t. & v.i.* **·phoned, ·phon·ing** To address or speak through a megaphone.

meg·a·pod (meg′ə·pod) *adj.* Having large feet, as certain Australian jungle birds.

Meg·a·ra (meg′ə·rə) An ancient city in east central Greece, capital of Megaris. — **Me·gar·i·an** (mə·gâr′ē·ən), **Me·gar·ic** (mə·gar′ik) *adj.*

Megarian school A school of philosophy founded at Megara about 400 B.C. by Euclid, a disciple of Socrates, holding that all being is unity and equal to the one true good. The school emphasized the art of disputation, from which it gained the name of *Eristic school*, thus formalizing the Socratic method.

Meg·a·ris (meg′ə·ris) A mountainous region of ancient Greece on the isthmus between the Peloponnesus and northern Greece.

meg·a·scope (meg′ə·skōp) *n.* A form of magic lantern for throwing enlarged images on a screen. — **meg′a·scop′ic** (-skop′ik) *adj.* — **meg′a·scop′i·cal·ly** *adv.*

meg·a·spo·ran·gi·um (meg′ə·spə·ran′jē·əm) *n. pl.* **·gi·a** (-jē·ə) *Bot.* A sporangium which bears only megaspores: also called *macrosporangium*.

meg·a·spore (meg′ə·spôr, -spōr) *n. Bot.* A large, asexual spore developed by certain seed plants, and giving rise always to a female gamete; the embryo sac of a seed plant: also called *macrospore*.

meg·a·spo·ro·phyll (meg′ə·spôr′ə·fil, -spō′rə-) *n. Bot.* A leaf or sporophyll which produces only megasporangia.

me·gass (mə·gas′, -gäs′) *n.* Bagasse. Also **me·gasse′**.

meg·a·there (meg′ə·thir) *n. Paleontol.* A gigantic, extinct, slothlike edentate (genus *Megatherium*), associated with the Pleistocene in North America. [<MEGA- + Gk. *thērion* wild animal]

MEGATHERE
(Fossils indicate length up to 20 feet)

meg·a·therm (meg′ə·thûrm) *n.* A macrotherm.

meg·a·tron (meg′ə·tron) *n.* A compact, sturdy type of vacuum tube designed to increase the range of wave frequencies and power in radio, television, and other electronic fields. [<MEGA- + (ELEC)TRON]

Megh·na (meg′nə) A river in East Pakistan, flowing 132 miles SW to the Bay of Bengal.

Me·gid·do (mə·gid′ō) An ancient city of NW Palestine at the western edge of the Plain of Jezreel.

me·gilp (mə·gilp′) See MAGILP.

me·grim (mē′grim) *n.* **1** A headache confined to one side of the head, characterized by nausea and vomiting; migraine. **2** *pl.* Dulness; depression of spirits. **3** A whim or fad. [<F *migraine*]

Me·hem·et A·li (mə·hem′et ä′lē, me·met′), 1769–1849, viceroy of Egypt 1805–48. Also known as *Mohammed Ali*.

Mei·ji (mā·jē) *n.* The reign, 1867–1912, of the Emperor Mutsuhito, 1852–1912, of Japan, regarded as a historic era. [<Japanese, lit., enlightened peace]

mei·kle (mē′kəl) See MICKLE.

Meik·le·john (mik′əl·jon), **Alexander**, born 1872, U.S. educator born in England.

Meil·hac (me·yàk′), **Henri**, 1831–97, French playwright.

Mei·nin·gen (mī′ning·ən) A town in the former state of Thuringia, SW East Germany.

Mein Kampf (mīn kämpf′) A book by Adolf Hitler, outlining his political philosophy and personal history, and setting forth his plans for the German domination of Europe. [<G, my battle]

mei·ny (mā′nē) *n.* **1** *Obs.* An army or retinue; attendants; household; crew. **2** *Scot.* A multitude or throng. Also **mei′nie**. [<OF *mesnee, meyné*, ult. <L *mansio* a dwelling. See MANSION.]

mei·o·sis (mī·ō′sis) *n.* **1** *Biol.* That process in the division of germ cells by which the number of chromosomes is reduced from the double or *diploid* number typical of somatic cells to the halved or *haploid* number characteristic of gametes: distinguished from *mitosis*: also spelled *maiosis*. 2 In rhetoric, understatement, often giving the effect of irony or humor, by representing a fact, thing, deed, etc., as being smaller than it really is: also spelled *miosis*. [<Gk. *meiōsis* lessening] —**mei·ot'ic** (-ot'ik) *adj.*

Meis·sen (mī'sən) A city on the Elbe in SE East Germany in the former state of Saxony.

Meissen ware A kind of porcelain or chinaware made in Meissen, Germany.

Meis·so·nier (mā·sô·nyā'), **Jean Louis Ernest,** 1815–91, French painter.

Meis·ter·sing·er (mīs'tər·sing'ər, *Ger.* mīs'tər·zing'ər) *n. pl.* **·sing·er** *German* One of the burgher poets and musicians of Germany in the 14th, 15th, and 16th centuries, the successors of the minnesingers: often called *mastersinger*. —**Die Meistersinger von Nürnberg** Title of a comic opera by Richard Wagner.

Meit·ner (mīt'nər), **Lise,** born 1878, German physicist.

Mé·ji·co (mā'hē·kō) The Spanish name for MEXICO.

Mek·ka (mek'ə) See MECCA.

Mek·nès (mek·nes') A city in NW Morocco: formerly *Mequinez*.

Me·kong (mā·kong') A river in SE Asia, flowing 2,600 miles south to the China Sea: Chinese *Lantsang*.

mel (mel) *n.* Honey; especially the pure, clarified honey used in the preparation of certain drugs. [<L]

mel·a·mine (mel'ə·mēn, -min) *n. Chem.* A transparent, colorless, crystalline compound, $C_3N_2(NH_2)_3$, the amide of cyanuric acid: combined with cellulose pulp and formaldehyde, it produces a synthetic resin of good qualities. [< *mel(am)*, a chemical compound + AMINE]

melan– Var. of MELANO–.

mel·an·cho·li·a (mel'ən·kō'lē·ə) *n. Psychiatry* Mental disorder characterized by excessive brooding and depression of spirits: typical of manic-depressive psychoses. [<L. See MELANCHOLY.] —**mel'an·cho'li·ac** *adj.* & *n.*

mel·an·chol·y (mel'ən·kol'ē) *adj.* 1 Morbidly gloomy; sad; dejected. 2 Suggesting or promoting sadness. 3 Somberly thoughtful; pensive. —*n.* 1 Low spirits; despondency; depression. 2 Melancholia. 3 Pensive contemplation; serious and sober reflection. 4 *Archaic* The dark, acrid, and viscous substance once believed to be secreted by the kidneys and to be responsible for gloomy dejection of spirits; one of the humors. [<F *melancolie* <L *melancholia* <Gk. < *melas, -anos* black + *cholē* bile] —**mel'an·chol'ic** *adj.* —**mel'an·chol'i·cal·ly** *adv.*

Me·lanch·thon (mə·langk'thən), **Philipp,** Grecized name of Philipp Schwarzert, 1497–1560, religious reformer and scholar; associate of Luther. Also **Me·lanc'thon.**

mel·a·ne·mi·a (mel'ə·nē'mē·ə) *n. Pathol.* A morbid excess of black pigment in the blood: noted chiefly in pernicious anemia. Also **mel'·a·nae'mi·a.** [<MELAN- + Gk. *haima* blood] —**mel'a·ne'mic** *adj.*

Mel·a·ne·sia (mel'ə·nē'zhə, -shə) The islands of the western Pacific south of the Equator, comprising one of the three main divisions of the Pacific Islands; total, 60,000 square miles.

Mel·a·ne·sian (mel'ə·nē'zhən, -shən) *n.* 1 One of the native people of Melanesia, having dark skins and thick, kinky beards and hair: believed to be a mixture of Papuan, Polynesian, and Malay stocks. 2 A branch of the Oceanic subfamily of the Austronesian family of languages, including the languages spoken in the Solomon, Loyalty, and Admiralty islands, Fiji, the New Hebrides, New Caledonia, etc., and the Micronesian group of the Caroline, Gilbert, and Marshall islands. —*adj.* Of or pertaining to Melanesia, its native inhabitants, or their languages.

mé·lange (mā·länzh') *n. French* A mixture or medley; also, a literary miscellany.

me·la·ni·an (mə·lā'nē·ən) *adj. Anthropol.* Having dark or black pigmentation: said of Negroes, Melanesians, etc. [<F *mélanien* <Gk. *melas, -anos* black]

me·lan·ic (mə·lan'ik) *adj.* 1 Relating to or resembling melanosis or melanism; melanoid. 2 Black; melanian.

mel·a·nin (mel'ə·nin) *n. Biochem.* The brownish-black pigment contained in various animal tissues, as the skin and hair: it is formed by the action of the enzyme tyrosinase upon tyrosine.

mel·a·nism (mel'ə·niz'əm) *n.* 1 Abnormal development of dark coloring matter in the skin, feathers, etc.: opposed to *albinism*. 2 Excessive darkness, as of the eyes, hair, skin, etc., due to extreme pigmentation. —**mel'a·nis'tic** *adj.*

mel·a·nite (mel'ə·nīt) *n.* A black variety of garnet.

melano– *combining form* Black; dark-colored: *melanosis*. Also, before vowels, *melan–*. [< Gk. *melas, melanos* black]

Mel·a·noch·ro·i (mel'ə·nok'rō·ī) *n. pl.* Caucasians having fair skins and dark hair. [< MELAN- + Gk. *ōchros* pale] —**Mel'a·noch'·roid** *adj.*

mel·an·o·crat·ic (mel'ən·ō·krat'ik) *adj. Geol.* Of or pertaining to those igneous rocks characterized by a predominance of dark or ferromagnesian minerals: opposed to *leucocratic*.

mel·a·noid (mel'ə·noid) *adj.* 1 Looking black or having a dark appearance. 2 Of the nature of melanosis.

mel·a·no·ma (mel'ə·nō'mə) *n. Pathol.* A black-pigmented tumor. [<MELAN- + -OMA]

mel·a·no·sis (mel'ə·nō'sis) *n. Pathol.* An organic disease in which pigment is deposited in the skin and other tissues; black degeneration. —**mel'a·not'ic** (-not'ik) *adj.*

mel·a·nous (mel'ə·nəs) *adj.* Having dark or black skin and hair: opposed to *xanthous*.

mel·an·tha·ceous (mel'ən·thā'shəs) *adj.* Of or pertaining to a former family *(Melanthaceae)* of monocotyledonous plants in the lily order, distinguished from lilies by the absence of bulbs: it included plants of the genera *Colchicum* and *Veratrum*. [<NL <Gk. *melas, -anos* black + *anthos* flower]

mel·a·phyre (mel'ə·fīr) *n.* Any igneous porphyry with a dark groundmass. [<F <Gk. *melas* black + F *(por)phyre* porphyry]

Mel·ba (mel'bə), **Dame Nellie,** 1866–1931, Australian operatic soprano: real name *Nellie Mitchell Armstrong*.

Melba toast Thinly sliced bread toasted until brown and crisp.

Mel·bourne (mel'bərn) A port and capital of Victoria, Australia.

Mel·bourne (mel'bərn), **Viscount,** 1779–1848, William Lamb, English statesman; prime minister 1834, 1835–41.

Mel·chi·or (mel'kē·ôr) One of the three Magi.

Mel·chis·e·dec (mel·kiz'ə·dek) In Old Testament history, a priest; king of Salem. *Gen.* xiv 18. Also **Mel·chiz'e·dek.**

meld (meld) *v.t.* & *v.i.* In pinochle and other card games, to announce or declare (a combination of cards in the hand), for inclusion in one's total score. —*n.* A group of cards to be declared, or the act of declaring them. [<G *melden* announce]

mel·der (mel'dər) *n. Scot.* The quantity of grain ground at one time; a grist.

Mel·e·a·ger (mel'ē·ā'jər) In Greek mythology, the son of Oeneus and Althaea, who slew the Calydonian boar and in a quarrel over the spoils killed his mother's brothers: Althaea took revenge by burning the log she had removed from the hearth at his birth, when it had been prophesied that her son would die after it was consumed.

mê·lée (mā'lā, mā·lā'; *Fr.* me·lā') *n.* A general hand-to-hand fight; an affray. [<F <OF *meslee*, var. of *medlee*. See MEDLEY.]

Me·le·gna·no (mā'lā·nyä'nō) A town in northern Italy SE of Milan; scene of French victory over Swiss, 1515: formerly *Marignano*.

me·li·a·ceous (mē'lē·ā'shəs) *adj.* Of or pertaining to a family *(Meliaceae)* of trees and shrubs of the order *Geraniales*, mainly native to the warm portions of Asia and America; the mahogany family, including the Spanish cedar (genus *Cedrela*). [<NL <Gk. *melia* an ash tree]

mel·ic (mel'ik) *adj.* Suitable for singing, or meant to be sung: said of poetry. In ancient Greek poetry it is the successor of the elegiac and iambic forms of verse and includes the Aeolian or single-voice lyric, and the Dorian or choral lyric. —*n.* Melic poetry. Compare ELEGIAC and IAMBIC. [<Gk. *melikos* < *melos* song]

Me·lil·la (mā·lē'lyä) A Spanish possession on the NW coast of Africa, constituting a port and enclave in eastern Morocco.

mel·i·lot (mel'ə·lot) *n.* Any one of several sweet-smelling, cloverlike herbs of the genus *Melilotus*, especially the **sweet clover** *(M. officinalis)* and the **Bokhara clover** *(M. alba)*. [<OF *melilot* <LL *melilotos* <Gk. *melilōtos* < *meli* honey + *lōtos* a lotus]

mel·i·nite (mel'ə·nīt) *n.* An explosive of great power and similar to lyddite, yielded by combining guncotton with picric acid. [<Gk. *mēlinos* yellow]

mel·io·rate (mēl'yə·rāt) *v.t.* & *v.i.* **·rat·ed, ·rat·ing** To improve, as in quality or condition; ameliorate. See synonyms under AMEND. [<LL *melioratus*, pp. of *meliorare* improve < *melior* better] —**mel'io·ra·ble** *adj.* —**mel'·io·ra'tive** *adj.* —**mel'io·ra'tor** *n.*

mel·io·ra·tion (mēl'yə·rā'shən) *n.* 1 A betterment. 2 *Ling.* An improvement or elevation in the meaning of a word, as in *nice* (formerly "foolish"): opposed to *pejoration*.

mel·io·rism (mēl'yə·riz'əm) *n.* 1 The improvement of society by bettering man's physical being and environment instead of by ethical or religious means. 2 A modified optimism, teaching that the world is neither the best nor the worst possible, but is susceptible of improvement through the increase of good as man evolves. Compare OPTIMISM and PESSIMISM. [<L *melior* better] —**mel'io·rist** *adj.* & *n.* —**mel'io·ris'tic** *adj.*

mel·ior·i·ty (mēl·yôr'ə·tē, -yor'-) *n.* The state of being better; superiority.

mel·is·mat·ic (mel'is·mat'ik) *adj. Music* Florid and ornate in phrasing. [<NL *melisma* song <Gk.]

Me·lis·sa (mə·lis'ə, *Ital.* mä·lēs'sä) A feminine personal name. Also *Fr.* **Mé·lisse** (mā·lēs') or **Mé·lite** (mā·lēt'). [<Gk., bee]

Mel·i·ta (mel'ə·tə) The ancient name for MALTA.

Mel·i·to·pol (mel'ə·tô'pəl) A city in southern Ukrainian S.S.R.

mell (mel) *v.t.* & *v.i. Obs.* 1 To mix; mingle. 2 To meddle. [<OF *meller*, var. of *mesler*. See MEDDLE.]

mel·lif·er·ous (mə·lif'ər·əs) *adj.* Producing or bearing honey. Also **mel·lif'ic.** [<L *mellifer* honey-bearing < *mel* honey + *ferre* bear]

mel·lif·lu·ous (mə·lif'lo͞o·əs) *adj.* 1 Sweetly or smoothly flowing; dulcet; honeyed. 2 Flowing like or as with honey. Also **mel·lif'lu·ent.** [<L *mellifluus* < *mel* honey + *fluere* flow] —**mel·lif'lu·ence, mel·lif'lu·ous·ness** *n.* —**mel·lif'lu·ous·ly, mel·lif'lu·ent·ly** *adv.*

mel·liph·a·gous (mə·lif'ə·gəs) *adj.* Feeding on honey, as certain animals and birds: often **mel·liv'o·rous** (-liv'ər·əs). Also **me·liph'a·gous.** [<Gk. *meli* honey + *phagein* to eat]

mel·lite (mel'īt) *n.* 1 Any medicated preparation containing honey. 2 The mineral honeystone. [<L *mel* honey + -ITE[1]]

Mel·lon (mel'ən), **Andrew William,** 1855–1937, U. S. banker, secretary of the treasury 1921–1932.

mel·lo·phone (mel'ō·fōn) *n.* A circular althorn. [<MELLOW + -PHONE]

mel·low (mel'ō) *adj.* 1 Soft by reason of ripeness; well-matured; not bitter or acid: *mellow* fruit; *mellow* wine. 2 Of a rich or delicate quality: *mellow* tints; *mellow* tones. 3 Companionable; jolly. 4 Made jovial by liquor. 5 Soft and friable, as soil. See synonyms under RIPE. —*v.t.* & *v.i.* To make or become mellow; ripen; soften. [ME *melwe*, ? <OE *melu* meal. Akin to Flemish *meluw* soft, tender.] —**mel'low·ly** *adv.* —**mel'low·ness** *n.*

me·lo·de·on (mə·lō'dē·ən) *n.* A small reed organ or harmonium. [<Gk. *melōdia* melody]

me·lo·di·a (mə·lō'dē·ə) *n.* An organ stop having wood pipes and a tone nearly like the clarabella; a stopped diapason. [<LL]

me·lod·ic (mə·lod'ik) *adj.* Pertaining to or containing melody; melodious. —**me·lod'i·cal·ly** *adv.*

add, āce, câre, pälm; end, ēven; it, īce; odd, ōpen, ôrder; to͝ok, po͞ol; up, bûrn; ə = a in *above*, e in *sicken*, i in *clarity*, o in *melon*, u in *focus*; yo͞o = u in *fuse*; oi, oil; ou, pout; ch, check; g, go; ng, ring; th, thin; th, this; zh, vision. Foreign sounds à, œ, ü, kh, ṅ; and ◆: see page xx. < from; + plus; ? possibly.

me·lod·ics (mə·lod′iks) *n. pl.* *(construed as singular)* The branch of musical science relating to the pitch of tones and the principles of melody.

me·lo·di·ous (mə·lō′dē·əs) *adj.* Agreeable to the ear; producing or characterized by melody; tuneful. **— me·lo′di·ous·ly** *adv.* **— me·lo′di·ous·ness** *n.*

mel·o·dize (mel′ə·dīz) *v.* **·dized, ·diz·ing** *v.t.* **1** To make melodious. **2** To compose melody for. **—** *v.i.* **3** To make melody or melodies. **— mel′o·diz′er, mel′o·dist** *n.*

mel·o·dra·ma (mel′ə·drä′mə, -dram′ə) *n.* **1** Originally, a drama with a romantic story or plot, sensational incidents, and usually including some music and song. **2** Any sensational and emotional drama, usually having a happy ending. **3** Behavior or language of a theatrical nature. [<F *mélodrame* <Gk. *melos* song + *drama* drama] **— mel′o·dram′a·tist** *n.*

mel·o·dra·mat·ic (mel′ə·drə·mat′ik) *adj.* Of, pertaining to, or like melodrama; sensational. **— mel′o·dra·mat′i·cal·ly** *adv.*

mel·o·dra·mat·ics (mel′ə·drə·mat′iks) *n. pl.* Melodramatic behavior.

mel·o·dy (mel′ə·dē) *n. pl.* **·dies** **1** Pleasing sounds or an agreeable succession of such sounds. **2** Musical sounds or quality, as in the words of a poem. **3** A poem written or suitable for being set to music. **4** *Music* **a** A succession of simple tones, usually in the same key, constituting, in combination, a rhythmic whole: distinguished as a formal element from *harmony* and *rhythm.* **b** The chief part or voice in a harmonic composition; the air. [<OF *melodie* <LL *melodia* <Gk. *melōidia* choral song <*melōidos* melodious <*melos* song + *aoidos* singer]

Synonyms: harmony, music, symphony, unison. *Harmony* is simultaneous; *melody* is successive; *harmony* is the correspondence of two or more notes sounded at once, *melody* the succession of a number of notes continuously following one another. A *melody* may be wholly in one part; *harmony* must be of two or more parts. Accordant notes of different pitch sounded simultaneously produce *harmony*; *unison* is the simultaneous sounding of two or more notes of the same pitch. Tones sounded at the interval of an octave are also said to be in *unison,* but this is not literally exact. *Music* may denote the simplest *melody* or the most complex and perfect *harmony.* A *symphony* (apart from its technical orchestral sense) is any pleasing consonance of musical sounds, vocal or instrumental. Compare METER², SONG, TUNE. *Antonyms*: discord, dissonance.

Me·lo·i·dae (mə·lō′ə·dē) *n. pl.* A family of coleopterous insects with plump cylindrical bodies; the blister beetles. [<NL <*meloē* oil beetle; ult. origin unknown] **— mel·oid** (mel′oid) *adj.* & *n.*

Mel·o·lon·thi·nae (mel′ə·lon·thī′nē) *n. pl.* A subfamily of beetles (family *Scarabaeidae*) including the cockchafers and June beetles. [<NL <Gk. *mēlolonthē* cockchafer] **— mel′o·lon′thine** (-thin, -thin) *adj.* & *n.*

mel·o·ma·ni·a (mel′ə·mā′nē·ə, -mān′yə) *n.* An excessive or morbid fondness for music. [<Gk. *melos* song + -MANIA] **— mel′o·ma′ni·ac** *n.*

mel·on (mel′ən) *n.* A trailing plant of the gourd family, or its fruit. There are two genera, the muskmelon and the watermelon, each with numerous varieties. [<F <LL *melo, melonis* <L *melopepo* <Gk. *mēlopepōn* apple-shaped melon <*mēlon* apple + *pepōn* melon]

mel·on·ite (mel′ən·īt) *n.* A reddish-white, granular nickel telluride, Ni_2Te_3; tellurnickel. [from *Melones* mine, Calif., where found]

Me·los (mē′los) See MILO.

Mel·pom·e·ne (mel·pom′ə·nē) The Muse of tragedy. [<Gk. *Melpomenē,* lit., the songstress <*melpein* sing]

Mel·rose (mel′rōz) A burgh of SE Scotland; site of a ruined Cistercian abbey, founded 1136.

melt (melt) *v.t.* & *v.i.* **melt·ed, melt·ed** (*Archaic* **mol·ten), melt·ing** **1** To reduce or change from a solid to a liquid state by heat; fuse. **2** To dissolve, as in water. **3** To disappear or cause to disappear; dissipate: often with *away.* **4** To blend by imperceptible degrees; merge. **5** To make or become softened in feeling or attitude. **—** *n.* **1** Something melted. **2** A single operation of fusing. **3** The amount of a single fusing. [OE *meltian.* Akin to MALT.] **— melt′a·ble** *adj.* **— melt′a·bil′i·ty** *n.* **— melt′er** *n.*

Synonyms (verb): dissolve, fuse, liquefy, thaw. *Antonyms*: congeal, freeze, harden, indurate, solidify.

melt·age (mel′tij) *n.* **1** The process of melting. **2** The amount resulting from melting.

melting point The temperature at which a specified solid substance becomes liquid.

melting pot **1** A vessel in which things are melted; crucible. **2** A country, city, or region in which immigrants of various racial and cultural backgrounds are assimilated.

mel·ton (mel′tən) *n.* A heavy woolen cloth with a short nap: used for overcoats. [after *Melton* Mowbray, England]

melt·wa·ter (melt′wô′tər, -wot′ər) *n.* The whitish water from melting glaciers.

Me·lun·geon (mə·lun′jən) *n.* One of a dark-skinned people of mixed white, Negro, and Indian stock, living in the mountains of Tennessee. [? <F *mélange* mixture]

mel·vie (mel′vē) *v.t. Scot.* To cover with meal.

Mel·ville (mel′vil), **Herman,** 1819–91, U.S. novelist and poet.

Melville Bay An inlet of Baffin Bay in NW Greenland.

Melville Island **1** The largest of the Parry Islands, in the Arctic Ocean, Northwest Territories, Canada; 16,503 square miles. **2** An Australian island comprising part of Northern Territory, Australia; 2,400 square miles.

Melville Lake A lake in SE Labrador; 120 miles long.

Melville Peninsula A peninsula in Northwest Territories, Canada, between the Gulf of Boothia and Foxe Basin; 24,156 square miles; 250 miles long.

Mel·vin (mel′vin) A masculine personal name. [OE, high protector]

mem (mem) *n.* The thirteenth Hebrew letter. See ALPHABET. [<Hebrew *mēm,* lit., water]

mem·ber (mem′bər) *n.* **1** A person belonging to an incorporated or organized body, society, etc.: a *Member* of Congress, a *member* of a club. **2** A limb or other functional organ of an animal body. **3** A part or element of a structural or composite whole, distinguishable from other parts or elements, as a part of a sentence, syllogism, period, or discourse, or any necessary part of a structural framework, as a tie rod, post, or strut in the truss of a bridge. **4** A subordinate classificatory part: A species is a *member* of a genus. **5** *Bot.* A part of a plant considered with reference to position and structure, but regardless of function. **6** *Math.* **a** Either side of an equation. **b** A set of figures or symbols forming part of a formula or number. **c** Any one of the items forming a series. See synonyms under PART, TERM. [<OF *membre* <L *membrum* limb]

mem·ber·ship (mem′bər·ship) *n.* **1** The state of being a member. **2** The members of an organization, collectively.

mem·brane (mem′brān) *n.* **1** A thin, pliable, sheetlike layer of animal or vegetable tissue serving as a cover, connection, or lining. **2** A piece of parchment. [<L *membrana,* lit., limb coating <*membrum* member] **— mem′bra·nous** (-brə·nəs), **mem′bra·na′ceous** (-nā′shəs) *adj.*

membrane bone *Anat.* A bone developed in membrane, as one of those of the vault of the skull.

Mem·el (mem′əl, *Ger.* mā′məl) **1** A territory of western Lithuanian S.S.R.; 1,026 square miles. **2** Its capital, a port on the Baltic Sea: Lithuanian *Klaipeda.* **3** The German name for the NEMAN.

me·men·to (mə·men′tō) *n. pl.* **·toes** or **·tos** **1** A hint or reminder to awaken memory; souvenir; memorial. **2** *Eccl.* Either of the two prayers in the canon of the mass in which the living and the departed are respectively mentioned. [<L, remember, imperative of *meminisse* remember]

memento mo·ri (môr′ī) *Latin* An emblem or reminder of death, as a skull, etc.: literally, remember that you must die.

Mem·ling (mem′ling), **Hans,** 1430–95, Flemish painter. Also **Mem·linc** (mem′lingk).

Mem·non (mem′non) **1** In Greek legend, a king of the Ethiopians killed by Achilles and made immortal by Zeus. **2** A statue of Amenhotep III at Thebes, Egypt, associated with Memnon by the Greeks and said to emit a musical note when touched by the sun at dawn. **— Mem·no·ni·an** (mem·nō′nē·ən) *adj.*

mem·o (mem′ō) *n. pl.* **mem·os** *Colloq.* A memorandum.

mem·oir (mem′wär) *n.* **1** An account of something deemed worthy of record; especially, one addressed to a public institution or scientific society. **2** *pl.* The reminiscences of a person, either general or relating to a particular period, published together. **3** A biographic memorial. See synonyms under HISTORY. [<F *mémoire* <L *memoria* memory. Doublet of MEMORY.] **— mem′oir·ist** *n.*

mem·o·ra·bil·i·a (mem′ə·rə·bil′ē·ə) *n. pl.* Things worthy of memory, or an account of them. [<L, neut. pl. of *memorabilis* memorable]

mem·o·ra·ble (mem′ər·ə·bəl) *adj.* Worthy to be remembered; noteworthy. [<L *memorabilis*] **— mem′o·ra·bil′i·ty, mem′o·ra·ble·ness** *n.* **— mem′o·ra·bly** *adv.*

mem·o·ran·dum (mem′ə·ran′dəm) *n. pl.* **·dums** or **·da** (-də) **1** Something to be remembered; hence, a brief note of a thing or things to be remembered. **2** *Law* A brief written outline of the terms of a transaction. **3** An informal letter. **4** A statement of goods sent from a consignor to a consignee. See synonyms under RECORD. [<L, a thing to be remembered]

me·mo·ri·al (mə·môr′ē·əl, -mō′rē-) *adj.* **1** Commemorating the memory of a deceased person or of any event. **2** Contained within one's memory: distinguished from *immemorial.* **—** *n.* **1** Something designed to keep in remembrance a person, event, etc. **2** A summary or presentation of facts usually made the ground of a petition or remonstrance. **3** *Law* A memorandum filed for record. See synonyms under HISTORY, RECORD, TRACE. [<OF <L *memorialis*]

Memorial Day Decoration Day.

me·mo·ri·al·ist (mə·môr′ē·əl·ist, -mō′rē-) *n.* **1** One who writes memoirs. **2** One who writes, signs, or presents a memorial.

me·mo·ri·al·ize (mə·môr′ē·əl·īz′, -mō′rē-) *v.t.* **·ized, ·iz·ing** **1** To commemorate. **2** To present a memorial to. Also *Brit.* **me·mo′ri·al·ise′.** **— me·mo′ri·al·i·za′tion** *n.*

mem·o·rize (mem′ə·rīz) *v.t.* **·rized, ·riz·ing** To commit to memory; learn by heart. See synonyms under LEARN. **— mem′o·ri·za′tion** *n.* **— mem′o·riz′er** *n.*

mem·o·ry (mem′ər·ē) *n. pl.* **·ries** **1** The mental process or faculty of representing in consciousness an act, experience, or impression, with recognition that it belongs to time past. **2** The experiences of the mind taken in the aggregate, and considered as influencing present and future behavior. **3** The accuracy and ease with which a person can retain and recall past experiences. **4** That which is remembered, as an act, event, person, or thing. **5** The period of time covered by the faculty of remembrance: beyond the *memory* of man. **6** The state of being remembered; posthumous reputation: The *memory* of Washington will endure. **7** That which reminds; a memorial; a memento. [<OF *memorie* <L *memoria* <*memor* mindful. Doublet of MEMOIR.]

Synonyms: recollection, remembrance, reminiscence, retrospect, retrospection. *Memory* is the faculty by which knowledge is retained or recalled; *memory* is a retention of knowledge within the grasp of the mind, while *remembrance* is the having what is known consciously before the mind. Either may be voluntary or involuntary. *Recollection* involves volition, the mind making a distinct effort to recall something, or fixing the attention actively upon it when recalled. *Reminiscence* is a half-dreamy *memory* of scenes or events long past; *retrospection* is a distinct turning of the mind back upon the past, bringing long periods under survey. *Antonyms*: forgetfulness, oblivion, obliviousness, oversight, unconsciousness.

Mem·phi·an (mem′fē·ən) *adj.* Of or pertaining to ancient Memphis, Egypt, or to its inhabitants.

Mem·phis (mem′fis) **1** A port on the Mississippi River in SW Tennessee. **2** The ancient capital of Egypt on the Nile above the apex of its delta.

Mem·phre·ma·gog (mem′frə·mā′gog), **Lake** A lake in northern Vermont and southern Quebec; 30 miles long.

mem·sah·ib (mem′sä·ib) *n. Anglo-Indian* A

European lady or mistress: a name given by native servants. [<MA'AM + Arabic *sāhib* master]

men[1] (men) *v.t.* & *v.i. Scot.* To mend.

men[2] (men) Plural of MAN.

men·ace (men'is) *v.* **·aced, ·ac·ing** *v.t.* **1** To threaten with evil or harm. **2** To make threats of. — *v.i.* **3** To make threats; appear threatening. See synonyms under THREATEN. — *n.* A threat; something which threatens; an impending evil. [<OF *manace* <L *minacia* <*minax, -acis* threatening <*minari* threaten] — **men'ac·er** *n.* — **men'ac·ing·ly** *adv.*

me·nac·me (mə·nak'mē) *n. Physiol.* The reproductive period of a woman's life, during which menstruation occurs. [<Gk. *mēn* month + *akmē* peak]

me·nad (mē'nad), **me·nad·ic** (mə·nad'ik) See MAENAD, etc.

mé·nage (mā·näzh', *Fr.* mā·nȧzh') *n.* **1** The persons of a household, collectively; a domestic establishment. **2** Household management. Also **me·nage'.** [<F <L *mansio, -onis* house. Doublet of MANSION.]

me·nag·er·ie (mə·naj'ər·ē) *n.* **1** A collection of wild animals kept for exhibition. **2** The enclosure in which they are kept. [<F]

Men·ai Strait (men'ī) A channel between the island of Anglesey and the mainland in NW Wales.

Me·nam (me·näm') See CHAO PHRAYA.

Me·nan·der (mə·nan'dər), 343?–291? B.C., Greek comic dramatist.

me·nar·che (mə·när'kē) *n. Physiol.* The commencement of menstrual function in women. [<Gk. *mēn* month + *archē* beginning]

Men·cius (men'shəs), 372?–289? B.C., Chinese philosopher: Latin form of Chinese *Meng-tse.*

Menck·en (meng'kən), **H(enry) L(ouis),** 1880–1956, U. S. author and editor.

mend (mend) *v.t.* **1** To make sound or serviceable again by repairing; patch. **2** To correct errors or faults in; reform; improve: *Mend your ways.* **3** To correct (some defect). — *v.i.* **4** To become better, as in health; improve. — *n.* **1** The act of repairing or patching. **2** A mended portion of a garment. — **on the mend** Recovering health; recuperating; convalescing. [Aphetic form of AMEND] — **mend'a·ble** *adj.* — **mend'er** *n.*

men·da·cious (men·dā'shəs) *adj.* **1** Addicted to lying; falsifying. **2** Characterized by deceit; false. [<L *mendax, -acis* lying] — **men·da'cious·ly** *adv.* — **men·da'cious·ness** *n.*

men·dac·i·ty (men·das'ə·tē) *n.* Lying; falsity. [<L *mendacitas, -tatis*]

Men·del (men'dəl), **Gregor Johann,** 1822–1884, Austrian monk and botanist; founder of the science of genetics.

Men·de·ley·ev (men'də·lā'əf), **Dmitri Ivanovich,** 1834–1907, Russian chemist; discovered the periodic law. Also **Men'de·lej'eff.**

men·de·le·vi·um (men'də·lē'vē·əm) *n. Chem.* The short-lived radioactive element (symbol Md) of atomic number 101 and mass number 256. [after Dmitri Ivanovich *Mendeleyev*]

Men·de·li·an (men·dē'lē·ən) *adj.* **1** Of or pertaining to Gregor Mendel. **2** Relating to or in accordance with Mendel's laws.

Men·del·ism (men'dəl·iz'əm) *n.* The theory of heredity as put forth by Mendel. Also **Men·de'li·an·ism.**

Mendel's laws *Genetics* Principles formulated by Gregor Mendel as a result of experiments in breeding garden peas. They state that certain contrasting characters of cross-bred parents, as color, height, etc., are inherited by the hybrid offspring through determining factors which act as units, and that subsequent cross-bred generations manifest these characters in varying combinations from dominant to recessive, each combination being present in a definite proportion of the total number of offspring.

Men·dels·sohn (men'dəl·sən, *Ger.* men'dəl·zōn), **Moses,** 1729–86, German-Jewish theologian and philosopher.

Men·dels·sohn-Bar·thol·dy (men'dəl·sən·bär·tōl'dē, *Ger.* men'dəl·zōn·bär·tôl'dē), **Felix,** 1809–47, German composer and musician; grandson of Moses Mendelssohn. Also **Men'·dels·sohn.**

Men·de·res (men'də·res') **1** A river in western Turkey in Asia, flowing SW to the Aegean: ancient *Meander.* **2** A river in NW Turkey in Asia, flowing NW across the plain of ancient Troy to the Dardanelles: ancient *Scamander.*

men·di·cant (men'də·kənt) *adj.* **1** Begging; depending on alms for a living. **2** Pertaining to or like a beggar. — *n.* **1** A beggar. **2** A begging friar. [<L *mendicans, -antis,* ppr. of *mendicare* beg <*mendicus* needy] — **men'di·can·cy, men·dic·i·ty** (men·dis'ə·tē) *n.*

mend·ing (men'ding) *n.* Articles to be mended.

Men·do·ci·no (men'də·sē'nō), **Cape** The westernmost point of California.

Men·do·za (men·dō'sä) A province of western Argentina; 58,239 square miles; capital, Mendoza.

Men·do·za (men·dō'thä), **Pedro de,** 1487?–1537, Spanish explorer who founded Buenos Aires and Asunción.

mends (mendz) *n. pl. Brit. Dial.* Amends. — **to the mends** *Brit. Dial.* Over and above.

Men·e·la·us (men'ə·lā'əs) In Greek legend, a son of Atreus, brother of Agamemnon, and a king of Sparta; after his wife Helen was abducted by Paris, Menelaus led the Greek princes against Troy, where he defeated Paris; reconciled with Helen, after eight years of wandering, he returned with her to Sparta.

Men·e·lik II (men'e·lik), 1844–1913, emperor of Abyssinia 1889–1910; defeated Italians at Aduwa, 1896.

me·ne, me·ne, tek·el, u·phar·sin (mē'nē mē'nē tek'əl yōō·fär'sin) *Aramaic* Numbered, numbered, weighed, (and) divided: in the Bible, the words that appeared on the wall at Belshazzar's feast, and which Daniel interpreted to mean that God had judged and doomed Belshazzar's kingdom. *Dan.* vi 24–28.

Me·nén·dez de A·vi·les (mā·nen'dāth thā ä·vē'läs), **Pedro,** 1519–74, Spanish captain, founder of St. Augustine, Florida.

Me·nes (mē'nēz) Traditionally the first Egyptian king; founder of the first dynasty of Egypt in the fifth millennium B.C. Also **Me·ni** (mē'nē).

Meng-tse (mung'dzu') See MENCIUS.

Meng-tze (mung'dzu') A city in SE Yunnan province, China.

men·ha·den (men·hād'n) *n.* A herringlike fish *(Brevoortia tyrannus)* found along the North Atlantic coast of the United States: it is the source of **menhaden oil,** used in industry, and of fertilizer. [Alter. of Algonquian *munnawhat* fertilizer]

men·hir (men'hir) *n.* A prehistoric sepulchral or battle monument consisting of a single tall stone, usually left rough. [<F <Celtic (Breton) *men* stone + *hir* long]

me·ni·al (mē'nē·əl, mēn'yəl) *adj.* **1** Pertaining or appropriate to servants. **2** Servile. See synonyms under BASE[2]. — *n.* **1** One doing servile work: generally in contempt. **2** Figuratively, a person of low or servile nature. [<AF <OF *meisniee, maisnie* household <LL *mansionata* <L *mansio* house] — **me'·ni·al·ly** *adv.*

Mé·nière's disease (mā·nyârz') *Pathol.* Progressive deafness of one ear, with vertigo, tinnitus, nausea, and vomiting. [after Prosper *Ménière,* 1799–1862, French physician]

me·nin·ges (mə·nin'jez) *n. pl.* of **me·ninx** (mē'·ningks) *Anat.* The membranes (the dura mater, pia mater, and arachnoid) enveloping the brain and spinal cord. [<NL <Gk. *mēninx, mēningos* membrane] — **me·nin'·ge·al** *adj.*

men·in·gi·tis (men'ən·jī'tis) *n. Pathol.* Inflammation of the enveloping membranes of an organ, especially those of the brain and spinal cord. Certain forms are caused by infection with diplococcus bacterium *(Neisseria intracellularis).* [<MENINGES + -ITIS] — **men'in·git'ic** (-jit'ik) *adj.*

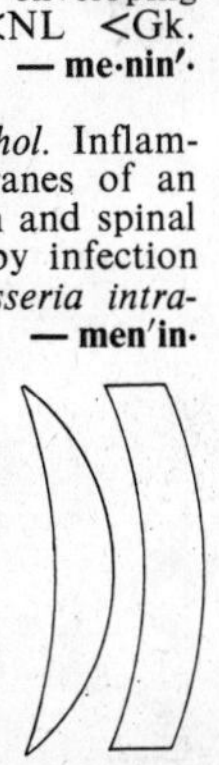

MENISCUS LENSES

me·nis·cus (mə·nis'kəs) *n. pl.* **·nis·cus·es** or **·nis·ci** (-nis'ī) **1** Any crescent-shaped body. **2** *Optics* A lens convex on one side and concave on the other. **3** *Anat.* A disklike body of fibro-cartilage found in some joints of the body exposed to concussion. **4** *Physics* The surface or upper part of a liquid column made convex or concave by capillarity. [<L <Gk. *mēniskos* crescent, dim. of *mēnē* the moon]

men·i·sper·ma·ceous (men'ē·spər·mā'shəs) *adj.* Designating a family *(Menispermaceae)* of mostly tropical, woody or herbaceous climbing plants having alternate leaves and small dioecious flowers, and yielding substances of narcotic and toxic properties; the moonseed family. [<NL *menispermaceae* <Gk. *mēnē* moon + *sperma* seed]

Men·non·ite (men'ən·īt) *n.* A member of a Protestant denomination that grew out of the Anabaptist movement in the 16th century, and still flourishes in Europe and the United States: named after Menno Simons, 1492–1559, a leader of the sect in the Netherlands. They are opposed to taking oaths, to military service, to theological learning, and to infant baptism.

me·no (mā'nō) *adv. Music* Less. [<Ital. <L *minus*]

me·nol·o·gy (mə·nol'ə·jē) *n.* **1** A calendar of the months; especially, one having a record of events by month. **2** A register or collection of lives of the saints arranged according to months and days of the month, as in the Greek Church. [<Gk. *mēn* month + -LOGY]

men·o·pause (men'ə·pôz) *n. Physiol.* Final cessation of the menses; change of life: opposed to *menarche.* [<Gk. *mēn* month + *pauein* cause to cease]

Me·nor·ca (mā·nôr'kä) The Spanish name for MINORCA.

men·or·rha·gi·a (men'ə·rā'jē·ə) *n. Pathol.* Excessive menstruation. [<Gk. *mēn* month + -RRHAGIA]

Me·not·ti (mə·not'ē), **Gian Carlo,** born 1911, U. S. composer and librettist.

men·sa (men'sə) *n. Dent.* The biting or chewing surface of a tooth. [<L, table]

Men·sa (men'sə) A southern constellation. See CONSTELLATION.

men·sal[1] (men'səl) *adj.* Belonging to or used at the table. [<L *mensalis* <*mensa* table]

men·sal[2] (men'səl) *adj.* Monthly. [<L *mensis* month]

mense (mens) *n. Scot.* & *Brit. Dial.* Dignified conduct or manner; decorum. [OE *mensk* <ON *mennska* humanity] — **mense'ful** *adj.* — **mense'less** *adj.*

men·ses (men'sēz) *n. pl. Physiol.* A periodical bloody flow from the uterus of a female mammal, resulting when an ovum is not fertilized, and occurring in women about once every lunar month; the menstrual flow. [<L, pl. of *mensis* month]

Men·she·vik (men'shə·vik) *n. pl.* **·vi·ki** (-vē'kē) or **·viks** A member of the conservative element in the Russian Social Democratic Party. Compare BOLSHEVIK, MAXIMALIST. Also **Men'she·vist.** [<Russian *menshe* smaller, minority] — **Men'she·vism** *n.*

mens le·gis (menz lē'jis) *Latin* The spirit of the law.

mens sa·na in cor·po·re sa·no (menz sā'nə in kôr'pə·rē sā'nō) *Latin* A sound mind in a sound body.

men·stru·al (men'strōō·əl) *adj.* **1** *Physiol.* Pertaining to the menses or to a menstruum. **2** Continuing a month; occurring monthly. Also **men'stru·ous.** [<L *menstrualis* <*menstruus* monthly <*mensis* month]

men·stru·ate (men'strōō·āt) *v.i.* **·at·ed, ·at·ing** To discharge the menses. [<L *menstruatus,* pp. of *menstruare* <*menstruus.* See MENSTRUAL.] — **men'stru·a'tion** *n.*

men·stru·um (men'strōō·əm) *n. pl.* **·stru·ums** or **·stru·a** (-strōō·ə) The medium in which a substance is dissolved; a solvent. [<L, neut. of *menstruus.* See MENSTRUAL.]

men·su·ra·ble (men'shər·ə·bəl) *adj.* **1** That can be measured. **2** Mensural. [<LL *mensurabilis* <*mensurare* measure <*mensura.* See MEASURE.] — **men'su·ra·bil'i·ty** *n.*

men·su·ral (men'shər·əl) *adj.* **1** Pertaining to measure. **2** *Music* Characterized by a fixed rhythm and measure.

men·su·rate (men'shə·rāt) *v.t.* **·rat·ed, ·rat·ing** *Rare* To measure the dimensions or quantity of. [<L *mensuratus,* pp. of *mensurare* <*mensura.* See MEASURE.]

men·su·ra·tion (men'shə·rā'shən) *n.* **1** The act, art, or process of measuring. **2** The branch of mathematical science that has to do with measurement, as of lines, surfaces, or volume. **3** The result of measuring; measure.

add, āce, câre, pälm; end, ēven; it, īce; odd, ōpen, ôrder; tŏŏk, pōōl; up, bûrn; ə = a in *above,* e in *sicken,* i in *clarity,* o in *melon,* u in *focus;* yōō = u in *fuse;* oi, oil; ou, pout; ch, check; g, go; ng, ring; th, thin; th, this; zh, vision. Foreign sounds ȧ, œ, ü, kh, ṅ; and ◆: see page xx. < from; + plus; ? possibly.

[<LL *mensuratio, -onis*] — **men′su·ra′tive** *adj.*
-ment *suffix of nouns* **1** The concrete result of; a thing produced by: *achievement*. **2** The instrument or means of: *atonement*. **3** The process or action of: *government*. **4** The quality, condition, or state of being: *astonishment*. [<F <L *-mentum*]
men·tal[1] (men′təl) *adj.* **1** Pertaining to the mind: contrasted with *corporeal*. **2** Effected by or due to the mind, especially without the aid of written symbols. [<F <LL *mentalis* <L *mens, mentis* mind] — **men′tal·ly** *adv.*
men·tal[2] (men′təl) *adj.* Of, pertaining to, or situated near the chin: the *mental* point. — *n.* A plate or scale of the chin, as in snakes. [<L *mentum* chin]
mental age See under AGE.
mental blindness A mental condition in which images conveyed by the optic nerves are not properly recognized: also called *mind-blindness, psychic blindness*.
mental deafness A form of deafness in which sounds and words are heard but cannot be interpreted: also called *mind-deafness, psychic deafness*.
mental deficiency *Psychiatry* Lack of one or more mental capacities and functions present in the normal individual, usually to the point of disqualifying from full participation in ordinary life; feeble-mindedness. The principal types, in order of increasing deficiency, are moronism, imbecility, and idiotism.
mental healing The curing of any disorder, ailment, or disease by concentrating the mind either directly on the healing forces in nature, or on the denial of the discomforts experienced.
mental hygiene The scientific study and rational application of all methods that will restore, preserve, promote, and improve mental health, especially in relation to the normal functioning of the personality as a whole.
men·tal·i·ty (men·tal′ə·tē) *n. pl.* **·ties 1** The sum of the mental faculties or powers; mental activity. **2** Cast or habit of mind.
Mental Science New Thought.
Men·ta·wai Islands (men·tä′wī) An Indonesian island group off the west coast of Sumatra; 2,354 square miles. Also **Men·ta′wei** (-wā).
men·tha·ceous (men·thā′shəs) *adj.* Designating or pertaining to a genus (*Mentha*) of odorous perennial herbs of the mint family, with opposite leaves and small flowers, including the peppermint, spearmint, etc. [<L *mentha* mint + -ACEOUS]
men·thane (men′thān) *n. Chem.* Any one of three isomeric, saturated hydrocarbons, $C_{10}H_{20}$, corresponding to cymenes: parent substance for several of the terpenes. [<MENTHOL + -ANE[2]]
men·thene (men′thēn) *n. Chem.* A colorless, liquid, oily hydrocarbon, $C_{10}H_{18}$, derived from the oil of peppermint. [<L *mentha* mint + -ENE]
men·thol (men′thôl, -thōl, -thol) *n. Chem.* A white, waxy, crystalline alcohol, $C_{10}H_{19}OH$, obtained from and having the odor of oil of peppermint: used as a flavoring agent, in perfumery, and in medicine as an anodyne for neuralgia and similar ailments. [<L *mentha* mint + -OL[1]]
men·tho·lat·ed (men′thə·lā′tid) *adj.* Treated with, containing, or impregnated with menthol.
men·ti·cide (men′tə·sīd) *n.* The undermining and destruction of a person's mental powers by deliberate intent and with the use of all available psychological means. [<L *mens, mentis* mind + -CIDE]
men·tion (men′shən) *v.t.* To speak of incidentally or briefly; refer to in passing; specify or name. See synonyms under ALLUDE, INFORM. — *n.* The act of mentioning; casual allusion; notice: used especially in the phrase *to make mention of*. [<OF <L *mentio, -onis* <*mens, mentis* mind] — **men′tion·a·ble** *adj.* — **men′tion·er** *n.*
Men·ton (män·tôn′) A resort town on the Italian border of the French Riviera. Italian **Men·to·ne** (mān·tō′nā).
men·tor (men′tər, -tôr) *n.* A wise and trusted teacher, guide, and friend; an elderly monitor or adviser. [<MENTOR] — **men·to′ri·al** (-tō′rē·əl, -tō′rē-) *adj.*
Men·tor (men′tər, -tôr) In the *Odyssey*, the sage guardian of Telemachus, appointed by Odysseus before he departed for the Trojan War. [<Gk., lit., adviser]
men·tum (men′təm) *n.* **1** The chin. **2** *Entomol.* The distal sclerite of the labium or lower lip of insects.
men·u (men′yōō, mān′-; *Fr.* mə·nü′) *n.* A bill of fare or the dishes included in it. [<F, small, detailed <L *minutus*. See MINUTE[2].]
Men·u·hin (men′yōō·in), **Yehudi**, born 1917, U. S. violinist.
me·ow (mē·ou′, myou) See MEW[2]. [Imit.]
me·per·i·dine hydrochloride (mə·per′ə·dēn, -din) A white, odorless, crystalline compound, $C_{15}H_{22}ClO_2$, used in medicine as an analgesic and sedative; Demerol.
Me·phis·to·phe·le·an (mə·fis′tə·fē′lē·ən) *adj.* Of, pertaining to, or like Mephistopheles; cynical; crafty; sardonic; fiendish. Also **Me·phis′to·phe′li·an.**
Meph·is·toph·e·les (mef′is·tof′ə·lēz) **1** In medieval legend, a devil to whom Faust sold his soul for wisdom and power. **2** A crafty, sardonic fiend; a diabolical person. Also **Me·phis·to** (mə·fis′tō).
me·phit·ic (mə·fit′ik) *adj.* Poisonous; pestilential; foul; noxious. Also **me·phit′i·cal.**
me·phi·tis (mə·fī′tis) *n.* **1** A noxious exhalation caused by the decomposition of organic remains. **2** A pestilential or deadly gas, as from a cave or mine. Also **me·phi′tism**. [<L]
Me·qui·nez (mä′kē·nāth′) The former name for MEKNÈS.
mer- Var. of MERO-.
Me·rak (mē′rak) The star Beta in the constellation Ursa Major; the smaller of the two stars composing the Pointers toward the polestar.
mer·can·tile (mûr′kən·til, -tīl) *adj.* **1** Pertaining to or characteristic of merchants. **2** Conducted or acting on business principles; commercial. [<F <Ital. <L *mercans, -antis*, pp. of *mercari* traffic. See MERCHANT.]
mercantile agency An institution which collects, records, and furnishes to regular clients full information about the financial standing, credit ratings, etc., of individuals and firms.
mercantile paper Negotiable instruments for the payment of money, given in course of business, as bills of exchange, promissory notes, etc.: also called *commercial paper*.
mercantile system A theory in political economy that wealth consists not in labor and its products, but in the quantity of silver and gold in a country, and hence that mining, the exportation of goods, and the importation of gold should be encouraged by the state.
mer·can·til·ism (mûr′kən·til·iz′əm) *n.* **1** The spirit or theory of mercantile life or trade in general. **2** The mercantile system. — **mer′can·til·ist** *n. & adj.* — **mer′can·til·is′tic** *adj.*
mer·cap·tan (mər·kap′tan) *n.* Thiol. [<G <Med. L *mer(curium) captan(s)* seizing mercury]
mer·cap·tide (mər·kap′tīd) *n.* The metal salt of a mercaptan or thiol, obtained by replacing the sulfur hydrogen constituent with a metal.
mer·cap·to (mər·kap′tō) *n.* Sulfhydryl. [See MERCAPTAN.]
Mer·ca·tor (mər·kā′tər, *Flemish* mer·kä′tôr), **Gerhard**, 1512–94, Flemish geographer and cartographer.

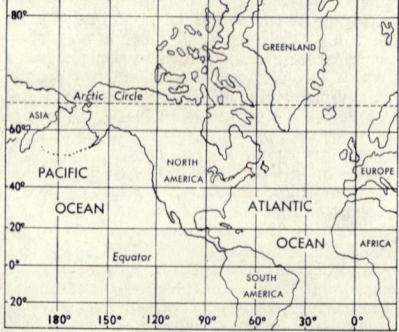

MERCATOR'S PROJECTION

Mercator's projection A system of making maps in which the meridians are represented by parallel straight lines, and the parallels of latitude by lines perpendicular to the meridians, and at increasing intervals, so as to preserve the actual ratio between the increments of longitude and latitude at every point. It is accurate at the equator, but areas become increasingly distorted toward the poles.
Merced River (mər·sed′) A river in central California, flowing 150 miles SW to the San Joaquin.
mer·ce·nar·y (mûr′sə·ner′ē) *adj.* **1** Influenced by desire for gain or reward; greedy; venal. **2** Serving for pay or profit; hired: *mercenary soldiers*. **3** Pertaining to or resulting from sordidness. See synonyms under VENAL[1]. — *n. pl.* **·nar·ies** A person working or serving only or chiefly for pay; a hired soldier in foreign service. See synonyms under AUXILIARY. [<L *mercenarius* <*merces* reward, hire] — **mer′ce·nar′i·ly** *adv.* — **mer′ce·nar′i·ness** *n.*
mer·cer (mûr′sər) *n. Brit.* **1** Formerly, a dealer in small wares. **2** A dealer in cloth or silks. [<F *mercier* <L *merx, mercis* wares]
mer·cer·ize (mûr′sə·rīz) *v.t.* **·ized, ·iz·ing** To treat (cotton fabrics) with caustic soda or potash, so as to increase their color-absorbing qualities and impart a silky gloss. [after John Mercer, 1791–1866, English inventor] — **mer′cer·i·za·tion** (mûr′sər·ə·zā′shən, -ī·zā′-) *n.*
mer·cer·y (mûr′sər·ē) *n. pl.* **·cer·ies** *Brit.* A mercer's wares or place of business. [<OF *mercerie*]
mer·chan·dise (mûr′chən·dīz, -dīs) *n.* **1** Anything movable customarily bought and sold for profit. **2** *Obs.* Mercantile dealings; commerce; trade; hence, gain or advantage. — *v.t. & v.i.* **·dised, ·dis·ing** To barter; trade; buy and sell. **2** To promote the sale of (an article) through advertising, etc. Also **mer′chan·dize.** [<F *marchandise*. See MERCHANT.] — **mer′chan·dis′er** *n.*
mer·chant (mûr′chənt) *n.* **1** A person who buys and sells commodities as a business or for profit; a trader. **2** A shopkeeper; storekeeper. — *adj.* Of or pertaining to merchants or merchandise; commercial. [<OF *marchant* <L *mercari* traffic, buy <*merx, mercis* wares]
mer·chant·a·ble (mûr′chən·tə·bəl) *adj.* That can be bought or sold.
merchant iron Wrought iron converted into marketable bars or rods of various sizes and shapes: often used for making hooks, chains, and rivets, and in reinforcing concrete.
mer·chant·man (mûr′chənt·mən) *n. pl.* **·men** (-mən) **1** A trading or merchant vessel. **2** *Archaic* A merchant.
merchant marine 1 All the vessels of a nation, collectively, both publicly and privately owned, engaged in commerce and trade. **2** The officers and men employed on these vessels.
mer·ci (mer·sē′) *interj. French* Thank you.
Mer·ci·a (mûr′shē·ə, -shə) An ancient Anglo-Saxon kingdom of central England, later annexed to Wessex.
Mer·ci·an (mûr′shē·ən, -shən) *adj.* Of or pertaining to Mercia, its people, or their dialect. — *n.* **1** An inhabitant of Mercia. **2** The dialect of Old English spoken in Mercia: the forerunner of the Midland dialects.
mer·ci beau·coup (mer·sē′ bō·kōō′) *French* Thank (you) very much.
Mer·cier (mer·syā′), **Désiré Joseph**, 1851–1926, Belgian cardinal.
mer·ci·ful (mûr′si·fəl) *adj.* **1** Full of mercy; compassionate. **2** Characterized by or indicating mercy. — **mer′ci·ful·ly** *adv.* — **mer′ci·ful·ness** *n.*
Synonyms: benignant, clement, compassionate, forgiving, gentle, gracious, humane, pitiful, pitying, tender, tender-hearted. The *merciful* man is disposed to withhold or mitigate the suffering even of the guilty; the *compassionate* man sympathizes with and desires to relieve actual suffering, while one who is *humane* would forestall and prevent the suffering which he sees to be possible. See CHARITABLE, GOOD, HUMANE, PROPITIOUS.
mer·ci·less (mûr′si·lis) *adj.* Having or showing no mercy. See synonyms under BARBAROUS, IMPLACABLE. — **mer′ci·less·ly** *adv.* — **mer′ci·less·ness** *n.*
mer·cu·ri·al (mər·kyoor′ē·əl) *adj.* **1** Pertaining to the god Mercury; hence, lively; volatile. **2** Of or relating to quicksilver. — *n.* A preparation containing mercury. [<L *mercurialis*

<*Mercurius* Mercury] **— mer·cu′ri·al·ly** *adv.* **— mer·cu′ri·al·ness** *n.*

mer·cu·ri·al·ism (mər·kyŏŏr′ē·əl·iz′əm) *n. Pathol.* The condition produced by excessive use of medicines containing mercury; mercury poisoning.

mer·cu·ri·al·ize (mər·kyŏŏr′ē·əl·īz′) *v.t.* **·ized, ·iz·ing 1** To make mercurial. **2** To treat with mercury. **— mer·cu′ri·al·i·za′tion** *n.*

mer·cu·ric (mər·kyŏŏr′ik) *adj. Chem.* Of, pertaining to, or containing mercury in its highest valence.

mercuric chloride Corrosive sublimate. Also **mercury chloride.**

mercuric oxide *Chem.* A compound, HgO, obtained from mercuric nitrate by heat, forming both red and yellow powders which, with lard, are used in making ointments for certain skin diseases.

mercuric sulfide *Chem.* A compound, HgS, found native as cinnabar, or artificially produced as a black or a vermilion powder by the action of hydrogen sulfide on mercury salts.

Mer·cu·ro·chrome (mər·kyŏŏr′ə·krōm) *n.* Proprietary name of an iridescent, green, crystalline mercury compound, $C_{20}H_8Br_2Na_2$-Hg, turning red in an aqueous solution, which is used as a germicide and as a moderately active local antiseptic. [<MERCURY + -CHROME]

mer·cu·rous (mər·kyŏŏr′əs) *adj. Chem.* Of, pertaining to, or containing mercury in its lowest valence: *mercurous* chloride, *mercurous* oxide, etc.

mer·cu·ry (mûr′kyə·rē) *n. pl.* **·ries 1** A heavy, silver-white metallic element (symbol Hg), liquid at ordinary temperatures; quicksilver. See ELEMENT. **2** The quicksilver in a thermometer or barometer, as indicating temperature, etc. **3** A messenger. **4** An Old World plant (genus *Mercurialis*), especially *M. annua,* the **annual** or **French mercury,** used in medicine, and *M. perennis,* the **perennial** (or **dog's**) **mercury,** which is poisonous.

Mer·cu·ry (mûr′kyə·rē) **1** In Roman mythology, the herald and messenger of the gods, god of commerce, eloquence, and skill, and patron of messengers, travelers, merchants, and thieves: identified with the Greek *Hermes.* **2** *Astron.* The planet of the solar system nearest the sun, from which its mean distance is about 36,000,000 miles. It is the smallest of the major planets, having a diameter of about 3,000 miles, and revolving about the sun in 88 of our days.

mer·cu·ry-va·por lamp (mûr′kyə·rē·vā′pər) A glass tube or bulb containing mercury vapor made luminous by the passage of an electric discharge. Its light is a source of ultraviolet rays.

Mer·cu·ti·o (mər·kyōō′shē·ō) In Shakespeare's *Romeo and Juliet,* a witty and brave young nobleman; friend of Romeo.

mer·cy (mûr′sē) *n. pl.* **·cies 1** The act of treating an offender with less severity than he deserves; also, forbearance to injure others when one has power to do so. **2** The act of relieving suffering, or the disposition to relieve it; compassion. **3** A providential blessing. [<OF <L *merces, mercedis* hire, payment, reward; with ref. to the heavenly reward for compassion]

Synonyms: benevolence, benignity, blessing, clemency, compassion, favor, forbearance, forgiveness, gentleness, grace, kindness, lenience, leniency, lenity, mildness, pardon, pity, tenderness. *Mercy* is the exercise of less severity than one deserves, or in a more extended sense, the granting of *kindness* or *favor* beyond what one may rightly claim. *Clemency* is a colder word than *mercy* signifying *mildness* and moderation in the use of power where severity would have legal sanction; it often denotes a habitual *mildness* of disposition on the part of the powerful, and is a matter rather of good nature or policy than of principle. *Leniency* or *lenity* denotes an easy-going avoidance of severity; these words are more general and less magisterial than *clemency. Grace* is *favor, kindness,* or *blessing* shown to the undeserving; *forgiveness, mercy,* and *pardon* are exercised toward the ill-deserving. *Pardon* remits the outward penalty which the offender deserves; *forgiveness* dismisses resentment or displeasure from the heart of the one offended. *Mercy* is also used in the wider sense of refraining from harshness or cruelty toward those who are in one's power without fault of their own; as, They besought the robber to have *mercy.* See LENITY, PITY. *Antonyms*: cruelty, hardness, harshness, implacability, justice, penalty, punishment, revenge, rigor, severity, sternness, vengeance.

mercy killing Euthanasia.

mercy seat 1 In ancient Jewish ritual, the golden lid of the ark of the covenant whence God gave his oracles to the high priest, and upon which was sprinkled the blood of the yearly atonement. **2** Figuratively, the throne of God.

mere[1] (mir) *adj.* **1** Such as is mentioned and no more; nothing but. **2** *Obs.* Absolute; entire; unqualified. See synonyms under PURE. [<L *merus* unmixed, bare]

mere[2] (mir) *n.* **1** A pond; pool. **2** *Scot.* The sea. [OE *mere*]

mere[3] (mir) *n. Brit.* A boundary line. [OE *gemǣre*]

-mere *combining form Zool.* A part or division: *blastomere.* [<Gk. *meros* part]

Mer·e·dith (mer′ə·dith), **George,** 1828–1909, English novelist and poet. **— Owen** See LYTTON.

mere·ly (mir′lē) *adv.* **1** Without including anything else; only; solely. **2** *Obs.* Absolutely; wholly. See synonyms under BUT[1].

mer·e·tri·cious (mer′ə·trish′əs) *adj.* **1** Deceitfully or artificially attractive; vulgar; tawdry. **2** Pertaining to a harlot; wanton. [<L *meretricius* <*meretrix, -icis* prostitute <*merere* earn, gain] **— mer′e·tri′cious·ly** *adv.* **— mer′e·tri′cious·ness** *n.*

Me·rezh·kov·ski (mi′rish·kôf′skē), **Dmitri,** 1865–1941, Russian novelist and critic. Also **Me′rej·kow′ski.**

mer·gan·ser (mər·gan′sər) *n.* A fish-eating duck (subfamily *Merginae*), with toothlike processes along the upper edge of the bill, and the head usually crested, as the hooded merganser (*Lophodytes cucullatus*) of North America. [<NL <L *mergus* diver <*mergere* plunge + *anser* goose]

merge (mûrj) *v.t. & v.i.* **merged, merg·ing** To combine or be combined so as to lose separate identity; blend. See synonyms under UNITE. [<L *mergere* dip, immerse] **— mer′gence** *n.*

Mer·gen·tha·ler (mûr′gən·thä′lər, *Ger.* mer′·gən·tä′lər), **Ottmar,** 1854–99, U. S. inventor of the Linotype, born in Germany.

merg·er (mûr′jər) *n.* **1** *Law* The extinguishment of a lesser estate, right, or liability in a greater one. **2** One who or that which merges. **3** A combination of a number of commercial interests or companies in one.

Mer·gui Archipelago (mər·gwē′) An island group in the Andaman Sea off the Tenasserim coast of Lower Burma, to which they belong.

Mé·ri·da (mā′rē·thä) A city in SE Mexico, capital of Yucatán state.

me·rid·i·an (mə·rid′ē·ən) *n.* **1** *Obs.* Noontime; midday. **2** The highest or culminating point of anything; the zenith: the *meridian* of life. **3** *Astron.* A great circle passing through the poles and zenith of the celestial sphere at any point; the celestial meridian. **4** *Geog.* **a** A great circle drawn from any point on the earth's surface and passing through both poles. **b** The half-circle so drawn between the poles, called a meridian of longitude, or geographic meridian. **5** A line on a surface of revolution in the same plane as its axis. **—** *adj.* **1** Of or pertaining to noonday: *meridian* heat. **2** Pertaining to or at the highest or culminating point; brightest: *meridian* fame. **3** Of or pertaining to a meridian. [<OF *meridien* <L *meridianus* <*meridies* noon, south <*medidies* <*medius* middle + *dies* day]

me·rid·i·o·nal (mə·rid′ē·ə·nəl) *adj.* **1** Of or pertaining to the meridian. **2** Relating to southern climates or people: *meridional* customs. **3** Approximating a direction north and south. **4** Situated or lying in the south; southerly. **—** *n.* An inhabitant of a southern country; specifically, a resident of southern France. [<OF <LL *meridionalis* southern] **— me·rid′·i·o·nal′i·ty** *n.* **— me·rid′i·o·nal·ly** *adv.*

Mé·ri·mée (mā·rē·mā′), **Prosper,** 1803–70, French novelist and historian.

me·ringue (mə·rang′) *n.* The beaten white of eggs sweetened, baked, and used to garnish pastry; also, pastry so garnished. [<F <G *meringe,* lit., cake of Mehringen (in Germany)]

me·ri·no (mə·rē′nō) *n. pl.* **·nos 1** A superior breed of sheep, originating in Spain, and having very fine, closely set, silky wool. See SHEEP. Also **Merino sheep. 2** The wool of this sheep. **3** A fabric made of merino wool. **4** A kind of knitted goods used for underwear. **—** *adj.* **1** Pertaining to merino sheep or their wool. **2** Made of merino wool. [<Sp., roving from pasture to pasture, shepherd, inspector of sheepwalks <Med. L *majorinus* steward <L *major* greater]

Mer·i·on·eth·shire (mer′ē·on′ith·shir) A county of NW Wales; 660 square miles; county town, Dolgelley. Also **Mer′i·on′eth.**

mer·i·stem (mer′ə·stem) *n. Bot.* Plant tissue in process of formation; vegetable cells in a state of active division and growth, as those at the apex of growing stems and roots. [<Gk. *meristos* divided <*merizein* divide <*meros* part] **— mer′i·ste·mat′ic** (-stə·mat′ik) *adj.*

me·ris·tic (mə·ris′tik) *adj.* Divided into parts; segmented. [<Gk. *meristos*]

mer·it (mer′it) *n.* **1** *Often pl.* The quality or fact of deserving, especially of deserving well; desert: Does his *merit* justify the reward? **2** Worth or excellence; quality: A man of *merit.* **3** That which deserves esteem, praise, or reward; a commendable act or quality: the *merit* of silence. **4** *pl.* The actual rights or wrongs of a matter considered exclusively of extraneous details or technicalities: to decide a case on its *merits.* **5** Reward, recompense, or, sometimes, punishment received or deserved; a token or award of excellence. See synonyms under WORTH. **—** *v.t.* To earn as a reward or punishment; deserve. [<OF *merite* <L *meritum* <*meritus,* pp. of *merere* deserve] **— mer′it·ed** *adj.* **— mer′it·ed·ly** *adv.*

mer·i·to·ri·ous (mer′ə·tôr′ē·əs, -tō′rē-) *adj.* Deserving of reward; praiseworthy. **— mer′i·to′ri·ous·ly** *adv.* **— mer′i·to′ri·ous·ness** *n.*

merit system A system adopted in the U. S. Civil Service whereby appointments and promotions are made on the basis of the merit and fitness of the appointee, ascertained through qualifying examinations.

merle (mûrl) *n.* The European blackbird (*Turdus merula*). Also **merl.** [<F <L *merula* blackbird]

mer·lin (mûr′lin) *n.* A small European falcon (*Falco columbarius aesalon*); also, a related American species, the pigeon hawk (*F. columbarius*). [<OF *esmerillon,* dim. of *esmeril* <OHG *smirl*]

Mer·lin (mûr′lin) In the Arthurian cycle and other medieval legends, a magician and prophet who built the Round Table for King Arthur. [<Med. L *Merlinus* <Welsh *Myrrdin,* lit., sea fortress]

mer·lon (mûr′lən) *n. Mil.* The solid part of a battlement, between the embrasures. See illustration under BATTLEMENT. [<F <Ital. *merlone,* aug. of *merlo* battlement]

mer·maid (mûr′mād′) *n.* A legendary marine creature having as its upper part the head and body of a lovely woman and as its lower part the scaled body and tail of a fish. Also **mer′·maid′en.** [<MERE[2] + MAID]

Mermaid Tavern A famous inn in London, England, frequented by Jonson, Raleigh, Shakespeare, and other celebrated Elizabethan writers.

mer·man (mûr′man′) *n. pl.* **·men** (-men′) A legendary marine creature, having as its upper part the head and body of a man and as its lower part the scaled body and tail of a fish. [<MERE[2] + MAN]

mero- *combining form* Part; partial; incomplete: *meroplankton.* Also, before vowels, *mer-.* [<Gk. *meros* a part, division]

mer·o·blast (mer′ə·blast) *n.* A meroblastic ovum. [<MERO- + Gk. *blastos* sprout]

mer·o·blas·tic (mer′ə·blas′tik) *adj. Biol.* Undergoing partial or incomplete segmentation, with formation of food yolk, as in the eggs of birds: opposed to *holoblastic.* **— mer′o·blas′ti·cal·ly** *adv.*

add,āce,câre,pälm; end,ēven; it,īce; odd,ōpen,ôrder; tŏŏk,pōōl; up,bûrn; ə = a in *above,* e in *sicken,* i in *clarity,* o in *melon,* u in *focus;* yōō = u in *fuse;* oi,oil; ou,pout; ch,check; g,go; ng,ring; th,thin; t͟h,this; zh,vision. Foreign sounds à,œ,ü,kh,ṅ; and ♦: see page xx. < from; + plus; ? possibly.

Mer·o·ë (mer′ō·ē) The ancient capital of Ethiopia on the Nile in northern Sudan; an archeological site of extensive ruins and groups of pyramids partly excavated.

mer·o·gen·e·sis (mer′ə·jen′ə·sis) *n. Biol.* Segmentation; reproduction by the formation of parts. [<MERO- + GENESIS] **—mer′o·ge·net′ic** (-jə·net′ik) *adj.*

Me·rom (mē′rəm), **Waters of** See LAKE HULA.

Mer·o·pe (mer′ə·pē) In Greek mythology, one of the Pleiades, not seen by the naked eye with the other six among the stars, supposedly having hidden herself from shame for loving a mortal: called the *lost Pleiad.*

mer·o·plank·ton (mer′ə·plangk′tən) *n. Biol.* Plankton found only at certain times or in certain seasons of the year.

-merous *suffix Zool.* Having (a specified number or kind of) parts: *trimerous, pentamerous* (often written 3-*merous,* 5-*merous,* etc.). [<Gk. *meros* a part, division]

Mer·o·vin·gi·an (mer′ə·vin′jē·ən, -jən) *adj.* Designating or pertaining to the first royal Frankish dynasty, founded by Clovis I in 486, and lasting until 751. —*n.* A member of the Merovingian dynasty. Also **Mer′o·win′gi·an.** [<L *Merovingi,* descendants of Merovaeus, a legendary Frankish king]

mer·o·zo·ite (mer′ə·zō′īt) *n. Zool.* One of the mature spores liberated in the sporulating stage of certain protozoa, as the parasite causing malaria (*Plasmodium*). [<MERO- + (SPORO)-ZO(A) + -ITE[1]]

Mer·ri·am (mer′ē·əm), **John Campbell,** 1869–1945, U.S. paleontologist.

Mer·ri·mac (mer′ə·mak) *n.* The first armored warship, a Confederate vessel; fought the *Monitor* at Hampton Roads, 1862: Confederate name *Virginia.*

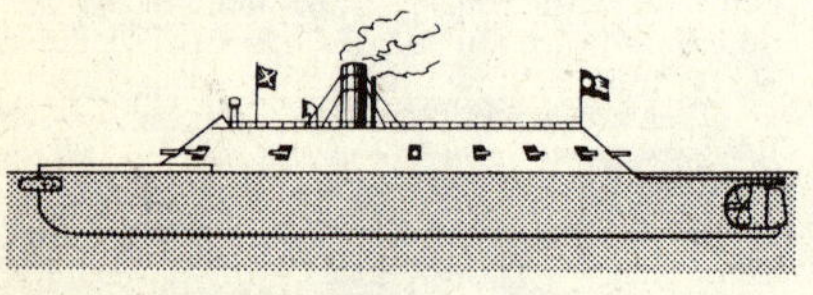

THE MERRIMAC: 35,000 TONS; DRAFT 22 FEET
Originally the U.S.S. *Merrimac;* later renamed C.S.S. *Virginia.*

Mer·ri·mac River (mer′ə·mak) A river in New Hampshire and Massachusetts, flowing 110 miles to the Atlantic. Also **Mer′ri·mack.**

mer·ri·ment (mer′i·mənt) *n.* The act of making merry; mirth; celebration. See synonyms under ENTERTAINMENT, HAPPINESS, LAUGHTER, SPORT.

mer·ry (mer′ē) *adj.* **·ri·er, ·ri·est 1** Inclined to mirth and laughter; full of fun; lively. **2** Of or pertaining to mirth or scenes of mirth; jovial and sportive; mirthful. **3** Inciting to mirth, cheerfulness, and gay spirits; fitted or calculated to enliven; exhilarating; bracing. **4** *Colloq.* Slightly tipsy; high. **5** *Obs.* Jibing; sarcastic. [OE *myrige* pleasant] **—mer′ri·ly** *adv.* **—mer′ri·ness** *n.*

Synonyms: blithe, blithesome, facetious, frolicsome, gay, glad, gladsome, gleeful, hilarious, jocose, jocund, jolly, jovial, joyous, lighthearted, lively, mirthful, sportive. See CHEERFUL, HAPPY, JOCOSE, VIVACIOUS, WANTON. *Antonyms:* see synonyms for SAD.

mer·ry-an·drew (mer′ē·an′drōō) *n.* A clown or buffoon.

Mer·ry del Val (mer′rē thel väl′), **Alfonso Marquis de,** 1864–1943, Spanish diplomat born in London. **—Rafael,** 1865–1930, cardinal; papal secretary of state 1903–1914: born in London.

mer·ry-go-round (mer′ē·gō·round′) *n.* **1** A revolving platform fitted with wooden horses, boatlike vehicles, etc., on which people ride for amusement, usually to music; a carousel. **2** A whirl, as of pleasure.

mer·ry·mak·ing (mer′ē·mā′king) *adj.* Frolicking. —*n.* Festivity; frolic. See synonyms under FROLIC, REVEL, SPORT. **—mer′ry·mak′er** *n.*

mer·ry·thought (mer′ē·thôt′) *n. Brit.* The wishbone of a fowl's breast.

Mer·sa Ma·trûh (mər·sä′ mə·trōō′) See MATRÛH.

Mer·sey (mûr′zē) A river between Cheshire and Lancashire, NW England, flowing 70 miles NW to the Irish Sea: its estuary is 16 miles long and forms Liverpool harbor.

Mer·sin (mər·sēn′) See IÇEL.

Mer·thyr Tyd·fil (mûr′thər tid′vil) A county borough in NE Glamorganshire, Wales.

Mer·zi·fon (mer′zē·fôn′) A town in north central Turkey in Asia: also *Marsivan.* Also **Mer′si·van′** (-vän′).

mes- Var. of MESO-.

me·sa (mā′sə, *Sp.* mā′sä) *n.* A high, broad, and flat tableland with sharp, usually rocky, slopes descending to the surrounding plain, common in the SW United States. [<Sp. <L *mensa* table]

Me·sa·bi Range (mə·sä′bē) A long, narrow range of iron-ore-producing hills in NE Minnesota.

mé·sal·li·ance (mā·zal′ē·əns, *Fr.* mā·zà·lyäns′) *n.* A marriage with one of inferior position; misalliance. [<F]

Mesa Verde National Park (mā′sə vûrd′) An area in SW Colorado containing ruins of prehistoric cliff-dwellings; 80 square miles; established as a national park in 1906.

mes·cal (mes·kal′) *n.* **1** A spineless cactus (*Lophophora williamsii*), native to the SW United States and northern Mexico. Its tops, which are often called **mescal buttons,** grow but little above the ground, contain a narcotic stimulating substance, and are chewed by the Indians, especially during the performance of religious ceremonies. **2** A mescal maguey. **3** An intoxicating liquor distilled from pulque. [<Sp. *mezcal* <Nahuatl *mexcalli*]

mes·ca·line (mes′kə·lēn, -lin) *n. Chem.* A white, crystalline alkaloid, $C_{11}H_{17}O_3N$, extracted from mescal buttons. It has narcotic and tetanic properties, and induces powerful color hallucinations: also spelled *mezcaline.* [<MESCAL]

mescal maguey Any plant from which the liquor mescal is obtained, especially the pulque agave (*Agave atrovirens*).

mes·dames (mā·däm′, *Fr.* mā·dàm′) Plural of MADAME.

mes·de·moi·selles (mād·mwà·zel′) *French* Plural of MADEMOISELLE.

me·seems (mē·sēmz′) *v., impersonal Archaic* It seems to me.

mes·en·ceph·a·lon (mes′en·sef′ə·lon) *n. Anat.* The central division of the brain; the midbrain. [<NL <MES- + ENCEPHALON] **—mes·en·ce·phal·ic** (mes·en′sə·fal′ik) *adj.*

mes·en·chyme (mes′eng·kim) *n. Biol.* The portion of the mesoderm that produces the connective tissues of the body, the blood vessels, lymphatic system, and heart. It is cellular in structure, and in some of the lower forms of life is the same as *mesoblast.* Also **mes·en·chy·ma** (mes·eng′kə·mə). [<NL *mesenchyma* <MES- + Gk. *en-* in + *chein* pour] **—mes·en′·chy·mal, mes′en·chym′a·tous** (-kim′ə·təs) *adj.*

mes·en·ter·i·tis (mes·en′tə·rī′tis) *n. Pathol.* Inflammation of the mesentery. [<MESENTER(ON) + -ITIS]

mes·en·ter·on (mes·en′tər·on) *n. pl.* **·ter·a** (-tər·ə) *Biol.* **1** The middle portion of the primitive intestinal cavity, lined with endoderm: distinguished from the buccal and anal parts, which are lined with ectoderm. **2** The midgut. [<MES- + ENTERON] **—mes·en′ter·on′ic** (-tə·ron′ik) *adj.*

mes·en·ter·y (mes′ən·ter′ē) *n. pl.* **·ter·ies** *Anat.* A fold of the peritoneum that invests an intestine and connects it with the abdominal wall; especially, the fold investing the small intestine. Also **mes′en·te′ri·um** (-tir′ē·əm). [<Med. L *mesenterium* <Gk. *mesenterion* <*mesos* middle + *enteron* intestine] **—mes′·en·ter′ic** *adj.*

Me·se·ta (mā·sā′tä) The entire interior of Spain, covering almost three fourths of the country and comprising an immense plateau with Madrid at its center.

mesh (mesh) *n.* **1** One of the open spaces between the cords of a net or the wires of a sieve: often expressed numerically in terms of a unit area: a 100-*mesh* screen. **2** *pl.* Such cords or wires collectively. **3** Anything that entangles or involves. **4** *Mech.* The engagement of gear teeth. —*v.t. & v.i.* **1** To make or become entangled, as in a net. **2** To make or become engaged, as gear teeth. [Cf. OE *max* a net and MDu. *maesche* a mesh] **—mesh′y** *adj.*

Me·shach (mē′shak) Babylonian captive. *Dan.* iii. See SHADRACH.

Me·shed (me·shed′) A walled city and Moslem shrine in NE Iran: also *Mashhad.* Also **Mesh·hed** (mesh·hed′).

mesh·work (mesh′wûrk′) *n.* A combination of meshes; network.

me·si·al (mē′zē·əl, mes′ē·əl) *adj.* Situated in or directed toward the middle: the *mesial* plane of the body. Also **me′si·an.** [<Gk. *mesos* middle] **—me′si·al·ly** *adv.*

mes·ic[1] (mes′ik, mē′zik) *adj. Bot.* Pertaining to or characterized by a medium moisture supply, as in certain plants. [<Gk. *mesos* middle]

mes·ic[2] (mes′ik, mē′sik) *adj. Physics* Of, pertaining to, characteristic of, or produced by mesons. [<MESON[2]]

me·sit·y·lene (mə·sit′ə·lēn, -lin) *n. Chem.* A colorless, liquid hydrocarbon, C_9H_{12}, made by heating acetone with concentrated sulfuric acid. [<*mesityl* a hypothetical organic radical (<Gk. *mesitēs* mediator + -YL) + -ENE]

mes·i·tyl oxide (mes′i·təl) *Chem.* A colorless hydrocarbon, $C_6H_{10}O$, used as a solvent for nitrocellulose and certain gums and resins. [See MESITYLENE]

mes·mer·ism (mes′mə·riz′əm, mez′-) *n.* **1** The theory, as exemplified by Franz Anton Mesmer, 1733–1815, that one person can produce in another an abnormal condition resembling sleep, during which the mind of the subject is passively responsive to the will of the operator: now identified with hypnotism. Compare ANIMAL MAGNETISM. **2** Personal magnetism. **—mes·mer·ic** (mes·mer′ik, mez-), **mes·mer′i·cal** *adj.* **—mes·mer′i·cal·ly** *adv.* **—mes′mer·ist** *n.*

mes·mer·ize (mes′mə·rīz, mez′-) *v.t.* **·ized, ·iz·ing** To hypnotize. Also *Brit.* **mes′mer·ise.** **—mes′mer·i·za′tion** *n.* **—mes′mer·iz′er** *n.*

mesn·al·ty (mē′nəl·tē) *n.* The estate of a mesne lord. Also **mesn·al·i·ty** (mē·nal′ə·tē). [<MF *mesnalte*]

mesne (mēn) *adj. Law* Being between two periods or extremes; intermediate; intervening. [<MF, alter. of AF *meen* <L *medianus* mean. Doublet of MEAN[3], MEDIAN, MIZZEN.]

mesne lord One holding lands as an intermediate between a superior lord and a subordinate tenant.

meso- *combining form* **1** Situated in the middle: *mesocarp.* **2** Intermediate in size or degree: *mesognathous.* Also, before vowels, *mes-.* [<Gk. *mesos* middle]

mes·o·blast (mes′ə·blast, mē′sə-) *n. Biol.* The middle germinal layer of the embryo. See MESENCHYME. [<MESO- + Gk. *blastos* sprout] **—mes′o·blas′tic** *adj.*

mes·o·carp (mes′ə·kärp, mē′sə-) *n. Bot.* The middle layer of a pericarp.

mes·o·ce·phal·ic (mes′ō·sə·fal′ik, mē′sō-) *adj. Anat.* **1** Intermediate in head form; having a cephalic index of from 76.0 to 80.9. **2** Having a medium cranial capacity. **3** Of or pertaining to the mesocephalon. Also **mes′o·ceph′a·lous** (-sef′ə·ləs). **—mes′o·ceph′a·ly** *n.*

mes·o·ceph·a·lon (mes′ə·sef′ə·lon, mē′sə-) *n.* The pons Varolii.

mes·o·crat·ic (mes′ə·krat′ik, mē′sə-) *adj. Geol.* Having the dark constituents slightly in excess of the light ones: said of certain igneous rocks.

mes·o·derm (mes′ə·dûrm, mē′sə-) *n.* **1** *Biol.* The middle germ layer of the embryo, from which are developed the muscular, vascular, and osseous systems. **2** *Bot.* The middle layer of the wall of a moss and capsule. **—mes′o·der′mal, mes′o·der′mic** *adj.*

mes·o·gas·tri·um (mes′ə·gas′trē·əm, mē′sə-) *n. Biol.* One of the two mesenteries in the stomach of an embryo; also, the region of the umbilicus. [<MESO- + Gk. *gaster* belly] **—mes′o·gas′tric** *adj.*

me·sog·na·thous (mə·sog′nə·thəs) *adj.* Having moderately projecting jaws; also, having a facial profile angle of 98° to 103°. Also **mes·og·nath·ic** (mes′əg·nath′ik). [<MESO- + -GNATHOUS] **—me·sog′na·thism, me·sog′na·thy** *n.*

mes·o·kur·to·sis (mes′ō·kər·tō′sis, mē′sō-) *n. Stat.* The symmetrical kurtosis characterizing the region near the mode of a normal probability curve. **—mes′o·kur′tic** (-kûr′tik) *adj.*

Mes·o·lith·ic (mes′ə·lith′ik, mē′sə-) *adj. Anthropol.* Pertaining to or describing that period of human culture immediately following the Magdalenian stage of the Paleolithic, characterized by small, delicately worked microliths and an economy transitional between food gathering and a settled agriculture. Also called *Epipaleolithic, Miolithic.* [<MESO- + LITH(O)- + -IC]

me·sol·o·gy (mə·sol′ə·jē) *n.* The study of the environment in its relations to organisms;

Mesolonghi / metacarpus

ecology. [<MESO- + -LOGY] — **mes′o·log′ic, mes′o·log′i·cal** (mes′ə-loj′i-kəl) *adj.*

Mes·o·lon·ghi (mes′ō-lông′gē) See MISSOLONGHI.

mes·o·mor·phic (mes′ə-môr′fik, mē′sə-) *adj.* **1** *Physics* Of or pertaining to a state of matter intermediate between the true liquid and the crystal; liquo-crystalline: also **mes′o·mor′phous.** **2** Designating a physical type developed predominantly from the mesodermal layer of the embryo; the muscular or athletic type. Compare ECTOMORPHIC, ENDOMORPHIC.

mes·on[1] (mes′on, mē′son) *n.* **1** The plane that divides the body longitudinally into two halves; the median or mesial plane. **2** *Music* Loosely, a tetrachord. [<NL <Gk. *mesos* middle]

mes·on[2] (mes′on, mē′son) *n. Physics* Any of a group of short-lived, unstable atomic particles having a mass intermediate between that of the electron and the proton. They are believed to be a product of cosmic-ray disintegration and may be electrically neutral or carry either a positive or negative charge. The principal types are the mu-meson and pi-meson. [<Gk. *mesos* middle]

mes·o·neph·ros (mes′ə-nef′rəs, mē′sə-) *n. Biol.* The middle of three tubular organs found in connection with the primitive genitourinary apparatus, and formed later than the pronephros; the mid-kidney or Wolffian body. It is the permanent kidney in some animals, as amphibians. [<NL <MESO- + Gk. *nephros* kidney] — **mes′o·neph′ric** *adj.*

mes·o·pause (mez′ō-pôz) *n.* A transition zone between the mesosphere and the exosphere, having an upper limit of about 600 miles.

mes·o·phyll (mes′ə-fil, mē′sə-) *n. Bot.* The soft, inner, parenchymatous tissue of a leaf; the cellular portion lying between the upper and lower epidermis. Also **mes′o·phyl, mes′o·phyl′lum.** [<MESO- + Gk. *phyllon* leaf]

mes·o·phyte (mes′ə·fit, mē′sə-) *n. Bot.* A plant requiring medium conditions of moisture and dryness, intermediate between a hydrophyte and a xerophyte. — **mes′o·phyt′ic** (-fit′ik) *adj.*

mes·o·plast (mes′ə-plast, mē′sə-) *n. Biol.* A cell nucleus. [<MESO- + Gk. *plastos* formed] — **mes′o·plas′tic** *adj.*

Mes·o·po·ta·mi·a (mes′ə·pə·tā′mē·ə) An ancient country of Asia comprising the region about the lower Tigris and the lower Euphrates, included in modern Iraq. [<Gk. <*mesos* middle + *potamos* river] — **Mes′o·po·ta′mi·an** *n. & adj.*

mes·or·rhine (mes′ə-rīn, -rin, mē′sə-) *adj.* Having a relatively broad, high-bridged nose. [<MESO- + Gk. *rhis, rhinos* nose]

mes·o·sere (mes′ə·sir, mē′sə-) *n. Bot.* The flora and major plant development of the Mesozoic era. [<MESO- + L *serere* sow, plant]

mes·o·sphere (mez′ō-sfir) *n.* A layer of the atmosphere lying between the ionosphere and exosphere.

mes·o·the·li·um (mes′ə·thē′lē·əm, mē′sə-) *n. Biol.* **1** The portion of the mesoderm and the tissues derived from it that in vertebrates forms two principal layers, visceral and parietal, and produces the epithelium of the peritoneum and pleurae, the striated muscles, etc. **2** Epithelium when mesoblastic in origin. [<NL <MESO- + (EPI)THELIUM] — **mes′o·the′li·al** *adj.*

mes·o·ther·mal (mes′ə·thûr′məl, mē′sə-) *adj.* Possessing or pertaining to medium warmth. Also **mes′o·ther′mic.**

mes·o·tho·rax (mes′ə·thôr′aks, -thō′raks, mē′sə-) *n. Entomol.* The middle one of the three segments of the thorax in insects, bearing the anterior wings and the middle legs. — **mes′o·tho·rac′ic** (-thō·ras′ik, -thō·ras′ik) *adj.*

mes·o·tho·ri·um (mes′ə·thôr′ē·əm, -thō′rē-, mē′sə-) *n. Physics* Either of two isotopes resulting from the radioactive disintegration of thorium, intermediate between thorium and radiothorium.

mes·o·tron (mes′ə·tron, mē′sə-) *n. Physics* Meson. [<MESO- + (ELEC)TRON]

Mes·o·zo·ic (mes′ə·zō′ik, mē′sə-) *n. Geol.* The era between the Paleozoic and the Cenozoic, including the Triassic, Jurassic, and Cretaceous periods: characterized by the dominance of the reptiles, the rise of flowering plants, and the beginnings of archaic mammals. — *adj.* Of or pertaining to this era. [<MESO- + -ZOIC]

mes·quite (mes·kēt′, mes′kēt) *n.* Either of two spiny, deep-rooted shrubs or small trees of the pea family, found in the southwestern United States, and extending southward to Peru. The **honey mesquite** (*Prosopis glandulosa* or *juliflora*) yields sweet algarroba pods used for cattle fodder; the **screw-pod mesquite** (*Strombocarpa odorata*), or screwbean, has edible spiral pods. Also spelled *mezquite, muskit.* Also **mes·quit′.** [<Sp. *mezquite* < Nahuatl *mizquitl*]

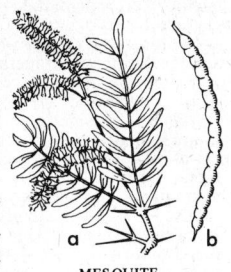
MESQUITE
a. Flower.
b. Fruit (edible).

mess (mes) *n.* **1** A quantity of food sufficient for one meal or for a particular occasion: a *mess* of beans; also, a portion of soft, partly liquid food, as pottage. **2** A number of persons who habitually take their meals together, as on board ship or in military units; also, a meal taken by them. **3** The sum or total of a haul of fish. **4** A state of disorder; especially, a condition of unclean confusion. **5** A confusing and embarrassing situation. **6** An unpleasant or unclean concoction; confused jumble. — *v.i.* **1** To busy oneself; dabble: often with *around* or *about.* **2** To make a mess; bungle: often with *up.* **3** To interfere; meddle: often with *around.* **4** To eat with a mess (def. 2). — *v.t.* **5** To make a mess of; muddle; botch: often with *up.* **6** To make dirty; befoul: often with *up.* **7** To provide meals for. [<OF *mes* <L *missus* course at a meal, orig. pp. of *mittere* send] — **mess′y** *adj.* — **mess′i·ly** *adv.* — **mess′i·ness** *n.*

mes·sage (mes′ij) *n.* **1** A communication, as of information, sent in any way. **2** A formal communication from a chief executive to a legislative body, not delivered in person: a *message* from the president to Congress. **3** An errand; the carrying out of a mission; a messenger's business. **4** An utterance divinely inspired; hence, any important communication embodying a truth, doctrine, principle, or advice. [<OF <Med.L *missaticum* < *missus,* pp. of *mittere*]

message center *Mil.* An agency attached to a headquarters or command post, and charged with the receipt, transmission, and delivery of messages.

Mes·sa·li·na (mes′ə·lī′nə), **Valeria,** executed A.D. 48, third wife of Emperor Claudius; notorious for profligacy.

mes·sa·line (mes′ə·lēn′, mes′ə·lēn) *n.* A lightweight, lustrous, twilled silk fabric.

Mes·sa·pi·an (mə·sā′pē·ən) *n.* **1** A member of an ancient people inhabiting SE Italy. **2** The Indo-European language of the Messapians, possibly related to ancient Illyrian. — *adj.* Of or pertaining to the Messapians or their language.

Mes·se·ne (me·sē′nē) A town in SW Peloponnesus, Greece, capital of ancient Messenia. Also **Mes·si′ni.**

mes·sen·ger (mes′ən·jər) *n.* **1** One sent with a message or on an errand of any kind; one employed to carry messages; specifically, a bearer of official dispatches. **2** A forerunner; herald. Also **mes′sa·ger** (-ə·jər). [ME *messanger, messager* <OF *messagier* < *message* MESSAGE: the *n* is non-historic]

Mes·se·ni·a (me·sē′nē·ə) A region of SW Peloponnesus, Greece, on the Ionian Sea, comprising an ancient country. Also **Mes·si′ni·a.**

Mes·ser·schmitt (mes′ər·shmit), **Wilhelm,** born 1898, German aircraft designer and manufacturer.

Mes·si·ah (mə·sī′ə) *n.* **1** The Anointed One; the Christ: the name for the promised deliverer of the Hebrews, assumed by Jesus, and given to him by Christians: with *the.* **2** Loosely, a looked-for liberator of a country or people. Also **Mes·si′as.** [<LL *Messias* <Gk. <Aramaic *měshīhā*, Hebrew *māshīah* anointed] — **Mes·si′ah·ship** *n.* — **Mes·si·an·ic** (mes′ē·an′ik) *adj.*

Mes·si·dor (me·sē·dôr′) See under CALENDAR (Republican).

mes·sieurs (mes′ərz, *Fr.* mā·syœ′) *n. pl.* of *Fr.* **mon·sieur** (mə·syœ′) Sirs; gentlemen: in English in the contracted form *Messrs.*, used as plural of *Mr.*

mes·sin (mes′in) *n. Scot.* A lap dog; hence, an insignificant person. Also **mes′san.**

Mes·si·na (mə·sē′nə, *Ital.* mäs·sē′nä) A port in NE Sicily on the **Strait of Messina,** the channel between Italy and Sicily, two miles wide at its narrowest point.

Mes·sines (me·sēn′) A village in western Belgium, scene of two battles in World War I, 1914 and 1917.

mess jacket A man's short, tailored jacket, usually white and terminating exactly at the waistline: worn on semiformal occasions.

mess kit A small, compactly arranged unit containing cooking and eating utensils: used by soldiers and campers.

mess·mate (mes′māt′) *n.* An associate at a mess, especially on board ship.

mes·suage (mes′wij) *n. Law* A dwelling house with its belongings, outhouses, garden, etc. [<OF *mesuage,* prob. alter. of *mesnage.* See MÉNAGE.]

Mes·ta (mes′tä) A river in SW Bulgaria and NE Greece, flowing 150 miles SE to the Aegean. *Greek* **Nes·tos** (nes′tos).

mes·tee (mes·tē′) *n.* The offspring of a white person and a quadroon; an octoroon: also spelled *mustee.* [<Sp. *mestizo* mongrel, hybrid]

mes·ti·zo (mes·tē′zō) *n. pl.* **·zos** or **·zoes** Any one of mixed blood; in Mexico and the western United States, a person of Spanish and Indian blood. Also **mes·te′so, mes·ti′no** (-nō). [<Sp. <LL *mixticius* <L *mixtus,* pp. of *miscere* mix] — **mes·ti′za** (-zə) *n. fem.*

Meš·tro·vić (mesh′trô·vich), ·**Ivan,** born 1883, Yugoslav sculptor.

met (met) Past tense and past participle of MEET[1].

met- Var. of META-.

meta- *prefix* **1** Changed in place or form; reversed; altered: *metamorphosis.* **2** *Anat. & Zool.* Behind; after; on the farther side of; later: often equivalent to *post-* or *dorso-*: *metathorax, metaplasis.* **3** With; alongside: *metabiosis.* **4** Beyond; over; transcending: *metaphysics, metapsychology.* **5** *Chem.* **a** A modification, usually polymeric, of. **b** A derivative of: *metaprotein.* **c** A derivative of an acid anhydride, formed by withdrawal of one or more water molecules: distinguished from *ortho-*: *metaphosphoric* acid. **d** A benzene derivative in which the substituted atoms or radicals occupy the positions 1, 3: abbr. *m-*. Compare ORTHO-, PARA-. See BENZENE RING. Also, before vowels and *h*, **met-**. [<Gk. < *meta* after, beside, with]

met·a·bi·o·sis (met′ə·bī·ō′sis) *n. Biol.* The condition of dependence of one organism upon another. [<META- + Gk. *bios* life] — **met′a·bi·ot′ic** (-ot′ik) *adj.*

met·a·bol·ic (met′ə·bol′ik) *adj.* **1** Of, pertaining to, or exhibiting metabolism: *metabolic* processes. **2** Pertaining to or undergoing change, transformation, or metamorphosis. Also **met′a·bol′i·cal.** [<Gk. *metabolikos*]

me·tab·o·lism (mə·tab′ə·liz′əm) *n. Biol.* The aggregate of all physical and chemical processes constantly taking place in living organisms, including those which use energy to build up assimilated materials (anabolism) and those which release energy by breaking them down (catabolism). Also **me·tab′o·ly** (-ə·lē). [<Gk. *metabolē* <*meta-* beyond + *ballein* throw]

me·tab·o·lite (mə·tab′ə·līt) *n.* A chemical product of metabolism.

me·tab·o·lize (mə·tab′ə·līz) *v.t. & v.i.* **·lized, ·liz·ing** To subject to or change by metabolism.

met·a·car·pal (met′ə·kär′pəl) *adj.* Of or pertaining to the metacarpus. — *n.* One of the bones of the metacarpus.

met·a·car·pus (met′ə·kär′pəs) *n. Anat.* The part of the fore- or thoracic limb between the carpus or wrist and the phalanges or bones of the finger. It consists in man of five bones. [<NL <Gk. *metakarpion* < *meta-* beyond + *karpos* wrist]

met·a·cen·ter (met′ə·sen′tər) *n. Physics* That point in a floating body slightly displaced from equilibrium through which the resultant upward pressure of the fluid always passes; the center of gravity of the unsubmerged portion of a floating body. Also **met′a·cen′tre.** — **met′a·cen′tric** *adj.*

met·a·chem·is·try (met′ə·kem′is·trē) *n.* The chemistry of elements and compounds which yield exceptionally large amounts of energy in relation to their mass. — **met′a·chem′i·cal** *adj.*

met·a·chro·ma·tism (met′ə·krō′mə·tiz′əm) *n.* An alteration in color; specifically, such alteration due to heating or cooling. — **met′a·chro·mat′ic** (-krō·mat′ik) *adj.*

Met·a·com·et (met′ə·kom′it) Indian name of American Indian chief, King Philip.

met·a·gal·ax·y (met′ə·gal′ək·sē) *n. pl.* **·ax·ies** *Astron.* The entire material universe, regarded especially as a system including all the galaxies.

met·age (mē′tij) *n.* **1** Measurement. **2** The price charged for measurement. **3** A general term for the tolls formerly exacted by the corporation of London over a part of the Thames above and below the city. [<METE[1]]

met·a·gen·e·sis (met′ə·jen′ə·sis) *n. Biol.* A type of reproduction in which a series of generations of unlike forms comes between the egg and the parent type; alternation of generations. — **met′a·ge·net′ic** (-jə·net′ik) *adj.*

me·tag·na·thous (mə·tag′nə·thəs) *adj. Ornithol.* Having the points of the mandibles crossing each other, as in the crossbill. [< META- + -GNATHOUS] — **me·tag′na·thism** *n.*

met·ag·nos·tic (met′ag·nos′tik) *adj.* Beyond the knowledge, whether of the sense or the understanding, of man as he is at present constituted; metaphysical. — *n.* A person holding a belief in the existence of a Supreme Being who transcends human knowledge. [<META- + Gk. *gnōstikos* knowing] — **met′ag·nos′ti·cism** *n.*

met·al (met′l) *n.* **1** An element that forms a base by combining with a hydroxyl group or groups. It is usually hard, heavy, lustrous, malleable, ductile, tenacious, and a good conductor of heat and electricity. **2** A composition of some metallic element; also, an alloy: generally with a qualifying word. **3** Cast iron while melted. **4** Broken stone for road surfaces or for railway ballast: also called **road metal.** **5** *Her.* Gold *(or)* or silver *(argent)* tincture. **6** Molten glass. **7** The weight of the projectiles that a warship's guns can throw at once. **8** *Printing* Type metal; also, composed type. **9** The constituent material of anything; essential quality. — **noble metal** A metal that does not readily oxidize in the open air, as gold, silver, and platinum. — **white metal** Any one of the various white alloys, such as pewter, used for making ornaments, small castings, etc; specifically, a soft, smooth, malleable, copper-zinc alloy of exceptional antifrictional properties used to form the bearing surface in the crankshaft and connecting-rod bearings in most internal-combustion engines. — *adj.* Consisting of or pertaining to metal. — *v.t.* **·aled** or **·alled, ·al·ing** or **·al·ling** To furnish or cover with metal. ◆ Homophone: *mettle.* [<OF <L *metallum* mine <Gk. *metallon.* Doublet of MEDAL.]

met·a·lin·guis·tics (met′ə·ling·gwis′tiks) *n.* An area of linguistic study concerned with the interrelationship of the structure and meaning of the language of a society and other aspects of its culture, such as the social system.

met·al·ist (met′l·ist) *n.* **1** One who works with or has special knowledge of metals. **2** An advocate of metallic money as against a paper currency. Also **met′al·list.**

met·al·ize (met′l·īz) *v.t.* **·ized, ·iz·ing** To turn into or treat with metal. Also **met′al·lize,** *Brit.* **met′al·ise.**

me·tal·lic (mə·tal′ik) *adj.* **1** Being, containing, yielding, or having the characteristics of a metal: a *metallic* voice; *metallic* luster. **2** Pertaining to a metal. — **me·tal′li·cal·ly** *adv.*

metallic soap A soapy, waxlike material made by combining the salts of certain metals, as lead or aluminum, with various fatty acids: used in the textile, varnish, and paint industries.

met·al·lif·er·ous (met′ə·lif′ər·əs) *adj.* Yielding or containing metal.

met·al·line (met′ə·lin, -līn) *adj.* **1** Relating to, having the properties of, or resembling metal. **2** Impregnated with metals or metallic salts.

met·al·log·ra·phy (met′ə·log′rə·fē) *n.* **1** The science that treats of metallic substances; also, a treatise on metals. **2** Microscopic study of the structure of metals and alloys. [<Gk. *metallon* mine, metal + -GRAPHY] — **me·tal·lo·graph·ic** (mə·tal′ə·graf′ik) *adj.*

met·al·loid (met′ə·loid) *n.* One of those nonmetallic elements that resemble the metals in some of their properties, as arsenic and antimony. — *adj.* **1** Resembling a metal. **2** Of, pertaining to, or having the properties of a metalloid. Also **met′al·loi′dal.**

me·tal·lo·ther·a·py (mə·tal′ō·ther′ə·pē) *n.* Medical treatment by the use of metals, especially metal salts. [<Gk. *metallon* mine, metal + THERAPY]

met·al·lur·gy (met′ə·lûr′jē) *n.* The art or science of extracting a metal or metals from ores, as by smelting, reducing, refining, alloying, electrolysis, etc. [<NL *metallurgia* <Gk. *metallourgos* working in mines <*metallon* mine + *-ergos* working] — **met′al·lur′gic** or **·gi·cal** *adj.* — **met′al·lur′gi·cal·ly** *adv.* — **met′·al·lur′gist** *n.*

met·al·work (met′l·wûrk′) *n.* **1** Articles made of metal. **2** Metalworking.

met·al·work·ing (met′l·wûr′king) *n.* The making or the business of making things out of metal. — **met′al·work′er** *n.*

met·a·math·e·mat·ics (met′ə·math′ə·mat′iks) *n.* That branch of mathematics which is concerned with the formalized and rigorously logical treatment of pure symbols, having regard only to internal consistency and the establishment of absolute proofs of the validity of a given set of axioms, postulates, theorems, etc., within a mathematical system. — **met′a·math′e·mat′i·cal** *adj.*

met·a·mer (met′ə·mər) *n. Chem.* A compound or substance exhibiting metamerism. [<META- + Gk. *meros* part] — **met′a·mer′ic** (-mer′ik) *adj.*

met·a·mere (met′ə·mir) *n. Biol.* One of the series of homologous segments that form the body of a chordate or articulate animal, as a worm; a somite. Also **me·tam·er·on** (mə·tam′·ər·on). [<META- + -MERE] — **met′a·mer′ic, me·tam′er·al** *adj.* — **met′a·mer′i·cal·ly** *adv.*

me·tam·er·ism (mə·tam′ə·riz′əm) *n.* **1** *Chem.* A variety of isomerism in which the compounds have not only the same percentage of composition, but also the same molecular weight. **2** *Biol.* Disposition in metameres; the state of being a metamere; also, an example of this. Also **me·tam′er·y.**

met·a·mor·phic (met′ə·môr′fik) *adj.* **1** Producing metamorphism. **2** Pertaining to, caused by, or exhibiting metamorphism. Also **met′a·mor′phous.**

met·a·mor·phism (met′ə·môr′fiz·əm) *n.* **1** *Geol.* The changes in the composition and texture of rocks caused by earth forces accompanied by heat, pressure, moisture, etc. **2** Any metamorphosis. [See METAMORPHOSIS]

met·a·mor·phop·si·a (met′ə·môr·fop′sē·ə) *n. Pathol.* A defect in vision which makes objects appear distorted. [<NL <Gk. *metamorphōsis* transformation + *ōps* eye]

met·a·mor·phose (met′ə·môr′fōz) *v.t.* **·phosed, ·phos·ing** **1** To change the form of; transmute. **2** To change by metamorphism. Also **met′a·mor′phize.** See synonyms under CHANGE. [<F *métamorphoser*]

met·a·mor·pho·sis (met′ə·môr′fə·sis) *n. pl.* **·pho·ses** (-fə·sēz) **1** A passing from one form or shape into another; transformation with or without change of nature: especially applied to change by means of witchcraft, sorcery, etc. **2** Complete transformation of character, purpose, circumstances, etc.; also, a person or thing metamorphosed. **3** *Biol.* A change in form, structure, or function in an organism resulting from development; transformation; specifically, the series of marked external changes through which an individual passes after leaving the egg and before attaining sexual maturity, as the larva, pupa, and imago of an insect. Compare METAGENESIS. **4** *Bot.* The varied development of plant organs of the same morphological value, such development resulting from their adaptations of different functions: also **met′a·mor′phy.** **5** *Pathol.* A morbid change of the elements of tissues into another form of structure. **6** The changes in form going on in living tissues, blood corpuscles, etc. [<L <Gk. *metamorphōsis* <*metamorphoein* transform <*meta-* beyond + *morphē* form]

met·a·neph·ros (met′ə·nef′ros) *n. Biol.* The posterior one of three similar tubular organs in connection with the genitourinary apparatus. It develops into the permanent kidney. [<NL <META- + Gk. *nephros* kidney]

met·a·phase (met′ə·fāz) *n. Biol.* The middle stage of mitotic cell division, during which the chromosomes split along the equatorial plane between the two poles of the spindle. [<META- + -PHASE]

met·a·phor (met′ə·fôr, -fər) *n.* A figure of speech in which one object is likened to another by speaking of it as if it were that other: distinguished from *simile* by not employing any word of comparison, such as "like" or "as." See synonyms under ALLEGORY, SIMILE. — **mixed metaphor** Figurative language in which incongruous, and often contradictory, metaphors are used; confusion of figurative with plain statement. [<F *métaphore* <L *metaphora* <Gk. <*metapherein* <*meta-* beyond, over + *pherein* carry] — **met′a·phor′·ic** (-fôr′ik, -for′ik) or **·i·cal** *adj.* — **met′a·phor′·i·cal·ly** *adv.*

met·a·phos·phate (met′ə·fos′fāt) *n. Chem.* A salt of metaphosphoric acid.

met·a·phos·phor·ic acid (met′ə·fos·fôr′ik,-for′·ik) *Chem.* The glacial phosphoric acid of commerce, HPO_3, usually sold in the form of transparent sticks. It is obtained by heating orthophosphoric acid.

met·a·phrase (met′ə·frāz) *v.t.* **·phrased, ·phras·ing** **1** To translate word for word. **2** To alter the wording of. — *n.* **1** A literal translation. **2** A phrase in response; retort. **3** A school exercise consisting in the rendering of a piece of poetry into prose or of prose into verse. [<Gk. *metaphrasis* <*metaphrazein* paraphrase <*meta-* beyond + *phrazein* phrase]

met·a·phrast (met′ə·frast) *n.* One who renders poetry into prose or prose into poetry, or changes the meter of verse. [<Gk. *metaphrastēs*] — **met′a·phras′tic** or **·ti·cal** *adj.*

met·a·phys·ic (met′ə·fiz′ik) *n.* Metaphysics. — *adj.* Metaphysical.

met·a·phys·i·cal (met′ə·fiz′i·kəl) *adj.* **1** Of or pertaining to metaphysics. **2** Treating of or versed in metaphysics. **3** Beyond or above the physical or experiential; pertaining to or being of the essential nature of reality; transcendental. **4** Dealing with abstractions; apart from, or opposed to, the practical. **5** Designating certain poets of the 17th century, notably Cowley and Donne, whose verses were characterized by complex, intellectualized imagery: term originating with Dr. Samuel Johnson. [See METAPHYSICS] — **met′a·phys′i·cal·ly** *adv.*

metaphysical healing Christian Science.

met·a·phy·si·cian (met′ə·fi·zish′ən) *n.* One skilled or versed in metaphysics. Also **met′a·phys′i·cist.**

met·a·phys·ics (met′ə·fiz′iks) *n. pl. (construed as singular)* **1** The systematic study or science of the first principles of being and of knowledge; the doctrine of the essential nature and fundamental relations of all that is real. **2** Speculative philosophy in the wide sense. **3** The principles of philosophy as applied to the methodology of any particular science. **4** Mental science in general; psychology. **5** In popular use, abstruse and bewildering discussion. Also *metaphysic.* [<Med. L *metaphysica* <Med. Gk. <*ta meta ta physika* the (works) after the physics; in ref. to Aristotle's ontological treatises, which came after his *Physics*]

met·a·pla·si·a (met′ə·plā′zhē·ə) *n. Biol.* The direct transformation of one kind of tissue into another, as cartilage into bone. [<NL <Gk. *meta-* beyond + *plassein* mold]

me·tap·la·sis (mə·tap′lə·sis) *n. Biol.* The period of completed growth in the life cycle of an organism; maturity. [<NL]

met·a·plasm (met′ə·plaz′əm) *n.* **1** *Biol.* The lifeless, non-protoplasmic material of a cell, as inclusions of fats and carbohydrates. **2** A reversal or change in the order of the letters or syllables of a word. [<L *metaplasmus* <Gk. *metaplasmos* <*meta-* beyond + *plassein* mold] — **met′a·plas′mic** *adj.*

met·a·po·di·um (met′ə·pō′dē·əm) *n. pl.* **·di·a** (-dē·ə) **1** The posterior part of the foot in

metaprotein

gastropods and pteropods. 2 The metacarpus and metatarsus of quadrupeds. Also **met′a·pode** (-pōd), **met′a·pod** (-pod).
met·a·pro·te·in (met′ə-prō′tē-in, -tēn) *n. Biochem.* A protein resulting from the action of acids and alkalis and soluble in weak solutions of either but not in solutions of neutral salts.
met·a·psy·chics (met′ə-sī′kiks) *n.* Parapsychology.
met·a·psy·chol·o·gy (met′ə-sī·kol′ə-jē) *n.* 1 Psychology restricted to philosophical speculations on the origin, structure, function, purpose, etc., of the mind. 2 *Psychoanal.* The investigation and study of mental processes from three points of view: the dynamic, topographical, and economic. — **met′a·psy′cho·log′i·cal** (-sī′kə·loj′i·kəl) *adj.*
met·a·psy·cho·sis (met′ə-sī·kō′sis) *n.* Interchange of mental influence or action without a recognized physical medium.
met·a·so·ma·to·sis (met′ə-sō′mə·tō′sis) *n. Geol.* That form of metamorphism by means of which a rock or mineral undergoes chemical change through the action of external materials. Also **met′a·so′ma·tism** (-sō′mə·tiz′əm). [<META- + SOMAT(O)- + -OSIS]
met·a·some (met′ə-sōm) *n. Geol.* A mineral which has developed individually within another mineral. [<META- + -SOME²]
met·a·sta·ble (met′ə-stā′bəl) *adj. Physics & Chem.* Denoting an apparent state of equilibrium, as in supersaturated solutions. — **met′a·sta·bil′i·ty** (-stə·bil′ə·tē) *n.*
me·tas·ta·sis (mə·tas′tə·sis) *n.*, *pl.* **·ses** (-sēz) 1 Change of one thing into another. 2 Metabolism. 3 *Pathol.* The transfer of a disease or its manifestations from one part of the body to another. 4 In rhetoric, a rapid change from one point to another. [<L <Gk. <*methistanai* place differently, change <*meta-* after + *histanai* place] — **met′a·stat′ic** (-stat′ik) *adj.* — **met′a·stat′i·cal·ly** *adv.*

tive changes, but without localized external lesions. 2 Parasyphilis. — **met′a·syph′i·lit′ic** (-sif′ə·lit′ik) *adj.*
met·a·tar·sal (met′ə·tär′səl) *adj.* Of or pertaining to the metatarsus. — *n.* One of the bones of the metatarsus. See illustration under FOOT.
met·a·tar·sal·gi·a (met′ə·tär·sal′jē·ə) *n. Pathol.* Neuralgia in the middle of the foot. [<METATARS(US) + -ALGIA]
met·a·tar·sus (met′ə·tär′səs) *n.*, *pl.* **·si** (-sī) *Anat.* The part of the hind or pelvic limb between the tarsus or ankle and the phalanges or bones of the toe. In man it consists of five bones. [<NL <META- + TARSUS]
met·a·the·ri·an (met′ə·thir′ē·ən) *n.* Any one of a subclass (*Metatheria*) of mammals whose young are born immature and are carried in a pouch until fully developed; one of the marsupials, as a kangaroo, opossum, or bandicoot. — *adj.* Of or pertaining to the *Metatheria*. [<META- + Gk. *thērion* beast]
me·tath·e·sis (mə·tath′ə·sis) *n.*, *pl.* **·ses** (-sēz) 1 The transposition of letters, syllables, or sounds in a word. 2 *Chem.* A substitution, as the replacing or exchange of one or more radicals or groups in a compound; double decomposition or mutual exchange. 3 *Surg.* The operation of removing a morbific substance from one place to another for relief, as by pushing a calculus lodged in the urethra back into the bladder. 4 Any change or reversal of conditions. [<LL <Gk. <*metatithenai* transpose <*meta-* over + *tithenai* place] — **met·a·thet·ic** (met′ə·thet′ik) or **·i·cal** *adj.*
met·a·tho·rax (met′ə·thôr′aks, -thō′raks) *n. Entomol.* The hindmost of the three segments of the thorax in insects, bearing the hind wings and the third pair of legs. — **met′a·tho·rac′ic** (-thô·ras′ik, -thō-) *adj.*
met·a·troph·ic (met′ə·trof′ik) *adj.* Saprophytic. [<META- + Gk. *trophikos* feeding, nursing]
me·tat·ro·phy (mə·tat′rə·fē) *n. Pathol.* A wasting away because of disordered nutrition. Also **met·a·tro·phi·a** (met′ə·trō′fē·ə). [<MET- + ATROPHY]
Me·tau·ro (mä·tou′rō) A river in central Italy, flowing 207 miles NE to the Adriatic; at its delta the Romans defeated the Carthaginians under Hasdrubal, 207 B.C. Ancient **Me·tau·rus** (me·tô′rus).
Me·tax·as (mə·tak′səs, *Greek* me′täk·säs′),

meteorology

Joannes, 1871-1941, Greek soldier and statesman.
met·a·xy·lem (met′ə·zī′ləm) *n. Bot.* The thick-walled cell portion in the woody tissues of plants, developed outside the primary xylem. [<META- + XYLEM]
met·a·zo·an (met′ə·zō′ən) *n.* Any member of a primary division (*Metazoa*) of the animal kingdom, whose cells become differentiated into at least an outer and an inner wall, including all animals higher than protozoans. Also **met′a·zo′on.** — *adj.* Of or pertaining to the metazoans: also **met′a·zo′ic**. [<META- + Gk. *zōion* animal]
Metch·ni·koff (mech′ni·kôf), **Elie**, 1845-1916, Russian physiologist and bacteriologist: also known as *Ilya Ilich Metchnikoff*.
mete¹ (mēt) *v.t.* **met·ed**, **met·ing** 1 To allot or distribute by measure; apportion: usually with *out*. 2 *Obs.* To measure. — *n. Obs.* Measure. ♦ Homophones: *meat*, *meet*. [OE *metan* measure]
mete² (mēt) *n. Obs.* A boundary line; limit; confine: usually in the phrase *metes and bounds*. ♦ Homophones: *meat*, *meet*. [<OF <L *meta* goal, boundary]
met·em·pir·i·cal (met′em·pir′i·kəl) *adj.* Lying beyond the bounds of experience, as intuitive principles; not derived from experience; transcendental; a priori: opposed to *empirical*. Also **met′em·pir′ic**.
met·em·pir·i·cism (met′em·pir′ə·siz′əm) *n.* The science of pure reason; metaphysics proper; hence, with some, transcendental philosophy. Also **met′em·pir′ics**.
me·temp·sy·cho·sis (mə·temp′sə·kō′sis, met′əm·sī-) *n.* Transmigration of souls from body to body. [<LL <Gk. *metempsychōsis* <*metempsychoein* <*meta-* over + *empsychoein* animate <*en-* in + *psychē* soul, life]
met·en·ceph·a·lon (met′ən·sef′ə·lən) *n.*, *pl.* **·la** (-lə) *Anat.* 1 The fifth cerebral vesicle of the brain and the parts derived therefrom, comprising the medulla oblongata and the posterior part of the roof of the fourth ventricle. 2 The part of the brain consisting of the cerebellum and pons Varolii. [<NL <MET- + ENCEPHALON] — **met′en·ce·phal′ic** (-sə·fal′ik) *adj.*
me·te·or (mē′tē·ər, -ôr) *n.* 1 *Astron.* A luminous phenomenon, produced by a small mass of matter from the celestial spaces which strikes the earth's atmosphere with great velocity and is dissipated by heat: when not very brilliant, called a *shooting star*. 2 A meteoroid. 3 *Obs.* Any phenomenon of the atmosphere: **aerial meteors** (winds, hurricanes, etc.), **aqueous meteors** (rain, snow, etc.), **igneous meteors** (lightning, shooting stars, etc.), **luminous meteors** (aurora, rainbow, etc.). [<Med. L *meteorum* <Gk. *meteōron* thing in the air <*meteōros* high in the air <*meta-* beyond + *eōra* suspension]
Meteor Crater A depression caused by a meteor in central Arizona; 600 feet deep, 4,000 feet in diameter: also *Diablo Crater*.
me·te·or·ic (mē′tē·ôr′ik, -or′ik) *adj.* 1 Relating to meteors. 2 Meteorological. 3 Transitorily brilliant: a *meteoric* career. Also **me′te·or′i·cal**. — **me′te·or′i·cal·ly** *adv.*
me·te·or·ite (mē′tē·ə·rīt′) *n.* 1 A fallen meteor; a mass of stone or iron that has fallen upon the earth from outer space. 2 A meteoroid. — **me′te·or·it′ic** (-ə·rit′ik) *adj.*
me·te·or·o·graph (mē′tē·ər·ə·graf′, -gräf′, mē′tē·ôr′ə-, -or′ə-) *n.* A self-recording instrument, frequently attached to a kite or balloon, by which several meteorological elements are plotted in the form of a diagram. [<F *météorographe*]
me·te·or·oid (mē′tē·ə·roid) *n. Astron.* One of innumerable small particles of matter moving through the celestial spaces, which, when they encounter the earth's atmosphere, form meteors or shooting stars.
me·te·or·ol·o·gy (mē′tē·ə·rol′ə·jē) *n.* 1 The science that treats of atmospheric phenomena, especially those that relate to weather. 2 The character of the weather and of atmospheric changes of any particular place. [<Gk. *meteōrologia* <*meteōros* high in the air + *logos* discourse] — **me′te·or′o·log′ic** (-ôr′ə·loj′ik), **me′te·or′o·log′i·cal** *adj.* — **me′te·or′o·log′i·cal·ly** *adv.* — **me′te·or·ol′o·gist** *n.*

METEOROLOGY SYMBOLS

Symbol	Meaning	Symbol	Meaning
,	Drizzle	↔	Duststorm; Sandstorm
•	Rain	<	Lightning
*	Snow	∩	Rainbow
≡	Fog	○	Smoke
∪	Frost	∞	Haze
▽	Showers	∧	Squalls
℞	Thunderstorms	0	Exceptional visibility
↔	Ice crystals or ice needles	§	Signs of tropical storm
▲	Sleet	○	Cloudless sky
⌒	Aurora	◐	Partly cloudy sky
∪	Lunar corona	◍	Cloudy sky
⊕	Solar corona	●	Overcast sky
⌒	Dew		

WIND DIRECTION

Symbol	Direction	Symbol	Direction
↓	North	↑	South
↓↙	NNE	↑↖	SSE
↙	NE	↖	SE
↘	NW	↗	SW
↓↘	NNW	↑↗	SSW
←	East	→	West
←↖	ESE	→↗	WSW
←↙	ENE	→↘	WNW

me·tas·ta·size (mə·tas′tə·sīz) *v.i.* **·sized**, **·siz·ing** *Pathol.* To shift or spread from one part of the body to another: said specifically of malignant growths.
met·as·then·ic (met′əs·then′ik) *adj. Biol.* Having the posterior or lower part of the body well developed, as a kangaroo. [<META- + STHENIC]
met·a·syph·i·lis (met′ə·sif′ə·ləs) *n. Pathol.* 1 Congenital syphilis, with typical degenera-

meteor shower *Astron.* Any of various displays of shooting stars which recur at definite intervals and are named after the constellation from which they appear to radiate, as the Leonids, Perseids, etc.

me·ter[1] (mē′tər) *n.* **1** An instrument, apparatus, or machine for measuring fluids, gases, electric currents, grain, etc., and recording the results obtained. **2** Any person or thing that measures; specifically, one of several officers appointed to measure certain commodities, as for tolls. —*v.t.* To measure or test by means of a meter. [<METE[1]]

me·ter[2] (mē′tər) *n.* **1** The fundamental unit of length in the metric system, originally defined as one ten-millionth of the distance on the earth's surface from the pole to the equator; 39.37 inches. In practice, it is the distance between two fiducial lines marked on the platinum-iridium International Prototype Meter deposited at Sèvres, France. See METRIC SYSTEM. **2** A measured verbal rhythm, the structure of verse; a definite arrangement of groups of syllables in a line, having a time unit and a regular beat; also, a specific arrangement of words, or a specific sequence of such lines in a stanza. **3** The character of a musical composition as being divisible into measures equal in time and length, and similar in rhythmic construction. Also spelled *metre.* [<F *mètre* <L *metrum* a measure <Gk. *metron*]

Synonyms: euphony, measure, rhythm, verse. *Euphony* is agreeable linguistic sound, however produced; *meter, measure,* and *rhythm* denote agreeable succession of sounds in the utterance of connected words; *euphony* may apply to a single word or even a single syllable; the other words apply to lines, sentences, paragraphs, etc.; *rhythm* and *meter* may be produced by accent only, as in English, or by accent and quantity combined, as in Greek or Italian; *rhythm* or *measure* may apply either to prose or to poetry, or to music, dancing, etc.; *meter* is more precise than *rhythm,* applies only to poetry, and denotes a measured *rhythm* with regular divisions into *verses,* stanzas, strophes, etc. A *verse* is strictly a metrical line, but the word is often used as synonymous with stanza. *Verse,* in the general sense, denotes metrical writing without reference to the thought involved; as, prose and *verse.* Compare MELODY, POETRY.

-meter *combining form* **1** That (instrument or unit) by which a thing is measured: *calorimeter.* **2** Measure according to or containing (the main element): *hexameter.* [<L *metrum* <Gk. *metron* measure]

me·ter·age (mē′tər·ij) *n.* **1** The act or result of measuring. **2** The charge for measurement.

met·es·trus (met·es′trəs) *n. Biol.* The period of sexual quiescence following estrus in female mammals: often spelled *metoestrus, metoestrum.* Also **met·es′trum.** [<NL <MET- + ESTRUS] —**met·es′trous** *adj.*

meth·ac·ry·late (meth·ak′rə·lāt) *n. Chem.* An ester of methacrylic acid.

meth·a·cryl·ic acid (meth′ə·kril′ik) *Chem.* A colorless acid, $C_4H_6O_2$, the esters of which are extensively used in the making of plastics. [<METH(YL) + ACRYLIC]

meth·ane (meth′ān) *n. Chem.* A colorless, inflammable gas, CH_4, the simplest of the saturated hydrocarbons, formed by decomposition of vegetable matter, or artificially by dry distillation of certain organic matter, or by chemical treatment of certain metal compounds, as aluminum carbide. It is an important component of illuminating gas: often called *marsh gas.* [<METH(YL) + -ANE[2]]

methane series *Chem.* A group of saturated aliphatic hydrocarbons having the general formula C_nH_{2n+2}, and identified by the ending *-ane*: also called *alkanes.*

meth·a·nol (meth′ə·nōl, -nol) *n. Chem.* Methyl alcohol; a colorless, volatile, inflammable liquid, CH_3OH, obtained by the destructive distillation of wood or by catalytic treatment of hydrogen and carbon monoxide: highly toxic and widely used in industry and the arts: also called *carbinol, wood alcohol.* [<METHANE + -OL[1]]

me·theg·lin (mə·theg′lin) *n.* A fermented drink made of water and honey; mead. [<Welsh *meddyglyn* <*meddyg* doctor (<L *medicus*) + *llyn* juice, liquor]

met·he·mo·glo·bin (met·hē′mə·glō′bin, -hem′ə-) *n. Biochem.* A stable, brown-red, crystalline compound formed by the decomposition of blood and by oxidation of hemoglobin. Also **met·hae′mo·glo′bin.** [<MET- + HEMOGLOBIN]

me·the·na·mine (mə·thē′nə·mēn, -min) *n. Chem.* An organic compound, $C_6H_{12}N_4$, crystallized from a mixture of formalin and ammonia: used in the vulcanization of rubber, in making synthetic resins, and as an antiseptic: also called *Urotropin, hexamethylenamine, hexamethylenetetramine.* [<*methene* (<METHYL + -ENE) + AMINE]

me·thinks (mē·thingks′) *v., impersonal Obs.* It seems to me; I think. [OE *me thyncth* < *thyncan* seem]

me·thi·o·nine (mə·thī′ə·nēn, -nin) *n. Biochem.* An amino acid, $C_5H_{11}O_2NS$, containing sulfur and closely related to cystine: obtained from various proteins. [<ME(THYL) + THIO- + -INE[2]]

metho- *combining form Chem.* Used to indicate the presence of a methyl group in a compound. Also, before vowels, **meth-**, as in *methane.* [<METHYL]

meth·od (meth′əd) *n.* **1** A general or established way or order of doing anything, or the means or manner by which it is presented or taught. **2** Orderly and systematic arrangement, as of ideas and topics, etc.; the design or plan of a speaker or of an author. **3** The arrangement of natural bodies according to their common characteristics. See synonyms under MANNER, RULE, SYSTEM. [<F *méthode* <L *methodus* <Gk. *methodos* <*meta-* after + *hodos* way]

me·thod·i·cal (mə·thod′i·kəl) *adj.* **1** Given to or characterized by orderly arrangement. **2** Arranged with method. Also **me·thod′ic.** —**me·thod′i·cal·ly** *adv.* —**me·thod′i·cal·ness** *n.*

meth·od·ist (meth′əd·ist) *n.* An observer of method or order. —**meth′od·ism** *n.* —**meth′od·is′tic, meth′od·is′ti·cal** *adj.*

Meth·od·ist (meth′əd·ist) *n.* A member of any one of the Protestant churches that have grown out of the religious movement begun at Oxford University in the first half of the 18th century, in which John Wesley as a leader was associated with Charles Wesley, George Whitefield, and others. The largest Methodist body in the United States is the **Methodist Church,** formed by the union of the **Methodist Episcopal Church,** the **Methodist Episcopal Church, South,** and the **Methodist Protestant Church.** —*adj.* Pertaining to, belonging to, like, or typical of this church or its doctrines. [<METHOD + -IST] —**Meth′od·ism** *n.* —**Meth′od·is′tic, Meth′od·is′ti·cal** *adj.*

meth·od·ize (meth′əd·īz) *v.t.* **·ized, ·iz·ing** To reduce or subject to method; systematize. Also *Brit.* **meth′od·ise.** See synonyms under REGULATE. —**meth′od·i·za′tion** *n.*

meth·od·ol·o·gy (meth′ə·dol′ə·jē) *n.* **1** The science of method or of arranging in due order. **2** The division of pure logic that treats of the methods of directing the means of thinking to the end of clear and connected thinking. [<Gk. *methodos* method + -LOGY]

me·thought (mē·thôt′) *v., impersonal Obs.* It seemed to me; I thought.

Me·thu·se·lah (mə·thōō′zə·lə) The son of Enoch; a Hebrew patriarch reputed to have lived 969 years. *Gen.* v 27.

meth·yl (meth′əl) *n. Chem.* An organic radical, CH_3, existing only in combination, as in methyl alcohol, etc. It is univalent, and forms esters with acids. [<METHYLENE] —**me·thyl·ic** (mə·thil′ik) *adj.*

methyl acetate *Chem.* A liquid product, $CH_3CO_2CH_3$, of wood alcohol or wood vinegar, used as a solvent or a flavoring material.

meth·yl·al (meth′əl·al) *n. Chem.* A colorless, volatile, inflammable liquid, $C_3H_8O_2$, with a pungent taste and the odor of chloroform, used in medicine and in the making of perfumery and artificial resins: also called *formal.* [<METHYL + AL(COHOL)]

methyl alcohol Methanol.

meth·yl·am·ine (meth′əl·am′ēn, -in) *n. Chem.* A colorless, gaseous, inflammable amine, CH_3NH_2, with a strong fishy odor, contained in the products of the decomposition of certain organic compounds, and also made synthetically from ammonia by replacement of hydrogen by methyl. [<METHYL + AMINE]

meth·yl·ate (meth′əl·āt) *v.t.* **·at·ed, ·at·ing** *Chem.* To mix or saturate with methyl or methanol. —*n.* A compound derived from methanol by replacing the hydrogen of the hydroxyl group with an element or radical of equal valence: potassium *methylate.* [<METHYL + -ATE] —**meth′yl·a′tion** *n.*

methylated spirit *Chem.* A mixture of ethyl alcohol with ten percent of methanol: used in the arts, unfit for drinking.

methyl chloride *Chem.* A gas, CH_3Cl, which on compression becomes a colorless, sweet-tasting liquid of ethereal odor yielding a poisonous vapor: used as a refrigerant and fire extinguisher.

meth·yl·cho·lan·threne (meth′əl·kō·lan′thrēn) *n. Biochem.* A yellow, crystalline hydrocarbon, $C_{21}H_{16}$, extracted from bile acids and also prepared synthetically: believed to be strongly carcinogenic. [<METHYL + CHOL(E)- + ANTHR(ACINE) + -ENE]

meth·yl·di·chlor·ar·sine (meth′əl·dī′klôr·är′·sēn, -sin) *n. Chem.* A colorless liquid compound, CH_3AsCl_2, adapted for chemical warfare as a vesicant and lung irritant. [<METHYL + DI- + CHLOR- + ARSINE]

meth·yl·ene (meth′əl·ēn) *n. Chem.* A bivalent organic radical, CH_2, derived from methane. [<F *méthyline* <Gk. *methy* wine + *hylē* wood]

methylene blue *Chem.* A dark, blue-green aniline dye, $C_{16}H_{18}N_3ClS{\cdot}3H_2O$, having medicinal properties in the treatment of diphtheria, malaria, etc., and as an antidote in cyanide poisoning: used also as a chemical indicator and bacteriological stain.

meth·y·lep·si·a (meth′ə·lep′sē·ə) *n.* A morbid craving for strong drink. [<Gk. *methy* wine + -LEPSIA]

meth·yl·gua·ni·dine (meth′əl·gwä′nə·dēn, -din) *n. Biochem.* A ptomaine, $C_2H_7N_3$, derived from creatinine and from certain foods, as spoiled fish. [<METHYL + GUANADINE]

methyl methacrylate *Chem.* A polymerized methacrylate forming transparent thermosetting plastics resembling glass in their optical properties. [See METHACRYLIC]

meth·yl·naph·tha·lene (meth′əl·naf′thə·lēn) *n. Chem.* A methyl compound, $C_{10}H_7CH_3$, occurring in two isomeric forms, one of which is used in the determination of cetane numbers. [<METHYL + NAPHTHALENE]

methyl orange *Chem.* An azo dyestuff used chiefly as an indicator in alkalimetry; a brilliant orange-yellow powder, readily soluble in water, which is colored red on the addition of acids.

meth·yl·ros·an·i·line (meth′əl·roz·an′ə·lēn, -lin) *n.* Gentian violet. [<METHYL + ROSANILINE]

me·tic·u·lous (mə·tik′yə·ləs) *adj.* Careful about trivial matters; finical; particular. [<F *méticuleux* <L *meticolosus* fearful <*metus* fear] —**me·tic′u·los′i·ty** (-los′ə·tē) *n.* —**me·tic′u·lous·ly** *adv.*

mé·tier (mā·tyā′) *n. French* Trade; profession.

mé·tif (mā·tēf′) *n. pl.* **·tifs** (-tēfs′) or **·tis** (-tēs′) *French* A person of mixed blood; especially, a person half French and half American Indian. Also **me·tiff′.**

mé·tis·sage (mā·tē·säzh′) *n. French* Miscegenation.

met·o·chy (met′ə·kē) *n. Entomol.* The relationship of mutual tolerance without mutual aid existing between ants and their neutral insect guests. [<Gk. *metochē* sharing <*metochein* share in]

met·o·don·ti·a·sis (met′ō·don·tī′ə·sis) *n. Dent.* **1** Abnormal or imperfect dentition. **2** Faulty teething. [<NL <MET- + Gk. *odontiaein* cut teeth + -IASIS]

met·oes·trus (met·es′trəs), **met·oes·trum** See METESTRUS.

Met·ol (met′ōl, -ol) *n.* A white, soluble powder, C_7H_9ON, derivative of cresol: used as a photographic developer: a trade name.

Me·ton·ic cycle (mə·ton′ik) A period of 19 Julian years, amounting to nearly 235 lunar revolutions, at the conclusion of which the phases of the moon recur at the same time of the year. [after *Meton,* Athenian astronomer of the fifth century B.C.]

met·o·nym (met′ə·nim) *n.* A word used as a substitute for another. [See METONYMY]

me·ton·y·my (mə·ton′ə·mē) *n.* A figure of speech that consists in the naming of a thing by one of its attributes, as "the crown prefers" for "the king prefers." [<L *metonymia* <Gk. *metōnymia* <*meta-* altered + *onyma* name] —**met·o·nym·ic** (met′ə·nim′ik), **met′o·nym′i·cal** *adj.* —**met′o·nym′i·cal·ly** *adv.*

met·o·pe¹ (met′ə·pē) *n. Archit.* 1 A square slab, sculptured or plain, between triglyphs in a Doric frieze. 2 Originally, the opening supposed to have been left by primitive Greek builders between the ends of adjoining ceiling beams. [< L *metopa* < Gk. *metopē* < *meta-* between + *opē* opening]

met·o·pe² (met′ə·pē) *n. Anat.* The face, forehead, or frontal surface. [Gk. *metōpon* forehead < *meta-* between + *ōps* eye] — **me·top·ic** (mə·top′ik) *adj.*

me·tox·e·ny (mə·tok′sə·nē) *n.* Heteroecism. [< Gk. *meta-* among, after + *xenos* host]

metr– Var. of METRO-.

me·tral·gi·a (mə·tral′jē·ə) *n. Pathol.* Pain in the womb. [< METR(O)-¹ + -ALGIA]

Met·ra·zol (met′rə·zōl, -zol) *n.* Proprietary name of a synthetic drug, $C_6H_{10}N_4$, used in medicine as a heart stimulant and in the shock treatment of certain nervous and mental disorders.

me·tre (mē′tər) See METER².

met·ric (met′rik) *adj.* 1 Pertaining to the meter as a unit of measurement or to the metric system. 2 Metrical (def. 1). [< L *metricus* < Gk. *metrikos*]

met·ri·cal (met′ri·kəl) *adj.* 1 Relating to meter or versification; composed in poetic measures; rhythmical. 2 Metric (def. 1). — **met′ri·cal·ly** *adv.*

metrical stress The emphasis required by the meter of a poem: opposed to *rhetorical stress.*

metric centner A quintal.

metric hundredweight A weight of 50 kilograms.

me·tri·cian (mə·trish′ən) *n.* 1 One versed in metrics. 2 A composer of verse. [< METR(O)-² + -ICIAN]

met·ri·cize (met′rə·sīz) *v.t.* **·cized, ·ciz·ing** 1 To adjust to the metric system. 2 To analyze or construe the meter of.

met·rics (met′riks) *n. pl.* (construed as singular) 1 The mathematical theory of measurement. 2 The science or art of metrical composition.

meters or skilled in metrical composition; a versemaker. [< Med. L *metrista*]

me·tri·tis (mə·trī′tis) *n. Pathol.* Inflammation of the womb. [< NL < METR(O)-¹ + -ITIS] — **me·trit′ic** (-trit′ik) *adj.*

Met·ro (met′rō) *n.* One of the underground railways of London; hence, any underground railway system, as in Paris, Madrid, etc., especially in Europe. [< METRO(POLITAN DISTRICT RAILWAY)]

metro-¹ *combining form* The uterus; pertaining to the uterus: *metropathic.* Also, before vowels, **metr-.** [< Gk. *metra* the uterus]

metro-² *combining form* Measure: *metrology.* Also, before vowels, **metr-.** [< Gk. *metron* a measure]

me·trol·o·gy (mə·trol′ə·jē) *n.* The science that treats of systems of weights and measures or of measure. [< METRO-² + -LOGY] — **me·tro·log·i·cal** (met′rə·loj′i·kəl) *adj.* — **me·trol′o·gist** *n.*

met·ro·nome (met′rə·nōm) *n.* An instrument for indicating and marking exact time in music, consisting usually of a reversed pendulum whose period of vibration is regulated by a shifting weight. [< METRO-² + Gk. *nomos* law] — **met′ro·nom′ic** (-nom′ik) *adj.*

me·tro·nym·ic (mē′trə·nim′ik, met′rə-) *adj.* Pertaining to or derived from the name of one's mother or female ancestors: also **me·tron·y·mous** (mə·tron′ə·məs). — *n.* 1 A name taken from the mother's side or derived from the maternal name: also **me′tro·nym.** Compare PATRONYMIC. 2 A metronymic designation. [< Gk. *mētrōnymikos* < *mētēr* mother + *onyma* name]

met·ro·path·ic (met′rə·path′ik) *adj. Pathol.* Of or pertaining to disorders of the uterus. [< METRO-¹ + Gk. *pathos* suffering] — **me·trop·a·thy** (mə·trop′ə·thē) *n.*

METRONOME

me·tror·rha·gi·a (mē′trə·rā′jē·ə, met′rə-) *n. Pathol.* Uterine hemorrhage; flooding; especially when not menstrual. [< METRO-¹ + -RRHAGIA]

-metry *combining form* The process, science, or art of measuring: *geometry.* [< Gk. *metria* < *metron* a measure]

Met·su (met′sü), **Gabriel**, 1630–67, Dutch painter. Also **Met′zu.**

Met·ter·nich (met′ər·nikh), **Prince**, 1773–1859, Klemens Wenzel Nepomuk Lothar von Metternich, Austrian statesman and diplomat.

met·tle (met′l) *n.* The stuff or material of which a thing is composed; especially, constitutional temperament or disposition; specifically, courage; ardor. See synonyms under COURAGE. — **on one's mettle** Aroused to one's utmost or best efforts. ◆ Homophone: *metal.* [Var. of METAL]

met·tle·some (met′l·səm) *adj.* Having courage or spirit; ardent; fiery. Also **met′tled.**

Metz (mets, *Fr.* mes) A city in NE France on the Moselle.

me·um (mē′əm) *pron. Latin* Mine; belonging to me: used in the phrase **meum and tuum,** mine and thine.

Meurthe (mûrt) A river in NE France, flowing 105 miles NW to the Moselle.

Meuse (myōōz, *Fr.* mœz) A river in western Europe, flowing 580 miles north from France to the North Sea: Dutch *Maas.*

mew¹ (myōō) *n.* 1 A cage for molting birds; an enclosure; pen for fattening or breeding; also, any place of concealment. 2 *pl.* A stable; specifically, the stables in London in which the royal horses are kept: so called because built on the site of the *mews* or cages of the royal hawks. — *v.t.* 1 To confine in or as in a mew; immure or conceal: often with *up.* 2 *Obs.* To change (feathers); molt. — *v.i.* 3 *Obs.* To molt. [< OF *muer* change < L *mutare*]

mew² (myōō) *v.i.* To cry as a cat. — *n.* The ordinary plaintive cry of a cat. Also spelled *miaou, miaow, meow.* [Imit.]

METRIC SYSTEM

LENGTH		EQUIVALENTS		WEIGHT OR MASS		EQUIVALENTS		CAPACITY		EQUIVALENTS	
		METRIC	U.S.			METRIC	U.S.			METRIC	U.S.
millimeter	(mm)	.001 m	.03937 in.	milligram	(mg)	.001 g	.0154 gr.	milliliter	(ml)	.001 l	.033 fl. oz.
centimeter	(cm)	.01 m	.3937 in.	centigram	(cg)	.01 g	.1543 gr.	centiliter	(cl)	.01 l	.338 fl. oz.
decimeter	(dm)	.1 m	3.937 in.	decigram	(dg)	.1 g	1.543 gr.	deciliter	(dl)	.1 l	3.38 fl. oz.
METER	(m)	1 m.	39.37 in.	GRAM	(g)	1 g	15.43 gr.	LITER	(l)	1 l	1.056 li. qts.
dekameter	(dkm)	10 m	10.93 yds.	dekagram	(dkg)	10 g	.3527 oz. av.	dekaliter	(dkl)	10 l	.283 bu.
hectometer	(hm)	100 m	328.08 ft.	hectogram	(hg)	100 g	3.527 oz. av.	hectoliter	(hl)	100 l	2.837 bu.
kilometer	(km)	1000 m	.6213 mi.	kilogram	(kg)	1000 g	2.2 lbs. av.	kiloliter	(kl)	1000 l	264.18 gal.

AREA		EQUIVALENTS		VOLUME		EQUIVALENTS	
		METRIC	U.S.			METRIC	U.S.
sq. millimeter	(mm²)	.000001 ca	.00155 sq. in.	cu. millimeter	(mm³)	.001 cm³	.016 minim
sq. centimeter	(cm²)	.0001 ca	.155 sq. in.	cu. centimeter	(cc, cm³)	.001 dm³	.061 cu. in.
sq. decimeter	(dm²)	.01 ca	15.5 sq. in.	cu. decimeter	(dm³)	.001 m³	61.023 cu. in.
CENTARE	(ca, m²)	1 m²	10.76 sq. ft.	STERE	(s, m³)	1 m³	1.307 cu. yds.
sq. dekameter	(dkm²)	100 ca	.0247 acre	cu. dekameter	(dkm³)	1000 m³	1307.942 cu. yds.
Also are	(a)	10,000 ca	2.47 acres	cu. hectometer	(hm³)	1,000,000 m³	1,307,942.8 cu. yds.
sq. hectometer	(hm²)	1,000,000 ca	.386 sq. mi.	cu. kilometer	(km³)	1,000,000,000 m³	0.24 cu. mile
Also hectare	(ha)						
sq. kilometer	(km²)						

Other prefixes occasionally used are: MICRO—one millionth MYRIA—10,000 times MEGA 1,000,000 times

metric system A decimal system of weights and measures having as fundamental units the gram and the meter. From the *gram* are derived measures of weight, and from the *meter* measures of length; measures of surface are based on the *square meter* and measures of capacity on the *liter.* The table shown above gives the basic and derived units of the metric system, together with abbreviations and a list of equivalent values in customary U.S. standards; originating in a report to the French Academy (1791) it was legalized in France, Nov. 2, 1801; now universally used in scientific measurement.

metric ton A unit of weight, equal to 1,000 kilograms, 0.984 long ton, or 1.023 short tons.

met·ri·fy (met′rə·fī) *v.t.* & *v.i.* **·fied, ·fy·ing** To write in meter; versify. [< METR(O)-¹ + -FY]

me·trist (mē′trist, met′rist) *n.* One versed in

me·trop·o·lis (mə·trop′ə·lis) *n.* 1 A chief city, either the capital or the largest or most important city of a state or country. 2 The seat of a metropolitan bishop. 3 In ancient Greece, the mother city of a colony. [< Gk. *mētropolis* city < *mētēr* mother + *polis* city]

met·ro·pol·i·tan (met′rə·pol′ə·tən) *n.* 1 An archbishop who exercises a limited authority over the bishops of the same ecclesiastical province. 2 A citizen of the mother city, as opposed to a colonist. 3 One who lives in a metropolis; also, one who has the manners and the ideas, or practices the customs, of a metropolis. — *adj.* Pertaining to a metropolis or to a metropolitan. [< LL *metropolitanus*]

Metropolitan France The chief member of the French Union, consisting of the Republic of France in western Europe.

mew³ (myōō) *n.* A European sea gull. Also **mew gull.** [OE *mæw*; imit.]

Me·war (mā·wär′) See UDAIPUR (def. 1).

mewl (myōōl) *v.i.* To cry, as an infant. — *n.* An infant's cry or crying. [Freq. of MEW²]

Mex·i·can (mek′sə·kən) *n.* 1 A native or inhabitant of Mexico. The inhabitants are chiefly of mestizo and native Indian blood, with a minority of white persons of Spanish ancestry. 2 The Nahuatl or Aztec language. — *adj.* Pertaining to Mexico, its inhabitants, or their language.

Mexican bean beetle A ladybird (*Epilachna varivestis*) with spotted wings, which feeds on the leaves and green pods of beans. For illustration see INSECTS (injurious).

Mexican broomroot Zacatón.

Mexican chickpea A leguminous herb of western Asia and tropical America (*Cicer* *arietinum*) with short inflated pods containing seeds used as food and in coffee blends: also called *garbanzo*.

Mexican hairless An ancient breed of small dog found in Mexico, hairless excepting a tuft on the head and a fuzzy growth toward the end of the tail: also called *biche*.

MEXICAN HAIRLESS (From 11 to 13 inches high at the shoulder)

Mexican saddle A heavy leather saddle, having a high pommel and cantle and, usually, wooden stirrups.

Mexican War See table under WAR.

Mex·i·co (mek'sə·kō) 1 A republic in southern North America between the Pacific and the Caribbean and the Gulf of Mexico, administratively divided into 29 states; 760,373 square miles. 2 Its capital, the largest city in the republic and capital of the **Federal District of Mexico** (573 square miles) in the center of the country: also **Mexico City**. 3 A state in central Republic of Mexico; 8,267 square miles; capital, Toluca. Spanish *Méjico*.

Mexico, Gulf of An arm of the Atlantic; nearly enclosed by the United States, Mexico, and Cuba; 700,000 square miles.

Mey·er (mī'ər), **Adolf**, 1866-1950, U.S. psychiatrist. — **Kuno**, 1858-1919, German Celtic scholar.

Mey·er·beer (mī'ər·bār), **Jakob**, 1791-1864, German composer.

Mey·er·hof (mī'ər·hôf), **Otto**, 1884-1951, German physiologist.

Meyn·ell (men'əl), **Alice**, *née* Thompson, 1847-1922, English poet and essayist.

mez·ca·line (mez'kə·lēn, -lin) See MESCALINE.

me·ze·re·on (mə·zir'ē·ən) n. 1 A low, Old World shrub (*Daphne mezereum*) with clusters of sweet-smelling, lilac-purple flowers: cultivated in the United States. 2 **Mezereum** (def. 1). [<Med. L <Arabic *māzarivān* camellia]

me·ze·re·um (mə·zir'ē·əm) n. 1 The dried bark of any of the *Daphne* species: used in medicine as an irritant and vesicant, and in the treatment of syphilis, rheumatism, and various skin diseases. 2 Mezereon (def. 1).

mez·qui·te (mes·kē'tā) See MESQUITE.

me·zu·zah (mə·zoo'zə, me·zoo'zä) n. pl. **·zoth** (-zōth) In Judaism, a parchment inscribed with the passages Deut. vi 4-9 and xi 13-21, to be rolled up in a case or tube and affixed to the doorpost of a dwelling, as the passages command. Also **me·zu'za**. [<Hebrew *mezūzāh* doorpost]

mez·za·nine (mez'ə·nēn, -nin) n. Archit. 1 A low-ceilinged story between two main ones, especially between the ground floor and the one above it. Also **mezzanine floor**, **mezzanine story**. 2 In a theater, the first balcony or the front rows of the first balcony. [<F <Ital. *mezzanino*, dim. of *mezzano* middle <L *medianus*]

mez·zo (met'sō, med'zō, mez'ō) adj. Half; medium; moderate. [<Ital. <L *medius* middle]

mez·zo-re·lie·vo (met'sō·ri·lē'vō) n. pl. **·vos** Half relief; a piece of sculpture in which the rounded figures project half-way from the background material; demi-relief. See RELIEF. [<Ital. *mezzo rilievo*. See MEZZO and RELIEF.]

mez·zo-ri·lie·vo (med'zō·rē·lyä'vō) n. pl. **mez·zi-ri·lie·vi** (-vē) Italian Mezzo-relievo.

mezzo soprano A voice lower than a soprano and higher than a contralto; also, a person possessing, or a part written for, such a voice.

mez·zo·tint (met'sō·tint', med'zō-, mez'ō-) n. 1 A method of copperplate engraving, producing an even gradation of tones, like a photograph. 2 An impression so produced. — v.t. To engrave in or represent by mezzotint. [<Ital. *mezzotinto* <*mezzo* middle + *tinto* painted] — **mez'zo·tint'er** n.

mho (mō) n. The practical unit of electrical conductance, being the reciprocal of the ohm. [<OHM reversed]

mi (mē) n. Music 1 The third note of the diatonic scale. 2 The note E. [<Ital. See GAMUT.]

Mi·am·i (mī·am'ē) n. A member of an Algonquian tribe of North American Indians formerly located between the Miami and Wabash rivers. [<N. Am. Ind.]

Miami A city on Biscayne Bay, southern Florida.

Miami Beach A resort city on an island between the Atlantic and Biscayne Bay, southern Florida.

Miami River A river in SW Ohio, flowing 160 miles SW to the Ohio River: also **Great Miami River**.

mi·a·na bug (mē·ä'nə) The tampan. [from *Miana*, an Iranian town]

mi·aou, mi·aow (mē·ou') See MEW[2].

mi·ar·o·lit·ic (mī·ar'ə·lit'ik) adj. Geol. Of, pertaining to, or designating a form of igneous rock containing small cavities into which fully developed crystals have projected. [<Ital. *miarola* a granite found near Baveno, Italy + Gk. *lithos* stone]

mi·as·ma (mī·az'mə, mē-) n. pl. **·mas** or **·ma·ta** (-mə·tə) Polluting exhalations; the poisonous effluvium once supposed to rise from putrid matter, swamps, and marshy ground, especially at night. Also **mi'asm**. [<NL <Gk., pollution <*miainein* stain, defile] — **mi·as'mal**, **mi·as·mat·ic** (mī'az·mat'ik), **mi·as'mic** adj.

mi·aul (mē·ôl', -oul') v.i. To cry as a cat; mew. — n. The mewing of a cat. [Imit.]

mib (mib) n. 1 A marble. 2 pl. The game of marbles. [? Alter. of MARBLE]

mi·ca (mī'kə) n. Any of a class of silicates, crystallizing in the monoclinic system and having eminently perfect basal cleavage, affording thin, tough laminae or scales, colorless to jet-black, transparent to translucent, and of widely varying chemical composition. The better grades are extensively used for electrical insulation; the transparent varieties are loosely called *isinglass*. [<L, crumb; infl. by *micare* glitter]

mi·ca·ceous (mī·kā'shəs) adj. 1 Of, pertaining to, consisting of, or containing mica. 2 Resembling mica; laminated; sparkling: also used figuratively: a *micaceous* literary style.

mica diorite A variety of diorite in which mica replaces hornblende.

Mi·cah (mī'kə) A masculine personal name. [<Hebrew, like Jehovah]

— **Micah** A Hebrew prophet of the eighth century B.C.; also, the book of the Old Testament bearing his name.

Mi·caw·ber (mə·kô'bər), **Wilkins** In Dickens's *David Copperfield*, an improvident family man, always waiting for "something to turn up."

mice (mīs) Plural of MOUSE.

mi·celle (mi·sel') 1 Biol. One of the theoretical structural units which are said to make up organized bodies. 2 Chem. The structural unit of a colloid, composed of an aggregate of molecules having crystalline properties and able to change in size without chemical alteration. Also **mi·cell'**, **mi·cel'la** (-sel'ə). [<NL *micella*, dim. of L *mica* crumb, grain] — **mi·cel'lar** adj.

Mi·chael (mī'kəl; Ger. mī'khä·el, Lat. mī'kə·el) A masculine personal name. Also Polish **Mi·chal** (mē'khäl), Russian **Mi·cha·il** (mē·kä·ēl'), Fr. **Mi·chel** (mē·shel'), Ital. **Mi·che·le** (mē·kā'lā). [<Hebrew, who is like God]

— **Michael** An archangel. Rev. xii 7. In Milton's *Paradise Lost*, he expels Adam and Eve from Paradise. In Moslem mythology, one of the four archangels, called the champion of the faith.

— **Michael I**, born 1921, king of Rumania 1927-30, 1940-47.

Mich·ael·mas (mik'əl·məs) The feast of St. Michael, Sept. 29; a quarterly rent day in England. [<MICHAEL + -MAS]

Michaelmas daisy 1 A European blue aster (*Aster tripolium*). 2 Any of several North American asters, wild and cultivated.

Mi·chel·an·ge·lo (mī'kəl·an'jə·lō), 1475-1564, Italian sculptor, painter, architect, and poet. Also **Mi'chael An'ge·lo**: full name **Mi·chel·an·ge·lo Buo·nar·ro·ti** (mē'kel·än'je·lō bwô'när·rô'tē).

Miche·let (mēsh·le'), **Jules**, 1798-1874, French historian.

Mi·chel·son (mī'kəl·sən), **Albert Abraham**, 1852-1931, U.S. physicist born in Germany.

Mich·i·gan (mish'ə·gən) A north central State of the United States; 57,980 square miles; capital, Lansing; entered the Union Jan. 26, 1837; nickname, *Wolverine State*: abbr. *Mich.* — **Mich'i·gan·ite'** (-īt'), **Mich'i·gan'der** (-gan'dər) n.

Michigan, Lake The only one of the Great Lakes entirely within the United States, lying between Michigan and Wisconsin; 22,400 square miles.

Mi·cho·a·cán (mē'chō·ä·kän') A maritime state of SW Mexico; 23,200 square miles; capital, Morelia.

Mick (mik) n. Slang An Irishman: a contemptuous usage. [<MICHAEL]

mick·ey finn (mik'ē fin') Slang A drugged drink. Also **mick'ey**, **Mick'ey Finn**. [Origin unknown]

Mic·kie·wicz (mits·kyä'vich), **Adam**, 1789-1855, Polish poet.

mick·le (mik'əl) Obs. or Scot. adj. 1 Large; great. 2 Much; many. — n. 1 A large amount or quantity; abundance. 2 By corruption, a small amount or quantity. Also spelled *meikle*. [OE *micel*]

Mic·mac (mik'mak) n. pl. **·mac** or **·macs** One of a tribe of North American Indians of Algonquian linguistic stock located in Nova Scotia, New Brunswick, and Newfoundland. [<N. Am. Ind., lit., allies]

mi·cra (mī'krə) Plural of MICRON.

mi·cra·ner (mī'krə·nər) n. Entomol. A dwarfed male ant. [<MICR(O)- + Gk. *anēr* male]

mi·cri·fy (mī'krə·fī) v.t. **·fied**, **·fy·ing** To make small or insignificant. [<MICR(O)- + -(I)FY]

micro- combining form 1 In the metric and other systems of measurement, the one-millionth part of (the specified unit):

microampere	microfarad	micromho
microangstrom	microhenry	micromicron
microbar	microhm	microphot
microcoulomb	microjoule	microvolt
microcurie	microlux	microwatt

2 An apparatus or instrument which enlarges in size or volume: *microphone*. 3 Exceptionally or abnormally small: *micro-organism*. 4 Microscopic; using or requiring a microscope:

microbotany	micropathology
microgeology	micropetrography
microhistology	micropetrology
micromechanics	microphysiography
micrometallurgy	microphysiology
micromineralogy	microzoology

Also, before vowels, sometimes **micr-**. [<Gk. *mikros* small]

mi·cro·a·nal·y·sis (mī'krō·ə·nal'ə·sis) n. The chemical analysis and identification of minute quantities. — **mi'cro·an'a·lyst** n. — **mi'cro·an'a·lyt'i·cal** (-an'ə·lit'i·kəl) adj. — **mi'cro·an'a·lyt'i·cal·ly** adv.

mi·cro·bar·o·graph (mī'krō·bar'ə·graf, -gräf) n. Meteorol. An instrument which records the lesser fluctuations of atmospheric pressure.

mi·crobe (mī'krōb) n. A microscopic organism; especially, a pathogenic bacterium. Also **mi·cro·bi·on** (mī·krō'bē·on). [<F <Gk. *mikros* small + *bios* life] — **mi·cro'bi·al**, **mi·cro'bi·an**, **mi·cro'bic** adj.

mi·cro·bi·cide (mī·krō'bə·sīd) n. Any substance or agent that destroys microbes; a germ-killer. [<MICROBE + -CIDE] — **mi·cro'bi·ci'dal** adj.

mi·cro·bi·ol·o·gy (mī'krō·bī·ol'ə·jē) n. The scientific study of the structure, development, function, and mode of action of micro-organisms, as bacteria, viruses, molds, etc., especially with regard to their significance in health and disease. — **mi'cro·bi·o·log·i·cal** (mī'krō·bī'ə·loj'i·kəl) adj. — **mi'cro·bi·ol'o·gist** n.

mi·cro·bi·o·ta (mī'krō·bī·ō'tə) n. pl. Ecol. The microscopic plant and animal organisms of a region. [<NL <MICRO- + BIOTA] — **mi'cro·bi·ot'ic** (-ot'ik) adj.

mi·cro·ceph·a·ly (mī'krō·sef'ə·lē) n. Abnormal smallness of the head; imperfect development of the cranium. Also **mi'cro·ce·pha'li·a** (-sə·fā'lē·ə). [<MICRO- + Gk. *kephalē* head] — **mi'cro·ce·phal'ic** (-sə·fal'ik), **mi'cro·ceph'a·lous** adj.

mi·cro·chem·is·try (mī'krō·kem'is·trē) n. The chemistry of very small objects or quantities, especially those requiring the use of the microscope. — **mi'cro·chem'i·cal** adj.

mi·cro·cli·mate (mī'krō·klī'mit) n. Meteorol. The climate of a very small area, as of a forest, meadow, lake, wheat field, etc., with special reference to local variations from the general climate of a region. Compare MACRO-

CLIMATE. — **mi·cro·cli·mat·ic** (mī'krō·klī·mat'ik) adj.
mi·cro·cline (mī'krə·klīn) n. A grayish, reddish, greenish, or green translucent potash feldspar, $KAlSi_3O_8$, crystallizing in the triclinic system. See FELDSPAR.
mi·cro·coc·cus (mī'krə·kok'əs) n. pl. ·coc·ci (-kok'sī) Any member of a genus (*Micrococcus*) of spherical bacteria occurring in irregular masses or in plates. They are generally Gram-positive, and feed either on living or on dead matter. [<NL]
mi·cro·cop·y (mī'krə·kop'ē) n. pl. ·cop·ies A reduced photographic copy, as of a letter, manuscript, etc.
mi·cro·cosm (mī'krə·koz'əm) n. 1 A little world; the world or universe on a small scale; hence, man, as if combining in himself all the elements of the great world: opposed to *macrocosm*. 2 A little community. Also **mi'·cro·cos'mos** (-koz'məs), **mi·cro·cos'mus**. [<F *microcosme* <LL *microcosmus* <Gk. *mikros kosmos*, lit., little world] — **mi'cro·cos'mic, mi'cro·cos'mi·cal, mi'cro·cos'mi·an** adj.
microcosmic salt *Chem.* A white crystalline salt, sodium ammonium hydrogen phosphate, $Na(NH_4)HPO_4·4H_2O$, originally derived from human urine: used to identify metallic oxides in lead tests, or as a reagent in blowpipe analysis.
mi·cro·crys·tal·line (mī'krō·kris'tə·lin, -lēn) adj. 1 Cryptocrystalline. 2 Having a crystalline structure distinguishable only under a microscope.
mi·cro·cyte (mī'krə·sīt) n. *Pathol.* A small red blood corpuscle found in cases of anemia.
mi·cro·de·tec·tor (mī'krō·di·tek'tər) n. A galvanometer.
mi·cro·dis·sec·tion (mī'krō·di·sek'shən) n. Dissection of tissue or other organic material under the microscope.
mi·cro·dont (mī'krə·dont) adj. Having unusually small teeth. Also **mi'cro·don'tous**. [<MICR(O)- + Gk. *odous, odontos* tooth]
mi·cro·film (mī'krə·film) n. 1 A photograph, as of a printed page, document, or other object, highly reduced for ease in transmission and storage, and capable of reenlargement or reading by suitable optical devices. 2 A microphotograph. — **mi'cro·film'ing** n.
mi·cro·fos·sil (mī'krə·fos'əl) n. The fossilized remains of an extremely minute organism, usually of submarine origin and of value in the study of early geological conditions and biological development.
mi·cro·ga·mete (mī'krō·gə·mēt', -gam'ēt) n. *Biol.* The male of two conjugating gametes: so called from its being the smaller one. Compare MACROGAMETE.
mi·cro·gram (mī'krə·gram) n. A unit of weight in the metric system, equal to one thousandth of a milligram. Also **mi'cro·gramme**.
mi·cro·graph (mī'krə·graf, -gräf) n. 1 A pantograph instrument for minute writing, drawing, or engraving. 2 A picture of an object as seen through a microscope. 3 An instrument for recording and photographically magnifying very small movements, as those of a diaphragm. — **mi'cro·graph'ic** adj.
mi·crog·ra·phy (mī·krog'rə·fē) n. 1 The description or study of microscopic objects. 2 The art or habit of writing very minutely. 3 Very minute handwriting, especially as a symptom of nervous disorder. Contrasted with *macrography*.
mi·cro·groove (mī'krə·groōv) n. 1 A groove or channel of exceptional fineness, especially as cut in the surface of the long-playing phonograph record. 2 A phonograph record having such grooves. — adj. Having microgrooves.
mi·cro·gyne (mī'krə·jīn) n. *Entomol.* A dwarfed queen or female ant. [<MICRO- + GYNE]
mi·cro·in·cin·er·a·tion (mī'krō·in·sin'ə·rā'shən) n. A method for the study of the inorganic constituents of cells and tissues by burning the organic materials and examining the residue under the microscope.
mi·cro·lec·i·thal (mī'krə·les'ə·thəl) adj. Alecithal. [<MICRO- + Gk. *lekithos* yolk]
mi·cro·li·ter (mī'krə·lē'tər) n. A metric unit of capacity, equal to one thousandth of a milliliter. Also **mi'cro·li'tre**.
mi·cro·lith (mī'krə·lith) n. *Anthropol.* A small, flint implement of the late Paleolithic, Mesolithic, and Neolithic culture periods. — **mi'cro·lith'ic** adj.
mi·crol·o·gy (mī·krol'ə·jē) n. 1 The branch of science that treats of microscopic objects or is dependent on microscopic investigations. 2 Undue attention to minute and unimportant matters. [<MICRO- + -LOGY]
mi·cro·mer·it·ics (mī'krō·mi·rit'iks) n. The study of the properties and behavior of finely divided matter, with special reference to practical applications, as in soil physics, metallurgy, etc. [<MICRO- + Gk. *meros* part]
mi·cro·me·te·or·ol·o·gy (mī'krō·mē'tē·ə·rol'ə·jē) n. The study of climatic conditions in very small areas or localities. Compare MACROMETEOROLOGY.
mi·crom·e·ter (mī·krom'ə·tər) n. 1 An instrument for measuring very small distances or dimensions. 2 A caliper or gage arranged to allow of minute measurements. [<MICRO- + METER] — **mi·cro·met'ri·cal** (mī'krō·met'ri·kəl), **mi'cro·met'ric** adj. — **mi'cro·met'ri·cal·ly** adv. — **mi·crom'e·try** n.
micrometer caliper A caliper or gage having a micrometer screw.
micrometer screw A screw with fine and very accurately cut threads, and a circular, graduated head, which shows the amount of advancement or retraction of the screw: used in making fine measurements, often to .0001 of an inch.

MICROMETER

mi·cro·mil·li·me·ter (mī'krō·mil'ə·mē'tər) n. 1 The one-millionth part of a millimeter, or the one-billionth part of a meter (1 x 10^{-9}m.). 2 One millimicron.
mi·cron (mī'kron) n. pl. ·cra (-krə) The one-thousandth part of a millimeter, or the one-millionth part of a meter (symbol μ). Also spelled *mikron*. [<NL <Gk. *mikron*, neut. of *mikros* small]
Mi·cro·ne·sia (mī'krə·nē'zhə, -shə) One of the three main divisions of Pacific islands, in the western Pacific north of the equator.
Mi·cro·ne·sian (mī'krə·nē'zhən, -shən) n. 1 A native of Micronesia. The Micronesians are a mixed race of Melanesian, Polynesian, and Malay stocks, but have shorter stature and darker skins than the Polynesians. 2 A group of Melanesian languages of the Austronesian family, spoken in the Caroline, Gilbert, and Marshall islands. — adj. Of or pertaining to Micronesia, its people, or their languages.
mi·cron·ize (mī'krən·īz) v.t. ·ized, ·iz·ing To comminute (a substance) into particles not more than a few microns in diameter. [<MICRON + -IZE] — **mi'cron·i·za'tion** n.
mi·cro·nu·cle·us (mī'krō·noō'klē·əs, -nyōō'-) n. pl. ·cle·i (-klē·ī) *Biol.* A small nucleus, especially the smaller of two nuclei, as in infusorians.
mi·cro·nu·tri·ent (mī'krō·noō'trē·ənt, -nyōō'-) n. A substance essential in the nourishment of animals and plants but only in small amounts or in low concentration, as certain minerals. — adj. Of or pertaining to such a substance.
mi·cro–or·gan·ism (mī'krō·ôr'gən·iz'əm) n. Any extremely small plant or animal organism, especially one visible only in an optical or electron microscope, as a bacterium, protozoan, or virus.
mi·cro·pa·le·on·tol·o·gy (mī'krō·pā'lē·ən·tol'ə·jē) n. The study of microscopic fossils in relation to their geologic and ecological environment, especially those forms found in core samples from submarine depths. — **mi'cro·pa·le·on·tol'o·gist** n. — **mi'cro·pa·le·on·to·log'ic** (-pā'lē·ən·tə·loj'ik) or **·i·cal** adj.
mi·cro·par·a·site (mī'krō·par'ə·sīt) n. A parasitic micro-organism.
mi·cro·phone (mī'krə·fōn) n. A device for amplifying sounds by the electromagnetic conversion of sound waves impinging upon a sensitive diaphragm. It forms the principal element of a telephone transmitter, and is the first component of any sound-reproducing system, as in radio broadcasting, sound films, etc. — **mi·cro·phon'ic** (-fon'ik) adj.
mi·cro·pho·to·graph (mī'krō·fō'tə·graf, -gräf) n. 1 A microscopic photograph of any object, as a writing, picture, etc. Compare PHOTOMICROGRAPH. 2 Microfilm. — **mi'cro·pho'to·graph'ic** adj. — **mi'cro·pho·tog'ra·phy** (-fə·tog'rə·fē) n.
mi·cro·pho·tom·e·ter (mī'krō·fō·tom'ə·tər) n. An instrument for the measurement of very small luminous intensities, and for the comparative study of spectral lines. — **mi'cro·pho'to·met'ric** (-fō'tə·met'rik) adj. — **mi'cro·pho·tom'e·try** n.
mi·cro·phys·ics (mī'krə·fiz'iks) n. 1 That branch of physics which investigates the structure, characteristics, and behavior of microscopic particles of matter. 2 Nucleonics.
mi·cro·phyte (mī'krə·fīt) n. A microscopic plant, generally parasitic. — **mi'cro·phy'tal, mi'cro·phyt'ic** (-fīt'ik) adj.
mi·cro·pla·si·a (mī'krə·plā'zhē·ə, -zē·ə) n. *Pathol.* A condition of arrested development; dwarfism.
mi·cro·print (mī'krə·print) n. A microphotograph reproduced in a print that may be examined or read by means of a magnifying glass.
mi·crop·si·a (mī·krop'sē·ə) n. *Pathol.* A defect in vision causing objects to appear unusually small. Also **mi·cro'pi·a** (-krō'pē·ə). [<NL <MICR(O)- + -OPSIS] — **mi·crop'tic** (-tik) adj.
mi·cro·pyle (mī'krə·pīl) n. 1 *Bot.* The aperture in the coats of a plant ovule through which the pollen tube penetrates. 2 *Biol.* An aperture in the vitelline membrane of the ovum, serving to admit a spermatozoon. [<MICRO- + Gk. *pylē* gate] — **mi'cro·py'lar** adj.
mi·cro·py·rom·e·ter (mī'krō·pī·rom'ə·tər) n. An optical instrument for observing the temperature, etc., of minute light- or heat-radiating bodies.
mi·cro·ra·di·om·e·ter (mī'krō·rā'dē·om'ə·tər) n. An instrument for the measurement of feeble radiations, especially in the infrared portion of the spectrum.
mi·cro·rock·et (mī'krō·rok'it) n. A miniature rocket used for testing a proposed full-scale model.
mi·cro·scope (mī'krə·skōp) n. An optical instrument consisting of a single lens or a combination of lenses for magnifying objects too small to be seen or clearly observed by the naked eye. [<NL *microscopium*]
mi·cro·scop·ic (mī'krə·skop'ik) adj. 1 Pertaining to a microscope or to microscopy. 2 Made with or as with a microscope. 3 Exceedingly minute; visible only under the microscope: opposed to *macroscopic*. See synonyms under LITTLE. Also **mi'cro·scop'i·cal**. — **mi'cro·scop'i·cal·ly** adv.
Mi·cro·sco·pi·um (mī'krə·skō'pē·əm) A southern constellation. See CONSTELLATION.
mi·cros·co·py (mī·kros'kə·pē, mī'krə·skō'pē) n. The art or practice of examining objects with a microscope. — **mi·cros'co·pist** n.
mi·cro·sec·ond (mī'krə·sek'ənd) n. One millionth of a second: used especially in timing the action of subatomic particles in nuclear physics.
mi·cro·seism (mī'krə·sīz'əm) n. *Geol.* A very slight tremor or vibration of the earth's crust, detectable only by a microseismometer. [<MICRO- + Gk. *seisma* shaking] — **mi'cro·seis'mic, mi'cro·seis'mi·cal** adj.
mi·cro·seis·mom·e·ter (mī'krō·sīz·mom'ə·tər, -sīs-) n. An apparatus for indicating the direction, duration, and intensity of microseisms. Also **mi'cro·seis'mo·graph** (-sīz'mə·graf, -gräf, -sīs'-). [<MICROSEISM + -(O)METER]
mi·cro·some (mī'krə·sōm) n. *Biol.* A minute corpuscle embedded in the protoplasm of an active cell. The great number of these corpuscles contributes to the granular appearance of protoplasm. Also **mi'cro·so'ma** (-sō'mə).
mi·cro·spe·cies (mī'krə·spē'shēz) n. *Biol.* A species having definite but slight variability; also, a species distributed within the confines of a limited area.
mi·cro·spec·tro·scope (mī'krō·spek'trə·skōp) n. A combination of the microscope and spectroscope for observing the absorptive spectrum of a minute body.
mi·cro·spo·ran·gi·um (mī'krō·spə·ran'jē·əm)

microspore

n. pl. **·gi·a** (-jē-ə) *Bot.* A sporangium producing or containing microspores, as in the pollen sac of the anther in seed plants.

mi·cro·spore (mī′krə-spôr, -spōr) *n. Bot.* A small, asexually produced spore in seed plants, male in function.

mi·cro·spo·ro·phyll (mī′krə-spôr′ə-fil, -spō′rə-) *n. Bot.* A sporophyll producing microsporangia.

mi·cros·to·mous (mī-kros′tə-məs) *adj.* Having an unusually small mouth. Also **mi·cro·stom·a·tous** (mī′krō-stom′ə-təs). [< MICRO- + -STOMOUS]

mi·cro·tome (mī′krə-tōm) *n.* An instrument for making very thin sections for microscopic observations. — **mi′cro·tom′ic** (-tom′ik) or **·i·cal** *adj.* — **mi·crot·o·mist** (mī-krot′ə-mist) *n.* — **mi·crot′o·my** *n.*

mi·cro·wave (mī′krə-wāv) *n.* A high-frequency electromagnetic wave having a wavelength of from 1 millimeter to about 30 centimeters.

mi·crur·gy (mī′krûr-jē) *n.* A highly refined precision technique for the microdissection, study, and investigation of living protoplasm, usually in the form of single cells, as bacteria, amebae, etc. [< MICR(O)- + -URGY] — **mi·crur·gic** (mī-krûr′jik) or **·gi·cal** *adj.* — **mi′crur·gist** *n.*

mic·tu·rate (mik′chə-rāt) *v.i.* **·rat·ed, ·rat·ing** To urinate. [< MICTURITION; an erroneous formation]

mic·tu·ri·tion (mik′chə-rish′ən) *n.* The act of urination. [< L *micturitus,* pp. of *micturire,* desiderative of *mingere* pass water]

mid[1] (mid) *adj.* **1** Middle. **2** *Phonet.* Produced with the tongue in a relatively midway position between high and low: said of certain vowels, as (ō) in *boat.* — *n. Archaic* The middle. [OE *midd*]

mid[2] (mid) *prep.* Amid; among: a poetical usage.

mid- *combining form* Middle; middle point or part of:

mid–act	midmonth
mid–African	midmonthly
midafternoon	midmorning
mid–April	mid–mouth
mid–arctic	mid–movement
mid–Asian	mid–nineteenth
mid–August	mid–November
midautumn	mid–ocean
midaxillary	mid–October
mid–block	mid–oestral
mid–breast	mid–orbital
mid–Cambrian	mid–Pacific
mid–career	mid–part
midcarpal	mid–period
mid–century	mid–periphery
mid–channel	mid–pillar
mid–chest	midpit
mid–continent	mid–Pleistocene
mid–course	mid–position
mid–court	midrange
mid–crowd	mid–refrain
mid–current	mid–region
mid–December	mid–Renaissance
mid–diastolic	mid–river
mid–dish	mid–road
middorsal	mid–sea
mid–eighteenth	midseason
mid–Empire	midsentence
mid–European	mid–September
midevening	mid–Siberian
midfacial	mid–side
mid–February	mid–sky
mid–field	mid–slope
mid–flight	mid–sole
midforenoon	midspace
mid–forty	mid–span
midfrontal	midstory
mid–hour	midstout
mid–ice	midstreet
mid–incisor	mid–stride
mid–Italian	mid–styled
mid–January	mid–sun
mid–July	mid–swing
mid–June	mid–tap
mid–lake	midtarsal
midleg	mid–Tertiary
mid–length	mid–thigh
mid–life	mid–thoracic
mid–line	mid–tide
mid–link	mid–time
mid–lobe	mid–totality
midmandibular	mid–tow
mid–March	mid–town
mid–May	mid–value

midventral
mid–volley
mid–walk
mid–wall
mid–watch
mid–water
midwintry
mid–world
midyear
mid–zone

mid-air (mid′âr′) *n.* The middle or midst of the air.

Mi·das (mī′dəs) In Greek legend, a king of Phrygia to whom Dionysus granted the power of turning whatever he touched into gold; when his food and even his daughter were thus transmuted, he prayed to have the power taken back.

mid-At·lan·tic (mid′ət-lan′tik) *n.* The middle part of the Atlantic Ocean.

mid-brain (mid′brān′) *n.* The mesencephalon.

mid-Can·a·da line (mid′kan′ə-də) A chain of radar stations in Canada along the 55th and 56th parallels, maintained by the Canadian government.

mid·day (mid′dā′) *n.* The middle of the day; noon.

mid·den (mid′n) *n.* **1** A kitchen midden. **2** *Brit. Dial.* A dunghill or heap of refuse. [ME *midding* < Scand. Cf. Dan. *mödding* < *mög* dung, muck + *dynge* heap (of dung).]

mid·dle (mid′l) *adj.* **1** Occupying a position equally distant from the extremes; mean. **2** Occupying any intermediate position. **3** *Gram.* In Greek and Sanskrit, designating a voice of the verb which indicates action by the subject for his own sake. See VOICE. — *n.* **1** The part or point equally distant from the extremities. ◆ Collateral adjective: *median.* **2** Something that is intermediate. **3** The middle voice. See synonyms under CENTER. — *v.t.* **·dled, ·dling** *Naut.* To fold or double in the middle. [OE *middel*]

middle age The time of life between youth and old age, commonly between 40 and 60. — **mid′dle-aged′** (-ājd′) *adj.*

Middle Ages The period in European history between classical antiquity and the Renaissance: usually regarded as extending from the downfall of Rome, in 476, to about 1450.

Middle America That part of Latin America north of South America and south of the United States; specifically, the six Central American states as well as Mexico, Cuba, Haiti, and the Dominican Republic.

Middle Atlantic States New York, New Jersey, and Pennsylvania.

mid·dle-break·er (mid′l-brā′kər) *n. Agric.* A lister. Also **mid′dle-bust′er** (-bus′tər).

middle C *Music* The note written on the first leger line above the bass staff and the first leger line below the treble staff; also, the corresponding tone or key.

middle class The class of a society that occupies a position between the laboring class and the very wealthy or the nobility. — **mid′dle-class′** (-klas′, -kläs′) *adj.*

Middle Congo See CONGO REPUBLIC.

Middle Dutch See under DUTCH.

middle ear *Anat.* The tympanum: occasionally applied to the tympanum, the mastoid cells, and the Eustachian tube.

Middle East See under EAST.

Middle English See under ENGLISH.

Middle Franconia An administrative division of western Bavaria, West Germany; 2,941 square miles.

Middle French See under FRENCH.

Middle High German See under GERMAN.

Middle Kingdom 1 The former Chinese Empire, regarded as occupying the center of the world: later considered as the 18 provinces of China proper. **2** A kingdom of ancient Egypt, 2400–1580 B.C., having first Heracleopolis, then Thebes, for its capital: also **Middle Empire.**

Middle Latin See MEDIEVAL LATIN under LATIN.

middle latitude The latitude midway between two places on the same hemisphere: also called **mean latitude.**

Middle Low German See under GERMAN.

mid·dle·man (mid′l-man′) *n. pl.* **·men** (-men′) **1** One who acts as an agent; one who buys in bulk from producers and resells. **2** The actor in the middle of a line of minstrel performers who propounds questions to the endmen; the interlocutor. **3** Any intermediary.

mid·dle·most (mid′l-mōst) *adj.* **1** Situated exactly in the middle. **2** Being in the middle or midst of; hence, very intimate. Also *midmost.*

middle passage The former slave trade route across the Atlantic from Africa to America or the West Indies.

mid·dle·piece (mid′l-pēs′) *n. Biol.* The part lying between the nucleus and the flagellum of a spermatozoon.

Middle Persian See under PERSIAN.

Mid·dles·brough (mid′lz-brə) A port and county borough in NE Yorkshire, England.

Mid·dle·sex (mid′l-seks) A county of SE England, comprising the NW part of London; 232 square miles.

Middle States The eastern States of the United States between New England and the South: New York, New Jersey, Pennsylvania, Delaware, and Maryland.

Middle Temple See INNS OF COURT.

Mid·dle·ton (mid′l-tən), **Arthur,** 1742–87, South Carolina patriot; signer of Declaration of Independence. — **Thomas,** 1570?–1627, English dramatist.

mid·dle·weight (mid′l-wāt) *n.* A boxer or wrestler weighing between 147 and 160 pounds.

Middle West That section of the United States between the Rockies and the Alleghenies and north of the Ohio River and the southern borders of Kansas and Missouri: also *Midwest.* Also **middle west.** — **Middle Western**

mid·dling (mid′ling) *adj.* **1** Of middle rank, condition, size, or quality. **2** In tolerable but not good health; in fair health. — *n.* **1** *pl.* The coarser parts of ground wheat, as distinguished from flour and bran: formerly used only for feed, but now manufactured into the best brand of flour, since it contains most gluten. **2** Pork or bacon cut from between the shoulder and ham of a hog. **3** A quality of cotton on which prices are based. [< MID + -LING[2]]

mid·dling·ly (mid′ling-lē) *adv.* Tolerably. Also *mid′dling.*

mid·dy (mid′ē) *n. pl.* **·dies** *Colloq.* A midshipman.

middy blouse A woman's or child's blouse not closely fitted and having a wide collar similar to that of a sailor's blouse.

Mid·gard (mid′gärd) In Norse mythology, the earth, middlemost of several worlds, and girdled by a great serpent: also spelled *Mithgarthr.* Also **Mid′garth** (-gärth). [< ON *Mithgarther* < *mithr* mid + *garthr* yard, house]

midge (mij) *n.* **1** A gnat or small fly; especially, a small, long-legged, dipterous insect of the family *Ceratopogonidae* that does not bite and has aquatic larvae. For illustration see INSECTS (injurious). **2** A small person; dwarf. [OE *mycge*]

midg·et (mij′it) *n.* **1** An extremely small person. **2** Anything very small of its kind. — *adj.* Small; diminutive. [Dim. of MIDGE]

mid·gut (mid′gut′) *n. Anat.* The primitive intestinal cavity, formed by the closure of the body walls of the embryo.

Mi·di (mē-dē′) *French* Southern France.

Mid·i·an·ite (mid′ē-ən-īt′) *n.* One of an ancient nomadic tribe of NW Arabia, descended from *Midian,* a son of Abraham. *Gen.* xxv 2. [< Hebrew *Midhyān* name of a son of Abraham] — **Mid′i·an·it′ish** *adj.*

mi·di·nette (mē′dē-net′) *n. fem. French* A young working girl, so called because she leaves the shop at noon for lunch.

mid-i-ron (mid′ī′ərn) *n.* An iron golf club with a loft intermediate between that of the cleek and that of the mashie.

mid·land (mid′lənd) *adj.* **1** In the interior country. **2** Surrounded by the land; mediterranean. — *n.* The interior of a country.

Mid·land (mid′lənd) *n.* The dialects of Middle English spoken in London and the Midland counties of England; especially, **East Midland,** the direct predecessor of Modern English.

Midland man Skeletal remains discovered with Neolithic artifacts in sand layers near Midland, Texas, in 1953, and believed to indicate the presence of man on the American continent at a period coeval with or possibly antedating the Folsom culture.

Mid·lands (mid′ləndz) *n. pl.* A region comprising the middle counties of England, generally coextensive with Anglo-Saxon Mercia.

Mid·lo·thi·an (mid-lō′thē-ən) A county in SE Scotland; 366 square miles; county town, Edinburgh.

mid·most (mid′mōst′) *adj.* Middlemost. — *n.*

The midmost part of anything. —*adv.* In the midst or middle. [OE *mydmest*]
mid·night (mid′nīt′) *n.* Twelve o'clock at night; the middle of the night. —*adj.* Pertaining to, occurring in, or like the middle of the night; dark.
midnight blue A very dark blue, almost black.
midnight sun The sun visible at midnight as a result of the fact that the latitude of the place from which it is viewed is greater than the polar distance of the sun, or above 66 degrees.
mid·noon (mid′nōōn′) *n.* The middle of the day; noon.
mid-point (mid′point′) *n.* A point at the middle.
Mid·rash (mid′rash, -räsh) *n. pl.* **Mid·rash·im** (mid·rash′im, -räsh′-) or **Mid·rash·oth** (mid·rash′ōth, -räsh′-) Jewish exegetical treatises on the Old Testament, dating from the 4th to the 12th century; specifically, the Haggadah. [< Hebrew, explanation]
mid·rib (mid′rib′) *n. Bot.* The primary vein, or rib, of a leaf, usually running from apex to base.
mid·riff (mid′rif) *n.* The diaphragm. [OE *midhrif* < *midd* mid + *hrif* belly]
mid·ship (mid′ship′) *adj. Naut.* At or pertaining to the middle of a vessel's hull.
mid·ship·man (mid′ship′mən) *n. pl.* **·men** (-mən) **1** A student training at the United States Naval Academy for commissioning as an officer. **2** *Brit.* In the Royal Navy: **a** An officer ranking between a naval cadet and the lowest commissioned officer. **b** Formerly, one of a class of youths performing minor duties on shipboard as training for commissioning as officers. [<*amidshipman*; so called from being amidships when on duty]
mid·ships (mid′ships) *adv.* Amidships. —*n. pl.* The midship timbers.
midst (midst) *n.* The central part; middle: often with the implication of being surrounded, beset, or hard pressed: in the *midst* of duties or dangers. See synonyms under CENTER. Compare synonyms for AMID. —**in our, your,** or **their midst** In the midst of, or among, us, you, or them. —*prep.* Amidst. —*adv.* In the middle. [OE *midd* + adverbial *-s* + intrusive or superlative *-t*]
mid·stream (mid′strēm′) *n.* The middle of a stream.
mid·sum·mer (mid′sum′ər) *n.* **1** The middle of summer. **2** Popularly, the time of the summer solstice, about June 21. —*adj.* Like, or occurring in, the middle of the summer.
mid-term (mid′tûrm′) *adj. & adv.* In the middle of the term. —*n.* **1** The middle of the term. **2** *Colloq.* A mid-term examination, as at a college.
mid-Vic·to·ri·an (mid′vik·tôr′ē·ən, -tō′rē-) *adj.* Of or characteristic of the middle period of Queen Victoria's reign in England, or the period approximately 1850–80. —*n.* **1** A British person belonging to this era. **2** A person of markedly Victorian ideas or tastes.
mid·way (mid′wā′, -wā′) *adj.* Being in the middle of the way or distance. —*n.* (mid′wā′) **1** The middle or the middle course. **2** The space, at a fair or exposition, assigned for the display of exhibits and along which the various amusements are situated. —*adv.* Half-way. [OE *midweg*]
Midway Islands Two islets in the North Pacific NW of Honolulu, belonging to the United States and under jurisdiction of the U.S. Navy; 2 square miles; scene of one of the decisive battles of World War II, June, 1942.
mid·week (mid′wēk′) *n.* The middle of the week. —*adj.* In the middle of the week. —**mid′week′ly** *adv.*
Mid·week (mid′wēk′) *n.* In the Society of Friends, Wednesday.
Mid·west (mid′west′) *n.* The Middle West. —**Mid′west′ern** *adj.*
Mid·west·ern·er (mid′wes′tər·nər) *n.* A person from the Middle West.
mid·wife (mid′wīf′) *n. pl.* **·wives** (-wīvz′) A woman who assists women in childbirth. [OE *mid* with + *wīf* wife]
mid·wife·ry (mid′wī′fər·ē, -wīf′rē) *n.* Obstetrics.
mid·win·ter (mid′win′tər) *n.* **1** The middle of winter. **2** Popularly, the winter solstice, about Dec. 21.
mid·years (mid′yirz′) *n. pl.* Examinations given in the middle of a school or college year.
mien (mēn) *n.* The external appearance or manner of a person; carriage; bearing. See synonyms under AIR[1], MANNER. ◆ Homophone: *mean.* [? Aphetic form of DEMEAN; infl. by F *mine* <Celtic (Breton) *min* beak, muzzle]
miff (mif) *Colloq. v.t.* To cause to be irritated or offended. —*v.i.* To take offense. —*n.* A huff. [Origin uncertain. Cf. G *muffen* sulk.]
mif·fy (mif′ē) *adj. Colloq.* **1** Supersensitive; easily offended. **2** Delicate: said of plants. —**mif′fi·ness** *n.*
might[1] (mīt) Past tense of MAY[1]. Both *may* and *might* are now considered subjunctives with present or future sense, the difference between the two being one of degree rather than of time. *May* implies a greater probability than *might*, the latter indicating possibility but less likelihood: He *might* be on time, but don't depend on it. As a request for permission, *might* is felt to be more hesitant in approach: *Might* we expect your reply by Tuesday? *Might* I come to dinner some evening?
might[2] (mīt) *n.* **1** Ability to do anything requiring force or power; strength. **2** The possession of great resources; intensity of will; ability. See synonyms under POWER. —**with might and main** With utmost endeavor; with one's whole strength. ◆ Homophone: *mite.* [OE *meaht, miht*]
might·i·ly (mī′tə·lē) *adv.* **1** With might; with great force, energy, or earnestness. **2** To a great degree; greatly; very much: I wanted *mightily* to go.
might·i·ness (mī′tē·nis) *n.* The state or quality of being mighty or powerful; great extent or degree.
might·y (mī′tē) *adj.* **might·i·er, might·i·est** **1** Possessed of might; powerful; strong. **2** Of unusual bulk, consequence, etc. **3** Momentous; wonderful: a *mighty* host. See synonyms under POWERFUL. —*adv. Colloq.* Very; to a great degree; very much: a *mighty* fine person.
mi·gnon (min′yon, *Fr.* mē·nyôn′) *adj.* Delicately small; dainty. [<F <Gmc. Cf. OHG *minna* love. Doublet of MINION.] —**mi·gnonne** (min′yon, *Fr.* mē·nyôn′) *adj. fem.*
mi·gnon·ette (min′yən·et′) *n.* A North African annual plant *(Reseda odorata)* having wedge-shaped leaves, greenish, fragrant flowers with fringed petals, and bladdery seed vessels open at the top. [<F]
mi·graine (mī′grān) *n.* A form of recurrent paroxysmal headache, usually confined to one side of the head, and associated with various gastric and nervous disturbances. See MEGRIM. [<F <LL *hemicrania* <Gk. *hēmikrania* < *hēmi* half + *kranion* skull]
mi·grant (mī′grənt) *adj.* Migratory. —*n.* A migratory bird or other animal or person. [<L *migrans, -antis,* ppr. of *migrare* roam, wander]
mi·grate (mī′grāt) *v.i.* **·grat·ed, ·grat·ing** **1** To move from one country, region, etc., to settle in another. **2** To move periodically from one region or climate to another, as birds or fish. See synonyms under EMIGRATE. [<L *migratus,* pp. of *migrare*] —**mi′gra·tor** *n.*
mi·gra·tion (mi·grā′shən) *n.* **1** The act of migrating. **2** The totality of persons or animals migrating, or the time occupied in migrating. **3** *Chem.* The removal or shifting of one or more atoms from one position in the molecule to another. **4** *Physics* The drift or movement of ions, under the influence of electromotive force, toward one or the other electrode. [<L *migratio, -onis*] —**mi·gra′tion·al** *adj.* —**mi·gra′tion·ist** *n.*
mi·gra·to·ry (mī′grə·tôr′ē, -tō′rē) *adj.* **1** Pertaining to migration. **2** Given to migrating. **3** Roving; nomadic.
migratory thrush The American robin.
Mi·guel (mē·gel′) Portuguese and Spanish form of MICHAEL. Also *Hungarian* **Mi·ha·ly** (mē′·hä·lē), *Sw.* **Mi·ka·el** (mē′kä·äl).
mi·ka·do (mi·kä′dō) *n.* The sovereign of Japan: a title used chiefly by foreigners. [<Japanese *mi* august + *kado* door]
mike (mīk) *n. Colloq.* A microphone. [Short for MICROPHONE]
Mike (mīk) Diminutive of MICHAEL.
Mi·khai·lo·vić (mē·khī′lô·vich), **Draja,** 1893–1946, Serb guerrilla leader in Yugoslavia in World War II; executed.
Mi·klós (mē′klōsh) Hungarian form of NICHOLAS.
mi·kron (mī′kron) See MICRON.
mil (mil) *n.* **1** A unit of length or diameter, equal to one thousandth of an inch or to 25.4001 microns. **2** A milliliter, or one cubic centimeter. **3** A monetary unit of Israel, equal to a thousandth part of the Israeli pound; also the coin having this value. —**artillery mil** A unit of angle measurement for artillery fire, equal to 1/6400 of the circumference of a circle. —**infantry mil** The angle subtended by the arc of a circle one thousandth the length of the radius: equal to 1.018 artillery mils. [<L *mille* thousand]
mi·la·dy (mi·lā′dē) *n.* An English noblewoman; a gentlewoman: a Continental term used of such a woman. Also **mi·la′di.** [<F <E *my lady*]
mil·age (mī′lij) See MILEAGE.
Mi·lan (mi·lan′, -län′, mil′ən) The second largest city of Italy, in Lombardy. *Italian* **Mi·la·no** (mē·lä′nō).
Mil·an·ese (mil′ən·ēz′, -ēs′) *adj.* Pertaining to Milan, in Italy. —*n. pl.* **·ese** A native or inhabitant of Milan; the people of Milan.
Mi·laz·zo (mē·lät′sō) A port in NE Sicily: site of ancient *Mylae.*
milch (milch) *adj.* Giving milk, as a cow. [OE *milc, meolc* milk]
milch·y (mil′chē) *adj.* Milky.
mild (mīld) *adj.* **1** Moderate in action or disposition. **2** Expressing kindness; calm. **3** Moderate in effect or degree. **4** Not of strong flavor. **5** *Metall.* Designating strong and tough but malleable steel containing only a small percentage of carbon. See synonyms under BLAND, CHARITABLE, PACIFIC. [OE *milde*] —**mild′ly** *adv.* —**mild′ness** *n.*
mild·en (mīl′dən) *v.t. & v.i.* To make or become mild.
mil·dew (mil′dōō, -dyōō) *n.* **1** Any of a family *(Erysiphaceae)* of ascomycetous parasitic fungi that attack a great variety of plants, as hops, cherries, roses, etc. **2** A fungous disease of plants, particularly one caused by a fungus that makes a superficial downy coating on the diseased part of the host plant. **3** Any mold on walls, food, clothing, etc. —*v.t. & v.i.* To affect or be affected with mildew. [OE *mildēaw* honeydew] —**mil′dew·y** *adj.*
Mil·dred (mil′drid) A feminine personal name. [OE, moderate power]
mile (mīl) *n.* A measure of distance (see STATUTE MILE below) in the United States, Great Britain, and Ireland, and in all British possessions; remotely derived from the ancient Roman mile, which was about 1,620 yards or 4,860 feet. —**air mile** A nautical mile by air. —**geographical, nautical,** or **sea mile** One sixtieth of a degree of the earth's equator, or 6,080.2 feet. That of the United States is now the international nautical mile, equal to 6,076.103 feet or 1,852 meters. —**statute mile** The legal mile of the United States, Great Britain, etc., 5,280 feet, or 1,609.35 meters. [OE *mil* <LL *milia, millia* <L *mille passuum* thousand paces]
mile·age (mī′lij) *n.* **1** The entire length or amount of anything that is or may be measured in miles, especially when stated in miles; aggregate number of miles of track, wire, etc., traversed, made, or used. **2** Compensation reckoned at so much a mile, allowed in lieu of expenses of travel. **3** The rate a mile paid by one traveling in a car, or using a car for any purpose. Also spelled *milage.*
mileage ticket A ticket entitling the holder to transportation for a specific number of miles.
mil·er (mī′lər) *n.* A person or horse trained to race a mile.
Miles (mīlz) A masculine personal name. [< Gmc., warrior, or L, soldier]
Mi·le·sian (mī·lē′zhən, -shən) *adj.* Pertaining to Miletus. —*n.* A native or citizen of Miletus.
Mi·le·sian (mī·lē′zhən, -shən) *n.* **1** In ancient Irish legend, one of the sons or people of Miledh, the mythical ancestor of all the Irish, who migrated from Asia Minor and wandered, via Spain, to Ireland; hence, one of the first Celtic settlers of Ireland. **2** An Irishman.

mile·stone (mīl'stōn') n. 1 A post, pillar, or stone set up to indicate distance in miles from a given point: also **mile'post'** (-pōst'). 2 An important event or turning point in a person's life.

Mi·le·tus (mī-lē'təs) An ancient port in western Asia Minor.

mil·foil (mil'foil) n. Yarrow. [< OF < L *millefolium* < *mille* thousand + *folium* leaf]

Mil·ford Ha·ven (mil'fərd hā'vən) A port in southern Pembroke, Wales.

Mil·haud (mē·yō'), **Darius**, born 1892, French composer.

mil·i·a·ri·a (mil'ē·âr'ē·ə) n. Pathol. An acute inflammatory disease of the sweat glands marked by an eruption of vesicles and papulae of the size of a pinpoint or larger; miliary fever; the millet-seed rash. [< NL, fem. of L *miliarius*. See MILIARY.]

mil·i·a·ri·sion (mil'ē·ə·rizh'ən) n. A silver coin introduced by Constantine and valued at one thousandth of the gold pound; the chief silver coin of the early Middle Ages. Also **mil'i·a·rense'** (-rens'), **mil'i·a·ren'sis** (-ren'sis). [< L *milia*, pl. of *mille* thousand]

mil·i·a·ry (mil'ē·er'ē, mil'yə·rē) adj. 1 Like millet seeds. 2 Accompanied by a rash of pimples the size of a millet seed. 3 Pathol. Designating a form of tuberculosis caused by the discharge into one or more organs of minute bacillary tubercles originating elsewhere in the body. [< L *miliarius* < *milium* millet]

Mil·i·cent (mil'ə·sənt) A feminine personal name. Also **Mil'li·cent**. [< Gmc., power to work]

mi·lieu (mē·lyœ') n. Surroundings; environment. [< F < OF *mi* middle (< L *medius*) + *lieu* place < L *locus*]

mil·i·tant (mil'ə·tənt) adj. 1 Pertaining to conflict with opposing powers or influences. 2 Of a warlike or combative tendency; aggressive. — n. A combative person; a soldier. [< L *militans, -antis*, ppr. of *militare* be a soldier] — **mil'i·tan·cy** n. — **mil'i·tant·ly** adv.

mil·i·ta·rism (mil'ə·tə·riz'əm) n. 1 A system emphasizing and aggrandizing the military spirit and the need of constant preparation for war. 2 A desire to foster the maintenance of a powerful military position. — **mil'i·ta·rist** n. — **mil'i·ta·ris'tic** adj. — **mil'i·ta·ris'ti·cal·ly** adv.

mil·i·ta·rize (mil'ə·tə·rīz') v.t. ·rized, ·riz·ing 1 To imbue with militarism. 2 To convert to a military system. — **mil'i·ta·ri·za'tion** (-rə·zā'shən, -rī-) n.

mil·i·tar·y (mil'ə·ter'ē) adj. 1 Pertaining to armed forces or to warfare; martial; warlike. 2 Done or carried on by force of arms: distinguished from *civil*. See synonyms under WARLIKE. — n. A body of armed men or soldiers; soldiery. See synonyms under ARMY. [< F *militaire* < L *militaris* < *miles, militis* soldier] — **mil'i·tar'i·ly** adv.

military academy A school for boys or young men combining military training and academic education; specifically, the U. S. Military Academy at West Point, New York.

military attaché An army officer attached to his country's embassy or legation in a foreign country.

military governor An army or navy officer serving (usually temporarily) as the civil governor of a state or territory under martial law.

military intelligence 1 Information, of whatever character and however obtained, that is of military value to a country in peace or war. 2 The branch or division of a government engaged in obtaining, interpreting, and using such information.

military mast A strong mastlike structure on a warship, designed to carry a turret, observation tower, etc.

military police *Mil.* A body of soldiers charged with police duties among troops.

mil·i·tate (mil'ə·tāt) v.i. ·tat·ed, ·tat·ing To have influence: with *against*, or, more rarely, *for*. [< L *militatus*, pp. of *militare* be a soldier < *miles, militis*]

mi·li·tia (mə·lish'ə) n. 1 A body of citizens enrolled and drilled in military organizations other than the regular military forces, and called out only in emergencies. 2 *U.S.* All able-bodied male citizens between eighteen and forty-five years of age not members of the regular military forces. — **organized militia** *U.S.* The National Guard, Organized Reserves, the Naval Reserve, and the Marine Corps Reserve. [< L, military service < *miles, militis*] — **mi·li'tia·man** (-mən) n.

mil·i·um (mil'ē·əm) n. Pathol. A skin disease characterized by small yellowish globules. [< L, millet; after the resemblance of the globules to millet seed]

Mil·i·um (mil'ē·əm) n. A genus of millet grass. [< NL]

milk (milk) n. 1 The opaque, whitish liquid secreted by the mammary glands of female mammals for the nourishment of their young; especially, cow's milk, drunk or used by human beings. ◆ Collateral adjectives: *galactic, lacteal*. 2 The sap of certain plants. 3 Any one of various emulsions. — v.t. 1 To draw milk from the mammary glands of. 2 To draw (milk). 3 To draw off as if by milking; extract: to *milk* sap from a tree. 4 To draw or extract something from: to *milk* someone of information. 5 To exploit; take advantage of. — v.i. 6 To yield milk. [OE *meolc, milc*]

milk adder The house snake.

milk–and–wa·ter (milk'ən·wô'tər, -wot'ər) adj. 1 Weak and vacillating; namby-pamby. 2 Mawkish; sentimental.

milk·ber·ry (milk'ber'ē) n. The snowberry.

milk·er (mil'kər) n. 1 One who or that which milks; specifically, a mechanical device for milking cows. 2 A domestic animal, especially a cow, that is milked or that gives milk.

milk fever 1 Pathol. A fever attending the secretion of milk about the third day after childbirth. 2 A similar disease of milk cows.

milk·fish (milk'fish') n. pl. ·fish or ·fish·es A large, toothless, silvery fish (genus *Chanos*) allied to the herring, especially *C. chanos* of South Pacific waters.

milk leg 1 Pathol. A painful white swelling of the leg of a parturient woman. 2 A chronic swelling of a horse's leg, caused by inflammation of the lymphatic vessels.

milk–liv·ered (milk'liv'ərd) adj. Cowardly; timorous.

milk·maid (milk'mād') n. A dairymaid.

milk·man (milk'man') n. pl. ·men (-men') 1 One who delivers milk from door to door. 2 A man who milks cows.

milk of magnesia See under MAGNESIA.

Milk River A river in Canada and Montana, flowing 625 miles east to the Missouri.

milk run *U.S. Slang* In the air force, a regularly repeated mission.

milk shake A drink made of chilled, flavored milk, and sometimes ice-cream, shaken or whipped until frothy.

milk sickness Pathol. A disease caused by drinking the milk or eating the dairy products of cows that have fed on certain poisonous plants: marked by vomiting, intestinal disturbances, and trembling.

milk snake The house snake. Also called *milk adder*.

milk·sop (milk'sop') n. An effeminate man; a sissy.

milk·stone (milk'stōn') See GALALITH.

milk sugar The sugar contained in milk; lactose.

milk tooth A tooth of the deciduous or first dentition.

milk·vetch (milk'vech') n. The ground plum: so called from the supposed increase in the secretion of milk by goats feeding upon it.

milk·weed (milk'wēd') n. 1 One of a genus (*Asclepias*) of plants having a milky juice. 2 Any of several similar or related plants.

milkweed butterfly The monarch butterfly.

milk white A bluish–white color, like the color of skimmed milk.

milk·wort (milk'wûrt') n. 1 Any of a genus (*Polygala*) of plants with varicolored, showy flowers: so called from the fancied property of increasing the secretion of milk in nursing women; specifically, the **orange milkwort** (*P. lutea*) of the SE United States. 2 The sea milkwort.

milk·y (mil'kē) adj. milk·i·er, milk·i·est 1 Containing or like milk. 2 Yielding milk. 3 Very mild; spiritless. 4 Containing young or spawn: said of oysters. Also spelled *milchy*. — **milk'i·ly** adv. — **milk'i·ness** n.

Milky Way *Astron.* A luminous band encircling the heavens, composed of distant stars and nebulae not separately distinguishable to the naked eye; the Galaxy.

mill¹ (mil) n. 1 A machine by means of which grain is ground for food. 2 Any one of various kinds of machines that transform raw material by other processes than grinding: often in combination: a *sawmill*; *planing mill*. 3 A machine for reducing to small or smaller proportions hard substances of any kind. 4 A building fitted up with the machinery requisite for a factory: often in combination: *powdermill*. 5 A hardened steel roller, bearing a design in relief, by which a printing plate or a die may be made by pressure. 6 A milling cutter. 7 *Slang* A pugilistic combat; set-to. 8 A raised or ridged edge or surface made by milling. 9 A machine for crushing or grinding vegetable substances in order to express the juice. — **to go through the mill** To receive or undergo the experiences or hardships needed to acquire a certain degree of skill or wisdom. — v.t. 1 To grind, shape, polish, roll, etc., in or with a mill. 2 To raise and indent the edge of (a coin). 3 To cause to move with a circular motion. 4 To beat or whip to a froth, as chocolate. 5 *Slang* To strike or thrash; beat. — v.i. 6 To move slowly in a circle, as cattle. 7 *Slang* To fight or box. [OE *myln, mylen* < LL *molina* < L *mola* millstone]

mill² (mil) n. 1 A thousandth part. 2 *U.S.* The thousandth part of a dollar, or the tenth part of a cent. [< L *mille* thousand]

Mill (mil), **James**, 1773–1836, Scottish historian, philosopher, and political economist. — **John Stuart**, 1806–73, English philosopher; political economist; son of the preceding.

Mil·lais (mi·lā'), **Sir John Everett**, 1829–96, English painter.

Mil·lay (mi·lā'), **Edna St. Vincent**, 1892–1950, U. S. poet.

mill·board (mil'bôrd', -bōrd') n. Heavy pasteboard used for the covers of books; imitation pressboard.

mill·cake (mil'kāk') n. 1 The by-product left after the oil has been extracted from linseed. 2 The cake formed by mixing and pressing together the materials of gunpowder previous to granulation.

mill car A railroad flat car upon which heavy hoisting apparatus is mounted.

mill cinder The slag from the puddling furnaces of a steel-rolling mill.

mill–dam (mil'dam') n. 1 A barrier thrown across a watercourse to raise its level sufficiently to turn a millwheel. 2 The pond formed by such a barrier.

milled (mild) adj. 1 Passed through, cut by, or mixed in a mill. 2 Having the edges fluted or grooved: said of coins.

mil·le·fi·o·ri (mil'ə·fē·ôr'ē, -ō'rē) n. Italian An ancient mosaic glass, later made in Venice, having a flowerlike pattern; literally, a thousand flowers.

mil·le·nar·i·an (mil'ə·nâr'ē·ən) adj. Of or pertaining to a thousand; specifically, relating to the millennium. — n. One who believes in the millennium. — **mil'le·nar'i·an·ism** n.

mil·le·nar·y (mil'ə·ner'ē) adj. Of or pertaining to a thousand; millenarian; millennial. — n. pl. ·nar·ies 1 The space of a thousand years; the millennium. 2 A millenarian. [< LL *millenarius* < L *milleni* a thousand each < *mille* thousand]

mil·len·ni·al (mi·len'ē·əl) adj. Of or pertaining to the millennium or to any period of a thousand years. — n. A thousandth anniversary. — **mil·len'ni·al·ist** n. — **mil·len'ni·al·ly** adv.

mil·len·ni·um (mi·len'ē·əm) n. pl. ·ni·a (-ē·ə) 1 A period of a thousand years. 2 The thousand years of the kingdom of Christ on earth. *Rev.* xx 1–5. 3 Hence, by extension, any period of happiness, beneficial government or the like. [< NL < *mille* thousand + *annus* year]

mil·le·ped (mil'ə·ped), **mil·le·pede** (mil'ə·pēd) See MILLIPEDE.

mil·le·pore (mil'ə·pôr, -pōr) n. *Zool.* A corallike hydrozoan (genus *Millepora*) with numerous cavities in the enclosing and often bulky limestone structure. Also **mil·lep·o·ra** (mə·lep'ər·ə). [< F *millépore* < *mille* thousand + *pore* pore]

mill·er (mil'ər) n. 1 One who keeps or tends a mill, particularly a gristmill. 2 A milling machine. 3 A mothmiller.

Mil·ler (mil'ər), **Arthur,** born 1915, U.S. playwright. — **Henry,** born 1891, U.S. author. — **Joaquin** Pseudonym of Cincinnatus Heine Miller, 1839-1913, U.S. poet. — **Joe,** 1684-1738, English comedian: real name *Josias Miller.* — **William,** 1782-1849, U.S. Adventist; founder of the Millerites.

Mille·rand (mēl·rän'), **Alexandre,** 1859-1943, French statesman; president 1920-24.

mil·ler·ite (mil'ər·īt) *n.* A metallic, brass-yellow, brittle nickel sulfide, NiS, crystallizing in the hexagonal system. [after W. H. *Miller,* 1801-80, English mineralogist]

Mil·ler·ite (mil'ər·īt) *n.* A follower of William Miller, who announced in 1831 that Christ's second coming and the end of the world would be in 1843. — **Mil'ler·ism** *n.*

miller's thumb A small, fresh-water fish (family *Cottidae*) with broad, flattened head and spiny fins.

Mil·les (mil'əs), **Carl,** 1875-1955, Swedish sculptor active in the United States.

mil·les·i·mal (mi·les'ə·məl) *adj.* Pertaining to thousandths. — *n.* A thousandth. [< L *millesimus* < *mille* thousand]

mil·let (mil'it) *n.* 1 A grass (*Panicum miliaceum*), or its seed, cultivated in the United States for forage, but in the Old World from earliest times, and still in some parts of Europe, as a cereal. 2 One of various other grasses, or their seed, as the **foxtail** or **Italian millet** (*Setaria italica*), **pearl millet** (*Pennisetum glaucum*), etc. [< F, dim. of *mil* < L *milium*]

Mil·let (mē·le'), **Jean François,** 1814-75, French artist.

mill finish A surface finish produced on sheet and plate steel characteristic of the ground finish that is present on the rollers used in fabrication.

mill·hand (mil'hand') *n.* A worker in a mill.

milli– *combining form* 1 A thousand: *millipede.* 2 In the metric and other systems of measurement, the thousandth part of (a specified unit):

milliampere	millicurie	millilux
milliangstrom	millifarad	milliphot
milliare	millihenry	millistere
millibar	millilambert	millivolt

[< L *mille* a thousand]

mil·liard (mil'yərd) *n.* A thousand millions; usually, in the United States, called a billion. [< F < Pg. *milhar* thousand]

mil·li·ar·y (mil'ē·er'ē) *adj.* Pertaining to a Roman mile. — *n. pl.* **·ar·ies** A Roman milestone, set up at intervals of 1,000 paces. [< L *milliarius*]

mil·lier (mē·lyā') *n.* A metric ton, 1,000 kilograms. [< F]

mil·li·gram (mil'ə·gram) *n.* The thousandth part of a gram. See METRIC SYSTEM. Also **mil'li·gramme.** [< F *milligramme*]

Mil·li·kan (mil'ə·kən), **Robert Andrews,** 1868-1955, U.S. physicist.

mil·li·li·ter (mil'ə·lē'tər) *n.* The thousandth part of a liter. See METRIC SYSTEM. Also **mil'li·li'tre.** [< F *millilitre*]

mil·li·me·ter (mil'ə·mē'tər) *n.* The thousandth part of a meter. See METRIC SYSTEM. Also **mil'li·me'tre.** [< F *millimètre*]

mil·li·mi·cron (mil'ə·mī'kron) *n.* A thousandth of a micron, or a millionth of a millimeter; also, ten millionth (symbol m, also μ).

mil·li·mol (mil'ə·mol) *n.* The thousandth part of a mol. Also **mil'li·mole** (-mōl). [< MILLI- + MOL]

mil·line (mil·līn') *n.* The rate of cost for placing an advertisement of one agate line before a million readers. [< MILL(ION) + LINE]

mil·li·ner (mil'ə·nər) *n.* 1 A person employed in making, trimming, or selling bonnets, women's hats, etc. 2 *Obs.* A dealer in small wares: a haberdasher. [< *Milaner* an inhabitant of Milan, in Italy; hence, a man from Milan who imported silks and the like]

mil·li·ner·y (mil'ə·ner'ē) *n. pl.* **·ies** 1 The articles made or sold by milliners. 2 The occupation or establishment of a milliner. [< MILLINER]

mill·ing (mil'ing) *n.* 1 The operating of a mill or mills, as in the grinding of meal or metals, the preparation of cloth, etc. 2 A milled surface, as of a coin, or the act or process of producing it: distinguished from *reeding.* 3 The slow, round-and-round motion of or as of a herd of cattle. [< MILL, v.]

mil·lion (mil'yən) *n.* 1 The cardinal number equivalent to ten hundred thousand or to a thousand thousand, or the symbols (1,000,000) representing it; ten to the sixth power. 2 Elliptically, a thousand thousand of the ordinary units of account, as dollars, francs, or pounds: He is worth *a million.* 3 An indefinitely great number. — **the million** The common people. [< OF < Ital. *millione* (now *milione*), aug. of *mille* thousand]

mil·lion·aire (mil'yən·âr') *n.* One whose possessions are valued at a million or more, as of dollars, pounds, etc. Also **mil'lion·naire'.** [< F *millionaire*]

mil·lion·fold (mil'yən·fōld') *adj.* A million times the quantity. — *adv.* In a millionfold proportion; a million times in amount: with the indefinite article *a,* construed as a plural.

mil·lionth (mil'yənth) *adj.* 1 Being last in a series of a million: an ordinal numeral. 2 Being one of a million equal parts: a *millionth part.* — *n.* One part in a million equal parts; one divided by one million.

mil·li·pede (mil'ə·pēd) *n.* A herbivorous, slow-moving arthropod (class *Diplopoda*) having a rounded body marked by numerous segments, from nearly all of which issue two pairs of appendages: also spelled **milleped, millepede.** Also **mil'li·ped** (-ped). [< L *millepeda* < *mille* thousand + *pes, pedis* foot]

mill·pond (mil'pond') *n.* 1 A body of water dammed up to run a mill. 2 A body of water adjoining a sawmill where logs are floated until sawn.

mill·race (mil'rās') *n.* The sluice through which the water runs to a millwheel.

mill·run (mil'run') *n.* 1 A millrace. 2 A certain amount of ore tested for content or quality by the process of milling. 3 The mineral yielded by the test.

mill-run (mil'run') *adj.* Average; just as it comes from the mill; not selected: also *run-of-the-mill.*

Mills bomb (milz) A highly explosive grenade weighing about 1 1/2 pounds and usually thrown by hand. Also **Mills grenade.** [after Sir Wm. *Mills,* 1856-1932, British inventor]

mill·stone (mil'stōn') *n.* 1 One of a pair of thick, heavy, stone disks for grinding something, as grain. 2 That which pulverizes or bears down. 3 A heavy or burdensome weight. *Matt.* xvii 6.

mill town A town whose center of activity is a mill or group of mills, and whose population is employed, for the most part, in the mill or mills.

mill·wheel (mil'hwēl') *n.* The waterwheel that drives a mill.

mill·work (mil'wûrk') *n.* Carpentry work delivered from a mill ready for installation or use.

mill·wright (mil'rīt') *n.* One who plans, builds, or fits out mills; also, a machinist who sets up shafting, etc.

Milne (miln), **A(lan) A(lexander),** 1882-1956, English playwright, novelist, and writer of children's books.

Mil·ner (mil'nər), **Sir Alfred,** 1854-1925, first Viscount Milner; British administrator in South Africa.

mi·lo (mī'lō) *n.* A non-saccharin sorghum similar to millet. [< Bantu *maili*]

Mi·lo (mī'lō) A renowned Greek athlete, about 520 B.C.

Mi·lo (mē'lō) The southwesternmost island in the Cyclades group; 61 square miles: also *Melos.* Also **Mī'los** (-lōs).

mil·o·maize (mil'ō·māz) *n.* See MILO.

mi·lord (mi·lôrd') *n.* An English nobleman; a gentleman: a Continental term used of such a man. [< F < E *my lord*]

Milque·toast (milk'tōst) See CASPAR MILQUETOAST.

mil·reis (mil'rās) *n.* 1 A former Brazilian monetary unit, equivalent to 1,000 reis. 2 A former Portuguese monetary unit, superseded in 1911 by the escudo. [< Pg., lit., a thousand reis]

milt¹ (milt) *n.* The spleen. [OE *milte* spleen]

milt² (milt) *n.* 1 The sperm of a fish. 2 The spermatic organs of a fish when filled with seminal fluid; the soft roe. — *v.t.* To impregnate (fish roe) with milt. [OE *milte*]

milt·er (mil'tər) *n.* 1 A male fish. 2 The milt of a fish.

Mil·ti·a·des (mil·tī'ə·dēz) Athenian general; defeated the Persians at Marathon, 490 B.C.

Mil·ton (mil'tən), **John,** 1608-74, English poet.

Mil·ton·ic (mil·ton'ik) *adj.* Of, pertaining to, or like the poet Milton or his works or style; sublime; majestic. Also **Mil·to'ni·an** (-tō'nē·ən).

Mil·town (mil'toun) *n.* Proprietary name for a synthetic drug used in pill form as a tranquilizer.

Mil·wau·kee (mil·wô'kē) A port on Lake Michigan in SE Wisconsin.

Mil·yu·kov (mē·lyōō'kôf), **Pavel Nikolaevich,** 1859-1943, Russian politician and historian.

mim (mim) *Brit. Dial.* Demure; precise. [Imit.]

mime (mīm) *n.* 1 A mimic play or farce or the dialog for this; a dramatic representation, akin to comedy, mimicking real persons or events: a favorite amusement among the Greeks and Romans. 2 An actor in a mime; hence, a mimic; clown, buffoon. — *v.* **mimed, mim·ing** *v.t.* To mimic; imitate. — *v.i.* To play the mimic; play a part with gestures and usually without words. [< L *mimus* < Gk. *mimos*] — **mim'er** *n.*

Mim·e·o·graph (mim'ē·ə·graf', -gräf') *n.* An apparatus in which a thin fibrous paper coated with paraffin is used as a stencil for reproducing copies of written or typewritten matter: a trade name. — *v.t.* To reproduce by means of a Mimeograph. [< Gk. *mimeisthai* imitate + -GRAPH]

mi·me·sis (mi·mē'sis, mī-) *n.* 1 Imitation or representation of the supposed speech, characteristic dialect, carriage, or gestures of an individual or a people, as in art and literature. 2 *Biol.* Mimicry. [< NL < Gk. *mimēsis* imitation]

mi·met·ic (mi·met'ik, mī-) *adj.* 1 Quick to mimic; imitative. 2 Relating to mimicry or mimesis. 3 Mimic (adj. def. 1). Also **mi·met'i·cal.** [< Gk. *mimētikos*] — **mi·met'i·cal·ly** *adv.*

mim·ic (mim'ik) *v.t.* **·icked, ·ick·ing** 1 To imitate the speech or actions of, as in ridicule. 2 To copy closely; ape. 3 To have or assume the color, shape, etc., of; simulate: Some insects *mimic* leaves. See synonyms under IMITATE. — *n.* 1 One who is given to mimicry; a mimic actor; buffoon. 2 A copy; imitation. — *adj.* 1 Of the nature of mimicry; imitative; mimetic: a *mimic* gesture. 2 Copying the real; simulated; mock: a *mimic* court. [< L *mimicus* < Gk. *mimikos* < *mimos* mime] — **mim'i·cal** *adj.* — **mim'ick·er** *n.*

mim·ic·ry (mim'ik·rē) *n. pl.* **·ries** 1 The act of imitating, especially for sport; also, a thing produced as a copy. 2 *Zool.* An imitative superficial resemblance in one animal to another or to its immediate environment, for purposes of concealment or protection. Compare PROTECTIVE COLORATION. See synonyms under CARICATURE.

Mi·mir (mē'mir) In Norse mythology, the giant who kept the **Mi·mis·brun·nen** (mē'mis·brōōn'ən), or well of wisdom, flowing from the root of Ygdrasil, and for a draft from which Odin bartered his eye. Also **Mī'mer.**

mi·mo·sa (mi·mō'sə, -zə) *n.* 1 Any plant of a large genus (*Mimosa*) of tropical herbs, shrubs, or trees of the bean family, with feathery, bipinnate foliage, and clusters of small flowers; especially, the sensitive plant. 2 A light yellow: the color of certain varieties of mimosa. [< NL < L *mimus* mime; from its supposed mimicry of animal life] — **mim·o·sa·ceous** (mim'ō·sā'shəs, mī'mō-) *adj.*

Min (min) 1 A river in Fukien province, China, flowing 350 miles SE to the China Sea. 2 A river in Szechwan province, China, flowing 500 miles south to the Yangtze.

min' (min) *v. & n. Scot.* Mind.

mi·na¹ (mī'nə) *n. pl.* **·nae** (-nē) or **·nas** An ancient Greek and Asian weight or sum of money, equal to 100 drachmas. [< L < Gk. *mna*]

mi·na² (mī'nə) See MYNA. Also **mī'nah.**

mi·na·cious (mi·nā'shəs) *adj.* Threatening; of a menacing character. [< L *minax, minacis.* See MENACE.] — **mi·na'cious·ly** *adv.* — **mi·na'cious·ness, mi·nac·i·ty** (mi·nas'ə·tē) *n.*

min·a·ret (min′ə·ret′) n. A high, slender tower attached to a Moslem mosque, surrounded by one or more balconies, from which is sounded the stated summons to prayer. [<Sp. *minarete* <Turkish *manārat* <Arabic *manārah* lamp, lighthouse <*minār* candlestick]

MINARET
A. Shown in relation to mosque.
B. Detail showing the balcony.

Mi·nas Basin (mī′nəs) The NE arm of the Bay of Fundy, Nova Scotia; connected by the **Minas Channel** (24 miles long) with the Bay of Fundy.
Mi·nas de Rí·o·tin·to (mē′näs thā rē′ō·tēn′tō) A town in SW Spain; site of copper mines since Roman times: also **Ríotinto**.
Mi·nas Ge·rais (mē′nəzh zhə·rīs′) An inland state in eastern Brazil; 224,701 square miles; capital, Bello Horizonte. Formerly **Minas Geraes**.
min·a·to·ry (min′ə·tôr′ē, -tō′rē) adj. Threatening, as with destruction or punishment. Also **min′a·to′ri·al**. [<OF *minatoire* <LL *minatorius* <*minatus*, pp. of *minari* threaten] — **min′a·to′ri·al·ly**, **min′a·to′ri·ly** adv.
mince (mins) v. **minced**, **minc·ing** v.t. 1 To cut or chop into small bits, as meat. 2 To subdivide minutely. 3 To diminish the force or strength of; moderate: He didn't *mince* words with her. 4 To say or do with affected primness or elegance. — v.i. 5 To walk with short steps or affected daintiness. 6 To speak or behave with affected primness. — n. 1 Mincemeat. 2 *Rare* An affectation. [<OF *mincier*; ult. <L *minuere* lessen, make smaller] — **minc′er** n. — **minc′ing·ly** adv.
minced oath (minst) A mild or weak oath: "Egad" is a *minced oath*.
mince·meat (mins′mēt′) n. 1 Meat chopped very fine. 2 A mixture of chopped meat, fruit, spices, etc., used in mince pie.
mince pie A pie made of mincemeat.
minc·ing-horse (min′sing·hôrs′) n. A heavy block of hardwood usually mounted on legs, used for chopping meats or vegetables.
minc·ing-spade (min′sing·spād′) n. A knife-edged spade used in whaling, for chopping or mincing blubber.
Min·cio (mēn′chō) A river in northern Italy, flowing 115 miles from Lake Garda to the Po. Ancient **Min·cius** (min′shəs).
mind (mīnd) n. 1 The aggregate of all conscious and unconscious processes originating in and associated with the brain, especially those pertaining to cognition, intelligence, and intellect. 2 Memory; remembrance: to bear in *mind*. 3 Opinion; way of thinking: to change one's *mind*. 4 Desire; inclination: to have a *mind* to leave. 5 Mental disposition, character, or temper: a liberal *mind*; a cheerful *mind*. 6 Intellectual power or capacity. 7 The faculty of cognition and intellect, as opposed to the will and emotions: a noble heart and a cultivated *mind*. 8 A person, regarded as having intellect: the great *minds* of our time. 9 Sanity; sound mentality: to lose one's *mind*. 10 *Philos*. The spirit of intelligence pervading the universe: opposed to *matter*. 11 In Christian Science, God: also **Divine Mind**. — **a month's mind** The monthly commemoration, usually the first, of a person's death. — **on one's mind** Occupying one's thoughts. — **to speak one's mind** To declare one's opinions freely or candidly. — **to take one's mind off** To turn one's thoughts from. — v.t. 1 To pay attention to; occupy oneself with: *Mind* your own business. 2 To be careful or wary concerning: *Mind* your step. 3 To give heed to, as commands; obey. 4 To care for; tend. 5 To be concerned about; regard with annoyance; dislike: Do you *mind* the noise? 6 To be aware of; notice; perceive. 7 *Colloq*. To remember: sometimes used reflexively. 8 *Archaic* To remind. 9 *Obs*. To intend; purpose. — v.i. 10 To pay attention; take notice; watch. 11 To be obedient. 12 To be concerned; care: I don't *mind*. 13 To be careful. [OE *gemynd*] — **mind′er** n.
Synonyms (noun): brain, consciousness, disposition, instinct, intellect, intelligence, reason, sense, soul, spirit, thought, understanding. *Mind* includes all the powers of sentient being apart from the physical factors in bodily faculties and activities; in a limited sense, *mind* is nearly synonymous with *intellect*, but includes *disposition*, or the tendency toward action, as appears in the phrase "to have a *mind* to work." The *intellect* is that assemblage of faculties which is concerned with knowledge, as distinguished from emotion and volition. *Understanding* is chiefly used of the reasoning powers: the *understanding* is distinguished by many philosophers from *reason* in that *reason* is the faculty of the high cognitions or a priori truth. *Thought*, the act, process, or power of thinking, is often used to denote the thinking faculty, and especially the *reason*. The *instinct* of animals is held to be of the same nature as the *intellect* of man, but inferior and limited; yet the apparent difference is very great. Human *instincts* denote tendencies independent of reasoning or instruction. As the seat of mental activity, *brain* is often used as a synonym for *mind*, *intellect*, *intelligence*. *Sense* is used as denoting clear mental action, good judgment, acumen; as, He is a man of *sense*, or, he showed good *sense*. *Consciousness* includes all that a sentient being perceives, knows, thinks, or feels, from whatever source. See GENIUS, SOUL, UNDERSTANDING. *Antonyms*: body, brawn, matter.
Min·da·na·o (min′də·nä′ō) The southernmost and second largest island in the Philippines; 36,537 square miles.
Mindanao Sea Part of the Pacific bounded by Mindanao to the south and by Leyte, Bohol, Cebu, and Negros to the north.
mind-blind·ness (mīnd′blīnd′nis) n. Mental blindness.
mind-cure (mīnd′kyoor′) n. 1 The treatment and cure of disease, especially of neuroses, by direct influence upon the mind of the patient and without the use of drugs. 2 Psychotherapy.
mind-deaf·ness (mīnd′def′nis) n. Mental deafness.
mind·ed (mīn′did) adj. 1 Disposed. 2 Having a specified kind of mind: evil-*minded*.
Min·del (min′dəl) See GLACIAL EPOCH.
mind·ful (mīnd′fəl) adj. Keeping in mind; heeding; having knowledge of; aware. See synonyms under THOUGHTFUL. — **mind′ful·ly** adv. — **mind′ful·ness** n.
mind·less (mīnd′lis) adj. 1 Devoid of intelligence. 2 Not giving heed or attention; careless. — **mind′less·ly** adv. — **mind′less·ness** n.
Min·do·ro (min·dô′rō) An island in the central Philippines; 3,759 square miles.
mind-read·ing (mīnd′rē′ding) n. 1 The ascertaining of the thoughts or purpose of some other mind by the interpretation of voluntary or involuntary muscle movements or facial expressions. 2 Telepathy. — **mind′-read′er** n.
Mind·szen·ty (mīnd′sen·tē) **Joseph**, born 1892, Hungarian Roman Catholic cardinal, imprisoned by the Nazis in 1944 and by the Communist regime in Hungary 1948-56.
mine[1] (mīn) n. 1 An excavation for digging out some useful product, as ore or coal. 2 Any deposit of such material suitable for excavation. 3 An underground tunnel dug beneath an enemy's fortifications, as for the placement of an explosive charge. 4 A case containing such a charge buried in the earth, or floating on or near, or anchored beneath, the surface of the water. 5 Any productive source of supply. 6 A burrow made by an insect. — **bounding mine** A land mine set just beneath the surface of the ground and designed to rise a few feet in the air before exploding its charge of shrapnel and fragments. — **land mine** A high-explosive or chemical mine, actuated by the weight of a person, troops, or vehicles. — v. **mined**, **min·ing** v.t. 1 To dig (coal, ores, etc.) from the earth. 2 To dig into (the earth, etc.) for coal, ores, etc. 3 To make by digging, as a tunnel; burrow. 4 To dig a tunnel under, as for placing an explosive mine; sap. 5 To attack or destroy by slow or secret means; undermine. 6 To place an explosive mine or mines in or under: to *mine* a harbor. — v.i. 7 To dig in a mine for coal, ores, etc.; work in a mine. 8 To make a tunnel, etc., by digging; burrow. 9 To place explosive mines. [<OF <Celtic. Cf. Irish *mein* vein of metal.] — **min′er** n.

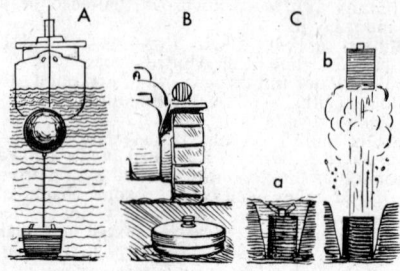

MINES
A. Marine mine anchored in harbor entrance.
B. Buried anti-tank mine.
C. Anti-personnel mine (a); inner canister rises and explodes (b).

mine[2] (mīn) pron. 1 The possessive case of *I* employed predicatively; belonging or pertaining to me: That book is *mine*. 2 The things or persons belonging or pertaining to me: His work is better than *mine*; Fortune has been good to me and *mine*. — **of mine** Belonging or relating to me; my: the double possessive. — *pronominal* adj. *Archaic* My: *mine* eyes. [OE *mīn*]
mine detector An electromagnetic instrument for detecting the presence and locating the position of mines.
mine field An area in water or on land systematically planted with mines.
mine layer An auxiliary naval vessel, of submarine or surface type, provided with special equipment for the laying of mines.
min·er·al (min′ər·əl) n. 1 A naturally occurring, homogeneous substance or material formed by inorganic processes and having a characteristic set of physical properties, a definite range of chemical composition, and a molecular structure usually expressed in crystalline forms. 2 Any inorganic substance, as ore, a rock, or a fossil. — **accessory minerals** Those components or minerals forming so small a part of a rock or occurring so rarely as not to be included in its description. — adj. 1 Pertaining to, consisting of, or resembling minerals. 2 Impregnated with mineral constituents. [<OF <Med. L *minerale*, neut. sing. of *mineralis* of a mine <*minera* a mine]
mineral butter *Chem*. Antimony trichloride, SbCl₃, a colorless, translucent, poisonous, crystalline substance: used as a mordant and reagent, and in the manufacture of certain dyes.
mineral carbon Graphite.
mineral charcoal A pulverulent substance showing patterned traces of vegetable origin but consisting mainly of carbon: it occurs in thin layers in bituminous coal.
min·er·al·ize (min′ər·əl·īz′) v. **·ized**, **·iz·ing** v.t. 1 To convert from a metal to a mineral, as iron to rust. 2 To convert to a mineral substance; petrify. 3 To impregnate with minerals or other inorganic substances. — v.i. 4 To observe, study, and collect minerals. Also **min·er·al·o·gize** (min′ə·ral′ə·jīz), *Brit*. **min′er·al·ise′**. — **min′er·al·i·za′tion** n.
min·er·al·iz·er (min′ər·əl·īz′ər) n. 1 An element that combines with a metal to form an ore, as sulfur. 2 A volatile or other substance, as boron or water, that facilitates the recrystallization of rocks.
mineral jelly See PETROLATUM.
mineral kingdom The great division of nature

which comprises all inorganic and non-living materials, as rocks, metals, minerals, etc. Compare ANIMAL KINGDOM, VEGETABLE KINGDOM.

min·er·al·o·gy (min′ə·ral′ə·jē) *n.* The science of minerals. — **min′er·a·log′i·cal** (-ər·ə·loj′i·kəl) *adj.* — **min′er·a·log′i·cal·ly** *adv.* — **min′·er·al′o·gist** *n.*

mineral oil Any of various oils derived from inorganic matter, especially petroleum and its products: used as a fuel, in medicine as a laxative, etc.

mineral pitch Asphalt.

mineral tar Maltha.

mineral water Any natural water containing one or more minerals in solution, especially one impregnated with salts or gases having therapeutic properties.

mineral wax Ozocerite.

mineral wool A fibrous, fluffy material resembling wool in appearance, and made by subjecting molten silicate, molten slag, or other similar materials to a steam blast and cooling rapidly to prevent crystallization: used as packing and as an insulator: also called *rock wool, slag wool.* Compare GLASS-WOOL.

Mi·ner·va (mi·nûr′və) In Roman mythology, the goddess of wisdom, handicraft, and technical dexterity: identified with the Greek *Athena.*

min·e·stro·ne (min′ə·strō′nē, *Ital.* mē′nä·strō′nä) *n. Italian* A thick vegetable soup containing vermicelli, barley, etc., in a meat broth.

mine sweeper 1 A naval vessel equipped for the detection, destruction, and removal of marine mines. 2 A heavy roller attached to the front of a tank for exploding land mines.

mine thrower A high-angle, muzzleloading gun, sometimes with a smooth bore, used for throwing large projectiles containing a heavy explosive charge: also called *trench mortar.*

Ming (ming) *n.* In Chinese history, the last ruling dynasty (1368–1644) of truly Chinese origin: noted for the artistic works, as porcelains, produced during its rule. [<Chinese, lit., luminous]

min·gle (ming′gəl) *v.* **min·gled, min·gling** *v.t.* 1 To mix or unite together; blend. 2 To join or combine; bring into close relation. — *v.i.* 3 To be or become mixed, united, or closely joined. 4 To enter into company; mix or associate, as with a crowd. 5 To take part; become involved, as in a dispute. See synonyms under MIX. [Freq. of OE *mengan*] — **min′gler** *n.*

Mi·nho (mē′nyōō) A river of Spain and Portugal, flowing 170 miles SW to the Atlantic Ocean. *Spanish* **Mi·ño** (mē′nyō).

Min·how (min′hō′) A former name for FOO-CHOW.

min·i·a·ture (min′ē·ə·chər, min′ə·chər) *n.* 1 A painting of small dimensions and delicate workmanship, usually a portrait, or the art of executing such paintings on ivory, metal, or vellum, in water colors or oils. 2 A portrayal of anything on a small scale; hence, reduced dimensions or extent. 3 *Obs.* Lettering in red; rubrication, as of medieval manuscripts. See synonyms under PICTURE. — *adj.* Much smaller than reality; on a very small scale. [<F <Ital. *miniatura* <Med. L *miniare* paint red < *minium* red lead, ? <Iberian; later infl. in meaning by L *minuere* lessen]

miniature camera *Photog.* 1 A small camera for taking pictures on film measuring 35 millimeters or less in width. 2 A subminiature camera.

min·i·a·tur·ize (min′ē·ə·chər·īz′, min′ə·chər·īz′) *v.t.* **·ized, ·iz·ing** To reduce the size of, as the parts of an instrument or machine. — **min′i·a·tur′i·za′tion** *n.*

Min·i·é ball (min′ē·ā, min′ē; *Fr.* mē·nyā′) A conical rifle ball with hollow base and a plug driven in by the explosion of the charge to expand the lead and fill the grooves of the rifling: formerly used in the Minié rifle, specially designed for it. [after Capt. Claude Étienne *Minié,* 1814–79, French inventor]

min·i·fy (min′ə·fī) *v.t.* **·fied, ·fy·ing** 1 To make small or less. 2 To lessen the worth or importance of; undervalue. [<L *minor* less + -FY]

min·i·kin (min′ə·kin) *adj. Obs.* Of small size or delicate form. — *n. Rare* Something very minute or delicate, as the smallest size of pin. [<Du. *minneken,* dim. of *minne* love]

min·im (min′im) *n.* 1 An apothecaries' fluid measure; roughly, one drop, or one sixtieth of a fluid dram: in the United States, 0.06 cubic centimeter; in England, 0.059 cubic centimeter. 2 *Music* A half note. 3 An extremely small creature; a pigmy. 4 A down-stroke in writing, as in forming the letter *n.* 5 *Printing* A size of type, about 7-point, between nonpareil and brevier. — *adj.* Extremely small. [<F *minime* <L *minimus* least, smallest. Doublet of MINIMUM.] — **min′i·mal** *adj.*

Min·i·mal·ist (min′ə·məl·ist) *n.* One belonging to the more conservative branch of the former Russian Social Democrats: distinguished from *Maximalist.*

min·i·mize (min′ə·mīz) *v.t.* **·mized, ·miz·ing** 1 To reduce to the smallest possible amount or degree. 2 To regard or represent at a minimum. Also *Brit.* **min′i·mise.** — **min′i·miz′er** *n.* — **min′i·mi·za′tion** *n.*

min·i·mum (min′ə·məm) *adj.* Consisting of or showing the least possible amount or degree; being a minimum. — *n. pl.* **·ma** (-mə) or **·mums** 1 The least possible quantity, amount, or degree. 2 A value of a function that is less than any value which immediately precedes and follows it. 3 The lowest degree, variation, etc., as of temperature, recorded. [<L, neut. of *minimus.* Doublet of MINIM.] — **min′i·mal** *adj.*

minimum wage 1 A wage fixed by law or agreement as the smallest amount an employer may offer an employee in a specific group. 2 A living wage.

min·ing (mī′ning) *n.* 1 The act, process, or business of extracting coal, ores, etc., from mines. 2 The act of laying explosive mines.

mining camp A temporary settlement around or near a mine.

min·ion (min′yən) *n.* 1 A servile favorite or follower: a term of contempt. 2 *Printing* A size of type, about 7-point. 3 A saucy girl or woman; minx. 4 *Obs.* One who is beloved; a paramour. — *adj. Rare* Dainty; delicate; fine. [<F *mignon.* Doublet of MIGNON.]

min·ish (min′ish) *v.t.* & *v.i. Obs.* To make or become less; diminish. [<F *menuisier* make small <L *minutus,* pp. of *minuere* lessen]

min·is·ter (min′is·tər) *n.* 1 The chief of an executive department of a government. 2 One commissioned to represent his government in diplomatic intercourse with another government. 3 One who is authorized to preach the gospel and administer the ordinances of the church; a clergyman. 4 One who acts in subservience to the will of another; a servant; agent. 5 One who promotes or dispenses. — *v.i.* 1 To give attendance or aid; provide for the wants or needs of someone. 2 To be conducive; contribute. — *v.i.* 3 To administer or apply. 4 *Archaic* To supply; furnish. See synonyms under SERVE. [<OF *ministre* <L *minister* an attendant < *minor* less; after L *magister* master < *magis* greater]

min·is·te·ri·al (min′is·tir′ē·əl) *adj.* 1 Of or pertaining to a minister of the gospel or the ministry; clerical. 2 Of or pertaining to a minister or executive staff in civil government. 3 *Law* Pertaining to an act or duty performed according to the mandate of legal authority, so that the agent is not accountable for its propriety or consequences: opposed to *judicial.* 4 Instrumental; causative. [<F *ministériel*] — **min′is·te′ri·al·ly** *adv.*

min·is·te·ri·al·ist (min′is·tir′ē·əl·ist) *n.* In English politics, one who supports the ministry.

minister plenipotentiary See PLENIPOTENTIARY.

minister resident See under RESIDENT.

min·is·trant (min′is·trənt) *adj.* Ministering. — *n.* One who ministers. [<L *ministrans, -antis,* ppr. of *ministrare* serve]

min·is·tra·tion (min′is·trā′shən) *n.* 1 The act of performing service as a minister. 2 Any religious ceremonial. [<L *ministratio, -onis*] — **min′is·tra′tive** *adj.*

min·is·try (min′is·trē) *n. pl.* **·tries** 1 The body of officials in charge of the administration of the departments of a government. In the United States it is selected by the president with the advice and consent of the Senate, and is called the *Cabinet.* 2 An executive department of government. 3 Ministers of the gospel collectively, or their office. 4 The act of ministering, or the state or office of a minister; ministration; service. [<L *ministerium*]

Min·i·track (min′i·trak′) *n.* A sensitive electronic system for tracking the paths of earth satellites by the timing of radio signals received by properly spaced ground stations: a trade name.

min·i·um (min′ē·əm) *n.* 1 A vivid opaque red lead oxide, Pb_3O_4: used chiefly as a pigment. 2 Cinnabar. [<L]

min·i·ver (min′ə·vər) *n.* 1 A fur or mixture of furs used in the Middle Ages for trimming. 2 Any white fur, as ermine. [<OF *menu vair,* lit., little spotted (fur) <L *minutus* small + *varius* variegated]

mink (mingk) *n.* 1 An amphibious, slender-bodied, carnivorous mammal (genus *Mustela*), related to the weasel and valued for its soft, thick, glossy, brown fur. 2 The fur of this mammal. [<Scand. Cf. Sw. *menk.*]

Min·kow·ski (ming·kôf′skē), **Hermann,** 1864–1909, Russian mathematician.

Min·ne·ap·o·lis (min′ē·ap′ə·lis) A city on the Mississippi in SE central Minnesota.

Min·ne·ha·ha (min′ē·hä′hä) In Longfellow's *The Song of Hiawatha,* a Dakota Indian maiden, the wife of Hiawatha.

min·ne·sing·er (min′ə·sing′ər) *n.* A lyric poet of medieval Germany. Compare TROUBADOUR. [<G < *minne* love + *singer* singer]

Min·ne·so·ta (min′ə·sō′tə) A State in the north central United States bordering on Canada and Lake Superior; 84,682 square miles; capital, St. Paul; entered the Union May 11, 1858; nickname, *Gopher State*: abbr. *Minn.* — **Min′ne·so′tan** *n.* & *adj.*

Min·ne·wit (min′ə·wit), **Peter** See MINUIT, PETER.

Min·nie (min′ē) A feminine personal name. [<G, love]

min·now (min′ō) *n.* 1 A small European cyprinoid fish (*Phoxinus phoxinus*). 2 One of various other small fishes; especially, in the United States, a fish of the carp family. 3 The young of various fishes. Also **min′nie.** [OE *myne* small fish; prob. infl. in meaning by F *menu* small]

Mi·no·an (mi·nō′ən) *adj.* 1 Of or pertaining to an advanced Bronze Age civilization that flourished in Crete from about 3000 to 1100 B.C. 2 Denoting two types of linear script using pictographic and syllabic signs, found on clay tablets at Minos's palace at Cnossus, Nestor's palace at Pylos, and other Greek and Cretan sites, and dating from the 17th to 13th centuries B.C.: Linear A is thought to be Akkadian on the basis of a tentative decipherment made in 1957; Linear B, deciphered in 1952, is an Achaean dialect of Greek.

mi·nor (mī′nər) *adj.* 1 Less in number, quantity, or extent: opposed to *major.* 2 Of secondary consideration. 3 *Music* **a** Higher than the corresponding major interval by a semitone. **b** Characterized by minor intervals, scales, or tones. **c** In a minor key; sad; plaintive. 4 *Logic* Designating a minor term. — *n.* 1 One below the age when full civil and personal rights can be exercised. 2 *Logic* A minor term or minor premise. See SYLLOGISM. 3 *Music* A minor key, interval, etc. 4 Hence, a pathetic or plaintive quality, as in literature or art. 5 *U.S.* A branch of study for degree candidates in colleges and universities, usually related to but requiring less time than a major; a secondary subject of study. 6 In English public schools, the younger of two namesakes; junior. ◆ Homophone: *miner.* [<L]

Mi·nor·ca (mi·nôr′kə) *n.* One of a breed of domestic fowls resembling Leghorns. [from *Minorca*]

Mi·nor·ca (mi·nôr′kə) Second largest of the Balearic islands; 264 square miles.

Mi·nor·can (mi·nôr′kən) *n.* 1 A native or inhabitant of Minorca. 2 A descendant of a group of 1500 Minorcans who migrated to Florida in 1769. — *adj.* Of or pertaining to Minorca or its people.

Mi·nor·ite (mī′nə·rīt) *n.* A Franciscan. Also

Mi′nor·ist. [<Med. L *(Fratres) Minores* Lesser (Brethren), original name of the order]
mi·nor·i·ty (mə·nôr′ə·tē, -nor′-, mī-) *n. pl.* **·ties 1** The smaller in number of two parts or parties: opposed to *majority.* **2** The state of being a minor; legal infancy. **3** A racial, religious, political, or national group smaller than and opposed to a larger group of which it is a part, as in voting. [<F *minorité*]
minor key *Music* A key or mode based on the minor scale and characterized by the use of the minor third, and producing a plaintive or mournful effect. See THIRD *n.*
minor league Any professional sports league not having the standing of a major league. **—mi′nor–league′** *adj.*
minor mode See under MODE.
minor scale See under SCALE.
minor suit In bridge, diamonds or clubs.
minor term *Logic* The subject of both the minor premise and the conclusion of a syllogism.
Mi·nos (mī′nəs, -nos) In Greek mythology, a king of Crete, son of Zeus and husband of Pasiphae, who became a judge of the lower world after his death. There is also historical evidence that Minos was an actual king of Crete. [<Gk. *Minōs*]
Mi·not (mī′nət), **George Richards,** 1885–1950, U.S. physician.
Min·o·taur (min′ə·tôr) In Greek mythology, the offspring, half man and half bull, of Pasiphae and a white bull sent by Poseidon to Minos, who confined it in the Labyrinth where it was annually fed human flesh until killed by Theseus. See ANDROGEUS. [<L *Minotaurus* <Gk. *Minōtauros* <*Minōs* Minos + *tauros* bull]
Minsk (mensk) The capital of Belorussian S.S.R. in western U.S.S.R.
min·ster (min′stər) *n.* A monastery church; in Great Britain, often a cathedral: often used in combination: *Axminster.* [OE *mynster* <LL *monasterium.* Doublet of MONASTERY.]
min·strel (min′strəl) *n.* **1** Originally, in the Middle Ages, a retainer whose business it was to play musical instruments for the entertainment of his lord. **2** A wandering musician who composed and sang to the harp; one of a class of vagrant musicians and mountebanks, repressed by Henry IV. **3** A performer in a minstrel show. **4** *Poetic* A poet; singer; musician. See synonyms under POET. [<OF *menestrel* <LL *ministerialis* servant, jester <L *minister* attendant]
minstrel show A comic variety show of songs, dances, jokes, etc., given by a company of performers in blackface.
min·strel·sy (min′strəl·sē) *n. pl.* **·sies 1** The occupation of a minstrel. **2** Ballads or lyrics collectively. **3** A troupe of minstrels. [<AF *menestralcie,* OF *menestralsie*]
mint[1] (mint) *n.* **1** A place where the coin of a country is manufactured. **2** Figuratively, an abundant supply of anything: used especially of money. **3** Figuratively, the source of a fabrication or invention. **—***v.t.* **1** To make (money) by stamping; coin. **2** To invent or fabricate, as a word. **—***adj.* Unused; in original condition: a *mint* stamp. [OE *mynet* coin <L *moneta* mint <*Moneta* epithet of Juno, whose temple at Rome was used as a mint. Doublet of MONEY.] **—mint′er** *n.*
mint[2] (mint) *n.* Any one of several aromatic herbs of the genus *Mentha,* of the family *Labiatae*; especially, spearmint and peppermint, used medicinally. [OE *minte* <L *menta, mentha* <Gk. *minthа*]
mint[3] (mint) *v.t. Scot.* To hint; intend.
mint·age (min′tij) *n.* **1** The act of minting, or that which is minted; coinage; also, figuratively, the act of fabricating or coining. **2** The duty paid for coining. **3** The authorized impression placed upon a coin.
mint julep A drink made of brandy or whisky mixed with crushed ice and sugar and flavored with fresh mint.
min·u·end (min′yōō·end) *n.* The number from which another is to be subtracted: opposed to *subtrahend.* [<L *minuendus* to be lessened, gerundive of *minuere* lessen]
min·u·et (min′yōō·et′) *n.* **1** A slow, stately dance for couples in triple measure: introduced in France in the 17th century. **2** A musical composition suited to this dance: often as a movement in a sonata or symphony. [<F *menuet,* dim. of *menu* small <L *minutus*]
Min·u·it (min′yōō·it), **Peter,** 1580?–1641, third governor of New Netherlands (New York): originally known as *Peter Minnewit.*
mi·nus (mī′nəs) *prep.* **1** Lessened, or requiring to be lessened by; less: 10 *minus* 5. **2** *Colloq.* Deprived of; lacking: *minus* a hat. **—***adj.* **1** Lying or reckoned in that one of two opposite directions arbitrarily assumed as negative. **2** Negative: A debt may be treated as a *minus* quantity. **—***n.* **1** A minus sign. **2** A minus quantity. [<L, neut. of *minor*]
mi·nus·cule (mi·nus′kyōōl) *n.* **1** A semi-uncial cursive script, developed by the monks out of the uncial in the 7th–9th centuries and forming the basis of the modern small Roman and Greek letters; hence, any small or lower-case letter. **2** Any very small thing. **—***adj.* **1** Of, pertaining to, like, or composed of minuscules. **2** Very small; miniature. [<L *minusculus,* dim. of *minor* less]
minus sign A sign (—) denoting subtraction, or reckoning or measuring in the negative direction.
min·ute[1] (min′it) *n.* **1** The 60th part of an hour; also, a moment; hence, any short, indefinite period of time. **2** The 60th part of a degree: a unit of angular measure indicated by the sign (′), and called a *minute* of arc. **3** A brief note or summary in writing of something to be remembered; memorandum; specifically, in the plural, an official record of the proceedings of any deliberative body. **—***v.t.* **·ut·ed, ·ut·ing 1** To make a minute or brief note of; record. **2** To time to the minute. [<F <Med.L *minuta (pars)* small (part), minute <L *minutus* small]
mi·nute[2] (mī·nōōt′, -nyōōt′, mi-) *adj.* **1** Exceedingly small; hence, unimportant; trifling. **2** Attending to small things; critically careful; very exact. [<L *minutus* small, little, orig. pp. of *minuere* lessen] **—mi·nute′ness** *n.*
Synonyms: circumstantial, comminuted, critical, detailed, diminutive, exact, fine, little, particular, precise, slender, tiny. That is *minute* which is of exceedingly limited dimensions, as a grain of dust. That which is broken up into *minute* particles is said to be *comminuted*; things may be termed *fine* which would not be termed *comminuted*; as, *fine* sand; *fine* gravel; but, in using an adverb, we say a substance is finely *comminuted.* An account extended to very *minute* particulars is *circumstantial, detailed, particular*; and examination so extended is *critical, exact, precise.* See FINE, LITTLE, PRECISE, SMALL. *Antonyms:* see synonyms for LARGE.
min·ute gun (min′it) A gun fired at intervals of a minute as a sign of distress, or of mourning as at a funeral.
min·ute hand (min′it) The hand of a timepiece that marks the minutes.
min·ute·ly[1] (min′it·lē) *adj. & adv.* At intervals of a minute.
mi·nute·ly[2] (mī·nōōt′lē, -nyōōt′-, mi-) *adv.* In a minute manner; very finely, closely, or exactly.
min·ute·man (min′it·man′) *n. pl.* **·men** (-men′) A man ready for service at a minute's notice: specifically applied to certain militiamen and armed citizens in the American Revolution.
min·u·the·sis (min′yə·thē′sis) *n. Physiol.* Diminished response of a sense organ due to continuous stimulation. [<LL <Gk. *minythēsis* a lessening <*minytheein* decrease]
mi·nu·ti·a (mi·nōō′shē·ə, -shə, -nyōō′-) *n. pl.* **·ti·ae** (-shi·ē) *Chiefly pl.* Small or unimportant details. [<L]
minx (mingks) *n.* A saucy, forward girl or woman. [Prob. <LG *minsk* impudent woman]
mi·o·car·di·a (mī′ə·kär′dē·ə) *n. Physiol.* Contraction of the heart; systole. [<NL <Gk. *meiōn* less + *kardia* heart]
Mi·o·cene (mī′ə·sēn) *adj.* Pertaining to or designating a geological epoch near the end of the Tertiary period, associated with a great development of modern mammals: also **Mi′o·cen′ic** (-sen′ik), See chart under GEOLOGY. **—***n.* The Miocene epoch or series. [<Gk. *meiōn* less + *kainos* recent]
Mi·o·lith·ic (mī′ə·lith′ik) *adj. Anthropol.* Mesolithic. [<Gk. *meiōn* less + -LITH[1] + -IC]
mi·o·sis (mī·ō′sis) *n.* **1** *Pathol.* **a** The period in the course of a disease when the symptoms begin to diminish. **b** Excessive contraction of the pupil of the eye: also spelled *myosis.* **2** Meiosis. [<Gk. *myein* close + -OSIS] **—mi·ot′ic** (-ot′ik) *adj.*
mi·o·ther·mic (mī′ə·thûr′mik) *adj.* Of or pertaining to those temperature conditions now prevailing on earth, especially as compared with past geological periods. [<Gk. *meiōn* less + *thermē* heat]
miq·ue·let (mik′ə·let) *n.* **1** A Spanish infantryman detailed for escort duty. **2** *Obs.* A partisan; a bandit of the Pyrenees. [<F <Sp. *miquelete*]
Mi·que·lon (mik′ə·lon, *Fr.* mē·kə·lôn′) An island off southern Newfoundland; 83 square miles. See under SAINT PIERRE.
mir (mir) *n.* A Russian local community, with common land apportioned by lot. [<Russian]
Mi·ra·beau (mē·rà·bō′), **Comte de,** 1749–91, Gabriel Honoré de Riquetti, French Revolutionary orator, writer, and statesman.
mi·ra·bi·le dic·tu (mi·rab′ə·lē dik′tōō, -tyōō) *Latin* Wonderful to relate.
mir·a·bil·i·a (mir′ə·bil′ē·ə) *n. pl. Latin* Miracles; wonders.
mir·a·cle (mir′ə·kəl) *n.* **1** Any wonderful or amazing thing, fact, or event; a wonder. **2** An event in the natural world, but out of its established order, possible only by the intervention of divine power. **3** A medieval dramatic representation of religious subjects: also **miracle play:** see MYSTERY[1] (def. 10). See synonyms under PRODIGY. [<F <L *miraculum* <*mirari* wonder <*mirus* wonderful]
mi·rac·u·lous (mi·rak′yə·ləs) *adj.* **1** Effected by direct divine agency; supernatural. **2** Surpassingly strange; wonderful. **3** Possessing the power to work miracles; wonder-working. See synonyms under SUPERNATURAL. [<F *miraculeux* <Med. L *miraculosus* <L *miraculum*] **—mi·rac′u·lous·ly** *adv.* **—mi·rac′u·lous·ness** *n.*
mir·a·dor (mir′ə·dôr′, -dōr′-) *n. Archit.* A balcony or oriel window. [<Sp. <*mirar* behold <L *mirari* wonder]
Mir·a·flo·res Lake (mir′ə·flôr′əs, -flō′rəs) An artificial lake in southern Canal Zone, linking Gaillard Cut with the Pacific section through **Miraflores Locks.**
mi·rage (mi·räzh′) *n.* **1** An optical illusion, as of an oasis or a sheet of water in the desert, or ships seen inverted in the air. It occurs when the lower strata of air are at a very different temperature from the higher strata, so that images are seen as by reflection. **2** Anything that falsely appears to be real. [<F <*se mirer* be reflected, look at oneself <L *mirari* wonder at]
Mi·ran·da (mi·ran′də) A feminine personal name. [<L, admirable]
—Miranda In Shakespeare's *Tempest,* the heroine; daughter of Prospero.
mir·bane oil (mûr′bān) Nitrobenzene. [Prob. a fanciful formation]
mire (mīr) *n.* Wet, yielding earth; swampy ground; deep mud or slush. **—***v.* **mired, mir·ing** *v.t.* **1** To cause to sink or become stuck in or as in mire. **2** To defile; soil. **—***v.i.* **3** To sink in mire; bog down. [<ON *mýrr* swampy ground] **—mir′y** *adj.* **—mir′i·ness** *n.*
Mir·i·am (mir′ē·əm) A feminine personal name. [<Hebrew, bitter]
—Miriam Sister of Moses and Aaron. *Ex.* xv 20.
Mi·rim (mē·rēn′) A tidewater lagoon in southern Brazil and eastern Uruguay; 110 miles long; 1,145 square miles. Spanish **Me·rín** (mā·rēn′).
mirk[1] (mûrk), **mirk·i·ly,** etc. See MURK, etc.
mirk[2] (mûrk) *Scot. adj.* Dark, gloomy. **—***n.* Darkness.
mir·mil·lon (mir·mil′ən) *n.* One of a class of ancient Roman gladiators armed with sword and shield and characterized by the figure of a fish on the helmet. See illustration under GLADIATOR. [<L *mirmillo* <Gk. *mormyros* sea fish]
Mi·ró (mē·rō′), **Joan,** born 1893, Spanish painter.
mir·ror (mir′ər) *n.* **1** An object having a nearly perfect reflecting surface; a looking-glass. **2** *Optics* A speculum. **3** Whatever reflects or clearly represents; an exemplar. **4** A crystal used by diviners. **—***v.t.* To reflect or show an image of, as in a mirror. [<OF *mirour* <LL *mirare* look at <L *mirari* wonder at, admire]
mir·ror·scope (mir′ər·skōp) *n.* **1** A mirror used to reflect a design so as to permit of rapid reproduction, as by artists. **2** A projector (def. 3).

mir·ror·stone (mir′ər·stōn′) n. Mica, especially of the muscovite variety.

mirth (mûrth) n. 1 Pleasurable feelings, or gaiety of spirits, manifested in jesting and laughter; social merriment; jollity. 2 *Obs.* Pleasure; joy. See synonyms under HAPPINESS, LAUGHTER, SPORT. [OE *myrth, myrgth* < *myrig* pleasant, merry]

mirth·ful (mûrth′fəl) adj. Merry. See synonyms under CHEERFUL, HAPPY, MERRY, VIVACIOUS. — **mirth′ful·ly** adv. — **mirth′ful·ness** n.

mirth·less (mûrth′lis) adj. Lacking mirth or merriment; joyless. — **mirth′less·ly** adv. — **mirth′less·ness** n.

mir·za (mir′zä) n. A Persian title, placed before a name to denote a scholar and after a name to denote a prince. [< Persian, contraction of *mīrzādah* < *mīr* prince (< Arabic *amīr* ruler) + *zādah* son of]

mis-[1] *prefix* Bad; amiss; wrongly. [OE *mis-* wrong; infl. in meaning by ME *mes-* MIS-[2]] *Mis-* may appear as a prefix in hyphemes or solidemes; for examples, see the list of self-explanatory words at the foot of the page.

mis-[2] *prefix* Bad; amiss; not: found with negative or depreciatory force in words borrowed from Old French; *misadventure, miscreant.* [< OF *mes-* < L *minus* less]

mis-[3] Var. of MISO-.

mis·ad·ven·ture (mis′əd·ven′chər) n. An unlucky chance; misfortune. See synonyms under ACCIDENT, MISFORTUNE. [< OF *mesaventure*]

mis·ad·ven·tured (mis′əd·ven′chərd) adj. *Obs.* Unfortunate.

mis·al·li·ance (mis′ə·lī′əns) n. An undesirable alliance, as marriage with one of inferior station or character. [< F *mésalliance*]

mis·an·thrope (mis′ən·thrōp, miz′-) n. One who entertains aversion to or distrust of his fellow men. Also **mis·an·thro·pist** (mis·an′thrə·pist). [< Gk. *misanthrōpos* hating mankind < *misein* hate + *anthrōpos* a man] — **mis′an·throp′ic** (-throp′ik) or **·i·cal** adj. — **mis′an·throp′i·cal·ly** adv.

mis·an·thro·py (mis·an′thrə·pē) n. Hatred or distrust of mankind.

mis·ap·ply (mis′ə·plī′) v.t. ·plied, ·ply·ing 1 To use or apply incorrectly or inefficiently. 2 To use or apply wrongfully or dishonestly.

mis·ap·pro·pri·ate (mis′ə·prō′prē·āt) v.t. ·at·ed, ·at·ing To use or take improperly or dishonestly; misapply. — **mis′ap·pro·pri·a′tion** n.

mis·be·come (mis′bi·kum′) v.t. ·came, ·come, ·com·ing To be unbecoming or not befitting to.

mis·brand (mis·brand′) v.t. To label or brand incorrectly or falsely.

mis·call (mis·kôl′) v.t. 1 To call by a wrong name; misname. 2 *Brit. Dial.* To revile; abuse.

mis·car·riage (mis·kar′ij) n. 1 *Med.* A premature delivery of a non-viable fetus; abortion. 2 Failure to reach or to bring to an expected result, destination, or conclusion.

mis·car·ry (mis·kar′ē) v.i. ·ried, ·ry·ing 1 To fail of an intended effect; go wrong. 2 To bring forth a fetus prematurely; have a miscarriage; abort.

mis·ce·ge·na·tion (mis′i·jə·nā′shən) n. *Biol.* Interbreeding of races, especially intermarriage or interbreeding between white and Negro or white and Oriental races. [< L *miscere* mix + *genus* race] — **mis′ce·ge·net′ic** (-jə·net′ik) adj.

mis·cel·la·ne·a (mis′ə·lā′nē·ə) n. pl. A miscellaneous collection; especially, literary miscellanies. [< L]

mis·cel·la·ne·ous (mis′ə·lā′nē·əs) adj. Consisting of several kinds; variously mixed; also, many-sided; promiscuous; varied. See synonyms under HETEROGENEOUS. [< L *miscellaneus* < *miscellus* mixed < *miscere* mix] — **mis′cel·la′ne·ous·ly** adv. — **mis′cel·la′ne·ous·ness** n.

mis·cel·la·nist (mis′ə·lā′nist, -lə·nist) n. A composer of miscellanies.

mis·cel·la·ny (mis′ə·lā′nē) n. pl. ·nies 1 *Often pl.* A collection of literary compositions on various subjects. 2 Any miscellaneous collection. [Anglicized var. of MISCELLANEA]

mis·chance (mis·chans′, -chäns′) n. An instance of ill-luck; a mishap. See synonyms under CATASTROPHE, MISFORTUNE. [< OF *mescheance*]

mis·chief (mis′chif) n. 1 Any occurrence attended with evil or injury; troublesome or damaging action or its result; damage; vexation. 2 Any annoying or vexatious action or course of conduct; a prank; also, the spirit or mood leading to such acts. 3 A prankish person. See synonyms under INJURY. [< OF *meschef* bad result < *meschever* come to grief < *mes-* mis- (< L *minus* less) + *chief* head, end < L *caput* head]

mis·chief-mak·er (mis′chif-mā′kər) n. One who causes mischief, especially by instigating ill-feeling between others. — **mis′chief-mak′ing** adj. & n.

mis·chie·vous (mis′chi·vəs) adj. 1 Inclined to mischief; of a prankish nature. 2 Injurious. See synonyms under BAD, MALICIOUS, NOISOME, PERNICIOUS. — **mis′chie·vous·ly** adv. — **mis′chie·vous·ness** n. [< AF *meschevous*]

mis·ci·ble (mis′i·bəl) adj. 1 Capable of being mixed. 2 Suitable for mixing. [< L *miscere* mix] — **mis′ci·bil′i·ty** n.

mis·col·late (mis′kə·lāt′) v.t. ·lat·ed, ·lat·ing In bookbinding, to assemble (sheets or signatures) in wrong sequence. — **mis′col·la′tion** n.

mis·col·or (mis·kul′ər) v.t. 1 To give a wrong color to. 2 To misrepresent. Also *Brit.* **mis·col′our.**

mis·con·ceive (mis′kən·sēv′) v.t. & v.i. ·ceived, ·ceiv·ing To conceive wrongly; misunderstand. — **mis′con·ceiv′er** n. — **mis′con·cep′tion** (-sep′shən) n.

mis·con·duct (mis′kən·dukt′) v.t. 1 To behave (oneself) improperly. 2 To mismanage. — n. (mis·kon′dukt) 1 Improper conduct; bad behavior. 2 *Law* Adultery. 3 Mismanagement.

mis·con·stru·al (mis′kən·stroo′əl) n. 1 An erroneous interpretation. 2 The act of putting a false meaning upon something said or done by another. 3 Misunderstanding or the result of a misunderstanding.

mis·con·strue (mis′kən·stroo′) v.t. ·strued, ·stru·ing To interpret erroneously; put a false or unwarranted meaning to; misunderstand. — **mis′con·struc′tion** (-struk′shən) n.

mis·cre·ance (mis′krē·əns) n. *Archaic* The condition or quality of adhering to a false faith; heresy. [< OF *mescreant* disbelieving, ppr. of *mescroire* < *mes-* mis- (< L *minus* less) + *croire* believe < L *credere*]

mis·cre·an·cy (mis′krē·ən·sē) n. *Archaic* 1 The condition or act of a miscreant; villainy. 2 Miscreance.

mis·cre·ant (mis′krē·ənt) n. 1 A vile wretch; an evil-doer. 2 *Archaic* An unbeliever; infidel. — adj. 1 Villainous; conscienceless. 2 *Archaic* Unbelieving; infidel. [< OF *mescreant* unbelieving]

mis·cue (mis·kyoo′) n. In billiards, a stroke spoiled in effect by a slipping of the cue. — v.i. ·cued, ·cu·ing 1 To make a miscue. 2 In the theater, to miss one's cue; to answer another's cue.

mis·deed (mis·dēd′) n. A wrong or improper act. See synonyms under OFFENSE, SIN. [OE *misdǣd*]

mis·de·mean·ant (mis′di·mē′nənt) n. One who is guilty of a misdemeanor or misconduct. — **first-class misdemeanant** In English law, one of a class of prisoners guilty of misdemeanor, but not subjected to the same prison regulations as a criminal, nor considered as a person convicted of a crime.

mis·de·mean·or (mis′di·mē′nər) n. 1 Misbehavior; a misdeed. 2 *Law* Any offense less than a felony. In England the distinction between a felony and a misdemeanor is still maintained. In the United States this distinction has, in most States, either been abolished or is treated in a manner that makes it of no practical value. Compare FELONY. Also *Brit.* **mis′de·mean′our.** See synonyms under OFFENSE.

mis·do (mis·doo′) v.t. & v.i. ·did, ·done, ·do·ing To do wrongly; bungle. [OE *misdon*] — **mis·do′er** n. — **mis·do′ing** n.

mis·doubt (mis·dout′) v.t. *Archaic* 1 To doubt; call in question. 2 To fear; suspect. — v.i. 3 To be in doubt; suspect. — n. 1 Doubt; wavering; irresolution. 2 Suspicion; apprehension.

mise (mīz) n. 1 *Law* The issue pleaded in a writ of right. 2 *Law* Expenses; specifically, the costs and charges in an action. 3 The adjustment of a dispute by arbitration or compromise. [< AF < OF *mis* put, laid out, pp. of *mettre* < L *mittere* send]

mise en scène (mēz än sen′) *French* The setting of a play on the stage; hence, general or visible surroundings.

Mi·se·no (mē·zā′nō), **Cape** A promontory in southern Italy between the Bay of Naples and the Gulf of Gaeta. Ancient **Mi·se·num** (mī·sē′nəm).

mi·ser (mī′zər) n. One who saves and hoards avariciously. [< L *miser* wretched]

mis·er·a·ble (miz′ər·ə·bəl, miz′rə-) adj. 1 Wretchedly unhappy. 2 Of mean quality; bad; valueless: sometimes expressing contempt. 3 Producing, proceeding from, or exhibiting misery; pitiable: *miserable* weather; a *miserable* groan. See synonyms under PITIFUL, SAD. 4 *Obs.* A miserable person. [< OF < L *miserabilis* pitiable < *miserari* pity < *miser* wretched] — **mis′er·a·ble·ness** n. — **mis′er·a·bly** adv.

mis·e·re·re (miz′ə·râr′ē, -rir′ē) n. In church stalls, a small wooden bracket affixed perpendicularly to the bottom of the seat and designed to afford support to the worshiper when the seat is turned up: also called *misericorde.*

Mis·e·re·re (miz′ə·râr′ē, -rir′ē) n. 1 The 51st psalm, so called from the opening words in the Latin version. 2 A musical setting of this psalm. [< L, imperative of *miserere* have mercy]

mis·er·i·corde (miz′ər·i·kôrd′, mi·zer′i·kôrd) n. 1 A small dagger used in the Middle Ages to give the death stroke to a fallen knight. 2 Formerly, a dispensation from fasting given to a member of a monastic order. 3 An apartment in a monastery serving as a refectory for monks who had received such dispensation. 4 Miserere. Also **mis′er·i·cord′.** [< OF < L *misericordia* < *misereri* have pity + *cor, cordis* heart]

mis·e·ri·cor·di·a (miz′ə·ri·kôr′dē·ə) *Latin* Pity; compassion; mercy.

mi·ser·ly (mī′zər·lē) adj. Of or like a miser; grasping; mean. See synonyms under AVARICIOUS. — **mi′ser·li·ness** n.

mis·er·y (miz′ər·ē) n. pl. ·er·ies 1 Extreme distress or suffering, especially as a result of poverty; wretchedness; also, a cause of wretchedness. 2 *Dial.* A cause of pain: a

misaccent	misadvise	misanalyze	misassay	misbandage	mischaracterization	miscoinage
misaccentuation	misaffection	misanswer	misassent	misbegin	mischaracterize	miscollocation
misachievement	misaffirm	misapparel	misassert	misbelove	mischarge	miscommand
misacknowledge	misagent	misappear	misassign	misbestow	mischristen	miscommit
misact	misaim	misappearance	misassociate	misbetide	miscipher	miscommunicate
misadapt	misalienate	misappellation	misassociation	misbias	miscitation	miscompare
misadaptation	misalinement	misappoint	misatone	misbill	miscite	miscomplain
misadd	misallegation	misappointment	misattribute	misbind	misclaim	miscomplaint
misaddress	misallege	misappraise	misauthorization	misbuild	misclaiming	miscompose
misadjust	misallotment	misappraisement	misauthorize	miscanonize	misclass	miscomputation
misadmeasurement	misallowance	misapprehensible	misauthorize	miscensure	misclassification	miscompute
misadministration	misalphabetize	misascribe	misaver	miscenter	misclassify	miscon
misadvice	misalter	misascription	misaward	mischallenge	miscoin	misconclusion

misery in the back. See synonyms under MISFORTUNE, PAIN. [<OF *miserie* <L *miseria* < *miser* wretched]

mis·fea·sance (mis·fē′zəns) *n. Law* The performance of a lawful act in an unlawful or culpably negligent manner. Compare MALFEASANCE, NONFEASANCE. [<OF *mesfaisance* < *mesfaire* do wrong < *mes-* mis- + *faire* do <L *facere*] — **mis·fea′sor** *n.*

mis·fire (mis·fir′) *n.* The failure to discharge or explode when desired: said of a firearm, explosive, or internal-combustion engine. — *v.i.* **·fired, ·fir·ing** To fail to explode or be fired.

mis·fit (mis·fit′) *v.t. & v.i.* **·fit·ted, ·fit·ting** To fail to fit or make fit. — *n.* **1** Something that fits badly. **2** (mis′fit′) A person who does not adjust or adapt himself readily to his surroundings. **3** The act or condition of fitting badly.

mis·for·tune (mis·fôr′chən) *n.* **1** Adverse or ill fortune. **2** An unlucky chance or occurrence; calamity.

Synonyms: adversity, affliction, bereavement, blow, calamity, chastening, chastisement, disappointment, disaster, distress, failure, hardship, harm, ill, misadventure, mischance, misery, mishap, reverse, ruin, sorrow, stroke, trial, tribulation, trouble, visitation. *Misfortune* is usually of lingering character or consequences, and such as the sufferer is not deemed directly responsible for; as, He had the *misfortune* to be born blind. Any considerable *disappointment, failure,* or *misfortune,* as regards outward circumstances, as loss of fortune, position, and the like, when long continued or attended with enduring consequences, constitutes *adversity*. For the loss of friends by death we commonly use *affliction* or *bereavement*. *Calamity* and *disaster* are used of sudden and severe *misfortunes*. We speak of the *misery* of the poor, the *hardships* of the soldier. *Affliction, chastening, trial,* and *tribulation* all suggest some disciplinary purpose of God with beneficent design. *Affliction* may be keen and bitter, but brief; *tribulation* is long and wearing. Compare ACCIDENT, ADVERSITY, BLOW, CATASTROPHE, LOSS. *Antonyms*: blessing, boon, comfort, consolation, gratification, happiness, joy, pleasure, prosperity, relief, success, triumph.

mis·give (mis·giv′) *v.* **·gave, ·giv·ing** *v.t.* To make fearful, suspicious, or doubtful: My heart *misgives* me. — *v.i.* To be apprehensive.

mis·giv·ing (mis·giv′ing) *n.* A feeling of doubt, premonition, or apprehension. See synonyms under ANXIETY, DOUBT, FEAR.

mis·guide (mis·gid′) *v.t.* **·guid·ed, ·guid·ing** To guide wrongly in action or thought; mislead. — **mis·guid′ance** *n.* — **mis·guid′er** *n.*

mi·shan·ter (mi·shan′tər) *n. Scot.* Misfortune; ill-luck; misadventure.

mis·hap (mis′hap, mis·hap′) *n.* An unfortunate accident; slight misfortune. See synonyms under ACCIDENT, CATASTROPHE, MISFORTUNE. [<MIS-[1] + HAP]

mish·mash (mish′mash′, -mosh′) *n.* A medley; hotch-potch. — *v.t.* To make a hotch-potch of; jumble. Also **mish′-mash′**. [Reduplication of MASH]

Mish·na (mish′nə) *n. pl.* **Mish·na·yoth** (mish′·nä·yōth′) **1** The first part of the Talmud, consisting of a collection of traditions and decisions made by Rabbi Juda, called the Holy (born about A.D. 150), summing up all previous rabbinical labors. **2** A paragraph of this collection. **3** The opinion of any notable expounder of the Jewish law; also, the sum and substance of his teachings. Also **Mish′nah.** [<Hebrew *mishnāh* repetition, oral law < *shānāh* repeat, teach] — **Mish·na·ic** (mish·nā′ik), **Mish′nic** or **·ni·cal** *adj.*

mis·in·form (mis′in·fôrm′) *v.t.* To give false or erroneous information to. — **mis′in·for·ma′tion** *n.* — **mis′in·form′er, mis′in·form′ant** *n.*

mis·in·ter·pret (mis′in·tûr′prit) *v.t.* To interpret wrongly; misunderstand. — **mis′in·ter′pre·ta′tion** *n.* — **mis′in·ter′pret·er** *n.*

Mi·sio·nes (mē·syō′nās) A national territory in NE Argentina; 11,514 square miles; capital, Posadas.

mis·join·der (mis·join′dər) *n. Law* The uniting of things or persons that should not be united: the *misjoinder* of parties in action: contrasted with *non-joinder*.

mis·kal (mis·käl′) *n.* An Oriental weight equivalent to 4.7 grams in Persia, and to 4.8 grams in Turkey. [<Arabic *mithqāl* weight]

Mis·kolc (mēsh′kōlts) A city in NE Hungary. Formerly **Mis′kolcz.**

mis·lay (mis·lā′) *v.t.* **·laid, ·lay·ing** To lay in a place not remembered; misplace. See synonyms under DISPLACE. — **mis·lay′er** *n.*

mis·lead (mis·lēd′) *v.t.* **·led, ·lead·ing** To direct wrongly; lead astray or into error. — **mis·lead′er** *n.* — **mis·lead′ing** *adj.* — **mis·lead′·ing·ly** *adv.*

mis·leared (mis·lird′) *adj. Scot.* Ill-bred or ill-taught.

mis·like (mis·lik′) *v.t.* **1** To dislike. **2** To displease. — *n.* Dislike; aversion; disapproval. — **mis·lik′er** *n.* — **mis·lik′ing** *n.* — **mis·lik′·ing·ly** *adv.*

mis·man·age (mis·man′ij) *v.t. & v.i.* **·aged, ·ag·ing** To manage badly or improperly. — **mis·man′age·ment** *n.* — **mis·man′ag·er** *n.*

mis·mar·riage (mis·mar′ij) *n.* An unhappy, incongruous, or inharmonious marriage.

mis·no·mer (mis·nō′mər) *n.* **1** A name wrongly applied; an inapplicable designation. **2** A misnaming; specifically, the giving of a wrong name to a person in a legal document. [<AF <OF *mesnommer* call by the wrong name < *mes-* wrongly + *nomer* <L *nominare* name]

miso- *combining form* Hating: *misogynist*. Also, before vowels, *mis-*. [<Gk. *misein* hate]

mis·og·a·my (mis·og′ə·mē) *n.* Hatred of marriage. [<MISO- + -GAMY] — **mis·og′a·mist** *n.*

mis·og·y·ny (mis·oj′ə·nē) *n.* Hatred of woman: opposed to *philogyny*. [<Gk. *misogynia* < *misein* hate + *gynē* woman] — **mis·og′y·nist** *n.* — **mis·og′y·nous** *adj.*

mis·ol·o·gy (mis·ol′ə·jē) *n.* Hatred of discussion or inquiry; aversion to enlightenment. [<Gk. *misologia* < *misein* hate + *logos* discourse] — **mis·ol′o·gist** *n.*

mis·o·ne·ism (mis′ō·nē′iz·əm, mī′sō-) *n.* Hatred of change, innovation or novelty. [<MISO- + Gk. *neos* new] — **mis′o·ne′ist** *n.*

mis·pick·el (mis′pik·əl) *n.* Arsenopyrite. [<G; ult. origin unknown]

mis·play (mis·plā′) *n.* In games, a wrong play; hence, any bad move.

mis·plead (mis·plēd′) *v.t. & v.i.* **·plead·ed** or **·pled, ·plead·ing** *Law* To plead erroneously.

mis·print (mis·print′) *v.t.* To print incorrectly. — *n.* (mis′print′, mis·print′) An error in printing.

mis·pri·sion (mis·prizh′ən) *n.* **1** *Law* Concealment of a crime, especially of treason or felony; also, loosely, contempt or high misdemeanor. **2** *Archaic* Misconception; misunderstanding. [<OF *mesprision* mistake < *mesprendre* do wrong, take amiss < *mes-* mis- + *prendre* take <L *prehendere*]

mis·rep·re·sent (mis′rep·ri·zent′) *v.t.* To give an incorrect or false representation of. See synonyms under PERVERT. — **mis′rep·re·sen·ta′tion** *n.* — **mis′rep·re·sen′ta·tive** *adj. & n.*

mis·rule (mis·rōōl′) *v.t.* **·ruled, ·rul·ing** To rule unwisely or unjustly; misgovern. — *n.* **1** Bad or unjust rule or government. **2** Disorder or confusion.

miss[1] (mis) *n.* **1** A young girl: chiefly colloquial, or in trade use: clothing for *misses*. **2** *Often cap.* A title used in speaking to an unmarried woman or girl: used without the name. [Contraction from MISTRESS]

miss[2] (mis) *v.t.* **1** To fail to hit or strike. **2** To fail to meet, catch, obtain, accomplish, see, hear, perceive, etc.: to *miss* the point. **3** To fail to attend, keep, perform, etc.: to *miss* church. **4** To overlook or fail to take advantage of, as an opportunity. **5** To discover or feel the loss or absence of. **6** To escape; avoid: He just *missed* being wounded. — *v.i.* **7** To fail to hit; strike wide of the mark. **8** To be unsuccessful; fail. — **to miss fire 1** To fail to discharge: said of firearms. **2** To be unsuccessful; fail. — *n.* **1** The act of missing; a failure to hit, find, attain, succeed, etc. **2** *Obs.* Loss; want. [OE *missan*]

Miss (mis) *n.* A title of address used with the name of a girl or an unmarried woman.

mis·sal (mis′əl) *n.* **1** The book containing the service for the celebration of mass throughout the year; a mass book. **2** Loosely, an illuminated black-letter or manuscript book of early date resembling the old mass books. ◆ Homophone: *missile*. [<Med. L *missale*, neut. of *missalis (liber)* mass (book) <LL *missa* mass]

mis·say (mis·sā′) *v.t. & v.i.* **·said, ·say·ing** *Archaic* To say (something) wrongly or incorrectly.

mis·sel thrush (mis′əl) A large European thrush (*Turdus viscivorus*) that feeds largely on mistletoe berries. Also **mis′sel.** [OE *mistel* mistletoe]

mis·shape (mis·shāp′) *v.t.* **·shaped, ·shaped** or **·shap·en, ·shap·ing** To shape badly; deform.

mis·shap·en (mis·shā′pən) *adj.* Shaped badly; deformed.

mis·sile (mis′əl) *n.* **1** Any object, especially a weapon, intended to be thrown or discharged; a projectile. **2** A guided missile. — *adj.* Such as may be thrown or hurled. ◆ Homophone: *missal*. [<L *missilis* < *missus*, pp. of *mittere* send]

mis·sile·man (mis′əl·mən) *n. pl.* **·men** (-mən) One who is skilled in the practical details of missilery, rocketry, and spacemanship.

mis·sile·ry (mis′əl·rē) *n.* The science and art of missiles, their design, construction, launching, range, and destructive capacities.

miss·ing (mis′ing) *adj.* Absent from the proper or accustomed place; lost; gone: said specifically of soldiers who are absent but whose fate has not been definitely ascertained. — **to turn up missing** To be absent; fail to arrive or be found.

missing link 1 Something lacking to make complete a chain or series. **2** *Biol.* A hypothetical form of life assumed to be intermediate in development between man and the anthropoid apes.

mis·sion (mish′ən) *n.* **1** The act of sending, or the state of being sent, as on some errand. **2** The sending forth of men with authority to preach or spread the gospel; authority so given by God or the church. **3** The business or service on which one is sent. **4** That which one is or feels destined to accomplish; the destined or chosen end of one's efforts. **5** An effort to spread, or the work of spreading, religious teaching. **6** A single field or locality covered by missionary work; the body of missionaries there established; a missionary station; also, any educational, religious, or welfare center for the underprivileged in a city. **7** *Eccl.* A regularly organized church and congregation not having the status of a parish in canon law; a quasi-parish. **8** The office of a foreign ambassador or envoy. **9** The persons sent on any errand or service. **10** *Mil.* A definite task assigned to an individual or unit of the armed services. **11** *Aeron.* A flight operation of a single aircraft or formation.

misconfer	**miscultivated**	**misdesire**	**misdraw**	**misenjoy**	**misexplain**	**misfond**
misconfident	**misculture**	**misdetermine**	**misdrive**	**misenrol**	**misexplanation**	**misform**
misconfiguration	**miscurvature**	**misdevise**	**miseat**	**misenter**	**misexplication**	**misformation**
misconjecture	**miscut**	**misdevoted**	**misecclesiastic**	**misentitle**	**misexposition**	**misframe**
misconjugate	**misdaub**	**misdevotion**	**misedit**	**misentry**	**misexpound**	**misgage**
misconjugation	**misdecide**	**misdiet**	**miseducate**	**misenunciation**	**misexpress**	**misgesture**
misconjunction	**misdecision**	**misdispose**	**miseducation**	**misevent**	**misexpression**	**misgraft**
misconsecrate	**misdeclaration**	**misdisposition**	**miseducative**	**misexample**	**misexpressive**	**misgrave**
misconsequence	**misdeclare**	**misdistinguish**	**miseffect**	**misexecute**	**misfashion**	**misground**
miscook	**misdefine**	**misdistribute**	**misencourage**	**misexecution**	**misfather**	**misgrow**
miscookery	**misdeliver**	**misdistribution**	**misendeavor**	**misexpectation**	**misfault**	**misgrown**
miscopy	**misdelivery**	**misdivide**	**misenforce**	**misexpend**	**misfield**	**misguess**
miscredulity	**misdentition**	**misdower**	**misengrave**	**misexpenditure**	**misfile**	**mis-hallowed**

missionary —*adj.* **1** Pertaining to or belonging to a mission. **2** Like the early Spanish architecture and simple furniture of the missions of the SW United States. — *v.t.* **1** To send on a mission. **2** To establish a mission in. [< L *missio, -onis* < *missus,* pp. of *mittere* send] — **mis′sion·al** *adj.*

mis·sion·ar·y (mish′ən·er′ē) *n. pl.* **·ar·ies 1** A person sent to propagate religion or to do educational or charitable work in some place where his church has no self-supporting local organization; hence, one who spreads any new system or doctrine. **2** A person sent on a mission; a messenger; ambassador. — *adj.* Pertaining to missions.

Missionary Baptist A Baptist who believes in or contributes to missionary work: opposed to *Primitive Baptist.*

Missionary Ridge A ridge of hills about 1,000 feet high in Tennessee and Georgia; a Civil War battleground, 1863.

Mis·sis·sip·pi (mis′ə·sip′ē) A State in the south central United States on the Gulf of Mexico; 47,716 square miles; capital, Jackson; entered the Union Dec. 10, 1817; nickname *Bayou State.* Abbr. *Miss.*

Mis·sis·sip·pi·an (mis′ə·sip′ē·ən) *adj.* **1** Relating to the Mississippi River or to the State. **2** *Geol.* Relating to the earliest of the two geological periods or systems in the American Carboniferous division of the Paleozoic era. See chart under GEOLOGY. — *n.* **1** One born or residing in Mississippi. **2** The Lower Carboniferous or Mississippian geological formation.

Mississippi River A river in the central United States, flowing 2,350 miles south to the Gulf of Mexico; from the headwaters of the Missouri River, flowing 3,892 miles to the Gulf of Mexico.

Mississippi scheme A plan formulated by the Scottish economist, John Law, in which a company was formed to exploit and monopolize all the land drained by the Mississippi, the Ohio, and the Missouri rivers. Subsequent mismanagement of minting rights which were granted the company led to wild speculation, panic, and inflation, causing the failure of the company in 1720. Also called **Mississippi bubble.**

Mississippi Sound An arm of the Gulf of Mexico bordering Louisiana, Mississippi, and Alabama.

mis·sive (mis′iv) *n.* That which is sent; especially a letter; a message in writing. — *adj.* Sent or designed to be sent. [< Med. L *missivus* < L *missus,* pp. of *mittere* send]

miss–lick (mis′lik) *n. Colloq.* A cut made by an ax or knife off the true line; hence, any mistake.

Mis·so·lon·ghi (mis′ə·lông′gē) A port in west central Greece; here Byron died, April 29, 1824, during the Greek struggle for independence from Turkish rule: Greek *Mesolonghi.*

Mis·sou·ri (mi·zŏor′ē) *n. pl.* **·ri** One of a tribe of North American Indians of the Siouan family, formerly inhabiting northern Missouri, now in Oklahoma with the Otoe.

Mis·sou·ri (mi·zŏor′ē, -zŏor′ə) A State in the west central United States west of the Mississippi; 69,674 square miles; capital, Jefferson City; entered the Union Aug. 10, 1821; nickname, *Ozark State* or *Show Me State;* abbr. *Mo.* — **to be from Missouri** To be on the alert against deception; be skeptical. — **Missou′ri·an** *adj. & n.*

Missouri River The longest river of the United States and chief tributary of the Mississippi, flowing 2,714 miles from the Rocky Mountains to the Mississippi River near St. Louis.

Missouri skylark A pipit of the North American plains (*Anthus spraguei*): also called *Sprague's pipit.*

Missouri Synod A leading conservative Lutheran body in the Lutheran Synodical Conference, first organized in 1847 and characterized by a strict adherence to traditional Lutheran doctrines: officially called *The Lutheran Church–Missouri Synod.*

mis·spell (mis·spel′) *v.t. & v.i.* **·spelled** or **·spelt, ·spell·ing** To spell incorrectly.

mis·state (mis·stāt′) *v.t.* **·stat·ed, ·stat·ing** To state wrongly or falsely.

mis·step (mis·step′) *n.* A false or wrong step; error.

mis·sus (mis′əz) *n. Colloq.* Mistress; wife. Also **mis′sis.** [Alter. of MISTRESS]

miss·y (mis′ē) *n. pl.* **miss·ies** Miss: a colloquial or diminutive form.

mist (mist) *n.* **1** An aggregation of fine drops of water in the atmosphere at or near the earth's surface, floating or falling very slowly: used either synonymously with *fog* or distinguished from it, as being less dense, or as consisting of drops large enough to fall perceptibly but slowly. **2** *Meteorol.* A very thin fog in which the horizontal visibility is greater than 1,100 yards. **3** Watery vapor condensed on and dimming a surface. **4** Any colloidal suspension of a liquid in a gas. **5** Anything that dims or darkens; that which obscures physical or mental vision. — *v.i.* **1** To be or become dim or misty; blur. **2** To rain in very fine drops. — *v.t.* **3** To make dim or misty; blur. [OE]

mis·take (mis·tāk′) *n.* An error in action, judgment, perception, or impression; a blunder. See synonyms under ERROR. — *v.* **·took, ·tak·en, ·tak·ing** *v.t.* **1** To understand wrongly; acquire a wrong conception of; misinterpret. **2** To take (a person or thing) to be another; fail to identify correctly: to *mistake* friends for enemies. — *v.i.* **3** To make a mistake. [< ON *mistaka*] — **mis·tak′a·ble** *adj.*

mis·tak·en (mis·tā′kən) *adj.* **1** Characterized by mistake; incorrect; wrong. **2** Being in error; wrong in opinion or judgment. See synonyms under ABSURD. — **mis·tak′en·ly** *adv.* — **mis·tak′en·ness** *n.*

Mis·tas·si·ni Lake (mis′tä·sē′nē) A lake in central Quebec, Canada; 840 square miles.

Mis·ter (mis′tər) *n.* **1** Master: a title of address prefixed to the name and to some official titles of a man: commonly written *Mr.: Mr.* Darwin; *Mr.* Chairman. **2** Official salutation in addressing a warrant officer, flight officer, or a cadet in the U.S. Military Academy at West Point, and, in some practice, officers below the rank of captain. In the Navy it is directed to those of all ranks below that of commander; in the Maritime Service it is applicable to all ranks below the captain of the ship. [Var. of MASTER]

mist·flow·er (mist′flou′ər) *n.* A tall perennial herb (*Eupatorium coelestinum*) of the composite family, with coarsely toothed leaves and clusters of light-blue or violet flowers.

Mis·ti (mēs′tē), **El** See EL MISTI.

mis·tle·toe (mis′əl·tō) *n.* **1** A European evergreen parasitic shrub (*Viscum album*) with yellowish-green leaves and inconspicuous flowers, succeeded by glutinous white berries: found on various deciduous trees. **2** An American plant (*Phoradendron flavescens*) related to this shrub: the State flower of Oklahoma. [OE *misteltan* mistletoe twig]

MISTLETOE

mis·took (mis·tŏok′) Past tense of MISTAKE. Also *Scot.* **mis·teuk** (-tyŏok′).

mis·tral (mis′trəl, *Fr.* mēs·tral′) *n.* **1** A cold, dry, and violent northwest Alpine wind blowing from the Ebro to the Gulf of Genoa and also through the southern provinces of France. **2** A worsted dress fabric with twisted warp and weft threads woven to give a nubbed effect. [< F, lit., master wind < L *magistralis* < *magister* master]

Mis·tral (mēs·tral′), **Frédéric Joseph Étienne,** 1830–1914, Provençal poet and lexicographer. — **Gabriela,** 1889–1957, Chilean poet; real name *Lucila Godoy de Alcavaga.*

mis·tress (mis′tris) *n.* **1** A woman in authority or control, or to whom service is rendered; a female head, chief, or owner, as of a country household, an institution, or an estate; also a female schoolteacher. **2** A woman who unlawfully, or without marriage, fills the place of a wife; also, a woman beloved and courted; a sweetheart. **3** A woman who is well skilled in or has mastered anything. **4** *Scot.* A married woman or wife. [< OF *maistresse,* fem. of *maistre.* See MASTER.]

Mis·tress (mis′tris) *n.* A title of address formerly applied to both married and unmarried women: now generally supplanted by *Mrs.* for married and *Miss* for unmarried women.

Mistress of the Adriatic Venice.
Mistress of the Seas Great Britain.
Mistress of the World Rome: when its empire embraced the known world.

mis·tri·al (mis·trī′əl) *n.* A trial of a lawsuit that is void because of errors; also, a trial of a lawsuit in which the jury cannot agree on a verdict.

mis·trust (mis·trust′) *n.* Lack of trust or confidence. — *v.t. & v.i.* To regard (someone or something) with suspicion or doubt; distrust. See synonyms under DOUBT, SUSPECT. — **mis·trust′er** *n.* — **mis·trust′ful** *adj.* — **mis·trust′ful·ly** *adv.* — **mis·trust′ful·ness** *n.*

mis·trust·ing·ly (mis·trus′ting·lē) *adv.* With mistrust; doubtingly.

mis·tryst (mis·trīst′) *v.t. Scot.* **1** To fail to keep an engagement with. **2** To perplex.

mist·y (mis′tē) *adj.* **mist·i·er, mist·i·est 1** Containing, characterized by, or accompanied by mist. **2** Dimmed or obscured by or as by mist; hence, lacking clearness or perspicuity; confused; indistinct; vague. See synonyms

mis–hear	misincite	misinter	mislive	misnatured	mispagination	misphrase
mis–hearer	misinclination	misinterment	mislocate	misnavigation	mispaint	mispoint
mis–heed	misincline	misintimation	mislocation	misnumber	misparse	mispoise
mis–hit	misinfer	misjoin	mislodge	misnurture	mispart	mispolicy
mis–hold	misinference	miskeep	mismark	misnutrition	mispassion	misposition
mishumility	misinflame	miskindle	mismeasure	misobservance	mispatch	mispossessed
misidentification	misingenuity	mislabel	mismeasurement	misobserve	mispen	mispractice
misidentify	misinspired	mislabor	mismeet	misoccupy	misperceive	misprejudiced
misimagination	misinstruct	mislanguage	mismenstruation	misopinion	misperception	mispresent
misimagine	misinstruction	mislearn	misminded	misordination	misperform	misprincipled
misimpression	misinstructive	mislie	mismingle	misorganization	misperformance	misproceeding
misimputation	misintend	mislight	mismotion	misorganize	mispersuade	misproduce
misimpute	misintention	mislikeness	misnarrate	mispage	misperuse	mispronounce

misunderstand under THICK. — **mist'i·ly** adv. — **mist'i·ness** n.

mis·un·der·stand (mis'un-dər-stand', mis·un'-) v.t. & v.i. **·stood**, **·stand·ing** To understand wrongly; misinterpret.

mis·un·der·stand·ing (mis'un-dər-stan'ding, mis·un'-) n. 1 A misapprehension; mistake as to meaning or motive. 2 A disagreement; dissension. See synonyms under QUARREL.

mis·un·der·stood (mis'un-dər-stood', mis·un'-) adj. 1 Not comprehended; wrongly interpreted. 2 Not appreciated at true worth.

mis·use (mis·yōōs') n. 1 Ill-treatment; violence; abuse. 2 Erroneous or improper use; misapplication. Also **mis·us'age**. — v.t. (mis·yōōz') **·used**, **·us·ing** 1 To use or apply wrongly or improperly. 2 To subject to ill-treatment; abuse; maltreat. See synonyms under ABUSE.

mis·us·er (mis·yōō'zər) n. 1 One who misuses. 2 Law Such a misuse or abuse of a privilege or franchise as should cause its forfeiture.

mis·ven·ture (mis·ven'chər) n. An ill venture; a misadventure.

Mi·tau (mē'tou) The German name for JELGAVA. Russian **Mi·ta·va** (mē'tə·və).

Mitch·ell (mich'əl), **John**, 1870–1919, U.S. labor leader. — **Maria**, 1818–89, U.S. astronomer. — **Silas Weir**, 1829–1914, U.S. physician and novelist. — **William**, 1879–1936, U.S. general; advocate of air power; called "Billy Mitchell."

Mitch·ell (mich'əl), **Mount** A peak in the Black Mountains, NW North Carolina; the highest point in the United States east of the Mississippi; 6,684 feet.

mite[1] (mīt) n. Any of various small arachnids (order *Acarina*) of both terrestrial and aquatic habits: many of them are parasitic on men, animals, plants, and stored grain, as the *itch mite*, *cheese mite*, etc. ◆ Homophone: *might*. [OE *mīte*] — **mit'y** adj.

mite[2] (mīt) n. 1 A very small amount or particle. 2 Any very small coin or sum of money: the widow's *mite*. Mark xii 42. See synonyms under PARTICLE. ◆ Homophone: *might*. [<Du. *mijt*]

mi·ter[1] (mī'tər) n. 1 A headdress worn by various church dignitaries, as popes, archbishops, bishops, and abbots: a tall ornamental cap terminating in two peaks; hence, the office or dignity of a bishop, etc. 2 The official headdress of the ancient Jewish high priest. 3 A headdress resembling a bishop's miter, worn in the 15th century by women. 4 The junction of two bodies at an equally divided angle, as at the corner of a picture frame: also **miter joint**. — v.t. 1 To confer a miter upon; raise to the rank of bishop. 2 To make or join with a miter joint. Also **mitre**. [<OF *mitre* <L *mitra* <Gk., belt, turban] — **mi'ter·er** n.

ECCLESIASTICAL MITER

miter box A box having a bottom and sides, but no top or ends, the sides having kerfs or sawguides in which wooden strips may be sawed to accurate miters.

mi·ter·wort (mī'tər·wûrt') n. Any of a genus (*Mitella*) of low, slender, mainly North American perennial herbs of the saxifrage family, having small miter-shaped flowers. Also **mi'tre·wort'**.

Mit·ford (mit'fərd), **Mary Russell**, 1787–1855, English author.

mith·er (mith'ər) n. Scot. Mother.

Mith·gar·thr (mith'gär·thər) See MIDGARD.

Mith·ra (mith'rə) 1 The ancient Persian god of light and truth. 2 In the Zoroastrian belief, a god often acting as the mediator between the Supreme God and man. Also

Mith'ras (-rəs). [<L <Gk. *Mithras* <OPersian *Mithra*] — **Mith·ra·ic** (mith·rā'ik) adj. — **Mith'ra·i·cism** n. — **Mith'ra·ism** n. — **Mith'ra·is'tic** adj.

Mith·ri·da·tes VI (mith'rə·dā'tēz), 132?–63 B.C., king of Pontus; defeated by Pompey: known as *Mithridates the Great*.

Mith·ri·dat·ic (mith'rə·dat'ik) adj. Of or pertaining to any of several kings of Pontus named Mithridates, especially Mithridates VI.

mith·ri·dat·ism (mith'rə·dā'tiz·əm) n. Immunity against poisons secured by the administration of gradually increasing doses: so called from King Mithridates VI of Pontus, who is said to have immunized himself by this method. [<obs. *mithridiate* an antidote against poison <LL *mithridatium* <*Mithridateus* of Mithridates] — **mith'ri·dat'ic** (-dat'ik) adj.

mi·ti·cide (mī'tə·sīd) n. A chemical agent destructive of mites: also called *acaricide*. [<*miti-* (<MITE[1]) + -CIDE]

mit·i·gate (mit'ə·gāt) v.t. & v.i. **·gat·ed**, **·gat·ing** To make or become milder, less harsh, or less severe; moderate. See synonyms under ABATE, ALLAY, ALLEVIATE, AMEND, PALLIATE, RELAX. [<L *mitigatus*, pp. of *mitigare* <*mitis* mild + *agere* do, drive] — **mit'i·ga·ble** adj. — **mit'i·gant** adj. & n. — **mit'i·ga·tion** n. — **mit'i·ga·tive** adj. — **mit'i·ga·tor** n. — **mit'i·ga·to·ry** (-gə·tôr'ē, -tō'rē) adj. & n.

mi·tis casting (mī'tis, mē'-) 1 The process of making castings of wrought iron of which the melting point has been lowered by the addition of a small amount of aluminum. 2 A casting so made. [<L *mitis* mild + CASTING]

mi·to·chon·dri·a (mī'tə·kon'drē·ə) n. Chondriosome. [<NL <Gk. *mitos* thread + *chondros* cartilage, granule]

mi·to·sis (mī·tō'sis) n. Biol. The series of changes in indirect cell division by which the chromatin of the nucleus is modified into a double set of chromosomes that splits longitudinally, one set going to each nuclear pole of the spindle before final division into two fully mature daughter cells. Compare MEIOSIS. [<NL <Gk. *mitos* thread + -OSIS] — **mi·tot·ic** (mī·tot'ik) adj. — **mi·tot'i·cal·ly** adv.

mi·trail·leur (mē·trà·yœr') n. 1 A soldier who operates a mitrailleuse. 2 A mitrailleuse. [<F <*mitrailler* fire grapeshot <*mitraille* grapeshot, small coins <OF *mitre*, *mite* small coin]

mi·trail·leuse (mē·trà·yœz') n. 1 A kind of breechloading machine-gun of grouped barrels for the rapid firing of small missiles. 2 Any machine-gun.

mi·tral (mī'trəl) adj. 1 Pertaining to or resembling a miter. 2 Of or pertaining to the mitral valve.

mitral valve Anat. A membranous valve between the left auricle and the left ventricle of the heart: it prevents the flow of blood into the auricle.

mi·tre (mī'tər) See MITER.

mitt (mit) n. 1 A glove, often of lace or knitwork, that does not extend over the fingers. 2 A mitten; specifically, in baseball, a heavy padded mitten used by the catcher and the first baseman. 3 pl. Slang The hands. [<MITTEN]

mit·ten (mit'n) n. 1 A covering for the hand, encasing the four fingers together and the thumb separately. 2 A mitt. 3 pl. Slang The hands; also, boxing gloves. [<F *mitaine*]

mit·ti·mus (mit'ə·məs) n. 1 Law An order by a magistrate committing a prisoner to jail. 2 A dismissal. [<L, we send <*mittere* send]

mitz·vah (mits'vä) n. pl. **·voth** (-vōth) A command of God; hence, the fulfilment of such a command, regarded as a special privilege; especially, a function of the synagog ceremonial or an act promoting the welfare of the Jews. Also **mits'vah**. [<Hebrew *mitzwah* commandment]

Mi·vart (mī'vərt, miv'ərt), **St. George Jackson**, 1827–1900, English biologist.

mix (miks) v. **mixed** or **mixt**, **mix·ing** v.t. 1 To put together in one mass or compound; blend. 2 To make by combining ingredients: to *mix* dough. 3 To combine or join: to *mix* business with pleasure. 4 To cause to associate or mingle: to *mix* social classes together. 5 To crossbreed. — v.i. 6 To be mixed or blended. 7 To associate; get along. 8 To take part; become involved. — **to mix up** 1 To blend thoroughly. 2 To confuse; bewilder. 3 To implicate or involve. — n. 1 The act or effect of mixing. 2 The proportion of certain substances or raw materials before their subjection to a fabricating or manufacturing process: a *mix* of cement. 3 Telecom. The correct blending of the sound input of two or more microphones. 4 Colloq. Confusion caused by blundering; a muddle; mess. [Back formation <MIXED] — **mix'a·ble**, **mix'i·ble** adj.

Synonyms (verb): amalgamate, associate, blend, combine, commingle, commix, compound, confuse, fuse, incorporate, join, meddle, mingle, unite. Compare COMPLEX, HETEROGENEOUS. Antonyms: see synonyms for SEPARATE.

mixed (mikst) adj. 1 Mingled in a body or mass; joined together; associated; blended: generally of different or even incongruous elements; a *mixed* mixture. 2 Containing persons of both sexes: a *mixed* school, *mixed* foursome, etc. 3 Mentally confused, as with liquor. 4 Law Designating statutes which concern both persons and property; also, designating property which is not altogether real nor personal, but a compound of both. 5 Phonet. Designating a vowel produced with neither front nor back articulation predominating; central. 6 Bot. Denoting inflorescence which combines cymose and racemose. Also **mixt**. [<F *mixte* <L *mixtus*, pp. of *miscere* mix] — **mix·ed·ly** (mik'sid·lē) adv. — **mix'ed·ness** n.

mixed economy A combination of laissez-faire with governmentally planned and/or controlled economy.

mixed marriage Marriage between persons of different religions or races.

mixed numbers See under NUMBER.

mixed train A train transporting both passengers and freight.

mix·en (mik'sən) n. Archaic A dunghill; compost heap. Also **mix'on**. [OE *mixen* <*meox* dung]

mix·er (mik'sər) n. 1 One who or that which mixes; a machine or device for mixing. 2 Colloq. A person with reference to his ability to mix socially or get along well in various groups.

mix·o·troph·ic (mik'sə·trof'ik) adj. Biol. Pertaining to or designating plants and animals capable of feeding on both inorganic and dead organic material, as certain flagellate protozoa. [<Gk. *mixis* mingling + *trophē* food]

mix·ture (miks'chər) n. 1 The act of mixing. 2 Something resulting from mixing; admixture. 3 Something added as an ingredient. 4 A pharmaceutical preparation consisting of an aqueous solution in which is suspended an insoluble compound: intended for internal use. 5 A commingling of two or more substances in varying proportions, in which the ingredients retain their individual chemical properties, and from which they may be separated, unaltered, by mechanical means. Compare COMPOUND. [<F <L *mixtura* <*miscere* mix]

mix·ty-max·ty (mik'stē-mak'stē) adj. Scot. Mingled confusedly or promiscuously.

---

misproportion	misrate	misreform	misresemblance	mis-sheathed	misstroke	misthread
misproposal	misread	misregulate	misresolved	mis-ship	misstyle	misthrive
mispropose	misrealize	misrehearsal	misresult	misshod	missuggestion	misthrow
misprovide	misreason	misrehearse	misreward	mis-sing	missuit	mistile
misprovidence	misreceive	misrelate	misrime	missolution	missummation	mistouch
misprovoke	misrecital	misrelation	misseason	missort	missuppose	mistranscribe
mispunctuate	misrecite	misreliance	misseat	missound	mis-sway	mistranscription
mispunctuation	misrecognition	misrely	mis-see	misspace	misswear	mistranslate
mispurchase	misrecognize	misrender	mis-seed	misspeak	missyllabication	mistranslation
mispursuit	misrecollect	misrepeat	missemblance	misstart	missyllabify	mistune
misqualify	misrefer	misreposed	mis-send	missteer	mistaught	mistutor
misquote	misreference	misreprint	mis-sense	misstop	mistend	misunion
misraise	misreflect	misrepute	missentence	misstrike	misterm	misyoke

mix–up (miks'up') *n.* 1 A confusion; muddle. 2 *Colloq.* A fight.

Mi·ya·za·ki (mē-yä-zä-kē) A city on SE Kyushu island, Japan.

Mi·zar (mī'zər, mē'-) The star Zeta, second from the end in the handle of the Dipper, constellation of Ursa Major: the faint star close to it is *Alcor.* [<Arabic *mi'zar* veil, cloak]

miz·zen (miz'ən) *n. Naut.* 1 A mizzenmast. 2 A triangular sail set on the mizzen. — *adj.* Of or pertaining to the mizzen or mizzenmast. Also **miz'en.** [<F *misaine* <Ital. *mezzana,* fem. of *mezzano* middle <L *medianus.* Doublet of MEDIAN, MEAN³, MESNE.]

miz·zen·mast (miz'ən·məst, -mast', -mäst') *n. Naut.* 1 The mast next abaft the mainmast. 2 The shorter of the two masts of a ketch or yawl.

Mjol·nir (myol'nir) *n.* In Norse mythology, Thor's terrible hammer. Also **Mjoll'nir, Mjöll·nir** (myœl'nir), **Mjol'ner.**

mks The meter–kilogram–second system of units for the measurement of physical quantities. It differs from the cgs system in using the international standards of length and mass instead of their submultiples, the centimeter and the gram.

mne·me (nē'mē) *n.* A hypothetical unit of memory assumed to exist in all animal cells and to function in behavior. [<Gk. *mnēmē* memory]

mne·mon·ic (ni·mon'ik) *adj.* Pertaining to, aiding, or designed to aid the memory. Also **mne·mon'i·cal.** [<Gk. *mnēmonikos* < *mnēmōn* mindful <*mnasthai* remember]

mne·mon·ics (ni·mon'iks) *n.* The science of memory improvement. Also **mne·mo·tech·nics** (nē'mō·tek'niks).

Mne·mos·y·ne (nē·mos'ə·nē, -moz'-) In Greek mythology, the goddess of memory and mother (by Zeus) of the Muses. [<L <Gk. *mnēmosynē* memory <*mnasthai* remember]

-mo *suffix Printing* Folded into a (specified) number of leaves: said of a sheet of paper: 12*mo* or *duodecimo.* Also shown by the symbol (°), as in 12°. [<L *-mo,* as in the phrase *in duodecimo* twelvefold]

mo·a (mō'ə) *n.* A large, flightless, extinct bird (family *Dinornithidae*) of New Zealand, having enormous legs with at least three toes; especially, the largest species (*Dinornis robustus*). [< native name]

Mo·ab (mō'ab) An ancient country in the upland area east of the Dead Sea.

Mo·ab·ite (mō'əb·īt) *n.* One of the descendants of Moab, son of Lot. *Gen.* xix 37. — **Mo·ab·it·ess** (mō'əb·it'is) *n. fem.*

Moabite Stone A stone slab with a Moabite inscription, dating from 850 B.C.; discovered, 1868.

moan (mōn) *n.* 1 A low mournful sound indicative of grief or pain. 2 A similar sound; the *moan* of the wind. 3 *Obs.* Lamentation; complaint. — *v.i.* 1 To utter moans of grief or pain. 2 To make a low, mournful sound, as wind in trees. — *v.t.* 3 To lament; bewail. [Cf. OE *mænan* lament, moan]

moat (mōt) *n.* A defensive ditch on the outside of a fortress wall. — *v.t.* To surround with or as with a moat. ◆ Homophone: *mote.* [<OF *mote* embankment]

mob¹ (mob) *n.* 1 A turbulent or lawless crowd or throng; a rabble. 2 The lowest class of people; the masses; populace. 3 *Slang* A gang, as of thieves. — *v.t.* **mobbed, mob·bing** 1 To attack in a mob; crowd around and annoy. 2 To crowd into, as a hall. [<L *mob(ile vulgus)* movable crowd] — **mob'ber** *n.* — **mob'bish** *adj.* — **mob'bish·ly** *adv.*
Synonyms (noun): canaille, crowd, masses, people, populace, rabble.

mob² (mob) *n.* A cap or headdress formerly worn by women and girls and usually tied under the chin. Also **mob'cap'.** [<Du. *mop* coif, cap]

mo·bile (mō'bəl, -bēl) *adj.* 1 Characterized by freedom of movement; movable. 2 Changing easily in expression or in state of mind; changeable. 3 Moving or flowing freely. 4 That can be easily and quickly moved, as military units. 5 Designating a mobile. — *n.* (mō'bēl) A form of sculpture arranged so that its movable parts, suspended or balanced on rods, wires, etc., describe kinetic rather than static patterns. [<F <L *mobilis* movable] — **mo·bil·i·ty** (mō·bil'ə·tē) *n.*
Synonyms: changeable, changing, expressive, fickle, movable, sensitive, variable, volatile. See ACTIVE. *Antonyms:* dull, fixed, immovable, inexpressive, still, stolid, unchanging, unvarying.

Mo·bile (mō·bēl') A port of Alabama, on **Mobile Bay,** an arm of the Gulf of Mexico in SW Alabama.

Mobile River A river in SW Alabama flowing 45 miles south from the confluence of the Tombigbee and Alabama rivers to Mobile Bay at Mobile.

mo·bi·lize (mō'bə·līz) *v.* **·lized, ·liz·ing** *v.t.* 1 To make ready for war, as an army, industry, etc. 2 To assemble for use; organize. 3 To make mobile; put into circulation or use. — *v.i.* 4 To get ready for war. Also *Brit.* **mo'bi·lise.** [<F *mobiliser*] — **mo·bi·li·za·tion** (-lə·zā'shən, -lī·zā'-) *n.*

Mö·bi·us surface (mœ'bē·ōōs) *Geom.* A surface both sides of which may be completely traversed without crossing either edge: made by joining the half-twisted ends of a rectangular strip of paper or other flexible material. Also **Möbius strip.** [after August Ferdinand *Möbius,* 1790–1868, German mathematician and astronomer]

mob law Lynch law.

mo·ble (mob'əl) *v.t. Archaic* To cover with a cap or mob. [<MOB²]

mob·oc·ra·cy (mob·ok'rə·sē) *n. pl.* **·cies** 1 Lawless control of public affairs by the mob or populace. 2 The mob considered as the dominant class. [<MOB¹ + -(O)CRACY]

mob·o·crat (mob'ə·krat) *n.* One who favors mobocracy; a demagog. — **mob'o·crat'ic, mob'o·crat'i·cal** *adj.*

mob·ster (mob'stər) *n. Slang* A gangster.

moc·ca·sin (mok'ə·sin) *n.* 1 A foot covering made of soft leather or buckskin: worn by North American Indians; also, a soft shoe or slipper. 2 A dark-colored, obscurely blotched, venomous snake (genus *Agkistrodon*) of the southern United States. *A. piscivorus* is the **water moccasin.** [< Algonquian *mohkisson*]

MOCCASIN
(Average length about 4 feet; largest specimens to 6 feet)

moccasin flower Any one of certain orchids of the genus *Cypripedium,* common in the United States, especially the showy ladyslipper (*C. reginae*), State flower of Minnesota.

mo·cha (mō'kə) *n.* 1 A choice coffee, originally grown in Arabia. 2 A rich, coffee-flavored icing, or a cake flavored with it. 3 A fine sheepskin leather used for making gloves. 4 A dark, dull, grayish-brown color. [from *Mocha*]

Mo·cha (mō'kə) A port of Yemen, in SW Arabia: also **Mokha.**

mock (mok) *v.t.* 1 To treat or address scornfully or derisively; hold up to ridicule. 2 To ridicule by imitation of action or speech; mimic derisively. 3 To deceive; delude. 4 To defy; make futile. 5 *Poetic* To imitate; counterfeit. — *v.i.* 6 To express or show ridicule, scorn, or contempt; scoff. — *adj.* Merely imitating the reality; sham. — *n.* An act of mocking; a jeer; mockery. [<OF *mocquer*] — **mock'er** *n.* — **mock'ing·ly** *adv.*
Synonyms (verb): banter, chaff, deride, flout, gibe, insult, jeer, taunt. See IMITATE, MISLEAD, SCOFF. Compare COUNTERFEIT. *Antonyms:* see PRAISE.

mock·er·y (mok'ər·ē) *n. pl.* **·er·ies** 1 Derisive or contemptuous mimicry. 2 A false show; sham. 3 A butt of ridicule. 4 Labor in vain. See synonyms under BANTER, SCORN.

mock–he·ro·ic (mok'hi·rō'ik) *adj.* Imitating or satirizing the heroic manner, style, attitude, or character. — *n.* 1 Any writing using the grand style as a comic expedient. 2 *pl.* Affectation of the grand manner in expressing trivialities.

mock·ing·bird (mok'ing·bûrd') *n.* 1 A bird (*Mimus polyglottos*) common in the southern and eastern United States, noted for its rich song and powers of imitating the calls of other birds. 2 One of various other birds that mock, as the catbird.

mocking thrush The thrasher.

mocking wren 1 The Carolina wren. 2 Bewick's wren. See under WREN.

mock moon A paraselene.

mock orange Any of a genus (*Philadelphus*) of ornamental shrubs of the saxifrage family: also called *syringa.*

mock sun A parhelion.

mock title See HALF-TITLE.

mock–tur·tle soup (mok'tûr'təl) Soup prepared from calf's head or other meat, and somewhat resembling green-turtle soup.

mock–up (mok'up') *n.* 1 A model, usually full-scale, of a proposed structure, machine, apparatus, etc. 2 An airplane, etc., constructed for purposes of study, testing, or training of personnel.

Moc·te·zu·ma (môk'tā·sōō'mä) See MONTEZUMA.

mo·dal (mōd'l) *adj.* 1 Of or denoting a mode or manner, especially a mode of grammar, a mode in music, or a mode of logical statement. 2 Characterized by form or manner without reference to matter or substance. 3 Pertaining to or designating a statistical mode. — **mo'dal·ly** *adv.*

mo·dal·i·ty (mō·dal'ə·tē) *n. pl.* **·ties** 1 Modal character; the fact or quality of being modal. 2 *Logic* The character of a proposition as expressing or asserting a sequence of necessity (including impossibility) or of contingency (including probability and possibility).

Mod·der (mod'ər) A river in Orange Free State, Union of South Africa, flowing 225 miles NW to the Riet.

mode (mōd) *n.* 1 Manner of being, doing, etc.; way; method. 2 Prevailing style; common fashion. 3 *Gram.* Mood. 4 *Music* A method of dividing an octave by placing the steps and half-steps of which it is composed in certain arbitrary relations. In the **major mode,** tones are arranged as given in the major scale; in the **minor mode,** as in the minor scale. 5 *Psychol.* A faculty or phenomenon of mind considered as a state of consciousness. 6 *Philos.* The manner of a thing's existence so far as it is not essential. 7 *Logic* **a** The style of the connection between the antecedent and the consequent of a proposition. **b** The arrangement of the propositions of a syllogism according to their quantity and quality. 8 *Stat.* That value, magnitude, or score which occurs the greatest number of times in a given series of observations: also called *norm.* 9 *Geol.* The actual mineral composition of a rock, expressed in percentages by weight: distinguished from *norm.* 10 A light bluish-gray color. See synonyms under MANNER, SYSTEM. [<L *modus* measure, manner]

mod·el (mod'l) *n.* 1 An object, usually in miniature, representing accurately something to be made or already existing; more rarely, a plan or drawing: a *model* of a building. 2 A person who poses for painters, sculptors, etc. 3 A thing or person to be imitated or patterned after; that which is taken as a pattern or an example. 4 A person employed to wear articles of clothing to display them to customers. 5 That which strikingly resembles something else; an approximate copy or image. — *v.* **·eled** or **·elled, ·el·ing** or **·el·ling** *v.t.* 1 To plan or fashion after a model or pattern. 2 To make a model of. 3 To fashion; make. 4 To display by wearing, as a coat or hat. — *v.i.* 5 To make a model or models. 6 To pose or serve as a model (defs. 2 and 4). 7 To assume the appearance of natural form. — *adj.* Serving or used as a model; suitable for a model; worthy to be imitated. [<F *modèle* <Ital. *modello*, dim. of *modo* <L *modus* measure, manner] — **mod′el·er, mod′el·ler** *n.*

Synonyms (noun): archetype, copy, design, ectype, example, facsimile, image, imitation, mold, original, pattern, prototype, replica, representation, type. A *pattern* must be closely followed in its minutest particulars by a faithful copyist; a *model* may allow a great degree of freedom. A sculptor may idealize his living *model*; his workmen must exactly copy in marble or metal the *model* he has made in clay. The *archetype* is the original form, actual or ideal, in accordance with which existing things are made; a *prototype* is either the original or an authenticated copy that has the authority of the original as a standard to which other objects of its kind must conform, even if the latter sense is comparatively rare. See EXAMPLE, IDEA, IDEAL. Compare FACSIMILE, DUPLICATE.

mod·el·ing (mod′l·ing) *n.* **1** In sculpture, the art of constructing in clay or wax a model afterward to be reproduced in plaster, stone, or metal. **2** In painting, the art of representing figures as if in natural relief or in solid form. Also **mod′el·ling.**

Model T An early model of automobile manufactured by Henry Ford in great numbers over a period of years, and considered by many to have established the automobile in the U. S.: also called *tin lizzie.*

Mo·de·na (mō′dä·nä) A city in north central Italy.

mod·er·ate (mod′ər·it) *adj.* **1** Keeping or kept within reasonable limits; not extreme, excessive, or radical; also, mild; temperate; calm; reasonable; gentle. **2** Not strongly partisan: said of political and religious parties, and their tenets or views. **3** Medium; fair; also, mediocre. **4** Slow in thought, speech, or action. **5** *Meteorol.* Designating a breeze (No. 4) or a gale (No. 7) on the Beaufort scale. See synonyms under GRADUAL, MODEST, SLOW, SOBER. — *n.* A person of moderate views, opinions, or practices; especially, a member of a political or religious party which is not strongly partisan. — *v.* (mod′ə·rāt) **·at·ed, ·at·ing** *v.t.* **1** To reduce the violence, severity, etc., of; make less extreme; restrain. **2** To preside over. — *v.i.* **3** To become less intense or violent; abate. **4** To act as moderator. See synonyms under ABATE, ALLAY, ALLEVIATE, TEMPER, TRANQUILIZE. [<L *moderatus*, pp. of *moderare* regulate < *modus* measure] — **mod′er·ate·ly** *adv.* — **mod′er·ate·ness** *n.* — **mod′er·a′tion** *n.*

mod·e·ra·to (mod′ə·rä′tō) *adj. & adv. Music* In moderate time; moderately. [<Ital.]

mod·er·a·tor (mod′ə·rā′tər) *n.* **1** One who restrains or regulates. **2** The presiding officer of a meeting; also, the presiding officer in Presbyterian and Congregational courts. **3** *Physics* A substance, as graphite or beryllium, used to control the rate of a nuclear chain reaction in an atomic-energy reactor. — **mod′er·a′tor·ship** *n.*

mod·ern (mod′ərn) *adj.* **1** Pertaining to the present or recent period; not ancient. **2** *Obs.* Commonplace; common; trite. — *n.* **1** A person of modern times, or modern views or characteristics: also **mod′ern·er.** **2** *Printing* A style of type face characterized by contrasting heavy down-strokes and thin cross-strokes. [<LL *modernus* recent <L *modo* just now] — **mod′ern·ly** *adv.* — **mo·der·ni·ty** (mo·dûr′nə·tē), **mod′ern·ness** *n.*

Synonyms (adj.): fresh, late, new, novel, recent. *Modern* history pertains to any period since the Middle Ages; *modern* literature, *modern* architecture, etc., are not strikingly remote from the styles and types prevalent today. That which is *late* is somewhat removed from the present, but not far enough to be called old. That which is *recent* is very close to the present, but not quite so sharply distinguished from the past as *new*; *recent* publications range over a longer time than *new* books. See NEW. *Antonyms*: see synonyms for ANCIENT.

mod·ern·ism (mod′ərn·iz′əm) *n.* **1** Something characteristic of modern as distinguished from former or classical times; a modern idiom or practice. **2** Modern character, methods, or mental attitude. **3** The humanistic tendency in religious thought to supplement old theological creeds and dogmas by new scientific and philosophical learning and thus to place emphasis on practical ethics and world-wide social justice: distinguished from *fundamentalism.* — **mod′ern·ist** *n.* — **mod′ern·is′tic** *adj.*

mod·ern·ize (mod′ərn·īz) *v.t. & v.i.* **·ized, ·iz·ing** To make or become modern in ideas, standards, methods, etc. — **mod′ern·i·za′tion** *n.* — **mod′ern·iz′er** *n.*

mod·est (mod′ist) *adj.* **1** Restrained by a sense of propriety or humility. **2** Characterized by reserve, propriety, or purity; decorous; chaste. **3** Free from excess; moderate. [<F *modeste* <L *modestus* moderate < *modus* measure] — **mod′est·ly** *adv.*

Synonyms: chaste, decent, decorous, humble, moderate, proper, pure, retiring, unassuming, unobtrusive, unostentatious, unpretending, unpretentious, virtuous. See HUMBLE.

mod·es·ty (mod′is·tē) *n.* Decent reserve and propriety; delicacy; decorum.

Synonyms: backwardness, bashfulness, coldness, constraint, coyness, diffidence, reserve, shyness, timidity, unobtrusiveness. *Bashfulness* is a shrinking from notice without assignable reason. *Coyness* is a half encouragement, half avoidance of offered attention, and may be real or affected. *Diffidence* is self-distrust; *modesty*, a humble estimate of oneself in comparison with others or with the demands of some undertaking. *Modesty* has also the specific meaning of a sensitive shrinking from anything indelicate. *Shyness* is a tendency to shrink from observation; *timidity*, a distinct fear of criticism, error, or failure. *Reserve* is holding oneself aloof from others, or holding back one's feelings from expression, or one's affairs from communication to others. Compare ABASH, PRIDE, RESERVE, TACITURN. *Antonyms*: abandon, arrogance, assumption, assurance, boldness, conceit, confidence, egotism, forwardness, frankness, freedom, haughtiness, impudence, indiscretion, loquaciousness, loquacity, pertness, sauciness, self-conceit, self-sufficiency, sociability.

mod·i·cum (mod′i·kəm) *n. pl.* **·cums** or **·ca** (-kə) **1** A moderate amount; a little. **2** A small thing or person. [<L < *modus* measure]

mod·i·fi·ca·tion (mod′ə·fə·kā′shən) *n.* **1** The act of modifying, or the state of being modified. **2** *Biol.* Variation in plants and animals, specifically by localized changes in an organism due to external influences and not inheritable. Compare MUTATION. **3** That which results from modifying. — **mod·i·fi·ca·to·ry** (mod′ə·fə·kā′tər·ē) *adj.*

mod·i·fi·er (mod′ə·fī′ər) *n.* **1** One who or that which qualifies, changes, limits, or varies. **2** *Gram.* A word, phrase, or clause that alters, restricts, or varies the application of another word or group of words, as an adjective or adverb. See also UNIT MODIFIER.

mod·i·fy (mod′ə·fī) *v.* **·fied, ·fy·ing** *v.t.* **1** To make somewhat different in form, character, etc.; vary. **2** To reduce in degree or extent; moderate. **3** *Gram.* To qualify the meaning of; restrict; limit. **4** *Ling.* To alter (a vowel) by umlaut. — *v.i.* **5** To be or become modified; change. See synonyms under CHANGE, TEMPER. [<F *modifier* <L *modificare* < *modus* measure + *facere* make] — **mod′i·fi′a·ble** *adj.*

Mo·di·glia·ni (mō′dē·lyä′nē), **Amedeo,** 1884–1920, Italian painter and sculptor.

mo·dil·lion (mō·dil′yən) *n. Archit.* An enriched block or horizontal bracket used in series under a Corinthian or Composite cornice: sometimes, with less ornament, under one of the Roman Ionic order. [<F *modillon* <Ital. *modiglione*]

mo·di·o·lus (mō·dī′ə·ləs) *n. pl.* **·li** (-lī) *Anat.* The central stem round which wind the passages of the cochlea of the internal ear. [<L, bucket on a water wheel < *modus* measure]

mod·ish (mō′dish) *adj.* Conformable to the current mode, fashion, or usage; stylish. — **mod′ish·ly** *adv.* — **mod′ish·ness** *n.*

mo·diste (mō·dēst′) *n.* A woman who makes or deals in fashionable articles, especially of women's dress or millinery. [<F]

Mo·djes·ka (mō·jes′kə), **Helena,** 1840–1909, Polish actress active in the United States.

Mo·djo·ker·to Man (mō′jō·kär′tō) A primitive hominid identified from the fossil skull of an infant found in 1931 in Pleistocene deposits near Modjokerto, Java: it bears anatomical resemblances both to Pithecanthropus and to Neanderthal man.

Mo·doc (mō′dok) *n.* A North American Indian of a small, nearly extinct tribe of Lutuamian linguistic stock, formerly living in California, now on reservations in Oregon and Oklahoma. See LUTUAMIAN.

mo·do et for·ma (mō′dō et fôr′mə) *Latin* In manner and form.

Mo·dred (mō′drid) In Arthurian legend, King Arthur's treacherous nephew (or son). During Arthur's absence Modred usurped the throne: after Arthur's return they killed each other in battle.

mod·u·lar (moj′o͝o·lər) *adj.* **1** Proportionate according to a module. **2** Of or pertaining to a modulus.

mod·u·late (moj′o͝o·lāt) *v.* **·lat·ed, ·lat·ing** *v.t.* **1** To vary the tone, inflection, or pitch of. **2** To regulate or adjust; temper; soften. **3** *Music* To change or cause to change to a different key. **4** To intone or sing. **5** *Electronics* To alter the frequency or amplitude of (a radio carrier wave). — *v.i.* **6** *Electronics* To alter the frequency or amplitude of a carrier wave. **7** *Music* To change from one key to another by using a transitional chord common to both. [<L *modulatus*, pp. of *modulari* regulate < *modulus* MODULE] — **mod′u·la·to′ry** (-lə·tôr′ē, -tō′rē) *adj.*

mod·u·la·tion (moj′o͝o·lā′shən) *n.* **1** The act of modulating, or the state of being modulated; specifically, a musical inflection of the voice; change in pitch. **2** *Music* A change from one key to another by the use of a transitional chord common to both. **3** *Telecom.* The process of varying the frequency, amplitude, intensity, or phase of a carrier wave so as to conform with a transmitted signal wave.

mod·u·la·tor (moj′o͝o·lā′tər) *n.* **1** One who or that which modulates. **2** *Telecom.* A tube or valve for effecting modulation. **3** A musical chart showing the relations of tones and scales.

mod·ule (moj′o͞ol) *n.* **1** A standard or unit of measurement. **2** *Archit.* A measure of proportion among the parts of a classical order, the size of the diameter or semidiameter of the base of a column shaft usually being taken as a unit. **3** *Mech.* A standard or unit of measurement for gearwheels, given as the pitch diameter in millimeters divided by the number of teeth in the wheel. **4** *Electronics* A small ceramic wafer for the control of circuits in a computer. **5** *Obs.* A mere image. [<L *modulus* < *modus* measure. Doublet of MOLD[1].]

mod·u·lor (moj′o͝o·lôr) *n.* A system of industrial design based upon the ideal proportions of the human body: units derived from the basic dimensions can be assembled to secure maximum harmony and utility. — *adj.* Of or characterized by such design.

mod·u·lus (moj′o͝o·ləs) *n. pl.* **·li** (-lī) **1** *Physics* A number, coefficient, or quantity that measures a force, function, or effect: *modulus* of elasticity: sometimes abbreviated to M or μ. See CONGRUENT. **2** *Math.* The logarithm of e to the base 10 ($\log_{10}e$): Napierian logarithms are multiplied by this factor to convert them to logarithms to the base 10. [<L, dim. of *modus* a measure]

mo·dus (mō′dəs) *n. Latin* Mode; manner.

modus op·er·an·di (op′ə·ran′dī) *Latin* A manner of operation.

modus vi·ven·di (vi·ven′dī) *Latin* A manner of living; especially, a temporary arrangement pending a final settlement.

moe (mō) *adj. & adv. Obs.* More.

mo·el·lon (mō′əl·on) *n.* **1** A form of rubble masonry used as a filling in the facing walls of a structure. **2** Dégras (def. 1). [<F, alter. of OF *moilon*; ? infl. by F *moelle* pith]

Moe·rae (mē′rē) See MOIRAI. Also **Mœ′ræ.**

Moe·ro (mwe·rō′) The French spelling of MWERU.

Moe·si·a (mē′shē·ə) An ancient country and former Roman province in SE Europe south of the Danube. Also **Mœ′si·a.**

Moe·so·goth (mē′sə·goth) *n.* A member of the Gothic tribe that settled in Moesia. — **Moe′so·goth′ic** *adj.*

mo·fette (mō·fet′) *n.* **1** A noxious emanation of gas from a fissure; a gas spring. **2** An opening in the earth from which noxious gas escapes, as from a volcano. Also **mof·fette′.** [<F <Ital. *mofetta* < *muffare* decay <G *muff* mold]

mo·fus·sil (mə·fus′əl) *n. Anglo-Indian* The

Mogadishu

country as distinguished from the residences and the towns.
Mog·a·di·shu (mog′ə-dish′ōō) The chief port and capital of Somalia, on the Indian Ocean. Italian **Mo·ga·di·scio** (mō′gä-dē′shō).
Mog·a·dor (mog′ə-dôr′, -dōr′; Fr. mô-gȧ-dôr′). A port on the Atlantic coast of SW Morocco.
mo·gen Da·vid (mō′gən dä′vid, duv′id) A mystic device formed by the intertwining of two equilateral triangles; the six-pointed star: used as a symbol of Judaism. Also called *star of David, shield of David, Solomon's seal*. [<Hebrew, star of David]
Mo·gi·lev (mō′gi·lef, *Russian* mô′gē·lyôf′) A city in SW Belorussian S.S.R., on the Dnieper. Also **Mo·hi·lev** (mō′hē·lôf′).
mo·gul (mō′gul, mō·gul′) n. 1 Any great or pretentious personage; autocrat: also **great mogul.** 2 A type of freight locomotive with three pairs of coupled drivers and one pair of leading truck wheels.
Mo·gul (mō′gul, mō·gul′) n. A Mongol; Mongolian; specifically, one of the Mongol conquerors of Hindustan, or a follower of Genghis Khan in the 13th century. Also **Mo·ghul** (″). — **the Great** or **Grand Mogul** The former Mongol emperor of Delhi. [<Persian *mugal* a Mongol].
Mo·hács (mō′häch) A city on the Danube in southern Hungary; scene of a decisive Turkish victory over the Hungarians, 1526.
mo·hair (mō′hâr) n. 1 The hair of the Angora goat. 2 A smooth, wiry fabric made of mohair filling and cotton warp: often called *brilliantine*. 3 A fabric of cut or uncut loops with cotton or wool back and mohair pile: used chiefly for upholstery. [Earlier *mocayare* <Arabic *mukhayyar*, infl. in form by *hair*]
Mo·ham·med (mō·ham′id) 570–632, Arabian religious and military leader; founder of Islam and author of the *Koran*: also spelled *Mahomet.* Also **Mo·hom′ed.**
— **Mohammed II,** 1430?–81, sultan of Turkey 1451–81; captured Constantinople, 1453.
Mohammed Ali See MEHEMET ALI.
Mo·ham·me·dan (mō·ham′ə·dən) adj. Pertaining to Mohammed or to his religion and institutions. — n. A follower of Mohammed or believer in Islam; a Moslem. Also *Mohometan, Muhammedan.*
Mo·ham·me·dan·ism (mō·ham′ə·dən·iz′əm) n. The religion founded by Mohammed; Islam.
mo·har·ra (mə·här′ə) See MOJARRA.
Mo·har·ran (mə·här′ən) See MUHARRAM.
Mo·ha·ve (mō·hä′vē) n. A member of a tribe of North American Indians of Yuman linguistic stock, formerly living along the Colorado River. — adj. Of or pertaining to this tribe. Also spelled *Mojave.*
Mohave Desert See MOJAVE DESERT.
Mo·hawk (mō′hôk) n. 1 One of a tribe of North American Indians of Iroquoian stock, one of the original tribes of the Five Nations, formerly ranging from the Mohawk River Valley, New York, to the St. Lawrence: now in Canada, New York, and Wisconsin. 2 The Iroquoian language of this tribe. 3 A Mohock. [<N. Am. Ind. Cf. Narragansett *mohowaicuck,* lit., they eat animate things, hence, eaters of human flesh; so named by enemy tribes.]
Mohawk River A river in central New York, flowing 140 miles SE to the Hudson.
Mo·he·gan (mō·hē′gən) n. One of a tribe of North American Indians of Algonquian linguistic stock, the eastern branch of the Mahican group: formerly occupying the region from the lower Connecticut and Thames rivers northward to Massachusetts. See MAHICAN. [<Algonquian *maingan* wolf]
Mo·hen·jo-Da·ro (mō·hen′jō·dä′rō) A site of Indus Valley civilization in NW Sind, West Pakistan.
Mo·hi·can (mō·hē′kən) n. One of a warlike tribe of North American Indians belonging to the Algonquian linguistic stock, and formerly dwelling along both banks of the Hudson. See MAHICAN.
Mo·hock (mō′hok, -hōk) n. One of a band of lawless rowdies, often aristocratic rakes, who frequented the streets of London early in the 18th century. Also **Mo′hawk.** [Var. of MOHAWK]
Mohs scale (mōz) *Mineral.* A qualitative scale

in which the hardness of a mineral is determined by its ability to scratch, or be scratched by, any one of 10 standard minerals arranged in the following increasing order of hardness: 1, talc; 2, gypsum; 3, calcite; 4, fluorite; 5, apatite; 6, feldspar; 7, quartz; 8, topaz; 9, corundum; 10, diamond. [after Friedrich *Mohs,* 1773–1839, German mineralogist who conceived it]
mo·hur (mō′hər) n. A former gold coin of India, equal in value to 15 rupees. [<Hind. *muhur*]
moi·dore (moi·dôr′, -dōr) n. A former Portuguese or Brazilian gold coin. [Pg. *moeda d'ouro* coin of gold <L *moneta* money + *aurum* gold]
moi·e·ty (moi′ə·tē) n. pl. **·ties** 1 A half. 2 A small portion. [<F *moitié* <L *medietas* < *medius* half]
moil (moil) v.i. To work hard; toil; drudge. — n. 1 A soiling; defilement; spot. 2 Confusion; vexation; trouble. 3 *Scot.* Toil; drudgery. [<OF *moillier, muiller* wet <L *mollis* soft; infl. in meaning by *toil*] — **moil′er** n. — **moil′ing·ly** adv.
Moi·rai (moi′rī) In Greek mythology, the three birth goddesses, identified with the Fates: also spelled *Moerae.*
moi·ré (mwä·rā′) adj. Having a wavelike or watered appearance, as certain fabrics. — n. 1 A corded silk or rayon fabric, having a wavy or watered pattern produced by passing the fabric between engraved cylinders which press the design into the material: also **moire** (mwär). 2 The finish or effect of this process on certain fabrics. [<F <*moirer* water silk <*moire* watered silk <MOHAIR]
Mo·ïse (mō·ēz′) French form of MOSES. Also *Ital.* **Mo·i·se** (mō′ē·zä′), *Sp.* **Mo·i·ses** (mō′ē·säs′, *Pg.* mō-ē·zesh′).
Mois·san (mwä·sän′), **Henri,** 1852–1907, French chemist.
moist (moist) adj. 1 Having slight sensible wetness; damp; humid. 2 Tearful: *moist* eyes. 3 Marked by the presence of pus, phlegm, etc. [<OF *moiste* <a fusion of L *musteus* dew + *mucidus* moldy <*mucus* mucus] — **moist′ly** adv. — **moist′ness** n.
mois·ten (mois′ən) v.t. & v.i. To make or become moist. — **mois′ten·er** n.
mois·ture (mois′chər) n. Slight sensible wetness; a small amount of liquid exuding from, diffused through, or resting on a substance; dampness. [<OF *moisteur*]
mo·jar·ra (mə·här′ə) n. A large salt-water fish (genus *Gerres*), similar to a bass, inhabiting mostly tropical waters: also spelled *moharra.* [<Sp.]
Mo·ja·ve (mō·hä′vē) See MOHAVE.
Mojave Desert An arid region comprising part of the Great Basin in southern California; 15,000 square miles: also *Mohave Desert.*
Mo·ji (mō·jē) A port on northern Kyushu island, Japan.
moke (mōk) n. *Slang* 1 A Negro: a derogatory usage. 2 A boring fellow. 3 One who performs on several musical instruments: usually called **musical moke.** 4 *Brit. Dial.* A donkey. [Origin unknown]
Mo·kha (mō′kə) See MOCHA.
Mo·ki (mō′kē) See MOQUI.
Mok·po (môk·pō) A port of SW Korea. *Japanese* **Mop·po** (mōp·pō).
mol (mōl) n. *Chem.* The gram-molecule; that weight of a substance, expressed in grams, which is equal numerically to its molecular weight: also spelled *mole.* [<G]
mo·la (mō′lə) n. pl. **·lae** (-lē) Mole[4]. [<L]
mo·lal (mō′ləl) adj. *Chem.* 1 Pertaining to the mol or gram-molecule. 2 Designating a solution which has a concentration equivalent to one mol of the solute in 1,000 grams of the solvent. Compare MOLAR[1]. [<MOL + -AL] — **mo·lal′i·ty** (mō·lal′ə·tē) n.
mo·lar[1] (mō′lər) adj. 1 *Physics* Pertaining to a mass; acting on or exerted by a mass, as force. 2 *Chem.* Having or containing a grammolecular weight or mol; specifically, denoting a solution containing one mol of solute to the liter. Compare MOLAL. [<MOL + -AR] — **mo·lar′i·ty** (mō·lar′ə·tē) n.
mo·lar[2] (mō′lər) n. A grinding tooth with flattened crown, situated behind the canine and incisor teeth. — adj. 1 Grinding; adapted

for grinding. 2 Pertaining to a molar. [<L *molaris* <*mola* mill]
mo·las·ses (mə·las′iz) n. A viscid, darkcolored liquor drained off from raw cane or beet sugar; treacle. [<Pg. *melaço* <L *mellaceus* honeylike <*mel* honey]
mold[1] (mōld) v.t. 1 To work into a particular shape or form; model; shape. 2 To shape or cast in or as in a mold; make on a mold. 3 In founding, to form a mold of or from. 4 To ornament with molding. See synonyms under BEND, GOVERN, INFLUENCE. [<n.] — n. 1 A matrix for shaping anything in a fluid or plastic condition: distinguished from *cast.* 2 Hence, that after which something else is patterned, or the thing that is molded. 3 Form; nature; also, kind; character. 4 The physical form; shape: now applied to the human form. 5 A molding, or number of moldings. See synonyms under MODEL. Also spelled *mould.* [<OF *modle* <L *modulus* < *modus* measure, limit. Doublet of MODULE.] — **mold′a·ble** adj. — **mold′er** n.

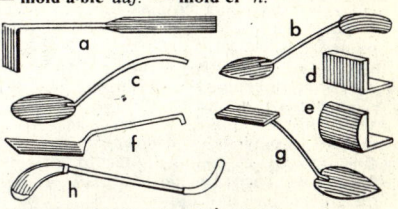

IRON-MOLDER'S TOOLS
a. Lifter. e. Half-round corner.
b. Taper and square. f. Yankee.
c. Oval or dog-tail. g. Heart and square.
d. Square corner. h. Flange and bead.

mold[2] (mōld) n. 1 Any fungous growth on food, clothing, walls, etc., especially such growths as form a woolly or furry coating on decaying vegetable matter or in moist, warm places. 2 Any of various fungi producing such growths. 3 Mustiness; decay. — v.t. & v.i. To become or cause to become moldy. Also spelled *mould.* [<obs. *moul* grow moldy <Scand. Cf. Dan. *mugle* grow moldy.]
mold[3] (mōld) n. 1 Earth that is fine and soft, and rich in organic matter. 2 The constituent material of anything; earthy material; matter. 3 The earth; ground; hence, a grave. — v.t. To cover with mold. Also spelled *mould.* [OE *molde* earth]
Mold (mōld) The county town of Flintshire, Wales.
Mol·dau (môl′dou) The German name for the VLTAVA.
Mol·da·vi·a (mol·dā′vē·ə) A historical province of eastern Rumania; 14,690 square miles. Also **Mol·do·va** (môl·dō′vä).
Mol·da·vi·an (mol·dā′vē·ən) adj. Of or relating to Moldavia. — n. 1 A native or naturalized inhabitant of Moldavia. 2 The Rumanian language of the Moldavians.
Moldavian Soviet Socialist Republic A constituent republic of SW European U.S.S.R.; 13,000 square miles; capital, Kishinev.
mol·da·vite (mol′də·vīt) n. 1 A dull-green natural glass resembling obsidian and thought to be of meteoritic origin. 2 A variety of ozocerite found in Moravia. [from the *Moldau,* near which it is found]
mold·board (mōld′bôrd′, -bōrd′) n. 1 *Agric.* The curved metal plate of a plow, by which the earth is turned over and pulverized. 2 A similar part of a machine for building roads: also spelled *mouldboard.* [<MOLD[3] + BOARD]
mold·er (mōl′dər) v.i. To decay gradually and turn to dust; crumble; waste away. — v.t. To cause to crumble. Also spelled *moulder.* See synonyms under DECAY. [Freq. of obs. *mold* crumble]
-mol·der combining form One who molds or fashions (a specific thing): *glass-molder, ironmolder.* [<MOLD[1]]
mold fungus A fungus which causes mold; specifically, any of an order (*Mucorales*) of phycomycetous fungi, as the common bread mold (*Rhizopus nigricans*).
mold·ing (mōl′ding) n. 1 The act of shaping with or as with a mold. 2 Anything made in or as in a mold. 3 *Archit.* a A more or less ornamental strip on some part of a structure.

b A cornice or other depressed or projecting decorative member on a surface or angle of any part of a building. Also spelled *moulding.*

mold·warp (mōld′wôrp) *n. Brit. Dial.* The European mole. Also spelled *molewarp, mouldwarp.* [OE *molde* soil + *weorpan* throw up]

mold·y (mōl′dē) *adj.* **mold·i·er, mold·i·est** Covered with mold; hence, old; musty: also spelled *mouldy.* **— mold′i·ness** *n.*

mole[1] (mōl) *n.* **1** A small permanent spot on the skin; a birthmark. **2** A stain or spot, as in a garment. [OE *māl*]

mole[2] (mōl) *n.* A small, insectivorous mammal (family *Talpidae*) with velvety fur, minute eyes, and very broad forefeet adapted for digging and forming extensive underground excavations. [ME *molle* <MLG. Cf. MDu. *molle.* Prob. related to MOLD[3].]

COMMON MOLE (About 5 1/2 inches long, with tail an inch long)

mole[3] (mōl) *n.* A jetty or breakwater, partially enclosing an anchorage or harbor. [<F *môle* <L *moles* great mass]

mole[4] (mōl) *n. Pathol.* A morbid mass formed in the womb by the degeneration of the partly developed ovum, and giving rise to false pregnancy: also spelled *mola.* [<F *môle* <L *mola* millstone, false conception]

mole[5] (mōl) See MOL.

Mo·lech (mō′lek) See MOLOCH.

mole cricket **1** A burrowing cricket (family *Gryllotalpidae*) with a soft, cylindrical body and broad, stout, molelike front legs, found in some sandy soils. For illustration see INSECTS (injurious). **2** Any of several related species.

mo·lec·u·lar (mə·lek′yə·lər) *adj.* **1** Of, pertaining to, or consisting of molecules. **2** Resulting from the action of molecules: *molecular* changes. [<NL *molecularis*]

molecular film A layer of a substance having a thickness of one molecule: also called *monolayer.*

molecular volume *Chem.* The molecular weight of a substance divided by its density under specified conditions, usually the boiling point and normal atmospheric pressure.

molecular weight *Chem.* The sum of the weights of the constituent atoms of a molecule; specifically, the weight of a molecule of any gas or vapor as compared with some standard gas, such as oxygen.

mol·e·cule (mol′ə·kyōōl) *n.* **1** *Physics* The smallest part of an element, substance, or compound that can exist freely in the solid, liquid, or gaseous state and still retain its composition and properties. **2** Any small particle. See synonyms under PARTICLE. [<F *molécule* <NL *molecula,* dim. of L *moles* mass]

mole·hill (mōl′hil′) *n.* A small heap or ridge of earth raised by a burrowing mole.

mole·skin (mōl′skin′) *n.* **1** The skin of a mole. **2** A heavy, twill fabric, usually cotton, having a thick, soft nap resembling moleskin: used chiefly for coats, jackets, etc.

mo·lest (mə·lest′) *v.t.* To annoy or harm by interference; disturb injuriously. See synonyms under ABUSE, PERSECUTE. [<OF *molester* <L *molestare* <*molestus* troublesome < *moles* mass, burden] **— mo·les·ta·tion** (mō′les·tā′shən, mol′es·tā′shən) *n.* **— mo·lest′er** *n.*

mole·warp (mōl′wôrp) See MOLDWARP.

Mol·fet·ta (môl·fet′tä) A port in Apulia, SE Italy, on the Adriatic.

Mo·lière (mô·lyâr′) Pseudonym of Jean Baptiste Poquelin, 1622–73, French dramatist.

moll (mol) *n. Slang* **1** A girl; a sweetheart. **2** The mistress of a thief or vagrant. **3** A prostitute. [<MOLL]

Moll (mol) Diminutive of Mary. [<*Molly,* dim. of MARY]

mol·lah (mol′ə) *n.* A Moslem complimentary title of address given to religious dignitaries; also, a judge. Also spelled *moolah, mullah.* [<Turkish *mulla* <Arabic *mawla* master, sir]

mol·le (mō·lā′) *n.* The sharp, astringent condiment extracted from the drupes of a tropical American pepper tree (*Schinus molle*).

mol·les·cent (mə·les′ənt) *adj.* Producing softness; softening. [<L *mollescens, -entis*] **— mol·les′cence** *n.*

mol·lient (mol′yənt) See EMOLLIENT.

mol·li·fy (mol′ə·fī) *v.t.* **·fied, ·fy·ing** **1** To make less angry; soothe; pacify; appease. **2** To reduce the violence or intensity of. See synonyms under ALLAY, TEMPER. [<F *mollifier* <LL *mollificare* <L *mollis* soft + *facere* make] **— mol′li·fi′a·ble** *adj.* **— mol′li·fi·ca′tion** *n.* **— mol′li·fi′er** *n.* **— mol′li·fy′ing·ly** *adv.*

mol·li·ti·es (mə·lish′i·ēz) *n. Pathol.* A softening of an organ or tissue. [<L]

Mol·lus·ca (mə·lus′kə) *n.* A large phylum of unsegmented, soft-bodied invertebrates, typically protected by a calcareous shell, and including snails, mussels, oysters, cuttlefish, squids, whelks, and limpets. [<NL]

mol·lus·coid (mə·lus′koid) *adj.* Like a mollusk. **—** *n.* One of the *Mollusca.* **— mol·lus·coi·dal** (mol′əs·koid′l) *adj.*

mol·lus·cum (mə·lus′kəm) *n. Pathol.* Any of various skin diseases, especially, **molluscum con·ta·gi·o·sum** (kən·tā′jē·ō′səm), caused by a filtrable virus and characterized by the formation of hard skin tubercles, usually on the face. [<NL]

mol·lusk (mol′əsk) *n.* Any member of the phylum *Mollusca.* Also **mol′lusc.** [<F *mollusque* <L *molluscus (nux)* soft, thin-shelled (nut) < *mollis* soft] **— mol·lus·can** (mə·lus′kən) *adj.* & *n.* **— mol·lus′cous** *adj.*

mol·ly·cod·dle (mol′ē·kod′l) *n.* Any excessively pampered or protected person; one who is coddled or coddles himself; a sissy. **—** *v.t.* **·dled, ·dling** To pamper; coddle. [<*Molly,* dim. of MARY, + CODDLE] **— mol′ly·cod′dler** *n.*

Mol·ly Ma·guire (mol′ē mə·gwīr′) **1** One of a secret society that terrorized inhabitants of the coal regions of eastern Pennsylvania (1867–77). **2** Originally, one of a secret society in Ireland organized (1843) to prevent evictions by terrorizing officers of the law. [So called because the members were often disguised as women]

Mol·nár (mōl′när), **Ferenc,** 1878–1952, Hungarian playwright.

mo·loch (mō′lok) *n.* A spiny Australian lizard (genus *Moloch*), resembling the horned toad. [<NL *Moloch,* genus name <MOLOCH]

Mo·loch (mō′lok) **1** In the Bible, a god of the Ammonites and Phoenicians to whom human sacrifices were offered. **2** Any system or principle involving merciless sacrifice: also spelled *Molech.* [<LL <Gk. <Hebrew *Mōlekh* a king]

Mo·lo·kai (mō′lə·kī′) An island in the central Hawaiian Islands; 260 square miles.

Mo·lo·tov (mô′lə·tôf) A former name for PERM.

Mo·lo·tov (mô′lə·tôf), **Vyacheslav Mikhailovich,** born 1890, U.S.S.R. statesman: original name *Skryabin.*

Molotov breadbasket A container filled with small incendiary bombs which are designed to scatter over a large area when dropped from aircraft.

Molotov cocktail See FRANGIBLE GRENADE under GRENADE.

molt (mōlt) *v.t.* & *v.i.* To cast off or shed (feathers, horns, skin, etc.) in preparation for replacement by new growth. **—** *n.* The molting process or season. Also spelled *moult.* [ME *mouten,* OE *bimūtian* exchange for <L *mutare* change] **— molt′er** *n.*

mol·ten (mōl′tən) Archaic past participle of MELT. **—** *adj.* **1** Reduced to fluid by heat; melted. **2** Made by molding; cast.

molten sea See BRAZEN SEA.

Molt·ke (môlt′kə), **Count Helmuth Karl Bernhard von,** 1800–91, Prussian field marshal. **— Helmuth Johannes Ludwig von,** 1848–1916, German general in World War I; nephew of preceding.

mol·to (môl′tō) *adv. Music* Much; very: *molto* adagio. [<Ital. <L *multum* much]

Mo·luc·ca Islands (mə·luk′ə) A widely scattered island group, comprising a province of Indonesia, between Celebes and New Guinea: 33,315 square miles: formerly *Spice Islands. Malay* **Ma·lu·ku** (mä·lōō′kōō). Also **Mo·luc′·cas.**

mo·ly (mō′lē) *n. pl.* **·lies** **1** A mythical plant of magic virtues, with a white flower and a black root: mentioned in the *Odyssey.* **2** A European wild garlic. **3** Molybdenum. [<L <Gk. *mōly*]

mo·lyb·date (mə·lib′dāt) *n. Chem.* A salt of molybdic acid.

mo·lyb·de·nite (mə·lib′də·nīt) *n.* A scaly, metallic, lead-gray, soft molybdenum disulfide, MoS_2: an important ore of molybdenum.

mo·lyb·de·num (mə·lib′də·nəm, mol′ib·dē′·nəm) *n.* A hard, heavy, silver-white, metallic element (symbol Mo) of the chromium group, occurring only in combination. It is used to harden steel. See ELEMENT. [<NL, alter. of L *molybdaena* lead, galena <Gk. *molybdaina* < *molybdos* lead]

mo·lyb·dic (mə·lib′dik) *adj. Chem.* Of, pertaining to, or containing molybdenum, especially in its higher valence.

mo·lyb·dous (mə·lib′dəs) *adj. Chem.* Of or pertaining to molybdenum, especially in its lower valence.

mom (mom) *n. U.S. Colloq.* Mother. [<MAMA]

Mom·ba·sa (mom·bä′sə, -bas′ə) A port in SE Kenya on **Mombasa Island** (7 square miles), separated from the mainland by Mombasa harbor.

mome (mōm) *n. Obs.* A stupid fellow; also, a buffoon. [? < MUM[1]]

mo·ment (mō′mənt) *n.* **1** A very short period of time; an instant; also, a point of time; definite period. **2** The present time. **3** Consequence or importance, as in influencing judgment or action. **4** *Stat.* The arithmetic mean of the deviations in a frequency distribution, each deviation being raised to the same power. **5** *Physics* **a** The product of a quantity and its distance to some significant related point: *moment* of area, *moment* of mass, etc. **b** The measure of a force with reference to its effect in producing rotation: also called *torque.* **6** The thing originating or causing; principle of movement or development; a moving force. See synonyms under WEIGHT. [<F <L *momentum* movement. Doublet of MOMENTUM.]

mo·men·tar·i·ly (mō′mən·ter′ə·lē, mō′mən·ter′·ə·lē) *adv.* **1** For a moment. **2** From moment to moment. **3** At any moment. Also **mo′ment·ly.**

mo·men·tar·y (mō′mən·ter′ē) *adj.* Lasting but a moment. See synonyms under TRANSIENT. **— mo′men·tar′i·ness** *n.*

mo·men·tous (mō·men′təs) *adj.* Of great importance; weighty. See synonyms under IMPORTANT, SERIOUS. **— mo·men′tous·ly** *adv.* **— mo·men′tous·ness** *n.*

mo·men·tum (mō·men′təm) *n. pl.* **·ta** (-tə) or **·tums** **1** *Mech.* The impetus of a moving body. **2** *Physics* The quantity of motion in a body as measured by the product of its mass and velocity. **3** An essential or constituent element. **4** *Music* An eighth rest. [<L. Doublet of MOMENT.] **— mo·men′tal** *adj.*

mom·ism (mom′iz·əm) *n.* Dominance of feminine values in a society, attributed to undue prolongation of maternal influence: a derogatory term. [<MOM + -ISM; coined by Philip Wylie, born 1902, U. S. author]

Momm·sen (mom′sən, -zən; *Ger.* mom′zən), **Theodor,** 1817–1903, German historian.

Mom·son lung (mom′sən) A respiratory device to aid persons to escape from a sunken submarine: invented by Rear Admiral Charles B. Momson, born 1896, U. S. Navy.

Mo·mus (mō′məs) In Greek mythology, the god of blame and mockery. [<L <Gk. *mōmos* blame, ridicule]

mon (mon) *n. Scot.* Man.

Mon (mōn) *n.* **1** One of the dominant native peoples of the Pegu region in Burma. **2** The Austro-Asiatic language of the Mons. Also *Peguan.*

mon– *combining form* Var. of MONO–.

Mo·na (mō′nə) The ancient name for the ISLE OF MAN. Also **Mo·na·pi·a** (mō·nä′pē·ə).

mon·ac·e·tin (mon·as′ə·tin) *n.* A colorless or pale-yellow, hygroscopic liquid, $C_5H_{10}O_4$, obtained by heating glycerol with glacial acetic acid: it is used as a solvent for basic dyes, and in making certain explosives: also spelled *monoacetin.* [<MON- + *acetin* a liquid ester of acetic acid]

mon·a·chism (mon′ə·kiz′əm) *n.* The monastic manner of life; monasticism. [<L *monachus* monk + -ISM] **— mon′a·chal** (-kəl) *adj.*

mon·ac·id (mon·as′id) See MONOACID.

Mon·a·co (mon′ə·kō, mə·nä′kō; *Fr.* mô·nà·kō′) An independent principality on the Mediterranean in SE France; 370 acres. **— Mon′a·can** *adj.*

mon·ad (mon′ad, mō′nad) *n.* **1** An indestructible unit; a simple and indivisible substance. **2** A minute, simple, single-celled organism, especially a flagellate infusorian. **3** In metaphysics, the one inseparable spirit in mankind manifesting itself in each person; also, the one inseparable spirit in nature. —*adj.* Of, pertaining to, or consisting of a monad: also **mo·nad'ic** or **·i·cal.** [<LL *monas, monadis* <Gk. *monas* a unit <*monos* alone]

mon·a·del·phous (mon'ə·del'fəs) *adj. Bot.* Having the stamens united by their filaments into a single set or tube, as in plants of the mallow family. Also **mon'a·del'phi·an.** [<MON- + Gk. *adelphos* brother]

mon·ad·ism (mon'ad·iz'əm, mō'nad-) *n.* A theory of monads in philosophy or physics. —**mon'ad·is'tic** *adj.*

mo·nad·nock (mə·nad'nok) *n. Geog.* An isolated hill or mass of rock rising above a peneplain. [from Mt. *Monadnock*]

Mo·nad·nock (mə·nad'nok), **Mount** An isolated peak in SW New Hampshire; 3,165 feet.

Mon·a·ghan (mon'ə·gən, -hən) A county of Ulster province, Ireland; 498 square miles; county seat, Monaghan.

Mo·na Li·sa (mō'nə lē'zə) A portrait by Leonardo da Vinci of a Neapolitan woman: also called *La Gioconda.*

mon a·mi (môn nà·mē') *French* My friend.

mo·nan·drous (mə·nan'drəs) *adj.* **1** Having one male or husband at a time. **2** *Bot.* Having one stamen to the flower. [<Gk. *monandros* <*mono-* single + *anēr, andros* male, man]

mo·nan·dry (mə·nan'drē) *n.* **1** The custom or practice of having only one husband at a time. **2** *Bot.* The condition of possessing only one perfect stamen, as in certain orchids. [<Gk. *monandria.* See MONANDROUS.]

mo·nan·thous (mə·nan'thəs) *adj. Bot.* Having but one flower: said of a peduncle or a whole plant. [<MON- + Gk. *anthos* flower]

Mo·na Passage (mō'nə) A strait between Hispaniola and Puerto Rico leading from the Atlantic to the Caribbean; about 80 miles wide.

mon·arch (mon'ərk) *n.* **1** A sovereign, as a king or emperor; in modern times, usually, a hereditary constitutional sovereign; originally, the sole ruler of a nation. **2** One who or that which surpasses others of the same kind. **3** A large, orange-brown butterfly (*Danaus menippe*) whose larva feeds on milkweed: also called *milkweed butterfly.* See synonyms under MASTER. [<LL *monarcha* <Gk. *monarchēs* <*monarchos* ruling alone <*monos* alone + *archein* rule] —**mo·nar·chal** (mə·när'kəl) *adj.* —**mo·nar'chal·ly** *adv.*

mo·nar·chi·an·ism (mə·när'kē·ən·iz'əm) *n.* A heretical doctrine of the second and third centuries which denied any real distinction between the persons of the Trinity. [<LL *monarchianus* <*monarchia* sovereignty of a single person <*monarcha* MONARCH] —**mo·nar'chi·an·is'tic** *adj.*

mo·nar·chi·cal (mə·när'ki·kəl) *adj.* Pertaining to, governed by, or favoring a monarch or monarchy. Also **mo·nar'chi·al, mo·nar'chic.** —**mo·nar'chi·cal·ly** *adv.*

mon·arch·ism (mon'ərk·iz'əm) *n.* Monarchical preferences or principles. —**mon'arch·ist** *n.* —**mon'arch·is'tic** *adj.*

mon·ar·chy (mon'ər·kē) *n. pl.* **·chies** **1** Government by a monarch; sovereign control. **2** A government or territory ruled by a monarch. —**absolute monarchy** A government in which the will of the monarch is positive law; a despotism. —**constitutional** or **limited monarchy** A monarchy in which the power and prerogative of the sovereign are limited by constitutional provisions. [<LL *monarchia* <Gk.]

Mo·nar·da (mə·när'də) *n.* A genus of aromatic American herbs of the mint family, with toothed leaves and large flowers in showy clusters; including the horsemint and Oswego tea. [<NL, after N. *Monardes,* 1493–1588, Spanish botanist]

mon·as (mon'əs, mō'nəs) *n.* A monad. [<Gk.]

mon·as·ter·y (mon'əs·ter'ē) *n. pl.* **·ter·ies** A dwelling place occupied in common by persons, especially monks, under religious vows of seclusion; also, the community of persons living in such a place. See synonyms under CLOISTER. [<LL *monasterium* <Gk. *monastērion* <*monastēs* a monk <*monazein* be alone <*monos* alone. Doublet of MINSTER.]

mo·nas·tic (mə·nas'tik) *adj.* **1** Pertaining to religious seclusion. **2** Characteristic of monasteries or their inhabitants; monkish. Also **mon·as·te·ri·al** (mon'əs·tir'ē·əl), **mo·nas'ti·cal.** —*n.* A monk or other religious recluse. [<F *monastique* <Med. L *monasticus* <Gk. *monastikos*] —**mo·nas'ti·cal·ly** *adv.*

mo·nas·ti·cism (mə·nas'tə·siz'əm) *n.* The monastic life; asceticism.

mon·a·tom·ic (mon'ə·tom'ik) *adj. Chem.* **1** Consisting of a single atom, as the molecules of certain elements. **2** Containing one replaceable or reactive atom. **3** Monovalent.

mon·ax·i·al (mon·ak'sē·əl) *adj.* Having but one axis; uniaxial.

mon·a·zite (mon'ə·zīt) *n.* A resinous, brownish-red or brown phosphate of the rare-earth metals, chiefly cerium, lanthanum, and didymium: an important source of thorium. [<G *monazit* <Gk. *monazein* be alone <*monos* alone]

mon cher (môn shâr') *French* My dear. —**ma chère** (mà shâr') *fem.*

Monck (mungk), **George** See MONK, GEORGE.

Monc·ton (mungk'tən) A city in SE New Brunswick, Canada.

Mon·day (mun'dē, -dā) *n.* The second day of the week. —**Black Monday** Easter Monday, 1360: so called on account of a sudden darkness followed by a violent hailstorm. [OE *mōn(an)dæg* day of the moon; trans. of L *lunae dies*]

monde (mônd) *n. French* The world; society. [<F <L *mundus* world]

Mond process (mond) *Metall.* A method for obtaining pure nickel by passing a stream of carbon monoxide over powdered nickel oxide and separating it from the resulting nickel-carbonyl vapor. [after Ludwig *Mond,* 1839–1909, English chemist born in Germany]

Mon·dri·an (môn'drē·än), **Piet,** 1872–1944, Dutch painter.

mo·ne·cious (mə·nē'shəs, mō-) See MONOECIOUS.

Mo·né·gasque (mō'nə·gask, *Fr.* mô·nā·gàsk') *French* A citizen of Monaco.

Mo·nel metal (mō·nel') A corrosion-resistant nickel alloy of copper, iron, and manganese, reduced from ore of the same composition: used for industrial equipment, machine parts, etc.: a trade name. [after Ambrose *Monel,* d. 1921, U. S. manufacturer]

Mo·net (mō·ne'), **Claude,** 1840–1926, French painter.

mon·e·tar·y (mon'ə·ter'ē, mun'-) *adj.* Pertaining to money, finance, or currency; consisting of money; pecuniary. See synonyms under FINANCIAL. [<L *monetarius* of a mint <*moneta* mint. See MINT.] —**mon'e·tar'i·ly** *adv.*

mon·e·tize (mon'ə·tīz, mun'-) *v.t.* **·tized, ·tiz·ing** **1** To legalize as money. **2** To give a standard value to (a metal) as currency. **3** To coin into money. Also *Brit.* **mon'e·tise.** [<L *moneta* mint, money] —**mon'e·ti·za'tion** *n.*

mon·ey (mun'ē) *n. pl.* **mon·eys** or **mon·ies** **1** Anything that serves as a common medium of exchange in trade, as coin or notes. ◆ Collateral adjective: *pecuniary.* **2** Legal tender for debts. **3** Purchasing power; credit; bank deposits, etc.; a denomination of value or unit of account. **4** Wealth; property. **5** *pl.* Cash payments or receipts. **6** A system of coinage. —**call money** Money loaned on security, or deposited in a bank, subject to repayment on demand of the lender. —**hard money** Metallic currency or specie. [<OF *moneie* <L *moneta.* Doublet of MINT.]

Synonyms: bills, bullion, capital, cash, coin, currency, funds, gold, notes, property, silver, specie. *Money* is the authorized medium of exchange; coined *money* is called *coin* or *specie.* What are termed in England *banknotes* are in the United States commonly called *bills*: a five-dollar *bill. Cash* is *specie* or *money* in hand, or paid in hand: the *cash* account; the *cash* price. In the legal sense, *property* is not *money,* and *money* is not *property*; for *property* is that which has inherent value, while *money,* as such, has but representative value, and may or may not have intrinsic value. *Bullion* is either *gold* or *silver* uncoined or the coined metal considered without reference to its coinage, but simply as merchandise, when its value as *bullion* may be very different from its value as *money.* The word *capital* is used chiefly of accumulated *property* or *money* invested in productive enterprises or available for such investment. Compare PROPERTY, WEALTH.

mon·ey·bags (mun'ē·bagz') *n. Slang* **1** A rich man. **2** Wealth.

mon·ey-chang·er (mun'ē·chān'jər) *n.* A person who changes money at a prescribed rate. Also **mon'ey-deal'er** (-dē'lər), **mon'ey-job'ber** (-job'·ər).

money cowry See COWRY.

mon·eyed (mun'ēd) *adj.* **1** Possessed of money; wealthy. **2** In the form of money. Also spelled *monied.*

mon·ey·er (mun'ē·ər) *n.* **1** A coiner of money; minter. **2** A broker; banker. [<OF *monoier* maker of coins <L *monetarius* mint master <*moneta* mint. See MINT.]

mon·ey-lend·er (mun'ē·len'dər) *n.* A person whose business is the lending of money at interest.

mon·ey-mak·ing (mun'ē·mā'king) *adj.* **1** Bent upon and successful in accumulating wealth. **2** Likely to bring in money; profitable. —*n.* The acquisition or procurement of money or wealth. —**mon'ey-mak'er** *n.*

money market The market in which money is the commodity bought and sold; the sphere of financial operations.

money of account A monetary denomination used in keeping accounts, but not represented by a coin, as the mill of the United States.

money order An order for the payment of a specified sum of money; specifically, such an order issued at one post office or telegraph office and payable at another.

mon·ey·wort (mun'ē·wûrt') *n.* A trailing herb (*Lysimachia nummularia*) of the primrose family with solitary yellow flowers and rounded leaves. [<MONEY + WORT; trans. of NL *Nummularia* <*nummus* a coin]

mon·ger (mung'gər, mong'-) *n.* **1** *Brit.* A dealer or trader: chiefly in compounds: *fishmonger.* **2** One who engages in discreditable matters: chiefly in compounds: a *scandalmonger.* [OE *mangere* <*mangian* traffic]

Mon·gol (mong'gəl, -gol, -gōl) *adj.* Of or pertaining to Mongolia or its inhabitants. —*n.* **1** A member of any of the native tribes of Mongolia; specifically, a Mongol (eastern Mongolia), a Buriat (Siberia), or an Eleut or Kalmuck (western Mongolia). **2** The Mongolian language of any of these peoples. **3** Any member of the Mongoloid race. [<Mongolian *mong* brave]

Mon·go·lia (mong·gō'lē·ə, mon-) A region of east central Asia south of Asiatic Russian S.F.S.R., east and north of China's Sinkiang-Uigur Autonomous Region, west of most of former Manchuria and the rest of NE China, and north of central China; about 1,000,000 square miles; divided into: (1) the **Mongolian People's Republic** (formerly *Outer Mongolia*), an independent country in the northern and western part; 590,966 square miles; capital, Ulan Bator; (2) *Inner Mongolia,* a region of northern China, including the former provinces of Ningsia, Suiyuan, Chahar, and Jehol and western (former) Manchuria, most of which region now comprises: (3) the *Inner Mongolian Autonomous Region,* an autonomous division of central northern China, including parts of western (former) Manchuria, and former Chahar and Jehol provinces and all of former Suiyuan province; over 400,000 square miles; capital, Huhehot (formerly *Kweisui*).

Mon·go·li·an (mong·gō'lē·ən, -gōl'yən, mon-) *adj.* **1** Of or pertaining to Mongolia, its people, or their languages; Mongol. **2** Exhibiting Mongolism. —*n.* **1** A native of Mongolia. **2** A subfamily of the Altaic languages, including the languages of the Mongols.

Mon·gol·ic (mong·gol'ik, mon-) *adj.* Of or peculiar to the Mongols; Mongolian. —*n.* Any of the Mongolian languages.

Mon·gol·ism (mong'gəl·iz'əm) *n.* A form of arrested physical and mental development characterized by slanting eyes and other facial resemblances to the Mongoloid ethnic type.

Mon·go·loid (mong'gə·loid) *adj.* **1** *Anthropol.* Of or pertaining to the so-called yellow race, with skin color varying from light yellow to

add,āce,câre,pälm; end,ēven; it,īce; odd,ōpen,ôrder; to͝ok,po͞ol; up,bûrn; ə = a in *above,* e in *sicken,* i in *clarity,* o in *melon,* u in *focus*; yo͞o = u in *fuse*; oi,oil; ou,pout; ch,check; g,go; ng,ring; th,thin; th,this; zh,vision. Foreign sounds à,œ,ü,kh,ṅ; and ◆: see page xx. < from; + plus; ? possibly.

mongoose (mong'gōōs, mung'-) n. pl. ·goos·es A small, ferretlike mammal (family *Viverridae*) which fearlessly attacks and kills venomous snakes; especially, the **Indian mongoose** (*Herpestes nyula*): often written *mungoose*. [<Marathi *muṅgūs*]

MONGOOSE
(The Indian mongoose has a body about 1 1/4 feet long, with tail 1 1/4 feet)

mon·grel (mung'grəl, mong'-) n. 1 The progeny of crossed breeding; sometimes restricted to the progeny of artificial varieties: distinguished from a *hybrid*; specifically, a dog of mixed breed. 2 Any incongruous mixture. —*adj.* Of mixed breed or origin; specifically, of a word or language made up of other words or languages: often a term of contempt. [< obs. *mong* mixture <OE *gemang* + *-rel*, dim. suffix]
'mongst (mungst) *prep.* Amongst: an aphetic form.
Mon·gu (mong'gōō) The capital of Barotseland, Northern Rhodesia.
mon·ied (mun'ēd) See MONEYED.
Mon·i·er–Wil·liams (mun'ē·ər·wil'yəmz, mon'-), **Sir Monier**, 1819–99, English Sanskrit scholar.
mon·i·ker (mon'ə·kər) n. *Slang* 1 A name; a signature. 2 A tramp's initials or other mark of identification. Also **mon'ick·er**. [Prob. blend of MONOGRAM and MARKER]
mon·i·li·a·sis (mon'ə·lī'ə·sis) n. *Pathol.* A disease caused by infection with any of various gas-forming fungi (family *Moniliaceae*) Also **mon'i·li'o·sis**. [<NL <L *monile* necklace + -IASIS; from the alternating swellings and constrictions caused by it]
mo·nil·i·form (mō·nil'ə·fôrm) *adj. Biol.* Resembling a string of beads; contracted or jointed at regular intervals so as to resemble a necklace. [<L *monile* necklace + -FORM]
mon·ish (mon'ish) *v.t. Obs.* To admonish. [See ADMONISH.]
mon·ism (mon'iz·əm, mō'niz·əm) n. 1 The doctrine of cosmology that attempts to explain the phenomena of the cosmos by one principle of being or ultimate substance: opposed to philosophical *dualism* and *pluralism*. 2 Any theory that refers many different facts to a single principle. 3 See MONOGENESIS (def. 1). [<NL *monismus* <Gk. *monos* single] — **mon'ist** n. — **mo·nis'tic** or **·ti·cal** *adj.* — **mo·nis'ti·cal·ly** *adv.*
mo·ni·tion (mō·nish'ən) n. 1 Friendly counsel given by way of warning and implying caution or reproof; admonition. 2 Indication; notice. 3 *Law* A summons or citation in civil law and admiralty practice. [<F <L *monitio, -onis* <*monitus*, pp. of *monere* warn] — **mon'i·tive** (mon'ə·tiv) *adj.*
mon·i·tor (mon'ə·tər) n. 1 One who advises or cautions. 2 A senior pupil placed in charge of a dormitory or class. 3 Something that warns or advises; a reminder. 4 *Nav.* An ironclad vessel having a low, flat deck and low freeboard, and fitted with a blister and with one or more turrets carrying heavy guns; specifically, the first vessel of the type, "**The Monitor.**" See MERRIMAC. 5 *Zool.* Any of several large carnivorous lizards (family *Varanidae*) of the Old World tropics; especially, the East Indian kabara-goya (*Varanus salvator*), which reaches a length of seven feet. 6 *Mining* A contrivance, consisting of nozzle and holder, whereby the direction of a stream can be readily changed. 7 *Telecom.* **a** A high-fidelity loudspeaker in the control-room of a radio studio, used to insure adequate sound transmission. **b** A receiver for listening to a station's broadcasts to check on quality and frequency of transmission, compliance with laws, material transmitted, etc. — *v.t. & v.i.* 1 *Telecom.* To listen to (a station, broadcast, etc.) with or as with a monitor. 2 To have charge of (a person or group) as a monitor. [<L <*monitus*. See MONITION.] — **mon'i·tor·ship'** n. — **mon'i·tress** n. *fem.*
mon·i·to·ri·al (mon'ə·tôr'ē·əl, -tō'rē-) *adj.* 1 Pertaining to a monitor or to instruction by monitors. 2 Monitory.
mon·i·to·ry (mon'ə·tôr'ē, -tō'rē) *adj.* Conveying monition; admonitory: a *monitory* look. — *n.* An ecclesiastical monition. [<L *monitorius.* See MONITION.]
monitory letter A papal letter of monition.
monk (mungk) n. 1 Formerly, a religious hermit. 2 One of a company of men vowed to separation from the world and to poverty, celibacy, and religious duties; a member of a monastic order. 3 *Printing* An area on a printed page or sheet containing too much ink: opposed to *friar.* [OE *munuc* <LL *monachus* <Gk. *monachos* <*monos* alone] — **monk'ish** *adj.* — **monk'ish·ly** *adv.*
Monk (mungk), **George**, 1608–70, Duke of Albemarle; English soldier. Also spelled *Monck*.
monk·er·y (mungk'ər·ē) n. pl. ·er·ies 1 Monastic life, ways, or beliefs: generally used opprobriously. 2 A monastery or its inmates.
mon·key (mung'kē) n. 1 Any of a group of primates (suborder *Anthropoidea*) having elongate limbs, hands and feet adapted for grasping, and a highly developed nervous system, including marmosets, baboons, and macaques, but not the anthropoid apes. 2 Any primate below man, especially the smaller arboreal forms, as lemurs and tarsiers. 3 A person regarded as a monkey, as a mischievous child. 4 One of various small articles or contrivances; especially, an iron block or ram with a catch, used in pile-driving by hoisting and dropping. — *v.i.* To play or trifle; fool; meddle: often with *with* or *around with*. — *v.t. Rare* To imitate or ape; mimic. [? <MLG *Moneke*, name of the son of Martin the Ape in *Reynard the Fox.* Cf. MF *monne,* Sp. *mona* a female ape.]
monkey bread See BAOBAB.
monkey business *Slang* Foolish tricks; deceitful or mischievous behavior; folly.
monkey chatter In electronics slang, garbled speech, music, or other sound signals in radio reception, caused by interference of adjoining frequencies with the carrier wave of the desired channel.
monkey cup An East Indian pitcherplant; any species of the genus *Nepenthes.*
monkey flower Any one of various figworts of the genus *Mimulus,* especially the cultivated species, *M. luteus* with yellow flowers, and *M. cardinalis* with red: so called from the gaping or grimacing appearance of the corolla.
monkey gaff *Naut.* A gaff attached to the mizzentopmast of a vessel for the display of signals.
monkey jacket 1 A short jacket of coarse material, worn especially by sailors. 2 *Slang* A dinner jacket.
monkey jar A large, undecorated earthenware jar, used for cooling drinking water.
monkey pot 1 The hard, woody, pot- or urn-shaped fruit of several tropical American trees of the family *Lecythidaceae,* especially *Lecythis ollaria* and *L. zabucajo*; also, the plant, the fruit of which has a circular lid which, when the nutlike seeds are ripe, separates with a cracking sound. 2 A barrel-shaped melting pot used in making flint glass.
monkey puzzle A large Chilean tree (*Araucaria araucana*) yielding a hard, durable, yellowish–white wood and edible seeds.
mon·key·shines (mung'kē·shīnz') n. pl. U.S. *Slang* Frolicsome tricks like a monkey's; pranks. [<MONKEY + SHINE, n. (def. 4)]
monkey wrench A wrench having an adjustable jaw for grasping a nut, bolt, or the like.
monk·fish (mungk'fish') n. pl. ·fish or ·fish·es 1 The angelfish. 2 The angler. [So called from a hoodlike protuberance suggesting a monk's cowl]
Mon–Khmer (mōn'kmer) n. A subfamily of the Austro-Asiatic family of languages, spoken chiefly in Indochina, including Mon, Khmer, and the Annamese dialects.
monk·hood (mungk'hŏŏd) n. 1 The character or condition of a monk. 2 Monks collectively. [OE *munukhade*]
monks·hood (mungks'hŏŏd') n. 1 A plant of the genus *Aconitum*, especially the poisonous *A. napellus*, having the upper sepal arched at the back like a hood. 2 Aconite.

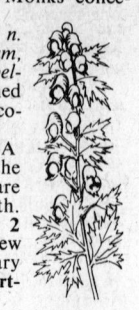

MONKS-HOOD
(Plants from 1 to 6 feet tall, varying with species)

Mon·mouth (mon'məth) 1 A county in western England on the border of Wales; 546 square miles; county town, Monmouth. Also **Mon'mouth·shire** (-shir). 2 A county in east central New Jersey; site of the Revolutionary War battle of **Monmouth Courthouse**, June 28, 1778.
Mon·mouth (mon'məth), **Duke of**, 1649–85, James Scott, an illegitimate son of Charles II of England, who claimed the throne and led an insurrection against James II; beheaded.
mono- *combining form* 1 Single; one. 2 *Chem.* Denoting the presence in a compound of a single atom, or an equivalent of the element or radical to the name of which it is prefixed: *monobasic.* Also, before vowels, **mon-**. [<Gk. <*monos* single, alone]
mon·o·ac·e·tin (mon'ō·as'ə·tin) See MONACETIN.
mon·o·ac·id (mon'ō·as'id) *adj. Chem.* Possessing one hydroxyl group that can replace the hydrogen of an acid: said of bases: also spelled *monacid.* Also **mon'o·a·cid'ic** (-ə·sid'ik).
mon·o·ba·sic (mon'ō·bā'sik) *adj. Chem.* Possessing but a single hydrogen atom replaceable by a metal or positive radical: applied to acids.
Mo·noc·a·cy River (mə·nok'ə·sē) A river in Pennsylvania and northern Maryland which rises near Gettysburg and flows 60 miles SW to the Potomac; scene of Civil War battle in 1864 near Frederick, Maryland.
mon·o·carp (mon'ō·kärp) n. *Bot.* A plant that yields fruit only once before dying.
mon·o·car·pel·lar·y (mon'ō·kär'pə·ler'ē) *adj. Bot.* Consisting of a single carpel.
mon·o·car·pic (mon'ō·kär'pik) *adj. Bot.* Bearing fruit only one time during its existence: said of annual or biennial plants and of some trees: a *monocarpic* palm. Also **mon'o·car'pal, mon'o·car'pi·an**.
mon·o·car·pous (mon'ō·kär'pəs) *adj. Bot.* Having a gynoecium composed of a single carpel.
Mo·noc·er·os (mə·nos'ər·əs) A southern constellation, the Unicorn. See CONSTELLATION. [<NL <Gk. *mono-* single + *keras* horn]
mon·o·cha·si·um (mon'ə·kā'zē·əm, -zhē-) n. *Bot.* A uniparous cyme; a cyme having only one lateral axis. [<NL <Gk. *mono-* single + *chasis* division] — **mon'o·cha'si·al** *adj.*
mon·o·chlo·ride (mon'ə·klôr'īd, -klō'rīd) n. *Chem.* A chloride which contains one chlorine atom in each molecule.
mon·o·chord (mon'ə·kôrd) n. 1 A single chord or string on which intervals can be marked in mathematical ratios, the vibrations giving the notes in the musical scale. 2 An acoustical instrument with one string and a movable bridge, used for measuring intervals. [<Med. L *monochordus* <Gk. *mono-* single + *chordē* string]
mon·o·chro·mat·ic (mon'ə·krō·mat'ik) *adj.* Of one color. Also **mon'o·chro'ic**. [<Gk. *monochrōmatikos*] — **mon'o·chro·mat'i·cal·ly** *adv.*
mon·o·chrome (mon'ə·krōm) n. A painting or the art of painting in a single color, or different shades of a single color. — *adj.* Monochromatic. [<Gk. *monochrōmos*] — **mon'o·chro'mic, mon'o·chro'mi·cal** *adj.* — **mon'o·chro'mist** n.
mon·o·cle (mon'ə·kəl) n. An eyeglass for one eye. [<F <LL *monoculus* one-eyed <Gk. *monos* single + L *oculus* eye] — **mon'o·cled** *adj.*
mon·o·cli·nal (mon'ə·klī'nəl) *adj. Geol.* Having an inclination in only one direction, or composed of rock strata so inclined. — *n.* A monocline. — **mon'o·cli'nal·ly** *adv.*
mon·o·cline (mon'ə·klīn) n. *Geol.* A stratum or fold of rocks inclined in only one direction. [<MONO- + Gk. *klinein* incline]
mon·o·clin·ic (mon'ə·klin'ik) *adj.* Pertaining to or designating a crystal system having two oblique axes and a third perpendicular to both.

mon·o·cli·nous (mon′ə-klī′nəs) *adj. Bot.* Containing both androecium and gynoecium in the same flower; bisexual; hermaphrodite. [<NL *monoclinus*]

mo·no·coque (mō-nō-kôk′) *n. French Aeron.* An airplane construction having a shell-shaped fuselage or nacelle in which the skin carries the main stresses.

mon·o·cot·y·le·don (mon′ə-kot′ə-lēd′n) *n.* Any of a great subclass (*Monocotyledones*) of seed plants (*Angiospermae*) bearing one cotyledon in the embryo, including palms, orchids, lilies, and grasses. Compare DICOTYLEDON. Also **mon′o·cot.** — **mon′o·cot′y·le′do·nous** *adj.*

mon·o·crat (mon′ə-krat) *n.* A person in favor of rule by a monarch: term used by Thomas Jefferson about 1790 to mean a Federalist sympathizer with England in the war between England and France. — **mon′o·crat′ic** *adj.* — **mo·noc·ra·cy** (mo-nok′rə-sē) *n.*

mo·noc·u·lar (mə-nok′yə-lər) *adj.* 1 One-eyed. 2 Of or pertaining to one eye. Also **mo·noc′u·lous.** [See MONOCLE]

mon·o·cul·ture (mon′ə-kul′chər) *n. Agric.* The use of a given tract of land for the intensive cultivation of only one crop or product, as cotton, wool, tobacco, etc.

mon·o·cy·cle (mon′ə-sī′kəl) *n.* A one-wheeled vehicle.

mon·o·cyte (mon′ə-sīt) *n. Physiol.* A large, white blood corpuscle with a horseshoe-shaped nucleus surrounded by clear cytoplasm.

mon·o·dac·ty·lous (mon′ə-dak′tə-ləs) *adj. Zool.* Having only one toe, finger, or claw. Also **mon′o·dac′tyl.** [<Gk. *monodaktylos* < *monos* single + *daktylos* finger]

MONOCYCLE

Mon·o·del·phi·a (mon′ə-del′fē·ə) See EUTHERIA.

mon·o·dra·ma (mon′ə-drä′mə, -dram′ə) *n.* A drama written for or acted by a single performer.

mon·o·dy (mon′ə-dē) *n. pl.* **·dies** 1 Any melancholy literary composition with a single emotional motive; especially, a poem on the death of a friend. Compare THRENODY. 2 In Greek tragedy, the lyric solo, usually of a somber character; hence, a dirge. 3 *Music* **a** A composition in which one vocal part predominates, or the style of such a composition; homophony: opposed to *polyphony*. **b** A song for a single voice, with instrumental accompaniment. 4 A monotonous sound; unvarying tone: the *monody* of waves. [<LL *monodia* <Gk. *monōidia* < *monōidos* singing alone < *monos* alone + *aedein* sing] — **mo·nod·ic** (mə-nod′ik), **mo·nod′i·cal** *adj.* — **mon·o·dist** *n.*

mo·noe·cious (mə-nē′shəs) *adj. Bot.* Having male and female organs in the same individual, as stamens and pistils in separate blossoms on the same plant. Also spelled *monecious, monoicous.* Also **mo·ne′cian.** [< MON- + Gk. *oikos* house]

mo·nog·a·mist (mə-nog′ə-mist) *n.* 1 One who has only one living spouse: opposed to *bigamist* and *polygamist.* 2 One who does not practice or believe in second marriage after the death of the first spouse: opposed to *digamist.* — **mo·nog′a·mis′tic** *adj.*

mo·nog·a·mous (mə-nog′ə-məs) *adj.* 1 Pertaining to monogamy: *monogamous* practices. 2 Having only one spouse; holding to monogamy. 3 Having or paired with but one mate, as certain birds. 4 *Biol.* Having flowers with the anthers united. Also **mon·o·gam·ic** (mon′ə-gam′ik). [<LL *monogamus*] — **mo·nog′a·mous·ly** *adv.*

mo·nog·a·my (mə-nog′ə-mē) *n.* 1 The principle or practice of single marriage: opposed to *bigamy* and *polygamy*, and to *digamy.* 2 The habit of pairing with or having but one mate. [<F *monogamie* <LL *monogamia* <Gk. < *monos* single + *gamos* marriage]

mon·o·gen·e·sis (mon′ə-jen′ə·sis) *n. Biol.* 1 Oneness of origin; the doctrine of the descent of all living organisms from a single cell. Compare POLYGENESIS. 2 Asexual reproduction. 3 Direct development of an ovum into an organism which resembles the parent. 4 Monogenism. [<NL]

mon·o·ge·net·ic (mon′ə-jə·net′ik) *adj.* 1 Pertaining to or exhibiting monogenesis or monogenism. 2 Asexual. 3 *Geol.* Resulting from one genetic process, as a group of mountains. 4 Giving only one color to textile fabrics: said of dyestuffs.

mon·o·gen·ic (mon′ə-jen′ik) *adj.* 1 *Biol.* Having but one method of reproduction, specifically asexual reproduction. 2 *Geol.* Having parts all of the same nature: said of certain rocks.

mo·nog·e·nism (mə-noj′ə-niz′əm) *n.* The doctrine that the whole human race is of one blood or species. [<MONO- + -GEN + -ISM] — **mo·nog′e·nist** *n.*

mo·nog·e·nous (mə-noj′ə-nəs) *adj.* Asexual. [<MONO- + -GENOUS]

mo·nog·e·ny (mə-noj′ə-nē) *n.* 1 Monogenesis. 2 Monogenism. ◆ Homophone: *monogyny.* [<MONO- + -GENY]

mon·o·gram (mon′ə-gram) *n.* 1 A character consisting of two or more letters interwoven into one, as the initials of several names. 2 A single character in writing, or a mark representing a word. [<LL *monogramma* <Gk. *monos* single + *gramma* letter] — **mon′o·gram·mat′ic** (-grə-mat′ik) *adj.*

mon·o·graph (mon′ə-graf, -gräf) *n.* A description or systematic exposition of one thing or class of things; a dissertation or treatise written in great detail. — **mo·nog·ra·pher** (mə-nog′rə-fər) *n.* — **mon′o·graph′ic** *adj.* — **mo·nog′ra·phy** *n.*

mo·nog·y·ny (mə-noj′ə-nē) *n.* The custom or practice of having only one wife at a time. Compare MONANDRY. ◆ Homophone: *monogeny.* [<MONO- + Gk. *gynē* woman] — **mo·nog′y·nist** *n.* — **mo·nog′y·nous** *adj.*

mon·o·hy·brid (mon′ō·hī′brid) *n. Biol.* A hybrid offspring of parents differing in one characteristic.

mon·o·hy·drate (mon′ō·hī′drāt) *n. Chem.* The union of a single molecule of water with an element or a compound.

mon·o·hy·dric (mon′ō·hī′drik) *adj. Chem.* Possessing a single hydroxyl radical.

mo·noi·cous (mə-noi′kəs) See MONOECIOUS.

mo·nol·a·try (mə-nol′ə·trē) *n.* Worship of some one of several gods: distinguished from *monotheism.* [<MONO- + -LATRY] — **mo·nol′a·ter, mo·nol′a·trist** *n.* — **mo·nol′a·trous** *adj.*

mon·o·lay·er (mon′ō·lā′ər) *n.* A molecular film.

mon·o·lith (mon′ə-lith) *n.* A single piece or block of stone fashioned or placed by art, particularly one notable for its size; any structure or sculpture in stone formed of a single piece, as a menhir or an obelisk. — **mon′o·lith′ic** *adj.*

mon·o·log (mon′ə-lôg, -log) *n.* 1 That which is spoken by one person alone; especially, a dramatic soliloquy, or a story or drama told or performed by one person; also, a lengthy speech by one person, occurring in the course of conversation. 2 A literary composition, or a poem, written as a soliloquy. Also **mon′o·logue.** [<F *monologue* <Gk. *monologos* speaking alone < *monos* alone + *logos* discourse] — **mon′o·log′ic** (-loj′ik) or **·i·cal** *adj.* — **mo·nol′o·gist** (mə-nol′ə-jist) *n.* — **mo·nol′o·gy** *n.*

mon·o·ma·ni·a (mon′ə-mā′nē·ə, -mān′yə) *n.* 1 A mental disorder characterized by obsession with one idea. 2 The unreasonable pursuit of one idea; a craze. [<NL] — **mon′o·ma′ni·ac** *n.* — **mon·o·ma·ni·a·cal** (-mə·nī′ə·kəl) *adj.*

mon·o·mer (mon′ə-mər) *n. Chem.* The structural unit of a polymer: either a single molecule or a substance consisting of identical molecules. [<MONO- + Gk. *meros* part]

mon·o·mer·ic (mon′ə·mer′ik) *adj.* 1 Having or consisting of a single part. 2 *Zool.* Derived from one segment or part. [<MONOMER]

mon·om·er·ous (mə-nom′ər·əs) *adj. Bot.* Having a single member in each whorl or circular series: said of a flower. Sometimes written *1-merous.* [<MONO- + -MEROUS]

mon·o·met·al·ism (mon′ō·met′əl·iz′əm) *n.* The theory or system of a single metallic standard in coinage. Also **mon′o·met′al·ism.** — **mon′o·met′al·ist** *n.*

mon·o·me·tal·lic (mon′ō·mə-tal′ik) *adj.* 1 Consisting of a single metal: a *monometallic* currency. 2 Using or favoring but one metal as a standard of value: a *monometallic* country.

mo·nom·e·ter (mə-nom′ə-tər) *n.* 1 A line of verse having one metrical foot. 2 Verse consisting of monometers. [<MONO- + METER²]

mon·o·met·ric (mon′ə-met′rik) *adj.* Isometric.

mo·no·mi·al (mō-nō′mē·əl) *adj.* Consisting of a single term: a *monomial* expression. — *n. Math.* An expression consisting of a single term. [<MONO- + *-nomial*, as in *binomial*; an irregular formation]

mon·o·mo·lec·u·lar (mon′ō·mə-lek′yə-lər) *adj. Chem.* Having a thickness of only one molecule.

mon·o·mor·phic (mon′ə-môr′fik) *adj.* 1 *Biol.* Of the same or an essentially similar type of structure; also, having the same form throughout successive stages of development. 2 *Bot.* Forming similar spores: said of fungi. Also **mon′o·mor′phous.**

Mo·non·ga·he·la River (mə-nong′gə-hē′lə) A river in West Virginia and western Pennsylvania, flowing 128 miles NE to the Allegheny River at Pittsburgh.

mon·o·nu·cle·o·sis (mon′ō·no͞o′klē·ō′sis, -nyo͞o-) *n. Pathol.* The presence in the blood of an abnormal number of leucocytes, as in glandular fever. [<NL <MONO- + NUCLE(US) + -OSIS]

mon·o·pet·al·ous (mon′ə·pet′əl·əs) *adj. Bot.* 1 Gamopetalous. 2 Having corollas actually consisting of a single laterally placed petal: applicable to a few flowers.

mo·noph·a·gous (mə-nof′ə·gəs) *adj.* Monotrophic. [<MONO- + -PHAGOUS]

mon·o·pho·bi·a (mon′ə·fō′bē·ə) *n. Psychiatry* Morbid fear of solitude.

mon·o·phon·ic (mon′ə·fon′ik) *adj.* Of or pertaining to a monody; monodic.

mon·oph·thong (mon′əf·thông, -thong) *n. Phonet.* 1 A vowel, or single vowel sound. 2 A vowel digraph, or two written vowels with a simple sound. [<MONO- + Gk. *phthongos* sound] — **mon′oph·thon′gal** *adj.*

mon·o·phy·let·ic (mon′ō·fī·let′ik) *adj.* 1 *Zool.* Of or pertaining to a single phylum. 2 *Biol.* Derived from one parent form.

mon·o·phyl·lous (mon′ə·fil′əs) *adj. Bot.* Having or composed of one leaf.

Mo·noph·y·site (mə-nof′ə·sīt) *n.* One of a Christian sect originating in the fifth century which affirms that Christ had but one nature, the divine alone or a single compounded nature, and not two natures so united as to preserve their distinctness. [<LGk. *monophysitēs* < *monos* single + *physis* nature] — **Mon·o·phy·sit·ic** (mon′ō·fi·sit′ik) *adj.* — **Mo·noph′y·sit′ism** *n.*

mon·o·plane (mon′ə-plān) *n. Aeron.* An airplane with only one supporting surface: distinguished from *biplane* and *triplane.*

mon·o·ple·gi·a (mon′ə·plē′jē·ə) *n. Pathol.* Paralysis of one part of the body. [<NL] — **mon′o·ple′gic** *adj.*

mon·o·pode (mon′ə-pōd) *n.* 1 Anything sustained by one foot; particularly, one of a fabulous Ethiopian race with only one leg. 2 A monopodium. — *adj.* One-footed: also **mon·o·po·di·al** (mon′ə-pō′dē·əl). [<LL *monopodius* <Gk. *monos* single + *pous, podos* foot]

mon·o·po·di·al (mon′ə·pō′dē·əl) *adj.* 1 Of or pertaining to a monopode. 2 *Bot.* Having but one main or primary axis, as ordinary plants. — **mon′o·po′di·al·ly** *adv.*

mon·o·po·di·um (mon′ə·pō′dē·əm) *n. pl.* **·di·a** (-dē·ə) *Bot.* A stem or axis of growth, as in the pine and other conifers, formed by the continued development of a terminal bud, all branches originating as lateral appendages. [<NL]

mo·nop·o·lize (mə-nop′ə-līz) *v.t.* **·lized, ·liz·ing** 1 To obtain or exercise a monopoly of. 2 To assume exclusive possession or control of: to *monopolize* one's time. Also *Brit.* **mo·nop′o·lise.** — **mo·nop′o·li·za′tion** *n.* — **mo·nop′o·liz′er** *n.*

mo·nop·o·ly (mə-nop′ə-lē) *n. pl.* **·lies** 1 The exclusive right or privilege of engaging in a particular traffic; especially, such control, as of a commodity, as allows prices to be raised. 2 A combination or company controlling a monopoly. 3 Exclusive possession or control of anything. 4 That which is the subject of a monopoly. 5 *Law* An exclusive license from the government for buying, selling, making, or using anything, and now granted only in case of patents and copyrights. [<L *monopolium* <Gk. *monopōlion* < *monos* alone + *pōlein* sell] — **mo·nop'o·lism** *n.* — **mo·nop'o·list** *n.* & *adj.* — **mo·nop'o·lis'tic** *adj.*

mon·o·pro·pel·lant (mon'ō·prə·pel'ənt) *n.* A liquid rocket propellant consisting of fuel and oxidizer mixed and ready for ignition in the combustion chamber: applied also to solid fuels.

mo·nop·so·ny (mə·nop'sə·nē) *n. Econ.* A condition of the market in which there is only one buyer for the product of a number of sellers. [<MON- + Gk. *opson* market]

mon·o·rail (mon'ō·rāl) *n.* 1 A single rail serving as a track for cars either suspended from it or balanced upon it by means of gyroscopes. 2 A railway using such a track.

mon·o·sac·cha·ride (mon'ə·sak'ə·rīd, -rid) *n. Biochem.* Any of a class of simple sugars which cannot be decomposed by hydrolysis, as glucose and fructose.

mon·o·sep·a·lous (mon'ə·sep'ə·ləs) *adj. Bot.* Having the sepals more or less united by their edges into a tube: also, more properly, *gamosepalous*: applied also to those rare cases in which the calyx actually consists of a single laterally placed sepal.

mon·o·sil·ane (mon'ə·sil'ān) *n. Chem.* Silane.

mon·o·some (mon'ə·sōm) *n. Genetics* The unpaired sex or X–chromosome.

mon·o·sper·mous (mon'ə·spûr'məs) *adj. Bot.* One–seeded. Also **mon'o·sper'mal.**

mon·o·stich (mon'ə·stik) *n.* A composition of one verse; especially, an epigram.

mon·o·stome (mon'ə·stōm) *adj. Zool.* Having a single sucker or mouth: also **mo·nos·to·mous** (mə·nos'tə·məs). — *n.* An animal with but a single mouth or sucker.

mo·nos·tro·phe (mə·nos'trə·fē) *n.* A metrical composition containing only one kind of strophe. [<MONO- + STROPHE] — **mon·o·stroph·ic** (mon'ə·strof'ik) *adj.*

mon·o·stroph·ics (mon'ə·strof'iks) *n. pl.* Monostrophic verses.

mon·o·sty·lous (mon'ə·stī'ləs) *adj. Bot.* Having only one style. [<MONO- + Gk. *stylos* pillar]

mon·o·syl·la·bism (mon'ə·sil'ə·biz'əm) *n.* The state or quality of being monosyllabic; addiction to the use of words having but one syllable.

mon·o·syl·la·ble (mon'ə·sil'ə·bəl) *n.* A word of one syllable. — **mon'o·syl·lab'ic** (-si·lab'ik) *adj.* — **mon'o·syl·lab'i·cal·ly** *adv.*

mon·o·the·ism (mon'ə·thē·iz'əm) *n.* The doctrine that there is but one God. [<MONO- + Gk. *theos* god] — **mon'o·the'ist** *n.* — **mon'o·the·is'tic** or **·ti·cal** *adj.* — **mon'o·the·is'ti·cal·ly** *adv.*

mon·o·ther·mi·a (mon'ə·thûr'mē·ə) *n. Med.* A condition of uniform temperature, as in the body. [<NL <Gk. *monos* single + *thermē* heat] — **mon'o·ther'mic** *adj.*

mon·o·tint (mon'ə·tint) *n.* A monochrome.

mon·o·tone (mon'ə·tōn) *n.* 1 Sameness of utterance or tone; monotony in the style of composition or speech, or something composed in such style. 2 A single musical tone unvaried in pitch or key; also, a chant in such a tone; an intoning.

mo·not·o·nous (mə·not'ə·nəs) *adj.* 1 Not varied in inflection, cadence, or pitch. 2 Tiresomely uniform. See synonyms under TEDIOUS. [<Gk. *monotonos*] — **mo·not'o·nous·ly** *adv.* — **mo·not'o·nous·ness** *n.*

mo·not·o·ny (mə·not'ə·nē) *n.* 1 The state or quality of being monotonous; irksome uniformity or lack of variety; also, sameness of tone; want of variety in cadence or inflection. 2 *Math.* **a** Continual increase or decrease of a quantity, function, etc. **b** Absence of either increase or decrease. [<Gk. *monotonia* < *monos* single + *tonos* tone] — **mon·o·ton·ic** (mon'ə·ton'ik) *adj.*

mon·o·treme (mon'ə·trēm) *n.* Any member of the lowest order of mammals *(Monotremata)*, without true teeth in the adult stage and having a single opening for the genitourinary and digestive organs, as in duckbills and echidnas. [<MONO- + Gk. *trēma* hole] — **mon'o·trem'a·tous** (-trem'ə·təs), **mon'o·tre'·mous** *adj.*

mo·not·ri·chous (mə·not'rə·kəs) *adj.* Having only one polar flagellum, as certain bacteria: also **mon·o·trich·ic** (mon'ə·trik'ik). [<MONO- + Gk. *thrix, trichos* hair]

mon·o·troph·ic (mon'ə·trof'ik) *adj.* Subsisting on or requiring only one kind of food; monophagous. [<MONO- + Gk. *trophē* food]

mo·not·ro·py (mə·not'rə·pē) *n.* The property, possessed by certain substances, tissues, or organs, of occurring in only one stable form. [<MONO- + -TROPY] — **mon·o·trop'ic** (mon'ə·trop'ik) *adj.*

mon·o·type (mon'ə·tīp) *n.* 1 *Biol.* The only representative of its kind, as a species of a genus or the like. 2 *Printing* A print from a metal plate on which a design, painting, etc., has been made.

Mon·o·type (mon'ə·tīp) *n. Printing* A machine which automatically casts and sets type in single characters or units: a trade name. Compare LINOTYPE.

mon·o·typ·ic (mon'ə·tip'ik) *adj.* 1 Containing but one representative; having only one type: a *monotypic* genus. 2 Being a monotype.

mon·o·va·lence (mon'ə·vā'ləns) *n. Chem.* Univalence. Also **mon'o·va'len·cy.** — **mon'o·va'lent** *adj.*

mon·ox·ide (mon·ok'sīd, mə·nok'-) *n. Chem.* A compound containing a single atom of oxygen in each molecule.

Mon·roe (mən·rō'), **James,** 1758–1831, president of the United States 1817–25.

Monroe Doctrine The doctrine, essentially formulated by President Monroe in his message to Congress (December 2, 1823), that any attempt by European powers to interfere in the affairs of the American countries or to acquire territory on the American continents would be regarded by the United States as an unfriendly act.

Mon·ro·vi·a (mən·rō'vē·ə) The capital of Liberia, a port on the Atlantic.

mons (monz) *n. pl.* **mon·tes** (mon'tēz) *Anat.* The rounded fatty eminence at the lower part of the abdomen, covered with hair in the adult: the **mons pu·bis** (pyōō'bis) of the male, or the **mons ven·er·is** (ven'ər·is) of the female. [<L, hill, mountain]

Mons (môṅs) A city in SW Belgium, capital of Hainaut province.

mon·sei·gneur (mon'sen·yûr', mon·sēn'yər; *Fr.* môṅ·se·nyœr') *n. pl.* **mes·sei·gneurs** (me·se·nyœr') My lord: a title given to princes of the church and formerly to the higher nobility of France. [<F < *mon* my + *seigneur* lord <L *senior* older. See SENIOR.]

mon·sieur (mə·syûr', *Fr.* mə·syœ') *n. pl.* **mes·sieurs** (mes'ərz, *Fr.* me·syœ') 1 The French title of courtesy for men, equivalent to *Mr.* and *sir*; abbreviated, M., *pl.* MM. It is capitalized when used with a proper name. 2 A Frenchman. [<F < *mon* + *sieur,* short for *seigneur.* See MONSEIGNEUR.]

mon·si·gnor (mon·sēn'yər, *Ital.* môn'sē·nyôr') *n. pl.* **·gnors** or *Ital.* **·gno·ri** (-nyō'rē) A title of honor of certain prelates and Roman Catholic officials, as of the papal court: abbreviated *Mgr.* Also **mon·si·gno·re** (môn'sē·nyō'rā). [<Ital. *monsignore* <F *monseigneur.* See MONSEIGNEUR.]

mon·soon (mon·sōōn') *n. Meteorol.* 1 A wind that blows more or less steadily along the Asiatic coast of the Pacific, in winter from the northeast **(dry monsoon),** in summer from the southwest **(wet monsoon).** 2 A trade wind. [<MDu. *monssoen* <Pg. *monção* <Arabic *mausim* season]

mon·ster (mon'stər) *n.* 1 A fabulous animal, compounded of various brute forms. 2 A being that is greatly malformed; anything hideous or abnormal in structure and appearance; a teratism. 3 One abhorred because of his unnatural or inhuman character. 4 A very large person, animal, or thing. See synonyms under PRODIGY. — *adj.* Extraordinary or enormous in size or numbers; huge. [<OF *monstre* <L *monstrum* divine omen, monster < *monere* warn]

mon·strance (mon'strəns) *n.* In Roman Catholic ritual, a sacred vessel in which the consecrated Host is exposed for adoration: also called *ostensorium, ostensory.* [<OF <Med. L *monstrantia* <L *monstrare* show]

mon·stros·i·ty (mon·stros'ə·tē) *n. pl.* **·ties** 1 Anything unnaturally huge, malformed, or distorted. 2 The character of being monstrous.

mon·strous (mon'strəs) *adj.* 1 Deviating greatly from the natural or normal; unnatural in form or structure. 2 Of extraordinary size or number; excessive; huge: a *monstrous* beast or multitude. 3 Inspiring abhorrence, hate, incredulity, etc., in a remarkable degree; hateful; hideous; incredible; intolerable: a *monstrous* cruelty. See synonyms under ABSURD, EXTRAORDINARY, FLAGRANT. — *adv. Archaic* Extremely. [<OF *monstreux* <L *monstrosus* < *monstrum.* See MONSTER.] — **mon'strous·ly** *adv.* — **mon'strous·ness** *n.*

mon·tage (mon·täzh') *n.* 1 A picture made by superimposing several different pictures so as to blend into one another, or so as to show figures upon a desired background; a composite picture; also, the process of composing such a picture. 2 In motion pictures, a swiftly run sequence of images or pictures illustrating a sequence of associated ideas; the dizzy revolving of several images around a central, focused image or picture to signify the passage of time or the like. 3 The section of a motion picture using this method. [<F]

Mon·ta·gu (mon'tə·gyōō), **Lady Mary Wortley,** 1689–1762, English writer.

Mon·ta·gue (mon'tə·gyōō) The family of Romeo, in Shakespeare's *Romeo and Juliet.*

Mon·taigne (mon·tān', *Fr.* môṅ·ten'y'), **Michel Eyquem de,** 1533–92, French essayist.

Mon·tan·a (mon·tan'ə) A State in the NW United States bordering on Canada; 147,138 square miles; capital, Helena; entered the Union Nov. 8, 1889; nickname, *Treasure State*: abbr. *Mont.* — **Mon·tan'an** *adj.* & *n.*

mon·tan·ic (mon·tan'ik) *adj.* Of or pertaining to mountains; mountainous. Also **mon·tane** (mon'tān). [<L *montan(us)* + -IC]

Mon·ta·nism (mon'tə·niz'əm) *n.* The doctrine of an ascetic Christian sect of the second century, founded by Montanus of Phrygia. — **Mon'ta·nist** *n.* — **Mon'ta·nis'tic** or **·ti·cal** *adj.*

mon·tan wax (mon'tən) A hydrocarbon wax of high melting point extracted from lignite, used as insulation and in making polishes, candles, phonograph records, etc. [<L *montanus* of the mountain]

Mon·tauk Point (mon'tôk) The easternmost point of New York, a promontory at the tip of the southern peninsula of Long Island.

Mont Blanc (mont blangk', *Fr.* môṅ bläṅ') The highest mountain of the Alps, on the French–Italian border; 15,780 feet.

Mont·calm (mont·käm', *Fr.* môṅ·kälm'), **Joseph Louis,** 1712–59, Marquis de Saint Véran, French general; fell in defense of Quebec against the British under Wolfe.

mont–de–pié·té (môṅ·də·pyā·tā') *n. French* A pawnshop authorized and controlled by the state to lend money to the poor; literally, mount of piety: originated in Italy in the 15th century. Also *Italian* **mon·te di pie·tá** (môn'·tā dē pyā·tä').

mon·te (mon'tē) *n.* 1 A Spanish gambling game of cards. 2 A table on which, or a place where, monte is played. 3 The money stacked in front of the dealer to pay off stakes: also **monte bank.** [<Sp., lit., mountain; in ref. to the pile of unplayed cards]

Mon·te Car·lo (mon'tē kär'lō, *Ital.* môn'tā kär'lō) A resort town in Monaco, on the Mediterranean.

Monte Cassino See CASSINO.

mon·teith (mon·tēth') *n.* An ornamental punch bowl. [from *Monteigh,* a surname]

Mon·te·mez·zi (môn'tā·med'dzē), **Italo,** 1875–1952, Italian composer.

Mon·te·ne·gro (mon'tə·nē'grō) A constituent republic of southern Yugoslavia; formerly a kingdom; 5,343 square miles; capital, Titograd. *Serbo–Croatian* **Cr·na Go·ra** (tsûr'nä gō'rä). — **Mon'te·ne'grin** *adj.* & *n.*

Mon·te·rey (mon'tə·rā') A city in western California on **Monterey Bay,** an inlet of the Pacific south of San Francisco.

mon·te·ro (mon·târ'ō) *n. pl.* **·ros** A huntsman's cap with a round crown and a flap. [<Sp. *montera* < *monte* hill]

Mon·ter·rey (mon'tə·rā', *Sp.* mōn'ter·rā') A city in NE Mexico, capital of Nuevo León state; captured by United States troops (September, 1846) in the Mexican War.

Mon·tes·pan (mon'tə·span', *Fr.* môṅ·tes·päṅ'),

Montesquieu

Marquise de, 1641-1707, Françoise Athénaïs de Rochechouart, mistress of Louis XIV of France.
Mon·tes·quieu (mon'təs·kyōō', *Fr.* môṅ·tes·kyœ'), **Baron de la Brède et de,** 1689-1755, Charles Louis de Secondat, French jurist and philosophical writer on history and government.
Mon·tes·so·ri (mon'tə·sôr'ē, -sō'rē; *Ital.* môn'·tes·sō'rē), **Maria,** 1870-1952, Italian educator.
Montessori method A system of teaching small children by training their sense perceptions, and by guiding rather than controlling their activity: devised in 1907 by Maria Montessori.
Mon·te·ver·di (mon'tə·vûr'dē, *Ital.* môn'tā·ver'·dē), **Claudio,** 1567-1643, Italian composer.
Mon·te·vi·de·o (mon'tə·vi·dā'ō, -vid'ē·ō; *Sp.* mōn'tā·vē·thā'ō) A port on the Rio de la Plata, capital of Uruguay.
Mon·te·zu·ma (mon'tə·zōō'mə), 1479?-1520, last Aztec emperor of Mexico; dethroned by Cortez: also *Moctezuma.*
Mont·fort (mont'fərt, *Fr.* môṅ·fôr'), **Simon de,** 1160?-1218, French crusader. — **Simon de,** 1208?-65, Earl of Leicester, English general and statesman; son of the preceding.
mont·gol·fi·er (mont·gol·fē'ər) *n.* A hot-air balloon. [after the *Montgolfier* brothers]
Mont·gol·fi·er (mont·gol·fē'ər, *Fr.* môṅ·gôl·fyā'), **Jacques Étienne,** 1745-99, and **Joseph Michel,** 1740-1810, French inventors, brothers, whose hot-air balloon was the first to make a successful ascent (1783).
Mont·gom·er·y (mont·gum'ər·ē) 1 The capital of Alabama, in the east central part of the State on the Alabama River. 2 A county in central Wales; 797 square miles; county town, Montgomery: also **Mont·gom'er·y·shire** (-shir).
Mont·gom·er·y (mont·gum'ər·ē, mən-), **Sir Bernard Law,** born 1887, first Viscount Montgomery of Alamein, English field marshal in World War II. — **Richard,** 1736?-75, general in the Continental Army; killed at Quebec.
month (munth) *n.* 1 A unit of time, originally equal to the interval between two new moons, afterward called a **lunar month,** and equal on the average to 29.53 days. 2 One of the 12 parts into which the calendar year is divided, called a **calendar month;** loosely, thirty days or four weeks. 3 The twelfth part of a solar year: also **solar month.** ◆ Collateral adjective: *mensal.* [OE *mōnath.* Akin to MOON.]
month·ly (munth'lē) *adj.* 1 Continuing a month, or done in a month. 2 Happening once a month. 3 Pertaining to the menses. — *adv.* Once a month. — *n. pl.* **·lies** 1 A periodical published once a month. 2 *pl.* The menses.
monthly meeting In the Society of Friends, a meeting held once a month, by two or more neighboring congregations, for worship and business; also, the organized unit composed of these congregations.
Mon·ti·cel·lo (mon'tə·sel'ō, -chel'ō) The estate and residence of Thomas Jefferson about 3 miles east of Charlottesville, Virginia.
Mont·mar·tre (môṅ·mär'tr') A northern district of Paris, famous for its night clubs and cafés.
Mont·mo·ren·cy River (mont'mə·ren'sē) A river in southern Quebec, Canada, flowing 60 miles south to the St. Lawrence River; near its mouth is **Montmorency Falls,** 275 feet high.
Mont·pel·ier (mont·pēl'yər) The capital of Vermont.
Mont·pel·lier (môṅ·pe·lyā') A city in southern France.
Mon·tre·al (mon'trē·ôl') A city in southern Quebec, Canada, on **Montreal Island** at the confluence of the St. Lawrence and Ottawa rivers. *French* **Mont·ré·al** (môṅ·rā·äl').
Mon·treux (môṅ·trœ') A resort in western Switzerland on the eastern shore of Lake Geneva.
Mon·trose (mon·trōz', mont·rōz'), **Marquis of,** 1612-50, James Graham, Scottish royalist leader; executed.
Mont Saint Mi·chel (môṅ saṅ mē·shel') A rocky islet, site of a famous ancient fortress and abbey, in **Mont Saint Michel Bay,** an inlet of the Gulf of St. Malo in NW France.

Mont·ser·rat (mont'sə·rat') 1 An island in the Leeward Islands, a British colony, member of The West Indies; 37.5 square miles. 2 A mountain in NE Spain, 4,054 feet; site of the 11th century Benedictine **Montserrat Monastery.**
mon·u·ment (mon'yə·mənt) *n.* 1 Something erected to perpetuate the memory of a person or of an event. 2 A notable structure, deed, production, etc., worthy to be considered as a memorial of the past, or of some event or person. 3 A stone or other permanent mark serving to indicate an angle or boundary. 4 A tomb. 5 *Obs.* A statue; effigy. [< F < L *monumentum* < *monere* remind]
mon·u·men·tal (mon'yə·men'təl) *adj.* 1 Pertaining to or like a monument. 2 Serving as a monument; memorial; impressive; conspicuous; enduring. 3 Spectacular; excessive: a *monumental fraud.* [< L *monumentalis*] — **mon'u·men'tal·ly** *adv.*
mon·u·men·tal·ize (mon'yə·men'təl·īz) *v.t.* **·ized, ·iz·ing** To establish a lasting record or memorial of.
mon·y (mon'ē) *adj. Scot.* Many. Also **mon'ie.**
-mony *suffix of nouns* The condition, state, or thing resulting from: *parsimony.* [< L *-monia, -monium*]
Mon·za (môn'tsä) A city in Lombardy, Italy.
mon·zo·nite (mon'zə·nīt) *n.* A coarse-grained igneous rock containing approximately equal amounts of orthoclase and plagioclase, with inclusions of colored silicates. [from Mount *Monzoni,* in the Tirol] — **mon'zo·nit'ic** (-nit'·ik) *adj.*
moo (mōō) *v.i.* To low as or like a cow. — *n.* The lowing noise made by a cow. [Imit.]
mooch (mōōch) *Slang v.t.* 1 To obtain without paying; beg; cadge: to *mooch* a drink. 2 To steal. — *v.i.* 3 To loiter about; skulk; sneak. Also spelled *mouch.* [Var. of dial. *miche* pilfer < OF *muchier* hide, skulk] — **mooch'er** *n.*
mood[1] (mōōd) *n.* 1 Temporary or capricious state of mind in regard to passion or feeling; humor; disposition. 2 *pl.* Fits of morose or sullen behavior; the state of being moody: to have *moods.* 3 *Obs.* Anger. See synonyms under FANCY, TEMPER. [OE *mōd*]
mood[2] (mōōd) *n.* 1 *Gram.* The manner in which the action or condition expressed by a verb is stated, whether as actual, doubtful, commanded, etc. The English moods proper are the indicative, subjunctive, and imperative. The infinitive is sometimes also classed as a mood. Certain verb-phrases are likewise loosely called moods, as those formed by *may, might, can, could* (potential), *should, would* (conditional), *must, ought* (obligative). Also *mode.* 2 *Logic* Mode. 3 *Music* Mode. [Var. of MODE]
mood·y (mōō'dē) *adj.* **mood·i·er, mood·i·est** 1 Given to or expressive of capricious moods. 2 Petulant; sullen; melancholy. See synonyms under MOROSE. — **mood'i·ly** *adv.* — **mood'i·ness** *n.*
Mood·y (mōō'dē), **Dwight Lyman,** 1837-99, U. S. evangelist. — **William Vaughn,** 1869-1910, U. S. poet and dramatist.
mool (mōōl) *Scot. n.* 1 Dirt; dust. 2 The grave. — *v.t.* 1 To crumble. 2 To bury.
moo·lah (mōō'lə) See MOLLAH.

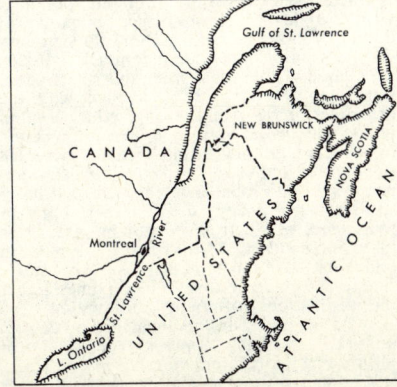

moo·ley (mōō'lē) *n.* A cow without horns. See MULEY.
moon (mōōn) *n.* 1 A celestial body revolving around the earth from west to east in a lunar month of 29.53 days or a sidereal month of 27.33 days; mean diameter, 2,160 miles; mean distance from the earth, 238,900 miles. In mass, the earth is 81.5 times greater than the moon; in volume, 49 times greater. 2 A satellite revolving about any planet. See under PLANET. 3 A lunar month. 4 Something resembling a moon or crescent. 5 Moonlight. — **dark of the moon** That period of time between the full moon and the new moon. — **man in the moon** The fancied appearance of a face in the disk of the full moon, caused by the shadows, lines, spots, etc., on its surface. — *v.i. Colloq.* To stare or wander about in an abstracted or listless manner. — *v.t.* To pass (time) thus. [OE *mōna.* Akin to MONTH.]

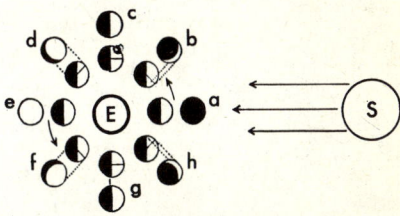

PHASES OF THE MOON
E. Earth. *S.* Sun.
a. New.   *d.* Gibbous.   *g.* 3rd Quarter.
b. Crescent.   *e.* Full.   *h.* Crescent.
c. 1st Quarter.   *f.* Gibbous.

Moon, Mountains of the See RUWENZORI.
moon·beam (mōōn'bēm') *n.* A ray of moonlight.
moon-blind (mōōn'blīnd') *adj.* 1 Purblind; feeble-sighted. 2 Moon-struck. 3 Affected with moonblindness.
moon-blind·ness (mōōn'blīnd'nis) *n.* 1 A form of blindness erroneously attributed to moonlight; nyctalopia. 2 A periodic inflammation of the eyes of a horse.
moon·calf (mōōn'kaf', -käf') *n.* 1 A stupid person; dolt; idiot. 2 A monster; deformity. [With ref. to the supposed bad influence of the moon]
mooned (mōōnd) *adj.* 1 Having the moon or crescent for a symbol, or bearing it as an emblem: *mooned* Diana. 2 Moonlike; crescent-shaped. 3 Spotted with moonlike marks.
Moo·ney (mōō'nē), **Thomas J.,** 1882-1942, U. S. labor leader: known as *Tom Mooney.*
moon-eye (mōōn'ī') *n.* 1 An eye affected with moonblindness. 2 Moonblindness.
moon-eyed (mōōn'īd') *adj.* 1 Having moon-eyes. 2 Moonblind (def. 3). 3 Having eyes wide open, as with amazement, awe, etc.
moon·fish (mōōn'fish') *n. pl.* **·fish** or **·fish·es** 1 Any of various carangoid fishes found on either coast of the western hemisphere, having a silvery or yellowish, much compressed body, especially *Vomer setipinnis:* also called *horsefish.* 2 The opah. 3 The Mexican top minnow (*Platypoecilus maculatus*).
moon-flow·er (mōōn'flou'ər) *n.* Any of a genus (*Calonyction*) of perennial climbing herbs of the morning-glory family; especially, *C. aculeatum* (formerly genus *Ipomoea*), bearing fragrant, white flowers which open at night.
moon-glade (mōōn'glād') *n.* The silvery reflection of moonlight on water.
moon·ish (mōō'nish) *adj.* Variable like the moon; affected by the moon; flighty; fickle; whimsical.
moon·light (mōōn'līt') *n.* The light of the moon. — *adj.* Pertaining to or illuminated by moonlight. — **moon'lit'** (-lit') *adj.*
moon·rise (mōōn'rīz') *n.* The appearing of the moon above the horizon, or the time when it appears.
moon·seed (mōōn'sēd') *n.* A North American climbing plant of the genus *Menispermum,* of the family *Menispermaceae:* so called from its crescent-shaped seeds.
moon·set (mōōn'set') *n.* The setting, or the time of setting, of the moon; specifically, the

---

add, āce, câre, pälm; end, ēven; it, īce; odd, ōpen, ôrder; tōōk, pōōl; ûp, bûrn; ə = a in *above,* e in *sicken,* i in *clarity,* o in *melon,* u in *focus;* yōō = u in *fuse;* oi, oil; ou, pout; ch, check; g, go; ng, ring; th, thin; ᵺ, this; zh, vision. Foreign sounds à,œ,ü,kh,ṅ; and ◆: see page xx. < from; + plus; ? possibly.

moonshine — **moray**

moon·shine (mōōn'shīn') *n.* 1 Moonlight. 2 Something visionary or unreal; pretence; nonsense. 3 *Colloq.* Smuggled or illicitly distilled spirits. — **moon'shin'y** *adj.*
moon·shin·er (mōōn'shī'nər) *n. Colloq.* An illicit distiller; smuggler.
moon·stone (mōōn'stōn') *n.* A whitish, cloudy feldspar, valued as a gemstone.
moon–struck (mōōn'struk') *adj.* Affected by or as by the moon; lunatic; deranged. Also **moon'–strick'en** (-strik'ən).
moon·wort (mōōn'wûrt') *n.* 1 The herb honesty. 2 Any fern of the genus *Botrychium*, especially *B. lunaria*. [Trans. of Med. L *lunaria* < L *luna* moon]
moon·y (mōō'nē) *adj.* **moon·i·er, moon·i·est** 1 Moon-struck. 2 Moonlit. 3 Like moonlight, or giving out light resembling moonlight. 4 Round or crescent-shaped. 5 Absent-minded; dreaming; vacant.
moop (mōōp) *v.i. Scot.* To live or associate: with *with*.
moor[1] (mōōr) *v.t.* 1 To secure (a ship, etc.) in one place by means of cables attached to shore, anchors, etc. 2 To secure in place; fix. — *v.i.* 3 To secure a ship in position; anchor. 4 To be secured by chains or cables. [Cf. OE *mǣrels* mooring rope]
moor[2] (mōōr) *n.* A tract of wasteland sometimes covered with heath, often elevated and frequently marshy and abounding in peat. 2 A tract of land on which game is preserved for shooting. [OE *mōr*] — **moor'ish** *adj.*
Moor (mōōr) *n.* 1 A person of mixed Berber and Arab blood, inhabiting Morocco and the southern Mediterranean coast. 2 Any North African native; specifically, a Saracen or a Spanish descendant of the Saracens. 3 A Mohammedan of India. [<F *more, maure* <L *Maurus* Moor, Mauritanian <Gk. *Mauros*] — **Moor'ish** *adj.*
moor·age (mōōr'ij) *n.* A mooring place.
moor·cock (mōōr'kok') *n.* The male of the red grouse.
Moore (mōōr, môr, mōr), **George,** 1852–1933, Irish novelist, critic, and dramatist. — **Henry,** born 1898, English sculptor. — **Sir John,** 1761–1809, British general. — **John Bassett,** 1860–1947, U.S. jurist. — **Thomas,** 1779–1852, Irish poet.
Mo·o·re·a (mō'ō·rā'ə) The second largest island in the Windward group of the Society Islands; 50 square miles.
moor·fowl (mōōr'foul') *n.* The red grouse.
moor·hen (mōōr'hen') *n.* 1 The female of the moorfowl. 2 The water hen (*Gallinula chloropus*). 3 The American coot.
moor·ing (mōōr'ing) *n.* 1 The act of securing a vessel. 2 *Chiefly pl.* The place where a vessel is moored. 3 *Chiefly pl.* Anything by which an object is fastened.
mooring mast The tower to which a dirigible or blimp is secured when not in flight: also **mooring tower.**
moor·land (mōōr'lənd) *n.* A moor or marsh. — *adj.* Having marshy properties. — **moor'land·er** *n.*
moor·wort (mōōr'wûrt') *n.* A low, smooth shrub (*Andromeda polifolia*) of the heath family, with narrow, thick, evergreen leaves, growing in wet bogs.
moor·y (mōōr'ē) *adj.* **moor·i·er, moor·i·est** Of the nature of moorland; marshy.
moose (mōōs) *n. pl.* **moose** 1 A large, heavily built mammal (*Alces americana*) of the deer family, found in northern North America: the male bears huge, palmate antlers. 2 The northern European elk. [<Algonquian (Massachuset) *moos* he strips off; because it eats the bark of trees]
moose·bird (mōōs'bûrd') *n.* The Canada jay.
moose·call (mōōs'kôl') *n.* A small horn or trumpet made of birch bark or other materials, used by hunters to lure moose within shooting distance. Also **moose horn.**

MOOSE
(Up to nearly 7 feet high at the shoulder; weight to 1,000 pounds)

moose·flow·er (mōōs'flou'ər) *n.* Trillium: also called *wakerobin*.
moose·grass (mōōs'gras', -gräs') *n.* Ground hemlock.
Moose·head Lake (mōōs'hed) The largest lake in Maine, source of the Kennebec River; 36 miles long; 120 square miles.
moose·wood (mōōs'wōōd') *n.* Leatherwood.
moot (mōōt) *n.* 1 In Anglo-Saxon times, the meeting of freemen and farmers for the discussion or debate of local affairs. 2 Hence, discussion or argument; especially, in modern usage, discussion of a mock lawsuit for the sake of practice. 3 The place where a meeting is held. — *v.t.* 1 To debate; discuss. 2 To argue (a case) in a moot court. — *adj.* Still open to discussion; intended for discussion: a *moot* point. [OE *mōt* assembly, court] — **moot'er** *n.*
moot court A court for the trial of a fictitious suit by law students.
mop[1] (mop) *n.* 1 A piece of cloth, or the like, attached to a handle: used for washing floors. 2 Any tangled bunch or mass, as of hair. — *v.t.* **mopped, mop·ping** To rub or wipe with or as with a mop. — **to mop up** 1 To take up with or as with a mop. 2 *Mil.* To wipe out remnants of enemy resistance in (captured areas). — **to. mop (up) the floor with** *Slang* To defeat easily and decisively. [Origin uncertain. Cf. F *mappe* napkin <L *mappa*.]
mop[2] (mop) *n.* 1 A wry mouth; grimace. 2 A pouting or petted young person; a young girl. [<*v.*] — *v.i.* To make a wry face; grimace. [Cf. Du. *mopper* pout]
mope (mōp) *v.* **moped, mop·ing** *v.i.* To be gloomy, listless, or dispirited. — *v.t.* To make gloomy or dispirited. — *n.* 1 One who mopes. 2 *pl.* Dejection; depression. [Prob. <Scand. Cf. Sw. *mopa* sulk, Dan. *maabe* be stupid, unaware.] — **mop'er** *n.* — **mop'ish** *adj.* — **mop'ish·ly** *adv.* — **mop'ish·ness** *n.*
mo·poke (mō'pōk) *n. Austral.* The frogmouth. [Imit. of the bird's note]
mop·pet (mop'it) *n.* 1 A rag doll or one made of cloth. 2 A child; youngster. [Dim. of ME *moppe* rag doll, ? <L *mappa* napkin]
mo·quette (mō·ket') *n.* A woolen fabric with a velvety pile, used for carpets and upholstery. [<F; ult. origin uncertain]
Mo·qui (mō'kē) *n.* A Hopi Indian: also spelled *Moki.*
mo·ra (môr'ə, mō'rə) *n. pl.* **mo·rae** (môr'ē, mō'rē) or **mo·ras** 1 In civil law, delay, especially if unjustifiable; default. 2 In prosody, a unit of meter, the common short foot, usually indicated by the breve. [<L, delay]
mo·ra·ceous (mō·rā'shəs, mô-) *adj.* Denoting or belonging to a family (*Moraceae*) of mostly tropical herbs, shrubs, and trees of the order *Urticales*, including the mulberry, common fig, hop, hemp, etc. [<L *morus* mulberry]
Mo·rad·a·bad (mō·rad'ə·bad, mō·räd'ä·bäd') A city in north central Uttar Pradesh State, northern India.
mo·raine (mō·rān', mō-) *n. Geol.* A ridge or heap of earth, stones, etc., carried by a glacier and deposited on adjacent ground, either along the course or at the edges of the glacier, as a **medial** or **lateral moraine,** or at its lower terminus, as a **terminal moraine.** [<dial. F] — **mo·rain'al, mo·rain'ic** *adj.*
mor·al (môr'əl, mor'-) *adj.* 1 Pertaining to character and behavior from the point of view of right and wrong, and obligation of duty; pertaining to rightness and duty in conduct. 2 Conforming to right conduct; actuated by a sense of the good, true, and right; good; righteous; virtuous. 3 Concerned with the principles of right and wrong; ethical: *moral* philosophy; *moral* values. 4 Acting or suited to act through man's intellect or sense of right: often opposed to *physical*: *moral* support. 5 *Logic* Probable as opposed to demonstrative: *moral* proof. 6 Of or pertaining to the science or doctrine of human nature as fitted for conduct. Most writers on modern philosophy use the term to cover the entire sphere of human conduct which comes under the distinctions of right and wrong. 7 Attempting or serving to inculcate or convey a moral; moralizing: a *moral* writer. 8 Of or influencing morals or morale: *moral* force. 9 Capable of understanding the difference between right and wrong: a *moral* agent. — *n.* 1 The lesson taught by a fable or the like. 2 *pl.* Conduct or behavior; ethics. [<F <L

moralis <*mos, moris* custom; in the pl. manners, morals] — **mor'al·ly** *adv.*
Synonyms (*adj.*): dutiful, ethical, excellent, faithful, good, honest, honorable, incorruptible, just, pious, religious, right, righteous, true, upright, virtuous, worthy. **Antonyms:** see synonyms for IMMORAL.
mo·rale (mə·ral', -räl', mō-) *n.* 1 State of mind with reference to confidence, courage, hope, etc.: used especially of a number of persons associated in some enterprise, as troops, workers, etc. 2 *Obs.* Morality. [<F. See MORAL.]
moral hazard In insurance, a risk resulting from doubt as to the honesty of the person insured.
moral insanity Mental deficiency amounting to the incapacity to distinguish between right and wrong, or characterized by compulsions to perform unsocial, irresponsible, or criminal acts: a legal term variously interpreted in different statutes.
mor·al·ism (môr'əl·iz'əm, mor'-) *n.* The belief in a morality divested of all religious character.
mor·al·ist (môr'əl·ist, mor'-) *n.* 1 A teacher of morals. 2 One who practices morality without religion. — **mor·al·is'tic** *adj.*
mo·ral·i·ty (mə·ral'ə·tē, mô-) *n. pl.* **·ties** 1 The doctrine of man's moral duties; ethics. 2 Virtuous conduct; rectitude; chastity. 3 The quality of being morally right. 4 A lesson inferred; a moral. 5 Conformity, or degree of conformity, to conventional rules, without or apart from inspiration and guidance by religion or other spiritual influences. 6 A form of allegorical drama of the 15th and 16th centuries in which the characters were personified virtues, vices, mental attributes, etc. See synonyms under VIRTUE. [<L *moralitas, -tatis* <*moralis*. See MORAL.]
mor·al·ize (môr'əl·īz, mor'-) *v.* **·ized, ·iz·ing** *v.i.* 1 To make moral reflections; talk about morality. — *v.t.* 2 To explain in a moral sense; derive a moral from. 3 To improve the morals of. Also *Brit.* **mor'al·ise.** — **mor·al·i·za'tion** *n.* — **mor'al·iz'er** *n.*
moral philosophy Ethics.
moral victory A defeat that is accounted a victory on the moral level, as when the righteousness of a defeated cause has been clearly established.
mo·rass (mə·ras', mô-, mō-) *n.* A tract of low-lying, soft, wet ground; marsh. [<Du. *moeras*, earlier *marasch* <OF *maresc* <Med. L *mariscus* <Gmc. Ult. akin to MARSH.]
mor·a·to·ri·um (môr'ə·tôr'ē·əm, -tō'rē-, mor'-) *n. pl.* **·ri·a** (-ē·ə) or **·ri·ums** *Law* An emergency act of legislation or a government edict authorizing a debtor or bank to suspend payments for a given period; also, the period during which it is, or is to be, in force. [<NL <LL *moratorius*. See MORATORY.]
mor·a·to·ry (môr'ə·tôr'ē, -tō'rē, mor'-) *adj.* Pertaining or intended to delay: particularly applied to legislation in the nature of a moratorium. [<LL *moratorius* <L *morari* delay <*mora* delay]
Mo·ra·tu·wa (mō'rə·tōō'wə) A port on the west coast of Ceylon.
Mo·ra·va (mō'rä·vä) 1 The Czech name for the river MARCH. 2 The Czech name for MORAVIA.
Mo·ra·vi·a (mō·rā'vē·ə, mō-) A central region of Czechoslovakia; 8,219 square miles; a former Austrian crownland: German *Mähren,* Czech *Morava.*
Mo·ra·vi·an (mō·rā'vē·ən, mō-) *adj.* Pertaining to Moravia or the Moravians. — *n.* 1 A native of Moravia. 2 One of a Christian sect founded by disciples of John Huss in Moravia (15th century), and now in Germany, Britain, and America. The sect is also known as the *Renewed Church of the United Brethren* or *Unity of the Brethren.*
Moravian Gap A mountain pass between the Carpathians and the Sudeten in central Europe.
Mo·rav·ská O·stra·va (mô'räf·skä ôs'trä·vä) A city of northern Moravia, Czechoslovakia: German *Mährisch–Ostrau:* also *Ostrava.*
mo·ray (môr'ā, mō·rā') *n.* A brightly colored, voracious, savage eel of the family *Muraenidae,* inhabiting tropical and subtropical waters, especially among coral reefs: also called *murry.* The Mediterranean moray (*Muraena helena*) is esteemed as a food fish. A related

species is the **banded moray** (*Gymnothorax waialuoe*) of Hawaiian waters. [? <Pg. *moreia* <L *muraena*]

Mor·ay (mûr'ē, *Scot.* mûr'ā) A county of NE Scotland; 476 square miles; county town, Elgin. Formerly *Elgin.* Also **Mor'ay·shire** (-shir).

mor·bid (môr'bid) *adj.* 1 Being in a diseased or abnormal state. 2 Caused by or denoting a diseased condition of body or mind. 3 Taking an excessive interest in matters of a gruesome or unwholesome nature; also, apprehensive; suspicious. 4 Grisly; gruesome: a *morbid* story. 5 Of or pertaining to disease; pathological: *morbid* anatomy. [<L *morbidus* < *morbus* disease] — **mor'bid·ly** *adv.* — **mor'bid·ness** *n.*

mor·bi·dez·za (môr'bē·ded'dzä) *n. Italian* In painting, delicacy or softness of flesh tints.

mor·bid·i·ty (môr·bid'ə·tē) *n.* 1 The condition of being morbid or diseased. 2 The rate of disease or proportion of diseased persons in a community: compare MORTALITY (def. 3).

mor·bif·ic (môr·bif'ik) *adj.* Producing disease. Also **mor·bif'i·cal.** — **mor·bif'i·cal·ly** *adv.*

mor·bil·li (môr·bil'ī) *n. pl.* The measles. [< Med. L, pl. of *morbillus,* dim. of *morbus* disease]

mor·bil·li·form (môr·bil'ə·fôrm) *adj.* Resembling measles. [<MORBILLI + -FORM]

mor·bil·lous (môr·bil'əs) *adj.* Relating to or affected with measles.

mor·ceau (môr·sō') *n. pl.* **·ceaux** (-sō) *French* A small bit or fragment; also, a short composition, as of music or poetry.

mor·da·cious (môr·dā'shəs) *adj.* Biting, or given to biting; hence, sarcastic. [<L *mordax, -acis* < *mordere* bite] — **mor·da'cious·ly** *adv.* — **mor·dac'i·ty** (-das'ə·tē) *n.*

mor·dan·cy (môr'dən·sē) *n.* Pungency; the quality of being biting or sarcastic.

mor·dant (môr'dənt) *n.* 1 Any substance, such as tannic acid or aluminum hydroxide, which, by combining with a dyestuff to form an insoluble compound (lake), serves to produce a fixed color in a textile fiber. 2 The corrosive used in etching. — *adj.* 1 Biting; pungent; sarcastic. 2 Acting as a mordant. — *v.t.* To treat or imbue with a mordant. [<OF]

Mor·de·cai (môr'də·kī) In the Old Testament, Esther's cousin, instrumental in saving the Jews from extermination. *Esth.* ii 15.

mor·dent (môr'dənt) *n. Music* The rapid alternation of a tone with the tone immediately below it or the character indicating it: a form of trill. Also **mor·den·te** (môr·den'tā). — **inverted mordent** The rapid alternation of a note with one above it; a pralltriller. [<G <Ital. *mordente,* ppr. of *mordere* bite <L]

Mor·dred (môr'drid) See MODRED.

Mord·vin·i·an Autonomous Soviet Socialist Republic (môrd·vin'ē·ən) An administrative division of central European Russian S.F.S.R.; 10,080 square miles; capital, Saransk: also **Mor·do·vi·an Autonomous S.S.R.** (môr·dō'vē·ən).

more (môr, mōr) *adj. superlative* **most** 1 Greater in amount, extent, or degree: comparative of *much: more* water. 2 Greater in number: comparative of *many.* 3 Greater in rank or dignity. 4 Added to some former number; extra. — *n.* 1 A greater quantity, amount, etc. 2 Something that exceeds something else. — *adv.* 1 To a greater extent or degree. 2 In addition; further; again: usually qualified by *any, never,* a numeral, etc.: I cannot walk any *more.* [OE *māra*]

More (môr, mōr), **Hannah,** 1745-1833, English religious writer. — **Paul Elmer,** 1864-1937, U. S. critic and essayist. — **Sir Thomas,** 1478?-1535, lord chancellor of England; author; beheaded; canonized, 1935.

Mo·re·a (mô·rē'ə, mō-) A former name for PELOPONNESUS.

Mo·reau (mô·rō'), **Jean Victor,** 1761-1813, French Republican general.

mo·reen (mə·rēn') *n.* A sturdy, ribbed, cotton, wool, or wool and cotton fabric, often with a watered or embossed finish, used for hangings and upholstery. [Prob. <MOIRE + -*een,* as in *velveteen*]

mo·rel[1] (mə·rel') *n.* An edible mushroom of the genus *Morchella,* somewhat resembling a sponge on the end of a stalk. [<MF *morille,* ult. <Gmc. Cf. OHG *morhila* little carrot.]

mo·rel[2] (mə·rel') *n.* The black nightshade (*Solanum nigrum*). Also **mo·relle'.** [<OF *morele* <Med. L *morella,* ? <LL *maurella,* a kind of plant]

Mo·re·li·a (mō·rā'lyä) A city of south central Mexico, capital of Michoacán state.

mo·rel·lo (mə·rel'ō) *n.* A variety of cultivated cherry, with a dark-red skin, flesh, and juice: used in cooking and preserving. [<Flemish *marelle,* aphetic var. of *amarelle* <Ital. *amarello,* dim. of *amaro* bitter <L *amarus*]

Mo·re·los (mō·rā'lōs) A state in southern Mexico; 1,916 square miles; capital, Cuernavaca.

more·o·ver (môr·ō'vər, mōr-) *adv.* Beyond what has been said; further; besides; likewise.

mo·res (môr'ēz, mō'rēz) *n. pl. Sociol.* 1 Those established, traditional customs or folkways regarded by a social group as essential to its preservation and welfare. 2 The accepted conventions of a group or community. [<L, pl. of *mos, moris* custom]

Mo·res·net (mô·rez·ne') A mining district on the Belgian-German frontier; ceded to Belgium under the Treaty of Versailles, 1919.

Mo·resque (mô·resk', mə-) *adj.* Moorish; decorated in the style of the Moors. — *n.* Decorative work, by means of interlacings, relief, etc., highly colored and gilded. Compare MORISCO (def. 3). [<F <Ital. *moresco* <L *Maurus* Mauritanian]

MORESQUE PANEL DECORATION

Mor·gain (môr'gān, -gən) See MORGAN LE FAY.

Mor·gan (môr'gən), **Daniel,** 1736-1802, American Revolutionary general. — **Sir Henry,** 1635?-88, Welsh leader of American buccaneers. — **John Hunt,** 1825-64, American Confederate cavalry leader. — **John Pierpont,** 1837-1913, U. S. banker, financier, and art collector. — **Thomas Hunt,** 1866-1945, U. S. zoologist.

mor·ga·nat·ic (môr'gə·nat'ik) *adj.* Designating a legitimate marriage between a male member of certain royal families of Europe and a woman of inferior rank, in which the titles and estates of the husband are not shared by the wife or their children. Also **mor'ga·nat'i·cal.** [<NL *morganaticus* <LL (*matrimonium ad*) *morganaticum* (wedding with) morning gift <OHG *morgengeba* morning gift (in lieu of a share in the estate)] — **mor'ga·nat'i·cal·ly** *adv.*

Morgan horse A breed of horse of Arabian strain descendent from the stallion *Justin Morgan* (died 1821): noted for its powerful frame, gentle disposition, and versatility. [after Justin *Morgan,* owner of the original horse]

mor·gan·ite (môr'gən·īt) *n.* A rose-pink variety of beryl, used as a semiprecious gemstone. [after John Pierpont *Morgan*]

Morgan le Fay (lə fā') In Arthurian legend, the fairy half-sister of Arthur. She appears in Carlovingian romance as *Fata Morgana.* [<OF *Morgain la fée* Morgan the fairy]

mor·gen (môr'gən) *n. pl.* **·gen** or **·gens** A land measure formerly used by the Dutch, and still employed in South Africa: equal to about two acres. [<Du. & G, morning; hence, area plowed in one morning]

Mor·gen·thau (môr'gən·thô), **Henry,** born 1891, U. S. secretary of the Treasury 1934-1945.

morgue (môrg) *n.* 1 A place where bodies of the dead are kept until identified or claimed. 2 In a newspaper editorial office, the department in charge of filed items and biographical material: used for obituary notices, etc. [<F, orig., the name of the Paris building used for this purpose]

Mo·ri·ah (mə·rī'ə) A hill in Jerusalem; site of Solomon's temple. II *Chron.* iii 1.

mor·i·bund (môr'ə·bund, -bənd, mor'-) *adj.* Dying; at the point of death. [<L *moribundus* < *mori* die] — **mor'i·bun'di·ty** *n.* — **mor'i·bund·ly** *adv.*

mo·rin (môr'in, mō'rin) *n.* A crystalline compound, $C_{15}H_{10}O_7$, obtained from old fustic: used as a yellow dyestuff and as a reagent for aluminum. [<F *morine* <L *morum* mulberry]

mo·ri·on[1] (môr'ē·on, mō'rē-) *n.* A kind of open helmet without vizor worn by men-at-arms: also spelled *morrion.* [<OF <Sp. *morrión* < *morra* crown of the head]

mo·ri·on[2] (môr'ē·on, mō'rē-) *n.* A dark, sometimes nearly black, variety of smoky quartz. [<F <LL, a misreading of L *mormorion*]

Mo·ris·co (mə·ris'kō) *adj.* Moorish. — *n. pl.* **·cos** or **·coes** 1 One of the Moors who remained in Spain after the conquest of Granada in 1492; a Moor. 2 A morris dance or dancer. 3 The Moresque style of architecture or decoration. [<Sp. < *moro* Moor]

Mo·ri·son (môr'ə·sən, mor'-), **Samuel Eliot,** born 1887, U. S. historian.

mo·ri·tu·ri te sa·lu·ta·mus (môr'i·tyŏŏr'ē tē sal'yŏŏ·tā'məs) *Latin* We (who are) about to die salute thee: salutation of the gladiators to the Roman emperor.

Mo·ritz (mō'rits) Danish, Swedish, and German form of MAURICE.

Mor·land (môr'lənd), **George,** 1763-1804, English painter.

Mor·ley (môr'lē), **Christopher,** 1890-1957, U. S. writer. — **John,** 1838-1923, first Viscount Morley of Blackburn, English statesman and man of letters.

Mor·mon (môr'mən) *n.* 1 A member of a religious sect officially styled "The Church of Jesus Christ of Latter-day Saints," founded in the U. S. by Joseph Smith in 1830: also **Mor'mon·ist, Mor'mon·ite.** 2 In Mormon belief, a prophet and sacred historian of the fourth century A.D. who wrote, on golden tablets, a history of an early American people. The tablets, called the **Book of Mormon,** the holy book of the Mormon faith, were found by Joseph Smith near Palmyra, New York, and translated and published by him in 1830. — **Mor'mon·ism** *n.*

Mormon Bible The Book of Mormon. See under MORMON.

Mormon Church The Church of Jesus Christ of Latter-day Saints.

Mormon State Nickname of UTAH.

Mormon trail The trail taken by the Mormons, 1847, from Iowa to Utah.

morn (môrn) *n. Poetic* The morning. [OE *morgen*]

Mor·nay (môr·nā'), **Philippe de,** 1549-1623, Seigneur du Plessis, French Huguenot leader; minister of Henry IV: also known as *Duplessis-Mornay.*

morn·ing (môr'ning) *n.* 1 The early part of the day; the time from midnight to noon, or from sunrise to noon; hence, any early stage. 2 The dawn: often personified as **Morning,** the goddess Eos or Aurora. — *adj.* Pertaining to or occurring in the early part of the day. ♦ Collateral adjective: *matutinal.* [<MORN; by analogy with EVENING]

morn·ing-glo·ry (môr'ning·glôr'ē, -glō'rē) *n. pl.* **·ries** 1 A twining plant (genus *Ipomoea*) with funnel-shaped flowers of various colors. 2 Any one of many convolvulaceous plants.

morning gun A gun fired at reveille on military posts as a signal for raising the flag.

Morning Prayer In the Anglican church, the order for public worship in the morning.

morning sickness Vomiting and nausea in the morning, common in early pregnancy.

morn·ing-star (môr'ning·stär') *n.* A weapon made of a metal ball set with spikes and chained to a handle; a mace.

morning star Any of the planets Jupiter, Mars, Saturn, Mercury, or especially Venus, when rising shortly before the sun.

Mo·ro (môr'ō, mō'rō) *n. pl.* **·ros** 1 A member of one of the Moslem tribes of the southern Philippines. 2 The Indonesian language of the Moros. [<Sp., a Moor]

MORNING-STAR

Mo·roc·co (mə·rok'ō) 1 A kingdom on the Atlantic and Mediterranean coasts of NW

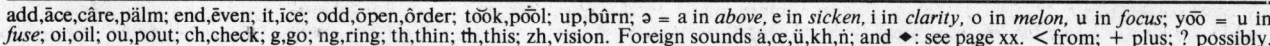

add, āce, câre, pälm; end, ēven; it, īce; odd, ōpen, ôrder; tōōk, pōōl; up, bûrn; ə = a in *above,* e in *sicken,* i in *clarity,* o in *melon,* u in *focus;* yōō = u in *fuse;* oi, oil; ou, pout; ch, check; g, go; ng, ring; th, thin; ṯẖ, this; zh, vision. Foreign sounds à, œ, ü, kh, ṅ; and ♦: see page xx. < from; + plus; ? possibly.

morocco leather Africa; 160,000 square miles; capital, Rabat: French *Maroc*, Spanish *Marruecos*. **2** A former name for MARRAKESH. *Arabic* El Maghreb el Aq·sa (al mä′greb el äg′sä). — **Mo·roc′can** *adj. & n.*

morocco leather A fine leather made from goatskin tanned with sumac. Also **mo·roc′co.**

Mo·ro Gulf (môr′ō, mō′rō) A large northern inlet of the Celebes Sea in SW Mindanao.

mo·ron (môr′on, mō′ron) *n.* A person whose mental capacity has been arrested during development and who represents mentally the condition of a child of from 8 to 12 years of age. [< Gk. *mōron*, neut. of *mōros* dull, sluggish] — **mo·ron·ic** (mō·ron′ik, mō-) *adj.* — **mo·ron′i·cal·ly** *adv.*

mo·ron·ism (môr′on·iz′əm, mō′ron-) *n.* The mildest degree of feeble-mindedness, rated above imbecility and idiocy. Also **mo·ron·i·ty** (mō·ron′ə·tē, mō-).

Mor·o·pus (môr′ə·pəs, mor′-) *n. Paleontol.* A genus of extinct North American chalicotheres of the Tertiary period having a skull similar to that of the horse, slender legs, and hoofs modified into three large claws. [< NL < Gk. *mōros* sluggish + *pous* foot]

mo·rose (mə·rōs′) *adj.* Having a surly temper; sullen; gloomy. [< L *morosus* < *mos, moris* manner, habit] — **mo·rose′ly** *adv.* — **mo·rose′ness, mo·ros·i·ty** (-ros′ə·tē) *n.*

Synonyms (adj.): acrimonious, bitter, churlish, crabbed, crusty, dogged, gloomy, gruff, ill-humored, ill-natured, moody, severe, sour, splenetic, sulky, sullen, surly. The *sullen* and *sulky* are discontented and resentful; *sullen* denotes more of pride, *sulky* more of resentful obstinacy. The *morose* are bitterly dissatisfied with the world in general, and disposed to vent their ill nature upon others. The *sullen* and *sulky* are for the most part silent; the *morose* growl out bitter speeches. A *surly* person is in a state of latent anger, resenting approach as intrusion, and ready to take offense at anything. See AUSTERE. Compare ACRIMONY. *Antonyms:* amiable, benignant, bland, complaisant, friendly, genial, gentle, good-natured, indulgent, kind, mild, pleasant, sympathetic, tender.

mor·phal·lax·is (môr′fə·lak′sis) *n. Biol.* The regeneration of parts of a living organism by the gradual transformation of other parts. [< Gk. *morphē* form + *allaxis* exchange]

mor·pheme (môr′fēm) *n. Ling.* The smallest lexical unit of a language, as a word, root, affix, or inflectional ending. *Man, run, pro-, -ess, -ing, ouch,* etc., are morphemes. [< F *morphème* < Gk. *morphē* form]

Mor·phe·us (môr′fē·əs, -fyoos) In Greek mythology, the god of dreams, son of Hypnos, god of sleep. [< L < Gk. *morphē* form; from the shapes he calls up in dreams] — **Mor′phe·an** *adj.*

mor·phic (môr′fik) *adj.* Morphologic.

-morphic *combining form* Having the form or shape of: *anthropomorphic.* [< Gk. *morphē* form]

mor·phine (môr′fēn) *n. Chem.* A bitter, white, crystalline, narcotic alkaloid, $C_{17}H_{19}NO_3 \cdot H_2O$, contained in opium and used for alleviating pain. Also **mor′phi·a** (-fē·ə). [< F < L *Morpheus* god of sleep]

mor·phin·ism (môr′fin·iz′əm) *n. Pathol.* A morbid condition of the system produced by an excessive dose or habitual use of morphine.

morpho- *combining form* Form; shape: *morpholysis.* Also, before vowels, **morph-**. [< Gk. *morphē* form]

mor·pho·gen·e·sis (môr′fō·jen′ə·sis) *n. Biol.* **1** The evolution of forms of structure. **2** The development of organic forms. Also **mor·phog·e·ny** (môr·foj′ə·nē). — **mor′pho·ge·net′ic** (-jə·net′ik) *adj.*

mor·phol·o·gy (môr·fol′ə·jē) *n.* **1** That branch of biology which treats of the form and structure of plants and animals. **2** The form of an organism considered apart from function. **3** *Geog.* The study of the forms of earth features. **4** *Ling.* **a** The branch of linguistics which deals with morphemes, their arrangement in words, and the changes they undergo in various grammatical constructions. **b** The arrangement, composition, and inflection of the morphemes of a language. [< MORPHO- + -LOGY] — **mor·pho·log·ic** (môr′fə·loj′ik) or **·i·cal** *adj.* — **mor′pho·log′i·cal·ly** *adv.* — **mor·phol′o·gist** *n.*

mor·phol·y·sis (môr·fol′ə·sis) *n.* The breakdown of form or structure. [< MORPHO- + -LYSIS]

mor·pho·sis (môr·fō′sis) *n. Biol.* The order or mode of formation of an organ or organism. [< Gk. *morphōsis* shaping < *morphē* form] — **mor·phot′ic** (-fot′ik) *adj.*

-morphous *combining form* Having a (specified) form: often equivalent to *-morphic: anthropomorphous.* [< Gk. *morphē* form]

Mor·rig·u (môr′ē·ōō) In Old Irish mythology, a war goddess, described as attending the battlefield in the form of a demoniacal raven. She helped the Tuatha De Danann defeat the Fomorians. Also **Mor·ri·gan** (môr′ē·ən).

mor·ri·on (môr′ē·on, mō′rē-) See MORION[1].

mor·ris (môr′is, mor′-) *n.* **1** An old English rustic dance, in which the performers took the part of Robin Hood and other characters in English folklore. **2** A dance by a single dancer with castanets. Also **mor′rice, morris dance.** [Earlier *morys, morish* Moorish]

Mor·ris (môr′is, mor′-), **Gouverneur,** 1752–1816, American statesman; financier. — **Robert,** 1734–1806, American statesman; signer of the Declaration of Independence. — **William,** 1834–96, English painter, craftsman, poet, and socialist; invented the **Morris chair,** an armchair with an adjustable back.

Mor·ris Jes·up (môr′is jes′əp, mor′is), **Cape** The world's northernmost point of land, at the northern extremity of Greenland; 440 miles from the North Pole.

Mor·ri·son (môr′ə·sən, mor′-), **Herbert Stanley,** born 1888, English Labour party leader and politician.

Mor·ri·son (môr′ə·sən, mor′-), **Mount** The highest mountain of Taiwan; 13,599 feet. *Chinese* Sin·kao (shin′kou′), *Japanese* Ni·i·ta·ka (nē·ē·tä·kä).

mor·ro (môr′ō, mor′ō; *Sp.* môr′rō) *n. pl.* **mor·ros** (môr′ōz, mor′-; *Sp.* môr′rōs) A round hill. [< Sp.]

Morro Castle A fort defending the entrance to Havana harbor, Cuba; bombarded, 1898.

mor·row (môr′ō, mor′ō) *n.* **1** The first day after the present or after a day specified; hence, any time following immediately after a specified event. **2** Formerly, morning: *good morrow.* — *adj.* Next succeeding, as a day. [OE *morgen* morning]

Mors (môrs) In Roman mythology, the god of death; identified with the Greek *Thanatos.* [< L, death]

Morse (môrs), **Philip McCord,** born 1903, U. S. physicist. — **Samuel Finley Breese,** 1791–1872, U. S. artist and inventor; constructed (1832–35) the first practical telegraph.

INTERNATIONAL MORSE CODE
LETTERS

a	·—	j	·———	s	···
b	—···	k	—·—	t	—
c	—·—·	l	·—··	u	··—
d	—··	m	——	v	···—
e	·	n	—·	w	·——
f	··—·	o	———	x	—··—
g	——·	p	·——·	y	—·——
h	····	q	——·—	z	——··
i	··	r	·—·		

NUMERALS

1	·————	4	····—	8	———··
2	··———	5	·····	9	————·
3	···——	6	—····	0	—————
		7	——···		

PUNCTUATION

Period	·—·—·—
Comma	——··——
Semicolon	—·—·—·
Question mark	··——··
Exclamation	——··——
Colon	———···
Apostrophe	·————·
Hyphen	—····—
Fraction bar	—··—·
Parenthesis	—·——·—
Quotation marks	·—··—·
Double dash	—···—

Morse code 1 A system of telegraphic signals invented by S. F. B. Morse, comprised of dots and dashes or short and long flashes representing the letters of the alphabet and used in transmitting messages. **2** International Morse code. Also **Morse alphabet.**

mor·sel (môr′səl) *n.* **1** A bit of food; bite. **2** A small piece of anything. [< OF, dim. of *mors* bite < L *morsus* < *mordere*]

Morse telegraph A telegraph employing the dot-and-dash or Morse code, recording the signals on a continuous paper strip. The first commercial system was set up between Baltimore and Washington in 1844.

mort[1] (môrt) *n. Obs.* Death. **2** A flourish on the hunting horn, announcing the death of game. Also **morte.** [< F *mort* < L *mors, mortis* < *mori* die]

mort[2] (môrt) *n. Brit. Dial.* A great quantity or number. [? < *mortal* (def. 6)]

mort[3] (môrt) *n. Brit. Dial.* A salmon in its third year. [Orig. unknown]

mor·tal (môr′təl) *adj.* **1** Subject to death; hence, pertaining to humanity; human. **2** Causing, or that may or will cause, death; fatal. **3** *Theol.* Incurring the penalty of eternal death, as a sin: opposed to *venial.* **4** Marking the end of life. **5** Subject to fatal injury, as a vital organ. **6** *Colloq.* Extreme: *a mortal fright;* also, long and tedious. **7** Deadly in malice or purpose; inveterate: *a mortal foe.* — *n.* Whatever is mortal or subject to death; a human being. — *adv. Colloq.* Very; exceedingly: *mortal tired.* [< OF < L *mortalis* < *mors, mortis* death]

mor·tal·i·ty (môr·tal′ə·tē) *n.* **1** The quality of being mortal. **2** Death. **3** Frequency of death; the proportion of deaths in a specified number of the population; the death rate. **4** Humanity; mankind. [< OF *mortalité* < L *mortalitas*]

mortality table A life table.

mor·tal·ly (môr′təl·ē) *adv.* **1** Fatally. **2** After the manner of a mortal. **3** Extremely.

mortal mind In Christian Science, nothing, claiming to be something, for Mind is immortal; a belief that life, substance, and intelligence are in and of matter: the opposite of Spirit and therefore the opposite of God, or good; the belief that man is the offspring of mortals; the belief that there can be more than one creator.

mor·tar[1] (môr′tər) *n.* **1** A strong bowl-like vessel in which substances are crushed or pounded with a pestle. **2** *Mil.* A piece of ordnance with a large bore for firing heavy shells at low muzzle velocity and great angles of elevation that they may drop upon the object aimed at. **3** *Mining* A tublike cast-iron receptacle with grated sides, in which ore is stamped. **4** Any of several devices for hurling pyrotechnic shells or bombs and also life lines. [OE *mortere* < L *mortarium*]

mor·tar[2] (môr′tər) *n.* **1** A building material prepared by mixing lime, plaster of Paris, or cement, with sand, water, and sometimes other materials. It is used in masonry, plastering, etc. **2** Loosely, a cement. — *v.t.* To plaster or join with mortar. [< OF *mortier* < L *mortarium*]

mor·tar·board (môr′tər·bôrd′, -bōrd′) *n.* **1** A square board with a handle, on which a mason holds mortar. **2** The academic cap: so called from the four-cornered piece forming its crown.

mortar hoe A hoe with openings in the blade, used for mixing mortar, cement, etc. See illustration under HOE.

mort·gage (môr′gij) *n.* **1** *Law* An estate in land created by conveyance coupled with a condition of defeasance on the performance of some stipulated condition, as the payment of money. **2** A lien upon land or other property as security for the performance of some obligation, to become void on such performance. **3** The act of conveying, or the instrument effecting the conveyance. **4** A state or condition of being pledged as security for a debt like that of a mortgage of property. — **first mortgage** A mortgage having precedence as a lien over all other mortgages. — *v.t.* **gaged, ·gag·ing 1** To make over or pledge (property) by mortgage. **2** To pledge. [< F, dead pledge]

mort·ga·gee (môr′gi·jē′) *n.* The person to whom a mortgage is given.

mort·ga·gor (môr′gi·jər) *n.* A person who gives a mortgage. Also **mort′gage·or, mort′gag·er.**

mor·tice (môr′tis) See MORTISE.

mor·ti·cian (môr·tish′ən) *n.* A funeral director; undertaker. [< L *mors, mortis* death + -ICIAN]

mor·ti·fi·ca·tion (môr′tə·fə·kā′shən) *n.* **1** The state of being humbled or depressed by disappointment or chagrin. **2** *Pathol.* The death

of one part of an animal body, as from gangrene, while the rest is still alive. 3 In religion, the act of subduing the passions and appetites by fasting, penance, or painful severities inflicted on the body. 4 That which mortifies or causes humiliation. See synonyms under CHAGRIN. [<LL *mortificatio, -onis*]
mor·ti·fy (môr′tə·fī) *v.* ·**fied**, ·**fy·ing** *v.t.* 1 To affect with humiliation, shame, or chagrin; humiliate. 2 To discipline or punish (the body, passions, etc.) by fasting, penance, or other ascetic practices. 3 *Pathol.* To cause mortification in (a part of the body). — *v.i.* 4 To practice ascetic self-discipline. 5 *Pathol.* To undergo mortification; become gangrenous. [<OF *mortifier* <LL *mortificare* <L *mors, mortis* death + *facere* make] — **mor′ti·fi′er** *n.* — **mor′ti·fy′ing·ly** *adv.*
Mor·ti·mer (môr′tə·mər), **Roger de**, 1287?-1330, Earl of March; Welsh rebel, favorite of Isabella, queen of Edward II of England.
mor·tise (môr′tis) *n.* 1 A space hollowed out, as in a timber, to receive a tenon or the like. 2 Figuratively, adhering power; firmness. — *v.t.* ·**tised**, ·**tis·ing** 1 To cut or make a mortise in. 2 To join by a tenon and mortise. Also spelled *mortice*. [<OF *mortaise* < Arabic *murtazza* joined, fixed in]

MORTISE AND TENON
a. Mortise.
b. Tenon.

mort·main (môrt′mān) *n. Law* The state of lands and tenements held by one that cannot alienate them, as a religious corporation; inalienable tenure or possession. The term is applied in some of the United States to statutes restricting the right of religious corporations to hold land. [<OF *mortemain* <Med. L *mortua manus* dead hand]
Mor·ton (môr′tən), **William Thomas Green**, 1819-68, U. S. dentist; pioneer in use of ether as an anesthetic.
mor·tu·ar·y (môr′choo·er′ē) *adj.* Pertaining to the burial of the dead; also, relating to or reminiscent of the dead. — *n. pl.* ·**ar·ies** 1 In old English law, a gift claimed by or given to a parish minister on the death of a parishioner. 2 A place for the temporary reception of the dead; dead house. [<L *mortuarius* belonging to the dead]
mor·u·la (môr′yoo·lə, -ōō-) *n. pl.* ·**lae** (-lē) *Biol.* The compact, spherical mass of cells formed by an ovum in the early stages of its development; the mulberry body. [<NL, dim. of L *morum* mulberry] — **mor′u·lar** *adj.*
mo·sa·ic (mō·zā′ik) *n.* 1 Inlaid work composed of bits of stone, glass, etc., forming a pattern or picture. 2 A piece of inlaid work of this kind, or anything resembling such work. 3 *Phot.* An assemblage of aerial photographs pieced together and joined at the margins so as to form a single, continuous picture of a terrain. 4 The plate covered with minute, photosensitive granules which is mounted in the image-scanning element of an electron television camera. — *adj.* Of, pertaining to, or resembling mosaic; tessellated; inlaid. — *v.t.* ·**icked**, ·**ick·ing** 1 To make as if by combining in a mosaic. 2 To decorate with mosaic. [<OF *mosaicq* <Med. L *musaicus* <Gk. *mouseios* of the Muses, artistic <*Mousa* Muse] — **mo·sa·i·cist** (mō·zā′ə·sist) *n.*
Mo·sa·ic (mō·zā′ik) *adj.* Of or pertaining to Moses or his laws. Also **Mo·sa′i·cal**. [<NL *Mosaicus* <L *Moses* Moses]
mosaic disease One of several destructive and infectious diseases of plants caused by a filtrable virus and characterized by a pale, mottled appearance of the foliage: tobacco *mosaic*; potato *mosaic*.
mosaic gold 1 An alloy of copper and zinc; ormolu varnish. 2 Stannic sulfide.
Mo·san·der (mōō·sän′dər), **Carl Gustav**, 1797-1858, Swedish chemist.
Mos·by (mōz′bē), **John Singleton**, 1833-1916, American Confederate soldier and author.
mos·ca·tel (mos′kə·tel′) See MUSCATEL (def. 2).
mos·chate (mos′kāt, -kit) *adj.* Having the odor of musk. [<NL *moschatus* <Med. L *moschus* musk]

mos·cha·tel (mos′kə·tel′, mos′kə·tel) *n.* A low perennial herb (*Adoxa moschatellina*) with greenish flowers and having a musky odor; muskroot. [<F *moscatelle* <Ital. *moscatella*]
Moś·cic·ki (môsh·tsits′kē), **Ignacy**, 1867-1946, Polish chemist; president of Poland 1926-39.
Mos·cow (mos′kou, -kō) The capital of the U.S.S.R. and of the Russian S.F.S.R., in western European Russian S.F.S.R. *Russian* **Mos·kva** (mos·kvä′).
Mose·ley (mōz′lē), **Henry Gwyn-Jeffreys**, 1887-1915, English physicist.
Mo·selle (mō·zel′) *n.* A light, dry wine made in the valley of the Moselle, chiefly in the region from Trier to Traben-Trarbach.
Mo·selle (mō·zel′) A river in NE France, Luxemburg, and western Germany, flowing 320 miles NE to the Rhine at Coblenz. *German* **Mo·sel** (mō′zəl).
Mo·ses (mō′zis, -ziz; *Ger., Sw.* mō′ses) A masculine personal name. [<Egyptian, child, son] — **Moses** In the Old Testament, the younger son of Amram and Jochebed, who led the Israelites out of Egypt into the Promised Land, received the Ten Commandments from God, and gave laws to the people; hence, a leader; legislator.
Mo·ses (mō′zis, -ziz), **Anna Mary**, born 1860, U. S. painter: called "Grandma Moses."
mo·sey (mō′zē) *v.i. U. S. Slang* 1 To saunter, or stroll; shuffle along. 2 To go away; move off. [? <VAMOOSE]
Mos·kva (mos·kvä′) A river in Russian S.F.S.R., flowing 315 miles to the Oka.
Mos·lem (moz′ləm) *n.* A believer in Islam; a follower of Mohammed: the name used by Mohammedan peoples themselves. — *adj.* Of or pertaining to Islam or its adherents. Also spelled **Muslem, Muslim**. [<Arabic *muslim* one who submits <*aslama* surrender (to God)] — **Mos·lem′ic** *adj.* — **Mos′lem·ism** *n.*
Mos·ley (mōz′lē), **Sir Oswald**, born 1896, English politician.
mosque (mosk) *n.* A Moslem temple of worship. Also **mosk**. [<F *mosquée* <Ital. *moschea* <Arabic *masjid* < *sajada* prostrate oneself, pray]
Mos·qui·ti·a (məs·kē′tē·ə, *Sp.* mōs·kē′tyä) An undeveloped region of Central America on the Caribbean coast of Honduras and Nicaragua. Also **Mosquito Coast**.
mos·qui·to (məs·kē′tō) *n. pl.* ·**toes** or ·**tos** 1 A two-winged, dipterous insect (family *Culicidae*), having (in the female) a long proboscis capable of puncturing the skin for extracting blood. The infections of malaria and yellow fever are spread by the bite of certain species. For illustration see INSECTS (injurious). 2 Any of various other gnats or flies inflicting a similar puncture. Also spelled *musquito*. [<Sp., dim. of *mosca* fly <L *musca*] — **mos·qui′tal** *adj.*
mosquito boat A patrol torpedo boat.
mosquito fleet An assemblage of small craft.
mosquito hawk 1 A nighthawk. 2 A dragonfly.
mosquito net A fine netting or gauze placed in windows, over beds, etc., to keep out mosquitoes.
mosquito netting A coarsely meshed fabric used to make mosquito nets.
moss[1] (môs, mos) *n.* 1 A delicate, bryophytic plant (class *Musci*), growing on the ground, decaying wood, rocks, trees, etc., having a stem and distinct leaves, and producing capsules which open by an operculum and contain spores unmixed with elaters. 2 Any of several other cryptogamous plants, as certain lichens, clubmosses, etc. — *v.t.* To cover with moss. [<MOSS[2]] — **moss′y** *adj.* — **moss′i·ness** *n.*
moss[2] (môs, mos) *n.* A bog; peat bog. [OE *mos*]
moss agate A variety of quartz containing mineral oxides, as manganese dioxide, arranged in mosslike forms.
moss·back (môs′bak′, mos′-) *n.* 1 An old fish or turtle on whose back is a growth of algae or the like. 2 One who is out of touch with the progress of the times; a conservative or reactionary person; especially, an extreme conservative in politics. 3 During the American Civil War, in the South, one who avoided conscription by hiding.

moss–backed (môs′bakt′, mos′-) *adj.* 1 Having moss or mosslike growth on the back. 2 Behind the times; reactionary.
moss·board (môs′bôrd′, -bōrd′, mos′-) *n.* A type of pasteboard made principally of peat moss and used in the preparation of surgical dressings.
moss·bunk·er (môs′bungk·ər, mos′-) *n.* The menhaden. Also **moss′bank·er**. [Alter. of Du. *marskanker*]
moss green Any of various shades of dull yellowish green.
moss·grown (môs′grōn′, mos′-) *adj.* Overgrown with moss; hence, very ancient.
moss hag *Scot.* A pit or slough in a moss or bog, where the moss or peat has been cut away.
mos·so (môs′sō) *adj. Music* Rapid; literally, moved. [<Ital., pp. of *movere* move <L]
moss pink A plant (*Phlox subulata*) of the eastern United States, occurring in several varieties which form mats close to the ground and have white, pink, or purplish flowers. Also **moss phlox**.
moss rose 1 A cultivated variety of the rose (*Rosa centifolia muscosa*) with a mossy calyx and stem. 2 Portulaca.
moss starch Lichenin.
moss–troop·er (môs′trōō′pər, mos′-) *n.* One of the marauders who infested the mossy marshes between Scotland and England prior to the union of the two kingdoms; hence, a bandit or undisciplined soldier.
most (mōst) *adj.* 1 Consisting of the greatest number: superlative of *many*. 2 Consisting of the greatest amount or quantity: superlative of *much*. 3 *Obs.* Greatest in size, rank, or age. — *n.* 1 The greater number; the larger part: the *most* of my belongings. 2 Greatest amount, value, or advantage; utmost degree, extent, or effect. — *adv.* 1 In the highest degree, or in the greatest number or quantity. 2 *Dial.* Almost. 3 Greatest, as in amount or degree: used with adjectives and adverbs to form the superlative degree. [OE *mǣst*]
-most *suffix* Most: added to adjectives, adverbs, and prepositions to form superlatives: *outmost*. [OE *-mest* <*-ma + -est*, superlative suffixes]
Mos·tar (môs′tär) A city in central Herzegovina, Yugoslavia.
most-fa·vored-na·tion clause (mōst′fā′vərd·nā′shən) A provision in many commercial treaties stipulating that the parties shall accord each other treatment as favorable as that granted to any other country.
most·ly (mōst′lē) *adv.* For the most part; principally.
Mo·sul (mō·sōōl′) The second largest city of Iraq, on the Tigris opposite the site of ancient Nineveh. Also **Mos·sul′**.
Mosz·kow·ski (môsh·kôf′skē), **Moritz**, 1854-1925, Polish composer born in Germany.
mot[1] (mō) *n.* A witty or pithy saying; bon mot. [<F, word <LL *muttum* uttered sound <*muttire* murmur]
mot[2] (mot) *n.* A bugle note, or its mark in music. [<F]
mote[1] (mōt) *v.* 1 *Archaic* May; might. 2 *Obs.* Must. ♦ Homophone: *moat*. [OE *mōt* it is permitted]
mote[2] (mōt) *n.* A minute particle; speck. ♦ Homophone: *moat*. [OE *mot* atom]
mo·tel (mō·tel′) *n.* A hotel for motorists, usually comprising private cabins and garage or parking facilities. [<MO(TOR) + (HO)TEL]
mo·tet (mō·tet′) *n. Music* A contrapuntal, polyphonic song of a sacred nature, usually unaccompanied. Also **mo·tet′to**. [<OF, dim. of *mot* word. See MOT[1].]
moth (môth, moth) *n. pl.* **moths** (môthz, môths, mothz, moths) Any of various typically nocturnal, lepidopterous insects (division *Heterocera*), distinguished from butterflies by their stout bodies and smaller, usually dull-colored wings, which fold laterally across the abdomen. The larvae of the gipsy moth, silkworm, etc., feed on plants; those of the clothes moth (family *Tineidae*) feed on textiles, clothing, and furs. [OE *moththe*]
moth ball A ball of camphor or naphthalene, for the protection of clothing, etc., from moths. — **in moth balls** In protective storage.

moth-ball (môth′bôl′, moth′-) *Mil. & Nav. adj.* Designating ships or military equipment laid up in reserve and covered or coated with protective materials. — *v.t.* To put in reserve and protective storage.

moth-eat-en (môth′ēt′n, moth′-) *adj.* Eaten by moths; hence, used up or worn out: also *mothy.*

moth-er[1] (muth′ər) *n.* 1 A female parent. 2 That which has produced or given birth to anything. 3 An abbess or other nun of rank or dignity. 4 An elderly woman or matron: a familiar title. — *v.t.* 1 To care for as a mother. 2 To bring forth as a mother; produce. 3 To admit or claim parentage, authorship, etc., of. — *adj.* 1 Native: *mother* tongue. 2 Holding a maternal relation. [OE *mōdor*]

moth-er[2] (muth′ər) *n.* A slimy film composed of bacteria and yeast cells that forms on the surface of fermenting liquids and is active in the production of vinegar: also called *mother-of-vinegar.* 2 Dregs; lees. — *v.i.* To become mothery, as vinegar. [Special use of MOTHER[1]] — **moth′er·y** *adj.*

Mother Car-ey's chicken (kâr′ēz) The petrel; especially, the storm petrel. [Alter. of L *mater cara* dear mother, an epithet of the Virgin Mary]

mother cell A cell which by division produces other cells.

Mother Goose 1 The imaginary narrator of a volume of folk tales, compiled in French by Charles Perrault in 1697. 2 The imaginary compiler of a collection of nursery rimes of English folk origin, first published in London about 1760 by John Newbery.

moth-er-hood (muth′ər-hood) *n.* The state of being a mother.

Mother Hub-bard (hub′ərd) 1 The main character in an old nursery rime. 2 A woman's loose, flowing gown, unconfined at the waist.

moth-er-in-law (muth′ər-in-lô′) *n. pl.* **moth-ers-in-law** 1 The mother of one's spouse. 2 *Brit. Dial.* A stepmother.

moth-er-land (muth′ər-land′) *n.* The land of one's ancestors; native land; mother country.

moth-er-less (muth′ər-lis) *adj.* Having no mother.

mother liquor The liquid remaining after the substances in solution have been deposited by crystallization or precipitation.

mother lode *Mining* 1 *Cap.* The great gold-bearing quartz vein in California, traced by its outcrop from Mariposa to Amador. 2 Any principal vein in a mining region.

moth-er-ly (muth′ər-lē) *adj.* 1 Resembling a mother: a *motherly* woman. 2 Pertaining to or becoming to a mother: *motherly* authority, *motherly* care. — *adv.* In the manner of a mother. — **moth′er·li·ness** *n.*

Mother of God A title officially given to the Virgin Mary at the Council of Ephesus in 431.

moth-er-of-pearl (muth′ər-əv-pûrl′) *n.* The hard, iridescent internal layer of certain shells, as of pearl oysters, abalones, and mussels; nacre.

Mother of Presidents The State of Virginia: seven of America's first twelve presidents were Virginians.

Mother of the Gods Rhea.

moth-er-of-vin-e-gar (muth′ər-əv-vin′ə-gər) See MOTHER[2].

Mother's Day A memorial day in honor of mothers, observed annually in the United States on the second Sunday in May. See FATHER'S DAY.

mother tongue 1 One's native language. 2 A language from which another language has sprung.

Moth-er-well and Wish-aw (muth′ər-wel ənd wish′ô) A burgh on the Clyde in Lanarkshire, Scotland; formerly two towns.

mother wit Inherent, natural, or native wit.

moth-er-wort (muth′ər-wûrt′) *n.* An Old World herb (*Leonurus cardiaca*) of the mint family, with lanceolate, toothed leaves, and small, purplish flowers: now common in the U. S. [So called because once thought to be valuable in the treatment of diseases of the womb]

moth-mill-er (môth′mil′ər, moth′-) *n.* 1 A pale moth with floury wings; a miller. 2 *Colloq.* Any moth.

moth-proof (môth′proof′, moth′-) *adj.* Resistant to the attack of moths. — *v.t.* **-proofed, -oof-ing** To render (textiles) resistant to s.

moth-y (môth′ē, moth′ē) *adj.* **moth·i·er, moth·i·est** Moth-eaten.

mo-tif (mō-tēf′) *n.* 1 The underlying idea or main feature or element in literary, musical, or artistic work. 2 In the decorative arts, a distinctive element of design. Also **motive** (mō′tiv). [<F]

mo-tile (mō′til) *adj.* 1 *Biol.* Having the power of spontaneous motion, as certain minute organisms. 2 Causing motion. [<L *motus,* pp. of *movere* move] — **mo·til′i·ty** *n.*

mo-tion (mō′shən) *n.* 1 Change of position in reference to an assumed point or center; a shifting movement. 2 The interaction of parts in a mechanism to produce a particular result. 3 A formal proposition in a deliberative assembly. 4 A significant movement of the limbs, eyes, etc.; a gesture: She made *motions* to him. 5 An impulse to action; incentive. 6 *Music* Melodic progression. 7 *Law* An application to a court to obtain an order or rule directing some act to be done. 8 A mechanism. 9 *Obs.* A puppet or puppet show. — **perpetual motion** Continuous mechanical motion, especially as applied to machines which are claimed to do useful work without the expenditure of equivalent amounts of work upon them. — *v.i.* To make a gesture of direction or intent, as with the hand. — *v.t.* To direct or guide by a gesture. [<F <L *motio, -onis* <*motus,* pp. of *movere* move] — **mo′tion·al** *adj.* — **mo′tion·less** *adj.* — **mo′tion·less·ly** *adv.*

Synonyms (noun): act, action, change, move, movement, passage, transit, transition. *Motion* may be either abstract or concrete, more frequently the former; *movement* is always concrete, that is, considered in connection with the thing that moves or is moved; thus we speak of the *movements* of the planets, but of the laws of planetary *motion;* of military *movements,* but of perpetual *motion. Motion* is *change* of place or position in space; *transition* is a passing from one point or position in space to another. *Move* is used chiefly of contests or competition, as in chess or politics: as, It is your *move;* a shrewd *move* of the opposition. We speak of mental or spiritual *acts* or processes, or of the laws of mental *action,* but a formal proposal of *action* in a deliberative assembly is termed a *motion. Action* is a more comprehensive word than *motion.* See ACT, TOPIC. *Antonyms:* quiescence, quiet, repose, rest.

motion picture 1 A sequence of pictures, each slightly different from the last, photographed by a special camera on a single strip of film, for projection on a screen, giving the optical illusion of continuous, ordered movement: also called *cinema, movie, moving picture.* 2 A photoplay. — **mo′tion-pic′ture** *adj.*

motion study The detailed observation and analysis of the different movements involved in the performance of a given repetitive task, with a view to lessening the fatigue and increasing the efficiency of the workers.

mo-ti-vate (mō′tə-vāt) *v.t.* **-vat-ed, -vat-ing** To provide with a motive; instigate; induce.

mo-ti-va-tion (mō′tə-vā′shən) *n.* Causative factor; incentive; drive. — **mo′ti·va′tion·al** *adj.*

motivation research The use of psychological or other social-science concepts and techniques in marketing, advertising, and public-opinion research to investigate the conscious or subliminal causes of consumer behavior, and the motives triggering action or decisions that bring about favorable or unfavorable response toward a product, package, advertising, or other communications.

mo-tive (mō′tiv) *n.* 1 That which incites to motion or action. 2 A predominant idea; design. — *adj.* 1 Having power to move; causing motion. 2 Relating to a motive or motives. — *v.t.* **-tived, -tiv-ing** 1 To motivate; prompt. 2 To relate to the leading idea or motif in a work of art, etc. [<OF *motif* <Med. L *motivus* <*motus,* pp. of *movere* move]

Synonyms (noun): consideration, ground, incentive, incitement, inducement, influence, reason. *Motive* may signify either a mental impulse, or something external that is an object of desire, and so an *inducement* or *incitement* to action. Compare CAUSE, IMPULSE, PURPOSE, REASON.

motive power 1 The power, or means of generating power, by which motion is imparted to an object, machine, etc. 2 Figuratively, an impelling force.

mo-tiv-i-ty (mō-tiv′ə-tē) *n.* The power of producing motion; motive energy.

mot juste (mō zhüst′) *French* The precise word; exactly the right expression.

mot-ley (mot′lē) *adj.* 1 Variegated in color; particolored. 2 Composed of heterogeneous elements. 3 Clothed in varicolored garments. — *n.* 1 A dress of various colors, such as was formerly worn by court jesters. 2 A jester or fool in motley garments. 3 A medley, as of colors. [ME *motteley;* origin uncertain]

Mot-ley (mot′lē), **John Lothrop,** 1814–77, U. S. historian.

mot-mot (mot′mot) *n.* One of a family of birds, the *Momotidae,* of the warmer parts of America, related to the kingfishers; a sawbill. The middle pair of tail feathers is usually elongated and racket-shaped. [Imit. of its note]

MOTLEY (*n.* def. 1)

mo-to-fa-cient (mō′tō-fā′shənt) *adj. Physiol.* Producing motion: said especially of muscles whose contraction results in a definite movement. [<L *motus* motion, moving <*movere* move + *faciens, -entis,* ppr. of *facere* make, cause]

mo-tor (mō′tər) *n.* 1 One who or that which produces motion, as a machine, nerve, etc. 2 An internal-combustion engine, especially one operating on gasoline. 3 A motorcar or motorcycle. 4 An electric motor. — **rotary motor** An internal-combustion engine having multiple radial cylinders rotating about a fixed crankshaft. — *adj.* 1 Causing, producing, or imparting motion. 2 Transmitting impulses from the nerve centers to the muscles. 3 Pertaining to consciousness of motion. — *v.i.* To travel or ride in an automobile. [<L <*motus,* pp. of *movere*] — **mo′tor·ing** *adj. & n.*

mo-tor-boat (mō′tər-bōt′) *n.* A boat propelled by an internal-combustion engine, or by an electric motor.

mo-tor-bus (mō′tər-bus′) *n.* A power-driven omnibus. Also **motor coach.**

mo-tor-cade (mō′tər-kād) *n.* A procession of motorcycles or automobiles. [<MOTOR + (CAVAL)CADE]

mo-tor-car (mō′tər-kär′) *n.* An automobile.

mo-tor-cy-cle (mō′tər-sī′kəl) *n.* A two-wheeled vehicle, sometimes having an attached sidecar with a third wheel, propelled by an internal-combustion engine. — *v.i.* **-cled, -cling** To travel or ride on a motorcycle. — **mo′tor·cy′clist** *n.*

motor drive A power unit consisting of an electric motor and auxiliaries, used to operate a machine or group of machines.

mo-tor-drome (mō′tər-drōm) *n.* An enclosure, course, or track where motor-driven vehicles of various kinds are tested in competition or otherwise.

motor generator A device for transforming electrical power by permanently connecting, usually on a common bedplate, a motor and a generator; a dynamotor.

motor impulse *Physiol.* The impulse transmitted by nerve fibers from the nerve centers to the muscles.

mo-tor-ist (mō′tər-ist) *n.* One who drives an automobile; one who travels by automobile.

mo-to-ri-um (mō-tôr′ē-əm, -tō′rē-) *n. Physiol.* 1 That portion of the nervous system which controls the motor apparatus. 2 Any center of a motor function. [<NL <L *motor* a mover]

mo-tor-ize (mō′tər-īz) *v.t.* **-ized, -iz-ing** 1 To equip with a motor or motors. 2 To equip with motor-propelled vehicles in place of horses and horse-drawn vehicles. — **mo′tor·i·za′tion** *n.*

mo-tor-man (mō′tər-mən) *n. pl.* **-men** (-mən) One who operates a motor, as on a street car or an electric locomotive.

motor oil A high-grade lubricating oil, designed for the exacting requirements of internal-combustion engines.

motor scooter See SCOOTER (def. 2).

mo-tor-ship (mō′tər-ship′) *n.* A vessel, as a

motor spirit passenger ship, of which the principal motive power is derived from an internal-combustion oil or gas engine.

motor spirit Any fuel adapted for spark-ignition, internal-combustion engines; specifically, coal or petroleum distillates blended with suitable additions, as alcohol.

motor transport *Mil.* Any motor vehicle used for transport only: distinguished from *combat vehicle.*

motor vehicle 1 Any vehicle operated by a motor or engine. 2 A vehicle adapted to be pulled by another, as a trailer.

Mott (mot), **John Raleigh,** 1865–1955, U.S. clergyman and leader in the Y.M.C.A. — **Lucretia,** 1793–1880, *née* Coffin, U.S. social reformer.

Mot·ta (môt'tä), **Giuseppi,** 1871–1940, Swiss lawyer and statesman.

motte (mot) n. *Dial.* A small growth of trees on a prairie. Also **mott.** [<Sp. *mata* clump, grove]

mot·tet·to (môt·tet'tō) n. *Italian* Motet.

Mot·teux (mô·tœ'), **Peter Anthony,** 1663–1718, English dramatist and translator, born in France.

mot·tle (mot'l) v.t. **·tled, ·tling** To mark with spots or streaks of different colors or shades; blotch. — n. 1 The spotted, blotched, or variegated appearance of any mottled surface, as of wood or marble. 2 One of a number of spots or blotches on any surface. [<MOTLEY]

mot·tled (mot'ld) adj. Marked with spots of different color or shades of color; blotched; variegated.

mot·to (mot'ō) n. pl. **·toes** or **·tos** 1 An expressive word or pithy sentence enunciating some guiding principle, rule of conduct, or the like; a phrase inscribed on something or prefixed to a literary composition as somehow indicative of its qualities. 2 A piece of paper printed with a motto or sentiment and wrapped around a small piece of candy; also, the piece of candy enclosed. See synonyms under ADAGE. [<Ital. <F *mot.* See MOT.]

mot·ty (mot'ē) adj. *Scot.* Abounding in motes. Also **mot'tie.**

mou' (moo) n. *Scot.* The mouth.

mouch (mooch) See MOOCH.

mou·choir (moo·shwâr') n. *French* A pocket handkerchief.

mou·die (moo'dē) n. *Scot.* A mole; also, a molecatcher. Also **mou'di·wort** (-wûrt).

moue (moo) n. A pouting grimace expressive of disdain or distaste. [<F]

mouf·lon (moo'flon) n. A hairy wild sheep, specifically, *Ovis musimon* of the mountains of Corsica and Sardinia, the males with very large and curved horns. Also **mouf'flon.** [<F <dial. Ital. *muffolo* <LL *muffione* <LL *mufro, -onis*]

mouil·lé (moo·yā') adj. Given a palatalized pronunciation, as *ll* in the French name *Villon.* [<F, pp. of *mouiller* moisten]

mou·jik (moo·zhēk') See MUZHIK.

Mouk·den (mook'den', mook'den, mook'-) See MUKDEN.

moul (mool) v.t. & v.i. *Obs.* or *Brit. Dial.* To make or become moldy.

mou·lage (moo·läzh') n. 1 A cast, in plaster of Paris or other similar material, of an object or of its impressed outlines on a surface: frequently used in criminal identification, as of footprints, tire marks, etc. 2 A synthetic, rubberlike, plastic material used in making casts. [<F]

mould (mōld), **moult** (mōlt), etc. See MOLD, etc.

mould goose The musk duck.

mould·warp (mōld'wôrp) See MOLDWARP.

mou·lin (moo·laṅ') n. A nearly vertical shaft in a glacier, formed by the surface water trickling through a crevice. [<F, mill <LL *molina.* See MILL.]

mou·line (moo·lēn') n. 1 The circular swing of a saber. 2 The drum of a winch, capstan, or the like; a windlass mechanism. 3 A form of turnstile. Also **mou·li·net** (moo'lē·net). [<F, dim. of *moulin* mill]

Mou·lins (moo·laṅ') The capital of Allier department, central France.

Moul·mein (mool·mān') A port on the Andaman Sea in Lower Burma, chief town of Tenasserim: also *Maulmain.*

mouls (moolz) *Scot.* See MOOLS.

moult (mōlt), **moult·ing,** etc. *Brit.* Molt, etc.

Moul·ton (mōl'tən), **Forest Ray,** 1872–1952, U.S. astronomer.

Moul·trie (mool'trē, moo'-, mōl'-), **William,** 1730–1805, American Revolutionary general; builder and defender (1776) of Fort Moultrie, Charleston Harbor, South Carolina.

mound[1] (mound) n. 1 A heap or pile of earth, either natural or artificial; hillock. 2 One of the earthworks built by the Mound Builders for burial or fortification. 3 In baseball, the slightly raised ground from which the pitcher pitches the ball. See synonyms under RAMPART. — v.t. 1 To fortify or enclose with a mound. 2 To heap up in a mound. [Origin uncertain]

mound[2] (mound) n. A jeweled ball or globe, often surmounted by a cross, forming part of the regalia of a king or emperor: an emblem of sovereignty; an orb. [<F *monde* <L *mundus* world]

MOUND[2]

Mound Builder One of the aboriginal people who built the burial mounds and fortifications found in the Mississippi basin and adjoining regions: the ancestors of the North American Indians dwelling in that region at the time of the first European explorers.

Mounds·ville (moundz'vil) A city in NW West Virginia; site of a conical Indian mound.

mount[1] (mount) v.t. 1 To ascend by climbing; go up, as stairs. 2 To climb upon; ascend and seat oneself on, as a horse. 3 To put on horseback. 4 To furnish with a horse or horses. 5 To set or place in an elevated position: to *mount* a plaque on a wall. 6 To place in position for use or operation, as a cannon or engine. 7 To put in or on a support, frame, etc., as for exhibition: to *mount* a butterfly. 8 To furnish, as a play, with scenery, costumes, etc. 9 To copulate with a female: said of male animals. 10 In microscopy: a To place or fix (a sample) on a slide. b To prepare (a slide) for examination. 11 *Mil. & Nav.* To carry or be equipped with: a ship *mounting* thirty-two guns. 12 To put on (clothing), especially for display. 13 *Mil.* To stand or post (guard). 14 *Mil.* To prepare for and begin: to *mount* an offensive. — v.i. 15 To rise or ascend; go up. 16 To increase in number, amount, or degree: often with *up*: The bills *mounted* up. 17 To get on horseback. 18 To get up on or on top of something. — n. 1 That upon or by which anything is prepared or equipped for use, exhibition, ornament, preservation, or examination. 2 The card, etc., upon which a drawing or photograph is mounted. 3 The parts and appliances by which a gun is attached to its carriage. 4 The glass slide and its adjuncts upon which a microscopic subject is secured for examination. 5 A saddle horse or other animal used for riding: by extension, a bicycle. 6 The act of riding a horse in a race; also, the privilege or opportunity of doing so. 7 A style of mounting. [<OF *monter* <L *mons, montis* mountain] — **mount'a·ble** adj. — **mount'er** n.

mount[2] (mount) n. 1 An elevation of the earth's surface; a mountain; a hill. When used as part of a proper name it usually precedes the specific application: *Mount Washington.* 2 *Obs.* A raised fortification commanding the surrounding country. 3 In palmistry, one of seven fleshy protuberances in the palm of the hand. [OE *munt* <L *mons* mountain]

moun·tain (moun'tən) n. 1 A natural elevation of the earth's surface, rising more or less abruptly to a small summit area, attaining an elevation greater than that of a hill, and standing either in a single mass or forming part of a series. 2 Any large heap or pile resembling this. 3 Something of great magnitude. — **the Mountain** A name given to the ultra-revolutionary party of the French National Assembly or Convention in 1793, from the fact that its members occupied the highest seats in the chamber. — adj. 1 Of, pertaining to, or living or growing on mountains. 2 Like or suggesting a mountain or mountains. [<OF *montaigne* <L *montanus* mountainous <*mons, montis* mountain]

mountain ash 1 Any of various American deciduous shrubs (genus *Sorbus*), having alternate simple or pinnate leaves, white flowers, and vivid red fruit, especially *S. americana.* 2 The rowan.

mountain avens A small evergreen plant (*Dryas octopetala*) of the rose family, growing in arctic and alpine regions.

mountain cat 1 The cougar. 2 The bobcat.

mountain chain A series of mountains connected and having some common characteristics.

mountain cork A variety of asbestos occurring in light, flexible sheets which will float on water. Also **mountain leather.**

mountain cranberry The mountain cowberry (*Vaccinium vitis-idaea minus*), having edible red berries, evergreen leaves, and pink or red flowers.

mountain dew *Slang* Illicitly distilled whiskey.

moun·tain·eer (moun'tən·ir') n. 1 An inhabitant of a mountainous district. 2 One who climbs mountains. — v.i. To climb mountains.

mountain goat The Rocky Mountain goat; a goat antelope.

mountain laurel In the eastern United States, the low-growing calicobush (*Kalmia latifolia*), an evergreen shrub with white or pink flowers; State flower of Connecticut and Pennsylvania. The foliage is poisonous to livestock.

ROCKY MOUNTAIN GOAT
(About 3 1/2 feet high at the shoulder)

mountain lion The puma or cougar.

moun·tain·ous (moun'tən·əs) adj. 1 Full of mountains. 2 Huge. — **moun'tain·ous·ly** adv.

mountain range 1 One of the components of a mountain chain, usually a group of more or less parallel ridges of similar origin, structure, etc. 2 A land area dominated by such a group of mountains, characterized by great variations in elevation above sea level.

mountain rat A pack rat.

mountain sheep The bighorn.

mountain sickness *Pathol.* A form of anoxemia accompanied by nausea, due to insufficient oxygen at high altitudes, especially on mountains.

Mountains of the Moon See RUWENZORI.

mountain specter The specter of the Brocken.

Mountain Standard Time See STANDARD TIME. Abbr. M.S.T.

Mount·bat·ten (mount·bat'n), **Lord Louis,** born 1900, English admiral in World War II: originally *Prince Louis Francis Battenberg.*

Mount Des·ert (dez'ərt, di·zûrt') An island in Acadia National Park off the SE coast of Maine; 100 square miles.

moun·te·bank (moun'tə·bangk) n. 1 A vendor of quack medicines at fairs, who usually mounts a platform or wagon to sell his wares. 2 Hence, a charlatan. See synonyms under QUACK. — v.i. To play the mountebank. [<Ital. *montimbanco* <*monta* mount + *in* on + *banco* bench]

mount·ed (moun'tid) adj. Elevated on or equipped with horses: *mounted* police.

Mount·ie (moun'tē) See MOUNTY.

mount·ing (moun'ting) n. 1 The act of mounting; elevation. 2 A mount, as of a picture. 3 The act of preparing for use, etc.

Mount McKinley National Park A region in south central Alaska; 3,030 square miles; established as a national park in 1917.

Mount Rainier National Park A region in the Cascade Mountains of west central Washington; 378 square miles; established as a national park in 1899.

Mount Rob·son Provincial Park (rob'sən) A national park in eastern British Columbia, Canada, near the Alberta border; 65 miles long, 10 to 20 miles wide.

Mount Ver·non (vûr'nən) The home and burial place of George Washington; 15 miles below

Mounty Washington, D.C., on the Potomac River.
Moun·ty (moun′tē) n. pl. **·ties** Colloq. A member of the Royal Canadian Mounted Police: also spelled **Mountie**.
mourn (môrn, mōrn) v.i. 1 To feel or express grief or sorrow. 2 To display the conventional signs of grief after someone's death; wear mourning. — v.t. 3 To grieve or sorrow for (someone dead). 4 To grieve over or lament (misfortune, failure, etc.); bewail; deplore. 5 To utter in a sorrowful manner. [OE murnan]
— Synonyms: bemoan, bewail, deplore, grieve, lament, regret, rue, sorrow. Mourning is thought of as prolonged; grief or regret may be transient. One may grieve or mourn, regret, rue, or sorrow without a sound; he bemoans with suppressed and often inarticulate sounds of grief; bewails with passionate utterance, whether of inarticulate cries or of spoken words; he laments in plaintive or pathetic words. One deplores with settled sorrow which may or may not find relief in words. One is made to rue an act by some misfortune resulting, or by some penalty or vengeance inflicted because of it. One regrets a slight misfortune or a hasty word; he sorrows over the death of a friend. Antonyms: exult, joy, rejoice, triumph.
mourn·er (môr′nər, mōr′-) n. 1 One who mourns; specifically, one who attends a funeral. 2 A penitent at a revival meeting.
mourner's bench U.S. A bench near the preacher reserved for penitents at a revival meeting; also called anxious seat.
mourn·ful (môrn′fəl, mōrn′-) adj. 1 Indicating or expressing grief. 2 Oppressed with grief. 3 Exciting sorrow. See synonyms under PITIFUL, SAD. — **mourn′ful·ly** adv. — **mourn′ful·ness** n.
mourn·ing (môr′ning, mōr′-) n. 1 The act of sorrowing or expressing grief; lamentation; sorrow. 2 The symbols or outward manifestations of grief, as the use of symbolical colors in dress, the draping of buildings or doors, and the half-masting of flags. — adj. Relating to or expressive of mourning. — **mourn′ing·ly** adv.
mourning cloak A brownish-black butterfly (Nymphalis antiopa) widely distributed in temperate regions: it has a row of dark spots just inside the yellow border on the upper side of the wings: also called Camberwell beauty.
mourning dove The Carolina turtledove (Zenaidura macroura), common in North America: so called for its plaintive note.
mourning paper Stationery with a black border.
mourning warbler A warbler of the eastern United States (Oporonis philadelphia): so called for its plaintive note.
mouse (mous) n. pl. **mice** (mīs) 1 One of various small rodents frequenting human habitations throughout the world, as the common **house mouse** (Mus musculus). ◆ Collateral adjective: murine. 2 Any of various similar animals, as the American **harvest mouse** (genus Reithrodontomys), or the **lemming mouse** (genus Synaptomys). 3 Slang A young woman. 4 Slang A discolored swelling of the eye, caused by a blow or bruise. 5 Naut. A swelling worked on a rope to prevent its slipping; also, a mousing. — v. (mouz) **moused**, **mous·ing** v.i. 1 To hunt or catch mice. 2 To hunt for something cautiously and softly; prowl. — v.t. 3 To hunt for, as a cat hunts mice. 4 Naut. To secure (a hook) with mousing. 5 Obs. To rend as a cat does a mouse. [OE mūs, pl. mȳs]
mouse·bane (mous′bān′) n. Aconite.
mouse·bird (mous′bûrd′) n. 1 An African bird (genus Colius) with a conical bill, long medial tail feathers, and soft plumage. 2 A shrike.
mouse deer The chevrotain.
mouse-ear (mous′ir′) n. 1 Any one of various plants whose leaves resemble the ears of a mouse; especially, the European hawkweed (Hieracium pilosella). 2 The forget-me-not.
mouse gray A medium shade of gray, the color of the fur of the common house mouse.
mous·er (mou′zər) n. 1 An animal that catches mice; especially, a cat. 2 A person who hunts stealthily.

ail (mous′tāl′) n. One of a genus rus) of plants of the crowfoot family: d from its slender spike.

mouse·trap (mous′trap′) n. A trap for catching mice.
mous·ing (mou′zing) n. 1 The act of hunting mice. 2 Naut. A lashing or shackle passed around the shank and point of a hook, to prevent its spreading or unhooking. — adj. 1 Given to catching mice; prowling. 2 Thorough; careful; patient, as a cat hunting a mouse.
mous·que·taire (moos′kə·târ′) n. 1 A member of one of the two companies of mounted musketeers forming the bodyguard of the French kings between 1622 and 1786. 2 A woman's cloth cloak with large buttons: in fashion about 1855. 3 A long glove of kid or silk, loose about the wrist: worn by women. [<F, musketeer]
mousse (moos) n. A light, frozen dessert made of whipped cream, white of egg, sugar, and flavoring extract, sometimes with the beaten yolks of eggs and gelatin; also, a similar dish made with finely ground meat, fish, or vegetables: lobster mousse. [<F]
mousse·line (moos′lēn′) n. 1 Fine French muslin. 2 A thin glass blown to resemble lace. [<F. See MUSLIN.]
mousse·line-de-soie (moos-lēn′də-swä′) n. French A plain-weave silk chiffon fabric, often figured; silk muslin.
Mous·sorg·sky (moo-sôrg′skē), **Modest Petrovich**, 1835–81, Russian composer: also Mussorgsky.
mous·tache (məs-tash′, mus′tash) See MUSTACHE.
Mous·te·ri·an (moo-stir′ē-ən) adj. Anthropol. Pertaining to or describing the culture stage of the Middle Paleolithic, represented in western Europe by artifacts of stone and other materials believed to indicate the social forms of the Neanderthal race of hunters. [<F moustérien<Le Moustier, a village in France where such remains were found]
mous·y (mou′sē, -zē) adj. **mous·i·er**, **mous·i·est** 1 Infested with or inhabited by mice. 2 Of, pertaining to, or like a mouse; having the color or smell of a mouse. 3 Like a mouse in appearance or manner; pallid; timid. Also **mous′ey**.
mouth (mouth) n. pl. **mouths** (mouthz) 1 The orifice at which food is taken into the body; the entrance to the alimentary canal; the cavity between the lips and throat. ◆ Collateral adjective: oral. 2 The human mouth as the channel of speech. 3 A wry face; grimace. 4 That part of a stream where its waters are discharged; also, the entrance to a harbor. 5 The opening for discharge in the muzzle of a firearm. 6 The slit in an organ pipe, from which the wind passes against the lip; also, the edge of the opening in a flute or similar instrument, against which the performer's breath is directed. 7 The opening of a vessel by which it is emptied or filled. 8 The entrance or opening into a cavity, mine, etc. 9 The space or opening between the jaws of a vice. — **down in** (or **at**) **the mouth** Disconsolate; dejected. — **to fix one's mouth for** To get ready for. — v.t. (mouth) 1 To utter in a forced or affected manner; declaim. 2 To seize or take in the mouth. 3 To caress or fondle with the mouth. 4 To accustom (a horse) to the bit. — v.i. 5 To speak in a forced or affected manner. 6 To distort the mouth; grimace. [OE mūth] — **mouth′er** (mou′thər) n.
mouthed (mouthd, moutht) adj. 1 Having a mouth: used in composition, to denote a characteristic of the mouth or of speech: a hard-mouthed horse. 2 Provided with a mouth.
mouth·ful (mouth′fool′) n. pl. **·fuls** (-foolz) 1 As much as can be or is usually put into or held in the mouth at one time. 2 A small quantity.

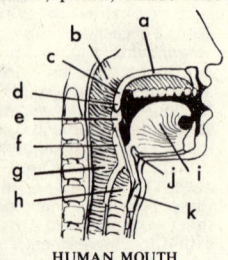

HUMAN MOUTH
a. Hard palate.
b. Pharynx.
c. Soft palate.
d. Uvula.
e. Tonsil.
f. Epiglottis.
g. Esophagus.
h. Trachea.
i. Tongue.
j. Hyoid bone.
k. Larynx.

mouth organ 1 A harmonica. 2 A set of panpipes.
mouth·piece (mouth′pēs′) n. 1 That part of any instrument, tool, etc., that is applied to the mouth. 2 One who speaks for others. 3 Slang A criminal lawyer.
mouth·y (mou′thē, -thē) adj. **mouth·i·er**, **mouth·i·est** 1 Garrulous. 2 Addicted to grimaces in speaking. — **mouth′i·ly** adv. — **mouth′i·ness** n.
mou·ton (moo′ton) n. Processed lambskin or sheepskin used in various types of apparel, especially coats. [<F, sheep]
mou·ton·née (moo′tə-nā′) French adj. Geog. Having the form of a sheep's back: said of rocks, etc. Also **mou′ton·néed**′.
mov·a·ble (moo′və·bəl) adj. 1 Capable of being moved in any way, as from one place, position, or posture to another; susceptible of transposition: movable property. 2 Capable of being moved in respect of time; recurring at varying intervals. See synonyms under MOBILE. — n. 1 Anything that can be moved; especially, anything that may be readily moved or is adapted for moving. 2 pl. House furniture of a movable nature; also, personal property, as distinguished from real or fixed property. Also **move′a·ble**. — **mov′a·ble·ness**, **mov·a·bil′i·ty** n. — **mov′a·bly** adv.
movable kidney A floating kidney.
move (moov) v. **moved**, **mov·ing** v.i. 1 To change place or position; pass or go from one place to another. 2 To change one's residence. 3 To make progress; advance; proceed. 4 To live or associate with; be active: to move in cultivated circles. 5 To operate or revolve; work: said of machines, etc. 6 Colloq. To depart; go or start: often with on. 7 To take action; begin to act. 8 To be disposed of by sale. 9 To make an application, appeal, or proposal: to move for adjournment. 10 To evacuate: said of the bowels. 11 In chess, checkers, etc., to make a move. — v.t. 12 To change the place or position of; carry, push, or pull from one place to another. 13 To set or keep in motion; stir or shake. 14 To rouse, influence, or impel to some action; prompt; actuate. 15 To affect with passion, sympathy, etc.; stir; excite. 16 To offer for consideration, action, etc.; propose, especially formally. 17 To cause (the bowels) to evacuate. See synonyms under ACTUATE, CARRY, CONVEY, INFLUENCE, PERSUADE, STIR[1]. — n. 1 The act of moving; movement. 2 An act in the carrying out of a plan. 3 In games, the changing of the place of a piece. — **to get a move on** To hurry; get going. See synonyms under MOTION. [<AF mover, OF moveir <L movere]
move·ment (moov′mənt) n. 1 The act of changing place or of moving in any way; any change of position. 2 A series of actions, incidents, or ethical impulses tending toward some end: the temperance movement. 3 An effect resembling motion, as in a picture. 4 Mech. A particular arrangement of related parts accomplishing a motion: the movement of a watch. 5 Music a The pace or speed at which a piece or section of music sounds best. b One of the sections of a larger work, as of a symphony. c Melodic progression. 6 Rhythm; meter. 7 The act of emptying the bowels; also, the state of being or the matter so emptied. 8 An elemental part of action in military or naval evolution or maneuver. See synonyms under ACT, MOTION.
mov·er (moo′vər) n. 1 One who or that which moves; specifically, one engaged in the business of moving household goods and other possessions. 2 A tenant farmer who moves away as soon as the soil is exhausted.
mov·ie (moo′vē) n. Colloq. 1 A motion picture. 2 A motion-picture theater. 3 pl. The motion-picture industry. 4 pl. A showing of motion pictures. [Contraction of moving picture]
mov·ie·go·er (moo′vē-gō′ər) n. Colloq. One who attends motion-picture showings regularly or frequently. — **mov′ie-go′ing** n.
mov·ing (moo′ving) adj. 1 Causing to move; impelling to act; influencing; instigating; persuading. 2 Exciting the susceptibilities; affecting; touching. — **mov′ing·ly** adv. — **mov′ing·ness** n.
moving picture A motion picture.
moving platform A platform operated by an endless belt or several such side by side,

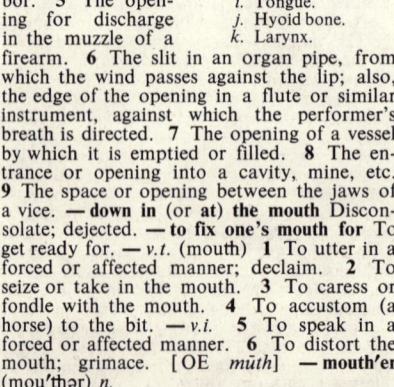

moving staircase An escalator.

mow[1] (mō) v. **mowed, mowed** or **mown, mowing** v.t. 1 To cut down, as grain, with a scythe or machine. 2 To cut the grain or grass of, as a field or lawn. 3 To cut down or kill rapidly or indiscriminately: with *down*: to *mow* down the enemy. —v.i. 4 To cut down grass or grain. [OE *māwan*]

mow[2] (mō, mou) v.i. To make faces; grimace. [< n.] — n. A grimace. [< OF *moue*]

mow[3] (mou) n. Hay or grain stored in a barn; also, the place of storage. —v.t. To store in a mow. [OE *mūga*]

mow·er (mō′ər) n. 1 One who mows. 2 A mowing machine.

mowing machine A farm machine with cutting blades used for mowing standing grass and other forage.

mown (mōn) adj. Cut down, as by mowing.

mox·a (mok′sə) n. 1 A cottony material, used, when ignited, as a counterirritant or cautery, prepared in China and Japan from certain species of *Artemisia*, especially *A. moxa*. 2 A substance for similar use obtained from other sources. 3 Any of the plants producing this material. [<Japanese *mogusa* a caustic < *moe kusa* burning herb]

moy·en âge (mwà·yen äzh′) *French* The Middle Ages.

Moy·zesz (moi′zesh) Polish form of MOSES. Also **Mo·zes** (*Dan.* mō′zes, *Hungarian* mō′zesh).

Mo·zam·bique (mō′zam·bēk′) 1 A Portuguese overseas province in SE Africa; 297,731 square miles; capital, Lourenço Marques: also *Portuguese East Africa*. 2 A port of northern Mozambique, on **Mozambique Channel**, a strait of the Indian Ocean between Madagascar and SE Africa. *Portuguese* **Mo·çam·bi·que** (mōō′səm·bē′kə).

Moz·ar·ab (mōz·ăr′əb) n. A member of a Christian congregation in Spain that maintained a modified form of its religion after the Moslem conquest. [<Sp. *Mozárabe* <Arabic *musta'rib* would-be Arab]

Mo·zart (mō′tsärt, -zärt), **Wolfgang Amadeus**, 1756–91, Austrian composer.

mo·zet·ta (mō·zet′ə) n. A hooded jacket worn by the prelates of the Roman Catholic Church. Also **moz·zet·ta**. [< Ital., fem. dim. of *mozzo* truncated]

Moz·za·rel·la (môd′dzä·rel′lä) n. A soft Italian curd cheese, originally made of buffalo's milk: used mainly in cooking. [< Ital.; ult. origin unknown]

Mrs. (mis′iz) n. A title prefixed to the name of a married woman: a contracted form of *Mistress*.

mu (myōō, mōō) n. 1 The twelfth letter in the Greek alphabet (M, μ): equivalent to *m*. As a numeral it denotes 40. 2 The micron (symbol μ).

muc- Var. of MUCO-.

much (much) adj. **more, most** 1 Great in quantity or amount. 2 *Obs.* Many in number. —n. 1 A considerable quantity. 2 A remarkable or important thing. —adv. 1 In a great degree. 2 For the most part; nearly. [ME *muchel*, OE *mycel*] —**much′ness** n.

mu·cic (myōō′sik) adj. *Chem.* Of, pertaining to, or designating a crystalline acid, $C_6H_{10}O_8$, formed by the oxidation of milk sugar, various gums, etc. [< F *mucique* < L *mucus* mucus]

mu·cid (myōō′sid) adj. Moldy; also, slimy; mucilaginous. [< L *mucidus* < *mucere* be moldy] — **mu′cid·ness** n.

mu·ci·lage (myōō′sə·lij) n. 1 A gummy or slimy substance obtained from the seeds, bark, or roots of various plants by infusion in water. 2 A solution of vegetable gum or mucus in water, especially when intended as an adhesive. [< F <LL *mucilago* <*mucere* be moldy, musty <*mucus* mucus]

mu·ci·lag·i·nous (myōō′si·laj′ə·nəs) adj. 1 Of, pertaining to, or like mucilage; soft, slimy, and viscid. 2 Producing mucilage, as glands. [< F *mucilagineux* — **mu′ci·lag′i·nous·ness** n.

mu·cin (myōō′sin) n. *Biochem.* A glycoprotein secreted by the mucous membranes. [< F *mucine* mucus] — **mu′cin·ous** adj.

muck (muk) n. 1 Moist manure containing decomposed vegetable matter. 2 A nasty mess; filth. 3 Vegetable mold combined with earth. —v.t. 1 To fertilize with manure. 2 *Colloq.* To make dirty; pollute. 3 To remove muck from. [ME *muk* <Scand. Cf. ON *myki* dung, *moka* shovel manure.]

Muck (mōōk), **Karl**, 1859–1940, German conductor, active in the United States.

muck·er (muk′ər) n. *Brit. Slang* A coarse, rude person. [Cf. G *mucker* a low person]

muck·le (muk′əl) *Scot. adj.* Much. —n. A large quantity.

muck·rake (muk′rāk′) v.i. **·raked, ·rak·ing** To search for or expose real or alleged corruption on the part of political officials, businessmen, etc. [Back formation <MUCKRAKER, used in 1906 by President Theodore Roosevelt, in allusion to the "man with the muckrake" in Bunyan's *Pilgrim's Progress*] — **muck′rak·er** n. — **muck′rak·ing** n.

muck rake A rake used in collecting muck or dung.

muck·worm (muk′wûrm′) n. 1 The larva of a beetle (*Ligyrus gibbosus*) common in dung heaps. 2 *Slang* A miser.

muck·y (muk′ē) adj. **muck·i·er, muck·i·est** Foul; nasty. — **muck′i·ly** adv. — **muck′i·ness** n.

muco- *combining form* Mucus; mucus and: *mucopurulent*: also, before vowels, **muc-**. Also **muci-**. [< L *mucus* mucus]

mu·coid (myōō′koid) adj. Like mucus. —n. *Biochem.* A compound glycoprotein similar to mucin, found in connective tissue, in cysts, in the vitreous humor, etc. [< MUC- + -OID]

mu·co·pro·te·in (myōō′kō·prō′tē·in, -tēn) n. *Biochem.* A glycoprotein combining a protein with a carbohydrate group.

mu·co·pu·ru·lent (myōō′kō·pyŏŏr′ə·lənt, -yə·lənt) adj. Relating to or consisting of both mucus and purulent matter.

Mu·co·ra·les (myōō′kō·rā′lēz) n. pl. An order of fungi (class *Phycomycetes*) which includes several mold species, as the common bread mold. [< NL]

mu·co·sa (myōō·kō′sə) n. pl. **·sae** (-sē) *Anat.* A mucous membrane. [< NL, fem. of L *mucosus* mucous] — **mu·co′sal** adj.

mu·cous (myōō′kəs) adj. 1 Secreting mucus. 2 Pertaining to or resembling mucus. Also **mu′cose** (-kōs). [< L *mucosus* slimy <*mucus*] — **mu·cos′i·ty** (-kos′ə·tē) n.

mucous membrane *Anat.* A membrane secreting or producing mucus, that lines passages communicating with the exterior, as the alimentary canal, air passages, etc.

mu·cro (myōō′krō) n. pl. **mu·cro·nes** (myōō·krō′nēz) *Biol.* A small, sharp process or part, as the point of a leaf; a spine. [< NL < L, point of a sword]

mu·cro·nate (myōō′krə·nāt) adj. *Biol.* Having a short, straight point, as a leaf, feather, etc.: also **mu′cro·nat′ed**. [< L *mucronatus* < *mucro* point of a sword]

mu·cus (myōō′kəs) n. 1 *Biol.* A viscid animal substance, as that secreted by the mucous membranes. 2 *Bot.* A gummy substance in plants. [< L]

mud (mud) n. 1 Wet and sticky earth; mire. 2 *Colloq.* Slander; abuse; detraction of character: to sling *mud* at someone. —v.t. **mud·ded, mud·ding** To soil or cover with mud. [? <MLG *mudde*]

mud boat 1 A scow or barge used in dredging: also **mud scow**. 2 A kind of low sledge with wide runners used for hauling logs over swampy ground.

mud·cap (mud′kap′) v.t. **·capped, ·cap·ping** 1 To cap with mud. 2 To cover (a charge of high explosive) with mud before detonating it above an exposed mass of rock. — **mud′-cap′ping** n.

mud·cat (mud′kat′) n. A large catfish of the Mississippi valley.

mud dauber 1 Any of various wasps (family *Sphecidae*) that build mud cells in which their larvae develop. For illustration see INSECTS (beneficial). 2 A swallow that builds a nest of mud.

mud·dle (mud′l) v. **·dled, ·dling** v.t. 1 To mix in confusion; jumble. 2 To confuse mentally; bewilder. 3 To make muddy or turbid; roil. 4 To stir up, as a drink. —v.i. 5 To act in a confused or ineffective manner. —to **muddle through** *Brit.* To achieve one's object despite one's own confusion and mistakes; succeed despite oneself. —n. 1 A muddy or dirty condition. 2 A mixed or confused condition, as of the mind; a mess. [<MUD + freq. suffix]

mud·dle·head·ed (mud′l·hed′id) adj. Mentally confused; addle-brained.

mud·dler (mud′lər) n. 1 A stick for churning or stirring liquids. 2 One who muddles.

mud drum An enclosed container placed at the bottom of any manufacturing or power apparatus, as a boiler, for the purpose of collecting insoluble waste matter, sludge, etc.

mud·dy (mud′ē) adj. **·di·er, ·di·est** 1 Bespattered with mud; turbid. 2 Mentally confused. 3 Consisting of mud; earthy; gross; impure. 4 Dull; cloudy: a *muddy* complexion, *muddy* weather. See synonyms under FOUL, THICK. —v.t. & v.i. **·died, ·dy·ing** To become or cause to become muddy. — **mud′di·ly** adv. — **mud′di·ness** n.

mud eel An eel-shaped amphibian having very small forelegs and no hind legs, that buries itself in the mud; especially, *Siren lacertina* of the swamps of the southern United States: also called *congo* and *siren*.

mud·fish (mud′fish′) n. pl. **·fish** or **·fish·es** A fish that inhabits the mud, as the bowfin, the common killifish, etc.

mud flat A low-lying strip of ground covered with mud, especially one between high and low tide.

mud·guard (mud′gärd′) n. A guard over the wheel of a vehicle to protect from splashing mud.

mud hen 1 The American coot (*Fulica americana*). 2 The Florida gallinule (*Gallenula chloropus*). 3 The clapper rail. See under RAIL[2].

mud·lark (mud′lärk′) n. *Brit. Colloq.* One who works or dabbles in mud; a street urchin.

mud·pot (mud′pot′) n. A geyser which ejects mud. Also **mud geyser**.

mud·pup·py (mud′pup′ē) n. pl. **·pies** 1 The hellbender. 2 A tailed amphibian with bushy, persistent, external gills, especially *Necturus maculosus*, found in the large lakes of North America.

MUDPUPPY (From 12 to 17 inches long)

mud·sill (mud′sil′) n. 1 The foundation timber of a structure placed directly on the ground. 2 *U.S. Dial.* A person of low social state or condition.

mud·sling·ing (mud′sling′ing) n. The practice of casting malicious slurs at an opponent, especially in a political campaign. — **mud′-sling′er** n.

mud·stone (mud′stōn′) n. A gray, sandy shale that readily decomposes into mud.

mud·suck·er (mud′suk′ər) n. A California fish related to the goby (*Gillichthys mirabilis*), much used as bait.

mud turtle Any of various turtles inhabiting muddy rivers in different parts of the world, especially the small common variety of the United States (genus *Kinosternon*).

mu·ez·zin (myōō·ez′in) n. In Moslem countries, a crier from a minaret or other part of the mosque who calls the faithful to prayer. Also **mu·ed′din** (-ed′in). [< Arabic *mu'adhdhin* < *adhana* call]

muff[1] (muf) v.t. & v.i. 1 To perform (some act) clumsily; blunder. 2 In baseball, to fail to hold (the ball) in attempting a catch. —n. 1 A bungling action; in baseball, etc., a failure to catch the ball. 2 A bungler. 3 A stupid fellow; dolt. [Origin unknown]

muff[2] (muf) n. A covering of fur or cloth, usually cylindrical, into which the hands are thrust from opposite ends to keep them warm. [< Du. *mof* < F *moufle*]

muf·fin (muf′in) n. A light, quick bread, baked in small cup-shaped tins, and usually eaten hot with butter; also, a small, flat yeast bread: also called *English muffin*. [Origin uncertain. Cf. OF *moufflet* soft (bread).]

muf·fin·eer (muf'in·ir') 1 A metal cruet with a perforated top for sprinkling salt or sugar on muffins. 2 A covered dish to keep muffins, etc., hot.

muffin stand A small tiered stand used in tea service for holding and passing cakes, sandwiches, etc.

muf·fle (muf'əl) v.t. ·fled, ·fling 1 To wrap up in a blanket, scarf, etc., as for warmth or concealment. 2 To prevent from seeing, hearing, or speaking by wrapping the head. 3 To deaden the sound of by or as by wrapping: to *muffle* a cry. 4 To deaden (a sound). — n. 1 Something used for muffling. 2 A clay oven for firing pottery without direct exposure to flame. 3 The naked upper lip and nose of ruminants and certain other mammals. [< OF *moufle* heavy leather or fur mitten]

muf·fler (muf'lər) n. 1 Anything used for wrapping up or muffling; specifically, a scarf worn about the neck; also, a veil or scarf worn by women. 2 A device to reduce noise, as from the exhaust of an internal-combustion engine. 3 A mitten. [< MUFFLE]

muf·ti[1] (muf'tē) n. In Moslem countries, an expounder of religious law. [< Arabic *āftā* expound the law]

muf·ti[2] (muf'tē) n. Civilian dress; plain clothes, especially when worn by one who normally wears a uniform. [< MUFTI[1]; prob. from the fact that a mufti is a civil official]

mug[1] (mug) n. 1 A drinking cup, usually cylindrical, with a handle and no lip. 2 That which is contained in a mug. [Cf. Sw. *mugg*, Norw. *mugga*]

mug[2] (mug) n. Slang 1 The human face or mouth. 2 A grimace. — v. mugged, mug·ging v.t. 1 To photograph (someone), especially for official purposes. 2 To assault from behind by throttling with an arm locked around the throat, usually so as to rob. — v.i. 3 To make faces; grimace exaggeratedly. [< MUG[1]; prob. from the fact that drinking mugs were often shaped to resemble a face] — **mug'ger** n.

mug·ger (mug'ər) n. A crocodile (*Crocodilus palustris*) of India and the East Indies, with a broad snout: it grows to a length of about 12 feet. Also **mug'gar**, **mug'gur**. [< Hind. *magar* < Skt. *makara* sea monster]

mug·ging (mug'ing) n. Slang Assault, often with the intention of robbery, usually by attacking the victim from behind and locking an arm about his throat. [< MUG[2]]

mug·gins (mug'inz) n. 1 One of several card games in which exposed cards are matched or suits are built. 2 A variant in the game of dominoes. 3 Brit. Slang A foolish person. [Prob. < *Muggins*, a surname used in arbitrary allusion to *mug* (slang) a card-sharper's dupe]

mug·gy (mug'ē) adj. ·gi·er, ·gi·est Warm, moist, and close; sultry. [< dial. E *mug* drizzle, prob. < ON *mugga* drizzling mist] — **mug'gi·ness** n.

mug·wort (mug'wûrt') n. An aromatic bitter herb (*Artemisia vulgaris*) of the composite family, sometimes used in folk medicine as a diaphoretic and emmenagog.

mug·wump (mug'wump) n. U.S. 1 A Republican who bolted the party candidate, James G. Blaine, in the presidential election of 1884. 2 Anyone who claims the right of independent action, especially in politics; an independent. [< Algonquian *mugquomp* great man, chief] — **mug'wump·er·y**, **mug'wump·ism** n.

Mu·ham·ma·dan (mōō·ham'ə·dən), etc. See MOHAMMEDAN, etc.

Mu·har·ram (mōō·har'əm) n. A Mohammedan month; also, the first ten days of the month, a period of lamentation. See CALENDAR (Mohammedan). Also spelled *Moharran*. [< Arabic *muharram* sacred, the first month of the year]

Mühl·bach (mül'bäkh), **Luise** Pen name of *Klara Mundt*, 1814–73, German novelist.

Muh·len·berg (myōō'lən·bûrg), **Frederick Augustus Conrad**, 1750–1801, American politician and clergyman; member of the Continental Congress. — **John Peter**, 1746–1807, American Revolutionary general.

Müh·len·berg (mü'lən·bûrg, Ger. mü'lən·berk), **Heinrich**, 1711–87, German clergyman, chief founder of Lutheranism in the United States.

muh·ly grass (myōō'lē) Any of a genus (*Muhlenbergia*) of mostly perennial, wiry grasses growing in the SW United States and Mexico, as the **ring muhly** (*M. torreyi*), and **spike muhly** (*M. wrightii*), valued as forage plants for livestock. [after Dr. G. H. E. Muhlenberg, 1753–1815, American botanist]

Mu·hu (mōō'hōō) An Estonian island in the Baltic; 79 square miles. Also *Russian* **Mukhu** (mōō'khōō).

muir (myōōr) n. *Scot.* A moor; heath.

Muir (myōōr), **John**, 1838–1914, U.S. naturalist.

mu·i·ra·pu·a·ma (mōō·ē·rä·pōō·ä'mä) n. The dried stems and roots of a Brazilian plant (*Liriosma ovata*), reputed to have properties as a nerve stimulant and aphrodisiac. [< Pg. < Tupian]

Muir Glacier (myōōr) A glacier in the St. Elias Mountains, Alaska; 350 square miles.

mu·jik (mōō·zhēk', mōō'zhik) See MUZHIK.

Muk·den (mōōk'den', mōōk'dən, mōōk'-) Former name of SHENYANG.

Mu·ker·ji (mōō'ker·jē'), **Dhan Gopal**, 1890–1936, East Indian writer, active in the United States.

muk·luk (muk'luk) n. 1 An Alaskan Eskimo boot of seal or other animal skin, made so that the fur is inside. 2 *pl.* Sport or lounging shoes of the soft moccasin type. [< Alaskan Eskimo *makliak*, *muklok* large seal]

mu·la·da (mōō·lä'thä) n. SW U.S. A drove of mules.

mu·lat·to (mə·lat'ō, myō·, -lä'tō) n. *pl.* ·toes A person having one white and one Negro parent; loosely, anyone having white and Negro blood. — *adj.* 1 Of or pertaining to a person of such descent. 2 Of a light-brown color. [< Sp. *mulato* of mixed breed < *mulo* mule < L *mulus*]

mul·ber·ry (mul'ber·ē, -bər·ē) n. *pl.* ·ries 1 The edible, berrylike fruit of a tree (genus *Morus*) whose leaves are valued for silkworm culture, especially the white mulberry (*M. alba*). 2 A deep purplish-red, the color of a mulberry. [ME *mulberie*, dissimilated var. of *murberie*, OE *morberie* < L *morum* mulberry + OE *berie* a berry]

mulberry body The morula.

mulch (mulch) n. Any loose material, as straw, placed about the stalks of plants to protect their roots. — v.t. To cover with mulch. [ME *molsh*. Cf. dial. G *molsch* soft, decaying.]

mulct (mulkt) v.t. 1 To punish (a person) by a fine or penalty. 2 To deprive of something fraudulently or deceitfully; cheat. — n. A fine, or similar penalty. [< L *mulctare* < *mulcta*, *multa* a fine]

mule[1] (myōōl) n. 1 A hybrid between the ass and horse, especially between a jackass and a mare, as distinguished from a hinny. 2 Any hybrid or cross, especially one that is sterile. 3 A spinning machine which draws, stretches, and twists at one operation: also called *spinning mule*, *jenny*, *spinning jenny*: also **mule'-jen'ny** (-jen'ē). 4 A person having the stubborn qualities of a mule. 5 A small electric engine or tractor for towing canal boats. [< F < L *mulus*]

mule[2] (myōōl) n. A backless lounging slipper. [< F < Du. *muil* < L *mulens* red slipper, ? < *mullus* red mullet]

mule deer The black-tailed deer (genus *Odocoileus*), of the western United States, having long ears.

mule–driv·er (myōōl'drī'vər) n. One who drives mules.

mule–skin·ner (myōōl'skin'ər) n. U.S. Colloq. A mule-driver.

mu·le·teer (myōō'lə·tir') n. A mule-driver. [< F *muletier* < *mulet* mule]

mule train A train of mules carrying packs; also, a train of heavy freight wagons drawn by mules.

mu·ley (myōō'lē, mōōl'ē, mōō'lē) *adj.* Hornless: said of cattle. — n. A hornless cow. Also spelled *mooley*, *mulley*. [< Irish *maol*, *moile* hornless, dismantled]

Mul·ha·cén (mōō·lä·thän') A mountain in SE Spain, in the Sierra Nevada; the highest peak in Spain; 11,411 feet. Also **Mu·ley-Ha·cén** (mōō·lā'ä·thän').

Mül·heim-an-der-Ruhr (mül'hīm·än·der·rōōr') A city on the Ruhr in western North Rhine–Westphalia, West Germany. Also **Mül'heim**.

Mul·house (mü·lōōz') A city in eastern France near the German border. *German* **Mül·hau·sen** (mül'hou·zən).

mu·li·eb·ri·ty (myōō'lē·eb'rə·tē) n. 1 The state of being a woman; womanhood; womanliness; hence, effeminacy. 2 The state of female puberty. [< LL *muliebritas*, *-tatis* < *muliebris* womanly < *mulier* woman]

mul·ish (myōō'lish) *adj.* Resembling a mule; stubborn. See synonyms under OBSTINATE. — **mul'ish·ly** adv. — **mul'ish·ness** n.

mull[1] (mul) v.t. To heat and spice, as wine or beer. [< MULSE]

mull[2] (mul) v.t. To ponder; cogitate: usually with *over*. [< obs. *mull* grind, OE *myl* dust]

mull[3] (mul) n. 1 A thin, soft, cotton, rayon, or silk dress goods. 2 A variety of soft, thin muslin used as a base for medicated ointments, as mulla. [Short for *mulmull* < Hind. *malmal*]

mull[4] (mul) n. A horn snuffbox. [Var. of *mill* (in a Scottish use); orig., a snuffbox in which tobacco could be ground to a powder by a mechanism]

Mull (mul) An island in NW Argyllshire, Scotland, the largest of the Inner Hebrides; 351 square miles.

mul·la (mul'ə) n. *Med.* An ointment having a base of lard and salt, spread on a piece of mull.

mul·lah (mul'ə, mōōl'ə) See MOLLAH.

mul·len (mul'ən) n. 1 A tall, stout, woolly herb (*Verbascum thapsus*) of the figwort family, the **great mullen**. 2 Any plant of the same genus, as the **moth mullen** (*V. blattaria*). Also **mul'lein**. [< AF *moleine* < OF *mol* soft < L *mollis*]

mull·er (mul'ər) n. 1 A pestlelike implement with which to mix paints. 2 A mechanical pulverizer or grinder. [< obs. *mull* pulverize, OE *myl* dust]

Mul·ler (mul'ər), **Hermann Joseph**, born 1890, U.S. geneticist.

Mül·ler (mül'ər), **(Friedrich) Max**, 1823–1900, English philologist and Orientalist born in Germany. — **Johannes Peter**, 1801–58, German physiologist.

mul·let[1] (mul'it) n. *pl.* ·lets or ·let 1 A food fish (family Mugilidae), usually greenish or copper-colored, with silvery sides, found on warm coasts. *Mugil cephalus* is the **striped mullet** of both coasts of the Atlantic. 2 A food fish (family Mullidae): often called *surmullet*. *Mullus barbatus* is the highly esteemed European **red mullet**. [< OF *mulet* < L *mullus* red mullet]

mul·let[2] (mul'it) n. *Her.* A star of five or more points. [< OF *molette* rowel]

mul·let–head (mul'it·hed') n. A fresh-water fish having a flat head.

mul·let–head·ed (mul'it·hed'id) *adj.* Slang Stupid.

mul·ley (mōōl'ē, mōō'lē) See MULEY.

mul·li·gan (mul'i·gən) n. Slang A stew, originally made by tramps, composed of odds and ends of food.

mul·li·ga·taw·ny (mul'i·gə·tô'nē) n. A strongly flavored soup of meat and curry. [< Tamil *milagu-tannīr* pepper water]

mul·li·grubs (mul'ə·grubz) n. Slang An acute colicky pain; colic; hence, the blues. [A grotesque arbitrary formation]

Mul·lin·gar (mul'in·gär') The county town of Westmeath, Ireland.

mul·lion (mul'yən) n. *Archit.* A vertical dividing piece between window lights or panels. — v.t. To furnish with or divide by means of mullions. [Prob. metathetic var. of earlier *monial* < OF *moinel*, *monial*; ult. origin unknown] — **mul'lioned** *adj.*

mul·lock (mul'ək) n. An accumulation of waste rock about a mine; refuse earth; a waste dump. [< obs. *mull* dust + -OCK] — **mul'lock·y** *adj.*

MULLIONS

Mu·lock (myōō'lok), **Miss** See CRAIK, DINAH MARIA.

mulse (muls) n. Wine heated and sweetened. [< L *mulsum*, pp. of *mulcere* sweeten]

Mul·tan (mōōl·tän') A city in southern Punjab, West Pakistan.

multi- *combining form* 1 Much; many; consisting of many; as in:

multiangular	multilobed
multiareolate	multilobular
multiarticulate	multilobulate
multiarticulated	multilocular
multiaxial	multiloculate
multiblade	multimedial
multibladed	multimetallic
multibranchiate	multimillion
multicamerate	multimolecular
multicapital	multinational
multicapitate	multinervate
multicapsular	multinodal
multicarinate	multinodous
multicellular	multinodular
multicentral	multiovular
multicentric	multiovulate
multicharge	multipartisan
multichord	multiperforate
multichrome	multiperforated
multiciliate	multipersonal
multiciliated	multipinnate
multicircuit	multipointed
multicoil	multipolar
multicolor	multiradial
multiconductor	multiradiate
multicore	multiradicate
multicorneal	multiramified
multicourse	multiramose
multicrystalline	multiramous
multidentate	multirate
multidenticulate	multireflex
multidenticulated	multirooted
multidigitate	multisaccate
multidimensional	multisaccate
multidirectional	multisegmental
multifaced	multisensitivity
multifaceted	multiseptate
multifactorial	multiserial
multifistular	multiserially
multiflagellate	multiseriate
multiflorous	multishot
multiflue	multispermous
multifocal	multispicular
multifoliate	multispiculate
multifurcate	multispindle
multigranulate	multispinous
multigyrate	multispiral
multihead	multispired
multihearth	multistaminate
multihued	multistoried
multi-infection	multistratified
multijet	multistriate
multijugate	multisulcate
multilaciniate	multisulcated
multilamellar	multisyllable
multilaminar	multiterminal
multilaminate	multititular
multilaminated	multitoed
multilighted	multituberculate
multilineal	multituberculated
multilinear	multitubular
multilingual	multivaned
multilobar	multivoiced
multilobate	multivolumed

2 Having more than two (or sometimes, more than one); as in:

multicostate	multimammae
multicuspid	multimammate
multicuspidate	multimotor
multicylinder	multinuclear
multielectrode	multinucleate
multiengine	multinucleolar
multiexhaust	multinucleolate
multiflow	multispeed

3 Many times over: *multimillionare.* 4 *Med.* Affecting many; as in:

multiarticular	multiganglionic
multifamilial	multiglandular

Also, before vowels, sometimes **mult–**. [<L *multus* much]

mul·ti·cip·i·tal (mul′tə·sip′ə·təl) *adj. Bot.* Many-headed: said of plants with many stems from one root. [<MULTI- + L *caput* head]
mul·ti·col·ored (mul′ti·kul′ərd) *adj.* Exhibiting or made up of many colors.
mul·ti·far·i·ous (mul′tə·fâr′ē·əs) *adj.* Having great diversity or variety. [<L *multifarius*] — **mul′ti·far′i·ous·ly** *adv.* — **mul′ti·far′i·ous·ness** *n.*

mul·ti·fid (mul′tə·fid) *adj. Bot.* Cut into many lobes or segments, as a leaf. Also **mul·tif·i·dous** (mul·tif′ə·dəs). [<MULTI- + -FID]
mul·ti·foil (mul′tə·foil) *n. Geom.* A plane figure made of the congruent arcs of circles which are symmetrically arranged along the sides of a regular polygon.
mul·ti·fold (mul′tə·fōld) *adj.* Many times doubled; manifold.
mul·ti·form (mul′tə·fôrm) *adj.* Having many forms, shapes, or appearances. See synonyms under COMPLEX. [<L *multiformis*] — **mul′ti·form′i·ty** *n.*
Mul·ti·graph (mul′tə·graf, -gräf) *n.* A typesetting and printing machine in which the type is moved from a typesetting drum to a printing drum: a trade name.
mul·ti·lat·er·al (mul′ti·lat′ər·əl) *adj.* 1 Having many sides. 2 *Govt.* Involving more than two nations: a *multilateral* trade agreement.
mul·ti·mil·lion·aire (mul′ti·mil′yən·âr′) *n.* A person having a fortune of several or many millions of dollars, pounds, francs, etc.
mul·ti·no·mi·al (mul′ti·nō′mē·əl) *adj.* Polynomial. [<MULTI- + (BI)NOMIAL]
mul·tip·a·ra (mul·tip′ə·rə) *n.* *pl.* **·rae** (-rē) A woman who has borne more than one child, or who is parturient the second time. [<NL <L *multiparus*]
mul·tip·a·rous (mul·tip′ə·rəs) *adj.* 1 Giving birth to many at one time. 2 Having borne more than one child. [<MULTI- + -PAROUS]
mul·ti·par·tite (mul′ti·pär′tīt) *adj.* 1 Divided into many parts. 2 *Govt.* Multilateral (def. 2).
mul·ti·pede (mul′tə·pēd) *n.* A many-footed animal or insect. — *adj.* Having many feet. Also **mul′ti·ped** (-ped).
mul·ti·phase (mul′tə·fāz) *adj.* Polyphase.
mul·ti·place (mul′tə·plās) *adj.* Having room for more than one: said especially of a military airplane having a crew of more than two.
mul·ti·plane (mul′tə·plān) *n. Aeron.* An airplane with two or more supporting surfaces, one above another.
mul·ti·ple (mul′tə·pəl) *adj.* 1 Containing or consisting of more than one; repeated more than once; manifold. 2 *Electr.* Having two or more conductors or pieces of apparatus, such as lamps, connected in parallel: a *multiple* circuit. — *n. Math.* The product of a given number and its factor. — **common multiple** Any number which is exactly divisible by two or more numbers, not including itself. — **lowest common multiple** The smallest number divisible by each of two or more numbers: 12 is the *least common multiple* of 2, 3, 4, and 6: often abbr. *L.C.M.* Also **least common multiple.** [<F <L *multiplex*]
multiple allele *Genetics* One of three or more alleles, only two of which may pass from the parents to a normal diploid offspring.
multiple cropping *Agric.* The successive cultivation in the same field and in the same year of two or more crops, of the same or different kinds.
multiple factors *Genetics* Two or more distinct genes which are believed to act as a unit or with cumulative effect in the transmission of certain plant and animal characteristics, as size, pigmentation, color of eyes, etc.
multiple fruit *Bot.* A fruit consisting of numerous smaller fruits, each developed from a single flower, as the pineapple: also called *collective fruit.*
multiple neuritis *Pathol.* Neuritis involving several nerves simultaneously.
multiple sclerosis *Pathol.* Sclerosis occurring in patches in the brain or spinal cord or both, and characterized by tremors, failure of coordination, and various nervous and mental symptoms.
multiple star *Astron.* A system of three or more stars revolving around a common gravitational center.
mul·ti·plet (mul′tə·plit) *n. Physics* Two or more spectral lines very close together in an atomic spectrum and associated with different energy characteristics of the atom. [<MULTIPLE]
mul·ti·plex (mul′tə·pleks) *adj.* 1 Multiple; manifold. 2 *Telecom.* Designating a system for the simultaneous transmission of two or more messages in either or both directions over the same wire, as in telegraphy or te-

lephony. 3 *Phot.* Designating a method based upon the stereoscopic principle: three cameras are used, together with auxiliary equipment designed to facilitate the construction of accurate maps. — *v.t. & v.i. Telecom.* To send (two or more messages) at the same time over the same wire. [<L *multus* much + stem of *plicare* fold]
mul·ti·pli·cand (mul′tə·plə·kand′) *n.* A number multiplied, or to be multiplied, by another. [<L *multiplicandus* to be multiplied, gerundive of *multiplicare.* See MULTIPLY.]
mul·ti·pli·cate (mul′tə·plə·kāt) *adj.* Consisting of or being many or more than one. [<L *multiplicatus,* pp. of *multiplicare* multiply]
mul·ti·pli·ca·tion (mul′tə·plə·kā′shən) *adj.* 1 The process of multiplying. 2 The process of finding the sum (the *product*) of a number (the *multiplicand*) repeated a given number of times (the *multiplier*). Opposed to *division.* [<OF <L *multiplicatio, -onis*]
mul·ti·pli·ca·tive (mul′tə·plə·kā′tiv) *adj.* Tending to multiply; indicating multiplication. — **mul′ti·pli·ca′tive·ly** *adv.*
mul·ti·plic·i·ty (mul′tə·plis′ə·tē) *n.* The condition of being manifold or various; hence, a large number. [<LL *multiplicitas, -tatis* <L *multiplex*]
mul·ti·pli·er (mul′tə·plī′ər) *n.* 1 One who or that which multiplies or increases in quantity, or causes something else to multiply or increase. 2 *Math.* The number by which a quantity is multiplied. 3 *Physics* An instrument or mechanical device for increasing or intensifying an effect. 4 *Electr.* An open spiral coil in a wireless telegraph receiver which has the effect of exalting the potential oscillations.
mul·ti·ply (mul′tə·plī) *v.* **·plied, ·ply·ing** *v.t.* 1 To increase the quantity, amount, or degree of; make more numerous. 2 *Math.* To perform the operation of multiplication upon. — *v.i.* 3 To become more in number, amount, or degree; increase. 4 *Math.* To perform the operation of multiplication. See synonyms under PROPAGATE. — *adv.* So as to be numerous; in many ways. [<OF *multiplier* <L *multiplicare* <*multiplex.* See MULTIPLEX.] — **mul′ti·pli′a·ble** *adj.*
mul·ti·pro·pel·lant (mul′tə·prə·pel′ənt) *n.* A rocket propellant consisting of two or more chemicals separately fed into the combustion chamber. — *adj.* Of or pertaining to such a propellant.
mul·ti·pur·pose (mul′tə·pûr′pəs) *adj.* Adapted to more than one use or type of service: a *multipurpose* gun.
mul·ti·range (mul′tə·rānj) *adj.* Having a wide range of operations or performance, as certain precision instruments.
mul·ti·sec·tion (mul′tə·sek′shən) *adj.* Having or occupying more than one section, compartment, etc.
multisection charge *Mil.* Propelling charge for artillery ammunition packed in separate powder bags to permit adjustments for range of fire.
mul·ti·stage (mul′tə·stāj) *adj.* 1 Having or characterized by a number of definite stages in the completion of a process or action. 2 Having several sections, each of which fulfils a given task before burnout: said especially of a rocket or ballistic missile.
mul·ti·tude (mul′tə·tōōd, -tyōōd) *n.* 1 The state of being many or numerous. 2 A large gathering; concourse. 3 A large number of things. See synonyms under ARMY, ASSEMBLY, COMPANY, THRONG. — **the multitude** The common people. [<OF <L *multitudo* <*multus* much, many]
mul·ti·tu·di·nous (mul′tə·tōō′də·nəs, -tyōō′-) *adj.* Consisting of a vast number. See synonyms under MANY. [<L *multitudo, -inis* <*crowd* + -OUS] — **mul′ti·tu′di·nous·ly** *adv.* — **mul′ti·tu′di·nous·ness** *n.*
mul·ti·va·lent (mul′ti·vā′lənt) *adj. Chem.* Having three or more valences. — **mul′ti·va′lence** *n.*
mul·ti·valve (mul′tə·valv) *n.* A shell with many valves. — *adv.* Having many valves. — **mul′ti·val′vu·lar** (-val′vyə·lər) *adj.*
mul·ti·verse (mul′tə·vûrs) *n. Philos.* The plurality of worlds as conceived in or projected by the mind: contrasted with *universe.* [<MULTI- + (UNI)VERSE]

mul·tiv·o·cal (mul·tiv′ə·kəl) *adj.* Having various meanings. — *n.* A word that has more than one signification. [<MULTI- + VOCAL]
mul·ti·vol·tine (mul′ti·vol′tin, -tēn) *adj. Entomol.* Having many broods of offspring in a year, as certain insects. [<MULTI- + Ital. *volta* turn]
Mult·no·mah Falls (mult·nō′mə) A waterfall in NW Oregon in a small tributary of the Columbia River; about 850 feet high.
mul·toc·u·lar (mul·tok′yə·lər) *adj.* 1 Having two or more eyes. 2 Having eyes divisible, like those of a fly, into facets. [<MULT(I)- + OCULAR]
mul·tum in par·vo (mul′təm in pär′vō) *Latin* Much in little.
mul·ture (mul′chər) *n.* 1 A grinding of grain. 2 The grain ground or the toll paid for the grinding. [<OF *moulture* <Med. L *molitura* <L *molere* grind]
mum[1] (mum) *v.i.* **mummed, mum·ming** *Obs.* To be silent. — *adj.* Silent; saying nothing. — *n.* Silence; *Mum's* the word. — *interj.* Hush! Be quiet! [Imit.]
mum[2] (mum) *v.i.* **mummed, mum·ming** To play or act in a mask, as at Christmas; be a mummer. Also **mumm**. [<MUM[1]. Cf. MDu. *mommen* mask, OF *momer* act in a dumb show.]
mum[3] (mum) *n.* A strong sweet beer, first brewed in Germany by Christian Mumme, 1492.
mum[4] (mum) Corruption of MADAM, MA'AM.
mum[5] (mum) *n. Colloq.* A chrysanthemum.
mum·ble (mum′bəl) *v.t.* & *v.i.* **·bled, ·bling** 1 To speak or utter in low, indistinct tones; mutter. 2 *Rare* To chew slowly and ineffectively, as with toothless gums. — *n.* A low, mumbling speech; mutter. [ME *momelen*, freq. of obs. *mum* make inarticulate sounds. Cf. G *mummeln*.] — **mum′bler** *n.* — **mum′bling·ly** *adv.*
mum·ble-the-peg (mum′bəl-thə-peg′) *n.* A boy's game played with a jackknife, which is tossed and flipped in various ways so as to stick into the ground: so called because the player who failed was originally required to draw a peg out of the ground with his teeth. Also **mum′ble-peg′, mum·ble-ty-peg** (mum′bəl-tē-peg′). [<MUMBLE, *v.* (def. 2) + PEG]
mum·bo jum·bo (mum′bō jum′bō) 1 Any object of superstitious homage; a fetish. 2 Incantation. [<MUMBO JUMBO]
Mum·bo Jum·bo (mum′bō jum′bō) Among certain tribes of the western Sudan, a village god or presiding genius, who protects the village from evil and terrifies the women and keeps them in subjection. [<Mandingo *mama dyambo*]
mu-mes·on (myoō′mes′on, -mē′son) *n.* An unstable, short-lived radioactive particle formed from the decay of a *pi*–meson and having a mass about 210 times that of the electron: a constituent of cosmic radiation. [<MU + MESON]
mum·mer (mum′ər) *n.* 1 One who acts or makes sport in a mask. 2 An actor.
mum·mer·y (mum′ər·ē) *n. pl.* **·mer·ies** 1 A masked performance. 2 Hypocritical parade of ritual. [<MF *mommerie* dumb show]
mum·mi·fy (mum′ə·fī) *v.* **·fied, ·fy·ing** *v.t.* To make a mummy of; preserve by drying. — *v.i.* To dry up; shrivel. Also **mummy**. [<F *momifier*] — **mum′mi·fi·ca′tion** *n.*
mum·my (mum′ē) *n. pl.* **·mies** 1 A body embalmed in the ancient Egyptian manner; also, any dead body which is very well preserved. 2 A person or thing that is dried up and withered. 3 *Obs.* The dried flesh of mummies; dead meat. — *v.t.* & *v.i.* **·mied, ·my·ing** To mummify. [<F *momie* <Med. L *mumia* <Arabic *mūmiyā* <Persian *mūm* wax] — **mum′mi·form** *adj.*
mummy cloth 1 The fabric in which a mummy is enwrapped. 2 A crêpelike fabric of cotton, silk, rayon, or wool.
mump·ish (mum′pish) *adj.* Sullen; sulky; morose; petulant. [<obs. *mump* mutter, ? <Du. *mompelen* mumble]
mumps (mumps) *n. pl.* (construed as singular) An acute, contagious, febrile disease characterized by inflammation of the facial glands, and occasional of the ovaries and testicles: also *Brit. Dial.* Man.
(ŏŏn) *v. Scot. & Brit. Dial.* Must.

Mu·na (moō′nə) An Indonesian island SE of Celebes; 659 square miles. *Dutch* **Moe·na** (moō′nə).
munch (munch) *v.t.* & *v.i.* To chew steadily and noisily. [ME *monchen, manchen*. Prob. ult. imit.] — **munch′er** *n.*
Mun·chau·sen (mun·chô′zən, mun′chou′zən), **Baron**, 1720–97, Hanoverian nobleman, whose extravagant stories of his exploits formed the basis of the *Tales of Munchausen*, collected by Rudolf Erich Raspe. Also **Münch·hau·sen** (münkh′hou′zən). — **Mun·chau′sen·ism** *n.*
Mün·chen-Glad·bach (mün′khən gläd′bäkh) A city in central western North Rhine-Westphalia, West Germany; the twin city of Rheydt.
Mun·cie (mun′sē) A city in central Indiana.
Mun·da (moōn′də) *n.* A subfamily of the Austro-Asiatic family of languages, spoken in central India and along the southern slope of the Himalayas.
mun·dane (mun′dān, mun·dān′) *adj.* Pertaining to the world; worldly. [<F *mondain* <L *mundanus* <*mundus* world] — **mun′dane·ly** *adv.* — **mun′dane·ness** *n.*
Mun·de·lein (mun′də·līn), **George William**, 1872–1939, U.S. cardinal.
mun·dic (mun′dik) *n. Brit. Dial.* Pyrite. [? <Celtic (Cornish) *maen tag* pretty stone]
mun·dun·go (mun·dung′gō) *n. Archaic* A black malodorous tobacco. Also **mun·dun′gus**. [Jocular alter. of Sp. *mondongo* tripe]
mun·go (mung′gō) *n.* The waste produced from hard-spun or felted cloth. Compare SHODDY. [? <*mung*, var. of obs. *mong* mixture]
mun·goose (mung′goōs) See MONGOOSE.
Mu·nich (myoō′nik) A city on the Isar, capital of Bavaria, SE West Germany. *German* **Mün·chen** (mün′khən).
mu·nic·i·pal (myoō·nis′ə·pəl) *adj.* 1 Pertaining to a town or city or its local government; also, having local self-government. 2 Pertaining to the internal government of a state or nation. [<L *municipalis* <*municipium* town possessing right of self-government <*municeps* free citizen <*munia* official duties + *capere* take] — **mu·nic′i·pal·ly** *adv.*
municipal borough See under BOROUGH.
municipal corporation An incorporated town.
mu·nic·i·pal·ism (myoō·nis′ə·pəl·iz′əm) *n.* Municipal government as opposed to central government; also, the theory of this. — **mu·nic′i·pal·ist** *n.*
mu·nic·i·pal·i·ty (myoō·nis′ə·pal′ə·tē) *n. pl.* **·ties** 1 An incorporated borough, town, or city. 2 In Cuba and some other Latin-American countries, an administrative area somewhat like a county. [<F *municipalité*]
mu·nic·i·pal·ize (myoō·nis′ə·pəl·īz) *v.t.* **·ized, ·iz·ing** 1 To place within municipal authority or transfer to municipal ownership. 2 To make a municipality of. — **mu·nic′i·pal·i·za′tion** *n.*
municipal ownership Ownership by a municipality: said especially of public services, as electricity, waterworks, etc.
mu·nif·i·cent (myoō·nif′ə·sənt) *adj.* Extraordinarily generous or bountiful; liberal. See synonyms under GENEROUS. [<L *munificens, -entis* <*munificus* <*munus* gift + *facere* make] — **mu·nif′i·cence, mu·nif′i·cen·cy** *n.* — **mu·nif′i·cent·ly** *adv.*
mu·ni·ment (myoō′nə·mənt) *n.* 1 That which supports or defends. 2 *Law* Any deed, record, or instrument by which title to property may be defended or evidenced: usually in the plural. 3 Any means of defending. [<OF <L *munimentum* fortification, support <*munire* fortify]
mu·ni·tion (myoō·nish′ən) *n.* 1 Ammunition and all necessary war materiel: usually in the plural. 2 The requisites for any undertaking. 3 *Obs.* A fort; stronghold. — *v.t.* To furnish with munitions. [<F <L *munitio, -onis* <*munire* fortify]
mun·nion (mun′yən) *n.* A mullion.
Mun·ro (mun·rō′), **Hector Hugh** See SAKI.
Mun·sell color system (mun·sel′) A system for the classification and identification of colors by means of reference to the three standard factors of hue, chroma (saturation), and value (lightness). [after Albert H. *Munsell*, 1858–1918, U.S. artist]
Mun·sey (mun′sē), **Frank Andrews**, 1854–1925, U.S. publisher.

Mun·ster (mun′stər) A province in southern Ireland; 9,317 square miles.
Mün·ster (mün′stər) A city in northern North Rhine-Westphalia, western West Germany. Also **Münster-in-West·fa·len** (-ēn·vest·fä′lən).
Mün·ster·berg (mün′stər·berkh), **Hugo**, 1863–1916, U.S. philosopher and psychologist born in Germany.
munt (munt) *v.t.* & *v.i. Scot.* To mount.
Mun·te·ni·a (mun·tē′nē·ə, *Rumanian* moōn·tā′nyä) The eastern region of Walachia, Rumania; 20,070 square miles: also *Greater Walachia*.
Mun·the (mun′te), **Axel**, 1857–1949, Swedish physician and writer.
munt·jac (munt′jak) *n.* Any of various small, short-legged deer (genus *Muntiacus*) of east Asia, the males having short, two-pronged horns on long pedicles; especially, *M. muntjak* of Java. Also **munt′iak**. [<Javanese *mĕnjañan*]
mu·on (myoō′on) *n. Physics* A mu-meson.
Muo·nio (mwô′nyô) A river rising in Lapland near the meeting point of the Norwegian, Swedish, and Finnish borders and flowing 180 miles SE and south along the Swedish-Finnish border, to its confluence with the Torne.
Mur (moōr) A river in Austria, Hungary, and Yugoslavia, flowing 300 miles east to the Drava. *Hungarian* **Mu·ra** (moō′rä).
mu·rae·na (myoō·rē′nə) *n.* An eel; moray. [<L *muraena*, a fish <Gk. *myraina* sea eel]
mu·ral (myoōr′əl) *adj.* 1 Pertaining to or supported by a wall. 2 Resembling a wall. — *n.* A painting or decoration on a wall. [<L *muralis* <*murus* wall]
mu·ral·ist (myoōr′əl·ist) *n.* A painter of murals.
Mu·ra·sa·ki (moō·rä·sä·kē), **Lady**, Japanese novelist and poet of the eleventh century.
Mu·rat (moō·rät′) A river in east central Turkey, flowing 380 miles west to the Euphrates. Also **Mu·rad** (-räd′).
Mu·rat (mü·rá′), **Joachim**, 1771–1815, French general; marshal of France 1804; king of Naples 1808–15; brother-in-law of Napoleon.
Mur·chi·son Falls (mûr′chə·sən) A series of three waterfalls in the lower Victoria Nile; 400 feet high.
Mur·chi·son River (mûr′chə·sən) A river in Western Australia, flowing 440 miles SW to the Indian Ocean.
Mur·cia (mûr′shə, moōr′-; *Sp.* moōr′thyä) 1 A region and former kingdom of southern Spain; 10,108 square miles. 2 Its former capital, a city in SE Spain.
mur·der (mûr′dər) *v.t.* 1 To kill (a human being) with premeditated malice. 2 To kill in a barbarous or inhuman manner; slaughter. 3 To spoil by bad performance, etc.; mangle; butcher. See synonyms under KILL. — *n.* The unlawful and intentional killing of one human being by another. — **murder will out** Murder cannot be concealed. [Fusion of OE *morthor* + OF *murdre*, both <Gmc.] — **mur′der·er** *n.* — **mur′der·ess** *n. fem.*
mur·der·ous (mûr′dər·əs) *adj.* 1 Pertaining to murder; destructive. 2 Given to murder. 3 Characterized by murder. See synonyms under SANGUINARY. — **mur′der·ous·ly** *adv.* — **mur′der·ous·ness** *n.*
mure (myoōr) *v.t.* **mured, mur·ing** To immure; confine. — *n. Obs.* A wall. [<F *murer*, ult. <L *murus* wall]
Mu·res (moō′resh) A river of NW Rumania and SE Hungary, flowing 550 miles from the Carpathians to the Tisza: *Hungarian Maros*. Also **Mu′resh**.
mu·rex (myoōr′eks) *n. pl.* **mu·ri·ces** (myoōr′ə·sēz) or **·rex·es** A rough-shelled marine gastropod (genus *Murex*) of warm seas, especially *M. trunculus* and *M. brandaris*, from whose large mucus gland a purple dye was obtained in ancient times. [<L, purple fish]
mu·rex·ide (myoō·rek′sīd) *n. Chem.* The ammonium hydrogen salt, $C_8H_8O_6N_6 \cdot H_2O$, of purpuric acid: formerly used to produce pink, purple, or red dyes, now displaced by aniline colors. [<MUREX + -IDE]
Mur·free (mûr′frē), **Mary Noailles** See CRADDOCK.
Mur·frees·bor·o (mûr′frēz·bûr′ō) A city in central Tennessee; site of the Civil War battle of Stones River, December 31, 1862, and January 2, 1863.
mur·geon (mûr′jən) *n. Scot.* A smirk; a grimace.

mu·ri·ate (myŏŏr′ē·āt) Former name for CHLORIDE.
mu·ri·at·ed (myŏŏr′ē·ā′tid) *adj.* 1 Salted; pickled. 2 *Archaic* Treated with or containing a chloride or hydrochloric acid.
mu·ri·at·ic (myŏŏr′ē·at′ik) *adj.* Hydrochloric. [< L *muriaticus* pickled < *muria* brine]
mu·ri·cate (myŏŏr′ə·kit) *adj. Biol.* Rough, with short, hard, tubercular excrescences. [< L *muricatus* murex-shaped, pointed < *murex*]
Mu·ri·el (myŏŏr′ē·əl) A feminine personal name. [< Gk., myrrh]
mu·ri·form (myŏŏr′ə·fôrm) *adj. Bot.* Regularly arranged like bricks in a wall: said of cells in plants. Also **mu′rine** (-ēn). [< L *murus* wall + -FORM]
Mu·ril·lo (myŏō·ril′ō, *Sp.* mōō·rē′lyō), **Bartolomé Esteban**, 1618–82, Spanish painter.
mu·rine (myŏŏr′īn, -in) *adj.* Of or pertaining to a family (*Muridae*) or a subfamily (*Murinae*) of rodents, embracing true mice and rats. — *n.* One of the *Murinae* or *Muridae*. [< L *murinus* < *mus, muris* a mouse]
murk (mûrk) *adj.* Murky; dark. — *n.* Darkness; gloom. Also spelled *mirk*. [< ON *myrkr* darkness] — **murk′ly** *adv.*
murk·y (mûr′kē) *adj.* **murk·i·er, murk·i·est** Darkened, thickened, or obscured; hazy; gloomy; obscure: also spelled *mirky*. See synonyms under DARK. — **murk′i·ly** *adv.* — **murk′i·ness** *n.*
Mur·man Coast (mŏŏr′män′) The Arctic coast of the Kola Peninsula in NW U.S.S.R.: also *Norman Coast*.
Mur·mansk (mŏŏr·mänsk′) A port of the western Murman Coast; world's largest city north of the Arctic Circle.
mur·mur (mûr′mər) *n.* 1 A low sound continually repeated. 2 A complaint uttered in a half-articulate voice. 3 An abnormal, rasping sound heard in certain morbid conditions: a *heart murmur*. — *v.i.* 1 To make a murmur. 2 To complain in a low tone; mutter. — *v.t.* 3 To utter in a low tone. See synonyms under BABBLE, COMPLAIN. [< OF < L] — **mur′mur·er** *n.* — **mur′mur·ing** *adj.* — **mur′mur·ing·ly** *adv.*
mur·mur·ous (mûr′mər·əs) *adj.* Characterized or accompanied by murmurs. — **mur′mur·ous·ly** *adv.* — **mur′mur·ous·ness** *n.*
Mu·ro·ran (mōō·rō·rän′) A port on SW Hokkaido island, Japan.
mur·phy (mûr′fē) *n. pl.* **·phies** *Slang* A potato. [from an Irish surname]
Mur·phy (mûr′fē), **Frank**, 1890–1949, U.S. jurist, associate justice of the Supreme Court 1940–49. — **William Parry**, born 1892, U.S. physician.
mur·ra (mûr′ə) *n. Latin* A material of ancient Rome which has been variously supposed to be Chinese jade, porcelain, iridescent glass, or artificially colored chalcedony. Also **mur′rha**.
mur·rain (mûr′in) *n.* 1 A malignant contagious fever affecting domestic animals, as anthrax. 2 Any plague or pestilence. See RINDERPEST. — *adj.* Affected with murrain. [< OF *morine* < L *mori* die]
Mur·ray (mûr′ē), **Gilbert**, 1866–1957, English classical scholar. — **Sir James Augustus Henry**, 1837–1915, Scottish philologist and lexicographer. — **John**, 1778–1843, English publisher; first published works of Byron, Jane Austen, etc. — **Lindley**, 1745–1826, U.S. grammarian. — **Philip**, 1866–1952, U.S. labor leader. — **William**, 1705–93, Earl of Mansfield; English jurist.
Murray River The principal river of Australia, forming part of the boundary between Victoria and New South Wales and flowing 1,600 miles to the Indian Ocean.
murre (mûr) *n. pl.* **murres** or **murre** 1 The foolish guillemot. 2 The razor-billed auk. Also **murr.** [Origin uncertain]
murre·let (mûr′lit) *n.* Any of certain small sea birds (family *Alcidae*) of the islands of the North Pacific. [Dim. of MURRE]
mur·rey (mûr′ē) *adj.* Of a purplish-red or mulberry color. — *n.* 1 *Her.* The tincture sanguine. 2 A dark purplish red. [< OF *moree* < L *morum* mulberry]
mur·rhine (mûr′in, -īn) *adj.* Of, pertaining to, or consisting of murra. Also **mur′rhine**. [< L *murrinus*]

murrine glass Glassware having a transparent ground with embedded flowers, ribbons, etc., of colored glass.
Mur·rum·bid·gee (mûr′əm·bij′ē) A river in New South Wales, Australia, rising in the Great Dividing Range and flowing 1,050 miles north to the Murray River.
mur·ry (mûr′ē) See MORAY.
mur·ther (mûr′thər) *n. Obs.* Murder. — **mur′ther·er** *n.*
Mur·vie·dro (mōōr·vyä′thrō) A former name for SAGUNTO.
mu·sa·ceous (myōō·zā′shəs) *adj.* Pertaining to or designating a family (*Musaceae*) of monocotyledonous plants including the common banana, proceeding from rootstocks, with stems composed of sheathing leafstalks and flowers bursting through spathes. [< NL *Musaceae* < *Musa*, genus name < Arabic *mawzah* banana]
Mus Al·lah (mōōs′ ä·lä′) See STALIN PEAK (def. 3).
mus·ca (mus′kə) *n. pl.* **mus·cae** (mus′sē) A fly; any of a genus (*Musca*) of dipterous insects of the family *Muscidae*, including the common housefly. [< L, fly] — **mus·cid** (mus′id) *adj.* & *n.*
Mus·ca (mus′kə) A southern constellation, the Fly. See CONSTELLATION. [< NL]
mus·ca·dine (mus′kə·din, -dīn) *n.* The fox grape or scuppernong (*Vitis* or *Muscadinia rotundifolia*) of the southern United States. [< Provençal *muscade*, fem. of *muscat*. See MUSCAT.]
mus·cae vol·i·tan·tes (mus′ē vol′ə·tan′tēz) Minute specks or motes apparently moving before the eye, caused by defects or impurities in the vitreous humor of the eye, etc. [< L, flying flies]
mus·ca·rine (mus′kə·rēn, -rin) *n. Chem.* A deliquescent, extremely poisonous, white, crystalline alkaloid, $C_5H_{13}O_3N$, found in certain fungi, as the fly agaric, and in putrefying fish. [< NL *muscarius* of flies < L *musca* fly]
mus·cat (mus′kat, -kət) *n.* 1 One of several varieties of musk-flavored Old World grapes. 2 A sweet, white wine made from such grapes. [< F < Provençal < Ital. *moscato* < LL *muscus* musk]
Mus·cat (mus·kat′) A port on the Gulf of Oman, capital of Muscat and Oman: also *Masqat*.
Muscat and Oman An independent sultanate of SE Arabia on the Gulf of Oman; 82,000 square miles; capital, Muscat: also *Oman*.
mus·ca·tel (mus′kə·tel′) *n.* 1 A rich, sweet wine made from the muscat grape. 2 The muscat grape: also spelled *moscatel*. Also **mus′ca·del′** (-del′). [< OF, dim. of *muscat*. See MUSCAT.]
mus·cle (mus′əl) *n.* 1 *Anat.* An organ composed of tissue arranged in bundles of fibers, by whose contraction bodily movements are effected. Two principal types are known: *striated* (striped), involved in voluntary movements, and *smooth* (unstriped), acting independently of the will. The heart muscle belongs anatomically between the two. 2 The tissue of the muscular organs. 3 Muscular strength. — *v.i.* **·cled, ·cling** To push in or ahead by sheer physical strength. ◆ Homophone: *mussel*. [< F < L *musculus*, lit., little

mouse, dim. of *mus*. Doublet of MUSSEL.]
mus·cle-bound (mus′əl·bound′) *adj.* Affected with a form of muscular hypertrophy characterized by diminution of elasticity in a muscle: caused by excessive exercise in training.
mus·cled (mus′əld) *adj.* Having or supplied with muscles.
muscle fiber *Physiol.* A muscle cell consisting of a soft contractile substance enclosed in a tubular sheath.
muscle plasma *Physiol.* The liquid that can be expressed from muscle tissue: it clots spontaneously and is sometimes injected intravenously as a restorative and stimulant.
muscle sense *Physiol.* The perception of muscular movement derived from the functioning of afferent nerves connected with muscle tissue, skin, joints, and tendons.
Muscle Shoals Rapids in the Tennessee River, NW Alabama; site of the Wilson Dam.
muscle spindle *Anat.* One of various groups of muscle fibers enclosed in a sheath of connective tissue and terminating in sensory organs.
muscle sugar Inositol.
mus·coid (mus′koid) *adj.* Mosslike. — *n.* A mosslike plant. [< L *muscus* moss + -OID]
mus·co·va·do (mus′kə·vā′dō) *n.* A raw brown sugar obtained by evaporating the juice of sugarcane and draining off the molasses. Also **mus′ca·va′da** (-də), **mus′co·vade** (-vād). [< Sp. *mascabado* unrefined, pp. of *mascabar* diminish, var. of *menoscabar* < *menos* (< L *minus*) + *cabo* head (< L *caput*)]
mus·co·vite (mus′kə·vīt) *n.* The most common and important white or potash mica, $KAl_2(OH)_2AlSi_3O_{10}$. [< earlier *Muscovy glass*]
Mus·co·vite (mus′kə·vīt) *n.* An inhabitant of Muscovy or ancient Russia; hence, a Russian. — *adj.* Of or pertaining to Muscovy or to Russia; Russian. — **Mus′co·vit′ic** (-vit′ik) *adj.*
Mus·co·vy (mus′kə·vē) *n. pl.* **·vies** A large greenish-black duck (*Cairina moschata*) of America from Mexico to Brazil, now widely domesticated. Also **muscovy duck.** [Alter. of MUSK DUCK]
Mus·co·vy (mus′kə·vē) *Archaic* Russia.
mus·cu·lar (mus′kyə·lər) *adj.* 1 Pertaining to or depending upon muscles. 2 Possessing strong muscles; powerful. 3 Accomplished by muscle or muscles. — **mus′cu·lar′i·ty** (-lar′ə·tē) *n.* — **mus′cu·lar·ly** *adv.*
mus·cu·la·ture (mus′kyə·lə·choor) *n.* 1 The disposition or arrangement of muscles in a part or organ. 2 The muscle system as a whole. Also **mus′cu·la′tion**. [< F]
muse¹ (myōōz) *n.* Something regarded as the source of artistic inspiration; the inspiring power of a poet or of poetry. [< MUSE]
muse² (myōōz) *v.t. & v.i.* **mused, mus·ing** To consider thoughtfully or at length; ponder; meditate. — *n.* 1 The act or state of musing; reverie. 2 Wonder. [< OF *muser*] — **mus′er** *n.* — **muse′ful** *adj.*
— *Synonyms (verb):* brood, cogitate, consider, contemplate, deliberate, dream, meditate, ponder, reflect, ruminate, stew, study, think. Compare REFLECTION.

THE MUSES—FROM A SARCOPHAGUS IN THE LOUVRE, PARIS
A. Clio. B. Thalia. C. Erato. D. Euterpe. E. Polyhymnia. F. Calliope. G. Terpsichore. H. Urania. I. Melpomene.

Muse (myōōz) In Greek mythology, any of the nine goddesses presiding over poetry, the arts, sciences, etc.: Calliope, Clio, Erato,

musette (myo͞o·zet′) *n.* **1** Any melody of soft, pastoral character written in imitation of bagpipe airs. **2** A small bagpipe formerly popular in France. **3** A variety of small oboe. **4** A small leather or canvas knapsack or wallet, used especially by soldiers, and carried by a strap worn over the shoulder: also **musette bag.** [< F, dim. of *muse* a bagpipe]

mu·se·um (myo͞o·zē′əm) *n.* **1** A place preserving and exhibiting works of nature, art, curiosities, etc.; also, any collection of such objects. **2** Any place where curiosities, freaks, monstrosities, etc., are exhibited. [< L < Gk. *mouseion* temple of the Muses < *Mousa*]

mush[1] (mush) *n.* **1** Thick porridge, made by boiling meal or flour in water or milk. **2** Anything soft and pulpy. **3** *Colloq.* Sentimentality. [Var. of MASH]

mush[2] (mush) *v.i.* In Alaska and the Canadian Arctic, to travel on foot, especially over snow with a dog sled. — *interj.* Get along! a call of the drivers of a dog team. [Prob. < *mush on,* alter. of F (Canadian) *marchons,* the cry of voyageurs and trappers to their dogs] — **mush′er** *n.*

mush·mel·on (mush′mel′ən) See MUSKMELON.

mush·room (mush′ro͞om, -ro͝om) *n.* **1** A large, rapidly growing fungus of the order *Agaricales,* consisting of an erect stalk and a caplike expansion: certain poisonous varieties are called *toadstools,* but the distinction is not scientifically correct. The best-known edible mushrooms are of the genus *Agaricus,* especially the **field mushroom,** *A. campestris.* **2** An object or excrescence of similar shape. — *v.i.* **1** To grow or spread rapidly, like a mushroom: The town *mushroomed* overnight. **2** To expand at one end into a mushroomlike shape: said of bullets. — *adj.* **1** Pertaining to or made of mushrooms. **2** Sudden in growth and rapid in decay; ephemeral; upstart. [< OF *mousseron* < *mousse* moss]

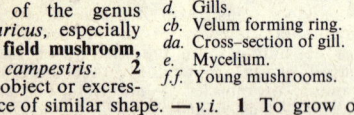

AGARIC MUSHROOM
a. Pileus, entire and cut.
b. Stipe in velum.
ca. Ruptured velum.
d. Gills.
cb. Velum forming ring.
da. Cross-section of gill.
e. Mycelium.
f.f. Young mushrooms.

mush·y (mush′ē) *adj.* **mush·i·er, mush·i·est** **1** Soft; pulpy. **2** *Colloq.* Sentimental; romantic. — **mush′i·ly** *adv.* — **mush′i·ness** *n.*

mu·sic (myo͞o′zik) *n.* **1** The science and art of the rhythmic combination of tones, vocal or instrumental, embracing melody and harmony. **2** A composition, or mass of compositions, conceived or executed according to musical rule or spirit. **3** Any rhythmic succession or combination of sounds, especially if pleasing to the ear; also, the sensations or emotions thus produced. **4** A band of musicians; an orchestra. See synonyms under MELODY. — **absolute music** Pure or abstract music wholly self-sufficient without representation or dependence on title, program, etc.: opposed to the pictorial or descriptive **program music.** — **to face the music** To take the consequences courageously and without complaint; accept facts. [< F *musique* < L *musica* < Gk. *mousikē* (*technē*) musical (art) < *Mousa* a Muse]

mu·si·cal (myo͞o′zi·kəl) *adj.* **1** Pertaining to music. **2** Capable of producing or appreciating music; fond of music. **3** Melodious — *n.* **1** A musical comedy. **2** *Colloq.* A musicale. — **mu′si·cal·ly** *adv.* — **mu′si·cal·ness** *n.*

musical chairs A marching game played to music, in which the contestants circle around a row of opposite-facing chairs, one less than the number of players. When the music stops, each marcher rushes for a chair, one person for whom there is no chair being eliminated. On each round, a chair is removed until finally only one remains, the contestant reaching it first being declared the winner. Also called "Going to Jerusalem."

musical comedy A kind of theatrical performance, characterized by music, songs, dances, jokes, and elaborate costumes, staging, and settings, usually based on a tenuous plot: also *musical.*

mu·si·cale (myo͞o′zə·kal′, *Fr.* mü·zē·kàl′) *n.* An informal concert or private recital: also *musical.* [< F]

music box A case containing a mechanism that reproduces melodies.

music hall **1** A public hall or building devoted to musical entertainments. **2** *Brit.* A vaudeville house.

music house A firm that publishes music.

mu·si·cian (myo͞o·zish′ən) *n.* One skilled in music; especially, a professional performer. [< OF *musicien*]

mu·si·cian·ly (myo͞o·zish′ən·lē) *adj.* Exhibiting musical taste, learning, or skill.

music of the spheres The harmony produced by the movements of the celestial spheres: a conception of Pythagorean philosophy.

mu·si·col·o·gist (myo͞o′zə·kol′ə·jist) *n.* One engaged in or versed in musicology.

mu·si·col·o·gy (myo͞o′zə·kol′ə·jē) *n.* The scientific and historical study of music as an art and as a craft.

music stand **1** A rack to hold sheet music for a performer. **2** A bandstand.

mus·ing (myo͞o′zing) *adj.* Thoughtful; dreamy; preoccupied. — **mus′ing·ly** *adv.*

musk (musk) *n.* **1** A soft, reddish-brown, powdery secretion of a penetrating odor, obtained from the preputial follicles or **musk bag** of the male musk deer. It is used by perfumers and in medicine. **2** A similar substance from some other animals, as the muskrat or civet. **3** *Chem.* Any of several organic compounds used to replace natural musk. **4** Muskroot. **5** The odor of musk. [< OF *musc* << LL *muscus* < Gk. *moschos* < Persian *mushk* (prob. akin to Skt. *mushka* testicle, dim. of *mus* a mouse)]

musk beaver The muskrat.

musk beetle A large European beetle (*Aromia moschata*) of a bronze-green color, and having a musky odor suggesting roses.

musk cat A civet. **2** *Obs.* A perfumed, effeminate man; a dandy.

musk deer A small hornless deer (*Moschus moschiferus*) of central and eastern Asia, of which the male has a musk-secreting gland.

musk duck **1** A muscovy. **2** An Australian duck (*Biziura lobata*), with a disklike appendage on the chin, and having a musky odor in the breeding season: also called *mould goose.*

MUSK DEER
(From 17 to 20 inches high; color brown speckled with gray or buff)

mus·keg (mus′keg) *n.* **1** A rocky basin filled by successive deposits of unstable material, as leaves, muck, and moss, incapable of sustaining much weight. **2** A swamp. Also spelled *maskeg.* [< Chippewa *muskig* grassy bog]

muskeg moss An absorbent, sterilized swamp moss (genus *Sphagnum*) used as a surgical dressing.

mus·kel·lunge (mus′kə·lunj) *n.* *pl.* **·lunge** *n.* A large North American pike (*Esox masquinongy*), valued as a game fish: also spelled *mascalonge, maskinonge.* [< Algonquian *mas·kinonge* < *mas* great + *kinonge* pike]

mus·ket (mus′kit) *n.* A smoothbore military hand gun; specifically, a hand gun for infantry, now superseded by the rifle. [< OF *mosquet* gun, hawk < Ital. *moschetto* hawk < L *musca* fly]

mus·ket·eer (mus′kə·tir′) *n.* A soldier armed with a musket; hence, a foot soldier. [< F *mousquetaire*]

mus·ket·ry (mus′kit·rē) *n.* **1** Muskets collectively. **2** The science of the operation of small arms.

Mus·kho·ge·an (mus·kō′gē·ən, mus′kō·gē′ən) *n.* One of the principal North American Indian linguistic stocks, well advanced in culture, including the Chickasaw, Choctaw, Creek, and Seminole tribes, formerly inhabiting the Gulf region of the SE United States. Also **Mus·ko·gi·an** (-kō′gē·ən).

mus·kit (mus′kit) See MESQUITE.

musk·mel·on (musk′mel′ən) *n.* **1** The juicy, edible, gourdlike fruit of a trailing herb (*Cucumis melo*) of the melon family; cantaloup. **2** The plant bearing this fruit. Also, *Colloq.,* mushmelon.

musk ox A shaggy, hollow-horned ruminant (*Ovibos moschatus*) combining the characteristics of the sheep and ox and emitting a strong odor of musk: now restricted to Greenland and the North American continent.

MUSK OX
(About 4 feet high at the shoulder)

musk·rat (musk′rat′) *n.* *pl.* **·rats** or **·rat** A North American aquatic rodent (*Ondatra zibethica*) yielding a valuable fur and secreting a substance with a musky odor: sometimes called *musquash.*

musk·root (musk′ro͞ot′, -ro͝ot′) *n.* The musky, spongy root of a plant (*Ferula sumbul*) of the parsley family, from Russian Turkestan: employed medicinally as a stimulant and antispasmodic. Also called *sumbul.*

MUSKRAT
(About 22 inches over-all in length)

musk rose A cultivated climbing rose (*Rosa moschata*) from Europe, with large white flowers in panicled clusters.

musk turtle A small turtle (*Sternotherus odoratus*) of the eastern United States and Canada, distinguished by bright-yellow lines on each side of the head.

musk·y (mus′kē) *adj.* **musk·i·er, musk·i·est** Like musk; smelling of musk. — **musk′i·ly** *adv.* — **musk′i·ness** *n.*

Mus·lem, Mus·lim (muz′ləm) See MOSLEM.

mus·lin (muz′lin) *n.* Any of several varieties of plain-weave cotton cloth ranging from thin batiste and nainsook to heavy sheetings such as longcloth and percale. [< F *mousseline* < Ital. *mussolino,* dim. of *mussolo* muslin < *Mussolo* Mosul, city in Iraq, where it was made]

mus·lin-kail (muz′lin·kāl′) *n.* *Scot.* A thin broth containing greens and shelled barley.

mu·so·pho·bi·a (myo͞o′sə·fō′bē·ə) *n.* A morbid fear of mice: a special form of zoophobia. [< L *mus* < Gk. *mys* mouse + PHOBIA] — **mu′so·pho′bic** *adj.*

mus·quash (mus′kwosh) *n.* The muskrat. [< N. Am. Ind., it is red]

mus·qui·to (mus·kē′tō) See MOSQUITO.

muss (mus) *v.t. U.S. Colloq.* To make messy or untidy; rumple; crumple: often with *up.* — *n. Colloq.* **1** A state of disorder or disturbance; mess. **2** Confused struggle or tumult; squabble. [Alter. of MESS]

mus·sel (mus′əl) *n.* **1** A small bivalve mollusk, especially the common edible mussel (*Mytilus edulis*). **2** One of several other fresh-water mollusks, of the genera *Anodonta, Unio,* and others, whose shells are made into buttons, etc. ♦ Homophone: *muscle.* [OE *musle* < L *musculus,* dim. of *mus* mouse. Doublet of MUSCLE.]

Mus·sel·shell River (mus′əl·shel) A river in central Montana, flowing 292 miles NE to the Missouri.

Mus·set (mü·se′), **(Louis Charles) Alfred de,** 1810–57, French poet, dramatist, and novelist.

Mus·so·li·ni (mo͞os′ə·lē′nē, *Ital.* mo͞os′sō·lē′nē), **Benito,** 1883–1945, Italian Fascist leader; premier 1922–43; executed: called "Il Duce" (the leader).

Mus·sul·man (mus′əl·mən) *n.* *pl.* **·mans** (-mənz) or **·men** (-mən) A Moslem; Mohammedan. — *adj.* Belonging or relating to the Moslems. [< Persian & Turkish *musulmān* < Arabic *muslimūn,* pl. of *muslim*]

muss·y (mus′ē) *U.S. Colloq. adj.* **muss·i·er, muss·i·est** Disarranged; rumpled; slightly soiled. — **muss′i·ly** *adv.* — **muss′i·ness** *n.*

must[1] (must) *v.* A defective verb now used only as an auxiliary followed by the infinitive without *to,* or elliptically with the infinitive understood, to express: **1** Obligation or compulsion: *Must* you go? I *must.* **2** Requirement: You *must* be healthy to be accepted. **3** Probability or supposition: You *must* be tired. **4** Conviction or certainty: War *must* follow. ♦ A past conditional is formed by placing the following verb in the perfect infinitive: *He must have gone.* —*n.* 1 Anything that is required or vital. 2 A news item or other material that must be printed: usually marked *must.* —*adj.* Important and essential: a *must* book. [OE *moste,* pt. of *mōtan* may]

must[2] (must) *n.* Mustiness; mold. —*v.t.* & *v.i.* *Obs.* To make or become musty. [Back formation <MUSTY[1]]

must[3] (must) *n.* 1 The expressed unfermented juice of the grape. 2 Unfermented potato pulp. [OE <L *mustum (vinum)* new wine]

must[4] (must) *n.* 1 A state of dangerous frenzy, related to sexual excitement, associated with adult male animals, especially elephants. 2 An elephant in this condition. —*adj.* Being in a state of must. Also spelled *musth.* [<Hind. *mast* drunk, lustful]

must[5] (must) *n. Scot.* Musk, especially in the form of hair powder.

mus·tache (məs·tash′, mus′tash) *n.* 1 The growth of hair on the upper lip of men: occasionally used in the plural, in reference to its two parts. 2 The hair or bristles growing near the mouth of an animal. 3 An old soldier: a brave old *mustache*: a gallicism. Also **mus·ta′chio** (-tä′shō): sometimes spelled *moustache.* [<F *moustache* <Ital. *mostaccio* face <Med. L *mustacia* <Gk. *mystax* upper lip <*mastax* mouth, jaws] —**mus·tached** (məs·tasht′) *adj.*

Mus·ta·fa (mōōs′tä·fä), **Kemal Pasha** See KEMAL.

Mus·tagh (mōō·stä′) See KARAKORAM.

mus·tang (mus′tang) *n.* 1 The wild horse of the American plains. 2 One of these horses broken to the saddle; a cow pony. [<Sp. *mesteño,* obs. *mestengo,* lit., belonging to a cattlemen's association, wild <*mesta* association, group <L *mixtus,* pp. of *miscere* mix]

mustang grape A vine *(Vitis candicans)* of the SW United States, having light-colored pungent berries or grapes.

mus·tard (mus′tərd) *n.* 1 Either of two species of *Brassica,* **white mustard** *(B. hirta)* and **black mustard** *(B. nigra),* both annual herbs with yellow flowers and pods of roundish seeds. 2 The pungent seed of the mustard, crushed and adapted for use as a condiment or as a medicinal rubefacient and diuretic. 3 A strong, dark-yellow color, the color of ground mustard. [<OF *moustarde* <L *mustum* must[3]; from once having been prepared with must]

mustard gas *Chem.* An oily amber liquid, dichlorethyl sulfide, $C_4H_8Cl_2S$, having an odor of mustard or garlic, and used in warfare because of its powerful blistering effect.

mustard oil A fixed oil of unpleasant odor extracted from mustard seeds and used in making soap, as a lubricant, etc.

mustard plaster A plaster of powdered mustard and flour used as a counterirritant and rubefacient.

must·ed (mus′tid) *adj. Scot.* Sprinkled with musk.

mus·tee (mus·tē′, mus′tē) See MESTEE.

mus·te·line (mus′tə·lin, -līn) *adj.* Pertaining to a family *(Mustelidae)* of fur-bearing, predacious mammals which includes weasels, skunks, badgers, otters, wolverines, and martens. [<L *mustelinus* <*mustela* weasel]

mus·ter (mus′tər) *v.t.* 1 To summon or assemble (troops, etc.), as for service, review, or roll call. 2 To collect or summon: often with *up*: to *muster* up one's courage. —*v.i.* 3 To come together or assemble, as troops for service, review, etc. —**to muster in** To enrol as military recruits. —**to muster out** To collect or assemble, as troops, for discharge from military service. See synonyms under CONVOKE. —*n.* 1 An assemblage, especially of troops for parade or review. 2 A muster roll. 3 A specimen; pattern; sample. 4 An imposing gathering; array. —**to pass muster** To pass inspection; hence, to be acceptable or accepted. [<OF *mostrer* exhibit <L *monstrare* show]

mus·ter-mas·ter (mus′tər·mas′tər, -mäs′-) *n.* An officer who inspects troops, their equipment, etc.

muster roll The official list or roll of officers and men in a military troop or a ship's crew.

musth (must) See MUST[4].

must·y[1] (mus′tē) *adj.* **must·i·er, must·i·est** 1 Having a moldy odor; ill-flavored; fetid; stale. 2 Without life, sparkle, or flavor. 3 Without life or energy; listless; apathetic. See synonyms under TRITE. [? Alter. of earlier *moisty* <MOIST] —**must′i·ly** *adv.* —**must′i·ness** *n.*

must·y[2] (mus′tē) *n.* Formerly, a cheap quality of snuff having a musty flavor.

mu·ta·ble (myōō′tə·bəl) *adj.* Capable of changing; liable to change; hence, fickle; unstable. See synonyms under FICKLE. [<L *mutabilis* <*mutare* change] —**mu′ta·ble·ness, mu′ta·bil′i·ty** *n.* —**mu′ta·bly** *adv.*

mu·ta·gen·ic (myōō′tə·jen′ik) *adj. Genetics* Having the power to produce mutations in plant or animal organisms, as X-rays. [<L *mutare* change + -GENIC]

Mu·tan·kiang (mōō′dän′jyäng′) A city in eastern Manchuria.

mu·tant (myōō′tənt) *n.* 1 That which admits of or undergoes mutation or change. 2 *Biol.* A plant or animal organism differing from its parents in one or more characteristics that are inheritable; a sport. [<L *mutans, -antis,* ppr. of *mutare* change]

mu·tate (myōō′tāt) *v.t.* & *v.i.* **·tat·ed, ·tat·ing** To undergo or subject to change, especially by mutation. [<L *mutatus,* pp. of *mutare* change] —**mu′ta·tive** *adj.*

mu·ta·tion (myōō·tā′shən) *n.* 1 The act or process of change; alteration; variation. 2 Modification of one vowel by another; umlaut. See UMLAUT. 3 *Biol.* A sudden, well-marked, transmissible variation in the organism of a plant or animal, especially as resulting from new combinations of genes and chromosomes and as distinguishable from cumulative evolutionary changes over a long period. 4 Change; hence, succession and serial succession; consecutive order. See synonyms under CHANGE. [<L *mutatio, -onis*] —**mu·ta′tion·al** *adj.*

mu·ta·tis mu·tan·dis (myōō·tā′tis myōō·tan′·dis) *Latin* The necessary changes having been made.

mutch (much) *n. Scot.* A woman's close-fitting cap; also, a man's nightcap; a child's cap.

mutch·kin (much′kin) *n. Scot.* A liquid measure containing three quarters of one imperial pint.

mute (myōōt) *adj.* 1 Uttering no word or sound; silent. 2 *Law* Refusing to plead upon arraignment. 3 Lacking the power of speech; dumb. 4 *Phonet.* Of or pertaining to a stop consonant. See synonyms under TACITURN. —*n.* 1 One who is silent; especially, a person who refuses or is unable to speak; a person who is dumb by reason of deafness or other infirmity: also called *deaf-mute.* 2 An undertaker's assistant. 3 *Law* A prisoner who refuses to plead on arraignment. 4 *Phonet.* A stop consonant, as (b), (p), (t), (d). 5 A device to silence, muffle, or deaden the tone of a musical instrument. —*v.t.* **mut·ed, mut·ing** To deaden or muffle the sound of (a musical instrument). [<L *mutus* dumb] —**mute′ly** *adv.* —**mute′ness** *n.*

mu·ti·cous (myōō′tə·kəs) *adj. Biol.* Without a point; unarmed; defenseless: said especially of certain plants and animals. Also **mu′tic, mu′ti·cate** (-kāt). [<L *muticus,* var. of *mutilus* docked, curtailed. See MUTILATE.]

mu·ti·late (myōō′tə·lāt) *v.t.* **·lat·ed, ·lat·ing** 1 To deprive (a person, animal, etc.) of a limb or essential part; maim. 2 To damage or injure by the removal of an important part or parts: to *mutilate* a speech. [<L *mutilatus,* pp. of *mutilare* maim <*mutilus* docked, maimed] —**mu′ti·la′tion** *n.* —**mu′ti·la′tive** *adj.* —**mu′ti·la′tor** *n.*

mu·ti·neer (myōō′tə·nir′) *n.* One who takes part in mutiny: also *Obs.* **mu·tine** (myōō′tin). —*v.i.* To mutiny. [<MF *mutinier* <*mutin.* See MUTINY.]

mu·ti·nous (myōō′tə·nəs) *adj.* Disposed to mutiny; seditious. See synonyms under REBELLIOUS, RESTIVE, TURBULENT. —**mu′ti·nous·ly** *adv.* —**mu′ti·nous·ness** *n.*

mu·ti·ny (myōō′tə·nē) *n. pl.* **·nies** 1 Rebellion against constituted authority; insubordination; especially, a revolt of soldiers or sailors against their officers or commander. 2 *Obs.* Tumult; discord; strife. See synonyms under REVOLUTION. —*v.i.* **·nied, ·ny·ing** To revolt against constituted authority, as in the army or navy; take part in a mutiny. [<F *mutiner* rebel <OF *mutin, meutin* riotous <*muete* riot <*motus,* pp. of *movere*]

mut·ism (myōō′tiz·əm) *n.* 1 Inability or refusal to utter articulate sounds: often associated with certain mental disorders. 2 Muteness. [<F *mutisme*]

Mu·tsu·hi·to (mōō·tsōō·hē·tō), 1852–1912, emperor of Japan 1867–1912.

mutt (mut) *n. Slang* 1 A cur; mongrel dog. 2 A stupid person; blockhead. Also **mut.** [<MUTT(ONHEAD)]

mut·ter (mut′ər) *v.i.* 1 To speak in a low, indistinct tone and with compressed lips, as in complaining or talking to oneself. 2 To complain; grumble. 3 To make a low, rumbling sound. —*v.t.* 4 To say in a low, indistinct tone. —*n.* An imperfect utterance; murmur. [Prob. imit. Cf. dial. G *muttern* and L *muttire.*] —**mut′ter·er** *n.* —**mut′ter·ing** *n.* & *adj.* —**mut′ter·ing·ly** *adv.*

mut·ton (mut′n) *n.* 1 The flesh of sheep as food. 2 Humorously, a sheep. [<F *mouton* <Celtic. Cf. O Irish *molt* ram, Welsh *mollt,* Breton *maout.*] —**mut′ton·y** *adj.*

mutton chop 1 A piece of mutton from the rib for broiling or frying. 2 *Printing* An em quad or em rule. —**mut′ton-chop′** (-chop′) *adj.*

mut·ton·chops (mut′n·chops′) *n. pl.* Burnsides or side whiskers shaped like mutton chops.

mut·ton·fish (mut′n·fish′) *n. pl.* **·fish** or **·fish·es** 1 The eelpout. 2 An abalone *(Haliotis iris)* said to taste like mutton. 3 The pargo or other snapper found from Florida to Brazil. 4 The mojarra.

mutton ham *Naut.* A small sail shaped like a leg of mutton, used on small fishing boats: also *mutton-leg.*

mut·ton·head (mut′n·hed′) *n. Slang* A stupid, dense person. [<MUTTON (def. 2) + HEAD; from the traditional stupidity of sheep] —**mut′ton-head′ed** (-hed′id) *adj.*

mut·ton-leg (mut′n·leg′) *n.* 1 A woman's dress sleeve, very full at the top: so called because shaped like a leg of mutton. 2 Mutton ham.

Mut·tra (mu′trə) The former name for MATHURA.

mu·tu·al (myōō′chōō·əl) *adj.* 1 Pertaining reciprocally to both of two; reciprocally related or bound; reciprocal in action or effect. 2 Shared or experienced alike; common. [<F *mutuel* <LL *mutualis* <L *mutuus* <*mutare* change] —**mu′tu·al·ly** *adv.* —**mu′tu·al′i·ty** *n.*

Synonyms: common, correlative, interchangeable, joint, reciprocal. That is *common* to which two or more persons have the same or equal claims, or in which they have equal interest or participation; that is *mutual* which is freely interchanged; that is *reciprocal* in respect to which one act or movement is met by a corresponding act or movement in return. *Antonyms*: detached, disconnected, dissociated, distinct, disunited, separate, separated, severed, sundered, unconnected, unreciprocated, unshared.

mutual fund An investment company which provides diversification and professional management for the pooled capital of the investors, paying out to its shareholders as mutual owners virtually all earnings including realized capital gains except for management fees and administrative costs, thus remaining tax free in its own operations. The shares of an **open-end mutual fund** are redeemable at the option of the stockholder at the net asset value of the portfolio calculated daily, and its shares are generally sold continually with a distribution charge on purchase. The shares are not listed on the stock exchange. A **closed-end mutual fund** has fixed capitalization, the shares not being offered or redeemable.

mutual insurance 1 A method of insurance based upon a reciprocal contract whereby various persons engage to indemnify each other against certain designated losses. 2 Popularly, the system of a company in which policyholders receive a certain share of the profits: also **mutual plan.**

add,āce,câre,pälm; end,ēven; it,īce; odd,ōpen,ôrder; tŏŏk,pōōl; up,bûrn; ə = a in *above,* e in *sicken,* i in *clarity,* o in *melon,* u in *focus*; yōō = u in *fuse*; oi,oil; ou,pout; ch,check; g,go; ng,ring; th,thin; *th*,this; zh,vision. Foreign sounds á,œ,ü,kh,ṅ; and ◆: see page xx. < from; + plus; ? possibly.

mu·tu·al·ism (myōō′chōō·əl·iz′əm) *n. Biol.* Symbiosis advantageous to both or all parties concerned.
mu·tu·al·ize (myōō′chōō·əl·īz) *v.t.* & *v.i.* **·ized, ·iz·ing 1** To make or become mutual. **2** To put (a firm or corporation) on the basis of majority employee or consumer ownership of common stock. — **mu′tu·al·i·za′tion** *n.*
mu·tule (myōō′chōōl) *n. Archit.* One of a series of rectangular blocks under a Doric corona, with dependent droplike ornaments called guttae. [<F <L *mutulus* modillion]
mu·zhik (mōō·zhĕk′, mōō′zhĕk) *n.* A Russian peasant in Czarist times: also spelled *moujik, mujik.* Also **mu·zjik′.** [<Russian]
muzz (muz) *v.i. Brit. Slang* **1** To idle; potter. **2** To study hard. [Origin unknown]
Muz·zey (muz′ē), **David Saville,** born 1870, U.S. historian.
muz·zle (muz′əl) *n.* **1** The projecting snout of an animal. **2** A guard or covering for an animal's snout to prevent biting or eating. **3** The front end of a firearm. — *v.t.* **·zled, ·zling 1** To put a muzzle on. **2** To restrain from speaking, expressing opinions, etc.; gag. [<OF *musel* <Med. L *musellum,* dim. of LL *musus* snout] — **muz′zler** *n.*
muz·zle·load·er (muz′əl·lō′dər) *n.* A firearm loaded through the muzzle. — **muz′zle·load′ing** *adj.*
muzzle velocity The velocity of a bullet or projectile at the instant of leaving the muzzle of a gun.
muz·zy (muz′ē) *adj. Colloq.* Muddled; hence, stupid.
MVD The secret police of Soviet Russia. [<Russian <*M(inisterstvo) V(nutrennikh) D(el)* Ministry of Internal Affairs]
Mwe·ru (mwā′rōō) A lake between SE Belgian Congo and Northern Rhodesia; 173 square miles: French *Moero.*
my (mī) *pronominal adj.* **1** The possessive case of the pronoun *I* employed attributively; belonging or pertaining to me: *my* house. **2** An adjective used in forms of address in customary phrases: *my* lord; also used in expressions of endearment: *my* boy. — *interj.* An exclamation of surprise: oh *my*! [OE *mīn*]
my– Var. of MYO–.
my·al·gi·a (mī·al′jē·ə) *n. Pathol.* Pain in a muscle; cramp. [<NL <MY- + -ALGIA] — **my·al′gic** *adj.*
my·a·sis (mī′ə·sis) See MYIASIS.
my·as·the·ni·a (mī′əs·thē′nē·ə) *n. Pathol.* Muscular debility, accompanied by general and usually progressive exhaustion but without marked sensory disturbance or atrophy. [<NL <MY- + ASTHENIA]
Myc·a·le (mik′ə·lē), **Mount** A peak in SW Turkey on the mainland opposite Samos; scene of a Greek naval victory over the Persians, 479 B.C.
my·ce·li·um (mī·sē′lē·əm) *n. Bot.* The thallus or vegetative portion of a fungus, consisting of threadlike tubes, or hyphae. See illustration under MUSHROOM. Also **my′cele** (-sēl). [<NL <Gk. *mykēs* mushroom] — **my·ce′li·al, my·ce′li·an** *adj.* — **my·ce′li·oid, my·ce·loid** (mī′sə·loid) *adj.*
My·ce·nae (mī·sē′nē) An ancient city in NE Peloponnesus, Greece; excavated 1876–77. Also **My·ce′næ.**
My·ce·nae·an (mī′sə·nē′ən) *adj.* Of or pertaining to Mycenae or to a civilization of which it was the center, existing in parts of Greece, Asia Minor, Sicily, and neighboring countries before the advance of the Hellenes, and thought to have been at its height about 1400 B.C.
–mycete *combining form Bot.* A member of a class of fungi: corresponding in use to class names in *–mycetes*: *Basidiomycete.* [See -MYCETES]
–mycetes *combining form Bot.* Used to form class names of fungi: *Basidiomycetes.* [<Gk. *mykēs, mykētos* fungus]
my·ce·to·gen·ic (mī·sē′tō·jen′ik) *adj.* Produced or caused by a fungus. Also **my·ce′to·ge·net′ic** (-jə·net′ik), **my′ce·tog′e·nous** (-toj′ə·nəs). [<Gk. *mykēs, mykētos* mushroom + -GENIC]
my·ce·to·ma (mī′sə·tō′mə) *n. pl.* **·to·ma·ta** (-tō′mə·tə) *Pathol.* A tumor or tumorlike growth caused by a fungus, as Madura foot. [<Gk. *mykēs, mykētos* fungus + -OMA]
my·ce·to·zo·an (mī·sē′tō·zō′ən) A myxomycete. [<Gk. *mykēs, mykētos* fungus + -ZOA]
myco– *combining form* Fungus: *mycology.* Also, before vowels, **myc–.** [<Gk. *mykēs* fungus]
my·co·bac·te·ri·um (mī′kō·bak·tir′ē·əm) *n. pl.* **·ri·a** (-ē·ə) One of a genus *(Mycobacterium)* of slender, typically aerobic bacteria difficult to stain, including the bacterium of tuberculosis and that of leprosy. [<MYCO- + BACTERIUM]
my·col·o·gy (mī·kol′ə·jē) *n.* **1** The science of fungi. **2** The fungous life of a region. [<MYCO- + -LOGY] — **my·co·log·ic** (mī′kə·loj′ik) or **·i·cal** *adj.* — **my·col′o·gist** *n.*
my·co·phile (mī′kə·fīl) *n.* A connoisseur of mushrooms.
my·cor·rhi·za (mī′kə·rī′zə) *n. Bot.* A subterranean hyphal mass or mycelium often found on the roots of certain trees, especially of the oak, heath, and pine families. Also **my′co·rhi′za.** [<MYCO- + Gk. *rhiza* root] — **my′cor·rhi′zic** *adj.*
my·co·sis (mī·kō′sis) *n. Pathol.* **1** A fungous growth within the body. **2** A disease or morbid condition caused by a fungous growth, as ringworm. [<MYC(O)- + -OSIS] — **my·cot′ic** (-kot′ik) *adj.*
my·co·tro·phism (mī′kō·trō′fiz·əm) *n. Bot.* The nutrition of the higher plants by the aid of fungi on their roots and in their leaves; nourishment by mycorrhiza. [<MYCO- + Gk. *trophē* nutrition] — **my′co·tro′phic** *adj.*
my·dri·a·sis (mi·drī′ə·sis, mī-) *n. Pathol.* An abnormal or prolonged dilatation of the pupil of the eye. [<LL <Gk.]
myd·ri·at·ic (mid′rē·at′ik) *adj. Med.* Relating to or causing dilatation of the pupil. — *n.* A drug that dilates the pupil, as belladonna.
my·e·len·ceph·a·lon (mī′ə·len·sef′ə·lon) *n. Anat.* The afterbrain; the posterior part of the rhombencephalon or that portion of the medulla oblongata lying behind the pons Varolii and cerebellum. [<MYEL(O)- + ENCEPHALON]
my·e·lin (mī′ə·lin) *n. Biochem.* A semisolid, fatlike substance that surrounds the axillary portion of medullated nerve fibers; the medullary sheath. Also **my′e·line** (-lēn, -lin). [<Gk. *myelos* marrow]
my·e·li·tis (mī′ə·lī′tis) *n. Pathol.* **1** Inflammation of the spinal cord. **2** Inflammation of the bone marrow. [<MYEL(O)- + -ITIS]
myelo– *combining form Anat.* The spinal cord. Also, before vowels, **myel–.** [<Gk. *myelon* marrow]
my·e·loid (mī′ə·loid) *adj. Anat.* **1** Pertaining to, from, or resembling marrow. **2** Of or pertaining to the spinal cord. [<MYEL(O)- + -OID]
My·ers (mī′ərz), **F(rederic) W(illiam) H(enry),** 1843–1901, English philosopher and writer on psychical research.
my·ia·sis (mī′yə·sis) *n. Pathol.* Any of various diseases caused by flies or maggots. [<NL <Gk. *myia* fly + -(O)SIS]
Myit·kyi·na (myi′chi′nä′) The capital of Kachin State, Upper Burma.
My·lae (mī′lē) The ancient name for MILAZZO; scene of a Roman naval victory over Carthage, 260 B.C. Also **My′læ.**
my·lo·nite (mī′lə·nīt) *n. Geol.* A hard, compact rock having a banded or streaky appearance, produced by the crushing and reforming of earth material under extreme pressure. [<Gk. *mylōn* mill]
my·na (mī′nə) *n.* One of various Oriental, starlinglike birds of the genera *Acridotheres* and *Eulabes. Eulabes religiosa,* the common myna of India, is often tamed and taught to speak words: sometimes spelled *mina, minah.* Also **my′nah.** [<Hind. *mainā*]

MYNA
(About 9 inches over-all)

myn·heer (mīn·hâr′, -hir′) *n.* Sir; Mr.: a Dutch title of address; hence, a Dutchman. [<Du. *mijn heer,* lit., my lord]
myo– *combining form* Muscle; of or pertaining to muscle: *myology.* Also, before vowels, *my–.* [<Gk. *mys, myos* a muscle]
my·o·car·di·al (mī′ō·kär′dē·əl) *adj. Anat.* Of or pertaining to the heart muscle. [<MYO- + Gk. *kardia* heart]
my·o·car·di·o·gram (mī′ō·kär′dē·ə·gram) *n.* The record made by a myocardiograph.
my·o·car·di·o·graph (mī′ō·kär′dē·ə·graf, -gräf) *n.* An instrument for registering the muscular action of the heart.
my·o·car·di·tis (mī′ō·kär·dī′tis) *n. Pathol.* Inflammation of the myocardium. [<NL] — **my′o·car·dit′ic** (-dit′ik) *adj.*
my·o·car·di·um (mī′ō·kär′dē·əm) *n. Anat.* The muscular tissue of the heart. [<NL <MYO- + Gk. *kardia* heart]
my·o·ge·net·ic (mī′ō·jə·net′ik) *adj. Physiol.* Originating in muscle or in muscle tissue. Also **my′o·gen′ic, my·og·e·nous** (mī·oj′ə·nəs).
my·o·gram (mī′ə·gram) *n.* The record made by a myograph.
my·o·graph (mī′ə·graf, -gräf) *n.* An instrument for recording and showing muscular movement.
my·og·ra·phy (mī·og′rə·fē) *n.* A scientific description of muscles.
my·oid (mī′oid) *adj.* Resembling a muscle or muscle tissue.
my·ol·o·gy (mī·ol′ə·jē) *n.* The study of the structure, functions, and diseases of the muscles. [<MYO- + -LOGY] — **my·o·log·ic** (mī′ə·loj′ik) or **·i·cal** *adj.* — **my·ol′o·gist** *n.*
my·o·ma (mī·ō′mə) *n. pl.* **·ma·ta** (-mə·tə) *Pathol.* A muscular tumor. [<MY- + -OMA] — **my·om′a·tous** (-om′ə·təs) *adj.*
my·o·mec·to·my (mī′ō·mek′tə·mē) *n. Surg.* The removal of a myoma. [<MYOMA + -ECTOMY]
my·op·a·thy (mī·op′ə·thē) *n. Pathol.* Disease of a muscle. Also **my·o·path·i·a** (mī′ō·path′ē·ə). [<MYO- + -PATHY] — **my′o·path′ic** *adj.*
my·ope (mī′ōp) *n.* One who is near-sighted. Also **my′ops** (-ops). [<F <LL *myops* <Gk. *myōps* <*myein* close + *ōps* an eye]
my·o·pi·a (mī·ō′pē·ə) *n. Pathol.* Defect in vision so that objects can be seen distinctly only when very near the eye; near-sightedness due to focusing of images in front of instead of on the retina. Also **my·o·py** (mī′ə·pē). [<NL] — **my·op′ic** (-op′ik) *adj.*
my·o·scope (mī′ə·skōp) *n. Med.* An instrument for observing the contraction of muscles.
my·o·sin (mī′ə·sin) *n. Biochem.* A globulin contained in the contractile muscular tissue. It is liquid during life, but coagulates after death. [<Gk. *mys, myos* muscle]
my·o·sis (mī·ō′sis) See MIOSIS.
my·o·so·tis (mī′ə·sō′tis) *n.* Any of a genus of plants *(Myosotis)* of the borage family; especially, *M. scorpioides,* having branched racemes of blue or pink flowers. Also **my′o·sote** (-sōt). [<NL <Gk., lit., mouse ear <*mys, myos* a mouse + *ous, ōtos* ear]
my·ot·ic (mī·ot′ik) *adj.* Of, pertaining to, or having miosis. — *n.* A drug causing contraction of the pupil of the eye.
My·ra (mī′rə) An ancient city of Lycia, Asia Minor. *Acts* xxvii 5.
myria– *combining form* **1** Very many; of great number: *myriapod.* **2** In the metric system, ten thousand: *myriagram.* Also, before vowels, **myri–.** [<Gk. *myrios* <*myrioi* ten thousand]
myr·i·ad (mir′ē·əd) *adj.* Composed of a very large indefinite number; innumerable. — *n.* **1** A vast indefinite number. **2** Ten thousand. [<Gk. *myrias, myriados* <*myrios* numberless]
myr·i·a·gram (mir′ē·ə·gram′), **myr·i·a·li·ter** (-lē′tər), **myr·i·a·me·ter** (-mē′tər) *n.* In the metric system, 10,000 grams, liters, or meters.
myr·i·a·pod (mir′ē·ə·pod) *n.* One of a class of arthropods *(Myriapoda)* whose bodies are made up of a varying number of segments, each of which bears one or two pairs of jointed appendages: includes the centipedes and millipedes. — **myr·i·ap·o·dan** (mir′ē·ap′ə·dən) *adj.* & *n.* — **myr′i·ap′o·dous** *adj.*
my·ri·ca (mi·rī′kə) *n.* The dried bark of the wax-myrtle root, used in medicine as an alterative and emetic. [<L <Gk. *myrikē* tamarisk]
my·ris·tic (mi·ris′tik) *adj.* **1** Of, pertaining to, or containing the principle of nutmeg. **2** *Chem.* Designating a white crystalline acid, $C_{14}H_{28}O_2$, contained in nutmeg butter and similar vegetable sources. [<NL *nux myristica* nutmeg genus <Gk. *myrizein* anoint]
myrmeco– *combining form* Ant; pertaining to ants: *myrmecophagous.* Also, before vowels, **myrmec–.** [<Gk. *myrmēx, myrmēkos* an ant]
myr·me·col·o·gy (mûr′mə·kol′ə·jē) *n.* The department of entomology that treats of ants. [<MYRMECO- + -LOGY] — **myr′me·co·log′i·cal** (-kə·loj′i·kəl) *adj.* — **myr′me·col′o·gist** *n.*
myr·me·coph·a·gous (mûr′mə·kof′ə·gəs) *adj.* Feeding on ants. [<MYRMECO- + -PHAGOUS]
myr·mi·don (mûr′mə·don, -dən) *n.* A faithful adherent; also, a follower or underling of rough or desperate character, who executes the commands of his master without question

or scruple; especially, a petty officer of the law. [<MYRMIDON]
Myr·mi·don (mûr′mə·don, -dən) *n.* One of a warlike people of ancient Thessaly, represented as followers of Achilles in the Trojan War. [<Gk. *Myrmidones*] — **Myr′mi·do′ni·an** (-dō′nē·ən) *adj.*
my·rob·a·lan (mī·rob′ə·lən, mi-) *n.* 1 Any of the prunelike fruits of several tropical plants (genus *Terminalia*), formerly much used by tanners and calico printers. 2 The weeping plum of the western Mediterranean region (*Phyllanthus emblica*). 3 The myrobalan plum of SW Asia (*Prunus cerasifera*), having a sweet, very juicy fruit. [<F <L *myrobalanum* <Gk. *myrobalanos* < *myron* juice, ointment + *balanos* nut]
My·ron (mī′rən) A Greek sculptor of the fifth century B.C. — **My·ron′ic** (-ron′ik) *adj.*
my·ro·sin (mī′rə·sin, mir′ə·sin) *n. Biochem.* An enzyme found in the seeds of black and white mustard: it hydrolyzes the glycosides. [<Gk. *myron* ointment]
myrrh (mûr) *n.* 1 An aromatic gum resin that exudes from several trees or shrubs of Arabia and Abyssinia: used in medicine. 2 Any shrub or tree that yields this gum, especially *Commiphora myrrha* and *C. abyssinica*. [OE *murra* <L *myrrha* <Gk. <Semitic. Cf. Arabic *myrr*, Hebrew *mōr*.]
myr·rhin (mûr′in, mir′-) *n. Biochem.* A resinous principle found in myrrh. Also **myr′rhine.**
myr·ta·ceous (mûr·tā′shəs) *adj.* Pertaining to or designating a family (*Myrtaceae*) of trees or shrubs, the myrtle family, widely distributed in tropical and semitropical countries, and including many valuable aromatic resin- and timber-producing genera, as *Pimenta*, *Eucalyptus*, *Caryophyllus*, etc. [<NL *Myrtaceae* <L *myrtus* a myrtle tree]
myr·tle (mûr′tal) *n.* 1 A tree or shrub of the genus *Myrtus*; especially, *M. communis* of southern Europe, originally from Asia. It is a bushy shrub or small tree with glossy evergreen leaves, fragrant white or rose-colored flowers, and black berries. 2 One of various other plants like the common myrtle. Among the ancients it was sacred to Venus. 3 The periwinkle (*Vinca minor*); also, the California laurel (*Umbellularia californica*). [<F *myrtille* bilberry <Med. L *myrtillus* myrtle, dim. of L *myrtus* <Gk. *myrtos* <Semitic. Cf. Persian *mûrd*.]
Myr·tle (mûr′tal) A feminine personal name. [<MYRTLE]
myrtle warbler A small insectivorous bird (*Dendroica coronata*) of North America, with blue-gray or black-and-yellow back.
my·sel (mī·sel′) *pron. Scot.* Myself; by myself. Also **my·sell′.**
my·self (mī·self′) *pron.* 1 I; me: the emphatic form of *I* and *me*, and reflexive of *me*: used in the nominative with *I* in apposition: I *myself* will see to it; I'll write it *myself*; also, in poetical nominative use as a simple subject: *Myself* hath often heard; also used as the object of a verb either direct or indirect: I deceived *myself* (reflexive); She invited Helen, Jeff, and *myself* (compound object); I got it for *myself* (object of a preposition). 2 Normal condition of mind or body: I feel *myself* again. [OE *mē sylf*]
My·si·a (mish′ē·ə) An ancient region of NW Asia Minor on the Aegean and the Sea of Marmara.
My·sore (mī·sôr′, -sōr′) 1 A constituent State of southern India on the southern Deccan Plateau; 74,326 square miles; capital, Bangalore. 2 The former dynastic capital of Mysore State. Also *Maisur*.
mys·ta·gog (mis′tə·gôg, -gog) *n.* An interpreter of religious mysteries; an initiator into mysteries; teacher. Also **mys′ta·gogue.** [<F <L *mystagogus* <Gk. *mystagōgos* < *mystēs* an initiate + *agōgos* leader < *agein*. See MYSTERY.] — **mys′ta·gog′ic** (-goj′ik) *adj.* — **mys′ta·go′gy** (-gō′jē) *n.*
mys·te·ri·ous (mis·tir′ē·əs) *adj.* Involved in or implying mystery; unexplained; obscure. [<L *mysterium*] — **mys·te′ri·ous·ly** *adv.* — **mys·te′ri·ous·ness** *n.*
Synonyms: abstruse, cabalistic, dark, enigmatic, hidden, incomprehensible, inexplicable, inscrutable, mystic, mystical, obscure, occult, recondite, secret, transcendental, unfathomable, unfathomed, unknown. That is *mysterious* in the true sense which is beyond human comprehension; that is *mystic* or *mystical* which has associated with it some *hidden* or *recondite* meaning. See DARK, SECRET. Antonyms: see synonyms for CLEAR.
mys·ter·y[1] (mis′tər·ē) *n. pl.* **·ter·ies** 1 Something unknown or incomprehensible in its nature; an inexplicable phenomenon. 2 Secrecy or obscurity: an event wrapped in mystery. 3 A secret: the *mysteries* of freemasonry. 4 Any affair or event so concealed or unexplained as to excite awe or curiosity: a murder *mystery*. 5 A literary or dramatic piece relating such an affair. 6 *Theol.* A truth known only through faith or revelation and incomprehensible to the human reason. 7 *Eccl.* **a** A sacrament, especially the Eucharist. **b** *pl.* The eucharistic elements. 8 *pl.* In classical antiquity, certain religious rites to which only selected worshipers were admitted. 9 *pl.* A cult practicing such rites. 10 A medieval dramatic performance based on Scriptural events or characters: also **mystery play**. [<L *mysterium* <Gk. *mystērion* secret worship, secret thing < *mystēs* an initiate into the mysteries < *myein* shut, shut the eyes]
mys·ter·y[2] (mis′tər·ē) *n. pl.* **·ter·ies** *Archaic* A trade; occupation. [<Med. L *misterium* <L *ministerium* service, office; infl. in form by L *mysterium*]
mys·tic (mis′tik) *adj.* 1 Pertaining to a mystery of the faith. 2 Spiritually symbolic. 3 Of or designating an occult or esoteric rite, practice, etc. 4 Of mysterious meaning or character; mysterious. — *n.* 1 One who professes a knowledge of spiritual truth or a feeling of union with the divine, reached through contemplation or intuition. 2 A practicer of occult or mystical rites. [<L *mysticus* <Gk. *mystikos* pertaining to secret rites < *mystēs* an initiate]
mys·ti·cal (mis′ti·kəl) *adj.* 1 Having a spiritual character or reality beyond the comprehension of human reason. 2 Relating to mystics or mysticism. 3 Mystic (defs. 2 and 3).
mys·ti·cism (mis′tə·siz′əm) *n.* 1 The belief that knowledge of divine truth or the soul's union with the divine is attainable by spiritual insight or ecstatic contemplation without the medium of the senses or reason. 2 Any theory advancing intense meditative and intuitive methods of thought or conduct. 3 Vague or obscure speculation involving confused or fanciful thinking.
mys·ti·fy (mis′tə·fī) *v.t.* **·fied, ·fy·ing** 1 To confuse, especially deliberately; perplex or bewilder; hoax. 2 To treat as obscure or mysterious. See synonyms under PERPLEX.
[<F *mystifier*] — **mys′ti·fi·ca′tion** *n.* — **mys′ti·fi′er** *n.* — **mys′ti·fy′ing** *adj.* — **mys′ti·fy′ing·ly** *adv.*
mys·tique (mis·tēk′) *n. French* The mythical and enigmatic character of a person, idea, or thing, as a focus of popular veneration: the *mystique* of Stalin.
myth (mith) *n.* 1 A story, presented as historical, dealing with the cosmological and supernatural traditions of a people, their gods, culture, heroes, religious beliefs, etc. 2 A popular fable or folk tale. 3 A parable; allegory. See synonyms under FICTION. [<LL *mythos* <Gk., word, speech, story]
myth·i·cal (mith′i·kəl) *adj.* 1 Pertaining to myth; legendary. 2 Fictitious. Also **myth′ic.** — **myth′i·cal·ly** *adv.*
myth·i·cize (mith′ə·sīz) *v.t.* **·cized, ·ciz·ing** To convert into or explain as a myth.
mytho– *combining form* Myth; myth and: mythography. Also, before vowels, **myth–**. [<Gk. *mythos* a legend]
my·thog·ra·phy (mi·thog′rə·fē) *n.* 1 The collecting of myths; descriptive mythology. 2 Expression of mythic characters or ideas in art form. [<MYTHO– + -GRAPHY] — **my·thog′ra·pher** *n.*
my·thol·o·gize (mi·thol′ə·jīz) *v.* **·gized, ·giz·ing** *v.i.* To narrate, classify, or explain myths. — *v.t.* To mythicize. Also *Brit.* **my·thol′o·gise.** [<F *mythologiser*] — **my·thol′o·gist, my·thol′o·giz′er** *n.*
my·thol·o·gy (mi·thol′ə·jē) *n. pl.* **·gies** 1 The myths and legends of a people concerning creation, gods, and heroes. 2 The scientific collection and study of myths; study of the beliefs of mankind; also, a volume of myths. [<LL *mythologia* <Gk., telling of tales < *mythos* legend + *logos* discourse] — **myth′o·log′i·cal** (mith′ə·loj′i·kəl), **myth′o·log′ic** *adj.* — **myth′o·log′i·cal·ly** *adv.*
myth·o·ma·ni·a (mith′ə·mā′nē·ə, -mān′yə) *n.* A compulsive tendency to tell lies.
myth·o·pe·ic (mith′ə·pē′ik) *adj.* Mythmaking; relating to a supposed stage of human culture when all natural phenomena are explained by myths. Also **myth′o·poe′ic.** [<Gk. *mythopoios* <*mythos* legend + *poieein* make] — **myth′o·pe′ist** *n.*
Myt·i·le·ne (mit′ə·lē′nē) 1 Lesbos. 2 A seaport on SE Lesbos, capital of the Aegean Islands division of Greece: formerly *Kastro*. Also **Myt′i·li′ni.**
myx·e·de·ma (mik′sə·dē′mə) *n. Pathol.* A disease associated with hypotrophy of the thyroid gland and characterized by dryness and wrinkling of the skin, swelling of the face, and progressive mental deterioration. Also **myx·oe·de′ma.** [<NL <MYX(O)– + EDEMA] — **myx′e·dem′ic** (-dem′ik), **myx′e·dem′a·tous** (-dem′ə·təs) *adj.*
myxo– *combining form* Slimy; like mucus. Also, before vowels, **myx–**. [<Gk. *myxa* mucus]
myx·o·ma (mik·sō′mə) *n. pl.* **·ma·ta** (-mə·tə) *Pathol.* A soft, elastic tumor composed of mucous tissue. [<NL <MYX(O)– + -OMA] — **myx·om·a·tous** (mik·som′ə·təs) *adj.*
myx·o·my·cete (mik′sō·mī·sēt′) *n.* One of the slime molds (*Myxomycetes*), a class of fungi exhibiting both plant and animal characteristics and classified by some authorities as *Mycetozoa*. They consist of masses of naked protoplasm with ameboid movements, and are chiefly saprophytic, living on dead and decaying matter. [<MYXO– + -MYCETE] — **myx′o·my·ce′tous** *adj.*

N

n, N (en) *n. pl.* **n's, N's** or **ns, Ns, ens** (enz) 1 The 14th letter of the English alphabet: from Phoenician *nun*, Greek *nu*, Roman *N*. 2 The sound of the letter *n*, a voiced, alveolar nasal. See ALPHABET. — *symbol* 1 *Printing* An en: see EN. 2 *Chem.* Nitrogen (symbol N). 3 *Math.* An indefinite number.
na (nä) *adv. Scot. & Brit. Dial.* No; not.
nab[1] (nab) *v.t.* **nabbed, nab·bing** *Colloq.* 1 To catch or arrest, as a criminal. 2 To take or seize suddenly. [Cf. Norw., Sw. *nappa* snatch]
nab[2] (nab) *n.* 1 *Geog.* A projecting part of a hill or rock; a peak, promontory, or summit. 2 *Mech.* A projection or spur on the bolt of a lock. [<ON *nabbi* a knoll]

Nabatean 842 **name**

Nab·a·te·an (nab'ə·tē'ən) n. 1 One of an ancient Arabic people dwelling east and southeast of Palestine, independent from 312 B.C. to A.D. 106, when they submitted to Roman rule. 2 The language of these people, belonging to the Aramaic subgroup of the Northwest Semitic languages.
Na·blus (nä·bloos') The chief town of Samaria, western Jordan, on the site of ancient *Sechem*, rebuilt by Hadrian as *Neapolis*.
na·bob (nā'bob) n. 1 A European who has become rich in India; hence, any rich man. 2 A nawab, viceroy, or governor in India under the old Mogul empire. [<Hind. *nawwāb* <Arabic *nuwwab*, pl. of *nā'ib* a deputy] —**na'bob·er·y** n. —**na'bob·ish** adj.
Na·both (nā'both) In the Bible, the owner of a vineyard which Ahab coveted and, with the aid of Jezebel, unlawfully procured. 1 *Kings* xxi.
nac·a·rat (nak'ə·rat) n. 1 Bright red-orange color. 2 A fine linen or crêpe fabric dyed this color. [<F, appar. <Sp. *nacarado* <*nacar* nacre]
na·celle (nə·sel') n. Aeron. 1 The basket suspended from a balloon. 2 The framework below the envelope of a dirigible balloon, which carries the motor, passengers, etc. 3 An enclosed shelter for housing the cargo or power plant and sometimes the personnel of an airplane. [<F]

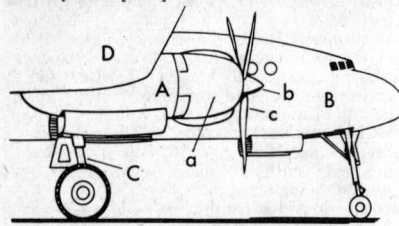

NACELLE
A. Nacelle. B. Fuselage.
 a. Engine cowling. C. Landing gear.
 b. Spinner. D. Wing.
 c. Propeller.

na·cre (nā'kər) n. Mother-of-pearl. [<F]
na·cre·ous (nā'krē·əs) adj. 1 Of, like, or producing nacre. 2 Iridescent; pearly.
na·da (nä'də) n. In East Indian music, the term for esthetically agreeable sound, as distinct from noise, grinding, clanging, etc.
na·dir (nā'dər, -dir) n. 1 The point of the celestial sphere directly beneath the place where one stands: opposed to *zenith*. 2 Figuratively, the lowest possible point: the *nadir* of melancholy. [<F <Arabic *naḍīr* (*es-semt*) opposite (the zenith)]
nae (nā) adj. *Scot.* No; none.
nae·thing (nā'thing) n. *Scot.* Nothing.
nae·void (nē'void), **nae·vus** (nē'vəs) See NEVOID, NEVUS.
Na·fud (na·food') A desert area in northern Saudi Arabia: also *Nefud*.
nag¹ (nag) v. **nagged, nag·ging** v.t. To torment with constant faultfinding, scolding, and urging. —v.i. To scold, find fault, or urge continually. —n. One who nags, especially a woman. [<Scand. Cf. Sw. *nagga* vex.] —**nag'ger** n. —**nag'ging** adj. —**nag'ging·ly** adv.
nag² (nag) n. 1 A pony or small horse. 2 An old or inferior horse. 3 *Archaic* A worthless person; jade. [ME *nagge*; origin uncertain]
Na·ga Hills (nä'gə) A series of hill ranges between NE India and western Burma.
na·ga·na (nə·gä'nə) n. A disease of cattle and horses caused by trypanosomes introduced into the blood by the tsetse fly. [<Zulu]
Na·ga·ri (nä'gə·rē) n. 1 Any one of a group of vernacular alphabets in India. 2 Devanagari.
Na·ga·sa·ki (nä'gä·sä·kē) A port on western Kyushu island, Japan; largely destroyed by an atomic bomb dropped by U. S. airmen on August 9, 1945.
Nä·ge·li (nā'gə·lē), **Karl Wilhelm von**, 1817-1891, Swiss botanist.
Na·gor·no-Ka·ra·bakh Autonomous Region (nä·gôr'nə·kä'rä·bäkh') An administrative division of SW Azerbaijan S.S.R.; 1,700 square miles.
Na·go·ya (nä·gō'yä) A city on southern Honshu island, Japan, on Ise Bay.
Nag·pur (näg'poor) A city in NE Bombay State, India, former capital of Madhya Pradesh State.

Nag·y·va·rad (nôd'y'·vä·rôd) Hungarian name for ORADEA.
Na·ha (nä·hä) Chief city of Okinawa island and headquarters of the U. S. military government of the Ryukyu Islands.
Na·hua (nä'wä) n. pl. **Na·hua** One of a group of civilized Indian peoples of Mexico and Central America belonging to the Uto-Aztecan linguistic stock, including the Aztecs, Toltecs, etc.
Na·hua·tl (nä'wät'l) n. The Uto-Aztecan language of the Aztecs, including many dialects. —adj. Designating or pertaining to this language.
Na·hua·tlan (nä'wät·lən) n. A branch of the Uto-Aztecan linguistic family of North American Indians, including the Aztec dialects. —adj. Of or pertaining to this linguistic branch.
Na·huel Hua·pí (nä·wel' wä·pē') A lake on the eastern slope of the Andes in western Argentina near the Chilean border; 210 square miles.
Na·hum (nā'əm, -hum) A Hebrew prophet; also, the book of the Old Testament containing his prophecies. [<Hebrew, consolation]
nai·ad (nā'ad, nī'-) n. pl. **·ads** or **·a·des** (-ə·dēz) 1 In classical mythology, one of the water nymphs who were believed to dwell in and preside over fountains, lakes, brooks, and wells. 2 *Bot.* A plant of the pondweed family (genus *Naias*). 3 The nymph stage in the life cycle of certain insects: applied especially to aquatic forms. [<Gk. *Naias, -ados.* Related to Gk. *naein* flow.]
nai·ant (nā'ənt) See NATANT.
Na·i·du (nä'i·dōō), **Sarojini**, 1879-1949, Hindu poet and reformer.
na·if (nä·ēf') adj. *French* Masculine form of NAIVE. Also **na·ïf'**.
naig (nāg) n. *Scot.* A nag; riding horse. Also **naig'ie.**
nail (nāl) n. 1 A thin horny plate on the end of a finger or toe. 2 A claw, talon, or hoof. 3 A slender piece of metal having a point and a head, and used for driving into or through wood, etc., as for fastening pieces together. 4 A measure of length, 2 1/4 inches. 5 A callosity on the inner side of a horse's leg. —**on the nail** *Colloq.* 1 Right away; immediately. 2 At the exact spot or moment. 3 Of immediate interest or importance; under discussion. —v.t. 1 To fasten or fix in place with a nail or nails. 2 To close up or shut in by means of nails. 3 To secure by decisive or prompt action: to *nail* a contract. 4 To fix firmly or immovably: Terror *nailed* him to the spot. 5 To succeed in hitting or striking. 6 *Colloq.* To catch or seize; intercept. 7 *Colloq.* To detect and expose, as a lie or liar. [OE *nægel*]
nail bed *Anat.* That portion of the true skin upon which the nail rests.
nail file A fine, flat file used for manicuring the fingernails.
nail·fold (nāl'fōld') n. *Anat.* The duplication of the skin that surrounds the edges of a nail; cuticle.
nail polish A lacquer applied to the nails to give a glossy finish, made from soluble cotton treated with various organic compounds, as toluene, ethyl acetate, ethanol, etc., and usually colored by the addition of dyes. Also **nail enamel.**
nain·sook (nān'sook, nan'-) n. A soft, lightweight cotton fabric, heavier than batiste; used for lingerie and infants' wear. [<Hind. *nainsukh*, lit., pleasure of the eye]
Nairn (nârn) A county in NE Scotland; 163 square miles; county town, Nairn. Also **Nairn·shire** (nârn'shir, -shər).
Nai·ro·bi (nī·rō'bi) The capital of Kenya in eastern Africa.
na·ive (nä·ēv') adj. 1 Ingenuous; artless; without sophistication. 2 Not consciously logical; uncritical. Also **na·ïve'**. See synonyms under CANDID. [<F, fem. of *naïf* <L *nativus*

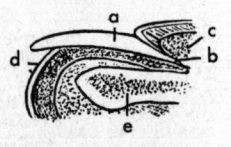

HUMAN FINGERNAIL
Longitudinal section
a. Nail.
b. Matrix.
c. Nailfold.
d. Epidermis.
e. Phalanx.

natural. Doublet of NATIVE.] —**na·ive'ly** adv. —**na·ive'ness** n.
na·ive·té (nä·ēv'tā', nä·ēv·tā') n. The state or quality of being naive. Also **na·ïve'té', na·ive·ty** (nä·ēv'tē).
Najd (näjd) See NEJD.
Na·jin (nä·jēn) A port of Northern Korea. *Japanese* **Ra·shin** (rä·shēn').
na·ked (nā'kid) adj. 1 Having no clothes or garments on; nude. 2 Having no covering, or lacking the usual covering. 3 Unsheathed; bare, as a sword. 4 Unsaddled, as a horse. 5 Having no defense or protection; exposed. 6 Being without means of sustenance, etc.; destitute; bare; also, stripped. 7 Being without concealment or excuse. 8 Without addition or adornment; plain; evident; mere. 9 Without some accessory, qualification, belonging, etc., which is customary or natural. 10 *Law* Having no consideration or inducement; unconfirmed; not validated. 11 *Bot.* **a** Not enclosed in an ovary or case. **b** Without a pericarp: said of seeds. **c** Destitute of leaves: said of stalks. **d** Having no hairs; smooth: said of leaves. 12 *Zool.* Lacking fur, hair, scales, or feathers. [OE *nacod*] —**na'ked·ly** adv. —**na'ked·ness** n.
naked eye The eye unaided by optical instruments.
Na·khi·che·van Autonomous S.S.R. (nä'khē·che·vän') 1 A republic forming part of the Azerbaijan S.S.R.; 2,100 square miles. 2 Its capital and chief city, on the Araxes.
Nak·tong (näk·tong) The largest river of Korea, flowing 326 miles to Korea Strait near Pusan.
Nal·chik (näl'chik) The capital of Kabardian Autonomous S.S.R.
nam (näm, nam) Past tense of NIM.
nam·a·ble (nā'mə·bəl) adj. 1 Capable of being named. 2 Memorable; worthy of being mentioned. Also **name'a·ble.** —**nam'a·bil'i·ty, name'a·bil'i·ty** n.
Na·man·gan (nä'män·gän') A city in eastern Uzbek S.S.R.
Na·ma·qua·land (nə·mä'kwə·land) A coastal region of SW Africa, divided by the Orange River into **Great Namaqualand** in South-West Africa and **Little Namaqualand** in NW Cape of Good Hope Province, Union of South Africa. Also **Na'ma·land.**
nam·ay·cush (nam'ə·kush, -ā-) n. The great lake trout (*Cristivomer namaycush*) of North America; the Mackinaw trout: also spelled *naymacush*. [<Algonquian (Cree) *namekus* trout]
nam·by-pam·by (nam'bē·pam'bē) adj. Weakly sentimental; insipid; inane. —n. pl. **-bies** 1 Writing, talk, or action of a feebly sentimental or finical character. 2 A person given to such talk or action. [<nickname of Ambrose Philips, 1671-1749, English poet; with ref. to his feeble, sentimental verse]
name (nām) n. 1 The distinctive appellation by which a person or thing is known. 2 A descriptive or arbitrary appellation; designation; title. 3 General reputation; eminence; fame. 4 A person, cause, thing, or class, or the claims of authority thereof, as represented by the name. 5 An opprobrious appellation. 6 A race or family, as having a common descent and patronymic. 7 A memorable person, character, or thing: great *names* in music. 8 Mere sound or simulation, in distinction from substance or reality: a wife in *name* only. —**by the name of** Named. —v.t. **named, nam·ing** 1 To give a name to; entitle; style; term. 2 To mention or refer to by name; cite. 3 To designate for some particular purpose or office; nominate; appoint. 4 To give the name of; identify: *Name* the capital of Peru. 5 To set or specify, as a price or requirement. [OE *nama*] —**nam'er** n.
Synonyms (noun): agnomen, appellation, cognomen, denomination, designation, epithet, style, title. *Name* in the most general sense includes all other words of this group; in the more limited sense a *name* is personal, an *appellation* is descriptive, a *title* is official. In the phrase William the Conqueror, king of England, William is the man's *name*, which belongs to him personally; Conqueror is the *appellation* which he won by his acquistion of England; king is the *title* denoting his royal rank. An *epithet* is given to mark some assumed characteristic, good or bad. *Designation* may be used in the sense of *appellation*.

but is far broader and more general in meaning. One's personal *name*, as John or Mary, is given in infancy, and is often called the given, or Christian, or first *name*. The *cognomen* is the family *name* which belongs to one by right of birth or marriage. In modern use, *style* is the legal *designation* by which a person or house is known in official or business relations. A *denomination* is a specific, and especially a collective name; the term is applied to a separate religious organization, also to money or notes of a certain value; as, The sum was in notes of the *denomination* of one thousand dollars. See TERM.

name-day (nām′dā′) *n.* The day of the saint for whom one is named.

name·less (nām′lis) *adj.* 1 Having no name; unnamed. 2 Having no fame or reputation; illegitimate; obscure; anonymous. 3 Not suitable or fit to be spoken of. 4 Not to be named; inexpressible; indescribable. — **name′less·ly** *adv.* — **name′less·ness** *n.*

name·lore (nām′lôr′, -lōr′) *n.* Lore relating to names.

name·ly (nām′lē) *adv.* That is to say; to wit; videlicet.

name·sake (nām′sāk) *n.* One who is named after or has the same name as another.

Nam·hoi (näm′hoi′) A city in central Kwangtung province, SE China: formerly *Fatshan*.

Nam·po (näm′pō′) See CHINNAMPO.

Nam Tso (näm′tsō′) The largest lake of Tibet, NW of Lhasa; elevation, 15,180 feet; 950 square miles: Mongolian *Tengri Nor*.

Na·mur (nä·mōōr′, *Fr.* nà·mür′) A city in south central Belgium. Flemish **Na·men** (nä′mən).

Nan (nan) Diminutive of ANN.

Nan (nän) A river in northern Thailand, flowing 500 miles south to the Chao Phraya.

Nan·chang (nän′chäng′) The capital city of Kiangsi province, SE China.

Nan·cy (nan′sē) Diminutive of ANN.

Nan·cy (nan′se, *Fr.* nän·sē′) A city of NE France, chief city of Lorraine.

Nan·da De·vi (nun′dä dā′vē) A mountain in northern Uttar Pradesh, India; 25,645 feet.

nane (nān) *adj.* & *pron. Scot.* None; no.

Nan·ga Par·bat (nung′gə pûr′bät) A mountain in NW Kashmir; 26,660 feet.

nan·keen (nan·kēn′) *n.* 1 A buff-colored Chinese cotton fabric. 2 *pl.* Clothes made of nankeen. Also **nan·kin′**. [from *Nanking*, where originally made]

Nan·king (nan′king′, nän′-) A port on the Yangtze, former capital of Kiangsu province; capital of China, 1928–37.

Nanking porcelain Any of various types of fine Chinese porcelain painted in blue on white: also called *blue-and-white*. Also *Nankeen porcelain*.

Nan Ling (nän′ ling′) A mountain chain in SE China, on the northern border of Kwangtung province.

Nan·nette (nan-et′) Diminutive of Ann.

Nan·ning (nän′ning′) A city of southern China, former capital of Kwangsi province.

nan·ny (nan′ē) *n. pl.* **·nies** *Colloq.* 1 A female goat: also **nanny goat**. 2 A child's nurse. [from *Nanny*, a personal name]

Nan·ny (nan′ē) Diminutive of ANN.

na·no·plank·ton (nā′nə·plangk′tən, nan′ə-) *n.* Floating plant and animal organisms of microscopic size. Also **nan′no·plank′ton** (nan′ə-). [< Gk. *nanos* dwarf + PLANKTON]

Nan·sei Islands (nän′-sā) See RYUKYU ISLANDS.

Nan·se·mond River (nan′sə·mond) A river in SE Virginia, flowing 25 miles north to the James River estuary near Hampton Roads.

Nan·sen (nän′sən), **Fridtjof**, 1861–1930, Norwegian Arctic explorer and naturalist.

Nan Shan (nän′ shän′) A mountain range in Tsinghai and Kansu provinces, China.

Nan·tas·ket Beach (nan·tas′kit) A resort village of eastern Massachusetts, on Massachusetts Bay south of Boston.

Nantes (nants, *Fr.* nänt) A city of western France, on the Loire. — **Edict of Nantes** An order granting freedom of conscience to Protestants, issued by Henry IV of France in 1598 and revoked in 1685 by Louis XIV.

Nan·tuck·et (nan·tuk′it) An island and summer resort off the SE coast of Massachusetts; 57 square miles. — **Nan·tuck′et·er** *n.*

Nan·tung (nän′tōōng′) A city of northern Kiangsu province, China: formerly *Tungchow*.

Na·o·mi (nā·ō′mē, nā′ə·mē, -mī) A feminine personal name. [< Hebrew, pleasant]
— **Naomi** In the Bible, the mother-in-law of Ruth. *Ruth* i 2.

na·os (nā′os) *n. Archit.* The principal chamber or body of an ancient Greek temple, usually containing a statue of the deity; a cella. [< Gk., a temple]

nap[1] (nap) *n.* A short sleep; doze. — *v.i.* **napped, nap·ping** 1 To take a nap; doze. 2 To be unprepared or off one's guard. [OE *hnappian* doze] — **nap′per** *n.*

nap[2] (nap) *n.* 1 The short fibers on the surface of flannel, etc., forming a soft surface lying smoothly in one direction. 2 A covering resembling this, as upon some plants or insects. — *v.t.* **napped, nap·ping** To raise a nap on. [< MDu. *noppe*]

nap[3] (nap) See NAPOLEON (def. 2).

na·palm (nā′päm) *n.* A jellied mixture of aluminum soap powder and oil or gasoline, used as an incendiary in bombs, flame-throwers, etc. [< *na*(phthenic) and *palm*(itic) acids, chemical compounds used in its manufacture]

nape (nāp) *n.* 1 The back of the neck, especially its upper part. ◆ Collateral adjective: *nuchal*. 2 The back of a fish next to the head. [Origin uncertain]

na·per·y (nā′pər·ē) *n. pl.* **·per·ies** An article of household linen, as napkins, tablecloths, etc., or such linen collectively. [< OF *naperie* < *nape*. See NAPKIN.]

Naph·ta·li (naf′tə·lī) The sixth son of Jacob. *Gen.* xxx 8.

naph·tha (naf′thə, nap′-) *n.* 1 A volatile mixture of low-boiling hydrocarbons between gasoline and benzine, obtained by distilling petroleum: used as a solvent, cleaning fluid, fuel, etc., and in the making of varnishes. 2 Petroleum. [< L < Gk., prob. < Persian *naft* petroleum]

naph·tha·lene (naf′thə·lēn, nap′-) *n. Chem.* A white, solid, aromatic hydrocarbon, $C_{10}H_8$, obtained from coal-tar distillates and crystallizing in white platelets: its derivatives are used in the making of dyestuffs: also called *tar camphor*. Also **naph′tha·line** (-lin, -lēn).

naph·thal·ic acid (naf·thal′ik, nap′-) Former name for PHTHALIC ACID.

naph·thene (naf′thēn, nap′-) *n. Chem.* Any of a group of saturated ring hydrocarbons having the general formula C_nH_{2n}, especially those obtained from Russian and Galician petroleum.

naph·thol (naf′thol, -thôl, nap′-) *n. Chem.* 1 Either of two isomeric compounds, the alpha and beta, $C_{10}H_7OH$, derived from naphthalene by replacing an atom of hydrogen by the hydroxyl group; specifically, the beta variety. They are used as antiseptics and in the manufacture of synthetic dyes. 2 Any one of a class of naphthalene derivatives containing the hydroxyl group. Also **naph′tol** (-tōl, -tol). [< NAPHTH(ALENE) + -OL[2]]

naph·tho·lism (naf′thə·liz′əm, nap′-) *n. Pathol.* Poisoning caused by excessive or prolonged use of naphthol.

Na·pier (nā′pēr, nə·pir′), **Sir Charles James**, 1782–1853, British general. — **John**, 1550–1617, Scottish mathematician; invented logarithms. — **Robert Cornelis**, 1810–80, Lord Napier of Magdala, British general.

Na·pier·i·an logarithms (nə·pir′ē·ən) *Math.* The logarithmic system employing the base *e*: (2.71828 . . .): also called *natural logarithms*. Also **Na·pe′ri·an**.

na·pi·form (nā′pə·fôrm) *adj. Bot.* Turnip-shaped; large above and small or slender below: a *napiform* rootstock. [< L *napus* turnip + -(I)FORM]

nap·kin (nap′kin) *n.* 1 A small cloth, as of linen, for use at table, etc. 2 *Scot.* A handkerchief. [ME *napekyn*, dim. OF *nape* < L *mappa* a cloth]

Na·ples (nā′pəlz) A port in SW Italy on the Bay of Naples, a semicircular inlet of the Tyrrhenian sea in SW Italy: Greek *Parthenope*, *Neapolis*. Italian **Na·po·li** (nä′pō·lē).

nap·less (nap′lis) *adj.* 1 Made without a nap. 2 Threadbare.

Naples yellow A semi-opaque, permanent pigment in various shades of yellow consisting of lead antimoniate. Inferior grades are mixtures of ocher, zinc oxide, or cadmium yellow.

Na·po (nä′pō) A river in NE Ecuador and north central Peru, flowing 550 miles SE to the Amazon.

na·po·le·on (nə·pō′lē·ən, *Fr.* nà·pô·lā·ôn′) *n.* 1 A former French gold coin, equivalent to 20 francs. 2 A card game: the highest bidder names trumps and, if he takes the tricks he has bid, receives from each adversary one chip for each trick; also, the taking of all the tricks in this game by one player: also called *nap*. 3 A pastry composed of layers of puff paste filled with cream or custard. [after *Napoleon* Bonaparte]

Na·po·le·on (nə·pō′lē·ən) A masculine personal name. Also *Fr.* **Na·po·lé·on** (nà·pô·lā·ôn′), *Ital.* **Na·po·le·o·ne** (nä·pō·lā·ō′nā). [< F < Gk., of the new city]
— **Napoleon** See BONAPARTE.

Na·po·le·on·ic (nə·pō′lē·on′ik) *adj.* 1 Belonging or relating to Napoleon Bonaparte, his conquests, etc. 2 Belonging or relating to Napoleon III.

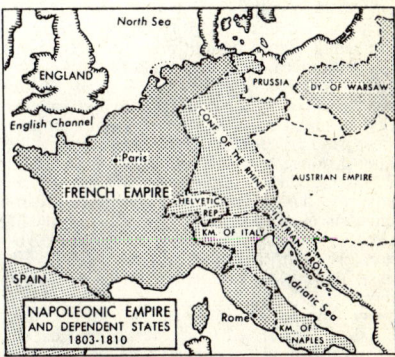

Napoleonic Wars See table under WAR.

nappe (nap) *n.* 1 *Geol.* A recumbent anticline, a portion of which has been thrust over other rocks. 2 *Engin.* The sheet of water overlying the top of a weir. 3 *Geom.* In a cone, one of the two conical surfaces divided by the vertex. [< F, a sheet]

nap·per (nap′ər) *n.* An implement or machine that raises a nap on fabrics; a napping machine; a gig.

nap·py[1] (nap′ē) *adj.* **·pi·er**, **·pi·est** Having or characterized by a nap, or abundance of nap or pile; shaggy.

nap·py[2] (nap′ē) *adj.* **·pi·er**, **·pi·est** 1 Inclined to fall asleep; drowsy. 2 Tending to produce drowsiness or intoxication; hence, slightly intoxicated. — *n. Scot.* Strong ale or beer.

nap·py[3] (nap′ē) *n. pl.* **·pies** A round earthen or glass dish with flat bottom and sloping sides. Also **nap′pie**. [OE *hnæp* bowl]

na·prap·a·thy (nə·prap′ə·thē) *n.* The treatment of disease by the manipulative correction of the disordered ligaments and connective tissues by which the disease is assumed to be caused. [< Czech *napra(va)* correction + -PATHY] — **nap′ra·path** *n.*

Nar·ba·da (nər·bud′ə) A river of central India, flowing about 775 miles west to the Gulf of Cambay: formerly *Nerbudda*.

Nar·bonne (när·bôn′) A town in southern France; formerly a port on the Gulf of the Lion: ancient **Nar·bo Mar·ti·us** (när′bō mär′shəs).

nar·ce·ine (när′sē·ēn, -in) *n. Chem.* A silky, bitter, crystalline alkaloid, $C_{23}H_{27}NO_8$, contained in the aqueous extract of opium from which the morphine has been separated. Also **nar·ce′ia**, **nar·ce·in**. [< L *narce* torpor (< Gk. *narkē*) + -INE[2]]

nar·cis·sism (när′sis·iz·əm) *n.* 1 *Psychoanal.* Sexual excitement or gratification derived from contemplation of the self: an arrested or regressive stage. 2 Self-love; excessive interest in or admiration for oneself. Also **nar·cism** (när′siz·əm). — **nar·cis′sist** *n.* — **nar′cis·sis′tic** *adj.*

nar·cis·sus (när·sis′əs) *n.* 1 One of a genus

(*Narcissus*) of bulbous flowering plants of the amaryllis family, including the daffodil and jonquil. 2 A flower or bulb of this genus. *N. poeticus* is the poet's narcissus.

Nar·cis·sus (när·sis′əs) In Greek mythology, a youth who caused the death of Echo by spurning her love: in punishment Nemesis caused him to pine away and die for love of his own image in a pool, and changed him into the narcissus.

narco- *combining form* Torpor; insensibility: *narcomania*. Also, before vowels, **narc-**. [<Gk. *narkē* numbness]

nar·co·lep·sy (när′kə·lep′sē) *n. Pathol.* A condition marked by an uncontrollable desire for sleep or by sudden attacks of drowsiness: sometimes associated with petit mal. — **nar′co·lep′tic** *adj.*

nar·co·ma·ni·a (när′kō·mā′nē·ə, -mān′yə) *n. Psychiatry* 1 A morbid craving to seek relief from pain, principally through the use of narcotics. 2 Psychotic alcoholism. — **nar′co·ma′ni·ac** *adj.*

nar·co·sis (när·kō′sis) *n.* Deep sleep or unconsciousness produced by a drug. Also **nar·co′ma** (-kō′mə).

nar·co·syn·the·sis (när′kō·sin′thə·sis) *n. Psychiatry* A condition of seminarcosis induced by certain drugs to aid in the treatment of abnormal mental conditions by encouraging the patient to talk freely about himself.

nar·cot·ic (när·kot′ik) *n.* 1 Any of various substances, as opium, morphine, and codeine, that in medicinal doses relieves pain, induces sleep, and in excessive or uncontrolled doses may produce convulsions, coma, and death. 2 An individual addicted to the use of narcotics. — *adj.* 1 Having the quality of causing narcosis or stupor. 2 Figuratively, causing sleep or dulness, as a book or sermon. Also **nar·cose** (när′kōs), **nar·cot′i·cal** *adj.* [<Gk. *narkōtikos* <*narkē* torpor] — **nar·cot′i·cal·ly** *adv.*

nar·co·tine (när′kə·tēn, -tin) *n.* An alkaloid, $C_{22}H_{23}O_7N$, derived from the aqueous extract of opium: used in medicine as an antispasmodic and tonic. Also **nar′co·tin** (-tin).

nar·co·tism (när′kə·tiz′əm) *n.* 1 Stupor due to narcotics. 2 Any method or influence inducing narcosis. 3 A morbid tendency to sleep. 4 Addiction to narcotics.

nar·co·tize (när′kə·tīz) *v.t.* **·tized**, **·tiz·ing** To bring under the influence of a narcotic; stupefy. — **nar′co·ti·za′tion** *n.*

nard (närd) *n.* 1 Spikenard the plant, oil, or ointment). 2 Any one of several aromatic plants or roots (mostly species of valerian) formerly used in medicine. [<OF *narde* <L *nardus* <Gk. *nardos*, prob. <Semitic] — **nard′ine** (när′dēn, -din) *adj.*

Na·ren·ta (nä·ren′tä) The Italian name for the NERETVA.

nar·es (nâr′ēz) *n. pl.* of **nar·is** (nâr′is) *Anat.* 1 Openings into the nose or nasal passages. 2 The nostrils. [<L, nostrils]

Na·rew (nä′ref) A river in western Belorussian S.S.R. and NE Poland, flowing 275 miles NW to the Vistula. Russian **Na·rev′**.

nar·ghi·le (när′gə·li) *n.* An Oriental tobacco pipe by which the smoke is drawn through water by means of a long tube. See HOOKAH. Also **nar′gi·le**, **nar′gi·leh**. [<Persian *nārgīleh* <*nārgīl* a coconut: because originally made of coconut shell]

nark (närk) *n. Brit. Slang* An informer; stool pigeon. [<Romany *nāk* a nose]

Nar·kom·vnu·del (när·kom·vnoo′dyel) *n.* People's Commissariat for Internal Affairs, U.S.S.R.: originally *Cheka*, later OGPU: abbreviated *NKVD*. [<Russian *Nar(odni) Kom(misariat) Vnu(trennikh) Del*]

Nar·ra·gan·sett (nar′ə·gan′sit) *n.* 1 One of a tribe of North American Indians of Algonquian stock, formerly inhabiting Rhode Island. 2 The Algonquian language of this tribe. 3 One of a breed of small, robust, surefooted horses, originally bred in Rhode Island and valued as saddle horses: also **Nar′ra·gan′sett**.

Nar·ra·gan·sett Bay (nar′ə·gan′sit) An inlet of the Atlantic Ocean in SE Rhode Island.

nar·rate (na·rāt′, nar′āt) *v.t.* **·rat·ed**, **·rat·ing** To tell or relate as a story; give an account of. See synonyms under RELATE. [<L *narratus*, pp. of *narrare* relate] — **nar·ra′tor** (-târ′, -tōr′ē) *n.*

nar·ra·tion (na·rā′shən) *n.* 1 The act of narrating the particulars of an event or series of events. 2 That which is narrated; narrative. See synonyms under HISTORY, REPORT.

nar·ra·tive (nar′ə·tiv) *n.* 1 An orderly, continuous account of an event or series of events. 2 The act or art of narrating. See synonyms under HISTORY, REPORT, STORY[1]. — *adj.* Pertaining to narration. — **nar′ra·tive·ly** *adv.*

nar·row (nar′ō) *adj.* 1 Having comparatively little distance from side to side. 2 Limited in extent or duration; circumscribed; small. 3 Illiberal; bigoted. 4 Limited in means or resources; straitened. 5 Niggardly; parsimonious. 6 Barely accomplished, attained, or sufficient: a *narrow* escape. 7 Scrutinizing closely. 8 *Phonet.* Tense. See synonyms under LITTLE, SCANTY, SMALL. — *v.t.* & *v.i.* To make or become narrower, as in width or scope. — *n.* 1 *Usually pl.* A narrow passage; strait; also, the narrowest part of an isthmus or cape. 2 A narrow part of a street, or of a valley or pass. — **The Narrows** 1 A strait connecting Upper New York Bay with Lower New York Bay between the western end of Long Island and Staten Island. 2 The narrowest part of the Dardanelles. [OE *nearu*] — **nar′row·ness** *n.*

nar·row–gage (nar′ō·gāj′) *adj.* 1 Denoting a width of railroad track less than the standard gage. 2 *Colloq.* Petty; illiberal; narrow-minded. — *n.* 1 A railroad having a gage narrower than 4 feet 8 1/2 inches. 2 A train for a narrow–gage railroad.

nar·row·ly (nar′ō·lē) *adv.* 1 With little breadth, width, or distance from side to side. 2 With small extent or duration; contractedly; restrictedly. 3 Barely; hardly.

nar·row–mind·ed (nar′ō–mīn′did) *adj.* Of contracted mental scope; also, illiberal or bigoted. — **nar′row–mind′ed·ly** *adv.* — **nar′row–mind′ed·ness** *n.*

nar·thex (när′theks) *n. Archit.* A porch, vestibule, or division of a church or basilica before the entrance proper. [<L <Gk. *narthēx*, orig. a plant with a hollow stalk]

Nar·va (när′vä) A city in NE Estonia.

Nar·vá·ez (när·vä′eth), **Pánfilo de**, 1480?–1528, Spanish general; defeated by Cortés in Mexico; explored Florida.

Nar·vik (när′vik) A port in northern Norway.

nar·whal (när′wəl, -hwəl) *n.* A large, arctic cetacean (*Monodon monoceros*) of the family Delphinidae, having in the male a long, straight, spiraled tusk: valued for its oil and ivory. Also **nar′wal**, **nar′whale′**. [<Dan. or Sw. *narhval*]

NARWHAL
a. Head of male showing tusk.
b. The female.
(From 12 to 16 feet in body length;
the spiraled tusk from 6 to 8 feet)

nar·y (nâr′ē) *adj. Dial.* Never a; not one: opposite of *ary*.

na·sal[1] (nā′zəl) *adj.* 1 Of or pertaining to the nose. 2 *Phonet.* Pronounced with the voiced breath passing partially or wholly through the nose, as in (m), (n), and (ng), and the French nasal vowels. — *n. Phonet.* A nasal sound. [<NL *nasalis* <L *nasus* the nose] — **na·sal·i·ty** (nā·zal′ə·tē) *n.*

na·sal[2] (nā′zəl) *n.* A nosepiece. [<OF *nasal*, *nasel* <L *nasus* nose]

nasal index A number expressing the ratio of the greatest breadth of the nose (multiplied by 100) to the length: it is greater when measured on the face than on the skull, and varies with age.

na·sal·ize (nā′zəl·īz) *v.* **·ized**, **·iz·ing** *v.t.* 1 To give a nasal sound to. — *v.i.* To pronounce oral sounds in the manner of nasals; talk through the nose. — **na′sal·i·za′tion** *n.*

nas·cent (nas′ənt, nā′sənt) *adj.* Beginning to exist or develop. [<L *nascens*, ppr. of *nasci* be born] — **nas′cence**, **nas′cen·cy** *n.*

nascent state *Chem.* The uncombined condition of an atom or radical when recently set free from a compound and ready to enter into combination with some other atom or radical. Also **nascent condition**.

nase·ber·ry (nāz′ber′ē) *n. pl.* **·ries** The plumlike fruit of the sapodilla. [<Sp. *níspero* medlar]

Nase·by (nāz′bē) A village in Northamptonshire, central England; site of decisive defeat of loyalists by Cromwell, 1645.

Nash (nash), **Ogden**, born 1902, U. S. poet and humorist. — **Thomas**, 1567–1601, English pamphleteer, poet, and dramatist: also **Nashe**.

Nash·u·a (nash′oo·ə) A city in southern New Hampshire, on the Merrimac River.

Nash·ville (nash′vil) The capital of Tennessee, on the Cumberland River.

Nashville warbler The common warbler of eastern North America (*Vermivora ruficapilla*), having an olive-green back and yellow breast.

na·si·on (nā′zē·on) *n. Anat.* The point at the root of the nose where the frontal and two nasal bones meet. [<NL <L *nasus* nose] — **na′si·al** *adj.*

Nas·myth (nāz′mith, nā′smith, naz′mith), **Alexander**, 1758–1840, Scottish portrait painter.

naso- *combining form* Nose; nasal and: *nasofrontal*. [<L *nasus* nose]

na·so·fron·tal (nā′zō·frun′təl) *adj. Anat.* Of or pertaining to the nasal and frontal bones.

na·so·phar·ynx (nā′zō·far′ingks) *n. Anat.* The upper part of the pharynx, above and behind the soft palate. — **na·so·pha·ryn′ge·al** (-fə·rin′jē·əl) *adj.*

na·so·scope (nā′zō·skōp) *n.* A small electric lamp for examining the nasal cavity.

Nas·sau (nas′ô) 1 (*Ger.* nä′sou) A former duchy in western Germany, now included in Hesse. 2 A port on New Providence Island, capital of the Bahama Islands.

Nassau Range A mountain system of west central New Guinea; highest peak, 16,400 feet.

Nas·ser (näs′ər, nas′-), **Gamal Abdel**, born 1918, prime minister of Egypt 1953–56; president 1956–58; president of the United Arab Republic, a federation of Egypt, Syria, and Yemen 1958–. Real surname *Abdel-Nasser*.

Nast (nast), **Thomas**, 1840–1902, U. S. political cartoonist born in Germany.

nas·tic (nas′tik) *adj. Bot.* Pertaining to or designating an automatic response in plants whose direction and character is determined by internal cellular pressure. [<Gk. *nastos* tight-pressed]

-nastic *combining form* Nastic toward or by: *epinastic*.

na·stur·tium (na·stûr′shəm) *n.* 1 A plant of the geranium family (genus *Tropaeolum*) with funnel-shaped flowers, commonly yellow, orange, scarlet, or crimson. 2 A rich yellow or reddish–orange color. [<L, cress <*nasus* the nose + *tortus*, pp. of *torquere* twist (from the pungent odor of the plant)]

nas·ty (nas′tē) *adj.* **·ti·er**, **·ti·est** 1 Filthy or offensively dirty. 2 Morally filthy; indecent. 3 Nauseating; disgusting to the senses; disagreeable: the *nasty* task of cleaning a chicken coop. 4 Difficult to handle or deal with; vexatious; annoying: a *nasty* turn of events. 5 Painful; serious; bad: a *nasty* cut. 6 Ill-natured: a *nasty* brat; a *nasty* trick. See synonyms under FOUL. [Cf. Sw. *naskug* filthy] — **nas′ti·ly** *adv.* — **nas′ti·ness** *n.*

-nasty *combining form* A generalized automatic response to a (specified) stimulus: *epinasty*. [<Gk. *nastos* close-pressed]

na·tal (nāt′l) *adj.* 1 Pertaining to one's birth; dating from birth. 2 *Poetic* Native. [<L *natalis* <*nasci* be born]

Na·tal (nä·täl′) Portuguese and Spanish forms of NOEL.

Na·tal (nə·tal′) 1 A province of eastern Union of South Africa; 35,284 square miles, including Zululand; capital, Pietermaritzburg. 2 A port in NE Brazil, capital of Rio Grande do Norte state.

na·tal·i·ty (nə·tal′ə·tē) *n.* The birth rate in a given community or place.

Na·tan (nä·tän′) Spanish form of NATHAN.

Na·ta·na·el (nä'tä·nä·el') Spanish form of NATHANIEL.

na·tant (nā'tənt) *adj. Bot.* Floating or swimming in water, as the leaves of certain aquatic plants. [<L *natans, -antis,* ppr. of *natare* swim]

na·ta·tion (nā·tā'shən) *n.* The art of swimming or floating. [<L *natatio, -onis* < *natare* swim] — **na·ta'tion·al** *adj.*

na·ta·to·ri·al (nā'tə·tôr'ē·əl, -tō'rē-) *adj.* Swimming, or adapted for swimming. Also **na'·ta·to'ry.**

na·ta·to·ri·um (nā'tə·tôr'ē·əm, -tō'rē-) *n. pl.* **·to·ri·ums** or **·to·ri·a** (-tôr'ē·ə, -tō'rē·ə) A swimming pool.

Natch·ez (nach'iz) *n.* One of a tribe of North American Indians of Muskhogean linguistic stock, formerly occupying the lower Mississippi Valley; overcome by the French in 1729–1730; later merged with Creek.

Natch·ez (nach'iz) A port on the Mississippi River in SW Mississippi.

na·tes (nā'tēz) *n. pl.* The buttocks. [<L]

Na·than (nā'thən; *Fr.* nà·täṅ', *Ger.* nä'tän) A masculine personal name. [<Hebrew, given; a gift]
— **Nathan** A prophet who called King David to account for the death of Uriah. II *Sam.* xii 1.

Na·than (nā'thən), **George Jean,** 1882–1958, U. S. editor and dramatic critic. — **Robert,** born 1894, U. S. novelist.

Na·than·a·el (nə·than'ē·əl, -than'yəl) A disciple of Jesus. *John* xxi 2.

Na·than·i·el (nə·than'ē·əl, -than'yəl; *Du., Ger.* nä·tä'nē·el; *Fr.* nà·tà·nyel') A masculine personal name. Also **Na·than·a·el** (nə·than'ē·əl, -than'yəl). [<Hebrew, gift of God]

nathe·less (nāth'lis, nath'-) *adv. Archaic* Nevertheless. Also **nath·less** (nath'lis).

na·tion (nā'shən) *n.* **1** A people as an organized body politic, usually associated with a particular territory and possessing a distinctive cultural and social way of life. **2** An aggregation of people of common origin and language. **3** A race; tribe; specifically, a tribe of American Indians or the territory occupied by them. See synonyms under PEOPLE. [<F <L *natio, -onis* breed, race < *nasci* be born]

Na·tion (nā'shən), **Carry Amelia,** 1846–1911, *née* Moore, Kansan temperance reformer: also *Carrie Nation.*

na·tion·al (nash'ən·əl) *adj.* **1** Belonging to a nation as a whole: opposed to *local.* **2** Of, pertaining to, or characteristic of a nation as distinguished from other nations. **3** Patriotic. **4** Authorized by a national government. — *n.* One who is a member of a nation. — **na'tion·al·ly** *adv.*

national bank **1** *U.S.* A commercial bank formerly authorized by statute to issue circulating notes secured by government bonds, and now exercising the ordinary functions of a bank of deposit and discount, with double liability of shareholders for all debts. **2** A bank associated with the national finances, such as the Bank of France.

national committee A committee which heads a political party of the United States: it consists of two delegates from each State serving for four years.

national convention In the United States, a convention of representatives of a political party elected at State and Territory primaries, held to decide upon party policy and to nominate candidates for president and vice president of the nation.

national debt The debt owed by any state; especially, the funded debt.

National Guard *U.S.* An organized land or air force maintained by a State, a Territory, or the District of Columbia, usually in conjunction with but not under the direct control of the U. S. Army or Air Force. Its units or personnel operate on a semiactive basis as part of the militia except in national emergencies or under special circumstances, when they may be called into Federal service.

National Guard of the United States Those members and units of the National Guard that have taken an oath of appointment in the Federal service and are thereby constituted as a component part of the United States Army.

na·tion·al·ism (nash'ən·əl·iz'əm) *n.* **1** Devotion to the nation as a whole; patriotism. **2** A system demanding national conduct of all industries. **3** A world order founded on the right of each nation to determine its policies unhindered by others: opposed to *internationalism.* **4** A demand for national independence. **5** A national custom, trait, etc. — **na'tion·al·ist** *adj.* & *n.* — **na'tion·al·is'tic** *adj.* — **na'tion·al·is'ti·cal·ly** *adv.*

na·tion·al·i·ty (nash'ən·al'ə·tē) *n. pl.* **·ties** **1** The quality of being national; national independence. **2** A nation. **3** A connection with a particular nation, as by citizenship. **4** Patriotism.

na·tion·al·ize (nash'ən·əl·īz) *v.t.* **·ized, ·iz·ing** **1** To place under the control or ownership of a nation. **2** To give a national character to. **3** To make into a nation. Also *Brit.* **na'tion·al·ise.** — **na'tion·al·i·za'tion** *n.* — **na'tion·al·iz'er** *n.*

National Labor Relations Board A U. S. government board established in 1935 to insure the rights of employees to self-organization and bargaining through the enforcement of codes detailing unfair labor practices and through the conducting of investigations and bargaining elections.

National Military Establishment See DEPARTMENT OF DEFENSE under DEFENSE.

national park A tract of U. S. government land withdrawn by special Act of Congress from settlement, occupancy, or sale, for the benefit and enjoyment of the public: preserved and maintained by the Federal government because of its historical interest, great natural beauty, or the value of its forests, wildlife, etc.

National Road See CUMBERLAND ROAD.

National Socialism The doctrines of the Nazi party. See NAZI.

na·tion-wide (nā'shən·wīd') *adj.* Throughout the entire nation.

na·tive (nā'tiv) *adj.* **1** Born or produced in a region or country in which one lives; indigenous: opposed to *foreign, exotic.* **2** Of or pertaining to one's birth or to its place or circumstances. **3** Natural rather than acquired; inborn; inherited. **4** Of or pertaining to natives; conferred by or peculiar to natives: usually applied to non-European peoples. **5** Natural to any one or any thing. **6** Plain, simple, unaffected, unadorned; untouched by art. **7** Occurring in nature in a pure state: *native* copper. **8** *Obs.* Related by birth; near; closely connected. — *n.* **1** One born in, or any product of, a given country or place; an aborigine. **2** Livestock common to a country or region. **3** In astrology, one born under a star or its aspect. [<F *natif* <L *nativus* < *nasci* be born. Doublet of NAIVE.] — **na'tive·ly** *adv.* — **na'tive·ness** *n.*

Synonyms (adj.): indigenous, innate, natal, natural, original. *Native* denotes that which belongs to one by birth; *natal* that which pertains to the event of birth; *natural* denotes that which rests upon inherent qualities of character or being. We speak of one's *native* country, or of his *natal* day; of *natural* ability, *native* genius. See INHERENT, NATURAL, PRIMEVAL, RADICAL. *Antonyms:* acquired, alien, artificial, assumed, foreign.

na·tive-born (nā'tiv·bôrn') *adj.* Born in the region or country specified.

Native States See INDIAN STATES.

na·tiv·ism (nā'tiv·iz'əm) *n.* **1** Partiality in favor of native-born citizens in preference to foreign-born. **2** The doctrine of innate ideas. — **na'tiv·ist** *n.* — **na'tiv·is'tic** *adj.*

na·tiv·i·ty (nā·tiv'ə·tē, nə-) *n. pl.* **·ties** **1** The coming into life or the world; birth. **2** A horoscope. **3** The condition of being born a serf or villein. **4** The condition of being a native. — **the Nativity** The birth of Jesus.

NATO (nā'tō) North Atlantic Treaty Organization.

na·tri·um (nā'trē·əm) *n.* Sodium: so called in pharmacy and formerly in chemistry. [<NL <F *natron* NATRON]

nat·ro·lite (nat'rə·līt, nā'trə-) *n.* A white or colorless, orthorhombic, hydrous sodium-aluminum zeolite occurring in prismatic, needlelike crystals: also called *needlestone.* [<NATRON + -LITE]

na·tron (nā'tron) *n.* A brittle, vitreous, white, alkaline, hydrous sodium carbonate, $Na_2CO_3 \cdot 10H_2O$, crystallizing in the monoclinic system. [<F <Sp. *natrón* <Arabic *naṭrūn, niṭrūn* <Gk. *nitron* niter]

nat·ty (nat'ē) *adj.* **·ti·er, ·ti·est** Smart or fine; trim; spruce; tidy. See synonyms under NEAT[1]. [? Akin to NEAT[1]] — **nat'ti·ly** *adv.* — **nat'ti·ness** *n.*

Na·tu·na (nə·to͞o'nə) See BUNGURAN ISLANDS.

nat·u·ral (nach'ər·əl) *adj.* **1** Of or pertaining to one's nature or constitution; innate; inborn; also, indigenous; native. **2** Of or pertaining to a particular nature; derived from nature; hence, exhibiting kindly feeling or affection. **3** Of or pertaining to nature; belonging or pertaining to the existing order of things: *natural* law; normal. **4** Coming within common experience; having to do with objects in the order of nature: opposed to *supernatural.* **5** Not forced or artificial; without affectation or exaggeration; lifelike. **6** Produced by nature; not artificial: a *natural* bridge. **7** Connected by ties of consanguinity; being such by birth: a *natural* brother. **8** Belonging to the inferior nature; not spiritual; animal. **9** Born out of wedlock; illegitimate. **10** *Music* Not sharped nor flatted: G *natural;* specifically, denoting the key of C, which is without flats or sharps in the signature. **11** *Math.* Designating an actual number in contradistinction to a logarithm: a *natural* sine, *natural* cosine, *natural* tangent, etc. See synonyms under INHERENT, NATIVE, NORMAL, PHYSICAL, RADICAL. — *n.* **1** *Music* **a** A note on a line or a space that is affected by neither a sharp nor a flat. **b** A character (♮) which acts upon a sharped degree of the staff as a flat and upon a flatted degree as a sharp. **2** In keyboard musical instruments, a white key. **3** One born without the usual powers of reason or understanding; a born fool. **4** *Colloq.* A person or thing admirably suited for some purpose, or obviously destined for success. [<F *naturel* <L *naturalis* < *natura* nature] — **nat'u·ral·ness** *n.*

Natural Bridge A rock-and-earth natural bridge over Cedar Creek in western Virginia; span, 90 feet; width, 50–100 feet; height, 215 feet.

Natural Bridges National Monument A region in SE Utah containing three natural bridges; 4 square miles; established, 1908.

natural gas Any gaseous hydrocarbon, consisting chiefly of methane, generated naturally in subterranean oil deposits: used as a fuel.

natural gender See under GENDER.

natural history The observation and study of the facts of the material universe as distinguished from man: commonly restricted to zoology, botany, geology, mineralogy, etc.

nat·u·ral·ism (nach'ər·əl·iz'əm) *n.* **1** Action or thought derived from or identified with exclusively natural desires and instincts. **2** In literature, art, etc.: **a** Adherence to observed nature; specifically, the principles of Zola, de Maupassant, and others who attempted to apply "scientific" objectivity to their treatment of life, without imposing judgments of value or avoiding what is considered ugly. **b** The qualities of a work of art resulting from such doctrines. **3** *Philos.* The doctrine that phenomena are derived from natural causes and can be explained by scientific laws: opposed to *supernaturalism.* **4** *Theol.* The doctrine that religion does not depend on supernatural revelation, but may be derived from the natural world.

nat·u·ral·ist (nach'ər·əl·ist) *n.* **1** One versed in natural sciences, as a zoologist or botanist. **2** One who holds the philosophical doctrines of naturalism.

nat·u·ral·is·tic (nach'ər·əl·is'tik) *adj.* **1** In accordance with nature; not conventional or ideal. **2** According to the doctrines of naturalism. **3** Pertaining to naturalists.

nat·u·ral·i·za·tion (nach'ər·əl·i·zā'shən, -ī·zā'-) *n.* The act or process of admitting an alien to citizenship.

naturalization papers Documents recording an alien's application for citizenship or verifying the conferment of citizenship.

nat·u·ral·ize (nach'ər·əl·īz) *v.* **·ized, ·iz·ing** *v.t.* **1** To confer the rights and privileges of citizenship upon, as an alien. **2** To adopt (a foreign word, custom, etc.) into the common use of

add,āce,câre,pälm; end, ēven; it,īce; odd,ōpen,ôrder; to͝ok,po͞ol; up,bûrn; ə = a in *above,* e in *sicken,* i in *clarity,* o in *melon,* u in *focus;* yo͞o = u in *fuse;* oi,oil; ou,pout; ch,check; g,go; ng,ring; th,thin; th,this; zh,vision. Foreign sounds à,œ,ü,kh,ṅ; and ◆: see page xx. < from; + plus; ? possibly.

a country or area. 3 To adapt (a foreign plant, animal, etc.) to the environment of a country or area. 4 To explain by natural laws: to *naturalize* a miracle. 5 To make natural; free from conventionality. —*v.i.* 6 To become as if native; adapt. Also *Brit.* **nat′u·ral·ise.** —**nat′u·ral·iz′er** *n.*

natural law A rule of conduct supposed to be inherent in man's nature and discoverable by reason alone.

natural logarithms Napierian logarithms.

nat·u·ral·ly (nach′ər·əl·ē) *adv.* 1 Without effort; spontaneously. 2 Without affectation or exaggeration. 3 As might have been expected; of course. 4 In a lifelike or natural manner.

natural philosophy 1 Natural history. 2 The physical sciences taken collectively.

natural resource Any of those sources of wealth provided by nature, as soil, forests, minerals, water supply, water power, and wild game.

natural sciences The sciences treating of the physical universe, taken collectively and in distinction from the mental and moral sciences and from abstract mathematics.

natural selection *Biol.* The process whereby individual variations of peculiarities that are of advantage in a certain environment tend to become perpetuated in the race; survival of the fittest.

na·ture (nā′chər) *n.* 1 The character, constitution, or essential traits of a person, thing, or class, especially if original rather than acquired. 2 The physical or psychic constitution or character of persons or things, whether native or acquired: often personified in poetry or figurative prose. 3 The entire material universe and its phenomena. 4 The system of natural existences, forces, changes, and events, regarded as distinguished from, or exclusive of, the supernatural: Man is included in *nature.* 5 The sum of physical or material existences and forces in the universe. 6 The constitution or inherited or habitual condition and tendencies of man. 7 *Theol.* The unregenerate state; character unchanged by grace. 8 *Obs.* Generative energy; genesis; birth. See synonyms under CHARACTER, SORT, TEMPER. [<F <L *natura* < *natus,* pp. of *nasci* be born]

-natured *combining form* Possessing a (specified) nature, disposition, or temperament: ill-*natured.* [<NATURE (def. 1)]

na·tur·op·a·thy (nā′chə·rop′ə·thē) *n. Med.* A system of therapy which avoids drugs in favor of such physical agencies as sunshine, air, water, exercise, etc. [<NATURE + -(O)PATHY] —**na·tur′o·path** (nə·chŏŏr′ə·path) *n.* —**na·tur′·o·path′ic** *adj.*

Nau·cra·tis (nô′krə·tis) An ancient Greek city in the Nile delta, Egypt. Also **Nau′kra·tis.**

naught (nôt) *n.* 1 Not anything; nothing. 2 A cipher; zero; the character 0. —*adj.* 1 Of no value or account. 2 *Obs.* Bad; wicked; also, poor in quality. —*adv.* Not in the least. Also spelled *nought.* [OE *nāwiht* < *nā* not + *wiht* thing]

naugh·ty (nô′tē) *adj.* **·ti·er, ·ti·est** 1 Perverse and disobedient; wayward; mischievous. 2 *Obs.* Corrupt; wicked. See synonyms under BAD. [<NAUGHT] —**naugh′ti·ly** *adv.* —**naugh′·ti·ness** *n.*

nau·ma·chy (nô′mə·kē) *n. pl.* **·chies** 1 In ancient Rome, a fight between ships for the amusement of the people; also, an artificial basin for the convenience of such battles. 2 A naval battle; especially, a mock sea fight. Also **nau·ma·chi·a** (nô·mā′kē·ə). [<L <Gk. *naumachia* sea fight < *naus* ship + *machē* fight]

Nau·pak·tos (nô·pak′təs, näf′päk·tôs) A town on the northern shore of the Gulf of Corinth, west central Greece: Italian *Lepanto*: also *Navpaktos.*

nau·path·i·a (nô·path′ē·ə) *n.* Seasickness. [<NL <Gk. *naus* a ship + *path-,* stem of *paschein* suffer]

nau·pli·us (nô′plē·əs) *n. pl.* **·pli·i** (-plē·ī) *Zool.* A larval stage of certain crustaceans, with body unsegmented, a median eye, and three pairs of legs which correspond to the anterior and posterior antennae and the mandibles of the adult. [<L, a kind of shellfish]

Na·u·ru (nä·ŏŏ′rŏŏ) A Pacific island just south of the equator and west of the Gilbert Islands, comprising a United Nations Trust Territory administered by Australia; 8 square miles: formerly *Pleasant Island.*

nau·sea (nô′zhə, -zē·ə, -shə, -sē·ə) *n.* 1 Sickness of the stomach, producing dizziness and an impulse to vomit. 2 A feeling of loathing in general. [<L <Gk. *nausia* < *naus* ship]

nau·se·ate (nô′zhē·āt, -zē, -shē-, -sē-) *v.t.* & *v.i.* **·at·ed, ·at·ing** To affect with or feel nausea or disgust. —**nau′se·a′tion** *n.*

nau·seous (nô′zhəs, -shəs) *adj.* 1 Nauseating; disgusting. 2 *Colloq.* Affected with nausea; queasy. —**nau′seous·ly** *adv.* —**nau′·seous·ness** *n.*

Nau·sic·a·a (nô·sik′ā·ə, -i·ə) In the *Odyssey,* the daughter of King Alcinous, who by the contrivance of Athena finds the shipwrecked Odysseus and guides him to her father, from whom he receives aid to return to Ithaca.

nautch (nôch) *n.* In India, an entertainment featuring dancing girls (**nautch girls**). [<Hind. *nāch* dance]

nau·ti·cal (nô′ti·kəl) *adj.* Pertaining to ships, seamen, or navigation. Also **nau′tic.** [<Gk. *nautikos* < *naus* ship] —**nau′ti·cal·ly** *adv.*

Synonyms: marine, maritime, naval, ocean, oceanic. *Marine* signifies belonging to the ocean, *maritime* bordering on or connected with the ocean; as, *marine* products; *marine* animals; *maritime* nations; *maritime* laws. *Naval* refers to the armed force of a nation on the sea, and on lakes and rivers; *nautical* denotes primarily anything connected with sailors, or with ships or navigation; as, a *nautical* almanac. *Oceanic* is especially applied to that which is connected with or suggestive of an *ocean.*

nautical mile See under MILE.

nau·ti·lus (nô′tə·ləs) *n. pl.* **·lus·es** or **·li** (-lī) 1 A cephalopod mollusk (genus *Nautilus*) of southern seas, having a shell whose spiral chambers are lined with mother-of-pearl; especially, the **chambered** or **pearly nautilus** (*N. pompilius*). 2 The argonaut. [<L < Gk. *nautilos* sailor]

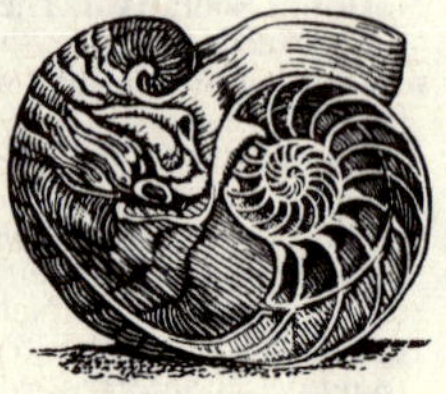

CHAMBERED NAUTILUS
Cross-section
(The shell up to 10 inches across)

Nautilus The first atomic-powered submarine, launched by the U. S. Navy: made the first undersea crossing of the North Pole on August 3, 1958.

Nav·a·ho (nav′ə·hō) *n. pl.* **·hos** or **·hoes** One of a tribe of North American Indians of Athapascan stock, now living on reservations in Arizona, New Mexico, and Utah. Also **Nav′a·jo.**

Nav·a·jo Mountain (nav′ə·hō) A peak in southern Utah; 10,416 feet.

na·val (nā′vəl) *adj.* 1 Pertaining to ships and a navy: distinguished from *civil.* 2 Having a navy; relating to the navy. See synonyms under NAUTICAL. ◆ Homophone: *navel.* [<L *navalis* < *navis* ship]

naval academy A school where men are trained as naval officers; specifically, the U. S. Naval Academy at Annapolis, Maryland.

naval auxiliary A launch or auxiliary vessel, as a tanker.

naval brass *Metall.* A type of brass containing a small percentage of tin to increase hardness and resistance to corrosion: used for marine fittings, etc.

naval stores Rosin and its products, as turpentine, pine oil, etc.; also, tar, pitch, asphalt, and other similar materials formerly or still used by shipbuilders.

na·var (nā′vär) *n.* A method of improving the efficiency of aircraft navigation by the use of an interlocking system of radar beacons operating from ground stations with receivers in aircraft. [<NAV(IGATION) A(ND) R(ANGING)]

Na·va·ri·no (nä′vä·rē′nō) The medieval name for PYLOS.

Na·varre (nə·vär′, *Fr.* nà·vàr′) 1 A region and former kingdom in northern Spain and SW France. 2 One of the Basque provinces, in northern Spain; 4,024 square miles; capital, Pamplona. *Spanish* **Na·var·ra** (nä·vär′rä).

nave[1] (nāv) *n. Archit.* The main body of a cruciform church, between the side aisles, and usually having a clerestory. ◆ Homophone: *knave.* [<OF <L *navis* ship]

nave[2] (nāv) *n.* 1 The central part or hub of a wheel. 2 *Obs.* The navel. ◆ Homophone: *knave.* [OE *nafu*]

na·vel (nā′vəl) *n.* 1 The depression on the abdomen where the umbilical cord was attached. 2 A central part or point. ◆ Homophone: *naval.* [OE *nafela* < *nafu* nave[2]]

na·veled (nā′vəld) *adj.* Having a navel; set as in a navel or hollow. Also **na′velled.**

navel orange An orange, usually seedless, having a small secondary fruit and a rind marked at the apex with a navel-like depression.

na·vel·wort (nā′vəl·wûrt′) *n.* 1 A low herb (genus *Omphalodes*) of the borage family, native in Europe and Asia, with alternate leaves and blue or white flowers resembling forget-me-nots: also **na′vel·seed′.** 2 A succulent herb (*Umbilicus pendulinus*) with yellowish or greenish tubular flowers. 3 Any of various other plants, as the pennywort or the water milfoil.

nav·i·cert (nav′ə·sûrt) *n. Brit.* A safe-conduct authorizing a merchant vessel of a friendly or neutral nation to pass through a naval blockade. [<L *navis* ship + CERT(IFICATE)]

na·vic·u·lar (nə·vik′yə·lər) *adj.* 1 Boat-shaped; scaphoid. 2 Pertaining to a boat. —*n. Anat.* 1 A bone on the radial side of the wrist; also, the bone in front of the astragalus in the foot: also **na·vic′u·la′re** (-lâr′ē). See illustration under FOOT. 2 A large bone behind the joint between the second and third phalanges of a horse's foot. [<LL *navicularis* <L *navis* ship]

na·vic·u·lar·thri·tis (nə·vik′yə·lär·thrī′tis) *n.* Inflammation of the navicular bone of the foot of a horse. Also **navicular disease.**

nav·i·gate (nav′ə·gāt) *v.* **·gat·ed, ·gat·ing** *v.t.* 1 To travel over, across, or on by ship or aircraft. 2 To steer; direct the course of. —*v.i.* 3 To travel by means of ship or aircraft. 4 To steer or manage a ship or aircraft. 5 To plot a course for a ship or aircraft. [<L *navigatus,* pp. of *navigare* < *navis* a boat + *agere* drive] —**nav·i·ga·ble** (nav′ə·gə·bəl) *adj.* —**nav′i·ga·bly** *adv.* —**nav′i·ga·bil′i·ty, nav′i·ga·ble·ness** *n.*

nav·i·ga·tion (nav′ə·gā′shən) *n.* 1 The act of navigating. 2 The art of ascertaining the position and directing the course of vessels at sea or of aircraft in flight. —**nav′i·ga′tion·al** *adj.*

nav·i·ga·tor (nav′ə·gā′tər) *n.* 1 One who navigates, or directs the course of a ship, aircraft, etc. 2 A person skilled in navigation.

Navigators' Islands See SAMOA.

Nav·pak·tos (näf′päk·tôs). See NAUPAKTOS.

nav·vy (nav′ē) *n. pl.* **·vies** *Brit.* A laborer on canals, railways, etc. [<NAVIGATOR, in obsolete sense of "one employed in digging canals"]

na·vy (nā′vē) *n. pl.* **·vies** 1 The entire marine military force of a country, under the control of a government department, and including vessels, men in the service, yards, etc. 2 The entire shipping of a country engaged in trade and commerce; the merchant marine. 3 A fleet of ships, as of merchantmen. —**United States Navy** The U. S. naval force administered by the Department of the Navy under the Department of Defense and including the Regular Navy, the Naval Reserve, the United States Marine Corps, and the United States Coast Guard when operating as a component of the Navy. [<OF *navie* <L *navis* ship]

navy bean The common, small, dried, white bean: so called from its common use in the U. S. Navy.

navy blue Any of various shades of dark blue. Also **navy.** —**na′vy-blue′** *adj.*

Navy Cross A decoration in the form of a bronze cross awarded by the U. S. Navy for extraordinary heroism or service in war.

navy gray A medium bluish-gray: adopted in World War II for the color of work uniforms of officers in the U. S. Navy. —**na′vy-gray′** *adj.*

navy yard A government-owned dockyard for the construction, repair, equipment, or care of warships.

na·wab (nə·wôb′) *n. Anglo-Indian* A Moslem ruler or viceroy in India; by courtesy, any person of rank and distinction. See NABOB. [< Hind. *nawwāb* nabob]

Nax·os (nak′səs) The largest island of the Cyclades group; 169 square miles. Formerly **Nax·i·a** (nak′sē·ə).

nay (nā) *adv.* **1** No: indicating negation. **2** Not only so, but also: He is a good, *nay,* an excellent man. — *n.* **1** A negative vote or voter: opposed to *yea.* **2** A negative; denial. [<ON *nei* < *ne* not + *ei* ever]

Na·ya·rit (nä′yä·rēt′) A maritime state in western Mexico; 10,547 square miles; capital, Tepic.

nay·ma·cush (nā′mə·kush) See NAMAYCUSH.

Naz·a·rene (naz′ə·rēn) *n.* **1** An inhabitant of Nazareth: applied specifically to Christ, **the Nazarene,** or disparagingly to the early Christians by opponents. **2** One of a sect of Jewish Christians (first to fourth century) that observed the Jewish ritual, but did not require its observance by Gentile Christians: they believed in the divinity of Christ and the apostleship of Paul. — *adj.* Of or pertaining to Nazareth or the Nazarenes. Also **Naz′a·re′an** (-rē′ən).

Naz·a·reth (naz′ə·rith) An ancient town in Lower Galilee, northern Israel; scene of Christ's childhood.

Naz·a·rite (naz′ə·rīt) *n.* **1** A Hebrew who had assumed certain vows, including total abstinence, leaving the hair uncut, and refraining from touching a dead body. *Numbers* vi. **2** Erroneously, a Nazarene. Also **Naz′i·rite.** [<Gk. *Nazaritēs* <Hebrew *nāzar* abstain] — **Naz′a·rit′ic** (-rit′ik), **Naz′i·rit′ic** *adj.*

Naze (nāz), **The 1** A cape at the southern extremity of Norway: Norwegian *Lindesnes.* **2** A headland on the NE coast of Essex, England.

Na·zi (nä′tsē, nat′sē, na′zē) *n.* A member of the National Socialist German Workers Party, founded in 1919 on fascist principles and dominant from 1933 to 1945 in Germany under the dictatorship of Hitler, where it followed the principles of extreme nationalism, racism, totalitarian direction of all cultural, political, and economic activity, and militarization, while urging a destiny of world leadership for Germany. — *adj.* Of or pertaining to the Nazis or their party. [<G, short for *Nationalsozialistische (Partei)* National Socialist (Party)]

Na·zi·fy (nä′tsə·fī, nat′sə-) *v.t.* **·fied, ·fy·ing** To subject to Nazi influence or control; cause to be Nazi-like. — **Na′zi·fi·ca′tion** *n.*

Na·zim·o·va (nə·zim′ə·və), **Alla,** 1879–1945, Russian actress active in the United States.

Na·zism (nä′tsiz·əm, nat′siz-) *n.* The doctrines or practices of the Nazi party. Also **Na·zi·ism** (nä′tsē·iz′əm, nat′sē-).

Ne·an·der·thal man (nē·an′dər·täl, -thôl, -thol; *Ger.* nā·än′dər·täl) A relatively advanced protohuman species *(Homo neanderthalensis)* first identified from fragments of a fossil skeleton discovered in 1856 in cave deposits of the Neander valley near Düsseldorf. It is regarded as typical of a race of ancient cave-dwellers who preceded modern man, developing the Mousterian stone culture in western Europe. [<G *Neanderthal* Neander valley]

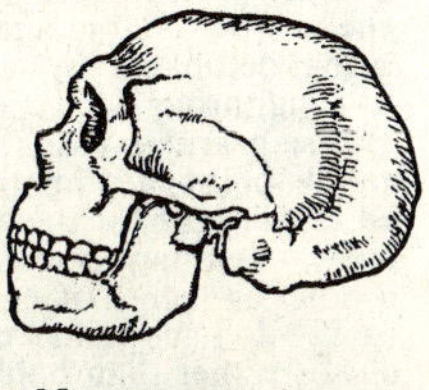

NEANDERTHAL SKULL

neap[1] (nēp) *adj.* Designating the tide occurring one or two days after the first and third quarters of the moon. — *n.* **1** A neap tide. **2** The lowest ebb. [OE *nēp-* in *nēpflod* low tide]

neap[2] (nēp) *n. U.S. Dial.* A wagon tongue. [? <Scand. Cf. dial. Norw. *neip* forked pole.]

Ne·ap·o·lis (nē·ap′ə·lis) **1** The Greek name for SHECHEM, Samaria. **2** A Greek name for NAPLES.

Ne·a·pol·i·tan (nē′ə·pol′ə·tən) *adj.* Of or pertaining to Naples. — *n.* A native or resident of Naples.

near (nir) *adj.* **1** Not distant in place, time, or degree; nigh; contiguous. **2** Closely related by blood or affection; familiar. **3** Closely touching one's interests. **4** In riding or driving, placed on the left: opposed to *off.* **5** Following or imitating closely; literal: a *near* copy; also, resembling or substituted for: *near* beer. **6** Penurious or miserly; stingy. **7** Short or speedy; tending to lessen a distance: a *near* way. **8** Avoiding by a narrow margin; close: a *near* escape. See synonyms under ADJACENT. — *adv.* **1** At little distance; not remote in place, time, or degree. **2** Nearly; almost; approximately. **3** In a close relation; intimately. **4** *Naut.* Close to the wind. **5** Stingily; parsimoniously. — *v.t.* & *v.i.* To come or draw near (to); approach. — *prep.* Close by or to. See synonyms under AT. [OE *nēar,* comp. of *nēah* NIGH] — **near′ness** *n.*

near beer Any imitation beer of little or no alcoholic content.

near·by (nir′bī′) *adj.* & *adv.* Close at hand; adjacent.

Ne·arc·tic (nē·ärk′tik) *adj.* Of or pertaining to a zoogeographic region including the northern part of the New World, or the realm embracing temperate and arctic North America with Greenland. [<NE(O)- + ARCTIC]

Near East See under EAST.

near-hand (nir′hand′) *Scot.* & *Brit. Dial. adj.* Close at hand; adjacent. — *adv.* **1** Nearby. **2** Nearly; almost.

Near Islands (nir) The westernmost group of the Aleutian Islands.

near·ly (nir′lē) *adv.* **1** Within a little; almost. **2** With a close regard to one's interest. **3** At no great distance; closely; narrowly. **4** Stingily.

near rime In prosody, a more or less radical substitute for rime, including such devices as assonance and consonance: also called *paraphone, half rime, oblique rime.*

near-sight·ed (nir′sī′tid) *adj.* Able to see distinctly at short distances only; myopic. — **near′-sight′ed·ly** *adv.* — **near′-sight′ed·ness** *n.*

neat[1] (nēt) *adj.* **1** Characterized by strict order, tidiness, and cleanliness. **2** Well-proportioned; trim; shapely. **3** Suited in character to a required purpose; hence, adroit; clever. **4** Clear of extraneous matter; free from admixture; undiluted: a glass of brandy *neat.* **5** Remaining after every deduction; net. [<AF *niet,* OF *net* <L *nitidus* shining] — **neat′ly** *adv.* — **neat′ness** *n.*

Synonyms: clean, cleanly, dapper, natty, nice, orderly, prim, spruce, tidy, trim. That which is *clean* is simply free from soil or defilement of any kind. Things are *orderly* in relation to other things; a room or desk is *orderly* when every article is in place; a person is *orderly* who habitually keeps things so. *Tidy* denotes that which conforms to propriety in general; an unlaced shoe may be perfectly *clean,* but is not *tidy. Neat* refers to that which is *clean* and *tidy,* with nothing superfluous, conspicuous, or showy; we speak of plain but *neat* attire; the same idea of freedom from the superfluous appears in the phrases "a *neat* speech," "a *neat* turn," "a *neat* reply," etc. A *clean* cut has no ragged edges; a *neat* stroke just does what is intended. *Nice* is stronger than *neat,* implying value and beauty; a cheap, coarse dress may be perfectly *neat,* but would not be termed *nice. Spruce* is applied to the show and affectation of neatness with a touch of smartness. *Trim* denotes a certain shapely and elegant firmness, often with suppleness and grace. *Prim* applies to a precise, formal, affected nicety. *Dapper* is *spruce* with the suggestion of smallness and slightness; *natty,* a diminutive of *neat,* suggests minute elegance, with a tendency toward the exquisite; as, a *dapper* man in a *natty* business suit. See BECOMING. *Antonyms*: dirty, disorderly, dowdy, negligent, rough, rude, slouchy, slovenly, soiled, untidy.

neat[2] (nēt) *n. Obs.* **1** Bovine cattle collectively. **2** A single bovine animal. — *adj.* Pertaining to bovine animals: *neat* cattle. [OE *nēat*]

'neath (nēth) *prep. Poetic* Beneath.

neat·herd (nēt′hûrd′) *n.* A herdsman.

neat's-foot oil (nēts′fo͝ot′) A pale yellow oil obtained by boiling the feet of neat cattle: used as a lubricant and softening agent for leather.

neb (neb) *n.* **1** The beak or bill, as of a bird; nose; snout. **2** The tip end of a thing; nib, as of a pen. **3** *Scot.* The face; also, the mouth. [OE *nebb*]

Neb·i·im (neb′i·ēm′, *Hebrew* nə·vē′im) *n. pl.* The second of the three divisions of the Hebrew scriptures known as the Prophets, as distinguished from the Law (Torah), etc.

Ne·bo (nē′bō), **Mount** A mountain in Moab from which Moses saw the Promised Land. *Deut.* xxxii 49. See PISGAH.

Ne·bras·ka (nə·bras′kə) A State in the north central United States; 77,237 square miles; capital, Lincoln; entered the Union Feb. 9, 1867; nickname, *Tree Planter State*: abbr. *Neb., Nebr.* — **Ne·bras′kan** *adj.* & *n.*

Neb·u·chad·nez·zar (neb′yo͞o·kəd·nez′ər) A Chaldean king of Babylon, 604?–561? B.C.; conquered Jerusalem and destroyed the Temple. *Dan.* i 1. Also **Neb′u·chad·rez′zar** (-rez′ər).

neb·u·la (neb′yə·lə) *n. pl.* **·lae** (-lē) or **·las 1** *Astron.* A luminous celestial body of cloudlike appearance and vast extent, composed of gaseous or stellar matter in various degrees of density. **2** *Pathol.* **a** A speck on the cornea; visual opacity. **b** Cloudiness of the urine. [<L, vapor, mist] — **neb′u·lar** *adj.*

nebular hypothesis *Astron.* A hypothesis that the solar system existed originally in the form of a nebula which, by cooling, condensing, and revolving, was formed into the sun and into rings of matter which later were consolidated into the planetary bodies.

neb·u·lize (neb′yə·līz) *v.t.* **·lized, ·liz·ing 1** To spray, as a wound or a morbid surface, with medicated liquid. **2** To reduce to a spray; atomize. — **neb′u·li·za′tion** *n.* — **neb′u·liz′er** *n.*

neb·u·lose (neb′yə·lōs) *adj.* Cloudlike; clouded.

neb·u·los·i·ty (neb′yə·los′ə·tē) *n.* **1** Nebulousness. **2** A misty or nebulous appearance; also, nebulous matter; a nebula.

neb·u·lous (neb′yə·ləs) *adj.* **1** Having its parts confused or mixed; hazy; indistinct: a *nebulous* idea. **2** Like a nebula. — **neb′u·lous·ly** *adv.* — **neb′u·lous·ness** *n.*

nec·es·sar·i·an·ism (nes′ə·sâr′ē·ən·iz′əm) *n.* The philosophical doctrine that acts of volition are predetermined by the force of inner motives; determinism; fatalism; necessity. Also **ne·ces′si·tar′i·an·ism.** — **nec′es·sar′i·an, ne·ces·si·tar·i·an** (nə·ses′ə·târ′ē·ən) *adj.* & *n.*

nec·es·sar·y (nes′ə·ser′ē) *adj.* **1** Being such in its nature or conditions that it must exist, occur, or be true; inevitable. **2** Absolutely needed to accomplish a desired result; essential; requisite. **3** Compulsory: a *necessary* action. **4** Being such that it must be believed. **5** *Archaic* Rendering useful and intimate service. — *n. pl.* **·sar·ies 1** That which is indispensable; an essential requisite: used usually in the plural: the *necessaries* of life. **2** That which is subject to the law of necessity: The *necessary* is opposed to the contingent. **3** A watercloset; privy. [<L *necessarius* necessary] — **nec′es·sar′i·ly** *adv.* — **nec′es·sar′i·ness** *n.*

Synonyms (adj.): compulsory, essential, indispensable, inevitable, needed, needful, required, requisite, unavoidable, undeniable. That which is *essential* belongs to the essence of a thing, so that the thing cannot exist in its completeness without it; that which is *indispensable* may be only an adjunct, but it is one that cannot be spared. That which is *requisite* is so in the judgment of the person requiring it. Food is *necessary,* death is *inevitable*; a *necessary* conclusion satisfies a thinker; an *inevitable* conclusion silences opposition. *Needed* and *needful* are more concrete than *necessary,* and respect an end to be attained, while *necessary* may apply simply to what exists; we speak of a *necessary* inference; *necessary* food is what one cannot live without, while *needful* food is that without which we cannot enjoy comfort, health, and strength. *Antonyms*: casual, contingent, needless, nonessential, optional, unnecessary, useless, worthless.

ne·ces·si·tate (nə·ses′ə·tāt) *v.t.* **·tat·ed, ·tat·ing 1** To make necessary, unavoidable, or certain. **2** To compel. See synonyms under COMPEL. [<Med. L *necessitatus,* pp. of *necessitare* compel < *necessitas, -tatis* necessity] — **ne·ces′si·ta′tion** *n.* — **ne·ces′si·ta′tive** *adj.*

ne·ces·si·tous (nə·ses′ə·təs) *adj.* Extremely needy; destitute; poverty-stricken. — **ne·ces′si·tous·ly** *adv.* — **ne·ces′si·tous·ness** *n.*

ne·ces·si·ty (nə·ses′ə·tē) *n. pl.* **·ties 1** The quality of being necessary: Food is a *necessity* for growth. **2** That which is unavoidable or necessary, as in physical, moral, or logical sequence; a state of things rendering something inevitable. 3 That which is indispensably requisite to an end desired: often in the plural: the *necessities* of life; also, the fact of being indispensable; indispensableness. 4 The condition of being in want; poverty. 5 The doctrine that all events are necessarily determined. It embraces physical determinism, or fatalism, and philosophical or rational determinism, thus precluding chance or free will. — **of necessity** Necessarily; unavoidably. [<L *necessitas, -tatis*]

Synonyms: compulsion, destiny, emergency, essential, exigency, extremity, fatality, fate, indispensability, indispensableness, need, requirement, requisite, unavoidableness, urgency, want. An *essential* is something, as a quality or element, that belongs to the essence of something else so as to be inseparable from it in its normal condition, or in any complete idea or statement of it. *Need* and *want* always imply a lack; but *necessity* simply denotes the exclusion of any alternative either in thought or fact. See PREDESTINATION, WANT. Compare NECESSARY. *Antonyms*: choice, contingency, doubt, doubtfulness, dubiousness, fortuity, freedom, option, possibility, uncertainty.

Nech·es River (nech′iz) A river in eastern Texas, flowing 416 miles SE to the head of Sabine Lake.

neck (nek) *n.* **1** *Anat.* **a** The part of an animal that connects the head with the trunk. **b** Any similarly constricted portion of an organ or part: the *neck* of the femur. ◆ Collateral adjective: *cervical.* **2** The narrowed part of an object, particularly if near one end. **3** Something likened to a neck, from its position, shape, etc. **4** The narrow part of a bottle. **5** A narrow passage of water connecting two larger bodies. **6** A peninsula, isthmus, or cape. **7** That part of a garment which is close to the neck. **8** That part of a stringed musical instrument of the banjo class between the head and the body, and bearing the frets, if any. **9** *Archit.* The upper part of the shaft of a column, immediately below the capital. **10** The diminished part of a shaft, axle, etc., where it rests in a bearing. — *v.i.* **1** *U.S. Slang* To make love by kissing and caressing. — *v.t.* **2** *U. S. Slang* To make love to (someone) in such a manner. **3** To behead or strangle, as a chicken. [OE *hnecca*]

neck and neck Keeping evenly abreast; keeping up with each other.

Neck·ar (nek′är) A river in Baden-Württemberg, southern West Germany, flowing 228 miles NE, NW, and west from the Black Forest to the Rhine at Mannheim.

neck·band (nek′band′) *n.* **1** The part of a garment that fits around the neck: the *neckband* of a shirt or dress. **2** A band around the neck.

neck·cloth (nek′klôth′) *n.* A folded cloth worn around the neck and collar; a cravat. Compare STOCK (def. 24).

Neck·er (nek′ər, *Fr.* ne·kâr′), **Jacques,** 1732–1804, French statesman; minister of finance of Louis XVI; born in Switzerland.

neck·er·chief (nek′ər·chif) *n.* A kerchief for the neck.

neck·ing (nek′ing) *n.* **1** *Archit.* An ornamental treatment, as a sculptured band, a hollow, etc., of the neck of a column; also, a gorgerin. **2** Any necklike stem. **3** *Slang* Kissing and caressing in lovemaking.

neck·lace (nek′lis) *n.* An ornament, as of precious stones, precious metal, beads, or the like, worn around the neck.

neck of the woods *U.S. Colloq.* A region or neighborhood.

neck-or-noth·ing (nek′ôr·nuth′ing) *adj.* Risking everything; desperate.

neck·tie (nek′tī′) *n.* **1** A band or scarf passing round the neck or collar and tying in front under the chin; any bow or tie worn under the chin. **2** *U.S. Slang* A halter. — **necktie sociable** or **party** *U.S. Slang* A lynching.

neck·wear (nek′wâr′) *n.* **1** Any article worn around the throat. **2** Ties, cravats, collars, mufflers, etc., collectively.

neck·yoke (nek′yōk′) *n.* **1** A yoke for the neck. **2** A crosspiece to connect the forward end of the tongue of a vehicle with the harness.

nec·rec·to·my (nek·rek′tə·mē) *n.* *Surg.* The removal of dead matter. [<NECR(O) + -ECTOMY]

ne·cre·mi·a (ne·krē′mē·ə) *n.* *Pathol.* A diminishing vitality of the blood. Also **ne·crae′mi·a.** [<NECR(O)- + -EMIA]

nec·ren·ceph·a·lus (nek′rən·sef′ə·ləs) *n.* *Pathol.* Softening of the brain. [<NECR(O)- + Gk. *enkephalos* brain]

necro- *combining form* Corpse; dead matter: *necropolis.* Also, before vowels, **necr-.** [<Gk. *nekros* a corpse]

nec·ro·bac·il·lo·sis (nek′rō·bas′ə·lō′sis) *n.* An infective disease of cattle, sheep, horses, elk, and swine due to invasion of the body by a micro-organism (*Actinomyces necrophorus*) which produces large areas of gangrenous and necrotic tissue.

nec·ro·bi·o·sis (nek′rō·bī·ō′sis) *n.* *Pathol.* Progressive decay and death of an organ or tissue. [<NECRO- + Gk. *bios* life] — **nec′ro·bi·ot′ic** (-ot′ik) *adj.*

nec·ro·gen·ic (nek′rə·jen′ik) *adj.* Originating or living in dead matter. Also **ne·crog·e·nous** (ne·kroj′ə·nəs).

ne·crol·o·gy (ne·krol′ə·jē) *n.* *pl.* **·gies** **1** A list of persons who have died in a certain place or time. **2** A treatise on or an account of the dead. **3** Formerly, a register of those for whose souls prayer was to be offered. [<NECRO- + Gk. *logos* a list, register] — **nec·ro·log·ic** (nek′rə·loj′ik), **nec′ro·log′i·cal** *adj.* — **nec′ro·log′i·cal·ly** *adv.* — **ne·crol′o·gist** *n.*

nec·ro·man·cy (nek′rə·man′sē) *n.* **1** Divination by means of pretended communication with the dead. **2** Black magic; sorcery. See synonyms under SORCERY. — **nec′ro·man′cer** *n.* — **nec′ro·man′tic** *adj.*

nec·ro·ma·ni·a (nek′rō·mā′nē·ə, -mān′yə) *n.* A morbid preoccupation with death or interest in dead persons. — **nec′ro·ma′ni·ac** (-nē·ak) *n.*

ne·croph·a·gous (ne·krof′ə·gəs) *adj.* Subsisting on carrion. [<NECRO- + -PHAGOUS]

nec·ro·phile (nek′rə·fīl, -fil) *n.* One who has a morbid attraction, usually of an erotic nature, to corpses. — **nec′ro·phil′i·a, ne·croph·i·lism** (ne·krof′ə·liz′əm), **ne·croph′i·ly** *n.*

nec·ro·pho·bi·a (nek′rō·fō′bē·ə) *n.* A morbid fear of death or of dead bodies. — **nec′ro·pho′bic** *adj.*

ne·crop·o·lis (ne·krop′ə·lis) *n.* A cemetery, especially one belonging to an ancient city. [<NECRO- + Gk. *polis* city]

nec·rop·sy (nek′rop·sē) *n.* *pl.* **·sies** An examination of a dead body; an autopsy. Also **ne·cros·co·py** (ne·kros′kə·pē). [<NECRO- + -OPSY]

ne·crose (ne·krōs′, nek′rōs) *v.t.* & *v.i.* **·crosed, ·cros·ing** To affect with or suffer from necrosis.

ne·cro·sis (ne·krō′sis) *n.* **1** *Pathol.* The death of a part of the body, as of a bone; mortification; gangrene. **2** *Bot.* A gradual decay in trees or plants. [<Gk. *nekrōsis* deadness] — **ne·crot′ic** (-krot′ik) *adj.*

necrotic enteritis An intestinal disorder of swine marked by extensive ulceration.

ne·crot·o·my (ne·krot′ə·mē) *n.* **1** The dissection of a dead body. **2** *Surg.* The excision of dead bone. [<NECRO- + -TOMY] — **nec·ro·tom·ic** (nek′rə·tom′ik) *adj.* — **ne·crot′o·mist** *n.*

nec·tar (nek′tər) *n.* **1** In Greek mythology, the drink of the gods. **2** Hence, any especially sweet drink: applied specifically to certain spiced or honeyed wines. **3** *Bot.* The saccharine substance secreted by some plants and forming the base of natural honey. [<L <Gk. *nektar*] — **nec·tar·e·an** (nek·târ′ē·ən) *adj.* — **nec·tar′e·ous** *adj.*

nec·tar·if·er·ous (nek′tə·rif′ər·əs) *adj.* Nectar- or honey-bearing.

nec·tar·ine (nek′tə·rēn′, nek′tə·rēn) *n.* A variety of peach having a smooth, waxy skin without down and a firm, aromatic pulp. — *adj.* Sweet and delicious.

nec·ta·ry (nek′tər·ē) *n.* *pl.* **·ries** **1** *Bot.* The organ or part of a plant that secretes nectar. **2** *Entomol.* One of the small, abdominal honey tubes of an aphid. — **nec·tar′i·al, nec·tar′e·al** (-târ′ē·əl) *adj.*

Ned (ned) Diminutive of EDWARD. Also **Ned′dy.** — **to raise Ned** *U.S. Colloq.* To start trouble or a disturbance.

ned·dy (ned′ē) *n.* *Brit. Dial.* **1** A donkey: usually as a proper or pet name. **2** A simpleton.

Ne·der·land (nā′dər·länt) The Dutch name for the NETHERLANDS.

née (nā) *adj.* Born: noting the maiden name of a married woman: Madame d'Arblay, *née* Burney. [<F, pp. of *naître* be born <L *nasci*]

nee·bour (nē′bər) *n.* *Scot.* A neighbor.

need (nēd) *v.t.* **1** To have need or want of; require. — *v.i.* **2** To be in want. **3** *Archaic* To be necessary: It *needs* not. **4** To be obliged or compelled: in this sense *need* is used as an uninflected auxiliary verb only in negative and interrogative sentences, followed by the infinitive without *to*: He *need* not go; *Need* he come? — *n.* **1** A lack of something requisite or desirable; hence, indigence; poverty: He was in *need.* **2** A situation of want or peril. **3** The thing needed. See synonyms under NECESSITY, POVERTY, WANT. ◆ Homophone: *knead.* [OE *nīed, nēd*] — **need′er** *n.*

need·ful (nēd′fəl) *adj.* **1** Needed; requisite; necessary. **2** Needy. — **need′ful·ly** *adv.* — **need′ful·ness** *n.*

Need·ham (nē′dəm), **John Turberville,** 1713–1781, English naturalist.

need·i·ness (nē′di·nis) *n.* The state of being in want; poverty.

nee·dle (nēd′l) *n.* **1** A small, slender, pointed instrument containing an eye at the head, or, in sewing machines, at the point, to carry thread through a fabric in sewing. **2** The straight rod, commonly of wire, bone, or wood, used in knitting; also, the hooked rod used in crocheting. **3** Any instrument or object shaped like or used as a needle, as a pinnacle of rock, or a leaf, such as that of the pine. **4** A straight wire, balanced and pivoted, as in a compass: a magnetic needle. **5** In a needle gun, the steel bolt that fires the cartridge. **6** A needle valve. **7** The sharp-pointed end of a hypodermic syringe. **8** *Colloq.* A hypodermic needle. **9** An obelisk. **10** A thin, sharp-pointed piece of steel, etc., often tipped with diamond, etc., to transmit the sound vibrations from a phonograph record. — *v.* **·dled, ·dling** *v.t.* **1** To sew or pierce with a needle. **2** *Colloq.* To heckle or annoy. **3** *Colloq.* To goad; prod. **4** *Colloq.* To increase the alcoholic content of: to *needle* the beer. — *v.i.* **5** To sew or work with a needle. [OE *nǣdl*]

nee·dle·bath (nēd′l·bath′, -bäth′) *n.* A form of shower bath in which the water is projected with force in fine jets.

nee·dle·fish (nēd′l·fish′) *n.* *pl.* **·fish** or **·fish·es** **1** One of the long, slender sea fishes of the family *Belonidae*, superficially resembling the fresh-water gar. **2** The pipefish.

nee·dle·ful (nēd′l·fōōl′) *n.* *pl.* **·fuls** The length of thread that can be suitably used in a needle at one time.

needle grass Feather grass.

needle gun A breechloading small arm firing a paper cartridge having the primer between the powder charge and the bullet, the primer being detonated by a needlelike firing pin passing through the powder: introduced in the Prussian army, 1842.

nee·dle·point (nēd′l·point′) *n.* **1** A sharp-pointed attachment for the leg of a drawing instrument; anything resembling the point of a needle, as spires of cathedrals or crystals of rock. **2** Lace made entirely with a sewing needle rather than bobbins, and worked with buttonhole and blanket stitches on a paper pattern. **3** A stitch used in needle tapestry, or embroidered needle tapestry itself.

need·less (nēd′lis) *adj.* Useless; not required. — **need′less·ly** *adv.* — **need′less·ness** *n.*

nee·dle·stone (nēd′l·stōn′) *n.* Natrolite.

needle valve *Mech.* **1** A valve having a conoidal opening closed by a plug of similar shape: designed to control accurately the flow of a liquid, as in a carbureter. **2** A valve with a conoidal plug fitting into a cylindrical opening: designed to give a large increase in opening with a slight increase in lift.

nee·dle·wom·an (nēd′l·wŏŏm′ən) *n.* *pl.* **·wom·en** (-wim′in) A seamstress.

nee·dle·work (nēd′l·wûrk′) *n.* **1** Work done with a needle; sewing; specifically, embroidery. **2** The business of sewing with a needle. — **nee′dle·work′er** *n.*

needs (nēdz) *adv.* Necessarily; indispensably: often with *must.*

need·y (nē′dē) *adj.* **need·i·er, need·i·est** Being in need, want, or poverty; necessitous. — **need′·i·ly** *adv.*

Ne·en·ga·tu (nye·eng′gə·tōō′) *n.* Tupi, the northern branch of the Tupian linguistic stock: widely used in the Amazon region.

neep (nēp) *n.* *Scot.* & *Brit. Dial.* A turnip.

ne'er (nâr) *adv.* Never: a contraction.

ne'er-do-well (nâr'do͞o·wel') *n.* A useless, unreliable person. —*adj.* Useless; good-for-nothing. Also **ne'er'-do-good'** (-go͝od'), *Scot.* **ne'er'-do-weel'** (-wēl').

Neer·win·den (nār·vin'dən) A village in eastern Belgium; scene of English defeat by the French, 1693, and Austrian victory over the French, 1793.

neeze (nēz) *v.i. Obs.* To sneeze. [Cf. ON *hnjosa* sneeze]

ne·far·i·ous (ni·fâr'ē·əs) *adj.* Wicked in the extreme; heinous. See synonyms under CRIMINAL, FLAGRANT, INFAMOUS, SINFUL. [< L *nefarius* < *nefas* a crime < *ne-* not + *fas* a divine command, right] —**ne·far'i·ous·ly** *adv.* —**ne·far'i·ous·ness** *n.*

Ne·fer·ti·ti (ne'fər·tē'tē) Egyptian queen of the 14th century B.C., wife of Ikhnaton: also *Nofretete.*

Ne·fud (ne·fo͞od') See NAFUD.

ne·gate (ni·gāt', nē'gāt) *v.t.* **·gat·ed, ·gat·ing** **1** To make ineffective; nullify. **2** To deny the existence of. [< L *negatus,* pp. of *negare* deny]

ne·ga·tion (ni·gā'shən) *n.* **1** The act of denying or of asserting the falsity of a proposition; denial in general. **2** Absence of anything affirmative or definite; voidness; nullity.

neg·a·tive (neg'ə·tiv) *adj.* **1** Containing contradiction or denial; expressing negation: opposed to *affirmative.* **2** Characterized by denial or refusal: a *negative* reply. **3** Exhibiting or characterized by absence of that which is essential to positive or affirmative character: the opposite of *positive.* **4** *Phot.* Exhibiting the reverse; showing dark for light and light for dark: a *negative* plate or film. **5** *Math.* **a** Denoting a direction or quality the opposite of another assumed as positive: usually denoted by the minus sign (-). **b** Less than zero; to be subtracted; minus: said of quantities. **6** *Geom.* Situated or measured downward from the axis of *X* or to the left of the axis of *Y.* **7** *Electr.* Denoting a type of electricity characterized by an excess of electrons on a charged body: opposed to *positive.* It is similar to that produced on a resinous object after rubbing with wool. **8** *Biol.* Describing a plant or animal response directed away from or in opposition to a stimulus. **9** *Med.* Indicating the absence of a suspected condition, or the absence of certain bacteria. See synonyms under PASSIVE. —*n.* **1** A proposition, word, or act expressing refusal or denial: My request received a *negative.* **2** The side of a question that denies; also, a negative decision. **3** The right to veto. **4** A photograph having the lights and shades reversed, used for printing positives. **5** *Gram.* A particle employing or expressing denial. The principal negative is *not.* **6** *Electr.* **a** Negative or frictional electricity. **b** The negative plate of a voltaic cell. **7** *Math.* A negative sign or quantity. —**double negative** *Gram.* The use of two negatives in the same statement, as in "I didn't see nobody." ◆ This usage is a descendant of a formation native to the Germanic languages and was regularly used in Old and Middle English to intensify negation. It survives in Modern English, but is now considered substandard on analogy with Latin, where a double negative becomes an affirmative. Such statements as "I am not unhurt," however, are standard English, and have the effect of weak affirmatives. —*v.t.* **·tived, ·tiv·ing** **1** To deny; contradict. **2** To refuse to sanction or enact; veto. **3** To prove to be false; disprove. **4** To neutralize; counteract. [< L *negativus* < *negare* deny] —**neg'a·tive·ly** *adv.* —**neg'a·tive·ness, neg'a·tiv'i·ty** *n.*

negative element **1** *Chem.* An element or atom having a tendency to attract electrons. **2** *Electr.* **a** That element or plate of a voltaic cell into which the current passes from the electrolyte. **b** The cathode of a vacuum tube.

neg·a·tiv·ism (neg'ə·tiv·iz'əm) *n.* **1** The beliefs or attitude of any negative thinker; atheism, agnosticism, etc. **2** The denial of traditional beliefs without proposing constructive substitutes. **3** *Psychol.* A type of behavior characterized by resistance to suggestion: when the subject fails to do what he is expected or asked to do, such behavior is known as **passive negativism**; when the subject does the opposite, **active** or **command negativism.** —**neg'a·tiv·ist** *n.* —**neg'a·tiv·is'tic** *adj.*

neg·a·to·ry (neg'ə·tôr'ē, -tō'rē) *adj.* Signifying negation.

Neg·ev (neg'ev, ne·gev') A triangular desert region in southern Israel; 4,700 square miles. Also **Neg·eb** (neg'eb, nə·geb').

neg·lect (ni·glekt') *v.t.* **1** To disregard; ignore. **2** To fail to give proper attention to or take proper care of: to *neglect* one's business. **3** To fail to do or perform through carelessness or oversight; leave undone. —*n.* **1** The act of neglecting, or the state of being neglected. **2** Habitual want of attention or care; negligence. [< L *neglectus,* pp. of *negligere* < *nec-* not + *legere* gather, pick up] —**neg·lect'a·ble** *adj.* —**neg·lect'er** *n.*

Synonyms (noun): carelessness, default, disregard, failure, heedlessness, inadvertence, inattention, indifference, neglectfulness, negligence, oversight, remissness, slackness, slight. *Neglect* is the failing to take such care, show such attention, pay such courtesy, etc., as may be rightfully or reasonably expected. *Negligence* may be used in almost the same sense, but with a slighter force; but it is often used to denote the quality or trait of character of which the act is a manifestation, or to denote the habit of neglecting that which ought to be done. *Negligence* in dress implies want of care as to its arrangement, tidiness, etc.; *neglect* of one's garments would imply leaving them exposed to defacement or injury, as by dust, moths, etc. See SLIGHT. *Antonyms:* see synonyms under CARE.

neg·lect·ful (ni·glekt'fəl) *adj.* Exhibiting or indicating neglect. See synonyms under INATTENTIVE. —**neg·lect'ful·ly, neg·lect'ing·ly** *adv.* —**neg·lect'ful·ness** *n.*

neg·li·gée (neg'li·zhā', neg'li·zhā; *Fr.* nā·glē·zhā') *n.* **1** A woman's soft, flowing, usually decorative dressing gown. **2** Any informal, careless, or incomplete attire. —*adj.* Of a woman, appearing careless in dress. [< F *négligée,* orig. pp. of *négliger* neglect]

neg·li·gence (neg'lə·jəns) *n.* **1** The act of neglecting. **2** An act or instance of neglect. **3** Disregard for outward appearances or for conventionalities. See synonyms under NEGLECT.

neg·li·gent (neg'lə·jənt) *adj.* **1** Apt to omit what ought to be done; neglectful. **2** Unconventional. See synonyms under INATTENTIVE. [< L *negligens, -entis,* ppr. of *negligere* NEGLECT] —**neg'li·gent·ly** *adv.*

neg·li·gi·ble (neg'lə·jə·bəl) *adj.* That can be disregarded; inconsiderable; trifling; of little importance or size. —**neg'li·gi·bil'i·ty, neg'li·gi·ble·ness** *n.* —**neg'li·gi·bly** *adv.*

ne·go·ti·a·ble (ni·gō'shē·ə·bəl, -shə·bəl) *adj.* **1** That can be negotiated. **2** *Law* Transferable to a third person, as for the payment of debts. **3** That can be managed, overcome, or successfully dealt with. —**ne·go'ti·a·bil'i·ty** *n.* —**ne·go'ti·a·bly** *adv.*

negotiable instruments Instruments, such as bills of exchange, notes, checks, drafts, bills of lading, etc., covered by the Negotiable Instrument Law in effect in most States of the United States.

ne·go·ti·ant (ni·gō'shē·ənt) *n.* One who negotiates; a negotiator.

ne·go·ti·ate (ni·gō'shē·āt) *v.* **·at·ed, ·at·ing** *v.i.* **1** To treat or bargain with others in order to reach an agreement. —*v.t.* **2** To procure, arrange, or conclude by mutual discussion: to *negotiate* an agreement. **3** To transfer for a value received; sell; assign, as a note or bond. **4** To surmount, cross, or cope with (some obstacle). See synonyms under TRANSACT. [< L *negotiatus,* pp. of *negotiari* trade < *negotium* business] —**ne·go'ti·a'tion** *n.* —**ne·go'ti·a'tor** *n.* —**ne·go'ti·a·to'ry** *adj.*

Ne·gress (nē'gris) *n.* Feminine of NEGRO: an offensive usage.

Ne·gril·lo (ni·gril'ō) *n. Anthropol.* Any of certain Negroid peoples of southern Africa; specifically, a Pygmy. See NEGRITO. [< Sp., dim. of *negro* black]

Ne·gri Sem·bi·lan (nā'grē sem·bē·län') A state in SW Federation of Malaya; 2,550 square miles; capital, Seremban.

Ne·gri·to (ni·grē'tō) *n. pl.* **·tos** or **·toes** *Anthropol.* **1** One belonging to any of the dwarfish Negroid peoples of southern Africa; a Negrillo. **2** One of the Pygmy peoples of southeast Asia, Malaya, and the Philippine Islands, by some authorities regarded as of Negroid descent. [< Sp., dim. of *negro* black] —**Ne·grit'ic** (-grit'ik) *adj.*

Ne·gro (nē'grō) *n. pl.* **·groes** **1** A member of the Negroid ethnic division of mankind; specifically, one belonging to the tribes inhabiting the Congo and Sudan regions of Africa. **2** One who is descended from the African Negro, full-blooded or of mixed descent. —*adj.* Of, pertaining to, like, for, or being, a Negro: *Negro* folklore; *Negro* school; a *Negro* entertainer, etc. [< Sp. < L *niger* black]

Ne·gro (nā'grō), **Río** See RÍO NEGRO.

Ne·groid (nē'groid) *adj.* **1** *Anthropol.* Pertaining to or characteristic of the so-called black race, having skin color varying from light brown to almost black, stature tall to dwarfish, curly or woolly hair, slight body hair, nose usually broad or flat, eyes brown or black, often with a vertical epicanthic fold, and some prognathism. **2** Resembling, related to, or characteristic of Negroes. —*n.* A person of Negro descent or having some Negro characteristics.

Ne·gro·ism (nē'grō·iz'əm) *n.* **1** Promotion of the rights and interests of Negroes. **2** An idiom, pronunciation, figure of speech, etc., peculiar to Negroes, especially to those of the southern United States.

ne·gro·phile (nē'grə·fīl, -fil) *n.* One not a Negro who is friendly to Negroes.

ne·gro·pho·bi·a (nē'grə·fō'bē·ə) *n.* Antipathy to or dislike of Negroes. —**ne'gro·phobe** *n.*

Neg·ro·pon·te (neg'rō·pôn'tā) The Italian name for EUBOEA.

Ne·gros (nā'grōs) One of the Visayan Islands, fourth largest of the Philippines; 4,905 square miles.

ne·gus (nē'gəs) *n.* A drink made of wine, water, and lemon juice, sweetened. [after Col. Francis *Negus,* died 1732, who concocted it]

Ne·gus (nē'gəs) *n.* The title of the kings of Abyssinia.

Ne·he·mi·ah (nē'hə·mī'ə) **1** A Hebrew statesman and historian. **2** A book of the Old Testament attributed to him, recounting the rebuilding of Jerusalem. [< Hebrew *Nechemiah* comforted of Jehovah]

Neh·ru (nā'ro͞o), **Ja·wa·har·lal** (jə·wä'hər·läl), born 1889, Indian nationalist leader; prime minister 1947–. —**(Pandit) Motilal,** 1861–1931, Indian nationalist; father of preceding.

neif (nēf) See NIEVE.

neigh (nā) *v.i.* To utter the cry of a horse; whinny. —*n.* A whinny. [OE *hnǣgan*]

neigh·bor (nā'bər) *n.* **1** One who lives near another. **2** One who is near by chance. **3** Friend; stranger: a colloquial and friendly term of address. **4** A fellow man. —*adj.* Close at hand; adjacent. —*v.t.* **1** To adjoin; live near to. **2** To bring near to or in close association. —*v.i.* **3** To be in proximity; lie close. **4** To live nearby; be neighborly. Also *Brit.* **neigh'bour.** [OE *nēahgebur* < *nēah* near + *gebur* farmer]

neigh·bor·hood (nā'bər·ho͝od) *n.* **1** The region near where one is or resides; vicinity. **2** The people collectively who dwell in the vicinity. **3** Nearness; the condition of standing in the relation of a neighbor. **4** Friendly relations; neighborliness. **5** A district considered with reference to a given characteristic. —**in the neighborhood of** About; near.

Synonyms: district, locality, vicinage, vicinity. See APPROXIMATION.

neigh·bor·ing (nā'bər·ing) *adj.* Adjacent; near.

neigh·bor·ly (nā'bər·lē) *adj.* Appropriate to a neighbor; sociable. See synonyms under AMICABLE, FRIENDLY. —**neigh'bor·li·ness** *n.*

Neil·son (nēl'sən), **William Allen,** 1869–1946, U. S. educator born in Scotland; president of Smith College 1917–39.

Neis·se (nī'sə) A river in Czechoslovakia, East Germany, and Poland, flowing 140 miles to the Oder: Polish *Nysa.*

Neis·ser (nī'sər), **Albert Ludwig Siegmund,** 1855–1916, German dermatologist; discoverer of the gonococcus.

neis·ser·o·sis (nī'sə·rō'sis) *n. Pathol.* Gonococcus infection. [after A. L. S. *Neisser*]

add,āce,câre,pälm; end,ēven; it,īce; odd,ōpen,ôrder; to͝ok,po͞ol; up,bûrn; ə = a in *above,* e in *sicken,* i in *clarity,* o in *melon,* u in *focus;* yo͞o = u in *fuse;* oi,oil; ou,pout; ch,check; g,go; ng,ring; th,thin; th,this; zh,vision. Foreign sounds à,œ,ü,kh,ṅ; and ◆: see page xx. < from; + plus; ? possibly.

neist (nēst) *adj., adv., & prep. Brit. Dial.* Next; nearest: also, *Scot., niest.*

nei·ther (nē′thər, nī-) *adj.* Not either. —*pron.* Not the one nor the other. —*conj.* 1 Not one nor the other: followed by correlative *nor*: He will *neither* go nor send. 2 Not at all: an intensive now replaced by *either* except in incorrect usage: He has no strength, nor sense *neither.* 3 Nor yet. [OE *nother*; infl. in form by EITHER]

Nejd (nejd) A province of Saudi Arabia comprising a viceroyalty of the country; 450,000 square miles; capital, Riyadh: also *Najd.*

nek·ton (nek′ton) *n.* The aggregate of marine organisms actively swimming on or near the surface of the sea. [<NL <Gk. *nektos* swimming] —**nek·ton′ic, nek·ter′ic** (-ter′ik) *adj.*

Ne·le·us (nē′lē·əs) In Greek legend, a son of Poseidon and king of Pylos, father of twelve sons, all of whom save Nestor were slain by Hercules.

Nell (nel), **Nel·lie, Nel·ly** (nel′ē) Diminutives of HELEN.

nel·son (nel′sən) *n.* A wrestling hold in which the arms are thrust under the opponent's armpits from behind, and the hands gripped at the back of his neck: also called **full nelson.** The **half, quarter,** and **three-quarter nelson** are variants of this fundamental hold.

Nel·son (nel′sən), **Viscount Horatio,** 1758–1805, English admiral; hero of Trafalgar: known as *Lord Nelson.*

Nel·son River (nel′sən) A river in NE Manitoba province, Canada, flowing 400 miles NE from Lake Winnipeg to Hudson Bay.

ne·lum·bo (ni·lum′bō) *n. pl.* **·bos** One of a genus *(Nelumbium)* of aquatic herbs of the waterlily family. *N. pentapetalum* is the water chinkapin and *N. nelumbo* is the sacred lotus of India. [<NL <Singhalese *nelumbu*]

Ne·man (nye′mən) A river in Belorussian S.S.R. and Lithuania, flowing 597 miles west to the Courland Lagoon: Polish *Nieman,* Lithuanian *Nemunas,* German *Memel.*

nem·a·thel·minth (nem′ə·thel′minth) *n.* One of a phylum or class *(Nemathelminthes)* of worms having a threadlike, unsegmented body with papillae or spines at the anterior extremity, as the nematodes and acanthocephalans; the roundworms. —*adj.* Pertaining to these worms. Also **nem′a·tel′minth** (-tel′-). [<NL <Gk. *nēma, -atos* thread + *helmins* a worm]

nemato- *combining form* Thread; filament: *nematocyst*: also, before vowels, **nemat-.** Also **nema-.** [<Gk. *nēma, -matos* thread]

nem·a·to·cyst (nem′ə·tō·sist′) *n. Zool.* One of the stinging cells in jellyfishes, polyps, and other hydrozoans, in the interior of which is coiled a long filament whose instantaneous release causes paralysis of the organism it touches: also called *lasso cell, nettle cell.* —**nem′a·to·cys′tic** *adj.*

nem·a·tode (nem′ə·tōd) *n.* Any of a phylum or class *(Nematoda)* of roundworms having a mouth and intestinal canal, some of which, as the hookworm, are intestinal parasites in man and other animals. [<NL <Gk. *nēma, -atos* a thread]

Ne·me·a (nē′mē·ə, nə·mē′ə) A valley in ancient Argolis, Greece; celebrated for the **Nemean games,** one of the four great pan-Hellenic festivals. —**Ne·me′an** *adj.*

Nemean lion In Greek legend, a fierce lion which Hercules strangled.

ne·mer·te·an (ni·mûr′tē·ən) *n.* One of a phylum or class *(Nemertea)* of flatworms, mostly marine and non-parasitic, with skin ciliated, proboscis retractile, and muscular, vascular, and nervous systems characteristically developed: often brilliantly colored. —*adj.* Of or pertaining to the *Nemertea.* [<NL <Gk. *Nēmertēs,* one of the Nereids] —**ne·mer′ti·an, ne·mer′tine, ne′mer·tin′e·an** *adj. & n.*

nem·e·sis (nem′ə·sis) *n.* Retributive justice; retribution. [<L <Gk. <*nemein* distribute]

Nem·e·sis (nem′ə·sis) In Greek mythology, the goddess of retributive justice or vengeance. [Prob. <*nemein* allot, distribute]

ne·mi·ne con·tra·di·cen·te (nem′ə·nē kon′trə·di·sen′tē) *Latin* No one speaking in opposition; hence, unanimously.

ne·mi·ne dis·sen·ti·en·te (nem′ə·nē di·sen′shē·en′tē) *Latin* No one dissenting; hence, unanimously.

ne·mo me im·pu·ne la·ces·sit (nē′mō mē im·pyōō′nē lə·ses′it) *Latin* No one attacks me with impunity: motto of Scotland.

nem·o·ral (nem′ər·əl) *adj.* Pertaining to a wood, grove, or the like. [<L *nemoralis* < *nemus, nemoris* a grove]

nem·o·rose (nem′ə·rōs) *adj. Bot.* Inhabiting groves or open woodland places: said especially of plants.

Ne·mu·nas (nye′mōō·näs) The Lithuanian name for the NEMAN.

neo- *combining form* 1 New; recent; a modern or modified form of: *Neo-Platonism.* 2 *Geol.* Denoting the most recent subdivision of a period: *Neocene.* Also, before vowels, usually **ne-.** [<Gk. <*neos* new]

ne·o·ars·phen·am·ine (nē′ō·ärs′fen·am′in, -fen·ə·mēn′) *n. Chem.* A modified compound of arsphenamine, $C_{13}H_{13}O_4N_2SAs_2Na$, used in the treatment of syphilis and certain other diseases.

Ne·o-Cath·o·lic (nē′ō·kath′ə·lik, -kath′lik) *adj.* 1 Of or pertaining to a school in the Anglican church in avowed sympathy with Roman Catholic doctrine and ritual. 2 In France, pertaining to a school of liberal Catholicism opposed to ultramontanism. —*n.* A member of either of these schools. —**Ne′o-Ca·thol′i·cism** (-kə·thol′ə·siz′əm) *n.*

Ne·o·cene (nē′ə·sēn) *adj. Geol.* Of or pertaining to the later of the two epochs into which the Tertiary period was at one time divided, or to the corresponding series of strata. —*n.* The Neocene epoch. [<NEO- + Gk. *kainos* new]

Ne·o-Chris·ti·an·i·ty (nē′ō·kris′chē·an′ə·tē) *n.* A rationalistic interpretation of Christianity. —**Ne′o-Chris′tian** *adj. & n.*

ne·o-clas·sic (nē′ō·klas′ik) *adj.* Of, pertaining to, or denoting neo-classicism. Also **ne′o-clas′si·cal.**

ne·o-clas·si·cism (nē′ō·klas′ə·siz′əm) *n.* 1 A revival of classical style in literature, art, etc. 2 In the later 17th and the 18th centuries, an esthetic and philosophical movement which sought to recover the classical spirit of order and moderation: characterized by close adherence to rules and conventional forms, restraint in the expression of emotion, and an emphasis on the typical and general rather than the individual or eccentric. —**ne′o-clas′si·cist** *n.*

ne·o·cul·tu·ra·tion (nē′ō·kul′chə·rā′shən) *n.* The creation and establishment of new cultural forms, especially as a result of transculturation.

Ne·o-Dar·win·ism (nē′ō·där′win·iz′əm) *n. Biol.* Darwinism as modified and extended by more recent students, who accept the theory of natural selection as sufficient to account for evolution, and deny, as in the case of Weismann especially, the inheritance of acquired characters. See NEO-LAMARCKISM, WEISMANNISM. —**Ne′o-Dar·win′i·an** (-där·win′ē·ən) *adj. & n.* —**Ne′o-Dar′win·ist** *n.*

ne·o·dym·i·um (nē′ō·dim′ē·əm) *n.* A metallic element (symbol Nd) forming rose-colored salts: found in combination with cerium and other elements of the lanthanide series. See ELEMENT. [<NEO- + (DI)DYMIUM]

Ne·o·gae·a (nē′ō·jē′ə) *n.* A zoogeographical region including the western hemisphere or New World. [<NEO- + Gk. *gaia* earth] —**Ne′o·gae′an** *adj.*

ne·o·ge·ic (nē′ō·jē′ik) *adj.* Of or belonging to the western hemisphere or New World: opposed to *gerontogeic.*

Ne·o·gene (nē′ō·jēn) *adj. Geol.* Of or pertaining to the Upper Tertiary and the Quaternary periods in the Cenozoic geological era: includes the Miocene, Pliocene, Pleistocene, and Holocene epochs. [<Gk. *neogenēs* newborn]

ne·o·gen·ic (nē′ō·jen′ik) *adj.* Newly formed: said especially of rocks and minerals. Also **ne′o·ge·net′ic** (-jə·net′ik).

Ne·o-He·bra·ic (nē′ō·hē·brā′ik) *n.* That form of the Hebrew language used in post-Biblical Jewish literature. —*adj.* Pertaining to post-Biblical Hebrew.

ne·o-im·pres·sion·ism (nē′ō·im·presh′ən·iz′əm) *n.* The doctrines and methods of a group of artists of the 19th century, based on a more strictly scientific practice of impressionist technique. Compare IMPRESSIONISM, POINTILLISM, POSTIMPRESSIONISM. —**ne′o-im·pres′sion·ist** *n. & adj.*

Ne·o-La·marck·ism (nē′ō·lə·märk′iz·əm) *n. Biol.* Lamarckism as revived, modified, and extended by students who hold to the inheritance of acquired habits as a potent influence in evolution. —**Ne′o-La·marck′i·an** *adj. & n.* —**Ne′o-La·marck′ist** *n.*

Ne·o-Lat·in (nē′ō·lat′n) *n.* 1 One of a group of peoples whose language is derived from Latin. 2 New Latin: see under LATIN.

ne·o·lith (nē′ə·lith) *n.* A Neolithic implement.

Ne·o·lith·ic (nē′ə·lith′ik) *adj. Anthropol.* Of or pertaining to the period of human culture following the Mesolithic: characterized by a great variety of polished stone implements and the development of new social forms based upon primitive techniques in weaving, spinning, and potterymaking, and the introduction of a settled agriculture exploiting many new domesticated plants. [<Gk. *neos* new + *lithos* stone]

ne·ol·o·gism (nē·ol′ə·jiz′əm) *n.* 1 A new word or phrase. 2 The use of new words or new meanings for old words. 3 A new doctrine in theology. —**ne·ol′o·gis′tic, ne·ol′o·gis′ti·cal** *adj.*

ne·ol·o·gist (nē·ol′ə·jist) *n.* 1 A person who invents or employs new words. 2 A person who adopts new views in theology.

ne·ol·o·gy (nē·ol′ə·jē) *n. pl.* **·gies** A neologism. [<NEO- + Gk. *logos* word] —**ne·o·log·i·cal** (nē′ə·loj′i·kəl) *adj.* —**ne′o·log′i·cal·ly** *adv.*

Ne·o-Mal·thu·sian (nē′ō·mal·thōō′zhən, -zē·ən) *n.* An advocate of birth control: so called from Malthus' belief that population is always checked at a level proportionate to the available means of subsistence. —*adj.* Pertaining to birth control. —**Ne′o-Mal·thu′sian·ism** *n.*

ne·o·morph (nē′ə·môrf) *n.* A newly acquired organ or part. Also **ne′o·mor′phism.**

ne·o·my·cin (nē′ə·mī′sin) *n.* An antibiotic related to streptomycin, used in the local treatment of certain skin and eye infections.

ne·on (nē′on) *n.* An inert gaseous element (symbol Ne) occurring in the atmosphere only to the extent of 1 or 2 parts per 100,000: first isolated in 1898: used in glow-discharge electric display lamps, called **neon lights.** See ELEMENT. [<NL <Gk. *neos* new]

ne·o·pho·bi·a (nē′ə·fō′bē·ə) *n.* Morbid fear of the new or unfamiliar. —**ne′o·phobe** *n.* —**ne′·o·pho′bic** *adj.*

ne·o·phre·ni·a (nē′ə·frē′nē·ə, -frēn′yə) *n. Psychiatry* A neurosis or other mental disorder occurring in childhood. [<NEO- + Gk. *phrēn* mind]

ne·o·phyte (nē′ə·fit) *n.* 1 A recent convert. 2 A novice of a religious or mystic order. 3 Any novice or beginner. 4 *Bot.* A new or recently introduced plant species; an exotic. [<Gk. *neophytos* novice]

ne·o·plasm (nē′ə·plaz′əm) *n. Pathol.* A growth or formation of tissue resulting from morbid action; a tumor.

ne·o·plas·ty (nē′ə·plas′tē) *n.* Plastic surgery for the restoration of old or the formation of new parts. —**ne′o·plas′tic** *adj.*

Ne·o-Pla·ton·ism (nē′ō·plā′tən·iz′əm) *n.* An Alexandrian system of philosophy of the third century, commingling Jewish and Christian ideas with doctrines of Plato and other Greek philosophers and Oriental mysticism. —**Ne′o-Pla·ton′ic** (-plə·ton′ik) *adj.* —**Ne′o-Pla′ton·ist** *n.*

ne·o·prene (nē′ə·prēn) *n. Chem.* A synthetic rubber obtained in a variety of types from chloroprene, and produced by the combination of hydrogen with acetylene gas. [<NEO- + (CHLORO)PRENE]

Ne·op·tol·e·mus (nē′op·tol′ə·məs) In Greek legend, a son of Achilles who fought in the Trojan War, killed Priam and Astyanax, sacrificed Polyxena on Achilles' grave, carried off Andromache, and later married Hermione: also known as *Pyrrhus.*

Ne·o·sal·var·san (nē′ō·sal′vər·san) *n.* Proprietary name for a brand of neoarsphenamine.

Ne·o-Scho·las·ti·cism (nē′ō·skə·las′tə·siz′əm) *n.* The revival in modern times of Scholasticism, specifically that of Thomas Aquinas, incorporating new elements to make it applicable to contemporary problems.

Ne·o·sho River (nē·ō′shō) A river in Kansas and Oklahoma, flowing 460 miles SE to the Arkansas River.

ne·o·style (nē′ə·stīl) *n.* A contrivance for making several copies of a document; a cyclostyle.

ne·o·ter·ic (nē′ə·ter′ik) *adj.* Recent in origin; new. Also **ne′o·ter′i·cal.** —*n.* One of modern times; a modern. [<Gk. *neōterikos* youthful] —**ne′o·ter′i·cal·ly** *adv.*

ne·o·ter·ism (nē·ot′ə·riz′əm) *n.* That which is new, modern, or recently introduced; the coinage of new words, or a newly coined word. — **ne·ot′er·ist** *n.* — **ne·ot′er·is′tic** *adj.*

Ne·o·trop·i·cal (nē′ō·trop′i·kəl) *adj.* Of, pertaining to, or designating the zoogeographical region of the New World that includes Central and South America and the adjacent islands.

ne·o·type (nē′ə·tīp) *n.* In systematics, any specimen of a plant or animal chosen to replace the original specimen when all type material has been lost or destroyed.

Ne·o·zo·ic (nē′ə·zō′ik) *adj. Geol.* 1 Of or pertaining to the Mesozoic and Cenozoic geological eras, as contrasted with the Paleozoic. 2 The Cenozoic era.

nep (nep) *n.* Small knots in cotton fiber produced by uneven growth of the plant or by friction in process machinery. [Cf. dial. E *nap* a knob, button]

Nep (nep) Contraction of NEW ECONOMIC POLICY: also written **NEP**.

Ne·pal (ni·pôl′) An independent kingdom between Tibet and India; 56,000 square miles; capital, Katmandu. — **Nep·a·lese** (nep′ə·lēz′, -lēs′) *adj. & n.*

ne·pen·the (ni·pen′thē) *n.* 1 A drug or potion supposed by the ancient Greeks to banish pain and sorrow. 2 Any agent causing oblivion. [<L <Gk. *nēpenthēs* free from sorrow < *nē-* not + *penthos* sorrow] — **ne·pen′the·an, ne·pen′thic** *adj.*

ne·pen·thes (ni·pen′thēz) *n.* One of a genus (*Nepenthes*) of mainly East Indian herbs or half-shrubby plants, the East Indian pitcher plants. [<NL <Gk. *nēpenthēs* NEPENTHE]

ne·per (nā′pər) *n.* A unit of power-level difference or of attenuation in electrical communication circuits: equal to 8.686 decibels. [after John *Napier*]

NEPENTHES (Pitchers from 4 to 10 inches long)

Neph·e·le (nef′ə·lē) In Greek legend, the wife of Athamas, mother of Phrixus and Helle, who after her husband set her aside for another, sent her children away on a ram with golden fleece. See PHRIXUS.

neph·e·line (nef′ə·lin) *n.* A colorless or variously colored hexagonal sodium aluminum silicate, NaAlSiO₄. Also **neph′e·lite**. [<Gk. *nephelē* cloud]

neph·e·lin·ite (nef′ə·lin·īt′) *n.* A dark-gray volcanic rock composed of the minerals nepheline, augite, and magnetite. Also **neph′e·lin·yte′**.

neph·e·lom·e·ter (nef′ə·lom′ə·tər) *n.* An instrument used for the measurement of light transmitted or scattered by translucent substances: also used for the determination of the quantity of matter suspended in a liquid. [<Gk. *nephelē* cloud + -METER] — **neph·e·lo·met·ric** (nef′ə·lō·met′rik) *adj.* — **neph′e·lom′e·try** *n.*

neph·ew (nef′yōō, *esp. Brit.* nev′yōō) *n.* 1 The son of a sister or a brother; by extension, a grandnephew. 2 An unlawfully begotten son: a euphemism. 3 *Obs.* A descendant; grandchild; also, a cousin. [<F *neveu* <L *nepos* grandson, nephew]

nepho- *combining form* Cloud; pertaining to the clouds: *nephology*. Also, before vowels, **neph-**. [<Gk. *nephos* cloud]

neph·o·gram (nef′ə·gram) *n. Meteorol.* A cloud picture made by a nephograph.

neph·o·graph (nef′ə·graf, -gräf) *n. Meteorol.* An electrically operated camera for photographing clouds, with special reference to their position in the sky.

ne·phol·o·gy (ne·fol′ə·jē) *n.* The branch of meteorology that treats of clouds. — **neph·o·log·i·cal** (nef′ə·loj′i·kəl) *adj.*

neph·o·scope (nef′ə·skōp) *n. Meteorol.* An instrument for indicating the direction and velocity of winds by observations of cloud drift.

ne·phral·gi·a (ni·fral′jē·ə, -jə) *n. Pathol.* Pain in the kidney or kidneys. [<NEPHR(O)- + -ALGIA]

ne·phrec·to·my (ni·frek′tə·mē) *n. Surg.* The excision of a kidney. [<NEPHR(O)- + -ECTOMY]

neph·ric (nef′rik) *adj.* Of, pertaining to, or situated near the kidneys; renal. [<Gk. *nephros* kidney]

ne·phrid·i·um (ni·frid′ē·əm) *n. pl.* **·phrid·i·a** (-frid′ē·ə) *Biol.* 1 One of the series of primitive excretory organs that afterward develop into uriniferous tubules, as in annelid worms, mollusks, and other invertebrates. 2 The embryonic tube which develops into the kidney in vertebrates. [<NL <Gk. *nephridios* pertaining to the kidneys] — **ne·phrid′i·al** *adj.*

neph·rism (nef′riz·əm) *n. Pathol.* General ill health due to chronic kidney disease.

neph·rite (nef′rīt) *n.* A very hard, compact, white to dark-green mineral: formerly worn as a remedy for diseases of the kidney. Compare JADE¹. [<G *nephrit* <Gk. *nephros* a kidney]

ne·phrit·ic (ni·frit′ik) *adj.* 1 Pertaining to, affecting, or affording relief to the kidneys. 2 Affected with nephritis. 3 Of the nature of nephrite. Also **ne·phrit′i·cal**. — *n.* Any medicine applicable to disease of the kidney.

ne·phri·tis (ni·frī′tis) *n. Pathol.* 1 Inflammation of the kidneys. 2 Bright's disease.

nephro- *combining form* A kidney; pertaining to the kidneys: *nephropathy*. Also, before vowels, **nephr-**. [<Gk. *nephros* kidney]

ne·phrog·e·nous (ni·froj′ə·nəs) *adj.* Originating in or caused by the kidney. Also **neph·ro·gen·ic** (nef′rə·jen′ik). [<NEPHRO- + -GENOUS]

neph·roid (nef′roid) *adj.* Shaped like a kidney.

ne·phrol·y·sis (ni·frol′ə·sis) *n.* 1 *Pathol.* The breakdown of kidney substance due to the action of poisons. 2 *Surg.* The separation of an inflamed kidney from morbid adhesions. [<NEPHRO- + -LYSIS] — **neph·ro·lyt·ic** (nef′rə·lit′ik) *adj.*

ne·phrop·a·thy (ni·frop′ə·thē) *n.* Any disease of the kidney. — **neph·ro·path·ic** (nef′rə·path′ik) *adj.*

ne·phro·sis (ni·frō′sis) *n. Pathol.* Disease of the kidney, especially any disease characterized by degenerative lesions of the renal tubules. — **ne·phrot·ic** (-frot′ik) *adj.*

ne·phrot·o·my (ni·frot′ə·mē) *n. Surg.* Incision of the kidney.

ne plus ul·tra (nē′ plus ul′trə) *Latin* The extreme or utmost point; hence, perfection; literally, nothing more beyond.

Ne·pos (nē′pos, nep′os), **Cornelius**, Roman historian of the first century B.C.

nep·o·tism (nep′ə·tiz′əm) *n.* Favoritism, especially governmental patronage, extended toward relatives. [<F *népotisme* <Ital. *nepotismo* <L *nepos* a grandson, nephew] — **nep′o·tist** *n.* — **nep′o·tis′tic** *adj.*

Nep·tune (nep′tōōn, -tyōōn) 1 In Roman mythology, son of Saturn and Ops, god of the sea: identified with the Greek *Poseidon*. 2 By personification, the ocean. 3 *Astron.* The eighth planet from the sun, discovered in 1846 by Galle of Berlin. Its mean distance from the sun is 2,793,000,000 miles; period of revolution, about 165 years; diameter, about 33,000 miles. It has two satellites. See PLANET.

Nep·tu·ni·an (nep·tōō′nē·ən, -tyōō′-) *adj.* 1 Of or pertaining to Neptune or his domain, the sea. 2 *Geol.* **a** Formed in or by the agency of water: said of rocks. **b** Of or pertaining to the theory of the aqueous origin of certain rocks: opposed to the *Plutonic* theory. — **Nep′tun·ist** *adj. & n.*

nep·tu·ni·um (nep·tōō′nē·əm, -tyōō′-) *n.* A radioactive element (symbol Np) of atomic number 93, artificially produced from uranium by neutron bombardment. Some isotopes give rise to plutonium by emission of a beta particle.

neptunium series *Physics* A sequence of radioactive elements beginning with plutonium of mass 241 and continuing through successive disintegrations to the stable isotope bismuth 209: named from its longest-lived member, neptunium 237, with a half-life of 2.2×10^6 years.

Ner·bud·da (nər·bud′ə) A former spelling of NARBADA.

Ne·re·id (nir′ē·id) *pl.* **Ne·re·i·des** (ni·rē′ə·dēz) or **Ne·re·ids** In Greek mythology, one of the fifty daughters of Nereus and Doris, sea nymphs who attend Poseidon.

ne·re·is (nir′ē·is) *n.* Any of a genus (*Nereis*) of burrowing annelid worms having a long, flattened body and a distinct head: common near the seashore: also called *clamworm*. [<NL <Gk., a Nereid]

Ne·ret·va (ne′ret·vä) A river in the Dinaric Alps, Herzegovina, and Dalmatia, flowing 135 miles NW to the Adriatic: Italian *Narenta*.

Ne·reus (nir′ōōs, -ē·əs) In Greek mythology, a sea god, father of the Nereides.

Ne·ri (nā′rē), **Saint Philip**, 1515–95, Italian priest; founded the Congregation of the Oratory; canonized, 1622.

ne·rit·ic (ni·rit′ik) *adj.* Of or pertaining to the coastline or to shallow water. [<L *nerita* mussel <Gk. *nēritēs*]

Nernst (nernst), **Walther Hermann**, 1864–1941, German physicist and chemist.

Ne·ro (nir′ō), 37–68, Nero Claudius Caesar Drusus Germanicus, Roman emperor 54–68: original name *Lucius Domitius Ahenobarbus*. — **Ne·ro′ni·an**, **Ne·ron·ic** (ni·ron′ik) *adj.*

ner·o·li (ner′ə·lē) *n.* The essential oil distilled from orange blossoms: an isomer of geraniol used in perfumery. [after Princess *Neroli*, an Italian noblewoman said to have discovered it]

Ner·va (nûr′və), **Marcus Cocceius**, 32?–98, Roman emperor 96–98.

ner·vate (nûr′vāt) *adj.* Provided with nerves or veins; having nerves.

ner·va·tion (nûr·vā′shən) *n.* The arrangement or disposition of nerves, as in plants and insects. Also **ner′va·ture** (nûr′və·chŏŏr).

nerve (nûrv) *n.* 1 A cordlike structure, composed of delicate filaments, by which impulses are transmitted to or from different parts of the body. ◆ Collateral adjective: *neural*. 2 A tendon: used only in the phrase, to strain every *nerve*. 3 Anything likened to a nerve, as a rib or vein of a leaf or of an insect's wing. 4 Active strength or vigor; coolness; intrepidity. 5 *pl.* Nervous excitability; a nervous attack. — *v.t.* **nerved**, **nerv·ing** To give strength, vigor, or courage to. [<L *nervus* sinew] — **nerv′al** *adj.*

nerve-block (nûrv′blok′) *n.* A method of surgical anesthesia in which sensation is cut off from definite nerves.

nerve canal *Dent.* The narrow cavity in a tooth for passage of the nerve to the pulp.

nerve cell *Physiol.* 1 The cell body of a neuron. 2 An individual cell of the nervous system.

nerve center *Anat.* An aggregation of nerve cells controlling a particular sense or function, as hearing, respiration, etc.

nerve fiber One of the essential threadlike units of which a nerve is composed.

nerve impulse *Physiol.* A wave of chemical and electrical change propagated along a nerve fiber and serving as a stimulus to body movements and activities.

nerve·less (nûrv′lis) *adj.* 1 Destitute of nerve or force; strengthless; unnerved. 2 Having no nerves. — **nerve′less·ly** *adv.* — **nerve′less·ness** *n.*

nerve net *Zool.* The primitive, reticulated nervous system of coelenterates, as in the hydra: its reactions to stimuli affect the entire organism.

nerve-rack·ing (nûrv′rak′ing) *adj.* Extremely irritating or exasperating to one's nerves. Also **nerve′-wrack′ing**.

ner·vi·duct (nûr′və·dukt) *n. Anat.* A passage in a bone for a nerve. [<L *nervus* a nerve + *ductus*, pp. of *ducere* lead]

ner·vine (nûr′vēn, -vin) *adj.* 1 Pertaining to the nerves. 2 Calming or quieting to the nerves. — *n.* Any medicine acting on the nerves.

nerv·ing (nûr′ving) *n.* A veterinary operation for the excision of a part of a nerve trunk when in a state of chronic inflammation.

ner·vos·i·ty (nûr·vos′ə·tē) *n.* The state of being nervous; the tendency or disposition to exhibit nervous tension.

nerv·ous (nûr′vəs) *adj.* 1 Affected or caused by the condition or action of the nerves: *nervous* prostration. 2 Easily disturbed or agitated, owing to weak nerves; excitable; timid. 3 Abounding in nerve or nerve force; vigorous; sinewy; nervy; also, highly strung. 4 Terse; crisp, as literary style. 5 Of or

pertaining to the nerves or nervous system. [< L *nervosus* sinewy] — **ner′vous·ly** *adv*. — **ner′vous·ness** *n*.
nervous prostration Neurasthenia.
nervous system *Biol*. The organized aggregate of all the nerve cells and nerve tissues of the higher animals, centralized in the spinal cord and brain of vertebrates. It has the functions of coordinating, controlling, and regulating responses to stimuli, directing behavior, and, in man, conditioning the phenomena of consciousness.
ner·vu·ra·tion (nûr′vyə·rā′shən) *n*. The arrangement or disposition of nervures.
ner·vure (nûr′vyoor) *n. Biol.* A principal vein, as on a leaf or an insect's wing. Also **ner′vule** (-vyool). [< F < L *nervus* sinew]
nerv·y (nûr′vē) *adj*. **nerv·i·er, nerv·i·est** 1 Exhibiting force or strength; sinewy. 2 Full of nerve or courage; brave. 3 *Slang* Displaying brazen assurance; cool; impudent. 4 *Brit*. Nervous; jumpy; excitable.
nes·cience (nesh′əns, -ē·əns) *n*. The state of not knowing; ignorance, especially that due either to the nature of the human mind or of external things. [< LL *nescientia* ignorance < *ne-* not + *scire* know] — **nes′cient** *adj. & n*. — **nes′cient·ist** *n*.
ness (nes) *n*. A promontory or cape: frequently used as a termination in the proper name of a headland: *Dungeness; Sheerness*. [OE *næs*]
-ness *suffix of nouns* 1 State or quality of being: *darkness*. 2 An example of this state or quality: to do someone a *kindness*. [OE *-nis(s), -nes(s)*]
Nes·sel·rode (nes′əl·rōd, *Russian* nyi′sil·rō′de), **Count Karl Robert**, 1780–1862, Russian diplomat.
Nes·sel·rode pudding (nes′əl·rōd) A custard made with preserved fruits and nuts, and flavored with rum: used as a pie filling. [after Count K. R. *Nesselrode*]
Ness·ler's reagent (nes′lərz) *Chem.* An aqueous solution of mercuric iodide, potassium iodide, and caustic potash: used in testing for ammonia. Also **Nessler's solution**. [after Julius *Nessler*, 1827–1905, German chemist]
Nes·sus (nes′əs) In Greek legend, a centaur who tried to abduct Deianira and was slain by her husband Hercules. The shirt of Nessus, steeped in his blood by Deianira, who believed it to be a charm to preserve her husband's love, caused the death of Hercules when he put it on.
nest (nest) *n*. 1 The habitation prepared by a bird for the hatching of its eggs and the rearing of its young. 2 The bed or home of

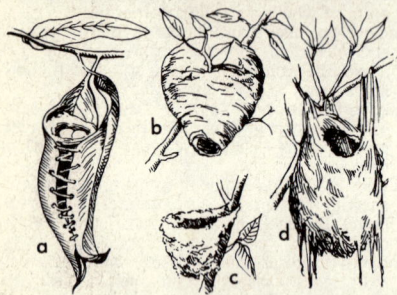

TYPES OF NESTS
a. Tailorbird.   c. Hummingbird.
b. Hornet.   d. Oriole.

certain fish, insects, turtles, mice, etc. 3 Any cozy place or abode; a retreat. 4 A haunt of anything bad, vulgar, or unpleasant; a den; also, those occupying it: a *nest* of brigands. 5 A series or set of similar things fitting into each other: a *nest* of bowls. 6 A connected set of small gearwheels, springs, or the like. 7 An isolated deposit of any ore or mineral in a rock. 8 A center of enemy resistance in battle: a machine-gun *nest*. — *v.t.* 1 To place in or as in a nest. 2 To pack or place one inside another. — *v.i.* 3 To build or occupy a nest. 4 To hunt for nests. [OE]
n'est-ce pas (nes pä′) *French* Isn't that so?
nest egg 1 A natural or artificial egg kept in a nest to attract a fowl. 2 Something laid by, as a sum of money, as a nucleus for future accumulation or for emergencies.

nest·er (nes′tər) *n. U.S. Dial.* A farmer seeking to settle on land used by cattlemen: a term of opprobrium.
nes·tle (nes′əl) *v*. **·tled, ·tling** *v.i.* 1 To lie closely or snugly; cuddle; snuggle. 2 To settle down in comfort and pleasure. 3 To lie as if sheltered; be half-hidden. 4 *Rare* To nest. — *v.t.* 5 To place or press lovingly or fondly. 6 To place in or as in a nest. [OE *nestlian*] — **nes′tler** *n*.
nest·ling (nest′ling, nes′-) *n*. A bird too young to leave the nest; hence, a young child. — *adj*. Recently hatched.
Nes·tor (nes′tər) 1 In Greek legend, a king of Pylos and one of the Argonauts, the oldest and wisest Greek chief in the Trojan War. 2 Any wise old man.
Nes·to·ri·an·ism (nes·tôr′ē·ən·iz′əm, -tō′rē-) *n. Theol.* The doctrine that Christ had two distinct natures, the divine and human, subsisting independently. — **Nes·to′ri·an** *n. & adj*.
Nes·to·ri·us (nes·tôr′ē·əs, -tō′rē-) Fifth century Syrian patriarch of Constantinople, condemned and banished as a heretic.
net[1] (net) *n*. 1 An open fabric, woven or tied with meshes, for the capture of fishes, birds, etc. ♦ Collateral adjective: *reticular*. 2 An openwork fabric, as lace. 3 Something constructed with meshes, as a tennis net. 4 In tennis and similar games, a returned ball which does not go over the net. — *v.t.* **net·ted, net·ting** 1 To catch in or as in a net; ensnare. 2 To make into a net. 3 To cover or enclose with a net. 4 In tennis, etc., to hit (the ball) into the net. — *adj*. 1 Manufactured or formed of netting, or resembling netting. 2 Captured or snared in a net. [OE]
net[2] (net) *adj*. 1 Free from everything extraneous; obtained after deducting all expenses. 2 Not subject to any discount or deduction. — *n*. A net profit, amount, weight, etc. — *v.t.* **net·ted, net·ting** To earn or yield as clear profit. [< F. See NEAT.]
net·ground (net′ground′) *n*. A foundation of net or meshes.
neth·er (neth′ər) *adj*. Situated at the lowest part; especially, pertaining to the parts beneath the heavens or the earth. [OE *neothera* under]
Neth·er·lands (neth′ər·ləndz) A country of NW Europe, first part of the Kingdom of the Netherlands; 12,425 square miles of land, 15,780 square miles with interior waters; capital, Amsterdam; seat of government, The Hague. Also, popularly, *Holland*. Dutch *Nederland*. — **Neth′er·land·er** *n*. — **Neth′er·land′ish** *adj*.
Netherlands, Kingdom of the A united kingdom comprising three equal and autonomous parts, the Netherlands, the Netherlands Antilles and Surinam, and the non-autonomous overseas territory of Netherlands New Guinea; 231,872 square miles; capital, Amsterdam.
Netherlands Antilles Part of the Kingdom of the Netherlands, comprising three islands north of Venezuela: *Aruba, Bonaire*, and *Curaçao*, each island largely autonomous; and a fourth largely autonomous unit comprising three islands in the Leeward Islands group: *Saba, Sint Eustatius*, and *Sint Maarten*; total, 336 square miles; capital, Willemstad, on Curaçao: also *Dutch West Indies, Netherlands West Indies*.
Netherlands East Indies Formerly, the island colonial possessions of the Netherlands in the Malay Archipelago: also *Dutch East Indies*. Also **Netherlands Indies**.
Netherlands Guiana See SURINAM.
Netherlands New Guinea A Netherlands overseas territory, comprising the western half of the island of New Guinea and several adjacent islands; 159,375 square miles; capital, Hollandia: also *Dutch New Guinea*.
Netherlands Timor See TIMOR.
Netherlands West Indies See NETHERLANDS ANTILLES.
neth·er·most (neth′ər·mōst) *adj*. Lowest.
neth·er·ward (neth′ər·wərd) *adv*. In a descending course; downward.
nether world The world of the shades or dead; specifically, the world of punishment after death; hell, conceived as being beneath the earth.
Né·thou (nā·too′), **Pic de** The French name for ANETO, PICO DE.
net knot *Biol*. A relatively large and thickened mass of chromatin in the nucleus of a cell.

ne·tsu·ke (nā·tsoo·kā) *n. Japanese* A small, carved or wrought toggle or button attached to a pipe case, etc.
Net·tie (net′ē) Diminutive of ANTOINETTE. Also **Net′ty**.
net·ting (net′ing) *n*. 1 A fabric of openwork; a net; network. 2 The act or operation of making net. 3 The act, practice, or right of using nets, as in fishing.
net·tle[1] (net′l) *n*. 1 An herb of the genus *Urtica*, with opposite leaves, inconspicuous, greenish, imperfect flowers, and minute stinging hairs. The stinging is due to the irritating watery juice discharged by the hairs when broken. 2 Any of the various plants of some other genus of the nettle family. 3 Any of various plants of the same or some other family, having some real or fancied resemblance to the nettle genus. 4 A condition of irritation. — *v.t.* **·tled, ·tling** 1 To sting as the nettle does. 2 To annoy or irritate; provoke. See synonyms under PIQUE[1]. [OE *netle*] — **net′tler** *n*.
net·tle[2] (net′l) *n. Naut.* A small rope made by tightly twisting two or three yarns. Also **net′tle·stuff** (-stuf′). [Var. of *knettle < knit*]
nettle cell A nematocyst.
nettle rash Urticaria. Also **nettle fever**.
net ton A short ton.
net·work (net′wûrk′) *n*. 1 A fabric of openwork; netting. 2 A system of interlacing lines, tracks, or channels. 3 Any complex arrangement of interconnected electrical circuits. 4 *Telecom.* A chain of broadcasting stations.
Neu·châ·tel (nœ′shə·tel′, *Fr*. nœ·shå·tel′) 1 A canton in western Switzerland in the Jura; 309 square miles. 2 Its capital, a town on the north shore of the Lake of Neuchâtel (24 miles long; 4 to 5 miles wide). German **Neu·en·burg** (noi′ən·boorkh).
Neuf·châ·tel (nœ′shə·tel′, *Fr*. nœ·shå·tel′) *n*. A soft, white cheese produced in Neufchâtel, a town in northern France.
Neuil·ly-sur-Seine (nœ·yē′sür·sen′) A suburb of NW Paris, France.
neuk (nyook) *n. Scot.* Nook; corner.
neume (nōom, nyōom) *n. Music* 1 One of various symbols in a system of notation first devised to aid rote singers of the Gregorian chants, indicating direction of the melody and later including the musical pitch and accents: also spelled **neum**. 2 *pl*. This system. [< LL *neuma* song < Gk. *pneuma* breath, sigh]
Neu-Meck·len·burg (noi′mek′lən·boorkh) A former name for NEW IRELAND.
Neu·pest (noi′pest) The German name for UJPEST.
Neu-Pom·mern (noi′pom′ərn) Former name of NEW BRITAIN.
Ne·u·quén (nā′oo·kān′) 1 A national territory of west central Argentina; 36,429 square miles; capital, Neuquén. 2 A river in this territory, flowing about 320 miles SE to the Río Negro.
neur– Var. of NEURO–.
neu·ral (noor′əl, nyoor′-) *adj*. 1 Of or pertaining to the nerves or nervous system: the *neural* axis. 2 Of, pertaining to, or situated on the side that contains the axis of the central nervous system; in vertebrates, the dorsal side. [< Gk. *neuron* nerve]
neu·ral·gi·a (nōo·ral′jē·ə, -jə, nyōō-) *n. Pathol*. An acute, paroxysmal pain along the course and over the local distribution of a nerve. [< NEUR– + -ALGIA] — **neu·ral′gic** *adj*.
neu·ras·the·ni·a (nōor′əs·thē′nē·ə, -thēn′yə, nyōōr′-) *n*. 1 *Pathol*. Derangement of the nervous system with depression of vital force; nervous prostration. 2 *Psychoanal*. A neurosis characterized by physical disorder, as headache, constipation, etc., originated by inadequate expression of the libido. [< NL < Gk. *neuron* a nerve + *asthenia* weakness] — **neu·ras·then′ic** (-then′ik) *adj. & n*.
Neu·rath (noi′rät), **Baron Konstantin von**, 1873–1956, German diplomat.
neu·ra·tion (nōo·rā′shən, nyōō-) *n*. Nervation.
neu·rax·is (nōo·rak′sis, nyōō-) *n. Anat.* The brain and spinal cord; the axon. [< NEUR– + AXIS] — **neu·rax′i·al** *adj*.
neu·rax·on (nōo·rak′son, nyōō-) *n. Anat.* The process of a nerve cell that forms the axial cylinder of a nerve; axon. [< NEUR– + AXON]
neu·rec·to·my (nōo·rek′tə·mē, nyōō-) *n. Surg.* The excision of a nerve. [< NEUR– + -ECTOMY]
neu·ren·ter·ic (nōor′ən·ter′ik, nyōor′-) *adj. Biol.* Of or pertaining to the neural and the

enteric tubes of the embryo: the *neurenteric canal,* the tube connecting the caudal end of the neural tube with the digestive tract of the embryo.

neu·ric (nŏŏr′ik, nyŏŏr′-) *adj.* 1 Of or pertaining to the nerves and nervous system; neural. 2 Having a nervous system.

neu·ri·lem·ma (nŏŏr′ə·lem′ə, nyŏŏr′-) *n. Anat.* The delicate sheath of a nerve fiber: also spelled *neurolemma.* Also **neu′ri·lem′a.** [<NL <Gk. *neuros* nerve + *eilēma* sheath]

neu·rine (nŏŏr′ēn, -in, nyŏŏr′-) *n. Biochem.* A poisonous sirupy ptomaine, $C_5H_{13}ON$, obtained from decomposing animal tissues.

neu·ri·tis (nōō·rī′tis, nyōō-) *n. Pathol.* Inflammation of a nerve. — **neu·rit′ic** (-rit′ik) *adj.*

neuro– *combining form* Nerve; pertaining to a nerve: *neurocyte*: also, before vowels, *neur–.* Also **neuri–.** [<Gk. *neuros* sinew, nerve]

neu·ro·blast (nŏŏr′ə·blast, nyŏŏr′-) *n. Biol.* 1 A cell with a large oval nucleus in the spinal cord of the early embryo. Prolongations of such cells form the nerve fibers. 2 A part of the nervous system of an insect resulting from histolysis in the larva. 3 An embryonic cell that develops into a nerve cell.

neu·ro·cele (nŏŏr′ə·sēl, nyŏŏr′-) *n. Anat.* The system of central communicating cavities (ventricles and passages) found in the spinal cord and brain. Also **neu′ro·coele.**

neu·ro·cyte (nŏŏr′ə·sīt, nyŏŏr′-) *n.* A nerve cell together with its processes.

neu·rog·li·a (nōō·rog′lē·ə, nyōō-) *n. Anat.* The supporting tissue of the central nervous system, composed of finely branched ectodermic cells with thin interlacing processes. [<NL <Gk. *neuro-* nerve + *glia* glue]

neu·roid (nŏŏr′oid, nyŏŏr′-) *adj.* Nervelike.

neu·ro·lem·ma (nŏŏr′ə·lem′ə, nyŏŏr′-) *n.* 1 The retina. 2 Neurilemma.

neu·rol·o·gy (nōō·rol′ə·jē, nyōō-) *n.* The science of the nervous system in health and disease. — **neu·ro·log·i·cal** (nŏŏr′ə·loj′i·kəl, nyŏŏr′-) *adj.* — **neu·rol′o·gist** *n.*

neu·rol·y·sis (nōō·rol′ə·sis, nyōō-) *n.* 1 The destruction of nerve tissue. 2 Liberation of a nerve from morbid adhesions. 3 Relief of nerve tension by stretching. 4 Nervous exhaustion through overstimulation. — **neu·ro·lyt·ic** (nŏŏr′ə·lit′ik, nyŏŏr′-) *adj.*

neu·ro·ma (nōō·rō′mə, nyōō-) *n. pl.* **·ma·ta** (-mə·tə) *Pathol.* A tumor developing from a nerve. [<NL <Gk. *neuros* nerve + *-ōma* a growth]

neu·ron (nŏŏr′on, nyŏŏr′-) *n.* 1 *Biol.* A nerve cell with all its processes and extensions. 2 *Entomol.* A vein or costa of an insect's wing. Also **neu′rone** (-ōn). [<NL <Gk. *neuros* nerve] — **neu·ron·ic** (nōō·ron′ik, nyōō-) *adj.*

neu·ro·path (nŏŏr′ə·path, nyŏŏr′-) *n. Psychiatry* One suffering from or subject to nervous disorders; a neurotic. — **neu′ro·path′ic, neu′ro·path′i·cal** *adj.* — **neu′ro·path′i·cal·ly** *adv.*

neu·ro·pa·thol·o·gy (nŏŏr′ō·pə·thol′ə·jē, nyŏŏr′-) *n.* The pathology of the nervous system. — **neu′ro·pa·thol′o·gist** *n.*

neu·ro·phys·i·ol·o·gy (nŏŏr′ō·fiz′ē·ol′ə·jē, nyŏŏr′-) *n.* The physiology of the nervous system. — **neu′ro·phys′i·o·log′i·cal** (-fiz′ē·ə·loj′i·kəl) *adj.* — **neu′ro·phys′i·ol′o·gist** *n.*

neu·ro·psy·chi·a·try (nŏŏr′ō·sī·kī′ə·trē, nyŏŏr′-) *n.* The study and treatment of diseases involving both neurological and mental factors; the pathology of nervous disorders combined with psychiatry. — **neu′ro·psy′chi·at′ric** (-sī′kē·at′rik) *adj.* — **neu′ro·psy·chi′a·trist** (-sī·kī′ə·trist) *n.*

neu·ro·psy·chol·o·gy (nŏŏr′ō·sī·kol′ə·jē, nyŏŏr′-) *n.* The study of the relationships existing between the mind and the nervous system. — **neu′ro·psy′cho·log′i·cal** (-sī′kə·loj′i·kəl) *adj.*

neu·ro·psy·cho·sis (nŏŏr′ō·sī·kō′sis, nyŏŏr′-) *n. Psychiatry* Mental disorder arising from a nervous disorder: often used interchangeably with *psychosis.*

neu·rop·ter (nōō·rop′tər, nyōō-) *n.* Any of an order *(Neuroptera)* of insects having four similar, net-veined wings, chewing mouth parts, and active carnivorous larvae, as antlions, etc. [<NL <Gk. *neuros* sinew, nerve + *pteron* wing] — **neu·rop′ter·al** *adj.* — **neu·rop′ter·an** *adj.* & *n.*

neu·rop·ter·oid (nōō·rop′tə·roid, nyōō-) *adj.* Like the *Neuroptera.* — *n.* A neuropteroid insect.

neu·rop·ter·ous (nōō·rop′tər·əs, nyōō-) *adj.* 1 Of or pertaining to the *Neuroptera.* 2 Having net-veined wings.

neu·ro·sis (nōō·rō′sis, nyōō-) *n. pl.* **·ses** (-sēz) *Psychiatry* A disorder of the psychic or mental functions without lesion of nerves and of less severity than a psychosis. — **neu·ro′sal** *adj.*

neu·rot·ic (nōō·rot′ik, nyōō-) *adj.* 1 Pertaining to or suffering from neurosis. 2 *Colloq.* Having a morbid nature or tendency. 3 Pertaining to a nerve or the nervous system. — *n.* 1 A person afflicted with neurosis. 2 Disease of the nerves.

neu·rot·o·my (nōō·rot′ə·mē, nyōō-) *n.* 1 *Surg.* The division or severing of a nerve, to relieve pain. 2 The dissection of the nervous system, as for study. [<NEURO- + -TOMY] — **neu·ro·tom·ic** (nŏŏr′ə·tom′ik, nyŏŏr′-) *adj.* — **neu·rot′o·mist** *n.*

Neu·satz (noi′zäts) The German name for NOVI SAD.

Neuse River (nōōs, nyōōs) A river in eastern North Carolina, flowing 300 miles SE to Pamlico Sound.

Neus·tri·a (nōōs′trē·ə, nyōōs′-) The western part of the Frankish Empire of the sixth century, comprising the region of modern France between the Meuse and the Loire.

neu·ter (nōō′tər, nyōō′-) *adj.* 1 *Gram.* **a** Neither masculine nor feminine in gender. Compare GENDER. **b** *Rare* Intransitive; neither active nor passive; middle: said of verbs in classical languages. 2 *Biol.* Sexless; having functionless or imperfectly developed sex organs, as certain plants and animals. 3 *Obs.* Neutral. — *n.* 1 An animal of no apparent sex, as a worker bee. 2 A eunuch. 3 A castrated animal. 4 *Gram.* **a** The neuter gender. **b** A word in this gender. 5 A neutral in warfare or other conflict. [<MF *neutre* <L *neuter* < *ne-* not + *uter* either]

neu·tral (nōō′trəl, nyōō′-) *adj.* 1 Refraining from interference in a contest; not taking the part of either or any belligerent: a *neutral* power. 2 Belonging to neither of two contestants; belonging to a neutral power: *neutral* forces. 3 Having no decided character; indefinite; middling. 4 Having no decided color; predominantly brownish or grayish. 5 *Biol.* Sexless; neuter. 6 *Bot.* Lacking pistils or stamens. 7 *Chem.* Lacking decided acid or alkaline qualities. 8 *Electr.* Neither positive nor negative. 9 *Phonet.* Pronounced with the tongue in a relaxed, mid-central position, as the *a* in *about.* — *n.* One who or that which is neutral. [<L *neutralis* < *neuter* neuter] — **neu′tral·ly** *adv.*

neu·tral·ism (nōō′trəl·iz′əm, nyōō′-) *n.* A political doctrine holding that abstention from alliance with ideologic or economic power blocs in international relations serves a country's best interests. — **neu′tral·ist** *n.*

neu·tral·i·ty (nōō·tral′ə·tē, nyōō-) *n. pl.* **·ties** 1 The state of being a neutral nation during a war. 2 The state of being neither good nor bad; indifference. 3 *Chem.* The condition of being neither acid nor basic. 4 The character of being neutral: the *neutrality* of a ship.

neu·tral·i·za·tion (nōō′trəl·ə·zā′shən, -ī·zā′-, nyōō′-) *n.* 1 Act of neutralizing or state of being neutralized. 2 *Ling.* The temporary suspension of a relevant feature in two phonemes, as /t/ and /d/ in *latter* and *ladder.*

neu·tral·ize (nōō′trəl·īz, nyōō′-) *v.t.* **·ized, ·iz·ing** 1 To counteract or destroy by an opposite force or influence; nullify; counterbalance. 2 To declare (a nation, area, etc.) to be neutral and not involved in hostilities. 3 *Chem.* To make neutral or inert. 4 *Electr.* To render electrically inert; void of electricity. 5 *Mil.* To render incapable of effective action. Also *Brit.* **neu′tral·ise.** — **neu′tral·iz′er** *n.*

neutral oil A light lubricating oil from petroleum, generally mixed with animal or vegetable oils.

neu·tri·no (nōō·trē′nō, nyōō-) *n. pl.* **·nos** *Physics* An atomic particle associated with the radioactive emission of beta rays, carrying no electric charge and having a mass comparable to that of the electron.

neu·tron (nōō′tron, nyōō′-) *n. Physics* An electrically neutral particle of the atom, having a mass approximately equal to that of the proton.

neu·tro·pe·ni·a (nōō′trə·pē′nē·ə, nyōō′-) *n. Pathol.* A blood disorder marked by a sharp reduction in the number of leucocytes; agranulocytosis. [<NEUTRO(PHILE) + Gk. *penia* dearth]

neu·tro·phile (nōō′trə·fīl, -fil, nyōō′-) *adj.* Stainable with neutral dyes: said of bacteria and leucocytes. Also **neu′tro·phil** (-fil), **neu′·tro·phil′ic.**

Neuve-Cha·pelle (nœv·shà·pel′) A village in northern France, the center of severe fighting in World War I, 1914–15.

Ne·va (nē′və, *Russian* nyi·vä′) A river in NW U.S.S.R., flowing 46 miles from Lake Ladoga to its delta at Leningrad on the Gulf of Finland.

Ne·vad·a (nə·vad′ə, -vä′də) A State in the western United States; 110,540 square miles; capital, Carson City; entered the Union Oct. 31, 1864; nickname, *Sagebrush State*: abbr. *Nev.* — **Ne·vad′an** *adj.* & *n.*

né·vé (nā·vā′) *n.* The consolidated snow on the summit of a mountain, composed of roundish grains, resembling sand: a transition stage in the formation of glacier ice. [< dial. F (Swiss), glacier, ult. <L *nix, nivis* snow]

nev·er (nev′ər) *adv.* 1 Not at any time: also used in composition to form adjectives: *never-ending.* 2 Not at all; positively not: used emphatically: *Never* fear. [OE *næfre* < *ne* not + *æfre* ever]

nev·er·more (nev′ər·môr′, -mōr′) *adv.* Never again.

Ne·vers (nə·vâr′) A city on the Loire in central France.

never so To an extent or degree beyond the actual or conceivable; no matter how: *never so* great.

nev·er·the·less (nev′ər·thə·les′) *conj.* & *adv.* None the less; however; yet. See synonyms under BUT[1], NOTWITHSTANDING.

Nev·in (nev′in), **Ethelbert,** 1862–1901, U.S. composer.

Ne·vis (nē′vis, nev′is) See ST. CHRISTOPHER, NEVIS, AND ANGUILLA.

Nev·ski (nev′skē, nef′-), **Alexander** See ALEXANDER NEVSKI.

ne·vus (nē′vəs) *n. pl.* **·vi** (-vī) A birthmark, or congenital mole on man or animal: also spelled *naevus.* [<L *naevus* blemish] — **ne′·void** (-void) *adj.*

new (nōō, nyōō) *adj.* 1 Recently come into existence or use; lately made; recently settled or recently opened to settlement: *new* country. 2 Lately discovered; also, recently become important or well known. 3 Beginning or recurring afresh; renewed: the *new* moon. 4 Changed in essence, constitution, force, etc.: usually for the better: I feel a *new* man. 5 Another; different from that heretofore known or used. 6 Recently come from any place or out of any condition. 7 Unaccustomed; unfamiliar: a horse *new* to the saddle. 8 Specifically, named for another: used in place names, to distinguish a place from its namesake: *New* Zealand, *New* Orleans. — *adv.* Newly; recently. [OE *nēowe*] — **new′ness** *n.*

Synonyms (adj.) fresh, modern, novel, recent, young, youthful. That which is *modern* has begun to exist in the present age, and still exists; *recent* denotes that which has come into existence within a comparatively brief period and may or may not be existing still. A *novel* contrivance is one that has never before been known. *Young* and *youthful* are applied to that which has life; that which is *young* is possessed of a comparatively new existence as a living thing, possessing actual youth; that which is *youthful* manifests the attributes of youth. Compare YOUTHFUL. *Fresh* applies to that which has newness or youth, but may deteriorate in time. *New* is opposed to *old, modern* to *ancient, recent* to *remote, young* to *old, aged,* etc.

New Amsterdam The capital of the Dutch colony of New Netherland; in 1664 renamed *New York.*

New·ark (nōō′ərk, nyōō′-) A city in NE New Jersey on **Newark Bay,** an estuary at the confluence of the Hackensack and Passaic rivers.

New Bedford A city on Buzzards Bay, SE Massachusetts.
new blue Ultramarine.
New·bolt (noo'bōlt, nyoo'-), **Sir Henry John**, 1862-1938, English poet and historian.
new·born (noo'bôrn', nyoo'-) *adj.* Just lately born; also, reborn.
New Britain The largest island in the Bismarck Archipelago; 14,600 square miles; chief city, Rabaul: formerly *Neu-Pommern.*
New Brunswick A province of SE Canada on the Bay of Fundy; 27,985 square miles; capital, Fredericton: abbr. *N.B.*
New·burg (noo'bûrg, nyoo'-) See À LA NEWBURG.
New Caledonia An island in the SW Pacific east of Australia; 8,548 square miles; with its dependencies, the Loyalty Islands, the Isle of Pines, and other smaller adjacent islands, it comprises a French overseas territory; total, 9,401 square miles; capital, Nouméa: French *Nouvelle Calédonie.*
New Castile A region of central Spain. *Spanish* Cas·til·la la Nue·va (käs·tē'lyä lä nwä'vä).
New·cas·tle-up·on-Tyne (noo'kas·əl·ə·pon'tin', nyoo'-) A port in SE Northumberland, England, and its county seat: also **Newcastle, Newcastle-on-Tyne. — to carry coals to Newcastle** To take goods to a place where they already abound; hence, to waste one's labor.
New Church See SWEDENBORGIANISM.
New·chwang (noo'chwäng') A town in SW Liutong province, Manchuria.
New·comb (noo'kəm, nyoo'-), **Simon**, 1835-1909, U.S. astronomer.
new·com·er (noo'kum'ər, nyoo'-) *n.* One who has recently arrived.
new deal 1 A dealing of cards with a new deck. **2** Any new system designed to do away with old ills and to promote reform.
New Deal The policies and principles of the administration under President Franklin D. Roosevelt, embracing various social, economic, and political measures through legislative and administrative change; also, the Roosevelt administration.
New Dealer A supporter of the measures advocated by the New Deal; hence, loosely, an adherent of President Roosevelt . rather than the Democratic party as such.
New Delhi The capital of India in Delhi State, an administrative center just SW of Delhi.
New Economic Policy The policy adopted in 1921 as a temporary measure by the government of the Soviet Union, permitting to a limited extent private ownership in minor industries, restoring free trade in agricultural production, substituting a labor tax for requisition of pay for services rendered in place of the uniform wage rate. Also *Nep, NEP.*
new·el (noo'əl, nyoo'-) *n.* **1** A post from which the steps of a winding stair radiate. **2** A post at the end of a stair or hand rail. [<OF *nouel* stone of a fruit <LL *nucale* <L *nux* nut]
New England The NE section of the United States, including Maine, New Hampshire, Vermont, Massachusetts, Rhode Island, and Connecticut. **— New Englander**

NEWEL

New England aster A perennial aster (*Aster novae-angliae*) having purple flowers and growing wild throughout eastern North America.
new·fan·gled (noo'fang'gəld, nyoo'-) *adj.* **1** Of new fashion: generally in depreciation: *newfangled* notions. **2** Disposed to value things for their novelty: *new-fangled* people. **— new'·fan'gled·ness** *n.*
new-fash·ioned (noo'fash'ənd, nyoo'-) *adj.* Made in a new style; recently become fashionable.
New Forest A partly wooded region in SW Hampshire, England; about 145 square miles.
New·found·land (noo'fənd·lənd, nyoo'-) *n.* A large dog of a breed originating in Newfoundland, characterized by a broad head, square muzzle, and thick, abundant, usually black coat.
New·found·land (noo'fənd·land', nyoo'-) The easternmost province of Canada, comprising the island of Newfoundland (42,734 square miles) and its dependency, Labrador, on the mainland; total, 152,734 square miles; capital, St. John's **— New·found'land·er** (-fənd'-) *n.*
New France The region discovered and settled by the French in North America, including Canada, the Great Lakes region, and Louisiana, from 1534 to 1763.
New·gate (noo'git, -gāt, nyoo'-) A prison in London, destroyed 1902.
New Georgia An island group in the British Solomon Islands; total, 2,000 square miles; also, the chief island of this group.
New Granada A Spanish viceroyalty in NW South America, later divided into the republics of Ecuador, Colombia, Panama, and Venezuela.
New Guin·ea (gin'ē) The world's second largest island, north of Australia in the Malay Archipelago; 304,200 square miles: also *Papua:* Indonesian **I·ri·an** (ē'rē·än); divided into Netherlands New Guinea, Territory of New Guinea, and Territory of Papua.
New Guinea, Territory of A United Nations Trust Territory, comprising NE New Guinea, the Bismarck Archipelago, and Bougainville and Buka in the Solomon Islands; about 93,000 square miles; administrative center, Port Moresby: became part of the Territory of Papua and New Guinea, 1949.
New Hampshire A State of the NE United States; 9,304 square miles; capital, Concord; entered the Union June 21, 1788, one of the thirteen original States; nickname, *Granite State:* abbr. *N.H.*
New Hanover A former name for LAVONGAI.
New Harmony A socialistic community in SW Indiana established in 1825 by Robert Owen.
New Haven A city on Long Island Sound, southern Connecticut; site of Yale University, founded 1701.
New Hebrides An island group in the SW Pacific north of New Caledonia and west of Fiji, comprising an Anglo–French condominium; 5,700 square miles; capital, Vila.
New High German See HIGH GERMAN under GERMAN.
New Ireland An island of the Bismarck Archipelago; 3,340 square miles: formerly *Neu-Mecklenburg.*
new·ish (noo'ish, nyoo'-) *adj.* Rather new.
New Jersey A State of the eastern United States on the Atlantic; 7,836 square miles; capital, Trenton; entered the Union Dec. 18, 1787, one of the thirteen original States; nickname, *Garden State:* abbr. *N.J.* **— New Jerseyite**
New Jerusalem The city of God; heaven. *Rev.* xxi 2.
New Jerusalem Church See SWEDENBORGIANISM.
New Latin See under LATIN.
New London A port in SE Connecticut on Long Island Sound; site of a United States naval base.
new·ly (noo'lē, nyoo'-) *adv.* **1** In a new or recent manner; lately. **2** In a different way; so as to be or appear new; afresh.
new·ly-wed (noo'lē·wed', nyoo'-) *n.* A person recently married.
New·man (noo'mən, nyoo'-), **Cardinal John Henry**, 1801-90, English author and theologian.
new·mar·ket (noo'mär·kit, nyoo'-) *n.* **1** A long, close-fitting coat for outdoor wear: also **Newmarket coat. 2** A game of cards resembling the game of *stops.*
New·mar·ket (noo'mär·kit, nyoo'-) A town in West Sussex, England; a horse-racing center since the 17th century.
New Mexico A State of the SW United States on the Mexican border; 121,666 square miles; capital, Santa Fé; entered the Union Jan. 6, 1912; nickname, *Sunshine State:* abbr. *N.M., New M., N. Mex.* **— New Mexican**
new moon That phase of the moon when it is directly between the earth and the sun, its disk then being invisible; also, the first visible crescent of the disk.
new-mown (noo'mōn', nyoo'-) *adj.* Recently cut or mown.
New Netherland The Dutch colony in North America, 1613-1664, near the mouth of the Hudson River; capital, New Amsterdam.
New Norwegian Landsmål.
New Order The system of regimentation imposed on Germany after 1933, and later on countries conquered or occupied by the Axis powers.
New Or·le·ans (ôr'lē·ənz, ôr·lēnz', ôr'lənz) A port of SE Louisiana on the Mississippi: a native of the city is sometimes known as an *Orleanian.* **— New Or·le·an·i·an** (ôr·lē·nē·ən or ôr·lē'nē-)
New·port (noo'pôrt, -pōrt, nyoo'-) **1** A city of SE Rhode Island on Narragansett Bay; site of a United States naval base. **2** A municipal borough of Hampshire, England, on the Isle of Wight. **3** A county borough and port of Monmouthshire, England, on the Severn estuary.
Newport News A city on the James River in SE Virginia.
New Providence Island The chief island of the Bahamas; 58 square miles.
New Quebec See UNGAVA.
new-rich (noo'rich', nyoo'-) *adj.* Newly rich; hence, showy; pretentious.
new rich Those who have recently acquired riches. Also, *French, nouveaux riches.*
news (nooz, nyooz) *n. pl.* (construed as singular) **1** Fresh information concerning something that has recently taken place. **2** A newspaper. **3** Anything new or strange. See synonyms under TIDINGS. [Trans. of OF *noveles* <LL *nova* new (things)]
news agency 1 A business concern that deals in and distributes newspapers and other periodicals. **2** An agency that sells news items to newspapers, etc.: also called **news bureau.**
New Sar·um (sâr'əm) See SALISBURY.
news·boy (nooz'boi', nyoo'-) *n.* A boy who sells or delivers newspapers.
news·cast (nooz'kast', -käst', nyoo'-) *n.* A radio news program. **—** *v.t.* & *v.i.* To broadcast (news). **— news'cast'er** *n.*
news-gath·er·er (nooz'gath'ər·ər, nyoo'-) *n.* One who collects news.
news·hawk (nooz'hôk', nyoo'-) *n. U.S. Colloq.* A journalist with a sharp eye for news.
New Siberian Islands An archipelago off the Arctic coast of the Yakut S.S.R.; total, 11,000 square miles.
news·man (nooz'man', -mən, nyoo'-) *n. pl.* **-men** (-men', -mən) A man who delivers or sells newspapers; also, a newspaper reporter.
news·mon·ger (nooz'mung'gər, -mong'-, nyoo'-) *n.* One who carries news about: especially, a gossip.
New South Wales A state of SE Commonwealth of Australia; 309,433 square miles; capital, Sydney.
New Spain The former Spanish possessions, comprising a viceroyalty in the SW United States, Central America north of Panama, the West Indies, and the Philippines.
news·pa·per (nooz'pā'pər, nyoo'-) *n.* A publication issued for general circulation at frequent intervals; a public print that circulates news, etc. **— news'pa'per·man** (-man', -mən) *n.*
news·print (nooz'print', nyoo'-) *n.* The thin, unsized paper on which the ordinary daily or weekly newspaper is printed.
news·reel (nooz'rēl', nyoo'-) *n.* A motion picture, usually of short duration, showing events of current interest.
news·stand (nooz'stand', nyoo'-) *n.* A stand or stall at which newspapers and periodicals are offered for sale.
New Style See GREGORIAN CALENDAR under CALENDAR.
New Sweden A Swedish colony in North America, 1638-55, on the Delaware River below Trenton.
news·worth·y (nooz'wûr'thē, nyoo'-) *adj.* Important enough to be written up in a newspaper; considered to be of general or current interest.
news·y (nooz'ē, nyoo'-) *adj. Colloq.* Full of news.
newt (noot, nyoot) *n.* One of various small, semiaquatic, salamander-like amphibians, chiefly of the genus *Triturus.* [Earlier *ewt, evet,* OE *efete;* in ME *an ewt* was taken as *a newt*]

NEWT
(From 3 1/2 to 20 inches in length, varying with the species)

New Territories A leased section of Hong Kong colony on the Chinese mainland north of Kowloon peninsula; 359 square miles.
New Testament 1 The promises of God to man as revealed in the life and teachings of

New Thought A modern religious philosophy stressing "God in man" and the power of right thinking over disease and failure: also called *Higher Thought, Mental Science, Practical Christianity.*

new·ton (noo'tən, nyoo'-) *n. Physics* A unit of force in the mks system, equal to 100,000 dynes, or that force which will impart to a mass of 1 kilogram an acceleration of 1 meter per second per second. [after Isaac *Newton*]

New·ton (noo'tən, nyoo'-), **Sir Isaac,** 1642–1727, English philosopher and mathematician; formulated the law of gravitation, the binomial theorem, and the elements of differential calculus. — **New·to·ni·an** (noo-to'nē-ən, nyoo-) *adj.*

New World The western hemisphere.

new year The year just begun or just about to begin.

New Year The first day of the year; in the Gregorian calendar, January 1. Also **New Year's Day.** Compare CALENDAR.

New Year's Eve The evening of December 31.

New York 1 A State of the NE United States on the Atlantic; 49,576 square miles; capital, Albany; entered the Union July 26, 1788; one of the thirteen original States; nickname, *Empire State;* abbr. *N.Y.* 2 A port at the mouth of the Hudson River in SE New York, divided into the five boroughs of the Bronx, Brooklyn, Manhattan, Queens, and Richmond, comprising the largest city of the United States; 365 square miles: also *Greater New York.*

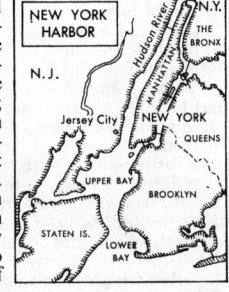

New York Bay An inlet of the Atlantic Ocean at the mouth of the Hudson River, forming **New York Harbor** and divided by the Narrows into **Upper New York Bay** and **Lower New York Bay.**

New York·er (yôr'kər) An inhabitant of New York; specifically, a native or resident of New York City.

New York State Barge Canal A waterway system of central New York, connecting the Hudson River with Lake Erie and with Lake Champlain; total length, 525 miles.

New Zea·land (zē'lənd) A self-governing member of the Commonwealth of Nations, comprising a group of islands, principally North Island and South Island, in the South Pacific SE of Australia; 103,416 square miles, excluding island territories; capital, Wellington.

New Zea·land·er (zē'lən·dər) *n.* A resident of New Zealand; formerly, a Maori.

Nex·ö (nek'sœ), **Martin Andersen,** 1869–1954, Danish novelist born in Germany.

next (nekst) *adj.* 1 Being nearest to, in time, space, order, rank, etc.; immediately succeeding or preceding. 2 Almost: *next* to impossible. See synonyms under ADJACENT, IMMEDIATE. — *adv.* In the nearest time, place, or rank; especially, immediately succeeding: when I *next* meet her. — *prep.* Nearest to. [OE *nēahst,* superl. of *nēah* nigh]

next–door (neks'dôr', -dōr') *adj.* In the next house: a *next-door* neighbor.

next door 1 The nearest or adjacent house. 2 In the next house: the lady *next door.*

next friend *Law* A person who, as the nearest friend, appears to prosecute an action in behalf of someone under legal disability, as a minor child, etc.

next of kin 1 *Law* The kindred of a person who would share in his estate according to the statutes of distribution. 2 The person most closely related to one.

nex·us (nek'səs) *n. pl.* **·us·es** or **·us** A bond or tie between the several members of a group or series; link. [<L <*nectere* tie]

Ney (nā), **Michel,** 1769–1815, French marshal under Napoleon; executed.

Ney·sha·pur (nā'shä·poor) See NISHAPUR.

Nez Per·cé (nez' pûrs', *Fr.* nā per·sā') One of a tribe of North American Indians of Shahaptian stock, formerly dwelling in Idaho, Oregon, and Washington. [<F, pierced nose]

Nga·mi (ngä'mē), **Lake** A marsh region of NW Bechuanaland Protectorate, formerly a lake covering 20,000 square miles.

Ngan·dong man (ngän'dong) Solo man.

ni·a·cin (nī'ə·sin) *n.* Nicotinic acid.

Ni·ag·a·ra Falls (nī·ag'ə·rə, -rə) 1 A city on the Niagara River, western New York. 2 See under NIAGARA RIVER.

Niagara River A river between Ontario province, Canada, and New York State, flowing 34 miles from Lake Erie to Lake Ontario; in its course occurs **Niagara Falls,** a cataract divided by Goat Island into the American Falls, 167 feet high and 1,000 feet wide, and Horseshoe Falls on the Canadian side, 160 feet high and 2,500 feet wide.

Nia·mey (nyä·mā') A port on the Niger, capital of the Republic of Niger.

Ni·as (nē'äs) An Indonesian island off the western coast of Sumatra; 1,569 square miles.

Ni·as·sus (nī·as'əs) Ancient name of NISH.

nib (nib) *n.* 1 A projecting, pointed part. 2 A beak of a bird; neb. 3 The point of a pen. 4 The point of anything. — *v.t.* **nibbed, nib·bing** 1 To furnish with a nib. 2 To sharpen or mend the nib or point of. [Var. of NEB]

nib·ble (nib'əl) *v.* **·bled, ·bling** *v.t.* 1 To eat (food) with small, quick bites. 2 To bite gently or cautiously, as bait. — *v.i.* 3 To bite off or eat little bits. 4 To take gentle or cautious bites: usually with *at.* — *n.* The act of nibbling; a little bite. [Cf. LG *nibbelen*] — **nib'bler** *n.*

Ni·be·lung (nē'bə·loong) *n. pl.* **·lungs** or **·lung·en** (-loong'ən) 1 In Teutonic mythology, one of the children of the mist, a dwarf people who held a magic ring and hoard of gold, which were wrested from them by Siegfried. 2 In the *Nibelungenlied,* one of the Burgundian kings.

Ni·be·lung·en·lied (nē'bə·loong'ən·lēt') The lay of the Nibelungs, a Middle High German epic poem written by an unknown author during the early 13th century, embodying legends of the Burgundian kings which were based on various compilations; the source of Wagner's operatic cycle, *The Ring of the Nibelung.*

nib·lick (nib'lik) *n.* A golf club with a slanted iron head for lifting the ball out of bunkers, long grass, etc. See illustration under GOLF CLUB. [Origin uncertain]

Ni·cae·a (nī·sē'ə) 1 An ancient town of Bithynia in Asia Minor on the site of modern Iznik. 2 An ancient name for NICE. — **Council of Nicaea** A council held at Nicaea, Asia Minor, A.D. 325, which condemned Arianism and promulgated the Nicene Creed. — **Ni·cae'an** *adj. & n.*

Nic·a·ra·gua (nik'ə·rä'gwə) A republic of Central America; 57,143 square miles; capital, Managua.

Nicaragua, Lake The largest lake of central America, in SW Nicaragua; 3,100 square miles.

Nic·a·ra·guan (nik'ə·rä'gwən) *adj.* Of or pertaining to Nicaragua. — *n.* A native of Nicaragua.

Ni·ca·ri·a (nyi·kä·rē'ə) A former name for ICARIA. Also **Ni'ka·ri·a.**

nic·co·lite (nik'ə·līt) *n.* A usually massive, brittle, metallic, pale copper-red nickel arsenide, NiAs, crystallizing in the hexagonal system: also called *copper nickel.* [<NL *niccolum* nickel]

nice (nīs) *adj.* **nic·er, nic·est** 1 Characterized by discrimination and judgment; acute; discerning. 2 Refined and pure in tastes or habits; refined; hence, overparticular; dainty; modest; fastidious; scrupulous. 3 Requiring careful consideration, discrimination, or treatment; delicate; subtle. 4 Exactly fitted or adjusted; accurate. 5 Delicately constructed; hence, easily disarranged or injured; fragile; tender. 6 Agreeable or pleasant in any way; a wide use. 7 Agreeable socially; respectable: *nice* people. See synonyms under CHOICE, FINE[1], NEAT[1], PRECISE, TASTEFUL. ◆ Homophone: *gneiss.* [<OF, stupid <L *nescius* ignorant < *ne-* not + *scire* know] — **nice'ly** *adv.* — **nice'ness** *n.*

Nice (nēs) A port of SE France: ancient *Nicaea,* Italian *Nizza.* — **Ni·çoise** (nē·swäz') *adj. & n.*

Ni·cene (nī'sēn, nī·sēn') *adj.* Pertaining to Nicaea, a town in Asia Minor: also spelled *Nicaean.*

Nicene Creed *Eccl.* 1 A Christian confession of faith, adopted against the Arian heresy by the first Council of Nicaea, A.D. 325. 2 A similar creed, later attributed to the Council of Constantinople, A.D. 381, and accepted by the Greek Church: also called *Constantinopolitan* or *Niceno-Constantinopolitan Creed.* 3 A modification of this, containing a clause referring to the Holy Spirit, adopted by the Council of Toledo, A.D. 589, and accepted by the Anglican, Roman Catholic, and various Protestant churches.

ni·ce·ty (nī'sə·tē) *n. pl.* **·ties** 1 The quality of being nice. 2 A delicate point or distinction; refinement of criticism; subtlety. 3 A rare or delicious thing; delicacy. 4 Fastidiousness. 5 *Obs.* Coyness; shyness. [<OF *niceté* folly <*nice.* See NICE.]

niche (nich) *n.* 1 A recessed space or hollow; specifically, a recess in a wall for a statue or the like. 2 Hence, any position specially adapted to its occupant. — *v.t.* **niched, nich·ing** To put in a niche. [<F <*nicher* nest, ult. <L *nidus* a nest]

nich·er (nikh'ər) See NICKER[2].

Nich·o·las (nik'ə·ləs) A masculine personal name. Also *Lat.* **Ni·co·la·us** (nik'ō·lā'əs), *Fr.* **Ni·co·las** (nē·kō·lä'), *Ital.* **Nic·co·lò** (nēk'kō·lô'), *Pg.* **Ni·co·láo** (nē'kō·lä'ōō), *Sp.* **Ni·co·lás** (nē'kō·läs'), *Ger.* **Ni·ko·laus** (nē'kō·lous), *Russian* **Ni·co·lai** (nē'kō·lī'). [<Gk., victory of the people]

NICHE

— **Nicholas, Saint** Fourth century prelate, bishop of Myra; patron saint of Russia, seamen, and children. In Dutch nursery lore, the Santa Claus who brings presents to children on Christmas Eve. See SANTA CLAUS.
— **Nicholas I,** 1796–1855, czar of Russia 1825–55.
— **Nicholas II,** 1868–1918, czar of Russia 1894–1917; executed.
— **Nicholas V,** 1397–1455, real name Tommaso Parentucelli, pope 1447–55.
— **Nicholas of Cusa,** 1401–64, German prelate and scholar.

Nich·ol·son (nik'əl·sən), **Sir Francis,** 1655–1728, British colonial administrator.

Ni·chrome (nī'krōm) *n.* An alloy of nickel, chromium, and iron, highly resistant to electricity, heat, and corrosion: a trade name.

nicht wahr (nikht vär') *German* Isn't that so?

Ni·ci·as (nish'ē·əs) Athenian general; killed at Syracuse, 413 B.C.

nick (nik) *n.* 1 A slight cut, chip, or indentation in the surface or edge of anything. 2 A score or tally: from the use of notched sticks for keeping tally. 3 A point of time; critical moment. 4 *Printing* A groove on the shank of a type character to aid in correct alinement. — *v.t.* 1 To make a nick or nicks in; notch. 2 To record or tally by making nicks. 3 To cut through or into. 4 *Slang* To cheat or trick. 5 To hit or catch at the exact moment; guess or understand exactly. 6 *Brit. Slang* To arrest; catch. [Origin uncertain] — **nick'er** *n.*

Nick (nik) The devil. Also **Old Nick.** [Nickname of NICHOLAS]

nick·el (nik'əl) *n.* 1 A hard, ductile, malleable, silver-white metallic element (symbol Ni) of the iron-cobalt group: widely used in metallurgy and the arts. See ELEMENT. 2 A five-cent coin of the United States, of a nickel-and-copper alloy. — *v.t.* To plate with nickel. [<Sw. <G *(kupfer)nickel,* lit., copper demon; because its ore looks like copper but contains none]

nick·el·bloom (nik'əl·bloom') *n.* Annabergite.

nickel carbonyl *Chem.* A yellow, volatile, highly poisonous liquid, $Ni(CO)_4$, obtained by passing carbon monoxide over finely divided nickel. Also **nickel tetracarbonyl.**

nick·el·ic (nik′əl·ik) *adj.* Of, pertaining to, or containing nickel, especially trivalent nickel.

nick·el·if·er·ous (nik′əl·if′ər·əs) *adj.* Containing nickel, as ore.

nick·el·o·de·on (nik′əl·ō′dē·ən) *n.* **1** An early type of motion-picture theater charging an admission fee of five cents. **2** An automatic slot machine, such as a cinematograph or a phonograph, which performs when a nickel is inserted. [<NICKEL (def. 2) + *odeon*, var. of ODEUM]

nick·el·ous (nik′əl·əs) *adj.* Of, pertaining to, or containing nickel, especially bivalent nickel.

nick·el-plate (nik′əl·plāt′) *v.t.* **-plat·ed, -plat·ing** To cover with nickel by electroplating. — **nick′el-plat′ed** *adj.*

nickel plate A thin layer of nickel deposited on the surface of objects by electrolysis.

nickel silver German silver.

nick·er[1] (nik′ər) *n.* A seed of the nickernut tree, used by children in games resembling marbles. Also **nick′er·nut.** [? <Du. *knikker* a marble]

nick·er[2] (nik′ər) *n.* & *v.i.* **1** Neigh. **2** Laugh. Also, *Scot.*, *nicher.* [Imit.]

nick·er·nut tree (nik′ər·nut) **1** A tropical American climbing shrub (*Caesalpinia crista*) with prickly, oval pods bearing seeds or nuts called *nickernuts* or *bonducnuts.* **2** The Kentucky coffee tree. [? <NICKER[1], from the use of its seeds as marbles]

nick·nack (nik′nak′) See KNICK-KNACK.

nick·name (nik′nām′) *n.* **1** A familiar name, sometimes a diminutive, as Tom for Thomas. **2** A descriptive or facetious name given to a person, place, or thing in derision, affection, or acclaim, as Longshanks, Honest Abe, Empire State, etc. — *v.t.* **·named, ·nam·ing 1** To give a nickname to. **2** To misname. [ME *ekename* a surname; later *an ekename* was taken as *a nickname*]

Nic·o·bar Islands (nik′ə·bär′) See under ANDAMAN AND NICOBAR ISLANDS. Also **Nic′o·bars′.**

Nic·o·de·mus (nik′ə·dē′məs) A ruler of the Jews and disciple of Christ. *John* iii 1.

Nic·o·lay (nik′ə·lā), **John George,** 1832–1901, U.S. biographer.

Ni·colle (nē·kôl′), **Charles Jean Henri,** 1866–1936, French physician and bacteriologist.

Nic·ol prism (nik′əl) *Optics* A set of two prisms of Iceland spar cemented together: used for producing plane-polarized light. Also **nic′ol.** [after Wm. *Nicol*, 1768?–1851, Scottish physicist]

Nic·ol·son (nik′əl·sən), **Harold,** born 1886, English biographer and diplomat.

Ni·cop·o·lis (ni·kop′ə·lis, nī-) An ancient city of southern Epirus, Greece.

Nic·o·si·a (nik′ə·sē′ə) The capital of Cyprus, in the north central part.

nic·o·tine (nik′ə·tēn, -tin) *n.* A poisonous, colorless, oily, liquid alkaloid, $C_{10}H_{14}N_2$, with a very acrid taste, contained in the leaves of tobacco. Also **nic′o·tin.** [after Jean *Nicot*, 1530–1604, French courtier, who introduced tobacco into France from Portugal] — **nic·o·tin·ic** (nik′ə·tin′ik) *adj.*

nicotinic acid The anti-pellagra factor of the vitamin B complex; a colorless, crystalline, water-soluble compound, $C_6H_5NO_2$, present in liver, kidney, muscle meats, fish, milk, and green vegetables, and also made by the oxidation of nicotine: also called *niacin.*

nic·o·tin·ism (nik′ə·tin·iz′əm) *n.* The morbid effects resulting from the excessive use of tobacco.

Nic·the·roy (nē′tə·roi′) A former spelling of NITERÓI.

nic·ti·tate (nik′tə·tāt) *v.i.* **·tat·ed, ·tat·ing** To wink. Also **nic′tate.** [<Med. L *nictitatus*, pp. of *nictitare*, freq. of L *nictare* wink]

nictitating membrane The third or lateral eyelid, in birds, crocodiles, etc., springing from the inner and anterior border of the eye.

nic·ti·ta·tion (nik′tə·tā′shən) *n.* **1** The act of winking. **2** *Pathol.* Rapid and involuntary winking due to nervous derangement. Also **nic·ta′tion.**

Ni·da·ros (nē′dä·rōs) A former name for TRONDHEIM.

nid·der·ing (nid′ər·ing) *n.* *Archaic* A coward. — *adj.* Cowardly; base. Also **nid′er·ing.** [Alter. of ME *nithing* a villain; misread in early text]

nide (nīd) *n.* A nest or brood of young pheasants. — *v.i.* **nid·ed, nid·ing** *Rare* To nest. [<L *nidus* nest]

ni·dic·o·lous (ni·dik′ə·ləs) *adj.* Remaining in the nest for some time after hatching: said of birds. [<L *nidus* nest + *colere* inhabit]

nid·i·fy (nid′ə·fī) *v.i.* **·fied, ·fy·ing** To build a nest. Also **nid′i·fi·cate′.** [<L *nidificare* <*nidus* a nest + *facere* make] — **nid′i·fi·ca′tion** *n.*

ni·dus (nī′dəs) *n.* *pl.* **·di** (-dī) **1** A place for the natural deposit of eggs, especially of insects, spiders, etc. **2** A place in an organism adapted to the development of some germ or parasite; hence, a center of infection. [<L]

Nie·buhr (nē′bŏŏr), **Barthold Georg,** 1776–1831, German historian, philologist, and critic. — **Reinhold,** born 1892, U.S. Protestant theologian.

niece (nēs) *n.* The daughter of a brother or sister; also, the daughter of a brother-in-law or sister-in-law. [<OF, ult. <L *neptis* niece, granddaughter]

ni·el·lo (nē·el′ō) *n.* *pl.* **·li** (-ī) **1** The art of decorating metal plates by incising designs upon them and then filling in the incised lines with a black alloy. **2** A work produced by this method: also **ni·el′lo-work′.** **3** A black alloy used in this work. [<Ital. <L *nigellus* blackish <*niger* black] — **ni·el′list** *n.*

Niel·sen (nēl′sən), **Carl,** 1865–1931, Danish composer.

Nie·men (nē′mən, *Polish* nye′men) The Polish name for the NEMAN.

Nie·möl·ler (nē′mœl·ər), **Martin,** born 1892, German anti-Nazi Protestant leader.

Nier·stein·er (nir′stī·nər, -shtī-) *n.* A full-bodied, white Rhine wine.

niest (nēst) See NEIST.

Nie·tzsche (nē′chə), **Friedrich Wilhelm,** 1844–1900, German philosopher.

Nie·tzsche·ism (nē′chi·iz′əm) *n.* The principles propounded in the philosophy of Nietzsche; especially, the doctrine of the development of the superman. Also **Nie′tzsche·an·ism.** — **Nie′tzsche·an** *adj.* & *n.*

Ni·eu·port (nē′ŏŏ·pōrt) A town in western Belgium; scene of Dutch victory over Spanish forces, 1600. Also **Ni′euw·poort.**

nieve (nēv) *n.* *Scot.* & *Brit. Dial.* The fist or hand. Also spelled *neif.*

nif·fer (nif′ər) *Scot.* *v.t.* To barter; exchange. — *n.* An exchange or barter.

Nif·l·heim (niv′əl·hām) The lowest of the nine worlds of Norse mythology, the world of fog or mist; the northern limit of cold and darkness. Also **Nif′el·heim.** [<ON <*nifl* fog + *heimr* world]

nif·ty (nif′tē) *adj.* **·ti·er, ·ti·est** *Slang* Stylish; pleasing.

Ni·ger (nī′jər, -gər) A river of western Africa, flowing 2,600 miles from near the Sierra Leone border through French West Africa and Nigeria to the Gulf of Guinea.

Ni·ger (nī′jər, -gər), **Republic of** An independent republic of the French Community in north central Africa; 458,976 square miles; capital, Niamey: formerly a French overseas territory.

Ni·ge·ri·a (nī·jir′ē·ə) An independent state of the British Commonwealth, in west Africa, including the northern part of the former British Cameroons; 356,093 square miles; capital, Lagos: formerly a British dependency. — **Ni·ge′ri·an** *adj.* & *n.*

Ni·ger seed (nī′jər, -gər) Ramtil.

nig·gard (nig′ərd) *n.* A parsimonious person. — *v.t.* & *v.i.* *Obs.* To act or treat in a niggardly manner. — *adj.* Niggardly. [Cf. ON *hnöggr* stingy]

nig·gard·ly (nig′ərd·lē) *adj.* Meanly covetous or avaricious; parsimonious; stingy. — *adv.* In the manner of a niggard. — **nig′gard·li·ness** *n.*

nig·ger (nig′ər) *n.* A Negro or member of any dark-skinned people: an offensive and vulgar term of contempt. [Earlier *neger* <F *nègre* <Sp. *negro* NEGRO]

nig·ger·head (nig′ər·hed′) *n.* **1** A stone or boulder, especially one having nodules. **2** The black-eyed Susan. **3** *Naut.* A spool about which a hauling rope is wound; a bollard.

nig·gle (nig′əl) *v.i.* **·gled, ·gling 1** To occupy oneself with trifles; be too precise. **2** To putter; trifle. [Cf. dial. Norw. *nigla*]

nig·gling (nig′ling) *n.* **1** Work that is too detailed or meticulous. **2** In art, overelaborate or too detailed a treatment. — *adj.* **1** Fussy; overelaborate. **2** Mean; petty; trite. **3** Troublesome; annoying. — **nig′gling·ly** *adv.*

nigh (nī) *adj.* **nigh·er, nigh·est** or, formerly, **next 1** Being close by; near in time or place. **2** On the left: used of a team. **3** Closely allied in kinship; intimate. **4** Most convenient; direct. See synonyms under ADJACENT. — *adv.* **1** Not remote in time or place; close by; near. **2** Almost; nearly. — *prep.* Close to; near. — *v.t.* & *v.i.* *Rare* To draw near; approach. [OE *neah, neh*]

night (nīt) *n.* **1** The period during which the sun is below the horizon from sunset to sunrise. ◆ Collateral adjective: *nocturnal.* **2** The close of day; evening. **3** A condition of darkness, or gloom; sorrow; misfortune. **4** Death. ◆ Homophone: *knight.* [OE *niht*]

night·bird (nīt′bûrd′) *n.* A bird that flies or sings by night.

night blindness Nyctalopia.

night-bloom·ing cereus (nīt′blōō′ming) See CEREUS.

night·cap (nīt′kap′) *n.* **1** A headcovering to be worn in bed. **2** *Colloq.* A drink of liquor taken just before going to bed.

night·chair (nīt′châr′) *n.* A commode (def. 3).

night·clothes (nīt′klōz′, -klōthz′) *n. pl.* Clothes to be worn in bed.

night club A restaurant open at night, providing entertainment, food, and drink for patrons.

night·crawl·er (nīt′krô′lər) *n.* *U.S. Colloq.* Any large earthworm that emerges at night. See NIGHTWALKER (def. 3).

night·fall (nīt′fôl′) *n.* The close of day.

night·glass (nīt′glas′, -gläs′) *n.* A spyglass or telescope arranged with concentrating lenses for use at night, especially at sea.

night·gown (nīt′goun′) *n.* A long, loose gown worn in bed: also called *night robe.* Also **night′dress′.**

night·hawk (nīt′hôk′) *n.* **1** An American goatsucker (genus *Chordeiles*) of nocturnal habits, related to the whippoorwill. **2** The nightjar. **3** One who works or stays up at night.

night heron A bird (genus *Nycticorax*) of somewhat nocturnal habits, having a comparatively short, stout bill. — **black-crowned night heron** One of two forms of night herons (*N. nycticorax*) found in North America.

night·in·gale (nī′tən·gāl, nī′ting-) *n.* **1** A small, Old World, migratory bird (genus *Luscinia*), of the thrush family (*Turdidae*), noted for the melodious night song of the male. **2** The bulbul. [OE *nihtegale*, lit., night-singer]

Night·in·gale (nī′tən·gāl, nī′ting-), **Florence,** 1820–1910, English nurse born in Italy; served in the military hospitals during the Crimean War; regarded as the founder of modern nursing service.

night·jar (nīt′jär′) *n.* A goatsucker, especially the common European species (*Caprimulgus europaeus*).

night latch A spring latch operated from the outside by a key and from the inside by a knob or the like. Also **night lock.**

night letter A lettergram sent during the night, usually at a reduced rate.

night·long (nīt′lông′, -long′) *adj.* Lasting through the night.

night·ly (nīt′lē) *adj.* **1** Pertaining to night or to every night; occurring at night or every night. **2** Dark; having the appearance of night. — *adv.* By night; every night.

night·mare (nīt′mâr) *n.* **1** A sensation of oppression or suffocation during sleep, with terrifying dreams and apparent inability to move or speak. **2** Hence, any oppressive influence. **3** An evil spirit formerly supposed to suffocate people during sleep; an incubus. [<NIGHT + MARE[2]] — **night′mar·ish** *adj.*

night owl 1 An owl especially nocturnal in habits. **2** Nighthawk (def. 3).

night raven 1 The nightjar. **2** The night heron.

night rider In the southern United States, any of a band of masked, mounted men who perform lawless acts of violence at night, generally to punish, intimidate, etc.

night robe A nightgown.

nights (nīts) *adv. Colloq.* At night.

night school A school that holds classes during the evening, especially for those who cannot attend day school.

night·shade (nīt′shād) *n.* **1** Any one of a genus (*Solanum*) of flowering plants; especially, the **common** or **black nightshade** (*S. nigrum*), a weedlike plant with white flowers and black berries, reputed poisonous, but used medicinally, and the **climbing** or **woody nightshade** (*S. dulcamara*). **2** The belladonna or **deadly nightshade.** **3** The henbane. —**enchanter's nightshade** A low, inconspicuous herb (genus *Circaea*) growing in damp woods.

BLACK NIGHTSHADE
a. Spray showing blossom *(b)* and fruit *(c)*.

night·shirt (nīt′shûrt′) *n.* A loose, shirtlike garment worn in bed, usually by men or boys.

night soil The contents of privies, cesspools, etc., usually removed at night.

night spot *U.S. Colloq.* A night club.

night·stick (nīt′stik′) *n.* A long, stout club carried by policemen, especially at night.

night sweats *Pathol.* Excessive sweating during sleep, often associated with phthisis.

night table A bedside table or stand.

night terrors *Med.* A disorder of children resembling nightmare, with fits of semiconscious screaming; pavor nocturnus.

night·tide (nīt′tīd′) *n.* Nighttime.

night·time (nīt′tīm′) *n.* The time from sunset to sunrise, or from dark to dawn.

night·walk·er (nīt′wô′kər) *n.* **1** One who walks in his sleep. **2** One who frequents the streets at night. **3** A large angleworm.

night watch **1** A guard for night duty. **2** A watch period of the night hours.

night watchman A person hired to keep watch and be on guard at night.

night·y (nī′tē) *n. pl.* **night·ies** *Colloq.* A nightgown.

ni·gres·cence (nī·gres′əns) *n.* The process of becoming black, or the blackness so produced. [<L *nigrescere* grow black < *niger* black] —**ni·gres′cent** *adj.*

nig·ri·fy (nig′rə·fī) *v.t.* ·fied, ·fy·ing To make black. [<L *nigrificare* < *niger* black + *facere* make] —**nig′ri·fi·ca′tion** *n.*

Ni·gri·ti·a (nī·grish′ē·ə) A former name for the region of the SUDAN.

Ni·gri·ti·an (nī·grish′ē·ən) *adj.* **1** Of or pertaining to Nigritia. **2** Negro. —*n.* A Negro.

nig·ri·tude (nig′rə·tōōd, -tyōōd) *n.* Blackness; darkness. [<L *nigritudo* < *niger* black]

ni·gro·sine (nig′rə·sēn, -sin) *n. Chem.* Any of a group of deep blue or black dyes obtained from aniline and its homologs. [<L *niger* black + -OS(E)[1] + -INE[2]]

ni·hil (nī′hil, nī′-) *n.* Nothing; nil. [<L]

ni·hil ad rem (nī′hil ad rem) *Latin* Nothing to the point.

ni·hil·ism (nī′əl·iz′əm, nī′hil-) *n.* **1** The doctrine that nothing exists or can be known; also, the rejection of religious and moral creeds, known as **ethical nihilism.** **2** A political doctrine holding that the existing structure of society should be destroyed; specifically, a movement in Russia in the 19th century advocating the overthrow of the social order and many revolutionary reforms, resulting in violence and terrorism. **3** Loosely, any revolutionary propaganda involving violence. **4** *Med.* The denial of the beneficial effects of medicines as remedial agents. **5** *Psychiatry* A sense of unreality; a delusion of non-existence. —**ni′hil·ist** *n.* —**ni′hil·is′tic** *adj.*

ni·hil·i·ty (nī·hil′ə·tē, nī-) *n.* Nothingness.

Ni·hon (nē′hōn′) See JAPAN.

Ni·i·ga·ta (nē′ē·gä′tä) A port on NW Honshu island, Japan.

Ni·i·hau (nē′ē·hä′ōō) An island in the NW Hawaiian Islands: 72 square miles.

Ni·jin·sky (nə·jin′skē), *Russian* ni·zhēn′skē), **Vaslav**, 1890-1950, Russian ballet dancer and choreographer.

Nij·me·gen (nī′mā·gən, *Du.* nī′mä·khən) A city on the Waal in eastern Netherlands; site of a peace conference, 1678: German *Nimwegen.*

Ni·ke (nī′kē) In Greek mythology, the winged goddess of victory: identified with the Roman Victoria.

Nik·ko (nēk′kō) A town and religious center on central Honshu island, Japan.

Ni·ko·la·ev (nē′kô·lä′yef) A port on the Bug estuary in SW Ukrainian S.S.R.: formerly *Vernoleninsk.*

Ni·ko·lay·evsk (nē′kô·lä′yefsk) A city near the mouth of the Amur in SE Russian S.F.S.R. Also **Ni′ko·lay′evsk-on-A·mur′** (-on-ä·mōōr′).

nil (nil) *n.* Nothing. [<L, contraction of *nihil* nothing]

nil des·pe·ran·dum (nil des′pə·ran′dəm) *Latin* Nothing to be despaired of; never despair.

Nile (nīl) The longest river in Africa, rising in Lake Victoria and flowing 3,485 miles north to the Mediterranean; between Lake Victoria and Lake Albert it is known as the **Victoria Nile**; between Lake Albert and the Sudan as the **Albert Nile** (Arabic *Bahr-el-Jebel*); between Lake No and Khartoum as the **White Nile** (Arabic *Bahr-el-Abiad*); at Khartoum it receives the **Blue Nile** (Arabic *Bahr-el-Azraq*), a tributary flowing 850 miles from Ethiopia, and is known as the **Nile** (Arabic *Bahr-en-Nil*) for the rest of its course. —**Battle of the Nile** A British naval victory over the French, near the mouth of the Nile on August 1, 1798.

Nile green Any of several light-green tints.

nil·gai (nil′gī) *n.* A large, short-maned Indian antelope (*Boselaphus tragocamelus*) with the hind legs much shorter than the fore: often spelled *nylghai, nylghau.* Also **nil′gau** (-gô), **nil′ghai, nil′ghau** (-gô). [<Persian *nīlgāu* < *nīl* blue + *gāu* cow]

Nil·gi·ri Hills (nil′gi·rē) A mountainous plateau district in Madras State, India; highest peak, 8,640 feet. Also **The Nil′gi·ris.**

nill (nil) *v.t. & v.i. Archaic* To be unwilling. [OE *nillan* < *ne-* not + *willan* will]

nil ni·si bo·num (nil nī′sī bō′nəm) *Latin* Short for DE MORTUIS NIL NISI BONUM.

Ni·lom·e·ter (nī·lom′ə·tər) *n.* A gage for measuring the height of water in the Nile. [<Gk. *Neilometrion* < *Neilos* Nile + *metron* measure] —**Ni·lo·met′ric** (nī′lə·met′rik) *adj.*

Ni·lot·ic (nī·lot′ik) *adj.* **1** Of, pertaining to, or characteristic of the Nile or the peoples native to the Nile basin. **2** Of or pertaining to a subfamily of Sudanic languages spoken by any of these peoples, including Dinka and Fula.

nim (nim) *v.t.* **nam** or **nimmed, no·men** or **nome, nim·ming** *Obs.* To take; steal. [OE *niman*]

nim·ble (nim′bəl) *adj.* ·bler, ·blest **1** Light and quick in motion or action; agile. **2** Intellectually alert or acute; keen; quick-witted. **3** Circulating freely, as money. **4** Indicating a ready mind; clever: a *nimble* answer. [OE *numel* quick at learning] —**nim′ble·ness** *n.* —**nim′bly** *adv.*

Synonyms: active, agile, alert, brisk, bustling, lively, prompt, quick, speedy, sprightly, swift. *Nimble* refers to lightness, freedom, and quickness of motion within a somewhat narrow range, with readiness to turn suddenly to any point; *swift* applies commonly to sustained motion over greater distances; a pickpocket is *nimble*-fingered, a dancer *nimble*-footed; an arrow, a race horse, or an ocean steamer is *swift.* We speak of *nimble* wit, *swift* destruction. *Alert,* which is a synonym for *ready,* sometimes comes near the meaning of *nimble* or *quick,* from the fact that the ready, wide-awake person is likely to be lively, quick, speedy. See ACTIVE, ALERT, SPRIGHTLY. *Antonyms:* clumsy, dilatory, dull, heavy, inactive, inert, slow, sluggish, unready.

nim·bus (nim′bəs) *n. pl.* ·bus·es or ·bi (-bī) **1** A dark, heavy, rain-bearing cloud: also **nim·bo·stra·tus** (nim′bō·strā′təs, -strat′əs). See CLOUD. **2** A halo or bright disk encircling the head, as of Jesus, saints, etc., in pictures, on medallions, etc. **3** A cloud of glory or surrounding aura of light in which the gods were supposed by the ancients to be clothed when appearing upon earth; hence, any atmosphere or aura of fame, glamour, etc., surrounding a person. [<L, rain cloud]

Nîmes (nēmz, *Fr.* nēm) A city in southern France. Formerly **Nismes** (nēm).

ni·mi·e·ty (ni·mī′ə·tē) *n.* Redundancy; excess. [<LL *nimietas, -tatis* < *nimis* too much] —**nim·i·ous** (nim′ē·əs) *adj.*

nim·i·ny-pim·i·ny (nim′ə·nē-pim′ə·nē) *adj.* Affectedly nice or delicate; effeminate. [Imit.]

Nim·itz (nim′its), **Chester William**, born 1885, U.S. admiral; in command of U.S. Pacific Fleet in World War II.

n'im·porte (nan·pôrt′) *French* It does not matter; it is of no importance.

Nim·rod (nim′rod) Grandson of Ham; a mighty hunter. *Gen.* x 8.

Nim·rud (nim·rōōd′) The modern name for the site of ancient KALAKH.

Nim·u·e (nim′ōō·ä) See under MERLIN.

Nim·we·gen (nim′vā·gən) The German name for NIJMEGEN.

Ni·na (nī′nə, nē′-) Diminutive of ANN.

Ni·ña (nē′nə, *Sp.* nē′nyä) *n.* One of the three ships of Columbus on his maiden voyage to America.

nin·com·poop (nin′kəm·pōōp) *n.* A foolish or silly person; simpleton. [Origin unknown]

nine (nīn) *n.* **1** The cardinal number preceding ten and following eight, or any of the symbols (9, ix, IX) representing this number. **2** Anything containing nine members or units, as a playing card with nine pips; specifically, a baseball team. —**the Nine** The Muses. —*adj.* Being or consisting of one more than eight; thrice three; novenary. [OE *nigon*]

nine·fold (nīn′fōld′) *adj. & adv.* Nine times as many or as great; nonuplicate.

nine-men's-mor·ris (nīn′menz-môr′is, -mor′-) *n.* A game played on a diagram marked out on the ground, or on a board. Each player, having five, nine, or twelve (according to the number playing) counters or pieces, endeavors to place three of them in a row, upon doing which he takes one of his opponent's pieces.

nine·pence (nīn′pəns) *n. Brit.* **1** The sum of nine pennies. **2** A coin of this value, no longer minted.

nine·pin (nīn′pin′) *n.* One of the pins in the game of ninepins.

nine·pins (nīn′pinz′) *n. pl.* (construed as singular) A game similar to tenpins, in which nine large wooden pins are employed.

nine·teen (nīn′tēn′) *adj.* Being nine more than ten. —*n.* The sum of ten and nine; also, its symbols (19, xix, XIX). —**nine′teenth′** (-tēnth′) *adj. & n.*

nine·ty (nīn′tē) *adj.* Being nine times ten. —*n. pl.* ·ties The sum of ten and eighty: a cardinal numeral; also, its symbols (90, xc, XC). —**nine′ti·eth** *adj. & n.* —**nine′ty·fold′** *adj. & adv.*

Nin·e·veh (nin′ə·və) An ancient city on the Tigris, capital of Assyria: Latin *Ninus.*—**Nin′e·vite** (-vīt) *n.*

Ning·po (ning′pō′) A treaty port in NE Chekiang province, China.

Ning·sia (ning′shyä′) **1** A former province of NW China in the region of Inner Mongolia; created 1928 from part of Kansu province; remerged with Kansu 1954: 106,150 square miles; capital Yinchwan. **2** See YINCHWAN. Also **Ning′hsia′.**

nin·ny (nin′ē) *n. pl.* ·nies A simpleton; dunce. [? Short for *an innocent*]

ni·non (nē′non) *n.* A kind of firm chiffon used for lingerie, neckwear, dresses, etc.: often called *triple voile.* [<F]

Ni·non (nē·nôn′) A French form of ANN.

ninth (nīnth) *adj.* **1** Next in order after the eighth: the ordinal of *nine.* **2** Being one of nine equal parts. —*n.* **1** One of nine equal parts; the quotient of a unit divided by nine. **2** *Music* **a** An interval of an octave and a second, or a note separated from another by this interval. **b** The two notes written or sounded together. —*adv.* In the ninth order, place, or rank: also, in formal discourse, **ninth'ly.**

Ni·nus (nī'nəs) **1** In Assyrian legend, the founder of Nineveh and husband of Semiramis. **2** The Latin name for NINEVEH.

Ni·o·be (nī'ə·bē) In Greek mythology, the daughter of Tantalus and wife of Amphion, whose children were killed by Apollo and Artemis after she had vaunted her superiority to their mother Leto: the weeping Niobe was turned by Zeus into a stone from which tears continued to flow.

NIOBE
After copy of a statue attributed to Scopas in the Uffizi, Florence.

ni·o·bi·um (nī·ō'bē·əm) *n.* A steel-gray, metallic element of rare occurrence (symbol Cb), valuable as an alloy metal: formerly called *columbium.*

ni·o·bous (nī·ō'bəs) *adj. Chem.* Denoting a compound which contains trivalent niobium.

Ni·o·brar·a River (nī'ə·brâr'ə) A river in eastern Wyoming and northern Nebraska, flowing 431 miles east to the Missouri.

Ni·os (nē'os) See Ios.

nip[1] (nip) *v.* **nipped, nip·ping** *v.t.* **1** To compress tightly between two surfaces or points; squeeze; bite. **2** To sever or remove by pinching or clipping, as shoots. **3** To check or destroy the growth or development of. **4** To affect painfully or injuriously; benumb: said of cold. **5** *Slang* To steal; pilfer. **6** *Slang* To snatch; take. —*v.i.* **7** *Brit. Colloq.* To move nimbly or rapidly: with *off, away, in,* etc. —*n.* **1** The act of compressing sharply. **2** A biting, pinching, or clipping off; also, whatever is pinched off; hence, a pinch. **3** A sudden blight, as by frost. **4** A sharp saying; cutting remark; gibe. [Cf. Du. *nijpen* nip]

nip[2] (nip) *n.* A small dram, especially of strong drink. —*v.t.* & *v.i.* To drink (liquor) in sips. [<earlier *nipperkin,* measure holding about a half pint]

Nip (nip) *n* & *adj. Slang* Nipponese: a contemptuous term.

ni·pa (nī'pə, nē'-) *n.* **1** Any of a genus (*Nipa*) of palms of tropical SE Asia. One species (*N. fruticans*) has feathery leaves, used for weaving, thatching, etc., and bunches of edible fruit. **2** An alcoholic beverage made from this palm. [<NL <Malay *nipah*]

nip and tuck A case of near equality, as between two runners; neck and neck.

niph·a·blep·si·a (nif'ə·blep'sē·ə) *n.* Snow blindness. Also **niph·o·typh·lo·sis** (nif'ə·tif·lō'sis). [<NL <Gk. *nipha* (accusative sing.) snow + *ablepsia* blindness]

Nip·i·gon (nip'ə·gon), **Lake** A lake in west central Ontario, Canada, north of Lake Superior; 1,870 square miles.

Nip·is·sing (nip'ə·sing), **Lake** A lake in east central Ontario, Canada, midway between Georgian Bay and the Ottawa River; 330 square miles.

nip·per[1] (nip'ər) *n.* **1** One who nips. **2** One of various pincers for nipping. **3** An incisor, as of a horse. **4** A great claw, as of a crab. **5** A heavy, padded, woolen mitten or glove, used by New England fishermen.

nip·per[2] (nip'ər) *n. Brit. Colloq.* A small boy.

nip·ping (nip'ing) *adj.* Pinching; biting; cutting; sarcastic. —**nip'ping·ly** *adv.*

nip·ple (nip'əl) *n.* **1** The pigmented cone-shaped process of the breast containing the milk duct; a pap; teat. **2** A protuberance to receive a percussion cap. **3** A small tubular pipefitting. **4** A tip, usually of rubber, for a nursing bottle. [Earlier *neble,* ? dim. of NEB]

Nip·pon (nip'on, ni·pon'; *Japanese* nēp·pōn) See JAPAN. —**Nip'pon·ese'** (-ēz', -ēs') *adj.* & *n.*

Nip·pur (ni·po͝or') An ancient Sumerian city of southern Babylonia, the earliest religious capital of the area.

nip·py (nip'ē) *adj.* **·pi·er, ·pi·est** **1** Biting; sharp; acid; sarcastic. **2** Active; vigorous; alert. —**nip'pi·ly** *adv.*

nir·va·na (nir·vä'nə, nər·van'ə) *n.* **1** In Hinduism, a "blowing out" of the spark of life; hence, spiritual reunion with Brahma; bliss. **2** In Buddhism, the ideal and goal of all religious effort: freedom from passion and delusion, and absorption of the individual into the supreme spirit; complete enlightenment. **3** The attainment of complete freedom from all mental, emotional, and psychic tension. [<Skt. *nirvāṇa* a blowing out <*nirvā* blow]

Nirvana principle *Psychoanal.* The tendency of mental life to keep psychic tension to the lowest possible degree: explained by Freud as the action of the death instinct, aiming toward the stability of the inorganic state.

Ni·san (nī'san, nis'ən; *Hebrew* nē·sän') A Hebrew month: also called *Abib.* See CALENDAR (Hebrew). Also **Nis'san.**

Ni·sei (nē·sā) *n. pl.* **·sei** or **·seis** An American citizen of immigrant Japanese parentage who was born and educated in the United States. Compare ISSEI, KIBEI.

Nish (nēsh) A city in SE central Serbia, Yugoslavia: ancient *Niassus. Serbo-Croatian* **Niš.** Formerly **Nis·sa** (nis'ə).

Ni·sha·pur (nē'shä·po͝or') A city in NE Iran: also *Neyshapur.*

Ni·shi·no·mi·ya (nē·shē·nō·mī·yä) A city on Osaka Bay, Honshu island, Japan.

ni·si (nī'sī) *conj. Law* Unless: used after the word *order, rule, decree,* etc., signifying that it shall become effective at a certain time, unless before the time named it shall have been modified or avoided. [<L <*ni* not + *si* if]

ni·si pri·us (nī'sī prī'əs) *Law* Literally, unless sooner; hence, a general designation suggestive of the trial of civil causes before a judge and jury.

Nis·sen hut (nis'ən) A portable, insulated structure of inverted U-sections for the shelter of troops, chiefly in arctic areas. Compare QUONSET HUT. [after P. N. *Nissen,* 1871–1930, Canadian Army engineer]

Ni·stru (nē'stro͞o) The Rumanian name for the DNIESTER.

ni·sus (nī'səs) *n. pl.* **·sus** *Latin* The exercise of power in acting or attempting to act; an effort, endeavor, or exertion.

nit (nit) *n.* **1** The egg of a louse or other insect. **2** The immature insect itself. **3** A small speck. **4** *Scot.* A nut. [OE *hnitu*] —**nit'ty** *adj.*

nite (nīt) *n.* A non-standard variant spelling of NIGHT: used in the trade press of the entertainment world, in advertising, and in copyright names.

ni·ter (nī'tər) *n.* **1** A crystalline white salt; saltpeter; potassium or sodium nitrate. **2** *Obs.* Natron. Also **ni'tre.** [<F *nitre* <L *nitrum* <Gk. *nitron* natron]

Ni·te·rói (nē'tə·roi') A city on Guanabara Bay, SE Brazil, capital of Rio de Janeiro state: formerly *Nictheroy.*

nit·id (nit'id) *adj.* **1** *Bot.* Shining; glossy, as many leaves and seeds. **2** *Obs.* Lustrous; bright, as metal. [<L *nitidus* <*nitere* shine]

ni·ton (nī'ton) *n.* Radon.

nitr– Var. of NITRO–.

ni·tra·mine (nī'trə·mēn, -min) *n. Chem.* Any of a class of compounds in which a nitro group is attached directly to a trivalent nitrogen atom: also *nitroamine.*

ni·trate (nī'trāt) *v.t.* **·trat·ed, ·trat·ing** *Chem.* To treat or combine with nitric acid or a compound; to change into a nitro derivative. —*n.* **1** A salt or ester of nitric acid: silver *nitrate.* **2** Sodium or potassium nitrate. [<L *nitratus* mixed with natron <*nitrum* niter] —**ni·tra'·tion** *n.*

ni·tric (nī'trik) *adj. Chem.* **1** Of, pertaining to, or obtained from nitrogen. **2** Containing nitrogen in the higher state of valence.

nitric acid *Chem.* A colorless, highly corrosive liquid, HNO_3, sometimes formed in the atmosphere in small quantities, but usually made by decomposing sodium or potassium nitrate with sulfuric acid.

nitric bacteria See NITROBACTERIUM.

nitric oxide *Chem.* A colorless, gaseous compound, NO, liberated when certain metals are dissolved in nitric acid.

ni·tride (nī'trīd, -trid) *n. Chem.* A compound of nitrogen with some more positive element, as boron, phosphorus, and any of the metals. Also **ni'trid** (-trid).

ni·tri·fi·ca·tion (nī'trə·fə·kā'shən) *n. Chem.* **1** The method or act of nitrifying. **2** The conversion of ammonium salts into nitrites and nitrates, especially by soil bacteria.

ni·tri·fi·er (nī'trə·fī'ər) *n. Chem.* A substance containing nitrogen that aids in the process of nitrification.

ni·tri·fy (nī'trə·fī) *v.t.* **·fied, ·fy·ing** *Chem.* **1** To combine with nitrogen. **2** To convert, by oxidation, into nitric or nitrous acid or into nitrates or nitrites. **3** To treat or impregnate (soil, etc.) with nitrates. [<F *nitrifier*] —**ni'·tri·fi'a·ble** *adj.*

ni·trile (nī'tril, -trēl, -trīl) *n. Chem.* Any of a group of cyanogen compounds, yielding ammonia upon saponification, and corresponding to the formula RCN, in which R is an organic radical. Also **ni'tril** (-tril).

ni·trite (nī'trīt) *n. Chem.* A salt of nitrous acid.

nitro– *combining form* **1** *Chem.* **a** Containing the univalent radical NO_2: *nitrophenol.* **b** Of or containing nitrogen in some other form: *nitroglycerin.* **2** Nitrifying: *nitrobacterium.* Also before vowels, *nitr–.* Also **nitri–.** [<L *nitrum* natron]

ni·tro·am·ine (nī'trō·am'ēn, -in) *n.* Nitramine.

ni·tro·bac·te·ri·um (nī'trō·bak·tir'ē·əm) *n. pl.* **·ri·a** (-ē·ə) Any of various soil bacteria involved in the process of nitrification, especially the nitrous group (genus *Nitrosomonas*), which converts ammonia into nitrites, and the nitric group (genus *Nitrobacter*), which oxidizes nitrites into nitrates. Also **nitric** or **nitrous bacterium.**

ni·tro·ben·zene (nī'trō·ben'zēn, -ben·zēn') *n. Chem.* A yellow, oily compound, $C_6H_5NO_2$, formed by the nitration of benzene: also called *mirbane oil.*

ni·tro·cel·lu·lose (nī'trō·sel'yə·lōs) *n.* Cellulose nitrate.

ni·tro·chlo·ro·form (nī'trō·klôr'ə·fôrm, -klō'·rə-) *n.* Chlorpicrin.

ni·tro·cot·ton (nī'trō·kot'n) *n.* Guncotton.

ni·tro·ga·tion (nī'trō·gā'shən) *n.* A method of increasing soil fertility by the addition to irrigation water of anhydrous ammonia in controlled amounts. [<NITRO- + (IRRI)GATION]

ni·tro·gen (nī'trə·jən) *n.* An odorless, colorless, gaseous element (symbol N) forming about four-fifths of the atmosphere by volume and playing a decisive role in the formation of compounds essential to life. See ELEMENT. [<NITRO- + -GEN]

nitrogen balance *Biochem.* The relation between the nitrogen intake and the nitrogen output of a human body. An excess of intake gives a plus balance, an excess of output a minus balance.

nitrogen cycle *Biol.* The sequence of physical and chemical processes by which atmospheric nitrogen is taken into the soil, utilized by plants and animals, and eventually returned to the atmosphere.

nitrogen equilibrium *Physiol.* That condition of the body in which the nitrogen intake and output are equal.

nitrogen fixation **1** The conversion of atmospheric nitrogen into nitrates by soil bacteria, either free-living or in symbiotic relations with the roots of certain leguminous plants. **2** The production of nitrogen compounds, as for fertilizers and explosives, by various electrochemical processes utilizing free nitrogen. —**ni'tro·gen-fix'ing** *adj.*

nitrogen iodide *Chem.* A chocolate-colored powder, $N_2H_3I_3$, explosive when dry.

ni·trog·en·ize (nī·troj'ən·īz, nī'trə·jən·īz') *v.t.* **·ized, ·iz·ing** To treat or combine with nitrogen.

nitrogen narcosis *Pathol.* A deranged, sometimes fatal nervous and mental condition resembling that of alcoholic intoxication, caused by the action of nitrogen inhaled by divers at excessive depths below the surface of the water.

ni·trog·e·nous (nī·troj'ə·nəs) *adj.* Pertaining to or containing nitrogen. Also **ni·tro·ge·ne·ous** (nī'trō·jē'nē·əs).

ni·tro·glyc·er·in (nī'trō·glis'ər·in) *n. Chem.* A colorless to pale-yellow, oily liquid, $C_3H_5(ONO_2)_3$, made by nitrating glycerol: an explosive and propellant, commonly combined, as with infusorial earth, to form dynamite and reduce the danger of its explosion by percussion. It is sometimes used in medicine. Also **ni'tro·glyc'er·ine.** [<NITRO- + GLYCERIN]

nitro group *Chem.* The univalent NO_2 radical.

ni·tro·gua·ni·dine (nī'trō·gwä'nə·dēn, -din, -gwan'ə-) *n. Chem.* Either of two colorless, crystalline nitro compounds, $N_4H_4CO_2$, the *alpha* form of which is used in the manufacture of a powerful, flashless, smokeless powder. [<NITRO- + GUANIDINE]

ni·tro·hy·dro·chlo·ric acid (nī'trō-hī'drə-klôr'ik, -klō'rik) Aqua regia.

ni·tro·jec·tion (nī'trō-jek'shən) *n.* The process of injecting anhydrous ammonia directly into the soil as a means of increasing soil fertility. [<NITRO- + (IN)JECTION]

ni·trol·ic (nī·trol'ik) *adj. Chem.* Noting a class of acids derived from nitroparaffin by the action of nitrous acid and having the general formula RCN₂O₃H.

ni·trom·e·ter (nī-trom'ə-tər) *n.* An apparatus or instrument used for the determination of nitrogen in some of its combinations when contained in mixtures. [<NITRO- + -METER]

ni·tro·par·af·fin (nī'trō-par'ə-fin) *n. Chem.* Any derivative of the methane series in which hydrogen has been replaced by a nitro group.

ni·tro·phe·nol (nī'trō-fē'nol, -nol) *n. Chem.* Any of a group of phenol compounds derived by the replacement of one or more hydrogen atoms by the nitro group: used in the making of dyestuffs.

ni·troph·i·lous (nī-trof'ə-ləs) *adj.* 1 Obtaining nutriment from nitrogenous soil. 2 Growing in a soil rich in nitrates. [<NITRO- + -PHILOUS]

ni·tro·phyte (nī'trə-fīt) *n.* A plant growing in a soil rich in nitrates.

ni·tros·a·mine (nī·trōs'ə-mēn, -min) *n. Chem.* Any of a group of organic compounds containing the bivalent radical N·NO. Also **ni·tros·o·a·min** (-min). [<NITROS(O)- + -AMINE]

nitroso- combining form *Chem.* Of or containing nitrosyl. Also, before vowels, **nitros-**. [<NL *nitrosus* <L, of natron <*nitrum* natron]

ni·tro·starch (nī'trō-stärch') *n. Chem.* An explosive compounded of starch and sulfuric and nitric acids.

ni·tro·syl (nī'trō-sil, nī'trə-sēl', nī'trə-sil) *n. Chem.* The univalent radical NO; known only in its combinations.

ni·trous (nī'trəs) *adj. Chem.* Of, pertaining to, or derived from nitrogen: especially applied to those compounds of nitrogen containing less oxygen than the nitric compounds. [<L *nitrosus* full of natron <*nitrum* natron]

nitrous acid *Chem.* An unstable compound, HNO₂, occurring only in solution.

nitrous bacterium Nitrobacterium.

nitrous oxide Laughing gas.

Nit·ti (nēt'tē), Francesco, 1868–1953, Italian economist and politician.

nit·wit (nit'wit') *n.* A silly or stupid person.

Ni·u·a·fo·o (nē·ōō'ə·fō'ō) The northernmost island of the Tonga group.

Ni·u·e (nē·ōō'ā) An island dependency of New Zealand SE of Samoa and east of Tonga; 100 square miles. Also *Savage Island.*

ni·val (nī'vəl) *adj.* Pertaining to the snow; also, growing under the snow. [<L *nivalis* < *nix, nivis* snow]

niv·e·ous (niv'ē-əs) *adj.* Snowy; like snow.

Ni·ver·nais (nē·ver·nā') A region and former province of central France.

Ni·vôse (nē·vōz') See CALENDAR (Republican).

nix¹ (niks) *n.* In Teutonic mythology, a water spirit appearing in male or female form, and sometimes appearing as part fish. [<G *nix*] — **nix'ie** *n.*

nix² (niks) *Slang n.* 1 Nothing. 2 No. — *adv.* No. — *interj.* Stop! Watch out!: an exclamation urging someone to stop saying or doing something. — *v.t.* To forbid or disagree with: He *nixed* our suggestions on the matter. [<G *nichts* nothing]

ni·zam (ni-zäm', -zam', nī-) *n. pl.* **zam** A Turkish regular soldier. [<Turkish *nizām* <Arabic *nazāma* govern]

Ni·zam (ni-zäm', -zam', nī-) *n.* The title of the native ruler of Hyderabad, India.

Ni·zam·ate (ni·zäm'āt, -zam'-, -it) *n.* The territory governed by the Nizam.

Nizam's Dominions A former name for HYDERABAD.

Nizh·ni Nov·go·rod (nēzh'nē nôv'gə·rot) A former name for GORKI.

Nizh·ni Ta·gil (nēzh'nē tä·gēl') A city in the central Urals in western Siberian Russian S.F.S.R.

Niz·za (nēt'tsä) The Italian name for NICE.

Njord (nyôrd) In Norse mythology, one of the Vanir, god of the winds and sea, father of Frey and Freya. Also **Njorth** (nyôrth).

NKVD See NARKOMVNUDEL.

no¹ (nō) *adv.* 1 Nay; not so: opposed to *yes.* 2 Not at all; not in any wise: used with comparatives: *no* better than the other. 3 *Scot. Not.* 4 Not: used to express an alternative after *or:* whether or *no.* — *adj.* Not any; not one: *No* seats are left. — *n. pl.* **noes** 1 A negative reply; a denial: He will not take *no* for an answer. 2 A negative vote or voter: The *noes* have it. [OE *na* < *ne* not + *a* ever]

no² (nō) *n.* The classical drama of Japan, traditionally tragic or noble in theme, requiring masks and elaborate costumes, stylized gesture, music, and dancing. Compare KABUKI. Also spelled *noh.*

No (nō), **Lake** A lake on the White Nile in south central Sudan; 40 square miles.

no·a (nō'ə) *adj.* In Polynesia, ordinary and generally accessible: opposite of *tabu.* [<Polynesian]

no-ac·count (nō'ə-kount') *adj. Slang* Worthless: also spelled *no-'count.*

No·a·chi·an (nō·ā'kē·ən) *adj.* Of or pertaining to Noah or to his times: the *Noachian* flood. Also **No·ach'ic** (-ak'ik), **No·ach'i·cal** (-ak'-).

No·ah (nō'ə, *Ger.* nō'ä) A masculine personal name. Also *Du.* **No·ach** (nō'äkh), *Fr.* **No·é** (nō·ā'), *Sw.* **No·a** (nō'ä). [<Hebrew, rest] — **Noah** In the Old Testament, a patriarch who, at the command of God, built an ark to save his family and "two of every sort of living thing" from the Flood. *Gen.* vi 14–22.

nob¹ (nob) *n.* 1 *Slang* The head. 2 In cribbage, the jack of trumps. [Var. of KNOB]

nob² (nob) *n. Brit. Slang* A person of social distinction; nobleman. [? Special use of NOB¹ (def. 1)]

no-ball (nō'bôl') *n.* In cricket, a ball unfairly bowled.

nob·ble (nob'əl) *v.t.* **bled, ·bling** *Brit. Slang* 1 To drug or disable (a horse) to prevent its winning a race. 2 To gain or influence by bribery or other underhand means. 3 To steal or cheat; swindle. 4 To catch. [Origin uncertain] — **nob'bler** *n.*

nob·by (nob'ē) *adj.* **·bi·er, ·bi·est** *Brit. Slang* Fit for a nob; hence, elegant or flashy; showy; stylish. [<NOB²]

No·bel (nō-bel'), **Alfred Bernhard,** 1833–96, Swedish industrialist and philanthropist; inventor of dynamite; founded by his will the **Nobel Prizes,** awarded annually to those whose work in physics, chemistry, medicine, literature, and furtherance of the world's peace is thought of most benefit to humanity.

Nob Hill A fashionable district of San Francisco, California.

No·bi·le (nō'bē·lā), **Umberto,** born 1885, Italian aeronautical engineer and Arctic explorer.

no·bil·i·ar·y (nō-bil'ē·er'ē, -bil'yə-rē) *adj.* Of or pertaining to the nobility. [<F *nobiliaire* <L *nobilis* noble]

nobiliary particle A preposition used as a prefix to a family name, indicating the noble birth of the person concerned, as *de* or *von.*

no·bil·i·ty (nō-bil'ə-tē) *n. pl.* **·ties** 1 The state of being noble, as in character or rank. 2 A class composed of nobles; in Great Britain, the peerage. 3 High-mindedness; magnanimity. 4 Great moral excellence. 5 Noble lineage. [<OF *nobilité* <L *nobilitas, -tatis* <*nobilis* noble]

no·ble (nō'bəl) *adj.* **·bler, ·blest** 1 Exalted in character or quality; excellent; worthy. 2 Characterized by or indicative of virtue or magnanimity; high-minded. 3 Of or pertaining to an aristocracy; of lofty lineage; aristocratic. 4 Imposing in appearance; magnificent; grand. 5 Precious; pure: said of minerals and metals. 6 Chemically resistant or inert, as helium. 7 In falconry, long-winged, as a true falcon: see IGNOBLE. See synonyms under AWFUL, GENEROUS, HIGH, ILLUSTRIOUS, IMPERIAL. — *n.* 1 A person having hereditary title, rank, and privileges; in England, a peer, a member of the Second Estate, as distinct from the clergy and commoners. 2 An old English gold coin weighing 120 grains (1351). [<F <L *nobilis* noble, well-known. Related to L *noscere* know.] — **no'bly** *adv.*

no·ble·man (nō'bəl-mən) *n. pl.* **·men** (-mən) A man of noble rank; in England, a peer.

noble metal *Chem.* A metal which strongly resists oxidation and the action of acids, especially gold and platinum.

no·ble·ness (nō'bəl-nis) *n.* The quality of being noble; nobility.

no·blesse (nō-bles') *n.* 1 *Obs.* Noble birth; nobleness. 2 The body of the nobility.

no·blesse o·blige (nō-bles' ō·blēzh') *French* Those of high birth, wealth, or social position should behave generously or nobly toward others; literally, nobility obliges.

no·ble·wom·an (nō'bəl-wŏŏm'ən) *n. fem. pl.* **·wom·en** (-wim'in) A woman of noble rank.

no·bod·y (nō'bod'ē, -bəd-ē) *pron.* Not anybody. — *n. pl.* **·bod·ies** A person of no importance or influence.

no·cent (nō'sənt) *adj. Rare* 1 Injurious; hurtful. 2 Guilty. [<L *nocens, -entis,* ppr. of *nocere* harm]

no·ci·as·so·ci·a·tion (nō'sē·ə·sō'sē·ā'shən, -sō'shē-) *n. Physiol.* The loss of nerve force resulting from overstimulation of nociceptors or from shock or exhaustion. [<L *nocere* harm + ASSOCIATION]

no·ci·cep·tor (nō'sē·sep'tər) *n. Physiol.* A sense organ or receptor which responds to and transmits painful stimuli. Compare BENECEPTOR. [<NL <L *nocere* injure + *-ceptor,* as in *receptor*] — **no'ci·cep'tive** *adj.*

nock (nok) *n.* 1 The notch on the butt end of an arrow. 2 The notch on the horn of a bow for securing the bowstring. — *v.t.* 1 To notch, as an arrow or bow. 2 To fit (an arrow) to the bowstring, as for shooting. [ME *nocke,* prob. <Scand. Cf. dial. Sw. *nokke* notch.]

noc·tam·bu·la·tion (nok·tam'byə·lā'shən) *n.* Somnambulism. Also **noc·tam'bu·lism.** [<L *nox, noctis* night + *ambulare* walk]

noc·tam·bu·list (nok·tam'byə·list) *n.* A somnambulist.

nocti- combining form By or at night: *noctiflorous.* Also, before vowels, **noct-**. [<L *nox, noctis* night]

noc·ti·flo·rous (nok'tə·flôr'əs, -flō'rəs) *adj. Bot.* Blooming at night. [<L *nox* night + *florere* flower]

noc·ti·lu·ca (nok'tə·lōō'kə) *n.* Any bioluminescent marine flagellate of the genus *Noctiluca,* found in warm seas, where its abundant presence gives the waves a brilliantly colored phosphorescent luminosity. [<NL <L *nox, noctis* night + *lucere* shine]

noc·ti·lu·cent (nok'tə·lōō'sənt) *adj. Meteorol.* Luminous by night: said especially of certain high altitude clouds visible at night by reflected sunlight.

noc·ti·pho·bi·a (nok'tə·fō'bē·ə) *n.* Nyctophobia.

noc·tu·id (nok'chōō·id) *n.* Any of a large family (Noctuidae) of medium-sized moths, especially those with stout bodies and shining eyes, as the army worm and the cutworm, whose larvae are very destructive pests. — *adj.* Of or pertaining to the Noctuidae. [<NL <L *noctua* night owl] — **noc'tu·id'e·ous** (-id'ē-əs), **noc·tu'id·ous, noc·tu·oid** (nok'chōō·oid) *adj.*

noc·tule (nok'chōōl) *n.* A large bat (*Nyctalus noctula*) of Europe.

noc·tur·nal (nok·tûr'nəl) *adj.* 1 Pertaining to night; occurring or active at night. 2 Seeking food by night, as animals. 3 Having blossoms that open by night. Opposed to *diurnal.* — **noc·tur'nal·ly** *adv.*

noc·turne (nok'tûrn) *n.* 1 In painting, a night scene. 2 *Music* A composition of a romantic, dreamy nature regarded as appropriate to night. [<F <L *nocturnus* nightly]

noc·u·ous (nok'yōō·əs) *adj.* Causing harm; noxious. [<L *nocuus* injurious < *nocere* harm] — **noc'u·ous·ly** *adv.* — **noc'u·ous·ness** *n.*

nod (nod) *v.* **nod·ded, nod·ding** *v.i.* 1 To make a brief forward and downward movement of the head, as in agreement, invitation, etc. 2 To let the head fall forward briefly and involuntarily, as when drowsy; be sleepy. 3 To be inattentive or careless; make an error. 4 To incline the top or upper part as if nodding: said of trees, flowers, etc. — *v.t.* 5 To bend (the head) forward and downward

briefly. **6** To express or signify by nodding: to *nod* approval. **7** To affect in a specified manner by nodding: He *nodded* me from the room. [Cf. G *notteln* shake] —**nod'der** *n.*
Nod (nod), **Land of** **1** The land east of Eden in which Cain settled after killing Abel. *Gen.* iv 16. **2** Sleep.
no·dal (nōd'l) *adj.* Of or pertaining to a node or nodes.
nod·dle (nod'l) *n. Colloq.* The head: a humorous use. —*v.t.* & *v.i.* **·dled** **·dling** To nod frequently. [<NOD]
nod·dy (nod'ē) *n. pl.* **·dies** **1** A dunce; a fool. **2** A light, two-wheeled, one-horse vehicle. **3** One of several terns (subfamily *Sterninae*) noted for their exceptional tameness and stolidity, especially *Anous stolidus* of the Atlantic coast. [<NOD]
node (nōd) *n.* **1** A knot or knob; swelling. **2** *Bot.* The joint or knob on the stem of a plant, from which leaves, buds, or other structures grow. **3** *Math.* A point at which a curve cuts or crosses itself. **4** *Astron.* Either of the two points at which the intersection of the planes of two orbits, especially those of a satellite and its primary, pierces the celestial sphere; specifically, the point where the orbit of a heavenly body intersects the ecliptic. The node encountered by a body in its northward passage is called its **ascending node**; in its southward passage, the **descending** or **setting node**. **5** *Anat.* A firm, flattened tumor on a bone, tendon, or the like. **6** The plot of a story or drama. **7** *Physics* A stationary point, line, or plane in a vibrating body where the amplitude of a wave is virtually zero. [<L *nodus* knot] —**no·dal** (nōd'l) *adj.*
nod·i·cal (nod'i·kəl, nō'di-) *adj. Astron.* Of or pertaining to the nodes: said especially of the revolution of a celestial body from a node back to the same node again.
no·dose (nō'dōs, nō·dōs') *adj.* Having nodes or knots; knobby. [<L *nodosus* full of knots <*nodus* a knot] —**no·dos·i·ty** (nō·dos'ə·tē) *n.*
no·dous (nō'dəs) *adj.* Knotty.
nod·u·lar (noj'ə·lər) *adj.* Relating to, shaped like, or containing nodules.
nod·ule (noj'ōōl) *n.* **1** A little knot, lump, or node. **2** *Bot.* A tubercle. [<L *nodulus*, dim. of *nodus* a knot]
nod·uled (noj'ōōld) *adj.* Having nodules.
nod·u·lose (noj'ə·lōs) *adj.* Having nodules. Also **nod'u·lous** (-ləs).
no·dus (nō'dəs) *n.* **1** A knot. **2** Node (def. 5). **3** A difficulty; complexity; knotty point. [<L]
No·el (nō'əl) A masculine personal name. Also *Fr.* No·ël (nō·el'). [<L, Christmas]
No·ël (nō·el') *n.* **1** Christmas. **2** A carol: also spelled *Nowel*. [<OF <L *natalis* birthday]
no·e·mat·ic (nō'ə·mat'ik) *adj.* Of or pertaining to mental operations and processes. [<Gk. *noēma* thought]
no·e·sis (nō·ē'sis) *n.* **1** Comprehension by the intellect alone. **2** Cognition, especially as applied to sources of self-evident knowledge. [<Gk. *noēsis* intelligence <*noeein* think]
no·et·ic (nō·et'ik) *adj.* Pertaining to or conceived by the mind; intellectual. —*n.* A person given to intellectual, especially abstract, reasoning. [<Gk. *noētikos* intelligent]
nog[1] (nog) *n.* **1** A peg or a square or oblong block of wood. **2** A wooden pin driven through a wall. [? Var. of ME *knag* a peg <LG]
nog[2] (nog) *n.* **1** A strong ale. **2** Eggnog. Also **nogg.** [<dial. E]
nog·gin (nog'in) *n.* **1** A small mug, or its contents. **2** A liquid measure equal to about a gill. **3** *Slang* A person's head. [Origin uncertain]
nog·ging (nog'ing) *n.* **1** Pieces of wood inserted in a masonry wall, to stiffen it, or upon which to nail finishing material. **2** Brick filling in the interstices of a frame wall. [<NOG[1]]

NOGGING
As seen in half-timbered gable.

no go An impasse; balk. Also **no-go** (nō'gō') *n.*
No·gu·chi (nō·gōō·chē), **Hi·deyo,** 1876–1928, Japanese bacteriologist active in the United States.
Noh (nō) See No.
no·how (nō'hou') *adv. Illit.* In no way; not by any means.
noil (noil) *n.* **1** Short-staple fibers combed out from long-staple during the combing process in preparing wool or cotton yarns. **2** Waste silk produced in the manufacture of spun silk. [Origin uncertain]
noise (noiz) *n.* **1** A sound of any kind, especially a loud or a disturbing sound. **2** In acoustics, the confused sound obtained by a number of discordant vibrations or undesirable random voltages in a radio or television channel. **3** *Obs.* Clamor; discussion; gossip. —*v.* **noised, nois·ing** *v.t.* **1** To spread by rumor or report: often with *about* or *abroad.* —*v.i.* **2** To be noisy; make a noise. **3** To talk in a loud and voluble manner. [<OF *noyse*; ult. origin uncertain]
Synonyms: blare, clamor, clatter, din, hubbub, jangle, outcry, racket, rattle, roar, tumult, uproar. See SOUND[1], TUMULT. *Antonyms*: calmness, noiselessness, peace, quiet, silence, stillness.
noise·less (noiz'lis) *adj.* Causing or making no noise; silent. —**noise'less·ly** *adv.* —**noise'·less·ness** *n.*
noise level In acoustics, that value of noise which may be expressed in decibels with reference to a specified frequency range.
noise·mak·er (noiz'mā'kər) *n.* **1** A horn, bell, or other device for making noise at celebrations. **2** One who or that which makes noise.
noi·some (noi'səm) *adj.* **1** Very offensive, particularly to the sense of smell. **2** Injurious; noxious. [<*noy*, aphetic var. of ANNOY + -SOME] —**noi'some·ly** *adv.* —**noi'some·ness** *n.*
Synonyms: deadly, deleterious, destructive, detrimental, fetid, foul, harmful, hurtful, insalubrious, mischievous, noxious, pernicious, pestiferous, pestilential, poisonous, unhealthful, unwholesome. *Noxious* is a stronger word than *noisome*, as referring to that which is injurious or *destructive*. *Noisome* denotes that which is disgusting, especially to the sense of smell; as, the *noisome* stench proclaimed the presence of *noxious* gases. *Antonyms*: beneficial, healthful, invigorating, salubrious, salutary, wholesome.
nois·y (noi'zē) *adj.* **nois·i·er, nois·i·est** **1** Making a loud noise. **2** Characterized by noise. —**nois'i·ly** *adv.* —**nois'i·ness** *n.*
Synonyms: blatant, blustering, boisterous, brawling, clamorous, obstreperous, riotous, tumultuous, turbulent, uproarious, vociferous. *Antonyms*: dumb, hushed, inaudible, mute, noiseless, quiet, silent, still.
no·lens vo·lens (nō'lenz vō'lenz) *Latin* Unwilling or willing; undecided; willy-nilly.
no·li-me-tan·ge·re (nō'lī·mē·tan'jə·rē) *n.* **1** A warning not to touch or meddle with. **2** Touch-me-not (def. 1). **3** The squirting cucumber. **4** A picture showing Jesus as he appeared to Mary Magdalene after his resurrection: so called from his words of warning to her. **5** Any person or thing not to be touched or interfered with. **6** *Pathol.* Rodent ulcer. [<L *noli me tangere* touch me not]
noll (nol, nōl) *n. Obs.* The head. [OE *hnoll* crown of the head]
nol·le pros·e·qui (nol'ē pros'ə·kwī) *Law* An entry of record in a civil or criminal case, to signify that the plaintiff or prosecutor will not press it. Abbr. *nol. pros.* [<L, be unwilling to prosecute]
no·lo con·ten·de·re (nō'lō kən·ten'də·rē) *Law* A plea by a defendant in a criminal action, which, while not an admission of guilt, has the same legal effect as regards the proceedings on the indictment. Such a plea does not debar a defendant from denying the truth of the charges in any other proceedings arising out of the same matter. [<L, I am unwilling to contend]
nol-pros (nol'pros') *v.t.* **-prossed, -pros·sing** *Law* To subject to a nolle prosequi. [Short for NOLLE PROSEQUI]
nom (nôṅ) *n. French* Name.
nom- Var. of NOMO-.
no·ma (nō'mə) *n. Pathol.* Gangrenous inflammation of the mouth, especially in young children. [<NL <Gk. *nomē* <*nomein* feed]
no·mad (nō'mad, nom'ad) *adj.* Nomadic. —*n.* A rover; one of an unsettled, wandering people, tribe, or race. [<L *nomas, -adis* <Gk. *nomas* <*nomein* pasture, feed] —**no'mad·ism** *n.*
no·mad·ic (nō·mad'ik) *adj.* **1** Pertaining to nomads; roaming. **2** Unsettled. Also **no·mad'·i·cal.** —**no·mad'i·cal·ly** *adv.*
no man's land **1** Waste or unowned land; specifically, a plot of land situated beyond the limits of the north wall of the City of London, where executions took place in the 14th century. **2** In warfare, the land between the opposing armies.
nom·arch (nom'ärk) *n.* The governor of a nomarchy or nome. [<Gk. *nomarchēs* < *nomos* province + *archein* rule]
nom·ar·chy (nom'är·kē) *n. pl.* **·chies** A province of modern Greece; a nome.
nom·bles (nom'bəlz, num'-) See NUMBLES.
nom·bril (nom'bril) *n. Her.* The navel point, between the fess point and the base point on an escutcheon. See illustration under ESCUTCHEON. [<F <OF *lombril*, ult. <L *umbilicus* navel]
nom de guerre (nôṅ də gâr') *French* An assumed name; a pseudonym; literally, a war name.
nom de plume (nom' də plōōm', *Fr.* nôṅ də plüm') A pen name; a writer's assumed name. [<F *nom* a name + *de* of + *plume* a pen]
nome[1] (nōm) *n.* A province or department of ancient Egypt or modern Greece. Also, *Greek*, *nomos.* [<Gk. *nomos* province]
nome[2] (nōm) Past participle of NIM.
Nome (nōm) A city in western Alaska; 12 miles west of **Cape Nome,** a promontory of the southern Seward Peninsula.
no·men (nō'mən) Past participle of NIM.
no·men·cla·tor (nō'mən·klā'tər) *n.* **1** One who assigns or announces names. **2** One who gives names; a scientific classifier. [<L *nomen* name + *calare* call]
no·men·cla·ture (nō'mən·klā'chər, nō·men'·klə-) *n.* A system of names, as used in any art or science, or by any recognized group, school, system, or authority. [<L *nomenclatura* list of names]
nom·i·nal (nom'ə·nəl) *adj.* **1** Of or pertaining to a name or names. **2** Existing in name only; not actual: a *nominal* peace. **3** So slight or inconsiderable as to be hardly worth naming; trifling: a *nominal* sum. **4** Consisting of or containing names: a *nominal* roll. **5** Assigned to a person by name, as stocks or shares. **6** *Gram.* **a** Pertaining to or like a noun or nouns. **b** Functioning as a noun or nouns. [<L *nominalis* <*nomen* name]
nom·i·nal·ism (nom'ə·nəl·iz'əm) *n. Philos.* The doctrine that abstract or generic conceptions, or universals, have no basis or representation in reality but are names only, and that only individual objects exist: opposed to *realism.* Compare CONCEPTUALISM. —**nom'i·nal·ist** *adj.* & *n.* —**nom'i·nal·is'tic** *adj.*
nom·i·nal·ly (nom'ə·nəl·ē) *adv.* In a nominal manner; in name only: opposed to *really, actually.*
nominal value The value named; face value.
nominal wages See REAL WAGES.
nom·i·nate (nom'ə·nāt) *v.t.* **·nat·ed, ·nat·ing** **1** To name as a candidate for elective office. **2** To appoint to some office or duty. **3** *Obs.* To name; entitle. —*adj.* **1** Nominated. **2** Having a legal or particular name. [<L *nominatus*, pp. of *nominare* <*nomen, -inis* a name] —**nom'i·na'tor** *n.*
nom·i·na·tion (nom'ə·nā'shən) *n.* **1** The act of nominating; the fact or condition of being nominated. **2** The power of appointment, as of a clergyman to a benefice. —**direct nomination** A method of nominating candidates for office by the direct votes of the people instead of by means of a representative convention.
nom·i·na·tive (nom'ə·nə·tiv, nom'ə·nā'-) *adj.* **1** *Gram.* Designating the case of the subject of a finite verb, or of a word agreeing with, or in apposition to the subject; in English grammar, subjective. **2** Appointed by nomination; nominated. **3** Bearing the name of a person, as an invitation or a share of stock. —*n. Gram.* **1** The nominative case. **2** A word in this case.
nom·i·nee (nom'ə·nē') *n.* **1** One who receives a nomination. **2** A designated person on whose life another's annuity depends.
no·mism (nō'miz·əm) *n.* Strict adherence to religious or moral law. [<Gk. *nomos* law] —**no·mis·tic** (nō·mis'tik) *adj.*
nomo- *combining form* Law; custom; usage: *nomocracy.* Also, before vowels, *nom-.* [<Gk. *nomos* law]
no·mo·gen·e·sis (nō'mə·jen'ə·sis) *n. Biol.* The doctrine which attributes all evolutionary change in plants and animals to the operation of predetermined and unchanging laws. —**no'·mo·ge·net'ic** (-jə·net'ik) *adj.*

no·mo·graph (nō′mə·graf, -gräf) *n.* **1** *Math.* A graph consisting of graduated scales for two or more interrelated variables, so arranged that a straight line joining given values of the two known variables will cut the scale of the third variable at the value sought; an isopleth. Also called **no′mo·gram.** **2** Any graphic representation of numerical relations.

no·mog·ra·phy (nō·mog′rə·fē) *n.* **1** The art of drafting laws, or a treatise on that art. **2** *Math.* A method for the graphic representation on a plane surface of the relations between two or more variables; the science and technique of graphic computation. — **no·mo·graph·ic** (nō′mə·graf′ik) or **·i·cal** *adj.*

no·mol·o·gy (nō·mol′ə·jē) *n.* **1** The science that treats of law and lawmaking. **2** The branch of any science that treats of the laws that explain its phenomena, as in biology, psychology, etc. — **nom·o·log·i·cal** (nom′ə·loj′i·kəl) *adj.* — **no·mol′o·gist** *n.*

no more 1 Dead; gone. **2** Nothing more: I'll say *no more.* **3** No longer: It rains *no more.* **4** Never again: She'll sing *no more.* **5** Not to any greater extent: We could *no more* see than the blind. **6** Neither: I did not speak, *no more* did he.

no·mos (nō′mos) See NOME.

nom·o·thet·ic (nom′ə·thet′ik) *adj.* **1** Giving or enacting laws. **2** Nomistic. **3** Pertaining to a science of universal or general laws. Also **nom·o·thet′i·cal.** [< Gk. *nomothetikos* < *nomos* law + *tithēnai* establish]

No·mu·ra (nō·moō′rä), **Kichisaburo,** born 1877, Japanese admiral and diplomat.

-nomy *combining form* The science or systematic study of: *astronomy, economy.* [< Gk. *nomos* law]

non- *prefix* Not. [< L *non* not]
◆ **Non-** is the Latin negative adverb adopted as an English prefix. It denotes in general simple negation or absence of, as in *non-attendance,* lack of attendance. Compare UN- and IN-, which are more commonly antithetical and emphatic. Numerous words beginning with *non-* are self-explaining, as in the following partial list:

non-absolute	non-conducive
non-absorbable	non-Congressional
non-absorbent	non-connivance
non-abstainer	non-connotative
non-acceptance	non-consent
non-accomplishment	non-consideration
non-acquaintance	non-consumption
non-acquiescence	non-contagious
non-action	non-contentiously
non-active	non-conversion
non-adherence	non-crystallized
non-adherent	non-currency
non-adjacent	non-deliquescent
non-admission	non-delivery
non-adult	non-development
non-aggression	non-discrimination
non-aggressive	non-distribution
non-agreement	non-divisibility
non-agricultural	non-doctrinal
non-alcoholic	non-efficient
non-alienating	non-elastic
non-alienation	non-electric
non-arrival	non-empirical
non-Aryan	non-entry
non-attached	non-equation
non-attendance	non-equilibrium
non-attention	non-essential
non-believing	non-eternal
non-belligerent	non-eternity
non-Biblical	non-European
non-budding	non-excusable
non-canonical	non-execution
non-Catholic	non-exempt
non-Christian	non-exercise
non-church	non-existence
non-citizen	non-existent
non-coincident	non-existing
non-collegiate	non-expert
non-colonial	non-explosive
non-combustible	non-extant
non-communicant	non-factual
non-Communist	non-fatal
non-competitive	non-fiction
non-complaisance	non-fictional
non-conception	non-financial
non-concurrence	non-fiscal
non-flammable	non-recognition
non-freedom	non-rectangular
non-freeman	non-regimented
non-fulfilment	non-reigning
non-Hellenic	non-relative
non-householder	non-reproductive
non-human	non-resemblance
non-identity	non-resinous
non-improvement	non-resistant
non-inflammable	non-resisting
non-inquiring	non-resonant
non-instruction	non-scheduled
non-interference	non-scientific
non-interrupted	non-sectarian
non-literary	non-sensitive
non-luminous	non-sentient
non-magnetic	non-separable
non-marital	non-sexual
non-material	non-sexually
non-materiality	non-society
non-member	non-specie
non-membership	non-spiritous
non-metalliferous	non-sporting
non-missionary	non-standard
non-modulated	non-submission
non-natural	non-subscriber
non-nitrogenous	non-subscribing
non-nutritive	non-supporter
non-observance	non-supporting
non-officially	non-syllabic
non-operative	non-sympathizer
non-orthodox	non-sympathy
non-oxidating	non-tax
non-parallel	non-technical
non-parasitic	non-terminating
non-paying	non-toxic
non-payment	non-truth
non-performance	non-understanding
non-pigmented	non-validity
non-poisonous	non-vascular
non-professional	non-venomous
non-profit	non-volatile
non-protection	non-volition
non-Protestant	non-voluntary
non-publication	non-vortical
non-punishment	non-voter
non-reactive	non-voting
non-recital	non-worker
	non-working

non-age (non′ij, nō′nij) *n.* The period of legal minority; immaturity. [< NON- + AGE]

non·a·ge·nar·i·an (non′ə·jə·nâr′ē·ən, nō′nə-) *adj.* Pertaining to the nineties in age. — *n.* A person between the ages of ninety and a hundred. [< L *nonagenarius* of ninety]

non·a·gon (non′ə·gon) *n.* A nine-sided polygon. [< L *nonus* ninth + Gk. *gōnia* angle]

non·ane (non′ān) *n. Chem.* A liquid hydrocarbon, C_9H_{20}, of the methane series. [< L *nonus* ninth; because ninth in the series]

no·na·no·ic acid (nō′nə·nō′ik) Pelargonic acid.

non-ap·pear·ance (non′ə·pir′əns) *n.* Failure to appear, especially failure to appear in court in answer to a summons.

non-bore·safe (non′bôr′sāf, -bōr′-) *adj. Mil.* Designating a type of fuze that does not have a safety device to prevent explosion of the burster charge of a projectile while it is in the bore of the gun.

nonce (nons) *n.* Present time or occasion. — **for the nonce** For the present time or occasion. [ME *for then ones* for the one (occasion), misread as *for the nones*]

nonce word A word coined for one occasion.

non·cha·lance (non′shə·ləns, non′shə·läns′) *n.* Jaunty indifference or unconcern.

non·cha·lant (non′shə·lənt, non′shə·länt′) *adj.* Without concern; casual; indifferent. [< F, orig. ppr. of *nonchaloir* < L *non calere* not be warm, be indifferent] — **non′cha·lant·ly** *adv.*

non-com (non′kom′) *Colloq. adj.* Non-commissioned. — *n.* A non-commissioned officer.

non-com·bat·ant (non′kom·bat′ənt, -kom′bə·tənt, -kum′-) *n.* **1** One who is not a combatant; especially, one attached to a military force but whose duties do not require that he fight, as a chaplain or medical officer. **2** Anyone not connected with the military service in time of war; a civilian.

non-com·mis·sioned (non′kə·mish′ənd) *adj.* Not holding a military commission.

non-commissioned officer See under OFFICER.

non-com·mit·tal (non′kə·mit′l) *adj.* Not committal; not having or expressing a decided opinion. — **non′-com·mit′tal·ly** *adv.*

non-com·pli·ance (non′kəm·pli′əns) *n.* Failure or neglect to comply. — **non′-com·pli′ant** *adj.* & *n.*

non com·pos men·tis (non kom′pəs men′tis) *Latin* Not of sound mind; mentally unbalanced: often shortened to **non compos.**

non-con·cur (non′kən·kûr′) *v.t.* **·curred, ·cur·ring** To reject, as a bill or resolution.

non-con·duc·tor (non′kən·duk′tər) *n.* A substance or material that offers resistance to the passage of some form of energy: a *non-conductor* of heat or electricity; an insulator or dielectric. — **non′-con·duct′ing** *adj.*

non-con·form·ist (non′kən·fôr′mist) *n.* One who does not conform to established usage, as in religion; specifically, a person, especially a Protestant clergyman, refusing to conform to the Book of Common Prayer where the Church of England is established by law; a dissenter. — **non′con·form′ing** *adj.* — **non′con·form′i·ty** (-fôr′mə·tē) *n.*

non con·stat (non kon′stat) *Latin* It does not follow.

non-co·op·er·a·tion (non′kō·op′ə·rā′shən) *n.* Refusal to cooperate; specifically, civil resistance to a government through disobedience, boycotting of institutions, etc. — **non′-co·op′er·a·tive** (kō·op′rə·tiv, -ə·rā′tiv) *adj.* — **non′-co·op′er·a′tion·ist** *n.* — **non′-co·op′er·a′tor** *n.*

non·de·script (non′di·skript) *adj.* Not distinctive enough to be described. — *n.* A person or thing of no particular type, kind, or character: often used disparagingly. [< NON- + L *descriptus,* pp. of *describere.* See DESCRIBE.]

non-dis·junc·tion (non′dis·jungk′shən) *n. Biol.* The failure of paired chromosomes to separate during cell mitosis.

non-dis·tinc·tive (non′dis·tingk′tiv) *adj.* **1** Not distinctive. **2** *Ling.* Non-relevant.

non·du·ty (non′doō′tē, -dyoō′-) *adj. Mil.* Designating the status of an enlisted man or officer who, for any reason, is not available for duty with his unit or command.

none (nun) *pron.* **1** Not one; no one. **2** No or not one specifically named person or thing: A bad book is better than *none.* **3** Not any: That is *none* of her business. **4** (*construed as pl.*) Not any (of the persons or things specified): *None* of them have their drawings finished; *None* of the apples are rotten. — *adv.* In no respect; not at all: *none* the worse for wear. — *adj. Archaic* Not one; no one; no: generally before a vowel: *none* other gods before me. [OE *nān* < *ne* not + *ān* one]

non-ef·fec·tive (non′i·fek′tiv) *adj.* **1** Not effective; inoperative. **2** Unfitted or unavailable for active service or duty in the army or navy: a *non-effective* officer. — *n.* A soldier or sailor unfitted for active service or duty, because of sickness, wounds, or the like.

non-e·go (non·ē′gō, -eg′ō) *n. pl.* **·gos 1** Whatever is not the self, or not of or pertaining to the conscious self; more especially, the object of the conscious ego as apposed to or set over against the ego. **2** The objective or material world.

non-en·ti·ty (non·en′tə·tē) *n. pl.* **·ties 1** A person or thing of no account; a nobody. **2** The negation of being; non-existence. [NON- + ENTITY]

nones (nōnz) *n. pl.* **1** The ninth day before the ides in the Roman calendar. **2** The canonical office, originally recited at 3 p.m., or the ninth hour by ancient Roman reckoning. [< OF < L *nonae* < *nonus* ninth < *novem* nine]

non est (non est) *Latin* It is not; it is wanting; it does not exist.

none·such (nun′such′) *n.* A person or thing having no equal; an unexampled thing: also spelled **nonsuch.**

none·the·less (nun′thə·les′) *adv.* In spite of everything; nevertheless. Also **none the less.**

non-Eu·clid·e·an (non′yoō·klid′ē·ən) *adj. Math.* Designating a geometry dealing with a space in which the axioms and postulates of Euclid do not necessarily hold.

non-ex·pend·a·ble (non′ik·spen′də·bəl) *adj.* Designating articles of public property for which there is responsibility and accountability: said especially of war materials which are not consumed or destroyed by the mere act of use.

non·fea·sance (non·fē′zəns) *n. Law* The nonperformance of some act which one is bound by legal or official duty to perform. Compare MALFEASANCE, MISFEASANCE. —**non·fea′sor** *n.*

non–fer·rous (non·fer′əs) *adj.* Pertaining to or designating any metal other than or not containing iron, as copper, tin, platinum, etc.

non–fi·nite (non·fī′nīt) *adj. Gram. & Logic* **1** Indefinite; at any unspecified time. **2** Limitless but finite, as a Möbius surface. **3** Endless; starting from the point specified and continuing indefinitely in one direction. **4** Infinite; without limits as to space or time. Example: *Continuing* and *extending*, as gerunds or verbal nouns, are called *non–finite* forms or infinitives in *–ing* by the older grammarians of English. **5** *Theol.* Eternal. Example: God *is* (now, ever was, and always will be). The word *is* here exhibits the eternal aspect of the *non–finite* verb. **6** Transcendent: where time is merely another dimension in the space–time continuum. Example: In the statement "Mass implies inertia," the verb *implies* exhibits the transcendent aspect of the *non–finite* tense, indicating that the relationship mass–energy is always true in the space–time multiverse.

non–frat·er·ni·za·tion (non·frat′ər·nə·zā′shən, -nī·zā′-) *n.* A policy pursued by the U.S. Army during and after World War II, forbidding the association of occupying military forces with civilians.

non–ha·la·tion film (non′hā·lā′shən) Film that has been opaqued to prevent reflection.

non–hy·gro·scop·ic (non·hī′grə·skop′ik) *adj.* Having little or no tendency to absorb moisture: a *non–hygroscopic* gunpowder.

no·nil·lion (nō·nil′yən) *n.* A cardinal number: in the French and American system of numeration, denoted by 1 followed by thirty ciphers; in the English system, 1 with fifty-four ciphers. [<F <L *nonus* ninth (< *novem* nine) + F *million* a million] —**no·nil′lionth** (-yənth) *adj.*

non–in·duc·tive (non′in·duk′tiv) *adj. Electr.* Not inductive: applied to a resistance that offers no greater opposition to a varying than to an unvarying current.

non–in·ter·course (non·in′tər·kôrs, -kōrs) *n.* No intercourse: commonly applied to a legal or diplomatic prohibition of commercial intercourse.

Non–Intercourse Act In U. S. history, an Act of Congress, passed in 1809, forbidding all commercial intercourse with England and France.

non–in·ter·ven·tion (non′in·tər·ven′shən) *n.* The state or condition of not interfering; refusal to intervene.

non–in·ter·ven·tion·ist (non′in·tər·ven′shən·ist) *n.* One who advocates a policy of non-intervention.

non–join·der (non·join′dər) *n. Law* An omission to join, as by a person who should be one of a party in an action. Compare MISJOINDER.

non·ju·ror (non·jŏŏr′ər) *n.* **1** A clergyman in England who refused to take the oath of allegiance to William and Mary after the revolution of 1688. **2** A Scottish Presbyterian who refused the oath abjuring the Stuart Pretenders. [<NON- + JUROR, in obs. sense "one who takes an oath"]

non–le·thal (non·lē′thəl) *adj.* Not capable of causing death: a *non–lethal* drug or chemical agent; non–toxic.

non li·cet (non lī′sit) *Latin* It is not lawful.

non li·quet (non lī′kwit, lik′wit) *Latin* The case is not clear.

non–met·al (non·met′l) *n. Chem.* Any element (as oxygen, nitrogen, carbon, sulfur, arsenic, and iodine) which has acid rather than basic properties, and is incapable of forming cations in solution.

non–me·tal·lic (non′mə·tal′ik) *adj.* **1** Not metallic. **2** Pertaining to a non–metal.

non–mor·al (non·môr′əl, -mor′-) *adj.* Having no relation to morals or to ethical conceptions and ideals; not moral or immoral. —**non′–mo·ral·i·ty** (-mô·ral′ə·tē, -mə-) *n.*

non–mo·tile (non·mō′til) *adj.* Incapable of motion of itself.

Non·ni (non′nē′) A river of central Manchuria, flowing 740 miles south to the Sungari.

non–nu·cle·at·ed (non·nōō′klē·ā′tid, -nyōō′-) *adj. Biol.* Not having a nucleus, as a cell.

non–ob·jec·tive art (non′əb·jek′tiv) Art that does not attempt to represent the recognizable form or effect of objects as they appear in nature.

non ob·stan·te (non ob·stan′tē) *Latin* Notwithstanding; in spite of.

non om·nis mo·ri·ar (non om′nis môr′ē·är) *Latin* My work will live; literally, I shall not wholly die.

non·pa·reil (non′pə·rel′) *adj.* Of unequaled excellence. —*n.* **1** Something of unequaled excellence. **2** *Printing* **a** A size of type between agate and minion: the former and now seldom used name for 6–point type. **b** A 6–point slug. **3** One of various birds of brilliant coloring of the southern United States, especially the painted bunting. **4** A variety of russet apple. [<OF <*non* not (<L) + *pareil* equal <LL *pariculus*, dim. of L *par* equal]

non–par·ous (non·par′əs) *adj.* Not having borne children.

non–par·tic·i·pat·ing (non′pär·tis′ə·pā′ting) *adj.* Not participating, nor conveying the right to participate, in the surplus or profits of an insurance company; pertaining to insurance in which the policyholders are not allowed to participate in the profits.

non–par·ti·san (non·pär′tə·zən) *adj.* Not pertaining or adhering to any established party.

Non–Partisan League A farmer's political organization formed in 1915. It controlled North Dakota for a time and extended its influence over 13 agricultural States with the object of fostering industrial democracy by organized political, financial, and economic cooperation.

non–pay (non′pā′) *adj.* Designating the status of an enlisted man or officer whose pay is canceled for periods of unauthorized absence.

non·plus (non·plus′, non′plus) *v.t.* **·plused** or **·plussed, ·plus·ing** or **·plus·sing** To bring to a nonplus; baffle; perplex. —*n.* A mental standstill; perplexity, especially as causing silence or indecision. [<L *non plus* no further <*non* not + *plus* more]

non pos·su·mus (non pos′ə·məs) *Latin* We cannot; we are not able.

non–pro·duc·tive (non′prə·duk′tiv) *adj.* Not producing: a labor term designating clerical workers, inspectors, etc. —**non′pro·duc′tive·ness** *n.*

non–pros (non·pros′) *v.t.* **–prossed, –pros·sing** *Law* To enter judgment against (a plaintiff who fails to prosecute). [Short for NON PROSEQUITUR]

non pro·se·qui·tur (non prō·sek′wi·tər) *Law* A judgment entered at common law against a plaintiff who fails to prosecute. Compare NOLLE PROSEQUI. [<L, lit., he does not prosecute]

non–rel·e·vant (non′rel′ə·vənt) *adj.* **1** Not relevant. **2** *Ling.* Denoting those features of a phoneme which do not function to differentiate it from other phonemes in a language, as aspiration in English.

non–rep·re·sen·ta·tion·al (non′rep′ri·zen·tā′·shən·əl) *adj.* Not representational; specifically, denoting a form of art that does not attempt to represent the recognizable form or effect of objects as they appear in nature.

non–res·i·dent (non·rez′ə·dənt) *adj.* Not resident in a place: a *non–resident* landlord. —*n.* One not permanently residing in, or systematically absent from, a particular place. —**non–res′i·dence, non–res′i·den·cy** *n.*

non–re·straint (non′ri·strānt′) *n.* Absence of restraint; especially, the treatment of insane persons without using a straitjacket or other physical restraint.

non·rig·id (non·rij′id) *adj. Aeron.* Denoting an airship whose form is maintained by the internal pressure in the gas chambers and ballonets.

non·sense (non′sens, -səns) *n.* **1** That which is without sense, or without good sense; meaningless or ridiculous language; absurd behavior. **2** Things of no importance. —**non·sen′si·cal** *adj.* —**non·sen′si·cal·ly** *adv.* —**non·sen′si·cal·ness** *n.*

non seq·ui·tur (non sek′wə·tər) The fallacy of irrelevant conclusion; an inference that does not follow from the premises. [<L, it does not follow]

non–sked (non′sked′) *Colloq. adj.* Non–scheduled: applied especially to passenger airplane service offered without scheduled flying times and at rates lower than those of the regular scheduled airlines. —*n.* Something not operating on or holding to a schedule. [Short for *non–scheduled*]

non·skid (non′skid′) *adj.* Having the surface treaded or corrugated to reduce skidding: said of tires.

non·stop (non′stop′) *adj.* Making, having made, or scheduled to make no stops: a *nonstop* flight; *nonstop* train.

non–stri·at·ed (non·strī′ā·tid) *adj.* Void of striations; without stripes, as muscle fibers.

non·such (nun′such′) See NONESUCH.

non·suit (non′sōōt′) *Law v.t.* To order the dismissal of the suit of. —*n.* **1** The abandonment of a suit. **2** A judgment dismissing a suit, when the plaintiff either abandons it or fails to establish a cause of action. [<AF *nonsute*, OF *nonsuite* <*non* not (<L) + *suite*. See SUIT.]

non–sup·port (non′sə·pôrt′, -pōrt′) *n.* Failure or neglect to provide for the support of dependents.

non–un·ion (non·yōōn′yən) *adj.* **1** Not belonging to a trade union. **2** Not employing or recognizing any trade union or its members. —*n.* Lack of union or joining: said specifically of broken bones that do not knit properly.

non–un·ion·ism (non·yōōn′yən·iz′əm) *n.* Non-adherence or opposition to the establishment or the principles of trade unions. —**non–un′ion·ist** *n.*

non·u·ple (non′yə·pəl) *adj.* Consisting of nine; having nine parts or members; ninefold; also, taken by nines. —*n.* A number or sum nine times as great as another. [<F <L *nonus* ninth; on analogy with *quadruple, quintuple*, etc.]

non·u·pli·cate (non·yōō′plə·kit) *adj.* **1** Nine-fold. **2** Raised to the ninth power. [<L *nonus* ninth; on analogy with *duplicate*]

non–us·er (non·yōō′zər) *n. Law* A continued omission to assert or exercise some right or privilege, whereby the right or privilege is lost.

non–vi·a·ble (non·vī′ə·bəl) *adj.* Having no capacity to survive independently: said especially of a fetus.

noo (nōō) *adv. Scot.* Now.

noo·dle[1] (nōōd′l) *n. Slang* **1** A simpleton: also **noo′dle·head′**. **2** The head. [Origin unknown]

noo·dle[2] (nōōd′l) *n.* A thin strip of dried dough, usually made with egg: used in soup, etc. [<G *nudel*]

nook (nŏŏk) *n.* A narrow and retired place, as in an angle; a recess, as in a garden. [ME *noke* corner, ? <Scand. Cf. dial. Norw. *nok* a hook.]

noon (nōōn) *n.* **1** That time of day when the sun is on the meridian; the middle of the day; in an exact sense, 12 o'clock in the daytime. **2** The highest point of any period or career: the *noon* of life. **3** Originally, the ninth hour after sunrise, or about 3 o'clock p.m; midway between 12 o'clock and sunset; hence, the canonical hour of nones. **4** *pl.* A noontime repast. [OE *non* <L *nona (hora)* ninth (hour)]

noon·day (nōōn′dā′) *n.* The middle of the day. —*adj.* Pertaining to midday.

noon·ing (nōō′ning) *n. Archaic* **1** A time of rest taken at noon. **2** A midday meal. **3** The time of noon.

noon mark Formerly, a mark made in some familiar place, on a floor, doorstep, window sill, etc., to indicate where a certain shadow fell at noon; hence, noon.

noon·tide (nōōn′tīd′) *n.* **1** The time of midday. **2** The period of culmination: the *noontide* of glory. **3** The position of the moon at midnight; midnight. —*adj.* Of, occurring at, or characteristic of noon: *noontide* glory. Also **noon′time′** (-tīm′).

noose (nōōs) *n.* **1** A loop furnished with a running knot, as in a hangman's halter or a snare; slipknot. **2** Anything that restricts one's freedom. —*v.t.* **noosed, noos·ing 1** To capture or secure with a noose. **2** To make a noose in or with. [<Provençal *nous* <L *nodus* a knot]

Noot·ka Sound (nōōt′kä, -kə) An inlet of the Pacific in western Vancouver Island, Canada.

no·pal (nō′pəl) *n.* **1** Any one of various cacti (especially genus *Nopalea*), as *N. cochenillifer*, used for rearing the cochineal insect. **2** A prickly pear. [<Sp. <Nahuatl *nopalli*]

nope (nōp) *adv. Slang* No.

nor[1] (nôr) *conj.* And not; likewise not. ◆ *Nor* is used, chiefly, as a correlative of a preceding negative, usually *neither* or *not*. It may be used, as for poetical effect, without a correlative: We sat still, *nor* stirred. In older writing and in poetry, it often appears as an introductory negative instead of *neither*: He heeded *nor* praise *nor* blame. [Contraction of ME *nother* neither]

nor[2] (nôr) *conj. Brit. Dial.* Than: He does better *nor* you.

nor- combining form *Chem.* A normal or a parent compound. [<NORMAL]

No·ra (nôr′ə, nō′rə) A feminine personal name; diminutive of ELEANOR, HONORA, LEONORA. Also **No′rah**.

no·ra·ghe (nə·rä′gä) See NURAGH.

Nord (nôr) The northernmost department of France, on the Belgian border; 2,229 square miles; capital, Lille.

Nor·dau (nôr′dou), **Max Simon**, 1849-1923, German physician, author, and Zionist leader born in Hungary: original surname *Südfeld*.

Nor·den·skjöld (nôr′dən·shœld), **Baron Nils Adolf Erik**, 1832-1901, Swedish Arctic explorer.

Nor·den·skjöld Sea (nôr′dən·shœld) A former name for the LAPTEV SEA.

Nord·hau·sen (nôrt′hou·zən) A city in the former state of Thuringia, western East Germany.

Nordhausen acid Fuming sulfuric acid. [from *Nordhausen*, where formerly made]

Nor·dic (nôr′dik) *adj. Anthropol.* Pertaining or belonging to the blond-haired subdivision of the Caucasian ethnic stock, inhabiting Scandinavia, Scotland, and England, and to other Germanic peoples of northwestern Europe. — *n.* A member of this subdivision. [<F *nordique* < *nord* north]

Nord·kyn (nôr′kün), **Cape**, A cape in northern Norway on the Arctic Ocean, the northernmost point of the European mainland. Compare NORTH CAPE.

Nore (nôr), **The** An anchorage, lighthouse, and sandbank in the Thames estuary, SE England.

nor′·east·er (nôr′ēs′tər) See NORTHEASTER.

Nor·folk (nôr′fək) 1 A county of eastern England, 2,054 square miles; county town, Norwich. 2 A port on the Elizabeth River south of Hampton Roads in SE Virginia; site of a United States naval base.

Norfolk Island An island dependency of Australia north of New Zealand; 13 square miles.

Norfolk jacket A loose-fitting jacket, with side pockets, belt and two box pleats at the back and front, worn in shooting, fishing, etc. Also **Norfolk coat**.

Nor·ge (nôr′gə) The Norwegian name for NORWAY.

nor·gine (nôr′jēn, -jin) *n.* Algin.

no·ri·a (nō′rē·ə) *n.* An undershot water wheel having buckets on its rim: utilized to raise water in the Levant, Spain, etc.: introduced from ancient Persia, and often called *Persian wheel*. [<Sp. <Arabic *nā′ūrah*]

NORIA

Nor·i·cum (nôr′i·kəm, nor′-) An ancient country and Roman province south of the Danube, corresponding to southern Austria.

Nor·land (nôr′lənd) *n. Poetic* Northland; also, northlander. — **Nor′land·er** *n.*

norm (nôrm) *n.* 1 A rule or authoritative standard; a model, type, pattern, or value considered as representative of a specified group. 2 *Psychol.* The average or median of performance in a given function or test, regarded as a standard for the group concerned. 3 *Stat.* The mode. 4 *Geol.* The theoretical standard of chemical composition of igneous rocks, expressed in terms of oxides: distinguished from *mode*. [<L *norma* rule]

Nor·ma (nôr′mə) A southern constellation. See CONSTELLATION.

Nor·ma (nôr′mə) A feminine personal name. [<L, pattern]

nor·mal (nôr′məl) *adj.* 1 In accordance with an established law or principle; conforming to a type or standard; regular; natural. 2 Constituting a standard; model. 3 *Math.* Of, pertaining to, or constituting a normal; perpendicular. 4 Average; mean. 5 *Chem.* Denoting an aliphatic hydrocarbon having a straight unbranched chain of carbon atoms: *normal* propyl alcohol. 6 *Biol.* Designating a condition not exposed to or modified by special experimental treatment: a *normal* animal. 7 *Psychol.* Well adjusted to the outside world; without undue mental tensions. — *n.* 1 *Math.* **a** A perpendicular; specifically, a perpendicular to a curve or curved surface; a straight line perpendicular to a tangent line or plane at the point of tangency. **b** The intercept, on the normal line, between the curve and either the X-axis or the center of curvature. 2 A common or natural condition. 3 A usual or accepted rule or process. 4 The average or mean value of observed quantities. 5 An abbreviated expression for normal temperature, volume, etc. [<L *normalis* < *norma* rule] — **nor′mal·i·ty** (nôr′mal′ə·tē) *n.* — **nor′mal·ly** *adv.* — **nor′mal·ness** *n. Synonyms* (adj.): common, natural, ordinary, regular, typical, usual. That which is *natural* is according to nature; that which is *normal* is according to the standard or rule which is observed or claimed to prevail in nature; a deformity may be *natural*, symmetry is *normal*; the *normal* color of the crow is black, while the *normal* color of the sparrow is gray, but one is as *natural* as the other. *Typical* refers to such an assemblage of qualities as makes the specimen, genus, etc., a type of some more comprehensive group, while *normal* is more commonly applied to the parts, qualities, etc., of a single object; the specimen was *typical*; color, size, and other characteristics *normal*. The *regular* is that which is steady and constant, as opposed to that which is fitful and changeable; the *normal* action of the heart is *regular*. That which is *common* or *usual* is shared by a great number of persons or things. See COMMON, GENERAL, NATURAL, USUAL. *Antonyms*: abnormal, exceptional, irregular, monstrous, rare, uncommon, unprecedented, unusual.

normal class *Mineral.* The class of highest symmetry in each crystal system; the holohedral class.

nor·mal·cy (nôr′məl·sē) *n.* The state or quality of being normal; normality. Compare NORMAL *n.* (def. 2).

normal distribution *Stat.* The frequency of occurrence of a given series of data for each change in an independent variable: usually represented by a bell-shaped curve.

NORMAL DISTRIBUTION CURVE

nor·mal·ize (nôr′məl·īz) *v.t.* **·ized**, **·iz·ing** 1 To make normal; reduce to a standard or normal state or form. 2 *Metall.* To heat (steel) above its critical range and hold it at a given temperature for a stated time before allowing it to cool in still air. — **nor′mal·i·za′tion** *n.* — **nor′mal·iz′er** *n.*

normal salt *Chem.* A salt which contains no hydrogen displaceable by a metal, as sodium carbonate.

normal school A school for the training of secondary school graduates to become teachers.

normal solution *Chem.* A solution of known strength used in volumetric analysis, made to contain one gram-equivalent of the solute in a liter of the solution.

normal spin *Aeron.* A tailspin which is continued voluntarily by the pilot.

normal value The average of values over a long period.

Nor·man (nôr′mən) *adj.* Pertaining to Normandy or to the Normans. — *n.* 1 A native of Normandy. See ANGLO-NORMAN, NORSEMAN, NORTHMAN. 2 Norman French. [<OF *Normans*, plural of *Normant* Northman]

Nor·man (nôr′mən) A masculine personal name. [<Gmc., northman]

Nor·man (nôr′mən) An erroneous name for PERCHERON.

Nor·man (nôr′mən), **Montagu Collet**, 1871-1950, first Baron Norman of St. Clare, English financier.

Norman architecture The form assumed by Romanesque architecture in Normandy and as developed in England: characterized by the round arch, barrel vault, and massive construction. Also **Norman style**.

Norman Coast See MURMAN COAST.

Norman Conquest The subjugation of England by William of Normandy after the Battle of Hastings in 1066.

Nor·man·dy (nôr′mən·dē) A region and former province of NW France, comprising Cotentin peninsula and the region to the SE and east.

Norman French The dialect of French spoken by the Norman conquerors in England: also called **Anglo-French**, **Anglo-Norman**.

Nor·man·ize (nôr′mən·īz) *v.t.* & *v.i.* To make or become Norman in style, customs, character, etc. — **Nor′man·i·za′tion** *n.*

nor·ma·tive (nôr′mə·tiv) *adj.* Pertaining to, based upon, or establishing a norm, especially a norm assumed to have the prescriptive value of a standard or rule of usage: *normative* grammar. Distinguished from *empirical*. [< NORM + -ATIVE]

normative science A department of knowledge which studies the phenomena and principles of human conduct with a view to establishing standards of value and norms of procedure, as politics, ethics, and esthetics: distinguished from *descriptive science*, *exact science*.

Norn (nôrn) *n. pl.* **Nor·nir** (nôr′nir) In Norse mythology, one of the three giant goddesses, Urd, Verdandi, and Skuld (past, present, and future), who disposed of the destinies of men and gods.

nor·nic·o·tine (nôr·nik′ə·tēn, -tin) *n. Chem.* A colorless, liquid alkaloid, $C_9H_{12}N_2$, found in the leaves of certain varieties of tobacco and having about half the toxicity of nicotine. [<NOR- + NICOTINE]

Nor·ris (nôr′is, nor′-), **Charles Gilman**, 1881-1945, U.S. novelist; brother of Frank. — **Frank**, 1870-1902, U.S. novelist and journalist. — **George William**, 1861-1944, U.S. statesman and legislator; senator from Nebraska 1913-43. — **Kathleen**, born 1880, U.S. novelist; wife of Charles.

Nor·ris Reservoir (nôr′is, nor′-) An artificial lake (53 square miles) formed by **Norris Dam**, the first major dam (1936) of the Tennessee Valley Authority.

Norr·kö·ping (nôr′chœ·ping) A port in SE Sweden.

Norse (nôrs) *adj.* 1 Scandinavian. 2 West Scandinavian, i.e., Norwegian, Icelandic, and Faroese. — *n.* 1 The Scandinavians or West Scandinavians collectively: with *the*. 2 The Scandinavian or North Germanic group of the Germanic languages; specifically, the language of Norway. 3 The West Scandinavian languages. — **Old Norse** The ancestor of the North Germanic languages, best represented in the literature of the period (before 1500) by Old Icelandic; Old Scandinavian. Abbr. ON [Prob. <Du. *Noorsch* a Norwegian, var. of *Noordsch* < *noord* north + *-sch* -ISH]

Norse·man (nôrs′mən) *n. pl.* **·men** (-mən) A Scandinavian of Viking times.

north (nôrth) *n.* 1 One of the four cardinal points of the compass; the direction on the left side of a person facing the rising sun, and opposite to the *south*. For this and other points of the compass, see illustration under COMPASS CARD. 2 Any region north of a given point. 3 *Poetic* The north wind. — *adj.* 1 Lying toward or in the north; northern; boreal. 2 Issuing from or inhabiting the north. 3 Facing or proceeding toward the north. — *adv.* Toward the north; northward. [OE]

North (nôrth) *n.* 1 That portion of the United States north of Maryland, the Ohio River, and

North

Missouri: the Free States opposed to the Confederacy (the South) in the Civil War. 2 The part of England north of the Humber.
North (nôrth), **Christopher** See WILSON, JOHN. — **Lord Frederick,** 1732-92, second earl of Guilford, British prime minister at the start of the American Revolution. — **Sir Thomas,** 1535-1603?, English author; translated Plutarch's *Lives*.
North Albanian Alps A mountain group at the southern end of the Dinaric Alps on the Albanian-Yugoslav border; highest peak, 8,714 feet.
North America The northern continent of the western hemisphere; 8,443,600 square miles. — **North American**
North American Indian An Indian of any of the tribes formerly inhabiting North America north of Mexico, now the United States and Canada.
North·amp·ton (nôr·thamp′tən) 1 A county in south central England; 914 square miles; county town, Northampton. Also **North·amp′·ton·shire** (-shir). 2 A city in western Massachusetts.
North Borneo A British crown colony comprising the northern part of Borneo and adjacent islands; 29,387 square miles; capital, Jesselton: also *British North Borneo*.
north–bound (nôrth′bound′) *adj.* Going northward. Also **north′bound′**.
North Brabant A province of southern Netherlands; 1,894 square miles; capital, s'Hertogenbosch.
north by east One point east of north on the mariner's compass. See COMPASS CARD.
north by west One point west of north on the mariner's compass. See COMPASS CARD.
North Cape A cape on a Norwegian island in the Barents Sea, popularly considered the northernmost point of Europe. Compare NORDKYN, CAPE.
North Car·o·li·na (kar′ə·lī′nə) A SE State of the United States on the Atlantic; 52,712 square miles; capital, Raleigh; entered the Union Nov. 21, 1789, one of the thirteen original States; nickname, *Tarheel State*: abbr. *N.C.* — **North Car′o·lin′i·an** (-lin′ē·ən)
North Channel A strait between Scotland and NE Ireland connecting the Irish Sea with the Atlantic; 13 miles wide at the narrowest point.
North·cliffe (nôrth′klif), **Viscount,** 1865-1922, Alfred Charles William Harmsworth, English newspaper publisher.
North Da·ko·ta (də·kō′tə) A north central State of the United States bordering on Canada; 70,665 square miles; capital, Bismarck; entered the Union Nov. 2, 1889; nickname, *Flickertail State*: abbr. *N. Dak.* — **North Da·ko′tan**
north·east (nôrth′ēst′, in nautical usage nôr·ēst′) *n.* That point on the mariner's compass midway between north and east; any region lying toward that point on the horizon. — *adj.* From the northeast. — *adv.* Toward the northeast. — **north′east′er·ly** *adj. & adv.* — **north′east′ern** *adj.* — **north′east′ward** *adj. & adv.* — **north′east′ward·ly, north′east′wards** *adv.*
northeast by east One point east of northeast on the mariner's compass. See COMPASS CARD.
northeast by north One point north of northeast on the mariner's compass. See COMPASS CARD.
north·east·er (nôrth′ēs′tər, in nautical usage nôr·ēs′tər) *n.* 1 A gale from the northeast. 2 A waterproof hat with sloping brim worn by fishermen and other mariners in stormy weather. Also spelled *nor'easter*.
Northeast Passage A water route from the Atlantic to the Pacific along the northern coast of Europe and Asia.
north·er (nôr′thər) *n.* A cold windstorm from the north; specifically, a wind blowing over Texas to the Gulf of Mexico. — **north′er·ly** *adj. & adv.* — **north′er·li·ness** *n.*
north·ern (nôr′thərn) *adj.* Pertaining to the north or the North. — *n.* 1 A northerner. 2 *Poetic* A north wind. [OE *northerne*] — **north′ern·most** *adj.*
Northern Car, The See CAR.
Northern Caucasia See CAUCASIA. — **Northern Caucasian**
Northern Caucasus Part of the Russian S.F.S.R north of the Caucasus: also *Ciscaucasia*.
Northern Cir·cars (sûr·kärz′) An historic region in the northern part of Madras, India.
Northern Cross The northern constellation Cygnus, so called from the cross formed by its principal stars.
Northern Crown Corona Borealis.
Northern Dvina See DVINA.
north·ern·er (nôr′thər·nər) *n.* One born or residing in the north.
North·ern·er (nôr′thər·nər) *n.* One from the North, as distinguished from a Southerner.
northern hemisphere The half of the earth north of the equator.
Northern Ireland See IRELAND.
Northern Kar·roo (kə·rōō′) The innermost and highest (up to 10,000 feet) plateau region of the Union of South Africa, in northern Cape of Good Hope Province, forming the watershed of the Union of South Africa: also *High Veld*.
northern lights The aurora borealis.
Northern Rhodesia See RHODESIA.
Northern Spy A large, yellow-and-red variety of apple.
Northern Territories A former British protectorate in western Africa, included since 1957 in Ghana.
Northern Territory A region of north central Australia; 523,620 square miles; capital, Darwin.
North Germanic See under GERMANIC.
North Holland A province of NW Netherlands; 1,016 square miles; capital, Haarlem.
north·ing (nôr′thing, -thing) *n.* 1 *Nav.* Difference of latitude, measured toward the north, between any position and the last one determined. 2 *Astron.* North declination. 3 Deviation toward the north.
North Island The northernmost of the principal islands of New Zealand; 44,281 square miles.
North Korea See under KOREA.
north·land (nôrth′lənd) *n.* A land in the north. — *adj.* Of or pertaining to a northern land or lands. [OE] — **north′land·er** *n.*
North·man (nôrth′mən) *n. pl.* **·men** (-mən) A Scandinavian; especially, in history, a Scandinavian of the Viking period. Compare NORMAN, NORSEMAN. [OE]
north–north·east (nôrth′nôrth′ēst′, in nautical usage nôr′nôr·ēst′) *adj., adv., & n.* Midway between north and northeast. See COMPASS CARD.
north–north·west (nôrth′nôrth′west′, in nautical usage nôr′nôr·west′) *adj., adv., & n.* Midway between north and northwest. See COMPASS CARD.
North·olt (nôrth′thōlt) A town in Middlesex, 11 miles NW of London: site of a major international airport.
North Os·se·tian Autonomous Soviet Socialist Republic (o·sē′shən) A constituent republic in southern European Russian S.F.S.R. on the northern slope of the Caucasus; 3,550 square miles; capital, Dzaudzhikau. Also **North Os·se′tia.**
North Platte River A river in northern Colorado, SE Wyoming, and western Nebraska, flowing 680 miles NE to join the South Platte, forming the Platte River.
North Pole The northern extremity of the earth's axis; the 90th degree of north latitude, from which all meridians are south. Its prolongation upward strikes the celestial sphere at a point a little more than 1 degree from Polaris.
North Rhine–West·pha·lia (rīn′west·fāl′yə) A state of the Federal Republic of Germany, in the western and west central parts; 13,108 square miles; capital Düsseldorf. German **Nord·rhein–West·fa·len** (nôrt′rīn·vest·fä′lən).
North Riding An administrative division of northern Yorkshire, England; 2,127 square miles.
North River The estuary of the Hudson River which flows between New York and New Jersey.
Nor·throp (nôr′thrəp), **John Howard,** born 1891, U.S. biochemist.
North Sea The arm of the Atlantic Ocean between Great Britain and the continent of Europe; 600 by 350 miles; formerly *German Ocean*.
North Star Polaris, the polestar.
North·um·ber·land (nôr·thum′bər·lənd) A county of northern England; 2,019 square miles; county town, Newcastle-upon-Tyne.
North·um·bri·a (nôr·thum′brē·ə) An ancient

nor'wester

Anglo-Saxon kingdom, extending from the Humber to the Firth of Forth.
North·um·bri·an (nôr·thum′brē·ən) *adj.* 1 Of the ancient English kingdom of Northumbria, its people, or their dialect. 2 Of the modern county of Northumberland in England, its people, or their dialect. — *n.* 1 A native or inhabitant of Northumbria. 2 The Old English dialect of these people. 3 A native or inhabitant of Northumberland. 4 The peculiarities of speech of modern Northumbrians.
North Vietnam 1 The Democratic Republic of Vietnam, comprising the former French protectorate of Tonkin and the northern part of the former Empire of Annam; 63,344 square miles; capital, Hanoi. 2 Tonkin alone, a former kingdom and French protectorate; 44,670 square miles; capital, Hanoi.
north·ward (nôrth′wərd) *adv.* Toward the north. Also **north′wards**. — *adj.* Directed or lying toward the north. — *n.* The northward direction or point of the compass. — **north′·ward·ly** *adj. & adv.*
north·west (nôrth′west′, in nautical usage nôr′west′) *n.* 1 That point on the mariner's compass lying midway between north and west; any region situated toward that point on the horizon. 2 The NW region of the United States when its western boundary was the Mississippi. 3 The NW part of the United States. 4 The NW part of Canada. — **Old Northwest** The Northwest Territory. — *adj.* From the northwest. — *adv.* Toward the northwest. — **north′west′er·ly** *adj. & adv.* — **north′west′ern** *adj.* — **north′west′ward** *adj. & adv.* — **north′west′ward·ly, north′west′wards** *adv.*
northwest by north One point north of northwest on the mariner's compass. See COMPASS CARD.
northwest by west One point west of northwest on the mariner's compass. See COMPASS CARD.
north·west·er (nôrth′wes′tər, in nautical usage nôr′wes′tər) *n.* A gale which blows from the northwest.
North–West Frontier Province A former province of West Pakistan on the Afghanistan border, included in West Pakistan province October, 1955; a province of British India, 1901-47; 13,560 square miles; with tribal areas and princely states, 39,259 square miles; capital, Peshawar.
Northwest Passage A water route from the Atlantic to the Pacific along the northern coast of America.
Northwest Semitic See under SEMITIC.
Northwest Territories A region and administrative unit of northern Canada east of Yukon Territory and north of Hudson Strait, Hudson Bay, and the provinces of Manitoba, Saskatchewan, Alberta, and British Columbia; 1,304,903 square miles including fresh water.
Northwest Territory A region awarded to the United States by Britain in 1783, extending from the Great Lakes to the Ohio River between Pennsylvania and the Mississippi: also *Old Northwest*.
Nor·ton (nôr′tən), **Charles Eliot,** 1827-1908, U.S. educator, editor, and author. — **Thomas,** 1532-84, English author.
Nor·ton Sound (nôr′tən) An arm of the Bering Sea on the southern shore of Seward Peninsula, western Alaska.
Nor·um·be·ga (nôr′əm·bē′gə) The old name of a region on the Atlantic coast of North America, mentioned in maps and writings of the 16th and 17th centuries.
Nor·way (nôr′wā) A kingdom of northern Europe, in the western part of the Scandinavian peninsula; 119,240 square miles; capital, Oslo: Norwegian *Norge*.
Norway maple A tall European maple (*Acer platanoides*), an excellent shade tree.
Norway pine The red pine (*Pinus resinosa*) of the eastern United States.
Norway spruce See under SPRUCE.
Nor·we·gian (nôr·wē′jən) *adj.* Of or pertaining to Norway, its inhabitants, or their language. — *n.* 1 A native of Norway. 2 The North Germanic language of Norway. See LANDSMÅL, RIKSMÅL. Abbr. *Norw.* [< Med. L *Norwegia*, *Norvegia* Norway < ON *Norvegr* < *northr* north + *vegr* way]
Norwegian Sea That part of the Atlantic off the coast of Norway.
nor'·west·er (nôr·wes′tər) *n.* An oilskin coat

worn by mariners in stormy weather. [Contraction of NORTHWESTER]
Nor·wich (nôr′ij, -ich, nor′-) A city, county borough, and county town of Norfolk, England.
nose (nōz) n. 1 That part of the face of an animal containing the nostrils and the organ of smell. ◆ Collateral adjectives: *nasal, rhinal.* 2 The power or sense of smelling; scent. 3 That which resembles a nose; a ship's prow; the frontal tapering end of a torpedo; a spout; nozzle, etc. 4 The working end of a tool; also, the threaded end of a lathe or milling-machine spindle. — **on the nose** *Slang* Exactly; precisely. — v. **nosed, nos·ing** v.t. 1 To perceive or discover by or with the sense of smell; scent. 2 To examine or touch with the nose; sniff. 3 To make (one's way) carefully and with the front end forward. — v.i. 4 To smell; sniff. 5 To pry; meddle. 6 To move, especially carefully. — **to nose out** To defeat by a small margin. — **to nose over** To turn over on its nose, as an airplane. [OE *nosu*, orig. the two nostrils]
nose·band (nōz′band′) n. That part of a bridle passing over the nose of a horse and attached to the cheek pieces.
nose·bleed (nōz′blēd′) n. 1 Bleeding from the nose; epistaxis. 2 Any of various plants, as the wakerobin, Indian paintbrush, or milfoil.
nose-dive (nōz′dīv′) n. 1 *Aeron.* A steep downward plunge of an airplane. 2 Any sudden descent or crash. — v.i. **dived, div·ing** To plunge downward.
nose·gay (nōz′gā′) n. A bouquet. [<NOSE + GAY, in obs. senses "a bright object, a pretty flower"]
nose-heav·y (nōz′hev′ē) adj. *Aeron.* Denoting the condition of an airplane in which the nose tends to sink when the longitudinal control is released unless corrected by trim controls: opposed to *tail-heavy.*
nose·piece (nōz′pēs′) n. 1 Any protective covering for the nose. 2 An attachment on a microscope to permit the use of two or more objectives without disturbing the focus. 3 The narrow band fitting across the nose in a pair of spectacles.
nose wheel *Aeron.* A third landing wheel attached under the nose of some types of airplane.
no-show (nō′shō′) n. One who makes a reservation for an airplane flight but fails to claim his seat at the time of take-off: used by airline offices.
nos·ing (nō′zing) n. 1 That part of a stair tread projecting beyond the riser; also, a shield for the edge of a stair tread. 2 *Archit.* A nose-shaped molding or dripstone. [<NOS(E) + -ING¹]
noso- combining form Disease: *nosogenesis.* [<Gk. *nosos* a disease]
no·so·gen·e·sis (nō′sə·jen′ə·sis) n. Pathogenesis.
no·so·ge·og·ra·phy (nō′sō·jē·og′rə·fē) n. The study of the geographical factors and distribution of diseases. [<NOSO- + GEOGRAPHY] — **no′so·ge′o·graph′ic** (-jē′ə·graf′ik) or **·i·cal** adj.
no·sog·ra·phy (nō·sog′rə·fē) n. A description and classification of diseases. — **no·sog′ra·pher** n.
no·sol·o·gy (nō·sol′ə·jē) n. 1 The branch of medical science that treats of the systematic classification of diseases. 2 A list or classification of this kind. 3 The special characteristics of a particular disease; also, opinions regarding it. [<NL *nosologia* <Gk. *nosos* a disease + -*logia* -LOGY] — **nos·o·log·i·cal** (nos′ə·loj′i·kəl) adj. — **no·sol′o·gist** n.
no·so·ma·ni·a (nō′sə·mā′nē·ə, -mān′yə) n. *Psychiatry* Extreme hypochondria; the morbid conviction of suffering from some particular disease. [<NOSO- + -MANIA]
no·so·pho·bi·a (nō′sə·fō′bē·ə) n. *Psychiatry* Morbid fear of sickness and disease. [<NOSO- + -PHOBIA]
nos·tal·gi·a (nos·tal′jē·ə, -jə) n. 1 Severe or poignant homesickness. 2 Any longing for something far away or long ago. 3 *Psychiatry* Prolonged, often morbid fixation of one's thoughts on home, family, and friends. [<Gk. *nostos* a return home + *algos* a pain] — **nos·tal′gic** adj.

nos·toc (nos′tok) n. Any of a genus (*Nostoc*) of fresh-water algae having a definite, globose or variously expanded, gelatinous or membranaceous thallus. They form greenish masses in fresh water, in damp places, and on stones. [<NL; coined by Paracelsus]
nos·tol·o·gy (nos·tol′ə·jē) n. The doctrines or science relating to the phenomena of extreme old age or second childhood; geriatrics. [<Gk. *nostos* a return home + -LOGY] — **nos·to·log·ic** (nos′tə·loj′ik) adj.
nos·to·ma·ni·a (nos′tə·mā′nē·ə, -mān′yə) n. *Psychiatry* Intense or excessive nostalgia. [<NL <Gk. *nostos* a return home + *mania* madness]
nos·top·a·thy (nos·top′ə·thē) n. *Psychiatry* An acute, often morbid fear of returning to one's home or to familiar scenes: the opposite of *nostalgia.* [<Gk. *nostos* a return home + -PATHY] — **nos·to·path·ic** (nos′tə·path′ik) adj.
Nos·tra·da·mus (nos′trə·dā′məs), 1503–66, French physician, astrologer, and prophet: original name *Michel de Notredame.*
nos·tril (nos′trəl) n. One of the anterior openings in the nose. [OE *nosthyrl* <*nos(u)* nose + *thyrel* a hole < *thurh* through]
nos·trum (nos′trəm) n. 1 A favorite remedy; patent medicine; quack recipe. 2 Anything savoring of quackery: *political nostrums.* [<L *nostrum*, neut. of *noster* our own; because prepared by those selling it]
nos·y (nō′zē) adj. *Colloq.* 1 Prying; snooping; inquisitive. 2 Stinking; malodorous.
not (not) adv. In no manner, or to no extent or degree: used in negation, prohibition, or refusal. [ME, contraction of NOUGHT]
not- Var. of NOTO-.
no·ta be·ne (nō′tə bē′nē) *Latin* Note well; take notice: abbr. *N.B.*
no·ta·bil·i·ty (nō′tə·bil′ə·tē) n. pl. **·ties** 1 Notableness. 2 A person of distinction.
no·ta·ble (nō′tə·bəl, also for def. 2 nō′tə·bəl) adj. 1 Worthy of note; remarkable; distinguished. 2 Eminently careful, thrifty, or skilful, as in housekeeping. — n. One who or that which is worthy of note, distinguished, or eminent. [<OF <L *notabilis* <*notare* note <*nota* a mark] — **no′ta·ble·ness** n. — **no′ta·bly** adv.
No·ta·ble (nō′tə·bəl) n. In France before the Revolution, one of the persons summoned by the king to a deliberative assembly in national crises.
no·ta·rize (nō′tə·rīz) v.t. **·rized, ·riz·ing** To attest to or authenticate as a notary. — **no′ta·ri·za′tion** n.
no·ta·ry (nō′tə·rē) n. pl. **·ries** 1 An officer empowered to authenticate contracts, administer oaths, take depositions, etc.: also **notary public.** 2 Formerly, a scrivener, or one who drew up legal papers. [<AF *notarie*, OF *notaire* <L *notarius* a shorthand writer, a clerk <*notare.* See NOTABLE.] — **no·tar·i·al** (nō·târ′ē·əl) adj.
no·ta·tion (nō·tā′shən) n. 1 The process of designating by figures, etc. 2 Any system of signs, figures, or abbreviations employed for convenience in any science or art, especially algebraic, arithmetical, or musical characters. [<L *notatio, -onis* <*notare.* See NOTABLE.] — **no·ta′tion·al** adj.
notch (noch) n. 1 A hollow cut or mark made in anything; a nick; indentation; especially, a mark or nick cut into the handle of a gun or other weapon to record each person killed. 2 A narrow, short defile. 3 *Colloq.* A degree: He is a *notch* above the others. — v.t. 1 To make a notch or notches in. 2 To record by means of notches; tally. [Prob. <ME *an oche* a notch <OF *oche, osche* <*oschier* notch] — **notch′er** n.
note (nōt) n. 1 That by which anything may be known; an outward sign. 2 A mark or character used in writing or printing to indicate or call attention to something: a *note* of interrogation (?) or exclamation (!). 3 A brief comment appended to text. 4 A brief record or summary; a memorandum. 5 A complete record or report: Make a *note* of that statement. 6 An official communication in writing from one government to another. 7 A brief letter; a billet. 8 *Logic* A distinctive mark or character of an object such as its qualities afford. 9 Notice; observation. 10

An account or bill. 11 High importance, estimation, or repute; distinction: something of *note.* 12 *Music* a An oval character in musical notation, either solid or formed in outline, used to indicate the length of a tone, and also, as placed on a staff, to point out, in conjunction with the signature, the pitch and relative position in a scale system. b Loosely, any musical sound: The first *notes* of

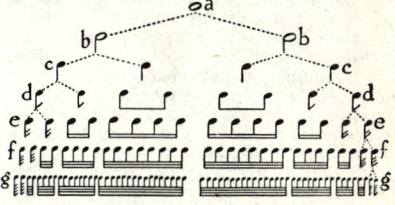

NOTES
a. Whole note.   d. Eighth note.
b. Half note.   e. Sixteenth note.
c. Quarter note.   f. Thirty-second note.
         g. Sixty-fourth note.

the fiddles were heard. c A key of the keyboard. 13 *Physics* Tone (def. 2): the preferable word in this sense. 14 A melodious or vocal sound, as of a bird; voice; tone. 15 Manner of speaking. 16 A signed promise by one party to another to pay a certain sum of money at a specified time: a promissory *note*; a bank *note.* 17 The general tone, coloring, or quality of a painting. See synonyms under REMARK, SIGN, SOUND. — v.t. **not·ed, not·ing** 1 To take notice or note of; observe; remark. 2 To set down, as in writing; make a note of. 3 To mention specially or separately in the course of writing. 4 To annotate. 5 To set down in musical notation. [<OF <L *nota* a mark, orig. pp. fem. of *noscere* know] — **not′er** n.
note·book (nōt′book′) n. 1 A book in which to enter memoranda. 2 A book in which notes of hand are registered; billbook.
not·ed (nō′tid) adj. Well known by reputation or report. See synonyms under EMINENT, ILLUSTRIOUS. — **not′ed·ly** adv.
note·less (nōt′lis) adj. 1 Not noted; unobserved; obscure. 2 Unmusical.
note of hand A promissory note.
note paper Paper for writing notes or letters.
note·wor·thy (nōt′wûr′thē) adj. Worthy of note; remarkable; significant. — **note′wor′thi·ly** adv. — **note′wor′thi·ness** n.
noth·ing (nuth′ing) n. 1 Not any being or existence; also, not any particular thing, act, or event; no thing: opposed to *thing, anything, something*: He has *nothing.* 2 A state of non-existence; nothingness; hence, insignificance or unimportance: to rise from *nothing.* 3 A person or thing of slight significance, consideration, or value; any trifle. 4 A cipher; zero; naught. — adv. In no degree; to no extent; not at all.
noth·ing·ness (nuth′ing·nis) n. 1 A state of non-existence. 2 Worthlessness; utter insignificance. 3 A trifle; nothing. 4 Unconsciousness; also, death.
no·tice (nō′tis) v.t. **·ticed, ·tic·ing** 1 To pay attention to or take cognizance of; remark or observe. 2 To treat courteously or with favor. 3 To mention; refer to or comment on. 4 To serve with a notice; notify. [<*n.*] — n. 1 The act of noticing or observing; attention: to take *notice* of. 2 Announcement; information; warning. 3 Respectful treatment; civility. 4 An order communicated to one; especially, a formal written or printed notification, instruction, or warning, as of the termination or intended termination of an agreement; also, a public communication openly displayed. 5 A short literary advertisement or review: a book *notice.* [<F *notitia* celebrity] — **no′tice·a·ble** adj. — **no′tice·a·bly** adv.
no·ti·fi·ca·tion (nō′tə·fə·kā′shən) n. 1 The act of notifying. 2 Notice given. 3 The writing that gives information.
no·ti·fy (nō′tə·fī) v.t. **·fied, ·fy·ing** 1 To give notice to; inform. 2 *Brit.* To give information of; make known. See synonyms under

ANNOUNCE, INFORM[1]. [<OF *notifier* <L *notificare* < *notus* known + *facere* make] —**no′·ti·fi′er** *n.*

no·tion (nō′shən) *n.* **1** A mental apprehension; an idea. **2** Loosely, an opinion; a hastily formed theory. **3** Intention; inclination. **4** *pl. U.S.* Miscellaneous, small, useful articles, such as ribbons, thread, pins, needles, hairpins, etc. See synonyms under IDEA, THOUGHT[1]. [<L *notio, -onis* < *noscere* know]

no·tion·al (nō′shən·əl) *adj.* **1** Pertaining to, expressing, or consisting of notions or concepts. **2** Existing in imagination only. **3** *U.S.* Given to entertaining pet ideas or hobbies; overfastidious. —**no′tion·al·ly** *adv.*

noto- *combining form* Back: *notochord.* Also, before vowels, **not-**. [<Gk. *nōton* back]

no·to·chord (nō′tə·kôrd) *n. Biol.* A cartilaginous, flexible rod of cells formed along the median line on the dorsal side of vertebrate embryos, in a situation afterwards occupied by the spinal column. It persists in the adult stage of certain primitive chordates, as lampreys and tunicates.

No·to·gae·a (nō′tə·jē′ə) *n.* A zoogeographical realm including the Australian and Neotropical regions. [<Gk. *notos* south + *gaia* earth] —**No′to·gae′al, No′to·gae′an, No′to·gae′ic** *adj.*

no·to·ri·e·ty (nō′tə·rī′ə·tē) *n. pl.* **·ties** **1** The character of being notorious. **2** Common knowledge or talk. **3** One who or that which is notorious. See synonyms under FAME. [<F *notoriété* <Med. L *notorius* making known. See NOTORIOUS.]

no·to·ri·ous (nō·tôr′ē·əs, -tō′rē-) *adj.* Being publicly known and the subject of general unfavorable remark. [<Med. L *notorius* < *notus* known, orig. pp. of *noscere* know] —**no·to′ri·ous·ly** *adv.*

Synonyms: egregious, evident, known, manifest, obvious, open, overt, patent, plain, undeniable, undenied, undisputed, unquestionable, well-known.

no·tor·nis (nō·tôr′nis) *n.* A ratite bird (genus *Notornis*) of New Zealand and neighboring islands, with rudimentary wings. [<NL <Gk. *notos* south + *ornis* bird]

No·tre Dame (nō′trə däm′, nō′tər dām′; *Fr.* nô′tr′ däm′) **1** *French* Our Lady (Mary, mother of Jesus). **2** A famous early Gothic cathedral in Paris, built 1163–1257.

no–trump (nō′trump′) *n.* In bridge, a bid or a declaration calling for play without a trump suit. —*adj.* Without a trump suit; denoting a hand suitable for playing without a trump suit. —**no′–trump′er** *n.*

Not·ta·way River (not′ə·wā) A river in western Quebec, Canada, flowing 205 miles NW to Hudson Bay.

Not·ting·ham (not′ing·əm) A county of north central England; 844 square miles; county town, Nottingham: shortened form **Notts.** Also **Not′ting·ham·shire** (-shir′).

no·tun·gu·late (nō·tung′gyə·lāt) *n. Paleontol.* Any member of an extinct order (*Notungulata* or *Notoungulata*) of herbivorous mammals of the Tertiary, whose principal habitat was South America. —*adj.* Of or pertaining to the *Notungulata.* Also **no′to·un′gu·late** (nō′tō·ung′gyə·lāt). [<NOTO- + UNGULATE]

not·with·stand·ing (not′with·stan′ding, -with-) *adv.* All the same; nevertheless: Though imprisoned, he escaped *notwithstanding.* —*prep.* In spite of: He left *notwithstanding* your orders. —*conj.* In spite of the fact that; although.

Synonym (prep.): despite. *Notwithstanding* simply states that circumstances shall not be or have not been allowed to withstand; *despite* refers primarily to personal and perhaps spiteful opposition; as, he failed *notwithstanding* his good intentions; or, he persevered *despite* bitter hostility.

Synonyms (conj.): although, but, howbeit, however, nevertheless, still, though, yet. *However* simply waives discussion and (like the archaic *howbeit*) says, "be that as it may, this is true"; *nevertheless* concedes the truth of what precedes, but claims that what follows is none the less true; *notwithstanding* marshals the two statements face to face, admits the one and its seeming contradiction to the other, while insisting that it cannot, after all, withstand the other. *Yet* and *still* are weaker than *notwithstanding,* while stronger than *but. Though* and *although* make as little as possible of the concession, dropping it, as it were, incidentally; as, "*though* we are guilty, thou art good"; to say "we are guilty, *but* thou art good," would make the concession of guilt more emphatic. See BUT[1].

notwithstanding that Although.

nou·gat (nōō′gət, *Fr.* nōō·gä′) *n.* A confection consisting usually of a honey or sugar paste mixed with chopped almonds, pistachios, etc. [<F <Provençal, ult. <L *nux, nucis* a nut]

nought (nôt) See NAUGHT.

Nou·mé·a (nōō·mē′ə, -mā′ə) A port, capital of New Caledonia.

nou·me·nal (nōō′mə·nəl, nou′-) *adj.* Of or pertaining to noumena or the noumenon: opposed to *phenomenal.* —**nou′me·nal·ly** *adv.* —**nou′me·nal·ism** *n.* —**nou′me·nal·ist** *n.*

nou·me·non (nōō′mə·non, nou′-) *n. pl.* **·me·na** (-mə·nə) *Philos.* **1** An object of intuition by the reason or understanding, as something transcending perception through the senses: opposed to *phenomenon.* **2** The unknown ground or cause of phenomena, regarded as necessarily assumed by the mind, but the real nature of which is wholly transcendent; the unknowable thing in itself. [<NL <Gk., orig. ppr. passive of *noeein* think]

noun (noun) *Gram. n.* **1** A word used as the name of a thing, quality, or action existing or conceived by the mind; a substantive. A **proper noun** is the name of an individual person, place, or thing, as *Paul, Nicole, Venice, Rover, U.S.S. Nautilus,* etc.; a **common noun** is the name an individual object has in common with others of its class, as *man, city, hill*; a **collective noun** is one expressing a collection or aggregate of individuals, as *assembly, army*; an **abstract noun** is one indicating a quality, as *goodness, beauty.* **2** Anything that can be used as subject, object, or appositive, as a substantive clause. —*adj.* Of or pertaining to a noun or nouns: also **noun′al·** [<AF, OF *nun* <L *nomen* name] —**noun′al·ly** *adv.*

nour·ish (nûr′ish) *v.t.* **1** To furnish material to sustain the life and promote the growth of (a living organism). **2** Hence, to support; maintain. **3** To furnish with knowledge; educate. See synonyms under CHERISH. [<OF *noriss-,* stem of *norir* <L *nutrire* nourish] —**nour′ish·a·ble** *adj.* —**nour′ish·er** *n.* —**nour′·ish·ing** *adj.*

nour·ish·ment (nûr′ish·mənt) *n.* **1** Nutriment. **2** The act of nourishing or the state of being nourished. **3** That which promotes growth in any way. See synonyms under FOOD, NUTRIMENT.

nous (nōōs, nous) *n. Philos.* **1** Mind, as employed in thinking, feeling, or willing. **2** The higher reason; emanation of the divine principle. [<Gk. *nous, noos* mind]

nous ver·rons (nōō ve·rôn′) *French* We shall see.

nou·veau riche (nōō·vō′ rēsh′) *French pl.* **nou·veaux riches** (nōō·vō′ rēsh′) One recently become rich; a parvenu.

nou·veau·té (nōō·vō·tā′) *n. French* A new thing; a novelty.

Nou·velle Ca·lé·do·nie (nōō·vel′ kà·lā·dō·nē′) The French name for NEW CALEDONIA.

no·va (nō′və) *n. pl.* **·vae** (-vē) or **·vas** *Astron.* A star which suddenly flares up in the heavens and fades away again to its former magnitude after a period of a few weeks or months. [<L *novus* new]

no·vac·u·lite (nō·vak′yə·līt) *n.* An extremely fine-grained sedimentary siliceous rock used for hones; whetstone. [<L *novacula* razor]

No·va Gô·a (nô′və gō′ə) A city in Gôa, capital of Portuguese India: also *Pangim, Panjim.*

No·va·lis (nō·vä′lis) Pseudonym of *Friedrich von Hardenberg,* 1772–1801, German poet.

No·va Lis·bo·a (nô′və lēzh·bō′ə) Capital-elect of Angola: formerly *Huambo.*

No·va·ra (nō·vä′rä) A city in Piedmont, northern Italy. Ancient **No·va′ri·a** (-rē·ä).

No·va Sco·tia (nō′və skō′shə) A maritime province of eastern Canada; 21,068 square miles; capital, Halifax: French *Acadia* (1605–1713). —**No′va Sco′tian** *adj.* & *n.*

No·va·tians (nō·vā′shənz) See CATHARI.

no·va·tion (nō·vā′shən) *n.* **1** *Law* A substitution of a new engagement, indebtedness, obligation, creditor, or debtor for an existing one. **2** A making anew; creation; inception. [<L *novatio* making new < *novare* make new < *novus* new]

No·va·ya Zem·lya (nô′və·yə zim·lyä′) An Arctic archipelago in European U.S.S.R. separating the Kara and Barents seas; 35,000 square miles.

nov·el (nov′əl) *n.* **1** A fictional prose narrative of considerable length, representing characters and events as if in real life by a plot or scheme of action of greater or less complexity. **2** The particular type of literature exemplified by fiction of this character: with the definite article: Dostoevsky is one of the fathers of the modern *novel.* **3** In Roman law, a new constitution or decree supplemental to a decree. **4** *Usually pl.* A novella. See synonyms under FICTION. —*adj.* Of recent origin; new, strange, or unusual. See synonyms under FRESH, MODERN, NEW. [Fusion of Ital. *novella* a novel and OF *novel* new, both <LL *novellus* <L *novus* new] —**nov′el·ly** *adj.*

nov·el·ette (nov′əl·et′) *n.* A short novel.

nov·el·ist (nov′əl·ist) *n.* A writer of novels.

nov·el·is·tic (nov′əl·is′tik) *adj.* Of, pertaining to, characteristic of, or found in novels. —**nov′el·is′ti·cal·ly** *adv.*

nov·el·ize (nov′əl·īz) *v.t.* **·ized, ·iz·ing** To put into the form of a novel. —**nov′el·i·za′tion** *n.*

no·vel·la (nō·vel′lä) *n. pl.* **·le** (-lā) *Italian* A short tale or narrative, usually with a moral, often of satirical nature: typified by the stories in Boccaccio's *Decameron.*

Nov·els (nov′əlz) *n. pl.* In civil law: **1** The amendments and supplementary laws to the Justinian Code decreed by Justinian and his immediate successors: in Latin, *Novellae Constitutiones.* **2** Similar decrees proclaimed by other Roman emperors.

nov·el·ty (nov′əl·tē) *n. pl.* **·ties** **1** The quality of being novel. **2** Something novel or unusual; especially, a small manufactured article or trinket for personal adornment: usually in the plural. **3** An innovation. See synonyms under CHANGE.

No·vem·ber (nō·vem′bər) The eleventh month of the year, containing 30 days. [<L *November* ninth month < *novem* nine]

no·ve·na (nō·vē′nə) *n.* In the Roman Catholic Church, a devotion consisting of a prayer said on nine successive days, asking for some special blessing. [<LL <L *novem* nine]

nov·e·nar·y (nov′ə·ner′ē) *adj.* Relating to the number nine. [<L *novenarius* < *novem* nine]

no·ven·ni·al (nō·ven′ē·əl) *adj.* Occurring every ninth year. [<L *novennis* < *novem* nine + *annus* year]

no·ver·cal (nō·vûr′kəl) *adj.* Of, pertaining to, or suitable for a stepmother. [<L *noverca* stepmother]

Nov·go·rod (nôv′gə·rot) A city of NW Russian S.F.S.R. on the Volkhov.

nov·ice (nov′is) *n.* **1** A beginner in any business or occupation; an untried or inexperienced person; tyro. **2** *Eccl.* **a** One who enters a religious house or community on probation. **b** One who has been recently converted. **3** In competitive games, etc., a person or animal entered in a class in which he or it has not already won an award. [<F <L *novicus* new < *novus*] —**nov′ice·hood** (-hōōd) *n.*

No·vi Sad (nô′vē säd′) A city on the Danube in NE Yugoslavia, capital of the autonomous province of Vojvodina, northern Serbia: German *Neusatz.*

no·vi·ti·ate (nō·vish′ē·it, -āt) *n.* **1** The state of being a novice. **2** *Eccl.* **a** The period of probation of a novice in a religious order. **b** The part of a monastic establishment inhabited by novices. Also **nov·ice·ship** (nov′·is·ship). **3** A novice. Also **no·vi′ci·ate.**

No·vo·cain (nō′və·kān) *n.* Proprietary name for a brand of procaine, used as a local anesthetic: less toxic than cocaine. Also **No′vo·caine.**

No·vo Kuz·netsk (nô′vō kōōz·netsk′) A former name for STALINSK.

No·vo·ros·siisk (nô′və·ro·sēsk′) A port on the Black Sea in SW European Russian S.F.S.R.

No·vo·si·birsk (nô′və·si·birsk′) A city in SW Asian Russian S.F.S.R. on the Ob.

no·vus ho·mo (nō′vəs hō′mō) *Latin* A new man; upstart; parvenu.

no·vus or·do se·clo·rum (nō′vəs ôr′dō sə·klôr′əm, -klō′rəm) *Latin* A new order of the ages: motto on the Great Seal of the United States.

now (nou) *adv.* **1** At once. **2** At or during the present time. **3** Nowadays. **4** In the immediate past: He said so just *now.* **5** In the immediate future: He is going just *now.* **6** In such circumstances; things being as they are: *Now* we can be sure of getting home. **7** At this point in the proceedings, narrative, etc.: The war was *now* virtually over. See synonyms under IMMEDIATELY, YET. —*conj.* Since; seeing that: *Now* the books have arrived, I must stay here and read them. —*n.* The present time, moment, or occasion. ◆ *Now* is often used as an expletive, as in command, remonstrance, etc.: Come *now*, don't make me insist! [OE *nū*]

now·a·days (nou'ə·dāz') *adv.* In the present time or age.

now and again Occasionally; from time to time. Also **now and then.**

no·way (nō'wā') *adv.* In no way, manner, or degree. Also **no'ways'.**

Now·el (nō·el') *Archaic* See NOËL.

no·where (nō'hwâr') *adv.* In no place; not anywhere. —*n.* No place. Also *U.S. Dial.* **no'wheres'.**

no·whith·er (nō'hwith'ər) *adv.* Toward no definite place.

no·wise (nō'wīz') *adv.* In no manner or degree.

nowt (nout) *n. Scot.* 1 An ox. 2 Figuratively, a stupid or clumsy person.

Nox (noks) In Roman mythology, the goddess of night: identified with the Greek *Nyx.*

nox·al (nok'səl) *adj. Law* Pertaining to damage inflicted by some person or animal in the ownership or legal custody of another. [<L *noxa* hurt <*nocere* hurt]

nox·ious (nok'shəs) *adj.* Causing, or tending to cause, injury to health or morals; pernicious. See synonyms under BAD, INIMICAL, NOISOME, PERNICIOUS. [<L *noxius* <*nocere* hurt] —**nox'ious·ly** *adv.* —**nox'ious·ness** *n.*

noy·ade (nwä·yàd') *n. French* Execution by drowning, especially as practiced during the Reign of Terror (1793–94) in Nantes, France.

Noyes (noiz), **Alfred,** 1880–1958, English poet.

Noy·on (nwà·yôn') A town in northern France; scene of Charlemagne's coronation; birthplace of Calvin.

noz·zle (noz'əl) *n.* 1 A projecting spout or pipe for discharge, as of a teapot, or the muzzle of a gun barrel, etc.; specifically, a rigid tube or vent, commonly tapering, at the end of a flexible tube, as a hose. 2 An inlet or outlet pipe. Also **noz'le.** [Dim. of NOSE]

nth (enth) *adj.* 1 Representing an ordinal equivalent to *n.* 2 Infinitely or indefinitely large or small: raised to the *nth* degree.

nu (no͞o, nyo͞o) *n.* The thirteenth letter in the Greek alphabet (N, *ν*): equivalent to English *n.* As a numeral it denotes 50. [<Gk. *ny*]

nu·ance (no͞o·äns', no͞o'äns, nyo͞o'-; *Fr.* nü·äns') *n.* A shade of difference in tone or color; hence, a slight degree of difference in anything perceptible to the mind. [<F <*nuer* shade, ult. <L *nubes* a cloud]

nub (nub) *n.* 1 A protuberance; knob. 2 The core of a matter; pith or point: the *nub* of the story. [Earlier *knub.* Related to KNOB.]

Nu·ba (no͞o'bä) *n.* 1 A Nubian. 2 One of a Negro tribe of the central Sudan, related to the Nubians. 3 The language of the Nuba peoples, related to the Sudanic languages: also called *Berberi.*

nub·bin (nub'in) *n. U.S.* An imperfectly developed ear of maize; hence, anything small and stunted. [<NUB]

nub·ble (nub'əl) *n. Dial.* 1 A protuberance; nub. 2 An island formed like a knob. [Dim. of NUB] —**nub'bly** *adj.*

nu·bi·a (no͞o'bē·ə, nyo͞o'-) *n.* A soft, light scarf or covering for the head, worn by women. [<L *nubes* cloud]

Nu·bi·a (no͞o'bē·ə, nyo͞o'-) A region and ancient country of NE Africa in the northern Sudan and southern Egypt, between the Red Sea and the Sahara.

Nu·bi·an (no͞o'bē·ən, nyo͞o'-) *adj.* Of or pertaining to Nubia, its people, or their language. —*n.* 1 A native of Nubia; specifically, a member of any of the Negroid tribes formerly ruling the territory between Egypt and Abyssinia. 2 The Sudanic language of the Nubians. 3 A Nubian horse or goat.

Nubian Desert A sandstone plateau in NE Sudan between the Nile valley and the Red Sea.

nu·bile (no͞o'bil, nyo͞o'-) *adj.* Of suitable age to marry; marriageable. [<L *nubilis* <*nubere* wed] —**nu·bil'i·ty** *n.*

nu·bi·lous (no͞o'bə·ləs, nyo͞o'-) *adj.* 1 Cloudy; foggy. 2 Obscure; indefinite. Also **nu'bi·lose** (-lōs). [<L *nubilus* <*nubes* cloud]

nu·cel·lus (no͞o·sel'əs, nyo͞o-) *n. pl.* **·li** (-ī) *Bot.* The body or essential part of a plant ovule, within which the embryo and its covering are developed. [<NL <L *nucella,* dim. of *nux, nucis* a nut] —**nu·cel'lar** *adj.*

nu·cha (no͞o'kə, nyo͞o'-) *n. pl.* **·chae** (-kē) The nape or back of the neck. [<LL <Arabic *nukhā'* spinal marrow] —**nu'chal** *adj.*

nu·cle·ar (no͞o'klē·ər, nyo͞o'-) *adj.* 1 Of, pertaining to, forming, of the nature of, or depending upon a nucleus or nuclei. Also **nu'cle·al.** 2 Of or employing the energy of the nucleus of the atom: *nuclear* weapons.

nuclear fission *Physics* See under FISSION.

nuclear physics 1 That branch of physics which investigates the structure, properties, and phenomena of the atomic nucleus. 2 Nucleonics.

nuclear plate *Biol.* Equatorial plate.

nu·cle·ase (no͞o'klē·ās, nyo͞o'-) *n. Biochem.* An enzyme which hydrolyzes nucleic acids.

nu·cle·ate (no͞o'klē·āt, nyo͞o'-) *adj.* Having a nucleus. Also **nu'cle·at'ed.** —*v.t.* & *v.i.* **·at·ed, ·at·ing** To form or gather into a nucleus. —**nu'cle·a'tion** *n.*

nu·cle·ic (no͞o·klē'ik, nyo͞o-) *adj. Biochem.* Designating a group of complex, non-crystalline acids present in organic nuclear material, as yeast, chromatin, the thymus gland, etc. They contain carbohydrates combined with phosphoric acids and bases derived from purine or pyrimidine.

nucleic acid *Biochem.* A complex acid derived from nuclein and nucleoproteins: it plays an important role in digestion and metabolism.

nu·cle·in (no͞o'klē·in, nyo͞o'-) *n. Biochem.* A colorless, amorphous protein containing nucleic acid, and found as a normal constituent of cell nuclei.

nu·cle·o·late (no͞o'klē·ə·lāt', nyo͞o'-) *adj.* Having nucleoli. Also **nu'cle·o·lat'ed.**

nu·cle·o·lus (no͞o·klē'ə·ləs, nyo͞o-) *n. pl.* **·li** (-lī) *Biol.* 1 A well-defined particle, easily affected by staining fluids, sometimes found within the nucleus of a cell. 2 A small nucleus. Also **nu'cle·ole** (-ōl). [<LL, dim. of *nucleus.* See NUCLEUS.] —**nu·cle'o·lar** *adj.*

nu·cle·on (no͞o'klē·on, nyo͞o'-) *n. Physics* Any of the particles composing the nucleus of an atom, as the proton, neutron, neutrino, etc. [<NL <NUCLEUS]

nu·cle·on·ics (no͞o'klē·on'iks, nyo͞o'-) *n.* The practical applications of nuclear physics in any field of science, engineering, and technology, especially in relation to the development of atomic energy. —**nu'cle·on'ic** *adj.*

nucleon number *Physics* Mass number.

nu·cle·o·plasm (no͞o'klē·ə·plaz'əm, nyo͞o'-) *n. Biol.* The more fluid part of the nucleus of a cell; karyoplasm. —**nu'cle·o·plas'mic** *adj.*

nu·cle·o·pro·te·in (no͞o'klē·ə·prō'tē·in, -tēn, nyo͞o'-) *n. Biochem.* Any of a class of substances found in the nuclei of plant and animal cells, and containing one or more protein molecules combined with nucleic acid.

nu·cle·o·side (no͞o'klē·ə·sīd', nyo͞o'-) *n. Biochem.* A glycoside derived from nucleic acid by removing the phosphoric acid from a nucleotide, leaving the carbohydrate in combination with the purine or pyrimidine derivative.

nu·cle·o·tide (no͞o'klē·ə·tīd', nyo͞o'-) *n. Biochem.* One of several compounds derived from nucleic acid by hydrolysis and consisting of phosphoric acid combined with a sugar and a purine or pyrimidine derivative.

nu·cle·us (no͞o'klē·əs, nyo͞o'-) *n. pl.* **·cle·i** (-klē·ī) 1 A center of development; central mass; kernel. 2 *Biol.* A complex, spheroidal body surrounded by a thin membrane and embedded in the protoplasm of most plant and animal cells. It contains the chromatin which is essential in the processes of heredity, and is the directive center of all the vital activities of the cell, as assimilation, metabolism, growth, and reproduction. 3 *Physiol.* A group of nerve cells within the nervous system from which the nerve fibers originate. 4 *Zool.* The apex, or earliest formed part of a shell; also, the central part, as of an operculum, around which additional parts are formed. 5 *Astron.* The starlike point seen in the head of a comet, and at the center of a nebula. 6 *Physics* The central core of an atom, believed to contain its effective mass and to have a positive charge balanced by the negative charge of the surrounding electrons. Its principal components are the proton and neutron. [<L, a kernel, dim. of *nux, nucis* a nut]

nu·clide (no͞o'klīd, nyo͞o'-) *n. Physics* An atomic nucleus considered apart from the extranuclear arrangement of electrons and identified according to the number and type of nucleons it contains; any species of nucleus.

nude (no͞od, nyo͞od) *adj.* 1 Without clothing or covering; naked; bare. 2 *Law* Naked; lacking an essential legal requisite. —*n.* 1 A nude figure, as in painting or sculpture. 2 The state of being unclad: to appear in the *nude.* 3 Any of several light beige or pinkish-beige tints. [<L *nudus* naked, bare] —**nude'ly** *adv.* —**nude'ness** *n.*

nudge (nuj) *v.* **nudged, nudg·ing** *v.t.* To touch or push gently, as with the elbow, in order to attract attention, convey a meaning, etc. —*v.i.* To give a nudge. —*n.* The act of nudging; a gentle push, as with the elbow. [? Akin to dial. Norw. *nugga* push]

nudi- *combining form* Naked; bare; without covering: *nudicaudate.* [<L *nudus* naked]

nu·di·branch (no͞o'di·brangk, nyo͞o'-) *n.* One of a suborder of variously colored marine gastropods (*Nudibranchiata*), lacking shells and true gills in the adult stage; a sea slug. [<NUDI- + Gk. *branchia* gills] —**nu'di·bran'chi·ate** (-brang'kē·it) *adj.* & *n.*

nu·di·cau·date (no͞o'di·kô'dāt, nyo͞o'-) *adj. Zool.* Having a naked or hairless tail, as rats and mice.

nu·di·cau·lous (no͞o'di·kô'ləs, nyo͞o'-) *adj. Bot.* Having naked or leafless stems. [<NUDI- + L *caulis* a stem]

nud·ism (no͞o'diz·əm, nyo͞o'-) *n.* The doctrine or practice of living in the state of nudity for hygienic reasons. —**nud'ist** *n.*

nu·di·ty (no͞o'də·tē, nyo͞o'-) *n. pl.* **·ties** 1 The state of being nude. 2 A naked part; anything unclad.

nu·dum pac·tum (no͞o'dəm pak'təm, nyo͞o'-) *Latin* A contract made without a consideration; literally, bare pact.

Nu·e·ces River (no͞o·ā'sās) A river in southern Texas, flowing 315 miles SE to **Nueces Bay,** an arm of Corpus Christi Bay.

Nue·vo Le·ón (nwā'vō lā·ōn') A state in NE Mexico; 25,136 square miles; capital, Monterrey.

nu·ga·to·ry (no͞o'gə·tôr'ē, -tō'rē, nyo͞o'-) *adj.* 1 Having no power; inoperative. 2 Having no worth or meaning; insignificant. See synonyms under USELESS. [<L *nugatorius* <*nugae* trifles, nonsense] —**nu'ga·to'ri·ly** *adv.* —**nu'ga·to'ri·ness** *n.*

nug·get (nug'it) *n.* A lump; specifically, a lump of precious metal, usually gold, found in a free state. [? dim. of dial. E *nug* lump]

nug·get·y (nug'it·ē) *adj.* 1 Found in the form of nuggets. 2 Nugget-shaped.

nui·sance (no͞o'səns, nyo͞o'-) *n.* 1 That which annoys, vexes, or harms. 2 *Law* That which by its use or existence works annoyance or damage to another. See synonyms under ABOMINATION. [<F <*nuire* harm <L *nocere*]

nuisance tax A small tax paid by the consumer and considered a nuisance by both the collector and the payer.

Nu·ku·a·lo·fa (no͞o'ko͞o·ä·lō'fä) Capital of the Tonga Islands.

Nu·ku Hi·va (no͞o'ko͞o hē'vä) The largest island of the Marquesas group; 46 square miles.

Nu·kus (no͞o·ko͞os') Capital of Kara-Kalpak Autonomous S.S.R.

null (nul) *adj.* 1 Of no legal force or effect; void: especially in the phrase **null and void.** 2 Having no existence. 3 Of no avail; useless; nugatory. 4 Lacking distinction or individuality; negative. 5 Zero. See synonyms under USELESS. —*n.* 1 Something that has no force or no meaning; a cipher. 2 *Telecom.* A cone of silence. [<L *nullus* no, none]

nul·lah (nul'ə) *n. Anglo-Indian* The dry bed of a small stream, or the stream itself; a gorge or ravine. [<Hind. *nālā*]

nul·li·fi·ca·tion (nul'ə·fə·kā'shən) *n.* The act of nullifying; in U.S. history, the claim of right by a State to refuse obedience to the laws of the United States, as by South Carolina in 1832. —**nul′li·fi·ca′tion·ist, nul′li·fi·ca′tor** *n.*

nul·li·fid·i·an (nul′ə·fid′ē·ən) *adj.* Having no religious faith. —*n.* One who has no religious faith. [<L *nullus* no + *fides* faith]

nul·li·fy (nul′ə·fī) *v.t.* **·fied, ·fy·ing 1** To bring to nothing; render ineffective or valueless. **2** To deprive of legal force or effect; make void; annul. See synonyms under ABOLISH, ANNUL, CANCEL. [<LL *nullificare* <*nullus* none + *facere* make] —**nul′li·fi′er** *n.*

nul·lip·a·ra (nu·lip′ər·ə) *n. pl.* **·a·rae** (-ə·rē) A woman who has never given birth to a child. Compara PRIMIPARA, MULTIPARA. [<L *nullus* none + *parere* bring forth] —**nul·li·par·i·ty** (nul′ə·par′ə·tē) *n.* —**nul·lip′a·rous** *adj.*

nul·li·pore (nul′ə·pôr, -pōr) *n. Bot.* A red-spored, coral-like, lime-secreting seaweed (family *Rhodophyceae*); a coralline. [<L *nullus* not any + *porus* pore] —**nul′li·po′rous** *adj.*

nul·li·ty (nul′ə·tē) *n. pl.* **·ties 1** The state of being null. **2** A nonentity. **3** *Law* A void act or instrument. [<F *nullité* <L *nullitas, -tatis* <*nullus* none]

Nu·man·ti·a (no͞o·man′shē·ə, nyo͞o-) An ancient ruined city on the Douro, north central Spain.

Nu·ma Pom·pil·i·us (no͞o′mə pom·pil′ē·əs, nyo͞o′-) The legendary second king of Rome, 715–675 B.C.

numb (num) *adj.* Destitute, wholly or partially, of the power of sensation or of motion; benumbed. —*v.t.* To make numb. [Orig. pp. of NIM; *b* added on analogy with *dumb, lamb*] —**numb′ly** *adv.* —**numb′ness** *n.*

Synonyms (adj.): benumbed, deadened, dull, insensible, narcotized, paralyzed, stupefied, torpid. *Antonyms*: feeling, impressionable, sensitive, sentient.

num·ber (num′bər) *n.* **1** One of a series of symbols or words used in classifying or arranging quantities; a numeral: Nine is a *number.* When a definite number is mentioned, the sign meaning number (#) is often used, followed by a numeral: R.F.D. #2: abbr. *no.,* or *No.,* from Latin *numero,* by number. See below for principal kinds of number. **2** A collection of units or individuals, whether large or small; an indefinite aggregation: often in the plural: a *number* of facts; large *numbers* of people. **3** *pl.* The science of numerals; arithmetic. **4** The character or quality of being numerous; Reliance is placed more on spirit than on *number.* **5** One of a numbered series, as of a periodical: the May *number* of "The Atlantic"; one of the parts of a literary, artistic, or musical work issued in parts. **6** One of the divisions or movements of a piece of music or of a musical or dancing program. **7** One of a numbered group. **8** *Often pl.* Poetic measure; rhythm; hence, verse or verses. **9** *Gram.* The form of inflection of a noun, pronoun, adjective, or verb, that indicates whether one thing or more is meant. English has the singular and the plural number. Greek and Sanskrit have in addition a dual number. See DUAL, PLURAL, SINGULAR. **10** *Colloq.* An article of merchandise numbered in a catalog; hence, any article, although unnumbered: This is our most popular *number.* —**by the numbers** *Mil.* A preparatory drill command to indicate that each subsequent movement is to be carried out step by step as its number is ordered. —**to get (or have) someone's number** *Colloq.* To have insight into a person's motives, character, etc. —*v.t.* **1** To determine the total number of; count; reckon. **2** To assign a number to; designate by a number or numbers. **3** To include as one of a collection or group. **4** To amount to; total: We *number* fifty men. **5** To set or limit the number of: Your days are *numbered.* —*v.i.* **6** To make a count; total. **7** To be included, as in a group. [<F *nombre* <L *numerus*] —**num′ber·er** *n.*

—**abstract number** Any number considered without reference to any particular object: distinguished from *concrete number.*

—**algebraic number** Any number which is the solution of an algebraic equation having integer coefficients.

—**amicable number** Either of two numbers, as 220 and 284, one of which is the sum of all the divisors of the other except itself.

—**cardinal number** Any number that directly expresses the number of digits under consideration as 1, 2, 3 . . . 8, etc.

—**composite number** Any integer exactly divisible by one or more integers other than itself or 1: opposed to *prime number.*

—**compound number** A number containing more than one unit or denomination, as feet and inches.

—**concrete number** A number applied to particular objects, as, four men; ten dollars: distinguished from *abstract number.*

—**defective number** A number which is greater than the sum of all its divisors except itself.

—**denominate number** A number expressing units of a specified kind, as, pounds, bushels, miles, etc.

—**irrational number** A number which cannot be expressed as an integer or the quotient of integers, as $\sqrt{2}$, $\sqrt{3}$, π, etc.

—**mixed number** A number, as 3 1/2, 5 3/4, which is the sum of an integer and a fraction.

—**ordinal number** A number that shows the order of a unit in a given series, as, first, second, third, etc.

—**perfect number** A number, as 6, 28, 496, which is equal to the sum of all its aliquots except itself. Compare *amicable number.*

—**polygonal number** The sum of an arithmetical progression that has the property of corresponding numerically to the number of points required to form a group of successively larger, regular polygons in accordance with a certain rule, as, 3, 6, 10, 15 . . .

—**prime number** A number divisible without remainder by no whole number except itself and unity: opposed to *composite number.*

—**Pythagorean numbers** Any set of three integers, as 3, 4, 5, satisfying the Pythagorean theorem.

—**rational number** A number which can be expressed as an integer or as a quotient of integers.

—**real number** Any rational or irrational number that does not contain an even root of a negative number.

—**sphenic number** A number product of three unequal prime factors.

—**square number** A number, as 1, 4, 9, 16, which is the square of some integer.

—**transcendental number** A number which is not an algebraic number, as π.

—**triangular number** A polygonal number which, apart from 1, is generated by making an array of dots in the form of an equilateral triangle, as 3, 6, 10, 15.

num·ber·less (num′bər·lis) *adj.* **1** Very numerous; innumerable; countless. **2** Having no number. See synonyms under INFINITE.

num·ber–one (num′bər·wun′) *adj.* The very finest.

number one *Colloq.* **1** Oneself. **2** Anything of the best quality. **3** *Brit.* A ship's officer ranking next below the captain, equivalent to the executive officer in the U. S. Navy.

Num·bers (num′bərz) The fourth book of the Pentateuch, giving the two censuses of Israel.

numbers pool A lottery, in which wagers are laid on the appearance of some particular, unpredictable number, as the last digits in the pari-mutuel racing totals of a given day: also called *policy racket.* Also **numbers game.**

numb·fish (num′fish′) *n. pl.* **·fish** or **·fish·es** The torpedo or electric ray.

num·bles (num′bəlz) *n. pl. Archaic* The entrails of a deer; especially, the edible organs, as heart, liver, etc.: also spelled *nombles.* [<OF *nombles,* ult. <L *lumbulus,* dim. of *lumbus* a loin]

numb·skull (num′skul′) See NUMSKULL.

nu·men (no͞o′mən, nyo͞o′-) *n. pl.* **·mi·na** (-mə·nə) In ancient Roman religion, a local divinity or presiding spirit. [<L]

nu·mer·a·ble (no͞o′mər·ə·bəl, nyo͞o′-) *adj.* That can be numbered.

nu·mer·al (no͞o′mər·əl, nyo͞o′-) *n.* **1** A symbol, character, or letter, alone or in combination with others, used to express a number. **2** A word that expresses number or is used in numerating or counting. —**Arabic numerals** The symbols, 1, 2, 3, 4, 5, 6, 7, 8, 9, 0, based on the decimal system and in general use since the tenth century. —**Roman numerals** The letters used until the tenth century as symbols in arithmetical notation. The basic letters are I(1), V(5), X(10), L(50), C(100), D(500), and M (1000), and intermediate and higher numbers are formed according to the following rules: Any symbol following another of equal or greater value adds to its value, as II = 2, XI = 11; any symbol preceding one of greater value subtracts from its value, as IV = 4, IX = 9, XC = 90; when a symbol stands between two of greater value, it is subtracted from the second and the remainder added to the first, as XIV = 14, LIX = 59. —*adj.* **1** Used in expressing or representing a number. **2** Pertaining to number. [<L *numeralis* <*numerus* number] —**nu′mer·al·ly** *adv.*

nu·mer·ar·y (no͞o′mə·rer′ē, nyo͞o′-) *adj.* Pertaining to numbers.

nu·mer·ate (no͞o′mə·rāt, nyo͞o′-) *v.t.* **·at·ed, ·at·ing 1** To enumerate; count. **2** To read, as a numerical expression, according to some system of numeration. [<L *numeratus,* pp. of *numerare* number <*numerus* a number]

nu·mer·a·tion (no͞o′mə·rā′shən, nyo͞o′-) *n.* **1** The act or art of reading or naming numbers, or a system of reading or naming them, especially those written decimally and according to the Arabic notation. Compare NOTATION. For numbers above and including 1,000,000,000 there are two systems in use: the French, used commonly in the United States, and the English. In the former, the above number is read *one billion*; in the latter, *one thousand million.* In general, in the former the successive names *billion, trillion,* etc., apply to the results obtained by multiplying 1,000 twice, thrice, etc., by itself; in the latter to those obtained by multiplying 1,000 by itself four times, six times, etc. **2** Enumeration.

nu·mer·a·tor (no͞o′mə·rā′tər, nyo͞o′-) *n.* **1** *Math.* In a common fraction, the term which stands above or to the left of the line and denotes how many of the parts of a unit (expressed by the denominator) are taken. **2** One who or that which numbers.

nu·mer·i·cal (no͞o·mer′i·kəl, nyo͞o-) *adj.* **1** Pertaining to or denoting number. **2** Numerable. **3** Represented by or consisting of numbers or figures, as in arithmetic, and not by letters, as in algebra. **4** *Math.* **a** Signifying that numbers have the place of letters: opposed to *literal.* **b** Designating a quantity considered opposed to *algebraic.* [<NL *numericus* <L *numerus* a number] —**nu·mer′i·cal·ly** *adv.*

nu·mer·ol·o·gy (no͞o′mə·rol′ə·jē, nyo͞o′-) *n.* **1** The science of numbers. **2** A system that purports to explain the occult influence of numbers, as those of the day of one's birth, the month in the year, and the year in the calendar, on life. —**nu·mer·o·log·i·cal** (no͞o′mər·ə·loj′i·kəl, nyo͞o′-) *adj.*

nu·mer·os·i·ty (no͞o′mə·ros′ə·tē, nyo͞o′-) *n.* **1** The state or condition of being numerous. **2** In symbolic logic, that property of a set, collection, or class which is defined by a cardinal number: the *numerosity* of a triplet, triad, or trilogy is 3.

nu·mer·ous (no͞o′mər·əs, nyo͞o′-) *adj.* Consisting of a great number of units; being many. See synonyms under FREQUENT, MANY. —**nu′mer·ous·ly** *adv.* —**nu′mer·ous·ness** *n.*

Nu·mid·i·a (no͞o·mid′ē·ə, nyo͞o-) An ancient kingdom and Roman province in northern Africa, roughly corresponding to modern Algeria.

Nu·mid·i·an (no͞o·mid′ē·ən, nyo͞o-) *adj.* Of or pertaining to ancient Numidia or its inhabitants. —*n.* **1** One of the ancient people of Numidia. **2** The Libyan dialect spoken by these people.

Numidian crane The demoiselle.

Numidian marble The yellow, pink, or red marbles found generally in northern Africa; especially those of Mauretania, first quarried by the Romans.

nu·mis·mat·ic (no͞o′miz·mat′ik, -mis-, nyo͞o′-) *adj.* Pertaining to or consisting of coins or medals. Also **nu′mis·mat′i·cal.** [<F *numismatique* <L *numisma, -atis* a coin <Gk. *nomisma* <*nomizein* sanction]

nu·mis·mat·ics (no͞o′miz·mat′iks, -mis-, nyo͞o′-) *n. pl. (construed as singular)* The science of coins and medals. Also **nu·mis·ma·tol·o·gy** (no͞o·miz′mə·tol′ə·jē, -mis′-, nyo͞o-). —**nu·mis′ma·tist, nu·mis′ma·tol′o·gist** *n.*

num·mu·lar (num′yə·lər) *adj.* **1** Of or pertaining to coins or money: also **num·ma·ry** (num′ər·ē). **2** Resembling coins: *nummular* sputa. Also **num′mu·lar′y, num′mu·lat′ed.** [<L *nummulus,* dim. of *nummus* a coin]

num·mu·la·tion (num′yə·lā′shən) *n.* The arrangement of red blood corpuscles in columns

nummulite 869 **nutritive**

like stacked-up coins, as seen under the microscope.
num·mu·lite (num′yə-līt) *n. Paleontol.* A large foraminifer of a nearly extinct family characteristic of the older Tertiary: preserved fossil forms show it as having a thin, coinlike shell. [<L *nummulus* small coin] — **num′mu·lit′ic** (-lit′ik) *adj.*
num·skull (num′skul) *n.* A blockhead; a dunce: also spelled *numbskull.*
nun[1] (nŭn, nōōn) *n.* The fourteenth Hebrew letter.

NUMMULITES

nun[2] (nun) *n.* 1 A woman devoted to a religious life, and living in a convent under vows of poverty, chastity, and obedience. 2 One of various birds, as the nunbird. 3 *Naut.* A conical or cone-shaped buoy made of metal: also **nun buoy**. [OE *nunne* <L *nonna*, fem. of *nonnus* an old man] — **nun′nish** *adj.*
Nun (nōōn) A principal outlet of the Niger in southern Nigeria.
nun·bird (nun′bûrd′) *n.* A South American bird (genus *Monasa*) having black plumage, usually with white about the head: also called *trappist.*
nunc di·mit·tis (nungk′ di·mit′is) *Latin* A dismissal; departure; permission to depart.
Nunc dimittis 1 The song or canticle of Simeon (*Luke* ii 29–32): so called from the first two words of the Latin version. 2 An English translation of this canticle. 3 A musical setting for this. [<L, now let depart]
nun·ci·a·ture (nun′shē·ə·chŏŏr) *n.* The office or term of office of a nuncio. [<Ital. *nunziatura* <*nunzio* NUNCIO]
nun·ci·o (nun′shē·ō) *n. pl.* ·**ci·os** 1 An ordinary ambassador of the pope to a foreign court or government: distinguished from *legate.* 2 Any messenger. Also **nun′ci·us** (-shē·əs). [<Ital. *nunzio* <L *nuntius* a messenger]
nun·cle (nung′kəl) *n. Dial.* An uncle. [<*an uncle*, taken as *a nuncle*]
nun·cu·pa·tive (nung′kyə·pā′tiv, nung·kyōō′pə·tiv) *adj. Law* Oral as distinguished from written: said especially of a will. Also **nun′cu·pa·to′ry** (-pə·tôr′ē, -tō′rē). [<LL *nuncupativus* <*nuncupare* call by name]
Nun·ea·ton (nun·ē′tən) A municipal borough of NE Warwick, England.
nun·na·tion (nu·nā′shən) *n.* The addition of the letter *n* to a word, as in the declension of Arabic nouns. [<Arabic *nūn*, the letter *n*]
nun·ner·y (nun′ər·ē) *n. pl.* ·**ner·ies** A convent for nuns. See synonyms under CLOISTER.
nun's-veil·ing (nunz′vā′ling) *n.* A soft, thin, untwilled woolen fabric, used for veiling and as a dress material.
Nu·per·caine (nōō′pər·kān, nyōō′-) Proprietary name for a white, crystalline, odorless compound, $C_{20}H_{29}N_3O_2$, used as a local anesthetic with an action similar to that of cocaine or procaine.
nup·tial (nup′shəl) *adj.* Pertaining to marriage or the marriage ceremony. See synonyms under MATRIMONIAL. [<L *nuptialis* <*nuptus*, pp. of *nubere* marry] — **nup′tial·ly** *adv.*
nuptial flight *Entomol.* The mating flight of many insects, as ants and gnats, during which conspicuous swarming may occur.
nup·tials (nup′shəlz) *n. pl.* (*construed as singular*) The marriage ceremony or state. See synonyms under MARRIAGE.
nu·ragh (nōō′räg) *n.* One of a class of prehistoric stone structures numerous in Sardinia: also spelled *noraghe.* Also **nu·ra′ghe** (-rä′gā). [<dial. Ital. *nuraghe*]
Nu·rem·berg (nōōr′əm·bûrg, nyōōr′-) A city of northern Bavaria, West Germany. *German* **Nürn·berg** (nürn′berkh).
Nu·ris·tan (nōōr′is·tan) A mountainous district of NE Afghanistan; 5,000 square miles: formerly *Kafiristan.*
nurl (nûrl) *v.t.* To mill or roughen, as the rim of a coin. [Var. of KNURL.]
nurse (nûrs) *n.* 1 A female servant who takes care of young children: in the case of one who suckles an infant, called a *wet-nurse*; otherwise, less frequently, a *drynurse.* 2 One who suckles a babe. 3 A person who cares

for the sick, wounded, or enfeebled, especially one who makes a profession of it. 4 One who or that which fosters, nurses, protects, or promotes. 5 One of various sharks, as the nursehound (genus *Ginglymostoma*). 6 *Entomol.* A sexually incomplete bee or ant, etc., whose duty it is to care for the young. — *v.* **nursed**, **nurs·ing** *v.t.* 1 To take care of, as in sickness or infirmity. 2 To feed (an infant) at the breast; suckle. 3 To feed and care for in infancy. 4 To promote the growth and development of; foster; cherish. 5 To use or operate carefully; preserve from injury, damage, or undue strain: to *nurse* a weak wrist. 6 To try to cure, as a cold, by taking care of oneself. 7 To clasp or hold carefully or caressingly; fondle. 8 In billiards, to keep (the balls) in a close group so as to score a series of caroms. — *v.i.* 9 To act or serve as a nurse. 10 To take nourishment from the breast. 11 To suckle an infant. See synonyms under CHERISH. [Earlier *nurice* <OF <L *nutricia* <L *nutrix* <*nutrire* nourish, foster] — **nurs′er** *n.*
nurse-maid (nûrs′mād′) *n.* A girl or woman employed to care for children.
nurs·er·y (nûr′sər·ē) *n. pl.* ·**er·ies** 1 A room in a house set apart for the occupation and use of children; also, a playroom. 2 A place where trees, shrubs, etc., are raised for sale or transplanting. 3 The place where anything is fostered, bred, or developed; hence, any condition that promotes growth. 4 *Obs.* The act of nursing; also, that which is nursed.
nurs·er·y·man (nûr′sər·ē·mən) *n. pl.* ·**men** (-mən) One who owns or manages a nursery for the cultivation of trees and shrubs.
nursery rime A simple story, riddle, proverb, etc., presented in rimed verse or jingle for children.
nursing bottle A small, graduated bottle fitted with a rubber nipple, for feeding infants.
nursing home A small private hospital.
nurs·ling (nûrs′ling) *n.* An infant; also, anything that is carefully tended or supervised. Also **nurse′ling**.
nur·ture (nûr′chər) *n.* 1 The act of nourishing. 2 That which nourishes or fosters; education. 3 *Biol.* The aggregate of environmental conditions and influences acting on an organism subsequent to birth. Compare NATURE. — *v.t.* ·**tured**, ·**tur·ing** 1 To feed or support; nourish; rear; foster. 2 To bring up or train; educate. [<OF *nurture*, var. of *nourriture* <LL *nutritura* <L *nutrire* nourish] — **nur′tur·er** *n.*
Synonyms (*noun*): breeding, discipline, education, instruction, schooling, teaching, training, tuition. *Breeding* and *nurture* include *teaching* and *training*, especially as directed by and dependent upon home life and personal association; *breeding* having reference largely to manners with such qualities as are deemed distinctively characteristic of high birth; *nurture* (literally *nourishing*) having more direct reference to moral qualities, not overlooking the physical and mental. See EDUCATION, CHERISH, TEACH.
Nu·sa Teng·ga·ra (nōō′sə teng·gä′rə) A province of Indonesia, comprising all of the Lesser Sunda Islands, exclusive of Portuguese Timor; 61,995 square miles; capital, Singaradja, on Bali.
nut (nut) *n.* 1 *Bot.* a A fruit consisting of a kernel or seed enclosed in a woody shell, as in the hazelnut, beechnut, or chestnut; also, the kernel of such fruit, especially when edible. b A hard, indehiscent, one-seeded pericarp resulting from a compound ovary. 2 *Mech.* A small block of metal having an internal screw thread so that it may be fitted upon a bolt, screw, or the like. 3 A person or matter difficult to deal with; a problem. 4 *Slang* The head. 5 *Slang* A crazy or irresponsible person. 6 The ridge at the upper end of the neck of stringed instruments, serving to elevate the strings; also, the adjustable end of a fiddle bow. — *v.i.* **nut·ted**, **nut·ting** To seek or gather nuts. [OE *hnutu*] — **nut′ter** *n.*
nu·tant (nōō′tənt, nyōō′-) *adj. Bot.* Nodding; drooping: said especially of flowers. [<L *nutans*, -*antis*, ppr. of *nutare* nod]
nu·ta·tion (nōō·tā′shən, nyōō′-) *n.* 1 *Astron.* The periodic inequalities in the motion of

the axis and pole of the earth around the pole of the ecliptic as a center. 2 *Bot.* A spontaneous rotatory movement, as of young growing parts of plants. 3 The act of nodding the head. [<L *nutatio*, -*onis* <*nutare* nod] — **nu·ta′tion·al** *adj.*
nut-crack·er (nut′krak′ər) *n.* 1 *Chiefly pl.* A device for cracking nuts. 2 One of certain crowlike birds (genus *Nucifraga*), as the common Old World nutcracker (*N. caryocatactes*), or **Clark's nutcracker** (*N. columbiana*) of the coniferous forests of western North America. 3 A nuthatch.
nut-gall (nut′gôl′) *n.* A nut-shaped gall, as on an oak tree; an oak apple.
nut-grass (nut′gras′, -gräs′) *n.* A perennial herb (*Cyperus rotundus*) of the sedge family bearing nutlike tubers: also called *cocograss.*
nut-hatch (nut′hach′) *n.* A small, short-tailed bird (family *Sittidae*) related to the titmouse, having a slender bill as long as the head and feeding on nuts and insects.
nut-let (nut′lit) *n.* 1 A diminutive nut. 2 The stone in a drupe.
nut-meg (nut′meg) *n.* 1 The aromatic kernel of the fruit of various tropical trees (genus *Myristica*), especially of the nutmeg tree (*M. fragrans*) of the Molucca Islands. 2 The tree itself. [ME *notemuge*, partial trans. of OF *nois mugue* <*nois* nut + *mugue* musk <L *muscus*]
nutmeg flower Fennelflower.
Nutmeg State Nickname of CONNECTICUT.
nut pick A small sharp-pointed instrument for picking out the kernels of nuts.
nu·tri·a (nōō′trē·ə, nyōō′-) *n.* 1 The coypu. 2 Its soft, brown fur, often dyed to resemble beaver. [<Sp., an otter <L *lutra*]
nu·tri·ent (nōō′trē·ənt, nyōō′-) *adj.* 1 Giving nourishment. 2 Conveying nutrition. — *n.* 1 Something that nourishes. 2 A drug or other substance which acts upon the nutritive processes of an organism. [<L *nutriens*, -*entis*, ppr. of *nutrire* nourish]
nutrient solution A solution containing, in correct proportions and strength, the various chemical substances required for plant growth: used in hydroponics.
nu·tri·ment (nōō′trə·mənt, nyōō′-) *n.* 1 That which nourishes; food. 2 That which promotes development. [<L *nutrimentum* <*nutrire* nourish] — **nu′tri·men′tal** *adj.*
Synonyms: aliment, food, meat, nourishment, provision, sustenance. *Nourishment* and *sustenance* apply to whatever can be introduced into the system as a means of sustaining life; we say of a convalescent: He is taking *nourishment. Aliment* is similar in meaning, but less frequent in use. *Nutriment* and *nutrition* have more of scientific reference to the vitalizing principles of various *foods;* thus, wheat is said to contain a great amount of *nutriment.* Compare FOOD.
nu·tri·tion (nōō·trish′ən, nyōō-) *n.* 1 The aggregate of all the processes by which food is assimilated, growth promoted, and waste repaired in living organisms. 2 Nutriment. See synonyms under FOOD. [<L *nutrire*] — **nu·tri′tion·al** *adj.* — **nu·tri′tion·al·ly** *adv.*
nu·tri·tion·ist (nōō·trish′ən·ist, nyōō-) *n.* One who specializes in the processes and problems of nutrition.

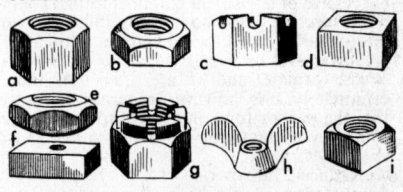

MECHANICAL NUTS
a. Hexagonal, soft.   e. Double-cupped.
b. Lock.   f. Joint, untapped.
c. Hexagonal, slotted.   g. Castle.
d. Square, plain.   h. Thumb.
     i. Square, chamfered.

nu·tri·tious (nōō·trish′əs, nyōō-) *adj.* Nourishing; promoting nutrition; trophic. — **nu·tri′tious·ly** *adv.* — **nu·tri′tious·ness** *n.*
nu·tri·tive (nōō′trə·tiv, nyōō′-) *adj.* 1 Having nutritious properties. 2 Of or relating to

nutrition. — **nu′tri·tive·ly** adv. — **nu′tri·tive·ness** n.
nuts (nuts) Slang adj. **1** Crazy; demented. **2** Madly in love: with about. **3** Extremely enthusiastic: with about. — interj. An exclamation of scorn, disapproval, etc. [<NUT]
nut·shell (nut′shel′) n. The shell of a nut. — **in a nutshell** In brief and concise statement or form.
Nut·tall (nut′ôl), **George Henry Falkiner**, 1862–1937, U.S. biologist.
nut·ty (nut′ē) adj. ·ti·er, ·ti·est **1** Abounding in nuts. **2** Having the flavor of nuts. **3** Slang Crazy; also, madly in love; very enthusiastic. — **nut′ti·ly** adv. — **nut′ti·ness** n.
nut·wood (nut′wood′) n. **1** Any tree bearing nuts, as walnut, hazel, hickory, etc. **2** The wood of such a tree.
nux vom·i·ca (nuks′ vom′i·kə) **1** The flattened, disklike, silky, poisonous seed of an Indian tree (Strychnos nux-vomica). It contains several alkaloidal poisons, principally strychnine and brucine. **2** The tree producing this fruit. [<Med. L <nux a nut + vomere vomit]
nuz·zle (nuz′əl) v. ·zled, ·zling v.i. **1** To root or dig with the nose or snout, as a hog does. **2** To nestle or snuggle; lie close. — v.t. **3** To rub with the nose; push the nose against. **4** To root up with the nose or snout. [Freq. of NOSE, v.]
ny·an·za (nī·an′zə) n. A sheet of water; lake; also, a river feeding a lake. [<Bantu]
Nya·sa (nyä′sä, nī·as′ə) **Lake** A lake in eastern Africa between Nyasaland and Mozambique; 11,000 square miles. Formerly **Nyas′sa**.
Nya·sa·land (nyä′sä·land, nī·as′ə-) A British protectorate in SE Africa; land area, 36,870 square miles; incl. water 47,870 square miles; capital, Zomba. See FEDERATION OF RHODESIA AND NYASALAND under RHODESIA.
nyc·ta·gi·na·ceous (nik′tə·ji·nā′shəs) adj. Bot. Of or pertaining to a family of plants (Nyctaginaceae) widely distributed in warm and tropical lands, including the bougainvillea; the four-o′-clock family. [<NL <Nyctago,

-inis, former genus name <Gk. nyx, nyktos night]
nyc·ta·lo·pi·a (nik′tə·lō′pē·ə) n. Pathol. Night blindness; a physical defect of the eyes in which one sees well by daylight, but poorly in the dark or in dim light: sometimes confused with day blindness or hemeralopia. Also **nyc·ta·lo′py**. [<NL nyx, nyktos night + alaos blind + ōps eye] — **nyc′ta·lop′ic** (-lop′ik) adj.
nyc·tan·thous (nik·tan′thəs) adj. Bot. Pertaining to or designating flowers which open at night. Also **nyc·ti·gam·ous** (nik′tə·gam′əs). [<NYCT(O) + Gk. anthos flower]
nyc·ti·trop·ism (nik·tit′rə·piz′əm) n. Bot. The changing of the position of the leaves of certain plants during the night. Also **nyc′ti·nas′ty** (-ti·nas′tē) — **nyc′ti·trop′ic** (-trop′ik) adj.
nycto- combining form Night; nocturnal: nyctophobia. Also, before vowels, **nyct-**. Also **nycti-**. [<Gk. nyx, nyktos night]
nyc·to·pho·bi·a (nik′tə·fō′bē·ə) n. Morbid fear of night or of darkness: also called noctiphobia, scotophobia. — **nyc′to·pho′bic** adj.
Nye (nī), **Edgar Wilson,** 1850–96, U.S. humorist: known as Bill Nye.
nyet (nyet) interj. Russian No.
Nyí·regy·há·za (nyē′redy′·hä′zō) A city in NE Hungary.
nyl·ghai (nil′gī), **nyl·ghau** (nil′gô) See NILGAI.
Ny·lon (nī′lon) n. A synthetic thermoplastic polyamide derivable from coal, air, and water, which may be formed into fibers, bristles, sheets, and other forms which, when drawn, are characterized by extreme toughness, elasticity, and strength: a trade name.
ny·lons (nī′lonz) n. pl. Stockings made of Nylon.
nymph (nimf) n. **1** In Greek and Roman mythology, a beautiful maiden belonging to a class of lesser divinities inhabiting groves, forests, fountains, springs, mountains, the ocean, etc. **2** Hence, an attractive girl; a lovely young woman. **3** Entomol. **a** The young of an insect which undergoes incomplete metamorphosis, at which stage the wing pads are first evident. **b** One of various nymphalid butterflies, as a fritillary. [<L nympha <Gk. nymphē nymph, bride] — **nymph′al, nym·phe·an** (nim·fē′ən) adj. — **nymph′ic, nymph′i·cal** adj.
nym·pha (nim′fə) n. pl. ·phae (-fē) **1** Anat. One of the inner folds of the mucous membrane of the female pudenda. **2** A nymph (def. 3a). [<L <Gk. nymphē a bride]
nym·phae·a·ceous (nim′fē·ā′shəs) adj. Pertaining to or designating a family (Nymphaeaceae) of aquatic, perennial herbs, the waterlily family, with a thick, horizontal rootstock, mainly peltate, floating or submersed leaves, and large solitary flowers living in fresh water. See LOTUS. [<NL <L nymphaea waterlily <Gk. nymphaia]
nym·pha·lid (nim′fə·lid) n. Any of a family (Nymphalidae) of medium to large butterflies, often brightly colored, including the emperor, tortoise-shell, and admiral butterflies. [<NL <L nympha a nymph]
nympho- combining form Nymph; bride: nymphomania. Also, before vowels, **nymph-**. [<Gk. nymphē a nymph]
nym·pho·lep·sy (nim′fə·lep′sē) n. **1** A kind of ecstasy or frenzy, said to have taken possession of one who looked upon a nymph. **2** An emotional state caused by an unrealizable desire. [<Gk. nympholeptos frenzied < nymphē a nymph + lambanein take] — **nym′pho·lept** (-lept) n. — **nym′pho·lep′tic** adj.
nym·pho·ma·ni·a (nim′fə·mā′nē·ə, -mān′yə) n. Psychiatry A morbid and ungovernable sexual desire in women. — **nym′pho·ma′ni·ac** adj. & n.
Ny·sa (nē′sä) The Polish name for the NEISSE.
nys·tag·mus (nis·tag′məs) n. Pathol. A spasmodic movement of the eyes, rotatory or from side to side. [<Gk. nystagmos drowsiness < nystazein nod in sleep, grow drowsy] — **nys·tag′mic** adj.
Nyx (niks) In Greek mythology, the goddess of night: identified with the Roman Nox.

O

o, O (ō) n. pl. **o's, O's,** or **os, Os,** or **oes** (ōz) **1** The 15th letter of the English alphabet: from Phoenician ayin, which was a consonant, through Greek omicron and omega, and Roman O. **2** Any sound of the letter o. See ALPHABET. — symbol **1** Math. Zero or naught: called also round O. **2** Chem. Oxygen (symbol O). **3** Anything shaped like an O; an oval or circle; a spot or spangle: Giotto's O. See appendix (ABBREVIATIONS).
O (ō) interj. **1** An exclamation prefixed to an expression of address, as a sign of the vocative, used especially in earnest appeal or exhortation, or in prayer to the Deity, to emphasize the feeling or passion conveyed by the words. A note of exclamation usually follows the vocative word, phrase, or clause: O Lord! O my countrymen! **2** An ejaculation expressive of a wish: an elliptical form: O stay! The object of desire sometimes follows in an interjectional or elliptical phrase, with for if a substantive, or that if a clause. **3** See OH. ◆ The forms O and oh are often used indiscriminately. It is, however, generally conceded that the proper form in the vocative use is O. — n. An exclamation or lamentation. Also spelled oh.
o- Reduced var. of OB-.
o′ prep. Of: one o'clock, man-o′-war, jack-o′-lantern.
O′ A descendant of: O′Conor: a patronymic prefix commonly used in Irish surnames, equivalent to the English and Scandinavian suffixes -son, -sen. Compare MAC, FITZ. [<Irish ó grandson, descendant]
oaf (ōf) n. **1** Originally, a misshapen bantling left in place of a pretty child supposed to be stolen by fairies; a changeling. **2** A simpleton; a stupid, lubberly person. [Earlier auf <ON alfr elf. Akin to ELF.]
oaf·ish (ō′fish) adj. Stupid; doltish. — **oaf′ish·ly** adv. — **oaf′ish·ness** n.
O·a·hu (ō·ä′hōō) The island in the north central Hawaiian Islands on which Honolulu is located; 589 square miles.
oak (ōk) n. **1** A hardwood, acorn-bearing tree or shrub (genus Quercus) of the beech family, valued for the hardness, strength, and durability of its timber. ◆ Collateral adjective: quercine. **2** The wood or timber of the oak. **3** One of various other plants having a resemblance or relation to the oak: Jerusalem oak. **4** A stout door: so called because usually made of oak. **5** Any of various shades of finished oak wood. **6** The leaves of the oak, as in a garland: used as a crown: in ancient Rome, the reward of a hero who saved the life of a fellow man. **7** Oaken woodwork or furniture. — **quartered oak** Oaken boards cut by quarter-sawing, and exhibiting a striking grain. — **to sport one's oak** To exclude visitors by closing the outer door of one's apartment: English university slang. [OE āc]
oak apple A gall produced on an oak by an insect: also called nutgall. Also **oak gall**.
oak·en (ō′kən) adj. Made of oak.
Oak·ham (ō′kəm) The county town of Rutland county, England.
Oak·land (ōk′lənd) A port on San Francisco Bay, California.
oak-leaf cluster (ōk′lēf′) A bronze decoration given to holders of certain U.S. military medals in recognition of acts meriting a second award of the same medal. It represents a small twig bearing 4 oak leaves and 3 acorns.
Oak·ley (ōk′lē), **Annie** See ANNIE OAKLEY.
Oak Ridge A town in eastern Tennessee; site of an atomic research center.
oak·um (ō′kəm) n. Hemp fiber obtained by untwisting old rope: used in calking, etc. [OE acuma, var. of acumba <a- off, without + cemban comb¹]
oar (ôr, ōr) n. **1** A wooden implement for propelling or, occasionally, for steering a boat, consisting of a long shaft with a blade at one end. **2** An oarsman. **3** An oarlike appendage in certain worms. — v.t. **1** To propel with or as with oars; row. **2** To make (one's way) or traverse (water) with or as with oars. — v.i. **3** To proceed by or as by rowing; row. ◆ Homophone: ore. [OE ār] — **oar′less** adj.
oared (ôrd, ōrd) adj. **1** Having oars for propulsion. **2** Having oarlike feet or swimming appendages.
oar·fish (ôr′fish′, ōr′-) n. pl. ·fish or ·fish·es Any of several fishes (genus Regalecus) of northern seas, with oarlike dorsal rays and a length of up to twenty feet.
oar·lock (ôr′lok′, ōr′-) n. A device on the side of a boat for keeping an oar in place; rowlock. [OE ārloc <ār oar + loc lock, enclosure]
oars·man (ôrz′mən, ōrz′-) n. pl. ·men (-mən) One who rows.
oars·man·ship (ôrz′mən·ship, ōrz′-) n. The art of rowing; skill in rowing.
oar·y (ôr′ē, ō′rē) adj. Having the form or use of an oar.
o·a·sis (ō·ā′sis, ō′ə·sis) n. pl. ·ses (-sēz) A small area in a waste or desert made fertile by groundwater or by surface irrigation. [<L <Gk. Oasis, a city in the Libyan Desert <Egyptian]
oast (ōst) n. A kiln for drying hops or malt. [OE āst a kiln]
oat (ōt) n. **1** Usually pl. A cereal grass (Avena sativa) extensively cultivated for its edible grain. **2** A musical pipe made from a stem of the oat; a shepherd's pipe; hence, a pastoral song. — **to feel one's oats 1** To feel lively; have a sense of vitality. **2** To feel important. — **to sow one's wild oats** To indulge in the follies or excesses to which youth is liable. [OE āte]
oat·cake (ōt′kāk′) n. A cake of oatmeal, usually rolled thin and baked hard. Also **oat cake**.
oat·en (ōt′n) adj. **1** Made of oats or oatmeal,

or of the straw of oats. **2** Sounded on a pipe made from a stem of oat.

Oates (ōts), **Titus,** 1649–1705, English impostor who fabricated a supposed Catholic conspiracy to massacre Protestants, burn London, and kill the king; convicted of perjury; pardoned.

oat·grass (ōt′gras′, -gräs′) *n.* **1** Any uncultivated kind of oats. **2** Any of various oatlike grasses.

oath (ōth) *n. pl.* **oaths** (ōthz) **1** A solemn attestation in support of a declaration or a promise, by an appeal to God or to some person or thing regarded as high and holy; also, the declaration or promise so supported. **2** *Law* Such an attestation or affirmation of the truth of a statement as renders liable to punishment for perjury one who wilfully thus asserts what is not true. **3** The form of words in which such attestation is made. **4** A frivolous and blasphemous use of the name of the Deity or of any sacred name or object, as in appeal or ejaculation. **5** An imprecation lightly or humorously used. [OE *āth*]

Synonyms: adjuration, affidavit, anathema, ban, blasphemy, curse, denunciation, execration, imprecation, malediction, profanity, reprobation, swearing, vow. In the highest sense, as in a court of justice, "an *oath* is a reverent appeal to God in corroboration of what one says"; an *affidavit* is a sworn statement made in writing in the presence of a competent officer; an *adjuration* is a solemn appeal to a person in the name of God to speak the truth. An *oath* is made to man in the name of God; a *vow* is usually made to God. In the lower sense, an *oath* may be mere *blasphemy* or profane *swearing*. *Anathema, curse, execration,* and *imprecation* are modes of invoking vengeance or retribution from a superhuman power upon the person against whom they are uttered. *Anathema* is a solemn ecclesiastical condemnation of a person or of a proposition. *Curse* may be just and authoritative; as, the *curse* of God; or, it may be wanton and powerless. *Execration* expresses most of personal bitterness and hatred; *imprecation* refers to the coming of the desired evil upon the person against whom it is uttered. *Malediction* is a general wish of evil, a less usual but very expressive word. Compare TESTIMONY. *Antonyms*: benediction, benison, blessing

oat·meal (ōt′mēl′) *n.* **1** The meal of oats. **2** Porridge made of it. Also **oat meal.**

Oa·xa·ca (wä·hä′kä) A state in SW Mexico on the Pacific; 36,355 square miles; capital, Oaxaca. Also **Oa·xa′ca de Juá·rez** (thā hwä′·räs).

Ob (ōb) A river of western Asian Russian S.F.S.R., flowing 2,500 miles NW to the **Ob Gulf,** an inlet of the Kara Sea 500 miles long.

ob– *prefix* **1** Toward; to; facing: *obvert.* **2** Against; in opposition to: *object, obstruct.* **3** Over; upon: *obliterate.* **4** Completely: *obdurate.* **5** Inversely: *obovate*: prefixed to adjectives in scientific Neo-Latin and English terms. Also: *o–* before *m,* as in *omit*; *oc–* before *c,* as in *occur; of–* before *f,* as in *offend; op–* before *p,* as in *oppress.* [<L *ob* toward, for, against]

O·ba·di·ah (ō′bə·dī′ə) A masculine personal name. [<Hebrew, servant of the Lord]

— **Obadiah** A minor prophet living in the sixth century B.C.; also, the book of the Old Testament containing his prophecies.

ob·bli·ga·to (ob′lə·gä′tō, *Ital.* ôb′blē·gä′tō) *adj.* **1** That cannot be dispensed with; necessary. **2** *Music* Referring to parts or accompaniments essential to the performance of a composition. — *n. pl.* **·tos** or **·ti** (-tē) *Music* A part or accompaniment, usually written for a single instrument. [<Ital. *obbligato, obligato* <L *obligatus,* pp. of *obligare.* See OBLIGE.]

ob·cor·date (ob·kôr′dāt) *adj. Bot.* Inversely heart-shaped, as the leaves of some plants. [<OB- + CORDATE]

ob·du·ra·cy (ob′dyə·rə·sē) *n.* Obstinacy; obdurateness.

ob·du·rate (ob′dyə·rit, -rāt) *adj.* **1** Unmoved by feelings of humanity or pity; inexorable. **2** Perversely impenitent. **3** Unyielding; stubborn. See synonyms under HARD, OBSTINATE. [<L *obduratus,* pp. of *obdurare* harden < *ob-* against + *durare* harden < *durus* hard] — **ob′·du·rate·ly** *adv.* — **ob′du·rate·ness** *n.*

o·be (ō′bē), **o·be·ah** (ō′bē·ə) See OBI.

o·be·di·ence (ō·bē′dē·əns, ə·bē′-) *n.* **1** Submission to command, prohibition, law, or duty. **2** The fact of being obeyed, or having subjects obedient to one. **3** *Eccl.* Sphere of authority, or those acknowledging it. See synonyms under ALLEGIANCE, SUBMISSION.

o·be·di·ent (ō·bē′dē·ənt, ə·bē′-) *adj.* Complying with or submitting to a behest, law, etc.; habitually yielding to authority; submissive; dutiful. See synonyms under DOCILE, OBSEQUIOUS. [<OF *obedient* <L *obediens, -entis,* ppr. of *obedire* OBEY] — **o·be′di·ent·ly** *adv.*

O·beid (ō·bād′), **El** Capital of Kordofan province, Sudan.

o·bei·sance (ō·bā′səns, ō·bē′-) *n.* An act of courtesy or reverence, consisting of bowing or a bending of the knee; a bow or courtesy; homage; deference. [<OF *obeissance* < *obeissant,* ppr. of *obeir* OBEY] — **o·bei′sant** *adj.*

ob·e·lisk (ob′ə·lisk) *n.* **1** A square shaft with pyramidal top, usually monumental. The Egyptian obelisks are always monolithic and slightly tapering. **2** *Printing* The dagger sign (†) used as a mark of reference; obelus. [<L *obeliscus* <Gk. *obeliskos,* dim. of *obelos* a spit, pointed pillar] — **ob′e·lis′cal, ob′e·lis′koid** *adj.*

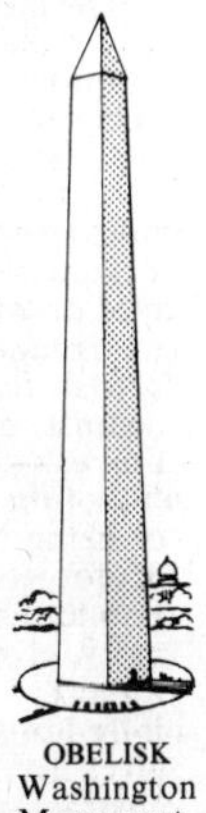

OBELISK
Washington Monument, Washington, D.C. (555 feet, 5 1/2 inches high)

ob·e·lize (ob′ə·līz) *v.t.* **·lized, ·liz·ing** To mark with an obelus. [<Gk. *obelizein* mark with an obelus < *obelos* an obelus]

ob·e·lus (ob′ə·ləs) *n. pl.* **·li** (-lī) **1** A critical mark, as — or ÷, used in ancient manuscripts to designate a suspected reading or to indicate a spurious passage. **2** *Printing* Obelisk. [<L <Gk. *obelos* a spit, obelisk, critical mark]

O·ber·am·mer·gau (ō′bər·äm′ər·gou) A village in Upper Bavaria, West Germany; noted for its Passion play, presented once a decade, performed by the villagers.

O·ber·hau·sen (ō′bər·hou′zən) A city in North Rhine-Westphalia, West Germany.

O·ber·land (ō′bər·länt) A mountainous region of central Switzerland, specifically the **Bernese Oberland** (*Bernese Alps*).

O·ber·on (ō′bə·ron) In medieval legend, folklore, and Shakespeare's *Midsummer Night's Dream,* the king of the fairies, husband of Titania.

o·bese (ō·bēs′) *adj.* Very corpulent. See synonyms under CORPULENT. [<L *obesus* fat, orig. pp. of *obedere* devour < *ob-* completely + *edere* eat] — **o·bese′ly** *adv.*

o·bes·i·ty (ō·bē′sə·tē, ō·bes′ə-) *n. Pathol.* An excessive accumulation of fat in the body; morbid corpulency. Also **o·bese′ness.**

o·bey (ō·bā′, ə·bā′) *v.t.* **1** To do the bidding of; be obedient to. **2** To carry into effect; execute, as a command. **3** To act in accordance with; be guided by: to *obey* the law. — *v.i.* **4** To be obedient. [<OF *obeir* <L *obedire,* var. of *oboedire* give ear, obey < *ob-* in the direction of + *audire* hear] — **o·bey′er** *n.*

Synonyms: comply, defer, keep, observe, submit, yield. See FOLLOW, KEEP, SERVE. *Antonyms*: contemn, defy, disobey, infringe, refuse, resist, violate. See synonyms for GOVERN.

ob·fus·cate (ob·fus′kāt, ob′fəs-) *v.t.* **·cat·ed, ·cat·ing** **1** To confuse or perplex; bewilder. **2** To darken or obscure. [<L *obfuscatus,* pp. of *obfuscare* darken, obscure < *ob-* against + *fuscare* darken < *fuscus* dark] — **ob′fus·ca′·tion** *n.*

o·bi[1] (ō′bē) *n.* **1** A kind of sorcery practiced by the Negroes of the West Indies and SE United States: a revival or survival of African magic rites, specializing in poisons and the power of terror. **2** A charm or fetish used in these magical practices. Also called *obe, obeah.* [Var. of *obeah* < native West African name] — **o′bi·ism** *n.*

o·bi[2] (ō′bē) *n.* A broad sash with a bow in the back, worn by Japanese women: also spelled *obe.* [<Japanese *ōbi*]

O·bi Islands (ō′bē) An Indonesian island group NW of Ceram; 1,069 square miles: also *Ombi.*

o·bi·it (ō′bē·it) *Latin* He (or she) died.

o·bit (ō′bit, ob′it) *n.* **1** The death or date of death of a person; also, an obituary. **2** A ceremony or service commemorating a death. [<OF <L *obitus* a going down, a death < *obire* go down, die < *ob-* down + *ire* go]

ob·i·ter dic·tum (ob′ə·tər dik′təm) *Latin pl.* **ob·i·ter dic·ta** (dik′tə) A remark by the way or in passing. See DICTUM.

o·bit·u·ar·y (ō·bich′ōō·er′ē) *adj.* Pertaining to the death of a person. — *n. pl.* **·ar·ies** A published notice of a death; a biographical sketch of one recently deceased. [<Med. L *obituarius* <L *obitus.* See OBIT.]

ob·ject[1] (əb·jekt′) *v.i.* **1** To offer arguments or opposition; dissent. **2** To feel or state disapproval; be averse. — *v.t.* **3** To offer as opposition or criticism; charge. See synonyms under OPPOSE. [<L *objectus,* pp. of *objicere* < *ob-* towards, against + *jacere* throw] — **ob·jec′tor** *n.*

ob·ject[2] (ob′jikt, -jekt) *n.* **1** Anything that lies within the cognizance of the senses; especially, anything tangible or visible; any material thing. **2** That which is affected or intended to be affected by feeling or action. **3** That on which one sets his mind as an end; purpose; aim. **4** *Gram.* A noun or pronoun to which the action of a transitive verb is directed, or which receives or endures the effect of this action. A **direct object** receives the direct action of the verb, as in the sentence "He ate the pie," *pie* is the direct object of *ate*; an **indirect object** receives the secondary action of the verb, as in "She gave him the pie," *him* is the indirect object of *gave.* **5** *Colloq.* A person of pitiable or ridiculous aspect; any sight that evokes laughter, disgust, pity, etc. See synonyms under AIM, DESIGN, PURPOSE, REASON. [<Med. L *objectum* something thrown in the way <L *objectus.* See OBJECT[1].]

object ball In billiards or pool, the ball which the player purposes to hit with his cue ball.

object glass *Optics* A lens or combination of lenses for focusing the rays of light passing through it; in a telescope or microscope, the lens nearest the object; in a camera or projector, the lens that makes the image of the object. Also **object lens.**

ob·jec·ti·fy (əb·jek′tə·fī) *v.t.* **·fied, ·fy·ing** To present, as in form or character, from an external viewpoint; make objective. [<OBJECT[2] + -(I)FY] — **ob·jec′ti·fi·ca′tion** *n.*

ob·jec·tion (əb·jek′shən) *n.* **1** The act of objecting. **2** An impediment raised; a dissenting argument; an adverse fact.

ob·jec·tion·a·ble (əb·jek′shən·ə·bəl) *adj.* Deserving of disapproval; offensive. — **ob·jec′·tion·a·bil′i·ty** *n.* — **ob·jec′tion·a·bly** *adv.*

ob·jec·tive (əb·jek′tiv) *adj.* **1** Of or belonging to an object; having the nature of an object or being that which is thought of or perceived, as opposed to that which thinks or perceives; outside the mind: opposed to *subjective.* **2** Directed to or pertaining to an object or end: an *objective* goal. **3** Having independent existence apart from experience or thought; substantive; self-existent. **4** Directing the mind or activity toward external things without reference to personal sensations; also, resulting from such direction; hence, representing things as they are; unbiased by thoughts, emotions, opinions, etc.: said of an artist, a writer, etc., or of his habits of thought. **5** Made up of objects represented precisely as they are, without idealization; realistic: said of a work of art, as a picture. **6** *Gram.* Denoting the case of the object of a transitive verb or of a preposition; accusative. — *n.* **1** *Gram.* **a** The objective or accusative case. **b** A word in this case. **2** *Optics* An object glass. **3** A result to be achieved or a point to be reached in any military action; the assigned goal of a mission. [<Med. L *objectivus* < *objectum.* See OBJECT[2].] — **ob·jec′tive·ly** *adv.*

ob·jec·tiv·ism (əb·jek′tə·viz′əm) *n.* **1** The power that enables an author or artist to treat subjects objectively, or apart from his own personality. **2** The tendency to give prominence to the facts of sense perception; the theory that human knowledge is based on the external world rather than within the ego. — **ob·jec′tiv·ist** *n.* — **ob·jec′tiv·is′tic** *adj.*

ob·jec·tiv·i·ty (ob′jek·tiv′ə·tē) *n.* 1 The state or relation of being objective. 2 Material reality. Also **ob·jec′tive·ness.**

ob·ject·less (ob′jikt·lis, -jekt-) *adj.* 1 Without aim; purposeless. 2 Having no corresponding object or concrete representation.

object lesson 1 A lesson in which the object to be known, or a representation of it, is shown to the eye. 2 The exemplification of a principle or moral in a concrete form or striking instance.

ob·jet d'art (ôb·zhe′ dȧr′) *French pl.* **ob·jets d'art** (ôb·zhe′) Any work of artistic value.

ob·jur·gate (ob′jər·gāt, əb·jûr′-) *v.t.* **·gat·ed, ·gat·ing** To rebuke severely; scold sharply; berate. [<L *objurgatus,* pp. of *objurgare* rebuke < *ob-* against + *jurgare* scold] **—ob′·jur·ga′tion** *n.* **—ob′jur·ga′tor** *n.* **—ob·jur′ga·to′ri·ly** *adv.* **—ob·jur′ga·to′ry** *adj.*

ob·lan·ce·o·late (ob·lan′sē·ə·lit, -lāt′) *adj. Bot.* Lance-shaped, but tapering toward the base, as the leaves of certain plants. [<OB- inversely + LANCEOLATE]

ob·late[1] (ob′lāt, ob·lāt′) *adj.* Flattened at the poles: opposed to *prolate.* [<NL *oblatus* < *ob-* against, inversely + L *(pro)latus* lengthened out] **—ob′late·ly** *adv.*

ob·late[2] (ob′lāt, ob·lāt′) *adj.* Consecrated; dedicated; devoted to a religious life. **—***n.* A person so devoted, as in a monastery or to certain religious work. [<Med. L *oblatus* <L, pp. to *offerre* OFFER]

ob·la·tion (ob·lā′shən) *n.* 1 The act of offering or anything offered in worship, especially the elements of the Eucharist. 2 Hence, any grateful and solemn offering. 3 In canon law, any property given to a church. [<OF, an offering, sacrifice <Med. L *oblatio, -onis* <L < *oblatus.* See OBLATE[2].] **—ob·la′tion·al** *adj.* **—ob·la·to·ry** (ob′lə·tôr′ē, -tō′rē) *adj.*

ob·li·gate (ob′lə·gāt) *v.t.* **·gat·ed, ·gat·ing** To bind or compel, as by contract, conscience, promise, etc. **—***adj.* (ob′lə·git, -gāt) 1 Bound or restricted. 2 *Biol.* Having only one life condition: distinguished from *facultative.* [<L *obligatus,* pp. of *obligare* OBLIGE]

ob·li·ga·tion (ob′lə·gā′shən) *n.* 1 The act of obligating or state of being obligated; also, the duty, promise, etc., by which one is bound. 2 The constraining power of conscience or law. 3 A requirement imposed by the customs of society or the laws of propriety and expediency; what one owes in return for a service, benefit, kindness, favor, etc. 4 A binding legal agreement, contract, bond, etc., bearing a penalty. 5 The condition of being indebted for an act of kindness, a service received, etc.; also, the kindness or service. See synonyms under DUTY. **—ob′li·ga′tor** *n.*

ob·li·ga·tive (ob′lə·gā′tiv) *adj.* Implying or expressing obligation: distinguished from *facultative.*

ob·lig·a·to·ry (ə·blig′ə·tôr′ē, -tō′rē, ob′lə·gə-) *adj.* 1 In civil or moral law, binding. 2 Of the nature of, or constituting a duty or obligation; imperative.

o·blige (ə·blīj′) *v.t.* **o·bliged, o·blig·ing** 1 To obligate; constrain. 2 To place under an obligation, as for a favor or kindness. 3 To do a favor or service for. See synonyms under ACCOMMODATE, BIND, COMPEL. [<OF *obliger, obligier* bind by oath or promise <L *obligare,* orig. tie around < *ob-* towards + *ligare* bind] **—o·blig′er** *n.*

ob·li·gee (ob′lə·jē′) *n.* One who is obliged; specifically, *Law,* the person in whose favor an obligation is entered into or incurred: opposed to *obligor.*

o·blig·ing (ə·blī′jing) *adj.* Disposed to do favors; accommodating; kind. See synonyms under GOOD, PLEASANT, POLITE.**—o·blig′ing·ly** *adv.* **—o·blig′ing·ness** *n.*

ob·li·gor (ob′lə·gôr′, ob′lə·gôr) *n. Law* The person who is bound to perform an obligation.

ob·lique (ə·blēk′, *in military usage* ə·blīk′) *adj.* 1 Deviating from the perpendicular or from a right line by any angle except a right angle; neither perpendicular nor horizontal; slanting. 2 Differing from a right angle; either acute or obtuse: said of angles. 3 Evasive; indirect; not straightforward; disingenuous. 4 Not in the direct line of descent; collateral. 5 *Bot.* Unequal-sided, as a leaf. 6 *Gram.* Having to do with cases other than the nominative and vocative. 7 *Anat.* Designating several muscles whose fibers run obliquely: the external *oblique* muscle of the abdomen. **—***n.* 1 One of the oblique muscles. 2 An oblique line. 3 A veering to the right or left less than ninety degrees, as in sailing. **—***v.i.* **·liqued, ·li·quing** 1 To deviate from the perpendicular; slant. 2 *Mil.* To march or advance in an oblique direction. [<L *obliquus* < *ob-* against, completely + *liquis* slanting, awry] **—ob·lique′ly** *adv.* **—ob·lique′ness** *n.*

ob·lique-an·gled (ə·blēk′ang′gəld) *adj.* Having the angles oblique: an *oblique-angled* triangle.

oblique case *Gram.* Any case other than the nominative or vocative.

oblique coordinates See CARTESIAN COORDINATE SYSTEM.

oblique rime Near rime.

oblique sailing Navigation along a course lying at an oblique angle to the meridian.

ob·liq·ui·ty (ə·blik′wə·tē) *n. pl.* **·ties** 1 Oblique quality or state. 2 Inclination from a vertical or horizontal line or plane; also, the amount or the angle of such inclination. 3 Deviation from right or moral principles or conduct. **—ob·liq′ui·tous** *adj.*

obliquity of the ecliptic *Astron.* The angle between the plane of the earth's equator and the plane of the ecliptic: equal to a mean of 23° 27′.

ob·lit·er·ate (ə·blit′ə·rāt) *v.t.* **·at·ed, ·at·ing** 1 To destroy utterly; leave no trace of. 2 To blot or wipe out; erase, as writing. See synonyms under ABOLISH, ANNUL, CANCEL. [<L *obliteratus,* pp. of *obliterare* blot out < *ob-* against, upon + *litera* a letter] **—ob·lit′er·a′·tion** *n.* **—ob·lit′er·a′tive** *adj.* **—ob·lit′er·a′tor** *n.*

ob·liv·i·on (ə·bliv′ē·ən) *n.* 1 The state or fact of being utterly forgotten. 2 The act or fact of forgetting completely; forgetfulness. 3 Public remission and pardon of offense; amnesty. [<OF <L *oblivio, -onis* < *oblivisci* forget]

ob·liv·i·ous (ə·bliv′ē·əs) *adj.* 1 Forgetful, or given to forgetfulness. 2 Lost in thought; abstracted. 3 Inducing forgetfulness. See synonyms under ABSTRACTED. **—ob·liv′i·ous·ly** *adv.* **—ob·liv′i·ous·ness** *n.*

ob·long (ob′lông, -long) *adj.* 1 Longer in one dimension than in another: applied most commonly to rectangular objects somewhat elongated. 2 Having one principal axis longer than the other or others. 3 *Bot.* Bluntly elliptical, as a leaf. **—***n.* A figure having greater length than breadth; a long rectangle. [<L *oblongus* somewhat long < *ob-* towards + *longus* long]

ob·lo·quy (ob′lə·kwē) *n. pl.* **·quies** 1 The state of one who is under odium or disgrace or spoken ill of. 2 Vilification; defamation; calumny. 3 *Obs.* A cause of disgrace or reproach. See synonyms under SCANDAL. [<LL *obloquium* a contradiction < *obloqui* < *ob-* against + *loqui* speak]

ob·nox·ious (əb·nok′shəs) *adj.* 1 Of a character to give offense or excite aversion; odious; objectionable. 2 *Law* Liable or answerable; amenable. 3 *Obs.* Subject; exposed: *obnoxious* to punishment. See synonyms under SUBJECT. [<L *obnoxiosus* < *obnoxius* exposed to harm, liable < *ob-* towards + *noxa* an injury] **—ob·nox′ious·ly** *adv.* **—ob·nox′ious·ness** *n.*

ob·nu·bi·la·tion (ōb·nōō′bə·lā′shən, -nyōō′-) *n. Psychiatry* A nebulous or clouded state of consciousness, as just before syncope or epileptic seizures. [<obs. *obnubilate* overcloud <L *obnubilatus,* pp. of *obnubilare* < *ob-* over + *nubilare* make cloudy < *nubila* clouds]

o·boe (ō′bō, ō′boi) *n.* A wooden double-reed wind instrument with a high, penetrating, melancholy tone. [<Ital. <F *hautbois* HAUTBOY] **—o′bo·ist** *n.*

ob·o·lus (ob′ə·ləs) *n. pl.* **·li** (-lī) 1 A weight and a silver coin of ancient Greece; one sixth of a drachma. 2 A medieval silver coin of Hungary and Bohemia. Also **ob·ol** (ob′əl). See DIOBOL. [<L <Gk. *obolos*]

ob·o·vate (ob·ō′vāt) *adj. Bot.* Inversely ovate, as certain leaves. [<OB- inversely + OVATE]

ob·o·void (ob·ō′void) *adj. Bot.* Solidly obovate, with the broader end upward or outward. [<OB- inversely + OVOID]

O·bre·gón (ō′brā·gōn′), **Alvaro,** 1880–1928, president of Mexico 1920–24, 1928; assassinated.

O'Bri·en (ō·brī′ən), **Edward Joseph,** 1890–1941, U.S. editor.

ob·scene (əb·sēn′, ob-) *adj.* 1 Offensive to chastity or decency. 2 Offensive to the senses; foul. See synonyms under FOUL, IMMODEST, VULGAR. [<F *obscène* <L *obscenus, obscaenus* ill-omened, filthy < *obs-,* var. of *ob-* towards + *caenum* filth] **—ob·scene′ly** *adv.*

ob·scen·i·ty (əb·sen′ə·tē, -sē′nə-, ob-) *n. pl.* **·ties** Obscene quality of act, thought, speech, or representation; gross indecency; unchaste action; lewdness. Also **ob·scene′ness.** See synonyms under INDECENCY.

ob·scur·ant (əb·skyŏŏr′ənt) *n.* One who obscures; specifically, one who opposes education, popular enlightenment, and freedom of thought. Also **ob·scur′ant·ist.** [<G *obskurant* <L *obscurans, -antis,* ppr. of *obscurare* darken] **—ob·scur′ant·ism** *n.* **—ob·scu·ra·tion** (ob′skyə·rā′shən) *n.*

ob·scure (əb·skyŏŏr′) *adj.* **·scur·er, ·scur·est** 1 Dim; dark; dusky; gloomy. 2 Not clear to the mind; vague; abstruse. 3 Faintly marked; hard to discern; undefined. 4 Remote or apart; hidden from view or notice; hence, little known; lowly: *obscure* birth. **—***v.t.* **·scured, ·scur·ing** 1 To darken or cloud; dim. 2 To hide from view; conceal. 3 To make unintelligible; confuse: to *obscure* an issue. 4 To make indefinite in sound, as a vowel. **—***n.* Indistinctness of outline or color. [<OF *obscur, oscur* <L *obscurus,* lit., covered over] **—ob·scure′ly** *adv.* **—ob·scure′ness** *n.*

Synonyms (adj.): abstruse, ambiguous, complicated, dark, difficult, dim, indistinct, involved, profound, unintelligible. That is *obscure* which the eye or mind cannot clearly see or understand. If the matter is *abstruse,* as if removed from the usual way of thinking, it is *difficult* to comprehend. The matter may be *complicated* by the intertwining of its many parts, or it may be so deep as to be *profound.* The expression of the thought may be *ambiguous,* as if looking in two ways, or *involved* and confused in form, or it may be *unintelligible* to the mind. Sometimes it is *dark, dim,* and *indistinct* by reason of lack of light or want of transparency.

ob·scu·ri·ty (əb·skyŏŏr′ə·tē) *n. pl.* **·ties** 1 The state or quality of being obscure. 2 Dimness; darkness. 3 Lack of distinctness or perspicuity. 4 The condition of being unknown to fame. 5 An unknown or obscure person, place, or thing.

ob·se·crate (ob′sə·krāt) *v.t.* **·crat·ed, ·crat·ing** *Rare* To supplicate; beseech. [<L *obsecratus,* pp. of *obsecrare* beseech < *ob-* on account of + *sacrare* make sacred < *sacer* sacred] **—ob′·se·cra′tion** *n.* **—ob′se·cra′to·ry** (-krə′tər·ē) *adj.*

ob·se·qui·ous (əb·sē′kwē·əs) *adj.* 1 Sycophantic or adulatory in manner; cringing; servile. 2 *Rare* Promptly obedient. [<L *obsequiosus* compliant < *obsequium* compliance < *obsequi* comply with < *ob-* towards + *sequi* follow] **—ob·se′qui·ous·ly** *adv.* **—ob·se′qui·ous·ness** *n.*

Synonyms: attentive, compliant, cringing, deferential, fawning, flattering, obedient, servile, slavish, submissive, sycophantic. See BASE[2], SUPPLE. *Antonyms:* independent, self-assertive, self-respecting. See synonyms for AUSTERE.

ob·se·quy (ob′sə·kwē) *n. pl.* **·quies** *Usually pl.* The last office for the dead; a funeral service. [<AF *obsequie* <Med. L *obsequiae,* pl., funeral, rites, fusion of LL *exequiae* funeral rites and L *obsequium* dutiful service. See OBSEQUIOUS.]

ob·serv·a·ble (əb·zûr′və·bəl) *adj.* 1 That can be observed; manifest. 2 Notable. 3 Customary; demanding observance. **—ob·serv′a·ble·ness** *n.* **—ob·serv′a·bly** *adv.*

ob·ser·vance (əb·zûr′vəns) *n.* 1 The act of observing, as a custom or ceremony; compliance, as with law or duty. 2 Any common custom, form, rite, etc. 3 Heedful attention; observation. 4 *Eccl.* a The rule or constitution of a religious order. b The order, or the house of such an order. 5 *Archaic* Obsequious compliance. See synonyms under FORM, SACRAMENT. [<OF <L *observantia* attention, reverence, < *observans, -antis,* ppr. of *observare* OBSERVE]

ob·ser·vant (əb·zûr′vənt) *adj.* 1 Carefully attentive; habitually noting. 2 Strict in observing rules; heedful of duties. 3 Obedient; attentive. [<F, orig. ppr. of *observer* OBSERVE] **—ob·ser′vant·ly** *adv.*

Ob·ser·van·tine (ob·zûr′vən·tin, -tēn) *n.* A member of the branch of the Franciscan order which observes the original rule strictly, especially with regard to the vow of poverty. Also **Ob·ser′vant.** [<F *Observantin* < *observant,* ppr. of *observer* OBSERVE]

ob·ser·va·tion (ob′zər·vā′shən) *n.* **1** The act, faculty, or habit of observing; the fact of being observed. **2** Scientific scrutiny of a natural phenomenon, for experiment, verification, or measurement and calculation; also, the record of such an examination and the data connected with it: an astronomical or meteorological *observation*: in this sense distinguished from *experimentation.* **3** Experience or knowledge acquired by observing. **4** An incidental remark. **5** *Obs.* Observance. See synonyms under REMARK. **— to take** (or **work out) an observation** *Naut.* To calculate the latitude and longitude from angular measurements of the altitude and position of the sun or other celestial body. **— ob′ser·va′tion·al** *adj.*

observation car A railway car with a rear section either open or glass-enclosed, used for passenger sight-seeing or for track inspection of the right-of-way.

observation post Any point, open or concealed, in which an observer may gather information of a specified nature; especially, in wartime, a station for directing gunfire, watching enemy action, etc.

ob·ser·va·to·ry (əb·zûr′və·tôr′ē, -tō′rē) *n. pl.* **·ries 1** A building designed and equipped for the systematic observation of natural phenomena: an astronomical *observatory.* **2** A tower built for obtaining a panoramic view. [<NL *observatorium* <L *observatus,* pp. of *observare* OBSERVE]

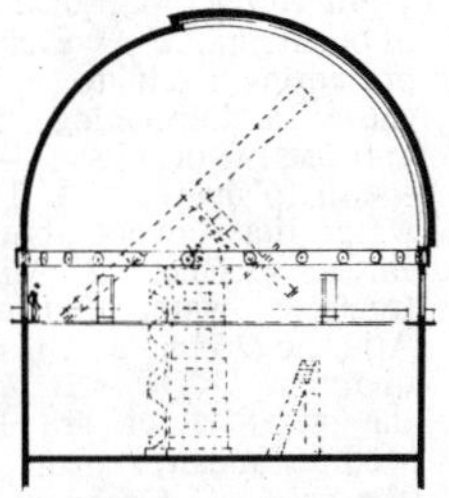

OBSERVATORY
Schematic plan.

ob·serve (əb·zûrv′) *v.* **·served, ·serv·ing** *v.t.* **1** To notice by the sense of sight; see. **2** To watch attentively; keep under surveillance: to *observe* enemy troop movements. **3** To make methodical observation of, as for scientific purposes: to *observe* sunspots. **4** To abide by the restrictions or provisions of: to *observe* a fast. **5** To celebrate or solemnize (an occasion), as with appropriate festivities or ceremony. **6** To say as a comment or opinion; mention. **—** *v.i.* **7** To make a remark; comment: often with *on* or *upon.* **8** To take notice. **9** To act as an observer. See synonyms under CELEBRATE, EXAMINE, FOLLOW, OBEY. [<OF *observer* <L *observare* watch < *ob-* towards + *servare* keep, watch] **— ob·serv′ing** *adj.* **— ob·serv′ing·ly** *adv.*

ob·serv·er (əb·zûr′vər) *n.* **1** One who observes; specifically, in modern warfare, one who observes the effect of artillery fire on the enemy, one who keeps a look-out for enemy aircraft, or one who makes meteorological observations. **2** A member of the crew of a military aircraft that makes reconnaissance flights.

ob·sess (əb·ses′) *v.t.* To occupy or trouble the mind of to an excessive degree; preoccupy; harass; haunt. [<L *obsessus,* pp. of *obsidere* besiege < *ob-* towards, against + *sedere* sit] **— ob·ses′sive** *adj.* **— ob·ses′sor** *n.*

ob·ses·sion (əb·sesh′ən) *n.* **1** A vexing or haunting, as by an evil spirit or morbidly dominant idea; the fact of being thus haunted; also, that which dominates or afflicts anyone in such manner. **2** *Psychiatry* A compulsive idea or emotion associated with the subconscious and exerting a more or less persistent influence upon conduct and behavior; also, the compulsion itself.

ob·sid·i·an (əb·sid′ē·ən, ob-) *n.* A glassy, volcanic rock, usually black and having the composition of rhyolite but containing few or no individualized crystals. [<L *obsidianus,* alter. of *obsianus,* after *Obsius,* a Roman said by Pliny to be its discoverer]

ob·sid·i·o·nal (əb·sid′ē·ə·nəl) *adj.* **1** Of or pertaining to a siege. **2** Pertaining to or contracted during trench warfare: *obsidional* infection. [<L *obsidionalis* < *obsidio, -onis* < *obsidere* besiege. See OBSESS.]

ob·so·les·cent (ob′sə·les′ənt) *adj.* Growing obsolete. [<L *obsolescens, -entis,* ppr. of *obsolescere* grow old < *ob-* against + *solere* be accustomed, use] **— ob′so·les′cence** *n.* **— ob′so·les′cent·ly** *adv.*

ob·so·lete (ob′sə·lēt) *adj.* **1** Gone out of use, as a word or phrase, a style, fashion, etc.; of a discarded type or fashion. **2** *Biol.* Imperfectly developed; atrophied; suppressed or lacking: said of markings, parts, organs, etc. [<L *obsoletus* grown old, worn out, pp. of *obsolescere.* See OBSOLESCENT.] **— ob′so·lete′ly** *adv.* **— ob′so·lete′ness** *n.* **— ob′so·let′ism** *n.*

Synonyms: ancient, antiquated, archaic, disused, obsolescent, old, out-of-date, rare. Some of the most *ancient* words are not *obsolete.* A word is *obsolete* which has quite gone out of use; a word is *archaic* or *obsolescent* which is falling out of use; *archaic* is also applied to a word which is *obsolete* in general usage but which survives in special texts, as the Bible, hymnals, poetry, etc.; a word is *rare* if there are few present instances of its use. See OLD. *Antonyms*: see synonyms for FRESH, MODERN, NEW.

ob·sta·cle (ob′stə·kəl) *n.* That which stands in the way; a hindrance or obstruction in either a physical or a moral sense. See synonyms under BARRIER, IMPEDIMENT. [<OF <L *obstaculum* < *obstare* stand before, withstand < *ob-* before, against + *stare* stand]

ob·ste·tri·cian (ob′stə·trish′ən) *n.* A medical and surgical specialist in childbirth.

ob·stet·rics (əb·stet′riks) *n.* The branch of medical science relating to pregnancy and childbirth; midwifery. [Orig. pl. of *obstetric, adj.,* <L *obstetricus* < *obstetrix, -icis* a midwife < *obstare.* See OBSTACLE.] **— ob·stet′ri·cal, ob·stet′ric** *adj.* **— ob·stet′ri·cal·ly** *adv.*

ob·sti·na·cy (ob′stə·nə·sē) *n. pl.* **·cies 1** Persistent and usually unreasonable adherence to one's own opinion or purpose; stubbornness. **2** The quality of being difficult to subdue or remedy: said especially of ailments. **3** Stubborn action.

ob·sti·nate (ob′stə·nit) *adj.* **1** Persistently and unreasonably resolved in a purpose or opinion; stubborn. **2** Hard to control or cure. [<L *obstinatus* stubborn < *obstinare* persist < *obstare.* See OBSTACLE.] **— ob′sti·nate·ly** *adv.* **— ob′sti·nate·ness** *n.*

Synonyms: contumacious, decided, determined, dogged, firm, fixed, headstrong, heady, immovable, indomitable, inflexible, intractable, mulish, obdurate, opinionated, persistent, pertinacious, refractory, resolute, resolved, stubborn, unflinching, unyielding. The *headstrong* person is not to be stopped in his own course of action, while the *obstinate* and *stubborn* is not to be driven to another's way. The *headstrong* act; the *obstinate* and *stubborn* may simply refuse to stir. *Stubborn* is the term most frequently applied to animals and inanimate things. *Refractory* implies more activity of resistance; the *stubborn* horse balks; the *refractory* animal plunges, rears, and kicks; metals that resist ordinary processes of reduction are termed *refractory.* One is *obdurate* who adheres to his purpose in spite of appeals that would move any tender-hearted person. *Contumacious* refers to a proud and insolent defiance of authority, as of the summons of a court. *Pertinacious* applies to that which is active and aggressive; *pertinacious* demand is contrasted with *obstinate* refusal. The *unyielding* conduct which we approve we call *decided, firm, inflexible, resolute*; that which we condemn we are apt to term *headstrong, obstinate, stubborn.* See INFLEXIBLE, PERVERSE. *Antonyms*: amenable, complaisant, compliant, docile, irresolute, obedient, pliable, pliant, submissive, teachable, tractable, undecided, wavering, yielding.

ob·sti·pant (ob′stə·pənt) *n. Med.* A drug or other substance that induces obstipation.

ob·sti·pa·tion (ob′stə·pā′shən) *n. Med.* Persistent or intractable constipation. [<L *obstipatio, -onis* a stopping up, ult. < *ob-* against + *stipare* press together]

ob·strep·er·ous (əb·strep′ər·əs) *adj.* Making a great disturbance; clamorous; boisterous; unruly. See synonyms under NOISY, TURBULENT. [<L *obstreperus* < *obstrepere* make noise against < *ob-* against + *strepere* roar] **— ob·strep′er·ous·ly** *adv.* **— ob·strep′er·ous·ness** *n.*

ob·struct (əb·strukt′) *v.t.* **1** To stop or impede movement through (a way or passage) by obstacles or impediments; barricade; choke; clog. **2** To block or retard the progress or way of; impede; check. **3** To come or be in front of so as to hide from sight. [<L *obstructus,* pp. of *obstruere* block up < *ob-* against + *struere* pile, build] **— ob·struct′er, ob·struc′tor** *n.* **— ob·struc′tive** *adj.* **— ob·struc′tive·ly** *adv.* **— ob·struc′tive·ness** *n.*

Synonyms: arrest, bar, barricade, check, choke, clog, embarrass, hinder, impede, interrupt, oppose, retard, stay, stop. To *obstruct* is literally to build up against: The road is *obstructed* by fallen trees. We may *hinder* one's advance by following and clinging to him; we *obstruct* his course by standing in his way. Anything that makes one's progress slower, whether from within or from without, *impedes, checks, hinders, retards,* or *stays*; an obstruction to one's progress is always from without. To *arrest* is to cause to stop suddenly; *obstructing* the way may have the effect of *arresting* progress. See CHECK, HINDER, OPPOSE. *Antonyms*: accelerate, advance, aid, clear, facilitate, forward, free, further, open, promote.

ob·struc·tion (əb·struk′shən) *n.* **1** A hindrance; obstacle. **2** The act of preventing progress; the state of being obstructed. See synonyms under BARRIER, IMPEDIMENT.

ob·struc·tion·ist (əb·struk′shən·ist) *n.* One who obstructs; especially, one who opposes reform, or who delays the progress of business, as in a legislature. **— ob·struc′tion·ism** *n.*

ob·stru·ent (ob′strōō·ənt) *adj. Med.* Causing obstruction or impediment, as of the stomach. **—** *n.* A medicine that obstructs. [<L *obstruens, -entis,* ppr. of *obstruere.* See OBSTRUCT.]

ob·tain (əb·tān′) *v.t.* **1** To gain possession of, especially by effort; acquire; get. **2** *Archaic* To arrive at; reach. **—** *v.i.* **3** To be customary or prevalent; hold good: Chivalry *obtained* until the Renaissance. **4** *Archaic* To succeed; prevail. [<OF *obtenir* <L *obtinere* < *ob-* against + *tenere* hold, keep] **— ob·tain′a·ble** *adj.* **— ob·tain′er** *n.* **— ob·tain′ment** *n.*

Synonyms: acquire, earn, gain, get, procure, receive, secure, win. When one *gets* the object of his desire, he is said to *obtain* it, whether he has *gained* or *earned* it or not. *Win* denotes contest, with a suggestion of chance or hazard; in popular language, a person is often said to *win* a lawsuit, but in legal phrase he is said to *gain* his suit, case, or cause. One *obtains* a thing commonly by some direct effort of his own, he *procures* it commonly by the intervention of someone else; he *secures* what has seemed uncertain or elusive, when he *gets* it firmly into his possession or under his control. Compare GAIN, GET, PURCHASE.

ob·tect (ob·tekt′) *adj. Entomol.* Covered with a hard chitinous case, as the pupa of most flies. Also **ob·tect′ed.** [<L *obtectus,* pp. of *obtegere* < *ob-* over + *tegere* cover]

ob·test (ob·test′) *v.t.* **1** To beseech; implore. **2** To invoke as a witness. **—** *v.i.* **3** To protest. [<L *obtestari* call to witness < *ob-* on account of + *testari* bear witness] **— ob·tes·ta·tion** (ob′tes·tā′shən) *n.*

ob·trude (əb·trōōd′) *v.* **ob·trud·ed, ob·trud·ing** *v.t.* **1** To thrust or force (oneself, an opinion, etc.) upon others without request or warrant. **2** To push forward or out; eject. **—** *v.i.* **3** To intrude oneself. [<L *obtrudere* < *ob-* towards, against + *trudere* thrust] **— ob·trud′er** *n.*

ob·tru·sion (əb·trōō′zhən) *n.* The act of obtruding or the thing obtruded; an instance of obtruding. [<LL *obtrusio, -onis* < *obtrusus,* pp. of *obtrudere* OBTRUDE]

ob·tru·sive (əb·trōō′siv) *adj.* Tending to obtrude; obtruding; pushing; intruding. See synonyms under MEDDLESOME. **— ob·tru′sive·ly** *adv.* **— ob·tru′sive·ness** *n.*

ob·tund (ob·tund′) *v.t.* To make blunt or dull; deaden, as pain. [<L *obtundere* blunt < *ob-* against + *tundere* beat] **— ob·tund′ent** *adj. & n.*

ob·tu·rate (ob′tyə·rāt, -tə-) *v.t.* **·rat·ed, ·rat·ing 1** To close or stop up (an opening or hole). **2** In ordnance, to close or seal (a gun breech)

add,āce,câre,pälm; end,ēven; it,īce; odd,ōpen,ôrder; tŏŏk,pōōl; up,bûrn; ə = a in *above,* e in *sicken,* i in *clarity,* o in *melon,* u in *focus*; yōō = u in *fuse*; oi,oil; ou,pout; ch,check; g,go; ng,ring; th,thin; th,this; zh,vision. Foreign sounds à,œ,ü,kh,ṅ; and ◆: see page xx. < from; + plus; ? possibly.

to prevent the escape of gas in firing. [<L *obturatus*, pp. of *obturare* stop up] —**ob′tu·ra′tion** *n.* —**ob′tu·ra′tor** *n.*

ob·tuse (əb·to͞os′, -tyo͞os′) *adj.* 1 *Bot.* Blunt or rounded at the extremity, as a leaf or petal: opposed to *acute.* 2 Dull intellectually or emotionally; stupid; insensible. 3 Heavy, dull, and indistinct, as a sound. See synonyms under BLUNT. [<L *obtusus* blunt, dulled, orig. pp. of *obtundere.* See OBTUND.] —**ob·tuse′ly** *adv.* —**ob·tuse′ness** *n.*

obtuse angle See under ANGLE.

ob·verse (ob·vûrs′, ob′vûrs) *adj.* 1 Turned toward or facing one: opposed to *reverse.* 2 Inverse; narrower at the base than at the apex: an *obverse* leaf. 3 Corresponding to something else as its counterpart. —*n.* (ob′·vûrs) 1 That side of a coin or medal upon which the face or main device is struck: opposed to *reverse*; that side of any object which is meant to be seen; the front as opposed to the back. 2 *Logic* The counterpart of any truth, fact, or statement. [<L *obversus*, pp. of *obvertere* OBVERT] —**ob·verse′ly** *adv.*

ob·ver·sion (ob·vûr′shən, -zhən) *n.* 1 A turning down or toward. 2 *Logic* A form of immediate inference in which the positive and negative or antecedent and consequent terms of a proposition are reversed in such a way that the converse or transverse forms can be legitimately inferred from the original proposition; conversion.

ob·vert (ob·vûrt′) *v.t.* 1 To turn the front, principal, or a different side of (a thing) toward another. 2 *Logic* To infer the obverse, or contradictory predicate of (a proposition). [<L *obvertere* <*ob-* towards, against, down + *vertere* turn]

ob·vi·ate (ob′vē·āt) *v.t.* **·at·ed, ·at·ing** To meet or provide for, as an objection or difficulty, by effective measures; clear away; prevent. [<L *obviatus*, pp. of *obviare* meet, withstand <*ob-* against + *via* a way] —**ob′vi·a′tion** *n.* —**ob′vi·a′tor** *n.*

ob·vi·ous (ob′vē·əs) *adj.* 1 Immediately evident without further reasoning or investigation; palpably true; manifest. 2 *Obs.* Standing or placed in the way. See synonyms under CLEAR, EVIDENT, MANIFEST, NOTORIOUS. [<L *obvius* in the way, obvious <*ob-* against + *via* a way] —**ob′vi·ous·ly** *adv.* —**ob′vi·ous·ness** *n.*

ob·vo·lute (ob′və·lo͞ot) *adj. Bot.* Overlapping: said of the margins of leaves or petals in vernation which are mutually infolded one within another. Compare CONVOLUTE. [<L *obvolutus*, pp. of *obvolvere* wrap around <*ob-* upon + *volvere* roll] —**ob′vo·lu′tion** *n.* —**ob′vo·lu′·tive** *adj.*

oc– Assimilated var. of OB–.

oc·a·ri·na (ok′ə·rē′nə) *n.* A small musical instrument in the shape of a sweet potato, usually of terra cotta, with a mouthpiece and finger holes. It yields soft, sonorous notes. [<Ital., dim. of *oca* a goose <L *auca*; so called with ref. to its shape]

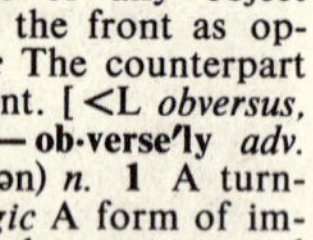

OCARINA

O'Ca·sey (ō·kā′sē), **Sean,** born 1880, Irish playwright.

Oc·cam (ok′əm) See OCKHAM.

oc·ca·sion (ə·kā′zhən) *n.* 1 A particular event, or juncture of events, considered simply as exciting notice or interest; especially, an important event or celebration. 2 An event or juncture of affairs that presents some reason, motive, or opportunity for action; an opportunity permitting or a reason requiring action; cause: no *occasion* for haste. 3 A determinative condition, as opposed to the main or principal cause. 4 A need or exigency. 5 *pl. Obs.* Needs. 6 *Obs.* Any matter of business requiring attention. See synonyms under CAUSE, OPPORTUNITY. —**by occasion of** In consequence of; by reason of. —**on occasion** On suitable opportunity; at need; now and then. —**to take occasion** To avail oneself of the opportunity. —*v.t.* To cause or bring about; cause accidentally or incidentally. See synonyms under MAKE, PRODUCE. [<L *occasio, -onis* a falling towards, an opportunity <*occidere* fall down <*ob-* towards, down + *cadere* fall]

oc·ca·sion·al (ə·kā′zhən·əl) *adj.* 1 Occurring at irregular intervals. 2 Belonging or suitable to some special occasion. 3 Happening casually or incidentally. See synonyms under INCIDENTAL.

oc·ca·sion·al·ism (ə·kā′zhən·əl·iz′əm) *n. Philos.* The doctrine that mind is not responsible for changes in the body, or vice versa, but that the Divine Spirit intervenes to produce the apparent interaction. —**oc·ca′sion·al·ist** *n.*

oc·ca·sion·al·ly (ə·kā′zhən·əl·ē) *adv.* In an occasional manner; more or less frequently at irregular times or intervals.

oc·ci·dent (ok′sə·dənt) *n.* The west: opposed to *orient.* [<OF <L *occidens, -entis* sunset, the west, orig. ppr. of *occidere* fall. See OCCASION.]

Oc·ci·dent (ok′sə·dənt) 1 The countries west of Asia; specifically, Europe. 2 The western hemisphere. Opposed to *Orient.*

oc·ci·den·tal (ok′sə·den′təl) *adj.* Of or belonging to the West, or the countries constituting the Occident: opposed to *oriental.* —*n.* One born or living in a western country. Also **Oc′ci·den′tal.**

Oc·ci·den·tal·ism (ok′sə·den′təl·iz′əm) *n.* The spirit, life, and culture of the people of the Occident. —**Oc′ci·den′tal·ist** *n.*

oc·ci·den·tal·ize (ok′sə·den′təl·īz) *v.t.* **·ized, ·iz·ing** To render occidental in spirit, culture, character, etc. —**oc′ci·den′tal·i·za′tion** *n.*

oc·cip·i·tal (ok·sip′ə·təl) *adj.* 1 Pertaining to the occiput. 2 Pertaining to the occipital bone. —*n.* The occipital bone. [<Med. L *occipitalis* <L *occiput, -itis* OCCIPUT]

occipital bone *Anat.* The hindmost bone of the skull between the parietal and temporal bones.

occipito– *combining form Anat.* Occipital; occipital and: *occipitofrontal,* pertaining to the occiput and the forehead. [<L *occiput* back of the head]

oc·ci·put (ok′sə·put, -pət) *n. pl.* **·cip·i·ta** (-sip′·ə·tə) *Anat.* The lower back part of the skull. [<L, back of the head <*ob-* against + *caput* head]

oc·clude (ə·klo͞od′) *v.* **·clud·ed, ·clud·ing** *v.t.* 1 To shut up or close, as pores or openings. 2 To shut in, out, or off. 3 *Chem.* To take up, either on the surface or internally, but without change of properties: Palladium *occludes* hydrogen. —*v.i.* 4 *Dent.* To meet so that the corresponding cusps fit closely together: said of the teeth of the upper and lower jaws. [<L *occludere* shut <*ob-* against, upon + *claudere* close] —**oc·clu′dent** *adj.* —**oc·clu·sion** (ə·klo͞o′zhən) *n.* —**oc·clu′sive** (-siv) *adj.*

oc·clud·er (ə·klo͞o′dər) *n. Optics* A device to shut off vision.

oc·cult (ə·kult′, ok′ult) *adj.* 1 Of, pertaining to, or designating those mystic arts involving magic, divination, astrology, alchemy, or the like. 2 Not divulged; secret. 3 Beyond human understanding; mysterious. See synonyms under MYSTERIOUS, SECRET. —*n.* Occult arts or sciences. —*v.t.* 1 To hide or conceal from view. 2 *Astron.* To hide or conceal by occultation. —*v.i.* 3 To become hidden or concealed from view. [<L *occultus*, pp. of *occulere* cover over, hide] —**oc·cult′ly** *adv.* —**oc·cult′ness** *n.*

oc·cul·ta·tion (ok′ul·tā′shən) *n.* 1 The act of occulting, or the state of being occulted. 2 *Astron.* Concealment of one celestial body by another interposed in the line of vision, as of a star or planet by the moon, or of a satellite by a planet. Compare ECLIPSE. 3 A disappearance from view or notice.

oc·cult·er (ə·kul′tər) *n.* A device used in a telescope to screen a light.

oc·cult·ism (ə·kul′tiz·əm) *n.* 1 The theory or practice of occult arts or sciences. 2 Belief in occult or supernatural powers. —**oc·cult′ist** *n.*

oc·cu·pan·cy (ok′yə·pən·sē) *n.* The act of occupying; a taking possession; also, the time during which anything is occupied. See synonyms under OCCUPATION.

oc·cu·pant (ok′yə·pənt) *n.* 1 One who occupies. 2 A tenant. [<L *occupans, -antis*, ppr. of *occupare.* See OCCUPY.]

oc·cu·pa·tion (ok′yə·pā′shən) *n.* 1 One's regular, principal, or immediate business. 2 The state of being busy. 3 The possession and holding of land by military force; the occupancy and holding of a nation by an army of another. 4 Occupancy. [<OF <L *occupatio, -onis* a seizing <*occupatus*, pp. of *occupare.* See OCCUPY.]

Synonyms: occupancy, possession, tenure, use. See BUSINESS, EXERCISE, WORK. *Antonyms:* dispossession, ejectment, eviction, resignation, vacating.

oc·cu·pa·tion·al (ok′yə·pā′shən·əl) *adj.* Of, pertaining to, or caused by, an occupation: *occupational* statistics; *occupational* diseases. —**oc′cu·pa′tion·al·ly** *adv.*

occupational therapy *Med.* The treatment of nervous, mental, or physical disabilities by means of work adapted to favor recovery and normal readjustment to external conditions.

oc·cu·py (ok′yə·pī) *v.t.* **·pied, ·py·ing** 1 To take and hold possession of, as by conquest. 2 To fill or take up (space or time): The estate *occupies* ten acres. 3 To inhabit; dwell in. 4 To hold; fill, as an office or position. 5 To busy or engage; employ: He *occupies* himself with trifles. [<OF *occuper* <L *occupare* seize, take possession of <*ob-* against + *capere* take] —**oc′cu·pi′er** *n.*

Synonyms: busy, employ, engage, fill, have, hold, keep, possess, preoccupy, use. See ENTERTAIN, HAVE, INTEREST, POSSESS.

oc·cur (ə·kûr′) *v.i.* **·curred, ·cur·ring** 1 To happen; come about. 2 To be found or met with; appear: Trout *occur* in this lake. 3 To present itself; come to mind: The theory just *occurred* to me. See synonyms under HAPPEN. [<L *occurrere* run to or against, befall <*ob-* towards, against + *currere* run]

oc·cur·rence (ə·kûr′əns) *n.* 1 The act or fact of occurring. 2 An event considered as simply presenting itself to notice without obvious cause; a happening. See synonyms under CIRCUMSTANCE, EVENT. —**oc·cur′rent** *Rare adj.*

o·cean (ō′shən) *n.* 1 The great body of salt water that covers about two thirds of the earth's surface. 2 Any one of the greater tracts of water that cover the globe: the Atlantic *Ocean.* 3 Any unbounded expanse or quantity. [<OF <L *oceanus* <Gk. *ōkeanos* the great outer sea (as opposed to the Mediterranean), orig. *Ōkeanos (potamos)* (the river of) Oceanus]

o·cean·ad (ō′shən·ad) *n.* An ocean-dwelling plant. [<OCEAN + -AD[1]]

ocean gray A light silvery-gray color used on ships of the U. S. Navy in World War II.

O·ce·an·i·a (ō′shē·an′ē·ə) The islands of the Pacific, including Melanesia, Micronesia, and Polynesia, sometimes including the Malay Archipelago and Australasia. Also **O′ce·an′·i·ca.** —**O′ce·an′i·an** *adj.* & *n.*

o·ce·an·ic (ō′shē·an′ik) *adj.* Relating to, or living in, the ocean; pelagic.

O·ce·an·ic (ō′shē·an′ik) *n.* A subfamily of the Austronesian family of languages, including the Melanesian languages of the Solomon Islands, Fiji, New Caledonia, the New Hebrides, etc., and the Micronesian group.

O·ce·a·nid (ō·sē′ə·nid) In Greek mythology, one of the 3,000 sea nymphs, daughters of Oceanus and Tethys. [<Gk. *Okeanis, -idos*]

Ocean Island A Pacific island west of the Gilbert group, comprising part of the Gilbert and Ellice Islands colony; 2 square miles.

o·cean–lane route (ō′shən·lān′) See LANE ROUTE.

o·ce·an·og·ra·phy (ō′shē·ən·og′rə·fē, ō′shən-) *n.* The branch of physical geography that treats of oceanic life and phenomena. [<OCEAN + -(O)GRAPHY] —**o′ce·an·og′ra·pher** *n.* —**o′ce·an·o·graph′ic** (-ə·graf′ik) or **·i·cal** *adj.* —**o′ce·an·o·graph′i·cal·ly** *adv.*

o·ce·an·oph·i·lous (ō′shē·ən·of′ə·ləs) *adj. Biol.* Living in the ocean, as a plant or animal. [<OCEAN + -(O)PHILOUS]

O·ce·a·nus (ō·sē′ə·nəs) 1 In Greek mythology, a Titan, husband and brother of Tethys, father of the Oceanids and all river gods. 2 In classical geography, the mighty stream said to encircle the habitable world.

oc·el·lat·ed (os′ə·lā′tid) *adj.* 1 Having an ocellus or ocelli (of color), as in the tail of a peacock. 2 Resembling an ocellus. 3 Spotted. Also **oc′el·late.** [<L *ocellatus* small-eyed <*ocellus.* See OCELLUS.] —**oc′el·la′tion** *n.*

o·cel·lus (ō·sel′əs) *n. pl.* **·li** (-ī) 1 *Biol.* A minute simple eye, as of many invertebrates. 2 A spot of color surrounded by a ring or rings of color as in the tail of a peacock. [<L, dim. of *oculus* eye] —**o·cel′lar** *adj.*

o·ce·lot (ō′sə·lot, os′ə-) *n.* A large Central and South American cat *(Felis pardalis)* of a prevailing yellowish- or reddish-gray, with black-edged blotches. [<F, short for Nahuatl *tlaocelotl* <*tlalli* a field + *ocelotl* a jaguar]

o·cher (ō′kər) *n.* 1 A native earth varying

ochlesis

from light yellow to deep orange or brown, and consisting of iron trioxide and water with varying proportions of clay in impalpable subdivision, largely used as a pigment. 2 Any metallic oxide occurring in an earthy or finely divided form. 3 A dark-yellow color derived from or compared to ocher. — **red ocher** A red, ferruginous native ocher: also called *Indian red, Venetian red*. Also **o′chre**. [< OF *ocre* < L *ochra* < Gk. *ōchra* yellow ocher < *ōchros* pale yellow] — **o′cher·ous, o·chre·ous** (ō′krē·əs), **o′cher·y, o′chry** *adj.*

och·le·sis (ok·lē′sis) *n. Pathol.* Any disease caused by overcrowding. [< NL < Gk. *ochlēsis* a disturbance < *ochlein* move, disturb < *ochlos* a crowd]

och·loc·ra·cy (ok·lok′rə·sē) *n. pl.* **·cies** Rule of the multitude; government by the populace; mob rule. [< MF *ochlocratie* < Gk. *ochlokratia* mob rule < *ochlos* a crowd + *krateein* rule] — **och′lo·crat** *n.* — **och·lo·crat·ic** (ok′lə·krat′ik) *adj.*

och·lo·pho·bi·a (ok′lə·fō′bē·ə) *n.* Morbid fear of crowds; demophobia. [< Gk. *ochlos* a crowd + -PHOBIA]

och·one (ə·khōn′) *interj.* Alas: a cry of grief. [< Irish *ochoin*]

o·chroid (ō′kroid) *adj.* Of the color of ocher.

Ochs (oks), **Adolph**, 1858-1935, U.S. newspaper publisher.

-ock suffix of nouns Small; little: now often without perceptible force: *hillock*. [OE *-oc, -uc*]

Ock·ham (ok′əm), **William of**, 1285?-1349?, English Franciscan and scholastic philosopher: known as the *Invincible Doctor*. Also spelled *Occam*.

Ockham's razor *Philos.* 1 A leading tenet of nominalism formulated by William of Ockham and stating that terms, concepts, and assumptions must not be multiplied beyond necessity. 2 The law of parsimony.

o'clock (ə·klok′) Of, according to, or by the clock.

Oc·mul·gee River (ōk·mul′gē) A river in central Georgia, flowing 225 miles east to the Altamaha.

O'Con·nell (ō·kon′əl), **Daniel**, 1775-1847, Irish nationalist statesman and orator: called "the Liberator."

O'Con·nor (ō·kon′ər), **Frank** Pseudonym of Michael O'Donovan, born 1903, Irish short-story writer. — **Thomas Power**, 1848?-1929?, Irish politician and journalist: known as *Tay Pay*.

o·co·til·lo (ō′kə·tēl′yō, *Sp.* ō′kō·tē′yō) *n.* The candlewood tree of California and Mexico. [< Sp., dim. of *ocote* Mexican pine < Nahuatl *ocotl*]

oc·re·a (ok′rē·ə, ō′krē·ə) *n. pl.* **oc·re·ae** (ok′ri·ē, ō′kri·ē) 1 *Bot.* a A stipule or combined pair of stipules forming a legging-shaped sheath about the stem of a plant. b A thin sheath around the seta of a moss: erroneously written *ochrea*. 2 *Ornithol.* A sheath, as the boot of a bird. [< L, a legging, a greave]

oc·re·ate (ok′rē·it, -āt, ō′krē-) *adj.* 1 Having ocreae. 2 *Ornithol.* Booted: said of the tarsi of certain birds.

oct-, octa- See OCTO-.

oc·ta·chord (ok′tə·kôrd) *n.* 1 A musical instrument with eight strings. 2 A diatonic scale of eight tones. [< LL *octochordos* < Gk. *oktachordos* < *okta-* eight + *chordē* a string] — **oc′ta·chor′dal** *adj.*

oc·tad (ok′tad) *n.* 1 A series of eight. 2 *Chem.* An atom, radical, or element that has a combining power of eight. 3 In ancient notation, a group of eight figures arranged similarly to successive powers of ten. [< L *octas, -adis* < Gk. *oktas, -ados* a group of eight < *oktō* eight] — **oc·tad′ic** *adj.*

oc·ta·gon (ok′tə·gon) *n.* A plane figure with eight sides and eight angles. [< L *octagonos* < Gk. *oktagōnos* < *okta-* eight + *gōnia* an angle]

oc·tag·o·nal (ok·tag′ə·nəl) *adj.* Eight-sided; eight-angled. — **oc·tag′o·nal·ly** *adv.*

oc·ta·he·dral *adj.* Having eight equal plane faces.

oc·ta·he·drite (ok′tə·hē′drīt) See ANATASE.

oc·ta·he·dron (ok′tə·hē′drən) *n. pl.* **·dra** (-drə) A solid figure bounded by eight plane faces. [< Gk. *oktaedron*, orig. neut. of *oktaedros* eight-sided < *okta-* eight + *hedra* a seat]

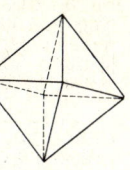

OCTAHEDRON

oc·tam·er·ous (ok·tam′ər·əs) *adj. Bot.* Having the parts in eights, as a flower with eight members in each set of organs: frequently written 8-merous. [< Gk. *oktamerēs* < *okta-* eight + *meros* a part]

oc·tam·e·ter (ok·tam′ə·tər) *adj.* In prosody, having eight measures or metrical feet. — *n.* A verse of eight feet. [< L *octameter* < Gk. *oktametros* < *okta-* eight + *metron* a measure]

oc·tan (ok′tən) *adj.* Recurring on the eighth day: an *octan* fever. [< F *octane, octaine*]

oc·tane (ok′tān) *n. Chem.* One of a class of isomeric hydrocarbons of the methane series which have the formula C_8H_{18}: especially, isooctane, correctly known as 2,2,4-trimethylpentane. — **high-octane gasoline** Gasoline of great efficiency, having a high octane number. [< OCT- + -ANE²]

octane number A measure of the anti-knock properties of gasoline. It is a number indicating the percentage, by volume, of 2,2,4-trimethylpentane in a mixture of this octane and normal heptane having the same knock-rating as the gasoline under test. Compare CETANE NUMBER.

oc·tan·gle (ok′tang·gəl) *n.* An octagon. [< LL *octangulus* < *octo* eight + *angulus* an angle]

oc·tan·gu·lar (ok·tang′gyə·lər) *adj.* Eight-angled and eight-sided; octagonal.

Oc·tans (ok′tanz) A southern constellation. See CONSTELLATION. Also **Oc′tant** (-tənt). [< NL. See OCTANT.]

oc·tant (ok′tənt) *n.* 1 An eighth part of a circle; an arc subtending an angle of 45 degrees. 2 *Astron.* The position in the heavens that is one eighth of a circle distant from conjunction or quadrature; one of the four positions of the moon midway between new or full moon and quarters. 3 An instrument resembling a sextant but having an arc of only 45 degrees: used for measuring the angular height of the sun, moon, and other celestial bodies as an aid in navigation. 4 *Geom.* One of the eight trihedral compartments of space formed by three planes with the three axes, *x*, *y*, and *z*, of a Cartesian coordinate system as edges. [< LL *octans* an eighth part < L *octo* eight] — **oc·tant′al** (-tan′təl) *adj.*

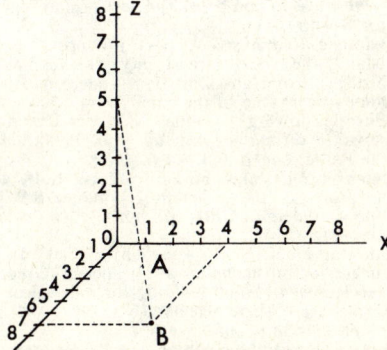

OCTANT OF CARTESIAN COORDINATE SYSTEM
(def. 4)
O. Origin. Coordinates of A: x = 4, y = 7, z = 5; of B: x = 4, y = 7.

oc·ta·pla (ok′tə·plə) *n.* 1 A Bible written or printed in eight languages or containing eight versions. 2 Any polyglot book in eight languages. [< Gk. *oktapla*, neut. pl. of *oktaploos* eightfold]

oc·tar·chy (ok′tär·kē) *n. pl.* **·chies** 1 A government headed by eight persons. 2 A country under eight rulers. 3 A group of eight allied governments. [< OCT- + -ARCHY]

oc·tave (ok′tiv, -tāv) *n.* 1 *Music* a The interval between any note and that given by twice as many or half as many vibrations in a second. b A note at this interval above or below any other, considered in relation to that other. c Two notes at this interval, sounded together; also, the resulting consonance. d An organ stop giving tones an octave higher than those normally corresponding to the keys played. 2 The eighth day from a feast day, beginning with the feast day as one; also, the lengthening of a festival so as to include a period of eight days. 3 Any group or series of eight. 4 In prosody, the first eight lines in an Italian sonnet, or a stanza of eight lines. — *adj.* 1 Composed of eight. 2 In prosody, composed of eight lines. 3 *Music* Producing tones an octave higher. Also **oc·ta·val** (ok·tā′vəl, ok′tə-) *adj.* [< OF < L *octava*, fem. of *octavus* eighth < *octo* eight]

Oc·ta·vi·a (ok·tā′vē·ə, *Ger., Sp.* ōk·tä′vē·ä) A feminine personal name. Also *Fr.* **Oc·ta·vie** (ôk·tà·vē′) or **Oc·tave** (ôk·tàv′). [See OCTAVIUS.]

— Octavia, died 11 B.C., sister of Augustus and wife of Mark Antony.

Oc·ta·vi·an (ok·tā′vē·ən) Augustus Caesar: so called from 45-27 B.C.

Oc·ta·vi·us (ok·tā′vē·əs) A masculine personal name. Also **Oc·ta·vi·a·nus** (ok·tā′vē·ā′nəs), **Oc·ta·vus** (ok·tā′vəs, ok′tə·vəs), *Fr.* **Oc·tave** (ôk·tàv′) or **Oc·ta·vien** (ôk·tà·vyan′), *Sp.* **Oc·ta·vio** (ôk·tä′vyō). [< L, eighth]

oc·ta·vo (ok·tā′vō, -tä′-) *n. pl.* **·vos** The page size (6 x 9 1/2 inches except where otherwise specified) of a book made up of printer's sheets folded into eight leaves; also, a book consisting of pages of this size: often called *eightvo* and written *8vo* or *8°*. — *adj.* In octavo; consisting of pages of this size. [< L (*in*) *octavo* (in) an eighth (of a sheet)]

oc·ten·ni·al (ok·ten′ē·əl) *adj.* 1 Recurring at intervals of eight years. 2 Occupying periods of eight years. [< LL *octennium* a period of eight years < *octo* eight + *annus* a year] — **oc·ten′ni·al·ly** *adv.*

oc·tet (ok·tet′) *n.* 1 A musical composition for eight parts or eight performers. 2 A choir of eight voices, or an orchestra of eight performers. 3 Any group of eight; especially, the first eight lines of an Italian sonnet; octave (def. 4). 4 *Physics* A group of eight electrons in the shell of an atom. Also **oc·tette′**. [< L *octo* eight; on analogy with *duet*]

oc·til·lion (ok·til′yən) *n.* A cardinal number: in the French and American systems, represented by a figure 1 with 27 ciphers annexed; in the English system by a figure 1 with 48 ciphers. [< MF < L *octo* eight; on analogy with *million*] — **oc·til′lionth** (-yənth) *adj. & n.*

octo- combining form Eight: *octopus*: also, before vowels, *oct-*. Also *octa-*. [< L *octo* and Gk. *oktō* eight]

Oc·to·ber (ok·tō′bər) 1 The tenth month of the year (the eighth of the Roman year), containing 31 days. 2 Ale or cider made in October. [< L, the eighth (month) < *octo* eight]

October Revolution See RUSSIAN REVOLUTION under REVOLUTION.

Oc·to·brist (ok·tō′brist) *n.* One of that faction in the Russian Duma which supported the czar in reform measures proposed in October, 1905. The proposals were only partly carried out and the party soon disappeared, its remnants supporting monarchical ideas, so that the term came to designate a reactionary.

oc·to·dec·i·mo (ok′tə·des′ə·mō) *n. pl.* **·mos** The page size (approximately 4 x 6 1/2 inches) of a book made up of printer's sheets folded into 18 leaves; also, a book consisting of pages of this size: often called *eighteenmo* and written *18mo* or *18°*. — *adj.* In octodecimo; consisting of pages of this size. [< L (*in*) *octodecimo* (in) an eighteenth (of a sheet)]

oc·to·ge·nar·i·an (ok′tə·jə·nâr′ē·ən) *adj.* Being eighty or from eighty to ninety years of age: also **oc·tog·e·nar·y** (ok·toj′ə·ner′ē). — *n.* A person between eighty and ninety years of age. [< L *octogenarius* < *octogeni* eighty each < *octoginta* eighty]

oc·to·he·dral (ok′tə·hē′drəl) See OCTAHEDRAL.

oc·to·lat·er·al (ok′tə·lat′ər·əl) *adj.* Eight-sided. [< OCTO- + LATERAL]

oc·to·nar·y (ok′tə·ner′ē) *adj.* 1 Of or pertaining to the number eight. 2 Having eight parts or members. — *n. pl.* **·nar·ies** 1 In prosody, an octave. 2 An ogdoad. [< L *octonarius* containing eight < *octoni* eight at a time < *octo* eight]

oc·to·pus (ok'tə·pəs) *n. pl.* **·pus·es** or **·pi** (-pī) or **oc·top·o·des** (ok·top'ə·dēz) **1** An eight-armed cephalopod (genus *Octopus*) as distinguished from the ten-armed species; a devilfish. **2** Figuratively, any organized power regarded as of far-reaching capacity for harm; specifically, a powerful business organization; a trust. [<NL <Gk. *oktapous* eight-footed < *okta-* eight + *pous* a foot]

OCTOPUS (Varies according to species, from a 6-inch to a 32-foot span over all)

oc·to·roon (ok'tə·ro͞on') *n.* A person who is one-eighth Negro: the offspring of a white person and a quadroon. [<L *octo* eight + (QUAD)ROON]

oc·to·sper·mous (ok'tə·spûr'məs) *adj. Bot.* Eight-seeded. [<OCTO- + -SPERMOUS]

oc·to·syl·lab·ic (ok'tə·si·lab'ik) *adj.* **1** Composed of eight syllables, as a line of verse. **2** Containing lines of eight syllables. —*n.* An octosyllabic line or verse. [<LL *octosyllabus* <L <Gk. *oktasyllabos* < *okta-* eight + *syllabē* a syllable]

oc·to·syl·la·ble (ok'tə·sil'ə·bəl) *n.* **1** A verse or line composed of eight syllables. **2** An eight-syllabled word.

oc·troi (ok'troi, *Fr.* ôk·trwà') *n.* **1** A government grant or privilege given a person or company; especially, a trade monopoly so conferred. **2** A tax levied at the gates of a city on articles of trade. **3** The gate where it is collected; also, the service of collection. [<F < *octroyer, ottroyer* grant <LL *auctorizare* <L *auctor* an author < *augere* increase]

oc·tu·ple (ok'to͞o·pəl, -tyo͞o-, ok·to͞o'pəl, -tyo͞o'-) *adj.* **1** Consisting of eight parts or copies. **2** Multiplied by eight. —*v.t.* **·pled, ·pling** To multiply by eight. —*n.* A number or sum eight times as great as another. [<L *octuplus* eightfold < *octo* eight] —**oc'tu·ply** *adv.*

oc·tu·pli·cate (ok·to͞o'plə·kit, -tyo͞o'-) *adj.* Eightfold. [<L *octuplicatus*, ult. < *octo* eight + *plic-*, stem of *plicare* fold]

oc·tyl (ok'til) *n. Chem.* The univalent radical C_8H_{17}.

oc·u·lar (ok'yə·lər) *adj.* Pertaining to, like, derived from, or connected with the eye; visual. —*n.* The lenses forming the eyepiece of an optical instrument. [<L *ocularis* < *oculus* eye] —**oc'u·lar·ly** *adv.*

oc·u·list (ok'yə·list) *n.* One skilled in treating diseases of the eye; an ophthalmologist. [<MF *oculiste* <L *oculus* eye]

oculo– *combining form* Eye; of or pertaining to the eye: *oculomotor.* Also, before vowels, **ocul–.** [<L *oculus* the eye]

oc·u·lo·mo·tor (ok'yə·lə·mō'tər) *Anat. adj.* Causing or connected with movement of the eye: the *oculomotor* nerve. —*n.* The oculomotor or third cranial nerve, which supplies most of the muscles that move the eyeball.

O·cyp·e·te (ō·sip'ə·tē) One of the Harpies.

od (od, ōd) *n.* A hypothetical force formerly supposed by some to pervade all nature: assumed to account for the phenomena of magnetism, chemical action, mesmerism, etc.: also *odyl.* [<G; coined by Baron Karl von Reichenbach, 1788–1869, German chemist] —**od·ic** (od'ik, ō'dik) *adj.*

Od (od) *n. & interj. Archaic* God: a euphemism used in oaths. Also **Odd.** [Alter. of GOD]

o·da·lisk (ō'də·lisk) *n.* A female slave or concubine in an Oriental harem. Also **o'da·lisque.** [<F *odalisque* <Turkish *ōdaliq* chambermaid < *ōdah* a chamber]

odd (od) *adj.* **1** Not even; leaving a remainder when divided by two. **2** Marked with an odd number. **3** Left over after a division. **4** Additional to any round number; thrown in or mentioned without exact enumeration: two hundred and *odd* miles; extra: an *odd* fork. **5** Occasional; casual: to work at *odd* jobs. **6** Peculiar; singular; queer; eccentric. **7** Single: an *odd* slipper. —*n.* In golf, an advantage of one or more strokes given to a less skilful player over his opponent. [<ON *odda-* odd, third < *oddi* a point of land, triangle, odd number] —**odd'ly** *adv.* —**odd'ness** *n.*

Synonyms (adj.): anomalous, bizarre, comical, droll, eccentric, extraordinary, fantastic, fantastical, grotesque, peculiar, quaint, queer, rare, strange, uncommon, unique, unmatched, unusual, whimsical. See QUEER, RARE[1]. *Antonyms*: common, conventional, customary, even, normal, ordinary, usual.

Odd Fellow A member of the Independent Order of Odd Fellows, a secret society for the mutual aid and benefit of the members: organized in the United States, 1819.

odd·i·ty (od'ə·tē) *n. pl.* **·ties** **1** Singularity. **2** An eccentricity. **3** Something odd or peculiar.

odd·ment (od'mənt) *n.* **1** That which is only an irregular and incidental and not an essential part of some course or system; something left over. **2** *Printing* A constituent part of a book other than the text, as the title page, index, etc. **3** *pl.* Small belongings; odds and ends.

odd–pin·nate (od'pin'āt, -it) *adj. Bot.* Pinnate with a single leaflet at the end, as the locust leaf.

odds (odz) *n. pl. (sometimes construed as singular)* **1** An equalizing allowance based on the apparent chances of success of an opponent or contestant. **2** The amount or proportions by which one's bet differs from that of another: The *odds* are three to two. **3** The balance of probability that something will happen or be found to be the case. **4** In a contest of any sort, a difference to the advantage of one side: the *odds* in one's favor. **5** In a contest, the advantage of one side over the opposing side. —**at odds** At variance; disagreeing. —**to give** (or **lay**) **odds** To offer to bet with someone on terms apparently favorable to him. —**to take odds** To agree to a wager on terms apparently favorable to the other person.

odds and ends Fragments; miscellaneous articles.

ode (ōd) *n.* **1** In classical prosody, a lyric poem intended to be sung or chanted, exemplified by the **Pindaric ode,** consisting of three stanzas (strophe, antistrophe, and epode), and the **Horatian ode,** consisting of one stanzaic form throughout. **2** In modern usage, a lyric poem, rimed or unrimed, of lofty tone, treating progressively one dignified theme, often in the form of an address. [<MF <LL *ode, oda* <Gk. *ōidē, aoidē* a song < *aeidein* sing] —**od·ic** (ō'dik) *adj.*

-ode[1] *combining form* Way; path: *anode, cathode.* [<Gk. *hodos* a way]

-ode[2] *suffix* Like; resembling; having the nature of: *phyllode.* [<Gk. *-ōdēs* < *eidos* form]

O·den·burg (ō'dən·bo͝orkh) The German name for SOPRON.

O·den·se (ō'thən·sə) A port on north Fyn Island, the third largest city of Denmark: birthplace of Hans Christian Andersen.

O·der (ō'dər) One of the chief rivers of central Europe, flowing 563 miles NE from Czechoslovakia through Germany and Poland to the Baltic: Czech and Polish *Odra.*

O·des·sa (ō·des'ə) A port on **Odessa Gulf,** an inlet of the Black Sea in Ukrainian S.S.R.

O·dets (ō·dets'), **Clifford,** born 1906, U.S. playwright.

o·de·um (ō'dē·əm) *n. pl.* **o·de·a** (ō'dē·ə) **1** A theater or music hall. **2** In ancient Greece and Rome, a roofed building for musical performances. Also **o·de·on** (ō'dē·on). [<LL <Gk. *ōideion* < *ōidē.* See ODE.]

O·din (ō'din) In Norse mythology, the supreme deity, god of war and founder of art and culture. Compare WODEN. Also spelled *Othin.*

o·di·ous (ō'dē·əs) *adj.* **1** Exciting hate, repugnance, or disgust. **2** Regarded with aversion or disgust. See synonyms under FOUL, INFAMOUS. [<AF <L *odiosus* < *odium.* See ODIUM.] —**o'di·ous·ly** *adv.* —**o'di·ous·ness** *n.*

o·di·um (ō'dē·əm) *n.* **1** The state of being odious; offensiveness; opprobrium. **2** A feeling of extreme repugnance, disgust, or hate. See synonyms under SCANDAL. [<L, hatred < *odisse* hate]

O·do·a·cer (ō'dō·ā'sər), 434?–493, German ruler of Italy 476–493: also *Ottokar.*

O·do·ar·do (ō'dō·är'dō) Italian form of EDWARD.

o·do·graph (ō'də·graf, -gräf) *n.* **1** A pedometer. **2** An odometer. **3** An automatic, portable mapmaking device designed to work from a moving vehicle. [<Gk. *hodos* a way + -GRAPH]

O·do·graph (ō'də·graf, -gräf) *n.* An instrument for automatically recording to correct scale the course of a ship at sea: a trade name.

o·dom·e·ter (ō·dom'ə·tər) *n.* An appliance for measuring distance traveled, as a mechanical registering attachment to the wheel of a vehicle: also spelled *hodometer.* [<Gk. *hodometros* < *hodos* a way, a road + *metron* a measure]

o·dom·e·try (ō·dom'ə·trē) *n.* Measurement by odometer.

O·don (ō·dôṅ') French form of OTTO. Also *Hungarian* **O·dön** (œ'dœn).

o·don·a·tous (ō·don'ə·təs) *adj.* Of or pertaining to an order *(Odonata)* of slender, long-bodied, generally large, predatory insects with four equal, net-veined wings, including the dragonflies and damsel flies. [<NL <Gk. *odous, odontos* a tooth]

o·don·tal·gi·a (ō'don·tal'jē·ə, -jə) *n. Pathol.* Toothache. [<Gk. < *odous, odontos* a tooth + *algos* a pain] —**o'don·tal'gic** *adj.*

odonto– *combining form* Tooth; of the teeth: *odontology.* Also, before vowels, **odont–.** [< Gk. *odous, odontos* a tooth]

o·don·to·blast (ō·don'tə·blast) *n. Anat.* A tooth cell that produces dentine. —**o·don'to·blas'tic** *adj.*

o·don·to·glos·sum (ō·don'tə·glos'əm) *n.* Any of a large genus *(Odontoglossum)* of tropical American epiphytic orchids with thick, fleshy leaves and large flowers with free, spreading sepals. [<NL <Gk. *odous, odontos* a tooth + *glossa* a tongue]

o·don·to·graph (ō·don'tə·graf, -gräf) *n.* **1** *Mech.* An instrument for correctly laying out gear teeth. **2** *Dent.* A device for showing irregularities occurring in the surface of tooth enamel.

o·don·toid (ō·don'toid) *adj.* **1** Toothlike. **2** Of or pertaining to the odontoid bone or process.

odontoid process *Anat.* A toothlike or peglike projection from the body of the axis or second vertebra of the neck, upon which the atlas rotates: found in mammals and birds. Also **odontoid peg.**

o·don·tol·o·gy (ō'don·tol'ə·jē) *n.* The body of scientific knowledge that relates to the structure, health, and growth of the teeth. [< ODONTO- + -LOGY] —**o·don·to·log·i·cal** (ō·don'tə·loj'i·kəl) *adj.* —**o·don'to·log'i·cal·ly** *adv.* —**o'don·tol'o·gist** *n.*

o·don·to·phore (ō·don'tə·fôr, -fōr) *n. Zool.* **1** A protrusible, ribbonlike organ covered with teeth for rasping, etc., and connected with the mouth of cephalous mollusks. **2** The radula, tongue, or lingual ribbon. [<Gk. *odontophoros* bearing teeth < *odous, odontos* tooth + *-phoros* bearing < *pherein* bear] —**o·don·toph·o·ral** (ō'don·tof'ər·əl) *adj.* —**o'don·toph'o·rine** (-rin, -rēn) *adj. & n.* —**o'don·toph'o·rous** *adj.*

o·dor (ō'dər) *n.* **1** That quality of a material substance that renders it perceptible to the sense of smell; scent. **2** Regard or estimation: to be in bad *odor.* **3** A perfume; incense. See synonyms under SAVOR, SMELL. Also *Brit.* **o'dour.** [<AF *odour,* OF *odor* <L *odor, -oris*] —**o'dored** *adj.*

o·dor·if·er·ous (ō'də·rif'ər·əs) *adj.* Diffusing an odor. [<L *odorifer* < *odor, -oris* an odor + *ferre* bear] —**o'dor·if'er·ous·ly** *adv.* —**o'dor·if'er·ous·ness** *n.*

o·dor·im·e·try (ō'də·rim'ə·trē) *n.* The measurement of the character, intensity, and permanence of odors. [<L *odor, -oris* an odor + -METRY] —**o'dor·im'e·ter** *n.*

o·dor·less (ō'dər·lis) *adj.* Having no odor. —**o'dor·less·ly** *adv.* —**o'dor·less·ness** *n.*

o·dor·ous (ō'dər·əs) *adj.* Having an odor; fragrant. —**o'dor·ous·ly** *adv.* —**o'dor·ous·ness** *n.*

Od·ra (ōd'rä) The Czech and Polish name for the ODER.

od·yl (od'il, ō'dil) *n.* An od. Also **od'yle.**

-odynia *combining form Med.* Pain; chronic pain in a (specified) part of the body: *osteodynia,* chronic pain in a bone. [<Gk. *odynē* pain]

O·dys·seus (ō·dis'yo͞os, -ē·əs) In Greek legend, king of Ithaca, one of the Greek leaders in the Trojan War, figuring in the *Iliad* and hero of the *Odyssey;* Ulysses.

Od·ys·sey (od'ə·sē) *n.* **1** An ancient Greek epic poem attributed to Homer, describing the wanderings of Odysseus during the ten years after the fall of Troy. 2 A long, wandering journey: often **od'ys·sey.** —**Od'ys·sey' an** *adj.*

oe– See also words beginning E–.

Oe·a (ē'ə) The Phoenician name for TRIPOLI. Also **Œ'a.**

oec·u·men·i·cal (ek'yōō·men'i·kəl, *Brit.* ē'kyōō·men'i·kəl) See ECUMENICAL.

Oe–Cus·se (ō·kōō'sē) See AMBENO.

oe·de·ma (i·dē'mə) See EDEMA.

Oed·i·pus (ed'ə·pəs, ē'də-) In Greek legend, the son of Laius and Jocasta, who, abandoned by them at birth because of an oracle and raised by the king of Corinth, eventually returned to Thebes. After guessing the riddle of the sphinx and accidentally killing his father, he unwittingly married his mother and became king of Thebes; discovering his relationship to Jocasta, he blinded himself and died in exile. Also **Œd'i·pus.** [<L <Gk. *Oidipous,* lit., swollen-footed]

Oedipus complex *Psychoanal.* A strong, typically unconcious attachment to the parent of opposite sex, with antagonism toward the other: regarded as a normal stage in the development of children but productive of various neurotic disorders when it persists unresolved into adult life: usually descriptive of a son's attachment to his mother. Compare ELECTRA COMPLEX.

Oeh·len·schlä·ger (œ'lən·shlā'gər), **Adam,** 1779–1850, Danish poet and dramatist: also spelled *Öhlenschläger.*

oeil–de–boeuf (œ'y'də·bœf') *n. pl.* **oeils–de–boeuf** (œ'y'-) A circular or oval window; a bull's-eye. [<F, lit., eye of an ox]

oeil·lade (œ·yàd') *n. French* A glance; an ogle; an amorous look.

Oe·ne·us (ē'nē·əs) In Greek mythology, a king of Calydon and father of Meleager, who neglected a sacrifice to Artemis, causing her to send a boar to devastate his country. See CALYDONIAN BOAR. Also **Œ'ne·us.**

oe·nol·o·gy (ē·nol'ə·jē) See ENOLOGY.

oe·no·mel (ē'nə·mel, en'ə-) *n.* 1 A beverage of wine and honey. 2 Hence, anything combining sweetness and strength. Also spelled *oinomel.* [<LL *oenomeli* <Gk. *oinomeli* < *oinos* wine + *meli* honey]

Oe·no·ne (ē·nō'nē) In Greek mythology, a nymph of Mount Ida; wife of Paris, who deserted her for Helen of Troy. Also **Œ·no'ne.**

o'er (ôr, ōr) *prep.* & *adv. Poetic* Over.

oer·sted (ûr'sted) *n.* The cgs unit of magnetic intensity, equal to a force of 1 dyne exerted on a unit magnetic pole. [after Hans C. *Oersted,* 1777–1851, Danish physicist]

oe·soph·a·gus (i·sof'ə·gəs), etc. See ESOPHAGUS, etc.

oes·trin (es'trin, ēs'-) See ESTROGEN.

oes·trum (es'trəm, ēs'-), **oes·trus** (es'trəs, ēs'-) See ESTRUM, ESTRUS.

Oe·ta (ē'tə) A mountain chain in east central Greece; highest peak, 7,063 feet. *Greek* **Oi·te** (oi'tā).

oeu·vre (œ'vr') *n. pl.* **oeu·vres** (œ'vr') *French* 1 A work, as of art or literature. 2 The totality of works, as of an author.

of (uv, ov; *unstressed* əv) *prep.* 1 Coming from; originating at or from: Anne *of* Cleves; an actor *of* noble birth. 2 Associated or connected with; included among: Is he *of* your party? 3 Located at: the Leaning Tower *of* Pisa. 4 Away or at a distance from: within six miles *of* home. 5 Named; specified as: the city *of* Newark; a fall *of* ten feet. 6 Characterized by: a man *of* strength. 7 With reference to; as to: quick *of* wit. 8 About; concerning: Good is said *of* him. 9 Because of: dying *of* pneumonia. 10 Possessing: a man *of* means. 11 Belonging to: the lid *of* a box. 12 Pertaining to: the majesty *of* the law. 13 Composed of: a ship *of* steel. 14 Containing: a glass *of* water. 15 Taken from; from the number or class of: six *of* the seven conspirators. 16 So as to be without: relieved *of* anxiety; despoiled *of* ornaments. 17 Proceeding from; produced by: the plays *of* Shakespeare; the work *of* a vanished hand. 18 Directed toward; exerted upon: the massacre *of* the innocents; a love *of* opera. 19 During or at a specified time or occasion: He came *of* a Sunday; *of* recent years. 20 Set aside for or devoted to: a program *of* Lieder. 21 Before; until: used in telling time: ten minutes *of* ten. 22 *Archaic* By: loved *of* all men. [OE, var. of *af, æf* away from]

of– Assimilated var. of OB–.

of course 1 In the usual order or procedure; naturally; as expected. 2 Doubtless; certainly.

off (ôf, of) *adj.* 1 Farther or more distant; remote: an *off* chance. 2 In a (specified) circumstance or situation: to be well *off.* 3 Not in accordance with the facts; wrong: Your reckoning is *off.* 4 Not in the usual health or condition; not up to standard: an *off* season for roses. 5 Not in existence; no longer considered active or effective: The deal is *off.* 6 Away from work; not on duty: He spent his *off* hours at the rink. 7 In riding or driving, on the right: opposed to *near*: Pass on the *off* side. 8 *Naut.* Seaward; farther from the coast. 9 In cricket, to the left of a bowler: said of the side of the field facing the batsman. —*adv.* 1 To a distance; so as to be away: My horse ran *off.* 2 To or at a (specified) future time: Your engagement is another week *off.* 3 To or at a (specified) distance: The inn is a mile *off.* 4 So as to be no longer in place, connection, etc.: Take *off* your hat. 5 So as to be no longer functioning, continuing, or in operation: Turn the lights *off.* 6 So as to be away from one's work, duties, etc.: to take the day *off.* 7 So as to be completed, exhausted, etc.: to kill *off* one's enemies; to drink *off* a draught. 8 So as to deviate from or be below what is regarded as standard: His game was *off.* 9 *Naut.* Away from land, a ship, the wind, etc.: Keep her four points *off.* —**off with . . . !** Take off! Remove!: *Off with* his head! —**off with you!** Go away! Leave! —**right** (or **straight**) **off** Forthwith; immediately. —**to be off** 1 To leave; depart. 2 *Colloq.* To be insane. —*prep.* 1 So as to be separated, detached, distant, or removed from (a position, source, etc.): Take your feet *off* the table; twenty miles *off* course. 2 Not engaged in or occupied with; relieved from: *off* duty. 3 Extending away or out from; no longer on: *off* Broadway. 4 So as to deviate from or below the (what is regarded as standard): to be *off* one's game. 5 On or from (the material or substance of): living *off* nuts and berries. 6 *Colloq.* No longer using, engaging in, or advocating: to be *off* drinking. 7 *Naut.* Opposite to and seaward of: the battle *off* the eastern cape. —*n.* 1 The state or condition of being off. 2 In cricket, the offside (of the field). [ME, orig. stressed var. of OF]

of·fal (ô'fəl) *n.* 1 Those parts of a butchered animal that are rejected as worthless. 2 Rubbish or refuse of any kind. See synonyms under WASTE. [ME *ofall* <OFF + FALL]

Of·fa·ly (of'ə·lē) A county in Leinster province, central Ireland; 771 square miles; county town, Tullamore.

off and on Now and then; intermittently.

off–beat (ôf'bēt', of'-; *for adj.* 2 *esp.* ôf'bēt', of'-) *n. Music* Any secondary or weak beat in a measure. —*adj.* 1 *Music* Having primary accents in 4/4 time on the second and fourth beats. 2 *Slang* Unconventional; out of the ordinary.

off·cast (ôf'kast', -käst', of'-) *adj.* Rejected; cast off. —*n.* Anything thrown away or rejected.

off chance A bare possibility.

off–col·or (ôf'kul'ər, of'-) *adj.* 1 Unsatisfactory in color, as a gem. 2 Bad, indelicate, or indecent by implication; of doubtful virtue. Also *Brit.* **off'–col'our.**

Of·fen·bach (ôf'ən·bäk, of'-; *Ger.* ôf'ən·bäkh) A city on the Main in southern Hesse, West Germany. Also **Offenbach–am–Main** (-äm·min').

Of·fen·bach (ôf'ən·bäk, of'-; *Fr.* ô·fen·bàk'), **Jacques,** 1819–80, French composer born in Germany.

of·fence (ə·fens') The usual British spelling of OFFENSE.

of·fend (ə·fend') *v.t.* 1 To give displeasure or offense to; displease; affront; anger. 2 To affect (the taste, eyes, etc.) with displeasure. 3 *Obs.* To transgress or violate. —*v.i.* 4 To give displeasure or offense; be offensive. 5 To commit an offense or crime; sin. See synonyms under AFFRONT, PIQUE[1]. [<OF *offendre* strike against <L *offendere* < *ob-* against + *fendere* hit, thrust] —**of·fend'er** *n.*

of·fense (ə·fens') *n.* 1 The act of offending; any sin, wrong, or fault. 2 That which injures the feelings or causes displeasure; that which provokes. 3 The state of being offended; umbrage; anger. 4 Assault or attack: a weapon of *offense.* 5 A cause of sin or stumbling. Also, *Brit., offence.* [<OF <L *offensa* a striking against, orig. pp. fem. of *offendere.* See OFFEND.]

Synonyms: affront, anger, crime, delinquency, displeasure, fault, indignity, insult, misdeed, misdemeanor, outrage, resentment, sin, transgression, trespass, umbrage. See ABOMINATION, ANGER, OUTRAGE, PIQUE[1], SIN[1]. *Antonyms*: see synonyms for APOLOGY. —**of·fense'less** *adj.*

of·fen·sive (ə·fen'siv) *adj.* 1 Serving, adapted, or intended to give offense; displeasing; annoying. 2 Causing unpleasant sensations; disagreeable. 3 Serving as a means of attack. 4 Injurious. See synonyms under FOUL, ROTTEN, VULGAR. —*n.* Aggressive methods, operations, or attitude: with the definite article. —**of·fen'sive·ly** *adv.* —**of·fen'sive·ness** *n.*

of·fer (ô'fər, of'ər) *v.t.* 1 To present for acceptance or rejection; tender. 2 To suggest for consideration or action; propose. 3 To present with solemnity or in worship; make an offering of. 4 To show readiness to do or attempt; propose or threaten: to *offer* battle. 5 To attempt to do or inflict; hence, to do or inflict: to *offer* insult or resistance. 6 To suggest as payment; bid. 7 To present for sale. —*v.i.* 8 To present itself; appear: No opportunity *offered.* 9 To make an offering in worship or sacrifice. —*n.* 1 The act of offering; a profferor proposal. 2 An attempt or endeavor to do something. See synonyms under PROPOSAL. [OE *offrian* offer a sacrifice <L *offerre* present < *ob-* before + *ferre* bring; infl. in meaning by OF *offrir* offer] —**of'fer·er, of'fer·or** *n.*

Synonyms (verb): adduce, allege, bid, exhibit, extend, present, proffer, propose, tender, volunteer. What one *offers* he brings before another for acceptance or rejection. *Proffer* is a more formal and deferential word, with a suggestion of contingency; as, to *proffer* one's services; the worshiper *offers,* but does not *proffer* sacrifice. See ALLEGE. *Antonyms*: alienate, divert, refuse, retain, retract, withdraw, withhold.

of·fer·ing (ô'fər·ing, of'ər-) *n.* 1 The act of making an offer. 2 That which is offered; sacrifice; any gift. 3 A contribution at a religious service.

of·fer·to·ry (ô'fər·tôr'ē, -tō'rē, of'ər-) *n. pl.* **·ries** *Eccl.* 1 *Usually cap.* A section of the eucharistic liturgy, usually following the saying of the creed, during which the bread and wine to be consecrated are offered, and the alms of the congregation are collected. 2 Any collection taken during a religious service; also, the part of a service when it is taken. 3 An antiphon, hymn, or anthem sung during the offertory. 4 A prayer of oblation said by the celebrant over the bread and wine to be consecrated. [<Med. L *offertorium* an offering <LL *offertus,* pp. of L *offerre.* See OFFER.]

off·hand (ôf'hand', of'-) *adv.* Without preparation; unceremonious or unceremoniously; extempore. —*adj.* Done, said, or made extemporaneously. Also **off'hand'ed.** —**off'hand'·ed·ly** *adv.*

of·fice (ô'fis, of'is) *n.* 1 A particular duty, charge, or trust; an employment undertaken by commission or authority; a post or position held by an official or functionary; specifically, a position of trust or authority under a government: the *office* of premier. 2 That which is performed, assigned, or intended to be done by a particular thing, or that which anything is fitted to perform; function; service. 3 A place, building, or series of rooms in which some particular branch of the public service is conducted: the Patent *Office,* Post *Office*; also, the persons conducting such business; specifically, the head of the department and his immediate assistants: The Executive *Office* serves the president. In the United States the term is applied to those branches of the government business ranking next to

the departments, the chiefs of which are not cabinet members; in Great Britain, to all branches of government business over which a secretary of state presides and to certain other departments having their chiefs in the cabinet: the War *Office*, Home *Office*. 4 A room or building in which a person transacts his business or carries on his stated occupation: distinguished from *shop, store, studio*, etc.: the lawyer's *office*. 5 *pl. Brit.* The outbuildings devoted to culinary or other domestic purposes. 6 The persons collectively, as an association or corporation, whose headquarters are in an office: The *office* has telegraphed me to return. 7 *Eccl.* A prescribed religious or devotional service, particularly that for the canonical hours, or the service itself: the divine *office*; specifically, the daily service of the breviary, morning and evening prayer, the mass, communion, etc. 8 A ceremony; rite. 9 *pl.* A proffered action of any kind; especially, a service: reinstated through the good *offices* of a friend. See synonyms under DUTY. [<AF <L *officium* a service, prob. <*opus* a work + *facere* do, make]
office boy A boy hired to run errands or do odd jobs in an office.
of·fice-hold·er (ôf′is-hōl′dər, of′-) *n.* One who holds an office under a government.
office hours The number of hours one works in an office; also, the hours an office is open for business.
of·fi·cer (ôf′ə·sər, of′ə-) *n.* 1 One elected or appointed to office, as in a company, a society, or an ecclesiastical body, or one filling some other semipublic position, as by appointment. 2 One appointed to a certain military or naval rank and authority, specifically by commission. 3 In the merchant marine, etc., the captain or any of the mates. 4 A member of the constabulary or police force. 5 In corporate bodies and other organizations, one who holds a position entailing certain duties, as secretary or treasurer, as distinguished from an employee. — **commissioned officer** An officer who receives a commission, ranking, in the U.S. Army, from second lieutenant to general, and in the U.S. Navy from ensign to admiral. — **line officer** An officer of a combat branch of the service; officer of the line. — **non-commissioned officer** An officer appointed from the military ranks by an authorized commanding officer. The non-commissioned grades rank from corporal through master sergeant. — **warrant officer** In the U.S. Army, Navy, Air Force, Marine Corps, and Coast Guard, an officer without a commission, but having authority by virtue of a certificate of warrant. His rank is superior to that of a non-commissioned officer. See table under GRADE. — *v.t.* 1 To furnish with officers. 2 To command; direct; manage. [<AF *officer*, OF *officier* <Med. L *officiarius* <L *officium* OFFICE]
officer of the day An officer, in a military body, responsible for a 24-hour period for the safety of the command, its property, maintenance of order, and performance of the guard in a post, camp, or station.
officer of the guard In a military body, an officer responsible to the officer of the day for the maintenance of discipline and performance of guard duty in a camp, post, or station.
of·fi·cial (ə·fish′əl) *adj.* 1 Pertaining to or holding an office or public trust. 2 Derived from the proper office or officer; authoritative. 3 In pharmacy, authorized to be used in medicine. 4 Formal; studied; ceremonious. — *n.* One holding an office or performing duties of a public nature. [<OF <L *officialis* <*officium* OFFICE] — **of·fi′cial·dom** *n.* — **of·fi′cial·ly** *adv.* — **of·fi′cial·ness** *n.*
of·fi·cial·ism (ə·fish′əl·iz′əm) *n.* 1 Official state, condition, or system. 2 Rigid adherence to official forms.
of·fi·ci·ant (ə·fish′ē·ənt) *n.* One who conducts or officiates at a religious service, office, or ceremony; celebrant. [<Med. L *officians, -antis*, ppr. of *officiare.* See OFFICIATE.]
of·fi·ci·ar·y (ə·fish′ē·er′ē) *n. pl.* **·ar·ies** A body of officials. — *adj.* Pertaining to or holding an office.
of·fi·ci·ate (ə·fish′ē·āt) *v.i.* **·at·ed, ·at·ing** 1 To act or serve as a priest or minister; conduct a service. 2 To perform the duties or functions of an office. 3 In sports, to act as a

referee, umpire, etc. [<Med. L *officiatus*, pp. of *officiare* perform divine service <L *officium* OFFICE] — **of·fi·ci·a′tion** *n.* — **of·fi′ci·a′tor** *n.*
of·fi·ci·nal (ə·fis′ə·nəl) *adj.* 1 Prepared and on hand, as drug preparations. 2 Employed in the arts or as a medicine. 3 Of any drug or medicine kept ready for sale. [<Med. L *officinalis* <L *officina* a workshop, contraction of *opificina* <*opifex* a workman <*opus* a work + *facere* do]
of·fi·cious (ə·fish′əs) *adj.* 1 Unduly forward in offering one's services. 2 Obtrusive and interfering; meddling with what is not one's concern. 3 *Obs.* Disposed to serve or oblige; friendly. See synonyms under MEDDLESOME. [<L *officiosus* obliging <*officium* OFFICE] — **of·fi′cious·ly** *adv.* — **of·fi′cious·ness** *n.*
off·ing (ô′fing, of′ing) *n. Naut.* 1 That part of the visible sea offshore but beyond anchorage. 2 A position some distance offshore. — **in the offing** 1 In sight and not very distant. 2 Ready or soon to happen, arrive, etc.
off·ish (ô′fish, of′ish) *adj.* Inclined to be distant in manner; aloof. — **off′ish·ness** *n.*
off-peak (ôf′pēk′, of′-) *adj.* Below the maximum: an *off-peak* load in a power plant.
off·print (ôf′print′, of′-) *v.t.* To reprint (an excerpt). — *n.* A reproduction or separate printing of an article or paragraph printed in some publication.
off·scour·ing (ôf′skour′ing, of′-) *n.* That which is scoured off; something vile or despised; refuse. See synonyms under WASTE.
off·set (ôf′set′, of′-) *n.* 1 Anything regarded or advanced as a counterbalance or equivalent; set-off. 2 *Geol.* A spur or branch from a range of mountains or hills. 3 *Bot.* A short lateral branch of a plant that takes root where it rests on the soil, thus serving for propagation. 4 A line drawn from a curved or irregular main line at right angles to an auxiliary line, to assist in measuring areas or in plotting. 5 A ledge or set-off in a wall; also, a fence spur set at right angles to the main fence. 6 A terrace; especially, a terrace on a hill or mountain side. 7 *Archit.* A comparatively thin place in the length of a wall; also, a recess below the general plane of a wall; a sunk panel. 8 A bend in a pipe bringing one part out of, but parallel with, the line of another part. 9 *Printing* **a** The smut or smear of a freshly printed sheet on the surface of the sheet in contact with it. **b** An impression made by the offset printing method. 10 A descendant; offspring; offshoot. — *adj.* 1 *Printing* Of or pertaining to offset printing. 2 *Archit.* Of or pertaining to a ledge, panel, frame or the like not flush with a surface into which it is set. — *v.* (ôf′set′, of′-) **·set, ·set·ting** *v.t.* 1 To compensate for, as by balancing or opposing with an equivalent; counterbalance. 2 To transfer (an impression) from one surface to another. 3 *Printing* **a** To print by offset printing. **b** To smudge or mark with an offset. 4 *Archit.* To make an offset in. — *v.i.* 1 To make an offset, as in printing. 6 To branch off; project as an offset. [<OFF + SET]
offset printing A method of printing from a lithographic surface to a rubber-surfaced cylinder, and thence onto the paper.
off·shoot (ôf′shoot′, of′-) *n.* 1 *Bot.* A side shoot or branch from the main stem of a plant. 2 Anything that branches off from the parent stock or is regarded as a side issue.
off·shore (ôf′shôr′, -shōr′, of′-) *adj.* 1 Moving or directed away from the shore. 2 Situated or occurring at some distance from the shore. — *adv.* 1 At a distance from the shore. 2 From or away from the shore.
off·side (ôf′sīd′, of′-) *adv.* 1 At or on the wrong side. 2 In football, out of play: said of a player in certain contingencies when he gets in front of the ball during a scrimmage, or when the ball has been last touched by his own side behind him. 3 In hockey, between the ball and the goal of the opposing side during a play.
off·spring (ôf′spring′, of′-) *n.* 1 That which springs from or is the progeny of any person, animal, or plant. 2 A child or children; issue. [OE *ofspring* <*of* of, off + *springan* spring]
off-stage (ôf′stāj′, of′-) *n.* The area behind or to the side of a stage, out of view of the audience. — *adj.* In or from this area: *off-*

stage dialog. — *adv.* To this area: He went *off-stage*.
off-white (ôf′hwīt′, of′-) *n.* Oyster-white.
O'Fla·her·ty (ō·flä′hər·tē), **Liam**, born 1896, Irish novelist and short-story writer.
oft (ôft, oft) *adv. Archaic & Poetic* Often. — *adj. Obs.* Frequent. [OE]
oft·en (ôf′ən, of′-) *adv.* On frequent or numerous occasions; repeatedly. — *adj. Obs.* Repeated; frequently occurring. [Var. of ME *ofte*, OE *oft*]
oft·en·times (ôf′ən·tīmz′, of′-) *adv.* Frequently; often. Also *Archaic* **oft-times** (ôf′tīmz′, of′-).
Og (og) The king of Bashan; conquered by the Israelites. *Josh.* xii 2.
O·ga·sa·wa·ra Ji·ma (ō·gä·sä·wä·rä jē·mä) The Japanese name for the BONIN ISLANDS.
Og·bo·mo·sho (og′bō·mō′shō) A city of SW Nigeria.
Og·den (og′dən) A city in northern Utah.
Og·den (og′dən), **Charles Kay**, 1889–1957, English psychologist, educator, and semanticist.
og·do·ad (og′dō·ad) *n.* 1 The number eight. 2 Anything constructed of eight parts, individuals, or members; any group of eight. [<LL *ogdoas, -adis* <Gk. *ogdoas, -ados* <*oktō* eight]
o·gee (ō′jē, ō·jē′) *n. Archit.* 1 A molding having in section a reverse or long S-curve. 2 Such a curve used in any construction. 3 An arch with two such curves meeting at the apex: called **ogee arch**. [Appar. alter. of OF *ogive* OGIVE]
O·gee·chee River (ō·gē′chē) A river in eastern Georgia, flowing SE to the Atlantic.
og·ham (og′əm) *n.* 1 An ancient British and Irish script of the fifth and sixth centuries having an alphabet of 20 characters (lines, dots, or notches) arranged along an arris, and ascribed to the god Ogma: found cut into the edges of tombstones and pillars. 2 An inscription in that script; also, any one of the characters used. Also **og′am**. [<Irish <OIrish *ogam*, after *Ogma*, its mythical inventor] — **o·gham·ic** (ō·gam′ik) *adj.*
O·gil·vie (ō′gəl·vē), **John**, 1797–1867, Scottish lexicographer.
o·give (ō′jīv, ō·jīv′) *n.* 1 *Archit.* **a** A diagonal rib of a vaulted arch or bay. **b** A pointed arch. 2 *Stat.* A frequency curve any of whose ordinates expresses a percentage or number of observations less than or more than the corresponding abscissa. 3 The tapering forward part of a projectile: also called *head*. [<OF *ogive, augive, orgive*, ? ult. <Arabic *auj* a summit] — **o·gi′val** *adj.*

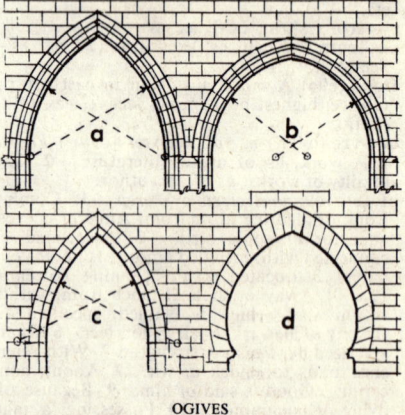

OGIVES
a. Equilateral.   b. Dropped.
c. Lancet.   d. Moorish.

o·gle (ō′gəl) *v.* **o·gled, o·gling** *v.t.* 1 To look at with admiring or impertinent glances. 2 To stare at; look at. — *v.i.* 3 To look or stare in an admiring or impertinent manner. — *n.* An amorous or coquettish look. [Prob. <LG *oegelen, ogelen*, freq. of *oegen* look at <*oege* an eye] — **o′gler** *n.*
O·gle·thorpe (ō′gəl·thôrp), **James Edward**, 1696–1785, British general; founder of the colony of Georgia.
OGPU (og′poo) *n.* Formerly, the Russian Soviet secret police: superseded by the *Narkomvnudel*, which was in turn replaced by the

O'Grady

MVD. Also called *Gay-Pay-Oo.* [< Russian *O(b'yedinennoye) G(osudarstvennoye) P(oliticheskoye) U(pravlenie)* Unified Government Political Administration]

O'Gra·dy (ō·grā′dē), **Standish James,** 1846-1928, Irish historian.

o·gre (ō′gər) *n.* **1** In fairy tales, a man-eating giant or monster. **2** A person regarded as resembling this. [< F; prob. coined by Perrault] — **o′gre·ish, o′grish** *adj.* — **o′gress** *n. fem.*

O·gyg·i·a (ō·jij′ē·ə) In the *Odyssey,* Calypso's island.

oh (ō) *interj.* **1** An exclamation expressing surprise, sudden emotion, etc. **2** See O (*interj.*).

O. Hen·ry (ō hen′rē) Pseudonym of *William Sydney Porter,* 1862-1910, U.S. short-story writer.

O'Hig·gins (ō·hig′ənz, *Sp.* ō·ē′gēns), **Bernardo,** 1778-1842, Chilean soldier and statesman: called "the Liberator of Chile."

O·hi·o (ō·hī′ō) A north central State of the United States, on Lake Erie; 41,222 square miles; capital, Columbus; entered the Union March 1, 1803; nickname, *Buckeye State*: abbr. *O.* — **O·hi′o·an** *adj. & n.*

Ohio River A river in the east central United States, flowing 981 miles SW from the junction of the Allegheny and Monongahela rivers in western Pennsylvania to the Mississippi.

Öh·len·schlä·ger (œ′lən·shlä′gər), **Adam** See OEHLENSCHLÄGER.

ohm (ōm) *n.* The unit of electrical resistance. The international ohm is the resistance at 0° C. of a uniform column of mercury having a mass of 14.4521 grams. It is equal to 1.000495 absolute ohms. [after G. S. *Ohm*] — **ohm′ic** *adj.*

Ohm (ōm), **Georg Simon,** 1787-1854, German physicist.

ohm·age (ō′mij) *n.* Electrical resistance of a conductor, expressed in ohms.

ohm·me·ter (ōm′mē′tər) *n.* A galvanometer having a dial or scale graduated to ohms and fractions of ohms, for measuring the resistance of a conductor.

o·ho (ō·hō′) *interj.* An exclamation expressing astonishment, exultation, etc.

O·hře (ôr′zhe) The Czech name for the EGER.

-oid *suffix* Like; resembling; having the form of: *ovoid, hydroid.* [< F *-oïdes* < L *-oides* < Gk. *-oeidēs* < *eidos* form]

-oidea *combining form Zool.* Used to form the names of classes or superfamilies: *Asteroidea, Echinoidea.* [< Gk. *-oeidēs* resembling < *eidos* form]

oil (oil) *n.* **1** A greasy or unctuous, generally combustible liquid, of vegetable, animal, or mineral origin, insoluble in water but sometimes soluble in alcohol, and always in ether: variously used as food, for lubricating, illuminating, and fuel, and in the manufacture of soap, candles, cosmetics, perfumery, etc. **2** Petroleum. **3** An oil paint; also, an oil painting. **4** Anything of an oily consistency. **5** Fawning or flattering speech; an apology or excuse. — *v.t.* **1** To smear, lubricate, or supply with oil. **2** To bribe; flatter. [< AF *olie,* OF *oile* < L *oleum* oil]

Oil may appear as a combining form in hyphemes or solidemes, or as the first element in two-word phrases; as in:

oil-bearing	oil-fired
oil box	oil-forming
oil-bright	oil-fueled
oil-burning	oil gage
oilcamp	oil gas
oil-carrying	oil groove
oil-containing	oil-harden
oil crane	oil-hardened
oil cup	oil-hardening
oil derrick	oil heater
oil-dispensing	oilhole
oil distiller	oil industry
oil-distributing	oil-insulated
oil drill	oil-laden
oil-driven	oil-ladened
oil engine	oil land
oil-fed	oil-lit
oil-filled	oil press
oil-finding	oil-producer
oil-finished	oil-producing

oilproof	oil-stained
oilproofing	oil stove
oil-pumping	oil tanker
oil refiner	oil tar
oil-refining	oil-tempered
oil-regulating	oil-testing
oil-rich	oil-thickening
oil-saving	oiltight
oil-secreting	oiltightness
oilskinned	oil tube
oil-smelling	oilway
oil-soaked	oil-yielding

oil·bird (oil′bûrd′) *n.* The guacharo.

oil burner **1** A furnace or heating unit that operates on oil fuel. **2** An atomizer for spraying oil into such a furnace.

oil·cake (oil′kāk′) *n.* The mass of compressed seeds of cotton, flax, etc., or coconut pulp from which oil has been expressed.

oil·can (oil′kan′) *n.* A can for holding lubricating or fuel oil.

oil·cloth (oil′klôth′, -kloth′) *n.* A cotton fabric coated on one side with a preparation of vegetable oils and pigments mixed with a clay filler, and sometimes ornamented with printed patterns: used for table, shelf, wall, or floor coverings, for bags, luggage, waterproofing, etc.

oil color A pigment ground in linseed or other oil, used chiefly by artists.

oil·er (oil′ər) *n.* **1** One who or that which oils; specifically, one who oils engines or machinery. **2** Any automatic device for oiling machinery. **3** A coat of oilskin. **4** A vessel for the transportation of oil.

oil field An oil-producing area, especially one under active exploitation.

oil gland **1** *Bot.* An oil-secreting gland, as in some plants. **2** *Ornithol.* The gland at the rump of a bird that secretes oil for the dressing of the plumage; the uropygial gland.

oil of lavender A yellow, fragrant oil distilled from the blossoms of lavender, especially the spike lavender: used in medicine and perfumery. Also called **oil of spike.**

oil of turpentine See under TURPENTINE.

oil of vitriol Sulfuric acid.

oil of wintergreen Wintergreen (def. 2).

oil painting **1** The art of painting in oils. **2** A painting done in pigments mixed in oil.

oil·paper (oil′pā′pər) *n.* Paper treated with oil for transparency and resistance against moisture and dryness.

oil pool An accumulation of petroleum in sedimentary rocks, usually associated with characteristic geological features.

Oil Rivers The Niger delta, Nigeria; governed by Britain as the **Oil Rivers Protectorate,** 1890-93.

oil shale A compact, typically laminated, brown or black sedimentary shale impregnated with petroleum in varying proportions.

oil·skin (oil′skin′) *n.* Cloth made waterproof with drying oil, or a garment of such material.

oil slick A smooth area on the surface of water caused by a film of oil.

oil·stone (oil′stōn′) *n.* A smooth stone, used when moistened with oil for sharpening tools, etc.

oil well A well or boring for petroleum.

oil·y (oi′lē) *adj.* **oil·i·er, oil·i·est** **1** Pertaining to or containing oil. **2** Smeared, rubbed, soaked, or coated with oil. **3** Smooth or deceitfully affable. — **oil′i·ly** *adv.* — **oil′i·ness** *n.*

oi·nol·o·gy (oi·nol′ə·jē) See ENOLOGY.

oi·no·mel (oi′nə·mel) See OENOMEL.

oint·ment (oint′mənt) *n.* A fatty preparation, with which a medicine has been incorporated; an unguent. [< OF *oignement,* ult. < L *unguentum* an unguent; infl. in form by obs. *oint* anoint < F, pp. of *oindre* < L *unguere*]

Oir·each·tas (er′əkh·thəs) *n. pl.* The combined legislatures of Ireland, consisting of the Dail Eireann (the representative assembly) and the Seanaid (the senate). [< Irish, assembly]

Oise (wäz) A river of Belgium and NE France, flowing 187 miles SW to the Seine.

O·je·da (ō·hā′thä), **Alonzo,** 1468?-1515?, Spanish explorer; accompanied Columbus on his second voyage.

O·jib·wa (ō·jib′wä) *n. pl.* **·wa** or **·was** **1** One of a tribe of North American Indians of Algonquian linguistic stock, formerly inhabiting the regions around Lake Superior; Chippewa. **2** The Algonquian language of this tribe. Also **O·jib′way.** [< Algonquian (Ojibwa) *Ojibwa* roast till puckered < *ojib* pucker + *ub-way* roast; with ref. to their puckered moccasin seams]

OK (ō′kā′) *interj., adj., & adv.* All correct; all right: expressing approval, agreement, etc. — *v.t.* (ō·kā′) **OK'd, OK'ing** To sign with an *OK*; endorse; approve. Also **O.K., o′kay′, o′keh′.** [< The Democratic *O.K.* Club, organized in 1840 to support President Martin Van Buren, nicknamed *Old Kinderhook,* from *Kinderhook, N.Y.,* his birthplace]

o·ka (ō′kə) *n.* **1** A weight used in Turkey, Egypt, Bulgaria, etc., equal to about 2.828 lbs. **2** A unit of capacity, equal to about one liter. [< Ital. *oca, occa* < Turkish *ōqah* < Arabic *ūqivah* < Gk. *oungia* an ounce < L *uncia*]

O·ka (o·kä′) **1** A river of the west central European Russian S.F.S.R., flowing 918 miles east to the Volga at Gorki. **2** A river in SW Buryat-Mongol S.S.R., flowing 500 miles north to the Angara.

O·ka·no·gan River (ō′kə·nō′gən) A river in northern Washington and southern British Columbia, Canada, flowing 115 miles south to the Columbia River.

o·ka·pi (ō·kä′pē) *n.* An African ruminant (*Okapia johnstoni*) related to the giraffe, but with a smaller body and a shorter neck. [< native Sudanic name]

O·ka·ya·ma (ō·kä·yä′mä) A port on SW Honshu island, Japan.

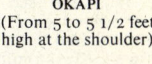
OKAPI
(From 5 to 5 1/2 feet high at the shoulder)

O·kee·cho·bee (ō′kē·chō′bē), **Lake** A lake in south central Florida; 730 square miles; 35 miles long, 30 miles wide.

Okeechobee Waterway See CROSS-FLORIDA WATERWAY.

O'Keeffe (ō·kēf′), **Georgia,** born 1887, U.S. painter.

O·ke·fi·no·kee Swamp (ō′kə·fə·nō′kē) A swamp in NE Florida and SE Georgia; 700 square miles. Also **O·kee·fe·no·kee** (ō′kē·fə·nō′kē).

O'Kel·ley (ō·kel′e), **Sean Thomas,** born 1883, Irish political leader; president of Ireland 1945-.

O·ken (ō′kən), **Lorenz,** 1779-1851, German naturalist and philosopher.

O·khotsk (ō·kotsk′, *Russian* o·khôtsk′), **Sea of** A NW arm of the Pacific west of Kamchatka and the Kurile Islands.

O·kie (ō′kē) *n. U.S. Slang* **1** An inhabitant of Oklahoma. **2** A migrant farmworker; originally, one from Oklahoma or other parts of the Dust Bowl forced by drought, mortgage foreclosure, etc., to leave his land or to seek work elsewhere. [Dim. of *Oklahoman*]

O·ki·na·wa (ō′ki·nä′wä) The largest of the Ryukyu Islands; 467 square miles; attacked and captured by United States forces, 1945, in World War II; subsequently administered by U.S. military government.

O·kla·ho·ma (ō′klə·hō′mə) A State in the south central United States; 69,919 square miles; capital, Oklahoma City; entered the Union Nov. 16, 1907; nickname, *Sooner State*: abbr. *Okla.* — **O'kla·ho′man** *adj. & n.*

Oklahoma City The capital of Oklahoma.

O·ko·van·go (ō′kō·vang′gō) A river in SW central Africa, flowing 1,000 miles south and east from central Angola to the **Okovanggo Basin,** a marsh in Bechuanaland: Portuguese *Cubango.* Also **Okovango.**

o·kra (ō′krə) *n.* **1** A tall annual herb (*Hibiscus* or *Abelmoschus esculentus*) of the mallow family. **2** Its green mucilaginous pods, used in soups and stews, or as a vegetable. **3** Gumbo. [< Ashanti]

O·ku·ma (ō·kōō′mä), **Marquis Shigenobu,** 1838-1922, Japanese statesman.

O·ku·si (ō·kōō′sē) See AMBENO.

---

add, āce, câre, pälm; end, ēven; it, īce; odd, ōpen, ôrder; tōōk, pōōl; up, bûrn; ə = a in *above,* e in *sicken,* i in *clarity,* o in *melon,* u in *focus;* yōō = u in *fuse;* oi, oil; ou, pout; ch, check; g, go; ng, ring; th, thin; ᴛʜ, this; zh, vision. Foreign sounds á, œ, ü, kh, ɴ; and ♦: see page xx. < from; + plus; ? possibly.

-ol[1] *suffix Chem.* Denoting an alcohol or phenol: *methanol, glycerol.* [<(ALCOH)OL]

-ol[2] *suffix Chem.* Var. of -OLE[1], as in *benzol.*

O·laf (ō′ləf; *Dan., Norw.* ō′läf, *Sw.* o͞o′läf) A masculine personal name; also, the name of a number of Scandinavian kings. Also **O·lav** (ō′lav, *Norw.* ō′läv, -läf), *Lat.* **O·la·us** (ō·lā′əs; *Dan.* ō·lä′o͝os, *Ger.* ō·lous′, *Sw.* o͞o·lä′o͝os). [<Scand., lit., left by his ancestors]

— **Olaf I,** 969?–1000, king of Norway 995?–1000: one of the heroes of the *Heimskringla*: known as *Olaf Trygvesson.*

— **Olaf II,** 995?–1030, king of Norway 1016?–1028; canonized; patron saint of Norway.

Ö·land (œ′länd) A Swedish island in the Baltic Sea; 519 square miles; chief town, Borgholm.

O·lav V (ō′läv, ō′läf), born 1903, king of Norway 1957–.

old (ōld) *adj.* **old·er** or **eld·er, old·est** or **eld·est** **1** Having lived or existed in a certain state for a long time: said of things liable to decay: an *old* elm; having lived beyond the middle period of life; aged: opposed to *young.* **2** Exhibiting discretion and judgment or deportment like a mature and experienced person. **3** Having some specified age: used after the noun expressing time or age: a child two months *old.* **4** Having been made, used, or known for a long time: opposed to *new, fresh, recent,* or *modern*; belonging to an early or remote period of history or development; ancient; antique: the *old* Greeks, *old* coins; also, belonging to a period long past or just preceding the present; not the latest; previous; former. **5** Belonging to the former of two or the earliest of several things: the *Old* Testament. **6** Long cultivated; not newly tilled: *old* land; not of this year's harvest: *old* corn. **7** Worthless on account of age or repeated use; shabby; worn-out: an *old* coat; also, stale; trite: an *old* joke. **8** Continued or established for a long time; known or used long; familiar: used often as an epithet of kindness or friendship: an *old* comrade. **9** Having had long experience or practice; hence, crafty; cunning: an *old* offender, an *old* hand at farming. **10** A general term of endearment or kindly familiarity: *old* boy. **11** In physical geography, in the later stages of a cycle of development: said of topographic forms, streams, etc. **12** Signifying the primeval character of the devil: the *old* enemy. **13** *Colloq.* More than enough; plentiful; great; wonderful: a great *old* racket. —*n.* **1** Past time: days of *old.* **2** A long time; long standing: my friend of *old.* [OE *ald*] —**old′ness** *n.*

Synonyms (adj.): aged, ancient, antiquated, antique, decrepit, elderly, gray, hoary, immemorial, obsolete, olden, patriarchal, remote, senile, time-honored, time-worn, venerable. That is termed *old* which has existed long, or which existed long ago. *Olden* is a statelier form of *old*, and is applied almost exclusively to time, not to places, buildings, persons, etc. As regards periods of time, the familiar are also the near; thus, the *old* times are not too far away for familiar thought and reference; the *olden* times are more remote, *ancient* times still further removed. *Aged* applies chiefly to long-extended human life. *Decrepit, gray,* and *hoary* refer to the effects of age on the body exclusively; *senile* upon the mind also; as, a *decrepit* frame, *senile* garrulousness. One may be *aged* and neither *decrepit* nor *senile. Elderly* is applied to those who have passed middle life, but scarcely reached *old* age. *Remote* primarily refers to space, but is extended to that which is far-off in time; as, at some *remote* period. See ANCIENT[1], OBSOLETE, PRIMEVAL. Compare ANTIQUE. *Antonyms*: compare synonyms for NEW, YOUTHFUL.

Old Bailey The former sessions house, or criminal court, in London; replaced (1906) by the Central Criminal Court.

Old Baldy A peak in southern Colorado, in the Sangre de Cristo Mountains; 14,125 feet: also *Baldy.*

Old·bridge (ōld′brij) A village in NE County Meath, Ireland; scene of the Battle of the Boyne, 1690.

Old Castile A region of central Spain. *Spanish* **Cas·til·la la Vie·ja** (käs·tē′lyä lä vyā′hä).

Old·cas·tle (ōld′kas·əl, -käs·əl), **Sir John** See LORD COBHAM.

Old Colony Plymouth Colony.

old country The native land of any emigrant, especially Europe.

Old Dominion Nickname of VIRGINIA.

old·en (ōl′dən) *adj.* Old; ancient. See synonyms under ANCIENT[1], OLD.

Old·en·burg (ōl′dən·bûrg, *Ger.* ōl′dən·bo͝orkh) **1** A former state of NW Germany, included after 1945 in Lower Saxony; 2,084 square miles. **2** Its former capital.

Old English See under ENGLISH.

old-fan·gled (old′fang′gəld) *adj.* Having a fondness for what is old-fashioned. [<OLD + FANGLED; on analogy with NEW-FANGLED]

old-fash·ioned (ōld′fash′ənd) *adj.* **1** Having the characteristics or customs of former times; antiquated; old-time. **2** Having the notions or ways of an old person. See synonyms under ANTIQUE.

old fo·gy (fō′gē) A person of extremely conservative or old-fashioned ideas. Also **old fo′gey.**

old-fo·gy·ish (ōld′fō′gē·ish) *adj.* Of, pertaining to, or like an old fogy; extremely conservative or behind the times. Also **old′-fo′gey·ish.**

Old French See under FRENCH.

Old Glory The flag of the United States.

old guard The conservative element in a community, political party, etc.

Old Guard The imperial guard formed by Napoleon I in 1804, composed of veterans of three campaigns. [Trans. of F *Vieille Garde*; so called in contrast with *Jeune Garde* Young Guard, formed 1810]

Old·ham (ōl′dəm) A county borough in SW Lancashire, England.

Old Harbour Bay A Caribbean inlet in southern Jamaica: also *Portland Bight.*

Old Hickory Nickname of Andrew Jackson.

Old High German See under GERMAN.

Old Icelandic See under ICELANDIC.

Old Irish See under IRISH.

Old Ironsides The U.S. frigate *Constitution*: in allusion to the hardness of her planking.

old·ish (ōl′dish) *adj.* Somewhat old.

old lady *Slang* **1** One's wife. **2** One's mother.

Old Latin See under LATIN.

old-light (ōld′līt′) *adj.* Favoring old principles; in Scottish ecclesiastical history denoting a party which favored union between church and state. —*n.* One who maintains old-light principles.

old-line (ōld′līn′) *adj.* Traditional; conservative; following a beaten path of thought or policy.

Old Line State Nickname of MARYLAND.

old maid An elderly single woman; a spinster. —**old′-maid′ish** *adj.*

old man *Slang* **1** One's father. **2** One's husband. **3** Any man in a position of authority, as one's employer, the captain of a vessel, etc.; especially, the senior officer on board a ship. **4** Old Mr. —: *Old Man* Brown. **5** A term of address, as to a friend. **6** Among North American Indians, a wise man.

old-man's beard (ōld′manz′) **1** Spanish moss. **2** The fringe tree. **3** The black gum.

Old Nick The devil.

Old Norse See under NORSE.

Old Northwest See NORTHWEST TERRITORY.

Old Orchard Beach A summer resort on the coast of SW Maine.

Old Persian See under PERSIAN.

Old Point Comfort A small peninsula at the north side of the entrance to Hampton Roads, SE Virginia.

Old Pretender See STUART, JAMES EDWARD.

Old Prussian See under PRUSSIAN.

old rose Any of various shades of grayish or purplish red.

Old Sar·um (sâr′əm) An ancient ruined city in SE Wiltshire, England.

Old Saxon See under SAXON.

Old Scandinavian See OLD NORSE under NORSE.

old sledge See SEVEN-UP.

Old South The South before the Civil War.

old squaw A sea duck *(Clangula hyemalis)* of the northern hemisphere: also called *oldwife.*

old·ster (ōld′stər) *n. Colloq.* A person of advanced years; an old or elderly person.

Old Stone Age The Paleolithic period of human culture.

old·style (ōld′stīl′) *n. Printing* A style of type first used in the 18th century, the downstrokes and the cross-strokes being of nearly the same thickness. Compare MODERN.

old-style (ōld′stīl′) *adj.* Of a former style; old-fashioned.

Old Style See GREGORIAN CALENDAR under CALENDAR.

Old Testament The first of the two main divisions of the Bible, containing the books of the old or Mosaic covenant, and including the historical books, the prophets, and the books of wisdom.

old-time (ōld′tīm′) *adj.* Of long standing.

old-tim·er (ōld′tī′mər) *n. Colloq.* **1** One who has been a member, resided in a place, or filled a position for a long time. **2** An old-fashioned person.

old·wife (ōld′wīf′) *n.* **1** Any of several fishes found in West Indian waters, as the parrotfish, the spot, alewife, menhaden, etc. **2** The old squaw.

old-wom·an·ish (ōld′wo͝om′ən·ish) *adj.* Characteristic of an old woman; fussy.

old-world (ōld′wûrld′) *adj.* **1** Of or pertaining to the Old World or eastern hemisphere; not American. **2** Prehistoric; antique.

Old World The eastern hemisphere.

-ole[1] *suffix Chem.* **1** Denoting a heterocyclic compound having five members in the ring and two hetero atoms: *pyrrole.* **2** Denoting certain aldehydes and ethers. Also spelled *-ol.* [<L *oleum* oil]

-ole[2] *suffix* Small; little: *nucleole, petiole.* [<L *-olus,* a diminutive suffix]

o·le·a·ceous (ō′lē·ā′shəs) *adj.* Designating a family *(Oleaceae)* of shrubs and trees; the olive family. It includes many widely distributed plants, including the lilac, jasmine, and ash. [<NL <L *olea* an olive tree]

o·le·ag·i·nous (ō′lē·aj′ə·nəs) *adj.* Pertaining to oil; oily. [<F *oléagineux* <L *oleaginus* pertaining to the olive <*olea* an olive tree] —**o′le·ag′i·nous·ly** *adv.* —**o′le·ag′i·nous·ness** *n.*

o·le·an·der (ō′lē·an′dər) *n.* An Old World evergreen ornamental shrub *(Nerium oleander)* with leathery leaves yielding a poisonous glycoside with medicinal properties, and clusters of fragrant, rose or white flowers. [<MF *oléandre* <Med. L *oleander* <LL *lorandrum,* ? alter. of L *rhododendron* RHODODENDRON]

o·le·as·ter (ō′lē·as′tər) *n.* An ornamental shrub or small tree *(Elaeagnus angustifolia)* of western Asia and southern Europe, with fragrant, yellow flowers; the Russian wild olive. [<L <*olea* an olive tree]

o·le·ate (ō′lē·āt) *n. Chem.* A salt or ester of oleic acid.

o·lec·ra·non (ō·lek′rə·non, ō′lə·krā′non) *n. Anat.* The curved process of the ulnar bone, marking its juncture with the humerus; the point of the elbow. [<Gk. *ōlekranon,* contraction of *ōlenokranon* head or point of the elbow <*ōlenē* elbow + *kranion* head, skull] —**o·lec′ra·nal** *adj.*

o·le·fi·ant (ō′lə·fī′ənt, ō·lē′fē·ənt) *adj.* Producing oil. [<F *(gaz) oléfiant* olefiant (gas), ppr. of *oléfier* make oil <L *oleum* oil + *facere* make]

olefiant gas Ethylene.

o·le·fin (ō′lə·fin) *n.* Alkene. [<OLEF(IANT) + -IN] —**o′le·fin′ic** *adj.*

o·le·ic (ō·lē′ik, ō′lē-) *adj.* Of, pertaining to, or derived from oil. [<L *oleum* oil + -IC]

oleic acid *Chem.* An oily compound, $C_{17}H_{33}$-CO_2H, contained as an ester in most mixed oils and fats, and obtained by saponification with an alkali.

o·le·in (ō′lē·in) *n. Chem.* A colorless liquid glyceride of oleic acid, the chief constituent of fatty oils: also *elain.* Also **o·le·ine** (ō′lē·in, ō′li·ēn). [<F *oléine* <L *oleum* oil]

O·lek·ma (o·lyek′mə) A river in southern Yakut Autonomous S.S.R., flowing 794 miles north to the Lena.

O·le·nek (o·lyi·nyôk′) A river in NW Yakut Autonomous S.S.R., flowing 1,500 miles north and east to the Laptev Sea.

o·le·o (ō′lē·ō) Short for OLEOMARGARINE.

oleo- *combining form* **1** Oil; of oil: *oleoresin.* **2** Olein; oleic: *oleomargarine.* [<L *oleum* oil]

o·le·o·graph (ō′lē·ə·graf′, -gräf′) *n.* **1** A chromolithograph imitating an oil painting. **2** The pattern assumed by a drop of oil placed on water. [<OLEO- + -GRAPH] —**o·le·og·ra·pher** (ō′lē·og′rə·fər) *n.* —**o′le·o·graph′ic** *adj.* —**o′le·og′ra·phy** *n.*

o·le·o·mar·ga·rine (ō′lē·ō·mär′jə·rin, -rēn, -gə-) *n.* Margarine. [<OLEO- + MARGARINE]

oleo oil Beef tallow, obtained as a yellow liquid consisting of olein with a small amount

oleoresin of palmitin: used in making margarine, soap, and as a base for some lubricants.

o·le·o·res·in (ō'lē·ō·rez'in) n. 1 A native compound of an essential oil and a resin. 2 A pharmaceutical preparation consisting of a fixed or volatile oil containing a resin and sometimes other active matter in solution.

O·lé·ron (dō·lā·rôn'), **Île d'** The largest island in the Bay of Biscay, belonging to France; 68 square miles.

ol·fac·tie (ol·fak'tē) n. Physiol. A unit of measurement of olfactory sensation, equal to the lowest perceptible concentration of a given scent. Also **ol·fac'ty**. [Coined by Dr. Zwaardemaker, 19th c. Dutch scientist, inventor of the olfactometer]

ol·fac·tion (ol·fak'shən) n. The act, sense, or process of smelling; scent. [<L olfactus. See OLFACTORY.]

ol·fac·tom·e·ter (ol'fak·tom'ə·tər) n. An instrument for measuring the keenness of the sense of smell. [<OLFACTO(RY) + -METER]

ol·fac·to·ry (ol·fak'tər·ē, -trē) adj. Pertaining to the sense of smell. — n. pl. ·ries 1 Usually pl. The organ of smell. 2 The capacity to smell. [<L olfactus, pp. of olfacere smell < olere have a smell + facere make]

Ol·ga (ol'gə, Russian ōl'gä) A feminine personal name. [<Russian <Scand., holy]

o·lib·a·num (ō·lib'ə·nəm) n. A gum resin; Oriental frankincense. [<Med. L <LL libanum <Gk. libanos <Arabic al-lubān the frankincense]

Ol·i·fants River (ol'ə·fənts) A river in the Union of South Africa and Mozambique, flowing 350 miles NE to the Limpopo river.

ol·i·garch (ol'ə·gärk) n. A ruler in an oligarchy. [<Gk. oligarches <oligos few + archein rule]

ol·i·gar·chy (ol'ə·gär'kē) n. pl. ·chies A form of government in which supreme power is restricted to a few. [Prob. <Med. L oligarchia <Gk. <oligarches. See OLIGARCH.] — **ol'i·gar'chic, ol'i·gar'chal, ol'i·gar'chi·cal** adj.

oligo- combining form Small; few; scanty: oligocythemia. Also, before vowels, **olig-**. [<Gk. oligos few]

Ol·i·go·cene (ol'ə·gō·sēn') n. Geol. The third in order of age of the geological epochs or series comprised in the Lower Tertiary system. — adj. Of or pertaining to the Oligocene. [<OLIGO- + Gk. kainos new, recent]

ol·i·go·chaete (ol'ə·gō·kēt') n. One of a class (Oligochaeta) of fresh-water and terrestrial, hermaphroditic annelid worms, including the earthworms, which lack a distinct head. — adj. Of or pertaining to the Oligochaeta. Also **ol'i·go·chete'**. [<NL <Gk. oligos few + chaitē bristle, mane; so called because it has a small number of bristly locomotive organs] — **ol'i·go·chae'tous** (-kē'təs) adj.

ol·i·go·clase (ol'ə·gō·klās') n. A massive, vitreous, whitish, triclinic soda-lime feldspar. [<OLIGO- + Gk. klasis a fracture <klaein break]

ol·i·go·chrom·e·mi·a (ol'ə·gō·krō·mē'mē·ə) n. Pathol. A deficiency of hemoglobin in the blood. [<NL oligochromaemia <Gk. oligos few + chrōma color + haima blood]

ol·i·go·cy·the·mi·a (ol'ə·gō·sī·thē'mē·ə) n. Pathol. A deficiency or diminution of the red blood corpuscles. Also **ol'i·go·cy·thae'mi·a**. [<NL oligocythaemia <Gk. oligos few + kytos hollow, a cell + haima blood]

ol·i·go·gen·ics (ol'ə·gō·jen'iks) n. Limitation in the number of children; birth control. [<OLIGO- + (EU)GENICS]

ol·i·go·phre·ni·a (ol'ə·gō·frē'nē·ə) n. Arrested mental development. [<NL <Gk. oligos little + phrēn mind] — **ol'i·go·phren'ic** (-fren'ik) adj.

ol·i·gop·o·ly (ol'ə·gop'ə·lē) n. pl. ·lies A form of monopoly in which the effective control of a market is exercised by a limited number of competitive sellers. [<OLIGO- + (MONO)POLY]

ol·i·gop·so·ny (ol'ə·gop'sə·nē) n. pl. ·nies A market condition in which the purchase of goods and services is restricted to a few buyers. [<OLIG(O)- + Gk. opsōneein buy victuals]

ol·i·go·syn·thet·ic (ol'ə·gō·sin·thet'ik) adj. 1 Based upon or derived from a few essential components. 2 Ling. Describing a language whose lexicon is composed of relatively few roots.

o·li·o (ō'lē·ō) n. pl. **o·li·os** 1 A miscellaneous collection, as of musical pieces or numbers; a medley. 2 An olla podrida; a seasoned meat and vegetable stew. [<Sp. olla. See OLLA.]

Ol·i·phant (ol'ə·fənt), **Margaret**, 1828–97, née Wilson, English novelist.

ol·i·va·ceous (ol'ə·vā'shəs) adj. Olive-green. [<NL olivaceus <L oliva an olive]

Ol·i·vant (ol'ə·vənt), **Alfred**, 1874–1927, English novelist.

ol·i·var·y (ol'ə·ver'ē) adj. 1 Like an olive, especially in shape. 2 Anat. Relating to the **olivary body**, an olive-shaped eminence containing a nucleus of gray matter, found at either side of the medulla oblongata. [<L olivarius belonging to olives <oliva an olive]

ol·ive (ol'iv) n. 1 An evergreen tree (Olea europaea) with leathery leaves, hard yellow wood, and an oily fruit. 2 The fruit of the olive tree. 3 A dull, medium yellowish-green color, like that of the unripe olive: also called **olive green**. 4 An olive branch. — adj. 1 Pertaining to the olive. 2 Having a dull yellowish-green color. [<OF <L oliva an olive]

Ol·ive (ol'iv) See OLIVIA.

olive branch 1 A branch of the olive tree, as an emblem of peace. 2 pl. offspring; children: alluding to Psalms cxxviii 3.

olive drab 1 Any of several shades of greenish-brown. 2 A woolen material of this color, used for uniforms by the United States Army. — **ol'ive-drab'** (-drab') adj.

olive oil Oil expressed from olives: used in making salad dressings, soap, etc.

Ol·i·ver (ol'ə·vər) A masculine personal name. Also Du., Ger., Sw. **O·li·vier** (ō'li·vēr', Fr. ō·lē·vyā'), Ital. **O·li·vie·ro** (ō'lē·vyā'rō), Sp. **O·li·vei·ro** (ō'lē·vā'rē·ō), **O·li·ve·ri·o** (ō'lē·vā·rē·ō). [Prob. <Gmc., army of the elves] — Oliver A paladin of Charlemagne's court.

Olives, Mount of A hill just east of Jerusalem: Matt. xvi 1. Also **Ol·i·vet** (ol'ə·vet, -vit).

O·liv·i·a (ō·liv'ē·ə, Du., Ger., Ital., Sw. ō·lē'vyä) A feminine personal name. Also Fr. **O·li·vie** (ō·lē·vē'). [<L oliva olive tree] — Olivia In Shakespeare's Twelfth Night, a countess courted by Orsino.

ol·i·vine (ol'ə·vēn, -vin) n. 1 Chrysolite. 2 Green garnet: used as a gem. [<L oliva an olive + -INE²]

ol·la (ol'ə, Sp. ô'lyä, ô'yä) n. 1 A wide-mouthed pot or jar, usually of earthenware. 2 An olla podrida. [<Sp. <L olla a pot]

ol·la po·dri·da (ol'ə pə·drē'də, Sp. ô'lyä pō·thrē'thä, ô'yä) 1 A dish of meat and vegetables, usually highly seasoned, cooked together. 2 Any heterogeneous mixture or miscellany. [<Sp., lit., a putrid pot <olla an olla + podrida putrid <L putridus]

ol·o·gy (ol'ə·jē) n. pl. ·gies Colloq. A science or branch of learning: a humorous term. [<-LOGY]

Ol·o·mouc (ô'lô·mōts) A city in north central Moravia, Czechoslovakia. German **Ol·mütz** (ôl'mūts).

Ol·sztyn (ôl'shtin) A city in NE Poland; formerly in East Prussia: German **Allenstein**.

Olt (ôlt) A river in Rumania, flowing 348 miles SW to the Danube. Also **O·tul** (ō'tōōl).

Ol·te·ni·a (ol·tē'nē·ə, Rumanian ōl·tā'nyä) The western part of Wallachia: also **Lesser Wallachia**.

Ol·tis (ol'tis) The ancient name for the LOT.

o·ly·koek (ō'lē·kōōk) n. U.S. Dial. A Dutch cake made like a cruller. [<Du. oliekoek a doughnut]

O·lym·pi·a (ō·lim'pē·ə) 1 An ancient city on a plain near Elis in western Peloponnesus, Greece; scene of the Olympic games. 2 A port on Puget Sound; capital of Washington.

O·lym·pi·ad (ō·lim'pē·ad) n. 1 The interval of four years between two successive celebrations of the Olympic games, by which intervals the ancient Greeks reckoned time: sometimes erroneously used to designate the games or their celebration. 2 The modern Olympic games. [<MF Olympiade <L Olympias, -adis <Gk. Olympias, -ados <Olympios OLYMPIAN]

O·lym·pi·an (ō·lim'pē·ən) adj. 1 Pertaining to the great gods of Olympus or to Mount Olympus itself; hence, of eminent, godlike power, excellence, or manner. 2 Pertaining to Olympia or to the Olympic games. Also **O·lym'pic**. — n. 1 One of the higher gods of Greek mythology, twelve in number, who dwelt on Mount Olympus. 2 A contestant in the Olympic games. 3 A resident or native of Olympia. [<LL Olympianus <L Olympius <Gk. Olympios <Olympia Olympia, the Olympic games]

Olympic games 1 Athletic games and races held at the chief ancient Pan-Hellenic festival, which was celebrated every four years at Olympia in honor of Zeus. See OLYMPIAD. 2 A modern revival of the old contests, held every four years at some city chosen for this event. The first of these games occurred at Athens in 1896. Also **Olympian games, Olympics**.

Olympic National Park A heavily forested region, 1,305 square miles; established 1938; in the **Olympic Mountains**, part of the Coast Range in NW Washington; highest peak, **Mount Olympus**, 7,954 feet.

Olympic Peninsula A peninsula bounded by Puget Sound, the Pacific, and Juan de Fuca Strait, including the **Olympic Mountains**.

O·lym·pus (ō·lim'pəs) 1 The highest mountain of Greece, between Thessaly and Macedonia on the Aegean, regarded in Greek mythology as the home of the Olympian gods; 9,570 feet. Also **Mount O·lym'pus**. 2 Any abode of gods; also, the sky; heaven.

O·lyn·thus (ō·lin'thəs) An ancient city in SE Macedonia; destroyed 348 B.C. by Philip of Macedon. — **O·lyn'thi·ac** (-thē·ak) adj.

Om (ōm) n. 1 In Hinduism, a mystic ejaculation representing the name of the Supreme Being, uttered on solemn occasions of invocation to Brahma. 2 In modern occultism, the spiritual essence. [<Skt.]

-oma suffix Med. Tumor; morbid growth: carcinoma. [<Gk. -ōma]

O·magh (ō'mə) The county town of County Tyrone, Northern Ireland.

O·ma·ha (ō'mə·hä, -hô) n. One of a Siouan tribe of North American Indians now living in Nebraska. [<Siouan (Omaha), lit., those going upstream]

O·ma·ha (ō'mə·hä, -hô) A city on the Missouri River in eastern Nebraska.

Omaha Beach A name given to that part of the Normandy coast where units of the United States Army landed on June 6, 1944, during the Allied invasion of France, World War II.

O·man (ō'man, ō·man', ō·män') 1 The coastal region of the eastern promontory (**Oman Promontory**) of the Arabian peninsula. 2 Muscat and Oman. See TRUCIAL OMAN.

Oman, Gulf of A NW arm of the Arabian Sea between the Oman section of the Arabian peninsula and Iran.

O·man (ō'mən), **Sir Charles William**, 1860–1946, English historian.

O·mar Khay·yám (ō'mär kī·äm', ō'mər), died 1123?, Persian poet and astronomer; author of the Rubáiyát.

o·ma·sum (ō·mā'səm) n. pl. ·sa (-sə) The manyplies or third stomach of a ruminant; the psalterium; paunch. [<NL <L, bullock's tripe]

O·may·yad (ō·mī'ad) See OMMIAD.

om·ber (om'bər) n. A gambling game played with 40 cards, popular in the 18th century; also, the player unsuccessful to win the pool in this game. Also **om'bre**. [<F ombre <Sp. hombre a man <L homo, hominis]

Om·bi Islands (om'bē) See OBI ISLANDS.

ombro- combining form Rain: ombrophilous. Also, before vowels, **ombr-**. [<Gk. ombros rain]

om·brom·e·ter (om·brom'ə·tər) n. A rain gage. [<OMBRO- + -METER]

om·broph·i·lous (om·brof'ə·ləs) adj. Bot. Relating to or characterizing plants able to withstand much rain. [<OMBRO- + -PHILOUS] — **om·bro·phile** (om'brə·fīl, -fil) n.

Om·dur·man (om'dŏŏr-män') A city on the White Nile, opposite Khartum in central Sudan.

-ome combining form Bot. Group; mass; body: caulome. [<Gk. -ōma. See -OMA.]

o·me·ga (ō·mē'gə, ō·meg'ə, ō'meg·ə) n. The twenty-fourth and last letter and seventh vowel in the Greek alphabet (Ω, ω); figuratively, the end; the last. It corresponds to English long o. As a numeral it denotes 800. Compare ALPHA, OMICRON. [<Gk. ō mega great o]

om·e·let (om'ə·lit, om'lit) n. A dish of eggs, etc., beaten together and cooked in a frying pan. Also **om'e·lette**. [<F omelette <OF amelette <alemette, lit., a thin plate < alemelle < la lemelle, lamelle <L lamella. See LAMELLA.]

o·men (ō'mən) n. A phenomenon or incident regarded as a prophetic sign. See synonyms under SIGN. — v.t. To foretell as or by an omen; presage; preshadow. [<L]

o·men·tum (ō·men'təm) n. pl. **·ta** (-tə) Anat. A free fold of the peritoneum passing between certain of the viscera. The **small omentum** passes from the lesser curvature of the stomach to the liver; the **great omentum** from the lower border of the stomach to the transverse colon, lying in front of the intestines like an apron. [<NL <L, a membrane enclosing the bowels] — **o·men'tal** adj.

o·mer (ō'mər) n. A Hebrew measure of capacity; the tenth of an ephah. [<Hebrew 'ōmer]

Om·fre·do (ōm·frā'dō) Italian form of HUMPHREY.

om·i·cron (om'ə·kron, ō'mə-) n. The fifteenth letter and fifth vowel of the Greek alphabet (Ο, ο): equivalent to English short o. As a numeral it denotes 70. Also **om'i·kron**. [<Gk. o mikron little o]

om·i·nous (om'ə·nəs) adj. 1 Of the nature of or marked by an omen or by a presentiment of evil; portentous; ill-omened: ominous fears. 2 Serving as an omen in general; prognostic. [<L ominosus <omen, ominis an omen] — **om'i·nous·ly** adv. — **om'i·nous·ness** n.

Om·ish (om'ish) See AMISH.

o·mis·si·ble (ō·mis'ə·bəl) adj. That can be omitted; subject to omission.

o·mis·sion (ō·mish'ən) n. 1 The act of omitting or state of being omitted or neglected. 2 Anything omitted or neglected. 3 Neglect or failure to do something that can and should be done; also, an instance of this. See synonyms under ERROR. [<L omissio, -onis <omissus, pp. of omittere OMIT] — **o·mis·sive** (ō·mis'iv) adj. — **o·mis'sive·ly** adv.

o·mit (ō·mit') v.t. **o·mit·ted, o·mit·ting** 1 To leave out; fail to include. 2 To fail to do, make, etc.; neglect or forbear. [<L omittere let go < ob- down, away + mittere send]

om·ma·tid·i·um (om'ə·tid'ē·əm) n. pl. **·tid·i·a** (-tid'ē·ə) Zool. One of the simple elements of a compound eye, as in arthropods. [<NL, dim. of Gk. omma, -atos eye] — **om'ma·tid'i·al** adj.

om·mat·o·phore (ə·mat'ə·fôr, -fōr) n. Zool. An eyestalk, as of a snail. [<NL ommatophorus <Gk. omma, -atos eye + -phoros bearing <pherein bear] — **om·ma·toph·o·rous** (om'ə·tof'ər·əs) adj.

Om·mi·ad (om'ē·ad) n. pl. **Om·mi·ads** or **Om·mi·a·des** (om'ē·ə·dēz) A member of a dynasty of early Moslem caliphs who ruled at Damascus 661–750, and in southern Spain 756–1031. Also spelled Omayyad. [after Omayyah, great-grandfather of the first caliph of the dynasty]

omni- combining form All; totally: omnipotent. [<L omnis all]

om·ni·a bo·na bo·nis (om'nē·ə bō'nə bō'nis) Latin All things (are) good to the good.

om·ni·a vin·cit a·mor (om'nē·ə vin'sit ā'môr) Latin Love conquers all things.

om·ni·bear·ing (om'nə·bâr'ing) n. The bearing of an omnirange.

om·ni·bus (om'nə·bəs, -bus) n. 1 A long passenger vehicle sometimes with two decks; a bus. 2 A printed anthology, either of works by a single author or of short stories, poems, etc., of the same general type: an omnibus of Westerns; a Conrad omnibus. 3 An omnibus box. — adj. Covering a full collection of objects or cases: an omnibus bill. [<F <L, for all, dat. pl. of omnis all]

omnibus bar A bus bar.

omnibus bill Any legislative bill or act, or section thereof, containing miscellaneous unrelated provisions.

omnibus box A large box or loge on a level with the stage of a theater.

om·ni·di·rec·tion·al (om'ni·di·rek'shən·əl) adj. Telecom. Capable of or adapted for operating equally well in all directions, as a radio transmitter or antenna.

om·ni·far·i·ous (om'nə·fâr'ē·əs) adj. Of all varieties, forms, or kinds. [<L omnifarius <omnis all + fari speak]

om·nif·er·ous (om·nif'ər·əs) adj. Producing all kinds. [<L omnifer <omnis all + ferre bear]

om·nif·ic (om·nif'ik) adj. All-creating. [<Med.L omnificus <L omnis all + facere make]

om·nip·a·rous (om·nip'ər·əs) adj. Producing or bearing all things. [<LL omniparus <L omnis all + parere produce]

om·nip·o·tence (om·nip'ə·təns) n. 1 Unlimited and universal power, as a divine attribute; hence, **Omnipotence** God. 2 Unlimited power within a certain sphere, or of a certain kind. Also **om·nip'o·ten·cy**.

om·nip·o·tent (om·nip'ə·tənt) adj. Almighty; not limited in authority or power. — **The Omnipotent** God. [<OF <L omnipotens, -entis <omnis all + potens, -entis able, powerful] — **om·nip'o·tent·ly** adv.

om·ni·pres·ence (om'nə·prez'əns) n. The quality of being everywhere present at the same time; ubiquity. [<Med.L omnipraesentia <omnipraesens, -entis <L omnis all + praesens, -entis present] — **om'ni·pres'ent** adj.

om·ni·range (om'nə·rānj) n. Aeron. A network of very-high-frequency radio beams emitted simultaneously in all directions from a system of ground stations, permitting aircraft pilots to chart their courses and positions anywhere within range of the network. [<L omnis all + RANGE]

om·nis·cience (om·nish'əns) n. 1 Infinite knowledge; an attribute of God; hence, **Omniscience** God. 2 Loosely, very extensive knowledge. Also **om·nis'cien·cy**.

om·nis·cient (om·nish'ənt) adj. Knowing all things; all-knowing. — **The Omniscient** God. [<NL omnisciens, -entis <L omnis all + sciens, -entis, ppr. of scire know] — **om·nis'cient·ly** adv.

om·ni·um-gath·er·um (om'nē·əm·gath'ər·əm) n. A miscellaneous collection; a medley. [<L omnium, genitive pl. of omnis all + GATHER + L -um, neut. suffix]

om·niv·o·rous (om·niv'ər·əs) adj. 1 Eating food of all kinds indiscriminately; hence, greedy. 2 Eating both animal and vegetable food: said of bears, crows, etc. [<L <omnis all + vorare devour] — **om·niv'o·rous·ly** adv. — **om·niv'o·rous·ness** n.

O·mo·lon (o'mə·lôn') A river in Asian Russian S.F.S.R., flowing 715 miles north to the Kolyma.

o·mo·pha·gi·a (o'mə·fā'jē·ə) n. The eating of raw flesh. Also **o·moph·a·gy** (ō·mof'ə·jē). [<NL <Gk. ōmophagia <ōmos raw + phagein eat] — **o·mo·phag·ic** (ō'mə·faj'ik), **o·moph·a·gous** (ō·mof'ə·gəs) adj. — **o·moph·a·gist** (ō·mof'ə·jist) n.

O'More (ō·môr'), **Rory** Name of three Irish rebel chieftains of the 16th century: also Rory O'more.

Om·pha·le (om'fə·lē) In Greek mythology, a Lydian queen in whose service Hercules, dressed as a woman, spun wool and did other womanly tasks for three years, in order to expiate a murder.

om·pha·los (om'fə·ləs) n. 1 A round stone in the temple of Apollo at Delphi, supposed by the ancient Greeks to mark the middle point of the earth. 2 The central boss of a shield. 3 A central point; hub. [<Gk., navel]

Omsk (ômsk) A city on the Irtish in western Asiatic Russian S.F.S.R.

O·mu·ta (ō·mōō'tä) A port on western Kyushu island, Japan. Also **O·mu·da** (ō·mōō'dä).

on (on, ôn) prep. 1 In contact with the upper surface of; above and supported by: lying on the ground. 2 In contact with any surface or part of: a blow on the head. 3 So as to be suspended from: a puppet on a string. 4 Directed or moving along the course of: Be on your way. 5 Near; adjacent to: the town on the river bank; the store on your right. 6 Within the duration of: He arrived on my birthday. 7 At the moment or point of: on the hour; at the time of: He withdrew on my speaking thus. 8 In a state or condition of: on fire; on record. 9 By means of; with the support of: on wheels; on all fours. 10 Using as a means of sustenance, activity, etc.: living on fruit. 11 Accumulated with; in addition to: thousands on thousands of them. 12 Sustained or confirmed by; with the authority of: committed on purpose; I swear to it on my honor. 13 In the interest or favor of: betting on a horse. 14 Concerning; about: a work on economics. 15 Engaged in; occupied or connected with: on a journey; on duty all night. 16 As a consequence or result of: making a profit on tips. 17 In accordance with or relation to; in terms of: measured on the Centigrade scale. 18 Directed, tending, or moving toward or against: making war on the enemy. 19 Following after: disease on the heels of famine. 20 Colloq. With; accompanying, as about one's person: Do you have five dollars on you? 21 Colloq. At the expense of; paid by: The joke is on them; drinks on the house. See synonyms under ABOVE, AT. — **to have something on** U.S. Colloq. To have knowledge, possess evidence, etc., against (a person). — adv. 1 In or into a position or condition of contact, adherence, covering, etc.: He put his hat on. 2 In the direction of something: He looked on while they played. 3 In advance; ahead, in space or time: a collision head on; later on. 4 In continuous course or succession: The music went on. 5 In or into operation, performance, or existence: to turn the electricity on. — **and so on** And like what has gone before; et cetera. — **on and on** Without interruption; continuously. — **to be on to** To be aware of or informed about (someone, something, etc.); understand. — adj. 1 In operation, progress, or application: The play is on; The brake is on. 2 Near; located nearer. 3 In cricket, indicating or pertaining to the side of the wicket and field where the batsman stands. — n. 1 The state or fact of being on. 2 In cricket, the on side of the field or wicket. [OE on, an]

On (on) The Egyptian name for HELIOPOLIS.

on·a·ger (on'ə·jər) n. pl. **·gers** or **·gri** (-grī) 1 A wild ass (Equus onager) of central Asia. 2 A medieval military engine by which stones were hurled with a slinglike device. [<L, a wild ass <Gk. onagros <onos an ass + agrios wild]

on·a·gra·ceous (on'ə·grā'shəs) adj. Pertaining to or designating a family (Onagraceae) of plants of temperate climates; the evening-primrose family. [<NL <L onagra, fem. of onager ONAGER]

o·nan·ism (ō'nən·iz'əm) n. 1 Withdrawal before orgasm; incomplete coitus. 2 Masturbation. [after Onan. See Gen. xxxviii 9.] — **o'nan·ist** n. — **o·nan·is'tic** adj.

once (wuns) adv. 1 One time, without repetition. 2 During some past time. 3 At any time; ever; also, at some future time. — adj. Former; formerly existing; quondam. — conj. As soon as; whenever. — n. One time. — **all at once** All of a sudden. — **at once** 1 Simultaneously. 2 Immediately. — **once for all** Finally. — **once in a while** Occasionally. — **this once** On this occasion only. [ME ones, OE anes, genitive of an one]

once-o·ver (wuns'ō'vər) n. Slang 1 A quick glance or survey. 2 A brief but comprehensive application of labor or study.

On·cid·i·um (on·sid'ē·əm) n. A large, varied genus of tropical American orchids with few leaves, and a loose raceme of striking flowers. They are among the most prized of cultivated orchids, O. papilio, the butterfly orchid, being one of the best-known. [<NL <Gk. onkos an arrow's barb; so called from the form of the lower petal]

on·col·o·gy (on·kol'ə·jē) n. The science of tumors. [<Gk. onkos bulk, a tumor + -LOGY]

on·come (on'kum') n. Scot. 1 A rainfall or snowfall. 2 A sudden attack of disease; also, a gathering or swelling.

on·com·ing (on'kum'ing) adj. Approaching. — n. An approach.

on·ding (on'ding') n. Scot. A fall of snow or rain.

on·do·gram (on'də·gram) n. The record made by an ondograph. [<F onde a wave (<L unda) + -(O)GRAM]

on·do·graph (on'də·graf, -gräf) *n.* An instrument by which electric wave forms, especially those of alternating currents, are recorded autographically. [<F *onde* a wave (<L *unda*) + -(O)GRAPH]

on·dom·e·ter (on·dom'ə·tər) *n.* A meter for registering the frequency of electromagnetic waves. [<F *onde* a wave (<L *unda*) + -(O)METER]

one (wun) *adj.* 1 Being a single individual or object; being a unit. 2 Being an individual or thing thought of as indefinite. 3 Designating a person, thing, or group as contrasted with another; this; that. 4 Single in kind; the same; closely united or alike. 5 Unitary. — *n.* 1 A single unit; the cardinal number preceding two; also, a symbol (1, i, I) representing this number. 2 A single thing or person. — *pron.* 1 Someone or something; anyone or anything. 2 One of certain persons or things already mentioned. — **all one** Of the same or of no consequence. — **at one** In harmony; the same. — **one another** Each other: said of an action or relation involving two or more persons or things reciprocally: They love *one another.* — **one by one** Singly and in succession. — **one day** Some indefinite day or period in the past or future. [OE *ān*]

-one *suffix Chem.* Denoting an organic compound of the ketone group: *acetone*. [<Gk. *-ōnē*, fem. patronymic]

O·ne·ga (ō·neg'ə), **Lake** The second largest lake in Europe, in NW European Russian S.F.S.R.; about 3,880 square miles.

one-horse (wun'hôrs') *adj.* 1 Drawn or adapted to be worked by one horse. 2 *Colloq.* Of inferior resources or capacity; small; unimportant: a *one-horse* town.

O·nei·da (ō·nī'də) *n.* A member of a tribe of North American Indians of Iroquoian stock.

Oneida Lake A lake in central New York; 20 miles long; 80 square miles.

O'Neill (ō·nēl'), **Eugene Gladstone**, 1888-1953, U.S. playwright.

o·nei·rism (ō·nī'riz·əm) *n. Psychol.* A psychic condition induced by or resembling dreams but prolonged into the waking period. Compare HYPNOPOMPIC. [<Gk. *oneiros* a dream]

oneiro- *combining form* Dream; of dreams: *oneiromancy*: also **oniro-**. Also, before vowels, **oneir-**. [<Gk. *oneiros* a dream]

o·nei·ro·crit·ic (ō·nī'rə·krit'ik) *adj.* Pertaining to or professing power to interpret dreams: also **o·nei·ro·crit'i·cal.** — *n.* One who interprets dreams. [<Gk. *oneirokritikos* pertaining to the interpretation of dreams <*oneiros* a dream + *kritikos* able to discern <*krinein* judge] — **o·nei·ro·crit'i·cal·ly** *adv.*

o·nei·rol·o·gy (ō·nī·rol'ə·jē) *n.* The study of dreams. [<ONEIRO- + -LOGY]

o·nei·ro·man·cy (ō·nī'rə·man'sē) *n.* Divination by means of dreams. — **o·nei'ro·man'cer, o·nei'ro·man'tist** *n.*

one·ness (wun'nis) *n.* 1 Singleness; unity; sameness. 2 Agreement; concord. 3 Quality of being unique. See synonyms under UNION.

one-night stand (wun'nīt') *U.S.* A town or theater in which a traveling show gives a performance on one night only; also, the performance itself.

on·er·ous (on'ər·əs) *adj.* 1 Burdensome or oppressive. 2 *Law* Legally liable for an obligation or subject to a burden: opposed to *gratuitous*. See synonyms under ARDUOUS, DIFFICULT. [<OF *onereus* <L *onerosus* <*onus, oneris* a burden] — **on'er·ous·ly** *adv.* — **on'er·ous·ness** *n.*

on·er·y (on'ər·ē) See ORNERY.

one·self (wun'self', wunz'-) *pron.* One's own self or personality; himself or herself. Also **one's self.**

one-sid·ed (wun'sī'did) *adj.* 1 Of or pertaining to but one side; hence, partial; unfair; inadequate. 2 Unequal-sided, as elm leaves. 3 Unilateral. — **one'-sid'ed·ly** *adv.* — **one'-sid'ed·ness** *n.*

one-step (wun'step') *n.* 1 A round dance consisting of a long step in two-four time. 2 The ragtime music for such a dance.

one-time (wun'tīm') *adj.* Former; quondam.

one-track (wun'trak') *adj. Colloq.* Limited to a single idea or purpose; undiversified: a *one-track* mind.

one-way (wun'wā') *adj.* 1 Moving in one direction only: *one-way* traffic. 2 Permitting traffic in one direction only: a *one-way* thoroughfare.

On-froi (ôn·frwà') French form of HUMPHREY.

Ong·jin (ông·jin) A city of central Korea. Japanese **O·shin** (ō·shin).

on·ie (on'ē) See ONY.

o·ni·o·ma·ni·a (ō'nē·ə·mā'nē·ə, -mān'yə) *n.* A propensity for the inordinate and indiscriminate buying of things. [<NL <Gk. *ōnios* for sale + *mania* madness]

on·ion (un'yən) *n.* 1 A field-grown edible bulb of an herb (*Allium cepa*) of the lily family; a succulent vegetable remarkable for its pungent odor and taste. 2 One of various allied plants. [<OF *oignon* <L *unio, -onis* unity, a pearl, an onion. Doublet of UNION.]

On·ions (un'yənz), **Charles Talbut**, born 1873, English philologist and lexicographer. — **Oliver**, 1873-1961, English novelist.

on·ion·skin (un'yən·skin') *n.* A thin, translucent paper.

oniro- See ONEIRO-.

on·look·er (on'lŏŏk'ər, ôn'-) *n.* One who looks on; a spectator.

on·look·ing (on'lŏŏk'ing, ôn'-) *adj.* 1 Looking on. 2 Looking forward.

on·ly (ōn'lē) *adv.* 1 Without another or others; singly. 2 In one manner or for one purpose alone. 3 In full; wholly. 4 Solely; merely; exclusively: limiting a statement to a single defined person, thing, or number. — *adj.* 1 Alone in its class; having no fellow or mate; sole; single; solitary: an *only* child. 2 Standing alone by reason of superior excellence. See synonyms under SOLITARY. — *conj.* Except that; but. [OE *ānlīc* <*ān* one + *-līc* -LY]

on·ly-be·got·ten (ōn'lē·bi·got'n) *adj.* Begotten as the sole issue or undisputed and incontestable heir: the *only-begotten* Son of God.

O·no·fre·do (ō'nō·frā'dō) Italian form of HUMPHREY.

on·o·mas·tic (on'ə·mas'tik) *adj.* 1 Of or pertaining to a name. 2 *Law* Designating a signature of an instrument the body of which is in another handwriting, or the instrument itself. [<Gk. *onomastikos* of naming <*onomastos* named <*onomazein* name <*onoma* a name]

on·o·ma·to·ma·ni·a (on'ə·mat'ə·mā'nē·ə, -mān'yə, -nom'ə-) *n.* A morbid dread of some particular word or name, or an irresistible impulse to repeat it. [<NL <Gk. *onoma, -atos* a name + *mania* madness]

on·o·ma·to·poe·ia (on'ə·mat'ə·pē'ə, ō·nom'ə·tə-) *n.* 1 The formation of words in imitation of natural sounds, as *crack, splash,* or *bow-wow*. 2 An imitative word. 3 The selection and use of such words. Also **on'o·mat'o·po·e'sis** (-pō·ē'sis), **on'o·mat'o·py** (-mat'ə·pē). [<L <Gk. *onomatopoiia* the making of words <*onoma, -atos* a name + *poiein* make] — **on'o·mat'o·poe'ic** or **·i·cal, on'o·mat'o·po·et'ic** (-pō·et'ik) *adj.* — **on'o·mat'o·poet'i·cal·ly** *adv.*

On·on·da·ga (on'ən·dô'gə, -dä'-) *n.* 1 One of a tribe of North American Indians of Iroquoian stock formerly living in New York and Ontario. 2 *Geol.* A limestone formation of the lower portion of the Devonian period. [<Iroquoian *ononta'gé*, lit., on top of the hill] — **On'on·da'gan** *adj.*

Onondaga Lake A salt lake in central New York; 5 miles long.

on·rush (on'rush', ôn'-) *n.* An onward rush or flow.

on·set (on'set', ôn'-) *n.* 1 An impetuous attack; assault, as of troops. 2 An attack, as of fever; seizure, as of passion. 3 A setting about; outset; start. See synonyms under ATTACK.

on·shore (on'shôr', -shōr', ôn'-) *adv. & adj.* To, toward, or near the shore.

on·slaught (on'slôt', ôn'-) *n.* A violent hostile assault. See synonyms under AGGRESSION, ATTACK. [Earlier *anslacht*, prob. <Du. *annslag* a striking at, attempt <*slagen* strike; refashioned after *draught, slaughter,* etc.]

on·stead (on'sted', ôn'-) *n. Dial. & Scot.* A farmhouse; homestead. Also **on'sted'**. [<ON + STEAD]

On·tar·i·o (on·târ'ē·ō) A province in SE Canada between the Great Lakes and Hudson Bay; 412,582 square miles; capital, Toronto: abbr. *Ont.* — **On·tar'i·an** *adj. & n.*

Ontario, Lake The smallest and easternmost of the five Great Lakes; 7,540 square miles.

on·to (on'tŏŏ, ôn'-) *prep.* 1 Upon the top of; to and upon: The cat jumped *onto* the table. 2 *Colloq.* Aware of; informed about: I'm *onto* your tricks. Also written **on to.**

onto- *combining form* Being; existence: *ontogeny*. Also, before vowels, **ont-**. [<Gk. *ōn, ontos,* ppr. of *einai* be]

on·tog·e·ny (on·toj'ə·nē) *n. Biol.* The history of the development of the individual organism: distinguished from *phylogeny*. Also **on·to·gen·e·sis** (on'tō·jen'ə·sis). [<ONTO- + -GENY] — **on·to·ge·net·ic** (on'tō·jə·net'ik), **on'to·gen'ic** (-jen'ik) *adj.* — **on'to·gen'ist** *n.*

on·to·log·i·cal (on'tə·loj'i·kəl) *adj.* Pertaining to ontology. Also **on'to·log'ic**. — **on'to·log'i·cal·ly** *adv.*

ontological argument The metaphysical a priori argument designed to prove that the real objective existence of God is necessarily involved in the existence of the very idea of God.

on·tol·o·gism (on·tol'ə·jiz'əm) *n.* The doctrine that man has an immediate and certain knowledge of God, and that this knowledge is the foundation and guaranty of all his knowledge: opposed to *psychologism*.

on·tol·o·gy (on·tol'ə·jē) *n.* The science of real being; the philosophical theory of reality; the doctrine of the universal and necessary characteristics of all existence. Compare METAPHYSICS, PHILOSOPHY. [<NL *ontologia* <Gk. *ōn, ontos* being + *-logia* <*logos* word, study] — **on·tol'o·gist** *n.*

o·nus (ō'nəs) *n.* A burden or responsibility; a duty. [<L]

onus pro·ban·di (ō'nəs prō·ban'dī) *Latin* The burden of proof; the responsibility of proving: generally resting upon the party holding the affirmative side of an issue.

on·ward (on'wərd, ôn'-) *adv.* 1 In the direction of progress; forward; ahead. 2 On in time. Also **on'wards.** — *adj.* Moving or leading forward or ahead.

on·y (ō'nē) *adj. & pron. Scot.* Any: also spelled *onie.*

on·y·choph·a·gy (on'ə·kof'ə·jē) *n.* Morbid or habitual biting of the nails. [<Gk. *onyx, onychos* a nail + -PHAGY] — **on'y·choph'a·gist** *n.*

on·yx (on'iks) *n.* A cryptocrystalline variety of quartz consisting of layers of different colors, usually in even planes; a variety of chalcedony. [<Gk., a nail, onyx]

onyx glass Cameo glass.

oo (ōō) *n. Scot.* Wool.

oo- *combining form* 1 Egg; pertaining to eggs: *oology*. 2 *Biol.* An ovum: *oogenesis*. [<Gk. *ōon* an egg]

o·o·cyte (ō'ə·sīt) *n. Biol.* 1 An egg which has not reached full development. 2 An immature female gamete, as in certain protozoans. [<OO- + -CYTE]

oo·dles (ōōd'lz) *n. pl. Slang* A great deal; many; more than plenty. [<dial. E (Irish) *oodle*, var. of HUDDLE, n.]

o·og·a·my (ō·og'ə·mē) *n. Biol.* The union of two gametes of different size and form, called egg and sperm cells. Compare ISOGAMY. [<OO- + -GAMY] — **o·og'a·mous** *adj.*

o·o·gen·e·sis (ō'ə·jen'ə·sis) *n. Biol.* The origin and development of the ovum. Also **o·og·e·ny** (ō·oj'ə·nē). — **o'o·ge·net'ic** (-jə·net'ik) *adj.*

o·o·go·ni·um (ō'ə·gō'nē·əm) *n. pl.* **·ni·a** (-nē·ə) 1 *Bot.* The female reproductive organ in thallophytic plants, a large spherical cell or sac within which the oospheres, or egg cells, are developed. 2 *Biol.* A cell whose divisions give rise to oocytes. Also **o·o·gone** (ō'ə·gōn'). [<NL <Gk. *ōon* an egg + *gonos* offspring]

o·o·lite (ō'ə·līt) *n.* A granular variety of limestone made up of nearly spherical concretions about some minute, preexisting particles, and resembling in texture the roe of a fish: used for building. [<F *oölithe* <Gk. *ōon* an egg + *lithos* a stone] — **o'o·lit'ic** (-lit'ik) *adj.*

O·o·lite (ō'ə·līt) *n. Geol.* The upper part of the Jurassic system in England.

o·ol·o·gy (ō·ol'ə·jē) *n.* The branch of ornithology that treats of eggs. [<OO- + -LOGY]

—**o·o·log·ic** (ō′ə·loj′ik), **o′o·log′i·cal** *adj.* — **o·ol′o·gist** *n.*

oo·long (o͞o′lông) *n.* A variety of dark tea that is partly fermented before being dried. [< dial. Chinese <Chinese *wu-lung* < *wu* black + *lung* a dragon]

oo·mi·ak (o͞o′mē·ak) See UMIAK.

o·o·my·cete (ō′ə·mi·sēt′) *n.* One of a subclass (*Oomycetes*) of fungi producing sexual and non-sexual spores: it includes the water molds and downy mildews. [<NL <Gk. *ōon* an egg + *mykēs, mykētos* a mushroom] —**o′o·my·ce′tous** *adj.*

o·o·phore (ō′ə·fôr, -fōr) *n.* An oophyte. [< OO- + -PHORE] —**o·o·phor·ic** (ō′ə·fôr′ik, -for′-) *adj.*

o·o·pho·rec·to·my (ō′ə·fə·rek′tə·mē) *n.* Ovariotomy. [<OOPHOR(O)- + -ECTOMY]

o·o·pho·ri·tis (ō′ə·fə·rī′tis) *n. Pathol.* Inflammation of an ovary or the ovaries, sometimes with inflammation of the Fallopian tubes. [<OOPHOR(O)- + -ITIS]

oophoro– *combining form* Ovary; ovarian. Also, before vowels, **oophor–**. [<Gk. *oophoros* egg-bearing]

o·o·phyte (ō′ə·fīt) *n. Bot.* The stage in the life history of mosses, ferns, and liverworts, during which sexual organs are developed: one of the examples of the alternation of generations; the gametophyte. [<OO- + -PHYTE] —**o′o·phyt′ic** (-fit′ik) *adj.*

oor[1] (o͝or) *adj. Scot.* Our.

oor[2] (o͝or) *n. Scot.* Hour.

oo·ra·li (o͞o·rä′lē) *n.* Urare.

Oor·du (o͝or′do͞o) See URDU.

oo·rie (o͞o′rē) *adj. Scot.* **1** Shivering; cold. **2** Bleak; desolate.

o·o·sperm (ō′ə·spûrm) *n.* **1** A fertilized ovum. **2** Oospore.

o·o·sphere (ō′ə·sfir) *n. Bot.* In algae and fungi, the egg cell prior to fertilization. [<OO- + SPHERE]

o·o·spore (ō′ə·spôr, -spōr) *n. Bot.* The fertilized and fully developed oosphere, produced within an oogonium. [<OO- + SPORE] —**o·o·spor·ic** (ō′ə·spôr′ik, -spor′ik), **o·os·po·rous** (o·os′pər·əs) *adj.*

Oost (ōst), **Jacob van,** 1600–71, Flemish religious painter. —**Jacob van,** 1639–1713, Flemish painter; son of the preceding.

oot (o͞ot) *adv. Scot.* Out.

o·o·the·ca (ō′ə·thē′kə) *n. pl.* **·cae** (-sē) **1** *Entomol.* The egg case of certain insects, as of a cockroach. **2** *Bot.* In ferns, a sporangium. [<NL <Gk. *ōon* an egg + *thēkē* a case] —**o′o·the′cal** *adj.*

ooze[1] (o͞oz) *v.* **oozed, ooz·ing** *v.i.* **1** To flow or leak out slowly or gradually, as through pores or small holes; seep; percolate. **2** To exude moisture. **3** To escape or disappear: His courage *oozed* away. —*v.t.* **4** To give off or exude. [<*n.*] —*n.* **1** A slow, gradual leak; gentle flow: the *ooze* of a small spring. **2** That which oozes. **3** An infusion or decoction of a tanniferous substance, such as oak bark, used in tanning. —*adj.* **1** Designating calfskin, sheepskin, goatskin, or other hide susceptible of a soft, velvety finish on the flesh side. **2** Denoting this kind of finish, or the process by which it is produced: *ooze* calf; *ooze* leather. [OE *wōs* sap, juice; infl. in meaning by OE *wāse* mire, dirt]

ooze[2] (o͞oz) *n.* **1** Slimy mud or moist, spongy soil. **2** A deposit of calcareous matter found on the ocean bottom and largely made up of the remains of foraminifers. **3** The fibers on the surface of unfinished cotton thread. **4** A piece of muddy or marshy ground; bog; fen. **5** Seaweed. [OE *wāse* mire, marsh]

oo·zy[1] (o͞o′zē) *adj.* **·zi·er, ·zi·est** Slowly leaking; gently dripping. —**oo′zi·ness** *n.*

oo·zy[2] (o͞o′zē) *adj.* **·zi·er, ·zi·est** Containing, composed of, or like mud or ooze; slimy. —**oo′zi·ly** *adv.* —**oo′zi·ness** *n.*

op– Assimilated var. of OB–.

o·pac·i·fi·er (ō·pas′ə·fī′ər) *n.* An agent used to increase the opacity of a substance to light or heat, especially as a safeguard against deterioration. [<*opacify* make opaque < OPACI(TY) + -FY]

o·pac·i·ty (ō·pas′ə·tē) *n. pl.* **·ties** **1** The state of being opaque; obscurity. **2** That which is opaque. **3** *Physics* Imperviousness to light or other forms of radiation. Compare TRANSMITTANCE. [<MF *opacité* <L *opacitas* < *opacus.* See OPAQUE.]

o·pah (ō′pə) *n.* A large fish (genus *Lampris*) of warm seas, with a compressed oviform body, long single dorsal and anal fins, and many-rayed ventrals: noted for the brilliancy of its colors. [<Ibo *úbà*]

o·pal (ō′pəl) *n.* An amorphous, variously colored, hydrous silica, softer and less dense than quartz. The **precious** (or **noble**) **opal** presents a peculiar play of delicate colors, and is highly valued as a gemstone. The **fire** (or **flame**) **opal** shows its various colors clearly disposed in streaks or rays: often called *girasol.* [<L *opalus* <Gk. *opallios* <Skt. *upala* a precious stone]

o·pal·esce (ō′pəl·es′) *v.i.* **·esced, ·esc·ing** To exhibit opalescence.

o·pal·es·cence (ō′pəl·es′əns) *n.* An iridescent play of pearly or milky colors, as in an opal. —**o′pal·es′cent, o′pal·esque′** (-esk′), **o′pal·ine** (-ēn, -in) *adj.*

o·paque (ō·pāk′) *adj.* **1** Impervious to light; not translucent. **2** Loosely, imperfectly transparent. **3** Impervious to reason; unintelligent. **4** Impervious to radiant heat, electric radiation, etc. **5** Having no luster; dull. **6** Unintelligible; obscure: an *opaque* style. **7** *Obs.* Dark; lying in shadow. See synonyms under DARK. —*n.* **1** Opacity; that which is opaque. **2** A pigment used for producing half-tones. [<L *opacus* shaded, darkened; refashioned after F *opaque*] —**o·paque′ly** *adv.* —**o·paque′ness** *n.*

ope (ōp) *v.t.* & *v.i.* **oped, op·ing** *Poetic* & *Archaic* To open. [ME, short for OPEN]

o·pen (ō′pən) *adj.* **1** Affording approach, view, passage, or access because of the absence or removal of barriers, restrictions, etc.; unobstructed: The new road is *open* for traffic; *open* country. **2** Public; unbounded; accessible to all: the *open* market; in *open* competition; the *open* sea. **3** Unconcealed; overt; not secret or hidden: *open* hostility. **4** Expanded; unfolded: an *open* flower. **5** Exposed; not enclosed or covered over; unprotected: an *open* car. **6** Ready for business, appointment, etc.: an *open* day in the schedule. **7** Not settled or decided; pending: an *open* account; an *open* question. **8** Ready and free for engagement, employment, etc.; available: The job is still *open.* **9** Ready to consider proof or argument; unbiased; receptive: often with *to*: an *open* mind; *open* to conviction. **10** Generous; liberal: He gives with an *open* hand. **11** *Phonet.* **a** Pronounced with a wide opening above the tongue; low: said of vowels, as the *a* in *father*: opposed to *close.* **b** Ending in a vowel or diphthong: said of a syllable. **12** Frank; ingenuous; not deceptive: *open* and aboveboard. **13** Eager or willing to receive: with *open* arms. **14** In hunting or fishing, without prohibition: *open* season. **15** Liable to attack, robbery, temptation, etc. **16** Having openings, holes, or perforations, as woven goods or needlework; porous. **17** Mild; free from fog, mist, or ice: an *open* winter; *open* weather; *open* water in northern seas. **18** *Printing* **a** Widely spaced, as a line on a page. **b** Widely leaded or containing many breaks; fat: said of composed or printed matter. **19** *Music* Not stopped by the finger, as a string, or having the top uncovered, as an organ pipe; also, produced by an open string or pipe: said of a note, tone, etc. **20** Unrestricted by union regulations in the employment of labor: an *open* shop. **21** *U.S. Colloq.* Not under control in the sale of intoxicants, gambling, or vice: an *open* town. **22** Out of doors. See synonyms under BLUFF[2], CANDID, EVIDENT, MANIFEST, NOTORIOUS, OVERT. —*v.t.* **1** To set open or ajar, as a door; unclose; unfasten. **2** To make passable; free from obstacles. **3** To make or force (a hole, passage, etc.). **4** To remove the covering, lid, etc., of: to *open* a package. **5** To expand, as for viewing; unroll; unfold, as a map. **6** To make an opening or openings into: to *open* an abscess. **7** To make or declare ready for commerce, use, etc.: to *open* a store. **8** To make or declare public or free of access, as a park; make available for settlement. **9** To make less compact; expand: to *open* ranks. **10** To make more receptive to ideas or sentiments; enlighten: to *open* the mind. **11** To bare the secrets of; divulge; reveal: to *open* one's heart. **12** To begin; commence, as negotiations. **13** *Law* To undo or recall (a judgment or order) so as to permit its validity to be questioned. —*v.i.* **14** To become open. **15** To come apart or break open; rupture: The wound *opened* again. **16** To come into view; spread out; unroll. **17** To afford access or view: The door *opened* on a courtyard. **18** To become receptive or enlightened. **19** To begin; be started: The season *opened* with a ball. **20** In the theater, to begin a season or tour. —*n.* Any wide space not enclosed, obstructed, or covered, as by woods, rocks, etc.; open land or water: usually with the definite article: in the *open.* [OE. Akin to UP.] —**o′pen·ly** *adv.* —**o′pen·ness** *n.*

o·pen–air (ō′pən·âr′) *adj.* **1** Out of doors; taking place in an open field or street: an *open-air* service. **2** Relating to the plein-air school of painters.

o·pen–and–shut (ō′pən·ənd·shut′) *adj.* Simple; obvious; easily determinable.

open chain *Chem.* An organic compound in which the carbon atoms are disposed in a chain open at both ends, as in aliphatic compounds.

open city In modern warfare, a city having no military installations, and not being used as a railroad junction for the passage of troops or materiel: presumably not subject to attack or bombardment.

open door **1** The policy of giving to all nations the same commercial privileges in a dependency, or recently conquered territory, as those exercised by the dominant country: used attributively in such phrases as **open-door policy, open-door principle,** etc. **2** Opportunity for free trade. **3** Admission to all without charge.

o·pen–end (ō′pən·end′) *adj.* Having a capitalization that is not fixed but subject to fluctuation as shares are sold or redeemed: A mutual fund is an *open-end* investment trust.

o·pen·er (ō′pən·ər) *n.* **1** An instrument for opening anything firmly closed: usually in combination: a can-*opener.* **2** A person who opens or is employed to open: usually in combination: a pew-*opener.* **3** In poker and similar games: **a** The player who opens the jackpot. **b** *pl.* Cards of sufficient value, as a pair of jacks, to enable the player to open a pot.

o·pen–eyed (ō′pən·īd′) *adj.* Having the eyes open; wary; watchful; also, amazed: in *open-eyed* wonder.

o·pen–faced (ō′pən·fāst′) *adj.* **1** Possessing a countenance suggestive of frankness, simplicity, and honesty. **2** Having a face uncovered by a casing, as a watch.

o·pen–hand·ed (ō′pən·han′did) *adj.* Giving freely; liberal. See synonyms under GENEROUS. —**o′pen–hand′ed·ly** *adv.* —**o′pen–hand′ed·ness** *n.*

o·pen–heart·ed (ō′pən·här′tid) *adj.* Disclosing the thoughts and intentions plainly; frank; candid. —**o′pen–heart′ed·ly** *adj.* —**o′pen–heart′ed·ness** *n.*

o·pen–hearth (ō′·pən·härth′) *adj. Metall.* **1** Having a shallow or open hearth: said of furnaces used in making steel by the Siemens-Martin process. **2** Made in a shallow or open hearth: said of steel.

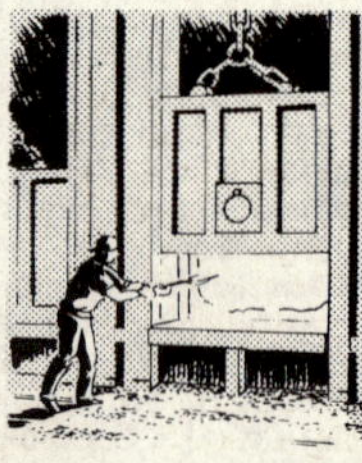

OPEN-HEARTH FURNACE

open house **1** A house extending hospitality to all who wish to come. **2** An occasion when a school, factory, institution, clubhouse, etc., is open to visitors, as for inspection, observation, etc.

o·pen·ing (ō′pən·ing) *n.* **1** The act of becoming open or of causing to be open. **2** Something that is open; a vacant or unobstructed space, as within barriers or boundaries; a hole, passage, or gap; a space. **3** A tract in a forest where trees are lacking or thinly scattered. **4** An aperture in a wall; especially, one for the admission of light or air. **5** The first part or stage, as of a period, act, or process; a beginning; prelude. **6** In chess, checkers, etc., a specific mode of beginning the game; the series of opening moves. **7** An opportunity for action, especially in business. See synonyms under BEGINNING, BREACH, ENTRANCE[1], HOLE, OPPORTUNITY.

o·pen-mind·ed (ō′pən·mīn′did) *adj.* Free from prejudiced conclusions; amenable to reason; receptive. **— o′pen-mind′ed·ly** *adv.*
o·pen-mouthed (ō′pən·mouthd′, -moutht′) *adj.* 1 Having the mouth open; gaping, as in wonder or surprise; greedy. 2 Noisy; clamorous.
open range An unfenced area of grazing country.
o·pen-ses·a·me (ō′pən·ses′ə·mē) *n.* An unfailing means or formula for opening secret doors and gaining entrance. [From the story of *Ali Baba and the Forty Thieves* where the door of the robbers' cave opened only at the magic conjuration *"open sesame"*]
open shop An establishment in which union labor and non-union labor are employed: opposed to *closed shop.*
open stove A stove having the firebox open to the room, resembling a fireplace; a Franklin stove.
open syllable See under SYLLABLE.
open timber A forest having no undergrowth.
o·pen·work (ō′pən·wûrk′) *n.* Any product of art or handicraft containing numerous small openings.
op·er·a (op′ər·ə, op′rə) *n.* 1 A form of drama in which music is a dominant factor, made up of arias, recitatives, choruses, etc., with orchestral accompaniment, scenery, acting, and sometimes dance: the principal types are **comic opera,** in which there is spoken dialog and the story ends happily; **grand opera,** a dramatic composition generally with a serious or tragic theme, of which the plot is elaborated as in a play and the dialog is set to music throughout; **light opera,** in which the plot has humorous situations, a happy ending, and some spoken dialog. 2 A particular musical drama or its music or libretto; also, its representation. 3 The theater in which operas are given. 4 Plural of OPUS. [<Ital. <L, a work, labor < *opus, operis* work] **— op·er·at·ic** (op′ə·rat′ik) *adj.* **— op′er·at′i·cal·ly** *adv.*
op·er·a·ble (op′ər·ə·bəl) *adj.* Capable of treatment by surgical operation. **— op′er·a·bil′i·ty** *n.*
o·pé·ra bouffe (ô·pā·rà bo͞of′) *French* A farcical comic opera. Also *Ital.* **o·pe·ra buf·fa** (ô′pā·rä bo͞of′fä).
o·pé·ra co·mique (ô·pā·rà kô·mēk′) *French* Comic opera.
opera glass A binocular telescope of small size, suitable for use at the theater. Also **opera glasses.**
opera hat A tall hat, the crown of which is extended by springs and is capable of being collapsed into an approximately flat form.
opera house A theater specially adapted for performance of operas; loosely, any theater.
op·er·and (op′ər·and) *n. Math.* Any quantity or symbol upon which an operation is performed: also called *faciend.* [<L *operandum,* neut. gerundive of *operari.* See OPERATE.]
op·er·ate (op′ə·rāt) *v.* **·at·ed, ·at·ing** *v.i.* 1 To act or function, especially with force or influence; work. 2 To bring about or produce the proper or intended effect. 3 To perform a surgical operation. 4 To deal in securities, stocks, etc., especially speculatively. 5 To carry on a military or naval operation: usually with *against.* —*v.t.* 6 To control the working or function of, as a machine. 7 To manage or conduct the affairs of: to *operate* a railroad. 8 To bring about or cause; effect. [<L *operatus,* pp. of *operari* work, have an effect < *opus, operis* a work] **— op′er·at′a·ble** *adj.*
op·er·a·tion (op′ə·rā′shən) *n.* 1 The act or process of operating; the exertion or action of any form of power or energy. 2 A method of exercising or applying force; a mode of action. 3 A single specific act or transaction, especially in the stock market. 4 A course or series of acts to effect a certain purpose; process. 5 The state of being in action: The machinery is in *operation.* 6 *Surg.* Any systematic manipulation upon the body, performed either with or without instruments, to restore disunited or deficient parts, to remove diseased or injured parts, or to extract foreign matter. 7 *Math.* **a** The act of making a change in the value or form of a quantity. **b** The change itself as indicated by symbols or rules: distinguished from the process by which such change is accomplished. 8 Some special kind of activity; manner of action; a vital or natural process of activity. 9 A military or naval campaign.
Synonyms: action, agency, effect, execution, force, influence, performance, procedure, result. *Operation* is *action* resulting in change, whether produced by the *agency* or *action* of an intelligent agent or of a material substance or *force*; as, military *operations*; the *operation* of a medicine. *Performance* and *execution* denote intelligent *action,* considered with reference to the actor or to that which he accomplishes; *performance* accomplishing the will of the actor, *execution* often the will of another. Compare ACT, EXERCISE. *Antonyms:* failure, inaction, ineffectiveness, inefficiency, inutility, powerlessness.
op·er·a·tion·al (op′ə·rā′shən·əl) *adj.* 1 Pertaining to an operation. 2 Organized or prepared to carry out assigned tasks, especially of a military character. 3 Fit or ready for some specified use. 4 In actual service, as a machine, aircraft, etc.
operations research The application of scientific method, engineering procedures, and technical skills to insure maximum efficiency in the conduct of planned operations in industry and government, both in war and in peace. Also called **operational research.**
op·er·a·tive (op′ər·ə·tiv, -ə·rā′tiv) *adj.* 1 Exerting force or influence. 2 Moving or working efficiently; effective. 3 Connected with operations: *operative* surgery. 4 Concerned with practical work, mechanical or manual. 5 Engaged in practical activity, as a workman or mechanic. —*n.* 1 A person employed as a skilled worker, as in a mill or factory, etc.; an artisan. 2 *Colloq.* A detective; one who works secretly. See synonyms under ARTIST. **— op′er·a·tive·ly** *adv.*
op·er·a·tor (op′ə·rā′tər) *n.* 1 One who operates; any skilled worker; specifically, one who works a telephone switchboard, one who receives or sends messages on a telegraph, or one who works a typesetting machine. 2 A broker who acts for others in trading in speculative securities. 3 The owner and director of a coal mine, oil field, or other large industrial organization. 4 *Math.* A symbol that briefly indicates a mathematical process. See synonyms under AGENT.
o·per·cu·lum (ō·pûr′kyo͞o·ləm) *n. pl.* **·la** (-lə) *Biol.* 1 A lid, cover, or lidlike part or organ, as of the orifice of the capsule in mosses, of certain capsules (as a pyxis) in flowering plants, of the hair follicles, etc. 2 A horny or shelly plate in many gastropods, serving to close the aperture when the animal is retracted. 3 In fishes, the gill cover; specifically, the hindmost and uppermost bone of the gill cover. 4 In the king crab, the plate that covers the abdominal limbs. 5 The labrum of certain dipterous insects. 6 A part of the cerebral cortex. Also **o·per·cele** (ō·pûr′sēl), **o·per·cule** (-kyo͞ol). [<L, a covering, lid < *operire* cover] **— o·per′cu·lar** *adj.* **— o·per′cu·late, o·per′cu·lat′ed** *adj.*
o·pe·re ci·ta·to (op′ə·rē sī·tā′tō) *Latin* In the work cited, or quoted: abbr. *op. cit.*
op·e·ret·ta (op′ə·ret′ə) *n.* A short, humorous opera with dialog. [<Ital., dim. of *opera* OPERA]
op·er·ose (op′ə·rōs) *adj.* Laborious; also, industrious. [<L *operosus*] **— op′er·ose′ly** *adv.* **— op′er·ose′ness** *n.*
O·phel·ia (ō·fēl′yə) A feminine personal name. Also *Fr.* **O·phé·lie** (ō·fā·lē′). [Prob. <Gk., a help]
— Ophelia In Shakespeare's *Hamlet,* the daughter of Polonius; in love with Hamlet.
oph·i·cleide (of′ə·klīd) *n.* A brass musical wind instrument producing fundamental tones and resembling a cornet, but having a greater number of finger levers: now usually replaced in orchestras by the tuba. [<F *ophicléide* <Gk. *ophis* a serpent + *kleis, kleidos* a key]
o·phid·i·an (ō·fid′ē·ən) *n.* One of a suborder of limbless reptiles *(Ophidia),* with mandibular rami connected only by an elastic ligament, and having no pectoral arch; a serpent; snake. —*adj.* Of or pertaining to the *Ophidia,* or to snakes; snakelike. [<NL <Gk. *ophis* a snake]
oph·i·dism (of′ə·diz′əm) *n.* Poisoning by the venom of a snake. Also **oph·i·di·a·sis** (of′ə·dī′ə·sis).
ophio– *combining form* Serpent; of or pertaining to serpents: *ophiolatry.* Also, before vowels, **ophi–.** [<Gk. *ophis* a serpent]
oph·i·ol·a·try (of′ē·ol′ə·trē) *n.* Serpent worship. [<OPHIO- + -LATRY] **— oph′i·ol′a·trous** *adj.*
oph·i·ol·o·gy (of′ē·ol′ə·jē) *n.* The branch of zoology that treats of serpents; herpetology. [<OPHIO- + -LOGY] **— oph′i·o·log′i·cal** (-ə·loj′i·kəl) *adj.* **— oph′i·ol′o·gist** *n.*
O·phir (ō′fər) In the Bible, a land rich in gold from which Solomon obtained his wealth. I *Kings* x 11.
oph·ite (of′īt, ō′fīt) *n.* A variety of greenish altered diabase of the late Mesozoic age occurring in the Pyrenees. [<L *ophites* <Gk. *ophitēs (lithos)* a serpentine (stone) < *ophis* a serpent]
o·phit·ic (ō·fit′ik) *adj. Mineral.* Characterized by feldspar crystals in a matrix of augite crystals.
Oph·i·u·chus (of′ē·yo͞o′kəs, ō′fē-) A northern constellation, the Serpent-holder or Doctor. Ras Alhagua is its principal star. See CONSTELLATION. [<L <Gk. *ophiouchos* < *ophis* a serpent + *echein* hold]
oph·thal·mi·a (of·thal′mē·ə) *n. Pathol.* Inflammation of the eye, its membranes, or its lids. Also **oph·thal′my.** [<LL <Gk. < *ophthalmos* an eye] **— oph·thal′mic** *adj.*
oph·thal·mi·tis (of′thal·mī′tis) *n. Pathol.* Inflammation of the eye, including the outer and internal structures.
ophthalmo– *combining form* Eye; pertaining to the eyes: *ophthalmology.* Also, before vowels, **ophthalm–.** [<Gk. *ophthalmos* the eye]
oph·thal·mol·o·gy (of′thal·mol′ə·jē) *n.* The science dealing with the structure, functions, and diseases of the eye. **— oph·thal′mo·log′ic** (-mə·loj′ik) or **·i·cal** *adj.* **— oph′thal·mol′o·gist** *n.*
oph·thal·mo·scope (of·thal′mə·skōp) *n.* An optical instrument having a concave mirror with a hole in its center, for illuminating and viewing the center of the eye. **— oph·thal′mo·scop′ic** (-skop′ik) or **·i·cal** *adj.* **— oph·thal·mos·co·py** (of′thal·mos′kə·pē) *n.*

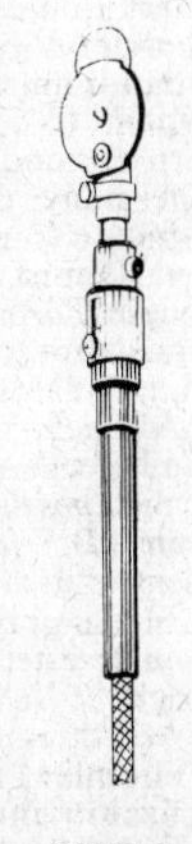
OPTHAL-MOSCOPE

–opia *combining form Med.* A (specified) defect of the eye, or condition of sight: *myopia.* Also spelled *-opy.* [<Gk. *-ōpia* < *ōps, ōpos* the eye]
o·pi·ate (ō′pē·it, -āt) *n.* 1 Medicine containing opium; a narcotic. 2 Something inducing sleep. —*adj.* Consisting of opium; tending to induce sleep. —*v.t.* (-āt) **·at·ed, ·at·ing** 1 To treat with opium or an opiate. 2 To deaden; dull. [<Med. L *opiatus,* pp. of *opiare* treat with opium <L *opium* OPIUM]
o·pine (ō·pīn′) *v.t.* & *v.i.* **o·pined, o·pin·ing** To hold or express as an opinion; think; conjecture: now usually humorous. [<MF *opiner* <L *opinari* think]
o·pin·ion (ə·pin′yən) *n.* 1 A conclusion or judgment held with confidence, but falling short of positive knowledge. 2 *Often pl.* A settled judgment or conviction on some subject, as religion or politics. 3 Favorable judgment or estimation; reputation. 4 Specifically, an estimate of the excellence or value of a person or a thing; also, a common or prevailing sentiment; public opinion. 5 *Law* The formal announcement of the conclusions of a court in a case before it; also, the conclusion of an attorney touching the merits of a submitted case: to take the *opinion* of counsel. [<OF <L *opinio, -onis* < *opinari* think]
Synonyms: belief, conviction, decision, determination, doctrine, estimate, idea, impression, judgment, notion, persuasion, sentiment, view. An *opinion* is a general conclusion held as probable, but without full certainty; a *conviction* is sustained by such evidence as removes all doubt from the believer's mind; a *persuasion* is a confident *opinion,* involving the heart as well as the intellect. In religion,

a *doctrine* is a statement of *belief* regarding a single point; a *creed* is a summary statement of *doctrines*. Such a system of *doctrines* is often called a faith; as, the Anglican or Lutheran faith. Compare BELIEF, FAITH, IDEA, THOUGHT[1].

o·pin·ion·at·ed (ə·pin′yən·ā′tid) *adj.* Unwarrantably attached to one's own opinion; obstinate. See synonyms under DOGMATIC, OBSTINATE. —**o·pin′ion·at′ed·ness** *n.*

o·pin·ion·a·tive (ə·pin′yən·ā′tiv) *adj.* **1** Opinionated. **2** Of the nature of opinion. **3** Relating to an opinion; doctrinal. —**o·pin′ion·a′·tive·ly** *adv.* —**o·pin′ion·a′tive·ness** *n.*

op·is·thog·na·thous (op′is·thog′nə·thəs) *adj.* Having receding jaws: opposed to *prognathous.* [<Gk. *opisthen* behind + *gnathos* jaw] —**op′is·thog′na·thism** (-nə·thiz′əm) *n.*

o·pis·tho·graph (ə·pis′thə·graf, -gräf) *n.* An ancient manuscript having writing on the back as well as on the front; also, a slab inscribed on both sides; especially, a mural slab whose back has been used for a later inscription, the front being turned to the wall. [<Gk. *opisthographos* written on the back or cover < *opisthen* behind + *graphein* write] —**o·pis′·tho·graph′ic** *adj.*

op·is·thot·o·nos (op′is·thot′ə·nəs) *n. Pathol.* A rigid muscular spasm of the neck and back, arching the body backward, as in tetanus. [<Gk. < *opisthen* behind + *tonos* tension < *teinein* stretch]

o·pi·um (ō′pē·əm) *n.* A milky exudation from the unripe capsules of the **opium poppy** (*Papaver somniferum*), containing a mixture of about 20 alkaloids, the most important of which is morphine. It is a powerful narcotic, with a sticky, gumlike body, bitter taste, and heavy odor. [<L <Gk. *opion* opium, dim. of *opos* vegetable juice]

opium den A room or place, usually illegal, for opium-smoking.

o·pi·um·ism (ō′pē·əm·iz′əm) *n. Pathol.* **1** Addiction to the use of opium. **2** A morbid condition due to such addiction.

op·o·del·doc (op′ə·del′dok) *n.* A camphorated soap liniment. [? <Gk. *opos* vegetable juice]

O·po·le (ô·pô′le) A port on the Oder in southern Poland; until 1945, in Upper Silesia, Germany. *German* **Op·peln** (ôp′əln).

O·por·to (ō·pôr′tō) A port of NW Portugal on the Douro, 3 miles from its mouth: Portuguese *Pôrto.*

o·pos·sum (ə·pos′əm, pos′əm) *n.* An American marsupial (genus *Didelphis*) of largely arboreal and nocturnal habits, having a prehensile tail and feet adapted for grasping: popularly called *possum.* The **common** (or **Virginia**) **opossum** (*D. virginiana*), which ranges from the central United States to Brazil, has a soft, whitish-gray pelage, with black ears and feet, and is esteemed as food. It is noted for its trick of feigning death, or *playing possum,* when threatened with danger. [<Algonquian (Virginian) *apasum,* lit., a white beast]

opossum shrimp A crustacean (family *Mysidae*) which carries its eggs in a pouch beneath the thorax.

Op·pen·heim (op′ən·hīm), **E(dward) Phillips,** 1866–1946, English novelist.

Op·pen·heim·er (op′ən·hī′mər), **J. Robert,** born 1904, U. S. physicist.

Op·per (op′ər), **Frederick Burr,** 1857–1937, U.S. cartoonist.

op·pi·dan (op′ə·dən) *adj.* Relating to a town; civic. —*n.* At Eton College in England, a student who boards in town: distinguished from *colleger.* [<L *oppidanus* a townsman, orig. adj. < *oppidum* a town]

op·pi·late (op′ə·lāt) *v.t.* **·lat·ed, ·lat·ing** *Med.* To block or obstruct; constipate. [<L *oppilatus,* pp. of *oppilare* stop up < *ob-* against + *pilare* ram down < *pilus* a pestle] —**op′pi·lant** *adj.* —**op′pi·la′tion** *n.*

op·po·nent (ə·pō′nənt) *n.* One who opposes another, as in battle or debate; antagonist. See synonyms under ENEMY. —*adj.* **1** Acting against something or someone; opposing. **2** *Anat.* Bringing one part, as of a muscle, into opposition to another. **3** Standing in front; opposite. [<L *opponens, -entis,* ppr. of *opponere* set against < *ob-* against + *ponere* place] —**op·po′nen·cy** *n.*

op·por·tune (op′ər·tōōn′, -tyōōn′) *adj.* Meeting some requirement; especially seasonable or timely. See synonyms under AUSPICIOUS, CONVENIENT. [<MF, fem. of *opportun* seasonable, exposed <L *opportunus* suitable, lit., at the port < *ob-* before + *portus* a harbor] —**op′·por·tune′ly** *adv.* —**op′por·tune′ness** *n.*

op·por·tu·nist (op′ər·tōō′nist, -tyōō′-) *n.* One who governs his course by opportunities or circumstances rather than by fixed principles or by regard for consistency or consequences. —**op′por·tu·nis′tic** (-tōō·nis′tik, -tyōō-) *adj.* —**op′por·tu′nism** *n.*

op·por·tu·ni·ty (op′ər·tōō′nə·tē, -tyōō′-) *n. pl.* **·ties** **1** A fit or convenient time; favorable occasion. **2** *Obs.* Opportuneness. **3** *Obs.* Importunity: an erroneous use.

Synonyms: convenience, occasion, opening, season. *Occasion* in the popular sense is a conjunction of circumstances which seems to require or inclines to or is fit for certain action; an *opportunity* is a conjunction of circumstances which makes certain action possible, with probability of success, advantage or gratification; as, I had *occasion* to interfere; I found an *opportunity* for a good investment.

op·pos·a·ble (ə·pō′zə·bəl) *adj.* **1** Capable of being placed opposite: said especially of the thumb. **2** That can be opposed. —**op·pos′a·bil′i·ty** *n.*

op·pose (ə·pōz′) *v.* **·posed, ·pos·ing** *v.t.* **1** To act or be in opposition to; resist; combat; fight. **2** To set in opposition or contrast: to *oppose* love to hatred. **3** To place before or in front. —*v.i.* **4** To act or be in opposition. [<OF *opposer, oposer* <L *ob-* against + OF *poser.* See POSE[1].] —**op·pos′er** *n.*

Synonyms: check, combat, confront, contradict, contravene, defy, face, object, obstruct, oppugn, resist, withstand. See CONTEND, CONTRAST, DISPUTE, HINDER[1], OBSTRUCT, REPEL. *Antonyms*: see synonyms for AID.

op·posed (ə·pōzd′) *adj.* **1** Set or placed in front or before; opposite. **2** Being in opposition, as in principle, meaning, purpose, etc. See synonyms under ALIEN, CONTRARY, INIMICAL, RELUCTANT.

op·pose·less (ə·pōz′lis) *adj.* Not to be opposed with effect; irresistible.

op·po·site (op′ə·zit) *adj.* **1** Situated or placed on the other side, or on each side, of an intervening space or thing; contrary in position: *opposite* ends of the room. **2** Facing or moving the other way; contrary: *opposite* directions. **3** Contrary in tendency or character; diametrically different: *opposite* opinions; the *opposite* sex. **4** *Bot.* **a** Arranged (as similar parts or organs) in pairs, so that the whole diameter of some intervening body separates them, as leaves on a stem. **b** Having one part or organ immediately before, or vertically over, another, as a stamen before a petal. See synonyms under CONTRARY. —*n.* **1** Something or someone that is opposite, opposed, or contrary. **2** An antonym. **3** *Obs.* An antagonist. —*adv.* In an opposite or complementary direction or position. —*prep.* **1** Across from; facing. **2** Complementary to, as in theatrical roles: He played *opposite* her. [<OF <L *oppositus,* pp. of *opponere.* See OPPONENT.] —**op′po·site·ly** *adv.* —**op′po·site·ness** *n.*

op·po·si·tion (op′ə·zish′ən) *n.* **1** The act of opposing or resisting; antagonism. **2** The state of being opposite or opposed; antithesis; also, a position confronting another or a placing in contrast. **3** That which is or furnishes an obstacle to some result: The stream flows without *opposition.* **4** *Often cap.* The political party opposed to the ministry or administration in power. **5** *Astron.* **a** The relative position of two heavenly bodies 180° apart in geometric longitude. **b** The position of a body opposite to the sun designated by the symbol ☍; as, ☍ ♂ ☉, *opposition* of Mars to the sun. **6** *Ling.* A state of contrast between any one phoneme and all the other phonemes in a language, as, /p/ is said to be in *bilateral opposition* to /b/ on the basis of the distinctive feature of voice, and in *multilateral opposition to* /d/ on the basis of the distinctive features of voice and place of articulation. **7** *Logic* The relation between two propositions which have the same subject and predicate but differ in quantity or quality or in both. See synonyms under AMBITION, ANTIPATHY, COLLISION, COMPETITION, EMULATION. —**in opposition** In the position of being opposed or hostile to a political party or measure: The Democratic party is *in opposition.* [<OF <L *oppositio, -onis* < *oppositus.* See OPPOSITE.] —**op′po·si′tion·al** *adj.* —**op′po·si′tion·ist** *n.* —**op′po·si′tion·less** *adj.*

op·pos·i·tive (ə·poz′ə·tiv) *adj.* Placed or capable of being placed in contrast. —**op·pos′i·tive·ly** *adv.* —**op·pos′i·tive·ness** *n.*

op·press (ə·pres′) *v.t.* **1** To burden or keep in subjugation by harsh and unjust use of force or authority; tyrannize. **2** To lie heavy upon physically or mentally; weigh down; depress; dispirit. **3** *Obs.* To crush or trample; overwhelm. See synonyms under ABUSE, LOAD, PERSECUTE. [<OF *oppresser, apresser* <Med. L *oppressare,* freq. of L *opprimere* crush < *ob-* against + *premere* press] —**op·pres′sor** *n.*

op·pres·sion (ə·presh′ən) *n.* **1** The act of oppressing. **2** Subjection to unjust hardships; tyranny. **3** Mental depression; languor; dulness of spirits. **4** A sense of weight or of constriction. **5** That which oppresses or is hard to bear; privation; hardship; cruelty.

op·pres·sive (ə·pres′iv) *adj.* **1** Characterized by, tending to, or disposed to practice oppression; burdensome; tyrannical. **2** Producing a sense of depression, physical or mental. See synonyms under HARD, HEAVY. —**op·pres′·sive·ly** *adv.* —**op·pres′sive·ness** *n.*

op·pro·bri·ous (ə·prō′brē·əs) *adj.* **1** Consisting of contemptuous abuse; imputing disgrace. **2** Shameful; disgraceful; odious; held in dishonor. [<OF *opprobrieus* <LL *opprobriosus* <L *opprobrium* OPPROBRIUM] —**op·pro′bri·ous·ly** *adv.* —**op·pro′bri·ous·ness** *n.*

op·pro·bri·um (ə·prō′brē·əm) *n.* **1** The state of being scornfully reproached; ignominy. **2** Reproach mingled with disdain. **3** A cause of disgrace or reproach. [<L, a disgrace < *opprobrare* reproach < *ob-* against + *probum* an infamy]

op·pugn (ə·pyōōn′) *v.t.* To assail or oppose with argument; call in question; controvert. See synonyms under OPPOSE. [<L *oppugnare* < *ob-* against + *pugnare* fight < *pugna* a fight] —**op·pugn′er** *n.*

op·pug·nant (ə·pug′nənt) *adj.* **1** Opposing in a hostile manner. **2** Combative. —**op·pug′·nan·cy, op·pug′nance** *n.*

Ops (ops) In Roman mythology, the wife of Saturn, goddess of the harvest: identified with the Greek *Rhea.*

-opsia *combining form Med.* A (specified) type or condition of sight: *macropsia.* Also spelled *-opsy.* [<NL <Gk. *opsis* sight]

-opsis *combining form Biol.* A thing having a (specified) appearance: often used in describing fruits: *caryopsis, coreopsis* [<Gk. *opsis* sight, appearance]

op·so·ma·ni·a (op′sə·mā′nē·ə, -mān′yə) *n.* A morbid craving for rich foods and delicate fare. [<NL <Gk. < *opson* cooked meat, dainties + *mania* madness]

op·son·ic (op·son′ik) *adj.* Of or pertaining to opsonin.

opsonic index *Bacteriol.* The ratio of the quantity of bacteria destroyed by phagocytes in the blood serum of any individual to that destroyed in a normal serum.

op·son·i·fy (op·son′ə·fī) *v.t.* **·fied, ·fy·ing** To render (bacteria) susceptible to phagocytosis by the action of opsonins. [<OPSON(IN) + -(I)FY] —**op·son′i·fi·ca′tion** *n.*

op·so·nin (op′sə·nin) *n. Bacteriol.* A component of blood serum which acts upon invading cells or bacteria, so as to assist in their absorption by the phagocytes. [<Gk. *opson* cooked meat + -IN]

op·son·ize (op′sən·īz) *v.t.* **·ized, ·iz·ing** To opsonify.

-opsy Var. of -OPSIS.

opt (opt) *v.i.* To choose; decide. [<F *opter* <L *optare* choose, wish]

op·ta·tive (op′tə·tiv) *adj.* **1** Expressing or indicative of desire or choice. **2** *Gram.* Denoting that mood in Greek and certain other languages which expresses wish or desire. —*n. Gram.* **1** The optative mood. **2** A word or construction in this mood. [<MF, fem. of *optatif* <LL *optativus* <L *optare* wish] —**op′·ta·tive·ly** *adv.*

op·tic (op′tik) *adj.* **1** Pertaining to the eye or to vision. **2** Optical. —*n. Colloq.* An eye. [<MF *optique* <Med. L *opticus* <Gk. *optikos* < *optos* seen < *ops-,* fut. stem of *horaein* see, behold]

op·ti·cal (op′ti·kəl) *adj.* **1** Pertaining to optics. **2** Of or pertaining to eyesight. **3** Designed to assist or improve vision. —**op′ti·cal·ly** *adv.*

optical activity The capacity to rotate the plane of polarization of light.
optical glass See under GLASS.
optic angle The angle formed at either eye by two lines drawn from the extremities of an object of vision, varying with the distances of the object beheld.
optic axis 1 One of the directions along which a ray of light undergoes no double refraction within a crystal. 2 The axis of the eye corresponding with the line of vision passing through the center of the lens and cornea.
optic disk Anat. The expanded portion of the optic nerve as it enters the retina.
op·ti·cian (op-tish'ən) n. One who makes or deals in optical goods. [<F opticien <Med. L optica OPTICS]
optic nerve Anat. The special nerve of vision, connecting the retina with the cerebral centers. See illustration under EYE.
op·tics (op'tiks) n. The science that treats of the phenomena of light, vision, and sight. [<OPTIC; trans. of Med. L optica <Gk. ta optika optical matters]
optic thalamus Anat. A large, ovoid mass of gray matter at the side of the third ventricle at the base of the brain, connected with the origin of the optic nerve.
op·ti·mal (op'tə·məl) adj. Of, pertaining to, indicating, or characterized by an optimum.
op·ti·me (op'tə·mē) n. In Cambridge University, England, one who has attained the second, **Senior optime**, or third, **Junior optime**, grade in mathematical honors. See WRANGLER (def. 2). [<L optimus best]
op·ti·mism (op'tə·miz'əm) n. 1 The doctrine that everything is ordered for the best; also, the doctrine that the universe is constantly tending toward a better state. 2 Disposition to look on the bright side of things: opposed to pessimism. [<F optimisme <L optimus best] — **op'ti·mist** n. — **op'ti·mis'tic** or ·ti·cal adj. — **op'ti·mis'ti·cal·ly** adv.
op·ti·mize (op'tə·mīz) v. ·mized, ·miz·ing v.t. 1 Technol. To plan or prepare plans for (industrial production) so that the methods for each process are coordinated into the most practical over-all operating procedures in order to attain the maximum efficiency in the economic end results. 2 To make the most of. — v.i. 3 To be optimistic. 4 To work toward obtaining an optimum.
op·ti·mum (op'tə·məm) n. pl. ·ma (-mə) or ·mums 1 The condition or degree producing the best result. 2 The combination of conditions, as of food, etc., that produces the best average result in the growth and development of organisms. 3 The most favorable degree, conditions, etc. — adj. Producing or conducive to the best results. [<L, neut. of optimus best]
op·tion (op'shən) n. 1 The right, power, or liberty of choosing; discretion; the exercise of such right, power, or liberty. 2 The purchased privilege of either buying or selling something at a specified price within a specified time. 3 A thing that is or can be chosen. See synonyms under ALTERNATIVE. [<MF <L optio, -onis <optare choose]
op·tion·al (op'shən·əl) adj. Depending on choice; elective. —n. A study or course to be chosen from two or more offered; an elective. — **op'tion·al·ly** adv.
op·tom·e·ter (op·tom'ə·tər) n. An optical instrument for measuring the range of vision of the eye, and its peculiarities as a refracting medium. [<OPT(IC) + -(O)METER]
op·tom·e·try (op·tom'ə·trē) n. Measurement of the powers of vision and correction of visual defects by optical means. — **op·tom'e·trist** n.
op·to·type (op'tə·tīp) n. The variously graded set of type specimens used in testing vision. [<opto- vision <Gk. optos visible) + TYPE]
op·u·lence (op'yə·ləns) n. 1 Wealth; affluence. 2 Luxuriance. Also **op'u·len·cy**. See synonyms under COMFORT.
op·u·lent (op'yə·lənt) adj. 1 Possessing great wealth. 2 Exuberant; profuse. [<L opulentus <opulens, -entis <ops, opis power, wealth] — **op'u·lent·ly** adv.
o·pun·ti·a (ō·pun'shē·ə) n. One of a large genus (Opuntia) of mainly tropical American cacti, the prickly pears, having a usually flattened, much-branched stem and tubular yellow, red, or purple flowers. [<NL <L Opuntia (herba) (a plant) native to Opus <Opus, Opuntis, a city in ancient Locris <Gk. Opous]
o·pus (ō'pəs) n. pl. **op·er·a** (op'ər·ə, op'rə) A literary or musical work or composition. [<L, a work]
o·pus·cule (ō·pus'kyool) n. A small or unimportant work. [<OF <L opusculum, dim. of opus a work]
-opy See -OPIA.
o·quas·sa (ō·kwas'ə) n. A small lake trout (Salvelinus oquassa) of Maine, with a bluish-black body. [from Oquassa Lake in Maine; ult. <Algonquian]

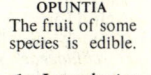
OPUNTIA
The fruit of some species is edible.

or[1] (ôr, unstressed ər) conj. 1 Introducing an alternative: stop or go; red or white. 2 Offering a choice of a series: Will you take milk or coffee or chocolate? 3 Introducing an equivalent: the culinary art or art of cookery. 4 Indicating uncertainty: He lives in Chicago or thereabouts. 5 Introducing the second alternative of a choice limited to two: with either or whether. It must be either black or white; I don't care whether he goes or not. 6 Poetic Either; whether: or in the heart or in the head. [ME, contraction of other, auther either, OE āther; infl. in meaning by OE oththe or]
or[2] (ôr) adv., prep., & conj. Obs. or Scot. Before; ere: chiefly in the phrase, **or ever**, before ever; ere. [OE ār, var. of ǣr earlier, before]
or[3] (ôr) n. Her. Gold: represented in engraving by a white surface powdered with dots. [<MF <L aurum gold]
-or[1] suffix The person or thing performing the action expressed in the root verb: competitor. See note under -ER[1]. [<AF -our, OF -or <L -or, -ator]
-or[2] suffix The quality, state, or condition of: favor. [<OF <L]
or·ach (ôr'əch, or'-) n. Any of various plants (genus Atriplex), especially the **garden orach** or **mountain spinach** (A. hortensis), a tall, hardy annual, formerly common in England as a pot herb. Also **or'ache**. [<AF arasche, OF arroche <L atriplex, -plicis <Gk. atraphaxys, -yos]
or·a·cle (ôr'ə·kəl, or'-) n. 1 The seat of the worship of some ancient divinity, as of Apollo at Delphi, where prophecies were given out by the priests in answer to inquiries. 2 A prophecy thus given. 3 The deity whose prophecies were given. 4 Hence, a person of unquestioned wisdom or knowledge, or something regarded as of infallible authority. 5 An infallible authority: often used ironically. 6 A wise saying. 7 In the Bible, a divine command or communication; also, the holy of holies in the temple. [<OF <L oraculum <orare speak, pray <os, oris mouth]
o·rac·u·lar (ō·rak'yə·lər, ō-) adj. 1 Pertaining to an oracle. 2 Obscure; enigmatical. 3 Prophetic. — **o·rac·u·lar'i·ty** n. — **o·rac'u·lar·ly** adv. — **o·rac'u·lar·ness** n.
O·ra·dea (ô·rä'dyä) A city in NW Rumania; German Grosswardein, Hungarian Nagyvárad. Also **Oradea Ma·re** (mä'rā).
o·ra e sem·pre (ō'rä ā sem'prā) Italian Now and always.
o·ral (ôr'əl, ō'rəl) adj. 1 Uttered through the mouth; consisting of spoken words. 2 Pertaining to or situated at or near the mouth. 3 Zool. Designating the side of the body on which the mouth is placed. 4 Of, pertaining to, or using speech. 5 Taken or administered by mouth. 6 Phonet. Pronounced through the mouth with the nasal passage closed; nonnasal: opposed to nasal. See synonyms under VERBAL. — n. An oral examination in a college or university. [<L os, oris mouth] — **o'ral·ly** adv.
O·ran (ō·ran', Fr. ô·rän') A port on **Oran Bay**, a semicircular inlet of the Mediterranean in NW Algeria.
o·rang (ō·rang') n. An orang-utan.
or·ange (ôr'inj, or'-) n. 1 A large, round, juicy fruit (technically a berry) of a low, much-branched, evergreen tree (genus Citrus), with a reddish-yellow rind enclosing membranous divisions and a refreshing, sweetish or subacid pulp. 2 Any of the trees yielding this fruit, as the Spanish sour orange (C. aurantium), the sweet orange (C. sinensis). 3 Any of many related species, such as the trifoliate orange (Poncirus trifoliata). 4 The kumquat. 5 The osage orange. 6 A reddish-yellow color; also, a pigment of this color. — **mandarin orange** A mandarin (def. 2). — adj. 1 Reddish-yellow. 2 Pertaining to an orange. [<OF <Provençal auranja (infl. by aur gold), earlier (n)aranja <Sp. naranja <Arabic nāranj <Persian nārang)
Or·ange (ôr'inj, or'-; Fr. ô·ränzh') 1 A former principality of western Europe, now a part of SE France, although the title has remained with the Dutch princes of Orange. 2 Its former capital, a city of Vaucluse department, SE France.
or·ange·ade (ôr'inj·ād') n. A beverage made of orange juice, sugar, and water. [<F]
orange blossom The white blossom of the orange tree: much worn by brides: State flower of Florida.
or·ange-crowned warbler (ôr'inj·kround', or'-) A small, greenish warbler (Vermivora celata) having an orange patch on its head: common in the United States.
Orange Free State A province of the east central Union of South Africa; 49,866 square miles; capital, Bloemfontein; a Boer republic until 1900; known as **Orange River Colony** from 1900 to 1910. Afrikaans **O·ran·je Vry·staat** (ō·rän'yə frā'stät).
Or·ange·ism (ôr'inj·iz'əm, or'-) n. The principles of the Orangemen; Irish Protestantism. Also **Or'ang·ism**. — **Or'ange·ist**, **Or'ang·ist** n.
Or·ange·man (ôr'inj·mən, or'-) n. pl. **·men** (-mən) A member of a secret society founded in northern Ireland in 1795 for the purpose of upholding Protestant ascendancy and succession in England: named in honor of William, Prince of Orange.
orange pekoe A finely sifted grade of black tea of India, Ceylon, and Java.
Orange Range A mountain system in central Netherlands New Guinea, the eastern section of the Snow Mountains; highest point, 15,585 feet.
Orange River A river in Basutoland, Union of South Africa, and South-West Africa, flowing 1,300 miles SW, NW, and west from NE Basutoland to the Atlantic and forming the southern boundaries of Orange Free State and South-West Africa.
or·ange·ry (ôr'inj·rē, or'-) n. pl. **·ries** A place for cultivating orange trees; an orange grove or greenhouse.
or·ange·wood (ôr'inj·wood', or'-) n. The fine-grained, yellowish wood of the orange tree: used in lathe work and in dentistry.
o·rang-u·tan (ō·rang'ə·tan, -ōō·tan) n. pl. **·tans** or **·tan** A large, anthropoid ape (genus Pongo or Simia) of Borneo and Sumatra, about 4 1/2 feet in height, having brownish-red hair, brown skin, small ears, doglike teeth, narrow lips, and long arms reaching to the ankles: also called **orang**. Also **o·rang'-ou·tang** (-ə·tang, -ōō·tang). [<Malay orañ utan <orañ a man + utan a forest]
o·ra pro no·bis (ō'rə prō nō'bis) Latin Pray for us.
o·rate (ô·rāt', ō·rāt', ō'rāt, ō·rāt') v.i. **o·rat·ed**, **o·rat·ing** To deliver an oration; talk oratorically or pompously; speechify: chiefly humorous. [<L oratus, pp. of orare. See ORATION.]
o·ra·tion (ō·rā'shən, ō·rā'-) n. 1 An elaborate public speech. 2 A graduation speech. See synonyms under SPEECH. [<L oratio, -onis <oratus, pp. of orare pray, speak <os, oris mouth. Doublet of ORISON.]
or·a·tor (ôr'ə·tər, or'-) n. 1 One who delivers an oration; an eloquent public speaker. 2 A high school or college student chosen to make a speech. 3 Law The complainant in a chancery proceeding; a petitioner in chancery. [<AF oratour <L orator <orare. See ORATION.] — **or'a·tor·ship** n. — **or'a·tress** n. fem.
or·a·to·ri·o (ôr'ə·tôr'ē·ō, -tō'rē·ō, or'-) n. pl. **·os** A musical composition, usually on a

sacred theme, for solo voices, chorus, and orchestra, dramatic in that it tells a connected story, though without scenery or acting. [<Ital., lit., a small chapel <L *oratorium.* See ORATORY[2].]

or·a·to·ry[1] (ôr′ə·tôr′ē, -tō′rē, or′-) *n. pl.* **·ries** **1** The art of public speaking; eloquence. **2** Eloquent language. See synonyms under SPEECH. [<L *oratoria (ars)* the oratorical (art), orig. fem. of *oratorius* < *orator* ORATOR] — **or′a·tor′i·cal** (-tôr′i·kəl, -tor′-) *adj.* — **or′a·tor′i·cal·ly** *adv.*

or·a·to·ry[2] (ôr′ə·tôr′ē, -tō′rē, or′-) *n. pl.* **·ries** **1** A place for prayer; a private chapel. **2** One of various congregations of priests in the Roman Catholic Church, who live together without vows, primarily for the purpose of teaching. [<LL *oratorium (templum)* (a temple) for prayer, orig. neut. of *oratorius* of prayer < *orator* one who prays < *oratus.* See ORATION.]

O·ra·zio (ō·rät′syō) Italian form of HORATIO.

orb (ôrb) *n.* **1** A rounded mass; a sphere or globe. **2** A circle or orbit; anything circular. **3** A sphere topped by a cross: symbolic of royal power. **4** *Obs.* The plane of the orbit or the orbit of a planet. — *v.t.* **1** To shape into a sphere or circle. **2** *Poetic* To enclose; encircle. [<L *orbis* a circle]

or·bic·u·lar (ôr·bik′yə·lər) *adj.* **1** Having the form of an orb or orbit. **2** Well-rounded. **3** *Bot.* Circular, as a leaf or petal. [<L *orbicularis* < *orbiculus,* dim. of *orbis* a circle] — **or·bic′u·lar′i·ty** (-lar′ə·tē) *n.* — **or·bic′u·lar·ly** *adv.*

or·bic·u·late (ôr·bik′yə·lit, -lāt) *adj.* Made into or taking the form of an orb or orbit; orbicular. Also **or·bic′u·lat′ed.** — **or·bic′u·late·ly** *adv.*
Synonyms: circular, globular, spherical, spheroidal.

or·bit (ôr′bit) *n.* **1** *Astron.* The path in space along which a heavenly body moves about its center of attraction. **2** *Anat.* One of the two cavities of the skull containing the eyes. **3** *Ornithol.* The eyelid and skin surrounding the eye of a bird. **4** *Physics* The assumed path of an electron around the atomic nucleus. **5** A range of influence or action: the *orbit* of imperialism. — *v.t.* To cause to move in an orbit, as an artificial satellite. — *v.i.* To move in an orbit. [<L *orbita* track of a wheel, an orbit < *orbis* a wheel, a circle] — **or′bi·tal** *adj.* — **or′bit·ing** *n.*

orbital cavity *Anat.* The bony socket enclosing and protecting the eyeball in the skull of vertebrates.

orbital decay The progressive spiraling change from elliptical to circular in the orbit of an artificial satellite whose velocity is gradually diminished by the residual air resistance encountered at perigee.

orbital index The maximum height of the orbital cavity multiplied by 100 and divided by the orbital width.

orb·y (ôr′bē) *adj.* **orb·i·er, orb·i·est** *Rare* Resembling an orb or rotating as an orb.

orc (ôrk) *n.* A grampus or some other cetacean: also spelled *ork.* [ME *orgue* <L *orca,* a kind of whale]

Or·ca·gna (ôr·kä′nyä), **Andrea,** 1308?–68, Florentine painter: original name *Andrea di Cione.*

or·ce·in (ôr′sē·in) *n. Chem.* A reddish-brown coloring matter, $C_{28}H_{24}N_2O_7$, obtained from orcinol by the action of aqueous ammonia and air. It is the tinctorial principle of archil. [<ORC(INOL) + -EIN]

or·chard (ôr′chərd) *n.* A plantation of trees grown for their products, as fruit, nuts, oils, etc.; also, the enclosure or ground containing them. [OE *orceard* < *ort-geard* a garden < *ort* (? <Med. L *ortus* a garden <L *hortus*) + *geard* a yard, enclosure]

or·chard·ist (ôr′chərd·ist) *n.* One who cultivates trees in orchards for their products. Also **or′chard·man** (-mən).

orchard oriole A common oriole *(Icterus spurius)* of eastern North America, smaller and having less brilliant plumage than the Baltimore oriole.

or·ches·tra (ôr′kis·trə) *n.* **1** A band of musicians playing together, especially a symphony orchestra; also, the instruments on which they play. **2** In theaters, the place immediately before the stage, occupied by the musicians; by extension, the main floor. **3** In the ancient Greek and Roman theaters, the approximately semicircular space from which the tiers of seats rose, in the Greek theater reserved for the chorus, and in the Roman theater reserved for the seats of the senators and other distinguished men. [<L <Gk. *orchēstra,* lit., a dancing space < *orcheesthai* dance] — **or·ches·tral** (ôr·kes′trəl) *adj.* — **or·ches′tral·ly** *adv.*

or·ches·trate (ôr′kis·trāt) *v.t.* & *v.i.* **·trat·ed, ·trat·ing** To compose or arrange (music) for an orchestra.

or·ches·tra·tion (ôr′kis·trā′shən) *n.* The arrangement of music for performance by an orchestra.

or·ches·tri·on (ôr·kes′trē·ən) *n.* A mechanical musical instrument, similar in action to a barrel organ, designed to imitate an orchestra. Also **or·ches·tri·na** (ôr′kis·trē′nə). [<ORCHESTRA]

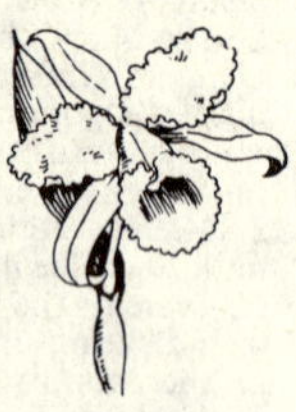
ORCHID FLOWER

or·chid (ôr′kid) *n.* **1** Any of a widely distributed family *(Orchidaceae)* of terrestrial or epiphytic monocotyledonous plants having thickened, bulbous roots and often very showy and distinctive flowers. **2** Any of various delicate, rosy-purple colors. [<NL <L *orchis,* an orchid <Gk., orig. a testicle; so called because of the shape of its tubers]

or·chi·da·ceous (ôr′ki·dā′shəs) *adj.* Of, pertaining to, or belonging to the orchid family.

orchido- *combining form* **1** *Bot.* Orchid; pertaining to orchids: *orchidology.* **2** *Med.* Orchio-. Also, before vowels, **orchid-.** [<ORCHID]

or·chid·ol·o·gy (ôr′ki·dol′ə·jē) *n.* The study and cultivation of orchids. — **or′chid·ol′o·gist** *n.*

or·chil (ôr′kil) *n.* **1** A purple or blue dye obtained from archil. **2** Archil. [<OF *orcheil, orchel* ARCHIL]

orchio- *combining form* Testicle; pertaining to the testicles. Also before vowels, **orchi-.** [<Gk. *orchis* a testicle]

or·chis (ôr′kis) *n.* Any plant of the genus *Orchis* having dense spikes of small flowers, frequently of striking shape and structure. [<NL <L, ORCHID]

or·chi·tis (ôr·kī′tis) *n. Pathol.* Inflammation of the testicle. Also **or·chei′tis, or·chi·di·tis** (ôr′ki·dī′tis). — **or·chit′ic** (-kit′ik) *adj.*

or·ci·nol (ôr′sə·nōl, -nol) *n. Chem.* A colorless crystalline compound, $C_7H_8O_2$, derived from certain lichens, as archil, and aloes. Also **or′cin.** [<Ital. *orc(ello)* archil + -IN + -OL[1]]

Or·cus (ôr′kəs) In Roman mythology: **1** The abode of the dead. **2** Pluto or Dis, the god of the underworld and lord of the dead.

Or·czy (ôr′tsē), **Baroness,** 1865–1947, English novelist born in Hungary. Also **Emmuska Orczy.**

or·dain (ôr·dān′) *v.t.* **1** To order or decree; enact; establish. **2** To predestine; destine: said of God, fate, etc. **3** To invest with ministerial or priestly functions. See synonyms under INSTALL, INSTITUTE. [<OF *ordener* <L *ordinare* order < *ordo, -inis* an order] — **or·dain′er** *n.*

or·deal (ôr·dēl′, -dē′əl, ôr′dēl) *n.* **1** A severe test of character or endurance; a trying course of experience. **2** A medieval form of judicial trial in which the accused was subjected to physical tests, as carrying or walking over burning objects or immersing the hand in scalding water, the result being considered a divine judgment of guilt or innocence. See synonyms under PROOF. [OE *ordāl, ordēl* a judgment < *or-* out + *dǣl* a deal; infl. in meaning by L *ordela* an ordeal <Gmc.]

ordeal bean Calabar bean.

or·der (ôr′dər) *n.* **1** Methodical and harmonious arrangement, as of successive things or as of military units in a formation. **2** Proper or working condition; available state. **3** A command or authoritative regulation. **4** *Law* Any direction of a court made to be entered of record in a cause, and not included in the final judgment. **5** A written commission or instruction to supply, purchase, or sell something. **6** Established use or customary procedure. **7** Established or existing state of things. **8** A class or body of persons united by some common bond, as for mutual insurance, protection, aid, social culture, etc.: the *Order* of Odd Fellows; a monastic or religious body: an *order* of mendicant friars. **9** A group of persons upon whom a government or sovereign has conferred an honor or dignity, and who are thus entitled to affix to their names designated initials and to wear specific insignia; also, the insignia worn as a sign of membership in such a group. **10** Social rank. **11** A class or kind of a common degree of excellence. **12** *Usually pl. Eccl.* **a** Any of the various grades or degrees of the Christian ministry: also **holy orders, sacred orders.** In the Anglican Church, there are three orders: bishops, priests, and deacons. The Greek Church recognizes in addition subdeacons and readers. In the Roman Catholic Church, there are seven orders: priests (including bishops), deacons, subdeacons (the **major orders**), acolytes, exorcists, readers, and doorkeepers (the **minor orders**). **b** The rank or position of an ordained clergyman. **c** The rite or sacrament of ordination. **d** A liturgical form for a service or the performance of a rite: the *order* of confirmation. **13** *Archit.* **a** The general character of a column and its parts as distinguishing a style of architecture; a style of architecture. Usually there are considered to be five orders of classical architecture — Doric, Ionic, Corinthian, Tuscan, and Composite. **b** A column with its entablature. **14** *Biol.* A taxonomic category ranking next below the class, and above the family. **15** *Math.* A number expressing the degree of complexity of an algebraic expression. **16** *Gram.* The sequence of words in a sentence or construction. **17** The position of the rifle as a result of the command *order arms.* **18** Any one of the ancient nine grades of angels. **19** *Obs.* Suitable care; preparation: usually in the phrase *to take order.* — **in order** **1** In accordance with rule; hence, apt; appropriate. **2** Neat; tidy. — **in order that** So that; to the end that. — **in order to** For the purpose of; to the end that. — **in short order** Quickly; without delay. — **on order** Ordered but not yet delivered. — **on the order of** Similar to. — **to order** According to the buyer's specifications: a shirt made *to order.* — *v.t.* **1** To give a command or direction to. **2** To command to go, come, etc.: They *ordered* him out of the city. **3** To give an order that (something) be done; prescribe. **4** To give an order for: to *order* a new suit. **5** To put in orderly or systematic arrangement; regulate. **6** To ordain: He was *ordered* deacon. — *v.i.* **7** To give an order or orders. — **to order arms** *Mil.* To bring a rifle perpendicularly against the right side, with the butt on the ground. See synonyms under DICTATE, REGULATE. [<OF *ordre* <L *ordo, -inis* a row, series, an order] — **or′der·er** *n.* — **or′der·less** *adj.*

Synonyms (noun): command, direction, directive, injunction, instruction, prohibition, requirement. *Instruction* implies more superiority of knowledge, *direction* more of authority; a teacher gives *instructions* to his pupils, an employer gives *directions* to his workmen; but the *instructions* of a superior regarding action are viewed as specific *commands.* A *directive* conveys all three of these — *instructions* for action, *directions* for procedure, and *command* for performance. *Order* is more absolute still; soldiers and railroad employees have simply to obey the *orders* of their superiors, without explanation or question. *Command* is a loftier word less frequent in common life; we speak of the *commands* of God. A *requirement* is imperative, but not always formal; it may be in the nature of things; as, the *requirements* of the position. *Prohibition* is wholly negative; it is a *command* not to do; *injunction* is now oftenest so used, especially as the *requirement* by legal authority that certain action be suspended or refrained from, pending final legal decision. Compare CLASS, FRAME, LAW[1], RULE, SORT, SYMMETRY, SYSTEM. *Antonyms:* allowance, consent, leave, liberty, license, permission, permit.

or·der·ing (ôr′dər·ing) *n.* **1** The act of directing, commanding, or disposing. **2** The act or process of arrangement, or the state of being arranged. **3** Right administration. **4** The act of ordination. **5** The act of arranging for procurement, purchase, or delivery of something.

or·der·ly (ôr′dər·lē) *adj.* **1** Having regard for arrangement; methodical; systematic. **2**

Peaceful. 3 Characterized by order. 4 Pertaining to orders. See synonyms under NEAT¹. — *n. pl.* **·lies** 1 A soldier or non-commissioned officer detailed to carry orders for superior officers. 2 A hospital attendant. — *adv.* According to the rules of order; methodically; regularly; properly. — **or′der·li·ness** *n.*

or·di·nal¹ (ôr′də·nəl) *adj.* 1 Denoting position in an order or succession. 2 Pertaining to an order, as of plants, animals, etc. — *n.* An ordinal number. See under NUMBER. [<LL *ordinalis* <L *ordo, -inis* an order]

or·di·nal² (ôr′də·nəl) *n. Eccl.* 1 A book of rites for clerical ordinations, episcopal consecrations, etc. 2 An ordo. [<Med. L *ordinale* <LL *ordinalis*. See ORDINAL¹.]

or·di·nance (ôr′də·nəns) *n.* 1 An authoritative rule; an order, decree, or law of a municipal body. 2 A religious rite or ceremony. 3 *Archit.* A system of arrangement. 4 A law or command of God, or a decree of fate. 5 *Obs.* Order, as arrangement, disposition, rank, position, array, provision, or preparation. 6 *Obs.* The act of devising, arranging, or contriving plans; a design or device. See synonyms under LAW¹, SACRAMENT. [<OF *ordenance* <Med. L *ordinantia* <L *ordinans, -antis*, ppr. of *ordinare* ORDAIN]

or·di·nand (ôr′də·nənd) *n.* A candidate for ordination. [<L *ordinandus*, gerundive of *ordinare* ORDAIN]

or·di·nant (ôr′də·nənt) *adj.* Exercising authority; ruling.

or·di·nar·i·ly (ôr′də·ner′ə·lē, ôr′də·nâr′ə·lē) *adv.* 1 In ordinary cases; commonly; usually: *Ordinarily*, he walks to work. 2 In the usual manner. 3 To the usual extent; normally.

or·di·nar·y (ôr′də·ner′ē) *adj.* 1 Of common or everyday occurrence; customary; usual. 2 According to an established order; regular; normal. 3 Common in rank or degree; of average merit or consequence; commonplace. 4 Having immediate or ex-officio jurisdiction, as a judge. See synonyms under COMMON, GENERAL, HABITUAL, NORMAL, USUAL. — *n. pl.* **·nar·ies** 1 That which is usual or common. 2 *Brit.* A meal provided regularly at a fixed price. 3 *Brit.* An eating house where such meals are served. 4 *Law* One who exercises jurisdiction in his own right, and not by delegation. 5 *U.S.* In some States, a judge exercising probate jurisdiction. 6 *Eccl.* **a** A rule or book prescribing the form for saying mass. **b** The practically unchangeable part of the mass: opposed to the *proper*. 7 An early type of bicycle with a large front wheel and a small rear wheel. 8 *Her.* A charge of the simplest kind, usually bounded between simple lines, as a chief, pale, fess, chevron, bend, cross, saltire, or quarter. — **in ordinary** 1 In actual and constant service. 2 *Naut.* Out of commission; laid up: said of a ship. [<L *ordinarius* regular, usual <*ordo, -inis* an order] — **or′di·nar′i·ness** *n.*

or·di·nate (ôr′də·nit) *adj.* 1 Characterized by order; regular. 2 *Biol.* Arranged in a regular row or rows, as spots on an insect's body or wings. — *n. Math.* 1 The distance of any point from the axis of abscissas, measured on a line parallel to the axis of ordinates in a coordinate system. 2 The line or number indicating such distance. [<L *ordinatus*, pp. of *ordinare* ORDAIN]

or·di·na·tion (ôr′də·nā′shən) *n.* 1 *Eccl.* The rite of consecration to the ministry. 2 The state of being ordained, regulated, or settled. 3 Arrangement of things in order; array. 4 Natural or proper order.

ord·nance (ôrd′nəns) *n. Mil.* 1 All kinds of weapons and their appliances used in war. 2 Cannon or artillery. [Contraction of ORDINANCE]

Ordnance Department A branch of the armed services responsible for the design, manufacture, procurement, maintenance, issue, and efficiency of ordnance.

or·do (ôr′dō) *n. pl.* **·di·nes** (-də·nēz) *Eccl.* 1 A book containing directions for the portions of the mass and the daily office which vary according to the calendar. 2 A book of rubrics for administering the sacraments and for other ceremonies. [<L, order]

or·don·nance (ôr′də·nəns, *Fr.* ôr·dô·näns′) *n.*

1 A right arranging of parts, as in a picture, so as to produce the best effect. 2 A law or ordinance; specifically, in French law, a code of laws on any subject. [<F *ordenance* ORDINANCE]

Or·do·vi·ces (ôr′də·vī′sēz, ôr·dov′ə·sēz) *n. pl.* An ancient Celtic tribe in Wales. [<L]

Or·do·vi·cian (ôr′də·vish′ən) *adj.* 1 *Geol.* Of, pertaining to, or designating a geological period of the Paleozoic era, following the Cambrian and preceding the Silurian. 2 Of or pertaining to the Ordovices. — *n.* An epoch of the Paleozoic era: sometimes called *Lower Silurian*. [<L *Ordovices* the Ordovices]

or·dure (ôr′jər, -dyŏŏr) *n.* Excrement; feces. [<OF <*ord* foul, nasty <L *horridus* HORRID]

Or·dzho·ni·kid·ze (ôr′jon·i·kēd′zə) A former name for DZAUDZHIKAU.

ore (ôr, ōr) *n.* 1 A natural substance containing an economically valuable metal, and sometimes forming part of a rock. 2 Loosely, a natural substance containing a non-metallic mineral: sulfur *ore*. ♦ Homophone: *oar.* [OE *ār, ǣr*, brass, copper; infl. in meaning by OE *ōra* unwrought metal]

ö·re (œ′rə) *n. pl.* **ö·re** A Scandinavian bronze coin, equaling 1/100 krone in Norway and Denmark, and 1/100 krona in Sweden. [< Norw., Dan., and Sw.]

O·re·ad (ôr′ē·ad, ō′rē-) *n.* 1 In classical mythology, a mountain nymph. 2 A sun-loving plant; a heliophyte. [<L *oreas, -adis* <Gk. *oreias, -ados* <*oros* a mountain]

Ö·re·bro (œ′rə·brōō′) A city in south central Sweden; the country's foremost paper-manufacturing center; site of many medieval buildings.

o·rec·tic (ō·rek′tik) *adj. Philos.* Of or pertaining to the appetites or desires; appetent; motive. Also **o·rec′tive**. [<Gk. *orektikos* appetitive < *orektos* stretched out, longed for < *oregein* stretch out, desire]

ore dressing *Metall.* The mechanical separation of valuable metals and minerals from the ores in which they occur.

o·reg·a·no (ō·reg′ə·nō) *n.* Origan. [<Sp. *orégano* <L *origanum* ORIGAN]

Or·e·gon (ôr′ə·gon, -gən, or′-) A State of the western United States on the Pacific; 96,981 square miles; capital, Salem; entered the Union Feb. 14, 1859; nickname, *Beaver State*: abbr. *Ore.* or *Oreg.* — **Or′e·go′ni·an** (-gō′nē·ən) *adj. & n.*

Oregon Country A name used between 1818 and 1846 for the region of the NW United States including the present States of Washington, Oregon, Idaho, and parts of Montana and Wyoming.

Oregon fir The Douglas fir. Also **Oregon pine.**

Oregon grape A thornless evergreen shrub (*Mahonia aquifolium*) of the barberry family, with dark-blue berry clusters resembling grapes: the State flower of Oregon.

Oregon jargon Chinook.

Oregon Trail A former route extending from the Missouri River about 2,000 miles NW to the Columbia River in Oregon; used by pioneers, 1804–1846.

ore house A building in which mined ore is stored. Also **ore shed.**

o·re·ide (ō′rē·id) See OROIDE.

O·rel (ō·rel′, *Russian* o·ryôl′) A city on the Oka in west central European Russian S.F.S.R.

Ore Mountains The English name for the ERZGEBIRGE.

O·ren·burg (ô′rən·bŏŏrkh) A former name for CHKALOV.

O·ren·se (ō·ren′sā) 1 A province of Galicia, NW Spain; 2,694 square miles. 2 Its capital, a city of great antiquity, occupied by the Romans and noted since their time for its warm sulfur springs.

O·re·o·pith·e·cus (ôr′ē·ō·pith′ə·kəs, ō′rē-) *n.* A primate of the Miocene or Pliocene epochs, identified in 1872 from a group of fossilized bones discovered in brown coal deposits in Tuscany: of uncertain evolutionary rank but possibly ancestral to the human family. Also called *mountain ape*. [<NL <Gk. *oros, oreos* a mountain + *pithēkos* an ape]

ore shoot That portion of an ore deposit which is exceptionally rich in metal content.

O·res·tes (ô·res′tēz, ō-) In Greek legend, the son of Agamemnon and Clytemnestra who, aided by his sister Electra, killed his mother and Aegisthus, her lover, to avenge his father's murder; he was then pursued by the Furies until Athena granted him absolution. In another version he expiated his guilt by carrying off an image of Artemis from Tauris to Greece with the aid of his sister Iphigenia. [<L <Gk. *Orestēs* <*oros, oreos* a mountain]

Ö·re·sund (œ′rə·sŏŏn) The strait between Zealand (Denmark) and Sweden, connecting the Kattegat with the Baltic; 87 miles long; average width 17 miles, at its narrowest, 2 1/2 miles: also *The Sound*.

o·rex·is (ō·rek′sis) *n.* Appetite; craving. [<Gk., a desire < *oregein*. See ORECTIC.]

Or·fa·ni (ôr·fä′nē), **Gulf of** See STRYMONIC GULF.

Or·fi·la (ôr·fē·lä′), **Matthieu Joseph Bonaventure**, 1787–1853, French chemist; founder of toxicology.

or·fray (ôr′frā) See ORPHREY.

or·gan (ôr′gən) *n.* 1 A musical wind instrument consisting of a collection of pipes made to sound by means of compressed air from bellows and played upon by means of keys: also **pipe organ**. 2 An electronic musical instrument not employing pipes or wind, designed to give the sounds and timbres of a pipe organ. 3 A musical instrument resembling or having some mechanism resembling the pipe organ: a reed *organ*; a barrel *organ*. 4 Any part of an organism, plant, or animal performing some definite function: the digestive *organs*. 5 An instrument or agency for communication of the views of a person or party; especially, a newspaper or periodical published in the interest of some political party or religious denomination. [Fusion of OE *organa* and OF *organe*, both <L *organum* an instrument, engine, organ <Gk. *organon* a tool, a musical instrument]

or·gan·dy (ôr′gən·dē) *n. pl.* **·dies** A very thin, crisp, transparent, cotton muslin, plain or figured, used for dresses, collars, cuffs, etc. Also **or′gan·die**. [<F *organdi*; ult. origin uncertain]

or·gan-grind·er (ôr′gən·grīn′dər) *n.* The player of a hand organ; specifically, a street musician playing a hand organ.

or·gan·ic (ôr·gan′ik) *adj.* 1 Of, pertaining to, or of the nature of animals and plants. 2 Affecting an organ or organs of an animal or plant: *organic* diseases. 3 Serving the purpose of an organ. 4 *Chem.* Of or pertaining to compounds containing carbon as an essential ingredient. 5 Inherent in or pertaining to the organization or fundamental structure; structural; constitutional. 6 Of or characterized by systematic coordination of parts; organized; systematized. 7 *Law* Designating that system of laws or principles forming the foundation of a government. 8 *Agric.* Pertaining to the use of compost, manure, peat moss, and other natural fertilizers in the cultivation of farms and gardens. Also **or·gan′i·cal**. See synonyms under RADICAL. — **or·gan′i·cal·ly** *adv.*

organic chemistry The branch of chemistry that relates to carbon compounds.

organic disease *Pathol.* A disease that affects some particular organ in its structure, as distinguished from its function.

or·gan·i·cism (ôr·gan′ə·siz′əm) *n.* 1 *Med.* The doctrine that all diseases are caused by specific lesions of one or more organs. 2 *Biol.* The theory of living processes as the result of the activity of all the organs considered as an autonomous, integrated system. 3 *Sociol.* The concept of society as an organism, of which beliefs, ideas, customs, etc., are component parts. — **or·gan′i·cist** *n.*

organic law 1 The law by which a government outlines and establishes its own political structure. 2 An act of Congress providing a form of government for a territory.

or·gan·ism (ôr′gən·iz′əm) *n.* 1 An animal or plant internally organized to maintain vital activities. 2 Anything that is analogous in structure and function to a living thing: the social *organism*. — **or′gan·is′mal** *adj.*

or·gan·ist (ôr′gən·ist) *n.* One who plays the organ. 2 In the Middle Ages, a singer who accompanied the plain song with another part. See ORGANUM (def. 2).

or·gan·i·za·tion (ôr′gən·ə·zā′shən, -ī·zā′-) *n.* **1** The act of organizing, or the state of being organized; also, that which is organized. **2** An animal or vegetable organism. **3** A number of individuals systematically united for some end or work: a military *organization.* **4** The officials, committeemen, etc., who control a political party: also called *machine.* **5** Any combination of parts. Also *Brit.* **or′·gan·i·sa′tion.**

or·gan·ize (ôr′gən·īz) *v.* **·ized, ·iz·ing** *v.t.* **1** To bring together or form as a whole or combination, as for a common objective. **2** To arrange systematically; order. **3** To furnish with organic structure. **4 a** To enlist (workers) in a trade union. **b** To unionize the workers of (a factory, etc.). — *v.i.* **5** To form or join an organization. Also *Brit.* **or′gan·ise.** See synonyms under INSTITUTE. [<Med. L *organizare* <L *organum* ORGAN] — **or′gan·iz′a·ble** *adj.* — **or′gan·iz′er** *n.*

organo- *combining form* **1** *Biol.* Related to an organ, or to the organs of the body: *organogenesis.* **2** *Chem.* Organic: *organometallic.* [<Gk. *organon* an instrument, organ]

organ of Corti A complex structure in the human ear directly involved with the perception of sound: discovered by Alfonso Corti.

or·ga·no·gen·e·sis (ôr′gə·nō·jen′ə·sis) *n. Biol.* The development of organs in animals and plants. — **or′ga·no·ge·net′ic** (-jə·net′ik) *adj.*

or·ga·nog·ra·phy (ôr′gə·nog′rə·fē) *n.* Scientific description of organs; descriptive organology. — **or′ga·no·graph′ic** (-nō·graf′ik) or **·i·cal** *adj.*

or·ga·no·lep·tic (ôr′gə·nō·lep′tik) *adj. Physiol.* **1** Having an effect upon one of the organs of sense, as taste or smell. **2** Of or pertaining to sense impressions and their evaluation: applied especially to laboratory tests of food, flavors, and odors. [<F *organoleptique* <Gk. *organon* organ + *lep-,* stem of *lambanein* take]

or·ga·nol·o·gy (ôr′gə·nol′ə·jē) *n.* The branch of biology that treats of organs of the body. — **or′ga·no·log′ic** (-nō·loj′ik) or **·i·cal** *adj.* — **or′ga·nol′o·gist** *n.*

or·ga·no·me·tal·lic (ôr′gə·nō·mə·tal′ik) *adj. Chem.* Designating or pertaining to a compound of metal and carbon.

or·ga·non (ôr′gə·non) *n. pl.* **·na** (-nə) or **·nons** A system of rules and principles considered as an instrument of guidance, as of knowledge or thought. Also *organum.* [<Gk., organ]

or·ga·no·sil·i·con (ôr′gə·nō·sil′ə·kon) *n. Chem.* Any of an important class of compounds or polymers containing silicon and carbon; a silicone.

or·ga·no·ther·a·py (ôr′gə·nō·ther′ə·pē) *n. Med.* The treatment of disease by remedies derived from animal organs. Also **or′ga·no·ther′a·peu′tics** (-ther′ə·pyōō′tiks).

organ pipe One of the long tubes of a pipe organ, in which a column of air is made to vibrate so as to produce a tone of definite pitch.

or·ga·num (ôr′gə·nəm) *n. pl.* **·na** (-nə) or **·nums** **1** An organon. **2** In medieval music, a part sung as an accompaniment to the melody or plain song at an interval of a fourth or a fifth above or below it; also, this method of part singing. [<L, ORGAN]

or·gan·za (ôr·gan′zə) *n.* A sheer, crisp fabric used for evening dresses, trimming, etc. [Origin uncertain]

or·gan·zine (ôr′gən·zēn) *n.* **1** A silk thread made of several single threads twisted together; thrown silk. **2** A fabric made of organzine. [<F *organsin* <Ital. *organzine*; ult. origin unknown]

or·gasm (ôr′gaz·əm) *n.* **1** Immoderate or extreme excitement or behavior. **2** *Physiol.* The acme of excitement at the culmination of the sexual act, followed by detumescence. [<F *orgasme* <Gk. *orgasmos* a swelling < *orgaein* swell, be excited] — **or·gas·tic** (ôr·gas′tik) *adj.*

or·geat (ôr′zhat, *Fr.* ôr·zhä′) *n.* A sirup made from barley water and sugar flavored with almonds, orange flowers, etc. [<MF <Provençal *orjat* < *ordi, orge* barley <L *hordeum*]

Or·get·o·rix (ôr·jet′ə·riks) Helvetian chief of the first century B.C. who opposed Julius Caesar.

or·gi·as·tic (ôr′jē·as′tik) *adj.* Pertaining to or resembling the Greek orgies; hence, marked by wild revelries. Also **or′gi·ac, or′gic.** [<Gk. *orgiastikos* < *orgiastēs* a celebrator of an orgy < *orgiazein* celebrate orgies]

or·gone (ôr′gōn) *n.* In the psychobiology of Wilhelm Reich, biological energy, accumulated from the environment and discharged gradually in all activity but suddenly in the orgasm: identified with the libido, as defined by Freud and, tentatively, with the ether of older physical theories.

or·gy (ôr′jē) *n. pl.* **·gies** **1** Wild or wanton revelry; a drunken carousal; debauch. **2** Any immoderate or excessive indulgence in something: an *orgy* of reading. **3** *pl.* The secret rites in honor of certain ancient Greek and Roman deities, as Dionysus, marked by ecstatic or frenzied songs and dances. [Earlier *orgies,* pl. <MF <L *orgia* <Gk., secret rites. Akin to WORK.]

or·i·bi (ôr′ə·bē, or′-) *n. pl.* **·bis** or **·bi** A small, dun-colored African antelope (genus *Ourebia*), about two feet high at the shoulder: also spelled *ourebi.* [<Afrikaans <Hottentot *arab*]

or·i·chalc (ôr′i·kalk, or′-) *n.* In ancient Greece, an alloy of copper and zinc, resembling gold; brass. Also **or′i·chalch.** [<L *orichalcum* <Gk. *oreichalkon,* lit., mountain copper < *oros, oreos* a mountain + *chalkos* copper, brass]

o·ri·el (ôr′ē·əl, ō′rē-) *n. Archit.* A bay window; especially, one built out from a wall and resting on a bracket or similar support. [<OF *oriol* a porch, gallery, ? <Med. L *oriolum*; ult. origin unknown]

o·ri·ent (ôr′ē·ənt, ō′rē-) *n.* **1** The east: opposed to *occident.* **2** The eastern sky; also, dawn; sunrise. **3** The iridescent luster of a pearl. — *v.t.* **1** To cause to face or turn to the east. **2** To place or adjust, as a map, in exact relation to the points of the compass. **3** To adjust according to first principles or recognized facts or truths; adapt (oneself) mentally to a situation. **4** To adjust in relation to something else: His experience *oriented* his ideas toward science. — *adj.* **1** Resembling sunrise; bright. **2** Ascending. [<OF <L *oriens, -entis* rising sun, east, orig. ppr. of *oriri* rise]

O·ri·ent (ôr′ē·ənt, ō′rē-) The East; Asia, especially eastern Asia: opposed to *Occident.*

o·ri·en·tal (ôr′ē·en′təl, ō′rē-) *adj.* **1** *Astron.* Eastern; appearing or being in the eastern sky: said of stars and planets. **2** Specially bright, clear, and pure: said of gems. **3** Noting a variety of precious corundum, especially sapphire, marked by colors suggestive of other gems: an *oriental* amethyst, *oriental* aquamarine, *oriental* emerald, *oriental* topaz. — **o′ri·en′tal·ly** *adv.*

O·ri·en·tal (ôr′ē·en′təl, ō′rē-) *adj.* **1** Of or pertaining to the Orient; eastern. **2** Magnificent; gorgeous; sumptuous: *Oriental* luxury. **3** Designating a zoogeographical realm or region which includes India, southern Asia, the East Indies, and the Philippine Islands. — *n.* An inhabitant of Asia; an Asian. Opposed to *Occidental.* Also **o′ri·en′tal.**

Oriental alabaster Alabaster (def. 2).

O·ri·en·tal·ism (ôr′ē·en′təl·iz′əm, ō′rē-) *n.* **1** An Oriental quality or character of thought, speech or manners, or the disposition to adopt such a quality or character. **2** Knowledge of or proficiency in Oriental languages, literature, etc. Also **o′ri·en′tal·ism.** — **O′ri·en′tal·ist** *n.*

O·ri·en·tal·ize (ôr′ē·en′təl·īz, ō′rē-) *v.t.* & *v.i.* **·ized, ·iz·ing** To make or become Oriental. Also *Brit.* **O′ri·en′tal·ise.**

Oriental rug A rug or carpet hand-woven in one piece in the Orient.

o·ri·en·tate (ôr′ē·en·tāt′, ō′rē-) *v.* **·tat·ed, ·tat·ing** *v.t.* To orient. — *v.i.* To face or turn eastward or in some specified direction; be oriented. [<F *orienter* <OF *orient* ORIENT]

o·ri·en·ta·tion (ôr′ē·en·tā′shən, ō′rē-) *n.* **1** The act of orienting, or the state of being oriented. **2** Position, or the determining of position, with relation to the points of the compass. **3** The determination or adjustment of one's position with reference to circumstances, ideals, etc. **4** *Psychol.* Awareness of one's own temporal, spatial, and personal relationships. **5** *Archit.* The construction of a church upon an east-west axis, so as to have the altar in the eastern end. **6** The homing instinct, as in pigeons. **7** *Chem.* The particular disposition of the constituent atoms in a compound, especially as determined by electrical forces.

O·ri·en·te (ôr′ē·en′tā, ō′rē-) **1** A province of eastern Cuba; 14,132 square miles; capital, Santiago de Cuba: formerly *Santiago de Cuba.* **2** A region and former province of Ecuador, comprising all its territory east of the Andes.

or·i·fice (ôr′ə·fis, or′-) *n.* A small opening into a cavity; an aperture. See HOLE. [<MF <LL *orificium* <L *os, oris* mouth + *facere* make]

or·i·flamme (ôr′ə·flam, or′-) *n.* **1** The red banner of the abbey of St. Denis, used as a battle standard by the kings of France until the 15th century. **2** *Her.* A blue banner charged with three fleurs-de-lis of gold. **3** Any flag or standard. Also spelled *auriflamme.* [<F <OF *oriflambe* <Med. L *auriflamma* <L *aurum* gold + *flamma* a flame]

or·i·gan (ôr′ə·gən, or′-) *n.* The wild marjoram (*Origanum vulgare*), an important source of carvacrol: its fragrant leaves are esteemed as a seasoning. Also spelled *oregano.* [<OF <L *origanum* <Gk. *origanon,* an herb like marjoram < *oros, oreos* a mountain + *ganos* brightness, joy]

Or·i·gen (ôr′ə·jen, -jən, or′-), 185?–254?, Alexandrian theologian and father of the Greek Church.

Or·i·gen·ism (ôr′ə·jen·iz′əm, -jən-, or′-) *n.* The system of religious and philosophical doctrine held by Origen, who taught a threefold interpretation of the Scriptures: literal, moral, and mystical.

or·i·gin (ôr′ə·jin, or′-) *n.* **1** The commencement of the existence of anything. **2** A primary source; cause. **3** Parentage; ancestry. **4** *Anat.* The relatively fixed point of attachment of a muscle. Compare INSERTION. **5** *Math.* **a** The point at which the axes of a Cartesian coordinate system intersect; the point where the ordinate and abscissa equal zero. **b** The point in a polar coordinate system where the radius vector equals zero. See illustration under QUADRANT, OCTANT. See synonyms under BEGINNING, CAUSE, SOURCE. [Appar. <OF *origine* <L *origo, -inis* a rise, beginning < *oriri* rise]

o·rig·i·nal (ə·rij′ə·nəl) *adj.* **1** Of or belonging to the beginning, origin, or first stage of existence of a thing. **2** Immediately produced by one's own mind and thought; not copied or produced by imitation. **3** Able to produce works requiring thought, without copying or imitating those of others; creative; inventive. See synonyms under AUTHENTIC, FIRST, NATIVE, PRIMEVAL, RADICAL, TRANSCENDENTAL. — *n.* **1** The first form of anything. **2** The language in which a book is first written. **3** A person of unique character or genius; also, an eccentric. **4** Originator; also, origin. See synonyms under IDEAL, MODEL.

o·rig·i·nal·i·ty (ə·rij′ə·nal′ə·tē) *n. pl.* **·ties** **1** The power of originating; inventiveness. **2** The quality of being original or novel.

o·rig·i·nal·ly (ə·rij′ə·nəl·ē) *adv.* **1** At the beginning. **2** In a new and striking manner.

original sin *Theol.* The natural corruption and depravity inherent in all mankind as a consequence of Adam's first sinful disobedience.

o·rig·i·nate (ə·rij′ə·nāt) *v.* **·nat·ed, ·nat·ing** *v.t.* To bring into existence; create; initiate. — *v.i.* To come into existence; have origin; arise. See synonyms under INSTITUTE, PRODUCE, PROPAGATE. — **o·rig′i·na′tion** *n.* — **o·rig′i·na′tive** *adj.* — **o·rig′i·na′tive·ly** *adv.* — **o·rig′i·na′tor** *n.*

o·ri·na·sal (ôr′ə·nā′zəl, or′-) *adj.* **1** *Anat.* Of or pertaining to the mouth and nose: the *orinasal* duct. **2** *Phonet.* Pronounced with the nasal and oral passages both open, as the French nasal vowels. — *n. Phonet.* An orinasal vowel. [<L *os, oris* mouth + NASAL]

O·ri·no·co (ôr′ə·nō′kō, or′-) A river in Venezuela, flowing about 1,700 miles NW and north to the Atlantic Ocean.

o·ri·ole (ôr′ē·ōl, ō′rē-) *n.* **1** Any of a family (*Oriolidae*) of black-and-yellow passerine birds of the Old World, related to the crows; the common **European** (or **golden**) **oriole** (*Oriolus oriolus*) is bright yellow with sharply contrasting black wings and tail, and builds a hanging nest. **2** One of various black-and-yellow American songbirds (family *Icteridae*) building a hanging nest; especially, the Baltimore oriole and the orchard oriole. [<OF *oriol* <Med. L *oriolus* <L *aureolus,* dim. of *aureus* golden < *aurum* gold]

O·ri·on (ō·rī′ən) **1** In Greek and Roman mythology, a giant hunter who pursued the

Pleiades and was killed by Diana: he was placed among the stars by her as a constellation. 2 *Astron.* An equatorial constellation, noted for its group of three stars in a line (the Sword Belt or Girdle), and for its two bright stars, Betelgeuse and Rigel. See CONSTELLATION.

O·ris·ka·ny (ō·ris′kə·nē) A village on the Mohawk River in central New York; scene of a battle in the Revolutionary War, 1777.

or·i·son (ôr′i·zən, or′-) *n. Usually pl.* A devotional prayer. See synonyms under PRAYER. [<OF *oreisun, orison* <L *oratio, -onis* a prayer. Doublet of ORATION.]

O·ris·sa (ō·ris′ə) A constituent State of east central India on the NW Bay of Bengal; 60,136 square miles; capital, Bhubaneswar.

O·ri·ya (ō·rē′yä) *n.* One of the major Indic languages of India, closely related to Bengali.

O·ri·za·ba (ō′rē·sä′bä) A city in Veracruz state, Mexico, SE of Pico de Orizaba.

Orizaba, Pico de An inactive volcano in Veracruz state, the highest point in Mexico; 18,700 feet: also *Citlatépetl.*

ork (ôrk) See ORC.

Or·khon (ôr′kon) A river in north central Mongolian People's Republic, flowing 700 miles east and north to the Selenga. Also **Or′hon** (-hon).

Ork·ney Islands (ôrk′nē) An island group and county of northern Scotland off the NE tip of the mainland; 376 square miles; capital, Kirkwall. Also **Ork′neys.**

Or·lan·do (ôr·lan′dō) A masculine personal name. See ROLAND.

Or·lan·do (ôr·län′dō), **Vittorio Emanuele,** 1860–1952, Italian statesman; premier 1917–1919.

orle (ôrl) *n. Her.* A subordinary bearing consisting of a band, half the width of a bordure, extending round the shield near the edge. [<F <OF *urle, ourle,* dim. <L *ora* a border]

Or·lé·a·nais (ôr·lā·á·ne′) A region and former province in north central France on both sides of the Loire.

Or·le·an·ism (ôr′lē·ən·iz′əm) *n.* Adherence to the Orléans family.

Or·le·an·ist (ôr′lē·ən·ist) *n.* A supporter of the Orléans branch of the French royal family descended from the younger brother of Louis XIV. —*adj.* Of or pertaining to Orleanists or Orleanism.

Or·lé·ans (ôr·lā·äṅ′) Name of a cadet branch of the Valois and Bourbon houses of France, many members of which have been prominent in French history from the 14th century. —**Louis Philippe Joseph d',** 1747–93, revolutionary and egalitarian; guillotined: known as *Philippe Égalité.*

Or·lé·ans (ôr·lā·äṅ′) A city of north central France, on the Loire.

Or·lon (ôr′lon) *n.* A synthetic fiber woven from an acrylic resin: it has high resistance to heat, light, and chemicals and is widely used as a textile material: a trade name.

or·lop (ôr′lop) *n. Naut.* The lowest deck of a ship, especially of a warship. [<Du. *overloop,* orig. a covering <*over* over + *loopen* run; so called because it covers the hold]

Or·ly (ôr·lē′) A SE suburb of Paris; site of an international airport.

or·mer (ôr′mər) *n.* 1 An ear shell; especially, *Haliotis tuberculata,* an edible univalve mollusk of the Channel Islands. 2 An abalone. [<dial. F (Channel Islands) <F *ormier,* contraction of *oreille de mer,* lit., ear of the sea; so called with ref. to its shape]

or·mo·lu (ôr′mə·lo͞o) *n.* Gilt or bronzed metallic ware, or lustrous bronze, used in decorating furniture, etc. [<F *or moulu,* lit., ground gold <*or* gold (<L *aurum*) + *moulu,* pp. of *moudre* grind <L *molere*]

Or·muz (ôr′muz, ôr·mo͞oz′) See HORMUZ.

Or·muzd (ôr′muzd) In Zoroastrian religion, the principle of good, source of light, and creator of the world: opponent of Ahriman, the evil deity: called *Ahura Mazda* in the Avesta. Also **Or′mazd** (-mäzd). [<Persian *Ormazd,* ult. <Avestan *Ahura-Mazda,* lit., wise lord]

or·na·ment (ôr′nə·mənt) *n.* 1 A part or an addition that contributes to the beauty or elegance of a thing. 2 Ornamentation in the abstract, or ornaments collectively. 3 Any thing or person considered as a source of honor or credit. 4 A mark of distinction; decoration. 5 *Music* A decorative note or notes not necessary to the melody; an appoggiatura. —*v.t.* (ôr′nə·ment) To adorn with ornaments; embellish. See synonyms under ADORN, GARNISH. [<OF *ornement* <L *ornamentum* equipment, ornament <*ornare* adorn] —**or′na·ment′er** *n.*

or·na·men·tal (ôr′nə·men′təl) *adj.* Serving to adorn. —*n.* An ornamental object, especially a plant meant for decorative purposes. —**or′·na·men′tal·ly** *adv.*

or·na·men·ta·tion (ôr′nə·men·tā′shən) *n.* 1 The act of adorning, or the state of being adorned. 2 Ornamental things collectively; that with which something is ornamented.

or·nate (ôr·nāt′) *adj.* 1 Ornamented to a marked degree; artistically elaborate, as a literary style. 2 Ornamented; decorated. [<L *ornatus,* pp. of *ornare* adorn] —**or·nate′ly** *adv.* —**or·nate′ness** *n.*

or·ner·y (ôr′nər·ē, ôrn′rē) *adj. U.S. Dial.* 1 Mean; low; also, unruly; stubborn. 2 Common; ordinary. [Alter. of ORDINARY] —**or′·ner·i·ness** *n.*

or·nis (ôr′nis) *n.* Avifauna. [<G <Gk., a bird]

or·nith·ic (ôr·nith′ik) *adj.* Of or pertaining to birds. [<Gk. *ornithikos* birdlike <*ornis, ornithos* a bird]

or·ni·thine (ôr′nə·thēn, -thin) *n. Biochem.* An amino acid, $C_5H_{12}O_2N_2$, found in the urine of birds; a product of arginine. [<Gk. *ornis, ornithos* a bird + -INE[2]]

or·ni·this·chi·an (ôr′nə·this′kē·ən) *Paleontol. adj.* Belonging to an order (*Ornithischia*) of birdlike, chiefly amphibious dinosaurs of the Jurassic and Cretaceous periods, including the iguanodon. —*n.* A member of this order. [<NL <Gk. *ornis, ornithos* a bird + *ischion* a hip]

ornitho- *combining form* Bird; of or related to birds: *ornithology.* Also, before vowels, **ornith-.** [<Gk. *ornis, ornithos* a bird]

Or·ni·tho·gae·a (ôr′nə·thō·jē′ə) *n.* A zoogeographical realm including New Zealand and Polynesia, characterized by extinct and existing avifauna not found elsewhere. Also **Or′·ni·tho·ge′a.** [<NL <Gk. *ornis, ornithos* a bird + *gaia* earth, land] —**Or′ni·tho·gae′an** or **·ge′an** *adj.*

or·ni·thoid (ôr′nə·thoid) *adj.* Resembling a bird or birds.

or·ni·thol·o·gy (ôr′nə·thol′ə·jē) *n.* The branch of zoology that treats of birds. [<ORNITHO- + -LOGY] —**or′ni·tho·log′ic** (-thə·loj′ik) or **·i·cal** *adj.* —**or′ni·tho·log′i·cal·ly** *adv.* —**or′ni·thol′o·gist** *n.*

or·ni·thoph·i·lous (ôr′nə·thof′ə·ləs) *adj. Bot.* Bird-loving: said of flowers that are adapted for or depend upon birds (usually hummingbirds) to transfer the pollen from the stamens to the stigma; bird-pollinated. [<ORNITHO- + -PHILOUS]

or·ni·tho·pod (ôr′nə·thō·pod, ôr·nī′-) *Paleontol. n.* One of an extinct order (*Ornithischia*) of bipedal dinosaurians of herbivorous habits. —*adj.* Of or pertaining to this order. [<NL *Ornithopoda* <Gk. *ornis, ornithos* a bird + *pous, podos* a foot]

or·ni·thop·ter (ôr′nə·thop′tər) *n.* A theoretical type of aircraft sustained and propelled by an upward and downward movement of the wings, as in the flight of a bird: also called *orthopter.* [<ORNITHO- + Gk. *pteron* a wing]

or·ni·tho·rhyn·chus (ôr′nə·thō·ring′kəs) *n. pl.* **·chi** (-kī) An egg-laying mammal with a ducklike bill; a duckbill. [<NL <Gk. *ornis, ornithos* a bird + *rhynchos* a beak] —**or′ni·tho·rhyn′chous** *adj.*

or·ni·tho·sis (ôr′nə·thō′sis) *n.* An infectious virus disease of turkeys, chickens, and other birds not of the parrot family: it resembles psittacosis and is transmissible to man.

oro-[1] *combining form* Mouth; oral: *oropharynx.* [<L *os, oris* the mouth]

oro-[2] *combining form Geol.* Mountain; of mountains: *orography.* [<Gk. *oros* a mountain]

or·o·ban·cha·ceous (ôr′ō·bang·kā′shəs, or′-) *adj.* Designating a genus (*Orobanche*) typical of a family (*Orobanchaceae*) of low, leafless, parasitic, yellowish or brownish herbs lacking chlorophyll; the broomrapes. [<NL <L, broomrape <Gk. *orobanchē* <*orobos* a kind of vetch + *anchein* throttle]

o·rog·e·ny (ô·roj′ə·nē, ō-) *n. Geol.* The process of mountain formation. [<ORO-[2] + -GENY] —**or·o·gen·ic** (ôr′ə·jen′ik, or′-) *adj.*

o·rog·ra·phy (ô·rog′rə·fē, ō-) *n.* The branch of physiography that treats of the development and relations of highlands and mountain ranges. [<ORO-[2] + -GRAPHY] —**or·o·graph·ic** (ôr′ə·graf′ik, or′-) or **·i·cal** *adj.*

o·ro·ide (ôr′ō·īd, ō′rō-) *n.* An alloy of copper, zinc, tin, and other metals, having a golden luster: also spelled *oreide.* [<F *or* gold (<L *aurum*) + Gk. *eidos* form]

o·rol·o·gy (ô·rol′ə·jē, ō-) *n.* The study of mountains. [<ORO-[2] + -LOGY] —**or·o·log·i·cal** (ôr′ə·loj′i·kəl, or′-) *adj.* —**o·rol′o·gist** *n.*

o·rom·e·ter (ô·rom′ə·tər, ō-) *n.* An aneroid barometer having, in addition to the usual scale, a second system of graduations giving elevations above sea level corresponding to barometric pressure; a mountain barometer. [<ORO-[2] + -METER] —**or·o·met·ric** (ôr′ə·met′·rik, or′-) *adj.*

O·ron·tes (ō·ron′tēz, ō-) A river of NW Syria and southern Turkey, flowing 240 miles north from the Anti-Lebanon mountains to the Mediterranean.

o·ro·phar·ynx (ôr′ō·far′ingks, ō′rō-) *n. pl.* **·pha·ryn·ges** (-fə·rin′jēz) or **·phar·ynx·es** *Anat.* That part of the pharynx behind the mouth; the pharynx proper. [<ORO-[1] + PHARYNX]

O·ro·si·us (ə·rō′shē·əs), **Paulus** Fifth century Spanish theologian and author.

o·ro·tund (ôr′ə·tund, ō′rə-) *adj.* 1 Full, clear, rounded, and resonant: said of the voice or utterance. 2 Pompous; inflated, as a manner of speech. —*n.* An orotund quality of voice: also **o′ro·tun′di·ty.** [<L *ore rotundo* with well-turned speech, lit., with a round mouth <*os, oris* mouth + *rotundus* round]

O·roz·co (ō·rōs′kō), **José Clemente,** 1883–1949, Mexican painter.

Or·pen (ôr′pən), **Sir William,** 1878–1931, Irish painter.

or·phan (ôr′fən) *n.* A child whose parents are dead. —*adj.* 1 Having lost one or (more commonly) both parents: said of a child. 2 Pertaining to a child so bereaved. —*v.t.* To bereave of parents or of a parent; make an orphan of. [<LL *orphanus* <Gk. *orphanos,* lit., bereaved] —**or′phan·hood** (-ho͝od) *n.*

or·phan·age (ôr′fən·ij) *n.* 1 The state of being an orphan. 2 An institution for orphans.

Orphans' Court In some States of the United States, a court having jurisdiction over the estates of deceased persons and guardianship of orphans; also, a probate court.

Or·phe·us (ôr′fē·əs) In Greek mythology, a musician who could charm beasts and even rocks and trees by his singing to the lyre. When his wife Eurydice died he was permitted to lead her back to earth from Hades provided he would not look at her: he failed in the test and was later killed by the Thracian women who were enraged at his mourning for Eurydice. —**Or′phe·an** *adj.*

Or·phic (ôr′fik) *adj.* 1 Belonging, relating, or similar to Orpheus or his works. 2 Of or pertaining to Orphism; hence, oracular; mysterious. Also **Or′phi·cal.** [<L *Orphicus* <Gk. *Orphikos* <*Orpheus* Orpheus]

Orphic mysteries Esoteric doctrines and rites practiced by the worshipers of Dionysus Zagreus, claiming Orpheus as their founder.

Or·phism (ôr′fiz·əm) *n.* The system of the Orphic mysteries.

or·phrey (ôr′frē) *n.* 1 Gold embroidery, or any costly embroidery. 2 A band of gold embroidery or other rich material put on certain ecclesiastical vestments. Also spelled *orfray.* [<OF *orfreis* <Med. L *aurifrisium* <L *auriphrygium* <*aurum* gold + *Phrygius* Phrygian]

or·pi·ment (ôr′pə·mənt) *n.* An easily cut, pearly, lemon-yellow, native arsenic trisulfide, As_2S_3, used as a pigment and as a dyestuff. [<OF <L *auripigmentum* gold pigment]

or·pine (ôr′pin) *n.* 1 An Old World species of stonecrop (*Sedum telephium*) with tuberous root, stout erect stem, ovate leaves, and white or purple flowers in dense tufts: naturalized in the United States. 2 A Mediterranean herb (*Telephium imperati*) of the pink family, with prostrate stems and white flowers in terminal clusters. Also **or'pin.** [<OF *orpin* < *orpiment* ORPIMENT]

Or·ping·ton (ôr'ping·tən) *n.* A variety of domestic fowl originating in Orpington, a village in Kent, England.

or·ra (ôr'ə, or'ə) *adj. Scot.* 1 Odd, in the sense of extra and occasional; incidental: an *orra* job. 2 Employed for odd jobs, as on a farm. 3 Composed of or belonging to the riff-raff; low; despicable: an *orra* gathering.

or·re·ry (ôr'ə·rē, or'-) *n. pl.* **·ries** A mechanical apparatus for exhibiting the relative motions and positions of the members of the solar system; a cosmoscope. [after the fourth Earl of *Orrery,* Charles Boyle, 1676–1731, English statesman for whom an early copy of the machine was made]

or·rhol·o·gy (ô·rol'ə·jē, ō-) *n.* Serology. [<Gk. *orrhos* serum + -LOGY]

or·ris (ôr'is, or'-) *n.* Any of the several species of *Iris* having a scented root, especially *I. florentina,* of which the dried rootstock is used in medicine, perfumery, etc. Also **or'rice.** [Prob. alter. of Ital. *ireos* <L *iris* IRIS]

or·seille (ôr·sāl') *n.* Archil. [<F <OF *orchel* ARCHIL]

Or·si·no (ôr·sē'nō) In Shakespeare's *Twelfth Night,* the duke of Illyria, who marries Viola, the heroine.

Or·so·la (ôr'sō·lä) Italian form of URSULA. Also *Du.* **Or·se·li·ne** (ôr'sə·lē'nə).

ort (ôrt) *n. Usually pl.* A worthless leaving, as of food after a meal. [Prob. <Du. *ooraete* remains of food]

Or·te·gal (ôr'tā·gäl'), **Cape** A headland of NW Spain.

Or·te·ga y Gas·set (ôr·tā'gä ē gä·set'), **José,** 1883–1955, Spanish philosopher and author.

Or·ten·si·a (ôr·ten'sē·ä) Italian form of HORTENSE.

or·tet (ôr'tet) *n. Bot.* The plant from which a clon is derived. [<L *ortus* an origin, a rising < *oriri* rise]

or·thi·con (ôr'thə·kon) *n.* A sensitive television camera tube which uses low-velocity electrons in scanning and can pick up scenes under all lighting conditions or by infrared radiations. Also **image orthicon.** [<ORTH(O)- + ICON(OSCOPE)]

ortho- *combining form* 1 Straight; upright; in line: *orthotropic.* 2 At right angles; perpendicular: *orthorhombic.* 3 Correct; proper; right: *orthography.* 4 *Med.* The correction of irregularities or deformities of: *orthodontia.* 5 *Chem.* a A compound, usually an acid, containing the greatest possible number of hydroxyl groups: distinguished from *meta-.* b A benzene derivative in which the substituted atoms or radicals occupy the positions 1, 2: abbr. *o-.* Compare META-, PARA-. See BENZENE RING. Also, before vowels, **orth-.** [<Gk. < *orthos* straight]

or·tho axis (ôr'thō) *Mineral.* That axis in a crystal of the monoclinic system which is perpendicular to the other two axes.

or·tho·bi·o·sis (ôr'thō·bī·ō'sis) *n.* Sound and correct living; living in accordance with proper hygienic principles.

or·tho·cen·ter (ôr'thō·sen'tər) *n. Geom.* The point at which the three altitudes of a triangle intersect.

or·tho·ce·phal·ic (ôr'thō·sə·fal'ik) *adj.* Having a head in which the ratio between the vertical and transverse diameters is from 70 to 75. Also **or'tho·ceph'a·lous** (-sef'ə·ləs). — **or'tho·ceph'a·ly** *n.*

or·tho·chro·mat·ic (ôr'thō·krō·mat'ik) *adj. Phot.* Maintaining natural relations of light and shade, especially by the use of films or plates treated to give correct values to the greens and yellows: also called *isochromatic.* — **or'tho·chro'ma·tism** (-krō'mə·tiz'əm) *n.*

or·tho·clase (ôr'thō·klās, -klāz) *n.* A massive, brittle, vitreous, varicolored potassium-aluminum silicate. [<ORTHO- + Gk. *klasis* a fracture < *klaein* break]

or·tho·clas·tic (ôr'thō·klas'tik) *adj.* Having right-angled cleavages, as orthoclase.

or·tho·cy·mene (ôr'thō·sī'mēn) *n. Chem.* One of the three isomeric forms of cymene.

or·tho·dome (ôr'thə·dōm) *n.* A domelike surface parallel to one of the axes in a monoclinic crystal. — **or'tho·dom'ic** (-dom'ik) *adj.*

or·tho·don·tia (ôr'thə·don'shə, -shē·ə) *n.* The branch of dentistry which is concerned with the prevention and correction of irregularities and faulty positions of the teeth. [<NL <Gk. *orthos* right, straight + *odous, odontos* a tooth] — **or'tho·don'tic** (-don'tik) *adj.* — **or'tho·don'tist** *n.*

or·tho·dox (ôr'thə·doks) *adj.* 1 Correct or sound in doctrine; holding the commonly accepted faith, established doctrines, etc.: opposed to *heterodox.* 2 Conforming to the Christian faith as formulated in the early ecumenical creeds. 3 Approved; accepted. [<MF *orthodoxe* <LL *orthodoxus* <Gk. *orthodoxos* having right opinion < *orthos* right + *doxa* opinion < *dokeein* think] — **or'·tho·dox'ly** *adv.*

Or·tho·dox (ôr'thə·doks) *adj.* Pertaining to the Greek Church.

Orthodox Church The Greek Church.

or·tho·dox·y (ôr'thə·dok'sē) *n. pl.* **·dox·ies** 1 Belief in established doctrine. 2 Agreement with accepted standards, established doctrines, ideas, etc. — **or'tho·dox'i·cal** *adj.*

or·tho·e·pist (ôr·thō'ə·pist, ôr'thō·ə·pist) *n.* An authority on pronunciation. — **or·tho·e·pis·tic** (ôr'thō·e·pis'tik) *adj.*

or·tho·e·py (ôr·thō'ə·pē, ôr'thō·ep'ē) *n.* 1 The art of correct pronunciation. 2 Pronunciation in general. [<Gk. *orthoepeia* correctness of diction < *orthos* right + *epos* a word] — **or·tho·ep·ic** (ôr'thō·ep'ik) or **·i·cal** *adj.*

or·thog·a·my (ôr·thog'ə·mē) *n.* 1 *Bot.* Immediate or direct self-fertilization of the ovary, as by the stamens of the same flower; autogamy. 2 Normal bisexual union. — **or·tho·gam·ic** (ôr'thō·gam'ik), **or·thog'a·mous** *adj.*

or·tho·gen·e·sis (ôr'thō·jen'ə·sis) *n. Biol.* The doctrine that the phylogenetic evolution of organisms takes place systematically in a definite direction and not accidentally in many directions; variation predetermined by the germ plasm. — **or'tho·ge·net'ic** (-jə·net'ik) *adj.*

or·thog·na·thous (ôr·thog'nə·thəs) *adj.* Having the lower jaw in line with the upper; having straight jaws. Also **or·thog·nath·ic** (ôr'thog·nath'·ik). [<ORTHO- + -GNATHOUS] — **or·thog'na·thism, or·thog'na·thy** *n.*

or·thog·o·nal (ôr·thog'ə·nəl) *adj.* Having or determined by right angles; perpendicular. [<F *orthogonale* < *orthogone* a right triangle <LL *orthogonium* <Gk. *orthogōnios* < *orthos* right + *gōnia* an angle] — **or·thog'o·nal·ly** *adv.*

or·tho·graph·ic (ôr'thə·graf'ik) *adj.* 1 Relating to orthography; also, correctly spelled. 2 Pertaining to right lines or angles. Also **or'tho·graph'i·cal.** — **or'tho·graph'i·cal·ly** *adv.*

orthographic projection A map projection in which the lines lie at right angles to the plane of projection.

or·thog·ra·phy (ôr·thog'rə·fē) *n. pl.* **·phies** 1 A mode or system of spelling, especially of spelling correctly or according to usage. 2 The science that treats of letters and spelling. 3 The art or act of drawing in correct architectural projection. — **or·thog'ra·pher, or·thog'ra·phist** *n.*

or·tho·hy·dro·gen (ôr'thō·hī'drə·jən) *n. Chem.* An unstable form of hydrogen in which the molecules consist of two hydrogen atoms spinning in the same direction, thus giving improperly alined poles. Compare PARAHYDROGEN. [<ORTHO- in line + HYDROGEN]

or·tho·ki·net·ic (ôr'thō·ki·net'ik) *adj.* Pertaining to or having movement in one direction, as molecules or particles.

or·tho·pe·dics (ôr'thə·pē'diks) *n.* The branch of surgery which is concerned with the correction of skeletal and spinal deformities, especially in children. Also **or'tho·pae'dics.** [<F *orthopédique* < *orthopédie* orthopedics <Gk. *orthos* right + *paideia* training of children < *pais, paidos* a child] — **or'tho·pe'dic** or **·di·cal** *adj.* — **or'tho·pe'dist** *n.*

or·thoph·o·ny (ôr·thof'ə·nē) *n.* The art of speaking correctly. [<ORTHO- + -PHONY] — **or·tho·phon·ic** (ôr'thə·fon'ik) *adj.*

or·tho·phos·phor·ic (ôr'thō·fos·fôr'ik, -for'-) *adj. Chem.* Designating common phosphoric acid, H_3PO_4, a colorless, sirupy liquid with acid taste: used in medicine and the arts.

or·tho·psy·chi·a·try (ôr'thō·sī·kī'ə·trē) *n.* The study, investigation, and treatment of incipient mental disorders and of the less extreme aberrations of behavior; mental hygiene. — **or'tho·psy'chi·at'ric** (-sī'kē·at'rik) or **·ri·cal** *adj.* — **or'tho·psy·chi'a·trist** *n.*

or·thop·ter (ôr·thop'tər) See ORNITHOPTER.

or·thop·ter·an (ôr·thop'tər·ən) *n.* Any of an order *(Orthoptera)* of insects with membranous hind wings and coriaceous, usually straight fore wings, including locusts, crickets, grasshoppers, cockroaches, etc. — *adj.* Of or pertaining to the *Orthoptera.* Also **or·thop'·ter·on.** [<NL <Gk. *orthos* straight + *pteron* a wing] — **or·thop'ter·al, or·thop'ter·ous** *adj.*

or·thop·tic (ôr·thop'tik) *adj.* 1 Of, pertaining to, or characterized by normal binocular vision. 2 Designating a method of correcting defective vision by muscular exercise of the eyes. [<ORTH(O)- + OPTIC]

or·thop·tics (ôr·thop'tiks) *n. pl. (construed as singular)* The treatment of defects in binocular vision and of poor visual habits, especially by training in eye movements.

or·tho·rhom·bic (ôr'thə·rom'bik) *adj.* Pertaining to a crystal system assumed to contain three unequal axes at right angles.

or·tho·scop·ic (ôr'thə·skop'ik) *adj.* 1 Having correct vision. 2 Constructed so as to correct optical distortion: an *orthoscopic* eyepiece of a telescope.

or·tho·stat·ic (ôr'thə·stat'ik) *adj.* Of or relating to an upright standing position.

or·thos·ti·chy (ôr·thos'tə·kē) *n. pl.* **·chies** *Bot.* A vertical row or rank: applied to an arrangement of organs on an axis, as leaves or flowers. [<ORTHO- + Gk. *stichos* a row] — **or'thos'ti·chous** *adj.*

or·tho·trop·ic (ôr'thə·trop'ik) *adj. Bot.* Growing vertically: said of developing plant organs that grow nearly vertically, either upward or downward. — **or·thot·ro·pism** (ôr·thot'rə·piz'əm) *n.*

or·thot·ro·pous (ôr·thot'rə·pəs) *adj. Bot.* Growing straight: said of an ovule in which the nucellus is straight. Also **or·thot'ro·pal.** [<NL *orthotropus* <Gk. *orthos* straight + *trepein* turn] — **or·thot'ro·py** *n.*

Ort·ler Range (ôrt'lər) A division of the Alps in northern Italy; highest peak, 12,792 feet.

or·to·lan (ôr'tə·lən) *n.* 1 An Old World bunting *(Emberiza hortulana)* reddish-green above with blackish spots, and with a greenish-gray head: highly esteemed as a table delicacy. 2 Any of several other birds, as the reedbird or bobolink of the United States. [<F <Provençal <Ital. *ortolano* a gardener <L *hortulanus* < *hortulus,* dim. of *hortus* a garden; so called because it frequents gardens]

O·ru·ro (ō·rōō'rō) A city of west central Bolivia.

Or·vie·to (ôr·vyā'tō) A city in central Italy: ancient *Urbs Vetus.*

-ory[1] *suffix of nouns* A place or instrument for (performing the action of the main element): *dormitory, lavatory.* [<OF *-oire, -orie* <L *-orium;* or directly <L]

-ory[2] *suffix of adjectives* Related to; like; resembling: *amatory, laudatory.* [<OF *-oire* <L *-orius;* or directly <L]

Or·yok·ko (ō·ryōk·kō) The Japanese name for the YALU.

o·ryx (ôr'iks, or'-, ō'riks) *n.* A long-horned African antelope (genus *Oryx*), as the gemsbok. [<NL <L <Gk., a pickax, a kind of antelope; so called from its pointed horns]

os[1] (os) *n. pl.* **o·ra** (ôr'ə, ō'rə) *Anat.* A mouth or opening into the interior of an organ. [<L]

os[2] (os) *n. pl.* **os·sa** (os'ə) *Anat.* A bone. [<L]

os[3] (ōs) *n. pl.* **o·sar** (ō'sär) *Geol.* A sinuous ridge of glacial sand and gravel, deposited by a stream flowing beneath; an esker: also spelled *ose.* [<Sw. *ås* a ridge, a chain of hills]

O·sage (ō'sāj) *n.* One of a tribe of North American Indians of southern Siouan stock, formerly inhabiting the region between the Missouri and Arkansas rivers: now living in Oklahoma. [<Siouan (Osage) *Wazhazhe* war people]

Osage orange A showy, spreading, moraceous tree *(Maclura pomifera)* native to Arkansas and adjacent regions, having alternate, entire, glossy leaves and a large inedible aggregate fruit somewhat like an orange in size and color; also, the fruit of this tree.

Osage River A river in eastern Kansas and west central Missouri, flowing 500 miles east to the Missouri.

O·sa·ka (ō·sä·kä) A port on SW Honshu island, on **Osaka Bay,** an arm of the Inland Sea.

Os·born (oz'bərn), **Henry Fairfield,** 1857–1935, U.S. paleontologist.

Os·borne (oz′bərn), **Thomas Mott,** 1859–1935, U. S. penologist.

Os·can (os′kən) *n.* 1 One of an ancient people who inhabited SW Italy. 2 The language of these people, belonging to the Osco-Umbrian branch of the Italic languages: also called *Samnite.* —*adj.* Of or pertaining to the Oscans or their language. [<L *Osci* people of Campania]

Os·car (os′kər) *n.* One of the small gold statuettes awarded annually (since 1928) by the Academy of Motion Picture Arts and Sciences for outstanding performances, productions, photography, etc., in motion pictures. [Said to be from the remark of an Academy secretary that the statuette resembled her uncle *Oscar*]

Os·car (os′kər, *Fr.* ô·skȧr′; *Ger., Sp.* ōs′kär; *Norw.* os′kär; *Russian, Sw.* ōs′kär) A masculine personal name. Also *Sw.* **Os·kar** (ōs′kär). [<Gmc., divine spear]

—**Oscar II,** 1829–1907, king of Sweden 1872–1907, and of Norway 1872–1905.

Os·ce·o·la (os′ē·ō′lə), 1804–38, Seminole chief.

os·cil·late (os′ə·lāt) *v.i.* **·lat·ed, ·lat·ing** 1 To swing back and forth; vibrate, as a pendulum. 2 To vary undecidedly; waver; fluctuate. 3 *Physics* To produce oscillations. See synonyms under FLUCTUATE, SHAKE. [<L *oscillatus,* pp. of *oscillare* swing <*oscillum* a swing]

os·cil·la·tion (os′ə·lā′shən) *n.* 1 The act or state of oscillating. 2 *Physics* **a** A single swing of an oscillating body between two extremes. **b** A continual fluctuation between extreme values of quantity or force, as in a high-frequency electric current, the maximum value of which constantly diminishes with a speed dependent upon the damping effect present. —**forced oscillation** The periodic oscillations of an electric system, as determined by forces applied from outside the system: also called *forced vibration.* —**sustained oscillation** Periodic oscillations imposed by forces outside of, but controlled by, the oscillating system, as a pendulum actuated by a clock mechanism. —**os′cil·la·to′ry** (-lə·tôr′ē, -tō′rē) *adj.*

os·cil·la·tor (os′ə·lā′tər) *n.* 1 One who or that which oscillates. 2 Any oscillating machine. 3 *Electronics* A device for producing electromagnetic oscillations of a specified frequency.

os·cil·lo·gram (ə·sil′ə·gram) *n.* A record made by an oscillograph. [<*oscillo-* <L *oscillare* oscillate + -GRAM]

os·cil·lo·graph (ə·sil′ə·graf, -gräf) *n.* 1 A device for making a visible representation of the oscillations of an alternating current, which are transmitted in the form of reflected light rays to a screen for observation, or to a moving photographic film for purposes of record. 2 A similar instrument used to locate the point of origin of remote sounds (as of enemy guns) by recording sound waves on a time scale.

os·cil·lo·scope (ə·sil′ə·skōp) *n.* A cathode-ray oscillograph of low voltage for recording wave forms on a fluorescent screen: used in radio and in sound-ranging devices.

os·cine[1] (os′in -īn) *n.* Any passerine bird of the suborder *Oscines,* having the most highly developed vocal ability among birds, as thrushes, sparrows, etc.: commonly called *singing birds.* —*adj.* Of or pertaining to the *Oscines.* [<NL <L *oscen, oscinis* a singing bird <*ob-* towards + *canere* sing]

os·cine[2] (os′ēn, -in) *n. Chem.* A decomposition product of scopolamine. [<G *oscin,* short for *hyoscin* hyoscine]

os·ci·tan·cy (os′ə·tən·sē) *n.* 1 The act of yawning or gaping. 2 Unusual drowsiness; dulness. Also **os′ci·tance.** [<L *oscitans, -antis,* ppr. of *oscitare* gape, yawn <*os* a mouth + *citare* move] —**os′ci·tant** *adj.*

os·ci·tate (os′ə·tāt) *v.i. Archaic* To yawn. [<L *oscitatus,* pp. of *oscitare.* See OSCITANCY.]

Os·co-Um·bri·an (os′kō·um′brē·ən) *n.* A branch of the Italic subfamily of Indo-European languages, comprising the ancient Oscan and Umbrian.

os·cu·lant (os′kyə·lənt) *adj. Biol.* 1 Intermediate in character between two groups of organisms; intergrading: an *osculant* genus or family. 2 Closely adherent. [<L *osculans, -antis,* ppr. of *osculari.* See OSCULATE.]

os·cu·lar (os′kyə·lər) *adj.* Of or pertaining to the mouth. [<L *osculum.* See OSCULATE.]

os·cu·late (os′kyə·lāt) *v.t. & v.i.* **·lat·ed, ·lat·ing** 1 To kiss. 2 To bring or come into close contact or union. 3 *Geom.* To touch so as to have three or more points in common, as two curves. 4 *Biol.* To have (characteristics) in common, as two genera or families. [<L *osculatus,* pp. of *osculari* kiss <*osculum* a little mouth, a kiss, dim. of *os* a mouth]

os·cu·la·tion (os′kyə·lā′shən) *n.* 1 The act of kissing; also, a kiss. 2 *Math.* A point on a curve where two branches have a common tangent but do not reverse direction, as at a cusp; a tacnode. —**os′cu·la·to′ry** (-lə·tôr′ē, -tō′rē) *adj.*

os·cu·lum (os′kyə·ləm) *n. pl.* **·la** (-lə) *Zool.* One of the comparatively large apertures in a sponge by which water with waste products is expelled. Also **os′cule** (-kyōōl). [<L, dim. of *os* a mouth]

ose (ōs) See OS[3].

-ose[1] *suffix of adjectives* 1 Full of or abounding in (what is indicated by the main element): *verbose.* 2 Like; resembling (the main element): *grandiose.* Compare -OUS. [<L *-osus*]

-ose[2] *suffix Chem.* 1 A carbohydrate: *glucose, cellulose.* 2 A derivative of a protein: *proteose.* [<(GLUC)OSE]

Ö·sel (œ′zəl) The Swedish name for the SAARE ISLANDS.

O-shaped (ō′shāpt′) *adj.* Having the round or oval shape of the letter O.

o·sier (ō′zhər) *n.* 1 Any of various species of willow (genus *Salix*), producing long, flexible shoots used in wickerwork, especially the European **velvet osier** (*Salix viminalis*). 2 One of the shoots of an osier. 3 A similar plant of some other genus or family, or its osierlike shoots, as the squawbush. —*adj.* Consisting of twigs of willow, etc. [<OF, prob. <Med. L *ausaria, osaria* a bed of willows]

O·si·jek (ô′sē·yek) A city in NE Croatia, Yugoslavia.

os in·nom·i·na·tum (os i·nom′ə·nā′təm) *Anat.* The innominate bone. [<NL]

O·si·ris (o·sī′ris) In Egyptian mythology, the god of the underworld and lord of the dead, husband of his sister Isis.

-osis *suffix of nouns* 1 The condition, process, or state of: *metamorphosis.* 2 *Med.* **a** A diseased or abnormal condition of: *melanosis.* **b** A formation of: *sclerosis.* [<L <Gk. *-ōsis*]

-osity *suffix of nouns* Forming nouns corresponding to adjectives in *-ose*: *verbosity, grandiosity.* [<F *-osité* <L *-ositas*; or directly <L]

Os·ler (ōs′lər, ōz′-), **Sir William,** 1849–1919, Canadian physician.

Os·lo (os′lō, oz′-; *Norw.* ōōs′lōō) The capital of Norway, on **Oslo Fiord,** an arm of the Skagerrak extending 80 miles inland: formerly *Christiania.*

Os·man (oz′mən, os′-; *Turkish* os·män′), 1259–1326, founder of the Ottoman Empire. Also *Othman.*

Os·man·li (oz·man′lē, os-) *n.* 1 An Ottoman Turk; one of the western branch of the Turkish peoples. 2 The language of the Ottoman Turks; Turkish. [<Turkish *Osmānli* of Osman <*Osmān* Osman <Arabic *'Othmān*]

Os·me·ña (ōs·mā′nyä), **Sergio,** born 1878, president of the Philippines 1944–46.

os·mes·the·sia (os′məs·thē′zhə, -zhē·ə) *n. Physiol.* A high susceptibility to odors. [<NL <Gk. *osmē* scent + *aisthēsis* perception]

os·mic (oz′mik, os′-) *adj. Chem.* 1 Of, pertaining to, or containing osmium, especially in its higher valence. 2 Designating a colorless to yellowish crystalline acid, OsO_4, used as a caustic and in medicine.

os·mi·dro·sis (os′mə·drō′sis, oz′-) *n. Pathol.* A condition in which the perspiration has an abnormally strong odor. [<Gk. *osmē* an odor + *drosos* dew, moisture]

os·mi·ous (oz′mē·əs, os′-) *adj. Chem.* Of, pertaining to, or containing osmium, especially in its lower valence.

os·mi·rid·i·um (os′mə·rid′ē·əm, oz′-) *n.* A varying isomorphous mixture of iridium and osmium, found native in flattened, metallic, tin-white, malleable grains, and used for pointing gold pens: also called *iridosmium.* [<OSM(IUM) + IRIDIUM]

os·mi·um (oz′mē·əm, os′-) *n.* A hard, brittle, extremely heavy, white, metallic element (symbol Os) of the platinum group. See ELEMENT. [<Gk. *osmē* an odor; so called with ref. to the pungent odor of one of its oxides]

Os·mond (oz′mənd) A masculine personal name. Also **Os′mund.** [<Gmc., protection of God]

os·mo·pho·bi·a (os′mə·fō′bē·ə, oz′-) *n. Psychiatry* A morbid fear of odors. [<Gk. *osmē* an odor + -PHOBIA]

os·mose (oz′mōs, os′-) *v.t. & v.i.* **·mosed, ·mos·ing** To subject to or to undergo the process of osmosis. [<*osmose, n.,* var. of OSMOSIS]

os·mo·sis (oz·mō′sis, os-) *n.* The diffusion of two miscible solutions through a semipermeable membrane in such a manner as to equalize their concentration: it is one of the essential processes of living matter, especially in its cellular forms: also called *diosmosis.* Also **os′mose.** Compare ENDOSMOSIS, EXOSMOSIS. [Earlier *osmose* <*-osmose* (as in *endosmose, exosmose*) <Gk. *ōsmos* a thrust, push] —**os·mot·ic** (oz·mot′ik, os-) *adj.* —**os·mot′i·cal·ly** *adv.*

os·mund (os′mənd, oz′-) *n.* Any of a genus (*Osmunda*) of showy ferns having pinnate fronds growing upright from a large crown, especially the royal fern (*O. regalis*). [<AF *osmunde* <Med. L *osmunda*; ult. origin unknown]

Os·na·brück (oz′nə·brŏŏk, *Ger.* ôs′nä·brük) A city in Lower Saxony, West Germany.

os·na·burg (oz′nə·bûrg) *n.* A tough, unbleached cotton cloth, often part waste, used for upholstery, for grain and cement sacks, and as a material for camouflage. [from *Osnaburg,* var. of OSNABRÜCK]

os·phre·sis (os·frē′sis) *n. Physiol.* The sense of smell. Also **os·phre′sia** (-frē′zhə). [<NL <Gk. *osphrēsis* <*osphrainesthai* smell] —**os·phret′ic** (-fret′ik) *adj.*

os·prey (os′prē) *n.* An American hawk (*Pandion haliaëtus*), brown above and white below, that preys upon fish. [Appar. <L *ossifraga* <*os, ossis* a bone + *frangere* break]

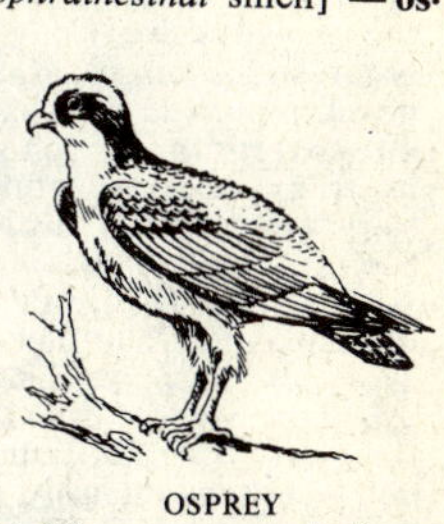

OSPREY
(From 22 to 24 inches in over-all length)

Os·sa (os′ə) A mountain in eastern Thessaly, Greece; 6,490 feet; in Greek mythology, the Titans attempted to scale Olympus by piling Pelion on Ossa.

os·sa·ture (os′ə·chər) *n. Anat.* The disposition and arrangement of the bones of the body. Compare MUSCULATURE. [<F, skeleton <L *os, ossis* a bone]

os·se·in (os′ē·in) *n. Biochem.* The soft protein substance of the bone that remains after the removal of mineral matter: also *ostein.* [<L *osse(us)* bony + -IN]

os·se·ous (os′ē·əs) *adj.* Pertaining to, of the nature of, or containing bones. [<L *osseus* bony <*os, ossis* a bone] —**os′se·ous·ly** *adv.*

Os·se·tia (o·sē′shə) A region in the central Caucasus mountains of the U.S.S.R., divided into the *North Ossetian S.S.R.* and the *South Ossetian Autonomous Region.*

Os·sian (osh′ən, os′ē·ən) A legendary Irish hero and bard of the third century, subject of a cycle of poems written by James MacPherson, published 1760–63, purporting to be translations from the original Ossian. —**Os·si·an·ic** (os′ē·an′ik) *adj.*

os·si·cle (os′i·kəl) *n. Anat.* 1 A small bone. 2 One of a chain of three small bones in the tympanic cavity of the ear. 3 One of various small, hard, nodular structures. [<L *ossiculum,* dim. of *os, ossis* a bone] —**os·sic·u·lar** (o·sik′yə·lər) *adj.*

Os·sietz·ky (ô·syet′skē), **Carl von,** 1889–1938, German pacifist.

os·sif·er·ous (o·sif′ər·əs) *adj.* Yielding or containing bones. [<L *os, ossis* a bone + -FEROUS]

os·si·frage (os′ə·frij) *n.* 1 The osprey. 2 The

add,āce,câre,pälm; end,ēven; it,īce; odd,ōpen,ôrder; tŏŏk,pōōl; up,bûrn; ə = a in *above,* e in *sicken,* i in *clarity,* o in *melon,* u in *focus;* yōō = u in *fuse;* oi,oil; ou,pout; ch,check; g,go; ng,ring; th,thin; th,this; zh,vision. Foreign sounds à,œ,ü,kh,ṅ; and ◆: see page xx. < from; + plus; ? possibly.

lammergeier. [<L *ossifraga.* See OSPREY.]

os·si·fy (os′ə·fī) *v.t.* & *v.i.* **·fied, ·fy·ing** 1 To convert or be converted into bones. 2 To make or become set, conventional, etc. [<L *os, ossis* a bone + -FY] **—os·sif·ic** (o·sif′ik) *adj.* **—os ′si·fi·ca′tion** *n.*

Os·si·ning (os′ə·ning) A village on the Hudson River in SE New York: formerly *Sing Sing.*

Os·so·li (ôs′sō·lē), **Marchioness** See MARGARET FULLER.

os·su·ar·y (os′o͞o·er′ē, osh′-) *n.* *pl.* **·ar·ies** A place for holding the bones of the dead; charnel house; grave mound. [<LL *ossuarium* <L *ossuarius* of, for bones <*os, ossis* a bone]

Os·ta·de (ôs′tä·də), **Adriaen van,** 1610–85, Dutch painter.

os·te·al (os′tē·əl) *adj.* Of, pertaining to, or like bone; osseous; bony. [<Gk. *osteon* a bone]

os·te·in (os′tē·in) See OSSEIN.

os·te·i·tis (os′tē·ī′tis) *n. Pathol.* Inflammation of bone. [<OSTE(O)- + -ITIS]

Os·tend (os′tend, os·tend′) A port of NW Belgium, on the North Sea. *French* **Os·tende** (ô·ständ′), *Flemish* **Oost·en·de** (ōst·en′də).

os·ten·si·ble (os·ten′sə·bəl) *adj.* Offered as real or having the character represented; seeming; professed or pretended. [<F <L *ostensus,* pp. of *ostendere* show <*ob-* against + *tendere* stretch] **—os·ten′si·bly** *adv.*

Synonyms: apparent, assigned, avowed, colorable, displayed, exhibited, expressed, plausible, professed, shown, specious. A man's *apparent* purpose or motive is what appears on the surface, with or without his own intent; his *ostensible* motive or purpose is that which is *assigned, avowed, displayed* by him; the word often implying that the *ostensible* may be only the pretended, a *specious* cover for a purpose or motive of a different sort. Compare synonyms for APPARENT. *Antonyms*: actual, genuine, real, true, veritable.

os·ten·sive (os·ten′siv) *adj.* Exhibiting; showing. **—os·ten′sive·ly** *adv.*

Os·ten·so (os′ten·sō), **Martha,** born 1900, U.S. novelist born in Norway.

os·ten·so·ri·um (os′tən·sôr′ē·əm, -sō′rē-) *n.* *pl.* **·ri·a** (-rē·ə) A monstrance. Also **os·ten·so·ry** (os·ten′sər·ē). [<Med. L <L *ostensus.* See OSTENSIBLE.]

os·ten·ta·tion (os′tən·tā′shən) *n.* 1 The act of displaying vauntingly; pretentious parade. 2 Public display. [<OF *ostentacion* <L *ostentatio, -onis* <*ostentatus,* pp. of *ostentare,* freq. of *ostendere* show] **—os′ten·ta′tious** *adj.* **—os′ten·ta′tious·ly** *adv.* **—os′ten·ta′tious·ness** *n.*

Synonyms: boast, boasting, display, flourish, pageant, pageantry, parade, pomp, pomposity, pompousness, show, vaunt, vaunting. *Ostentation* is an ambitious showing forth of whatever is thought adapted to win admiration or praise; *ostentation* may be without words; as, the *ostentation* of wealth in luxuriously equipped cars; when in words, *ostentation* is rather in manner than in direct statement; as, the *ostentation* of learning. *Boasting* is in direct statement, and is louder and more vulgar than *ostentation.* There may be great *display* or *show* with little substance; *ostentation* suggests something substantial to be shown. *Pomp* is some material demonstration of wealth and power, as in grand and stately ceremonial, etc., considered as worthy of the person or occasion in whose behalf it is manifested; *pomp* is the noble side of that which as *ostentation* is considered arrogant and vain. *Pageant* and *pageantry* are inferior to *pomp,* denoting spectacular *display* designed to impress the public mind. *Parade* is an exhibition as of troops in camp going through military evolutions, and suggests a lack of immediate occasion or demand; hence, in the more general sense, a *parade* is an uncalled-for exhibition, and so used is a more disparaging word than *ostentation. Pomposity* and *pompousness* are the affectation of *pomp.* See PRIDE. *Antonyms*: diffidence, modesty, quietness, reserve, retirement, shrinking, timidity.

osteo- *combining form* Bone; pertaining to bone or the bones: *osteoblast.* Also, before vowels, **oste-.** [<Gk. *osteon* a bone]

os·te·o·blast (os′tē·ə·blast′) *n.* A bone-forming cell.

os·te·oc·la·sis (os′tē·ok′lə·sis) *n.* 1 *Surg.* The operation of breaking a bone to correct a deformity or of rebreaking to remedy a bad setting. 2 The gradual absorption of bony tissue by osteoclasts. [<NL <Gk. *osteon* a bone + *klasis* a fracture <*klaein* break]

os·te·o·clast (os′tē·ə·klast′) *n.* 1 *Surg.* An instrument for effecting osteoclasis. 2 *Anat.* A large multinucleate cell found in the marrow of bones and concerned in the absorption of bony tissue during the formation of canals, cavities, etc. [<G *osteoklast* <Gk. *osteon* a bone + *klastos* broken <*klaein* break] **—os′·te·o·clas′tic** *adj.*

os·te·oid (os′tē·oid) *adj.* Resembling bone; bony.

os·te·ol·o·gy (os′tē·ol′ə·jē) *n.* The science that treats of the bones of the skeleton and of the properties of osseous tissue. **—os′te·o·log′i·cal** (-ə·loj′i·kəl) *adj.* **—os′te·o·log′i·cal·ly** *adv.* **—os′te·ol′o·gist** *n.*

os·te·o·ma (os′tē·ō′mə) *n.* *pl.* **·ma·ta** (-mə·tə) *Pathol.* A tumor consisting of bony substance; a morbid outgrowth from bone or from cartilage. [<OSTE(O)- + -OMA]

os·te·o·ma·la·ci·a (os′tē·ō′mə·lā′shē·ə) *n.* *Pathol.* Softening of the bones, with progressive osseous deformity and exhaustion: caused by calcium deficiency. [<NL <Gk. *osteon* a bone + *malakia* softness]

os·te·o·my·e·li·tis (os′tē·ō·mī′ə·lī′tis) *n. Pathol.* Inflammation of the bone marrow. [<OSTEO- + MYEL(O)- + -ITIS]

os·te·op·a·thy (os′tē·op′ə·thē) *n.* A system of healing based on a theory that most diseases are caused by the pressure of displaced bones on nerves, etc., and that therapy consists in manipulation of the parts affected or involved. [<OSTEO- + -PATHY] **—os′te·o·path′, os′te·op′a·thist** *n.* **—os′te·o·path′ic** (-ə·path′ik) *adj.*

os·te·o·phyte (os′tē·ə·fīt′) *n. Pathol.* A bony excrescence. **—os′te·o·phyt′ic** (-fit′ik) *adj.*

os·te·o·plas·ty (os′tē·ə·plas′tē) *n. Surg.* 1 An operation to remedy loss of bone. 2 The restoration to its place of a bone temporarily removed. **—os′te·o·plas′tic** *adj.*

os·te·o·scle·ro·sis (os′tē·ō·sklə·rō′sis) *n. Pathol.* A morbid condition marked by a hardening and increased density of bone. **—os′te·o·scle·rot′ic** (-sklə·rot′ik) *adj.*

os·te·o·sis (os′tē·ō′sis) *n. Pathol.* The abnormal formation of bony tissue. [<OSTE(O)- + -OSIS]

os·te·o·tome (os′tē·ə·tōm′) *n. Surg.* An instrument for dividing or cutting bone.

os·te·ot·o·my (os′tē·ot′ə·mē) *n. Surg.* The operation of dividing a bone, especially beneath the integuments, as to remedy deformity. **—os′te·ot′o·mist** *n.*

Ös·ter·reich (œs′tə·rīkh) The German name for AUSTRIA.

Os·ti·a (os′tē·ə, *Ital.* ô′styä) An ancient city and port of Rome at the mouth of the river Tiber.

os·ti·ar·y (os′tē·er′ē) *n.* *pl.* **·ar·ies** 1 In the Roman Catholic Church, a cleric belonging to the lowest of the minor orders. 2 A doorkeeper. Also **os′ti·ar′i·us** (-âr′ē·əs). [<L *ostiarius* doorkeeper <*ostium* a door]

os·ti·na·to (ôs′tē·nä′tō) *n. Music* A melodic phrase persistently reiterated in the same voice and pitch. [<Ital. *(basso) ostinato,* lit., obstinate (bass)]

os·ti·ole (os′tē·ōl) *n.* 1 A small opening. 2 *Zool.* Any one of the small inhalant orifices of a sponge. [<L *ostiolum,* dim. of *ostium* a door] **—os′ti·o′lar** *adj.*

os·tler (os′lər) See HOSTLER.

ost·mark (ôst′märk′) See MARK² (def. 1).

os·to·sis (os·tō′sis) *n.* Ossification. [<OST(EO)- + -OSIS]

Ost·preus·sen (ôst′proi′sən) The German name for EAST PRUSSIA.

os·tra·cism (os′trə·siz′əm) *n.* 1 Exclusion, as from society or common privileges, by general consent. 2 In ancient Greece, banishment by popular vote. [<Gk. *ostrakismos* <*ostrakizein* OSTRACIZE]

os·tra·cize (os′trə·sīz) *v.t.* **·cized, ·ciz·ing** 1 To shut out or exclude by ostracism. 2 In ancient Greece, to exile by ostracism. Also *Brit.* **os′·tra·cise.** See synonyms under BANISH. [<Gk. *ostrakizein* <*ostrakon* a potsherd, shell, voting tablet]

Os·tra·va (ôs′trä·vä) See MORAVSKÁ OSTRAVA.

os·tre·o·dy·na·mom·e·ter (os′trē·ō·dī′nə·mom′·ə·tər) *n.* An instrument for detecting the movements of an oyster in its shell without disturbing its normal activities: used especially in relation to water pollution. [<Gk. *ostreon* an oyster + DYNAMOMETER]

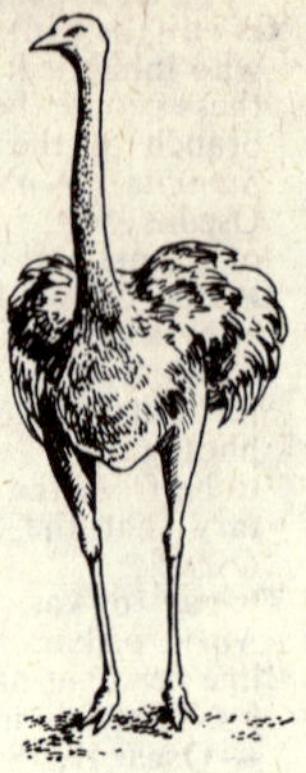

OSTRICH (The adult male about 8 feet tall)

os·trich (ôs′trich, os′-) *n.* 1 A large, two-toed bird (genus *Struthio*) of Africa and Arabia, with aborted wings. Its long, powerful legs give it great speed. The plumage of the male is black, with white plumes at the ends of the wings and tail, much esteemed for ornamental purposes. 2 A rhea. [<OF *ostruce, ostruche* <LL *avistruthius* <L *avis* a bird + LL *struthio* an ostrich <Gk. *strouthiōn* <*strouthos* a sparrow, an ostrich]

Os·tro·goth (os′trə·goth) *n.* A member of the eastern branch of the Goths, who established a kingdom in Italy from 493 to 555. See VISIGOTH. [<LL *Ostrogothus* <*ostro-* eastward (prob. <OHG *oster*) + *Gothus* a Goth] **—Os′tro·goth′ic** *adj.*

Ost·wald (ôst′vält), **Wilhelm,** 1853–1932, German chemist.

Os·ty·ak (os′tē·ak) *n.* 1 One of a Finno-Ugric people inhabiting western Siberia and the Ural Mountains. 2 The Ugric language of these people. Also **Os′ti·ak.**

Os·we·go (os·wē′gō) A port of entry on Lake Ontario in central New York.

Oswego tea A species of mint (*Monarda didyma*) with a showy head of bright-red flowers, growing in wet places in the eastern United States: also called *bee balm.*

Oś·wię·cim (ôsh·vyan′chim) See AUSCHWITZ.

ot- Var. of OTO-.

O·ta·hei·te (ō′tə·hē′tē, -hā′-) A former name for TAHITI.

o·tal·gi·a (ō·tal′jē·ə) *n. Pathol.* Neuralgia of the ear; earache. Also **o·tal′gy.** [<NL <Gk. *ōtalgia* <*ous, ōtos* an ear + *algos* a pain] **—o·tal′gic** *adj.*

O·ta·ru (ō·tä·ro͞o) A port on SW Hokkaido island, Japan.

o·ta·ry (ō′tər·ē) *n.* *pl.* **·ries** or **·ry** An eared seal. [<NL *Otaria,* genus name <Gk. *ōtaros* large-eared <*ous, ōtos* an ear] **—o·tar·i·an** (ō·târ′ē·ən) *adj.*

O tem·po·ra! O mo·res! (ō tem′pər·ə ō môr′ēz, mō′rēz) *Latin* O the times! O the manners!

O·thel·lo (ō·thel′ō, ə-) In Shakespeare's tragedy of the same name, a Moor of Venice who smothers his wife, Desdemona, in a jealous rage inspired by the treachery of Iago, and who later kills himself after learning of her innocence.

oth·er (uth′ər) *adj.* 1 Different from the one specified; not the same. 2 Being over and above; additional. 3 Second: noting the remaining one of two persons or things. 4 Opposite; contrary: the *other* side. 5 Alternate: every *other* day. **—the other day (night,** etc.) A day (night, etc.) not long ago; recently. *—pron.* 1 A different person or thing. 2 The second of two; the opposite one. *—adv.* Otherwise: with *than.* [OE *ōther*] **—oth′er·ness** *n.*

oth·er·guess (uth′ər·ges′) *Obs. adj.* Of another sort; other. *—adv.* In a different manner.

oth·er·where (uth′ər·hwâr′) *adv. Archaic* or *Dial.* In some other place; elsewhere.

oth·er·while (uth′ər·hwīl′) *adv. Archaic* At another time. Also **oth′er·whiles′.**

oth·er·wise (uth′ər·wīz′) *adv.* 1 In a different manner or by other means. 2 In all other circumstances or conditions. 3 In all other respects: an *otherwise* sensible writer. *—adj.* Different: How could such notions be *otherwise* than useless?

other world The unseen world; the life after death; the future state.

oth·er·world·ly (uth′ər·wûrld′lē) *adj.* 1 Of or characteristic of an ideal world, especially heaven. 2 Concerned with the hereafter to the neglect of the present. **—oth′er·world′li·ness** *n.*

O·thin (ō′thin) See ODIN.

Oth·man (oth′mən, *Arabic* o͞oth·män′) See OSMAN.

O·tho (ō′thō, *Du., Ger., Sw.* ō′tō) Dutch, Swedish, and German form of OTTO. Also *Fr.* **O·thon** (ô·tôn′).

o·tic (ō′tik, ot′ik) *adj.* Pertaining to or situated near the ear. [< Gk. *ōtikos* < *ous, ōtos* ear]
-otic *suffix of adjectives* 1 *Med.* Of, related to, or affected by: corresponding to nouns in *-osis: psychotic, sclerotic.* 2 Causing or producing: *narcotic.* [< Gk. *-otikos*, suffix of adjectives]
o·ti·ose (ō′shē·ōs, -tē-) *adj.* 1 Being at rest or ease; having nothing to do. 2 Characterized by indolence or easy negligence. 3 Futile; useless. [< L *otiosus* < *otium* leisure] — **o′ti·ose′ly** *adv.* — **o′ti·os′i·ty** (-os′ə-tē) *n.*
Otis (ō′tis), **James**, 1725–83, American patriot and orator.
o·ti·tis (ō·tī′tis) *n. Pathol.* Inflammation of the mucous membrane of the ear. [< OT(O)- + -ITIS]
oto- *combining form* Ear; pertaining to the ear: *otoscope.* Also, before vowels, *ot-*. [< Gk. *ous, ōtos* the ear]
o·to·co·ni·a (ō′tō·kō′nē·ə) *n. Anat.* A dustlike substance consisting of minute crystalline otoliths forming part of the contents of the statocyst: also called *ear dust.* [< NL < F *otoconic* < Gk. *ous, ōtos* an ear + *konis* sand, dust]
o·to·cyst (ō′tə·sist) *n. Anat.* 1 An auditory vesicle, as in many invertebrates. 2 The similar vesicle contained in the embryo of a vertebrate.
O·toe (ō′tō) *n.* One of a Siouan tribe of North American Indians living in southeastern Nebraska.
o·to·lar·yn·gol·o·gy (ō′tō·lar′ing·gol′ə·jē) *n.* The branch of medicine which treats of the ear and throat. — **o′to·lar′yn·gol′o·gist** *n.*
o·to·lith (ō′tə·lith) *n. Anat.* 1 One of the concretions of calcium carbonate and calcium phosphate found in the internal ear of vertebrates and in the auditory organ of many invertebrates. 2 An ear bone. [< OTO- + -LITH]
o·tol·o·gy (ō·tol′ə·jē) *n.* The science of the ear and its diseases. — **o·to·log·i·cal** (ō′tə·loj′i·kəl) *adj.* — **o·tol′o·gist** *n.*
O·to·nio (ō·tō′nyō) Spanish form of OTTO.
o·to·rhi·no·lar·yn·gol·o·gy (ō′tō·rī′nō·lar′ing·gol′ə·jē) *n.* The branch of medicine dealing with the ear, nose, and larynx in health and disease. [< OTO- + RHINO- + LARYNGO- + -LOGY] — **o′to·rhi′no·lar′yn·gol′o·gist** *n.*
o·to·scope (ō′tə·skōp) *n. Med.* An instrument for viewing or examining the interior of the ear; especially, an ear speculum.
O·tran·to (ō·trän′tō) A port of SE Italy on the **Strait of Otranto**, a strait between the Adriatic and Ionian seas (43 miles wide).
ot·tar (ot′ər) See ATTAR.
ot·ta·va (ōt·tä′vä) *n. Ital.* An octave.
ottava ri·ma (rē′mä) In English prosody, a stanza of eight-, ten-, or eleven-syllable iambic lines with the rime scheme *ababcbcc:* used by Byron in *Don Juan.* [< Ital., octave rime]
Ot·ta·vi·a (ōt·tä′vyä) Italian form of OCTAVIA.
Ot·ta·vi·o (ōt·tä′vyō) Italian form of OCTAVIUS. Also **Ot·ta·vi·a·no** (ōt·tä·vyä′nō).
Ot·ta·wa (ot′ə·wə) *n.* One of a tribe of North American Indians of Algonquian stock, originally inhabiting the region around Georgian Bay, Lake Huron, Ontario, later migrating to the region around Lake Superior and Lake Michigan. [< dial. F *otaua, otawa* <Algonquian (Cree) *atawea* a trader]
Ot·ta·wa (ot′ə·wə) The capital of Canada in SE Ontario on the **Ottawa River**, which flows 696 miles SE through Ontario and Quebec to the St. Lawrence.
ot·ter (ot′ər) *n.* 1 A weasel-like, web-footed carnivore (genus *Lutra*) inhabiting streams and lakes, and feeding upon fish. The common otter (*L. canadensis*) is about two feet long, exclusive of the flattened, oarlike tail, and furnishes a valuable, dark-brown fur. 2 The sea otter. [OE *oter*. Akin to WATER.]

OTTER

Ot·ter·burn (ot′ər·bûrn) A village in Northumberland, England; scene of a Scots defeat of the English, 1388. See CHEVY CHASE.
otter hound A breed of hound used in England for hunting otters: strongly built, with a close black-and tan coat, a large broad head, and floppy ears.
otter shrew An insectivorous, aquatic animal (family *Potamogalidae*) of central and western Africa.
Ot·to (ot′ō; *Dan., Du.* ot′ō; *Ger.* ôt′ō; *Sw.* ō′tō) A masculine personal name. Also *Ital.*
Ot·to·ne (ōt·tō′nā). [< Gmc., rich]
— **Otto I**, 912–973, king of Germany 936–973; Holy Roman Emperor 962–973. Called "Otto the Great."
ot·to·man (ot′ə·mən) *n.* 1 An upholstered, backless and armless seat or sofa. 2 An upholstered footrest. 3 A heavy corded silk or rayon fabric used for coats and trimmings. [< F *ottomane*, orig. fem. of *Ottoman* OTTOMAN]
Ot·to·man (ot′ə·mən) *n. pl.* **·mans** A Turk. — *adj.* Pertaining to the Turks. [< F < Ital. *Ottomano* < Med. L *Ottomanus* < Arabic *'Uthmāni, 'Othmāni* of Osman < *'Othmān* Osman]
Ottoman Empire A former empire (1300–1919) of the Turks in Asia Minor, NE Africa, and SE Europe; capital, Constantinople: also *Turkish Empire.*

OTTOMAN EMPIRE 1683–1913

Ot·way (ot′wā), **Thomas**, 1652–85, English dramatist.
oua·ba·in (wä·bä′in) *n. Chem.* A white, lustrous, extremely poisonous, crystalline glycoside, $C_{29}H_{44}O_{12}$, from the seeds of two South African trees (*Strophanthus glaber* and *Acokanthera ouabaio*): used in medicine as a cardiac stimulant. It is a constituent of Zulu arrow poison. [< F *ouabaio*, a South African tree [< Somali native name) + -IN]
Ouach·i·ta River (wosh′ə·tô, wôsh′-) A river in SW Arkansas and NE Louisiana, flowing 605 miles SE to the Red River: also *Washita.*
Oua·daï (wä·dī′) The French name for WADAI.
Oua·ga·dou·gou (wä′gə·dōō′gōō) The capital of the Republic of the Upper Volta.
oua·na·niche (wä′nə·nēsh′, *Fr.* wä·nä·nēsh′) *n.* A small Canadian salmon (*Salmo salar ouananiche*), identified with the landlocked salmon of Maine. [< dial. F (Canadian) <Algonquian (Cree) *wananish*]
ou·bli·ette (ōō′blē·et′) *n.* A secret dungeon with an entrance only through the top. [< OF *oublier* forget]
ouch[1] (ouch) *interj.* An exclamation indicating sudden pain.
ouch[2] (ouch) *n.* 1 The setting of a jewel. 2 An ornament, as a clasp or brooch, of gold. — *v.t.* To ornament with or as with ouches. [< AF *nouche* <LL *nusca* <OHG *nuscka, nuscha* a buckle, clasp, appar. ult. <Celtic; in ME *a nouche* became *an ouche*]
Ou·de·naar·de (ou′də·när′də) A town on the Scheldt in eastern Flanders, Belgium; scene of an English victory over the French, 1708. French **Aude·nar·de** (ōd·närd′).
Oude Rijn (oud rīn) See RHINE.
Oudh (oud) An historic region of east central Uttar Pradesh, northern India; 24,071 square miles.
Ou·di·not (ōō·dē·nō′), **Nicolas**, 1767–1847, Duc de Reggio, French general and marshal of France.
Oues·sant (dwe·sän′), **Île d'** The French name for USHANT.
ought[1] (ôt) *v.* A defective verb now used only as an auxiliary followed by the infinitive with *to*, or elliptically with the infinitive understood, to express: 1 Obligation or moral duty: He *ought* to keep his promises. 2 Advisability or expediency: You *ought* to be careful. 3 Probability or expectation: He *ought* to be here tomorrow. ◆ A past is formed by placing the following verb in the perfect infinitive, as in He *ought* to have been there. ◆ Homophone: aught. [OE *ãhte*, past tense of *ãgan* owe, possess]
— Synonym: should. *Ought* is the stronger word, holding most closely to the sense of moral obligation, or sometimes of imperative logical necessity; *should* may have the sense of moral obligation or may apply merely to propriety or expediency, as in the proverb, "The liar *should* have a good memory"; that is, he will need it. *Ought* is sometimes used as indicating what the mind deems to be logical in view of all the conditions; as, These goods *ought* to go into that space; *should* in such connections would be correct, but less emphatic. Compare DUTY.
ought[2] (ôt) *n.* Aught; anything. [Var. of AUGHT]
oui (wē) *interj. French* Yes.
Oui·da (wē′də) Pen name of *Louise de la Ramée*, 1839–1908, English novelist.
oui·ja (wē′jə) *n.* A device consisting of a board inscribed with the alphabet and other characters and a planchette, the pointer of which is thought to spell out mediumistic communications. [< F *oui* yes + G *ja* yes]
oui·sti·ti (wis′ti·tē) *n.* The common marmoset (genus *Callithrix* or *Hapale*) of South America, with tufted ears and a long, banded tail. [< F; name coined by Buffon, imit. of the animal's cry]
ounce[1] (ouns) *n.* 1 A unit of weight; one sixteenth of a pound avoirdupois, or 28.349 grams; one twelfth of a pound troy, or 31.1 grams. Abbr. *oz.* 2 A fluid ounce. 3 A small quantity. — **fluid ounce** 1 *U.S.* One sixteenth of a pint; 29.5737 cubic centimeters; 480 minims. 2 *Brit.* 28.413 cubic centimeters; 480 minims. [< OF *unce* <L *uncia* twelfth part (of a pound or foot), orig. a unit. Doublet of INCH.]
ounce[2] (ouns) *n.* 1 A feline carnivorous mammal (*Panthera uncia*) of central Asia, about the size of a leopard, having long fur and a long, thick tail; the snow leopard. 2 Some similar American cat, as the jaguar. [< OF *l'once*, var. of *lonce* the lynx <L *lyncea* < *lynx, lyncis* LYNX]
ounce metal An alloy of copper with five percent each of tin, zinc, and lead: formerly made by adding one ounce of each minor metal to one pound of copper.
ouphe (ōōf) *n. Archaic* A goblin or elf. [Var. of OAF]
our (our) *pronominal adj.* Belonging or pertaining to us: the possessive case of the pronoun *we* employed attributively. [OE *ūre*, earlier *ūser*, genitive of US]
-our See -OR.
ou·rang (ōō·rang′) *n.* The orang-utan. [Var. of ORANG]
ou·ra·nog·ra·phy (ōōr′ə·nog′rə·fē) See URANOGRAPHY.
Ou·ra·nos (ōōr′ə·nəs) See URANUS (def. 1).
ou·ra·ri (ōō·rä′rē) *n.* Curare. [Var. of WOORALI <Tupian]
Ourcq (ōōrk) A river in northern France, flowing 50 miles SW to the Marne; scene of a battle of World War I, 1918.
ou·re·bi (ōō′rə·bē) See ORIBI.
Our Lady The Virgin Mary.
Our Lord Jesus Christ.
ouro- See URO-.
ours (ourz) *pron.* 1 Belonging or pertaining to us: That dog is *ours*: the possessive case of *we* used predicatively. 2 The things or persons belonging or pertaining to us: their country and *ours.* — **of ours** Belonging or relating to us; our: the double possessive. [ME *ures* <*ure* OUR]
our·self (our·self′) *pron.* Myself: only in formal or regal style.
our·selves (our·selvz′) *pron. pl.* We or us.
-ous *suffix of adjectives.* 1 Full of; having: *studious, glorious.* 2 *Chem.* Having a lower valence than that indicated by the suffix *-ic: nitrous.* [< OF <L *-osus*]

Ouse (o͞oz) **1** A river in North Riding, Yorkshire, England, flowing 61 miles SE to the Humber. **2** A river in Northamptonshire and Lincolnshire, England, flowing 156 miles NE to the Wash: also **Greater Ouse.**

ou·sel (o͞o′zəl) See OUZEL.

oust (oust) *v.t.* To force from possession or occupancy; turn out; eject. See BANISH. [<AF *ouster* take away, ? <L *obstare* obstruct < *ob-* against + *stare* stand]

oust·er (ous′tər) *n. Law* The act of putting one out of possession or occupancy; dispossession.

out (out) *adv.* **1** Away from the inside or center: to go *out*; to branch *out.* **2** Away from a specified or usual place: to set *out* from Paris. **3** From a receptacle or source: to pour *out* wine. **4** So as to free of undesired parts or refuse: to thresh *out* grain; to sweep *out* a room. **5** From among others: to pick *out* a dress. **6** Into the charge or care of another or others: to hire *out* laborers; to deal *out* cards. **7** So as to project or be extended: to stretch *out.* **8** Into extinction or inactivity: The flame went *out*; The excitement died *out.* **9** To a result or conclusion: to fight it *out*; to find *out.* **10** Completely; fully: tired *out.* **11** Into existence or outward manifestation: An epidemic broke *out*; The sun came *out.* **12** Into blossom or leaf. **13** Into public notice or circulation: to bring *out* a new edition. **14** Aloud: to call *out.* **15** On strike. **16** Into disagreement; at odds: to be put *out* over trifles. **17** *Colloq.* Into unconsciousness: to pass *out.* **18** In baseball, cricket, etc., so as to be retired from active or leading play: to strike *out.* —*adj.* **1** External; exterior; outer. **2** Away from one's home, place of work, or other place regarded as a base: to be *out* on maneuvers. **3** Away at a distance: to be *out* in California. **4** Exposed or bare, as by rents in the clothing: *out* at the knees. **5** Visible; manifest: The stars are *out.* **6** Made public; disclosed: The truth is *out.* **7** In blossom or leaf. **8** Removed; displaced: The stain is *out.* **9** Mistaken; in error: *out* in one's calculations. **10** Extinguished; exhausted: The fire is *out.* **11** Finished; at an end: before the week is *out.* **12** At odds; in disagreement. **13** At a financial loss; in default: to be *out* five dollars. **14** Not in effective operation: The machine is *out.* **15** *Colloq.* Unconscious: He's *out.* **16** In baseball, cricket, etc., no longer in active or leading play. —*prep.* **1** From within; forth from: *out* the door. **2** Outside; on the outside of: the view *out* this window. —*n.* **1** Something that is out. **2** An escape; a way to dodge responsibility or involvement: He had an *out.* **3** A person not in office or position of power; specifically, in the plural, the party not in power. **4** In baseball, retirement of a batter or base runner. **5** In tennis, a return of the ball, which, untouched by the opponent, falls outside the court. **6** *Printing* Matter in the copy omitted from the composed type. —*v.t.* To drive out; expel. —*v.i.* To come or go out; be revealed: Murder will *out.* —*interj.* Go out! away! begone! [OE *ūt*]

out– *combining form* **1** Living or situated outside; external; away from the center; detached: *outlying, outpatient.* **2** Going forth; issuing; outward: *outbound, outstretch.* **3** Used to denote the time, place, or result of the action expressed by the root verb: *outcome, outcry.* **4** Excessive; surpassing; more; beyond. Dissyllabic nouns with this prefix are pronounced with an almost even stress on each syllable, the first slightly more emphatic. In dissyllabic verbs the stress is usually strongly upon the second element; but when their participles are used as adjectives or nouns, the stress becomes even or, for emphasis, shifts to the first syllable.

Out– is widely used in sense 4 to form compounds, as in the following list:

outact	**outbeg**	**outbluster**
outambush	**outbeggar**	**outboast**
outargue	**outbellow**	**outbrag**
outbabble	**outblaze**	**outbrazen**
outbake	**outbleat**	**outbribe**
outbalance	**outbless**	**outbuild**
outbanter	**outbloom**	**outbully**
outbargain	**outblossom**	**outburn**
outbark	**outbluff**	**outcaper**
outbawl	**outblunder**	**outcavil**
outbeam	**outblush**	**outcharm**

outchatter	**outmatch**	**outsprint**
outcheat	**outpaint**	**outstare**
outchide	**outpass**	**outstate**
outclimb	**outperform**	**outstay**
outcomplete	**outpity**	**outstrain**
outcompliment	**outplan**	**outstride**
outcrawl	**outplod**	**outstrive**
outcrow	**outpoison**	**outstudy**
outcurse	**outpopulate**	**outstunt**
outdance	**outpractice**	**outsuffer**
outdare	**outpraise**	**outsulk**
outdazzle	**outpray**	**outswagger**
outdodge	**outpreach**	**outswear**
outdrink	**outpreen**	**outswindle**
outeat	**outprice**	**outtalk**
outecho	**outproduce**	**outtask**
outfable	**outpromise**	**outthieve**
outfast	**outpush**	**outthreaten**
outfawn	**outqueen**	**outthrob**
outfeast	**outquestion**	**outthwack**
outfight	**outquibble**	**outtower**
outflatter	**outquote**	**outtrade**
outfool	**outrace**	**outtravel**
outfrown	**outrank**	**outtrick**
outgabble	**outrave**	**outtrot**
outgain	**outreason**	**outtrump**
outgallop	**outredden**	**outtyrannize**
outgamble	**outring**	**outvalue**
outglare	**outrival**	**outvaunt**
outglitter	**outroar**	**outvenom**
outglow	**outrun**	**outvociferate**
outgnaw	**outsail**	**outvoice**
outgrin	**outsatisfy**	**outvote**
outguess	**outsavor**	**outwait**
outhasten	**outscold**	**outwalk**
outhowl	**outscorn**	**outwallop**
outhumor	**outscream**	**outwar**
outhyperbolize	**outshame**	**outwarble**
outinvent	**outshout**	**outwaste**
outjazz	**outshriek**	**outwatch**
outjinx	**outsin**	**outweary**
outjockey	**outsing**	**outweep**
outjourney	**outsit**	**outwhirl**
outleap	**outskirmish**	**outwile**
outlighten	**outslander**	**outwill**
outlinger	**outsleep**	**outwish**
outlove	**outsmile**	**outwrangle**
outluster	**outsnore**	**outwrestle**
out–maneuver	**outsoar**	**outwriggle**
outmarch	**outsophisticate**	**outyelp**
outmaster	**outsparkle**	**outyield**

out-and-out (out′ənd·out′) *adj.* Thoroughgoing; unqualified; genuine. —*adv.* Unqualifiedly; genuinely.

out·bid (out·bid′) *v.t.* **·bid, ·bid·den** or **·bid, ·bid·ding** To bid more than; offer a higher price than.

out·board (out′·bôrd′, -bōrd′) *Naut. adj.* Situated on the outside of a vessel, as a motor for temporary attachment to the stern of a small boat. —*adv.* Away from the center.

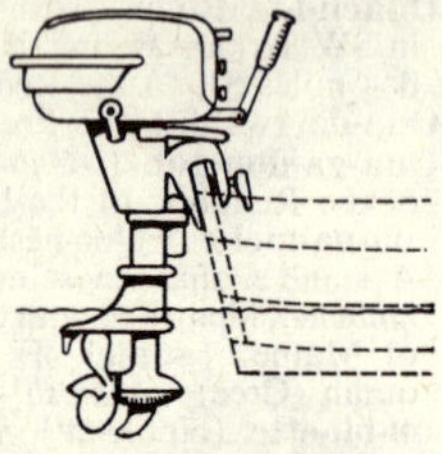

OUTBOARD MOTOR
Showing method of attachment.

out·bound (out′·bound′) *adj.* Outward bound.

out·brave (out·brāv′) *v.t.* **·braved, ·brav·ing** **1** To surpass in bravery. **2** To stand in defiance of. **3** To excel or surpass in splendor.

out·break (out′brāk′) *n.* A sudden and violent breaking forth: said of passion or of disease affecting large numbers of people at once. See synonyms under TUMULT. —*v.i.* (out′·brāk′) **·broke, ·brok·en, ·break·ing** To burst out; break forth.

out·breed (out′brēd′) *v.t.* **·bred, ·breed·ing** *Biol.* To breed or mate (individuals) belonging to stocks or families not closely related: opposed to *inbreed.* —**out′breed′ing** *n.*

out·build·ing (out′bil′ding) *n.* A smaller building appurtenant to a main building and generally separate from it; specifically, a chicken house, woodshed, smokehouse, etc.

out·burst (out′bûrst′) *n.* A bursting out; a violent manifestation, especially of passion.

out·by (out′bī′) *adv. Scot.* At a distance; outlying; also, outdoors. Also **out′bye′.**

out·cast (out′kast′, -käst′) *n.* **1** One who is cast out from home or country; one rejected and despised. **2** *Scot.* A quarrel; disagreement. —*adj.* Rejected as unworthy or useless; cast out; forlorn.

out·caste (out′kast′, -käst′) *n.* In India, a person who has forfeited his caste; a pariah.

out·class (out·klas′, -kläs′) *v.t.* To exceed decisively in skill, quality, or powers.

out·come (out′kum′) *n.* The consequence or visible result. See synonyms under CONSEQUENCE, END, EVENT, PRODUCT.

out·crop (out′krop′) *n. Geol.* **1** The exposure at or above the surface of the ground of any stratum, vein, etc. **2** The rock so exposed. —*v.i.* (out·krop′) **·cropped, ·crop·ping** **1** To crop up or out. **2** To appear above the ground, as rocks.

out·cross (out·krôs′, -kros′) *Biol. v.t.* To mate (individuals) of the same breed but of different strains. —*n.* The act of so mating, or its result. —**out′cross′ing** *n.*

out·cry (out′krī′) *n. pl.* **·cries** **1** A vehement or loud cry, as of distress, alarm, or opposition. **2** A public auction. See synonyms under NOISE. —*v.t.* (out·krī′) **·cried, ·cry·ing** To surpass in noise or crying; cry down.

out·curve (out′kûrv′) *n.* **1** In baseball, a pitched ball that curves away from the batter: opposed to *incurve.* **2** A small projection in a coastline.

out·date (out·dāt′) *v.t.* **·dat·ed, ·dat·ing** To make obsolete or out of date.

out·dat·ed (out·dā′tid) *adj.* Made obsolete; antiquated; old-fashioned.

out·dis·tance (out·dis′təns) *v.t.* **·tanced, ·tanc·ing** **1** To outrun; outstrip. **2** To surpass completely; outdo.

out·do (out·do͞o′) *v.t.* **·did, ·done, ·do·ing** To exceed in performance; surpass; excel. See synonyms under SURPASS. —**out·do′er** *n.*

out·done (out·dun′) *adj. U.S. Colloq.* Tired; exasperated; also, puzzled.

out·door (out′dôr′, -dōr′) *adj.* Being or done in the open air; belonging or occurring outside the house: *outdoor* sports. Also *out-of-door.*

out·doors (out·dôrz′, -dōrz′) *adv.* Outside of the doors; out of the house; in the open air. —*n.* The world beyond the house; the open air. Also *out-of-doors.* —**all outdoors** *Colloq.* The whole world.

out·en (out′n) *prep. Dial.* Out of.

out·er (ou′tər) *adj.* **1** Being on the exterior side; external. **2** Farther from a center or from something regarded as the inside. —*n.* **1** In rifle practice, the part of a target outside the rings. **2** A shot that strikes this part. Compare INNER. —**out′er·most′** *adj.* & *adv.*

Outer Hebrides See HEBRIDES.

Outer Mongolia See MONGOLIA.

outer space The space beyond the extreme limits of the earth's atmosphere; interplanetary and interstellar space.

out·face (out·fās′) *v.t.* **·faced, ·fac·ing** **1** To face or stare down. **2** To defy or confront fearlessly or impudently.

out·fall (out′fôl′) *n.* **1** The place where a river, culvert, or conduit discharges; mouth. **2** The discharged matter.

out·field (out′fēld′) *n.* **1** In baseball, cricket, etc., the players who take their positions in the outer part of the field, or the field occupied by them; specifically, in baseball, right, left, and center field, or all the field beyond the bases. **2** *Scot.* Arable land continually cropped without being manured. **3** A field not situated near a house. **4** A bordering region or domain. —**out′field′er** *n.*

out·fit (out′fit′) *n.* **1** A fitting out or equipment. **2** The expenses occasioned by and incidental to a journey. **3** All the garments and incidentals of a person's costume. **4** The tools or equipment for any particular occupation, calling, or trade; a kit: a painter's *outfit.* **5** Mental acquirements suitable to any intellectual purpose. **6** Any expedition or party, with its proper equipment; hence, any industry, or any group of persons unified in a common undertaking; specifically, the cowboys, horses, wagons, etc., working on a certain ranch. —*v.t.* & *v.i.* **·fit·ted, ·fit·ting** To provide with or acquire an outfit. —**out′fit′ter** *n.*

out·flank (out·flangk′) *v.t.* To get around and in back of the flank of (an opposing force or army); turn the flank of; flank.

out·flow (out′flō′) *n.* **1** That which flows out, or the process of flowing out. **2** An outlet.

out·foot (out·fo͝ot′) *v.t.* **1** To exceed or surpass, as in running or dancing. **2** *Naut.* To sail faster than; outsail.

out·gen·er·al (out·jen′ər·əl) *v.t.* ·aled or ·alled, ·al·ing or ·al·ling To surpass in generalship; out-maneuver.
out·go (out·gō′) *v.t.* ·went, ·gone, ·go·ing To go farther than; exceed or outstrip. —*n.* (out′·gō′) *pl.* ·goes 1 That which goes out; cost or outlay: opposed to *income.* 2 An outgoing. 3 An exit. See synonyms under EXPENSE. —**out′go′er** *n.*
out·go·ing (out′gō′ing) *adj.* 1 Going out; leaving. 2 Friendly; sympathetic. —*n.* 1 The act of going out; departure. 2 That which goes out. 3 *Usually pl.* An expenditure; outlay.
out–group (out′grōōp′) *n. Sociol.* Persons not included in an in-group.
out·grow (out·grō′) *v.t.* ·grew, ·grown, ·grow·ing 1 To surpass in growth. 2 To grow too large for. 3 To lose or get rid of in the course of time or growth: to *outgrow* a habit.
out·growth (out′grōth′) *n.* That which grows out of something else; an excrescence. See synonyms under CONSEQUENCE.
out·guard (out′gärd′) *n.* An outlying guard or post; an advanced picket.
out·gush (out′gush′) *n.* A gushing out.
out·haul (out′hôl′) *n. Naut.* A rope for extending a sail along a spar.
out–Her·od (out·her′əd) *v.t.* 1 To outdo (Herod as depicted in the old mystery plays): usually, to **out-Herod Herod.** 2 To surpass (anyone) in cruelty or excess.
out·house (out′hous′) *n.* An outbuilding; specifically, a privy.
out·ing (ou′ting) *n.* 1 The act of going out; a holiday excursion; short pleasure trip; airing. 2 The distance out at sea: the farthest *outing.* —*adj.* Of, pertaining to, or suitable for an outing, as various garments and fabrics.
outing flannel A soft, lightweight cotton fabric, usually napped on both sides: used chiefly for sleeping garments, children's underwear, etc.
out·land (out′land′) *n.* Land lying beyond the limits of occupation or cultivation. [OE *ūtland*] —**out′land′er** *n.*
out·land·ish (out·lan′dish) *adj.* 1 Of strange or barbarous aspect or action. 2 Situated in an unfamiliar spot; remote. 3 Not native. Also **out′land.** See synonyms under RUSTIC. [OE *ūtlandisc*] —**out·land′ish·ly** *adv.* —**out·land′ish·ness** *n.*
out·last (out·last′, -läst′) *v.t.* To last longer than.
out·law (out′lô′) *n.* 1 A person who is deprived of the protection or benefit of the law; a proscript. 2 One who habitually breaks or defies the law; a freebooter; a person having a price on his head. 3 A vicious horse. —*v.t.* 1 To declare an outlaw; proscribe. 2 To prohibit; ban. 3 To deprive of legal force or protection, as contracts or debts. [OE *ūtlaga* <ON *útlagi*]
out·law·ry (out′lô′rē) *n. pl.* ·ries The state, fact, or process of outlawing or being outlawed.
out·lay (out·lā′) *v.t.* ·laid, ·lay·ing To expend (money). —*n.* (out′lā′) A laying out or disbursing; hence, that which is disbursed; expenditure. See synonyms under EXPENSE, PRICE.
out·ler (ōōt′lər) *n. Scot.* An animal not housed during the night or during winter; hence, a person out of work.
out·let (out′let) *n.* 1 A passage or vent for escape or discharge; an egress; specifically, in commerce, a market for the sale of any commodity. 2 The act of letting out. 3 *Electr.* That point in a wiring system at which the current is taken to supply fixtures, lamps, motors, etc.
out·li·er (out′lī′ər) *n.* 1 One whose residence is not in the same place as his office or business. 2 One who or that which is beyond or excluded from the main body. 3 A person who camps or lies out in the forest, prairie, or other deserted place. 4 *Geol.* An exposed mass of rock surrounded by older rock strata which have been worn away; opposed to *inlier.*
out·line (out′līn′) *n.* 1 *Often pl.* A preliminary sketch showing the principal features of a thing; general plan. 2 The bordering line that serves to define a figure; hence, a sketch made of such lines without shading; also, the art of making such sketches. See synonyms under SKETCH. —*v.t.* ·lined, ·lin·ing 1 To draw the outline of; sketch. 2 To describe in general terms; give the main points of.
out·live (out·liv′) *v.t.* ·lived, ·liv·ing 1 To live longer than (another). 2 To live through; survive.
out·look (out′lŏŏk′) *n.* 1 The expanse in view; hence, the condition or prospect of a thing. 2 Distance of view; hence, foresight. 3 Vigilance. 4 A place where watch is kept. 5 The watch; sentinel.
out·ly·ing (out′lī′ing) *adj.* 1 Situated apart from the main body; distant; extrinsic. 2 Outside the boundary.
out–man (out·man′) *v.t.* –manned, –man·ning 1 To surpass in number of men. 2 To excel in manliness.
out·mo·ded (out·mō′did) *adj.* Out of fashion; not in current style.
out·most (out′mōst′) *adj.* Outermost.
out·ness (out′nis) *n.* 1 The quality or condition of being outside; externality. 2 The quality of being interested in external things.
out of 1 From or beyond the inside of; from among. 2 Beyond the limits, reach, scope, or proper position of; not in or included in: *out of* sight. 3 Without: *out of* breath. 4 Influenced, inspired, or caused by: *out of* pity; *out of* respect for him.
out of commission Completely out of order; not working; laid aside.
out–of–date (out′əv·dāt′) *adj.* Old-fashioned; archaic.
out–of–door (out′əv·dôr′, -dōr′) See OUTDOOR.
out–of–doors (out′əv·dôrz′, -dōrz′) See OUTDOORS.
out of sorts 1 Indisposed or unwell. 2 Dissatisfied or unhappy. 3 *Printing* Without sorts of particular letters, figures, etc.; having insufficient sorts.
out–of–the–way (out′əv·thə·wā′) *adj.* 1 Remotely situated; difficult to reach; secluded. 2 Different from what is common; out of the common range; unusual; singular; eccentric.
out of the woods Clear of doubts and difficulties; safe after peril or hazard.
out page *Printing* The first page of a complete signature when folded.
out parish A parish situated in the country or somewhat distant from a city parish.
out part An outer or remote part.
out·pa·tient (out′pā′shənt) *n.* A patient, not an inmate, treated at a hospital or dispensary.
out·play (out·plā′) *v.t.* To play better than; defeat.
out·point (out·point′) *v.t.* 1 To score more points than. 2 *Naut.* To sail closer to the wind than.
out·post (out′pōst′) *n.* 1 A detachment of troops stationed at a distance from the main body as a guard against surprise. 2 The station occupied by them.
out·pour (out·pôr′, -pōr′) *v.t.* & *v.i.* To pour out. —*n.* (out′pôr′, -pōr′) A free outflow; outpouring. —**out′pour′er** *n.* —**out′pour′ing** *n.*
out·pull (out·pŏŏl′) *v.t.* To pull more strongly than.
out·put (out′pŏŏt′) *n.* 1 The quantity put out or produced in a specified time; amount or rate of production, collective or individual, as from a mine or mines, or from a furnace or a country. 2 That which is excreted from the body by the lungs, skin, or kidneys. 3 The electric power of a dynamo; also, the energy or power given by a machine. See synonyms under PRODUCT.
out·rage (out′rāj′) *n.* 1 An act of shocking violence or cruelty; a gross infringement of morality or decency; a gross insult. 2 Something that violates the feelings or the proprieties. 3 *Obs.* Violent rage; a dangerous display of passion. —*v.t.* ·raged, ·rag·ing 1 To commit outrage upon; wrong or abuse grossly; violate; offend. 2 To rape. See synonyms for VIOLATE. [<OF *ultrage,* ult. <L *ultra* beyond; infl. in meaning by RAGE]

Synonyms (noun): abuse, affront, indecency, indignity, injury, insult, offense, violence. An *outrage* combines *insult* and *injury.* See INJURY, OFFENSE, VIOLENCE. Compare synonyms for AFFRONT.
out·ra·geous (out·rā′jəs) *adj.* 1 Of the nature of an outrage; heinous; atrocious. 2 Heedless of authority or decency. 3 Exceeding bounds. See synonyms under FLAGRANT, INFAMOUS, VIOLENT. —**out·ra′geous·ly** *adv.* —**out·ra′geous·ness** *n.*
ou·trance (ōō·träns′) *n. French* The utmost extremity; the bitter end.
out·range (out·rānj′) *v.t.* ·ranged, ·rang·ing 1 To surpass in range; have a greater range than. 2 To go beyond the range of.
ou·tré (ōō·trā′) *adj. French* Deviating from conventional usage; strikingly odd; exaggerated.
out·reach (out·rēch′) *v.t.* 1 To reach or go beyond; surpass. 2 To reach out; extend. —*v.i.* 3 To reach out. —*n.* (out′rēch′) The act of reaching out.
ou·tre·mer (ōō′trə·mer′, *Fr.* ōō·tr′·mâr′) *n.* The region beyond the sea; foreign lands. —*adv.* Beyond the sea. [<F <*outre* beyond (<L *ultra*) + *mer* the sea <L *mare*]
out·ride (out·rīd′) *v.t.* ·rode, ·rid·den, ·rid·ing To ride faster, farther, or better than.
out·rid·er (out′rī′dər) *n.* 1 A mounted servant who rides in advance of a carriage. 2 One who rides along the edge of a herd of cattle to prevent stampeding or straying. 3 One who rides out or forth.
out·rig·ger (out′rig′ər) *n.* 1 A part built or arranged to project beyond a natural outline, as of a vessel or machine. 2 A projecting contrivance terminating in a boatlike float, braced to the side of a canoe to prevent capsizing. 3 *Naut.* a A spar for extending a sail or rope farther than the beam of the vessel would otherwise permit. b A boom swung out from a vessel, to which to secure boats. 4 A bracket projecting from the side of a narrow rowboat or shell, and provided with a rowlock for an oar or scull. 5 A boat or shell equipped with such a bracket. 6 *Aeron.* A projecting contrivance to support various components of an airplane: also called *tail boom.*
out·right (out′rīt′) *adj.* 1 Free from reserve or restraint; positive; downright. 2 Complete; entire. 3 Going straight on. —*adv.* (out′rīt′) 1 Without reservation or limitation; entirely; utterly; openly. 2 Without delay.
out·root (out·rōōt′, -rŏŏt′) *v.t.* To root out; eradicate.
out·run·ner (out′run′ər) *n.* 1 An attendant who runs before or beside a carriage. 2 The leading dog in a dog team.
out·sell (out·sel′) *v.t.* ·sold, ·sell·ing 1 To sell more readily or for a higher price than. 2 To sell more goods than.
out·sen·try (out′sen′trē) *n. pl.* ·tries An outer sentry.
out·sert (out′sûrt) *n. Printing* A folded sheet placed around a folded section of printed matter: opposed to *insert.*
out·set (out′set′) *n.* A first entrance on any business, journey, or the like; a setting out; beginning; start; opening. Also **out′set′ting.** See synonyms under BEGINNING.
out·shine (out·shīn′) *v.* ·shone, ·shin·ing *v.t.* 1 To shine brighter than. 2 To surpass, as in wit or finery. —*v.i.* 3 To shine forth. —**out·shin′er** *n.*
out·shoot (out·shōōt′) *v.* ·shot, ·shoot·ing *v.t.* 1 To excel in shooting. 2 To shoot out or beyond; project. —*v.i.* 3 To shoot out; project. —*n.* (out′shōōt′) 1 A projection; branch; bud, as of a plant. 2 A rushing forth, as of water. 3 In baseball, an outcurve.
out–shop (out′shop′) *v.t.* –shopped, –shop·ping In railroading, to turn out (equipment) from a repair shop or factory.
out·side (out′sīd′) *n.* 1 The external part of a thing; the side or part that forms or adjoins the surface. 2 The part or side that is seen; hence, superficial appearance. 3 The space beyond a bounding line or surface; outer region: opposed to *inside.* —**at the outside** At the farthest, longest, or most, as in an estimate. —*adj.* 1 Pertaining to, located on, or restricted to the outside; exterior. 2 Originating or situated beyond the outside limits; foreign. 3 Reaching the limit; extreme: an *outside* estimate. 4 Slight; inconsequential: There is only an *outside* possibility. —*adv.* 1 On or to the outside; externally. 2 Beyond the outside limits of. 3 In the open air; outdoors. —*prep.* (out′sīd′) 1 On or to the exterior of: *outside* the park; *outside* the box. 2 Beyond the limit of: Don't tell it *outside*

the club. 3 *Colloq.* Except: No one knows *outside* yourself.
outside of 1 *U.S. Colloq.* Except; besides: No one came *outside of* me. 2 Outside.
out·sid·er (out′sī′dər) *n.* 1 One who is outside; an intruder. 2 A race horse whose chance of winning is slight.
outside roll *Aeron.* A roll executed while flying in a negative angle–of–attack range, which begins and ends with the airplane on its back.
out·sight (out′sīt′) *n. Rare* 1 Observation of that which is without. 2 The power of noticing external things: opposed to *insight.*
out sister A member of a cloistered order of nuns who attends to the business of the order, or convent, with the outside world.
out·size (out′sīz′) *n.* A size, as of clothing, footwear, etc., that is larger than the regular sizes.
out·skirt (out′skûrt′) *n. Often pl.* A place on the skirts or border; outer verge.
out·smart (out·smärt′) *v.t. U.S. Colloq.* To outwit; fool.
out·sole (out′sōl′) *n.* The outside or lower sole of a boot or shoe: distinguished from *insole.* See illustration under SHOE.
out·span (out·span′) *v.t.* & *v.i.* **·spanned, ·span·ning** In South Africa, to unharness or unyoke (animals). — *n.* The act or the place of outspanning. [<Afrikaans *uitspannen* <*uit-* out + *spannen* stretch]
out·speak (out·spēk′) *v.* **·spoke, ·spo·ken, ·speak·ing** *v.t.* 1 To outdo in speaking. 2 To say openly or boldly. — *v.i.* 3 To speak out. — **out·speak′er** *n.*
out·spent (out·spent′) *adj.* Completely spent or wearied; tired out.
out·spo·ken (out′spō′kən) *adj.* 1 Bold or free of speech; frank. 2 Spoken boldly or frankly. — **out′spo′ken·ly** *adv.* — **out′spo′ken·ness** *n.*
out·spread (out·spred′) *v.t.* & *v.i.* **·spread, ·spread·ing** To spread out; extend.
out·stand (out·stand′) *v.* **·stood, ·stand·ing** *v.i.* 1 To stand out; project. 2 *Naut.* To put to sea; sail. — *v.t.* 3 *Archaic* To stay or last beyond. 4 *Dial.* To withstand. — **out·stand′er** *n.*
out·stand·ing (out·stan′ding) *adj.* 1 Standing prominently forth; salient; conspicuous; preeminent. 2 Still standing, as a debt unpaid or not due.
out·stretch (out·strech′) *v.t.* 1 To stretch out; expand; extend. 2 To extend beyond. — **out·stretched′** *adj.*
out·strip (out·strip′) *v.t.* **·stripped, ·strip·ping** 1 To leave behind; outrun, as in a race. 2 To excel; surpass. See synonyms under LEAD[1], SURPASS.
out·stroke (out′strōk′) *n.* An outward stroke, as the thrust of an engine's piston toward the crankshaft. Compare INSTROKE.
out·tell (out·tel′) *v.t.* **·told, ·tell·ing** To declare; say openly.
out·turn (out′tûrn′) *n.* 1 Output; product. 2 In commerce, the quantity, condition, or quality of goods actually turned out and delivered.
out·ward (out′wərd) *adj.* 1 Of or pertaining to the exterior of an object; outer; external; outside: *outward* show. 2 Tending to the outside; directed outward: an *outward* course. 3 Derived or added from without; not inherent; extraneous; extrinsic: *outward* grace. 4 Relating to the physical or bodily as opposed to the mental aspect; external: His *outward* attitude belies his inward feeling. 5 Of or pertaining to the world or the outer man; not spiritual; carnal; corporeal. — *adv.* 1 To or in the direction of the outside; away from an inner place. 2 On the surface; superficially. 3 Away from port or home. Also **out′wards.** — *n.* External form; outside appearance. [OE *ūtweard*] — **out′ward·ly** *adv.* — **out′ward·ness** *n.*
out·wash (out′wäsh′, -wôsh′) *n.* Detritus and waste materials carried away by the water of melting glaciers.
out·wear (out·wâr′) *v.t.* **·wore, ·worn, ·wear·ing** 1 To wear or stand use better than; outlast. 2 To wear out, as by constant use. 3 To exhaust; use up. 4 To outlive; outgrow.
out·weigh (out·wā′) *v.t.* 1 To weigh more than. 2 To exceed in importance, value, etc.
out·wit (out·wit′) *v.t.* **·wit·ted, ·wit·ting** To defeat by superior ingenuity or cunning; overreach. See synonyms under BAFFLE, DECEIVE.
out·work[1] (out·wûrk′) *v.t.* **·worked** or **·wrought, ·work·ing** 1 To work faster or better than; excel in working. 2 To work out; complete.
out·work[2] (out′wûrk′) *n. Mil.* Any outer defense, as beyond the ditch of a fort. See synonyms under RAMPART.
out·worn (out·wôrn′) Past participle of OUTWEAR.
out·wrought (out·rôt′) Alternative past tense and past participle of OUTWORK.
ou·zel (o͞o′zəl) *n.* 1 One of various European thrushes, as the blackbird (*Turdus merula*), the **ring ouzel** (*T. torquatus*). 2 The related dipper or **water ouzel** (*Cinclus aquaticus*). Also spelled *ousel.* [OE *ōsle*]
o·va (ō′və) Plural of OVUM.
o·val (ō′vəl) *adj.* 1 Having the figure of the plane longitudinal section of an egg, usually rounded at one end and tapering at the other. 2 Ellipsoidal. — *n.* A figure or body of such form or outline. [<NL *ovalis* <L *ovum* egg] — **o′val·ly** *adv.* — **o′val·ness** *n.*
o·var·i·ot·o·my (ō·vâr′ē·ot′ə·mē) *n. Surg.* The excision of either or both ovaries, or of an ovarian tumor: also called *oophorectomy.* [OVARY + -TOMY]
o·va·ri·tis (ō′və·rī′tis) *n. Pathol.* Inflammation of the ovary. [<OVARY + -ITIS]
o·var·i·um (ō·vâr′ē·əm) *n. pl.* **o·var·i·a** (ō·vâr′·ē·ə) An ovary.
o·va·ry (ō′və·rē) *n. pl.* **·ries** 1 *Biol.* The genital gland of female animals in which are produced the essential reproductive elements or ova. In the higher vertebrates there are two. 2 *Bot.* In angiospermous plants, that portion of the pistil or gynoecium in which the ovules are contained. [<NL *ovarium* <L *ovum* an egg] — **o·var′i·an, o·var′i·al** *adj.*
o·vate (ō′vāt) *adj. Bot.* Egg–shaped: said of leaves. [<L *ovatus* <*ovum* egg] — **o′vate·ly** *adv.*
o·va·tion (ō·vā′shən) *n.* 1 A spontaneous acclamation of popularity; enthusiastic reception of a popular personage. 2 In ancient Rome, a secondary triumphal honor. [<L *ovatio, -onis* a rejoicing <*ovare* rejoice, exult] — **o·va′tion·al** *adj.*
ov·en (uv′ən) *n.* 1 An enclosed chamber in which substances are heated or cooked: used also for baking, annealing, etc., as in a kiln or assaying furnace. 2 A furnace. [OE *ofen*]
ov·en·bird (uv′ən·bûrd′) *n.* 1 A bird that builds a domed nest, as a South American passerine bird (genus *Furnarius*), whose nests are oven–shaped structures of clay. 2 The American golden–crowned thrush (*Seiurus aurocapillus*): also called *teacher bird.*
o·ver (ō′vər) *prep.* 1 In or to a place or position above; higher than: the sky *over* our heads. 2 So as to pass or extend across; from one end or side of to the other: the plane flying *over* the lake; walking *over* the bridge. 3 On the other side of: lying *over* the ocean. 4 Upon the surface or exterior of: Oil was smeared *over* the axle. 5 Here and there upon or within; throughout all parts of: traveling *over* land and sea. 6 So as to rise above, cover, or submerge: The mud is now *over* my boots. 7 So as to close or cover: a cloth tied *over* the mouth of the jar. 8 During the continuance of; through: a diary kept *over* the years. 9 Up to the end of and beyond: Stay with us *over* Christmas. 10 More than; in excess of, as in amount, degree, number, extent, etc.: *over* a million dollars in assets. 11 In preference to: chosen *over* all other contenders. 12 In higher rank, authority, power, etc., than: They want a strong man *over* them. 13 Upon, as an effect: His influence *over* her is profound. 14 Concerning; with regard to: time wasted *over* trifles. 15 While engaged in or partaking of: falling asleep *over* Shakespeare; a bargain made *over* a bottle of wine. — **over all** From one end or aspect to the other. — **over and above** In addition to; besides. — *adv.* 1 Above; on high. 2 So as to close, cover, or be covered: The pond froze *over.* 3 To pass above from one of two sides or places to the other; across an intervening space, brim, or edge. 4 At or on the other side; at a distance in a specified direction or place: *over* in Europe; They're playing music *over* there. 5 From one side, opinion, or purpose to another: to be won *over* to a point of view. 6 From one person, condition, or custody to another: to make property *over* to someone. 7 From beginning to end; all through; completely: I'll think the matter *over.* 8 With the upper surface downwards; from an upright position, especially so as to invert, reverse, or transpose: to turn one's hand *over*; to topple *over.* 9 With repetition; again: He added his figures *over.* 10 So as to overflow: The cup ran *over.* 11 So as to constitute a surplus; in excess; beyond: enough to have some left *over.* 12 Beyond a stated time; until later: Plan to stay *over.* — **all over with** Finished. — **over again** Once more; afresh. — **over against** Opposite to; as contrasted with; in front of. — **over and over** Repeatedly. — **over there** *Colloq.* In Europe: a phrase popularized in the United States during World War I. — *adj.* 1 Finished; complete. 2 On the other side; having got across. 3 Outer; superior; upper. 4 In excess or addition; extra. — *n.* 1 Something remaining or in addition. 2 In cricket: **a** The succession of four to six balls bowled during a turn at one end of the wicket. **b** The part of the game in which this occurs. 3 *Mil.* A shot hitting or exploding beyond the target. — *v.t.* & *v.i.* To go or pass over. [OE *ofer*]
over– *combining form* 1 Above; on top of; superior: *overlord.* 2 Passing above; going beyond the top or limit of: *overarch, overflow.* 3 Moving or causing to move downward, as from above: *overthrow, overturn.* 4 Excessively; excessive; too much.
Over– is widely used in def. 4 to form compounds, as in the list beginning at the foot of the page.
o·ver·act (ō′vər·akt′) *v.t.* & *v.i.* To act with exaggeration.
o·ver·age (ō′vər·ij) *n.* In commerce, an amount of money or goods in excess of that which is listed as being on hand.
o·ver–age (ō′vər·āj′) *adj.* Past the age of usefulness; too old to be of service: *over–age* guns.
o·ver–all (ō′vər·ôl′) *adj.* 1 From one extremity to the other: said of dimensions measured. 2 Including or covering everything.
o·ver·alls (ō′vər·ôlz′) *n. pl.* 1 Loose, coarse trousers, often with suspenders and a piece extending over the breast, worn over the clothing as protection from soiling. 2 *Brit.* Waterproof leggings.

overabound	**overargue**	**overbig**	**overbrown**	**overcaution**	**overcold**	**overconscious**
overabstemious	**overassert**	**overbitter**	**overbrush**	**overcautious**	**overcolor**	**overconsciousness**
overabundance	**overassertion**	**overblame**	**overbrutal**	**overcautiously**	**overcommend**	**overconservatism**
overabundant	**overassertive**	**overblithe**	**overbulky**	**overcentralization**	**overcompetitive**	**overconservative**
overaccentuate	**overassess**	**overboastful**	**overburdensome**	**overcharitable**	**overcomplacency**	**overconsiderate**
overaccumulation	**overassessment**	**overbold**	**overbusy**	**overcheap**	**overcomplacent**	**overconsideration**
overactive	**overassumption**	**overbookish**	**overbuy**	**overcherish**	**overcomplete**	**overconsume**
overactivity	**overattached**	**overborrow**	**overcapacity**	**overchildish**	**overcomplex**	**overconsumption**
overadvance	**overattentive**	**overbravely**	**overcaptious**	**overchill**	**overcompliant**	**overcontented**
overambitious	**overbake**	**overbred**	**overcaptiousness**	**overcivil**	**overconcentration**	**overcontribute**
overanalyze	**overbanked**	**overbreed**	**overcareful**	**overcivilized**	**overconcern**	**overcook**
overanxiety	**overbarren**	**overbright**	**overcareless**	**overclean**	**overcondense**	**overcool**
overanxious	**overbashful**	**overbrilliant**	**overcaring**	**overclever**	**overconfidence**	**overcoolly**
overapprehensive	**overbelief**	**overbroaden**	**overcarry**	**overclose**	**overconfident**	**overcopious**
overapt	**overbet**	**overbroil**	**overcasual**	**overcloseness**	**overconscientious**	**overcorrect**

o·ver·arch (ō′vər·ärch′) *v.t.* & *v.i.* To form an arch over (something).

o·ver·awe (ō′vər·ô′) *v.t.* ·awed, ·aw·ing To subdue or restrain by awe. See synonyms under ABASH.

o·ver·bal·ance (ō′vər·bal′əns) *v.* ·anced, ·anc·ing *v.t.* 1 To exceed in weight, importance, etc. 2 To cause to lose balance. — *v.i.* 3 To lose one's balance. — *n.* Excess of weight or value.

o·ver·bear (ō′vər·bâr′) *v.* ·bore, ·borne, ·bear·ing *v.t.* 1 To crush or bear down by physical weight or force. 2 To prevail over; domineer. — *v.i.* 3 To bear too much fruit; be too fruitful. See synonyms under SUBDUE.

o·ver·bear·ing (ō′vər·bâr′ing) *adj.* 1 Arrogant; dictatorial. 2 Overwhelming; crushing. See synonyms under ARBITRARY, DOGMATIC, IMPERIOUS. — **o′ver·bear′ing·ly** *adv.* — **o′ver·bear′ing·ness** *n.*

o·ver·bid (ō′vər·bid′) *v.t.* & *v.i.* ·bid, ·bid·den or ·bid, ·bid·ding 1 To outbid (someone). 2 To bid more than the fair value of (something).

o·ver·bite (ō′vər·bīt′) *n.* *Dent.* Projection of the upper incisor teeth in front of the lower in attempted occlusion.

o·ver·blow (ō′vər·blō′) *v.t.* ·blew, ·blown, ·blow·ing 1 To blow down, over, or away. 2 To cover by blowing, as with snow or sand.

o·ver·blown (ō′vər·blōn′) *adj.* Too productive of flowers; also, past flowering; past full bloom.

o·ver·board (ō′vər·bôrd′, -bōrd′) *adv.* Over the side of or out of a boat or ship.

o·ver·borne (ō′vər·bôrn′, -bōrn′) Past participle of OVERBEAR.

o·ver·build (ō′vər·bild′) *v.t.* ·built, ·build·ing 1 To build over: to *overbuild* a ravine. 2 To erect more buildings within (an area) than are needed. 3 To build, as a house, on too elaborate a scale.

o·ver·bur·den (ō′vər·bûr′dən) *v.t.* To load with too much weight.

o·ver·by (ō′vər·bī′) *adv.* *Scot.* & *Brit. Dial.* Nearby; a little way on; across the way.

o·ver·cap·i·tal·ize (ō′vər·kap′i·təl·īz′) *v.t.* ·ized, ·iz·ing 1 To invest capital in to an extent not warranted by actual prospects. 2 To affix an unjustifiable or unlawful value to the nominal capital of (a corporation). 3 To estimate the value of (a property, company, etc.) too highly. — **o′ver·cap′i·tal·i·za′tion** *n.*

o·ver·cast (ō′vər·kast′, -käst′, ō′vər·kast′, -käst′) *v.* ·cast, ·cast·ing *v.t.* 1 To overcloud; darken. 2 To sew, as the edge of a fabric, with long wrapping stitches so as to prevent raveling. — *adj.* 1 Clouded, as the sky. 2 Sewn with a blanket stitch. — *n.* *Meteorol.* A cloud or clouds covering more than nine tenths of the sky.

o·ver·charge (ō′vər·chärj′) *v.t.* ·charged, ·charg·ing 1 To charge (someone) too high a price. 2 To load or fill to excess; overburden. 3 To exaggerate. — *n.* An excessive charge.

o·ver·check (ō′vər·chek′) *n.* A checkrein passing over a horse's head between the ears to draw the bit upward.

o·ver·clothes (ō′vər·klōz′, -klōthz′) *n. pl.* Outer garments.

o·ver·cloud (ō′vər·kloud′) *v.t.* To cover with clouds; darken or make gloomy.

o·ver·coat (ō′vər·kōt′) *n.* An extra outdoor coat worn over a suit; a greatcoat; topcoat.

o·ver·come (ō′vər·kum′) *v.t.* 1 To get the better of in any conflict or struggle; defeat; conquer. 2 To prevail over or surmount, as difficulties, obstacles, etc. 3 To affect violently so as to render helpless, as by emotion, sickness, etc. — *v.i.* 4 To gain mastery; win. See synonyms under BEAT, CONQUER, REPRESS, SUBDUE. [OE *ofercuman*] — **o′ver·com′er** *n.*

o·ver·com·pen·sate (ō′vər·kom′pən·sāt) *v.* ·sat·ed, ·sat·ing *v.i.* 1 To make too great a compensation, as in balancing the arms of a scales. 2 *Psychol.* & *Psychoanal.* To engage in overcompensation. — *v.t.* 3 To make too great a compensation for.

o·ver·com·pen·sa·tion (ō′vər·kom′pən·sā′shən) *n.* 1 *Psychol.* More than the necessary or normal adjustments in a given situation. 2 *Psychoanal.* The cultivation of attitudes and forms of behavior designed to compensate in an exaggerated manner for the fact or feeling of inferiority. — **o′ver·com·pen′sa·to·ry** (-kəm·pen′sə·tôr′ē, -tō′rē) *adj.*

o·ver·con·trol (ō′vər·kən·trōl′) *v.t.* & *v.i.* ·trolled, ·trol·ling *Aeron.* To move the controls of (an aircraft) so as to compensate excessively for a previously performed, incorrect movement.

o·ver·crop (ō′vər·krop′) *v.i.* ·cropped, ·crop·ping *Agric.* To exhaust by continuous cropping, as land.

o·ver·de·ter·mined (ō′vər·di·tûr′mind) *adj.* *Psychoanal.* Having several factors contributory to a given condition or state of mind: an *overdetermined* neurosis. — **o′ver·de·ter′mi·na′tion** (-tûr′mə·nā′shən) *n.*

o·ver·de·vel·op (ō′vər·di·vel′əp) *v.t.* 1 To develop excessively. 2 *Phot.* To develop (a plate or film) to too great a degree. — **o′ver·de·vel′op·ment** *n.*

o·ver·do (ō′vər·doō′) *v.* ·did, ·done, ·do·ing *v.t.* 1 To do excessively; carry too far; exaggerate. 2 To overtax the strength of; exhaust: usually used passively or reflexively. 3 To cook too much, as meat. 4 *Poetic* To surpass; outdo. — *v.i.* 5 To do too much. [OE *oferdōn*]

o·ver·dose (ō′vər·dōs′) *v.t.* ·dosed, ·dos·ing To dose to excess. — *n.* (ō′vər·dōs′) An excessive dose.

o·ver·draft (ō′vər·draft′, -dräft′) *n.* 1 The act of overdrawing an account, as at a bank. 2 The amount by which a check or draft exceeds the sum against which it is drawn. 3 A current of air passing over, not through, the ignited fuel in a furnace. Also **o′ver·draught′** (-draft′, -dräft′).

o·ver·draw (ō′vər·drô′) *v.t.* ·drew, ·drawn, ·draw·ing 1 To draw against (an account) beyond one's credit. 2 To draw or strain excessively, as a bow. 3 To exaggerate a representation of. — *n.* (ō′vər·drô′) The act of overdrawing; an overdraft.

o·ver·drive (ō′vər·drīv′) *v.t.* ·drove, ·driv·en, ·driv·ing 1 To push too hard or too far; overwork. 2 To drive too hard. — *n.* (ō′vər·drīv′) *Mech.* A gearing device which turns a drive shaft at a speed greater than that of the engine, thus decreasing power output: opposed to *underdrive*.

o·ver·due (ō′vər·doō′, -dyoō′) *adj.* 1 Remaining unpaid after becoming due. 2 Past due: an *overdue* plane or train.

o·ver·dye (ō′vər·dī′) *v.t.* ·dyed, ·dye·ing 1 To dye with too much color. 2 To dye with a second color.

o·ver·ex·pose (ō′vər·ik·spōz′) *v.t.* ·posed, ·pos·ing 1 To expose excessively. 2 *Phot.* To expose (a film or plate) too long. — **o′ver·ex·po′sure** (-spō′zhər) *n.*

o·ver·fall (ō′vər·fôl′) *n.* 1 A rapid sea current formed by the peculiarities of the bottom, or by winds, tide, etc.; a race. 2 A sudden drop in the bottom of the sea. 3 A catch basin for overflow, as from a canal.

o·ver·flow (ō′vər·flō′) *v.* ·flowed, ·flown, ·flow·ing *v.i.* 1 To flow or run over the brim or bank, as water, rivers, etc. 2 To be filled beyond capacity; spill over. 3 To superabound. — *v.t.* 4 To flow over the brim or bank of. 5 To flow or spread over; cover. 6 To fill beyond capacity; cause to overflow. See synonyms under INUNDATE. — *n.* (ō′vər·flō′) 1 The act of overflowing. 2 That which flows over. 3 A flood. 4 A surplus. 5 A passage or outlet for liquid. [OE *oferflōwan*]

o·ver·flow·ing (ō′vər·flō′ing) *adj.* Running over the brim or edge; copious; abundant. — *n.* Overflow; copiousness.

o·ver·gar·ment (ō′vər·gär′mənt) *n.* An outer garment.

o·ver·glaze (ō′vər·glāz′) *v.t.* ·glazed, ·glaz·ing To glaze over; apply an overglaze to. — *n.* A decoration or second glaze applied to pottery.

o·ver·grow (ō′vər·grō′) *v.* ·grew, ·grown, ·grow·ing *v.t.* 1 To grow over; cover with growth. 2 To grow too big for; outgrow. — *v.i.* 3 To grow or increase excessively; grow too large. — **o′ver·grown′** *adj.*

o·ver·growth (ō′vər·grōth′) *n.* 1 Luxuriant or excessive growth. 2 A growth upon or over something.

o·ver·hand (ō′vər·hand′) *adj.* 1 In baseball, delivering the ball, or delivered, as the ball, with the hand well above the level of the elbow or shoulder. 2 Made by carrying the thread over and over, as a seam. 3 With the hand above the object which it holds, seizes, or throws. 4 Striking downward. — *adv.* In an overhand manner. — *v.t.* In sewing, to overcast. — *n.* 1 An overhand stroke or delivery in baseball or tennis; also, a ball so served or delivered. 2 The act or method of such delivery or stroke. 3 A kind of knot. See illustration under KNOT.

o·ver·hand·ed (ō′vər·han′did) *adj.* Overhand (def. 1).

o·ver·hang (ō′vər·hang′) *v.* ·hung, ·hang·ing *v.t.* 1 To hang or project over (something); jut over. 2 To impend over; threaten. 3 To adorn with hangings. — *v.i.* 4 To hang or jut over something. — *n.* 1 An overhanging portion of a structure, as of a roof, the bow of a ship, etc.; also, the amount or degree of such projection. 2 *Aeron.* **a** One half the distance in span of any two main supporting surfaces of an airplane: when the upper surface is the greater, it is called positive overhang. **b** The distance from the outer strut attachment to the edge of a wing.

o·ver·haul (ō′vər·hôl′) *v.t.* 1 To examine carefully, as for needed repairs; turn over or take apart for this purpose. 2 To catch up with; gain on. 3 *Naut.* **a** To slacken (a rope) by hauling in the opposite direction. **b** To prepare (a tackle) for use by separating the blocks. See synonyms under EXAMINE. — *n.* (ō′vər·hôl′) 1 A thorough inspection or examination. 2 Examination and complete repair. Also **o′ver·haul′ing**.

o·ver·head (ō′vər·hed′) *adj.* 1 Placed or working above or aloft: an *overhead* railway; working from above; working downward: an *overhead* valve. 2 Chosen by random sampling; average: an *overhead* sample. 3 Situated or working overhead; also, passing over the head. 4 Denoting such general expenditure in a financial or industrial enterprise as cannot be attributed to any one department or product, excluding cost of materials, labor, and

overcorrupt	overcured	overdelicate	overdilute	overeasy	overenthusiastic	overexpenditure
overcostly	overcurious	overdelicately	overdiscipline	overeat	overesteem	overexpress
overcount	overcuriousness	overdemand	overdiscourage	overeducate	overestimate	overexuberant
overcourteous	overdaintiness	overdemocratic	overdistant	overelaborate	overestimation	overfacile
overcovetous	overdainty	overdepress	overdiversification	overelaboration	overexcelling	overfaithful
overcoy	overdance	overdepressive	overdiversify	overelate	overexcitable	overfamiliar
overcram	overdare	overdesirous	overdogmatic	overelegant	overexcite	overfanciful
overcredit	overdazzle	overdestructive	overdominate	overembellish	overexcitement	overfast
overcredulous	overdear	overdestructiveness	overdoubt	overemotional	overexercise	overfastidious
overcriticize	overdecorate	overdevoted	overdramatic	overemphasis	overexert	overfastidiousness
overcull	overdecorative	overdevotion	overdrink	overemphasize	overexertion	overfasting
overcultivate	overdeepen	overdiffuse	overdry	overemphatic	overexpand	overfat
overcultivation	overdeeply	overdignified	overeager	overenjoy	overexpansion	overfatigue
overcunning	overdeliberate	overdiligence	overearnest	overenrich	overexpect	overfatten
overcunningly	overdeliberation	overdiligent	overeasily	overenter	overexpectant	overfavor

add,āce,câre,pälm; end,ēven; it,īce; odd,ōpen,ôrder; tŏŏk,pōōl; up,bûrn; ə = a in *above*, e in *sicken*, i in *clarity*, o in *melon*, u in *focus*; yōō = u in *fuse*; oi,oil; ou,pout; ch,check; g,go; ng,ring; th,thin; th,this; zh,vision. Foreign sounds à,œ,ü,kh,ń; and ◆: see page xx. < from; + plus; ? possibly.

selling; fixed charges; in transportation, bond interest and other expenses previous to operating expenses, taxes, etc. — *n.* 1 General expenditure applicable to all departments of a business, as light, heat, taxes, etc. 2 *Naut.* Ceiling of a cabin or hold. — *adv.* (ō′vər·hed′) 1 Above one's head; aloft. 2 So as to be submerged; over one's head.

overhead valve *Mech.* A valve that is located in the upper part of the combustion chamber of an engine, above the piston face. See VALVE-IN-HEAD ENGINE.

o·ver·hear (ō′vər·hir′) *v.t.* **·heard, ·hear·ing** To hear (something said or someone speaking) without the knowledge or intention of the speaker. — **o′ver·hear′er** *n.*

o·ver·hours (ō′vər·ourz′) *n. pl.* 1 Time outside and in addition to the assigned or usual number of hours; overtime. 2 Unduly long hours of employment.

O·ver·ijs·sel (ō′vər·ī′səl) A province of eastern Netherlands; 1,255 square miles; capital, Zwolle. Also **O′ver·ys′sel.**

o·ver·is·sue (ō′vər·ish′o͞o, -yo͞o) *v.t.* **·sued, ·su·ing** To issue in excess of a legal or authorized amount, or in excess of ability to meet the demands thus created: to *overissue* stock, notes, or bonds. — *n.* (ō′vər·ish′o͞o, -yo͞o) An excessive or unauthorized issue.

o·ver·joy (ō′vər·joi′) *v.t.* To delight or please greatly. See synonyms under RAVISH.

o·ver·kill (ō′vər·kil′) *n. U. S. Mil.* The surplus of nuclear weapons beyond the number considered necessary to demolish all key enemy targets: a neologism.

o·ver·lade (ō′vər·lād′) *v.t.* **·lad·ed, ·lad·ed** or **·lad·en, ·lad·ing** To overload: now used chiefly in the past participle.

o·ver·lain (ō′vər·lān′) Past participle of OVERLIE.

o·ver·land (ō′vər·land′) *adj.* Journeying or accomplished by or principally by land. — *adv.* Across, over, or via land. — *n.* An overland stage or train.

overland mail A stagecoach line carrying mail and passengers semiweekly from St. Louis to San Francisco: operated from 1858 until superseded by the railroad in 1869.

overland route A transcontinental route, especially one crossing North America from coast to coast.

overland stage A stagecoach used on any overland route.

o·ver·lap (ō′vər·lap′) *v.t. & v.i.* **·lapped, ·lap·ping** To lie or extend partly over or upon (another or one another); lap over. — *n.* (ō′vər·lap′) 1 The state, condition, or extent of overlapping; also, the part that overlaps. 2 *Geol.* The extension of younger rock strata beyond the limits of the older underlying strata.

o·ver·lay (ō′vər·lā′) *n.* 1 *Printing* A piece of paper placed on the tympan of a press to make the impression heavier at the corresponding part of the form, or to compensate for a depression in the form. 2 Anything that overlies, covers, or partly covers something. 3 Ornamental work overlaid on wood, as with veneers, etc. 4 A sheet of transparent material carrying information of a special or confidential nature to supplement the details of the map on which it is laid. 5 *Scot.* A cravat. — *v.t.* (ō′vər·lā′) **·laid, ·lay·ing** 1 To spread something over, as with a decorative pattern or layer. 2 To lay or place over or upon something else. 3 *Printing* To put an overlay upon. 4 To overburden; weigh down.

o·ver·leaf (ō′vər·lēf′) *adv. & adj.* On the reverse side of a leaf (of paper).

o·ver·leap (ō′vər·lēp′) *v.t.* 1 To leap over or across. 2 To omit; overlook. 3 To leap farther than; outleap. — **to overleap oneself** To miss one's purpose by going too far. [OE *oferhlēapan*]

o·ver·learn·ing (ō′vər·lûr′ning) *n. Psychol.* A degree of learning beyond what is necessary for immediate or adequate use in prescribed situations.

o·ver·lie (ō′vər·lī′) *v.t.* **·lay, ·lain, ·ly·ing** 1 To lie over or upon. 2 To suffocate by lying upon, as a baby.

o·ver·live (ō′vər·liv′) *v.* **·lived, ·liv·ing** *v.t.* To live longer than; survive. — *v.i.* To survive; live too long. [OE *oferlibban*]

o·ver·load (ō′vər·lōd′) *v.t.* To load excessively; overburden. — *n.* (ō′vər·lōd′) 1 An excessive burden. 2 *Electr.* A circuit-breaker.

o·ver·look (ō′vər·lo͝ok′) *v.t.* 1 To fail to see or notice; miss. 2 To disregard purposely or indulgently; ignore. 3 To look over or see from a higher place. 4 To afford a view of: The castle *overlooks* the harbor. 5 To supervise; oversee. 6 To examine or inspect. 7 To look upon or bewitch with the evil eye. See synonyms under PARDON. — *n.* (ō′vər·lo͝ok′) 1 The act of looking over, as from a height; an inspection; survey. 2 Oversight; neglect. 3 The jack bean: so called because believed by West Indian Negroes to serve as a watchman.

o·ver·lord (ō′vər·lôrd′) *n.* 1 In Saxon times, a superior lord or chief who outranked and held authority over other lords. 2 Hence, one who holds supremacy over another. — **o′ver·lord′ship** *n.*

o·ver·loup (ō′vər·lo͞op′) *n. Scot.* A leap over; hence, a trespass.

o·ver·ly (ō′vər·lē) *adv.* To an excessive degree; too much; too.

o·ver·man[1] (ō′vər·mən) *n. pl.* **·men** (-mən) 1 An overseer. 2 An umpire. 3 (ō′vər·man′) A superman.

o·ver·man[2] (ō′vər·man′) *v.t.* **·manned, ·man·ning** To provide with more men than necessary: The ship was *overmanned.*

o·ver·mas·ter (ō′vər·mas′tər, -mäs′-) *v.t.* To overcome; overpower. — **o′ver·mas′ter·ing** *n. & adj.*

o·ver·match (ō′vər·mach′) *v.t.* To be more than a match for; surpass. — *n.* (ō′vər·mach′) 1 One who or that which is superior in strength, skill, etc. 2 A contest in which one party overmatches the other.

o·ver·ma·ture (ō′vər·mə·cho͝or′) *adj.* Denoting the state of a forest in which, as a result of age, the growth of the trees has almost entirely ceased and degeneration has started.

o·ver·much (ō′vər·much′) *adj.* Exceeding what is necessary or proper; too much. — *adv.* In too great a degree. — *n.* An excess; too much.

o·ver·night (ō′vər·nīt′) *adj.* 1 Of or belonging to or done during the previous evening; lasting all night: an *overnight* visit. 2 For use in nighttime travel or for short visits: an *overnight* bag. — *adv.* 1 During or through the night. 2 On the previous evening. — *n.* (ō′vər·nīt′) The previous evening.

o·ver·pass (ō′vər·pas′, -päs′) *v.t.* 1 To pass across, over, or through; cross. 2 To surpass or exceed. 3 To overlook; disregard. 4 To transgress. — *n.* (ō′vər·pas′, -päs′) An elevated section of highway crossing other lines of travel.

o·ver·pay (ō′vər·pā′) *v.t.* **·paid, ·pay·ing** 1 To pay more than (a sum due). 2 To pay (someone) too much. 3 To reward too highly. — **o′ver·pay′ment** *n.*

o·ver·per·suade (ō′vər·pər·swād′) *v.t.* **·suad·ed, ·suad·ing** To persuade (someone) to an action or view, especially against his judgment or inclination. — **o′ver·per·sua′sion** *n.*

o·ver·play (ō′vər·plā′) *v.t.* 1 To play or act (a part or role) to excess; overdo; exaggerate. 2 To outplay; surpass. 3 In golf, to send (the ball) beyond the putting green.

o·ver·plus (ō′vər·plus′) *n.* That which remains after a certain part has been used or set aside; surplus; excess. See synonyms under EXCESS. [Partial trans. of OF *surplus* < *sur-* over + *plus* more]

o·ver·pow·er (ō′vər·pou′ər) *v.t.* 1 To gain supremacy over; subdue. 2 To render wholly helpless or ineffective; overcome. 3 To supply with more power than necessary. See synonyms under CONQUER, REPRESS, SUBDUE. — **o′ver·pow′er·ing** *adj.* — **o′ver·pow′er·ing·ly** *adv.*

o·ver·pres·sure (ō′vər·presh′ər) *n.* Atmospheric pressure greater than normal resulting from any violent blast or explosion, as from an atomic bomb.

o·ver·print (ō′vər·print′) *v.t.* To print additional material of another color on (sheets already printed). — *n.* (ō′vər·print′) 1 Anything printed over another impression. 2 Any word, symbol, etc., printed on a stamp which changes its value or use.

o·ver·prize (ō′vər·prīz′) *v.t.* **·prized, ·priz·ing** To value too highly.

o·ver·pro·duce (ō′vər·prə·do͞os′, -dyo͞os′) *v.t.* **·duced, ·duc·ing** To produce too much of or so as to exceed demand.

o·ver·pro·duc·tion (ō′vər·prə·duk′shən) *n.* Production in excess of demand, or of the possibility of profitable sale.

o·ver·proof (ō′vər·pro͞of′) *adj.* Containing a larger proportion of alcohol than proof spirit; said of alcoholic liquors.

o·ver·pro·por·tion (ō′vər·prə·pôr′shən, -pōr′-) *v.t.* To make or depict in excess of true proportions. — **o′ver·pro·por′tion·ate** *adj.* — **o′ver·pro·por′tion·ate·ly** *adv.* — **o′ver·pro·por′tioned** *adj.*

o·ver·rate (ō′vər·rāt′) *v.t.* **·rat·ed, ·rat·ing** To rate or value too highly; to credit with undue merit; overestimate.

o·ver·reach (ō′vər·rēch′) *v.t.* 1 To reach over or beyond. 2 To spread over; cover. 3 To defeat (oneself), as by trying too hard or being too clever. 4 To miss by stretching or reaching too far. 5 To get the advantage of; outwit; cheat. — *v.i.* 6 To reach too far. 7 To cheat. 8 To hit a toe of the hind foot against the heel of the forefoot: said of horses, etc. See synonyms under DECEIVE. — **o′ver·reach′er** *n.* — **o′ver·reach′ing** *n.*

o·ver·ride (ō′vər·rīd′) *v.t.* **·rode, ·rid·den, ·rid·ing** 1 To ride over or across. 2 To trample down; suppress. 3 To disregard summarily, as if trampling down; supersede; prevail over. 4 To ride (a horse) to exhaustion. 5 *Surg.* To slide over (the corresponding fragment), as one end of a fractured bone. [OE *oferrīdan*]

o·ver·rule (ō′vər·ro͞ol′) *v.t.* **·ruled, ·rul·ing** 1 To decide against or nullify by superior authority; set aside; invalidate. 2 To disallow the arguments of (someone). 3 To have control over; rule. 4 To influence, as to another opinion or course of action; prevail over. — **o′ver·rul′ing** *adj.* — **o′ver·rul′ing·ly** *adv.*

o·ver·run (ō′vər·run′) *v.* **·ran, ·run, ·run·ning** *v.t.* 1 To spread or swarm over, especially harmfully, as vermin or invaders do; ravage; invade; infest. 2 To run or flow over; overflow. 3 To spread rapidly across or throughout: said of fads, ideas, etc. 4 To run beyond; pass the limit of. 5 *Printing* **a** To shift (words, lines of type, etc.) from one line, page, or column to another. **b** To rearrange (matter) in this way. 6 *Archaic* To run faster than; outrun. — *v.i.* 7 To run over; overflow. 8 To pass the usual or desired limit. — *n.* (ō′vər·run′) An instance of overrunning; the amount or extent of overrunning.

o·ver·seas (ō′vər·sēz′) *adv.* Beyond the sea;

overfear
overfearful
overfeatured
overfed
overfeminine
overfierce
overflatten
overflourish
overfluent
overfond
overfondle
overfondness
overfoolish
overfoul
overfrail
overfrank
overfraught
overfree
overfreedom
overfreely
overfrequency
overfrequent
overfrighten
overfruitful
overfull
overfullness
overfunctioning
overfurnish
overgamble
overgeneralize
overgenerous
overgenial
overgentle
overgifted
overglad
overgloomy
overglorious
overgoad
overgodly
overgracious
overgrasping
overgrateful
overgratify
overgraze
overgreasy
overgreediness
overgreedy
overgrieve
overhandle
overhappy
overharass
overharden
overhardy
overharsh
overhasty
overhate
overhaughty
overheartily
overhearty
overheavy
overhelpful
overhigh
overhold
overholy
overhomely
overhonest
overhonor
overhope
overhot
overhotly
overhuman
overhumanize
overhurriedly
overhysterical
overidealism
overidealistic
overidle
overillustrate
overimaginative
overimitate
overimitative
overimpress
overimpressible
overinclination
overinclined
overindividualistic
overindulge
overindulgence
overindulgent
overindustrialize
overinflate
overinflation
overinfluential
overinsistent
overinsolent
overinstruct
overinsure
overintellectual
overintense
overinterest
overinventoried
overinvest
overirrigate
overirrigation
overjealous

abroad. — *adj.* Coming from or for use beyond the sea; foreign. Also **o′ver·sea′**.

overseas cap A garrison cap.

o·ver·see (ō′vər·sē′) *v.t.* **·saw, ·seen, ·see·ing** 1 To direct as supervisor; superintend. 2 To survey; watch. 3 To examine; peruse. [OE *ofersēon*]

o·ver·se·er (ō′vər·sē′ər) *n.* 1 A person who oversees; especially, one who superintends laborers at their work. 2 *Brit.* A parish officer who administrates relief funds: also **overseer of the poor**. See synonyms under MASTER, SUPERINTENDENT.

o·ver·sell (ō′vər·sel′) *v.t.* **·sold, ·sell·ing** 1 To sell to excess. 2 To sell more of (a stock, etc.) than one can deliver or provide a margin for.

o·ver·set (ō′vər·set′) *v.* **·set, ·set·ting** *v.t.* 1 To overcome or disorder mentally or physically; disconcert. 2 *Printing* **a** To set too much (type or copy) in a given space. **b** To set too much type in. 3 *Rare* To cause to overturn; capsize. — *v.i.* 4 To overturn; fall over; spill. — *n.* (ō′vər·set′) 1 A turning over; upset. 2 *Printing* Excess of composed type.

o·ver·sew (ō′vər·sō′, ō′vər·sō′) *v.t.* **·sewed, ·sewed** or **·sewn, ·sew·ing** To sew overhand, especially with close stitches.

o·ver·shade (ō′vər·shād′) *v.t.* **·shad·ed, ·shad·ing** To overshadow.

o·ver·shad·ow (ō′vər·shad′ō) *v.t.* 1 To render unimportant or insignificant by comparison; loom above; dominate. 2 To throw a shadow over; dim; obscure. [OE *ofersceadwian*]

o·ver·shine (ō′vər·shīn′) *v.t.* **·shone, ·shin·ing** 1 To shine over or upon; illumine. 2 To excel in some respect; outshine. [OE *ofersćīnan*]

o·ver·shoe (ō′vər·shōō′) *n.* A shoe worn for protection over another: usually of rubber or felt.

o·ver·shoot (ō′vər·shōōt′) *v.* **·shot, ·shoot·ing** *v.t.* 1 To shoot or go over or beyond. 2 To go beyond; exceed, as a limit. 3 To drive or force (something) beyond the proper limit. — *v.t.* 4 To shoot or go over or beyond the mark. 5 To go too far.

o·ver·shot (ō′vər·shot′) *adj.* 1 Surpassed in any way. 2 Projecting, as the upper jaw beyond the lower jaw. 3 Driven by water flowing over from above: an *overshot* wheel.

overshot wheel A water wheel with buckets that are filled by water from a race over the top, the weight and impetus of the water turning the wheel.

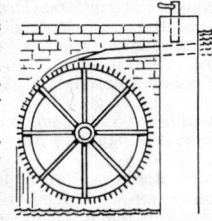
OVERSHOT WHEEL

o·ver·sight (ō′vər·sīt′) *n.* 1 An error due to inattention; an inadvertent mistake or omission. 2 Watchful supervision; superintendence.

Synonyms: care, charge, command, control, direction, inspection, management, superintendence, supervision, surveillance, watch, watchfulness. *Oversight* strictly implies constant personal presence; *superintendence* requires only so much of presence or communication as to know that the superintendent's wishes are carried out; the superintendent of a railroad will personally oversee very few of its operations; the railroad company has supreme *direction* of all its affairs without *super-*intendence or *oversight*. But a person may look over a matter in order to survey it carefully in its entirety, or he may look over it with no attention to the thing itself because his gaze and thought are concentrated on something beyond; *oversight* has thus two contrasted senses, in the latter sense denoting inadvertent error or omission, and in the former denoting watchful *supervision*. *Control* is chiefly used with reference to restraint or the power of restraint. *Surveillance* signifies watching with something of suspicion. See CARE, ERROR, NEGLECT.

o·ver·signed (ō′vər·sīnd′) *n.* The person whose name appears at the head of an article, document, report, etc.: distinguished from *undersigned*.

o·ver·size (ō′vər·sīz′) *adj.* Of a larger size than necessary or larger than normal. — *n.* An exceptionally large size. — **o′ver·sized′** *adj.*

o·ver·skirt (ō′vər·skûrt′) *n.* A skirt or drapery worn over the skirt of a dress.

o·ver·sleep (ō′vər·slēp′) *v.* **·slept, ·sleep·ing** *v.i.* To sleep too long. — *v.t.* To sleep beyond (a specified time).

o·ver·soul (ō′vər·sōl′) *n.* The spiritual being or element of the universe individualized in or uniting together and influencing human souls; the absolute unity, in which subject and object, knower and known, are one: a concept in Emerson's transcendentalist philosophy.

o·ver·spend (ō′vər·spend′) *v.* **·spent, ·spend·ing** *v.t.* 1 To spend more than; exceed. 2 *Archaic* To use up; exhaust. — *v.i.* 3 To spend more than one can afford.

o·ver·spread (ō′vər·spred′) *v.t.* **·spread, ·spread·ing** To spread or extend over; cover completely. [OE *ofersprǣdan*]

o·ver·square (ō′vər·skwâr′) *adj. Mech.* Having the cylinder bore greater than the piston stroke: said of engines.

o·ver·state (ō′vər·stāt′) *v.t.* **·stat·ed, ·stat·ing** To state in too strong terms; exaggerate. — **o′ver·state′ment** *n.*

o·ver·stay (ō′vər·stā′) *v.t.* To stay beyond the limits or duration of.

o·ver·step (ō′vər·step′) *v.t.* **·stepped, ·step·ping** To step over or go beyond; exceed (some limit or restriction). [OE *ofersteppan*]

o·ver·strung (ō′vər·strung′) *adj.* 1 Strung too tensely; too sensitive. 2 *Music* Having two sets of strings crossing obliquely, as the lower-bass strings in a piano.

o·ver·stuffed (ō′vər·stuft′) *adj.* Excessively stuffed; especially, as furniture, completely covered with deep upholstery; heavily upholstered.

o·ver·sup·ply (ō′vər·sə·plī′) *n. pl.* **·plies** An excessive supply. — *v.t.* (ō′vər·sə·plī′) **·plied, ·ply·ing** To supply in excess.

o·vert (ō′vûrt, ō·vûrt′) *adj.* 1 Open to view; outwardly manifest. 2 *Law* Done with criminal intent. [<OF, pp. of *ovrir* open] — **o′vert·ly** *adv.*

Synonyms: see EVIDENT, MANIFEST, NOTORIOUS, OPEN. *Antonyms:* contemplated, hidden, intended, meditated, secret.

o·ver·take (ō′vər·tāk′) *v.t.* **·took, ·tak·en, ·tak·ing** 1 To catch up with. 2 To come upon suddenly. See synonyms under CATCH.

o·ver-the-count·er (ō′vər·thə·koun′tər) *adj.* Not sold on the floor of a stock exchange: said of stocks, bonds, etc.

o·ver·throw (ō′vər·thrō′) *v.t.* **·threw, ·thrown, ·throw·ing** 1 To throw over or down; upset. 2 To bring down or remove from power by force; defeat; ruin. See synonyms under ABOLISH, CONQUER, DEMOLISH, EXTERMINATE, RUIN, SUBVERT. — *n.* (ō′vər·thrō′) 1 The act of overthrowing; destruction; demolition. 2 In cricket and baseball, a throwing of the ball over and beyond the player at whom it is aimed. 3 In cricket, a run obtained as the result of such a throw.

o·ver·thrust (ō′vər·thrust′) *adj. Geol.* Characterized by or belonging to earlier and originally lower rock strata, which by faulting are pushed over later and originally higher strata. Compare FAULT, THRUST.

o·ver·time (ō′vər·tīm′) *v.t.* **·timed, ·tim·ing** *Phot.* To expose too long, as a plate or film. — *n.* (ō′vər·tīm′) Time used in working beyond the specified hours. — *adj.* During or for extra working time: *overtime* pay. — *adv.* Beyond the stipulated time.

o·ver·tone (ō′vər·tōn′) *n.* 1 *Music* A harmonic: so called because it is heard with and above the fundamental tone produced by a musical instrument. 2 The color of the light reflected by a painted surface. 3 The associations, connotations, implications, etc., of language, thoughts, etc. [<G *oberton*]

o·ver·top (ō′vər·top′) *v.t.* **·topped, ·top·ping** 1 To rise above the top of; tower over. 2 To surpass; excel.

o·ver·topped (ō′vər·topt′) *adj.* Having the crown shaded from above by other contiguous trees: said of a tree.

o·ver·trade (ō′vər·trād′) *v.i.* **·trad·ed, ·trad·ing** To trade beyond one's capital or the requirements of the market.

o·ver·trick (ō′vər·trik′) *n.* In card games, a trick more than game or than the number bid.

o·ver·trump (ō′vər·trump′) *v.t.* In card games, to trump (a player or a trump played) with a higher trump.

o·ver·ture (ō′vər·chər) *n.* 1 *Music* **a** An instrumental prelude, as to an opera or other large work. **b** An orchestral piece of varying form, usually illustrating a dramatic or graphic theme. 2 A proposal intended to lead to further negotiations by expressing willingness to make terms; also, the proposal submitted. See synonyms under PROPOSAL. — *v.t.* **·tured, ·tur·ing** 1 To offer as an overture or proposal. 2 To introduce with or as an overture. [<OF <*ouvert*. See OVERT.]

o·ver·turn (ō′vər·tûrn′) *v.t.* 1 To turn or throw over; capsize; upset. 2 To destroy the power of; overthrow; defeat; ruin. — *v.i.* 3 To turn over; capsize; upset. See synonyms under DEMOLISH, SUBVERT. — *n.* (ō′vər·tûrn′) 1 The act of overturning or the state of being overturned; an upset; overthrow. 2 A subversion or destruction. 3 Turnover (def. 5).

o·ver·volt·age (ō′vər·vōl′tij) *n.* 1 *Electr.* Excess voltage in a circuit. 2 *Chem.* The minimum potential at which a particular electrolytic reaction will take place at an appreciable rate on an electrode.

o·ver·watch (ō′vər·woch′, -wôch′) *v.t.* 1 To watch over. 2 To weary with watching.

o·ver·wear (ō′vər·wâr′) *v.t.* **wore, ·worn, ·wear·ing** 1 To wear out. 2 To outgrow.

o·ver·wea·ry (ō′vər·wir′ē) *adj.* Overtired; exhausted. — *v.t.* **·ried, ·ry·ing** To tire to excess.

o·ver·weath·er (ō′vər·weth′ər) *adj. Aeron.* Denoting flight conditions or activities which are not affected by storm, overcast, or other meteorological phenomena.

o·ver·ween·ing (ō′vər·wē′ning) *adj.* Characterized by presumptuous pride or conceit; arrogant; excessive; exaggerated. — *n.* Overconfidence; presumption. [OE *oferwēnan*

overjocular	overlawful	overlogical	overmeasure	overmortgage	overnumerous	overparticular
overjoyful	overlax	overloud	overmeek	overmournful	overobedient	overpassionate
overjoyous	overlaxness	overlove	overmellow	overmultiply	overobese	overpatient
overjudicious	overlearnedness	overloyal	overmelt	overnarrow	overoblige	overpatriotic
overjust	overlet	overluscious	overmerciful	overnear	overobsequious	overpensive
overkeen	overlewd	overlustiness	overmerrily	overneat	overobsequiousness	overpert
overkind	overliberal	overlusty	overmerry	overneglect	overoffensive	overpessimistic
overknowing	overliberality	overluxuriant	overmighty	overnegligence	overofficious	overpiteous
overlabor	overlighted	overluxurious	overmild	overnegligent	overoften	overplain
overlade	overlinger	overmagnify	overminutely	overnervous	overoptimistic	overplausible
overlarge	overliterary	overmany	overminuteness	overnervousness	overornamented	overplease
overlate	overliveliness	overmasterful	overmix	overnew	overpainful	overplentiful
overlaudatory	overlively	overmasterfulness	overmodest	overnimble	overpamper	overplenty
overlaunch	overload	overmature	overmoist	overnotable	overpartial	overplump
overlavish	overlofty	overmeanness	overmoisten	overnourish	overpartiality	overpointed

add, āce, câre, pälm; end, ēven; it, īce; odd, ōpen, ôrder; tōōk, pōōl; up, bûrn; ə = a in *above*, e in *sicken*, i in *clarity*, o in *melon*, u in *focus*; yōō = u in *fuse*; oi, oil; ou, pout; ch, check; g, go; ng, ring; th, thin; ᵺ, this; zh, vision. Foreign sounds à, œ, ü, kh, ṅ; and ◆: see page xx. < from; + plus; ? possibly.

< *ofer-* over + *wēnan* think] — **o′ver·ween′ing·ly** *adv.*

o·ver·weigh (ō′vər·wā′) *v.t.* **1** To outweigh; overbalance. **2** To overburden; oppress.

o·ver·weight (ō′vər·wāt′) *n.* **1** Excess of weight, as beyond the legal or customary amount: to give *overweight.* **2** Greater weight; preponderance; also, more than normal weight. — *adj.* Being more than the usual or permitted weight. — *v.t.* (ō′vər·wāt′) To weigh down; overburden.

o·ver·whelm (ō′vər·hwelm′) *v.t.* **1** To bury or submerge completely, as with a wave or flood. **2** To overcome or defeat by or as by irresistible force or numbers; crush; render helpless. **3** *Obs.* To overthrow. See synonyms under BURY[1], HIDE[1], INUNDATE, INVOLVE, SUBDUE.

o·ver·whelm·ing (ō′vər·hwel′ming) *adj.* Crushing by reason of force, weight, or numbers; irresistible. — **o′ver·whelm′ing·ly** *adv.*

o·ver·wind (ō′vər·wīnd′) *v.t.* **·wound, ·wind·ing** **1** To wind too far or too tightly, as a watch. **2** *Electr.* To wind (the magnet of a motor) in order to produce a maximum magnetism with a smaller current than is normally required.

o·ver·word (ō′vər·wûrd′) *n.* A word repeated; also, the burden of a song; refrain. See OVERWORD.

o·ver·work (ō′vər·wûrk′) *v.* **·worked** or **·wrought, ·work·ing** *v.t.* **1** To cause to work too hard; exhaust with work or use. **2** To work on or elaborate excessively: to *overwork* an argument. **3** To work up or excite excessively. — *v.i.* **4** To work too hard; do too much work. — *n.* (ō′vər·wûrk′) **1** Work done in overtime, or in excess of the stipulated amount. **2** Excessive work.

o·ver·write (ō′vər·rīt′) *v.t.* & *v.i.* **·wrote, ·writ·ten, ·writ·ing** **1** To write over other writing. **2** To write in too elaborate or labored a style. **3** To write too much about (a subject) or at too great length.

o·ver·wrought (ō′vər·rôt′) *adj.* **1** Worked up or excited excessively; overstrained: *overwrought* feelings. **2** Worked all over, as with embroidery. **3** Worked too hard. **4** Too elaborate; overdone. [<OVERWORK]

ovi– *combining form* Egg; of or pertaining to eggs: *oviparous.* Also **ovo–**. [<L *ovum* an egg]

o·vi·bos (ō′və·bos) *n.* The musk ox. [<NL <L *ovis* sheep + *bos* cow]

Ov·id (ov′id), 43 B.C.–A.D. 17, Roman poet: full name *Publius Ovidius Naso.* — **O·vid·i·an** (ō·vid′ē·ən) *adj.*

o·vi·duct (ō′vi·dukt) *n. Anat.* The passage by which the ova are conveyed from the ovary to the uterus, as the Fallopian tube.

O·vie·do (ō·vyā′thō) A city of NW Spain.

o·vif·er·ous (ō·vif′ər·əs) *adj.* Bearing or holding eggs. [<OVI- + -FEROUS]

o·vi·form (ō′vi·fôrm) *adj.* Having the form of an egg or ovum.

o·vine (ō′vīn, ō′vin) *adj.* Of or pertaining to a sheep; sheeplike. — *n.* An ovine animal. [<L *ovinus* < *ovis* sheep]

o·vip·a·ra (ō·vip′ər·ə) *n. pl.* Animals that lay eggs. [<NL <L *oviparus* laying eggs < *ovum* egg + *-parus* < *parere* bring forth]

o·vip·a·rous (ō·vip′ər·əs) *adj. Biol.* Producing eggs or ova that mature and are hatched outside the body: contrasted with *ovoviviparous*: opposed to *viviparous.* — **o·vip′a·rous·ly** *adv.* — **o·vip′a·rous·ness** *n.*

o·vi·pos·it (ō′vi·poz′it) *v.i.* **1** *Biol.* To lay eggs. **2** *Entomol.* To deposit eggs by means of an ovipositor. [<OVI- + L *positus,* pp. of *ponere* place] — **o·vi·po·si·tion** (ō′vi·pə·zish′·ən) *n.*

o·vi·pos·i·tor (ō′vi·poz′ə·tər) *n. Entomol.* The tubular organ at the extremity of the abdomen in many insects, by which the eggs are deposited: sometimes modified as a sting, as in bees and wasps.

o·vi·sac (ō′vi·sak) *n. Anat.* The closed capsule in which ova are developed within the ovary; a Graafian vesicle. [< OVI- + SAC]

o·vo·glob·u·lin (ō′və·glob′yə·lin) *n. Biochem.* A globulin from egg yolk.

o·void (ō′void) *adj.* Egg-shaped: also **o·voi′dal.** — *n.* An egg-shaped body.

o·vo·lo (ō′və·lō) *n. pl.* **·li** (-lē) *Archit.* A convex molding: usually, a quarter of a circle or ellipse. [<Ital., dim. of *ovo* egg <L *ovum*]

o·vo·tes·tis (ō′və·tes′tis) *n. Zool.* The bisexual reproductive gland of certain gastropods, as the snail. [< *ovo-,* var. of OVI- + TESTIS]

o·vo·vi·vip·a·rous (ō′vō·vī·vip′ər·əs) *adj. Zool.* Producing eggs that are incubated and hatched within the parent's body, as some reptiles and fishes. [<OVO- + VIVIPAROUS] — **o′vo·vi·vip′a·rous·ly** *adv.* — **o′vo·vi·vip′a·rous·ness** *n.*

o·vu·late (ō′vyə·lāt) *v.i.* **·lat·ed, ·lat·ing** To produce ova; discharge ova from an ovary. [<NL *ovulum,* dim. of L *ovum* an egg]

o·vu·la·tion (ō′vyə·lā′shən) *n. Biol.* The formation and discharge of ova; the period when this occurs.

o·vule (ō′vyōōl) *n.* **1** *Bot.* The rudimentary seed of a plant; the body within the ovary which, upon fertilization, becomes the seed. **2** A small ovum. [<F, dim. of L *ovum* an egg] — **o′vu·lar, o·vu·lar·y** (ō′vyə·ler′ē) *adj.*

o·vum (ō′vəm) *n. pl.* **o·va** (ō′və) **1** *Biol.* **a** A cell formed in the ovary; an egg. **b** An ovule. **2** *Archit.* An egg-shaped ornament. [<L]

owe (ō) *v.* **owed** (*Obs.* **ought**), **ow·ing** *v.t.* **1** To be indebted to the amount of; be obligated to pay or repay. **2** To be obligated to render or offer: to *owe* an apology. **3** To have or possess by virtue of gift, labor, etc.: with *to*: He *owes* his success to his own efforts. **4** To cherish (a certain feeling) toward another: to *owe* a grudge. **5** *Obs.* To own; have. — *v.i.* **6** To be in debt. [OE *āgan*]

Ow·en (ō′in) A masculine personal name. [< Welsh, young warrior]

Owen, Robert, 1771–1858, Welsh social reformer. — **Wilfred,** 1893–1918, English poet.

Owen Falls A waterfall (65 feet high) in the Victoria Nile in SE Uganda; site of a dam 2,725 feet long, 85 feet high, built in 1949.

Ow·ens River (ō′inz) A river in eastern California, flowing 120 miles SE and supplying water to the city of Los Angeles.

Owen Stanley Range A mountain range in SE New Guinea; highest peak, 13,240 feet.

ow·er·word (ou′ər·wûrd′) *n. Scot.* The burden of a song; refrain.

ow·ing (ō′ing) *adj.* Due; yet to be paid: six dollars *owing.*

owing to Attributable to; on account of; in consequence of.

owl (oul) *n.* **1** A predatory nocturnal bird of the order *Strigiformes,* having large eyes and head, short, sharply hooked bill, long powerful claws, and a circular facial disk of radiating feathers. Of the North American owls, prominent species are the circumpolar **snowy owl** (genus *Nyctea*), the **great horned owl,** the **barred owl** *(Syrnium varium),* the **great gray owl** (genus *Scotiaptex*), **screech owl** (genus *Otus*), and the **long-eared** and **short-eared owls** (genus *Asio*). **2** One of a breed of domestic pigeons having an owl-like head and a prominent frill. **3** A person with nocturnal habits. **4** A person of solemn appearance, etc. [OE *ūle*]

owl·et (ou′lit) *n.* **1** A small owl: the European owlet *(Athene noctua),* or a similar Oriental species *(A. brama)* of India. **2** A young owl.

owl·ish (ou′lish) *adj.* **1** Like an owl; grave. **2** Nocturnally active. **3** *Brit. Dial.* Stupid. — **owl′ish·ly** *adv.* — **owl′ish·ness** *n.*

owl's-clo·ver (oulz′klō′vər) *n.* Any of a genus *(Orthocarpus)* of herbs of the figwort family of western North and South America, especially a California species *(O. purpurascens)* with crimson or purple flowers.

own (ōn) *adj.* **1** Belonging to oneself; peculiar; particular; individual: following the possessive (usually a possessive pronoun) as an intensive to express ownership, interest, or individual peculiarity with emphasis, or to indicate the exclusion of others: my *own* horse; his *own* idea; It is my *own*: in this sense often with ellipsis of the noun. **2** Being of the nearest degree: *own* cousin. — **to come into one's own** **1** To obtain possession of one's property. **2** To receive one's reward; come into one's rightful position. — **to hold one's own** **1** To maintain one's place or position. **2** To keep up with one's work, or remain undefeated. — **on one's own** Entirely dependent on one's self for support or success. — *v.t.* **1** To have or hold as one's own; have as a belonging; possess. **2** To admit or acknowledge. — *v.i.* **3** To confess. See synonyms under ACKNOWLEDGE, AVOW, CONFESS, HAVE. [OE *āgen,* orig. pp. of *āgan* owe, possess] — **own′a·ble** *adj.*

own·er (ō′nər) *n.* One who has the legal title or right to or has possession of a thing. See synonyms under MASTER. — **own′er·less** *adj.*

own·er·ship (ō′nər·ship) *n.* The state of being an owner; proprietorship; also, legal title. ◆ Collateral adjective: *allodial.* See synonyms under PROPERTY.

owre-hip (our′hip) *n. Scot.* A way of striking a blow with a hammer swung from the hip.

owse (ouz) *n. Scot.* Ox.

O·wy·hee River (ō·wī′hē) A river in SE Oregon, flowing 300 miles north to the Snake River.

ox (oks) *n. pl.* **ox·en** (ok′sən) **1** An adult castrated male of a domestic bovine quadruped. **2** A bovine quadruped, as a buffalo, bison, or yak; specifically, the common domesticated *Bos taurus,* or the zebu or Indian ox *(Bos indicus).* ◆ Collateral adjective: *bovine.* [OE *oxa*]

oxa– *combining form Chem.* Denoting the presence of oxygen, especially as replacing carbon in a ring compound. Also, before vowels, **ox–**, as in *oxazine.* [<OXYGEN]

ox·a·late (ok′sə·lāt) *n. Chem.* A salt or ester of oxalic acid.

ox·al·ic (ok·sal′ik) *adj.* Pertaining to or derived from the oxalis or sorrel. [<F *oxalique* <L *oxalis* OXALIS]

oxalic acid *Chem.* A white, crystalline, poisonous compound, $C_2H_2O_4 \cdot 2H_2O$, found extensively in plant tissues as oxalates, and made artificially in various ways, as by decomposing sugar with nitric acid: used in bleaching and in dyeing, and for removing ink stains from paper, linen, etc.

ox·a·lis (ok′sə·lis) *n.* A plant of a large, widely distributed genus *(Oxalis)* of mostly stemless herbs of the wood-sorrel family, with purple, rose, or white flowers; wood sorrel. [<L <Gk. < *oxys* sharp, acid]

ox·a·lu·ri·a (ok′sə·lŏŏr′ē·ə) *n. Pathol.* Excess of calcium oxalate in the urine. [< OXAL(ATE) + -URIA]

ox·a·lyl (ok′sə·lil) *n. Chem.* The bivalent OCCO radical: *Oxalyl* chloride, $C_2O_2Cl_2$, is a colorless, fuming, poisonous liquid obtained from oxalic acid.

ox·blood (oks′blud′) *n.* A monochrome glaze

overpolemical	**overproficient**	**overpunish**	**overreflective**	**overrigorous**	**overscrupulous**	**overshort**
overpolish	**overprolific**	**overpunishment**	**overrelax**	**overripe**	**overseason**	**overshorten**
overponderous	**overprominent**	**overquick**	**overreliant**	**overroast**	**overseasoned**	**overshrink**
overpopular	**overprompt**	**overquiet**	**overreligious**	**overrough**	**oversecure**	**oversick**
overpopulous	**overpromptness**	**overquietness**	**overrepresent**	**overrude**	**oversensible**	**oversilent**
overpositive	**overprosperous**	**overrapturize**	**overreserved**	**oversad**	**oversensitive**	**oversimple**
overpowerful	**overprotect**	**overrash**	**overresolute**	**oversalt**	**oversententious**	**oversimplicity**
overpraise	**overprotract**	**overrational**	**overrestrain**	**oversalty**	**oversentimental**	**oversimplification**
overprecise	**overproud**	**overrationalize**	**overretention**	**oversanguine**	**overserious**	**oversimplify**
overpreciseness	**overprovide**	**overreadiness**	**overreward**	**oversaturate**	**overservile**	**overskeptical**
overpreoccupation	**overprovision**	**overready**	**overrich**	**oversaturation**	**oversettled**	**overslander**
overpress	**overprovocation**	**overrealism**	**overrife**	**oversaucy**	**oversevere**	**overslow**
overpresumptuous	**overprovoke**	**overrealistic**	**overrighteous**	**overscare**	**overseverely**	**oversmall**
overprocrastination	**overpublic**	**overrefinement**	**overrighteousness**	**overscented**	**overseverity**	**oversmooth**
overproductive	**overpublicity**	**overreflection**	**overrigid**	**overscrub**	**oversharp**	**oversoak**

ox·bow (oks'bō') n. 1 A bent piece of wood in an ox yoke, that forms a collar for the ox. 2 A bend in a river shaped like this.

Ox·en·stiern (ok'sən·stirn), **Count Axel**, 1583–1654, Swedish statesman. Also **Ox·en·stier·na** (ok'sən·stir'nä).

ox·eye (oks'ī') n. 1 Any of several plants of various genera of the composite family, especially any species of *Buphthalmum*, with large yellow heads. 2 The oxeye daisy. 3 One of various birds, as the least sandpiper. 4 An oval dormer window.

ox-eyed (oks'īd') adj. Having large, calm eyes like those of an ox.

oxeye daisy An erect perennial weed (*Chrysanthemum leucanthemum*) with oblong leaves and solitary white flowers with yellow centers.

Ox·ford (oks'fərd) 1 A county of south central England; 749 square miles. Also **Ox'ford·shire** (-shir). 2 Its county town, on the Thames; site of **Oxford University**, established in the 12th century.

oxford gray 1 A very dark gray. 2 A woolen fabric of this color.

Oxford Group movement See BUCHMANISM.

Oxford movement See TRACTARIANISM.

Oxford shoe A low, laced shoe tied at the instep. Also **Oxford tie**.

Oxford unit That quantity of penicillin which, when dissolved in 50 cubic centimeters of meat extract, will completely inhibit growth in a test strain of *Staphylococcus aureus*. Also called *Florey unit*.

ox·heart (oks'härt) n. A variety of sweet cherry.

ox·i·dase (ok'si·dās) n. *Biochem.* One of many oxidizing ferments found widely distributed in plant and animal tissues. [<OXID(E) + -ASE] — **ox'i·da'sic** adj.

ox·i·date (ok'sə·dāt) v.t. & v.i. ·dat·ed, ·dat·ing To oxidize.

ox·i·da·tion (ok'sə·dā'shən) n. *Chem.* 1 The act of uniting or of causing a substance to unite with oxygen. 2 The state of being so united. 3 Any changes in an element or a compound that result in addition to it of a negative radical or a relative decrease of the positive constituent. 4 The process by which the atoms of an element lose electrons: opposed to *reduction*. — **ox'i·da'tive** adj.

ox·ide (ok'sīd, -sid) n. *Chem.* Any binary compound of oxygen either with an element or with an organic radical, as iron rust. Also **ox·id** (ok'sid). [<F *ox(ygène)* oxygen + (ac)ide acid]

ox·i·dize (ok'sə·dīz) v. ·dized, ·diz·ing v.t. *Chem.* 1 To unite with oxygen; cause the oxidation of; rust. 2 To add an electronegative element or radical to, or to decrease by an electropositive element or radical. 3 To remove electrons from (an element or compound). 4 To change (an element) to a higher valence: to *oxidize* ferrous iron to ferric iron. — v.i. 5 *Chem.* To become oxidized. Also *Brit.* **ox'i·dise**. — **ox'i·diz'a·ble** adj.

ox·i·diz·er (ok'sə·dī'zər) n. That which oxidizes or produces oxidation, as an oxygen compound that frees its oxygen easily.

ox·ime (ok'sēm, -sim) n. *Chem.* One of a series of compounds containing the group C-NOH, formed by the action of hydroxylamine on aldehydes, ketones, and ketonic compounds. Also **ox·im** (ok'sim). [<OX(YGEN) + IM(IDE)]

ox·lip (oks'lip') n. 1 A species of primrose (*Primula elatior*), closely resembling the cowslip. 2 A hybrid primrose.

Ox·o·ni·an (ok·sō'nē·ən) adj. Of or pertaining to Oxford, England, or to its university. — n. A student or graduate of Oxford University. [<LL *Oxonia* Oxford]

ox·o·ni·um compound (ok·sō'nē·əm) *Chem.* Any of a class of organic compounds containing a basic oxygen atom combined with a mineral acid.

ox·peck·er (oks'pek'ər) n. An African bird of the starling family (genus *Buphagus*) that devours the parasites on oxen, etc.

ox·tail (oks'tāl') n. The tail of an ox, especially when skinned for use in soup.

ox·ter (ok'stər) n. *Brit. Dial. & Scot.* The armpit.

ox·tongue (oks'tung') n. 1 Any of various plants having rough, tongue-shaped leaves, as the European alkanet or bugloss. 2 A short broadsword.

Ox·us River (ok'səs) The ancient name for the AMU DARYA.

oxy-[1] *combining form* 1 Sharp; pointed; keen: *oxytone*. 2 Acid: *oxygen*. [<Gk. *oxys* sharp, keen]

oxy-[2] *combining form Chem.* 1 Oxygen; of or containing oxygen, or one of its compounds: *oxyphyte*. 2 An oxidation product of: *oxysulfide*. 3 Containing the hydroxyl group: *oxyacid*. [<OXYGEN]

ox·y·a·cet·y·lene (ok'sē·ə·set'ə·lēn) adj. Designating or pertaining to a mixture of acetylene and oxygen, used to obtain high temperatures, as in welding and blowpipe analysis.

ox·y·ac·id (ok'sē·as'id) n. *Chem.* 1 An acid containing oxygen: contrasted with *hydracid*. 2 A hydroxy acid.

ox·y·cal·ci·um (ok'si·kal'sē·əm) adj. Of, pertaining to, or containing oxygen and calcium; especially, designating the action of the oxyhydrogen flame on lime, as in the calcium light.

ox·y·ceph·a·ly (ok'si·sef'ə·lē) n. The condition of having the skull conical in the upper frontal region. Also **ox'y·ce·pha'li·a** (-sə·fā'lē·ə). [<OXY-[1] + Gk. *kephalos* head] — **ox'y·ce·phal'ic** (-sə·fal'ik), **ox'y·ceph'a·lous** adj. — **ox'y·ceph'a·lism** n.

ox·y·gen (ok'sə·jin) n. A colorless, tasteless, and inodorous gaseous element (symbol O), occurring free in the atmosphere, of which it forms 21 percent by volume. It is slightly heavier than air, sparingly soluble in water, a very abundant and active element, essential to combustion and in the respiration of plants and animals. See ELEMENT. [<F *oxygène* < *oxy-* OXY-[1] + *-gène* -GEN; so called because formerly considered essential to all acids]

oxygen acid An oxyacid.

ox·y·gen·ate (ok'sə·jən·āt') v.t. ·at·ed, ·at·ing To treat, combine, or impregnate with oxygen; oxidize. — **ox'y·gen·a'tion** n.

ox·y·gen·ic (ok'sə·jen'ik) adj. Of, pertaining to, resembling, or containing oxygen. Also **ox·yg·e·nous** (ok·sij'ə·nəs).

ox·y·gen·ize (ok'sə·jən·īz') v.t. ·ized, ·iz·ing To oxidize; oxygenate.

oxygen point *Physics* The boiling point of liquid oxygen at standard atmospheric pressure, -182.97° C.: one of the fixed points of the international temperature scale.

oxygen tent A tentlike chamber placed over a patient's head and shoulders and supplied with oxygen for the purpose of facilitating his respiration.

ox·y·gon (ok'sə·gon) n. *Geom.* A triangle with three acute angles. [<OXY-[1] + Gk. *gōnia* corner] — **ox·yg'o·nal** (ok·sig'ə·nəl) adj.

ox·y·he·mo·glo·bin (ok'si·hē'mə·glō'bin) n. *Biochem.* A combination of oxygen and hemoglobin, formed in the red blood corpuscles of the pulmonary capillaries.

ox·y·hy·dro·gen (ok'si·hī'drə·jən) adj. Of, pertaining to, or using a mixture of oxygen and hydrogen, especially for the production of very high temperatures. — n. A mixture of oxygen and hydrogen.

oxyhydrogen blowpipe A blowpipe in which jets of oxygen and hydrogen are combined in order to obtain very high temperatures: used in welding.

ox·y·mel (ok'sə·mel) n. A mixture of honey and vinegar boiled to a sirup. [<L *oxymel* <Gk. *oxymeli* <*oxys* acid + *meli* honey]

ox·y·mo·ron (ok'si·môr'on, -mō'ron) n. pl. **·mo·ra** (-môr'ə, -mō'rə) A figure of speech consisting of that form of antithesis in which, for emphasis or in an epigram, contradictory terms are brought sharply together, as in the phrase, "O heavy lightness, serious vanity!" [<Gk. *oxymōron*, neut. of *oxymōros* <*oxys* keen + *mōros* foolish]

ox·yn·tic (ok·sin'tik) adj. *Physiol.* Acid-secreting: said especially of certain cells of the stomach. [<Gk. *oxynein* sharpen]

ox yoke A yoke consisting of a heavy piece of wood which lies over the necks of the oxen and from which depend two oxbows.

ox·y·phyte (ok'si·fīt) n. *Bot.* A plant adapted to soil which lacks oxygen.

OX YOKE
With 2 oxbows.

Ox·y·rhyn·cus (ok'sə·ring'kəs) An excavation site of Upper Egypt near the Nile. Also **Ox'y·rhyn'chus**.

ox·y·salt (ok'si·sôlt) n. *Chem.* A salt of an acidic oxide.

ox·y·sul·fide (ok'si·sul'fīd) n. *Chem.* A compound of a sulfide with an oxide in which part of the sulfur has been replaced by oxygen. Also **ox'y·sul'phide**.

ox·y·toc·ic (ok'si·tos'ik, -tō'sik) adj. *Med.* Bringing on or hastening parturition. — n. A medicine efficacious in hastening parturition. [<OXY-[1] + Gk. *tokos* birth]

ox·y·to·cin (ok'si·tō'sin) n. 1 *Med.* Any drug which stimulates movements of the uterus, as ergotine. 2 *Physiol.* A hormone of the posterior lobe of the pituitary gland, believed to facilitate uterine contractions during childbirth.

ox·y·tone (ok'si·tōn) adj. 1 Having the acute accent on the last syllable. 2 Causing a preceding word to take an acute accent. — n. A word thus accented. [<Gk. *oxytonos* <*oxys* sharp + *tonos* pitch] — **ox'y·ton'i·cal** (-ton'i·kəl) adj.

oy (oi) n. *Scot.* A grandchild. Also **oye**.

o·yer (ō'yər, oi'ər) n. *Law* 1 A hearing or trial of causes; an assize; formerly, in pleading, a petition by a party to an action, praying that he might hear read to him a deed held by the opposite party; in modern practice, the production of such a document, or a copy thereof, by the party holding it. 2 Oyer and terminer: a contracted form. [<AF *oyer* <OF *oir*, *oyr*, ult. <L *audire* hear]

oyer and ter·min·er (tûr'mə·nər) 1 *Brit.* A court composed of two or more judges of assize held at least twice a year in each county. 2 *U.S.* A court of higher criminal jurisdiction. [<F, to hear and determine]

o·yez (ō'yes, ō'yez) *interj.* Hear! hear ye! an introductory word to call attention to a proclamation, as by a court crier: usually thrice

oversoft	overspeed	overstretch	oversuspiciously	overtenderness	overtrim	overvigorous
oversoftness	overspeedily	overstrict	oversweet	overtense	overtrust	overviolent
oversolemn	oversqueamishness	overstrident	oversystematic	overtension	overtrustful	overwarm
oversolicitous	overstale	overstriving	oversystematize	overthick	overtruthful	overwarmed
oversoon	overstaring	overstrong	overtalkative	overthin	overunionized	overwary
oversoothing	overstately	overstudious	overtalkativeness	overthoughtful	overurbanization	overwealthy
oversophisticated	oversteadfast	overstudiousness	overtame	overthrifty	overurge	overwet
oversophistication	oversteadfastness	oversublime	overtart	overthrong	over-use	overwilling
overspacious	oversteady	oversubtle	overtask	overthrust	overvaluable	overwily
oversparingly	overstiff	oversubtlety	overtaxation	overtight	overvaluation	overwise
overspecialization	overstimulate	oversufficient	overteach	overtimid	overvariety	overworry
overspecialize	overstimulation	oversuperstitious	overtechnical	overtimorous	overvehement	overworship
overspeculate	overstir	oversure	overtedious	overtinseled	overventilate	overyoung
overspeculation	overstout	oversusceptible	overtenacious	overtire	overventuresome	overyouthful
overspeculative	overstress	oversuspicious	overtender	overtrain	overventurous	overzealous

add, āce, câre, pälm; end, ēven; it, īce; odd, ōpen, ôrder; tŏŏk, pōōl; up, bûrn; ə = a in *above*, e in *sicken*, i in *clarity*, o in *melon*, u in *focus*; yōō = u in *fuse*; oi, oil; ou, pout; ch, check; g, go; ng, ring; th, thin; ŧh, this; zh, vision. Foreign sounds à, œ, ü, kh, ń; and ◆: see page xx. < from; + plus; ? possibly.

repeated. Also **o'yes.** [<OF *oyez*, imperative of *oir*. See OYER.]
oys·ter (ois'tər) *n.*
1 A bivalve of the genus *Ostrea* or the family *Ostreidae*, found in salt and brackish water and moored by the shell to stones, other shells, etc., as the common edible species of Europe and America, respectively. **2** Some other analogous bivalve, as the pearl oyster (*Pinctada margaritifera*). **3** The morsel of dark meat found in the hollow of the bone on both sides of the back of a fowl. **4** Some delicacy; titbit; prize. **5** *Slang* A very uncommunicative person. — *v.i.* To gather or farm oysters. [<OF *oistre* <L *ostrea* <Gk. *ostreon*]
oyster bed A place where oysters breed or are grown.
oyster catcher A shore bird of the genus *Haematopus*; especially, the American *H. palliatus*, about 20 inches long, having black-and-white plumage and red feet and bill. It feeds mainly upon small mollusks caught between tide marks.

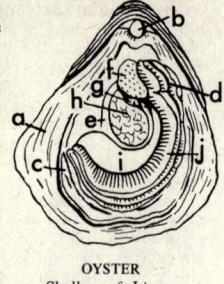

OYSTER
a. Shell. f. Liver.
b. Hinge. g. Heart.
c. Mantle. h. Adductor.
d. Palpi. i. Stomach.
e. Anus. j. Gills.

oyster crab A smooth-bodied crab (*Pinnotheres ostreum*) living symbiotically in the mantle of the oyster.
oyster cracker A small biscuit or hard, salted cracker served with oysters.
oys·ter·ing (ois'tər·ing) *n.* The gathering or farming of oysters.
oys·ter·man (ois'tər·mən) *n. pl.* **·men** (-mən)
1 A man who dredges for, raises, or sells oysters. **2** A vessel engaged in the oyster trade.
oyster plant 1 Salsify. **2** The sea lungwort.
oyster planting The placing of small oysters on submerged artificial beds for propagation.
oy·ster-root (ois'tər·root', -root') *n.* Salsify.
oyster seed See SEED OYSTER.
oyster white Any of several very light gray tints: also called *off-white*.
O·zark Mountains (ō'zärk) The hilly uplands in SW Missouri, NW Arkansas, and NE Oklahoma; highest point, 2,500 feet. Also **Ozark Plateau.**
O·zarks (ō'zärks), **Lake of the** An artificial lake in the Osage River, central Missouri; 130 miles long.
Ozark State Nickname of MISSOURI.
o·ze·na (ō·zē'nə) *n. Pathol.* An evil-smelling ulceration of the nasal cavities. Also **o·zae'na.** [<L *ozaena* <Gk. *ozaina* <*ozein* smell] — **o·ze'nic, o·zae'nic** *adj.*
o·zo·ce·rite (ō·zō'sə·rīt, -sə-, ō'zō·sir'īt) *n.* A waxy, translucent mixture of natural paraffins, or hydrocarbons. It is colorless to white when pure, but otherwise often leek-green, yellowish, brownish-yellow, or brown. It is used extensively as a purified paraffin. [<Gk. *ozein* smell + *keros* wax]

o·zo·na·tion (ō'zō·nā'shən) *n.* The act or process of producing or treating with ozone.
o·zone (ō'zōn) *n.* A blue gas with a pungent odor like that of chlorine, formed variously, as by the passage of electricity through the air, and regarded as an allotropic form of oxygen containing three atoms in the molecule (O_3). It is unstable and is a powerful oxidizing agent, being much more active than ordinary oxygen: employed for bleaching oils, waxes, ivory, flour, and starch, and for sterilizing drinking water. [<F <Gk. *ozein* smell] —
o·zon·ic (ō·zon'ik, ō·zō'nik), **o·zo·nous** (ō'zə·nəs) *adj.*
ozone paper A filter paper coated with a mixture of starch and potassium iodide, which turns blue when exposed to the action of ozone.
ozonic ether A solution in ether of hydrogen peroxide and alcohol.
o·zo·nide (ō'zō·nīd) *n. Chem.* Any of a group of unstable, sometimes violently explosive, organic compounds containing ozone held in a double bond.
o·zo·nize (ō'zō·nīz) *v.t.* **·nized, ·niz·ing 1** To treat or charge with ozone. **2** To convert (oxygen) into ozone.
o·zo·niz·er (ō'zō·nī'zər) *n.* An apparatus for generating ozone.
o·zon·o·sphere (ō·zon'ə·sfir) *n.* A narrow layer in the stratosphere at a height of about 20 miles and containing an unusual concentration of ozone formed by the action of ultraviolet solar radiation on oxygen. Also called **ozone layer.** [<*ozono*- (<OZONE) + (STRATO)SPHERE]

P

p, P (pē) *n. pl.* **p's, P's** or **ps, Ps, pees** (pēz)
1 The 16th letter of the English alphabet: from Phoenician *pe*, Greek *pi*, Roman *P.* **2** The sound of the letter *p*, the voiceless bilabial stop. See ALPHABET. — *symbol* **1** *Chem.* Phosphorus (P). **2** *Genetics* The parental generation: followed by a subscript numeral, as P_1, P_2, to indicate the first, second, etc., parental generation. — **to mind one's P's and Q's** To be careful of one's behavior.
pa (pä) *n. Colloq.* Papa.
Paan (pän) A town on the Yangtze, placed in the Tibetan Autonomous District in 1950: formerly *Batang*: also **Baan.**
Paar·de·berg (pär'də·bûrg, -berkh) A village of western Orange Free State, Union of South Africa; scene of a British victory in the Boer War, 1900.
Paa·si·ki·vi (pä'si·kē've), **Juhoo Kusti,** 1870–1956, president of Finland 1946–56.
Pa·bia·ni·ce (pä·byä·nē'tse) A city in central Poland SW of Łódź.
Pa·blo (pä'blō) Spanish form of PAUL.
pab·u·lum (pab'yə·ləm) *n.* Any substance affording nourishment; aliment. See synonyms under FOOD. — **pab'u·lar** *adj.*
pac (pak) *n.* A leather moccasin having a wide sole that turns up and is sewed to the upper; also, a heavy half-boot or legged moccasin of felt or leather worn by lumbermen in the winter. [<N. Am. Ind.]
pa·ca (pä'kə, pak'ə) *n.* A large seminocturnal rodent (genus *Cuniculus*) of Central and South America, brownish with white spots. [<Pg. or Sp. <Tupian]
pace[1] (pās) *n.* **1** A step in walking; also, the progress made in one such movement. **2** A conventional measure of length approximating the average length of stride in walking: usually 3 feet, but sometimes 3.3 feet, making 5 paces to the rod. The **Roman pace** was measured from the point where the heel of one foot left the ground to the point where it descended in the next stride, and was 5 Roman feet, equal to about 58.1 inches, a thousand such double strides making a mile. Such a double step is now called a **geometrical pace**, reckoned at 5 feet. The U.S. Army **regulation pace**

is 30 inches, quick time; 36 inches, double time. **3** The manner or speed of movement in going on the legs; gait; carriage and action, especially of a horse. **4** Rate of speed, as in movement or work: often applied to a fast or ruinous life: the *pace* that kills; also, the speed with which a baseball pitcher delivers the ball. **5** A gait of a horse, etc., in which both feet on the same side are lifted and moved forward at once. — **to put (one) through his paces** To test the abilities, speed, etc., of. — *v.* **paced, pac·ing** *v.t.* **1** To walk back and forth across: to *pace* the floor. **2** To measure by paces. **3** To set or make the pace for. **4** To train to a certain gait or pace. — *v.i.* **5** To walk with slow or regular steps. **6** To move at a pace (def. 5). [<F *pas* <L *passus* step < *pandere* stretch. Doublet of PASS *n.*]
pa·ce[2] (pā'sē) *adv.* & *prep.* With the permission (of): used to express courteous disagreement. [<L, ablative of *pax, pacis* peace, pardon]
paced (pāst) *adj.* **1** Having a particular pace: used in compounds: slow-*paced*. **2** Measured in paces or by pacing. **3** Done behind or with the help of a pacemaker.
pace·mak·er (pās'mā'kər) *n.* One who makes or sets the pace for another in a race. — **pace'mak'ing** *n. & adj.*
pac·er (pā'sər) *n.* **1** A pacing horse: usually five-gaited. **2** One who paces, or measures by paces. **3** A pacemaker.
pa·cha (pə·shä', päsh'ə), **pa·cha·lic** (pə·shä'lik) See PASHA, PASHALIK.
pa·chi·si (pə·chē'zē, pä-) *n.* **1** A game of East Indian origin resembling backgammon. **2** A similar game played elsewhere. Also spelled *parcheesi, parchesi, parchisi.* [<Hind. *pach(ch)īsī*, lit., of twenty-five (the highest throw)]
Pach·mann (päkh'män), **Vladimir de,** 1848–1933, Russian pianist.
pach·ou·li (pach'ōō·lē, pə·chōō'lē) See PATCHOULI.
Pa·chu·ca (pä·chōō'kä) The capital of Hidalgo state, Mexico.
pachy- *combining form* Thick; massive: *pachyderm.* [<Gk. *pachys* thick]
pach·y·ceph·a·ly (pak'i·sef'ə·lē) *n.* Exceptional thickness of the skull. Also **pach'y·ceph'a·li·a** (-sə·fā'lē·ə). [<PACHY- + Gk. *kephalē* head] — **pach'y·ce·phal'ic** (-sə·fal'ik), **pach'y·ceph'a·lous** *adj.*
pach·y·derm (pak'ə·dûrm) *n.* **1** Any of certain thick-skinned, non-ruminant ungulates; especially, an elephant, hippopotamus, or rhinoceros: formerly included in the obsolete order *Pachydermata*, which embraced also the horse, pig, tapir, etc. **2** A stolid, thick-skinned, insensitive person. [<F *pachyderme* <Gk. *pachydermos* <*pachys* thick + *derma* skin] — **pach'y·der·ma·tous, pach'y·der'mous** *adj.*
pach·y·me·ni·a (pak'ə·mē'nē·ə) *n. Pathol.* Any thickening of the skin or of a membrane. [<NL <Gk. *pachys* thick + *hymēn* a membrane] — **pach'y·men'ic** (-men'ik) *adj.*
pach·y·rhi·zid (pak'ə·ri'zid) *n.* A yellowish-green, poisonous glycoside, $C_{30}H_{18}O_8(OCH_3)_2$, extracted from the seeds of a tropical leguminous vine (*Pachyrhizus angulatus*). [< PACHY- + Gk. *rhiza* root]
pach·y·san·dra (pak'ə·san'drə) *n.* A hardy, evergreen, spurgelike plant (*Pachysandra terminalis*) of the box family, much cultivated for ground cover in shady spots: also called *shade grass.* [<NL, with thick stamens <Gk. *pachys* thick + *anēr, andros* man, a male]
pa·cif·ic (pə·sif'ik) *adj.* Pertaining to the making of peace; inclined or leading to peace or conciliation; peaceable; calm. Also **pa·cif'i·cal.** [<F *pacifique* <L *pacificus* peacemaking < *pax, pacis* peace + *facere* make] — **pa·cif'i·cal·ly** *adv.*
Synonyms: calm, conciliating, conciliatory, gentle, meek, mild, peaceable, peaceful, placid, quiet, smooth, still, tranquil, unruffled, waveless. *Antonyms:* belligerent, contentious, controversial, enraged, exasperated, exasperating, fighting, furious, harsh, hateful, hostile, irritated, irritating, provoked, provoking, quarrelsome, stormy, tumultuous, turbulent, warlike.
Pa·cif·ic (pə·sif'ik) *adj.* Pertaining to the Pacific Ocean.
pa·cif·i·cate (pə·sif'ə·kāt) *v.t.* **·cat·ed, ·cat·ing** To pacify. [<L *pacificatus*, pp. of *pacificare* make peace, pacify <*pax, pacis* peace + *facere* make] — **pac·i·fi·ca·tion** (pas'ə·fə·kā'-

Pacific Islands

shən) n. — pa·cif'i·ca'tor n. — pa·cif'i·ca·to'ry (-kə·tôr'ē, -tō'rē) adj.
Pacific Islands A United Nations Trust Territory (1947) administered by the United States, comprising the Marshall Islands, Marianas (except Guam), and the Caroline Islands; 680 square miles; headquarters, Agana on Guam; permanent capital, Truk.
pa·ci·fi·co (pä·sē'fē·kō) n. A peaceable person; a neutral; specifically, a native of Cuba or the Philippines who did not oppose the Spaniards. [<Sp. <L pacificus. See PACIFIC.]
Pacific Ocean An ocean between the American continents and Asia and Australia, extending from the Arctic to the Antarctic Ocean; 70,000,000 square miles: divided by the equator into the **North Pacific Ocean** and the **South Pacific Ocean**.
Pacific slope That region of North America west of the Continental Divide.
Pacific Standard Time See STANDARD TIME. Abbr. *P.S.T.*
pac·i·fi·er (pas'ə·fī'ər) n. 1 One who or that which pacifies; a peacemaker. 2 A rubber nipple attached to a round guard, used to quiet a fretful baby. 3 A rubber ring used to relieve the teething discomfort of a baby.
pac·i·fist (pas'ə·fist) n. One who opposes military ideals, war, or military preparedness, and proposes that all international disputes be settled by arbitration. — **pac'i·fism** n. — **pac'i·fis'tic** adj.
pac·i·fy (pas'ə·fī) v.t. ·fied, ·fy·ing 1 To bring peace to; end war or strife in. 2 To allay the anger or agitation of; appease; calm. See synonyms under ALLAY, TEMPER, TRANQUILIZE. [<F *pacifier* <L *pacificare* <*pacificus*. See PACIFIC.]
Pa·cin·i·an body (pə·sin'ē·ən) *Anat.* One of the flattened, oval end organs of nerves, found especially in the hands, feet, and mesentery: after **Filippo Pa·ci·ni** (pä·chē'nē), 1812–1883, Italian anatomist. Also **Pacinian corpuscle**.
pack[1] (pak) n. 1 A bundle or large package, especially one to be carried on the back of a man or animal; a collection of anything; heap. 2 A full set of like or associated things usually handled collectively, as cards. 3 A number of dogs or wolves that hunt together; hence, any gang or band, especially one existing for criminal purposes. 4 A large area of floating broken ice: also **ice pack**. 5 Face pack. 6 A wrapping of sheets or blankets about a patient, used in certain water-cure treatments: a wet *pack*, cold *pack*, etc. 7 *Obs.* A lewd or low person: usually with *naughty*. 8 A parachute, fully assembled and folded for use. 9 The quantity of something, as vegetables or other foods, put in containers for preservation at one time or in a season. See synonyms under FLOCK, LOAD. — v.t. 1 To make a pack or bundle of. 2 To place compactly in a trunk, box, etc., for storing or carrying. 3 To fill compactly, as for storing or carrying: to *pack* a suitcase. 4 To put up for preservation or sale: to *pack* fruit. 5 To compress tightly; crowd together. 6 To fill completely or to overflowing; cram. 7 To cover, fill, or surround so as to prevent leakage, damage, etc.: to *pack* a piston rod. 8 To load with a pack; burden. 9 To carry or transport on the back or on pack animals. 10 To carry or wear habitually: to *pack* a gun. 11 To send or dispatch summarily: with *off* or *away*. 12 To treat with a pack (def. 6). 13 *Slang* To be able to inflict: He *packs* a wallop. — v.i. 14 To place one's clothes and belongings in trunks, boxes, etc., for storing or carrying. 15 To allow of being stowed or packed. 16 To crowd together; form a pack or packs. 17 To settle in a hard, firm mass. 18 To leave in haste: often with *off* or *away*. See synonyms under JAM. — **to send packing** To send away or dismiss summarily. [ME *pakke*, appar. <LG *pak*]
pack[2] (pak) v.t. To arrange, select, or manipulate to one's own advantage: to *pack* a jury. [? Var. of PACT]
pack[3] (pak) adj. Scot. Intimate: usually in the phrase *pack an' thick*. — **pack'ly** adv. — **pack'ness** n.
pack·age (pak'ij) n. 1 The process or act of packing; also, that which is packed, as for transportation; something wrapped up or bound together; a packet or parcel. 2 A box, case, crate, or other receptacle in which goods are packed. — v.t. ·aged, ·ag·ing To bind or tie into a package or bundle. [<PACK[1] + -AGE]
pack animal An animal, as a horse or mule, used to carry packs or burdens.
pack·er (pak'ər) n. 1 One who packs; specifically, one who makes a business of packing goods for transportation or preservation. 2 One who cures and packs wholesale provisions: a pork-*packer*. 3 One who transports goods on pack animals. 4 One who manipulates or packs, as a convention, jury, cards, etc. 5 Any of certain machines or devices for packing commodities.
pack·et (pak'it) n. 1 A small package; parcel. 2 A steamship for conveying mails, passengers, and freight at stated times; especially, one plying up and down a coast or on a canal: also **packet boat**. — v.t. To make into a packet or parcel. [<AF *pacquet*, dim. of PACK[1]]
pack·ing (pak'ing) n. 1 The act or operation of filling an empty space, putting up for transportation, etc. 2 The canning or putting up of meat, fish, fruit, etc., for market, home consumption, etc. 3 The substance used in adjusting or protecting the article packed. 4 A greasy or other material for closing a joint. 5 A fibrous or porous substance for holding oil by absorption and assisting in the lubrication of a journal, etc. 6 A mechanism or device for making a leak-proof fit, as between a piston head and its cylinder.
packing box 1 A stout box in which goods are packed: also **packing case**. 2 A stuffing box.
packing effect *Physics* The crowding together of the component particles of the atomic nucleus resulting from the mass defect.
packing fraction *Physics* A measure of the binding force of the nucleus of an atom; the difference between the isotopic weight of an atom and the mass number, divided by the mass number, the quotient usually being multiplied by 10,000.
packing house An establishment devoted to packing provisions, especially beef, pork, and oysters.
packing press A press used in baling cotton, hay, or the like.
pack·man (pak'mən) n. pl. ·men (-mən) A peddler.
pack rat A common North American rat (genus *Neotoma*) which feeds chiefly on seeds, nuts, fruit, and green vegetation: so called from its habit of carrying off provisions: also called *wood rat* and *mountain rat*.
pack·sack (pak'sak') n. A canvas or leather traveling sack for blankets, etc.: usually carried across the shoulders.
pack·sad·dle (pak'sad'l) n. A saddle for a pack animal, to which the packs are fastened so as to balance evenly.
pack·thread (pak'thred') n. Strong thread or twine used for doing up packages.
pack·wax (pak'waks') See PAXWAX.
pact (pakt) n. An agreement; compact. [<OF *pact* <L *pactum* agreement <*pactus*, pp. of *paciscere* agree]
pac·ta con·ven·ta (pak'tə kən·ven'tə) *Latin* Stipulations agreed upon.
Pac·to·lus (pak·tō'ləs) A river of ancient Lydia; the gold washed from its sands was a traditional source of Croesus' gold.
pad[1] (pad) n. 1 A cushion; also, any stuffed, cushionlike thing serving to protect from jarring, friction, etc. 2 A soft saddle. 3 A number of sheets of paper packed and gummed together at the edge; a tablet. 4 A large floating leaf of an aquatic plant: a lily *pad*. 5 A soft cushionlike enlargement of skin on the under surface of the toes of many animals. 6 The foot of a fox, otter, etc. 7 The footprint of an animal. 8 A pulvillus. — v.t. **pad·ded**, **pad·ding** 1 To stuff, line, or protect with pads or padding. 2 To lengthen (speech or writing) by inserting unnecessary matter. 3 To expand (an expense account) by recording non-existent expenditures. [Origin unknown]
pad[2] (pad) v.i. **pad·ded**, **pad·ding** 1 To travel by walking; tramp; trudge. 2 To move with soft, almost noiseless footsteps. — n. A dull, padded sound, as of a footstep. [Related to PAD[3] (path). Cf. LG *padden* tread.]
pad[3] (pad) n. 1 An easy-paced road horse: also **pad horse**. 2 A footpad; a highwayman. 3 *Brit. Dial.* A path; road. [<LG *pad* path]
Pa·dang (pä·däng') An Indonesian port on the western coast of Sumatra.
Pa·dauk wood (pə·dôk') Amboina wood: also *Padouk wood*.
pad·ding (pad'ing) n. 1 The act of stuffing or forming a pad. 2 That of which a pad is made. 3 Matter used in writing to fill space.
Pad·ding·ton (pad'ing·tən) A metropolitan borough of NW London.
pad·dle (pad'l) n. 1 A broad-bladed implement resembling a short oar, used without a rowlock in propelling a canoe or small boat. 2 The distance covered during one trip in a canoe over a given time. 3 A paddle board. 4 A straight iron tool for stirring ore in a furnace. 5 A bat or pallet, as used in tempering clay. 6 A scoop for stirring and mixing, as used in glassmaking. 7 A paddle-shaped implement for inflicting bodily punishment. 8 A limb or appendage of service in swimming; a flipper. 9 The snout of the paddlefish. 10 The act of paddling. 11 A flat instrument with which clothes are beaten while being washed in a stream. — v. **·dled**, **·dling** v.i. 1 To move a canoe, etc., on or through water by means of a paddle; ply a paddle. 2 To row gently or lightly. 3 To swim with short, downward strokes, as ducks do. 4 To play in water with the hands or feet; dabble; wade. — v.t. 5 To propel by means of a paddle or paddles. 6 To convey by paddling. 7 To beat with a paddle; spank. 8 To stir. — **to paddle one's own canoe** To be independent; get along without help. [ME *padell* small spade, prob. var. of *patel* shallow pan <L *patella*] — **pad'dler** n.
paddle board One of the broad, paddlelike boards set on the circumference of a paddle wheel or water wheel.
paddle boat A boat propelled by paddle wheels.
paddle box The housing or box over a paddle wheel: usually with semicircular upper outline.
pad·dle·fish (pad'l·fish') n. pl. ·fish or ·fish·es A large fish (*Polyodon spathula*) of the sturgeon family, having a scaleless body with inferior mouth and spatuliform snout, found in the Mississippi Valley streams.
paddle wheel A wheel having projecting floats or boards for propelling a vessel.

PADDLE WHEEL
As seen on a Mississippi River stern-wheeled steamboat.

pad·dling (pad'ling) n. 1 The act of propelling by paddle. 2 A beating or spanking.
pad·dock[1] (pad'ək) n. 1 A pasture lot or enclosure for exercising horses, adjoining a stable. 2 A grassed enclosure at a racecourse where horses are walked about and saddled before a race. 3 In Australia, any enclosed piece of land, whether tilled or untilled. — v.t. To confine, as horses, in a paddock. [Alter. of dial. E *parrock*, OE *pearruc* enclosure. Akin to PARK.]
pad·dock[2] (pad'ək) n. *Scot.* A toad or frog.
pad·dy[1] (pad'ē) n. pl. ·dies The ruddy duck. Also *paddywhack*. [from PADDY, proper name]
pad·dy[2] (pad'ē) n. pl. ·dies Rice in the husk, whether gathered or growing. [<Malay *pādī* rice in the straw]
Pad·dy (pad'ē) n. *Slang* An Irishman: a nickname for PATRICK. [See PATRICK.]
paddy wagon *U.S. Slang* A patrol wagon.
pad·dy·whack (pad'ē·hwak') n. 1 *Brit. Dial.* A fit or rage of temper. 2 *Brit. Dial.* A beating or thrashing. 3 *U.S.* The ruddy duck. [<PADDY + WHACK]
Pa·de·rew·ski (pä'də·ref'skē), Ignace Jan,

1860–1941, Polish pianist, composer, and statesman; first premier of the Polish Republic 1919.

pa·di·shah (pä′di·shä) *n.* **1** Lord protector; chief ruler: a title of the shah of Iran. **2** Formerly, the title of the sultan of Turkey. **3** Formerly, the title of the British sovereign as emperor of India. Also **pad·shah** (päd′shä). [<Persian *Pādshāh* < *pati* master + *shāh* king]

pad·le (päd′l) *n. Scot.* A garden hoe.

pad·lock (pad′lok′) *n.* A detachable lock, designed to hang on the object fastened, having a shackle attached at one end, and devised so as to fasten through a staple. —*v.t.* To fasten with or as with a padlock. [ME *padlocke*, ? < *pad* a basket + LOCK[1]]

pad·nag (pad′nag′) *n. Scot.* A horse ridden with a pad instead of a saddle; hence, an ambling nag.

Pa·douk wood (pə·dōōk′) Amboina wood: also *Padauk wood.*

Pad·raic (pôth′rig) Irish form of PATRICK. Also **Pad′raig.**

pa·dre (pä′drā) *n.* **1** Father: a title used in Italy, Spain, and Spanish America in addressing or speaking of priests, and in India for all clergymen. **2** An army or navy chaplain. [<Ital., Sp., and Pg. <L *pater, patris* father]

pa·dro·ne (pä·drō′nā) *n. pl.* **·nes** (-nās) or **·ni** (-nē) **1** Master: an appellation of an Italian house proprietor or employer of labor. **2** The master of a small vessel in the Mediterranean trade. [<Ital. <L *patronus* PATRON] —**pa·dro′nism** *n.*

Pad·u·a (paj′ōō·ə, pad′yōō·ə) A city in NE Italy west of Venice. *Italian* **Pa·do·va** (pä′dō·vä).

pad·u·a·soy (paj′ōō·ə·soi′) *n.* A strong, rich silk fabric, originally made at Padua; also, an article made of it. [Prob. alter. of F *pou-de-soie* a silk fabric; infl. in form by *Padua say* serge from Padua]

Pa·dus (pä′dəs) The ancient name for the Po.

pae·an (pē′ən) *n.* **1** A choral ode; originally, a song of praise honoring Apollo. **2** Hence, any song of joy or exultation. Also spelled *pean.* [<L <Gk. *paian* a hymn addressed to Paian, the god Apollo]

paed–, paedo– See PEDO–.

paed·e·rast (ped′ər·ast), etc. See PEDERAST, etc.

pae·do·gen·e·sis (pē′dō·jen′ə·sis) *n.* The reproduction of young in the larval stage, as in the axolotl and certain dipterous insects.

pae·on (pē′on) *n.* In Greek and Latin prosody, a foot of four syllables, one long and three short in any order. [<L <Gk. *paiōn* <*Paiōn* Apollo; because used in hymns to Apollo]

Paes·tum (pes′təm) An ancient Greek city in southern Italy; on the site of modern *Pesto.* Also **Pæs′tum.**

pa·gan (pā′gən) *n.* **1** One who is neither a Christian, a Jew, nor a Moslem; a heathen. **2** In early Christian use, an idol-worshiper; a non-Christian. **3** An irreligious person. —*adj.* Pertaining to pagans; heathenish; idolatrous. [<LL *paganus* heathen <L, orig., a rural villager < *pagus* the country] —**pa′gan·dom** *n.* —**pa′gan·ish** *adj.* —**pa′gan·ism** *n.*

Pa·ga·ni·ni (pä′gä·nē′nē), **Niccolò,** 1782–1840, Italian violinist and composer.

pa·gan·ize (pā′gən·īz) *v.t.* & *v.i.* **·ized, ·iz·ing** To make or become pagan.

page[1] (pāj) *n.* **1** A male servant or attendant; specifically, in chivalry, a lad or young man in training for knighthood, or a youth of gentle parentage attending a royal or princely personage. **2** A boy whose duty it is to attend upon legislators while in session. **3** A boy in livery, employed in a hotel, club, theater, or private house to perform light duties. —*v.t.* **paged, pag·ing** **1** To seek or summon (a person) by calling his name, as a hotel page does. **2** To wait on as a page. [<OF <Ital. *paggio* <Med. L *pagius,* ? <Gk. *paidion,* dim. of *pais, paidos* child]

page[2] (pāj) *n.* **1** One side of a leaf of a book, letter, manuscript, etc.; also, the type for printing on one side: abbreviated p., *pl.* pp. **2** Hence, any source or record of knowledge. —*v.t.* **paged, pag·ing** To mark the pages of with numbers. [<F <L *pagina* leaf of a book, written page < *pag-,* stem of *pangere* fasten] —**pag′ing** *n.*

Page (pāj), **Thomas Nelson,** 1853–1922, U.S. novelist. —**Walter Hines,** 1855–1918, U.S. diplomat and publisher.

pag·eant (paj′ənt) *n.* **1** A community outdoor celebration presenting scenes from local history and tradition. **2** An imposing exhibition or spectacular parade devised for a public ceremony or celebration. **3** A theatrical spectacle; hence, unsubstantial display. **4** Hangings having scenic enrichment. **5** Originally, a traveling car or float having a stage for presenting mystery plays; hence, any pompous or showy object or decoration designed for public parades. See synonyms under OSTENTATION, SPECTACLE. [<Med. L *pagina* a framework, ? <L *pegma* <Gk. *pēgma* a framework, scaffold < *pēgnynai* fasten]

pag·eant·ry (paj′ən·trē) *n. pl.* **·ries** Pageants collectively; ceremonial splendor or display; pomp; showy quality. See synonyms under OSTENTATION.

Pag·et (paj′it), **Sir James,** 1814–99, English surgeon and pathologist.

pag·i·nal (paj′ə·nəl) *adj.* Consisting of, or pertaining to, the pages of a book; also, page for page. [<LL *paginalis* <L *pagina* leaf of a book, page]

pag·i·nate (paj′ə·nāt) *v.t.* **·nat·ed, ·nat·ing** To number the pages of (a book) consecutively. [<L *pagina* page + -ATE[1]]

pag·i·na·tion (paj′ə·nā′shən) *n.* **1** The numbering of the pages, as of a book. **2** The system of figures and marks used in paging.

pa·go·da (pə·gō′də) *n.* In the countries of the Far East, a sacred tower or temple, usually pyramidal and profusely adorned. Also **pag·od** (pag′əd, pə·god′). [<Pg. *pagode,* prob. <Persian *butkadah* idol-temple < *but* idol + *kadah* house, ? ult. <Skt. *bhagavati* divine]

Pa·go Pa·go (päng′ō päng′ō) The only port of call in American Samoa, on Tutuila island: also *Pango Pango.* Also **Pa′go·pa′go.**

pa·go·plex·i·a (pä′gō·plek′sē·ə) *n. Pathol.* **1** Frostbite. **2** Chilblain. [<NL <Gk. *pagos* frost + *plēxis* stroke < *plēssein* strike]

Pa·gu·ri·dae (pə·gyōōr′ə·dē) *n. pl.* A family of crustaceans containing the hermit crabs. [<NL <L *pagurus* a kind of crab <Gk. *pagouros* < *pagos* a fixed thing (< *pēgnynai* fasten) + *oura* a tail] —**pa·gu′ri·an** (-ē·ən), **pa·gu·rid** (pə·gyōōr′id, pag′yə·rid) *adj.* & *n.*

pah (pä, pa) *interj.* Bah! faugh! an exclamation of contemptuous disgust.

Pa·hang (pä·häng′) The largest state in the Federation of Malaya; 13,873 square miles; capital, Kuala Lipis.

pah·la·vi (pä′lə·vē) *n.* A gold coin of Iran, equivalent to 100 rials, adopted in 1927. [<Persian, after Rhiza Khan *Pahlavi,* shah of Persia]

Pah·la·vi (pä′lə·vē) *n.* A literary form of Middle Persian, in use from the third to the seventh century: preserved in the Zoroastrian sacred writings, transliterated in Semitic characters: also spelled *Pehlevi.* [<Persian *Pahlavī* Parthian <OPersian *Parthava* Parthia]

pa·ho·e·ho·e (pä·hō′ə·hō′ə) *n.* A variety of lava having a smooth, ropy, or corded appearance. [<Hawaiian, lit., smooth, polished]

Pai (pī) A river in Hopeh province, China, flowing 300 miles SE to the Gulf of Chihli: Chinese *Pei Ho.*

paid (pād) Past tense and past participle of PAY.

pai·deu·tics (pī·dōō′tiks, -dyōō′-) *n.* The theory or the art of instruction; pedagogy. [<Gk. *paideutikos* of teaching < *paideuein* teach < *pais, paidos* a child]

pai·dle (päd′l) *v.t.* & *v.i. Scot.* To paddle.

pai·dol·o·gy (pī·dol′ə·jē) *n.* Pedology. [<Gk. *pais, paidos* a child + -LOGY]

paik (pāk) *n. Scot.* A blow; a beating. —*v.t.* To beat or strike.

pail (pāl) *n.* **1** A cylindrical vessel for carrying liquids, etc., properly having a bail as a handle. **2** The amount carried in this vessel. ◆ Homophone: *pale.* [OE *paegel* a gill, wine measure; infl. by OF *paelle* frying pan, liquid measure <L *patella* a small pan] —**pail′ful′** *n.*

pail·lasse (pal·yas′, pal′yas) *n.* A mattress of straw, excelsior, or the like: also spelled *palliase.* [<F < *paille* straw <L *palea* chaff]

pail·lette (pal·yet′) *n.* **1** A bit of metal or colored foil, used in enamel painting. **2** A spangle; one of a hanging bunch of spangles. [<F, dim. of *paille* straw] —**pail·let′ted** *adj.*

pain (pān) *n.* **1** The sensation or feeling resulting from or accompanying some injury, derangement, overstrain, or obstruction of the physical powers; any distressing or afflicting emotion, or such emotions in general; grief: opposite of *pleasure.* **2** *pl.* Care, trouble, effort, or exertion expended on anything: used often as singular: with much *pains.* **3** *pl.* The pangs of childbirth. —**on** (or **upon** or **under**) **pain of** With the penalty of (some specified punishment). —**to take pains** To be careful; to make an effort. —*v.t.* To cause pain to; hurt or grieve; disquiet. —*v.i.* To cause pain. See synonyms under HURT, PIQUE. ◆ Homophone: *pane.* [<OF *peine* <L *poena* <Gk. *poinē* a penalty]

Synonyms (noun): ache, affliction, agony, anguish, discomfort, distress, misery, pang, paroxysm, suffering, throe, torment, torture, trouble, twinge, uneasiness, woe, wretchedness. *Pain* is the most general term of this group, including all the others; *pain* is a disturbing sensation from which nature revolts, resulting from some injurious external interference (as from a wound, a bruise, a harsh word, etc.), or from some lack of what one needs, craves, or cherishes (as, the *pain* of hunger or bereavement), or from some abnormal action of bodily or mental functions (as, the *pains* of disease, envy, or discontent). *Ache* is lingering *pain,* more or less severe; *pang,* a *pain* short, sharp, intense, and perhaps repeated. We speak of the *pangs* of hunger or of remorse. *Throe* is a violent *pain. Paroxysm* applies to the alternately recurring and receding *pain,* which comes as though in waves; the *paroxysm* is the rising of the wave. *Torment* and *torture* are intense and terrible *sufferings.* Compare ADVERSITY, AFFLICTION, AGONY, INJURY, SUFFERING. *Antonyms*: comfort, delight, ease, enjoyment, peace, rapture, relief, solace.

painch (pänch) *n. Scot.* Paunch.

Paine (pān), **Robert Treat,** 1731–1814, American jurist; signer, for Massachusetts, of the Declaration of Independence. —**Thomas,** 1737–1809, American patriot, author, and political philosopher born in England.

pained (pānd) *adj.* **1** Hurt (physically or mentally); distressed. **2** Showing pain: a *pained* expression.

pain·ful (pān′fəl) *adj.* **1** Giving or attended with pain; distressing. **2** Requiring labor, effort, or care; arduous. **3** Affected with pain: said of the body or of some part of it. **4** *Archaic* Painstaking; laborious. See synonyms under TROUBLESOME. —**pain′ful·ly** *adv.* —**pain′ful·ness** *n.*

pain–kill·er (pān′kil′ər) *n. U.S. Colloq.* A medicine that relieves pain.

pain·less (pān′lis) *adj.* Free from pain; causing no pain. —**pain′less·ly** *adv.* —**pain′less·ness** *n.*

pains·tak·ing (pānz′tā′king) *adj.* Taking pains; careful; assiduous. —*n.* Diligent and careful endeavor. —**pains′tak′ing·ly** *adv.*

paint (pānt) *n.* **1** A color or pigment, either dry or mixed with oil, water, etc. **2** A cosmetic, as rouge. **3** A film, layer, or coat of pigment applied to the surface of an object. —*v.t.* **1 a** To make a representation of in paints or colors. **b** To make, as a picture, by applying paints or colors. **2** To describe or depict vividly, as in words or thoughts. **3** To cover or coat with or as with paint. **4** To decorate with or as with paints: The setting sun *paints* the clouds red. **5** To apply cosmetics to. **6** To apply (medicine, etc.) with or as with a swab. —*v.i.* **7** To cover or coat something with paint. **8** To practice the art of painting; paint pictures. **9** To apply cosmetics to the face, etc. [<OF *peint,* pp. of *peindre* <L *pingere* paint]

paint·brush (pānt′brush′) *n.* The painted cup.

paint brush A brush for applying paint.

paint·ed (pān′tid) *adj.* **1** Covered or coated with paint. **2** Depicted in colors; existing merely in semblance. **3** Marked with bright or varied colors.

painted beauty A brightly colored butterfly (*Vanessa virginiensis*) having dusky orange wings with two eyelike spots on the under side of each.

painted bunting A brilliantly colored finch (*Passerina ciris*) widely distributed in the southern United States. Also **painted finch.**

painted cup Any of several showy North American flowers (*genus Castilleja*) of the figwort family, especially the Wyoming painted cup (*C. linariaefolia*) with vivid scarlet bracts and calyxes and yellow corollas: the State flower of Wyoming. Also called *Indian paintbrush.*

Painted Desert A large arid area in northern

Arizona, extending SE along the Little Colorado River.
painted goose The emperor goose.
painted lady The thistle butterfly.
paint·er[1] (pān′tər) n. 1 One whose occupation is painting; specifically, one who covers surfaces with a preservative or decorative coat of paint. 2 An artist who portrays scenes or objects in colors.
paint·er[2] (pān′tər) n. Naut. A rope with which to fasten a boat by its bow. [Prob. <OF pentoir a rope for hanging things <L pendere hang]
paint·er[3] (pān′tər) n. U.S. Dial. The puma, or cougar. [Var. of PANTHER]
Paint·er (pān′tər), **Theophilus Shickel**, born 1889, U.S. zoologist.
painters' colic Pathol. A form of lead poisoning characterized by sharp abdominal pains, slow pulse, and increased arterial tension.
paint horse In the western United States, a pied or spotted horse; a pinto.
paint·ing (pān′ting) n. 1 The act, art, or employment of laying on paints with a brush. 2 The art of representing objects on a surface by means of pigments. 3 A picture.
paint·y (pān′tē) adj. **paint·i·er**, **paint·i·est** 1 Of, belonging to, or covered with paint. 2 Heavily daubed with paint: said of a painting.
pair (pâr) v.t. 1 To bring together or arrange in a pair or pairs; match; couple; mate. — v.i. 2 To come together as a couple or pair. 3 To marry or mate. — **to pair off** 1 To separate into couples. 2 To arrange by pairs. — n. 1 In general, two persons or things of a kind, joined, related, correspondent, or associated; a couple; brace. 2 A single thing having two like or correspondent parts dependent on each other: a pair of scissors: in this sense, always linked with a singular verb when one object is counted. 3 A married couple; two animals mated. 4 In legislative bodies, two opposed members who agree to abstain from voting, and so offset each other. 5 A set of like or equal things making a whole: now restricted in use. 6 In some games of cards, two cards of the same denomination: a pair of queens. 7 Mech. A combination of two elements forming a unit in the mutual production or constraint of motion, as a piston and cylinder, a screw and nut, etc. 8 A racing shell for two oarsmen. ♦ Current usage calls for pair in the plural after a numeral of two or more, as, four pairs of shoes, though colloquially the singular is often used, as, four pair of shoes. ♦ Homophones: pare, pear. [<F paire <L paria, neut. plural of par equal]
pair-oar (pâr′ôr′, -ōr′) n. A boat in which two men sitting one behind the other pull one oar each.
pair production Physics The instantaneous conversion of a photon into an electron and positron by its passage through a strong electric field.
pairt (pärt) Scot. See PART.
pair-trick (pâr′trik′) n. Scot. Partridge.
pais (pā) n. The country; the people; especially, the people from whom a jury is selected: also spelled pays. [<OF, the country]
Pais·ley (pāz′lē) adj. Made of or resembling a certain patterned woolen fabric made in Paisley, a suburb of Glasgow, Scotland. — n. 1 The Paisley fabric. 2 A Paisley shawl: designed to imitate a Kashmir shawl.
Pai·ute (pī·yōōt′) n. One of a tribe of North American Indians of Shoshonean stock, living in SW Utah. Also spelled Piute.
pa·ja·mas (pə·jä′məz, -jam′əz) n. pl. 1 Loose trousers of silk or cotton, worn by both men and women in the Orient. 2 Similar trousers with coats to match, used as nightwear. Also,

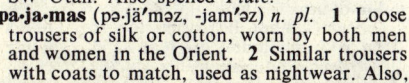
TYPE OF PAISLEY DESIGN

Brit., pyjamas. [<Hind. pājāmā <Persian pāi a leg + jāmah a garment] — **pajama** adj.
Pa·ki·stan (pä′ki·stän′), **Islamic Republic of** An independent republic remaining within the Commonwealth of Nations; 365,907 square miles; capital, Karachi; consisting of two sections separated by 900 miles of Indian territory: West Pakistan; 311,406 square miles; and East Pakistan, coextensive with East Bengal; 54,501 square miles. — **Pa′ki·sta′ni** adj. & n.
pak·tong (pak′tong) n. A Chinese alloy of zinc, nickel, and copper. [< dial. Chinese (Cantonese) <Chinese peh t'ung < peh white + t'ung copper]
pal (pal) n. Colloq. A mate; chum; confederate. [<Romany pal brother, mate, ult. <Skt. bharātṛ. Akin to BROTHER.]
Pal (päl) Hungarian form of PAUL.
pa·la·bra (pä·lä′brä) n. Spanish A word; speech.
pal·ace (pal′is) n. 1 A royal residence, or the official residence of some high dignitary, as of a bishop. 2 Any splendid residence or stately building. 3 A large building or room used as a place of public entertainment: an oyster palace. See synonyms under HOUSE. [<OF palais <L palatium, orig., the Palatine Hill at Rome, on which stood the palace of the Caesars]
Pa·la·cio Val·dés (pä·lä′thyō väl·thās′), Armando, 1853-1938, Spanish novelist.
pal·a·din (pal′ə·din) n. Any of the twelve peers of Charlemagne; hence, a paragon of knighthood. [<F <L palatinus of the palace <palatium. See PALACE.]
palae-, palaeo- See words beginning PALE-, PALEO-.
Pal·a·me·des (pal′ə·mē′dēz) In Greek legend, one of the Greek warriors in the Trojan War.
pal·an·quin (pal′ən·kēn′) n. A type of covered litter used as a means of conveyance in the Orient, borne by poles on the shoulders of two or more men. Also **pal′an·keen′**. [<Pg. palanquim <Javanese pĕlangki <Skt. palyaṅka, var. of paryaṅka bed]
pal·at·a·ble (pal′it·ə·bəl) adj. 1 Agreeable to the taste or palate; savory. 2 Acceptable. — **pal′at·a·bil′i·ty** — **pal′at·a·ble·ness** n. — **pal′at·a·bly** adv.
pal·a·tal (pal′ə·təl) adj. 1 Pertaining to the palate. 2 Phonet. a Produced by placing the front (not the tip) of the tongue near or against the hard palate, as y in English yoke, ch in German ich. b Produced with the blade of the tongue near the hard palate, as ch in child, j in joy. — n. Anat. 1 A bone of the palate. 2 Phonet. A palatal sound. [<F <L palatum palate]
pal·a·tal·ize (pal′ə·təl·īz′) v.t. & v.i. **·ized**, **·iz·ing** Phonet. To change to a palatal sound, as (s) to (sh) in censure, (t) to (ch) in nature. — **pal′a·tal·i·za′tion** n.
pal·ate (pal′it) n. 1 Anat. The roof of the mouth. The **hard** (or **bony**) **palate**, or anterior part, has a bony skeleton; the **soft palate**, or posterior division, is composed of muscular tissue and mucous membrane. 2 The sense of taste; relish: so used originally from the false notion that the palate is the organ of taste. 3 Intellectual taste; mental relish. ♦ Homophones: palet, palette, pallet, pallette. [<OF palat <L palatum palate]
pa·la·tial (pə·lā′shəl) adj. Of, like, or befitting a palace; magnificent. [<L palatium PALACE + -AL] — **pa·la′tial·ly** adv.
pa·lat·i·nate (pə·lat′ə·nāt, -nit) n. 1 A political division ruled over by a prince possessing certain prerogatives of royalty within his own domain. 2 The office of a count palatine or of an elector palatine.
Pa·lat·i·nate (pə·lat′ə·nāt, -nit) n. A native or resident of the Palatinate.
Palatinate, the A region west of the Rhine, formerly a state of the German Empire, administered (1837-1945) by Bavaria; incorporated (1945) into the Rhineland-Palatinate: also Rhine Palatinate. German Pfalz (pfälts).
pal·a·tine[1] (pal′ə·tīn, -tin) adj. Of or pertaining to the palate. — n. Anat. Either of the two bones forming the hard palate. [<F palatin <L palatum palate]
pal·a·tine[2] (pal′ə·tīn, -tin) adj. 1 Pertaining to a royal palace or its officials. 2 Possessing royal prerogatives; exercising or endowed with

regal rights within a certain domain: a count or county palatine. — n. 1 A high judicial functionary in medieval France and Germany; hence, by the delegation of sovereign power, a lord exercising sovereign power over a province; a vassal enjoying the exercise of royal privileges over his territory. See COUNT PALATINE. 2 The ruler of a palatinate or county palatine. 3 A fur tippet worn by women over the shoulders. — **the Palatine** or **the Palatine Hill** The central hill of the seven on which ancient Rome was built. [<F palatin <L palatinus <palatium. See PALACE.]
Pal·a·tine (pal′ə·tīn, -tin) adj. Of or pertaining to the Palatinate. — n. An inhabitant of the Palatinate.
Pa·lau Islands (pä·lou′) An island group in the western Caroline Islands; 188 square miles: also Pelew Islands.
pa·lav·er (pə·lav′ər) n. 1 Empty talk, especially that intended to flatter or deceive. 2 A profuse parley; hence, public discussion or conference: a term originated by the Portuguese explorers of Africa. — v.t. To flatter; cajole. — v.i. To talk idly and at length. See synonyms under BABBLE. [<Pg. palavra word, speech <LL parabola a story, word <L, comparison. Doublet of PARABLE, PARABOLA, PAROLE.] — **pa·lav′er·er** n. — **pa·lav′er·ing** adj. & n.
Pa·la·wan (pä·lä′wän) An island in the SW Philippines; 4,550 square miles.
pa·lay (pä·lī′) n. A natural rubber extracted by the natives of Madagascar from wild plants of the genus Cryptostegia: also spelled pulay. [<Tamil]
pale[1] (pāl) n. 1 Originally, a pointed stick of wood for driving into the ground; a stake; a paling; a fence picket. 2 A fence enclosing a piece of ground; hence, any boundary or limit. 3 That which is enclosed within bounds, literally or figuratively: the social pale. 4 Her. An ordinary consisting of a vertical band through the middle of the shield, occupying one third of its width. — **English pale** 1 The varying portion of Irish territory which the Anglo-Normans conquered and governed for several centuries after their invasion of Ireland in the 12th century: also **the Pale**. 2 Formerly, the territory of Calais in France. — v.t. **paled**, **pal·ing** To enclose with pales; fence in. ♦ Homophone: pail. [<OF pal <L palus a stake]
pale[2] (pāl) adj. 1 Of a whitish or ashen appearance; pallid; wan. 2 Of a very light shade of any color; lacking in brightness or intensity of color. — v.t. & v.i. To make or turn pale; blanch. ♦ Homophone: pail. [<OF palle <L pallidus. Doublet of PALLID.] — **pale′ly** adv. — **pale′ness** — **pal′ish** adj. Synonyms (adj.): ashy, bloodless, cadaverous, colorless, ghastly, marble, pallid, wan, white. See GHASTLY. Antonyms: blushing, flaming, florid, flushed, purple, red, roseate, rosy, rubicund, ruddy.
pa·le·a (pā′lē·ə) n. pl. **·le·ae** (-li·ē) Bot. 1 A chafflike bract. 2 One of the chaffy inner scales subtending a flower in the grass spikelet. [<NL <L, chaff] — **pa′le·a′ceous** (-ā′shəs) adj.
Pa·le·arc·tic (pā′lē·ärk′tik) adj. Designating a zoogeographical realm which embraces Europe, North Africa, and Asia north of the tropic of Cancer: also spelled Palaearctic.
pa·le·eth·nol·o·gy (pā′lē·eth·nol′ə·jē) n. Ethnology dealing with prehistoric man: also spelled palaeethnology. — **pa′le·eth′no·log′ic** (-eth′nə·loj′ik) or **·i·cal** adj. — **pa′le·eth·nol′o·gist** n.
pale·face (pāl′fās′) n. A white person: a term allegedly originated by North American Indians.
Pa·lem·bang (pä′lem·bäng′) The largest city of Sumatra, in the SE part.
Pa·len·que (pä·leng′kā) A village in NE Chiapas state, Mexico, on the site of extensive Mayan ruins.
paleo- combining form 1 Ancient; old: paleography. 2 Primitive: paleolithic. Also, before vowels, **pale-**. Also palaeo-. [<Gk. palaios old, ancient]
pa·le·o·bi·o·chem·is·try (pā′lē·ō·bī′ō·kem′is·trē) n. The study of the chemical composition of extinct plant and animal organisms as shown by their fossil remains.

pa·le·o·bot·a·ny (pā′lē-ō-bot′ə-nē) *n.* The study of fossil plants. — **pa′le·o·bot·an′ic** (-bə-tan′ik) or **·i·cal** *adj.* — **pa′le·o·bot′a·nist** *n.*

Pa·le·o·cene (pā′lē-ə-sēn′) *n. Geol.* The oldest epoch of the Cenozoic era, preceding the Eocene.

pa·le·o·e·col·o·gy (pā′lē-ō-ē-kol′ə-jē) *n.* The study of the environment of plant and animal organisms living in past geologic periods. — **pa′le·o·ec′o·log′i·cal** (-ek′ə-loj′i-kəl) *adj.*

Pa·le·o·gene (pā′lē-ə-jēn′) *n. Geol.* The Eogene.

pa·le·o·ge·og·ra·phy (pā′lē-ō-jē-og′rə-fē) *n.* The study and description of earth features in past geologic periods. — **pa′le·o·ge′o·graph′ic** (-jē′ə-graf′ik) *adj.*

pa·le·og·ra·phy (pā′lē-og′rə-fē) *n.* 1 An ancient mode of writing; ancient writings collectively. 2 The science of describing or deciphering ancient writings. — **pa′le·o·graph** (pā′lē-ə-graf′, -gräf′) *n.* — **pa′le·og′ra·pher** *n.* — **pa′le·o·graph′ic** or **·i·cal** *adj.*

pa·le·o·hy·drol·o·gy (pā′lē-ō-hī-drol′ə-jē) *n.* The study of ancient systems of irrigation, water supply, and the like. — **pa′le·o·hy′dro·log′ic** (-hī′drə-loj′ik) *adj.* — **pa′le·o·hy·drol′o·gist** *n.*

pa·le·o·lith (pā′lē-ō-lith′) *n.* A chipped stone object or implement of the Paleolithic period of human culture.

Pa·le·o·lith·ic (pā′lē-ō-lith′ik) *adj. Anthropol.* Of, pertaining to, or associated with a period of human culture contemporaneous with the Pleistocene epoch and followed by the Mesolithic period. It is characterized by stone implements of increasing technical refinement; cave paintings, many in vivid color; sculptured forms and bas-reliefs. Also called *Old Stone Age*. The principal stages, reading from the earliest, are as follows:

LOWER PALEOLITHIC:	UPPER PALEOLITHIC:
Abbevillian	Chatelperronian
Clactonian	Aurignacian
Acheulean	Gravettian
Levalloisian	Solutrean
MIDDLE PALEOLITHIC:	Magdalenian
Mousterian	

paleolithic man *Anthropol.* Any type of human being belonging to the Paleolithic period, such as the Neanderthal or Cro-Magnon men.

pa·le·ol·o·gy (pā′lē-ol′ə-jē) *n.* The study of antiquity or antiquities; archeology: also spelled *palaeology*. [<PALEO- + -LOGY] — **pa′le·o·log′i·cal** (-ə-loj′i-kəl) *adj.* — **pa′le·ol′o·gist** *n.*

pa·le·on·tog·ra·phy (pā′lē-on-tog′rə-fē) *n.* The description of fossils; descriptive paleontology: also spelled *palaeontography*. [< PALEO- + Gk. *ōn, ontos* being, ppr. of *einai* be + -GRAPHY] — **pa′le·on′to·graph′ic** (-on′tə-graf′ik) *adj.*

pa·le·on·tol·o·gy (pā′lē-on-tol′ə-jē) *n.* The science that treats of the ancient life of the globe or of fossil organisms, either plants or animals: also spelled *palaeontology*. — **pa′le·on′to·log′ic** (-on′tə-loj′ik) or **·i·cal** *adj.* — **pa′le·on·tol′o·gist** *n.*

pa·le·o·pa·thol·o·gy (pā′lē-ō-pə-thol′ə-jē) *n.* The study of pathological conditions in fossil or extinct organisms.

pa·le·o·pe·dol·o·gy (pā′lē-ō-pə-dol′ə-jē) *n.* The study of ancient or fossil soils.

pa·le·o·pho·bi·a (pā′lē-ō-fō′bē-ə) *n.* An immoderate hostility toward or fear of the past, considered as a trend in contemporary culture and as a possible factor in the origin and development of certain types of mental disorder. — **pa′le·o·pho′bic** *adj.*

pa·le·o·psy·chol·o·gy (pā′lē-ō-sī-kol′ə-jē) *n.* The investigation of mental phenomena traceable to or persisting from an earlier stage in evolution.

pa·le·o·tem·per·a·ture (pā′lē-ō-tem′pər-ə-chər, -prə-chər) *n.* The condition of heat and cold prevailing on the earth and in the oceans in past geologic eras: a determinative factor in the study of extinct forms of life.

Pa·le·o·zo·ic (pā′lē-ō-zō′ik) *adj. Geol.* Of or pertaining to the era following the Pre-Cambrian and below the Mesozoic. — *n.* The Paleozoic era or group. Also spelled *Palaeozoic*. [< PALEO- + Gk. *zōē* life]

pa·le·o·zo·ol·o·gy (pā′lē-ō-zō-ol′ə-jē) *n.* The branch of paleontology that treats of fossil animals. — **pa′le·o·zo′o·log′i·cal** (-zō′ə-loj′i-kəl) *adj.*

Pa·ler·mo (pä-ler′mō) The capital of Sicily, a port on the NW coast: ancient *Panormus*.

Pal·es·tine (pal′əs-tīn) 1 In Biblical times, a territory on the eastern coast of the Mediterranean, the country of the Jews: Old Testament *Canaan*: also *Holy Land*. 2 Parts of this territory, not including Syria or Jordan, placed (1920) under British mandate by the League of Nations: 10,434 square miles; capital, Jerusalem: divided (1947) by the United Nations into independent Arab and Jewish states. See ISRAEL, JORDAN. — **Pal′es·tin′i·an** (-tin′ē-ən), **Pal′es·tin′e·an** *adj. & n.*

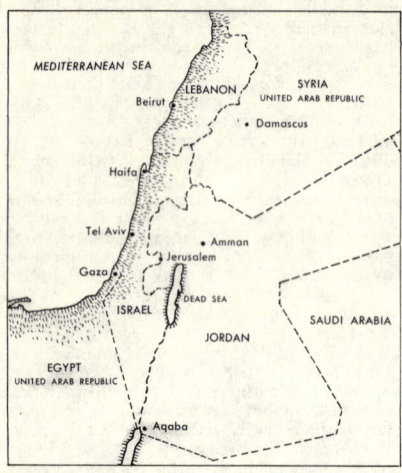

pa·les·tra (pə-les′trə) *n. pl.* **·trae** (-trē) 1 In ancient Greece, a school of athletics; also, the wrestling court in a public gymnasium; hence, any school for youth. 2 A gymnasium. Also spelled *palaestra*. [<L *palaestra* <Gk. *palaistra* < *palaiein* wrestle] — **pa·les′tral, pa·les′tri·an** *adj.*

pa·les·tric (pə-les′trik) *adj.* Of or pertaining to wrestling; athletic.

Pa·les·tri·na (pä′les-trē′nä), **Giovanni Pierluigi da**, 1526?–94, Italian composer.

pal·et (pal′it) *n.* A palea. ◆ Homophones: *palate, palette, pallet, pallette*. [<L *palea* chaff + -ET]

pal·e·tot (pal′ə-tō, pal′tō) *n.* A loose overcoat or outer garment. [<F <OF *palletoc, ?* < *palle* cloak + *toque* hood, cap]

pal·ette (pal′it) *n.* 1 A thin tablet, with a hole for the thumb, upon which artists lay and mix their colors for painting. 2 An arrangement of colors placed on the tablet. ◆ Homophones: *palate, palet, pallet, pallette*. [<F, dim. of *pale* shovel <L *pala* spade]

palette knife A thin, flat knife with flexible blade, usually offset from the handle, for mixing and applying oil colors.

Pa·ley (pā′lē), **William**, 1743–1805, English theologian and philosopher.

pal·frey (pôl′frē) *n.* A saddle horse, especially a woman's saddle horse. [<OF *palefrei* <LL *palafredus*, alter. of *paraveredus* <Gk. *para* beside, extra + LL *veredus* post horse]

Pal·grave (pôl′grāv), **Francis Turner**, 1824–1897, English poet and anthologist.

Pa·li (pä′lē) *n.* The sacred language of the early Buddhist writings, comprised of various Indic dialects: preserved in Ceylon and still surviving in the religious literature of Burma and Thailand. [<Skt. *pāli*, short for *pālibhāsā* language of the canonical texts < *pāli* line, canon + *bhāsā* language]

pal·i·kar (pal′i-kär) *n.* A Greek soldier in the struggle against Turkey for Grecian independence. [<Modern Gk. *palikari* lad <Gk. *pallax, pallakos* youth]

pa·limp·sest (pal′imp·sest) *n.* A parchment, etc., written upon two or three times, the earlier writing having been wholly or partially erased to make room for the next. — *adj.* Rewritten or superinscribed. [<L *palimpsestus* <Gk. *palimpsēstos*, lit., scraped again < *palin* again + *psēein* rub]

pal·in·drome (pal′in·drōm) *n.* A word or words that read the same forward or backward, as "Madam, I'm Adam," or "radar." [<Gk. *palindromos* a running back again < *palin* again + *dromos* a running]

pal·in·dro·mi·a (pal′in·drō′mē-ə) *n. Pathol.* The recurrence or worsening of a diseased condition; a relapse. [<NL <Gk. *palin* again

+ *dromos* a running] — **pal′in·drom′ic** (-drom′ik) *adj.*

pal·ing (pā′ling) *n.* 1 One of a series of upright pales forming a fence; also, such pales or pickets collectively. 2 A fence made of pales or pickets; hence, a limit or enclosure. 3 The act of erecting a fence with pales.

pal·in·gen·e·sis (pal′in·jen′ə-sis) *n.* 1 A new or second birth into a higher or better life or being; a regeneration; theory or belief that souls are continually reborn. 2 *Biol.* The development of an individual in which the ethnic or group history of its ancestors is repeated: opposed to *cenogenesis*. 3 *Entomol.* The metamorphosis of an insect. 4 *Obs.* The supposed generation of an animal from an organism on which it is parasitic or from decaying animal matter. [<NL <Gk. *palin* again + GENESIS] — **pal′in·ge·net′ic** (-jə·net′ik) *adj.*

pal·i·node (pal′i·nōd) *n.* 1 A poem retracting the matter of an earlier poem; metrical recantation. 2 Hence, any retraction or recantation. [<MF *palinod* <LL *palinodia* <Gk. *palinōidia* < *palin* again + *ōidē* a song]

Pal·i·nu·rus (pal′i·nyoor′əs) In Vergil's *Aeneid*, the helmsman of Aeneas, who fell asleep, fell overboard, and was drowned as the Trojans approached the western coast of Italy.

pal·i·sade (pal′ə·sād′) *n.* 1 A fence or fortification made of strong timbers set in the ground. 2 *pl.* An extended cliff or precipice of rock, usually along the bank of a river. — *v.t.* **·sad·ed, ·sad·ing** To enclose or fortify with a palisade. [<MF *palissade* < *palisser* enclose with pales <L *palus* a stake]

Pal·i·sades (pal′ə·sādz′), **The** The ridge of cliffs extending about 25 miles along the western bank of the Hudson River in New York and New Jersey, included in **Palisades Interstate Park**, a chain of recreational areas, 47,070 acres.

Pa·lis·sy (pȧ·lē·sē′), **Bernard**, 1509?–89, French potter; creator of a type of highly colored ware decorated with reliefs of animals, fish, etc., known as **Palissy ware**.

Palk Strait (pôk) An inlet of the Bay of Bengal between India and the northern coast of Ceylon; 85 miles long.

pall[1] (pôl) *n.* 1 A covering, usually of black cloth, thrown over a coffin or over a tomb; figuratively, that which brings deep sorrow or fear; also, metaphorically, a dark, heavy covering: a *pall* of smoke. 2 *Eccl.* **a** A chalice cover, consisting of a square piece of cardboard faced on both sides with lawn or linen. **b** An altar cloth. **c** A prelate's pallium. — **heraldic pall** The Y-shaped bearing resembling a pallium. — *v.t.* To cover with or as with a pall. ◆ Homophone: *pawl*. [OE *paell* a cloak <L *pallium* a pallium]

HERALDIC PALL
From the arms of the See of Canterbury.

pall[2] (pôl) *v.i.* 1 To become insipid or uninteresting. 2 To have a dulling or displeasing effect: with *on*. — *v.t.* 3 To satiate; cloy; disgust. ◆ Homophone: *pawl*. [Appar. aphetic var. of APPAL]

Pal·la·di·an (pə·lā′dē-ən) *adj.* Belonging to or characteristic of the goddess Pallas or the Palladium; hence, characterized by wisdom or learning. [<L *Palladius* <Gk. *Palladios* of Pallas (Athena)]

Pal·la·di·an (pə·lā′dē-ən) *adj. Archit.* Pertaining to or in the Renaissance style of Andrea Palladio: a modification of the classic Roman style.

pal·lad·ic (pə·lad′ik) *adj. Chem.* Pertaining to or designating compounds containing tetravalent palladium.

Pal·la·di·o (päl·lä′dyō), **Andrea**, 1518–80, Italian architect.

pal·la·di·um[1] (pə·lā′dē-əm) *n. pl.* **·di·a** (-dē-ə) Any object considered essential to the safety of a community or organization; a safeguard. [<PALLADIUM]

pal·la·di·um[2] (pə·lā′dē-əm) *n.* A rare metallic element (symbol Pd) occurring in combination with platinum, iridium, and rhodium: it is silver-white, malleable, ductile, does not tarnish in the air, and in the spongy state is capable of absorbing hydrogen in large quantities. See ELEMENT. [<NL, after *Pallas*, an

asteroid discovered contemporaneously with the element]
Pal·la·di·um (pə-lā′dē-əm) In Greek and Roman legend, a statue of Pallas Athena in Troy on the preservation of which the city depended for safety. [<L <Gk. *palladion*, neut. of *palladios* of Pallas (Athena)]
pal·la·dous (pə-lā′dəs) *adj. Chem.* Of, pertaining to, or containing palladium, especially in its lower valence.
pal·lah (pä′lə) See IMPALA.
Pal·las (pal′əs) 1 In Greek mythology, a name of Athena, the goddess of wisdom: often **Pallas Athena**. 2 The second largest asteroid.
pall-bear·er (pôl′bâr′ər) *n.* One who attends a coffin at a funeral. [<PALL[1] + BEARER]
pal·les·the·sia (pal′is-thē′zhə, -zhē·ə) *n. Physiol.* Sensitiveness to vibration, as of the skin or a bony prominence to a vibrating tuning fork. [<Gk. *pallein* quiver + ESTHESIA] — **pal′les·thet′ic** (-thet′ik) *adj.*
pal·let[1] (pal′it) *n.* 1 *Mech.* A click or pawl used to convert a reciprocating into a rotary motion, or the reverse, as in a feed motion; also, the lip or point of a pawl. 2 A paddle for mixing and shaping clay for crucibles, etc. 3 A tool used in gilding the backs of books or for taking up gold leaf. 4 A movable platform for the storage or transportation of goods. 5 A painter's palette. ◆ Homophones: *palate, palet, palette, pallette.* [<F *palette.* See PALETTE.]
pal·let[2] (pal′it) *n.* 1 A small, mean bed of mattress, usually of straw. 2 A blanket laid on the floor for a bed. ◆ Homophones: *palate, palet, palette, pallett.* [<OF *paillet* < *paille* straw <L *palea* chaff]
pal·let·ize (pal′it-īz) *v.t.* ·ized, ·iz·ing To load or store (goods) on pallets. See PALLET[1] (def. 4). [<PALLET[1] (def. 4) + -IZE]
pal·lette (pal′it) *n.* One of the plates in a suit of armor protecting the armpits. ◆ Homophones: *palate, palet, palette, pallet.* [<F *palette.* See PALETTE.]
pal·liasse (pal·yas′, pal′yas) See PAILLASSE.
pal·li·ate (pal′ē·āt) *v.t.* ·at·ed, ·at·ing 1 To cause (a crime, fault, etc.) to appear less serious or offensive; extenuate. 2 To relieve the symptoms or effects of without curing, as a disease; alleviate; mitigate. [<LL *palliatus*, pp. of *palliare* cloak <*pallium* a cloak] — **pal·li·a′tion** *n.* — **pal′li·a′tor** *n.*
— **Synonyms:** cloak, conceal, cover, excuse, extenuate, hide, mitigate, screen, veil. *Cloak*, from the French, and *palliate*, from the Latin, are the same in original signification, but have diverged in meaning; to *cloak* a sin is to attempt to *hide* it from discovery; to *palliate* it is to attempt to hide some part of its blameworthiness. Either to *palliate* or to *extenuate* is to admit the fault; but *extenuate* seeks especially to lessen the culpability involved; hence we speak of *extenuating* circumstances, since circumstances, while they cannot change the inherent wrong of an act, may yet lessen the blameworthiness of him who does it. In reference to diseases, to *palliate* is to diminish their violence, or partly to relieve the sufferer. See ALLAY, ALLEVIATE, HIDE. *Antonyms*: see synonyms for AGGRAVATE.
pal·li·a·tive (pal′ē·ā′tiv) *adj.* Having a tendency to palliate. — *n.* That which serves to palliate. — **pal′li·a·tive·ly** *adv.*
pal·lid (pal′id) *adj.* Of a pale or wan appearance; feeble in color. See synonyms under PALE[2]. [<L *pallidus* < *pallere* be pale. Doublet of PALE[2].] — **pal′lid·ly** *adv.* — **pal′lid·ness** *n.*
pal·li·um (pal′ē·əm) *n. pl.* ·li·a (-ē·ə) 1 A himation, the distinctive ancient Greek mantle, later adopted by the Romans as a Hellenism. 2 A vestment of the pope, archbishops, and metropolitans in the Roman Catholic Church, and of patriarchs in the Eastern Church; a pall. The Roman pallium is a yokelike band of white wool, with pendants on the breast and back, and is adorned with crosses. 3 *Zool.* The mantle of a brachiopod, mollusk, or bird. 4 *Anat.* The brain mantle or cerebral cortex, which is developed from the anterior vesicle, including the central white substance and the cortical gray. [<L]
pall-mall (pel′mel′) *n.* 1 A game formerly played in England and France by driving a wooden ball along an alley and through a raised iron ring by means of a mallet. 2 The mallet used in this game. 3 An alley or long space for playing the game. It gave its name to one of the streets of London, **Pall Mall** (pel′ mel′, pal′ mal′), noted for its numerous clubs. [<MF *pallemaille* <Ital. *pallamaglio* < *palla* a ball + *maglio* a mallet <L *malleus* a hammer]
pal·lor (pal′ər) *n.* The state of being pale or pallid; paleness. [<L < *pallere* be pale]
palm[1] (päm) *n.* 1 The hollow inner surface of the hand between the wrist and the base of the fingers. ◆ Collateral adjective: *thenar.* 2 The breadth (3 or 4 inches) or the length (about 8 1/2 inches) of the hand used as a linear measure. 3 That which covers the palm, as part of a glove or mitten. 4 The flattened, palmate portion of an antler, as of a moose. 5 The flat expanding end of any armlike projection; specifically, the blade of an oar. 6 A shield attached to the palm of the hand, used by sailmakers in pushing a needle through heavy canvas. — **to grease the palm** To give a bribe. — *v.t.* 1 To hide (cards, dice, etc.) in or about the hand, as in sleight of hand. 2 To handle or touch with the palm. — **to palm off** To pass off or impose fraudulently. [<F *paume* <OF *paulme* <L *palma* a hand; refashioned after L]
palm[2] (päm) *n.* 1 Any of a large and varied family (*Palmaceae*) of tropical trees or shrubs usually having an unbranched columnar trunk topped by a crown of large palmate or pinnate leaves. 2 A leaf or branch of the palm, used as a symbol of victory or joy. 3 Hence, supremacy; triumph; the reward or symbol of victory or preeminence. [OE *palm* <L *palma* a palm tree, orig., a hand; so called because its leaves are hand-shaped] — **pal·ma·ceous** (pal-mā′shəs) *adj.*
Pal·ma (päl′mä) 1 A port on Majorca, capital of the Balearic Islands. Also **Palma de Mallor·ca** (thä mä·lyôr′kä, mä·yôr′kä). 2 An island in the NW Canary group, 280 square miles: also **La Pal′ma**.
pal·ma-Chris·ti (päl′mä-kris′tē) *n.* The castor-oil plant.
pal·mar (pal′mər) *adj.* Pertaining to, like, or situated near or in the palm. [<L *palmaris* < *palma* a hand]
Pal·mas (päl′mäs), **Las** 1 One of two Spanish provinces in the Canary Islands; 1,583 square miles. 2 Its capital. Also **Las Palmas de Gran Ca·na·ri·a** (thä grän kä·nä′rē·ä).
pal·mate (pal′māt) *adj.* 1 Resembling an open hand. 2 Broad and flat, with fingerlike projections, as the antlers of the moose, or some corals. 3 *Bot.* Having lobes (usually five) that diverge from the apex of the petiole, as a leaf. 4 *Zool.* Webbed, as a bird's foot. Also **pal′mat·ed.** [<L *palmatus*, pp. of *palmare* mark with the palm of the hand < *palma* a hand] — **pal′mate·ly** *adv.*
pal·ma·tion (pal-mā′shən) *n.* 1 The state or quality of being palmate. 2 Any division of a palmate structure.
Palm Beach (päm) A resort town in SE Florida.
Palm Beach cloth A lightweight summer fabric of cotton warp and mohair filling: a trade name.
palm civet A long-tailed arboreal civet (family *Viverridae*) of Asia and Africa, especially the common *Paradoxurus hermaphroditus* of India.
palm·er (pä′mər) *n.* A medieval pilgrim who had visited Palestine and brought back a palm branch. [<AF <Med. L *palmarius* <L *palma* palm tree + -*arius* -ARY]
Palm·er (pä′mər), **Daniel David,** 1845–1913, U.S. founder of chiropractic. — **George Herbert,** 1842–1933, U.S. scholar and educator.
Palmer Archipelago An island group off the NW coast of Palmer Peninsula: also *Antarctic Archipelago.*
Palmer Peninsula A region of Antarctica, extending 800 miles toward South America: claimed (as *Graham Land*) by Great Britain as part of the Falkland Islands: 63–70° S., 53–68° W.
Palm·er·ston (pä′mər·stən), **Viscount,** 1784–1865, Henry John Temple, British statesman.
Palm·er·ston (pä′mər·stən) 1 One of the Cook Islands; one square mile: also *Avaran.* 2 The former name for DARWIN, Australia.
Palmerston North A city in southern North Island, New Zealand.
palm·er·worm (pä′mər·wûrm′) *n.* The caterpillar of a tineid moth, especially the *Dichomeris ligulella* which skeletonizes apple leaves.
pal·mette (pal·met′) *n.* A conventional carved or painted ornament in ancient art, resembling the palm leaf. [<F, dim. of *palme* a palm tree]
pal·met·to (pal-met′ō) *n. pl.* ·tos or ·toes Any one of various fan palms, especially the cabbage palm of the southern United States. [<Sp. *palmito*, dim. of *palma* a palm tree <L; ending infl. by Ital. -*etto*, dim. suffix]
Palmetto State Nickname of SOUTH CAROLINA.
palmi- combining form Palm. [<L *palma* palm]
pal·mi·ped (pal′mi·ped) *adj.* Web-footed, as a swimming bird. — *n.* A swimming bird. [<L *palmipes, palmipedis* < *palma* palm + *pes, pedis* foot]
palm·ist (pä′mist) *n.* One who practices palmistry.
palm·is·try (pä′mis·trē) *n.* The art of reading the past life or future of a person by the lines and marks in the palm of the hand. [ME *palmestrie* < *palme* palm + -*estrie*, prob. <OF *maistrie* mastery <L *magister* master]
pal·mi·tate (pal′mə·tāt) *n. Chem.* A salt or ester of palmitic acid. [<PALMITIC + -ATE[3]]
pal·mit·ic (pal·mit′ik) *adj.* Of, pertaining to, or derived from the palm, or especially, from palm oil. [<F *palmitique* <L *palma* a palm tree]
palmitic acid *Biochem.* A crystalline fatty acid, $C_{15}H_{31}CO_2H$, contained in numerous animal and vegetable fats and fixed oils, principally as glycerides: used in making candles and soaps.
pal·mi·tin (pal′mə·tin) *n. Chem.* A colorless crystalline compound, glyceryl palmitate, $(C_{15}H_{31}COO)_3C_3H_5$, contained in those natural fats that yield palmitic acid on saponification, and especially in palm oil: also called *tripalmitin.* [<F *palmitine*]
pal·mi·toyl (pal′mə·toil) *n. Chem.* The univalent radical, $CH_3(CH_2)_{14}CO$, from palmitic acid. [<PALMIT(IC ACID) + -(O)YL]
palm oil 1 A yellow or reddish fat or butter obtained from the fruit of several varieties of palm, especially the African oil palm (*Elaeis guineensis*): used in the manufacture of soap, candles, and for coloring and scenting ointments. 2 *Slang* Money given as a bribe or tip.
palm sugar Sugar made from palm sap.
Palm Sunday The Sunday before Easter, being the last Sunday in Lent and the first day in Holy Week: so called in commemoration of Christ's triumphal entry into Jerusalem, when palm branches were strewn before him. John xii 13.
palm·y (pä′mē) *adj.* **palm·i·er, palm·i·est** 1 Marked by prosperity; flourishing. 2 Abounding in or resembling palms.
pal·my·ra (pal·mī′rə) *n.* An East Indian palm (*Borassus flabellifer*) with a cylindrical stem 50 to 100 feet in height bearing a crown of large fan-shaped leaves. [<Pg. *palmeira* a palm tree <L *palma*; infl. by Gk. *Palmyra* PALMYRA]
Pal·my·ra (pal·mī′rə) 1 An ancient, ruined city in central Syria, NE of Damascus. 2 A northern atoll of the Line Islands, comprising part of the Territory of Hawaii; one square mile.
pal·nut (pôl′nut′, pal′-) *n. Mech.* A thin steel nut having a shallow, concave bottom face which by deformation under stress causes the nut to exert a binding grip on the bolt. [Origin unknown]
Pa·lo Al·to (pä′lō äl′tō) A battlefield of the Mexican War (1846) in southern Texas near Brownsville.
Pal·o·mar (pal′ə·mär), **Mount** A mountain of southern California, NW of San Diego; 6,126 feet; site of **Mount Palomar Observatory**, having the world's largest reflecting telescope.
pal·o·mi·no (pal′ə·mē′nō) *n. pl.* ·nos A golden-brown or yellow horse. [<Am. Sp., orig. a

dove-colored horse <Sp. *palomillo,* dim. of *paloma* a dove <LL *palumbus* <L *palumbes* a ring dove]

pa·loo·ka (pə·lo͞o′kə) *n. U.S. Slang* An inferior or bungling athlete; a lout; lummox. [? <Sp. *peluca,* a term of reproof, lit., a wig]

Pa·los (pä′lōs) A village, formerly a port, in SW Spain, where Columbus embarked in 1492. Also **Palos de la Fron·te·ra** (thä lä frōn·tā′rä).

Pa·louse River (pə·lo͞os′) A river in northern Idaho and SE Washington, flowing 140 miles west to the Snake River.

palp (palp) *n.* A palpus. [<F *palpe* <L *palpus.* See PALPUS.]

pal·pa·ble (pal′pə·bəl) *adj.* **1** That may be touched or felt. **2** Readily perceived; obvious. **3** That may be perceived by touch, or by any of the other senses. See synonyms under EVIDENT, MANIFEST. [<LL *palpabilis* <L *palpare* touch] **—pal′pa·bil′i·ty, pal′pa·ble·ness** *n.* **—pal′pa·bly** *adv.*

pal·pate (pal′pāt) *v.t.* **·pat·ed, ·pat·ing** To feel or examine by touch, especially for medical diagnosis. —*adj.* Having a palpus or sense organ. [<L *palpatus,* pp. of *palpare* touch]

pal·pa·tion (pal·pā′shən) *n. Med.* The process of examining or exploring the body by means of touch; a digital exploration.

pal·pe·bra (pal′pi·brə) *n. pl.* **·brae** (-brē) An eyelid. [<NL <L, an eyelid] **—pal′pe·bral** *adj.*

pal·pi·tate (pal′pə·tāt) *v.i.* **·tat·ed, ·tat·ing 1** To quiver; tremble. **2** *Med.* To beat more rapidly than normal; flutter: said especially of the heart. [<L *palpitatus,* pp. of *palpitare* tremble, freq. of *palpare* touch]

pal·pi·ta·tion (pal′pə·tā′shən) *n.* Rapid and irregular pulsation.

pal·pus (pal′pəs) *n. pl.* **·pi** (-pī) *Zool.* A feeler; especially, one of the jointed sense organs attached to the mouth parts of arthropods: also called *palp.* [<NL <L *palpus* a feeler <*palpare* touch]

pals·grave (pôlz′grāv, palz′-) *n.* In German history, one having charge of a royal or the imperial court or household; also, one of an order of nobility; one of the hereditary rulers of the Palatinate. [<MDu. *paltsgrave* <*palts* a palace (ult. <L *palatium*) + *grave* count] **—pals′gra·vine** (-grə·vēn) *n. fem.*

pal·sy (pôl′zē) *n.* **1** Paralysis. **2** Any impairment or loss of sensation or of ability to control movement. —*v.t.* **·sied, ·sy·ing** To paralyze. [<OF *paralisie* <L *paralysis.* Doublet of PARALYSIS.] **—pal′sied** *adj.*

pal·ter (pôl′tər) *v.i.* **1** To speak or act insincerely; equivocate; lie. **2** To treat something lightly; trifle. **3** To haggle or quibble. [Cf. dial. E *palt* a piece of coarse or dirty cloth] **—pal′ter·er** *n.*

pal·try (pôl′trē) *adj.* **·tri·er, ·tri·est** Having little or no worth or value; trifling; trivial; contemptible; petty. See synonyms under BASE, CHILDISH, INSIGNIFICANT, LITTLE, PITIFUL. [<dial. E, rags, rubbish <*palt* a piece of coarse or dirty cloth] **—pal′tri·ly** *adv.* **—pal′tri·ness** *n.*

pa·lu·dal (pə·lo͞od′l) *adj.* Pertaining to a marsh; swampy. [<L *palus, paludis* marsh]

pal·u·dism (pal′yə·diz′əm) *n. Pathol.* The morbid condition observed in those who live among marshes; malaria.

pal·y[1] (pā′lē) *adj.* Lacking brilliance; pale; pallid.

pal·y[2] (pā′lē) *adj. Her.* Divided palewise, the number of such divisions (always even) being specified. [<OF *palé* a row of stakes <*pal* PALE[1]]

pal·y·nol·o·gy (pal′ə·nol′ə·jē) *n.* The scientific study of pollen and other spores, their dispersal and application; pollen analysis. [<Gk. *palynein* strew <*pallein* brandish + -LOGY]

pam (pam) *n.* **1** In the game of loo, the knave of clubs. **2** A game resembling napoleon, wherein the highest trump is the knave of clubs. [<Short for F *pamphile* knave of clubs <Gk. *Pamphilos,* a personal name; lit., beloved of all]

Pa·mir (pä·mir′) An elevated region of central Asia, chiefly in the Tadzik S.S.R., central Asia; a range of the Pamir-Alai, with high plains lying at 11,000 to 13,000 feet, and rising to Stalin Peak, 24,590 feet, the highest in the U. S. S. R.; known as "the roof of the world." Also **The Pamirs.**

Pa·mir-A·lai (pä·mir′a·lī′) A principal mountain system of Central Asia, in the U. S. S. R., China, and Afghanistan.

Pam·li·co Sound (pam′li·kō) A body of water between eastern North Carolina and its offshore islands; 80 miles long; greatest width, 25 miles.

Pam·pa (päm′pä), **La** See LA PAMPA.

pam·pas (pam′pəz, *Sp.* päm′päs) *n. pl.* The great open treeless plains south of the Amazon river, extending from the Atlantic to the Andes. [<Sp. *pampa* <Quechua, plain]

pam·pas grass (pam′pəs) A tall, ornamental, reedlike grass (*Cortaderia selloana*) native to South America, having very large, thick, silvery panicles.

pam·pe·an (pam′pē·ən, pam·pē′ən) *adj.* Of or pertaining to the pampas or to their native inhabitants. —*n.* An Indian of the pampas; a pampero.

pam·per (pam′pər) *v.t.* **1** To treat too indulgently; gratify the whims or wishes of; coddle. **2** *Obs.* To feed with rich food; glut. [Appar. freq. of obs. *pamp* cram] **—pam′per·er** *n.*

Synonyms: caress, coddle, glut, indulge, pet, spoil. See CARESS, INDULGE. *Antonyms:* deny, discipline, harden, inure, starve, stint.

pam·pe·ro (päm·pā′rō) *n. pl.* **·ros 1** A strong, cold, dry, southwest wind of the Argentine pampas, generally advancing with a well-marked and very black cloud front. **2** An Indian of the pampas. [<Sp. <Quechua *pampa* plain + Sp. *-ero* -ARY[1]]

pam·phlet (pam′flit) *n.* **1** A printed work stitched or pasted, but not permanently bound. **2** A brief treatise or essay, printed and published without a binding, and usually on a subject of current interest. [<OF *Pamphilet,* popular title of a 12th c. Latin love poem, *Pamphilus, seu de Amore*]

pam·phlet·eer (pam′flə·tir′) *n.* One who writes pamphlets: sometimes a term of contempt. —*v.i.* To write and issue pamphlets.

Pam·phyl·i·a (pam·fil′ē·ə) An ancient country and Roman province in southern Asia Minor. *Acts* ii 10.

Pam·plo·na (päm·plō′nä) A city of northern Spain; capital of Navarre. Also **Pam·pe·lu·na** (päm′pä·lo͞o′nä).

Pa·mun·key River (pə·mung′kē) A river in eastern Virginia, flowing 90 miles SE to the York River; scene of Civil War battles, 1864.

pan[1] (pan) *n.* **1** A wide, shallow vessel, usually metallic or earthen, for domestic use, as in holding liquids or in cooking. **2** A vessel, either open or closed, for boiling and evaporating. **3** A natural or artificial depression in the earth for evaporating brine. **4** A circular sheet-iron dish with sloping sides, in which gold is separated. **5** The powder cavity of a flintlock. **6** The skull; brainpan. **7** Any natural depression in the earth containing water or mud. **8** Hardpan. **9** Either of the two receptacles on a pair of scales or a balance. —*v.* **panned, pan·ning** *v.t.* **1** To separate (gold) by washing gold-bearing earth in a pan. **2** To wash (earth, gravel, etc.) for this purpose. **3** To cook and serve in a pan. **4** *Colloq.* To criticize severely. **5** *Colloq.* To obtain; secure. —*v.i.* **6** To search for gold by washing earth, gravel, etc., in a pan. **7** To yield gold, as earth. **—to pan out** *U.S.* To result or turn out; transpire. [OE *panne,* ? <LL *panna* <L *patina* a pan or dish <*patere* stand open]

pan[2] (pan) *n.* **1** The leaf of the climbing pepper (*Piper betle*) used with the nuts of the betel palm as a masticatory in the East Indies. **2** The masticatory obtained from this leaf. [<Hind. *pān* a betel leaf <Skt. *parṇa* a feather, a leaf]

pan[3] (pan) *v.t.* **panned, pan·ning** To move (a motion-picture or television camera) across a scene in order to secure a panoramic effect. [<PANORAMA]

Pan (pan) In Greek mythology, a horned, goat-footed god of forests, flocks, and shepherds: identified with the Roman *Faunus.*

pan- *combining form* **1** All; every; the whole: *panchromatic.* **2** Comprising, including, or applying to all: usually capitalized when preceding proper nouns or adjectives, as in:

Pan-African	**Pan-Asian**	**Pan-Slav**
Pan-Arab	**Pan-Islam**	**Pan-Slavic**
Pan-Arabian	**Pan-Islamic**	**Pan-Slavonic**
Pan-Asia	**Pan-Russian**	**Pan-Syrian**

pan·a·ce·a (pan′ə·sē′ə) *n.* **1** A remedy for all diseases; a cure-all. **2** An herb credited with remarkable healing virtues; formerly, allheal. [<L <Gk. *panakeia* a universal remedy <*panakēs* all-healing <*pan,* neut. of *pas* all + *akos, akeos* cure] **—pan′a·ce′an** *adj.*

pa·nache (pə·nash′, -näsh′) *n.* A plume or bunch of feathers, especially as an ornament on a helmet. [<F <Ital. *pennacchio* <*penna* a feather <L]

PANACHE

pa·na·da (pə·nä′də, -nā′-) *n.* A dish made of crackers or bread soaked with boiling water, sweetened and eaten with milk, or flavored with wine, etc. [<Sp. <Ital. *pane* bread <L *panis*]

Pan·a·ma (pan′ə·mä, -mô; *Sp.* pä′nä·mä′) **1** A republic on the Isthmus of Panama; 28,575 square miles (excluding Canal Zone). **2** Its capital, near the Pacific terminus of the Panama Canal.

Panama, Isthmus of An isthmus connecting North and South America and separating the Atlantic and Pacific; 30 miles wide at its narrowest point: formerly *Isthmus of Darien.*

Panama Canal A ship canal across the Isthmus of Panama, extending about 40 miles from Colón on the SE Atlantic (Caribbean) to Panama on the Pacific; completed (1914) by the United States on the leased territory of Canal Zone.

Panama Canal Zone See CANAL ZONE.

Panama fever **1** Yellow fever. **2** Malaria.

Panama hat A hat woven from the young leaves of the jipijapa tree of Central and South America.

Pan·a·ma·ni·an (pan′ə·mā′nē·ən, -mä′-) *adj.* Of or pertaining to the Isthmus of Panama or its inhabitants: also **Pa·nam·ic** (pə·nam′ik). —*n.* A native or naturalized inhabitant of Panama. Also **Pan·a·man** (pan′ə·män′).

Pan American Including or pertaining to the whole of America, both North and South, or to all Americans. Also **Pan′-A·mer′i·can.**

Pan American Conference Any one of various conferences of delegates from the republics of North and South America.

Pan American Highway A projected system of roads, totaling 15,714 miles, to link the nations of the western hemisphere.

Pan-A·mer·i·can·ism (pan′ə·mer′ə·kən·iz′əm) *n.* The advocacy of a political union or alliance, or of closer political and economic cooperation, among the republics of the western hemisphere; also, the life of the American peoples as represented in republican forms of government and tending toward such a union.

Pan American Union A bureau (formerly the *International Bureau of the American Republics*) established in Washington, D.C., in 1890, by the 21 American republics to promote mutual peace, commerce, and friendship.

Pan·a·mint Mountains (pan′ə·mint) A range in SE California; highest point, 11,045 feet.

Pan-An·gli·can (pan′ang′gli·kən) *adj.* Pertaining to or including all branches or members of the Anglican church.

pan·a·tel·a (pan′ə·tel′ə) See PANETELA.

Pan·ath·e·nae·a (pan·ath′ə·nē′ə) *n. pl.* Ancient Athenian festivals celebrated yearly in midsummer, with special magnificence (the **greater Panathenaea**) every fifth year: held in honor of Athena, founded by Pisistratus, and said to commemorate the union of Attica under Theseus. The **lesser Panathenaea,** held in the other four years, were founded by Erechtheus. [<Gk. *Panathēnaia* (*hiera*) (festival of) Athena <*pan,* neut. of *pas* all + *Athēnē* Athena] **—Pan·ath′e·nae′an, Pan·ath′e·na′ic** (-nā′ik) *adj.*

pan·at·ro·phy (pan·at′rə·fē) *n. Pathol.* Atrophy involving many or all parts of the body.

Pa·nay (pä·nī′) An island of the central Philippines; 4,446 square miles; chief town, Iloilo.

pan-broil (pan′broil′) *v.t.* & *v.i.* To cook (meat) in a heavy frying pan placed over direct heat, using little or no fat.

pan·cake (pan′kāk′) *n.* **1** A thin battercake

fried in a pan or baked on a griddle: also **pan cake.** 2 *Aeron.* An abrupt or violent landing effected by an airplane which levels off and settles rapidly on a steep flight path. — *v.i.* **·caked, ·cak·ing** To level off and decelerate an airplane so that it drops to the ground with little forward movement.

pan·chro·mat·ic (pan′krō·mat′ik) *adj. Phot.* Sensitive to all the colors of the spectrum in proportion to their respective visual luminosities: said of an emulsion, film, or plate. — **pan·chro′ma·tism** (-krō′mə·tiz′əm) *n.*

pan·cra·ti·um (pan·krā′shē·əm) *n. pl.* **·ti·a** (-shē·ə) An ancient Greek contest of athletes, including both boxing and wrestling. [<L <Gk. *pan,* neut. of *pas* all + *kratos* strength] — **pan·crat′ic** (-krat′ik) *adj.*

pan·cre·as (pan′krē·əs, pang′-) *n. Anat.* A gland connecting with the alimentary canal; in vertebrates, a large racemose gland behind the peritoneum, between the lower part of the stomach and the vertebrae of the loins, and emptying into the duodenum by one or more small ducts; the sweetbread. [<NL <Gk. *pankreas* sweetbread < *pan,* neut. of *pas* all + *kreas* flesh] — **pan′cre·at′ic** (-at′ik) *adj.*

pancreatic juice *Biochem.* A colorless fluid, containing certain enzymes, secreted by the pancreas and forming an important factor in digestion by emulsifying fats.

pan·cre·a·tin (pan′krē·ə·tin, pang′-) *n.* 1 *Biochem.* One of the active ferments of the pancreatic juice, or a mixture of them. 2 A commercial digestant extract of the pancreas of the ox or hog. [<Gk. *pankreas, -atos* pancreas + -IN]

pan·da (pan′də) *n. pl.* **·das** or **·da** 1 A small raccoonlike carnivore (*Ailurus fulgens*) of the southeastern Himalayas, with long red-dish-brown fur and ringed tail; the red bearcat. 2 The great panda (*Ailuropoda melanoleuca*) of Tibet and China, a mammal of bearlike appearance, with black-and-white coat and rings around the eyes: also *giant panda.* [Prob. <Nepalese]

GIANT PANDA
(About the size of a large bear)

Pan·da·nus (pan·dā′nəs) *n.* A genus of trees and shrubs of southeastern Asia (family *Pandanaceae*), characterized by stiltlike aerial roots and the spiral arrangement of their long gracefully recurved leaves; a screwpine. [<NL <Malay *pandan* conspicuous] — **pan·da·na·ceous** (pan′də·nā′shəs) *adj.*

Pan·da·rus (pan′də·rəs) In the *Iliad,* son of Lycaon and leader of the Lycians in the Trojan War; in medieval legend, Chaucer's *Troilus and Criseyde,* and Shakespeare's *Troilus and Cressida,* a go-between who procures Cressida for Troilus. Also **Pan·dar** (pan′dər).

Pan·de·an (pan·dē′ən) *adj.* Pertaining to the god Pan.

Pandean pipes A primitive wind instrument made of graduated reeds; a panpipe.

pan·dect (pan′dekt) *n.* 1 An encyclopedic treatise; a complete digest of some department of knowledge. 2 Any complete system of law. [<F *pandecte* <L *pandecta* <Gk. *pandektēs* an all-receiver < *pan,* neut. of *pas* all + *dechesthai* take]

Pan·dects (pan′dekts) *n. pl.* A compilation or digest from the decisions of Roman jurists, made by direction of the Emperor Justinian about A.D. 533: also called *The Digest.* See CORPUS JURIS CIVILIS.

pan·dem·ic (pan·dem′ik) *adj.* 1 Pertaining to or affecting all the people. 2 *Med.* Widely epidemic. — *n.* A pandemic disease. [<Gk. *pandēmos* pertaining to all the people < *pan,* neut. of *pas* all + *dēmos* people]

pan·de·mo·ni·um (pan′də·mō′nē·əm) *n.* 1 The abode of all demons; the infernal regions; as used by Milton, **Pandemonium,** the palace of Satan in Hell. 2 Hence, any place or gathering remarkable for disorder and uproar. 3 A fiendish or riotous uproar. Also **pan′·dae·mo′ni·um.** [<NL <Gk. *pan,* neut. of *pas* all + *daimōn* an evil spirit]

pan·der (pan′dər) *n.* 1 A go-between in sexual intrigues; a procurer; pimp. 2 One who ministers to the passions or base desires of others. Also **pan′der·er.** — *v.i.* To act as a pander. — *v.t.* To act as a pander for. [after *Pandarus,* with ref. to his role in medieval legend] — **pan′der·age, pan′der·ism** *n.* — **pan′·der·ess** *n. fem.*

Pan·dit (pun′dit), **Vijaya Lakshmi,** born 1900, Indian diplomat and administrator; sister of Jawaharlal Nehru.

Pan·do·ra (pan·dôr′ə, -dō′rə) In Greek mythology, the first mortal woman, sent to earth by the gods as punishment for the theft of fire by Prometheus. She brought with her a box (**Pandora's box**) containing all human ills. When in curiosity she opened the lid, the ills escaped into the world, leaving only Hope, which had been at the bottom of the box. See EPIMETHEUS. [<Gk., all-gifted]

pan·dore (pan′dôr, -dōr) *n.* A bandore. Also **pan·do′ra, pan·du·ra** (pan·dŏŏr′ə, -dyŏŏr′ə). [<F *pandore* <L *pandura.* See BANDORE.]

pan·dour (pan′dŏŏr) *n.* 1 A member of a force of Croatian foot soldiers, noted for their brutality: organized locally in 1741, and later incorporated in the Austrian army. 2 Any inhuman or marauding soldier. Also **pan′door.** [<F *pandour* <Serbo-Croatian *pandūr* constable, ult. <Med. L *banderius* follower of a banner]

pan·dow·dy (pan·dou′dē) *n. pl.* **·dies** A deep-dish pie or pudding made of baked sliced apples. Also **apple pandowdy.** [Cf. obs. dial. E (Somerset) *pandoulde* a custard]

pan·du·rate (pan′dyə·rāt) *adj. Bot.* Fiddle-shaped, as certain leaves. Also **pan·du·ri·form** (pan·dŏŏr′ə·fôrm, -dyŏŏr′-). [<L *pandura* a lute + -ATE[1]]

pan·dy (pan′dē) *n. pl.* **·dies** *Scot. & Dial.* A stroke on the palm of the hand with a cane or strap, as a punishment. — *v.t.* **·died, ·dy·ing** To punish thus. [<L *pande (palmam)* extend (your hand), imperative of *pandere* extend]

pane[1] (pān) *n.* 1 A piece or compartment, particularly if flat and rectangular. 2 A piece of window glass filling one opening in a frame. 3 A flat surface, as on an object having several sides: the *pane* of a tower, nut, or brilliant-cut diamond. 4 A panel, or space between timbers; also, a bay. ◆ Homophone: *pain.* [<OF *pan(e)* <L *pannus* a piece of cloth]

pane[2] (pān) *n.* The peen of a hammer. ◆ Homophone: *pain.* [Cf. F *panne* peen of a hammer]

pan·e·gyr·ic (pan′ə·jir′ik) *n.* A formal public eulogy, either written or spoken; encomium; laudation in general. See synonyms under EULOGY, PRAISE. — *adj.* Elaborately eulogistic or laudatory: also **pan′e·gyr′i·cal.** [<F *panégyrique* <L *panegyricus* <Gk. *panēgyrikos* of an assembly < *panēgyris* an assembly < *pan,* neut. of *pas* all + *agyris,* Aeolic form of *agora* gathering] — **pan′e·gyr′i·cal·ly** *adv.* — **pan′e·gyr′ist** *n.*

pan·e·gy·rize (pan′ə·jə·rīz′) *v.* **·rized, ·riz·ing** *v.t.* To deliver or write a panegyric upon; eulogize. — *v.i.* To make panegyrics.

pan·el (pan′əl) *n.* 1 A rectangular piece set in or as in a frame, as in a door, or sunken below it, as a window pane; hence, any such piece, even if raised above the general plane. 2 A bordered member to which the effect of framing is given by moldings or by working away material from the solid plane. 3 One or more pieces of a different fabric and color inserted lengthwise in the skirt of a woman's dress. 4 A tablet of wood, used as the surface for an oil painting; also, the picture on such a tablet. 5 A picture very long for its width, in a simple frame, or with no frame at all. 6 A size of photograph longer than it is wide, usually about 8 1/2 × 4 inches. 7 A face on a hewn stone. 8 A section of a book cover having a framed effect; also, a subdivision of the back of a bound book, between two bands. 9 *Law* The official list of persons summoned for jury duty; the body of persons composing a jury. 10 *Aeron.* **a** One of the construction units forming the wing surface of an airplane. **b** The unit of fabric of which the outer covering of a balloon or the canopy of a parachute is made. 11 An upright board of insulating material sustaining the controlling devices of an electric circuit. 12 An array of dials, gages, and instruments for the operation of an aircraft, automobile, or other complex apparatus. 13 A small group of persons assembled for judging, discussion, etc. — *v.t.* **·eled** or **·elled, ·el·ing** or **·el·ling** 1 To fit, furnish, or adorn with panels. 2 To divide into panels. 3 *Scot. Law* To indict. [<OF *panel* a piece of cloth <Med. L *panellus,* dim. of L *pannus* a cloth]

panel fence A kind of worm fence made in sections.

pan·el·ing (pan′əl·ing) *n.* 1 Work in panels; panels collectively. 2 The introduction or use of panelworking. Also **pan′el·ling.**

pan·el·ist (pan′əl·ist) *n.* A person serving on a panel of judges or debaters.

pan·el·work (pan′əl·wûrk′) *n.* 1 Wainscoting; any work using or introducing panels. 2 Panelworking.

pan·el·work·ing (pan′əl·wûrk′ing) *n. Mining* A method of working a colliery by dividing it into large rooms separated by very wide masses of coal.

pa·nem et cir·cen·ses (pā′nəm et sər·sen′sēz) *Latin* Bread and circuses; food and amusements.

pan·e·tel·a (pan′ə·tel′ə) *n.* A long, slender, cylindrical-shaped cigar: also spelled *panatela.* Also **pan′e·tel′la.** [<Sp.]

pan fish Any little fish that can be fried whole.

pan-fry (pan′frī′) *v.t.* **-fried, -fry·ing** To fry in a frying pan.

pang[1] (pang) *n.* A sudden and poignant pain; keen transient agony; hence, a throe of mental anguish. See synonyms under AGONY, PAIN. [? Alter. of ME *prange* a prong, point]

pang[2] (pang) *Scot. v.t.* To cram; squeeze. — *adj.* Crammed full.

pan·ga·my (pang′gə·mē) *n. Biol.* Indiscriminate or random mating. — **pan·gam·ic** (pan·gam′·ik), **pan′ga·mous** *adj.* — **pan′ga·mous·ly** *adv.*

pan·gen·e·sis (pan·jen′ə·sis) *n. Biol.* The theory, advanced by Charles Darwin, that all the cells of an organism throw off very minute gemmules, **pangens,** which circulate through the body and develop buds or germ cells which have the power of reproduction and contain the units of heredity which they transmit from parent to offspring. — **pan′ge·net′ic** (-jə·net′ik) *adj.*

Pan-Ger·man·ism (pan′jûr′mən·iz′əm) *n.* Originally, the theory of, or organized effort toward, political union of all Teutonic peoples; later, the German doctrine of world domination by aggression and ideology. — **Pan′-Ger′man** *adj. & n.* — **Pan′-Ger·man′ic** (-jər·man′ik) *adj.* — **Pan′-Ger′man·ist** *n.*

Pan·gim (puṅ·zhēṅ′) See NOVA GOA.

pan·go·lin (pang·gō′lin) *n.* A heavily armored, typically long-tailed edentate mammal (genus *Manis*) of Asia and Africa; the scaly ant-eater. [<Malay *peng-goling* a roller < *gōling* roll, in ref. to its power of rolling itself up]

Pang·o Pang·o (päng′ō päng′ō) See PAGO PAGO.

pan·han·dle[1] (pan′han′dəl) *v.i.* **·dled, ·dling** *U.S. Colloq.* To beg, especially on the street. [Back formation from PANHANDLER a beggar < PAN[1] (used to receive alms) + HANDLE, *v.*] — **pan′han′dler** *n.*

pan·han·dle[2] (pan′han′dəl) *n.* 1 A narrow strip of land attached to a larger region: from its resemblance to the handle of a pan. 2 *Usually cap.* A region of this shape in either Texas or West Virginia.

Panhandle State Nickname of WEST VIRGINIA.

Pan-Hel·le·nism (pan′hel′ə·niz′əm) *n.* The aspiration for the political union of all Greeks, or the effort to accomplish such union; also, that which pertains to universal Greek interests and ideas. — **Pan′-Hel·len′ic** (-he·len′ik) *adj.* — **Pan′-Hel′le·nist** *n.*

pan·ic (pan′ik) *adj.* 1 Of the nature of or resulting from sudden and infectious terror. 2 Of or pertaining to the Greek god Pan as the cause of fear: *panic* flight. — *n.* 1 A sudden, unreasonable, overpowering fear, especially when affecting a large number simultaneously. 2 Sudden and overpowering alarm or distrust in financial or commercial circles, precipitating mercantile and banking failures. See synonyms under ALARM, FEAR,

pan ice Loose fragments of ice detached from ice floes and drifting along the seacoast.

panic grass A North American grass (*Panicum capillare*) used for forage: also called *witchgrass*. [< L *panicum* a kind of millet < *panis* bread + GRASS]

pan·i·cle (pan′i·kəl) *n. Bot.* A loose compound flower cluster, produced by irregular branching. [< L *panicula*, dim. of *panus* a swelling, an ear of millet] — **pan′i·cled** *adj.*

pan·ic-strick·en (pan′ik-strik′ən) *adj.* Overcome by panic. Also **pan′ic-struck′**.

pa·nic·u·late (pə·nik′yə·lāt, -lit) *adj.* Arranged or borne in panicles; panicled. Also **pa·nic′u·lat·ed**. — **pa·nic′u·late·ly** *adv.*

Pan·ja·bi (pun·jä′bē) See PUNJABI.

pan·jan·drum (pan·jan′drəm) *n.* An imaginary character of exaggerated importance or great pretensions; hence, a pompous personage in a small place. [Coined by Samuel Foote, English dramatist and actor, in 1755]

Pan·jim (puṅ·zhēn′) See NOVA GOA.

Panj·nad (punj′näd) The combined waters of five rivers of the Punjab, West Pakistan, flowing 50 miles SW to the Indus.

Pank·hurst (pangk′hûrst), **Emmeline**, 1858–1928, née Goulden, English suffragist.

pan·mix·i·a (pan·mik′sē·ə) *n.* Indiscriminate interbreeding; unrestricted mating in a mixed population. [< NL < Gk. *pan*, neut. of *pas* all + *mixis* a mixture]

Pan-mun·jom (pän·mŏon′jom) A village in southern Korea south of Kaesong; site of truce talks, 1951.

panne satin (pan) Silk or rayon satin with an unusually high luster because of a special finish. [< F *panne*, a type of soft cloth + SATIN]

panne velvet Silk or rayon velvet with a flattened pile, lustrous and lightweight.

pan·nier (pan′yər) *n.* 1 One of a pair of baskets adapted to be slung on both sides of a beast of burden. 2 A basket for carrying a load on the back. 3 A light framework for extending a woman's dress at the hips; also, a skirt or overskirt extended at the hips. [< MF < L *panarium* bread basket < *panis* bread]

pan·ni·kin (pan′ə·kin) *n.* A small saucepan or tin cup. [Dim. of PAN¹]

Pan·no·ni·a (pə·nō′nē·ə) An ancient Roman province south and west of the Danube in modern Hungary and Yugoslavia.

pa·no·cha (pə·nō′chə) *n.* 1 A coarse Mexican sugar. 2 A kind of candy made from brown sugar, milk, and usually containing chopped nuts: also spelled *penuche, penuchi*. Also **pa·no′che**. [< Am. Sp. < L *panucula, panicula*. See PANICLE.]

pan·o·ply (pan′ə·plē) *n. pl.* **·plies** 1 The complete equipment of a warrior. 2 Hence, any complete covering that protects or magnificently arrays. [< Gk. *panoplia* full armor < *pan*, neut. of *pas* all + *hopla* arms] — **pan′o·plied** *adj.*

pan·op·tic (pan·op′tik) *adj.* Inclusive of all that is visible in one view. Also **pan·op′ti·cal**.

pan·o·ra·ma (pan′ə·ram′ə, -rä′mə) *n.* 1 A series of pictures representing a continuous scene, arranged to unroll and pass before the spectator. 2 A complete view in every direction; also, a complete or comprehensive view of a subject or of constantly passing events. 3 A cyclorama. [< PAN- + Gk. *horama* sight < *horaein* see] — **pan′o·ram′ic** *adj.* — **pan′o·ram′i·cal·ly** *adv.*

panoramic sight A sight constructed on the periscope principle, intended for use by marksmen.

Pan·or·mus (pə·nôr′məs) The ancient name for PALERMO.

pan·pipes (pan′pīps′) *n. pl.* A wind instrument formed of short hollow tubes (originally reeds) fastened together in proper order to produce a scale.

pan·psy·chism (pan·sī′kiz·əm) *n.* The doctrine that the universe as a whole, and every physical part of it also, has a psychic aspect. [< PAN- + Gk. *psychē* soul, breath] — **pan·psy′chic** *adj.* — **pan·psy′chist** *n.*

pan·so·phism (pan′sə·fiz′əm) *n.* Profession to universal wisdom. — **pan′so·phist** *n.*

pan·so·phy (pan′sə·fē) *n.* Complete or comprehensive knowledge; a system embracing all human knowledge. [< PAN- + Gk. *sophia* wisdom] — **pan·soph′ic** (pan·sof′ik), **pan·soph′i·cal** *adj.*

pan·sy (pan′zē) *n. pl.* **·sies** 1 A species of garden violet (*Viola tricolor hortensis*) having blossoms of a variety of colors of great beauty. 2 A bright, deep-purple color, the color of some pansies. 3 *Slang* An effeminate or homosexual man. [< MF *pensée* thought, orig. pp. of *penser* think]

pant (pant) *v.i.* 1 To breathe rapidly or spasmodically; gasp for breath. 2 To emit smoke, steam, etc., in loud puffs. 3 To gasp with desire; yearn: with *for* or *after*. 4 To beat or pulsate rapidly; throb, as the heart. — *v.t.* 5 To breathe out or utter gaspingly. See synonyms under PUFF. — *n.* A short or labored breath; a gasp; also, a quick or violent heaving, as of the breast; a throb, as of the heart. [Appar. < OF *pantoisier* gasp, ult. < L *phantasia* a nightmare] — **pant′er** *n.*

pant- Var. of PANTO-.

Pan·ta·gru·el (pan·tag′rōō·el, pan′tə·grōō′əl; *Fr.* pän·tȧ·grü·el′) In François Rabelais' satirical romance *Gargantua and Pantagruel*, the giant son of Gargantua, characterized by his broad, somewhat cynical good humor. — **Pan·ta·gru·e·li·an** (pan·tag′rōō·el′ē·ən) *adj.*

Pan·ta·gru·el·ism (pan·tag′rōō·əl·iz·əm, pan′tə·grōō′əl·iz·əm) *n.* The theories and practices of Pantagruel; good-natured cynicism. — **Pan·ta·gru·el·ist** *n.*

pan·ta·lets (pan′tə·lets′) *n. pl.* Long ruffled drawers, formerly worn by women and children. Also **pan′ta·lettes′**. [Dim. of PANTALOON]

pan·ta·loon (pan′tə·lōōn′) *n.* 1 In pantomimes, an absurd old man on whom the clown plays tricks. 2 *pl.* Trousers; formerly, a tight-fitting garment for the legs. [< F *pantalon* < Ital. *pantalone* a clown <*Pantalone*, nickname for a Venetian, after *Pantaleone*, a popular Venetian saint]

Pan·ta·loon (pan′tə·lōōn′) In early Italian comedies, an old dotard wearing a certain kind of tight-fitting trousers.

Pan·tar (pän′tär) One of the Lesser Sunda Islands in the Alor group; 281 square miles.

pan·tech·ni·con (pan·tek′ni·kon) *n. Brit.* 1 *Obs.* A place for the exhibition and sale of manufactured articles. 2 A van for moving furniture. [< PAN- + Gk. *technikon*, neut. of *technikos* belonging to the arts < *technē* art, craft]

Pan·tel·le·ri·a (pän′tel·le·rē′ä) An Italian island off the SW coast of Sicily; 32 square miles; ancient *Cossyra*.

Pan-Teu·ton·ism (pan·tōōt′n·iz′əm, -tyōō′-) *n.* The doctrine advocating a political union, or a union of interests, of all Teutonic peoples. — **Pan′-Teu·ton′ic** (-tōō·ton′ik, -tyōō-) *adj.*

pan·the·ism (pan′thē·iz′əm) *n.* 1 The form of monism that identifies mind and matter, the finite and the infinite, making them manifestations of one universal or absolute being; the doctrine which holds that the self-existent and self-developing universe, conceived as a whole, is God. 2 The worship of all gods. — **pan′the·ist** *n.* — **pan′the·is′tic, pan′the·is′ti·cal** *adj.* — **pan′the·is′ti·cal·ly** *adv.*

pan·the·on (pan′thē·on) *n.* 1 All the gods of a people, collectively. 2 A mausoleum or temple resembling the Roman Pantheon, and commemorating the great. [< L *pantheon* < Gk. *pantheion* a temple consecrated to all the gods < *pan*, neut. of *pas* all + *theos* a god]

Pan·the·on (pan′thē·on) A circular temple at Rome, dedicated to all the gods: built by Agrippa, rebuilt by Hadrian: after A.D. 609, a Christian church (Santa Maria della Rotunda).

pan·ther (pan′thər) *n.* 1 A leopard, especially the black leopard of southern Asia. 2 Some other large feline carnivore, as the North American puma or cougar; also, a jaguar. [< OF *pantère* < L *panthera* < Gk. *panthēr*] — **pan′ther·ess** *n. fem.*

pant·ies (pan′tēz) *n. pl.* A woman's or child's underpants. Also **pant′ie**.

pan·tile (pan′tīl′) *n.* A tile displaying a curved cross-section, making laps on each side with adjacent tiles of reverse form.

panto- combining form All; every: *pantoscope*. Also, before vowels, **pant-**. [< Gk. *pantos*, genitive of *pas* all]

pan·to·base (pan′tə·bās) *adj. Aeron.* Designating an aircraft capable of landing on or taking off from any kind of terrain, as mud, ice, snow, water, etc.

pan·to·fle (pan′tə·fəl, pan·tof′əl, -tōō′fəl) *n.* A slipper. Also **pan′to·fle**. [< MF *pantoufle* <Ital. *pantufola, pantofola*, ? <Med. Gk. *pantophellos* cork shoe, lit. whole cork <Gk. *pas, pantos* all + *phellos* cork]

pan·to·graph (pan′tə·graf, -gräf) *n.* 1 An instrument for copying a drawing, diagram, or map, either on the same scale or with reduction or enlargement. 2 *Electr.* A trolley whose current-collecting member is borne on a jointed, quadrilateral frame. — **pan′to·graph′ic** or **·i·cal** *adj.* — **pan·tog′ra·phy** (pan·tog′rə·fē) *n.*

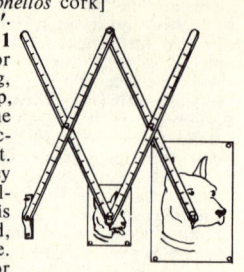
PANTOGRAPH (def. 1)

pan·tol·o·gy (pan·tol′ə·jē) *n.* A system comprehending all departments of human knowledge. — **pan·tol′o·gist** *n.*

pan·to·mime (pan′tə·mīm) *n.* 1 A series of actions, as gestures and postures, used to express ideas or convey information. 2 Any play in which the actors express their meaning by action without dialog. 3 An ancient, classical play or part of a play in which the actor used gestures or movement only, while the chorus sang. 4 *Brit.* A play relating some popular story accompanied with burlesque, gorgeous scenery, and music: a Christmastide production. — *v.t.* & *v.i.* **·mimed**, **·mim·ing** To act or express in pantomime. [< F < L *pantomimus* a pantomimist <Gk. *pantomimos* an imitator of all <*pan*, neut. of *pas* all + *mimos* an imitator] — **pan′to·mim′ic** (-mim′ik) or **·i·cal** *adj.* — **pan′to·mi′mist** (-mī′mist) *n.*

pan·to·scope (pan′tə·skōp) *n. Phot.* A very wide-angled lens.

pan·to·scop·ic (pan′tə·skop′ik) *adj.* Giving a broad scope or sweep of vision, as in a panoramic camera.

pan·to·then·ic (pan′tə·then′ik) *adj. Biochem.* Designating an acid, $C_9H_{17}NO_5$, of the vitamin B complex, widely distributed in many plant and animal tissues. It is obtained as a pale-yellow, viscous oil forming a calcium salt, and is also made synthetically. [< Gk. *pantothen* from every side]

pan·toum (pän·tōōm′) *n.* A verse form of Malay origin, consisting of a series of quatrains in which the second and fourth lines of each quatrain recur as the first and third of the next, and in which the second and fourth lines of the final quatrain repeat the first and third lines of the first. Also **Pan·tun′** (-tōōn′). [< F <Malay *pantun*]

pan·try (pan′trē) *n. pl.* **·tries** A room or closet in which to keep provisions, dishes, table linen, etc. [< AF *panetrie*, OF *paneterie* bread room <Med. L *panetaria* < *panetarius* baker < L *panis* bread]

pants (pants) *n. pl.* Trousers; drawers. [Short for PANTALOONS]

pan·ty·waist (pan′tē·wāst′) *n.* 1 A child's waist with buttons on which to fasten short pants. 2 *Slang* An effeminate young man.

Pá·nu·co (pä′nōō·kō) A river in NE central Mexico, flowing 100 miles NE to the Gulf of Mexico.

pa·nung (pä′nung) *n.* A long, broad strip of cloth worn by Siamese men and women as a loincloth or skirt. [< Siamese < *phā* cloth + *nŭṅ* one]

Pan·urge (pan·ûrj′) In Rabelais' *Pantagruel*, the boon companion of the hero.

Pa·nyush·kin (pə·nyōōsh′kin), **Aleksandr**, born 1905, U.S.S.R. diplomat.

Pan·za (pan′zə, *Sp.* pän′thä), **Sancho** See SANCHO PANZA.

pan·zer (pan′zər, *Ger.* pän′tsər) *adj.* Armored; also, using armored tanks or mechanized troops: a *panzer* attack.

Pan·zer (pän′tsər) *n. German* Armor-plating; literally, coat of mail.

Panzer division *Mil.* An armored division; especially, a division of tanks.

Pão de A·çú·car (pouń thi ə·soō'kər) The Portuguese name for SUGAR LOAF MOUNTAIN.

Pa·o·li (pä'ō·lē), **Pasquale di**, 1725–1807, Corsican patriot.

Pa·o·li·na (pä'ō·lē'nä) Italian form of PAULINE.

Pa·o·lo (pä'ō·lō) Italian form of PAUL.

Pa·o·lo (pä'ō·lō) See under FRANCESCA DA RIMINI.

Pao·shan (bou'shän') A city on the Burma Road in western Yunnan, China.

Pao·ting (bou'ting') The capital of Hopeh province, China.

pap[1] (pap) *n.* 1 A teat; nipple. 2 A hill or other object having a conical shape. [ME *pappe*, appar. <Scand. Cf. dial. Sw. *papp*.]

pap[2] (pap) *n.* 1 Any soft food for babes. 2 *Slang* The fees, favors, and privileges of public office. [ME *pappe*, ? <MLG. Cf. MDu. *pap*, LG *pappe*.]

pa·pa[1] (pä'pə, pə·pä') *n.* Father. [<F <Gk. *papas*, a child's word]

pa·pa[2] (pä'pä) *n.* 1 The bishop of Rome; the pope. 2 In the Greek Church, the patriarch of Alexandria; also, a parish priest. [<Med. L <Gk. *papas* father. Doublet of POPE.]

pa·pa·cy (pä'pə·sē) *n. pl.* **·cies** 1 The dignity, office, or jurisdiction of the pope of Rome. 2 The succession of popes and the administration of affairs in the Roman Catholic Church. 3 The tenure of office of the pope. [<Med. L *papatia* < *papa*. See PAPA[2].]

Pa·pa·cy (pä'pə·sē) *n.* The Roman Catholic system of church government.

pa·pa·in (pə·pā'in, pä'pə·in) *n. Biochem.* A vegetable enzyme, resembling the trypsin of the pancreatic juice, contained in the milk of the papaya: used in medicine as a digestive. [<PAPA(YA) + -IN]

pa·pal (pä'pəl) *adj.* 1 Pertaining to the papacy or the pope. 2 Assuming supreme authority. 3 Pertaining to the Roman Catholic Church. [<OF <Med. L *papalis* belonging to the pope < *papa*. See PAPA[2].]

papal delegate A representative of the pope in a country in which a papal nuncio is not in residence: also called *internuncio*.

Papal States See STATES OF THE CHURCH.

Pa·pav·er·a·ce·ae (pə·pav'ə·rā'sē·ē) *n. pl. Bot.* A family of widely distributed herbs and shrubs with polypetalous, typically showy flowers and a milky juice yielding several important narcotic alkaloids; the poppy family. [<NL <L *papaver* poppy + *-aceus* of the nature of] — **pa·pav'er·a'ceous** (-rā'shəs) *adj.*

pa·pav·er·ine (pə·pav'ə·rēn, -ər·in, pə·pā'və-) *n. Chem.* A white, odorless, tasteless, crystalline alkaloid, $C_{20}H_{21}O_4N$, obtained from opium: used in medicine chiefly as the hydrochloride. [<L *papaver* poppy + -INE[2]]

pa·pav·er·ous (pə·pav'ər·əs) *adj.* Having the properties of the poppy.

pa·paw (pä·pô', pô'pô) *n.* 1 A small, deciduous, North American tree (*Asimina triloba*, family *Annonaceae*) bearing edible fruit. 2 The fruit. Also spelled *pawpaw*. [<Sp. *papayo* <Cariban]

pa·pa·ya (pä·pä'yä, pə·pä'yə) *n.* 1 The yellow, melonlike fruit of a tropical American evergreen tree (*Carica papaya*), valued for its nutritious and palatable qualities. The fruit may be eaten raw, cooked, or pickled. 2 The tree. [<Sp. <Cariban]

Pa·pe·e·te (pä'pā·ā'tā, pə·pē'tē) A port on Tahiti, capital of the Society Islands and of French Oceania.

Pa·pen (pä'pən), **Franz von**, born 1879, German diplomat.

pa·per (pä'pər) *n.* 1 A substance made from fibrous cellulose material, as rags, wood, or bark, treated with various chemicals and formed into thin sheets or strips for writing, printing, wrapping and a wide variety of other uses in industry and the arts. 2 A sheet or a web of such material. 3 A printed or written instrument or document. 4 A printed journal; newspaper. 5 A written discourse; essay. 6 Written or printed pledges or promises to pay which are negotiable, as bills of exchange, notes, etc.: called *commercial paper*. 7 A package containing in a paper wrapping a limited amount or number: a *paper* of pins. 8 Wallpaper, paperhangings. 9 *Slang* A marked playing card. 10 *pl.* Small strips of paper on which the hair is twisted to be curled: also *curl papers*. 11 *pl.* A ship's papers, as invoices, etc. See SHIP'S PAPERS. 12 *pl.* Personal documents or identification, etc. 13 Something having a similar appearance to paper, as papyrus or papier-mâché. 14 Collectively, free orders of admission to a place of amusement; also, an audience so admitted. — **first papers** Papers declaring intention of becoming a citizen of the United States: filed by an applicant for naturalization as the first step in the process. — **second papers** Popular name for a certificate of naturalization. — *v.t.* 1 To put paper on; cover with wallpaper. 2 To fold or enclose in paper. 3 To write or describe on paper. 4 To issue free tickets of admission to (a place of amusement) to *paper* the house. — *adj.* 1 Made of paper. 2 Enrolled, described, or stated on paper; existing only on paper. [<AF *papir*, OF *papier* <L *papyrus*. Doublet of PAPYRUS.] — **pa'per·y** *adj.*

pa·per·back (pä'pər·bak') *adj.* Having a paper cover or binding. — *n.* A book so bound.

paper birch See under BIRCH.

pa·per·bound (pä'pər·bound') *adj.* Paperback.

pa·per·er (pä'pər·ər) *n.* One who applies paper; a paperhanger.

paper farmer A farmer educated in agricultural theory, but inexperienced.

pa·per·hang·ing (pä'pər·hang'ing) *n.* 1 The act or process of covering walls, ceilings, etc., with paper. 2 *pl.* Wallpaper. — **pa'per·hang'er** *n.*

pa·per·knife (pä'pər·nīf') *n. pl.* **·knives** (-nīvz') A blade of bone or other hard substance, for cutting folded leaves, creasing paper, etc.

paper money 1 Currency consisting of paper on which certain fixed values are printed, as banknotes, government notes, etc. 2 Negotiable commercial papers, as promissory notes, bills of exchange, etc.

paper nautilus The argonaut: a cephalopod named for its thin, paperlike shell.

paper spar A form of calcite occurring in thin laminae or sheets.

paper wasp Any of various wasps, as the hornets, yellow jackets, and certain other social wasps, that build nests of a material resembling paper.

pa·per·weight (pä'pər·wāt') *n.* A small, heavy object, often ornamental, used to keep loose papers in place by its weight.

pa·per·work (pä'pər·wûrk') *n.* Work involving the preparation or handling of reports, letters, correction of papers, etc.

pa·pes·cent (pə·pes'ənt) *adj. Med.* Having the qualities of or containing pap or milky fluid. [<PAP[2] + -ESCENT]

pap·e·te·rie (pap'ə·trē, *Fr.* pȧp·trē') *n.* A box or case containing writing materials. [<F <*papetier* a papermaker, ult. <*papier* paper]

Paph·la·go·ni·a (paf'lə·gō'nē·ə) An ancient country and Roman province in northern Asia Minor on the Black Sea.

Pa·phos (pā'fos) An ancient city in SW Cyprus, considered sacred to Aphrodite. — **Pa'phi·an** (-fē·ən) *adj. & n.*

pa·pier-mâ·ché (pā'pər·mə·shā', *Fr.* pȧ·pyā'mȧ·shā') *n.* A tough plastic material made from paper pulp containing an admixture of size, paste, oil, resin, etc., or from sheets of paper glued and pressed together. [<F *papier* PAPER + *mâché*, pp. of *mâcher* chew <L *masticare*]

pa·pil·i·o·na·ceous (pə·pil'ē·ō·nā'shəs) *adj. Bot.* 1 Butterfly-shaped; composed of five petals of a certain shape and arrangement. 2 Belonging to the bean or pea family (*Leguminosae*), in which the flowers have irregular corollas. [<NL *papilionaceus* <L *papilio, -onis* butterfly]

Pa·pil·i·on·i·dae (pə·pil'ē·on'ə·dē) *n. pl.* A family of butterflies, including mostly large species with a tail-like lobe on each hind wing; the swallowtail butterflies. [<NL <L *papilio, -onis* butterfly] — **pa·pil'i·on'id** *adj. & n.*

pa·pil·la (pə·pil'ə) *n. pl.* **·lae** (-ē) 1 *Anat.* a The nipple of the mammary glands. b Any small nipplelike process of connective tissue, as on the tongue or the epidermal layer of the skin, or at the root of a developing tooth, hair, feather, etc. 2 *Bot.* A small, elongate, nipple-shaped protuberance on a flower or leaf. [<L, dim. of *papula* a swelling, pimple] — **pap·il·lar·y** (pap'ə·ler'ē) *adj.*

pap·il·lo·ma (pap'ə·lō'mə) *n. pl.* **·ma·ta** (-mə·tə) *Pathol.* A morbid growth on the skin, consisting of small tumors composed of and covered by the normal skin, as corns, warts, or mucous tubercles. [<PAPILLA]

pap·il·lon (pap'ə·lon) *n.* A breed of toy spaniel descended from the European 16th century dwarf spaniel, having large fringed ears resembling the wings of a butterfly, a plumy tail, and a thick, solidly colored coat. [<F, butterfly <L *papilio, -onis*]

pap·il·lose (pap'ə·lōs) *adj.* Papillary; also, pimply; warty. Also **pap'il·lous**. [<NL *papillosus* <L *papilla*] — **pap'i·los'i·ty** (-los'ə·tē) *n.*

pap·il·lote (pap'ə·lōt) *n. French* 1 A small strip of paper used in curling the hair. 2 A small paper encircling the end of the bone of a chop or cutlet when served.

pa·pist (pā'pist) *n.* An adherent of the papacy; a Roman Catholic: usually an opprobrious use. [<F *papiste* <Med. L *papa*. See PAPA[2].]

pa·pis·ti·cal (pə·pis'ti·kəl) *adj.* Of or pertaining to the doctrines of the Roman Catholic Church or to the papal system: used in disparagement. Also **pa·pis'tic**.

pa·pis·try (pā'pis·trē) *n.* The religion or ceremonial of the papist; Roman Catholicism: used disparagingly.

pa·poose (pa·poōs') *n.* A North American Indian infant. Also **pap·poose'**. [<Algonquian (Narraganset) *papoos* child]

Pap·pen·heim (päp'ən·hīm), **Count Gottfried zu**, 1594–1632, German general.

pap·pus (pap'əs) *n. pl.* **pap·pi** (pap'ī) *Bot.* The peculiar limb on the calyx of a floret of the composite family, consisting either of a downy tuft of hairs, as in thistles, or of teeth, scales, bristles, or awns. [<NL <Gk. *pappos* grandfather] — **pap'pose** (-ōs), **pap'pous** (-əs) *adj.*

pap·py[1] (pap'ē) *adj.* **·pi·er**, **·pi·est** Resembling pap; pulpy; soft.

pap·py[2] (pap'ē) *n. pl.* **·pies** Papa; father. [Dim. of PAPA[1]]

pa·pri·ka (pa·prē'kə, pap'rə·kə) *n.* A condiment made from the ripe fruit of a mild variety of red pepper (*Capsicum frutescens*). Also **pa·pri'ca**. [<G Magyar, red pepper <Gk. *peperi* pepper]

Pap·u·a (pap'yoō·ə, pä'poō·ä') New Guinea.

Pap·u·a (pap'yoō·ə, pä'poō·ä'), **Territory of** An Australian territory in SE New Guinea; with its dependencies in adjacent islands, 90,540 square miles; capital, Port Moresby: formerly *British New Guinea*: became part of the Territory of Papua and New Guinea 1949.

Papua and New Guinea, Territory of A United Nations Trust Territory administered by Australia and comprising the formerly separate Territory of Papua and Territory of New Guinea; established 1949; 183,600 square miles; capital, Port Moresby.

Pap·u·an (pap'yoō·ən, pä'poō·ən) *adj.* Of or pertaining to the Island of Papua or New Guinea, or to the Papuan peoples. — *n.* One of any of the dark peoples inhabiting the Melanesian Archipelago from Fiji westward to the Aru Islands, including New Guinea or Papua.

pap·u·la (pap'yə·lə) *n. pl.* **·lae** (-lē) *Pathol.* An isolated pimple. Also **pap'ule** (-yoōl). [<L, a pimple]

pap·y·ra·ceous (pap'ə·rā'shəs) *adj.* Made of papyrus; papery.

pa·py·rus (pə·pī'rəs) *n. pl.* **·ri** (-rī) 1 The writing paper of the ancient Egyptians, made from the papyrus plant. 2 A manuscript written on this material. 3 A perennial rushlike plant of the sedge family (*Cyperus papyrus*) having stems 6 to 10 feet high. [<L <Gk. *papyros*. Doublet of PAPER.]

par (pär) *n.* 1 Equality of value; equivalence;

PAPAYA
(The tree from 15 to 25 feet in height)

parity; specifically, equality between nominal and actual value. Shares of stock are said to be **at** (or **up to**) **par** when exchangeable at face value in money, **above par** when the market price is greater, and **below par** when less than the nominal value. 2 An accepted standard with which to compare variations. 3 In golf, the number of strokes allotted to a round or hole on the basis of faultless play. —**mint par** The reduction of the monetary unit of one country to expression in terms of that of another: also **par of exchange.** —**on a par** On a level. [<L, equal]

par- *prefix* Per-: used in a few words from French: *pardoner.* [<F <L *per-* <*per* through]

pa·ra (pä·rä′, pä′rä) *n.* 1 A Turkish copper coin. 2 A small coin of Yugoslavia. [<Turkish *pārah* <Persian, a piece]

Pa·rá (pä·rä′) 1 An estuary of the Amazon; 200 miles long; 40 miles wide at its mouth. 2 A state in NE Brazil; 469,778 square miles; capital, Belém. 3 A former name for BELÉM.

para-[1] *prefix* 1 Beside; near by; along with: *paradigm.* 2 Beyond; aside from; amiss: *paradox.* 3 *Chem.* **a** An isomeric or polymeric modification of: *paraldehyde.* **b** A modification of or a compound similar to (not necessarily isomeric or polymeric): *paramorphine.* **c** A benzene derivative in which the substituted atoms or radicals occupy the positions 1, 4: *paradichlorobenzene*: abbr. *p*-. Compare META-, ORTHO-. See BENZENE RING. 4 *Med.* **a** A diseased or abnormal condition: *paranoia.* **b** Accessory to: *parasympathetic.* **c** Similar to but not identical with a true condition or form: *paratyphoid.* Also, before vowels and *h*, usually **par-**. [<Gk. <*para* beside]

para-[2] *combining form* Shelter or protection against: *parasol.* [<Ital. *parare* defend]

par·a-a·mi·no·ben·zo·ic acid (par′ə·ə·mē′nō·ben·zō′ik, -am′ə·nō-) *Biochem.* A colorless crystalline compound, $C_7H_7NO_2$, forming part of the vitamin B complex. Present in yeast and also made synthetically: source of several local anesthetics.

Par·a·bell·um (par′ə·bel′əm) *n.* The European name for the Luger pistol. [<G *(Pistole) Parabellum* (pistol) for war <Gk. *para* for + L *bellum* war]

par·a·bal·loon (par′ə·bə·lōōn′) *n.* A radar antenna in the form of an inflated sphere of Fiberglas partly coated with a thin metal layer forming the reflector. [<PARA-[2] + BALLOON]

par·a·bi·o·sis (par′ə·bī·ō′sis) *n.* 1 *Biol.* A fusion of two individuals, congenital or by surgery, resulting in mutual physiological intimacy, as in Siamese twins. 2 *Physiol.* A temporary suppression of the irritability and conductivity of a nerve. —**par′a·bi·ot′ic** (-ot′ik) *adj.*

par·a·blast (par′ə·blast) *n. Biol.* The yolk of a meroblastic egg. —**par′a·blas′tic** *adj.*

par·a·ble (par′ə·bəl) *n.* A comparison; simile; specifically, a short narrative making a moral or religious point by comparison with natural or homely things: the New Testament *parables.* See synonyms under ALLEGORY. [<OF *parabole* <LL *parabola* allegory, speech <L, comparison <Gk. *parabolē* a placing side by side, a comparison <*para-* beside + *ballein* throw. Doublet of PALAVER, PARABOLA, PAROLE.]

pa·rab·o·la (pə·rab′ə·lə) *n. Math.* The locus of a point moving in a plane so that its distances from a fixed point (focus) and a fixed straight line (directrix) are equal; the curve formed by the edges of a plane when cutting through a right circular cone at an angle parallel to one of its sides. [<Med. L <Gk. *parabolē.* Doublet of PALAVER, PARABLE, PAROLE.]

pa·rab·o·le (pə·rab′ə·lē) *n.* A rhetorical comparison; a formal simile. [<Gk. *parabolē.* See PARABLE.]

par·a·bol·ic (par′ə·bol′ik) *adj.* 1 Pertaining to a parable. 2 Pertaining to or having the form of a parabola. Also **par′a·bol′i·cal.** [<LL *parabolicus* <LGk. *parabolikos* figurative <*parabolē.* See PARABLE.] —**par′a·bol′i·cal·ly** *adv.*

parabolic spiral *Math.* The polar curve traced by a point moving so that the square of its distance from the pole is proportional to its polar angle: also called *Fermat's spiral.*

pa·rab·o·lize (pə·rab′ə·līz) *v.t.* **·lized, ·liz·ing** 1 To relate in parable form. 2 *Math.* To give the form of a parabola to.

pa·rab·o·loid (pə·rab′ə·loid) *n. Math.* A surface or solid generated by the rotation of a parabola about its axis. —**pa·rab′o·loi′dal** *adj.*

par·a·ca·se·in (par′ə·kā′sē·in, -sēn) *n. Biochem.* A form of casein produced by the action of rennin.

Pa·ra·cel Islands (pä′rä′sel′) A Chinese island group SE of Hainan in the South China Sea.

Par·a·cel·sus (par′ə·sel′səs) Pseudonym of Theophrastus von Hohenheim, 1493–1541, Swiss physician and alchemist. —**Par′a·cel′·si·an** *adj.*

par·a·cen·es·the·sia (par′ə·sen′is·thē′zhə, -zhē·ə) *n. Psychiatry* Any perversion or abnormality of the general sense of well-being, as an obsession or mania.

par·a·cen·tric (par′ə·sen′trik) *adj.* Directed to or from the center: said of motion. Also **par′·a·cen′tri·cal.**

par·a·chor (par′ə·kôr) *n. Chem.* A constant expressing the relationship between the surface tension and density of a given substance or compound. [<PARA-[1] + Gk. *chōros* space]

par·a·chor·dal (par′ə·kôr′dəl) *adj. Anat.* Situated beside the notochord: said of the paired cartilaginous plates at the base of the developing cranium.

par·a·chro·ma·tism (par′ə·krō′mə·tiz′əm) *n. Pathol.* Abnormal perception of colors; color blindness. —**par′a·chro·mat′ic** (-krō·mat′ik) *adj.*

par·a·chute (par′ə·shōōt) *n.* 1 A large, expanding, umbrella-shaped apparatus for retarding the speed of a body descending through the air, especially from an airplane. 2 *Zool.* A lateral extension of the skin in flying squirrels, etc., enabling them to glide through the air. —**pilot parachute** *Aeron.* A small parachute whose release serves to open the canopy of the main parachute quickly and with minimum danger of jamming or tearing. —*v.* **·chut·ed, ·chut·ing** *v.t.* To land (troops, materiel, etc.) by means of parachutes. —*v.i.* To descend by parachute. [<F <*para* PARA-[2] + *chute* fall]

parachute fabric A plain-woven, very firm fabric used for making parachutes: made of silk or nylon if for humans, of rayon if for cargo, bombs, etc.

parachute flare A flare attached to a parachute and released from an aircraft to illuminate the terrain.

parachute troop See PARATROOP.

par·a·chut·ist (par′ə·shōō′tist) *n.* A person, specifically a soldier, trained and equipped to drop by parachute.

par·a·clete (par′ə·klēt) *n.* One called to the aid of another, especially in legal process; an advocate; hence, **Paraclete,** the Holy Spirit as the helper or comforter. [<OF *paraclet* < LL *paracletus* <LGk. *paraklētos* a comforter, advocate <*parakalein* call to one's aid < *para-* to + *kalein* call]

par·ac·me (par·ak′mē) *n.* 1 *Biol.* The stage of degeneration following maturity. 2 *Pathol.* The period of decline or remission, as of a disease. [<Gk. *parakmē* <*para-* beyond + *akmē* the highest point]

par·a·cy·mene (par′ə·sī′mēn) *n.* A common form of cymene.

pa·rade (pə·rād′) *n.* 1 A marshaling and maneuvering of troops for display or official inspection; a review. 2 A ceremonious procession. 3 A ground where military reviews are held. 4 A promenade or public walk. 5 A setting forth or arrangement of persons or things for display. 6 Pompous show; ostentation. See synonyms under OSTENTATION, SPECTACLE. —*v.* **·rad·ed, ·rad·ing** *v.t.* 1 To walk or march through or about: to *parade* the streets. 2 To display or show off ostentatiously; flaunt. 3 To cause to assemble for military parade or review. —*v.i.* 4 To march formally or with display. 5 To walk in public for the purpose of showing oneself. 6 To assemble in military order for inspection or review. See synonyms under FLAUNT. [<MF <Sp. *parada* a stopping place, exercise ground <LL *parare* adorn, prepare <L] —**pa·rad′er** *n.*

parade rest A formal ceremonial position of rest, exactly prescribed with or without rifle, in which soldiers stand without moving or speaking, as distinguished from the informal positions of *rest* and *at ease.*

par·a·di·chlo·ro·ben·zene (par′ə·dī·klôr′ō·ben′·zēn, -klō′rō-) *n. Chem.* A white, volatile, crystalline compound, $C_6H_4Cl_2$, widely used as an insecticide.

par·a·digm (par′ə·dim, -dīm) *n.* 1 *Gram.* An ordered list or table of all the inflected forms of a word or class of words, as of a particular declension, conjugation, etc. 2 Any pattern or example. [<LL *paradigma* <Gk. *paradeigma* a pattern <*para-* beside + *deiknynai* show] —**par′a·dig·mat′ic** (-dig·mat′ik) *adj.*

par·a·dise (par′ə·dīs) *n.* 1 The intermediate place or state where the souls of the saved await the resurrection. 2 Heaven, the ultimate abode of righteous souls; also, the abode of the deceased faithful of Islam. 3 Any region or state of surpassing delight. 4 In the Near East, a park or pleasure ground. [<F *paradis* <L *paradisus* <Gk. *paradeisos* park <OPersian *pairidaēza* an enclosure, park <*pairi* around + *daēza* wall] —**par′a·di·sa′ic** (-di·sā′ik) or **·i·cal, par′a·dis′i·ac** (-dis′ē·ak) or **par·a·di·si·a·cal** (par′ə·di·sī′ə·kəl) *adj.*

Par·a·dise (par′ə·dīs) The garden of Eden.

Paradise Lost and **Paradise Regained** Epic poems by John Milton depicting the fall and redemption of man.

par·a·dos (par′ə·dos) *n.* An embankment, as behind a trench, for protection against gunfire from the rear. [<F <*para-* PARA-[2] + *dos* back]

par·a·dox (par′ə·doks) *n.* 1 A statement, doctrine, or expression seemingly absurd or contradictory to common notions or to what would naturally be believed, but in fact really true. 2 A statement essentially absurd and false. See synonyms under RIDDLE[2]. [<F *paradoxe* <L *paradoxum* <Gk. *paradoxos, -on* incredible <*para-* contrary to + *doxa* opinion <*dokeein* think] —**par′a·dox′·i·cal** *adj.* —**par′a·dox′i·cal′i·ty** (-kal′ə·tē), **par′a·dox′i·cal·ness** *n.* —**par′a·dox′i·cal·ly** *adv.*

par·a·drop (par′ə·drop) *n.* The dropping of supplies, equipment, and the like by parachute, especially over terrain not adapted to landings by aircraft. [<PARA(CHUTE) + DROP]

par·aes·the·sia (par′is·thē′zhə, -zhē·ə), etc. See PARESTHESIA, etc.

par·af·fin (par′ə·fin) *n. Chem.* 1 A translucent, waxy, solid mixture of hydrocarbons, indifferent to most chemical reagents: it is a constituent of peat, soft coal, and shale, but is derived principally from the distillation of petroleum. 2 Any saturated aliphatic hydrocarbon of the methane series having the formula C_nH_{2n+2}. —*v.t.* To treat or impregnate with paraffin. Also **par′af·fine** (-fin, -fēn). [<G <L *parum* too little + *affinis* related to; so named because it has little affinity for other bodies]

paraffin wax Solid paraffin.

par·a·form (par′ə·fôrm) *n. Chem.* A white crystalline powder, $(CH_2O)_3$, obtained by concentrating formaldehyde solutions: used as a disinfectant. Also **par′a·for·mal′de·hyde** (-fôr·mal′də·hīd).

par·a·frag bomb (par′ə·frag) A demolition bomb designed to be dropped by parachute.

par·a·gen·e·sis (par′ə·jen′ə·sis) *n. Mineral.* The formation of minerals in contact in such manner as to affect the development of the individual crystals. Also **par′a·ge·ne′si·a** (-jə·nē′sē·ə). —**par′a·ge·net′ic** (-jə·net′ik) *adj.*

par·a·go·ge (par′ə·gō′jē) *n.* The addition of an inorganic sound or syllable at the end of a word without a change in meaning, as in *amongs-t, whils-t,* etc. [<LL <Gk. *paragōgē* <*para-* beyond + *agōgē* a leading <*agein* lead] —**par′a·gog′ic** (-goj′ik) *adj.*

par·a·gon (par′ə·gon) *n.* 1 A model of excellence. 2 *Printing* A size of type: about 3 1/2 lines to the inch: 20-point. 3 A round pearl of exceptional size. —*v.t. Archaic* or *Poetic* 1 To match; equal. 2 To surpass. 3 To compare with. 4 To hold up as a paragon. [<OF <Ital. *paragone* a touchstone, prob. <Gk. *parakonaein* sharpen one thing against another <*para-* beside + *akonē* whetstone]

pa·rag·o·nite (pə·rag′ə·nīt) *n.* A scaly, pearly, variously colored, translucent mica containing sodium instead of potassium, and found massive. [<Gk. *paragōn,* ppr. of *paragein* lead astray + -ITE[1]]

par·a·graph (par′ə·graf, -gräf) *n.* 1 A short passage in a written or printed discourse, begun on a new and usually indented line. 2 A short article, complete and unified; especially, in a newspaper, a short article, item, or comment. 3 A mark (¶) used to indicate where a paragraph is to be begun, or as a reference mark. —*v.t.* 1 To arrange in or into paragraphs. 2 To comment on or express in a paragraph. [<OF *paragraphe* <LL *paragraphus* <Gk. *paragraphos,* orig. a short line in a text marking a break in sense <*para-* beside + *graphein* write] —**par′a·graph′er, par′a·graph′ist** *n.* —**par′a·graph′ic, par′a·graph′i·cal** *adj.*

par·a·graph·i·a (par′ə·graf′ē·ə) *n. Psychiatry* A symptom of mental disorder in which the patient writes wrong letters or words. [<NL <Gk. *para-* beside + *graphein* write]

Par·a·guay (par′ə·gwā, -gwī; *Sp.* pä·rä·gwī′) A republic in south central South America; 157,047 square miles; capital, Asunción. — **Par′a·guay′an** *adj.* & *n.*

Paraguay A river in south central South America, flowing 1,300 miles south to the Paraná.

Paraguay tea Maté.

par·a·he·li·ot·ro·pism (par′ə·hē′lē·ot′rə·piz′əm) *n. Bot.* A manifestation of irritability in motile leaves when exposed to bright sunlight, whereby they assume such a position that their surfaces are parallel to the direction of the incident rays; diurnal sleep of leaves. — **par′a·he′li·o·trop′ic** (-ə·trop′ik) *adj.*

Pa·ra·hi·ba (pä′rä·ē′bä) A former name for JOÃO PESSOA and former spelling of PARAÍBA. Also **Pa′ra·hy′ba.**

par·a·hy·dro·gen (par′ə·hī′drə·jən) *n. Chem.* A form of hydrogen which has been stabilized by treating orthohydrogen with hydrous ferric oxide acting as a catalyst: an important liquid fuel for rockets and missiles. [<PARA-[1] (def. 1) + HYDROGEN]

par·a·hyp·no·sis (par′ə·hip·nō′sis) *n. Psychiatry* Abnormal sleep, suggesting the effects of but not necessarily due to hypnosis, as in somnambulism. —**par′a·hyp·not′ic** (-not′ik) *adj.*

Pa·ra·í·ba (pä′rä·ē′bä) 1 A state of NE Brazil; 21,831 square miles; capital, João Pessoa. 2 **Paraíba do Sul** (thōō sōōl′) The chief river of Rio de Janeiro state, SE Brazil, rising in SE São Paulo state and flowing about 600 miles SW and NE to the Atlantic, near Campos. 3 **Paraíba do Norte** (thōō nôr′ti) A river of Paraíba state, NE Brazil, rising on the Pernambuco state border, and flowing about 180 miles NE to the Atlantic, near João Pessoa.

par·a·keet (par′ə·kēt) *n.* 1 Any of certain small parrots, having a long, wedge-shaped tail, as the crimson rosella of Australia (*Platycercus elegans*). 2 The Carolina parrot (*Conuropsis carolinensis*) of the southern United States: now extinct. Also spelled *paroquet, parrakeet, parrakeeto, parroquet.* [<OF *paroquet,* ? <Ital. *parrochetto,* dim. of *parroco* parson]

par·a·ki·ne·sia (par′ə·ki·nē′zhə, -zhē·ə) *n. Pathol.* Clumsy and unnatural movements, caused by impairment of motor functions. Also **par′a·ki·ne′sis** (-nē′sis). [<NL <Gk. *para-* beside + *kinēsis* movement] —**par′a·ki·net′ic** (-net′ik) *adj.*

par·a·la·li·a (par′ə·lā′lē·ə) *n. Pathol.* Any defect in the faculty of speech. [<NL <Gk. *paralaleein* talk at random <*para-* beside + *laleein* talk]

par·al·de·hyde (pə·ral′də·hīd) *n. Chem.* A colorless, transparent liquid, $C_6H_{12}O_3$, derived by the action of sulfuric acid on acetaldehyde: used as a hypnotic.

par·a·leip·sis (par′ə·līp′sis) *n.* In rhetoric, a pretended suppression of what is really said; a feigned omission, as in "not to mention his insufferable conceit"; apophasis. Also **par′a·lep′sis** (-lep′sis), **par′a·lip′sis** (-lip′sis). [<Gk., an omission <*paraleipein* <*para-* beside + *leipein* leave]

par·al·lax (par′ə·laks) *n.* 1 Such difference in the apparent position of an object, specifically a star or other heavenly body, as would appear if it were viewed from two points. It is **diurnal** or **geocentric parallax** when due to the change of place of the observer caused by the earth's rotation; **annual** or **heliocentric parallax** when the observer's change of place is due to the earth's motion around the sun. 2 Any apparent displacement of an object due to an observer's position. [<MF *parallaxe* <Gk. *parallaxis* a change <*parallassein* alter <*para-* beside + *allassein* change] —**par′al·lac′tic** (-lak′tik) or **·ti·cal** *adj.* —**par′·al·lac′ti·cal·ly** *adv.*

par·al·lel (par′ə·lel) *adj.* 1 Not meeting or intersecting, however far extended: said of straight lines or planes. 2 In projective geometry, meeting at infinity. 3 Having lines or surfaces lying in the same or about the same direction. 4 Having a like course; conforming in action. 5 Essentially alike; similar. 6 *Music* Separated by the same interval: *parallel* fifths. 7 Having sides parallel to one another. 8 *Electr.* Connected between like terminals, as a group of cells, condensers, etc. —*n.* 1 A line extending in the same direction and being equidistant at all points from another line. 2 Essential likeness. 3 A comparison tracing similarity, as between persons. 4 A counterpart. 5 Any person or thing ranked as equal to another; a match. 6 A trench dug parallel to the outline of a fortification. 7 A degree of latitude. —*v.t.* **·leled** or **·lelled, ·lel·ing** or **·lel·ling** 1 To place in parallel; make parallel. 2 To be, go, or extend parallel to. 3 To furnish with a parallel or equal; find a parallel to. 4 To be a parallel to; correspond to. 5 To compare; liken. See synonyms under COMPARE. [<MF *parallele* <L *parallelus* <Gk. *parallēlos* <*para* beside + *allēlos* one another]

parallel bars Two horizontal crossbars, parallel to each other and supported a few feet from the ground by upright posts: used for gymnastic exercises.

par·al·lel·e·pi·ped (par′ə·lel′ə·pī′pid, -pip′id) *n.* A prism with six faces, each of which is a parallelogram. Also **par′al·lel′o·pi′ped, par′·al·lel′e·pip′e·don, par′al·lel′o·pip′e·don** (-pip′ə·don, -pī′pə-). [<Gk. *parallēlepipedon* <*parallēlos* parallel + *epipedon* a plane surface <*epi-* upon + *pedon* ground]

par·al·lel·ism (par′ə·lel·iz′əm) *n.* 1 Parallel position. 2 Essential likeness; similarity; analogy. 3 Correspondence or similarity of construction in successive passages or clauses, especially in Hebrew poetry. 4 *Philos.* The opinion that the relation between physical and mental processes is one of concomitant or parallel variation, and not of cause and effect. —**psychophysical parallelism** 1 *Philos.* The theory that related mental and physical events, although occurring simultaneously, are separate and distinct phenomena: opposed to *interactionism.* 2 *Psychol.* The theory that changes in mental processes are accompanied, although not necessarily in a causal relation, by parallel changes in neural processes.

parallel motion *Mech.* Motion of a machine element, as a piston, reproduced by a linkwork system and recorded as a straight line parallel to the original motion.

par·al·lel·o·gram (par′ə·lel′ə·gram) *n.* 1 *Geom.* A four-sided plane figure whose opposite sides are parallel and equal, including the *square, rectangle, rhombus,* and *rhomboid.* 2 Any area or object having such form. [<F *parallélogramme* <L *parallelogrammum* <Gk. *parallēlogrammon,* orig. adj., bounded by parallel lines <*parallēlos* parallel + *grammē* a line]

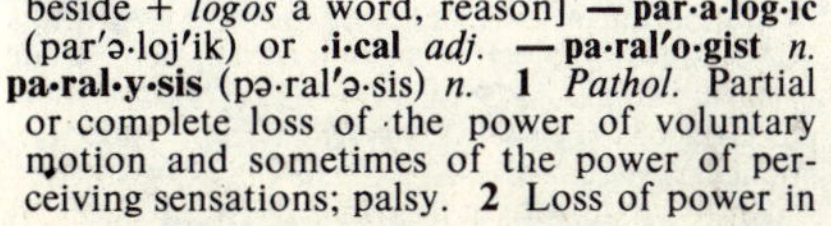

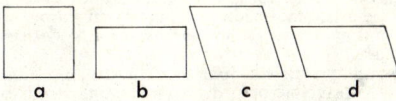

PARALLELOGRAMS
a. Square. *b.* Rectangle. *c.* Rhombus. *d.* Rhomboid.

pa·ral·o·gism (pə·ral′ə·jiz′əm) *n.* A fallacy in reasoning; reasoning contrary to logic or logical form; hence, any false reasoning. See synonyms under SOPHISTRY. [<F *paralogisme* <LL *paralogismus* <Gk. *paralogismos* <*paralogizesthai* reason falsely, ult. <*para-* beside + *logos* a word, reason] —**par·a·log·ic** (par′ə·loj′ik) or **·i·cal** *adj.* —**pa·ral′o·gist** *n.*

pa·ral·y·sis (pə·ral′ə·sis) *n.* 1 *Pathol.* Partial or complete loss of the power of voluntary motion and sometimes of the power of perceiving sensations; palsy. 2 Loss of power in general. [<L <Gk. <*paralyein* disable <*para-* beside + *lyein* loosen, untie. Doublet of PALSY.]

pa·ral·y·sis a·gi·tans (pə·ral′ə·sis aj′ə·tanz) *Latin* Shaking palsy.

par·a·lyt·ic (par′ə·lit′ik) *adj.* 1 Pertaining to or affected with paralysis. 2 Subject to paralysis. —*n.* A person subject to or suffering from paralysis. [<OF *paralytique* <L *paralyticus* <Gk. *paralytikos* <*paralyein.* See PARALYSIS.]

par·a·lyze (par′ə·līz) *v.t.* **·lyzed, ·lyz·ing** 1 To bring about paralysis in; make paralytic. 2 To render powerless, ineffective, or inactive. See synonyms under WEAKEN. [Appar. <F *paralyser*] —**par′a·lyz′er** *n.* —**par′a·ly·za′·tion** *n.*

par·a·mag·net·ic (par′ə·mag·net′ik) *adj. Physics* 1 Capable of being attracted by a magnet, as iron. 2 Having a magnetic permeability greater than that of a vacuum: distinguished from *diamagnetic.* —**par′a·mag′net** (-mag′nit) *n.* —**par′a·mag′net·ism** *n.*

Par·a·mar·i·bo (par′ə·mar′i·bō) A port, capital of Surinam.

par·a·mat·ta (par′ə·mat′ə) *n.* A kind of light, twilled, dress goods with cotton warp and filling of combed merino wool: also spelled *parramatta.* [from *Paramatta,* a town in N. So. Wales, Australia]

par·a·me·cin (par′ə·mē′sin) *n.* A powerful protein transmitted by inheritance within certain strains of paramecium and usually lethal to other strains of this organism.

par·a·me·ci·um (par′ə·mē′shē·əm, -sē·əm) *n. pl.* **·ci·a** (-shē·ə, -sē·ə) A ciliate infusorian (genus *Paramecium*) having a flattened elongate body, and feeding by a primitive oral groove. *P. caudatum* is the slipper animalcule found in stagnant water. [<NL <Gk. *paramēkēs* oblong, oval]

pa·ram·e·ter (pə·ram′ə·tər) *n. Math.* Any given constant or element whose values characterize one or more of the variables entering into a system of expressions, functions, etc.: A road gradient is a *parameter* of the performance of an automobile. [<NL *parametrum* <Gk. *para-* beside + *metron* a measure] — **par·a·met·ric** (par′ə·met′rik) *adj.*

par·am·ne·sia (par′am·nē′zhə, -zhē·ə) *n. Psychiatry* 1 Distortion and falsification of memory. 2 A condition in which the proper meaning or use of words cannot be remembered.

pa·ra·mo (pä′rə·mō, par′ə-) *n. pl.* **·mos** A treeless alpine plain in tropical South America. [<Sp. *páramo,* appar. <S. Am. Ind.]

par·a·mor·phine (par′ə·môr′fēn) *n.* Thebaine.

par·a·mor·phism (par′ə·môr′fiz·əm) *n. Mineral.* The alteration of one mineral to another having the same chemical composition but other molecular structure and physical properties. [<PARA-[1] + Gk. *morphē* a form + -ISM] — **par′a·mor′phic, par′a·mor′phous** *adj.*

par·a·mount (par′ə·mount) *adj.* 1 Having the highest title. 2 Superior to all others; supremely controlling. —*n.* A supreme lord; highest ruler. [<AF *paramont* above <OF *par* by (<L *per*) + *à mont* up, above <L *ad montem* to the hill] —**par′a·mount·ly** *adv.* —**par′a·mount·cy** *n.*

Synonyms (adj.): chief, eminent, foremost, preeminent, principal, superior, supreme. *Antonyms*: inferior, minor, secondary, subordinate.

par·a·mour (par′ə·mōōr) *n.* A lover, especially one who unlawfully takes the place of a husband or wife. [<OF *par amour* with love <*par* by, with (<L *per* through) + *amour* love <L *amor*]

Pa·ra·ná (pä′rä·nä′) 1 A state of southern Brazil; 77,717 square miles; capital, Curitiba. 2 An important river of SW Brazil and NE Argentina, forming the eastern and southern boundaries of Paraguay and flowing 2,050 miles south to the Rio de la Plata: above the confluence of the Paraguay, called the **Al·to Paraná** (äl′tō). 3 A city in eastern Argentina on the Paraná.

Pa·ra·na·í·ba (pä′rä·nä·ē′bä) A river of east central Brazil, flowing 500 miles SW to the Paraná. Formerly **Pa′ra·na·hi′ba.**

par·a·neph·ric (par′ə·nef′rik) *adj. Anat.* Found in the tissue beside the kidneys.

par·a·neph·ros (par′ə·nef′ros) *n. Anat.* A capsule of the suprarenal gland. [<NL <Gk. *para-* beside + *nephros* a kidney]

pa·rang (pä′räng) *n.* A short, heavy sheath knife with a straight edge, used especially by the Dyaks of Borneo for chopping and as a weapon. [<Malay *pārang*]

par·a·noi·a (par′ə·noi′ə) *n. Psychiatry* A chronic, often progressive, mental disorder or psychosis, characterized by monomania, systematized delusions of persecution, and sometimes hallucinations. Also **par′a·noe′a** (-nē′ə). [<NL <Gk., madness <*paranoos* distraught <*para-* beside + *noos, nous* mind]

par·a·noi·ac (par′ə·noi′ak) *adj.* Relating to or affected by paranoia. —*n.* One affected by paranoia. Also **par′a·no′ic** (-nō′ik), **par′a·noe′ac** (-nē′ak).

par·a·noid (par′ə·noid) *adj.* Resembling or suggestive of paranoia. —*n.* A person affected by paranoia.

par·a·no·sis (par′ə·nō′sis) *n. Psychoanal.* The primary advantage obtained by a patient from subjective exploitation of his illness. Compare EPINOSIS. [<NL <Gk. *para-* beside + *nosos* sickness] —**par′a·no′sic** *adj.*

par·a·nymph (par′ə·nimf) *n.* In ancient Greece, a groomsman or bridesmaid; specifically, the best man who went with a bridegroom to fetch home the bride, or the maiden who conducted the bride to the bridegroom. [<L *paranymphus* <Gk. *paranymphos* best man or bridesmaid <*para-* near, beside + *nymphē* bride]

par·a·pet (par′ə·pit, -pet) *n.* **1** A low wall about the edge of a roof, terrace, bridge, fortification, etc. **2** A breastwork. See synonyms under BARRIER. [<F <Ital. *parapetto* <*para-* PARA-² + *petto* breast <L *pectus* breast] —**par′a·pet·ed** *adj.*

par·aph (par′əf) *n.* A flourish made with the pen at the end of a signature, often as a protection against forgery; a rubric. —*v.t.* To affix a paraph to; sign, especially with initials; initial. [<OF *paraphe* <Med. L *paraphus*, var. of L *paragraphus*. See PARAGRAPH.]

par·a·pher·na·li·a (par′ə·fər·nā′lē·ə, -nāl′yə, -fə-) *n. pl.* **1** Trappings or accessories of equipment or adornment; furnishings, especially for ceremonious occasions; the parts of any outfit, apparatus, or equipment. **2** *Law* Formerly, the personal articles reserved to a wife over and above her dower. [<Med. L *paraphernalia (bona)* a wife's own (goods) <L *parapherna* <Gk. <*para-* beside + *phernē* dower <*pherein* carry]

pa·ra·phi·a (pə·rā′fē·ə) *n. Pathol.* An abnormal or perverted sense of touch. [<NL <Gk. *para-* beside + *aphē* touch]

par·a·phone (par′ə·fōn) *n.* Near rime. [< PARA-¹ + Gk. *phōnē* sound]

par·a·phrase (par′ə·frāz) *n.* A restatement of the meaning of a passage, work, etc. —*v.t.* & *v.i.* **·phrased, ·phras·ing** To express in or make a paraphrase. [<F <L *paraphrasis* <Gk. <*paraphrazein* tell the same thing in other words <*para-* beside + *phrazein* tell] —**par′a·phras′er, par′a·phrast** (-frast) *n.* —**par′a·phras′tic** or **·ti·cal** *adj.* —**par′a·phras′ti·cal·ly** *adv.*

pa·raph·y·sis (pə·raf′ə·sis) *n. pl.* **·ses** (-sēz) *Bot.* A sterile unicellular or pluricellular filament or narrow plate accompanying sexual or sporogenous organs found in certain mosses and other cryptogams. [<NL <Gk. *para* beside, subsidiary + *physis* growth]

par·a·ple·gi·a (par′ə·plē′jē·ə) *n. Pathol.* Paralysis of the lower half of the body, due to disease or injury of the spinal cord. Also **par′a·ple′gy** (-plē′jē). [<NL <Gk. *paraplēgia, paraplēxia* a stroke on one side <*paraplēssein* strike at the side <*para-* beside + *plēssein* strike] —**par·a·ple·gic** (par′ə·plē′jik, -plej′ik) *adj.* & *n.*

par·a·prax·is (par′ə·prak′sis) *n. Psychoanal.* Any faulty action, blunder, or lapse, as a slip of the tongue, failure of memory, etc. [<NL <Gk. *paraprassein* <*para-* beside + *prassein* do]

par·a·psy·chol·o·gy (par′ə·sī·kol′ə·jē) *n.* The investigation of extrasensory perception and the sporadic phenomena supposedly associated with it, such as telepathy, clairvoyance, telekinesis, prevision, dreams that prove prophetic, experiences of déjà vu, and poltergeist activity; metapsychics.

par·a·psy·cho·sis (par′ə·sī·kō′sis) *n. Psychiatry* A psychosis characterized by abnormal or unnatural thinking.

par·a·qui·none (par′ə·kwi·nōn′) *n.* Quinone.

Pa·rá rubber (pä·rä′) Rubber obtained from the tropical American rubber tree *(Hevea brasiliensis)*. [<Pará, Brazil]

par·a·sang (par′ə·sang) *n.* An ancient Persian measure of length, varying from 2 to 4 miles. [<L *parasanga* <Gk. *parasangēs* <OPersian]

par·a·sceve (par′ə·sēv, par′ə·sē′vē) *n.* **1** The day before the Jewish Sabbath, on which preparation is made for the Sabbath; also, what is then prepared. **2** In the Roman Catholic Church, Good Friday. [<L, day of preparation, day before the Sabbath <Gk. *paraskeuē* preparation <*para-* beside, against + *skeuē* equipment]

par·a·se·le·ne (par′ə·si·lē′nē) *n. pl.* **·nae** (-nē) *Meteorol.* A luminous spot appearing on a lunar halo; a mock moon. [<NL <Gk. *para* beside, subsidiary + *selēnē* moon]

par·a·shah (par′ə·shä) *n. pl.* **·shoth** (-shōth) One of the fifty-four sections or lessons into which the Pentateuch is divided for weekly readings throughout the annual cycle, or one of the smaller sections read on festivals. [<Hebrew *pārāshāh* a division <*pārash* divide]

par·a·site (par′ə·sīt) *n.* **1** *Biol.* An animal or a plant that lives on or in another organism at whose expense it obtains nourishment and shelter. **2** An obsequious sycophant who lives at another's expense. **3** In ancient Greece and Rome, one who secured a welcome at the tables of the rich by means of fawning and flattery. [<L *parasitus* <Gk. *parasitos*, lit., one who eats at another's table <*para-* beside + *sitos* food] —**par′a·sit′ic** (-sit′ik) or **·i·cal** *adj.* —**par′a·sit′i·cal·ly** *adv.*

parasite drag *Aeron.* That portion of the drag of an aircraft exclusive of the drag of the wings: also called *head resistance.* Also **parasite resistance.** Compare INDUCED DRAG.

par·a·sit·i·cide (par′ə·sit′ə·sīd) *n.* Any agent that destroys parasites. —*adj.* Efficacious for destroying parasites: also **par′a·sit′i·ci′dal.**

par·a·sit·ism (par′ə·sī′tiz·əm) *n.* **1** The condition or conduct of a fawner or sycophant. **2** The state or condition of being parasitic. **3** *Biol.* Destructive symbiosis. **4** *Med.* Disease, especially of the skin, caused by parasites.

par·a·si·tol·o·gy (par′ə·sī·tol′ə·jē) *n.* The scientific study of parasites and parasitism. —**par′a·si′to·log′i·cal** (-sī′tə·loj′i·kəl) *adj.* —**par′a·si·tol′o·gist** *n.*

par·a·sol (par′ə·sôl, -sol) *n.* A small, light sunshade carried by women. [<MF <Ital. *parasole* <*para-* PARA-² + *sole* sun]

par·a·sphe·noid (par′ə·sfē′noid) *n. Anat.* A bone forming the floor of the cranium in some vertebrates.

par·a·sym·pa·thet·ic (par′ə·sim′pə·thet′ik) *adj. Anat.* Designating that part of the autonomic nervous system originating in the cranial and sacral regions of the spinal cord. Its functions include constriction of the pupil, dilation of the blood vessels and salivary glands, slowing of the heart, and stimulation of the digestive and genitourinary systems: also called *cranio-sacral.*

par·a·syn·ap·sis (par′ə·si·nap′sis) *n. Biol.* In meiosis, the side-by-side conjugation of chromosomes. Also **par′a·syn′de·sis** (-sin′də·sis).

par·a·syn·the·sis (par′ə·sin′thə·sis) *n. Gram.* The principle or process of forming words by both combination and derivation; especially, the creation of a derivative word by compounding with a particle, as in *downfallen.* —**par′a·syn·thet′ic** (-sin·thet′ik) *adj.* —**par′a·syn′the·ton** (-thə·ton) *n.*

par·a·syph·i·lis (par′ə·sif′ə·lis) *n. Pathol.* A morbid sequela of syphilis, but not itself syphilitic. —**par′a·syph′i·lit′ic** (-sif′ə·lit′ik) *adj.*

par·a·tax·ic (par′ə·tak′sik) *adj. Psychol.* Pertaining to or characterized by behavior in which one or more individuals seek adjustment to a group situation in terms of private meanings or images projected upon other people or surroundings. [<PARA-¹ + Gk. *taxis* order]

par·a·tax·is (par′ə·tak′sis) *n. Gram.* Independent arrangement of clauses, phrases, etc., without connectives, as in "I die, I faint, I fail." Opposed to *hypotaxis.* [<Gk., lit., a placing side by side <*paratassein* place side by side <*para-* beside + *tassein* arrange] —**par′a·tac′tic** (-tak′tik) or **·ti·cal** *adj.* —**par′a·tac′ti·cal·ly** *adv.*

par·a·thy·mi·a (par′ə·thī′mē·ə) *n. Psychiatry* Disordered emotional response; perversion of mood. [<NL <Gk. *para-* beside + *thymos* mind]

par·a·thy·roid (par′ə·thī′roid) *adj. Anat.* **1** Lying near the thyroid gland. **2** Pertaining to or designating one of several (usually four) small, bean-shaped glands found typically in pairs on the inner side near the back of each lobe of the thyroid.

par·a·troop (par′ə·trōōp) *n.* A military offensive force, with equipment, trained to land in hostile territory from an airplane by parachutes: also called *parachute troop.* [<PARA(CHUTE) + TROOP] —**par′a·troop′er** *n.*

par·a·ty·phoid (par′ə·tī′foid) *Pathol. adj.* Resembling typhoid fever but not responding to the tests for that disease. —*n.* Paratyphoid fever.

par·a·u·nit (par′ə·yōō′nit) *n.* A unit of paratroops. [<PARA(CHUTE) + UNIT]

par·a·vane (par′ə·vān) *n.* **1** A torpedo-shaped underwater device equipped with sharp projecting teeth for cutting the moorings of sunken mines. **2** A similar device loaded with high explosives for use against submarines. [<*para-*¹ + VANE]

par·boil (pär′boil′) *v.t.* **1** To boil partially. **2** To make uncomfortable with heat. [<OF *parboillir* <LL *parabullire* boil thoroughly <*per-* through + *bullire* bubble]

par·buck·le (pär′buk′əl) *n.* **1** A purchase made by looping a rope in the middle to aid in rolling casks, etc., up or down an incline, or in furling a sail by rolling the yards. **2** A sling made by passing both ends of a rope through its bight. —*v.t.* **·led, ·ling** To hoist or lower by means of a parbuckle. [Earlier *parbunkle*; origin unknown]

Par·cae (pär′sē) *n. pl.* In Roman mythology, the three Fates: also called *Destinies.* Also **Par′cæ.**

par·cel (pär′səl) *n.* **1** Anything wrapped up; a package; bundle. **2** An integral part. **3** A group or lot composed of an indefinite number or quantity (of people or animals). **4** A distinct portion of land. **5** A separated part of anything; an indefinite number. —*v.t.* **·celed** or **·celled, ·cel·ing** or **·cel·ling** **1** To divide or distribute in parts or shares: usually with *out.* **2** To make up into a parcel or parcels. **3** *Naut.* To wrap or cover with canvas strips, as a rope. —*adj.* & *adv.* Half or part; partly; partially. [<F *parcelle* <L *particula*, dim. of *pars, partis* part]

parcel post A postal service for the carriage and delivery of parcels not exceeding a specified size and weight: begun in the United States January 1, 1913. Also **parcels post.**

par·ce·nar·y (pär′sə·ner′ē) See COPARCENARY.

par·ce·ner (pär′sə·nər) *n. Law* A coparcener; coheir. [<AF *parcener*, OF *parçonier* <Med. L *partionarius* <L *partitionarius* <*partitio, -onis* share. See PARTITION.]

parch (pärch) *v.t.* **1** To make extremely dry; shrivel with heat. **2** To dry (corn, peas, etc.) by exposing to great heat; roast slightly. **3** To dry up or shrivel with cold. —*v.i.* **4** To become extremely dry; shrivel with heat. [ME *parchen, perchen*,? ult. <L *persiccare* <*per-* thoroughly + *siccare* dry]

par·chee·si (pär·chē′zē), **par·che·si, par·chi·si** See PACHISI.

parch·ment (pärch′mənt) *n.* **1** The skin of sheep, goats, and other animals prepared and polished with pumice stone for writing. **2** An imitation parchment made by treating paper with sulfuric acid and water: also called **vegetable parchment.** **3** A formal writing on parchment. **4** A college graduation diploma. **5** A light-tan or cream color, the color of parchment. [<OF *parchemin* <LL *Pergamena (charta)* (paper) of Pergamum <Gk. *Pergamon* the city of Pergamum; because it was used there instead of papyrus]

pard¹ (pärd) *n. Archaic* A leopard; panther. [<OF *parde* <L *pardus* <Gk. *pardos* <*pardalis*, ? <Persian *pars* a panther]

pard² (pärd) *n. Slang* A partner; mate; chum. [Short for PARDNER]

par·di (pär·dē′) *adv.* & *interj. Archaic* By God; verily: formerly a form of profanity: also spelled *perdie, perdy.* Also **par·dee′, par·die′, par·dy′.** [<OF *par dé* by God <L *per* by + *deus* God]

pard·ner (pärd′nər) *n. U.S. Dial.* Chum; friend. [Alter. of PARTNER]

par·don (pär′dən) *v.t.* 1 To remit the penalty of (a crime, insult, etc.). 2 To release from punishment; forgive for an offense. 3 To grant courteous allowance for or to: *Pardon* my French. —*n.* 1 The act of pardoning; remission of penalty incurred. 2 A waiving, by sovereign prerogative, of the execution of the penal sanctions of the violated law: distinguished from *justification.* 3 Courteous forbearance; acquittal of blame: used in making polite excuses. 4 *Law* Remission of guilt; also, an official warrant declaring the act of pardon. 5 An indulgence. [<OF *pardun* <*pardonner* <LL *perdonare* grant <*per-* through + *donare* give] —**par′don·a·ble** *adj.* —**par′don·a·bly** *adv.*

Synonyms (verb): absolve, acquit, condone, exculpate, excuse, forgive, overlook, release, remit. *Forgive* has reference to feelings, *pardon* to consequences; hence, the executive may *pardon,* but has nothing to do officially with *forgiving.* To *pardon* is the act of a superior, implying the right to punish; to *forgive* is the privilege of the humblest person who has been wronged or offended. In law, to *remit* the whole penalty is equivalent to *pardoning* the offender; but a part of a penalty may be *remitted* and the remainder inflicted, as where the penalty includes both fine and imprisonment. To *condone* is to put aside a recognized offense by some act which restores the offender to forfeited right or privilege, and is the act of a private individual, without legal formalities. To *excuse* is to *overlook* some slight offense, error, or breach of etiquette; *pardon* is often used by courtesy in nearly the same sense. Compare ABSOLVE, MERCY. *Antonyms*: castigate, chasten, chastise, condemn, convict, correct, doom, punish, recompense, scourge, sentence, visit.

Synonyms (noun): absolution, acquittal, amnesty, forbearance, forgiveness, mercy, oblivion, remission. *Acquittal* is a release from a charge, after trial, as not guilty. *Pardon* is a removal of penalty from one who has been adjudged guilty. *Acquittal* is the adjudging one to be not guilty, as by the decision of a court, commonly of a jury; *pardon* is the act of the executive. An innocent man may demand *acquittal,* and need not plead for *pardon. Oblivion* signifies overlooking and virtually forgetting an offense, so that the offender stands before the law in all respects as if it had never been committed. *Amnesty* conveys the same idea. *Pardon* affects individuals; *amnesty* and *oblivion* are said of great numbers. *Pardon* is oftenest applied to the ordinary administration of law; *amnesty* or *oblivion,* to national and military affairs. An *amnesty* is issued after war, insurrection, or rebellion. *Absolution* is a religious word (compare synonyms for ABSOLVE). *Remission* is a discharge from penalty; as, the *remission* of a fine. *Antonyms*: penalty, punishment, retaliation, retribution, vengeance.

par·don·er (pär′dən·ər) *n.* 1 One who pardons. 2 In the Middle Ages, a layman commissioned to collect offerings for which indulgences were promised.

par·don·nez-moi (pȧr·dôn′ā·mwȧ′) *French* Pardon me.

pare (pâr) *v.t.* **pared, par·ing** 1 To cut off the covering layer or part of. 2 To cut off or trim away (a covering layer or part): often with *off* or *away.* 3 To reduce or diminish, especially gradually or little by little. ◆ Homophone: *pair, pear.* [<F *parer* prepare, trim <L *parare*] —**par′er** *n.* —**par′ing** *adj.* & *n.*

Pa·ré (pȧ·rā′), **Ambroise,** 1510?–90, French surgeon; father of modern surgery.

pa·re·cious (pə·rē′shəs), etc. See PAROECIOUS, etc.

par·e·gor·ic (par′ə·gôr′ik, -gor′ik) *n.* 1 A medicine that assuages pain. 2 A camphorated tincture of opium. [<LL *paregoricus* <Gk. *parēgorikos, parēgoros* soothing <*para-* beside + *-agoros* speaking <*agora* assembly, market place]

pa·rei·ra bra·va (pə·râr′ə brä′və, brā′və) A drug obtained from the roots of a tropical American plant of the moonseed family (*Chondrodendron tomentosum*): used in treating chronic disorders of the urinary passages. [<Brazilian Pg., a wild vine]

par·en·ceph·a·lon (par′ən·sef′ə·lon) *n.* The cerebellum. [<NL <Gk. *para-* beside + *enkephalon* brain <*en-* in + *kephalē* head] —**par·en·ce·phal·ic** (par·en′sə·fal′ik) *adj.*

pa·ren·chy·ma (pə·reng′ki·mə) *n. Biol.* 1 The soft cellular substance of glandular and other organs. 2 The proper substance of an organ, excluding the connective tissue and the like. 3 *Bot.* The thin-walled, soft cell tissue of higher plants, as found in stem pith and in the pulp of fruits. Also **pa·ren′chyme** (-kīm). [<NL <Gk., lit., something poured in beside <*para-* beside + *enchyma* infusion] —**par·en·chym·a·tous** (par′eng·kim′ə·təs) *adj.*

par·ent (pâr′ənt) *n.* 1 A father or a mother. 2 Any organism that generates another; a producer. 3 Cause; occasion. [<OF <L *parens, -entis* parent, ancestor, orig. ppr. of *parere* beget]

par·ent·age (pâr′ən·tij) *n.* 1 The relation of parent to child, of producer to the produced. 2 Relation of cause to effect. 3 Descent or derivation from parents; extraction; lineage; origin. 4 Derivation from any source. 5 Parenthood.

pa·ren·tal (pə·ren′təl) *adj.* 1 Pertaining to or characteristic of a parent. 2 *Genetics* Pertaining to or designating that generation from whose crossbreeding hybrids are produced. —**pa·ren′tal·ly** *adv.*

Par·en·ta·li·a (par′ən·tā′lē·ə) *n. pl. Latin* In ancient Rome, an annual feast in commemoration of the dead and veneration of ancestors.

par·en·ter·al (par·en′tər·əl) *adj. Med.* Pertaining to or designating a mode of assimilation other than through the alimentary canal, as intravenous or subcutaneous. [<Gk. *para-* beside + *enteron* intestine]

pa·ren·the·sis (pə·ren′thə·sis) *n. pl.* **·ses** (-sēz) 1 *Gram.* A word, phrase, or clause inserted in a sentence that is grammatically complete without it, separated usually by commas, dashes, or upright curves. 2 Either or both of the upright curves ( ) so used. 3 Hence, any intervening episode or incident; interval. [<Med. L <Gk. <*parentithenai* put in beside <*para-* beside + *en-* in + *tithenai* place]

pa·ren·the·size (pə·ren′thə·sīz) *v.t.* **·sized, ·siz·ing** 1 To insert as a parenthesis. 2 To insert parentheses in. 3 To place within parentheses (def. 2).

par·en·thet·i·cal (par′ən·thet′i·kəl) *adj.* 1 Pertaining to a parenthesis. 2 Abounding in parentheses. 3 Thrown in; episodical. Also **par′en·thet′ic.** —**par′en·thet′i·cal·ly** *adv.*

par·ent·hood (pâr′ənt·ho͝od) *n.* The condition or relation of a parent.

Par·ent-Teach·er Association (pâr′ənt·tē′·chər) *U.S.* Any of several local organizations composed of parents and public-school teachers, seeking mutual cooperation in the guidance of school children.

pa·re·sis (pə·rē′sis, par′ə·sis) *n. Pathol.* Partial paralysis affecting muscular motion but not sensation. —**general paresis** General paralysis accompanied by dementia, caused by syphilitic degeneration of the brain. [<NL <Gk., a letting go <*parienai* let go <*para-* beside + *hienai* let go]

par·es·the·sia (par′is·thē′zhə, -zhē·ə) *n. Pathol.* Abnormal or perverted sense of touch; a sensation of itching, tingling, or prickling of the skin: also spelled *paraesthesia.* Also **par′es·the′sis.** —**par′es·thet′ic** (-thet′ik) *adj.*

pa·ret·ic (pə·ret′ik, -rē′tik) *adj.* Pertaining to or afflicted with paresis. —*n.* One who suffers from paresis. —**pa·ret′i·cal·ly** *adv.*

Pa·re·to (pä·rā′tō), **Vilfredo,** 1848–1923, Italian economist and sociologist.

pa·re·u (pä′rā·o͞o) *n.* A rectangular, figured cotton cloth worn as a skirt or loincloth by the natives of southern Pacific islands. [<Tahitian]

par ex·cel·lence (pär ek′sə·läns, *Fr.* pȧr ek·se·läṅs′) Of the highest excellence; beyond comparison; preeminently. [<F, by way of excellence]

par ex·em·ple (pȧr eg·zäṅ′pl′) *French* For example; for instance.

par·fait (pär·fā′) *n.* A frozen dessert or confection made with eggs, sugar, whipped cream, and fruit or other flavoring. [<F, lit., perfect <L *perfectus.* See PERFECT.]

par·fleche (pär·flesh′) *n.* 1 Rawhide, usually of buffalo skin, which has been freed of hair and dried on a stretcher. 2 An article, as a shield, made from such a hide. Also **par·flesh′.** [<dial. F (Canadian) <F *par-* PARA-[2] + *flèche* an arrow]

par·get (pär′jit) *n.* 1 Gypsum. 2 Plaster suitable for lining chimneys. 3 Pargeting. —*v.t.* **·get·ed** or **·get·ted, ·get·ing** or **·get·ting** To cover or adorn with parget or pargeting. [Appar. <OF *pargeter, parjeter* throw over a surface <*par-* all over + *jeter* throw <L *per-* thoroughly + *jactare,* freq. of *jacere* throw]

par·get·ing (pär′jit·ing) *n.* 1 Plastering; specifically, ornamental stuccowork or plasterwork in relief. 2 Parget (def. 2).

par·go (pär′gō) *n. pl.* **·gos** or **·go** The American muttonfish. [<Sp. *pargo* <L *pagrus, phagrus* <Gk. *phagros* a sea bream]

par·he·lic circle (pär·hē′lik, -hel′ik) A band of light or halo, passing through the sun and parallel to the horizon: it is an effect of solar reflection from the vertical faces of ice crystals in the atmosphere. Also **parhelic ring.**

par·he·li·on (pär·hē′lē·ən) *n. pl.* **·li·a** (-lē·ə) One of two bright solar images appearing on the parhelic circle on either side of the sun; a mock sun or sundog. Also **par·he′li·um.** [<L *parelion* <Gk. *parēlion* <*para-* beside + *hēlios* sun] —**par·he′lic, par·he·li·a·cal** (pär′hi·lī′ə·kəl) *adj.*

pari– *combining form* Equal: *parisyllabic.* [<L *par, paris* equal]

Pa·ria (pä′ryä), **Gulf of** An inlet of the Caribbean between Trinidad and the coast of Venezuela; enclosed on the north by **Paria Peninsula,** a promontory of NE Venezuela extending 75 miles east.

pa·ri·ah (pə·rī′ə, par′ē·ə) *n.* 1 One of low caste (but not lowest or outcast) people of southern India and Burma, employed as servants. 2 A person of low caste or no caste. 3 A social outcast. [<Tamil *paraiyar,* pl. of *paraiyon,* lit., (hereditary) drummer <*parai* a large festival drum]

Par·i·an (pâr′ē·ən) *adj.* 1 Of, from, or pertaining to Paros or the white marble mined there. 2 Resembling the marble of Paros. —*n.* 1 A native or inhabitant of Paros. 2 Ware of Parian marble: also **Parian biscuit.**

Pa·ri·cu·tín (pä′rē·ko͞o·tēn′) A volcano in Michoacán state, Mexico; 8,200 feet; first erupted February, 1943.

pa·ri·es (pâr′i·ēz) *n. pl.* **pa·ri·e·tes** (pə·rī′ə·tēz) *Usually pl. Biol.* The wall of any cavity in the body, as of any organ. [<L, a wall]

pa·ri·e·tal (pə·rī′ə·təl) *adj.* 1 *Biol.* Of, pertaining to, or forming the walls of any cavity in the body. 2 Of or pertaining to a wall. 3 Pertaining to the care of or residence within walls or precincts, as of a college. 4 *Bot.* Pertaining to or borne on a wall: said especially of the placentae or ovules borne on the wall of the ovary of a plant. —*n.* A parietal bone.

parietal bone *Anat.* Either of two bones between the occipital and frontal bones that form a part of the top and sides of the cranium.

parietal lobe *Anat.* That portion of the hemispheres of the brain that lies between the central sulcus and the occipital lobe.

pa·ri·e·tes (pə·rī′ə·tēz) Plural of PARIES.

pa·ril·lin (pə·ril′in) *n. Biochem.* A white crystalline saponin of variable composition, contained in sarsaparilla root, and to which the drug owes its medicinal qualities: often called *smilacin.* Also **pa·ril·lic acid** (pə·ril′ik). [<Sp. *parrilla,* dim. of *parra* a vine + -IN]

par·i·mu·tu·el (par′i·myo͞o′cho͞o·əl) *n.* 1 A system of betting at races in which those who have bet on the winners share in the total amount wagered: also **parimutuel.** 2 A machine for recording bets under this system; a totalisator. [<F, a stake or mutual wager]

par·ing (pâr′ing) *n.* 1 The act of cutting off the surface or edge of. 2 The part pared off.

pa·ri pas·su (pâr′ī pas′o͞o, par′ē) *Latin* With equal pace; of the same speed.

par·i·pin·nate (par′i·pin′āt) *adj. Bot.* Equally or abruptly pinnate: said of leaves.

par·is (par′is) *n.* The herb-Paris. [<L *pars, paris* equal; infl. by PARIS]
Par·is (par′is) In Greek mythology, a son of Priam and Hecuba who carried off Helen, wife of Menelaus, thus causing the Trojan War. See APPLE OF DISCORD.
Par·is (par′is, *Fr.* pȧ·rē′) A port on the Seine, 111 miles from its mouth, and capital of France: ancient *Lutetia.*
Pa·ris (pä·rēs′), **Gaston,** 1839–1903, French philologist.
Paris Basin The chief depression of north and north central France, at the center of which is Paris.
Paris blue Prussian blue.
Paris green A poisonous compound prepared from copper acetate and arsenic trioxide, used largely as an insecticide and pigment.
par·ish (par′ish) *n.* **1** *Eccl.* In the Anglican, Roman Catholic, and some other churches, a district, usually part of a diocese, with its own church, and in charge of a priest or other clergyman. **2** *U.S.* **a** A religious congregation, comprising all those who worship at the same church. **b** The district in which they live. **3** *Brit.* A political subdivision of a county, often corresponding to an ecclesiastical parish. **4** In Louisiana, a civil district corresponding to a county. **5** The people of a parish, in any of the above senses. ◆ Collateral adjective: *parochial.* [<OF *paroche, paroisse* <LL *parochia* <Gk. *paroikia,* orig. a neighborhood, later a diocese < *para-* beside + *oikeein* dwell] —**pa·rish′ion·al** *adj.*
pa·rish·ion·er (pə·rish′ən·ər) *n.* A member of a parish.
Pa·ri·sian (pə·rizh′ən, -riz′ē·ən) *adj.* Of or pertaining to the city of Paris. —*n.* A native or resident of Paris.
par·i·syl·lab·ic (par′i·si·lab′ik) *adj.* Having the same number of syllables. Also **par′i·syl·lab′i·cal.**
par·i·tor (par′ə·tər) *n. Obs.* An officer of an ecclesiastical court who issues summonses. [Aphetic var. of APPARITOR]
par·i·ty[1] (par′ə·tē) *n.* **1** Equality, as of condition or rank; like state or degree; equivalent position; equal value. **2** The equivalence in legal weight and quality of the legal tender of one class of money to another. **3** Par (def. 1). **4** Equality between the currency or prices of commodities of two countries or cities. **5** Perfect analogy; close resemblance. **6** *U.S.* A level for farm prices which gives to the farmer the same purchasing power that he averaged during each year of a chosen base period, originally the five years of farm prosperity prior to World War I. **7** *Physics* That property of a wave whereby its function is symmetrically unchanged by inversion in a coordinate system (**even parity**), or changed only in sign (**odd parity**). See synonyms under ANALOGY, SYMMETRY. [<L *paritas* equality < *pars* equal]
par·i·ty[2] (par′ə·tē) *n. Med.* Fitness or ability to bear offspring. [<L *parere* bear + -ITY]
par·i·vin·cu·lar (par′ə·ving′kyə·lər) *adj. Zool.* Designating a bivalve that has an elongated semicylindrical ligament. [<PARI- + VINCULUM]
park (pärk) *n.* **1** In English law, a tract of enclosed land stocked with wild beasts of the chase, and held through royal grant or by immemorial prescription. **2** A tract of land for public use in or near a city, usually laid out with walks, drives, and recreation grounds. **3** An open square or plaza in a city, usually containing shade trees and seats. **4** A large area of country containing natural curiosities reserved by the government for public enjoyment: a national *park.* **5** A plateaulike valley between mountain ranges: used most frequently in Colorado and Wyoming. **6** *Mil.* An enclosure where guns, trucks, wagons, animals, etc., are placed for safety; also, the objects thus enclosed: an artillery *park;* also, a complete train of cannon, including equipment, ammunition, gunners, etc., for an army in the field. —*v.t.* **1** To place or leave (an automobile, etc.) standing for a time, as on the street. **2** *U.S. Colloq.* To place; set: *Park* your hat on the table. **3** To assemble or mass together: to *park* artillery. **4** To enclose in or as in a park. —*v.i.* **5** To park an automobile, etc. [<OF *parc* a game preserve, ult. <Gmc. Akin to PADDOCK.]

Park (pärk), **Mungo,** 1771–1806, Scottish African explorer.
par·ka (pär′kə) *n.* **1** An outer garment of undressed skins worn by Eskimos. **2** A similar woolen garment, sometimes fur-lined, with attached hood: worn for skiing and other winter sports. Also **par′kee** (-kē). [<Aleut]

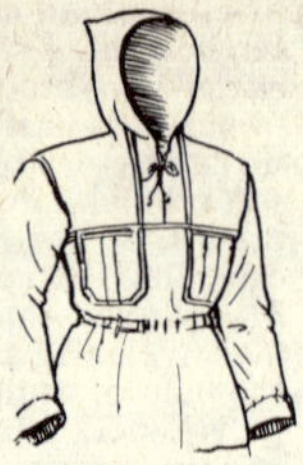
HOODED PARKA

Park Avenue A residential street running north and south in Manhattan borough of New York City: regarded as a symbol of wealth and fashion.
Par·ker (pär′kər), **Dorothy,** born 1893, *née* Rothschild, U. S. writer. —**Sir (Horatio) Gilbert,** 1862–1932, Canadian author and dramatist. —**Theodore,** 1810–60, U. S. minister, and New England abolitionist leader.
Parkes (pärks), **Sir Henry,** 1815–96, Australian statesman.
park·ing (pär′king) *n.* **1** Parks collectively, or ground resembling a park, as a strip of sward in a street. **2** The forming into a park, as of cannon. **3** The act of leaving a vehicle in a public place: No *parking* on this street.
Park·man (pärk′mən), **Francis,** 1823–93, U. S. historian.
Park Range A part of the Rocky Mountains in western Colorado; highest peak, 14,284 feet.
park·way (pärk′wā′) *n.* **1** A wide thoroughfare adorned with spaces planted with turf and trees. **2** A specially constructed track for the use of automobiles.
par·lance (pär′ləns) *n.* **1** Manner of speech; language; phrase: common *parlance,* legal *parlance.* **2** *Archaic* Talk; conversation. [<OF < *parler* speak <LL *parabolare* < *parabola.* See PARABLE.]
par·lan·do (pär·län′dō) *adj. & adv. Music* Declamatory in style; in recitative. Also **par·lan′·te** (-tā). [<Ital. ppr. of *parlare* speak <LL *parabolare* < *parabola.* See PARABLE.]
Par·la·to·ri·a (pär′lə·tôr′ē·ə, -tō′rē·ə) *n.* A genus of scale insects (family *Coccidae*), including many species injurious to fruits, especially the date-palm scale (*P. blanchardi*). [<NL, after Felipe *Parlatore,* 1816–77, Italian physician and botanist]
par·lay (pär·lā′, pär′lē) *v.t. & v.i.* To place (an original bet and its winnings) on a later race, contest, etc. —*n.* Such a bet. [Alter. of earlier *paroli* <F <Ital., a grand cast at dice < *paro* equal <L *par*]
par·le·men·taire (pär′lə·men·târ′, *Fr.* pȧr·lə·mäṅ·târ′) *n.* The agent of a field commander authorized to enter enemy lines to negotiate with the enemy commander. [<F < *parlement.* See PARLIAMENT.]
par·ley (pär′lē) *n.* **1** An oral conference, as with an enemy; a discussion of terms. **2** Mutual discourse. See synonyms under CONVERSATION. —*v.i.* To hold a conference, especially with an enemy. Also *Obs.* **parle** (pärl). [<F *parlée,* fem. pp. of *parler* speak <LL *parabolare* < *parabola.* See PARABLE.]
par·lia·ment (pär′lə·mənt) *n.* A meeting or assembly for consultation and deliberation; a legislative body; a national legislature, especially when composed of various estates. Also *Obs.* **par′le·ment.** [<OF *parlement* speaking < *parler* speak <LL *parabolare* < *parabola.* See PARABLE.]
Par·lia·ment (pär′lə·mənt) *n.* **1** The supreme legislature of Great Britain and Northern Ireland, composed of the three estates of the realm—the Lords Spiritual, the Lords Temporal, and the Commons—together with, in a strict legal sense, the sovereign. **2** The legislature in any of Great Britain's self-governing colonies or dominions. **3** In France, before the French Revolution, one of several tribunals of justice. **4** The legislative assembly of Scotland until 1707, or that of Ireland until 1800. —**Long Parliament** The British Parliament which first assembled in 1640 and finally dissolved by its own consent in 1660: after the enforced expulsion (*Pride's Purge*) of some of its members in 1648, it was known as the **Rump Parliament.**
par·lia·men·tar·i·an·ism (pär′lə·men·târ′ē·ən·iz′əm) *n.* The system of government, developed in England, in which the tenure of the cabinet is dependent on the will of the majority in the lower house: also called **parliamentary system.**
par·lia·men·ta·ry (pär′lə·men′tər·ē) *adj.* **1** Pertaining to, characterized by, or enacted by a parliament. **2** According to the rules of Parliament; admissible in a deliberative assembly. —**par′lia·men·tar′i·an** *adj. & n.*
parliamentary law See under LAW.
par·lor (pär′lər) *n.* **1** A room for reception of callers or entertainment of guests; drawing-room. **2** A room in an inn, hotel, etc., for private conversation, appointments, etc. **3** *U.S.* Formerly, a smartly furnished room for the performance of personal or professional services: a tonsorial *parlor:* a genteelism. Also *Brit.* **par′lour.** [<AF *parlur,* OF *parleor* <Med. L *parlatorium* < *parlare* <LL *parabolare* speak < *parabola.* See PARABLE.]
parlor car A railway car fitted with luxurious chairs, and run as a day coach; a drawing-room car.
par·lous (pär′ləs) *Archaic adj.* **1** Dangerous or exciting; perilous. **2** Shrewd; venturesome; waggish; mischievous. —*adv.* Exceedingly; very. [Var. of PERILOUS] —**par′lous·ly** *adv.*
Par·ma (pär′mä) **1** A province and former duchy of north central Italy. **2** Its capital, a city near the Apennines.
Par·men·i·des (pär·men′ə·dēz) Greek philosopher of the fifth century B.C.
Par·me·san (pär′me·zan′) *adj.* Of or pertaining to Parma. Also **Par·mese** (pär·mēs′). [<F <Ital. *parmigiano* belonging to Parma]
Parmesan cheese A hard, dry cheese, originally made in Parma: usually grated and served on soups, spaghetti, etc.
Par·na·í·ba (pär′nä·ē′bä) A river in NE Brazil, flowing 750 miles NE to the Atlantic. Also **Par′na·hy′ba.**
Par·nas·si·an (pär·nas′ē·ən) *adj.* **1** Belonging or relating to Parnassus or the Parnassians. **2** Of or pertaining to poetry. —*n.* A representative or member of a school of poetry founded in France during the last half of the 19th century, which emphasized the technical aspect of the art of poetry: so called from the title of its first collection of poems, *Le Parnasse contemporain* (1866). [<L *Parnas(s)ius*] —**Par·nas′si·an·ism** *n.*
Par·nas·sus (pär·nas′əs) **1** A mountain north of the Gulf of Corinth in central Greece; 8,062 feet: formerly *Liákoura:* anciently regarded as sacred to Apollo and the Muses. *Greek* **Par·nas′sos. 2** A domain of poetry or of literature. **3** A collection of poems or literary works.
Par·nell (pär·nel′), **Charles Stewart,** 1846–91, Irish statesman.
pa·ro·chi·al (pə·rō′kē·əl) *adj.* **1** Pertaining to, supported by, or confined to a parish. **2** Hence, narrow; provincial; restricted in scope: *parochial* ideas. [<OF <LL *parochialis* < *parochia.* See PARISH.]
pa·ro·chi·al·ism (pə·rō′kē·əl·iz′əm) *n.* **1** Government or control by a vestry or parochial board. **2** Narrowness of view; provincialism.
parochial school See under SCHOOL.
par·o·dy (par′ə·dē) *n. pl.* **·dies 1** A literary composition imitating and ridiculing some serious work; a comical imitation, especially of a poem; a travesty. **2** Any burlesque imitation of something serious. **3** A poor imitation. See synonyms under CARICATURE. —*v.t.* **·died, ·dy·ing** To make a parody of; travesty. [<Gk. *parōidia* a burlesque poem or song < *para-* beside + *ōidē* a song, poem] —**pa·rod·ic** (pə·rod′ik) or **·i·cal** *adj.* —**par′o·dist** *n.*
pa·roe·cious (pə·rē′shəs) *adj. Bot.* Having the male and female sexual organs of plants developed side by side or in the same inflorescence, as in many bryophytes: also spelled *parecious.* Also **pa·roi′cous.** [<Gk. *paroikos* dwelling side by side < *paroikeein* < *para-* beside + *oikeein* dwell] —**pa·roe′cious·ly** *adv.* —**pa·roe′cious·ness** *n.* —**pa·roe′cism** (-siz·əm) *n.*
pa·rol (pə·rōl′) *Law n.* **1** Something spoken or said; specifically, in the legal phrase **by parol,** by word of mouth. **2** The pleadings filed in an action. —*adj.* **1** Given or expressed by word of mouth; oral. **2** Written but not under seal. Also **pa·role′.** [<AF, var. of OF *parole* <L *parabola* word. See PARABLE.]

pa·role (pə·rōl′) *n.* **1** A pledge of honor by a prisoner of war that he will not seek to escape or will not serve against his captors until exchanged; also, the condition of being under parole. **2** The release of a prisoner from jail prior to the expiration of his term on his own recognizance. **3** The watchword used only by officers of the guard and of the day: distinguished from *countersign.* **4** *Law* Parol. **5** Word of honor. —*v.t.* **·roled, ·rol·ing** To release (a prisoner) on parole. [<F *parole (d'honneur)* word (of honor) <OF <L *parabola.* Doublet of PALAVER, PARABLE, PARABOLA.]

pa·role d'hon·neur (pȧ·rôl′ dô·nœr′) *French* Word of honor.

par·o·no·ma·si·a (par′ə·nō·mā′zhē·ə, -zē·ə) *n.* Any use for effect of words similar in sound, but differing in meaning; a play on words, especially one in which the similarity of sound is the prominent characteristic. Compare PUN. [<L <Gk. <*paronomazein* alter slightly in meaning <*para-* beside + *onoma* name] —**par′o·no·ma′si·al, par′o·no·mas′tic** (-mas′tik) or **·ti·cal** *adj.* —**par′o·no·mas′ti·cal·ly** *adv.*

par·o·nym (par′ə·nim) *n.* A word having the same root as another; a cognate word. [<Gk. *parōnymon,* orig. neut. of *parōnymos* derivative <*para-* beside + *onyma,* Aeolic var. of *onoma* name] —**pa·ron·y·mous** (pə·ron′ə·məs), **par′o·nym′ic** (-nim′ik) *adj.*

par·o·quet (par′ə·ket) See PARAKEET.

par·o·ral (par·ôr′əl, -ō′rəl) *adj.* Adjacent to the mouth or oral region.

Pa·ros (pä′ros) An island of the central Cyclades, Greece; 77 square miles.

pa·rot·ic (pə·rot′ik, -rō′tik) *adj.* Situated near the ear: the *parotic* region. [<NL *paroticus* <Gk. *para-* beside + *otikos* of the ear <*ous, ōtos* ear]

pa·rot·id (pə·rot′id) *Anat. adj.* **1** Situated near the ear. **2** Designating one of the paired salivary glands in front of and below the external ear in mammals. —*n.* A salivary gland below the ear. [<F *parotide* <L *parotis, -idis* <Gk. *parōtis, -idos* a tumor near the ear <*para-* beside + *ous, ōtos* ear]

par·o·ti·tis (par′ə·tī′tis) *n. Pathol.* Inflammation and swelling of the parotid gland; mumps. Also **par′o·ti·di′tis** (-ti·dī′tis). —**par′o·tit′ic** (-tit′ik) *adj.*

pa·ro·toid (pə·rō′toid) *Biol. adj.* **1** Resembling a parotid gland. **2** Designating a cutaneous gland situated behind the eye and above the tympanum in anurous amphibians. —*n.* A parotoid gland.

-parous *suffix* Giving birth to; bearing; producing: *oviparous.* [<L *-parus* <*parere* beget]

par·ox·ysm (par′ək·siz′əm) *n.* **1** *Pathol.* A periodic attack of disease; a fit. **2** Sudden and violent excitement or emotion, as of anger. **3** A convulsion of any kind. See synonyms under AGONY, PAIN. [<MF *paroxysme* <Med. L *paroxysmus* irritation <Gk. *paroxysmos* <*paroxynein* goad <*para-* beside, beyond + *oxynein* goad <*oxys* sharp]

par·ox·ys·mal (par′ək·siz′məl) *adj.* **1** Relating to, of the nature of, accompanied or characterized by a paroxysm. **2** Resulting from convulsive action of natural forces, as a volcanic eruption, flood, etc. Also **par′ox·ys′mic.** —**par′ox·ys′mal·ly** *adv.*

par·ox·y·tone (par·ok′sə·tōn) *adj.* Having the acute accent on the penultimate syllable. —*n.* A word thus accented, as *ly′kos.* [<NL *paroxytonus* <Gk. *paroxytonos* <*para-* beside, past + *oxytonos* OXYTONE]

par·quet (pär·kā′, -ket′) *n.* **1** The main-floor space behind the orchestra of a theater; sometimes, the whole lower floor. **2** Parquetry. —*v.t.* **·queted** (-kād′, -ket′id), **·quet·ing** (-kā′·ing, -ket′ing) To make of or ornament with parquetry. Also **par·quette** (pär·ket′). [<F <OF *parchet* a small compartment, dim. of *parc.* See PARK.]

parquet circle The section of theater seats at the rear of the parquet and under the balcony.

par·quet·ry (pär′kit·rē) *n.* Wooden mosaic, used especially for floor surfaces. [<F *parqueterie*]

par·quine (pär′kēn) *n. Chem.* A yellowish, bitter, crystalline alkaloid, $C_{21}H_{39}O_8N$, extracted from the bark of certain solanaceous plants, especially *Cestrum parqui*: similar in action to strychnine and atropine. [<NL *parqui,* name of a species + -INE[2]]

parr (pär) *n.* A young salmon before its first migration seaward. [? <dial. E (Scottish)]

Parr (pär), **Catherine,** 1513–48, sixth and last wife of Henry VIII of England.

par·ra·keet (par′ə·kēt), **par·ra·kee·to** (-tō) See PARAKEET.

par·ra·mat·ta (par′ə·mat′ə) See PARAMATTA.

Par·ran (par′ən), **Thomas,** born 1892, U.S. physician.

par·rel (par′əl) *n.* **1** A chimneypiece or the ornaments of a chimneypiece collectively. **2** *Naut.* A sliding hoop, rope, or chain by which a yard is attached to a mast. Also **par′·ral.** [ME *parail,* aphetic var. of *aparail* equipment. See APPAREL.]

par·ri·cide (par′ə·sīd) *n.* **1** The murder of a parent, or of an ancestor. **2** One who has committed such a crime. [<F <L *paricidium* a killing of a relative, and *paricida* a killer of a relative] —**par′ri·ci′dal** *adj.* —**par′ri·ci′dal·ly** *adv.*

Par·ring·ton (par′ing·tən), **Vernon Louis,** 1871–1929, U. S. literary historian.

Par·rish (par′ish), **Maxfield,** born 1870, U.S. artist.

Par·ris Island (par′is) One of the Sea Islands, NE of Savannah, Georgia; site of a United States Marine Corps training camp.

par·ritch (par′ich) *n. Scot.* Porridge of oatmeal. Also **par′ridge** (-ij).

par·ro·quet (par′ə·ket) See PARAKEET.

par·rot (par′ət) *n.* **1** Any of certain birds of warm regions of the order *Psittaciformes,* having a hooked bill, paired toes, and usually brilliant plumage, including the macaws, parakeets, cockatoos, lories, and related genera. Some parrots are noted for their ability to simulate human laughter and speech. ◆ Collateral adjective: *psittacine.* **2** A person who repeats or imitates without understanding. —*v.t.* To repeat or imitate by rote or without understanding. [? <F *perrot,* var. of *Pierrot,* dim. of *Pierre* Peter, a personal name] —**par′rot·er** *n.*

parrot fever Psittacosis.

parrot fish **1** Any of many small fishes of the family *Scaridae,* inhabiting warm seas: so called from their vivid coloring and beaklike jaws. **2** A labroid fish of the genus *Labrichthys,* especially the parrot perch *(L. psittacula)* of Australasia.

par·ry (par′ē) *v.* **·ried, ·ry·ing** *v.t.* **1** To ward off, as a thrust in fencing. **2** To avoid or evade. —*v.i.* **3** To make a parry. —*n. pl.* **·ries** **1** A defensive movement, as in fencing. **2** An evasion or diversion in a contest of wits. [Prob. <F *parez,* imperative of *parer* ward off <Ital. *parare* defend <L, prepare]

Par·ry (par′ē), **Sir William Edward,** 1790–1855, English Arctic explorer.

Par·ry Islands (par′ē) An arctic archipelago in west central Franklin district, Northwest Territories, including Melville, Bathurst, and numerous smaller islands.

parse (pärs) *v.t.* **parsed, pars·ing** **1** To describe (a sentence) grammatically by giving the form, function, etc., of each of its components. **2** To describe (a word) as to its part of speech, form, and relation to the other elements in a sentence. [Prob. <L *pars, partis* part] —**pars′er** *n.*

par·sec (pär′sek) *n. Astron.* A unit of length used in expressing the distance of stars. One parsec is almost exactly 206,265 times the mean distance of the earth from the sun, or 19.2 trillion miles, or 3.26 light years. A star is at a distance of one parsec from the earth if its annual parallax amounts to one second of arc (1″). [<PAR(ALLAX) + SEC(OND)[1] (def. 2)]

Par·see (pär′sē, pär·sē′) *n.* A Zoroastrian; especially, an adherent of the old Persian religion whose ancestors fled to India about the eighth century on account of Mohammedan persecution. Also **Par′si.** [<Persian *Pārsī* a Persian <*Pārs* Persia] —**Par′see·ism, Par′si·ism, Par′sism** *n.*

Par·si·fal (pär′si·fäl, -fəl) An opera by Richard Wagner based on the *Parzifal* of Wolfram von Eschenbach. See PERCEVAL.

par·si·mo·ni·ous (pär′sə·mō′nē·əs) *adj.* Niggardly; penurious. See synonyms under AVARICIOUS, SCANTY. —**par′si·mo′ni·ous·ly** *adv.* —**par′si·mo′ni·ous·ness** *n.*

par·si·mo·ny (pär′sə·mō′nē) *n.* Undue sparingness in the expenditure of money; stinginess. See synonyms under FRUGALITY. [<L *parsimonia* <*parcere* spare]

pars·ley (pärs′lē) *n.* A cultivated umbelliferous herb *(Petroselinum latifolium* or *crispum)* with aromatic, finely divided leaves and greenish-yellow flowers, used as a garnish and for flavoring soups. [Fusion of OF *peresil* and OE *petersilie,* both <LL *petrosilium,* alter. of L *petroselinum* <Gk. *petroselinon* <*petra* rock + *selinon* parsley]

pars·nip (pärs′nip) *n.* A European herb *(Pastinaca sativa)* of the parsley family, with a large, sweetish, edible root, widely cultivated as a vegetable and as fodder. The root of the wild plant is acrid and poisonous. [Alter. of ME *passenep,* ? <OF *pasnaie* <L *pastinaca* <*pastinare* dig up; infl. in form by OE *nǣp* turnip <L *napus*]

par·son (pär′sən) *n.* **1** The clergyman of a parish or congregation; a minister. **2** Specifically, a beneficed clergyman of the Anglican Church, having full charge of a parish; a rector. [<OF *persone* <Med. L *persona* a rector. Doublet of PERSON.] —**par·son·i·cal** (pär·son′i·kəl), **par·son′ic** *adj.*

par·son·age (pär′sən·ij) *n.* **1** A clergyman's dwelling, especially a free official residence provided for a pastor; in England, a rectory. **2** *Scot.* A tax paid for the maintenance of a parson. **3** The benefice of a parson.

Par·sons (pär′sənz), **Sir Charles Algernon,** 1854–1931, English engineer; invented the steam turbine. —**William Barclay,** 1859–1932, U. S. engineer; designed and built the subway system of New York, 1894–1904.

par·son's-nose (pär′sənz·nōz′) *n. Colloq.* The rump of a fowl: also called *pope's-nose.*

part (pärt) *n.* **1** A certain portion or amount of anything; a piece; segment; fraction. **2** *Math.* One of certain fractional portions or components of a thing; an aliquot division; a submultiple: a fifth *part.* **3** An essential portion of a body or an organism; a member. **4** *Usually pl.* A portion of territory; region; quarter: in foreign *parts.* **5** So much as is allotted or belongs to one; an individual share, as of duty, business, or performance: If he'll do his *part,* we'll win. **6** The role or lines assigned to an actor in a play; occasionally, a role played in actual life. **7** A side, cause, or party opposed to another. **8** *Usually pl.* A component or quality of mind or character; talent; intellectual gift or faculty: That man is a person of *parts.* **9** The melody intended for a single voice or instrument in a concerted piece; also, the written or printed copy for the performer's use. **10** A section of a book, poem, or play; also, a portion of a literary work issued at intervals, at a fixed price. **11** A parting or division of the hair. —**for my part** As far as I am concerned. —**in good (or ill) part** With a good (or a bad) grace. —**in part** Partly. —**part and parcel** An essential part: an emphatic phrase. —**principal part** One of the inflected forms of a verb from which all other inflected forms may be derived. In English, the principal parts of a verb are the infinitive *(go, walk),* the past tense *(went, walked),* and the past participle *(gone, walked).* In this dictionary, the past, past participle, and present participle are shown *(gave, given, giving)*; in cases where the past tense and past participle are identical, only the one is shown *(behaved, behaving).* —**to take part** To participate; share or cooperate: usually with *in.* —**to take someone's part** To support someone in a contest or disagreement; side with someone. —*v.t.* **1** To divide or break (something) into parts. **2** To sever or discontinue (a relationship or connection): to *part* company. **3** To separate by being or coming between; keep or move apart: The referee *parted* the two men. **4** To comb (the hair) so as to leave a dividing line on the sides or elsewhere on the scalp. **5** To separate (mingled substances) chemically or mechanically: to *part* gold and silver. **6** *Archaic* To divide into shares or portions. **7** *Obs.* To depart from; leave. —*v.i.* **8** To become divided or broken into parts; come

add,āce,câre,pälm; end,ēven; it,īce; odd,ōpen,ôrder; to͝ok,po͞ol; up,bûrn; ə = a in *above,* e in *sicken,* i in *clarity,* o in *melon,* u in *focus*; yo͞o = u in *fuse*; oi,oil; ou,pout; ch,check; g,go; ng,ring; th,thin; th,this; zh,vision. Foreign sounds à,œ,ü,kh,ṅ; and ◆: see page xx. < from; + plus; ? possibly.

apart; divide. 9 To go away from each other; cease associating; separate. 10 To depart; leave. — **to part from** To separate from; leave. — **to part with** 1 To give up; relinquish. 2 To part from. — *adv.* In some degree; to some extent; partly. [<OF <L *pars, partis* part]

Synonyms (noun): atom, component, constituent, division, element, fraction, fragment, ingredient, member, particle, piece, portion, section, segment, share, subdivision. *Part* is the general word, including all the others. A *fragment* is the result of breaking, rending, or disruption of some kind, while a *piece* may be smoothly or evenly separated and have completeness in itself. *Division* and *fraction* are always regarded as in connection with the total; *divisions* may be equal or unequal; a *fraction* is one of several equal *parts* into which the whole is supposed to be divided. A *portion* is a *part* viewed with reference to some one who is to receive it or some special purpose to which it is to be applied; a *share* is a *part* to which one has or may acquire a right in connection with others; a *particle* is an exceedingly small *part*. A *component, constituent, ingredient,* or *element* is a *part* of some compound or mixture; an *element* is necessary to the existence, as a *component* or *constituent* is necessary to the completeness of that which it helps to compose; an *ingredient* may be foreign or accidental. A *subdivision* is a *division* of a *division*. We speak of a *segment* of a circle. Compare BRANCH, PARTICLE, PIECE, PORTION. *Antonyms*: see synonyms for AGGREGATE.

par·take (pär·tāk′) *v.* **·took, ·tak·en, ·tak·ing** *v.i.* 1 To take part or have a share; participate: with *in.* 2 To receive or take a portion or share: with *of.* 3 To have something of the quality or character; bear a trace: with *of*: replies *partaking* of insolence. — *v.t.* 4 To take or have a part in; share. [Back formation <*partaker*, var. of *part-taker*, trans. of L *particeps* <*pars, partis* a part + *capere* take] — **par·tak′er** *n.*

par·tan (pär′tan) *n. Scot.* A crab.

part·ed (pär′tid) *adj.* 1 Situated or placed apart; separated; cloven. 2 *Bot.* Cut almost but not quite to the base, as certain leaves. 3 Having or divided into parts. 4 *Archaic* Departed; dead.

par·terre (pär·târ′) *n.* 1 A flower garden having beds arranged in a pattern. 2 A level plot or space. 3 The part of a theater on the main floor under the balcony and behind the parquet. [<MF <*par terre* on (the) ground <L *per* through, all over + *terra* land]

par·the·no·gen·e·sis (pär′thə·nō·jen′ə·sis) *n.* 1 *Biol.* Reproduction by means of unfertilized eggs, as in many rotifers and polyzoans. 2 *Entomol.* Production of a new individual from a virgin female without intervention of a male, as in plant lice and some hymenopters. 3 *Bot.* Reproduction from unfertilized seeds or spores, as in many algae and fungi. Also **par′the·nog′e·ny** (-noj′ə·nē). [<Gk. *parthenos* virgin + GENESIS] — **par′the·no·ge·net′ic** (-jə·net′ik), **par′the·no·gen′ic** *adj.* — **par′the·no·ge·net′i·cal·ly** *adv.*

Par·the·non (pär′thə·non) The Doric temple of Athena Parthenos on the Acropolis at Athens, now largely in ruins; built under the supervision of Phidias during the administration of Pericles, and dedicated 438 B.C.

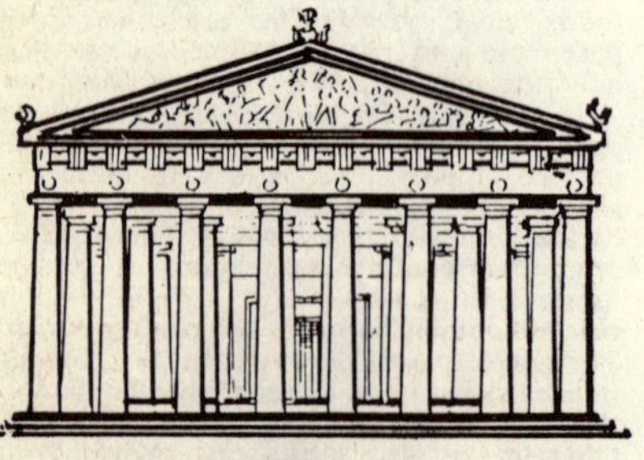

THE PARTHENON
Temple of Athena, on the Acropolis at Athens, representative of classical Greek architectural style.

Par·the·no·pae·us (pär′thə·nō·pē′əs) See SEVEN AGAINST THEBES. Also **Par′the·no·pæ′us.**

Par·then·o·pe (pär·then′ə·pē) 1 In Greek legend, one of the sirens, who, unable to charm Odysseus by her singing, cast herself into the sea. 2 The Greek name for NAPLES. [<Gk. *parthenos* virgin + *ōps* face]

Par·the·nos (pär′thə·nos) *n.* A virgin; epithet of several Greek goddesses, especially of Athena. [<Gk.]

Par·thi·a (pär′thē·ə) An ancient kingdom occupying what is now NE Iran.

Par·thi·an (pär′thē·ən) *n.* An inhabitant of Parthia. — *adj.* Of or pertaining to Parthia or the Parthians. — **Parthian shot** Any aggressive remark or action made in leaving or fleeing, after the manner of Parthian cavalry who shot at their enemies while retreating or pretending to retreat.

par·tial (pär′shəl) *adj.* 1 Pertaining to, constituting, or involving a part only. 2 Favoring one side; prejudiced; biased. 3 Having a special liking: usually with *to.* [<OF *parcial* <LL *partialis* <L *pars, partis* a part] — **par′·tial·ly** *adv.*

partial fraction See under FRACTION.

par·ti·al·i·ty (pär′shē·al′ə·tē) *n.* 1 The state of being partial. 2 Unfairness; bias. 3 A particular fondness; predilection. Also **par′·tial·ness** (-shəl·nis). See synonyms under RELISH.

partial tone *Music* A harmonic. — **upper partial tone** An overtone or harmonic; one of the accessory tones generated by the fundamental.

part·i·ble (pär′tə·bəl) *adj.* Divisible. — **part′i·bil′i·ty** *n.*

par·ti·ceps cri·mi·nis (pär′ti·seps krim′i·nis) *Latin* A sharer in crime; accomplice.

par·tic·i·pant (pär·tis′ə·pənt) *adj.* Sharing; taking part in. — *n.* One who participates; a sharer. [<L *participans, -antis,* ppr. of *participare.* See PARTICIPATE.]

par·tic·i·pate (pär·tis′ə·pāt) *v.* **·pat·ed, ·pat·ing** *v.i.* To take part or have a share in common with others; partake: with *in.* — *v.t. Rare* To take or have a part or share of; partake of. [<L *participatus,* pp. of *participare* <*particeps, -cipis* a partaker <*pars, partis* a part + *capere* take] — **par·tic′i·pa′tion, par·tic′i·pance** *n.* — **par·tic′i·pa′tor** *n.*

par·ti·cip·i·al (pär′tə·sip′ē·əl) *Gram. adj.* Having the nature, form, or use of a participle; characterized by, consisting of, or based on a participle: a *participial* adjective, a *participial* meaning. — *n.* A word derived from a verb, essentially a noun or adjective, but having the syntactical use of the verb, as a gerund, gerundive, supine, or infinitive. — **par′ti·cip′i·al·ly** *adv.*

par·ti·ci·ple (pär′tə·sip′əl) *n. Gram.* A verbal derivative that may function as both a verb and an adjective. The **present participle** ends in *-ing* and the **past participle** commonly in *-d, -ed, -en, -n,* or *-t.* — **dangling participle** A participle that modifies the wrong substantive, as in "*Opening* the door, the *room* looked large" instead of "*Opening* the door, *I* saw that the room looked large." [<OF, var. of *participe* <L *participium* a sharing, partaking <*participare.* See PARTICIPATE.]

par·ti·cle (pär′ti·kəl) *n.* 1 A minute part, piece, or portion of matter. 2 Any very small amount or slight degree: without a *particle* of truth. 3 *Physics* One of the elementary components of an atom, as an electron, proton, neutron, meson, etc. 4 *Gram.* **a** A short, uninflected part of speech, as a preposition, an interjection, an article, and especially a conjunction. **b** A prefix or suffix. **c** A small part of a sentence or composition, as a clause. 5 In the Roman Catholic Church, the small Host used for lay communicants; also, a fragment of a consecrated Host. [<L *particula,* dim. of *pars, partis* a part]

Synonyms: atom, element, grain, iota, jot, mite, molecule, scintilla, scrap, shred, tittle, whit. A *particle* is a very small part of any material substance; as, a *particle* of sand or dust; it is a general term, not accurately determinate in meaning. *Atom* etymologically signifies that which cannot be cut or divided, and was formerly considered the smallest conceivable *particle* of matter, regarded as absolutely homogeneous and as having but one set of properties. A *molecule* is made up of *atoms,* and is regarded as separable into its constituent parts. *Element* in chemistry denotes, without reference to quantity, a substance regarded as simple, that is, one incapable of being resolved into simpler substances without losing its specific physico-chemical properties; the *element* gold may be represented by an ingot or by a *particle* of gold dust. In popular language, an *element* is any essential constituent; the ancients believed that the universe was made up of the four *elements,* earth, air, fire, and water; a storm is spoken of as a manifestation of the fury of the *elements.* Compare synonyms for PART. *Antonyms*: aggregate, entirety, mass, quantity, sum, total, whole.

par·ti·col·ored (pär′tē·kul′ərd) *adj.* Having various colors; variegated; also, diversified. Also *party-colored.* [<F *parti,* pp. of *partir* divide + COLORED]

par·tic·u·lar (pər·tik′yə·lər) *adj.* 1 Specifying or comprising a part; separate: a *particular* act. 2 Peculiar or pertaining to a specified person, thing, time, or place; not common or general; private; specific: my *particular* hobby. 3 Specially noteworthy: of *particular* importance. 4 Comprising all details or circumstances; circumstantial: a *particular* description. 5 Marked by, requiring, or giving minute attention; exact in performance or requirement; precise; also, nice in taste; fastidious: *particular* in dress. 6 *Law* Separate or separable; being apart from others; special; limited; specific. 7 *Logic* Including some, not all, of a class: opposed to *subalternant* or *universal*: "Some trees are oaks" is a *particular* proposition. — *n.* 1 A separate matter or item, as of a class or number. 2 An individual instance; a single or separate case; a given fact that may be brought under or be the ground of a generalization. [<OF *particulier* <L *particularis* concerning a part <*particula.* See PARTICLE.]

Synonyms (adj.): accurate, appropriate, characteristic, circumstantial, definite, detailed, distinct, distinctive, especial, exact, individual, peculiar, separate, single, special. See MINUTE[2], PRECISE, SQUEAMISH.

par·tic·u·lar·ism (pər·tik′yə·lə·riz′əm) *n.* 1 Exclusive attachment to the interests of one's particular state, party, people, or religion. 2 Care or regard for particulars; attention to details. 3 The theological doctrine of the election of particular individuals to grace and salvation; particular election. — **par·tic′u·lar·ist** *n.* — **par·tic′u·lar·is′tic** *adj.*

par·tic·u·lar·i·ty (pər·tik′yə·lar′ə·tē) *n. pl.* **·ties** 1 The state, character, or quality of being particular; exactitude in description; circumstantiality; strict or careful attention to detail; fastidiousness. 2 Something that is particular; a circumstance or detail; also, a characteristic; peculiarity.

par·tic·u·lar·ize (pər·tik′yə·lə·rīz′) *v.* **·ized, ·iz·ing** *v.t.* To speak of or treat individually or in detail. — *v.i.* To give particulars; be specific. — **par·tic′u·lar·i·za′tion** *n.* — **par·tic′u·lar·iz′er** *n.*

par·tic·u·lar·ly (pər·tik′yə·lər·lē) *adv.* 1 With specific reference; distinctly: a fact *particularly* mentioned. 2 In an unusually great degree; in an especial manner: *particularly* difficult. 3 Part by part; in detail. 4 Severally; personally.

par·tic·u·late (pər·tik′yə·lāt) *adj.* 1 Of, pertaining to, having, or characterized by particles. 2 *Genetics* Designating the inheritance of specific characters from either or both of the parents. [<Med. L *particulatus,* pp. of *particulare* divide into particles <*particula.* See PARTICLE.]

par·tim (pär′tim) *adv. Latin* Partly: said of taxonomic synonyms that in part include the same things.

part·ing (pär′ting) *adj.* 1 Of or pertaining to a parting or going away, often in death. 2 Departing; declining. 3 Capable of being parted. 4 Separating; severing; dividing. — *n.* 1 The act of separating, or the state of being separated; division. 2 A leave-taking; a departure; especially, a final separation. 3 *Metall.* The separation of metals in an alloy; specifically, the separation of gold and silver by acid in assaying. 4 A place, line, or surface of separation or division. 5 Something that parts or separates.

parting strip A strip or piece of thin wood or metal separating contiguous parts.

par·ti pris (pȧr·tē′ prē′) *French* An opinion formed beforehand; prejudice.

par·ti·san (pär′tə·zən) *adj.* 1 Relating to or unreasonably devoted to a party or faction. 2 Pertaining to or carried on by partisans or irregular troops. — *n.* 1 An adherent and upholder of an individual or of a party or cause; especially, a blind or fanatical adherent or devotee. 2 A member of a body of detached or irregular troops; a guerilla. See synonyms under ADHERENT. Also **par′·ti·zan.** [<F <Ital. *partigiano, partisano* < *parte* a part <L *pars, partis*] — **par′ti·san·ship** *n.*

par·tite (pär′tīt) *adj.* 1 Divided into or composed of parts: used in composition: *bipartite, tripartite.* 2 *Bot.* Cleft nearly to the base, as a leaf. [<L *partitus,* pp. of *partire* divide]

par·ti·tion (pär·tish′ən) *n.* 1 Division; separation. 2 A dividing line or boundary. 3 A wall or other barrier dividing one part or apartment from another. 4 An internal wall separating cells or cavities. 5 *Law* The division of property, especially of lands, among co-owners, either by agreement or by judicial decree; also, the dividing of lands held by tenants in common into separate parcels, so that they may be held in severalty. 6 *Math.* The representation of a positive whole number as the sum of whole numbers in all possible ways; also, any one of such ways. 7 *Logic* **a** The form of analysis that systematically unfolds the properties or attributes of a concept. **b** The process of explanation that exhibits the theme by means of its attributes. 8 A compartment; apartment; department; division. — *v.t.* 1 To divide into parts, segments, etc. 2 To separate by a partition: with *off.* 3 To divide, as property, into shares or portions; apportion. [<OF *particion* <L *partitio, -onis* < *partire* divide] — **par·ti′tion·er** *n.*

par·ti·tion·ment (pär·tish′ən·mənt) *n.* 1 The act of partitioning, as property. 2 A compartment; partition.

par·ti·tive (pär′tə·tiv) *adj.* 1 Separating into integral parts or into distinct divisions. 2 *Gram.* Denoting a part as distinct from the whole. Example: *Of them* is the partitive genitive in the sentences "Many of them were there" and "They couldn't do it for the life of them." — *n. Gram.* A partitive word or case. [<F *partitif* <L *partitivus* < *partitus,* pp. of *partire* divide] — **par′ti·tive·ly** *adv.*

part·let (pärt′lit) *n.* A garment, frequently ruffled, covering the throat and bust, worn, especially by women, in the 16th century. [Var. of obs. *patlet* <OF *patelette* band of stuff, dim. of *pat* paw, flap]

part·ly (pärt′lē) *adv.* In some part; in some degree; partially.

part music Music with two or more melodies written in harmony and sung or played by two or more performers: said especially of vocal music.

part·ner (pärt′nər) *n.* 1 One who takes part or is associated with another or others; a sharer. 2 One of two or more persons associated by contract for the carrying on of a commercial, manufacturing, or other undertaking with their joint capital, labor, or skill. 3 One of two persons united in some enterprise, as marriage, a dance, or a game. 4 *pl. Naut.* Framing pieces surrounding a mast to strengthen and relieve the deck from strain. See synonyms under ACCESSORY, ASSOCIATE. — **secret** or **sleeping partner** One who is inactive and unknown in the business. — **silent partner** Strictly, one who is inactive, though he may be known to be a partner. The terms *silent partner* and *dormant partner* are often interchanged. — *v.t.* 1 To make a partner or partners. 2 To be or act as the partner of. [Var. of PARCENER; infl. by PART]

part·ner·ship (pärt′nər·ship) *n.* 1 The state of being a partner or partners; joint interests or ownership; also, the group of persons so associated. 2 *Law* An association founded on a contract between two or more persons to combine their money, effects, labor, or skill, or any or all of them, in lawful commerce or business, and to share the profit and bear the loss in certain proportions; a co-partnership. 3 Fellowship (def. 6). See synonyms under ALLIANCE, ASSOCIATION.

part of speech One of the eight traditional classes of words in English; namely: noun, pronoun, verb, adjective, adverb, conjunction, preposition, and interjection.

par·took (pär·took′) Past tense of PARTAKE.

part owner One of two or more persons who own a thing in common, but not as partners; especially, one of the joint owners of a ship.

par·tridge (pär′trij) *n.* 1 Any of certain small, plump-bodied, Old World, gallinaceous game birds of genera *Perdix, Alectoris* (synonym *Caccabis*), etc. 2 Any of certain other similar birds, often so called, as the ruffed grouse of the northern United States and the bobwhite of the South. Compare QUAIL¹. 3 A tinamou of the South American pampas. [<OF *perdriz,* var. of *perdix* <L *perdix, -icis* <Gk. *perdix, -ikos* a partridge]

PARTRIDGE
(The ruffed grouse; 16 to 18 inches in length)

par·tridge·ber·ry (pär′trij·ber′ē) *n. pl.* **·ries** 1 A small, trailing evergreen herb (*Mitchella repens*) of the madder family, with dark-green leaves, fragrant white flowers, and a scarlet double berry; also, the berry. 2 The wintergreen or its berry.

partridge hawk The goshawk: also called *dove hawk.*

part song A song composed of three or more parts; specifically, a secular choral piece without accompaniment.

part-time (pärt′tīm′) *adj.* For, during, or by part of the time: a *part-time* student.

part time A part of the time.

par·tu·ri·ent (pär·tyoor′ē·ənt, -toor′-) *adj.* Bringing forth or about to bring forth young; pertaining to childbirth: used also figuratively of plans, ideas, etc. [<L *parturiens, -entis,* ppr. of *parturire* be in labor, desiderative of *parere* bring forth] — **par·tu′ri·en·cy** *n.*

par·tu·ri·fa·cient (pär·tyoor′ə·fā′shənt, -toor′-) *Med. adj.* Promoting parturition. — *n.* A medicine promotive of parturition. [<L *parturire* be in labor + -FACIENT]

par·tu·ri·tion (pär·tyoo·rish′ən, -chōō-) *n.* The act of bringing forth young; delivery; childbirth. [<L *parturitio, -onis* < *parturire* be in labor]

par·ty (pär′tē) *n. pl.* **·ties** 1 A body of persons united for some common purpose, as political ascendency; a political organization; also, partisanship. 2 A social company or gathering: a tea *party.* 3 *Mil.* A small company or detachment of soldiers: a firing *party.* 4 *Law* One of the persons named on the record in an action either as plaintiff or defendant; a person interested, as in a contract, deed, suit, etc.: a *party* to a suit. 5 One concerned in or privy to a matter: He was a *party* to the affair. 6 *Colloq.* A person. 7 *Obs.* A cause or interest; side. See synonyms under SECT. — *adj.* 1 Of or pertaining to a political party: *party* platforms. 2 Divided into or consisting of parts, or of different parties; composite. 3 *Her.* Divided; parted: said of a shield. [<OF *partie,* orig., fem. pp. of *partir* divide <L *partire* < *pars, partis* a part]

par·ty-col·ored (pär′tē·kul′ərd) See PARTI-COLORED.

party line 1 A telephone line or circuit serving two or more subscribers: also **party wire.** 2 A boundary line between the properties of two or more owners. 3 A belief or principle of a political party regarded as an essential conviction of every loyal member.

party wall A wall erected on a line between adjoining properties, and used in common.

pa·rure (pə·roor′, *Fr.* pá·rür′) *n.* A set of ornaments, especially of trimmings for a costume or of jewels to be worn together. Also **pa·ru′ra** (-roor′ə). [<F <L *paratura* preparation < *parare* make ready]

par·ve·nu (pär′və·nōō, -nyōō) *n.* One who has suddenly attained wealth or position beyond his birth or worth, as by accident or fortune; an upstart. — *adj.* 1 Being a parvenu. 2 Like or characteristic of a parvenu. [<F, orig., pp. of *parvenir* arrive <L *pervenire*]

par·vis (pär′vis) *n.* 1 An enclosed or raised area in front of a church. 2 A portico or colonnade before a church. [<F <L *paradisum* paradise; later, the court in front of St. Peter's, Rome]

par·vo·line (pär′və·lēn, -lin) *n. Chem.* An oily liquid, $C_9H_{13}N$, obtained either as a ptomaine in decaying flesh or as a product of the destructive distillation of certain shales and coals. Also **par′vo·lin** (-lin). [<L *parvus* small + (QUIN)OLINE; so named because of its low volatility]

Par·zi·val (pär′tsi·fäl) See PERCEVAL.

pas (pä) *n.* 1 A step. 2 A dance. 3 Right of going before; precedence. [<F, a step]

Pas·a·de·na (pas′ə·dē′nə) A city in SW California.

Pa·sar·ga·dae (pə·sär′gə·dē) A ruined city of ancient Persia in south central Iran; capital of Cyrus the Great. Also **Pa·sar′ga·de.**

Pa·say (pä′sī) A city in Luzon on Manila Bay, Philippines.

Pas·cal (pas·kal′, pas′kəl; *Fr.* pás·kál′), **Blaise,** 1623–62, French mathematician and philosopher.

pasch (pask) *n.* The feast of the Passover; also, Easter. Also **pas·cha** (pas′kə). [<OF *pasche* <L *pascha* <Gk. <Aramaic *paskhā* <Hebrew *pesakh* a passing over, the Passover <*pāsakh* pass over]

pas·chal (pas′kəl) *adj.* Pertaining to the Jewish Passover or to Easter: *paschal* sacrifice. — *n.* 1 A paschal candle or candlestick. 2 The celebration of the Passover; also, the paschal lamb; the paschal supper. [<OF *pascal* <LL *paschalis* <L *pascha* PASCH]

paschal flower The pasqueflower.

paschal lamb The lamb eaten at the feast of the Passover.

Paschal Lamb Jesus Christ.

pas·cu·al (pas′kyōō·əl) *adj. Ecol.* Of or pertaining to plants growing in pastures and grassy commons. [<OF <Med. L *pascualis* <L *pascuum* a pasture < *pascere* feed]

Pas-de-Ca·lais (pä·də·ká·le′) 1 A department of northernmost France, on the English Channel and the Strait of Dover; 2,607 square miles; capital, Arras. 2 The French name for the STRAIT OF DOVER.

pas de deux (pä də dœ′) *French* A dance or ballet figure for two persons.

pas du tout (pä dü tōō′) *French* Not at all.

pash¹ (pash) *Obs.* or *Dial. v.t.* To strike violently; dash to pieces. — *n.* A crushing blow. [ME *passchen,* prob. imit.]

pash² (pash) *n. Scot.* The head. [Prob. <PASH¹, with ref. to blows on the head]

pa·sha (pə·shä′, pash′ə, pä′shə) *n.* A Turkish honorary title placed after the name, formerly given to generals, governors of provinces, etc.: also spelled **pacha.** [<Turkish *pāshā* < *bāsh* head]

pa·sha·lik (pə·shä′lik) *n.* The province or jurisdiction of a pasha: also spelled *pachalic.* Also **pa·sha′lic.** [<Turkish *pāshālik*]

Pash·to (push′tō) See PUSHTU.

Pa·siph·a·e (pə·sif′ē·ē) In Greek mythology, the wife of Minos and by him mother of Phaedra and Ariadne, and of the Minotaur by a white bull sent to Minos by Poseidon.

pasque·flow·er (pask′flou′ər) *n.* Any of several plants (genus *Anemone*) with showy white, red, or purple flowers, blooming about Easter; especially, the daneflower or campana (*A. pulsatilla*) of the Old World, or *A. ludoviciana,* the State flower of South Dakota. Also **pasch′·flow′er.** [<Earlier *passeflower* <F *passefleur* < *passer* excel + *fleur* flower; infl. in form by F *pasque* Easter]

pas·quil (pas′kwil) *n.* A pasquinade. [<Med. L *pasquillus* <Ital. *pasquillo,* dim. of *pasquino.* See PASQUIN.]

pas·quin (pas′kwin) *n.* 1 A pasquinade. 2 A pasquinader. [<Ital. *pasquino,* orig., a disinterred statue at Rome on which satirical verses were pasted]

pas·quin·ade (pas′kwin·ād′) *n.* An abusive or coarse personal satire posted in a public place; a malicious squib. — *v.t.* **·ad·ed, ·ad·ing** To attack or ridicule in pasquinades; lampoon. — **pas′quin·ad′er** *n.*

pass (pas, päs) *v.* **passed** (*Rare* **past**), **passed** or **past, pass·ing** *v.t.* **1** To go by or move past and leave behind. **2** To go across, around, over, or through. **3** To permit to go unnoticed or unmentioned. **4** To undergo; experience: to *pass* a bad night. **5** To undergo successfully, as a test; meet the requirements of. **6** To go beyond or exceed; surpass: It *passes* comprehension. **7** To cause to go or move: to *pass* one's eyes over a book; to *pass* a rope through a pulley. **8** To cause to go or move past: to *pass* troops in review. **9** To cause or allow to advance or proceed: They *passed* him through their ranks. **10** To cause or allow to elapse; spend: to *pass* the night at an inn. **11** To give approval to; sanction; allow. **12** To enact, as a law. **13** To be approved by: The bill *passed* the senate. **14** To omit paying (a dividend). **15** To cause to go from person to person; put in circulation; transmit: *Pass* the word. **16** To utter or pronounce, especially judicially, as judgment or sentence. **17** To excrete (waste). **18** To pledge, as one's word. **19** To perform a pass (*n.* def. 5) on or over. **20** In sports, to transfer (the ball, etc.) to another player on the same side. **21** *Law* To transfer or assign ownership of to another by will, deed, etc. — *v.i.* **22** To go or proceed; move. **23** To have course or direction; extend: The road *passed* under a bridge. **24** To go away; depart. **25** To come to an end; disappear. **26** To elapse or go by; be spent: The day *passed* slowly. **27** To die. **28** To go by; move past in or as in review. **29** To go from person to person; obtain currency; circulate. **30** To be mutually given and received, as greetings or recriminations. **31** To go or change from one condition, circumstance, etc., to another; alter: to *pass* from hot to cold. **32** To take place; happen; occur. **33** To be allowed or tolerated; go unheeded or unpunished. **34** To undergo a test, examination, etc., successfully; meet the requirements. **35** To be approved, ratified, enacted, etc. **36** To obtain or force passage; make a way: They shall not *pass*! **37** To be excreted or voided. **38** *Law* **a** To give or pronounce judgment, sentence, etc.: with *on* or *upon.* **b** To sit in inquest: with *on* or *upon.* **c** To adjudicate: with *between.* **d** To be transferred or assigned to another by will, deed, etc. **39** In sports, to transfer the ball, etc., to another player on the same side. **40** In fencing, to make a pass or thrust; lunge. **41** In card games, to decline to make a play, bid, etc. — **to pass away** **1** To come to an end; disappear. **2** To die. **3** To allow (time) to elapse. — **to pass for** To be accepted as, usually incorrectly. — **to pass off** **1** To come to an end; disappear. **2** To give out or circulate as genuine; palm off. **3** To be emitted, as a vapor. — **to pass out** **1** To distribute. **2** *Colloq.* To faint. — **to pass over** To fail to notice or consider. — **to pass up** *Colloq.* To reject or fail to take advantage of, as an offer or opportunity. — *n.* **1** A way or opening that affords a passage; a place through which one can pass; a gap in a mountain range through which a road may be or has been made; a defile; waterway. **2** Permission or a permit to pass; a ticket; passport: a *pass* through an army's lines. **3** A state of affairs; crisis. **4** The successful undergoing of an examination, test, or inspection; in a university, a degree gained without honors. **5** A movement of a hand, wand, or the like, as in mesmeric manipulation; transference of objects in sleight-of-hand tricks, magic, etc. **6** A movement made in attempting to stab or strike; a thrust; lunge; also, figuratively, a verbal thrust; a witty sally. **7** In football, hockey, lacrosse, etc., the action of passing the ball between players, in the course of the game. **8** In court tennis, a ball so served that it strikes the penthouse or the floor of the court between the main wall and the pass line. **9** In baseball, a base on balls. **10** *Mil.* Authority in writing given a soldier to be absent from duty or station for a specified period. — **forward pass** In football, the throwing or passing of the ball toward the opponent's goal. — **lateral pass** In football, a pass which does not travel towards either goal. — **to bring to pass** To cause to be fulfilled, accomplished, or realized. — **to come to pass** To happen; come about; be realized. — **to make a pass at** **1** To attempt to hit. **2** *Slang* To attempt to caress. [<OF *passer* <L *passus* a step; *n.*, doublet of PACE]

pass·a·ble (pas′ə·bəl, päs′-) *adj.* **1** Capable of being passed in any sense; capable of being penetrated or traversed. **2** Fairly good or acceptable; not open to great objection; moderate; mediocre; tolerable. **3** Fit for general circulation. — **pass′a·bly** *adv.*

pas·sade (pə·säd′) *n.* **1** *Obs.* A forward thrust in fencing; a pass made by advancing the body: also **pas·sa·do** (pə·sä′dō). **2** In horsemanship, a moving of a horse back and forth over the same ground. [<F <Ital. *passata* or Provençal *passada* <LL *passare* pass <L *passus* a step]

pas·sage¹ (pas′ij) *n.* **1** The act of passing; a passing by, through, or over; transition from one state or condition to another. **2** A journey by conveyance, as by a vessel; a voyage: a stormy *passage.* **3** Hence, conveyance on a journey; right of transportation, especially on a ship; also, money paid for conveyance. **4** A way or channel by which a person or thing may pass; a way through or over. **5** Any corridor, hall, or gallery affording communication between apartments in a building: also called *passageway.* **6** Liberty or power of passing; free entrance, exit, or transit. **7** A separate portion of a discourse, treatise, or writing; a clause, verse, paragraph, or similar division. **8** The course of a legislative measure through the various stages of debate and action; especially, its enactment by the final vote or approval by the supreme authority. **9** A part of a train of events; a series of incidents; episode. **10** A navigable route; especially, a channel connecting large bodies of water. **11** A personal encounter; a fight or a dispute: a *passage* with swords. **12** Migration, especially of birds; a migratory flight. **13** An evacuation of the bowels. **14** *Music* **a** A portion of a musical composition. **b** A run or series of short notes. **15** *Obs.* Departure; hence, death. See synonyms under CAREER, MOTION, ROAD, WAY. — *v.i.* **·saged, ·sag·ing** **1** To make a journey. **2** To fence physically or verbally. [<OF <*passer.* See PASS.]

pas·sage² (pas′ij) *v.* **·saged, ·sag·ing** *v.t.* In equitation, to cause (a horse) to sidle or walk sidewise. — *v.i.* In equitation: **a** To move sidewise; sidle. **b** To cause a horse to move sidewise. — *n.* A sidewise movement made by a horse in which diagonal pairs of feet are raised alternately. [<F *passager,* alter. of *passéger* <Ital. *passeggiare* walk <L *passus* a step]

pas·sage·way (pas′ij·wā′) *n.* A way affording passage; especially, a way made or kept open for walking, between rooms, localities, etc., as a hall or corridor, gangway, lane, etc.

Pas·sa·ic (pə·sā′ik) A city in NE New Jersey on the **Passaic River,** a river flowing 80 miles south and east past Newark to Newark Bay.

Pas·sa·ma·quod·dy Bay (pas′ə·mə·kwod′ē) An inlet of the Bay of Fundy between SE Maine and SW New Brunswick.

pas·sant (pas′ənt) *adj. Her.* Walking and looking toward the dexter, with the dexter fore paw raised: said of a beast. [<F, ppr. of *passer.* See PASS.]

pass·book (pas′bo͝ok′, päs′-) *n.* **1** A bankbook. **2** A book given by a merchant to a customer, showing all items bought on credit.

pas·sé (pa·sā′, pas′ā; *Fr.* pà·sā′) *adj.* Past the prime; faded; also, out-of-date; old-fashioned. [<F, orig., pp. of *passer.* See PASS.] — **pas·sée′** *adj. fem.*

passed (past, päst) *adj.* **1** Having passed an examination for promotion. **2** Unpaid, as dividends.

passed ball In baseball, a pitched ball that the batsman fails to hit, which passes by the catcher and enables a runner to advance a base.

passe·garde (pas′gärd′, päs′-) *n.* In medieval armor, a projecting piece or ridge on the shoulder to turn the point of a lance. [Earlier *passguard,* appar. <PASS + GUARD: refashioned after the French]

pas·sel (pas′əl) *n. Dial.* A parcel (def. 3).

passe·men·terie (pas·men′trē, *Fr.* päs·mäṅ·trē′) *n.* Trimming for dresses, as beaded lace, tinsel, etc. [<MF <*passement* lace <*passer* PASS]

pas·sen·ger (pas′ən·jər) *n.* **1** A person who travels in a conveyance. **2** *Rare* A traveler; passer-by: a foot *passenger.* [<OF *passager* (with intrusive *-n*), orig. passing <*passage.* See PASSAGE.]

passenger pigeon The wild pigeon of North America (*Ectopistes migratorius*): now extinct.

passe par·tout (pas pär·to͞o′, Fr. päs pȧr·to͞o′) **1** A light picture frame consisting of a glass and a pasteboard back put together with strips of decorative tape pasted around the edge; also, the pasteboard mat of a picture. **2** Strips of gummed paper; tape. **3** A master key. [<F, orig. a key <*passe,* imperative of *passer* PASS + *partout* everywhere]

passe·pied (päs′pyā′) *n. Music* **1** A quick, lively French dance of the 17th century. **2** The music for this dance, sometimes used as the movement of a suite. [<MF, lit., pass the foot <*passe,* imperative of *passer* PASS + *pied* foot <L *pes, pedis*]

pass·er-by (pas′ər·bī′, päs′-) *n. pl.* **pas·sers-by** A person who passes by.

Pas·ser·i·for·mes (pas′ər·i·fôr′mēz) *n. pl.* An order of birds, including all singing birds, and more than half of the living birds of various sizes ranging from crows and jays to sparrows and titmice. [<NL <L *passer, -eris* a sparrow + *forma* form]

pas·ser·ine (pas′ər·ēn, -in) *adj.* **1** Pertaining to the *Passeriformes.* **2** Resembling or characteristic of a sparrow. — *n.* One of the *Passeriformes.* [<L *passer* sparrow + -INE¹]

Pas·se·ro (päs′sä·rō), **Cape** The NE tip of an islet in the Ionian Sea off SE Sicily.

pas seul (pä sœl′) *French* A dance or ballet figure by a single person.

Pass·field (pas′fēld), **Baron** See under WEBB.

pas·si·ble (pas′ə·bəl) *adj.* Capable of feeling or of suffering. [<F <LL *passibilis* <L *passus,* pp. of *patiri* suffer] — **pas′si·bil′i·ty, pas′si·ble·ness** *n.*

pas·si·flo·ra·ceous (pas′i·flô·rā′shəs) *adj.* Of or pertaining to a family (*Passifloraceae*) of climbing vines and erect herbs abundant in tropical America, including the passionflower. [<NL *passiflora,* genus of the passionflower <L *flos passionis* flower of the Passion]

pas·sim (pas′im) *adv. Latin* Here and there; in various passages: a reference.

pass·ing (pas′ing, päs′-) *adj.* **1** Going by or away. **2** Transitory; fleeting. **3** Happening or occurring; current. **4** Done, said, found, used, or given in or as in passing; cursory. **5** Indicating fulfilment of requirements for advancement; satisfactory: a *passing* grade. **6** *Obs.* Surpassing. See synonyms under TRANSIENT. — *n.* **1** A going by or away; hence, dying. **2** An act of passing or passage. **3** A means of passing, as a ford. — **in passing** Incidentally; in the course of discussion. — *adv.* In a surpassing degree or manner; exceedingly.

passing bell A bell tolled to announce a person's death.

passing note A note or tone foreign to the harmony: used in passing from one chord to another.

pas·sion (pash′ən) *n.* **1** Intense or overpowering emotion. **2** An eager outreaching of mind toward some special object, as art, travel, etc.; fervid devotion. **3** Ardent affection for one of the opposite sex; love; also, the object of such feeling. **4** A fit of intense and furious anger; rage. **5** Any transport of excited feeling; violent agitation. **6** A strong impulse tending to physical indulgence. **7** *pl.* Inordinate appetites; sexual desires. **8** The state or condition of being acted upon; subjecting to external force, as opposed to acting or doing: the philosophic sense. **9** The endurance of some painful infliction; suffering. **10** *Obs.* Some painful disease. See synonyms under ANGER, APPETITE, ENTHUSIASM, FEELING, LOVE, VIOLENCE, WARMTH. [<OF *passiun* <L *passio, -onis* suffering <*passus,* pp. of *patiri* suffer]

Pas·sion (pash′ən) *n.* **1** The sufferings of Christ, especially in the agony of the garden and on the cross; also, their representation in art. **2** That part of the Gospels which relates the Passion and death of Christ.

pas·sion·al (pash′ən·əl) *adj.* Of, pertaining to, or characterized by passion, especially amorous: *passional* poetry. — *n.* A book descriptive of the sufferings of saints and martyrs.

pas·sion·ate (pash′ən·it) *adj.* **1** Capable of or inclined to strong passion; susceptible of vehement emotion; excitable. **2** Easily moved to anger; quick-tempered. 3 Expressing or displaying some passion; characterized by passion; intense; ardent. 4 Of a strong, ardent quality or excessive degree: said of feeling and emotion. See synonyms under ARDENT, HOT, IMPETUOUS, VIOLENT. [<Med. L *passionatus,* ult. <L *passio, -onis.* See PASSION.] **—pas′sion·ate·ly** *adv.* **—pas′sion·ate·ness** *n.*

pas·sion·flow·er (pash′ən·flou′ər) *n.* Any of various climbing vines or shrubs (genus *Passiflora*) of tropical America, with showy flowers and sometimes edible berries: so called from the fancied resemblance of certain parts of the flower to the instruments of the crucifixion. [Trans. of Med. L *flos passionis*]

pas·sion·less (pash′ən·lis) *adj.* Unimpassioned; calm.

Passion play A mystery or drama representing the Passion of Christ.

Passion Sunday *Eccl.* The second Sunday before Easter.

Passion Week *Eccl.* 1 The week that begins with Passion Sunday. 2 Formerly, Holy Week.

pas·si·vate (pas′ə·vāt) *v.t.* **·vat·ed, ·vat·ing** *Chem.* To remove minute impurities from the surface of (stainless iron or steel) by submerging in a pickling solution, for the purpose of preventing rust. [<PASSIVE (def. 4) + -ATE[1]]

pas·sive (pas′iv) *adj.* 1 Acted upon or receiving impressions from external agents or causes; being the object rather than the subject of action; moved by or as by external force or influence. 2 In a state of rest or quiescence; not vitally or mentally active; unresponsive. 3 Unresisting; submissive; *passive* obedience. 4 *Chem.* Characterized by a disinclination to enter into combination; inactive; inert. 5 *Gram.* Designating a voice of the verb which indicates that the subject is being acted upon, as, *was killed* is in the passive voice in *Caesar was killed by Brutus*: opposed to *active.* 6 Not bearing interest: said of bonds which yield the holder a profit or benefit, while no rate percent is named; having reference to a debt which, by agreement, is non-interest-bearing; of the nature of a liability. 7 *Med.* Designating certain abnormal conditions marked by relaxation of blood vessels and tissues, indicating impaired vitality and reaction. 8 Not provided with, or not making use of, motive power. **—***n. Gram.* 1 The passive voice. 2 A verb or construction in this voice. [<L *passivus* <*passus,* pp. of *pati* suffer] **—pas′sive·ly** *adv.* **—pas′sive·ness** *n.*

Synonyms (adj.): inactive, inert, negative, patient, quiescent, submissive, suffering. *Antonyms*: active, positive, resistant.

passive resistance Opposition to constituted authority that does not offer open violence, but resorts instead to voluntary fasting, refusal to obey laws, etc.

pas·siv·i·ty (pa·siv′ə·tē) *n.* 1 Passiveness. 2 The suspension or abeyance of the rational functions and the reduction of the physical functions to the lowest possible degree. 3 Resistance to oxidation and chemical reagents, as iron which has been immersed in strong nitric acid.

pass·key (pas′kē′, päs′-) *n.* 1 A latch key or night key. 2 A skeleton key; master key.

pass·o·ver (pas′ō′vər, päs′-) *n.* The sacrifice offered at the paschal feast; the paschal lamb.

Pass·o·ver (pas′ō′vər, päs′-) *n.* 1 A Jewish feast commemorating the night when the Lord, smiting the first-born of the Egyptians, "passed over" the houses of the children of Israel. *Ex.* xii. 2 By extension, the entire festival of seven days following the paschal supper; the feast of unleavened bread.

pass·port (pas′pôrt′, -pōrt′, päs′-) *n.* 1 An official warrant certifying the citizenship of the bearer and affording protection to him when traveling abroad; a safe conduct. 2 A permit to travel or convey goods through a foreign country. 3 A documentary permission for a ship to proceed on a voyage. 4 A means or authority to pass; that which empowers one to arrive at anything: a *passport* to success. 5 That which gives the privilege or right to enter into some sphere of action. See NAVICERT. [<F *passeport* <*passer* pass + *port* harbor]

pas·sus (pas′əs) *n. pl.* **·sus** or **·sus·es** A part or canto, as of a poem. [<Med. L <L, a stop <*pandere* stretch]

pass·word (pas′wûrd′, päs′-) *n.* A word identifying one as entitled to pass; a watchword.

Pas·sy (pȧ·sē′), **Paul Édouard,** 1859–1940, French phoneticist; chief originator of the International Phonetic Alphabet.

past (past, päst) *adj.* 1 Belonging to time gone by; hence, accomplished or ended. 2 In the usage of some societies, having completed a full term and been succeeded by another person: a *Past* Master in Masonry. 3 *Gram.* Denoting a tense or construction which refers to time or action belonging to the past. **—***n.* 1 Time gone by; an antecedent period; former days collectively. 2 One's antecedents or record, especially if disreputable or kept secret. 3 *Gram.* **a** The past tense. **b** A verb or construction in this tense. **—***adv.* In such manner as to go by and beyond. **—***prep.* 1 Beyond in time; at a later period than; after: It is now *past* noon. 2 Beyond in place or position; farther than: walking *past* the house. 3 Beyond the reach, scope, power, or influence of: The matter is *past* hope. 4 Beyond in amount or degree: He couldn't count *past* ten. [Orig. pp. of PASS]

Pas·ta·za (päs·tä′sä) A river in Ecuador and Peru flowing 400 miles SE to the Marañón.

paste[1] (pāst) *n.* 1 An adhesive mixture, usually of flour and water: used for joining or affixing paper articles and the like, and in bookbinding, etc. 2 A mixture of flour and water, often with other materials, for culinary purposes; dough. 3 Any doughy or moist plastic substance; anything of the consistency of paste, as for consumption or application: usually with a qualifying word: fish *paste;* almond *paste.* 4 A vitreous composition for making imitation gems. 5 A confection made of fruit juices, sugar, gum, etc. 6 A mixture of clay for making stoneware or porcelain. **—***v.t.* **past·ed, past·ing** 1 To stick or fasten with paste or the like. 2 To cover by applying pasted material. **—***adj.* Made of paste; artificial. [<OF <LL *pasta* <Gk. *pastē* barley porridge]

paste[2] (pāst) *Slang v.t.* **past·ed, past·ing** To strike, as with the fist; beat. **—***n.* A hard blow. [<BASTE[3]]

paste·board (pāst′bôrd′, -bōrd′) *n.* 1 Paper pulp compressed, or paper pasted together and rolled into a stiff sheet. 2 A board on which dough for pastry is rolled. 3 *Colloq.* A visiting card; also, a playing card; playing cards generally. **—***adj.* Made of or resembling pasteboard; hence, thin and flimsy; also, sham.

pas·tel[1] (pas·tel′, pas′tel) *n.* 1 A picture drawn with colored crayons. 2 The art of drawing such pictures. 3 A hard crayon made of pipe clay and a pigment, mixed with gum water. 4 A sketchy poetic study in prose. **—***adj.* 1 Of or pertaining to a pastel. 2 Having a delicate, soft, or slightly grayish tint. [<MF <Ital. *pastello* PASTEL[2]] **—pas′tel·ist** or **pas′tel·list** *n.*

pas·tel[2] (pas′tel) *n.* Woad, either as a dye or as a plant. [<MF <Ital. *pastello,* dim. of *pasta* <LL. See PASTE[1].]

past·er (pās′tər) *n.* 1 One who pastes. 2 A strip of gummed paper, to cover a portion of a circular, an election ballot, or the like; a sticker.

pas·tern (pas′tərn) *n.* 1 That part of a horse's foot that is between the fetlock and the coffin joint. 2 A hobble for a horse's foot. [<OF *pasturon* <*pasture* a tether for a grazing animal]

pastern bone Either the proximal or first phalanx, **great pastern bone,** or the median or second phalanx, **small pastern bone,** of a horse's foot.

pastern joint The joint between the pastern bones.

Pas·teur (päs·tœr′), **Louis,** 1822–95, French chemist and bacteriologist.

Pas·teu·rel·la (pas′tə·rel′ə) *n.* A genus of Gram-negative, rod-shaped, aerobic bacteria parasitic in animals and man: *P. pestis* is the plague bacillus, and *P. tularensis* is the cause of tularemia. [<NL, after Louis *Pasteur*]

pas·teur·ism (pas′tə·riz′əm) *n. Med.* A method of progressive inoculation developed by Pasteur for the prevention or cure of certain diseases, as hydrophobia.

pas·teur·i·za·tion (pas′tər·ə·zā′shən, -chər-) *n.* A process of arresting or preventing fermentation in liquids, as beer, milk, wine, etc., by heating from 60° to 70° C., so as to destroy the vitality of the ferment: originally proposed by Pasteur. Also *Brit.* **pas′teur·i·sa′tion.**

pas·teur·ize (pas′tə·rīz, -chə·rīz) *v.t.* **·ized, ·iz·ing** 1 To treat (milk, beer, etc.) by pasteurization. 2 To treat by pasteurism. Also *Brit.* **pas′teur·ise.**

pas·teur·iz·er (pas′tə·rī′zər, -chə-) *n.* 1 One who pasteurizes. 2 An apparatus for pasteurizing liquids.

pas·tic·cio (päs·tē′chō) *n.* A work of art, music, or literature made up of fragments from various sources, as from other works, connected so as to form a complete work; medley. [<Ital., a paste <Med. L *pasticium* <LL *pasta.* See PASTE[1].]

pas·tiche (pas·tēsh′, päs-) *n.* A pasticcio, especially one imitating or satirizing the style of other works of art or artists. [<F <Ital. *pasticcio* PASTICCIO]

pas·tille (pas·tēl′, -til′) *n.* 1 A compound of aromatic substances with niter for fumigating. 2 A troche; lozenge. 3 A flavored confection. 4 A small paper disk coated with a chemical that changes color on exposure to X-rays: used to determine X-ray dosages. 5 Pastel[1] (def. 3). Also **pas·til** (pas′til). [<MF <L *pastillus* a little loaf, a lozenge, ? dim. of *pasta* PASTE[1]]

pas·time (pas′tīm′, päs′-) *n.* Something that serves to make time pass agreeably; recreation; sport. See synonyms under ENTERTAINMENT, SPORT. [<PASS, *v.* + TIME; trans. of F *passe-temps*]

past·i·ness (pās′tē·nis) *n.* The appearance, feeling, or consistency of paste.

past master 1 One who has held the office of master in certain social and benevolent organizations. 2 One who has thorough experience in something; an adept.

Pas·to (päs′tō) A city in SW Colombia at the foot of the active **Pasto Volcano** (also *Galeras*) 13,996 feet.

pas·tor (pas′tər, päs′-) *n.* 1 A Christian minister who has a church or congregation under his official charge. 2 *Obs.* A shepherd. [<AF *pastour,* OF *pastur* <L *pastor, -oris* a shepherd, lit., a feeder <*pascere* feed]

pas·tor·age (pas′tər·ij, päs′-) See PASTORATE.

pas·tor·al (pas′tər·əl, päs′-) *adj.* 1 Pertaining to the life of shepherds and rustics; rural in spirit or sentiments: a *pastoral* poem. 2 Pertaining to a pastor and his work. See synonyms under RUSTIC. **—***n.* 1 A poem dealing with rural matters; a bucolic; an idyl. 2 A picture illustrating rural scenes. 3 *Eccl.* A letter from a pastor to his flock. 4 A simple melody in 6/8 time in a rustic style; also, a complete symphony portraying a series of pastoral scenes. 5 A book or treatise on the cure of souls. 6 *pl.* The pastoral epistles. 7 A crozier or pastoral staff. [<L *pastoralis* <*pastor.* See PASTOR.] **—pas′tor·al·ism** *n.* **—pas′tor·al·ist** *n.* **—pas′tor·al·ly** *adv.*

pas·to·ra·le (pas′tə·rä′lē, päs′-, pas′tə·ral′) *n.* A cantata or operetta on a rustic theme; also, a piece of instrumental music simple and idyllic in character. [<Ital., lit., pastoral <L *pastoralis.* See PASTORAL.]

pastoral epistles In the New Testament, the three epistles addressed to Timothy and Titus, and ascribed to St. Paul: so called because they deal almost entirely with the office of a Christian pastor.

pastoral staff *Eccl.* A staff, usually curved like a shepherd's crook, borne as an emblem of authority by or before a bishop, abbot, etc.

pas·tor·ate (pas′tər·it, päs′-) *n.* 1 The office or jurisdiction of a pastor. 2 The duration of a pastoral charge. 3 Pastors collectively.

pas·to·ri·um (pas·tôr′ē·əm, -tō′rē·əm, päs-) *n.* A parsonage.

pas·tor·ship (pas′tər·ship, päs′-) *n.* The place, dignity, or work of a pastor.

past participle See under PARTICIPLE.

past perfect *Gram.* The verb tense indicating

an action completed prior to the occurrence of some other past action or before some specified past time. Example: He *had finished* before the bell rang, or, He *had finished* by last Friday. Also called *pluperfect.*

pas·tra·mi (pə·strä′mē) *n.* Smoked beef, heavily seasoned and usually cut from the shoulder. [<Yiddish <Magyar]

pas·try (pās′trē) *n. pl.* **·tries** Articles of food made with a crust of shortened dough, as pies. [Appar. <PAST(E)[1] + -RY]

pas·tur·age (pas′chər·ij, päs′-) *n.* 1 Grass and herbage for cattle. 2 Ground used or suitable for grazing. 3 The business or right of grazing cattle. [<OF <*pasturer* feed <*pasture.* See PASTURE.]

pas·ture (pas′chər, päs′-) *n.* 1 Ground for the grazing of domestic animals. 2 Grass or herbage that cattle or other grazing domestic animals eat. —*v.t.* **·tured, ·tur·ing** 1 To lead to or put in a pasture to graze. 2 To graze on (grass, land, etc.). 3 To provide pasturage for (cattle, etc.): said of land. [<OF <L *pastura,* lit., feeding <*pastus,* pp. of *pascere* feed] —**pas′tur·a·ble** *adj.* —**pas′tur·er** *n.*

past·y[1] (pās′tē) *adj.* **past·i·er, past·i·est** Like paste.

past·y[2] (päs′tē, *Brit.* pas′tē, päs′tē) *n. pl.* **past·ies** A pie, as of meat, enclosed and baked in a crust of pastry, especially relished by Cornish miners. [<OF *pastée* <LL *pasta.* See PASTE[1].]

pat[1] (pat) *v.* **pat·ted, pat·ting** *v.t.* 1 To touch or tap lightly with something flat, especially with the hand in caressing, soothing, etc. 2 To shape or mold by a pat or pats. 3 To strike or tap with lightly sounding steps, as in running. —*v.i.* 4 To tap or strike gently. 5 To run or walk with light steps. —*n.* 1 A light, caressing stroke; a gentle tap. 2 The sound of patting or pattering. 3 A small, molded mass, as of butter. —*adj.* 1 Exactly suitable in time or place; fitting; apt. 2 Satisfactory; needing no change: a *pat* hand in a card game. —*adv.* 1 Firm; steadfast: to stand *pat.* 2 In a fit or convenient manner; aptly; also, perfectly; unforgettably: to know one's lesson *pat.* [ME *patte;* prob. imit.] —**pat′ness** *n.* —**pat′ter** *n.*

pat[2] (pat) *n. Scot.* A pot.

pa·ta·gi·um (pə·tā′jē·əm) *n. pl.* **·gi·a** (-jē·ə) 1 *Zool.* The wing membrane of a bat; also, a parachute, as of a flying squirrel, and the membranous expansion of a bird's wing. 2 *Entomol.* One of two scalelike appendages on the front dorsal part of the thorax of some moths. [<NL <L, gold edging of a tunic <Gk. *patageion*]

Pat·a·go·ni·a (pat′ə·gō′nē·ə) A region at the southern extremity of South America; now divided between Chile and the Argentine; the name is, however, usually applied only to the Argentine portion; 300,000 square miles. —**Pat′a·go′ni·an** *adj.* & *n.*

pat·a·mar (pat′ə·mär) *n.* A coasting vessel with upward-arched keel and great stem and stern rake: used from Bombay to Ceylon: also spelled *pattemar.* [<Pg. <Malayalam *pattamāri*]

Pa·tan (pä′tən) A city in central Nepal.

Pa·tap·sco River (pə·tap′skō) A river in central Maryland, flowing 65 miles SE to Chesapeake Bay.

patch (pach) *n.* 1 A small piece of material, especially of cloth, used to repair a garment, etc. 2 Something resembling a patch, as a piece of courtplaster or the like, applied to the skin to hide a blemish or to set off the complexion; a beauty spot. 3 A small piece of ground; also, the plants growing on it: a *patch* of corn; a small area in a larger expanse. 4 A piece of cloth or other material worn over an injured eye. 5 Any small part of a surface not agreeing with the general character or appearance of the whole. 6 A shred or scrap. —*v.t.* 1 To put a patch or patches on. 2 To repair, make whole, or put together, especially hurriedly or crudely: often with *up* or *together.* 3 To make of patches, as a quilt. [ME *pacehe;* origin uncertain] —**patch′a·ble** *adj.* —**patch′er** *n.*

patch·head (pach′hed′) *n.* The surf scoter. Also **patchhead coot.**

patch·ou·li (pach′ŏŏ·lē, pə·chōō′lē) *n.* 1 An East Indian herb *(Pogostemon heyneamus* or *patchouly)* of the mint family. 2 A perfume obtained from it. Also spelled *pachouli.* Also **patch′ou·ly.** [<F <Tamil *paccilai* <*paccu* green + *ilai* a leaf]

patch test *Med.* A test for allergy in which an area of unbroken skin is covered by a small patch of linen or blotting paper impregnated with the suspected substance: upon removal of the patch the skin reaction is noted.

patch·work (pach′wûrk′) *n.* 1 A fabric made of patches of cloth, as for quilts, etc. 2 Work made up of heterogeneous materials; work done hastily or carelessly; a jumble.

patch·y (pach′ē) *adj.* **patch·i·er, patch·i·est** 1 Abounding in patches; resembling patchwork; hence, lacking in proper effect; incongruous: a *patchy* architectural design. 2 Peevish; irritable. —**patch′i·ly** *adv.* —**patch′i·ness** *n.*

pate (pāt) *n.* The top of the head, especially a human head; often, the head in reference to brains; intellect: usually derogatory. [ME; origin uncertain]

pâte (pät) *n. French* Paste; specifically, porcelain paste.

pâ·té (pä·tā′) *n.* A little pie or pasty; a patty. [<F <OF *pasté* <LL *pasta.* See PASTE[1].]

pâté de foie gras (də fwä grä′) *French* A paste of fat goose liver.

pa·tel·la (pə·tel′ə) *n. pl.* **·lae** (-ē) 1 *Anat.* The flat, movable, oval bone in front of the knee joint; kneecap. 2 A small pan or dish. [<L, dim. of *patina* a pan, dish <*patere* lie open] —**pa·tel′lar** *adj.* —**pa·tel·late** (pə·tel′āt, -it) *adj.*

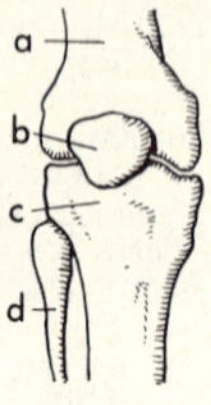

PATELLA
a. Femur.
b. Patella.
c. Tibia.
d. Fibula.

pa·tel·li·form (pə·tel′i·fôrm) *adj.* 1 Having the form of a patella or kneepan, or of a flattened cone. 2 Having the shape of a limpet shell. [<NL *patelliformis* <L *patella* a patella + *forma* form]

pat·en (pat′n) *n.* 1 A plate; especially, a plate for the eucharistic bread, or one held beneath the chin of the person receiving it. 2 A thin, metallic plate. [<OF *patène* <L *patena, patina* a pan <*patere* lie open]

pa·ten·cy (pāt′n·sē) *n.* 1 The condition of being patent or evident. 2 The state of being open, spread, enlarged, or without obstruction.

pat·ent (pat′nt) *n.* 1 A government protection to an inventor, securing to him for a specific time the exclusive right of manufacturing, exploiting, using, and selling an invention; the right granted. 2 Hence, any official document securing a right. 3 A government grant or franchise of land; also, land so granted; the official certificate of such a grant. 4 That which is protected by a patent or its distinctive marks or features. —*v.t.* 1 To obtain a patent on (an invention). 2 *Rare* To grant a patent for or to. —**pa·tent** (pāt′nt *for defs. 1, 4, 5;* pat′nt *for defs. 2, 3, 6*) *adj.* 1 Manifest or apparent to everybody. 2 Protected or conferred by letters patent. 3 Open for general inspection or use: letters *patent.* 4 Expanded; spreading widely, as leaves from the stem of a plant. 5 Open; unobstructed, as an intestine. 6 Designating grades of flour, usually those of superior quality. See synonyms under EVIDENT, MANIFEST, NOTORIOUS, OPEN. ◆ In British English, pāt′nt is the usual pronunciation, except in *Patent Office* and *letters patent.* [<F <L *patens, -entis,* ppr. of *patere* lie open] —**pat′ent·a·bil′i·ty** *n.* —**pat′ent·a·ble** *adj.*

pat·en·tee (pat′n·tē′) *n.* One who holds a patent.

pat·ent leather (pat′nt) Leather finished with a glossy, black, varnishlike coat; lacquered leather.

patent log *Naut.* A torpedo-shaped device with projecting rotary fins: when trailed on a braided line from the stern of a vessel it records the distance traveled by means of an attached mechanism. Also called *screw log, taffrail log.*

pa·tent·ly (pāt′nt·lē, pat′nt-) *adv.* Manifestly.

Patent Office A bureau of the U. S. Department of Commerce where applications for patents are examined and patents are issued.

pat·en·tor (pat′n·tər) *n.* One who grants a patent, as of land: correlative of *patentee.*

patent right 1 An exclusive right conferred by a government grant. 2 The exclusive privilege, for a limited time, to the use and control of an invention and its manufacture.

pa·ter (pā′tər) *n.* Father. [<L]

Pa·ter (pā′tər), **Walter Horatio,** 1839–94, English essayist and critic.

pat·er·a (pat′ər·ə) *n. pl.* **·er·ae** (-ər·ē) 1 A shallow, platelike vessel of earthenware, silver, etc., used by the Romans in pouring libations. 2 In architecture and cabinetwork, a small, dishlike ornament, often in bas-relief. [<L <*patere* lie open] —**pat′er·i·form′** *adj.*

pa·ter·fa·mil·i·as (pā′tər·fə·mil′ē·əs) *n.* 1 The father of a family or master of a house. 2 In Roman law, an independent person; the head of a family. [<L <*pater* father + *familias,* archaic genitive of *familia* family, household]

pa·ter·nal (pə·tûr′nəl) *adj.* 1 Pertaining to a father; fatherly. 2 Derived from, related through, or connected with one's father; hereditary. [<LL *paternalis* <L *paternus* fatherly <*pater* a father] —**pa·ter′nal·ly** *adv.*

pa·ter·nal·ism (pə·tûr′nəl·iz′əm) *n.* The care or control of a country, community, group of employees, etc., in a manner suggestive of a father looking after his children. —**pa·ter′·nal·is′tic** *adj.* —**pa·ter′nal·is′ti·cal·ly** *adv.*

pa·ter·ni·ty (pə·tûr′nə·tē) *n.* 1 The condition of being a father. 2 Parentage on the male side; descent from a father. 3 Origin in general; authorship. [<OF *paternité* <L *paternitas, -tatis* <*paternus.* See PATERNAL.]

Pa·ter·nò (pä′ter·nō′) A town in eastern Sicily on the site of ancient Hybla.

pa·ter·nos·ter (pā′tər·nos′tər) *n.* 1 The Lord's Prayer: the prayer taught to his disciples by Jesus. *Matt.* vi 9–13. Also **Pater Noster.** 2 A recitation of this prayer. 3 A bead of the rosary, indicating that a paternoster is to be recited. 4 Any formula repeated in a low voice. [<L *pater noster* our father, the opening words of the prayer in Latin]

pa·ter pa·tri·ae (pā′tər pā′tri·ē) *Latin* Father of his country.

Pat·er·son (pat′ər·sən) A city on the Passaic River in NE New Jersey.

path (path, päth) *n. pl.* **paths** (pathz, päthz, paths, päths) 1 A walk or way, as one beaten by the foot, used by men or animals. 2 Any road, track, or course. 3 Hence, course or way of life or action. See synonyms under ROAD, WAY. [OE *pæth*]

Pa·than (pə·tän′, pət·hän′) *n.* An Afghan; specifically, one of a people of Afghanistan of Indo-Iranian stock and Moslem religion, who settled in India and on its NW frontier. [<Hind. *Paṭhān* <Afghan *Pĕṣtana,* pl. of *Pĕṣtūn* an Afghan]

pa·thet·ic (pə·thet′ik) *adj.* Of the nature of or expressing sadness, pity, tenderness, etc.; arousing compassion. Also **pa·thet′i·cal.** See synonyms under PITIFUL. [<LL *patheticus* <Gk. *pathētikos* sensitive <*path-,* stem of *paschein* suffer] —**pa·thet′i·cal·ly** *adv.* —**pa·thet′i·cal·ness** *n.*

path·find·er (path′fīn′dər, päth′-) *n.* 1 One skilled in leading or finding a way; especially, one who opens new trails into unknown regions; an explorer; also, one who opens new fields, as in science, philosophy, art. 2 An aircraft carrying flares to light targets in enemy territory for raiding bombers.

-pathia See -PATHY.

path·less (path′lis, päth′-) *adj.* Trackless; untrodden: the *pathless* forest —**path′less·ness** *n.*

patho- *combining form* Suffering; disease: *pathogenesis.* Also, before vowels, **path-.** [<Gk. *pathos* suffering]

path·o·gen (path′ə·jən) *n.* Any disease-producing bacterium or micro-organism. Also **path′o·gene** (-jēn).

path·o·gen·e·sis (path′ə·jen′ə·sis) *n. Med.* The production or development of any morbid or diseased condition: also called *nosogenesis.* Also **pa·thog·e·ny** (pə·thoj′ə·nē).

path·o·gen·ic (path′ə·jen′ik) *adj. Med.* Productive of or pertaining to the production of disease. Also **path′o·ge·net′ic** (-jə·net′ik).

path·o·log·i·cal (path′ə·loj′i·kəl) *adj.* 1 Pertaining to pathology. 2 Related to, involving, concerned with, or caused by disease. Also **path′o·log′ic.** —**path′o·log′i·cal·ly** *adv.*

pa·thol·o·gist (pə·thol′ə·jist) *n.* One skilled in pathology.

pa·thol·o·gy (pə·thol′ə·jē) *n. pl.* **·gies** 1 The branch of medical science that treats of morbid conditions, their causes, nature, etc. 2 The sum of the morbid conditions, processes, and results in the course of a disease. [<PATHO- + -LOGY]

path·o·mi·me·sis (path′ō·mi·mē′sis, -mī·mē′-)

n. Psychiatry The simulation of a disease or of its symptoms, found in hysteria. Also **path′o·mim′i·cry** (-mim′i·krē). — **path′o·mi·met′ic** (-mi·met′ik, -mī·met′-) *adj.*

path·o·mor·phism (path′ə·môr′fiz·əm) *n. Pathol.* Any abnormality of bodily structure and appearance: said of extreme physical types. [<PATHO- + -MORPH(O)- + -ISM] — **path′o·mor′phic** *adj.*

path·o·pho·bi·a (path′ə·fō′bē·ə) *n.* A morbid fear of disease or of being sick. — **path′o·pho′bic** *adj.*

pa·thos (pā′thos) *n.* **1** The quality, attribute, or element, in events, speech, or art, that rouses emotion or passion, especially the tender emotions, as compassion or sympathy; also, tender or sorrowful feeling. **2** In art, the quality of the contingent and evanescent phenomena of life, as the facts of personality, individuality, human passion, or emotion, that the artist's conception embodies or concretely expresses: opposed to the quality of the ideal or *ethos.* **3** *Obs.* Suffering. See synonyms under FEELING. [<Gk., suffering < *path-*, stem of *paschein* suffer]

path·way (path′wā′, päth′-) *n.* A path; a footway.

-pathy *combining form* **1** Suffering; affection: *sympathy.* **2** *Med.* Disease, or the treatment of disease: *psychopathy.* Also spelled *-pathia.* [<Gk. *-patheia* < *pathos* suffering]

Pat·i·a·la (put′ē·ä′lə) **1** A former princely state now included in Punjab State, India. **2** Its former capital, also formerly capital of Patiala and East Punjab States Union.

Patiala and East Punjab States Union A former constituent state of NW India, consisting of several detached areas surrounded by or bordering on Punjab State; became part of Punjab State, November 1, 1956; 10,099 square miles; capital, Patiala: also *Pepsu.*

pa·tience (pā′shəns) *n.* **1** The quality or habit of enduring without complaint. **2** The exercise of sustained endurance and perseverance. **3** Forbearance toward the faults or infirmities of others. **4** Tranquil waiting or expectation. **5** Ability to await events without perturbation. **6** Any solitaire card game. **7** *Obs.* Permission or sufferance. [< OF *pacience* <L *patientia* < *patiens.* See PATIENT.]

Synonyms: calmness, composure, endurance, forbearance, fortitude, leniency, long-suffering, resignation, submission, sufferance. *Endurance* hardens itself against suffering, and may be merely stubborn; by modifiers it may be made to have a passive force, as when we speak of "passive *endurance*"; *fortitude* is *endurance* animated by courage; *patience* is not so hard as *endurance* nor so self-effacing as *submission. Submission* is ordinarily and *resignation* always applied to matters of great moment, while *patience* may apply to slight worries and annoyances. *Patience* may also have an active force denoting uncomplaining steadiness in doing, as in tilling the soil. Compare INDUSTRY, SUBMISSION. *Antonyms*: see synonyms for ANGER, IMPATIENCE.

pa·tient (pā′shənt) *adj.* **1** Possessing quiet, uncomplaining endurance under distress or annoyance; long-suffering. **2** Tolerant, tender, and forbearing. **3** Capable of tranquilly awaiting events. **4** Capable of bearing: with *of*: *patient* of hunger. **5** Persevering. — *n.* **1** A person undergoing treatment for disease or injury. **2** Anything passively affected; the object of external impressions or actions: opposed to *agent.* See synonyms under CHARITABLE, PASSIVE. [<OF *pacient* <L *patiens, -entis,* ppr. of *patiri* suffer] — **pa′tient·ly** *adv.*

pat·i·na[1] (pat′ə·nə) *n. pl.* **·nae** (-nē) An earthenware or metal bowl or basin used as a domestic utensil by the Romans; a patella. [<L < *patere* be open]

pat·i·na[2] (pat′ə·nə) *n.* **1** A green rust or aerugo that covers ancient bronzes, copper coins, medals, etc. **2** An aspect of the surface of stone implements, giving evidence of antiquity. **3** Any surface of antique appearance. Also **pa·tine** (pə·tēn′). [<Ital. <F *patine,* prob. <L *patina* a plate; with ref. to the tarnish on a copper dish]

Pa·ti·ño (pä·tē′nyō), **Simón,** 1868?–1947, Bolivian industrialist and diplomat.

pa·ti·o (pä′tē·ō, pat′ē·ō; *Sp.* pä′tyō) *n. pl.* **·ti·os** **1** The open inner court of a Spanish or Spanish-American dwelling. **2** *U.S.* The enclosed outdoor terrace of a ranch-type dwelling. [<Sp., prob. <L *patere* lie open]

pa·tis·se·rie (pə·tis′ər·ē, *Fr.* pä·tēs·rē′) *n.* A pastry shop; also, one in which light luncheons are served. [<F *pâtisserie* < *pâtissier* a pastry cook <Med. L *pasticerius* < *pasticium* a pastry <LL *pasta.* See PASTE[1].]

Pat·more (pat′môr, -mōr), **Coventry Kersey Dighton,** 1823–96, English poet.

Pat·mos (pat′mos, -mōs, pät′-) An Aegean island in the northern Dodecanese off the western coast of Turkey: place of St. John's exile. *Rev.* i 9. *Italian* **Pat·mo** (pät′mō).

Pat·na (put′nə) A city on the Ganges, capital of Bihar State, India.

pat·ois (pat′wä, *French* pȧ·twȧ′) *n. pl.* **pat·ois** (pat′wäz, *French* pȧ·twȧ′) A dialect, especially one that is provincial or illiterate. See synonyms under LANGUAGE. [<OF, a village, later, rustic speech <Med. L *patriensis* a native <L *patria* native land]

pa·to·la (pə·tō′lə) *n.* An East Indian silk fabric, used especially for native wedding garments. [<Skt. *paṭola*]

Pa·tos (pä′to͞os), **Lagoa dos** A tidal lagoon in SE Río Grande do Sul, Brazil; 3,917 square miles; 150 miles long.

Pa·tras (pə·träs′, pat′rəs) A port in northern Peloponnesus, Greece, on the **Gulf of Patras** (ancient *Gulf of Calydon*), an inlet of the Ionian Sea west of the Gulf of Corinth. *Greek* **Pa·trai** (pä′tre).

patri- *combining form* Father: *patricide.* [<L *pater, -tris* father]

pa·tri·arch (pā′trē·ärk) *n.* **1** The leader of a family or tribe who rules by paternal right. **2** One of the earliest fathers of the human race, from Adam to Noah: in full: **antediluvian patriarch.** **3** One of the fathers of the Hebrew race, Abraham, Isaac, or Jacob. **4** One of the twelve sons of Jacob considered as the progenitors of the tribes of Israel. **5** A venerable man; especially the founder of a religion, order, etc. **6** In later Jewish history, the title of the president of the Sanhedrin in Syria and Babylon. **7** *Eccl.* **a** In the primitive Christian church, any of the bishops of Antioch, Alexandria, Rome, Constantinople, or Jerusalem. **b** In the Roman Catholic Church, a prelate inferior only to the pope and the cardinals, appointed as head of one of the ancient eastern patriarchates or of some modern Uniat churches. **c** In the Greek Orthodox Church, any of the bishops of Constantinople, Alexandria, Antioch, or Jerusalem, sometimes also a prelate of other cities. The bishop of Constantinople, the highest ranking dignitary in the Greek Church, is titled the **ecumenical patriarch.** **d** The title of the heads of other Eastern churches, as the Coptic, Armenian, Jacobite, or Nestorian churches. **8** In the Mormon Church, one of the superior order of priests, with special authority and jurisdiction in bestowing blessings. [<OF *patriarche* <L *patriarcha* <Gk. *patriarchēs* head of a family < *patria* family, clan + *archein* rule]

pa·tri·ar·chal (pā′trē·är′kəl) *adj.* **1** Of or pertaining to a patriarch; governed by a patriarch: a *patriarchal* see. **2** Of the nature of a patriarchy. **3** Of or belonging to the patriarchs. **4** Having the nature or character of a patriarch; venerable. — **pa′tri·ar′chal·ly** *adv.*

pa·tri·ar·chate (pā′trē·är′kit) *n.* **1** The office, dominion, or residence of a patriarch. **2** A patriarchal system of government.

pa·tri·ar·chy (pā′trē·är′kē) *n. pl.* **·chies** **1** A patriarchate. **2** A system of government in which the father or the male heir of his choice rules.

Pa·tri·cia (pə·trish′ə) A feminine personal name. [<L. See PATRICK.]

pa·tri·cian (pə·trish′ən) *adj.* **1** Pertaining to the aristocracy. **2** Noble or aristocratic. **3** Of or pertaining to the Roman aristocracy; also, relating to patricians of the Italian republics, German free cities, etc. — *n.* **1** An aristocrat; specifically, a member of the hereditary aristocracy that, for the first four centuries of her history, monopolized the government and priesthood of Rome. **2** Any one of the upper classes. **3** An honorary title bestowed by the later Roman emperors. **4** In medieval history, one of the upper class in the Italian republics, German free cities, etc. [<OF *patricien* <L *patricius* belonging to the senatorial class < *pater, -tris* a senator, lit., a father] — **pa·tri′cian·ly** *adv.*

pa·tri·ci·ate (pə·trish′ē·it, -āt) *n.* **1** The patricians as a class; the nobility. **2** The rank, dignity, or term of office of a patrician.

pat·ri·cide (pat′rə·sīd) *n.* **1** The killing of a father. **2** One who slays a father; a parricide. — **pat′ri·ci′dal** *adj.*

pat·rick (pat′rik) *n. Scot.* A partridge.

Pat·rick (pat′rik) A masculine personal name. Also *Fr.* **Pa·trice** (pȧ·trēs′), **Pa·tri·ci·o** (*Pg.* pə·trē′sē·o͞o, *Sp.* pä·trē′thyō), **Pa·tri·ci·us** (*Du.* pä·trē′sē·o͞os, *Lat.* pə·trish′əs), *Ital.* **Pa·tri·zi·o** (pä·trē′tsē·ō), *Greek* **Pa·tri·zi·us** (pä·trē′tsē·əs). [<L, patrician]

— **Patrick, Saint,** 389?–461?, apostle to and patron saint of Ireland.

pat·ri·cli·nous (pat′rə·klī′nəs) *adj.* Showing hereditary characteristics inclining toward the paternal side: opposed to *matriclinous.* Also **pat′ro·cli′nous.** [<PATRI- + Gk. *klinein* lean]

pat·ri·lin·e·al (pat′rə·lin′ē·əl) *adj.* Derived from or descending through the male line. Compare MATRILINEAL.

pat·ri·lo·cal (pat′rə·lō′kəl) *adj. Anthropol.* Describing that form of residence in clan societies in which a married couple lives in the husband's community. Compare MATRILOCAL. [<PATRI- + L *locus* a place]

pat·ri·mo·ny (pat′rə·mō′nē) *n. pl.* **·nies** **1** An inheritance from a father or an ancestor; also, any inheritance. **2** An endowment, as of a church. [<OF *patrimoine* <L *patrimonium* < *pater, -tris* a father] — **pat′ri·mo′ni·al** *adj.* — **pat′ri·mo′ni·al·ly** *adv.*

pa·tri·ot (pā′trē·ət, -ot) *n.* One who loves his country and zealously guards its welfare; especially, a defender of popular liberty. [<F *patriote* <LL *patriota* a fellow countryman <Gk. *patriōtēs,* < *patris* fatherland]

pa·tri·ot·ic (pā′trē·ot′ik) *adj.* Characterized by patriotism; intended for the public good. — **pa′tri·ot′i·cal·ly** *adv.*

pa·tri·ot·ism (pā′trē·ə·tiz′əm) *n.* Devotion to one's country.

Patriots' Day The day of the battle of Lexington, April 19, 1775: its anniversary is observed as a legal holiday in Maine and Massachusetts.

pa·tris·tic (pə·tris′tik) *adj.* Of or pertaining to the fathers of the Christian church or to their writings. Also **pa·tris′ti·cal.** [<L *pater, -tris* father + -IST + -IC] — **pa·tris′ti·cal·ly** *adv.*

Pa·tro·clus (pə·trō′kləs) In the *Iliad,* a Greek soldier and friend of Achilles in the Trojan War: wearing Achilles' armor, he was mistaken for him and killed by Hector.

pa·trol (pə·trōl′) *v.t. & v.i.* **·trolled, ·trol·ling** To walk or go through or around (an area, town, etc.) for the purpose of guarding or inspecting. — *n.* **1** One or more soldiers, policemen, etc., patrolling a district. **2** A reconnaissance or safety group sent out from a security detachment or by the main body in air, ground, or naval warfare. **3** The act of patrolling. **4** A division of a troop of Boy Scouts. [<MF *patrouille* a night watch < *patrouiller,* alter. of *patouiller,* orig. paddle in mud, ? ult. < *patte* a paw, foot] — **pa·trol′ler** *n.*

patrol boat A small, ruggedly built boat for patrolling harbors and coastal waters.

patrol bomber A large navy bombing plane adapted for long-range patrol activities.

patrol car A squad car.

pa·trol·man (pə·trōl′mən) *n. pl.* **·men** (-mən) One who patrols; specifically, a policeman assigned to a beat.

patrol torpedo boat A small, rugged, highly maneuverable vessel, lightly armed and equipped with torpedoes for rapid action against enemy shipping: also called *PT boat.*

patrol wagon A police wagon or truck for the conveyance of prisoners, etc.

pa·tron (pā′trən) *n.* **1** One who protects, fosters, countenances, or supports some person or thing; a protector or benefactor. **2** A regular customer. **3** A saint regarded as the peculiar protector of some special person, country, cause, etc.; a tutelary saint; also, the canonized founder of a religious order. **4** In Greek and Roman religion, a tutelary deity; protector of some city, cause, occupation, etc. **5** One in the position of father, guardian, or helper, as one who sponsors a concert or charitable entertainment, one who champions a cause. **6** In ancient Rome, a master who freed his slave and sustained toward him a legal relation analogous to that of father. [<OF *patrun* <L *patronus* protector < *pater, -tris* father] —**pa′tron·al** *adj.*

pa·tron·age (pā′trən·ij, pat′rən-) *n.* **1** Special countenance; guardianship. **2** An uncalled-for distribution of favors, or an overly condescending manner. **3** The right to control the distribution of offices, etc., in the public service; also, the offices, etc., so distributed. **4** The financial support given by customers to commercial enterprises; the customers themselves, as a group.

pa·tron·ess (pā′trən·is, pat′rən-) *n.* A female patron; a matron who promotes and assists in the management of a social event.

pa·tron·ize (pā′trən·īz, pat′rən-) *v.t.* **·ized, ·iz·ing 1** To act as a patron toward; give support or protection to. **2** To treat in a condescending manner. **3** To trade with as a regular customer; frequent. —**pa′tron·iz′er** *n.* —**pa′·tron·iz′ing·ly** *adv.*

pat·ro·nym·ic (pat′rə·nim′ik) *adj.* Formed after one's father's name. —*n.* **1** A name derived from an ancestor; a family name. **2** A name formed by the addition of a prefix or suffix to a proper name: Fitzhugh, son of Hugh; Johnson, son of John. Also **pat′ro·nym** *n.* [<Gk. *patronymos* (< *patēr, -tros* father + *onyma, onoma* name) + -IC] —**pat′·ro·nym′i·cal·ly** *adv.*

pa·troon (pə·trōōn′) *n.* **1** Formerly, a holder of entailed estates, chiefly in New York and New Jersey, with manorial rights, under old Dutch law. **2** *Obs.* The master of a vessel; also, the steersman. [<Du. <F *patron* <L *patronus.* See PATRON.]

pat·te·mar (pat′ə·mär) See PATAMAR.

pat·ten (pat′n) *n.* A shoe having a thick, wooden sole; a clog. [<OF *patin,* prob. < *patte* a paw, foot]

pat·ter[1] (pat′ər) *v.i.* **1** To make a succession of light, sharp sounds. **2** To move with light, quick steps. —*v.t.* **3** To cause to patter. —*n.* Pattering, or the sound of pattering. [Freq. of PAT[1]]

pat·ter[2] (pat′ər) *v.t.* & *v.i.* To speak or say glibly or rapidly; mumble or recite (prayers, etc.) mechanically or indistinctly. —*n.* **1** Glib and rapid talk, as used by comedians, etc. **2** Patois or dialect; any professional jargon. **3** Rapid speech set to music. [Short for PATERNOSTER; from the rapid repetition of the prayer] —**pat′ter·er** *n.*

pat·tern (pat′ərn) *n.* **1** An original or model proposed for or worthy of imitation: Ancient Athens was a *pattern* of democracy. **2** Anything shaped or designed to serve as a model or guide in making something else: a *pattern* for a coat. **3** Any decorative design or figure, or such design worked on something: a vase with a geometrical *pattern.* **4** Arrangement of natural or accidental markings: the *pattern* of a butterfly's wings. **5** The stylistic composition or design of a work of art: the *pattern* of Hardy's novels. **6** A complex of integrated parts functioning as a whole: the behavior *pattern* of a five-year-old; *patterns* of American culture. **7** In gunnery, the distribution of shot or shots about a target. **8** A representative example; sample or instance: a book of *patterns.* **9** *U.S.* Material in sufficient quantity to make a garment, especially a dress. **10** *Obs.* Something made after a model or prototype; a copy. See synonyms under EXAMPLE, IDEA, IDEAL, MODEL, SIGN. —*v.t.* **1** To make after a model or pattern: with *on, upon,* or *after.* **2** To decorate or furnish with a pattern. [Alter. of PATRON]

pattern bombing The covering of an entire target area, according to a definite plan, by bombs dropped simultaneously by a formation of aircraft.

Pat·ter·son (pat′ər·sən), **Robert Porter,** 1891–1952, U.S. lawyer and statesman.

Pat·ti (pät′tē), **Adelina,** 1843–1919, Baroness Cederstrom, Italian soprano born in Madrid.

Pat·ti·son (pat′ə·sən), **Mark,** 1813–84, English scholar and author.

pat·tle (pat′l) *n. Scot.* A plow staff.

Pat·ton (pat′n), **George Smith,** 1885–1945, U.S. Army officer in World War II.

pat·tu (put′ōō) *n.* An East Indian homespun wool or tweed, used for shawls, etc. [<Hind. *paṭṭu*]

pat·ty (pat′ē) *n. pl.* **·ties** A small pie. [Alter. of F *pâté.* See PÂTÉ.]

patty shell A small puff-paste shell in which to serve creamed meat, fish, vegetables, or fruit.

pat·u·lous (pach′ōō·ləs) *adj.* **1** Spreading; gaping. **2** *Bot.* Spreading slightly, as a calyx. **3** Having a wide aperture. [<L *patulus* standing open < *patere* lie open] —**pat′u·lous·ly** *adv.* —**pat′u·lous·ness** *n.*

Pau (pō) A city of SW France.

pau·cis ver·bis (pô′sis vûr′bis) *Latin* In few words.

pau·ci·ty (pô′sə·tē) *n.* Smallness of number or quantity; fewness; also, scarcity; insufficiency. [<L *paucitas, -tatis* <*paucus* few]

paugh·ty (pôkh′tē) *adj. Scot.* Haughty; insolent.

Pau·ker (pou′kər), **Ana,** 1894?–1960, *née* Rabinsohn, Rumanian Communist leader.

Paul (pôl; *Fr.* pôl, *Dan., Ger., Sw.* poul) **1** A masculine personal name. Also *Lat.* **Pau·li·nus** (pô·lī′nəs), *Pg.* **Pau·lo** (pou′lōō), *Lat.* **Pau·lus** (pô′ləs). **2** Appellation of five popes. [<L, small]

—**Paul** The apostle to the Gentiles, a Hebrew who, before his conversion, was called *Saul of Tarsus*; author of various New Testament books; died about A.D. 67. Also *Paul the Apostle, Saint Paul.*

—**Paul I,** 1754–1801, emperor of Russia 1798–1801; assassinated.

—**Paul I,** born 1901, king of Greece 1947–.

—**Paul III,** 1468–1549, pope 1534–49; real name Alessandro Farnese; excommunicated Henry VIII; approved Jesuit order.

Paul and Virginia An idyllic romance by Bernardin de St. Pierre; also, the juvenile lovers around whom the story centers.

Paul Bunyan The famous hero lumberjack of American folklore, of superhuman size and strength and credited with amazing feats.

Paul·ding (pôl′ding), **James Kirk,** 1778–1860, U.S. author.

paul·dron (pôl′drən) *n.* A detachable piece of plate armor to protect the shoulder. [Aphetic var. of OF *espauleron* < *espaule* shoulder; with intrusive *-d*]

Pau·li (pou′lē), **Wolfgang,** born 1900, U.S. physicist born in Austria.

pau·lin (pô′lin) *n.* A sheet of heavy canvas, usually waterproof. [Short for TARPAULIN]

Paul·ine (pô′lēn, -lin) *adj.* **1** Relating to the apostle Paul, his teachings, or writings. **2** Characterized by the assumed trend of Paul in his theological thinking. —**Paul′in·ism** *n.* —**Paul′in·ist** *n.*

Pau·line (pô·lēn′; *Fr.* pō·lēn′, *Ger.* pou·lē′nə) A feminine personal name. Also **Pau·li·na** (pô·lī′nə; *Pg., Sp.* pou·lē′nä). [<L, of Paul]

Pau·ling (pô′ling), **Linus Carl,** born 1901, U.S. chemist.

Pau·li·nus (pô·lī′nəs), **Saint,** died 644, a Roman missionary to England in 601; archbishop of York.

Paul·ist (pô′list) *n.* A member of the Congregation of the Missionary Priests of St. Paul the Apostle, founded in New York in 1858.

pau·low·ni·a (pô·lō′nē·ə) *n.* Any of a genus (*Paulownia*) of Chinese trees of the figwort family, with heart-shaped leaves and panicles of handsome, fragrant, purple flowers. [<NL, after Anna *Paulovna,* daughter of Czar Paul I]

Paul Pry 1 A comedy by John Poole; also, the inquisitive title character. **2** An inquisitive person.

Pau·lus (pou′lōōs), **Friedrich von,** 1890–1957 German field marshal.

Pa·u·mo·tu Archipelago (pä′ōō·mō′tōō) A former name for TUAMOTU ARCHIPELAGO.

paunch (pônch) *n.* **1** The abdomen; the belly; also, a potbelly. **2** The first stomach of a ruminant. [<AF *panche* <L *pantex, -ticis* belly, bowels] —**paunch′y** *adj.* —**paunch′i·ness** *n.*

pau·per (pô′pər) *n.* One dependent on charity; a destitute person who receives, or is entitled to receive, public charity; any very poor person. [<Med. L (*in forma*) *pauperis* (in the character) of a poor man < *pauper* poor; orig. a legal phrase. Doublet of POOR.]

pau·pered (pô′pərd) *adj.* Made a pauper.

pau·per·ism (pô′pə·riz′əm) *n.* **1** Poverty. **2** Paupers collectively. See synonyms under POVERTY.

pau·per·ize (pô′pər·īz) *v.t.* **·ized, ·iz·ing** To make a pauper of.

pauper's oath An oath that one is destitute and incapable of supporting oneself: sometimes required when making a plea for public assistance.

pau·ro·me·tab·o·lous (pô′rō·mə·tab′ə·ləs) *adj. Entomol.* Having a gradual or incomplete metamorphosis, as in the grasshopper and certain other insects. Also **pau′ro·met′a·bol′ic** (-met′ə·bol′ik). [<Gk. *pauros* small + *metabolos* changing]

Pau·sa·ni·as (pô·sā′nē·əs) Greek traveler and geographer of the second century A.D.

pause (pôz) *v.i.* **paused, paus·ing 1** To cease action or utterance temporarily; stop; hesitate; delay. **2** To dwell or linger: with *on* or *upon*: to *pause* on a word. See synonyms under CEASE, STAND. [<*n.*] —*n.* **1** A ceasing of action; an intermission; rest; stop. **2** A holding back because of doubt or irresolution; suspense; hesitation. **3** A momentary cessation in speaking or music for the sake of meaning or expression. **4** A character or sign indicating such cessation, as most marks of punctuation, a break, or a paragraph, or, in music, a hold or a rest. **5** A calculated interval of silence in a meter, or the place at which the voice naturally pauses in reading a verse. See CAESURA. See synonyms under RESPITE, REST[1]. [<MF <L *pausa* a stop <Gk. *pausis* < *pauein* stop] —**paus′er** *n.*

pav·an (pav′ən) *n.* **1** A slow, stately dance of the 16th and 17th centuries. **2** The music for this dance. Also **pav·ane** (pav′ən, *Fr.* pȧ·vȧn′). [<MF *pavane* <Sp. *pavana,* prob. < *pavo* a peacock <L *pavo, -onis*]

pave (pāv) *v.t.* **paved, pav·ing** To cover or surface with asphalt, gravel, concrete, macadam, etc., as a road. —**to pave the way (for)** To make preparation (for); lead up to. [<F *paver* <L *pavire* ram down]

pa·vé (pä·vā′) *n.* **1** A street pavement. **2** The close setting of jewels in which no metal shows. [<F, orig. pp. of *paver.* See PAVE.]

Pa·vel (pä′vyel) Polish and Russian form of PAUL.

pave·ment (pāv′mənt) *n.* **1** A hard, solid, surface covering or flooring for a road or footway, usually resting immediately on the ground. **2** A paved road or footway. **3** The material with which a surface is paved. [<OF <L *pavimentum* a rammed floor < *pavire* ram down]

pav·er (pā′vər) *n.* **1** One who lays pavement. **2** A paver's rammer.

Pa·vi·a (pä·vē′ä) A city in northern Italy on the Ticino; scene of a victory of Charles V over Francis I of France, 1525.

pa·vil·ion (pə·vil′yən) *n.* **1** A movable or open structure for temporary shelter or dwelling; a large tent; summerhouse. **2** A related or connected part of a principal building, especially such a structure appropriated to amusement: the dancing *pavilion.* **3** A canopy. **4** The external ear. **5** The sloping surface of a brilliant-cut gem between the girdle and the culet. **6** A detached building for patients, as at a hospital. —*v.t.* **1** To provide with a pavilion or pavilions. **2** To shelter by a pavilion. [<OF *pavillon* <L *papilio, -onis* a butterfly, tent]

pav·ing (pā′ving) *n.* **1** The laying of a pavement. **2** A pavement; also, the material used for pavement.

pav·ior (pāv′yər) *n.* A pavement layer. Also, *esp. Brit.* **pav′iour.** [Alter. of PAVER]

pav·is (pav′is) *n.* A large, medieval shield protecting the whole body of the soldier bearing it. [<OF *pavais,* appar. <Ital. *Pavia* Pavia, where these shields were made]

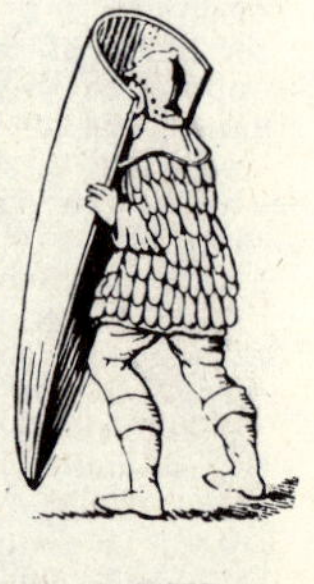

PAVISER

pav·is·er (pav′is·ər) *n.* A soldier bearing a pavis.

Pav·lov (päv′lôf), **Ivan Petrovich,** 1849–1936, Russian physiologist.

Pav·lo·va (päv·lō′və), **Anna,** 1885–1931, Russian ballet dancer.

Pa·vo (pā′vō) A southern constellation, the

pavonine | 927 | **Peacock**

Peacock; its principal star, also called Peacock, is used by navigators. See CONSTELLATION. [<L, a peacock]
pav·o·nine (pav′ə-nīn, -nin) *adj.* 1 Resembling or characteristic of the peacock. 2 Iridescent like the tail of a peacock. [<L *pavoninus* like a peacock < *pavo, -onis* a peacock]
pa·vor noc·tur·nus (pā′vər nok·tûr′nəs) *Latin* Night terrors.
paw (pô) *n.* 1 The foot of an animal having nails or claws. 2 A clumsy human hand. — *v.t.* & *v.i.* 1 To strike or scrape with the feet or paws: to *paw* the air; to *paw* at the ground. 2 *Colloq.* To handle rudely or clumsily; maul. [<OF *powe*, prob. <Gmc.] — **paw′er** *n.*
pawk·y (pô′kē) *adj. Scot.* **pawk·i·er, pawk·i·est** Cunning; sly; humorous. — **pawk′i·ly** *adv.* — **pawk′i·ness** *n.*
pawl (pôl) *n. Mech.* A hinged or pivoted member shaped to engage with ratchet teeth, either to drive a ratchet wheel or to stop its reverse motion; a click or detent. ◆ Homophone: *pall.* [? <Welsh, a pole, prob. ult. <L *palus* a stake]

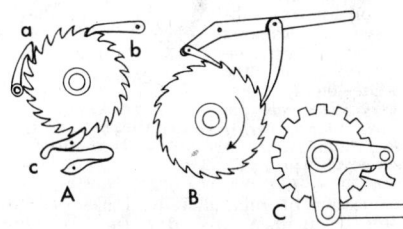

PAWLS
A. Types of pawls: *a.* Hook; *b.* Straight gravity; *c.* Spring.
B. Double pawl.
C. Reversible double pawl.

pawn¹ (pôn) *n.* 1 A chessman of lowest rank, that moves on file but captures diagonally. 2 Hence, any insignificant person used at another's will. [<AF *poun*, OF *paon*, var. of *peon, pedon* a foot soldier <LL *pedo, -onis* <L *pes, pedis* a foot. Doublet of PEON.]
pawn² (pôn) *n.* 1 Something pledged as security for a loan; especially, personal property pledged to secure a loan. 2 The condition of being held as a pledge for money loaned. 3 The act of pawning. — *v.t.* 1 To give (personal property) as security for a loan. 2 To risk or stake; pledge: to *pawn* one's life. [<OF *pan*, prob. <L *pannus* a cloth; infl. in meaning by MDu. *pand* a pledge <Gmc.] — **pawn′a·ble** *adj.* — **pawn′age** *n.*
pawn·brok·er (pôn′brō′kər) *n.* One engaged in the business of lending money on pledged personal property. — **pawn′brok′ing** *n.*
pawn·ee (pô·nē′) *n. Law* One with whom goods have been left in pawn; a pawnbroker.
Paw·nee (pô·nē′) *n.* A member of one of four tribes of North American Indians of Caddoan linguistic stock, formerly inhabiting the region between the Arkansas River and the Platte River, Nebraska: now living in Oklahoma.
pawn·er (pô′nər) *n.* One who pawns personal property. Also **pawn′or.**
pawn·shop (pôn′shop′) *n.* The place of business of a pawnbroker.
pawn ticket A certificate given by a pawnbroker for goods pawned.
paw·paw (pô′pô) See PAPAW.
Paw·tuck·et (pô·tuk′it) A city in NE Rhode Island.
Pax (paks) In Roman mythology, the goddess of peace: identified with the Greek *Irene.*
pax vo·bis·cum (paks vō·bis′kəm) *Latin* Peace be with you.
pax·wax (paks′waks) *n.* A strong fibrous band extending from the dorsal vertebrae to the occiput, and supporting the head in many mammals, as horses: also spelled *packwax.* [Alter. of dial. *fax-wax* <OE *feax* hair + *weaxan* grow]
pay¹ (pā) *v.* **paid, pay·ing** *v.t.* 1 To give to (someone) what is due for a debt, purchase, etc.; recompense; remunerate. 2 To give (money, etc.) for a purchase, service rendered, etc. 3 To provide or hand over the amount of; discharge, as a debt, bill, etc. 4 To yield as return or recompense. 5 To afford profit or benefit to: It wouldn't *pay* me to do it. 6 To defray, as expenses. 7 To requite, as for an insult. 8 To render or give, as a compliment, attention, etc. 9 To make, as a call or visit. — *v.i.* 10 To make recompense or payment. 11 To afford compensation or profit; be worthwhile: It *pays* to be honest. — **to pay back** To repay. — **to pay off** 1 To pay the entire amount of (a debt, mortgage, etc.). 2 To pay the wages of and discharge. 3 To gain revenge upon or for. 4 *Colloq.* To afford full return; be fully effective. 5 *Naut.* To turn or cause to turn to leeward. — **to pay out** 1 To disburse or expend. 2 *Naut.* To let out by slackening, as a rope or cable. — **to pay up** To make full payment of. — *n.* 1 That which is given as a recompense or to discharge a debt; compensation; wages. 2 The act of paying or the state of being paid. 3 Whatever compensates for labor or loss; an equivalent. 4 Requital; reward; also, punishment; retaliation. 5 A person considered from the point of view of his ability to pay or his promptness or slowness in paying. 6 A worthwhile yield of metal in a vein of ore. 7 Retribution. See synonyms under SALARY. — *adj.* 1 Of or pertaining to payments, persons who pay, or services paid for: *pay* day, *pay* students, a *pay* library, etc. 2 Yielding enough metal to be worth mining: *pay* dirt, *pay* quartz, etc. [<OF *payer* pay, appease <L *pacare* appease < *pax, pacis* peace] — **pay′er** *n.* Synonyms (*verb*): compensate, defray, discharge, indemnify, liquidate, recompense, remunerate, requite, reward, satisfy, settle. See SETTLE. Antonyms: default, repudiate.
pay² (pā) *v.t.* **paid** or **payed, pay·ing** To coat with pitch or other waterproof composition, as the seams of a vessel, etc. [<AF *peier* <L *picare* < *pix, picis* pitch]
pay·a·ble (pā′ə·bəl) *adj.* 1 Due and unpaid. 2 Capable of being discharged by payment; that can or will be paid. 3 Likely to be profitable; specifically, likely to yield a surplus, as a mine. — **pay′a·bly** *adv.*
pay dirt Soil containing enough metal, especially gold, to be profitable to mine.
pay·ee (pā·ē′) *n.* A person to whom money has been or is to be paid.
pay·load (pā′lōd′) *n.* That part of a cargo producing revenue.
pay·mas·ter (pā′mas′tər, -mäs′-) *n.* One who has charge of the paying of employees.
pay·ment (pā′mənt) *n.* 1 The act of paying. 2 Pay; requital; recompense. 3 Punishment. See synonyms under SALARY.
Payne (pān), **John Howard,** 1791–1852, U.S. dramatist and song writer; author of *Home, Sweet Home.*
pay·nim (pā′nim) *n. Archaic* A non-Christian; a pagan; heathen; especially, a Moslem. [<OF *paienime* <LL *paganismus* heathenism < *paganus.* See PAGAN.]
pay-off (pā′ôf′, -of′) *n.* 1 Payment; specifically, the time or act of payment of wages to employees; also, the time or act of paying an employee in full and discharging him. 2 *Colloq.* Any settlement; the end; reward or punishment. 3 *Colloq.* The climax of an incident or narrative.
pay·roll (pā′rōl′) *n.* A list of those entitled to receive pay, with the amounts due them; also, the total sum of money needed to make the payments. Also **pay roll.**
pays (pās) See PAIS.
Paz (päz, *Sp.* päs), **La** See LA PAZ.
pe (pā) *n.* The seventeenth Hebrew letter. See ALPHABET.
pea (pē) *n. pl.* **peas** or **pease** 1 A climbing annual leguminous herb (*Pisum sativum*) having pinnate leaves. 2 Its edible seed. 3 The seed of any one of various other plants of the same family, as the chickpea or cowpea. 4 A grade of anthracite coal less than an inch in size; also, a lump of this. [<PEASE, incorrectly taken as a plural]
pea·bird (pē′bûrd′) *n.* The Baltimore oriole.
Pea·bod·y (pē′bod′ē), **George,** 1795–1869, U.S. philanthropist.
pea·bod·y bird (pē′bod·ē) The white-throated sparrow.

peace (pēs) *n.* 1 A state of quiet or tranquillity; freedom from disturbance or agitation; calm; repose. 2 Specifically, absence or cessation of war. 3 General order and tranquillity; freedom from riot or violence. 4 A state of reconciliation after strife or enmity; peaceable or friendly relations; agreement; concord. 5 Freedom from mental agitation or anxiety. 6 Spiritual content. See synonyms under REST. — **to hold** (or **keep**) **one's peace** To be silent. — *v.i. Obs.* except as an imperative. To be or become quiet or silent. ◆ Homophone: *piece.* [<OF *pais* <L *pax, pacis*]
Peace may appear as a combining form in hyphemes or solidemes, or as the first element in two-word phrases; as in:

peace-abiding	peace-lover
peace-bearer	peace-loving
peace-breaker	peace-minded
peace-breaking	peacemonger
peace-bringer	peace movement
peace conference	peace offer
peace congress	peace party
peace-destroying	peace plan
peace-giving	peace-preserver
peace-inspiring	peace-restoring
peace-keeper	peace-seeker
peace-keeping	peacetime

peace·a·ble (pē′sə·bəl) *adj.* 1 Inclined to peace. 2 Peaceful; tranquil. — **peace′a·ble·ness** *n.* — **peace′a·bly** *adv.*
peace·ful (pēs′fəl) *adj.* 1 Exempt from war, riot, or commotion; undisturbed. 2 Averse to strife. 3 Inclined to or used in peace. See synonyms under CALM, PACIFIC. — **peace′ful·ly** *adv.* — **peace′ful·ness** *n.*
peace·mak·er (pēs′mā′kər) *n.* One who effects, or seeks to effect, a reconciliation between unfriendly parties. — **peace′mak′ing** *n.* & *adj.*
peace offering An offering made for the sake of peace or reconciliation; specifically, an offering prescribed by Levitical law.
peace officer Any conservator of the peace, especially a justice of the peace, sheriff, constable, or policeman.
peace pipe The calumet.
Peace River (pēs) A river in western Canada, flowing 1,050 miles NE to the Slave River in NE Alberta.
peach¹ (pēch) *n.* 1 The fruit of the peach tree (*Prunus persica*), a fleshy, juicy, edible drupe. 2 The tree, widely cultivated in many varieties. 3 The orange-yellow color of the fruit. 4 *Slang* Any person or thing particularly beautiful, pleasing, or excellent. [<OF *peche, pesche* <L *persica* <L *Persicum (malum)* Persian (apple) <Gk. *persikos*]
peach² (pēch) *v.i. Slang* To inform against an accomplice; turn informer. — *v.t. Obs.* To impeach; inform against. [Aphetic var. of APPEACH]
peach-bloom (pēch′blōōm′) *n.* 1 A monochrome glaze in Chinese porcelain in various tones of pinkish-red: also **peach′blow′** (-blō′). 2 A kind of ware thus glazed or tinted. 3 A delicate pink color.
peach·y (pē′chē) *adj.* **peach·i·er, peach·i·est** 1 Resembling a peach, especially in color or downiness. 2 *Slang* Delightfully pleasant. — **peach′i·ness** *n.*
pea-coat (pē′kōt′) *n.* A peajacket. [<PEA-(JACKET) + COAT]
pea·cock (pē′kok′) *n.* The male of a gallinaceous crested bird (genus *Pavo*), which has the tail coverts enormously elongated, erectile, and marked with ocelli or eyelike spots, and the neck and breast of an iridescent greenish-blue. ◆ Collateral adjective: *pavonine.* — *v.i.* To strut vainly; make a display. [OE *pēa, pāwa* a peacock (< L *pavo*) + COCK¹]
Pea·cock (pē′kok) A

PEACOCK
(Body length about 30 to 36 inches; tail up to 6 feet)

add, āce, câre, pälm; end, ēven; it, īce; odd, ōpen, ôrder; tōōk, pōōl; up, bûrn; ə = a in *above*, e in *sicken,* i in *clarity,* o in *melon,* u in *focus;* yōō = u in *fuse;* oi, oil; ou, pout; ch, go; ng, ring; th, thin; ŧħ, this; zh, vision. Foreign sounds á, œ, ü, kh, ń; and ◆: see page xx. < from; + plus; ? possibly.

Peacock — **pecten**

star of 2.12 magnitude in the constellation Pavo. See STAR.
Pea·cock (pē′kok), **Thomas Love**, 1785–1866, English novelist.
peacock blue A vivid greenish blue, the color of the blue in a peacock's feathers.
peacock copper Bornite.
pea·cock·ish (pē′kok·ish) adj. Vain; pretentious. Also **pea′cock·y**.
peacock moth A large, dark-gray moth (*Saturnia pyri*), with wings fringed with white and having a black ocellus with brown, violet, and red crescents: common in southern Europe and western Asia.
pea·fowl (pē′foul) n. A peacock or peahen. [< obs. *pea*, OE *pēa* a peacock + FOWL]
peag (pēg) n. Wampum. [< Algonquian (Massachuset) *piak*, pl. of *pi* a strung bead of shell money]
pea green Any of several shades of light yellowish green.
pea·hen (pē′hen) n. A female peafowl. [OE *pēa* a peacock + *henne* hen]
pea·jack·et (pē′jak·it) n. A short coat of thick woolen cloth, worn by seamen. [Prob. <obs. *pee* a coat of coarse wool <MDu. *pie* + JACKET]
peak[1] (pēk) n. 1 A projecting point or edge; an end terminating in a point; summit: the *peak* of a roof. 2 A mountain with pointed summit; a conspicuous or precipitous mountain. 3 *Naut.* **a** The after upper corner of a fore-and-aft sail. **b** The upper end of a gaff. **c** The sharply narrowed part of the hull or hold of a vessel at the bow or stern, called respectively the *forepeak* and *afterpeak*. 4 A point formed on the forehead by the growth or cut of the hair: a widow's *peak*. See synonyms under SUMMIT. — *v.t. Naut.* To raise to or almost to a vertical position, as a gaff or yard. ◆ Homophones: peek, pique. [OE *pīc*; def. 2 infl. by Sp. *pico* a beak, peak; *v.* infl. by *peak* vertically, aphetic var. of *apeak* <F *à pic* to a peak]
peak[2] (pēk) *v.i.* To become sickly, weak, or dispirited. ◆ Homophones: peek, pique. [Origin unknown]
peak·ed (pē′kid, pēkt) adj. Having a thin or sickly appearance. [<PEAK[2]]
peak load *Electr.* The maximum power load consumed or produced by a generating unit or group of units during a specified time.
peal (pēl) n. 1 A prolonged, sonorous sound, as of a bell, trumpet, or thunder. 2 A set of large bells attuned to the major scale. 3 The change rung on a chime, usually a scale or part of a scale. — *v.t. & v.i.* To sound with a peal or peals; ring out; resound. See synonyms under ROAR. ◆ Homophone: peel. [ME *pele*, aphetic var. of *apele* <OF *apeler*. See APPEAL.]
Peale (pēl) Name of a family of American artists, especially, **Charles Willson**, 1741–1827, who painted many portraits of George Washington; his brother **James**, 1749–1831, and his son **Rembrandt**, 1778–1860.
pe·an[1] (pē′ən) See PAEAN.
pean[2] (pēn) See PEEN.
pea·nut (pē′nut′) n. 1 The nutlike seed or seed pod of an annual herbaceous vine (*Arachis hypogaea*) of the pea family, ripening underground from the pistillate flowers, which bury themselves after fertilization. 2 The plant. 3 A small or insignificant person.
peanut brittle A hard candy containing roasted peanuts.
peanut butter A spread resembling butter in consistency, made from ground, roasted peanuts.
peanut oil Oil from the seeds of the peanut.
pear (pâr) n. 1 The juicy, edible, fleshy fruit of a tree (*Pyrus communis*) of the rose family, widely cultivated in many varieties. 2 The tree. ◆ Homophones: pair, pare. [OE *pere* <LL *pera*, *pira* <L *pira*, orig. pl. of *pirum* <a pear]
pearl (pûrl) n. 1 A lustrous, calcareous concretion deposited in layers around a central nucleus in the shells of various mollusks, and largely used as a gem. 2 Something like or likened to such a jewel in form, luster, value, etc. 3 Nacre or mother-of-pearl; also, the color of nacre, or **pearl blue. 4** The color of a pearl, a delicate gray: also **pearl gray. 5** *Printing* A size of type, smaller than agate, 5 points. — *adj.* 1 Pertaining to, consisting of, made of, or resembling pearl or mother-of-pearl: a *pearl* button; a *pearl* ring. 2 Shaped like a pearl: *pearl* barley. — *v.i.* 1 To seek or fish for pearls. 2 To form beads like pearls. — *v.t.* 3 To adorn or set with or as with pearls. 4 To color or shape like pearls. 5 To make into small round grains, as barley. ◆ Homophone: purl. [<OF *perle* <Med. L *perla*; ult. origin unknown]
Pearl (pûrl), **Raymond**, 1879–1940, U. S. biologist.
pearl ash Crude potassium carbonate.
pearl barley Barley reduced to a round shotlike form by pearling: used in soups. Also **pearled barley**.
pearl·er (pûr′lər) n. 1 A diver for or trader in pearls. 2 A boat engaged in pearling.
Pearl Harbor An inlet on the southern coast of Oahu, near Honolulu, Hawaii; site of a U. S. naval base, attacked by Japanese, December 7, 1941.
pearl·ing (pûr′ling) n. 1 The process of removing the outer coat of grain, as in making pearl barley. 2 The business of fishing for pearls.
pearl·ite (pûr′līt) n. Perlite.
pearl millet The East Indian millet, a tall cereal grass (*Pennisetum glaucum*) having edible seeds: used in the United States as a forage grass.
Pearl River A river in central Mississippi, flowing 480 miles SW to the Gulf of Mexico and forming part of the boundary between Mississippi and Louisiana. See CANTON RIVER.
pearl·y (pûr′lē) adj. **pearl·i·er**, **pearl·i·est** Adorned with, yielding, or resembling pearls; margaric.
pear·main (pâr′mān) n. A variety of apple. [<OF *permain*, *parmain*, lit., of Parma, Italy]
Pearse (pîrs), **Padraic**, 1879–1916, Irish educator and patriot; executed.
Pear·son (pîr′sən), **Karl**, 1857–1936, English mathematician and scientist.
peart (pîrt, pûrt) adj. *Dial.* In good health and spirits; active; lively. [Var. of PERT] — **peart′ly** adv.
Pea·ry (pîr′ē), **Robert Edwin**, 1856–1920, U.S. Arctic explorer; first to reach the North Pole, April 6, 1909.
Peary Land A region of northern Greenland along the Arctic Ocean, the world's northernmost land region.
peas·ant (pez′ənt) n. 1 In Europe, a small farmer; a farm laborer; any rustic workman. 2 *Obs.* A rascal; base character; scamp. [<AF *paisant* <OF *païs* country <L *pagensis* (*ager*) (territory) of a canton <*pagus* a district]
peas·ant·ry (pez′ən·trē) n. 1 The peasant class; a body of peasants. 2 Rusticity.
peas·cod (pēz′kod) n. A pea pod. Also **pease′·cod**. [<PEAS(E) + COD[2]]
pease (pēz) n. pl. Peas collectively. [OE *pise* <L *pisa* <L, orig. pl. of *pisum* <Gk. *pison* <*pisos* pulse, pease]
pea-shoot·er (pē′shoo′tər) n. A toy blowgun through which small pellets, as dried peas, are blown.
peat[1] (pēt) n. 1 A substance consisting of partially carbonized vegetable material, chiefly mosses, found usually in bogs. 2 A block of this substance, pressed and dried for fuel. [<Med. L *peta* a piece of peat, ? <*petia* a fragment]
peat[2] (pēt) n. *Obs.* 1 A pet; a favorite woman or girl. 2 A favorite; a minion. [Cf. MDu. *pete* a goddaughter]
peat bog A marsh with an accumulation of peat.
peat·man (pēt′mən) See PETEMAN.
peat moss 1 A moss that enters largely into the composition of peat. 2 *Brit.* A peat bog.
peat·y (pē′tē) adj. **peat·i·er**, **peat·i·est** Resembling or containing peat.
pea·vy (pē′vē) n. pl. **·vies** An iron-pointed lever fitted with a movable hook and used for handling logs. Also **pea′vey**. [after Joseph Peavey, its inventor]
peb·ble (peb′əl) n. 1 A small, rounded fragment of rock, its form being due to attrition of water, ice, etc. 2 Quartz crystal; also, a lens made of it. 3 Leather that has been pebbled. 4 Pebbleware. — *v.t.* **·bled**, **·bling** 1 To impart a rough grain to (leather). 2 To pave, cover, or pelt with pebbles. [OE *pabol(stān)* a pebble(stone)] — **peb′bly** adj.
peb·ble·stone (peb′əl·stōn′) n. 1 A pebble. 2 A material consisting of a mass of pebbles. [See PEBBLE.]

peb·ble·ware (peb′əl·wâr′) n. A ware having different-colored clays in the paste.
pe·can (pi·kan′, -kän′, pē′kan) n. 1 A large hickory (*Carya illinoensis*) of the central and southern United States, with olive-shaped, thin-shelled nuts. 2 The nut borne by this tree, containing a sweet, oily kernel. [Earlier *paccan* <Algonquian (Cree) *pakan*]
pec·ca·ble (pek′ə·bəl) adj. Capable of sinning. [<Med. L *peccabilis* <L *peccare* sin] — **pec′ca·bil′i·ty** n.
pec·ca·dil·lo (pek′ə·dil′ō) n. pl. **·los** or **·loes** A slight or trifling sin; a fault. See synonyms under FOIBLE. [<Sp. *pecadillo*, dim. of *pecado* a sin <L *peccatum*, orig. pp. of *peccare* sin]
pec·cant (pek′ənt) adj. 1 Guilty of sin; sinful; offending. 2 Corrupt and offensive; diseased. 3 Violating some rule or principle. [<L *peccans*, *-antis*, ppr. of *peccare* sin] — **pec′can·cy** n. — **pec′cant·ly** adv.
pec·ca·ry (pek′ər·ē) n. pl. **·ries** Either of two pugnacious hoglike ungulates of Central and South America, secreting an oily, musky substance, the **collared peccary** (*Pecari angulatus*), or the **white-lipped peccary** (*Tayassus pecari*). [<Sp. *pecari* <Cariban *pakira*]

PECCARY
(From 1 1/4 to 1 1/2 feet high at the shoulder)

pec·ca·vi (pe·kā′vī, -kä′vē) n. Latin A confession of guilt; literally, I have sinned.
pech (pekh) v.i. *Scot. & Brit. Dial.* To pant; puff. — **pech′in** n.
pech·an (pekh′an) n. *Scot.* The gullet; stomach.
Pe·chen·ga (pə·cheng′gə) A port of Russian S.F.S.R. on the Barents Sea, near the Norwegian border. Finnish **Pet·sa·mo** (pet′sə·mō).
Pe·cho·ra (pə·chôr′ə) A river of NE European U.S.S.R., flowing 1,110 miles north from the Urals to the Barents Sea.
peck[1] (pek) v.t. 1 To strike with the beak, as a bird does, or with something pointed. 2 To make by striking thus: to *peck* a hole in a wall. 3 To pick up, as food, with the beak. — *v.i.* 4 To make strokes with the beak or with something pointed. — *n.* 1 A quick, sharp blow, as with a beak or something pointed. 2 A mark, dent, or hole made by such a blow. [Var. of PICK] — **peck′er** n.
peck[2] (pek) n. 1 A measure of capacity: the fourth of a bushel, 8 quarts, or 8.8 liters. See DRY MEASURE. 2 A vessel for measuring a peck. 3 *Slang* A great quantity. [<OF *pek*, a measure of oats for horses; ult. origin uncertain]
peck·er·wood (pek′ər·wood′) n. *U.S. Dial.* 1 A woodpecker. 2 A poor white.
peck·or·der (pek′ôr′dər) n. A hierarchy of social privilege and status among the members of a flock of chickens, established by the enforced right of the more aggressive hens or cocks to peck at, harass, and dominate all those lower in the scale.
Peck·sniff (pek′snif), **Seth** An unctuous, canting hypocrite in Dickens' *Martin Chuzzlewit*.
Peck·sniff·i·an (pek·snif′ē·ən) adj. Suggestive of Seth Pecksniff; hypocritical; insincere.
Pe·co·ra (pə·kôr′ə, -kō′rə), **Ferdinand**, born 1882, U.S. jurist.
Pe·cos Bill (pā′kōs) A legendary cowboy of the American West, who was raised by coyotes and performed many fantastic feats, such as digging the Rio Grande.
Pecos River (pā′kəs, -kōs) The principal tributary of the Rio Grande, flowing 926 miles from north central New Mexico to western Texas.
Pécs (pāch) A city of SW Hungary: German *Fünfkirchen*.
pec·tase (pek′tās) n. *Biochem.* An enzyme obtained from fruits which combines with pectin to yield pectic acid. [<PECT(IN) + (DIAST)ASE]
pec·tate (pek′tāt) n. *Chem.* A salt or ester of pectic acid. [<PECT(IC) + -ATE[3]]
pec·ten (pek′tən) n. pl. **·ti·nes** (-tə·nēz) *Zool.* 1 A comb, or comblike part or process; specifically, in birds and reptiles, a vascular pigmented membrane of the eyeball. 2 A scallop. 3 *Anat.* The pubic bone. [<L, a comb, a scallop <*pectere* comb, dress the

hair] — **pec′ti·nate, pec′ti·nat′ed** *adj.* — **pec′ti·na′tion** *n.*

pec·tic (pek′tik) *adj.* Of, pertaining to, or derived from pectin. [<Gk. *pēktikos* < *pēktos* congealed < *pēgnynai* make solid]

pectic acid *Chem.* Any of a group of compounds derived from pectin by the hydrolysis of methyl ester.

pec·tin (pek′tin) *n. Biochem.* Any of a class of compounds of high molecular weight contained in the cell walls of various fruits and vegetables, as apples, lemons, or carrots: it is the basis of fruit jellies. [<PECT(IC) + -IN]

pec·tize (pek′tīz) *v.t.* & *v.i.* **·tized, ·tiz·ing** To coagulate. [<Gk. *pēktos* congealed + -IZE]

pec·to·ral (pek′tər·əl) *adj.* **1** Of or pertaining to the breast or thorax. **2** As if proceeding from the breast or inner consciousness; more especially, of an emotional character: *pectoral* theology. **3** Adapted to, efficacious in, or designed for relieving or curing diseases of the lungs or chest. — *n.* **1** An ornament worn on the breast; especially, the **pectoral cross** worn on the breast by bishops, abbots, etc. **2** A pectoral organ, fin, or muscle. **3** Any medicine for ailments of the chest. [<L *pectoralis* < *pectus, -oris* the breast]

pectoral arch or **girdle** *Anat.* **1** The arch formed by the collar bone and shoulder blade in man. **2** That part of the skeleton with which the forelimbs of a vertebrate animal are articulated.

pectoral fin *Zool.* One of the anterior paired fins of fishes, homologous with the anterior limb of higher vertebrates.

pectoral sandpiper An American sandpiper (*Pisobia melanotos*), occasional in Europe, having a buff–gray breast with dusky streaks: also called *grass snipe.*

pec·u·late (pek′yə·lāt) *v.t.* & *v.i.* **·lat·ed, ·lat·ing** To steal or appropriate wrongfully (funds, especially public funds) entrusted to one's care; embezzle. [<L *peculatus*, pp. of *peculari* embezzle < *peculium.* See PECULIUM.] — **pec′u·la′tion** *n.* — **pec′u·la′tor** *n.*

pe·cul·iar (pi·kyo͞ol′yər) *adj.* **1** Having a character exclusively its own; unlike anything else or anything of the same class or kind; specific; particular. **2** Singular; odd; strange. **3** Select or special; separate; distinguished. **4** Belonging particularly or exclusively to one. — *n.* **1** A person or thing that is peculiar; formerly, any private possession. **2** A member of the sect known as the Peculiar People. See synonyms under EXTRAORDINARY, ODD, PARTICULAR, QUEER, RARE[1]. [<MF *peculier* <L *peculiaris* < *peculium.* See PECULIUM.] — **pe·cul′iar·ly** *adv.*

pe·cu·li·ar·i·ty (pi·kyo͞o′lē·ar′ə·tē) *n. pl.* **·ties** **1** A characteristic. **2** The quality of being peculiar. See synonyms under CHARACTERISTIC.

Peculiar People **1** A denomination of Christians, founded in England in 1838, who hold that sinless perfection is immediately obtainable by those willing to seek and accept it. **2** In the Scripture, the Jews, as being God's chosen people and separated from the rest of mankind. *Deut.* xxvi 18.

pe·cu·li·um (pi·kyo͞o′lē·əm) *n.* In Roman law, property that a slave, a wife, or a child was permitted to hold as his own. [<L, private property, orig. one's cattle < *pecus* cattle, money]

pe·cu·ni·ar·y (pi·kyo͞o′nē·er′ē) *adj.* **1** Consisting of or relating to money; monetary. **2** Having a monetary penalty; entailing a fine. See synonyms under FINANCIAL. [<L *pecuniarius* < *pecunia* money < *pecus* cattle]

ped–[1] Var. of PEDI–.

ped–[2] Var. of PEDO–.

–ped Var. of –PEDE.

ped·a·gog (ped′ə·gog, -gôg) *n.* **1** A schoolmaster; especially, a pedantic, narrow–minded teacher. **2** In ancient Greece and Rome, a slave who attended children to school. Also **ped′a·gogue.** [<OF *pedagoge* <L *paedagogus* <Gk. *paidagōgos* < *pais, paidos* a child + *agōgos* a leader < *agein* lead]

ped·a·gog·ic (ped′ə·goj′ik, -gō′jik) *adj.* **1** Pertaining to the science or art of teaching. **2** Of or belonging to a pedagog; affected with a conceit of learning. Also **ped′a·gog′i·cal.** — **ped′a·gog′i·cal·ly** *adv.*

ped·a·gog·ics (ped′ə·goj′iks) *n. pl.* (*construed as singular*) The science and art of teaching; pedagogy.

ped·a·gog·ism (ped′ə·gog′iz·əm, -gôg′-) *n.* The nature, character, or business of teachers or a teacher. Also **ped′a·gogu′ism.**

ped·a·go·gy (ped′ə·gō′jē, -goj′ē) *n.* The science or profession of teaching; also, the theory or the teaching of how to teach.

ped·al (ped′l) *adj.* **1** Of or pertaining to a foot, feet, or a footlike part. **2** Of or pertaining to a pedal. — *n. Mech.* A lever operated by the foot, differing from a treadle in that it is usually applied only to musical instruments, cycles, sewing machines, and light machinery. — *v.t.* & *v.i.* **·aled** or **·alled, ·al·ing** or **·al·ling** To move or operate by working pedals; use the pedals (of). ◆ Homophone: *peddle.* [<L *pedalis* < *pes, pedis* the foot]

pedal curve *Math.* The curve traced by the foot of a perpendicular drawn from a fixed point to a variable tangent to a given curve: the curve so traced is called the pedal of the curve with respect to the point.

ped·al·fer (ped·al′fər) *n.* A type of soil characterized by a downward shifting of alumina and iron oxide, with an absence of carbonate of lime accumulations. Compare PEDOCAL. [<Gk. *ped(on)* ground + AL(UMINA) + L *fer(rum)* iron]

ped·a·lier (ped′ə·lir′) *n.* A pedal keyboard for a pianoforte. [<F *pédalier* < *pédale* <L *pedalis.* See PEDAL.]

pedal point *Music* A tonic or dominant note sustained (usually in the bass) while the other parts proceed with varying harmonies. Also **pedal note.**

ped·ant (ped′ənt) *n.* **1** A scholar who makes needless and inopportune display of his learning, or who insists upon the importance of trifling points of scholarship. **2** *Obs.* A schoolmaster; teacher. [<F *pédant* <Ital. *pedante*, prob. <Med. L *paedagogans, -antis,* ppr. of *paedagogare* teach <L *paedagogus.* See PEDAGOG.] — **pe·dan·tic** (pi·dan′tik) *adj.* — **pe·dan′ti·cal·ly** *adv.*

ped·ant·ry (ped′ən·trē) *n. pl.* **·ries** **1** Ostentatious display of knowledge. **2** Undue and slavish adherence to forms or rules.

ped·ate (ped′āt) *adj.* **1** Resembling or having the functions of a foot; having feet. **2** *Bot.* Palmately divided or parted, the lateral divisions being subdivided: said especially of leaves. [<L *pedatus* having feet < *pes, pedis* a foot] — **ped′ate·ly** *adv.*

pedati– *combining form Bot.* Pedately: *pedatifid.* [<L *pedatus* having feet < *pes, pedis* foot]

pe·dat·i·fid (pi·dat′ə·fid, -dā′tə-) *adj. Bot.* Having the subdivisions of a simple leaf, which is pedately nerved, extending half–way to the base.

pe·dat·i·sect (pi·dat′ə·sekt, -dā′tə-) *adj. Bot.* Pedate and cleft almost to the midrib: said of leaves. Also **pe·dat′i·sect′ed** (-sek′tid).

ped·dle (ped′l) *v.* **·dled, ·dling** *v.i.* **1** To travel about selling small wares. **2** To occupy oneself with trifles; piddle. — *v.t.* **3** To carry about and sell in small quantities. **4** To sell or dispense in small quantities. ◆ Homophone: *pedal.* [Appar. back formation <ME *ped(d)ler(e)* a peddler; infl. by PIDDLE]

ped·dler (ped′lər) *n.* One who peddles; a hawker: also spelled *pedlar, pedler.* [ME *pedlere,* ? alter. of *pedder* a peddler < *ped* a basket] — **ped′dler·y** *n.*

ped·dling (ped′ling) *adj.* Small; trifling; piddling.

–pede *combining form* Footed: *centipede.* Also spelled *–ped,* as in *quadruped.* [<L *pes, pedis* foot]

Pe·der (pā′thər) Danish form of PETER.

ped·er·ast (ped′ə·rast, pē′də-) *n.* One addicted to pederasty: also spelled *paederast.*

ped·er·as·ty (ped′ə·ras′tē, pē′də-) *n.* Sodomy, especially as practiced between men and boys. [<NL *paederastia* <Gk. *paiderastia* < *paiderastēs* a lover of boys < *pais, paidos* a boy + *erastēs* a lover < *eraein* love] — **ped′er·as′tic** *adj.* — **ped′er·as′ti·cal·ly** *adv.*

pe·des (pē′dēz) *Latin* Plural of PES.

ped·es·tal (ped′is·təl) *n.* **1** A base or support for a column, statue, or vase. **2** Hence, any foundation, base, or support, either material or immaterial. — **to put on a pedestal** To hold in high estimation; to put in the position of an idol or hero. [<MF *pédesta;* <Ital. *piedestallo* < *piè di stallo* < *piè, pied* foot (<L *pes, pedis*) + *di* of (<L *de*) + *stallo* a stall, standing place <OHG *stal*]

pedestal table A table whose top is supported by a central column with three or more spreading feet.

pe·des·tri·an (pə·des′trē·ən) *adj.* **1** Moving on foot; walking; pertaining to walking. **2** Pertaining to common people; plebeian. **3** Hence, commonplace, prosaic, or dull, as prose or mechanical verse. — *n.* One who journeys or moves from place to place on foot; a walker. [<L *pedester, -tris* on foot < *pes, pedis* a foot] — **pe·des′tri·an·ism** *n.*

pedi–[1] *combining form* Foot; related to the foot or feet: *pedicure.* Also, before vowels, *ped–.* [<L *pes, pedis* foot]

pedi–[2] *combining form* Pedo–. [<Gk. *pais, paidos* a child]

pe·di·a·tri·cian (pē′dē·ə·trish′ən, ped′ē-) *n.* A specialist in children's diseases. Also **pe′di·at′·rist.**

pe·di·at·rics (pē′dē·at′riks, ped′ē-) *n.* That branch of medicine dealing with the diseases and hygienic care of children: also spelled *paediatrics.* [<Gk. *pais, paidos* a child + -IATRICS] — **pe′di·at′ric** *adj.*

ped·i·cel (ped′ə·səl) *n.* **1** *Bot.* **a** The stalk supporting a single flower in an inflorescence composed of flowers arranged upon a common peduncle. **b** A small or delicate support of various special organs, as of a sporangium in ferns or a capsule in mosses. **2** *Entomol.* A stalk or supporting part, as the second segment of the antenna of an insect, between the scape and the funicle. **3** *Zool.* An ambulacral sucker in an echinoderm. [<NL *pedicellus,* dim. of L *pediculus,* dim. of *pes, pedis* a foot] — **ped′i·cel′lar** (-sel′ər) *adj.*

ped·i·cel·late (ped′ə·sə·lit, -lāt′) *adj.* Borne on or having a pedicel.

pe·dic·u·lar (pə·dik′yə·lər) *adj.* Of, pertaining to, or infested with lice. [<L *pedicularis* < *pediculus,* dim. of *pedis* a louse]

pe·dic·u·late (pə·dik′yə·lit, -lāt) *adj.* Pertaining or belonging to an order (*Pediculati*) of teleost fishes having a spinous dorsal fin with the front ray adapted as a lure, and including the broad, soft–bodied angler fish. [<L *pediculus* footstalk + -ATE[1]]

pe·dic·u·lo·sis (pə·dik′yə·lō′sis) *n. Pathol.* The condition of being infested with lice; lousiness; phthiriasis. [<L *pediculus* louse + -OSIS] — **pe·dic′u·lous** *adj.*

ped·i·cure (ped′i·kyo͝or) *n.* **1** The care of the feet; the surgical treatment of corns, bunions, etc. **2** A chiropodist. **3** The cosmetic treatment of the feet and toenails. — *v.t.* **·cured, ·cur·ing** To treat (the feet) for corns, bunions, etc. [<L *pes, pedis* a foot + *cura* care] — **ped′·i·cur′ist** *n.*

ped·i·form (ped′i·fôrm) *adj.* Resembling a foot in shape.

ped·i·gree (ped′ə·grē) *n.* **1** A line of ancestors; lineage. **2** A list or table of descent and relationship; a genealogical register, especially of an animal of pure breed. [<MF *pié de grue* a crane's foot; from a three line mark denoting succession in pedigrees]

ped·i·greed (ped′ə·grēd) *adj.* Having a pedigree; of notable ancestry.

PEDIMENT
Pediment of the U. S. Supreme Court, Washington, D.C.

ped·i·ment (ped′ə·mənt) *n. Archit.* **1** A broad triangular part above a portico or door. **2** Any similar piece with a long base surmounting a door, screen, bookcase, etc. [Earlier *periment,* prob. alter. of PYRAMID; infl. in form by *pes, pedis* a foot] — **ped′i·men′tal** (-men′təl) *adj.*

ped·i·palp (ped′i·palp) *n. Zool.* **1** One of the second pair of appendages at the sides of the mouth in arachnids, terminally pincerlike, as in scorpions; long and leglike, as in solpugids;

or leglike with the terminal joint serving to convey the semen in copulation, as in male spiders. 2 One of the *Pedipalpi.* —**ped'i·pal'pous** *adj.*

Ped·i·pal·pi (ped'i·pal'pī) *n. pl.* An order of arachnids with segmented abdomen; whip scorpions. [<NL <L *pes, pedis* foot + *palpus* feeler]

ped·lar (ped'lər), **ped·ler,** etc. See PEDDLER.

pedo– *combining form* Child; children; offspring: also, before vowels, *ped–,* as in *pedagogy.* Also spelled *paedo–.* [<Gk. *pais, paidos* a child]

pe·do·bap·tism (pē'dō·bap'tiz·əm) *n.* Infant baptism, and the system that sustains it. —**pe'do·bap'tist** *n.*

ped·o·cal (ped'ō·kal) *n.* A type of soil characterized by an accumulation of carbonates of calcium or of calcium and magnesium. Compare PEDALFER. [<Gk. *pedon* ground + CAL(CIUM)]

pe·do·gen·e·sis (pē'dō·jen'ə·sis) *n. Entomol.* Reproduction in the sexually immature or larval stage, as in certain insects.

pe·dol·o·gy[1] (pi·dol'ə·jē) *n.* The scientific study of the development and behavior of children. [<PEDO- + -LOGY] —**ped·o·log·i·cal** (ped'ə·loj'i·kəl) *adj.* —**ped'o·log'i·cal·ly** *adv.* —**pe·dol'o·gist** *n.*

pe·dol·o·gy[2] (pi·dol'ə·jē) *n.* The science that treats of the origin, nature, and properties of soils, especially in their more fundamental aspects. [<Gk. *pedon* ground + -LOGY] —**pe·dol'o·gist** *n.*

pe·dom·e·ter (pi·dom'ə·tər) *n.* An instrument that measures distance traveled by recording the number of steps taken by the person who carries it. [<F *pédomètre* <L *pes, pedis* foot + Gk. *metron* measure]

ped·o·sphere (ped'ə·sfir) *n.* The soil-bearing layer of the earth's surface. [<Gk. *pedon* ground + -SPHERE]

ped·re·gal (ped'rə·gäl') *n.* In Mexico and SW United States, a rough, rocky tract of land, especially in a volcanic region; a lava field. [<Sp. <*piedra* a stone]

pe·dro (pē'drō) *n. pl.* **·dros** 1 The five of trumps in the card game of cinch. 2 A game of pitch in which the five of trumps counts five. [<Sp. *Pedro* Peter]

Pe·dro (pā'thrō) Portuguese and Spanish form of PETER.

ped·ule (pej'o͞ol) *n.* A leg covering of flexible leather, flannel, or other material, worn in ancient and medieval times. [<L, neut. of *pedulis* of or for the feet <*pes, pedis* a foot]

pe·dun·cle (pi·dung'kəl) *n.* 1 *Bot.* The general stalk or support of an inflorescence. 2 *Anat.* A stalk or stem, as for the attachment of an organ or organism: the *peduncles* of the brain. [<NL *pedunculus* a footstalk, dim. of *pes, pedis* a foot] —**pe·dun'cled, pe·dun'cu·lar** *adj.*

pe·dun·cu·late (pi·dung'kyə·lit, -lāt) *adj.* Borne on or having a peduncle. Also **pe·dun'cu·lat'ed.**

pee·been (pē'bēn) *n.* A large hardwood evergreen tree (*Syncarpia hillii*) of the myrtle family, native to Australia: also called *turpentine tree.*

Pee·bles (pē'bəlz) A county in SE Scotland; 347 square miles; county town, Peebles: also *Tweeddale.* Also **Pee'bles·shire** (-shir).

Pee Dee River (pē dē) A river in southern North Carolina and NE South Carolina, flowing 233 miles SE to Winyah Bay.

peek (pēk) *v.i.* To look furtively, slyly, or quickly; peep. —*n.* A peep, glance. ◆ Homophones: *peak, pique.* [ME *piken,* ? var. of *kiken* peer; infl. by PEEP[2]]

peek–a–boo (pē'kə·bo͞o') *n.* A children's game in which one hides (one's face) and calls out "peek-a-boo!" or "Bo-peep!"

peel[1] (pēl) *n.* The natural coating of certain kinds of fruit, as oranges and lemons; skin; rind. —*v.t.* 1 To strip off the bark, skin, etc., of. 2 To strip off; remove. —*v.i.* 3 To lose bark, skin, etc. 4 To come off: said of bark, skin, etc. 5 *Slang* To undress. —**to keep one's eye peeled** *Colloq.* To keep watch; be alert. —**to peel off** *Aeron.* To veer off from a flight formation so as to dive or prepare for a landing. ◆ Homophone: *peal.* [Var. of earlier *pill* a skin, covering; infl. by F *peler* strip of skin]

peel[2] (pēl) *n.* 1 A broad, thin, long-handled, shovel-like implement used by bakers in moving bread, etc., about an oven. 2 The blade or broad part of an oar. ◆ Homophone: *peal.* [<OF *pele* <L *pala* a spade]

peel[3] (pēl) *n.* 1 A square stronghold or tower of the 16th century, especially on the borders of Scotland and England: also **peel'–house'** (-hous'). 2 *Obs.* A stake; stockade; palisade. ◆ Homophone: *peal.* [<AF *pel* <L *palus* a stake]

PEEL
Vaulted lower story often used as a stable; living quarters above.

Peel (pēl) A port and resort on the west coast of the Isle of Man; site of an ancient castle.

Peel (pēl), **Sir Robert,** 1788–1850, English statesman; introduced police reforms.

Peele (pēl), **George,** 1558?–97?, English playwright and poet.

peel·ing (pē'ling) *n.* Something peeled off; a strip of rind, bark, skin, or outer layer.

Peel River (pēl) A river in Northwest Territories, Canada, flowing 365 miles north to the Mackenzie River.

peel·er (pē'lər) *n. Brit. Slang* A policeman. [after Sir Robert *Peel*]

peen (pēn) *n.* The end of a hammer head opposite the face: usually shaped for indenting, riveting, chipping, etc., as when straight, pointed, conical, hemispherical, or wedge-shaped: ball-*peen,* cross-*peen,* etc. See illustration under HAMMER. —*v.t.* To beat, bend, or shape with the peen. Also spelled *pane, pean, pein.* [Appar. var. of PANE[2], ? infl. by Scand. Cf. Norw. *paenn* the sharp end of a hammer.]

peenge (pēnj) *v.i. Scot. & Brit. Dial.* To complain.

peep[1] (pēp) *v.i.* 1 To utter the small, sharp cry of a young bird or chick; chirp; cheep. 2 To speak in a weak, small voice. —*n.* 1 The cry of a chick or small bird, or of a young frog; chirp. 2 A small sandpiper; especially, the least sandpiper; sandpeep. [ME *pepen,* var. of *pipen* PIPE]

peep[2] (pēp) *v.i.* 1 To look through a small hole, from concealment, etc.; look furtively or quickly; peek. 2 To begin to appear; be just visible. —*v.t.* 3 *Rare* To cause to stick out slightly. —*n.* 1 A furtive look; a glimpse or glance. 2 An aperture or crevice through which one may look; peephole. 3 The earliest appearance: the *peep* of day. [ME *pepen.* ? Akin to ME *piken* PEEK.]

peep[3] (pēp) *n. Slang* A jeep; also, sometimes, a larger command car. [Alter. of JEEP]

peep·er[1] (pē'pər) *n.* An animal that peeps or makes a chirping sound, especially a very young chick or any of several tree frogs.

peep·er[2] (pē'pər) *n.* 1 One who peeps or peeks; a spying person. 2 *Slang* An eye.

peep·hole (pēp'hōl') *n.* An aperture, as a hole or crack, through which one may peep; also, a small window in a door.

peep·ing–tom (pē'ping·tom') *n.* An overly inquisitive or pruriently prying person, especially one who peeps in at windows.

Peeping Tom of Coventry In British legend, a curious tailor who peeped at Lady Godiva during her ride through Coventry and was struck blind.

peep·show (pēp'shō') *n.* An exhibition of pictures or other objects viewed through a small orifice fitted with a magnifying lens.

peep sight *Mil.* An adjustable plate on the breech of a gun or cannon, having in its center a small orifice through which an aim can be taken with great accuracy by centering the front sight therein.

pee·pul (pē'pəl) See PIPAL.

peer[1] (pir) *v.i.* 1 To look narrowly or searchingly, as in an effort to see clearly. 2 To come partially into view: The sun *peers* over the horizon. 3 *Poetic* To appear. ◆ Homophone: *pier.* [Cf. obs. *pear, pere,* aphetic var. of APPEAR]

peer[2] (pir) *n.* 1 An equal, as in natural gifts or in social rank. 2 An equal before the law. 3 A noble; especially, a member of a hereditary legislative body. In the United Kingdom, a duke, marquis, earl, viscount, or baron; also, an archbishop or a bishop having a seat in the House of Lords. Until 1922 peers were of three classes: **Peers of the United Kingdom,** all of whom sit in the House of Lords, **Peers of Scotland,** and **Peers of Ireland.** 4 *Obs.* A companion; mate; associate; also, rival. See synonyms under ASSOCIATE. —**House of Peers** *Brit.* The House of Lords. ◆ Homophone: *pier.* [<OF *per* <L *par* equal]

peer[3] (pir) *n. Scot.* 1 A pear. 2 A peg top. ◆ Homophone: *pier.*

peer·age (pir'ij) *n.* 1 In England, the office or rank of a peer of the realm, or nobleman. 2 Peers collectively; the nobility. 3 A book containing a genealogical list of the nobility.

peer·ess (pir'is) *n.* A woman who holds a title of nobility, either in her own right or by marriage with a peer.

peer·less (pir'lis) *adj.* Of unequaled excellence. —**peer'less·ly** *adv.* —**peer'less·ness** *n.*

peer of the blood royal A member of the royal family who is entitled to sit as a member of the House of Lords.

peer of the realm One of the lords of Parliament.

peer·y (pir'ē) *n. pl.* **peer·ies** *Scot.* A child's top spun with a string. Also **peer'ie.**

pees·weep (pēz'wēp') *n. Scot.* The pewit or lapwing.

peet·weet (pēt'wēt) *n.* The spotted sandpiper. [Imit.]

peeve (pēv) *v.t.* **peeved, peev·ing** *Colloq.* To make peevish. —*n. Colloq.* A complaint; grievance. [Back formation <PEEVISH]

peeved (pēvd) *adj.* Vexed; discontented; disagreeable.

pee·vish (pē'vish) *adj.* 1 Irritable or querulous; fretful; cross. 2 Showing or marked by petulant discontent and vexation. See synonyms under FRETFUL. [ME *pevische;* origin unknown] —**pee'vish·ly** *adv.* —**pee'vish·ness** *n.*

pee·wee (pē'wē) *n.* 1 The pewee. 2 *Colloq.* Anything or anyone especially small or diminutive. —*adj.* Tiny; insignificant. [Prob. <Algonquian (Massachuset) *pewe* little]

peg (peg) *n.* 1 A wooden pin used for fastening articles together or for holding fast the end of a string and adjusting its tension in a musical instrument. 2 A projecting wooden pin upon which something may be fastened or hung, or which may serve to mark a boundary. 3 Hence, a reason or excuse for an action: a *peg* to hang an argument upon. 4 A degree or step, as in rank or estimation. 5 *Brit.* A drink of brandy and soda or of whisky and soda. 6 *Colloq.* A leg, often one of wood. —**to take (one) down a peg** To lower the self-esteem of (a person), as by humiliating. —*v.* **pegged, peg·ging** *v.t.* 1 To drive or force a peg into; fasten with pegs. 2 To mark or designate with pegs. 3 To strike or pierce with a peg or sharp instrument. 4 *Colloq.* To throw: to *peg* stones. —*v.i.* 5 To work or strive hard and perseveringly: usually with *away.* 6 In croquet, to hit a peg. 7 In cribbage, etc., to mark the score with pegs. [ME *pegge* <LG; cf. MDu. *pegge*]

Peg (peg), **Peg·gy** (peg'ē) Diminutive of MARGARET.

Peg·a·sus (peg'ə·səs) 1 In Greek mythology, a winged horse, sprung from the blood of Medusa, a blow of whose hoof caused the fountain of poetic inspiration, Hippocrene, to spring from Mount Helicon; hence, poetic inspiration. See BELLEROPHON. 2 *Astron.* A northern constellation. See CONSTELLATION. [<L <Gk. *Pēgasos,* after *pēgai* the springs of Ocean, where Medusa was killed]

peg leg *Colloq.* 1 An artificial leg of rodlike or tapering shape. 2 A person with such a leg.

peg·ma·tite (peg'mə·tīt) *n.* A very coarse-grained granitic rock composed chiefly of orthoclase, quartz, and mica (usually muscovite); graphic granite: often occurs in veins or dikes. [<Gk. *pēgma, -atos* a solid mass + -ITE[1]] —**peg'ma·tit'ic** (-tit'ik) *adj.*

peg–top (peg'top') *adj.* Having the shape of a peg top: applied especially to trousers wide at the hip and narrow at the ankle.

peg top A child's wooden spinning top, pear-shaped and having a sharp metal peg.

Pe·gu (pe·go͞o') An administrative division of Lower Burma east of the Irrawaddy; 20,223 square miles; capital, Rangoon.

Pe·gu·an (pe·go͞o'ən) *n.* Mon.

Peh·le·vi (pā'lə·vē) See PAHLAVI.

pei·gnoir (pān·wär′, pān′wär) *n.* A loose dressing robe worn by women; a bathrobe; a negligée. [<F <*peigner* comb <L *pectinare* < *pecten* a comb]

Pei Ho (bā′ hō′) The Chinese name for the PAI.

pein (pēn) See PEEN.

pei·no·ther·a·py (pī′nō·ther′ə·pē) *n. Med.* The treatment of disease by severe fasting and starvation; the hunger cure. [<Gk. *peina* hunger + THERAPY]

Pei·ping (bā′ping′) A former name for PEKING.

Pei·pus (pī′pəs) A lake on the boundary between Estonia and the U.S.S.R.; 1,356 square miles: Russian *Chudskoe.* Estonian **Peip·si** (pāp′sē).

Pei·rai·evs (pē′rē·efs′) The Greek name for PIRAEUS.

Peirce (pûrs), **Charles Sanders**, 1839–1914, U.S. mathematician and logician; founder of philosophical pragmatism.

Peirse (pirs), **Sir Richard Edmund**, born 1892, British air marshal.

Pei·xot·to (pā·shō′tō), **Ernest Clifford**, 1869–1940, U.S. painter and illustrator.

pe·jo·ra·tion (pē′jə·rā′shən, pej′ə-) *n.* **1** A worsening; deterioration. **2** *Ling.* A degeneration or lowering in the meaning of a word, as in *silly* (formerly "blessed"): opposed to *melioration*.

pe·jo·ra·tive (pē′jə·rā′tiv, pej′ə-, pi·jôr′ə·tiv, -jor′-) *adj.* Giving a deteriorating effect or meaning, as to the sense of a word. — *n.* A word expressing depreciation. [<L *pejoratus*, pp. of *pejorare* make worse < *pejor* worse] — **pe′jo·ra·tive·ly** *adv.*

Pe·ka·long·an (pē′kä′long′än) An Indonesian port on the north central coast of Java.

pek·an (pek′ən) *n.* A North American carnivore; the fisher. [<dial. F (Canadian) <Algonquian *pékané*]

pe·kin (pē′kin′) *n.* A silk fabric, usually figured or striped. [<F *pékin* <*Pékin* Peking, China]

Pe·king (pē′king′, Chinese bā′jing′) A city in northern Hopeh province, the capital of the People's Republic of China; from 1928 to 1949 known as *Peiping*.

Pe·king·ese (pē′king·ēz′) *n. pl.* **·ese** **1** A native or inhabitant of Peking. **2** The dialect spoken in Peking. **3** (pē′kə·nēz′) A Pekingese dog. — *adj.* Of or pertaining to Peking. Also **Pe′kin·ese′**.

Pekingese dog A variety of the Chinese (or Pekingese) pug, with long silky hair, especially upon the ears, diminutive snub nose, and short legs.

PEKINGESE DOG
(From 7 to 10 inches high at the shoulder)

Peking lacquer Chinese carved lacquer.

Peking man *Paleontol.* Sinanthropus.

pe·koe (pē′kō, *Brit.* pek′ō) *n.* A superior kind of black tea, made from the downy tips of the young buds of the tea plant. [<dial. Chinese (Amoy) *pek-ho* <*pek* white + *ho* down]

pel·age (pel′ij) *n.* The coat or covering of a mammal, as of fur, wool, etc. [<MF <OF *peil, pel* hair <L *pilus*]

pe·la·gi·an (pə·lā′jē·ən) *adj.* Pelagic. — *n.* A deep-sea animal. [<L *pelagius* of the sea <Gk. *pelagios* <*pelagos* the sea]

Pe·la·gi·an·ism (pə·lā′jē·ən·iz′əm) *n. Theol.* The body of doctrines held by the followers of Pelagius, who denied original sin, confined grace to forgiveness, and affirmed that man's unaided will is capable of spiritual good. — **Pe·la′gi·an** *n. & adj.*

pe·lag·ic (pə·laj′ik) *adj.* **1** Of, pertaining to, or inhabiting the sea far from land; oceanic. **2** Living on or near the surface of the ocean. **3** Conducted or operating on the open sea: *pelagic* sealing or sealers. [<L *pelagicus* <Gk. *pelagikos* <*pelagos* the sea]

Pe·la·gie Islands (pə·lā′jē) A Sicilian island group in the Mediterranean between Tunis and Malta; 10 square miles.

Pe·la·gi·us (pə·lā′jē·əs), 360?–420, British monk. See PELAGIANISM.

pelargonic acid *Chem.* A colorless compound, $CH_3(CH_2)_7 \cdot COOH$, liquid at ordinary temperatures, obtained as an ester from oil of pelargonium: also called *nonanoic acid*.

pel·ar·go·ni·um (pel′är·gō′nē·əm) *n.* Any of a large genus (*Pelargonium*) of strong-scented, ornamental evergreen herbs or shrubs, generally known in cultivation as *geraniums*, having entire lobed or dissected leaves, and handsome, variously colored flowers. [<NL <Gk. *pelargos* a stork < *pelos* blackish + *argos* shining] — **pel′ar·gon′ic** (-gon′ik) *adj.*

Pe·las·gi (pə·laz′jī) *n. pl.* A primitive, seafaring people who inhabited the coasts of Greece, Asia Minor, Crete, Thrace, etc.: mentioned by ancient Greek writers as the pre-Greek inhabitants of the eastern Mediterranean region. Also **Pe·las′gi·ans**. — **Pe·las′gi·an** *adj. & n.* — **Pe·las′gic** *adj.*

Pe·le (pā′lā) In Polynesian mythology, the goddess of volcanoes.

Pe·lée (pə·lā′), **Mont** A volcano in northern Martinique; 4,429 feet; last eruption, 1902.

pel·er·ine (pel′ə·rēn′) *n.* A waist-length cape worn by women. [<F *pèlerine*, fem. of *pèlerin*. See PILGRIM.]

Pe·le's hair (pā′lāz) Volcanic glass drawn out into long, fine threads by ejected driblets of fused lava; capillary volcanic glass. [after PELE]

Pe·leus (pēl′yōos, pē′lē·əs) In Greek legend, a king of the Myrmidons and father of Achilles.

Pe·lew Islands (pē·lōō′) See PALAU ISLANDS.

pelf (pelf) *n.* Money; wealth: often implying ill-gotten gains. See synonyms under WEALTH. [<AF *peufe*, OF *pelfre* spoil; ult. origin uncertain]

Pe·li·as (pē′lē·əs, pel′ē·əs) In Greek mythology, a son of Poseidon and king of Iolcus, who sent his nephew Jason to get the Golden Fleece; after Jason's return, Medea caused Pelias' death by persuading his daughters to cut their father into little pieces and stew him as a means of restoring his youth.

pel·i·can (pel′i·kən) *n.* A large, gregarious, fish-eating, web-footed bird (genus *Pelecanus*) of warm regions, having a distensible membranous pouch on the lower jaw for the temporary storage of fish. [OE *pellican* <LL *pelicanus, pelecanus* <Gk. *pelekan,* ? < *pelekys* an ax; in ref. to its bill]

Pelican State Nickname of LOUISIANA.

Pel·i·des (pel′ə·dēz) Patronymic of Achilles. See PELEUS.

Pe·li·on (pē′lē·on) A mountain range on the coast of SE Thessaly; highest peak, 5,252 feet. See OSSA.

WHITE PELICAN
(From 4 to 5 feet high according to species)

pe·lisse (pə·lēs′) *n.* A long outer garment or cloak: originally one of fur or lined with fur. [<F <Med. L *pellicia* <L, a garment of skins or fur < *pellis* skin]

pe·lite (pē′līt) *n.* A sedimentary rock composed of clay, quartz particles, or rock flour: also called *argillite.* [<Gk. *pelos* clay + -ITE¹] — **pe·lit·ic** (pi·lit′ik) *adj.*

Pel·la (pel′ə) A city of ancient Greece; birthplace of Alexander the Great and capital of Macedonia.

pel·la·gra (pə·lā′grə, -lag′rə) *n. Pathol.* A disease characterized by gastric disturbance, skin eruptions, and nervous derangement; endemic in many parts of the world, and known to be caused by nutritional deficiencies. [<NL prob. <Ital. *pelle agra* rough skin; ? infl. by Gk. -*agra* <*agra* a seizure, as in *podagra* gout, lit., a seizure in the feet] — **pel·la′grous** *adj.*

pel·la·grin (pə·lā′grin, -lag′rin) *n.* A sufferer from pellagra.

Pel·le·as (pel′ē·əs), **Sir** In Arthurian legend, one of the Knights of the Round Table.

pel·let (pel′it) *n.* **1** A small round ball or imitation projectile, as of wax, paper, bread, etc. **2** A small shot. **3** A very small pill. **4** A slingstone; also, a bullet; cannonball. — *v.t.* **1** To make into pellets. **2** To strike with pellets. [<OF *pelote* a ball <Med. L *pelota, pilota* <L *pila*]

pel·le·tier·ine (pel′ə·tir′in, -ēn) *n. Chem.* A sirupy liquid alkaloid, $C_8H_{15}ON$, from the roots of the pomegranate tree: its salts are powerful anthelminthics. [after Bertrand *Pelletier,* 1761–1797, French chemist]

pel·li·cle (pel′i·kəl) *n.* A thin skin, film, or layer. [<L *pellicula,* dim. of *pellis* skin] — **pel·lic·u·lar** (pə·lik′yə·lər) *adj.*

pel·li·to·ry (pel′ə·tôr′ē, -tō′rē) *n. pl.* **·ries** Any of various diffuse or tufted weedlike herbs of the nettle family (genus *Parietaria*); especially, the European **wall pellitory** (*P. officinalis*), which grows on old walls. [Alter. of earlier *paretarie* <AF *paritarie* <L (*herba*) *parietaria* wall (plant) < *parietarius* of a wall < *paries, -etis* a wall]

pellitory of Spain A perennial herb (*Anacyclus pyrethrum*) of the composite family, with procumbent stems, dissected leaves, and white flowers: cultivated for its pungent roots, and used in medicine.

pell-mell (pel′mel′) *adv.* **1** In a confused or promiscuous way or manner; indiscriminately; higgledy-piggledy. **2** With a headlong rush. — *adj.* Devoid of order or method. — *n.* A confused crowd or mixture; a medley; disorder. Also **pell′mell′**. [<F *pêle-mêle* <OF *pesle-mesle,* varied reduplication < *mesler* mix]

pel·lo·tine (pel′ə·tēn, -tin) *n. Chem.* A white crystalline alkaloid, $C_{13}H_{19}O_3N$, from the mescal cactus: used as a sedative. [< *pellotte,* var. of PEYOTE, + -INE²]

pel·lu·cid (pə·lōō′sid) *adj.* **1** Permitting to a certain extent the passage of light; translucent; limpid. **2** Transparent; clear; understandable: a *pellucid* style. See synonyms under CLEAR, TRANSPARENT. [<L *pellucidus* < *perlucere* < *per-* through + *lucere* shine] — **pel·lu′cid·ly** *adv.* — **pel·lu′cid·ness, pel·lu·cid′i·ty** (pel′ōō·sid′ə·tē) *n.*

Pel·ly River (pel′ē) A river in south central Yukon, Canada, flowing 330 miles NW to a confluence with the Lewes River, forming the Yukon River.

pe·lon (pə·lōn′) *adj.* Hairless: said of animals. [<Am. Sp. <Sp. *pelón* bald]

pelon dog The Mexican hairless dog.

pel·o·phyte (pel′ō·fīt) *n.* A plant subsisting in clayey soil. [<Gk. *pēlos* clay + -PHYTE]

Pe·lop·i·das (pə·lop′i·dəs), died 364 B.C., Theban general.

Pel·o·pon·ne·sian (pel′ə·pə·nē′shən, -zhən) *adj.* Of or pertaining to the Peloponnesus. — *n.* A native or inhabitant of the Peloponnesus.

Peloponnesian War See table under WAR.

Pel·o·pon·ne·sus (pel′ə·pə·nē′səs) The southern peninsula of Greece between the Ionian and Aegean seas; 8,400 square miles: formerly *Morea.* Also **Pel′o·pon·nese′**. Greek **Pel′o·pon·ne′sos**.

Pe·lops (pē′lops) In Greek mythology, the son of Tantalus who was restored to life by Demeter after his father had killed him and served his flesh to the gods.

pe·lo·ri·a (pə·lôr′ē·ə, -lō′rē·ə) *n. Bot.* Reversion of an irregular flower form, by abnormal development of complementary irregularities or by the loss of the irregular parts. Also **pel·o·rism** (pel′ə·riz′əm). [<NL <Gk. *pelōros* monstrous <*pelōr* a monster] — **pe·lo′ri·ate, pe·lor·ic** (pə·lôr′ik, -lor′-) *adj.*

pe·lo·rus (pə·lôr′əs, -lō′rəs) *n.* **1** *Aeron.* A circular plate having its movable rim graduated in degrees: used to determine the actual or relative bearings of objects. **2** *Nav.* An instrument for correcting errors in the compass by stellar observations. [after *Pelorus,* said to have been Hannibal's pilot]

pe·lo·ta (pe·lō′tə) *n.* A game similar to handball, popular among Basques, Spaniards, and Spanish Americans: it is played in a court with a hard rubber ball and a long wickerwork gauntlet (cesta) attached to the player's wrist. See JAI ALAI. [<Sp., lit., a ball, aug. of *pella* <L *pila*]

Pe·lo·tas (pə·lō′təs) A port in southern Río Grande do Sul state, Brazil.

pelt¹ (pelt) *n.* **1** An undressed fur skin; raw hide; also, a garment made of skins. **2** *Slang* The human skin. [Prob. back formation <AF *pelterie.* See PELTRY.]

pelt² (pelt) *v.t.* **1** To strike repeatedly with or as with missiles or blows. **2** To throw or hurl

(missiles). 3 To assail with words. — v.i. 4 To beat or descend with violence. 5 To move rapidly; hurry. — n. 1 A blow, as one given by something thrown. 2 A steady or swift pace: especially in the expression **full pelt**. [ME *pelten*, ? var. of *pulten* thrust <L *pultare*, freq. of *pellere* beat, drive] — **pelt'er** n.
pel·ta (pel'tə) n. In ancient Greece, a light, leather-covered shield. [<Gk. *peltē* a shield]
pel·tast (pel'tast) n. In ancient Greece, a soldier who bore a pelta; a lightly armed soldier. [<L *peltasta* <Gk. *peltēs* <*peltē* a shield]
pel·tate (pel'tāt) adj. 1 Shield-shaped. 2 Bot. Attached to the stalk at or near the center of the lower surface, as a leaf. Also **pel'tat·ed**. See SCUTATE. [<L *peltatus* armed with a shield <Gk. *peltē* a shield] — **pel'tate·ly** adv.
pel·try (pel'trē) n. pl. **·ries** 1 Pelts collectively. 2 A pelt. 3 A place for keeping or storing pelts. [<AF *pelterie* <OF *peletier* a furrier < *pel* a skin <L *pellis*]
pel·vic (pel'vik) adj. Of or pertaining to the pelvis.
pelvic arch Anat. That part of the skeleton in vertebrates to which the hind limbs (in man, the lower limbs) are attached. Also **pelvic girdle**.
pel·vim·e·ter (pel·vim'ə·tər) n. Med. An instrument used in pelvimetry. [<PELVI(S) + -METER]
pel·vim·e·try (pel·vim'ə·trē) n. Med. The measurement of the size and capacity of the pelvis, especially by X-rays prior to childbirth. — **pel·vi·met·ric** (pel'vi·met'rik) adj.
pel·vi·ra·di·og·ra·phy (pel'vi·rā'dē·og'rə·fē) n. Med. X-ray examination of the pelvis. [<PELVI(S) + RADIOGRAPHY]
pel·vis (pel'vis) n. pl. **·ves** (-vēz) 1 A basin or basinlike structure. 2 Anat. a The part of the skeleton that forms a bony girdle joining the lower or hind limbs to the body: composed, in man, of two hip bones and the sacrum. b The hollow interior portion of the kidney, into which the uriniferous tubules empty: formed by the expanded part of the ureter. [<NL <L, a basin]

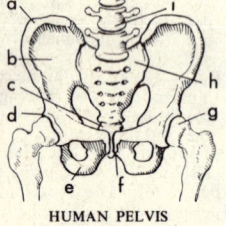

HUMAN PELVIS
a. Crest of ilium.
b. Ilium.
c. Coccyx.
d. Socket of thigh bone.
e. Ischium.
f. Pubic symphysis.
g. Head of femur.
h. Sacrum.
i. Lumbar vertebrae.

Pem·ba (pem'bə) An island in the Indian Ocean off the NE coast of Tanganyika, in the Zanzibar protectorate; 380 square miles.
Pem·broke (pem'brŏok) 1 A maritime county of SW Wales; 614 square miles: also **Pem'broke·shire** (-shir). 2 Its county town.
pem·mi·can (pem'ə·kən) n. 1 Lean venison cut into strips, dried, pounded into paste with fat and a few berries, and pressed into cakes. 2 A similar concentrated and nutritious food made from beef and dried fruits: used by Arctic explorers, etc. Also **pem'i·can**. [< Algonquian (Cree) *pimekan* < *pime* fat]
pem·phi·gus (pem'fə·gəs, pem·fī'-) n. Pathol. A skin disease characterized by watery vesicles successively formed on various parts of the body; water blebs; bladdery fever. Also **pem'phix** (-fiks). [<NL <Gk. *pemphix*, -*igos* a breath, a pustule]
pen[1] (pen) n. 1 An instrument for writing with a fluid ink: formerly made of a quill, now usually of metal and fitted to a holder; by extension, pen and holder together. 2 Quality of penmanship or of composition. 3 A writer; also, the profession of writing. 4 Bot. The midrib of a leaf. 5 Zool. The internal shell of a cuttlefish. 6 Ornithol. A feather; quill; also, in the plural, wings. — v.t. **penned**, **pen·ning** To write with a pen; indite. [<OF *penne* a pen, a feather <LL *penna* <L, a feather] — **pen'ner** n.
pen[2] (pen) n. 1 A small enclosure, as for pigs; also, the animals contained in a pen collectively. 2 Any small place of confinement, as in a police court. 3 Slang A penitentiary. — v.t. **penned** or **pent**, **pen·ning** To enclose in or as in a pen; confine. [OE *penn*]
pen[3] (pen) n. A female swan. [Origin unknown]

pe·nal (pē'nəl) adj. 1 Pertaining to punishment or its means or place. 2 Liable, or rendering liable, to punishment. 3 Enacting or prescribing punishment. [<OF *penal* <L *penalis*, *poenalis* <*poena* a penalty <Gk. *poinē* a fine]
penal code See under CODE.
pe·nal·ize (pē'nəl·īz, pen'əl-) v.t. **·ized**, **·iz·ing** 1 To subject to a penalty, as for a violation. 2 To declare, as an action, subject to a penalty. Also Brit. **pe'nal·ise**. — **pe'nal·i·za'tion** n.
pen·al·ty (pen'əl·tē) n. pl. **·ties** 1 The consequences, as suffering, detriment, etc., that follow the transgression of laws. 2 Judicial punishment for crime or violation of the law. 3 Law a A sum of money fixed by a statute as a fine or mulct for a violation of its provisions. b A sum of money paid and stipulated to be forfeited in case of the non-performance of the conditions of a contract. 4 A handicap imposed for a violation of rules or regulations of a game. [<Med. L *poenalitas*, -*tatis* <L *poenalis* PENAL]
pen·ance (pen'əns) n. 1 Eccl. A sacramental rite involving contrition, confession to a priest, the acceptance of penalties, and absolution. 2 A feeling of sorrow for sin or fault, evinced by some outward act; repentance. 3 A penalty, suffering, mortification, or act of piety, imposed or voluntarily undertaken as an atonement or outward sign of repentance for sin. 4 The performance of a penitential act or acts. — **to do penance** To perform an act or acts of penance; to repent of one's sins. — v.t. **pen·anced**, **pen·anc·ing** To impose a penance upon. [<OF <L *paenitentia*. Doublet of PENITENCE.]
pe·nang (pē·nang') n. A heavy cotton fabric resembling percale. [? from *Penang*]
Pe·nang (pē·nang'), **Settlement of** A civil division of the NW Federation of Malaya under British protection; 400 square miles; capital, George Town; comprising **Penang Island** (formerly *Prince of Wales Island*); 110 square miles; and Province Wellesley on the mainland; 290 square miles.
Pe·na·tes (pə·nā'tēz) In the ancient Roman religion, the household gods: associated with the *Lares*. [<L, prob. akin to *penus* the inmost part of the temple of Vesta]
pence (pens) Plural of PENNY.
pen·cel (pen'səl) n. A small pennon or streamer; a pennoncel: also spelled *pensil*, *pensile*. ◆ Homophone: *pencil*. [<AF, alter. of *penoncel*. See PENNONCEL.]
pen·chant (pen'chənt, Fr. pän·shän') n. A strong leaning or inclination in favor of something. [<F, orig. ppr. of *pencher* incline, ult. <L *pendere* hang]
pen·cil (pen'səl) n. 1 A long, pointed strip of graphite, colored chalk, slate, etc., often encased in wood: used for writing or drawing. 2 A small, finely pointed paint brush: also **hair pencil**. 3 A set of rays diverging from or converging upon a given point. 4 Skill, as in drawing or painting; the painter's art. 5 A small stick of any substance having caustic or styptic properties. 6 An eyebrow pencil. — v.t. **·ciled** or **·cilled**, **·cil·ing** or **·cil·ling** To mark, write, or draw with or as with a pencil. ◆ Homophone: *pencel*. [<OF *pincel* <L *penicillum* a paint brush, double dim. of *penis* a tail; infl. by *pen*[1]] — **pen'cil·er** or **pen'cil·ler** n.
pen·ciled (pen'səld) adj. 1 Marked with fine lines, with or as if with a finely pointed pencil. 2 Having pencils, or lines or rays. Also **pen'cilled**.
pend (pend) v.i. 1 To await or be in process of adjustment or settlement. 2 Dial. To hang; depend. [<MF *pendre* hang <L *pendere*]
pen·dant (pen'dənt) n. 1 Anything that hangs or depends from something else, either for ornament or for use. 2 Something attached to another thing as an ending; an appendix. 3 A parallel; one of a pair. 4 Archit. A hanging ornament, as a boss or knot, particularly in late Perpendicular work, on ceilings, roofs, etc. 5 The stem of a watchcase and the ring by which it is attached to a chain. 6 A suspended chandelier; also, an electrical fitting hanging from a ceiling lamp, chandelier, etc., by which to switch on and off a light. Also spelled *pendent*. — adj. Pendent. [<OF, orig. ppr. of *pendre*. See PEND.]
pen·dent (pen'dənt) adj. 1 Hanging loosely; drooping downward; suspended. 2 Projecting or overhanging. 3 Undetermined; pending; incomplete. Also spelled *pendant*. — n. Pendant. [Var. of PENDANT, refashioned after L] — **pen'dent·ly** adv.
pen·den·te li·te (pen·den'tē lī'tē) Latin Pending or during suit.
pen·den·tive (pen·den'tiv) n. Archit. 1 The vaulting that serves to connect an angle of a square area enclosed by four arches with a dome that rests upon the arches. 2 The principle or system of such vaulting and use of the dome. [<MF *pendentif* <L *pendens*, -*entis*, ppr. of *pendere* hang]

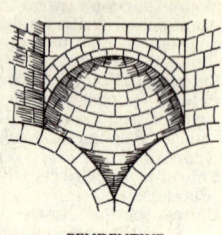

PENDENTIVE

pend·ing (pen'ding) adj. 1 Remaining unfinished or undecided. 2 Imminent; impending. — prep. 1 During the continuance of. 2 Awaiting; until: The court adjourned *pending* the jury's verdict.
pen·drag·on (pen·drag'ən) n. A supreme head, ruler, or chief: in early Britain, a title conferred in times of danger. [<Welsh, a chief leader in war <*pen* head + *dragon* a war chief <L *draco*, -*onis*, orig. a dragon (used as a military standard)] — **pen·drag'on·ish** adj. — **pen·drag'on·ship** n.
pen·du·lous (pen'jōō·ləs) adj. Hanging, especially so as to swing. [<L *pendulus* hanging < *pendere* hang] — **pen'du·lous·ly** adv. — **pen'du·lous·ness** n.
pen·du·lum (pen'jōō·ləm, -də-) n. 1 A body suspended from a fixed point, and free to swing to and fro. 2 Such a device, consisting of rod and bob, and serving, by oscillation under the forces of gravity plus momentum, to regulate the rate of running of a clock. [<NL <L, neut. of *pendulus*. See PENDULOUS.]
Pe·nel·o·pe (pə·nel'ə·pē) A feminine personal name. Also Fr. **Pé·né·lo·pe** (pā·nā·lōp'). [<Gk., a weaver] — **Penelope** In the *Odyssey*, the faithful wife of Odysseus, who, during her husband's absence after the fall of Troy, kept her many suitors in check under pretext of having to complete a shroud she was weaving for Laertes, her father-in-law: the moment of her choice among them was indefinitely postponed, since every night she unraveled all the work she had done during the previous day. See TELEGONUS.
pe·ne·plain (pē'nə·plān', pē'nə·plān') n. Geol. A region of faint or low relief, the penultimate result of long-continued action of denudation on a once larger land mass, whose ultimate result is a base-leveled plain. Also **pe'ne·plane'**. [<L *paene* almost + PLAIN]
pen·e·tra·li·a (pen'ə·trā'lē·ə) n. pl. 1 The inmost parts of anything, but especially of a house or temple; a sanctuary; shrine. 2 Secret things. [<L, orig. neut. pl. of *penetralis* innermost < *penetrare*. See PENETRATE.]
pen·e·trance (pen'ə·trəns) n. Genetics A measure of the frequency with which a given gene will show its effects, expressed as a percentage of the total number of cases observed. [<L *penetrans*, -*antis*, ppr. of *penetrare*. See PENETRATE.]
pen·e·trate (pen'ə·trāt) v. **·trat·ed**, **·trat·ing** v.t. 1 To force a way into or through; pierce; enter. 2 To spread or diffuse itself throughout; permeate. 3 To perceive the meaning of; understand. 4 To affect or move profoundly. — v.i. 5 To enter or pass through something. 6 To have effect on the mind or emotions. See synonyms under PIERCE. [<L *penetratus* < *penetrare* put within (*penitus* inside)] — **pen'e·tra·ble** (-trə·bəl) adj. — **pen'e·tra·bil'i·ty** n. — **pen'e·trant** adj. & n.
pen·e·trat·ing (pen'ə·trā'ting) adj. Tending or having power to penetrate; acute; discerning.

PENDULUM
a. Bob (adjustable on rod)
b. Rod.
c. Pallets.
d. Escape wheel.
e. Second hand.
f. Weight.

penetration

See synonyms under ACUTE, ASTUTE, KNOWING, SHARP. — **pen′e·trat′ing·ly** adv. — **pen′e·trat′ing·ness** n.
pen·e·tra·tion (pen′ə·trā′shən) n. 1 The act or power of penetrating physically. 2 Ability to penetrate mentally; acuteness; sagacity. 3 The depth to which a bullet or other projectile sinks in a target. See synonyms under ENTRANCE[1].
pen·e·tra·tive (pen′ə·trā′tiv) adj. Tending or having power to penetrate, physically or mentally; insinuating and pervasive; pungent: a *penetrative* odor; acute; discerning: *penetrative* wisdom. See synonyms under ASTUTE. — **pen′e·tra′tive·ly** adv. — **pen′e·tra′tive·ness** n.
pen·e·trom·e·ter (pen′ə·trom′ə·tər) n. 1 An instrument for indicating the quality and measuring the strength of X-rays. 2 A device for testing the hardness of relatively plastic substances under given conditions. [< L *penetrare* PENETRATE + -METER]
Pe·ne·us (pē·nē′əs) The chief river of Thessaly, Greece, flowing 135 miles SE to the Aegean: formerly *Salambria*. Greek **Pi·nei′os**.
Peng·hu Islands (pung′hōō′) An island group in Formosa Strait, comprising a district of the Chinese province of Taiwan; 49 square miles: formerly *Pescadores*.
pen·gö (peng′gœ′) n. pl. **·gö** or **·gös** A former monetary unit of Hungary; a coin equivalent to 100 filler.
pen·guin (pen′gwin, peng′-) n. 1 A web-footed, flightless, aquatic bird (genus *Spheniscus*) of the southern hemisphere, with flipperlike wings, short legs, and plantigrade feet. 2 Originally, the great auk. 3 *Aeron.* An airplane with an engine of low motive power, so as to be incapable of flight: used in the early training of aviators. [Cf. F *pingouin, penguyn* the great auk]
pen·hold·er (pen′hōl′dər) n. A handle with a device for inserting a metallic pen; also, a rack for pens.
pen·i·cil·late (pen′ə·sil′it, -āt) adj. 1 Pencil-shaped. 2 *Biol.* Bordered or tufted with fine hairs resembling a hair pencil. Also **pen′i·cil·li·form**. [< L *penicillus* a pencil + -ATE[1]] — **pen′i·cil′late·ly** adv. — **pen′i·cil·la′tion** (-si·lā′shən) n.
pen·i·cil·lin (pen′ə·sil′in) n. A powerful antibacterial substance found in the mold fungus *Penicillium*: prepared in several forms for the treatment of a wide variety of infective conditions. [< PENICILL(IUM) + -IN]
pen·i·cil·li·um (pen′ə·sil′ē·əm) n. pl. **·li·a** (-ē·ə) Any member of a genus (*Penicillium*) of ascomycetous fungi characterized by feltlike masses of tubular hyphae, and growing on decaying fruits, ripening cheese, etc. *P. notatum* is the principal source of penicillin. [< NL < L *penicillus* a pencil; so called because of resemblance of its tufts to small paint brushes]
pe·nin·su·la (pə·nin′sə·lə, -syə-) n. A piece of land almost surrounded by water, and connected with the mainland by an isthmus. [< L *paeninsula* < *paene* almost + *insula* an island] — **pe·nin′su·lar** adj.
Peninsula, the 1 The Iberian Peninsula, comprising Spain and Portugal. 2 The region between the James and York rivers in SE Virginia; scene of several battles of the Civil War, 1862.
pe·nin·su·lar·i·ty (pə·nin′sə·lar′ə·tē, -syə-) n. The state or quality of being a peninsula; hence, narrowness of views; provincialism; bigotry. Compare INSULARITY.
Peninsula State A nickname for FLORIDA.
pe·nis (pē′nis) n. pl. **·nes** (-nēz) The male copulatory organ. [< L, orig. a tail] — **pe′ni·al** (-nē·əl), **pe′nile** (-nil, -nīl) adj.
pen·i·tence (pen′ə·təns) n. The state of being penitent; sorrow for sin, with desire to amend and atone; contrition. See synonyms under CONTRITION. [< OF *penitence* < L *paenitentia* < *paenitens*, ppr. of *paenitere* repent. Doublet of PENANCE.]

pen·i·tent (pen′ə·tənt) adj. Affected by a sense of one's own guilt, and resolved on amendment; repentant; contrite. — n. 1 One who is penitent. 2 One who confesses his sins to a priest and submits himself to the penance prescribed. — **pen′i·tent·ly** adv.
pen·i·ten·tial (pen′ə·ten′shəl) adj. 1 Pertaining to or expressing penitence. 2 Pertaining to penance or punishment. — n. 1 *Eccl.* A book of rules relating to penance and the reconciliation of penitents. 2 A penitent. — **pen′i·ten′tial·ly** adv.
pen·i·ten·tia·ry (pen′ə·ten′shər·ē) n. pl. **·ries** 1 A prison, especially one operated by a state or government as a place of confinement and correction for those convicted of serious crimes. 2 One who prescribes or superintends penances; also, something that has to do with penances; specifically, in the Roman Catholic Church, an office, having at its head a cardinal (called the **Grand Penitentiary**), for deciding questions of conscience, absolution, special dispensation, etc. — adj. 1 Pertaining to penance. 2 Relating to or used for the punishment and discipline of criminals. 3 Rendering the offender liable to imprisonment in a penitentiary. [<Med. L *poenitentiarius* < L *poenitentia* PENITENCE]
Pen·ki (bun′chē′) A city in west central Liaotung province, Manchuria.
pen·knife (pen′nīf′) n. pl. **·knives** (-nīvz′) A small pocket knife: formerly used for making or sharpening quill pens.
pen·man (pen′mən) n. pl. **·men** (-mən) 1 A person considered with regard to his handwriting; also, a teacher of penmanship, or one skilled in penmanship. 2 A writer.
pen·man·ship (pen′mən·ship) n. 1 The art of writing. 2 Handwriting; calligraphy.
Penn (pen), **William**, 1644–1718, English Quaker; founder of Pennsylvania.
pen·na (pen′ə) n. pl. **·nae** (-ē) *Ornithol.* A feather; plume; especially, a quill feather of wing or tail. [<NL < L, a feather] — **pen·na·ceous** (pə·nā′shəs) adj.
pen name An author's assumed name; pseudonym; nom de plume.
pen·nant (pen′ənt) n. 1 A long, narrow flag displayed on a commissioned naval vessel; also, a triangular flag flown as a signal. 2 A small flag peculiar in shape, color or design, flown during a public function. 3 A flag awarded to the winners in some sports leagues; also, the championship thus symbolized. 4 *Music* The hook distinguishing notes shorter than quarter notes. [< PENNON, infl. by PENDANT]
pen·nate (pen′āt) adj. Having wings or feathers: usually in composition: *longipennate*. Also **pen′nat·ed**. [< L *pennatus* winged < *penna* a feather]
Pen·nell (pen′əl), **Joseph**, 1857–1926, U.S. etcher and illustrator.
pen·ni·less (pen′i·lis) adj. Being without even a penny; poverty-stricken.
Pen·nine Alps (pen′īn, -in) A SW division of the Alps on the Swiss-Italian border; highest peak, 15,203 feet.
Pen·nine Chain (pen′īn, -in) A long hill range, called "the backbone of England," extending south from the Cheviot Hills on the Scottish border to Derbyshire and Staffordshire; highest point, 2,930 feet.
pen·non (pen′ən) n. 1 A small, pointed or swallow-tailed flag, borne by medieval knights on their lances and displaying a personal device. 2 A wing. 3 A banner or flag of any sort. [< OF *penon* a streamer, feather of an arrow < L *penna* a feather]
pen·non·cel (pen′ən·sel) n. A small, narrow pennon: also spelled *penoncel*. Also **pen′non·celle**. [< OF *penoncel*, dim. of *penon*. See PENNON.]
Penn·syl·va·nia (pen′səl·vā′nē·ə, -vān′yə) An eastern State of the United States; 45,333 square miles; capital, Harrisburg; entered the Union Dec. 12, 1787; one of the thirteen original States; nickname, *Keystone State* or *Quaker State*; abbr. *Pa., Penn.*, or *Penna.* Official name: *Commonwealth of Pennsylvania*.
Pennsylvania Avenue A principal street of Washington, D.C., which runs in part from the Capitol to the White House.
Penn·syl·va·ni·a-Dutch (pen′səl·vā′nē·ə-duch′,

-vān′yə-) adj. 1 Pertaining to the Pennsylvania Dutch. 2 Denoting a style of furniture, pottery, etc., made by these people, characterized by carved or gaily colored decorations of flowers, fruits, etc.
Pennsylvania Dutch 1 Descendants of immigrants from the Palatinate, SW Germany, and Switzerland who settled in Pennsylvania in the 17th and 18th centuries. 2 The language spoken by these people: a High German dialect with an admixture of English. Also **Pennsylvania German**.
Penn·syl·va·ni·an (pen′səl·vā′nē·ən, -vān′yən) adj. 1 Belonging to or relating to the State of Pennsylvania. 2 *Geol.* Belonging to or denoting a Paleozoic period between the Mississippian and the Permian periods of the Carboniferous. — n. 1 A native or inhabitant of Pennsylvania. 2 *Geol.* The Pennsylvanian period or system.
pen·ny (pen′ē) n. pl. **pen·nies** or **pence** (pens) 1 A cent. 2 A bronze, formerly copper, English coin, one twelfth of a shilling; originally, a silver coin, weighing 22 1/2 grains and first issued in England in the eighth century, and coined until 1343. 3 Money in general: to turn an honest *penny*. [OE *penning, penig, pending*]
-penny *combining form* Costing (a specified number of) pennies: formerly designating the cost of nails per hundred, but now denoting their length, beginning at 1 inch for twopenny nails and increasing by quarter-inches up to tenpenny, thereafter irregularly. [< PENNY]
pen·ny·a·line (pen′ē·ə·līn′) adj. Cheap; inferior: said of writing.
pen·ny·a·lin·er (pen′ē·ə·līn′ər) n. A literary drudge; a hack writer.
penny ante A poker game in which the ante is limited to one cent.
penny dreadful *Brit. Colloq.* A cheap book or magazine containing popular, usually sensational fiction.
pen·ny·roy·al (pen′ē·roi′əl) n. 1 A low, erect, branching, strong-scented American herb (*Hedeoma pulegioides*) of the mint family, yielding the oil of pennyroyal used in medicine. 2 A species of European mint (*Mentha pulegium*) resembling the American pennyroyal in taste, odor, and uses. [Alter. of earlier *pulyole ryale* <AF *puliol real* <L *pulegium fleabane* + *real* royal]
pen·ny·weight (pen′ē·wāt′) n. The twentieth part of an ounce in troy weight, or 1.55 grams.
penny wheep *Scot.* Small beer or ale. Also **penny whip**.
pen·ny·wise (pen′ē·wīz′) adj. Unduly economical in small matters: usually in the phrase **penny-wise and pound-foolish**, economical in small matters, but wasteful in large ones. — **pen′ny·wis′dom** (-wiz′dəm) n.
pen·ny·wort (pen′ē·wûrt′) n. Any one of various plants with round or peltate leaves, as the several species of *Hydrocotyle*, of the parsley family, the navelwort, and the American gentian (*Obolaria virginica*), with funnel-shaped, white, pink, or purple flowers.
pen·ny·worth (pen′ē·wûrth′) n. 1 As much as can be bought for a penny. 2 The amount given or received for money paid; a bargain. 3 A small amount; trifle.
Pe·nob·scot (pə·nob′skot) n. One of a tribe of North American Indians of the Algonquian confederacy of 1749.
Penobscot River A river in central Maine, flowing 350 miles south to **Penobscot Bay**, an inlet of the Atlantic Ocean.
pe·nol·o·gy (pē·nol′ə·jē) n. The science that treats of the punishment and prevention of crime and of the management of prisons and reformatories: also spelled *poenology*. [<L *poena* a penalty + -LOGY] — **pe·no·log·i·cal** (pē′nə·loj′i·kəl) adj. — **pe·nol′o·gist** n.
pen·on·cel (pen′ən·sel) See PENNONCEL.
pen·point (pen′point′) n. The point of a pen; especially, the metal nib for insertion into a holder.
Pen·rhyn (pen′rin) An atoll of the Manihiki group in the South Pacific; 4,000 acres: also *Tongareva*.
Pen·sa·co·la (pen′sə·kō′lə) A port on **Pensacola Bay**, an arm of the Gulf of Mexico in NW Florida; site of a U.S. naval and air base.

EMPEROR PENGUIN (From 3 to 3 1/2 feet tall; the largest of the flightless, swimming birds)

pen·se·mon (pen·sē′mən) See PENSTEMON.
pen·sil (pen′sil), **pen·sile** (pen′sil) See PENCEL.
pen·sile (pen′sil) *adj.* 1 Pendent and swaying; pendulous; suspended. 2 Hanging loosely: a *pensile* nest. 3 Constructing pensile nests: said of birds. [<L *pensilis* hanging down < *pensus,* pp. of *pendere* hang] — **pen′·sile·ness, pen·sil′i·ty** *n.*
pen·sion[1] (pen′shən) *n.* 1 A periodical allowance to an individual or his representative on account of some meritorious work or service; especially, an allowance made by a government to a veteran soldier or to his widow or children. 2 *Obs.* A payment; specifically, a payment made to one not a servant to retain his good will, or to a man of science or letters to enable him to carry on his work. See synonyms under SUBSIDY. — *v.t.* 1 To grant a pension to. 2 To dismiss with a pension: with *off.* [<OF <L *pensio, -onis* payment < *pensus,* pp. of *pendere* weigh, pay] — **pen′sion·a·ble** *adj.*
pen·sion[2] (pen′shən, *Fr.* pän·syôn′) *n. French* A boarding school; also, a boarding house.
pen·sion·ar·y (pen′shən·er′ē) *adj.* 1 Living by means of a pension; pensioned. 2 Consisting of a pension: a *pensionary* provision. — *n. pl.* **·ar·ies** 1 A pensioner. 2 A hireling: often used in a contemptuous sense.
pen·sion·er (pen′shən·ər) *n.* 1 One who receives a pension; hence, one who is dependent on the bounty of another. 2 In Cambridge University, England, and Dublin University, Ireland, a student who pays his own expenses: at Oxford University, England, called *commoner.* 3 A boarder, as in a convent or school. 4 *Archaic* One of the gentlemen-at-arms comprising the royal bodyguard within the palace: instituted by Henry VIII. 5 *Obs.* A paid soldier; a mercenary.
pen·sive (pen′siv) *adj.* 1 Engaged in or addicted to serious or quiet reflection; thoughtful with a touch of sadness. 2 Expressive of, suggesting, or causing sad thoughtfulness. [<OF *pensif, pensive* < *penser* think] — **pen′sive·ly** *adv.* — **pen′sive·ness** *n.*
pen·ste·mon (pen·stē′mən) *n.* Any member of a North American genus (*Penstemon*) of perennial or shrubby plants of the figwort family, with opposite leaves and variously colored flowers; the beard tongues: also spelled *pensemon, pentstemon.* [Var. of NL *pentstemon* <Gk. *pente* five + *stēmōn* thread, stamen; so called from the rudimentary fifth stamen of this genus]
pen·stock (pen′stok′) *n.* 1 A conduit from a millrace to a water-wheel gate. 2 A sluice or floodgate, controlling the discharge of water, as from a pond, or of sewage. 3 A fire hydrant. 4 A penholder.
pent (pent) *adj.* Penned up or in; closely confined. [Pp. of *pend,* obs. var. of PEN[2], *v.*]
penta- *combining form* Five: *pentahedron.* Also, before vowels, **pent-.** [<Gk. *pente* five]
pen·ta·cle (pen′tə·kəl) *n.* 1 A figure composed of five straight lines, making a star that includes a pentagon. 2 In magic, a circle containing certain mystical figures and symbols; a pentagram: also spelled *penticle.* Also **pen·tal·pha** (pen·tal′fə). [<Med. L *pentaculum,* ult. <Gk. *pente* five]

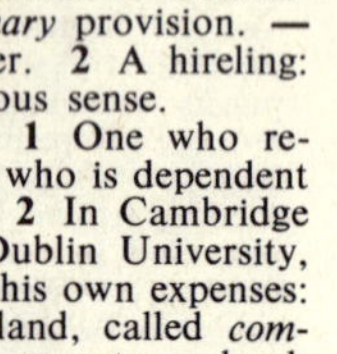

PENTACLE

pen·tad (pen′tad) *n.* 1 The number five; a group of five things. 2 A period of five years. 3 *Chem.* An atom, radical, or element with a combining power of five. — *adj.* Having a combining power of five. [<Gk. *pentas, -ados* a group of five < *pente* five]
pen·ta·dac·tyl (pen′tə·dak′til) *n.* An animal having five fingers or toes. — *adj.* Having five fingers or toes. [<L *pentadactylus* <Gk. *pentadaktylos* < *pente* five + *dactylos* a finger] — **pen′ta·dac′ty·lous** *adj.*
pen·ta·gon (pen′tə·gon) *n.* A figure with five angles and five sides. [<L *pentagonum* <Gk. *pentagōnon* < *pente* five + *gonia* an angle] — **pen·tag·o·nal** (pen·tag′ə·nəl) *adj.* — **pen·tag′o·nal·ly** *adv.*
Pentagon, the A five-sided building in Arlington, Virginia, housing the Department of Defense and other military and naval installations and government offices.
pen·ta·gram (pen′tə·gram) *n.* A figure having five points or lobes; specifically, a pentacle. [<Gk. *pentagrammon,* neut. of *pentagrammos* having five lines < *pente* five + *grammē* a line]
pen·ta·he·dron (pen′tə·hē′drən) *n. pl.* **·dra** (-drə) A solid bounded by five plane faces. — **pen′ta·he′dral** *adj.*
pen·tam·er·ous (pen·tam′ər·əs) *adj.* 1 Composed of or having five similar parts. 2 *Bot.* Five-parted, as a corolla. Also **pen·tam′er·al.**
pen·tam·e·ter (pen·tam′ə·tər) *n.* 1 A line of verse of five metrical feet; especially, English iambic pentameter. 2 Verse comprised of pentameters; heroic verse. 3 In classical prosody, the second line of an elegiac distich: a hexameter with third and sixth feet lacking one long syllable. [<L <Gk. *pentametros* (a verse) of five measures < *pente* five + *metron* a measure]
pen·tane (pen′tān) *n. Chem.* Any one of three isomeric, volatile, liquid hydrocarbons of the alkane series, C_5H_{12}, two of which are contained in petroleum and similar compounds. They differ from one another in behavior with reagents.
pentane lamp A lamp burning pentane under standardized conditions, and used in photometric work: also called *Harcourt lamp.*
Pen·tap·o·lis (pen·tap′ə·lis) A name for any of several groups of five ancient cities in Italy, North Africa, and Asia Minor; the most important was that of the chief cities of Cyrenaica: Apollonia, Arsinoë, Berenice, Cyrene, and Ptolemaïs.
pen·tar·chy (pen′tär·kē) *n. pl.* **·chies** 1 A government administered by five joint rulers; also, a group of five such rulers. 2 An association of five kingdoms, each ruled separately. [<Gk. *pentarchia* < *pente* five + *archein* rule] — **pen·tar′chi·cal** *adj.*
pen·ta·stich (pen′tə·stik) *n.* A stanza of five lines, or a poem containing five lines. [<NL *pentastichus* <Gk. *pentastichos* of five lines < *pente* five + *stichos* a row, line]
pen·ta·style (pen′tə·stīl) *Archit. adj.* Having five columns in front. — *n.* A pentastyle portico or other edifice. [<PENTA- + Gk. *stylos* a pillar]
Pen·ta·teuch (pen′tə·to͞ok, -tyo͞ok) *n.* The first five books of the Bible taken collectively. [<LL *pentateuchus* <Gk. *pentateuchos (biblos)* (the book) of five books < *pente* five + *teuchos* a book, orig. an implement, vessel] — **Pen′ta·teuch′al** *adj.*
pen·tath·lon (pen·tath′lən) *n.* 1 In ancient Greece, an athletic contest of five events — leaping, running, wrestling, throwing the discus, and hurling the spear (earlier, boxing) — that occurred all on the same day between the same contestants. 2 In the modern Olympic games, a contest comprising a 200-meter running race, a 1,500-meter running race, throwing the discus, throwing the javelin, and a running broad jump. [<Gk. < *pente* five + *athlon* a contest] — **pen·tath′lete** (-lēt) *n.* — **pen·tath′lic** *adj.*
pen·ta·va·lent (pen′tə·vā′lənt, pen·tav′ə-) *adj. Chem.* Having a valence of five. Also *quinquevalent.* [<PENTA- + L *valens, -entis,* ppr. of *valere* be strong, have power]
Pen·te·cost (pen′tə·kôst, -kost) *n.* 1 A Jewish festival occurring fifty days after the Passover; Shabuoth. 2 The feast of Whitsunday, commemorating the descent of the Holy Ghost upon the apostles on the Jewish Pentecost. *Acts* ii. [<LL *pentecoste* <Gk. *pentēkostē (hēmera)* the fiftieth (day) < *pentēkonta* fifty] — **pen′te·cos′tal** *adj.*
Pen·tel·i·kon (pen·tel′i·kon) A mountain NW of Athens, Greece; 3,640 feet. Also **Pen·del′·i·kon** (-del′i·kon), *Latin* **Pen·tel′i·cus** (-tel′i·kəs).
Pen·the·si·le·a (pen′thə·si·lē′ə) In Greek legend, a queen of the Amazons who aided the Trojans against the Greeks and was killed by Achilles.
pent·house (pent′hous′) *n.* 1 An apartment or dwelling on the roof of a building. 2 A shed or roof with a single slope affixed to the wall of another building. 3 A small building, generally one-storied, adjoined to the wall of another building; an annex. 4 A canopy or awning projecting above a doorway or window. [Alter. of *pentice,* ME *pentis,* aphetic form of OF *apentis, apendis* <LL *appendicium,* lit., an appendage <L *appendere* APPEND]
pen·ti·cle (pen′ti·kəl) See PENTACLE (def. 2).
pen·tom·ic (pen·tom′ik) *adj.* Referring to a U. S. Army division designed primarily for use in nuclear warfare, having as its basic elements five self-contained battle groups of high mobility, supported by atomic weapons. [<Gk. *pente* five + (AT)OMIC]
pen·to·san (pen′tə·san) *n. Biochem.* One of a group of polysaccharides, found in foods and plant juices, which yield pentoses on hydrolysis. [<PENTOS(E) + -AN]
pen·tose (pen′tōs) *n. Biochem.* Any of an unfermentable class of simple sugars derived from woods, gums, fruits, and some animal tissues, and having five carbon atoms in the molecule. [<Gk. *pent(e)* five + -OSE[2]]
pent·ste·mon (pent·stē′mən) See PENSTEMON.
pent-up (pent′up′) *adj.* Confined; repressed: *pent-up* emotions.
pe·nu·che (pə·no͞o′chē), **pe·nu·chi** See PANOCHA.
pe·nuch·le (pē′nuk·əl), **pe·nuck·le** See PINOCHLE.
pe·nult (pē′nult, pi·nult′) *n.* The syllable next to the last in a word. Also **pe·nul·ti·ma** (pi·nul′tə·mə). [Short for *penultima* <L *paenultima (syllaba)* next to the last (syllable) < *paene* almost + *ultimus* last]
pe·nul·ti·mate (pi·nul′tə·mit) *adj.* 1 Being the last but one. 2 Of or belonging to the last syllable but one. — *n.* A syllable or member of a series that is last but one. [<L *paene* almost + ULTIMATE, on analogy with L *paenultimus* next to the last]
pe·num·bra (pi·num′brə) *n. pl.* **·brae** (-brē) or **·bras** 1 A margin of a shadow within which the rays of light from an illuminating body are partly but not wholly intercepted. 2 *Astron.* **a** The partial shadow between the umbra, or region of total eclipse, and the region of unobstructed light. **b** The dark fringe around the central part of a sunspot. 3 In painting, the blending point, or line between light and shade. [<NL <L *paene* almost + *umbra* a shadow] — **pe·num′bral, pe·num′brous** (-brəs) *adj.*
pe·nu·ri·ous (pə·no͝or′ē·əs, -nyo͝or′-) *adj.* 1 Excessively sparing or saving in the use of money; parsimonious. 2 Affording or yielding little; scanty. See synonyms under AVARICIOUS. — **pe·nu′ri·ous·ly** *adv.* — **pe·nu′ri·ous·ness** *n.*
pen·u·ry (pen′yə·rē) *n.* Extreme poverty or want. See synonyms under POVERTY. [<OF *penurie* <L *penuria, paenuria* want]
Pe·nu·ti·an (pə·no͞o′tē·ən, -shən) *n.* A postulated family of northwestern North American Indian languages, including the Chinook and Shahaptian linguistic stocks.
Pen·za (pen′zä) A city in south central European Russian S.F.S.R.
Pen·zance (pen·zans′) A port and municipal borough in Cornwall, England; the westernmost town in England.
pe·on (pē′ən) *n.* 1 In Latin America: **a** A laborer; servant. **b** Formerly, a debtor kept in virtual servitude until he had worked out his debt. 2 In India: **a** A foot soldier. **b** A messenger, attendant, or orderly. **c** A native police officer or constable. [<Sp. *peón* <LL *pedo, -onis* a foot soldier. Doublet of PAWN[1].]
pe·on·age (pē′ən·ij) *n.* The condition of a peon, or the system of employing this form of labor. Also **pe′on·ism.**
pe·o·ny (pē′ə·nē) *n. pl.* **·nies** 1 A plant of the crowfoot family (genus *Paeonia*) having large, handsome, crimson, rose, or white flowers. 2 Its flower. [Fusion of OE *peonie* and AF *pione,* both from L *paeonia* <Gk. *paiōnia* < *Paion* Paeon, the physician of the gods]
peo·ple (pē′pəl) *n. pl.* **·ple** or (*for def.* 1) **·ples** 1 The aggregate of human beings living under the same government, speaking the same language, or being of the same blood: a general term, used when the technical terms *race, tribe, nation,* or *language* would be misleading: the *people* of England. 2 Ethnologically, a body of human beings belonging to the same linguistic stock and having the same culture. 3 The whole body of persons composing a state or nation, or that part of the population invested with political rights; the enfranchised: the *people* of the state. 4 Persons collectively: taking a verb in the plural: *people* say; also, bodies of persons classified according to their collective occupation or interest: literary *people.* 5 The commonalty, as distinguished from the titled, the rich, or the learned; the populace: with *the.* 6 Those who are connected with one as subjects, attendants, kinfolk, etc.; formerly, all the Negro

slaves belonging to one family. 7 Animals collectively: the ant *people.* 8 Human beings; also, a collection or company. —**chosen people** The Israelites. —**good people** In Ireland, fairies: also **little people.** —*v.t.* **·pled, ·pling** To fill with inhabitants; populate. [<AF *people, poeple* <L *populus* the populace] —**peo′pler** *n.*

Synonyms (noun): commonwealth, community, folk, nation, population, race, state, tribe. A *community* is the aggregate of persons inhabiting any territory in common and having common interests; a *commonwealth* is such a body of persons having a common government, especially a republican government; as, the *commonwealth* of Massachusetts. A *community* may be very small; a *commonwealth* is ordinarily of considerable extent. A *people* is the aggregate of any public *community,* either in distinction from their rulers or as including them; a *race* is a division of mankind in the line of origin and ancestry; the *people* of the United States includes members of almost every *race.* The term *people* is used ethnologically to mean *folk* having the same linguistic and cultural origins, the same customs, traditions, and beliefs, and usually the same geographic distribution: as distinguished from political affiliations or physical origins. The *population* of a country is simply the aggregate of persons residing within its borders, without reference to *race,* organization, or allegiance; unnaturalized residents form part of the *population,* but not of the *nation,* possessing none of the rights and being subject to none of the duties of citizens. In American usage *state* signifies one *commonwealth* of the federal union known as the United *States. Tribe* is now almost wholly applied to primitive *peoples* with primitive political organization; as, the Indian *tribes*; nomadic *tribes.* Compare MOB, STATE.

People's party A political organization formed in the United States in 1891, its platform being increase in currency, free coinage of silver, public control of railways, an income tax, and limitation of ownership of land: also called *Populist party.*

Pe·o·ri·a (pē·ôr′ē·ə, -ō′rē·ə) A city in NW central Illinois.

pep (pep) *Slang n.* Vim; energy; sprightliness; activity; punch; snap; vigor; ginger. —*v.t.* **pepped, pep·ping** To inspire with energy or pep: usually with *up.* [Short for PEPPER] —**pep′pi·ness** *n.* —**pep′py** *adj.*

Pe·phre·do (pə·frē′dō) See GRAEAE.

Pep·in the Short (pep′in), 714?–768, king of the Franks, 751–768; father of Charlemagne. Also *Pippin.*

pep·los (pep′ləs) *n.* In ancient Greece, an elaborate shawl or upper garment worn by women: also *peplum.* Also **pep′lus.** [<Gk.]

pep·lum (pep′ləm) *n. pl.* **·la** (-lə) 1 A short over-skirt, ruffle, or flounce attached to a blouse or coat at the waist, and extending down over the hips. 2 A peplos. [<L <Gk. *peplos* a peplos]

pe·po (pē′pō) *n. pl.* **·pos** The fleshy fruit of the gourd family, with hardened rind and numerous enclosed seeds, as the squash, cucumber, pumpkin, melon, etc. Also **pe·pon·i·da** (pi·pon′ə·də), **pe·po·ni·um** (pi·pō′nē·əm). [<L, a pumpkin <Gk. *pepōn (sikyos)* a ripe (gourd)]

pep·per (pep′ər) *n.* 1 A pungent aromatic condiment consisting of the dried immature berries of the pepper plant, entire or powdered. It is usually black, but when the outer coating of the seeds is removed, the product is **white pepper.** 2 Any plant yielding pepper; especially, a tropical climbing shrub *(Piper nigrum)* of the pepper family *(Piperaceae),* native to India, now widely distributed. 3 Any plant of the genus *Capsicum,* or its product, entire or powdered: red *pepper* or Cayenne *pepper.* 4 *Colloq.* Spiciness; pungency; raciness. —*v.t.* 1 To sprinkle or season with pepper. 2 To sprinkle like pepper. 3 To shower, as with missiles; spatter; pelt. —*v.i.* 4 To discharge missiles at something. [OE *pipor,* ult. <L *piper* <Gk. *peperi* < an Oriental source. Cf. Skt. *pippali* a peppercorn.]

pep·per-and-salt (pep′ər·ən·sôlt′) *adj.* Mixed white and black, so closely intermingled as to present a finely speckled grayish appearance: said of cloth. —*n.* A pepper-and-salt cloth, or a garment made of this cloth.

pep·per·box (pep′ər·boks′) *n.* 1 A cylindrical container with a perforated lid for sprinkling pepper. 2 A quick-tempered person.

pep·per·corn (pep′ər·kôrn′) *n.* 1 A berry of the pepper plant. 2 Anything trifling or insignificant.

pep·per·grass (pep′ər·gras′, -gräs′) *n.* Any cress of the genus *Lepidium,* especially *L. sativum,* a garden salad, and *L. virginicum,* the wild peppergrass or tonguegrass. Also **pep′per·weed′** (-wēd′). —**California peppergrass** A herbaceous annual *(Brassica japonica)* sometimes used in salads.

pep·per·idge (pep′ər·ij) *n.* The tupelo, sour-gum, or blackgum tree. [Var. of dial. E *pip(p)eridge* a barberry tree <Med. L *berberis, barbaris* barberry]

pep·per·mint (pep′ər·mint′) *n.* 1 A pungent aromatic herb *(Mentha piperita),* used in medicine and confectionery. 2 An oil or other preparation from peppermint. 3 A confection, usually disk-shaped, flavored with peppermint.

pep·per·pot (pep′ər·pot′) *n.* 1 A pepperbox. 2 A West Indian stew of meat or fish with okra, chilis, and other vegetables, flavored with cassareep, Cayenne pepper, and the like. 3 In Pennsylvania, a soup of tripe and dough balls highly seasoned with pepper.

pep·per·root (pep′ər·rōōt′, -rŏŏt′) *n.* Crinkleroot.

pepper tree 1 A Tasmanian and Australian shrub *(Drimys aromatica)* with small, greenish-yellow flowers and globular berries sometimes used as a substitute for pepper. 2 The Peruvian mastic *(Schinus molle),* whose seeds are used as a spice known as mollé and whose fruit yields an intoxicating beverage.

pep·per·wort (pep′ər·wûrt′) *n.* 1 Any plant of the pepper family. 2 Any species of peppergrass.

pep·per·y (pep′ər·ē) *adj.* 1 Pertaining to or like pepper; pungent. 2 Quick-tempered; hasty; stinging. See synonyms under HOT. —**pep′per·i·ness** *n.*

pep·sin (pep′sin) *n.* 1 *Biochem.* A proteolytic enzyme secreted by the gastric juices of the stomach. 2 A medicinal preparation obtained from the stomachs of various animals, as the pig and the calf, used to aid digestion. 3 A similar enzyme found in the cells of certain plants. Also **pep′sine.** [<G <Gk. *pepsis* digestion <*peptein* digest]

pep·sin·ate (pep′sin·āt) *v.t.* **·at·ed, ·at·ing** To treat, make up, or prepare with pepsin.

pep·sin·o·gen (pep·sin′ə·jən) *n. Biochem.* The inactive form of pepsin, found in the stomach mucosa and converted into pepsin in a slightly acid solution.

Pep·su (pep′sōō) An acronym for PATIALA AND EAST PUNJAB STATES UNION. Also **PEPSU.**

pep·tic (pep′tik) *adj.* 1 Of, pertaining to, or promotive of digestion. 2 Of, pertaining to, or producing pepsin. 3 Able to digest: opposed to *dyspeptic.* —*n.* An agent that promotes digestion. [<Gk. *peptikos* able to digest <*peptein* digest]

pep·tide (pep′tīd, -tid) *n. Biochem.* Any combination of amino acids in which the carboxyl group of one is joined with the amino group of another. [<PEPT(ONE) + -IDE]

pep·tize (pep′tīz) *v.t.* **·tized, ·tiz·ing** *Chem.* To bring about or increase the colloidal dispersion of (a substance); especially, to convert (a sol) into a gel.

pep·tone (pep′tōn) *n. Biochem.* Any of the soluble protein compounds into which the albuminous substances contained in food are converted when acted upon by pepsin, by acids and alkalis, by putrefaction, etc. [<G *pepton* <Gk., neut. of *peptos* digested, cooked <*peptein* digest] —**pep·ton·ic** (pep·ton′ik) *adj.*

pep·to·nize (pep′tə·nīz) *v.t.* **·nized, ·niz·ing** 1 To change into peptones. 2 To subject, as food, to the action of a peptone or other proteolytic enzyme. 3 To subject to partial predigestion. —**pep′to·ni·za′tion** *n.*

Pepys (pēps, peps, pep′is), **Samuel,** 1633–1703, English diarist. —**Pepys′i·an** *adj.*

Pe·quot (pē′kwot) *n.* One of a tribe of North American Indians of Algonquian stock, formerly inhabiting southern New England. Also **Pe′quod** (-kwod).

per (pûr) *prep.* 1 By; by means of; through: used in commercial and business English: *per* bearer. 2 To or for each: ten cents *per* yard. 3 By the; every: especially in Latin phrases: *per diem.* [<L, through, by]

per- *prefix* 1 Through; throughout: *pervade, perennial.* 2 Thoroughly; completely: *perturb.* 3 Away: *pervert, peremptory.* 4 Very: *perfervid.* 5 *Chem.* **a** Denoting the higher degree of valence in two similar compounds: *barium peroxide* as distinguished from *barium monoxide.* **b** Indicating a relatively large amount of the compound or radical named: *perchloric acid,* contrasted with *chloric acid.* ◆ The prefix occurs in other forms in *pardon, paramour, pellucid,* etc. [<L *per* through, by means of; in some words <OF or F]

Pe·ra (pā′rä) A suburb of Istanbul; formerly the foreign quarter.

per·ac·id (pûr′as′id) *n. Chem.* Any of a class of acids containing more than the normal proportion of oxygen, as perboric, or persulfuric acids.

per·a·cid·i·ty (pûr′ə·sid′ə·tē) *n.* Excessive acidity, as of the stomach.

per·ad·ven·ture (pûr′əd·ven′chər) *adv.* Perchance; it may be; perhaps; not improbably: often preceded by *if* or *unless.* —*n.* Possibility of failure, miscarriage, or error; doubt; question. [<OF *par aventure* by chance; infl. in form by L *adventura* chance]

Pe·rae·a (pə·rē′ə) In ancient geography, a region east of the Jordan, corresponding to Gilead, and sometimes including Bashan.

Pe·rak (pā′rak, *Malay* pe′rä) A state in the Federation of Malaya on the Strait of Malaya; 7,980 square miles; capital, Taiping.

per·am·bu·late (pə·ram′byə·lāt) *v.* **·lat·ed, ·lat·ing** *v.t.* 1 To walk through or over; traverse. 2 To walk through or around so as to inspect, survey, etc. —*v.i.* 3 To walk about; stroll. [<L *perambulatus,* pp. of *perambulare* <*per-* through + *ambulare* walk]

per·am·bu·la·tion (pə·ram′byə·lā′shən) *n.* 1 The act of perambulating; specifically, an annual survey of boundaries. 2 The district or jurisdiction within which one perambulates or surveys.

per·am·bu·la·tor (pə·ram′byə·lā′tər) *n.* 1 One who perambulates. 2 A rolling chair. 3 A baby carriage. 4 A surveyor's measuring wheel, constructed on the principle of the odometer. —**per·am′bu·la·to′ry** *adj.*

per an·num (pûr an′əm) *Latin* By the year.

per·bo·rate (pər·bôr′āt, -bō′rāt) *n. Chem.* A salt of perboric acid, as sodium *perborate.*

per·bo·ric (pər·bôr′ik, -bō′rik) *adj. Chem.* Denoting an acid, HBO_3, known only from salts formed by the action of hydrogen peroxide on borates.

per·cale (pər·kāl′, -kal′) *n.* A closely woven cotton fabric without gloss, in solid colors or prints. [<F, prob. <Persian *pergālah* a rag]

per·ca·line (pûr′kə·lēn′) *n.* A glossy cotton cloth, usually dyed in a solid color: used chiefly as lining. [<F, dim. of *percale* PERCALE]

per cap·i·ta (pûr kap′ə·tə) *Latin* For each person; literally, by heads.

per·ceive (pər·sēv′) *v.t.* & *v.i.* **·ceived, ·ceiv·ing** 1 To become aware of (something) through the senses; see, hear, feel, taste, or smell. 2 To come to understand; apprehend with the mind. [<AF *perceivre,* OF *perçoivre* <L *percipere* seize, perceive <*per-* thoroughly + *capere* take] —**per·ceiv′a·ble** *adj.* —**per·ceiv′a·bly** *adv.* —**per·ceiv′er** *n.*

Synonyms: apprehend, comprehend, conceive, know, understand. We *perceive,* primarily, what is presented through the senses. We *apprehend* what is presented to the mind, whether through the senses or by any other means. That which we *apprehend* we catch, as with the hand; that which we *conceive* we are able to analyze and recompose in our mind; that which we *comprehend* we, as it were, grasp around, take together, seize, embrace wholly within the mind. Compare APPREHEND, KNOW, KNOWLEDGE, LEARN. *Antonyms*: ignore, lose, misapprehend, misconceive, miss, overlook.

per·cent (pər·sent′) *n.* **1** Number of parts in or to every hundred, often specified: fifty *percent* of the people. **2** Amount or quantity commensurate with the number of units in proportion to one hundred: ten *percent* of fifty is five: (symbol, %). **3** *pl.* Securities bearing a certain percentage of interest. Also **per cent., per cent** [Short for L *per centum* by the hundred]

per·cent·age (pər·sen′tij) *n.* **1** Rate per hundred, or proportion in a hundred parts. **2** A proportion of what is under consideration; a part considered in its quantitative relation to the whole. **3** In commerce, the allowance, commission, duty, or interest on a hundred. **4** *Colloq.* Advantage; profit.

per·cen·tile (pər·sen′tīl, -til) *n. Stat.* Any of 100 points measured within the range of a plotted variable, each of which denotes that percentage of the total cases lying below it in value: thus, 1, 2, 3, etc., percent of the cases are in the first, second, third, etc., percentile. Also called *centile.* — *adj.* **1** Of or pertaining to a percentile. **2** Having to do with percentage.

per·cept (pûr′sept) *n. Psychol.* **1** The object of knowledge as mentally presented in sense perception. **2** Immediate knowledge derived from perceiving. [<L *perceptum* (a thing) perceived, orig. neut. pp. of *percipere* PERCEIVE, on analogy with *concept*]

per·cep·ti·ble (pər·sep′tə·bəl) *adj.* That may be seen or apprehended; perceivable; cognizable; evident. See synonyms under EVIDENT. — **per·cep′ti·bil′i·ty, per·cep′ti·ble·ness** *n.* — **per·cep′ti·bly** *adv.*

per·cep·tion (pər·sep′shən) *n.* **1** The act, power, process, or product of perceiving; knowledge through the senses of the existence and properties of matter and the external world. **2** Cognition of fact or truth in general by the activity of thinking: moral *perception*; apprehension; knowledge. **3** *Psychol.* **a** The faculty or power of acquiring immediate and fundamental knowledge through the senses: often called *sense perception.* **b** The process of acquiring such knowledge. **c** The mental product so obtained, often called the *percept.* **4** Any insight or intuitive judgment that implies unusual discernment of fact or truth. **5** *Law* The taking into possession, as of crops or profits. See synonyms under KNOWLEDGE, SENSATION, UNDERSTANDING. [<OF <L *perceptio, -onis* a receiving <*percipere* PERCEIVE] — **per·cep′tion·al** *adj.*

per·cep·tive (pər·sep′tiv) *adj.* **1** Perceiving, or having the power of perception. **2** Pertaining to perception; perceptional. — **per·cep′tive·ly** *adv.* — **per·cep·tiv·i·ty** (pûr′sep·tiv′ə·tē), **per·cep′tive·ness** *n.*

per·cep·tu·al (pər·sep′chōō·əl) *adj.* Pertaining to or involving the power or act of perceiving. — **per·cep′tu·al·ly** *adv.*

Per·ce·val (pûr′sə·vəl) A knight of Arthur's Round Table, type of high chivalry and purity, who together with Galahad achieved the Grail. Also spelled *Percival, Percivale.*

perch[1] (pûrch) *n.* **1** A staff, pole, or slat, variously used, especially as a roost for poultry, etc.; any place on which birds alight or rest; hence, any elevated seat or situation. **2** A measure: one rod (16.5 feet), or, in surveying, a square rod; also, in stonework, a variable measure, usually about 25 cubic feet. **3** A bracket or corbel; a console. **4** A frame on which cloth is examined for imperfections. **5** A pole set to mark a shallow place in navigable water. **6** A pole connecting the fore gear and hind gear of a spring carriage; a reach. — *v.i.* **1** To alight or sit on or as on a perch; roost. — *v.t.* **2** To set on or as on a perch. **3** To examine (cloth) on a perch. [<OF *perche* <L *pertica* a pole]

perch[2] (pûrch) *n.* **1** A small, spiny-finned, predaceous, fresh-water fish (genus *Perca*); especially, the common **European perch** (*P. fluviatilis*), and the American **yellow perch** (*P. flavescens*). **2** One of various other similar or related fishes, including many marine forms. [<OF *perche* <L *perca* <Gk. *perkē*, ? <*perknos* dark-colored]

per·chance (pər·chans′, -chäns′) *adv.* **1** In a possible case; peradventure; perhaps. **2** *Obs.* By chance. [<AF *par chance* by chance <*par* (<L *per* through) + *chance* CHANCE]

Perche (persh) A region of NW France, mostly in western Maine and southern Normandy.

perch·er (pûr′chər) *n.* **1** One who or that which perches. **2** A perching or insessorial bird. **3** A person who perches or examines cloth.

Per·che·ron (pûr′chə·ron, -shə-) *adj.* Belonging or originating in Perche: said of a breed of large, usually dapple-gray or black draft horses. The name *Norman* or *Percheron-Norman* is erroneously applied to this breed. — *n.* A horse of the Percheron breed. [<F, from Perche]

PERCHERON (14 hands high at the withers)

per·chlo·rate (pər·klôr′āt, -klō′rāt) *n. Chem.* A salt of perchloric acid.

per·chlo·ric (pər·klôr′ik, -klō′rik) *adj. Chem.* Pertaining to or designating a colorless, liquid, unstable acid, $HClO_4$, formed when potassium perchlorate is distilled with sulfuric acid.

per·chlo·ride (pər·klôr′īd, -klō′rīd) *n. Chem.* A chloride having a larger proportion of chlorine than any other chloride of the same series: iron *perchloride*, $FeCl_3$. Also **per·chlo′rid** (-klôr′id, -klō′rid).

per·chlo·ron (pər·klôr′on, -klō′ron) *n. Chem.* Calcium hypochlorite, $Ca(ClO)_2 \cdot 4H_2O$, a bleaching agent with a high chlorine content.

per·cip·i·ent (pər·sip′ē·ənt) *adj.* **1** Having the power of perception. **2** Perceiving rapidly or keenly. — *n.* One who or that which perceives. [<L *percipiens, -entis,* ppr. of *percipere* PERCEIVE] — **per·cip′i·ence** or **·en·cy** *n.*

Per·ci·val, Per·ci·vale (pûr′sə·vəl) See PERCEVAL.

per·coid (pûr′koid) *adj.* Of or pertaining to an order (*Percomorphi*) of spiny-finned teleost fishes, including the fresh-water perches, the sunfishes, mackerels, tunas, blennies, and many others; perchlike. — *n.* One of the *Percomorphi.* Also **per·coi′de·an.** [<L *perca* PERCH[2] + -OID]

per·co·late (pûr′kə·lāt) *v.t.* & *v.i.* **·lat·ed, ·lat·ing** To pass or cause to pass through fine interstices; filter; strain; permeate. — *n.* **1** That which has percolated; a filtered liquid. **2** A liquid containing the soluble portion of a drug through which it has passed. [<L *percolatus,* pp. of *percolare* <*per-* through + *colare* strain <*colum* a strainer] — **per′co·la′tion** *n.*

per·co·la·tor (pûr′kə·lā′tər) *n.* **1** One who or that which percolates, as a filter. **2** A filtering coffee pot.

per con·tra (pûr kon′trə) *Latin* On the contrary.

per·cur·rent (pər·kûr′ənt) *adj. Bot.* Extending from one end to another or from base to apex, as the veins of certain leaves. [<L *percurrens, -entis,* ppr. of *percurrere* <*per-* through + *currere* run]

per·cuss (pər·kus′) *v.t.* **1** To strike or tap quickly or forcibly. **2** *Med.* To test or treat by percussion. [<L *percussus,* pp. of *percutere* strike <*per-* through + *quatere* shake] — **per·cus′sor** *n.*

per·cus·sion (pər·kush′ən) *n.* **1** The sharp striking of one body against another; sudden collision, especially such as causes a shock or sound. **2** The act of striking the percussion cap in a firearm. **3** The shock or vibration produced by collision; the impression of sound upon the ear. **4** *Med.* A light, quick tapping of the finger tips upon the back, chest, or abdomen, for determining, by the resonance, the condition of the organ beneath. **5** Those musical instruments, collectively, whose tone is produced by striking or hitting, as the timpani, glockenspiel, piano, etc. — *adj.* Pertaining to or operating by percussion; percussive: *percussion* cap, *percussion* lock. — **per·cus′sive** (-kus′iv) *adj.* — **per·cus′sive·ly** *adv.* — **per·cus′sive·ness** *n.*

percussion cap A percussion primer.

percussion figure *Mineral.* The figure assumed by the various cracks in a crystal or mineral made by the impact of a dull point against it: also called *strike figure.*

percussion fuze A fuze at the head of a shell that causes explosion by impact.

percussion instruments Musical instruments played by striking, as cymbals, drums, piano, etc.

percussion lock The hammer of a firearm.

percussion primer A small cap of thin metal, containing mercury fulminate, or other detonator, used in ammunition to explode the propelling charge.

Per·cy (pûr′sē) A masculine personal name. [<*Percy*, an English surname]

Per·cy (pûr′sē), **Sir Henry,** 1364–1403, English soldier; in Shakespeare's *Richard II* and *Henry IV*: known as *Hotspur.* — **Thomas,** 1729–1811, English antiquary; editor of *Reliques of Ancient English Poetry.*

Per·di·do (per·thē′thō), **Monte** A peak in the central Pyrenees of NE Spain near the French border; 10,997 feet. *French* **Mont Per·du** (môn per·dü′).

per·die, per·dy (pər·dē′) See PARDI.

per di·em (pər dē′əm, dī′əm) **1** By the day. **2** An allowance (of money) for expenses each day. [<L]

per·di·tion (pər·dish′ən) *n.* **1** *Theol.* Future misery or eternal death as the condition of the wicked; hell. **2** *Obs.* Utter destruction or ruin. **3** *Obs.* Lessening; diminution. See synonyms under RUIN. [<OF *perdicium* <L *perditio, -onis* <*perdere* destroy, lose <*per-* through, away + *dare* give]

per·du (pər·dōō′) *adj.* Hidden; concealed. — *n.* *Obs.* A soldier on a perilous assignment. Also **per·due′.** [<F *perdue,* orig. pp. fem. of *perdre* lose <L *perdere.* See PERDITION.]

per·du·ra·ble (pər·dŏŏr′ə·bəl, -dyŏŏr′-) *adj.* Very durable; lasting. [<OF <LL *perdurabilis* <L *perdurare* <*per-* through + *durare* endure <*durus* hard] — **per·du·ra·bil·i·ty** (pûr′dŏŏr·ə·bil′ə·tē, -dyŏŏr-) *n.* — **per′du·ra·bly** *adv.*

père (pâr) *n. French* Father: used after a surname to distinguish father from son: Dumas *père.*

per·e·gri·nate (per′ə·gri·nāt′) *v.* **·nat·ed, ·nat·ing** *v.i.* To travel from place to place. — *v.t.* To travel through or along. — *adj. Obs.* Of foreign birth or manners; traveled; foreign. [<L *peregrinatus,* pp. of *peregrinari* travel abroad <*peregrinus.* See PEREGRINE.] — **per′e·gri·na′tion** *n.* — **per′e·gri·na′tor** *n.*

per·e·grine (per′ə·grin) *adj.* **1** Coming from foreign regions. **2** Foreign. **3** Upon a pilgrimage; on one's travels. — *n.* The peregrine falcon. Also **per′e·grin.** [<L *peregrinus* foreign <*pereger* traveling <*per-* through + *ager, agri* a field, land]

peregrine falcon A widely distributed falcon (*Falco peregrinus*) generally blackish-blue above and whitish below, streaked with black in the typical form, and with black cheek patches: formerly much used in falconry on account of its courage and speed; the duck hawk. See FALCON.

pe·rei·ra bark (pə·rā′rə) The bark of a tropical American tree (*Geissospermum vellosii*), valued for its medicinal properties. [<NL *pereira,* former genus name, after J. *Pereira,* 1804–53, English medical professor]

pe·rei·rine (pə·rā′rēn, -rin) *n. Chem.* An amorphous alkaloid, $C_{20}H_{26}N_2O$, contained in pereira bark, and used in medicine as an antipyretic tonic. [<PEREIR(A BARK) + -INE[2]]

per·emp·to·ry (pə·remp′tər·ē, per′əmp·tôr′ē, -tō′rē) *adj.* **1** Not admitting of debate or appeal; decisive; absolute. **2** Positive in judgment or opinion; dogmatic. **3** Intolerant of opposition; dictatorial. **4** *Law* That which destroys, puts an end to, or precludes debate or discussion; final; positively fixed. See synonyms under ARBITRARY. [<AF *peremptorie* <L *peremptorius* destructive <*peremptor* a destroyer <*perimere* destroy <*per-* entirely + *emere* buy, take] — **per·emp′to·ri·ly** *adv.* — **per·emp′to·ri·ness** *n.*

per·en·ni·al (pə·ren′ē·əl) *adj.* **1** Continuing or enduring through the year or through many years. **2** Hence, unfailing; unceasing: *perennial* courage. **3** Growing continually; surviving more than one year. **4** *Bot.* Lasting more than two years. See synonyms under ETERNAL, PERPETUAL. — *n.* A plant that grows for three or more years, usually blossoming and fructifying annually. [<L *perennis* <*per-* through + *annus* a year] — **per·en′ni·al·ly** *adv.*

Pé·rez Gal·dós (pā′rāth gäl·thōs′), **Benito,** 1843–1920, Spanish novelist and dramatist.

per·fect (pûr′fikt) *adj.* **1** Having all the qualities, excellences, or elements that are requisite

to its nature or kind; without defect or lack; consummated; supremely excellent; complete. **2** Thoroughly versed or informed; completely skilled: a *perfect* soldier. **3** Closely correspondent; accurately reproducing: a *perfect* replica. **4** Thoroughly effectual; meeting the requirements of the occasion: a *perfect* antidote; a *perfect* answer. **5** *Colloq.* Excessive in degree; very great: She has a *perfect* horror of spiders. **6** *Bot.* Having the essential organs, stamens, and pistils: said of flowers. **7** *Gram.* Denoting the tense of the verb expressing completed action in the past. Some grammarians note in English a *present perfect, past perfect* (or *pluperfect*), and a *future perfect* tense, a *conditional perfect,* and a *perfect infinitive* and *participle.* **8** *Music* **a** Of a character not altered by inversion: said of interval: a *perfect* fifth or octave. **b** Complete: a *perfect* cadence. **9** *Obs.* Assured; positive. —*n. Gram.* The perfect tense, or a verb in this tense. —*v.t.* (pər·fekt′) **1** To bring to perfection; complete; finish. **2** To make thoroughly skilled or accomplished: to *perfect* oneself in an art. [<OF *parfit* <L *perfectus,* pp. of *perficere* accomplish <*per-* thoroughly + *facere* do, make] —**per·fect′er** *n.* —**per·fect′i·bil′i·ty** *n.* —**per·fect′i·ble** *adj.* —**per′·fect·ly** *adv.*

Synonyms (*adj.*): absolute, accurate, blameless, complete, completed, consummate, correct, entire, faultless, finished, holy, ideal, immaculate, infallible, sinless, spotless, stainless, unblemished, undefiled. That is *perfect* to which nothing can be added and from which nothing can be taken without impairing its excellence, symmetry, or worth; as, a *perfect* flower; a copy of a document is *perfect* when it is *accurate* in every particular; a vase may be called *perfect* when *entire* and *unblemished,* even if not artistically *faultless*; the best judges never pronounce a work of art *perfect,* because they see always *ideal* possibilities not yet attained; even the *ideal* is not *perfect,* by reason of the imperfection of the human mind; a human character faultlessly *holy* would be morally *perfect* but finite. That which is *absolute* is free from admixture (as *absolute* alcohol) and from imperfection or limitation. See CORRECT, IMPLICIT, INNOCENT, PURE, RADICAL, RIPE. *Antonyms*: bad, blemished, corrupt, corrupted, defaced, defective, deficient, deformed, fallible, faulty, imperfect, incomplete, inferior, insufficient, marred, meager, perverted, poor, ruined, short, spoiled, worthless.

per·fec·tion (pər·fek′shən) *n.* **1** The state or condition of being perfect; supreme excellence; also, an embodiment of this: also **per′fect·ness.** **2** A particular quality that is supreme. **3** The highest degree of a thing: the *perfection* of rudeness. **4** The act or process of perfecting; the fact of having been perfected.

per·fec·tion·ism (pər·fek′shən·iz′əm) *n. Philos.* The theory that moral perfection may be attained, or has been attained, by men: variously held and taught by different sects and schools. Also **per·fect′ism.**

per·fec·tion·ist (pər·fek′shən·ist) *n.* **1** One who demands an exceedingly high degree of excellence in the performance, behavior, etc., of himself or in that of others. **2** One who adheres to the theory of perfectionism.

per·fec·tive (pər·fek′tiv) *adj.* **1** Tending to make perfect. **2** *Gram.* Denoting an aspect of the verb expressing the completion of an action: opposed to *imperfective.* —**per·fec′·tive·ly** *adv.* —**per·fec′tive·ness** *n.*

perfect number See under NUMBER.

per·fec·to (pər·fek′tō) *n.* A cigar shaped to taper at either end, and of medium size. [<Sp., perfect <L *perfectus.* See PERFECT.]

perfect pitch Absolute pitch (def. 2).

per·fer·vid (pər·fûr′vid) *adj.* Very or excessively fervid; glowing; intensely zealous. [<NL *perfervidus* <L *per-* thoroughly + *fervidus* FERVID]

per·fid·i·ous (pər·fid′ē·əs) *adj.* **1** Characterized by or guilty of perfidy; treacherous. **2** Involving a breach of faith; contrary to loyalty and truth. [<L *perfidiosus* <*perfidia* PERFIDY] —**per·fid′i·ous·ly** *adv.* —**per·fid′i·ous·ness** *n.*

Synonyms: deceitful, disloyal, double-faced, faithless, false, forsworn, perjured, traitorous, treacherous, two-faced, unfaithful, untrue, untrustworthy. *Antonyms*: faithful, honest, incorruptible, staunch, steadfast, true, trustworthy, trusty.

per·fi·dy (pûr′fə·dē) *n. pl.* **·dies** The act of violating faith or allegiance; treachery; faithlessness. [<MF *perfidie* <L *perfidia* treachery <*per-* through, away + *fides* faith]

per·fo·li·ate (pər·fō′lē·it, -āt) *adj. Bot.* Growing so that the stem passes, or seems to pass, through it: said of a leaf. The condition is brought about by the union of the basal lobes of a clasping leaf. [<NL *perfoliatus* <L *per-* through + *folium* a leaf] —**per·fo′li·a′tion** *n.*

per·fo·rate (pûr′fə·rāt) *v.t.* **·rat·ed, ·rat·ing** **1** To make a hole or holes through, by or as by stamping or drilling. **2** To pierce with holes in rows or patterns, as sheets of stamps, etc. See synonyms under PIERCE. —*adj.* (-rit) Perforated. [<L *perforatus,* pp. of *perforare* <*per-* through + *forare* bore] —**per′fo·ra·ble** *adj.* —**per′fo·ra′tive, per′fo·ra·to′ry** *adj.* —**per′fo·ra′tor** *n.*

per·fo·rat·ed (pûr′fə·rā′tid) *adj.* Pierced with a hole or holes, especially in lines or patterns, as sheets of stamps to facilitate tearing.

per·fo·ra·tion (pûr′fə·rā′shən) *n.* **1** A perforating or state of being perforated. **2** A hole or series of holes drilled in or stamped through something, especially in lines or patterns.

per·force (pər·fôrs′, -fōrs′) *adv.* By force; by or of necessity; necessarily. [<OF *par force* <*par* through, by (<L *per-*) + *force* FORCE]

per·form (pər·fôrm′) *v.t.* **1** To carry out in action; execute; do: to *perform* an operation. **2** To act in accord with the requirements or obligations of; fulfil; discharge, as a duty or command. **3** To act (a part) or give a performance of (a play, piece of music, etc.). —*v.i.* **4** To carry through to completion an action, undertaking, etc. **5** To give an exhibition or performance, as of a role in a play, singing, etc.: The actress will *perform* tomorrow. See synonyms under ACCOMPLISH, EFFECT, EXECUTE, MAKE, TRANSACT. [<AF *parfourmer,* OF *parfournir* accomplish entirely <*par-* thoroughly (<L *per-*) + *fournir* accomplish; infl. in form by OF *former* form] —**per·form′·a·ble** *adj.*

per·form·ance (pər·fôr′məns) *n.* **1** The act of performing; also, the thing done; execution; completion; action; achievement. **2** A representation before spectators; an exhibition of feats; any entertainment: two *performances* daily. See synonyms under ACT, EXERCISE, OPERATION, PRODUCTION, WORK.

per·form·er (pər·fôr′mər) *n.* **1** One who performs or acts; one who carries a part upon the stage or in any performance, as an actor, musician, or acrobat. **2** One who carries out his promise or does his duty. See synonyms under AGENT.

per·frig·er·a·tion (pər·frij′ə·rā′shən) *n.* Frostbite. [<L *perfrigeratus,* pp. of *perfrigerare* <*per-* thoroughly + *frigerare* cool <*frigus* cold]

per·fume (pûr′fyo͞om, pər·fyo͞om′) *n.* **1** A pleasant odor, as from flowers; fragrance. **2** A fragrant substance, usually a volatile liquid, prepared to emit a pleasant odor; scent. See synonyms under SMELL. —*v.t.* (pər·fyo͞om′) **·fumed, ·fum·ing** To fill or impregnate with a fragrant odor; scent. [<F *parfum* <Ital. *perfumare,* lit., impregnate with smoke <*per-* through (<L) + *fumare* smoke <*fumus* smoke]

per·fum·er (pər·fyo͞o′mər) *n.* **1** One who makes or deals in perfumes. **2** One who or that which perfumes.

per·fum·er·y (pər·fyo͞o′mər·ē) *n. pl.* **·er·ies** **1** Perfumes in general, or a specific perfume. **2** A place where perfumes are manufactured.

per·func·to·ry (pər·fungk′tər·ē) *adj.* Done merely for the sake of getting through; mechanical and without interest; half-hearted; negligent; superficial; careless. [<LL *perfunctorius* negligent <*perfunctor* one who performs an act <*perfungi* get through with <*per-* through + *fungi* perform] —**per·func′·to·ri·ly** *adv.* —**per·func′to·ri·ness** *n.*

per·fuse (pər·fyo͞oz′) *v.t.* **·fused, ·fus·ing** **1** To overspread, suffuse, or sprinkle with a liquid, color, etc.; permeate. **2** To spread, as a liquid, over or through something; diffuse. [<L *perfusus,* pp. of *perfundere* <*per-* through, all over + *fundere* pour out] —**per·fu′sion** (-zhən) *n.* —**per·fu′sive** (-siv) *adj.*

Per·ga·mum (pûr′gə·məm) **1** An ancient kingdom of western Asia Minor, later a Roman province. **2** Its capital, a Greek city on the site of modern Bergama. Also **Per′ga·mon, Per′ga·mos** (-məs), **Per′ga·mus.**

per·go·la (pûr′gə·lə) *n.* An arbor; specifically, an arbor or trelliswork of a structural nature, covered with vegetation or flowers; a covered walk. [<Ital., an arbor <L *pergula* a projecting roof, arbor <*pergere* go forward <L *per-* through + *regere* keep straight]

PERGOLA

Per·go·le·si (per′gō·lā′zē), **Giovanni Battista,** 1710–36, Italian composer.

per·haps (pər·haps′) *adv.* It may be; possibly. [<PER + *happes, haps,* pl. of HAP[1]]

pe·ri (pir′ē) *n.* In Persian mythology, a fairy or elf descended from the disobedient angels, doing penance until readmitted into paradise. [<Persian *pari, peri* a demon, fairy]

peri- *prefix* **1** Around; encircling; all about: *periphery.* **2** Situated near; close; adjoining: *perihelion.* [<Gk. <*peri* around]

Per·i·an·der (per′ē·an′dər), died 585 B.C., tyrant of Corinth, 625–585 B.C.; one of the seven wise men of Greece.

per·i·anth (per′ē·anth) *n. Bot.* The combined calyx and corolla of a flower when so alike as to be indistinguishable: sometimes called *perigonium.* [Appar. <F *périanthe* <NL *perianthium* <Gk. *peri-* around + *anthos* a flower] —**per′i·an′the·ous** *adj.*

per·i·apt (per′ē·apt) *n. Obs.* A charm to protect the wearer from disease or misfortune; an amulet. [<MF *périapte* <Gk. *periapton* <*peri-* around + *aptos* fastened <*aptein* fasten]

per·i·as·tron (per′ē·as′tron) *n. Astron.* That point in the orbit of either member of a double star when the stars are closest to each other: opposed to *apastron.* [<NL <Gk. *peri-* around + *astron* a star]

per·i·blem (per′ə·blem) *n. Bot.* A sheath of meristematic tissue surrounding the plerome of a plant and giving rise to the primary cortex tissue. [<G <Gk. *periblēma* a covering <*periballein* put on, around <*peri-* around + *ballein* throw]

per·i·blep·sis (per′ə·blep′sis) *n.* The wild, intense, staring expression of a delirious or insane person. [<NL <Gk. *peri-* around + *blepsis* an act of sight <*blepein* look]

per·i·car·di·al (per′ə·kär′dē·əl) *adj.* Of or pertaining to the pericardium. Also **per′i·car′·di·ac** or **·di·an.**

per·i·car·di·ec·to·my (per′ə·kär′dē·ek′tə·mē) *n. Surg.* Removal of the pericardium. [<*pericardi-* (<PERICARDIUM) + -ECTOMY]

per·i·car·di·ot·o·my (per′ə·kär′dē·ot′ə·mē) *n. Surg.* Incision of the pericardium.

per·i·car·di·tis (per′ə·kär·dī′tis) *n. Pathol.* Inflammation of the pericardium.

per·i·car·di·um (per′ə·kär′dē·əm) *n. pl.* **·di·a** (-dē·ə) *Anat.* A membranous bag that surrounds and protects the heart. [<NL <Gk. *pericardion* (the membrane) around the heart <*peri-* around + *kardia* heart]

per·i·carp (per′ə·kärp) *n. Bot.* The wall of the ripened ovary of a flower, constituting the germ of the fruit. [<NL *pericarpium* <Gk. *pericarpion* a husk <*peri-* around + *karpos* a fruit] —**per′i·car′pi·al** *adj.*

per·i·chon·dri·um (per′ə·kon′drē·əm) *n. pl.* **·dri·a** (-drē·ə) *Anat.* The vascular membrane that envelops the surface of a cartilage between the joints. [<NL <Gk. *peri-* around + *chondros* a cartilage] —**per′i·chon′dri·al** *adj.*

Per·i·cle·an (per′ə·klē′ən) *adj.* Pertaining to, characteristic of, or named after Pericles, or the period of his supremacy, the age when Greek art, literature, philosophy, and statesmanship are considered to have been at their height.

add,āce,câre,pälm; end,ēven; it,īce; odd,ōpen,ôrder; to͝ok,po͞ol; up,bûrn; ə = a in *above,* e in *sicken,* i in *clarity,* o in *melon,* u in *focus*; yo͞o = u in *fuse*; oi,oil; ou,pout; ch,check; g,go; ng,ring; th,thin; th,this; zh,vision. Foreign sounds à,œ,ü,kh,ṅ; and ♦: see page xx. <from; + plus; ? possibly.

PERIODIC TABLE OF ELEMENTS

The atomic number will be found in the upper left corner of each box; the atomic weight (1954), correct to two decimal places, in the lower right (numbers in the table in parentheses are mass numbers of the most stable isotopes); and the symbol in the lower left.

Period	Group 1	Group 2	Group 3	Group 4	Group 5	Group 6	Group 7	Group 8			Group 0
1	1 H 1.00										2 He 4.00
2	3 Li 6.94	4 Be 9.01	5 B 10.82	6 C 12.01	7 N 14.00	8 O 16.00	9 F 19.00				10 Ne 20.18
3	11 Na 22.99	12 Mg 24.32	13 Al 26.98	14 Si 28.09	15 P 30.97	16 S 32.06	17 Cl 35.45				18 Ar 39.94
4A	19 K 39.10	20 Ca 40.08	21 Sc 45.96	22 Ti 47.90	23 V 50.95	24 Cr 52.01	25 Mn 54.94	26 Fe 55.85	27 Co 58.94	28 Ni 58.69	
4B	29 Cu 63.54	30 Zn 65.38	31 Ga 69.72	32 Ge 72.60	33 As 74.91	34 Se 78.96	35 Br 79.91				36 Kr 83.70
5A	37 Rb 85.48	38 Sr 87.63	39 Y 88.92	40 Zr 91.22	41 Nb 92.91	42 Mo 95.95	43 Tc (99)	44 Ru 101.10	45 Rh 102.91	46 Pd 106.70	
5B	47 Ag 107.88	48 Cd 112.41	49 In 114.76	50 Sn 118.70	51 Sb 121.76	52 Te 127.61	53 I 126.91				54 Xe 131.30
6A	55 Cs 132.91	56 Ba 137.36	57-71 Lanthanide Series	72 Hf 178.60	73 Ta 180.95	74 W 183.92	75 Re 186.31	76 Os 190.20	77 Ir 192.2	78 Pt 195.23	
6B	79 Au 197.00	80 Hg 200.61	81 Tl 204.39	82 Pb 207.21	83 Bi 209.00	84 Po 210.00	85 At (210)				86 Rn 222.00
7	87 Fr (223)	88 Ra 226.05	89-103 Actinide Series								
Lanthanide Series	57 La 138.92	58 Ce 140.13	59 Pr 140.92	60 Nd 144.27	61 Pm (145)	62 Sm 150.43	63 Eu 152.0	64 Gd 156.9	65 Tb 158.93	66 Dy 162.46	67 Ho 164.94
	68 Er 167.2	69 Tm 169.94	70 Yb 173.04	71 Lu 174.99							
Actinide Series	89 Ac 227.0	90 Th 232.05	91 Pa 231	92 U 238.07	93 Np (237)	94 Pu (242)	95 Am (243)	96 Cm (245)	97 Bk (245)	98 Cf (248)	99 Es
	100 Fm	101 Md	102 No	103 ?							

Per·i·cles (per′ə·klēz), died 429 B.C., Athenian statesman and general.
— **Pericles, Prince of Tyre** Hero of Shakespeare's play of that name.

per·i·cline (per′ə·klīn) *n.* One of the varieties of albite found in the Swiss Alps in the form of white twinned crystals. [<Gk. *periklinēs* sloping all around <*peri-* around + *klinein* lean; with ref. to the large inclination between the terminal and lateral faces of the crystals]

per·i·cra·ni·um (per′ə·krā′nē·əm) *n. pl.* **·ni·a** (-nē·ə) *Anat.* The periosteum of the external surface of the cranium. [<NL <Gk. *perikranion* <*perikranios (chitōn)* (the membrane) under the skin of the skull <*peri-* around + *kranion* skull] — **per′i·cra′ni·al** *adj.*

per·i·cy·cle (per′ə·sī′kəl) *n. Bot.* The outer portion of the central or fibrovascular cylinder in plants, capable of active growth. [<Gk. *perikyklos* all around, spherical <*perikykloein* <*peri-* around + *kykloein* encircle] — **per′i·cy′clic** *adj.*

per·i·den·tal (per′ə·den′təl) *adj.* Periodontal.

per·i·derm (per′ə·dûrm) *n. Bot.* The outer bark; the tissue produced by the cork cambium layer in plants. — **per′i·der′mal** or **·der′mic** *adj.*

pe·rid·i·um (pə·rid′ē·əm) *n. pl.* **·i·a** (-ē·ə) *Bot.* The outer coat or coats of an angiocarpous fungus, forming a complete investment of the fructification, as in puffballs. [<NL <Gk. *pēridion*, dim. of *pēra* a leather bag] — **pe·rid′·i·al** *adj.*

per·i·dot (per′ə·dot) *n.* A yellowish-green gem variety of olivine. [<F *péridot* <OF *peritot*; ult. origin uncertain] — **per′i·dot′ic** *adj.*

per·i·do·tite (per′ə·dō′tīt) *n.* A granular igneous rock composed essentially of olivine or chrysolite. [<PERIDOT + -ITE[1]]

per·i·gee (per′ə·jē) *n. Astron.* The point in the orbit of the moon where it is nearest the earth: opposed to *apogee.* [<MF *périgée* <Med. L *perigaeum* <Gk. *perigeion*, orig. neut. of *perigeios* close around the earth <*peri-* around + *gē* earth] — **per′i·ge′al, per′i·ge′·an** *adj.*

per·i·gon (per′ə·gon) *n. Geom.* An angle equal to two straight angles or 360 degrees.

per·i·go·ni·um (per′ə·gō′nē·əm) *n. pl.* **·ni·a** (-nē·ə) The perianth. [<NL <Gk. *peri-* around + *gonos* offspring, a seed]

Pé·ri·gord (pā·rē·gôr′) A former division of Guienne, SW France.

Per·i·gor·di·an (per′ə·gôr′dē·ən) *adj. Anthropol.* Pertaining to either of two extensions of the Aurignacian culture stage: the Lower Perigordian (Chatelperronian) and the Upper Perigordian (Gravettian). [from *Périgord*]

Pé·ri·gueux (pā·rē·gœ′) A city of SW France, former capital of Périgord.

pe·rig·y·nous (pə·rij′ə·nəs) *adj. Bot.* Situated around the ovary: said of parts of a flower, as the stamens, in which the ovary is nearly or quite free and surrounded by a cup formed by the torus or by the adnation of two or more of the floral organs, upon which the other parts seem to be inserted. [<NL *perigynus* <Gk. *peri-* around + *gynē* female] — **pe·rig′y·ny** *n.*

per·i·he·li·on (per′ə·hē′lē·ən) *n. pl.* **·li·a** (-lē·ə) *Astron.* The point in the orbit of a planet or comet where it is nearest the sun: opposed to *aphelion.* [<NL *perihelium* <Gk. *peri-* close about + *hēlios* the sun; refashioned after Greek]

per·il (per′əl) *n.* Exposure to the chance of injury, loss, or destruction; danger; jeopardy; risk. See synonyms under DANGER, HAZARD. — *v.t.* **·iled** or **·illed, ·il·ing** or **·il·ling** To expose to danger; imperil. [<OF *péril* <L *periculum* trial, danger]

per·il·ous (per′əl·əs) *adj.* Full of, involving, or attended with peril; hazardous. See synonyms under PRECARIOUS. — **per′il·ous·ly** *adv.* — **per′·il·ous·ness** *n.*

Pe·rim (pə·rim′) An island dependency of Aden Colony in the strait Bab el Mandeb; 5 square miles. *Arabic* **Ba·rim** (bä·rim′).

pe·rim·e·ter (pə·rim′ə·tər) *n.* **1** The bounding line of any figure of two dimensions. **2** The sum of the sides of a plane figure. **3** An instrument for testing the scope of the field of vision. [<L *perimetros* <Gk. <*peri-* around + *metron* a measure] — **per·i·met·ric** (per′ə·met′rik) or **·ri·cal** *adj.* — **per′i·met′ri·cal·ly** *adv.*

pe·rim·e·try (pə·rim′ə·trē) *n.* Measurement of the scope of vision by use of a perimeter.

per·i·morph (per′ə·môrf) *n.* A mineral that encloses another. Compare ENDOMORPH. [<PERI- + Gk. *morphē* a form] — **per′i·mor′phic** or **·phous** *adj.* — **per′i·mor′phism** *n.*

per·i·ne·phri·tis (per′ə·nə·frī′tis) *n. Pathol.* Inflammation of the cellular and fibrous tissues around the kidney. [<NL <Gk. *peri-* around + *nephros* a kidney + -ITIS] — **per′·i·ne·phrit′ic** (-frit′ik) *adj.*

per·i·neph·ri·um (per′ə·nef′rē·əm) *n. Anat.* The capsule of adipose tissue that invests the kidney. [<NL <Gk. *peri-* around + *nephros* a kidney] — **per′i·neph′ral** or **·ri·al** or **·ric** *adj.*

per·i·ne·um (per′ə·nē′əm) *n. pl.* **·ne·a** (-nē′ə) *Anat.* **1** The region of the body at the lower end of the trunk, between the genital organs and the rectum. **2** The entire region at the outlet of the pelvis, comprising the anus and the internal genitals. Also **per′i·nae′um.** [<LL <Gk. *perinaion, perineos*] — **per′i·ne′al** *adj.*

per·i·neu·ri·tis (per′ə·no͞o·rī′tis, -nyo͞o-) *n. Pathol.* Inflammation of the perineurium.

per·i·neu·ri·um (per′ə·no͝or′ē·əm, -nyo͝or′-) *n. pl.* **·ri·a** (-ē·ə) *Anat.* The connective tissue investing one of the bundles of fibers composing a nerve. [<NL <Gk. *peri-* around + *neuron* a nerve] — **per′i·neu′ri·al** *adj.*

per in·ter·im (pər in′tər·im) *Latin* In the meantime.

pe·ri·od (pir′ē·əd) *n.* **1** A definite portion of time marked and defined by some recurring event or phenomenon. **2** A lapse of time; a series of years; an age; era; also, a stage of life. **3** The concluding limit of any sequence of years, events, acts, or phenomena; termination. **4** The present day or time: with *the.* **5** *Astron.* The time of revolution of a heavenly body about its primary. **6** *Med.* A special phase or epoch distinguishable in the course of a disease: the *period* of augmentation; also, the menses. **7** A dot (.) placed on the line: used as a mark of rhetorical punctuation after every complete declarative sentence, after most abbreviations, as LL.D., pp., after titles, headings, and sideheads, and often after Roman numerals. The same mark serves also as a decimal point. **8** A sentence in which completion of the sense is suspended till the close. **9** *Geol.* One of the divisions of geologic time, intermediate between the shorter *epoch* and the longer *era*: the Cretaceous *period.* **10** *Music* A group of measures arranged in two or more phrases and comprising a complete musical statement. **11** *Math.* **a** The interval between the equal recurring values of a dependent variable. **b** Any one of similar groups into which a number is divided, as when a root is to be extracted: in numeration or in recurring decimals. **c** The length of the smallest subinterval in the graph of the function of a real variable. **12** *Physics* The time that elapses between two successive similar phases of a vibration. **13** The completion or end of a cycle, event, or series of events. **14** *Obs.* A particular occasion or moment. See synonyms under END, TIME. [<OF *periode* <L *periodus* <Gk. *periodos* a going around, a rounded surface <*peri-* around + *hodos* a way]

per·i·o·date (pə·rī′ə·dāt) *n. Chem.* A salt of periodic acid. [<PERIOD(IC ACID) + -ATE[3]]

period furniture Furniture in a style characteristic of any given period.

pe·ri·od·ic (pir′ē·od′ik) *adj.* **1** Pertaining to or of the nature of a period; characterized by periods; recurring after a definite interval; cyclic. **2** *Gram.* Belonging to a sentence that is grammatically complete. **3** In rhetoric, pertaining to or expressed in complete sentences: pertaining especially to a style in which several clauses hang upon one principal statement or sentence; hence, rhetorically elaborate. See also PERIODIC SENTENCE. **4** *Math.* **a** Of or pertaining to curves with ordinates repeated at equal distances along the abscissa. **b** Of or pertaining to the function of a real variable such that a graph of the function is identical within each subinterval.

per·i·od·ic acid (pûr′ī·od′ik) *Chem.* A compound, $HIO_4 \cdot 2H_2O$, containing iodine combined with oxygen at its highest valence. [<PER- (def. 5) + IODIC]

pe·ri·od·i·cal (pir′ē·od′i·kəl) *adj.* **1** Pertaining to publications, as magazines, etc., that appear at fixed intervals of more than one day; also, published at regular intervals. **2** Periodic (def. 1). — *n.* A publication, usually a weekly, monthly, or quarterly magazine, appearing at regular intervals. — **pe′ri·od′i·cal·ly** *adv.*

pe·ri·o·dic·i·ty (pir′ē·ə·dis′ə·tē) *n.* The quality of being periodic or of recurring at definite intervals of time, as in sunspots, an electric current, or the symptoms of a disease.

periodic law *Chem.* The statement that the physicochemical properties of the elements are functionally related to their atomic numbers and recur periodically when the elements are arranged in the order of these numbers.

periodic sentence A sentence that is not grammatically complete until the end; specifically, one of several rhetorical clauses so constructed as to suspend completion of both sense and structure until the close.

periodic spiral *Chem.* A complex graphic diagram of the elements arranged in a series of curves so as to illustrate the various relationships of properties, chemical behavior, etc., as expressed in the periodic law.

periodic system *Chem.* A classification of the elements in accordance with the relationships formulated by the periodic law.

periodic table *Chem.* A table in which the elements are arranged in physicochemical groups as determined, formerly by their atomic weights, now by atomic numbers. See opposite page.

per·i·o·dide (pə·rī′ə·dīd) *n. Chem.* An iodide having a larger proportion of iodine than any other iodide of the same series. Also **per·i′o·did** (-did).

per·i·o·don·tal (per′ē·ə·don′təl) *adj. Anat.* Occurring or situated around a tooth: the *periodontal* membrane lining the cement of a tooth: also called *peridental.* [<PERI- + -ODONT(O)- + -AL]

per·i·o·ma·ni·a (per′ē·ə·mā′nē·ə, -mān′yə) *n.* Dromomania. [<Gk. *peraioein* carry across, pass over + MANIA]

per·i·os·te·um (per′ē·os′tē·əm) *n. Anat.* A tough, fibrous, two-layered vascular membrane that surrounds and nourishes the bones. [<NL <L *periosteon* <Gk., neut. of *periosteos* around the bones <*peri-* around + *osteon* a bone] — **per′i·os′te·al, per′i·os′te·ous** *adj.*

per·i·os·ti·tis (per′ē·əs·tī′tis) *n. Pathol.* Inflammation of the periosteum. [<PERIOST(EUM) + -ITIS] — **per′i·os·tit′ic** (-tit′ik) *adj.*

per·i·o·tic (per′ē·ō′tik, -ot′ik) *Anat. adj.* Surrounding the inner ear; specifically, relating to the bony structure or capsule enclosing the labyrinth. — *n.* A periotic bone.

per·i·pa·tet·ic (per′i·pə·tet′ik) *adj.* **1** Walking about; moving from place to place. **2** Rambling, as of speech. — *n.* One given to walking about. [<OF *peripatetique* <L *peripateticus* <Gk. *peripatētikos* given to walking about <*peripatētēs* one who walks about <*peri-* around + *pateein* walk]

Per·i·pa·tet·ic (per′i·pə·tet′ik) *adj.* Pertaining to the philosophy of Aristotle, who lectured to his disciples while walking in the Lyceum at Athens. — *n.* A disciple of Aristotle; an adherent of his teachings.

pe·riph·er·al (pə·rif′ər·əl) *adj.* **1** Of or pertaining to a periphery. **2** Distant from the center; hence, distal; external. Also **per′i·pher′ic** or **·i·cal** (per′ə·fer′i·kəl). — **pe·riph′·er·al·ly** *adv.*

pe·riph·er·y (pə·rif′ər·ē) *n. pl.* **·er·ies** **1** The outer surface. **2** The surface of the body. **3** Circumference. **4** *Geom.* The sum of the sides of any polygon. **5** A surrounding region, country, or area. [<OF *periferie* <LL *peripheria* <Gk. *periphereia* circumference <*peripherēs* moving around <*peri-* around + *pherein* carry]

per·i·phrase (per′i·frāz′) *n.* Periphrasis.

pe·riph·ra·sis (pə·rif′rə·sis) *n. pl.* **·ses** (-sēz) Circumlocution, or an instance of it. See synonyms under CIRCUMLOCUTION. [<L <Gk. <*periphrazein* <*peri-* around + *phrazein* speak]

per·i·phras·tic (per′ə·fras′tik) *adj.* **1** Of the nature of or involving periphrasis; employing indirect words; circumlocutory. 2 *Gram.* Denoting a construction in which a phrase is substituted for an inflected form of similar function, as, *the hat of John* for *John's hat.* Also **per'i·phras'ti·cal.** — **per'i·phras'ti·cal·ly** *adv.*

periphrastic conjugation A conjugation formed by simple verbs with the aid of auxiliaries, instead of by inflection of the verb itself, as, *he did run* for *he ran.*

periphrastic genitive A genitive case formed not by inflection, but by a preposition, as in English by *of.*

per·i·plo·cin (per'ə·plō'sin) *n.* A powerful glycoside, $C_{30}H_{49}O_{12}$, extracted from the bark of the silk vine in the form of a yellow, bitter, amorphous powder: it resembles digitalis in action and is sometimes used in the treatment of certain heart conditions. [<NL *Periploca (graeca)* silk vine [<Gk. *periplokē* a twining (<*peri-* around + *plekein* twine) + -IN]

pe·rip·ter·al (pə·rip'tər·əl) *Archit. adj.* Having a detached row of columns extending around the cella: said especially of a temple. — *n.* A peripteral temple; peristyle: also **pe·rip'·ter, pe·rip'ter·os** (-tər·os). [<MF *périptère* <Med. L *peripteron* <Gk., neut. of *peripteros* winged about <*peri-* around + *pteron* a wing]

pe·rique (pə·rēk') *n.* A dark, strongly flavored tobacco grown in Louisiana. [<Creole, prob. alter. of slang E *prick* a phallus; so called when made into a carotte]

per·i·sarc (per'ə·särk) *n. Zool.* The chitinous excretion by which the soft parts of a hydroid colony are invested. [<PERI- + Gk. *sarx, sarkos* flesh] — **per'i·sar'cal, per'i·sar'cous** *adj.*

per·i·scope (per'ə·skōp) *n.* 1 An instrument consisting of a revolving prism capable of reflecting light rays down a vertical tube: used to guide submarine boats or to watch an enemy from a trench. 2 A special variety of photographic objective; a periscopic or wide-angled lens. — **per'i·scop'ic** or **·i·cal** (-skop'i·kəl) *adj.*

PERISCOPE
Showing the principle of reflection.

per·ish (per'ish) *v.i.* 1 To suffer a violent or untimely death. 2 To be destroyed; pass from existence. See synonyms under DIE. [<OF *periss-,* stem of *perir* <L *perire* <*per-* away + *ire* go]

per·ish·a·ble (per'ish·ə·bəl) *adj.* Liable to perish; mortal; liable to speedy decay, as fruit in transportation. — **per'ish·a·ble·ness, per'ish·a·bil'i·ty** *n.* — **per'·ish·a·bly** *adv.*

per·ish·a·bles (per'ish·ə·bəlz) *n. pl.* Goods liable to speedy decay: used chiefly of foods in transit.

per·i·sperm (per'ə·spûrm) *n. Bot.* Tissue surrounding the embryo sac in an ovule, in which nutrient material is stored. [<F *périsperme* <NL *perispermum* <Gk. *peri-* around + *sperma* seed] — **per'i·sper'mic** *adj.*

per·i·sphere (per'ə·sfir) *n. Physics* That portion of space within which the magnetic, electrical, or gravitational fields of an object produce observable effects.

per·i·spom·e·non (per'ə·spom'ə·non) *Gram. adj.* In Greek, having the circumflex accent on the final syllable. — *n.* A perispomenon word. [<Gk. *perispōmenon,* neut. ppr. passive of *perispaein* mark with the circumflex < *peri-* around + *spaein* draw]

pe·ris·so·dac·tyl (pə·ris'ō·dak'til) *adj.* 1 Odd-toed. 2 Of or pertaining to an order of ungulates (*Perissodactyla*) with an odd number of digits and an enlarged cecum, including horses, tapirs, rhinoceroses, etc. — *n.* An ungulate mammal belonging to this order. Also **pe·ris'so·dac'tyle.** [<NL <Gk. *perissos* odd, uneven + *dactylos* finger, toe] — **pe·ris'·so·dac·tyl'ic, ·dac'ty·lous** *adj.* — **pe·ris'so·dac'·ty·lism** *n.*

per·i·stal·sis (per'ə·stôl'sis, -stal'-) *n. pl.* **·ses** (-sēz) *Physiol.* A contractile muscular movement of any hollow organ of the body, as of the alimentary canal and intestines, whereby the contents are gradually propelled toward the point of expulsion. Compare SYSTALTIC. [<NL <Gk. *peristaltikos* <*peristellein* surround <*peri-* around + *stellein* place] — **per'i·stal'tic** *adj.*

pe·ris·ta·sis (pə·ris'tə·sis) *n. Biol.* The total environment of an individual organism, including all its vital processes. [<NL <Gk., an environment <*peri-* around + *stasis* a standing <*histanai* stand]

per·i·stome (per'ə·stōm) *n.* 1 *Bot.* The fringe of delicate teeth, generally some multiple of four, around the mouth of the capsule of mosses. 2 *Zool.* The parts that surround the mouth; specifically, the lip or margin of the mouth of a univalve; the space between the mouth and the tentacles of a sea anemone. 3 *Entomol.* The oval margin of the face or border of the mouth in a dipterous insect. Also **pe·ris·to·ma** (pə·ris'tə·mə), **per'i·sto'·mi·um.** [<NL *peristoma* <Gk. *peri-* around + *stoma* a mouth]

per·i·style (per'ə·stīl) *n. Archit.* 1 A system of columns about a building or an internal court. 2 An area or space so enclosed. [<MF *péristyle* <L *peristylum* <Gk. *peristylon,* neut. of *peristylos* surrounded by a colonnade <*peri-* around + *stylos* a pillar] — **per'i·sty'lar** *adj.*

per·i·the·ci·um (per'ə·thē'shē·əm, -sē·əm) *n. pl.* **·ci·a** (-shē·ə, -sē·ə) *Bot.* A closed or narrow-mouthed receptacle containing the fructification in certain fungi, especially the powdery mildews. [<NL <Gk. *peri-* around + *thēkē* a case] — **per'i·the'ci·al** *adj.*

per·i·to·ne·um (per'ə·tə·nē'əm) *n. pl.* **·ne·a** (-nē'ə) *Anat.* A serous membrane that lines the abdominal cavity in mammals and is reflected as a more or less complete investment over the viscera. In the higher vertebrates the peritoneum forms a completely closed sac, except in females, where the Fallopian tubes open into the cavity. Also **per'i·to·nae'um.** [<LL *peritonaeum* <Gk. *peritonaion* <*peritonos* stretched round <*peri-* around + *teinein* stretch] — **per'i·to·ne'al** or **·nae'al** *adj.*

per·i·to·ni·tis (per'ə·tə·nī'tis) *n. Pathol.* Acute inflammation of the peritoneum. [<*periton-* (<PERITONEUM) + -ITIS]

pe·rit·ri·cha (pə·rit'rə·kə) *n. pl.* Bacteria having flagella all around the body. [<NL <Gk. *peri-* around + *thrix, trichos* hair] — **pe·rit'·ri·chous** *adj.*

per·i·vis·cer·al (per'i·vis'ər·əl) *adj. Anat.* Situated about the viscera: the *perivisceral* cavity.

per·i·wig (per'ə·wig) *n.* A wig; peruke. [Earlier *perwyke,* alter. of *perruck* <MF *perruque* PERUKE]

per·i·win·kle[1] (per'ə·wing'kəl) *n.* 1 A small marine snail of the genus *Littorina,* especially the edible **European periwinkle** (*L. littorea*), now common on the east coast of the United States, or the **American periwinkle** (*L. palliata*). 2 Any of various other small univalves. [OE *pinewinclan, winewinclan,* pl., ? <L *pinna* a mussel (<Gk.) + OE *wincla* a shellfish; ? infl. in form by PERIWINKLE[2]]

per·i·win·kle[2] (per'ə·wing'kəl) *n.* A plant of the genus *Vinca* (family *Apocynaceae*); especially, either of two European trailing shrubs, *V. minor* and *V. major,* with shining, evergreen, opposite leaves, and blue, or sometimes white, flowers. They are commonly called *myrtle* or *creeping myrtle* in the United States. [OE *peruince* <L *pervinca* <*vinca pervinca,* prob. <*pervincire* <*per-* thoroughly + *vincire* bind]

periwinkle blue A medium mauve blue, the color of a periwinkle flower.

per·jure (pûr'jər) *v.t.* **·jured, ·jur·ing** 1 To make (oneself) guilty of perjury. 2 To find guilty of or involved in perjury: usually in the passive: if they are *perjured.* [<OF *parjurer* <L *perjurare* <*per-* through, badly + *jurare* swear] — **per'jur·er** *n.*

per·jured (pûr'jərd) *adj.* Guilty of perjury; having sworn falsely; forsworn: a *perjured* witness. — **per'jured·ly** *adv.*

per·ju·ry (pûr'jə·rē) *n. pl.* **·ries** *Law* The wilful and voluntary giving of false testimony or the withholding of material facts or evidence, in regard to a matter or thing material to the issue or point of inquiry in a legal document or while under oath lawfully administered in a judicial proceeding.

perk[1] (pûrk) *v.i.* 1 To recover one's spirits or vigor: with *up.* 2 To carry oneself or lift one's head jauntily. — *v.t.* 3 To raise quickly or smartly, as the ears: often with *up.* 4 To make (oneself) trim and smart in appearance: often with *up* or *out.* — *adj.* Holding up the head smartly or jauntily; pert: also **perk'y.** [ME *perken,* ? <AF *perquer* perch, roost] — **perk'i·ly** *adv.* — **perk'i·ness** *n.*

perk[2] (pûrk) *v.i. Colloq.* To percolate. [Short for PERCOLATE]

Per·kin (pûr'kin), **Sir William Henry,** 1838–1907, English chemist; founder of aniline dye industry, 1856.

Per·kins (pûr'kinz), **Frances,** born 1882, U. S. social worker and administrator; secretary of labor 1933–45.

Per·lis (pûr'lis) The smallest state in the Federation of Malaya, on the Strait of Malacca; 310 square miles; capital, Kangar.

per·lite (pûr'līt) *n.* An acid, igneous, glassy rock of the composition of obsidian, but divided into small spherical bodies by the stress developed by its contraction on cooling: also spelled *pearlite.* [<F <*perle* a pearl] — **per·lit'ic** (-lit'ik) *adj.*

Perm (pûrm) A city on the Kama in western Asiatic Russian S.F.S.R.: formerly *Molotov.*

per·ma·frost (pûr'mə·frôst, -frost) *n.* That part of the earth's surface in arctic regions which is permanently frozen. [<PERMA(NENT) + FROST]

Perm·al·loy (pûr'mə·loi) *n.* Any of a group of iron and nickel alloys with small quantities of other metals, characterized by high magnetic permeability: a trade name. [<PERM(EABLE) + ALLOY]

per·ma·nence (pûr'mə·nəns) *n.* The state of being permanent; durability; fixity. [<Med. L *permanentia* <L *permanens.* See PERMANENT.]

per·ma·nen·cy (pûr'mə·nən·sē) *n. pl.* **·cies** 1 Permanence. 2 Something permanent.

per·ma·nent (pûr'mə·nənt) *adj.* Continuing in the same state or without essential change; durable; fixed; stable: opposed to *temporary.* — *n.* A permanent wave. [<L *permanens, -entis* <*permanere* stay to the end <*per-* through + *manere* remain] — **per'ma·nent·ly** *adv.*

Synonyms (adj.): abiding, changeless, constant, durable, enduring, fixed, immutable, invariable, lasting, perpetual, persistent, stable, steadfast, unchangeable, unchanging. *Durable* is said almost wholly of material substances that resist wear; *lasting* is said of either material or immaterial things. *Permanent* is a word of wider meaning; a thing is *permanent* which is not liable to change; as, a *permanent* color. Buildings upon a farm are called *permanent* improvements. *Enduring* is applied to that which resists both time and change; as, *enduring* fame. See PERPETUAL. *Antonyms:* see synonyms for TRANSIENT.

Permanent Court of Arbitration See under COURT.

Permanent Court of International Justice See under COURT.

permanent set *Physics* A deformation of a rigid body or material that persists after the stress has been removed.

permanent wave An artificial wave mechanically or chemically produced on growing hair and lasting several months.

per·man·ga·nate (pər·mang'gə·nāt) *n. Chem.* A dark-purple salt of permanganic acid.

per·man·gan·ic (pûr'man·gan'ik) *adj. Chem.* Of, pertaining to, or designating an acid, $HMnO_4$, which is a powerful oxidizing agent in aqueous solutions.

per·me·a·bil·i·ty (pûr'mē·ə·bil'ə·tē) *n.* 1 The quality or condition of being permeable. 2 *Physics* The property of being easily traversed by magnetic lines of force; susceptibility to magnetization. 3 *Aeron.* The measure of the rate of diffusion of a gas per unit area of a balloon fabric under standard conditions: generally given in liters per square meter per 24 hours.

per·me·a·ble (pûr'mē·ə·bəl) *adj.* 1 Allowing passage, especially of fluids. 2 Designating a type of protective clothing treated to resist penetration by vapors and gases, but not by liquids. [<L *permeabilis* <*permeare* PERMEATE] — **per'me·a·bly** *adv.*

per·me·ance (pûr'mē·əns) *n. Electr.* Permeation; the reciprocal of the reluctance of a magnetic circuit.

per·me·ate (pûr'mē·āt) *v.* **·at·ed, ·at·ing** *v.t.* 1 To spread thoroughly through; pervade. 2 To pass through the pores or interstices of. — *v.i.* 3 To spread itself. [<L *permeatus,* pp. of *permeare* <*per-* through + *meare* pass] — **per'me·ant** *adj.* — **per'me·a'tion** *n.* — **per'·me·a'tive** *adj.*

per men·sem (pûr men'səm) *Latin* By the month.

Per·mi·an (pûr′mē·ən) *Geol.* *n.* The latest period of the Paleozoic era, following the Pennsylvanian and succeeded by the Triassic; also, the Permian rock system. —*adj.* Relating to this period. [after *Perm,* a former E. Russian province]

per mill (pûr mil′) In, into, or by the thousand. Also **per mil.** [<L *per* by + *mille* thousand]

per·mil·lage (pər·mil′ij) *n.* Proportion or rate per thousand; the number of thousandth parts.

per·mis·si·ble (pər·mis′ə·bəl) *adj.* That can be permitted; allowable. —**per·mis′si·bil′i·ty** *n.* —**per·mis′si·bly** *adv.*

per·mis·sion (pər·mish′ən) *n.* The act of permitting or allowing; license granted; formal authorization; consent. [<L *permissio, -onis* <*permissus,* pp. of *permittere* PERMIT]

Synonyms: allowance, authority, authorization, consent, leave, liberty, license, permit. *Authority* is rightful power conferred and limited by law; in a more general sense, *authority* is applied to any conceded power of control. *Permission* justifies another in acting without interference or censure, and usually implies some degree of approval. A *permit* is a special authorization, generally given in writing. A *license* is *permission* granted rather than *authority* conferred; the sheriff has *authority* (not *permission* or *license*) to make an arrest. *Consent* is *permission* by the concurrence of wills in two or more persons, a mutual approval or acceptance of something proposed. Compare synonyms for ALLOW. *Antonyms*: denial, hindrance, objection, opposition, prevention, prohibition, refusal, resistance.

per·mis·sive (pər·mis′iv) *adj.* **1** That permits; granting permission. **2** That is permitted; optional. —**per·mis′sive·ly** *adv.* —**per·mis′sive·ness** *n.*

per·mis·so·ry (pər·mis′ər·ē) *adj.* **1** Pertaining to or of the nature of permission. **2** *Law* Arising from or founded on permission; authorized; licensed.

per·mit (pər·mit′) *v.* **·mit·ted, ·mit·ting** *v.t.* **1** To allow the doing of; consent to. **2** To give (someone) leave or consent; authorize. **3** To afford opportunity for: His answer *permits* no misinterpretation. —*v.i.* **4** To afford possibility or opportunity. —*n.* (pûr′mit) Permission or warrant; especially, a written authorization to do something. [<L *permittere* <*per-* through <*mittere* send, let go] —**per·mit′ter** *n.*

Synonyms (verb): allow, authorize, empower, let, license, suffer, tolerate. See ALLOW, ENDURE. Compare synonyms for PERMISSION. *Antonyms*: disallow, forbid, prohibit, refuse.

per·mit·tiv·i·ty (pûr′mə·tiv′ə·tē) *n.* *Electr.* Specific inductive capacity of a dielectric.

per·mut·a·ble (pər·myōō′tə·bəl) *adj.* Capable of being changed or of undergoing change or interchange.

per·mu·ta·tion (pûr′myōō·tā′shən) *n.* **1** The act of permuting; change; transformation. **2** *Math.* **a** Change in the order of sequence of elements or objects in a series; especially, the making of all possible changes of sequence, as *abc, acb, bac, bca,* etc. **b** Any one of the arrangements thus made: distinguished from *combination.* [<OF *permutacion* <L *permutatio, -onis* <*permutatus,* pp. of *permutare.* See PERMUTE.]

per·mute (pər·myōōt′) *v.t.* **·mut·ed, ·mut·ing** To subject to permutation, especially, to change the order of. [<L *permutare* <*per-* thoroughly + *mutare* change]

Per·nam·bu·co (pûr′nəm·byōō′kō, *Pg.* per′·nəm·bōō′kōō) **1** A state in NE Brazil on the Atlantic; 37,458 square miles; capital, Recife. **2** Recife.

per·ni·cious (pər·nish′əs) *adj.* **1** Having the power of destroying or injuring; tending to kill or hurt; very injurious; deadly. **2** Malicious; wicked. [<OF *pernicieux* <L *perniciosus* <*pernicies* destruction <*per-* thoroughly + *nex, necis* death] —**per·ni′cious·ly** *adv.* —**per·ni′cious·ness** *n.*

Synonyms: bad, baneful, deadly, destructive, evil, harmful, hurtful, injurious, mischievous, noxious, perverting, ruinous. *Pernicious* is stronger than *injurious*; that which is *injurious* is capable of doing harm; that which is *pernicious* is likely to be *destructive.* See BAD, INIMICAL, NOISOME. *Antonyms*: advantageous, beneficial, favorable, good, healthful, helpful, profitable, salutary, serviceable, wholesome.

pernicious anemia *Pathol.* A morbid condition characterized by a diminution in the number of red blood corpuscles, abnormalities in the composition of the blood, and progressive disturbances in the muscular, nervous, and gastrointestinal systems.

per·nick·e·ty (pər·nik′ə·tē) See PERSNICKETY.

Pe·rón (pā·rōn′), **Juan Domingo,** born 1895, Argentine president 1946–55; deposed 1955.

per·o·ne·al (per′ə·nē′əl) *adj.* *Anat.* Of, pertaining to, or near, the fibula. [<NL *peronaeus* <*perone* the fibula <Gk. *peronē,* orig. a pin <*peirein* pierce]

Pé·ronne (pā·rôn′) A town on the Somme in northern France; scene of several battles during World War I.

per·o·rate (per′ə·rāt) *v.i.* **·rat·ed, ·rat·ing** **1** To speak at length; harangue. **2** To sum up or conclude a speech.

per·o·ra·tion (per′ə·rā′shən) *n.* The concluding portion of an oration; the recapitulation and summing up of an argument. [<L *peroratio, -onis* <*peroratus,* pp. of *perorare* speak to the end <*per-* through + *orare* speak]

per·ox·i·dase (pə·rok′sə·dās) *n.* *Biochem.* Any of a class of enzymes which, in the presence of hydrogen peroxide, accelerate the oxidation of various compounds. [<PEROXID(E) + -ASE]

per·ox·ide (pə·rok′sīd) *n.* *Chem.* **1** An oxide having a larger proportion of oxygen than any other oxide of the same series: distinguished from *protoxide.* **2** Hydrogen peroxide. Also **per·ox′id** (-sid). —*v.t.* **·id·ed, ·id·ing** To treat with peroxide; bleach, as hair, with peroxide.

per·pend[1] (pər·pend′) *v.t.* & *v.i.* *Obs.* To ponder; consider. [<L *perpendere* <*per-* thoroughly + *pendere* weigh]

per·pend[2] (pûr′pənd) *n.* In masonry, a stone header extending through a wall so that one end appears on each side of it. Also **perpend stone, per′pent** (-pənt). [Var. of *parpen* <OF *parpain,* ? <*par-* through (<L *per-*) + *pan (de mur)* a side (of a wall); infl. in form by PEND]

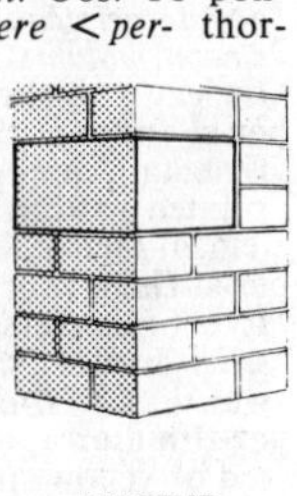

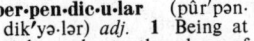

PERPEND

per·pen·dic·u·lar (pûr′pən·dik′yə·lər) *adj.* **1** Being at right angles to the plane of the horizon; upright or vertical. **2** *Math.* Meeting a given line or plane at right angles. See synonyms under RIGHT. —*n.* **1** A perpendicular line. **2** An appliance or instrument used to indicate the vertical line from any point; a plumb rule. **3** A line at right angles to another line or to a plane. **4** A vertical line or vertical face; loosely, any steep incline or face. **5** Perpendicular position. **6** Moral uprightness. [<OF *perpendiculer* <L *perpendicularis* <*perpendiculum* a plumb line <*per-* thoroughly + *pendere* hang] —**per′pen·dic′u·lar′i·ty** (-lar′ə·tē) *n.* —**per′pen·dic′u·lar·ly** *adv.*

Perpendicular architecture A late style of Gothic architecture in England from the end of the 14th century through the 16th century: characterized by accentuation of vertical lines in its tracery.

per·pe·trate (pûr′pə·trāt) *v.t.* **·trat·ed, ·trat·ing** To do, perform, or commit (a crime, etc.). [<L *perpetratus,* pp. of *perpetrare* carry through <*per-* through + *patrare* effect] —**per′pe·tra′tion** *n.* —**per′pe·tra′tor** *n.*

per·pet·u·al (pər·pech′ōō·əl) *adj.* **1** Continuing unlimited in time. **2** Incessant; ceaseless. **3** *Bot.* Being in bloom during all or nearly all the year, as certain hybrid flowers. —*n.* Any perennial plant; also, any of certain perpetual hybrid roses. [<OF *perpetuel* <L *perpetualis* <*perpetuus* <*per-* through + *petere* seek] —**per·pet′u·al·ly** *adv.* —**per·pet′u·al·ness** *n.*

Synonyms (adj.): ceaseless, constant, continual, continuous, endless, enduring, eternal, incessant, interminable, lasting, perennial, permanent, sempiternal, unceasing, unending, unfailing, unintermitted, uninterrupted. See CONTINUAL, ETERNAL, PERMANENT. *Antonyms*: see synonyms for TRANSIENT.

perpetual calendar See under CALENDAR.

perpetual motion See under MOTION.

per·pet·u·ate (pər·pech′ōō·āt) *v.t.* **·at·ed, ·at·ing** To make perpetual or enduring. [<L *perpetuatus,* pp. of *perpetuare* perpetuate <*perpetuus* PERPETUAL] —**per·pet′u·a′tion** *n.* —**per·pet′u·a′tor** *n.*

per·pe·tu·i·ty (pûr′pə·tōō′ə·tē, -tyōō′-) *n.* *pl.* **·ties** **1** The quality or state of being perpetual. **2** Something that has perpetual existence or worth. **3** Unending or unlimited time. **4** *Law* A limitation rendering property inalienable; also, the property so limited. **5** In annuities, a perpetual annuity, or the number of years' purchase to be given for it; the number of years in which the simple interest of a sum becomes equal to the principal. [<OF *perpetuité* <L *perpetuitas* <*perpetuus* PERPETUAL]

Per·pi·gnan (per·pē·nyän′) A city of southern France near the Spanish border and the Mediterranean.

per·plex (pər·pleks′) *v.t.* **1** To cause to hesitate, as from doubt; confuse, as with difficult problems; puzzle. **2** To make complicated, intricate, or confusing. [Back formation from PERPLEXED]

Synonyms: bewilder, bother, complicate, confound, confuse, distract, embarrass, entangle, harass, involve, mystify, pose, puzzle, trouble. *Antonyms*: clarify, disentangle, elucidate, explain, simplify.

per·plexed (pər·plekst′) *adj.* **1** Confused; embarrassed. **2** Of a complicated character; involved. [Appar. alter. of obs. *perplex,* adj., intricate <L *perplexus* involved <*per-* thoroughly + *plexus,* pp. of *plectere* plait] —**per·plex′ed·ly** *adv.* —**per·plex′ed·ness** (-plek′sid·nis) *n.*

per·plex·ing (pər·plek′sing) *adj.* Confusing; puzzling; embarrassing; intricate. —**per·plex′ing·ly** *adv.*

per·plex·i·ty (pər·plek′sə·tē) *n.* *pl.* **·ties** **1** Mental difficulty owing to doubt, confusion, etc. **2** That which perplexes; also, an instance of bewilderment. **3** The quality of being intricate or complicated; entanglement.

Synonyms: amazement, astonishment, bewilderment, confusion, distraction, disturbance, doubt, embarrassment. *Perplexity* is the drawing of the thoughts or faculties by turns in different directions or toward contrasted or contradictory conclusions; *confusion* is a state in which the mental faculties are thrown into chaos, so that the clear and distinct action of perception, memory, reason, and will, is lost; *bewilderment* is akin to *confusion,* but is less overwhelming, and more readily recovered from. *Perplexity* has not the unsettling of the faculties implied in *confusion,* nor the overwhelming of the faculties implied in *amazement* or *astonishment.* With an excitable person, *bewilderment* may deepen into *confusion* that will make him unable to think clearly or even to see or hear distinctly. *Amazement* results from the sudden and unimagined occurrence of great good or evil or the sudden awakening of the mind to unthought-of truth. *Astonishment* often produces *bewilderment,* which the word was formerly understood to imply. See AMAZEMENT, ANXIETY, CARE, DOUBT.

per·qui·site (pûr′kwə·zit) *n.* Any incidental profit from service beyond salary or wages; hence, any privilege or benefit claimed as due. [<Med. L *perquisitum* an acquisition <L, a thing diligently sought, orig. pp. neut. of *perquirere* <*per-* thoroughly + *quaerere* seek]

Per·rault (pe·rō′), **Charles,** 1628–1703, French author and compiler of fairy tales.

Per·rin (pe·raṅ′), **Jean,** 1870–1942, French physicist.

per·ron (per′ən, *Fr.* pe·rôṅ′) *n.* *Archit.* A flight of external steps and a platform before the entrance door of a building. [<OF <L *petra* stone]

per·ry (per′ē) *n.* A fermented drink made from the expressed juice of pears. [<OF *peré* <LL *pera.* See PEAR.]

Perry (per′ē), **Bliss,** 1860–1954, U.S. author and critic. —**Matthew Calbraith,** 1794–1858,

U.S. commodore; opened Japan to commerce in 1852. — **Oliver Hazard**, 1785–1819, U.S. naval commander; defeated British on Lake Erie, Sept. 10, 1813; brother of the preceding. — **Ralph Barton**, 1876–1956, U.S. philosopher and educator.

per·salt (pûr′sôlt′) *n. Chem.* A salt formed by combination of a negative radical or ion with a metal at a high, or its highest, state of oxidation.

per sal·tum (pər sôl′təm) *Latin* By a leap; without intermediate degrees.

perse (pûrs) *adj.* Grayish-blue. —*n.* A grayish blue. [<OF *pers* <LL *persus*]

per se (pûr sē′, sā′) *Latin* By itself, himself, or herself; intrinsically; in or of its own nature, without reference to its relations.

per·se·cute (pûr′sə·kyōōt) *v.t.* **·cut·ed, ·cut·ing** 1 To harass with cruel or oppressive treatment, especially because of race, religion, or opinions. 2 To annoy or harass persistently. [<OF *persecuter* <L *persecutus*, pp. of *persequi* pursue <*per-* thoroughly + *sequi* follow] —**per′se·cu′tive** *adj.* —**per′se·cu′tor** *n.*
Synonyms: afflict, distress, harass, harry, molest, oppress, torment, worry. See ABUSE. *Antonyms*: advance, advocate, aid, assist, befriend, cherish, countenance, encourage, favor, help, indulge, support, sustain, tolerate.

per·se·cu·tion (pûr′sə·kyōō′shən) *n.* 1 The act or process of persecuting; cruel oppression. 2 Any period characterized by systematic oppression, infliction of torture, death, or the like, on account of religious belief. —**per′se·cu′tion·al, per·se·cu·to·ry** (pûr′sə·kyōō′tər·ē) *adj.*

Per·se·ids (pûr′sē·idz) *n. pl. Astron.* The meteors belonging to the group that has its radiant point in the constellation Perseus, which appear about August 12 of each year. Also **Per·se·i·des** (pər·sē′ə·dēz). [<NL *Perseis*, pl. *Perseides* <Gk. *Perseïs*, pl. *Perseïdes*, a daughter of Perseus]

Per·seph·o·ne (pər·sef′ə·nē) In Greek mythology, the daughter of Zeus and Demeter, abducted by Pluto and made queen of the kingdom of the dead, but allowed to return to the earth for a third of each year: identified with the Roman *Proserpine.*

Per·sep·o·lis (pər·sep′ə·lis) An ancient, ruined capital of Persia NE of Shiraz in SW central Iran.

Per·seus (pûr′syōōs, -sē·əs) 1 In Greek mythology, the son of Zeus and Danae, slayer of Medusa and savior and husband of Andromeda. 2 In the Apocrypha, the last king of Macedonia; died about 164 B.C. I *Mac.* viii 5. 3 A northern constellation. See CONSTELLATION. [<L <Gk.]

per·se·ver·ance (pûr′sə·vir′əns) *n.* 1 The act or habit of persevering; persistence. 2 *Theol.* In Calvinism, the continuance in grace and certain salvation of those whom God effectually calls, accepts in Christ, and sanctifies by his spirit. [<OF *perseverance* <L *perseverantia* steadfastness <*perseverans, -antis*, pp. of *perseverare* PERSEVERE]
Synonyms: constancy, indefatigableness, persistence, persistency, resolution, steadiness, steadfastness, tenacity. See INDUSTRY. *Antonyms*: caprice, fickleness, fitfulness, inconstancy, irresolution, levity, unsteadiness, vacillation, volatility.

per·sev·er·a·tion (pər·sev′ə·rā′shən) *n. Psychol.* 1 The continual repetition of an activity or mental state. 2 The spontaneous recurrence in the mind of the same idea, phrase, tune, mental image, etc., irrespective of associative factors.

per·se·vere (pûr′sə·vir′) *v.i.* **·vered, ·ver·ing** To persist in any purpose or enterprise; continue striving in spite of difficulties, etc. [<OF *perseverer* <L *perseverare* <*perseverus* very strict <*per-* thoroughly + *severus* strict]
Synonyms: continue, endure, persist. *Persevere* is almost uniformly employed in the good and high sense of holding to a worthy course against all difficulty, danger, hindrance, or opposition; *persist* is often used of an annoying or perverse adherence to a demand or purpose that might well be abandoned. See INSIST, PERSIST. Compare OBSTINATE. *Antonyms*: see synonyms for CEASE.

per·se·ver·ing (pûr′sə·vir′ing) *adj.* Persistent of purpose. —**per′se·ver′ing·ly** *adv.*

Per·shing (pûr′shing), **John Joseph**, 1860–1948, U.S. general; commander in chief of the American Expeditionary Force of World War I.

Per·sia (pûr′zhə, -shə) The former name for IRAN. See PERSIAN EMPIRE.

Per·sian (pûr′zhən, -shən) *adj.* Of or pertaining to ancient Persia or modern Iran, its people, its language, or its architecture. —*n.* 1 A native or inhabitant of Persia or Iran. 2 The Iranian language of the Persians: historically divided into **Old Persian**, recorded in the cuneiform inscriptions of Darius I and his successors, and closely related to the language of the Avesta; **Middle Persian**, chiefly represented by Pahlavi, a literary language written in a Semitic alphabet, used in the sacred writings of the Zoroastrian religion from the third to the seventh century; and **Modern Persian**, containing many Arabic loan words and written in Arabic script. 3 A fine silk used formerly for linings. 4 *pl.* Persian blinds. [<OF *persien*, ult. <L *Persia* Persia <Gk. *Persis* <OPersian *Pārsa*]

Persian blinds Outside window shutters of thin, movable slats fastened in a frame.

Persian carpet A hand-woven Oriental carpet with connected design, the warp and filling of silk, wool, or cotton, and the pile of wool.

Persian Empire An empire of SW Asia, extending from the Indus to the Mediterranean: founded by Cyrus the Great (sixth century B.C.) and destroyed by Alexander the Great (331 B.C.).

Persian Gulf An arm of the Arabian Sea between Iran and Arabia; 90,000 square miles.

Persian lamb The young of certain sheep of central Asia, especially of the karakul; also, its skin, used as a fur; astrakan. See KARAKUL.

Persian wheel A noria.

per·si·car·y (pûr′sə·ker′ē) *n. pl.* **·car·ies** The knotweed; especially, the lady's-thumb (*Polygonum persicano*). [<NL *persicaria* <L *persicum (malum)* a peach; from the likeness of its leaves to those of the peach tree]

per·si·enne (pûr′zē·en′, *Fr.* per·syen′) *n.* 1 An Oriental cambric or muslin with colored printed pattern. 2 *pl.* Persian blinds. [<F, fem. of *persien* PERSIAN]

per·si·flage (pûr′sə·fläzh′) *n.* A light, flippant style of conversation or writing. [<F <*persifler* banter <L *per-* through + F *siffler* whistle <L *sifilare*]

per·sim·mon (pər·sim′ən) *n.* 1 The orange-red or yellow, plumlike fruit of an American tree of the ebony family (genus *Diospyros*), very astringent in taste until exposed to frost. 2 The tree, its hard blackish wood, or its tonic and astringent bark. [<Algonquian. Cf. Cree *pasiminan* dried fruit.]

Per·sis (pûr′sis) The ancient name for FARS.

per·sist (pər·sist′, -zist′) *v.i.* 1 To continue firmly in some course, state, etc., especially despite opposition or difficulties. 2 To be insistent, as in repeating a statement. 3 To continue to exist; endure. [<L *persistere* <*per-* thoroughly + *sistere* stand]
Synonyms: continue, endure, insist, last, persevere, remain, stay. As applied to duration, *last* is applied chiefly to things, *endure* to either persons or things. That *remains* or *stays* which is simply let alone; that which *endures* or *persists* does so against opposing forces. A man *insists* upon his demand, *persists* in his refusal. See under INSIST, PERSEVERE. *Antonyms*: see synonyms for CEASE.

per·sis·tence (pər·sis′təns, -zis′-) *n.* 1 The act of persisting in any course or enterprise; the quality of being persistent; perseverance; fixed adherence to a resolve, course of conduct, etc. 2 The continuance of an effect longer than the cause that first produced it. Also **per·sis′ten·cy.** See synonyms under INDUSTRY, PERSEVERANCE.

per·sis·tent (pər·sis′tənt, -zis′-) *adj.* 1 Firm and persevering in a course or resolve. 2 Enduring; permanent; continuous. 3 *Bot.* Not falling away; remaining for a long time or after the neighboring parts have reached maturity, as the calyx or petals in certain flowers. 4 *Zool.* Retained throughout life, as the gills of fishes and some amphibians. Compare DECIDUOUS. See synonyms under INDEFATIGABLE, INFLEXIBLE, OBSTINATE, PERMANENT. [<L *persistens, -entis*, ppr. of *persistere*. See PERSIST.] —**per·sis′tent·ly** *adv.*

Per·si·us (pûr′shē·əs, -shəs), A.D. 34–62, Roman satirist.

per·snick·e·ty (pər·snik′ə·tē) *adj. Colloq.* 1 Unduly fastidious; fussy; overprecise. 2 Demanding minute care or pains. Also *pernickety.* [<dial. E, ? alter. of PARTICULAR] —**per·snick′e·ti·ness** *n.*

per·son (pûr′sən) *n.* 1 A human being as including body and mind; an individual. 2 The body of a human being or its characteristic appearance and condition. 3 *Law* Any human being, corporation, or body politic having legal rights and duties. 4 *Theol.* One of the three individualities in the triune God; hypostasis. 5 *Gram.* **a** A modification of the pronoun and verb that distinguishes the speaker (**first person**), the person or thing spoken to (**second person**), and the person or thing spoken of (**third person**). **b** Any of the forms or inflections indicating this, as *I* or *we, you, he, she, it.* 6 An individual. 7 Superciliously, a common individual. 8 A part acted on the stage. —**in person** Present in the flesh; present and acting for oneself. [<F *personne* <L *persona* mask for actors. Doublet of PARSON.]

per·so·na (pər·sō′nə) *n. pl.* **·nae** (-nē) 1 Literally, person; specifically, a character in a drama, novel, etc.: dramatis *personae.* 2 In Jung's analytic psychology, the conscious artificial or masked personality complex developed by an individual, in contrast to his innate personality characteristics, for purposes of concealment, defense, deception, or adaptation to his environment. [<L, a person, orig. a mask]

per·son·a·ble (pûr′sən·ə·bəl) *adj.* Attractive in person; of good appearance. —**per′son·a·bly** *adv.*

per·son·age (pûr′sən·ij) *n.* 1 A man or woman of importance or rank. 2 An assumed character; an impersonation. 3 A character in fiction, history, etc.; a character in a play. [<OF <L *persona* a person]

per·so·na gra·ta (pər·sō′nə grä′tə, grā′tə) *Latin* An acceptable person.

per·son·al (pûr′sən·əl) *adj.* 1 Pertaining to or characteristic of a particular person; not general or public: a purely *personal* matter. 2 Belonging or relating to or constituting a person or persons, as distinguished from things; characteristic of human beings or free agents. 3 Performed by or done to the person directly concerned; done in person: a *personal* service. 4 Affecting or relating to one individually: *personal* habits. 5 Of or pertaining to the body or appearance: *personal* beauty. 6 Directly characterizing an individual; hence, concerning one's character or conduct, often in the sense of disparaging. 7 *Law* Appertaining to the person; movable: *personal* effects. 8 *Gram.* Denoting or indicating the person: *personal* pronouns. —*n.* 1 *Law* A movable article of property; chattel. 2 A paragraph or advertisement of personal reference or application. [<OF <LL *personalis* <L *persona* a person]

personal effects Goods, articles, and items of property having a more or less intimate relation to the person of the possessor.

personal equation 1 In precision observations and measurements, any deviation from a correct or standard value caused by variations in the technical skill or personal qualities of the observer. 2 Any individual characteristic which influences attitudes, judgments, or the full use of the reasoning powers in any situation.

per·son·al·i·ty (pûr′sən·al′ə·tē) *n. pl.* **·ties** 1 That which constitutes a person; also, that which distinguishes and characterizes a person; personal existence. 2 Anything said of a person, especially if disparaging: usually in the plural: offensive *personalities.* 3 A person, especially one of exceptional qualities. —**double** or **multiple personality** *Psychiatry* A condition in which two or more relatively distinct sets of experiences and behavior patterns reveal themselves alternately in the same individual. See synonyms under CHARACTER. [<OF *personalité* <Med. L *personalitas, -tatis* <LL *personalis* of a person]

per·son·al·ize (pûr′sən·əl·īz′) *v.t.* **·ized, ·iz·ing** 1 To make personal. 2 To personify.

per·son·al·ly (pûr′sən·əl·ē) *adv.* 1 In proper person; not through an agent. 2 With reference to one's own personality. 3 In a personal manner.

personal pronoun See under PRONOUN.

personal property Property that may attend the person of the owner; movables.

per·son·al·ty (pûr′sən·əl·tē) *n. pl.* **·ties** Personal property. [<AF *personaltie* <Med. L *personalitas.* See PERSONALITY.]

per·so·na non gra·ta (pər·sō′nə non grä′tə, grā′tə) *Latin* A person not acceptable.

per·son·ate (pûr′sən·āt) *v.t.* **·at·ed, ·at·ing 1** To act the part of, as a character in a play. **2** To personify, as in poetry, art, etc. **3** *Law* To impersonate with intent to deceive. See synonyms under IMITATE. — *adj.* **1** *Bot.* Masklike; masked: said specifically of a gamopetalous, two-lipped corolla in which the mouth of the tube is closed by an inflated projection of the throat. **2** Impersonated; feigned. [<L *personatus* masked <*persona* a mask] —**per′·son·a′tion** *n.* —**per′son·a′tive** *adj.* —**per′son·a′tor** *n.*

per·son·i·fi·ca·tion (pər·son′ə·fə·kā′shən) *n.* **1** The figurative endowment of inanimate objects or qualities with personality or human attributes. **2** Striking or typical exemplification of a quality or attribute in one's person; embodiment: She was the *personification* of joy. **3** The emblematic representation of an abstract quality or idea by a human figure. **4** Impersonation.

per·son·i·fy (pər·son′ə·fī) *v.t.* **·fied, ·fy·ing 1** To think of or represent as having life or human qualities. **2** To represent (an abstraction or inanimate object) as a person; symbolize. **3** To be the embodiment of; typify: He *personifies* honor. [<F *personnifier* <L *persona* a mask, person + *facere* make] —**per·son′i·fi′er** *n.*

per·son·nel (pûr′sə·nel′) *n.* **1** Persons collectively. **2** The persons employed in a business or in military service. **3** The collective characteristics of such a body of persons. — *adj.* Of or pertaining to personnel; directing personnel. [<F <OF *personal* PERSONAL]

per·spec·tive (pər·spek′tiv) *n.* **1** The art or theory of representing, by a drawing made on a flat or curved surface, solid objects or surfaces conceived of as not lying in that surface; delineation of objects as they appear to the eye. **2** The art of conveying the impression of depth and distance; representation of scenes as they appear to the eye, by means of correct drawing, shading, etc. **3** The effect of distance upon the appearance of objects, by means of which the eye judges spatial relations. **4** The relative importance of facts or matters from any special point of view; also, their presentation with just regard to their proportional importance. **5** A distant view; vista; prospect: also figuratively. **6** A picture giving the illusion of a scene of nature. —**aerial perspective** The art of indicating the relative distances of objects by gradations of tone and color. —**linear perspective** The art or method of producing an appearance of distance by means of converging lines. —*adj.* Pertaining to the art of perspective; also, drawn in perspective. [<Med. L *perspectiva (ars)* optical (art) <LL *perspectivus* optical <L *perspectus*, pp. of *perspicere* <*per-* through + *specere* look] —**per·spec′tive·ly** *adv.*

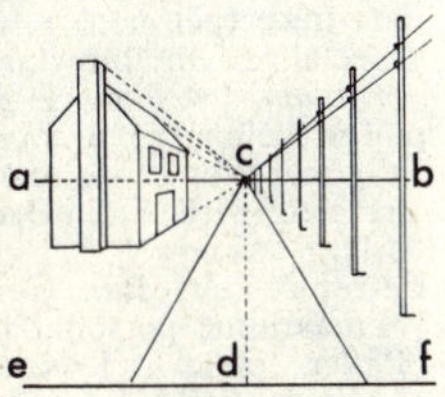

PERSPECTIVE
ab. Horizon.
c. Vanishing point (point of sight).
dc. Line of sight.
ef. Ground line.

per·spi·ca·cious(pûr′spə·kā′shəs)*adj.* **1** Keenly discerning or understanding. **2** Quick-eyed; sharp-sighted. See synonyms under ACUTE, ASTUTE, SAGACIOUS. [<L *perspicax, -acis* sharp-sighted <*perspicere.* See PERSPECTIVE.] —**per′spi·ca′cious·ly** *adv.* —**per′spi·ca′cious·ness** *n.*

per·spi·cac·i·ty (pûr′spə·kas′ə·tē) *n.* Keenness in mental penetration or discernment. See synonyms under ACUMEN.

per·spi·cu·i·ty (pûr′spə·kyo͞o′ə·tē) *n.* Clearness of expression or style; lucidity.
Synonyms: clearness, distinctness, explicitness, intelligibility, lucidity, plainness. *Antonyms*: ambiguity, cloudiness, confusion, incomprehensibility, indistinctness, intricacy, obscurity, unintelligibility, vagueness.

per·spic·u·ous (pər·spik′yo͞o·əs) *adj.* Having the quality of perspicuity; clear; lucid. See synonyms under CLEAR, PLAIN. [<L *perspicuus* clear, transparent <*perspicere.* See PERSPECTIVE.] —**per·spic′u·ous·ly** *adv.* —**per·spic′u·ous·ness** *n.*

per·spi·ra·tion (pûr′spə·rā′shən) *n.* **1** The exuding of the saline fluid secreted by the sweat glands of the skin. **2** The saline fluid excreted; sweat. —**per·spir·a·to·ry** (pər·spī′rə·tôr′ē, -tō′rē) *adj.*

per·spire (pər·spīr′) *v.* **·spired, ·spir·ing** *v.i.* To give off perspiration through the pores of the skin; sweat. — *v.t.* To give off through pores; exude. [<L *perspirare* breathe, blow constantly <*per-* through + *spirare* breathe] —**per·spir′a·ble** *adj.*

per·suade (pər·swād′) *v.t.* **·suad·ed, ·suad·ing 1** To move (a person, etc.) to do something by arguments, inducements, pleas, etc. **2** To induce to a belief; convince. [<L *persuadere* <*per-* thoroughly + *suadere* advise] —**per·suad′a·ble** *adj.* —**per·suad′er** *n.*
Synonyms: allure, coax, convince, dispose, entice, impel, incite, incline, induce, influence, lead, move, urge, win. Of these words *convince* alone has no direct reference to moving the will, denoting an effect upon the understanding only; one may be *convinced* of truth that has no manifest connection with duty or action, as of a mathematical proposition. To *persuade* is to bring the will of another to a desired decision by some influence exerted upon it short of compulsion; one may be *convinced* that the earth is round; he may be *persuaded* to travel around it; but persuasion is so largely dependent upon conviction that it is commonly held to be the orator's work first to *convince* in order that he may *persuade.* *Coax* is a slighter word than *persuade*, seeking the same end by shallower methods, largely by appeal to personal feeling, with or without success; as, a child *coaxes* a parent to buy him a toy. One may be *induced* by means not properly included in persuasion, as by bribery or intimidation; he is *won* over chiefly by personal influence. See ACTUATE, BEND, CONVINCE, INFLUENCE. *Antonyms*: deter, discourage, dissuade, hinder, repel, restrain.

per·sua·si·ble (pər·swā′sə·bəl) *adj.* Open to persuasion; persuadable. —**per·sua′si·bil′i·ty, per·sua′si·ble·ness** *n.*

per·sua·sion (pər·swā′zhən) *n.* **1** The act of persuading or of using persuasive methods. **2** The state of being persuaded; settled opinion; conviction. **3** A settled belief; accepted creed; hence, a party, sect, or denomination. **4** *Colloq.* Sort; kind: the male *persuasion.* See synonyms under COUNSEL. [<L *persuasio, -onis* <*persuasus*, pp. of *persuadere* PERSUADE]

per·sua·sive (pər·swā′siv) *adj.* Having power or tendency to persuade. —*n.* That which persuades or tends to persuade. —**per·sua′sive·ly** *adv.* —**per·sua′sive·ness** *n.*

per·sul·fate (pər·sul′fāt) *n. Chem.* A salt of persulfuric acid. Also **per·sul′phate.**

per·sul·fu·ric (pûr′sul·fyo͝or′ik) *adj. Chem.* Designating an acid, $H_2S_2O_8$, formed by the electrolysis of concentrated sulfuric acid. Also **per′sul·phu′ric.**

pert (pûrt) *adj.* **1** Disrespectfully forward or free; saucy. **2** *Dial.* Of fine appearance; comely; sprightly. See synonyms under IMPUDENT. [Aphetic var. of APERT] —**pert′ly** *adv.* —**pert′ness** *n.*

per·tain (pər·tān′) *v.i.* **1** To have reference; relate. **2** To belong as an adjunct, function, quality, etc.: the house and lands that *pertain* thereto. **3** To be fitting or appropriate: the joys that *pertain* to youth. [<OF *partenir* <L *pertinere* extend <*per-* through + *tenere* hold]
Synonyms: appertain, belong, concern, regard, relate.

pertaining to Having to do with or characteristic of; belonging or relating to.

Perth (pûrth) **1** An eastern midland county of Scotland; 2,493 square miles; county town, Perth. Also **Perth′shire** (-shir). **2** The capital of Western Australia, near the SW coast.

per·ti·na·cious (pûr′tə·nā′shəs) *adj.* **1** Tenacious of purpose; stubbornly adhering to a pursuit or opinion; also, perversely or doggedly persistent. **2** Continuing without abatement; incessant. See synonyms under INFLEXIBLE, OBSTINATE, URGENT. [<L *pertinax, -acis* <*per-* thoroughly, very + *tenax, -acis* tenacious] —**per′ti·na′cious·ly** *adv.*

per·ti·nac·i·ty (pûr′tə·nas′ə·tē) *n.* **1** Persistent tenacity of purpose; unyielding adherence. **2** Dogged perseverance; obstinacy.

per·ti·nent (pûr′tə·nənt) *adj.* Related to or properly bearing upon the matter in hand; relevant. See synonyms under APPROPRIATE. [<OF *partenant*, ppr. of *partenir* PERTAIN; refashioned after L *pertinens, -entis*] —**per′ti·nence, per′ti·nen·cy** *n.* —**per′ti·nent·ly** *adv.*

per·turb (pər·tûrb′) *v.t.* To disquiet or disturb greatly; alarm; agitate. [<OF *perturber* <L *perturbare* <*per-* thoroughly + *turbare* disturb <*turba* turmoil] —**per·turb′a·ble** *adj.*

per·tur·ba·tion (pûr′tər·bā′shən) *n.* **1** The state of being perturbed, or the act of perturbing. **2** *Astron.* Deviation in the motion of a heavenly body, caused by the attraction of some other body than that round which it moves. **3** A cause of disquiet or disturbance. Also **per·turb·ance** (pər·tûr′bəns). [<OF *perturbacion* <L *perturbatio, -onis* <*perturbare* PERTURB]

per·tus·sis (pər·tus′is) *n.* **1** *Pathol.* Whooping cough. **2** Any violent convulsive or spasmodic cough. [<NL <L *per-* thoroughly, very great + *tussis* a cough] —**per·tus′sal** *adj.*

Pe·ru (pə·ro͞o′, *Sp.* pā·ro͞o′) A republic in western South America on the Pacific; 533,916 square miles; capital, Lima.

Pe·ru·gia (pā·ro͞o′jä) A city in central Italy. **Pe·ru·gia** (pā·ro͞o′jä), **Lake of** See TRASIMENO.

Pe·ru·gi·no (pā′ro͞o·jē′nō), 1446–1523, Italian painter; real name *Pietro Vannucci.*

pe·ruke (pə·ro͞ok′) *n.* A periwig. [<MF *perruque* <Ital. *perruca*, ? ult. <L *pilus* hair]

per·ul·ti·mate (pər·ul′tə·mit) *adj.* Designating a magnitude or condition that cannot be exceeded: a *perultimate* yield of crops. [<PER- (def. 2) + ULTIMATE]

pe·ruse (pə·ro͞oz′) *v.t.* **·rused, ·rus·ing 1** To read carefully or attentively. **2** To read. **3** To examine; scrutinize. [<PER- + USE, *v.*] —**pe·rus′a·ble** *adj.* —**pe·ru′sal** *n.* —**pe·rus′er** *n.*

Pe·ru·vi·an (pə·ro͞o′vē·ən) *adj.* Of or pertaining to Peru or the Peruvians. —*n.* **1** A native or inhabitant of Peru. **2** An Indian of any of the Quechuan tribes of the ancient Inca empire.

Peruvian bark Jesuits' bark.

Pe·ruz·zi (pā·ro͞o′tsē), **Baldassare**, 1481–1536, Italian architect and painter.

per·vade (pər·vād′) *v.t.* **·vad·ed, ·vad·ing** To pass or spread through every part of; be diffused throughout; permeate. [<L *pervadere* <*per-* through + *vadere* go] —**per·va′sion** (-zhən) *n.*

per·va·sive (pər·vā′siv) *adj.*Thoroughly penetrating or permeating. [<L *pervasus*, pp. of *pervadere* PERVADE] —**per·va′sive·ly** *adv.* —**per·va′sive·ness** *n.*

per·verse (pər·vûrs′) *adj.* **1** Wrong or erring; different or varying from the correct or normal; also, unreasonable. **2** Thwarting or refractory. **3** Disposed to vex; petulant. [<OF *pervers* <L *perversus* turned the wrong way, orig. pp. of *pervertere.* See PERVERT.] —**per·verse′ly** *adv.* —**per·verse′ness** *n.* —**per·ver′sive** *adj.*
Synonyms: contrary, factious, fractious, froward, intractable, obstinate, petulant, stubborn, ungovernable, untoward, wayward, wilful. *Perverse* signifies wilfully wrong or erring, unreasonably set against right, reason, or authority. The *stubborn* or *obstinate* person will not do what another desires or requires; the *perverse* person will do anything contrary to what is desired or required of him. The *petulant* person frets, but may comply; the *perverse* individual may be smooth or silent, but is wilfully *intractable. Wayward* refers to a *perverse* disregard of morality and duty; *froward* is now almost obsolete; *untoward* is rarely heard except in certain phrases; as, *untoward* circumstances. Compare OBSTINATE. *Antonyms*: accommodating, amenable, complaisant, compliant, genial, governable, kind, obliging.

add,āce,câre,pälm; end,ēven; it,īce; odd,ōpen,ôrder; to͝ok,po͞ol; up,bûrn; ə = a in *above*, e in *sicken*, i in *clarity*, o in *melon*, u in *focus*; yo͞o = u in *fuse*; oi,oil; ou,pout; ch,check; g,go; ng,ring; th,thin; th,this; zh,vision. Foreign sounds à,œ,ü,kh,ṅ; and ♦: see page xx. < from; + plus; ? possibly.

per·ver·sion (pər·vûr′zhən, -shən) *n.* 1 The act of perverting, or the state of being perverted. 2 *Pathol.* A deviation from the normal in structure or function. 3 *Psychiatry* Deviation from the normal in sexual desires or activities.

per·ver·si·ty (pər·vûr′sə·tē) *n. pl.* **·ties** 1 The state or quality of being perverse. 2 Perverse nature or behavior. 3 An instance of perverseness.

per·vert (pər·vûrt′) *v.t.* 1 To turn to an improper use or purpose; misapply. 2 To distort the meaning of; misconstrue. 3 To turn from approved opinions or conduct; lead astray; corrupt. — *n.* (pûr′vûrt) 1 An apostate; renegade: opposed to *convert.* 2 *Psychiatry* One affected with or addicted to sexual perversion. [<F *pervertir* <L *pervertere* turn around, over < *per-* away + *vertere* turn] — **per·vert′er** *n.* — **per·vert′i·bil′i·ty** *n.* — **per·vert′i·ble** *adj.* — **per·vert′i·bly** *adv.*

Synonyms (verb): corrupt, distort, falsify, garble, misquote, misrepresent, misstate, stretch, twist. See ABUSE. *Antonyms*: correct, quote, rectify, restore.

per·vert·ed (pər·vûr′tid) *adj.* 1 Turned from the right purpose; misused. 2 Wilfully sinful; wicked; vicious. — **per·vert′ed·ly** *adv.*

per·vi·ous (pûr′vē·əs) *adj.* Capable of being penetrated; permeable. [<L *pervius* having a way through < *per-* through + *via* way] — **per′vi·ous·ly** *adv.* — **per′vi·ous·ness** *n.*

pes (pēz) *n. pl.* **pe·des** (pē′dēz) *Zool.* 1 The distal segment of the hind limb of a vertebrate, composed of tarsus, metatarsus, and phalanges. 2 A footlike organ, appearance, or part. 3 In prosody, the name for each of the first two quatrains of a sonnet. [<L, a foot]

Pe·sach (pā′säkh) *n.* The feast of the Passover. Also **Pe′sah.** See JEWISH HOLIDAYS. [<Hebrew *pesakh,* lit., a passing over < *pāsakh* pass over]

pe·sade (pə·säd′, -zād′, -zäd′) *n.* The act or position of a saddle horse in rearing. [<F, alter. of *posade* <Ital. *posata* a pause < *posare* pause <L *pausare* halt < *pausa* a stop]

Pe·sa·ro (pā′zä·rō) A port on the Adriatic in The Marches, central Italy.

Pes·ca·do·res (pes′kä·dō′räs) A former name for the PENGHU ISLANDS.

Pes·ca·ra (pās·kä′rä) A port on the Adriatic in south central Italy.

pe·se·ta (pə·sā′tə, *Sp.* pā·sā′tä) *n.* A Spanish monetary unit; a silver coin equivalent to 100 centesimos. [<Sp., dim. of *pesa* a weight]

Pe·sha·war (pə·shä′wər, pā·shä·vər) 1 A commissioners' division of northern West Pakistan, near Afghanistan; 27,563 square miles. 2 The chief city of this division, anciently, a Greco-Buddhist cultural center; capital of North-West Frontier Province, British India, 1901–47; capital of North-West Frontier Province, West Pakistan, 1947–55.

Pe·shi·to (pə·shē′tō) *n.* The oldest Syriac version of the Bible. Also **Pe·schi′to, Pe·shit′ta** (-shē′tä), **Pe·shit·to.** [<Syriac *p'shīṭ(t)to,* lit. plain, simple]

pes·ky (pes′kē) *adj.* **·ki·er, ·ki·est** *Colloq.* 1 Annoying; troublesome; plaguy. 2 Darned; damned: a euphemism. [< dial. E, prob. alter. of *pesty* <PEST] — **pes′ki·ly** *adv.*

pe·so (pā′sō) *n. pl.* **·sos** 1 A monetary unit of Cuba, the Philippines, Mexico, and certain other Latin-American countries, equal to 100 centavos. 2 An old Spanish coin equal to 8 reales; the Spanish dollar or piece-of-eight. [<Sp., orig. a weight <L *pensum,* orig. pp. neut. of *pendere* weigh]

pes·sa·ry (pes′ə·rē) *n. pl.* **·ries** *Med.* 1 An instrument used to remedy a uterine displacement. 2 A medicated suppository for use in the vagina. 3 A contraceptive device worn over or in the uterine cervix. [<Med. L *pessarium* <L *pessum* <Gk. *pessos* an oval stone]

pes·si·mism (pes′ə·miz′əm) *n.* 1 A disposition to take a gloomy view of affairs: opposed to *optimism.* 2 Cynicism. 3 A theory of cosmology that regards the cosmos, or the world and life, or some main constituent thereof, as essentially evil, or (in its extreme form) as the worst possible world. [<L *pessimus* worst + -ISM, on analogy with *optimism*] — **pes′si·mist** *n.* — **pes′si·mis′tic** or **·ti·cal** *adj.* — **pes′si·mis′ti·cal·ly** *adv.*

pest (pest) *n.* 1 A pernicious or vexatious person or thing, especially a destructive or injurious insect. 2 A virulent epidemic; pestilence. [<MF *peste* <L *pestis* a plague]

Pes·ta·loz·zi (pes′tä·lôt′tsē), **Johann Heinrich,** 1746–1827, Swiss educational reformer who held that development was the chief end of education. — **Pes′ta·loz′zi·an** *adj.* & *n.*

pes·ter (pes′tər) *v.t.* To harass with petty and persistent annoyances; bother; plague. [Aphetic var. of obs. *impester* entangle <OF *empestrer, empasturer,* orig. hobble a grazing horse < *em-* (<L *in-* in) + LL *pastorium* foot shackles <L *pastus,* pp. of *pascere* feed] — **pes′ter·er** *n.*

pest·hole (pest′hōl′) *n.* A breeding place for pestilence.

pest·house (pest′hous′) *n.* A public hospital where patients suffering from infectious or pestilential diseases are treated and kept isolated.

pes·ti·cide (pes′tə·sīd) *n.* A chemical or other substance effective in the destruction of such plant and animal pests as fungi, bacteria, insects, and the like. — **pes′ti·ci′dal** *adj.*

pes·tif·er·ous (pes·tif′ər·əs) *adj.* 1 Carrying pestilence. 2 Threatening or bringing danger or evil. 3 *Colloq.* Annoying; disagreeable. See synonyms under NOISOME. [<L *pestiferus* bringing plague < *pestis* a plague + *ferre* bear] — **pes·tif′er·ous·ly** *adv.* — **pes·tif′er·ous·ness** *n.*

pes·ti·lence (pes′tə·ləns) *n.* 1 Any wide-spread and fatal infectious or contagious malady. 2 Figuratively, a noxious or malign doctrine, influence, etc.

pes·ti·lent (pes′tə·lənt) *adj.* 1 Tending to produce malignant zymotic disease; pestilential. 2 Having a malign influence or effect. 3 Making trouble; causing irritation; vexatious. [<L *pestilens, -entis* < *pestis* a plague] — **pes′ti·lent·ly** *adv.*

pes·ti·len·tial (pes′tə·len′shəl) *adj.* 1 Having the nature of or breeding pestilence. 2 Morally harmful or pernicious; baneful. See synonyms under NOISOME.

pes·tle (pes′əl) *n.* 1 An implement used for braying, bruising, or mixing substances, as in a mortar. 2 A vertical moving bar employed in pounding, as in a stamp mill, etc. — *v.t.* & *v.i.* **·tled, ·tling** To pound, grind, or mix with or as with a pestle. [<OF *pestel* <L *pistillum* < *pistus,* pp. of *pinsere* pound]

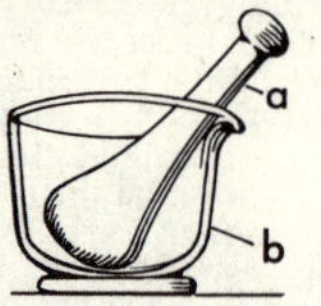

PESTLE
a. Pestle.
b. Mortar.

Pe·sto (pā′stō) A village on the site of ancient PAESTUM.

Pest·szent·er·zsé·bet (pesht′sent·er′zhā·bet) A city in north central Hungary SE of Budapest.

pet[1] (pet) *n.* 1 A tame, fondled animal. 2 Any loved and cherished creature; also, a favorite: teacher's *pet.* — *adj.* 1 Being a pet; indulged and fondled: a *pet* cat. 2 Regarded as a favorite; cherished: my *pet* hobby. — *v.* **pet·ted, pet·ting** — *v.t.* 1 *Rare* To treat as a pet; indulge. 2 To stroke or caress. — *v.i.* 3 *U.S. Slang* To make love by kissing and caressing. See synonyms under PAMPER, CARESS. [< dial. E (Scottish), ? back formation <PETTY, in affectionate use]

pet[2] (pet) *n.* A fit of pique or ill temper; peevish mood. [< obs. *to take the pet* take offence, sulk; origin uncertain]

Pé·tain (pā·tan′), **Henri Philippe,** 1856–1951, French marshal; chief of state 1940–44; convicted of treason 1945.

pet·al (pet′l) *n. Bot.* One of the leaves or subordinate parts of a corolla. [<NL *petalum* <L, a metal plate <Gk. *petalon* a thin plate, leaf, orig. neut. of *petalos* outspread < *petannynai* expand] — **pet′aled** or **pet′alled** *adj.*

-petal *combining form* Seeking: *centripetal.* [<L *petere* seek]

pet·a·lif·er·ous (pet′ə·lif′ər·əs) *adj.* Bearing petals. [<PETAL + -(I)FEROUS]

pet·a·line (pet′ə·lin, -līn) *adj.* Of, pertaining to, or like a petal.

pet·al·ism (pet′l·iz′əm) *n.* A form of ostracism in use among the Greeks of ancient Syracuse, who wrote on olive leaves their votes to banish for five years a citizen obnoxious to his fellow citizens. [<Gk. *petalismos* < *petalon* a leaf]

pet·a·lo·dy (pet′ə·lō′dē) *n. Bot.* A metamorphosis of other organs, as sepals or stamens, into petals, as in cultivated roses and double flowers. [<Gk. *petalōdēs* leaflike < *petalon* a leaf + *eidos* a form] — **pet′a·lod′ic** (-lod′ik) *adj.*

pet·al·oid (pet′l·oid) *adj.* Like or consisting of petals.

pet·al·ous (pet′l·əs) *adj.* Petaled; provided with petals.

pe·tard (pi·tärd′) *n.* 1 An explosive device formerly used for making breaches, etc., as in walls. 2 A small paper bomb used in pyrotechnics to imitate the sound of musketry. [<MF *pétard* < *péter* break wind <OF *pet* a fart <L *peditum,* orig. pp. neut. of *pedere* break wind]

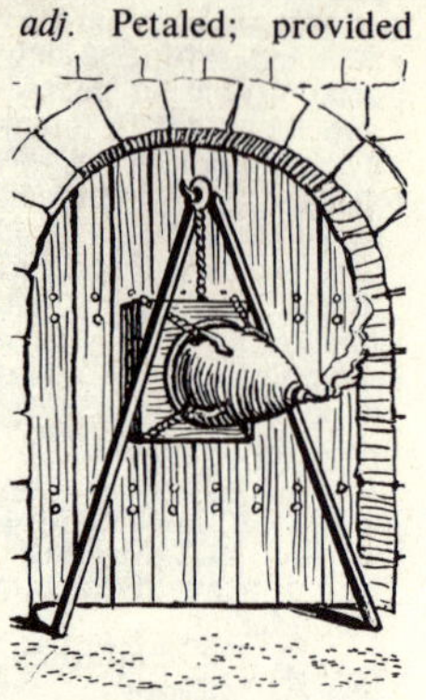

PETARD

pet·a·sus (pet′ə·səs) *n.* 1 A hat, typically with broad brim and low crown, worn by heralds and travelers of the ancient Greece. 2 The winged hat of the god Mercury. Also **pet′a·sos.** [<L <Gk. *petasos* < *petannynai* spread out]

pet·cock (pet′kok′) *n. Mech.* A small cock, as at the end of a steam cylinder or on a pipe or pump, for testing or draining. [? < obs. *pet* a fart + COCK[1] a valve]

Pete (pēt) Diminutive of PETER.

pe·te·chi·a (pə·tē′kē·ə) *n. pl.* **·chi·ae** (-ki·ē) *Pathol.* One of a number of small purplish spots on the skin, which appear in certain severe fevers. [<NL <Ital. *petecchia* a freckle; ult. origin unknown] — **pe·te′chi·al** *adj.*

pete·man (pēt′mən) *n. Slang* A criminal who specializes in blowing safes: also spelled *peatman.* [< slang E *pete(r)* a safe + MAN]

pe·ter (pē′tər) *v.i. Colloq.* To diminish gradually and then cease or disappear; become exhausted: with *out.* [Orig. U.S. mining slang; origin unknown]

Pe·ter (pē′tər; *Du., Ger., Norw., Sw.* pā′tər) A masculine personal name. Also *Hungarian* **Pé·ter** (pā′tər). [<Gk., a stone]

— **Peter, Saint** A Galilean fisherman, one of the Twelve Apostles, reputed author of two epistles of the New Testament; also called "Simon Peter."

— **Peter I,** 1672–1725, czar of Russia 1682–1725, known as **Peter the Great.**

— **Peter II,** born 1923, king of Yugoslavia 1934–45.

— **Peter the Hermit,** 1050?–1115?, French monk; preacher of the First Crusade.

Pe·ter·bor·ough (pē′tər·bûr′ō, -bər·ə) A city in NE Northamptonshire, England.

Peterborough, Soke of See SOKE OF PETERBOROUGH.

Pe·ter·kin (pē′tər·kin) Diminutive of PETER.

Pe·ter·mann Peak (pā′tər·män) A mountain in eastern Greenland; 9,645 feet.

Peter Pan 1 In J. M. Barrie's play, *Peter Pan* (1904), the little boy "who never grew up." 2 A statue, in Kensington Gardens, London, symbolizing perpetual youth; hence, any fully grown person of youthful or childish enthusiasm.

Pe·ters (pā′tərs), **Karl,** 1856–1918, German explorer and colonial statesman in Africa.

Pe·ters·burg (pē′tərz·bûrg) A port on the Appomattox River in SE Virginia; scene of Civil War battles, 1864, 1865.

pe·ter·sham (pē′tər·shəm) *n.* 1 A heavy, rough, tufted woolen cloth. 2 Formerly, a heavy greatcoat of such cloth. [after Viscount *Petersham,* who introduced it]

Peter's pence 1 Voluntary contributions raised by Roman Catholics for the pope since 1860. 2 The tax of a penny for every house, once paid by the English to support the English hospice in Rome; also, a like tribute paid by them and other peoples to aid the pope: so called because collected on St. Peter's Day: also called *hearth money.* Also **Peter pence.**

pet·i·ole (pet′ē·ōl) *n.* 1 *Bot.* The footstalk of a leaf. 2 *Zool.* A stalk or peduncle. [<L *petiolus* a stem, fruitstalk, orig. dim. of *pes, pedis* a foot] — **pet′i·o·lar** *adj.* — **pet′i·o·late′, pet′i·o·lat′ed** *adj.*

pet·it (pet′ē) *adj.* Small; lesser; minor; trivial: used in legal phrases: *petit larceny*: also spelled *petty*. [<OF, small, ? <Celtic *pit* something pointed, thin]

Pe·tit (pə·tē′), **Alexis Thérèse**, 1791–1820, French physicist. See DULONG AND PETIT'S LAW.

pe·tite (pə·tēt′) *adj. fem.* Diminutive; little. [<F, fem. of *petit*. See PETIT.]

Pe·tite Terre (pə·tēt târ′) An island group comprising a dependency of Guadeloupe; 1.2 square miles.

pe·ti·tion (pə·tish′ən) *n.* 1 A request, supplication, or prayer; a solemn or formal supplication. 2 A formal request, written or printed, addressed to a person in authority and asking for some grant or benefit, the redress of a grievance, etc. 3 *Law* A formal application in writing made to a court, requesting judicial action concerning some matter therein set forth. 4 That which is requested or supplicated. — *v.t.* 1 To make a petition to; entreat. 2 To ask for. — *v.i.* 3 To make a petition. See synonyms under ASK, PRAY. [<OF *peticiun* <L *petitio, -onis* <*petere* seek; refashioned after L] — **pe·ti′tion·ar′y** *adj.* — **pe·ti′tion·er** *n.*
Synonyms (noun): appeal, application, craving, entreaty, pleading, prayer, request, supplication. See PRAYER. *Antonyms*: command, demand, denial, exaction, refusal, requirement.

pe·ti·ti·o prin·cip·i·i (pə·tish′ē·ō prin·sip′ē·ī) *Logic* Begging the question; assuming in the premise that which is to be proved. [<L]

petit juror A member of a petit jury: also spelled *petty juror*.

petit jury See under JURY.

petit larceny The theft of property of less than such amount as may be fixed by statute: the distinction between petit and grand larceny has been almost wholly abolished: also spelled *petty larceny*. See LARCENY.

pe·tit mal (pə·tē′ mäl′) *Pathol.* A form of epileptic seizure characterized by a momentary loss of memory or consciousness and a brief interval of helplessness. See GRAND MAL.

pet·i·to·ry (pet′ə·tôr′ē, -tō′rē) *adj.* Soliciting or solicited by petition. [<LL *petitorius* <L *petitor* a candidate <*petere* seek]

pet·it point (pet′ē) A fine needle-tapestry stitch: also called *tent stitch*.

pet·its fours (pet′ē fôrz′, fōrz′; *Fr.* pə·tē′fōōr′) Small cakes, often elaborately iced. [<F, lit., little ovens]

pe·tits pois (pə·tē′ pwä′) *French* Specially selected small peas.

Pe·tö·fi (pe′tœ·fi), **Sándor**, 1822–49, Hungarian poet.

Pe·tra (pē′trə) A ruined ancient city of SW Jordan.

Pe·trarch (pē′trärk), **Francesco**, 1304–74, Italian scholar and poet.

Pe·trar·chan sonnet (pi·trär′kən) See under SONNET.

pet·rel (pet′rəl) *n.* A long-winged, black-and-white sea bird (order *Procelariiformes*); Mother Carey's chicken. [Earlier *pitteral,* ? a dim. of PETER; from its seeming to walk on the water like St. Peter. Matt. xiv 29]

PETREL (Body from 7 to 16 inches long; the storm petrel, Mother Carey's chicken, 5 to 6 inches)

pe·tres·cent (pə·tres′ənt) *adj.* Petrifying or tending to petrify. [<L *petra* a rock + -ESCENT]

Pe·trie (pē′trē), **Sir (William Matthew) Flinders**, 1853–1942, English Egyptologist.

pet·ri·fac·tion (pet′rə·fak′shən) *n.* 1 Partial or entire replacement of the material of an organism by mineral matter: also **pe·tres′cence, pe·tres′cen·cy.** 2 Anything petrified. Also **pet′ri·fi·ca′tion.** [<PETRIFY, on analogy with *satisfaction, stupefaction,* etc.] — **pet′ri·fac′tive** *adj.*

Petrified Forest National Monument A region of eastern Arizona containing petrified flora; 133 square miles; established 1906.

pet·ri·fy (pet′rə·fī) *v.* **·fied, ·fy·ing** *v.t.* 1 To convert (organic material) into a substance of stony character. 2 To make fixed and unyielding; deaden; harden. 3 To daze or paralyze with fear, surprise, etc. — *v.i.* 4 To become stone or a stony substance. [<MF *pétrifier* <L *petra* a rock + *facere* make] — **pe·trif′ic** (pə·trif′ik) *adj.*

petro- combining form Rock; stone: *petroglyph.* Also, before vowels, **petr-.** [<Gk. *petra* a rock and *petros* a stone]

pet·ro·chem·is·try (pet′rō·kem′is·trē) *n.* The chemistry of petroleum and its derivatives, especially the natural and synthetic hydrocarbons. — **pet′ro·chem′i·cal** *adj.* & *n.*

pet·ro·glyph (pet′rə·glif) *n.* A primitive figure or legend cut in rock. [<F *pétroglyphe* <Gk. *petra* a rock + *glyphē.* See GLYPH.] — **pet′ro·glyph′ic** *adj.*

Pet·ro·grad (pet′rə·grad, *Russian* pe·trō·grät′) A former name for LENINGRAD.

pet·ro·graph (pet′rə·graf, -gräf) *n.* A prehistoric carving or inscription on a rock.

pe·trog·ra·phy (pə·trog′rə·fē) *n.* The systematic description and classification of rocks. — **pe·trog′ra·pher** *n.* — **pet·ro·graph·ic** (pet′rə·graf′ik) or **·i·cal** *adj.* — **pet′ro·graph′i·cal·ly** *adv.*

pet·rol (pet′rəl) *n.* 1 *Brit.* Gasoline. 2 *Obs.* Petroleum. [<OF *petrole* <Med. L *petroleum* PETROLEUM]

PETROGRAPH
From Australia.

pet·ro·la·tum (pet′rə·lā′təm) *n.* A fatty semi-solid mixture of the paraffin hydrocarbons, obtained from petroleum, used in preparing ointments, and internally: often called *mineral jelly.* [<NL <PETROLEUM]

pe·tro·le·um (pə·trō′lē·əm) *n.* An oily, liquid mixture of numerous hydrocarbons, chiefly of the paraffin series, found in many widely scattered subterranean deposits, and extensively used for heat and light. A number of very important substances are obtained by the fractional distillation of petroleum, such as petroleum ether, gasoline, naphtha, benzine, kerosene, paraffin, etc. [<Med. L <L *petra* a rock (<Gk.) + *oleum* oil]

petroleum naphtha Ligroin.

petroleum spirit Benzine.

Pe·tro·le·um V. Nas·by (pə·trō′lē·əm vē naz′bē) Pseudonym of David Ross Locke.

pe·trol·ic (pə·trol′ik) *adj.* Of or pertaining to petroleum.

pe·trol·o·gy (pə·trol′ə·jē) *n.* The science of the origin, structure, constitution, and characteristics of rocks: a branch of geology. — **pet·ro·log·ic** (pet′rə·loj′ik) or **·i·cal** *adj.* — **pe·trol′o·gist** *n.*

pet·ro·nel (pet′rə·nəl) *n.* A 15th century firearm about the size of a large horse pistol, fired with the stock resting against the breast. [<MF *petrinal,* dial. var. of *poitrinal,* adj., pectoral < *poitrine* the chest, ult. <L *pectus*]

Pe·tro·ni·us (pə·trō′nē·əs), **Gaius**, died A.D. 66?, Roman author, called "Arbiter of Elegance."

Pet·ro·pav·lovsk (pet′rō·päv·lôfsk′, *Russian* pe′tro·päv′ləfsk) A city in Kazakh S.S.R., on the Ishim.

Pe·tró·po·lis (pə·trō′pōō·lēs) A city in Rio de Janeiro state, Brazil.

pet·rous (pet′rəs, pē′trəs) *adj.* 1 Hard, like stone. 2 *Anat.* Pertaining to or situated near the hard portion of the temporal bone. Also **pe·tro·sal** (pə·trō′səl). [<L *petrosus* rocky <L *petra* rock <Gk.]

Pe·tro·vitch (pe·trō′vich), **George** See KARAGEORGE.

Pet·ro·za·vodsk (pet′rə·zä·vôtsk′) A city on Lake Onega, capital of Karelo-Finnish S.S.R. Finnish **Pe·tro·skoi** (pe′trō·skoi).

Pe·trus (*Lat.* pē′trəs, *Du., Sw.* pā′trəs) See PETER. Also *Greek* **Pe·tros** (pe′tros).

pet·ti·coat (pet′ē·kōt) *n.* 1 A skirt or loose garment depending from the waist; especially, a woman's underskirt. 2 One who wears a petticoat; hence, a woman. 3 An electric insulator shaped like an inverted cup, for use on high-tension wires. — *adj.* Of, pertaining to, or influenced by, women: *petticoat* politics. [Earlier *petty coat* <PETTY + COAT]

pet·ti·fog (pet′ē·fog, -fôg) *v.i.* **·fogged, ·fog·ging** To be a pettifogger. [Appar. back formation <PETTIFOGGER]

pet·ti·fog·ger (pet′ē·fog′ər, -fôg′ər) *n.* An inferior lawyer, especially one chiefly employed on mean or petty cases, or resorting to small or tricky methods. [Earlier *petty fogger* <PETTY + obs. *fogger* a trickster for gain, prob. <FUGGER] — **pet′ti·fog′ger·y** *n.*

pet·tish (pet′ish) *adj.* Capriciously ill-tempered; testy. See synonyms under FRETFUL. [Prob. <PET² + -ISH] — **pet′tish·ly** *adv.*

pet·ti·toes (pet′ē·tōz′) *n. pl.* The aborted toes at the back of a pig's foot. [Earlier sense "giblets" <F *petit oie* goose giblets, lit., a little goose; later mistakenly understood as *petty toes*]

pet·tle (pet′l) *n. Scot.* A plowstaff. Also **pet′tul.**

pet·to (pet′ō) *n.* The breast. — **cardinal in petto** A cardinal appointed, but not yet formally announced. — **in petto** Within one's own breast; to oneself. [<Ital., the breast <L *pectus*]

pet·ty (pet′ē) *adj.* **·ti·er, ·ti·est** 1 Having little worth, importance, position, or rank; trifling; inferior: also spelled *petit.* 2 Having little scope or generosity; narrow-minded. 3 Mean; spiteful. See synonyms under CHILDISH, INSIGNIFICANT, LITTLE, SMALL. — *n.* A small amount of money advanced from a week's wages. [<F *petit* small. See PETIT.] — **pet′ti·ly** *adv.* — **pet′ti·ness** *n.*

petty cash Money used for the purchase of small items.

petty jury, petty larceny, etc. See PETIT JURY, etc.

petty officer In the navies of the United States and Great Britain, an enlisted man comparable in rank with a non-commissioned officer of the army.

pet·u·lance (pech′ōō·ləns) *n.* 1 Fretfulness; peevishness; temporary ill-humor. 2 *Obs.* Insolence; pertness. Also **pet′u·lan·cy.** See synonyms under IMPATIENCE.

pet·u·lant (pech′ōō·lənt) *adj.* 1 Displaying or characterized by capricious fretfulness; peevish. 2 *Obs.* Saucily rude; insolently wanton; pert. See synonyms under FRETFUL, PERVERSE. [<OF *petulant* <L *petulans, -antis* forward, ult. <*petere* seek, assail] — **pet′u·lant·ly** *adv.*

pe·tu·ni·a (pə·tōō′nē·ə, -tyōō′-) *n.* A plant of a tropical American genus (*Petunia*) of herbs of the nightshade family, with funnel-shaped, fragrant flowers, in various shades of red, purple, and white. [<NL <obs. E *petun* tobacco <F <Guarani *petün*; so called because of its close relation to tobacco]

pe·tun·tze (pe·tōōn′tse, *Chinese* bī′dun′dzu′) *n.* A variety of feldspar that is mixed with kaolin, and used by the Chinese in the manufacture of porcelain. Also **pe·tun′tse.** [<Chinese *pai-tun-tze* <*pai* white + *tun* stone]

peu à peu (pœ à pœ′) *French* Little by little.

peu de chose (pœ də shōz′) *French* A small matter; a trifle.

pew (pyōō) *n.* A bench for seating people in church, frequently with a kneeling rack attached; formerly, a boxlike quadrangle, usually raised on a low platform, with seats on three sides for a family. [ME *puwe,* appar. <OF *puye* a parapet <L *podia,* pl. of *podium* a height, a balcony <Gk. *podion* a base, dim. of *pous, podos* a foot]

pew·age (pyōō′ij) *n.* Rent paid for a pew or pews, or income derived from the rental of pews.

pe·wee (pē′wē) *n.* 1 A small olive-green flycatcher, especially the eastern wood pewee (*Myiochanes virens*) of North America; the pewee flycatcher. 2 The phoebe. Also spelled *peewee.* [Imit.]

pe·wit (pē′wit, pyōō′it) *n.* 1 A pewee. 2 The lapwing. 3 The black-headed gull (*Larus ridibundus*). 4 The spotted sandpiper. [Imit.]

pew·ter (pyōō′tər) *n.* 1 An alloy, usually of tin and lead, formerly much used for tableware. 2 Pewter vessels collectively. — *adj.* Made of pewter. [<OF *peutre, pialtre,* prob. <Ital. *peltro*; ult. origin unknown]

pew·ter·er (pyōō′tər·ər) *n.* A smith who works in pewter.

Pey·e·ri·an (pī·ir′ē·ən) *adj. Anat.* Pertaining to or designating certain glands (called **Peyer's patches**) in the lower part of the small intestine. [after J. K. *Peyer*, 1653–1712, Swiss anatomist]

pe·yo·te (pā·ō′tē, *Sp.* pā·yō′tā) *n.* A powerful intoxicant and narcotic drug obtained from the dried upper part of the mescal cactus found in Mexico and Texas. Also **pe·yo′tl** (-yōt′l). [<Am. Sp., the mescal cactus <Nahuatl *peyotl*, lit., a caterpillar; so called because of the down at its center]

pfen·nig (pfen′ikh) *n. pl.* **·nigs** or **pfen·ni·ge** (pfen′i·ge) A small bronze coin of Germany, equivalent to 1/100 of a mark: formerly called *reichspfennig*. [<G, a penny]

Pforz·heim (pfôrts′hīm) A city in Baden-Württemberg, West Germany.

***p*H** *Chem.* Denoting the negative logarithm of the hydrogen-ion concentration, in grams per liter, of a solution: used in expressing relative acidity and alkalinity. Thus, the *p*H of pure water is 7, indicating a concentration of 10^{-7}, or 0.0000001 gram per liter, usually regarded as neutral. Decreasing acidity means decreasing hydrogen-ion concentration, or a rise in the absolute value of *p*H. [<P(OTENTIAL OF) H(YDROGEN)]

Phae·a·cia (fē·ā′shə) In Homeric legend, an island visited by Ulysses after the fall of Troy. Also **Phæ·a′cia**. — **Phae·a′cian** *adj.* & *n.*

Phae·dra (fē′drə) In Greek mythology, a daughter of Minos and Pasiphaë, and wife of Theseus: she fell in love with her stepson Hippolytus and killed herself because he spurned her, indirectly causing his death. Also **Phæ′dra.**

Phae·drus (fē′drəs) Roman fabulist of the first century A.D.; 97 of his stories are extant.

phae·no·gam (fē′nə·gam) *n.* A phanerogam. [<NL *Phaenogama (Vegetabilia)* flowering (plants) <Gk. *phainein* show + *gamos* marriage]

phae·o·phy·ce·an (fē′ə·fī′sē·ən, -fish′ən) *n.* One of a family *(Phaeophyceae)* of brown algae. — *adj.* Of or pertaining to the *Phaeophyceae.* [<NL <Gk. *phaios* dusky + *phykos* seaweed]

Pha·e·thon (fā′ə·thon) In Greek mythology, the son of Helios, who borrowed his father's chariot of the sun, and would have set heaven and earth on fire by his careless driving if Zeus had not slain him with a thunderbolt.

pha·e·ton (fā′ə·tən, *esp. Brit.* fā′tən) *n.* 1 A light four-wheeled boxless carriage, having one or two seats, open at the sides, and sometimes having a top. 2 An open two-seated automobile. [<F <*phaéton* <L *Phaethon* Phaethon]

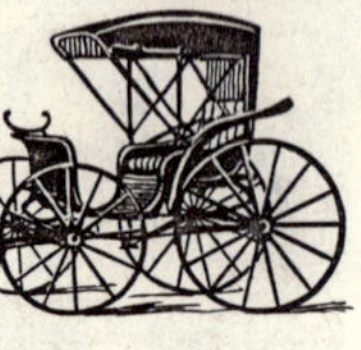
AMERICAN TWO-SPRING PHAETON *(def. 1)*

-phage *combining form* One who or that which eats or consumes: *bacteriophage.* [<Gk. *phagein* eat]

phag·e·de·na (faj′ə·dē′nə) *n. Pathol.* 1 An eating, sloughing ulcer; hospital gangrene. 2 Leishmaniasis of the skin. 3 Ravenous hunger; bulimia. Also **phag′e·dae′na**. [<L *phagedaena* an eating ulcer <Gk. *phagedaina* <*phagein* eat] — **phag′a·den′ic** (-den′ik) *adj.*

phago- *combining form* Eating: *phagocyte.* Also, before vowels, **phag-**. [<Gk. *phagein* eat]

phag·o·cyte (fag′ə·sīt) *n. Physiol.* A leucocyte that takes into its substance and digests bacteria and other noxious matters in the blood and tissues of the body. — **phag′o·cyt′ic** (-sit′ik) or **·i·cal** *adj.*

phagocytic index A number expressing the average number of bacteria ingested by a single leucocyte under specified conditions.

phag·o·cy·to·sis (fag′ə·sī·tō′sis) *n.* The destruction and absorption of bacteria or microorganisms by phagocytes.

pha·gol·y·sis (fə·gol′ə·sis) *n.* The dissolution or destruction of phagocytes. Also **phag·o·cy·tol·y·sis** (fag′ə·sī·tol′ə·sis). — **phag′o·lyt′ic** (-lit′ik) *adj.*

phag·o·ma·ni·a (fag′ə·mā′nē·ə, -mān′yə) *n.* A morbid or uncontrollable desire to eat.

-phagous *combining form* Consuming; tending to eat: *anthropophagous.* [<Gk. *phagein* eat]

-phagy *combining form* The consumption or eating of: *geophagy.* Also **-phagia**. [<Gk. *-phagia* <*phagein* eat]

pha·i·no·pep·la (fā·ī′nō·pep′lə, fā′ə-) *n.* A small flycatcher of the western United States *(Phainopepla nitens)* with a slender crest and, in the male, conspicuous white wing patches. [<NL <Gk. *phainein* show + *peplos* a peplos]

phal·ange (fal′ənj, fə·lanj′) *n.* A phalanx of the fingers or toes. See illustration under FOOT. [<F <L *phalanx, phalangis* a line of battle]

pha·lan·ge·al (fə·lan′jē·əl) *adj.* Of, pertaining to, or resembling the phalanges of the fingers and toes. Also **pha·lan′gal** (-gəl), **pha·lan′ge·an.**

pha·lan·ger (fə·lan′jər) *n.* Any one of a family *(Phalangeridae)* of small marsupials of Australia and Papua, having long tails, often prehensile. [<NL <*phalanges*, pl. of *phalanx* phalanx (def. 3); in ref. to the peculiarly constructed phalanges of its hind feet]

pha·lan·ges (fə·lan′jēz) Plural of PHALANX.

phal·an·ster·y (fal′ən·ster′ē) *n. pl.* **·ster·ies** 1 The building inhabited by a community of Fourierites; also, such a community. 2 Any group or community of individuals. [<F *phalanstère* <*phalan(x)* a phalanx (<L) + *(mona)stère* <LL *monasterium* a monastery] — **phal′an·ste′ri·an** (-stir′ē·ən) *adj.* & *n.* — **phal′an·ste′ri·an·ism** *n.*

pha·lanx (fā′langks, *esp. Brit.* fal′angks) *n. pl.* **pha·lan·ges** (fə·lan′jēz) or **pha·lanx·es** 1 In ancient Greece, a marching order of heavy infantry, with close ranks and files, joined shields, and spears overlapping. 2 Any massed or compact body or corps, such as a group of Fourierites. 3 *Anat.* One of the bones articulating with the joints of the fingers or toes. See synonyms under ARMY. [<L *phalanx, phalangis* <Gk. *phalanx, phalangos* a line of battle]

phal·a·rope (fal′ə·rōp) *n.* A migratory swimming bird (family *Phalaropodidae*), breeding in northern regions, resembling the sandpiper, but having the body depressed, the toes bordered by lateral webs, and the plumage close underneath. [<F <NL *Phalaropus*, the genus name <Gk. *phalaris* coot + *pous* foot]

phal·lin (fal′in) *n.* The characteristic hemolytic poison of the deathcup fungus. [<NL *phall(oides)*, species name of the deathcup (<L *phallus* a phallus) + -IN]

phal·lism (fal′iz·əm) *n.* Worship of the generative power in nature as symbolized by the phallus; phallic worship, as in the Dionysiac festivals of ancient Greece. Also **phal′li·cism.** [<PHALL(US) + -ISM] — **phal′list, phal′li·cist** *n.*

phal·lus (fal′əs) *n. pl.* **·li** (-ī) 1 A figure of the male generative organ, used in many systems of religion, especially in the Orient, as a symbol of the generative power of nature. 2 The generative organ of the male or the clitoris of the female. 3 *Psychoanal.* The sexually immature penis. [<L <Gk. *phallos* penis] — **phal′lic** or **phal′li·cal** *adj.*

Pha·nar (fä·när′) A section of Istanbul, formerly the residence of privileged Greek families known as **Pha·nari·ots** (fə·nar′ē·ots): also *Fanar.*

-phane *combining form* That which resembles or is similar to (a specified substance or material): *cellophane.* [<Gk. *-phanēs* <*phainein* appear]

pha·ner·ic (fə·ner′ik) *adj. Geol.* Clearly visible to the naked eye, as the textures of certain igneous rocks. Also **phan·ic** (fan′ik).

phanero- *combining form* Visible: *phanerophyte.* Also, before vowels, **phaner-**. [<Gk. *phaneros* visible <*phainein* appear]

phan·er·o·crys·tal·line (fan′ər·ō·kris′tə·lin, -līn) *adj. Mineral.* Obviously crystalline: opposed to *cryptocrystalline.*

phan·er·o·gam (fan′ər·ə·gam′) *n. Bot.* A flowering, seed-producing plant: also called *phaenogam.* Compare CRYPTOGAM. [<F *phanérogame* <Gk. *phaneros* visible (<*phainein* show) + *gamos* marriage] — **phan′er·o·gam′ic, phan·er·og·a·mous** (fan′ə·rog′ə·məs) *adj.*

Phan·er·o·ga·mi·a (fan′ər·ə·gā′mē·ə) *n. pl.* One of the two primary divisions into which Linnaeus divided all plants, embracing those with flowers having stamens and pistils; flowering plants: distinguished from *Cryptogamia*, or flowerless plants. [<NL <Gk. *phaneros* visible + *gamos* marriage]

phan·er·o·phyte (fan′ər·ə·fīt′) *n. Bot.* A plant having aerial buds.

phan·tasm (fan′taz·əm), etc. See FANTASM, etc.

phan·tas·ma·go·ri·a (fan·taz′mə·gôr′ē·ə, -gō′rē·ə) See FANTASMAGORIA.

phan·tom (fan′təm) *n.* 1 Something that exists only in appearance. 2 An apparition; specter; illusion. 3 The visible representative of an abstract state or incorporeal person. — *adj.* Illusive; ghostlike: a *phantom* ship. Also spelled *fantom.* [<OF *fantosme* <L *phantasma* <Gk., an appearance <*phantazein* make visible <*phantos* visible <*phainein* show. Doublet of FANTASM.]

phantom section In mechanical drawing, cross-hatching superimposed on an external view of an object, assembly, or structure to show interior construction and details, often eliminating the need for an additional drawing or view.

phantom word A spurious word that exists only through the error of a lexicographer, writer, or printer, as one resulting from a false etymology or wrong attribution of meaning. Also called *ghost word.*

-phany *combining form* Appearance; manifestation: *epiphany, theophany.* [<Gk. *-phaneia* <*phainein* appear]

Phar·a·mond (far′ə·mənd) A legendary king of the Franks during the fifth century.

Phar·aoh (fâr′ō, fā′rō, fâr′ē·ō) *n.* Any one of the monarchs of ancient Egypt. [OE *Pharaon* <LL *Pharao, -onis* <Gk. *Pharaō* <Hebrew *Par′ōh* <Egyptian *pr-′ōh* the great house] — **Phar′a·on′ic** (-ē·on′ik) or **·i·cal** *adj.*

Pharaoh's serpent A stick or pellet of mercuric thiocyanate which, when ignited, glows and swells up, developing a long strip of ash which curls out like a serpent.

phar·i·sa·ic (far′ə·sā′ik) *adj.* 1 Pertaining to the Pharisees. 2 Observing the form, but neglecting the spirit, of religion; self-righteous; hypocritical. Also **phar′i·sa′i·cal.** [<LL *pharisaicus* <Gk. *pharisaikos* <*pharisaios* PHARISEE] — **phar′i·sa′i·cal·ly** *adv.* — **phar′i·sa′i·cal·ness** *n.*

phar·i·sa·ism (far′ə·sā·iz′əm) *n.* The principles and practices of the Pharisees; hence, formality, self-righteousness, censoriousness, or hypocrisy. Also **phar′i·see·ism** (-sē·iz′əm).

Phar·i·see (far′ə·sē) *n.* 1 One of an ancient, exclusive Jewish sect that paid excessive regard to tradition and ceremonies, and in so doing led its members, by their sense of superior sanctity, to separate themselves from the other Jews. 2 Hence, a formal, sanctimonious, hypocritical person. [OE *fariseus*, infl. by OF *pharise*, both <L *pharisaeus* <Gk. *pharisaios* <Aramaic *perīshayā*, pl. of *perīsh* <Hebrew *pārūsh* separated <*parash* cleave]

phar·ma·ceu·tic (fär′mə·sōō′tik) *adj.* Pertaining to, using, or relating to pharmacy or the pharmacopoeia: also **phar·ma·cal** (fär′mə·kəl). — *n.* A drug. Also **phar′ma·ceu′ti·cal.** [<L *pharmaceuticus* <Gk. *pharmakeutikos* of drugs <*pharmakeutēs* druggist <*pharmakeuein* give drugs <*pharmakon* a drug] — **phar′ma·ceu′ti·cal·ly** *adv.* — **phar′ma·ceu′tist** *n.*

phar·ma·ceu·tics (fär′mə·sōō′tiks) *n.* The science of pharmacy.

phar·ma·cist (fär′mə·sist) *n.* A qualified druggist; pharmaceutist.

pharmaco- *combining form* A drug; of or pertaining to drugs: *pharmacology.* Also, before vowels, **pharmac-**. [<Gk. *pharmakon* a drug]

phar·ma·co·dy·nam·ics (fär′mə·kō·dī·nam′iks) *n.* The experimental science of the action and effects of drugs.

phar·ma·cog·no·sy (fär′mə·kog′nə·sē) *n.* The knowledge of drugs, especially their origin, structure, and chemical constitution. [<NL *pharmacognosia* <Gk. *pharmakon* a drug + *gnōsis* a knowing, knowledge] — **phar′ma·cog′no·sist** *n.*

phar·ma·col·o·gy (fär′mə·kol′ə·jē) *n.* The science of the action of medicines, their nature, preparation, administration, and effects: includes materia medica and therapeutics. — **phar′ma·co·log′ic** (-kə·loj′ik) or **·i·cal** *adj.* — **phar′ma·co·log′i·cal·ly** *adv.* — **phar′ma·col′o·gist** *n.*

phar·ma·co·ma·ni·a (fär′mə·kō·mā′nē·ə, -mān′yə) *n.* A morbid craving for drugs.

phar·ma·co·poe·ia (fär′mə·kə·pē′ə) *n.* 1 A book, usually published by authority, containing standard formulas and methods for the

preparation of medicines, drugs, and other remedial substances. 2 A collection of drugs. [<NL <Gk. *pharmakopoiïa* art of making drugs < *pharmakon* a drug + *poieein* make] — **phar'ma·co·poe'ial** *adj.* — **phar'ma·co·poe'·ist** *n.*

phar·ma·cy (fär'mə·sē) *n. pl.* **·cies** 1 The art or business of compounding, preserving, and identifying drugs, and of compounding and dispensing medicines. 2 A drugstore. [<OF *farmacie* <LL *pharmacia* <Gk. *pharmakeia* < *pharmakeus* a druggist < *pharmakon* a drug]

pha·ros (fâr'os, fā'rôs, fā'-) *n.* A lighthouse; beacon. [<L <Gk. <*Pharos* Pharos]

Pha·ros (fâr'os, fā'rôs, fā'-) A peninsula of Alexandria, Egypt; formerly an island, the site of an ancient lighthouse.

Phar·sa·la (fär'sä·lä) A city of southern Thessaly, Greece; chief town of the ancient district of **Phar·sa·li·a** (fär·sā'lē·ə) and scene of Caesar's victory over Pompey, 48 B.C.: also *Farsala.* Ancient **Phar·sa·lus** (fär·sā'ləs).

pha·ryn·ge·al (fə·rin'jē·əl, far'in·jē'əl) *adj.* Of or pertaining to the pharynx. Also **pha·ryn'gal** (-gəl). [<NL *pharyngeus* < *pharynx, -yngis* PHARYNX]

phar·yn·gi·tis (far'in·jī'tis) *n. Pathol.* Inflammation of the pharynx, as in diphtheria and malignant sore throat. [<NL < *pharynx, -yngis* pharynx]

pharyngo– *combining form* The throat; related to the throat: *pharyngoscope.* Also, before vowels, **pharyng–.** [<Gk. *pharynx* throat]

phar·yn·gol·o·gy (far'ing·gol'ə·jē) *n.* The science of the pharynx and its diseases.

pha·ryn·go·scope (fə·ring'gə·skōp) *n.* An apparatus for examining the pharynx. — **phar·yn·gos·co·py** (far'ing·gos'kə·pē) *n.*

phar·yn·got·o·my (far'ing·got'ə·mē) *n. Surg.* The operation of making an incision into the pharynx.

phar·ynx (far'ingks) *n. pl.* **pha·ryn·ges** (fə·rin'jēz) or **phar·ynx·es** *Anat.* The part of the alimentary canal between the palate and the esophagus, serving as a passage for air and food. [<NL *pharynx, -yngis* <Gk. *pharynx, -yngos* throat]

phase (fāz) *n.* 1 The view that anything presents to the eye; any one of varying distinctive manifestations of an object. 2 *Astron.* One of the appearances or forms presented periodically by the moon and planets. 3 *Physics* **a** In an oscillatory motion, the position and character of a wave at any instant: often measured as an angle, the whole period being regarded as a circle, or 360°. **b** The instant when the maximum, zero, or other relative value is attained by any cyclical system, as sound or light waves, an alternating electric current, etc. **c** Any of the homogeneous forms of a given substance that may occur alone, or exist independently as a component of a larger heterogeneous system, as ice in water, water vapor in fog, etc. 4 *Biol.* **a** One of the distinct stages in the reduction or division process of a cell. **b** Any characteristic or decisive stage in the growth, development, or life pattern of an organism. Also **pha·sis** (fā'sis). ◆ Homophone: *faze.* [<NL *phasis* <Gk., an appearance < *phainein* make appear] — **pha·sic** (fā'zik) *adj.*

phase meter An instrument for measuring the difference in phase between two alternating oscillations of the same frequency.

phase modulation *Electronics* Modulation of a carrier wave by varying its phase in accordance with the amplitude or pitch of the transmitted signal.

phase rule *Physics* A mathematical generalization of the equilibrium relations between two or more phases of a material system, according to which the degrees of freedom possible to a given heterogeneous system equal the number of components in the system, less the number of phases, plus 2.

–phasia *combining form Med.* Defect or malfunction of speech: *dysphasia.* Also **–phasy.** [<Gk. *-phasia* < *phanai* speak]

pha·sine (fā'sēn, -sin) *n. Biochem.* A poisonous substance isolated as a white amorphous powder from soya and certain other beans: it acts as a strong agglutinating agent on the red blood cells of man and other animals. Also **pha'sin** (-sin). [<NL *Phas(eolus),* a genus of beans (<L, a kidney bean, dim. of *phaseolus* <Gk. *phasēlos,* a kind of bean) + -INE[2]]

phat (fat) See FAT *adj.* (def. 4).

pheas·ant (fez'ənt) *n.* 1 A long-tailed gallinaceous bird of *Phasianus* or related genus, noted for the gorgeous plumage of the male: native to Asia, but long semidomesticated elsewhere and bred in game preserves. 2 One of various other birds, as the ruffed grouse. [< AF *fesant* <L *Phasianus* <Gk. *Phasianos (ornis)* the Phasian (bird) <*Phasis* the Phasis, a river in the Caucasus, where it was first found]

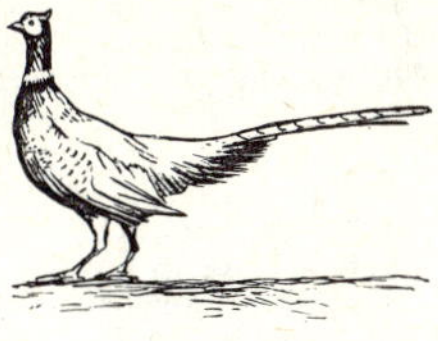

RING–NECKED PHEASANT
(From 31 to 36 inches long over–all)

pheasant eye 1 A low, hardy European annual *(Adonis annua)* having crimson or scarlet flowers with dark centers: cultivated in the United States. 2 The garden pink.

pheas·ant·ry (fez'ənt·rē) *n.* A place for keeping pheasants.

Phe·be (fē'bē) See PHOEBE.

Phei·dip·pi·des (fī·dip'ə·dēz) Greek runner sent from Athens to Sparta to secure help against the Persians, 490 B.C.: sometimes confused with the runner, whose name is not preserved, who carried news of victory at Marathon to Athens. Also *Phidippides.*

phel·lo·derm (fel'ə·dûrm) *n. Bot.* A layer of green parenchymatous tissue sometimes developed on the inner side of a layer of cork. [<Gk. *phellos* cork + *derma* skin] — **phel'·lo·der'mal** *adj.*

phel·lo·gen (fel'ə·jən) *n. Bot.* The active meristematic tissue out of which cork is developed. [<Gk. *phellos* cork + -GEN] — **phel'lo·ge·net'ic** (-jə·net'ik), **phel'lo·gen'ic** *adj.*

Phelps (felps), **William Lyon,** 1865–1943, U. S. educator and critic.

phen– Var. of PHENO–.

phe·na·caine (fē'nə·kān) *n.* A white, odorless, crystalline substance, $C_{18}H_{25}O_3N_2Cl$, used as a quick-acting local anesthetic. [<PHEN- + A(CETO)- + (CO)CAINE]

Phe·nac·e·tin (fə·nas'ə·tin) *n.* Proprietary name for a brand of acetophenetidin.

phen·a·cite (fen'ə·sīt) *n.* A brittle, vitreous, colorless, transparent to subtranslucent beryllium silicate, Be_2SiO_4, sometimes used as a gemstone. [<Gk. *phenax, -akos* a cheat + -ITE[1]; so called because mistaken for quartz]

phen·a·kis·to·scope (fen'ə·kis'tə·skōp) *n.* A disk bearing a series of representations of an object in successive phases of motion, revolved before a mirror to give the illusion of continuous motion. Compare ZOETROPE. [<Gk. *phenakistēs* a cheat + -SCOPE]

phe·nan·threne (fə·nan'thrēn) *n. Chem.* A crystalline isomer, $C_{14}H_{10}$, of anthracene; a coal-tar product used in the synthesis of drugs and dyes. [<PHEN(YL) + ANTHR(AC)ENE]

phen·a·zine (fen'ə·zēn, -zin) *n. Chem.* A yellowish basic compound, $C_{12}H_8N_2$, on which many dyestuffs are based. Also **phen'a·zin** (-zin). [<PHEN(YL) + AZ(O)- + -INE[2]]

phe·net·i·dine (fə·net'ə·dēn, -din) *n. Chem.* Any one of three isomeric liquid derivatives, $C_8H_{11}ON$, of phenetole, especially the *para* form, which is the base of acetophenetidin. Also **phe·net'i·din** (-din). [<PHEN(OL) + ET(HYL) + (AM)ID(O)- + -INE[2]]

phen·e·tole (fen'ə·tōl, -tol) *n. Chem.* An aromatic liquid compound, $C_6H_5OC_2H_5$, the ethyl ether of phenol. Also **phen'e·tol.** [< PHEN(YL) + ET(HYL) + -OLE]

phen·ic acid (fen'ik) See PHENOL.

Phe·ni·cia (fə·nē'shə, -nish'ə), **Phe·ni·cian,** etc. See PHOENICIA, etc.

phe·nix (fē'niks) See PHOENIX.

pheno– *combining form Chem.* Related to benzene; a derivative of benzene: *phenobarbital.* Also, before vowels, *phen–.* [<PHENYL]

phe·no·bar·bi·tal (fē'nō·bär'bə·tal, -tôl) *n. Chem.* A white, odorless, slightly bitter, crystalline powder, $C_{12}H_{12}O_3N_2$, a derivative of barbituric acid: used as a sedative and hypnotic. Also **phe'no·bar'bi·tone** (-tōn). [< PHENO- + BARBITAL]

phe·no·cryst (fē'nō·krist) *n. Geol.* A mineral constituent of a rock, occurring in well-defined crystals embedded in a fine-grained groundmass. [<F *phénocryste* <Gk. *phainein* show + *krystallos* a crystal] — **phe'no·crys'tic** *adj.*

phe·nol (fē'nōl, -nol) *n. Chem.* 1 Any one of a series of aromatic hydroxyl derivatives of benzene or its homologs. 2 A white crystalline compound, $C_6H_5{\cdot}OH$, with a burning taste and characteristic odor, derived from coal-tar oil by distillation and used as a disinfectant: popularly called *carbolic acid,* formerly *phenic acid.* Phenol is a powerful caustic poison. [<Gk. *phaino-* shining (< *phainein* show) + -OL[1]; so called because derived from coal tar, a by-product of illuminating gas]

phe·no·late (fē'nə·lāt) *n. Chem.* A salt of phenol.

phe·nol·ic (fi·nol'ik, -nō'lik) *adj. Chem.* Of, pertaining to, derived from, or containing phenol: *phenolic* resins, a large and important class of synthetic plastics made from an aldehyde-phenol base.

phe·nol·o·gy (fi·nol'ə·jē) *n.* The study of the periodic phenomena of plant life and animal behavior in relation to seasonal changes, climatic and other ecological factors. [Contraction of PHENOMENOLOGY, with a restricted application] — **phe·no·log·ic** (fē'nə·loj'ik) or **·i·cal** *adj.* — **phe'no·log'i·cal·ly** *adv.* — **phe·nol'o·gist** *n.*

phe·nol·phthal·ein (fē'nōl·thal'ēn, fē'nolf·thal'·ē·in) *n.* A whitish or yellowish-white crystalline compound, $C_{20}H_{14}O_4$, obtained by treating phenol with phthalic anhydride. Because its brilliant red alkaline solutions are readily decolorized by acid, it is valuable as an indicator in acid-base titrations: used also in medicine as a laxative.

phe·nom·e·na (fi·nom'ə·nə) Plural of PHENOMENON: erroneously used as a singular.

phe·nom·e·nal (fi·nom'ə·nəl) *adj.* 1 Pertaining to phenomena. 2 Extraordinary or marvelous. — **phe·nom'e·nal·ly** *adv.*

phe·nom·e·nal·ism (fi·nom'ə·nəl·iz'əm) *n.* The metaphysical opinion that no realities of which the human mind can have knowledge underlie phenomena. Compare POSITIVISM. — **phe·nom'e·nal·ist** *n.* — **phe·nom'e·nal·is'tic** *adj.* — **phe·nom'e·nal·is'ti·cal·ly** *adv.*

phe·nom·e·nol·o·gy (fi·nom'ə·nol'ə·jē) *n.* 1 The scientific investigation or description of phenomena. 2 *Philos.* The general doctrine of phenomena, as distinguished from ontology. [<PHENOMENO(N) + -LOGY] — **phe·nom'e·no·log'i·cal** (-nə·loj'i·kəl) *adj.* — **phe·nom'e·no·log'i·cal·ly** *adv.*

phe·nom·e·non (fi·nom'ə·non) *n. pl.* **·na** (-nə) 1 Something visible or directly observable, as an appearance, action, change, or occurrence of any kind, as distinguished from the force by which, or the law in accordance with which, it may be produced. 2 Any unusual occurrence; an inexplicable fact; marvel; prodigy. 3 Any fact, appearance, or occurrence in consciousness; that which is apprehended by the senses, in contrast with or in opposition to that which really exists, or to things in themselves: contrasted with *noumenon.* 4 A symptom of disease. [<LL *phaenomenon* <Gk. *phainomenon* an appearance, orig. ppr. passive neut. of *phainein* show]

phe·no·thi·a·zine (fē'nō·thī'ə·zēn, -zin) *n. Chem.* A light-yellow crystalline compound, $C_{12}H_9NS$, prepared by combining diphenylamine and sulfur in the presence of iodine: used as an insecticide and as a remedy against livestock parasites.

phe·no·type (fē'nə·tīp) *n. Biol.* A type or strain of organisms distinguishable from others by some visibly manifested group of characters, as contrasted with genetic constitution. Compare GENOTYPE. [<F *phéno-* (<Gk. *phaino-* < *phainein* show, appear) + -TYPE] — **phe'no·typ'ic** (-tip'ik) or **·i·cal** *adj.* — **phe'no·typ'i·cal·ly** *adv.*

phen·yl (fen'əl, fē'nəl) *n. Chem.* The univalent radical C_6H_5, regarded as the basis of numerous aromatic derivatives. [< obs. *phene* benzene + -YL]

phen·yl·a·mine (fen'əl·ə·mēn', -am'in, fē'nəl-) *n.* Aniline.

phen·yl·ene (fen'əl·ēn, fē'nəl-) *n. Chem.* A

add,āce,câre,pälm; end,ēven; it,īce; odd,ōpen,ôrder; to͝ok,po͞ol; up,bûrn; ə = a in *above,* e in *sicken,* i in *clarity,* o in *melon,* u in *focus;* yo͞o = u in *fuse;* oi,oil; ou,pout; ch,check; g,go; ng,ring; th,thin; th,this; zh,vision. Foreign sounds à,œ,ü,kh,ń; and ◆: see page xx. < from; + plus; ? possibly.

bivalent radical, C_6H_4, contained in certain benzene derivatives.
phew (fyōō, fōō) *interj.* An exclamation of disgust or surprise.
phi (fī, fē) *n. Greek* The twenty-first letter in the Greek alphabet (Φ,φ): corresponding to English *ph* and *f.* As a numeral it denotes 500.
phi·al (fī′əl) See VIAL.
Phi Be·ta Kap·pa (fī bā′tə kap′ə, bē′tə) An American honorary society founded in 1776 at William and Mary College, with chapters in many of the nation's colleges, and with membership based on conditions of high academic standing. The name is formed from the initials of the Greek phrase *Philosophia biou kybernētēs* (Philosophy the guide of life).
Phid·i·as (fid′ē·əs), 500?–432? B.C., Greek sculptor and architect; designed the Parthenon. —**Phid′i·an** *adj.*
Phi·dip·pi·des (fī·dip′ə·dēz) See PHEIDIPPIDES.
Phil (fil) Diminutive of PHILIP.
phil- Var. of PHILO-.
-phil Var. of -PHILE.
phil·a·beg, phil·i·beg (fil′ə·beg) See FILIBEG.
Phil·a·del·phi·a (fil′ə·del′fē·ə) **1** A city on the Delaware River in SE Pennsylvania. **2** An ancient city of Lydia, Asia Minor, on the site of modern Alashehir. —**Phil′a·del′phi·an** *adj. & n.*
Philadelphia Chippendale Fine, usually richly carved mahogany furniture in the Chippendale style, made in Philadelphia in the 18th century.
Philadelphia lawyer An unusually sharp lawyer, especially one adept in phrasing legal technicalities: originally a tribute to the high caliber of the Philadelphia bar, now implying over-shrewd trickery.
Phi·lae (fī′lē) An island in the Nile river, near Aswan, Upper Egypt; submerged half the year by the backwater from the Aswan Dam; site of the temples of Isis and Hathor. Also **Phi′læ.**
phi·lan·der (fi·lan′dər) *v.i.* To make love without serious intentions: said of a man. [<*n.*] —*n.* A male flirt or suitor: also **phi·lan′der·er.** [<Gk. *philandros* <*phileein* love + *anēr, andros* man; from its use as a proper name for a lover in drama] —**phi·lan′der·ing** *n.*
Phi·lan·der (fi·lan′dər) A masculine personal name. [<Gk., loving man or mankind]
phi·lan·thro·py (fi·lan′thrə·pē) *n. pl.* **·pies** Disposition or effort to promote the happiness or social elevation of mankind; desire, effort, or beneficence, as by making donations, intended to mitigate social evils and increase social comfort; comprehensive benevolence, but often specific in its objects; literally, love of man. See synonyms under BENEVOLENCE. [<LL *philanthropia* <Gk. *philanthrōpia* < *phileein* love + *anthropos* man] —**phil·an·throp·ic** (fil′ən·throp′ik) or **·i·cal** *adj.* —**phil′·an·throp′i·cal·ly** *adv.* —**phi·lan′thro·pist** *n.*
phi·lan·thro·pize (fi·lan′thrə·pīz) *v.* **·pized, ·piz·ing** *v.t.* To deal with philanthropically. —*v.i.* To act as a philanthropist.
phi·lat·e·ly (fi·lat′ə·lē) *n.* The study and collection of postage stamps, stamped envelopes, wrappers, etc.; stamp collecting. [<F *philatélie* <Gk. *philos* loving + *ateleia* exemption from tax; with ref. to prepaid postage] —**phil·a·tel·ic** (fil′ə·tel′ik) or **·i·cal** *adj.* —**phil′a·tel′·i·cal·ly** *adv.* —**phi·lat′e·list** *n.*
-phile *combining form* One who supports or is fond of; one devoted to: *bibliophile.* [<Gk. *-philos* loving <*phileein* love]
Phi·le·mon (fi·lē′mən) A masculine personal name. Also *Fr.* **Phi·lé·mon** (fē·lā·môṅ′). [< Gk., loving thought]
—**Philemon** In Greek mythology, the husband of Baucis.
—**Philemon** A Greek of Colossae, converted to Christianity by Paul; also, the epistle addressed by Paul to him, forming one of the books of the New Testament.
phil·har·mon·ic (fil′här·mon′ik, -ər·mon′-) *adj.* Fond of harmony or music; often, **Philharmonic,** used in the names of musical societies. [<F *philharmonique* <Ital. *filarmonico* <Gk. *philos* loving + *harmonikos* HARMONIC]
Phil·hel·lene (fil·hel′ēn) *n.* **1** One who loves Greece or the Greeks. **2** A sympathizer with the modern Greeks in their effort (1821–29) to throw off the Turkish yoke and revive the Greek nation. Also **Phil·hel′le·nist.** [<Gk. *philellēn* loving Greeks <*phileein* love + *Hellēn* a Greek] —**Phil·hel·len·ic** (fil′he·len′ik) *adj.* —**Phil·hel′le·nism** *n.*
-philia *combining form* **1** A tendency toward: *hemophilia.* **2** A morbid affection or fondness for: *necrophilia.* Also spelled *-phily.* [<Gk. *-philia* <*phileein* love]
Phil·ip (fil′ip) A masculine personal name. Also *Ger.* **Phi·lipp** (fē′lip), *Fr.* **Phi·lippe** (fē·lēp′), **Phi·lip·pus** (*Du.* fē·lip′əs, *Lat.* fi·lip′əs). [<Gk., a lover of horses]
—**Philip** One of the seven deacons of the early Christian church. *Acts* viii 5.
—**Philip, King,** died 1676, American Indian chief; made war on New England colonists, **King Philip's War,** 1675–76: Indian name *Metacomet.*
—**Philip, Prince,** born 1921, consort of Queen Elizabeth II of Great Britain; third duke of Edinburgh (1947).
—**Philip, Saint** One of the Twelve Apostles.
—**Philip II,** 382–336 B.C., king of Macedon 359–336; conqueror of Thessaly and Greece; father of Alexander the Great.
—**Philip II,** 1165–1223, king of France 1180–1223; went on Third Crusade; took English provinces in France from King John: called "Philip Augustus."
—**Philip II,** 1527–98, king of Spain 1556–98; sent the Armada against England.
—**Philip IV,** 1268–1314, king of France 1285–1314; supported Clement V at Avignon: called "Philip the Fair."
—**Philip V,** 237–179 B.C., king of Macedon 220–179 B.C.; defeated by Romans.
—**Philip V,** 1683–1746, king of Spain 1700–1746; founder of Bourbon line in Spain.
—**Philip the Bold,** 1342–1404, duke of Burgundy; conquered Flanders.
—**Philip the Good,** 1396–1467, duke of Burgundy; acquired the Netherlands.
Phi·lip·pa (fi·lip′ə) A feminine personal name; feminine of PHILIP. Also *Fr.* **Phi·lip·pine** (fē·lē·pēn′), *Ger.* **Phi·lip·pi·ne** (fē′lē·pē′nə).
Phi·lippe·ville (fil′ip·vil, *Fr.* fē·lēp·vēl′) A port of NE Algeria on the Mediterranean.
Phi·lip·pi (fi·lip′ī) An ancient town in northern Macedonia, Greece; scene of the defeat of Brutus and Cassius by Octavius and Anthony, 42 B.C., and of St. Paul's first preaching in Europe. *Acts* xvi 12. —**Phi·lip′pi·an** *adj. & n.*
Phi·lip·pi·ans (fi·lip′ē·ənz) In the New Testament, an epistle of St. Paul addressed to Christians at Philippi.
phi·lip·pic (fi·lip′ik) *n.* An impassioned speech characterized by invective: from the **Philippics,** a series of twelve speeches in which Demosthenes denounced Philip of Macedon. [<L *Philippicus* <Gk. *Philippikos* of Philip]
Phil·ip·pine Islands (fil′ə·pēn) An archipelago of 7,083 islands SE of China and NE of Borneo; 114,830 square miles; ceded by Spain to the United States, 1898, for $20,000,000; a commonwealth since 1935; seized by Japan, 1942–44, in World War II; since 1946 the **Republic of the Philippines;** capital, Quezon City. Also **Phil′ip·pines.** A native of the Philippines is known as a *Filipino.* —**Phil′ip·pine** *adj.*
Philippine Scouts The regiments of U. S. troops formerly maintained in the Philippines, consisting of native enlisted personnel and American officers.
Philippine Sea A part of the western Pacific between the Philippines and the Marianas.
Phil·ip·pop·o·lis (fil′i·pop′ə·ləs) An ancient name for PLOVDIV.
Phil·ips (fil′ips), **Ambrose,** 1675?–1749, English poet and dramatist.
Phi·lis·ti·a (fi·lis′tē·ə) An ancient region on the Mediterranean, SW Palestine. *Ps.* lx 8.
Phi·lis·tine (fi·lis′tin, -tēn, -tīn, fil′əs-) *n.* **1** One of a warlike race of ancient Philistia. I *Sam.* xvii 23. **2** An ignorant, narrow-minded person, devoid of culture and indifferent to art. [<F *Philistin* <LL *Philistinus,* pl. *Philistini* <Gk. *Philistinoi, Palaistinoi* <Hebrew *p'lishtim*]
Phi·lis·tin·ism (fi·lis′tin·iz′əm) *n.* Blind conventionalism; devotion to low aims.
Phil·lips (fil′ips), **Stephen,** 1868–1915, English poet and dramatist. —**Wendell,** 1811–84, U. S. orator and abolitionist.
Phil·lis (fil′is) See PHYLLIS.
philo- *combining form* Loving; fond of: *philomath.* Also, before vowels, *phil-.* [<Gk. *philos* loving <*phileein* love]
Phil·oc·te·tes (fil′ək·tē′tēz) In Greek legend, a Greek hero, heir to the bow and poisoned arrows of Hercules, who killed Paris with one of them in the tenth year of the siege of Troy.
phil·o·den·dron (fil′ə·den′drən) *n.* Any of a genus (*Philodendron*) of tropical American climbing plants, with thick, glossy, evergreen leaves: cultivated as an ornamental house plant. [<NL <Gk., neut. of *philodendros* fond of trees <*philos* fond + *dendron* a tree]
phi·log·y·ny (fi·loj′ə·nē) *n.* Fondness for or devotion to women: opposed to *misogyny.* [<Gk. *philogynia* <*philos* fond + *gynē* a woman] —**phi·log′y·nist** *n.* —**phi·log′y·nous** *adj.*
Phi·lo Ju·dae·us (fī′lō jōō·dē′əs) Jewish Platonist philosopher of the first century.
phi·lol·o·gy (fi·lol′ə·jē) *n.* **1** The scientific study of written records (chiefly literary works of art), in order to set up accurate texts and determine their meaning, often in terms of linguistic and cultural history. **2** Linguistics. **3** In popular use, etymology. **4** Formerly, literary scholarship, especially classical scholarship. [<F *philologie* <L *philologia* <Gk. <*philologos* fond of argument, words <*philos* fond + *logos* a word] —**phil·o·log·ic** (fil′ə·loj′ik) or **·i·cal** *adj.* —**phil′o·log′i·cal·ly** *adv.* —**phi·lol′o·gist, phi·lol′o·ger, phil′o·log** or **·logue** (-lôg, -log) *n.*
phil·o·math (fil′ə·math) *n.* One who loves learning; a scholar. [<Gk. *philomathēs* fond of learning <*philos* fond + *math-,* root of *manthanein* learn] —**phil′o·math′ic** or **·i·cal** *adj.* —**phi·lom·a·thy** (fi·lom′ə·thē) *n.*
phil·o·mel (fil′ə·mel) *n.* In poetic usage, the nightingale. Also **phil′o·me′la.** [<F *philomèle* <L *philomela* <Gk. *philomēla,* ? <*philos* fond of + *melos* a song]
Phil·o·me·la (fil′ə·mē′lə) In Greek mythology, a princess of Athens who was raped by her brother-in-law Tereus, who then tore out her tongue; when, in revenge, she and her sister Procne killed his son Itylus, the gods changed Tereus into a hoopoe, Procne into a swallow, and Philomela into a nightingale.
phi·lom·e·try (fi·lom′ə·trē) *n.* The study and collecting of postal meter impressions on mail matter: a branch of philately. [<PHILO- + *(postal) meter,* on analogy with PHILATELY] —**phi·lom′e·trist** *n.* —**phil·o·met·ric** (fil′ə·met′rik) *adj.*
phil·o·pe·na (fil′ə·pē′nə) *n.* **1** A game in which anyone finding a nut with twin kernels shares it with another person. The one who first says *philopena* when next they meet receives a forfeit from the other. **2** The twin kernels shared. **3** The gift made as a forfeit. Often spelled *fillipeen.* Also **phil′lip·pine, phil′li·peen′er.** [Appar. <Du. *phillipine,* alter. of G *vielliebchen* very dear <*viel* much + *liebchen,* dim. of *liebe* love]
Phil·o·poe·men (fil′ə·pē′mən), 252?–183 B.C., Greek patriot; advocate of unity among Greek city-states; called "Last of the Greeks."
phil·o·pro·gen·i·tive (fil′ə·prō·jen′ə·tiv) *adj.* **1** Fond of offspring or of children in general. **2** Prolific. [<PHILO- + L *progenitus,* pp. of *progignere* beget] —**phil′o·pro·gen′i·tive·ly** *adv.* —**phil′o·pro·gen′i·tive·ness** *n.*
phi·los·o·pher (fi·los′ə·fər) *n.* **1** A student of or specialist in philosophy. **2** A man of practical wisdom; one who schools himself to calmness and patience under all circumstances, as originally enjoined by the Stoic philosophy. [<AF *philosophre,* var. of OF *filosofe* <L *philosophus* <Gk. *philosophos* a lover of wisdom <*philos* loving + *sophos* wise]
philosopher's stone An imaginary stone or substance having the property of transmuting the baser metals into gold: sought by the alchemists.
phil·o·soph·i·cal (fil′ə·sof′i·kəl) *adj.* **1** Pertaining to or founded on the principles of philosophy. **2** Proper to or characteristic of a philosopher. **3** Self-restrained and serene; rational; thoughtful; calm. **4** *Archaic* Pertaining to or used in the study of natural philosophy or physics. Also **phil′o·soph′ic.** —**phil′o·soph′i·cal·ly** *adv.* —**phil′o·soph′i·cal·ness** *n.*
phi·los·o·phism (fi·los′ə·fiz′əm) *n.* Unsound or pretended philosophy; sophistry.
phi·los·o·phist (fi·los′ə·fist) *n.* One who affects philosophy; a would-be philosopher.
phi·los·o·phis·tic (fi·los′ə·fis′tik) *adj.* Of the nature of philosophism; characteristic of a philosophist. Also **phi·los′o·phis′ti·cal.**

phi·los·o·phize (fi·los'ə·fīz) v.i. ·phized, ·phiz·ing To speculate like a philosopher; seek ultimate causes and principles; moralize. — **phi·los'o·phiz'er** n.

phi·los·o·phy (fi·los'ə·fē) n. pl. ·phies 1 The love of wisdom as leading to the search for it; hence, knowledge of general principles — elements, powers, or causes and laws — as explaining facts and existences. 2 The general laws that furnish the rational explanation of anything: the *philosophy* of banking. 3 The calm judgment and equable temper resulting from study of causes and laws; practical wisdom; fortitude, as in enduring reverses and suffering. 4 Reasoned science; a scientific system; as (formerly), natural *philosophy*, now natural science. 5 A philosophical system or treatise. 6 The sciences as formerly studied in the universities. [< OF *filosofie, philosophie* < L *philosophia* < Gk. < *philosophos*. See PHILOSOPHER.]

philosophy of the Academy Platonism: so called because Plato taught in the Academy, near Athens: also **intuitional philosophy**.

philosophy of the garden Epicureanism: so called because Epicurus taught in a garden at Athens.

philosophy of the Lyceum Aristotelianism: so called because Aristotle taught it in the Lyceum at Athens: also **empirical philosophy**.

philosophy of the Porch Stoicism: so called because Zeno taught it in the porch of the Poecile in Athens.

-philous combining form Loving; fond of: *anemophilous*. [< Gk. *-philos*. See -PHILE.]

phil·ter (fil'tər) n. A charmed draft supposed to have power to excite sexual love; a love potion; hence, any magic potion. — v.t. To charm with a philter. Also **phil'tre**. ♦ Homophone: *filter*. [< MF *philtre* < L *philtrum* < Gk. *philtron* a love potion < *phileein* love]

-phily Var. of -PHILIA.

phi·mo·sis (fi·mō'sis) n. *Pathol.* The abnormal constriction of the opening of the prepuce, preventing the uncovering of the glans penis. [< NL < Gk. *phimōsis* a muzzling < *phimos* a muzzle] — **phi·mot'ic** (-mot'ik) adj.

Phin·e·as (fin'ē·əs) A masculine personal name. Also **Phin'e·has** (-həs), Fr. **Phi·né·as** (fē·nā·ä'). [Prob. < Egyptian, mouth of brass]

Phi·neus (fi·nyōōs', fi-, fin'ē·əs) In Greek mythology, a Thracian king who for some offense against the gods was punished by having the Harpies defile or snatch away his food.

Phin·ti·as (fin'tē·əs) Pythias.

Phips (fips), **Sir William**, 1651–95, first royal governor of Massachusetts. Also **Phipps**.

phiz (fiz) n. *Slang* Visage; face. [Short for *phiznomy*, obs. var. of PHYSIOGNOMY]

phle·bi·tis (fli·bī'tis) n. *Pathol.* Inflammation of the inner membrane of a vein. [< NL < Gk. *phleps, phlebos* a vein + *-itis* -ITIS] — **phle·bit'ic** (-bit'ik) adj.

phlebo- combining form Venous: *phlebotomy*. Also, before vowels, **phleb-**. [< Gk. *phleps, phlebos* a vein]

phleb·o·scle·ro·sis (fleb'ō·skli·rō'sis) n. *Pathol.* Thickening and hardening of the walls of the veins. [< PHLEBO- + Gk. *sklērōsis* a hardening < *sklēros* hard] — **phleb'o·scle·rot'ic** (-rot'ik) adj.

phle·bot·o·mize (fli·bot'ə·mīz) v.t. ·mized, ·miz·ing To treat by phlebotomy.

phle·bot·o·my (fli·bot'ə·mē) n. *Surg.* The practice of opening a vein for letting blood as a remedial measure; bloodletting. [< OF *flebothomie* < L *phlebotomia* < Gk., the opening of a vein < *phleps, phlebos* a vein + *temnein* cut] — **phleb·o·tom·ic** (fleb'ə·tom'ik) or **·i·cal** adj. — **phle·bot'o·mist** n.

Phleg·e·thon (fleg'ə·thon, flej'-) In Greek mythology, the river of fire, one of the five rivers surrounding Hades. [< Gk., lit., blazing]

phlegm (flem) n. 1 A viscid, stringy mucus secreted in the air passages or in the stomach, especially when produced as a morbid discharge. 2 Inflammation; cold, undemonstrative temper; self-possession. 3 *Obs.* One of the four natural humors (the cold and moist) in ancient physiology. See synonyms under APATHY. [< OF *fleume, flemme* < L *phlegma* the clammy humor of the body < Gk., inflammation < *phlegein* blaze; refashioned after Gk.]

phleg·ma·si·a do·lens (fleg·mā'zhē·ə dō'lənz) *Pathol.* Milk leg. [< L < *phlegmasia* an inflammation + *dolens, -entis* painful, ppr. of *dolere* feel pain]

phleg·mat·ic (fleg·mat'ik) adj. Sluggish; indifferent; not easily moved or excited. Also **phleg·mat'i·cal**. [< OF *fleumatique* < L *phlegmaticus* < Gk. *phlegmatikos* < *phlegma, -matos*. See PHLEGM.] — **phleg·mat'i·cal·ly** adv.

phlegm·y (flem'ē) adj. 1 Relating to, resembling, or containing phlegm. 2 Phlegmatic in temperament.

phlo·em (flō'em) n. *Bot.* The complex plant tissue composed of sieve tubes with associated cells, and forming part of the vascular system serving for the conduction of the sap; the bast; leptome. Compare XYLEM. [< G < Gk. *phloos* bark]

phlo·gis·tic (flō·jis'tik) adj. 1 Pertaining to phlogiston or to the theory of its existence. 2 Inflammatory; inflamed. [< Gk. *phlogistos* inflammable. See PHLOGISTON.]

phlo·gis·ton (flō·jis'tən) n. The fiery principle formerly assumed to be a necessary constituent of all combustible bodies, and to be given up by them in burning. [< NL < Gk., neut. of *phlogistos* inflammable < *phlogizein* set on fire < *phlox, phlogos* a flame < *phlegein* burn]

phlog·o·pite (flog'ə·pīt) n. A yellowish-brown to brownish-red monoclinic magnesium mica. [< G *phlogopit* < Gk. *phlogōpos* fiery; so called from its appearance]

phlo·go·sis (flə·gō'sis) n. *Pathol.* 1 Inflammation. 2 Erysipelas. [< NL < Gk. *phlogōsis* an inflammation < *phlox, phlogos* a flame] — **phlo·gosed'** (-gōzd', -gōst'), **phlo·got'ic** (-got'ik) adj.

phlor·i·zin (flôr'ə·zin, flor'-, flə·rī'-) n. *Chem.* A bitter crystalline glycoside, $C_{21}H_{24}O_{10}$, contained in the root bark of the apple, pear, plum, and cherry tree: used in medicine as a tonic. Also **phlo·rid·zin** (flə·rid'zin). [< Gk. *phloios* bark + *rhiza* a root + -IN]

phlox (floks) n. Any plant or flower of a North American genus (*Phlox*) of herbs of the Polemoniaceae family, with opposite leaves and clusters of showy flowers in various shades of red, purple, white, or variegated. [< NL < L, a kind of flower < Gk. *phlox* a wallflower, lit., a flame < *phlegein* burn]

phlyc·te·na (flik·tē'nə) n. pl. ·nae (-nē) *Pathol.* A small blister containing watery or serous fluid. Also **phlyc·tae'na**. [< NL < Gk. *phlyk·taina* a blister < *phlyein* swell]

PERENNIAL PHLOX
(Plant from 2 to 6 feet tall)

Phnôm·penh (pə·nôm'pen') See PNOM-PENH.

-phobe combining form One who fears or has an aversion to: *Anglophobe*. [< Gk. *-phobos* fearing < *phobeesthai* fear]

pho·bi·a (fō'bē·ə) n. 1 A morbid, compulsive, and persistent fear of any specified type of object, stimulus, or situation. 2 Any strong aversion or dislike. [< L < Gk. < *phobos* fear] — **pho'bic** (-bik) adj.

-phobia combining form *Psychiatry* Aversion to; morbid fear or dislike of. [< Gk. *phobos* fear]

In the following list each entry denotes a morbid fear or dislike of the thing or situation indicated by the translation of the first part of the word:

acrophobia	heights
agoraphobia	open spaces
ailurophobia	cats
algophobia	pain
androphobia	men
astraphobia	thunder and lightning
autophobia	being alone; self
bathophobia	depth
claustrophobia	closed space
cynophobia	dogs (rabies)
demophobia	crowds
dromophobia	crossing streets
genophobia	sex
gynophobia	women
haptephobia	being touched
hemophobia	blood
hydrophobia	water
hypnophobia	falling asleep
musophobia	mice
mysophobia	contamination
neophobia	the new
nyctophobia	night, darkness
ophidiophobia	snakes
photophobia	light
sitophobia	eating; food
taphephobia	being buried alive
thanatophobia	death
toxicophobia	poison
xenophobia	strangers, foreigners
zoophobia	animals

pho·ca (fō'kə) n. pl. ·cae (-sē) One of a genus (*Phoca*) of seals (family Phocidae), including the typical earless seals of temperate waters. [< NL < L < Gk. *phōkē* a seal] — **pho'cine** (-sīn, -sin) adj. — **pho'coid** (-koid) adj. & n.

Pho·cae·a (fō·sē'ə) An ancient Ionian port on the Aegean, later becoming an important maritime state in western Asia Minor. Also **Pho·cæ'a**.

Pho·ci·on (fō'shē·on), 402?–317 B.C., Athenian general and patriot; executed.

Pho·cis (fō'sis) A region of central Greece on the north shore of the Gulf of Corinth.

phoe·be (fē'bē) n. An American flycatcher (*Sayornis phoebe*) with grayish-brown plumage and slightly crested head: common throughout the eastern United States. Also **phoebe bird**. [Imit. of its cry; infl. in form by PHOEBE]

Phoe·be (fē'bē) 1 A feminine personal name. 2 In Greek mythology, a Titaness, mother of Leto. 3 A name for Artemis as goddess of the moon. 4 *Poetic* The moon. 5 Saturn's ninth satellite. Also spelled *Phebe*. Also **Phœ'be**. (-bē, bright)

Phoe·bus (fē'bəs) 1 In Greek mythology, Apollo as god of the sun. 2 *Poetic* The sun. Also **Phœ'bus**.

Phoe·ni·cia (fə·nē'shə, -nish'ə) An ancient country of western Asia at the eastern end of the Mediterranean in modern Syria and Lebanon, comprising a group of city-states that flourished around 1200 B.C.: also *Phenicia*.

Phoe·ni·cian (fə·nē'shən, -nish'ən) adj. Of or pertaining to ancient Phoenicia, its people, or its language. — n. 1 One of the people of ancient Phoenicia or any of its colonies, as Carthage: ethnically belonging to the Canaanite branch of the Semitic peoples. 2 The Northwest Semitic language of these people. Also spelled *Phenician*.

phoe·nix (fē'niks) n. 1 In Egyptian mythology, a legendary bird of great beauty, unique of its kind, which was supposed to live for 500 or 600 years in the Arabian Desert and then consume itself by fire, rising again from its ashes young and beautiful to live through another cycle: a symbol of immortality. 2 A person of matchless beauty or excellence; a paragon. Also spelled *phenix*. [OE *fenix* < Med. L *phenix* < L *phoenix* < Gk. *phoinix* the phoenix]

PHOENIX
Described by Herodotus as golden-winged with eagle-like red body.

Phoe·nix (fē'niks) *Astron.* A southern constellation. See CONSTELLATION.

Phoe·nix (fē'niks) The capital of Arizona.

Phoenix Islands A group of eight islands, comprising a district of the Gilbert and Ellice Islands colony of Great Britain; the islands of Canton and Enderbury are administered jointly with the United States; total, 11 square miles; headquarters on Canton.

phon (fon) n. *Physics* The unit of loudness level of a sound, numerically equal to the sound-pressure level in decibels, relative to a pressure of 0.0002 microbar, of a simple tone of 1,000 cycles per second which is judged by the listener as equal in loudness. [< Gk. *phōnē* a voice]

phon- Var. of PHONO-.

pho·nate (fō'nāt) v.t. ·nat·ed, ·nat·ing To make

articulate sounds. [<Gk. *phōnē* + -ATE[1]] — **pho·na'tion** *n.*

pho·nau·to·graph (fō·nô'tə·graf, -gräf) *n.* 1 An apparatus designed to record the vibrations of sounds. It was the forerunner of Edison's phonograph. 2 A writing or tracing produced by the mechanical use of sound vibrations. Also **pho·nau'to·gram.** [<Gk. *phōnē* sound + *autos* self + -GRAPH] — **pho·nau'to·graph'ic** or **·i·cal** *adj.* — **pho·nau'to·graph'i·cal·ly** *adv.*

phone[1] (fōn) *n.* & *v. Colloq.* Telephone. [Short for TELEPHONE]

phone[2] (fōn) *n.* A sound used in human speech. [<Gk. *phōnē* a sound]

-phone *combining form* Voice; sound: used in names of musical instruments and other sound-transmitting devices: *saxophone, microphone.* [<Gk. *phōnē* voice]

pho·neme (fō'nēm) *n.* A class of acoustically similar sounds in a language, usually written with the same phonetic symbol, which differ non-relevantly as conditioned by environment; the smallest unit in the sound system of a language, functioning to distinguish one morpheme from another. The contrasting phonemes /t/ and /p/ distinguish the words *tin* and *pin*, whereas the varying pronunciations of *t* in *tip, stop* and *pit* are not recognized by speakers of English and are considered members of the one phoneme /t/. See ALLOPHONE. [<F *phonème* a sound <Gk. *phōnēma* < *phōnē* a voice, sound]

pho·ne·mic (fə·nē'mik) *adj.* 1 Of or referring to phonemes: the *phonemic* pattern of a language. 2 Involving distinctive speech sounds: a *phonemic* difference. — **pho·ne'mi·cal·ly** *adv.*

pho·ne·mics (fə·nē'miks) *n.* The study of the phonemic system of a language.

pho·nen·do·scope (fə·nen'də·skōp) *n. Med.* An amplifying stethoscope. [<PHON- + END(O)- + -SCOPE]

pho·net·ic (fə·net'ik) *adj.* 1 Of or pertaining to phonetics, or to speech sounds and their production. 2 Representing articulate sounds or speech; specifically, designating the representation of each speech sound by a distinct character, or by a distinctive spelling or mark: *phonetic* alphabet, *phonetic* spelling. Also **pho·net'i·cal.** [<Gk. *phōnētikos* < *phōnē* sound] — **pho·net'i·cal·ly** *adv.*

pho·ne·ti·cian (fō'nə·tish'ən) *n.* An authority on phonetics. Also **pho·net·i·cist** (fə·net'ə·sist), **pho'ne·tist.**

phonetic law A description of a pattern of sound-changes occurring under given conditions in a language or group of languages, as Grimm's law.

pho·net·ics (fə·net'iks) *n.* 1 The branch of linguistics which deals with the analysis, description, and classification of the sounds of speech, including **articulatory phonetics,** the study of the physiological processes involved in speech production, by means of which the sounds of a language are recorded and described, and **acoustic phonetics,** the study of the physical attributes of speech sounds by the use of laboratory instruments. 2 The system of sounds of a language: the *phonetics* of American English.

pho·ney (fō'nē) See PHONY.

-phonia See -PHONY.

phon·ic (fon'ik, fō'nik) *adj.* 1 Pertaining to or of the nature of sound. 2 Caused or accompanied by sound-articulation.

phon·ics (fon'iks, fō'niks) *n. pl. (construed as singular)* 1 The phonetic rudiments used in teaching reading and pronunciation. 2 Acoustics.

phono- *combining form* Sound; speech; voice: *phonograph.* Also, before vowels, *phon-.* [< Gk. *phōnē* a voice]

pho·no·chem·is·try (fō'nō·kem'is·trē) *n.* The study of chemical reactions as induced or affected by sound waves. — **pho'no·chem'i·cal** *adj.*

pho·no·deik (fō'nə·dēk) *n.* An instrument for making sound waves visible by converting them into light waves reflected from a vibrating mirror. [<PHONO- + Gk. *deiknynai* show]

pho·no·gram (fō'nə·gram) *n.* 1 The tracing produced by a phonograph, from which articulate sounds are reproduced; a phonograph record. 2 A graphic character symbolizing an articulate sound, word, syllable, etc. 3 A telephone message taken down on paper and delivered, like a telegram. — **pho'no·gram'ic** or **·gram'mic** *adj.* — **pho'no·gram'i·cal·ly** or **·gram'mi·cal·ly** *adv.*

pho·no·graph (fō'nə·graf, -gräf) *n.* An apparatus for recording and reproducing sounds, as speech, music, etc.

pho·no·graph·ic (fō'nə·graf'ik) *adj.* 1 Pertaining to or produced by a phonograph. 2 Pertaining to or written in phonography. 3 Relating to the representation of articulate sound. — **pho'no·graph'i·cal·ly** *adv.*

pho·nog·ra·phy (fə·nog'rə·fē) *n.* 1 The art or science of writing by sound; especially, the art of representing words according to a system of sound elements that reduces their graphic reproduction to the simplest form: a style of shorthand which owes its principal development to Isaac Pitman, of Bath, England, upon whose alphabet the majority of the existing stenographic systems are based. 2 The art of representing articulate sounds by marks or letters. 3 The art of making or using phonographs; the mechanical recording and reproduction of sounds or speech. — **pho·nog'ra·pher, pho·nog'ra·phist** *n.*

pho·no·lite (fō'nə·līt) *n.* A grayish-green compact igneous rock composed essentially of orthoclase, nephelite, and augite; clinkstone. — **pho'no·lit'ic** (-lit'ik) *adj.*

pho·nol·o·gy (fə·nol'ə·jē) *n.* 1 The study of the sound system of a language. 2 The historical study of the sound-changes that have taken place in a language. 3 The phonetic or phonemic pattern of a language. — **pho·no·log·ic** (fō'nə·loj'ik) or **·i·cal** *adj.* — **pho'no·log'i·cal·ly** *adv.* — **pho·nol'o·gist** *n.*

pho·no·ma·ni·a (fō'nə·mā'nē·ə, -mān'yə) *n.* Homicidal mania. [<Gk. *phonos* murder + -MANIA]

pho·nom·e·ter (fə·nom'ə·tər) *n.* An instrument for measuring the intensity of sounds or the frequency of sound vibrations. — **pho·nom'e·try** *n.*

pho·no·scope (fō'nə·skōp) *n.* An instrument for observing, testing, or exhibiting the properties of musical strings or other sounding bodies.

pho·no·type (fō'nə·tīp) *n.* 1 A writing or printing alphabet having a distinct character for each simple sound of speech. 2 A production written or printed in such characters. — **pho'no·typ'ic** (-tip'ik) or **·i·cal** *adj.* — **pho'no·typ'i·cal·ly** *adv.*

pho·no·typ·y (fō'nə·tī'pē) *n.* The art or practice of representing every elementary sound of articulate speech by a mark or letter of its own; phonetic transcription. — **pho'no·typ'ist** *n.*

pho·ny (fō'nē) *U.S. Slang adj.* **·ni·er, ·ni·est** Fake; false; spurious; counterfeit. — *n.* 1 Something fake or not genuine. 2 One who impersonates another; an impostor. Also spelled *phoney.* [< slang E *fawney* a gilt brass ring used in a fraud <Irish *fain(n)e* a ring]

-phony *combining form* A (specified) type of sound or sounds: *cacophony.* Also *-phonia,* as in *aphonia.* [<Gk. *phōnē* sound, voice]

Phor·cus (fôr'kəs) In Greek mythology, the leader of the Tritons and father of the Graeae.

-phore *combining form* A bearer or producer of: *semaphore.* [<Gk. *-phoros* < *pherein* bear]

pho·re·sis (fə·rē'sis) *n. Chem.* The passage of ions through a membrane by the action of an electric current. [<NL <Gk. *phorēsis* a being borne < *phoreein,* freq. of *pherein* bear]

phor·e·sy (fôr'ə·sē, for'-) *n. Biol.* An inter-relationship between small and large organisms by which the smaller, by attachment to the bodies of the larger, are transported or dispersed, as mites on the bodies of certain insects. [<Gk. *phorēsis.* See PHORESIS.]

-phorous *combining form* Bearing or producing: found in adjectives corresponding to nouns in *-phore.* [See -PHORE]

phos·gene (fos'jēn) *n. Chem.* Carbonyl chloride, $COCl_2$, a colorless, highly toxic gas with a suffocating odor: used in organic chemistry and as a chemical warfare agent. [<Gk. *phōs* light + *gen-,* root of *gignesthai* be born]

phos·ge·nite (fos'jə·nīt) *n.* A white, adamantine carbonate and chloride of lead, $Pb_2Cl_2CO_3$, crystallizing in the tetragonal system. [< PHOSGENE + -ITE[1]]

phosph- Var. of PHOSPHO-.

phos·pha·tase (fos'fə·tās) *n. Biochem.* An enzyme found in various animal tissues, as the kidneys and intestines, which hydrolyzes phospholipids. [<PHOSPHAT(E) + -ASE]

phos·phate (fos'fāt) *n.* 1 *Chem.* A salt or ester of phosphoric acid. Phosphates, especially those of calcium, are necessary to the growth of plants, which absorb them in the form of soluble salts. 2 *Agric.* Any fertilizer valued for its phosphoric acid. 3 A beverage of carbonated water, variously flavored, containing small amounts of phosphoric acid. [<F]

phos·phat·ic (fos·fat'ik) *adj.* 1 Relating to the phosphates. 2 Containing some phosphate.

phos·pha·tide (fos'fə·tīd, -tid) *n.* A phospholipid. [<PHOSPHAT(E) + -IDE]

phos·pha·tize (fos'fə·tīz) *v.t.* **·tized, ·tiz·ing** 1 To treat with phosphates. 2 To reduce to a phosphate. — **phos'pha·ti·za'tion** *n.*

phos·pha·tu·ri·a (fos'fə·tŏŏr'ē·ə, -tyŏŏr'-) *n. Pathol.* Excess of phosphates in urine. [<NL < *phosphatum* phosphate + Gk. *ouron* urine] — **phos'pha·tu'ric** *adj.*

phos·phene (fos'fēn) *n. Physiol.* The spectrum or luminous image made by pressing the eyeball: due to mechanical excitement of the retina, and seen internally opposite the point of pressure. [<Gk. *phōs* light + *phainein* make appear]

phos·phide (fos'fīd, -fid) *n. Chem.* A binary compound of phosphorus with a more positive element: calcium *phosphide,* Ca_3P_2. Also **phos'phid** (-fid).

phos·phine (fos'fēn, -fin) *n. Chem.* 1 A colorless, gaseous, highly toxic hydride of phosphorus, PH_3, with an odor resembling that of putrid fish. 2 One of a class of compounds derived from phosphine by replacing the hydrogen with alkyl radicals. 3 An acridine dye. Also **phos'phin** (-fin).

phos·phite (fos'fīt) *n.* A salt of phosphorous acid.

phospho- *combining form* Phosphorus; of or containing phosphorus, or any of its compounds: *phospholipid.* Also, before vowels, *phosph-.* [<PHOSPHORUS]

phos·pho·cre·a·tine (fos'fō·krē'ə·tēn, -tin) *n. Biochem.* An organic compound, $C_4H_{10}O_5N_3$, present in muscle tissue, to which it supplies the energy for contraction. [<PHOSPHO(RIC ACID) + CREATINE]

phos·pho·lip·id (fos'fō·lip'id) *n. Biochem.* Any of a group of fatty substances widely distributed in plant and animal tissue, as cephalin and lecithin: they contain nitrogen and phosphoric acid. Also called *phosphatide.* [<PHOSPHO- + LIPID]

phos·pho·ni·um (fos·fō'nē·əm) *n. Chem.* The univalent radical PH_4 regarded as a base. It resembles the radical ammonium, and forms crystalline halides. [<PHOSPH(ORUS) + (AMM)ONIUM]

phos·pho·pro·te·in (fos'fō·prō'tē·in, -tēn) *n. Biochem.* Any of a class of conjugated proteins containing phosphoric acid combined with the hydroxy group of certain amino acids, as the casein of milk.

phos·phor (fos'fər) *n. Chem.* 1 Phosphorus. 2 Any of a class of substances that will emit light under the action of certain chemicals or radiations. — *adj.* Phosphorescent. [<L *phosphorus* PHOSPHORUS]

Phos·phor (fos'fər) *n.* The morning star, especially Venus, as the harbinger of day. Also **Phos'phore.** [<L *Phosphorus* the morning star <Gk. *phōsphoros.* See PHOSPHORUS.]

phosphor bronze An alloy of copper and tin containing small amounts of phosphorus, noted for its toughness, durability, and high tensile strength.

phos·phor·esce (fos'fə·res') *v.i.* **·esced, ·esc·ing** To glow with a faint light unaccompanied by sensible heat. [? Back formation <PHOSPHORESCENT]

phos·phor·es·cence (fos'fə·res'əns) *n.* 1 The emission of light without sensible heat, or the light so emitted. 2 The property of continuing to shine in the dark after exposure to light, shown by many mineral substances: distinguished from *fluorescence.* 3 *Biol.* The property of producing a faint light, shown by infusorians, fireflies, etc.

phos·phor·es·cent (fos'fə·res'ənt) *adj.* Exhibiting phosphorescence. [<PHOSPHORUS + -ESCENT]

phos·phor·et·ed (fos'fə·ret'id) *adj. Chem.* Combined with phosphorus: *phosphoreted* hydrogen, PH_3. Also **phos'phor·et'ted, phos'phu·ret'ed** (-fyə·ret'id) or **·ret'ted.** [<NL *phosphoretum* a phosphide]

phos·phor·ic (fos·fôr′ik, -for′-) *adj.* 1 *Chem.* Pertaining to or derived from phosphorus, especially in its highest valence. 2 Phosphorescent.
phosphoric acid *Chem.* One of three oxyacids of phosphorus known respectively as *metaphosphoric acid* (HPO₃), *orthophosphoric acid* (H₃PO₄), and *pyrophosphoric acid* (H₄P₂O₇).
phos·phor·ism (fos′fə·riz′əm) *n.* *Pathol.* Chronic phosphorus poisoning.
phos·phor·ite (fos′fə·rīt) *n.* 1 A massive fibrous variety of apatite. 2 Phosphate rock in general.
phos·phor·o·scope (fos′fər·ə·skōp′) *n.* An apparatus for measuring the duration of phosphorescent light after the source is withdrawn.
phos·pho·rous (fos′fər·əs, fos·fôr′əs, -fō′rəs) *adj. Chem.* Of, pertaining to, resembling, containing, or derived from phosphorus, especially in its lower valence.
phosphorous acid *Chem.* A crystalline acid, H₃PO₃, with a garlic taste, obtained by the oxidation of phosphorus.
phos·pho·rus (fos′fər·əs) *n.* 1 A soft, yellowish, non-metallic element (symbol P) that readily combines with oxygen, exhibiting a phosphorescent glow at a low temperature, and inflaming violently when heated very slightly, as by friction. White phosphorus is exceedingly active and highly poisonous; red phosphorus is not. 2 Any phosphorescent substance. [< NL < L *Phosphorus* the morning star < Gk. *phôsphoros* (*astēr*), lit., light-bringing < *phôs* a light + *phoros* bearing < *pherein* bear]
phot (fot, fōt) *n.* The cgs unit of illumination, equal to one lumen per square centimeter. [< Gk. *phōs, phōtos* a light]
pho·tic (fō′tik) *adj.* 1 Relating to light and the production of light. 2 Designating those underwater regions which are penetrated by sunlight: the *photic* zone.
Pho·ti·us (fō′shē·əs, -shəs), 816?-891?, a patriarch of Constantinople who refused to recognize papal jurisdiction and was therefore excommunicated.
pho·to (fō′tō) *n. pl.* **-tos** *Colloq.* A photograph. [Short for PHOTOGRAPH]
photo- *combining form* 1 Light; of, pertaining to, or produced by light: *photometer*. 2 Photograph; photographic: *photoengrave*. [< Gk. *phōs, phōtos* light]
pho·to·ac·tin·ic (fō′tō·ak·tin′ik) *adj.* Capable of emitting actinic radiation.
pho·to·ar·chive (fō′tō·är′kiv) *n.* A collection of photographs assembled and classified for purposes of study and research.
pho·to·bath·ic (fō′tō·bath′ik) *adj.* Of or pertaining to sea depths to which sunlight reaches. [< PHOTO- + Gk. *bathos* depth]
pho·to·bi·og·ra·phy (fō′tō·bī·og′rə·fē) *n.* Photography that reveals the subject's personality through the use of characteristic poses or suggestive settings; portrait photography. [< PHOTO- (def. 2) + BIOGRAPHY]
pho·to·bi·ot·ic (fō′tō·bī·ot′ik) *adj.* 1 Living in the light. 2 Requiring light for life or development.
pho·to·cell (fō′tō·sel′) *n.* A photoelectric cell.
pho·to·chem·is·try (fō′tō·kem′is·trē) *n.* The branch of chemistry dealing with chemical reactions produced or influenced by light. — **pho′to·chem′i·cal** *adj.*
pho·to·chro·mog·ra·phy (fō′tō·krō·mog′rə·fē) *n.* The art of reproducing on a printing press photographic images in several colors.
pho·to·chro·my (fō′tō·krō′mē) *n.* Color photography. [< PHOTO- (def. 2) + CHROM- + -Y¹]
pho·to·chron·o·graph (fō′tō·kron′ə·graf, -gräf) *n.* 1 An instrument for taking pictures at minute, regular, timed intervals, of a body in motion. 2 A picture so taken. 3 A chronograph adapted for use in photographing a moving body, as a star in transit. — **pho′to·chron·og′ra·phy** (-krə·nog′rə·fē) *n.*
pho·to·com·pos·er (fō′tō·kəm·pō′zər) *n.* Any machine or apparatus which composes printed matter by photographic means.
pho·to·com·po·si·tion (fō′tō·kom′pə·zish′ən) *n.* The composing of printed matter by photographic means rather than directly from movable type.
pho·to·con·duc·tion (fō′tō·kən·duk′shən) *n. Physics* The property, possessed by many substances, of exhibiting increased electrical conductivity when subjected to light waves. — **pho′to·con·duc′tive** (-duk′tiv) *adj.*
pho·to·cur·rent (fō′tō·kûr′ənt) *n.* An electric current produced by the action of light or by the photoelectric effect.
pho·to·di·a·gram (fō′tō·dī′ə·gram) *n.* A diagrammatic view, as of a factory, planned community, etc., superimposed upon an actual photograph of the area.
pho·to·dis·in·te·gra·tion (fō′tō·dis·in′tə·grā′shən) *n. Physics* A breaking down by the action of radiant energy: said especially of the atomic nucleus.
pho·to·dra·ma (fō′tō·drä′mə, -dram′ə) *n.* A motion picture or photoplay. — **pho′to·dra·mat′ic** (-drə·mat′ik) *adj.*
pho·to·dy·nam·ic (fō′tō·dī·nam′ik, -di·nam′-) *adj.* Of, pertaining to, or operating by the energy of light.
pho·to·dy·nam·ics (fō′tō·dī·nam′iks, -di·nam′-) *n.* The study of the action and influence of light on plants and animals.
pho·to·e·lec·tric (fō′tō·i·lek′trik) *adj.* Of or pertaining to the electrical effects due to the action of light, as in the emission of electrons from gaseous, liquid, or solid bodies when subjected to radiation of suitable wavelength. Also **pho′to·e·lec′tri·cal.**
photoelectric cell A vacuum tube, one of whose electrodes gives off electrons under the action of light: incorporated in electrical circuits as a controlling, testing, and counting device: also called *phototube, electric eye,* or *photocell.*
pho·to·e·lec·tron (fō′tō·i·lek′tron) *n.* An electron emitted from a metal surface when exposed to suitable radiation.
pho·to·e·lec·tro·type (fō′tō·i·lek′trə·tīp) *n.* 1 An electrotype produced by a photomechanical process. 2 A picture printed from such a block.
pho·to·en·grave (fō′tō·in·grāv′) *v.t.* **·graved**, **·grav·ing** To reproduce by photoengraving. — **pho′to·en·grav′er** *n.*
pho·to·en·grav·ing (fō′tō·in·grā′ving) *n.* 1 The act or process of producing by the aid of photography a relief block or plate for printing. 2 A plate or picture so produced.
photo finish The finish of a game or race, as in horse-racing, in which the two leads are so close as they cross the finish line that only the accuracy of a photograph can determine the winner.
pho·to·flash bulb (fō′tō·flash′) An electric bulb containing aluminum or magnesium which, on the passage of a current, ignites and gives an incandescent light of brief duration: used in photography.
pho·to·flood lamp (fō′tō·flud′) An electric lamp operating at excess voltage to give high illumination, as in taking photographs and motion pictures.
pho·to·gel·a·tin (fō′tō·jel′ə·tin) *adj.* Characterized by the use of gelatin: said of a photographic process.
pho·to·gen (fō′tō·jen) *n. Biol.* An organism that generates light; a phosphorescent plant or animal.
pho·to·gene (fō′tō·jēn) *n.* 1 *Physiol.* An impression on the retina, continuing after the object producing it has disappeared; an afterimage. 2 *Biol.* Photogen.
pho·to·gen·ic (fō′tō·jen′ik) *adj.* 1 Of or pertaining to the action of light; generating or producing light. 2 Producing phosphorescence; phosphorescent, as fireflies. 3 Having certain characteristics and qualities, as coloration, form, etc., suitable for being photographed. [< PHOTO- + -GENIC; def. 3 coined from PHOTO(GRAPH), on analogy with *pathogenic, eugenic,* etc.] — **pho′to·gen′i·cal·ly** *adv.*
pho·to·ge·ol·o·gy (fō′tō·jē·ol′ə·jē) *n. Geol.* The study of geological formations and processes by means of aerial photographs taken on a uniform scale. — **pho′to·ge′o·log′i·cal** (-jē′ə·loj′i·kəl) *adj.*
pho·to·go·ni·om·e·ter (fō′tō·gō′nē·om′ə·tər) *n.* A device for studying the phenomena of crystal X-ray diffraction and X-ray spectra.
pho·to·gram·me·try (fō′tō·gram′ə·trē) *n.* The art and technique of making surveys or maps by means of photographs. [< *photogram*, var. of PHOTOGRAPH + -METRY]

pho·to·graph (fō′tō·graf, -gräf) *n.* A picture taken by photography. — *v.t.* 1 To take a photograph of. — *v.i.* 2 To practice photography. 3 To undergo photographing. See synonyms under PICTURE.
pho·tog·ra·pher (fə·tog′rə·fər) *n.* One who makes a business of or is expert in photography.
pho·to·graph·ic (fō′tə·graf′ik) *adj.* 1 Pertaining to, used in, or produced by photography. 2 Like a photograph; vividly depicted. Also **pho′to·graph′i·cal.** — **pho′to·graph′i·cal·ly** *adv.*
pho·tog·ra·phy (fə·tog′rə·fē) *n.* 1 The process of forming and fixing an image of an object or objects by the chemical action of light and other forms of radiant energy on photosensitive surfaces. 2 The art or business of producing and printing photographs.
pho·to·gra·vure (fō′tə·grə·vyoor′, -grāv′yər) *n.* 1 The process of producing an intaglio plate for printing in which there are no sharp incised lines, but minute depressions, the deep parts producing the shadows, and the high parts showing white. 2 A picture produced from such a plate. [< F]
pho·to·he·li·o·graph (fō′tə·hē′lē·ə·graf′, -gräf) *n.* A telescopic photographic instrument, variously constructed, for taking pictures of the sun, as during an eclipse.
pho·to·jour·nal·ism (fō′tō·jûr′nəl·iz′əm) *n.* A form of journalism in which a story or news item is recounted largely or entirely by means of photographs. — **pho′to·jour′nal·ist** *n.*
pho·to·ki·net·ic (fō′tō·ki·net′ik) *adj. Biol.* Capable of movement under the influence of light, as certain plants. — **pho′to·ki·ne′sis** (-ki·nē′sis) *n.*
pho·to·lith·o·graph (fō′tō·lith′ə·graf, -gräf) *v.t.* To reproduce by photolithography. — *n.* A picture produced by photolithography.
pho·to·li·thog·ra·phy (fō′tō·li·thog′rə·fē) *n.* The art or operation of producing on stone, largely by photographic means, a printing surface from which impressions may be taken by a lithographic process. — **pho′to·lith′o·graph′ic** (-lith′ə·graf′ik) *adj.*
pho·tol·y·sis (fō·tol′ə·sis) *n.* Chemical or biological decomposition due to the action of light. [< NL < Gk. *phôs, phōtos* a light + *lysis* a loosening < *lyein* loosen] — **pho·to·lyt·ic** (fō′tə·lit′ik) *adj.*
pho·to·map (fō′tō·map′) *n.* A map composed of one or more aerial photographs, laid off into a grid, contour lines, etc.
pho·to·me·chan·i·cal (fō′tō·mi·kan′i·kəl) *adj.* Pertaining to a process, illustration, plate, etc., produced by any one of a variety of methods, by which photography is brought to the aid of the etcher or engraver. — **pho′to·me·chan′i·cal·ly** *adv.*
pho·tom·e·ter (fō·tom′ə·tər) *n.* 1 Any instrument for measuring the intensity of light or comparing the intensity of two lights. 2 A device for determining the proper duration of exposure in photography.
pho·tom·e·try (fō·tom′ə·trē) *n.* 1 The art of measuring the intensity of light, especially by means of the photometer. 2 The branch of optics that treats of such measurement. — **pho·to·met·ric** (fō′tə·met′rik) *adj.* — **pho·tom′e·trist** *n.*
pho·to·mi·cro·graph (fō′tō·mī′krə·graf, -gräf) *n.* A photograph of a microscopic object taken through a microscope. Compare MICROPHOTOGRAPH.
pho·to·mi·crog·ra·phy (fō′tō·mī·krog′rə·fē) *n.* The art or process of making photographs of minute objects, as by a camera attached to a microscope.
pho·to·mon·tage (fō′tō·mon·täzh′, -môn-) *n.* The process of montage with photographs; also, a picture produced by this process.
pho·to·mu·ral (fō′tō·myoor′əl) *n.* A photograph enlarged to a size suitable for wall decoration.
pho·ton (fō′ton) *n. Physics* A quantum of radiant energy moving with the velocity of light and an energy proportional to its frequency. [< PHOTO- + (ELECTR)ON] — **pho′ton′ic** *adj.*
Pho·ton (fō′ton) *n.* A keyboard-operated machine assembly for the composition of printed matter by means of high-speed photography and photoelectric action geared to

a matrix disk bearing the required characters in a series of concentric fonts, thus eliminating the use of metal type: a trade name.

pho·to·neu·tron (fō′tō·nōō′tron, -nyōō′-) *n.* *Physics* A neutron emitted in a photonuclear reaction.

pho·to·nu·cle·ar (fō′tō·nōō′klē·ər, -nyōō′-) *adj.* *Physics* Of, pertaining to, or designating a reaction initiated in an atomic nucleus by a photon.

pho·to·off·set (fō′tō·ôf′set, -of′-) *n.* Offset printing from a metal surface on which the text or design has been imprinted by photography.

pho·to·pe·ri·od (fō′to·pir′ē·əd) *n.* The length of day most favorable to the growth of a specified plant. —**pho′to·pe′ri·od′ic** (-pir′ē·od′ik) *adj.*

pho·toph·i·lous (fō·tof′ə·ləs) *Biol. adj.* Light-loving; possessing positive phototaxis. —*n.* A photophilous organism.

pho·to·pho·bi·a (fō′tə·fō′bē·ə) *n.* **1** Aversion to or intolerance of light. **2** *Pathol.* Morbid sensitivity of the eye to light.

pho·to·pia (fō·tō′pē·ə) *n.* Vision under lighting conditions which permit color discrimination; daylight vision. [<NL <Gk. *phōs, phōtos* a light + *ōps, ōpos* an eye] —**pho·top·ic** (fō·top′ik) *adj.*

pho·to·play (fō′tō·plā′) *n.* **1** The representation of a play in motion pictures. **2** A play arranged for a motion-picture performance.

pho·to·pro·ton (fō′tō·prō′ton) *n.* *Physics* A proton resulting from a photonuclear reaction.

pho·to·re·cep·tor (fō′tō·ri·sep′tər) *n.* *Physiol.* A nerve receptor sensitive to light stimuli.

pho·to·sen·si·tive (fō′tō·sen′sə·tiv) *adj.* Sensitive to light. —**pho′to·sen′si·tiv′i·ty** *n.*

pho·to·shock (fō′tō·shok′) *n.* *Psychiatry* A method of treating certain forms of mental disorder by the application of controlled flashes of light used in connection with appropriate drugs.

pho·to·spec·tro·scope (fō′tō·spek′trə·skōp) *n.* A spectrograph.

pho·to·sphere (fō′tə·sfir) *n.* *Astron.* The visible shining surface of the sun, or, more rarely, of a fixed star. —**pho′to·spher′ic** (-sfer′ik) *adj.*

pho·to·sta·ble (fō′tə·stā′bəl) *adj.* Unaffected by or resistant to the influence of light.

pho·to·stat (fō′tə·stat) *v.t. & v.i.* **·stat·ed** or **·stat·ted**, **·stat·ing** or **·stat·ting** To make a reproduction (of) with a Photostat. —*n.* The reproduction so produced. —**pho′to·stat′ic** *adj.*

Pho·to·stat (fō′tə·stat) *n.* A camera designed to reproduce documents, drawings, etc., directly, prints being made from the primary negative: a trade name.

pho·to·sur·vey (fō′tō·sûr′vā) *n.* A survey, as of industrial processes or social phenomena, illustrated and documented by photographs.

pho·to·syn·the·sis (fō′tō·sin′thə·sis) *n.* *Biol.* **1** The synthesis of chemical compounds by means of light and other forms of radiant energy. **2** The process by which plants form carbohydrates from carbon dioxide and water through the agency of sunlight acting upon chlorophyll. —**pho′to·syn·thet′ic** (-sin·thet′ik) *adj.* —**pho′to·syn·thet′i·cal·ly** *adv.*

pho·to·tax·is (fō′tō·tak′sis) *n.* *Biol.* The assumption by an organism of a definite position with reference to the direction of the incident ray of light, called **negative phototaxis** when the movement is away from the light, and **positive phototaxis** when the movement is toward the light. Also **pho′to·tax′y**. [<NL <Gk. *phōs, phōtos* a light + *taxis* an arrangement] —**pho′to·tac′tic** (-tak′tik) *adj.* —**pho′·to·tac′ti·cal·ly** *adv.*

pho·to·tel·e·graph (fō′tō·tel′ə·graf, -gräf) *v.t. & v.i.* To transmit by phototelegraphy. —*n.* Something so transmitted.

pho·to·te·leg·ra·phy (fō′tō·tə·leg′rə·fē) *n.* **1** The electrical transmission of messages, photographs, etc., by facsimile. **2** Telephotography. —**pho′to·tel′e·graph′ic** (-tel′ə·graf′ik) *adj.* —**pho′to·tel′e·graph′i·cal·ly** *adv.*

pho·to·tel·e·scope (fō′tō·tel′ə·skōp) *n.* A telescope provided with a photographic apparatus, adapted to photographing the heavenly bodies. —**pho′to·tel′e·scop′ic** (-tel′ə·skop′ik) *adj.*

pho·to·ther·a·py (fō′tō·ther′ə·pē) *n.* Treatment of diseases, especially diseases of the skin, by the application of light. Also **pho′to·ther′a·peu′tics** (-ther′ə·pyōō′tiks). —**pho′to·ther′a·peu′tic, pho′to·the·rap′ic** (-thə·rap′ik) *adj.*

pho·to·ther·mic (fō′tō·thûr′mik) *adj.* Denoting the thermic activity of the light rays.

pho·tot·o·nus (fō·tot′ə·nəs) *n.* *Biol.* **1** The influence of light upon the movement and the growth of plants. **2** The condition thus induced. **3** Increased irritability or motility induced by exposure to light. [<NL <Gk. *phōs, phōtos* a light + *tonos* tension. See TONE.] —**pho·to·ton·ic** (fō′tə·ton′ik) *adj.*

pho·to·to·pog·ra·phy (fō′tō·tə·pog′rə·fē) *n.* The art and technique of preparing topographic maps with the aid of photographs, as in the multiplex system or by data provided by aerial photographs. —**pho′to·top′o·graph′ic** (-top′ə·graf′ik) *adj.*

pho·to·tran·sis·tor (fō′tō·tran·zis′tər) *n.* A very small disk of germanium which produces a multiplied photocurrent by transistor action.

pho·to·trop·ic (fō′tə·trop′ik) *adj.* *Biol.* Turning toward the light; heliotropic. —**pho′to·trop′i·cal·ly** *adv.*

pho·tot·ro·pism (fō·tot′rə·piz′əm) *n.* *Biol.* The effect of light on the direction of growth of plant and animal organisms.

pho·tot·ro·py (fō·tot′rə·pē) *n.* **1** Phototropism. **2** The color alteration observed in some substances after exposure to light.

pho·to·tube (fō′tō·tōōb′, -tyōōb′) *n.* A photoelectric cell.

pho·to·type (fō′tə·tīp′) *n.* **1** A relief plate made for printing by photography. **2** The process by which it is produced. **3** A picture printed from such a plate. —**pho′to·typ′ic** (-tip′ik) *adj.*

pho·to·ty·pog·ra·phy (fō′tō·tī·pog′rə·fē) *n.* Any photomechanical process of engraving in relief that may be reproduced in connection with type on a printing press. —**pho′to·ty′po·graph′ic** (-tī′pə·graf′ik) *adj.*

pho·to·typ·y (fō′tō·tī′pē) *n.* The production or use of phototypes.

pho·to·vol·ta·ic (fō′tō·vol·tā′ik) *adj.* Capable of producing an electromotive force under the action of light; photoelectric.

pho·to·zin·cog·ra·phy (fō′tō·zing·kog′rə·fē) *n.* Photoengraving which uses a sensitized zinc plate.

phrais·in (frā′zin) *adj.* *Scot.* Flattering.

phrase (frāz) *n.* **1** An expression, consisting usually of but a few words, denoting a single idea or forming a separate part of a sentence; specifically, a group of two or more associated words, not containing a subject and predicate: distinguished from *clause*. See synonym below. **2** A concise, sententious expression. **3** Characteristic mode of expression; peculiar habit of language; phraseology. **4** *Music* A fragment of a melody having well-determined motion and repose, but incomplete sense. —*v.t. & v.i.* **phrased, phras·ing** **1** To express or be expressed in words or phrases. **2** *Music* To execute or divide (notes) into phrases by accentuation. [<LL *phrasis* diction <Gk., speech <*phrazein* point out, tell] —**phras′al** *adj.*

Synonym (noun): clause. A *clause* is a short sentence forming a distinct part of a composition, or in more extended use a distinct and separable statement forming part of a legal or state document, as of a will, an indictment, etc.; a *phrase* is a group of words conveying a single idea, and forming a distinct part of a sentence. In grammar, a *clause* is a simple sentence which is combined with some other sentence or sentences, so as to form a complex or compound sentence. A simple sentence standing alone is not, in grammatical use, called a *clause*, but every *clause* of a complex or compound sentence is a simple sentence. Thus, the *clause* always contains a subject and predicate. A *phrase* does not contain a subject and predicate, but it may include as many words as a *clause*. See DICTION, TERM.

phra·se·o·gram (frā′zē·ə·gram′) *n.* A symbol or combination of stenographic signs standing for a phrase. [<PHRASE + -(O)GRAM; on analogy with PHRASEOLOGY]

phra·se·o·graph (frā′zē·ə·graf′, -gräf′) *n.* A phrase having a symbol or phraseogram. —**phra′se·o·graph′ic** *adj.*

phra·se·ol·o·gist (frā′zē·ol′ə·jist) *n.* **1** One who pays much attention to phraseology; a maker of phrases. **2** One who collects phrases.

phra·se·ol·o·gy (frā′zē·ol′ə·jē) *n.* **1** The choice and arrangement of words and phrases in expressing ideas; diction; style. **2** A compilation or handbook of phrases. See synonyms under DICTION. [<NL *phraseologia*, irregularly formed <Gk. *phrasis, -eōs* speech + *logos* a word] —**phra′se·o·log′i·cal** (-ə·loj′·i·kəl) *adj.*

phras·ing (frā′zing) *n.* **1** The rendering of phrases. **2** *Music* Grouping and accentuation of the sounds in a melody. **3** Manner or form of verbal expression.

phra·try (frā′trē) *n.* *pl.* **·tries** **1** In ancient Athens, a clan or subdivision of a phyle. **2** Any similar tribal unit among primitive peoples, as a tribe composed of several totemic clans among North American Indians. [<Gk. *phratria* <*phratēr* clansman, brother] —**phra′tric** *adj.*

phre·at·ic (frē·at′ik) *adj.* Of or pertaining to underground waters, especially those at or below the water table and accessible through wells. [<Gk. *phrear, phreatos* well]

phre·net·ic (frə·net′ik) *adj.* **1** Of, pertaining to, or suffering from brain fever. **2** Excessively excited; frantic. Also **phre·net′i·cal.** —*n.* A madman. Also spelled *frenetic*. [<OF *frenetike* <L *phreneticus* <LGk. *phrenētikos*, var. of Gk. *phrenetikos* afflicted with delirium <*phrenitis* PHRENITIS] —**phre·net′i·cal·ly** *adv.*

phren·ic (fren′ik) *adj.* **1** Of or pertaining to the mind. **2** *Anat.* Of or pertaining to the diaphragm; diaphragmatic: the *phrenic* nerve. [<NL *phrenicus* <Gk. *phrēn, phrenos* diaphragm, mind]

phre·ni·tis (fri·nī′tis) *n.* *Pathol.* **1** Brain fever. **2** Wild delirium; frenzy. **3** Inflammation of the diaphragm. [<LL <Gk., delirium < *phrēn, phrenos* diaphragm, mind]

phreno– *combining form* **1** Mind; brain: *phrenotropic*. **2** Diaphragm; of or related to the diaphragm. Also, before vowels, **phren–**. [<Gk. *phrēn, phrenos* the diaphragm (thought to be the seat of intellect)]

phre·nol·o·gy (fri·nol′ə·jē) *n.* The doctrine that the conformation of the human skull, its shape and protuberances, indicate the position and degree of development of separate parts of the brain which control the various mental faculties and characteristics; loosely, character analysis by interpreting cranial formations. [<Gk. *phrēn, phrenos* mind + -LOGY] —**phren·o·log·ic** (fren′ə·loj′ik) or **·i·cal** *adj.* —**phren′o·log′i·cal·ly** *adv.* —**phre·nol′o·gist** *n.*

phren·o·sin (fren′ə·sin) *n.* *Biochem.* A cerebroside found in brain substance and isolated as a yellowish-white powder, $C_{48}H_{93}O_9N$: also called *cerebrin*. [<Gk. *phrēn, phrenos* mind, on analogy with *myosin*]

phren·o·trop·ic (fren′ə·trop′ik) *adj.* *Med.* Acting upon, influencing, or affecting the mind, as certain drugs.

Phrix·us (frik′səs) In Greek legend, son of Athamas and Nephele who, with his sister Helle, escaped his stepmother on a ram with golden fleece: only Phrixus arrived safely in Colchis, where he sacrificed the ram. See GOLDEN FLEECE.

Phryg·i·a (frij′ē·ə) An ancient country in Asia Minor divided into **Greater Phrygia,** the central part, and **Lesser Phrygia,** the NW section. *Acts* xvi 6.

Phryg·i·an (frij′ē·ən) *adj.* Of or pertaining to Phrygia or to its people. —*n.* **1** One of a prehistoric European people who migrated to Phrygia via Thrace and settled there. **2** The Indo-European language of this people, known from a few inscriptions.

Phrygian cap An ancient headdress. See LIBERTY CAP.

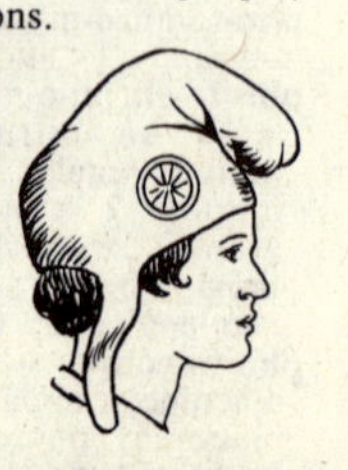

PHRYGIAN CAP

Phry·ne (frī′nē) A beautiful Athenian courtesan of the fourth century B.C.; supposedly the model for sculptures by Praxiteles.

phthal·e·in (thal′ē·in, -ēn, fthal′-) *n.* *Chem.* Any one of a series of organic compounds formed, with elimination of water, by the combination of a phenol with phthalic acid or its anhydride. Some compounds of phthalein are coloring matters. Also **phthal′e·ine.** [<PHTHAL(IC) + *-ein*, var. of -IN]

phthal·ic (thal′ik, fthal′-) *adj.* *Chem.* Of, pertaining to, or derived from naphthalene. [Short for NAPHTHALIC]

phthalic acid *Chem.* One of three aromatic crystalline compounds, $C_8H_6O_4$, derived variously, as by the oxidation of naphthalene: formerly called *alizaric acid, naphthalic acid.*

phthal·in (thal'in, fthal'-) *n. Chem.* Any of several colorless crystalline compounds obtained by reducing phthalein. Also **phthal'ine.** [<PHTHAL(EIN) + -IN]

phthal·o·cy·a·nin (thal'ō·sī'ə·nin, fthal'-) *n. Chem.* Any of a group of organic dyestuffs related to porphyrin and yielding blue and green pigments. [<*phthalo-* (<PHTHALIC) + CYANIN]

phthi·ri·a·sis (thi·rī'ə·sis, fthi-) *n.* Pediculosis; lousiness. [<L <Gk. *phtheiriasis* <*phtheiriaein* be lousy <*phtheir* a louse]

phthis·ic (tiz'ik) *n.* **1** *Pathol.* A wasting disease of the lungs. **2** Asthma; difficulty of breathing. [<OF *tisike* <L *phthisicus* <Gk. *phthisikos* consumptive <*phthisis* PHTHISIS]

phthis·i·cal (tiz'i·kəl) *adj.* **1** Relative to or affected with disease of the lungs; consumptive. **2** Asthmatic; wheezy. Also **phthis'·ick·y.**

phthis·i·o·pho·bi·a (tiz'ē·ə·fō'bē·ə) *n.* Morbid dread of phthisis. [<NL <Gk. *phthisis* phthisis + *phobeein* fear]

phthi·sis (thī'sis, fthī'-) *n. Pathol.* **1** Pulmonary consumption; tuberculosis of the lungs; less frequently, tuberculosis of some other part. **2** Progressive emaciation; any continuous destruction of tissue. [<L <Gk., a wasting away <*phthiein* decay]

-phyceae *combining form Bot.* Seaweed: used in the names of various classes of algae: *Rhodophyceae.* [<Gk. *phykos* seaweed]

phyco- *combining form* Seaweed; of or related to seaweed: *phycology.* [<Gk. *phykos* seaweed]

phy·co·e·ryth·rin (fī'kō·ə·rith'rin) *n. Biochem.* The accessory pigment associated with chlorophyll in red algae. [<PHYCO- + Gk. *erythros* red + -IN]

phy·col·o·gy (fī·kol'ə·jē) *n.* The branch of botany dealing with seaweeds or algae.

Phy·co·my·ce·tes (fī'kō·mī·sē'tēz) *n. pl.* A class of fungi, both saprophytic and parasitic, resembling algae, but destitute of chlorophyll, including the water molds and downy mildews. [<NL <Gk. *phykos* seaweed + *mykēs, -ētos* a mushroom] —**phy'co·my·ce'·tous** *adj.*

phy·co·phae·in (fī'kō·fē'in) *n.* Fucoxanthin. [<PHYCO-+ Gk. *phaios* dusky + -IN]

Phyfe (fīf), **Duncan,** 1768?–1854, American cabinetmaker, noted for the excellence and beauty of his furniture; born in Scotland.

phy·lac·ter·y (fi·lak'tər·ē) *n. pl.* **·ter·ies 1** A charm or amulet worn on the person; specifically, among the Jews, a strip or strips of cowhide parchment inscribed with passages of Scripture (*Ex.* xiii 8–10, 11–16; *Deut.* vi 4–9, xi 13–22) and enclosed in a black calfskin case, having thongs for binding it on the forehead or around the left arm in memory of the early Israelitish history and of the duty to observe the law, or sometimes to serve as an amulet: also **phyl·ac·te·ri·um** (fil'ak·tir'·ē·əm). **2** An inscribed scroll represented in medieval art as held in the hands, or issuing from the mouths, of angels. **3** A reminder. [<LL *phylacterium* <Gk. *phylaktērion* a safeguard <*phylaktēr* a guard <*phylassein* guard]

phy·le (fī'lē) *n. pl.* **·lae** (-lē) In ancient Athens, a political subdivision. [<Gk. *phylē* a tribe]

phy·let·ic (fī·let'ik) *adj.* **1** Pertaining to a phyle or clan. **2** Of or pertaining to a phylum. Also **phy·lic** (fī'lik). [<Gk. *phyletikos* <*phyletēs* a tribesman <*phylē* a tribe]—**phy·let'·i·cal·ly** *adv.*

Phyl·lis (fil'is) A feminine personal name. [<Gk., green leaf]

—**Phyllis** In Greek mythology, a maiden who, believing herself deserted by her betrothed, hanged herself and was changed to an almond tree.

—**Phyllis** A country girl in Vergil's *Eclogues*; hence, a poetic name for a rustic maiden: often written *Phillis.*

phyl·lite (fil'īt) *n.* A lustrous schistose rock containing small particles of mica. [<Gk. *phyllon* a leaf + -ITE[1]]

phyl·li·um (fil'ē·əm) *n.* Any of a genus (*Phyllium*) of green, flattened, leaflike insects (family *Phasmatidae*); a leaf insect. [<NL <Gk. *phyllion,* dim. of *phyllon* a leaf]

phyllo- *combining form* Leaf; pertaining to a leaf: *phyllotaxis.* Also, before vowels, **phyll-.** [<Gk. *phyllon* a leaf]

phyl·lo·clade (fil'ə·klād) *n. Bot.* A flattened branch or stem performing the functions of a leaf in certain plants, as the cacti. Also **phyl'lo·clad** (-klad). [<NL *phyllocladium* <Gk. *phyllon* a leaf + *klados* a branch]

phyl·lode (fil'ōd) *n. pl.* **phyl·lo·di·a** (fi·lō'dē·ə) *Bot.* A petiole that develops into a flattened expansion, thus taking the place, structure, and function of a leaf. [<F <NL *phyllodium* <Gk. *phyllōdēs* leaflike <*phyllon* a leaf + *eidos* form] —**phyl·lo'di·al** *adj.*

phyl·lo·dy (fil'ə·dē) *n.* Frondescence.

phyl·loid (fil'oid) *adj.* Resembling a leaf; foliaceous. [<NL *phylloides* <Gk. *phyllon* a leaf + *eidos* form]

phyl·lome (fil'ōm) *n. Bot.* The leaf or its equivalent; foliage: one of the four members that make up a perfect plant. [<NL *phylloma* <Gk. *phyllōma* foliage, leaves <*phyllon* a leaf] —**phyl·lom·ic** (fi·lom'ik) *adj.*

phyl·lo·phore (fil'ə·fôr, -fōr) *n. Bot.* The budding summit of a stem, especially a palm stem on which leaves are developed. [<Gk. *phyllophoros* bearing leaves <*phyllon* a leaf + *pherein* bear] —**phyl·loph·o·rous** (fi·lof'ər·əs) *adj.*

phyl·lo·pod (fil'ə·pod) *adj.* **1** Having leaflike feet. **2** Of or pertaining to the *Phyllopoda.* —*n.* One of the *Phyllopoda.*

Phyl·lop·o·da (fi·lop'ə·də) *n. pl.* A division of crustaceans (subclass *Entomostraca*), with the body elongated, a shell or bivalve carapace, and at least four pairs of flattened, leaflike swimming feet which also function as gills, as the fresh-water fairy shrimp. [<NL <Gk. *phyllon* a leaf + *pous, podos* a foot] —**phyl·lop'o·dan** *n. & adj.*

phyl·lo·qui·none (fil'ə·kwi'nōn) *n.* Vitamin K_1.

phyl·lo·tax·is (fil'ə·tak'sis) *n. Bot.* **1** The arrangement of leaves upon a stem. **2** The laws governing this arrangement. Also **phyl'lo·tax'y.** [<NL <Gk. *phyllon* a leaf + *taxis* arrangement <*tassein* arrange] —**phyl'lo·tac'·tic** (-tak'tik) *adj.*

-phyllous *combining form* Having (a specified kind or number of) leaves: *monophyllous.* [<Gk. *phyllon* a leaf]

phyl·lox·e·ra (fil'ək·sir'ə, fi·lok'sər·ə) *n.* A minute aphis or plant louse (family *Phylloxeridae*), especially the grape phylloxera (*Dactylosphaera vitifoliae*), which is very destructive to grape vines. For illustration see INSECTS (injurious). [<NL <Gk. *phyllon* a leaf + *xēros* dry]

phylo- *combining form* Tribe; race; species: *phylogeny.* Also, before vowels, **phyl-.** [<Gk. *phylē, phylon* a tribe]

phy·log·e·ny (fi·loj'ə·nē) *n. Biol.* The history of the evolution of a species or group; tribal or racial history. Compare ONTOGENY. Also **phy·lo·gen·e·sis** (fī'lə·jen'ə·sis). [<G *phylogenie* <Gk. *phylon* a race + *-geneia* birth, origin <*gen-,* root of *gignesthai* be born]—**phy'lo·ge·net'ic** (-jə·net'ik), **phy'lo·gen'ic** *adj.* —**phy'lo·ge·net'i·cal·ly** *adv.*

phy·lum (fī'ləm) *n. pl.* **·la** (-lə) *Biol.* A great division of animals or plants ranking next below a kingdom and above a class, of which the members are believed to have a common evolutionary ancestor. [<NL <Gk. *phylon* a race <*phyein* produce]

-phyre *combining form Geol.* In petrography, a porphyritic rock: *granophyre.* [<F *porphyre* porphyry]

physi- Var. of PHYSIO-.

phys·ic (fiz'ik) *n.* **1** Medicine in general. **2** A cathartic; a purge. **3** *Archaic* The art or practice of medicine; the medical profession. **4** *Obs.* Physics. —*v.t.* **phys·icked, phys·ick·ing 1** To treat with medicine or, especially, a cathartic; purge. **2** To cure or relieve. [<OF *fisique* <L *physica* <Gk. *physikē* (*epistēmē*) (the knowledge) of nature <*physis* nature < *phyein* produce]

phys·i·cal (fiz'i·kəl) *adj.* **1** Relating to the material universe or to the physical sciences. **2** Pertaining to material things, as opposed to *mental, moral,* or *spiritual*; especially, relating to the human body apart from the mind or spirit; material; corporeal. **3** Of or pertaining to the phenomena treated of in physics. **4** Accessible to the senses; external: *physical* characteristics of a mineral; *physical* changes. [<Med. L *physicalis* <L *physica.* See PHYSIC.] —**phys'i·cal·ly** *adv.*

Synonyms: bodily, corporal, corporeal, material, natural, sensible, tangible, visible. Whatever is composed of or pertains to matter may be termed *material*: *physical* applies to *material* things considered as parts of a system or organic whole; hence, we speak of *material* substances, *physical* forces. *Bodily, corporal,* and *corporeal* apply primarily to the human body; *bodily* and *corporal* both denote pertaining or relating to the body; *corporeal* signifies of the nature of or like the body; *corporal* is now almost wholly restricted to signify applied to or inflicted upon the body; we speak of *bodily* sufferings, *bodily* presence, *corporal* punishment, the *corporeal* frame. See NATURAL (def. 8). *Antonyms*: hyperphysical, immaterial, intangible, intellectual, invisible, mental, moral, spiritual, unreal, unsubstantial.

physical anthropology 1 The study of the physical characteristics of man during the course of his evolution from the primate stock, and of the genetic relations between ethnic groups. **2** Anthropometry.

physical chemistry The branch of chemistry that deals with the physical properties of substances, especially in their quantitative relations to energy transformations and chemical change.

physical education Training and development of the human body by means of athletics and other exercises; also, education in hygiene.

physical geography Geography dealing with the natural features of the earth, as vegetation, land forms, drainage, ocean currents, climate, etc.

physical sciences The sciences that treat of the structure, properties, and energy relations of matter apart from the phenomena of life, as physics, astronomy, chemistry, geology, mineralogy, meteorology, etc.

physical therapy The science of treating disability, injury, and disease by external physical means, such as electricity, heat, light, massage, exercise, etc.: also called *physiotherapy.*

phy·si·cian (fi·zish'ən) *n.* **1** One legally authorized to practice medicine; a doctor. **2** One engaged in the general practice of medicine as distinguished from a surgeon. **3** Any healer.

phys·i·cist (fiz'ə·sist) *n.* A student of or specialist in physics.

phys·i·co·chem·i·cal (fiz'i·kō·kem'i·kəl) *adj.* **1** Of or pertaining to the physical and chemical properties of matter. **2** Pertaining to physical chemistry. [<*physico-* (<PHYSICAL) + CHEMICAL]

phys·ics (fiz'iks) *n.* The science that treats of matter and energy and of the laws governing their reciprocal interplay under conditions susceptible to precise observation, experimental control, and exact measurement. Physics generally includes the subjects of mechanics, heat, light and sound, electricity and magnetism, and radiation, but not the phenomena peculiar to living matter or to chemical change.

physio- *combining form* Nature; related to natural functions or phenomena: *physiology.* Also, before vowels, *physi-.* [<Gk. *physis* nature]

phys·i·oc·ra·cy (fiz'ē·ok'rə·sē) *n.* The doctrine of François Quesnay, who taught that society should be governed by a natural order inherent in itself, that land and its unmanufactured products are the only true wealth, the precious metals being a false standard, that the proper source of state revenue is direct taxation of land; and maintained the right of freedom of trade, person, opinion, and property. [<F *physiocratie*] —**phys'i·o·crat'** (-ə·krat') *n.* —**phys'i·o·crat'ic** *adj.*

phys·i·og·no·my (fiz'ē·og'nə·mē, *esp. Brit.* fiz'ē·on'ə·mē) *n. pl.* **·mies 1** The face or features as revealing character or disposition.

2 The outward look of a thing. 3 The art of reading character by the lineaments of the face or form of the body. [<OF *fiznomie* <Med. L *phisnomia* <Gk. *physiognōmonia* the judging of a man's nature (by his features) < *physis* nature + *gnōmōn, -onos* a judge] — **phys'i·og·nom'ic** (-og·nom'ik) or **·i·cal** *adj.* — **phys'i·og·nom'i·cal·ly** *adv.* — **phys'i·og'no·mist** *n.*

phys·i·og·ra·phy (fiz'ē·og'rə·fē) *n.* 1 A description of nature. 2 The study of the development of the features of the earth's surface; physical geography. [<PHYSIO- + -GRAPHY] — **phys'i·og'ra·pher** *n.* — **phys'i·o·graph'ic** (-ə·graf'ik) or **·i·cal** *adj.* — **phys'i·o·graph'i·cal·ly** *adv.*

phys·i·o·log·i·cal (fiz'ē·ə·loj'i·kəl) *adj.* Pertaining to the functions of living organisms. Also **phys'i·o·log'ic.** — **phys'i·o·log'i·cal·ly** *adv.*

phys·i·ol·o·gy (fiz'ē·ol'ə·jē) *n.* 1 The branch of biology that treats of the vital phenomena manifested by animals or plants; the science of organic functions, as distinguished from *anatomy* and *morphology.* 2 The aggregate of organic processes: the *physiology* of the frog. [<F *physiologie* <L *physiologia* <Gk., natural philosophy < *physiologos* a speaker on nature < *physis* nature + *logos* a word] — **phys'i·ol'o·gist** *n.*

phys·i·o·ther·a·py (fiz'ē·ō·ther'ə·pē) See PHYSICAL THERAPY.

phy·sique (fi·zēk') *n.* The physical structure, organization, or appearance of a person. [<F, orig. adj., physical <L *physicus* <Gk. *physikos* natural < *physis.* See PHYSIC.]

phy·so·clis·tous (fī'sō·klis'təs) *adj. Zool.* Having no connection between the air bladder and the digestive tract, as in most teleost fishes. [<NL *Physoclisti,* a genus of fishes <Gk. *physa* a bladder + *kleistos* closed < *kleiein* close]

phy·so·stig·mine (fī'sō·stig'mēn, -min) *n. Chem.* A white, tasteless, toxic alkaloid, $C_{15}H_{21}N_3O_2$, derived from the Calabar bean; used as a miotic: also called *eserine.* [<NL *Physostigma,* genus of the Calabar bean (<Gk. *physa* a bladder + *stigma* a mark) + -INE[2]]

phy·sos·to·mous (fī·sos'tə·məs) *adj. Zool.* Having the air bladder united by a duct with the intestinal canal, as in certain teleost fishes. Also **phy·so·stom·a·tous** (fī'sō·stom'ə·təs). [<NL *physostomus* <Gk. *physa* a bladder + *stoma* a mouth]

-phyte *combining form* A (specified) kind of plant; a plant having a (specified) habitat: *thallophyte, hydrophyte.* [<Gk. *phyton* a plant]

phy·tin (fī'tin) *n. Biochem.* A calcium-magnesium salt isolated as a white, odorless powder from the seeds of various plants, as sunflowers, hemp, peas, etc.

phyto- *combining form* Plant; of or related to vegetation: *phytogenesis.* Also, before vowels, **phyt-.** [<Gk. *phyton* a plant]

phy·to·ben·thon (fī'tō·ben'thon) *n. Bot.* Plant organisms growing at the bottom of seas, lakes, and other large bodies of water. Compare GEOBENTHOS. [<PHYTO- + Gk. *benthos* the depth of the sea]

phy·to·gen·e·sis (fī'tō·jen'ə·sis) *n.* The science of the generation, origin, and development of plants. Also **phy·tog·e·ny** (fī·toj'ə·nē). — **phy'to·ge·net'ic** (-jə·net'ik) or **·i·cal** *adj.* — **phy'to·ge·net'i·cal·ly** *adv.*

phy·to·gen·ic (fī'tō·jen'ik) *adj.* 1 Phytogenetic. 2 Of plant origin, as coal and some other biogenic formations. Also **phy·tog·e·nous** (fī·toj'ə·nəs).

phy·to·ge·og·ra·phy (fī'tō·jē·og'rə·fē) *n.* That department of geography which deals with the distribution of plants; plant geography: also called *geobotany.*

phy·tog·ra·phy (fī·tog'rə·fē) *n.* Descriptive botany. [<NL *phytographia* <Gk. *phyton* a plant + *graphein* write]

phy·toid (fī'toid) *adj.* Plantlike.

phy·tol·o·gy (fī·tol'ə·jē) *n.* Botany. [<NL *phytologia* <Gk. *phyton* a plant + *logos* a word, study] — **phy·to·log·ic** (fī'tə·loj'ik) or **·i·cal** *adj.*

phy·to·pa·thol·o·gy (fī'tō·pə·thol'ə·jē) *n.* 1 The study of the diseases of plants and their control. 2 The pathology of diseases which are caused by fungi, bacteria, and other plant organisms.

phy·toph·a·gous (fī·tof'ə·gəs) *adj. Biol.* Feeding on plants; herbivorous. [<PHYTO- + -PHAGOUS]

phy·to·plank·ton (fī'tō·plangk'tən) *n.* Free-floating aquatic plants.

phy·tos·te·rol (fī·tos'tə·rōl, -rol) *n. Biochem.* Any of various sterols found in and isolated from plant organisms, as ergosterol, sitosterol, etc. [<PHYTO- + -STEROL]

phy·to·tron (fī'tə·tron) *n.* A large-scale field laboratory for the study of plant growth under artificially produced climatic conditions ranging from the tropical to the arctic. [< PHYTO- + Gk. *-tron,* instrumental suffix]

pi[1] (pī) *n.* 1 The sixteenth letter in the Greek alphabet (Π,π): corresponding to English *p.* As a numeral it denotes 80. 2 This letter used to designate the ratio of the circumference of a circle to its diameter, 3.14159 + ; also, this ratio. [Def. 2 <Gk. *p(eriphereia)* periphery]

pi[2] (pī) *n. Printing* Type that has been thrown into disorder; hence, any jumble or disorder. — *v.t.* **pied, pie·ing** To jumble or disorder, as type. Also spelled *pie.* [Var. of PIE[1]]

Pia·cen·za (pyä·chen'tsä) A city on the Po in northern Italy: ancient *Placentia.*

pi·ac·u·lar (pī·ak'yə·lər) *adj.* 1 Expiatory; having power to atone. 2 Requiring expiation; criminal. [<L *piacularis* < *piaculum* an expiation < *piare* appease]

piaffe (pyaf) *v.i.* **piaffed, piaf·fing** To perform or move by performing the piaffer. [<F *piaffer* paw the ground, lit., strut; ult. origin uncertain]

piaf·fer (pyaf'ər) *n.* In equitation, a movement in which the horse lifts one forefoot and the opposite hind foot in unison and slowly places them forward, backward, or to the side. [<F *piaffer.* See PIAFFE.]

pi·a ma·ter (pī'ə mā'tər) *Anat.* The delicate inner vascular membrane that invests the brain and spinal cord: it is overlaid by the arachnoid and dura mater. [<Med. L, trans. of Arabic *umm raqīqah* a tender mother]

pi·an·ism (pē·an'iz·əm, pē'ə·niz'əm) *n.* 1 Arrangement of music for the pianoforte. 2 Performance on the piano; the technique of piano playing.

pi·a·nis·si·mo (pē'ə·nis'i·mō, *Ital.* pyä·nēs'sē·mō) *adj.* & *adv. Music* Very soft or softly; a musical direction: abbr. *pp.* or *ppp.* — *n.* A musical movement played very softly. [<Ital. <L *planissimus,* superl. of *planus.* See PIANO[2].]

pi·an·ist (pē·an'ist, pē'ə·nist) *n.* One who plays on the piano; specifically, an expert or a professional performer.

pi·an·o[1] (pē·an'ō) *n. pl.* **·an·os** A stringed musical instrument having felt-covered hammers, operated from a manual keyboard, which strike upon steel wires to produce the tones; a pianoforte. — **concert grand piano** The largest size of grand piano, used for concert performances. — **grand piano** A horizontal, harp-shaped piano, having three or more strings to each note, and action without springs. — **square piano** A piano having a horizontal rectangular case and horizontally strung wires. — **upright piano** A piano in which the case is upright, with the strings vertical and overstrung to save space. [< Ital., short for *pianoforte* PIANOFORTE]

pi·a·no[2] (pē·ä'nō, *Ital.* pyä'nō) *adv.* With slight force; softly: a direction to a singer or player of a musical instrument: abbr. *p.* — *adj.* Performed or to be performed with slight force; soft: a *piano* passage. — *n.* A passage of music rendered softly and lightly. [<Ital. <L *planus* flat, soft (of sound)]

pi·an·o·for·te (pē·an'ə·fôr'tē, -fôr'-, -fôrt', -fōrt') *n.* A piano. [<Ital. < *piano e forte* soft and strong]

Pi·a·no·la (pē'ə·nō'lə) *n.* A small, portable, cabinetlike, piano-playing mechanism: a trade name.

Pi·a·rist (pī'ə·rist) *n.* One of a Roman Catholic monastic order the members of which are known as Regular Clerks of the Scuole Pie, an institute of instruction, founded in Rome about 1600. [<NL *(patres scholarum) piarum* (fathers of the) religious (schools) <L *pius* dutiful, pious]

pi·as·sa·va (pē'ə·sä'və) *n.* 1 A coarse, stiff fiber obtained from the leafstalks of two Brazilian palms, *Attalea funifera* and *Leopoldinia piassaba*: used for making ropes, brooms, brushes, etc. 2 Either of these palms. Also **pi'a·sa'ba** (-bə), **pi'as·sa'ba, pi'a·sa'va.** [<Pg. *piassava, piassaba* <Tupian *piaçába*]

pi·as·ter (pē·as'tər) *n.* 1 A Turkish and Egyptian coin and monetary unit. 2 The Spanish peso or dollar. Also **pi·as'tre.** [<F *piastre* <Ital. *piastra,* lit., a plate of metal, short for *piastra d'argento* a plate of silver, ult. <L *emplastrum* PLASTER]

Pi·au·í (pē'ou·ē') 1 A river in NE Brazil, flowing 250 miles north to the Canindé river. 2 A state in NE Brazil; 97,150 square miles; capital, Teresina. Formerly **Pi'au·hy'.**

Pia·ve (pyä'vā) A river in NE Italy, flowing 137 miles SE to the Adriatic; scene of World War I battles, 1917–18.

pi·az·za (pē·az'ə, *Ital.* pē·at'tsä) *n.* 1 A veranda or porch. 2 In Europe, especially in Italy, a plaza. 3 A covered outer walk or gallery. [<Ital., a square, market place, ult. <L *platea* a broad street <Gk. *plateia (hodos).* Doublet of PLACE, PLAZA.]

pi·bal (pī'bal) *n. Meteorol.* An observation on conditions in the atmosphere as reported by or from a pilot balloon. [<PI(LOT) + BAL(LOON)]

pi·broch (pē'brokh) *n.* A martial air played on the bagpipe. [< dial. E (Scottish) <Scottish Gaelic *piobaireachd* playing the bagpipe < *piobair* a piper < *piob* a pipe <PIPE]

pi·ca[1] (pī'kə) *n.* A size of type; 12-point; 1/6 inch; also, a standard unit of measurement, as for leads or pages. See POINT SYSTEM.— **small pica** A size of type; about six and a half lines to the inch; 11-point. — **two-line** or **double pica** Type having a depth of body of two lines of pica; 24-point. ◆ Homophone: *pika.* [<Med. L, a pie[4]; ? because used in printing pies]

pi·ca[2] (pī'kə) *n. Pathol.* A morbid appetite for unusual or unfit food, as clay, chalk, ashes, etc., showing itself especially in hysteria, pregnancy, and chlorosis. ◆ Homophone: *pika.* [<L *pica* a magpie, ? trans. of Gk. *kissa, kitta* a magpie, a craving for strange food; with ref. to the bird's omnivorousness] — **pi'cal** *adj.*

pi·ca·cho (pē·kä'chō) *n. pl.* **·chos** *SW U.S.* An isolated peak of a hill or butte. [<Am. Sp. <Sp. *pico* a peak]

pic·a·dor (pik'ə·dôr, *Sp.* pē'kä·thôr') *n.* 1 In bullfighting, a horseman armed with a lance, whose function is to enrage the bull. 2 A clever debater; one with ready wit. [<Sp., lit., a pricker < *picar* prick, pierce < *prica.* Akin to PIKE[1].]

Pi·card (pē·kär'), **Jean,** 1620–82, French astronomer.

Pic·ar·dy (pik'ər·dē) A region and former province of northern France. — **Pic'ard** *adj.* & *n.*

pic·a·resque (pik'ə·resk') *adj.* Pertaining to picaroons or rogues: specifically applied to the **picaresque novel,** a form having a slight plot consisting of episodes loosely connected by the hero, a rogue; originated in Spain in the 17th century, popular in France and England in the 18th century, and still used occasionally. [<Sp. *picaresco* roguish < *pícaro* a rogue; ult. origin uncertain]

pic·a·roon[1] (pik'ə·rōōn') *n.* 1 One who lives by cheating or robbery: a wrecker or pirate; rogue; adventurer. 2 A pirate vessel. [<Sp. *picaron,* aug. of *pícaro* a rogue]

pic·a·roon[2] (pik'ə·rōōn') *n.* A piked pole used by log drivers: also spelled *pickaroon.* [? <MF *piqueron* a spur, dim. of *pique* a pike[1]]

Pi·cas·so (pē·kä'sō), **Pablo,** born 1881, Spanish painter and sculptor active in France.

pic·a·yune (pik'i·yōōn') *adj. U.S.* Little; worthless; mean. — *n.* 1 A former small Spanish-American coin; a half-real. 2 *U.S.* A person or thing of trifling value. [<F *picaillon* an old Piedmontese coin, a farthing <Provençal *picaioun, picalhoun,* dim. of *picalo* money] — **pic'a·yun'ish** *adj.*

Pic·ca·dil·ly (pik'ə·dil'ē) A famous thoroughfare of western London, running from Piccadilly Circus to Hyde Park Corner.

pic·ca·lil·li (pik'ə·lil'ē) *n.* A highly seasoned relish of chopped vegetables. [? <PICKLE[1]]

Pic·card (pē·kär'), **Auguste,** born 1884, and his twin brother **Jean Felix,** Swiss physicists and aeronauts.

pic·co·lo (pik'ə·lō) *n. pl.* **·los** 1 A small flute with tones an octave higher than those of the ordinary flute. 2 An organ stop of similar tone. [<Ital., small] — **pic'co·lo'ist** *n.*

pice (pīs) *n.* A copper coin of British India:

1/4 anna. [<Hind. *paisā*, ? <*pāi*, *paī* a quarter <Skt. *pad, padī*]
pic·e·ous (pis′ē·əs, pī′sē-) *adj*. **1** Relating to or resembling pitch; inflammable. **2** Pitch-black. [<L *piceus* <*pix, picis* pitch]
Piche·gru (pēsh·grü′), **Charles,** 1761–1804, French Revolutionary general.
pich·i·ci·a·go (pich′ə·sē·ä′gō, -ā′gō) *n. pl.* **·gos** A small burrowing armadillo (*Chlamydophorus truncatus*) of South America. [<Am. Sp. *pichiciego* <Guarani *pichey* the little armadillo + Sp. *ciego* blind <L *caecus*]
pich·u·rim (pich′ə·rim) *n*. One of the aromatic cotyledons of the seed of a South American tree (genus *Ocotea* or *Nectandra*) of the laurel family, resembling in taste and smell both sassafras and nutmeg: used medicinally and for flavoring. Also **pichurim bean.** [<Tupian]
pick[1] (pik) *v.t.* **1** To choose; select; cull, as from a group or number. **2** To detach; pluck, as with the fingers or beak: to *pick* a flower. **3** To gather or harvest: to *pick* cotton. **4** To prepare by removing the feathers, hulls, leaves, etc.: to *pick* a chicken. **5** To remove extraneous matter from (the teeth, etc.) with the fingers, a pointed instrument, etc. **6** To pull apart, as rags. **7** To break up, penetrate, or indent with or as with a pointed instrument. **8** To form or make in this manner: to *pick* a hole. **9** To seek or point out too critically: to *pick* flaws. **10** To seek or bring on purposely; provoke: to *pick* a quarrel. **11** To remove the contents of by stealth: to *pick* a pocket or purse. **12** To open (a lock) by means other than the key, as by a piece of wire or metal. **13 a** To pluck (the strings) of a musical instrument. **b** To play: to *pick* a banjo. — *v.i.* **14** To work with a pick. **15** To eat daintily or without appetite; nibble. **16** To make careful selection of: to *pick* and choose. **17** To pilfer: to *pick* and steal. See synonyms under CHOOSE. — **to pick at 1** To touch or toy with. **2** To eat without appetite. **3** *U.S. Colloq.* To nag at. — **to pick off 1** To remove by picking. **2** To shoot with careful and deliberate aim. — **to pick on** *Colloq.* To single out for attention, duty, etc.; tease; annoy. — **to pick one's way** (or **steps**) To advance by careful selection of one's course or actions. — **to pick out 1** To choose or select. **2** To distinguish (something) from its surroundings. **3** To produce the notes of (a tune) singly or slowly, as by ear. — **to pick over** To examine carefully or one by one. — **to pick to pieces 1** To pull apart. **2** To destroy the arguments or claims of by critical or carping analysis. — **to pick up 1** To take up, as with the hand. **2** To take up or receive into a group, vehicle, etc.: We *picked up* more passengers in Hoboken. **3** To get or acquire casually or by chance. **4** To gain speed; accelerate. **5** To recover spirits, health, etc.; improve. **6** To be able to perceive or receive, as a distant radio station. **7** *Colloq.* To make the acquaintance of, casually or informally. — *n*. **1** Right of selection; choice; hence, the best. **2** The quantity of certain crops that are picked by hand. **3** A blow, as with a spear. **4** The act of picking. See synonyms under ALTERNATIVE. [ME *piken, pikken,* OE *pican, pician* (assumed), infinitive of OE *picung* pricking, infl. by OF *piquer* pierce. Akin to PIKE[1].]

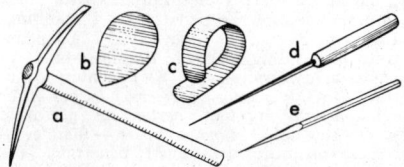

TYPES OF PICKS
a. Pickax.   *c.* Guitar thumb pick.
b. Mandolin pick.   *d.* Ice pick.
e. Quill toothpick.

pick[2] (pik) *n*. **1** A double-headed, pointed metal tool mounted on a strong wooden handle, as a pickax: used for breaking ground, etc. **2** Any of various implements for picking, as an ice pick, toothpick, or a picklock.

3 A plectrum, as for a stringed instrument. [Appar. var. of PIKE[1]]
pick[3] (pik) *n*. **1** In weaving, the blow that drives a loom shuttle. **2** A thread: the number of picks to the inch determines the relative value of cotton cloth or muslin. [<dial. E *pick, v.,* var. of PITCH[2] throw]
pick-a-back (pik′ə·bak′) *adv*. On the back or shoulders: also spelled *piggy-back.* [Earlier *a pickback, a pickpack.* Cf. dial. E *pick* throw, toss.]
pick·a·dil (pik′ə·dil) *n*. A standing collar, usually with a scalloped edge, as worn in the 17th century. Also **pick′a·dill.** [Var. of *piccadil* <MF *piccadilles* pieces fastened on edge of a doublet's collar <dim. of Sp. *picado,* pp. of *picar* prick]
pick·a·nin·ny (pik′ə·nin′ē) *n. pl.* **·nies** A little child; specifically, a Negro child: a condescending or contemptuous term. Also **pic′·ca·nin′ny.** [Alter. of Sp. *pequenino,* dim. of *pequeño* little, small]
pick·a·roon (pik′ə·rōōn′) See PICAROON.
pick·ax (pik′aks′) *n*. A pick or mattock with one arm of the head edged like a chisel and the other pointed; also, one with both arms pointed. Also **pick′axe′.** [Alter. of ME *pikoys* <OF *picois,* ? <*pic* a pike[1]; infl. in form by *ax*]
pick·ed[1] (pik′id, pikt) *adj*. **1** Having spines or prickles. **2** Sharp-pointed, as a stick. [< PICK[2], *n.*]
pick·ed[2] (pikt) *adj*. **1** Carefully selected; chosen for a purpose. **2** Cleaned by picking out refuse, stalks, etc., as cotton. **3** Caused intentionally or sought out, as a quarrel. See synonyms under CHOICE. [Orig. pp. of PICK[4], *v.*]
picked-o·ver (pikt′ō′vər) *adj*. **1** Handled; left after the best have been removed. **2** Left after the undesirable ones have been removed, as berries.
Pick·ens (pik′inz), **Andrew,** 1739–1817, American Revolutionary general.
pick·er[1] (pik′ər) *n*. **1** One who or that which picks. **2** A machine for loosening up fibrous material. **3** A tool like a graver used by electrotypers. [<PICK[1]]
pick·er[2] (pik′ər) *n*. In a loom, the part that strikes the shuttle. [< dial. E *pick, v.* See PICK[3].]
pick·er·el (pik′ər·əl) *n*. **1** A North American fresh-water fish (family *Esocidae*), a pike; especially, one of the smaller species. *Esox reticulatus* is the common pond pickerel of the eastern United States. **2** A young pike. [Dim. of PIKE[1], ? <AF]
pickerel frog A frog (*Rana palustris*) of the eastern United States: also called *marsh frog.*
pick·er·el·weed (pik′ər·əl·wēd′) *n*. A perennial aquatic herb (*Pontederia cordata*), growing in the shallows of lakes of the United States and Canada, or other species of the same genus.
Pick·er·ing (pik′ər·ing), **Timothy,** 1745–1829, American Revolutionary general. — **William Henry,** 1858–1938, U.S. astronomer.
pick·et (pik′it) *n*. **1** A pointed stick, tent peg, bar, fence paling, or stake. **2** *Mil.* A soldier or detachment of soldiers posted to guard a camp, army, etc. **3** A person stationed by a labor organization outside a place affected by a strike. — *v.t.* **1** To fence or fortify with pickets or pointed stakes. **2** *Mil.* **a** To guard by means of a picket. **b** To post as a picket. **3** To station pickets outside of. **4** To tie to a picket, as a horse. — *v.i.* **5** To act as a picket (def. 3). [<F *piquet* a pointed stake < *piquer* pierce] — **pick′et·er** *n*.
picket fence A fence made of upright pickets.
picket pin A long iron pin or wooden stake driven into the ground and having a swivel loop at the upper end: used for tethering horses.
Pick·ett (pik′it), **George Edward,** 1825–75, American Confederate general.
Pick·ford (pik′fərd), **Mary,** born 1893, U.S. motion-picture actress born in Canada: real name *Gladys Smith.*
pick glass A magnifying glass for determining the thread count of fabrics.
pick·ing (pik′ing) *n*. **1** The act of picking; also, that which is or may be picked. **2** *pl.* That which is left to be picked up or gleaned:

scanty *pickings.* **3** Pilfering, or that which is pilfered. **4** *Usually pl.* That which is taken by questionable means; spoils.
pick·le[1] (pik′əl) *n*. **1** A preserving, flavoring liquid, as brine or vinegar, sometimes spiced, for meat, fish, vegetables, etc. **2** One of certain objects preserved or flavored in pickle, as a cucumber or onion. **3** Diluted acid used in cleaning metal castings, etc. **4** *Colloq.* An embarrassing condition or position. **5** *Colloq.* A mischievous child. — *v.t.* **·led, ·ling 1** To preserve with brine or vinegar. **2** To immerse in diluted acid, as castings, for cleansing. [Appar. <MDu. *pekel, peeckel*; ult. origin uncertain]
pick·le[2] (pik′əl) *n. Scot.* A grain of corn; a small quantity.
pickled finish A finish having the effect of a cloudy white patina over light-toned wood: originally produced on old painted furniture by removing the plaster base of the paint with vinegar, or exposing a surface which had been bleached with lime.
pick·lock (pik′lok′) *n*. **1** A special implement, as a bent wire, for opening a lock; a false key. **2** A burglar.
pick·maw (pik′mô′) *n. Dial.* The laughing gull. [Prob. <dial. E *pick* PITCH[2] + obs. E *maw* a gull[1]]
pick-me-up (pik′mē·up′) *n. Colloq.* A drink, especially an alcoholic drink, taken to renew one's energy or spirits.
pick·pock·et (pik′pok′it) *n*. One who steals from pockets.
pick·thank (pik′thangk′) *n. Archaic & Dial.* A flatterer; sycophant. [Earlier *pick a thank(s)*]
pick·up (pik′up′) *n*. **1** Acceleration, as in the speed of an automobile, engine, etc. **2** The electromagnetic vibrating device holding the needle in a phonograph or similar sound-reproducing apparatus. **3** *Telecom.* **a** In radio, the location of microphones in relation to program elements. **b** The system for broadcasting material gathered outside the studio. **c** In television, the scanning of an image by the electron beam. **d** The scanning apparatus. **4** *Colloq.* Gain; improvement; renewal of prosperity, etc. **5** *Slang* A stranger with whom a casual acquaintance is made, usually in a public place and for the purposes of lovemaking. **6** A small, usually open, truck for light loads.
Pick·wick (pik′wik), **Mr. (Samuel)** In Dickens's *Pickwick Papers,* the president of the Pickwick Club, a stout, good-hearted man, fond of travel, and distinguished for blundering simplicity.
Pick·wick·i·an (pik·wik′ē·ən) *adj*. Relating to or characteristic of Mr. Pickwick.
Pickwickian sense A technical or esoteric sense; not the common sense: usually said of a word or a phrase.
pic·nic (pik′nik) *n*. **1** An outdoor party, usually held in the countryside, during which a meal is eaten. **2** *Slang* An easy or pleasant time or experience. — *v.i.* **·nicked, ·nick·ing** To have or attend a picnic. Also **pick′nick.** [<F *pique-nique,* ? reduplication of *piquer* pick, peck] — **pic′nick·er** *n*.
Pi·co (pē′kō) An active volcano on **Pico Island** (167 square miles) in the central Azores; 7,613 feet.
Pico de Aneto See ANETO, PICO DE.
Pi·co del·la Mi·ran·do·la (pē′kō del′lä mē·rän′dō·lä), **Count Giovanni,** 1463–94, Italian humanist scholar.
pic·o·line (pik′ə·lēn, -lin) *n. Chem.* Any of three isomeric liquid compounds, C_6H_7N, contained in coal tar, naphtha, bone oil, etc., and homologous with pyridine. Also **pic′o·lin** (-lin). [<L *pix, picis,* PITCH[1] + -OL[2] + -INE[2]]
pi·cot (pē′kō) *n*. A small thread loop on ornamental edging, sometimes having knots or stitches added. — *v.t. & v.i.* To sew with this edging. [<F, dim. of OF *pic* a point]
pic·o·tee (pik′ə·tē′) *n*. A variety of carnation, having white or light-colored petals edged with scarlet or other strong color. [<F *picotée,* pp. fem. of *picoter* mark with pricks or dots < *picot* PICOT]
picot stitch A loop stitch.
pic·quet (pi·kā′, -ket′) See PIQUET.
pic·rate (pik′rāt) *n. Chem.* One of the salts

picric or esters of picric acid, exploding when heated or struck.

pic·ric (pik′rik) adj. Of, pertaining to, or having an exceedingly bitter taste. [<Gk. pikros bitter]

picric acid Chem. A yellow crystalline compound, $C_6H_2(NO_2)_3OH$, obtained variously, as by the action of nitric acid on phenol: used in dyeing and as an ingredient in certain explosives.

pic·rite (pik′rīt) n. An olivine-augite peridotite containing some magnetite or ilmenite, biotite, and brown hornblende. [<Gk. pikros bitter + -ITE[1]]

picro- combining form Bitter: picrol. Also, before vowels, **picr-**. [<Gk. pikros bitter]

pic·rol (pik′rol, -rol) n. Chem. A colorless, bitter, crystalline compound, $C_6H_3O_5I_2SK$, used in medicine as an antiseptic. [<PICR(O)- + -OL[2]]

pic·ro·tox·in (pik′rə·tok′sin) n. Chem. A bitter, odorless, crystalline compound, $C_{30}H_{34}O_{13}$, contained in and forming the poisonous principle of the fishberry: it resembles strychnine in action. — **pic′ro·tox′ic** adj.

Pict (pikt) n. One of an ancient people of uncertain origin who inhabited Britain and the Scottish Highlands, and waged war on the Romans: conquered in 846 by the Scots. [<LL Picti, ? <L pictus, pp. of pingere paint; with ref. to their being painted or tattooed]

Pict·ish (pik′tish) n. The language of the Picts, of undetermined relationship. — adj. Of or pertaining to the Picts.

pic·to·graph (pik′tə·graf, -gräf) n. A picture representing an idea: the earliest form of record. [<picto- pictorial <L pictus, pp. of pingere paint) + -GRAPH] — **pic′to·graph′ic** adj. — **pic·to·graph′i·cal·ly** adv. — **pic·tog·ra·phy** (pik·tog′rə·fē) n.

Pic·tor (pik′tər) A southern constellation. [<L, lit., a painter <pictus, pp. of pingere paint]

pic·to·ri·al (pik·tôr′ē·əl, -tō′rē-) adj. 1 Pertaining to or concerned with pictures. 2 Representing in or as if in pictures; graphic. 3 Containing or illustrated by pictures. — n. An illustrated publication. See synonyms under GRAPHIC. [<LL pictorius <L pictor. See PICTOR.] — **pic·to·ri·al·ly** adv.

pic·ture (pik′chər) n. 1 A surface representation of an object or scene, as by a painting, drawing, engraving, or photograph; also, a mental image. 2 A vivid or graphic verbal delineation. 3 A striking resemblance to another person, object, or general idea: She is the picture of her mother; the very picture of despair. 4 A tableau vivant: also called **living picture**. 5 A visual image or scene produced by the working of physical laws or their use, as in the lens. 6 A motion picture. — **to be in the picture** To belong to the group or the occasion; also, to be successful. — v.t. **·tured, ·tur·ing** 1 To give visible representation to; draw, paint, etc. 2 To describe graphically; depict verbally. 3 To form a mental image of. [<L pictura <pictus, pp. of pingere paint]

— Synonyms (noun): cartoon, copy, delineation, drawing, engraving, image, likeness, miniature, painting, photograph, print, representation, resemblance, semblance, similitude, sketch. See IMAGE, SKETCH.

picture gallery A room or hall for the exhibition of pictures; also, the pictures, collectively, exhibited.

picture hat A woman's hat with a very wide brim which frames the face: often trimmed with plumes, as hats seen in certain famous paintings, especially those of Gainsborough.

picture ratio In television, the ratio of the length of the received image to the width.

pic·tur·esque (pik′chə·resk′) adj. 1 Having pictorial quality; like or suitable for a picture; especially, having a striking, irregular beauty, quaintness, or charm. 2 Abounding in striking or original expression or imagery; figurative; richly graphic. See synonyms under BEAUTIFUL, GRAPHIC, ROMANTIC. [<F pittoresque <Ital. pittoresco <pittore a painter <L pictor. See PICTOR.] — **pic·tur·esque′ly** adv. — **pic·tur·esque′ness** n.

picture writing 1 The use of pictures or pictorial symbols in writing. 2 A writing so made.

pic·tur·ize (pik′chə·rīz) v.t. **·ized, ·iz·ing** To make a picture or motion picture of; present pictorially. — **pic′tur·i·za′tion** n.

pic·ul (pik′ul) n. A varying commercial weight, usually about 100 catties, or 133 1/3 pounds: used in China, Japan, Thailand, and other countries of Asia. [<Malay pikul a man's load]

pid·dle (pid′l) v. **·dled, ·dling** v.t. 1 To trifle; dawdle: usually with away. — v.i. 2 To trifle; dawdle. 3 To urinate. [? Freq. of PISS, ? infl. in form by PUDDLE]

pid·dling (pid′ling) adj. Unimportant; trivial; trifling.

pid·dock (pid′ək) n. A bivalve mollusk (genus Pholas) with an elongated shell, which burrows in clay and sand; especially, P. castata of the southern United States. [Cf. OE puduc a wart]

pidg·in (pij′ən) n. A mixed language, such as bêche-de-mer, combining the vocabulary and grammar of dissimilar languages and providing a simplified, mutually intelligible form of communication for use in commerce: distinguished from a creolized language in that it is used only as an additional, auxiliary language. [<Pidgin English, alter. of BUSINESS]

Pidgin English A jargon composed of English and local native elements, used as the language of commerce between natives and foreigners in areas of China, Melanesia, Northern Australia, and West Africa. Also called **Pidgin**.

pie[1] (pī) n. 1 A baked food consisting of one or two layers or crusts of pastry with a filling of fruit, vegetables, or meat. 2 Pi^2. 3 Slang Anything very good or very easy. 4 Slang Political graft. — **to have** (or **put**) **one's finger in the pie** To have a share in an activity or project; hence, to meddle. [ME pie, pye, ? <PIE[2]; with ref. to the variety of objects collected by magpies and of foods baked in pies]

pie[2] (pī) n. A magpie, or a related bird. [<OF <L pica a magpie]

pie[3] (pī) n. A copper coin of India of smallest value, worth a third of a pice. [<Hind. pā′ī <Skt. pad, padi a fourth]

pie[4] (pī) n. 1 In the pre-Reformation English church, a book of rules and directions for services on days when two or more feasts concur. 2 An index; a register. Also spelled **pye**. [<LL pica <L, a magpie; ? because its pages resembled the bird's black-and-white plumage]

pie·bald (pī′bôld) adj. Having spots, especially of white and black. — n. A spotted or mottled animal, especially a horse. [<PIE[2] + BALD; because like a magpie's plumage]

piece (pēs) n. 1 A small portion considered as forming or having formed a distinct part of a whole. 2 A portion or quantity existing as an individual entity or mass: a piece of paper; a piece of music. 3 An object considered as forming one of a class or group: a piece of furniture, luggage, etc. 4 A definite quantity or thing in which an article is manufactured or sold. 5 An instance; specimen; example: a piece of luck. 6 A firearm. 7 A coin: a fifty-cent piece. 8 A literary composition. 9 A drama; play. 10 A picture. 11 A musical composition. 12 Dial. A short time, space, or distance: to walk a piece. 13 Archaic or Dial. A person; individual. 14 Any of the figures used in the game of chess; technically, any man but the pawns. 15 One of the disks or counters used in checkers, backgammon, etc. — **a piece of one's mind** Criticism or censure frankly expressed. — **of a piece** 1 Of the same kind, sort, or class. 2 Of one piece; undivided. — **to go to pieces** 1 To fall apart. 2 To lose moral or emotional self-control. — **to speak one's piece** To voice one's opinions. — adj. Of, made of, or by the piece. — v.t. **pieced, piec·ing** 1 To add or attach a piece or pieces to, as for enlargement. 2 To unite or reunite the pieces of, as in mending. 3 To unite (parts or fragments) into a whole. ♦ Homophone: peace. [<OF pece <LL pettia (assumed), prob. ult. <Celtic. Cf. Welsh peth little.]

pièce de résistance (pyes də rā·zē·stäns′) French The principal or most important work in a collection, as of art, poems, etc; also, the most substantial dish of a dinner; literally, piece of resistance.

pièce d'occasion (pyes dô·kä·zyôn′) French Something for a special occasion.

piece goods Dry-goods; fabrics, usually sold by the piece, as shirtings and sheetings.

piece·meal (pēs′mēl′) adj. Made up of pieces. — adv. 1 Piece by piece; gradually. 2 In pieces. [ME pece-mele <pece PIECE + -mele, OE -maelum <mael a measure; partial trans. of OE styccemaelum in pieces]

piec·er (pē′sər) n. One who or that which pieces; especially, one who ties broken threads in a spinning mill.

piece·work (pēs′wûrk′) n. Work done, or paid for, by the piece or quantity. — **piece′work·er** n.

pie chart Stat. A graph in the form of a circle each of whose sectors is proportional in area and subtended angle to a component part of an entire series.

piec·ing (pē′sing) n. Pieces of cloth, especially those collected and saved to be sewed together, as for a quilt.

Pieck (pēk), **Wilhelm**, born 1876, German politician; president of the German Democratic Republic (East Germany) 1949–.

pie-crust table (pī′krust) A small table having a top, usually round, with a raised scalloped edge.

pied (pīd) adj. Spotted; piebald; mottled. [<PIE[2]]

pied-à-terre (pyä·dà·târ′) n. French Temporary lodging; literally, foot on the ground.

pied·mont (pēd′mont) adj. At the foot of a mountain or mountain range: a piedmont plain. [<Piedmont, Italy <L Pedimontium <pes, pedis a foot + mons, montis a mountain]

Pied·mont (pēd′mont) 1 A region of northern and NW Italy; 9,817 square miles; capital, Turin. Italian **Pie·mon·te** (pyā·mōn′tā). 2 A region of the eastern United States extending from New Jersey to Alabama east of the Appalachian Mountains; approximately 80,000 square miles. — **Pied′mont·ese′** (-ēz′, -ēs′) adj. & n.

Pied Piper of Hamelin In medieval legend, a piper who rid the town of Hamelin of its rats by leading them with his music into the river: when not rewarded as promised, he led the town's children to a hill into which they disappeared.

PIER
Steamship pier and dock.

pie plant The garden rhubarb, much used for pies.

pier (pir) n. 1 A plain, detached mass of masonry, usually serving as a support: the pier of a bridge. 2 An upright projecting portion of a wall. 3 A mole or jetty, or projecting wharf. 4 A solid portion of a wall between window openings, etc. ♦ Homophone: peer. [ME per <Med. L pera, ? ult. <Gk. petra rock]

pierce (pirs) v. **pierced, pierc·ing** v.t. 1 To pass into or through; penetrate, in the manner of a pointed object, weapon, etc.; puncture; stab. 2 To make an opening or hole in, into, or through: Many windows pierced the old walls. 3 To make or cut (an opening or hole) in or through something. 4 To force a way into or through: to pierce the wilderness. 5 To affect sharply or deeply, as with emotion, pain, etc. 6 To penetrate as if stabbing: Lightning pierced the night sky. 7 To penetrate as if seeing; perceive or understand: to pierce a mystery. — v.i. 8 To enter; penetrate. [<OF percer, percier, ? <pertuisier <L pertusus, pp. of pertundere perforate <per- through + tundere beat] — **pierc′er** n.

— Synonyms: bore, drill, enter, penetrate, perforate, puncture, stab, transfix.

Pierce (pirs), **Franklin**, 1804–69, president of the United States 1853–57.

pierc·ing (pir′sing) adj. Penetrating by or as if by a sharp-pointed instrument; cutting; keen; poignant; shrill, as a look or cry. See synonyms under ACUTE, BLEAK, SHARP. — n. Penetration. — **pierc′ing·ly** adv. — **pierc′ing·ness** n.

pier glass A large, high mirror intended to stand against a pier and thus fill the space between two openings in the wall.

Pi·e·ri·a (pī·ir′ē·ə) A coastal region of ancient Macedon, at the base of Mount Olympus:

Pierian Spring

legendary birthplace of the nine Muses. — **Pi·e′ri·an** *adj.*
Pierian Spring A spring in Pieria, supposed to give poetic inspiration to those who drank from it.
Pi·er·i·des (pī·er′ə·dēz) 1 In Greek mythology, the nine Muses. 2 The nine daughters of Pierus, who were vanquished by the Muses in a musical contest and changed into magpies.
pi·er·i·dine (pī·er′ə·dīn, -din) *adj.* Of or pertaining to a family (*Pieridae*) of butterflies, including species of predominantly white or yellow color. [<NL *Pierdinae*, subfamily name <*Pieris*, genus name <Gk., a Muse]
Pie·ro del·la Fran·ces·ca (pyä′rō del′lä fränches′kä), 1418?-92, Italian painter. Also **Piero de·i Fran·ces·chi** (dā′ē fränches′kē).
Pierre (pyâr) French form of PETER. Also *Du.* **Pie·ter** (pē′tər), *Ital.* **Pie·tro** (pyä′trō).
Pierre (pir) The capital of South Dakota, on the Missouri River.
Pier·rot (pye·rō′) *n.* Originally, a stock character of French pantomimes, usually taking the part of valet, wearing white pantaloons and loose white jacket with big buttons; now, **pierrot**, a white-faced buffoon dressed in Pierrot costume. [<F, dim. of *Pierre* Peter]
Piers Plow·man (pirz plou′mən) The chief character in the 14th century allegorical poem, *Vision of Piers Plowman*, ascribed to William Langland.
pier table A low table occupying the space between two wall openings, usually combined with a pier glass.
Pi·er·us (pī′ər·əs) In Greek mythology, a king of Thrace, father of the nine Pierides.
pi·et (pī′ət) *n.* 1 The magpie. 2 *Scot.* The water ouzel. [ME *piot*, dim. of PIE²]
Pie·tà (pyä·tä′) *n.* In painting, sculpture, etc., a representation of Mary mourning over the body of Christ in her lap. [<Ital., lit., piety <L *pietas, -tatis* PIETY]
pi·et·ed (pī′it·id) *adj. Scot.* Piebald.
Pie·ter·ma·ritz·burg (pē′tər·mär′its·boŏrkh) The capital of Natal province, Union of South Africa: also *Maritzburg*.
pi·e·tism (pī′ə·tiz′əm) *n.* 1 Piety or godliness; devotion, as distinguished from insistence on religious creeds or forms. 2 Affected or exaggerated piety. — **pi′e·tist** *n.* — **pi′e·tis′tic** *adj.*
Pi·e·tism (pī′ə·tiz′əm) *n.* A movement in the Lutheran Church in Germany during the later 17th century, advocating a revival of the devotional ideal. — **Pi′e·tist** *n.* — **Pi·e·tis′tic** *adj.*
pi·e·ty (pī′ə·tē) *n.* 1 Reverence toward God or the gods; religious devoutness. 2 Religiousness in general. 3 Filial honor and obedience as due to parents, superiors, or country. See synonyms under RELIGION. [<OF *piete* <L *pietas, -tatis* dutifulness < *pius* dutiful. Doublet of PITY.]
piezo- *combining form* Pressure; related to or produced by pressure: *piezometer*. [<Gk. *piezein* press]
pi·e·zo·chem·is·try (pī·ē′zō·kem′is·trē) *n.* The study of chemical reactions under the influence of high pressures.
pi·e·zo·e·lec·tric·i·ty (pī·ē′zō·i·lek′tris′ə·tē, -ē′lek-) *n.* Electricity or electric phenomena resulting from pressure upon certain bodies, especially crystals. — **pi·e′zo·e·lec′tric** or **·tri·cal** *adj.*
pi·e·zom·e·ter (pī′ə·zom′ə·tər) *n.* 1 An instrument for determining pressure; specifically, an apparatus for measuring the compressibility of liquids. 2 An attachment for a sounding line that denotes by the compression of air in a tube the depth of water to which the appliance descends. 3 A similar instrument used in ascertaining the sensitiveness of the skin to pressure. — **pi′e·zo·met′ric** (-zō·met′rik) or **·ri·cal** *adj.* — **pi′e·zom′e·try** *n.*
pif·fle (pif′əl) *Colloq. v.i.* **·fled, ·fling** To talk nonsensically; babble. — *n.* Nonsense. [? OE *pyffan* puff + -LE]
pig (pig) *n.* 1 A hog or hoglike animal, especially when small or young; also, its flesh (pork). ◆ Collateral adjective: *porcine*. 2 An oblong mass of metal, especially iron or lead, just run from the smelter and cast in a rough mold, usually in sand; also, the mold or

trough. 3 Pig iron or iron pigs in general. 4 *Colloq.* A person regarded as like a pig, especially one who is filthy, gluttonous, or grasping. 5 *Scot.* An earthen article or vessel. 6 *Colloq.* A railroad locomotive: also called *hog.* — *v.i.* **pigged, pig·ging** 1 To bring forth pigs. 2 To act or live like pigs: with *it*. [ME *pigge,* ? OE *picga,* as in *pic(g)bred* food for hogs]
pig bed The bed of sand into which iron is run in casting pigs.
pig·boat (pig′bōt′) *n. Colloq.* A submarine.
pi·geon (pij′ən) *n.* 1 Any of a widely distributed family (*Columbidae*) of birds, of arboreal and terrestrial habit, as the rock pigeon (*Columba livia*); also, a dove. 2 *Slang* One easily swindled. [<OF *pijon* <LL *pipio, -onis* a young chirping bird <L *pipire* chirp]
pigeon breast *Pathol.* A deformity in which the chest is flattened from side to side and the breast bone pressed forward and outward. — **pi′geon-breast′ed** *adj.*
pigeon hawk The American merlin.
pig·eon-heart·ed (pij′ən·här′tid) *adj.* Timid; fearful.
pig·eon·hole (pij′ən·hōl′) *n.* 1 A hole for pigeons to nest in, especially in a compartmented pigeon house. 2 A small compartment, as in a desk, for filing papers. — *v.t.* **·holed, ·hol·ing** 1 To place in a pigeonhole; file. 2 To file away and ignore. 3 To place in categories; classify mentally.
pig·eon-liv·ered (pij′ən·liv′ərd) *adj.* Very mild or weak-spirited; meek.
pig·eon-toed (pij′ən·tōd′) *adj.* Having the toes turned inward; toeing in.
pig·eon-wing (pij′ən·wing′) *n.* 1 A fancy dance step. 2 A figure in skating, outlining the shape of a pigeon's spread wing.
pig·fish (pig′fish′) *n. pl.* **·fish** or **·fish·es** 1 A salt-water fish that makes a grunting noise; especially, a grunt, as the sailor's-choice (*Orthopristis chrysopterus*), common off the South Atlantic coast of the United States. 2 A sculpin. 3 A sea robin.
pig·ger·y (pig′ər·ē) *n. pl.* **·ger·ies** A place for keeping or raising pigs.
pig·gin (pig′in) *n.* A small wooden vessel having one stave projecting above the rim for a handle; also, a pitcher. [? Dim. of dial. E *pig* a crock]
pig·gish (pig′ish) *adj.* Like a pig; greedy; dirty; selfish. — **pig′gish·ly** *adv.* — **pig′gish·ness** *n.*
pig·gy (pig′ē) *n.* A little pig. Also **pig′gie.**
pig·gy-back (pig′ē-bak′) *adv.* Pick-a-back.
pig·gy-back·ing (pig′ē-bak′ing) *n. U.S.* Transshipment of loaded truck bodies on railway flat cars.
pig-head·ed (pig′hed′id) *adj.* Stupidly obstinate. — **pig′-head′ed·ly** *adv.* — **pig′-head′ed·ness** *n.*
pig iron Crude iron poured from a blast furnace into variously shaped molds or pigs of sand or the like.
pig·ment (pig′mənt) *n.* 1 Any of a class of finely powdered, insoluble coloring matters suitable for making paints, enamels, oil colors, etc. 2 Any substance that imparts color to animal or vegetable tissues, as chlorophyll. 3 Any substance used for coloring. [<L *pigmentum* <*pingere* paint. Doublet of PIMENTO.]
pig·men·tar·y (pig′mən·ter′ē) *adj.* Producing, secreting, or containing pigment, as a cell.
pig·men·ta·tion (pig′mən·tā′shən) *n.* 1 Coloration. 2 *Biol.* Deposition of pigment by cells.
pigment cell *Biol.* A cell secreting or containing pigment, as an epithelial cell of the iris or a connective-tissue cell; a chromatophore.
pig·my (pig′mē) *adj.* Diminutive; dwarfish: also **pig·me′an.** — *n. pl.* **·mies** A dwarf. Also spelled *pygmy.* [<L *pygmaeus* <Gk. *pygmaios* dwarfish, a dwarf < *pygmē*, the length from elbow to knuckles]
Pig·my (pig′mē) See PYGMY.
pig·nus (pig′nəs) *n. pl.* **·no·ra** (-nər·ə) *Law* A contract of pawn of personal property; also, personal property pawned. [<L, a pledge, a pawn]
pig·nut (pig′nut′) *n.* 1 The fruit of a species of hickory (*Carya glabra*) common in the United States. 2 The tree. 3 St. Anthony's nut. 4 The Old World earthnut.

pig·pen (pig′pen′) *n.* A pen or sty where pigs are kept.
pig·skin (pig′skin′) *n.* 1 The skin of a pig. 2 Something made of this skin, as a saddle or football.
pig-stick·ing (pig′stik′ing) *n.* The hunting of wild boars with spears.
pig·sty (pig′stī′) *n. pl.* **·sties** A sty or pen for pigs.
pig·tail (pig′tāl′) *n.* 1 The tail of a pig. 2 *Colloq.* A cue or plait of hair; also, one who wears a cue. 3 A twist of tobacco. — **pig′tailed′** *adj.*
pig·weed (pig′wēd′) *n.* 1 Any of several American goosefoots; especially, the white pigweed (*Chenopodium album*). 2 The common purslane. 3 One of several amaranths, as the redroot (*Amaranthus retroflexus*).
pi·ka (pī′kə) *n.* A small mammal (family *Ochotonidae*) mostly of North America and Asia; a tailless hare. *Ochotona princeps* is the little chief hare or cony of the Rocky Mountains. ◆ Homophone: *pica.* [<Tungusic *peeka*]
pike¹ (pīk) *n.* A long pole having a metal spearhead, used by foot soldiers in medieval warfare. — *v.t.* **piked, pik·ing** To run through or kill with a pike. [<MF *pique* <*piquer* pierce <OF *pic* PIKE⁵]
pike² (pīk) *n.* 1 A slender, long-snouted, voracious, spiny-finned food fish (family *Esocidae*), widely distributed in fresh waters of Europe, Asia, and America. 2 Some other fish resembling it, as the garpike. [Appar. short for *pikefish* <PIKE⁵ + FISH; with ref. to its pointed snout]

PIKE HEADS

pike³ (pīk) *n.* 1 A turnpike road. 2 A tollbar. — *v.i. Slang* **piked, pik·ing** To go in haste. [Short for TURNPIKE]
pike⁴ (pīk) *n.* A mountain peak or pointed hill. [? <ON *pik* a pointed mountain. Akin to OE *piic* PIKE⁵.]
pike⁵ (pīk) *n.* A spike or sharp point, as the central spike in a buckler. [OE *piic, pic*, prob. <OF, ? <L *picus* a woodpecker]
Pike (pīk), **Zebulon Montgomery,** 1779-1813, U. S. general and explorer.
piked (pīkt, pī′kid) *adj.* Having a pike; pointed. [<PIKE⁵]
pike·man (pīk′mən) *n. pl.* **·men** (-mən) One of a body of soldiers armed with pikes, as in the 16th and 17th centuries. [<PIKE¹ + MAN]
pike perch A pikelike percoid fish, as the walleyed pike or sauger.
pik·er (pī′kər) *n. U.S. Slang* 1 One who bets or speculates in a small, niggardly way. 2 One who does anything in a small, niggardly way. [Appar. from *Pike* County, Missouri, whose inhabitants were considered lazy, poor, suspicious, etc.]
Pike's Peak A mountain in central Colorado; 14,110 feet.
pike·staff (pīk′staf′, -stäf′) *n. pl.* **·staves** (-stāvz′) 1 A piked staff, formerly carried by pilgrims, travelers, etc. 2 The wooden handle of a pike. [<PIKE⁵ + STAFF¹]
pi·lar (pī′lər) *adj.* Of, pertaining to, or covered with hair. [<NL *pilaris* <L *pilus* hair]
pi·las·ter (pi·las′tər) *n. Archit.* A rectangular column, with capital and base, engaged in a wall. [<F *pilastre* <Ital. *pilastro* <L *pila* a column]
Pi·late (pī′lət), **Pontius** A Roman official; procurator of Judea A.D. 26-36?; delivered Jesus to be crucified.
Pi·la·tus (pē·lä′toŏs), **Mount** A peak 7 miles SW of the Lake of Lucerne, Switzerland; 6,998 feet.
pi·lau (pi·lou′, -lō′) *n.* An Oriental dish of boiled rice, raisins, spice, and some kind of meat or fowl. Also **pi·laf** (pi·läf′), **pi·laff′, pi·law** (pi·lô′). [<Turkish *pilaw* <Persian *pilaw*]
pil·chard (pil′chərd) *n.* 1 A herringlike food fish (*Sardinia pilchardus*) of European Atlantic and Mediterranean waters.

PILASTER

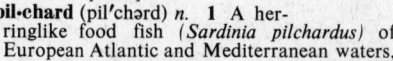

the sardine. 2 The California sardine. Also *Obs.* **pil′cher, pil′cherd.** [Earlier *pilcher,* ? <Scand. Cf. Norw. *pilk* an artificial bait.]

Pil·co·ma·yo (pēl′kō·mä′yō) A river in south central South America, flowing 700 miles from central Bolivia to the Paraguay, and forming part of the boundary between Argentina and Paraguay.

pile[1] (pīl) *n.* 1 A quantity of anything gathered or thrown together in one place; a heap. 2 *Electr.* Any of various devices for generating an electric current by means of superimposed plates of different metals in contact with a suitable liquid: a galvanic *pile,* voltaic *pile.* 3 A funeral pyre. 4 A large accumulation or number of something. 5 A massive building or group of buildings. 6 A pyramid. 7 A great quantity, especially of money; a fortune. 8 *Physics* A reactor. — **to make one's pile** To amass a fortune. — *v.* **piled, pil·ing** *v.t.* 1 To make a heap or pile of: often with *up.* 2 To cover or burden with a pile or piles: to *pile* a plate with food. — *v.i.* 3 To form a heap or pile. 4 To proceed or go in a confused mass: with *in, on, off, out,* etc. See synonyms under HEAP. — **to pile up** 1 To accumulate. 2 *Colloq.* To reduce or become reduced to a pile or wreck: He *piled* the ship *up* on a reef. [<OF <L *pila* a pillar, pier]

pile[2] (pīl) *n.* 1 A heavy timber pointed at one end, forced into the earth to form a foundation; a spile. 2 An arrowhead. 3 Formerly, a pointed stake. 4 *Obs.* A javelin. — *v.t.* **piled, pil·ing** 1 To drive piles into, as for a foundation. 2 To furnish or strengthen with piles. [OE *pil* a dart, pointed stake <L *pilum* a heavy javelin]

pile[3] (pīl) *n.* 1 Hair collectively; fur. 2 The manner in which hair is laid or set. 3 A fiber, as of cotton. 4 The cut or uncut loops which make the surface of certain fabrics, as velvets, plushes, corduroys, etc. [<L *pilus* hair] — **piled** *adj.*

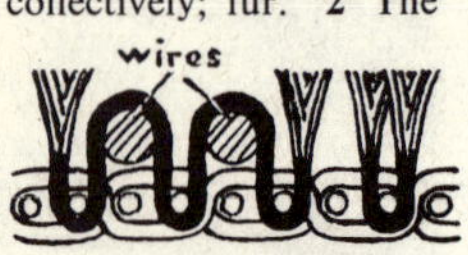

CROSS-SECTION OF PILE WEAVE

pi·le·at·ed (pī′lē·ā′tid, pil′ē-) *adj.* 1 *Bot.* Provided with a pileus or cap. 2 *Ornithol.* Having the feathers of the pileum elongated or conspicuous; crested: the *pileated* woodpecker. Also **pi′le·ate.** [<L *pileatus* capped <*pileus* a felt cap]

pile driver 1 A machine for driving piles. In the ordinary forms, a heavy weight, raised by a small hoisting engine and sliding between upright guides, falls on the head of the pile by the force of gravity: also **pile engine.** 2 A boat or barge carrying a pile driver.

pi·le·ous (pī′lē·əs) See PILOSE.

piles (pīlz) *n. pl.* Hemorrhoids. [<LL *pilae,* pl. of *pila* a ball]

pi·le·um (pī′lē·əm, pil′ē-) *n. pl.* **·le·a** (-lē·ə) *Ornithol.* The top of the head of a bird, from the base of the bill to the nape and above the eyes. [<L, var. of *pileus* a felt cap]

pi·le·us (pī′lē·əs, pil′ē-) *n. pl.* **·le·i** (-lē·ī) 1 *Bot.* The cap or expanded umbrella-shaped portion of a mushroom. See illustration under MUSHROOM. 2 In ancient Rome, a brimless round felt cap. [<L, a felt cap]

pile·wort (pīl′wûrt′) *n.* 1 An Old World crowfoot *(Ranunculus ficaria),* producing tuberous roots and yellow flowers. 2 An American species of fireweed *(Erechtites hieracifolia).* 3 The princess feather. [Trans. of Med. L *ficaria* <L *ficus* a fig, piles; with ref. to its reputed ability to cure hemorrhoids]

pil·fer (pil′fər) *v.t.* & *v.i.* To steal in small quantities. See synonyms under STEAL. [<OF *pelfrer* rob <*pelfre* plunder] — **pil′fer·er** *n.*

pil·fer·ing (pil′fər·ing) *n.* Petty thieving.

pil·gar·lic (pil·gär′lik) *n.* 1 *Obs.* A person made bald by disease. 2 *Dial.* A person regarded with mock pity or contempt. [<PILL[2] + GARLIC; from the appearance of a bald head]

pil·grim (pil′grim) *n.* 1 One who journeys to some sacred place from religious motives. 2 Any wanderer or wayfarer. [ME *pelegrim* <OF *pelegrin* (assumed) <L *peregrinus.* See PEREGRINE.]

Pil·grim (pil′grim) *n.* One of the English Puritans who founded Plymouth Colony in 1620.

pil·grim·age (pil′grə·mij) *n.* 1 A long journey, especially one made to a shrine or sacred place. 2 Man's life as a journey through the world. See synonyms under JOURNEY. [<OF *pelrimage, pelerinage* <*peleriner* go as a pilgrim <*pelerin,* var. of *pelegrin* PILGRIM]

Pilgrim Fathers The founders of Plymouth Colony, Massachusetts, in 1620.

Pilgrim's Progress A religous allegory in two parts (1678 and 1684) by John Bunyan, depicting the life journey of Christian from the City of Destruction to the Celestial City.

pi·li (pē·lē′) *n.* 1 An edible nut of a Philippine tree *(Canarium ovatum),* considered a delicacy after roasting. 2 The tree. [<Tagalog]

pili- *combining form* Hair; related to the hair. [<L *pilus* a hair]

pi·lif·er·ous (pī·lif′ər·əs) *adj.* Bearing or tipped with hairs.

pil·ing (pī′ling) *n.* 1 Piles collectively. 2 A structure formed of piles. 3 The act or process of driving piles. [<PILE[2]]

pill[1] (pil) *n.* 1 A medicinal substance put up in a pellet, convenient for swallowing whole. 2 Hence, a disagreeable necessity. 3 *Slang* A person difficult to bear with; a bore. 4 *Colloq.* A baseball or a golf ball. — *v.t.* 1 To form into pills. 2 To dose with pills. 3 *Slang* To blackball. [Prob. <MDu. *pille* <L *pilula,* dim. of *pila* a ball]

pill[2] (pil) *v.t.* & *v.i. Obs.* 1 To pillage. 2 To peel off; scale. 3 To make or become bald. [OE *pylian,* prob. <L *pilare* make hairless, plunder; infl. by OF *piller* plunder and OF *peler* peel]

pil·lage (pil′ij) *n.* 1 The act of pillaging; open robbery, as in war. 2 Spoil; booty. See synonyms under PLUNDER. — *v.* **·laged, ·lag·ing** *v.t.* 1 To strip of money or property by open violence, especially in war; loot. 2 To take as loot. — *v.i.* 3 To take plunder. See synonyms under STEAL. [<OF <*piller* plunder <LL *pillare,* var. of L *pilare* plunder] — **pil′lag·er** *n.*

pil·lar (pil′ər) *n.* 1 A firm, upright, separate support; column or shaft. 2 Something resembling a column in form or use. 3 One who or that which strongly supports a work or cause. — **from pillar to post** From one thing to another; from one predicament to another; hither and thither. — *v.t.* To adorn or support with or as with pillars. [<OF *piler* <LL *pilare* <L *pila* a pillar]

pillar box *Brit.* A box, supported by a pillar or post in the street, in which letters, etc., may be placed to be collected by a postman. Also **pillar post.**

Pillars of Hercules Two promontories on opposite sides of the eastern entrance to the Strait of Gibraltar: identified with *Gibraltar* and *Jebel Musa.*

pill·box (pil′boks′) *n.* 1 A small box for pills. 2 A small, round, concrete emplacement for a machine-gun, antitank gun, etc.

pill bug A small isopod crustacean (family *Armadillidiidae*), found under logs, etc.: so called because they roll into tiny pills when disturbed.

pil·lion (pil′yən) *n.* A pad on a horse's back, behind the saddle, on which a second person may ride: formerly used by women. [Appar. <Scottish Gaelic *pillean,* dim. of *pell* a cushion <L *pellis* a skin]

pil·li·winks (pil′ē·wingks) *n.* An old instrument of torture; thumbscrew: also called *pinnywinkle.* [ME *pyrwykes*; origin unknown]

pil·lo·ry (pil′ər·ē) *n. pl.* **·ries** Formerly, a framework in which an offender was fastened by the neck and wrists and exposed to public scorn. — *v.t.* **·ried, ·ry·ing** 1 To set in the pillory. 2 To hold up to public scorn or ridicule. [<OF *pellori, pilori,* appar. < dial. OF (Gascon) *espilori* <Provençal *espitlori,* ? < Catalan *espitlera* a little window, peephole]

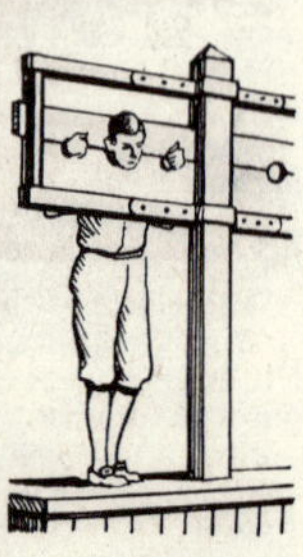
PILLORY

pil·low (pil′ō) *n.* 1 A case of cloth stuffed with some yielding material, or inflated with air, used as a support for the head, as in sleeping. 2 Any body rest. 3 One of various supporting blocks or devices, as a journal bearing. — *v.t.* 1 To rest on or as on a pillow. 2 To act as a pillow for. — *v.i.* 3 To recline as on a pillow. [OE *pyle, pylu,* ult. <L *pulvinus* a cushion] — **pil′low·y** *adj.*

pillow block *Mech.* A block or other device resting on firm foundations and designed to support a journal or shaft; a bearing.

pillow case A covering drawn over a pillow. Also **pillow slip.**

pillow lace Bobbin lace.

pillow sham A decorative covering or spread to be laid over a bed pillow.

pi·lo·car·pine (pī′lō·kär′pēn, -pin, pil′ō-) *n. Chem.* A colorless to yellow, poisonous, liquid or crystalline alkaloid, $C_{11}H_{16}N_2O_2$, contained in jaborandi: its salts are used in medicine. Also **pi′lo·car′pin** (-pin). [<NL *Pilocarpus,* genus of the jaborandi + -INE[2]]

pi·lose (pī′lōs) *adj.* Covered with hair, especially with fine and soft hair; hairy: also spelled *pileous, pilous.* [<L *pilosus* <*pilus* hair] — **pi·los·i·ty** (pī·los′ə·tē) *n.*

pi·lot (pī′lət) *n.* 1 A helmsman; specifically, one duly qualified by training and licensed by law to conduct vessels in and out of port. 2 Any guide. 3 One who controls the operation of an airplane, dirigible, or other aircraft. 4 The cowcatcher of a locomotive. — *v.t.* 1 To act as the pilot of; steer. 2 To guide or conduct through difficulties, intricate dealings, etc. 3 To serve as pilot on, over, or upon. [<MF *pillotte, pilot* <Ital. *pilota,* ? <*pedota,* ult. <Gk. *pēda* a rudder, orig. pl. of *pēdon* an oar]

pi·lot·age (pī′lət·ij) *n.* 1 The act of piloting a vessel or aircraft. 2 The fee for such service.

pilot balloon A small balloon sent up before dispatching a large one, to show the direction and velocity of the wind.

pilot bread Ship biscuit. Also **pilot biscuit.**

pilot cell A selected cell of a storage battery whose voltage, temperature, etc., are considered to indicate the condition of the whole battery.

pilot engine A locomotive preceding and piloting a train.

pilot fish 1 An oceanic fish *(Naucrates ductor),* often seen in warm latitudes in company with sharks; the banded pilot. 2 A whitefish of North American waters *(Prosopium quadrilaterale).*

pilot house An enclosed structure, usually in the forward part of a vessel, containing the steering wheel and compass.

pi·lot·ing (pī′lət·ing) *n.* 1 The occupation of a pilot. 2 The branch of navigation that has to do with steering vessels in and out of ports or along coasts, or finding a ship's position by knowledge of landmarks, buoys, etc.

pilot lamp A small electric light used to indicate whether the power in a given circuit, motor, control unit, etc., is on or off.

pi·lot·less plane (pī′lət·lis) An aircraft designed and equipped to operate without a pilot.

pilot light A minute jet of gas kept burning beside an ordinary burner, for igniting the latter as soon as the gas is turned on. Also **pilot burner.**

pilot parachute See under PARACHUTE.

pilot plant An experimental assembly of various units of machinery and other equipment for the purpose of testing the value of new production methods.

pilot snake 1 The copperhead. 2 A black snake *(Elaphe obsoleta)* of the eastern United States.

pilot whale The blackfish.

pi·lous (pī′ləs) See PILOSE.

Pil·sen (pil′zən) A city of western Bohemia, Czechoslovakia: Czech *Plzeň.*

Pil·sud·ski (pēl·sōōt′skē), **Joseph,** 1867–1935, Polish general; first president of Poland 1918–1921.

Pilt·down (pilt′doun) A locality of eastern Sussex, England.

Piltdown man Eoanthropus.

pil·ule (pil′yōōl) *n.* A little pill; pellet. [<L *pilula.* See PILL[1].] — **pil′u·lar** *adj.*

Pi·man (pē′mən) *n.* A branch of the Uto-Aztecan linguistic family of North American Indians of southern Arizona and northern Mexico. — *adj.* Of or pertaining to this linguistic branch.

pim·e·lo·sis (pim′ə·lō′sis) *n. Pathol.* Conversion into fat; fatty degeneration; obesity. [<Gk. *pimelē* fat + -OSIS] — **pim′e·lot′ic** (-lot′ik) *adj.*

pi·men·to (pi·men'tō) *n. pl.* **·tos** 1 The dried, unripe, aromatic berries of the West Indian allspice; also, the spice made from these berries, or the tree producing them. 2 The Spanish páprika or the sweet pepper from which it is made; pimiento. [<Sp. *pimienta* pepper <Med. L *pigmentum* a spiced drink, spice <L, a paint, juice of plants. Doublet of PIGMENT.]
pimento cheese Cheese made from processed Neufchâtel curds, cream cheese, or cheddar with pimientos added.
pi-mes·on (pī'mes'on, -mē'son) *n. Physics* A short-lived, highly unstable, radioactive particle produced by the impact of fast-moving protons on atomic nuclei. It has a mass about 275 times that of the electron, may be either positive or negative in charge, and decays spontaneously into mu-mesons and neutrinos.
pi·mien·to (pi·myen'tō) *n. pl.* **·tos** The sweet pepper, of which the fruit is used as a relish and for stuffing olives. [<Sp. <*pimienta*. See PIMENTO.]
pim·o·la (pi·mō'lə) *n.* An olive which has been stuffed with a sweet red pepper. [<PIM(IENTO) + OL(IVE) + -a]
pimp (pimp) *n.* A pander. — *v.i.* To act as a pimp. [Prob. <F *pimpant* seductive, ppr. of *pimper* dress elegantly; ult. origin uncertain]
pim·per·nel (pim'pər·nel) *n.* A plant of the primrose family (genus *Anagallis*) usually with red flowers, as the common **scarlet pimpernel** (*A. arvensis*). [<OF *pimprenele, piprenelle* <Med. L *pipinella*, ? <a dim. of L *bipennis* two-winged <*bi-* two + *penna* a feather]
pimp·ing (pim'ping) *adj. Colloq.* Puny; mean; sickly. [< dial. E. Akin to Du. *pimpel* a weak man.]
pim·ple (pim'pəl) *n.* 1 A minute swelling or small elevation of the skin, with an inflamed base. 2 Any small protuberance. [ME *pimplis* pimples. Cf. OE *pipligende* afflicted with herpes.] — **pim′pled, pim′ply** *adj.*
pin (pin) *n.* 1 A short stiff piece of wire, with a sharp point and a round, usually flattened head, used in fastening together parts of clothing, sheets of paper, etc. 2 An ornamental device mounted on a pin or having a pin as a clasp: often serving to fasten parts of the dress in addition to its use as a decoration: frequently a badge. 3 A peg or bar of metal or

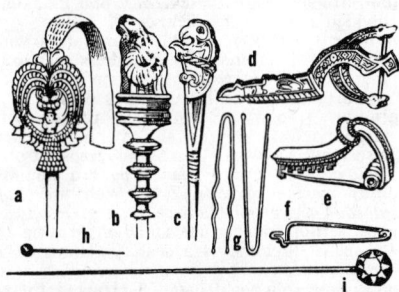

PINS OF VARIOUS TYPES
a. Greek. *f.* Safety pin.
b. Roman. *g.* Hairpins.
c. Early French. *h.* Round-headed.
d. Russian. *i.* Hatpin.
e. Scandinavian.

wood used for a fastening or support, as the thole of a boat, the bolt of a door, a peg serving to stop a hole or to fasten two beams together, or to keep a wheel from slipping from an axle, or one of the pegs to which the strings of a musical instrument are fastened. 4 Anything like a pin, as a hairpin or clothespin. 5 *Usually pl.* A wooden club turned in long, oval, or cylindrical shape, set up as a mark or target in various bowling or ball-throwing games; a skittle. 6 *pl. Colloq.* Legs. 7 A belaying-pin; a rolling pin; a thole pin. 8 The merest trifle. 9 The cylindrical part of a key forward of the stem that enters the lock. 10 *Obs.* A peg showing the center of a target. — *v.t.* **pinned, pin·ning** 1 To fasten with a pin or pins. 2 To transfix with a pin, spear, etc. 3 To seize and hold firmly: to *pin* an opponent against a wall. 4 To force (someone) to make a definite statement, abide by a promise, etc.: usually with *down*. [OE *pinn* a peg, ? ult. <L *pinna* a point, pinnacle]
pi·ña (pē'nyä) *n.* 1 The pineapple. 2 A sweet drink prepared from the pineapple. Also **pi′na** (-nä). [<Sp., a pineapple, orig., a pine cone <L *pinea*. See PINEAL.]
Pi·na·ce·ae (pī·nā'si·ē) *n. pl.* A family of widely distributed coniferous trees and shrubs having needlelike leaves and bearing hard, woody cones, as the pine, cedar, redwood, larch, hemlock, etc.; the pine family. [<NL <L *pinus* a pine] — **pi·na′ceous** (-shəs) *adj.*

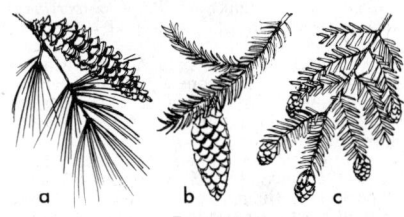

PINACEAE
a. White pine. *b.* Red spruce. *c.* Hemlock.

piña cloth A material for scarfs, handkerchiefs, etc., made from the fibers of the pineapple leaf. It is soft, transparent, and pale yellow.
pin·a·fore (pin'ə·fôr, -fōr) *n.* A sleeveless apron, especially for protecting the front of a child's dress. [<PIN + AFORE]
pi·nang (pi·nang') *n.* The betel palm, or its fruit. [<Malay, an areca, a betel nut]
Pi·nar del Rí·o (pē·när' thel rē'ō) A province in western Cuba; 5,211 square miles; capital, Pinar del Río.
pi·nas·ter (pī·nas'tər, pi-) *n.* An Old World pine (*Pinus pinaster*) common in the Mediterranean region. [<L, a wild pine <*pinus* a pine]
pin·ball (pin'bôl') *n.* A game in which a ball, spring-propelled to the top of an inclined board, contacts in its descent any of various numbered pins, holes, etc., the contacts so made comprising the player's score.
pince-nez (pans'nā', pins'-, *Fr.* pańs·nā') *n.* Eyeglasses held upon the nose by a spring. [<F, lit., pinch-nose <*pincer* pinch + *nez* nose]
pin·cers (pin'sərz) *n. pl.* (sometimes construed as singular) 1 An instrument having two handles and a pair of jaws working on a pivot, used for holding objects. 2 *Zool.* A nipperlike organ, as the chela of a lobster. Also spelled **pinchers**. [ME *pinsours*, appar. <AF *pincer*, OF *pincier* pinch]
pinch (pinch) *v.t.* 1 To squeeze between two hard edges, or surfaces, a finger and thumb, etc. 2 To bind or compress painfully: This collar *pinches* my neck. 3 To affect with pain or distress: The cold *pinched* his fingers. 4 To contract or make thin, as from cold or hunger. 5 To reduce in means; distress, as for lack of money; straiten. 6 To move by means of a pinchbar. 7 *Slang* To capture or arrest. 8 *Slang* To steal. 9 *Naut.* To sail (a vessel) close-hauled. — *v.i.* 10 To squeeze; hurt. 11 To be careful with money; be stingy. 12 *Mining* Of veins, to become narrow; also, to disappear: with *out*. — **to pinch pennies** To be economical or stingy. — *n.* 1 The act of pinching. 2 Painful pressure of any kind. 3 A case of emergency. 4 So much of a loose substance as can be taken between the finger and thumb. 5 A narrow or tapering section on a vein of rock or fissure of earth. 6 A pinchbar. 7 *Slang* A theft. 8 *Slang* An arrest or raid. [<AF *pincher*, OF *pincier*, prob. <Gmc.] — **pinch′er** *n.*
pinch·bar (pinch'bär') *n.* A crowbar with a short projection and a heel or fulcrum at the end so that it may be used to pry forward heavy objects.
pinch·beck (pinch'bek) *n.* 1 An alloy of copper, zinc, and tin, forming a cheap imitation of gold. 2 Anything spurious or pretentious. — *adj.* Made of pinchbeck; spurious. [after Christopher *Pinchbeck*, 1670?-1732, English inventor]
pinch·bug (pinch'bug') *n.* A stag beetle.

pinch effect *Physics* The constriction produced in a plasma jet or vaporized electrical conductor when forced through a narrow orifice by the combined action of magnetohydrodynamic and thermal forces applied in the generating chamber.
pinch·ers (pin'chərz) See PINCERS.
pinch-hit (pinch'hit') *v.i.* **-hit, -hit·ting** 1 In baseball, to go to bat in place of a regular player, as when a hit is needed. 2 *Colloq.* To substitute for another in an emergency. [<PINCH an emergency + HIT]
pinch hitter One who pinch-hits.
Pin·chot (pin'shō), **Gifford**, 1865-1946, U. S. conservationist and politician.
Pinck·ney (pingk'nē), **Charles Cotesworth**, 1746-1825, American soldier and patriot.
pin-cush·ion (pin'koosh'ən) *n.* A small cushion into which pins are stuck when they are not being used.
Pin·dar (pin'dər), 522?-443 B.C., Greek lyric poet.
Pin·dar·ic (pin·dar'ik) *adj.* Of or pertaining to Pindar. — *n.* A Pindaric ode; any ode written in the complex style of Pindar.
pin·der (pin'dər) See PINNER[2].
pind·ling (pind'ling) *adj.* 1 Dwindling; delicate; sickly. 2 Trifling. 3 *Dial.* Peevish. [Var. of PIDDLING]
Pin·dus (pin'dəs) The mountain range of northern Greece, between Epirus and Thessaly; highest point, 8,650 feet.
pine[1] (pīn) *n.* 1 A cone-bearing tree (genus *Pinus*) having needle-shaped evergreen leaves growing in clusters; especially, the American **white pine** (*P. strobus*), the long-leaved southern **Georgia** or **yellow pine** (*P. palustris*), the **loblolly** or **oldfield pine** (*P. taeda*), the **red pine** (*P. resinosa*) of the eastern United States, and a **nut pine** (*P. cembroides*) of the Pacific States. 2 The wood of any pine tree. 3 *Colloq.* The pineapple. [Fusion of OE *pin* and OF *pin*, both <L *pinus* a pine tree]
pine[2] (pīn) *v.* **pined, pin·ing** *v.i.* 1 To grow thin or weak with longing, grief, etc. 2 To have great desire or longing: with *for*. — *v.t.* 3 *Archaic* To grieve for. [OE *pīnian* torment, ult. <L *poena* a punishment]
pin·e·al (pin'ē·əl) *adj.* 1 Shaped like a pine cone: the *pineal* body. 2 Pertaining to the pineal body. [<F *pinéal* <L *pinea* a pine cone <*pinus* a pine tree]
pineal body *Anat.* A small, reddish-gray, vascular, conical body of rudimentary glandular structure found behind the third ventricle of the brain, embraced by its two peduncles, but not a part of the brain, and having no known function. Also **pineal gland**.
pineal eye *Biol.* The pineal body which in certain reptiles emerges as an eyelike structure.
pine·ap·ple (pīn'ap'əl) *n.* 1 A tropical American plant (*Ananas comosus*) having spiny, recurved leaves and a cone-shaped fruit consisting of the inflorescence clustering densely around a fleshy axis tipped with a rosette of spiked leaves. 2 The edible fruit of this plant. 3 *Slang* A bomb. 4 In decoration, an ornament frequently in the form of a finial resembling either a pineapple or a pine cone: used especially on furniture. [OE *pīnæppel* a pine cone <*pīn* a pine + *æppel* an apple; so called because the fruit resembles a pine cone]

PINEAPPLE
Fruit in crown of plant.

pine cone The cone-shaped fruit of the pine tree. The pine cone and tassel compose the floral emblem of Maine.
pine·drops (pīn'drops') *n.* 1 Beechdrops. 2 A stout brownish-purple saprophytic plant (*Pterospora andromedea*) with terminal clusters of nodding white flowers.
pine grosbeak A north American finch (*Pinicola enucleator*) having a slate-gray plumage which in the male is tinged with red.
Pine·hurst (pīn'hûrst) A winter resort in central North Carolina.

pine knot 1 A knot in pine wood. 2 Any person or thing as tough as a pine knot.

Pi·nel (pē·nel′), **Philippe,** 1745–1826, French physician.

pine mouse The little reddish-brown vole *(Pitymys pinetorum)* of the pine barrens of the eastern United States.

pi·nene (pī′nēn) *n. Chem.* Either of two isomeric terpenes, $C_{10}H_{16}$, the principal constituent of turpentine, and an ingredient of many essential oils, such as oil of juniper, eucalyptus, etc. [<PIN(E)[1] + -ENE]

pine needle The needle-shaped leaf of the pine tree.

Pi·ner·o (pi·nâr′ō, -nir′ō), **Sir Arthur Wing,** 1855–1934, English dramatist.

pin·er·y (pī′nər·ē) *n. pl.* **·er·ies** 1 A hothouse for growing pineapples. 2 A pine forest, especially one where lumbering is carried on. 3 A large collection of pine trees.

Pines, Isle of See ISLE OF PINES.

pine·sap (pīn′sap′) *n.* A low, fragrant plant *(Hypopitys latisquama)*, whitish or reddish, parasitic on roots or living on dead vegetable material: native of the north temperate zone. [<PINE[1] + SAP[2]]

pine siskin A finch *(Spinus pinus)* of North America having streaked plumage. Also **pine finch.**

pine squirrel The American red squirrel *(Sciurus hudsonicus)*.

pine tar A dark, viscous tar obtained by the destructive distillation of the wood of pine trees: used in the treatment of skin ailments.

pine-tree shilling (pīn′trē′) A famous silver coin of Massachusetts from 1652 to about 1684, stamped with the image of a tree resembling a pine tree.

Pine-Tree State Nickname of MAINE.

pi·ne·tum (pī·nē′təm) *n. pl.* **·ta** (-tə) A plantation of pine trees. [<L, a pine grove <*pinus* a pine tree]

pine warbler A small, olive-green warbler *(Dendroica pinus* or *D. vigosi)* common in the pine forests of the eastern United States.

pine·y (pī′nē) See PINY.

Pine·y (pī′nē) *n. U.S.* A poor white living in a pine woods area of New Jersey or the South. Also **Pin′er.**

pin feather *Ornithol.* 1 A rudimentary feather. 2 A feather just beginning to grow through the skin.

pin·fish (pin′fish′) *n. pl.* **·fish** or **·fish·es** A sparoid fish *(Lagodon rhomboides)* common on the Atlantic coast of the southern United States.

pin·fold (pin′fōld′) *n.* A pound for stray animals; especially, a cattle pound. —*v.t.* To shut in a pinfold; confine. [OE *pundfald.* See POUND[2], FOLD[2]; infl. in form by *pyndan* enclose.]

ping (ping) *n.* 1 The sound made by a bullet striking an object. 2 The sound made by a bullet as it cuts the air. —*v.i.* To make this sound. [Imit.]

Ping (ping) A river in northern Thailand, flowing 300 miles south to the Chao Phraya.

Ping-pong (ping′pong′, -pông′) *n.* The game of table tennis: a trade name.

pin·grass (pin′gras′, -gräs′) *n.* Alfileria.

pin·guid (ping′gwid) *adj.* Containing or resembling oil or fat; unctuous. [<L *pinguis* fat] —**pin·guid′i·ty** *n.*

Ping·yu·an (ping′yü·än′) A former province of north central China, constituted, 1949; repartitioned among other provinces, 1952; 20,072 square miles; former capital, Sinsiang.

pin·head (pin′hed′) *n.* 1 A small minnow. 2 *Slang* A brainless or stupid person; a fool.

pin head 1 The head of a pin. 2 Any small object. —**pin-head·ed** (pin′hed′id) *adj.*

pin·hole (pin′hōl′) *n.* A minute puncture made by or as by a pin.

pinhole camera A camera of simple design and construction, usually home-made, consisting of a box having a small aperture functioning as a lens at one end, the image being projected on the film at the other end.

pin·ion[1] (pin′yən) *n.* 1 The wing of a bird. 2 A feather; a quill. 3 The outer segment of a bird's wing, bearing the flight feathers. 4 The anterior border of the wing of an insect. —*v.t.* 1 To cut off one pinion or bind the wings of (a bird) so as to prevent flight. 2 To cut or bind (the wings) of a bird. 3 To bind or hold the arms of (someone) so as to render helpless. 4 To shackle; confine. [<OF *pignon* a streamer, a feather <L *penna, pinna* a feather] —**pin′ioned** *adj.*

pin·ion[2] (pin′yən) *n. Mech.* A toothed wheel (or sometimes, in watches, a ribbed wire) driving or driven by a larger cogwheel. [<F *pignon,* orig. a battlement <OF *pinun* <L *pinna,* orig. a pinnacle]

pin·ite[1] (pin′īt, pī′nīt) *n.* A hydrous potassium-aluminum silicate. [<G *pinit* <*Pin(i),* a mine in Saxony + *-it* -ITE[1]]

pi·nite[2] (pī′nīt) *n. Chem.* A white, very sweet, crystalline substance, $C_6H_6(OH)_5{\cdot}OCH_3$, from the resin of the sugar pine *(Pinus lambertiana)*: also **pi′ni·tol** (-ni·tôl, -tol). [<F <L *pinus* a pine]

pink[1] (pingk) *n.* 1 A pale hue of crimson. 2 A flower of any one of several garden plants (genus *Dianthus*) with narrow grasslike leaves and fragrant flowers, as the common pink *(D. plumarius)*. 3 The plant itself. 4 A flower or plant of some other genus, including the **moss pink.** 5 A type of excellence or perfection: the *pink* of politeness. 6 A red-colored coat; especially, a scarlet hunting coat; also, one who wears such a coat. 7 *Often cap.* A person who holds somewhat radical economic and political opinions. —**in the pink** *Colloq.* In the best possible condition or degree. —*adj.* 1 Having the color called pink; pale rose. 2 Fashionably dainty. [? Short for obs. *pink eye* a small eye (< obs. *pink* small + EYE), trans. of F *oeillet* a pink (flower), orig. dim. of *oeil* an eye]

pink[2] (pingk) *v.t.* 1 To prick or stab with a pointed weapon. 2 To decorate, as cloth or leather, with a pattern of holes. 3 To cut or finish the edges of (cloth) with a notched pattern, as to prevent raveling or for decoration. 4 *Brit.* To adorn; deck. [ME *pynken,* prob. a nasalized form of *pikken* PICK[1]]

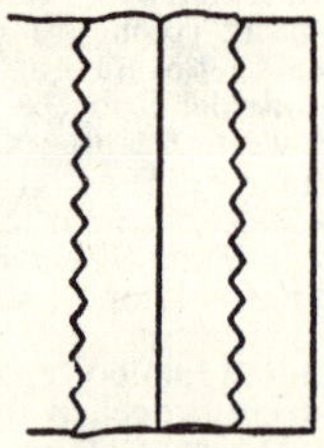

PINKED SEAM
(v.t. def. 3)

pink[3] (pingk) *n. Naut.* 1 A sailing vessel with a narrow stern, originally flat-bottomed with bulging sides. 2 A small sailing vessel with a narrow stern, used for fishing and coasting: also called *pink-stern*: also **pink′ie, pink′y.** [<MDu. *pincke* a small sea-going ship; ult. origin unknown]

pink[4] (pingk) *v.i. Brit. Dial.* To draw in; fade. [Cf. Du. *pinken* blink, glimmer]

Pink·er·ton (pingk′ər·tən), **Allan,** 1819–84, U. S. private detective born in Scotland.

pink·eye (pingk′ī′) *n.* 1 A febrile contagious keratitis of sheep, with inflammation of the mucous membrane lining the eyelids. 2 *Pathol.* An acute, contagious conjunctivitis in man, marked by redness of the eyeball. 3 A variety of white potato having pink eyes or buds.

pink·ie (pingk′ē) *n. U.S.* The little finger. Also **pink′y.** [Dim. of obs. *pink* small]

pink·ing (pingk′ing) *n.* 1 The act or process of pinking fabrics. 2 The act of stabbing, as with a rapier. [Orig. ppr. of PINK[2]]

pinking shears Shears with serrated blades for scalloping the edges of fabrics.

pink·ish (pingk′ish) *adj.* Somewhat pink.

pin knot See CAT-EYE.

pink rhododendron The California rosebay *(Rhododendron macrophyllum)* common on the Pacific coast: State flower of Washington.

pink·root (pingk′rōōt′, -rŏŏt′) *n.* 1 The root of any of several perennial herbs (genus *Spigelia*), especially the **Carolina pinkroot** with bright red flowers *(S. marilandica)*, a well-known anthelminthic: also called *wormroot.* 2 A plant yielding this root.

Pink·ster (pingk′stər) *n.* Whitsuntide: formerly observed as a day of revelry in New York. [<Du. *pinkster,* ult. <Gothic *paintekuste* <Gk. *pentēkostē* PENTECOST]

pinkster flower An American shrub *(Azalea nudiflora* or *Rhododendron nudiflorum)* with showy, flesh-colored to dark-red flowers, blooming at Whitsuntide: also spelled *pinxter flower.* Also called *wild honeysuckle.*

pink-stern (pingk′stûrn′) See PINK[3] (def. 2).

pink tea *U.S. Colloq.* 1 A women's social gathering or tea, at which the decorations or refreshments are exceptionally dainty. 2 Any innocuous occasion.

pink·y (pingk′ē) *adj.* Small and blinking: said of eyes. [Prob. <obs. *pink* small]

pin money An allowance for incidentals; formerly, an allowance made by a husband to his wife for her personal expenses.

pin·na (pin′ə) *n. pl.* **pin·nae** (pin′ē) 1 *Bot.* A single leaflet of a pinnate leaf. 2 *Anat.* The auricle of the ear. 3 *Zool.* A wing, fin, or the like. [<NL <L *pinna, penna* a feather] —**pin′·nal** *adj.*

pin·nace (pin′is) *n. Naut.* 1 A six- or eight-oared boat, carried by men-of-war; any ship's boat. 2 Formerly, a small schooner-rigged vessel used as a tender, scout, etc. [<OF *pinasse* <Ital. *pinaccia,* prob. <L *pinus* a pine]

pin·na·cle (pin′ə·kəl) *n.* 1 A small turret or tall ornament, as on a parapet. 2 Anything resembling a pinnacle; a high or topmost point, as a mountain peak; summit. —*v.t.* **·cled, ·cling** 1 To place on or as on a pinnacle. 2 To furnish with a pinnacle; crown. See synonyms under SUMMIT. [<OF *pinacle* <LL *pinnaculum,* dim. of L *pinna* a wing, a pinnacle]

pin·nate (pin′āt, -it) *adj.* 1 *Bot.* Having the shape or arrangement of a feather: said of compound leaves or leaflets arranged on each side of a common axis: when terminated by a single leaf, *odd-pinnate*; when lacking a terminal leaf, *abruptly pinnate.* 2 Having winglike parts or appendages. Also **pin′nat·ed.** [<L *pinnatus* feathered <*pinna* a feather, wing] —**pin′nate·ly** *adv.* —**pin·na′tion** *n.*

pinnati- *combining form* 1 *Bot.* Feathered; resembling a feather: *pinnatifid.* 2 *Zool.* Pinni-. [<L *pinnatus* feathered <*pinna* feather]

pin·nat·i·fid (pi·nat′ə·fid) *adj. Bot.* Cleft in a pinnate manner, with the incisions half-way down or more and the lobes or sinuses narrow.

pin·nat·i·lo·bate (pi·nat′ə·lō′bāt) *adj. Bot.* Pinnately lobed.

pin·nat·i·par·tite (pi·nat′ə·pär′tīt) *adj. Bot.* Pinnately parted.

pin·nat·i·ped (pi·nat′ə·ped) *adj. Ornithol.* Having lobed membranes to the toes, as certain birds.

pin·nat·i·sect (pi·nat′ə·sekt) *adj. Bot.* Pinnately divided as far as the rachis.

pin·ner[1] (pin′ər) *n.* 1 One who fastens with pins. 2 A pinafore. 3 A headdress with long flaps at each side, worn by women in the 18th century; also, a cloth band for a dress.

pin·ner[2] (pin′ər) *n. Obs.* An officer who impounded stray animals; a poundkeeper: also spelled *pinder.* [Var. of obs. *pinder* <obs. *pind* enclose, OE *pyndan* <*pund* a pound[2]]

pinni- *combining form Zool.* Web; fin: *pinniped.* [<L *pinna* feather]

pin·ni·grade (pin′ə·grād) *adj. Biol.* Moving by means of flippers, as a seal. [<PINNI- + L *gradi* walk]

pin·ni·ped (pin′ə·ped) *adj.* 1 Having finlike locomotive organs. 2 Of or pertaining to a suborder *(Pinnipedia)* of aquatic carnivorous mammals, the seals and walruses. —*n.* A fin-footed carnivorous mammal, as a walrus, seal, etc. [<NL *Pinnipes, -pedis* <L, wing-footed, fin-footed <*pinna* a wing, fin + *pes, pedis* a foot] —**pin′ni·pe′di·an** (-pē′dē·ən) *adj.* & *n.*

pin·nu·la (pin′yə·lə) *n. pl.* **·lae** (-lē) 1 *Ornithol.* A barb of a feather. 2 Pinnule. [<NL <L, dim. of *pinna* a feather] —**pin′nu·lar** *adj.* —**pin′nu·late, pin′nu·lat′ed** *adj.*

pin·nule (pin′yōōl) *n.* 1 *Zool.* A small, detached fin, as in mackerel; a finlike appendage. 2 *Bot.* One of the smaller or ultimate divisions of a pinnate leaf or frond; a secondary pinna. [<NL *pinnula* PINNULA]

pin·ny·win·kle (pin′ē·wing′kəl) See PILLIWINKS.

pin oak An oak tree *(Quercus palustris)* common in the eastern United States, forming a conical head with long pendulous branches, the leaves of which turn a bright scarlet in autumn.

pi·noch·le (pē′nuk·əl, -nok-) *n.* A card game resembling bezique played with a double pack of 48 cards, ranking as follows: ace, ten, king, queen, jack, and nine: also spelled *penuchle, penuckle.* Also **pi′noc·le.** —**check pinochle** A four-handed, partnership card game, based on bridge and pinochle, with a double-scoring

pinole system of points and checks. [Origin unknown]

pi·no·le (pi·nō'lā) n. SW U.S. A meal ground from corn, mesquite beans, or other plant seeds, and roasted. [<Am. Sp. <Nahuatl *pinolli*]

pi·ñon (pin'yən, pēn'yōn; Sp. pē·nyōn') n. 1 The edible seed of any of various pines of the Pacific coast of the United States; especially, the New Mexican piñon (*Pinus cembroides*). 2 The tree: also spelled *pinyon*. [<Sp., a pine nut <L *pinea* a pine cone <*pinus* a pine tree]

pin·point (pin'point') n. 1 The point of a pin. 2 Something extremely small. —v.t. To locate or define precisely: to pinpoint a target, argument, etc.

pins and needles A tingling or prickling sensation in some part of the body, as the fingers or toes; paresthesia. —**on pins and needles** Uneasy; nervous.

pin·scher (pin'shər) n. One of a breed of large, short-haired dogs, originally bred in Germany, usually black, brown, or gray-blue, with rust markings: usually called **Doberman pinscher**, after its first breeder. [<G, a terrier]

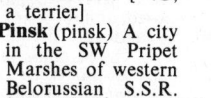

DOBERMAN PINSCHER (From 26 to 28 inches high at the shoulder)

Pinsk (pinsk) A city in the SW Pripet Marshes of western Belorussian S.S.R.

pin·son (pin'sən) n. A thin, light shoe or slipper. [? <OF *pinçon* toe-piece of a horseshoe, dim. of *pince* toe of a hoof]

pint (pīnt) n. 1 A dry and liquid measure of capacity equal to half a quart or (liquids only) four gills or 0.832 British pint. 2 Scot. A measure equivalent to three English pints. [<OF *pinte*, prob. <MDu.; ult. origin uncertain]

pin·ta (pin'tə, Sp. pēn'tä) n. Pathol. A tropical skin disease caused by a treponeme and characterized by patches of discoloration: also called *carate*. [<Sp., lit., a colored spot <LL *pincta* <L *picta*, orig. pp. fem. of *pingere* paint]

Pin·ta (pin'tə, Sp. pēn'tä) n. One of the three ships of Columbus on his initial voyage to America.

pin·ta·do (pin·tä'dō) n. pl. ·dos A large, spotted fish (*Scomberomorus maculatus*) of the mackerel family, found in tropical Atlantic waters of North and South America; the Spanish mackerel. [<Pg., lit., painted, pp. of *pintar* paint, ult. <L *pingere* paint]

pin·tail (pin'tāl') n. 1 A duck (*Dafila acuta*) of the northern hemisphere, the male of which has a long, sharp tail. 2 A sharp-tailed grouse (*Pediaecetes phasianellus*) of northern North America. 3 The ruddy duck.

pin·ta·no (pin·tä'nō) n. pl. ·nos or ·no A tropical fish (genus *Abudefduf*) having a green body marked with dark bands; found in the neighborhood of coral reefs. Also called *cow pilot*. [Prob. <Sp. *pinta*. See PINTA.]

pin·tle (pin'tl) n. A pin upon which anything pivots, as one of the metal braces or hooks upon which a rudder swings, the pin of a hinge of a gunlock, etc. [OE *pintel* penis. Akin to PIN.]

pin·to (pin'tō) adj. SW U.S. Piebald; pied, as an animal. —n. pl. ·tos 1 A pied animal: said especially of a horse, or Western pony. 2 A kind of spotted bean (*Phaseolus vulgaris*) of the southwestern United States: also **pinto bean**. [<Am. Sp. <Sp., lit., painted, ult. <L *pingere* paint]

Pintsch gas (pinch) A fuel and illuminating gas made by the destructive distillation of oil. [after Richard *Pintsch*, 1840-1919, German inventor]

Pin·tu·ric·chio (pēn·tōō·rēk'kyō), 1454-1513, Italian painter; original name *Bernardino di Betti*.

pin-up (pin'up') U.S. Colloq. n. That which is affixed to a board or wall for scrutiny or perusal; specifically, a clipping or photograph, usually of an attractive young woman. —adj. Designating a photograph, clipping, or drawing used in this manner, or a person who models such a picture: She's my pin-up girl.

pin·weed (pin'wēd') n. Any one of various perennial herbs (genus *Lechea*), with very small greenish or purplish flowers.

pin·wheel (pin'hwēl') n. 1 A firework that revolves when ignited, forming a wheel of fire. 2 A wheel with pins set in its face in place of cogs on the periphery. 3 A child's toy resembling a windmill, made of colored paper, revolving on a pin attached to a stick.

pin·worm (pin'wûrm') n. A nematode worm (*Enterobius vermicularis*) parasitic in the lower intestines and rectum of man, especially of children.

pinx·it (pingk'sit) Latin He (or she) painted (it).

pinx·ter flower (pingk'stər) See PINKSTER FLOWER.

pin·y (pī'nē) adj. Pertaining to, suggestive of, or covered with pines: also spelled *piney*.

pin·yon (pin'yən) See PIÑON.

pinyon jay A small, short-tailed, uncrested corvine bird (*Cyanocephalus cyanocephalus*) of the Rocky Mountain region having a dark plumage and a long, sharp bill.

Pin·zón (pēn·thōn'), **Martín Alonzo**, 1441-1493, Spanish navigator; commanded the *Pinta* of Columbus' fleet. — **Vicente Yáñez**, 1460-1524, commanded the *Nina*; discovered Brazil; brother of the preceding.

pi·on (pī'on) n. A pi-meson. [<PI + (MES)ON]

pi·o·neer (pī'ə·nir') n. 1 One of the first explorers, settlers, or colonists of a new country or region. 2 One of the first investigators or developers in a new field of research, enterprise, etc. 3 Mil. An engineer who goes before the main body building roads, bridges, etc. See synonyms under HERALD. —v.t. 1 To prepare (a way, etc.). 2 To prepare the way for. 3 To be a pioneer of. —v.i. 4 To act as a pioneer. [<F *pionnier* <OF *paonier*, orig. a foot soldier <*peon*, *pion*]

Pio·tr·ków (pyō'tər·kōōf) A city in central Poland SE of Łódź. Russian **Pe·tro·kov** (pet'rə·kôf, Russian pye'trō·kôf').

pi·ous (pī'əs) adj. 1 Actuated by reverence for a Supreme Being; religious; godly. 2 Marked by a reverential spirit. 3 Practiced in the name of religion. 4 Obs. Exhibiting filial respect and affection; filial. See synonyms under GOOD, MORAL. [<L *pius* dutiful, devout] —**pi'ous·ly** adv. —**pi'ous·ness** n.

Pi·oz·zi (pē·ōz'ē, Ital. pyōt'tsē), **Hester Lynch Thrale**, 1741-1821, English writer and friend of Dr. Johnson: best known as *Mrs. Thrale*.

pip¹ (pip) n. 1 The seed of an apple, orange, etc. 2 Slang An admirable person or thing. [Short for PIPPIN]

pip² (pip) n. 1 A spot, as on a playing card, domino, or die. 2 Any of the small buds of the lily-of-the-valley. 3 Any dormant rootstock of several flowering plants, as anemones and peonies. 4 One of the diamond-shaped sections on the rind of a pineapple. 5 A sharp sound or luminous signal produced mechanically or electronically, as in radar. [<earlier *peep*; orig. unknown]

pip³ (pip) v. pipped, pip·ping v.t. To break through (the shell), as a chick in the egg. —v.i. To peep; chirp. [Prob. var. of PEEP¹]

pip⁴ (pip) n. 1 A contagious disease of fowls in which mucus forms in the throat or a scale on the tongue. 2 Slang A mild human ailment: used humorously; also, a grouch; an ill temper. [<MDu. *pippe* <LL *pipita* <L *pituita* mucus, the pip]

pip·age (pī'pij) n. 1 Pipes collectively; a system of pipes. 2 The carriage of oil, gas, water, etc., through pipes.

pi·pal (pē'pəl) n. The sacred fig tree of India (*Ficus religiosa*): also spelled *peepul*. [<Hind. *pīpal* <Skt. *pippala*]

pipe (pīp) n. 1 An apparatus, usually a small bowl with a hollow stem, for smoking tobacco, opium, or other narcotic. 2 Enough tobacco to fill the bowl of a pipe. 3 A long conducting passage of wood, metal, tiling, etc., for conveying a fluid. 4 A single tube or long, hollow case: when part of a line of piping, often called a *piece of pipe*. 5 Any hollow or tubular part in an animal or plant body. 6 Music a A tubular wind instrument, such as the flageolet. b pl. The bagpipe. 7 The voice; also, a bird's note or call. 8 A large cask for wine; also, a liquid measure of half a tun. 9 Metall. A conical cavity in the head of a steel ingot, made by an escape of gas while the metal was cooling. 10 Slang An easy college course. 11 A boatswain's whistle. 12 Mining An elongated, usually vertical or highly inclined body of mineral or rich ore: also called a *chimney*. —v. **piped**, **pip·ing** v.i. 1 To play on a pipe. 2 To make a shrill sound. 3 Naut. To signal the crew by means of a boatswain's pipe. 4 Metall. To form conical cavities in hardening, as ingots. —v.t. 5 To convey by or as by means of pipes. 6 To provide with pipes. 7 To play, as a tune, on a pipe. 8 To utter shrilly or in a high key. 9 Naut. To call to order by means of a boatswain's pipe. 10 To lead, entice, or bring by piping. 11 To trim, as a dress, with piping. —**to pipe down** Slang To become silent; stop talking or making noise. —**to pipe up** 1 To start playing or singing: *Pipe up the band!* 2 To speak out, especially in a shrill voice. [OE *pīpe*, ult. <L *pipare* cheep. Doublet of FIFE.]

pipe-clay (pīp'klā') v.t. To whiten with pipe clay.

pipe clay A white clay used for pottery, especially pipes, and, formerly, for whitening military accouterments: also called *terra alba*.

pipe dream Any wish, plan, or groundless hope; a daydream.

pipe·fish (pīp'fish') n. pl. ·fish or ·fish·es 1 One of a family (*Syngnathidae*) of slender marine and fresh-water fishes having a straight, tubelike snout and bodies enclosed in a series of bony rings. 2 A sea horse.

pipe·fit·ter (pīp'fit'ər) n. One who joins pipes together; a plumber.

pipe·fit·ting (pīp'fit'ing) n. 1 A piece of pipe used to connect two or more pipes together. 2 The work of a pipefitter.

pipe·line (pīp'līn') n. 1 A line of pipe, as for the transmission of water, oil, etc. 2 A channel for the transmission of information, usually private or secret. —v.t. 1 To convey by pipeline. 2 To furnish with a pipeline.

pipe of peace The calumet.

pipe organ An organ having pipes: distinguished from a *reed organ*.

pip·er (pī'pər) n. 1 One who lays pipes. 2 One who plays upon a pipe, especially a bagpipe.

Pi·per·a·ce·ae (pī'pə·rā'si·ē, pip'ə-) n. pl. A family of tropical aromatic herbs and shrubs, including the common pepper plant of Asia and South America. [<NL <L *piper* pepper] —**pi·per·a·ceous** (-rā'shəs) adj.

pi·per·a·zine (pi·per'ə·zēn) n. Chem. A crystalline substance, $C_4H_{10}N_2$, formed by the action of aniline on ethylene bromide: used in the treatment of rheumatism and gout. [<PIPER(INE) + AZ- + -INE²]

pi·per·i·dine (pi·per'ə·dēn) n. Chem. A colorless liquid alkaloid, $C_5H_{11}N$, with a strong odor and caustic taste, contained in piperine and also made synthetically. [<L *piper* pepper + -ID(E) + -INE²]

pip·er·ine (pip'ər·ēn, -in) n. Chem. A colorless crystalline alkaloid, $C_{17}H_{19}NO_3$, contained in pepper and made synthetically: used in medicine as an antipyretic. [<L *piper* pepper + -INE²]

pip·er·o·nal (pip'ər·ə·nal') n. Chem. A white crystalline aldehyde, $C_8H_6O_3$, derived from benzene: used in making perfumes. [<G <*piper(in)* piperine]

pipe stem 1 The stem of a tobacco pipe. 2 pl. Thin, skinny legs. —**pipe-stem** (pīp'stem') adj.

pipe·stone (pīp'stōn') n. An indurated red clay much valued by the American Indians for making tobacco pipes.

pi·pette (pī·pet', pi-) n. 1 A small tube, often graduated, for removing small portions of a fluid. 2 A funnel-like can used in applying liquid decoration. Also **pi·pet'**. [<F, dim. of *pipe* pipe]

pip·ing (pī'ping) adj. 1 Playing on the pipe. 2 Hissing or sizzling: *piping hot*. 3 Having a shrill sound. 4 Characterized by peaceful rather than martial music. —n. 1 The act of one who pipes. 2 The music of pipes; hence, a wailing or whistling sound. 3 A system of pipes, as for drainage. 4 A narrow strip of cloth folded on the bias, used for trimming dresses. 5 A cordlike decoration of icing on a cake.

pip·it (pip′it) *n.* 1 One of various lark-like singing birds (genus *Anthus*) widely distributed in North America; especially, the common American pipit (*A. spinoletta*); a titlark. 2 The Missouri skylark (*A. spraguei*): also called *Sprague's pipit*. [Prob. imit.]

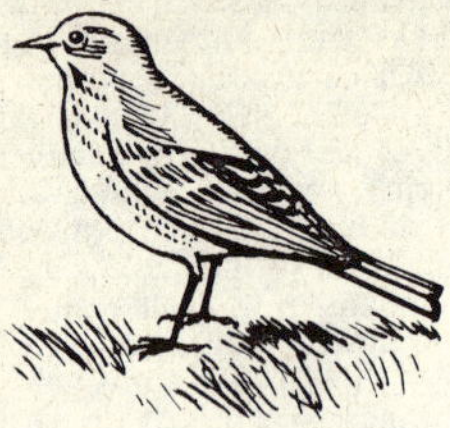

AMERICAN PIPIT
(About 6 inches long)

pip·kin (pip′kin) *n.* 1 A small earthenware jar. 2 A piggin. [? Dim. of PIPE]

pip·pin (pip′in) *n.* 1 An apple of many varieties. 2 A seed; pip. 3 *Slang* An admirable person or thing. [<OF *pepin* seed of a fruit; origin uncertain]

Pip·pin (pip′in) Pepin the Short.

pip·sis·se·wa (pip·sis′ə·wə) *n.* A low-growing evergreen (genus *Chimaphila*) of the heath family, with white or pink flowers and thick leaves, used in medicine as an astringent and diuretic. [<Algonquian. Cf. Cree *pipisisikweu*, lit., it breaks it (gallstone) into pieces.]

pip–squeak (pip′skwēk′) *n.* 1 A petty and contemptible person or thing. 2 A small, insignificant person. [Orig. imit. name for a small German high-velocity shell employed in World War I]

pip·y (pī′pē) *adj.* **pip·i·er, pip·i·est** 1 Pipelike; tubular; containing pipes. 2 Piping; thin and shrill, or reedlike, in sound.

pi·quant (pē′kənt) *adj.* 1 Having an agreeably pungent or tart taste. 2 Interesting; tart; racy; also, charmingly lively. 3 *Obs.* Stinging; sharp. [<F, orig. ppr. of *piquer* sting] —**pi′quan·cy** *n.* —**pi′quant·ly** *adv.*

pique[1] (pēk) *n.* A feeling of irritation or resentment. —*v.t.* **piqued, pi·quing** 1 To excite resentment in. 2 To stimulate or arouse; provoke. 3 To pride (oneself): with *on* or *upon.* ◆ Homophones: *peak, peek.* [<MF <*piquer* sting, prick]

Synonyms (noun): displeasure, grudge, irritation, offense, resentment, umbrage. *Pique* signifies primarily a prick or a sting, as of a nettle; the word denotes a sudden feeling of mingled pain and anger, usually transient, arising from some neglect or *offense*, real or imaginary. *Umbrage* is a deeper and more persistent *displeasure* at being overshadowed or subjected to any treatment that one deems unworthy of oneself. *Resentment* rests on more solid grounds, and is deep and persistent. See ANGER. *Antonyms*: approval, complacency, contentment, delight, gratification, pleasure, satisfaction.

Synonyms (verb): affront, annoy, chafe, displease, fret, goad, irritate, nettle, offend, pain, provoke, rouse, stimulate, sting, stir, urge, vex, wound. See ANGER.

pique[2] (pēk) *n.* In piquet, the scoring of 30 points in one hand before the other side scores at all. —*v.t.* To win a pique from. ◆ Homophones: *peak, peek.* [<F *pic* a mountain peak]

pi·qué (pē·kā′) *n.* A fabric of cotton, rayon, or silk, with raised cord or welts, called wales, running lengthwise in the fabric. [<F, lit., quilted, orig. pp. of *piquer* prick, backstitch]

pi·quet (pē·ket′, *Fr.* pē·ke′) *n.* A two-handed game of cards in which the cards below the seven are excluded: also spelled *picquet.* [<F, ? dim. of *pique* a spade in cards]

pi·ra·cy (pī′rə·sē) *n. pl.* **·cies** 1 Robbery on the high seas. 2 The unauthorized publication, reproduction, or use of another's invention, idea, or literary creation. [<Med. L *piratia* <Gk. *peirateia* <*peiratēs* a pirate]

Pi·rae·us (pī·rē′əs) A town in SE Greece, 5 miles SW of Athens and its ancient port; the leading port and industrial city of Greece: Greek *Peiraievs.*

pi·ra·gua (pi·rä′gwə) *n.* 1 A dug-out canoe. 2 A flat-bottomed boat with two masts: used in the Caribbean Sea. Also called *pirogue.* [<Sp. <Cariban, a dug-out]

Pi·ran·del·lo (pē′rän·del′lō), **Luigi,** 1867–1936, Italian dramatist and novelist.

pi·ra·nha (pi·rä′nyə) *n.* A caribe. Also **pi·ra′ya** (-rä′yä). [<Pg. (Brazilian) <Tupian, toothed fish <*piro* a fish + *sainha* a tooth]

pi·rate (pī′rit) *n.* 1 A rover and robber on the high seas. 2 A vessel engaged in piracy. 3 A person who appropriates without right the work of another. See synonyms under ROBBER. —*v.t.* & *v.i.* **·rat·ed, ·rat·ing** 1 To practice or commit piracy (upon). 2 To publish or appropriate (the work, ideas, etc., of another) wrongfully or illegally; plagiarize. [<L *pirata* <Gk. *peiratēs* <*peiraein* attack] —**pi·rat·ic** (pī·rat′ik) or **·i·cal** *adj.* —**pi·rat′i·cal·ly** *adv.*

Pi·rith·o·us (pī·rith′ō·əs) In Greek mythology, a king of the Lapithae who, with his friend Theseus, attempted to carry off Persephone from Hades; he was punished by Pluto by being bound to a rock. See LAPITHAE.

Pir·ma·sens (pir′mä·zens) A city in Rhineland–Palatinate, West Germany.

pirn (pûrn) *n. Scot.* 1 A small spindle. 2 Yarn on a shuttle. 3 A spinning-wheel bobbin. 4 A fishing-rod reel.

pi·rogue (pi·rōg′) *n.* A piragua. [<F <Cariban *piragua*]

pir·ou·ette (pir′ōō·et′) *n.* A rapid whirling upon the toes in dancing. —*v.i.* **·et·ted, ·et·ting** To make a pirouette. [<F, lit., a spinning top, prob. <dial. F *piroue* a top; ult. origin uncertain]

Pi·sa (pē′zə, *Ital.* pē′sä) A city on the Arno river in north central Italy; celebrated for its Leaning Tower. —**Pi′san** *adj.* & *n.*

pis al·ler (pē zà·lā′) *French* The last shift; last resource; literally, to go worse.

Pi·sa·no (pē·sä′nō), **Andrea,** 1270–1348?, Italian goldsmith, sculptor, and architect. —**Giovanni,** 1245?–1320?, Italian sculptor and architect. —**Nicola,** 1225?–78?, Italian sculptor; father of the preceding.

pis·ca·ry (pis′kər·ē) *n. pl.* **·ries** 1 *Law* The right of fishing in waters that belong to another: now usually in the phrase **common of piscary.** 2 A fishing place; fishery. [<Med. L *piscaria* fishing rights, orig. neut. pl. of L *piscarius* of fishing <*piscis* a fish]

pis·ca·tol·o·gy (pis′kə·tol′ə·jē) *n.* The science of fishing. [<L *piscatus*, pp. of *piscari* fish + -LOGY]

pis·ca·tor (pis·kā′tər) *n.* An angler; fisherman. [<L]

pis·ca·to·ri·al (pis′kə·tôr′ē·əl, -tō′rē-) *adj.* 1 Pertaining to fishes or fishing. 2 Engaged in fishing. Also **pis′ca·to′ry.** [<L *piscatorius* <*piscator* a fisherman <*piscatus*, pp. of *piscari* fish] —**pis′ca·to′ri·al·ly** *adv.*

Pis·ces (pis′ēz, pī′sēz) *n. pl.* 1 A class of vertebrates: the true fishes. 2 The twelfth sign of the zodiac. See ZODIAC. 3 *Astron.* A zodiacal constellation south of Pegasus and Andromeda; the Fish or Fishes. See CONSTELLATION. [<L, pl. of *piscis* a fish]

Pisces Aus·tri·nus (ô·strī′nəs) *Astron.* A zodiacal constellation. See CONSTELLATION.

pisci– *combining form* Fish; of or related to fish: *piscivorous.* Also, before vowels, **pisc–.** [<L *piscis* a fish]

pis·ci·cul·ture (pis′i·kul′chər) *n.* The hatching and rearing of fish. —**pis′ci·cul′tur·al** *adj.* —**pis′ci·cul′tur·ist** *n.*

pis·ci·na (pi·sī′nə, -sē′-) *n. pl.* **·nae** (-nē) *Eccl.* A stone basin with a drain in which the priest washes the chalice after administering communion. [<Med. L <L, lit., a fish pond, basin <*piscis* fish] —**pis·ci·nal** (pis′ə·nəl) *adj.*

pis·cine (pis′īn, -in) *adj.* Of, pertaining to, or resembling a fish or fishes. [<L *piscis* a fish + -INE[1]]

pis·civ·o·rous (pi·siv′ər·əs) *adj.* Feeding or subsisting on fish.

Pis·gah (piz′gə), **Mount** A mountain in Jordan, NE of the Dead Sea: in the Old Testament, the peak from which Moses beheld the Promised Land; highest peak, Mount Nebo.

pish (pish) *interj.* An exclamation of contempt. —*v.t.* & *v.i.* To use this exclamation (to).

Pi·sid·i·a (pi·sid′ē·ə) An ancient country, later a Roman province, of south central Asia Minor. —**Pi·sid′i·an** *adj.*

pi·si·form (pī′sə·fôrm) *adj.* 1 Shaped like a pea. 2 *Anat.* Pertaining to a pea-shaped bone on the inner or ulnar side of the carpus. [<NL *pisiformis* <L *pisum* a pea + *forma* form]

Pi·sis·tra·tus (pī·sis′trə·təs, pi-), 600?–527 B.C., Athenian tyrant and statesman.

pis·mire (pis′mīr) *n.* An ant. [ME *pissemyre* <*pisse* urine + *myre, mire* an ant; with ref. to urinous smell of an anthill]

pis·mo clam (piz′mō) A clam (*Tivela stultorum*) of the southwestern coast of North America. [from *Pismo* beach, California]

pi·so·lite (pī′sə·līt) *n.* A coarse concretionary limestone, composed of globules with a distinct pisiform structure. [<NL *pisolithus* <Gk. *pisos* a pea + *lithos* a stone] —**pi′so·lit′ic** (-lit′ik) *adj.*

piss (pis) *n.* Urine. [<*v.*] —*v.i.* To urinate. —*v.t.* To discharge as or with the urine. [<OF *pissier*; prob. orig. imit.]

Pis·sar·ro (pē·sà·rō′), **Camille,** 1831?–1903, French painter.

pis·ta·chi·o (pis·tä′shē·ō, -tash′ē·ō) *n. pl.* **·chi·os** 1 The edible nut of a small tree (genus *Pistacia*) of western Asia and the Levant. 2 The tree. 3 The flavor produced by, or a delicacy flavored with the pistachio nut. 4 A delicate shade of green, the color of the pistachio nut. Also **pis·tache′** (-täsh′). [<Ital. *pistacchio* <L *pistacium* <Gk. *pistakion* <*pistakē* a pistachio tree, prob. <OPersian *pistah* a pistachio nut]

pis·ta·reen (pis′tə·rēn′) *n.* An old Spanish coin formerly worth about 20 cents: used in the West Indies and United States in the 18th century. [Appar. alter. of Sp. *peseta* a peseta]

pis·til (pis′til) *n. Bot.* The seed-bearing organ of flowering plants, composed of the ovary, with its contained ovules, and the stigma, usually with a style. [<F <L *pistillum* a pestle]

pis·til·late (pis′tə·lit, -lāt) *adj. Bot.* 1 Having a pistil. 2 Having pistils and no stamens. Also **pis′til·lar′y** (-ler′ē).

Pis·to·ia (pēs·tō′yä) A city in north central Italy; where pistols are said to have been first manufactured.

pis·tol (pis′təl) *n.* A small firearm having a stock to fit the hand, and a short barrel: now either the revolver or automatic type fired from one hand. —*v.t.* **·toled** or **·tolled, ·tol·ing** or **·tol·ling** To shoot with a pistol. [<F *pistole* <Ital. *pistola*, prob. ult. <PISTOIA]

Pis·tol (pis′təl) In Shakespeare's *Merry Wives of Windsor* and *Henry IV*, a despicable follower of Falstaff.

pis·tole (pis·tōl′) *n.* An obsolete gold coin of varying value, formerly current in Europe. [<F, short for MF *pistolet* a pistol. Earlier called an *écu*, lit., a shield; the name was changed in humorous allusion to the coin's debasement in value.]

pis·to·leer (pis′tə·lir′) *n.* One who fires a pistol; formerly, a soldier carrying a pistol. Also **pis′to·lier′.**

pis·tol–whip (pis′təl·hwip′) *v.t.* **–whipped** or **–whipt, –whip·ping** To strike or beat with the barrel of a pistol.

pis·ton (pis′tən) *n.* 1 *Mech.* A disk fitted to slide in a cylinder, as in a steam engine, and connected with a rod for receiving the pressure of or exerting pressure upon a fluid in the cylinder. 2 A valve in a wind instrument for altering the pitch of tones. [<F <Ital. *pistone* a piston, var. of *pestone* a large pestle <*pestare* pound <LL *pistare*, freq. of L *pinsere* pound]

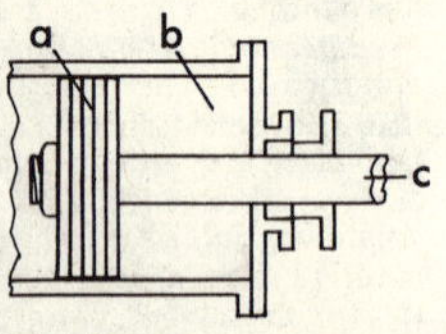

STEAM-ENGINE PISTON
a. Piston. *b.* Cylinder. *c.* Piston rod.

piston ring *Mech.* An adjustable ring, usually of cast iron, fitted within a groove on the piston body and designed to prevent leakage of the fluid by expansion against the cylinder wall.

piston rod *Mech.* A rod attached to a piston at one end and to a cross-head or crankpin at the other: used to impart motion.

pit[1] (pit) *n.* 1 A natural or artificial cavity in the ground, especially when relatively wide and deep. 2 A pitfall for snaring animals; snare. 3 An abyss so deep that one cannot return from it, as the grave. 4 Hell. 5 Great distress or trouble. 6 The main floor of the auditorium of a theater, especially, in Great Britain, that portion under the first balcony; also, that part of the audience occupying this portion of the theater. Compare ORCHESTRA, PARQUET. 7 An enclosed space in which animals trained for combat are pitted against each other. **8** Any natural cavity or depression in the body: the *armpit*, the *pit* of the stomach. **9** An indention like that made by a smallpox pustule; any slight depression or excavation. **10** A thin spot in the cell walls of some plants. **11** That part of the floor of an exchange where a special line of trading is done: the wheat *pit*. **12** A mining excavation, or the shaft of a mine. — *v.* **pit·ted, pit·ting** *v.t.* **1** To mark with dents, pits, or hollows. **2** To put, bury, or store in a pit. **3** To match as antagonists; set in opposition. — *v.i.* **4** To become marked with pits. [OE *pytt* <L *puteus* a well]

pit^2 (pit) *n.* The kernel of certain fruits, as the plum. — *v.t.* **pit·ted, pit·ting** To remove pits from, as fruits. [<Du. <MDu. *pitte* kernel, pith. Akin to PITH.]

pi·ta (pē′tə) *n.* **1** The fiber of the century plant and other allied species of *Agave*: used for making paper, cordage, etc. **2** The plant yielding the fiber. **3** The fiber obtained from several kinds of yucca. [<Sp. <Quechua, a fine thread made from vegetable fiber]

pit–a–pat (pit′ə·pat′) *v.i.* **-pat·ted, -pat·ting** To move or sound with a succession of light, quick steps or pulsations. — *n.* A tapping or succession of taps, steps, or similar sounds. — *adv.* With a rapid succession of light steps, beats or taps; flutteringly. Also spelled *pit-ty-pat*. [Imit.]

Pit·cairn Island (pit′kârn) A British colony in the South Pacific, settled in 1790 by mutineers of the British ship *Bounty*; 2 square miles; with dependencies, 18.5 square miles.

pitch^1 (pich) *n.* **1** A thick, viscous, dark substance obtained by boiling down tar from the residues of distilled turpentine, etc., used in coating seams. **2** Any of a class of hydrocarbon residues obtained from the refining of fats, oils, and greases: linseed *pitch*, cottonseed *pitch*, etc. **3** The resinous sap of pines. **4** Bitumen or asphaltum, especially when unrefined. — *v.t.* To smear, cover, or treat with or as with pitch. [OE *pic* <L *pix, picis* pitch]

pitch^2 (pich) *v.t.* **1** To erect or set up (a tent, camp, etc.). **2** To throw or hurl; toss; fling. **3** To set the level, angle, degree, etc., of. **4** To put in a definite place or position. **5** To set in order; arrange: obsolete except in *pitched battle*. **6** In baseball, to deliver (the ball) to the batter. **7** *Music* To set the pitch or key of. **8** In card games, to determine or announce (the trump suit) by leading a card of that suit. — *v.i.* **9** To fall or plunge forward or headlong. **10** To lurch. **11** To rise and fall alternately at the bow and stern, as a ship: to *pitch* and roll. **12** To incline downward; slope. **13** To encamp; settle. **14** To decide, especially at random: with *on* or *upon*. **15** In baseball, to deliver the ball to the batter; act as pitcher. — **to pitch in** *Colloq.* To start vigorously. — **to pitch into** To attack; assail. — *n.* **1** Point or degree of elevation or depression; especially, the extreme top or bottom point; hence, the ultimate reach. **2** The degree of descent of a declivity; also, a descent, slope, or inclination to the horizon. **3** In building, the inclination of a roof. **4** *Aeron.* **a** An angular displacement about an axis parallel to the lateral axis of an aircraft. **b** The distance an aircraft advances along its flight path for one revolution of the propeller. **5** *Mech.* **a** The amount of advance of a screw thread in a single turn. **b** The distance between two corresponding points on the teeth of a gearwheel. **6** *Physics* The dominant frequency of a sound wave perceived by the ear, ranging from a low tone of about 20 cycles per second to a maximum high approaching 30,000 cycles. **7** *Music* The acuteness or gravity of all the tones of a given instrument with reference to some standard. The pitch of an instrument is expressed by the vibrations per second of some one of its notes, generally middle C, treble C, or the A between them. The high pitch, known as **concert pitch**, has about 450 vibrations for middle A. In 1859 a commission of French musicians and scientists determined the pitch of A′ as 435 (true C′′ 522, equal temperament C′′ 517) which is known as **French, international,** or **low pitch.** The present standard or **philharmonic pitch** has 440 vibrations for middle A. **8** In games, the act of pitching; a throw; specifically, in baseball, the delivery of the ball by the pitcher; also, the place of pitching or the distance pitched. **9** *Geol.* The inclination or dip of a rock stratum or vein of ore. **10** The act of dipping or plunging downward; the pitching of a ship: correlative of *scend*. **11** A game of cards; seven–up. **12** A location or station for a vender, on a sidewalk, etc. **13** A short, steep stretch of a mountain climb. **14** An attempt to sell or persuade: to make a *pitch*. — **auction pitch** A variety of the game of pitch in which the privilege of pitching is sold at auction by the player entitled to it. — **full pitch** In cricket, bowled to hit the wicket before touching the ground. [ME *picchen*; origin uncertain]

pitch–and–toss (pich′ən·tôs′, -tos′) *n. Brit.* A game played by pitching pennies at a line.

pitch–black (pich′blak′) *adj.* Intensely black; as dark as pitch.

pitch·blende (pich′blend′) *n.* A black or brown uranium oxide with a luster resembling that of pitch: the chief source of uranium and radium. See URANINITE. [<G *pechblende* <*pech* pitch^1 + *blende* blende]

pitch–dark (pich′därk′) *adj.* Very dark; as black as pitch.

pitch·er^1 (pich′ər) *n.* One who pitches; specifically, in baseball, the player who delivers the ball to the batter. [<PITCH^2]

pitch·er^2 (pich′ər) *n.* **1** A vessel with a spout and a handle, used for holding liquids to be poured out. **2** A peculiar form of leaf suggestive of a pitcher. [<OF *pichier, picher* <LL *bicarium* a jug <Gk. *bikos* a wine jar]

pitch·er·plant (pich′ər·plant′, -plänt′) *n.* A plant having leaves arranged in the form of pitchers, urns, or goblets which often function as insect traps; especially, the common American pitcherplant (genus *Sarracenia*).

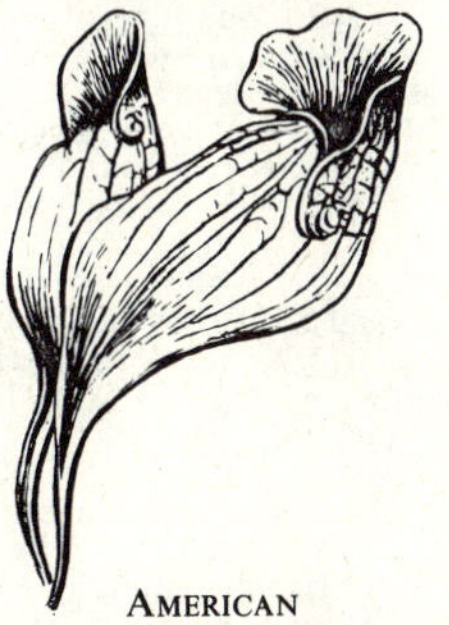

AMERICAN PITCHERPLANT
(From 6 inches to 4 feet tall according to variety)

pitch·fork (pich′fôrk′) *n.* **1** A large fork with which to handle hay, straw, etc. **2** A tuning fork. — *v.t.* To lift and throw with or as with a pitchfork. [<PITCH^2 + FORK]

Pitch Lake A deposit of natural asphalt in SW Trinidad; 114 acres; greatest depth, 285 feet.

pitch·man (pich′mən) *n. pl.* **·men** (-mən) *Slang* One who sells small articles from a temporary stand, as at a fair, etc.; a sidewalk vender.

pitch pine Any of several American pines that yield pitch; especially, *Pinus rigida* and the longleaf Georgia pine (*P. palustris*).

pitch ratio *Aeron.* The ratio of the pitch of an aircraft propeller to its diameter.

pitch·stone (pich′stōn′) *n.* An acid volcanic glass having a resinous luster and containing more water than obsidian. [Trans. of G *pechstein* <*pech* pitch^1 + *stein* stone]

pitch·y (pich′ē) *adj.* **pitch·i·er, pitch·i·est** **1** Resembling pitch; pitchlike; intensely dark. **2** Full of or daubed with pitch. — **pitch′i·ly** *adv.* — **pitch′i·ness** *n.*

pit·e·ous (pit′ē·əs) *adj.* **1** Exciting pity, sorrow, or sympathy. **2** *Archaic* Affected with or feeling pity. See synonyms under PITIFUL. [<OF *pitos, piteus*, ult. <L *pietas, -tatis*. See PIETY.] — **pit′e·ous·ly** *adv.* — **pit′e·ous·ness** *n.*

pit·fall (pit′fôl′) *n.* A pit contrived for entrapping wild beasts or men; hence, any hidden danger. [ME *pitfalle, putfal* <PIT^1 + *falle, fal* <OE *fealle* a trap]

pith (pith) *n.* **1** *Bot.* The cylinder of soft, spongy tissue in the center of the stems and branches of certain plants. **2** *Ornithol.* The spongy substance of the interior of the shaft of a feather. **3** The marrow of bones or of the spinal cord. **4** Concentrated force; vigor; substance; hence, the essential part; quintessence; gist. — *v.t.* **1** To destroy the central nervous system or spinal cord of, as a frog, by passing a wire through the vertebral column. **2** To remove the pith from, as a plant stem. **3** To kill (cattle) by severing the spinal cord. [OE *pitha*. Akin to PIT^2.]

pith·e·can·thrope (pith′ə·kan′thrōp) *n. Paleontol.* A member of the genus *Pithecanthropus*.

Pith·e·can·thro·pus (pith′ə·kan′thrə·pəs, -kan·thrō′pəs) *n. pl.* **·pi** (-pī) *n. Paleontol.* The type genus of two small–brained Pleistocene primates transitional between ape and man: *P. erectus*, represented by a fossil cranium, femur, and other fragments discovered near Trinil, central Java, in 1891; and *P. robustus*, based upon skeletal remains found in the same area about 1938. Also called *Java man, Trinil man.* [<NL <Gk. *pithēkos* an ape + *anthropos* a man] — **pith′e·can′thro·pine** (-pēn, -pin) *adj.*

pith·less (pith′lis) *adj.* Having no pith; lacking force.

pith·y (pith′ē) *adj.* **pith·i·er, pith·i·est** **1** Consisting of pith; like pith. **2** Forcible; effective. See synonyms under TERSE. — **pith′i·ly** *adv.* — **pith′i·ness** *n.*

pit·i·a·ble (pit′ē·ə·bəl) *adj.* **1** Arousing or meriting pity or compassion; pathetic. **2** Insignificant; contemptible. See synonyms under PITIFUL. [<OF *piteable* <*piteer, pitier* pity <*pitie* PITY] — **pit′i·a·ble·ness** *n.* — **pit′i·a·bly** *adv.*

pit·i·ful (pit′i·fəl) *adj.* **1** Calling forth pity or compassion; miserable; wretched. **2** Calling forth a feeling of contempt, because of littleness, meanness, or the like; contemptible. **3** *Archaic* Full of pity; compassionate. — **pit′i·ful·ly** *adv.* — **pit′i·ful·ness** *n.*

Synonyms: abject, base, contemptible, despicable, lamentable, miserable, mournful, moving, paltry, pathetic, piteous, pitiable, sorrowful, touching, woeful, wretched. *Pitiful* originally signified full of pity; as, "the Lord is very *pitiful* and of tender mercy"; but this usage is now archaic, and the meaning in question is appropriated by such words as merciful and compassionate. *Pitiful* and *pitiable* now refer to what may be deserving of pity, *pitiful* being used chiefly for that which is merely an object of thought, *pitiable* for that which is brought directly before the senses; as, a *pitiful* story; a *pitiable* condition. Since pity, however, always implies weakness or inferiority in that which is pitied, *pitiful* and *pitiable* are often used, by an easy transition, for what might awaken pity, but does awaken contempt; as, a *pitiful* excuse; He presented a *pitiable* appearance. *Piteous* is now rarely used in its earlier sense of feeling pity, but in its derived sense applies to what really excites the emotion; as, a *piteous* cry. See MERCIFUL. Compare HUMANE, MERCY, PITY. *Antonyms*: august, beneficent, commanding, dignified, exalted, glorious, grand, great, lofty, mighty, noble, superb.

pit·i·less (pit′i·lis) *adj.* Having no pity or mercy; ruthless. See synonyms under IMPLACABLE. — **pit′i·less·ly** *adv.* — **pit′i·less·ness** *n.*

pit·man (pit′mən) *n.* **1** *pl.* **·men** (-mən) One who works in a pit, as in sawing, mining, etc.; especially, in mining, the man who has charge of the underground machinery. **2** *pl.* **·mans** (-mənz) A rod that connects a rotary with a reciprocating part; a connecting rod.

Pit·man (pit′mən), **Sir Isaac**, 1813–97, English educator; inventor of a system of shorthand.

Pi·to·cin (pi·tō′sin) *n.* Proprietary name for a brand of oxytocin.

pi·ton (pi·ton′, *Fr.* pē·tôn′) *n.* An iron spike, with an eye or ring in one end, that can be driven into a crack in rock or ice: used in mountaineering as a hold or support for hand or foot, or for karabiner and rope. See KARABINER. [<F, ? <Sp., a little horn; ult. origin unknown]

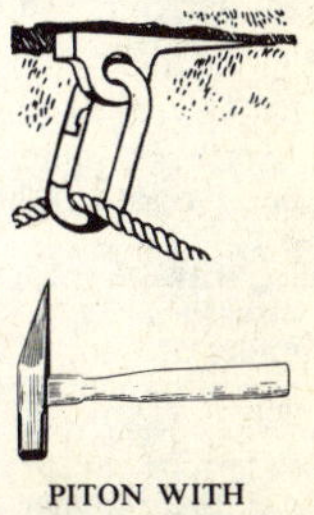

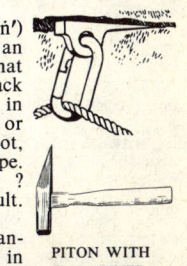

PITON WITH KARABINER AND PITON HAMMER

piton hammer A short–handled hammer for driving in pitons.

Pi·tot tube (pē·tō′) **1** A device for measuring the velocity of a fluid flow, consisting of a narrow

bent tube with its opening against the current and its upper portion above the surface of the fluid. 2 Any similar device for measuring pressure or pressure differences. [after Henri *Pitot*, 1695–1771, French hydraulic engineer]

pit saw A two-handled saw for cutting logs over the mouth of a pit, one man standing in the pit.

Pitt (pit), **William,** 1708–78, first Earl of Chatham, English statesman. — **William,** 1759–1806, English prime minister 1784–1801, 1804–1806; son of the preceding: called "William Pitt the Younger."

pit·tance (pit′əns) *n.* 1 A small allowance of money. 2 Any meager income or remuneration. [<OF *pitance*, orig. a monk's food allotment, pity <Med. L *pietantia* <L *pietas, -tatis.* See PIETY.]

pit·ter-pat·ter (pit′ər·pat′ər) *n.* A rapid series of light sounds or taps. [Varied reduplication of PATTER[1]]

Pitts·burgh (pits′bûrg) A city at the confluence of the Allegheny, Monongahela, and Ohio rivers in SW Pennsylvania.

Pittsburg Landing A village in SW Tennessee; scene of the Civil War battle of Shiloh, 1862.

pit·ty-pat (pit′ē·pat′) See PIT-A-PAT.

pi·tu·i·tar·y (pi·to͞o′ə·ter′ē, -tyo͞o′-) *adj.* 1 Secreting mucus; mucous. 2 Of or pertaining to the pituitary gland. — *n. pl.* **·tar·ies** 1 The pituitary gland. 2 Any of various preparations made from extracts of the anterior or posterior lobe of the pituitary gland. [<L *pituitarius* pertaining to mucus <*pituita* phlegm <*sputus*, pp. of *spuere* spit]

pituitary gland *Anat.* A small, rounded body at the base of the brain in vertebrates, consisting of an anterior and a posterior lobe, which secretes hormones having a wide range of effects upon the growth, metabolism, and other functions of the body; the hypophysis cerebri. Also **pituitary body.**

pi·tu·i·tous (pi·to͞o′ə·təs, -tyo͞o′-) *adj.* Containing, due to, resembling, or discharging mucus. [<L *pituitosus* <*pituita* phlegm. See PITUITARY.]

pit viper See under VIPER.

pit·y (pit′ē) *n. pl.* **pit·ies** 1 The feeling of grief or pain awakened by the misfortunes or sorrows of others; compassion. 2 That which arouses compassion; misfortune. — *v.t.* & *v.i.* **pit·ied, pit·y·ing** To feel compassion or pity (for). [<OF *pitet, pitié* <LL *pietas, -tatis.* Doublet of PIETY.] — **pit′i·er** *n.* — **pit′y·ing·ly** *adv.*

Synonyms (noun): commiseration, compassion, condolence, mercy, sympathy, tenderness. *Pity* is a feeling of grief or pain aroused by the weakness, misfortunes, or distresses of others, joined with a desire to help or relieve. *Sympathy* (feeling or suffering with) implies some degree of equality, kindred, or union; *pity* is for what is weak or unfortunate; hence, *pity* is often resented where *sympathy* would be welcomed. We have *sympathy* with one in joy or grief, in pleasure or pain, *pity* only for those in suffering or need. *Pity* may be only in the mind, but *mercy* does something for those who are its objects. *Compassion*, like *pity*, is exercised only with respect to the suffering or unfortunate, but combines with the tenderness of *pity* the dignity of *sympathy* and the active quality of *mercy*. *Commiseration* is as tender as *compassion*, but more remote and hopeless; we have *commiseration* for sufferers whom we cannot reach or cannot relieve. *Condolence* is the expression of *sympathy*. See MERCY. *Antonyms*: barbarity, brutality, cruelty, ferocity, hard-heartedness, harshness, inhumanity, mercilessness, pitilessness, rigor, ruthlessness, severity, sternness, truculence.

pit·y·ri·a·sis (pit′i·rī′ə·sis) *n.* 1 *Pathol.* A skin disease in which the epidermis sheds thin scales as dandruff. 2 A disease of domestic animals characterized by dry scales. [<NL <Gk. <*pityron* bran, scale]

più (pyo͞o) *adv. Music* More: *più* allegro, faster; *più* forte, louder; *più* lento, slower; *più* piano, softer. [<Ital. *più* <L *plus*]

Pi·us (pī′əs) 1 A masculine personal name. 2 Appellation of 12 popes. [<L, dutiful, devout]

— **Pius II,** 1405–64, real name Aeneas Silvius Piccolomini, pope 1458–64; diplomat, humanist, and historian.

— **Pius IV,** 1499–1565, real name Giovanni Angelo Medici, pope 1559–65; issued the *Tridentine Creed.*

— **Pius V,** 1504–72, real name Michele Ghislieri, pope 1566–72; promoted the Counter Reformation.

— **Pius VII,** 1742–1823, real name Luigi Barnaba Chiaramonti, pope 1800–23; crowned Napoleon I as emperor of France, was later imprisoned by him at Fontainebleau.

— **Pius IX,** 1792–1878, real name Giovanni Maria Mastai-Ferretti, pope 1846–78; lost temporal power to Victor Emmanuel, 1870.

— **Pius X,** 1835–1914, real name Giuseppi Melchiore Sarto, pope 1903–14; canonized in 1954.

— **Pius XI,** 1857–1939, real name Achille Ratti, pope 1922–39; signed treaty with Mussolini establishing Vatican City as a sovereign state and regulating the position of the Roman Catholic Church in Italy.

— **Pius XII,** 1876–1958, real name Eugenio Pacelli, pope 1939–58.

Pi·ute (pi·o͞ot′) See PAIUTE.

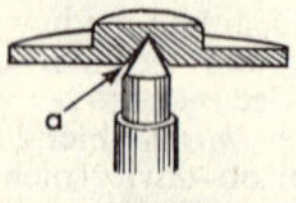

PIVOT
a. Bearing point.

piv·ot (piv′ət) *n.* 1 *Mech.* Something, typically a pin or a short shaft, upon which a related part turns, oscillates, or rotates: often a short cylindrical bearing, fixed on only one end, for carrying or rotating a swinging part. 2 Something on which an important matter hinges or turns; a turning point. 3 *Mil.* In wheeling troops, the soldier, officer, or point upon which the line turns. — *v.t.* To place on, attach by, or provide with a pivot or pivots. — *v.i.* To turn on a pivot; swing. [<F. Cf. Ital. *pivolo* a peg.] — **piv′ot·al** *adj.* — **piv′ot·al·ly** *adv.*

pix (piks) *n. pl. U. S. Slang* 1 Motion pictures. 2 Photographs. [Short for PICTURES]

pix·i·lat·ed (pik′sə·lā′tid) *adj.* 1 Affected by the pixies; mentally unbalanced; fey. 2 *Slang* Drunk. [Prob. alter. of dial. E (Cornish) *pixy-led* bewitched]

pix·y (pik′sē) *n. pl.* **pix·ies** A fairy or elf: also spelled *pyxie.* Also **pix′ie.** [<dial. E *pixey, pisky* <Scand. Cf. dial. Sw. *pysk, pyske* a small fairy, dwarf.]

Pi·zar·ro (pi·zär′ō, *Sp.* pē·thär′rō), **Francisco,** 1475?–1541, Spanish conqueror of Peru.

piz·za (pēt′sə, *Ital.* pēt′tsä) *n.* An Italian food comprising a doughy crust overlaid with a mixture of cheese, tomatoes, spices, etc., and baked. [<Ital., ? <dial. Ital. *picca* a pie]

piz·ze·ri·a (pēt′sə·rē′ə) *n.* A place where pizzas are prepared, sold, and eaten. [<Ital. <*pizza* a pizza]

piz·zi·ca·to (pit′sə·kä′tō, *Ital.* pēt′tsē·kä′tō) *Music adj.* Plucked: a direction to the performer that the notes for a bowed instrument are to be played with the fingers. — *n. pl.* **·ti** (-tē) A musical movement or phrase played by plucking the strings. [<Ital., orig. pp. of *pizzicare* pluck, pinch <*pizzare* <*picciare* peck <*picco* a beak]

PK See PSYCHOKINESIS.

pla·ca·ble (plā′kə·bəl, plak′ə-) *adj.* Appeasable; yielding; forgiving. [<OF <L *placabilis* <*placare* appease] — **pla′ca·bil′i·ty, pla′ca·ble·ness** *n.* — **pla′ca·bly** *adv.*

plac·ard (plak′ärd) *n.* 1 A printed or written paper publicly displayed, as a proclamation or poster. 2 A tag or plate bearing the owner's name.

— **pla·card** (plə·kärd′, plak′ärd) *v.t.* 1 To announce by means of placards. 2 To post placards on or in. 3 To display as a placard. [<OF *plackart* <*plaquier* plaster, lay flat <MFlemish *placken* bedaub, plaster]

pla·cate (plā′kāt, plak′āt) *v.t.* **·cat·ed, ·cat·ing** To appease the anger of; pacify. [<L *placatus*, pp. of *placare* appease] — **pla′cat·er** *n.* — **pla·ca′tion** *n.*

pla·ca·to·ry (plā′kə·tôr′ē, -tō′rē, plak′ə-) *adj.* Tending or intended to placate or appease. Also **pla′ca·tive.**

place (plās) *n.* 1 A particular point or portion of space, especially that part of space occupied by or belonging to a thing under consideration; a definite locality or location. 2 An occupied situation or building; space regarded as abode or quarters; an estate, town, military post, etc. 3 An open space or square in a city; also, a court or street. 4 Position in relative order; hence, station in life; degree; rank. 5 An office, appointment, or employment; also, rank, position, or station. 6 Room for occupation; hence, reception; welcome; lodgment; seat. 7 Room; stead; hence, precedence: One thing gives *place* to another. 8 A particular passage, as in a book; a text; a topic. 9 The second position among the first three competitors in a horse race. 10 The position of a figure in relation to the other figures of a given arithmetical series or group. — **in place** 1 In its natural position; also, in a suitable place, situation, job, etc. 2 In situ. — **in place of** In substitution or exchange for; instead of. — **out of place** Removed from or not situated in the natural or appropriate place, order, or relation; unsuitable; inappropriate; ill-timed. — **to go places** *Slang* To rise to success. — **to take place** To happen; occur. — *v.* **placed, plac·ing** *v.t.* 1 To put in a particular place or position. 2 To put or arrange in a particular relation or sequence. 3 To find a place, situation, home, etc., for. 4 To appoint to a post or office. 5 To identify; classify: Historians *place* him in the time of Nero. 6 To arrange for the satisfaction, handling, or disposition of: to *place* an order for a garbage truck. 7 To bestow or entrust: I *place* my life in your hands. 8 To invest, as funds. 9 To emphasize or resonate tones of (the voice) consciously, as in singing or speaking. — *v.i.* 10 In racing, to finish among the first three contestants; especially, to finish second. See synonyms under PUT, SET. ◆ Homophone: *plaice.* [<OF, ult. <L *platea* a wide street <Gk. *plateia (hodos)* <*platys* wide. Doublet of PIAZZA, PLAZA.]

Synonyms (noun): locality, location, part, position, post, room, site, situation, space, spot, station. See SCENE.

pla·ce·bo (plə·sē′bō) *n. pl.* **·bos** or **·boes** 1 In the Roman Catholic Church, the opening antiphon of the vespers for the dead. 2 *Med.* Any harmless substance given to humor a patient or as a test in controlled experiments on the effects of drugs. 3 Anything said to flatter or please. [<L *placebo* I shall please <*placere* please]

place kick In football, a kick for a goal in which the ball is placed on the ground for kicking.

place·man (plās′mən) *n. pl.* **·men** (-mən) An office-holder: often derogatory.

place·ment (plās′mənt) *n.* 1 The act of placing or the state of being placed. 2 In football, the putting of the ball in position for a place kick from the field; also, the kick itself.

pla·cen·ta (plə·sen′tə) *n. pl.* **·tas** or **·tae** (-tē) 1 *Anat.* In higher mammals, the vascular, spongy organ of interlocking fetal and maternal tissue by which the fetus is nourished in the uterus. 2 *Bot.* The part of the ovary that supports the ovules. [<NL *placenta (uterina)* (uterine) cake <L, a cake <Gk. *plakoeis, -oentos* a flat cake <*plax, plakos* a flat plate] — **pla·cen′tal, plac·en·tar·y** (plas′ən·ter′ē, plə·sen′tər·ē) *adj.*

pla·cen·tate (plə·sen′tāt) *adj.* Having a placenta.

plac·en·ta·tion (plas′ən·tā′shən) *n.* 1 *Biol.* **a** The process of fetal attachment to the uterus. **b** The type of placenta or manner of its construction. 2 *Bot.* The way in which the seeds are arranged in the pericarp of a plant, or the manner in which the placentas are attached.

Pla·cen·tia (plə·sen′shə) Ancient name for PIACENZA.

Pla·cen·tia Bay (plə·sen′shə) An inlet of the Atlantic extending NE 100 miles into SE Newfoundland.

plac·er[1] (plā′sər) *n.* One who or that which places.

plac·er[2] (plas′ər) *n. Mining* 1 An alluvial or glacial deposit of sand, gravel, etc., containing gold or other mineral in particles large enough to be obtained by washing. 2 Any place where deposits are washed for valuable minerals. [<Am. Sp. *placer* a deposit <Sp. *plaza* a place, ult. <L *platea.* See PLACE.]

placer digging 1 The act of obtaining minerals from deposits by washing. 2 A place where placer mining is carried on.

pla·cet (plā′sit) *n. Latin* 1 Literally, it pleases; permission given by authority; sanction. 2 A vote of assent, as by a council: expressed by saying the word *placet.*

pla·ce·ta (plä·thā′tä) *n. Spanish* A small garden

adjoining a building. Also **pla·ci·ta** (-thē′tä).

plac·id (plas′id) *adj.* Having a smooth, unruffled surface, as a sheet of still water; unruffled; calm. See synonyms under CALM, PACIFIC. [<L *placidus* pleasing <*placere* please] — **pla·cid·i·ty** (plə·sid′ə·tē), **plac′id·ness** *n.* — **plac′id·ly** *adv.*

plack (plak) *n.* A small copper coin, formerly current in Scotland. [Prob. <Flemish *placke,* a small coin of Brabant and Flanders <*plak* a flat disk. Doublet of PLAQUE.]

plack·et (plak′it) *n.* **1** The opening or slit in the upper part of a petticoat or skirt: also **placket hole. 2** A pocket in a woman's skirt. [? Var. of *placat,* var. of *placard,* in obs. sense of "a breastplate, top of a skirt"]

plac·oid (plak′oid) *adj.* Platelike, as the hard, spiny scales resembling teeth found on sharks and rays. — *n.* A fish having platelike scales; an elasmobranch. [<Gk. *plax, plakos* a flat plate + -OID]

pla·fond (plà·fôn′) *n. Archit.* **1** A flat or arched ceiling, decorated with painting or carving. **2** The under side of a projecting member (cornice, soffit, balcony, etc.); the under face of an architrave between columns, or of a staircase. **3** A painting on a ceiling. [<F <MF *platfond* <*plat* flat + *fond* bottom <L *fundus*]

pla·gal (plā′gəl) *adj. Music* Designating a cadence in which the tonic chord is preceded by the major or minor chord of the subdominant. [<Med. L *plagalis* <*plaga* a plagal mode <Med. Gk. *plagios (ēchos)* <Gk., oblique <*plagos* a side]

pla·gia·rism (plā′jə·riz′əm, -jē·ə-) *n.* The act of plagiarizing, or something plagiarized. — **pla′gia·rist** *n.* — **pla′gia·ris′tic** *adj.*

pla·gia·rize (plā′jə·rīz, -jē·ə-) *v.* **·rized, ·riz·ing** *v.t.* **1** To appropriate and pass off as one's own (the writings, ideas, etc., of another). **2** To appropriate and use passages, ideas, etc., from. — *v.i.* **3** To commit plagiarism. Also *Brit.* **pla′gia·rise.** — **pla′gia·riz′er** *n.*

pla·gia·ry (plā′jər·ē, -jē·ər·ē) *n.* **1** Plagiarism, the act or its result. **2** A plagiarist. [<L *plagiarius* a kidnapper, a plagiarist <L *plagium* a kidnapping <Gk. *plagios* oblique, treacherous]

plagio- *combining form* Oblique; slanting: *plagiotropism.* Also, before vowels, **plagi-.** [<Gk. *plagios* slanting]

pla·gi·o·clase (plā′jē·ə·klās′) *n.* Feldspar consisting chiefly of the silicates of sodium, calcium, and aluminum, and crystallizing in the triclinic system. [<PLAGIO- + Gk. *klasis* a cleavage] — **pla′gi·o·clas′tic** (-klas′tik) *adj.*

pla·gi·ot·ro·pism (plā′jē·ot′rə·piz′əm) *n. Bot.* Oblique geotropism, under the influence of which certain plant organs grow at an angle from the vertical. — **pla·gi·o·trop·ic** (plā′jē·ə·trop′ik), **pla′gi·ot′ro·pous** *adj.*

plague (plāg) *n.* **1** Anything troublesome or harassing, producing mental distress; affliction. **2** A pestilence or epidemic disease of man or animals, occurring in many forms and usually intensely malignant and contagious: bubonic *plague,* pulmonary *plague.* **3** Any great natural evil or calamity. **4** *Colloq.* Nuisance; bother. See synonyms under ABOMINATION. — *v.t.* **plagued, pla·guing 1** To harass or torment; vex; annoy. **2** To afflict with plague or disaster. [<OF *plage, plague* <LL *plaga* a pestilence <L, a stroke, prob. <dial. Gk. (Doric) *plaga* a stroke <*plag-,* root of *plēssein* strike]

pla·guy (plā′gē) *Colloq. adj.* Characterized by vexation or annoyance; troublesome. — *adv.* Vexatiously; intolerably. — **pla′gui·ly** *a*

plaice (plās) *n.* **1** A European flounder (*Pleuronectes platessa*). **2** One of various American flatfishes, as the summer flounder (*Paralichthys dentatus*). ◆ Homophone: *place.* [<OF *plaïz* <LL *platessa,* prob. <Gk. *platys* broad]

plaid (plad) *adj.* Having a tartan pattern; checkered. — *n.* An oblong woolen scarf of tartan or checkered pattern, worn in the Scottish Highlands as a cloak fastened over one shoulder; also, any fabric of this pattern. [< Scottish Gaelic *plaide* a blanket <*peallaid* a sheepskin <*peall* <L *pellis* a skin] — **plaid′ed** *adj.*

plain[1] (plān) *adj.* **1** Having no noticeable elevation or depression; flat; smooth. **2** Presenting few difficulties; easy. **3** Clear; understandable: *plain* English; also, straightforward; guileless. **4** Lowly in condition or station; unlearned. **5** Having no conspicuous ornamentation; unadorned; unvariegated; in the case of cloths, not figured or twilled. **6** Homely. **7** Not rich, as food. — *n.* An expanse of level, treeless land; a prairie. ◆ Homophone: *plane.* [<OF <L *planus* flat; *n.,* doublet of PLAN] — **plain′ly** *adv.* — **plain′·ness** *n.*

Synonyms (adj.): clear, distinct, explicit, intelligible, perspicuous, straightforward, transparent, unadorned, unambiguous, unequivocal. That is *clear* which offers no impediment to vision—is not dim, dark, or obscure. *Transparent* refers to the medium through which a substance is seen, *clear* to the substance itself, without reference to anything to be seen through it; we speak of a stream as *clear* when we think of the water itself; we speak of it as *transparent* with reference to the ease with which we see objects at the bottom. *Plain* is level to the thought, so that one goes straight on without difficulty or hindrance; as, *plain* language; a *plain* statement; a *clear* explanation. *Perspicuous* is often equivalent to *plain,* but *plain* never wholly loses the meaning of *unadorned,* so that we can say the style is *perspicuous* even if highly ornate, when we could not call it at once ornate and *plain.* See APPARENT, BLANK, CLEAR, EVIDENT, EXPLICIT, HORIZONTAL, LEVEL, MANIFEST, NOTORIOUS, RUSTIC, SMOOTH. *Antonyms*: see synonyms for EQUIVOCAL, OBSCURE.

plain[2] (plān) *v.i.* **1** *Dial.* To complain. **2** *Obs.* To mourn. ◆ Homophone: *plane.* [<OF *plaign-,* stem of *plaindre* <L *plangere* beat the breast, lament]

plain-clothes man (plān′klōz′, -klōthz′) A member of a police force not in uniform; specifically, a detective.

plain-deal·ing (plān′dē′ling) *adj.* Dealing frankly and sincerely. — *n.* Frankness; straightforwardness.

plain·dress (plān′dres′) *n.* A radiotelegraph message carrying the address either in plain text or in a cipher different from that used for the message. Compare CODRESS. [<PLAIN + (AD)DRESS]

plain-laid (plān′lād′) *adj.* Consisting of strands twisted together in the ordinary way: a *plain-laid* rope.

Plains Indian A member of any of the tribes of American Indians formerly inhabiting the Great Plains of North America, belonging variously to the Algonquian, Athapascan, Caddoan, Kiowan, Siouan, and Uto-Aztecan linguistic stocks, but having in common the nomadic culture of the plains and dependence on the buffalo: also called *Buffalo Indian.*

plains·man (plānz′mən) *n. pl.* **·men** (-mən) A dweller on the plains.

Plains of Abraham See ABRAHAM, PLAINS OF.

plain people *U.S.* The Amish, Mennonites, and Dunkers: so called from their plain dress.

plain song The old ecclesiastical chant, having simple melody, not governed by strict rules of time, but by accentuation of the words. Also **plain chant.** [Trans. of Med. L *cantus planus*]

plain-spo·ken (plān′spō′kən) *adj.* Plainly or frankly uttered: a *plain-spoken* promise; also, habitually frank.

plain·stanes (plān′stānz′) *n. Scot.* A pavement; flagstones, as opposed to cobbles. Also **plain′·stones′** (-stōnz′).

plaint (plānt) *n.* **1** Audible utterance of sorrow or grief; lamentation; a complaint. **2** In English law, a writ setting forth a grievance and asking redress. [<OF *plainte* <Med. L *plancta* <L, pp. fem. of *plangere* lament]

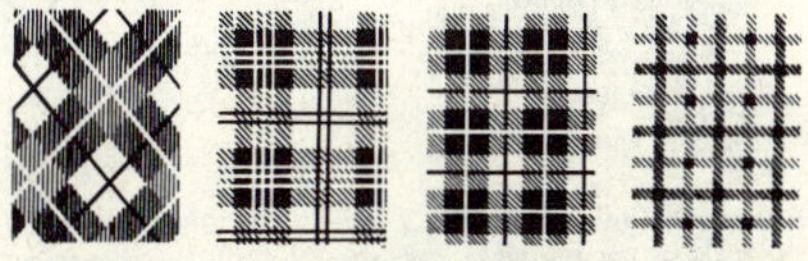

Argyle Blanket Tartan Tattersall

PLAIDS

plain text In cryptography, the original text of a message to be converted into or reconverted from a code or cipher cryptogram: also called *clear text.*

plain·tiff (plān′tif) *n.* The party that begins an action at law; the complaining party in an action. [<OF *plaintif, plaintive* plaintive <L *planctus.* See PLAINT.]

plain·tive (plān′tiv) *adj.* Expressing a subdued sadness; mournful. [<OF, fem. of *plaintif*] — **plain′tive·ly** *adv.* — **plain′tive·ness** *n.*

plait (plāt, *Brit.* plat) *v.t.* **1** To braid. **2** To pleat. **3** To make by pleating or braiding. — *n.* **1** A braid, especially of hair. **2** A pleat. [<OF *pleit* <L *plicitum* a folded thing, orig. pp. neut. of *plicare* fold]

plan (plan) *n.* **1** An arrangement of means or steps for the attainment of some object; a scheme; method; design. **2** A drawing showing the proportion and relation of parts, as of a building; any outline sketch; draft. **3** A mode of action. **4** One of a number of hypothetical planes perpendicular to the line of vision in which the size of the pictured object is increased or diminished proportionately to the distance from the eye at which they are interposed. See PERSPECTIVE. See synonyms under DESIGN, IDEA, PROJECT, PURPOSE, SKETCH. — *v.* **planned, plan·ning** *v.t.* **1** To form a scheme or method for doing, achieving, etc. **2** To make a plan of, as a building; design. **3** To have as an intention or purpose. — *v.i.* **4** To make plans. [<OF, a plane (surface), a ground plan <Ital. *plano* <L *planus* flat. Doublet of PLAIN[1], *n.*] — **plan′·ner** *n.*

Synonyms (verb): concoct, contrive, design, devise, invent, plot, project, propose, purpose, scheme, sketch. Compare BREW, PROPOSE.

pla·nar (plā′nər) *adj.* Of or pertaining to a plane; lying in one plane; flat. [<L *planaris* <*planum* a plane <*planus* flat]

pla·nar·i·an (plə·nâr′ē·ən) *n. Zool.* Any turbellarian, chiefly aquatic flatworm, usually dark-colored, having a body covered with cilia: a few species dwell in moist places upon land, and others are parasitic in or upon holothurians. [<NL *Planaria,* genus name of the flatworm <L, fem. of *planarius* flat <*planus*]

planch (planch, plänch) *n.* A plank; board. Also **planche.** [<OF *planche* <LL *planca*]

planch·et (plan′chit) *n.* A piece of metal ready to receive an impression. [Dim. of PLANCH, in sense "a flat plate of metal"]

plan·chette (plan·chet′, -shet′) *n.* **1** A small board, usually resting on a vertical pencil and two casters; believed by some to spell out messages, as on a ouija board, when the fingers are rested lightly upon it, independently of the volition of the persons touching it: used in the investigation of psychic phenomena. [<F, dim. of *planche* a plank]

Planck (plängk), **Max,** 1858–1947, German physicist; developed the quantum theory.

Planck's constant *Physics* The quantum of action (symbol *h*); a universal constant having the value of approximately 6.624 x 10^{-27} erg second. For any specified radiation, the magnitude of the energy emitted is given by the product *hv,* where *v* is the frequency of the radiation in cycles per second.

Plan·çon (plän·sôn′), **Pol,** 1854–1914, French bass singer.

plane[1] (plān) *n.* **1** *Geom.* A surface such that a straight line joining any two of its points lies wholly within the surface. **2** Hence, any flat or uncurved surface. **3** A grade of development; stage; level, as of thought, knowledge, rank, etc. **4** *Aeron.* A supporting surface of an airplane: often used in combination: *monoplane.* **5** An airplane. — *adj.* **1** Lying in a plane; level; flat. **2** Having a flat surface; dealing only with flat surfaces: *plane* geometry. See synonyms under HORIZONTAL, LEVEL, SMOOTH. ◆ Homophone: *plain.* [Var. of PLAIN[2], *n.*; refashioned after L *planus* flat]

plane[2] (plān) *n.* **1** A tool used for smoothing boards or other surfaces of wood. **2** A trowel-like tool for striking off clay that projects above the mold. — *v.* **planed, plan·ing** *v.t.* **1** To make smooth or even with or as with a plane. **2** To remove with or as with a plane. — *v.i.* **3** To use a plane. **4** To do the work of a plane. ◆ Homophone: *plain.* [<MF *plane* <OF *plain* <LL *plana* a plane <*planare* plane <L *planus* flat]

plane (plān) *n.* A plane tree. ◆ Homophone: *plain.* [<OF *plane* <*plasne* <L *platanus* <Gk. *platanos* <*platys* broad; because of its broad leaves]

plane⁴ (plān) *v.i.* **planed, plan·ing** 1 To rise partly out of the water, as a power boat when driven at high speed. 2 To glide; soar. 3 To travel by airplane. ◆ Homophone: *plain.* [<F *planer* <*plan* a plane <L *planus* flat]

plan·er (plā'nər) *n.* 1 A machine for planing wood or metal. 2 A smooth wooden block used for leveling a form of type, etc. 3 One who or that which planes. [<PLANE²]

planer tree A small tree (*Planera aquatica*) allied to and resembling the elms, but with nutlike wingless fruit and small ovate leaves, growing in wet places in the southern United States; the water elm. [after J.J. *Planer*, 1743–1789, German botanist]

plane sailing *Nav.* A system for ascertaining a vessel's position on the supposition that the earth's surface is plane, not spherical.

plane-sheer (plān'shir) See PLANK-SHEER.

plan·et (plan'it) *n.* 1 *Astron.* One of the non-self-luminous bodies of the solar system revolving around the sun as their center of motion. Those within the Earth's orbit, Mercury and Venus, are called **inferior planets**; those beyond it, the **superior planets**, are Mars, the asteroids or planetoids known collectively as **minor planets**, Jupiter, Saturn, Uranus, Neptune, and Pluto. 2 In ancient astronomy, one of the seven heavenly bodies (the Sun, Moon, Mercury, Venus, Mars, Jupiter, and Saturn) that have an apparent motion among the fixed stars. 3 One of these bodies considered in relation to its supposed influence on human beings and their affairs. [<OF *planete* <LL *planeta* <Gk. (*asteres*) *planētai* wandering (stars) <*planaesthai* wander] at the same time rotating axially. 5 In astrology, under the influence or domination of some of the planets. [<LL *planetarius* an astrologer <*planeta* PLANET]

plan·e·tes·i·mal (plan'ə·tes'ə·məl) *Astron. adj.* Of or pertaining to the small, solid, planetary bodies of space. — *n.* Any of such bodies resembling a meteorite in composition and revolving around a larger mass as a planet revolves around the sun. [<PLANET + (IN-FINIT)ESIMAL]

planetesimal hypothesis *Astron.* The hypothesis that the solar system developed from large masses of planetesimals which, by the crossing of orbits, gravity, and electric attraction, coalesced gradually, without extreme intensity of heat.

plan·et·fall (plan'it·fôl') *n.* The descent of a rocket or artificial satellite to the surface of a planet.

plan·e·toid (plan'ə·toid) *n.* An asteroid. — **plan'e·toi'dal** *adj.*

plane tree Any tree of the genus *Platanus* characterized by broad, lobed leaves and spreading growth, as the sycamore or buttonwood. [See PLANE³]

plan·et·struck (plan'it·struk') *adj.* Affected by the influence of planets; blasted; moon-struck. Also **plan'et-strick'en** (-strik'ən).

planet wheel One of the smaller wheels in an epicyclic train.

plan·gent (plan'jənt) *adj.* Dashing noisily; resounding; reverberating, as the sound of bells; plaintive, as certain qualities of the human voice. [<L *plangens, -entis,* ppr. of *plangere* lament, mourn] — **plan'gen·cy** *n.* — **plan'gent·ly** *adv.*

plan·gor·ous (plang'gər·əs) *adj.* Wailing; moaning; lamenting. [<L *plangor* lamentation <*plangere.* See PLANGENT.]

TABLE OF MAJOR PLANETS

Name	Symbol	Distance from sun: millions of miles	Mean diameter: miles	Period of sidereal revolution	Period of rotation	No. of satellites	Mass: Earth considered as 1.	Escape velocity: in miles per second.
Mercury	☿	36	3,000	88 days	88 days?	0	0.0543	2
Venus	♀	67	7,600	225 "	20–30 d.	0	0.8136	6.3
Earth	⊕	93	7,918	365.25 d.	23 h. 56 m.	1	1.0000	6.95
Mars	♂	142	4,200	687 days	24 h. 37 m.	2	0.1069	3.1
Jupiter	♃	483	87,000	12 years	9 h. 50 m.	12	318.35	37.
Saturn	♄	886	72,000	29.5 "	10 h. 14 m.	9	95.3	22.
Uranus	♅	1780	33,200	84 "	10 h. 45 m.	5	14.58	13.
Neptune	♆	2790	31,000	165 "	15 h. 48 m.	2	17.26	15.
Pluto	♇	3670	4,000	248 "	?	0	.1?	?

plane·ta·ble (plān'tā'bəl) *n.* A surveying instrument used in mapping in the field. [<PLANE¹ + TABLE]

plan·e·tar·i·um (plan'ə·târ'ē·əm) *n. pl.* **·tar·i·ums** or **·tar·i·a** 1 An apparatus for exhibiting the features of the heavens as they exist at any time and for any place on earth, consisting of a suitably mounted projector installed in a room having a circular dome. 2 An apparatus or model representing the planetary system. [<NL <LL *planetarius* PLANETARY]

plan·e·tar·y (plan'ə·ter'ē) *adj.* 1 Of or pertaining to a planet or the planets; the *planetary* bodies. 2 Mundane; terrestrial. 3 Having the character anciently ascribed to the planets; wandering; erratic: a *planetary* career. 4 *Mech.* Pertaining to or noting a type of gearing in which one or more small wheels mesh with the toothed circumference of a larger wheel, around which they revolve,

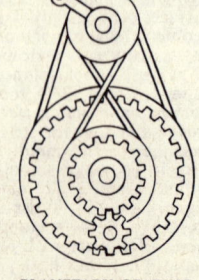

PLANETARY GEARING

pla·ni·cop·ter (plā'ni·kop'tər) *n.* A convertiplane. [<(AIR)PLAN(E) + -i- + (HELI)COPTER]

pla·ni·form (plā'nə·fôrm, plan'ə-) *adj.* Having the surfaces nearly flat.

pla·nim·e·ter (plə·nim'ə·tər) *n.* An instrument for measuring the area of any plane surface, however irregular, by moving a pointer around its boundary and reading the indications of a scale. [<F *planimètre*] — **pla·ni·met·ric** (plā'nə·met'rik, plan'ə-) or **·ri·cal** *adj.* — **pla·nim'e·try** *n.*

PLANIMETER

plan·ish (plan'ish) *v.t.* To condense, smooth, toughen, or polish, as metal, by hammering, rolling, etc. [<MF *planiss-,* stem of *planir* flatten < *plan* flat <L *planus* flat]

plan·i·sphere (plan'ə·sfir) *n.* A plane projection of the sphere; especially, a polar projection of the heavens on a chart, which shows the stars visible at a given place and time. [<OF *planisphère* <Med. L *planisphaerium* <L *planus* flat + *sphaera* a sphere]

plank (plangk) *n.* 1 A broad piece of sawed timber, thicker than a board. 2 Timber when sawed into planks. 3 Anything that sustains or upholds; a support. 4 One of the principles of a political platform. — **to walk the plank** To walk off a plank projecting from the side of a ship: a method once used by pirates for executing prisoners. — *v.t.* 1 To cover, furnish, or lay with planks. 2 To broil or bake and serve on a plank, as fish. 3 *Colloq.* To put down emphatically or forcibly. 4 *Colloq.* To pay: with *out, down,* etc. [<AF *planke* <LL *planca* a board, slab, prob. <Gk. *plax, plakos* flat]

plank·ing (plangk'ing) *n.* 1 The act of laying planks; also, anything made of planks. 2 Planks collectively.

plank-sheer (plangk'shir') *n. Naut.* A timber extending around a vessel's deck, covering and fastening the timberheads: also called *plane-sheer.* [Var. of *planesheer,* alter. of *plancher* <OF *planchier* planking, floor < *planche* PLANCH]

plank·ton (plangk'tən) *n. Biol.* The floating, weakly swimming or drifting plant or animal organic life of the sea, as distinguished from the coastal or the bottom forms: used also of analogous life forms in fresh-water lakes. Compare BENTHOS. [<G <Gk., neut. of *planktos* drifting, wandering < *plazesthai* wander] — **plank·ton·ic** (plangk·ton'ik) *adj.*

plano-¹ *combining form* Roaming; wandering; *planoblast.* Also, before vowels, **plan-**. [<Gk. *planos* wandering]

plano-² *combining form* Flat; level; plane: *plano-convex:* also, before vowels, **plan-**. Also **plani-**. [<L *planus* flat]

plan·o·blast (plan'ə·blast) *n. Zool.* The free-swimming or medusa form of a hydroid. [<PLANO-¹ + Gk. *blastos* a sprout]

pla·no·con·cave (plā'nō·kon'kāv) *adj.* Flat or plane on one side and concave on the other. See illustration under LENS. [<PLANO-² + CONCAVE]

pla·no·con·i·cal (plā'nō·kon'i·kəl) *adj.* Plane on one side and conical on the other. [<PLANO-² + CONICAL]

pla·no·con·vex (plā'nō·kon'veks) *adj.* Flat or plane on one side and convex on the other. See illustration under LENS. [<PLANO-² + CONVEX]

pla·nom·e·ter (plə·nom'ə·tər) *n.* A device for gaging a plane surface, such as a surface plate used in metalworking. [<PLANO-² + -METER] — **pla·nom'e·try** *n.*

plant (plant, plänt) *n.* 1 A living organism belonging to the vegetable, as distinguished from the animal kingdom, having typically rigid cell walls, promoting an indefinite growth of tissue, and characterized by growth from the synthesis of simple, usually inorganic food materials from soil, water, and air or, in some cases, from other organisms. 2 Loosely, one of the smaller forms of vegetable life, in distinction from shrubs and trees. 3 A set of machines, tools, apparatus, etc., necessary to conduct a manufacturing enterprise or other business: a chemical *plant:* often including the buildings and grounds, or, in case of a railroad, the rolling stock, but not including the material or product; hence, the permanent appliances needed for any institution, as a post office, a college, etc. 4 A sapling; a slip or cutting from a tree or bush. 5 *Slang* A trick; dodge; imposition; swindle. 6 A person placed in a theater audience to encourage applause, speak lines, or contribute to the action of a play. 7 An apparently trivial passage early in a story or play that later becomes important in shaping the outcome of the action. — *v.t.* 1 To set in the ground for growing. 2 To furnish with plants or seed: to *plant* a field. 3 To set or place firmly; put in position. 4 To found; establish. 5 To introduce into the mind; implant, as an idea or principle. 6 To introduce into a country, as a breed of animal. 7 To deposit (fish or spawn) in a body of water. 8 To stock, as a river. 9 To bed (oysters). 10 *Slang* To deliver, as a blow. 11 *Slang* To place or station for purposes of deception, observation, etc.: to plant evidence. 12 *Slang* To hide; bury. [OE *plante* <L *planta* a sprout, something planted] Synonyms (*verb*): seed, set, sow. We *set* or *set out* slips, cuttings, young trees, etc., but we may also be said to *plant* them; we plant

corn, potatoes, etc., which we put in definite places, as in hills; we *sow* wheat or other small grains and seeds which are scattered in the process. Notwithstanding the fact that by modern agricultural machinery the smaller grains are almost as precisely *planted* as corn, the old word for broadcast scattering is retained. Land is *seeded* to grass. See SET. *Antonyms*: eradicate, extirpate, uproot.

Plan·tag·e·net (plan·taj′i·net) A patronymic of the English sovereigns from Henry II (1154) to the accession of the House of Tudor (1485): from the sprig of broom (in Medieval Latin, *planta genista*) worn by Geoffrey of Anjou, founder of the line. See table under ENGLAND.

plan·tain[1] (plan′tin) *n.* An annual or perennial herb (genus *Plantago*) widely distributed in temperate regions; especially, the **common** or **greater plantain** *(P. major)* with large, ovate, or oval, ribbed leaves. [<OF <L *plantago, -ginis* <*planta* sole of the foot; with ref. to its broad, flat leaves]

plan·tain[2] (plan′tin) *n.* A tropical perennial herb *(Musa paradisiaca)*; also, its edible, bananalike fruit. [Earlier *plantan* <Sp. *plátano, plántano* <Cariban *balatanna*; infl. in form by PLANTAIN[1]]

plan·tar (plan′tər) *adj.* Pertaining to the sole of the foot. See illustration under FOOT. [<L *plantaris* <*planta* sole of the foot]

plant association *Ecol.* A group of plants (including several species) in a certain area, needing similar nourishment and growing conditions, and taking on a similar aspect.

plan·ta·tion (plan·tā′shən) *n.* **1** Any place that is planted; especially, a farm or estate of many acres in the southern United States planted in cotton, tobacco, rice, or sugarcane, and formerly worked by slave labor. **2** A colony. **3** An oyster bed or oyster farm. **4** A grove cultivated for its wood. **5** The act of planting. **6** In Maine, a minor civil division. [<L *plantatio, -onis* a planting <*plantare* plant]

plant·er (plan′tər) *n.* **1** One who plants. **2** An early settler or colonizer. **3** An owner of a plantation. **4** An agricultural implement for dropping seed in soil. **5** A decorative container in which shrubs and flowers are planted, especially outdoors.

plan·ti·grade (plan′tə·grād) *adj.* Walking on the whole sole of the foot, as men, bears, etc. —*n.* A plantigrade animal. Compare DIGITIGRADE. [<F <NL *plantigradus* <L *planta* the sole of the foot + *gradi* walk]

plant louse **1** A small insect (family *Aphididae*) which infests plants and sucks the juices from leaves and stalks; an aphid. See illustration under INSECTS (injurious). **2** A leaping insect (family *Psyllidae*) of similar habits.

plan·u·la (plan′yə·lə) *n. pl.* **·lae** (-lē) *Zool.* The free-moving, ciliated embryo of certain coelenterates, as the hydroids. [<NL <L, dim. of *planus* flat] —**plan′u·lar, plan′u·late** (-lit, -lāt) *adj.*

plaque (plak) *n.* **1** A plate, disk, or slab of metal, porcelain, ivory, etc., artistically ornamented, as for wall decoration. **2** A brooch, etc. [<F <MDu. *placke* flat disk, tablet. Doublet of PLACK.]

plash[1] (plash) *n.* A slight splash. —*v.t.* & *v.i.* To splash lightly, as water. [Prob. imit.] —**plash′y** *adj.*

plash[2] (plash) *n.* A small pool. [OE *plæsc* a pool]

plash[3] (plash) *v.t.* **1** To bend down and interweave, as twigs or branches, so as to form a hedge or arbor. **2** To form or trim (a hedge) in this manner. [<OF *plaissier* <L *plectere* weave] —**plash′er** *n.*

-plasia *combining form* Growth; development; formative action: *heteroplasia*: also spelled *-plasy.* Also **-plasis.** [<Gk. *plasis* a molding <*plassein* make, form]

-plasm *combining form Biol.* The viscous material of an animal or vegetable cell: *protoplasm.* [<Gk. *plasma* figure, form <*plassein* mold, make]

plas·ma (plaz′mə) *n.* **1** The liquid portion of nutritive animal fluids, as blood, lymph, or intercellular fluid. **2** The clear, fluid portion of blood, freed from blood cells and used for transfusions. **3** The viscous material of a cell; protoplasm. **4** A green, translucent variety of chalcedony, used among the Romans as a gem. **5** *Physics* That region in a gas-discharge tube which is rendered nearly neutral by the presence of approximately equal numbers of positive ions and electrons. [<LL, a molded thing <Gk. *plasma* <*plassein* mold, form] —**plas·mat·ic** (plaz·mat′ik), **plas′mic** *adj.*

plas·ma·gene (plaz′mə·jēn) *n.* A counterpart of the gene present in the cytoplasm of the cell and subject to environmental influences: believed to play a significant role in the processes of inheritance. [<PLASMA + *-gene,* var. of -GEN]

plasma gun *Physics* A magnetohydrodynamic apparatus for the generation of plasma and its forcible expulsion as a plasma jet.

plasma jet *Physics* A beam of plasma ejected from a specially constructed generator which utilizes the pinch effect in forming a brilliantly luminous jet of extremely high energy and temperature.

plasmo- *combining form* Plasma; of or pertaining to plasma: *plasmolysis.* Also, before vowels, **plasm-.** [See -PLASM]

Plas·mo·chin (plaz′mə·kin) *n.* Proprietary name of a white to pale-yellow, tasteless powder, $C_{19}H_{20}N_3O$, synthesized from quinoline, used as an antimalarial drug.

plas·mo·di·um (plaz·mō′dē·əm) *n. pl.* **·di·a** (-dē·ə) **1** A mobile, naked, slimy mass of protoplasm resulting from the fusion of ameboid organisms, typical of the slime molds. **2** A malaria parasite. [<NL <Gk. *plasma* + *eidos* form]

Plas·mo·di·um (plaz·mō′dē·əm) *n.* A genus of protozoan blood parasites (class *Sporozoa*) which includes the causative agents of malaria in man and animals, especially *P. vivax, P. malariae,* and *P. falciparum.*

plas·moid (plaz′moid) *n. Physics* A small particle of plasma ejected from a plasma gun and capable of reacting as a unit under the influence of a wide range of magnetic, electrical, and thermal forces.

plas·mol·y·sis (plaz·mol′ə·sis) *n. Biol.* The process of withdrawing water from the protoplasm of a cell, resulting in shrinkage of the cell body.

plas·mo·lyze (plaz′mə·līz) *v.t.* & *v.i.* **·lyzed, ·lyz·ing** To subject to or undergo plasmolysis.

plas·mo·some (plaz′mə·sōm) *n. Biol.* The acid-staining nucleolus in the nucleus of a cell. [<PLASMO- + Gk. *sōma* a body]

Plas·sey (plas′ē) A village north of Calcutta in West Bengal, India; scene of Clive's victory over the nawab of Bengal, 1757.

-plast *combining form* An organized living particle or cell: *protoplast.* [<Gk. *plastos* formed <*plassein* form]

plas·te·in (plas′tē·in, -tēn) *n. Biochem.* Any of a class of proteins formed in the presence of pepsin during digestion. [<G *plastein* <Gk. *plastos* formed]

plas·ter (plas′tər, pläs′-) *n.* **1** A composition of lime, sand, and water for coating walls and partitions. **2** Calcined gypsum for making sculptor's casts, etc.; plaster of Paris. **3** A viscid substance spread on linen, silk, etc., and applied to some part of the body: used for healing purposes. —*v.t.* **1** To cover or overlay with or as with plaster. **2** To apply a plaster to, as a boil or part of the body. **3** To apply like plaster or a plaster: to *plaster* posters on a fence. **4** To cause to adhere or lay flat like plaster. **5** *Slang* To strike with great force or effect. [OE <LL *plastrum* <L *emplastrum* <Gk. *emplastron, emplaston* <*en-* on + *plassein* daub, mold; defs. 1, 2, and 4 reborrowed in ME from cognate OF *plastre*] —**plas′ter·er** *n.* —**plas′ter·ing** *n.*

plaster cast **1** A cast or model of a person or object made by molding plaster of Paris. **2** *Surg.* An application of gauze stiffened with plaster of Paris, applied to an injured or broken part of the body to prevent movement, allow knitting of bones, etc.

plas·ter·ing (plas′tər·ing, pläs′-) *n.* **1** Applying or working with plaster. **2** A coating of plaster.

plaster of Paris Calcined gypsum: mixed with water it sets readily and is useful in making molds, casts, bandages, etc.

plas·ter·y (plas′tər·ē, pläs′-) *adj.* Like plaster; viscid.

plas·tic (plas′tik) *adj.* **1** Giving form or fashion to matter. **2** Capable of being molded; pliable. **3** Pertaining to modeling or molding; sculptural. **4** *Surg.* Efficacious or instrumental in recreating or remodeling injured or destroyed protoplasm; also, capable of being thus renewed. —*n.* **1** Anything moldable; specifically, any material, natural or synthetic, which may be fabricated into a variety of shapes, usually by the application of heat and pressure. **2** *Chem.* One of a class of organic compounds synthesized from hydrocarbons, proteins, cellulose, or resins, capable of being molded, extruded, cast, or otherwise fabricated into various shapes: usually in the plural, **plas′tics.** [<L *plasticus* <Gk. *plastikos* moldable <*plastos* formed <*plassein* form, mold] —**plas′ti·cal·ly** *adv.*

-plastic *combining form* Growing; developing; forming: *cytoplastic.* [<Gk. *plastikos* plastic, moldable]

plas·tic·i·ty (plas·tis′ə·tē) *n.* **1** Plastic quality. **2** *Physics* The ability of certain bodies to exhibit a continuous change of shape under suitable distorting forces. **3** Capacity for mental or spiritual molding.

plas·ti·cize (plas′tə·sīz) *v.t.* & *v.i.* **·cized, ·ciz·ing** To make or become plastic.

plas·ti·ciz·er (plas′tə·sī′zər) *n.* **1** That which functions to make a substance plastic. **2** *Chem.* Any of a class of substances adapted to preserve the softness and flexibility of materials to which they are added.

plastic surgery Surgery that deals with the restoration or healing of lost, wounded, or deformed parts of the body.

plas·tid (plas′tid) *n. Biol.* **1** An elementary organism, as a cell. **2** Any permanent organ of the cell situated in the cytoplasm. [<G *plastiden,* pl. <Gk. *plastides,* fem. pl. of *plastēs* a molder <*plastos.* See PLASTIC.]

plas·ti·sol (plas′tə·sôl, -sol) *n.* A suspension of finely divided resin particles in a plasticizer: useful in the application of plastic coatings to various surfaces. [<PLASTI(C) + SOL[4]]

plas·to·mer (plas′tə·mər) *n. Chem.* Any of a group of polymerized, thermosetting plastics, as Celluloid or the acrylic resins. [<PLAST(IC) + (POLY)MER]

plas·tom·e·ter (plas·tom′ə·tər) *n.* An instrument for measuring the plasticity of a substance or material. [<*plasto-* (<PLASTICITY) + -METER]

plas·tron (plas′trən) *n.* **1** An ornamental addition to the front of a woman's dress, reaching from the throat to the waist. **2** A leather shield worn on the breast by fencers. **3** *Zool.* The under or ventral part of the shell or armor of a turtle or tortoise: also **plas′trum.** **4** The starched bosom of a man's shirt. **5** An iron breastplate worn under a coat of mail. [<F, orig. a breastplate <Ital. *piastrone,* aug. of *piastra* a breastplate] —**plas′tral** *adj.*

-plasty *combining form Med.* An operation in plastic surgery involving: **a** A (specified) part of the body: *osteoplasty.* **b** Tissue from a (specified) source: *zooplasty.* **c** A (specified) process or formation: *neoplasty.* [<Gk. *-plastia* formation <*plastos.* See -PLAST.]

-plasy See -PLASIA.

plat[1] (plat) *v.t.* **plat·ted, plat·ting** To plait or braid. —*n.* A plait. [ME *platten,* var. of *playten* <OF *pleit* PLAIT]

plat[2] (plat) *n.* **1** A small piece of ground; a plot. **2** A plotted map, chart, or plan. —*v.t.* To make a plot or plan of. [Var. of PLOT; infl. in form by obs. *plat* a flat thing or area <OF. See PLATE.]

plat- Var. of PLATY-.

Pla·ta (plä′tä), **La** See LA PLATA (def. 1).

Pla·ta (plä′tä), **Río de la** The estuary of the Paraná and Uruguay rivers, extending 170 miles SE between Uruguay and Argentina to the Atlantic where it is 140 miles wide; scene of a British naval engagement with Germany in World War II, 1939. Also **River Plate.**

Pla·tae·a (plə·tē′ə) An ancient city NW of Athens, Greece; scene of a Spartan and Athenian victory over the Persians, 479 B.C. Also **Pla·tae′ae** (-tē′ē). —**Pla·tae′an** *adj.*

add,āce,câre,pälm; end,ēven; it,īce; odd,ōpen,ôrder; to͝ok,po͞ol; up,bûrn; ə = a in *above,* e in *sicken,* i in *clarity,* o in *melon,* u in *focus;* yo͞o = u in *fuse;* oi,oil; ou,pout; ch,check; g,go; ng,ring; th,thin; th,this; zh,vision. Foreign sounds à,œ,ü,kh,ṅ; and ◆: see page xx. <from; + plus; ? possibly.

plat·an (plat′ən) *n.* The plane tree. Also **plat′·ane** (-ən). [<L *platanus* PLANE[3]]

plate (plāt) *n.* **1** A flat, extended, rigid body of metal or any material of slight but even thickness. **2** Metal in sheets. **3** A shallow vessel, formerly often of wood or pewter, now usually of crockery, in which food is served or from which it is eaten at table. **4** Articles of household service, as goblets, tea sets, etc., made originally of precious metals, but now largely of base metal coated with precious metals. **5** A portion of food served at table; a dish; a plateful; also, a whole course served on one plate. **6** A cup or other article of silver or gold offered as a prize in a race or other contest. **7** A piece of flat metal bearing a design or inscription, either for use in that form, as in a doorplate or coffin plate, or intended for reproduction by stamping, printing, or otherwise, as in a bookplate; also, an impression from a plate of the latter kind. **8** An electrotype or stereotype. **9** A horizontal timber laid on a wall to receive a framework. **10** *Dent.* A piece of metal, vulcanite, or plastic, fitted to the mouth and holding one or more artificial teeth. **11** Plate armor. **12** A thin part of the brisket of beef. **13** *Phot.* A sensitized sheet of glass, metal, or the like, for taking photographs. **14** In baseball, the home base, a flat, pentagonal figure, 12 inches in diameter and usually of hard white rubber level with the surface of the diamond. **15** *Biol.* A lamina; a lamella. **16** A dish like a table plate used in taking up collections, as in churches; also, a collection. **17** A hinge. See illustration under HINGE. **18** The principal anode in a vacuum tube. **19** *Obs.* A piece of silver money. — *v.t.* **plat·ed, plat·ing 1** To coat with a thin layer of gold, silver, etc. **2** To cover or sheathe with metal plates for protection. **3** In papermaking, to give a high gloss to (paper) by pressure between metal plates. **4** *Printing* To make an electrotype or stereotype from. [<OF, a plate of metal, orig. fem. of *plat* flat <LL *plattus,* prob. <Gk. *platys* broad, flat]

plate armor *Mil.* **1** Defensive armor of strong metallic plates for protecting ships or fortifications against artillery. **2** Formerly, defensive armor for the person made of overlapping plates, in distinction from chain or mail. See ARMOR.

pla·teau (pla·tō′, *esp. Brit.* plat′ō) *n. pl.* **·teaus** or **·teaux** (-tōz′) **1** An extensive stretch of elevated and comparatively level land; tableland; mesa. **2** A broad, low stand for table decorations; also, a decorative plaque. **3** *Psychol.* A relatively level portion in the curve indicating a subject's rate of learning; also, the condition it typifies. **4** *Mil.* A device for making a rough preliminary setting on certain gunsights. [<F <OF *platel* a flat piece of metal or wood, orig. dim. of *plat.* See PLATE.]

plat·ed (plā′tid) *adj.* **1** Provided with plates of metal, as for defense. **2** Coated with a layer of silver, tin, etc. **3** Having one kind of yarn on the face and another on the back: said of certain fabrics.

plate·ful (plāt′fo͝ol′) *n. pl.* **·fuls** The quantity that fills a plate.

plate glass See under GLASS.

plate hinge A hinge with one long, narrow plate as the movable unit. See illustration under HINGE.

plate·let (plāt′lit) *n.* **1** A small, platelike object. **2** *Physiol.* One of the small, disk-shaped bodies found in blood and thought to aid in the process of clotting. [Dim. of PLATE]

plat·en (plat′n) *n. Mech.* **1** The part of a printing press, typewriter, or the like, on which the paper is supported to receive the impression. **2** In a machine tool, the adjustable table that carries the work. [<OF *platine* a flat piece, metal plate < *plat.* See PLATE.]

plat·er (plā′tər) *n.* **1** One who plates articles with a layer of gold, silver, etc. **2** One who makes or works upon metallic plates. **3** An inferior race horse.

plate rail A shelflike molding around a room, for holding ornamental plates or bric-a-brac.

plat·form (plat′fôrm) *n.* **1** Any floor or flat surface raised above the adjacent level, as a stage for public speaking, a raised walk upon which passengers alight from railroad cars. **2** A projecting stage at the end of a car or similar vehicle. **3** A formal scheme of principles put forth by a religious, political, or other body; also, the document stating the principles of a political party. **4** The business of public speaking. [<MF *plateforme* < *plate* flat + *forme* form]

platform car A flat car.

platform scale A scale for weighing heavy objects, having a platform on which the load may stand.

pla·tie (plā′tē) *n. Scot.* A small plate.

pla·til·la (plə·til′ə) *n.* A kind of white linen fabric, originally of Silesian manufacture. [<Sp., appar. orig. dim. of *plata* silver]

pla·ti·na (plat′ə·nə, plə·tē′nə) *n.* **1** Platinum. **2** A white, brittle alloy of zinc and copper. [<NL <Sp., platinum, orig. dim. of *plata* silver]

plat·i·nate (plat′ə·nāt) *n. Chem.* A salt of platinic acid. [<PLATIN(IC) + -ATE[3]]

plat·ing (plā′ting) *n.* **1** A layer of metal of varying thickness: silver *plating.* **2** A sheathing of metal plates, or plate armor for protection. **3** The act or process of sheathing or coating something with plates or metal.

pla·tin·ic (plə·tin′ik) *adj. Chem.* Of, pertaining to, or containing platinum, especially in its higher valence. [<PLATIN(UM) + -IC]

plat·i·nif·er·ous (plat′ə·nif′ər·əs) *adj.* Containing or yielding platinum. [<PLATIN(UM) + -(I)FEROUS]

plat·in·i·rid·i·um (plat′in·i·rid′ē·əm) *n.* A whitish to gray native alloy of iridium, platinum, and other allied metals. [<PLATIN(UM) + IRIDIUM]

plat·i·nize (plat′ə·nīz) *v.t.* **·nized, ·niz·ing** To coat or combine with platinum, especially by electroplating.

platino– *combining form* Platinum; of, related to, or containing platinum: *platinocyanic.* Also, before vowels, **platin–.** [<PLATINUM]

plat·i·no·cy·an·ic (plat′ə·nō·sī·an′ik) *adj. Chem.* Of, pertaining to, or derived from compounds containing platinum and cyanogen.

plat·i·no·cy·a·nide (plat′ə·nō·sī′ə·nīd, -nid) *n. Chem.* A cyanide of platinum and some other element or radical. Also **plat′i·no·cy′a·nid** (-nid).

plat·i·noid (plat′ə·noid) *adj.* Like platinum. — *n.* **1** An alloy of German silver and 1 or 2 percent of tungsten, used in the manufacture of resistance coils and other electrical appliances. **2** A platinum metal.

plat·i·no·type (plat′ə·nō·tīp′) *n. Phot.* **1** A process in which the positive is obtained by a deposit of finely precipitated platinum in combination with iron salts. **2** A positive print obtained by the foregoing process.

plat·i·nous (plat′ə·nəs) *adj. Chem.* Of, pertaining to, or containing platinum, especially in its lower valence. [<PLATIN(UM) + -OUS]

plat·i·num (plat′ə·nəm) *n.* **1** A whitish, steel-gray, malleable and ductile metallic element (symbol Pt), usually found native, and also in combination. It is very infusible and resistant to most acids, has a high electrical resistance, and is widely used as a catalyst, for jewelry, and in dental work. See ELEMENT. **2** A color resembling that of platinum, but having a slightly bluish tone. [<NL, alter. of Sp. *platina* PLATINA]

platinum black Finely divided metallic platinum in the form of black powder made by reduction of platinum salts: used as a catalyst.

platinum blond A very light, almost white, blond.

plat·i·tude (plat′ə·to͞od, -tyo͞od) *n.* **1** A flat, dull, or commonplace statement; an obvious truism. **2** Dulness; triteness. [<F, flatness < *plat* flat]

plat·i·tu·di·nize (plat′ə·to͞o′də·nīz, -tyo͞o′-) *v.i.* **·nized, ·niz·ing** To utter platitudes.

plat·i·tu·di·nous (plat′ə·to͞o′də·nəs, -tyo͞o′-) *adj.* **1** Of the nature of platitude; insipid; flat. **2** Abounding in or given to platitudes.

Pla·to (plā′tō), 427?–347? B.C., Greek philosopher.

Pla·ton·ic (plə·ton′ik) *adj.* Of, pertaining to, or characteristic of Plato or of Platonism; academic; theoretical. Also **Pla·ton′i·cal.** — **Pla·ton′i·cal·ly** *adv.*

Platonic love Love which is purely spiritual, or devoid of sensual feeling.

Platonic year See PRECESSION OF THE EQUINOXES.

Pla·to·nism (plā′tə·niz′əm) *n.* **1** The philosophy of Plato; specifically, the doctrine that objects are copies or images of eternal ideas, that these ideas are the ultimate metaphysical realities and therefore the object of true knowledge. **2** A tenet or maxim of the Platonic philosophy. **3** The doctrine or practice of Platonic love. — **Pla′to·nist** *n.*

Pla·to·nize (plā′tə·nīz) *v.* **·nized, ·niz·ing** *v.t.* To make Platonic; idealize. — *v.i.* To conform to Platonism in views or utterance.

pla·toon (plə·to͞on′) *n.* **1** A subdivision of a company, troop, or other military unit, commanded by a lieutenant. **2** A company of people; set. [<F *peloton* ball, group of men, dim. of *pelote* a ball]

Platt·deutsch (plät′doich′) *n.* The Low German vernacular of the north of Germany. [<G]

Plat·ten·see (plät′ən·zā) The German name for BALATON, LAKE.

plat·ter (plat′ər) *n.* **1** An oblong shallow dish on which meat or fish is served. **2** *Colloq.* A phonograph record. [<AF *plater* < *plat* a dish]

Platte River (plat) A river in Nebraska formed by the confluence of the North Platte and the South Platte and flowing 310 miles east to the Missouri River.

plat·ting (plat′ing) *n.* **1** The process of weaving by hand. **2** Any fabric made by coarse weaving, as a straw hat.

Platt National Park A federal park in southern Oklahoma containing mineral springs; 912 acres.

Platts·burg (plats′bûrg) A city in NE New York, on Lake Champlain; scene of an American naval victory over the British (1814) in the War of 1812.

platy– *combining form* Flat: *platyrrhine.* Also, before vowels, *plat–.* [<Gk. *platys* flat]

plat·y·hel·minth (plat′ē·hel′minth) *n. Zool.* Any of a phylum (*Platyhelminthes*) of soft-bodied, bilaterally symmetrical, flattened or sometimes cylindrical worms, including many parasitic tapeworms. [<PLATY- + Gk. *helmins* worm]

plat·y·pus (plat′ə·pəs) *n. pl.* **·pus·es** or **·pi** (-pī) A burrowing egg-laying and aquatic monotrematous mammal (*Ornithorhynchus anatinus*) of Australia, with a ducklike bill; the duckbill. [<PLATY- + Gk. *pous* foot]

PLATYPUS
(Body about 18 inches in length)

plat·yr·rhine (plat′ə·rīn, -rin) *adj.* **1** Having a broad nose, with widely separated nostrils. **2** *Zool.* Designating a group of monkeys (the *Platyrrhini*) inhabiting the New World. — *n.* A broad-nosed person or monkey. Also **plat′yr·rhin′i·an** (-rin′ē·ən). [<PLATY- + Gk. *rhis, rhinos* nose]

plau·dit (plô′dit) *n.* An expression of applause; praise bestowed. See synonyms under APPLAUSE. [Short for L *plaudite,* 2nd pl. imperative of *plaudere* applaud]

Plau·en (plou′ən) A city in the former state of Saxony, south central East Germany. Also **Plauen im Vogt·land** (im fōkht′länt).

plau·si·ble (plô′zə·bəl) *adj.* **1** Seeming likely to be true, but open to doubt; specious. **2** Superficially trustworthy; endeavoring or calculated to gain trust or confidence: a *plausible* speaker. **3** *Colloq.* Apparently believable; credible. See synonyms under OSTENSIBLE. [<L *plausibilis* deserving applause] — **plau′·si·bil′i·ty, plau′si·ble·ness** *n.* — **plau′si·bly** *adv.*

plau·sive (plô′siv) *adj.* **1** Manifesting praise; applauding. **2** *Obs.* Plausible.

Plau·tus (plô′təs), **Titus Maccius,** 254?–184 B.C., Roman comic dramatist.

play (plā) *v.i.* **1** To engage in sport or diversion; amuse oneself; frolic; gambol. **2** To take part in a game of skill or chance; gamble. **3** To act in a way which is not to be taken seriously. **4** To act or behave in a specified manner: to *play* false. **5** To deal carelessly; behave lightly or insincerely: with *with.* **6** To make love sportively. **7** To move quickly or irregularly as if frolicking: lights *playing* along a wall. **8** To discharge or be discharged freely or continuously: a fountain *playing* in the square. **9** To perform on a musical instrument. **10** To give forth musical sounds;

sound: The bugles are *playing.* **11** To be performed or exhibited: *Hamlet* is *playing* tonight. **12** To act on or as on a stage; perform. **13** To move freely or loosely, especially within limits, as part of a mechanism. —*v.t.* **14** To engage in (a game, etc.). **15** To pretend to be; imitate in play: to *play* cowboys and Indians. **16** To perform sportively or wantonly: to *play* a trick. **17** To oppose in a game or contest. **18** To move or employ (a piece, card, etc.) in a game. **19** To employ (someone) in a game as a player. **20** To cause; bring about: to *play* hob. **21** To perform upon (a musical instrument). **22** To perform or produce, as a piece of music, a play, etc. **23** To act the part of on or as on the stage; assume the character of: to *play* the fool. **24** To perform or act in: to *play* Chicago. **25** To cause to move quickly or irregularly: to *play* lights over a surface. **26** To put into or maintain in action; wield; ply. **27** In angling, to let (a hooked fish) tire itself by maintaining pressure on the line. **28 a** To bet. **b** To bet on. —**to play at 1** To take part in. **2** To pretend to be doing; do half-heartedly. —**to play down** To treat as being of little importance; minimize. —**to play into the hands of** To act to the advantage of (a rival or opponent). —**to play off 1** To oppose against one another. **2** To decide (a tie) by playing one more game. —**to play on 1** To take unscrupulous advantage of (another's hopes, emotions, etc.) for one's own advantage. **2** To continue: The band *played on.* —**to play out 1** To come to an end; be exhausted. **2** To continue to the end; finish. —**to play the game** To behave in a fair manner. —**to play up** *Colloq.* To emphasize. —**to play up to** *Colloq.* To try to win the favor of by flattery, etc. —*n.* **1** Action without special aim or for amusement: opposed to *work.* **2** Exercise or action for recreation or diversion; sport; jest; fun; competitive trial of skill for amusement. **3** Gambling. **4** Manner of contending in a game; also, a move in a game: rough *play,* a fine *play,* sword *play.* **5** A dramatic composition; also, a dramatic representation; especially, a public theatrical exhibition. **6** Action without specified or special hindrance; freedom of movement. **7** Manner of acting toward others; dealing. **8** Active operation. **9** Light, quick, fitful movement. **10** Length of stroke, as of a piston. —**to make a play for** *Colloq.* To attempt to ingratiate oneself with. [OE *plegan*] —**play′a·ble** *adj.*

pla·ya (plä′yä) *n.* A plain with a hard clayey surface intermittently covered by a shallow lake. [<Sp.]

play·back (plā′bak′) *n.* In motion pictures and radio, the reproduction of a sound record from a disc, for testing.

play·bill (plā′bil′) *n.* A bill or program advertising or giving the cast of a play.

play·book (plā′bo͝ok′) *n.* **1** A book containing the script of a play. **2** A book of plays.

play·boy (plā′boi′) *n.* **1** *Colloq.* An irresponsible pleasure-seeker. **2** One who assumes a role for his own advantage or glory; a pretender: from the central character in J. M. Synge's *The Playboy of the Western World.*

play-by-play (plā′bī·plā′) *adj.* Dealing with each play consecutively: a *play-by-play* report.

play·day (plā′dā′) *n.* A holiday.

played out 1 Performed until finished. **2** Used up; exhausted: originally employed by gamblers.

play·er (plā′ər) *n.* **1** One who takes part in a game; also, one who specializes in a game: a tennis *player.* **2** One who performs on the dramatic stage; an actor. **3** A performer on a musical instrument. **4** One who works without a purpose or makes idle pretensions; also, an idler; a trifler. **5** A gambler. **6** An automatic device for playing a musical instrument; specifically, a mechanical device for playing a piano: also **player piano.**

play·fel·low (plā′fel′ō) *n.* An associate in games; a playmate.

play·ful (plā′fəl) *adj.* Frolicsome; merry; jocose. See synonyms under WANTON. —**play′ful·ly** *adv.* —**play′ful·ness** *n.*

play·go·er (plā′gō′ər) *n.* A frequenter of the theater.

play·ground (plā′ground′) *n.* A ground used for playing games; space set aside for recreation.

play·house (plā′hous′) *n.* **1** A theater. **2** A small house for children to play in.

playing card One of a pack of cards used in playing various games, the pack usually consisting of 52 cards divided into four suits (spades, hearts, diamonds, clubs) of 13 cards each.

play·mate (plā′māt′) *n.* A companion in sports, games, or recreation; especially, a child's companion in play.

play-off (plā′ôf′, -of′) *n.* In sports, a decisive game or contest, especially after a tie.

play·thing (plā′thing′) *n.* A thing to play with; a toy.

play·time (plā′tīm′) *n.* Time allowed for or given up to play or amusement.

play upon words Words used with double meaning; a pun.

play·wright (plā′rīt′) *n.* A writer of plays.

pla·za (plä′zə, plaz′ə; *Sp.* plä′thä) *n.* An open square or market place, especially in a Spanish or Spanish-American town. [<Sp. <L *platea.* Doublet of PIAZZA, PLACE.]

plea (plē) *n.* **1** An act of pleading, or that which is pleaded; an appeal; entreaty; prayer: a *plea* for aid. **2** An excuse; pretext or justification: necessity, the tyrant's *plea.* **3** *Law* **a** An allegation made by either party in a cause; a pleading. **b** In common-law practice, a defendant's answer of fact to the plaintiff's declaration, known in the United States as the *answer.* **c** In equity, a special answer, showing a reason why the writ should be dismissed, delayed, or barred. **d** A suit or action: usually in the plural. See synonyms under APOLOGY. —**Common Pleas** The Court of Common Pleas. See under COURT. —**special plea** A plea to prevent action which, while admitting the plaintiff's allegations, avoids them by setting up new matter. [<OF *plait* <L *placitum* opinion, orig. pp. of *placere* seem right, please]

pleach (plēch) *v.t.* To plait (vines or twigs) together, as in forming a hedge or arbor; interweave. [<AF *plechier,* OF *plaissier* <L *plectere* weave]

pleached (plēcht) *adj.* Interwoven; covered with interwoven branches.

plead (plēd) *v.* **plead·ed** (*Colloq.* or *Dial.* **pled**), **plead·ing** *v.i.* **1** To make earnest entreaty; implore; beg. **2** *Law* **a** To advocate a case in court. **b** To file a pleading. —*v.t.* **3** To allege as an excuse or defense: to *plead* insanity. **4** *Law* To discuss or maintain (a case) by argument. [<OF *plaider* <*plait.* See PLEA.] —**plead′a·ble** *adj.* —**plead′er** *n.*

Synonyms: advocate, argue, ask, beg, beseech, entreat, implore, press, solicit, urge. To *plead* for one is to employ argument or persuasion, or both, in his behalf, with earnestness or importunity; similarly one may *plead* for himself or for a cause, etc.; *plead* also takes a direct object, as to *plead* a case; in legal usage, *pleading* is argumentative; in popular usage, *pleading* always implies some appeal to the feelings. One *argues* a case solely on rational grounds with fair consideration of both sides; he *advocates* one side for the purpose of carrying it, and under the influence of motives that may range all the way from cold self-interest to the highest and noblest impulses; he *pleads* a cause, or *pleads for* a person, with still more intense feeling. *Beseech, entreat,* and *implore* imply impassioned earnestness, with direct and tender appeal to personal considerations. *Press* and *urge* imply determined or perhaps authoritative insistence. *Solicit* is a weak word denoting merely an attempt to secure one's consent or cooperation, sometimes by sordid or corrupt methods. See ALLEGE.

plead·ing (plē′ding) *n.* **1** The act of making a plea or argument in behalf of someone or something; specifically, the oral advocacy of a cause in court. **2** *Law* The art, science, or system of preparing the formal written statements of the parties to an action, leading to the joinder of issue; also, any one of such statements: collectively called the *pleadings* in a case. See synonyms under PETITION. —**plead′ing·ly** *adv.*

pleas·ance (plez′əns) *n.* **1** A secluded garden; pleasure ground. **2** *Archaic* The feeling of being pleased; that which pleases. Also **pleas′·aunce.** [<OF *plaisance* <*plaisant.* See PLEASANT.]

pleas·ant (plez′ənt) *adj.* **1** Giving or promoting pleasure; pleasing; agreeable. **2** Conducive to merriment; gay. [<F *plaisant* <L *placens,* ppr. of *placere* please] —**pleas′ant·ly** *adv.* —**pleas′ant·ness** *n.*

Synonyms: agreeable, attractive, good-natured, kind, kindly, obliging, pleasing, pleasurable. That is *pleasing* from which pleasure is received, or may readily be received, without reference to any action or intent in that which confers it; as, a *pleasing* picture; a *pleasing* landscape. Whatever has active qualities adapted to give pleasure is *pleasant;* as, a *pleasant* breeze; a *pleasant* (not a *pleasing*) day. As applied to persons, *pleasant* always refers to a disposition ready and desirous to please, and in this sense is near akin to *kind,* but *kind* refers to act or intent, while *pleasant* stops with the disposition. *Pleasant* keeps always something of the sense of actually giving pleasure, and thus surpasses the meaning of *good-natured;* there are *good-natured* people who by reason of rudeness and ill-breeding are not *pleasant* companions. A *pleasing* face has good features, complexion, expression, etc.; a *pleasant* face indicates a *kind* heart and an *obliging* disposition, as well as *kindly* feelings in actual exercise. See AGREEABLE, AMIABLE, ATTRACTIVE, COMFORTABLE, DELIGHTFUL, GOOD, VIVACIOUS. *Antonyms:* arrogant, austere, crabbed, disagreeable, displeasing, dreary, forbidding, gloomy, glum, grim, harsh, hateful, ill-humored, ill-natured, offensive, repellent, repulsive, unkind, unpleasant.

Pleasant Island A former name for NAURU.

pleas·an·try (plez′ən·trē) *n. pl.* **·tries 1** The spirit of playful and jocose companionship; playfulness. **2** A playful, amusing, or good-natured remark, jest, or trick. See synonyms under SPORT, WIT[1].

please (plēz) *v.* **pleased, pleas·ing** *v.t.* **1** To give pleasure to; be agreeable to; gratify. **2** To be the wish or will of: May it *please* you. —*v.i.* **3** To give satisfaction or pleasure. **4** To have the will or preference; wish: Go when you *please.* See synonyms under ENTERTAIN, INDULGE, REJOICE. [<OF *plaisir* <L *placere* please]

pleas·ing (plē′zing) *adj.* Affording pleasure or satisfaction. See synonyms under AGREEABLE, AMIABLE, LOVELY, PLEASANT. —**pleas′·ing·ly** *adv.* —**pleas′ing·ness** *n.*

pleas·ur·a·ble (plezh′ər·ə·bəl) *adj.* Affording gratification; pleasant. See synonyms under DELIGHTFUL, PLEASANT. —**pleas′ur·a·ble·ness** *n.* —**pleas′ur·a·bly** *adv.*

pleas·ure (plezh′ər) *n.* **1** An agreeable sensation or emotion; gratification; enjoyment. **2** Sensual or mental gratification. **3** Amusement in general; diversion. **4** One's preference; choice. See synonyms under COMFORT, ENTERTAINMENT, HAPPINESS, SPORT. —*v.* **·ured, ·ur·ing** *v.t.* To give or afford pleasure to; please; gratify. —*v.i.* To take pleasure; delight. [<OF *plaisir.* See PLEASE.]

pleasure principle *Psychoanal.* The concentration of the ego on securing a maximum gratification of instincts with a minimum of pain and effort. Compare REALITY PRINCIPLE.

pleat (plēt) *n.* A fold of cloth doubled on itself and pressed or sewn in place. —*v.t.* To make a pleat or pleats in. Also *plait.* [Var. of PLAIT]

pleat·er (plē′tər) *n.* **1** One who pleats. **2** A sewing-machine attachment for pleating; a ruffler.

pleb (pleb) *n.* **1** A plebeian. **2** A plebe. ◆

plebe (plēb) *n.* **1** *U.S.* A member of the lowest class in the academies at West Point and Annapolis. **2** *Obs.* Plebs. [Short for PLEBEIAN]

ple·be·ian (pli·bē′ən) *adj.* Pertaining to the common people, originally to the common people of ancient Rome; hence, common. —*n.* One of the common people. [<L *plebeius* <*plebs* the common people] —**ple·be′·ian·ism** *n.* —**ple·be′ian·ly** *adv.* —**ple·be′ian·ness** *n.*

pleb·i·scite (pleb′ə·sīt, -sit) *n.* An expression of the popular will by means of a vote by the

add,āce,câre,pälm; end,ēven; it,īce; odd,ōpen,ôrder; to͝ok,po͞ol; up,bûrn; ə = a in *above,* e in *sicken,* i in *clarity,* o in *melon,* u in *focus;* yo͞o = u in *fuse;* oi,oil; ou,pout; ch,check; g,go; ng,ring; th,thin; th,this; zh,vision. Foreign sounds à,œ,ü,kh,ṅ; and ◆: see page xx. < from; + plus; ? possibly.

whole people, usually resorted to in important changes, as those dealing with the constitution, sovereignty, etc. [<F *plébiscite* <L *plebiscitum* <*plebs* people + *scitum* decree <*scire* know] —**ple·bis·ci·tar·y** (plə·bis′ə·ter′ē) *adj.*

plebs (plebz) *n.* 1 The lower order of the ancient Roman people. 2 The populace. [<L, common people]

ple·cop·ter·an (plə·kop′tər·ən) *n.* Any of an order (*Plecoptera*) of soft-bodied, flattened insects of which the nymphs are aquatic: the adults usually have a pair of long caudal appendages, and two pairs of wings folding flat over the abdomen; a stone fly. [<Gk. *plekein* twine + *pteron* wing]

plec·tog·nath (plek′tog·nath) *n.* Any of an order or suborder (*Plectognathi*) of teleost fishes having spiny bodies, generally inedible and often poisonous flesh, and including a large number of odd forms, as the trigger-fishes, swellfishes, globefishes, etc. —*adj.* Of or pertaining to the *Plectognathi.* [<Gk. *plektos* twisted + *gnathos* jaw]

plec·trum (plek′trəm) *n.* *pl.* **·tra** (-trə) A small implement with which the player on a lyre, zither, etc., picks or strikes the strings to set them in vibration. Also **plec′tron** (-tron). [<L <Gk. *plēktron* spur <*plēssein* strike]

pled (pled) *Dial.* or *Colloq.* Past tense and past participle of PLEAD.

pledge (plej) *v.t.* **pledged, pledg·ing** 1 To give or deposit as security for a loan, etc.; pawn. 2 To bind by or as by a pledge. 3 To promise solemnly, as assistance. 4 To offer (one's word, life, etc.) as a guaranty or forfeit. 5 To drink a toast to. 6 To promise to join (a fraternity). 7 To accept (someone) as a pledge (def. 5). [<*n.*] —*n.* 1 A guaranty for the performance of an act, contract, or duty. 2 A formal promise to do or not to do something; especially, a vow to abstain from intoxicating liquors. 3 The drinking of a health or to good cheer. 4 A pawn of personal property; also, the property delivered. 5 One who has promised to join a fraternity but who has not yet been formally inducted. [<OF *plege* security, prob. <Gmc.]

pledg·ee (plej·ē′) *n.* The person to whom anything is pledged.

pledg·er (plej′ər) *n.* One who gives a pledge.

pledg·or (plej′ər) *n.* *Law* A pledger. Also **pledge′or**.

pledg·et (plej′it) *n.* 1 A little plug. 2 A compressed wad of lint, cotton, etc., as for a wound. 3 An oakum string used in calking. [Origin unknown]

-plegia *combining form Med.* A (specified) kind of paralysis, or paralytic condition: *hemiplegia.* Also **-plegy.** [<Gk. <*plēgē* a stroke]

Plei·ad (plē′əd, plī′ad) *n.* *pl.* **Plei·a·des** (plē′ə·dēz, plī′-) 1 One of the Pleiades. 2 One of any cluster of brilliant persons, usually seven.

Plei·a·des (plē′ə·dēz, plī′-) 1 In Greek mythology, the seven daughters of Atlas (Maia, Electra, Taygeta, Alcyone, Celaeno, Sterope, and Merope), who were set by Zeus among the stars: also called *Atlantides.* 2 *Astron.* A loose cluster of many hundred stars in the constellation Taurus, six of which are visible to ordinary sight and represent the daughters of Atlas, of whom the seventh, Merope, is known as the **Lost Pleiad.**

plein-air (plān′âr′) *adj.* Characterizing the work of a school of French impressionist painters concerned with the representation of objects seen under brilliant sunlight, and other outdoor effects. [<F, open-air] —**plein-air·ism** (plān′âr′iz·əm) *n.* —**plein′-air′ist** *n.*

Plei·o·cene (plī′ə·sēn) See PLIOCENE.

plei·o·syl·lab·ic (plī′ə·si·lab′ik) *adj.* Having more than one syllable; especially, having two or three syllables. See POLYSYLLABLE. [<Gk. *pleiōn* more + *syllabē* syllable]

Pleis·to·cene (plīs′tə·sēn) *Geol. n.* 1 The epoch following the Pliocene; the first of the two epochs of the Quaternary period, characterized by ice sheets over much of the northern hemisphere; the glacial epoch of northern Asia, Europe, and North America. 2 The rock series of this period. —*adj.* Pertaining to this epoch or to its rock series. [<Gk. *pleistos* most + *kainos* new]

ple·na·ry (plē′nə·rē, plen′ə-) *adj.* 1 Full in all respects or requisites; entire; absolute; also, complete, as embracing all the parts or members: *plenary* authority. 2 Having full powers: *plenary* jurisdiction. 3 Fully or completely attended; consisting of the full number of members: said of an assembly. [<L *plenus* full] —**ple′na·ri·ly** *adv.* —**ple′na·ri·ness** *n.*

plenary indulgence In the Roman Catholic Church, the remission of all temporal penalties incurred by sin.

ple·nip·o·tent (plə·nip′ə·tənt) *adj.* Possessing full power or authority.

plen·i·po·ten·ti·ar·y (plen′i·pə·ten′shē·er′ē, -shə·rē) *adj.* Possessing or conferring full powers. —*n. pl.* **·ar·ies** 1 A person fully empowered, especially an ambassador, minister, or envoy, invested with full powers by a government. 2 Specifically, a diplomatic representative of the second class ranking next below an ambassador, accredited by the sovereign or head of one state to that of another: full title *envoy extraordinary and minister plenipotentiary.* [<L *plenus* full + *potens* powerful]

plen·ish (plen′ish) *v.t. Scot.* & *Brit. Dial.* To fill or stock; replenish.

ple·nism (plē′niz·əm) *n.* The doctrine that space is a plenum. —**ple′nist** *n.*

plen·i·tude (plen′ə·to͞od, -tyo͞od) *n.* The state of being full, complete, or abounding; also, that which fills to repletion. [<L *plenitudo* <*plenus* full]

plen·te·ous (plen′tē·əs) *adj.* 1 Characterized by plenty; amply sufficient. 2 Yielding an abundance. See synonyms under AMPLE, PLENTIFUL. [<OF *plentieus, plentivous* <*plenté* PLENTY] —**plen′te·ous·ly** *adv.* —**plen′te·ous·ness** *n.*

plen·ti·ful (plen′ti·fəl) *adj.* 1 Existing in great quantity; abundant. 2 Yielding or containing plenty; affording ample supply. —**plen′ti·ful·ly** *adv.* —**plen′ti·ful·ness** *n.*

Synonyms: abounding, abundant, adequate, affluent, ample, bounteous, bountiful, complete, copious, enough, exuberant, full, generous, large, lavish, liberal, luxuriant, overflowing, plenteous, profuse, replete, rich, sufficient, teeming. *Plentiful* is used of supplies, as of food, water, etc. We may say a *copious* rain; but *copious* can also be applied to thought, language, etc., where *plentiful* cannot well be used. *Affluent* and *liberal* both apply to riches, resources; *liberal*, with especial reference to giving or expending. *Affluent*, referring especially to riches, may be used of thought, feeling, etc. Neither *affluent, copious*, nor *plentiful* can be used of time or space; a field is sometimes called *plentiful*, with reference to its productiveness. *Complete* expresses not excess or overplus, and yet not mere sufficiency, but harmony, proportion, fitness to a design, or ideal. *Ample* and *abundant* may be applied to any subject and mean more than *enough. Lavish* and *profuse* imply a decided excess. We rejoice in *abundant* resources, and honor *generous* hospitality; *lavish* or *profuse* expenditure suggests extravagance and wastefulness. *Luxuriant* is used especially of that which is *abundant* in growth; as, a *luxuriant* crop. Compare ADEQUATE, AMPLE, ENOUGH. *Antonyms*: deficient, drained, exhausted, impoverished, inadequate, insufficient, mean, miserly, narrow, niggardly, poor, scanty, scarce, scrimped, short, small, sparing, stingy, straitened.

plen·ty (plen′tē) *n.* 1 The state of being abundantly sufficient, or of having an abundance, particularly of necessaries and comforts: to live in peace and *plenty.* 2 As much as can be required; an abundance or sufficiency: now generally without the article: *plenty* of water; I have *plenty.* See synonyms under COMFORT, WEALTH. —*adj.* Existing in abundance; plentiful. —*adv. Colloq.* In a sufficient degree: The house is *plenty* large enough. [<OF *plenté* <L *plenitas, -tatis* <*plenus* full]

ple·num (plē′nəm) *n.* *pl.* **·nums** or **·na** (-nə) 1 Fullness of matter in space; that state of things in which space is considered as fully occupied by matter, especially by absolutely continuous matter. 2 Space so considered: opposed to *vacuum.* 3 Any condition of fullness or plethora, or that which produces it. 4 An enclosed body of gas under greater than normal pressure. 5 A completely attended meeting, as of an association or legislative body. 6 Fullness. —*adj.* Pertaining to or utilizing fullness (of air, etc.) [<L <*plenus* full]

ple·och·ro·ism (plē·ok′rō·iz′əm) *n. Mineral.* The property exhibited by double-refracting colored crystals of showing different colors when the transmitted light is viewed along different axes. [<Gk. *pleōn* more + *chroos* color + -ISM] —**ple·o·chro·ic** (plē′ə·krō′ik) *adj.*

ple·o·mor·phism (plē′ə·môr′fiz·əm) *n.* 1 *Mineral.* The ability of a substance to crystallize in two or more distinct fundamental forms, embracing *dimorphism* and *trimorphism.* 2 *Biol.* The occurrence of several independent stages in the life cycle of an organism or plant. Also **ple′o·mor′phy.** [<Gk. *pleōn* more + *morphē* form] —**ple′o·mor′phic, ple′o·mor′phous** *adj.*

ple·o·nasm (plē′ə·naz′əm) *n.* 1 The use of needless words; redundancy; tautology or any instance of it; a redundant word or phrase. 2 A superabundance of parts, as in an organism. [<Gk. *pleonasmos* <*pleōn* more] —**ple′o·nas′tic** (-nas′tik) *adj.* —**ple′o·nas′ti·cal·ly** *adv.*

ple·o·nex·i·a (plē′ə·nek′sē·ə) *n.* 1 *Pathol.* A tendency of the hemoglobin of the blood to retain oxygen, yielding less than normal amounts to the tissues of the body. 2 *Psychiatry* A morbid greediness. [<NL <Gk., greediness] —**ple′o·nec′tic** (-nek′tik) *adj.*

ple·o·pod (plē′ə·pod) *n. Zool.* An abdominal limb of a crustacean; a swimmeret. [<Gk. *pleein* swim + *pous, podos* foot]

ple·rome (plē′rōm) *n. Bot.* That part of the actively growing tissue of the apex of the stem and root out of which the stele arises. Also **ple′rom** (-rəm). [<Gk. *plērōma* fullness <*plērēs* full]

ple·si·o·saur (plē′sē·ə·sôr) *n. Paleontol.* Any of an extinct genus (*Plesiosaurus*) of fish-eating marine reptiles which flourished in the Jurassic period, having a small head, long neck, and limbs modified into swimming paddles. Also **ple′si·o·sau′rus.** [<Gk. *plēsios* near + *sauros* lizard]

PLESIOSAUR
(Up to 50 feet in length)

ples·sor (ples′ər) *n.* Plexor.

pleth·o·ra (pleth′ər·ə) *n.* 1 A state of excessive fullness; superabundance; excess. 2 *Bot.* An excess of juices. 3 *Pathol.* Superabundance of blood in the whole system or in an organ or part. [<Med. L <Gk. *plēthōrē* fullness <*plēthein* be full]

ple·thor·ic (ple·thôr′ik, -thor′-, pleth′ə·rik) *adj.* 1 Affected or characterized by plethora. 2 Excessively full; overloaded; turgid; inflated. See synonyms under CORPULENT. —**ple·thor′i·cal·ly** *adv.*

ple·thys·mo·graph (ple·thiz′mə·graf, -gräf, -this′-) *n.* An instrument for recording variations in size of parts of the body, especially as caused by the circulation of the blood. [<Gk. *plēthysmos* enlargement (<*plēthyein* increase) + -GRAPH]

pleu·ra (plo͝or′ə) *n.* *pl.* **pleu·rae** (plo͝or′ē) *Anat.* The serous membrane that infolds the lungs and is reflected upon the walls of the thorax and upon the diaphragm. [<Gk. *pleura* side] —**pleu′ral** *adj.*

pleu·ri·sy (plo͝or′ə·sē) *n. Pathol.* Inflammation of the pleura, commonly attended with fever, pain in the chest, difficult breathing, exudation, etc. [<OF *pleurisie* <L *pleurisis* <Gk. *pleuritis* <*pleura* side] —**pleu·rit·ic** (plo͝o·rit′ik) *adj.*

pleurisy root 1 Butterfly weed. 2 Its root, formerly used in treating pleurisy.

pleuro- *combining form* 1 Of or pertaining to the side: *pleurodont.* 2 *Med.* Of, related to, or affecting the pleura: *pleurotomy.* Also, before vowels, **pleur-.** [<Gk. *pleura* side and *pleuron* rib]

pleu·ro·dont (plo͝or′ə·dont) *Zool. adj.* Having teeth attached to the inner side of the jaw, as in an iguanoid lizard. —*n.* A lizard with pleurodont dentition. [<PLEUR(O)- + -ODONT]

pleu·ron (plo͝or′on) *n.* *pl.* **pleu·ra** (plo͝or′ə)

1 *Entomol.* The lateral wall of a thoracic segment in insects. 2 *Zool.* In crustaceans, the lateral process of an abdominal segment. [<NL <Gk., a rib]

pleu·ro·pneu·mo·ni·a (plŏŏr′ō·nōō·mō′nē·ə, -mōn′yə, -nyōō-) *n. Pathol.* Pleurisy combined with pneumonia.

pleu·rot·o·my (plŏŏ·rot′ə·mē) *n. Surg.* The operation of making an incision into the pleural cavity, for drawing off effused liquids.

pleus·ton (plōōs′tən) *n. Bot.* That type of vegetation which consists of large aquatic plants floating on the water. [<NL <Gk. *pleus-*, stem of *pleein* swim]

Plev·en (plev′ən) A city in northern Bulgaria; surrendered by the Turks after a siege, 1877. Also **Plev·na** (plev′nä).

plex·i·form (plek′sə·fôrm) *adj.* Having the form of a plexus; complicated.

Plex·i·glas (plek′si·glas, -gläs) *n.* A thermoplastic acrylic resin used in the fabrication of transparent objects, as windows for airplane gun turrets, gages, etc.: a trade name.

plex·im·e·ter (plek·sim′ə·tər) *n.* A plate to be placed against the body to receive the blows in percussion. [<Gk. *plēxis* a stroke + -METER] — **plex·i·met·ric** (plek′si·met′rik) *adj.* — **plex·im′e·try** *n.*

plex·or (plek′sər) *n. Med.* An instrument used like a hammer in percussion of the chest: also called *plessor.* [<Gk. *plessein* strike; on analogy with *flexor*]

plex·us (plek′səs) *n. pl.* **·us·es** or **·us** 1 A network or interlacement; a complication of parts. 2 *Anat.* An interlacement of cordlike structures, as blood vessels or nerves. [<L, braid]

pli·a·ble (plī′ə·bəl) *adj.* 1 Easily bent or twisted; flexible. 2 Easily persuaded or controlled. See synonyms under DOCILE, SUPPLE. — **pli′a·bil′i·ty, pli′a·ble·ness** *n.* — **pli′a·bly** *adv.*

pli·an·cy (plī′ən·sē) *n.* The state or quality of being pliant; pliability: opposed to *rigidity.*

pli·ant (plī′ənt) *adj.* 1 Capable of being bent or twisted with ease; supple; lithe. 2 Easily yielding to influence; tractable. [<OF, ppr. of *plier.* See PLY.] — **pli′ant·ly** *adv.*

pli·ca (plī′kə) *n. pl.* **·cae** (-sē) 1 A fold of membrane, skin, or the like, as between the fingers. 2 *Zool.* A ridge, as on the outer wall of the body whorl in a shell, or on the wing covers of some beetles. 3 *Pathol.* A disease affecting the hair, causing it to become matted and agglutinated. [<Med. L <L *plicare* fold]

pli·cate (plī′kāt) *adj.* Plaited; folded in plaits like a fan, as a leaf. Also **pli′cat·ed.** — **pli′cate·ness** *n.* — **pli′cate·ly** *adv.*

pli·ca·tion (plī·kā′shən) *n.* A folding, or that which is folded; a fold. Also **plic·a·ture** (plik′ə·chŏŏr).

pli·er (plī′ər) *n.* 1 One who or that which plies. 2 *pl.* Small pincers for bending, holding, or cutting.

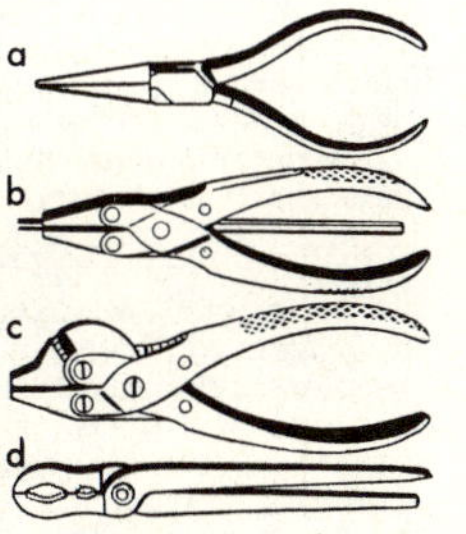

VARIOUS TYPES OF PLIERS
a. Round-nose.
b. Flat-nose, showing wire running through.
c. Flat-nose, with wire-cutting attachment.
d. Gas-fitter's.

plight[1] (plīt) *n.* A condition, state, or case: usually distressed or complicated. [<OF *ploit,* var. of *pleit.* See PLAIT.]

plight[2] (plīt) *n.* A solemn engagement; betrothal; a pledge subject to forfeiture. — *v.t.* 1 To pledge (one's word, faith, etc.). 2 To promise, as in marriage; betroth: She is *plighted* to a judge. — **to plight one's troth** 1 To pledge one's solemn word. 2 To promise oneself in marriage. [OE *pliht* peril] — **plight′er** *n.*

Plim·soll (plim′sol, -səl), **Samuel,** 1824–98, English statesman; secured Parliamentary reforms, 1876, against overloading of ships.

Plimsoll line A mark painted on the outside of a British vessel's hull to show how deeply she may be loaded; load line. Also called **Plimsoll mark.** [after Samuel *Plimsoll*]

plinth (plinth) *n. Archit.* 1 The slab, block, or stone on which a column, pedestal, or statue rests. 2 A thin course, as of slabs, usually projecting: also **plinth course.** [<L *plinthus* <Gk. *plinthos* a brick]

Plin·y (plin′ē) Anglicized name of two Roman authors: **Pliny the Elder,** 23–79, Gaius Plinius Secundus, naturalist, and his nephew, **Pliny the Younger,** 62–113, Gaius Plinius Caecilius Secundus, statesman.

Pli·o·cene (plī′ə·sēn) *Geol. adj.* Of or pertaining to the latest epoch of the Tertiary period, following the Miocene and succeeded by the Pleistocene. — *n.* The Pliocene epoch or rock series. Also spelled *Pleiocene.* [<Gk. *pleiōn* more + *kainos* new] — **Pli′o·cen′ic** (-sen′ik) *adj.*

Pli·o·film (plī′ə·film) *n.* A flexible, transparent rubber sheeting, used for raincoats, umbrellas, etc.: a trade name.

plod (plod) *v.* **plod·ded, plod·ding** *v.i.* 1 To walk heavily or laboriously; trudge. 2 To work in a steady, laborious manner; drudge. — *v.t.* 3 To walk along heavily or laboriously. — *n.* 1 A tiring walk; tramp; act or duration of plodding. 2 The sound of a heavy step, as of a horse. [Imit.] — **plod′ding** *adj.* — **plod′ding·ly** *adv.*

plod·der (plod′ər) *n.* One who plods; a drudge; also, a slow but persevering person.

Plo·eș·ti (plō·yesht′, -yesh′tē) A city of south central Rumania; chief center of the Rumanian petroleum industry. Also **Plo·esh′ti.**

-ploid *combining form Biol.* In cytology and genetics, having a (specified) number of chromosomes: *diploid.* Corresponding nouns end in **-ploidy.** [<Gk. *-ploos,* as in *diploos* twofold]

plop (plop) *v.t. & v.i.* **plopped, plop·ping** To drop with a sound like that of a pebble striking the water without making a splash. — *n.* The act or sound of plopping. — *adv.* Suddenly with the sound of plop: They fell *plop* into the river. [Imit.]

plo·sion (plō′zhən) *n. Phonet.* The sudden release of breath after closure of the oral passage in the articulation of a stop consonant, as after the *p* in *pat*: also *explosion.* [<EXPLOSION]

plo·sive (plō′siv) *Phonet. adj.* Designating a speech sound produced by a total blockage of the breath stream followed by an explosive release, as (p) and (t) before vowels. — *n.* A consonant so produced; a stop. Also *explosive.*

plot (plot) *n.* 1 A piece or patch of ground set apart: also called *plat.* 2 A chart or diagram, as of a building, for showing certain data; also, a surveyor's map. 3 A secret plan to accomplish some questionable purpose; conspiracy. 4 The series of incidents forming the plan of action of a story, play, or poem. — *v.* **plot·ted, plot·ting** *v.t.* 1 To make a map, chart, or plan of, as a ship's course, a building, etc. 2 To plan for secretly: to *plot* an enemy's ruin. 3 To arrange the plot of (a novel, etc.). 4 *Math.* **a** To represent graphically the position of (a measured value) by a point located with reference to its coordinates on plotting paper. **b** To draw (a curve) through a series of such points. — *v.i.* 5 To form a plot; conspire. [OE]

Plo·ti·nus (plō·tī′nəs), 205?–270?, Roman philosopher born in Egypt.

plot·ter (plot′ər) *n.* 1 One who plots or contrives; a conspirator. 2 A maker of a plot or map. 3 A contrivance, as for plotting coordinates.

plotting paper Paper which has been ruled into small squares for plotting curves, and making diagrams.

plot·ty (plot′ē) *n. Scot.* A hot, spiced beverage.

plough (plou) See PLOW.

Plov·div (plôv′dēf) The second largest city of Bulgaria, in the south central part: ancient *Philippopolis.*

plov·er (pluv′ər, plō′vər) *n.* 1 A shore bird (family *Charadriidae*), especially of *Charadrius* or related genus, with long, pointed wings and a short tail, especially the **American golden plover** (*Pluvialis dominica*). 2 Any of certain related shore birds, as the ruddy turnstone and the **upland plover** (*Bartramia longicauda*). [<AF, ult. <L *pluvia* rain]

plow (plou) *n.* 1 An implement (usually drawn by horses or oxen, or by mechanical power) for cutting, turning over, stirring, or breaking up the soil. 2 Any implement that operates like a plow: often in combination: a *snowplow*; also, any one of various furrowing or grooving tools. 3 Figuratively, agriculture. — *v.t.* 1 To turn up the surface of (land) with a plow. 2 To make or form (a furrow, ridge, etc.) by means of a plow. 3 To furrow or score the surface of: Shot *plowed* the field. 4 To dig out or remove with a plow: with *up* or *out.* 5 To move or cut through (water): to *plow* the waves. 6 To pluck (def. 7). — *v.i.* 7 To turn up soil with a plow. 8 To undergo plowing in a specified way, as land. 9 To move or proceed as a plow does: usually with *through* or *into.* 10 To advance laboriously; plod. — **to plow under** *U.S.* To put from sight by or as by plowing in such a way as to cover with soil; obliterate. Also spelled *plough.* [OE *plōh*] — **plow′a·ble** *adj.* — **plow′er** *n.*

Plow The group of seven stars commonly called *Charles's Wain* or *the Dipper*; sometimes also *Ursa Major.* Also **Plough.**

plow beam The horizontal projecting part of a plow frame, whose front end is attached to the swingletree. See illustration under SWINGLETREE.

plow·boy (plou′boi′) *n.* A boy who drives or guides a team in plowing; hence, a young rustic. Also **plough′boy′.**

plow·man (plou′mən) *n. pl.* **·men** (-mən) One who plows; a cultivator; hence, a rustic. Also **plough′man.**

plow·share (plou′shâr′) *n.* The blade of a plow. Also **plough′share′.**

plow·staff (plou′staf′, -stäf′) *n.* The handle of a plow. Also **plough′staff′.**

ploy[1] (ploi) *v.i. Mil.* To diminish front; maneuver from line into column: opposite of *deploy.* [<DEPLOY] — **ploy′ment** *n.*

ploy[2] (ploi) *n. Scot.* Sport; merrymaking.

pluck (pluk) *v.t.* 1 To pull out or off; pick: to *pluck* a flower. 2 To pull with force; snatch or drag: with *off, away,* etc. 3 To pull out the feathers, hair, etc., of: to *pluck* a chicken. 4 To give a twitch or pull to, as a sleeve. 5 To cause the strings of (a musical instrument) to sound by such action. 6 *Slang* To rob; swindle. 7 *Brit. Slang* To reject (a candidate) for failure to pass an examination. — *v.i.* 8 To give a sudden pull; tug: with *at.* — **to pluck up** To rouse or summon (one's courage). — *n.* 1 Confidence and spirit in the face of difficulty or danger; courage. 2 The heart, liver, windpipe, and lungs of an animal. 3 A sudden pull; twitch. 4 The act of plucking or state of being plucked; also, the person plucked. See synonyms under COURAGE. [OE *pluccian*] — **pluck′er** *n.*

pluck·y (pluk′ē) *adj.* **pluck·i·er, pluck·i·est** Brave and spirited; courageous. — **pluck′i·ly** *adv.* — **pluck′i·ness** *n.*

plug (plug) *n.* 1 Anything, as a piece of wood or a cork, used to stop a hole; a wedge or peg driven into anything. 2 A spark plug. 3 *Electr.* A device containing conducting material, as projecting prongs, for inserting in an outlet, etc., so as to complete a circuit or make contact. 4 A flat cake of pressed or twisted tobacco. 5 Any worn-out or useless thing, particularly a horse past his prime: often with *old.* 6 *U.S. Slang* A man's high silk hat: also **plug hat.** 7 *Slang* Mention of a product, song, etc., as on a radio or television program, to give it publicity; an advertisement. 8 *Geol.* The hard core of igneous rock which fills the neck of a volcano. 9 The discharge outlet from a water main: also called *hydrant.* 10 *Mech.* The cylindrical part of a cylinder lock which contains the keyhole and is turned by the key. 11 In angling, a type of lure, usually cylindrical and with several hooks attached, similar to a spoon. — *v.* **plugged, plug·ging** *v.t.* 1 To stop or close, as a hole, by inserting a plug: often with *up.* 2 To insert as a plug. 3 *Slang* To shoot a bullet into. 4 *U.S. Slang* To advertise frequently or insistently; publicize. — *v.i.* 5 *Colloq.* To work doggedly; persevere. — **to plug in** To insert the plug of (a lamp, etc.) in an electrical outlet. [<MDu. *plugge*] — **plug′ger** *n.*

plug-ug·ly (plug′ug′lē) *n. pl.* **·lies** *U.S. Slang* A city ruffian; gangster; a street rowdy.

plum (plum) *n.* **1** The edible drupaceous fruit of any one of various trees of the genus *Prunus,* especially *P. domestica,* the **European** or **garden plum. 2** The tree. **3** The plumlike fruit of any one of various other trees having an edible drupe; also, the tree bearing such fruit. **4** A raisin, especially as used in cooking: *plum* pudding. **5** The best part of anything; a choice piece or portion; a desirable post or appointment. **6** *Brit. Slang* A sum of £100,000 sterling; a handsome fortune, or the possessor of it. **7** Any of various shades of dull reddish purple or purplish red. **8** A sugarplum; anything resembling a plum, as in shape or flavor. ◆ Homophone: *plumb.* [OE *plume* <LL *pruna.* Doublet of PRUNE[1].]

plum·age (plōō′mij) *n.* **1** The feathers that cover a bird, collectively. **2** Gaudy costume; adornment. [<F <*plume* plume]

plu·mate (plōō′māt) *adj.* Resembling plumage or feathers. [<L *plumatus* feathered]

plumb (plum) *n.* **1** A lead weight on the end of a line used by masons, carpenters, etc., to find the exact perpendicular; a plumb bob; a plummet. **2** A plummet or nautical sounding lead; a sinker on a fishing line, etc. **—off** (or **out of**) **plumb** Not exactly vertical; not in alinement. **—***adj.* **1** True, accurate, and upright; vertical or perpendicular; hence, figuratively, upright in principle. **2** *Colloq.* Sheer; complete: also **plum. —***adv.* **1** In a line perpendicular to the plane of the horizon; vertically. **2** *Colloq.* With exactness; correctly; exactly; completely; entirely: also **plum. —***v.t.* **1** To test the perpendicularity of with a plumb. **2** To make vertical; straighten: usually with *up.* **3** To test the depth of; sound. **4** To reach the lowest level or extent of; fathom: to *plumb* the depths of despair. **5** To seal with lead. ◆ Homophone: *plum.* [<F *plomb* <L *plumbum* lead]

plumb- Var. of PLUMBO-.

plum·ba·go (plum·bā′gō) *n. pl.* **·gos 1** Graphite: used for pencils, crucibles, lubricating, and in electroplating to coat non-conducting surfaces, as gutta-percha. **2** A drawing made with a lead-pointed instrument. **3** Any of a genus *(Plumbago)* of hardy, shrubby plants cultivated for their showy blue, white, or purplish flowers: also called *leadwort.* [<L *plumbum* lead] **—plum·bag′i·nous** (-baj′ə·nəs) *adj.*

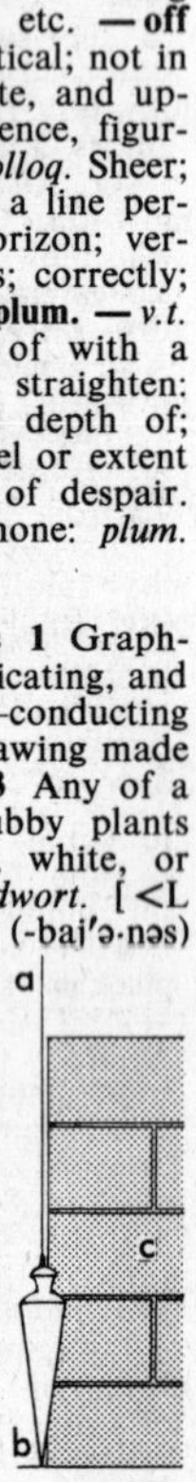

PLUMB BOB
a. Plumb line.
b. Bob.
c. Wall.

plumb bob The weight used at the end of a plumb line.

plum·be·ous (plum′bē·əs) *adj.* **1** Resembling lead; heavy. **2** Lead-colored. [<L *plumbeus* <*plumbum* lead]

plumbeous vireo A blue-headed vireo *(Vireo solitarius plumbeus),* of a leaden-gray color, common from northern Nevada to Mexico.

plumb·er (plum′ər) *n.* One who makes a business of plumbing.

plumb·er·y (plum′ər·ē) *n. pl.* **·er·ies 1** The business of plumbing. **2** A plumber's place of business. **3** Leadwork.

plum·bic (plum′bik) *adj. Chem.* Of, pertaining to, or containing lead, especially in its higher valence.

plum·bif·er·ous (plum·bif′ər·əs) *adj.* Containing or yielding lead.

plumb·ing (plum′ing) *n.* **1** The art or trade of putting into buildings the tanks, pipes, etc., for water, gas, and sewage. **2** The pipe system of a building. **3** The act of sounding for depth, etc., with a plumb line.

plum·bism (plum′biz·əm) *n. Pathol.* Chronic lead poisoning.

plumb line 1 A cord by which a weight is suspended to test the perpendicularity or depth of something. **2** A plumb bob and its cord together. **3** A sounding line.

plumbo- *combining form* Lead; of or containing lead. Also, before vowels, *plumb-,* as in *plumbiferous.* [<L *plumbum* lead]

plum·bous (plum′bəs) *adj. Chem.* Of, pertaining to, or containing lead, especially in its lower valence.

plumb rule A narrow rule furnished with a plumb line or a cross level, with which masons and carpenters test the verticality of their work.

plum·bum (plum′bəm) *n.* Lead: so called in pharmacy and old chemistry. [<L]

plum duff A suet and flour pudding with raisins, currants, etc., boiled in a cloth bag.

plume (plōōm) *n.* **1** A feather, especially when long and ornamental. **2** A large feather or tuft of feathers worn as an ornament, especially on a helmet; a panache. **3** *Her.* Three feathers, unless more are specified. **4** A featherlike form or part; the plumose appendage of a seed. **5** Plumage. **6** A decoration of honor; a prize. **—***v.t.* **plumed, plum·ing 1** To adorn, dress, or furnish with or as with plumes. **2** To smooth or dress (itself or its feathers); preen. **3** To congratulate or pride (oneself): with *on* or *upon.* [<F <L *pluma* small soft feather]

PLUME *(def. 3)*
Prince of Wales.

plumed partridge The mountain quail *(Oreortyx picta)* of the western United States.

plume·let (plōōm′lit) *n.* **1** A plumule. **2** A little plume.

plu·mi·ped (plōō′mə·ped) *adj.* Having feathered feet. **—***n.* A plumiped bird, as an owl. Also **plu′mi·pede** (-pēd). [<L *pluma* feather + -PED]

plum·met (plum′it) *n.* **1** A piece of lead or heavy substance, attachable to a line for making soundings, adjusting walls to the vertical, etc.; a plumb bob; hence, a standard of truth or rectitude. **2** A plumb rule. **3** A weight; especially an oppressive weight. **—***v.i.* To drop straight down; plunge. [<OF *plommet,* dim. of *plom* lead]

plum·my (plum′ē) *adj.* **·mi·er, ·mi·est 1** Full of plums. **2** *Colloq.* Full of desirable things; profitable.

plu·mose (plōō′mōs) *adj.* **1** Bearing feathers or plumes. **2** Having fine processes on opposite sides, like the vane of a feather. **3** Resembling plumes. [<L *plumosus* <*pluma* feather] **—plu′mose·ly** *adv.* **—plu·mos·i·ty** (plōō·mos′ə·tē) *n.*

plump[1] (plump) *adj.* Swelled out or enlarged to the full; somewhat fat. See synonyms under ROUND[1]. **—***v.t.* & *v.i.* To make or become plump: often with *up* or *out.* **—***n. Archaic* A closely united group; a cluster or clump. [<MDu. *plomp*] **—plump′ly** *adv.* **—plump′·ness** *n.*

plump[2] (plump) *v.i.* **1** To fall suddenly or heavily; drop with full impact. **2** To give one's complete support: with *for.* **—***v.t.* **3** To drop or throw down heavily or all at once. **4** To utter bluntly or abruptly: often with *out.* **—***n.* The act of plumping or falling; the sound made by the impact of a falling object. **—***adj.* Containing no reservation or qualification; blunt; downright. **—***adv.* **1** With a sudden impact or fall into or as into water; in a sudden or forcible manner; also, unexpectedly. **2** Directly; without hesitation, circumlocution, or qualification; bluntly. [<MDu. *plompen*] **—plump′ly** *adv.*

plump·er[1] (plum′pər) *n.* **1** A heavy fall or drop. **2** Votes cast all for one candidate instead of for several; also, a person so voting. **3** *Brit. Slang* An unqualified lie.

plump·er[2] (plum′pər) *n.* A disk or padding placed in the mouth, as by persons who have lost their teeth, to distend the cheek and give it an appearance of plumpness.

plum pudding A boiled pudding made with flour, raisins, suet, currants, spices, etc.

plu·mule (plōō′myōōl) *n.* **1** *Ornithol.* A feather having the barbs soft and free; a downy feather. **2** *Bot.* The rudimentary or first bud of a plant embryo; the first bud of a germinating plant above the cotyledons. [<L *plumula,* dim. of *pluma* feather]

plum·y (plōō′mē) *adj.* **plum·i·er, plum·i·est 1** Covered with feathers. **2** Adorned with plumes.

plun·der (plun′dər) *v.t.* **1** To rob of goods or property by open violence, as in war; pillage; loot. **2** To despoil by robbery or fraud. **3** To take as plunder. **—***v.i.* **4** To take plunder; steal. See synonyms under STEAL. **—***n.* **1** That which is taken by plundering; booty. **2** The act of plundering or robbing. **3** *U.S. Colloq.* Personal belongings or goods, etc. **4** Political booty. [<G *plündern*] **—plun′der·er** *n.*

Synonyms (noun): booty, pillage, prey, rapine, robbery, spoil.

plun·der·age (plun′dər·ij) *n.* **1** Pillage; specifically, in maritime law, the embezzlement of goods on board a ship. **2** The goods so obtained.

plunge (plunj) *v.* **plunged, plung·ing** *v.t.* **1** To thrust or force suddenly into a fluid, penetrable substance, hole, etc. **2** To force into some condition or state: to *plunge* a nation into debt. **—***v.i.* **3** To dive, jump, or fall into a fluid, chasm, etc. **4** To move suddenly or with a rush: to *plunge* through a door. **5** To move violently forward and downward, as a horse or ship. **6** To descend abruptly or steeply, as a road or cliff. **7** *Colloq.* To gamble or speculate heavily and recklessly. See synonyms under IMMERSE. **—***n.* **1** The act of plunging; a leap; dive. **2** A sudden and violent motion, as of a breaking wave. **3** A place, tank, or pool for diving or swimming. **4** An exceptionally heavy bet or speculation; an extravagant or reckless expenditure. [<OF *plunjer,* ult. <L *plumbum* lead]

plung·er (plun′jər) *n.* **1** One who or that which plunges; a heavy or reckless speculator. **2** *Mech.* Any appliance having or adapted for a plunging motion, as the piston of a force pump. **3** A cavalryman. **4** A small boat.

plunk (plungk) *Colloq. v.t.* **1** To pluck, as a banjo or its strings; strum. **2** To place or throw heavily and suddenly: with *down.* **—***v.i.* **3** To emit a twanging sound. **4** To fall heavily or suddenly; plump. **—***n.* **1** A heavy blow, or its sound. **2** *Slang* A dollar. [Imit.]

plu·per·fect (plōō·pûr′fikt) See PAST PERFECT.

plu·ral (plŏŏr′əl) *adj.* **1** Containing, consisting of or designating more than one. **2** *Gram.* Denoting more than one (in languages that have dual number, such as Sanskrit and Greek, more than two): opposed to *singular.* **—***n.* *Gram.* The plural number, or a word in this number. [<L *pluralis* <*plus* more] **—plu′·ral·ly** *adv.*

◆ English nouns regularly form their plurals by adding *s* or *es* to the singular; most nouns ending in *f* change the *f* to *v* and add *es*; as wol*f,* wol*ves*; hal*f,* hal*ves.* Nouns ending in *y* change it to *ies* if it is preceded by a consonant: bod*y,* bod*ies*; or merely add an *s* if it is preceded by a vowel; as donke*y,* donke*ys.* Some nouns of Old English origin have an irregular plural in *en,* as, child, child*ren*; or by a vowel change; as, mouse, *mice*; goose, *geese*; man, *men*; tooth, *teeth.* A few nouns retain the singular form unchanged in the plural; as, *deer, hose, moose, series, sheep, species, vermin.* Some such nouns, especially the names of animals, have also an alternative plural regularly formed: as, fish, *fish* or *fishes. Fish* is the usual collective plural; *fishes* is used to indicate more than one genus, variety, species, etc. Many words of foreign derivation retain the plural form peculiar to the languages from which they are severally derived; as, addendum, *addenda*; antithesis, *antitheses*; crisis, *crises*; datum, *data,* etc. Many nouns of this class have also a plural of the regular English form; as, appendix, *appendixes* or *appendices*; beau, *beaus* or *beaux*; cherub, *cherubs* or *cherubim*; focus, *focuses* or *foci*; index, *indexes* or *indices,* etc. Compounds commonly form the plural regularly by adding *s* or *es* to the complete word; as, armful, *armfuls*; cut-throat, *cut-throats*; football, *footballs*; teaspoonful, *teaspoonfuls.* If the last element of the compound forms its plural irregularly, the same form usually appears in the plural of the compound; as, footman, *footmen.* Nouns that end in *-man,* but are not compounds, form the plural regularly by adding *s,* as Mussulman, *Mussulmans.* Hyphenated compounds in which the principal word forms the first element change that element to form the plural; as, father-in-law, *fathers-in law*; son-in-law, *sons-in-law.*

plu·ral·ism (plŏŏr′əl·iz′əm) *n.* **1** The condition of being plural. **2** *Eccl.* The holding of more than one office, or, in the Anglican church, of more than one ecclesiastical living, at one time. **3** *Philos.* The doctrine that there is a plurality of ultimate substances, as spirit and matter: opposed to *monism.* **4** A political theory that denies the concept of sovereignty and emphasizes the importance of organizations and groups within the state.

plu·ral·ist (plŏŏr′əl·ist) *n.* **1** One who holds more than one ecclesiastical benefice at the same time. **2** Anyone who holds a plurality

plu·ral·i·ty (ploo·ral'ə·tē) *n. pl.* **·ties** 1 The state of being plural. 2 The larger portion or greater number; majority. 3 In U.S. politics, the greatest of more than two numbers, whether it is or is not a majority of the whole; also, the excess of the highest number of votes cast for any one candidate over the next highest number. 4 *Eccl.* Pluralism; also, one of the livings held by a pluralist. 5 Polygamy.

plu·ral·ize (ploor'əl·īz) *v.t.* **·ized, ·iz·ing** 1 To make plural. 2 To express in the plural.

pluri– *combining form* More; many; several: *pluriaxial.* [< L *plus, pluris* more]

plu·ri·ax·i·al (ploor'ē·ak'sē·əl) *adj.* 1 Having more than one axis. 2 *Bot.* Denoting plants whose flowers grow on secondary shoots.

plus (plus) *prep.* 1 Added to or to be added to: Three *plus* two equals five: opposed to *minus*. 2 Increased by: salary *plus* commission. — *adj.* 1 Being or indicating more than nothing; above zero; positive. 2 Electrified positively. 3 *Colloq.* Possessing (something) in addition: used predicatively: He was *plus* a new hat. 4 Extra; supplemental: *plus* value. 5 *Colloq.* Denoting a value higher than ordinary in a specified grade: B *plus*. 6 *Bot.* Designating a form of sexual differentiation in certain plants: the *plus* strain of heterothallic fungi. — *n. pl.* **plus·es** 1 The plus sign. 2 An addition; an extra quantity. 3 A positive quantity. — *adv. Electr.* Positively. [< L, more]

plus–fours (plus'fôrz', -fōrz') *n.* Knickerbockers, cut very full and bagging below the knees. [Orig. tailor's cant; because they were four inches longer than ordinary knickerbockers]

plush (plush) *n.* A pile fabric of silk, rayon, or mohair having a deeper pile than velvet. — *adj.* 1 Of or made of plush. 2 *Slang* Luxurious. [< F *pluche, peluche* < L *pilus* hair] — **plush'y** *adj.*

plus sign The symbol (+) signifying addition or a positive quantity: opposed to *minus sign*.

Plu·tarch (ploo'tärk) A.D. 46?–120?, Greek moralist and biographer.

plu·tar·chy (ploo'tär·kē) *n. pl.* **·chies** Government by the rich. [< Gk. *ploutos* wealth + *archein* rule]

Plu·to (ploo'tō) 1 In Greek and Roman mythology, the god of the dead: identified with the Greek *Hades* and the Roman *Dis*. 2 *Astron.* The ninth planet of the solar system in order of distance from the sun, invisible to the naked eye: discovered 1930. See PLANET. [< L < Gk. *Ploutōn*]

plu·toc·ra·cy (ploo·tok'rə·sē) *n. pl.* **·cies** 1 A class in a community that controls the government by its wealth; the wealthy classes. 2 Plutarchy. [< Gk. *ploutokratia* < *ploutos* wealth + *kratein* rule]

plu·to·crat (ploo'tə·krat) *n.* One who has or exercises power by virtue of his wealth; one of a plutocracy. — **plu'to·crat'ic** or **·i·cal** *adj.* — **plu'to·crat'i·cal·ly** *adv.*

plu·to·ma·ni·a (ploo'tə·mā'nē·ə, -mān'yə) *n.* An excessive desire for great wealth. [< Gk. *ploutos* wealth + -MANIA]

Plu·to·ni·an (ploo·tō'nē·ən) *adj.* 1 Pertaining to Pluto and the lower world; hence, subterranean. 2 *Geol.* Of or pertaining to the Plutonic theory of rock formation. Also **Plu·ton'ic** (-ton'ik). [< L *Plutonius* < Gk. *Ploutonios* like Pluto]

plu·ton·ic (ploo·ton'ik) *adj. Geol.* Deeply subterranean in original position; crystallized, probably from a fused condition: said of igneous rocks: distinguished from *volcanic*. [< L *Pluto, -onis*]

Plutonic theory *Geol.* The doctrine that the principal phenomena of rock structure are chiefly due to igneous agency: distinguished from *Neptunian theory*.

plu·to·ni·um (ploo·tō'nē·əm) *n.* A radioactive element (symbol Pu), of atomic number 94, occurring in several isotopes, of which Pu-239, formed as a decay product of neptunium, was produced by fission during research on the atomic bomb: an important source of atomic energy. [< NL < *Pluto* (the planet)]

Plu·tus (ploo'təs) In Greek mythology, the god of riches, blinded by Zeus so that his gifts should be distributed without discrimination.

plu·vi·al (ploo'vē·əl) *adj.* 1 Pertaining to rain; rainy. 2 Arising from the action of rain. [< L *pluvialis* < *pluvia* rain]

pluvio– *combining form* Rain; pertaining to rain: *pluviometer*. Also, before vowels, **pluvi–**. [< L *pluvia* rain]

plu·vi·om·e·ter (ploo'vē·om'ə·tər) *n.* An instrument for measuring the depth of rainfall. [< PLUVIO– + -METER] — **plu'vi·om'e·tric** (-ə·met'rik) or **·ri·cal** *adj.* — **plu'vi·om'e·try** *n.*

Plu·vi·ôse (ploo'vē·ōs, *Fr.* plü·vyôs') See under CALENDAR (Republican).

plu·vi·ous (ploo'vē·əs) *adj.* Pertaining to rain; rainy. Also **plu'vi·ose** (-ōs). [< L *pluviosus*]

ply[1] (plī) *v.* **plied, ply·ing** *v.t.* To bend; mold; shape. — *v.i. Obs.* To bend or yield. — *n. pl.* **plies** 1 A web, layer, fold, or thickness, as in a carpet, cloth, etc. 2 A strand, turn, or twist of rope, yarn, thread, etc.: used in combination to mean a (certain) number of folds, twists, or strands: three-*ply* yarn. 3 A bent or bias; inclination to one side, as of the mind. [< F *plier* < L *plicare*]

ply[2] (plī) *v.* **plied, ply·ing** *v.t.* 1 To use in working, fighting, etc.; wield; employ. 2 To work at; be engaged in: He *plies* the trade of shoemaker. 3 To subject to repeated action, as by offering unwanted gifts, asking questions insistently, etc.: to *ply* a person with drink; to *ply* one with requests. 4 To strike or assail persistently: He *plied* the donkey with a whip. 5 To traverse regularly: ferryboats that *ply* the river. — *v.i.* 6 To make regular trips; sail: usually with *between*. 7 To work steadily; do one's or its work. 8 *Poetic* To proceed; steer. 9 *Naut.* To beat; tack. [Aphetic var. of APPLY]

ply·er (plī'ər) *n.* 1 A plier. 2 *pl.* A balance of crossed timbers used in raising and lowering a drawbridge.

Plym·outh (plim'əth) 1 A city of eastern Massachusetts; first settlement (*Plymouth Colony*) in New England; site of Plymouth Rock. 2 A port on **Plymouth Sound**, an inlet of the English Channel between Cornwall and Devon in SW England.

Plymouth Colony The colony on the shore of Massachusetts Bay founded by the Pilgrim Fathers who sailed from Plymouth, England, in 1620.

Plymouth Rock 1 The rock at Plymouth, Massachusetts, on which the Pilgrim Fathers are said to have stepped when landing from the *Mayflower* in 1620. 2 One of a breed of domestic fowls of large size, with small single comb and buff, white, black, or gray barred plumage.

ply·wood (plī'wood') *n.* Laminated wood consisting of an odd number of sheets or plies tightly glued together, the grains of adjoining layers usually being at right angles to each other: widely used as a structural and building material.

Pl·zeň (pul'zen'y') The Czech name for PILSEN.

pne·o·ste·no·sis (nē'ō·stə·nō'sis) *n. Pathol.* A condition marked by any obstruction of air entering or leaving the respiratory tract. [< Gk. *pneein* breathe + STENOSIS]

pneu·ma (noo'mə, nyoo'-) *n.* The breath of life; the soul or spirit. [< Gk.]

pneu·mat·ic (noo·mat'ik, nyoo'-) *adj.* 1 Pertaining to the science of pneumatics. 2 Describing machines or devices that make use of compressed air: a *pneumatic* engine. 3 Pertaining to or containing air or gas, especially compressed air: a *pneumatic* tire. Also **pneu·mat'i·cal.** — *n.* A pneumatic tire. [< L *pneumaticus* < Gk. *pneumatikos* < *pneuma* breath < *pneein* breathe] — **pneu·mat'i·cal·ly** *adv.*

pneu·mat·ics (noo·mat'iks, nyoo'-) *n.* The branch of physics that treats of the mechanical properties of air and other gases, such as their pressure, elasticity, and density, and also of pneumatic mechanisms.

pneumato– *combining form* 1 Air: *pneumatophore*. 2 Breath; breathing: *pneumatometer*. 3 Spirit; spirits: *pneumatology*. Also, before vowels, **pneumat–**. [< Gk. *pneuma, pneumatos* air, spirit, breath]

pneu·ma·tog·ra·phy (noo'mə·tog'rə·fē, nyoo'-) *n.* Spirit writing. [< PNEUMATO– + -GRAPHY]

pneu·ma·tol·o·gy (noo'mə·tol'ə·jē, nyoo'-) *n.* 1 The doctrine of the nature and operation of spirit, or a treatise on that science; the science of spiritual beings or existence. 2 The science of the beliefs of men touching a world of spirits. 3 The science dealing with the physiology of air or gases. 4 *Obs.* Pneumatics. — **pneu'ma·to·log'ic** (-tə·loj'ik) or **·i·cal** *adj.* — **pneu'ma·tol'o·gist** *n.*

pneu·ma·tol·y·sis (noo'mə·tol'ə·sis, nyoo'-) *n. Geol.* The process of forming minerals during the later stages in the consolidation of molten rock-magmas under the influence of the gases which are then present.

pneu·ma·to·lyt·ic (noo'mə·tō·lit'ik, nyoo'-) *adj.* Of, pertaining to, formed by, or characteristic of pneumatolysis.

pneu·ma·tom·e·ter (noo'mə·tom'ə·tər, nyoo'-) *n.* An instrument for measuring the volume of air exhaled or inhaled at one breath; a spirometer. — **pneu'ma·tom'e·try** *n.*

pneu·ma·to·phore (noo'mə·tə·fôr', -fōr', nyoo'-) *n.* 1 *Zool.* The air-containing sac of a siphonophore. 2 *Bot.* A root structure found on certain tropical swamp trees: it contains lenticels and is supposed to act as a respiratory organ. — **pneu'ma·toph'o·rous** (-tof'ər·əs) *adj.*

pneu·ma·to·ther·a·py (noo'mə·tō·ther'ə·pē, nyoo'-) *n. Med.* The treatment of disease by rarefied or condensed air; also, the use of gases for the relief of pain, asphyxiation, etc.

pneu·mec·to·my (noo·mek'tə·mē, nyoo'-) *n. Surg.* The operation of removing lung tissue or a part of the lung. [< PNEUM(O)– + -ECTOMY]

pneumo– *combining form* Lung; related to the lungs; respiratory: *pneumobacillus*: also **pneumono–**. Also, before vowels, **pneum–**. [< Gk. *pneumon, pneumonos* a lung]

pneu·mo·ba·cil·lus (noo'mō·bə·sil'əs, nyoo'-) *n. pl.* **·cil·li** (-sil'ī) A bacillus (*Klebsiella pneumoniae*) found in infections of the respiratory tract.

pneu·mo·coc·cus (noo'mō·kok'əs, nyoo'-) *n. pl.* **·coc·ci** (-kok'sī) Any of a group of bacteria (genus *Diplococcus*) which inhabit the respiratory tract of man and animals, especially *D. pneumoniae*, the causative agent of lobar pneumonia. — **pneu'mo·coc'cal, pneu'mo·coc'cous, pneu'mo·coc'cic** (-kok'sik) *adj.*

pneu·mo·co·ni·o·sis (noo'mō·kon'ē·ō'sis, nyoo'-) *n. Pathol.* Any of various lung disorders resulting from the inhalation of dust or other minute particles. Also **pneu'mo·no·con'i·o'sis** (noo'mō·nō-, nyoo'-). [< PNEUMO– + Gk. *konia* dust + -OSIS]

pneu·mo·dy·nam·ics (noo'mō·dī·nam'iks, nyoo'-) *n.* Pneumatics.

pneu·mo·ec·ta·sis (noo'mō·ek'tə·sis, nyoo'-) *n. Pathol.* Emphysema of the lungs. [< PNEUMO– + Gk. *ektasis* extension, swelling < *ekteinein* stretch out < *ek*– out + *teinein* stretch]

pneu·mo·e·de·ma (noo'mō·i·dē'mə, nyoo'-) *n. Pathol.* An abnormal accumulation of fluid in the intercellular cavities of the lungs. [< PNEUMO– + EDEMA]

pneu·mo·gas·tric (noo'mō·gas'trik, nyoo'-) *adj.* 1 Of or pertaining to the lungs and the stomach. 2 Of or pertaining to the vagus. — *n.* The vagus.

pneu·mo·graph (noo'mə·graf, -gräf, nyoo'-) *n.* An instrument which records movements of the chest in breathing.

pneu·mo·nec·to·my (noo'mə·nek'tə·mē, nyoo'-) *n. Surg.* The total removal of a lung. [< PNEUMON(O)– + -ECTOMY]

pneu·mo·ni·a (noo·mōn'yə, nyoo-) *n. Pathol.* An infectious disease characterized by inflammation of the lung tissue. The two principal types are *bronchopneumonia*, involving the bronchi and parenchyma of the lungs; and *lobar* or *croupous pneumonia*, involving one or more lobes of the lungs. [< NL < Gk. *pneumonia* < *pneumōn* lung < *pneein* breathe]

pneu·mon·ic (noo·mon'ik, nyoo-) *adj.* 1 Affected with pneumonia; pertaining to pneumonia. 2 Pulmonary. [< NL *pneumonicus* < Gk. *pneumonikos*]

pneumono– *combining form* Pneumo–.

pneu·mo·tho·rax (noo'mō·thôr'aks, -thō'raks,

nyōō′-) *n.* An accumulation of air or gas within the pleural cavity: sometimes artificially induced to collapse the lung in tuberculosis.

Pnom-Penh (nom′pen′, pnōōm·pen′y′) A city on the Mekong, capital of Cambodia; also *Phnompenh.* Also **Pnom Penh.**

Pnyx (niks) The place in ancient Athens where the people met to deliberate and vote upon public affairs.

Po (pō) The largest river in Italy, flowing 405 miles from the Alps to the Adriatic: ancient *Padus.*

po·a·ceous (pō·ā′shəs) *adj. Bot.* Of or pertaining to a large, widely distributed family *(Poaceae)* of annual or perennial herbs, the grasses. The inflorescence is spicate, racemose, or paniculate, with very small flowers, generally perfect or staminate. The fruit is a seedlike grain *(caryopsis),* having a starchy endosperm. The grasses producing food grains are known as cereals. Formerly called *Gramineae* or *Graminaceae.* [<Gk. *poa* grass + -ACEOUS]

poach[1] (pōch) *v.t.* To cook (eggs without their shells, fish, etc.) in boiling water, milk, or other liquid until coated. [<OF *pochier* put in a pocket <*poche* pocket; from the "pocketed" position of the egg yolk]

poach[2] (pōch) *v.i.* **1** To trespass on another's property, etc., especially for the purpose of taking game or fish. **2** To take game or fish unlawfully. **3** To become soft and muddy by being trampled: said of land. **4** To sink into mud or soft earth while walking. —*v.t.* **5** To trespass on, as for taking game or fish. **6** To take (game or fish) unlawfully. **7** To make muddy or tear up (land, etc.) by trampling. **8** To reduce to a uniform consistency by mixing with water, as clay. [<OF *pochier* thrust one's fingers into <LG *poken* poke] —**poach′·er** *n.*

poach·y (pō′chē) *adj.* Easily trodden into holes by cattle; soft and miry. [<POACH[2] (def. 3)] —**poach′i·ness** *n.*

Po·ca·hon·tas (pō′kə·hon′təs), 1595?–1617, American Indian princess; daughter of Powhatan, a Virginian chief; she reputedly saved the life of Captain John Smith.

po·chard (pō′chərd, -kərd) *n.* A sea duck (genus *Nyroca*) having the head and neck reddish, found in America, Europe, and South Africa. *N. ferina* is the **common pochard** of the Old World; *N. americana,* the **American pochard** or redhead. [Origin uncertain]

pock[1] (pok) *n.* **1** A pustule in an eruptive disease, as in smallpox; a pockmark. **2** *Obs.* Smallpox. [OE *poc*]

pock[2] (pok) *n. Scot.* A bag; pouch: also spelled *poke.*

pock·et (pok′it) *n.* **1** A small bag or pouch; especially, a pouch attached to a garment, as for carrying money. **2** Hence, money; pecuniary means or interests. **3** A cavity, opening, or receptacle. **4** *Mining* A cavity containing gold or other ore; also, an accumulation of alluvial gold in one spot. **5** One of the pouches in a billiard or pool table, into which the balls are driven. **6** A bin for holding grain, coal, etc., for storage. **7** A glen among mountains. **8** In horse-racing, a position in which a horse is behind the leading horse or horses, and is kept from going past by others at the side. **9** An air pocket. —**in one's pocket** **1** On terms of intimacy as close to one as one's pocket. **2** Under one's influence or control. —*adj.* **1** Diminutive, as if pocketable. **2** Pertaining to, for, or carried in a pocket: *pocket* lining, *pocket* knife. —*v.t.* **1** To put into or confine in a pocket. **2** To appropriate as one's own, especially dishonestly, as profits or funds. **3** To enclose as if in a pocket. **4** To accept or endure without open resentment or reply, as an insult. **5** To conceal or suppress: *Pocket* your pride. **6** To retain without signing. See POCKET VETO. **7** In billiards, etc., to drive (a ball) into a pocket. [<AF *pokette, poquette,* dim. of OF *poque, poche* bag, pouch] —**pock′et·a·ble** *adj.* —**pock′et·er** *n.*

pock·et·book (pok′it·bo͝ok′) *n.* **1** A small book or case for carrying money and papers in the pocket; wallet. **2** A woman's purse or handbag. **3** A notebook or other book for the pocket. **4** Money or pecuniary resources.

pocket borough In England before the Reform Bill of 1832, a Parliamentary borough owned or controlled by a single individual or family; hence, any constituency controlled by a boss.

pocket edition An edition or copy of a book small enough to be carried in the pocket.

pock·et·ful (pok′it·fo͝ol′) *n. pl.* **·fuls** As much as a pocket will hold.

pocket knife A knife, having one or more blades which fold into the handle, for carrying in the pocket; a penknife.

pocket money Money for occasional expenses; spending money.

pocket veto *U.S.* The act of a chief executive who, where the legislative session will end within the period allowed for returning a measure with his signature or veto, simply retains ("pockets") it until the session adjourns and thus causes it to fail without a direct veto.

pock·et·y (pok′it·ē) *adj.* **1** Characterized by pockets: said of a lode or a placer. **2** Characterized by air pockets.

pock·mark (pok′märk′) *n.* A pit or scar left on the skin by smallpox or similar diseases. —**pock′marked′** *adj.*

pock·y (pok′ē) *adj.* **pock·i·er, pock·i·est** **1** Pertaining to, resembling, or affected with smallpox; pockmarked. **2** Syphilitic.

po·co (pō′kō) *adv. Music* Slightly; a little. [<Ital.]

po·co a po·co (pō′kō ä pō′kō) *Music* Little by little; gradually. [<Ital.]

po·co·cu·ran·te (pō′kō·ko͞o·ran′tē, *Ital.* pō′kō·ko͞o·rän′tā) *adj.* Indifferent; not caring. —*n.* An indifferent person. [<Ital. *poco curante* caring (but) little <*poco* (<L *paucus*) little + *curante,* ppr. of *curare* care <L] —**po′co·cu·ran′te·ism** or **·ran′tism** *n.*

pod[1] (pod) *n.* **1** A seed vessel or capsule of a plant; a legume. **2** Any dry and many-seeded dehiscent fruit. —*v.i.* **pod·ded, pod·ding** **1** To fill out like a pod. **2** To produce pods. [Origin uncertain]

pod[2] (pod) *n.* A flock or collection of animals, especially of seals, whales, or walruses. [Origin unknown]

pod[3] (pod) *n. Mech.* **1** The lengthwise groove in certain augers, bits, and gimlets. **2** An auger so grooved. [Origin unknown]

-pod *combining form* **1** One who or that which has (a specified number or kind of) feet: *arthropod.* **2** A (specified kind of) foot: *pleopod.* Also **-pode.** [<Gk. *pous, podos* a foot]

-poda *combining form Zool.* Plural of -POD: used in names of phyla, orders, classes, etc.: *Arthropoda.* [<NL <Gk. *pous, podos* a foot]

po·dag·ra (pō·dag′rə, pod′ə·grə) *n. Pathol.* Gout in the foot. [<L <Gk. <*pous, podos* foot + *agra* seizure] —**po·dag′ral, po·dag′ric** *adj.*

po·des·ta (pō·des′tə, *Ital.* pō′des·tä′) *n.* **1** A chief magistrate in the medieval Italian republics. **2** One of the governors of the Lombard cities appointed by Frederick I. **3** A subordinate municipal judge in Fascist Italy. [<Ital. *podestà* <L *potestas* power]

podg·y (poj′ē) *adj.* **podg·i·er, podg·i·est** Dumpy and fat. [Var. of PUDGY] —**podg′i·ness** *n.*

po·di·a·try (pə·dī′ə·trē, pō-) *n.* The study and treatment of diseases of the feet. [<Gk. *pous, podos* foot + -IATRY] —**po·di′a·trist** *n.*

po·dis·mos (pō·dis′məs) *n.* In ancient Greece, any of several military dances performed in full battle dress, simulating pursuit and victory. [<Gk.]

po·di·um (pō′dē·əm) *n. pl.* **·di·a** (-dē·ə) **1** *Archit.* **a** A solid basement or pedestal supporting a structure, as a Roman temple. **b** The parapet surrounding the arena of an ancient amphitheater or circus, and hence also the platform or path behind or above it. **2** A dais, platform, or stage; especially, the platform for the conductor of an orchestra. **3** *Zool.* A foot, or any footlike structure. [<L <Gk. *podion,* dim. of *pous, podos* foot]

-podium *combining form* A footlike part: *pseudopodium.* [<Gk. *podion,* dim. of *pous, podos* a foot]

Po·do·li·a (pō·dō′lē·ə) A region of SW Ukrainian S.S.R.

Po·dolsk (po·dôlsk′) A city 23 miles south of Moscow in Russian S.F.S.R.

pod·o·phyl·lin (pod′ə·fil′in) *n.* A bitter, resinous substance obtained from the dried root of *Podophyllum,* the May apple, used in medicine as a purgative. [<NL *Podophyllum,* generic name of the May apple <Gk. *pous, podos* foot + *phyllon* leaf]

-podous *combining form* -footed: used in adjectives corresponding to nouns in *-pod* and *-poda*: *arthropodous.* [<-POD + -OUS]

Po·dunk (pō′dungk) *n.* One of a tribe of North American Indians of Algonquian stock, formerly inhabiting parts of Connecticut and Massachusetts.

Po·dunk (pō′dungk) *n.* Any small town regarded as typically dull and non-progressive. [? from *Podunk,* Massachusetts <N. Am. Ind.]

po·du·rid (pō·do͞or′id, -dyo͞or′-) *n.* Any of a widely distributed family *(Poduridae)* of primitive insects which includes the springtails. [<NL *Podura* <Gk. *pous, podos* foot + *oura* tail; from their ability to leap by sudden extensions of their infolded tails]

pod·zol (pod′zol) *adj.* Of, pertaining to, or designating a major soil type of northern regions developed principally under forest conditions and characterized by a strongly acid, infertile humus underlying a thin mat of leaves and decayed vegetation. Also **pod·zol′ic.** —*n.* Podzol soil. Also **pod′sol** (-sol). [<Russian, ashlike, salty <*sol′* salt]

pod·zol·i·za·tion (pod′zol·ə·zā′shən, -ī·zā′-) *n.* The process or processes by which a soil develops podzol characteristics.

Poe (pō), **Edgar Allan,** 1809–49, U.S. poet, critic, and short story writer.

po·e·chore (pō′ə·kôr, -kōr) *n. Ecol.* The semiarid regions of the steppes. [<Gk. *poa* grass + *chōra* region] —**po′e·chor′ic** (-kôr′ik, -kor′ik) *adj.*

po·em (pō′əm) *n.* **1** A composition in verse, either in meter or in free verse, characterized by the imaginative treatment of experience and a heightened use of language more intensive than ordinary speech. **2** Any composition in verse. **3** Any composition characterized by intensity and beauty of language or thought: a prose *poem.* **4** Any experience which produces an effect upon the mind similar or likened to that of a poem: a *poem* in stone. See synonyms under POETRY, SONG. [<F *poème* <L *poema* <Gk. *poiēma,* lit., anything made <*poieein* make]

poe·nol·o·gy (pē·nol′ə·jē) See PENOLOGY.

po·e·sy (pō′ə·sē, -zē) *n. pl.* **·sies** **1** *Poetic* Poetry taken collectively. **2** *Poetic* The art or faculty of writing poetry. **3** *Obs.* A poem. **4** *Obs.* A motto or conceit, as one engraved on jewelry. See synonyms under POETRY, SONG. [<OF *poesie* <L *poesia* <Gk. *poiēsis* <*poieein*]

po·et (pō′it) *n.* **1** One who writes poems. **2** One especially endowed with imagination, the power of rhythmical expression, and the creative faculty or power of artistic construction. [<OF *poete* <L *poeta* <Gk. *poiētēs* <*poieein* make] —**po′et·ess** *n. fem.*

Synonyms: bard, minnesinger, minstrel, rimer, rimester, singer, troubadour. Compare synonyms for POETRY.

po·et·as·ter (pō′it·as′tər) *n.* An inferior poet; a mere rimer or writer of mediocre verse. [<NL]

po·et·ic (pō·et′ik) *adj.* **1** Pertaining to poetry; having the nature or quality of or expressed in poetry: a *poetic* theme. **2** Pertaining to, befitting, or characteristic of a poet: *poetic* fire. **3** Fit to be described in poetry; of a nature to evoke poetic expression: a *poetic* incident or scene. **4** Having or showing the sensibility, feelings, faculty, etc., of a poet. **5** Celebrated or recounted in poetry or verse. Also **po·et′i·cal.** —*n.* Poetics. [<F *poétique* <L *poeticus* <Gk. *poiētikos*]

poetic justice The distribution of rewards to the good and punishment to the evil as often represented in literature; ideal justice.

poetic license The departure from the rules of diction, pronunciation, or from what is generally regarded as fact, for the sake of rime, meter, or an over-all enhancement of effect.

po·et·ics (pō·et′iks) *n. pl. (construed as singular)* **1** The principles and nature of poetry or, by extension, of any art: the *poetics* of music. **2** A treatise on poetry. Also *poetic.*

po·et·ize (pō′it·īz) *v.* **·ized, ·iz·ing** *v.i.* **1** To write poetry. —*v.t.* **2** To turn into or describe by means of poetry; express in poetic form. **3** To make poetic. —**po′et·iz′er** *n.*

poet laureate *pl.* **poets laureate** **1** The poet officially invested with the title of laureate by the crown of England, an officer of the royal household receiving a salary and formerly expected to write for public occasions. **2** In former times, a poet publicly crowned with laurel in recognition of his merits, usually by a sovereign. **3** A poet acclaimed as the most eminent in a locality.

po·et·ry (pō'it·rē) *n.* **1** The writing of poems; the art by which the poet projects feeling and experience onto an imaginative plane, in rhythmical words, to stir the imagination and the emotions. **2** The quality or effect of a poem manifested in any work of literature. **3** That which resembles poetry: Dancing is the *poetry* of motion. **4** A work or works metrically composed; verse or poems collectively; also, metrical composition in general: a book of *poetry.* [<OF *poetrie* <LL *poetria* <L *poeta* poet]

Synonyms: meter, numbers, poem, poesy, rime, song, verse. In ordinary usage, *poetry* is both imaginative and metrical. *Poetry* often exists without *rime*; it may exist without regular *meter*, as in free verse; substitution may be made for *meter*, as in the Hebrew parallelism; *poetry* may be expressed in a way beautiful, lyrically comic, or sharply satiric, but it must involve, besides the artistic form, the exercise of the fancy or imagination to heighten, intensify, and integrate feeling or experience. Failing this, there may be *verse, rime,* and *meter,* but not *poetry.* In a very wide sense *poetry* may be anything rhythmical; as, the *poetry* of motion. There is much in literature that is beautiful and sublime in thought and artistic in construction, which is yet not *poetry,* in the strict sense, because quite devoid of the rhythmical element, and the patterned arrangement and economy of words; the dividing line between poetry and "the other music of prose" is hard to draw. Compare METER², SONG. *Antonym*: prose.

pog·a·mog·gan (pog'ə·mog'ən) *n.* A war club consisting of a stone or antler secured to the end of a wooden handle, used as a weapon and as a ceremonial symbol by the Plains Indians and also by the Algonquians around the Great Lakes. [<Algonquian]

Po·ga·ny (pō·gä'nē, *Hungarian* pō'gän·y'), **Willy,** 1882–1955, U. S. illustrator, mural painter, and designer, born in Hungary.

po·gey bait (pō'gē) *U. S. Slang* Any confection, as candy bars, etc.

po·go·ni·a (pə·gō'nē·ə, -gōn'yə) *n.* One of a genus (*Pogonia*) of widely distributed terrestrial orchids, especially an American species, *P. ophioglossoides*, having fragrant, rose-pink flowers. [<NL <Gk. *pōgōn* a beard]

pog·o·nip (pog'ə·nip) *n. Meteorol.* A cold fog containing particles of ice, characteristic of the Sierra Nevada mountains and valleys; a frost fog. [<Shoshonean (Paiute)]

po·go stick (pō'gō) A stiltlike toy, with a spring at the base and fitted with two projections for the feet, on which a person may stand and propel himself in a series of hops.

po·grom (pō'grəm, pō·grom') *n.* An officially instigated local massacre, especially one directed against the Jews. [<Russian, destruction]

po·gy (pō'gē, pog'ē) *n. pl.* **·gies** or **·gy** The menhaden. [<N. Am. Ind. *pauhagen*]

poh (pô) *interj.* Pshaw! bah! an expletive signifying disgust or contempt. [Imit.]

Po·hai (bō'hī'), **Strait of** See CHIHLI, STRAIT OF.

Po·hang (pō·häng) A town in southern Korea on the Sea of Japan. *Japanese* **Ho·ko** (hō·kō).

poi (poi, pō'ē) *n.* A native Hawaiian food made from the ground root of the taro. [<Hawaiian]

-poietic *combining form* Making; producing; creating: *hemapoietic.* [<Gk. *poiētikos* forming <*poieein* make]

poign·ant (poin'yənt, poi'nənt) *adj.* **1** Severely painful or acute to the spirit; keenly piercing; bitter; severe: *poignant* grief; a *poignant* retort. **2** Sharp or stimulating to the taste; pungent; biting. See synonyms under VIOLENT. [<OF, ppr. of *poindre* prick <L *pungere*] **—poign'·an·cy** *n.* **—poign'ant·ly** *adv.*

poi·ki·lo·ther·mal (poi'kə·lō·thûr'məl) *adj. Zool.* Variable in body temperature, as cold-blooded animals: distinguished from *homothermal.* [<Gk. *poikilos* variegated + THERMAL]

poi·lu (pwä·lü') *French adj.* Hairy; bearded. **—***n.* A French soldier; originally, an experienced French soldier of World War I.

Poin·ca·ré (pwaṅ·kȧ·rā'), **Jules Henri,** 1854–1912, French mathematician and author. **—Raymond,** 1860–1934, French statesman.

poin·ci·a·na (poin'sē·ā'nə, -an'ə) *n.* **1** One of a small genus of tropical trees or shrubs (*Poinciana,* family *Leguminosae*), especially the flower-fence. **2** The royal poinciana (*Delonix regia*), a tropical tree with bright orange and scarlet flowers and large flat pods. [<NL, after M. de *Poinci,* a 17th century governor of the West Indies]

poind (poind) *v.t. Scot.* **1 a** To seize and sell (the property of a debtor) to satisfy a debt. **b** To distrain the property of. **2** To impound.

poin·set·ti·a (poin·set'ē·ə) *n.* Any of a genus (*Euphorbia*) of American plants of the spurge family, with large showy bracts and inconspicuous flowers, especially an ornamental evergreen hothouse shrub (*E. pulcherrima*) from Mexico, with richly colored, red, leaflike bracts. [after J. R. *Poinsett,* 1779–1851, U. S. statesman]

POINSETTIA
Flower in bracts.

point (point) *n.* **1** The sharp end of a thing, particularly of anything that tapers so as to be very small and keen at the extremity: the *point* of a needle or a thorn. **2** *pl.* The extremities of a horse. **3** An object, as a tool or instrument, having a sharp or tapering end, as a needle, etching tool, etc. **4** A tapering tract of land extending into water; a promontory; cape: *Point* Judith. **5** A prominent feature or peculiarity; typical attribute; salient quality; essential physical characteristic: the *points* of a thoroughbred horse. **6** That to which effort is directed, on which attention is fixed, or to which especial importance is attached; the precise subject of discussion; aim; gist; purport: the *point* of a story. **7** A particular place, location, or position. **8** A position considered as one of a series; a unit of fluctuation, as of count in a game: to gain a *point.* **9** A precise grade, limit, or degree attained or determined, especially in temperature. **10** A particular juncture in the course of events. **11** Any single item or particular; detail. **12** A vital step or division of an argument or discourse; a proposition; head: to note every *point*; to contest *point* by *point.* **13** In schools and colleges, a unit of credit equal to a certain number of hours of academic work. **14** An indivisible portion of time; a particular moment. **15** The moment when something is about or likely to be done or to take place; verge: on the *point* of starting; at the *point* of death. **16** Point lace. **17** A cord or strap by which a thing is fastened, as a rope for reefing sails. **18** In 16th and 17th century costume, a ribbon or string with an aglet on one end, used to fasten together two pieces of clothing. **19** A mark made by or as by the end of a pointed instrument or tool; a prick; puncture; dot. **20** Any mark of punctuation; especially, among printers, a period; stop; end. **21** *Ling.* A vowel point as used in Hebrew. **22** *Music* A dot or other mark to designate time, or formerly tone; also, a short tune or strain; also, such tune when played on an instrument as a signal. **23** A decimal point. **24** Point system (def. 5). **25** The attitude of a pointer or setter when it finds game: The dog came to a *point.* **26** In fencing, a thrust; also, in dancing, the act of pointing the foot downward. **27** A trifle; punctilio: a mere *point.* **28** In cricket, a fielder stationed the nearest to the right of the wicket and slightly in advance of it; also, the position thus occupied. **29** *pl.* In baseball, the positions occupied by the pitcher and catcher. **30** The leading group of an advanced guard. **31** One of the 32 divisions of the compass. See POINT OF THE COMPASS. **32** That which is conceived to have position, but not parts or dimensions, as the extremity of a line. **33** A unit of variation in price of shares, stocks, etc., in the stock market; also, a rumor on which speculation is made; a tip. **34** A fixed place from which position and distance are reckoned. **35** A spot or place which is regarded as having position only, without extent, as a locality. **36** The tail of an animal: used in the phrase *heads and points.* **37** *Electr.* Any of a set of contacts determining the direction of current flow in a circuit. See synonyms under CIRCUMSTANCE, END, TOPIC. **—at** (or **on, upon**) **the point of** On the verge of. **—beside the point** Irrelevant. **—in point** Pertinent. **—in point of** In the matter of; as regards. **—to make a point of** To treat as vital or essential. **—to see the point** To understand the purpose of a course of action; get the important meaning of a story, joke, etc. **—to stretch a point** To make an exception. **—to the point** Relevant. **—***v.t.* **1** To direct or aim, as a finger or weapon. **2** To indicate; direct attention to: often with *out*: to *point* the way; to *point* out errors. **3** To give force or point to, as a meaning or remark: often with *up.* **4** To shape or sharpen to a point. **5** To punctuate, as writing. **6** To mark or separate with points, as decimal fractions: with *off.* **7** In hunting, to indicate the presence or location of (game) by standing rigid and directing the muzzle toward it: said of dogs. **8** In masonry, to fill and finish the joints of (brickwork) with mortar. **9** *Ling.* To mark with a vowel point. **—***v.i.* **10** To call attention or indicate direction by or as by extending the finger: usually with *at* or *to.* **11** To direct the mind: Everything *points* to your being wrong. **12** To be directed; have a specified direction; tend; face: with *to* or *toward.* **13** To point game: said of hunting dogs. **14** *Med.* To come to a head, as an abscess. **15** *Naut.* To sail close to the wind. See synonyms under ALLUDE. [Fusion of OF *pointe* a sharp point (<Med. L *puncta* <L *punctus*) + OF *point* prick, dot, moment <L *punctum,* neut. of *punctus,* pp. of *pungere* prick]

point alphabet The alphabet of the point system for the blind.

Point Bar·row (bar'ō) See BARROW, POINT.

point-blank (point'blangk') *adj.* **1** Aimed directly at the mark; in gunnery, fired horizontally without allowing for dropping. **2** Hence, direct; plain: a *pointblank* question. **—***n.* A shot with direct aim. **—***adv.* In a horizontal line; hence, directly; without circumlocution.

point d'ap·pui (pwaṅ dȧ·pwē') *French* Point of support; base.

Point de Galle (pwaṅ də gäl) A former name for GALLE.

point d'es·prit (pwaṅ des·prē') *French* **1** Net or tulle with dots. **2** Lace with the small oval or square dots first used in Normandy lace.

point-de·vice (point'di·vīs') *adj.* Scrupulously neat; precise; finical. **—***adv.* Precisely; exactly. Also **point'-de-vise'.** [ME (*at point*) *devis,* i.e., (at an) exact (point) <OF *devis* exact]

Pointe-à-Pi·tre (pwaṅ·tȧ·pē'tr') A city on Grande-Terre, principal port and commercial center of Guadeloupe.

point·ed (poin'tid) *adj.* **1** Having a point. **2** Piquant; pungent; epigrammatic; to the point. **3** Aimed at a particular person; emphasized; conspicuous. See synonyms under ACUTE, SHARP. **—point'ed·ly** *adv.* **—point'ed·ness** *n.*

pointed arch A narrow, pointed arch used in medieval architecture in Europe, characteristic of the Gothic style: also called *Gothic arch.*

pointed architecture The European architecture of the Middle Ages characterized by its consistent use of the pointed arch, with details to correspond.

Pointe-Noire (pwaṅt·nwȧr') The capital of Middle Congo since 1950.

point·er (poin'tər) *n.* **1** One who or that which points. **2** A hand or index finger, as on a clock or scale. **3** A long tapering rod used in class rooms to point out things on wall maps, charts, diagrams, etc. **4** One of a breed of dogs trained to scent and point out game. **5** A useful bit of information; hint. **6** *Nav.*

add,āce,câre,pälm; end, ēven; it,īce; odd,ōpen,ôrder; to͝ok,po͞ol; up,bûrn; ə = a in *above,* e in *sicken,* i in *clarity,* o in *melon,* u in *focus*; yo͞o = u in *fuse*; oi,oil; ou,pout; ch,check; g,go; ng,ring; th,thin; t͟h,this; zh,vision. Foreign sounds ȧ,œ,ü,kh,ṅ; and ◆: see page xx. < from; + plus; ? possibly.

Pointers

One whose business is to bring the gun or turret to its proper elevation. Compare TRAINER. **7** The cowboy who rides at the head of the herd in a cattle drive.
Point·ers (poin'tərz) n. pl. Astron. Two stars, Alpha and Beta in the constellation Ursa Major, whose connecting line points nearly to the North Star: called *Dubhe* and *Merak*.
pointes (points) n. pl. In ballet, dancing on tiptoe. [<F]
Point Four The fourth point in President Truman's Inaugural Address, January 20, 1949, in which he recommended "a bold new program for making the benefits of our scientific advances and industrial progress available for the improvement and growth of underdeveloped areas."
poin·til·lism (pwan'tə·liz'əm) n. A French neo-impressionist method of producing effects of light by placing small spots of varying hues on a surface in close proximity, the eye blending them together. [<F *pointillisme* < *pointiller* mark with dots] — **poin'til·list** n.
point·ing (poin'ting) n. **1** The act of sharpening or bringing to a point. **2** *Punctuation*. See under PUNCTUATION. **3** In sculpture, the making of a plaster or clay model with points or marks at intervals and the transferring of these points to the surface of a stone block as an aid in reproducing the model accurately. **4** *Archit.* **a** The process of treating joints in masonry, slating, or tiling, by filling interstices, smoothing out, etc., to finish or repair and to weatherproof. **b** The removal of the thin top layer of mortar between courses of brick and masonry, to replace it with a more moisture-resistant compound. **5** In milling, the rubbing off of the points of grain.
Point Judith A promontory and lighthouse at the western entrance of Narragansett Bay, Rhode Island.
point lace Needlepoint (def. 2). — **point-laced** (point'lāst') adj.
point·less (point'lis) adj. Without a point; dull; also, having no significance: a *pointless* remark. See synonyms under BLUNT, FLAT¹. — **point'less·ly** adv. — **point'less·ness** n.
point of honor Something that vitally affects one's honor.
point of order In parliamentary language, a question of procedure under the rules.
point of the compass One of the 32 equidistant directions or division points marked on the card of the mariners' compass, or a corresponding point in the horizon, or a vertical plane passing through the horizon and one of such points. See COMPASS CARD.
point of view The relative position from which one sees an object, a proposition, or the like. Compare STANDPOINT.
point system 1 *Printing* A standard system of sizes for type bodies, 996 points of which are equal to 35 centimeters, one point being .0138 inch (or approximately 1/72 inch), as adopted by the Typefounders' Association of the United States. **2** Any system of raised letters for the blind, as braille, in which the alphabet is formed of groups of raised dots or points. **3** An academic system of allowing students to progress according to points or credits earned in individual subjects. **4** Any method of rating based on the accumulation of points.
point target A particular structure, object, or installation selected for direct gunfire or bombing. Compare AREA TARGET.
poise¹ (poiz) v. poised, pois·ing v.t. **1** To bring into or hold in balance; maintain in equilibrium. **2** To hold; support, as in readiness. **3** *Rare* To weigh. — v.i. **4** To be balanced or suspended; hover. — n. **1** The state or quality of being balanced; equilibrium; equipoise; also, indecision; suspense. **2** Equanimity; repose; dignity, as in bearing or carriage. **3** A balance weight or counterpoise. **4** Any position that indicates suspended motion. [<OF *il poise, peise*, 3rd person sing. of *peser* <L *pensare*, intens. of *pendere* weigh] — **pois'er** n.
poise² (poiz) n. pl. poise The unit of viscosity in the cgs system, equal to 1 dyne-second per square centimeter. [after Jean Marie *Poiseuille*, 1797?–1869, French physiologist]
poi·son (poi'zən) n. **1** Any substance which, introduced into an organism in relatively small amounts, acts chemically upon the tissues to produce serious injury or death. **2** *Physics* Any substance or material which, by absorbing neutrons, prevents fission in an atomic reactor. **3** Anything that tends to taint or destroy character or to mislead, corrupt, or pervert. — v.t. **1** To administer poison to; kill or injure with poison. **2** To put poison into or on. **3** To affect wrongfully; corrupt; pervert: to *poison* one's mind. — adj. Killing; venomous; corrupting. [<OF <L *potio, -onis* a drink, poisonous draft. Doublet of POTION.] — **poi'son·er** n.
poison dogwood, poison elder Poison sumac.
poison gas Any of a class of toxic chemical agents, usually a liquid having high vapor pressure, employed in warfare for the purpose of disabling or killing enemy personnel.
poison hemlock See under HEMLOCK.
poison ivy A climbing shrub (*Toxicodendron radicans* or *Rhus toxicodendron*), a species of sumac with three broadly ovate, variously notched, sinuate or cut-lobed leaflets and whitish berries: poisonous to many persons by touch.
poison oak 1 A species of poison sumac, especially *Toxicodendron quercifolium*. **2** A species of poison ivy (*T. rydlergii*) common in the western United States.
poi·son·ous (poi'zən·əs) adj. **1** Containing or being a poison. **2** Having the effect of a poison; toxic; vitiating. See synonyms under NOISOME. — **poi'son·ous·ly** adv. — **poi'son·ous·ness** n.
poison sumac 1 A handsome shrub or small tree (*Toxicodendron vernix* or *Rhus vernix*), growing in swamps in the United States and Canada. It has smooth, entire leaflets, and loose panicles of smooth greenish-yellow drupes. The whole plant is poisonous to taste or touch. **2** Poison ivy.
Poi·tiers (pwä·tyā') A city of west central France; formerly, capital of Poitou.
Poi·tou (pwä·tōō') A region and former province of west central France.
poi·trel (poi'trəl) n. The armor formerly used to protect the breast of a war horse. [<OF *poitral* <L *pectorale* breastplate <*pectus* breast]
poke¹ (pōk) v. poked, pok·ing v.t. **1** To push or prod, as with the elbow; jab: to *poke* a person in the ribs. **2** To make by or as by thrusting: to *poke* a hole. **3** To thrust or push in, out, through, from, etc.: to *poke* one's head from a window. **4** To stir (a fire, etc.) by prodding: often with *up*. — v.i. **5** To make thrusts, as with a stick or weapon: often with *at*. **6** To intrude or meddle. **7** To go or look curiously; pry. **8** To appear or show: logs *poking* above the surface. **9** To proceed slowly; dawdle; putter. — **to poke one's nose into** To meddle in. — **to poke fun at** To ridicule, especially slily. — n. **1** A push; prod. **2** A yokelike collar with long projections to prevent animals from passing through fences. **3** One who moves sluggishly; a dawdler. **4** *Colloq.* A punch. [ME *poken*. Cf. LG, Du. *poken* push.]
poke² (pōk) n. A pocket, or small bag. See POCK². [<OF <Gmc. Cf. ON *poki* and MDu. *poke*.]
poke³ (pōk) n. The pokeweed. [<earlier *pocan* <Algonquian (Virginian) *pakon* weed used for staining <*pak* blood. Akin to PUCCOON.]
poke⁴ (pōk) n. A large bonnet with projecting front. Also **poke bonnet**.
poke·ber·ry (pōk'ber'ē) n. pl. ·ries **1** A berry of the pokeweed. **2** The plant.
pok·er¹ (pō'kər) n. **1** One who or that which pokes. **2** An iron rod for poking a fire.
pok·er² (pō'kər) n. Any of several games of cards in which the players bet on the value of the cards, usually five, dealt to them, and he whose hand contains the group of highest value wins the entire sum wagered, provided he has not dropped out of the betting. The groups usually recognized, in the ascending order of value, are the *pair, two pairs, three of a kind, straight, flush, full hand or house, four of a kind, straight flush*. [Origin uncertain. Cf. G *pochspiel*, lit., bragging game <*pochen* brag.]
pok·er·face (pō'kər·fās') n. A face that reveals nothing: so called from the controlled and inscrutable faces of professional poker-players.
pok·er·ish (pō'kər·ish) adj. **1** Stiff or unbending, as a poker. **2** Ghastly; unearthly. — **pok'er·ish·ly** adv.

Polaris

poke·weed (pōk'wēd') n. A stout perennial herb of the United States (*Phytolacca americana*), having dark-purple berries and a root used in medicine: often called *inkberry*. Also **poke'root'** (-rōōt', -rŏŏt'). [See POKE³]
poke welding A method of electric welding in which manual pressure is applied to one electrode only.
pok·ing (pō'king) adj. **1** Drudging; servile; mean. **2** Projecting.
pok·y (pō'kē) adj. pok·i·er, pok·i·est **1** Lacking life or spirit; dull; slow. **2** Shabby. **3** Cramped; stuffy. Also **poke'y**.
Po·la (pō'lä) The Italian name for PULA.
po·lac·ca (pō·lak'ə) n. A two- or three-masted Mediterranean vessel. Also **po·la·cre** (pō·lä'kər). [<Ital.]
Po·lack (pō'lok, -lak) n. *Slang* A Pole; especially, an immigrant from Poland: a contemptuous term. — adj. Polish. [<Polish *polak*]
Po·land (pō'lənd) A republic of north central Europe; 120,359 square miles; capital, Warsaw: Polish *Polska*.
Poland China An American mixed breed of large pigs, similar to Berkshires.
Po·land·er (pō'lən·dər) n. A Pole.
po·lar (pō'lər) adj. **1** Pertaining to the poles of a sphere, as of the earth. **2** Coming from or found near the North or South Pole. **3** Pertaining to the poles of a magnet or other center of attraction or repulsion. **4** Exhibiting ionization. **5** Having or proceeding from a point of radiation. **6** Attracting; guiding. **7** *Math.* **a** Of or pertaining to a coordinate system of representing equations graphically whereby a point is located by its linear distance from the pole and by the angle subtended by a line from the point to the pole and the polar axis. **b** Of or pertaining to a curve or an equation traced or traceable by means of such coordinates. [<Med. L *polaris* < *polus* pole]

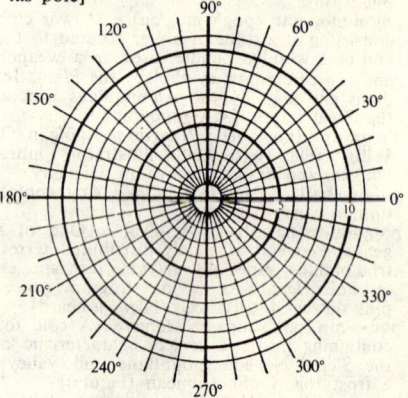

POLAR COORDINATE SYSTEM
The polar axis is the horizontal axis.

polar angle *Math.* In a polar coordinate system, the angle subtended between the polar axis and a line drawn from a point to the pole.
polar axis *Math.* A fixed line directed from the pole in the polar coordinate system from which angles are measured in a counterclockwise direction.
polar bear A large, amphibious, white bear of arctic regions (*Thalarctus maritimus*).
polar body *Biol.* One of the two spherical bodies that separate from the ovum at the time of its maturation.
polar circles The Arctic and Antarctic circles.
polar compound *Chem.* Any of a class of compounds which will conduct an electric current when either fused or in solution, as most inorganic acids, bases, and salts.
polar distance Codeclination.
polar front *Meteorol.* The line or surface of discontinuity separating an air mass originating in polar regions from one of tropical origin.
po·lar·im·e·ter (pō'lə·rim'ə·tər) n. **1** An instrument for measuring the rotation of the plane of polarization or the proportion of polarized light in a beam. **2** A form of polariscope. [<L *polaris* polar + -METER]
Po·lar·is (pō·lâr'is) The polestar or North Star: Alpha in the constellation Ursa Minor. See under STAR. [<L]

po·lar·i·scope (pō·lar′ə·skōp) *n.* An optical instrument for exhibiting or measuring the polarization of light. [<L *polaris* polar + -SCOPE]

po·lar·i·ty (pō·lar′ə·tē) *n.* 1 The quality of having opposite poles. 2 *Physics* That quality of a body by which it exhibits certain properties related to a line of direction through its mass, the properties at one end of this line being of opposite or contrasting nature to the properties at the other end, as in a magnet. 3 The quality of being attracted to one pole and repelled from the other.

po·lar·i·za·tion (pō′lər·ə·zā′shən, -ī·zā′-) *n.* 1 The act of polarizing, or state of being polarized; bestowal or gaining of polarity. 2 *Physics* A condition of radiant energy, most noticeable in light, in which its vibrations assume a definite form or direction when subjected to special influences. Light may be polarized by reflection, at an angle which differs for different substances, or by transmission, as through most crystals or solutions. If light thus treated be examined by subjecting it to such reflection or transmission a second time, it is found that in certain positions of the reflector or crystal it will pass most easily, while in the positions at right angles to these it will be totally quenched, and in intermediate positions it will pass partially. The plane of polarization is altered or rotated by the passage of light through suitable media; this is called **rotary polarization**, which takes two directions, right-handed and left-handed. 3 *Electr.* A change in the potential of the electrode of a cell due to the accumulation upon it of dissociation products liberated by the current. — **angle of polarization** or **polarizing angle** That angle of reflection from a plane surface at which light is polarized. — **plane of polarization** The plane in which the light vibrations occur when polarized.

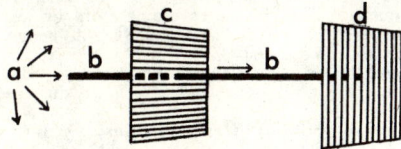

POLARIZATION PRINCIPLE
Light source at *a* emits rays, a beam of which *b* impinges on the first polarizing medium *c* whose axis is placed so as to allow the beam to pass through. The second medium *d* with its axis at right angles to that of *c*, arrests the beam, preventing it from emerging on the other side.

po·lar·ize (pō′lə·rīz) *v.t.* **·ized**, **·iz·ing** 1 To develop polarization in; give polarity to. 2 To give a special meaning or direction to. Also *Brit.* **po′lar·ise**. — **po′lar·iz′a·ble** *adj.* — **po′lar·iz′er** *n.*
polar lights The aurora borealis or the aurora australis.
Po·lar·oid (pō′lə·roid) *n.* A material composed of a sheet of specially prepared plastic between layers of glass and having the property of polarizing and thus reducing the intensity of the light passing through it: a trade name.
Polar Regions The areas within the Arctic and Antarctic circles.
polar star The polestar.
pol·der (pōl′dər) *n.* A tract of marshy land, lower than the sea, which has been diked and reclaimed to cultivation. Also **pol′der·land′**. [<Du.]
pole[1] (pōl) *n.* 1 Either of the extremities of an axis or sphere. 2 One of two points where the axis of rotation, as of the earth, meets the surface. 3 Either of the Polar Regions of the earth; also, either of the two extremities of the earth's axis, called the North *Pole* and the South *Pole*. See CELESTIAL POLE. 4 *Physics* One of the two points at which opposite physical qualities are concentrated; especially, a point (usually one of two) of maximum intensity of electric or magnetic force. 5 The polestar. 6 *Biol.* The differentiated extremities of an ovum or other cell. 7 *Physiol.* The point of a nerve cell where a process has its origin. 8 *Math.* In polar coordinate and spherical coordinate systems, that point where all radius vectors equal zero. ◆ Homophone: *poll*. [<L *polus* <Gk. *polos* pivot, pole <*pelein* be in motion]

pole[2] (pōl) *n.* 1 A long slender piece of wood or metal, commonly tapering and more or less rounded; a Maypole, beanpole, the mast of a vessel, etc. 2 The tongue of a vehicle. 3 In linear and surface measure, a perch or rod. 4 A fishing rod. — *v.* **poled**, **pol·ing** *v.t.* 1 To propel, push, or strike with a pole. 2 To support on poles, as growing beans. 3 To push a boat, raft, etc., with a pole. ◆ Homophone: *poll*. [OE *pal* <L *palus* stake]

Pole (pōl) *n.* A native or inhabitant of Poland.
Pole (pōl), **Reginald**, 1500–58, English statesman; archbishop of Canterbury.
pole-ax (pōl′aks′) *n.* A medieval weapon consisting of an ax, or a combined ax and pick, set on a long pole; a battle-ax. — *v.t.* To strike or fell with a pole-ax. Also **pole′-axe′**. [ME *pollax* <*pol* poll[1] + AX]
pole bean Any variety of climbing bean supported by poles.
pole·cat (pōl′kat′) *n.* 1 One of certain European carnivores (genus *Mustela*) of the weasel family, noted for a fetid odor when irritated or alarmed. 2 *U.S.* A skunk. [<F *poule* pullet + CAT; from its predacity]
pole fence A fence made of horizontal unsplit poles.
pole horse A horse hitched beside the pole, as distinguished from a leader.
pole jump See POLE VAULT.
pole line A line of telephone or telegraph poles.
po·lem·ic (pō·lem′ik) *adj.* Pertaining to controversy; disputatious. Also **po·lem′i·cal**. — *n.* 1 A controversy; also, the speeches, papers, etc., comprising this. 2 One who engages in controversy. [<Gk. *polemikos* warlike <*polemos* war]
po·lem·i·cist (pō·lem′ə·sist) *n.* One skilled or engaged in polemics. Also **pol·e·mist** (pol′ə·mist).
po·lem·ics (pō·lem′iks) *n.* The art or practice of disputation; especially, the use of aggressive argument to refute errors of doctrine.
pol·e·mo·ni·a·ceous (pol′ə·mō′nē·ā′shəs) *adj. Bot.* Designating or belonging to a family (*Polemoniaceae*) of herbs (rarely shrubs or small trees) including many ornamental garden species; the phlox family. [<Gk. *polemōnion* kind of plant]
pol·er (pō′lər) *n.* 1 The draft animal harnessed nearest the pole of a cart or wagon; a wheeler. 2 One who poles a boat.
pole·star (pōl′stär′) *n.* 1 *Astron.* The North Star; Polaris; Alpha in Ursa Minor. 2 That which governs, guides, or directs; an attracting or controlling principle.
pole-vault (pōl′vôlt′) *v.i.* To perform a pole vault. — **pole′-vault′er** *n.*
pole vault A vault or jump with a long pole, usually over a light horizontal bar: an athletic field event.
po·lice (pə·lēs′) *n.* 1 A body of civil officers, especially in a city, organized under authority to maintain order and enforce law; constabulary. 2 The whole system of internal regulation of a state, or the local government of a city or town; that department of government that maintains and enforces law and order, and prevents, detects, or deals with crime. 3 The cleansing or keeping clean of a camp or garrison; also, the soldiers detailed for the duties of policing in camp. — *v.t.* **·liced**, **·lic·ing** 1 To protect, regulate, or maintain order in (a city, country, etc.) with or as with police. 2 *U.S.* To make clean or orderly, as a military camp. [<F <LL *politia* governmental administration <Gk. *politeia* polity <*politēs* citizen <*polis* city]

police court A municipal court where minor criminal cases are tried. Its jurisdiction corresponds with that of a justice of the peace.
police dog See GERMAN SHEPHERD DOG.
po·lice·man (pə·lēs′mən) *n. pl.* **·men** (-mən) A member of a police force.
police power The broad authority of a state to limit private rights to the extent necessary to promote the peace, good order, morals, health, and safety of the general community.
police state A country whose citizens are rigidly supervised by a national police, often working secretly.
police station The headquarters of a community police force, to which arrested persons are taken and from which policemen operate.
po·lice·wom·an (pə·lēs′wŏŏm′ən) *n. pl.* **·wom·en** (-wim′in) A woman member of a police force.
pol·i·clin·ic (pol′i·klin′ik) *n.* The dispensary of a hospital, or that part of it in which out-patients are treated. Compare POLYCLINIC. [<G *poliklinik* <Gk. *polis* city + G *klinik* clinic]
pol·i·cy[1] (pol′ə·sē) *n. pl.* **·cies** 1 Prudence or sagacity in the conduct of affairs. 2 A course or plan of action, especially of administrative action. 3 Any system of management based on self-interest as opposed to equity; finesse in general; artifice. 4 *Obs.* Political science; government. See synonyms under POLITY. [<OF *policie* <L *politia* <Gk. *politeia*. See POLICE.]
pol·i·cy[2] (pol′ə·sē) *n. pl.* **·cies** 1 A written contract of insurance. 2 A gambling game in which certain numbers (12 or 13) are drawn from a possible 78, bets being made as to what combinations will appear; also, any variation of this game. [<F *police* <Ital. *polizza*, aphetic alter. of LL *apodixa*, *apodissa* receipt <Gk. *apodeixis* proof <*apodeiknynai* make known]
pol·i·cy·hold·er (pol′ə·sē·hōl′dər) *n.* One who holds a policy of insurance.
policy racket Numbers pool.
Po·lil·lo Islands (pō·lē′yō) A Philippine island group off the eastern coast of central Luzon; approximately 295 square miles.
po·li·o (pō′lē·ō) *n. Colloq.* Poliomyelitis.
po·li·o- *combining form Med.* Of or pertaining to the gray matter of the brain, or the spinal cord: *polioencephalitis*. [<Gk. *polios* gray]
pol·i·o·en·ceph·a·li·tis (pol′ē·ō·en·sef′ə·lī′tis) *n. Pathol.* Inflammation of the gray matter of the brain. Also **pol′i·en·ceph′a·li′tis**. [<POLIO- + ENCEPHALITIS]
pol·i·o·my·e·li·tis (pol′ē·ō·mī′ə·lī′tis, pō′lē-) *n.*

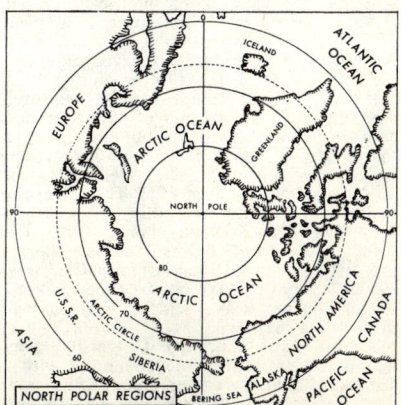

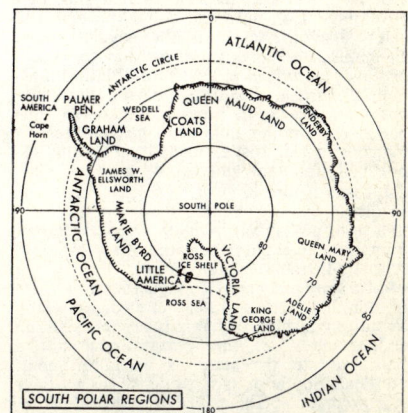

Pathol. An acute, communicable disease caused by infection with a virus, occurring especially in children, and characterized by inflammation of the gray matter of the spinal cord, followed by paralysis and atrophy of various muscle groups: also called *infantile paralysis.* [<NL <POLIO- + Gk. *myelos* marrow + -ITIS]

pol·ish (pol′ish) *n.* **1** Smoothness or glossiness of surface; finish. **2** A substance used to produce a bright, smooth, or glossy surface; a varnish. **3** Refinement of manner or style. **4** The process of polishing. — *v.t.* **1** To make smooth or lustrous, as by rubbing. **2** To make complete; finish; perfect. **3** To free from crudeness; make refined or elegant: to *polish* the mind. — *v.i.* **4** To take a gloss; shine. **5** To become elegant or refined. — **to polish off 1** To do or finish completely or quickly. **2** To dispose of; overwhelm. — **to polish up** To make better; improve. [<OF *poliss-*, stem of *polir* <L *polire* make smooth] — **pol′ish·er** *n.*

Po·lish (pō′lish) *adj.* Pertaining to Poland, its inhabitants, or their language. — *n.* The West Slavic language of the Poles.

Polish Corridor A strip of land in NW Poland, extending to the Baltic between Germany and East Prussia 1919–1939; part of Germany prior to 1919, and from 1939 to 1945; now a part of Poland.

pol·ished (pol′isht) *adj.* **1** Made smooth by polishing. **2** Naturally smooth and glossy. **3** Refined and polite. See synonyms under FINE[1], POLITE, SMOOTH.

Po·lit·bu·ro (po·lit′byŏŏr′ō) *n.* The leading policy-forming committee of the Communist party in the U. S. S. R. until 1952, when it was replaced by the Presidium. [<Russian *polit(icheskoe) buro*]

po·lite (pə·līt′) *adj.* **1** Exhibiting in manner or speech a considerate regard for others; courteous; also, cultivated: *polite* society. **2** Finished and elegant in style. [<L *politus*, pp. of *polire* polish] — **po·lite′ly** *adv.* — **po·lite′ness** *n.*

Synonyms: accomplished, civil, complaisant, courteous, courtly, cultivated, cultured, elegant, genteel, gracious, obliging, polished, urbane, well-behaved, well-bred, well-mannered. A man may be *civil* with no consideration for others, simply because self-respect forbids him to be rude; but one who is *polite* has at least some care for the opinions of others, and if *polite* in the highest and truest sense, he cares for the comfort and happiness of others in the smallest matters. *Civil* is a colder and more distant word than *polite*; *courteous* is fuller and richer, dealing often with greater matters, and is used only in the good sense. *Courtly* suggests that which befits a royal court, and is used of external grace and stateliness without reference to the prompting feeling. *Genteel* refers to an external elegance, which may be showy and superficial, and the word is thus inferior to *polite* or *courteous. Urbane* refers to a politeness that is genial and successful in giving others a sense of ease and cheer. *Polished* refers to external elegancies of speech and manner without reference to spirit or purpose; as, a *polished* gentleman or a *polished* scoundrel; *cultured* refers to a real and high development of mind and soul, of which the external manifestation is the smallest part. *Complaisant* denotes a disposition to please or favor. *Antonyms*: awkward, bluff, blunt, boorish, brusk, clownish, coarse, discourteous, ill-behaved, ill-bred, ill-mannered, impertinent, impolite, impudent, insolent, insulting, raw, rude, rustic, uncivil, uncouth, unpolished, untaught, untutored.

po·li·tesse (pô·lē·tes′) *n. French* Politeness; civility.

Po·li·tian (pō·lish′ən), **Angelo**, 1454–94, Italian humanist and poet: real name *Ambrogini.* Also **Po·li·zia·no** (pō′lē·tsyä′nō).

pol·i·tic (pol′ə·tik) *adj.* **1** Sagacious and wary in planning; artful; crafty; shrewd, especially in statesmanship. **2** Wisely adapted to an end; specious. **3** *Rare* Pertaining to public polity, or to the state or its government; political. See BODY POLITIC. [<OF *politique* <L *politicus* <Gk. *politikos* civic < *politēs* citizen] — **pol′i·tic·ly** *adv.*

Synonyms: artful, crafty, cunning, diplomatic, discreet, judicious, prudent, sagacious, shrewd, wary, wily, wise. *Antonyms*: see synonyms under IMPRUDENT.

po·lit·i·cal (pə·lit′i·kəl) *adj.* **1** Pertaining to public policy; concerned in the administration of government: a *political* system: distinguished from *civil.* **2** Belonging to the science of government; treating of polity or politics: *political* principles. **3** Having an organized system of government; administering a polity. **4** Pertaining to or connected with a party or parties controlling or seeking to control government in a state: *political* methods. [<L *politicus*] — **po·lit′i·cal·ly** *adv.*

political economist A person skilled in political economy.

political economy Economics.

political science The science of the form and principles of civil government, and the extent and manner of its intervention in public and private affairs; politics.

pol·i·ti·cian (pol′ə·tish′ən) *n.* **1** One engaged in politics, especially professionally. **2** *U.S.* One who engages in politics for personal or partisan aims rather than for reasons of principle; also, a political schemer or opportunist. **3** *Brit.* One skilled in the science of government or politics; a statesman. **4** The white-eyed vireo: so called because it feathers its nest with bits of newspaper or whatever comes easily. [<F *politicien*]

po·lit·i·cize (pə·lit′ə·sīz) *v.* **·cized, ·ciz·ing** *v.i.* To take part in or discuss politics. — *v.t.* To make political.

po·lit·i·co (pə·lit′i·kō) *n. pl.* **·cos** A politician. [<Sp. *político*]

pol·i·tics (pol′ə·tiks) *n.* **1** The science of civil government. **2** Political affairs in a party sense; party intrigues, etc. **3** One's political sentiments: construed as plural. — **to play politics** To speak or act for political reasons; hence, to scheme for an advantage.

pol·i·ty (pol′ə·tē) *n. pl.* **·ties 1** The form or method of government of a nation, state, church, etc. **2** Any community living under some definite form of government. [<OF *politie*, var. of *policie.* See POLICY[1].]

Synonym: policy. *Polity* is the permanent system of government of a state, a church, or a society; *policy* is the method of management with reference to the attainment of certain ends; the national *polity* of the United States is republican; each administration has a *policy* of its own. *Policy* is often used as equivalent to expediency; as, Many think honesty to be good *policy. Polity* in ecclesiastical use serves a valuable purpose in distinguishing that which relates to administration and government from that which relates to faith and doctrine. See LEGISLATION.

Polk (pōk), **James Knox,** 1795–1849, president of the United States 1845–49.

pol·ka (pōl′kə, pō′-) *n.* **1** A round dance of Bohemian origin in common time, with three steps to every second measure. **2** Music for such a dance: a lively Bohemian or Polish tune in 2/4 time. — *v.i.* **·kaed, ·ka·ing** To dance the polka. [<F <Czech *pulka* half (step)]

polka dot 1 One of a series of spots of various sizes and spacing on a textile fabric. **2** A pattern made up of such spots.

poll[1] (pōl) *n.* **1** The head; hence, a person; also, the top or back of the head; crown. **2** A list of persons. **3** The voting at an election; the votes thus registered or voted; also, the place where they are registered or voted: used in the United States in the plural. **4** A poll tax. **5** The blunt or round end of a hammer or ax. **6** A survey of public opinion on a given subject, usually obtained from a sample group. — *v.t.* **1** To receive (a specified number of votes). **2** To enrol, as for taxation or voting; register. **3** To cast at the polls. **4** To canvass in a poll (def. 6). **5** To cut off or trim, as hair, horns, etc.; clip; shear. **6** To cut off or trim the hair, horns, top, etc., of: to *poll* cattle; to *poll* a tree. — *v.i.* **7** To vote at the polls; cast one's vote. ◆ Homophone: *pole.* [<MDu. *polle* top of the head] — **poll′er** *n.*

poll[2] (pol) *n.* In Cambridge University, England, a student who contents himself with a degree, without trying for honors. Such students are called collectively *the poll.* [<Gk. *(hoi) polloi* (the) many]

Poll (pol) *n.* A parrot. Also **Poll parrot, Pol′ly.**

pol·lack (pol′ək) *n.* A gadoid food fish (genera *Pollachius* and *Theragra*), resembling the true cod, but with the lower jaw projecting and barbel obsolete. *P. pollachius* is the common **European pollack,** *P. virens*, the **green pollack** or coalfish of the North Atlantic, and *T. chalcogramma*, of the North Pacific. Also spelled *pollock.* [Origin uncertain]

pol·lard (pol′ərd) *n.* **1** A tree shorn of its top so that it puts out a dense head of slender shoots. **2** An animal that has lost its horns. — *v.t.* To convert into a pollard. [<POLL[1]]

polled (pōld) *adj.* **1** Shorn of the head or top. **2** Shorn of the hair; bald.

poll·ee (pōl·ē′) *n.* A person whose opinion is polled.

pol·len (pol′ən) *n.* The fine yellowish powder formed within the anther of the flowering plant; the fecundating element in seed plants. [<L, fine flour]

pollen count A measure of the relative concentration of pollen grains in the atmosphere at a given locality and date: usually expressed in the number of grains of a specified variety of pollen per cubic yard.

poll evil An ulcerous abscess on a horse's poll, usually resulting from a bruise.

pol·lex (pol′eks) *n. pl.* **pol·li·ces** (pol′ə·sēz) The first or radial digit of the hand or forelimb of a vertebrate; the thumb. [<L] — **pol′li·cal** *adj.*

pol·li·ce ver·so (pol′ə·sē vûr′sō) *Latin* With thumbs reversed or extended downward: used among the Romans to denote that a defeated gladiator be killed. (The exact signals of mercy and punishment are matters of dispute.)

pol·li·nate (pol′ə·nāt) *v.t.* **·nat·ed, ·nat·ing** To supply or convey pollen to. Also **pol′len·ate.**

pol·li·na·tion (pol′ə·nā′shən) *n.* The transfer of pollen from anthers to stigmas. Also **pol′len·a′tion.**

pol·li·nif·er·ous (pol′ə·nif′ər·əs) *adj.* **1** Producing pollen. **2** Bearing or carrying pollen. Also **pol′len·if′er·ous.**

pol·lin·i·um (pə·lin′ē·əm) *n. pl.* **·i·a** (-ē·ə) A mass or body of pollen grains more or less coherent; a pollen mass. [<NL <L *pollen, pollinis* fine flour]

pol·li·no·sis (pol′ə·nō′sis) *n. Pathol.* Hay fever. [<NL <L *pollen, pollinis* dust + -OSIS]

Pol·li·o (pol′ē·ō), **Gaius Asinius,** 75 B.C.–A.D. 5, Roman orator and politician.

pol·li·wog (pol′ē·wog) *n.* A tadpole. Also **pol′ly·wog.** [ME *polwygle.* Cf. POLL[1], WIGGLE.]

pol·lock (pol′ək) See POLLACK.

Pol·lock (pol′ək), **Channing,** 1880–1946, U.S. dramatist, novelist, and lecturer. — **Sir Frederick,** 1845–1937, English jurist. — **Jackson,** 1912–1956, U.S. painter.

poll·ster (pōl′stər) *n.* One who takes public opinion polls. Also **poll′ist.**

poll tax A tax on a person, as distinguished from that on property, especially as a prerequisite for voting.

pol·lut·ant (pə·lōō′tənt) *n.* **1** That which pollutes. **2** Any of various noxious chemicals and refuse materials which impair the purity of water, soil, or the atmosphere.

pol·lute (pə·lōōt′) *v.t.* **·lut·ed, ·lut·ing** To make unclean or impure; dirty; corrupt; profane. [<L *pollutus*, pp. of *polluere* make unclean] — **pol·lut′ed** *adj.* — **pol·lut′ed·ly** *adv.* — **pol·lut′ed·ness** *n.* — **pol·lut′er** *n.* — **pol·lu′tion** *n.*

Synonyms: abuse, contaminate, corrupt, debauch, defile, degrade, deprave, dishonor, infect, ravish, soil, stain, taint, violate, vitiate. See CORRUPT, DEFILE[1], VIOLATE. *Antonyms*: clarify, clean, cleanse, clear, filter, fine, purge, purify, redeem, refine, renew, restore.

Pol·lux (pol′əks) See CASTOR AND POLLUX.

Pol·ly (pol′ē) **1** Mary; a familiar nickname used instead of *Molly.* **2** A parrot.

Pol·ly·an·na (pol′ē·an′ə) *n.* A person who always finds good in everything: so called from the heroine of stories by Eleanor H. Porter, 1868–1920.

po·lo (pō′lō) *n.* **1** A game played on horseback, usually with a light wooden ball and long-handled mallets. **2** A similar game played on ice or roller skates. [Cf. Tibetan *pulu* ball] — **po′lo·ist** *n.*

Po·lo (pō′lō), **Marco,** 1254?–1324?, Venetian traveler and author.

polo coat A tailored coat of camel's hair or material imitating camel's hair.

pol·o·naise (pol′ə·nāz′, pō′lə-) *n.* **1** A garment

for women, consisting of a waist and an overskirt in one piece. 2 A stately marchlike Polish dance, or the music for it. 3 A kind of antique Oriental carpet with a silk pile. [<F]

po·lo·ni·um (pə·lō′nē·əm) *n.* A radioactive element (symbol Po) produced by the disintegration of various uranium minerals: discovered in 1898 by Pierre and Marie Curie. See ELEMENT. [<NL <Med. L *Polonia* Poland]

Po·lo·ni·us (pə·lō′nē·əs) In Shakespeare's *Hamlet*, the chamberlain to the king and father of Ophelia and Laertes.

Pol·ska (pôl′skä) The Polish name for POLAND.

Pol·ta·va (pol·tä′və) A city of NE Ukrainian S.S.R.; scene of Peter the Great's victory over the Swedes under Charles XII, 1709.

pol·ter·geist (pōl′tər·gīst) *n.* A ghost or spirit reputed to make its presence known by any kind of clatter, as knockings and the noises of moving objects. [<G <*poltern* make a noise + *geist* spirit]

pol·troon (pol·tro͞on′) *n.* 1 A mean-spirited coward; dastard. 2 A lazy idler; sluggard. — *adj.* Cowardly; contemptible. [<F *poltron* <Ital. *poltrone* cowardly, sluggish <*poltro* bed] **— pol·troon′er·y** *n.*

poly- *combining form* 1 Many; several; much: *polygamy, polygon.* 2 Excessive; abnormal: *polydactylism.* [<Gk. *polys* much, many]

pol·y·am·ide (pol′ē·am′id) *n. Chem.* A polymer derived from compounds containing amine and carboxyl groups: used in the making of various synthetic fibers.

pol·y·an·dry (pol′ē·an′drē) *n.* 1 The civil condition of having more than one husband. 2 A social order that includes a plurality of husbands. 3 *Bot.* Having 20 or more stamens. [<POLY- + Gk. *anēr, andros* a man] **— pol′y·an′drous** *adj.*

pol·y·an·thus (pol′ē·an′thəs) *n.* 1 A variety of primrose (*Primula polyantha*), with many-flowered umbels. 2 A widely distributed, fragrant-flowered narcissus (*Narcissus tazetta*). [<POLY- + Gk. *anthos* flower]

pol·y·ar·chy (pol′ē·är′kē) *n. pl.* **·chies** Government by several persons of whatever class. **— pol′y·ar′chic** or **·chi·cal** *adj.*

pol·y·a·tom·ic (pol′ē·ə·tom′ik) *adj. Chem.* 1 Having more than one atom in the molecule. 2 Containing or capable of combining with several replaceable atoms.

po·ly·ba·sic (pol′ē·bā′sik) *adj. Chem.* Containing two or more atoms of hydrogen replaceable by a base or basic radicals: said of certain acids.

pol·y·ba·site (pol′ē·bā′sīt) *n.* A metallic, iron-black ore of silver crystallizing in the monoclinic system. [<G *polybasit*]

Po·lyb·i·us (pə·lib′ē·əs), 205?–120? B.C., Greek historian.

pol·y·brid (pol′i·brid) *n. Bot.* A hybrid plant derived from the crossing of two particular genera, species, or varieties. [<POLY- + (HY)BRID]

Pol·y·carp (pol′i·kärp), **Saint,** 69?–155?, bishop of Smyrna; martyred.

pol·y·car·pel·lar·y (pol′i·kär′pə·ler′ē) *adj. Bot.* Made up of many carpels.

pol·y·car·pous (pol′i·kär′pəs) *adj. Bot.* 1 Having the fruit composed of two or more distinct carpels. 2 Fruiting many times. Also **pol′y·car′pic.**

pol·y·chae·tous (pol′i·kē′təs) *adj. Zool.* 1 Having several setae. 2 Of or pertaining to a class (*Polychaeta*) of annelids, including most marine worms. Also **pol′y·chae′tal, pol′y·chae′tan.** [<POLY- + Gk. *chaitē* hair] **— pol′y·chaete** *adj.* & *n.*

pol·y·cha·si·um (pol′i·kā′zē·əm, -zhē·əm) *n. pl.* **·si·a** (-zē·ə, -zhē·ə) *Bot.* A form of cymose inflorescence in which, below each flower, more than two secondary branches are given off from the main axis. [<NL <POLY- + Gk. *chasis* division]

pol·y·cho·ry (pol′i·kôr′ē, -kō′rē) *n.* In ballet, the use of contrapuntal patterns of line and rhythm for the movements of two groups of dancers. [<POLY- + Gk. *choros* dance]

pol·y·chrome (pol′i·krōm) *adj.* Done in several or many colors. — *n.* An association of several colors, as in decoration.

pol·y·chro·mic (pol′i·krō′mik) *adj.* Exhibiting many colors or changes of color. Also **pol′y·chro·mat′ic** (-krō·mat′ik) **pol′y·chro′mous.**

pol·y·chro·my (pol′i·krō′mē) *n.* The art of decorating or executing in several or many colors, as in ancient statuary and architecture.

pol·y·clin·ic (pol′i·klin′ik) *n.* 1 An institution furnishing clinical instruction in all kinds of diseases. 2 A general hospital in which all forms of diseases are treated. Compare POLICLINIC.

Pol·y·cli·tus (pol′i·klī′təs) A Greek sculptor of the fifth century B.C. Also **Pol′y·cle′tus** (-klē′təs). **— Pol′y·cli′tan** or **·cle′tan** *adj.*

pol·y·con·ic (pol′i·kon′ik) *adj.* Of, relating to, or based on many cones.

polyconic projection A type of map projection in which the parallels of latitude are arcs of circles which are not concentric and the meridians, except the central one, are curved lines.

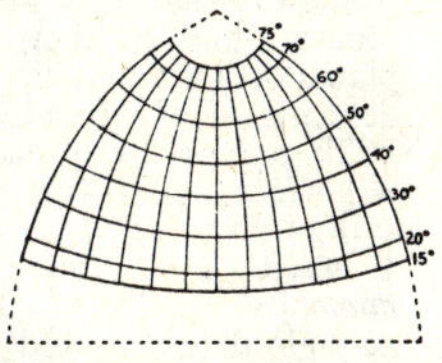

POLYCONIC PROJECTION

Po·lyc·ra·tes (pə·lik′rə·tēz), died 522 B.C., tyrant of Samos; crucified.

pol·y·dac·tyl (pol′i·dak′til) *adj.* Having an abnormally large number of fingers or toes; many-fingered or many-toed: also **pol′y·dac′ty·lous.** — *n.* A polydactyl animal. **— pol′y·dac′tyl·ism** *n.*

pol·y·dem·ic (pol′i·dem′ik) *adj. Ecol.* Occurring or dwelling in two or more regions: said of plants and animals. [<POLY- + Gk. *dēmos* region]

Pol·y·deu·ces (pol′i·do͞o′sēz, -dyo͞o′-) Pollux. See CASTOR AND POLLUX.

Pol·y·do·rus (pol′i·dôr′əs, -dō′rəs) Greek sculptor of the first century B.C.

pol·y·em·bry·o·ny (pol′ē·em′brē·ō′nē, -brē·ə·nē) *n.* 1 *Bot.* The production of two or more viable embryos in a seed. 2 *Zool.* The production of two or more offspring from a single fertilized ovum, as identical twins in man.

pol·y·er·gic (pol′ē·ûr′jik) *adj.* Capable of accomplishing many tasks; energetically versatile. [<POLY- + Gk. *ergon* work]

pol·y·es·ter fiber (pol′ē·es′tər) *Chem.* A synthetic fiber of high tensile strength made by the esterification of ethylene glycol and other organic compounds.

pol·y·eth·y·lene (pol′ē·eth′ə·lēn) *n. Chem.* A tough, flexible thermoplastic resin, C_2H_4, made by the polymerization of ethylene: used in the making of moistureproof plastics having high electrical resistance.

Pol·y·euc·tus (pol′ē·yo͞ok′təs) Greek sculptor of the third century B.C.

po·lyg·a·la (pə·lig′ə·lə) *n.* 1 Any of a large genus (*Polygala*) of herbs and shrubs, natives of temperate and subtropical regions, and distinguished by simple, entire leaves, sometimes dotted, and showy magenta, purple, or white flowers; especially, the North American fringed polygala (*P. paucifolia*). 2 The milkwort. [<POLY- + Gk. *gala* milk]

po·lyg·a·mous (pə·lig′ə·məs) *adj.* 1 Of, pertaining to, or characterized by polygamy. 2 Mating with more than one of the opposite sex. 3 *Bot.* Bearing male, female, and bisexual or hermaphrodite flowers on the same plant. [<POLY- + -GAMOUS] **— po·lyg′a·mous·ly** *adv.*

po·lyg·a·my (pə·lig′ə·mē) *n.* 1 The condition of having more than one wife or husband at the same time. 2 The state of having more than one mate. Compare MONOGAMY. **— po·lyg′a·mist** *n.*

pol·y·gen·e·sis (pol′i·jen′ə·sis) *n. Biol.* The doctrine that organisms originate from cells of different kinds. Compare MONOGENESIS. **— pol′y·ge·net′ic** (-jə·net′ik), **pol′y·gen′ic, po·lyg·e·nous** (pə·lij′ə·nəs) *adj.*

pol·y·glot (pol′i·glot) *adj.* 1 Expressed in several tongues. 2 Speaking several languages. — *n.* 1 A book giving versions of the same text, as of the Scriptures, in several languages. 2 One who speaks or writes several languages. [<Gk. *polyglōttos*]

Pol·yg·no·tus (pol′ig·nō′təs) Greek painter of the fifth century B.C. **— Pol′yg·no′tan** *adj.*

pol·y·gon (pol′i·gon) *n. Geom.* A closed, usually plane, figure bounded by straight lines or arcs, especially more than four; a figure having many sides and angles. [<LL *polygonum* <Gk. *polygōnon*]

pol·y·go·na·ceous (pol′i·gə·nā′shəs) *adj. Bot.* Designating a family (*Polygonaceae*) of apetalous, widely distributed herbs, vines, shrubs, and trees; the buckwheat family, including the sorrels. [<POLYGONUM]

po·lyg·o·nal (pə·lig′ə·nəl) *adj.* Constituting or having the form of a polygon; having many angles: a *polygonal* figure. Also **po·lyg′o·nous.** **— po·lyg′o·nal·ly** *adv.*

polygonal number See under NUMBER.

po·lyg·o·num (pə·lig′ə·nəm) *n.* Any of a large and widely distributed genus (*Polygonum*) of annual or perennial herbs. The common smartweed, the prince's-feather, and the bistort are among the best-known species. Also **po·lyg′o·ny.** [<NL <L *polygonos* <Gk. *polygonon* knotgrass <*poly-* many + *gony* knee; from its many joints]

pol·y·graph (pol′i·graf, -gräf) *n.* 1 A device for reproducing a drawing or writing many times; a copy pad. 2 A mechanism for multiplying a drawing or writing. 3 A versatile or prolific author. 4 A collection of different treatises or books. 5 *Med.* A device for recording variations in the heartbeat and in respiratory movements: sometimes used as a lie detector. Compare PSYCHOGALVANOMETER. [<Gk. *polygraphos*] **— pol′y·graph′ic** or **·i·cal** *adj.*

po·lyg·ra·phy (pə·lig′rə·fē) *n.* 1 The use of a polygraph. 2 The art of writing in or of interpreting various ciphers.

po·lyg·y·nous (pə·lij′ə·nəs) *adj.* 1 Of, pertaining to, or practicing polygyny. 2 *Bot.* Having many styles.

po·lyg·y·ny (pə·lij′ə·nē) *n.* The marriage, mating, or cohabitation of one male with more than one female. [<POLY- + Gk. *gynē* woman]

pol·y·he·dral (pol′i·hē′drəl) *adj.* Of or pertaining to a polyhedron.

polyhedral angle *Geom.* The angle formed by three or more planes passing through a point; an angle at a vertex of a solid.

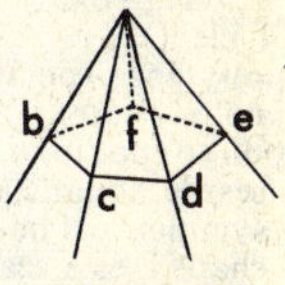

POLYHEDRAL ANGLE
Lateral angles *bac, bad,* etc., form angle at vertex *a.*

pol·y·he·dron (pol′i·hē′drən) *n. pl.* **·dra** (-drə) or **·drons** *Geom.* A solid bounded by plane faces, especially by more than four. [<NL <Gk. *polyedros* many-sided <*polys* many + *hedra* side]

Pol·y·hym·ni·a (pol′i·him′nē·ə) The Muse of sacred song. Also **Po·lym·ni·a** (pə·lim′nē·ə).

pol·y·mer (pol′i·mər) *n. Chem.* 1 Any of two or more polymeric compounds. 2 Any compound formed by polymerization, especially one of higher molecular weight than the parent substance. [<POLY- + Gk. *meros* part]

pol·y·mer·ic (pol′i·mer′ik) *adj. Chem.* 1 Of, pertaining to, or manifesting polymerism. 2 Having the same chemical composition but different molecular weights and different properties, as acetylene and benzene.

po·lym·er·ism (pə·lim′ə·riz′əm, pol′i·mə-) *n. Chem.* The property possessed by several compounds of having identical percentage composition but different molecular weights.

po·lym·er·i·za·tion (pə·lim′ər·ə·zā′shən, pol′·i·mər·ə-) *n. Chem.* The process of changing the molecular arrangement of a compound so as to form new compounds having the same percentage composition as the original, but of different (usually greater) molecular weight and different properties. The method may be *linear,* by the successive addition of small structural units to form a chain; *cyclic,* by the formation of rings; or *cross-linked,* by a three-dimensional fusion of either linear or cyclic elements.

po·lym·er·ize (pə·lim′ə·rīz, pol′i·mə·rīz′) *v.t.* & *v.i.* **·ized, ·iz·ing** To subject to or undergo polymerization. Also *Brit.* **po·lym′er·ise.**

po·lym·er·ous (pə·lim′ər·əs) *adj.* 1 *Biol.* Consisting of many parts. 2 *Bot.* Having many parts or members in each whorl or series.

pol·y·morph (pol′i·môrf) *n.* A substance or organism that exhibits polymorphism. [<Gk. *polymorphos* <*poly-* many + *morphē* form]

add,āce,câre,pälm; end,ēven; it,īce; odd,ōpen,ôrder; to͝ok,po͞ol; up,bûrn; ə = a in *above,* e in *sicken,* i in *clarity,* o in *melon,* u in *focus;* yo͞o = u in *fuse;* oi,oil; ou,pout; ch,check; g,go; ng,ring; th,thin; th,this; zh,vision. Foreign sounds à,œ,ü,kh,ṅ; and ◆: see page xx. < from; + plus; ? possibly.

pol·y·morph·ism (pol′i·môr′fiz·əm) *n.* 1 *Zool.* The property of assuming or passing through several forms, as an animal exhibiting seasonal changes in coloration. 2 *Mineral.* The occurrence in a mineral of two or more distinct crystal forms of identical chemical composition.

pol·y·mor·phous–per·verse (pol′i·môr′fəs·pər·vûrs′) *adj. Psychoanal.* Designating the generalized sexual potentialities of an individual, especially of a young child.

Pol·y·ne·sia (pol′i·nē′zhə, -shə) The islands of Oceania in the central and SE Pacific, extending east of Melanesia and Micronesia from the Hawaiian Islands to New Zealand; total, about 10,000 square miles.

Pol·y·ne·sian (pol′i·nē′zhən, -shən) *n.* 1 One of the native brown-skinned people of Polynesia: believed to be either of Malay stock originally stemming from a Caucasian strain of Asia, or of mixed Melanesian, Malay, and Caucasian stock. 2 A subfamily of the Austronesian family of languages spoken by these people. — *adj.* Of or pertaining to Polynesia, its people, or their languages.

pol·y·neu·ri·tis (pol′i·nŏŏ·rī′tis, -nyŏŏ-) *n. Pathol.* Simultaneous inflammation of many peripheral nerves.

pol·y·neu·rop·a·thy (pol′i·nŏŏ·rop′ə·thē, -nyŏŏ-) *n. Pathol.* Any morbid condition which affects several nerves at once, as alcoholism or vitamin deficiency. [<POLY- + NEURO- + -PATHY] — **pol′y·neu′ro·path′ic** (-nŏŏr′ə·path′·ik, -nyŏŏr′-) *adj.*

Pol·y·ni·ces (pol′i·nī′sēz) In Greek legend, a son of Oedipus and Jocasta. See SEVEN AGAINST THEBES.

pol·y·no·mi·al (pol′i·nō′mē·əl) *adj.* Of, pertaining to, or consisting of many names or terms. — *n.* 1 *Math.* An expression, as in algebra, containing two or more terms. 2 *Biol.* A scientific name consisting of more than two terms. [<POLY- + (BI)NOMIAL]

pol·y·nu·cle·ar (pol′i·nŏŏ′klē·ər, -nyŏŏ′-) *adj.* Having many nuclei. Also **pol′y·nu′cle·ate** (-klē·it).

pol·y·ose (pol′ē·ōs) *n.* Polysaccharide. [< POLY- + -OSE[2]]

pol·yp (pol′ip) *n.* 1 *Zool.* **a** A many-tentacled, sessile aquatic coelenterate having a radially symmetrical body typically cylindrical or cup-shaped, as a sea anemone or coral. **b** A single unit of a colonial organism. 2 *Pathol.* A polypus. [<MF *polype* <L *polypus* a cuttlefish, a polypus <Gk. *polypous* <*poly-* many + *pous* a foot]

pol·y·par·y (pol′i·per′ē) *n. pl.* **·par·ies** *Zool.* The solid calcareous or chitinous stock of a colony of polyps, especially of coral. Also called *polypidom.* [<NL *polyparium* <L *polypus* a polypus]

pol·y·pep·tide (pol′i·pep′tīd) *n. Chem.* A compound formed by the union of two or more amino acids.

pol·y·pet·al·ous (pol′i·pet′əl·əs) *adj. Bot.* Having the petals free and distinct. [<NL *polypetalus* <Gk. *poly-* many + *petalon* a leaf, a petal]

pol·y·pha·gi·a (pol′i·fā′jē·ə) *n.* 1 Excessive craving for food; voracity. 2 *Zool.* The practice of eating many kinds of food. [<NL <Gk. <*polyphagos* eating to excess <*poly-* much + *phagein* eat] — **pol′y·pha′gi·an** *n.* & *adj.*

pol·y·phag·ic (pol′i·faj′ik) *adj.* Eating many things; subsisting on various kinds of food. Also **po·lyph·a·gous** (pə·lif′ə·gəs).

pol·y·phase (pol′i·fāz) *adj. Electr.* Having or producing several phases, as an alternating current.

pol·y·phe·mus (pol′i·fē′məs) *n.* 1 An animal, or sometimes a person, having but one eye. 2 A large American silkworm moth *(Telea polyphemus)* having a conspicuous ocellus on each hind wing. [<NL <L, POLYPHEMUS]

Pol·y·phe·mus (pol′i·fē′məs) In Homer's *Odyssey,* the Cyclops who imprisoned Odysseus and his companions in a cave, from which they escaped after blinding him in his sleep. [<L <Gk. *Polyphēmos,* a Cyclops, lit., many-voiced <*poly-* many + *phēmē* a voice]

pol·y·phon·ic (pol′i·fon′ik) *adj.* 1 *Phonet.* Representing more than one sound or combination of sounds, as some written characters. 2 Consisting of many sounds or voices. 3 *Music* **a** Designating or involving the simultaneous and harmonious combination of two or more independent parts or melodies. **b** Denoting an instrument, as a piano, by which two or more sounds may be produced simultaneously. Also **po·lyph·o·nous** (pə·lif′ə·nəs). [<Gk. *polyphōnos* having many tones < *poly-* many + *phōnē* a voice, sound]

polyphonic prose A poem set down on the page as prose: closer to the rhythms of prose than to those of verse and employing such devices as rime, assonance, and alliteration to produce poetic effects.

po·lyph·o·ny (pə·lif′ə·nē, pol′i·fō′nē) *n.* 1 Multiplicity of sounds, as in an echo. 2 *Phonet.* The representation by one written character or sign of more than one sound. 3 Counterpoint. [<Gk. *polyphōnia* a variety of tones or speech <*polyphōnos.* See POLYPHONIC.]

pol·y·phy·le·sis (pol′i·fi·lē′sis) *n. Biol.* The separate and distinct origin of a species of plants or animals from more than one line of descent. Also **pol·y·phy·ly** (pol′i·fī′lē). [<NL <Gk. *polyphylos* of many tribes <*poly-* many + *phylē* a clan] — **pol′y·phy·let′ic** (-let′ik) *adj.*

po·lyp·i·dom (pə·lip′ə·dəm) *n.* A polypary. [< L *polypus* a polypus + *domus* house <Gk. *domos*]

pol·y·ploid (pol′i·ploid) *adj. Genetics* Having more than two basic chromosome sets in the body cells. — *n.* A polyploid cell. [<POLY- + -PLOID]

pol·y·pod (pol′i·pod) *adj.* 1 Having many feet. 2 *Zool.* Pertaining to many-footed organisms. — *n.* A myriapod. [<POLY- + -POD]

pol·y·po·dy (pol′i·pō′dē) *n. pl.* **·dies** 1 Any one of a genus *(Polypodium)* of widely distributed ferns, typically epiphytic, and having naked sori. 2 The possession of many legs or abdominal attachments. [<NL <L <Gk. *polypodion* a kind of fern, dim. of *polypous, -podos* many-footed; so called from its many root branches]

pol·y·pous (pol′i·pəs) *adj.* 1 Having many feet or roots. 2 Pertaining to or resembling a polyp. 3 *Pathol.* Pertaining to, afflicted with, or resembling polypi.

pol·yp·tych (pol′ip·tik) *n.* An altarpiece or panel having more than three folds or leaves. [< Gk. *polyptychos* having many folds <*poly-* many + *ptyx, ptychos* a fold]

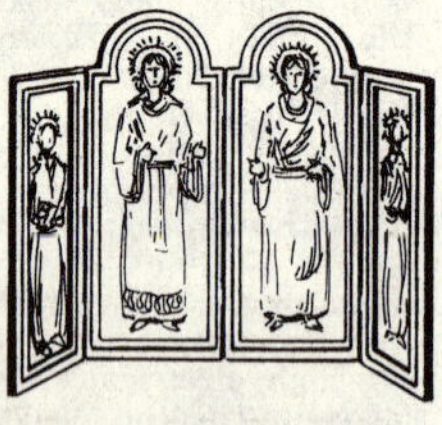
POLYPTYCH

pol·y·pus (pol′i·pəs) *n. pl.* **·pi** (-pī) *Pathol.* 1 A smooth growth of hypertrophied mucus found in mucous membrane, as in the nasal passages, bladder, rectum, etc. 2 A tumor. [<NL <L, a polypus]

pol·y·sac·cha·rid (pol′i·sak′ə·rid, -rīd) *n. Chem.* Any of a class of carbohydrates of high molecular weight, formed by the union of three or more monosaccharide molecules: they include starch, dextrin, inulin, cellulose, mucilage, and glycogen. Also **pol′y·sac′cha·rose** (-rōs).

pol·y·sep·al·ous (pol′i·sep′əl·əs) *adj. Bot.* Having sepals free and unconnected.

pol·y·soap (pol′i·sōp) *n. Chem.* Any of a class of detergent compounds formed by the polymerization of many soap molecules into one large molecule having properties useful in the study of organic colloids, especially proteins. [<POLY(MERIZED) + SOAP]

pol·y·sperm (pol′i·spûrm) *n. Bot.* A tree bearing a many-seeded fruit. [<Gk. *polyspermos* abounding in seed <*poly-* many + *sperma* a seed]

pol·y·sper·my (pol′i·spûr′mē) *n. Bot.* The condition of having numerous seeds in the fruit. Also **pol′y·sper′mi·a.** [<Gk. *polyspermia* < *polyspermos.* See POLYSPERM.] — **pol′y·sper′·mal** or **·sper′mic** or **·sper′mous** *adj.*

pol·y·ste·lic (pol′i·stē′lik) *adj. Bot.* Consisting of more than one stele or internal vascular cylinder. [<POLY- + STELE[2]]

pol·y·sty·rene (pol′i·stī′rēn) *n. Chem.* A thermoplastic polymer of styrene, C_8H_8; a clear, colorless, water-resistant resin: much used in the making of plastics, housewares, light fixtures, electrical components, surface coatings, etc. [<POLY(MER) + STYRENE]

pol·y·sul·fide (pol′i·sul′fīd, -fid) *n. Chem.* A binary compound containing more than one atom of sulfur in the molecule.

pol·y·syl·la·ble (pol′i·sil′ə·bəl) *n.* A word of several syllables, especially of more than three. [<Med. L *(vox) polysyllaba* (a) many-syllabled (word), fem. of *polysyllabus* polysyllabic <Gk. *polysyllabos* <*poly-* many + *syllabē* a syllable] — **pol′y·syl·lab′ic** (-si·lab′·ik) or **·i·cal** *adj.* — **pol′y·syl′la·bism** or **·syl·lab′i·cism** (pol′i·si·lab′ə·siz′əm) *n.*

pol·y·syn·de·ton (pol′i·sin′də·ton) *n.* Repetition of connectives or conjunctions for rhetorical effect, as, "east and west and south and north": distinguished from *asyndeton.* [<NL <Gk. *poly-* much + *syndetos* bound together <*syndeein* <*syn-* together + *deein* bind]

pol·y·syn·thet·ic (pol′i·sin·thet′ik) *adj. Ling.* Describing a language, such as Eskimo, or certain of the American Indian languages, in which the subject, object, verb, etc., of a sentence are combined into a single word and have no existence as separate elements. [<Gk. *polysynthetos* much compounded < *poly-* much + *syntithenai* <*syn-* together + *tithenai* put]

pol·y·tech·nic (pol′i·tek′nik) *adj.* Embracing many arts: also **pol′y·tech′ni·cal.** — *n.* A school of applied science and the industrial arts. [<F *polytechnique* <Gk. *polytechnos* skilled in many arts <*poly-* many + *technē* an art]

pol·y·the·ism (pol′i·thē·iz′əm) *n.* The belief in and worship of more gods than one. [<F *polythéisme* <Gk. *polytheos* of many gods <*poly-* many + *theos* a god] — **pol′y·the′ist** *n.* — **pol′y·the·is′tic** or **·is′ti·cal** *adj.*

Pol·y·thene (pol′ə·thēn) *n.* Polyethylene: a trade name. [Contraction of POLYETHYLENE]

pol·y·top·ic (pol′i·top′ik) *adj.* Occurring at two or more distinct places. [<Gk. *poly-* many + *topikos* of a place <*topos* a place]

pol·y·troph·ic (pol′i·trof′ik, -trō′fik) *adj.* Obtaining nourishment from several sources, as certain pathogenic bacteria. [<Gk. *polytrophos* highly nourished <*poly-* much + *trephein* feed]

pol·y·typ·ic (pol′i·tip′ik) *adj.* Existing in many types or forms. Also **pol′y·typ′i·cal.** [<POLY- + Gk. *typikos* <*typos* a type]

pol·y·u·ri·a (pol′i·yŏŏr′ē·ə) *n. Pathol.* Excessive urination. [<NL <Gk. *poly-* much + *ouron* urine] — **pol′y·u′ric** *adj.*

pol·y·va·lent (pol′i·vā′lənt) *adj.* 1 *Bacteriol.* Designating a type of vaccine containing antigens derived from two or more different strains of micro-organisms. 2 *Chem.* Multivalent. — **pol′y·va′lence** *n.*

pol·y·vi·nyl (pol′i·vī′nil) *adj. Chem.* Designating any of a group of polymerized vinyl derivatives extensively used in the production of many high-quality resins: *polyvinyl* acetate, *polyvinyl* chloride. [<POLY(MERIZED) + VINYL]

Po·lyx·e·na (pə·lik′sə·nə) In Greek legend, a daughter of Priam, betrothed to Achilles and after his death sacrificed to appease his shade by Neoptolemus.

pol·y·zo·ar·i·um (pol′i·zō·âr′ē·əm) *n. pl.* **·ar·i·a** (-âr′ē·ə) *Zool.* The entire colony of a compound bryozoan, or its supporting skeleton. Also **pol·y·zo′a·ry** (-zō′ə·rē). [<NL <*polyzoa* a bryozoan <Gk. *poly-* many + *zōion* an animal]

pol·y·zo·ic (pol′i·zō′ik) *adj. Zool.* 1 Of or pertaining to the *Bryozoa.* 2 Denoting a spore which produces many sporozoites. [< NL *polyzoa.* See POLYZOARIUM.]

pom·ace (pum′is) *n.* 1 The substance of apples or like fruit crushed by grinding. 2 Fish scrap. 3 The cake left after the expression of oil from castor beans. [<Med. L *pomacium* cider <L *pomum* an apple]

pomace fly A fruit fly.

po·ma·ceous (pō·mā′shəs) *adj.* 1 Relating to or made of apples. 2 Of or pertaining to a pome, or to trees of the rose family that produce pomes. [<NL *pomaceus* <L *pomum* an apple]

po·made (pō·mād′, -mäd′) *n.* A perfumed dressing for the hair or an ointment for the scalp. — *v.t.* **·mad·ed, ·mad·ing** To anoint with pomade. [<MF *pommade* <Ital. *pomata* <*pomo* an apple, fruit <L *pomum*]

po·man·der (pō′man·dər, pō·man′dər) *n.* A perfume ball, or perfumed powder, formerly worn as an amulet; also, a box for carrying such perfume. [Earlier *pomamber* <OF *pomme*

d'ambre apple of amber < *pomme* an apple (< L *pomum* + *ambre* amber]
pome (pōm) *n. Bot.* A fleshy, many-celled fruit with a core, as an apple, quince, pear, or the like. [< OF, an apple < L *pomum*, orig. a fruit]
pome·gran·ate (pom′gran·it, pŭm′, pəm·gran′·it) *n.* 1 The fruit of a tropical Asian and African tree (*Punica granatum*), about the size of an orange and having a hard rind and subacid red pulp with many seeds. 2 The tree. [< OF *pome grenate* < *pome* an apple (< L *pomum*) + *grenate* < LL *granata* < L *granatum*, orig. neut. of *granatus* very seedy < *granum* a grain, a seed]
pom·e·lo (pom′ə·lō) *n. pl.* **·los** A small variety of the shaddock; grapefruit. [Prob. < POME; infl. in form by Du. *pompelmoes* a pompelmous]
Pom·e·ra·ni·a (pom′ə·rā′nē·ə) A region of north central Europe along the Baltic, extending from Stralsund to the Vistula and including a former Prussian province with its capital at Stettin and the former Free City of Danzig; now a part of Poland except for a small area in NE East Germany: German *Pommern*.
Pom·e·ra·ni·an (pom′ə·rā′nē·ən) *adj.* Relating to Pomerania or its inhabitants. — *n.* 1 A native or inhabitant of Pomerania. 2 A small dog with pointed ears and muzzle, a bushy tail turned over the back, and long, straight, silky coat varying in color: believed to have originated in Pomerania.

POMERANIAN
(From 7 to 10 inches high; weight, 3 to 7 pounds)

Pom·e·rel·ia (pom′ə·rel′yə) A region of northern Poland on the Baltic and the Gulf of Danzig. German **Pom·mer·el·len** (pô′mer·el′ən), Polish **Po·mo·rze** (pô·mô′zhe).
po·mi·cul·ture (pō′mi·kul′chər) *n.* Fruit culture. [< *pomi-* (< L *pomum* an apple, fruit) + CULTURE]
po·mif·er·ous (pō·mif′ər·əs) *adj.* Pome-bearing. [< L *pomifer* < *pomum* an apple, fruit + *ferre* bear]
pom·mel (pum′əl, pom′-) *v.t.* **·meled** or **·melled**, **·mel·ing** or **·mel·ling** To beat with or as if with the pommel of a sword or with the fists. See synonyms under BEAT. — *n.* 1 A knob at the front of a saddle or on the hilt of a sword. 2 The butt of a firearm. Also spelled *pummel*. [< OF *pomel* a rounded knob, dim. of *pome*. See POME.]
Pom·mern (pôm′ərn) The German name for POMERANIA.
po·mol·o·gy (pō·mol′ə·jē) *n.* The science of fruits and the art of fruit culture. [< NL *pomologia* < L *pomum* an apple, fruit + *-logia* -LOGY] — **po·mo·log′i·cal** (pō′mə·loj′i·kəl) *adj.* — **po′mo·log′i·cal·ly** *adv.* — **po·mol′o·gist** *n.*
Po·mo·na (pə·mō′nə) In Roman mythology, the goddess of fruit and fruit trees.
Po·mo·na Island (pə·mō′nə) Largest of the Orkney Islands, Scotland; about 189 square miles: also *Mainland*.
pomp (pomp) *n.* 1 Magnificent or ostentatious display, especially in costume, equipage, etc. 2 *Obs.* A grand procession; pageant. See synonyms under OSTENTATION. [< OF *pompe* < L *pompa* < Gk. *pompē* a sending, pomp *pempein* send]
pom·pa·dour (pom′pə·dôr, -dōr, -dōōr) *n.* 1 A style of arranging the hair by brushing it up from the forehead in a manner reminiscent of an 18th century style. 2 A style of bodice with low, square neck. — *adj.* Characterizing anything made fashionable by the Marquise de Pompadour: *pompadour* silk. [after Marquise de *Pompadour*]
Pom·pa·dour (pôn′pà·dōōr′), **Marquise de**, 1721–64, Jeanne Antoinette Poisson, mistress of Louis XV of France.
pom·pa·no (pom′pə·nō) *n. pl.* **·nos** 1 A highly prized carangoid food fish (genus *Trachinotus*) of warm seas, especially *T. carolinus*, found off the coasts of the South Atlantic States.

2 A food fish of the American Pacific coast (*Rhombus simillimus*). [< Sp. *pámpano*]
Pom·pe·ii (pom·pā′ē) An ancient city of southern Italy SE of Naples; destroyed by an eruption of Vesuvius, A.D. 79. — **Pom·pe·ian** (pom·pā′ən, -pē′ən) *adj. & n.*
pom·pel·mous (pom′pəl·mōōs) *n.* An East Indian variety of shaddock. [< Du. *pompelmoes*, ? < Du. *pompoen* a pumpkin + older Malay *limoes* a shaddock < Pg., pl. of *limão* a lemon, citron]
Pom·pey (pom′pē) Anglicized name of Gnaeus Pompeius Magnus, 106–48 B.C., Roman general, statesman, and triumvir; rival of Julius Caesar: known as *Pompey the Great*.
pom·pho·ly·he·mi·a (pom′fə·li·hē′mē·ə) *n. Pathol.* An abnormal accumulation of gas bubbles in the blood, as in caisson disease. [< NL < Gk. *pompholyx* a bubble (< *pomphos* a blister) + *haima* blood]
pom·pom (pom′pom) *n.* A rapid-fire, automatic cannon used especially as an anti-aircraft weapon. [From the sound made by the charge when fired]
pom·pon (pom′pon, *Fr.* pôn·pôn′) *n.* 1 In millinery, a tuft or ball, as of feathers or ribbon. 2 The colored ball of wool on the front of a shako, or on top of a sailor's cap. 3 A variety of chrysanthemum or dahlia having a small, compact, globe-shaped flower head. [< F < MF *pomper* exhibit pomp < OF *pompe* pomp]
pom·pous (pom′pəs) *adj.* 1 Marked by assumed stateliness; overbearing; ostentatious. 2 Magnificent; marked by ceremonious or impressive display. — **pom·pos·i·ty** (pom·pos′ə·tē) *n.* — **pom′pous·ness** *n.* — **pom′pous·ly** *adv.*
Po·na·pe (pō′nə·pā) One of the most important of the eastern Caroline Islands; 129 square miles: formerly *Ascension*.
Pon·ce (pôn′sā) A port in southern Puerto Rico.
Ponce de Le·ón (pons′ də lē′ən, *Sp.* pôn′·thā thā lā·ôn′), **Juan**, 1460–1521, Spanish discoverer of Florida.
pon·cho (pon′chō) *n. pl.* **·chos** 1 A South American cloak like a blanket with a hole in the middle for the head. 2 A similar garment, waterproofed or rubberized, and used as a raincoat. [< Araucan *poncho*, *pontho*]
pond (pond) *n.* A body of still water, smaller than a lake. [ME *ponde*, var. of POUND²]
pon·der (pon′dər) *v.t.* To weigh in the mind; consider carefully. — *v.i.* To meditate; reflect. See synonyms under CONSIDER, DELIBERATE, EXAMINE, MUSE. [< OF *ponderer* < L *ponderare* < *pondus, ponderis* a weight] — **pon′der·a·ble** *adj.* — **pon′der·a·bil′i·ty** *n.* — **pon′der·er** *n.*
pon·der·ous (pon′dər·əs) *adj.* 1 Having great weight; also, huge; bulky. 2 Heavy to the extent of dulness; lumbering; labored. See synonyms under HEAVY. [< OF *pondereux* < L *ponderosus* < *pondus, ponderis* a weight] — **pon·der·os·i·ty** (pon′dər·os′ə·tē), **pon′der·ous·ness** *n.* — **pon′der·ous·ly** *adv.*
Pon·di·cher·ry (pon′di·cher′ē) A former French settlement in SE Madras, India; a free city 1947–54; incorporated into India, November 1, 1954; 192 square miles. French **Pon·di·ché·ry** (pôn·dē·shā·rē′).
pond·lil·y (pond′lil′ē) *n. pl.* **·lil·ies** The water-lily.
Pon·do·land (pon′dō·land′) A district of eastern Cape of Good Hope Province, Union of South Africa; 4,000 square miles.
pond scum Any of a group of free-floating, fresh-water, green algae (*Spirogyra* and related genera).
pond·weed (pond′wēd′) *n.* Any of various submersed or partially floating perennial aquatic plants (genus *Potamogeton*) common in the Old and the New World.
pone¹ (pōn) *n.* 1 Bread made of cornmeal, sometimes with milk and eggs: also *corn pone*. 2 A small cake or patty of cornbread. [< Algonquian (Virginian), bread < *ápan* something baked]
pone² (pōn) *n.* In card games, the player at the dealer's right. [< L, imperative sing. of *ponere* place]
po·nent (pō′nənt) *adj.* Affirmative, constructive; positing: term used in logic. [< L *ponens, -entis*, ppr. of *ponere* place]
pon·gee (pon·jē′) *n.* A thin, natural, un-

bleached silk with a knotty, rough weave, originally made in China from the product of wild silkworms. [? Alter of dial. Chinese *pen chi* home loom < Chinese *pun ki*]
pon·iard (pon′yərd) *n.* A small dagger, especially one with a slender triangular or square blade. — *v.t.* To stab with a poniard. [< MF *poignard* < *poing* a fist < L *pugnus*]
pons (ponz) *n. pl.* **pon·tes** (pon′tēz) *Latin* A bridge: in Latin phrases.
pons as·i·no·rum (ponz as′·i·nôr′əm, -nō′rəm) *Latin* Asses' bridge.
pons Va·ro·li·i (ponz və·rō′·lē·ī) *Anat.* The organ containing the commissural fibers which connect the cerebrum, cerebellum, and medulla oblongata. Also **pons**. [< NL, bridge of Varoli; after Costanzo *Varoli*, 1542–75, Italian anatomist]

PONIARDS
a. Knife.
b. Japanese.
c. Senegalese.

Pon·selle (pon·sel′), **Rosa**, born 1897, U.S. soprano.
Pon·ta Del·ga·da (pôn′tə del·gä′də) 1 Easternmost of the three districts of the Azores; 326 square miles. 2 Its capital, a port and chief city of the Azores, on São Miguel.
Pont·char·train (pon′chər·trān′), **Lake** A shallow lake in SE Louisiana, joining with the Mississippi at New Orleans by a canal; about 40 by 25 miles.
Pon·te·fract (pom′frit, pum′-, pon′ti·frakt) A municipal borough of West Riding, Yorkshire; site of an 11th century castle where Richard II died. Also **Pom·fret** (pom′frit).
Pon·te·ve·dra (pôn′tā·vä′thrä) An Atlantic province of NW Spain; 1,427 square miles; capital, Pontevedra.
Pon·ti·ac (pon′tē·ak), died 1769, Ottawa Indian chief who made war on the British.
Pon·ti·ac (pon′tē·ak) An industrial city in SE Michigan, 24 miles NW of Detroit.
pon·ti·a·nak (pon′tē·ä′näk) *n.* 1 A grayish-white gum resin obtained from the jelutong tree of Borneo: used as a friction compound on belting, etc. 2 A variety of copal from various species of dammar pine (genus *Agathis*): used in varnishes. [after *Pontianak*]
Pon·ti·a·nak (pon′tē·ä′näk) An Indonesian port on the west coast of Borneo.
Pon·tic (pon′tik) *adj.* Of or pertaining to the Black Sea or adjacent regions. [< L *Ponticus* < Gk. *Pontikos* < *Pontos* the Black Sea, *Pontus* < *pontos* open sea]
pon·ti·fex (pon′tə·feks) *n. pl.* **pon·tif·i·ces** (pon·tif′ə·sēz) A member of the highest priestly college of ancient Rome, the Pontifical College, which had supreme jurisdiction in religious matters. [< L *pontifex, -ficis* < Osco-Umbrian *puntis* a sacrificial offering + L *facere* make; infl. in form by L *pons, pontis* a bridge]
pon·tiff (pon′tif) *n.* 1 The pope; also, any bishop. 2 A pontifex of ancient Rome. [< F *pontife* < L *pontifex* a pontifex] — **pon·tif′ic** *adj.*
pon·tif·i·cal (pon·tif′i·kəl) *adj.* 1 Of, pertaining to or appropriate for a pontiff. 2 Having the pomp or dogmatism sometimes ascribed to a pontiff; hence, haughty; pompous; dogmatic. [< L *pontificalis* < *pontifex* a pontifex]
pon·tif·i·cate (pon·tif′i·kit, -kāt) *n.* 1 The office of a pontiff. 2 A pope's term of office. — *v.i.* (-kāt) **·cat·ed**, **·cat·ing** 1 To perform the offices of a pontiff. 2 To act or speak pompously or dogmatically.
pon·til (pon′til) *n.* An iron rod used in glassmaking to shape hot glass; a punty. [< F, appar. < Ital. *pontello, puntello*, dim. of *punto* a point < L *punctus*]
pontil mark The slight excrescence or scar left on a finished glass article after detaching it from the pontil: also spelled *punty mark*.
pon·tine (pon′tin) *adj.* Of or pertaining to a bridge or bridges. [< L *pons, pontis* a bridge + -INE²]
Pon·tine Marshes (pon′tin, -tīn) A plain SE of Rome in west central Italy; about 300 square miles; formerly a swamp.

Pon·tius (pon'shəs, -tē-əs) See PILATE.
pon·ton (pon'tən) n. A pontoon. [<OF. See PONTOON.]
pon·to·nier (pon'tə·nir') n. 1 A soldier in charge of pontoons. 2 A builder of pontoon bridges. [<OF pontonnier <Med. L pontonarius <L ponto, -onis a pontoon]
pon·toon (pon·tōōn') n. 1 A flat-bottomed boat, an airtight cylinder, or the like, used in the construction of floating bridges, to support the roadway. 2 A bridge so supported: in the United States Army, usually ponton. 3 A float or a raft to ferry goods across water. 4 A float on the landing gear of a hydroplane. [<OF ponton <L ponto, pontonis <pons, pontis a bridge]

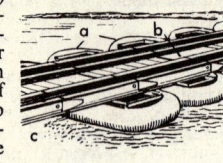

PONTOON BRIDGE
a. Pontoons.
b. Locking bridge sections.
c. Shore.

pontoon bridge A bridge supported on pontoons: distinguished from fixed bridge. Also **ponton bridge**.
Pon·top·pi·dan (pôn·tôp'i·dän), Henrik, 1857–1943, Danish novelist.
Pon·tus (pon'təs) An ancient country, later a Roman province, on the Black Sea in NE Asia Minor.
po·ny (pō'nē) n. pl. ·nies 1 A very small horse, especially one of a small breed; specifically, an Indian pony. 2 Anything small of its kind; specifically, a pony engine. 3 U.S. Slang A translation used in the preparation of foreign language lessons; a crib; trot. 4 In British racing slang, the sum of 25 pounds. 5 Colloq. A very small glass, for spirits, beer, etc. — v.t. & v.i. ·nied, ·ny·ing U.S. Slang 1 To translate (lessons) with the aid of a pony or trot. 2 To pay (money) that is due: with up. [Var. of dial. E (Scottish) powney, prob. <OF poulenet, dim. of poulain a foal, colt <LL pullanus <L pullus a young animal]
pony engine U.S. A small locomotive for use in railroad yards.
pony express In 1860–61, a postal system by which mail was relayed from Missouri to California by riders mounted on swift ponies; also, the rider. See HORSE–POST.
poo (pōō) v.t. Scot. To pull.
pooch (pōōch) n. Slang A dog, especially a small mongrel. [? <dial. E and obs. pooch, var. of POUCH; ? with ref. to appetite]
pood (pōōd) n. A Russian weight equivalent to 36.1 pounds avoirdupois. [<Russian pud <LG pund ult. <L pondo a pound]
poo·dle (pōōd'l) n. One of a breed of dogs of high intelligence, with long, curly, usually white or black hair. [<G pudel <LG, short for pudelhund <pudeln splash in water; with ref. to its being a water dog]
pooh (pōō) interj. Bah! hoh!: an exclamation of disdain: also spelled poh.
Pooh–Bah (pōō'bä') n. Colloq. One who fills many offices inefficiently: from a character in Gilbert and Sullivan's The Mikado.
pooh–pooh (pōō'pōō') v.t. To reject or speak of disdainfully. [Reduplication of POOH]
pook (pōōk) v.t. Scot. To pluck; pick.
pool¹ (pōōl) n. 1 A small body, usually of fresh water, as a spring. 2 A deep place in a stream. 3 Any small, isolated body of liquid: a pool of blood; a puddle. [OE pōl]
pool² (pōōl) n. 1 A collective stake in a gambling game. 2 A combination, generally formed to overcome the effects of excessive competition, whereby companies or corporations agree to fix rates or prices and divide the collective profits pro rata; also, any combination formed for a speculative operation, as in stocks or the like, or the common fund raised for that purpose. 3 Any of various games played on a six-pocket billiard table, in which the object is to drive balls numbered from 1 to 15 into the pockets. 4 A combining of efforts or resources, as for a purpose or the benefit of the contributors. — v.t. To combine in a mutual fund or pool, as to satisfy a mutual need, finance an enterprise, etc. — v.i. To form a pool. [<F poule a stake, orig. a hen <L pulla; infl. in form by POOL¹]
Poole (pōōl) A municipal borough of SE Dorset, England, on the English Channel.

pool·room (pōōl'rōōm', -rōōm') n. A place equipped for the playing of pool, billiards, etc.
pool table A billiard table with six pockets, one at each corner and one in the middle of each long side.
poon (pōōn) n. Any of various East Indian trees (genus Calophyllum), especially C. tacamahaca, from which is obtained the bitter **poon** (or puna) **oil**, used chiefly as an illuminant, and the hard light **poon wood**, formerly used for making masts, etc.: also spelled puna. [<Singhalese pūna]
Poo·na (pōō'nə) A city of central Bombay State, India.
poop¹ (pōōp) Naut. n. 1 A short deck built over the after part of the spar deck of a vessel of war; hence, generally, the stern of a vessel: also poop deck. 2 A cabin covered by the poop deck: also poop cabin. — v.t. 1 To break over the stern or poop of: said of a wave. 2 To take (a wave) over the stern. [<OF pupe, pope <Ital. poppa <L puppis]
poop² (pōōp) U.S. Slang v.t. To bring to exhaustion; weary: usually used passively; He was pooped by the long climb. — v.i. To stop; cease or withdraw. [Cf. MDu. poepen blow, toot]
Po·o·pó (pō'ō·pō'), Lake A lake in west central Bolivia; 12,106 feet above sea level; 60 miles long, 20 to 30 miles wide; about 8 feet deep.
poo·quaw (pōō'kwä) n. A quahaug. [<Algonquian poquaw hock a tightly closed shell]
poor (pōōr) adj. 1 Lacking means of comfortable subsistence; indigent; needy. 2 Lacking in good qualities, or the qualities that render a thing valuable; specifically, lacking in abundance or quality: a poor crop; of inferior workmanship or quality: a poor watch; deficient in vigor; feeble: poor health; lean; thin; feeble from ill feeding: That animal is poor; lacking in fertility; sterile: poor soil. 3 Wanting in strength or spirit; cowardly. 4 Devoid of elegance or refinements; uncomfortable: poor surroundings. 5 Deserving of pity; unhappy; wretched: the poor dog. 6 Devoid of merit; unsatisfactory: a very poor speaker. See synonyms under BAD¹, BASE², HUMBLE, MEAGER, SCANTY. [<OF povre <L pauper. Doublet of PAUPER.] — poor'ness n.
Poor Clare A member of a religious order founded in 1212 by St. Clare of Assisi and following a rule prescribed by St. Francis of Assisi; a Franciscan nun.
poor farm A farm where paupers are cared for at public expense.
poor·house (pōōr'hous') n. A public establishment maintained as a dwelling for paupers.
poor·ly (pōōr'lē) adv. 1 With poor results. 2 Imperfectly; badly. 3 In the manner of the poor. 4 In a spiritless manner. — adj. Colloq. Poor in health; somewhat ailing.
Poor Richard Richard Saunders, the imaginary author of wise precepts in almanacs issued by Benjamin Franklin from 1732 to 1757.
poor soldier The friarbird of Australia: so called from its cry.
poor–spir·it·ed (pōōr'spir'it·id) adj. Having little spirit or courage; cowardly. See synonyms under BASE². — poor'–spir'it·ed·ness n.
poor·tith (pōōr'tith) n. Scot. Poverty.
poor white In the southern United States, one of a class of poverty–stricken white farmers or laborers, contemptuously called **poor white trash**.
pop¹ (pop) v. popped, pop·ping v.i. 1 To make a sharp, explosive sound. 2 To burst open or explode with such a sound. 3 To move or go suddenly or quickly: with in, out, etc. 4 To protrude; bulge: His eyes popped. 5 In baseball, to bat the ball into the air so that an opposing player can catch it, thus retiring the batter: with up or out. — v.t. 6 To cause to burst or explode, as corn by heating. 7 To thrust or put suddenly: with in, out, etc.: He popped his head out of the window. 8 To fire (a gun, etc.). 9 To shoot; also, to hit. 10 In baseball, to bat (the ball) into the air. — **to pop the question** Colloq. To make a proposal of marriage. — n. 1 A sharp explosive sound; a small report: the pop of a pistol. 2 The shot of a firearm. 3 An artificial, unintoxicating, variously flavored drink containing carbon dioxide: so called because its bottle opens with an explosive sound. — adv. Like, or with the sound of a pop; suddenly; unexpectedly. [Imit.]

pop² (pop) n. Slang Papa. [Short for poppa, var. of PAPA]
pop³ (pop) n. A concert of popular or light classical music. — adj. Featuring popular or light classical music: a pop concert; a pop orchestra. Also pops. [Short for POPULAR]
pop·corn (pop'kôrn') n. A variety of maize, the kernels of which explode when heated, forming large white balls; also, the corn after popping.
pope (pōp) n. 1 Often cap. The bishop of Rome, the visible head of the Roman Catholic Church, accounted by that church the vicar of Christ and successor of St. Peter. He is elected by the college of cardinals, usually from their own number. 2 Any person having, or thought to have, similar great authority. 3 In the Greek Church, a parish priest. [OE papa <LL <LGk. papas a bishop, father <Gk. pappas father]
Pope (pōp), **Alexander**, 1688–1744, English poet and satirist. — **John**, 1822–92, U.S. general in the Civil War.
pope·dom (pōp'dəm) n. The office or dominion of a pope; papacy.
Pope Joan An old card game, a variety of newmarket. [after an alleged female pope, central figure of a legend dating from the middle of the ninth century]
pop·er·y (pōp'ər·ē) n. The religion of the Roman Catholic Church with all its doctrines and practices: an opprobrious term.
pope's–nose (pōps'nōz') See PARSON'S–NOSE.
pop–eyed (pop'īd') adj. Having bulging or protruding eyes; hence, amazed; filled with wonder.
pop·gun (pop'gun') n. A tube with a piston that compresses air, expelling a pellet with a pop.
pop·in·jay (pop'in·jā) n. 1 A coxcomb. 2 The figure of a bird, formerly used as a mark in archery, and later for firearms. 3 Archaic A parrot. [<OF papegai <Med. Gk. papagas a parrot <Arabic babhagā; infl. in form by AF gai, OF geai a jay]
pop·ish (pō'pish) adj. Pertaining to popes or popery: used opprobriously. — **pop'ish·ly** adv. — **pop'ish·ness** n.
pop·lar (pop'lər) n. 1 Any of a genus (Populus) of dioecious trees and bushes of the willow family, widely distributed in the northern hemisphere; especially, the **white** or **silver poplar** (P. alba) or the **Lombardy poplar** (P. nigra). 2 The wood of any of these trees. 3 Any one of several trees in some way resembling a poplar: the **Queensland poplar** (Homalanthus populifolius) of tropical Australia, and the **western**, **white**, or **yellow poplar** of the United States, more properly called tuliptree. [<OF poplier <L populus]
pop·lin (pop'lin) n. A durable plain-weave silk, cotton, rayon, or wool fabric, having cross ribs made of warp threads finer than the woof or filling threads: used for dresses, upholstery, etc. [<F popeline, papeline <Ital. papalina papal; with ref. to Avignon, a papal residence where the fabric was originally made]
pop·lit·e·al (pop·lit'ē·əl, pop'li·tē'əl) adj. Of or pertaining to the back part of the leg behind the knee. Also **pop·li·tae·al** (pop'li·tē'əl).
pop·lit'ic. [<NL (musculus) popliteus popliteal (muscle) <L poples, poplitis ham]
Po·po·cat·e·pet·l (pō'pō·kät'ə·pet'l, pō·pō'kä·tä'pet'l) A dormant volcano in central Mexico 45 miles SE of Mexico City in Puebla state; crater 250 feet deep, 2,000 feet across, and over a mile in circumference; 17,887 feet high.
pop·o·ver (pop'ō'vər) n. A very light egg muffin: so named from its rising over the dish in which it is baked.
Pop·pae·a Sa·bi·na (po·pē'ə sə·bī'nə), died A.D. 65?, wife of Nero.
pop·per (pop'ər) n. 1 Anything that pops or makes an explosive noise, as a popgun, firecracker, etc. 2 A container or device for popping corn.
pop·pet (pop'it) n. 1 Mech. A poppet head or a poppet valve. 2 A dainty little person; pretty child; darling: a pet name. 3 One of several small bits of wood on a boat's gunwale to support the rowlocks. [Earlier form of PUPPET]
poppet head A pulley frame over a mine shaft, bearing the hoisting gear.
poppet valve Mech. A disk valve mounted on a stem and having a reciprocating motion in

the direction of the longitudinal axis of the stem.

pop·pied (pop′ēd) *adj.* **1** Abounding in or adorned with poppies. **2** Caused by or as by the poppy; causing sleep: a *poppied* drink. **3** Drowsy as with opium.

popping crease In cricket, a line 4 feet in front of and parallel to the wicket, marking the limit of the batsman's position.

pop·ple[1] (pop′əl) *v.i.* **·pled, ·pling** To have a heaving motion; ripple; bubble, as agitated water. — *n.* Rippling or bubbling water; bubbling, or its sound. [Prob. imit.]

pop·ple[2] (pop′əl) *n. Dial.* Poplar. [<L *populus*]

pop·py (pop′ē) *n. pl.* **·pies 1** Any plant of the genus *Papaver*, typical of a widely distributed family *(Papaveraceae)* having lobed or toothed leaves and vivid red, violet, orange, or white flowers; especially, the **opium poppy** *(P. somniferum)*, the **oriental poppy** *(P. orientale)*, the **Iceland poppy** *(P. nudicaule)*, the **mission poppy** *(P. californicum)*, etc. ◆ Collateral adjective: *papaverous.* **2** The medicinal extract from such a plant. **3** The bright scarlet color of certain poppy blossoms: also **poppy red.** See CALIFORNIA POPPY. [OE *popæg, papoeg* <L *papaver*]

pop·py·cock (pop′ē·kok) *n. Colloq.* Pretentious talk; humbug; nonsense. [<colloq. Du. *pappekak*, lit., soft dung]

pop·py·head (pop′ē·hed′) *n.* A small, carved wooden finial, particularly at the end of a church pew.

pop·u·lace (pop′yə·lis) *n.* The body of the common people; the masses. See synonyms under MOB[1]. [<MF <Ital. *popolaccio, popolazzo* <L *populus*]

pop·u·lar (pop′yə·lər) *adj.* **1** Pertaining to the people at large: *popular* demonstrations or government. **2** Widely approved or admired: a *popular* officer. **3** Suitable for the common people; easily comprehended: *popular* lectures. **4** Prevalent among the people: *popular* errors. **5** Suited to the means of the people: *popular* prices. **6** Of folk origin: the *popular* ballad. **7** Used by the people; current; colloquial: said also of many words on the borderline between slang and reputable usage. **8** Plebeian; vulgar; common. See synonyms under COMMON, GENERAL. [<L *popularis* <*populus* the people] — **pop′u·lar·ly** *adv.*

popular etymology A folk etymology.

pop·u·lar·i·ty (pop′yə·lar′ə·tē) *n.* The condition of being popular, especially of possessing the confidence and favor of the people or of a set of people.

pop·u·lar·ize (pop′yə·lə·rīz′) *v.t.* **·ized, ·iz·ing** To make popular. Also *Brit.* **pop′u·lar·ise′.** — **pop′u·lar·i·za′tion** *n.* — **pop′u·lar·iz′er** *n.*

pop·u·late (pop′yə·lāt) *v.t.* **·lat·ed, ·lat·ing 1** To furnish with inhabitants; people. **2** To inhabit. [<Med. L *populatus*, pp. of *populare* <L *populus* the people]

pop·u·la·tion (pop′yə·lā′shən) *n.* **1** The whole number of people in a place or given area; also, any specific portion of that number: the foreign *population* of New York. **2** The act or process of populating or furnishing with inhabitants; the multiplying of inhabitants. **3** *Biol.* The total number of individual organisms being studied by statistical or biometric methods. See synonyms under PEOPLE. [<LL *populatio, -onis* <L *populus* the people]

Pop·u·list (pop′yə·list) *adj.* Of or pertaining to the Populist or People's party. — *n.* A member of the People's party. [<L *populus* the people] — **Pop′u·lism** *n.* — **Pop′u·lis′tic** *adj.*

Populist party See PEOPLE'S PARTY.

pop·u·lous (pop′yə·ləs) *adj.* Containing many inhabitants; thickly settled. [<L *populosus* <*populus* the people] — **pop′u·lous·ly** *adv.* — **pop′u·lous·ness** *n.*

por·bea·gle (pôr′bē·gəl) *n.* A large voracious shark *(Lamna nasus)* of northern waters, sometimes 10 feet long. [<dial. E (Cornish); ult. origin unknown]

porce·lain (pôrs′lin, pōrs′-, pôr′sə-, pōr′-) *n.* A white, hard, translucent ceramic ware, usually glazed, existing in many varieties, according to its composition and method of manufacture; china; chinaware. It is made from pure clay to which a little of the more fusible feldspar is added. [<OF *porcelaine* <Ital. *porcellana*, orig. a cowry] — **por·ce·la·ne·ous** (pôr′sə·lā′nē·əs, pōr′-) or **por′cel·la′ne·ous** *adj.*

porch (pôrch, pōrch) *n.* **1** A covered structure forming an entrance to a building, outside and with a separate roof, or as a recess in the interior as a kind of vestibule; a veranda. **2** An ancient covered walk or portico. Compare LOGGIA. — **the Porch** The Stoic school of philosophy in ancient Athens, named from the Stoa Poecile, or Painted Porch. See STOIC. [<OF *porche* <L *porticus* a colonnade <*porta* a gate. Doublet of PORTICO.]

por·cine (pôr′sīn, -sin) *adj.* Pertaining to, like, or characteristic of swine. [<F, fem. of *porcin* <L *porcinus* <*porcus* a hog]

por·cu·pine (pôr′kyə·pīn) *n.* A large, hystricomorphic rodent, having coarse hair thickly interspersed with erectile quill-like spines used for defense. *Hystrix cristata* is the common porcupine of the Mediterranean region; *Erethizon dorsatum* is the common Canada porcupine of eastern North America: also called *hedgehog.* [<OF *porc espin*, lit., a spiny hog <*porc* a hog (<L *porcus*) + *espin* a thorn <L *spina*]

CANADA PORCUPINE
(From 30 to 35 inches long in body length)

porcupine ant-eater An echidna.

porcupine fish A globefish.

porcupine grass A tall grass *(Stipa spartea)* of the western United States, yielding good forage and hay, but having long, stiff, sharp awns which twist through the wool into the flesh of sheep.

Porcupine River A river in northern Yukon and NE Alaska, flowing 525 miles north to the Yukon River.

pore[1] (pôr, pōr) *v.i.* **pored, por·ing 1** To gaze steadily or intently. **2** To study or read with care and application: with *over*: to *pore* over one's accounts. **3** To meditate; ponder. [ME *pouren*; origin unknown]

pore[2] (pôr, pōr) *n.* **1** A small orifice or opening, especially a minute perforation in a membrane or tissue, as in the skin. **2** A minute interstice between the molecules of a body. **3** Any inlet or means of absorption or communication. [<OF *pore, porre* <L *porus* <Gk. *poros*]

por·gy (pôr′gē) *n. pl.* **·gies 1** A sparoid, perchlike, salt-water food fish *(Pagrus pagrus)* of the Mediterranean and North Atlantic: often called **red porgy. 2** Any of various other fishes, as the scup, sailor's-choice, or pinfish. [? Var. of PARGO; infl. by *pogy*]

po·rif·er·ous (pô·rif′ər·əs, pō-) *adj.* **1** Bearing or having pores. **2** Of or pertaining to a phylum *(Porifera)* of primitive, aquatic, chiefly marine animals, having bodies perforated by pores which lead to an internal cavity, and living attached to rocks, shells, and other supports; the sponges. [<NL *porifer* <L *porus* a pore + *ferre* bear]

po·rism (pôr′iz·əm, pō′riz-) *n. Math.* One of an ancient class of geometrical propositions intermediate between theorems and problems that asserted a relation between variables or affirmed the possibility of finding conditions under which a problem would become indeterminate. [<L *porisma* a corollary, a problem <Gk. <*porizein* carry, deduce <*poros* a way, a voyage]

pork (pôrk, pōrk) *n.* **1** The flesh of swine used as food. **2** A swine or swine collectively. **3** *U.S. Slang* Government money, distinctions, favors, etc., obtained by a representative for his constituency: a form of political patronage. [<OF *porc* <L *porcus* a hog]

pork barrel 1 A barrel in which pork is pickled and kept. **2** *U.S. Slang* A Federal appropriation for some local enterprise that will favorably impress a representative's constituents.

pork·er (pôr′kər, pōr′-) *n.* A pig or hog, especially regarded as a source of pork.

pork·pie (pôrk′pī, pōrk′-) *n.* **1** A thick-crusted pie with pork filling. **2** A man's hat with a low, flat crown.

pork·wood (pôrk′wŏŏd′, pōrk′-) *n.* **1** The brown, coarse-grained wood of a small tree *(Torrubia longifolia)* with small flowers in cymes, found in southeastern Florida and tropical America. **2** The tree.

pork·y (pôr′kē, pōr′-) *adj.* **pork·i·er, pork·i·est 1** Of or like pork. **2** Obese; fat.

por·nog·ra·phy (pôr·nog′rə·fē) *n.* **1** The expression or suggestion of the obscene in speaking, writing, etc.; licentious art or literature. **2** Originally, a description of prostitutes and of prostitution as related to public hygiene. [<Gk. *pornographos* writing of harlots <*pornē* a harlot + *graphein* write] — **por·no·graph·ic** (pôr′nə·graf′ik) *adj.*

po·ros·co·py (pô·ros′kə·pē, pō-) *n.* The study of the character and arrangement of the sweat pores, especially as shown on fingerprints: used in identification. [<*poro-* (<Gk. *poros* a pore) + -SCOPY] — **po·ro·scop·ic** (pôr′ə·skop′ik, pō′rə-) or **·i·cal** *adj.*

po·ros·i·ty (pô·ros′ə·tē, pō-) *n.* **1** The property of being porous; porousness. **2** A porous part or structure. [<Med. L *porositas, -tatis* <*porosus* <L *porus* a pore]

po·rous (pôr′əs, pō′rəs) *adj.* Having pores. — **po′rous·ly** *adv.* — **po′rous·ness** *n.*

por·phy·rin (pôr′fə·rin) *n. Biochem.* Any of a class of organic pigments derived from the breakdown of hemoglobin and chlorophyll, and consisting of four pyrrole nuclei. [Short for *(hemato)porphyrin* <HEMATO- + Gk. *porphyra* the purple whelk and its dye + -IN]

por·phy·rit·ic (pôr′fə·rit′ik) *adj.* **1** Pertaining or relating to porphyry. **2** *Mineral.* Containing well-defined, relatively large crystals in a fine-grained, glassy base or groundmass. Also **por′phy·rit′i·cal.** [<Med. L *porphyriticus* <L *porphyrites* porphyry <Gk. *porphyritēs (lithos)*, lit., (a) purplelike (stone) <*porphyros* purple]

Por·phy·ro·gen·i·tus (pôr′fə·rō·jen′i·təs) See CONSTANTINE VII.

por·phy·roid (pôr′fə·roid) *n.* A greenish, grayish, or reddish crystalline and perfectly schistose rock, containing porphyritic crystals of feldspar.

por·phy·ry (pôr′fə·rē) *n. pl.* **·ries** An igneous rock that has a groundmass enclosing crystals of feldspar or quartz. [<OF *porfire* <Med. L *porphyreus* <Gk. *porphyros* purple <*porphyra* the purple whelk and its dye]

Por·phy·ry (pôr′fə·rē) Anglicized name of Malchus Porphyrius, 233–304?, Neo-Platonic philosopher of Syrian origin; a disciple of Plotinus; opposed Christianity.

por·poise (pôr′pəs) *n. pl.* **·poises** or **·poise 1** A gregarious cetacean, of the genus *Phocaena*, without a distinct beak; especially, *P. phocaena* of the North Atlantic and Pacific, from 5 to 6 feet long, blackish above and white below. **2** Any small cetacean; popularly, the common dolphin or the bottlenose. [<OF *porpeis, porpois*, lit., hog fish <L *porcus* a hog + *piscis* a fish]

COMMON PORPOISE
(Smaller than the 7 to 8 foot common dolphin)

por·ridge (pôr′ij, por′-) *n.* **1** A soft food made by boiling meal or flour in water or milk until it becomes thick. **2** A broth or stew of vegetables, sometimes containing meat. [Alter. of POTTAGE; infl. in form by OF *poree* vegetable soup]

por·rin·ger (pôr′in·jər, por′-) *n.* A small, shallow dish, having straight sides and sometimes ears; a porridge dish. [Earlier *pottanger* <MF *potager* a soup bowl; infl. in form by PORRIDGE]

Por·se·na (pôr′sə·nə), **Lars** A semilegendary Etruscan king of the sixth century B.C. who marched against Rome to restore the Tarquins. Also **Por·sen·na** (pôr·sen′ə).

Por·son (pôr′sən), **Richard,** 1759–1808, English classical scholar.

port[1] (pôrt, pōrt) *n.* **1** A harbor or haven; hence, a place of customary entry and exit for vessels, as for commerce. **2** *Law* Any place designated as a point at which persons or merchandise may enter or pass out of a country, under specified supervision: also *port of entry.*

[Fusion of OE and OF, both <L *portus* a harbor]

port[2] (pôrt, pōrt) *n.* **1** An opening in the side of a ship, as for a gun, light and air, or for the passage of cargo. **2** A gate, portal, door, or other entrance. **3** An orifice for the passage of a motive fluid, as air, gas, etc.: a steam *port*; exhaust *port*. [Prob. fusion of OE and OF, both <L *porta* a gate, door]

port[3] (pôrt, pōrt) *n.* **1** The way in which one bears or carries himself; mien; external manner: a majestic *port*. **2** The position of a rifle when ported. See synonyms under AIR[1]. — **high port** *Mil.* The position in which a soldier carries his rifle, diagonally across his body, while running or jumping. — *v.t.* **1** *Mil.* To carry, as a rifle, saber, or other weapon, diagonally across the body and sloping to the left shoulder. **2** To carry. [<OF *porte* < *porter* carry <L *portare*]

port[4] (pôrt, pōrt) *Naut. n.* The left side of a vessel as one looks from stern to bow: formerly called *larboard*: opposed to *starboard*. — *v.t.* To put or turn to the port or larboard side: to *port* the helm. —*adj.* Left; larboard: *port* side. [Prob. <PORT[1]]

port[5] (pôrt, pōrt) *n.* A sweet variety of wine, usually of a dark-red color. [Short for *Oporto wine*, from *Oporto*, Portugal; so called because orig. shipped from there]

port·a·ble (pôr'tə·bəl, pōr'-) *adj.* **1** That can be readily carried or moved. **2** *Obs.* Endurable; supportable. [<LL *portabilis* <L *portare* carry] — **port'a·ble·ness, port'a·bil'i·ty** *n.* — **port'a·bly** *adv.*

port·age (pôr'tij, pōr'-) *n.* **1** The act of transporting, especially canoes, boats, and goods, from one navigable water to another. **2** The route over which such transportation is made, or that which is transported. **3** The charge for transportation. [<F <OF <Med. L *portaticum* <L *portare* carry]

Por·ta (pôr'tä), **Giacomo della,** 1541–1604, Italian architect and sculptor. — **Giambattista della,** 1538?–1577, Italian physicist.

por·tal (pôr'təl, pōr'-) *n.* **1** A passage for gaining entrance; door; gate; especially, one that is grand and imposing. **2** The architectural composition that includes the entrances and porches of a large church or similar building. **3** Any opening or entrance resembling or suggesting the portal of an edifice: often in the plural. See synonyms under ENTRANCE[1]. — *adj.* **1** Pertaining to or entering at a port or gate. **2** *Anat.* Pertaining to or arranged like the **portal vein,** which conveys blood from the intestines and other abdominal viscera to the liver, there subdividing into capillaries. [<OF, a gate <Med. L *portale* a city gate, a porch, orig. neut. of *portalis* <L *porta* a gate]

Por·tal (pôr·tàl'), **Baron Antoine,** 1742–1832, French anatomist.

por·tal-to-por·tal pay (pôr'təl·tə·pôr'təl, pōr'-) A wage computed on the full time spent on mine or factory property from arrival to departure, not on actual working time.

por·ta·men·to (pôr'tə·men'tō, pōr'-; *Ital.* pôr'·tä·men'tō) *n. pl.* **·ti** (-tē) *Music* **1** A slur or glide from one note to another, sounding all the intervening tones. **2** Loosely, à legato passage or effect. [<Ital., lit., a carrying < *portare* carry <L]

port·ance (pôr'təns, pōr'-) *n. Archaic* Personal carriage; deportment; mien. [<MF, a carrying, support < *porter* carry <L *portare*]

Port A·pra (ä'prä) See APRA HARBOR.

port arms *Mil.* **1** A command to carry a rifle, saber, or other weapon at the port. **2** The position of the weapon when so carried. [< PORT[3], *v.* + ARMS]

Port Arthur A city of southern Manchuria at the tip of Liaotung Peninsula; site of a major naval base, operated jointly by the Soviet Union and the Chinese People's Republic after 1945: Chinese *Lüshun*, Japanese *Ryojun*.

por·ta·tive (pôr'tə·tiv, pōr'-) *adj.* **1** Of or pertaining to carrying; capable of carrying. **2** Portable. [<OF, fem. of *portatif*, lit., portable <L *portatus*, pp. of *portare* carry]

Port-au-Prince (pôrt'ō·prins', pōrt'-; *Fr.* pôr·tō·praṅs') A port, capital of Haiti.

port authority Any official body having charge of the coordination of all rail and water traffic of a port.

Port Blair The capital of the Andaman and Nicobar Islands, a port on SE South Andaman Island.

Port Castries See CASTRIES.

port·cul·lis (pôrt·kul'is, pōrt-) *n.* A grating made of strong bars of wood or iron that can be let down suddenly to close the portal of a fortified place. [<OF *porte coleice* < *porte* a gate (<L *porta*) + fem. of *coleis* sliding <L *colare* strain, filter]

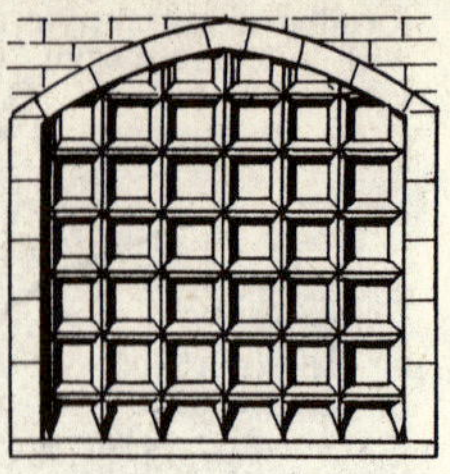

MEDIEVAL PORTCULLIS

Port du Sa·lut (pôrt də sə·lōōt', sə·lōō') A creamy, compact cheese with a flavor similar to that of Gouda.

Porte (pôrt, pōrt) *n.* The former Turkish government: with *the*: officially called **the Sublime Porte.** [<F *(la Sublime) Porte* (the High) Gate, trans. of Turkish *Babi Ali*, the chief office of the Ottoman Empire]

porte-co·chère (pôrt'kō·shâr', pōrt'-; *Fr.* pôrt·kô·shâr') *n.* **1** A large gateway for vehicles, leading into a courtyard. **2** A porch at the door of a building for sheltering persons entering or leaving carriages. [<F < *porte* a gate (<L *porta*) + *cochère*, fem. adj. < *coche* a coach]

por·tée (pôr·tē', -tā', pōr-) *adj.* Towed, carried, or transported by vehicles: said of artillery, cavalry units, etc. Also **por·té'** (-tē', -tā'). [<F, pp. fem. of *porter* carry <L *portare*]

Port Elizabeth A port of SE Cape of Good Hope Province, Union of South Africa.

porte-mon·naie (pôrt'mun'ē, pōrt'-; *Fr.* pôrt·mô·ne') *n. French* A pocketbook for money; especially, a small purse with clasps.

por·tend (pôr·tend', pōr-) *v.t.* **1** To warn of as an omen; presage; forebode. **2** *Obs.* To mean; signify. See synonyms under AUGUR. [<OF *portendre* stretch forth <L *portendere*, var. of *protendere* < *pro-* forth + *tendere* stretch]

Por·te·ño (pôr·tā'nyō) *n.* A native or inhabitant of Buenos Aires.

por·tent (pôr'tent, pōr'-) *n.* **1** Anything that portends what is to happen, especially a momentous or calamitous event. **2** The quality of portending; ominous significance. **3** A prodigy; marvel. [<L *portentum* < *portendere*. See PORTEND.]

por·ten·tous (pôr·ten'təs, pōr-) *adj.* **1** Full of portents of ill; ominous. **2** Of strange and ill-boding character, as if supernatural; monstrous; prodigious. See synonyms under AWFUL, FRIGHTFUL. — **por·ten'tous·ly** *adv.* — **por·ten'tous·ness** *n.*

por·ter[1] (pôr'tər, pōr'-) *n.* **1** One who carries things; especially, a man who carries travelers' luggage, etc., for hire, as in a hotel or at a railroad station. **2** *U.S.* An attendant in a Pullman car. [<OF *porteour* <L *portator* < *portatus*, pp. of *portare* carry]

por·ter[2] (pôr'tər, pōr'-) *n.* **1** A keeper of a door or gate. **2** One who waits at a door to carry messages. [<AF, OF *portier* <LL *portarius* <L *porta* a gate, a door]

por·ter[3] (pôr'tər, pōr'-) *n.* A dark-brown, heavy, English malt liquor resembling ale. [Short for *porter's beer* <PORTER[1]; so called because formerly drunk chiefly by porters]

Por·ter (pôr'tər, pōr'-), **Cole,** born 1893, U.S. composer and lyricist. — **David,** 1780–1843, U. S. commodore. — **David Dixon,** 1813–91, U. S. admiral; son of David Porter. — **Jane,** 1776–1850, English novelist. — **Noah,** 1811–1892, U. S. educator and editor. — **William Sydney** See O. HENRY.

por·ter·age (pôr'tər·ij, pōr'-) *n.* **1** The business of a porter. **2** The cost of carriage by a porter.

por·ter·house (pôr'tər·hous', pōr'-) *n.* **1** A place where porter, ale, etc., are retailed. **2** A restaurant; chophouse. **3** A choice cut of beefsteak including a part of the tenderloin, usually next to the sirloin: also **porterhouse steak.** [<PORTER[3] + HOUSE]

port·fo·li·o (pôrt·fō'lē·ō, pōrt-) *n. pl.* **·li·os** **1** A portable case for holding drawings, writing materials, documents, etc. **2** The position or office of a minister of state or member of a government. **3** A list of investments. [Earlier *porto folio* <Ital. *portafoglio* < *portare* carry (<L) + *foglio* a leaf <L *folium*]

port·hole (pôrt'hōl', pōrt'-) *n.* **1** A small opening in a ship's side. **2** Hence, an embrasure; loophole for shooting through. **3** The entrance to a port in an engine. See PORT[2].

Port Hudson A village on the east bank of the Mississippi in Louisiana; scene of a Union victory in the Civil War, 1863.

Port Huron A city on the St. Clair River and Lake Huron, SE Michigan.

Por·tia (pôr'shə, -shē·ə, pōr'-) The heroine of Shakespeare's *The Merchant of Venice*. She acts the part of a lawyer and defeats Shylock's claim for a pound of Antonio's flesh.

por·ti·co (pôr'ti·kō, pōr'-) *n. pl.* **·coes** or **·cos** An open space or ambulatory with roof upheld by columns; a porch. [<Ital. <L *porticus*. Doublet of PORCH.] — **por'ti·coed** *adj.*

por·tière (pôr·tyâr', pōr-; *Fr.* pôr·tyâr') *n.* A curtain for a doorway, used either instead of a door or as an ornament. Also **por·tiere'.** [<F < *porte* a door <L *porta*]

por·tion (pôr'shən, pōr'-) *n.* **1** A part of a whole, whether separated from it or not. **2** An allotment; share; especially, the quantity of any kind of food usually served to one person. **3** The part of an estate coming to an heir. **4** A dowry (def. 1). **5** One's fortune or destiny. — *v.t.* **1** To divide into shares for distribution; parcel: often with *out*. **2** To give a dowry to; dower. **3** To assign; allot. [<OF *porcion* <L *portio, -onis*] — **por'tion·a·ble** *adj.* — **por'tion·less** *adj.*

Synonyms (noun): part, proportion. When any whole is divided into *parts*, any *part* that is allotted to some person, subject, or purpose is called a *portion*, whether or not the division may be by some fixed rule or relation. But when we speak of a *part* as a *proportion*, we think of the whole as divided according to some rule or scale, so that the different *parts* bear a contemplated and intended relation or ratio to one another; thus, the *portion* allotted to a child by will may not be a fair *proportion* of the estate. See PART.

por·tion·er (pôr'shən·ər, pōr'-) *n.* One who divides in shares or holds a share or shares.

Port Jackson An inlet on the southern shore of New South Wales, Australia, forming the harbor of Sydney.

Port Jin·nah (jin'ə) A port of SW East Bengal, East Pakistan: also *Chalna Anchorage*.

Port·land (pôrt'lənd, pōrt'-) **1** A port on Casco Bay, SW Maine. **2** A port of entry on the Willamette River, NW Oregon.

Portland Bight See OLD HARBOR BAY.

Portland cement See under CEMENT.

Portland Race A dangerous, swift current off the coast of Dorset, SW England.

Port Lou·is (lōō'is, lōō'ē) A port, capital of Mauritius.

port·ly (pôrt'lē, pōrt'-) *adj.* **·li·er, ·li·est** **1** Somewhat corpulent; stout. **2** Of a stately appearance and carriage; impressive, especially on account of size. See synonyms under CORPULENT. [<PORT[3] + -LY] — **port'li·ness** *n.*

Port-Ly·au·tey (pôrt'lē·ō·tā', *Fr.* pôr·lyō·tā') A river port of NW French Morocco ten miles from the Atlantic: formerly *Kénitra*.

Port Ma·hón (mə·hōn', *Sp.* mä·ōn') A former name for MAHÓN.

port·man·teau (pôrt·man'tō, pōrt-) *n. pl.* **·teaus** or **·teaux** (-tōz) **1** Originally, a case for carrying clothing, etc., behind a saddle. **2** An oblong leather suitcase, hinged at the back, and fitted with catches, straps, and a lock, and with handles by which it can be carried. [<MF < *porter* carry (<L *portare*) + *manteau* a coat <OF *mantel* a mantle]

portmanteau word A word arbitrarily formed of two distinct words, as *chortle*, from *chuckle* and *snort*; *cyclotron*, from *cycle* and *electron*; a telescope word; a blend. [Coined by Lewis Carroll]

Port Mores·by (môrz'bē, mōrz'-) A port on the SE coast of New Guinea; administrative center of the Territory of Papua and New Guinea.

Pôr·to (pôr'tōō) The Portuguese name for OPORTO.

Pôr·to A·le·gre (pôr'tōō ä·le'grə) A port of southern Brazil, capital of Rio Grande do Sul state.

Por·to Bel·lo (pôr'tō bel'ō) A port NE of

Colón on the Caribbean coast of Panama. Also **Por·to·be·lo** (pôr′tō-bā′lō).

port of call A port where vessels put in for supplies, repairs, discharge or taking on of cargo, etc.

port of entry A place, whether on the coast or inland, designated as a point at which persons or merchandise may enter or pass out of a country under the supervision of customs and other proper authorities.

Port–of–Spain (pôrt′əv-spăn′, pōrt′-) A port of NW Trinidad, capital of Trinidad and Tobago colony. Also **Port of Spain**.

Por·to–No·vo (pôr′tō-nō′vō, pōr′-) A port in western Africa, capital of Dahomey.

Porto No·vo (nō′vō) A port in SE Madras State, India.

Porto Ri·co (rē′kō) The former official name of PUERTO RICO.

Porto San·to (sän′tō) Northernmost island of Madeira; 16 square miles.

Pôr·to Vê·lho (pôr′tōō ve′lyōō) Capital of Guaporé territory, western Brazil.

Port Philip Bay A bay on the southern coast of Victoria, Australia, forming the harbor of Melbourne.

por·trait (pôr′trit, pōr′-, -trāt) n. 1 A likeness of an individual, especially of the face, produced by an artist in oils, water color, etc., or by photography. 2 Hence, a vivid description of something or someone having existence. [<MF, orig. pp. of *portraire* <OF *pourtraire* PORTRAY]

por·trait·ist (pôr′trā·tist, pōr′-) n. One who makes portraits; a portrait painter or photographer.

por·trai·ture (pôr′tri·chər, pōr′-) n. 1 A representation of an object. 2 The act or art of portraying; especially, the art or practice of making portraits. 3 Portraits or pictures collectively. [<OF <*pourtrait*, pp. of *pourtraire* PORTRAY]

por·tray (pôr·trā′, pōr-) v.t. 1 To represent by drawing, painting, etc.; delineate. 2 To describe in words; depict verbally. 3 To represent, as in a play; act. See synonyms under IMITATE. [<OF *pourtraire* <Med. L *protrahere* <L, draw forth < *pro-* forth + *trahere* draw] — **por·tray′a·ble** adj. — **por·tray′er** n.

por·tray·al (pôr·trā′əl, pōr-) n. 1 The act of portraying by any method of depiction or delineation: the *portrayal* of a character on the stage. 2 The making of a likeness of persons, places, or things; picturing. 3 A portrait.

por·tress (pôr′tris, pōr′-) n. A woman porter or doorkeeper. Also **por′ter·ess**.

Port–Roy·al (pôrt′roi′əl, pōrt′-; *Fr.* pôr·rwȧ·yȧl′) A Cistercian abbey SW of Paris, France; noted as a Jansenist center in the 17th century; suppressed, 1709. Also **Port–Royal–des–Champs** (-dā·shäṅ′).

Port Royal 1 A town and naval station in Jamaica, British West Indies; destroyed by earthquake, 1692. 2 A town in southern South Carolina on **Port Royal Island**, one of the Sea Islands. 3 The former name for Annapolis Royal.

Port Sa·id (sä·ēd′) A port on the Mediterranean end of the Suez Canal, Egypt.

Ports·mouth (pôrts′məth, pōrts′-) 1 A port and the chief naval station of Great Britain, in Hampshire, England. 2 A port and naval station in New Hampshire; site of the signing of the **Treaty of Portsmouth**, ending the Russo–Japanese war, Sept. 5, 1905. 3 A port in SE Virginia, site of a U. S. naval base.

Por·tu·gal (pôr′tə·gəl, pōr′-; *Pg.* pôr′tōō·gäl′) A republic of SW Europe in the western Iberian Peninsula; 34,222 square miles; including the Azores and Madeira islands, 35,419 square miles; capital Lisbon: ancient *Lusitania*.

Por·tu·guese (pôr′chə·gēz′, -gēs′, pōr′-) adj. Pertaining to Portugal, its inhabitants, or their language. — n. 1 A native or inhabitant of Portugal. 2 The people of Portugal collectively: with *the*. 3 The Romance language of Portugal and Brazil. [<Pg. *Portuguez* <*Portugal* <*Portucal* <Med. L *Portus Cale* Oporto]

Portuguese East Africa A former name of MOZAMBIQUE.

Portuguese Guin·ea (gin′ē) A Portuguese overseas province on the coast of western Africa; 13,944 square miles; capital, Bissau.

Portuguese India A Portuguese overseas province on the west coast of India, comprising the territories of Gôa, Damão, and Diu with their dependencies in western India; 1,537 square miles; capital, Nova Gôa in Gôa.

Portuguese man–of–war A pelagic siphonophore (genus *Physalia*) of warm seas, having long, stinging tentacles hanging down from a bladderlike float: also *man–of–war*.

Portuguese Timor See TIMOR.

Portuguese West Africa See ANGOLA.

por·tu·lac·a (pôr′chə·lăk′ə, pōr′-) n. Any plant of a genus (*Portulaca*) of low, fleshy herbs of the purslane family, with scattered leaves, ephemeral flowers of many colors which open only in sunshine, and a globular pod. [<L, purslane] — **por′tu·la·ca′ceous** (-lə·kā′shəs) adj.

po·sa·da (pō·sä′thä) n. Spanish An inn.

pose[1] (pōz) n. 1 The position of the whole or part of the body, especially such a position assumed for or represented by an artist, photographer, etc.: the *pose* of the head. 2 Hence, a mental attitude; attitudinizing for effect. See synonyms under ATTITUDE. — v. **posed, pos·ing** v.i. 1 To assume or hold an attitude or position, as for a portrait. 2 To affect poses; attitudinize. 3 To represent oneself: to *pose* as an expert. — v.t. 4 To cause to assume an attitude or position, as an artist's model. 5 To state or propound; put forward, as a theory or problem. [<F <*poser* put down, rest, place; fusion of L *pausare* lie down and *pos-*, stem of *ponere* lay down, put]

pose[2] (pōz) v.t. **posed, pos·ing** 1 To puzzle or confuse by asking a difficult question. 2 *Obs.* To question closely. See synonyms under PERPLEX. [Aphetic var. of obs. *appose*, var. of OPPOSE]

Po·sei·don (pō·sī′dən) In Greek mythology, brother of Zeus and husband of Amphitrite, god of the sea and of horses: identified with the Roman *Neptune*. — **Po′sei·do′ni·an** (-dō′nē·ən) adj.

Po·sen (pō′zən) The German name for POZNAŃ.

pos·er[1] (pō′zər) n. One who poses; one who strikes affected attitudes. [<POSE[1], v.]

pos·er[2] (pō′zər) n. A question or problem that baffles. [<POSE[2], v.]

po·seur (pō·zœr′) n. One who assumes or affects a particular attitude to make an impression on others. [<F <*poser* POSE[1], v.]

pu·sie (pō′zē) n. *Scot.* A posy.

po·sied (pō′zēd) adj. 1 Inscribed with a posy, as a ring. 2 With many posies or bunches of flowers.

pos·it (poz′it) v.t. To lay down or assume as a fact; affirm; postulate. Compare INFER. — n. That which is posited. [<L *positus*, pp. of *ponere* place]

po·si·tion (pə·zish′ən) n. 1 The manner in which a thing is placed; also, the place of its location. 2 Disposition of the parts of the body, especially with reference to therapeutic, surgical, or obstetric procedures; posture. 3 Relative social standing; high rank: Wealth commands *position*. 4 Employment or job: He lost his *position*. 5 The act of positing a principle or proposition, or the proposition posited; also, ground of argument; hence, the attitude assumed with reference to a subject; point of view: my *position* on the labor question. 6 *Music* The arrangement of the notes of a chord, as in voice parts. 7 In ancient prosody, the situation of a short vowel before two consonants or their equivalent, causing prolonged utterance: In "texunt," the vowels are long by *position*. See synonyms under ATTITUDE, CIRCUMSTANCE, PLACE. — v.t. 1 To place in a particular or appropriate position. 2 *Rare* To locate. [<OF <L *positio, -onis* <*positus*, pp. of *ponere* place] — **po·si′tion·al** adj.

position light *Aeron.* Any of several variously colored lights used on an aircraft to indicate its position and path of motion.

pos·i·tive (poz′ə·tiv) adj. 1 That is or may be directly affirmed; real; actual; existing: opposed to *negative*. 2 Inherent in a thing and of itself, regardless of its relations to other things; absolute: opposed to *relative*. 3 Openly and plainly expressed; explicit; express; emphatic: opposed to *implied* or *inferred*: a *positive* denial. 4 Imperative: opposed to *discretionary*. 5 Dependent on authority, agreement, or convention: opposed to *natural*: *positive* law. 6 Not admitting of doubt or denial; incontestable: *positive* proof. 7 Free from doubt or hesitation; confident; certain; also, overconfident; dictatorial. 8 *Philos.* Pertaining to positivism (def. 2). 9 Noting one of two opposite directions, qualities, properties, etc., which is taken as primary, or as indicating increase or progression. 10 *Math.* Greater than zero; plus: said of quantities. 11 *Electr.* Having a relatively high potential: the *positive* electrode of a cell; specifically, designating the kind of electricity exhibited by a glass object when rubbed with silk. 12 *Physics* Having a deficiency of electrons: said of atoms which yield electrons. 13 *Biol.* Noting the response of an organism toward a stimulus: a *positive* tropism. 14 *Bacteriol.* Noting the presence of a specified condition or organism: a *positive* bacterial culture. 15 *Mech.* Operated by mechanical power, not by springs or gravity; operated or communicating power through intermediate inelastic parts that are under exact control. 16 Noting the north-seeking pole of a magnet and the corresponding (south) pole of the earth. 17 *Phot.* Having the lights and shades in their natural relation, as in a photograph. 18 *Gram.* Denoting the simple, uncompared degree of the adjective or adverb. See synonyms under DOGMATIC, RADICAL, SURE. — n. 1 That which is capable of being directly and certainly affirmed. 2 *Philos.* In positivism, that which is cognizable by the senses. 3 *Phot.* A picture giving the lights and shades as in nature; a print from a negative. 4 *Gram.* The positive degree of an adjective or adverb; also, a word in this degree, as *good, glad*. 5 *Electr.* A positive plate, pole, etc. See synonyms under CERTAINTY. [<OF, fem. of *positif* <L *positivus* <*positus*. See POSITION.] — **pos′i·tive·ly** adv. — **pos′i·tive·ness** n.

positive rays *Physics* Canal rays.

pos·i·tiv·ism (poz′ə·tiv·iz′əm) n. 1 A way of thinking that regards nothing as ascertained or ascertainable beyond the facts of physical science or of sense. 2 A system of philosophy elaborated by Auguste Comte, holding that man can have no knowledge of anything but actual phenomena and facts and their interrelations, rejecting all speculation concerning ultimate origins or causes. Compare HUMANITARIANISM. 3 Certitude, or the claim of certitude, in knowledge. — **pos′i·tiv·ist** n. — **pos′i·tiv·is′tic** adj.

pos·i·tron (poz′ə·tron) n. *Physics* A positively charged particle of an atom, with a mass equal to that of the electron. [<POSI(TIVE) + (ELEC)TRON]

pos·i·tron·i·um (poz′ə·trō′nē·əm) n. *Physics* An unstable, short-lived atomic entity consisting of a positron and an electron subject to mutual annihilation, with conversion of mass into energy. [<NL <POSITRON + *-ium*, suffix of names of elements]

po·sol·o·gy (pō·sol′ə·jē) n. The branch of medicine that treats of the dosages of drugs. [<F *posologie* <Gk. *posos* how much + *logos* word, study] — **pos·o·log·ic** (pos′ə·loj′ik) or **-i·cal** adj.

pos·se (pos′ē) n. 1 A posse comitatus. 2 A force of men; squad. 3 *Law* Possibility: chiefly in the phrase *in posse* (capable of being): distinguished from *in esse*. [<Med. L, power, armed force <L, be able]

pos·se com·i·ta·tus (pos′ē kom′ə·tā′təs) The body of men that a sheriff or other peace officer calls or may call to his assistance in the discharge of his official duty, as to quell a riot or make an arrest. [<Med. L, power of the county <*posse* a posse + *comitatus* a county <*comes, -itis* a count]

pos·sess (pə·zes′) v.t. 1 To have as property; own. 2 To have as a quality, attribute, etc.: to *possess* a conscience. 3 To enter and exert control over; dominate: often used passively: He was *possessed* by a devil; The idea *possessed* him. 4 To maintain control over (oneself, one's mind, etc.): *Possess* yourself in patience. 5 To put in possession, as of property, news, etc.: with *of.* 6 To have knowledge of; gain mastery of, as a language. 7 To imbue or impress, as with wonder or an idea: with *with.* 8 *Obs.* To seize; gain. See synonyms under HAVE, OCCUPY. [<OF *possessier* <L *possessus,* pp. of *possidere* possess]

pos·sessed (pə·zest′) *adj.* 1 Having; owning: *possessed* of a ready tongue. 2 Calm; cool: to be *possessed* in time of danger. 3 Controlled by or as if by evil spirits; beyond self-control; frenzied. —**like all possessed** *U.S. Colloq.* As if driven by the devil; frenziedly.

pos·ses·sion (pə·zesh′ən) *n.* 1 The act or state of possessing. 2 A thing possessed or owned. 3 *pl.* Property; wealth. 4 The state of being possessed, as by evil spirits. 5 Self-possession. See synonyms under OCCUPATION, PROPERTY, WEALTH. Compare POSSESS.

pos·ses·sive (pə·zes′iv) *adj.* 1 Pertaining to or expressive of possession. 2 *Gram.* Designating a case of the noun or pronoun that denotes possession, origin, or the like. In English, this is formed in nouns by adding *'s* to the singular and to irregular plurals: *John's* book; *men's* souls; the *boss's* office; and a simple apostrophe to the regular plural and sometimes to singulars and proper names ending in a sibilant: *boys'* shoes; *Dickens'* (or *Dickens's*) writings; *James'* (or *James's*) brother. See also -'s[1]. Pronouns in the possessive case have special forms, as *my, mine, his, her, hers, its, our, ours, your, yours, their, theirs, whose.* By some grammarians possessive nouns and pronouns are called *possessive adjectives.* —*n. Gram.* 1 The possessive case. 2 A possessive form or construction. —**double possessive** A redundant possessive. Example: a book *of Mike's.*

pos·ses·sive·ness (pə·zes′iv·nis) *n.* Strong or excessive concern with one's own possessions.

pos·ses·so·ry (pə·zes′ər·ē) *adj.* 1 Pertaining to or having possession. 2 *Law* Proceeding from or depending upon possession.

pos·set (pos′it) *n.* A drink of hot milk curdled with liquor, sweetened and spiced. [ME *poshote, possot;* origin unknown]

pos·si·bil·i·ty (pos′ə·bil′ə·tē) *n. pl.* **·ties** 1 The fact or state of being possible. 2 A possible thing. See synonyms under ACCIDENT, EVENT. [<OF *possibilite* <L *possibilitas* <*possibilis* POSSIBLE]

pos·si·ble (pos′ə·bəl) *adj.* 1 That may be or may become true: opposed to *actual:* said of a thing, an event, or a statement. 2 That may be true in some contingency; imaginably true: sometimes used to denote extreme improbability: opposed to *certain, necessary, impossible.* [<OF <L *possibilis* <*posse* be able <*potis* able + *esse* be] —**pos′si·bly** *adv.*

pos·sum (pos′əm) *n. Colloq.* An opossum. —**to play possum** To pretend; deceive; feign ignorance or inattention; dissemble: from the fact that the opossum feigns death when threatened. [Short for OPOSSUM]

pos·sum·haw (pos′əm·hô′) *n.* The bearberry. [<POSSUM + HAW[2]]

post[1] (pōst) *n.* 1 An upright piece of timber or other material used as a support, a point of attachment, etc., as in a building. 2 A central projection in a lock for receiving the tube of a key. 3 A line or post serving to mark the starting or finishing point of a racecourse. —*v.t.* 1 To put up (a poster, etc.) in some public place. 2 To fasten posters upon; placard. 3 To announce by or as by a poster: to *post* a reward. 4 To denounce thus: to *post* one as a coward. 5 To publish the name of on a list. 6 To publish the name of (a ship) as lost or overdue. See synonyms under SET. [OE <L *postis* a door post]

post[2] (pōst) *n.* 1 Any fixed place or station, occupied or for occupation; especially, a place occupied by a detachment of troops; also, the garrison of such a station; the limits of a sentry's beat; the beat or position to which a policeman is assigned. 2 *U.S.* A local unit of a veterans' organization. 3 An office or employment; a position, as of trust or emolument; situation; especially, a public office. 4 A trading post or settlement. 5 *Brit.* One of the two bugle calls known respectively as **first post** and **last post.** The latter corresponds to *taps* in the army of the United States. See synonyms under PLACE. —*v.t.* 1 To assign to a particular position or post; station, as a sentry. 2 To appoint to a military or naval command. [<MF *poste* a post, a station <Ital. *posto* <LL *postum,* contraction of L *positum,* pp. neut. of *ponere* place]

post[3] (pōst) *n.* 1 **a** A rider or courier who travels over a fixed route or between stations on such a route carrying letters, dispatches, etc. **b** Any of the series of stations furnishing relays of men and horses on such a route. 2 An established system, especially a government system, for transporting the mails; also, the aggregate of mail matter transported from one place to another at one time; the mail; by extension, a post office: Has the *post* come in? Put your letter in the *post.* 3 A size of writing paper, 16 by 20 inches: so called because it bore a postman's horn for watermark. —*v.t.* 1 *Brit.* To place in a mailbox or post office; mail. 2 To inform: He *posted* us on the latest news. 3 In bookkeeping: **a** To transfer (items or accounts) to the ledger. **b** To make the proper entries in (a ledger). —*v.i.* 4 To travel with post horses. 5 To travel with speed; hasten. 6 In horseback riding, to rise from the saddle in rhythm with a horse's gait when trotting. —*adv.* By post horses; hence, rapidly. [<MF *poste* <Ital. *posta,* orig. a station <LL, contraction of L *posita,* pp. fem. of *ponere* place]

post- *prefix* 1 After in time or order; following: *postdate, postwar.* 2 Chiefly in scientific terms, after in position; behind: *postorbital.* [<L *post-* <*post* behind, after]

Post (pōst), **Emily,** 1873–1960, *née* Price, U. S. columnist and writer on social etiquette.

post·age (pōs′tij) *n.* 1 The charge levied on mail matter. 2 The act of going by post. [<POST[3] (def. 2) + -AGE]

postage stamp A small, printed label issued and sold by a government to be affixed to letters, parcels, etc., in payment of postage.

pos·tal (pōs′təl) *adj.* Pertaining to the mails or to mail service. —*n.* A postal card.

postal card A card, issued officially, for carrying a written or printed message through the mails under government stamp: also *postal.* Compare POSTCARD.

postal currency An emergency stamp money, used during the Civil War in the United States (1862–65). Also **postage currency.**

Postal Union An aggregation of countries, organized in 1874, agreeing to deliver foreign mail: officially designated *Universal Postal Union.*

post–bel·lum (pōst′bel′əm) *adj.* Coming or occurring after the war, especially the Civil War. [<L, after the war <*post* after + *bellum* a war]

post box A mailbox.

post·ca·non·i·cal (pōst′kə·non′i·kəl) *adj.* Occurring later than the writing of the Scripture canon.

post captain 1 Formerly, in the British Navy, a captain of three years' standing. 2 Formerly, in the U.S. Navy, a senior captain.

post·card (pōst′kärd′) *n.* 1 A postal card. 2 An unofficial card of any regulation size transmissible under postal regulations through the mails on prepayment of the same postage as a postal card.

post–chaise (pōst′shāz′) *n.* A traveling carriage.

post·clas·si·cal (pōst′klas′i·kəl) *adj.* Being or occurring between the Greek and Latin classical and the medieval writers. Also **post′·clas′sic.**

post·com·mun·ion (pōst′kə·myōōn′yən) *adj.* Coming after communion: a *postcommunion* prayer. —*n.* The part of the Eucharist which follows the distribution of the elements.

post·date (pōst′dāt′) *v.t.* **·dat·ed, ·dat·ing** 1 To assign or affix a date later than the actual date to (a check, document, etc.). 2 To follow in time.

post·di·lu·vi·al (pōst′di·lōō′vē·əl) *adj.* Coming after the deluge.

post·di·lu·vi·an (pōst′di·lōō′vē·ən) *n.* One living after the deluge. —*adj.* Postdiluvial.

post·ed (pōs′tid) *adj.* Possessed of the latest information or news: Keep me *posted.* [<POST[3], *v.* (def. 2)]

pos·teen (pos·tēn′) *n.* An Indian garment made of sheepskin with the fleece left on: also *postin.* [<Persian *pōstīn* of leather <*pōst* a skin]

post·er[1] (pōs′tər) *n.* 1 A placard or bill used for advertising, public information, etc., to be posted on a wall or other surface. 2 A billposter. [<*post*[1], *v.*]

post·er[2] (pōs′tər) *n.* 1 One who travels post. 2 A post horse. [<POST[3], *v.*]

poste res·tante (pōst res·tänt′, *Fr.* pôst res·tänt′) *French* The department of a post office that has charge of mail matter to be held until called for.

pos·te·ri·or (pos·tir′ē·ər) *adj.* 1 Situated behind or toward the hinder part: opposed to *anterior.* 2 Coming after another in a series; especially, subsequent in point of time; later: in this sense opposed to *prior.* 3 *Bot.* Situated or growing on the side next the parent axis: the *posterior* side of an axillary flower. 4 *Zool.* In the direction of the tail; caudal. 5 *Anat.* Dorsal. —*n. Often pl.* The buttocks. [<L *posterior,* comp. of *posterus* following <*post* after]

pos·te·ri·or·i·ty (pos·tir′ē·ôr′ə·tē, -or′ə-) *n.* The state of being posterior or later in point of time: opposed to *priority.*

pos·te·ri·or·ly (pos·tir′ē·ər·lē) *adv.* 1 Subsequently. 2 Behind.

pos·ter·i·ty (pos·ter′ə·tē) *n.* 1 The stock that proceeds from a progenitor; a person's descendants; also, succeeding generations, taken collectively: the *posterity* of Adam. 2 Posteriority. [<OF *posterite* <L *posteritas* <*posterus.* See POSTERIOR.]

pos·tern (pōs′tərn, pos′-) *n.* 1 A back gate or door; a private entrance, especially a small gate beside a large one in a fortified place. 2 A covered passage closed by a gate and leading from a bastion to the ditch. —*adj.* Situated at the back; private: a *postern* gate. [<OF *posterne, posterle* <L *posterus.* See POSTERIOR.]

Post Exchange An establishment for the sale of merchandise and services to military personnel: abbr. *PX.*

post·ex·il·i·an (pōst′eg·zil′ē·ən) *adj.* Pertaining to that period of Jewish history subsequent to the Babylonian exile (605 to 536 B.C.). Also **post′ex·il′ic.**

post·fix (pōst′fiks′) *v.t.* To add at the end of a word, as a letter, syllable, etc.: opposed to *prefix.* —*n.* (pōst′fiks′) That which is so added; a suffix. [<POST- + (AF)FIX]

post·gla·cial (pōst′glā′shəl) *Geol. adj.* Later than the glacial epoch; specifically, formed since the disappearance of the Pleistocene continental glaciers. —*n.* A sedimentary deposit resulting from the retreat of a continental glacier.

post·grad·u·ate (pōst′graj′ōō·it, -āt) *adj.* Of or pertaining to studies pursued after receiving a first degree; graduate. —*n.* One who pursues or has completed a postgraduate course.

post·haste (pōst′hāst′) *adj.* Done with speed; instant. —*n.* Great haste or speed like that of the post. —*adv.* With utmost speed; hurriedly. [Appar. <*Haste, post, haste,* an old direction written on letters]

post horse A horse kept at a post-house for postriders or for hire to travelers.

post–house (pōst′hous′) *n.* A house where post horses were kept for relay; also, formerly, a post office.

post·hu·mous (pos′chōō·məs) *adj.* 1 Born after the father's death: said of a child. 2 Published after the author's death, as a book. 3 Arising or continuing after a person's death: a *posthumous* reputation. [<LL *posthumus* <L *postumus* latest, last, superl. of *posterus.* See POSTERIOR.] —**post′hu·mous·ly** *adv.*

pos·tiche (pôs·tēsh′) *adj.* 1 Added after the completion of the work: said especially of a superadded and inappropriate architectural ornament. 2 Spurious; artificial. —*n.* 1 Pretense; sham. 2 An imitation; artificial substitute. Also **pos·tique′** (-tēk′). [<F <Ital. *posticcio* counterfeit <LL *appositicius* <L *appositus.* See APPOSITE.]

pos·ti·cous (pos·tī′kəs) *adj. Bot.* Hinder; posterior. [<L *posticus* <*post* after]

pos·til (pos′til) *n.* A marginal note; especially, one written on the margin of the Scriptures; also, a series of Scriptural comments. [<OF *postille* <Med. L *postilla* a gloss on the gospel, ? <L *post illa (verba textus)* after those (words of the text) <*post* after + *illa* those]

pos·til·ion (pōs·til′yən, pos-) *n.* A rider of one of the near horses of a team drawing a vehicle, with or without a coachman. Also **pos·til′lion.** [<MF *postillon* <Ital. *postiglione* <*posta* a post, station]

post·im·pres·sion·ism (pōst'im·presh'ən·iz'əm) n. The methods, theories, or practice of a group of painters of the late 19th century who emphasized the subjective prerogatives of the artist as opposed to the literal or idealistic representation of academic painting and the supposed objectivity of impressionism. Cézanne, Van Gogh, and Gauguin are considered its chief exponents. — **post'im·pres'sion·ist** n. & adj. — **post'im·pres·sion·is'tic** adj.

pos·tin (pos·tēn', -tin') See POSTEEN.

post·li·min·i·um (pōst'li·min'ē·əm) n. In international law, a right (Latin jus postliminii), derived from Roman law, whereby persons or things taken in war by the enemy are restored to their former civil condition or previous ownership upon their coming again under the power of the nation to which they belonged. Also **post·lim·i·ny** (pōst·lim'ə·nē). [< L < post after, behind + limen, liminis a threshold]

post·lude (pōst'lōōd) n. An organ voluntary concluding a church service. See PRELUDE. [<POST- + (PRE)LUDE]

post·man (pōst'mən) n. pl. **·men** (-mən) A letter-carrier; mail-carrier; formerly, a courier.

post·mark (pōst'märk') n. The stamp of a post office on mail matter handled there, sometimes also serving to cancel stamps, and giving the name of the office and the day (in large cities also the hour) of mailing or arrival. — v.t. To stamp with a postmark.

post·mas·ter (pōst'mas'tər, -mäs'-) n. 1 An official having charge of a post office. 2 One who provides horses for posting. — **post'mis'tress** (-mis'tris) n. fem.

postmaster general pl. **postmasters general** The executive head of the postal service of a government.

post·me·rid·i·an (pōst'mə·rid'ē·ən) adj. Pertaining to the afternoon. Also **post'me·rid'i·o·nal**. [< L postmeridianus < post- after + meridianus MERIDIAN]

post me·ri·di·em (pōst mə·rid'ē·əm) After midday: abbr. p.m. or P.M. [< L]

post·mil·len·ni·al (pōst'mi·len'ē·əl) adj. Of or pertaining to a period after the millennium. Also **post'mil·len'ni·an**.

post·mil·len·ni·al·ism (pōst'mi·len'ē·əl·iz'əm) n. Theol. The tenet that Christ's second coming will follow the millennium: opposed to premillennialism. Also **post·mil·le·nar·i·an·ism** (pōst'mil'ə·nâr'ē·ən·iz'əm). — **post'mil·len'ni·al·ist** n.

post–mor·tem (pōst·môr'təm) n. Expert examination of a human body after death for pathological or judicial purposes; an autopsy. [< L, after death < post after + mors, mortis death]

post mor·tem (môr'təm) Latin After death.

post·na·tal (pōst·nāt'l) adj. Occurring after birth.

post note A promissory note issued by a bank and payable at a fixed time after its date.

post·nup·tial (pōst·nup'chəl) adj. Happening or occurring after marriage; made after marriage: a postnuptial settlement.

post–o·bit (pōst·ō'bit) adj. Made or done after death; taking effect after death: also **post–o·bit'u·ar'y** (-ō·bich'ōō·er'ē). — n. A bond given to secure payment by the obligor of a sum of money on the death of a designated person, generally one from whose estate he has expectations: also **post–obit bond**. [Contraction of POST OBITUM]

post o·bi·tum (ō'bi·təm) Latin After death.

post office 1 That branch of the civil service of a government charged with carrying and delivering the mails. 2 An office for the receipt, transmission, and delivery of mails, and for the transaction of business connected with the same. 3 Any town or place having a post office. 4 A kissing game. — **post–of·fice** (pōst'ôf'is, -of'-) adj.

Post Office Department An executive department of the U.S. government since 1872 (originally established in 1789), headed by the Postmaster General, which maintains and operates the postal system.

post·op·er·a·tive (pōst·op'ər·ə·tiv, -ə·rā'-) adj. Surg. Occurring after an operation.

post·or·bi·tal (pōst·ôr'bi·təl) adj. Anat. Situated behind the orbit or socket of the eye. — n. 1 A bone of some reptiles at the posterior part of the orbit. 2 A scale behind the orbit, as in snakes.

post–paid (pōst'pād') adj. Having postage prepaid.

post–par·tum (pōst'pär'təm) adj. Med. After childbirth: a postpartum fever. [<POST- + L partus childbirth < parere bear]

post·pone (pōst·pōn') v.t. **·poned**, **·pon·ing** 1 To put off to a future time; defer; delay. 2 To subordinate. [< L postponere < post- after + ponere put] — **post·pon'a·ble** adj. — **post·pone'ment** n. — **post·pon'er** n.

— Synonyms: adjourn, defer, delay, procrastinate. Adjourn signifies literally to put off to another day, and, hence, to any future time. A deliberative assembly may adjourn to another day or to another hour of the same day, and resume business, where it left off, as if there had been no interval; or it may adjourn to a definite later date or, when no day can be fixed, to meet at the call of the president or other officer. In common usage, to adjourn a matter is to hold it in abeyance until it may be more conveniently or suitably attended to; in such use defer and postpone are close synonyms of adjourn; defer is simply to lay or put aside temporarily; to postpone is strictly to lay or put aside until after something else occurs, or is done, known, obtained, or the like; but postpone is often used without such limitation. Adjourn, defer, and postpone all imply definite expectation of later consideration or action; delay is much less definite, while procrastinate is hopelessly vague. One who procrastinates gives no assurance that he will ever act. Compare HINDER, PROCRASTINATE. Antonyms: act, complete, consummate, dispatch, do, expedite, hasten, hurry, quicken.

post·po·si·tion (pōst'pə·zish'ən) n. 1 The act of placing after or state of being placed behind. 2 Gram. A word placed after another word, as an enclitic; especially, a suffixed element which functions as a preposition, as -de in Greek oikade homeward. [< L postpositus, pp. of postponere. See POSTPONE.]

post position The place, in relation to the inner rail, occupied by a horse at the start of a race.

post·pos·i·tive (pōst·poz'ə·tiv) Gram. adj. Appended to something; suffixed; enclitic. — n. An appended word; a postposition. [< L postpositus. See POSTPOSITION.]

post·pran·di·al (pōst·pran'dē·əl) adj. After-dinner.

post·rid·er (pōst'rī'dər) n. A person who journeys by relays of horses.

post road A road built and maintained for the transportation of mail, formerly having post-houses at specified distances.

post·script (pōst'skript') n. 1 A supplemental addition to a written or printed document. 2 Something added to a letter after the writer's signature: abbr. P.S. [< L postscriptum, pp. of postscribere write after]

post terminal A point or port of destination to which goods are transshipped after being delivered by an oceanic carrier: usually applied to additional rates for extra haulage.

post town 1 A town furnishing relays of post horses. 2 A town containing a post office.

pos·tu·lant (pos'chə·lənt) n. 1 One who or that which presents a request. 2 Eccl. An applicant for admission into a religious order or the sacred ministry. Compare NOVICE. [< F < L postulans, -antis, ppr. of postulare demand] — **pos'tu·lant·ship** n.

pos·tu·late (pos'chə·lit) n. 1 A position claimed or basis of argument laid down as well known or too plain to require proof; a self-evident truth. 2 Geom. A self-evident statement regarding the possibility of a geometrical construction: distinguished from axiom. 3 A condition precedent that must be assumed to explain or account for a thing: Peace is a postulate of prosperity. 4 A hypothesis; an unproved assumption. — v.t. (pos'chə·lāt) **·lat·ed**, **·lat·ing** 1 To claim; demand; require. 2 To set forth as self-evident or already known: to postulate the existence of matter. 3 To assume the truth or reality of, especially as a basis for discussion: His theory postulates the validity of an older theory. See synonyms under ASSUME. [< L postulatus, pp. of postulare demand] — **pos'tu·la·tor** n.

pos·tu·la·tion (pos'chə·lā'shən) n. 1 The act of postulating or supposing something as not needing proof; the assumption of a thing as a fact or truth. 2 Eccl. The election or presentation of a person to an office notwithstanding some disqualification.

pos·tu·la·tum (pos'chə·lā'təm) n. pl. **·ta** (-tə) A postulate. [< L]

pos·ture (pos'chər) n. 1 The visible disposition, either natural or assumed, of the several parts of a material thing, and especially of a living thing, with reference to each other; attitude; pose; in art, the position of a figure with regard to its members. 2 Situation as connected with or resulting from a relation of parts; state: the posture of national affairs. 3 Mental or spiritual attitude or condition. See synonyms under ATTITUDE. — v.t. & v.i. **·tured**, **·tur·ing** To place in or assume a posture; pose. [< F < L postura < positura, pp. of ponere place] — **pos'tur·al** adj. — **pos'tur·er**, **pos'tur·ist** n.

pos·tur·ize (pos'chə·rīz) v.t. & v.i. **·ized**, **·iz·ing** To posture.

post–war (pōst'wôr') adj. After a war.

po·sy (pō'zē) n. pl. **·sies** 1 A bunch of flowers, or a single flower; a bouquet; nosegay. 2 Generally, a brief inscription or motto, originally one in verse; especially, one inscribed on a ring or other trinket. [Contraction of POESY]

pot (pot) n. 1 A round earthen, metal, or glass vessel for culinary and other domestic purposes. 2 A metal drinking cup; mug. 3 The contents of a pot; hence, liquor; drink. 4 The amount of stakes wagered or played for; the pool, as in poker. 5 Colloq. A large sum of money. 6 A chimney pot. 7 Scot. A deep pit. 8 In fishing, the circular part of a net; also, a basketlike trap for catching lobsters, eels, fish, etc. 9 A pot shot. 10 A crucible. — v. **pot·ted**, **pot·ting** v.t. 1 To put into a pot. 2 To preserve, as meat, in pots or jars. 3 To cook in a pot; stew. 4 To shoot (game) for food rather than for sport. 5 To shoot or kill with a pot shot. 6 Colloq. To secure, capture, or win; bag. — v.i. 7 To take a pot shot; shoot. [OE pott]

po·ta·ble (pō'tə·bəl) adj. Suitable for drinking: said of water. — n. Something drinkable; a drink. [< F < L potabilis < potare drink]

po·tage (pō·täzh') n. French Any thick soup.

pot·ash (pot'ash') n. 1 Potassium hydroxide: also called caustic potash or potassa. 2 The crude potassium carbonate obtained by leaching the ashes of plants: when purified it is called pearl ash. 3 The oxide of potassium, K_2O. 4 Potash water. Also, in pharmaceutical use, **po·tas·sa** (pə·tas'ə). [Earlier potashes, pl., after Du. potaschen; from being prepared in iron pots]

potash feldspar Orthoclase.

potash water An artificial mineral water containing potassium bicarbonate and charged with carbon dioxide. Also **potassic water**.

po·tas·sa (pə·tas'ə) n. Potassium hydroxide.

po·tas·si·um (pə·tas'ē·əm) n. A bluish-white, highly reactive, metallic element (symbol K). It is never found free in nature, but its many salts are of great practical value, as in fertilizers, gunpowder, dyeing, and medicine. See ELEMENT. [< NL < POTASSA] — **po·tas'sic** adj.

potassium arsenite Chem. A white, hygroscopic, very poisonous mixture of potassium, arsenic, hydrogen, and oxygen, used mostly in solution, as in Fowler's mixture.

potassium carbonate Chem. A white, strongly alkaline compound, K_2CO_3, prepared from wood ashes and also from potassium sulfate obtained from salt beds. It is largely used in the manufacture of soft soap and glass. Also called potash.

potassium chlorate Chem. A colorless crystalline salt, $KClO_3$, used in the manufacture of matches, explosives, etc.

potassium chloride Chem. A colorless crystalline salt, KCl, occurring naturally in large mineral deposits in Germany, and also in certain giant kelps of the Pacific coast; sylvite.

potassium cyanide *Chem.* An intensely poisonous, white, crystalline compound, KCN, used in photography, in electrometallurgy, and as a reagent.
potassium dichromate *Chem.* A reddish crystalline salt, $K_2Cr_2O_7$, used in the arts as an oxidizing agent and in making sensitive coatings for photographs.
potassium hydroxide *Chem.* A whitish deliquescent solid, KOH, yielding a strong caustic solution: used in saltmaking, electroplating, as a chemical reagent, etc. Also called *potash*.
potassium nitrate Niter.
potassium permanganate *Chem.* A purplered crystalline salt, $KMnO_4$, used as an oxidizing agent in antiseptics and deodorizing substances.
potassium sulfate *Chem.* A salt, K_2SO_4, used in the manufacture of glass and alum, and in the crude state as a component of fertilizers: derived from kainite.
po·ta·tion (pō·tā′shən) *n.* 1 The act of drinking; a drink. 2 A drinking bout. [<OF <L *potatio*, *-onis* < *potatus*, pp. of *potare* drink]
po·ta·to (pə·tā′tō) *n. pl.* **·toes** 1 One of the edible, farinaceous tubers of a plant (*Solanum tuberosum*) of the nightshade family. 2 The plant. 3 The sweet potato. [<Sp. *patata* < Arawakan (Taino) *batata* sweet potato]
potato beetle 1 The Colorado beetle (*Leptinotarsa decemlineata*), yellowish, with ten longitudinal black stripes on the wing covers. Both the adult and the larva feed on the leaves of the potato, tomato, and similar plants, and are among the world's greatest agricultural pests: also **potato bug**. For illustration see also INSECTS (injurious). 2 Any of several beetles feeding on the foliage of the potato, especially *Lema trilineata*, with three longitudinal black stripes on the wing covers.
POTATO BEETLE
First described in 1824; widespread by 1874.
(About 3/8 inch long; 1/4 inch wide)
potato chip A very thin slice of potato fried crisp and salted.
potato rot A disease of the potato caused by a mildew (genus *Phytophthora*).
po·ta·to·ry (pō′tə·tôr′ē, -tō′rē) *adj.* Pertaining to potation; given or addicted to drinking: a *potatory* club. [<L *potatorius* < *potator* drinker < *potare* drink]
potato stone A quartz geode resembling a potato.
pot-au-feu (pô·tō·fœ′) *n.* French A variety of beef stew.
Pot·a·wat·o·mi (pot′ə·wot′ə·mē) *n.* One of a tribe of North American Indians of Algonquian stock, formerly inhabiting the western shores of Lake Michigan.
pot·bel·ly (pot′bel′ē) *n. pl.* **·lies** A protuberant belly. — **pot′bel′lied** *adj.*
pot·boil·er (pot′boi′lər) *n. Colloq.* A literary or artistic work produced simply to obtain the means of subsistence. — **pot′boil′ing** *n.*
pot·boy (pot′boi′) *n.* In a public house, a boy or young man who cleans the pots, serves customers, etc.
pot cheese Cottage cheese.
pot companion A boon companion; fellow toper.
po·teen (pō·tēn′) *n.* In Ireland, illicitly manufactured whisky: also spelled *potheen*, *potteen*. [<Irish *poitín*, dim. of *poite* pot]
Po·tem·kin (pō·tem′kin, Russian pô·tyôm′kin), **Prince Grigory Alexandrovich**, 1739-91, Russian field marshal and favorite of Catherine the Great.
po·ten·cy (pōt′n·sē) *n. pl.* **·cies** 1 The quality of being potent; inherent ability; mental, moral, or physical power. 2 The power of effecting particular results: the *potency* of a drug or liquor. 3 In homeopathy, the efficacy of a drug as increased by dilution or attenuation; also, the degree to which such attenuation has been carried. 4 Power arising from external circumstances; authority; hence, power of the prime minister; hence, power to move or influence. 5 Capacity to respond to certain influences; latent power. Also **po′tence**. [<L *potentia*]
po·tent (pōt′nt) *adj.* 1 Physically powerful; able to accomplish material results; efficacious: a *potent* drug. 2 Morally powerful; of a character to influence; convincing: a

potent argument. 3 Having great authority: a *potent* prince. 4 Sexually competent; able to procreate. See synonyms under POWERFUL. [<L *potens*, *-entis*, ppr. of *posse* be able, have power <*potis* able + *esse* be] — **po′tent·ly** *adv.* — **po′tent·ness** *n.*
po·ten·tate (pōt′n·tāt) *n.* One having great power or sway; a sovereign. [<LL *potentatus*]
po·ten·tial (pə·ten′shəl) *adj.* 1 Possible but not actual. 2 Having capacity for existence, but not yet existing. 3 *Physics* Existing by virtue of position: distinguished from *kinetic*. 4 *Gram.* Indicating possibility or power. See POTENTIAL MOOD. 5 Having force or power. — *n.* 1 Anything that may be possible; a possible development. 2 *Gram.* The potential mood. 3 *Physics* A condition at a point in space, due to local attraction or repulsion, such that a mass, electric charge, etc., at that point becomes capable of doing work. 4 *Electr.* The charge on a body as referred to another charged body or to a given standard, as the earth, considered as having zero potential. [<LL *potentialis*] — **po·ten′tial·ly** *adv.*
po·ten·ti·al·i·ty (pə·ten′shē·al′ə·tē) *n. pl.* **·ties** 1 Inherent capacity for development or accomplishment; capability; power; efficiency. 2 Potential quality or being; possibility. [<Med. L *potentialitas*, *-tatis*]
potential mood *Gram.* The verb phrase made up by means of the auxiliaries *may*, *can*, *could*, *must*, *should*, or *would*, with an infinitive, and expressing power, liberty, or possibility: I *could* go; it *may* be.
po·ten·til·la (pō′tən·til′ə) *n.* Any plant of a large genus (*Potentilla*) of herbs or, rarely, of shrubs of the rose family, the cinquefoils or five-fingers, having compound leaves and solitary or cymose flowers with a many-bracted calyx. Many are in cultivation for their profuse, showy flowers. [<NL <L *potens*, *-entis*, ppr. of *posse* be able + *-illa*, dim. suffix]
po·ten·ti·om·e·ter (pə·ten′shē·om′ə·tər) *n.* An apparatus for measuring electromotive force or difference of potential. [<L *potenti(a)* potency + -METER]
po·ten·tize (pōt′n·tīz) *v.t.* **·ized**, **·iz·ing** In homeopathy, to render potent, as drugs, by attenuation. — **po′tent·iz′er** *n.*
poth·e·car·y (poth′ə·ker′ē) *n. Scot. & Brit. Dial.* Apothecary.
po·theen (pō·thēn′) See POTEEN.
poth·er (poth′ər) *n.* Excitement mingled with confusion; bustle; fuss. — *v.t. & v.i.* To worry; bother. [Origin uncertain]
pot herb Any plant or herb, especially greens, cooked by boiling, or used to flavor boiled foods.
pot·hole (pot′hōl′) *n.* 1 A pot-shaped cavity in a rock, as that worn by loose stone gyrated in an eddy. 2 A deep hole, as in a road.
pot·hook (pot′hook′) *n.* 1 A curved or hooked piece of iron for lifting or hanging pots. 2 A curved mark or elementary stroke used in teaching penmanship; also, a scrawl, or, popularly, any curved stroke in stenography.
pot·house (pot′hous′) *n.* An alehouse; saloon. — *adj.* Of or pertaining to a pothouse; vulgar: *pothouse* politics.
pot·hunt·er (pot′hun′tər) *n.* 1 One who kills game for food rather than for sport: usually a contemptuous use. 2 One who engages in a competition simply to win the prizes offered. — **pot′hunt′ing** *adj. & n.*
po·tiche (pō·tēsh′) *n.* A vase having an elongated round body, a cylindrical neck, and a detached cover. [<F]
Pot·i·dae·a (pot′ə·dē′ə) An ancient Macedonian city on the Chalcidice peninsula, near the Aegean Sea. Also **Pot′i·dæ·a**.
po·tion (pō′shən) *n.* A draft, as a large dose of liquid medicine: often used of a magic or poisonous draft. [<F <L *potio*, *-onis* < *potare* drink. Doublet of POISON.]
Pot·i·phar (pot′ə·fär, -fər) An officer of Pharaoh, who bought Joseph as a slave. *Gen.* xxxix 1.
pot·latch (pot′lach) *n.* 1 Among American Indians of the northern Pacific coast: a A gift. b *Often cap.* A winter festival. 2 A ceremonial feast in which gifts are exchanged and property destroyed in a competitive show of wealth. Also **pot′lach**, **pot′lache**. [<Chinook *patshatl* gift]
pot·lead (pot′led′) *n.* Graphite, especially as

used on the bottoms of racing vessels to reduce friction.
pot-lead (pot′led′) *v.t.* To coat with potlead.
pot liquor The liquid left in a pot after cooking greens and meat (usually pork or bacon) together.
pot luck Whatever may chance to be in the pot; hence, a meal or food not prepared for guests: usually in the phrase **to take pot luck**.
pot marigold The calendula.
pot metal 1 Cast iron suitable for making pots. 2 A copper-and-lead alloy formerly used for large pots. 3 A kind of glass colored throughout while still in a molten state.
Po·to·mac River (pə·tō′mək) A river forming the boundaries between Maryland, West Virginia, and Virginia, and flowing 287 miles from the Allegheny Mountains near Cumberland, Md., to Chesapeake Bay about 70 miles SE of Washington, D.C. — **Army of the Potomac** The chief Union army in the Civil War, commanded at first by General George McClellan.
po·to·ma·ni·a (pō′tə·mā′nē·ə, -mān′yə) *n.* Delirium tremens; dipsomania. [<Gk. *potos* drunk + -MANIA]
po·tom·e·ter (pō·tom′ə·tər) *n.* An instrument for measuring the amount of moisture absorbed by a plant, as determined by the amount lost in transpiration. [<Gk. *poton* drink + -METER]
Po·to·sí (pō′tō·sē′) A city of south central Bolivia; 13,255 feet above sea level.
pot·pie (pot′pī′) *n.* A pie, baked in a deep dish, containing meat and vegetables and having only a top crust; also, meat stewed with dumplings.
pot·pour·ri (pot-poor′ē, *Fr.* pō·pōō·rē′) *n.* 1 A ragout of meats and vegetables; a stew. 2 A mixture of dried sweet-smelling flower petals used to perfume a room; also, a small covered jar for containing such a mixture. 3 A medley of musical airs. 4 A literary production composed of miscellaneous parts; an anthology. [<F, lit., rotten pot. See OLLA PODRIDA.]
pot roast Meat braised and cooked in a pot until tender, often with vegetables.
Pots·dam (pots′dam, *Ger.* pôts′dám) A city in East Germany, capital of the former state of Brandenburg; scene of a United Nations conference, July-August, 1945.
pot·sherd (pot′shûrd) *n.* A bit of broken crockery. Also **pot′shard** (-shärd). [<POT + SHARD]
pot shot 1 A shot fired to kill, without regard to the rules of sports. 2 A shot fired, as from ambush, at a person or animal within easy range. 3 A random shot.
pot·stone (pot′stōn′) *n.* Steatite.
pott (pot) *n.* A size of paper, varying in size according to use, but generally about 15 1/2 × 12 1/2 inches. [Var. of POT; so named from having once borne the watermark of a pot]
pot·tage (pot′ij) *n.* 1 A thick broth or stew. 2 A porridge. [<F *potage* < *pot* pot]
pot·ted (pot′id) *adj.* 1 Placed or kept in a pot. 2 Cooked or preserved in a pot. 3 *Slang* Drunk.
pot·teen (po·tēn′) See POTEEN.
pot·ter[1] (pot′ər) *v.t. & v.i.*, *n. Brit.* Putter.
pot·ter[2] (pot′ər) *n.* 1 One who makes earthenware or porcelain. 2 One who pots meats, vegetables, etc. [OE *potere*]
Pot·ter (pot′ər), **Paul**, 1625-54, Dutch painter.
potter's field A piece of ground appropriated as a burial ground for the destitute and the unknown. *Matt.* xxvii 7.
potter's flint Finely pulverized quartz mixed with porcelain to impart strength and rigidity and to reduce shrinkage: used also in enamel mixtures.
potter's wheel A horizontal rotating disk used by potters for holding and manipulating prepared clay.
potter wasp A digger wasp (genus *Eumenes*) which constructs vaselike cells of mud as a nest, especially the North

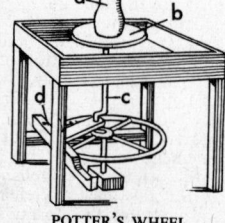
POTTER'S WHEEL
a. Molding clay.
b. Rotating wheel.
c. Shaft.
d. Treadle.

American potter wasp, *E. fraterna*. For illustration see INSECTS (beneficial).

pot·ter·y (pot′ẽr·ē) *n. pl.* **·ter·ies** 1 A factory where potters' ware is made. 2 The manufacture of earthenware or porcelain. 3 Clay ware molded and hardened. ◆ Collateral adjective: *fictile*. [<F *poterie* < *potier* a potter < *pot* a pot]

pot·ting (pot′ing) *n.* 1 The preserving of articles of food in pots for future use. 2 The placing of buds, bulbs, or plants in pots.

pot·tin·ger (pot′in·jər) *n.* 1 *Obs.* A maker of pottage; a cook. 2 *Scot.* & *Brit. Dial.* A porridge dish. [See PORRINGER]

pot·tle (pot′l) *n.* 1 A drinking vessel, pot, or tankard holding about half a gallon. 2 An old liquid measure of half a gallon. 3 A small vessel or basket for holding fruit. [<OF *potel*, dim. of *pot* pot]

pot·to (pot′ō) *n. pl.* **·tos** 1 A small, slow-moving lemur (genus *Perodicticus*) of tropical Africa, having a rudimentary tail and large hands and feet. 2 The kinkajou. [<West African native name]

Pott's disease *Pathol.* Caries or tuberculosis of the vertebrae, causing angular curvature of the spine: first described scientifically by Percival Pott, 1714–88, English surgeon.

pot·ty (pot′ē) *adj. Brit. Colloq.* 1 Insignificant. 2 Slightly drunk; hence, a little silly. [Prob. < POT, in the phrase *go to pot* deteriorate]

pot·val·iant (pot′val′yənt) *adj.* Courageous from drink. — **pot′-val′ian·cy, pot′-val′ian·try, pot′-val′or** (-val′ər) *n.*

pot·wal·lop·er (pot′wol′ə·pər, pot′wol′-) *n.* 1 *Slang* One employed to clean or wash pots, etc.; a scullion. 2 *Brit.* Formerly, by the requirements of some boroughs before 1832, a parliamentary voter who was a householder (not a tenant), having his own fireplace as a qualification for suffrage. [< POT + WALLOP (def. 5)]

pouch (pouch) *n.* 1 A small bag or sack, or something serving a similar purpose, as a pocket or a purse. 2 *Zool.* A saclike part for temporarily containing food, as in gophers and pelicans; also, a marsupium. 3 *Bot.* Any pouchlike cavity, as the silique of the mustard plant. 4 A leather receptacle for carrying small-arms ammunition; also, a wooden cartridge box. 5 An inner mailbag. — *v.t.* 1 To put in or as in a pouch; pocket. 2 To fashion or arrange in pouchlike form. 3 To swallow. — *v.i.* 4 To take on a pouchlike shape; form a pouchlike cavity. [<OF *poche*, var. of *poke, poque* bag] — **pouch′y** *adj.*

pouched (poucht) *adj.* Having pouches or sacs; characterized by pouches.

pouched rat 1 A rodent with cheek pouches; especially, a pocket gopher. 2 A kangaroo rat. 3 Any of certain ratlike rodents of Africa having large cheek pouches (genera *Cricetomys* and *Saccostomus*).

pouf (poōf) *n.* 1 A hair arrangement in high rolled puffs, popular in the 18th century. 2 Any puffed part of a dress. 3 An upholstered tabouret for one or more persons. [<F, a puff]

Pough·keep·sie (pə·kip′sē) A city on the Hudson River in SE New York.

pou·laine (poō·lān′) *n.* 1 The long pointed toe of a medieval shoe. 2 A shoe with such a toe. [<OF (*soulier à la*) *poulaine* (shoe in the) Polish (fashion)]

pou·lard[1] (poō·lärd′) *n.* A pullet having the ovaries removed to produce abnormal growth and fattening and superior quality; hence, a fat pullet. Compare CAPON. [<F *poularde* < *poule* pullet]

pou·lard[2] (poō·lärd′) *n.* A variety of spring or winter wheat closely related to durum, having broad leaves, thick culms, and hard, starchy kernels. Also **poulard wheat**. [<POULARD[1]; so named because suitable only for stock feed]

Poul·sen (poōl′sən), **Valdemar**, 1869–1942, Danish electrical engineer and inventor.

poult (pōlt) *n.* A young turkey, chicken, etc. [Contraction of ME *pulet* pullet]

poul·ter (pōl′tər) *n. Obs.* A poulterer. [<OF *pouletier* < *poulet* pullet]

poul·ter·er (pōl′tər·ər) *n.* A dealer in poultry. [<POULTER + -ER[2]]

poulter's measure A verse form consisting of alternating lines of twelve and fourteen syllables: so called from the poulterer's custom of sometimes giving fourteen eggs to the dozen.

poul·tice (pōl′tis) *n.* A mollifying remedy of a moist, mealy nature, applied to inflamed surfaces. — *v.t.* **·ticed, ·tic·ing** To cover or treat with a poultice. [<L *pultes*, pl. of *puls* porridge]

poul·try (pōl′trē) *n.* Domestic fowls, generally or collectively, as hens, ducks, etc. [<OF *pouleterie* < *poulet* fowl]

pounce[1] (pouns) *v.i.* **pounced, pounc·ing** To swoop or spring in or as in seizing prey: with *on, upon*, or *at*. — *n.* 1 A talon or claw. 2 The act of pouncing; a sudden leap, swoop, spring, or seizure. [Origin uncertain] — **pounc′er** *n.*

pounce[2] (pouns) *v.t.* 1 To perforate with holes in decorative patterns; scallop; pink. 2 To emboss (metalwork) with a design hammered on the reverse side. [<OF *poinçonnier, ponchonner < poinçon, poinchon* a puncheon]

pounce[3] (pouns) *n.* 1 A powder formerly used to absorb excess of ink, as on a manuscript. 2 A finely pulverized substance used in transferring designs. 3 *Obs.* A perfumed powder used as a cosmetic. — *v.t.* To sprinkle, smooth, or rub with pounce. [<F *ponce* <L *pumex, pumicis* pumice]

pounce box 1 A box with perforated lid formerly used for dusting out pounce as a perfume; a perfume box. 2 A box formerly used for dusting powder or sand on freshly written paper. Also **poun·cet** (poun′sit) **box**.

pound[1] (pound) *n.* 1 A variable unit of weight (symbol lb.): the avoirdupois pound is 16 ounces, 7,000 grains, or 453.59 grams; the troy pound, 12 ounces, 5,760 grains, or 373.24 grams. 2 An English money of account, equal to 20 shillings; specifically, a pound sterling (symbol £). See SOVEREIGN. 3 A former Scottish coin worth about twenty pence: more fully **pound Scots**. [OE *pūnd* <L *pondo*, orig. ablative of *pondus* weight]

pound[2] (pound) *n.* 1 A place, enclosed by authority, in which stray or trespassing cattle and distrained cattle or goods are left till redeemed; also, a similar enclosure for stray dogs. 2 An enclosed shelter for cattle or sheep. 3 A trap for wild animals. 4 An area or place in which to catch or stow fish; a poundnet. — *v.t.* To confine in or as in a pound; impound; restrain. [OE *pund(fald)* pinfold] — **pound′er, pound′keep′er** (-kē′pər), **pound′mas′ter** (-mas′tər, -mäs′-) *n.*

pound[3] (pound) *v.t.* 1 To strike heavily and repeatedly, as with a hammer; beat. 2 To reduce to a pulp or powder by beating; pulverize; triturate. — *v.i.* 3 To strike heavy, repeated blows: with *on, at*, etc. 4 To move or proceed heavily or vigorously. 5 To rise and fall heavily, as a ship in rough water. 6 To throb heavily or resoundingly. — *n.* 1 A heavy blow; thump; thud. 2 The act of pounding. See synonyms under BEAT. [OE *punian*] — **pound′er** *n.*

Pound (pound), **Sir Dudley**, 1877–1943, English admiral in World War II. — **Ezra Loomis**, born 1885, U.S. poet. — **Louise**, 1872–1958, U.S. linguist. — **Roscoe**, born 1870, U.S. jurist; brother of Louise.

pound·age[1] (poun′dij) *n.* 1 A rate on the pound sterling. 2 Formerly, in England, a subsidy to the crown on each pound of merchandise exported or imported.

pound·age[2] (poun′dij) *n.* 1 The charges for the redemption of impounded cattle. 2 The act of impounding cattle.

pound·al (poun′dl) *n. Physics* The unit of force in the foot-pound-second system, which, acting on the mass of a pound, imparts to it an acceleration of one foot per second per second.

pound cake A rich cake having ingredients equal in weight, as a pound each of flour, butter, and sugar, with eggs added.

pound·er (poun′dər) *n.* 1 Anything weighing a pound: The trout's a *pounder*. 2 A person or thing weighing, having, or having a certain relation to, a given number of pounds: used only in compounds: The baby is an eight-*pounder*.

pound-fool·ish (pound′foo̅′lish) *adj.* 1 Extravagant with large sums, but watching small sums closely: penny-wise and *pound-foolish*. 2 Having little capacity for business.

pound·net (pound′net′) *n.* A weir or arrangement of nets supported upon stakes to form a trap for fish.

pound party *U.S.* A social gathering to which each guest brings a pound of something, usually food, to be given to an individual or a charitable cause.

pour (pôr, pōr) *v.t.* 1 To cause to flow in a continuous stream, as water, sand, etc. 2 To send forth, emit, or utter profusely or continuously: The radio *poured* forth music. — *v.i.* 3 To flow in a continuous stream; gush. 4 To rain heavily. 5 To move in great numbers; swarm: The northern hordes *poured* over Italy. — *n.* A pouring, flow, or downfall. [Origin unknown] — **pour′er** *n.* — **pour′ing·ly** *adv.*

pour·boire (poor·bwär′) *n. French* A gratuitous gift of money as a tip; literally, for drink.

pour le mé·rite (poor lə mā·rēt′) *French* For merit.

pour·par·ler (poor·pàr·lā′) *n. French* A preliminary or informal conference or consultation.

pour·point (poor′point′, *Fr.* poor·pwan′) *n.* A quilted cloth doublet worn in the 14th and 15th centuries. [<F, prob. orig. pp. of *pourpoindre* perforate]

pour point 1 *Physics* The lowest temperature at which a liquid, especially a fuel oil, will flow under prescribed conditions. 2 *Metall.* The temperature at which molten metal is cast.

pousse-ca·fé (poos·kà·fā′) *n. French* A drink, commonly a mixture of cordials and brandy in successive layers, served after the coffee at dinner.

pous·sette (poo·set′) *n.* A dance figure in which a couple or couples swing round and round while holding hands. — *v.i.* **·set·ted, ·set·ting** To perform a poussette. [<F, dim. of *pousse* a push < *pousser* push]

pous·sie (poos′ē) *n. Scot.* Pussy; also, a hare.

Pous·sin (poō·san′), **Nicolas**, 1594–1665, French painter.

pou sto (poō stō′, pou′) A place to stand on; hence, a foundation for operations in any line of endeavor. [<Gk. *pou stō* where I may stand: from the alleged saying of Archimedes on his discovery of the lever, "Give me a place where I may stand and I will move the earth."]

pout[1] (pout) *v.i.* 1 To thrust out the lips, especially in ill humor. 2 To be sullen; sulk. 3 To swell out; protrude. — *v.t.* 4 To thrust out (the lips, etc.). 5 To utter with a pout. — *n.* A pushing out of the lips as in pouting; hence, a fit of ill humor. [Cf. Sw. *puta* be swollen]

pout[2] (pout) *n.* 1 One of various fresh-water catfishes having a pouting appearance. 2 The eelpout. [OE (*æle*)*pūte* eelpout]

pout·er[1] (pou′tər) *n.* 1 One who or that which pouts. 2 A breed of pigeon having the habit of puffing out the crop.

pout·er[2] (pou′tər) *v.t.* & *v.i. Scot.* To poke; stir.

pou·ther (poō′thər) *v.* & *n. Scot.* Powder. Also spelled *powther*.

pou·try (poō′trē) *n. Scot.* Poultry.

pov·er·ty (pov′ər·tē) *n.* 1 The state of being poor or without competent subsistence; need; penury. 2 The condition that relates to the absence or scarcity of requisite resources or elements. 3 A lack or meagerness of supply; dearth. [<OF *povreté* <L *paupertas* < *pauper* poor]

Synonyms: beggary, destitution, distress, indigence, mendicancy, need, pauperism, penury, privation, want. *Poverty* denotes a condition below that of easy, comfortable living; *privation* denotes a condition of painful lack of what is useful or desirable; *indigence* is lack of ordinary means of subsistence; *destitution* is lack of the comforts, and even of the necessaries of life; *penury* is cramping *poverty*; *pauperism* is such *destitution* as throws one upon public charity for support; *beggary* and *mendicancy* denote *poverty* that appeals for indiscriminate private charity.

poverty grass Any of certain grasses, especially *Aristida divaricata*, having little or no nutriment and found growing in old fields too poor for cultivation.

pov·er·ty-strick·en (pov′ər·tē·strik′ən) *adj.* Suffering from poverty; destitute.

pow (pou) *n. Scot.* The poll; head.

pow·der (pou′dər) *n.* **1** A finely ground or comminuted mass of free particles formed from a solid substance in the dry state; dust. **2** A pulverized cosmetic preparation for toilet use. **3** A medicine in the form of powder. **4** An explosive dry powder, as gunpowder. — **flashless powder** Moisture–resistant smokeless powder which does not flash when exploded: used as a propelling charge in shells. — *v.t.* **1** To reduce to powder; pulverize. **2** To sprinkle or cover with or as with powder. **3** To sprinkle with small objects or ornaments. — *v.i.* **4** To be reduced to powder. **5** To use powder as a cosmetic. [<OF *poudre* <L *pulvis, pulveris* dust] — **pow′der·er** *n.*

powder blue **1** Pulverized smalt having a deep–blue color: used as laundry bluing. **2** Its deep–blue color. **3** A valuable porcelain glaze. **4** The color of this glaze, a soft medium blue.

powder flask A metallic or other flask for carrying gunpowder.

powder horn The hollow horn of an ox or cow, formerly fitted with a cover and used by hunters or soldiers for holding gunpowder.

powder metallurgy The science and technique of manufacturing objects from finely powdered metals and alloys, shaped under controlled pressures and temperatures.

powder puff A soft pad used to apply powder to the skin.

Powder River **1** A river in NE Oregon, flowing 110 miles NE to the Snake River. **2** A river in Wyoming and Montana, flowing 486 miles north to the Yellowstone River.

pow·der·y (pou′dər·ē) *adj.* **1** Consisting of or like fine powder or dust. **2** Covered with or as with powder; mealy; dusty. **3** Capable of being easily powdered or crumbled; friable.

pow·er (pou′ər) *n.* **1** Ability to act; potency; specifically, the property of a substance or being that is manifested in effort or action, and by virtue of which that substance or being produces change, moral or physical. **2** Potential capacity. **3** Strength or force actually put forth. **4** The right, ability, or capacity to exercise control; legal authority, capacity, or competency, particularly, authority to do some act in relation to lands, as to create estates therein or charges thereon; also, a legal instrument or document conferring it. See POWER OF APPOINTMENT. **5** Any agent that exercises power, as in control or dominion; a military or naval force; an important and influential sovereign nation. **6** Great or telling force or effect. **7** *Colloq.* A great number or quantity. **8** Religious frenzy, especially as exemplified in exhortation: believed to be by possession of the Holy Spirit. **9** Any form of energy available for doing work; specifically, energy developed by mechanical or electrical means. **10** *Physics* The time rate at which energy is transferred, or converted into work. **11** *Math.* **a** The product of a number multiplied by itself a given number of times. **b** An exponent. **12** *Optics* Magnifying capacity, as of a lens. **13** *pl.* The sixth of the nine grades or orders of angels. — *v.t.* To provide with means of propulsion. [<OF *poeir,* ult. <L *posse* be able]

Synonyms: ability, capacity, efficacy, efficiency, energy, force, might, potency, puissance, strength. *Power* is the most general term of this group of words, including every quality, property, or faculty by which any change, effect, or result is, or may be, produced, as, the *power* of the legislature to enact laws, or of the executive to enforce them; the *power* of an acid to corrode a metal; the *power* of a polished surface to reflect light. *Ability* is nearly coextensive with *power,* but does not reach its positiveness and vigor, *ability* often implying latent, as distinguished from active, *power. Power* and *ability* include *capacity,* which is *power* to receive; but *ability* is often distinguished from *capacity,* as *power* that may be manifested in doing, as *capacity* is in receiving. *Efficacy* is *power* to produce effects; *efficiency* is effectual agency, competent *power. Energy* is *power* both actual and potential; *force* is *power* enough to overcome resistance. *Puissance* is a poetic or literary synonym. See ABILITY, CAUSE, GENIUS, WEIGHT. *Antonyms*: feebleness, helplessness, imbecility, impotence, inability, incapability, incapacity, inertness, powerlessness, weakness.

power boat A motorboat.

power dive *Aeron.* A descent in which the engine increases the acceleration due to gravity.

power drill A motor–operated drill.

pow·er·ful (pou′ər·fəl) *adj.* **1** Possessing great force; very efficient; strong. **2** Having great intensity or energy. **3** Exercising great authority, or manifesting high qualities; mighty. **4** Having great effect on the mind; convincing. — *adv. Colloq.* Very; exceedingly. — **pow′er·ful·ly** *adv.*

Synonyms (adj.): able, cogent, commanding, controlling, effective, effectual, efficacious, efficient, forceful, forcible, influential, mighty, potent, puissant, robust, strong, sturdy, valid, vigorous.

pow·er·house (pou′ər·hous′) *n.* **1** *Electr.* A station where electricity is generated. **2** *Slang* A person or thing of great might or force.

pow·er·less (pou′ər·lis) *adj.* **1** Destitute of power; unable to accomplish an effect; impotent. **2** Without authority. — **pow′er·less·ly** *adv.* — **pow′er·less·ness** *n.*

power loading *Aeron.* The gross weight of an aircraft divided by its rated engine power.

power loom A loom operated by power–driven machinery.

power of appointment *Law* Authority conferred, as by power of attorney, deed, or will, to appoint or designate a person or persons to make disposition of an estate or interest in the property of another.

power of attorney *Law* **1** The authority or power to act conferred upon an agent. **2** The instrument or document by which that power or authority is conferred or guaranteed. See under ATTORNEY.

power pack A compact assemblage of electrical units to provide requisite steady power, as in radio communication from an airplane.

power plant Any source of power, together with its housing, installations and accessory equipment: the *power plant* of an airplane.

power politics The use or threatened use of superior force to exact international concessions.

power train Drive train.

Pow·ha·tan (pou′hə·tan′) *n.* **1** A confederacy of Algonquian Indian tribes of Virginia (1607–1705) comprising about thirty tribes, organized by Wahunsonacock, commonly known as Powhatan, chief of the Powhatan tribe. **2** One of a tribe of North American Indians of Algonquian stock, formerly inhabiting a part of eastern Virginia.

pow·ney (pou′nē) *n. Scot.* A pony.

pow·ter (pou′tər) *n.* The pouter pigeon.

pow·ther (po͞o′thər) See POUTHER.

pow·wow (pou′wou′) *U.S. n.* **1** A North American Indian medicine man, priest, or magician. **2** The ceremony of a medicine man involving a dance, feast, or other demonstration, to cure the sick or effect success in hunting, war, etc. **3** Hence, magic; witchcraft. **4** An Indian council. **5** *Colloq.* Any meeting for conference. — *v.i.* To hold a deliberative council. [<Algonquian (Massachuset) *pauwaw,* lit., he dreams]

Pow·ys (pō′is) Name of three English authors, brothers: **John Cowper,** born 1872; **Llewelyn,** 1884–1939; **Theodore Francis,** 1875–1953.

pox (poks) *n.* **1** Any disease characterized by eruptions of a purulent nature: chicken *pox.* **2** Syphilis. [Var. of *pocks,* pl. of POCK]

Po·yang (pō′yäng′) A lake in northern Kiangsi province, eastern China; about 1,070 square miles at low water, 3,600 square miles when it receives the flood waters of the Yangtze in summer.

poy·ou (poi′o͞o) *n.* The six–banded armadillo (*Dasypus sexcinctus*) of Argentina and Brazil. [<Guarani (*tatu*)*–po–yu* (armadillo) with a yellow band <*po* band + *yu* yellow]

Poz·nań (pôz′nän·y′) A city of western Poland: German *Posen.*

Po·zsony (pô′zhôn′y′) The Hungarian name for BRATISLAVA.

poz·zuo·la·na (pot′swə·lä′nə, *Ital.* pōt′tswô·lä′nä) *n.* A volcanic ash, first collected at Pozzuoli, used in making hydraulic cement: also made artificially. Also **poz′zo·la′na** (pot′sə-). [<Ital., from *Pozzuoli*] — **poz′zuo·lan′ic** (-lan′ik) *adj.*

Poz·zuo·li (pōt·tswô′lē) A town on the site of an ancient city SW of Naples, Italy. Ancient **Pu·te·o·li** (pyo͞o·tē′ə·lī).

P–P factor *Biochem.* The pellagra–preventive factor of the vitamin B complex; nicotinic acid.

praam (präm) *n.* A Baltic flat–bottomed barge. Also spelled *pram.* [<Du. <Slavic. Cf. Polish *pram* boat.]

prac·tic (prak′tik) *adj. Obs.* Practical.

prac·ti·ca·ble (prak′tə·kə·bəl) *adj.* **1** That can be put into practice; feasible. **2** That can be used for an intended purpose; usable. — **prac′ti·ca·bil′i·ty, prac′ti·ca·ble·ness** *n.* — **prac′ti·ca·bly** *adv.*

prac·ti·cal (prak′ti·kəl) *adj.* **1** Pertaining to or governed by actual use and experience or action, as contrasted with ideals and speculations. **2** Trained by or derived from practice or experience. **3** Having reference to useful ends to be attained; applicable to use. **4** Manifested in practice. **5** Being such to all intents and purposes; virtual. [<obs. *practic* <obs. F *practique* <LL *practicus* <Gk. *praktikos* fit for doing <*prassein* do] — **prac′ti·cal′i·ty** (-kal′ə·tē), **prac′ti·cal·ness** *n.*

Practical Christianity New Thought.

practical joke A joke involving action instead of wit or words; a prank or trick.

prac·ti·cal·ly (prak′tik·lē) *adv.* **1** In a practical manner. **2** To all intents and purposes; in fact or effect; virtually.

practical nurse A nurse with practical experience in the care of the sick, but who is not a registered nurse.

prac·tice (prak′tis) *v.* **·ticed, ·tic·ing** *v.t.* **1** To make use of habitually or often: to *practice* economy. **2** To apply in action; make a practice of: *Practice* what you preach. **3** To work at or pursue as a profession: to *practice* law. **4** To do or perform repeatedly in order to acquire skill or training; rehearse. **5** To instruct, as pupils, by repeated exercise or lessons. — *v.i.* **6** To repeat or rehearse something in order to acquire skill or proficiency: to *practice* for a concert. **7** To work at or pursue a profession: He *practiced* for twenty years. **8** *Rare* To conspire; scheme. Also **prac′tise.** — *n.* **1** Any customary action or proceeding regarded as individual; habit. **2** An established custom or usage. **3** The act or process of executing or accomplishing; doing or performance: distinguished from *theory.* **4** The regular prosecution of a business pursuit requiring education; professional business. **5** Frequent and repeated exercise in any matter. **6** *pl.* Stratagems or schemes for bad purposes; tricks. **7** A rule or method in arithmetic to facilitate multiplying quantities in different denominations. **8** The rules by which legal proceedings are governed. [<OF *practiser* <*practiquer* <Med. L *practicare* <LL *practicus.* See PRACTICAL.] — **prac′tic·er** *n.*

Synonyms (noun): drill, exercise. *Exercise* is action with a view to employing, maintaining, or increasing power, or merely for enjoyment; *practice* is systematic *exercise* with a view to the acquirement of facility and skill; a person takes a walk for *exercise,* or takes time for *practice* on the piano. *Practice* is also used of putting into action and effect what one has learned or holds as a theory; as, the *practice* of law or medicine. Educationally, *practice* is the voluntary and persistent attempt to make skill a *habit*; as, *practice* in penmanship. *Drill* is systematic, rigorous, and commonly enforced *practice* under a teacher or commander. See CUSTOM, EXERCISE, HABIT, MANNER.

prac·ticed (prak′tist) *adj.* **1** Expert by practice; skilled by use or habit; experienced. **2** Acquired by practice. Also **prac′tised.**

prac·ti·tion·er (prak·tish′ən·ər) *n.* **1** One who practices an art or profession. **2** A Christian Science healer. [<earlier *practician* <OF *practicien,* ult. <L *practica* practice]

prae– See PRE–.

prae·ci·pe (pres′i·pē, prē′si-) See PRECIPE.

prae·di·al (prē′dē·əl), **prae·fect** (prē′fekt), etc. See PREDIAL, PREFECT, etc.

prae·mu·ni·re (prē′myo͞o·nī′rē) *n.* **1** In English law, the offense of introducing an alien power within the realm; specifically, the offense of maintaining the papal power in England; also, certain other grave offenses. **2** The writ or process issued as the initial step in the prosecution of the offense of praemunire. [<Med. L *praemunire* (*facias*) (see that you)

warn, a legal phrase; from a confusion of L *praemunire* protect with *praemonere* warn]

prae·no·men (prē·nō′mən) *n. pl.* **·nom·i·na** (-nom′ə·nə) The name prefixed to an ancient Roman family name to mark the individual, corresponding to the modern Christian name: also spelled *prenomen.* [<PRAE- + L *nomen* name]

prae·pos·tor (prē·pos′tər) *n.* A prepositor: also spelled *prepostor.* Also **prae·pos′i·tor.** [< Med. L *praepositor* one who puts another in charge <L *praepositus,* pp. of *praeponere.* See PREPOSITION.]

praeter- See PRETER-.

prae·tex·ta (prē·teks′tə) *n. pl.* **·tae** (-tē) An ordinary white toga with a purple border or stripe, worn by free-born Roman boys until they assumed the toga virilis at 14–16 years, and by girls until they were married. It was also the distinctive mark of the Roman curule magistrates, censors, state priests (when performing their functions), and emperors. [<L, lit., woven before, fringed, fem. of *praetextus,* pp. of *praetexere*]

prae·tor (prē′tər), **prae·to·ri·al** (prē·tôr′ē·əl, -tō′rē-), **prae·to·ri·an** (prē·tôr′ē·ən, -tō′rē-), etc. See PRETOR, etc.

Pra·ga (prä′gä) A suburb of Warsaw, Poland, on the east bank of the Vistula.

prag·mat·ic (prag·mat′ik) *adj.* **1** Pertaining to the accomplishment of duty or of business; specifically, relating to the civil affairs of a sovereign state. **2** Pertaining to or occupied with the scientific evolution of causes and effects; philosophical: said especially of history: the *pragmatic* method. **3** Pragmatical; practical. **4** Of or pertaining to the philosophy of pragmatism. [<L *pragmaticus* active or skilled in practical affairs <Gk. *pragmatikos* <*pragma, pragmatos* a thing done, an affair <*prassein* do, perform]

prag·mat·i·cal (prag·mat′i·kəl) *adj.* **1** Inclined to be officious or meddlesome; self-important; busy. **2** Relating to or engrossed with everyday business; practical; hence, commonplace. — **prag·mat′i·cal·ly** *adv.* — **prag·mat′i·cal·ness** *n.*

prag·mat·i·cism (prag·mat′ə·siz′əm) *n. Philos.* The pragmatism of C. S. Peirce, renamed by him to distinguish it from the teachings of William James and others, and referring to his philosophy that concepts are predictions of facts to be found and consequences to result should specified action be taken.

pragmatic sanction An imperial or royal edict or decree operating as a fundamental law. The most famous of these edicts was that of Charles VI of Austria in 1724, which admitted heirs in the female line to the Austrian succession.

prag·ma·tism (prag′mə·tiz′əm) *n. Philos.* The doctrine that thought or ideas have value only in terms of their practical consequences, and that results are the sole test of the validity or truth of one's beliefs. — **prag′ma·tist** *n.*

Prague (präg) The capital of Czechoslovakia, on the Vltava, in central Bohemia. *German* **Prag** (präkh), *Czech* **Pra·ha** (prä′hä).

pra·hu (prä·hōō′) *n.* A proa. [<Malay *prâu*]

Prai·ri·al (pre·rē·äl′) See under CALENDAR (Republican).

prai·rie (prâr′ē) *n.* A level or rolling tract of treeless land covered with coarse grass and generally of rich soil, especially as in parts of the western United States. [<F, a large meadow <Med. L *prataria* <L *pratum* meadow]

prairie chicken See under GROUSE. Also **prairie hen.**

prairie cock The cock of the plains.

prairie dog A burrowing rodent (genus *Cynomys*) of the plains of North America; specifically, *C. ludovicianus,* which lives in large communities and is very destructive to vegetation. Also **prairie squirrel.**

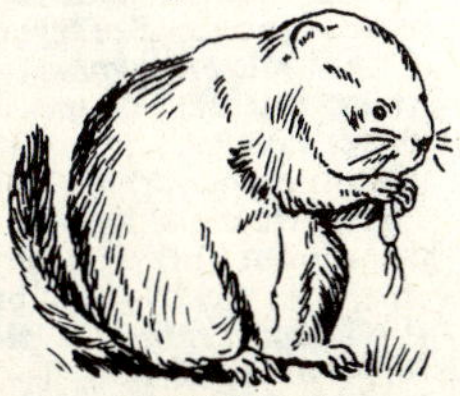

PRAIRIE DOG
(Body from 12 to 15 inches long; tail, 3 to 4 inches)

prairie owl **1** The burrowing owl. **2** The short-eared owl.

Prairie Provinces The provinces of Manitoba, Saskatchewan, and Alberta, in western Canada.

prairie schooner A covered wagon.

prairie state One of the States of the prairie regions of the Western and Middle Western United States.

Prairie State Nickname of ILLINOIS.

prairie wolf A coyote: distinguished from *timber wolf.*

praise (prāz) *v.t.* **praised, prais·ing** **1** To express approval and commendation of; applaud; eulogize. **2** To express adoration of; glorify (God, etc.). — *n.* **1** Commendation expressed, as of a person for his virtues, or concerning meritorious actions; utterance of approval; honor given; also, applause. **2** Thanksgiving for blessings conferred; laudation to God; worship expressed in song. **3** The object, ground, reason, or subject of praise. ◆Homophone: *prase.* [<OF *preisier* <LL *pretiare* prize <L *pretium* price. See PRICE.] — **prais′er** *n.*

Synonyms (verb): adore, applaud, approve, bless, celebrate, commend, eulogize, extol, flatter, glorify, honor, laud, magnify, worship. See PUFF. *Antonyms*: see synonyms for ASPERSE, BLAME.

Synonyms (noun): acclaim, acclamation, adulation, applause, approbation, approval, commendation, compliment, encomium, eulogy, flattery, laudation, panegyric, plaudit, sycophancy. *Praise* is the hearty approval of an individual, or of a multitude considered individually, and is expressed by spoken or written words; *applause,* the spontaneous outburst of many at once. *Applause* is expressed by stamping of feet, clapping of hands, waving of handkerchiefs, etc., as well as by the voice; *acclamation* is the spontaneous and hearty approval of many at once, and strictly by the voice alone. One is chosen moderator by *acclamation* when he receives a practically unanimous viva voce vote; he could not be nominated by *applause. Acclaim* is the more poetic term for *acclamation*; as, a nation's *acclaim. Plaudit* is a shout of *applause,* and is commonly used in the plural; as, the *plaudits* of a throng. *Applause* is also used in the general sense of *praise. Approbation* is a milder and more qualified word than *praise*; *praise* is always uttered, *approbation* may be silent. The industry and intelligence of a clerk win his employer's *approbation*; his decision in a special instance receives his *approval. Praise* is always understood as genuine and sincere, unless the contrary is expressly stated; *compliment* is a light form of *praise* that may or may not be sincere; *flattery* is often insincere. Compare APPLAUSE, EULOGY. *Antonyms*: abuse, animadversion, blame, censure, condemnation, contempt, denunciation, disapprobation, disapproval, disparagement, obloquy, reproach, reproof, repudiation, scorn, slander, vilification, vituperation.

praise·wor·thy (prāz′wûr′thē) *adj.* Worthy of praise; commendable. — **praise′wor′thi·ly** *adv.* — **praise′wor′thi·ness** *n.*

Pra·ja·dhi·pok (prä·jä′di·pôk), 1893–1941, king of Siam 1925–35; abdicated.

Pra·krit (prä′krit) *n.* The popular dialects or any one of the vernaculars of northern and central India, arising from or connected with Sanskrit, and forming a link between Sanskrit and the modern Indic languages. [<Skt. *prakṛtā* natural, common, lit., created before <*pra-* before + *kṛ* do. Cf. SANSKRIT.]

pra·line (prä′lēn, prā′-) *n.* A crisp confection made of pecans or other nuts browned in boiling sugar. [<F, after Marshal Duplessis-Praslin, 1598–1675, whose cook invented it]

prall·tril·ler (präl′tril·ər) *n. Music* An inverted mordent. [<G, lit., elastic trill]

pram[1] (pram) *n. Brit. Colloq.* A baby carriage. [Short for PERAMBULATOR]

pram[2] (präm) See PRAAM.

prance (prans, präns) *v.* **pranced, pranc·ing** *v.i.* **1** To move proudly with high steps, as a spirited horse; spring from the hind legs. **2** To ride gaily, proudly, or insolently, as on a prancing horse. **3** To move in an arrogant or elated manner; swagger. **4** To gambol; caper. — *v.t.* **5** To cause to prance. — *n.* The act of prancing; a high step; a caper. [ME *prauncen,* ? <Scand. Cf. dial. Dan. *pranse* walk proudly.] — **pranc′er** *n.*

pran·di·al (pran′dē·əl) *adj.* Of or pertaining to a meal, especially a dinner. [<L *prandium* breakfast or lunch]

prank[1] (prangk) *v.t.* To decorate gaudily; deck with showy ornaments. — *v.i.* To make an ostentatious show. [Cf. Du. *pronken,* G *prunken* make a show of]

prank[2] (prangk) *n.* A mischievous or frolicsome act. See synonyms under FROLIC, SPORT. — *v.i.* To play pranks or tricks. [Origin uncertain] — **prank′ish** *adj.*

pranked (prangkt) *adj.* Decorated; dressed up: often with *out* or *up.*

prase (prāz) *n.* An olive-green, translucent quartz, usually cryptocrystalline. ◆ Homophone: *praise.* [<F <L *prasius* light green <Gk. *prason* leek; with ref. to its color]

pra·se·o·dym·i·um (prā′zē·ō·dim′ē·əm, prā′sē-) *n.* A yellowish-white metallic element (symbol Pr) of the lanthanide series, having olive-green salts. See ELEMENT. [<NL <Gk. *prasios* light green + (DI)DYMIUM]

prate (prāt) *v.* **prat·ed, prat·ing** *v.i.* To talk idly and at length; chatter. — *v.t.* To utter idly or emptily. See synonyms under BABBLE. — *n.* Idle talk; prattle. [<Cf. MDu. & MLG *praten* chatter, ON *prata* talk] — **prat′er** *n.* — **prat′ing·ly** *adv.*

prat·fall (prat′fôl′) *n. Slang* A fall on the buttocks.

prat·in·cole (prat′ing·kōl, prā′tin-) *n.* Any one of a genus (*Glareola*) of Old World shore birds having long, pointed wings and deeply forked tail. [<L *pratum* meadow + *incola* inhabitant]

pra·tique (pra·tēk′, prat′ik; *Fr.* prà·tēk′) *n.* Intercourse or correspondence; especially, privilege granted to the master of a vessel to land passengers after compliance with sanitary inspection or quarantine. [<F]

prat·tle (prat′l) *v.* **·tled, ·tling** *v.i.* To talk foolishly or like a child; prate. — *v.t.* To utter in a foolish or childish way: to *prattle* secrets. See synonyms under BABBLE. — *n.* **1** Childish speech; babble. **2** Idle or foolish talk. [Freq. of PRATE] — **prat′tler** *n.*

Prav·dinsk (präv′dēnsk) **1** A town on the Volga in Russian S.F.S.R. **2** A city in the former German province of East Prussia, now in Russian S.F.S.R.: formerly *Friedland.*

prawn (prôn) *n.* An edible shrimplike decapod (suborder *Natantia*) occurring in a variety of genera and species, especially numerous in tropical and temperate waters, principally marine. [ME *prane*; origin unknown]

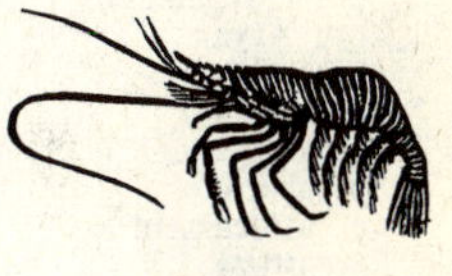

PRAWN
(Up to 6 inches in length)

prax·is (prak′sis) *n.* Exercise or discipline for a specific purpose; practical application of rules as distinguished from theory. [<NL <Gk. <*prassein* accomplish, do]

Prax·it·e·les (prak·sit′ə·lēz) Greek sculptor of the fourth century B.C.

pray (prā) *v.i.* **1** To address prayers to a deity, idol, etc.; say prayers. **2** To make entreaty; beg. — *v.t.* **3** To address by means of prayers; say prayers to. **4** To ask (someone) earnestly; entreat. **5** To ask for by prayers or entreaty. **6** To effect by prayer. ◆ Homophone: *prey.* [<OF *preier* <LL *precare* <L *precari* ask, pray <*prex, precis* prayer]

Synonyms: ask, beg, beseech, bid, conjure, entreat, implore, importune, invoke, petition, request, supplicate. See ASK.

prayer (prâr) *n.* **1** The act of offering reverent petitions, especially to God. **2** The act of beseeching earnestly; entreaty. **3** *Often pl.* A religious service of which prayer is the most prominent part: evening *prayers.* **4** Communion with God and recognition of His presence, as in praise, thanksgiving, intercession, etc. **5** A form of words appropriate to prayer. **6** A memorial or petition. **7** *Law* The request in a bill in equity for the specific relief sought by the complainant; also, the part of the bill

add,āce,câre,pälm; end,ēven; it,īce; odd,ōpen,ôrder; tŏŏk,pōōl; up,bûrn; ə = a in *above,* e in *sicken,* i in *clarity,* o in *melon,* u in *focus*; yōō = u in *fuse*; oi,oil; ou,pout; ch,check; g,go; ng,ring; th,thin; th,this; zh,vision. Foreign sounds à,œ,ü,kh,ṅ; and ◆: see page xx. <from; + plus; ? possibly.

prayer book in which the request is made. — **common prayer** The prescribed form of public worship of the Anglican Church as contained in the *Book of Common Prayer*. [<OF *preiere* < Med. L *precaria* <L *precarius* obtained by prayer < *precari*. See PRAY.]
  Synonyms: adoration, devotion, invocation, litany, orison, petition, request, suit, supplication. See PETITION.
prayer book A book of ritual prescribed for conducting divine service.
prayer·ful (prâr′fəl) *adj.* Inclined or given to prayer; devotional. — **prayer′ful·ly** *adv.* — **prayer′ful·ness** *n.*
prayer wheel A wheel, cylinder, or vertical drum containing written prayers, which is revolved to make the prayers efficacious: used by the Buddhists of Tibet. Also **praying wheel.**
praying mantis The mantis.
pre- *prefix* **1** Before in time or order; prior to; preceding; as in:

preaccusation	preconfiguration
preacquaint	preconfirm
preacquaintance	preconnection
preacquire	preconnubial
preact	preconsent
preaction	preconsideration
preadapt	preconsign
preadaptation	preconstitute
preadjust	preconstruction
preadjustment	preconsult
preadministration	preconsultation
preadmit	preconsume
preadmonish	precontract
preadmonition	precontrive
preadvertise	preconviction
preadvertiser	pre-cool
preadvise	precorrupt
preadviser	precounsel
preaestival	pre-Darwinian
preallege	predecision
preannounce	prededication
preannouncement	predeliberation
preannouncer	predemand
preantiquity	predescribe
preapperception	predesign
preappoint	predeterminable
preappointment	predevised
preapproval	predirect
prearm	prediscipline
prearrange	prediscovery
prearrangement	preelect
pre-Aryan	preelection
preassigned	preembodiment
preassume	preemploy
preassurance	preenact
preassure	preengage
preattachment	preengagement
preattune	pre-epic
preaver	preestablish
preavowal	preestablishment
prebaptize	preexamination
prebasal	preexamine
prebasilar	preexist
preboding	preexistence
preboil	preexistent
prebranchial	preexpose
pre-British	preform
prebronchial	preglacial
prebuccal	preheat
precalculable	preheater
precalculate	preinhabitation
precalculation	preinstruct
precancerous	preintimation
pre-Carboniferous	preknowledge
pre-Centennial	pre-Paleozoic
precerebellar	pre-Reformation
pre-Christian	pre-Renaissance
pre-Christianize	prerequire
precited	prerevolutionary
preclassical	pre-Roman
precogitate	preselect
precogitation	preshadow
precognition	pre-Shakespearian
precognizable	preshow
precognizant	presuccess
precollection	presurmise
pre-Columbian	pre-Tertiary
precompose	pretribal
precomputation	pretypify
precompute	preunite
preconcession	pre-Victorian
preconclusion	prewarm
precondemn	prewarn
precondemnation	

2 Before in position; anterior: chiefly in scientific terms; as in:

preabdomen	precardiac	prerectal
preanal	precerebral	prerenal
preaortic	precostal	preretinal
preauricular	prepatellar	prevertebral

3 Preliminary to; preparing for; as in:

| precollege | prelegal | premedical |
| preflight | prelexical | pre-military |

[<L *prae-* < *prae* before]
preach (prēch) *v.i.* **1** To deliver a sermon, as on a religious topic or a text of Scripture. **2** To give advice or instruction, especially persistently and intrusively. — *v.t.* **3** To advocate or recommend urgently: to *preach* temperance. **4** To proclaim; expound upon: to *preach* the gospel. **5** To deliver (a sermon, etc.). [<OF *prechier* <L *praedicare* proclaim < *prae-* before + *dicare* make known]
preach·er (prē′chər) *n.* One who preaches; specifically, a clergyman. [<OF *prechor*]
preach·i·fy (prē′chə·fī) *v.i.* **·fied**, **·fy·ing** *Colloq.* To preach or discourse tediously. — **preach′i·fi·ca′tion** *n.*
preach·ing (prē′ching) *n.* **1** The act or practice of delivering sermons. **2** The style of a preacher. **3** The doctrine preached.
Preaching Friars See DOMINICAN.
preach·ment (prēch′mənt) *n.* A preaching or moral lecture; especially, a wearisome exhortation. [<OF *prechement*]
preach·y (prē′chē) *adj.* **preach·i·er, preach·i·est** Given to or resembling preachments; marked by sanctimony or cant: not a complimentary term.
pre·ad·am·ite (prē·ad′əm·īt) *adj.* Existing before Adam; relating to the preadamites. Also **pre·a·dam·ic** (prē′ə·dam′ik), **pre·ad·am·it′ic** (-it′ik). — *n.* **1** One who or that which existed before Adam or before man. **2** One holding that there were men on the earth before Adam.
pre·ag·o·nal (prē·ag′ə·nəl) *adj.* Immediately preceding the death agony. [<PRE- + AGON(Y) + -AL¹]
pre·am·ble (prē′am·bəl) *n.* **1** A statement introductory to and explanatory of what follows; the introductory portion of a writing or speech: used chiefly of formal resolutions. **2** *Law* An introductory clause in a constitution, contract, or other instrument. [<F *préamble* <Med. L *praeambulum*, orig. neut. of L *praeambulus* walking before <L *praeambulare* precede < *prae-* before + *ambulare* walk] — **pre·am′bu·lar′y** *adj.*
pre·ax·i·al (prē·ak′sē·əl) *adj. Biol.* Situated on that side of the axis of a limb or body that is in front.
preb·end (preb′ənd) *n.* **1** A stipend allotted to an ecclesiastic from the revenues of a cathedral or conventual church in consideration of his officiating and serving therein; also, the land or tithe yielding the stipend, the tenure of which is a benefice. **2** A prebendary. [<OF *prebende* <Med. L *praebenda*, lit., things to be furnished, neut. pl. of L *praebendus*, gerundive of *praebere* supply < *prae-* in front of, before + *habere* have] — **preb′·en·dal** *adj.*
preb·en·dar·y (preb′ən·der′ē) *n. pl.* **·dar·ies** A person, as a canon, who receives a stated income from the revenues of a cathedral. [<Med. L *prebendarius*]
Pre·ble (preb′əl), **Edward**, 1761–1807, U.S. naval officer; captain of the *Constitution*.
Pre-Cam·bri·an (prē·kam′brē·ən) *adj. Geol.* Of or pertaining to all geological time and rock formations preceding the Cambrian. — *n.* Pre-Cambrian rocks.
pre·can·cel (prē·kan′səl) *v.t.* **·celed** or **·celled**, **·cel·ing** or **·cel·ling** To cancel (stamps) before use on mail. — *n.* A stamp so canceled.
pre·car·i·ous (pri·kâr′ē·əs) *adj.* **1** Subject to continued risk; that may be taken away at another's pleasure or by accident; uncertain. **2** Subject or leading to danger; hazardous. **3** Not firmly established; untrustworthy; without foundation. [<L *precarius*. See PRAYER.] — **pre·car′i·ous·ly** *adv.* — **pre·car′i·ous·ness** *n.*
  Synonyms: doubtful, dubious, equivocal, hazardous, insecure, perilous, risky, unassured, uncertain, unsettled, unstable, unsteady. *Uncertain* is applied to things that human knowledge cannot certainly determine or that human power cannot certainly control; *precarious* originally meant dependent on the will or pleasure of another; now it also means dependent on chance or hazard; one holds office by a *precarious* tenure, or land by a *precarious* title; the strong man's hold on life is *uncertain*, the invalid's is *precarious*. *Antonyms*: assured, certain, firm, immutable, incontestable, settled, stable, steady, strong, sure, undoubted, unquestionable.
pre-cast (prē′kast′, -käst′) *adj.* Receiving a finished shape before being put to final use: *pre-cast* concrete blocks.
prec·a·tive (prek′ə·tiv) *adj.* Expressing entreaty; supplicatory. Also **prec′a·to·ry**. [<L *precativus*]
pre·cau·tion (pri·kô′shən) *n.* **1** Prudent forethought, as against danger, etc. **2** A provision made for some emergency. See synonyms under CARE. [<F *précaution* <LL *praecautio, -onis* <L *praecautus*, pp. of *praecavere* guard against beforehand < *prae-* before + *cavere* take care]
pre·cau·tion·ar·y (pri·kô′shən·er′ē) *adj.* **1** Of or pertaining to precaution. **2** Expressing, advising, or using precaution. Also **pre·cau′tion·al**.
pre·cau·tious (pri·kô′shəs) *adj.* Exercising care; precautional. — **pre·cau′tious·ly** *adv.* — **pre·cau′tious·ness** *n.*
pre·cede (pri·sēd′) *v.* **·ced·ed**, **·ced·ing** *v.t.* **1** To go before in order, place, rank, time, etc. **2** To preface; introduce. — *v.i.* **3** To have or take precedence. [<F *précéder* <L *praecedere* go before < *prae-* before + *cedere* go]
  Synonyms: head, herald, lead. See LEAD¹. *Antonyms*: see synonyms for FOLLOW.
prec·e·dence (pri·sēd′ns, pres′ə·dəns) *n.* The act or right of preceding, or the state of being precedent; priority in place, time, or rank. Also **pre·ce′den·cy**.
  Synonyms: antecedence, ascendency, lead, leadership, preeminence, preference, priority, superiority, supremacy. *Antonyms*: inferiority, subjection, subjugation, subordination.
prec·e·dent (pres′ə·dənt) *n.* **1** Previous usage or established mode of precedure. **2** An antecedent. **3** A judicial decision taken as furnishing a rule for subsequent decisions. — **pre·ce·dent** (pri·sēd′nt) *adj.* Former; previous; preceding. [<F *précédent* <L *praecedens, -entis*, ppr. of *praecedere*. See PRECEDE.]
  Synonyms (noun): antecedent, case, example, instance, pattern, warrant. A *precedent* is an authoritative *case*, *example*, or *instance*. *Cases* decided by irregular or unauthorized tribunals are not *precedents* for the regular administration of law. See ANTECEDENT, CAUSE, EXAMPLE.
prec·e·den·tial (pres′ə·den′shəl) *adj.* Of the nature of a precedent; preliminary; having social priority.
pre·ced·ing (pri·sē′ding) *adj.* Going before, as in time, place, or rank; earlier; foregoing; immediately antecedent: The citation was on the *preceding* page.
pre·cent (pri·sent′) *v.i.* To act as precentor. [Back formation <PRECENTOR]
pre·cen·tor (pri·sen′tər) *n.* The leader of the musical part of a church service. [<LL *praecentor* <L *praecinere* sing before < *prae-* before + *canere* sing] — **pre·cen·to·ri·al** (prē′sen·tôr′ē·əl, -tō′rē-) *adj.* — **pre·cen′tor·ship** *n.*
pre·cept (prē′sept) *n.* **1** A prescribed rule of conduct or action; instruction or direction regarding a given course or action; especially, a maxim in morals; distinguished from *counsel*. **2** *Law* A judicial command in writing; writ; process. See synonyms under ADAGE. [<OF <L *praeceptum*, pp. of *praecipere* give rules, instruct < *prae-* before + *capere* receive, take]
pre·cep·tive (pri·sep′tiv) *adj.* Consisting of precepts; didactic.
pre·cep·tor (pri·sep′tər) *n.* A teacher; instructor; specifically, the principal of a school. [<L *praeceptor*] — **pre·cep·to·ri·al** (prē′sep·tôr′ē·əl, -tō′rē-) *adj.*
pre·cep·to·ry (pri·sep′tər·ē) *adj.* Preceptive; mandatory. — *n. pl.* **·ries** A place of instruction; specifically, a religious house of the Knights Templars. [<Med. L *praeceptoria*]
pre·cep·tress (pri·sep′tris) *n.* A woman preceptor; governess.

pre·ces·sion (pri·sesh′ən) *n.* The act of preceding or coming in advance of time or of other persons or things. [<LL *praecessio, -onis* < *praecessus,* pp. of *praecedere.* See PRECEDE.]

pre·ces·sion·al (pri·sesh′ən·əl) *adj.* Pertaining to or of the nature of precession.

precession of the equinoxes *Astron.* A slow rotary motion of the equinoctial points on the ecliptic from east to west, causing the time between successive equinoxes to be appreciably shorter than it would otherwise be: caused by the combined attractive forces of the moon, sun, and planets upon the equatorial protuberance of the earth and completing a full cycle in about 26,000 years, a period known as the *Platonic* or *great year.*

pre·cinct (prē′singkt) *n.* **1** A place definitely marked off by fixed lines; also, the boundary of a designated place. **2** A minor territorial or jurisdictional district. **3** An election district of a town, township, county, etc. **4** A police subdivision of a city or town, or its police station. **5** *Brit.* The immediate neighborhood of a church or temple. **6** *pl.* Neighborhood; environs. [<LL *praecinctum* boundary, orig. neut. of *praecinctus,* pp. of *praecingere* gird about <*prae-* before + *cingere* gird]

pre·ci·os·i·ty (presh′ē·os′ə·tē) *n.* Extreme fastidiousness or affected refinement, as in speech, style, or taste. [<OF *preciosité* <L *pretiositas* <*pretiosus.* See PRECIOUS.]

pre·cious (presh′əs) *adj.* **1** Highly priced or prized, as for rarity, or for intrinsic, exchangeable, or other value; valuable. **2** Beloved; dear. **3** Good-for-nothing; undeserving: used ironically. **4** *Colloq.* Very considerable; surpassing: a *precious* scoundrel. **5** Overnice; fastidious: a *precious* writer. See synonyms under CHOICE, EXCELLENT, GOOD, RARE[1]. [<OF *precios* <L *pretiosus* <*pretium* price] —**pre′cious·ly** *adv.* —**pre′cious·ness** *n.*

precious garnet Pyrope.

prec·i·pe (pres′i·pē, prē′si-) *n. Law* **1** A written order directing the issuance of a specified writ. **2** *Brit.* Formerly, a writ commanding a defendant, in the alternative, to do some particular thing, or to show cause for not doing it. Also spelled *praecipe.* [<L *praecipe,* lit., admonish, imperative of *praecipere* admonish, instruct. See PRECEPT.]

prec·i·pice (pres′i·pis) *n.* **1** A high, steep place; the brink of a cliff. **2** A perilous situation. [<F *précipice* <L *praecipitium* <*praeceps* headlong <*prae-* before + *caput* head]

pre·cip·i·ta·ble (pri·sip′ə·tə·bəl) *adj.* Capable or susceptible of being precipitated: a *precipitable* salt.

pre·cip·i·tance (pri·sip′ə·təns) *n.* **1** The quality of being precipitant; rashness. **2** An instance of this. Also **pre·cip′i·tan·cy.**

pre·cip·i·tant (pri·sip′ə·tənt) *adj.* **1** Rushing or falling headlong; moving onward quickly and heedlessly: *precipitant* speed. **2** Rash in thought or action; overhasty; impulsive; precipitate; sudden; abrupt. —*n. Chem.* Any substance, as a reagent, that when added or applied to a solution results in the formation of a precipitate. [<L *praecipitans, -antis,* ppr. of *praecipitare.* See PRECIPITATE.] —**pre·cip′i·tant·ly** *adv.*

pre·cip·i·tate (pri·sip′ə·tāt, -tit) *adj.* **1** Rushing down headlong; moving or moved speedily or hurriedly. **2** Wanting due deliberation; hasty; rash. **3** Done prematurely; hurried; undeliberated. **4** Sudden and brief, as a disease. See synonyms under IMPETUOUS. —*v.* (pri·sip′ə·tāt) **·tat·ed, ·tat·ing** *v.t.* **1** To bring about before expected or needed; hasten the occurrence of: to *precipitate* a quarrel. **2** To throw headlong; hurl from or as from a height. **3** *Meteorol.* To cause (vapor, etc.) to condense and fall as dew, rain, etc. **4** *Chem.* To separate (a constituent) in solid form, as from a solution. —*v.i.* **5** *Meteorol.* To fall as condensed vapor. **6** *Chem.* To separate and settle, as a substance held in solution. **7** To fall headlong; rush. —*n.* (pri·sip′ə·tāt, -tit) *Physics* A deposit of solid matter formed in a solution by the action of chemical reagents or by certain physical forces, as low temperature. [<L *praecipitatus,* pp. of *praecipitare* <*praeceps.* See PRECIPICE.] —**pre·cip′i·tate·ly** *adv.* —**pre·cip′i·tate·ness** *n.* —**pre·cip′i·ta′tive** *adj.* —**pre·cip′i·ta′tor** *n.*

pre·cip·i·ta·tion (pri·sip′ə·tā′shən) *n.* **1** The act of casting down; the state of being thrown downward. **2** Headlong or rash haste or hurry; precipitancy; hastening; acceleration. **3** A falling, flowing, or rushing down with violence or rapidity. **4** *Chem.* The process of rendering insoluble and so separating any of the constituents of a solution, as by reagents; also, the precipitate. **5** *Meteorol.* The deposition of moisture from the atmosphere upon the general surface of the earth. **6** Materialization, as of spirits. [<F *précipitation*]

pre·cip·i·tin (pri·sip′ə·tin) *n. Biochem.* An antibody produced in the blood serum by inoculation with foreign protein and capable of providing immunity against specific bacteria; coagulin. [<PRECIPIT(ATE) + -IN]

pre·cip·i·tin·o·gen (pri·sip′ə·tin′ə·jen) *n. Biochem.* The antigen which reacts with the blood to form precipitin. [<PRECIPITIN + -(O)GEN]

pre·cip·i·tous (pri·sip′ə·təs) *adj.* **1** As steep as or consisting of a very precipice; very steep. **2** Headlong and downward in motion. **3** Headlong in disposition; precipitate; hasty. See synonyms under STEEP[1]. [<MF *précipiteux*] —**pre·cip′i·tous·ly** *adv.* —**pre·cip′i·tous·ness** *n.*

pré·cis (prā·sē′, prā′sē) *n. pl.* **·cis** (-sēz′, -sēz) A concise, brief summary of the ideas and point of view of a book, article, or document. [<F]

pre·cise (pri·sīs′) *adj.* **1** Sharply or clearly determined; strictly accurate; exact. **2** No more and no less than. **3** Noting or confined to a certain thing; particular; identical. **4** Scrupulously observant of rule; punctilious. [<F *précis* <L *praecisus,* pp. of *praecidere* cut off short <*prae-* before + *caedere* cut] —**pre·cise′ly** *adv.* —**pre·cise′ness** *n.*

Synonyms: accurate, careful, correct, definite, distinct, exact, explicit, faultless, flawless, minute, nice, particular, perfect, rigid, right, scrupulous, strict. *Accurate, correct, definite, exact, precise, nice,* all denote absolute conformity to some standard or truth. *Accurate* indicates conformity secured by *scrupulous* care. An *accurate* measurement or account can be verified and found true in all particulars. *Careful* carries less sharp certainty. *Exact* indicates that which is worked out to the utmost limit of requirement in every respect; *precise* refers to a like conformity or to an excessive *exactness. Exact* and *precise* are often interchangeable; but *precise* has often an invidious meaning, denoting excessive care of petty details; we speak of the martinet as insufferably *precise,* not insufferably *exact. Correct* applies to a required or enforced correspondence with a standard. This is especially seen in the use of the verb; the printer *corrects* the proof. That is *correct* which is free from fault or mistake. *Nice* denotes a very fine and discriminating exactness, and refers to intellectual distinctions oftener than to material measurements. Compare CORRECT, MINUTE[2]. *Antonyms*: careless, doubtful, erroneous, false, faulty, inaccurate, inexact, loose, mistaken, misty, nebulous, untrue, vague, wrong.

pre·ci·sian (pri·sizh′ən) *n.* One who adheres punctiliously to rules and forms: a term especially applied in a religious sense to the Puritans.

pre·ci·sion (pri·sizh′ən) *n.* The quality of being precise; accuracy of limitation, definition, or adjustment. [<F *précision* <L *praecisio, -onis*] —**pre·ci′sion·ist** *n.*

pre·clin·i·cal (prē·klin′i·kəl) *adj.* **1** *Med.* In the period of disease before the appearance of symptoms sufficient for diagnosis. **2** Pertaining to medical studies which precede practical study of patients.

pre·clude (pri·klōōd′) *v.t.* **·clud·ed, ·clud·ing** **1** To render impossible or ineffectual by antecedent action; prevent. **2** To shut out; exclude. [<L *praecludere* <*prae-* before + *cludere* shut] —**pre·clu′sion** (-klōō′zhən) *n.* —**pre·clu′sive** (-klōō′siv) *adj.* —**pre·clu′sive·ly** *adv.*

Synonyms: obviate, prevent. To *obviate* is to *prevent* by interception and making unnecessary; to *preclude,* to close or shut in advance, is to *prevent* by anticipation or by logical necessity; walls and bars *precluded* the possibility of escape; a supposition is *precluded*; a necessity or difficulty is *obviated.* Compare PROHIBIT, SHUT.

pre·co·cial (pri·kō′shəl) *adj. Ornithol.* Of or pertaining to birds whose young are able to run about as soon as they are hatched. [See PRECOCIOUS]

pre·co·cious (pri·kō′shəs) *adj.* **1** Developing before the natural season. **2** Unusually forward or advanced, especially mentally. **3** *Bot.* Flowering or ripening early, as certain plants. [<OF *precoce* <L *praecox, praecocis* <*praecoquere* cook or ripen beforehand <*prae-* before + *coquere* cook] —**pre·co′cious·ly** *adv.* —**pre·co′cious·ness, pre·coc′i·ty** (-kos′ə·tē) *n.*

pre·con·ceive (prē′kən·sēv′) *v.t.* **·ceived, ·ceiv·ing** To conceive in advance; form an idea or opinion of beforehand.

pre·con·cep·tion (prē′kən·sep′shən) *n.* **1** An idea or opinion formed or conceived in advance. **2** A prejudice or misconception; bias. —**pre′con·cep′tion·al** *adj.*

pre·con·cert (prē′kən·sûrt′) *v.t.* To arrange in advance, as by agreement. —*n.* (prē·kon′sûrt) Previous arrangement.

pre·co·nize (prē′kə·nīz) *v.t.* **·nized, ·niz·ing** **1** To announce the appointment of (a new bishop) in public consistory: said of the pope. **2** To proclaim or extol publicly. [<LL *praeconizare* proclaim <L *praeco, praeconis* crier, herald]

pre·con·scious (prē·kon′shəs) *n. Psychoanal.* That area of the psyche containing mental processes of which the individual is unaware at any given time but which are more or less readily available to consciousness: formerly called *foreconscious.*

pre·crit·i·cal (prē·krit′i·kəl) *adj. Med.* Preceding the crisis (of a disease).

pre·cur·sive (pri·kûr′siv) *adj.* Going before as a precursor or harbinger; premonitory; preliminary. Also **pre·cur′so·ry** (-sər·ē).

pre·cur·sor (pri·kûr′sər) *n.* One who or that which precedes and gives intimation of a coming event. See synonyms under HERALD. [<L *praecursor* <*praecursus,* pp. of *praecurrere* run before <*prae-* before + *currere* run]

pre·da·cious (pri·dā′shəs) *adj.* Predatory. Also **pre·da′ceous.** [<L *praeda* prey] —**pre·da′cious·ness, pre·dac′i·ty** (-das′ə·tē) *n.*

pre·date (prē·dāt′) *v.t.* **·dat·ed, ·dat·ing** **1** To date before the actual time. **2** To precede in time.

pred·a·to·ry (pred′ə·tôr′ē, -tō′rē) *adj.* **1** Characterized by or undertaken for plundering. **2** Addicted to pillaging. **3** Constituted for living by preying upon others, as a beast or bird; raptorial. [<L *predatorius* <*praeda* prey] —**pred′a·to′ri·ly** *adv.* —**pred′a·to′ri·ness** *n.*

pre·de·cease (prē′di·sēs′) *v.t.* **·ceased, ·ceas·ing** To die before: She *predeceased* her husband by five years.

pred·e·ces·sor (pred′ə·ses′ər) *n.* **1** One who goes or has gone before another in point of time, as an early settler, a previous incumbent of an office, etc. **2** An ancestor. [<OF *predecesseur* <LL *praedecessor* <*prae-* before + *decessor* retiring official <*decessus* pp. of *decedere* go away. See DECEASE.]

pre·des·ig·nate (prē·dez′ig·nāt) *v.t.* **·nat·ed, ·nat·ing** **1** To designate beforehand. **2** *Logic* To begin (a proposition) with a designation of quantity, as *some, many,* etc. —**pre·des′ig·na′tion** *n.*

pre·des·ti·nar·i·an (prē·des′tə·nâr′ē·ən) *adj.* **1** Pertaining to predestination. **2** Holding the doctrine of predestination. —*n.* A believer in theological predestination; also, a fatalist. —**pre·des′ti·nar′i·an·ism** *n.*

pre·des·ti·nate (prē·des′tə·nit, -nāt) *adj.* **1** Designed for some special fate. **2** Foreordained by divine decree, as to salvation. —*n.* One who is predestined, as to salvation. —*v.t.* (-nāt) **·nat·ed, ·nat·ing** **1** To destine or decree beforehand; foreordain. **2** *Theol.* To foreordain by divine decree or purpose. [<L *praedestinatus,* pp. of *praedestinare* <*prae-* before + *destinare* destine]

pre·des·ti·na·tion (prē·des′tə·nā′shən) *n.* **1** The act of predestinating, or the state of being predestinated; destiny; fate. **2** *Theol.* The foreordination of all things by God, including

add,āce,câre,pälm; end,ēven; it,īce; odd,ōpen,ôrder; tŏŏk,pōōl; up,bûrn; ə = a in *above,* e in *sicken,* i in *clarity,* o in *melon,* u in *focus*; yōō = u in *fuse*; oi,oil; ou,pout; ch,check; g,go; ng,ring; th,thin; th,this; zh,vision. Foreign sounds à,œ,ü,kh,ṅ; and ◆: see page xx. <from; + plus; ? possibly.

the future bliss or sorrow of men. See CALVINISM. [<LL *predestinatio, -onis*]

pre·des·tine (prē·des′tin) *v.t.* **·tined, ·tin·ing** To predestinate.

pre·de·ter·mi·nate (prē′di·tûr′mə·nit, -nāt) *adj.* Decided or decreed beforehand.

pre·de·ter·mine (prē′di·tûr′min) *v.t.* **·mined, ·min·ing 1** To determine beforehand; decide in advance. **2** To foreordain. **3** To imbue with an antecedent tendency. — **pre′de·ter′mi·na′tion** *n.*

pre·di·al (prē′dē·əl) *adj.* **1** Consisting of land; composed of farms or landed property; real. **2** Relating to the country. **3** Attached to the soil; belonging to real estate, or resulting from tenancy of farms. Also spelled *praedial.* [<OF <Med. L *praedialis* <L *praedium*]

pred·i·ca·ble (pred′i·kə·bəl) *adj.* That may be predicated or affirmed. — *n.* **1** Anything ascribable. **2** *Logic* A property or attribute affirmable of a class. [<F *prédicable* <L *praedicabilis* <*praedicare.* See PREACH.] — **pred′i·ca·bil′i·ty, pred′i·ca·ble·ness** *n.*

pre·dic·a·ment (pri·dik′ə·mənt) *n.* **1** A trying, embarrassing, puzzling, or amusing situation or plight. **2** A specific state, position, or situation. **3** *Logic* A class or kind distinguished by definite marks; a category. [<LL *praedicamentum* that which is predicated < *praedicare.* See PREACH.]

pred·i·cant (pred′i·kənt) *adj.* Preaching. — *n.* A preacher. [<L *praedicans, -antis,* ppr. of *praedicare.* See PREACH.]

pred·i·cate (pred′i·kāt) *v.* **·cat·ed, ·cat·ing** *v.t.* **1** To declare; affirm; proclaim. **2** To state or affirm concerning the subject of a proposition. **3** To affirm as a quality or attribute of something. **4** To imply or connote. **5** *U.S.* To found or base (an argument, proposition, etc.): with *on* or *upon.* — *v.i.* **6** To make a statement or affirmation. See synonyms under AFFIRM. — *adj.* (-kit) **1** Predicated. **2** *Gram.* Belonging, relating to, or of the nature of a predicate: a *predicate* adjective. — *n.* (-kit) **1** *Gram.* The word or words in a sentence that express what is affirmed or denied of a subject, as, in the sentence, "Life is short," "is short" is the *predicate.* **2** A quality or property inherent in or asserted to belong to a thing. **3** *Logic* In a proposition, that which is stated about a subject. [<L *praedicatus,* pp. of *praedicare* make known] — **pred′i·ca′tive** *adj.* — **pred′i·ca′tive·ly** *adv.*

predicate adjective *Gram.* An adjective which describes the subject of a copulative verb, as, He is *sad*; The water turned *green,* etc.

predicate noun *Gram.* A noun which designates or identifies the subject of a copulative verb, as, He was *king*; The water became *ice.*

pred·i·ca·tion (pred′i·kā′shən) *n.* **1** The act of publicly setting forth or proclaiming. **2** The act of predicating or asserting. **3** *Logic* The assertation of something of or concerning a subject; assertion. **4** Something predicated; a predicate. [<L *praedicatio, -onis*] — **pred′i·ca′tion·al** *adj.*

pred·i·ca·to·ry (pred′i·kə·tôr′ē, -tō′rē) *adj.* **1** Of or pertaining to a preacher or preaching. **2** Proclaimed. [<LL *praedicatorius*]

pre·dict (pri·dikt′) *v.t.* To make known beforehand; prophesy; foretell. — *v.i.* To make a prediction; prophesy. See synonyms under AUGUR, PROPHESY. [<L *praedictus,* pp. of *praedicere* speak beforehand <*prae-* before + *dicere* say] — **pre·dict′a·ble** *adj.*

pre·dic·tion (pri·dik′shən) *n.* **1** The act of foretelling. **2** The thing foretold; a prophecy; forecast. [<L *praedictio, -onis*] — **pre·dic′tive** *adj.* — **pre·dic′tive·ly** *adv.*

pre·dic·tor (pri·dik′tər) *n.* **1** One who or that which predicts. **2** *Mil.* A mechanism used in connection with anti-aircraft guns for automatically determining the position, speed, and course of approaching aircraft.

pre·di·gest (prē′di·jest′, -dī-) *v.t.* To treat (food) by a process of partial digestion before introduction into the stomach; peptonize. — **pre′di·ges′tion** *n.*

pre·di·lec·tion (prē′də·lek′shən, pred′ə-) *n.* A favorable prepossession or predisposition; partiality; preference: with *for.* See synonyms under FANCY, INCLINATION, RELISH. [<F *prédilection* <Med. L *praedilectio, -onis* <*praedilectus,* pp. of *praedilegere* prefer <L *prae-* before + *diligere* love, choose]

pre·dis·pose (prē′dis·pōz′) *v.t.* **·posed, ·pos·ing 1** To give a tendency or inclination to; make susceptible or liable: Exhaustion *predisposes* one to sickness. **2** To dispose beforehand. **3** To dispose of beforehand; bequeath. — **pre·dis·po·si·tion** (prē′dis·pə·zish′ən) *n.*

pre·dom·i·nance (pri·dom′ə·nəns) *n.* **1** The state or quality of being predominant. **2** Superiority; ascendance; preponderance. Also **pre·dom′i·nan·cy.**

pre·dom·i·nant (pri·dom′ə·nənt) *adj.* Superior in power, influence, effectiveness, number, or degree; prevailing over others. [<F *prédominant*] — **pre·dom′i·nant·ly** *adv.*

Synonyms: ascendent, chief, commanding, controlling, dominant, prevailing, prevalent, regnant, sovereign, superior, supreme. *Antonyms*: accessory, complementary, contributory, inferior, subordinate, subsidiary, unimportant.

pre·dom·i·nate (pri·dom′ə·nāt) *v.i.* **·nat·ed, ·nat·ing 1** To have governing influence or control; be in control: often with *over.* **2** To be superior to all others, as in power, height, number, etc.; prevail; preponderate. [<Med. L *predominatus*] — **pre·dom′i·nat′ing·ly** *adv.* — **pre·dom′i·na′tion** *n.*

pree (prē) *v.t. Scot.* To test, especially by tasting; also, to kiss.

preef (prēf) *n. Scot.* Proof.

pre·em·i·nent (prē·em′ə·nənt) *adj.* **1** Supremely eminent; distinguished above all others; transcendent; supreme. **2** Extraordinary in degree; outstanding; conspicuous; superlative. See synonyms under PARAMOUNT. [<L *praeëminens, -entis,* ppr. of *praeëminere* be prominent <*prae-* before + *eminere* stand out, project] — **pre·em·i·nent·ly** *adv.* — **pre·em′i·nence** *n.*

pre·empt (prē·empt′) *v.t.* **1** To acquire or appropriate beforehand. **2** To secure by preemption; occupy (public land) so as to acquire by preemption. [Back formation < PREEMPTION] — **pre·emp′tor** *n.* — **pre·emp′to·ry** (-tər·ē) *adj.*

pre·emp·tion (prē·emp′shən) *n.* **1** The right or act of purchasing before others. **2** Public land obtained by exercising this right. [< Med. L *praeëmptio, -onis* <L *prae-* before + *emptus,* pp. of *emere* buy]

pre·emp·tive (prē·emp′tiv) *adj.* Pertaining to or capable of preemption.

preemptive bid In auction or contract bridge, a bid of a high number of tricks in order to shut out probable bids of an opponent.

preen[1] (prēn) *v.t.* **1** To trim and dress with the beak, as birds their feathers. **2** To dress or adorn (oneself) carefully; primp; prink. — *v.i.* **3** To primp; prink. [Prob. var. of PRUNE[3]]

preen[2] (prēn) *n. Scot.* A pin; a brooch. — *v.t.* To sew; pin.

pre·ex·il·i·an (prē′eg·zil′ē·ən) *adj.* In Jewish history, pertaining to or denoting a period prior to the Babylonian exile (sixth century B.C.). Also **pre′ex·il′ic.**

pre·fab (prē′fab′) *n.* A prefabricated structure or part: also used attributively.

pre·fab·ri·cate (prē·fab′rə·kāt) *v.t.* **·cat·ed, ·cat·ing 1** To fabricate or build beforehand. **2** To manufacture in standard sections that can be rapidly assembled. — **pre·fab′ri·ca′tion** *n.*

pref·ace (pref′is) *n.* **1** A brief explanation or address to the reader at the beginning of a book or other publication. **2** Any introductory speech, writing, etc. — *v.t.* **·aced, ·ac·ing 1** To introduce or furnish with a preface. **2** To serve as a preface for. [<OF <L *praefatio* <*praefatus,* pp. of *praefari* utter beforehand, premise <*prae-* before + *fari* speak]

Pref·ace (pref′is) *n. Eccl.* **1** The prayer of thanksgiving, ending with the Sanctus, which introduces the canon of the mass. **2** The corresponding section in other eucharistic liturgies.

pref·a·to·ry (pref′ə·tôr′ē, -tō′rē) *adj.* Of the nature of a preface; introductory. Also **pref′a·to′ri·al.** — **pref′a·to′ri·ly** *adv.*

pre·fect (prē′fekt) *n.* **1** In ancient Rome, any of various civil and military officials, as certain magistrates, governors, and commanders. **2** Any magistrate, chief official, etc.; specifically, in France, the chief administrator of a department, or the head of the Paris police. **3** In Roman Catholic schools, the dean. **4** *Brit.* A senior pupil charged with maintaining order and discipline among other pupils. Also spelled *praefect.* [<OF <L *praefectus,* orig. pp. of *praeficere* set over <L *prae-* before + *facere* make, do]

pre·fec·ture (prē′fek·chər) *n.* **1** The office, jurisdiction, or province of a prefect. **2** The official building for his use. [<L *praefectura*] — **pre·fec′tur·al** *adj.*

pre·fer (pri·fûr′) *v.t.* **·ferred, ·fer·ring 1** To hold in higher regard or esteem; like better. **2** To give priority to, as one creditor or form of securities over others. **3** To advance or promote, as in status or rank. **4** To offer, as a suit or charge, for consideration or decision. See synonyms under CHOOSE, PROMOTE. [<F *préferer* <L *praeferre* carry, set in front <*prae-* before + *ferre* bear] — **pre·fer′rer** *n.*

pref·er·a·ble (pref′ər·ə·bəl) *adj.* To be preferred; more desirable; worthy of choice. — **pref′er·a·ble·ness, pref′er·a·bil′i·ty** *n.* — **pref′er·a·bly** *adv.*

pref·er·ence (pref′ər·əns) *n.* **1** The act of preferring; estimation or choice of one thing or person over another; also, the privilege of making such choice. **2** The state of being preferred. **3** That which is preferred; an object of favor or choice. **4** A priority of payment given by an insolvent debtor to one or to a certain class of his creditors over others; also, priority of payment by operation of law. **5** Promotion; preferment. **6** The granting of special advantage over others to one country or group of countries in international trade. See synonyms under ALTERNATIVE, PRECEDENCE. [<F *préférence* <L *praeferentia,* orig. neut. pl. of *praeferens, -entis,* ppr. of *praeferre.* See PREFER.]

pref·er·en·tial (pref′ə·ren′shəl) *adj.* **1** Indicating or arising from preference or partiality. **2** Possessing or giving priority or preference, as in tariffs or railroad charges. — **pref′er·en′tial·ism** *n.* — **pref′er·en′tial·ly** *adv.*

preferential shop A shop that gives precedence to union members when hiring employees, usually by agreement with a union.

preferential voting A form of voting in which an order of choice of candidates may be signified by a voter on his ballot.

pre·fer·ment (pri·fûr′mənt) *n.* **1** The act of preferring. **2** The state of being preferred. **3** The act of promoting or appointing to higher office; advancement; promotion. **4** A superior post or dignity: said especially of ecclesiastical rank.

pre·ferred (pri·fûrd′) *adj.* **1** Having the first claim: *preferred* bonds or stock. **2** Having gained promotion. **3** Chosen by preference.

pre·fig·ur·a·tion (prē·fig′yə·rā′shən) *n.* **1** Antecedent representation by types, figures, etc. **2** A prototype. — **pre·fig·ur·a·tive** (prē·fig′yər·ə·tiv) *adj.* — **pre·fig′ur·a·tive·ly** *adv.* — **pre·fig′ur·a·tive·ness** *n.*

pre·fig·ure (prē·fig′yər) *v.t.* **·ured, ·ur·ing 1** To represent in advance; serve as an indication or suggestion of; foreshadow. **2** To imagine or picture to oneself beforehand. [<LL *praefigurare*]

pre·fix (prē′fiks) *n.* **1** *Gram.* A non-separable syllable, or syllables, affixed to the beginning of a word to modify or alter the meaning, as *pre-* in prefix, *be-* in behead, *dis-* in disagree, *re-* in renew, *post-* in postwar, *un-* in unhorse, etc. **2** Something placed before, as a title before a noun. Compare SUFFIX. — *v.t.* (prē·fiks′) **1** To put or attach before or at the beginning; add as a prefix: opposed to *postfix.* **2** *Obs.* To arrange or settle beforehand. [< OF *prefixer* <L *praefixus,* pp. of *praefigere* <*prae-* before + *figere* fix] — **pre′fix·al** *adj.* — **pre′fix·al·ly** *adv.* — **pre·fix·ion** (prē·fik′shən) *n.*

pre·flight training (prē′flīt) *Aeron.* Preliminary ground instruction in aviation.

pre·flo·ra·tion (prē′flə·rā′shən) *n. Bot.* The disposition of flowers within the flower bud; estivation. [<PRE- + L *flos, floris* flower + -ATION]

pre·fo·li·a·tion (prē·fō′lē·ā′shən) *n. Bot.* The disposition of leaves within the bud; venation.

pre·for·ma·tion (prē′fôr·mā′shən) *n.* **1** The act of preforming; the state of being formed in advance. **2** *Biol.* An early theory of generation according to which an organism exists fully preformed in the germ, developing only by increase in size. Compare EPIGENESIS.

Pre·gel (prā′gəl) A river in the former German province of East Prussia, now in Russian

S.F.S.R., flowing 78 miles west to the Vistula Lagoon below Kaliningrad.

Pre·gl (prā′gəl), **Fritz,** 1869–1930, Austrian chemist.

preg·na·ble (preg′nə·bəl) *adj.* Weak enough to be conquered; likely to yield when attacked, as a fort. [<OF *prenable* <*prendre* take <L *prehendere* seize] —**preg′na·bil′i·ty** *n.*

preg·nan·cy (preg′nən·sē) *n.* 1 The state of being with young or with child. 2 *Obs.* Quickness of intelligence.

preg·nan·di·ol (preg·nan′dē·ôl, -ol) *n. Biochem.* A complex organic compound of the sterol group, found in the urine of pregnant women and chemically related to progesterone. [< *pregnane,* a sterol from the urine of pregnant women + *diol,* a glycol]

preg·nant (preg′nənt) *adj.* 1 Carrying a growing fetus in the uterus; with child; impregnated; gestating. 2 Carrying great weight or significance; full of meaning or contents; leading to important results. 3 Fruitful; prolific; teeming with ideas; imaginative; inventive. 4 In rhetoric and logic, implying more than is expressed. [<L *praegnans, -antis,* ult. < *prae-* before + *gnasci* be born] —**preg′nant·ly** *adv.*

pre·hen·si·ble (pri·hen′sə·bəl) *adj.* Capable of being apprehended or grasped. [<L *prehensus,* pp. of *prehendere* seize]

pre·hen·sile (pri·hen′sil) *adj.* Adapted for grasping or holding; formed to grasp or coil around and cling to objects, as the tail of a monkey. [<F *préhensile*] —**pre·hen·sil·i·ty** (prē′hen·sil′ə·tē) *n.*

pre·hen·sion (pri·hen′shən) *n.* The act of grasping, physically or mentally. [<L *prehensio, -onis*]

pre·his·tor·ic (prē′his·tôr′ik, -tor′-) *adj.* Of or belonging to a period before that covered by written history. Also **pre′his·tor′i·cal.** —**pre′·his·tor′i·cal·ly** *adv.*

pre·his·to·ry (prē·his′tə·rē) *n.* The history of the development of mankind based on archeological and ethnological findings; the period of history preceding written records.

prehn·ite (pren′īt) *n.* A light-green, gray, or white hydrous silicate of calcium and aluminum: similar in composition and occurrence to zeolite. [after Col. van *Prehn,* 18th century Dutch colonist]

pre·ig·ni·tion (prē′ig·nish′ən) *n.* Ignition of the charge in the cylinder of an internal-combustion engine previous to the completion of the compression stroke, often the result of faulty ignition timing.

pre·judge (prē·juj′) *v.t.* **·judged, ·judg·ing** To judge before or without proper inquiry; pass judgment on hastily or beforehand. [<F *préjuger* <*praejudicare* <*prae-* before + *judicare* judge] —**pre·judg′er** *n.* —**pre·judg′·ment,** *Brit.* **pre·judge′ment** *n.*

prej·u·dice (prej′ŏŏ·dis) *n.* 1 A judgment or opinion, favorable or unfavorable, formed beforehand or without due examination; a mental decision based on other grounds than reason or justice; especially, a premature or adversely biased opinion. 2 Detriment arising from a hasty and unfair judgment; injury; harm. —**in** (or **to**) **the prejudice of** To the injury or detriment of. —**without prejudice** *Law* Without detriment to any right that previously existed: usually applied to the dismissal of a bill in equity without consideration of the merits; or to the reservation, express or implied, of all rights in favor of one who offers to compromise a claim or litigation, in case his offer is rejected. —*v.t.* **·diced, ·dic·ing** 1 To affect or influence with a prejudice; bias. 2 To affect injuriously or detrimentally; damage; impair. [<OF <L *praejudicium* <*prae-* before + *judicium* judgment]

Synonyms (noun): bias, preconception, predilection, prepossession, unfairness. A *prejudice* or *prepossession* is grounded often on feeling, fancy, associations, etc. A *prepossession* is always favorable, a *prejudice* usually unfavorable, unless the contrary is expressly stated. See INJURY. *Antonyms*: certainty, conclusion, conviction, demonstration, evidence, reason, reasoning.

prej·u·di·cial (prej′ŏŏ·dish′əl) *adj.* Having power or tendency to prejudice or injure; injurious; detrimental. —**prej′u·di′cial·ly** *adv.*

prel·a·cy (prel′ə·sē) *n. pl.* **·cies** 1 The system of church government by prelates: often a hostile term for episcopacy. 2 The dignity or function of a prelate. 3 Prelates collectively. [<AF *prelacie* <Med. L *praelatia* <*praelatus.* See PRELATE.]

prel·ate (prel′it) *n.* One of a higher order of clergy, as a bishop or abbot. [<OF *prelat* <L *praelatus,* pp. to *praeferre* set over. See PREFER.] —**prel′ate·ship** *n.* —**pre·lat·ic** (pri·lat′ik) or **·i·cal** *adj.*

prel·a·tism (prel′ə·tiz′əm) *n.* 1 Prelacy; episcopacy. 2 Prelatic partisanship.

prel·a·tist (prel′ə·tist) *n.* One who supports the prelacy; an advocate of High Church government: sometimes used contemptuously.

prel·a·ture (prel′ə·chər) *n.* Prelacy (defs. 2 and 3).

pre·lect (pri·lekt′) *v.i.* To lecture; discourse. [<L *praelectus,* pp. of *praelegere* read before <*prae-* before + *legere* read] —**pre·lec′tion** *n.* —**pre·lec′tor** *n.*

pre·li·ba·tion (prē′lī·bā′shən) *n.* 1 A preliminary offering. 2 A tasting beforehand or by anticipation; anticipation. [<LL *praelibatio, -onis* <*prae-* before + *libatio* a libation]

pre·lim·i·nar·y (pri·lim′ə·ner′ē) *adj.* Antecedent or introductory to the main discourse, proceedings, or business; prefatory; preparatory. —*n. pl.* **·ries** 1 An initiatory step; a preparatory act. 2 A preliminary examination. See synonyms under ANTECEDENT. [<PRE- + L *liminaris* pertaining to a threshold < *limen, liminis* threshold] —**pre·lim′i·nar′i·ly** *adv.*

pre·lit·er·ate (prē·lit′ər·it) *adj.* Without written records; prehistoric: said especially of the earliest human cultures.

prel·ude (prel′yōōd, prē′lōōd) *n.* 1 *Music* **a** An independent instrumental composition of moderate length, in a free style suggesting improvisation. **b** An opening piece at the start of a church service; a voluntary. **c** The overture of an opera. **d** An opening strain or movement at the beginning of a musical composition, usually introducing the theme of the whole work. 2 Any introductory or opening performance or event, or that which foreshadows a coming event. —*v.* **·ud·ed, ·ud·ing** *v.t.* 1 To introduce with a prelude. 2 To serve as a prelude to. —*v.i.* 3 To serve as a prelude. 4 To provide or play a prelude. [<F *prélude* <Med. L *praeludium* <L *praeludere* play before <*prae-* before + *ludere* play] —**pre·lud·er** (pri·lōō′dər, prel′yə·dər) *n.* —**pre·lu·di·al** (pri·lōō′dē·əl) *adj.*

pre·lu·sion (pri·lōō′zhən) *n.* That which serves as a prelude. [<L *praelusio, -onis* <*praelusus,* pp. of *praeludere.* See PRELUDE.]

pre·lu·sive (pri·lōō′siv) *adj.* Having the character of a prelude; indicating beforehand. Also **pre·lu′so·ry** (-sər·ē). [<L *praelusus.* See PRELUSION.] —**pre·lu′sive·ly, pre·lu′so·ri·ly** *adv.*

pre·ma·ture (prē′mə·chŏŏr′, -tŏŏr′, -tyŏŏr′) *adj.* Existing, happening, matured or developed before the natural period; done before the proper time; untimely. [<L *praematurus* <*prae-* before + *maturus* ripe, seasonable] —**pre′ma·ture′ly** *adv.* —**pre′ma·tu′ri·ty, pre′·ma·ture′ness** *n.*

pre·max·il·la (prē′mak·sil′ə) *n. pl.* **·max·il·lae** (-mak·sil′ē) *Anat.* One of the two bones set between the maxillae in front of the vertebrate jaw. [<NL] —**pre·max·il·lar·y** (prē·mak′sə·ler′ē) *adj.*

pre·med·i·tate (prē·med′ə·tāt) *v.t. & v.i.* **·tat·ed, ·tat·ing** To plan or consider beforehand. [<L *praemeditatus,* pp. of *praemeditari* think over <*prae-* before + *meditari.* See MEDITATE.] —**pre·med′i·tat′ed·ly** *adv.* —**pre·med′i·ta′tive** *adj.* —**pre·med′i·ta′tor** *n.*

pre·med·i·ta·tion (prē·med′ə·tā′shən) *n.* The considering and planning of a subsequent act; deliberate intention and plan to do a certain thing, especially to commit a crime.

pre·mi·er (prē′mē·ər, *esp. Brit.* prem′yər) *adj.* 1 First in rank or position; principal: the *premier* place, *premier* officer. 2 First in order of occurrence; earliest; specifically, first in order of creation; senior: the *premier* duke of England. —*n.* (prē′mē·ər, pri·mir′; *Brit.* prem′yər) The head of government; the prime minister of England, France, etc. [<F <L *primarius* <L *primus* first] —**pre′mi·er·ship′** *n.*

pre·mière (pri·mir′, *Fr.* prə·myâr′) *adj.* First. —*n.* 1 The leading lady in a theatrical company. 2 The first public presentation of a play, etc. [<F]

pre·mil·le·nar·i·an (prē′mil·ə·nâr′ē·ən) *adj.* Existing or occurring before the millennium: also **pre′mil·len′ni·al** (-mi·len′ē·əl). —*n.* One who believes in premillennialism: also **pre′·mil·len′ni·al·ist.**

pre·mil·len·ni·al·ism (prē′mi·len′ē·əl·iz′əm) *n.* The doctrine that the millennium is to be introduced by the personal return of Christ: opposed to *postmillennialism.*

prem·ise (prem′is) *n.* 1 A proposition laid down, proved, supposed, or assumed, that serves as a ground for argument or for a conclusion; a judgment leading to another judgment as a conclusion. 2 *Logic* Either of the two propositions in a syllogism from which, their truth being granted, the conclusion necessarily follows. 3 *pl. Law* **a** Foregoing statements; facts previously stated. **b** That part in a deed that sets forth the date, names of parties, the land or thing conveyed or granted, the consideration, and all other matters down to the phrase "to have and to hold." 4 *pl.* A distinct portion of real estate; land or lands; land with its appurtenances, as buildings: He lingered about the *premises.* Also **prem′iss.** —**major premise** *Logic* The premise in which the predicate of the conclusion of a syllogism, called the **major term,** is contained; the first proposition of a syllogism. —**minor premise** *Logic* The premise in which the subject of the conclusion of a syllogism, called the **minor term,** is contained; the second proposition of a syllogism.

—**pre·mise** (pri·mīz′, prem′is) *v.* **·mised, ·mis·ing** *v.t.* 1 To stay or state beforehand, as by way of introduction or explanation. 2 To state or assume as a premise or basis of argument. 3 *Obs.* To send in advance. —*v.i.* 4 To make a premise. [<OF *premisse* <Med. L *praemissa,* orig. fem. of L *praemissus,* pp. of *praemittere* send before <*prae-* before + *mittere* send]

pre·mi·um (prē′mē·əm) *n.* 1 A reward or prize for a superior performance or production in competition. 2 A price paid for a loan; a sum offered or given to secure a loan, either a sum in addition to interest, a bonus, or the interest itself. 3 The rate or price at which stocks, shares, or money are valued in excess of their nominal or par value: bank shares at a *premium* of five percent. 4 The amount paid for insurance, as admission fees, annual dues, periodical payments, etc., according to the kind of insurance secured. 5 Any object offered free to those who purchase goods to a certain value, as a set of books given free as an inducement to subscribe to a magazine. 6 A fee for instruction in a trade or a profession. See synonyms under SUBSIDY. —**at a premium** Above par; hence, valuable and in demand. [<L *praemium,* ult. <*prae-* before + *emere* buy]

pre·mo·lar (prē·mō′lər) *n. Anat.* One of the teeth situated before the molars and behind the canines. Compare BICUSPID. —*adj.* Situated in front of or appearing before the molar teeth.

pre·mon·ish (pri·mon′ish) *v.t.* To admonish in advance; forewarn. [<PRE- + MONISH]

pre·mo·ni·tion (prē′mə·nish′ən, prem′ə-) *n.* 1 An actual warning of something yet to occur. 2 A presentiment not based on information received; an instinctive foreboding. [<OF *premonicion* <LL *praemonitio, -onis* <*praemonitus,* pp. of *praemonere* premonish <*prae-* before + *monere* warn] —**pre·mon·i·to·ry** (pri·mon′ə·tôr′ē, -tō′rē) *adj.* —**pre·mon′i·to′·ri·ly** *adv.*

pre·morse (pri·môrs′) *adj. Biol.* Terminating abruptly, as if bitten or broken off: a *premorse* root. [<L *praemorsus,* pp. of *praemordere* bite off <*prae-* before + *mordere* bite]

pre·mun·dane (prē·mun′dān) *adj.* Antemundane.

pre·name (prē′nām′) *n.* A forename; Christian name. [Trans. of L *praenomen*]

pre·na·tal (prē·nāt′l) *adj.* Before birth: *prenatal* care or health. —**pre·na′tal·ly** *adv.*

pre·no·men (prē·nō′mən) See PRAENOMEN.

pre·nom·i·nate (prē·nom′ə·nāt) *Obs. v.t.* To mention or name beforehand. —*adj.* Named beforehand.

pre·no·tion (prē·nō′shən) *n.* A preconception; a generalization with slight basis of fact or experience.

prent (prent) *v.* & *n. Scot.* Print.

pren·tice (pren′tis) *n.* An apprentice. Also **'pren′tice.** [Aphetic var. of APPRENTICE]

pre·oc·cu·pa·tion (prē·ok′yə·pā′shən) *n.* **1** The act of occupying before others, or the state of being or having a prior occupant: also **pre·oc′cu·pan·cy. 2** The state of being pre-occupied, as in mind, attention, or inclination; prepossession. **3** Something that preoccupies. [<L *praeoccupatio, -onis*] —**pre·oc′cu·pant** *n.*

pre·oc·cu·pied (prē·ok′yə·pīd) *adj.* **1** Engrossed in thought or business; abstracted. **2** Previously occupied. **3** Already in use, as a scientific name. See synonyms under ABSTRACTED.

pre·oc·cu·py (prē·ok′yə·pī) *v.t.* **·pied, ·py·ing 1** To engage fully; engross, as the mind. **2** To occupy or take possession of in advance of another or others. See synonyms under OCCUPY. [<L *praeoccupare*]

pre·or·dain (prē′ôr·dān′) *v.t.* To foreordain. —**pre·or·di·na·tion** (prē′ôr·də·nā′shən) *n.*

prep (prep) *adj. Colloq.* Preparatory: a *prep* school, student, etc.

prep·a·ra·tion (prep′ə·rā′shən) *n.* **1** The act, process, or operation of preparing. **2** An act or proceeding designed to bring about some event; a precaution; provision: *preparations* for war or for a journey. **3** The fact or state of being prepared; readiness. **4** Something made or prepared, as a compound, composition, etc.: medicinal or chemical *preparations.* **5** Preliminary study; training, as for college or business. **6** *Music* The previous introduction, as an integral part of a chord, of a note which is then continued into a following dissonance; also, the note so treated. **7** *Eccl.* Devotional exercises introducing an office, as that of the Eucharist. [<OF <L *praeparatio, -onis*]

pre·par·a·tive (pri·par′ə·tiv) *adj.* Serving or tending to prepare. —*n.* **1** That which is preparatory. **2** An act of preparation. —**pre·par′a·tive·ly** *adv.*

pre·par·a·tor (pri·par′ə·tər) *n.* One who prepares subjects for scientific purposes, as specimens for dissection or objects for preservation in collections. [<LL *praeparator*]

pre·par·a·to·ry (pri·par′ə·tôr′ē, -tō′rē) *adj.* **1** Serving as a preparation. **2** Occupied in preparation: a *preparatory* scholar. —*adv.* As a preparation: *Preparatory* to writing, I will consider this: also **pre·par′a·to′ri·ly.**

preparatory school A school in which students are prepared for admission to a college or university.

pre·pare (pri·pâr′) *v.* **·pared, ·par·ing** *v.t.* **1** To make ready, fit, or qualified; put in readiness. **2** To provide with what is needed; outfit; equip: to *prepare* an expedition. **3** To bring to a state of completeness, as a meal, lesson, or prescription. **4** *Music* To introduce by a preliminary note or notes. —*v.i.* **5** To make preparations; get ready. [<F *préparer* <L *praeparare* <*prae-* before + *parare* make ready] —**pre·par·ed·ly** (pri·pâr′id·lē) *adv.* —**pre·par′er** *n.*

pre·par·ed·ness (pri·pâr′id·nis, -pârd′-) *n.* Readiness; especially, a condition of military readiness for war.

pre·pay (prē·pā′) *v.t.* **·paid, ·pay·ing** To pay or pay for in advance. —**pre·pay′ment** *n.*

pre·pense (pri·pens′) *adj.* Premeditated; considered beforehand: chiefly in the phrase malice *prepense.* [<OF *purpensé,* pp. of *purpenser* <*pur-* (<L *pro-*) ahead + *penser* think <L *pensare*] —**pre·pense′ly** *adv.*

pre·pon·der·ance (pri·pon′dər·əns) *n.* Superiority in weight, influence, force, quantity, etc. Also **pre·pon′der·an·cy.**

pre·pon·der·ant (pri·pon′dər·ənt) *adj.* Having such superior force, weight, importance, efficacy, quantity, or number as to overbalance something else or all other things of a class; predominant. —**pre·pon′der·ant·ly** *adv.*

pre·pon·der·ate (pri·pon′də·rāt) *v.i.* **·at·ed, ·at·ing 1** To be of greater weight. **2** To incline downward or descend, as the scale of a balance. **3** To be of greater power, importance, quantity, etc.; predominate; prevail. [<L *praeponderatus,* pp. of *praeponderare* <*prae-* before + *ponderare* weigh <*pondus, ponderis* weight] —**pre·pon′der·a′tion** *n.*

prep·o·si·tion (prep′ə·zish′ən) *n. Gram.* **1** In some languages, a word functioning to indicate the relation of a substantive (the object of the preposition) to another substantive, a verb, or an adjective: one of the eight traditional parts of speech. Some English prepositions are *by, for, from, in, to, with.* A preposition is usually placed before its object (whence its name), and together they constitute a prepositional phrase which serves as an adjectival or an adverbial modifier: He sat *beside* the fire; sick *at* heart; a man *of* honor. There is a close relationship between certain prepositions and adverbs, and the same word may have either function, depending on the context: We saw it *through* (adverb); It sailed out *through* the window (preposition). **2** Any word or construction that functions in a similar manner: He telephoned *in reference to* (equals *about*) your letter. —**inseparable preposition** A preposition so closely connected with a verb as to have all the force of a compound: to *laugh at.* —**participal preposition** A participle used without direct connection with a subject, so that it has the force of a preposition: They spoke to him *concerning* that affair. —**postpositive preposition** A preposition in postposition; also, a suffix added to a noun and serving as a preposition: Hope soars heaven*ward.* [<L *praepositio, -onis* <*praepositus,* pp. of *praeponere* place before <*prae-* before + *ponere* place]

prep·o·si·tion·al (prep′ə·zish′ən·əl) *adj.* Pertaining to, formed with, or having the character or force of prepositions. —**prep′o·si′tion·al·ly** *adv.*

pre·pos·i·tive (prē·poz′ə·tiv) *adj.* **1** Prefixed. **2** *Gram.* Placed before the word governed or qualified. —*n. Gram.* A prepositive word or particle. [<L *praepositivus* <*praepositus.* See PREPOSITION.]

pre·pos·i·tor (prē·poz′ə·tər) *n. Brit.* A pupil or student in a school or college who directs or oversees others; monitor. [Alter. of L *praepositus.* See PREPOSITION.] —**pre·pos·i·to·ri·al** (prē·poz′ə·tôr′ē·əl, -tō′rē-) *adj.*

pre·pos·sess (prē′pə·zes′) *v.t.* **1** To preoccupy to the exclusion of other ideas, beliefs, etc.; prejudice; bias. **2** To impress or influence beforehand or at once, especially favorably. **3** *Rare* To take possession of in advance of others, as land.

pre·pos·sess·ing (prē′pə·zes′ing) *adj.* Inspiring a favorable opinion from the beginning. —**pre′pos·sess′ing·ly** *adv.*

pre·pos·ses·sion (prē′pə·zesh′ən) *n.* **1** The state of being prepossessed; a previous impression of a particular person or thing; a preconceived liking; bias. **2** Prior possession. See synonyms under INCLINATION, PREJUDICE.

pre·pos·ter·ous (pri·pos′tər·əs) *adj.* Contrary to nature, reason, or common sense; strikingly or utterly absurd or impracticable. See synonyms under ABSURD, EXTRAORDINARY, RIDICULOUS. [<L *praeposterus* the last first, inverted <*prae-* before + *posterus* last] —**pre·pos′ter·ous·ly** *adv.* —**pre·pos′ter·ous·ness** *n.*

pre·pos·tor (prē·pos′tər) *n.* A prepositor. [Alter. of PREPOSITOR]

pre·po·ten·cy (pri·pō′tən·sē) *n.* **1** The quality of superior potency; preponderance of influence or efficiency. **2** *Biol.* The pronounced capacity of one parent, strain, or breed to transmit its own characteristics to the offspring. Also **pre·po′tence.** [<PREPOTENT]

pre·po·tent (pri·pō′tənt) *adj.* **1** Endowed with prevailing potency; predominant. **2** Having potential power or efficacy; possessing power to shape or influence what comes after. **3** Pertaining to or exhibiting prepotency. Also **pre·po·ten·tial** (prē′pə·ten′shəl). [<L *praepotens, -entis* very powerful] —**pre·po′tent·ly** *adv.*

pre·puce (prē′pyōōs) *n. Anat.* The loose skin that covers the glans of the penis; the foreskin. [<F *prépuce* <L *praeputium*] —**pre·pu·tial** (pri·pyōō′shəl) *adj.*

Pre-Raph·a·el·ite (prē·raf′ē·ə·līt, -rā′fē-) *n.* **1** A follower or adherent of the Pre-Raphaelite Brotherhood, a society of artists, formed in England, 1847–49, by D. G. Rossetti, W. Holman-Hunt, John Millais, and others, stressing the truth to nature and delicacy of poetic sentiment that supposedly characterized Italian art before the time of Raphael. **2** Any modern artist with similar or related aims. **3** Any Italian painter before the time of Raphael. —*adj.* **1** Before the time of Raphael. **2** Of or pertaining to the Pre-Raphaelite Brotherhood or its followers. —**Pre-Raph′a·el·it′ism** *n.*

pre·req·ui·site (prē·rek′wə·zit) *adj.* Required as an antecedent condition; necessary to something that follows. —*n.* A necessary antecedent condition.

pre·rog·a·tive (pri·rog′ə·tiv) *n.* **1** An indefeasible and unquestionable right belonging to a person or body of persons by virtue of position or relation, and exercised without control or accountability; specifically, a hereditary or official right: the royal *prerogative.* **2** Hence, any characteristic and generally recognized privilege peculiar to a person or class: It is a woman's *prerogative* to change her mind. **3** Precedence; preeminence. See synonyms under RIGHT. —*adj.* Of or pertaining to a prerogative; possessing or held by prerogative. [<OF <L *praerogativa* right of voting first <*praerogatus,* pp. of *praerogare* ask before another <*prae-* before + *rogare* ask]

prerogative court 1 *Brit.* Formerly, a court having jurisdiction of testamentary matters, as an archbishop's court in which were involved effects up to five pounds in each of two or more dioceses of the archiepiscopal province. **2** *U.S.* A court held in New Jersey by the chancellor sitting as ordinary in probate matters, and for the determination of appeals from the Orphan's Court.

pre·sa (prā′sä) *n. Music* A sign 𝄋, +, or ✻ in fugues or canons, where the voices are successively to take up the theme. [<Ital., lit., a taking, orig. fem. of *preso,* pp. of *prendere* take <L *prehendere*]

pres·age (pres′ij) *n.* **1** An indication of something to come; prophetic token; portent; omen. **2** A prophetic impression; presentiment; foreboding. **3** Prophetic meaning or import; prediction; foresight. See synonyms under SIGN.

—**pre·sage** (pri·sāj′) *v.* **·saged, ·sag·ing** *v.t.* **1** To give a presage or portent of; betoken; foreshadow. **2** To have a presentiment of. **3** To predict; foretell. —*v.i.* **4** To make a prediction; prophesy. See synonyms under AUGUR. [<F *présage* <L *presagium* omen <*praesagire* perceive beforehand <*prae-* before + *sagire* be aware of. Akin to SAGACIOUS.] —**pre·sage′ment** *n.* —**pre·sag′er** *n.*

pres·by·cu·sis (prez′bi·kyōō′sis, pres′-) *n. Pathol.* Impairment of hearing due to advancing years or old age. Also **pres·by·a·cu·sis** (prez′bē·ə·kyōō′sis, pres′-). [<NL <Gk. *presbys* old + *akousis* hearing]

pres·by·o·phre·ni·a (prez′bē·ə·frē′nē·ə, pres′-) *n. Psychiatry* Failure of mental powers due to old age, characterized by confabulation and loss of memory. [<NL <Gk. *presbys* old + *phrēn* mind] —**pres′by·o·phren′ic** (-fren′ik) *adj.*

pres·by·o·pi·a (prez′bē·ō′pē·ə, pres′-) *n. Pathol.* Long-sightedness, especially that incident to old age and due to rigidity of the crystalline lens, which renders accommodation difficult for near objects. [<NL <Gk. *presbys* old + -OPIA] —**pres′by·op′ic** (-op′ik) *adj.*

pres·by·ter (prez′bə·tər, pres′-) *n.* **1** In the early church, one of the elders of a church. **2** *Eccl.* **a** In hierarchical churches, a priest. **b** In Presbyterian churches, an ordained clergyman (a **teaching elder**); also, a layman who is a member of the governing body of a congregation (a **ruling elder**). [<LL <Gk. *presbyteros* an elder. Doublet of PRIEST.]

pres·byt·er·ate (prez·bit′ər·it, -ə·rāt, pres-) *n.* **1** The office or dignity of a presbyter or elder. **2** The order or the body of presbyters.

pres·by·te·ri·al (prez′bə·tir′ē·əl, pres′-) *adj.* Pertaining to a presbytery or a presbyter. Also **pres·byt′er·al.** —**pres·byt′er·al·ly** *adv.*

Pres·by·te·ri·an (prez′bə·tir′ē·ən, pres′-) *n.* **1** One who believes in the government of the church by presbyters. **2** A member of any of various Protestant churches, mostly Calvinist in doctrine, and holding to the government of the church by presbyters. —*adj.* Of or pertaining to the Presbyterian Church, its form of government, or its doctrines. [<LL *presbyterium* presbytery + -IAN] —**Pres′by·te′ri·an·ism** *n.*

pres·by·ter·y (prez′bə·ter′ē, pres′-) *n.* *pl.* **·ter·ies** **1** In the Presbyterian Church, a court having the ecclesiastical and spiritual rule and oversight of a given district; the district so represented; also, presbyters collectively. **2** The system of church government by presbyters: distinguished from the *Independent* system and *prelacy.* **3** That part of a church set apart for the clergy. **4** In the Roman Catholic Church, the residence of the priest. [<OF *presbiterie* <LL *presbyterium* assembly of elders <Gk. *presbyterion* <*presbyteros* elder]

pre–school (prē·skōōl′) *adj.* For or designating a child past infancy but under school age.

pre·sci·ence (prē′shē·əns, presh′ē-) *n.* Knowledge of events before they take place. See synonyms under WISDOM. [<OF <L *praescientia,* orig. neut. pl. of *praesciens, -entis,* ppr. of *praescire* know beforehand <*prae-* before + *scire* know]

pre·sci·ent (prē′shē·ənt, presh′ē-) *adj.* Having prescience; foreknowing; also, far-seeing. [<F <L *presciens, -entis.* See PRESCIENCE.] **—pre′sci·ent·ly** *adv.*

pre·scind (pri·sind′) *v.t.* **1** To set apart in thought; consider separately. **2** To cut off; remove. **—***v.i.* **3** To withdraw the attention: with *from.* [<L *praescindere* cut off in front <*prae-* before + *scindere* cut]

Pres·cott (pres′kət), **William,** 1726–95, American officer; commanded at Bunker Hill in the Revolutionary War. **—William Hickling,** 1796–1859, U.S. historian.

pre·scribe (pri·skrīb′) *v.* **·scribed, ·scrib·ing** *v.t.* **1** To set down as a direction or rule to be followed; ordain; enjoin. **2** *Med.* To order the use of (a medicine, treatment, etc.) as a remedy. **3** *Law* To render invalid by lapse of time. **—***v.i.* **4** To lay down laws or rules; give directions. **5** *Law* **a** To assert a title to something on the basis of prescription: with *for* or *to.* **b** To become invalid or unenforceable by lapse of time. See synonyms under DICTATE, SET. [<L *praescribere* write beforehand <*prae-* before + *scribere* write] **—pre·scrib′er** *n.*

pre·script (prē′skript) *n.* A prescription or direction, as a rule of conduct. **—***adj.* (pri·skript′, prē′skript) Prescribed as a rule or model; laid down. [<L *praescriptus,* pp. of *praescribere.* See PRESCRIBE.]

pre·scrip·ti·ble (pri·skrip′tə·bəl) *adj.* Derived from or acquirable by prescription; depending on prescriptive right. **—pre·scrip′ti·bil′i·ty** *n.*

pre·scrip·tion (pri·skrip′shən) *n.* **1** The act of prescribing, directing, or dictating. **2** That which is prescribed or appointed, as a rule or precept; a prescript. **3** *Med.* **a** A physician's formula for compounding and administering a medicine. **b** The remedy so prescribed. **c** A formula issued by a licensed oculist or optometrist giving directions for the grinding of eyeglass lenses. **4** *Law* A title to property, or a mode of acquiring title to property, founded on uninterrupted possession; a mode of losing a right or title by failure to assert it within a given time; the period after which a neglected right or title cannot be asserted; also, the period, if any, after which prosecution for a crime is barred. **5** Old or continued custom, particularly when considered authoritative. **6** A claim based on long usage. [<L *praescriptio, -onis*]

pre·scrip·tive (pri·skrip′tiv) *adj.* **1** Making strict requirements or rules: *prescriptive* grammar. **2** Sanctioned by custom or long use: a *prescriptive* right to grumble. **3** *Law* Acquired by immemorial use; based on prescription: a *prescriptive* title. [<LL *praescriptivus*] **—pre·scrip′tive·ly** *adv.*

preselector gearbox An automobile transmission which allows the manual selection of a gear ratio in advance of its actual use, the gear being automatically engaged by the actuation of the clutch.

pres·ence (prez′əns) *n.* **1** The state or fact of being present: opposed to *absence.* **2** Situation face to face; close approach or vicinity within view or access. **3** Something invisible but near and sensible, as a spiritual being. **4** Personal appearance; bearing. **5** Personal qualities collectively; self; personality: used also absolutely of a sovereign. **6** *Obs.* A distinguished assembly, as before a prince or exalted personage. **7** Formerly, the room or apartment in which a high dignitary or ruler received assemblies: also **presence chamber.** [<OF <L *praesentia,* orig. neut. pl of *praesens, -entis,* ppr. of *praeesse.* See PRESENT.]

presence of mind Full command of one's faculties; coolness, alertness, and readiness of resource in a situation of sudden danger, embarrassment, etc.

pres·ent[1] (prez′ənt) *adj.* **1** Being in a place or company referred to or contemplated; being at hand: opposed to *absent.* **2** Now going on; current; not past or future. **3** Actually in mind. **4** Immediately impending or actually coming on; not delayed; instant. **5** *Gram.* Relating to or signifying what is going on at the time being: the *present* tense, *present* participle. **6** Ready at hand; prompt in emergency: a *present* wit, a *present* aid. See synonyms under IMMEDIATE. **—***n.* **1** Present time; now; the time being. **2** *Gram.* The present tense; also, a verbal form denoting it. **3** A present matter or affair; a question under consideration. **4** *pl. Law* Present writings: term for the document in which the word occurs: Know all men by these *presents.* **—at present** Now. **—for the present** For the time being. [<OF <L *praesens, -entis* being in front of or at hand, ppr. of *praeesse* <*prae-* before + *esse* be]

pre·sent[2] (pri·zent′) *v.t.* **1** To bring into the presence or acquaintance of another; introduce, especially to a superior: The ambassador was *presented* to the king. **2** To exhibit to view or notice; display. **3** To suggest to the mind: This *presents* a problem. **4** To put forward for consideration or action; submit, as a petition. **5** To make a gift or presentation of or to, usually formally. **6** *Archaic* To represent on the stage; act. **7** *Law* **a** To offer, as a charge, for judicial action or inquiry. **b** To bring a charge or indictment against.
—pres·ent (prez′ənt) *n.* That which is presented or given; a gift; donation. [<OF *presenter* <L *praesentare* set before <*praesens, -entis* present. See PRESENT[1].] **—pre·sent′er** *n.*

pre·sent·a·ble (pri·zen′tə·bəl) *adj.* **1** Fit to be presented; in suitable condition or attire for company. **2** Capable of being offered, exhibited, or bestowed. **—pre·sent′a·bil′i·ty, pre·sent′a·ble·ness** *n.* **—pre·sent′a·bly** *adv.*

present arms A command requiring a soldier to salute by holding his gun or other weapon vertically, in front of and close to his body. Correct position for the gun is muzzle up and trigger facing forward.

pres·en·ta·tion (prez′ən·tā′shən, prē′zən-) *n.* **1** The act of presenting or proffering for acceptance, approval, etc.; especially, the formal offering of a complimentary gift. **2** *Rare* That which is bestowed; a present. **3** The act of introducing or bringing to notice; formal introduction, especially to a superior: *presentation* at court. **4** *Eccl.* The nomination of a clergyman to a living; also, the right of such nomination. **5** The manner of bringing into view, as a play, thought, or case; way of putting; exhibition; representation; also, that which is represented. **6** The fact or process of being present in consciousness; also, the object of consciousness, without added reference. **7** *Med.* The position of the fetus at birth: designated by the part that is first presented to the touch at the mouth of the womb: breech *presentation,* etc. **8** The condition of being placed in a certain position or direction, with regard to something else, or to an observer. **9** Presentment; the offering of a negotiable instrument for payment. [<OF *presentacion* <L *praesentatio, -onis*]

pres·en·ta·tion·al (prez′ən·tā′shən·əl, prē′zən-) *adj.* Relating to or composed of presentations.

pres·en·ta·tion·al·ism (prez′ən·tā′shən·əl·iz′·əm, prē′zən-) *n. Philos.* The doctrine that man has an immediate perception of all the elemental forms of entity, as space, time, substance, and power; natural realism. Also **pres′·en·ta′tion·ism.** **—pres′en·ta′tion·ist** *adj.* & *n.*

pre·sen·ta·tive (pri·zen′tə·tiv) *adv.* **1** Having to do with the mental awareness or knowledge of an activity, power, or object: distinguished from *representative*: a *presentative* judgment. **2** Having the right to present to a benefice; also, admitting of the presentation of a clergyman. **—pre·sen′ta·tive·ness** *n.*

pres·ent–day (prez′ənt·dā′) *adj.* Of the present time; current.

pres·en·tee (prez′ən·tē′) *n.* **1** One who is presented, as to a benefice or at court. **2** The recipient of a gift.

pre·sen·ti·ment (pri·zen′tə·mənt) *n.* A prophetic sense of something to come; a foreboding. See synonyms under ANTICIPATION. [<MF <L *praesentire* perceive beforehand <*prae-* before + *sentire* feel]

pre·sen·tive (pri·zen′tiv) *adj.* Conveying or embodying (as nouns, adjectives, and most verbs) a distinct and complete conception, whether of an object, act, or quality: distinguished from *symbolic.* **—***n.* A presentive word. **—pre·sen′tive·ly** *adv.* **—pre·sen′tive·ness** *n.*

pres·ent·ly (prez′ənt·lē) *adv.* **1** After a little time; shortly. **2** *Archaic & Dial.* At once; immediately. See synonyms under IMMEDIATELY. ◆ *Presently* in the sense "at once" has not been in use in literary English since the 17th century, though it has persisted in dialectal use in both England and the United States.

pre·sent·ment (pri·zent′mənt) *n.* **1** The act of presenting; also, the state or manner of being presented; presentation. **2** That which is represented or exhibited; a representation or picture; semblance. **3** *Law* A report made by a grand jury, concerning some wrongdoing, and presented to the court; also, the finding and setting forth of charges in an indictment by a grand jury; an indictment. **4** The presentation of a negotiable instrument for payment. **5** *Philos.* The mental images of a perception or idea.

present participle See under PARTICIPLE.

present perfect *Gram.* The verb tense expressing an action completed by the present time: By now he *has finished* the task.

present tense The tense marking present time: I *go, do go, am going.*

pre·ser·va·tive (pri·zûr′və·tiv) *adj.* Serving or tending to preserve. **—***n.* That which serves or tends to preserve; a substance that preserves; a safeguard. [<F *préservatif, -ive* <Med. L *praeservativus*]

pre·serve (pri·zûrv′) *v.* **·served, ·serv·ing** *v.t.* **1** To keep in safety; protect from destruction, loss, death, or detriment; guard: May the gods *preserve* you. **2** To keep intact or unimpaired; maintain: to *preserve* appearances. **3** To prepare (food) for future consumption, as by boiling with sugar or salting. **4** To keep from decomposition or change, as by chemical treatment: to *preserve* a specimen in alcohol. **5** To keep for one's private hunting or fishing: to *preserve* foxes; to *preserve* a wood. **—***v.i.* **6** To make preserves, as of fruit. **7** To maintain a game preserve. **—***n.* **1** *Usually pl.* Fruit which has been cooked, usually with sugar, to prevent its fermenting. **2** Something preserved or which preserves. **3** A place set apart for one's own private use, or in which game or fish are protected for purposes of sport. [<OF *preserver* <LL *praeservare* <L *prae-* before + *servare* keep] **—pre·serv′a·bil′i·ty** *n.* **—pre·serv′a·ble** *adj.* **—pres·er·va·tion** (prez′ər·vā′shən) *n.* **—pre·serv′er** *n.*
Synonyms (verb): conserve, defend, guard, keep, maintain, protect, save, secure, sustain, uphold. See KEEP, RETAIN. *Antonyms*: abandon, lavish, lose, neglect, scatter, spend, spoil, waste.

pre–shrunk (prē′shrungk′) *adj.* Shrunk during manufacture to minimize later shrinkage: *pre-shrunk* cotton.

pre·side (pri·zīd′) *v.i.* **·sid·ed, ·sid·ing** **1** To sit in authority, as over a meeting; be in charge of an assembly, government, etc.; act as chairman or president. **2** To exercise direction or control. [<F *présider* <L *praesidere* sit in front of, protect, guard <*prae-* before + *sedere* sit] **—pre·sid′er** *n.*

pres·i·den·cy (prez′ə·dən·sē) *n.* *pl.* **·cies** **1** The office, function, or term of office of a president. **2** *Often cap.* The office of president of the United States. **3** *Often cap.* Formerly, any of the three original provinces of British India: Bengal, Madras, and Bombay. **4** *Brit.* An administrative subdivision. **5** In the Mormon

president

Church, a local administrative council of three men; also, the highest governing body of the church (**First Presidency**), consisting of the president and his two counselors. [< Med. L *praesidentia*]

pres·i·dent (prez′ə-dənt) *n.* **1** One who is chosen to preside over an organized body; specifically, the chief executive of a republic, as of the United States. See list of presidents of the United States. **2** The chairman of the meetings and chief executive officer of a department of the government, a corporation, society, board of trade or trustees, or similar body. **3** The chief officer of a college or university. **4** The chairman of a meeting conducted under parliamentary rules. [< F *président* < L *praesidens, -entis*, ppr. of *praesidere*. See PRESIDE.] — **pres·i·den·tial** (prez′ə-den′shəl) *adj.* — **pres′i·dent·ship**′ *n.*

THE PRESIDENTS OF THE UNITED STATES

Number — Name	Birthplace—Inaugurated: year	Age
1 George Washington	Westmoreland Co., Va. 1789	57
2 John Adams	Quincy, Mass. 1797	61
3 Thomas Jefferson	Shadwell, Va. 1801	57
4 James Madison	Port Conway, Va. 1809	57
5 James Monroe	Westmoreland Co., Va. 1817	58
6 John Quincy Adams	Quincy, Mass. 1825	57
7 Andrew Jackson	Union Co., N.C. 1829	61
8 Martin Van Buren	Kinderhook, N.Y. 1837	54
9 William H. Harrison	Berkeley, Va. 1841	68
10 John Tyler	Greenway, Va. 1841	51
11 James K. Polk	Little Sugar Creek, N.C. 1845	49
12 Zachary Taylor	Orange Co., Va. 1849	64
13 Millard Fillmore	Summerhill, N.Y. 1850	50
14 Franklin Pierce	Hillsboro, N.H. 1853	48
15 James Buchanan	Cove Gap, Pa. 1857	65
16 Abraham Lincoln	Hardin Co., Ky. 1861	52
17 Andrew Johnson	Raleigh, N.C. 1865	56
18 Ulysses S. Grant	Point Pleasant, O. 1869	46
19 Rutherford B. Hayes	Delaware, O. 1877	54
20 James A. Garfield	Cuyahoga Co., O. 1881	49
21 Chester A. Arthur	Fairfield, Vt. 1881	50
22 Grover Cleveland	Caldwell, N.J. 1885	47
23 Benjamin Harrison	North Bend, O. 1889	55
24 Grover Cleveland	Caldwell, N. J. 1893	55
25 William McKinley	Niles, O. 1897	54
26 Theodore Roosevelt	New York, N.Y. 1901	42
27 William H. Taft	Cincinnati, O. 1909	51
28 Woodrow Wilson	Staunton, Va. 1913	56
29 Warren G. Harding	Corsica, O. 1921	55
30 Calvin Coolidge	Plymouth, Vt. 1923	51
31 Herbert C. Hoover	West Branch, Ia. 1929	55
32 Franklin D. Roosevelt	Hyde Park, N.Y. 1933	51
33 Harry S Truman	Lamar, Mo. 1945	60
34 Dwight D. Eisenhower	Denison, Tex. 1953	62
35 John F. Kennedy	Brookline, Mass. 1961	43

pre·sid·i·al (pri·sid′ē-əl) *adj.* Of or having a garrison or a garrisoned post.

[< F *présidial* < LL *praesidialis* < *praesidium* a garrison, fort]

pre·sid·i·o (pri·sid′ē-ō) *n.* *pl.* **·sid·i·os 1** A garrisoned post; fort; fortified settlement. **2** A Spanish penal settlement in a foreign country. — **the Presidio** A U.S. military reservation in San Francisco. [< Am. Sp.]

pre·sid·i·um (pri·sid′ē-əm) *n.* Any of several executive committees in the U.S.S.R. serving as the permanent organ of a larger governmental body. [< L *praesidium*]

Pre·sid·i·um (pri·sid′ē-əm) *n.* **1** The administrative cabinet of the Soviet Union, headed by the premier, which exercises executive powers between sessions of the Supreme Soviet. **2** The supreme policy-making committee of the Communist party of the Soviet Union, headed by the party secretary. See POLITBURO

pre·sig·ni·fy (prē·sig′nə·fī) *v.t.* **·fied, ·fy·ing** To signify or give token of in advance; presage; foreshadow.

press[1] (pres) *v.t.* **1** To act upon by weight or pressure: to *press* a button. **2** To compress so as to extract the juice: to *press* grapes. **3** To extract by pressure, as juice. **4** To exert pressure upon so as to smooth, shape, make compact, etc. **5** To smooth or shape by heat and pressure, as clothes; iron. **6** To embrace closely; hug. **7** To force or impel; drive. **8** To distress or harass; place in difficulty: I am *pressed* for time. **9** To urge persistently; importune; entreat: They *pressed* me for an answer. **10** To advocate persistently; insist on; emphasize. **11** To put forward insistently: to *press* a gift on a friend. **12** To urge onward; hasten. **13** *Obs.* To crowd. — *v.i.* **14** To exert pressure; bear heavily. **15** To advance forcibly or with speed: *Press* on! **16** To press clothes, etc. **17** To crowd; cram. **18** To be urgent or importunate. See synonyms under IMPRESS[1], JAM, PLEAD, PUSH. — *n.* **1** A dense throng. **2** The act of crowding together or of straining forward. **3** Hurry or pressure of affairs; urgency: the *press* of business. **4** A movable upright closet or case in which clothes, books, etc., are kept: a linen *press*. **5** An apparatus or machine by which pressure is applied, as for making wine, compressing bulky substances for packing, etc.; a printing press. **6** Newspapers or periodical literature collectively, or the body of persons collectively, as editors, reporters, etc., engaged upon such publications; also, printed literature in the abstract. **7** The art, process, or business of printing. **8** The place of business in which a printing press is set up and where printing is carried on: the Clarendon *Press*; to go to *press*. See synonyms under THRONG. [< OF *presser* < L *pressare*, freq. of *premere* (pp. *pressus*) press]

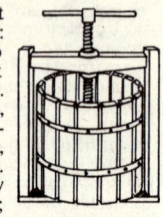
PRESS
Cider, wine, and fruit press.

press[2] (pres) *v.t.* **1** To force into military or naval service; impress. **2** To put to use in a manner not intended or desired. — *n.* A commission to impress men into the public service; also, the impressment of men. [< *obs. prest* engage for military service by payment of earnest money < OF *prester* lend < L *praestare* guarantee, furnish money for < *prae-* before + *stare* stand; influenced in form and meaning by *press*[1]]

press agent A person employed, as in a theatrical company, to advance the interests of his client by advertisements and other notices; a publicity agent for any person or business.

press·board (pres′bôrd′, -bōrd′) *n.* **1** A wooden board placed between sheets in a standing press. **2** An ironing board. — **imitation pressboard** Millboard. — **electrical pressboard** Fullerboard.

Press·burg (pres′bŏŏrk) The German name for BRATISLAVA.

press conference An interview granted by a celebrity, government official, etc., to a number of journalists at the same time.

press·er (pres′ər) *n.* **1** One who or that which presses. **2** *Mech.* Any machine or apparatus exerting pressure, as by a spring; a presser foot. **3** One who cleans and presses clothes. **4** One who operates a press: a cotton *presser*.

presser foot A footpiece in a sewing machine to hold the fabric down to the feed plate.

press·gang (pres′gang′) *n.* A detachment of men detailed to press men into naval or military service. Also **press gang**.

press·ing (pres′ing) *adj.* **1** Demanding immediate attention; urgent; important. **2** Importunate. — **press′ing·ly** *adv.*

press·man[1] (pres′mən) *n.* *pl.* **·men** (-mən) **1** A man who has charge of a press, as a printing press. **2** A man who presses clothes. **3** *Brit.* A member of the press; journalist.

press·man[2] (pres′mən) *n.* *pl.* **·men** (-mən) *Obs.* A member of a pressgang.

press·mark (pres′märk′) *n.* **1** A mark in a book to point out its particular place in a book press or bookcase of a library. **2** A mark, as a number or letter on the margin of a newspaper, showing on which press it was printed.

press money The king's shilling. See under SHILLING.

press of canvas or **sail** *Naut.* The maximum spread of sail that can be carried with safety under wind pressure.

press·or (pres′ər) *adj. Physiol.* Increasing the functional activities of an organ: a *pressor* nerve, the stimulating of which raises the arterial blood pressure: opposed to *depressor*. [< PRESS[1]]

press·pahn (pres′pän) *n.* Fullerboard. [< G]

press proof 1 The last proof taken before printing. **2** A proof taken on a press.

press release A bulletin, prepared by a press agent, public relations department, or other official representative, announcing an event, development in a business, newsworthy decision, etc.

press·room (pres′rŏŏm′, -rŏŏm′) *n.* A room containing the presses of a printing concern.

pres·sure (presh′ər) *n.* **1** The act of pressing, or the state of being pressed. **2** *Physics* Any force which acts against an opposing force; a thrust, stress, or strain between opposed masses, uniformly distributed over the surfaces in contact: steam *pressure*; the *pressure* of gas in a confined space. **3** Electromotive force. **4** An impelling or constraining moral force; compulsory motive: bringing *pressure* to bear. **5** Exigent demand on one's time or strength; urgency: the *pressure* of business. **6** The oppressive influence or depressing effect of something hard to bear; weight, as of grief or trouble; onerousness: *pressure* of taxation; *pressure* of calamity. **7** A printed character; stamp; an impression. — **fluid pressure** Pressure of a fluid or resembling that of a fluid, being invariable and uniform in all directions. — *v.t.* **·sured, ·sur·ing** *Colloq.* To compel, as by forceful persuasion or influence. [< OF < L *pressura* < *pressus*, pp. of *premere* press]

pressure cabin *Aeron.* An enclosed compartment in an airplane, supplied with air maintained at or near sea-level pressure to provide sufficient oxygen for crew and passengers at high altitudes.

pressure cooker An airtight receptacle for the cooking of food at high temperature under pressure; an autoclave.

pressure gage An instrument for measuring the pressure of a gas or liquid, and for indicating it by a pointer on a graduated dial; a manometer. Also **pressure gauge**.

pressure gradient *Meteorol.* The decrease in barometric pressure per unit of horizontal distance along the course in which the pressure decreases most rapidly.

pressure group An organized minority group which seeks, through propaganda and lobbying, to influence legislators and public opinion in behalf of its own special interests, or to defeat restrictive legislation.

pres·sur·ize (presh′ə·rīz) *v.t.* **·ized, ·iz·ing 1** To subject to high pressure. **2** *Aeron.* To maintain normal atmospheric pressure in (the cabin or cockpit of) an airplane at high altitudes. — **pres′sur·i·za′tion** *n.*

press·work (pres′wûrk′) *n.* **1** The operating, adjustment, or management of a printing press. **2** The work done by the press. **3** Cabinetwork made up of cross veneers glued together and pressed while hot.

prest[1] (prest) *adj. Obs.* Ready; prepared at hand; daring; also, tidy; neat. [< OF < L *praesto*, dative of *praestus* ready, at hand]

prest[2] (prest) *n.* **1** An advance or loan; also,

ready money. 2 Press money: also **prest money.** [<OF <*prester* lend <L *praestare.* See PRESS².]

Pres·teigne (pres·tēn′) The county town of Radnorshire, Wales.

pres·ter (pres′tər) *n. Obs.* A priest; presbyter. [<OF *prestre* <LL *presbyter.* See PRIEST.]

Pres·ter John (pres′tər jon′) A legendary medieval Christian priest, traditionally a powerful and wealthy king of central Asia or Abyssinia.

pres·ti·dig·i·ta·tion (pres′tə·dij′ə·tā′shən) *n.* The practice of sleight of hand; jugglery; legerdemain. [<F <*preste* (<Ital. *presto* <LL *praestus*) nimble + L *digitus* finger] **—pres′·ti·dig′i·ta′tor** *n.*

pres·tige (pres·tēzh′, pres′tij) *n.* Authority or importance based on past achievements or reputation; ascendency based on recognition of power; renown. [<F <L *praestigium* illusion, juggler's trick, spell <*praestringere* bind fast <*prae-* before + *stringere* bind]

pres·tis·si·mo (pres·tis′i·mō, *Ital.* pres·tēs′sē·mō) *adj. & adv. Music* As fast as possible; in very quick time. [<Ital., superlative of *presto.* See PRESTO.]

pres·to (pres′tō) *adv. & adj.* 1 *Music* In fast time. 2 At once; speedily. **—***n. Music* A movement, passage, or phrase performed in fast tempo. [<Ital. <L *praesto.* See PREST¹.]

Pres·ton (pres′tən) A county borough and river port in central Lancashire, England.

Pres·tone (pres′tōn) *n.* Ethylene glycol, used as an anti-freeze mixture: a trade name.

Pres·ton·pans (pres′tən·panz′) A burgh in East Lothian, Scotland, east of Edinburgh on the Firth of Forth; scene of a Scottish victory over the English, 1745.

pre·sum·able (pri·zo͞o′mə·bəl) *adj.* That may be assumed or presumed; reasonable. See synonyms under APPARENT, LIKELY, PROBABLE. **—pre·sum′a·bly** *adv.*

pre·sume (pri·zo͞om′) *v.* **·sumed, ·sum·ing** *v.t.* 1 To take upon oneself without warrant or permission; dare; venture: usually with the infinitive: Do you *presume* to address me? 2 To take for granted; assume to be true until disproved: I *presume* you are right. 3 To indicate the probability of; seem to prove: The receipt for this month *presumes* preceding payments. **—***v.i.* 4 To act or proceed presumptuously or overconfidently. 5 To make excessive demands; rely too heavily: with *on* or *upon*: He *presumes* on my good nature. See synonyms under ASSUME. [<OF *presumer* <L *praesumere* take first <*prae-* before + *sumere* take] **—pre·sum·ed·ly** (pri·zo͞o′mid·lē) *adv.* **—pre·sum′er** *n.*

pre·sump·tion (pri·zump′shən) *n.* 1 Blind or overweening confidence or self-assertion. 2 A passing beyond the ordinary bounds of good breeding, respect, or reverence; offensively forward or arrogant conduct or expression; effrontery. 3 The act of forming a judgment on probable grounds, awaiting further evidence; also, the judgment so formed, or a ground or reason for it. 4 That which may be logically or legally assumed to be true until disproved: the *presumption* of guilt. 5 *Law* The inference of a fact on proof of circumstances that usually or necessarily attend such a fact. See synonyms under ARROGANCE, ASSURANCE, IMPUDENCE, PROBABILITY, TEMERITY. [<OF *presomption* <L *praesumtio, -onis* <*praesumptus,* pp. of *praesumere.* See PRESUME.]

pre·sump·tive (pri·zump′tiv) *adj.* Creating or resting upon a presumption; affording reasonable grounds for belief. [<F *présomptif*] **—pre·sump′tive·ly** *adv.*

presumptive heir See HEIR PRESUMPTIVE.

pre·sump·tu·ous (pri·zump′cho͞o·əs) *adj.* 1 Unduly confident or bold; audacious; arrogant; insolent. 2 Exhibiting, characterized by, or founded on presumption; presuming unduly, as upon success or the forbearance of others; foolhardy. [<OF *presumptuoux* <LL *praesumptiosus*] **—pre·sump′tu·ous·ly** *adv.* **—pre·sump′tu·ous·ness** *n.*

pre·sup·pose (prē′sə·pōz′) *v.t.* **·posed, ·pos·ing** 1 To imply or involve as a necessary antecedent condition. 2 To take for granted; assume to start with. [<F *présupposer*] **—pre·sup·po·si·tion** (prē′sup·ə·zish′ən) *n.*

pre·tend (pri·tend′) *v.t.* 1 To assume or display a false appearance of; feign: to *pretend* friendship for an enemy. 2 To claim or assert falsely: He *pretended* that there was gold on his property. 3 To feign in play; make believe. **—***v.i.* 4 To make believe, as in play or for the purpose of deception: She is only *pretending* when she says that. 5 To put forward a claim: with *to.* [<OF *pretendre* <L *praetendere* spread out before <*prae-* before + *tendere* spread out]

Synonyms: affect, assume, counterfeit, feign, profess, sham, simulate. See ASSUME, MASK¹.

pre·tend·ed (pri·ten′did) *adj.* Alleged; asserted; professed. **—pre·tend′ed·ly** *adv.*

pre·tend·er (pri·ten′dər) *n.* 1 One who advances a claim or title; a claimant; specifically, a claimant of a throne who is an heir of a deposed dynasty. 2 In English history, the son and grandson of James II, the former being known in literature as the **Pretender** or the **Old Pretender,** and the latter as the **Young Pretender.** 3 A hypocrite. See synonyms under HYPOCRITE.

pre·tense (pri·tens′, prē′tens) *n.* 1 That which is pretended; a pretext; a ruse or wile. 2 The act or state of pretending, or of being a pretender or claimant; specifically, a false assumption of a character or condition; hence, affectation; ostentation. 3 Any act of simulation. 4 A right or title asserted. 5 An intention, aim, or effort. Also *Brit.* **pre·tence′.** [<AF *pretensse* <Med. L *praetensus,* alter. of L *praetentus,* pp. of *praetendere.* See PRETEND.]

Synonyms: affectation, air, assumption, cloak, color, disguise, dissimulation, excuse, mask, pretension, pretext, ruse, seeming, semblance, show, simulation. A *pretense,* in the unfavorable and usual sense, is something advanced or displayed for the purpose of concealing the reality. A person makes a *pretense* of something for the credit or advantage to be gained by it; he makes what is allowed or approved a *pretext* for doing what would be opposed or condemned; a tricky schoolboy makes a *pretense* of doing an errand which he does not do, or he makes the actual doing of an errand a *pretext* for playing truant. A *ruse* is something employed to blind or deceive so as to mask an ulterior design, and enable a person to gain some end that he would not be allowed to approach directly. A *pretension* is a claim that is or may be contested; the word is now commonly used in an unfavorable sense. See DISGUISE, HYPOCRISY. *Antonyms*: actuality, candor, fact, guilelessness, honesty, ingenuousness, openness, reality, simplicity, sincerity, truth.

pre·ten·sion (pri·ten′shən) *n.* 1 A claim put forward, whether true or false. 2 Affectation; display. 3 A bold or presumptuous assertion. See synonyms under PRETENSE.

pre·ten·tious (pri·ten′shəs) *adj.* Characterized by pretension; making an ambitious outward show; ostentatious. [<F *prétentieux*] **—pre·ten′tious·ly** *adv.* **—pre·ten′tious·ness** *n.*

preter- *prefix* Beyond; past; more than: *preternatural.* [<L *praeter* beyond <*prae* before]

pret·er·hu·man (prē′tər·hyo͞o′mən) *adj.* Beyond what is human.

pret·er·it (pret′ər·it) *adj.* 1 *Gram.* Signifying past time or completed past action. 2 *Rare* Belonging to the past; bygone. **—***n. Gram.* The tense that expresses absolute past time; the past tense. Also **pret′er·ite.** [<OF *preterit* <L *praeteritus* past, pp. of *praeterire* go past <*praeter-* beyond + *ire* go]

pret·er·i·tion (pret′ə·rish′ən) *n.* 1 The act of passing over or omitting. 2 The omission or passing by of a natural heir without mention by a testator in his will. 3 In the doctrine of predestination, the passing-by of the non-elect. [<LL *praeteritio, -onis* <L *praeteritus.* See PRETERIT.]

pre·ter·i·tive (pri·ter′ə·tiv) *adj. Gram.* 1 Used to indicate past actions or states: said of tenses of verbs. 2 Employed only in a past tense or past tenses: said of certain verbs.

pre·ter·mit (prē′tər·mit′) *v.t.* **·mit·ted, ·mit·ting** 1 To fail or cease to do; neglect; omit. 2 To let pass without noticing; overlook; disregard. [<L *praetermittere* let go by <*praeter-* beyond + *mittere* send] **—pre′ter·mis′sion** (-mish′ən) *n.*

pre·ter·nat·u·ral (prē′tər·nach′ər·əl) *adj.* Diverging from or exceeding the common order of nature; inexplicable in terms of the known facts and laws of science, but not outside the universal natural order: distinguished from *supernatural.* See synonyms under SUPERNATURAL. **—pre′ter·nat′u·ral·ism** *n.* **—pre′ter·nat′·u·ral·ly** *adv.*

pre·text (prē′tekst) *n.* 1 A fictitious reason or motive advanced to conceal a real one. 2 A specious excuse or explanation. See synonyms under PRETENSE. [<F *prétexte* <L *praetextus.* See PRAETEXTA.]

pre·tor (prē′tər) *n.* An ancient Roman city magistrate having charge of the administration of justice: also spelled *praetor.* [<L *praetor* <*praeire* go before. See PRETERITE.] **—pre·to·ri·al** (pri·tôr′ē·əl, -tō′rē-) *adj.* **—pre′·tor·ship** *n.*

Pre·to·ri·a (pri·tôr′ē·ə, -tō′rē·ə) Capital of Transvaal province and administrative capital of the Union of South Africa, in south central Transvaal.

pre·to·ri·an (pri·tôr′ē·ən, -tō′rē-) *adj.* Of or pertaining to a pretor; pretorial. **—***n.* A pretor or ex-pretor. Also spelled *praetorian.*

Pre·to·ri·an (pri·tôr′ē·ən, -tō′rē-) *adj.* Denoting the imperial bodyguard of the Caesars. **—***n.* A soldier of the imperial bodyguard of the Caesars. Also spelled *Praetorian.*

Pretorian Guard 1 The bodyguard of the Roman emperors, organized by Augustus to take the place of the old *cohors praetoria,* or bodyguard of the general, and disbanded by Constantine the Great. 2 A member of the Pretorian Guard.

Pre·to·ri·us (prə·to͞o′rē·o͞os), **Andries Wilhelmus,** 1799–1853, and his son **Marthinus,** 1819–1901, South African Dutch colonizers.

pret·ti·fy (prit′i·fī) *v.t.* **·fied, ·fy·ing** To make pretty; embellish overmuch. [<PRETTY + -FY]

pret·ty (prit′ē) *adj.* **·ti·er, ·ti·est** 1 Characterized by delicacy, gracefulness, or proportion rather than by striking beauty; pleasing; attractive. 2 Decent; good; sufficient: often used ironically as a term of deprecation: A *pretty* mess you've made of it! 3 *Colloq.* Considerable; rather large in size or degree. 4 Sweet; precious: a diminutive of endearment: *pretty* girl. 5 Characterized by effeminacy; affected; foppish. 6 *Scot.* Bold; vigorous; athletic. 7 *Obs.* Strong; able; cunning. See synonyms under BEAUTIFUL. **—***adv.* 1 Moderately; somewhat; to a fair extent: He looked *pretty* well. 2 Very; quite: He's grown *pretty* fast. 3 *Dial.* Prettily; finely. **—sitting pretty** *Colloq.* In good circumstances. **—***n.* A pretty thing or person. [OE *prættig* tricky, cunning] **—pret′ti·ly** *adv.* **—pret′ti·ness** *n.*

pret·zel (pret′səl) *n.* A glazed salted biscuit baked in the form of a loose knot. [<G *brezel*]

Preus·sen (proi′sən) The German name for PRUSSIA.

pre·vail (pri·vāl′) *v.i.* 1 To gain mastery; be victorious; triumph: with *over* or *against.* 2 To be effective or efficacious; succeed. 3 To use persuasion or influence successfully: with *on, upon,* or *with.* 4 To be or become a predominant feature or quality; be prevalent. 5 To have general or wide-spread use or acceptance; be in force. See synonyms under SUCCEED. [<OF *prevaloir* <L *prevalere* <*prae-* before + *valere* be strong]

pre·vail·ing (pri·vā′ling) *adj.* 1 Current; prevalent. 2 Having effective power or influence; efficacious. See synonyms under PREDOMINANT, USUAL. **—pre·vail′ing·ly** *adv.* **—pre·vail′ing·ness** *n.*

prev·a·lent (prev′ə·lənt) *adj.* 1 Predominant. 2 Of wide extent or frequent occurrence; common. 3 Efficacious; effective. See synonyms under GENERAL, PREDOMINANT, USUAL. [<L *praevalens, -entis,* ppr. of *praevalere* PREVAIL] **—prev′a·lence** *n.* **—prev′a·lent·ly** *adv.*

pre·var·i·cate (pri·var′ə·kāt) *v.i.* **·cat·ed, ·cat·ing** To speak or act in a deceptive, ambiguous, or evasive manner; quibble; lie. [<L *praevaricatus,* pp. of *praevaricare,* lit., walk crookedly <*prae-* before + *varicare* straddle <*varicus* straddling <*varus* crooked] **—pre·var′i·ca′·tor** *n.*

pre·var·i·ca·tion (pri·var′ə·kā′shən) *n.* 1 The act of prevaricating. 2 Misleading or equivocal statement. 3 A trick. See synonyms under DECEPTION, SOPHISTRY.

pré·ve·nance (prā·və·näṅs′) *n. French* Prevenience.

pre·ven·ience (pri·vēn′yəns) *n.* The act or state of going before; anticipation.

pre·ven·ient (pri·vēn′yənt) *adj.* 1 Preceding or preventing. 2 Anticipatory; expectant. [<L *praeveniens, -entis,* ppr. of *praevenire.* See PREVENT.]

pre·vent (pri·vent′) *v.t.* 1 To keep from happening, as by previous measures or preparations; preclude; thwart. 2 To keep from doing something; forestall; hinder. 3 *Obs.* To anticipate; precede. [<L *praeventus;* pp. of *praevenire* precede, come before, anticipate <*prae-* before + *venire* come] **—pre·vent′a·ble** or **pre·vent′i·ble** *adj.* **—pre·vent′a·bil′i·ty** or **pre·vent′i·bil′i·ty** *n.* **—pre·vent′er** *n.*

Synonyms: anticipate, forestall. The original sense of *prevent,* to go or come before, act in advance of, now practically obsolete, was still in good use when the authorized version of the Bible was made, as appears in such passages as "Thou *preventest* him with the blessings of goodness" (that is, by sending the blessings before the desire is formulated or expressed), *Ps.* xxi 3. *Anticipate* is now the only single word usable in this sense; to *forestall* is to take or act in advance in one's own behalf and to the prejudice or hindrance of another. But to *anticipate* is very frequently used in the favorable sense; as, his thoughtful kindness *anticipated* my wish (that is, met the wish before it was expressed); or one *anticipates* a payment (by making it before the time). For the present use of *prevent,* see synonyms for HINDER[1], PRECLUDE, PROHIBIT.

pre·ven·tion (pri·ven′shən) *n.* 1 The act of preventing. 2 A hindrance; obstruction. 3 A preventive.

pre·ven·tive (pri·ven′tiv) *adj.* Intended or serving to ward off harm, diseases, etc.: *preventive* medicine. **—***n.* That which prevents or hinders, as a medicine to ward off disease; a precautionary measure. Also **pre·vent·a·tive** (pri·ven′tə·tiv). **—pre·ven′tive·ly** *adv.* **—pre·ven′tive·ness** *n.*

pre·verb (prē′vûrb′) *n.* A verbal prefix, as *be-* in *behave.*

pre·ver·nal (pri·vûr′nəl) *adj.* 1 Prior to spring. 2 *Bot.* Flowering in the early spring, as certain trees and plants.

pre·view (prē′vyōō′) *n.* 1 An advance showing, as of a motion picture, a fashion show, etc., to invited guests before it is presented publicly. 2 In motion pictures, the showing of scenes or parts of scenes to advertise a coming picture.

pre·vi·ous (prē′vē·əs) *adj.* 1 Being or taking place before something else in time or order; antecedent; prior to. 2 *Colloq.* Acting, occurring, or speaking too soon; premature. [<L *praevius* going before <*prae-* before + *via* way, road] **—pre′vi·ous·ly** *adv.* **—pre′vi·ous·ness** *n.*

Synonyms: antecedent, anterior, earlier, foregoing, former, precedent, preceding, preliminary, prior. *Antecedent* may denote simple priority in time, implying no direct connection between that which goes before and that which follows; as, the striking of one clock may be always *antecedent* to the striking of another with no causal connection between them. *Antecedent* and *previous* may refer to that which goes or happens at any distance in advance, *preceding* is limited to that which is immediately or next before; an *antecedent* event may have happened at any time before; the *preceding* transaction is the one completed just before the one with which it is compared; a *previous* statement or chapter may be in any part of the book that has gone before; the *preceding* statement or chapter comes next before without an interval. *Foregoing* is used only of that which is spoken or written; as, the *foregoing* statements. *Anterior,* while it can be used of time, is coming to be employed chiefly with reference to place; as, the *anterior* lobes of the brain. *Prior* bears exclusive reference to time, and commonly where that which is first in time is first also in right; as, a *prior* demand. *Former* is used of time, or of position in written or printed matter, not of space in general. We say *former* times, a *former* chapter, etc. *Former* has a close relation, or sharp contrast, with something following; the *former* always implies the latter, even when not fully expressed. Compare ANTECEDENT. *Antonyms:* after, concluding, consequent, following, hind, hinder, hindmost, later, latter, posterior, subsequent, succeeding.

Previous Examination See LITTLE GO.

previous question In parliamentary practice, a motion to avoid or secure a vote at once. In the British Parliament, the motion is put to prevent a speedy vote on a measure. In the United States House of Representatives it is used to end debate and secure an immediate vote. Compare CLOSURE.

previous to 1 Antecedent to; being before. 2 Before: *previous* being loosely used for *previously.*

pre·vise (prē·vīz′) *v.t.* **·vised, ·vis·ing** 1 To see beforehand; foresee. 2 To notify beforehand; forewarn. [<L *praevisus,* pp. of *praevidere* foresee <*prae-* before + *videre* see]

pre·vi·sion (prē·vizh′ən) *n.* 1 The act or power of foreseeing; prescience; foresight. 2 A prophetic or anticipatory vision. See synonyms under ANTICIPATION. [<F *prévision*]

pre·vo·ca·tion·al (prē′vō·kā′shən·əl) *adj.* Of or pertaining to the training given or requisite in schools of a lower grade than the vocational schools.

Pré·vost (prā·vō′), **Marcel,** 1862–1941, French novelist.

Pré·vost d'Ex·iles (prā·vō′ deg·zēl′), **Antoine François,** 1679–1763, French novelist: known as *Abbé Prévost.*

pre·vue (prē′vyōō) *n.* A preview. [<F *prévue,* fem. pp. of *prévoir* foresee]

pre·war (prē′wôr′) *adj.* Of or pertaining to a condition, arrangement, time, etc., before a war.

prex·y (prek′sē) *n. Slang* A college president. Also **prex.**

prey (prā) *n.* 1 Any animal seized by another for food. 2 Booty; plunder; pillage. 3 Anything made the victim of that which is hostile or evil. 4 The act of preying; depredation; robbery. See synonyms under PLUNDER. **—***v.i.* 1 To seek or take prey for food: Cats *prey* on birds. 2 To take booty; plunder. 3 To make a victim of someone, as by cheating. 4 To exert a wearing or harmful influence: His losses *preyed* on his mind. ◆ Homophone: *pray.* [<OF *preie* <L *praeda* booty] **—prey′er** *n.*

Pri·am (prī′əm) In Greek legend, the son of Laomedon, husband of Hecuba, and father of fifty sons including Hector and Paris; he was the last king of Troy and was killed during its capture at the end of the Trojan War.

Pri·a·pe·an (prī′ə·pē′ən) *adj.* Of or pertaining to Priapus; phallic.

pri·a·pus (prī·ā′pəs) *n.* A phallus. [<PRIAPUS]

Pri·a·pus (prī·ā′pəs) In Greek and Roman mythology, the god of male procreative power, son of Dionysos and Aphrodite. [<L <Gk. *Priapos*]

Prib·i·lof Islands (prib′i·lof) A group of four Alaskan islands in the SE Bering Sea; major breeding ground of the Alaska fur seal.

price (prīs) *n.* 1 An equivalent given or asked in exchange; valuation; cost (to the buyer). 2 Anything given or done to obtain something: Death is the *price* of glory. 3 The quality of possessing value; worth; especially, high value. 4 A bribe or anything used for a bribe. 5 A reward for the capture or death of. **—beyond price** 1 So valuable that no adequate price can be set; priceless. 2 Unbribable. **—market price** The price that something will bring in the open market. **—to set a price on one's head** To offer a reward for the capture of a person, dead or alive. **—***v.t.* **priced, pric·ing** 1 To ask the price of. 2 To set a price on; value; appraise. [<OF *pris* <L *pretium.* Related to PRAISE.]

Synonyms (noun): charge, cost, expenditure, expense, outlay, value, worth. The *cost* of a thing is all that has been expended upon it, whether in discovery, production, refinement, decoration, transportation, or otherwise, to bring it to its present condition in the hands of its present possessor; the *price* of a thing is what the seller asks for it. *Price* always implies that an article is for sale; what a man will not sell he declines to put a *price* on. *Value* is the estimated equivalent for an article, whether the article is for sale or not; the market *value* is what something would bring if it were for sale in the open market; the intrinsic *value* is the inherent *worth* of the article considered by itself alone; the market *value* of an old and rare volume may be very great, while its intrinsic *value* may be practically nothing. *Value* has always more reference to others' estimation (literally, what the thing will avail with others) than *worth,* which regards the thing in and by itself; thus, intrinsic *value* is a weaker expression than intrinsic *worth. Charge* has especial reference to services, *expense* to outlays; as, the *charges* of a lawyer or physician; traveling *expenses,* etc.

Price may appear as a combining form in hyphemes or solidemes, or as the first element in two-word phrases:

price adjustment	**price–making**
price administration	**price–manipulation**
price boom	**price notice**
price–control	**price reduction**
price cut	**price–ruling**
price–fixer	**price–stabilizer**
price freeze	**price–stabilizing**
price history	**price–support**
price–level	**price–supporting**
price–maintenance	**price tag**

price cutting The act of reducing the price of an article to one below the figure at which it is usually advertised or sold.

price–fix·ing (prīs′fik′sing) *n.* 1 The establishment and maintenance of a scale of prices agreed upon within specified groups of producers or distributors. 2 The establishing by government action of maximum or minimum or fixed prices for certain goods and services. 3 The fixing by a manufacturer or producer of the price at which retailers must sell his product. **—***adj.* Of or pertaining to price–fixing.

price·less (prīs′lis) *adj.* 1 Beyond price or valuation; invaluable. 2 *Colloq.* Wonderfully amusing or absurd.

price·list (prīs′list′) *n.* A catalog of goods in which the prices are named.

prick (prik) *v.t.* 1 To pierce slightly, as with a sharp point; puncture. 2 To affect with sharp mental pain; sting; spur. 3 To mark, outline, or indicate by or as by punctures. 4 *Obs.* To urge on with or as with a spur; goad. 5 In farriery: **a** To drive a nail into the quick of (a horse's hoof), causing lameness. **b** To nick (a horse's tail). 6 To transplant, as young plants, preparatory to later planting. **—***v.i.* 7 To have or cause a stinging or piercing sensation. 8 *Archaic* To ride at full speed; go at a gallop. **—to prick up one's (or its) ears** 1 To raise the ears erect. 2 To listen attentively. **—***n.* 1 The act of pricking; the state or sensation of being pricked. 2 A mental sting or spur: the *prick* of conscience. 3 That which pricks; a slender, sharp-pointed thing, as a thorn or pointed weapon. 4 A mark made by a sharp, pointed instrument; puncture; dot. 5 The footprint of an animal, as a rabbit or deer. 6 *Archaic* A goad or spur. [OE *prica* sharp point] **—prick′er** *n.*

prick·et (prik′it) *n.* 1 A buck of the second year. 2 A sharp point upon which to stick a candle; hence, a candlestick. [Dim. of PRICK]

prick·ing (prik′ing) *n.* 1 The act of puncturing with a sharp point, or the resulting sensation. 2 The laming of a horse by improper shoeing. 3 The nicking of a horse's tail.

pricking wheel A toothed wheel mounted on a handle, used by saddlers to mark equidistant places for stitch holes, or by dressmakers in copying patterns. Also **prick wheel.**

prick·le (prik′əl) *n.* 1 A small, sharp point, as on the bark of a plant. 2 A prickling or stinging sensation. **—***v.* **·led, ·ling** *v.t.* 1 To prick; pierce. 2 To cause a prickling or stinging sensation in. **—***v.i.* 3 To have a prickling or stinging sensation; tingle. [OE *pricel*]

prick·ly (prik′lē) *adj.* 1 Furnished with prickles. 2 Stinging, as if from a prick or sting: a *prickly* sensation.

prickly ash A prickly shrub or tree (*Zanthoxylum americanum*) of the rue family, with pungent and aromatic bark.

prickly heat *Pathol.* A summer rash of bright red pimples, with heat, itching, and pricking as if by needles; miliaria.

prickly pear 1 A flat-stemmed cactus (genus *Opuntia*) bearing a pear-shaped and often prickly fruit. **2** The fruit itself.
prickly poppy A weedlike annual (*Argemone mexicana*) of the poppy family, with prickly stem and leaves, showy yellow flowers, and yellow juice.
prick punch A pointed steel punch for marking reference points on metal.
prick·song (prik′sông′, -song′) *n. Archaic* **1** Music pricked down or written. **2** Counterpoint.
pride (prīd) *n.* **1** An undue sense of one's own superiority; inordinate self-esteem; arrogance or superciliousness; conceit. **2** A proper sense of personal dignity and worth; honorable self-respect. **3** That of which one is justly proud; a cause of exultation. **4** The acme of excellence. **5** Consciousness of youth or power; high spirits; mettle. **6** *Obs.* Sexual desire. **7** *Archaic* Ostentatious splendor; display. **8** A group or company: said only of lions. — *v.t.* **prid·ed, prid·ing** To take pride in (oneself) for something: with *on* or *upon*. [OE *prȳte < prūt* proud]
Synonyms (noun): conceit, ostentation, self-complacency, self-conceit, self-esteem, self-exaltation, self-respect, vainglory, vanity. *Conceit* and *vanity* are associated with weakness, *pride* with strength. *Conceit* may be founded upon nothing, *pride* is founded upon something that one is, or has, or has done; *vanity*, too, is commonly founded on something real, but far slighter than would afford foundation for *pride*. *Vanity* is eager for admiration and praise and seeks them; *pride* could never solicit admiration or praise. *Conceit* is stronger than *self-conceit*. *Self-conceit* is ridiculous; *conceit* is offensive. *Self-respect* is a thoroughly worthy feeling; *self-esteem* is a more generous estimate of one's own character and abilities than the rest of the world is ready to allow. *Vainglory* is more pompous and boastful than *vanity*. Compare synonyms for ARROGANCE, EGOTISM, OSTENTATION, RESERVE. *Antonyms:* humility, lowliness, meekness, modesty, self-abasement.
Pride (prīd), **Thomas**, died 1658, English general; one of the judges who condemned Charles I.
pride·ful (prīd′fəl) *adj.* Full of pride; haughty; disdainful.
pride of China The azedarach tree.
Pride's Purge The expulsion of Royalist and Presbyterian members from the House of Commons in 1648, conducted by Thomas Pride.
Prid·win (prid′win) King Arthur's shield.
prie–dieu (prē·dyœ′) *n.* A small desk arranged to support a book or books and with a footpiece on which to kneel; a praying desk. [<F, pray God]
pri·er (prī′ər) *n.* One who pries.
priest (prēst) *n.* **1** One especially consecrated to the service of a divinity, and serving as mediator between the divinity and his worshipers in sacrifice, worship, prayer, teaching, etc. **2** In the Anglican, Greek, and Roman Catholic churches, a clergyman in the second order of the ministry, ranking next below a bishop, and having authority to administer the sacraments. **3** Any ordained clergyman or pastor; an official minister of any religious system: distinguished from *layman*. **4** In the early Christian church, an elder or presbyter. **5** One who performs functions or duties similar to those of a priest. — **parish priest** The priest in charge of a parish; specifically, in the Roman Catholic Church, a priest exercising personal jurisdiction in a parish, all members of which are obliged to apply to him for the ministrations of the church: distinguished from *rector* and *curate*. [OE *prēost*, ult. <L *presbyter*. Doublet of PRESBYTER.]

PRIE-DIEU

priest·craft (prēst′kraft′, -kräft′) *n.* **1** Priestly arts and wiles: an invidious term. **2** The knowledge and skill of priests.
priest·ess (prēs′tis) *n.* A woman or girl who exercises priestly functions or who performs sacred rites.
priest·hood (prēst′hood) *n.* **1** The priestly office or character. **2** The priestly order; priests collectively. [OE *prēosthad*]
Priest·ley (prēst′lē), **J(ohn) B(oynton),** born 1894, English author. — **Joseph,** 1733–1804, English philosopher and chemist; discoverer of oxygen.
priest·ly (prēst′lē) *adj.* **1** Of or pertaining to a priest or the priesthood; sacerdotal. **2** Suitable to or befitting a priest. — **priest′li·ness** *n.*
priest–rid·den (prēst′rid′n) *adj.* Completely under the influence or domination of priests.
prig¹ (prig) *n.* A formal and narrow-minded person who assumes superior virtue, wisdom, or learning; pedant. [Origin unknown]
prig² (prig) *v.* **prigged, prig·ging** *v.t. Brit. Slang* To steal. — *v.i. Scot & Brit. Dial.* To bargain; haggle.
prig·gish (prig′ish) *adj.* Like a prig; conceited. — **prig′gish·ly** *adv.* — **prig′gish·ness** *n.*
prig·gism (prig′iz·əm) *n.* The characteristics or manners of a prig.
prill (pril) *n.* **1** A small metal particle formed in assay work. **2** A spherical pellet about the size of buckshot. — *v.t.* To convert into prills for some purpose or use. [? <Cornish]
prim (prim) *adj.* Minutely or affectedly precise and formal; stiffly proper and neat. See synonyms under NEAT¹. — *v.* **primmed, prim·ming** *v.i.* To fix the face or mouth in a precise or prim expression; be prim. — *v.t.* To fix in a precise or prim manner. [Prob. <OF *prim* first, prime, fine, delicate <L *primus* first] — **prim′ly** *adv.* — **prim′ness** *n.*
Pri·ma·cord (prī′mə·kôrd) *n.* A flexible tube of fabric or lead, filled with high explosive and used as a primer or bursting charge: a trade name.
pri·ma·cy (prī′mə·sē) *n. pl.* **·cies 1** The state of being first, as in rank or excellence. **2** The office or province of a primate; archbishopric: also **pri′mate·ship** (-mit·ship). [<OF *primacie* <Med. L *primatia* <LL *primas, primatis* one of the first. See PRIMATE.]
pri·ma don·na (prē′mə don′ə) **1** A leading female singer, as in an opera company. **2** *Colloq.* A temperamental or vain person. [<Ital., lit., first lady]
pri·ma fa·ci·e (prī′mə fā′shi·ē, fā′shē) *Latin* At first view; so far as at first appears.
prima–facie evidence Evidence which, if unexplained or uncontradicted, would establish the fact alleged.
pri·mage (prī′mij) *n.* An allowance in addition to wages, formerly paid by a shipper to the master of a vessel, now paid to the owner of the vessel as an addition to freight charges, for care in loading or unloading goods in port. [<PRIME + -AGE; after Med. L *primagium*]
pri·mal (prī′məl) *adj.* **1** Being at the beginning or foundation; first; original. **2** Most important; chief. See synonyms under PRIMEVAL. [<Med. L *primalis*]
primal cut Any one of the cuts into which a side of beef may be divided for sale at wholesale. These cuts are: hindquarter, trimmed full loin, round sirloin, short loin, flank, flank steak, kidney, hanging tender, forequarter, cross-cut chuck, triangle, arm chuck, rib, short plate, brisket, fore shank, back, regular chuck.
pri·ma·quine (prī′mə·kwīn) *n.* An antimalarial drug synthesized from chemicals derived from corn and coal tar. [<PRIM(E) + A(MINO)- QUIN(OLIN)E]
pri·ma·ri·ly (prī′mer·ə·lē, -mər·ə·lē, *emphatic* prī·mâr′ə·lē) *adv.* In the first place; originally; essentially.
pri·ma·ry (prī′mer·ē, -mər·ē) *adj.* **1** First in time or origin; primitive; original. **2** First in a series or sequence. **3** First in degree, rank, or importance; chief. **4** Constituting the fundamental or original elements of which a whole is comprised; basic; elemental: the *primary* forces of life. **5** Of the first stage of development; elementary; lowest: *primary* school. **6** *Ornithol.* Of or pertaining to the principal flight feathers of a bird's wing. **7** *Geol.* Paleozoic. **8** *Electr.* Of, pertaining to, or noting an inducing current or its circuit: a *primary* coil. **9** *Chem.* **a** Having some characteristic in the first degree, as an initial replacement, substitution, etc. **b** In organic compounds, denoting a radical in which a carbon atom is directly joined to only one other carbon atom. **c** Denoting a compound containing such a radical. — *n. pl.* **·ries 1** That which is first in rank, dignity, or importance, as a primary planet in distinction from a satellite. **2** A primary meeting or balloting of the voters belonging to one political party in an election district to nominate candidates. **3** *Ornithol.* One of the large flight feathers of the pinion or hand bones of a bird's wings. See synonyms under FIRST, PRIMEVAL. — **direct primary election** A primary election in which candidates for office are nominated directly by the voters and not by a convention or by a body of delegates. [<L *primarius* < *primus* first]
primary cell *Electr.* A cell which cannot be efficiently recharged after use owing to an irreversible electrochemical reaction.
primary colors See under COLOR.
pri·mate (prī′mit, -māt) *n.* **1** The prelate highest in rank in a nation or province. **2** Any of an order (*Primates*) of mammals, including the tarsiers, lemurs, marmosets, monkeys, apes, and man. [<OF *primat* <LL *primas, primatis* of the first <L *primus*] — **pri·ma′tial** (prī-mā′shəl) *adj.*
pri·ma·tol·o·gy (prī′mə·tol′ə·jē) *n.* The branch of zoology which treats of the origin, structure, evolution, and development of primates. — **pri·ma·tol′o·gist** *n.*
pri·ma·ve·ra (prē′mä·vä′rä) *n.* A tropical American tree (*Cybistax donnell-smithi*) of the bignonia family, yielding a creamy-white or yellowish wood resembling satinwood; erroneously called *white mahogany.* [<Sp., lit., spring <L *prima vera*, pl. of *primum ver* earliest spring]
prime¹ (prīm) *adj.* **1** First in rank, dignity, or importance; chief. **2** First in value or excellence; of excellent quality; first-rate. **3** First in time or order; original; primitive; primeval. **4** *Math.* Divisible by no whole number except itself and unity: said of a number. Two or more numbers are said to be *prime* to each other when they have no common factor but unity. **5** Having or pertaining to the strength and vigor of fresh maturity; blooming. **6** Original; not derived; first: opposed to *secondary*. **7** Marked with the sign (′). See synonyms under EXCELLENT, PRIMEVAL. — *n.* **1** The period of fresh, full vigor, beauty, and power succeeding youth and preceding age; formerly, youth. **2** The period of full perfection in anything. **3** The beginning of anything, as of the day; dawn; spring. **4** The best of anything; a prime grade. **5** A prime number. **6** A mark or accent (′) written above and to the right of a letter or figure; also, an inch, a minute, etc., as indicated by that sign, used in indicating and measuring degrees. **7** *Music* The tonic; the interval of unison; also, a note in unison with another. — *v.* **primed, prim·ing** *v.t.* **1** To prepare; make ready for some purpose. **2** To put a primer into (a gun, mine, etc.) preparatory to firing. **3** To pour water into (a pump) so as to displace air and promote suction. **4** To cover (a surface) with sizing, a first coat of paint, etc. **5** To supply beforehand with facts, information, etc.: to *prime* a witness. — *v.i.* **6** To carry water along with the steam into the cylinder: said of a steam boiler or engine. **7** To make something ready, as for firing, pumping, etc. [<OF <L *primus*] — **prime′ly** *adv.* — **prime′ness** *n.*
prime² (prīm) *n.* **1** The first canonical hour succeeding lauds; first of the day hours. **2** The office recited at this time. [OE *prīm* <LL *prima (hora)* first (hour)]
prime conductor *Electr.* The conductor of a frictional machine which collects and retains the positive electricity.
prime cost The direct cost of obtaining or producing something; the cost of labor and material, exclusive of capital and management expenses.

prime meridian A meridian from which longitude is reckoned: now, generally, the one that passes through Greenwich, England, but formerly that of the local capital, as, in the United States, Washington, D.C.: in France, Paris; etc.

prime minister The chief of the cabinet or ministry; in Great Britain, the principal minister of the sovereign. Compare PREMIER.

prime ministry The office of a prime minister.

prime mover 1 An original or chief force in an undertaking. 2 That which is regarded as an original or natural source of the energy required to perform work or develop power, as muscular force, wind, the motion of water, etc. 3 An object or machine used to convert natural forces to productive power, as a turbine, water wheel, windmill, or the like. 4 In Aristotelian philosophy, the first cause of all movement, which does not itself move.

prime number See under NUMBER.

prim·er[1] (prim′ər) *n.* 1 An elementary textbook; especially, a beginning reading book. 2 Originally, a small prayer book or the like. 3 *Printing* Either of two sizes of type, **great primer** (18-point) and **long primer** (10-point). [<Med. L *primarius*]

prim·er[2] (prī′mər) *n.* 1 Any device, as a cap, tube, etc., used to detonate the main charge of a gun, mine, etc. 2 One who or that which primes.

prim·er[3] (prim′ər, prī′mər) *adj. Obs.* First; original; primary. [<OF <L *primarius* <*primus* prime]

pri·me·ro (pri·mâr′ō) *n.* An old gambling card game. [<Sp.]

pri·me·val (prī·mē′vəl) *adj.* Belonging to the first ages; primitive in time; primary. [<L *primaevus* youthful <*primus* first + *aevum* age] — **pri·me′val·ly** *adv.*

Synonyms: aboriginal, ancient, autochthonic, immemorial, indigenous, native, old, original, primal, primary, prime, primitive, primordial, pristine. *Aboriginal* signifies pertaining to the earliest known inhabitants of a country in the widest sense, including not merely human beings, but animals and plants. *Primeval* signifies strictly belonging to the first ages, earliest in time, but often only the earliest of which man knows or conceives. *Prime* and *primary* may signify either first in time, or first in importance; *primary* has also the sense of elementary or preparatory; we speak of a *prime* minister, a *primary* school. *Primal* is chiefly poetic, in the sense of *prime*; as, the *primal* curse. *Primordial* is first in an order of existence or development; as, a *primordial* leaf. *Primitive* frequently signifies having the original characteristics of that which it represents, as well as standing first in time; as, the *primitive* church, or early characteristics without remoteness in time. *Primeval* simplicity is the simplicity of the earliest ages; *primitive* simplicity may be found in retired villages now. *Pristine* is used almost exclusively in a good sense of that which is *original* and perhaps *ancient*; as, *pristine* purity, innocence, vigor. *Immemorial* refers solely to time, independently of quality, denoting, in legal phrase, that "whereof the memory of man runneth not to the contrary." Compare synonyms for ANCIENT, FIRST, OLD. *Antonyms*: adventitious, exotic, foreign, fresh, late, modern, new, novel, recent.

pri·mi·ge·ni·al (prī′mə·jē′nē·əl) *adj.* Being the first or first-born; primal; primitive; original. [<L *primigenius* first, original]

pri·mine (prī′min) *n. Bot.* The outermost and last-developed integument of an ovule; also, the inner integument as being formed first. Compare SECUNDINE. [<L *primus* first]

prim·ing (prī′ming) *n.* 1 That with which anything is primed. 2 A combustible composition used to ignite an explosive charge. 3 The ground or first layer of paint laid on a surface that is to be painted.

pri·mip·a·ra (prī·mip′ər·ə) *n. pl.* **·a·rae** (-ər·ē) A woman pregnant for the first time or one who has borne just one child. [<L <*primus* first + *parere* give birth to] — **pri·mi·par·i·ty** (prī′mi·par′ə·tē) *n.* — **pri·mip′a·rous** *adj.*

prim·i·tive (prim′ə·tiv) *adj.* 1 Pertaining to the beginning or origin; first; earliest; primary. 2 Resembling the manners or style of long ago; old-fashioned; simple; plain. 3 *Geol.* Of, belonging to, or characterized by the earliest geological period: said especially of the crystalline, unstratified, and massive rocks, the oldest known. 4 *Anthropol.* Of or pertaining to the beginning or earliest anthropological forms or civilizations: *primitive* man, *primitive* weapons. 5 *Biol.* a Being or occurring at an early stage of development or growth; first-formed; rudimentary. b Not much changed by evolution: a *primitive* species. 6 *Ling.* Standing in original relation, as a word from which a derivative is made; radical: opposed to *derived.* 7 *Theol.* Adhering to strictly traditional interpretation of doctrine and Scripture: the *primitive* church. — *n.* 1 *Ling.* A primary or radical word; also, a word from which another is derived. 2 *Math.* A form in algebra or geometry from which another is derived. 3 An artist, or a work of art, belonging to a very early period of art, or to the earliest phase of an art development or movement; also, a work of any period resembling or imitating such art, or an artist producing it: often characterized by simplicity or a childlike quality. See synonyms under FIRST, PRIMEVAL, RADICAL. [<L *primitivus* <*primus* first] — **prim′i·tive·ly** *adv.* — **prim′i·tive·ness, prim′i·tiv′i·ty** *n.*

Primitive Baptist A member of a branch of the Baptist church (separate since 1835) holding to Calvinistic and antimission doctrines: also called *Hardshell Baptist.*

prim·i·tiv·ism (prim′ə·tiv·iz′əm) *n.* Belief in or adherence to primitive forms and customs.

pri·mo·gen·i·tor (prī′mə·jen′ə·tər) *n.* An earliest ancestor; a forefather. [<Med. L <L *primus* first + *genitor* a father]

pri·mo·gen·i·ture (prī′mə·jen′ə·chər) *n.* 1 The state of being the first-born child of the same parents. 2 The right of the eldest son to inherit the property, title, etc., of a parent, to the exclusion of all other children. See ULTIMOGENITURE. [<Med. L *primogenitura* <L *primus* first + *genitura* birth <*genitus*, pp. of *gignere* beget]

prim′·o·mo (prē·mō′mō) *Italian* First man; leading actor or singer.

pri·mor·di·al (prī·môr′dē·əl) *adj.* 1 First in order or time; original; elemental. 2 *Biol.* First in order of appearance in the growth or development of an organism. — *n.* An elementary principle. See synonyms under FIRST, PRIMEVAL, TRANSCENDENTAL. [<LL *primordialis* <*primordius* original <*primordium* beginning <*primus* first + *ordiri* begin a web] — **pri·mor′di·al·ly** *adv.*

pri·mor·di·al·ism (prī·môr′dē·əl·iz′əm) *n.* The survival or persistence of primitive arts and customs.

primp (primp) *v.t.* & *v.i.* To prink; dress up, especially with superfluous attention to detail. [Akin to PRIM]

prim·rose (prim′rōz) *n.* 1 An early-blossoming perennial herb (genus *Primula*) with tufted basal leaves and variously colored flowers. 2 The flower. 3 The evening primrose. 4 A pale-yellow color, named for the common English primrose: a term indiscriminately applied to various yellow pigments. — *adj.* 1 Pertaining to a primrose; of primrose color. 2 Flowery; gay. [Alter. of ME *primerole* <OF <Med. L *primula,* fem. dim. of L *primus* first; infl. by *rose*]

PRIMROSE (*def. 1*) (The wild species from 1 to 18 inches tall)

primrose path The life of worldly or sensual pleasures.

prim·sie (prim′zē) *adj. Scot.* Demure; prim.

prim·u·la·ceous (prim′yə·lā′shəs) *adj. Bot.* Designating or belonging to a family (*Primulaceae*) of herbs widely distributed in the northern hemisphere, including the pimpernel, cyclamen, and loosestrife. [<NL <Med. L *primula* a primrose]

pri·mum mo·bi·le (prī′məm mō′bi·lē) 1 A prime mover. 2 *Astron.* In the Ptolemaic cosmology, the tenth and outermost of the concentric spheres of the universe, regarded as causing all the other spheres to repeat its own revolution around the earth once in 24 hours. [<L, first moving thing]

pri·mus in·ter pa·res (prī′məs in′tər pâr′ēz) *Latin* First among equals.

prince (prins) *n.* 1 A non-reigning male member of a royal family. 2 A male monarch or sovereign. 3 *Brit.* The son of a sovereign or of a son of the sovereign. 4 One of a high order of nobility. 5 The ruler of a small state; head of a principality. 6 A chief or leader, or one of the highest rank of the class to which he belongs: a merchant *prince.* See synonyms under MASTER. [<OF <L *princeps* first, principal <*primus* first + stem of *capere* take]

Prince Albert A long, double-breasted frock coat.

Prince Albert National Park A park in central Saskatchewan, Canada; 1,496 square miles; established 1927.

prince consort The husband of a reigning female sovereign.

Prince Edward Island An island in the Gulf of St. Lawrence off eastern New Brunswick, comprising a maritime province of Canada; 2,184 square miles; capital, Charlottetown.

Prince Island The English name for PRINCIPE. See SÃO TOMÉ E PRINCIPE.

prince·kin (prins′kin) *n.* A little or inferior prince.

prince·ling (prins′ling) *n.* 1 A young prince. 2 A subordinate prince. Also **prince′let** (-lit).

prince·ly (prins′lē) *adj.* **·li·er, ·li·est** 1 Like or characteristic of a prince; liberal; generous. 2 Belonging to, ruled by, or suitable for a prince. 3 Having the rank of a prince. See synonyms under KINGLY. — *adv.* In a princely manner. — **prince′li·ness** *n.*

Prince of Darkness Satan.

Prince of Peace Jesus Christ.

Prince of Wales The eldest son or male heir apparent of the British sovereign: he is born Duke of Cornwall, and becomes Prince of Wales only by creation.

Prince of Wales, Cape The westernmost point of the American continent, at the tip of Seward Peninsula, Alaska, on the Bering Strait 100 miles NW of Nome.

Prince of Wales Island 1 A former name for PENANG ISLAND. 2 An island in the Arctic Ocean, in south central Franklin district of Northwest Territories, Canada; 13,736 square miles; the north magnetic pole was located on it in 1948. 3 The largest island of the Alexander Archipelago, SE Alaska, in the North Pacific west of Ketchikan; 2,231 square miles; 135 miles long, 45 miles wide.

Prince of Wales plumes In furniture and decoration, a motif of three ostrich feathers tied with a bowknot.

Prince Rupert A port in western British Columbia, Canada.

prin·ce's-feath·er (prin′siz·feth′ər) *n.* A tall, hardy plant (*Polygonum orientale*) with plume-like inflorescences and dark-crimson flowers, growing wild in eastern North America.

prin·cess (prin′sis) *n.* 1 A non-reigning female member of a royal family. 2 The consort of a prince. 3 A female sovereign. 4 *Brit.* The daughter of a sovereign or of a son of the sovereign. [<F *princesse*]

prin·cesse (prin·ses′, prin′sis) *adj.* Designating a woman's close-fitting garment cut in a single piece from shoulder to flared hem. Also **prin′cess.** [<F, princess]

princess feather A tall, graceful plant (*Amaranthus hybridus hypochondriacus*) with many-branched panicles of showy flowers.

princess royal The eldest daughter of a sovereign.

Prince·ton (prins′tən) A borough of central New Jersey; scene of an American victory in the Revolutionary War (1777) and seat of Princeton University, founded 1746.

Prince William Sound An inlet of the Gulf of Alaska in southern Alaska; 100 miles across.

prin·ci·pal (prin′sə·pəl) *adj.* First in rank, character, or importance; chief. See synonyms under FIRST, PARAMOUNT. — *n.* 1 One who takes a leading part; one concerned directly and not as an auxiliary; one who is a leader or chief in some action. 2 *Law* a The actor in a crime, or one present aiding and abetting. b The employer of one who acts as an agent. c One primarily liable for whom another has become surety. d The most important thing, or part of a given property, to which other things or parts are incidental. e The capital or body of an estate. 3 One who is at the head of some body; a chief; one in authority; a presiding officer, as of a society. 4 The head teacher or master in a public or private school. 5 The chief executive of some colleges and universities in Great Britain. 6 Property or capital, as opposed to interest or income. 7 A rafter extending to the ridge pole; a principal rafter. 8 *Music* **a** The chief metal organ stop, an octave higher in pitch than the other diapasons. **b** The subject of a fugue: distinguished from *answer.* See synonyms under CHIEF, MASTER. ◆ Homophone: *principle.* [<F <L *principalis* < *princeps* chief] — **prin′·ci·pal·ly** *adv.* — **prin′ci·pal·ship′** *n.*

prin·ci·pal·i·ty (prin′sə·pal′ə·tē) *n. pl.* **·ties** 1 The territory of a reigning prince, or one that gives to a prince a title of courtesy. 2 *pl.* Powers or powerful influences, as celestial or demoniacal powers; in the celestial hierarchy of Dionysius, the seventh of the nine emanations from the Divine. 3 *Obs.* The state of being supreme or preeminent; sovereignty.

principal part See under PART.

Prin·ci·pe (prin′si·pē, *Pg.* prēn′sē·pə) See under SÃO TOMÉ E PRINCIPE.

prin·cip·i·um (prin·sip′ē·əm) *n. pl.* **·cip·i·a** (-sip′ē·ə) 1 Beginning; origin; first principle. 2 *pl.* Fundamentals. [<L]

prin·ci·ple (prin′sə·pəl) *n.* 1 A general truth or law, basic to other truths: the *principle* of self-government. 2 A settled law or rule of personal conduct: He followed the *principle* of the Golden Rule. 3 That which is inherent in anything, determining its nature; essential character; essence. 4 A source or cause from which a thing proceeds; fundamental cause. 5 An established mode of action or operation in natural phenomena: the *principle* of Archimedes. 6 *Chem.* An essential constituent of a compound or substance that gives character to it. 7 Moral standards collectively: a man of *principle.* See synonyms under DOCTRINE, LAW[1], REASON. ◆ Homophone: *principal.* [<L *principium* a beginning]

prin·cock (prin′kok) *n. Obs.* A coxcomb. Also **prin′cox** (-koks). [Prob. <PRIM + COCK[1]]

Prings·heim (prinks′hīm), **Ernst,** 1859–1917, German physicist. — **Nathanael,** 1823–94, German biologist.

prink (pringk) *v.t.* To dress or adorn (oneself) for show. — *v.i.* To dress oneself showily or fussily. [Prob. alter. of PRANK[1] under infl. of PREEN] — **prink′er** *n.*

print (print) *n.* 1 An impression with ink from type, plates, etc.; printed characters collectively; printed matter. 2 Anything printed from an engraved plate or lithographic stone; a proof; a printed picture or design. 3 A newspaper, pamphlet, or the like. 4 An impression or mark made upon or sunk into a substance by pressure; imprint. 5 A reproduction from such an impression. 6 Any fabric stamped with a design by means of dyes used on engraved rollers, wood blocks, or screens. 7 Any tool or device bearing a pattern or design, or that upon which it is impressed. 8 *Phot.* A positive picture made from a negative. 9 Newsprint. — **in print** 1 Printed; also, for sale in printed form: opposed to *out of print.* 2 *Obs.* In an exact or formal manner. — **India print** Muslin printed, specifically handblocked, with the native patterns and glowing colors of India. — **out of print** No longer on sale, the edition being exhausted. See synonyms under MARK[1], PICTURE. — *v.t.* 1 To mark, as with inked type, a stamp, die, etc. 2 To stamp or impress (a mark, seal, etc.) on or into a surface. 3 To fix as if by impressing: The scene is *printed* on my memory. 4 To produce (a book, newspaper, etc.) by the application of inked type, plates, etc., to paper or similar material. 5 To cause to be put in print; publish: The newspaper *printed* the story. 6 To write in letters similar to those used in print: Please *print* your name and address. 7 *Phot.* To produce (a positive picture) by transmitting light through a negative onto a sensitized surface. — *v.i.* 8 To be a printer. 9 To take or give an impression in printing. 10 To form letters similar to printed ones. See synonyms under IMPRESS[1]. [<OF *preinte, priente,* fem. of pp. of *preindre* <L *premere* press] — **print′a·ble** *adj.*

print·er (prin′tər) *n.* 1 One engaged in the trade of typographical printing; one who sets type or runs a printing press; specifically, a compositor. 2 One who owns a printing establishment and employs printers. 3 One who prints, stamps, impresses, or transfers copies of anything as a business.

printer's devil A printer's apprentice. See under DEVIL.

print·er·y (prin′tər·ē) *n. pl.* **·er·ies** 1 A place where cotton goods, as calico, are printed. 2 A printing office.

print·ing (prin′ting) *n.* 1 The making and issuing of matter for reading by means of type and the printing press. 2 Presswork. 3 The act of reproducing a design upon a surface by any process. 4 That which is printed.

printing machine 1 Any machine for printing, as on cotton cloth. 2 A printing press.

printing press A mechanism for printing from an inked surface as of type, plates, wood blocks, etc., operating by pressure, either against a flat bed, as in the platen press, or against a series of revolving cylinders, as in the rotary press.

print·less (print′lis) *adj.* Making, bearing, or retaining no print or impression.

pri·or (prī′ər) *adj.* Preceding in time, order, or importance. See synonyms under ANTECEDENT, ANTERIOR. — **prior to** Before: The theater closed *prior to* our arrival. — *n.* 1 A monastic officer next in rank below an abbot. 2 Formerly, an Italian magistrate. [<L, earlier, superior] — **pri′or·ate** (-it), **pri′or·ship** *n.*

Pri·or (prī′ər), **Matthew,** 1664–1721, English poet and diplomat.

pri·or·ess (prī′ər·is) *n.* A woman holding a position corresponding to that of a prior; a nun next in rank below an abbess.

pri·or·i·ty (prī·ôr′ə·tē, -or′-) *n. pl.* **·ties** 1 Antecedence; precedence: opposed to *posteriority.* 2 A first right established on emergency or need: Defense plants have *priorities* on steel in time of war. 3 A certificate giving this right to a manufacturer or contractor; hence, a restriction on the use of a commodity or service.

pri·or·y (prī′ər·ē) *n. pl.* **·or·ies** A monastic house presided over by a prior or prioress. See synonyms under CLOISTER. [<OF *priorie*]

Pri·pet (prē′pet) A river in NE Ukrainian S.S.R. and southern Belorussian S.S.R., flowing about 500 miles east through the **Pripet Marshes,** a swampy region of 33,500 square miles, to the Dnieper, 50 miles north of Kiev. *Russian* **Pri·pyat** (pryē′pyət·y′), *Polish* **Pry·peć** (prē′pech).

Pris·cian (prish′ən) Latin grammarian of the fifth century A.D.: full name *Priscianus Caesariensis.*

Pris·cil·la (pri·sil′ə; *Du.* pri·sil′ä, *Ital.* prē·sil′lä) A feminine personal name. Also *Fr.* **Pris·cille** (prē·sēl′). [<L, somewhat ancient] — **Priscilla** In Longfellow's poem *The Courtship of Miles Standish,* the Puritan maiden, Priscilla Mullens, courted by John Alden as proxy for Standish. Later she married Alden.

Pris·cil·li·an (pri·sil′ē·ən, -sil′yən), died 385, bishop of Avila, Spain; burned at the stake for sorcery. — **Pris·cil′li·an·ist, Pris·cil′li·an·ite′** *n.*

prise (prīz) *n.* & *v.t.* Prize[2]; lever.

prism (priz′əm) *n.* 1 *Geom.* A solid whose bases or ends are any similar equal and parallel plane figures, and whose lateral faces are parallelograms. 2 *Optics* An instrument consisting of such a solid, usually having triangular ends and made of glass or other translucent substance, its refracting surfaces making an angle with each other. 3 Any

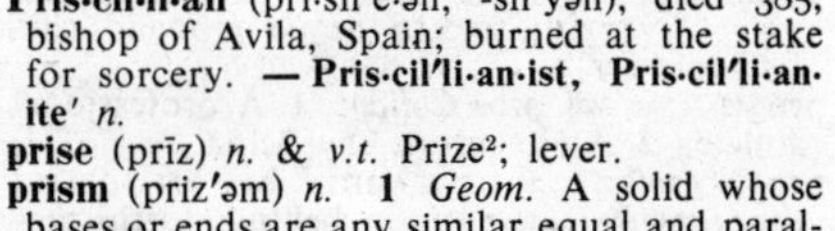

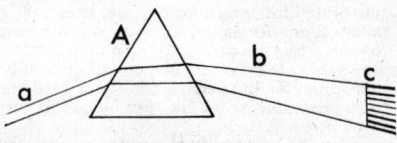

PRISM
A beam of white light *(a)*, entering the prism at A, is refracted according to wavelengths and, on emerging at *b*, produces its spectrum at *c*.

medium that resolves a seemingly simple matter into its elements. 4 The spectrum. 5 *Mineral.* A crystal form consisting of three or more intersecting planes whose intersections are parallel and vertical. — **Nicol prism** A prism of calcite (Iceland spar) so cut that light emerging from it is polarized in a definite plane: used in polarizing microscopes, etc. [<LL *prisma* <Gk., something sawed < *prixein* saw]

pris·mat·ic (priz·mat′ik) *adj.* 1 Refracted or formed by a prism. 2 Resembling the spectrum; exhibiting rainbow tints. 3 Pertaining to or shaped like a prism. 4 Orthorhombic. Also **pris·mat′i·cal.** [<Gk. *prisma, prismatos*] — **pris·mat′i·cal·ly** *adv.*

pris·moid (priz′moid) *n.* A body resembling a prism in form. — **pris·moi·dal** (priz·moid′l) *adj.*

pris·on (priz′ən) *n.* A place of confinement; specifically, a public building for the safekeeping of persons in legal custody; a penitentiary. — *v.t.* To imprison. [<F *prisoun* <L *praehensio, -onis* seizure < *praehensus,* pp. of *praehendere*]

pris·on-breach (priz′ən·brēch′) *n.* The escape of a prisoner, against the will of his custodian, from the place where he is held in lawful custody. Also **pris′on-break′ing** (-brā′king).

pris·on·er (priz′ən·ər, -nər) *n.* 1 One who is confined in a prison or whose liberty is forcibly restrained; one held in custody; a captive; specifically, in law, a person confined in a prison by virtue of an order of arrest or of a legal committal. 2 A person confined to a place or position through some cause over which he has not control: A sick man is a *prisoner* to his bed. [<OF *prisonier*]

prisoner of war A combatant or person in arms taken by the enemy either by capture or surrender during war.

prisoner's base A game played in various forms and popular in England as early as the 14th century. Opposing players occupy opposite bases, the object being to touch a player of the opposite side while he is away from his base, when he either joins his captor's side or is confined at another goal called a prison.

prison fever Malignant typhus: so called from its former prevalence in prisons: also called *ship fever.*

pris·sy (pris′ē) *adj.* **·si·er, ·si·est** Effeminate; overprecise; prim. — *n.* A person who acts, dresses, or speaks very meticulously. [Blend of PRIM or PRECISE + SISSY]

Priš·ti·na (prēsh′ti·na) A city of south central Yugoslavia, capital of an autonomous region included in Serbia as its SW part; a 12th century capital of Serbia; included in Albania 1941–44.

pris·tine (pris′tēn, -tin; *Brit.* pris′tīn) *adj.* Of or pertaining to the earliest state or time; primitive; untouched. See synonyms under FIRST. [<L *pristinus* primitive]

prith·ee (prith′ē) *interj. Archaic* I pray thee.

pri·va·cy (prī′və·sē) *n. pl.* **·cies** 1 The condition of being private; seclusion; retirement. 2 A matter that is or should be private. 3 The state of being secret; avoidance of display or publicity; secrecy. 4 A place of seclusion; retreat. See synonyms under RETIREMENT, SECLUSION, SOLITUDE.

Pri·vat·do·zent (prē·vät′dō·tsent′) *n. German* A lecturer or tutor recognized by a university but unsalaried and dependent on his student fees. Also **Pri·vat′do·cent′.**

pri·vate (prī′vit) *adj.* 1 Removed from public view; retired; secluded; confidential; secret: a *private* parlor, a *private* agreement. 2 Personal or unofficial, as opposed to public; hence, without rank: a *private* citizen, *private* property, a *private* soldier. 3 Not common or general; special: a *private* interpretation. 4 *Obs.* Privy. See synonyms under SECRET. — *n.* 1 A soldier in the ranks. See table under GRADE. 2 *pl.* The private parts; genitals. 3 Privacy. — **in private** In secret; privately. See synonyms under SECRET. [<L *privatus* apart from the state, orig. pp. of *privare* set apart < *privus* single, one's own. Doublet of PRIVY.] — **pri′vate·ly** *adv.* — **pri′vate·ness** *n.*

private enterprise 1 Business owned and operated by private individuals, as opposed to government-owned operations. 2 An economic system based upon private ownership and operation of business. Also called *free enterprise.*

pri·va·teer (prī′və·tir′) *n.* 1 A vessel owned and officered by private persons, but carrying on maritime war under letters of marque. 2 The commander or one of the crew of a privateer: also **pri′va·teers′man.** —*v.i.* To cruise in or as a privateer. —**pri′va·teer′ing** *n.*

private first class A soldier ranking next above a private and below a corporal. See table under GRADE.

private nurse A nurse in exclusive attendance on one patient, whether in a hospital or at home.

private school See under SCHOOL.

pri·va·tion (prī·vā′shən) *n.* 1 The state of lacking something necessary or desirable; especially, want of the common comforts of life. 2 Deprivation. 3 *Logic* The absence from an object of what ordinarily or naturally belongs to objects of that kind. 4 *Eccl.* Suspension or degradation from office, as of a priest. See synonyms under LOSS, POVERTY, WANT. [<OF <L *privatio, -onis* <*privare.* See PRIVATE.]

priv·a·tive (priv′ə·tiv) *adj.* 1 Causing privation, want, or destitution; depriving. 2 *Gram.* Altering a word so as to express a negative instead of a positive meaning; also, denoting negation: *privative* particles (such prefixes and suffixes as *a-, an-, in-, -less*). 3 *Logic* Noting or denoting negation or privation. —*n.* 1 That which has its only reality in the absence of something; a negative conception. 2 *Gram.* A prefix indicating negation; an adjective indicating the absence of that which is ordinarily or naturally inherent. [<L *privativus*] —**priv′a·tive·ly** *adv.* —**priv′a·tive·ness** *n.*

priv·et (priv′it) *n.* 1 An ornamental, bushy European shrub (*Ligustrum vulgare*) with white flowers and black berries, used for hedges: naturalized in the United States. 2 Any other plants of the same genus. 3 The swamp privet, an oleaceous tree (*Forestiera acuminata*) of the southern United States. [Earlier *primet*; prob. infl. by *private* because of its screening effect]

priv·i·lege (priv′ə·lij) *n.* 1 A special or peculiar benefit, favor, or advantage; a right or immunity enjoyed only under special conditions; a prerogative, franchise, or permission: the *privileges* of the rich. 2 A special right or power conferred on or possessed by one or more individuals, in derogation of the general right; also, the law or grant conferring it. 3 An exemption, by virtue of one's office or station, from burdens or liabilities to which others are subject: the *privilege* of a member of Congress. 4 A fundamental or specially important legal or political right: the *privilege* of voting. 5 A form of contract used by speculators, but not recognized by the exchanges, giving the holder the privilege of putting (tendering to) or calling for, or either (in which latter case the privilege is called a *straddle*), a certain number of shares of a certain stock, or a specified quantity, as of grain or provisions, under specified conditions as to time and price. Compare OPTION. 6 An advantage. See synonyms under RIGHT. —*v.t.* **·leged, ·leg·ing** 1 To grant a privilege to. 2 To exempt or free: with *from.* [<OF <L *privilegium* a piece of special legislation < *privus* one's own + *lex, legis* law]

priv·i·leged (priv′ə·lijd) *adj.* Having or invested with a privilege; enjoying a peculiar right or immunity.

priv·i·ly (priv′ə·lē) *adv.* Privately; secretly.

priv·i·ty (priv′ə·tē) *n. pl.* **·ties** 1 Knowledge shared with another or others regarding a private matter: usually implying consent or concurrence. 2 *Law* **a** A mutual or successive relationship to the same rights of property. **b** A participation in interest. **c** A relation to another founded on common knowledge. 3 *Obs.* Privacy; secrecy; a secret. [<OF *privité* <L *privus* one's own]

priv·y (priv′ē) *adj.* 1 Participating with another or others in the knowledge of a secret transaction: with *to*: *privy* to the plot. 2 *Archaic* Removed from publicity; clandestine; secret: a *privy* meeting. 3 Designed for individual or private use; personal: a *privy* purse, *privy* chamber. —*n. pl.* **priv·ies** 1 One who is concerned with another in a matter affecting the interests of both: *privies* in contract, *privies* in estate. 2 A small room or outhouse for evacuation and disposal of feces. See WATERCLOSET. [<F *privé* <L *privatus.* Doublet of PRIVATE.]

Privy Council In Great Britain, the sovereign's ordinary council. Since the duties of government were assumed by the cabinet, the political importance of the Privy Council has largely disappeared.

privy council 1 A body similar to the Privy Council in some British colonies and dominions. 2 A term used by British writers for an analogous body in other countries.

privy councilor 1 A member of a privy council. Also **privy councillor** or **counsellor.**

privy seal In Great Britain, the seal used by the king on papers which later pass under the great seal: also affixed to such documents as do not demand the great seal.

prix fixe (prē fēks′) *French* Table d'hôte: literally, fixed price.

prize[1] (prīz) *n.* 1 That which is offered or won as an honor and reward for superiority or success, as in a contest; an award. 2 Anything to be striven for; a desirable acquisition; also, anything offered or won in a scheme of chance. —*adj.* 1 Offered or awarded as a prize: a *prize* medal. 2 Having drawn a prize; entitled to a prize. 3 Highly valued or esteemed. —*v.t.* **prized, priz·ing** 1 To value highly; regard as very valuable. 2 To estimate the value of; appraise. See synonyms under APPRECIATE, ESTEEM. [Var. of PRICE]

prize[2] (prīz) *n.* 1 In international law, property, as a vessel and cargo, captured by a belligerent at sea in conformity with the laws of war. 2 The act of capturing; also, the person or thing captured. 3 A lever or pry; also, the hold or purchase of a lever: also spelled *prise.* —*v.t.* **prized, priz·ing** 1 To seize as a prize, as a ship. 2 To raise or force with a lever; pry: also spelled *prise.* [<F *prise* something taken, booty, orig. fem. of pp. of *prendre* take <L *praehendere* seize]

prize court A court sitting for the adjudication of prize causes. In the United States the federal courts have exclusive jurisdiction as prize courts.

prize crew A crew put on board a captured vessel by the captor, to navigate and carry her into port.

prize fight A fight between pugilists for a wager or prize, generally limited to a specified number of rounds. —**prize fighter** —**prize fighting**

prize money The proceeds of the sale of a maritime prize, distributable among the officers and crew of the vessel making the capture: abolished in the United States in 1899.

priz·er (prī′zər) *n.* 1 An appraiser. 2 *Archaic* A contestant for a prize, as in athletics.

prize ring A roped enclosure, 16 or 24 feet square, within which pugilists fight; also, with the definite article, professional pugilism.

pro[1] (prō) *n. pl.* **pros** 1 An argument or vote in favor of something: in the phrase *pros and cons.* 2 One who votes for or favors a proposal: usually in the plural. —*adv.* In behalf of; in favor of; for: to argue *pro* and con. [<L *pro* for]

pro[2] (prō) *n. pl.* **pros** *Colloq.* 1 A professional athlete. 2 An expert in any field.

pro–[1] *prefix* 1 Forward; to or toward the front from a position behind; forth: *produce,* to lead forth; *project,* to throw forth. 2 Forth from its place; away: *profugate,* to flee away. 3 To the front of; forward and down: *prolapse,* to slip forward and down. 4 Forward in time or direction: *proceed,* to go forward. 5 In front of: *prohibit,* to hold in front of. 6 In behalf of: *prolocutor.* 7 In place of; substituted for: *procathedral, proconsul.* 8 In favor of: *pro-Russian.* [<L *pro-* <*pro* before, forward, for]

pro–[2] *prefix* 1 Prior; occurring earlier in time: *prognosis.* 2 Situated in front; forward; before: *prognathous.* [<Gk. *pro-* <*pro* before, in front]

pro·a (prō′ə) *n.* A swift Malaysian vessel, sailing equally well in either direction, having a sharp stem and stern, a flat lee side, a single outrigger, and a lateen sail. Also *prahu.* [<Malay *prāu*]

prob·a·bil·ism (prob′ə·bəl·iz′əm) *n. Philos.* 1 The doctrine that certainty is unattainable, but that belief and action must be governed by probability. 2 The doctrine that, as long as the existence, interpretation, or application of a law remains truly doubtful, one may follow his own inclination, on the ground that a doubtful law cannot impose a certain obligation. [<L *probabilis*] —**prob′a·bil·ist** *n.* —**prob′a·bil·is′tic** *adj.*

prob·a·bil·i·ty (prob′ə·bil′ə·tē) *n. pl.* **·ties** 1 The state or quality of being probable; likelihood; also, a probable event or statement. 2 *Stat.* The ratio of the chances favoring an event to the total number of chances for and against it. [<F *probabilité* <L *probabilitas, -tatis* <*probabilis.* See PROBABLE.]

Synonyms: chance, credibility, likelihood, likeliness, presumption, verisimilitude. *Antonyms*: doubt, dubiousness, impossibility, improbability, inconceivability, inconceivableness, unlikelihood.

prob·a·ble (prob′ə·bəl) *adj.* 1 Having more evidence than the contrary, but not proof; likely to be true or to happen, but leaving room for doubt. 2 That renders something worthy of belief, but falls short of demonstration: *probable* evidence. [<OF <L *probabilis* <*probare* prove, test]

Synonyms: credible, likely, presumable, reasonable. See APPARENT, LIKELY. *Antonyms*: doubtful, dubious, improbable, incredible, questionable, unlikely.

probable cause A state of facts to warrant the belief that an accused person committed the crime charged.

prob·a·bly (prob′ə·blē) *adv.* In all probability; so far as the evidence shows; presumably.

pro·bands (prō′bandz) *n. Genetics* The original cases constituting the starting point of studies of a specific tainted family. [<L *probandus* to be proved, gerundive of *probare*]

pro·bang (prō′bang) *n. Med.* A slender, flexible rod, tipped with sponge, ball, button, or other attachment, especially used for the insertion of remedies into, or the removal of an obstruction from, the esophagus or larynx; also, a larger form for the relief of choking cattle. [Earlier *provang,* ? blend of obs. *provet* a probe + *fang* catch; infl. in form by *probe*]

pro·bate (prō′bāt) *adj.* 1 Of or pertaining to a probate court. 2 Pertaining to making proof: *probate* proceedings. —*n.* 1 Formal, legal proof, as of a will. 2 The right or jurisdiction of proving wills. Compare PROBATE COURT under COURT. —*v.t.* **·bat·ed, ·bat·ing** To secure probate of, as a will. [<L *probatus,* pp. of *probare* prove]

pro·ba·tion (prō·bā′shən) *n.* 1 A proceeding designed to test character, qualifications, etc., as of candidates for holy orders; examination; trial; novitiate. 2 In criminal administration, a method of allowing a person convicted of a minor offense to go at large under suspension of sentence, but usually under the supervision of a probation officer. 3 The period throughout which a trial or examination extends. 4 The act of proving; also, proof. [<L *probatio, -onis*] —**pro·ba′tion·al, pro·ba′tion·ar′y** *adj.*

pro·ba·tion·er (prō·bā′shən·ər) *n.* 1 One on probation or trial; a novice. 2 A candidate for membership in a church. 3 A convicted criminal or delinquent allowed to be at large but under the supervision of the convicting court and its probation officer.

probation officer A person delegated by the magistrate of a municipal criminal court to supervise an offender on suspended sentence.

pro·ba·tive (prō′bə·tiv) *adj.* 1 Serving to prove or test. 2 Pertaining to probation; proving. Also **pro·ba·to·ry** (prō′bə·tôr′ē, -tō′rē). [<L *probativus*]

probe (prōb) *v.* **probed, prob·ing** *v.t.* 1 To explore with a probe. 2 To investigate or examine thoroughly. —*v.i.* 3 To penetrate; search. —*n.* 1 *Med.* An instrument for exploring cavities, the course of wounds, etc. 2 That which proves or tests. 3 *U.S.* An examination; a searching investigation or inquiry, especially into crime or wrongdoing. [<L *probare* <*probus* good, proper. Doublet of PROVE.] —**prob′er** *n.*

pro·bi·ty (prō′bə·tē, prob′ə-) *n.* Virtue or integrity tested and confirmed; strict honesty. See synonyms under VIRTUE. [<F *probité* <L *probitas* <*probus* good, honest]

prob·lem (prob′ləm) *n.* 1 A perplexing question demanding settlement, especially when difficult or uncertain of solution; also, any puzzling circumstance or person. 2 *Math.* A proposition in which some operation or construction is required, as to bisect an angle; anything proposed to be worked out. See synonyms under RIDDLE[2]. —*adj.* 1 Presenting and dealing with a problem, especially a moral, sociological, or emotional problem: *problem* drama. 2 Being a problem, especially in point of behavior, maladjustment, etc.: a *problem* child. [<L *problema* <Gk. *problēma* something thrown forward (for discussion) <*pro-* forward + *ballein* throw]

prob·lem·at·ic (prob′ləm·at′ik) *adj.* Constituting or involving a problem; questionable; contingent. Also **prob′lem·at′i·cal.** [<Gk. *problēmatikos*] — **prob′lem·at′i·cal·ly** *adv.*

pro bo·no pub·li·co (prō bō′nō pub′li·kō) *Latin* For the public good; for the benefit of the public.

pro·bos·cid·i·an (prō′bə·sid′ē·ən) *n.* Any of an order (*Proboscidea*) of ungulates with columnar legs and a snout bearing a proboscis, consisting of the elephants and certain extinct related mammals, as the mammoth, mastodon, etc. —*adj.* 1 Pertaining or belonging to the *Proboscidea.* 2 Of, having, or pertaining to a proboscis.

pro·bos·cis (prō·bos′is) *n. pl.* **·bos·cis·es** or **·bos·ci·des** (-bos′ə·dēz) 1 *Zool.* A long flexible snout, as in the tapir; specifically, the trunk of an elephant. 2 *Entomol.* One of various tubular structures protruding or capable of being protruded from the front of the head of certain insects, as the combined mouth parts adapted for sucking in bees, or in certain dipterous insects, as the mosquito, the sheath and needlelike organs for piercing. 3 A human nose, especially when unusually large or prominent: a humorous use. [<L <Gk. *proboskis* <*pro-* before + *boskein* feed]

pro·caine (prō·kān′, prō′kān) *n.* A white crystalline compound, $C_{13}H_{20}O_2N_2$, used in its hydrochloride form as a local anesthetic. [<PRO-[1] + (CO)CAINE]

pro·cam·bi·um (prō·kam′bē·əm) *n. Bot.* The nascent tissue giving rise to the vascular bundle of plants. [<NL <PRO-[1] + CAMBIUM] — **pro·cam′bi·al** *adj.*

pro·carp (prō′kärp) *n. Bot.* A one– or several–celled female sexual organ in certain algae, which on fertilization becomes a sporocarp. [<PRO-[1] + -CARP]

pro·ca·the·dral (prō′kə·thē′drəl) *n.* A church or edifice used temporarily as a cathedral.

pro·ce·den·do (prō′sə·den′dō) *n. pl.* **·dos** *Law* A writ issued by a superior court to an inferior, remitting a cause that had been brought up on insufficient grounds, and commanding the inferior court to proceed to its determination. [<L, oblique case of *procedendum*, gerundive of *procedere* proceed]

pro·ce·dure (prə·sē′jər) *n.* 1 A manner of proceeding or acting; also, an act or a special course of action. 2 The methods or forms of conducting a business, collectively. 3 *Law* The methods of conducting judicial proceedings as distinguished from the legal definition and recognition of rights. 4 A course of action; a proceeding. 5 The manner of carrying on parliamentary affairs. See synonyms under OPERATION. [<F *procédure*] — **pro·ce′du·ral** *adj.*

pro·ceed (prə·sēd′) *v.i.* 1 To go on or forward, especially after a stop or interruption. 2 To begin and carry on an action or process: He *proceeded* to strike her about the head. 3 To issue or come, as from some cause, source, or origin: with *from.* 4 *Law* To institute and carry on legal proceedings. [<OF *proceder* <L *procedere* go forward <*pro-* forward + *cedere* go] — **pro·ceed′er** *n.*

pro·ceed·ing (prə·sē′ding) *n.* 1 An act or course of action; a transaction or procedure: an outrageous *proceeding.* 2 The action of issuing forth; emanation. 3 *pl.* The records or minutes of the meetings of a society, etc. 4 *Law* **a** Any action instituted in a court: a judicial *proceeding.* **b** Any of the various steps taken in a cause by either party: a *proceeding* by writ of error. See synonyms under ACT, TRANSACTION.

pro·ceeds (prō′sēdz) *n. pl.* The useful or material results of an action or course; also, that which accrues therefrom; the amount derived from the disposal of goods, work, or the use of capital; return; yield. See synonyms under HARVEST, PRODUCT, PROFIT.

proc·e·leus·mat·ic (pros′ə·lo͞os·mat′ik) *adj.* 1 In prosody, composed of four short syllables, or pertaining to feet so composed. 2 Animating or inciting, as a song. —*n.* A metrical foot of four short syllables. [<Gk. *prokeleusmatikos* <*prokeleusma* incitement <*prokeleuein* incite <*pro-* before + *keleuein* rouse]

pro·ce·phal·ic (prō′sə·fal′ik) *adj. Anat.* Of or pertaining to the anterior part of the head: a *procephalic* lobe of an invertebrate.

proc·ess (pros′es, *esp. Brit.* prō′ses) *n.* 1 A course or method of operations in the production of something: a metallurgical *process.* 2 A forward movement; progressive or continuous proceeding; passage; advance; course. 3 Any judicial writ or order issued at the commencement or during the progress of an action, as summons, citation, subpoena, or execution; especially, a writ issued to bring a defendant into court; also, the whole course of proceedings in a cause, civil or criminal, from beginning to end. 4 *Biol.* An accessory outgrowth or prominence of an organism. 5 *Physiol.* The fibrous prolongation from the body of the nerve cell (neuron) that carries the outgoing nervous impulse. 6 In patent law, a means of effecting a result otherwise than by mechanism, as by chemical action. 7 *Phot.* Any of the modern methods of producing relief printing surfaces by photography and mechanical or chemical means. —*adj.* 1 Produced by a special method: *process* butter; *process* cheese. 2 Pertaining to, for, or made by, a mechanical or chemical photographic process: a *process* illustration. —*v.t.* 1 To treat or prepare by a special method. 2 *Law* **a** To issue or serve a process on. **b** To proceed against. [<L *processus* progress, orig. pp. of *procedere.* See PROCEED.]

processing tax A tax imposed by the government on the processing of various farm products.

pro·ces·sion (prə·sesh′ən) *n.* 1 An array, as of persons or vehicles, arranged in succession and moving in a formal manner; a parade: a funeral *procession*; also, any continuous course: the *procession* of the stars. 2 The act of proceeding or issuing forth: the *procession* of the Holy Ghost from the Father. 3 A litany or hymn sung by persons moving in orderly array; a processional. —*v.i.* To march in procession. [<OF]

Synonyms: cavalcade, column, cortège, train. *Antonyms*: herd, mob, rabble, rout.

pro·ces·sion·al (prə·sesh′ən·əl) *adj.* Of or pertaining to or moving in a procession. —*n.* 1 A book containing the services in a religious procession. 2 A hymn sung during a religious procession. — **pro·ces′sion·al·ly** *adv.*

process printing Color printing from halftone plates each of which carries one of the primary colors, red, yellow, and blue, with sometimes a fourth plate for black.

process server A person, as a deputy sheriff, who serves summonses or processes.

pro·cès–ver·bal (prô·se′ver·bàl′) *n. pl.* **·baux** (-bō′) In French law, a detailed statement in writing made by an official relating to the commission of a crime within his jurisdiction; hence, any official report. [<F, lit., verbal process]

pro·chein (prō′shen) *adj. Law* Nearest in time, relation, or degree. Also **pro·chain** (prō′shān, *Fr.* prô·shaṅ′). [<F *prochain* <L *proximus* next]

pro·claim (prō·klām′) *v.t.* 1 To announce or make known publicly or officially; declare. 2 To make plain; manifest: His manner *proclaimed* his innocence. 3 To outlaw, prohibit, or restrict by proclamation. See synonyms under ANNOUNCE, AVOW, PUBLISH. [<OF *proclamer* <L *proclamare* <*pro-* before + *clamare* call] — **pro·claim′er** *n.*

proc·la·ma·tion (prok′lə·mā′shən) *n.* 1 The act of proclaiming. 2 That which is proclaimed; a public authoritative announcement. [<OF *proclamacion*]

pro·clit·ic (prō·klit′ik) *adj.* Attached to or dependent on a following word: said of monosyllables attached so closely as to have no separate accent. Compare ENCLITIC, ATONIC. —*n.* A proclitic word. [<NL *procliticus* <Gk. *proklinein* lean forward; formed on analogy of ENCLITIC]

pro·cliv·i·ty (prō·kliv′ə·tē) *n. pl.* **·ties** Natural disposition or tendency; propensity: usually with *to*: a *proclivity* to grumble. See synonyms under APPETITE, DESIRE, INCLINATION. [<L *proclivitas* <*proclivus* downward <*pro-* before + *clivus* slope]

Proc·ne (prok′nē) In Greek mythology, an Athenian princess whom the gods transformed into a swallow after she killed her son. Compare PHILOMELA.

pro·con·sul (prō·kon′səl) *n.* 1 In ancient Rome, an official, usually an ex–consul, who exercised consular authority over a province or an army. 2 A governor of a dependency, especially a British one; a viceroy. [<L] — **pro·con′su·lar** (-sə·lər) *adj.* — **pro·con′su·late** (-sə·lit), **pro·con′sul·ship** *n.*

Pro·con·sul (prō·kon′səl) *n. Paleontol.* An extinct ape related to Dryopithecus. [<NL]

Pro·co·pi·us (prō·kō′pē·əs), 500?–565?, Byzantine historian.

pro·cras·ti·nate (prō·kras′tə·nāt) *v.* **·nat·ed, ·nat·ing** *v.i.* To put off taking action until a future time; be dilatory. —*v.t.* To defer or postpone. [<L *procrastinatus*, pp. of *procrastinare* <*pro-* forward + *crastinus* pertaining to the morrow <*cras* tomorrow] — **pro·cras′ti·na′tor** *n.*

Synonyms: adjourn, defer, delay, postpone. See POSTPONE. *Antonyms*: accelerate, dispatch, drive, expedite, hasten, hurry, press, quicken, urge.

pro·cras·ti·na·tion (prō·kras′tə·nā′shən) *n.* The act, tendency, or habit of procrastinating; dilatoriness; delay.

pro·cre·ant (prō′krē·ənt) *adj.* Effecting, conducive to, or connected with procreation or reproduction; generating; fruitful. [<L *procreans, -antis*]

pro·cre·ate (prō′krē·āt) *v.t.* **·at·ed, ·at·ing** 1 To engender or beget (offspring). 2 To originate; produce. See synonyms under PROPAGATE. [<L *procreatus*, pp. of *procreare* <*pro-* before + *creare* create] — **pro′cre·a′tion** *n.* — **pro′cre·a′tor** *n.*

pro·cre·a·tive (prō′krē·ā′tiv) *adj.* Possessed of generative power; reproductive; pertaining to procreation.

Pro·crus·te·an (prō·krus′tē·ən) *adj.* 1 Pertaining to or characteristic of Procrustes. 2 Hence, ruthlessly or violently forcing to conform.

Pro·crus·tes (prō·krus′tēz) In Greek mythology, an Attic giant, killed by Theseus, who tied travelers to an iron bed and amputated or stretched their limbs until they fitted it. [<L <Gk. *Prokroustēs* <*prokrouein* stretch out <*pro-* thoroughly + *krouein* beat]

pro·cryp·tic (prō·krip′tik) *adj. Biol.* 1 Having protective or imitative coloration: said of certain animals, insects, etc. 2 Having the power to adapt coloration to environment, as chameleons. [<PRO-[1] + CRYPTIC]

procto– *combining form Med.* Related to or affecting the rectum or anus: *proctology.* Also, before vowels, **proct–.** [<Gk. *proktos* the anus]

proc·tol·o·gy (prok·tol′ə·jē) *n.* The branch of medicine which treats of the anatomy, physiology, and diseases of the rectum. [<PROCTO- + -LOGY] — **proc·to·log·i·cal** (prok′tə·loj′i·kəl) *adj.* — **proc·tol′o·gist** *n.*

proc·to·plas·ty (prok′tə·plas′tē) *n.* Plastic surgery of the rectum and anus. [<PROCTO- + -PLASTY]

proc·tor (prok′tər) *n.* 1 An agent acting for another; attorney; proxy; specifically, a practitioner in an admiralty, ecclesiastical, or probate court. 2 A university or college official charged with maintaining order, supervising examinations, etc. —*v.t.* & *v.i.* To supervise (an examination). [ME *proketour, procutour*, contraction of L *procurator* PROCURATOR] — **proc·to·ri·al** (prok·tôr′ē·əl, -tō′rē-) *adj.* — **proc′tor·ship** *n.*

proc·to·scope (prok′tə·skōp) *n.* A surgical instrument for examining the interior of the rectum. — **proc·tos·co·py** (prok·tos′kə·pē) *n.*

Proc·u·lus (prok′yə·ləs), 412?–485, Greek Neo–Platonist and religious commentator. Also **Pro·clos** (prō′kləs, prok′ləs).

pro·cum·bent (prō·kum′bənt) *adj.* 1 *Bot.* Lying on the ground; trailing: said of certain vines and trailing plants. 2 Leaning forward or lying down or on the face; prone; prostrate. [<L *procumbens, -entis,* ppr. of *procumbere* lean forward <*pro-* forward + *cubare* lie down]

pro·cur·a·ble (prō·kyŏŏr′ə·bəl) *adj.* That can be procured.

proc·u·ra·cy (prok′yər·ə·sē) *n. pl.* **·cies** The management of another's affairs; the office or service of a procurator or proctor.

pro·cur·ance (prō·kyŏŏr′əns) *n.* The process of procuring. Also **pro·cur′al.**

proc·u·ra·tion (prok′yə·rā′shən) *n.* **1** The act of procuring. **2** *Law* **a** The function of an attorney; an agency; a proxy. **b** A power of attorney. [<F <L *procuratio, -onis*] —**proc′u·ra·to′ry** (-rə·tôr′ē, -tō′rē) *adj.*

proc·u·ra·tor (prok′yə·rā′tər) *n.* **1** A person authorized and employed to act for and manage the affairs of another. **2** In ancient Rome, one who had charge of the imperial revenues; an imperial collector, especially in a province; a provincial administrator; a viceroy. **3** The public magistrate of some Italian cities. [<L <*procurare.* See PROCURE.] —**proc′u·ra·to′ri·al** (-rə·tôr′ē·əl, -tō′rē-) *adj.* —**proc′u·ra′tor·ship** *n.*

pro·cure (prō·kyŏŏr′) *v.* **·cured, ·cur·ing** *v.t.* **1** To obtain by some effort or means; acquire. **2** To bring about; cause. **3** To obtain (women) for the gratification of the lust of others. —*v.i.* **4** To be a procurer or procuress. See synonyms under GAIN, GET, OBTAIN, PROVIDE, PURCHASE. [<F <L *procurare* look after <*pro-* on behalf of + *curare* attend to <*cura* care]

pro·cure·ment (prō·kyŏŏr′mənt) *n.* **1** The act of procuring; obtainment; attainment. **2** The act of effecting or causing to be effected.

pro·cur·er (prō·kyŏŏr′ər) *n.* One who procures for another, as to gratify lust; a pander. [<AF *procurour* <L *procurator*] —**pro·cur′ess** *n. fem.*

Pro·cy·on (prō′sē·on) *n.* The most conspicuous star in the constellation Canis Minor; magnitude, 0.5. See STAR. [<L <Gk. *Prokyōn* <*pro-* before + *kyōn* dog]

prod (prod) *v.t.* **prod·ded, prod·ding** **1** To punch or poke with or as with a pointed instrument. **2** To arouse mentally; urge; goad. —*n.* **1** Any pointed instrument used for prodding; a goad. **2** A thrust or punch with or as with a prod; a poke. **3** Hence, a reminder. [Origin unknown] —**prod′der** *n.*

prod·i·gal (prod′ə·gəl) *adj.* **1** Addicted to wasteful expenditure, as of money, time, or strength; extravagant. **2** Yielding in profusion; bountiful. **3** Lavish; profuse. —*n.* One who is wasteful or profligate; a spendthrift. See synonyms under IMPROVIDENT. [<OF <Med. L *prodigalis* <L *prodigus* wasteful <*prodigere* drive forth, get rid of <*pro-* forward + *agere* drive] —**prod′i·gal·ly** *adv.*

prod·i·gal·i·ty (prod′ə·gal′ə·tē) *n. pl.* **·ties** Extravagance; wastefulness; lavishness; also, bounteousness. See synonyms under EXCESS. [<OF *prodigalité*]

pro·di·gious (prə·dij′əs) *adj.* **1** Enormous or extraordinary in size, quantity, or degree; vast; excessive. **2** Marvelous; amazing. **3** *Obs.* Of the nature of a prodigy. See synonyms under IMMENSE. [<L *prodigiosus*] —**pro·dig′ious·ly** *adv.* —**pro·dig′ious·ness** *n.*

prod·i·gy (prod′ə·jē) *n. pl.* **·gies** **1** Something so extraordinary as to excite wonder and admiration. **2** A person or thing of remarkable qualities or powers: an infant *prodigy.* **3** Something out of the ordinary course of nature; a monstrosity. **4** *Archaic* A portent. [<L *prodigium*]

Synonyms: marvel, monster, miracle, portent, wonder.

pro·drome (prō′drōm) *n. Pathol.* A sign of approaching disease; a premonitory symptom. [<F <L *prodromus* <Gk. *prodromos* forerunner <*pro-* before + *dromos* a running] —**prod·ro·mal** (prod′rə·məl) *adj.*

pro·duce (prə·dōōs′, -dyōōs′) *v.* **·duced, ·duc·ing** *v.t.* **1** To bring forth or bear; yield, as young or a natural product. **2** To bring forth by mental effort; compose, write, etc.: to *produce* a book. **3** To bring about; cause to happen or be: His words *produced* a violent reaction. **4** To bring to view; exhibit; show: to *produce* evidence. **5** To manufacture; make. **6** To bring to performance before the public, as a play. **7** To extend or lengthen, as a line. **8** *Econ.* To create (anything with exchangeable value). —*v.i.* **9** To yield or generate an appropriate product or result.
—**prod·uce** (prod′ōōs, -yōōs, prō′dōōs, -dyōōs) *n.* That which is produced; a product; specifically, farm products collectively. See synonyms under HARVEST, PRODUCT, WEALTH. [<L *producere* lead forward <*pro-* forward + *ducere* lead] —**pro·duc′i·ble** *adj.*

Synonyms (verb): bear, breed, cause, create, effect, engender, furnish, generate, make, manufacture, occasion, originate, propagate, yield. See ALLEGE, EFFECT, PROVIDE.

pro·duc·er (prə·dōō′sər, -dyōō′-) *n.* **1** One who produces. **2** One who cultivates or makes things for sale and use in distinction from the user or consumer. **3** That which produces or generates. **4** An apparatus for manufacturing producer gas.

producer gas A combustible gas formed by driving air and steam over burning coke: used for heating and to drive engines for power.

producers' goods *Econ.* Goods having indirect use, as tools or raw materials used in making other goods: opposed to *consumers' goods.*

prod·uct (prod′əkt, -ukt) *n.* **1** Anything produced or obtained as a result of some operation or work, as by generation, growth, labor, study, or skill. **2** *Math.* The result obtained by multiplication. **3** *Chem.* Any substance resulting from chemical change. Compare EDUCT. [<L *productus,* pp. of *producere.* See PRODUCE.]

Synonyms: crop, effect, fruit, harvest, outcome, output, proceeds, produce, production, result, return, yield. See HARVEST, WORK.

pro·duc·tile (prə·duk′til) *adj.* Capable of being extended or drawn out. [<PRO-[1] + DUCTILE]

pro·duc·tion (prə·duk′shən) *n.* **1** The act or process of producing. **2** In political economy, a producing for use, involving the creating or increasing of economic wealth: in contradistinction to *consumption* (by use). **3** That which is produced or made; any tangible result of industrial, artistic, or literary labor. [<F <L *productio, -onis* a prolongation]

Synonyms: composition, performance, work. See PRODUCT, WORK.

pro·duc·tive (prə·duk′tiv) *adj.* **1** Producing or tending to produce; fertile; creative, as of artistic things. **2** Producing or tending to produce profits or increase in quantity, quality, or value: *productive* labor. **3** Causing; resulting in: with *of.* See synonyms under FERTILE, PROFITABLE. [<Med. L *productivus* <LL, fit for production] —**pro·duc′tive·ly** *adv.* —**pro·duc·tiv·i·ty** (prō′·duk·tiv′ə·tē), **pro·duc′tive·ness** *n.*

pro·em (prō′əm) *n.* An introductory statement; preface; prelude. [<OF *proeme* <L *prooemium* <Gk. *prooimion* an overture <*pro-* before + *oimē* way of a song, lay] —**pro·e·mi·al** (prō·ē′mē·əl) *adj.*

pro et con (prō′ et kon′) *Latin* For and against.

prof·a·na·tion (prof′ə·nā′shən) *n.* **1** The act of profaning; abuse or dishonoring of sacred things; desecration. **2** Abusive or improper treatment of anything; misuse. [<F <LL *profanatio, -onis*]

pro·fane (prə·fān′) *v.t.* **·faned, ·fan·ing** **1** To treat (something sacred) with irreverence or abuse; desecrate; pollute. **2** To put to an unworthy or degrading use; debase. See synonyms under VIOLATE. —*adj.* **1** Manifesting irreverence, disrespect, or undue familiarity toward the Deity or sacred things; blasphemous. **2** Secular: opposed to *sacred.* **3** Not initiated into the inner mysteries; hence, vulgar; common. [<F *profaner* <L *profanare* <*profanus* before or outside the temple, hence, unsacred <*pro-* before + *fanum* temple] —**pro·fan·a·to·ry** (prə·fan′ə·tôr′ē, -tō′rē) *adj.* —**pro·fane′ly** *adv.* —**pro·fan′er** *n.*

Synonyms (adj.): blasphemous, godless, impious, irreligious, sacrilegious, secular, temporal, unconsecrated, ungodly, unhallowed, unholy, unsanctified, wicked, worldly. *Antonyms:* consecrated, devout, godly, holy, pious, religious, reverent, sacred, sanctified, spiritual.

pro·fan·i·ty (prə·fan′ə·tē) *n. pl.* **·ties** **1** The state of being profane. **2** Profane speech or action. Also **pro·fane′ness** (-fān′nis). See synonyms under OATH.

pro·fa·num vul·gus (prō·fā′nəm vul′gəs) *Latin* The common herd.

pro·fert (prō′fərt) *n. Law* The formal allegation in a pleading or on the record that the pleader produces in court an instrument on which an action or defense is founded. [<L, he brings forward]

pro·fess (prə·fes′) *v.t.* **1** To declare openly; avow; affirm. **2** To assert, usually insincerely; make a pretense of: to *profess* remorse. **3** To declare or affirm faith in: to *profess* Taoism. **4** To claim skill or learning in; have as one's profession: to *profess* the law. **5** To receive into a religious order. —*v.i.* **6** To make open declaration; avow; offer public affirmation. **7** To take the vows of a religious order. See synonyms under ACKNOWLEDGE, AVOW, PRETEND. [<OF *professe,* fem. of *profes* bound by a vow <L *professus,* pp. of *profiteri* avow, confess <*pro-* before + *fateri* confess]

pro·fess·ed·ly (prə·fes′id·lē) *adv.* **1** By open profession; avowedly. **2** Pretendedly.

pro·fes·sion (prə·fesh′ən) *n.* **1** An occupation that properly involves a liberal education or its equivalent, and mental rather than manual labor; especially, one of **the three learned professions,** law, medicine, or theology. **2** Hence, any calling or occupation other than commercial, manual, etc., involving special attainments or discipline, as editing, music, teaching, etc.; also, the collective body of those following such vocation. **3** The act of professing or declaring; declaration; avowal: *professions* of good will. **4** That which is avowed or professed; a declaration; a faith; also, a pretense: His *professions* are not trustworthy. See synonyms under BUSINESS. [<F]

pro·fes·sion·al (prə·fesh′ən·əl) *adj.* **1** Connected with, preparing for, engaged in, appropriate, or conforming to a profession: *professional* courtesy, a *professional* soldier, a *professional* job. **2** Of or pertaining to a special occupation, often for gain: opposed to *amateur:* a *professional* ball game or player. —*n.* **1** One who pursues as a business some vocation or occupation. **2** A person who engages for money to compete in sports: opposed to *amateur.* **3** One skilled in a profession. —**pro·fes′sion·al·ly** *adv.*

pro·fes·sion·al·ism (prə·fesh′ən·əl·iz′əm) *n.* **1** The methods, manner, or spirit of a profession; also, its practitioners. **2** The practice of some profession as a business: opposed to *amateurism.*

pro·fes·sor (prə·fes′ər) *n.* **1** A teacher of the highest grade in a university or college, or in an institution where professional or technical studies are pursued; usually, an officer holding a chair in some particular branch of higher instruction. **2** One who professes skill and offers instruction in some sport or art: a *professor* of gymnastics. **3** One who makes open declaration of his opinions or sentiments; specifically, one who avows a religious faith. [<L, a public teacher <*professus.* See PROFESS.]

pro·fes·sor·ate (prə·fes′ər·it) *n.* The position of a professor.

pro·fes·so·ri·al (prō′fə·sôr′ē·əl, -sō′rē-, prof′ə-) *adj.* Of or pertaining to a professor; pedagogic; academic. —**pro′fes·so′ri·al·ly** *adv.*

pro·fes·so·ri·ate (prō′fə·sôr′ē·it, -sō′rē-, prof′·ə-) *n.* Professors collectively, as in a college; professorship.

pro·fes·sor·ship (prə·fes′ər·ship) *n.* The office and duties of a professor; the state of being a professor.

prof·fer (prof′ər) *v.t.* To offer for acceptance. —*n.* The act of proffering, or that which is proffered; a tender; offer. [<AF *proffrir,* OF *poroffrir* <*por-* (<L *pro-*) in behalf of + L *offerre.* See OFFER.] —**prof′fer·er** *n.*

pro·fi·cien·cy (prə·fish′ən·sē) *n. pl.* **·cies** An advanced state of attainment in some knowledge, art, or skill; expertness.

pro·fi·cient (prə·fish′ənt) *adj.* Thoroughly versed, as in an art or science; skilled; expert. —*n.* An expert in any branch of skill or knowledge; an adept. See synonyms under SKILFUL. [<L *proficiens, -entis,* ppr. of *proficere* make progress, go forward <*pro-* forward + *facere* do] —**pro·fi′cient·ly** *adv.*

pro·file (prō′fīl, *esp. Brit.* prō′fēl) *n.* **1** An outline, or contour; a drawing in outline. **2** *Archit.* The outline of a perpendicular section of a building, fort, etc., or the contour of an architectural member, as a base or cornice. **3** A drawing showing the outline of a human face or figure as seen from the side. **4** A short biographical sketch vividly presenting the most striking characteristics of a personality. **5** A vertical section of soil extending from the surface through all its levels to the underlying parent material. — *v.t.* **·filed, ·fil·ing 1** To draw a profile of; outline. **2** To write a profile of. [<Ital. *profilo, proffilo* outline <*proffilare* draw in outline <L *pro-* forward + *filum* thread, line]

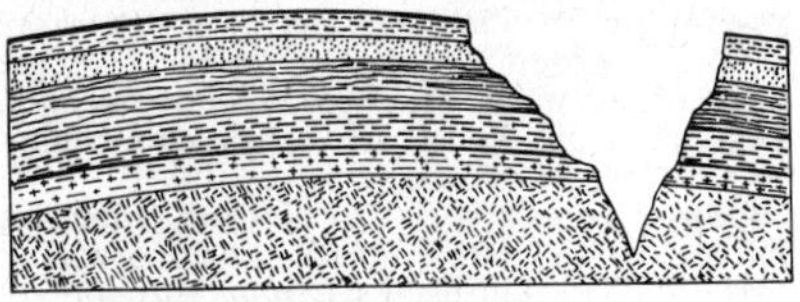

PROFILE OF GRAND CANYON AND KAIBAB PLATEAU

profile drag *Aeron.* The difference between the total wing drag of an airplane and the induced drag.

prof·it (prof′it) *n.* **1** Any accession of good—physical, mental, or moral—from labor or exertion; benefit; return. **2** *Often pl.* Excess of returns over outlay or expenditure: a business yielding fair *profits.* **3** The return from the employment of capital after deducting the amount paid for raw material and for wages, real or estimated rent, interest, insurance, etc. **4** That part of the amount received for goods which exceeds the sum originally paid for them with or without all secondary expenses involved. **5** The income of invested property without counting its increased value by any actual rise in the market. **6** In invested capital, the ratio of the increment to the actual amount of capital for a given year. —**gross profit** The profit apparent on the face of a transaction or business; the excess of receipts from sales over expenditures for purchase: opposed to **net profit,** the surplus remaining after all necessary deductions, as for interest, transportation, bad debts, etc. —*v.i.* **1** To be of advantage or benefit. **2** To derive gain or benefit. —*v.t.* **3** To be of profit or advantage to. ◆ Homophone: *prophet.* [<OF <L *profectus,* pp. of *proficere* go forward. See PROFICIENT.]

Synonyms (noun): advantage, avail, benefit, emolument, expediency, gain, good, improvement, proceeds, receipts, return, returns, service, utility, value. The *returns* or *receipts* include all that is received from any outlay or investment; the *profit* is the excess (if any) of the *receipts* over the outlay; hence, in government, morals, etc., the *profit* is what is really good, helpful, useful, valuable. *Utility* is chiefly used in the sense of some immediate or personal and generally some material *good. Advantage* is that which gives one a vantage ground, either for coping with competitors or with difficulties, needs, or demands; as, to have the *advantage* of a good education; it is frequently used of what one has beyond another or secures at the expense of another; as, to have the *advantage* in argument, or to take *advantage* in a bargain. *Gain* is what one secures beyond what he previously possessed. *Benefit* is anything that does one good. *Emolument* is *profit, return,* or *value* accruing through official position. *Expediency* has respect to *profit* or *advantage,* real or supposed, considered apart from or perhaps in opposition to right, in actions having a moral character. See UTILITY. *Antonyms*: damage, detriment, disadvantage, harm, hurt, injury, loss, ruin, waste.

prof·it·a·ble (prof′it·ə·bəl) *adj.* Bringing profit or gain; remunerative; advantageous. —**prof′it·a·ble·ness** *n.* —**prof′it·a·bly** *adv.*

Synonyms: advantageous, beneficial, desirable, expedient, gainful, lucrative, productive, remunerative, useful. See EXPEDIENT, GOOD, USEFUL. Compare synonyms for PROFIT. *Antonyms*: detrimental, disadvantageous, disastrous, fruitless, harmful, hurtful, undesirable, unproductive, unprofitable, worthless.

profit and loss In bookkeeping, an account in the ledger in which profits are entered on the creditor side and losses on the debtor side. —**prof′it-and-loss′** (prof′it·ənd·lôs′, -los′) *adj.*

prof·i·teer (prof′ə·tir′) *v.i.* To seek or obtain excessive profits. —*n.* One who is given to making excessive profits, especially to the detriment of others. —**prof′i·teer′ing** *n.*

prof·it·less (prof′it·lis) *adj.* Resulting in no gain or benefit; unprofitable.

prof·it–shar·ing (prof′it·shâr′ing) *n.* A system of remuneration by which workmen are given a percentage, according to wages, of the net profits of a business. —*adj.* Of or related to profit–sharing.

prof·li·ga·cy (prof′lə·gə·sē) *n.* **1** Corruptness of morals; viciousness of character or conduct. **2** Great extravagance; wastefulness; overabundance. Also **prof′li·gate·ness** (-git·nis, -gāt′nis).

prof·li·gate (prof′lə·git, -gāt) *adj.* **1** Lost or insensible to principle, virtue, or decency; abandoned to vice. **2** Recklessly extravagant; in great profusion. —*n.* **1** A depraved or dissolute person. **2** A reckless spendthrift. See synonyms under IMMORAL. [<L *profligatus,* pp. of *profligare* strike to the ground, destroy <*pro-* forward + *fligere* dash] —**prof′li·gate·ly** *adv.*

prof·lu·ent (prof′lo͞o·ənt) *adj.* Flowing smoothly or plentifully; fluent. [<L *profluens, -entis,* ppr. of *profluere* flow along <*pro-* before + *fluere* flow] —**prof′lu·ence** *n.*

pro for·ma (prō fôr′mə) *Latin* As a matter of form.

pro·found (prə·found′) *adj.* **1** Intellectually deep; thorough; exhaustive: *profound* learning. **2** Reaching to, arising from, or affecting the depth of one's nature or of any matter: *profound* respect. **3** Situated far below the surface; deep; unfathomable. **4** Bent low: said of a bow. —*n.* **1** A fathomless depth; an abyss. **2** The ocean; the deep. See synonyms under OBSCURE, WISE. [<OF *profond* <L *profundus* <*pro-* very + *fundus* deep] —**pro·found′ly** *adv.* —**pro·found′ness** *n.*

pro·fun·di·ty (prə·fun′də·tē) *n. pl.* **·ties 1** The state or quality of being profound, in any sense; depth. **2** A deep place or thing. **3** A profound or abstruse statement, theory, or the like. See synonyms under WISDOM. [<OF *profundité* <LL *profunditas*]

pro·fuse (prə·fyo͞os′) *adj.* **1** Giving or given forth lavishly; liberal; extravagant; prodigal. **2** Copious; overflowing: *profuse* gratitude, *profuse* vegetation. [<L *profusus,* pp. of *profundere* pour forth <*pro-* forward + *fundere* pour] —**pro·fuse′ly** *adv.* —**pro·fuse′·ness** *n.*

pro·fu·sion (prə·fyo͞o′zhən) *n.* **1** A lavish supply or condition; plenty: a *profusion* of ornaments. **2** The act of pouring forth or supplying in great abundance; prodigality: *profusion* in giving. See synonyms under EXCESS. [<F]

prog (prog) *Dial. v.i.* **progged, prog·ging** To prowl about for food or plunder. —*n.* Food obtained by begging. [? Blend of PROD and BEG] —**prog′ger** *n.*

pro·gen·i·tor (prō·jen′ə·tər) *n.* A forefather or parent. [<L <*progenitus,* pp. of *progignere* beget <*pro-* forth + *gignere* beget] —**pro·gen′i·tor·ship′** *n.*

prog·e·ny (proj′ə·nē) *n. pl.* **·nies** Offspring. [<L *progenies* <*progignere.* See PROGENITOR.]

pro·ge·ri·a (prō·jir′ē·ə) *n. Pathol.* Retarded development with premature senility. [<NL <PRO-¹ + Gk. *gēras* old age]

pro·ges·ta·tion·al (prō′jes·tā′shən·əl) *adj. Med.* **1** Promoting gestation. **2** Designating those substances and processes which are active in the menstrual cycle or during pregnancy.

pro·ges·ter·one (prō·jes′tə·rōn) *n. Biochem.* A hormone from the corpus luteum: isolated as a white, crystalline compound, $C_{21}H_{30}O_2$, and also made synthetically. It is active in preparing the uterus for reception of the fertilized ovum. [<PRO-¹ + GE(STATION) + STER(OL) + -ONE]

pro·ges·tin (prō·jes′tin) *n. Biochem.* **1** Any substance which promotes the gestational activity of the corpus luteum and uterus after fertilization of the ovum. **2** Progesterone. [<PRO-¹ + GEST(ATION) + -IN]

pro·glot·tid (prō·glot′id) *n. pl.* **·glot·ti·des** (-glot′ə·dēz) *Zool.* One of the segments or joints of a tapeworm, in which the reproductive organs develop. Also **pro·glot′tis.** [<NL *proglottis, proglottidis* <Gk. *proglossis* tip of the tongue; from their shape] —**pro·glot′tic** *adj.*

prog·na·thous (prog′nə·thəs, prog·nā′-) *adj.* Having projecting jaws: opposed to *opisthognathous.* Also **prog·nath·ic** (prog·nath′ik). [<PRO-² + -GNATHOUS] —**prog·na·thism** (prog′nə·thiz′əm), **prog·na·thy** (prog′nə·thē) *n.*

prog·no·sis (prog·nō′sis) *n.* **1** *Med.* A prediction or conclusion in regard to the course and termination of a disease. **2** Any prediction or forecast; foreknowledge. [<NL <Gk. *prognōsis* <*pro-* before + *gignōskein* know]

prog·nos·tic (prog·nos′tik) *adj.* Betokening something future by signs or symptoms; relating to prognosis. —*n.* **1** A sign of some future occurrence; an omen. **2** *Med.* A symptom indicative of the course of a disease. See synonyms under SIGN. [<Med. L *prognosticum* omen <Gk. *prognōstikon* <*prognōsis.* See PROGNOSIS.]

prog·nos·ti·cate (prog·nos′tə·kāt) *v.t.* **·cat·ed, ·cat·ing 1** To foretell (future events, etc.) by present indications. **2** To indicate beforehand; foreshadow: Events *prognosticate* war. See synonyms under AUGUR, PROPHESY. [<Med. L *prognosticatus,* pp. of *prognosticare* <L *prognosticum.* See PROGNOSTIC.] —**prog·nos′ti·ca′tor** *n.*

prog·nos·ti·ca·tion (prog·nos′tə·kā′shən) *n.* **1** The act of prognosticating; prediction. **2** That which foretokens.

pro·gram (prō′gram, -grəm) *n.* **1** A list giving in order the items, turns, selections, etc., making up an entertainment; also, the selections, etc., collectively. **2** Any prearranged plan or course of proceedings; a prospectus. **3** *Electronics* A sequence of instructions set up on the control panels of an electronic computer as guides in the performance of a desired operation or group of operations. **4** A preface, or prefatory statement. **5** *Obs.* A public proclamation; official edict or decree. Also **pro′gramme.** —*v.t.* **·gramed** or **·grammed, ·gram·ing** or **·gram·ming** To arrange in an appropriate sequence the separate items of (a program, set of instructions, etc.). [<F *programme* <LL *programma* public announcement <Gk. <*prographein* write in public <*pro-* before + *graphein* write] —**pro·gram·mat·ic** (prō′grə·mat′ik) *adj.*

program music Descriptive music; music having a subject and intended to suggest or convey a succession of moods, scenes, or incidents: distinguished from *absolute music.*

Pro·gre·so (prō·grā′sō) A port of entry in Yucatán state, SE Mexico.

prog·ress (prog′res, *esp. Brit.* prō′gres) *n.* **1** A moving forward in space; movement forward nearer a goal. **2** Advancement toward maturity or completion; gradual development, as of mankind or civilization; improvement. **3** A journey of state, as of a monarch over his realm.

—**pro·gress** (prə·gres′) *v.i.* **1** To move forward or onward. **2** To advance toward completion or fuller development. [<OF *progres* <L *progressus,* pp. of *progredi* go forward <*pro-* forward + *gradi* walk]

Synonyms (noun): advance, advancement, attainment, development, growth, improvement, increase, proficiency, progression. *Attainment, development,* and *proficiency* are more absolute than the other words of the group, denoting some point of advantage or of comparative perfection reached by forward or onward movement; we speak of *attainments* in scholarship, *proficiency* in music or languages, the *development* of new powers or organs; *proficiency* includes the idea of skill. *Advance* denotes a forward movement or the point gained by forward movement; *progress* (Latin *progredior,* walk forward) is steady and constant forward movement, admitting of pause, but not of retreat. Compare ATTAIN. *Antonyms*: check, decline, delay, retreat, recession, retrogression, stay, stop, stoppage.

pro·gres·sion (prə·gresh′ən) *n.* **1** The act of progressing; advancement. **2** *Math.* A sequence of numbers or quantities each of which is derived from the preceding by a constant law. See ARITHMETIC PROGRESSION, GEOMETRIC PROGRESSION, SERIES (def. 2). **3** *Music* a An advance from one tone or chord to another. b A sequence or succession of tones or chords. 4 Course or lapse of time; passage. See synonyms under PROGRESS. [<L *progressio, -onis*] — **pro·gres′sion·al** *adj.* — **pro·gres′sion·ism** *n.*

pro·gres·sion·ist (prə·gresh′ən·ist) *n.* 1 One who believes that society is progressing toward perfection. 2 An evolutionist.

prog·ress·ist (prog′res·ist, prō′gres-) *n.* 1 A progressionist. 2 A member of any party devoted to some scheme of progress.

pro·gres·sive (prə·gres′iv) *adj.* 1 Moving forward; advancing: *progressive* movement; also, moving forward gradually or step by step. 2 Aiming at or characterized by progress. 3 Spreading from one part to others; increasing: said of a disease: *progressive* paralysis. 4 Striving for or favoring progress or reform, especially social, political, educational or religious: a *progressive* party, *progressive* schools. 5 *Gram.* Designating an aspect of the verb which expresses the action as being in progress at some time in the past, present, or future: formed with any tense of the auxiliary *be* and the present participle; as, He *is speaking*; he *had been speaking*; he *was to have been speaking*; he *will be speaking*. See synonyms under GRADUAL. — *n.* 1 One who believes in progress or progressive methods; especially, one who favors or promotes reforms or changes, as in politics or religion; a radical: opposed to *conservative* or *reactionary*. 2 *Gram.* A progressive verb form. — **pro·gres′sive·ly** *adv.* — **pro·gres′sive·ness** *n.* — **pro·gres′siv·ism** *n.* — **pro·gres′siv·ist** *n.*

Progressive party 1 A political party formed under the leadership of Theodore Roosevelt in 1912, which sought political and labor reforms and social security legislation. 2 A political party seeking labor and agricultural reforms, formed in 1924 under the leadership of Robert M. LaFollette. 3 A political party formed in 1948, which nominated Henry A. Wallace for president on a platform advocating full employment and a modification of the then current U.S. foreign policy, particularly in respect to the U.S.S.R.

pro·hib·it (prō·hib′it) *v.t.* 1 To forbid, especially by authority or law; interdict. 2 To prevent or hinder. [<L *prohibitus,* pp. of *prohibere* <*pro-* before + *habere* have] — **pro·hib′it·er** *n.*

Synonyms: debar, disallow, forbid, hinder, inhibit, interdict, preclude, prevent. *Debar* is said of persons, *disallow* of acts; one is *debarred* from anything when shut off by authority or necessity; an act is *disallowed* by the authority that might have allowed it. *Forbid* is less formal and more personal, *prohibit* more official and judicial, with the implication of readiness to use force; a parent *forbids* a child to take part in some game or to associate with certain companions; the opium trade is now *prohibited* by the leading nations of the world. Many things are *prohibited* by law which cannot be wholly *prevented,* as gambling and prostitution; on the other hand, things may be *prevented* which are not *prohibited,* as the services of religion, the payment of bets or military conquest. Compare ABOLISH, HINDER, PREVENT, SHUT. *Antonyms*: allow, authorize, command, direct, empower, enjoin, let, license, order, permit, require, sanction, suffer, tolerate, vouchsafe, warrant.

pro·hi·bi·tion (prō′ə·bish′ən) *n.* 1 The act of prohibiting, preventing, or stopping; also, a decree or order forbidding anything; an interdiction. 2 The forbidding of the manufacture, transportation, and sale of alcoholic liquors as beverages: instituted in the United States effective January 16, 1920. See synonyms under BARRIER, ORDER. [<L *prohibitio, -onis*]

Prohibition Amendment The Eighteenth Amendment to the Constitution of the United States, ratified January 1919, prohibiting the manufacture, sale, or transportation of intoxicating liquors for beverage purposes: repealed in 1933. Compare VOLSTEAD ACT.

Pro·hi·bi·tion·ist (prō′ə·bish′ən·ist) *n.* 1 One who believes in prohibition. 2 One who favors the prohibition by law of the manufacture and sale of alcoholic liquors as beverages.

Prohibition party A political party advocating the prohibition by law of the manufacture and sale of alcoholic liquors as beverages.

pro·hib·i·tive (prō·hib′ə·tiv) *adj.* Prohibiting or tending to prohibit. Also **pro·hib′i·to′ry** (-tôr′ē, -tō′rē). — **pro·hib′i·tive·ly** *adv.*

proj·ect (proj′ekt) *n.* 1 Something proposed or mapped out in the mind, as a course of action; a plan. 2 In schools, a problem involving the theory of the subject matter, given to a student or group of students to be worked out in practice.

— **pro·ject** (prə·jekt′) *v.t.* 1 To cause to extend forward or out. 2 To throw forth or forward, as missiles. 3 To visualize as an external reality: to *project* an image of one's destiny. 4 To cause (an image, shadow, etc.) to fall on a surface. 5 To propose or plan. 6 *Math.* a To make a projection of (a solid, etc.) on a plane. b To reproduce (a figure) by drawing lines from a vertex through every point (of the figure) to the corresponding point of the reproduction. — *v.i.* 7 To extend forward or out; protrude. See synonyms under PLAN, THROW. [<L *projectus,* pp. of *projicere* throw out, cause to protrude <*pro-* before + *jacere* throw]

Synonyms (noun): contrivance, design, device, invention, plan, purpose, scheme.

pro·jec·tile (prə·jek′təl) *adj.* 1 Projecting, or impelling forward. 2 Capable of being or intended to be projected or shot forth. 3 Protrusile. — *n.* 1 A body projected or thrown forth by force. 2 *Mil.* A missile for discharge from a gun or cannon. [<F]

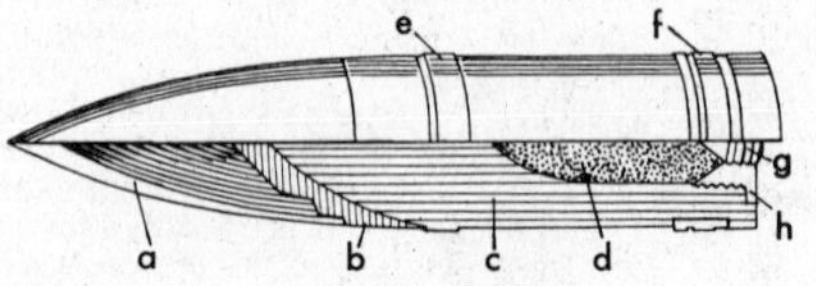

ARMOR–PIERCING PROJECTILE

a. Windshield. *b.* Armor–piercing cap. *c.* Body. *d.* Bursting charge. *e.* Bourrelet. *f.* Copper rotating band. *g.* Fuze. *h.* Plug.

pro·jec·tion (prə·jek′shən) *n.* 1 The act of projecting; a jutting, throwing, or shooting out or forth. 2 That which projects; a prominence; projecting part or subject. 3 A scheme; project. 4 A system of lines drawn on a given fixed plane, as in a map, which represents, point for point, a corresponding system of imaginary lines on a given terrestrial or celestial datum surface: when used in delineating part of the earth's surface, called a *map projection.* 5 *Psychol.* The process or result of externalizing or objectifying a perception or mental image: compare INTROJECTION. 6 The exhibiting of motion pictures or lantern slides upon a screen. [<F <L *projectio, -onis*] — **pro·jec′tion·al** *adj.*

pro·jec·tion·ist (prə·jek′shən·ist) *n.* 1 One who projects. 2 The operator of motion–picture and sound–reproducing equipment.

projection printing *Phot.* The process of enlarging a photograph by projecting the original negative onto sensitized paper by appropriate adjustment of light source and lens.

projection test *Psychol.* Any of various tests for the determination of personality traits and concealed motivations, as by theatrical performances, the completion of sentences or designs, the interpretation of ink blots, and the like.

pro·jec·tive (prə·jek′tiv) *adj.* 1 Pertaining to, treating of, or derived by projection: *projective* geometry, a *projective* figure. 2 *Geom.* Such as may be derived from one another by projection, as two plane figures. — **pro·jec′tive·ly** *adv.*

projective geometry A branch of geometry which investigates the properties of figures by means of projections in two or three dimensions, including the study of corresponding forms of various dimensions.

pro·jec·tor (prə·jek′tər) *n.* 1 One who devises projects; a schemer; a promoter. 2 That which projects something. 3 A mirror or combination of lenses for projecting a beam of light. 4 An apparatus for throwing illuminated images or motion pictures upon a screen. 5 A device for throwing grenades, bombs, etc.

pro·jet (prô·zhe′) *n. French* A plan or outline; specifically, a draft of a proposed treaty or law.

Pro·kof·iev (prô·kôf′yəf), **Sergei,** 1891–1953, Russian composer. Also **Pro·kof′ieff.**

Pro·ko·pyevsk (prô·kô′pyifsk) A city in the Kuznetsk Basin of Russian S.F.S.R.

pro·lac·tin (prō·lak′tin) *n. Biochem.* A hormone from the anterior lobe of the pituitary gland, believed to be active in initiating lactation. [<PRO-[1] + LACT- + -IN]

pro·la·mine (prō′lə·mēn, -min) *n. Biochem.* Any of a group of simple proteins that are insoluble in pure water or absolute alcohol, as gliadin from wheat. Also **pro′la·min** (-min). [<PROL(INE) + AM(MONIA) + -INE[2]]

pro·lapse (prō·laps′) *v.i.* **·lapsed, ·laps·ing** *Pathol.* To fall out of place, as an organ or part. — *n.* Prolapsus. [<L *prolapsus,* pp. of *prolabi* fall forward <*pro-* forward + *labi* glide, fall]

pro·lap·sus (prō·lap′səs) *n. Pathol.* The falling down of an organ, as the womb, from its normal position. [<L]

pro·late (prō′lāt) *adj.* 1 Extended lengthwise. 2 Lengthened toward the poles, as a spheroid generated by the revolution of an ellipse around its long axis: opposed to *oblate.* [<L *prolatus,* pp. to *proferre* extend, carry forward <*pro-* forward + *ferre* carry]

pro·leg (prō′leg) *n. Entomol.* One of the abdominal legs of insect larvae, as of caterpillars. [<PRO-[1] + LEG]

pro·le·gom·e·non (prō′lə·gom′ə·non) *n. pl.* **·na** (-nə) *Often pl.* An introductory remark or remarks; a preface. [<Gk., neut. passive ppr. of *prolegein* say beforehand <*pro-* before + *legein* say] — **pro′le·gom′e·nous** *adj.*

pro·lep·sis (prō·lep′sis) *n. pl.* **·ses** (-sēz) 1 Anticipation. 2 A rhetorical figure consisting in the anticipation, and answering or nullifying beforehand, of objections or opposing arguments. 3 The use of an adjective or a noun as an objective predicate in anticipation of the result of the verbal action: to shoot a person *dead.* 4 An error by which a date earlier than the true date is assigned to an event. [<L <Gk. *prolēpsis* anticipation <*prolambanein* take beforehand <*pro-* before + *lambanein* seize, take] — **pro·lep′tic** (-tik) or **·ti·cal** *adj.*

pro·le·tar·i·an (prō′lə·târ′ē·ən) *adj.* 1 Formerly, of or pertaining to the lower classes of society. 2 Of or pertaining to proletarians or the proletariat. — *n.* 1 Formerly, a person of the lowest or poorest class. 2 A laborer; a wageworker. [<L *proletarius* <*proles* offspring; so called because, being propertyless, they served the state only by having children] — **pro′le·tar′i·an·ism** *n.*

pro·le·tar·i·at (prō′lə·târ′ē·ət) *n.* 1 Formerly, the indigent classes collectively of a community; the lower classes. 2 Wageworkers collectively, regarded as the creators of wealth; workingmen. 3 *Bot.* Self–pollinated plants having a small or limited reserve of food materials. [<F *prolétariat.* See PROLETARIAN.]

pro·le·tar·y (prō′lə·ter′ē) *n. pl.* **·tar·ies** In ancient Rome, one of the lowest or poorest class, regarded as contributing to the state nothing but offspring. [See PROLETARIAN]

pro·let·cult (prō·let′kōōlt) *n.* A group formed in Russia at the time of the 1917 revolution, originally to develop a proletarian culture as an educational instrument; later transformed into a literary movement. [< Russian *Prolet(arskaya) Kult(ura)* proletarian culture]

pro·li·cide (prō′lə·sīd) *n.* The crime of killing one's own child, before or after birth; infanticide. [<L *proles* offspring + -CIDE]

pro·lif·er·ate (prō·lif′ə·rāt) *v.t.* & *v.i.* **·at·ed, ·at·ing** To produce, reproduce, or grow, especially with rapidity, as cells in tissue formation. [<PROLIFER(OUS) + -ATE[1]] — **pro·lif′er·a′tion** *n.* — **pro·lif′er·a′tive** *adj.*

pro·lif·er·ous (prō·lif′ər·əs) *adj.* 1 Producing offspring freely. 2 Producing branchlets, as a coral. 3 *Bot.* Having an excessive development of parts; developing buds, branches, and flowers from unusual places; bearing progeny in the way of offshoots, buds, etc. [<Med. L *prolifer* <L *proles, prolis* offspring + *ferre* bear]

pro·lif·ic (prō·lif′ik) *adj.* 1 Producing abundantly, as offspring or fruit; fertile. 2 Producing results abundantly; creative: a *prolific* writer. See synonyms under FERTILE. [<F

prolifique <Med. L *prolificus* <L *proles, prolis* offspring + stem of *facere* make] — **pro·lif′·i·ca·cy** (-i·kə·sē), **pro·lif′ic·ness** *n.* — **pro·lif′·i·cal·ly** *adv.*

pro·line (prō′lēn, -lin) *n. Biochem.* An amino acid, $C_5H_9O_2N$, found in proteins. [Contraction of *pyrroline* <PYRROLE + -INE[2]]

pro·lix (prō′liks, prō·liks′) *adj.* 1 Unduly long and verbose, as an address. 2 Indulging in long and wordy discourse; tedious: a *prolix* orator. [<F *prolixe* <L *prolixus* extended <*pro-* before + stem of *liquere* flow] — **pro·lix·i·ty** (prō·lik′sə·tē), **pro′lix·ness** *n.* — **pro′·lix·ly** *adv.*

pro·loc·u·tor (prō·lok′yə·tər) *n.* 1 One who speaks for another; a spokesman or advocate. 2 The presiding officer of a convocation; specifically, the speaker or chairman of the lower house of convocation in the Church of England. [<L <*prolocutus,* pp. of *proloqui* declare, speak for <*pro-* in behalf of + *loqui* talk]

Pro·loc·u·tor (prō·lok′yə·tər) *n. Brit.* In the House of Lords, the lord chancellor. [See PROLOCUTOR]

pro·log (prō′lôg, -log) *n.* A prefatory statement to a poem, discourse, or performance; specifically, an introduction, often in verse, spoken or sung by an actor before a play or opera; hence, any anticipatory act or event. — *v.t.* To introduce with a prolog or preface. Also **pro′logue.** [<OF *prologue* <L *prologus* <Gk. *prologos* <*pro-* before + *logos* discourse]

pro·log·ize (prō′lôg·īz, -log-) *v.i.* **·ized, ·iz·ing** To make or utter a prolog. Also **pro′logu·ize.** — **pro′log·iz′er** or **pro′logu·iz′er** *n.*

pro·long (prə·lông′, -long′) *v.t.* To extend in time or space; continue; lengthen. See synonyms under INCREASE, PROTRACT. Also **pro·lon′gate** (-lông′gāt, -long′-). [<OF *prolonguer* <L *prolongare* <*pro-* forth + *longus* long] — **pro·long′er** *n.* — **pro·long′ment** *n.*

pro·lon·ga·tion (prō′lông·gā′shən, -long-) *n.* 1 The act of prolonging. 2 That by which anything is increased; an extension. [<F]

pro·longe (prō·lonj′, *Fr.* prô·lôṅzh′) *n. Mil.* A rope having a hook at one end and a toggle at the other: used for drawing a gun carriage. [<F <*prolonger* <OF *prolonguer.* See PROLONG.]

pro·lu·sion (prō·lo͞o′zhən) *n.* 1 That which is introductory to the principal effort or performance; a preliminary attempt; a prolog; prelude. 2 An essay written as a test of the writer's powers, or as preliminary to a more elaborate treatise. [<L *prolusio, -onis* prelude <*prolusus,* pp. of *proludere* play beforehand <*pro-* before + *ludere* play]

prom (prom) *n. U.S. Colloq.* A formal college or school dance or ball: short for *promenade.*

prom·e·nade (prom′ə·nād′, -näd′) *n.* 1 A walk for amusement or exercise, or as part of a formal or social entertainment. 2 A ceremonious parade on horseback or in a vehicle. 3 A place for promenading. 4 A concert or ball opened with a formal march; also, the march. — *v.* **·nad·ed, ·nad·ing** *v.i.* 1 To take a promenade. — *v.t.* 2 To take a promenade through or along. 3 To take or exhibit on or as on a promenade; parade. [<F <*promener* take for a walk <L *prominare* drive forward <*pro-* before + *minare* drive (cattle)] — **prom′·e·nad′er** *n.*

promenade deck The deck above the shelter deck in merchant vessels.

Pro·me·the·an (prə·mē′thē·ən) *adj.* 1 Of, pertaining to, or like Prometheus. 2 Creative or life-bringing.

Pro·me·theus (prə·mē′thē·əs) In Greek mythology, a Titan who stole fire from heaven for mankind and as a punishment was chained to a rock, where an eagle daily devoured his liver, which was made whole again at night: he was released by Hercules.

pro·me·thi·um (prə·mē′thē·əm) *n.* The rare radioactive element of atomic number 61, separated from uranium fission products and belonging to the lanthanide series: new name for the element formerly known as *illinium.* [<NL <PROMETHEUS]

prom·i·nence (prom′ə·nəns) *n.* 1 The state of being prominent; conspicuousness; fame. 2 That which is prominent; a protuberance. 3 *Astron.* One of the great tongues of flame shooting out from the sun's surface, seen during total eclipses: also **solar prominence.** Also **prom′i·nen·cy.**

prom·i·nent (prom′ə·nənt) *adj.* 1 Jutting out; projecting; protuberant. 2 Conspicuous in position, character, or importance. 3 Eminent. See synonyms under EMINENT, IMPORTANT. [<L *prominens, -entis,* ppr. of *prominere* project] — **prom′i·nent·ly** *adv.*

pro·mis·cu·i·ty (prō′mis·kyo͞o′ə·tē, prom′is-) *n.* 1 Condition or state of being promiscuous; indiscriminate or confused mixture. 2 Promiscuous sexual union.

pro·mis·cu·ous (prə·mis′kyo͞o·əs) *adj.* 1 Composed of individuals or things confusedly mingled. 2 Unrestricted in distribution or application; exercised or shared without discrimination. 3 Indiscriminate; not fastidious, especially in sexual relations. 4 *Colloq.* Lacking plan or purpose; casual; irregular. [<L *promiscuus* mixed <*pro-* thoroughly + stem of *miscere* mix] — **pro·mis′cu·ous·ly** *adv.* — **pro·mis′cu·ous·ness** *n.*

prom·ise (prom′is) *n.* 1 An assurance given by one person to another that the former will or will not do a specified act. 2 Reasonable ground for hope or expectation, especially of future excellence or satisfaction: a youth of great *promise.* 3 Something promised; the fulfilment or obtainment of that which is promised. See synonyms under CONTRACT. — *v.* **·ised, ·is·ing** *v.t.* 1 To engage or pledge by a promise: used with the infinitive or a clause: He *promised* that he would do it. 2 To make a promise of (something) to someone. 3 To give reason for expecting: The sky *promised* rain. 4 *Colloq.* To assure (someone). — *v.i.* 5 To make a promise. 6 To give reason for expectation: often with *well* or *fair.* [<F *promesse* <L *promissum,* pp. of *promittere* send forward <*pro-* forth + *mittere* send] — **prom′is·er** *n.*

Promised Land See LAND OF PROMISE.

prom·is·ee (prom′is·ē′) *n. Law* One to whom a promise is made.

prom·is·ing (prom′is·ing) *adj.* Giving promise of good results or development. See synonyms under AUSPICIOUS. — **prom′is·ing·ly** *adv.*

prom·is·or (prom′is·ôr) *n. Law* One who makes a promise.

prom·is·so·ry (prom′ə·sôr′ē, -sō′rē) *adj.* 1 Containing or of the nature of a promise; expressing an engagement to pay: a *promissory* note. 2 Indicating what is to be required or to take place after the signing of an insurance contract. Compare WARRANTY. [<Med. L *promissorius*]

promissory note A written promise by one person to pay another unconditionally a certain sum of money at a specified time: also called *note of hand.*

prom·on·to·ry (prom′ən·tôr′ē, -tō′rē) *n. pl.* **·ries** 1 A high point of land extending into the sea; headland. 2 *Anat.* A rounded projection or part. [<LL *promontorium* <L *promunturium,* ? <*prominere.* See PROMINENT.]

Promontory Point A peninsula extending 20 miles south into Great Salt Lake, NW Utah.

pro·mote (prə·mōt′) *v.t.* **·mot·ed, ·mot·ing** 1 To contribute to the progress, development, or growth of; further; encourage. 2 To advance to a higher position, grade, or honor. 3 To work in behalf of; advocate actively: to *promote* social reforms. 4 In education, to advance (a pupil) to the next higher school grade. [<L *promotus,* pp. of *promovere* move forward <*pro-* forward + *movere* move]

Synonyms: advance, aid, assist, elevate, encourage, exalt, excite, foment, forward, foster, further, help, prefer, raise. We *promote* a person by *advancing, elevating,* or *exalting* him to a higher position or dignity. A person *promotes* a scheme or an enterprise which others have projected or begun, and which he *encourages, forwards, furthers,* especially when he acts as the agent of the prime movers of the enterprise. One who *excites* a quarrel originates it; to *promote* a quarrel is strictly to *foment* it, the one who *promotes* keeping himself in the background. See ABET, ENCOURAGE, QUICKEN, SERVE. *Antonyms*: see synonyms for ABASE, ALLAY.

pro·mot·er (prə·mō′tər) *n.* 1 One who or that which promotes. 2 One who assists (by securing capital, etc.) in promoting a financial or commercial enterprise, or who makes this his regular business. 3 *Chem.* A substance used to increase the action of a catalyst. See synonyms under AGENT, AUXILIARY.

pro·mo·tion (prə·mō′shən) *n.* 1 Advancement or preferment in honor, dignity, rank, or grade. 2 Furtherance; encouragement. 3 The act of promoting. 4 The state of being promoted. — **pro·mo′tion·al** *adj.*

pro·mo·tive (prə·mō′tiv) *adj.* Tending to promote.

prompt (prompt) *v.t.* 1 To incite to action; instigate. 2 To suggest or inspire (an act, thought, etc.). 3 To remind of what has been forgotten or of what comes next; give a cue to. — *v.i.* 4 To give help or suggestions. See synonyms under ACTUATE, ENCOURAGE, INFLUENCE, STIR[1]. — *adj.* 1 Acting, or ready to act, at the moment; quick to respond or decide; punctual. 2 Done or rendered with readiness or alacrity; taking place at the appointed time. See synonyms under ACTIVE, ALERT, NIMBLE. — *n.* 1 A term of credit allowed for the payment of a debt as stated in a prompt-note. 2 An act of prompting; also, the information imparted by prompting; a reminder. [<OF <L *promptus* brought forth, hence, at hand, pp. of *promere* <*pro-* forth + *emere* take] — **prompt′ly** *adv.* — **prompt′ness** *n.*

prompt·book (prompt′bo͝ok′) *n.* An annotated script of a play used by a prompter or director.

prompt·er (promp′tər) 1 In a theater, one who follows the lines and prompts the actors. 2 One who or that which prompts.

promp·ti·tude (promp′tə·to͞od, -tyo͞od) *n.* The quality, habit, or fact of being prompt; promptness. [<F]

prompt-note (prompt′nōt′) *n.* In commerce, a note or memorandum delivered to a purchaser of merchandise as a reminder, and containing a statement of the sum due, day of payment, etc.

pro·mul·gate (prō·mul′gāt, prom′əl·gāt) *v.t.* **·gat·ed, ·gat·ing** To make known or announce officially and formally; put into effect by public proclamation, as a law or dogma. See synonyms under ANNOUNCE, PUBLISH, SPREAD. [<L *promulgatus,* pp. of *promulgare* make known, prob. alter. of *provulgare* <*pro-* forth + *vulgus* the people] — **pro·mul·ga·tion** (prō′mul·gā′shən, prom′əl-) *n.* — **pro·mul·ga·tor** (prō·mul′gā·tər, prom′əl-) *n.*

pro·mulge (prō·mulj′) *v.t.* **·mulged, ·mulg·ing** *Archaic* To promulgate. [<L *promulgare*]

pro·my·ce·li·um (prō′mī·sē′lē·əm) *n. pl.* **·li·a** (-lē·ə) *Bot.* A short-jointed filament, developed on the germination of certain smut or rust spores, and which gives rise to sporidia. — **pro′my·ce′li·al** *adj.*

pro·na·os (prō·nā′os) *n.* In ancient Greece, a portico or vestibule of a temple. [<Gk. <*pro-* before + *naos* temple]

pro·nate (prō′nāt) *v.t.* **·nat·ed, ·nat·ing** To place in a position of pronation. [<L *pronatus,* pp. of *pronare* bow <*pronus* prone]

pro·na·tion (prō·nā′shən) *n. Physiol.* 1 The act or movement of turning the palm of the hand, or the corresponding surface of the forelimb, downward or backward. 2 The position of a limb so turned: opposed to *supination.*

pro·na·tor (prō·nā′tər) *n. pl.* **pro·na·to·res** (prō′nə·tôr′ēz, -tō′rēz) *Anat.* A muscle of the forearm by which pronation is effected.

prone (prōn) *adj.* 1 Lying flat, especially with the face, front, or palm downward; prostrate: opposed to *supine.* 2 Leaning forward or downward; also, moving or sloping sharply downward. 3 Mentally inclined or predisposed: with *to.* See synonyms under ADDICTED, SUBJECT. [<L *pronus* prostrate <*pro-* before] — **prone′ly** *adv.* — **prone′ness** *n.*

pro·neph·ros (prō·nef′ros) *n. Anat.* The primordial kidney, the anterior of three similar tubular organs found in connection with the genitourinary apparatus of typical vertebrates. [<NL <Gk. *pro-* before + *nephros* kidney] — **pro·neph′ric** *adj.*

prong (prông, prong) *n.* 1 A pointed end of an instrument; a tine of a fork. 2 Any pointed

prongbuck / **proof spirit**

and projecting part, as the end of an antler, etc. — *v.t.* To prick or stab with or as with a prong. [Cf. LG *prange* a pointed stick, Du. *prangen* pinch.]

prong·buck (prông'buk, prong'-) *n.* The male of the pronghorn.

prong·horn (prông'hôrn', prong'-) *n. pl.* **·horns** or **·horn** A ruminant (*Antilocapra americana*) of western North America, having deciduous branched horns; the Rocky Mountain antelope: not a true antelope.

PRONGHORN
(About 3 feet high at the shoulder)

pro·nom·i·nal (prō-nom'ə-nəl) *adj.* Of, pertaining to, like, or having the nature of a pronoun. [<LL *pronominalis* <L *pronomen*. See PRONOUN.] — **pro·nom'i·nal·ly** *adv.*

pronominal adjective The possessive case of a personal pronoun used attributively: *my, your, his, her, its, our, their, whose,* and, poetically, *mine* and *thine.*

pro·noun (prō'noun) *n.* A word used as a substitute for a noun, as *he, she, that.* [<OF *pronom* <L *pronomen* <*pro-* in place of + *nomen* name, noun]

— **adjective pronoun** Any pronoun used like an adjective; as, *that* boy, *this* house, *which* man. Any demonstrative pronoun, any indefinite pronoun (except *none*), and any interrogative and relative pronoun (except *who*) may be used as an adjective pronoun.

— **demonstrative pronoun** A pronoun that directly points out its antecedents.

	Singular	Plural
	this	these
	that	those

The same forms are used for all genders, persons, and cases.

— **indefinite pronoun** A pronoun that represents an object indefinitely or generally. The principal indefinite pronouns are *another, any, both, each, either, neither, none, one, other, some, such. None* and *any* are both singular and plural.

— **interrogative pronoun** A pronoun that is used to ask a question.

	Subjective	Possessive	Objective
Singular	who	whose	whom
and	which	whose, of which	which
Plural	what	of what	what

Of what occurs in such sentences as *Of what are you speaking? What are you speaking of?*

— **personal pronoun** A pronoun that shows by its form the person speaking, the person spoken to, or the person or thing spoken of.

	Singular			Plural		
	Subjective	Possessive	Objective			
1st person	I	my *or* mine	me			
2nd person	you (thou)	your *or* yours (thy *or* thine)	you (thee)			
3rd person						
masculine	he	his	him			
feminine	she	her *or* hers	her			
neuter	it	its	it			
Plural:						
1st person	we	our *or* ours	us			
2nd person	you (ye)	your *or* yours	you			
3rd person	they	their *or* theirs	them			

— **reflexive pronoun** A pronoun formed by adding *-self* or *-selves* to the oblique cases of the personal pronoun. They serve as an intensive: I, *myself,* was there; or a reference back to a personal pronoun where the same person is both subject and object: He hit *himself.*

	Singular	Plural
1st person	myself	ourselves
2nd person	yourself	yourselves
3rd person	himself, herself, itself	themselves

— **relative pronoun** A pronoun that relates to an antecedent and introduces a qualifying clause: We found a boatman *who* ferried us.

	Subjective	Possessive	Objective
	who	whose	whom
	which	of which	which
	what	of what	what
	that		that

Sometimes *as* and *but* are regarded as relative pronouns: Such men *as* survived the accident;

There is not a man *but* remembers that day. ♦ The relative pronouns *who* (with its inflected forms *whose* and *whom*), *which,* and *what,* are identical in form with the interrogative pronouns but they undergo shifts of meaning in interrogative use, often being indefinite and general in reference: *What* (if anything or of all possible things) is he talking about? But: He is talking about *what* (specifically) he knows best. These pronouns, when used to introduce an indirect question, are by nature both relative and interrogative: They asked *what* he wanted; *whom* he preferred as a colleague; *which* party he belonged to. Similarly, *that* is not only a demonstrative but a relative pronoun and it makes a specific and limiting reference in either use.

pro·nounce (prə-nouns') *v.* **·nounced, ·nounc·ing** *v.t.* 1 To utter or deliver officially or solemnly; proclaim. 2 To assert; declare, especially as one's judgment: The judge *pronounced* her insane. 3 To give utterance to; articulate (words, etc.). 4 To articulate in a prescribed manner. 5 To indicate the sound of (a word) by phonetic symbols. — *v.i.* 6 To make a pronouncement or assertion. 7 To articulate words; speak. See synonyms under ASSERT, SPEAK. [<OF *pronuncier* <L *pronunciare* <L *pronuntiare* proclaim <*pro-* forth + *nuntiare* announce] — **pro·nounce'a·ble** *adj.* — **pro·nounc'er** *n.*

pro·nounced (prə-nounst') *adj.* Of marked character; decided.

pro·nounce·ment (prə-nouns'mənt) *n.* The act of pronouncing; a formal declaration or announcement.

pro·nounc·ing (prə-noun'sing) *adj.* Pertaining to or serving as a guide in pronunciation.

pron·to (pron'tō) *adv. U.S. Slang* Quickly; promptly; instantly. [<Sp. <L *promptus.* See PROMPT.]

pro·nu·cle·us (prō-n$\overline{oo}$'klē-əs, -ny$\overline{oo}$'-) *n. pl.* **·cle·i** (-klē-ī) *Biol.* The nucleus of either the spermatozoon or ovum, the union of which forms the nucleus of the fertilized ovum. [<NL <Gk. *pro-* before + L *nucleus* a kernel]

pro·nu·mer·al (prō-n$\overline{oo}$'mər-əl, -ny$\overline{oo}$'-) *n. Math.* A letter or symbol that stands for a number, as *x, y,* and *z* in the equation $3x + 2y - z = 13$.

pro·nun·ci·a·men·to (prə-nun-sē-ə-men'tō, -shē-ə-) *n. pl.* **·tos** A public announcement; proclamation; manifesto. [<Sp. *pronunciamiento,*

lit., a pronouncement <L *pronuntiare.* See PRONOUNCE.]

pro·nun·ci·a·tion (prə-nun-sē-ā'shən) *n.* The act or manner of pronouncing words; articula-

tion. [<L *pronunciatio, -onis* <*pronuntiatus,* pp. of *pronuntiare.* See PRONOUNCE.]

proof (pr$\overline{oo}$f) *n.* 1 The act or process of proving, in any sense; specifically, the establishment of a fact by evidence or a truth by other truths. 2 A trial of strength, truth, fact, or excellence, etc.; a test. 3 Evidence and argument sufficient to induce belief. 4 *Law* Anything that serves to convince the mind of the truth or falsity of a fact or proposition, including facts and admissions of parties, which are properly called *evidence,* and presumptions either of fact or of law, and citations of law. 5 The state or quality of having successfully undergone a proof or test; impenetrability; also, impenetrable armor. 6 The standard of strength of alcoholic liquors: see PROOF SPIRIT. 7 *Printing* A printed trial sheet showing the contents or condition of matter in type or of a plate, or the like, either with or without marked corrections. 8 In engraving and etching, a trial impression taken from an engraved plate, stone, or block; also, a perfect impression from such a plate, etc., when finished, and usually before the title or inscription has been added. 9 *Phot.* A trial print from a negative. 10 *Math.* A process to check a computation by using its result; also, a demonstration. 11 Anything proved true; experience. 12 In philately, an experimental printing of a stamp. — *adj.* 1 Employed in or connected with proving or correcting. 2 Capable of resisting successfully; firm; impenetrable: with *against: proof against* bribes. 3 Of standard alcoholic strength, as liquors. [<OF *prueve* <LL *proba* <*probare* PROVE]

Synonyms (noun): attestation, certification, confirmation, demonstration, essay, evidence, fact, ordeal, test, testimony, trial. See CERTAINTY, DEMONSTRATION, TESTIMONY. *Antonyms:* assertion, conjecture, disproof, failure, fallacy, fancy, hypothesis, imagination, likelihood, possibility, presumption, probability, refutation.

-proof *combining form* 1 Impervious to; able to withstand; not damaged by: *waterproof, bombproof.* 2 Protected against: *mothproof, stormproof.* 3 As strong as: *armorproof.* 4 Resisting; showing no effects of: *joyproof, panicproof.* Adjectives formed with *-proof* may also be used as verbs. [<PROOF, *adj.*]

proof·read (pr$\overline{oo}$f'rēd') *v.t. & v.i.* **·read** (-red'), **·read·ing** (-rē'ding) To read and correct (printers' proofs).

PROOFREADER'S MARKS

Symbols in the column headed MARGIN are used only in the outer margins of the proof: the symbols used within the body of the text are given in the TEXT column.

MARGIN		TEXT	MARGIN		TEXT
l.c.	Set in lower-case type	circled or /	✕	Broken letter: examine	circled
Cap.	Set in capitals	underscored	*tr.*	Transpose matter marked	⌐⌐
s.c.	Set in small capitals	underscored	*eq. #*	Equalize spacing	∨∨∨ ∧∧∧∧
C+s.c.	Set in caps and small caps		⌐	Move to left to point marked	⌐
l.f.	Set in lightface type	circled	⌐	Move to right to point marked	⌐
b.f.	Set in boldface type	underscored			
Rom.	Set in roman type	circled	⌐⌐	Raise to point marked	⌐⌐
Ital.	Set in italic type	underscored	⌐⌐	Lower to point marked	⌐⌐
⊙	Insert period	∧	⌣	Push down space	
⁏?	Insert colon; semicolon	∧ ∧	⊂	Close up	⊂
∧	Insert comma	∧	¶	Begin new paragraph	
∨	Insert apostrophe	∧	No ¶	Run matter on, not a paragraph	⌣
/?/	Insert interrogation mark	∧	=	Aline type	=
/!/	Insert exclamation mark	∧	*Stet.*	Retain words crossed out	.....
/=/	Insert hyphen	∧	⌒	Take out and close up	⌐
⌣	Insert quotation marks	∧	‖	Line up matter	‖
∨	Insert superior figure or letter	∧	*Out*	Omission here; see copy	
⌃	Insert inferior figure or letter	∧	⌐	Move this to left	
⌐	Insert one em-dash	∧	⌐	Move this to right	
⌐	Take out (delete matter marked)	/	*Qu.?*	Query: is this right?	∧
w.f.	Wrong font	circled or /	*Sp.*	Spell out	circled
			#	Insert space	∧

proof·read·er (pr$\overline{oo}$f'rē'dər) *n.* One whose business is to read and mark the errors in printers' proofs. — **proof'read'ing** *n.*

proof spirit An alcoholic liquor that contains

a standard amount of alcohol: in the United States, half its volume of alcohol, with a specific gravity of 0.7939 at 60° F., which is rated *100-proof.*

prop[1] (prop) *v.t.* **propped, prop·ping** **1** To support or keep from falling by or as by means of a prop. **2** To lean or place: usually with *against.* **3** To support; sustain. —*n.* That which sustains an incumbent weight; a buttress; stay. [<MDu. *proppe* a vine prop, a support]
Synonyms: bolster, brace, buttress, shore, stay, support, sustain. See SUPPORT.

prop[2] (prop) *n. Colloq.* On a theater stage, any adjunct except the scenery or the costumes of the actors; a property. [Short for PROPERTY]

prop[3] (prop) *n. Colloq.* A propeller.

pro·pae·deu·tic (prō′pə·do͞o′tik, -dyo͞o′-) *adj.* Pertaining to or of the nature of preliminary instruction; relating to or introductory to an art or science: also **pro′pae·deu′ti·cal.** —*n.* A preparatory or introductory subject or course. [<Gk. *propaideuein* teach beforehand <*pro-* before + *paideuein* teach <*pais, paidos* a child]

pro·pae·deu·tics (prō′pə·do͞o′tiks, -dyo͞o′-) *n.* The body of principles or rules introductory to an art or science.

prop·a·ga·ble (prop′ə·gə·bəl) *adj.* That can be propagated; capable of being disseminated or spread abroad, as principles, etc. [<L *propagare* PROPAGATE + -ABLE]

prop·a·gan·da (prop′ə·gan′də) *n.* **1** Any institution or scheme for propagating a doctrine, or system. **2** Effort directed systematically toward the gaining of public support for an opinion or course of action. **3** The tenets, views, etc., put forward by propaganda. [<PROPAGANDA]

Prop·a·gan·da (prop′ə·gan′də) *n.* A society of cardinals, the overseers of foreign missions; also, the College for the Propagation of the Faith, founded by Pope Urban VIII, in 1627, for the education of missionary priests; Sacred Congregation *de Propaganda Fide.* Also **College of Propaganda.** [<Ital. <NL *(congregatio de) propaganda (fide)* (the congregation for) propagating (the faith) <L, gerund of *propagare.* See PROPAGATE.]

prop·a·gan·dism (prop′ə·gan′diz·əm) *n.* The art, practice, or system of using propaganda. —**prop′a·gan′dist** *n.*

prop·a·gan·dize (prop′ə·gan′dīz) *v.* **·dized, ·diz·ing** *v.t.* **1** To subject to propaganda. **2** To spread by means of propaganda. —*v.i.* **3** To carry on or spread propaganda.

prop·a·gate (prop′ə·gāt) *v.* **·gat·ed, ·gat·ing** *v.t.* **1** To cause (animals, plants, etc.) to multiply by natural reproduction; breed. **2** To reproduce (itself). **3** To spread abroad or from person to person; diffuse; disseminate. **4** To transmit through a medium; extend the action of: to *propagate* heat. **5** *Obs.* To increase. —*v.i.* **6** To multiply by natural reproduction; have offspring; breed. [<L *propagatus,* pp. of *propagare* slip or layer a plant, multiply <*propago* a slip for transplanting <*pro-* forth + *pag-,* root of *pangere* fasten] —**prop′a·ga′tive** *adj.* —**prop′a·ga′tor** *n.*
Synonyms: beget, breed, engender, generate, increase, multiply, originate, procreate. See PRODUCE, SPREAD. *Antonyms*: annihilate, destroy, eradicate, exterminate, extirpate.

prop·a·ga·tion (prop′ə·gā′shən) *n.* **1** The act of propagating; reproduction. **2** Dissemination; diffusion.

prop·a·gule (prop′ə·gyo͞ol) *n. Bot.* A bud, shoot, or other plant part which vegetatively propagates the species. Also **pro·pag·u·lum** (prō·pag′yə·ləm). [<NL *propagulum,* dim. of L *propago.* See PROPAGATE.]

pro·pane (prō′pān) *n. Chem.* A gaseous hydrocarbon of the methane series, C_3H_8, obtained from petroleum and also made synthetically. [<PROP(YL) + (METH)ANE]

pro·par·ox·y·tone (prō′pə·rok′sə·tōn) *adj.* Having an acute accent on the antepenult. —*n.* A word with an acute accent on the antepenult. [<Gk. *proparoxytonos* <*pro-* before + *paroxytonos* paroxytone]

pro pa·tri·a (prō pā′trē·ə) *Latin* For one's country.

pro·pel (prə·pel′) *v.t.* **·pelled, ·pel·ling** To cause to move forward or ahead; drive or urge forward. See synonyms under DRIVE, PUSH, SEND. [<L *propellere* drive before one <*pro-* forward + *pellere* drive]

pro·pel·lant (prə·pel′ənt) *n.* **1** That which propels. **2** *Mil.* An explosive which, upon ignition, propels a projectile from a gun. **3** A solid or liquid fuel which serves to propel a rocket, guided missile, or the like.

pro·pel·lent (prə·pel′ənt) *adj.* Propelling; able to propel. [<L *propellens, -entis,* ppr. of *propellere.* See PROPEL.]

pro·pel·ler (prə·pel′ər) *n.* **1** One who or that which propels. **2** Any device for propelling a craft through water or air; especially, one having blades mounted at an angle on a power-driven shaft and producing a thrust by their rotary action on the fluid.

pro·pend (prō·pend′) *v.i. Obs.* To be disposed in favor; tend. [<L *propendere* hang forward, be inclined or favorable <*pro-* forward + *pendere* hang]

pro·pene (prō′pēn) *n.* Propylene. [<PROP(YL) + -ENE]

pro·pe·no·ic acid (prō′pə·nō′ik) See under ACRYLIC. [<PROPEN(E) + (BENZ)OIC]

pro·pense (prō·pens′) *adj. Obs.* Having a propensity; prone. [<L *propensus,* pp. of *propendere.* See PROPEND.] —**pro·pense′ly** *adv.*

pro·pen·si·ty (prə·pen′sə·tē) *n. pl.* **·ties** **1** Natural disposition to or for; tendency. **2** *Obs.* A liking for; partiality. See synonyms under APPETITE, DESIRE, INCLINATION. [<L *propensus.* See PROPENSE.]

prop·er (prop′ər) *adj.* **1** Having special adaptation or fitness; specially suited; applicable; appropriate. **2** Conforming to a standard; becoming; seemly; correct. **3** Naturally belonging to a person or thing; particular; peculiar. **4** Understood in the most correct sense; strictly so called: commonly following the noun modified. **5** *Gram.* Belonging to an individual person, family, place, or the like: a *proper* noun: opposed to *common.* **6** *Archaic* Belonging to or affecting oneself; own. **7** *Her.* Represented in the natural color. **8** *Eccl.* Appointed for special use: the *proper* psalms for Christmas. **9** *Archaic* Of becoming form or appearance. **10** *Archaic* Good; excellent; pleasant. **11** *Obs.* Respectable; worthy; honest. See synonyms under APPROPRIATE, BECOMING, CONVENIENT, CORRECT, GOOD, MODEST. —*n.* A collection of prayers; specifically, that portion of the breviary or missal containing the prayers and collects suitable to special occasions. [<OF *propre* <L *proprius* one's own] —**prop′er·ness** *n.*

proper fraction See under FRACTION.

prop·er·ly (prop′ər·lē) *adv.* In a proper manner; suitably; rightly.

proper noun See under NOUN.

prop·er·tied (prop′ər·tēd) *adj.* Owning property.

Pro·per·tius (prō·pûr′shəs), **Sextus,** 50?–14? B.C., Roman poet.

prop·er·ty (prop′ər·tē) *n. pl.* **·ties** **1** Any object of value that a person may lawfully acquire and hold; anything that may be owned; stocks, land, etc.; any possession. **2** Ownership or dominion; the legal right to the possession, use, enjoyment, and disposal of a thing; a valuable legal right or interest in or to particular things. **3** Whatever belongs or pertains to any object, as a distinguishing quality or characteristic; a peculiarity. **4** In the theater, any portable article, except scenery, which is not personally owned by the actors, but which is used by them in the performance, as flowers, books, dishes, etc. **5** A characteristic attribute of a body or substance under stated conditions, especially in relation to the senses, as color, odor, hardness, density, etc. **6** Any typical mode of action or behavior observed in natural phenomena: a *property* of radiation. [<OF *proprieté* <L *proprietas, -tatis* <*proprius* one's own. Doublet of PROPRIETY.]
Synonyms: chattels, estate, goods, means, money, ownership, possessions, resources, right, wealth. See ATTRIBUTE, CHARACTERISTIC, MONEY, WEALTH.

pro·phase (prō′fāz) *n. Biol.* One of the preparatory changes in the mitosis of the cell, during which the chromatin of the nucleus is formed into longitudinally split chromosomes. [<PRO-[2] before + PHASE]

proph·e·cy (prof′ə·sē) *n. pl.* **·cies** **1** A prediction made under divine influence and direction; loosely, any prediction. **2** Discourse delivered by a prophet under divine inspiration: the common Biblical sense. **3** A book of prophecies. **4** *Obs.* Public interpretation of Scripture; preaching. [<OF *profecie* <LL *prophetia* <Gk. *prophēteia* <*prophētēs* <*pro-* before + *phanai* speak]

proph·e·sy (prof′ə·sī) *v.* **·sied, ·sy·ing** *v.t.* **1** To utter or foretell with or as with divine inspiration. **2** To predict (a future event). **3** To point out beforehand. —*v.i.* **4** To speak by divine influence, or as a medium between God and man. **5** To foretell the future; make predictions. **6** To explain or teach religious subjects; preach. [<OF *prophecier* <*profecie* PROPHECY] —**proph′e·si′er** *n.*
Synonyms: augur, divine, foretell, predict, prognosticate. *Prophesy* differs from *predict* by assuming a claim to supernatural or divine inspiration. To *prognosticate* is to *predict* from observed signs, indications, or conditions. To *prophesy* in the Scriptural sense is to utter religious truth under divine inspiration, not necessarily to *foretell* future events, but to warn, exhort, comfort, etc. See AUGUR. *Antonyms*: chronicle, recall, recite, recollect, record, remember.

proph·et (prof′it) *n.* **1** One who delivers divine messages or interprets the divine will. **2** One who foretells the future; especially, an inspired predictor. **3** A religious leader. **4** An interpreter or spokesman for any cause. **5** A mantis. —**the Prophet** According to Islam, Mohammed. —**the Prophets** The Old Testament books written by the prophets. ◆ Homophone: *profit.* [<Gk. *prophētēs* <*pro-* before + *phanai* speak] —**proph′et·ess** *n. fem.* —**proph′et·hood** (-ho͝od) *n.*

pro·phet·ic (prə·fet′ik) *adj.* **1** Of or pertaining to a prophet or prophecy; vatic. **2** Pertaining to or involving prediction or presentiment; predictive. Also **pro·phet′i·cal.** —**pro·phet′i·cal·ly** *adv.* —**pro·phet′i·cal·ness** *n.*

pro·phy·lac·tic (prō′fə·lak′tik, prof′ə-) *adj.* Operating to ward off something, especially disease; preventive. —*n.* A prophylactic medicine or appliance. [<Gk. *prophylaktikos* <*prophylassein* be on guard <*pro-* before + *phylassein* guard]

pro·phy·lax·is (prō′fə·lak′sis, prof′ə-) *n.* Preventive treatment for disease. [<NL <Gk. *pro-* before + *phylaxis* a guarding]

pro·pine (prō·pīn′) *Scot. v.t.* To offer, as a gift; propose. —*n.* An offering; pledge.

pro·pin·qui·ty (prō·ping′kwə·tē) *n.* **1** Nearness in place or time. **2** Kinship. See synonyms under APPROXIMATION. [<OF *propinquité* <L *propinquitas, -tatis* <*propinquus* near]

pro·pi·o·nate (prō′pē·ə·nāt′) *n. Chem.* An organic compound containing the radical CH_3CH_2COO. [<PRO(TO)- + Gk. *piōn* fat + -ATE[3]]

pro·pi·on·ic (prō′pē·on′ik, -ō′nik) *adj. Chem.* Designating a colorless, liquid acid, $C_3H_6O_2$, occurring in nature, as in beet root molasses, and also produced variously by synthesis. It is the first member in the series of fatty acids.

pro·pi·theque (prō′pə·thēk′) *n.* Sifaka. [<F <NL *Propithecus* <Gk. *pro-* before + *pithēkos* an ape]

pro·pi·ti·ate (prō·pish′ē·āt) *v.t.* **·at·ed, ·at·ing** To cause to be favorably disposed; appease; conciliate. [<L *propitiatus,* pp. of *propitiare* render favorable, appease <*propitius* PROPITIOUS] —**pro·pi·ti·a·ble** (prō·pish′ē·ə·bəl) *adj.* —**pro·pi′ti·at′ing·ly** *adv.* —**pro·pi′ti·a′tive** *adj.* —**pro·pi′ti·a′tor** *n.*

pro·pi·ti·a·tion (prō·pish′ē·ā′shən) *n.* **1** The act of propitiating. **2** That which propitiates.
Synonyms: atonement, expiation, reconciliation, satisfaction. *Atonement* (at-one-ment), originally denoting *reconciliation,* or the bringing into agreement of those who have been estranged, is now chiefly used, as in theology, in the sense of some offering, sacrifice, or suffering sufficient to win forgiveness or make up for an offense. *Expiation* is the enduring of the full penalty of a wrong or crime. *Propitiation* is an offering, action, or sacrifice that makes the governing power propitious toward the offender. *Satisfaction*

denotes the rendering a full legal equivalent for the wrong done. *Propitiation* appeases the lawgiver; *satisfaction* meets the requirements of the law. *Antonyms*: alienation, condemnation, estrangement, offense, penalty, punishment, reprobation.

pro·pi·ti·a·to·ry (prō·pish′ē·ə·tôr′ē, -tō′rē) *adj.* Pertaining to or causing propitiation. —*n. pl.* **·ries 1** A propitiation. **2** In Jewish antiquity, the mercy seat regarded as symbolizing the merciful presence of Jehovah.

pro·pi·tious (prō·pish′əs) *adj.* **1** Kindly disposed; gracious. **2** Attended by favorable circumstances; auspicious. [<OF *propicius* <L *propitius* favorable, prob. <*pro-* before, forward + *petere* seek] —**pro·pi′tious·ly** *adv.* —**pro·pi′tious·ness** *n.*

Synonyms: auspicious, benign, benignant, clement, favorable, friendly, gracious, kind, kindly, merciful. That which is *auspicious* is of *favorable* omen; that which is *propitious* is of favoring influence or tendency; as, an *auspicious* morning; a *propitious* breeze. *Propitious* applies to persons, implying *kind* disposition and *favorable* inclinations, especially toward the suppliant; *auspicious* is not used of persons. See AUSPICIOUS. *Antonyms*: adverse, antagonistic, forbidding, harsh, hostile, ill-disposed, inauspicious, repellent, unfavorable, unfriendly, unpropitious.

prop·o·lis (prop′ə·lis) *n.* A resinous, adhesive substance elaborated by bees to serve as a cementing material. [<L <Gk. <*pro-* before + *polis* a city]

pro·pone (prə·pōn′) *v.t.* **·poned, ·pon·ing** *Scot.* To propose or propound; put forward.

pro·po·nent (prə·pō′nənt) *n.* **1** One who makes a proposal or puts forward a proposition; one who propounds a thing. **2** *Law* One who presents a will for probate. **3** One who advocates or supports a cause or doctrine. [<L *proponens, -entis,* ppr. of *proponere* set forth <*pro-* forth + *ponere* put]

Pro·pon·tis (prə·pon′tis) The ancient name for the SEA OF MARMARA.

pro·por·tion (prə·pôr′shən, -pōr′-) *n.* **1** Relative magnitude, number, or degree, as existing between parts, a part and a whole, or different things. **2** Fitness and harmony; symmetry. **3** A proportionate or proper share; any share or part. **4** An equality or identity between ratios. **5** *Math.* That rule by which, when three numbers are given, a fourth can be found having the same ratio to the third as the second has to the first: also called *the rule of three,* three of the four terms being always given, and the unknown being called a fourth proportional. See synonyms under ANALOGY, PORTION, SYMMETRY. —*v.t.* **1** To adjust properly as to relative magnitude, amount, or degree: to *proportion* one's expenses to one's means. **2** To form with a harmonious relation of parts. [<OF *proporcion* <L *proportio, -onis* <*pro-* before + *portio, -onis* a share] —**pro·por′tion·a·ble** *adj.* —**pro·por′tion·a·bly** *adv.* —**pro·por′tion·er** *n.*

pro·por·tion·al (prə·pôr′shən·əl, -pōr′-) *adj.* **1** Of, pertaining to, or being in proportion. **2** *Math.* **a** Constituting the terms of a proportion: said of four quantities: The numbers 2, 3, and 8, 12 are *proportional.* **b** Varying so that corresponding values form a proportion. —*n.* Any quantity or number in proportion to another or others. —**pro·por′tion·al·ly** *adv.* —**pro·por′tion·al′i·ty** (-al′ə·tē) *n.*

proportional representation A system of election by which political parties secure legislative representation in a government in proportion to voting strength.

pro·por·tion·ate (prə·pôr′shən·it, -pōr′-) *adj.* Being in due proportion; proportional. —*v.t.* (-āt) **·at·ed, ·at·ing** To make proportionate. —**pro·por′tion·ate·ly** *adv.* —**pro·por′tion·ate·ness** *n.*

pro·por·tion·ment (prə·pôr′shən·mənt, -pōr′-) *n.* The act of placing or putting things in proportion; arrangement; distribution.

pro·po·sal (prə·pō′zəl) *n.* **1** An offer proposing something to be accepted or adopted. **2** An offer of marriage. **3** Something proposed, as a scheme or plan.

Synonyms: bid, offer, overture, proposition. An *offer* or *proposal* puts something before one for acceptance or rejection, *proposal* being the more formal word; a *proposition* sets forth truth (or what is claimed to be truth) in formal statement. The *proposition* is for consideration, the *proposal* for action; as, a *proposition* in geometry, a *proposal* of marriage; but *proposition* is often used nearly in the sense of *proposal* when it is a matter for deliberation; as, a *proposition* for the surrender of a fort. A *bid* is commercial and often verbal; as, a *bid* at an auction. An *overture* opens negotiation or conference, and the word is especially used of some movement toward reconciliation; as, *overtures* of peace. See synonyms under DESIGN. *Antonyms*: acceptance, decision, denial, refusal, rejection, repulse.

pro·pose (prə·pōz′) *v.* **·posed, ·pos·ing** *v.t.* **1** To put forward for acceptance or consideration. **2** To nominate, as for admission or appointment. **3** To intend; purpose. **4** To suggest the drinking of (a toast or health). —*v.i.* **5** To form or announce a plan or design. **6** To make an offer, as of marriage. [<OF *proposer* <*pro-* forth (<L) + *poser.* See POSE[1].] —**pro·pos′er** *n.*

Synonym: purpose. In its most frequent use, *propose* differs from *purpose* in that what we *purpose* lies in our own mind as a decisive act of will, a determination; what we *propose* is offered or stated to others. In this use of the word, what we *propose* is open to deliberation, as what we *purpose* is not. In another use of the word one *proposes* something to or by himself which may or may not be stated to others. In this latter sense *propose* is nearly identical with *purpose.* See PLAN, PURPOSE.

prop·o·si·tion (prop′ə·zish′ən) *n.* **1** A scheme or proposal offered for consideration or acceptance. **2** *U.S. Colloq.* Any matter or person to be dealt with: a tough *proposition.* **3** *Colloq.* An indecent or immodest proposal. **4** A subject or statement presented for discussion. **5** *Logic* A statement in which something (the *subject*) is affirmed or denied in terms of something else (the *predicate*), the two being related usually by a copula. In the propositions, *Grass is green* and *Grass is not red,* grass in each case is the subject and green and red are the predicates respectively. **6** *Math.* A statement of a truth to be demonstrated (a *theorem*) or of an operation to be performed (a *problem*). See synonyms under PROPOSAL. —*v.t. Colloq.* To make an improper suggestion to. [<OF <L *propositio, -onis* a setting forth <*propositus,* pp. of *proponere.* See PROPONENT.] —**prop′o·si′tion·al** *adj.* —**prop′o·si′tion·al·ly** *adv.*

pro·pos·i·tus (prō·poz′i·təs) *n. pl.* **·ti** (-tī) *Law* The person from whom a line of descent is reckoned. [<L. See PROPOSITION.]

pro·pound (prō·pound′) *v.t.* To put forward for consideration, solution, etc. See synonyms under AFFIRM, ANNOUNCE. [Earlier *propone* <L *proponere* set forth. See PROPONENT.] —**pro·pound′er** *n.*

pro·pre·tor (prō·prē′tər) *n.* In ancient Rome, an officer, especially a governor of a province, having the authority of a pretor without pretorian rank. Also **pro·prae′tor.** [<L *propraetor* <*pro praetore* (one acting) for the pretor <*pro* for + *praetor* a pretor]

pro·pri·e·tar·y (prə·prī′ə·ter′ē) *adj.* **1** Pertaining to a proprietor; subject to exclusive ownership. **2** Designating an article, as a therapeutic device or medicine, protected as to name, composition, or process of manufacture by copyright, patent, secrecy, or other means. —*n. pl.* **·tar·ies 1** A proprietor or proprietors collectively. **2** Proprietorship. [<LL *proprietarius* <*proprietas* PROPERTY]

proprietary colony A colony organized under a royal grant of territory with full administrative powers to a private person or persons. Maryland, Pennsylvania, and Delaware remained proprietary colonies until the Revolution.

pro·pri·e·tor (prə·prī′ə·tər) *n.* A person having the exclusive title to anything. —**pro·pri′e·tor·ship′** *n.* —**pro·pri′e·tress** *n. fem.*

pro·pri·e·ty (prə·prī′ə·tē) *n. pl.* **·ties 1** The character or quality of being proper; especially, accordance with recognized usage, custom, or principles; becomingness; fitness; correctness. **2** *Obs.* An exclusive right of possession; also, a possession or property owned. —**the proprieties** The methods or standards of good society. [<OF *proprieté.* Doublet of PROPERTY.]

pro·pri·o·cep·tor (prō′prē·ə·sep′tər) *n. Physiol.* One of the sensory receptors situated within the body which are responsive to internal stimuli, as the muscles, joints, and tendons. [<NL <L *proprius* one's own + (RE)CEPTOR] —**pro′pri·o·cep′tive** *adj.*

prop root *Bot.* The supporting root of a plant, growing into the soil from above ground, as in corn.

prop·to·sis (prop·tō′sis) *n. Med.* A forward displacement; bulging, as of the eyeball. [<LL <Gk. *proptōsis* a falling forward <*propiptein* fall forwards <*pro-* forward + *piptein* fall]

pro·pul·sion (prə·pul′shən) *n.* **1** The act or operation of propelling. **2** An impulse given or received. [<F <L *propulsus,* pp. of *propellere* PROPEL] —**pro·pul′sive** (-siv) *adj.*

pro·pyl (prō′pil) *n. Chem.* The univalent radical, C_3H_7, derived from propane. [<PROP(IONIC) + -YL]

prop·y·lae·um (prop′ə·lē′əm) *n. pl.* **·lae·a** (-lē′ə) *Usually pl.* A structure forming an imposing entrance or gateway before an ancient temple; more widely, a porch or vestibule. [<L <Gk. *propylaion* <*pro-* before + *pylē* a gate]

pro·pyl·ene (prō′pə·lēn) *n. Chem.* A gaseous hydrocarbon, C_3H_6, obtained from propane and as a by-product in petroleum refining. Also called *propene.* [<PROPYL + -ENE]

prop·y·lite (prop′ə·līt) *n.* A variety of andesite which has been altered by the action of hot water. [<Gk. *propylon* PROPYLON + -ITE[1]; so called because thought of as opening the Tertiary epoch]

prop·y·lon (prop′ə·lon) *n. pl.* **·la** (-lə) A monumental gateway placed before the principal entrance of an important building of ancient Egypt, as a temple. [<L <Gk. <*pro-* before + *pylē* a gate]

pro ra·ta (prō rā′tə, rat′ə, rä′tə) In proportion: The loss was shared *pro rata.* [<L *pro rata (parte)* according to the calculated (share)]

pro·rate (prō·rāt′, prō′rāt′) *v.t.* & *v.i.* **·rat·ed, ·rat·ing** To distribute or divide proportionately. [<PRO RATA] —**pro·rat′a·ble** *adj.* —**pro·ra′tion** *n.*

prore (prôr) *n. Obs.* A prow. [<MF <L *prora* a prow]

pro·ro·ga·tion (prō′rə·gā′shən) *n.* **1** The act of proroguing, as a session of the British Parliament. **2** The act of prolonging or extending in time; also, continuance; prolongation. [<OF *prorogacion* <L *prorogatio, -onis* <*prorogatus,* pp. of *progrogare* PROROGUE]

pro·rogue (prō·rōg′) *v.t.* **·rogued, ·ro·guing 1** To discontinue a session of (an assembly, especially the British Parliament). **2** *Obs.* To put off or postpone. **3** *Obs.* To protract or prolong. [<MF *proroguer* <L *prorogare* prolong <*pro-* forth + *rogare* ask]

pro·sa·ic (prō·zā′ik) *adj.* **1** Lacking in those qualities that impart animation or interest; unimaginative; commonplace; dull. **2** Pertaining to or having the form of prose. Also **pro·sa′i·cal.** [<LL *prosaicus* <L *prosa* prose] —**pro·sa′ic·ness** *n.*

pro·sa·i·cism (prō·zā′ə·siz′əm) *n.* Prosaic character.

pro·sa·ism (prō′zā·iz′əm) *n.* A prosaic expression, phrase or style. [<F *prosaïsme* <L *prosa* prose]

pro·sce·ni·um (prō·sē′nē·əm) *n. pl.* **·ni·a** (-nē·ə) **1** In a modern theater or similar building, that part of the stage between the curtain or drop scene and the orchestra, sometimes including the curtain and its arch. **2** In the ancient theater, the wall that formed a background for the actors. [<L <Gk. *proskēnion* <*pro-* before + *skēnē* a stage, orig. a tent]

pro·scribe (prō·skrīb′) *v.t.* **·scribed, ·scrib·ing 1** To denounce or condemn; prohibit; interdict. **2** To outlaw or banish. **3** In ancient Rome, to publish the name of (one condemned or exiled). [<L *proscribere* <*pro-* before + *scribere* write] —**pro·scrib′er** *n.*

pro·scrip·tion (prō·skrip′shən) *n.* The act of proscribing, or state of being proscribed; interdiction; ostracism; outlawry. [<L *proscriptio, -onis* <*proscriptus,* pp. of *proscribere* PROSCRIBE] —**pro·scrip′tive** *adj.* —**pro·scrip′tive·ly** *adv.* —**pro·scrip′tive·ness** *n.*

prose (prōz) *n.* **1** Speech or writing without metrical structure: opposed to *verse* or *poetry.* **2** Commonplace or tedious discourse. **3** *Eccl.* A hymn of irregular meter sometimes sung in the eucharistic liturgy after the gradual; a sequence. 4 A proser. —*adj.* Pertaining to prose; not poetic; hence, tedious. —*v.t.* & *v.i.* **prosed, pros·ing** To write or speak in prose. [<OF <L *prosa (oratio)* straightforward (discourse) <*prorsus* <*pro-* forward + *versus,* pp. of *vertere* turn]

pro·se·cre·tin (prō′si·krē′tin) *n. Biochem.* The inactive form of secretin converted into the active form by stomach acids. [<PRO-² before + SECRETIN]

pro·sect (prō·sekt′) *v.t.* To dissect for purposes of anatomical demonstration and instruction. [Back formation <*prosector* an anatomist <LL <L *prosectus,* pp. of *prosecare* cut up <*pro-* before + *secare* cut] —**pro·sec′tion** (-sek′shən) *n.* —**pro·sec′tor** *n.*

pros·e·cute (pros′ə·kyōōt) *v.* **·cut·ed, ·cut·ing** *v.t.* 1 To go on with so as to complete; pursue to the end: to *prosecute* an inquiry. 2 To carry on or engage in, as a trade or profession. 3 *Law* **a** To bring suit against for redress of wrong or punishment of crime. **b** To seek to enforce or obtain, as a claim or right, by legal process. —*v.i.* 4 To begin and carry on a legal proceeding. See synonyms under PUSH. [<L *prosecutus,* pp. of *prosequi* pursue <*pro-* before + *sequi* follow]

prosecuting attorney The attorney empowered to act in behalf of the government, whether state, county, or national, in prosecuting for penal offenses.

pros·e·cu·tion (pros′ə·kyōō′shən) *n.* 1 The act or process of prosecuting. 2 *Law* **a** The instituting and carrying forward of a judicial proceeding to obtain some right or to redress and punish some wrong. **b** The institution and continuance of a criminal proceeding. **c** The party instituting and conducting it.

pros·e·cu·tor (pros′ə·kyōō′tər) *n.* 1 One who prosecutes, in any sense. 2 *Law* **a** One who institutes and carries on a suit, especially a criminal suit. **b** A prosecuting attorney.

pros·e·lyte (pros′ə·līt) *n.* One brought over to any opinion, belief, sect, or party, especially from one religious belief to another. See synonyms under CONVERT. —*v.* **·lyt·ed, ·lyt·ing** *v.i.* To make proselytes. —*v.t.* To make a convert of. [<LL *proselytus* <Gk. *prosēlytos* a convert to Judaism, orig. a newcomer <*prosēlyth-,* stem of *proserchesthai* approach]

pros·e·lyt·ism (pros′ə·lit′iz·əm, -li·tiz′əm) *n.* The making of converts to a religion, sect, or party, or the state of being thus converted. —**pros′e·lyt·ist** *n.*

pros·e·lyt·ize (pros′ə·lit·īz′) *v.t.* & *v.i.* **·ized, ·iz·ing** To proselyte. Also *Brit.* **pros′e·lyt·ise′.**

pros·en·ceph·a·lon (pros′en·sef′ə·lon) *n. Anat.* The forebrain. [<NL <Gk. *pros-* near, before + *encephalon* the brain. See ENCEPHALON.] —**pros′en·ce·phal′ic** (-sə·fal′ik) *adj.*

pros·en·chy·ma (pros·eng′ki·mə) *n. Bot.* Plant tissue composed of elongated, pointed, typically thick-walled cells, as distinguished from the parenchyma. [<NL <Gk. *pros-* toward, near + *enchyma* an infusion. See ENCHYMA.] —**pros·en·chym·a·tous** (pros′eng·kim′ə·təs) *adj.*

prose poem A prose work which resembles poetry either in style, structure, or emotional content.

pros·er (prō′zər) *n.* A dull or tedious writer or talker; a bore.

Pros·er·pine (pros′ər·pīn, prō·sûr′·pə·nē) In Roman mythology, the daughter of Ceres and wife of Pluto: identified with the Greek *Persephone.* Also **Pro·ser·pi·na** (prō·sûr′pə·nə).

pro·sim·i·an (prō·sim′ē·ən) *adj. Zool.* Designating any member of a suborder or group *(Prosimii)* of widely distributed early primates, as lemurs, indris, lorises, and tarsiers, characterized by small size, primitive brain development, and extensive adaptive radiation. —*n.* Any primate of this group. [<NL <Gk. *pro-* before + L *simia* an ape]

pro·sit (prō′sit) *Latin* Literally, may it benefit (you): a toast used in drinking health.

pro·slav·er·y (prō·slā′vər·ē, -slāv′rē) *adj.* In United States history, advocating Negro slavery or the policy of non-interference with it. —*n.* The advocacy of slavery.

pros·o·dem·ic (pros′ə·dem′ik) *adj. Med.* Transmitted from one person to another: said of diseases which spread by contact with affected individuals. [<*proso-* forward (<Gk. *prosō*) + (EPI)DEMIC]

pros·o·dist (pros′ə·dist) *n.* One versed in prosody.

pros·o·dy (pros′ə·dē) *n.* The science of poetical forms, including quantity and accent of syllables, meter, and versification and metrical composition. [<L *prosodia* the accent of a syllable <Gk. *prosōidia* a song sung to music <*pros-* to + *ōidē* a song] —**pro·sod·ic** (prō·sod′ik) or **·i·cal, pro·so·di·ac** (prō·sō′·dē·ak), **pro·so·di·al** (prō·sō′dē·əl) *adj.*

pro·so·po·pe·ia (prō·sō′pə·pē′ə) *n.* 1 A rhetorical figure in which the speaker impersonates another. 2 Personification. Also **pro·so′po·poe′ia.** [<L <Gk. *prosōpopoiia* <*prosōpon* a face, person + *poieein* make]

pros·pect (pros′pekt) *n.* 1 A future probability based on present indications. 2 A scene spread out before one's eyes; an extended view. 3 The direction in which anything faces; an exposure; outlook. 4 A prospective buyer. 5 The act of observing; sight; survey. 6 *Mining* **a** An indication of the presence of mineral ore. **b** A place having promising signs of the presence of mineral ore. **c** The sample or specimen of mineral obtained by washing a small portion of ore or dirt. 7 A consideration of the future; foresight. See synonyms under SCENE. —*v.t.* & *v.i.* To explore (a region) for gold, oil, etc. [<L *prospectus* a look-out, view <*prospicere* look forward <*pro-* forward + *specere* look]

pro·spec·tive (prə·spek′tiv) *adj.* 1 Being still in the future; anticipated; expected. 2 Looking toward or concerned with the future; anticipatory. —**pro·spec′tive·ly** *adv.*

pros·pec·tor (pros′pek·tər) *n.* One who searches or examines a region for mineral deposits or precious stones.

pro·spec·tus (prə·spek′təs) *n.* 1 A paper containing information of a proposed literary, commercial, or industrial undertaking. 2 A summary; outline. [<L. See PROSPECT.]

pros·per (pros′pər) *v.i.* To be prosperous; thrive; flourish. —*v.t.* To render prosperous: God *prospers* the Republic. See synonyms under FLOURISH, SUCCEED. [<OF *prosperer* <L *prosperare* cause to succeed or prosper <*prosper, prosperus* favorable, prosperous]

pros·per·i·ty (pros·per′ə·tē) *n.* The state of being prosperous; attainment of the object desired; material well-being; success.

Pros·per·o (pros′pər·ō) In Shakespeare's *Tempest,* the banished Duke of Milan.

pros·per·ous (pros′pər·əs) *adj.* 1 Successful; flourishing. 2 Favoring or tending to success; auspicious. 3 Promising; favorable. See synonyms under AUSPICIOUS, FORTUNATE, HAPPY, WELL. [<MF *prospereus* <OF *prospere* <L *prosper, prosperus* favorable] —**pros′·per·ous·ly** *adv.* —**pros′per·ous·ness** *n.*

pros·tate (pros′tāt) *adj.* 1 *Anat.* Designating a partly muscular gland at the base of the bladder around the urethra in male mammals. 2 Standing in front. —*n.* The prostate gland. [<Med. L *prostata* <Gk. *prostatēs* one who stands before <*proïstania* <*pro-* before + *histanai* set] —**pro·stat·ic** (prō·stat′ik) *adj.*

pros·ta·tec·to·my (pros′tə·tek′tə·mē) *n. Surg.* Excision of the prostate gland.

prostato- *combining form Med.* The prostate gland; of or related to the prostate: *prostatotomy.* Also, before vowels, **prostat-.** [<Gk. *prostatēs.* See PROSTATE.]

pros·ta·tot·o·my (pros′tə·tot′ə·mē) *n. Surg.* An incision into the prostate gland. [<PROSTATO- + -TOMY]

pros·the·sis (pros′thə·sis) *n.* 1 The addition of a letter or syllable to a word, especially at the beginning, as *yclept, bewail*: also spelled *prothesis.* 2 *Surg.* The fitting of artificial parts to the body, as a glass eye, a false tooth, etc. 3 Replacement or substitution of parts. [<L <Gk., addition <*prostithenai* add <*pros-* toward, besides + *tithenai* place, put] —**pros·thet·ic** (pros·thet′ik) *adj.*

pros·thet·ics (pros·thet′iks) *n.* The branch of surgery or dentistry which specializes in artificial parts and organs. [<Gk. *prostheti-kos* additional <*prosthetos* added, put on <*prostithenai.* See PROSTHESIS.] —**pros·the·tist** (pros′thə·tist) *n.*

pros·tho·don·ti·a (pros′thə·don′shē·ə, -shə) *n.* Dental prosthetics. [<NL <Gk. *prosthesis* addition + *odous, odontos* tooth]

pros·tho·don·tist (pros′thə·don′tist) *n.* A dentist who specializes in dental prosthetics.

pros·ti·tute (pros′tə·tōōt, -tyōōt) *n.* 1 A woman who practices prostitution; a harlot; whore. 2 Any base hireling; a corrupt person. —*v.t.* **·tut·ed, ·tut·ing** 1 To apply to base or unworthy purposes: to *prostitute* one's talent. 2 To offer (oneself or another) for lewd purposes, especially for hire. See synonyms under ABUSE. —*adj.* 1 Openly devoted to lewdness or promiscuity, as a woman. 2 Surrendered to base purposes. [<L *prostitutus,* pp. of *prostituere* expose publicly, prostitute <*pro-* before + *statuere* cause to stand] —**pros′ti·tu′tor** *n.*

pros·ti·tu·tion (pros′tə·tōō′shən, -tyōō′-) *n.* 1 The act or business of prostituting; the offering, by a woman, of her body for purposes of intercourse with men for hire. 2 The act of hiring or devoting to base purposes, as one's honor, talents, resources, etc.

pros·trate (pros′trāt) *adj.* 1 Lying prone, or with the face to the ground; hence, figuratively, brought low in mind or spirit. 2 Lying at the mercy of another; defenseless. 3 *Bot.* Trailing along the ground; procumbent. —*v.t.* **·trat·ed, ·trat·ing** 1 To bow or cast (oneself) down, as in adoration or pleading. 2 To throw flat; lay on the ground. 3 To overthrow or overcome; reduce to weakness or helplessness. [<L *prostratus,* pp. of *prosternere* lay flat <*pro-* before + *sternere* stretch out] —**pros′tra·tor** *n.*

pros·tra·tion (pros·trā′shən) *n.* 1 The act of prostrating in any sense. 2 Exhaustion of body or mind; great dejection or depression.

pro·style (prō′stīl) *adj. Archit.* Having a range of detached columns in front, but no columns on the sides or back of the building; also, constituting such a portico: a *prostyle* temple. [<L *prostylus* <Gk. *prostylos* <*pro-* before + *stylos* a pillar]

pros·y (prō′zē) *adj.* **pros·i·er, pros·i·est** 1 Like mere prose; prosaic. 2 Dull; tedious; commonplace. —**pros′i·ly** *adv.* —**pros′i·ness** *n.*

prot- Var. of PROTO-.

pro·tac·tin·i·um (prō′tak·tin′ē·əm) *n.* A radioactive element (symbol Pa) intermediate between thorium and uranium: its disintegration by the loss of an alpha particle gives rise to actinium. [<PROT- + ACTINIUM]

pro·tag·o·nist (prō·tag′ə·nist) *n.* The actor who played the chief part in a Greek drama; hence, a leader in any enterprise or contest. [<Gk. *prōtagōnistēs* <*prōtos* first + *agōnistēs* a contestant, an actor]

Pro·tag·o·ras (prō·tag′ər·əs), 481?–411 B.C., Greek philosopher.

pro·ta·mine (prō′tə·mēn, -min) *n. Biochem.* One of a class of strongly basic simple proteins, uncoagulable by heat, soluble in ammonia, and yielding a few amino acids when hydrolyzed. Also **pro′ta·min** (-min). [<PROT- + -AMINE]

pro·ta·no·pi·a (prō′tə·nō′pē·ə) *n. Pathol.* Color blindness marked by inability to distinguish between red and green; red blindness. [<NL <Gk. *prōtos* first + *an-* not + *ōps, ōpos* an eye] —**pro·ta·nope** (prō′tə·nōp) *n.*

prot·a·sis (prot′ə·sis) *n.* 1 In a conditional sentence, the clause (usually introductory) that contains the condition or antecedent: distinguished from *apodosis.* 2 The introductory or subordinate clause in a sentence not conditional. 3 In classical drama, the introductory part of a play. [<LL <Gk., a hypothesis <*pro-* before + *teinein* stretch]

pro·te·an (prō′tē·ən, prō·tē′an) *adj.* Readily assuming different forms or various aspects; changeable. —*n. Biochem.* Any of a group of derived proteins which are the first product of protein hydrolysis. [<PROTEUS]

pro·te·ase (prō′tē·ās) *n. Biochem.* An enzyme that digests proteins. [<PROTE(OLYSIS) + -ASE]

pro·tect (prə·tekt′) *v.t.* 1 To shield or defend from attack, harm, or injury; guard; defend. 2 *Econ.* To assist (domestic industry) by means of protective tariffs. 3 In commerce, to provide funds to guarantee payment of (a draft, etc.). See synonyms under CHERISH,

KEEP, PRESERVE, SHELTER. [<L *protectus,* pp. of *protegere* protect <*pro-* before + *tegere* cover] — **pro·tect′ing** *adj.* — **pro·tect′ing·ly** *adv.*

pro·tec·tant (prə·tek′tənt) *n.* That which protects from or guards against damage, disease, or injury; especially, a germicide, insecticide, fungicide, or the like.

pro·tect·ed (prə·tek′tid) *adj.* Shielded from harm; cared for; guarded.

pro·tec·tion (prə·tek′shən) *n.* **1** The act of protecting; a protected condition; that which protects. **2** Specifically, a system aiming to protect the industries of a country by governmental action, as by imposing duties. See PROTECTIVE TARIFF. **3** A safe-conduct; passport. **4** *U.S. Slang* Security purchased under threat of violence from racketeers; also, the money so paid. See synonyms under DEFENSE, REFUGE, SHELTER.

pro·tec·tion·ism (prə·tek′shən·iz′əm) *n.* The economic doctrine or system of protection. — **pro·tec′tion·ist** *n.*

pro·tec·tive (prə·tek′tiv) *adj.* **1** Affording or suitable for protection; sheltering; defensive; specifically, in political economy, insuring or intended to insure protection to home industries: a *protective* tariff. **2** Providing or alleging to provide protection: *protective* custody. — *n.* Something that protects; specifically, an aseptic covering for a wound. — **pro·tec′tive·ly** *adv.* — **pro·tec′tive·ness** *n.*

protective coloration *Biol.* Any coloration of a plant or animal that makes it almost indistinguishable from its natural or habitual environment, and thus safe from detection by its enemies.

protective tariff *Econ.* A tariff that is intended to insure protection of domestic industries against foreign competition: opposed to *free trade.*

pro·tec·tor (prə·tek′tər) *n.* **1** One who protects; a defender. **2** In English history, one appointed as a regent of the kingdom during minority or incapacity of the sovereign. Also **pro·tect′er.** — **pro·tec′tress** *n. fem.*

Pro·tec·tor (prə·tek′tər) *n.* The official title of the chief ruler during the Commonwealth: in full, **Lord Protector.** The title was borne by Oliver Cromwell, 1653–58, and by Richard Cromwell, 1658–59.

pro·tec·tor·ate (prə·tek′tər·it) *n.* **1** A relation of protection and partial control by a strong nation over a weaker power. **2** A country or region under the protection of another. **3** The office, or period of office, of a protector of a kingdom. Also **pro·tec′tor·ship.**

Pro·tec·tor·ate (prə·tek′tər·it) *n.* The English government during the time of the Cromwells, 1653–59.

pro·tec·to·ry (prə·tek′tər·ē) *n. pl.* **·to·ries** An institution for the care and education of homeless or destitute children.

pro·tect·o·scope (prə·tek′tə·skōp) *n. Mil.* A device resembling the periscope, to permit tank gunners to observe around their protective shields without exposing themselves to gunfire. [<PROTECT + -(O)SCOPE]

pro·té·gé (prō′tə·zhā, *Fr.* prô·tā·zhā′) *n.* One specially cared for by another who is older or more powerful. [<F, pp. of *protéger* <L *protegere* PROTECT] — **pro′té·gée** *n. fem.*

pro·te·in (prō′tē·in, -tēn) *n. Biochem.* Any of a class of highly complex nitrogenous organic compounds occurring naturally in all living matter, and forming an essential part of animal food requirements. They are composed principally of amino acids in varying combinations, and are usually classified as: *simple* (hydrolyzed only by enzymes or acids into alpha-amino acids or their derivatives); *conjugated* (simple proteins combined with non-proteins in a form other than a salt); *derived* (obtained by the action of heat, enzymes, or reagents upon naturally occurring proteins). Also **pro′te·id** (-id). [<G <Gk. *prōteios* chief <*prōtos* first; so called because the chief constituent of living matter]

pro tem·po·re (prō tem′pə·rē) *Latin* For the time being; temporary: abbr. *pro tem.*

pro·tend (prō·tend′) *v.i. Psychol.* To exhibit protensity. [<L *protendere* stretch forth < *pro-* forth + *tendere* stretch]

pro·ten·si·ty (prō·ten′sə·tē) *n. Psychol.* The temporal attribute of a sensation or other mental phenomenon: the psychic analog of duration. [<L *protensus,* pp. of *protendere.* See PROTEND.] — **pro·ten′sive** *adj.*

pro·te·ol·y·sis (prō′tē·ol′ə·sis) *n. Biochem.* The change or splitting up of proteins into simpler products during digestion. [<NL <*proteo-* (<PROTEIN) + Gk. *lysis* a loosening <*lyein* loosen] — **pro′te·o·lyt′ic** (-ə·lit′ik) *adj.*

pro·te·ose (prō′tē·ōs) *n. Biochem.* Any of a group of derived proteins formed naturally in the process of digestion and produced artificially, as by treating the corresponding proteins with dilute mineral acids. [<PROTE(IN) + -OSE[2]]

Prot·er·o·zo·ic (prot′ər·ə·zō′ik) *Geol. adj.* Pertaining to or designating the geological era following the Archeozoic and succeeded by the Paleozoic. — *n.* The Proterozoic era. [<*protero-* (<Gk. *proteros* former, anterior) + -ZOIC]

Pro·tes·i·la·us (prō·tes′ə·lā′əs) In the *Iliad,* the husband of Laodamia and first of the Greeks killed at Troy.

pro·test (prō′test) *n.* **1** The act of protesting; a solemn or formal objection or declaration. **2** A formal notarial certificate attesting the fact that a note or bill of exchange has been presented for acceptance or payment and that it has been refused. **3** In maritime law, a written declaration by the master of a vessel stating that an injury to the vessel or the cargo was not owing to the neglect or misconduct of the master. **4** A formal statement in writing made by a person called upon by public authority to pay a sum of money, as an import duty or a tax, in which he declares that he does not concede the legality of the claim. — *v.* (prə·test′) *v.t.* **1** To assert earnestly or positively; state formally, especially against opposition or doubt. **2** To make a protest against; object to: I *protested* his actions. **3** To declare formally that payment of (a promissory note, etc.) has been duly submitted and refused. — *v.i.* **4** To make solemn affirmation. **5** To make a protest; object. See synonyms under AFFIRM, ASSERT, AVOW. [<OF *protester* <L *protestari* <*pro-* forth + *testari* affirm, give evidence <*testis* a witness] — **pro·test′er** *n.*

prot·es·tant (prot′is·tənt, prə·tes′-) *n.* One who makes a protest. [<MF <L *protestans, -antis,* ppr. of *protestari* PROTEST]

Prot·es·tant (prot′is·tənt) *n.* **1** A member of one of those bodies of Christians that adhere to Protestantism, as opposed to Roman Catholicism: a use opposed by some Anglicans. **2** In the 17th century, a Lutheran or Anglican. **3** Originally, one of those German princes who, at the second Council of Spires, April 19, 1529, protested against the decree of the majority representing the Roman Catholic states which involved a virtual submission to the authority of the Roman Catholic Church. — *adj.* Pertaining to Protestants or Protestantism.

Protestant Episcopal Church A religious body in the United States which is descended from the Church of England, but has been organized as a separate and independent body since 1789.

Prot·es·tant·ism (prot′is·tənt·iz′əm) *n.* **1** The principles and common system of doctrines taught by Luther, and by the evangelical churches since. Its positive and formal principle is that nothing that is not taught in the Holy Scriptures, the authoritative rule of faith and practice in the church, enters as an essential element into the Christian system. **2** The ecclesiastical system founded upon this faith; also, Protestants, collectively. **3** The state of being a Protestant.

prot·es·ta·tion (prot′is·tā′shən) *n.* **1** The act of protesting; also, that which is protested. **2** A formal declaration of dissent. **3** Any solemn or urgent avowal.

Pro·test·er (prə·tes′tər) *n.* A Scotsman who protested against the union of the Presbyterians and the Royalists in 1650. Also **Pro·tes′tor.**

pro·test·ing·ly (prə·tes′ting·lē) *adv.* In such a manner as to protest.

Pro·te·us (prō′tē·əs, -tyōōs) In Greek mythology, a sea god who had the power of assuming different forms. — **Pro′te·an** *adj.*

pro·tha·la·mi·on (prō′thə·lā′mē·on, -ən) *n. pl.* **·mi·a** (-mē·ə) A song celebrating a marriage; an epithalamium. Also **pro′tha·la′mi·um** (-mē·əm). [<NL <Gk. *pro-* before + *thalamos* a bridal chamber; coined by Spenser on analogy with *epithalamion*]

pro·thal·li·um (prō·thal′ē·əm) *n. pl.* **·li·a** (-ē·ə) *Bot.* The first or false thallus formed on the germination of the asexually produced spores in pteridophytes; a delicate, evanescent cellular structure bearing the sexual organs. Also **pro·thal′lus.** [<NL <Gk. *pro-* before + *thallion,* dim. of *thallos* a shoot] — **pro·thal′li·al** *adj.* — **pro·thal′line** (-thal′in, -in) *adj.*

proth·e·sis (proth′ə·sis) *n.* **1** Prosthesis (def. 1). **2** In the Greek Orthodox Church, a service by which the elements are prepared for consecration in the Eucharist. [<LL <Gk., a placing before, or in public <*protithenai* set before <*pro-* before + *tithenai* place] — **pro·thet·ic** (prō·thet′ik) *adj.* — **pro·thet′i·cal·ly** *adv.*

pro·thon·o·tar·y (prō·thon′ə·ter′ē, prō′thə·nō′·tər·ē) *n. pl.* **·tar·ies** **1** A chief clerk; specifically, in the Roman Catholic Church, one of the seven (formerly twelve) ecclesiastics at Rome who keep the registry of important pontifical proceedings, or one having the title and some of the associated privileges. **2** In some States of the United States, a probate officer. Also spelled *protonotary.* [<LL *protonotarius* <LGk. *prōtonotarios* <Gk. *prōtos* first + L *notarius.* See NOTARY.] — **pro·thon′o·tar′i·al** (-târ′ē·əl) *adj.*

prothonotary warbler A North American warbler (*Protonotaria citrea*) the male of which is noted for the brilliant yellow to orange coloring of its head and under parts, with bluish-gray wings and tail.

pro·tho·rax (prō·thôr′aks, -thō′raks) *n. pl.* **·rax·es** or **·tho·ra·ces** (-thôr′ə·sēz, -thō′rə-) *Entomol.* The anterior segment of the thorax of an insect. [<NL <Gk. *pro-* in front + *thorax* thorax] — **pro·tho·rac·ic** (prō′thô·ras′·ik, -thō-) *adj.*

pro·throm·bin (prō·throm′bin) *n. Biochem.* The inactive precursor of thrombin: it is converted into thrombin by the action of calcium and thromboplastin, and is essential to the process of blood-clotting: also called *thrombogen.* [<NL <Gk. *pro-* before + *thrombos* a clot]

pro·tist (prō′tist) *n. Biol.* **1** Any unicellular organism, whether animal or plant. **2** Formerly, any member of a large division (*Protista*) including all single-celled organisms. [<NL <Gk. *prōtista,* neut. pl. of *prōtistos* the very first, superl. of *prōtos* first] — **pro·tis′tan** *adj.* & *n.* — **pro·tis′tic** *adj.*

pro·ti·um (prō′tē·əm) *n. Chem.* The hydrogen isotope of atomic mass 1 (symbol, H^1): sometimes so called in distinction from deuterium and tritium. [<NL <Gk. *prōtos* first. Cf. PROTO- (def. 3).]

proto- *combining form* **1** First in rank or time; chief; typical: *protomartyr.* **2** Primitive; original: *prototype.* **3** *Chem.* **a** Designating the first or lowest member of a series; having the least amount (of an element or radical): *protoxide.* **b** Denoting the parent form or source of: *protoactinium.* Also, before vowels, *prot-.* [<Gk. *prōto-* <*prōtos* first]

pro·to·ac·tin·i·um (prō′tō·ak·tin′ē·əm) *n.* Protactinium.

Pro·to·coc·cus (prō′tə·kok′əs) *n.* The typical genus of a family (*Chlorophyceae*) of green algae, growing on damp walls, rocks, and trunks of trees. [<PROTO- + COCCUS]

pro·to·col (prō′tə·kol) *n.* **1** The preliminary draft of an official document, as a treaty; specifically, the preliminary draft or report of the negotiations and conclusions arrived at by a diplomatic conference, having the force of a treaty when ratified. **2** The rules of diplomatic and state etiquette and ceremony. — *v.i.* To write or form protocols. [<OF *prothocole* <Med. L *protocollum* <LGk. *prōtokollon* the first glued sheet of a papyrus roll <*prōtos* first + *kolla* glue]

pro·to·derm (prō′tə·dûrm) *n.* Dermatogen.

pro·to·gene (prō′tə·jēn) *n.* The hypothetical prototype of the gene, assumed to have been formed from complex carbon compounds at the time when life evolved from inorganic matter.

pro·to·gram (prō′tə·gram) *n.* An acronym.

Pro·to·hip·pus (prō′tō·hip′əs) *n. Paleontol.* A genus of extinct, three-toed horses of the Miocene period. [<NL <Gk. *prōtos* first + *hippos* a horse]

pro·to·hu·man (prō′tō·hyōō′mən) *adj.* **1** Anterior to or more primitive than man. **2** *Paleontol.* Of, pertaining to, or describing any of several hominoid primates regarded as being at an earlier stage of development than *Homo sapiens.* —*n.* Any primate antedating modern man in evolutionary characteristics. Principal types are:

Africanthropus	Meganthropus
Australopithecus	Oreopithecus
Dryopithecus	Pithecanthropus
Gigantopithecus	Sinanthropus

pro·to·lith·ic (prō′tō·lith′ik) *adj.* Pertaining to the earliest period of the stone age; eolithic.

pro·to·mar·tyr (prō′tō·mär′tər) *n.* The first martyr or victim in any cause. [<OF *prothomartir* <Med. L *protomartyr* <Gk. *prōtomartyr* <*prōtos* first + *martyr* a witness]

pro·to·mor·phic (prō′tō·môr′fik) *adj. Biol.* Of or pertaining to, or having the most primitive or elementary form or structure. —**pro′to·morph** *n.*

pro·ton (prō′ton) *n. Physics* **1** The positively charged nucleus of the atom of the light isotope of hydrogen (symbol, H^1), constituting its principal mass. **2** One of the elementary particles in the nucleus of an atom, having a unitary positive charge and a mass of approximately 1.672×10^{-24} gram. The atomic number of an element is equivalent to the number of protons in its nucleus. [<NL <Gk. *prōton,* neut. of *prōtos* first]

pro·to·ne·ma (prō′tə·nē′mə) *n. pl.* **·ne·ma·ta** (-nē′mə·tə) *Bot.* An early stage in the development of the prothallium of ferns; a green confervoid or filamentous structure developed from the spore in mosses, on which the leafy plant arises as a lateral or terminal shoot. Also **pro′to·neme** (-nēm). [<NL <Gk. *prōtos* first + *nēma* a thread]

pro·ton·o·tar·y (prō·ton′ə·ter′ē, prō′tə·nō′tər·ē) See PROTHONOTARY.

pro·ton–pro·ton reaction (prō′ton·prō′ton) A thermonuclear chain reaction which is assumed to provide stellar energy by means of the fusion of 4 protons to make a helium nucleus, with a residue of 2 protons returned to the cycle. Compare CARBON CYCLE.

proton synchrotron *Physics* A bevatron.

pro·to·path·ic (prō′tə·path′ik) *adj. Physiol.* Pertaining to or designating primary sensibility, responsive only to gross, typically painful stimuli: distinguished from *epicritic.* [<PROTO- + -PATHIC]

pro·to·phyte (prō′tə·fīt) *n. Bot.* **1** Any single-celled plant. **2** A member of a former division (*Protophyta*) embracing only the lowest and simplest plants. [<NL <Gk. *prōtos* first + *phyton* a plant]

pro·to·plasm (prō′tə·plaz′əm) *n. Biol.* **1** The physicochemical basis of living matter, a viscid, grayish, translucent, colloidal substance of granular structure and complex composition that forms the essential part of plant and animal cells. **2** The cytoplasm of the cell, as distinguished from the nuclear material. [<G *protoplasma* <Gk. *prōtos* first + *plasma.* See PLASMA.] —**pro′to·plas′mic** or **·plas′mal** or **·plas·mat′ic** *adj.*

pro·to·plast (prō′tə·plast) *n. Biol.* **1** That which is first formed; the original or primordial cell. **2** The parent pair or one of the parent pair of the first-formed individuals of a species. **3** The protoplasmic contents of a cell. **4** A plastid. [<F *protoplaste* <LL *protoplastus* <Gk. *protoplastos* formed first <*prōtos* first + *plastos* formed <*plassein* form] —**pro′to·plas′tic** *adj.*

pro·to·ste·le (prō′tə·stē′lē, -stēl) *n. Bot.* The dense central cylinder of roots and young stems, and, in various pteridophytes, the axes. [<PROTO- + STELE[2]] —**pro′to·ste′lic** *adj.*

Pro·to·the·ri·a (prō′tə·thir′ē·ə) *n. pl.* A subclass of primitive, egg-laying mammals; the monotremes, as the duckbill. [<NL <Gk. *prōtos* first + *thēria,* pl. of *thērion,* dim. of *thēr* a beast]

pro·to·troph·ic (prō′tə·trof′ik, -trō′fik) *adj. Biol.* Capable of assimilating only simple inorganic substances: said of the earliest forms of life.

pro·to·type (prō′tə·tīp) *n.* **1** *Biol.* A primitive or ancestral organism; an archetype: opposed to *ectype.* **2** A first or original model on which subsequent forms are to be based. **3** An accepted standard to which all others must conform. See synonyms under EXAMPLE, IDEAL, MODEL. [<MF <NL *prototypon* <Gk. *prōtotypon,* orig. neut. sing. of *prōtotypos* original <*prōtos* first + *typos* a model] —**pro′to·typ′al** (-tī′pəl), **pro′to·typ′ic** (-tip′ik), **pro′to·typ′i·cal** *adj.*

pro·tox·ide (prō·tok′sīd, -sid) *n. Chem.* An oxide containing the lowest proportion of oxygen for a given series: contrasted with *peroxide*: iron *protoxide* (ferrous oxide). Also **pro·tox′id** (-sid). [<PROT- + OXIDE]

Pro·to·zo·a (prō′tə·zō′ə) *n. pl.* A phylum of the animal kingdom embracing microscopic organisms consisting of a single cell, and reproducing typically by binary fission. They are largely aquatic, and include many parasitic forms. [<NL <Gk. *prōtos* first + *zōia,* pl. of *zōion* an animal] —**pro′to·zo′an** *adj.* & *n.* —**pro′to·zo′ic** *adj.* —**pro′to·zo′on** *n.*

pro·to·zo·ol·o·gy (prō′tō·zō·ol′ə·jē) *n.* The study or science of unicellular organisms. —**pro′to·zo′o·log′i·cal** (-zō′ə·loj′i·kəl) *adj.* —**pro′to·zo·ol′o·gist** *n.*

pro·tract (prō·trakt′) *v.t.* **1** To extend in time; prolong. **2** In surveying, to draw or map by means of a scale and protractor; plot. **3** *Zool.* To protrude or extend: opposed to *retract.* [<L *protractus,* pp. of *protrahere* extend <*pro-* forward + *trahere* draw] —**pro·trac′tive** *adj.*

Synonyms: continue, delay, elongate, extend, lengthen, prolong. To *protract* is to cause to occupy a longer time than is usual, expected, or desirable. We *protract* a negotiation which we are slow to conclude; *delay* may be used either of the beginning or of any stage in the proceedings; we may *delay* a person as well as an action, but *protract* is not used of persons. *Elongate* is used only of material objects or extension in space; *protract* is rarely, except in mathematics, used of concrete objects or extension in space; we *elongate* a line, *protract* a discussion. *Protract* has usually an unfavorable sense; *continue* is neutral, applying equally to the desirable or the undesirable. Compare HINDER. *Antonyms*: abbreviate, abridge, conclude, contract, curtail, hasten, hurry, limit, reduce, shorten.

pro·tract·ed (prō·trak′tid) *adj.* Unduly or unusually extended or prolonged.

protracted meeting A series of religious, usually revival, meetings, held morning, afternoon, and evening, and sometimes continued for several days.

pro·tract·er (prō·trak′tər) *n.* **1** One who or that which protracts. **2** A protractor.

pro·trac·tile (prō·trak′til) *adj.* Capable of being protracted or protruded; protrusile.

pro·trac·tion (prō·trak′shən) *n.* **1** The act of drawing out or lengthening in time; the act of delaying the termination of anything. **2** In prosody, the irregular lengthening of a syllable ordinarily short. **3** The making of a surveyor's plot on paper.

pro·trac·tor (prō·trak′tər) *n.* **1** An instrument for measuring and laying off angles. **2** A tailor's adjustable pattern. **3** *Anat.* A muscle that extends a limb or moves it forward. **4** *Surg.* An instrument for extracting foreign bodies from a wound.

pro·trude (prō·trōōd′) *v.t.* & *v.i.* **·trud·ed, ·trud·ing** To push or thrust out; project outward. [<L *protrudere* <*pro-* forward + *trudere* thrust]

pro·tru·sile (prō·trōō′sil) *adj.* Adapted to being thrust out, often rapidly, as the tongue of an ant-eater. Also **pro·tru′si·ble.** [<L *protrusus,* pp. of *protrudere* PROTRUDE + -ILE]

pro·tru·sion (prō·trōō′zhən) *n.* **1** The act of protruding, or the state of being protruded. **2** The part or object protruded. [<F <L *protrusus.* See PROTRUSILE.]

pro·tru·sive (prō·trōō′siv) *adj.* **1** Tending to protrude; protruding. **2** Pushing or driving forward. —**pro·tru′sive·ly** *adv.* —**pro·tru′sive·ness** *n.*

pro·tu·ber·ance (prō·tōō′bər·əns, -tyōō′-) *n.* **1** Something that protrudes; a knob; prominence. **2** The state of being protuberant. Also **pro·tu′ber·an·cy, pro·tu′ber·a′tion.**

pro·tu·ber·ant (prō·tōō′bər·ənt, -tyōō′-) *adj.* Swelling out beyond the surrounding surface; bulging. [LL *protuberans, -antis,* ppr. of *protuberare* bulge out <L *pro-* forth + *tuber* a swelling] —**pro·tu′ber·ant·ly** *adv.*

pro·tu·ber·ate (prō·tōō′bə·rāt, -tyōō′-) *v.i.* **·at·ed, ·at·ing** To be protuberant; bulge out. [<LL *protuberatus,* pp. of *protuberare.* See PROTUBERANT.]

pro·tyle (prō′tīl, -til) *n.* The hypothetical primitive material of the universe; a substance of which all existing elements have been supposed to be modifications. Also **pro′tyl** (-til). [<Gk. *prōtos* first + *hylē* timber, matter]

proud (proud) *adj.* **1** Actuated by, possessing, or manifesting pride; arrogant; haughty; also, self-respecting. **2** Sensible of honor and personal elation: generally followed by *of* or by a verb in the infinitive. **3** High-mettled, as a horse; spirited. **4** Proceeding from or inspired by pride. **5** Being a cause of honorable pride, as a distinction or achievement. **6** *Obs.* Bold; fearless; daring. See synonyms under HAUGHTY, HIGH. [OE *prūt, prūd* <OF *prud, prod,* prob. ult. <L *prodesse* be of value] —**proud′ly** *adv.*

proud flesh *Pathol.* A granulated growth resembling flesh in a wound or sore. [So called because of its swelling up]

Prou·dhon (prōō·dôn′), **Pierre Joseph,** 1809–1865, French socialist, philosophical anarchist, and writer on politics and economics.

Proust (prōōst), **Joseph Louis,** 1754–1826, French chemist. —**Marcel,** 1871–1922, French novelist.

proust·ite (prōōs′tīt) *n.* An adamantine ruby-red sulfide of silver and arsenic, crystallizing in the rhombohedral system. [<F, after J. L. *Proust,* its discoverer]

Prout's hypothesis (prouts) *Chem.* A hypothesis that all atomic weights are simple multiples of the atomic weight of hydrogen. [after William *Prout,* 1785–1850, English chemist]

prove (prōōv) *v.* **proved, proved** or **prov·en, prov·ing** *v.t.* **1** To show to be true or genuine, as by evidence or argument. **2** To determine the quality or genuineness of; test: to *prove* a gun. **3** To establish the authenticity or validity of, as a will. **4** *Math.* To verify the accuracy of (a calculation or demonstration) by an independent process. **5** *Printing* To take a proof of or from. **6** *Archaic* To learn by experience; undergo. —*v.i.* **7** To be shown to be by the result or outcome; turn out to be: His hopes *proved* vain. **8** *Archaic* To make trial. See synonyms under CONFIRM. [OF *prouver* <L *probare* test, try. Doublet of PROBE.] —**prov′a·ble** *adj.* —**prov′er** *n.*

pro·vec·tion (prō·vek′shən) *n. Ling.* A transfer of the final consonant of one word to the beginning of the next word, as in *a newt,* the old form of which was *an ewt.* [<LL *provectio, -onis* <*provectus,* pp. of *provehere* advance <*pro-* forward + *vehere* carry]

prov·en (prōō′vən) Alternative past participle of PROVE: the less common form. —*adj.* Proved; established; verified.

prov·e·nance (prov′ə·nəns) *n.* Provenience; origin. [<F <*provenant,* ppr. of *provenir* come forth <L *provenire* <*pro-* forth + *venire* come]

Pro·ven·çal (prō′vən·säl′, *Fr.* prō·vän·sȧl′) *n.* **1** A native or resident of Provence, France. **2** The Romance language of Provence: developed from *langue d'oc,* and used especially in the 12th and 13th centuries in the lyric literature of the troubadours. —*adj.* Of or pertaining to Provence, its inhabitants, or their language. [<MF, of Provence <L *provincialis* <(*nostra*) *provincia* (our) province, i.e., Provence]

Pro·vence (prô·väns′) A region and former province of SE France.

prov·en·der (prov′ən·dər) *n.* Food for cattle; especially, dry food, as hay; rarely, provisions generally. See synonyms under FOOD. —*v.t.* To provide with food, as cattle. [<OF *provendre, provende* an allowance of food <L *praebenda.* See PREBEND.]

pro·ve·ni·ence (prō·vē′nē·əns, -vēn′yəns) *n.* The origin or source of a thing: used especially in the fine arts and archeology. [<L *proveniens, -entis,* ppr. of *provenire.* See PROVENANCE.]

prov·erb (prov′ərb) *n.* **1** A pithy saying, especially one condensing the wisdom of experience; adage; saw; maxim. **2** An enigmatical saying: to speak in a *proverb.* **3** Something proverbial; a typical example; byword. [<OF *proverbe* <L *proverbium* <*pro-* before + *verbum* a word]

Synonyms: adage, aphorism, apothegm, axiom, byword, dictum, maxim, motto, precept, saw, saying, truism. The *proverb* or *adage* gives homely truth in condensed, practical form; the latter especially gains authority by long usage. An *aphorism* is a summary statement of a general truth. An *apothegm* is a sententious statement. A *dictum* is a statement of some person or school, on whom it depends for authority. A *saying* is impersonal, current among the people. A *saw* is a *saying* that is old, but somewhat worn and tiresome. *Precept* is a command or a rule for behavior; a *motto* or *maxim* is a brief statement of cherished truth, the *maxim* being more uniformly and directly practical. A *byword* is a *saying* used reproachfully or contemptuously. Compare ADAGE, AXIOM.

pro·ver·bi·al (prə·vûr′bē·əl) *adj.* 1 Of the nature of, pertaining to, or like a proverb: *proverbial* brevity. 2 Supplying the subject for a proverb; being the object of general remark, especially as a typical case; well-known; notorious. —**pro·ver′bi·al·ly** *adv.*

Prov·erbs (prov′ərbz) An Old Testament didactic poetical book of moral sayings and instructions.

pro·vide (prə·vīd′) *v.* **·vid·ed, ·vid·ing** *v.t.* 1 To supply or furnish. 2 To afford; yield. 3 To prepare, make ready, or procure beforehand. 4 To set down as a condition; stipulate. —*v.i.* 5 To take measures in advance: with *for* or *against*. 6 To furnish means of subsistence: usually with *for*. 7 To make a stipulation. [<L *providere* foresee <*pro-* before + *videre* see. Doublet of PURVEY.]

Synonyms: arrange, cater, furnish, prepare, procure, produce, supply. *Antonyms*: alienate, divert, lose, misemploy, mismanage, neglect, overlook, scatter, squander, waste.

pro·vid·ed (prə·vī′did) *conj.* On condition: with *that* expressed or understood: He will get the loan *provided* he offers good security. See synonyms under BUT. [Orig. pp. of PROVIDE]

prov·i·dence (prov′ə·dəns) *n.* 1 The care exercised by the Supreme Being over the universe. 2 An event or circumstances ascribable to divine interposition. 3 The exercise of foresight and care for the future; prudent economy. See synonyms under FRUGALITY, PRUDENCE. [<OF <L *providentia* <*providens, -entis*, ppr. of *providere*. See PROVIDE.]

Prov·i·dence (prov′ə·dəns) God; the Deity.

Prov·i·dence (prov′ə·dəns) The capital of Rhode Island, a port of entry on Narragansett Bay.

Providence Plantations Original name of the colony established by Roger Williams (1636) in Rhode Island.

prov·i·dent (prov′ə·dənt) *adj.* Exercising foresight; economical; anticipating and making ready for future wants or emergencies. See synonyms under THOUGHTFUL. —**prov′i·dent·ly** *adv.*

prov·i·den·tial (prov′ə·den′shəl) *adj.* Resulting from or exhibiting the action of God's providence. —**prov′i·den′tial·ly** *adv.*

pro·vid·er (prə·vī′dər) *n.* One whose income supports a family: He's a good *provider*.

pro·vid·ing (prə·vī′ding) *conj.* Provided; in case that.

prov·ince (prov′ins) *n.* 1 A considerable country incorporated with a kingdom or empire and subject to the central administration without having itself any voice in that administration. 2 Any large administrative division of a country with a permanent local government: the *provinces* of the Roman Empire, the *Provinces* of the Dominion of Canada or of the Union of South Africa, the United *Provinces* of Agra and Oudh. The word is often loosely used in the plural to denote those regions that lie at a distance from the capital; specifically, in Great Britain, the whole country except London. 3 A comprehensive department or sphere of knowledge or activity: the *province* of chemistry. 4 A definite sphere of action, especially one authoritatively assigned or properly belonging to a person: The *province* of the judge is to apply the laws. 5 *Ecol.* A zoogeographical area less than a region, having its own special flora, fauna, and types of mankind. [<OF <L *provincia* an official duty or charge, a province]

Province Welles·ley (welz′lē) See PENANG, SETTLEMENT OF.

pro·vin·cial (prə·vin′shəl) *adj.* 1 Pertaining to or characteristic of a province. 2 Confined to a province; rustic; hence, local, as a word or idiom; also, narrow; uncultured; illiberal: said of people. —*n.* A native or inhabitant of a province; one who is provincial, in any sense. —**pro·vin′ci·al′i·ty** (-shē·al′ə·tē) *n.* —**pro·vin′cial·ly** *adv.*

pro·vin·cial·ism (prə·vin′shəl·iz′əm) *n.* The quality of being provincial; a provincial custom or peculiarity, especially of speech.

proving ground A site used for testing new weapons, equipment, scientific theories, etc.

pro·vi·sion (prə·vizh′ən) *n.* 1 Measures taken or means made ready in advance; the act of taking such measures. 2 *pl.* Food or a supply of food; victuals. 3 Something provided or prepared, as against future need. 4 A stipulation or requirement; the part of an agreement, instrument, etc., referring to one specific thing. 5 Appointment to a see or benefice not yet vacant, including designation, institution, and installation; especially, such appointment when made by the pope, before a vacancy, so as to set aside nomination by the ordinary patron. 6 *pl.* Medieval English statutes by which certain important matters were provided for: the *provisions* of Oxford. See synonyms under NUTRIMENT, STOCK. —*v.t.* To provide with food or provisions. [<OF <L *provisio, -onis* a foreseeing <*provisus*, pp. of *providere*. See PROVIDE.] —**pro·vi′sion·er** *n.*

pro·vi·sion·al (prə·vizh′ən·əl) *adj.* Provided for a present service or temporary necessity: a *provisional* army; adopted tentatively or for lack of something better. —**pro·vi′sion·al·ly** *adv.*

provisional government A temporary government established to provide for a present situation or emergency, to be superseded later by a permanent government.

pro·vi·sion·ar·y (prə·vizh′ən·er′ē) *adj.* 1 Providing or intended to provide for some future occasion or want; provident; also, containing the statement of a provision. 2 Provisional.

pro·vi·so (prə·vī′zō) *n.* *pl.* **·sos** or **·soes** A conditional stipulation; a clause, as in a contract or statute, limiting, modifying, or rendering conditional its operation. [<Med. L *proviso (quod)* it being provided (that), ablative neut. sing. pp. of L *providere*. See PROVIDE.]

pro·vi·so·ry (prə·vī′zər·ē) *adj.* 1 Containing or made dependent on a proviso; conditional. 2 Provisional. —**pro·vi′so·ri·ly** *adv.*

pro·vi·ta·min (prō·vī′tə·min) *n. Biochem.* Any of various substances believed to promote the formation of vitamins, as carotene (**provitamin A**) or ergosterol. [<*pro-* undeveloped (<L, before) + VITAMIN]

prov·o·ca·tion (prov′ə·kā′shən) *n.* 1 The act of provoking. 2 An incitement to action; stimulus; something that stirs to anger. [<OF <L *provocatio, -onis* <*provocatus*, pp. of *provocare* PROVOKE]

pro·voc·a·tive (prə·vok′ə·tiv) *adj.* Serving to provoke; stimulating. —*n.* That which provokes or tends to provoke. —**pro·voc′a·tive·ly** *adv.* —**pro·voc′a·tive·ness** *n.*

pro·voke (prə·vōk′) *v.t.* **·voked, ·vok·ing** 1 To stir to anger or resentment; irritate; vex. 2 To arouse or stimulate to some action. 3 To stir up or bring about: to *provoke* a quarrel. 4 To induce or cause; elicit: to *provoke* a smile. 5 *Obs.* To call forth; summon. [<OF *provoker* <L *provocare* challenge <*pro-* forth + *vocare* call] —**pro·vok′ing** *adj.* —**pro·vok′ing·ly** *adv.* —**pro·vok′ing·ness** *n.*

prov·ost (prov′əst) *n.* 1 A person having charge or authority over others. 2 The chief magistrate of a Scottish city, corresponding to the English mayor: in Edinburgh, Dundee, Glasgow, and Aberdeen called **Lord Provost.** 3 In some English and American colleges, the head of the faculty. 4 The head of a collegiate chapter or a cathedral; a dean. 5 (prō′vō) Provost marshal. [Fusion of OE *profost, prafost* and AF. OF *provost*, both <LL *propositus* <L *praepositus* a prefect, orig. pp. of *praeponere* <*prae-* before + *ponere* place] —**prov′ost·ship** *n.*

pro·vost court (prō′vō) A summary military court for trying those (especially civilians in a theater of war) charged with minor offenses committed within areas controlled by the army. They are usually guided by the rules of evidence. Their jurisdiction is concurrent with that of courts martial. The military commission is resorted to in like situations for graver offenses such as espionage.

pro·vost guard (prō′vō) A company of soldiers detailed for police duty under the provost marshal.

pro·vost marshal (prō′vō) A military or naval officer exercising police functions.

pro·vost sergeant (prō′vō) A non-commissioned officer who supervises the work and duties of the military police.

prow[1] (prou) *n.* 1 The fore part of a vessel's hull or of an airship; the bow. 2 Any pointed projection. 3 *Poetic* A ship. [<MF *prove* <Provençal *proa* <L *prora* <Gk. *prōira*]

prow[2] (prou) *adj. Archaic* Brave; valiant: a *prow* knight. [<OF *prou* brave <LL *prode*, back formation <L *prodesse* be of use <*pro-*, *prod-* for + *esse* be]

prow·ess (prou′is) *n.* 1 Strength, skill, and courage in battle. 2 A daring and valiant deed. [<OF *prouesse, proece* <*prou* PROW[2]]

Synonyms: bravery, courage, gallantry, heroism, intrepidity, strength, valor. *Bravery, courage, heroism*, and *intrepidity* may be silent, spiritual, or passive; they may be exhibited by a martyr at the stake. *Courage* is a nobler word than *bravery*, involving more of the deep, spiritual, and enduring elements of character; it applies to matters to which *valor* and *prowess* cannot, as submission to a surgical operation, or the facing of censure or detraction for conscience' sake. *Prowess* and *valor* imply both daring and doing. *Valor* meets odds or perils with courageous action, doing its utmost to conquer at any risk or cost; *prowess* has power and ability adapted to the need; dauntless *valor* is often vain against superior *prowess*. Compare synonyms for BRAVE, COURAGE, FORTITUDE. *Antonyms*: cowardice, cowardliness, effeminacy, fear, timidity.

prowl (proul) *v.t. & v.i.* To roam about stealthily, as in search of prey or plunder. —*n.* A roaming about for prey. [ME *prollen* search; ult. origin uncertain] —**prowl′er** *n.*

prowl car *U.S.* A police patrol car.

prox·i·mal (prok′sə·məl) *adj.* 1 *Anat.* Relatively nearer the central portion of the body or point of origin: opposed to *distal*. 2 Proximate. —**prox′i·mal·ly** *adv.*

prox·i·mate (prok′sə·mit) *adj.* Being in immediate relation with something else; next. See synonyms under IMMEDIATE. [<LL *proximatus*, pp. of *proximare* come near <L *proximus* nearest, superl. of *prope* near] —**prox′i·mate·ly** *adv.*

prox·im·i·ty (prok·sim′ə·tē) *n.* The state or fact of being near or next; nearness. [<MF *proximité* <L *proximitas, -tatis* <*proximus*. See PROXIMATE.]

proximity fuze A complete miniature radio set placed in the nose of a projectile or bomb, capable of detonating the charge by simple proximity to the target: also called *VT fuze*.

prox·i·mo (prok′sə·mō) *adv.* In or of the next or coming month: opposed to *ultimo*. Abbr. *prox.* [<L *proximo (mense)* in the next (month), ablative of *proximus*. See PROXIMATE.]

prox·y (prok′sē) *n.* *pl.* **prox·ies** A person empowered by another to act for him, the office or right so to act, or the instrument conferring it. [Contraction of PROCURACY]

prude (prōōd) *n.* A person who makes an affected display of modesty and propriety, especially in matters relating to sex. [<F, prob. back formation <*prudefemme* an excellent woman <OF *prou, prode* honest, upright + *feme* a woman]

pru·dence (prōōd′ns) *n.* The quality of being prudent; sagacity; economy; discretion.

Synonyms: care, carefulness, caution, circumspection, consideration, discretion, forecast, foresight, forethought, frugality, judgment, judiciousness, providence, wisdom. *Care* may respect only the present; *prudence* and *providence* look far ahead and sacrifice the present to the future, *prudence* watching, saving, guarding, *providence* planning, doing, preparing, and perhaps expending largely to meet the future demand. *Frugality* is in many cases one form of *prudence*. *Foresight* merely sees the future, and may even lead to the recklessness and desperation to which *prudence* and *providence* are strongly opposed. *Forethought* is thinking of the future, a *consideration* of what might arise. See CARE, FRUGALITY, WISDOM. Antonyms: folly, heedlessness, improvidence, imprudence, indiscretion, rashness, recklessness, thoughtlessness.

pru·dent (prōō′dnt) *adj.* 1 Habitually careful to avoid errors and in following the most politic and profitable course; cautious; worldly-wise. 2 Exercising sound judgment; sagacious; judicious. 3 Characterized by practical wisdom or discretion; not extravagant. 4 Decorously discreet: a *prudent* maiden. [<OF <L *prudens, -entis* knowing, skilled, contraction of *providens*. See PROVIDENCE.] — **pru′·dent·ly** *adv.*
Synonyms: careful, cautious, circumspect, considerate, discreet, economical, frugal, judicious, politic, provident, sagacious, thoughtful, thrifty, wary, wise. See POLITIC. Compare synonyms for PRUDENCE. *Antonyms:* audacious, daring, desperate, foolhardy, foolish, imprudent, indiscreet, rash, reckless, spendthrift, thoughtless, unwary.

pru·den·tial (prōō·den′shəl) *adj.* 1 Proceeding from or marked by prudence. 2 Exercising prudence and wisdom officially. — **pru·den′tial·ly** *adv.*

prud·er·y (prōō′dər·ē) *n. pl.* **·er·ies** Primness; extreme priggishness; also, prudish action or language.

prud·ish (prōō′dish) *adj.* Showing prudery; prim. — **prud′ish·ly** *adv.* — **prud′ish·ness** *n.*

pru·i·nose (prōō′i·nōs) *adj. Biol.* Having the surface characterized by a secretion or outgrowth so as to appear frosted; powdery, as the bloom on a cabbage leaf, or the floury appearance of the body of some cicadas and beetles. [<L *pruinosus* frosty < *pruina* hoarfrost]

prune¹ (prōōn) *n.* 1 The dried fruit of any of several varieties of the common plum. 2 A plum. 3 *Slang* A stupid or uninteresting person. [<OF <LL *pruna* <L *prunum* <Gk. *proumnon, prounon* a plum. Doublet of PLUM.]

prune² (prōōn) *v.t.* & *v.i.* **pruned, prun·ing** 1 To trim or cut superfluous branches or parts (from) so as to improve growth, appearance, etc. 2 To cut off (superfluous branches or parts). See synonyms under ABBREVIATE. [<OF *prooignier, proignier,* ? < *provaignier* cut < *provain* a slip <L *propago*; prob. infl. in form by *rooignier* cut off, ult. <L *rotundus* round] — **prun′er** *n.*

prune³ (prōōn) *v.t.* **pruned, prun·ing** *Archaic* To dress up; preen. [<OF *poroindre* anoint < *por-* (<L *pro* before) + *oindre* anoint <L *ungere*]

pru·nel·la (prōō·nel′ə) *n.* 1 A strong woolen cloth used for the uppers of shoes. 2 A similar twilled heavy dress fabric. 3 *pl.* Shoes made partly of prunella. Also **pru·nel′lo** (-nel′ō). [<F *prunelle* a sloe, dim. of *prune* plum, prune; prob. so called from its dark color]

pru·nelle (prōō·nel′) *n.* 1 A small yellow prune, usually packed with the stone and skin removed. 2 A plum-flavored liqueur. [<F, dim. of *prune.* See PRUNE¹.]

pru·nif·er·ous (prōō·nif′ər·əs) *adj.* Plum-bearing. [< *pruni-* plum (<L *prunum*) + -FEROUS]

pru·ri·ent (prŏŏr′ē·ənt) *adj.* 1 Impure in thought and desire; lewd. 2 Having lustful cravings or desires. 3 Longing; desirous. [<L *pruriens, -entis,* ppr. of *prurire* itch, long for] — **pru′ri·ence, pru′ri·en·cy** *n.* — **pru′·ri·ent·ly** *adv.*

pru·ri·go (prŏŏ·rī′gō) *n. Pathol.* A chronic inflammatory skin disease marked by eruption and severe itching. [<L, an itching, lasciviousness < *prurire* itch] — **pru·rig′i·nous** (-rij′ə·nəs) *adj.*

pru·ri·tus (prŏŏ·rī′təs) *n. Pathol.* Itching. [<L < *prurire* itch] — **pru·rit′ic** (-rit′ik) *adj.*
pruritus hi·e·ma·lis (hī′ə·mā′lis) Frost itch.

Pru·sa (prōō′sä) An ancient name for BRUSA.
Prus·sia (prush′ə) A former state, the largest and most important, of northern Germany, 113,410 square miles; capital, Berlin; formally dissolved, Feb. 1947; territory divided between East and West Germany, Poland, and Russian S.F.S.R.: German *Preussen*.

Prus·sian (prush′ən) *adj.* 1 Of or pertaining to Prussia, its inhabitants, or their language. 2 Characteristic of the Junkers of Prussia; militaristic; overbearing. — *n.* 1 A native or naturalized inhabitant of Prussia. 2 The old language of Prussia, belonging to the Baltic branch of the Balto-Slavic subfamily of Indo-European languages: extinct since the 17th century, and often called *Borussian*: also **Old Prussian.**

Prussian blue 1 *Chem.* Any one of a group of cyanogen compounds formed from ferrous sulfate and potassium ferrocyanide: formerly much used in dyeing. 2 A deep, strong, blue pigment with a coppery sheen, obtained from these compounds: used in oil painting but impermanent on alkali surfaces, as fresco: also called **Paris blue** (formerly called **Berlin blue**). Heat changes it to **Prussian brown.** [So called because discovered accidentally in Berlin, 1704, by H. de Diesbach, a colormaker]

Prus·sian·ism (prush′ən·iz′əm) *n.* The practices or policies of the Prussian ruling class during its leadership of Germany, characterized by militarism and *esprit de corps.*

prus·si·ate (prush′ē·āt, -it, prus′-) *n. Chem.* 1 A salt of prussic acid; also, a cyanide. 2 A ferrocyanide or a ferricyanide.

prus·sic (prus′ik) *Chem. adj.* Hydrocyanic. — *n.* Prussic acid. [<F *prussique* < *Prusse* Prussia + *-ique* -IC; so called because derived from *Prussian blue*]

prussic acid Hydrocyanic acid.

Prut (prŏŏt) A river forming the boundary between SW U.S.S.R. and Rumania and flowing 530 miles from the Carpathians in SW Ukrainian S.S.R. to the Danube. Formerly **Pruth.**

pry¹ (prī) *v.i.* **pried, pry·ing** To look or peer carefully, curiously, or slyly; snoop. — *n. pl.* **pries** 1 A sly and searching inspection. 2 One who pries; an inquisitive, prying person. [ME *prien*; ult. origin unknown] — **pry′·ing** *adj.* & *n.* — **pry′ing·ly** *adv.*

pry² (prī) *v.t.* **pried, pry·ing** 1 To raise, move, or open by means of a lever; prize. 2 To obtain by effort. — *n.* A lever, as a bar, stick, or beam; also, leverage. [Back formation <PRIZE², *v.,* mistaken as a 3rd person sing.]

pry·er (prī′ər) See PRIER.

Prynne (prin), **William,** 1600–69, English Presbyterian lawyer, pamphleteer, and statesman.

Prze·myśl (pshe′mish·əl) A city in SE Poland near the Ukrainian S.S.R. border; scene of several battles in World War I, 1915.

psalm (säm) *n.* A sacred song or lyric, especially one of those contained in the Old Testament Book of Psalms; a hymn. See synonyms under SONG. — *v.t.* To celebrate or praise in psalms; hymn. [Fusion of OE *sealm, psalm* and OF *salme, psaume,* both <LL *psalmus* <Gk. *psalmos* a song sung to the harp, lit., a twanging < *psallein* twitch]

psalm·ist (sä′mist) *n.* 1 A maker or composer of psalms. 2 In the early Christian church, one of the minor clergy who led the singing; a precentor. — **the Psalmist** King David, as the traditional author of many of the Scriptural psalms.

psalm·o·dy (sä′mə·dē, sal′-) *n. pl.* **·dies** 1 The use of psalms in divine worship; psalm-singing. 2 A collection of psalms. [<LL *psalmodia* <Gk. *psalmōidia* singing to the harp < *psalmōidos* a psalmist < *psalmos* a psalm + *ōidē* a song] — **psalm′o·dist** *n.*

Psalms (sämz) A lyrical book of the Old Testament, containing 150 hymns, many ascribed to David. Also **Book of Psalms.**

psal·ter (sôl′tər) *n.* 1 The psalms appointed to be read or sung at any given service. 2 In the Roman Catholic Church, a rosary of 150 beads, equaling the number of the Psalms. [OE *psaltere, saltere* <L *psalterium* a psaltery] — **psal·te·ri·an** (sȯl·tir′ē·ən, sal′-) *adj.*

Psal·ter (sôl′tər) *n.* 1 The Book of Psalms; specifically, the version of Psalms in the Book of Common Prayer. 2 The Latin version of Psalms used in the Roman Catholic breviary. Also **Psal′ter·y.**

psal·te·ri·um (sȯl·tir′ē·əm, sal-) *n. pl.* **·te·ri·a** (-tir′ē·ə) The manyplies, or third stomach of a ruminant. [<L, a psaltery; so called because its many folds make it resemble the instrument] — **psal·te′ri·al** *adj.*

psal·ter·y (sôl′tər·ē) *n. pl.* **·ter·ies** An ancient stringed musical instrument, similar to a dulcimer but played by plucking with the fingers or a plectrum. [<OF *sautere, psalterie* <L *psalterium* <Gk. *psalterion* < *psallein* twitch, twang]

PSALTERY
Twelfth century.

psam·mite (sam′īt) *n.* Fine-grained sandstone. [<F <Gk. *psammos* sand + *-ite* -ITE¹]

psam·mit·ic (sa·mit′ik) *adj. Geol.* 1 Composed of material in the form of rounded grains of sand: contrasted with *gritty.* 2 Specifically, having the texture of fine sand: said of detrital deposits or fragmental rocks: contrasted with *psephitic.*

psel·lism (sel′iz·əm) *n.* Imperfect articulation; stammering. [<Gk. *psellismos* stammering < *psellizein* stammer]

pse·phite (sē′fīt) *n.* A conglomerate of small pebbles; fragmental rock. [<Gk. *psēphos* a pebble + -ITE¹]

pse·phit·ic (sē·fit′ik) *adj. Geol.* Having the texture of coarse sand: said of detrital deposits or fragmental rocks: contrasted with *psammitic.*

pseph·ol·o·gy (sef·ol′ə·jē) *n.* The study and statistical analysis of the elective process and its results. [<Gk. *psēphos* pebble used in voting, the vote itself + -LOGY; coined by R. B. McCallum of Oxford University in 1952] — **pseph·ol′o·gist** *n.*

pseu·dax·is (sōō·dak′sis) *n.* A sympodium. [<PSEUD(O)- + AXIS]

pseu·de·pig·ra·pha (sōō′də·pig′rə·fə) *n. pl.* Spurious writing; especially, spurious religious writings, falsely ascribed to Scriptural characters or times and not considered as canonical by any branch of the Christian church. [<Gk., neut. pl. of *pseudepigraphos* with a false title < *pseudēs* false + *epigraphein.* See EPIGRAPH.] — **pseu·de·pig·raph·ic** (sōō′dep·i·graf′ik) or **·i·cal, pseu′de·pig′ra·phous** *adj.*

pseu·do (sōō′dō) *adj.* Pretended; sham.

pseudo- combining form 1 False; pretended: *pseudonym.* 2 Counterfeit; not genuine: *pseudepigrapha.* 3 Closely resembling; serving or functioning as: *pseudopodium.* 4 Illusory; apparent: *pseudoaquatic.* 5 Abnormal; erratic: *pseudocarp.* Also, before vowels, **pseud-.** [<Gk. < *pseudēs* false]

pseu·do·a·quat·ic (sōō′dō·ə·kwat′ik, -kwot′-) *adj.* Not really aquatic, but native to or found in wet places.

pseu·do·bulb (sōō′dō·bulb′) *n. Bot.* A swollen, bulblike internode at the base of the stem in many orchids.

pseu·do·carp (sōō′dō·kärp) *n. Bot.* A false fruit; an often conspicuous portion of a fructification which consists of other parts besides the pericarp and seeds, as the apple, checkerberry, and mulberry. [<PSEUDO- + -CARP] — **pseu′do·car′pous** *adj.*

pseu·do·clas·sic (sōō′dō·klas′ik) *adj.* Emulating classic style; pretending to be classic; wrongly classed as classic.

pseu·do-Is·i·dore (sōō′dō·iz′ə·dôr, -dōr) The unknown author or compiler of the "False Decretals." See DECRETALS. — **pseu′do-Is·i·do′ri·an** *adj.*

pseu·do·morph (sōō′dō·môrf) *n.* 1 An irregular or false form. 2 *Mineral.* A mineral having the external crystalline form of another mineral. [<PSEUDO- + -MORPH] — **pseu′do·mor′phic** *adj.* — **pseu′do·mor′phism** *n.* — **pseu′·do·mor′phous** *adj.*

pseu·do·nym (sōō′də·nim) *n.* A fictitious name; pen name. [<F <Gk. *pseudonymon,* orig. neut. of *pseudonymos* having a false name < *pseudēs* false + *onoma, onyma* a name] — **pseu·don·y·mous** (sōō·don′ə·məs) *adj.* — **pseu·don′y·mous·ly** *adv.* — **pseu·don′y·mous·ness, pseu·do·nym′i·ty** *n.*

pseu·do·pod (sōō′də·pod) *n.* 1 A pseudopodium. 2 An organism with pseudopodia;

a rhizopod. —**pseu·dop·o·dal** (so͞o·dop′ə·dəl) *adj.*

pseu·do·po·di·um (so͞o′də·pō′dē·əm) *n. pl.* **·di·a** (-dē·ə) **1** *Zool.* A process formed by the temporary extension of the protoplasm of a cell or of a unicellular animal, serving for taking in food, for locomotion, etc. **2** *Bot.* A false pedicel in certain mosses. Also **pseu′do·pode** (-pōd). [<NL <Gk. *pseudēs* false + *podion.* See PODIUM.]

pshaw (shô) *interj.* & *n.* An exclamation of annoyance, disapproval, disgust, or impatience. —*v.t.* & *v.i.* To exclaim *pshaw* at (a person or thing).

psi[1] (sī, psī, psē) *n.* The twenty-third letter in the Greek alphabet (Ψ, ψ): equivalent to English *ps.*

psi[2] (sī) *n.* Pounds per square inch: a unit of pressure.

psi·lan·thro·py (sī·lan′thrə·pē) *n.* The doctrine of the mere humanity of Christ. Also **psi·lan′·thro·pism.** [<Gk. *psilanthrōpos* merely human <*psilos* mere, bare + *anthrōpos* man] —**psi·lan·throp·ic** (sī′lan·throp′ik) *adj.*

psi·lom·e·lane (sī·lom′ə·lān) *n.* A submetallic, iron-black to steel-gray, hydrous oxide of manganese, found massive. [<Gk. *psilos* bare + *melan,* neut. of *melas* black]

Psi·lo·ri·ti (psē′lô·rē′tē) See MOUNT IDA.

psi·lo·sis (sī·lō′sis) *n.* **1** Sprue. **2** Alopecia. [<NL <Gk. *psilōsis* a stripping bare <*psiloein* strip bare, make bald] —**psi·lot′ic** (-lot′ik) *adj.*

Psit·ta·ci·for·mes (sit′ə·si·fôr′mēz) *n. pl.* An order of climbing, arboreal birds, characterized by sharp, recurved beaks, typically brilliant plumage, and the ability to articulate: including the parrots, macaws, and cockatoos. [<NL <Gk. *psittakos* a parrot + L *forma* form]

psit·ta·cine (sit′ə·sīn, -sin) *adj.* Of or pertaining to parrots. [<L *psittacinus* <*psittacus* a parrot <Gk. *psittakos*]

psit·ta·co·sis (sit′ə·kō′sis) *n.* An acute, infectious, wasting disease of parrots and related birds, caused by a filtrable virus: transmitted to man, it causes fever and nausea, with complications resembling influenza and typhoid fever: also called *parrot fever.* [<NL <Gk. *psittakos* a parrot + *-osis* -OSIS]

Pskov (pskôf) A city of NW European Russian S.F.S.R. near the border of the Estonian S.S.R. on the southern end of **Lake Pskov,** the southern arm of Lake Peipus.

pso·as (sō′əs) *n. Anat.* Either of two muscles of the interior of the pelvis, arising from the spine and constituting the loins. [<NL <Gk., acc. pl. of *psoa* the muscle of the loins]

Pso·cop·ter·a (sō·kop′tər·ə) See CORRODENTIA.

pso·ra (sôr′ə, sō′rə) *n. Pathol.* **1** The itch, or some similar skin disease, as scabies. **2** Psoriasis. [<L <Gk. *psōra* an itch] —**pso′ric** *adj.* & *n.*

pso·ra·le·a (sə·rā′lē·ə) *n.* A scented herb or shrub (genus *Psoralea*) of the bean family, especially the common breadroot. [<NL <Gk. *psōraleos* scabby]

pso·ri·a·sis (sə·rī′ə·sis) *n. Pathol.* A noncontagious, inflammatory skin disease, chronic or acute, characterized by reddish patches and white scales. [<NL <Gk. *psōriaein* have an itch <*psōra* an itch] —**pso·ri·at·ic** (sôr′ē·at′ik, sō′rē-) *adj.*

psych– See PSYCHO–.

psy·chal·gi·a (sī·kal′jē·ə) *n. Psychiatry* Mental suffering; morbid depression; pain in the mind, as distinguished from *somatalgia.* [<NL <Gk. *psychē* mind + *algos* a pain, an affliction]

psy·chas·the·ni·a (sī′kas·thē′nē·ə) *n. Psychiatry* A morbid mental state characterized by mental fatigue, obsessive anxiety, phobias, tics, etc. [<NL <Gk. *psychē* mind + *astheneia* debility, weakness] —**psy′chas·then′ic** (-then′ik) *adj.* & *n.*

psy·che (sī′kē) *n.* **1** The human soul; the mind; the intelligence. **2** *Psychoanal.* The aggregate of all the psychic components constituting a human individual, sometimes considered as an entity functioning apart from or independently of the body. **3** A knot of hair coiled at the back of the head by women in imitation of an ancient Greek style of hairdressing: also **Psyche knot.** [<Gk. *psychē* <*psychein* breathe, blow]

Psy·che (sī′kē) In Greek and Roman mythology, a maiden beloved by Eros, who, after many tribulations caused by the jealousy of Venus, is united with her lover and accorded a place among the gods as a personification of the soul.

psy·chi·a·trist (sī·kī′ə·trist) *n.* A medical doctor specializing in the practice of psychiatry. Also **psy·chi′a·ter.**

psy·chi·a·try (sī·kī′ə·trē) *n.* The branch of medicine that treats disorders of the mind or psyche, especially psychoses, but also neuroses. [<PSYCH- + -IATRY] —**psy·chi·at·ric** (sī′kē·at′rik) or **·ri·cal** *adj.*

psy·chic (sī′kik) *adj.* **1** Pertaining to the mind or soul; mental, as distinguished from physical and physiological. **2** *Psychol.* Pertaining to or designating those mental phenomena which are, or appear to be, independent of normal sensory stimuli and which cannot be fully explained in terms of the known data of experimental science, as clairvoyance, telepathy, and extrasensory perception. Compare PARAPSYCHOLOGY. **3** Caused by, proceeding from, associated with, or attributed to a nonmaterial or occult agency. **4** Sensitive to mental or occult phenomena. Also **psy′chi·cal.** —*n.* **1** A person sensitive to mental or extrasensory phenomena; especially, a spiritualistic medium. **2** The field of extrasensory phenomena: with *the.* [<Gk. *psychikos* <*psychē* soul] —**psy′chi·cal·ly** *adv.*

psychic blindness Hysterical blindness.

psychic deafness Hysterical deafness.

psy·chics (sī′kiks) *n.* Psychic research.

psycho– *combining form* Mind; soul; spirit: *psychosomatic.* Also, before vowels, *psych–.* [<Gk. *psychē* spirit, soul]

psy·cho·a·nal·y·sis (sī′kō·ə·nal′ə·sis) *n.* **1** The doctrine that mental life and all forms of behavior may be interpreted in terms of reciprocally acting forces largely governed by the dynamic interplay of conflicting drives and processes originating in the unconscious. **2** A system of psychotherapy originated and developed by Freud which seeks to alleviate mental and nervous disorders by the technical analysis of controlling factors persistently repressed in, and manifested through, the unconscious. Also **psy′cha·nal′y·sis.** —**psy′cho·an′a·lyt′ic** (-an′ə·lit′ik) or **·i·cal** *adj.* —**psy′cho·an′a·lyt′i·cal·ly** *adv.*

psy·cho·an·a·lyst (sī′kō·an′ə·list) *n.* One who practices psychoanalysis.

psy·cho·an·a·lyze (sī′kō·an′ə·līz) *v.t.* **·lyzed, ·lyz·ing** To treat by psychoanalysis. Also *Brit.* **psy′cho·an′a·lyse.**

psy·cho·bi·ol·o·gy (sī′kō·bī·ol′ə·jē) *n.* **1** The study of the mind and of mental processes in relation to anatomy, physiology, and the nervous system, with special reference to the influence of the environment. **2** Psychology in its biological aspects. Also called *biopsychology.* —**psy′cho·bi′o·log′i·cal** (-bī′ə·loj′i·kəl) *adj.* —**psy′cho·bi·ol′o·gist** *n.*

psy·cho·dra·ma (sī′kō·drä′mə, -dram′ə) *n.* A form of psychotherapy in which the patient acts out, occasionally before an audience, situations involving his problems.

psy·cho·dy·nam·ics (sī′kō·dī·nam′iks) *n.* The study of mental processes in action. —**psy′·cho·dy·nam′ic** *adj.*

psy·cho·gal·va·nom·e·ter (sī′kō·gal′və·nom′ə·tər) *n.* A type of galvanometer designed to indicate and record electrical disturbances in the body caused by or associated with various forms of emotional stress: often used as a lie detector.

psy·cho·gen·e·sis (sī′kō·jen′ə·sis) *n.* **1** The development of the individual soul; the science of the origin of psychic life. **2** Genesis or specific change due to vitality of the organism, as opposed to external influences. Also **psy·chog·e·ny** (sī·koj′ə·nē). —**psy′cho·ge·net′ic** (-jə·net′ik) *adj.* —**psy′cho·ge·net′i·cal·ly** *adv.*

psy·cho·gen·ic (sī′kō·jen′ik) *adj.* Having mental origin, or being affected by mental actions and states.

psy·chog·no·sis (sī′kog·nō′sis) *n.* The close study and diagnosis of mental states. [<PSYCHO- + -GNOSIS] —**psy′chog·nos′tic** (-nos′·tik) *adj.*

psy·cho·graph (sī′kə·graf, -gräf) *n.* **1** A chart graphically representing the personality traits of an individual: also **psy′cho·gram** (-gram). **2** A description of the personality traits of an individual, especially in literary form. —**psy′·cho·graph′ic** *adj.*

psy·chog·ra·phy (sī·kog′rə·fē) *n.* **1** Involuntary or unconscious writing, as by a medium. **2** The making of a psychograph. [<PSYCHO- + -GRAPHY]

psy·cho·ki·ne·sis (sī′kō·ki·nē′sis) *n.* The alleged power of controlling the chance behavior of physical objects, as cards, dice, etc., by the direct influence upon them of emotional states, strong desire, or other psychic factors: related to extrasensory perception. Abbr. *PK.*

psy·cho·log·i·cal (sī′kə·loj′i·kəl) *adj.* **1** Of or pertaining to psychology. **2** Of or in the mind. **3** Suitable for affecting the mind: the *psychological* moment. Also **psy′cho·log′ic.** —**psy′cho·log′i·cal·ly** *adv.*

psy·chol·o·gism (sī·kol′ə·jiz′əm) *n.* Idealistic philosophy as opposed to sensationalism. Compare ONTOLOGISM.

psy·chol·o·gist (sī·kol′ə·jist) *n.* A student of or a specialist in psychology.

psy·chol·o·gize (sī·kol′ə·jīz) *v.i.* **·gized, ·giz·ing** **1** To study psychology. **2** To theorize psychologically.

psy·chol·o·gy (sī·kol′ə·jē) *n.* **1** The science of the human mind in any of its aspects, operations, powers, or functions. **2** The systematic investigation of mental phenomena, especially those associated with consciousness, behavior, and the problems of adjustment to the environment. **3** The aggregate of the emotions, traits, and behavior patterns regarded as characteristic of an individual or type: the *psychology* of a fanatic. [<NL *psychologia* <Gk. *psychē* soul + -LOGY]

psy·chom·e·try (sī·kom′ə·trē) *n.* **1** The science of the measurement of psychophysical processes, especially of their accuracy or duration in time; mental testing: also **psy·cho·met·rics** (sī·kō·met′riks). **2** Divination by physical contact or proximity of the properties of things touched or approached. —**psy·chom′e·trist** *n.*

psy·cho·mo·tor (sī′kō·mō′tər) *adj. Physiol.* Of or pertaining to muscular movements resulting from or caused by compulsive mental processes.

psy·cho·neu·ro·sis (sī′kō·no͝o·rō′sis, -nyo͝o-) *n. pl.* **·ses** (-sēz) *Psychiatry* A nervous disorder originating in disturbed psychic or mental functions, usually without or independent of organic symptoms: characterized by anxiety, phobias, compulsions, obsessions, etc. —**psy′·cho·neu·rot′ic** (-rot′ik) *adj.* & *n.*

psy·cho·nom·ics (sī′kə·nom′iks) *n.* The psychological study of the development of the individual in relation to his environment. Also **psy·chon·o·my** (sī·kon′ə·mē). [<PSYCHO- + Gk. *nomos* law] —**psy′cho·nom′ic** *adj.*

psy·cho·path (sī′kō·path) *n.* One subject to or afflicted by mental instability.

psy·cho·path·ic (sī′kō·path′ik) *n.* A psychopath. —*adj.* Of or characterized by psychopathy.

psy·cho·pa·thol·o·gy (sī′kō·pə·thol′ə·jē) *n.* The pathology of the mind. —**psy′cho·path′o·log′i·cal** (-path′ə·loj′i·kəl) *adj.* —**psy′cho·pa·thol′o·gist** *n.*

psy·chop·a·thy (sī·kop′ə·thē) *n.* **1** Mental disorder, especially as apart from disease of the brain, and typified by emotional immaturity and instability, moral deficiency, and perversions. **2** Psychotherapy. [<PSYCHO- + -PATHY]

psy·cho·phar·ma·col·o·gy (sī′kō·fär′mə·kol′ə·jē) *n.* The branch of pharmacology which investigates the structure, properties, and uses of drugs acting primarily on the nervous system and serving to modify various types of human behavior, as chlorpromazine, reserpine, mescaline, etc.

psy·cho·phys·ics (sī′kō·fiz′iks) *n.* The science of the relations between mental and physical phenomena. —**psy′cho·phys′i·cal** *adj.* —**psy′·cho·phys′i·cist** *n.*

psy·cho·phys·i·ol·o·gy (sī′kō·fiz′ē·ol′ə·jē) *n.* The physiology of mental processes.

psy·cho·sis (sī·kō′sis) *n. pl.* **·ses** (-sēz) *Psychiatry* A mental disorder, severe in character, often involving disorganization of the total personality, with or without organic disease. ◆ Homophone: *sycosis.* [<NL <Gk. *psychōsis* a giving of life <*psychoein* animate <*psychē* a soul]

psy·cho·so·mat·ic (sī′kō·sō·mat′ik) *adj.* **1** Of or pertaining to the interrelationships of mind and body, with especial reference to disease. **2** Designating a branch of medicine which investigates the reciprocal influences of body

psy·cho·tech·ni·cian (sī'kō·tek·nish'ən) *n.* One skilled in psychotechnics.
psy·cho·tech·nics (sī'kō·tek'niks) *n.* The direct application of psychological principles and methods to practical ends, especially in the management of large industrial, commercial, and business enterprises. — **psy'cho·tech'ni·cal** *adj.*
psy·cho·ther·a·py (sī'kō·ther'ə·pē) *n.* The treatment of nervous and mental disorders, especially by psychological methods, as hypnosis, re-education, psychoanalysis, etc. Also **psy'cho·ther'a·peu'tics** (-ther'ə·pyoo'tiks). — **psy'cho·ther'a·peu'tic** *adj.* — **psy'cho·ther'a·pist** *n.*
psy·chot·ic (sī·kot'ik) *n.* One suffering from a psychosis. — *adj.* Of or characterized by a psychosis.
psychro- *combining form* Cold: *psychrophobia*. [<Gk. *psychros* cold]
psy·chrom·e·ter (sī·krom'ə·tər) *n.* An instrument for measuring the vapor tension and relative humidity of the air, consisting of two thermometers, the bulb of one being kept moist. [<PSYCHRO- + -METER]
psy·chro·pho·bi·a (sī'krō·fō'bē·ə) *n.* A morbid fear of, or sensibility to, cold. [<PSYCHRO- + -PHOBIA]
psy·chro·ther·a·py (sī'krō·ther'ə·pē) *n.* Medical treatment by the use of cold.
psyl·li·um (sil'ē·əm) *n.* 1 A plantain of Asia Minor (*Plantago psyllium*). 2 Its small, reddish-brown seeds, resembling flaxseed in medicinal properties, used as a mild laxative. [<L <Gk. *psyllion* <*psylla* a flea; so called because supposed to destroy fleas]
Ptah (ptä, ptäkh) In ancient Egyptian religion, the chief divinity of ancient Memphis, the creator of gods and men.
ptar·mi·gan (tär'mə·gən) *n. pl.* **·gans** or **·gan** A grouse (genus *Lagopus*) of the northern hemisphere, with the winter plumage chiefly pure white, and with feathered toes; especially, the **willow ptarmigan** (*L. lagopus*) of the Arctic, and the **white-tailed ptarmigan** (*L. leucurus*) of the Rocky Mountains. [< Scottish Gaelic *tarmachan*; excrescent *p* prob. due to false analogy with Gk. *pteron* wing.]
PT boat A patrol torpedo boat.
pter·i·dol·o·gy (ter'i·dol'ə·jē) *n.* The department of botany that treats of ferns. [<Gk. *pteris, pteridos* a fern + -LOGY] — **pter'i·do·log'i·cal** (-də·loj'i·kəl) *adj.* — **pter'i·dol'o·gist** *n.*
pter·i·do·phyte (ter'i·dō·fīt') *n.* Any of a phylum (*Pteridophyta*) of flowerless plants comprising the ferns, clubmosses, and their allies. [<NL <Gk. *pteris, pteridos* a fern + *phyton* a plant] — **pter'i·do·phyt'ic** (-fit'ik), **pter'i·doph'y·tous** (-dof'ə·təs) *adj.*
ptero- *combining form* Wing; feather; plume; resembling wings: *pterodactyl*. Also, before vowels, **pter-**. [<Gk. *pteron* wing]
pter·o·dac·tyl (ter'ə·dak'til) *n. Paleontol.* 1 Any of a genus (*Pterodactylus*) of extinct flying reptiles which flourished in the Jurassic period, characterized by a large, birdlike skull, long jaws, and flying membrane somewhat like that of a bat. 2 A pterosaurian. [<NL <Gk. *pteron* a wing + *daktylos* a finger]

PTERODACTYL
(American Cretaceous: wing span about 6 feet)

pter·o·pod (ter'ə·pod) *n.* One of a subclass or order (*Pteropoda*) of gastropods having the middle region of the foot expanded into winglike lobes or fins; a sea butterfly. — *adj.* 1 Having the foot expanded into swimming lobes. 2 Of or pertaining to the *Pteropoda*: also **pte·rop·o·dan** (tə·rop'ə·dən). [<NL <Gk. *pteron* a wing + *pous, podos* a foot]
pter·o·sau·ri·an (ter'ə·sôr'ē·ən) *n. Paleontol.* One of an extinct order (*Pterosauria*) of flying reptiles, including pterodactyls, of the Mesozoic, with external digits long and developed to support a flying membrane. Also **pter'o·saur**. — *adj.* Of or pertaining to the *Pterosauria*. [<NL <Gk. *pteron* a wing + *sauros* a lizard]
-pterous *combining form* Having (a specified number or kind of) wings: *dipterous*. [<Gk. *pteron* wing]
pter·y·goid (ter'ə·goid) *adj.* 1 Having the form of a wing; winglike. 2 *Anat.* Pertaining to, or situated near the winglike processes of the sphenoid. Also **pter'y·goi'dal, pter'y·goi'de·an**. — *n. Anat.* A pterygoid bone, plate, process, or muscle. [<Gk. *pteryx, pterygos* a wing + -OID]
ptis·an (tiz'ən) *n.* 1 A slightly medicinal decoction or tea of herbs: also spelled *tisane*. 2 The juice of grapes drained off without pressure. 3 A decoction of barley water. [<OF *ptisane, tisane* <L *ptisana* barley groats, a drink made from them <Gk. *ptisanē* peeled barley < *ptissein* peel]
Ptol·e·ma·ic (tol'ə·mā'ik) *adj.* Of or pertaining to Ptolemy, the astronomer, or to the Ptolemies, the Egyptian kings. [<Gk. *Ptolemaïkos*]
Ptolemaic system The ancient astronomical system of Ptolemy, which assumed that the earth was the central body around which the sun, planets, and celestial bodies revolved: this system was accepted till replaced in the 16th century by the Copernican system.
Ptol·e·ma·is (tol'ə·mā'is) The New Testament name for ACRE.
Ptol·e·ma·ist (tol'ə·mā'ist) *n.* A believer in or adherent of the Ptolemaic system.
Ptol·e·my (tol'ə·mē) Second century A.D. astronomer, mathematician, and geographer of Alexandria: full name *Claudius Ptolomaeus*.
Ptol·e·my (tol'ə·mē) Name of 14 kings of Egypt, of whom the most noted are:
— **Ptolemy I**, 367?–283? B.C., king 323–285; a general of Alexander the Great; founded the dynasty: called "Soter."
— **Ptolemy II**, 309–246 B.C., king 285–46; patron of literature and the arts: called "Philadelphus."
— **Ptolemy III**, 282?–221 B.C., king 246–21; conquered much of the Seleucid dominions; built many temples: called "Euergetes."
pto·maine (tō'mān, tō·mān') *n. Biochem.* Any of a class of basic organic chemical compounds derived from decomposing or putrefying animal or vegetable protein. They bear some resemblance to the alkaloids, and some of them are poisonous. Also **pto'main**. [<Ital. *ptomaina* <Gk. *ptōma* a corpse]
ptomaine poisoning Botulism.
pto·sis (tō'sis) *n. Pathol.* The permanent drooping of the upper eyelid, due to paralysis of the lifting muscle of the lid. [<NL <Gk. *ptōsis* a falling < *piptein* fall] — **pto·tic** (tō'tik) *adj.*
pty·a·lin (tī'ə·lin) *n. Biochem.* An amylase contained in the saliva of man and other mammals; the enzyme of saliva which converts starch into dextrin and maltose. [<Gk. *ptyalon* saliva + -IN]
pty·a·lism (tī'ə·liz'əm) *n.* Abnormal flow of saliva; salivation. [<Gk. *ptyalon* saliva + -ISM]
pu' (pōō) *v.t. Scot.* To pull.
pub (pub) *n. Brit. Slang* A public house; an inn; tavern. [Short for *public house*]
pu·ber·ty (pyoo'bər·tē) *n.* The period in life at which a person of either sex becomes functionally capable of reproduction: in civil law, usually the age of 14 years in males and 12 in females. [<OF *puberte* <L *pubertas* < *pubes, puberis* an adult]
pu·bes (pyoo'bēz) *n.* 1 *Anat.* The part of the lower central hypogastric region that is covered with hair in the adult; the pubic region. 2 The hair that appears on the body at puberty; specifically, the hair on the pubic region. 3 *Biol.* Pubescence. [<L, pubic hair, groin]
pu·bes·cence (pyoo·bes'əns) *n.* 1 The state or quality of being pubescent (def. 1). 2 *Biol.* A covering or growth of soft, fine hairs or down, especially that upon certain plants.
pu·bes·cent (pyoo·bes'ənt) *adj.* 1 Arriving or having arrived at puberty. 2 *Biol.* Covered with hairs, especially fine, soft, short hairs; hairy or downy, as leaves, etc. [<MF <L *pubescens, -entis,* ppr. of *pubescere* become

downy, attain puberty < *pubes*. See PUBES.]
pu·bic (pyoo'bik) *adj.* Of or pertaining to the region in the lower part of the abdomen: the *pubic* bones.
pu·bis (pyoo'bis) *n. pl.* **·bes** (-bēz) *Anat.* Either of the two bones which join with a third to form an arch on either ventral side of the pelvis. [<NL, short for L *os pubis* pubic bone < *pubes*. See PUBES.]
pub·lic (pub'lik) *adj.* 1 Of, pertaining to, or affecting the people at large or the community: distinguished from *private* or *personal.* 2 Open to all; maintained by or for the public: *public* parks; participated in by the people: a *public* demonstration. 3 For the use of the public; specifically, for hire: a *public* cab, hall, etc. 4 Done or made in public or without concealment; well-known; open; notorious: a *public* scandal. 5 Occupying an official or professional position; acting before or for the community: a *public* speaker. See synonyms under COMMON, GENERAL. — *n.* The people collectively, or in general, of a particular locality or nation; also, all those persons who may be grouped together for any given purpose: the church-going *public*. [<OF <L *publicus,* alter. of *poplicus* (through infl. of *pubes* an adult) < *poplus, populus* people]
pub·lic-ad·dress system (pub'lik-ə·dres') A complete assembly of sound-reproducing apparatus for the proper amplification and recording of programs for performance in public places.
pub·li·can (pub'lə·kən) *n.* 1 In England, the keeper of a public house. 2 In ancient Rome, one who farmed or collected the public revenues. [<OF *publicain* <L *publicanus* a tax farmer, tax gatherer < *publicum* public revenue, orig. neut. of *publicus* PUBLIC]
pub·li·ca·tion (pub'lə·kā'shən) *n.* 1 The act of publishing or offering to public notice; notification to people at large orally or by writing or print; promulgation; proclamation. 2 In the law of libel and slander, the communication of a defamation to a third person. 3 That which is published; any printed work placed on sale or otherwise distributed or offered for distribution. See PUBLISH. [<OF *publicacion* <L *publicatio, -onis* < *publicatus,* pp. of *publicare* PUBLISH]
public debt The national debt.
public domain Lands owned by a state or national government; public lands. — **in the public domain** Available for unrestricted use: said of material on which copyright or patent right has expired.
public enemy 1 Any government with which a nation is at open war. 2 A person, especially a criminal, regarded as a menace to the public.
Public Health Service *U.S.* A Federal agency under the Surgeon General, which, as a constituent organization of the Department of Health, Education, and Welfare, is responsible for protecting and improving the health of the nation.
public house 1 An inn, tavern, or hotel. 2 In England, a place licensed to sell intoxicating liquors; a saloon.
pub·li·cist (pub'lə·sist) *n.* 1 A writer on international law or on topics of public interest. 2 A public-relations man or publicity agent. [<F *publiciste* <L *(jus) publicum* public (law), neut. of *publicus* PUBLIC]
pub·lic·i·ty (pub·lis'ə·tē) *n.* 1 The state of being public, or the act or fact of making or becoming public; exposure; notoriety: opposed to *secrecy.* 2 Advertising; advance information, or personal news intended to promote the interests of individuals, institutions, causes, etc., especially that appearing in print. 3 The attention or interest of the public gained by any method.
pub·li·cize (pub'lə·sīz) *v.t.* **·cized, ·ciz·ing** To give publicity to; advertise.
public library 1 A library maintained for the use of the public. 2 The building in which it is contained.
pub·lic·ly (pub'lik·lē) *adv.* 1 In an open or public manner; openly. 2 In the name or with the consent and concurrence of the public.
pub·lic·ness (pub'lik·nis) *n.* 1 The state or quality of being public or of belonging to the public. 2 Publicity.

public opinion The prevailing ideas, beliefs, and aims of the people, collectively: in politics, considered as a massed power or entity.
public relations 1 The activities and techniques utilized by public and private organizations and enterprises to establish favorable attitudes and responses in their behalf on the part of the general public or of special groups: included are analysis of attitudes, appraisal of procedures and policies, recommendations for internal change, and effective presentation of the organization's purposes and objectives. 2 The public conduct of the affairs of an organization with regard to its reputation and standing and to public opinion. 3 The state of the relationship between the general public and an institution of any kind.
public school See under SCHOOL.
public servant A government official.
public service 1 Official employment under the government, especially in the civil departments. 2 The radio or television broadcasting of announcements of civic interest, such as appeals for donations to charitable organizations.
pub·lic–ser·vice corporation (pub′lik-sûr′vis) Any corporation operating a public utility, as a railroad, gas, electric, or water company, etc.
public spirit Active, enlightened interest in and concern for matters that affect the welfare of the community. — **pub·lic–spir·it·ed** (pub′-lik·spir′it·id) adj.
public utility A business organization or industry which performs some public service, as the supplying of water or electric power, and is subject to particular governmental regulations; a public-service corporation.
public works Permanent architectural or engineering works or improvements built with public money, as post offices, museums, canals, harbors, parks, playgrounds, roads, bridges, etc.
pub·lish (pub′lish) v.t. 1 To make known or announce publicly; promulgate; proclaim. 2 To print and issue (a book, magazine, map, etc.) to the public. 3 Law To communicate (a defamation) to a third person. [<OF publier, puplier <L publicare make public <publicus PUBLIC] — **pub′lish·a·ble** adj.
Synonyms: advertise, announce, blazon, bruit, communicate, declare, disclose, divulge, impart, proclaim, promulgate, reveal, spread, tell. See ANNOUNCE, SPREAD. Antonyms: conceal, cover, hide, hush, suppress, withhold.
pub·lish·er (pub′lish·ər) n. One who publishes; especially, one who makes a business of publishing books or periodicals.
Puc·ci·ni (p○○t·chē′nē), **Giacomo**, 1858–1924, Italian operatic composer.
puc·coon (pə·kō○n′) n. 1 Any of several North American herbs (genus Lithospermum) of the borage family, yielding a red or yellow dye; especially, the **hoary puccoon** (L. canescens), with orange-yellow flowers, of which the root yields a red dye. 2 The pigment or dye made from these plants. 3 The bloodroot. [<Algonquian (Virginian) puccoon, pakon <pak blood]
puce (pyō○s) adj. Of a dark-brown or purplish-brown. [<F, flea color, a flea <L pulex, -icis a flea]
pu·celle (pyō○·sel′, Fr. pü·sel′) n. A virgin; maid: obsolete except in the phrase **La Pucelle**, Joan of Arc, the Maid of Orleans. [<OF pucele, pulcella <LL pulcella a young girl; ult. origin uncertain]
puck[1] (puk) n. 1 An evil sprite or hobgoblin. 2 In English folklore, **Puck**, a mischievous elf or goblin: also called Robin Goodfellow; specifically, in Shakespeare's A Midsummer Night's Dream, a mischievous fairy servant of Oberon. [OE púca a goblin] — **puck′ish** adj.
puck[2] (puk) n. The hard rubber disk used in playing hockey. [<dial. E, strike. Akin to POKE[1].]
puck·a (puk′ə) adj. Anglo-Indian Made of good materials; substantial; hence, genuine; superior: also spelled pukka. [<Hind. pakkā substantial, lit., cooked, ripe]
puck·er (puk′ər) v.t. & v.i. To gather or draw up into small folds or wrinkles. — n. 1 A wrinkle, or group of wrinkles. 2 Colloq. Agitation; perplexity; confusion. [Appar. freq. of POKE[1].] — **puck′er·y** adj.
pud·ding (pŏŏd′ing) n. 1 A sweetened and flavored dessert of soft food, usually farinaceous. 2 A skin or gut filled with seasoned minced meat, blood, or the like, and usually boiled or broiled. [ME poding, orig. sausage, black pudding, prob. <OF bodin, boudin]
pud·dle (pud′l) n. 1 A small pool of dirty water. 2 Pudding (def. 2). — v.t. ·dled, ·dling 1 Metall. To convert (molten pig iron) into wrought iron by melting and stirring in the presence of oxidizing substances. 2 To mix (clay, etc.) with water so as to obtain a watertight paste. 3 To line, as canal banks, with such a mixture. 4 To make muddy; stir up. [ME podel, appar. dim. of OE pudd a ditch] — **pud′dly** adj.
pud·dle·ball (pud′l·bôl′) n. A ball of heated iron fresh from the puddling furnace.
pud·dle·bar (pud′l·bär′) n. A bar into which a puddleball is rolled or hammered.
pud·dler (pud′lər) n. 1 One who puddles. 2 A device for stirring fused metal. 3 A puddling furnace.
pud·dling (pud′ling) n. 1 Metall. The operation or business of making wrought iron from pig iron in a puddling furnace. 2 Puddled clay for lining the banks of canals, etc.; puddle. 3 The operation of lining a canal with such clay.
puddling furnace A reverberatory furnace for puddling pig iron.
pu·den·cy (pyō○′dən·sē) n. Shame; modesty; also, prudishness. [<LL pudentia <L pudens, -entis, ppr. of pudere be ashamed]
pu·den·dum (pyō○·den′dəm) n. pl. ·da (-də) 1 The vulva. 2 pl. The external genitals of either sex. [<L, neut. of pudendus (something) to be ashamed of, gerundive of pudere be ashamed] — **pu′dic, pu·den′dal** adj.
pudg·y (puj′ē) adj. **pudg·i·er, pudg·i·est** Short and thick; fat. [?<dial. E (Scottish) <pud belly] — **pudg′i·ly** adv. — **pudg′i·ness** n.
Pue·bla (pwä′blä) A state in SE Mexico; 13,124 square miles; capital, Puebla.
pueb·lo (pweb′lō for def. 1, pwä′blō for defs. 2 and 3) n. pl. ·los 1 A communal adobe or stone building or group of buildings of the Indians of the SW United States. 2 A town or village of Indians or Spanish Americans, as in Mexico. 3 In the Philippines, a municipality: originally the civilian quarter of a Spanish community. [<Sp., a town, people <L populus]

HOPI INDIAN PUEBLO

Pueb·lo (pweb′lō) n. A member of one of the Indian tribes of Mexico and the SW United States, representing several linguistic stocks, as Zuñi, Uto-Aztecan, etc., but having in common the pueblo culture.
pu·er·ile (pyō○′ər·il, Brit. pyō○′ə·rīl) adj. Pertaining to or characteristic of childhood; juvenile; hence, immature; weak; silly: a puerile suggestion. See synonyms under CHILDISH, YOUTHFUL. [<MF puéril <L puerilis <puer, pueri a boy] — **pu′er·ile·ly** adv. — **pu′er·ile·ness** n.
pu·er·il·ism (pyō○′ər·il·iz′əm) n. Childishness, especially as indicative of mental disorder.
pu·er·il·i·ty (pyō○′ə·ril′ə·tē) n. pl. ·ties 1 Puerile state; childishness. 2 A childish act or expression.
pu·er·per·al (pyō○·ûr′pər·əl) adj. Med. Pertaining to, resulting from, or following childbirth: puerperal fever. [<L puerperus parturient <puer a boy + parere bring forth]
Puer·to A·ya·cu·cho (pwer′tō ä′yä·kō○′chō) Capital of Amazonas territory, on the Orinoco in southern Venezuela.
Puerto Bar·ri·os (bär′ryōs) A port of eastern Guatemala.
Puerto Ca·bel·lo (kä·bā′yō) A port on the Caribbean in northern Venezuela.
Puerto Li·món (lē·mōn′) See LIMÓN.
Puerto Me·xi·co (mā′hē·kō) A former name of COATZACOALCOS.

Puerto Montt (mônt) A port of south central Chile.
Puerto Ri·co (rē′kō) The easternmost island of the Greater Antilles, ceded to the United States by Spain in 1898; since 1952 a commonwealth; 3,423 square miles; capital, San Juan: former official name, Porto Rico. — **Puer′to–Ri′can** adj. & n.
puff (puf) n. 1 A breath emitted suddenly and with force; a sudden emission, as of air, smoke, or steam; a whiff. 2 A light, air-filled piece of pastry: a cream puff. 3 A light ball, tuft, wad, or pad for dusting powder on the hair or skin; a powder puff. 4 A loose roll of hair in a coiffure, or a light cushion over which it is rolled. 5 A quilted bed coverlet, usually filled with cotton, wool, or down; a comforter. 6 In dressmaking, a part of a fabric so gathered as to produce a loose, fluffy distention. 7 A public expression of fulsome praise, as in a newspaper or advertisement. 8 A puffball. — v.i. 1 To blow in puffs, as the wind. 2 To breathe hard, as after violent exertion. 3 To emit smoke, steam, etc., in puffs. 4 To smoke a cigar, etc., with puffs. 5 To move, act, or exert oneself while emitting puffs: with away, up, etc. 6 To swell as with air or pride; dilate: often with up or out. — v.t. 7 To send forth or emit with short puffs or breaths. 8 To move, impel, or stir up with or in puffs. 9 To smoke, as a pipe or cigar, with puffs. 10 To swell or distend: He puffed his cheeks with pride. 11 To praise fulsomely; advertise in a puff (def. 7). 12 To arrange (the hair) in a puff. [ME puf <puffen, OE pyffan]
Synonyms (verb): blow, compliment, flatter, inflate, pant, praise, swell. Compare SWELL. Antonyms: belittle, contract, disparage, shrink, shrivel.
puff adder 1 A large, sluggish, venomous African viper (Bitis arietans), with variously colored chevron-and-crescent markings and a habit of violently puffing out its breath. 2 The American hognose snake.
puff·ball (puf′bôl′) n. A globular fungus (genus Lycoperdon) that puffs out its dustlike spores when broken open. Some species are edible.
puff·er (puf′ər) n. 1 One who puffs. 2 A plectognath fish that inflates its body with air; a globefish. 3 The little harbor porpoise (Phocaena phocaena) of the North Atlantic and Pacific oceans.
puff·er·y (puf′ər·ē) n. pl. ·er·ies 1 The act or practice of puffing. 2 Fulsome public praise or commendation.
puf·fin (puf′in) n. 1 A sea bird allied to the auk and murre (family Alcidae), with deep compressed bill and thick naked skin at the corner of the mouth; especially, the common puffin (Fratercula arctica) of the North Atlantic; the Labrador auk. 2 The Pacific coast sea parrot (Lunda cirrhata). [Prob. < PUFF; with ref. to its puffed-out beak or the plumpness of its young]

PUFFIN
(Body from 12 to 15 inches long)

puff paste A short flaky paste for fine pastry.
puff·y (puf′ē) adj. **puff·i·er, puff·i·est** 1 Swollen with air or any soft matter; soft; bloated. 2 Inflated in manner; bombastic. 3 Blowing in puffs. — **puff′i·ly** adv. — **puff′i·ness** n.
pug[1] (pug) n. 1 Clay ground and worked with water, for molding pottery or bricks. 2 A machine in which clay is ground and mixed or tempered: also **pug mill**. — v.t. **pugged, pug·ging** 1 To knead or work (clay) with water, as in brickmaking. 2 To fill in with clay, etc. 3 To fill in or cover with mortar, felt, etc., to deaden sound. [<dial. E, ?<pug punch]
pug[2] (pug) n. 1 A breed of dog characterized by a short, square body, upturned nose, curled tail, and short, smooth coat. 2 A pug nose. [Prob. alter. of PUCK]
pug[3] (pug) n. Anglo-Indian An animal's footprint; trail. — v.t. **pugged, pug·ging** To track,

pug as game, by pugs; trail. [<Hind. *pag* a foot]
pug⁴ (pug) *n. Slang* A professional pugilist. [Short for PUGILIST.]
Pu·get Sound (pyōō′jit) An inlet of the Pacific in NW Washington, extending 100 miles south from Juan del Fuca Strait to Olympia.
pugh (pōō, pōō) *interj.* An exclamation of contempt or disgust.
pu·gi·lism (pyōō′jə·liz′əm) *n.* The art or practice of boxing or fighting with the fists, as in the prize ring. [<L *pugil* a boxer]
pu·gi·list (pyōō′jə·list) *n.* One who fights with his fists; a boxer; specifically, a prize fighter. — **pu′gi·lis′tic** *adj.*
pug·na·cious (pug·nā′shəs) *adj.* Disposed or inclined to fight; quarrelsome. [<L *pugnax, -acis* < *pugnare* fight < *pugnus* a fist] — **pug·na′cious·ly** *adv.*
pug·nac·i·ty (pug·nas′ə·tē) *n.* The quality of being pugnacious; quarrelsome disposition; combativeness. Also **pug·na′cious·ness** (-nā′shəs·nis).
pug nose A thick, short nose, tilted upward at the end. [<PUG² + NOSE] — **pug-nosed** (pug′nōzd′) *adj.*
pug·ree (pug′rē) *n. Anglo-Indian* A light scarf wound round a hat to keep off the sun; also, a turban worn by natives of India. Also **pug′gree, pug′gry.** [<Hind. *pagri* a turban]
puir (pür) *adj. Scot.* Poor.
puis·ne (pyōō′nē) *adj. Law* Junior as to rank; younger; inferior: a *puisne* judge. — *n.* One who is of inferior rank or younger; a junior associate. — Homophone: *puny*. [<OF *puisne* < *puis* afterwards (<L *postea* < *post* after) + *ne* born <L *natus*]
pu·is·sance (pyōō′ə·səns, pyōō·is′əns, pwis′- əns) *n.* The power to accomplish or achieve, especially against resistance; potency. [<OF]
pu·is·sant (pyōō′ə·sənt, pyōō·is′ənt, pwis′ənt) *adj.* Powerful; mighty. See synonyms under POWERFUL. [<OF <L *posse* be able] — **pu′is·sant·ly** *adv.*
puke (pyōōk) *v.t. & v.i.* **puked, puk·ing** To vomit or cause to vomit. — *n.* Vomit, or the act of vomiting. [Cf. LG *spucken* spew, spit <L *spuere*]
puk·ka (puk′ə) See PUCKA.
Pu·la (pōō′lä) A port of NW Croatia, Yugoslavia; formerly in Italy: Italian *Pola*.
Pu·las·ki (pōō·las′kē, pə-; *Polish* pōō·läs′kē) **Count Casimir,** 1748?-79, Polish soldier and American Revolutionary general; killed at Savannah.
pu·lay (pə·lī′) See PALAY.
pul·chri·tude (pul′krə·tōōd, -tyōōd) *n.* Beauty; grace; physical charm. [<L *pulchritudo, -inis* < *pulcher* beautiful]
pul·chri·tu·di·nous (pul′krə·tōō′də·nəs, -tyōō′-) *adj.* Beautiful; lovely; especially, having physical beauty.
pule (pyōōl) *v.i.* **puled, pul·ing** To cry plaintively, as a child; whimper; whine. [Cf. F *piauler* <MF *pioler* chirp] — **pul′er** *n.*
pu·lex (pyōō′leks) *n.* One of a genus (*Pulex*) of fleas, including the human flea (*P. irritans*). 2 Any flea. [<L, a flea]
pu·li (pōō′lē) *n. pl.* **pu·lik** (pōō′lik) A breed of working dog, of medium height, white, gray, or black in color, with a long, wavy coat: used in Hungary for sheepherding. [Hungarian]
pu·li·cene (pyōō′lə·sēn) *adj.* Of, pertaining to, or abounding with fleas. [<L *pulex, -icis* a flea]
pul·ing (pyōō′ling) *n.* A plaintive cry; whining. — *adj.* Whimpering; whining. — **pul′ing·ly** *adv.*
Pul·itz·er (pyōō′lit·sər, pōōl′it-), **Joseph,** 1847-1911, U.S. journalist and publisher, born in Hungary.
Pulitzer Prize One of several annual awards for outstanding work in American journalism and literature: established by Joseph Pulitzer.
pul·kha (pul′kə) *n.* A canoe-shaped traveling sledge, drawn by one reindeer: used in Lapland. [<Lapp *pulkke*]
pull (pōōl) *v.t.* 1 To apply force to so as to cause motion toward or after the person or thing exerting force; drag; tug. 2 To draw or remove from a natural or fixed place: to *pull* a tooth or plug. 3 To give a pull or tug to. 4 To pluck, as a fowl. 5 To draw asunder; tear; rend: with *to pieces, apart,* etc. 6 To strain so as to cause injury: to *pull* a ligament. 7 In sports, to strike (the ball) so as to cause it to curve obliquely from the direction in which the striker faces. 8 *Slang* To put into effect; carry out: often with *off:* to *pull* off a prank. 9 *Slang* To make a raid on; arrest. 10 *Slang* To draw out so as to use: to *pull* a knife. 11 *Printing* To make or obtain by impression from type: to *pull* a proof. 12 In boxing, to deliver (a punch, etc.) with less than one's full strength. 13 In horse-racing, to rein in or otherwise restrain (a horse) so as to prevent its winning. 14 In rowing: **a** To operate (an oar) by drawing toward one. **b** To propel or transport by rowing. **c** To be propelled by: The gig *pulls* four oars. — *v.i.* 15 To use force in hauling, dragging, moving, etc. 16 To move: with *out, in, away, ahead,* etc. 17 To drink deeply: to *pull* at a bottle. 18 To inhale deeply: to *pull* at a cigar. 19 To row. See synonyms under DRAW. — **to pull for** 1 To strive in behalf of. 2 *Colloq.* To declare one's allegiance to. — **to pull oneself together** To regain one's composure. — **to pull out** *Aeron.* To return to level flight after a dive, as an airplane. — **to pull through** 1 Succeed. 2 To survive. — **to pull up** To come to a halt. — **to pull up with** To advance to a position even with. — *n.* 1 The act of pulling; the exertion of force to draw something toward one. 2 Something that is pulled; specifically, the handle of a doorbell, drawer, cabinet, or the like. 3 An impression made by pulling the lever of a hand press. 4 A long swallow, or a deep puff, as on a pipe or cigar. 5 Exercise in rowing: a *pull* on the river. 6 The exertion expended in climbing a mountain; hence, any steady, continuous effort. 7 *Slang* A means of influencing those in power: political *pull*; influence to one's advantage. 8 Attraction: These ads have *pull*. 9 The action of restraining a horse by pulling on the reins; specifically, in horse-racing, the dishonest checking of a horse so that he may be defeated. 10 In sports, a stroke causing the ball to curve to the left. [OE *pullian* pluck] — **pull′er** *n.*
pull·back (pōōl′bak′) *n.* 1 That which keeps or holds back; a restraint or drawback. 2 A device for drawing or holding something back, as part of a dress, or a window.
pull-doo (pōōl′dōō) *n.* The coot. [<F *poule d'eau* a water hen]
pul·let (pōōl′it) *n.* A young hen, or one not fully grown. [<OF *polete, poulet,* dim. of *poule* a hen <L *pullus* a chicken, young animal]
pul·ley (pōōl′ē) *n.* 1 A wheel grooved to receive a rope, and usually mounted in a block, used to increase the mechanical advantage of an applied force and to transmit or change the direction of power by means of a flexible belt or rope. 2 A block with its pulleys or tackle. 3 A wheel driving, carrying, or being driven by a belt: often called *belt pulley*. [<OF *polie* <Med. L *poleia,* prob. ult. <Gk. *polos* a pivot, axis]

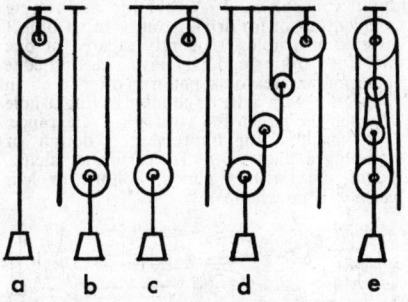

PULLEYS
a. Single fixed.   c. Fixed and runner.
b. Single runner.   d. First system.
                 e. Second system.

Pull·man (pōōl′mən) *n.* A sleeping-car or chair car on a passenger train: a trade name. Also **Pullman car.** [after George M. *Pullman,* 1831-97, U. S. inventor]
pull-out (pōōl′out′) *n. Aeron.* The maneuver of an airplane in passing from a dive to horizontal flight.
pull-o·ver (pōōl′ō′vər) *adj.* Donned by being drawn over the head. — *n.* A garment so donned, as a sweater or shirt.
pul·lu·late (pul′yə·lāt) *v.i.* **·lat·ed, ·lat·ing** 1 To germinate; bud. 2 To breed in abundance; swarm; teem. [<L *pullulatus,* pp. of *pullulare* sprout < *pullulus,* dim. of *pullus* a young animal] — **pul′lu·la′tion** *n.* — **pul′lu·la′tive·ly** *adv.*
pull-up (pōōl′up′) *n. Aeron.* The passing of an airplane from approximately level flight into a short climb.
pul·mom·e·ter (pul·mom′ə·tər) *n.* An instrument for determining lung capacity by measuring the quantity of air in a single respiration; a spirometer. [<L *pulma* lung + -METER] — **pul·mom′e·try** *n.*
pul·mo·nar·y (pul′mə·ner′ē) *adj.* 1 Pertaining to or affecting the lungs. 2 Having lunglike organs. [<L *pulmonarius* < *pulmo, -onis* lung]
pulmonary artery *Anat.* An artery which conveys (venous) blood from the right ventricle of the heart to the lungs. In man it divides into the right and left pulmonary arteries, leading respectively to the right and left lungs.
pulmonary vein *Anat.* One of four veins which return arterial blood from the lungs to the left side of the heart.
pul·mo·nate (pul′mə·nāt, -nit) *adj.* 1 Having lunglike organs. 2 Of or pertaining to an order of gastropods (*Pulmonata*), including most land snails, slugs, and fresh-water snails, which have lunglike organs. — *n.* One of the *Pulmonata*. [<NL *pulmonatus* <L *pulmo, -onis* lung]
pul·mon·ic (pul·mon′ik) *adj.* 1 Pertaining to or affecting the lungs; pulmonary. 2 Pertaining to pneumonia. — *n.* 1 A medicine for lung disease. 2 One affected by lung disease. [<MF *pulmonique* <L *pulmo, -onis* lung]
Pul·mo·tor (pul′mō′tər, pōōl′-) *n.* An apparatus for producing artificial respiration in asphyxiated persons by forcing oxygen into the lungs: a trade name. [<L *pul(mo)* lung + MOTOR]
pulp (pulp) *n.* 1 A moist, soft, slightly cohering mass of matter, usually organic, as chyme, or the soft, succulent part of fruit. 2 A mixture of wood fibers or rags, reduced to a pulpy consistency, and forming the basis from which paper is made. 3 *pl.* Magazines printed on rough, unglazed, wood-pulp paper, and usually having contents of a cheap, sensational nature: distinguished from *slicks*. 4 Powdered ore mixed with water; slime. 5 A pulplike organ or part. 6 *Dent.* The soft tissue of vessels and nerves that fills the central cavity of a tooth. — *v.t.* 1 To reduce to pulp. 2 To remove the pulp or envelope from. — *v.i.* 3 To be or become of a pulpy consistency. [<MF *pulpe* <L *pulpa* flesh, pulp of fruit, pith] — **pulp′less** *adj.*
pulp·ous (pul′pəs) *adj.* Resembling pulp; pulpy.
pul·pit (pōōl′pit) *n.* 1 An elevated stand or desk for a preacher in a church. 2 The office or work of preaching; hence, the clergy as a class. 3 An elevated platform usually boxed in and variously used: the harpooner's *pulpit* on a whaling vessel. — *adj.* Of or pertaining to the pulpit: *pulpit* oratory. [<L *pulpitum* a scaffold, stage, platform]
pul·pit·eer (pōōl′pə·tir′) *n.* A preacher: used depreciatively.
pulp·wood (pulp′wōōd′) *n.* The soft wood of certain trees, as the spruce, used in the manufacture of paper.
pulp·y (pul′pē) *adj.* **pulp·i·er, pulp·i·est** 1 Consisting of or resembling pulp. 2 Of a soft, juicy consistency; succulent. — **pulp′i·ly** *adv.* — **pulp′i·ness** *n.*
pul·que (pul′kē, pōōl′-; *Sp.* pōōl′kä) *n.* A fermented drink made from various species of agave, especially from the juice of the maguey. [<Sp., prob. <Nahuatl]
pul·sate (pul′sāt) *v.i.* **·sat·ed, ·sat·ing** 1 To move or throb with rhythmical impulses, as the pulse or heart. 2 To vibrate; quiver. [<L *pulsatus,* pp, of *pulsare,* freq. of *pellere* (pp. *pulsus*) beat]
pul·sa·tile (pul′sə·til) *adj.* 1 Pulsatory. 2 That must be struck in order to produce sound; specifically, in music, percussive.
pul·sa·til·la (pul′sə·til′ə) *n.* The dried herb of

pul·sa·tion the pasqueflower, used as a sedative and alterative; also, the plant itself. [< Med. L, dim. of L *pulsata*, pp. fem. of *pulsare* beat, strike; with ref. to the beating of the flower by the wind]

pul·sa·tion (pul·sā′shən) *n.* 1 A throbbing or vibrating. 2 A single throb or heartbeat.

pul·sa·tive (pul′sə-tiv) *adj.* Pulsating; throbbing; pulsatile. — **pul·sa·tive·ly** *adv.*

pul·sa·tor (pul·sā′tər) *n.* A machine which operates by pulsation, as a pneumatic rock drill operated by puffs of air.

pul·sa·to·ry (pul′sə-tôr′ē, -tō′rē) *adj.* Of or pertaining to pulsation; having rhythmical movement; throbbing; beating; pulsatile.

pulse¹ (puls) *n.* 1 *Physiol.* The rhythmic beating of the arteries due to the successive contractions of the heart, especially as felt in pressing upon the radial artery at the wrist. 2 Any throbbing characterized by a short, quick, regular stroke or motion; pulsation. 3 *Telecom.* A brief surge of electrical or electromagnetic energy, usually transmitted as a signal in communication. 4 Any movement, drift, or tendency indicative of general opinion, feeling, or sentiment. — *v.i.* **pulsed, puls·ing** To manifest a pulse; pulsate; throb. [< OF *pous* < L *pulsus* (*venarum*) the beating (of the veins), orig. pp. of *pellere* beat] — **pulse′less** *adj.*

pulse² (puls) *n.* Leguminous plants collectively, as peas, beans, etc., or their edible seeds. [< OF *pols* < L *puls* pottage of meal or pulse]

pulse-jet (puls′jet′) *adj. Aeron.* Designating a type of jet engine equipped in front with movable vanes which intermittently take in air to develop power in rapid bursts rather than continuously. Also **pul′so-jet′**.

pulse repeater *Electronics* A transponder.

pul·sim·e·ter (pul·sim′ə·tər) *n.* An instrument for indicating and registering the frequency, force, and variations of the pulse; a sphygmograph. [< *pulsi-* (< PULSE¹) + -METER]

pul·som·e·ter (pul·som′ə·tər) *n.* 1 A device for pumping liquids by steam pressure, operating without pistons and consisting of two pear-shaped chambers connected by valves; a vacuum pump. 2 A pulsimeter. [< *pulso-* (< PULSE¹) + -METER]

Pul·tusk (po͞o′o͞o-to͞osk′) A town in east central Poland, north of Warsaw.

pul·ver·a·ble (pul′vər-ə-bəl) *adj.* Pulverizable.

pul·ver·a·ceous (pul′və-rā′shəs) *adj.* Having a powdery surface; pulverulent. [< L *pulvis, pulveris* a powder + -ACEOUS]

pul·ver·ize (pul′və-rīz) *v.* **·ized, ·iz·ing** *v.t.* 1 To reduce to powder or dust, as by grinding or crushing. 2 To demolish; annihilate. — *v.i.* 3 To become reduced to powder or dust. Also *Brit.* **pul′ver·ise**. [< MF *pulveriser* < LL *pulverizare* < L *pulvis*, a powder, dust] — **pul′ver·iz·a·ble** *adj.* — **pul′ver·i·za′tion** *n.* — **pul′ver·iz′er** *n.*

pul·ver·u·lent (pul·ver′yə·lənt) *adj.* 1 Consisting of, reducible or reduced to, fine powder or dust. 2 Dusty; powdery. [< L *pulverulentus* dusty < *pulvis, pulveris* a powder, dust] — **pul·ver′u·lence** *n.*

pul·vil·lus (pul·vil′əs) *n. pl.* **·vil·li** (-vil′ī) *Entomol.* One of a pair of adhesive pads between the claws of an insect's foot, as the paired positions of a fly's foot. [< L, contraction of *pulvinulus*, dim. of *pulvinus* a cushion]

pul·vi·nate (pul′və-nāt) *adj.* 1 Cushion- or pillow-shaped. 2 Having a pulvinus. 3 Swelling out like a pillow: said of a convex frieze. Also **pul′vi·nat′ed**. [< L *pulvinatus* < *pulvinus* a cushion] — **pul′vi·nar** *adj.*

pul·vi·nus (pul·vī′nəs) *n. pl.* **·ni** (-nī) *Bot.* The enlargement or swelling at the base of the leaves and leaflets of many leguminous and other plants, through which the sensitive movements are rendered possible. [< L, a cushion]

pu·ma (pyo͞o′mə) *n.* An American carnivore (*Felis couguar*) ranging from Canada to Patagonia, of a reddish-tawny color, about 4 feet in length, exclusive of the tail; the cougar: also called *mountain lion*. [< Sp. < Quechua]

PUMA
(From 2 to 2 1/2 feet at the shoulder)

pum·ice (pum′is) *n.* Spongy or cellular volcanic lava, used as an abrasive and polishing material, especially when powdered: also **pumice stone**. — *v.t.* **·iced, ·ic·ing** To smooth, polish, or clean with pumice. [< OF *pomis, pumis* < L *pumex, pumicis*] — **pu·mi·ceous** (pyo͞o-mish′əs) *adj.*

pum·mel (pum′əl) See POMMEL.

pump¹ (pump) *n.* A mechanical device for raising, circulating, exhausting, or compressing a liquid or gas by drawing or pressing it through apertures and pipes. — *v.t.* 1 To raise with a pump, as water or other liquid. 2 To remove the water, etc., from. 3 To inflate with air by means of a pump. 4 To propel, discharge, force, etc., from or as if from a pump: The heart *pumps* blood. 5 To cause to operate in the manner of a pump or pump handle. 6 To question or obtain information from persistently or subtly: to *pump* a witness. 7 To obtain (information) in such a manner. — *v.i.* 8 To work a pump; raise water or other liquid with a pump. 9 To move up and down like a pump or pump handle. [< MDu. *pompe*, prob. < Sp. *bomba*; prob. ult. imit.] — **pump′er** *n.*

pump² (pump) *n.* A low-cut slipper without a fastening, having either a high or a low heel. [? < F *pompe* pomp]

pum·per·nick·el (pum′pər-nik′əl) *n.* A coarse, dark, sour bread made from unsifted rye. [< G, Westphalian rye bread, orig. a lout, a peasant]

pump gun A repeating shotgun operated by a sliding handle.

pump·kin (pump′kin, pung′-) *n.* 1 A large trailing vine (*Cucurbita pepo*) with heart-shaped leaves. 2 Its large, round, edible, yellow fruit. 3 In Europe, the winter squash (*C. maxima*) or any of its varieties. [Earlier *pompion* < MF *pompon, popon* < L *pepo, peponis* < Gk. *pepōn* a melon, lit., ripe, cooked by the sun]

pump·kin·seed (pump′kin-sēd′, pung′-) *n.* 1 The seed of a pumpkin. 2 A small freshwater sunfish, especially the common North American sunfish (*Lepomis gibbosus*). 3 The butterfish.

pump-prim·ing (pump′prī′ming) *n.* 1 Any device or method for priming a pump, usually the application of a little water to wet the valve. 2 Government spending for the purpose of stimulating business.

pun (pun) *n.* The witty use of two words having the same or similar sounds but different meanings, or of two different, more or less incongruous meanings of the same word. — *v.* **punned, pun·ning** *v.i.* 1 To make a pun or puns. — *v.t.* 2 To treat as a pun. 3 To affect in a specified manner by puns. [? < Ital. *puntiglio* a fine point, a verbal quibble. See PUNCTILIO.] — **pun′ning·ly** *adv.*

pu·na¹ (po͞o′nä) See POON.

pu·na² (po͞o′nä) *n.* 1 A cold, arid region at high altitudes, as in the Andes. 2 Mountain sickness; illness caused by rarefaction of the air; soroche. [< Sp. < Quechua]

punch¹ (punch) *n.* 1 A tool for perforating or indenting, or for driving out or in an object inserted in a hole: frequently tapered at one end. The working end may have a cutting edge enclosing an area or a pattern: often used in connection with a die or counter having a hole in which the punch fits with slight clearance. 2 A machine for impressing a design or stamping a die. — *v.t.* To perforate, shape, indent, etc., with a punch. [Short for ME *punchon* a puncheon¹]

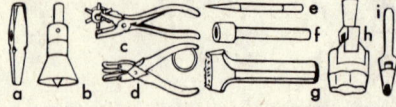

PUNCHES
a. Blacksmith's square. d. Ticket.
b. Center. e, g. Stamping.
c. Revolving belt. f, h, i. Cutting.

punch² (punch) *v.t.* 1 To strike sharply, especially with the fist. 2 To poke with a stick; prod. 3 *Western U.S.* To drive (cattle). — *n.* 1 A swift blow with the fist; also, a thrust or nudge. 2 *Slang* Hence, vitality; effectiveness; force; directness: an editorial with *punch*. [Prob. var. of POUNCE²]

punch³ (punch) *n.* A beverage having wine or spirits, milk, tea, or fruit juices as a basic ingredient, sweetened, sometimes spiced, and diluted with water: variously called milk punch, fruit punch, rum punch, etc., from its principal ingredient. [< Hind. *pānch* < Skt. *panchan* five; from the five original ingredients: arrack, tea, sugar, water, and lemon]

Punch (punch) The quarrelsome, grotesque hero of a comic puppet show, **Punch and Judy**. — **pleased as Punch** Extremely pleased; highly gratified. [Short for PUNCHINELLO]

Punch (punch) An English illustrated humorous weekly journal, founded in 1841.

punch-drunk (punch′drungk′) *adj.* 1 Suffering from the effects of repeated blows so as to be groggy, slow in movement, etc.: said of prize fighters. 2 Confused; dazed.

pun·cheon¹ (pun′chən) *n.* 1 An upright supporting timber. 2 A punch or perforating tool, especially one for chipping stone or for stamping figures on plate or other material. 3 A broad, heavy piece of roughly dressed timber, having one flat, hewed side. [< OF *poinçon, poinchon* a punch, ult. < L *punctus*, pp. of *pungere* prick]

pun·cheon² (pun′chən) *n.* 1 A liquor cask of variable capacity, from 72 to 120 gallons. 2 A liquor measure of varying amount: mostly of wine, 84 gallons. [< OF *ponçon, poinchon*; ult. same as PUNCHEON¹]

punch·er (pun′chər) *n.* 1 One who or that which punches. 2 A cowboy; cowpuncher.

Pun·chi·nel·lo (pun′chə-nel′ō) *n. pl.* **·los** or **·loes** 1 A character in an Italian burlesque or puppet show, the original of the English Punch. 2 Hence, **punchinello**, any similarly comic or grotesque character; buffoon. [Earlier *polichinello* < dial. Ital. (Neapolitan) *Polcenella*]

punching bag An inflated or stuffed ball, usually suspended, that is punched with the fists for exercise.

punch press A machine equipped with dies for cutting or forming metal.

punc·tate (pungk′tāt) *adj.* 1 Covered or studded with dots, points, or minute depressions; dotted. 2 Pointed. Also **punc′tat·ed**. [< NL *punctatus* < L *punctum* a point] — **punc·ta′tion** *n.*

punc·til·i·o (pungk-til′ē-ō) *n. pl.* **·til·i·os** 1 A nice point of etiquette. 2 Preciseness in the observance of etiquette or ceremony. [< Sp. *puntillo* < Ital. *puntiglio*, dim. of *punto* a point < L *punctum*]

punc·til·i·ous (pungk-til′ē-əs) *adj.* 1 Very nice or exact in the observance of forms of etiquette, etc. 2 Of or pertaining to precise etiquette. [< F *pointelleux* < *pointille* < Ital. *puntiglio* small point] — **punc·til′i·ous·ly** *adv.* — **punc·til′i·ous·ness** *n.*

punc·tu·al (pungk′cho͞o-əl) *adj.* 1 Exact as to appointed time; acting or arriving promptly; prompt. 2 Done or made precisely at an appointed time, as a work or payment. 3 Punctilious; exact. 4 Consisting of or confined to a point as related to space. [< Med. L *punctualis* < L *punctus* a pricking, a point] — **punc′tu·al·ly** *adv.*

punc·tu·al·i·ty (pungk′cho͞o-al′ə-tē) *n. pl.* **·ties** The quality, characteristic, or habit of being punctual, in any sense.

punc·tu·ate (pungk′cho͞o-āt) *v.* **·at·ed, ·at·ing** *v.t.* 1 To divide or mark with punctuation. 2 To interrupt at intervals. 3 To emphasize. — *v.i.* 4 To use punctuation. [< Med. L *punctuatus*, pp. of *punctuare* < L *punctus* a point] — **punc′tu·a′tor** *n.*

punc·tu·a·tion (pungk′cho͞o-ā′shən) *n.* The use of points or marks in written or printed matter, to indicate the separation of the words into sentences, clauses, and phrases, and to aid in the better comprehension of the meaning and grammatical relation of the words; also, the marks so used. See also under PRINTING. — **punc′tu·a′tive** *adj.* The chief punctuation points are:

period	.
colon	:
semicolon	;
comma	,
interrogation point	?
(question mark)	
exclamation point	!
parentheses	( )
brackets	[ ]
dash (em-dash)	—
(en-dash)	–
hyphen	-
quotation marks	" "
virgule (virgil)	/

punc·ture (pungk′chər) *v.* **·tured, ·tur·ing** *v.t.* 1 To pierce with a sharp point. 2 To make

puncturevine by pricking, as a hole. 3 To cause to collapse: to *puncture* a reputation. — *v.i.* 4 To be pierced or punctured. See synonyms under PIERCE. — *n.* 1 A small hole, as in a pneumatic tire, made by piercing with something sharp-pointed. 2 A minute depression; pit. 3 The act of puncturing. [<LL *punctura* a prick, puncture <L *punctus,* pp. of *pungere* prick] — **punc′tur·a·ble** *adj.*

punc·ture·vine (pungk′chər·vīn′) *n.* A low-growing weed *(Tribulus terrestris)* of the caltrop family, having sharp divergent spines which often damage automobile tires: common in the western United States.

pun·dit (pun′dit) *n.* A learned Brahman, especially one versed in Sanskrit lore and in the science, laws, and religion of the Hindus; hence, any learned man. [<Hind. *paṇḍit* <Skt. *paṇḍita,* lit., learned, skilled]

pung (pung) *n. U.S. Dial.* A low box sled for one horse. [Short for *tom pung,* prob. alter. of TOBOGGAN]

pun·gent (pun′jənt) *adj.* 1 Having or causing sharp pricking, stinging, piercing, or acrid effects upon the senses. 2 Affecting the mind or feelings, as by sharp points, so as to cause pain; piercing; sharp. 3 Caustic; keen; racy: *pungent* sarcasm. 4 Terminating in a hard sharp point, as a pine needle. See synonyms under BITTER, HOT, RACY. [<L *pungens, -entis,* ppr. of *pungere* prick] — **pun′gence** or **pun′gen·cy** *n.* — **pun′gent·ly** *adv.*

Pu·nic (pyōō′nik) *adj.* Of or pertaining to ancient Carthage or the Carthaginians, who were regarded by the Romans as treacherous; hence, faithless; untrustworthy. — *n.* The Northwest Semitic language of the Carthaginians, a dialect of Phoenician. [<L *punicus* <*poenicus* <*Poenus* a Carthaginian, a Phoenician <Gk. *Phoinix, -ikos*]

Punic Wars See table under WAR.

pun·ish (pun′ish) *v.t.* 1 To subject (a person) to pain, confinement, or other penalty for a crime or fault. 2 To subject the perpetrator of (an offense) to a penalty: to *punish* forgery. 3 To use roughly; injure; hurt. 4 To make heavy inroads upon; deplete, as a stock of food. See synonyms under AVENGE, CHASTEN, REQUITE. [<OF *puniss-,* stem of *punir* <L *punire* punish <*poenire* <*poena* a punishment, penalty, fine] — **pun′ish·er** *n.*

pun·ish·a·ble (pun′ish·ə·bəl) *adj.* Deserving of or liable to punishment: said of offenders or offenses. — **pun′ish·a·bil′i·ty** *n.*

pun·ish·ment (pun′ish·mənt) *n.* 1 Penalty imposed, as for transgression of law. ♦ Collateral adjective: *penal.* 2 Any ill suffered in consequence of wrongdoing. 3 The act of punishing. 4 *Colloq.* Rough handling, as in a pugilistic encounter, a naval engagement, etc.

pu·ni·tive (pyōō′nə·tiv) *adj.* 1 Pertaining to or inflicting punishment. 2 *Law* Of a character to punish or vindicate. Also **pu′ni·to′ry** (-tôr·ē, -tō′rē). [<Med. L *punitivus* <L *punitus,* pp. of *punire* PUNISH] — **pu′ni·tive·ly** *adv.* — **pu′ni·tive·ness** *n.*

Pun·jab (pun′jäb, pun·jäb′) 1 A region of NW India and West Pakistan; 148,610 square miles. 2 A former province of British India in this region, divided in 1947 between Punjab State, India, and West Pakistan; 99,089 square miles; former capital, Lahore. 3 A State of India in this region; 47,456 square miles; capital, Chandigarh. 4 A former province of West Pakistan in this region, a part of West Pakistan province since October, 1955; 63,134 square miles; former capital, Lahore.

Punjab Hill States A former political agency in NW India, under British rule, consisting of 22 princely states which came to be part of India: one in Uttar Pradesh, two in Punjab, and the rest in Himachal Pradesh; from 1936 to 1947-48 included in Punjab States; 11,375 square miles; headquarters, Simla.

Pun·ja·bi (pun·jä′bē) *n.* 1 A native of the Punjab. 2 The Sanskritic language of the Punjab, belonging to the Indic branch of the Indo-Iranian languages: also spelled *Panjabi.*

Punjab States A former political agency in NW India, under British rule, consisting of 14 princely states (of which two were included in West Pakistan, three in Himachal Pradesh, India, and the rest in Punjab, India) and also, after 1936, the Punjab Hill States;

38,146 square miles; headquarters, Lahore.

punk¹ (pungk) *n.* 1 Wood decayed through the action of some fungus, and useful as tinder; touchwood. 2 An artificial preparation that will smolder without flame. [<Algonquian (Lenape) *punk, ponk* fine ashes]

punk² (pungk) *n.* 1 *U.S. Slang* Rubbish; nonsense; anything worthless. 2 *U.S. Slang* A petty hoodlum. 3 *Obs.* A prostitute. — *adj. U.S. Slang* Worthless; useless. [Origin uncertain]

pun·ka (pung′kə) *n.* A fan; especially, a rectangular strip of cloth, etc., swung from the ceiling and moved by a servant or by machinery. Also **pun′kah.** [<Hind. *paṅkhā* a fan <Skt. *pakshaka* <*paksha* a wing]

pun·ky (pung′kē) *n. pl.* **·kies** A minute, annoying, bloodsucking gnat or midge (genus *Culicoides*): also called *sand fly.* Also **pun′key, pun′kie.** [<Du. *punki* <Algonquian (Lenape) *punk, ponk,* orig. fine ashes]

pun·ster (pun′stər) *n.* One who puns; one addicted to punning. Also **pun′ner.**

punt¹ (punt) *n.* A flat-bottomed, square-ended boat, usually with a seat in the middle and a well or seat at one or each end, for use in shallow waters, and propelled with a pole. — *v.t.* 1 To propel (a boat) by pushing with a pole against the bottom of a shallow stream, lake, etc. 2 To convey in a punt. — *v.i.* 3 To go or hunt in a punt. [OE <L *ponto, -onis* a punt, a pontoon <*pons, pontis* a bridge] — **punt′er** *n.*

PUNT

punt² (punt) *v.i.* To gamble or bet, especially against a bank, as at faro, roulette, or baccarat. [<F *ponter* <*ponte* a point <L *punctum*] — **punt′er** *n.*

punt³ (punt) *n.* In football, a kick made by dropping the ball from the hands and kicking it before it strikes the ground. — *v.t.* In football, to propel (the ball) with a punt. — *v.i.* In football, to make a punt. [Prob. var. of BUNT] — **punt′er** *n.*

Pun·ta A·re·nas (pōōn′tä ä·rā′näs) A port of southern Chile, on the Strait of Magellan.

pun·til·la (pun·til′ə) *n.* Lacework, lace edging, or lace design with points. [<Sp., dim. of *punto* a point <L *punctum*]

pun·to (pun′tō) *n.* A hit or thrust in fencing. [<Ital., a point <L *punctum*]

pun·ty (pun′tē) See PONTIL.

pu·ny (pyōō′nē) *adj.* **·ni·er, ·ni·est** 1 Weak and insignificant; of small and feeble development or importance; petty. 2 *Obs.* Puisne; born later; younger. See synonyms under SMALL. ♦ Homophone: *puisne.* [<OF *puisne.* See PUISNE.] — **pu′ni·ly** *adv.* — **pu′ni·ness** *n.*

pup (pup) *n.* 1 A puppy (def. 1). 2 A young seal. — *v.i.* **pupped, pup·ping** To bring forth pups. [Short for PUPPY]

pu·pa (pyōō′pə) *n. pl.* **·pae** (-pē) 1 *Entomol.* The quiescent stage in the development of an insect that undergoes a complete metamorphosis, following the larval and preceding the adult stage; also, an insect in such a stage. 2 *Zool.* A similar developmental state in some echinoderms, as holothurians. [<NL <L, a girl, doll, puppet] — **pu′pal** *adj.*

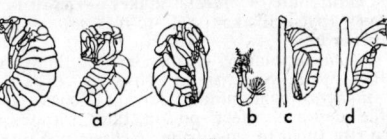
PUPAE
a. Three pupal stages of a bumblebee.
b. Aquatic pupa of a gnat.
c. Suspended pupa of a butterfly.
d. Girdled pupa of a butterfly.

pu·pate (pyōō′pāt) *v.i.* **·pat·ed, ·pat·ing** To enter upon or undergo the pupal condition. — **pu·pa′tion** *n.*

pu·pil¹ (pyōō′pəl) *n.* 1 A person of either sex or of any age under the care of a teacher; scholar; learner. 2 In civil law, a minor who is under the age of puberty and has a guardian. See synonyms under SCHOLAR. [<

OF *pupille,* orig. an orphan, ward <L *pupillus,* dim. of *pupus* a boy and *pupilla,* dim. of *pupa* a girl]

pu·pil² (pyōō′pəl) *n. Anat.* The contractile opening in the iris of the eye, through which light reaches the retina. [<L *pupilla* a figure reflected in the eye, the pupil of the eye, dim. of *pupa.* See PUPA.]

pu·pil·age (pyōō′pəl·ij) *n.* The state or period of being a pupil. Also **pu′pil·lage.**

pu·pi·lar·i·ty (pyōō′pə·lar′ə·tē) *n.* In Scots law, the interval between birth and the age of 14 in males and 12 in females. Also **pu′pil·lar′i·ty.** [<OF *pupillarité* <L *pupillaris* pertaining to an orphan <*pupillus, pupilla.* See PUPIL¹.]

pu·pi·lar·y (pyōō′pə·ler′ē) *adj.* Of or pertaining to a pupil or a ward. Also **pu′pil·lar′y.**

Pu·pin (pyōō·pēn′, *Hungarian* pōō·pēn′), **Michael Idvorsky,** 1858-1935, U. S. physicist and inventor born in Hungary.

pu·pip·a·rous (pyōō·pip′ə·rəs) *adj.* Of or pertaining to a division *(Pupipara)* of dipterous insects in which the young are born ready to pupate, as bat ticks, sheep ticks, etc. [<NL <PUPA + L *parere* bring forth]

pup·pet (pup′it) *n.* 1 A small figure of a human being, that by means of strings or wires is made to perform mock drama; a marionette. 2 A person slavishly subject to the will of another; a tool. 3 A doll. — *adj.* 1 Of or pertaining to puppets or mummery. 2 Performing the will of an unseen power; not autonomous: a *puppet* state or government. [<OF *poupette* <L *pupa* a girl, doll, puppet]

pup·pet·eer (pup′i·tir′) *n.* A person who manipulates puppets.

pup·pet·ry (pup′it·rē) *n.* The performances of puppets or the manipulation of puppets; mummery.

puppet show A mock drama, with puppets for the actors.

pup·py (pup′ē) *n. pl.* **·pies** 1 The young of a canine mammal, as of a dog; a pup. 2 A conceited and forward young man; a silly fop. [<OF *poupee, popee* <L *pupa* a girl, doll] — **pup′py·ish** *adj.*

puppy love Adolescent love; sentimental, temporary infatuation.

pup tent A shelter tent.

pur (pûr) See PURR.

Pu·ra·cé (pōō′rä·sā′) An active volcano in SW Colombia; 15,420 feet; last major eruption, 1869.

Pu·ra·na (pōō·rä′nə) *n.* Any of a number of Hindu scriptures in the form of verse dialogs, coming next in order after the Vedas, dealing mainly with theogony and cosmogony, especially with the god Vishnu and his incarnations. There are 18 Puranas and 18 Upa Puranas or subordinate works. [<Skt. *purāna,* lit., ancient <*purā* of old]

pur·blind (pûr′blīnd′) *adj.* 1 Afflicted with dimness of vision; near-sighted. 2 Having little or no insight or understanding. 3 *Obs.* Totally blind. [ME *pur blind* <*pur* (<OF, plain) + *blind* blind] — **pur′blind′ly** *adv.* — **pur′blind′ness** *n.*

Pur·cell (pûr′səl), **Henry,** 1658?-95, English composer.

Pur·chas (pûr′chəs), **Samuel,** 1575?-1626, English author and compiler.

pur·chas·a·ble (pûr′chəs·ə·bəl) *adj.* That can be purchased; hence, venal; corrupt. — **pur′chas·a·bil′i·ty** *n.*

pur·chase (pûr′chəs) *v.t.* **·chased, ·chas·ing** 1 To acquire by paying money or its equivalent; buy. 2 To obtain by exertion, sacrifice, flattery, etc. 3 *Law* To acquire (property) by means other than descent or inheritance. 4 To move, hoist, or hold by a mechanical purchase. — *n.* 1 The act of purchasing; acquisition by giving an equivalent in money or other exchange, or by exertion, risk, etc. 2 That which is purchased; especially, that which is bought with money. 3 A mechanical hold or grip. 4 A device that gives a mechanical advantage, as a tackle or lever. 5 Leverage. 6 Any means of increasing influence or advantage. 7 *Law* The act of acquiring property by payment of a price or value; hence, any lawful mode of acquiring property other than by inheritance or descent or by the mere operation of law. 8 Value; worth, especially as measured by the annual income, expressed in terms of years to indicate the period at the end of which the income received from a property will have covered the price paid for it: to buy at ten years' *purchase.* **9** A small territorial division in New Hampshire, originally made when the land was sold in lots to individuals by the State. **10** *Obs.* A seeking; also, attempt; endeavor. [<AF *purchacer,* OF *porchacier* seek for <*pur-, por-* for (<L *pro-*) + *chacier* CHASE] **—pur'chas·er** *n.*

Synonyms (verb): acquire, buy, get, obtain, procure, secure. *Buy* and *purchase* are close synonyms, in numerous cases freely interchangeable, but with the difference usually found between words of Anglo-Saxon and French or Latin origin. The Anglo-Saxon *buy* is used for all the concerns of common life, the French *purchase* is often restricted to transactions of more dignity; yet *buy* is commonly more emphatic, and also appeals more strongly to the feelings. One may either *buy* or *purchase* fame, favor, honor, pleasure, etc., but we speak of victory or freedom as dearly *bought. Antonyms*: barter, exchange, sell.

pur·dah (pûr'də) *n. Anglo-Indian* **1** A curtain or screen, especially one used to seclude women; also, the state of seclusion so secured. **2** The material of which a curtain is made. [<Urdu *pardah* <Persian]

pure (pyo͝or) *adj.* **1** Free from mixture or contact with that which weakens, impairs, or pollutes; containing no foreign or vitiating material. **2** Free from adulteration; clear; clean; hence, genuine; stainless: *pure* food, *pure* motives. **3** Free from moral defilement; innocent; chaste; unsullied; also, free from coarseness; refined: a *pure* life, *pure* language. **4** Free from foreign or imported elements: said especially of language and works of art. **5** *Music* Mathematically correct as to intervals; free from harsh quality in tone; also, correct in form or style; finished. **6** *Philos.* Considered apart from its attributes or from concrete experience; abstract; also, a priori. **7** *Phonet.* Having a single, unvarying tone or sound: said of vowels. **8** Theoretical; concerned with fundamental research, as distinguished from practical application: said of sciences. **9** *Genetics* Breeding true with respect to one or more characters; homozygous. **10** Nothing but; real; sheer: *pure* mischief, *pure* luck. [<OF *pur* <L *purus* clean, pure] **—pure'ness** *n.*

Synonyms: absolute, chaste, classic, classical, clean, clear, continent, fair, genuine, guileless, guiltless, holy, immaculate, incorrupt, innocent, mere, perfect, real, sheer, simple, spotless, stainless, true, unadulterated, unblemished, uncorrupted, undefiled, unmingled, unmixed, unpolluted, unspotted, unstained, unsullied, untainted, untarnished, upright, virtuous. Material substances are called *pure* in the strict sense when free from foreign admixture of any kind; as, *pure* oxygen; the word is often used to signify free from any defiling or objectionable admixture (the original sense); we speak of water as *pure* when it is bright, clear, and refreshing, even if it contains mineral salts in solution; in the medical and chemical sense, only distilled water *(aqua distillata)* is *pure.* In moral and religious use *pure* denotes positive excellence of a high order; one is *innocent* who knows nothing of evil and has experienced no touch of temptation; one is *pure* who, with knowledge of evil and exposure to temptation, keeps heart and soul *unstained. Virtuous* refers primarily to right action, *pure* to right feeling; as, "Blessed are the *pure* in heart: for they shall see God." *Matt.* v 8. See FINE[1], INNOCENT, MODEST, VIRTUOUS. *Antonyms*: adulterated, defiled, dirty, filthy, gross, impure, indecent, indelicate, lewd, mixed, obscene, polluted, stained, sullied, tainted, tarnished, unchaste, unclean; see also synonyms for FOUL, IMMODEST.

pure·blood (pyo͝or'blud) *n.* **1** An individual descended from a long line of ancestors of the same ethnic or racial stock: said especially of American Indians. **2** A purebred animal. **—pure'-blood'ed** *adj.*

pure·bred (pyo͝or'bred') *adj.* Bred from stock having had no admixture for many generations: said especially of livestock. **—***n.* (pyo͝or'·bred') A purebred animal.

pure culture *Bacteriol.* A culture or medium for the isolation and cultivation of microorganisms of a particular kind, as those of anthrax, diphtheria, etc.

pu·rée (pyo͝o·rā', pyo͝or'ā; *Fr.* pü·rā') *n.* A thick soup, usually of vegetables, boiled and strained. [<F <OF, pp. fem. of *purer* strain <L *purare* purify <*purus* pure]

pure line *Genetics* A strain of plants or animals which through self-fertilization, continued inbreeding, or other means, exhibit a high degree of stability in one or more genetic characteristics.

pure·ly (pyo͝or'lē) *adv.* **1** So as to be free from admixture, taint, or any harmful substance. **2** Chastely; innocently. **3** Merely.

pur·fle (pûr'fəl) *v.t.* **·fled, ·fling** To decorate, as with a wrought or flowered border; border. **—***n.* A richly ornamented border: also **pur'·fling.** [<OF *porfiler, pourfiler* <*por-, pour-* for (<L *pro-*) + *fil* a thread <L *filum*]

pur·ga·tion (pûr·gā'shən) *n.* A purging; catharsis. [<OF *purgacion* <L *purgatio, -onis* < *purgatus,* pp. of *purgare* PURGE]

pur·ga·tive (pûr'gə·tiv) *adj.* Efficacious in purging; cathartic. **—***n.* A cathartic.

pur·ga·to·ri·al (pûr'gə·tôr'ē·əl, -tō'rē-) *adj.* **1** Pertaining to purgatory. **2** Tending to purge from sin; expiatory. Also **pur'ga·to'ri·an.**

pur·ga·to·ry (pûr'gə·tôr'ē, -tō'rē) *n. pl.* **·ries** **1** In Roman Catholic theology, a state or place where the souls of those who have died penitent are made fit for paradise by expiating venial sins and undergoing any punishment remaining for previously forgiven sins. **2** Any place or state of temporary banishment, suffering, or punishment. [<AF *purgatorie,* OF *purgatoire* <Med. L *purgatorium* <L *purgatorius* cleansing <*purgare* PURGE]

purge (pûrj) *v.* **purged, purg·ing** *v.t.* **1** To cleanse of what is impure or extraneous; purify. **2** To remove (impurities, etc.) in cleansing: with *away, off,* or *out.* **3** To rid (a group, nation, etc.) of elements regarded as undesirable or inimical, especially by killing. **4** To remove or kill (a person or persons) in such a manner. **5** To cleanse or rid of sin, fault, or defilement. **6** *Med.* **a** To cause evacuation of (the bowels, etc.). **b** To induce evacuation of the bowels of. **7** *Law* To clear of accusation, suspicion, or guilt. **—***v.i.* **8** To become clean or pure. **9** *Med.* To have or induce evacuation. **—***n.* **1** The act or operation of purging, in any sense. **2** That which purges; specifically, a medicine causing active evacuation of the bowels; a cathartic; also, its administration or operation. [<OF *purgier* <L *purgare* cleanse <*purigare* <*purus* pure] **—purg'er** *n.* **—purg'ing** *n.*

Pu·ri (po͞o'rē) A port and Hindu pilgrimage center on the Bay of Bengal in SW Orissa, India: also *Jaganath.*

pu·ri·fi·ca·tion (pyo͝or'ə·fə·kā'shən) *n.* **1** The act or operation of purifying: said of things physical or spiritual. **2** The act or observance of formal cleansing from ceremonial defilement. ◆ Collateral adjective: *lustral.*

pu·ri·fy (pyo͝or'ə·fī) *v.* **·fied, ·fy·ing** *v.t.* **1** To make pure or clean; rid of extraneous or noxious matter. **2** To free from sin or defilement. **3** To free of foreign or debasing elements, as a language. **—***v.i.* **4** To become pure or clean. [<OF *purifier* <L *purificare* <*purus* pure + *facere* make] **—pu·rif·i·ca·to·ry** (pyo͞o·rif'ə·kə·tôr'ē, -tō'rē-) *adj.* **—pu'·ri·fi'er** *n.*

Synonyms: clarify, clean, cleanse, filter, refine, wash. See AMEND, CHASTEN, CLEANSE. *Antonyms*: contaminate, corrupt, debase, defile, deprave, infect, poison, taint, vitiate.

Pu·rim (po͝or'im, pyo͝or'im; *Hebrew* po͞o·rēm') A Jewish festival commemorating the defeat of Haman's plot to massacre the Jews (*Esth.* ix 26), observed about the first of March. [<Hebrew *pūrīm,* pl. of *pūr* a lot]

pu·rine (pyo͝or'ēn, -in) *n. Biochem.* A white, crystalline compound, $C_5H_4N_4$, which is closely related to uric acid in structure. Also **pu·rin** (pyo͝or'in). [<G *purin* <L *purus* pure + NL *uricum* uric acid + *-in* -INE[2]]

purine group *Biochem.* An important group of organic compounds widely distributed in nature and related to purine, as caffeine, xanthine, uric acid, etc.

pur·ism (pyo͝or'iz·əm) *n.* Extreme strictness in regard to the use of words, or an instance of it. **—pur'ist** *n.* **—pu·ris'tic** *adj.*

Pu·ri·tan (pyo͝or'ə·tən) *n.* **1** One of a group or party of English Protestants (1599) who advocated simpler forms of creed and ritual in the established church, freedom of conscience and worship, and condemned all laxity of morals. Many of them emigrated to the American colonies in the 17th century, especially to the Massachusetts Bay colony. **2** One who is scrupulously strict, or censorious and exacting in his religious life: often not capitalized. **—***adj.* Of or pertaining to the Puritans or their beliefs or customs. [<LL *puritas* purity <L *purus* pure + -AN; orig. used by opponents to suggest a resemblance to the *Cathari* (lit., purists)] **—Pu'ri·tan'ic** *adj.*

pu·ri·tan·i·cal (pyo͝or'ə·tan'i·kəl) *adj.* Governed by the Puritan code; rigidly scrupulous in religious observance and morals; strict. **—pu'·ri·tan'i·cal·ly** *adv.* **—pu'ri·tan'i·cal·ness** *n.*

Pu·ri·tan·ism (pyo͝or'ə·tən·iz'əm) *n.* **1** The spirit, doctrines, and practices of the Puritans. **2** Religious and moral scrupulousness and austerity. **3** The New England character and spirit.

pu·ri·ty (pyo͝or'ə·tē) *n.* **1** The character or state of being pure, in any sense, as freedom from dirt or foreign or adulterating matter; cleanness; moral cleanness; innocence; freedom from sinister or improper design; absence of admixture. **2** Saturation: said of a color. **3** The use of no foreign words, phrases, or idioms; use of words with only the precise form, connection, and meaning assigned to them by good usage. See synonyms under INNOCENCE, VIRTUE.

Pur·kin·je (po͝or'kin·ye), **Johannes Evangelista,** 1787–1869, Czech physiologist.

Purkinje cell *Physiol.* One of the large, flask-shaped ganglion cells interposed as a single layer between the two layers of gray matter in the cerebellar cortex of the brain. [after J. E. *Purkinje*]

purl[1] (pûrl) *v.i.* **1** To whirl; turn. **2** To flow with a bubbling sound; ripple. **3** To move in eddies. **—***n.* **1** A circling movement of water; an eddy. **2** A gentle, continued murmur, as of a rippling stream. ◆ Homophone: *pearl.* [Cf. Norw. *purla* gush out, bubble up]

purl[2] (pûrl) *v.t.* **1** To purfle. **2** In knitting, to make (a stitch) backward. **3** To edge with lace, embroidery, etc. **—***v.i.* **4** To do edging with lace, etc. [<*n.*] **—***n.* **1** An edge of lace, embroidery, etc.; in lacework, a spiral of gold or silver wire. **2** In knitting, the inversion of the knit stitch giving a horizontal rib effect. ◆ Homophone: *pearl.* [Earlier *pyrle,* orig. twisted gold or silver thread <*pyrl* twist; ult. origin unknown]

pur·lieu (pûr'lo͞o) *n.* **1** *pl.* The outlying districts or outskirts of any place. **2** A place in which one is free to come and go; a haunt. **3** Formerly, ground unlawfully taken for a royal forest, but afterward disafforested and restored to its rightful owners. [<AF *puralee* <OF <*puraler* go through <*pur-* through (<L *per-*) + *aler* go; infl. in form by MF *lieu* a place]

pur·lin (pûr'lin) *n.* One of several horizontal timbers supporting rafters. Also **pur'line** (-lin). [ME *purlyn,* prob. <OF]

pur·loin (pûr·loin') *v.t.* & *v.i.* To steal; filch. See synonyms under ABSTRACT, STEAL. [<AF *purloignier,* OF *porloignier* remove, put far off <*pur-, por-* for (<L *pro-*) + *loing, loin* far <L *longe*] **—pur·loin'er** *n.*

pur·ple (pûr'pəl) *n.* **1** A color of mingled red and blue, between crimson and violet; in ancient times, the color obtained from the murex, properly a crimson. **2** Cloth or a garment of this color, worn formerly by sovereigns, especially the emperors of Rome; hence, royal power or dignity; preeminence in rank or wealth. **3** The office of a cardinal: from the official red hat and robes; also, the episcopal dignity: from its purple insignia. **—***v.t.* & *v.i.* **·pled, ·pling** To make or become purple. **—***adj.* **1** Of the color of purple. **2** Hence, imperial; regal. **3** Conspicuously brilliant or ornate: said of language. [Alter. of ME *purpre,* OE *purpure,* the color purple <L *purpura,* orig. the shellfish yielding Tyrian purple dye, the dye, or cloth dyed with it <Gk. *porphyra*]

purple finch The rose-breasted American finch *(Carpodacus purpureus).*

pur·ple-fringed orchid (pûr'pəl·frinjd') A terrestrial orchid of North America (genus

Habenaria) with fragrant, purple, lilac, or, rarely, white flowers.
Purple Heart A decoration of honor of the **Order of the Purple Heart** in the form of a purple enameled heart surrounded by a gold-colored border and bearing the head of George Washington in gold-colored relief: established by George Washington in 1782, revived 1932: awarded to members of the armed forces or to citizens of the United States honorably wounded in action, or as a result of enemy action.
purple medic Lucerne.
purple of Cassius A rich and powerful pigment obtained from a mixture of stannic, stannous, and gold chlorides: used chiefly in miniature painting and enamel painting.

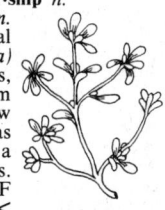
PURPLE HEART

purple osier Red osier.
pur·plish (pûr'plish) *adj.* Somewhat purple.
pur·port (pər·pôrt', -pōrt', pûr'pôrt, -pōrt) *v.t.* 1 To have or bear as its meaning; signify; imply. 2 To claim or profess (to be), especially falsely. See synonyms under IMPORT. — *n.* (pûr'pôrt, -pōrt) 1 That which is conveyed or suggested to the mind as the meaning or intention; import; significance. 2 The substance of a statement, etc., given in other than the exact words. See synonyms under PURPOSE. [<AF, OF *purporter* extend <*purforth* (<L *pro-*) + *porter* carry <L *portare*] — **pur·port'ed·ly** *adv.*
pur·pose (pûr'pəs) *v.t. & v.i.* ·posed, ·pos·ing To have the intention of doing or accomplishing (something); intend; aim; design. — *n.* 1 The idea or ideal kept before the mind as an end of effort or action; plan; design; aim. 2 The particular thing to be effected or attained; practical advantage or result; consequence; use: words to little *purpose.* 3 Settled resolution; determination; constancy. 4 Purport; intent, as of spoken or written language. 5 A proposition; proposal; question at issue. — **on purpose** With previous design; intentionally. [<OF *porposer,* var of *proposer.* See PROPOSE.]
Synonyms (noun): aim, design, determination, drift, end, intent, intention, meaning, motive, object, plan, project, purport, resolution, resolve, view. Compare AIM, CAUSE, DESIGN, END, IDEA, PLAN, PROJECT, REASON, SERVICE. *Antonyms:* See synonyms for ACT.
Synonyms (verb): design, determine, intend, mean, propose, resolve. See PROPOSE.
pur·pose·ful (pûr'pəs·fəl) *adj.* Having, or marked by, purpose; intentional; important; significant. — **pur'pose·ful·ly** *adv.* — **pur'pose·ful·ness** *n.*
pur·pose·less (pûr'pəs·lis) *adj.* Having no definite design or use; aimless. See synonyms under FAINT. — **pur'pose·less·ly** *adv.*
pur·pose·ly (pûr'pəs·lē) *adv.* For a purpose; intentionally; deliberately; on purpose.
pur·po·sive (pûr'pə·siv) *adj.* 1 Pertaining to, having, or indicating purpose. 2 Functional. — **pur'po·sive·ly** *adv.* — **pur'po·sive·ness** *n.*
pur·pu·ra (pûr'pyŏŏ·rə) *n. Pathol.* A disease characterized by especially livid spots on the skin caused by extravasated blood. [<L. PURPLE.]
pur·pure (pûr'pyŏŏr) *n.* Purple: one of the colors or tinctures used in heraldic description. [OE]
pur·pu·ric (pûr·pyŏŏr'ik) *adj.* 1 Of or pertaining to a purple tint. 2 Relating to or resembling purpura.
pur·pu·rin (pûr'pyŏŏ·rin) *n. Chem.* A red crystalline compound, $C_{14}H_8O_5$, contained in madder, largely used in dyeing: also prepared synthetically. Also **pur'pu·rine** (-rin). [<L *purpura* purple + -IN]
purr (pûr) *n.* An intermittent murmuring sound, such as a cat makes when pleased. — *v.i.* To make such a sound. — *v.t.* To express by or as by purring. Also spelled *pur.* [Imit.]
purse (pûrs) *n.* 1 A small bag or pouch of leather or the like, often having the mouth drawn together with a drawstring; especially, one for carrying money; hence, anything for carrying money on the person. 2 Available resources or means; a treasury: the public *purse.* 3 A sum of money offered as a prize or tendered as a gift, as for a contest or charitable collection. — *v.t.* **pursed, purs·ing** 1 To contract into wrinkles or folds like the mouth of a purse; pucker: to *purse* the lips. 2 *Rare* To place in a purse. [OE *purs* <LL *bursa* <Gk. *byrsa* a skin, a hide]
purse·pride (pûrs'prīd') *n.* Arrogance due to the possession of wealth. — **purse'proud'** (-proud') *adj.*
purs·er (pûr'sər) *n.* An officer having charge of the accounts, etc., of a vessel; formerly, a naval paymaster. — **purs'er·ship** *n.*
purs·lane (pûrs'lin, -lān) *n.* A procumbent fleshy annual plant (*Portulaca oleracea*) of gardens and waste places, with reddish-green stem and leaves and small yellow flowers: used in Europe as a salad, but regarded as a weed in the United States. Also spelled *pussley.* [<OF *porcelaine* <L *porcilaca* < *portulaca*]
PURSLANE
pur·su·ance (pər·sōō'əns) *n.* The act of pursuing; a following after or following out; prosecution: usually in the phrase *in pursuance of.*
pur·su·ant (pər·sōō'ənt) *adj.* Done in accordance with or by reason of something; conformable. — *adv.* In accordance; agreeably; conformably: usually with *to:* also **pur·su'ant·ly.**
pur·sue (pər·sōō') *v.* ·sued, ·su·ing *v.t.* 1 To follow in an attempt to overtake or capture; chase. 2 To seek or attain or gain: to *pursue* fame. 3 To advance along the course of; keep to the direction or provisions of, as a path, plan, or system. 4 To apply one's energies to or have as one's profession or chief interest: to *pursue* one's studies. 5 To follow persistently; harass; worry. — *v.i.* 6 To follow. 7 To continue. See synonyms under FOLLOW. [<AF *pursuer,* OF *porsievre* <LL *prosequere* <L *prosequi* <*pro-* forth + *sequi* follow] — **pur·su'a·ble** *adj.* — **pur·su'er** *n.*
pur·suit (pər·sōōt') *n.* 1 The act of pursuing; a chase. 2 That which is followed as a continued employment; a business; vocation. See synonyms under HUNT. [<AF *pursuete,* OF *porsuite, poursuite* <*porsievre* PURSUE]
pursuit plane *Mil.* A powerful, speedy, highly maneuverable airplane, heavily armed, but with short range, designed to intercept, pursue, and attack enemy aircraft: also called *fighter plane.*
pur·sui·vant (pûr'swi·vənt) *n.* 1 An attendant upon a herald; an officer of the third and lowest rank in the College of Heralds, performing similar duties to a herald. 2 *Obs.* A follower; especially, a military attendant of the king. [<OF *porsivant,* ppr. of *porsievre* pursue]
purs·y (pûr'sē) *adj.* **purs·i·er, purs·i·est** Short-breathed; asthmatic; hence, fat. See synonyms under CORPULENT. [Earlier *pursive* <AF *pursif,* OF *polsif* <*polser* pant, gasp] — **purs'i·ness** *n.*
pur·te·nance (pûr'tə·nəns) *n. Obs.* Appurtenance; specifically, the inwards of an animal. [<AF *purtinaunt,* OF *partenant.* See PERTINENT.]
pu·ru·lent (pyŏŏr'ə·lənt, -yə·lənt) *adj.* Consisting of or secreting pus; suppurating. [<L *purulentus* <*pus, puris* pus] — **pu'ru·lence** or **·len·cy** *n.* — **pu'ru·lent·ly** *adv.*
Pu·rus (pōō·rōōs') A river in SE Peru and western Brazil, flowing 2,100 miles NE to the Amazon.
pur·vey (pər·vā') *v.t. & v.i.* To furnish or provide (provisions, etc.). [<AF *purveier,* OF *porveier* <L *providere.* Doublet of PROVIDE.]
pur·vey·ance (pər·vā'əns) *n.* 1 The act of purveying. 2 That which is purveyed or supplied; provisions. 3 A former prerogative of royalty, abolished in 1660, enabling a monarch to buy goods at an appraised value, and also to enforce personal service.
pur·vey·or (pər·vā'ər) *n.* 1 One who furnishes supplies for living, especially for the table; a caterer. 2 Formerly, an officer who, by exaction or otherwise, made provision for the king's household.
pur·view (pûr'vyōō) *n.* 1 Extent, sphere, or scope of anything, as of official authority. 2 Range of view, experience, or understanding; outlook. 3 *Law* The body or the scope or limit of a statute. [<AF *purveu* provided, OF *porveu,* pp. of *porveier* PURVEY; orig. in AF legal phrases *purveu est* it is provided and *purveu que* provided that]
pus (pus) *n. Med.* A secretion from inflamed tissues, as in healing wounds, usually viscid or creamy, and consisting of modified leucocytes and other cells in a liquid plasma: the result of suppuration. [<L. Akin to PUTRID.]
Pu·san (pōō·sän) A port on Korea Strait in SE Korea: Japanese *Fusan.*
Pu·sey (pyōō'zē), **Edward Bouverie,** 1800–82, English theologian.
Pu·sey·ism (pyōō'zē·iz'əm) *n.* Tractarianism. [after E. B. Pusey] — **Pu'sey·ite** *n.*
push (pŏŏsh) *v.t.* 1 To exert force upon or against (an object) for the purpose of moving. 2 To force (one's way), as through a crowd, jungle, etc. 3 To press forward, prosecute, or develop with vigor and persistence: to *push* trade with South America. 4 To urge, advocate, or promote vigorously and persistently: to *push* a new product. 5 To bear hard upon; distress; harass: I am *pushed* for time. — *v.i.* 6 To exert steady pressure against something so as to move it. 7 To move or advance vigorously or persistently. 8 To exert great effort. — *n.* 1 A propelling or thrusting pressure; repulsion as opposed to attraction or pull; a shove. 2 *Colloq.* An extremity; exigency: at a *push* for money. 3 Determined activity; energy. 4 Anything pushed to cause action; a pushbutton. 5 *Slang* The crowd; a number of friends or associates: He fooled the whole *push*; also, an influential clique. 6 *Austral. Slang* A body of larrikins. [<OF *pousser, polser* <L *pulsare.* See PULSATE.]
Synonyms: crowd, drive, expedite, force, hasten, impel, importune, press, propel, prosecute, shove, thrust, urge. See DRIVE, HUSTLE, JAM[1]. *Antonyms:* see DRAW.
push·ball (pŏŏsh'bôl') *n.* A game, played with a ball 6 feet in diameter and weighing 48 pounds, in which each of two sides tries to push the ball across the opponent's goal.
push·but·ton (pŏŏsh'but'n) *n.* A button or knob which, on being pushed, opens or closes a circuit in an electric system, thereby turning on or off a light, ringing a bell, etc.
push·cart (pŏŏsh'kärt') *n.* A two- or four-wheeled cart pushed by hand: used by fruit venders, peddlers, hawkers, etc.
push·er (pŏŏsh'ər) *n.* 1 One who or that which pushes; especially, an active, energetic person. 2 *Aeron.* An airplane with the propeller in the rear of the wings. 3 *U.S. Slang* One who sells illegally, especially one who sells narcotics to addicts.
push·ing (pŏŏsh'ing) *adj.* 1 Possessing business enterprise and energy. 2 Possessing aggressiveness; impertinent. — **push'ing·ly** *adv.*
Push·kin (pŏŏsh'kin) A city south of Leningrad in Russian S.F.S.R.: formerly *Tsarskoe Selo.*
Push·kin (pŏŏsh'kin), **Alexander Sergeyevich,** 1799–1837, Russian poet.
push·o·ver (pŏŏsh'ō'vər) *n. Slang* A susceptible person; an easy mark; also, anything done or that can be done with little or no effort.
push·pin (pŏŏsh'pin') *n.* 1 A children's game in which pins are pushed over each other. 2 A sharp pin with a large head, inserted by thumb pressure into a bulletin board, drawing board, etc., for mounting and holding in place papers, drawings, etc.
Push·tu (push'tōō) *n.* The Iranian language of the dominant peoples of Afghanistan; Afghan: also spelled *Pashto.* Also **Push'to** (-tō).
pu·sil·la·nim·i·ty (pyōō'səl·ə·nim'ə·tē) *n.* Faintheartedness; indecision; cowardice. Also **pu'·sil·lan'i·mous·ness.**
pu·sil·lan·i·mous (pyōō'sə·lan'ə·məs) *adj.* 1 Lacking strength of mind, courage, or spirit; mean-spirited; cowardly. 2 Characterized by weakness of purpose or lack of courage.

[<LL pusillanimis <L pusillus very little + animus mind] — **pu·sil′lan′i·mous·ly** *adv.*
Synonyms: cowardly, dastardly, faint–hearted, feeble, mean-spirited, recreant, spiritless, timid, timorous, weak. *Antonyms:* see synonyms for BRAVE.
puss[1] (pŏŏs) *n.* **1** A cat. **2** A child or young woman: a term of affection. [Cf. Du. *poes,* LG *puus,* a name for a cat]
puss[2] (pŏŏs) *n. Slang* The mouth; face. [Irish *pus* mouth, lips]
puss·ley (pŏŏs′lē) *n.* Purslane. Also **puss′ly**. [Alter. of PURSLANE]
puss moth A common European moth (*Cerura vinula*) with grayish wings and two rows of black spots on the abdomen.
pus·sy[1] (pŏŏs′ē) *n. pl.* **·sies 1** Puss; a cat: a diminutive. **2** A fuzzy catkin, as of a willow, a birch, etc. [Dim. of PUSS[1]]
pus·sy[2] (pŭs′ē) *adj.* Full of pus.
pus·sy·foot (pŏŏs′ē·fŏŏt′) *v.i.* **1** To move softly and stealthily, as a cat does. **2** To act or proceed without committing oneself or revealing one's intentions.
pus·sy willow (pŏŏs′ē) **1** A small American willow (*Salix discolor*) with silky catkins in early spring: also called *glaucous willow*. **2** One of various other willows bearing catkins in early spring.
pus·tu·lant (pus′chŏŏ·lənt) *adj.* Causing pustules. — *n.* A medicine that causes pustules.

PUSSY WILLOW

pus·tu·lar (pus′chŏŏ·lər) *adj.* **1** Proceeding from or marked by pustules: a *pustular* eruption. **2** Pustulate. Also **pus′tu·lous**.
pus·tu·late (pus′chŏŏ·lāt) *v.t. & v.i.* **·lat·ed, ·lat·ing** To form into or become pustules. — *adj.* (-lāt, -lit) Covered with pustules or pustulelike elevations. [<L *pustulatus,* pp. of *pustulare* blister <*pustula* a pustule]
pus·tu·la·tion (pus′chŏŏ·lā′shən) *n.* **1** The formation of pustules; a pustular eruption. **2** A pustule.
pus·tule (pus′chōōl) *n.* **1** *Pathol.* A small, circumscribed elevation of the skin with an inflamed base containing pus. **2** Any elevation resembling a pimple or a blister. [<L *pustula*]
put[1] (pŏŏt) *v.* **put, put·ting** *v.t.* **1** To bring into or set in a specified or implied place or position; lay: *Put* the book on the table. **2** To bring into a specified state, condition, or relation: to *put* a prisoner to death. **3** To apply; bring to bear: *Put* your back into it! **4** To impose: to *put* a tariff on bicycles. **5** To ascribe or attribute, as the wrong interpretation on a remark. **6** To place according to one's estimation: I *put* the time at five o'clock. **7** To throw with a pushing motion of the arm: to *put* the shot. **8** To incite; prompt: Who *put* him up to it? **9** To bring forward for debate, answer, consideration, etc.: to *put* a question. **10** To subject: Let's *put* it to a vote. **11** To express in words; state: That's *putting* it mildly. **12** To risk; bet: I'll *put* six dollars on that horse. — *v.i.* **13** To go; proceed: to *put* to sea. — **to put about** *Naut.* To change to the opposite tack; change direction. — **to put aside (or away or by) 1** To place in reserve; save. **2** To thrust aside; discard. — **to put down 1** To repress; crush. **2** To degrade; demote. **3** To write. — **to put forth 1** To extend, as the arm or hand. **2** To grow, as shoots or buds. **3** To exert. **4** To set out; leave port. — **to put forward** To advance; urge, as a claim. — **to put in 1** *Naut.* To enter a harbor or place of shelter. **2** To interpolate; interpose. **3** *Colloq.* To devote; expend, as time. **4** To advance (a claim, etc.). **5** To submit, as an application. — **to put off 1** To delay; postpone. **2** To discard. — **to put on 1** To don. **2** To bring into action; turn on. **3** To simulate; pretend. **4** To give a representation of; stage. — **to put out 1** To extinguish. **2** To expel; eject. **3** To disconcert; embarrass. **4** To inconvenience. **5** To put forth. **6** In baseball, to retire (a batter or base runner). — **to put over 1** To place in command or charge. **2** *Colloq.* To accomplish successfully. — **to put one (or something) over on** *Colloq.* To deceive or dupe. — **to put through 1** To bring to successful completion. **2** To cause to perform. — **to put up 1** To erect; build. **2** To preserve or can. **3** To wager. **4** To provide (money, capital, etc.). **5** To sheathe, as a weapon. — **to put upon** To deceive; cheat. — **to put up with** To endure; tolerate. — *n.* **1** The act of putting, as a cast or throw. **2** A contract by which one person, in consideration of money paid to another, acquires the privilege of selling or delivering to the latter within a certain time some article named, as wheat or cotton, or shares at a stipulated price: opposed to *call.* — *adj. Colloq.* Fixed; settled as fixed: My hat won't stay *put.* [Fusion of OE *putian* place, *potian* thrust, and *pȳtan* push, prob. all <Scand. Cf. Dan. *putte.*]
Synonyms (verb): deposit, lay, place, set. *Put* is the most general term for bringing an object to some point or within some space, however exactly or loosely; we may *put* a horse in a pasture, or *put* a bullet in a rifle or into an enemy. *Place* denotes more careful movement and more exact location; as, to *place* a crown on one's head, or a garrison in a city. To *lay* is to *place* in a horizontal or recumbent position; to *set* is to *place* or adjust in a certain place or position; we *lay* a cloth, and *set* a dish upon a table. To *deposit* is to *put* in a place of security for future use; as, to *deposit* money in a bank; in the original sense, to *lay* down is also common; as, the stream *deposits* sediment; insects *deposit* eggs. Compare SET.
put[2] (put) *n. Colloq.* One of the sharp sounds made by a gasoline engine, machine-gun, etc. [Imit.]
pu·ta·men (pyōō·tā′min) *n. pl.* **·tam·i·na** (-tam′ə-nə) *Bot.* The hard bony stone of certain fruits, as the cherry. [<L, waste, a husk <*putare* cleanse, prune] — **pu·tam′i·nous** *adj.*
put and take A game of chance in which the players add to or take from a pool.
pu·ta·tive (pyōō′tə·tiv) *adj.* Supposed; reported; reputed. [<MF *putatif* <LL *putativus* <L *putatus,* pp. of *putare* think] — **pu′ta·tive·ly** *adv.*
Put-in-Bay (pŏŏt′in·bā′) A harbor on South Bass Island in Lake Erie near the Canadian border in northern Ohio; site of Perry's defeat of the British in a naval battle (1813) in the War of 1812.
put·log (pŏŏt′lôg, -log, put′-) *n.* A crosspiece in a scaffolding, its inner end resting in a hole in the wall and its outer on a ledger. [Earlier *putlock* <*put,* pp. of PUT[1]]
Put·nam (put′nəm), **Israel,** 1718–90, American Revolutionary general.
put-off (pŏŏt′ôf′, -of′) *n.* An evasion; excuse.
put-out (pŏŏt′out′) *n.* The act of causing an out, as of batter or base runner in baseball.
put-put (put′put′) *n. Slang* A gasoline engine; especially, one used in propelling a small boat. [Imit.]
pu·tre·fac·tion (pyōō′trə·fak′shən) *n.* **1** The progressive chemical decomposition of organic matter, as by the agency of anaerobic bacteria, with the production of evil-smelling compounds. **2** The state of being putrefied. **3** Putrescent or putrefied matter. [<OF <L *putrefactio, -onis* <*putrefacere* PUTREFY]
pu·tre·fac·tive (pyōō′trə·fak′tiv) *adj.* **1** Of or pertaining to putrefaction. **2** Producing putrefaction.
pu·tre·fy (pyōō′trə·fī) *v.t. & v.i.* **·fied, ·fy·ing 1** To decay or cause to decay with fetid odor; rot; decompose. **2** To make or become gangrenous. [<L *putrefacere* <*putrere* decay (<*puter* rotten) + *facere* make] — **pu′tre·fi′er** *n.*
Synonyms: corrupt, decay, decompose, rot. See CORRUPT, DECAY. *Antonyms:* disinfect, embalm, freshen, preserve, purify, vitalize, vivify.
pu·tres·cence (pyōō·tres′əns) *n.* **1** The state of undergoing putrefaction. **2** Something that is putrescent.
pu·tres·cent (pyōō·tres′ənt) *adj.* **1** Becoming putrid; undergoing putrefaction. **2** Pertaining to putrefaction. [<L *putrescens, -entis,* ppr. of *putrescere* grow rotten, inceptive of *putrere* be rotten. See PUTREFY.]
pu·tres·ci·ble (pyōō·tres′ə·bəl) *adj.* Liable to putrefy. — *n.* A substance that decomposes at a certain temperature in contact with air and moisture: generally containing nitrogen. — **pu·tres′ci·bil′i·ty** *n.*
pu·tres·cine (pyōō·tres′ēn, -in) *n. Biochem.* A colorless, ill-smelling ptomaine, $C_4H_{12}N_2$, resulting from the bacterial decomposition of animal tissues.
pu·trid (pyōō′trid) *adj.* **1** Being in a state of putrefaction; decomposed or decomposing; rotten: *putrid* meat. **2** Indicating or produced by putrefaction: a *putrid* smell. **3** Rotten; corrupt. See synonyms under BAD[1], ROTTEN. [<L *putridus* <*putrere.* See PUTREFY.] — **pu·trid′i·ty** *n.* — **pu′trid·ness** *n.*
Putsch (pŏŏch) *n.* An outbreak or rebellion; an attempted coup d'état. [<G <dial. G (Swiss), lit., a push, blow]
putt (put) *n.* In golf, a light stroke made on a putting green to place the ball in or near the hole. — *v.t. & v.i.* To strike (the ball) with such a stroke. [Var. of PUT[1]]
put·tee (put′ē, pu·tē′) *n.* A strip of cloth wound spirally about the leg from knee to ankle, as used by soldiers, sportsmen, etc.; also, a leather gaiter strapped around the leg. Also **put′ty.** [<Hind. *paṭṭī* a bandage <Skt. *paṭṭa* a strip of cloth]
put·ter[1] (put′ər) *n.* **1** One who putts: He is a poor *putter.* **2** An upright, stiff-shafted golf club used on the putting green. [<PUTT]
put·ter[2] (put′ər) *v.i.* To act, work, or proceed in a dawdling or ineffective manner; trifle. — *v.t.* To waste or spend (time, etc.) in dawdling or puttering. [Var. of POTTER[1]]
put·ti·er (put′ē·ər) *n.* One who putties; a glazier. [<PUTTY]
put·ting (pŏŏt′ing) *n.* The action of the verb *to put,* as *putting* the shot.
putting green (put′ing) In golf, the smooth ground within twenty yards of the hole; also, a place set aside for putting practice. [<PUTT]
put·ty (put′ē) *n.* **1** Whiting mixed with linseed oil to the consistency of dough: used for filling holes or cracks in wood surfaces, securing panes of glass in the sash, making relief ornaments, etc. **2** Fine lime mortar for filling cracks, finishing, etc. — **iron putty** Ferric oxide mixed with boiled linseed oil: used in making pipe–joint connections. — **red-lead putty** Red and white lead mixed with boiled linseed oil, used mainly for cementing pipe joints. — *v.t.* **·tied, ·ty·ing** To fill, stop, fasten, etc., with putty. [<OF *potee* calcined tin, lit., a potful <*pot* a pot]
putty knife A knife with a spatulalike blade, used by glaziers in puttying window glass, etc.
putty powder Tin oxide, or tin and lead oxide, used for polishing glass, metals, etc.
put·ty-root (put′ē·rōōt′, -rōŏt′) *n.* An American orchid (*Aplectrum hyemale*) with a scape bearing a loose raceme of brownish flowers produced yearly. [So called from a sticky substance found in its bulbs]
Pu·tu·ma·yo (pōō′tōō·mä′yō) A river in Ecuador, Colombia, and Peru, flowing about 1,000 miles SE to the Amazon, forming the greater part of the boundary between Colombia and Peru: called *Içá* in its lower courses in Brazil.
put-up (pŏŏt′up′) *adj. Colloq.* Prearranged or contrived in an artful manner: a *put-up* job.
Pu·vis de Cha·vannes (pü·vē′ də shà·vàn′), **Pierre,** 1824–98, French painter.
puy (pwē) *n.* A conical hill of volcanic origin. [<F <OF *pui, poi* a hill <L *podium* a height]
Puy-de-Dôme (pwē·də·dôm′) An extinct volcano of the Massif Central in central France; site of an observatory and a ruined temple of Mercury; 4,806 feet.
Pu-yi (pōō′yē′), **Henry,** born 1906, last Manchu emperor of China 1908–12; abdicated; puppet emperor of Manchukuo 1934–45, under name *Kang Te*; abdicated.
puz·zle (puz′əl) *v.* **·zled, ·zling** *v.t.* **1** To confuse or perplex; mystify. **2** To solve by investigation and study, as something perplexing: with *out.* — *v.i.* **3** To be perplexed or confused. See synonyms under PERPLEX. — **to puzzle over** To attempt to understand or solve. — *n.* **1** A thing difficult to understand or explain; perplexing problem; an enigma or problem. **2** Something, as a toy, purposely arranged so as to require time, patience, and ingenuity to solve its intricacies. **3** The state of being puzzled; a quandary; perplexity. See synonyms under RIDDLE[2]. — **cross-word puzzle** A pattern of white and black spaces, of which the white spaces are to be filled with letters that form words, vertically, horizontally, or diagonally, to agree with accompanying definitions. [Related to ME *poselet* confused; ult. origin unknown]

puz·zle·ment (puz′əl-mənt) *n.* State of being nonplused; perplexity.

puz·zler (puz′lər) *n.* One who or that which puzzles; a knotty question.

PX A military post exchange or general store. [<P(OST) (E)X(CHANGE)]

py– Var. of PYO–.

Pya·ti·gorsk (pyä′ti-gôrsk′) A city in the northern Caucasus, Russian S.F.S.R.

pyc·nid·i·um (pik·nid′ē-əm) *n. pl.* **·nid·i·a** (-nid′ē-ə) *Bot.* A spore-bearing receptacle found in certain fungi. [<NL <Gk. *pyknos* thick + *-idion,* dim. suffix] — **pyc·nid′i·al** *adj.*

pyc·nom·e·ter (pik·nom′ə-tər) *n.* A specific-gravity bottle or flask. [<Gk. *pyknos* dense, thick + -METER]

pyc·no·spore (pik′nə-spôr, -spōr) *n. Bot.* A conidium developed within a pycnidium. [Contraction of *pycnidiospore* <PYCNIDIUM + SPORE]

Pyd·na (pid′nə) An ancient city in south central Macedonia, Greece; scene of the final Roman victory over Macedonia, 168 B.C.

pye (pī) See PIE[4].

py·e·li·tis (pī′ə-lī′tis) *n. Pathol.* Inflammation of the pelvis and calices of the kidneys. [<NL <Gk. *pyelos* the pelvis, orig. a trough + *-itis* -ITIS] — **py′e·lit′ic** (-lit′ik) *adj.*

py·e·lo·gram (pī′ə·lō-gram′) *n.* A picture taken by pyelography. [<*pyelo-* <Gk. *pyelos* a trough, pelvis + -GRAM]

py·e·log·ra·phy (pī′ə·log′rə-fē) *n.* The technique of making X-rays of the ureter and the kidney by the use of a radiopaque dye. [<*pyelo-* <Gk. *pyelos* a trough, pelvis + -GRAPHY] — **py′e·lo·graph′ic** (-lō-graf′ik) *adj.*

py·e·mi·a (pī-ē′mē-ə) *n. Pathol.* A poisonous infection of the blood, due to the absorption of vitiated pus or pyogenic micro-organisms into the circulation: it causes suppuration marked by multiple abscesses, phlebitis, high fever, etc. Also **py·ae′mi·a.** [<NL <Gk. *pyon* pus + *haima* blood] — **py·e′mic** *adj.*

py·et (pī′it) *n. Scot.* The magpie.

py·gid·i·um (pī·jid′ē-əm) *n. pl.* **·gid·i·a** (-jid′ē-ə) *Entomol.* The terminal or posterior segment, as of an insect; a caudal shield. [<NL <Gk. *pygidion,* dim. of *pygē* rump] — **py·gid′i·al** *adj.*

Pyg·ma·li·on (pig·mā′lē-ən, -māl′yən) In Greek mythology, a sculptor of Cyprus, who fell in love with his statue, Galatea, which Aphrodite later brought to life.

pyg·my (pig′mē) See PIGMY.

Pyg·my (pig′mē) *n. pl.* **·mies** 1 A member of a Negroid people of equatorial Africa, ranging in height from four to five feet. 2 Any of the Negrito peoples of the Philippines, Andaman Islands, and Malaya. 3 In the *Iliad,* one of a race of dwarfs. [<Gk. *pygmaeus.* See PIGMY.]

py·ic (pī′ik) *adj.* Of or pertaining to pus; purulent. [<PY- + -IC]

py·in (pī′in) *n. Biochem.* A protein compound contained in pus. [<PY- + -IN]

py·ja·mas (pə·jä′məz, -jam′əz) See PAJAMAS.

pyke (pīk) *v.t. Scot.* To pick.

pyk·nic (pik′nik) *adj.* Characterized by plump contours and a broad, stocky build; fat; squat. — *n.* A person of this physical type. [<Gk. *pyknos* thick, compact]

pyk·no·phra·si·a (pik′nə·frā′zhē-ə, -zhə) *n. Pathol.* A thickening of speech. [<NL <Gk. *pyknos* thick + *phrasis* speech]

Pyl·a·des (pil′ə-dēz) In Greek legend, a nephew of Agamemnon and friend of Orestes, whose sister Electra he married.

pyle (pīl) *n. Scot.* A grain.

Pyle (pīl), **Howard,** 1853–1911, U.S. illustrator, painter, and writer.

py·lon (pī′lon) *n.* 1 *Archit.* A monumental structure constituting an entrance to an Egyptian temple or other large edifice, consisting of a central gateway, flanked on each side by a truncated pyramidal tower. 2 A stake marking the course in an airdrome or turning point in an aerial race. 3 One of the tall, mastlike metal structures from whose summits high-tension wires are carried across open country. 4 *Surg.* An artificial leg, usually temporary. [<Gk. *pylōn* a gateway <*pylē* a gate]

py·lo·rec·to·my (pī′lə-rek′tə-mē) *n. Surg.* Excision of the pylorus. [<PYLOR(US) + -ECTOMY]

py·lo·rus (pī·lôr′əs, -lō′rəs, pi-) *n. pl.* **·ri** (-rī) *Anat.* The opening between the stomach and the duodenum, surrounded by circular muscle fibers; also, the adjoining portion of the stomach. [<LL <Gk. *pylōros* a gatekeeper <*pylē* a gate + *ouros* a watcher] — **py·lor′ic** (-lôr′ik, -lor′ik) *adj.*

Py·los (pī′los) 1 A port of SW Peloponnesus, Greece, at the southern entrance to **Pylos Bay.** 2 An ancient city, 4 miles NW of modern Pylos; said to be the seat of Nestor. Medieval *Navarino:* also *Pilos.* Latin **Py·lus** (pī′ləs).

Pym (pim), **John,** 1584–1643, English statesman and orator.

pyo– *combining form* Pus; of or related to pus: *pyorrhea.* Also, before vowels, *py–.* [<Gk. *pyon* pus]

py·o·gen·e·sis (pī′ō·jen′ə-sis) *n.* 1 *Pathol.* The formation or secretion of pus; suppuration. 2 The doctrine or theory of the origin, source, and process of the generation of pus. — **py′o·gen′ic** *adj.*

py·oid (pī′oid) *adj.* Resembling pus; purulent. [<PY- + -OID]

Pyong·yang (pyông·yäng) A city in northern Korea, capital of the Democratic People's Republic of Korea: *Japanese* **Hei·jo** (hā·jō).

py·or·rhe·a (pī′ə-rē′ə) *n. Pathol.* A discharge of pus with a continuous flow; especially, **pyorrhea al·ve·o·la·ris** (al′vē·ō·lā′ris), a loosening of the teeth accompanied by progressive inflammation of their lining membrane; Riggs's disease. Also **py′or·rhoe′a.** [<NL <Gk. *pys, pyos* pus + *rheein* flow] — **py′or·rhe′al** *adj.*

py·o·sis (pī·ō′sis) *n.* Suppuration. [<NL <Gk. *pyōsis* <*pys, pyos* pus]

pyr– Var. of PYRO–.

py·ra·can·tha (pī′rə·kan′thə, pir′ə-) *n.* The firethorn. [<L <Gk. *pyrakantha* <*pyr, pyros* fire + *akantha, -ēs* a thorn]

py·ral·i·did (pi·ral′ə-did) *adj.* Of or pertaining to a family (*Pyralididae*) of small or medium-sized moths of slender build and broad hind wings, including many groups sometimes classified as separate families. — *n.* A moth belonging to this family. Also **pyr·a·lid** (pir′ə-lid). [<NL <L *pyralis, -idis* a winged insect supposed to live in fire <Gk. < *pyr, pyros* a fire] — **py·ral′i·dan** *adj. &. n.*

pyr·a·mid (pir′ə-mid) *n.* 1 *Archit.* A solid structure of masonry with a square base and triangular sides meeting in an apex. Such structures were used as tombs or temples. The pyramids of Egypt, raised over the sepulchral chambers of kings, are the best examples. The most interesting group is at Giza, near Cairo. The pyramids of Mexico

THE PYRAMIDS AT GIZA

served as temples. The largest is the pyramid near Cholula on the Pueblo plateau, in central Mexico. 2 Something in pyramidal form. 3 *Geom.* A solid consisting of a polygonal base and triangular sides, the apices of the triangles coming together at the vertex. 4 *Mineral.* A crystal form consisting of three or more similar planes having a common point of intersection. 5 *Physiol.* One of various pyramidal or conical structures found in animal organisms. 6 *Anat.* A small bony projection in the cavity of the tympanum. 7 Any tree trained in pyramidal form. 8 The operations involved in pyramiding. — *v.t. & v.i.* 1 To arrange or form in the shape of a pyramid. 2 To buy or sell (stock) with paper profits shown by the change in price of stock already purchased or sold, without any additional deposit of money being made, and to continue so buying or selling on each movement in price. [<F *pyramide* <L *pyramis, -idis* <Gk. *pyramis, -idos,* prob. <Egyptian *pi-mar* a pyramid]

py·ram·i·dal (pi·ram′ə-dəl) *adj.* Of or shaped like a pyramid. Also **pyr·a·mid·ic** (pir′ə-mid′ik), **pyr′a·mid′i·cal.** — **py·ram′i·dal·ly** *adv.*

py·ram·i·da·lis (pi·ram′i·dā′lis) *n. pl.* **·les** (-lēz) *Anat.* Any one of several conical or triangular muscles; especially, the flat triangular muscle arising from the pubis and inserted into the linea alba. [<NL <LL, pyramidal; with ref. to its shape]

Pyr·a·mus and This·be (pir′ə-məs, thiz′bē) In classical legend, two Babylonian lovers: believing Thisbe slain by a lion, Pyramus killed himself, and Thisbe, finding his body, took her own life.

py·ran (pī′ran, pi·ran′) *n. Chem.* Either of two isomeric cyclic compounds, C_5H_6O, each having in its ring 5 carbon atoms: the parent forms of certain carbohydrates, alkaloids, and other physiologically active compounds. [<PYRONE]

py·ra·nom·e·ter (pī′rə·nom′ə-tər) *n.* An instrument for measuring sky radiation or radiation from the earth, especially at night. [<PYR- + -ANO + -METER]

py·rar·gy·rite (pī·rär′jə·rīt) *n.* A metallic, black sulfide of antimony and silver, Ag_3SbS_3, crystallizing in the rhombohedral system. [<PYR- + Gk. *argyros* silver + -ITE[1]]

pyre (pīr) *n.* 1 A heap of combustibles arranged for burning a dead body. 2 Any pile or heap of combustible material. [<L *pyra* a hearth, funeral pile <Gk. < *pyr* a fire]

py·rene[1] (pī′rēn) *n. Chem.* A tetracyclic hydrocarbon, $C_{16}H_{10}$, contained in that portion of coal-tar oil boiling above 360° C. [<PYR- + -ENE]

py·rene[2] (pī′rēn) *n. Bot.* The stone of a drupe; also, any nutlet; putamen. [<NL *pyrena* <Gk. *pyrēn* fruit stone]

Py·rene (pī′rēn) *n.* Carbon tetrachloride, prepared for use as a chemical fire extinguisher: a trade name.

Pyr·e·nees (pir′ə-nēz) A mountain chain between France and Spain, extending about 270 miles from the Bay of Biscay to the Mediterranean; highest point, Pico de Aneto, 11,168 feet. — **Pyr′e·ne′an** *adj.*

py·re·noid (pī·rē′noid) *adj.* Having the form of a fruit stone. — *n.* 1 *Bot.* A small, colorless mass of protein substance of a crystalline form, appearing in the chloroplasts of green algae. 2 *Zool.* A transparent body in the chromatophores of certain protozoa. [<Gk. *pyrēnoeidēs* <*pyrēn* a fruit stone + *eidos* a form, shape] — **py·re·no·de·an** (pī′ri·nō′dē-ən) *adj.*

Py·re·no·my·ce·tes (pī′rē·nō·mī·sē′tēz) *n. pl.* A large class of fungi characterized by the forcible expulsion of ascospores from the perithecium, including many parasitic species, as ergot. [<NL <Gk. *pyrēn* a fruit stone + *mykēs, mykētis* a mushroom]

py·reth·rum (pī·reth′rəm, -rē′thrəm) *n.* 1 The dried and powdered roots of the pellitory used in medicine as a sialogog and rubefacient. 2 The powdered flowers of a chrysanthemum (*Chrysanthemum cinerariaefolium*), used medically as an ointment, and as an insecticide. [<L, feverfew <Gk. *pyrethron* < *pyr* fire]

py·ret·ic (pī·ret′ik) *adj.* 1 Affected with or relating to fever; febrile. 2 Remedial in fevers. — *n.* A febrifuge. [<NL *pyreticus* <Gk. *pyretos* a fever < *pyr* fire]

pyr·e·tol·o·gy (pir′ə·tol′ə·jē, pī′rə-) *n.* The department of medical science that treats of fevers. [<Gk. *pyretos* a fever + -LOGY] — **pyr′e·tol′o·gist** *n.*

py·re·to·ther·a·py (pir′ə·tō·ther′ə·pē, pī′rə-) *n.* Medical treatment by the artificial induction of fever by electricity, bacterial infection, etc.; fever therapy. [<Gk. *pyretos* (a fever) + THERAPY]

Py·rex (pī′reks) *n.* A type of heat-resisting glass having a high silica content, with additions of soda, aluminum, and boron: a trade name.

py·rex·i·a (pī·rek′sē-ə) *n. Pathol.* An abnormal elevation of bodily temperature; fever. [<NL <Gk. *pyrexis* <*pyressein* be feverish

<*pyretos.* See PYRETIC.] —**py·rex′i·al** *adj.* —**py·rex′ic** *adj.*

pyr·ge·om·e·ter (pīr′jē·om′ə·tər, pir′-) *n.* A pyranometer. [<PYR- + GEO- + -METER]

pyr·he·li·om·e·ter (pīr·hē′lē·om′ə·tər, pir-) *n. Astron.* An instrument for measuring the quantity and rate of solar radiation by its thermal effects on a silvered disk or other sensitive surface. [<PYR- + HELIO- + -METER]

Pyr·i·ben·za·mine (pir′ə·ben′zə·mēn, -min) *n.* Proprietary name of an antihistamine drug, $C_{16}H_{21}N_3HC$, used in the treatment of certain allergies.

pyr·i·dine (pir′ə·dēn, -din) *n. Chem.* A colorless, liquid, nitrogenous compound, C_5H_5N, with a pungent, noxious odor, obtained by the distillation of coal tar and bone oil and also made synthetically: used in organic syntheses, as a disinfectant, antiseptic, alcohol denaturant, and asthma remedy. [< PYR(ROLE) + -ID(E) + -INE[2]] —**py·rid·ic** (pī·rid′ik) *adj.*

pyr·i·dox·ine (pir′ə·dok′sēn, -sin) *n. Biochem.* A factor of the vitamin B complex known to prevent dermatitis in rats, vitamin B_6, a water-soluble compound, $C_8H_{10}NO_3$, occurring in cereal grains, vegetable oils, legumes, yeast, meats, and fish: also made synthetically. [PYRID(INE) + OX(Y)-[2] + -INE[2]]

pyr·i·form (pir′ə·fôrm) *adj.* Pear-shaped. [< NL *pyriformis* <Med. L *pyrum* a pear (<L *pirum*) + L *forma* form]

py·rim·i·dine (pī·rim′ə·dēn, -din, pir′ə·mə·dēn′, -din′) *n. Chem.* An organic compound, $C_4H_4N_2$, resulting from the acid hydrolysis of a nucleic acid; a constituent of thiamine. Also **py·rim′i·din** (-din). [<G *pyrimidin* <*pyridin* pyridine]

py·rite (pī′rīt) *n. pl.* **py·ri·tes** (pī·rī′tēz) A metallic, pale brass-yellow, opaque, isometric iron disulfide, FeS_2; fool's gold; iron pyrites. [<L *pyrites* <Gk. *pyritēs* flint <*pyritēs (lithos)* fire (stone) <*pyr* fire] —**py·rit′ic** (-rit′ik) or **·i·cal** *adj.*

py·ri·tes (pī·rī′tēz) *n. pl.* The common name for various metallic sulfides: copper *pyrites.* Compare CHALCOPYRITE, PYRITE.

py·ro (pī′rō) *n.* Pyrogallol: so called in photography. [Short for PYROGALLOL]

pyro- *combining form* **1** Fire; heat: *pyromania.* **2** *Chem.* Denoting actual or hypothetical derivation by the action of heat; specifically, in certain inorganic acids, indicating derivation from two molecules of an ordinary acid by the elimination of one molecule of water: $2H_3AsO_4$ (arsenic acid) — H_2O = $H_4As_2O_7$ (*pyroarsenic* acid). **3** *Geol.* Resulting from the action of fire or heat: *pyrolusite.* Also, before vowels, *pyr-.* [<Gk. *pyr, pyros* fire]

py·ro·cat·e·chol (pī′rə·kat′ə·kōl, -chōl, -kol, pir′ə-) *n. Chem.* A white crystalline phenol compound, $C_6H_6O_2$, contained in various barks, originally obtained when catechin was subjected to dry distillation: used in photography as a developer and in medicine. Also **py′ro·cat′e·chin** (-kin, -chin). [<PYRO- + CATECH(U) + (PHEN)OL]

py·ro·cel·lu·lose (pī′rə·sel′yə·lōs) *n.* A form of guncotton used as a propellant in smokeless powder. Also **py′ro·cot′ton** (-kot′n).

Py·ro·ce·ram (pī′rō·sə·ram′) *n.* A strongly heat-resistant, crystalline ceramic material formed from glass and characterized by extreme hardness, great tensile strength, and high dielectric properties: a trade name.

py·ro·chem·i·cal (pī′rə·kem′i·kəl) *adj.* Pertaining to chemical changes induced or effected by high temperature.

py·ro·clas·tic (pī′rə·klas′tik) *adj. Geol.* Formed from or consisting of the fragmentary or comminuted ejecta of volcanic or igneous eruptions: said of rocks or their composition. [<PYRO- + CLASTIC]

py·ro·con·duc·tiv·i·ty (pī′rə·kon′duk·tiv′ə·tē) *n.* Conductivity of an electric current dependent upon or improved by the application of heat. —**py′ro·con·duc′tive** (-kən·duk′tiv) *adj.*

py·ro·crys·tal·line (pī′rə·kris′tə·lin, -līn, pir′ə-) *adj.* Crystallized from materials in a state of fusion: *pyrocrystalline* masses.

py·ro·e·lec·tric (pī′rō·i·lek′trik, pir′ō-) *adj.* **1** Of or pertaining to pyroelectricity. **2** Manifesting pyroelectricity; developing poles when heated. —*n.* A substance that becomes polar when heated.

py·ro·e·lec·tric·i·ty (pī′rō·i·lek′tris′ə·tē, -ē′lek-, pir′ō-) *n.* **1** Electrification or electric polarity developed in certain minerals by a change in temperature. **2** The branch of science treating of this phenomenon.

py·ro·gal·late (pī′rə·gal′āt, pir′ə-) *n.* A salt of pyrogallol.

py·ro·gal·lic (pī′rə·gal′ik, pir′ə-) *adj. Chem.* **1** Of, pertaining to, or derived by heat from gallic acid. **2** Pertaining to or designating pyrogallol.

py·ro·gal·lol (pī′rə·gal′ōl, -ol, -gə·lōl′, pir′ə-) *n. Chem.* A white, crystalline, poisonous compound, $C_6H_3(OH)_3$, obtained by heating gallic acid: used to reduce silver and mercury salts to a metallic state, as a photographic developer, as a dye, and in certain medical preparations. Also **pyrogallic acid.** [<PYRO- + GALL(IC) + (PHEN)OL]

py·ro·gen·ic (pī′rə·jen′ik, pir′ə-) *adj.* **1** Causing or produced by heat. **2** Caused by or inducing fever. **3** Igneous. Also **py·rog·e·nous** (pī·roj′ə·nəs, pi-).

py·rog·nos·tics (pī′rog·nos′tiks, pir′əg-) *n. pl.* The characteristics of a mineral as shown by heat of varying intensity produced with a blowpipe. [<PYRO- + Gk. *gnostikos* knowing]

py·rog·ra·phy (pī·rog′rə·fē, pi-) *n.* The art or process of producing a design, as on wood or leather, by a red-hot point or fine flame. —**py·ro·graph** (pī′rə·graf, -gräf, pir′ə-) *n.* —**py·rog′ra·pher** *n.* —**py′ro·graph′ic** *adj.*

py·ro·gra·vure (pī′rō·grə·vyŏŏr′, pir′ō-) *n.* **1** The art or process of producing a design on wood by pyrography. **2** A picture thus made.

py·ro·lig·ne·ous (pī′rə·lig′nē·əs, pir′ə-) *adj.* Pertaining to that which is derived from wood by heat, specifically by dry distillation. [<PYRO- + LIGNEOUS]

pyroligneous acid Crude acetic acid as derived from wood by distillation; wood vinegar.

py·rol·o·gy (pī·rol′ə·jē, pi-) *n.* **1** The scientific examination of materials by heat; blowpipe analysis. **2** The branch of physics that treats of heat. [<PYRO- + -LOGY] —**py·ro·log·i·cal** (pī′rə·loj′i·kəl, pir′ə·loj′i·kəl) *adj.*

py·ro·lu·site (pī′rə·lōō′sīt, pī·rol′yə·sīt) *n.* A soft, metallic, iron-black or steel-gray manganese dioxide, MnO_2, of great value in the arts, and used in the manufacture of oxygen, chlorine, etc. [<G *pyrolusit* <Gk. *pyr, pyros* a fire + *lousis* a washing (<*louein* wash) + G *-it* -ITE[1]]

py·rol·y·sis (pī·rol′ə·sis) *n. Chem.* Decomposition by the application of or as a result of heat. [<NL <Gk. *pyr, pyros* a fire + *lysis* a loosing <*lyein* loosen] —**py·ro·lit·ic** (pī′rə·lit′ik, pir′ə-) *adj.*

py·ro·mag·net·ic (pī′rō·mag·net′ik, pir′ō-) *adj.* Of, pertaining to, or produced by the changes in magnetic intensity caused by change of temperature.

py·ro·man·cy (pī′rə·man′sē, pir′ə-) *n.* Divination by fire. [<PYRO- + -MANCY]

py·ro·ma·ni·a (pī′rə·mā′nē·ə, -mān′yə, pir′ə-) *n.* A morbid propensity to set things on fire. —**py′ro·ma′ni·ac** (-ak) *adj.* & *n.* —**py·ro·ma·ni·a·cal** (pī′rō·mə·nī′ə·kəl, pir′ō-) *adj.*

py·ro·man·tic (pī′rə·man′tik, pir′ə-) *adj.* Of or pertaining to pyromancy. —*n.* One who professes to divine by means of fire.

py·rom·e·ter (pī·rom′ə·tər) *n.* An instrument for measuring high degrees of heat, as caused by electrical resistance, degree of incandescence, expansion, radiation, etc. —**py·ro·met·ric** (pī′rə·met′rik, pir′ə-) or **·ri·cal** *adj.* —**py·rom′e·try** *n.*

py·ro·mor·phite (pī′rə·môr′fīt, pir′ə-) *n.* A resinous, variously colored phosphate and chloride of lead, found in masses or crystals; green lead ore. [<G *pyromorphit* <Gk. *pyr, pyros* a fire + *morphos* form]

py·rone (pī′rōn, pī·rōn′) *n. Chem.* A cyclic compound, $C_5H_4O_2$, existing in two isomeric forms: it yields yellow dyestuffs. [<G *pyron*]

py·rope (pī′rōp) *n.* A variety of deep-red garnet: also called *precious garnet.* [<OF *pirope* <L *pyropus* gold-bronze <Gk. *pyrōpos,* lit., fiery-eyed <*pyr, pyros* a fire + *ōps, ōpos* eye, face]

py·ro·phor·ic (pī′rə·fôr′ik, -for′ik) *adj.* **1** Fire-bearing; spontaneously combustible. **2** Designating materials which are easily and quickly inflammable, as finely divided metals on exposure to air. [<Gk. *pyrophoros* <*pyr, pyros* a fire + *pherein* carry] —**py′ro·phore** (-fôr, -fōr) *n.*

py·ro·phos·phate (pī′rə·fos′fāt, pir′ə-) *n.* A salt of pyrophosphoric acid.

py·ro·phos·phor·ic acid (pī′rō·fos·fôr′ik, -for′ik, pir′ō-) *Chem.* An acid, $H_4P_2O_7$, obtained by heating orthophosphoric acid to about 255° C.

py·ro·pho·tom·e·ter (pī′rō·fō·tom′ə·tər, pir′ō-) *n.* A pyrometer used to determine high temperatures by means of the luminosity of a substance.

py·ro·phyl·lite (pī′rə·fil′īt, pir′ə-) *n.* A compact, soft, variously colored, hydrous aluminum silicate, $HAlSi_2O_6$, used in making slate pencils. [<PYRO- + PHYLL(O)- + -ITE[1]]

py·ro·sis (pī·rō′sis) *n. Pathol.* Heartburn; acid dyspepsia, accompanied by a burning sensation and belching of an acrid fluid. [<NL <Gk. *pyrōsis* a burning <*pyroein* burn <*pyr* fire]

py·ro·stat (pī′rə·stat, pir′ə-) *n.* A thermostat; specifically, one for the higher temperatures. [<PYRO- + -STAT]

py·ro·sul·fate (pī′rə·sul′fāt, pir′ə-) *n.* A salt of pyrosulfuric acid; a disulfate.

py·ro·sul·fu·ric acid (pī′rō·sul·fyŏŏr′ik, pir′ō-) *Chem.* A brown, fuming liquid, $H_2SO_4SO_3$, obtained by adding liquid sulfuric oxide to strong sulfuric acid: also called *disulfuric acid.*

py·ro·tech·nic (pī′rə·tek′nik, pir′ə-) *adj.* Pertaining to fireworks or their manufacture. Also **py′ro·tech′ni·cal.**

py·ro·tech·nics (pī′rə·tek′niks, pir′ə-) *n.* **1** The art of making or using fireworks: also **py′ro·tech′ny.** **2** A display of fireworks. **3** An ostentatious display, as of oratory. [Earlier *pyrotechny* <F *pyrotechnie* <Gk. *pyr, pyros* fire + *technē* an art; infl. in form by *pyrotechnic*] —**py′ro·tech′nist** *n.*

py·rot·ic (pī·rot′ik, pi-) *adj.* Caustic. —*n.* A caustic substance or remedy. [<NL *pyroticus* <Gk. *pyrōtikos* <*pyroein* burn <*pyr* fire]

py·ro·tox·in (pī′rə·tok′sin, pir′ə-) *n. Biochem.* Any one of a number of toxic substances found in the body as a result of bacterial action and inducing a rise of bodily temperature, or symptoms of fever.

py·rox·ene (pī′rok·sēn) *n.* **1** A monoclinic mineral, usually in short, prismatic crystals, composed principally of calcium and magnesium: next to feldspar, the most frequent component of igneous rocks. **2** Any member of the pyroxene group, as diopside and augite: they are essentially metasilicates. [<F *pyroxène* <Gk. *pyr, pyros* fire + *xenos* a stranger; because at first considered alien to igneous rocks] —**py·rox·en·ic** (pī′rok·sen′ik) *adj.*

py·rox·e·nite (pī·rok′sə·nīt) *n.* A granitoid igneous rock composed mostly of pyroxene, but without olivine. [<PYROXENE + -ITE[1]]

py·rox·y·lin (pī·rok′sə·lin) *n. Chem.* A cellulose nitrate mixture soluble in ether, alcohol, and organic solvents, less explosive than guncotton, and widely used in making Celluloid, lacquers, adhesives, etc. Also **py·rox′y·line** (-lēn, -lin). [<F *pyroxyline* <Gk. *pyr, pyros* fire + *xylon* wood + -INE[2]]

Pyr·rha (pir′ə) In Greek mythology, the daughter of Epimetheus and wife of Deucalion.

pyr·rhic[1] (pir′ik) *n.* A foot in ancient prosody composed of two short syllables. —*adj.* Of, pertaining to, or composed of pyrrhics. [<L *(pes) pyrrhicius* a pyrrhic (foot) <Gk. *(pous) pyrrichios* warlike, martial]

pyr·rhic[2] (pir′ik) *adj.* In Greek antiquity, pertaining to a martial dance in which the movements necessary to assail and avoid an enemy were imitated. —*n.* The pyrrhic dance. [<L *pyrrhicius*<Gk. *pyrrhichios* <*pyrrhichē* a war-dance <*Pyrrhichos* Pyrrhichus, a Greek said to have invented it]

Pyrrhic victory A victory gained at a ruinous loss, such as that of Pyrrhus over the Romans at Heracles Asculum, 279 B.C. [after *Pyrrhus*]

Pyr·rho·nism (pir′ə·niz′əm) *n.* A system of philosophy taught by Pyrrho of Elis, 365?–275? B.C., founder of the first and inspirer of subsequent skeptical schools of Greek philosophy; skepticism. [<L *Pyrrhoneus* pertaining to Pyrrho <Gk. *Pyrrhōn* Pyrrho]

pyr·rho·tite (pir′ə·tīt) *n.* A metallic, bronze-colored, magnetic iron sulfide, FeS; magnetic pyrites. Also **pyr′rho·lite** (-līt), **pyr′rho·tine** (-tīn). [<Gk. *pyrrhotēs* redness (<*pyrrhos* flame-colored <*pyr* a fire) + -ITE[1]]

pyr·rhu·lox·i·a (pir′ə·lok′sē·ə) *n.* A grosbeak of the western United States *(Pyrrhuloxia*

Pyrrhus

sinuata) with a slender, gray-and-red body and parrotlike bill. [<NL <*Pyrrhula*, a genus of Fringillidae (dim. <Gk. *pyrrhos* fiery < *pyr* fire) + *Loxia*, genus of the crossbills < Gk. *loxos* oblique]

PYTHAGOREAN THEOREM
Sum of squares *acde* and *bhkc* equals square *abgf* (a^2 + b^2 = ab^2)

Pyr·rhus (pir′əs), 318?–272 B.C., king of Epirus; aided Tarentum against the Romans.
Pyr·rhus (pir′əs) In Greek legend, Neoptolemus.
pyr·role (pi·rōl′, pir′ōl) *n. Chem.* A colorless, poisonous, weakly basic, liquid compound, C$_4$H$_4$NH, having an odor of chloroform, obtained from bone oil, coal tar, and by synthesis, and occurring in many natural substances, as chlorophyll and hemoglobin. Also **pyr·rol′**. [<G *pyrrol* (<Gk. *pyrros* reddish < *pyr* fire) + -ol -OLE[1]]
pyr·rol·i·dine (pi·rōl′ə·dēn, -din) *n. Chem.* A colorless nitrogenous compound, C$_4$H$_9$NO, with a mild ammonia odor, found in tobacco and carrot leaves. [<PYRROL(E) + -ID(E) + -INE[2]]
py·ru·vic acid (pī·rōō′vik, pi-) *Chem.* A colorless to pale-yellow ketone compound, C$_3$H$_4$O$_3$, obtained by the distillation of a mixture of racemic acid and potassium bisulfate. [<PYR- + L *uva* grape + -IC]
Py·thag·o·ras (pi·thag′ər·əs) Greek philosopher of the sixth century B.C. — **Py·thag′o·re′an** *adj. & n.*
Py·thag·o·re·an·ism (pi·thag′ə·rē′ən·iz′əm) *n.* The mystical philosophy taught by Pythagoras, its central idea being that number is the essence of all things and the metaphysical principle of rational order in the universe.

The leading theological doctrine was metempsychosis. [<L *Pythagoreus* <Gk. *Pythagoreios* <*Pythagoras* Pythagoras]
Pythagorean numbers See under NUMBER.
Pythagorean theorem *Math.* The theorem that the sum of the squares of the legs of a right triangle is equal to the square of the hypotenuse.
Pyth·i·a (pith′ē·ə) In ancient Greece, the priestess of the Pythian Apollo at Delphi, who was believed to be inspired by the god when seated on a tripod over the rock sacred to him, and to utter his oracles. — **Pyth′ic** *adj.*
Pyth·i·ad (pith′ē·ad) *n.* The period from one celebration of the ancient Greek Pythian games to another. [<Gk. *Pythias, -ados* <(*hiera*) *Pythia* the Pythian (games), neut. pl. of *Pythios* Pythian]
Pyth·i·an (pith′ē·ən) *adj.* 1 Relating to Delphi, to Apollo's temple there, its oracle, or priestess. 2 Relating to the Pythian games. — *n.* 1 A native or inhabitant of Delphi; specifically, the priestess of Apollo. 2 An epithet of the Delphic Apollo. [<L *Pythius* <Gk. *Pythios* <*Pytho*, older name for Delphi]
Pythian games Games held every four years in ancient Greece, of which musical contests were a feature.
Pyth·i·as (pith′ē·əs) See DAMON AND PYTHIAS. Also *Phintias*.
py·tho·gen·e·sis (pī′thō·jen′ə·sis, pith′ō-) *n.* Generation from or because of filth. [<Gk. *pythein* rot + GENESIS] — **py′tho·gen′ic, py′tho·ge·net′ic** (-jə·net′ik) *adj.*
pythogenic fever *Pathol.* 1 Typhoid fever. 2 Any fever due to filth. Also **pythogenetic fever.**
py·thon (pī′thon, -thən) *n.* 1 A large, non-venomous serpent (genus *Python*) that crushes its prey in its folds. 2

PYTHON (From 3 to 32 feet in length)

Any non-venomous serpent related to the boas. 3 A soothsayer or soothsaying spirit: from the tradition that the Python delivered oracles at Delphi; also, a ventriloquist. [<L <Gk. *Python* Python <*Pytho*. See PYTHIAN.]
Py·thon (pī′thon, -thən) In Greek mythology, a monstrous serpent which haunted the caves of Parnassus and was killed by Apollo near Delphi.
py·tho·ness (pī′thə·nis, pith′ə-) *n.* 1 The priestess of the Delphic oracle. 2 Any woman supposed to be possessed of the spirit of prophecy; a witch. [<OF *phitonise* <Med. L *phitonissa* <LL *pythonissa* <Gk. *Pytho* a familiar spirit, orig. Delphi]
py·thon·ic (pī·thon′ik, pi-) *adj.* 1 Of, pertaining to, or resembling pythons or a python. 2 Inspired; prophetic.
py·u·ri·a (pī·yoor′ē·ə) *n. Pathol.* The presence of pus in the urine. [<NL <Gk. *pyon* pus + *ouron* urine]
pyx (piks) *n.* 1 A vessel or casket, usually of precious metal, in which the Host is preserved. 2 A receptacle for coins selected for trial at the British mint: short for **pyx chest.** [<L *pyxis* a box <Gk. < *pyxos* a box tree. Doublet of BOX.]
pyx·ie[1] (pik′sē) See PIXY.
pyx·ie[2] (pik′sē) *n.* A creeping shrub (*Pyxidanthera barbulata*) with numerous solitary white or rose-colored flowers: it is the flowering moss or pixie of the pine barrens of New Jersey and North Carolina. [Prob. short for NL *Pyxidanthera*, the genus name <Gk. *pyxos* the box tree + *antheros* flowery]
pyx·is (pik′sis) *n. pl.* **pyx·i·des** (pik′sə·dēz) 1 A box or pyx; especially, an ancient form of ornamental jewel case or toilet box. 2 An emollient ointment. 3 *Bot.* A capsule or seed vessel with transverse dehiscence, the upper portion separating as a lid, as in the common purslane: also **pyx·id·i·um** (pik·sid′ē·əm). [<L. See PYX.]

Q

q, Q (kyōō) *n. pl.* **q's, Q's** or **qs, Qs, ques** (kyōōz) 1 The 17th letter of the English alphabet, from Phoenician *Q'oph* and Greek *koppa*, which was present in five eastern Greek alphabets, obsolete in the late alphabets of Elis and Athens, but survived in the Chalcidian and Boeotian, whence it passed into the Italian, or Roman Q. 2 The sound of the letter *q*. In English *q* is always followed by *u* and is pronounced *kw*, as in quack, queen, quest, quote, conquest, equal, etc. In some words borrowed from French, however, it retains its French pronunciation of *k*, as in appliqué, conquer, coquette, pique, piquant, toque. Final *-que* is always pronounced as *k*, as in antique, oblique, physique, unique, etc. See ALPHABET.
Qair·wan (kīr·wän′) See KAIROUAN.
Qan·da·har (kän·dä·här′) See KANDAHAR.
Qaz·vin (käz·vēn′) See KAZVIN.
Qa·tar (kä′tär) An independent Arab sheikdom under British protection, containing the whole **Qatar Peninsula** on the Persian Gulf coast of the Arabian Peninsula; about 8,000 miles; capital, Doha.
Qat·ta·ra Depression (kä·tä′rä) A desert basin in the Libyan Desert of northern Egypt; 7,500 square miles. Also **Qat′ta′rah.**
Q-boat (kyōō′bōt′) *n.* A merchant vessel having masked guns: used in World War I as a decoy for submarines. Also **Q′-ship.**
Qe·na (kā′nə) A city on the east bank of the Nile, Upper Egypt. Also **Qi·na** (kē′nə).
Qishm (kish′əm) The largest island in the Persian Gulf, at the entrance of the Strait of Hormuz, southern Iran; 70 by 7 to 20 miles. Also **Qeshm** (kesh′əm).
Qi·shon (kī′shon, kish′on) A river in NW

Israel, flowing 45 miles NW to the Bay of Acre.
qua (kwā, kwä) *adv.* In the capacity of; by virtue of being; in so far as. [<L, ablative sing. fem. of *qui* who]
quack[1] (kwak) *v.i.* To utter a harsh, croaking cry, as a duck. — *n.* The sound made by a duck, or a similar croaking noise. [Imit.]
quack[2] (kwak) *n.* 1 A pretender to medical knowledge or skill. 2 A charlatan. — *adj.* Of or pertaining to quacks or quackery; ignorantly or falsely pretending to cure. — *v.i.* To play the quack. [Short for QUACKSALVER] — **quack′ish** *adj.* — **quack′ish·ly** *adv.*
Synonyms (*noun*): charlatan, empiric, humbug, impostor, mountebank. Antonyms: adept, expert, master.
quack·er·y (kwak′ər·ē) *n. pl.* **·er·ies** Ignorant or fraudulent practice. Also **quack′hood, quack′ism.**
quack·grass (kwak′gras′, -gräs′) *n.* Couch-grass.
quack·sal·ver (kwak′sal′vər) *n.* A quack. [< MDu. < *quacken* quack[1] + *salf* a salve]
quad[1] (kwod) *n. Colloq.* A quadrangle of a college or prison. [Short for QUADRANGLE]
quad[2] (kwod) *n. Printing* A quadrat. [Short for QUADRAT]
quad[3] (kwod) See QUOD.
quad·ra·ge·nar·i·an (kwod′rə·jə·nâr′ē·ən) *adj.* Forty years old or relating to this age. — *n.* A person forty years old. [<L *quadragenarius* < *quadrageni* forty each < *quadraginta* forty]
Quad·ra·ges·i·ma (kwod′rə·jes′ə·mə) *n. Obs.* The forty fast days before Easter; Lent. [<L, fortieth]
quad·ra·ges·i·mal (kwod′rə·jes′ə·məl) *adj.* 1 Of or pertaining to the number forty, especially to the forty days of Lent. 2 Used during or appropriate to Lent; Lenten.
Quadragesima Sunday The first Sunday in Lent.
quad·ran·gle (kwod′rang·gəl) *n.* 1 *Geom.* A plane figure having four sides and four angles. 2 A court, square or oblong, as within a public building. 3 A tract of land as represented by the United States Geological Survey on one of its atlas sheets. [<L *quadrangulum* < *quattuor* four + *angulus* angle] — **quad·ran′gu·lar** *adj.*
quad·rant (kwod′rənt) *n.* 1 The quarter of a circle, or of its circumference. 2 An instrument having a graduated arc of 90°, with a movable radius for measuring angles on it; especially, a nautical instrument for measuring the altitude of the sun. 3 *Geom.* In a Cartesian coordinate system, any of the four sections formed by the intersection of the axes: beginning with the upper right-hand quadrant where the ordinate and abscissa are positive, they are called the **first, second, third,** and **fourth quadrants,** in counterclockwise order. See illustration on page 1030. 4 A device or machine-part having the shape of, or suggesting the quadrant of a circle. [<L *quadrans, -antis* a fourth

QUADRANT (def. 2)

quadrat | 1030 | **quake**

part < *quattuor* four] — **quad·ran·tal** (kwod·ran'təl) *adj.*

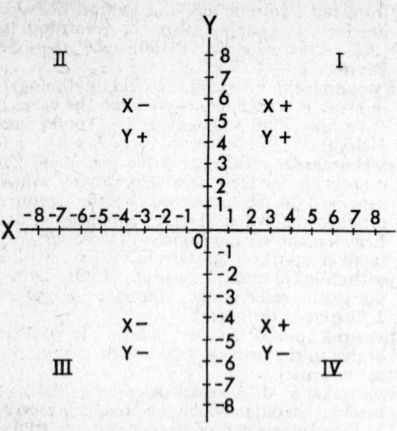

CARTESIAN QUADRANT

quad·rat (kwod'rət) *n.* 1 *Printing* A piece of type metal lower than the letters, used for spacing: abbreviated *quad.* 2 *Ecol.* A square area of varying size laid down in a plant association or formation to estimate the number of plants enclosed, or to determine the character of successional changes. [See QUADRATE.]
quad·rate (kwod'rāt, -rit) *n.* 1 *Zool.* A bone or cartilaginous element suspending the lower jaw in certain vertebrates below the mammals. 2 In astrology, an aspect of two heavenly bodies in which they are distant from each other 90°. 3 A cubical or square object, or an object resembling a cube. — *adj.* 1 Square; four-sided, as a muscle. 2 Distant from each other 90°: said of two heavenly bodies. 3 Of or pertaining to the quadrate bone or cartilage. — *v.* (-rāt) **·rat·ed, ·rat·ing** *v.i.* To correspond or agree: with *with.* — *v.t.* To cause to conform; bring in accordance with. [< L *quadratus,* pp. of *quadrare* square < *quattuor* four]
quad·rat·ic (kwod·rat'ik) *adj.* 1 Pertaining to or resembling a square. 2 Relating to a quadratic equation. — *n. Math.* 1 An equation of the second degree. It is a **pure, simple,** or **incomplete quadratic** when it contains only the second power of the variable, as $ax^2 + c = 0$; a **complete** or **adfected quadratic** when it contains also the first power, as $ax^2 + bx + c = 0$. 2 A formula, $x = \frac{-b \pm \sqrt{b^2 - 4ac}}{2a}$, for computing the roots of the standard quadratic equation, $ax^2 + bx + c = 0$. 3 *pl.* The part of algebra that treats of quadratic equations.
quad·ra·ture (kwod'rə·chər) *n.* 1 The act or process of squaring. 2 The finding in square measure of the area of any surface, especially one bounded by a curve. 3 *Astron.* **a** The relative position of two heavenly bodies that are 90° apart as viewed from the center of a third body. **b** Either intersection of an orbit with a line whose ends terminate in the curve drawn perpendicular to the major axis through the focus. **c** The moon in either of the two phases when half its disk is visible.
quad·ren·ni·al (kwod·ren'ē·əl) *adj.* 1 Occurring once in four years. 2 Comprising four years. — **quad·ren'ni·al·ly** *adv.*
quad·ren·ni·um (kwod·ren'ē·əm) *n. pl.* **·ren·ni·a** (-rēn'ē·ə) A space or period of four years. Also **quad·ren·ni·um** (kwod'rē·en'ē·əm). [< L]
quadri- *combining form* Var. of QUADRI-.
Also, before vowels, **quadr-**: also **quadru-**. [< L *quattuor* four]
quad·ric (kwod'rik) *adj. Math.* Of the second degree: applied especially where there are more than two variables. — *n.* A quantic of the second degree.
quadric curve *Math.* A curve with an algebraic Cartesian equation of the second degree.
quad·ri·cen·ten·ni·al (kwod'ri·sen·ten'ē·əl) *n.* A four-hundredth anniversary. — *adj.* Of or pertaining to such an anniversary.
quad·ri·ceps (kwod'rə·seps) *n. Anat.* The extensor of the leg. [< QUADRI- + L *caput* head] — **quad'ri·cip'i·tal** (-sip'ə·təl) *adj.*
quadric surface *Math.* A surface with an algebraic Cartesian equation of the second degree.
quad·ri·fid (kwod'rə·fid) *adj. Bot.* Four-cleft;

divided into four segments, as a flower petal.
quad·ri·ga (kwod·rī'gə) *n. pl.* **·gae** (-jē) In ancient Rome, a two-wheeled chariot to which four horses were harnessed abreast. [< L < *quattuor* four + *jugum* yoke]
quad·ri·lat·er·al (kwod'rə·lat'ər·əl) *adj.* Formed or bounded by four lines; four-sided. — *n.* 1 *Geom.* **a** A figure bounded by four straight lines terminated at four angles. **b** A figure formed of four infinite straight lines, having six intersections. 2 A space or area defended by four enclosing fortresses. [< L *quadrilaterus* < *quattuor* four + *latus, lateris* side]

QUADRILATERAL
abdc. Quadrilateral.
ad., bc. Diagonals.
ebfc. Quadrilateral.
a., b., c., d., f. Vertices.
ai., bh., ei. Diagonals.
g., hh., i. Centers.

quad·ri·lin·gual (kwod'rə·ling'gwəl) *adj.* 1 Consisting of, or knowing, four languages. 2 Written in four languages.
qua·drille[1] (kwə·dril') *n.* 1 A square dance for four couples and having five figures. 2 Music for such a dance. [< F < Sp. *cuadrilla* little square < L *quadrum* < *quattuor* four]
qua·drille[2] (kwə·dril') *n.* A card game for four persons, played with a deck of 40 cards and popular in the 18th century. [< F < Sp. *cuartillo* < *cuarto* fourth < L *quartus*]
quadrille paper See GRAPH PAPER.
quad·ril·lion (kwod·ril'yən) *n.* 1 In the French and United States system of numeration, a thousand trillions, or 1 followed by 15 ciphers. 2 In the English system, a million trillions, or 1 followed by 24 ciphers. [< F < *quatre* four + (m)*illion*]
quad·ri·mes·ter (kwod'rə·mes'tər) *n.* A period of four months; one third of a year. [< L *quadrimestris* of four months] — **quad'ri·mes'tral** *adj.*
quad·ri·no·mi·al (kwod'rə·nō'mē·əl) *n. Math.* An algebraic expression having four terms.
quad·ri·par·tite (kwod'rə·pär'tīt) *adj.* 1 Consisting of or embracing four parts. 2 Having four parties, as an agreement or contract. [< L *quadripartitus* < *quattuor* four + *partitus* divided] — **quad·ri·par·ti·tion** (-tish'ən) *n.*
quad·ri·syl·la·ble (kwod'rə·sil'ə·bəl) *n.* A word of four syllables. — **quad·ri·syl·lab·ic** (-si·lab'ik) *adj.*
quad·ri·va·lent (kwod'rə·vā'lənt) *adj. Chem.* Having a valence of four, as carbon; tetravalent. [< QUADRI- + L *valens, -entis,* ppr. of *valere* be worth] — **quad'ri·va'lence** or **·len·cy** *n.*
quad·riv·i·al (kwod·riv'ē·əl) *adj.* 1 Having four radiating ways. 2 Leading to or going in four directions: *quadrivial* streets. [< L *quadrivius* < *quattuor* four + *via* way]
quad·riv·i·um (kwod·riv'ē·əm) *n. pl.* **·i·a** (-ē·ə) In the Pythagorean system, the four sciences, geometry, astronomy, arithmetic, and music, making up with the trivium the seven liberal arts. Compare TRIVIUM. [< L, a place where four roads meet]
quad·roon (kwod·roon') *n.* A person having one-fourth Negro and three-fourths white blood. [< Sp. *cuarteron* < *cuarto* fourth]
quadru- Var. of QUADRI-.
quad·ru·ma·nous (kwod·roo'mə·nəs) *adj.* 1 Four-handed; having four feet resembling hands. 2 Of or pertaining to a former order (*Quadrumana*) of mammals now classed with man under the primates. [< QUADRU- + L *manus* hand] — **quad·ru·mane** (kwod'roo·mān) *n.*
quad·ru·ped (kwod'roo·ped) *n.* An animal having four feet; especially, a four-footed mammal. — *adj.* Having four feet. [< L *quadrupes, -pedis* < *quattuor* four + *pes* foot] — **quad·ru·pe·dal** (kwod·roo'pə·dəl, kwod'roo·ped'l) *adj.*
quad·ru·ple (kwod'roo·pəl, kwod·roo'pəl) *adj.* Consisting of four; having four parts or members; fourfold; also, taken by fours. — *n.* A number or sum four times as great as another. — *v.t.* & *v.i.* **·pled, ·pling** To multiply by four; make or become four times larger. — *adv.* Fourfold. [< L *quadruplus*]

quad·ru·plet (kwod'roo·plit, kwod·roo'-) *n.* 1 A compound or combination of four things or objects. 2 One of four offspring born of the same mother at one birth.
quadruple time *Music* Measure or time having four beats; four-two, four-four, or four-eight time.
quad·ru·plex (kwod'roo·pleks, kwod·roo'-) *adj.* 1 Fourfold. 2 Pertaining to or designating a telegraph system such that four messages, two in each direction, may be sent simultaneously over one wire. — *n.* A sending instrument used in quadruplex telegraphy. [< L < *quattuor* four + stem of *plicare* fold]
quad·ru·pli·cate (kwod·roo'plə·kit, -kāt) *adj.* 1 Fourfold. 2 Raised to the fourth power. — *v.t.* (-kāt) **·cat·ed, ·cat·ing** To multiply by four; quadruple. — *n.* One of four like things. — **quad·ru'pli·ca'tion** *n.* — **quad·ru'pli·cate·ly** *adv.*
quae·re (kwē'rē) *n.* Literally, seek; inquire: an annotation inserted, as in law reports. [< L. See QUERY.]
quaes·tor (kwes'tər, kwēs'-) *n.* Any of a number of public officials in ancient Rome; originally, one of two magistrates who inquired into and punished capital crimes; later, one who took charge of the public treasury and expenditure: also spelled *questor.* [< L < *quaerere* seek; inquire] — **quaes·to·ri·al** (kwes·tôr'ē·əl, -tō'rē-, kwēs'-) *adj.* — **quaes'tor·ship** *n.*
quaff (kwaf, kwof, kwôf) *v.t.* & *v.i.* To drink, especially copiously or with relish. — *n.* A drink; swallow. [< earlier *quaft,* ? blend of QUENCH and DRAUGHT] — **quaff'er** *n.*
quag (kwag, kwog) *n.* A quagmire. [< obs. *quag, v.,* blend of QUAKE and SAG]
quag·ga (kwag'ə) *n.*
1 A South African equine mammal (*Equus quagga*) intermediate between the ass and the zebra and resembling the latter: now extinct. 2 A zebra: an erroneous use. [< native Hottentot name]

QUAGGA
(About 11 hands high at the withers)

quag·gy (kwag'ē, kwog'ē) *adj.* Yielding to or quaking under the foot, as soft, wet earth; boggy.
quag·mire (kwag'mīr', kwog'-) *n.* 1 Marshy ground that gives way under the foot; bog. 2 A difficult situation. [< QUAG + MIRE] — **quag'mired', quag'mir'y** *adj.*
qua·haug (kwô'hôg, -hog, kwə·hôg', -hog') *n.* The round, thick-shelled clam (*Venus mercenaria*) of the Atlantic coast of North America. The young are called cherrystone clams. Also **qua'hog**: sometimes spelled *cohog, quohog.* [< Algonquian (Narraganset) *poquauhock*]
Quai d'Or·say (kā dôr·sā') 1 A quay on the left bank of the Seine in Paris, toward which the French Foreign Office faces. 2 The French Foreign Office.
quaigh (kwäkh) *n. Scot.* A small cup or drinking vessel. Also **quaich.**
quail[1] (kwāl) *n.* 1 An Old World migratory game bird (*Coturnix coturnix*) similar to the partridge, having a very short tail. 2 Any of various small American game birds related to the partridge (family *Perdicadae*), especially the bobwhite and the California quail (*Lophortyx californica*). See BOBWHITE. 3 *Obs.* A prostitute. [< OF *quaille,* prob. < Gmc.]
quail[2] (kwāl) *v.i.* To shrink with fear; lose heart or courage. [ME *quailen*; origin uncertain]
quaint (kwānt) *adj.* 1 Combining an antique appearance with a pleasing oddity, fancifulness, or whimsicalness. 2 Hence, pleasingly odd or old-fashioned; fanciful. 3 *Obs.* Curiously wrought; hence, ornamental. 4 *Obs.* Crafty. See synonyms under ANTIQUE, ODD, QUEER. [< OF *cointe* < L *cognitus* known] — **quaint'ly** *adv.* — **quaint'ness** *n.*
quake (kwāk) *v.i.* **quaked, quak·ing** 1 To shake, as with violent emotion or cold; quiver. 2 To shake or tremble, as earth during an earthquake. — *n.* A shaking, tremulous motion, quickly repeated; a shaking or shuddering. [OE *cwacian* shake]
Synonyms (verb): quaver, quiver, shake, shiver, shudder, tremble, vibrate, waver. See SHAKE.

Quak·er (kwā′kər) *n.* A member of the Society of Friends: originally a term of derision, and still not used within the society. See SOCIETY OF FRIENDS. [<QUAKE, *v.*; with ref. to their founder's admonition to them to tremble at the word of the Lord] **— Quak′er·ess** *n. fem.* **— Quak′er·ish** *adj.* **— Quak′er·ish·ly** *adv.*

quaker buttons The dried, ripe seeds of nux vomica.

Quaker City A nickname of PHILADELPHIA.

Quaker gun A dummy gun, as one made of wood: from the Friends' doctrine of non-resistance.

Quak·er·ism (kwā′kə·riz′əm) *n.* The beliefs or practices of the Quakers.

quaker ladies Bluets.

Quak·er·ly (kwā′kər·lē) *adj.* Like the Quakers. *— adv.* After the manner of the Quakers.

Quaker meeting **1** Any meeting of the Society of Friends for worship, in which, following their usage, they remain silent until the Spirit moves some member to speak or pray aloud. **2** Hence, any gathering at which silence prevails.

Quaker State Nickname of PENNSYLVANIA.

quak·y (kwā′kē) *adj.* **quak·i·er, quak·i·est** Shaky; tremulous. **— quak′i·ly** *adv.* **— quak′i·ness** *n.*

qual·i·fi·ca·tion (kwol′ə·fə·kā′shən) *n.* **1** The act of qualifying, or the state of being qualified. **2** That which fits a person or thing for something. **3** A restriction; mitigation. See synonyms under ABILITY.

qual·i·fied (kwol′ə·fīd) *adj.* **1** Competent or fit, as for public office. **2** Restricted or modified in some way. See synonyms under COMPETENT. **— qual′i·fied′ly** *adv.*

qual·i·fy (kwol′ə·fī) *v.* **·fied, ·fy·ing** *v.t.* **1** To make fit or capable, as for an office, occupation, or privilege. **2** To make legally capable, as by the administration of an oath. **3** To limit or restrict, as by conditions or exceptions. **4** To attribute a quality to; describe; characterize or name. **5** To make less strong or extreme; soften; moderate. **6** To change the strength or flavor of. **7** *Gram.* To modify. *— v.i.* **8** To be or become qualified or fit; meet the requirements, as for entering a race. See synonyms under CHANGE. [<MF *qualifier* <Med. L *qualificare* <L *qualis* of such a kind + *facere* make] **— qual′i·fi′a·ble** *adj.* **— qual′i·fi′er** *n.*

qual·i·ta·tive (kwol′ə·tā′tiv) *adj.* Of or pertaining to quality: distinguished from *quantitative.* [<LL *qualitativus* <L *qualitas* quality] **— qual′i·ta′tive·ly** *adv.*

qualitative analysis *Chem.* The process of finding how many and what elements or ingredients are present in a substance or compound.

qual·i·ty (kwol′ə·tē) *n. pl.* **·ties** **1** That which makes a being or thing such as it is; a distinguishing element or characteristic. **2** The characteristics of anything regarded as determining its value, place, worth, rank, position, etc., or the condition of a thing as so determined; character; kind; when unqualified, peculiar excellence. **3** A moral trait or characteristic. **4** Degree of excellence; relative goodness; grade: high *quality* of fabric. **5** Capability of producing specific effects. **6** Particular character or part; capacity; function. **7** *Archaic* Social rank; persons of rank, collectively. **8** *Music* That which distinguishes sounds of the same pitch and intensity from different sources, as from different instruments; timbre. **9** *Logic* The character of a proposition or judgment as asserting or denying. **10** *Philos.* An essential property or attribute. **11** *Phonet.* The character of a vowel sound as determined by the resonance of the oral cavity. See synonyms under ATTRIBUTE, CHARACTERISTIC. [<L *qualitas, -tatis* < *qualis* of such a kind]

qualm (kwäm, kwôm) *n.* **1** A feeling of sickness. **2** A twinge of conscience; moral scruple. **3** A sensation of fear or misgiving. [OE *cwealm* death]

qualm·ish (kwä′mish, kwô′-) *adj.* **1** Feeling or affected with qualms. **2** Likely to produce qualms. See synonyms under SQUEAMISH. Also **qualm′y.** **— qualm′ish·ly** *adv.* **— qualm′ish·ness** *n.*

quam·ash (kwom′ash, kwə·mash′) *n.* Camas.

quan·da·ry (kwon′dər·ē, -drē) *n. pl.* **·da·ries** A state of hesitation or perplexity; predicament. [Origin uncertain]

quand même (käṅ mem′) *French* Notwithstanding; even though; nevertheless.

quan·dong (kwon′dong) *n.* **1** A small Australian tree of the sandalwood family (*Fusanus acuminatus*). **2** Its edible drupaceous fruit, used as a preserve. Also **quan′dang.** [< native Australian name]

quant (kwant, kwont) *n. Brit.* A punting pole with a flange at the end to prevent its sinking in the mud. *— v.t. & v.i.* To propel or be propelled with a quant. [? <L *contus* a boat pole]

quan·ta (kwon′tə) Plural of QUANTUM.

quan·tic (kwon′tik) *n. Math.* A rational homogeneous function of two or more variables, usually containing only positive integers.

Quan·ti·co (kwon′ti·kō) A town on the Potomac in northern Virginia; site of a United States Marine Corps base.

quan·ti·fy (kwon′tə·fī) *v.t.* **·fied, ·fy·ing** **1** To determine the quantity of. **2** *Logic* To express the quantity of explicitly, as by using *all, some,* or *none.* [<Med. L *quantificare* <L *quantus* how great + *facere* make] **— quan′ti·fi·ca′tion** *n.* **— quan′ti·fi′er** *n.*

quan·tim·e·ter (kwon·tim′ə·tər) *n. Med.* A dosimeter.

quan·ti·ta·tive (kwon′tə·tā′tiv) *adj.* **1** Of or pertaining to quantity. **2** Having to do with quantities only: distinguished from *qualitative.* [<LL *quantitativus* <L *quantitas* quantity] **— quan′ti·ta′tive·ly** *adv.* **— quan′ti·ta′tive·ness** *n.*

quantitative analysis *Chem.* The process of finding the amount or percentage of each element or ingredient present in a material or compound.

quan·ti·ty (kwon′tə·tē) *n. pl.* **·ties** **1** The condition of being much. **2** That property of a thing which admits of exact measurement and numerical statement. **3** An object regarded as possessing a certain determinable magnitude, as of length, size, mass, volume, or number. **4** *Electr.* The strength of a current, as opposed to intensity or potential. **5** In prosody, the relative period of time, regarded as short or long, required to pronounce a syllable. **6** *Music* The duration of a musical note. **7** A specified, or indefinite, number of persons or things. **8** *Logic* The extent of a general term or proposition as applying to the whole or to a part of a class. Considered with reference to quantity, propositions are *universal,* as "all men are mortal," and *particular,* as "some men are honest," while with reference to conceptions quantity relates either to their extension, or to their intension or comprehension. **9** Considerable bulk or amount. [<OF *quantité* <L *quantitas, -tatis* < *quantus* how much, how large]

quan·tize (kwon′tīz) *v.t.* **·tized, ·tiz·ing** *Physics* To express (an energy relationship) in terms of quanta or in accordance with the quantum theory. **— quan′ti·za′tion** *n.*

Quan·trill (kwon′tril), **William Clarke,** 1837–1865, American Confederate guerrilla commander.

quan·tum (kwon′təm) *n. pl.* **·ta** (-tə) **1** An object that has quantity or is concrete. **2** A certain amount; also, a prescribed or a sufficient quantity. **3** *Physics* A fundamental unit of energy or action as provided for in the quantum theory. [<L, neuter of *quantus* how much]

quantum liquid *Physics* Helium in the superfluid condition: so called because it confirms the quantum theory that even at temperature of absolute zero molecular motion does not completely cease.

quantum number *Physics* A number indicating any of the energy levels possible in an atom under specified conditions.

quantum state Energy level.

quantum theory *Physics* The theory that energy is not a smoothly flowing continuum but is manifested by the emission from radiating bodies of discrete particles or *quanta,* the values of which are expressed as the product of Planck's constant, *h,* and the frequency, *ν,* of the given radiation.

Qua·paw (kwä′pô) *n.* One of a tribe of North American Indians of Siouan stock, formerly living in Arkansas, now in Oklahoma: also called *Arkansas.*

quar·an·tine (kwôr′ən·tēn, kwor′-) *n.* **1** The enforced isolation for a fixed period of persons, ships, or goods arriving from places infected with contagious disease, or of any persons who have been exposed to such infection. **2** A place designated for the enforcement of such interdiction. **3** The enforced isolation of any person or place infected with contagious disease; loosely, any enforced isolation. **4** A period of forty days. *— v.t.* **·tined, ·tin·ing** To subject to or retain in quarantine; isolate by or as by quarantine. [<Ital. *quarantina* <L *quadraginta* forty]

Quarles (kwôrlz, kworlz), **Francis,** 1592–1644, English poet.

Quar·ne·ro (kwär·ne′rō), **Gulf of** See VELIKI KVARNER.

quar·rel[1] (kwôr′əl, kwor′-) *n.* **1** An unfriendly, angry, or violent dispute. **2** A falling out or contention; breach of amity. **3** The cause for dispute. *— v.i.* **·reled** or **·relled, ·rel·ing** or **·rel·ling** **1** To engage in a quarrel; dispute; contend; fight. **2** To break off a mutual friendship; fall out; disagree. **3** To find fault; cavil. [<F *querelle* <L *querela* complaint] **— quar′rel·er** or **quar′rel·ler** *n.*

Synonyms (noun): affray, altercation, bickering, brawl, breach, broil, contention, contest, controversy, disagreement, discussion, dispute, dissension, feud, fracas, fray, fuss, jangle, jar, misunderstanding, quarreling, rupture, scene, squabble, strife, tumult, variance, wrangle. A *quarrel* is in word or act, or both, and is often slight and transient, as we speak of childish *quarrels*; but *quarrel* may denote the cause or ground of *contention* or *strife,* and so be deep and enduring. *Contention* and *strife* may be in word or deed; *contest* ordinarily involves some form of action. *Controversy* is commonly in words; *strife* extends from verbal *controversy* to the *contests* of armies. An *affray* or *broil* may arise at a street corner; the *affray* always involves physical force; the *brawl* or *broil* may be confined to violent language. See ALTERCATION, FEUD[1]. *Antonyms*: accord, amity, concord, harmony, pacification, peace, reconcilement, reconciliation.

quar·rel[2] (kwôr′əl, kwor′-) *n.* **1** A dart or arrow with a four-edged head, formerly used with a crossbow. **2** A graver, stonemason's chisel, glazier's diamond, or other tool having a several-edged point. [<OF <LL *quadrellus,* dim. of L *quadrum* a square < *quattuor* four]

quar·rel·some (kwôr′əl·səm, kwor′-) *adj.* Inclined to quarrel; contentious. **— quar′rel·some·ly** *adv.* **— quar′rel·some·ness** *n.*

quar·ri·er (kwôr′ē·ər, kwor′-) *n.* A workman in a stone quarry.

quar·ry[1] (kwôr′ē, kwor′ē) *n. pl.* **·ries** **1** A beast or bird hunted, seized, or killed, as in the chase; game; prey: now chiefly poetical. **2** Anything hunted, slaughtered, or eagerly pursued. **3** *Obs.* A heap of slaughtered game. [<OF *cuirée* <L *corium* hide]

quar·ry[2] (kwôr′ē, kwor′ē) *n. pl.* **·ries** An excavation from which stone is taken by cutting, blasting, or the like. *— v.t.* **·ried, ·ry·ing** **1** To cut, dig, or take from or as from a quarry. **2** To establish a quarry in. [<Med. L *quareia, quareria* <LL *quadraria* place for squaring stone < *quadrare.* See QUADRATE.]

quar·ry[3] (kwôr′ē, kwor′ē) *n. pl.* **·ries** **1** A square or lozenge. **2** A small square or lozenge-shaped pane of glass, tile, etc. **3** In archery, a quarrel. [<OF *quarré* <L *quadratus.* See QUADRATE.]

quart[1] (kwôrt) *n.* **1** A measure of capacity; the fourth part of a gallon, or two pints. In the United States, the dry quart is equal to 1.10 liters and the liquid quart is equal to 0.946 liter. **2** A vessel of such capacity. [<F *quarte* <L *quartus* fourth]

quart[2] (kärt) *n.* **1** In fencing, a quarte. **2** In piquet, a sequence of four cards of the same suit: called **quart major** if they are the highest four. [<F *quarte.* See QUART[1]]

quar·tan (kwôr′tən) *adj.* Pertaining to the fourth in a series; especially, occurring every fourth day. *— n. Pathol.* A malarial fever caused by the parasite *Plasmodium malariae,* in which the paroxysms recur every fourth

quarte 1032 **quasi-**

day, or 72 hours, reckoning inclusively. [<F *quartaine* <L *quartanus* < *quartus* fourth]
quarte (kärt, *Fr.* kȧrt) *n.* In fencing, a thrust or parry: the fourth regular position: also spelled *carte*. [<F]
quar·ter (kwôr′tər) *n.* 1 One of four equal parts into which anything is or may be divided; a fourth part; specifically, the fourth of a hundredweight; eight bushels; a fourth of a ton (of grain); the fourth of a yard, or a span; a fourth of a pound; a fourth of a mile; fifteen minutes or the fourth of an hour, or the moment with which it begins or ends. 2 A fourth of a year or three months; hence, a term of school. 3 A limb of a quadruped with the adjacent parts; also, a haunch of venison. 4 In the United States and Canada, a coin of the value of 25 cents. 5 *Astron.* Either of two phases of the moon: the first quarter, between the new and full moon; or the last quarter, between the full moon and the new. 6 *Music* A quarter note. 7 *Nav.* One of the four principal points of the compass or divisions of the horizon; also, a point or direction of the compass. 8 The place, origin, or source from which anything comes. 9 A particular division or district; a locality. 10 Usually *pl.* Proper or assigned station, position, or place, as of officers and crew on a warship. 11 *pl.* A place of lodging or residence, especially temporary shelter; specifically, a group of cabins provided for the Negroes on a Southern plantation. 12 A region embracing one fourth, or about one fourth, of a space; one of four corresponding localities or parts. 13 The side of a horse's hoof, just in front of the heel; also that part of a boot or shoe from the middle of the heel to the line of the ankle bone. 14 *Naut.* a The upper part of a vessel's side from the after part of the main chains to the stern. b That part of a yard outside the slings. 15 *Her.* Any of four equal divisions into which a shield is divided, or an ordinary occupying such a division. 16 Mercy shown to a vanquished foeman by sparing his life; clemency. 17 One of the four periods into which a game, as football, is divided. — **at close quarters** Close by; at close range. — *adj.* 1 Being one of four equal parts. 2 Having one fourth of a standard value. — *v.t.* 1 To divide into four equal parts. 2 To divide into a number of parts or pieces. 3 To cut the body of (an executed person) into four parts: He was hanged, drawn, and *quartered*. 4 To range from one side to the other of (a field, etc.) while advancing: The dogs *quartered* the field. 5 To furnish with quarters or shelter; lodge, station, or billet. 6 *Her.* a To divide (a shield) into quarters by vertical and horizontal lines. b To bear or arrange (different coats of arms) quarterly upon a shield or escutcheon. 7 *Mech.* To mark or place at intervals of a quarter, especially of a quarter of a circle. — *v.i.* 8 To be stationed or lodged. 9 To range from side to side of an area, as dogs in hunting. 10 *Naut.* To blow on a ship's quarter: said of the wind. [<OF <L *quartarius* < *quartus* fourth]
quar·ter·age (kwôr′tər·ij) *n.* 1 A quarterly allowance or payment. 2 Board and lodging; quarters, especially for troops, a work gang, etc.; also, the cost of lodging or shelter.
quar·ter·back (kwôr′tər·bak′) *n.* In American football, one of the backfield, who often calls the signals.
quar·ter·crack (kwôr′tər·krak′) *n.* A crack on the inner quarter of a horse's forehoof. Compare SANDCRACK.
quar·ter·day (kwôr′tər·dā′) *n.* One fourth of a day.
quarter day Any of the days of the year when quarterly payments are due. Quarter days for the U.S. government are the first days of January, April, July, and October; for England, Lady Day (March 25), Midsummer Day (June 24), Michaelmas (September 29), and Christmas (December 25).
quar·ter·deck (kwôr′tər·dek′) *n. Naut.* The rear part of a ship's upper deck, reserved for officers.
quar·tered (kwôr′tərd) *adj.* 1 Divided into four quarters. 2 *Her.* Divided into quarterings. 3 Quarter-sawed: *quartered* oak. 4 Lodged; stationed; also, having quarters.
quar·ter·fi·nal (kwôr′tər·fī′nəl) *n.* A competition immediately preceding the semifinal in sporting events; also, one of four competitions in a tournament, the winners of which play in the two semifinals. — *adj.* Next to the semifinal. — **quar′ter·fi′nal·ist** *n.*
quar·ter·foil (kwôr′tər·foil′) See QUATREFOIL.
Quarter horse A breed of horse descendent from the thoroughbred stallion *Janus* imported from England in 1756: first known as a racing breed, now widely popular as a ranch horse and cow pony. [From the quarter-of-a-mile path over which it was raced by the early settlers of Virginia]
quar·ter·hour (kwôr′tər·our′) *n.* Fifteen minutes. — **quar′ter·hour′ly** *adj.*
quar·ter·ing (kwôr′tər·ing) *adj.* 1 *Naut.* a Blowing against or being on the quarter. b Blowing from any point between beam and stern: a *quartering* wind. c Sailing so as to have the wind on the quarter. 2 Set or being at right angles. — *n.* 1 A dividing or marking off into quarters. 2 *Her.* a The grouping of two or more coats of arms in compartments on one shield, to indicate family alliances, etc. b Any of the coats which are quartered on the shield, or the quarter containing it. 3 Quarters, or the assigning of quarters, as for soldiers.

QUARTERING

quar·ter·ly (kwôr′tər·lē) *adj.* 1 Containing or being a fourth part. 2 Occurring at intervals of three months. — *n. pl.* **·lies** A publication issued once every three months. — *adv.* 1 Once in a quarter of a year. 2 In or by quarters.
quar·ter·mas·ter (kwôr′tər·mas′tər, -mäs′-) *n.* 1 Usually *cap.* The officer on an Army post who is responsible for carrying out the functions required of the Quartermaster Corps. 2 On shipboard, a petty officer who assists the navigator.
Quartermaster Corps A branch of the U.S. Army which is responsible for the supply of food, fuel, clothing, and related items of equipment.
Quartermaster General In the U.S. Army, the major general who is at the head of the Quartermaster Corps.
quar·tern (kwôr′tərn) *n.* 1 A fourth part of certain measures or weights, as of a peck or pound; a gill. 2 A four-pound loaf of bread. [<OF *quarteron* < *quarte* a fourth part <L *quartus* fourth]
quar·ter·ni·on (kwôr·tûr′nē·ən) *n. Printing* A gathering of four sheets, each folded into pages, usually four to a sheet, to make a section of a book, pamphlet, etc.
quarter note *Music* A note having one fourth the value of a semibreve. See illustration under NOTE.
quar·ter·phase (kwôr′tər·fāz′) *adj. Electr.* Diphase.
quar·ter·sawed (kwôr′tər·sôd′) *adj.* Sawed lengthwise into quarters, as a log, or sawed from quartered timber.
quar·ter·sec·tion (kwôr′tər·sek′shən) *n.* In the system of land surveying adopted by the governments of the United States and Canada, a tract of land half a mile square, containing one fourth of a square mile, or 160 acres.
quar·ter·ses·sions (kwôr′tər·sesh′ənz) *n.* A court held quarterly. In England and Scotland it tries many indictable offenses, hears appeals from the petty sessions, and exercises a minor civil jurisdiction.
quar·ter·staff (kwôr′tər·staf′, -stäf′) *n. pl.* **·staves** (-stāvz′) A stout, iron-tipped staff about 6 1/2 feet long, formerly used in England as a weapon; also, the use of, or exercise with, the quarterstaff.
quar·ter·tone (kwôr′tər·tōn′) *n.* 1 In photoengraving, a coarse zinc halftone plate having 65 lines or less to the inch: used in newspaper printing. 2 *Music* Half of a semitone; also, any interval less than a semitone: also **quarter tone**.
quar·tet (kwôr·tet′) *n.* 1 A composition for four voices or instruments. 2 The set of four persons who render such compositions. 3 A stanza of four lines. 4 Any group or set of four things of a kind. Also **quar·tette**′. [< Ital. *quartetto* < *quarto* fourth]
quar·tic (kwôr′tik) *Math. adj.* Denoting a quantic function of the fourth degree. — *n.* Such a function.
quar·tile (kwôr′tīl, -til) *n.* 1 In astrology, a quadrate. 2 *Stat.* That portion of a frequency distribution which comprises an exact fourth of the total observed cases. — *adj.* Of or pertaining to a quartile. [<LL *quartilis* <L *quartus* fourth]
quar·to (kwôr′tō) *adj.* Having four leaves or eight pages to the sheet: a *quarto* book. — *n. pl.* **·tos** A book or pamphlet whose pages are of the size of the fourth of a sheet: often written *4to* or 4°. [<L (*in*) *quarto* (in) fourth]
quartz (kwôrts) *n.* Silicon dioxide, SiO_2, a hard, vitreous, widely distributed mineral occurring in many varieties, sometimes massive, as jasper and chalcedony: sometimes in colorless and transparent or diversely colored forms crystallizing in the hexagonal system. [<G *quarz*]
quartz crystal A thin section of pure quartz, accurately ground so as to vibrate at the required frequency in radio transmission; a piezoelectric oscillator. Also **quartz plate**.
quartz·if·er·ous (kwôrt·sif′ər·əs) *adj.* Consisting of or containing quartz.
quartz·ite (kwôrt′sīt) *n.* A massive or schistose metamorphic rock formed by the induration of sandstone through the deposition of secondary quartz about each grain.
quartz lamp A mercury-vapor lamp enclosed in a quartz tube, which transmits ultraviolet wavelengths.
quash[1] (kwosh) *v.t. Law* To make void or set aside, as an indictment; annul. See synonyms under ANNUL, CANCEL. [<OF *quasser* <LL *cassare* < *cassus* empty]
quash[2] (kwosh) *v.t.* To put down or suppress forcibly or summarily. [<OF *quasser* <L *quassare*, freq. of *quatere* shake]
quasi- *prefix* 1 (With nouns) Resembling; not genuine, as in:

quasi-accident	quasi-injury
quasi-adult	quasi-insight
quasi-approval	quasi-integrity
quasi-artist	quasi-invasion
quasi-attack	quasi-kindred
quasi-authority	quasi-lament
quasi-bargain	quasi-luxury
quasi-blunder	quasi-market
quasi-certificate	quasi-method
quasi-characteristic	quasi-miracle
quasi-comprehension	quasi-neutrality
quasi-conquest	quasi-owner
quasi-conservative	quasi-pleasure
quasi-consultation	quasi-poem
quasi-dependence	quasi-protection
quasi-despair	quasi-purity
quasi-development	quasi-reality
quasi-difference	quasi-recreation
quasi-distress	quasi-refusal
quasi-endorsement	quasi-remedy
quasi-escape	quasi-repair
quasi-faith	quasi-scholar
quasi-farmer	quasi-tradition
quasi-friend	quasi-triumph
quasi-guarantee	quasi-victory
quasi-handicap	quasi-worship
quasi-illness	quasi-zeal

2 (With adjectives) Nearly; almost, as in:

quasi-absolute	quasi-grateful
quasi-amiable	quasi-hereditary
quasi-beneficial	quasi-human
quasi-classic	quasi-humorous
quasi-colloquial	quasi-important
quasi-comic	quasi-infinite
quasi-complex	quasi-internal
quasi-conservative	quasi-jocose
quasi-continuous	quasi-medical
quasi-converted	quasi-natural
quasi-devoted	quasi-normal
quasi-eligible	quasi-official
quasi-equal	quasi-practical
quasi-evil	quasi-private
quasi-exempt	quasi-probable
quasi-explicit	quasi-righteous
quasi-financial	quasi-similar
quasi-forgotten	quasi-spiritual
quasi-formidable	quasi-stylish
quasi-genteel	quasi-sufficient

quasi–tangible
quasi–theatrical
quasi–typical
quasi–valid
quasi–vital
quasi–willing

3 *Law* Superficially resembling but intrinsically different, as in:

quasi–corporation
quasi–delict
quasi–deposit
quasi–entail
quasi–legislative
quasi–partner

[<L, as if]

qua·si–con·tract (kwā'sī·kon'trakt, -zī-, kwä'sē-) *n.* An obligation to do something, enforceable by a contract remedy, but imposed by operation of law regardless of the consent of the defendant.

qua·si–ju·di·cial (kwā'sī·jōō·dish'əl, -zī-, kwä'sē-) *adj.* Exercising functions of a judicial nature as a guide for official action, as a committee investigating facts and drawing conclusions from them.

quas·si·a (kwosh'ē·ə, kwosh'ə) *n.* **1** The wood of either of two tropical American trees (*Picrasma excelsa* or *Quassia amara*). **2** The bitter principle of this wood, used in medicine as a tonic and anthelmintic. **3** The tree itself. [<NL, after Graman *Quassi*, a Surinam Negro who discovered its use in 1730]

quas·sin (kwos'in, kwas'-) *n. Chem.* A white, crystalline, intensely bitter amaroid, $C_{22}H_{30}O_6$, contained in quassia wood.

quatch·grass (kwoch'gras', -gräs') *n.* Couchgrass.

quate (kwāt) *adj. Scot.* Quiet.

qua·ter·na·ry (kwə·tûr'nə·rē) *adj.* **1** Consisting of four things. **2** Fourth in order. —*n. pl.* **·ries 1** The number four; a group of four things. **2** *Math.* A quantic function having four variables. [<L *quaternarius* < *quaterni* by fours]

Qua·ter·na·ry (kwə·tûr'nə·rē) *adj. Geol.* Of, pertaining to, or designating a geological period and system of the Cenozoic era, following the Tertiary and still continuing. —*n.* The Quaternary system or period.

qua·ter·ni·on (kwə·tûr'nē·ən) *n.* **1** A set, system, or file of four. **2** *Math.* **a** An operator or factor that changes one vector into another: so called because expressible as the sum of four quantities. **b** The form of the calculus of vectors based on and making use of the quaternion operator. [<LL *quaternio, -onis* < *quattuor* four]

Quath·lam·ba (kwät·läm'bä) See DRAKENSBERG.

qua·torze (kə·tôrz') *n.* In piquet, four cards of the same denomination, higher than a nine, held in one hand and counting 14 points. [<F, *fourteen*]

quat·rain (kwot'rān) *n.* A stanza of four lines. [<F < *quatre* four]

qua·tre (kä'tər, Fr. kà'tr') *n. French* **1** Anything, as a card or a domino, marked with four spots or pips. **2** The number four; four.

Qua·tre–Bras (kä'tr'·brä') A village in Belgium, SE of Brussels; scene of an English victory by Wellington over French forces under Marshal Ney in the Waterloo campaign of the Napoleonic Wars, 1815.

quat·re·foil (kat'ər·foil', kat'rə-) *n.* **1** *Bot.* A leaf or flower with four leaflets or petals. **2** *Archit.* An ornament with four foils or lobes. Sometimes spelled *quarterfoil*. [<OF *quatre* four + *foil* leaf]

quat·tro·cen·to (kwät'trō·chen'tō) *n.* The 15th century as connected with the revival of art and literature (especially in Italy). —*adj.* Of or pertaining to the quattrocento. [<Ital., four hundred < *quattro* four + *cento* hundred]

qua·ver (kwā'vər) *v.i.* **1** To tremble or shake: said usually of the voice. **2** To produce trills or quavers in singing or in playing a musical instrument. —*v.t.* **3** To utter or sing in a tremulous voice. See synonyms under QUAKE, SHAKE. —*n.* **1** A quivering or tremulous motion. **2** A shake or trill, as in singing. **3** An eighth note. [Freq. of

QUATREFOILS

obs. *quave*, ME *cwafian* tremble] —**qua'ver·y** *adj.*

quay (kē) *n.* A wharf or artificial landing place where vessels unload. ◆ Homophone: *key.* [<F]

quay·age (kē'ij) *n.* **1** Wharfage; quay dues. **2** Space for quays; quays collectively.

quean (kwēn) *n.* **1** A brazen or ill-behaved woman; harlot; prostitute. **2** *Scot.* A young or unmarried woman; a girl. [OE *cwene* prostitute]

quea·sy (kwē'zē) *adj.* **·si·er, ·si·est 1** Sick at the stomach. **2** Nauseating; also, caused by nausea. **3** Easily nauseated; hence, fastidious; squeamish. **4** Requiring to be carefully treated; delicate; ticklish. **5** Uncertain; hazardous. [Cf. Norw. *kveis* nausea] —**quea'si·ly** *adv.* —**quea'si·ness** *n.*

Que·bec (kwi·bek') **1** A province in eastern Canada; 523,860 square miles; formerly *Lower Canada:* abbr. *Que.* or *P.Q.* **2** Its capital, a port on the St. Lawrence River; captured from the French under Montcalm by Wolfe, Sept. 13, 1759.

que·bra·cho (kā·brä'chō) *n. pl.* **·chos 1** Any of several tropical American trees producing a medicinal bark, especially the **white quebracho** (*Aspidosperma quebracho-blanco*), a Chilean tree whose bark is used as a febrifuge and for diseases of the respiratory organs; also, **red quebracho** (*Schinopsis lorentzii*), a tree whose heartwood is rich in tannin. **2** The wood or bark of any of these trees. [<Sp., var. of *quiebrahacha*, lit., ax-breaker < *quebrar* break + *hacha* ax]

Quech·ua (kech'wä) *n.* **1** One of a tribe of South American Indians which dominated the Inca empire prior to the Spanish conquest. **2** The language of the Quechuas, still spoken as a mother tongue in parts of Peru and Ecuador: also called *Incan.* Also spelled *Kechua.*

Quech·uan (kech'wən) *adj.* Of or pertaining to the Quechua or their language. —*n.* Quechua. Also spelled *Kechuan.*

queen (kwēn) *n.* **1** The wife of a king. **2** A female sovereign or monarch. **3** A woman preeminent in a given sphere. **4** The most powerful piece in chess, capable of moving any number of squares in a straight line. **5** A playing card bearing a conventional picture of a queen in her robes. **6** *Entomol.* The single fully developed female in a colony of social insects, as bees, ants, etc.: distinguished from workers, soldiers, and unproductive females. —*v.t.* **1** To make a queen of. **2** In chess, to make a queen of (a pawn) by moving it to the eighth row. —*v.i.* **3** To reign as or play the part of a queen: usually with *it.* [OE *cwēn* woman, queen]

Queen Anne's lace The wild carrot (*Daucus carota*), having filmy white flowers in umbels.

Queen Anne style 1 *Archit.* A style prevalent in England in the early 18th century, or a style similar to it used in the United States in the latter part of the 19th century, characterized by the use of red brickwork on which relief ornaments are carved, and by plain, unpretentious design. **2** A type of furniture characterized by much upholstery and marquetry.

Queen Anne's War See WAR OF THE SPANISH SUCCESSION in table under WAR.

Queen Charlotte Sound A bay of the Pacific in British Columbia between Vancouver Island and **Queen Charlotte Islands**, an archipelago (3,970 square miles) of western British Columbia; narrowing to **Queen Charlotte Strait**, 60 miles long and 16 miles wide, the northern end of the channel separating Vancouver Island from the mainland.

queen consort The wife of a reigning king, who does not share his sovereignty.

queen dowager The widow of a king who has reigned in his own right.

queen·ly (kwēn'lē) *adj. & adv.* Like a queen; stately; reginal. See synonyms under IMPERIAL. —**queen'li·ness** *n.*

Queen Mary Coast Part of Antarctica on the Indian Ocean, west of Wilkes Land.

Queen Maud Land Part of Antarctica south of Africa, claimed by Norway, 1939; made a dependency of Norway, 1949.

Queen Maud Mountains A range extending

south of the Ross Shelf Ice, Antarctica; rising over 13,000 feet.

queen mother A queen dowager who is mother of a reigning sovereign.

queen of the meadows An Old World meadowsweet (*Filipendula ulmaria*), naturalized in the United States.

queen of the prairie A tall perennial herb (*Filipendula rubra*) common to American meadows and prairies.

queen olive A large variety of Spanish olive.

queen·post (kwēn'pōst') *n.* One of two upright suspending or sustaining posts or compression members in a truss.

queen regent 1 A queen who rules in behalf of another. **2** A queen who rules in her own right: also **queen regnant**.

Queens (kwēnz) The easternmost borough of New York City, located on Long Island; 108 square miles.

Queen's Bench See under COURT.

Queens·ber·ry Rules (kwēnz'ber·ē) See MARQUIS OF QUEENSBERRY RULES.

queen's counsel See KING'S COUNSEL.

queen's–de·light (kwēnz'di·līt') *n.* A smooth, erect perennial (*Stillingia sylvatica*) of the spurge family, with alternate leaves and a medicinal root.

queen's English See KING'S ENGLISH under ENGLISH.

queen's evidence See STATE'S EVIDENCE.

Queens·land (kwēnz'lənd) The second largest state of the Commonwealth of Australia, in the NE part; 670,500 square miles; capital, Brisbane.

queen's metal An alloy of tin, antimony, bismuth, and lead, softer than Britannia metal: used for ornamental purposes.

queen snake A small water snake (*Natrix lebris*) of the central and eastern United States.

Queens·town (kwēnz'toun) A former name for COBH.

queen's ware Fine, glazed, cream-colored English earthenware; specifically, cream-colored Wedgwood: named for Queen Charlotte by Josiah Wedgwood, 1761.

queer (kwir) *adj.* **1** Being out of the usual course of events in minor respects; singular; odd. **2** Of questionable character; open to suspicion; mysterious. **3** *Slang* Counterfeit. —*n. Slang* Counterfeit money. —*v.t. U.S. Slang* To jeopardize or spoil. [<G *quer* oblique] —**queer'ly** *adv.* —**queer'ness** *n.*

— Synonyms (adj.): anomalous, bizarre, crotchety, curious, droll, eccentric, erratic, fantastic, funny, grotesque, laughable, ludicrous, mysterious, odd, peculiar, quaint, ridiculous, singular, strange, unique, unusual, whimsical. *Odd* is unmatched, as an *odd* shoe, and so uneven, as an *odd* number. *Singular* is alone of its kind; as, the *singular* number. What is *singular* is *odd*, but what is *odd* may not be *singular*, as, a drawerful of *odd* gloves. A *strange* thing is something either unnatural or extraordinary. A *singular* coincidence is one the happening of which is unusual; a *strange* coincidence is one the cause of which is hard to explain. That which is *peculiar* belongs especially to a person as his own; in its ordinary use there is the implication that the thing *peculiar* to one is not common to the majority. *Eccentric* is off center, and so off or aside from the ordinary and normal course; as, genius is commonly *eccentric. Eccentric* is a higher and more respectful word than *odd* or *queer. Erratic* signifies wandering, a stronger and more censorious term than *eccentric. Queer* is aside from the common in a way that is comical or perhaps slightly *ridiculous* or *mysterious. Quaint* denotes that which is pleasingly *odd* and fanciful, often with something of the antique; as, the *quaint* architecture of medieval towns. That which is *funny* is calculated to provoke laughter; that which is *droll* is more quietly amusing. That which is *grotesque* in the material sense is irregular or misshapen in form or outline or ill-proportioned so as to be somewhat *ridiculous*; the French *bizarre* is practically equivalent to *grotesque*. See ODD. Antonyms: common, customary, familiar, natural, ordinary, regular, usual.

quell (kwel) *v.t.* **1** To put down or suppress

by force; extinguish. 2 To quiet; allay, as pain. [OE *cwellan* kill] —**quell′er** *n.*

Quel·part (kwel′pärt) A former name for CHEJU.

quel·que chose (kel′kə shōz′) *French* A trifle; something.

quench (kwench) *v.t.* 1 To put out or extinguish, as a fire. 2 To put an end to; cause to cease. 3 To slake or satisfy (thirst). 4 To suppress or repress, as emotions. 5 To cool, as heated iron or steel, by thrusting into water or other liquid. [ME *cwenken*] —**quench′·a·ble** *adj.* —**quench′er** *n.*

quench·less (kwench′lis) *adj.* Incapable of being quenched; insatiable; irrepressible. —**quench′less·ly** *adv.* —**quench′less·ness** *n.*

que·nelle (kə·nel′) *n. French* A ball of savory paste made of minced meat, as chicken, veal, or fish, with bread crumbs and egg, usually poached.

Quen·tin (kwen′tin) A masculine personal name. [<L, fifth]

quer·ce·tin (kwûr′sə·tin) *n. Biochem.* A yellow crystalline compound, $C_{15}H_{10}O_7 \cdot H_2O$, found in the bark of the American oak and in the rind of certain fruits: used as a base for dyestuffs. [Prob. <L *quercus* oak + -IN] —**quer·cet·ic** (kwər·set′ik, -sē′tik) *adj.*

quer·cine (kwûr′sin, -sīn) *adj.* Of or pertaining to oaks. [<LL *quercinus* <L *quercus* oak]

quer·cit·rin (kwûr′sit′rin) *n. Biochem.* A yellow crystalline glycoside, $C_{21}H_{20}O_{11} \cdot 2H_2O$, contained in quercitron bark. Also **quer′cit′·rine.**

quer·cit·ron (kwûr′sit·ron) *n.* 1 The crushed and powdered inner bark of the American black oak *(Quercus velutina)*, used in dyeing and tanning. 2 The yellow dye made therefrom. 3 The dyer's oak *(Q. coccinea)*. [<L *quercus* oak + CITRON]

Quer·cus (kwûr′kəs) *n.* A genus of hardwood trees and shrubs of the beech family, widely distributed in north temperate regions; the oaks. [<NL <L, an oak]

Que·ré·ta·ro (kā·rā′tä·rō) A state in central Mexico; 4,432 square miles; capital, Querétaro.

que·ri·da (kā·rē′dä) *n. fem. SW U.S.* A beloved; a darling. [<Sp.]

que·rist (kwir′ist) *n.* An inquirer; questioner.

querl (kwûrl) *v.t.* & *n. U.S. Dial.* Curl; twist: also spelled *quirl.* [? <G]

quern (kwûrn) *n.* 1 An old form of hand mill for grinding grain. 2 A small hand mill for grinding spices. [OE *cweorn*]

quer·u·lous (kwer′ə·ləs, -yə·ləs) *adj.* 1 Disposed to complain or be fretful; faultfinding. 2 Indicating or expressing a complaining or whining disposition. 3 Quarrelsome. [<LL *querulosus* <L *querulus* <*queri* complain] —**quer′u·lous·ly** *adv.* —**quer′u·lous·ness** *n.*

que·ry (kwir′ē) *v.* **·ried, ·ry·ing** *v.t.* 1 To inquire into; ask about. 2 To ask questions of; interrogate. 3 To express doubt concerning the correctness or truth of, especially, as in printing, by marking with a query. —*v.i.* 4 To have or express doubt; question. See synonyms under INQUIRE, QUESTION. [<*n.*] —*n. pl.* **·ries** 1 An inquiry, or a memorandum of an inquiry, to be answered; a question. 2 A doubt; interrogation: often indicated, as in printing, by the interrogation point (?). See synonyms under INQUIRY, QUESTION. [<L *quaere*, imperative sing. of *quaerere* ask]

Ques·nay (ke·nā′), **François,** 1694–1774, French physician and economist.

quest (kwest) *n.* 1 The act of seeking; a looking for something; a search, as an adventure or expedition in medieval romance; also, the person or persons making the search. 2 *Rare* An inquest. —*v.i.* 1 To go on a quest. 2 To make a search. 3 To search for game; also, to bay on the trail of game: said of hunting dogs. —*v.t.* 4 To search for; seek. [<OF *queste* <L *quaesitus*, pp. of *quaerere* ask, seek] —**quest′er** *n.*

ques·tion (kwes′chən) *n.* 1 An interrogative sentence calling for an answer; an inquiry. 2 A subject of inquiry or debate; a matter to be decided; a point at issue; problem. 3 A subject of dispute; a controversy; difference: A *question* rose about it. 4 A proposition under discussion in a deliberative assembly. 5 Objection raised or entertained; doubt: a statement accepted without *question.* 6 Interrogation; the act of asking or inquiring. —*v.t.* 1 To put a question or questions to; interrogate. 2 To be uncertain of; doubt. 3 To make objection to; challenge; dispute. —*v.i.* 4 To ask a question or questions. [<AF *questiun* <L *quaestio, -onis* <*quaerere* ask] —**ques′tion·er** *n.*

Synonyms (noun): doubt, inquiry, inquisition, interrogation, interrogatory, investigation, query. An *inquiry* seeks information for the benefit of the inquirer; a *question* may do the same, or may have the intent to perplex, confuse, or entrap the one of whom it is asked; one makes *inquiry* as to his way; we speak of idle or frivolous *questions* rather than of idle or frivolous *inquiries.* A *query* is a *question* more or less vaguely formulated and indefinite in purpose, often amounting to no more than a suspense of judgment. An *interrogation* or *interrogatory* is a formal *inquiry. Interrogatory* has a special legal use, denoting an *inquiry* in writing by order of a court, to be answered under oath. An *investigation* is an elaborate search for truth or fact, not only by *questions,* but by every other means of procuring information; an *inquisition* is an *investigation* which is either unwarranted, unduly minute, or in some other way offensive. See DOUBT, INQUIRY, TOPIC.

Synonyms (verb): ask, challenge, dispute, doubt, inquire, interrogate, investigate, query, quiz. To *ask* is to seek information, favor, or aid; *inquire, question, interrogate,* respect only the obtaining of information. To *interrogate* is to *examine* formally or officially, commonly by a series of questions. One may *inquire* casually and indifferently; he *questions* intently and resolutely. *Question* also has nearly the meaning of *challenge*; as, "I *question* the truth of that statement—the value of that article," or the like. See INQUIRE.

ques·tion·a·ble (kwes′chən·ə·bəl) *adj.* 1 Liable to be called in question; debatable; open to question or to suspicions; dubious; suspicious: *questionable* motives. 2 Of doubtful meaning; difficult to decide. 3 *Obs.* Capable of being questioned or inquired of. See synonyms under EQUIVOCAL. —**ques′tion·a·ble·ness, ques′·tion·a·bil′i·ty** *n.* —**ques′tion·a·bly** *adv.*

ques·tion·ar·y (kwes′chən·er′ē) *adj.* Of the nature of an examination; interrogatory. —*n. pl.* **·ar·ies** A questionnaire.

ques·tion·less (kwes′chən·lis) *adj.* Unquestionable; indubitable; also, unquestioning. —**ques′tion·less·ly** *adv.*

question mark An interrogation point (?).

ques·tion·naire (kwes′chə·nâr′) *n.* A written or printed form comprising a series of questions submitted to a number of persons in order to obtain data for a survey or report. [<F]

ques·tor (kwes′tər) See QUAESTOR.

Quet·ta (kwet′ə) 1 A commissioners' division of west central West Pakistan, near the border of Afghanistan; 35,027 square miles. 2 The leading city of this division, formerly capital of the former province of Baluchistan, included, October, 1955, in the province of West Pakistan.

quet·zal (ket·säl′) *n. pl.* **·zal·es** (-sä′lās) 1 A trogon *(Pharomacrus mocinno)* of brilliant plumage, the national symbol of Guatemala, and anciently regarded as a deity by the Mayas, whose chiefs alone were permitted to wear its plumes. 2 A silver coin, the monetary unit of Guatemala. Also **que·zal** (kā·säl′). [<Sp. <Nahuatl]

QUETZAL
(About 4 feet long including tail)

Quet·zal·co·a·tl (ket·säl′kō·at′l) A traditional god and heroic figure of the Aztecs.

queue (kyōō) *n.* 1 A pendent braid of hair on the back of the head; a pigtail. 2 A line of persons or vehicles waiting in the order of their arrival. —*v.i.* **queued, queu·ing** *Brit.* To form such a line: usually with *up.* Also spelled *cue.* [<F <OF *coe, coue* <L *cauda* a tail]

que vou·lez–vous (kə vōō·lā·vōō′) *French* What do you want? What can you expect?: an expression of indifference or cynicism.

quey (kwā) *n. Scot.* A young cow; heifer.

Que·zal·te·nan·go (kā·säl′tā·näng′gō) The second largest city of Guatemala, in the SW part of the western highlands.

Que·zon (kā′zon, *Sp.* kā′sôn, -thôn), **Manuel Luis,** 1878–1944, Filipino statesman; first president of the Philippines, 1935–1944. Also **Que·zon y Mo·li·na** (kā′sôn ē mō·lē′nä).

Que·zon City (kā′sôn, -thôn) The capital (since 1948) of the Philippines, in southern Luzon, NE of Manila.

quib·ble (kwib′əl) *n.* 1 An evasion of a point or question; an equivocation. 2 *Rare* A pun. —*v.i.* **·bled, ·bling** To use quibbles; evade the truth or the point in question. [<obs. *quib* <L *quibus,* ablative pl. of *qui* who, which; with ref. to its use in legal documents] —**quib′bler** *n.*

Qui·be·ron (kēb·rôn′) A town at the southern end of **Quiberon Peninsula,** which projects seven miles into the Bay of Biscay from Brittany, France, nearly enclosing **Quiberon Bay,** scene of a British naval victory over the French, 1759.

Qui·ché (kē·chā′) *n.* 1 An Indian of a tribe of Mayan linguistic stock inhabiting Guatemala. 2 The Mayan language of this tribe.

quick (kwik) *adj.* 1 Done or occurring in a short time; expeditious; brisk; rapid; swift; speedy. 2 Characterized by rapidity or readiness of movement or action; nimble; prompt. 3 Sharp; steep, as a curve. 4 Alert; sensitive; perceptive: a *quick* ear; *quick* wit. 5 Responding readily to impressions; excitable; hasty. 6 Having life; living: opposed to *dead*: an archaic use. 7 Pregnant; with child. 8 Burning briskly; fiery. 9 Shifting; moving: said of soil or sand. 10 Refreshing; bracing. See synonyms under ACTIVE, ALIVE, CLEVER, IMPETUOUS, NIMBLE, SWIFT[1], VIVID. —*n.* 1 That which has life; those who are alive: chiefly in the phrase **the quick and the dead.** 2 The living flesh; any vital or tender part; especially, the tender flesh under a nail; hence, the feelings: cut to the *quick.* 3 A hedge plant; quickset. —*adv.* Quickly; rapidly. [OE *cwic* alive]

quick assets Assets which are readily convertible into cash; liquid assets.

quick bread Any bread, biscuits, etc., whose leavening agent makes immediate baking possible.

quick–break (kwik′brāk′) *n. Electr.* A current switch equipped with a spring or other device to permit rapid contact–opening independent of the operator.

quick·en (kwik′ən) *v.t.* 1 To cause to move more rapidly; hasten or accelerate. 2 To make alive or quick; give or restore life to. 3 To excite or arouse; stimulate: to *quicken* the appetite. —*v.i.* 4 To move or act more quickly; become more rapid. 5 To come or return to life; revive. 6 To reach the stage of pregnancy at which the motions of the fetus first become perceptible: said of the mother. 7 To begin to manifest signs of life: said of the fetus. —**quick′en·er** *n.*

Synonyms: accelerate, advance, dispatch, drive, expedite, facilitate, further, hasten, hurry, promote, speed, urge. To *quicken* is to increase speed, move or cause to move more rapidly, as through more space or with a greater number of motions in the same time. To *accelerate* is to increase the speed of action or of motion. A motion whose speed increases upon itself is said to be *accelerated,* as the motion of a falling body, which becomes swifter with every second of time. To *accelerate* any work is to *hasten* it toward a finish. To *dispatch* is to do and be done with, to get a thing off one's hands. To *dispatch* an enemy is to kill him outright and quickly; to *dispatch* a messenger is to send him in haste; to *dispatch* a business is to bring it quickly to an end. To *promote* a cause is in any way to bring it forward, *advance* it in power, prominence, etc. To *speed* is really to secure swiftness; to *hasten* is to attempt it, whether successfully or unsuccessfully. *Hurry* always indicates something of confusion. To *facilitate* is to *quicken* by making easy; to *expedite* is to *quicken* by removing hindrances. *Antonyms*: check, clog, delay, drag, hinder, impede, obstruct, retard.

quick fire The firing of quick successive shots:

quick-firing (kwik′fīr′ing) *adj.* Able to fire shots rapidly and continuously.

quick-freeze (kwik′frēz′) *v.t.* -froze, -fro·zen, -freez·ing To subject (food) to rapid refrigeration for storing at or below freezing temperatures. — **quick′-fro′zen** *adj.*

quick·grass (kwik′gras′, -gräs′) *n.* Couchgrass.

quick·ie (kwik′ē) *n. U.S. Slang* Anything done hastily, as by short cuts or makeshift methods.

quick·lime (kwik′līm′) *n.* Unslaked lime. See LIME[1].

quick·ly (kwik′lē) *adv.* In a quick manner; rapidly; soon.

quick march A march in quick time; quickstep.

quick·match (kwik′mach′) *n.* A cord impregnated with black powder and used as a fast-burning fuse for flares, fireworks, etc.

quick·ness (kwik′nis) *n.* 1 The state or quality of being quick; speed; celerity; liveliness; readiness. 2 Acuteness of perception or sensibility; sharpness; keenness.

quick·sand (kwik′sand′) *n.* A bed of sand so water-soaked as readily to engulf any person or animal that attempts to move or rest upon it.

quick·set (kwik′set′) *n.* 1 A hedge plant, especially hawthorn. 2 A hedge made of it. — *adj.* Composed of quickset.

quick·sil·ver (kwik′sil′vər) *n.* 1 Metallic mercury: widely used in metallurgy, industry, and the arts. All of its compounds are poisonous. 2 An amalgam of tin, used for the backs of mirrors. [Trans. of L *argentum vivum*.]

quick·step (kwik′step′) *n.* A march or dance written in a rapid tempo; also, a quick march.

quick-tem·pered (kwik′tem′pərd) *adj.* Easily angered.

quick time A marching step of 120 paces a minute, each pace of 30 inches: used in military drills and ceremonies.

quick·wa·ter (kwik′wô′tər, -wot′ər) *n.* A stream or that part of a stream having a decided current.

quick-wit·ted (kwik′wit′id) *adj.* Having a ready wit or quick discernment; keen; alert. See synonyms under CLEVER, SAGACIOUS. — **quick′-wit′ted·ly** *adv.* — **quick′-wit′ted·ness** *n.*

qui·cun·que vult (kwī·kung′kwē vult) *Latin* Whosoever will. See ATHANASIAN CREED.

quid[1] *n.* 1 A small portion of chewing tobacco. 2 A cud, as of a cow. [Var. of CUD]

quid[2] (kwid) *n. Brit. Slang* In England, a pound sterling, or a sovereign. [Origin uncertain]

Quid·de (kvid′ə), **Ludwig**, 1858–1941, German historian and pacifist.

quid·di·ty (kwid′ə·tē) *n. pl.* **·ties** 1 The essence of a thing. 2 A subtle or trifling distinction or objection; cavil. [<LL *quidditas, -tatis* <L *quid* which, what]

quid·nunc (kwid′nungk′) *n.* One who seeks or affects to know all that is going on; an inquisitive busybody. [<L *quid nunc* what now]

quid pro quo (kwid′ prō kwō′) *Latin* 1 Something for something; an equivalent in return. 2 Formerly, one medicine used in place of another; hence, a substitution.

quién sa·be (kyen sä′vä) *Spanish* Literally, "who knows?": used to mean, in reply to a question, "I do not know," or "I do not care to say."

qui·es·cent (kwī·es′ənt) *adj.* 1 Being in a state of repose or inaction; quiet; still. 2 Resting free from anxiety, emotion, or agitation. 3 *Phonet.* In Semitic languages, having no sound; silent. See synonyms under PASSIVE. [<L *quiescens, -entis*, ppr. of *quiescere* be quiet] — **qui·es′cence** *n.* — **qui·es′cent·ly** *adv.*

qui·et (kwī′ət) *adj.* 1 Being in a state of repose; still; calm; motionless. 2 Free from turmoil, strife, or alarm; tranquil; peaceful. 3 Silent. 4 Gentle or mild of disposition. 5 Undisturbed by din or bustle; retired; secluded: a *quiet* nook. 6 Restful to the eye; soft in hue; hence, not showy or obtrusive; as dress. See synonyms under CALM, PACIFIC, SEDATE, SOBER. — *n.* 1 The condition or quality of being free from motion, disturbance, noise, etc.; peace; calm. See synonyms under REST[1]. — *v.t. & v.i.* To make or become quiet: often with *down.*

See synonyms under ALLAY, REPRESS, SETTLE, TRANQUILIZE. — *adv.* In a quiet or peaceful manner. [<OF *quiete* <L *quietus* <*quies* rest, repose. Doublet of COY.] — **qui′et·ly** *adv.* — **qui′et·ness** *n.*

qui·et·ism (kwī′ə·tiz′əm) *n.* The doctrine that spiritual exaltation is attained by self-abnegation and withdrawing the soul from outward activities, fixing it in passive religious contemplation; especially, mystic meditation or introspection, as cultivated by certain devotees in the 17th century; also a state of quiet; quietude.

qui·et·ist (kwī′ə·tist) *n.* 1 An advocate or practicer of quietism. 2 One who seeks or enjoys quiet.

qui·e·tude (kwī′ə·tōōd, -tyōōd) *n.* A state or condition of calm or tranquillity; repose; rest.

qui·e·tus (kwī·ē′təs) *n.* 1 A silencing or suppressing; death; repose. 2 A final discharge or quittance; a settlement. 3 A killing blow. [<L *quietus est* he is quiet]

quill[1] (kwil) *n.* 1 *Ornithol.* One of the large, strong flight feathers or tail feathers of a bird. 2 A pen made from a feather; hence, any pen. 3 The hollow, horny stem of a feather; a calamus. 4 Such a stem used for a receptacle or measure, as for a drug, or as a plectrum for playing a stringed instrument. 5 *Zool.* One of the large, sharp spines of a porcupine or hedgehog. 6 A piece of cane or reed used as a musical pipe. 7 A slow-burning fuse made formerly of the quill of a feather filled with powder. 8 A piece of bark rolled into cylindrical form: a cinnamon *quill.* 9 A quill toothpick. 10 *Mech.* A hollow shaft, with or without openings, designed to revolve on a solid shaft when the clutches are engaged. 11 In weaving, a spindle or bobbin; pirn. 12 A fluted, rounded ridge, or cylindrical fold, as in a ruff or ruffle. — *v.t.* 1 To make or iron (a garment or fabric) with rounded plaits or ridges. 2 To wind (thread or yarn) on a quill or quills. — *v.i.* 3 To wind thread or yarn on a quill or quills. [Cf. LG *quiele* a quill of a feather]

quil·lai (ki·lī′) *n.* A large Chilean evergreen tree, the soapbark tree, whose alkaline inner bark (quillai bark) is used as medicine and as a substitute for soap: also spelled *cullay.* [<Sp. <Araucanian]

quill·back (kwil′bak′) *n.* A carplike fish (*Carpiodes velifer*) common in the Mississippi Valley.

quill·driv·er (kwil′drī′vər) *n. Colloq.* 1 One who writes; a literary hack. 2 Formerly, a clerk; copyist.

Quil·ler-Couch (kwil′ər-kōōch′), **Sir Arthur Thomas**, 1863–1944, English author and editor.

quil·let (kwil′it) *n. Obs.* A quibble; subtlety; nice distinction. [?Alter. of L *quidlibet* what you please <*quid* what + *libet* it pleases]

quill·pig (kwil′pig′) *n. U.S. Colloq.* A porcupine.

quill·wort (kwil′wûrt′) *n.* A small plant (genus *Isoetes*), found in marshes, pond edges, etc., consisting of a cormlike stem sending up a tuft of quill-like leaves.

Quil·mes (kēl′mäs) A city on the Río de la Plata SE of Buenos Aires, Argentina.

quilt (kwilt) *n.* 1 A bedcover made by stitching together firmly two layers of cloth or patchwork with some soft and warm substance (as wool or cotton) between them. 2 Any bedcover, especially if thick. 3 A quilted skirt or other quilted article. 4 *Obs.* A mattress. — *v.t.* 1 To stitch together (two pieces of material) with a soft substance between. 2 To stitch in ornamental patterns or crossing lines. 3 To sew up or secure between two layers. 4 To pad or line with something soft. — *v.i.* 5 To make a quilt or quilted work. [<OF *cuilte* <L *culcita*]

quilt·ing (kwil′ting) *n.* 1 The act or process of making a quilt, or of stitching as in making a quilt. 2 Material for quiltwork. 3 A quilting bee or party.

quilting bee A social gathering of the women of a community for working on a quilt or quilts. Also **quilting frolic, quilting party.**

quin·a·crine (kwin′ə·krēn) *n.* Atabrine.

qui·na·ry (kwī′nə·rē) *adj.* Consisting of or containing five parts or elements; arranged by

fives, or in sets or groups of five. — *n. pl.* **·ries** A number, body, group, or system of five; something composed of five like parts. [<L *quinarius* <*quini* five each]

qui·nate (kwī′nāt, kwin′āt) *adj.* 1 Arranged in five. 2 *Bot.* Having five similar parts together, as the five leaflets of the Virginia creeper. [<L *quini* five each <*quinque* five]

quince (kwins) *n.* 1 The hard, acid, applelike, yellowish fruit, used for preserves, of a small deciduous Asian tree (*Cydonia oblonga*) of the rose family. 2 The tree. [Orig. pl. of obs. *coyn* <OF *cooin* <L *cotoneum*, var. of (*malum*) *cydonium* (apple) of Cydonia <Gk. *Kydōnia*, a town in Crete]

quin·cun·cial (kwin·kun′shəl) *adj.* 1 Arranged in the form of a quincunx. 2 *Bot.* Arranged in a set of five, as leaves. Also **quin·cunx′ial** (-kungk′shəl). — **quin·cun′cial·ly** *adv.*

quin·cunx (kwin′kungks) *n.* 1 An arrangement of five things, as trees, in a square having one in each corner and one in the center. 2 A disposition of such squares repeated indefinitely. 3 A quincuncial arrangement, as of flower parts. [<L *quincunx* five twelfths <*quinque* five + *uncia* twelfth part]

Quin·cy (kwin′sē), **Josiah**, 1744–75, American statesman and Revolutionary leader.

Quin·cy (kwin′sē) A city in eastern Massachusetts south of Boston.

quin·dec·a·gon (kwin·dek′ə·gon) *n. Geom.* A figure, especially a plane figure, with fifteen sides and fifteen angles. [<L *quindecim* fifteen + Gk. *gōnia* angle]

quin·de·cen·ni·al (kwin′di·sen′ē·əl) *n.* A fifteenth anniversary. — *adj.* Of or pertaining to the fifteenth anniversary. [<L *quindecim* fifteen + *annus* year]

quin·dec·i·mal (kwin·des′ə·məl) *adj.* Fifteen. [<L *quindecim* fifteen]

quin·ic (kwin′ik) *adj.* Of, pertaining to, or derived from quinine.

quinic acid *Chem.* A white crystalline compound, $C_7H_{12}O_6$, contained in cinchona bark, coffee beans, etc.

quin·i·dine (kwin′ə·dēn, -din) *n. Chem.* A white crystalline alkaloid, $C_{20}H_{24}N_2O_2$, isomeric with quinine, contained in certain cinchona barks. It is used in medicine to regulate the heartbeat, and its sulfate is officinal.

qui·nine (kwī′nīn, *esp. Brit.* kwi·nēn′) *n. Chem.* A white, amorphous or slightly crystalline, very bitter alkaloid, $C_{20}H_{24}N_2O_2$, contained in cinchona barks. Its salts, as the hydrochlorate, sulfate, and others, are largely used in medicine on account of their tonic and antipyretic qualities, especially in malarial affections of all kinds. Also **quin·in** (kwin′in). [<earlier *quina* (<Sp. <Quechua *kina* bark) + -INE[2]]

quin·nat (kwin′at) *n.* A salmon (*Oncorhynchus tschawytscha*) of the coasts of the North Pacific: also called *Chinook salmon.* [<Chinook *ikwaná*]

quin·oid (kwin′oid) *adj.* Having a quinone nucleus.

qui·noi·dine (kwi·noi′dēn, -din) *n. Chem.* A brown, resinous, amorphous compound, consisting chiefly of uncrystallizable products of cinchona bark: a cheap substitute for real quinine. Also **qui·noi′din** (-din).

quin·o·line (kwin′ə·lēn, -lin) *n. Chem.* 1 A colorless liquid compound, C_9H_7N, with a tarry odor, obtained variously, as by distilling quinine, cinchonine, or by the destructive distillation of coal and bones. 2 Any of a class of quinoline derivatives, among which are many dyes and medicinal compounds. Also **quin′o·lin** (-lin).

qui·none (kwi·nōn′, kwin′ōn) *n. Chem.* Either of two isomeric compounds obtained from benzene and its homologs; especially, a golden-yellow crystalline compound, $C_6H_4O_2$, with pungent odor, formed variously, as by the oxidation of quinic acid, aniline, etc.: also called *paraquinone.*

qui·non·i·mine (kwi·non′ə·mēn, -min) *n. Chem.* A crystalline organic compound, C_6H_5NO, derived from a quinone through replacement of an oxygen atom by an imine. See INDOPHENOL.

qui·o·noid (kwin′ə·noid) *adj.* Resembling or like quinone.

quin·ox·a·line (kwin·ok′sə·lēn, -lin) *n. Chem.*

A fully basic, white, crystalline compound, $C_8H_6N_2$, used in organic synthesis.

quin·qua·ge·nar·i·an (kwin′kwə·jə·nâr′ē·ən) *adj.* Being fifty years old; relating to this age. — *n.* A person fifty years old.

quin·qua·gen·a·ry (kwin′kwə·jen′ər·ē) *adj.* 1 Consisting of or containing fifty. 2 Denoting a group or set of fifty. [<L *quinquagenarius* < *quinquageni* fifty each]

quin·qua·ges·i·ma (kwin′kwə·jes′ə·mə) *adj.* Fiftieth. — *n.* A period of fifty days. [<L *quinquagesima (dies)* fiftieth (day)]

quin·qua·ges·i·mal (kwin′kwə·jes′ə·məl) *adj.* Of, pertaining to, comprising, or containing fifty.

Quinquagesima Sunday The fiftieth day before Easter; the Sunday before Lent; Shrove Sunday.

quinque– *combining form* Five: *quinquefoliate.* Also, before vowels, **quinqu–**. [<L *quinque* five]

quin·que·fo·li·ate (kwin′kwə·fō′lē·it, -āt) *adj.* *Bot.* Five–leaved. Also **quin′que·fo′li·o·late′** (-fō′lē·ə·lāt′). [<QUINQUE- + L *foliatus* < *folium* leaf]

quin·quen·ni·al (kwin·kwen′ē·əl) *adj.* Occurring every five years, or once in five years; also, lasting five years. — *n.* 1 A fifth anniversary or its celebration. 2 A quinquennium. [<L *quinque* five + *annus* year]

quin·quen·ni·um (kwin·kwen′ē·əm) *n.* A period of five years. [<L]

quin·que·va·lent (kwin′kwə·vā′lənt) *adj.* *Chem.* Having a valence or combining value of five; pentavalent. See VALENCE. — **quin′que·va′·lence** *n.*

quin·sy (kwin′zē) *n.* *Pathol.* Inflammation of the tonsils and the adjoining tissues, especially when suppurative. [<Med. L *quinancia* <Gk. *kynanchē* a dog's collar < *kyōn* dog + *anchein* choke]

quint (kwint) *n.* 1 A fifth. 2 A set of five. 3 The E string of a violin. 4 *Colloq.* A quintuplet. 5 In piquet, a sequence of five of the same suit: if of the five highest cards, called a **quint major.** 6 An organ stop giving tones a fifth above those of the keys that are pressed. [<L *quintus* < *quinque* five]

quin·tain (kwin′tin) *n.* *Obs.* An object set up to be tilted at; also, the place for the sport. [OF *quintaine* <L *quintana* street in a camp < *quintus* fifth]

quin·tal (kwin′təl) *n.* A measure of weight, a hundredweight; in the metric system, 100 kilograms: also called *metric centner.* See METRIC SYSTEM. [<OF <Arabic *qinṭar*]

quin·tan (kwin′tən) *adj.* Recurring on every fifth day, reckoning inclusively: a *quintan* fever. — *n.* A quintan fever. [<L *quintanus* < *quintus* fifth]

Quin·ta·na Ro·o (kēn·tä′nä rō′ō) A territory in the eastern part of the Yucatán peninsula, Mexico; 19,625 square miles; capital, Chetumal.

Quin·te·ro (kēn·tā′rō), **Alvarez** See ALVAREZ QUINTERO.

quin·tes·sence (kwin·tes′əns) *n.* 1 An extract from anything, containing in concentrated form its most essential principle. 2 The purest and most essential part, manifestation, or embodiment of anything. 3 *Philos.* In the doctrine of the Pythagoreans, the fifth or celestial essence, ether, above the four elements of earth, air, fire, and water. [<F <L *quinta essentia* fifth essence] — **quin′tes·sen′tial** *adj.*

quin·tet (kwin·tet′) *n.* 1 A musical composition arranged for five voices or instruments; also, the five persons performing it. 2 Any group of five; anything arranged for a set of five performers, as in a game. Also **quin·tette′.** [<Ital. *quintetto* < *quinto* fifth]

quin·tic (kwin′tik) *Math. adj.* Denoting a quantic function of the fifth degree. — *n.* Such a function.

quin·tile (kwin′til) *n.* 1 In astrology, the aspect of planets separated by 72°, or the fifth part of the zodiac. 2 *Stat.* a That part of a frequency distribution containing one fifth of the total observations or cases. b The point marking such a part. [<L *quintus* fifth, on analogy with *quartile*]

Quin·til·i·an (kwin·til′ē·ən, -til′yən), A.D. 35?–95?, Roman rhetorician: full name *Marcus Fabius Quintilianus.*

quin·til·lion (kwin·til′yən) *n.* In the French system of numeration, almost universally followed in the United States, 1 followed by 18 ciphers; in the English system, 1 followed by 30 ciphers. [<L *quintus* fifth + MILLION] — **quin·til′lionth** (-yənth) *adj.* & *n.*

quin·tin (kwin′tin) *n.* *Rare* A fine linen fabric. Also **quin′tain** (-tin). [from *Quintin,* a town in Brittany]

quin·tu·ple (kwin′tōō·pəl, -tyōō-, kwin·tōō′pəl, -tyōō′-) *v.t.* & *v.i.* **·pled, ·pling** To multiply by five; make or become five times as large. [< *adj.*] — *adj.* 1 Consisting of five united or of five parts. 2 Multiplied by five. — *n.* A number or a sum five times as great as another. [<F <L *quintuplex* < *quintus* fifth + *plic-*, stem of *plicare* fold]

quin·tu·plet (kwin′tōō·plit, -tyōō-, kwin·tōō′·plit, -tyōō′-) *n.* 1 Five things of a kind used or occurring together. 2 One of five born of the same mother at one birth.

quin·tu·pli·cate (kwin·tōō′plə·kit, -tyōō′-) *adj.* 1 Fivefold. 2 Raised to the fifth power. — *v.t.* & *v.i.* (-kāt) **·cat·ed, ·cat·ing** To multiply by five; quintuple. — *n.* (-kit) One of five identical things. [<L *quintuplex, -icis* + -ATE[1]] — **quin·tu′pli·cate·ly** *adv.* — **quin·tu′·pli·ca′tion** *n.*

quip (kwip) *n.* 1 A sarcastic or sharp jest, remark, or retort; gibe; also, a clever or witty sally without sarcasm. 2 A quibble. 3 An odd, fantastic action or object. — *v.i.* **quipped, quip·ping** To make a witty remark; jest. [Earlier *quippy* <L *quippe* indeed] — **quip′pish** *adj.*

quip·ster (kwip′stər) *n.* One who makes quips.

qui·pu (kē′pōō, kwip′·ōō) *n.* An aboriginal Peruvian device for recording and conveying information, consisting of a series of varicolored and knotted strings tied at one end to a thicker cord. The order, color, and knots of the strings were used like elements of a written language. Also **quip′pu.** [<Quechua *quipu* knot]

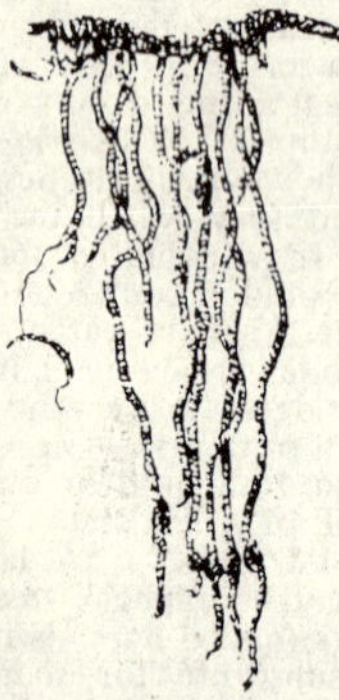

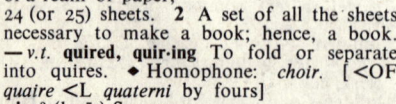

PERUVIAN QUIPU

quire[1] (kwīr) *n.* 1 The twentieth part of a ream of paper; 24 (or 25) sheets. 2 A set of all the sheets necessary to make a book; hence, a book. — *v.t.* **quired, quir·ing** To fold or separate into quires. ◆ Homophone: *choir.* [<OF *quaire* <L *quaterni* by fours]

quire[2] (kwīr) See CHOIR.

Quir·i·nal (kwir′ə·nəl) 1 One of the seven hills on which Rome stands, containing the **Quirinal palace,** formerly a papal residence; after 1870 the official residence of the kings of Italy. 2 Figuratively, the monarchical regime of Italy, as distinguished from the *Vatican,* or papal government. — *adj.* Pertaining to or situated on the Quirinal.

Qui·ri·no (kē·rē′nō), **Elpidio,** 1890–1956, president of the Philippines 1948–54.

Qui·ri·nus (kwi·rī′nəs) An ancient Italic god of war: ultimately identified with the deified *Romulus.*

Qui·ri·tes (kwi·rī′tēz) *n. pl.* The citizens of ancient Rome in their civil as distinguished from their military or political capacity. [<L]

quirk (kwûrk) *n.* 1 A short or sharp turn; twist. 2 A quaint turn of the fancy; bright retort; hence, a personal peculiarity; caprice. 3 An artful turn for evasion or subterfuge; quibble. 4 A sudden curve or flourish, especially in drawing or writing. 5 *Archit.* a A small groove in, beside, or between moldings or beads. b A molding or bead having a groove on one or both edges. See synonyms under WHIM. [Origin uncertain] — **quirk′y** *adj.*

quirl (kwûrl) See QUERL.

quirt (kwûrt) *n.* A short–handled riding whip with a braided rawhide lash. — *v.t.* To strike with a quirt. [<Mexican Sp. *cuarta*]

qui s'ex·cuse s'ac·cuse (kē seks·küz′ sà·küz′) *French* He who excuses himself accuses himself.

quish (kwish) See CUISH.

quis·ling (kwiz′ling) *n.* One who betrays his country to the enemy and is then given political power by the conquerors. [after Vidkun *Quisling,* 1887–1945, Norwegian Nazi party leader and traitor] — **quis′ling·ism** *n.*

quit (kwit) *v.* **quit** or **quit·ted, quit·ting** *v.t.* 1 To cease or desist from; discontinue. 2 To give up; renounce; relinquish. 3 To go away from; leave. 4 To let go of (something held). 5 *Archaic* To acquit (oneself). — *v.i.* 6 To stop; cease; discontinue. 7 To leave; depart. 8 *Colloq.* To resign from a position, etc. See synonyms under ABANDON, CEASE, END, REQUITE. — *adj.* Released, relieved, or absolved from something, as a duty, obligation, encumbrance, or debt; clear; free; rid. — *n.* The act of quitting. — **to be quits** To be even (with another). — **to cry quits** To declare to be even, or that neither has the advantage; declare (oneself) willing to stop competing. [<OF *quiter* <LL *quietare* set free <L *quies* rest, repose]

quitch·grass (kwich′gras′, -gräs′) *n.* Couch-grass. Also **quitch.**

quit·claim (kwit′klām′) *n.* *Law* A full release and acquittance given by one to another in regard to a certain demand, suit, or right of action: also **quit′claim′ance.** — *v.t.* To relinquish or give up claim or title to; release from a claim. [<QUIT + CLAIM]

quitclaim deed A conveyance, in the nature of a release, of all the maker's interest in the land in question, but not professing that the title is valid, nor containing any warranty or covenants for title.

quite (kwīt) *adv.* 1 To the fullest extent; without limitation or reservation; fully; totally: *quite* dead. 2 *Colloq.* To a great or considerable extent; noticeably; very: *quite* ill. [ME; var. of QUIT, *adj.*]

Qui·to (kē′tō) The capital of Ecuador; 9,343 feet above sea level in the Andes of north central Ecuador.

quit–quit (kwit′kwit′) *n.* The honey creeper.

quit–rent (kwit′rent′) *n.* A fixed rent formerly paid by a freeholder, whereby he was released from feudal services.

quit·tance (kwit′ns) *n.* 1 Discharge or release, as from a debt or obligation; acquittance. 2 Something given or tendered by way of requital; repayment. [<F < *quiter* QUIT]

quit·ter[1] (kwit′ər) *n.* One who quits needlessly; a shirker; slacker; coward.

quit·ter[2] (kwit′ər) *n.* 1 A fistulous sore on the hoof of a horse or any solid–hoofed animal: also **quit′ter·bone′** (-bōn′), **quit′tor.** 2 Purulent matter. [? <OF *quiture* a cooking]

qui va là? (kē vä lä′) *French* Who goes there?: a watchword.

quiv·er[1] (kwiv′ər) *v.i.* To shake with a slight, tremulous motion; vibrate; tremble. See synonyms under QUAKE, SHAKE. — *n.* The act or fact of quivering; a trembling or shaking. [Prob. related to QUAVER]

quiv·er[2] (kwiv′ər) *n.* A portable case or sheath for arrows; also, its contents. [<AF *quiveir,* OF *coivre* <Gmc.]

quiv·er[3] (kwiv′ər) *adj.* *Obs.* Brisk; active; nimble. [OE *cwifer-,* found in *cwiferlice* zealously]

Qui·vi·ra (kē·vē′rä) A land in the central United States, sought and found by Coronado in 1541: often identified with Kansas.

qui vive? (kē vēv′) *French* Literally, who lives?: as used by French sentinels, "Who goes there?" — **to be on the qui vive** To be on the look–out; be wide–awake.

quix·ot·ic (kwik·sot′ik) *adj.* Pertaining to or like Don Quixote, the hero of a Spanish romance ridiculing knight–errantry; hence, ridiculously chivalrous or romantic; having high but impractical sentiments, aims, etc.; extravagant; visionary. See synonyms under IMAGINARY. — **quix·ot′i·cal·ly** *adv.* — **quix·ot·ism** (kwik′sə·tiz′əm) *n.*

quiz (kwiz) *n.* 1 The act of questioning; specifically, an oral or written examination of a class or individual. 2 Something or someone odd or ridiculous; an eccentric. 3 One given to quizzing. 4 A hoax; practical joke. — *v.t.* **quizzed, quiz·zing** 1 To examine by asking questions; question. 2 *Brit.* To make fun of; ridicule. See synonyms under QUESTION. [Origin unknown] — **quiz′zer** *n.*

quiz program A television or radio program in which selected contestants or a panel of experts try to answer questions presented by the master of ceremonies.

quiz·zi·cal (kwiz′i·kəl) *adj.* 1 Addicted to quizzing or chaffing; bantering. 2 Queer; odd. — **quiz′zi·cal·ly** *adv.*

quizzing glass A monocle or single eyeglass.
Qum (koom) A city in north central Iran. Also **Qom** (kōm).
quo' (kwō) v. Scot. Quoth.
quod (kwod) n. Brit. Slang A prison. Also spelled quad.
quod e·rat de·mon·stran·dum (kwod er′ət dem′ən·stran′dəm) Latin Which was to be demonstrated: abbreviated Q.E.D.
quod·li·bet (kwod′li·bet) n. 1 A debatable or nice point; subtlety; especially, a scholarly dissertation on such a subject. 2 Music A fantasia or medley, usually humorous. [<L, anything at all] — **quod·li·bet·ic** (kwod′li·bet′ik) or **-i·cal** adj.
quod vi·de (kwod vī′dē) Latin Which see: usually abbreviated to q.v. and used in parentheses after a word by way of reference.
quo·hog (kwô′hôg, -hog, kwə·hôg′, -hog′) n. A quahaug.
quoin (koin, kwoin) n. 1 A large square ashlar or stone at the angle of a wall. 2 An external angle of a building. 3 A vertical, angular, ornamental projection from a wall face. 4 A wedge-shaped stone of an arch. 5 A block cut obliquely at the bottom to support a vertical column or pilaster on an inclined plane. 6 An internal angle, as of a room; a corner. 7 A wedge, or wedgelike piece. 8 Printing A wedge, or pair of wedges, by which to lock up type in a chase or galley. — v.t. To fasten or provide with a quoin or quoins. [Var. of COIN]

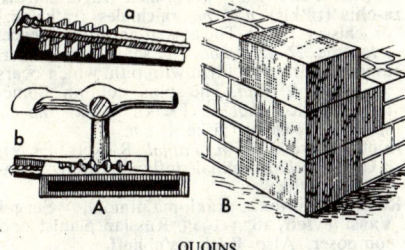

QUOINS

A. Printer's metal quoins.
  a. Single quoin.
  b. Pair of quoins ready for locking with key.
B. Quoins of dressed stone.

quoit (kwoit, esp. Brit. koit) n. 1 A disk of iron or other material with a round hole in the center to be thrown over a stake: used in the game of quoits. 2 pl. A game played by throwing these disks at a short stake. — v.t. To pitch as a quoit. [ME coyte; origin unknown]
quo ju·re (kwō joor′ē) Latin By what right? By what law?
quo·mo·do (kwō·mō′dō) Latin adv. In what manner? How? — n. The means; manner.
quon·dam (kwon′dəm) adj. Having been formerly; former. [<L]
Quon·set hut (kwon′sit) A portable structure resembling the Nissen hut, designed for use by the U.S. armed services: a trade name. [from Quonset, a town in Rhode Island where first made]

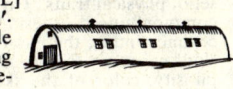

QUONSET HUT

quo·rum (kwôr′əm, kwō′rəm) n. 1 Such a number of members of any deliberative or corporate body as is necessary for the legal transaction of business: commonly, a majority. 2 Formerly, in England, certain designated justices of the peace without the presence of some one of whom the others could not act: now applied loosely to all justices. 3 A select or chosen body. [<L, of whom < qui who]
quo·ta (kwō′tə) n. A proportional part or share required for making up a certain number or quantity; proportionate contribution. [<Med. L quota (pars) how great (a part) <L quotus how great]
quot·a·ble (kwō′tə·bəl) adj. Suitable for quotation. — **quot′a·bil′i·ty** n.
Quo Tai-chi (gwō′ tī′chē′), 1889?–1952, Chinese diplomat.
quo·ta·tion (kwō·tā′shən) n. 1 The act of quoting. 2 The words quoted or cited; a passage from a book or writing, cited or adduced. 3 A price quoted or current, as of securities, etc.: the quotations for wheat. [<Med. L quotatio, -onis < quotare. See QUOTE.] — **quo·ta′tion·al** adj. — **quo·ta′tion·al·ly** adv. — **quo·ta′tion·ist** n.
quotation mark One of the marks placed at the beginning and end of a quoted word or passage. In English usage, one or two inverted commas (',") mark the beginning of a quotation, and, correspondingly, one or two apostrophes (',") the close, the single marks usually being used to set off a quotation within a quotation.
quote (kwōt) v. **quot·ed**, **quot·ing** v.t. 1 To repeat or reproduce the words of. 2 To repeat or cite (a rule, author, etc.), as for authority or illustration. 3 In commerce: a To state (a price). b To give the current or market price of. 4 Printing To enclose within quotation marks. — v.i. 5 To make a quotation, as from a book. — n. A quotation; also, a quotation mark. [<Med. L quotare distinguish by number <L quot how many] — **quot′a·ble** adj. — **quot′er** n. — **quote′·wor′thy** (-wûr′thē) adj. — **quot′ing·ly** adv.
— Synonyms (verb): cite, excerpt, extract, paraphrase, plagiarize, recite, repeat. To quote is to give an author's words, either exactly, as in direct quotation, or in substance, as in indirect quotation; to cite is, etymologically, to call up a passage, as a witness is summoned. In citing a passage its exact location by chapter, page, or otherwise must be given, so that it can be promptly called into evidence; in quoting, the location may or may not be given, but the words or substance of the passage must be given. To paraphrase is to state an author's thought more freely than in indirect quotation, keeping the substance of his thought and his order of statement, but changing the language and style, and perhaps expanding by explanation, inference, etc. To plagiarize is to quote without credit, appropriating another's words or thought as one's own. To recite or repeat is usually to quote orally, but recite is applied in legal phrase to a particular statement of facts which is not a quotation.
quoth (kwōth) v.t. Said or spoke; uttered: the imperfect tense of the obsolete verb queth, used only in the first and third persons, the nominative always following the verb, as quoth he. [OE cwæth, pt. of cwethan say]
quo·tha (kwō′thə) interj. Archaic Indeed! forsooth!: usually in slight contempt. [<quoth he]
quo·tid·i·an (kwō·tid′ē·ən) adj. Recurring or occurring every day. — n. A fever whose paroxysms return every day. [<L quotidianus daily]
quo·tient (kwō′shənt) n. Math. The result obtained by division; a number indicating how many times one number or quantity is contained in another. [<L quotiens how often < quot how many]
quo war·ran·to (kwō wô·ran′tō, wo-) Latin Literally, by what warrant; a judicial writ commanding a person to show by what authority he exercises an office or franchise never granted or forfeited by some fault. In England, and generally in the United States, this writ has given way to an **information in the nature of a quo warranto**, criminal in form, but in substance civil.

R

r, R (är) n. pl. **r's, R's** or **rs, Rs, ars** (ärz) 1 The 18th letter of the English alphabet: from Phoenician resh, Greek rho, Roman R. 2 The sound of the letter r. See ALPHABET. — symbol 1 Chem. An organic radical. 2 Math. Ratio. 3 Electr. Resistance. — **the three R's** Reading, writing, and arithmetic (regarded humorously as spelled reading, 'riting, and 'rithmetic); hence, the essential elements of a primary education.
Ra (rä) The supreme Egyptian deity, the sun-god, usually represented as a hawk-headed man crowned with the solar disk and uraeus: also spelled Re. [<Egyptian Rā the sun]
Raab (räb) The German name for GYÖR, Hungary.
ra·ban·na (rə·ban′ə) n. A textile fabric of raffia, made in Madagascar and used for draperies,

RA

curtains, and the like. [<Malagasy rebana]
Ra·bat (rä·bät′) A port on the Atlantic, capital of Morocco.
ra·ba·to (rə·bä′tō, -bä′-) See REBATO.
Ra·baul (rə·boul′, rä′boul) The chief city of New Britain; formerly the administrative center of the Territory of New Guinea.
Rab·bath Am·mon (rab′əth am′ən) The Old Testament name for AMMAN.
rab·bet (rab′it) n. 1 A recess or groove in or near the edge of one piece of wood or other material to receive the edge of another piece. 2 A joint so made. 3 A rabbet plane. — v. **·bet·ed**, **·bet·ing** v.t. 1 To cut a rectangular groove in. 2 To unite in a rabbet. 3 To be jointed by a rabbet. Also spelled rebate. ◆ Homophone: rabbit. [<OF rabat < rabattre beat down. See REBATE.]

RABBET JOINTS

rabbet joint A joint between two edges, as of timbers, each of which is partly cut away so that their faces are flush.
rabbet plane A plane for cutting a rectangular groove, as in or near the edge of a plank.
rab·bi (rab′ī) n. pl. **·bis** Master, teacher: a Jewish title for those distinguished for learning, authoritative teachers of the Law, and appointed spiritual heads of a community. Also **rab′bin** (-in). [OE <L <Gk. rhabbi <Hebrew rabbī my master <rab great, master + -i my (pronominal suffix)]
rab·bin·ate (rab′in·āt) n. 1 The office or term of office of a rabbi. 2 Rabbis collectively. [<Med. L rabbinus rabbi]
Rab·bin·ic (rə·bin′ik) n. The language or dialect of the rabbis; especially, the Hebrew language as used in Biblical and Talmudic exegesis by Jewish scholars of the late ancient and early medieval periods.
rab·bin·i·cal (rə·bin′i·kəl) adj. Pertaining to the rabbis or to their opinions, language, or writings. Also **rab·bin′ic**. — **rab·bin′i·cal·ly** adv.
rab·bin·ism (rab′in·iz′əm) n. 1 The teachings or doctrines of the rabbis. 2 A rabbinical phrase, expression, or idiom.
rab·bin·ist (rab′in·ist) n. One among the Jews who adhered to the Talmud and the

traditions of the rabbis, in opposition to those who rejected the traditions. Also **rab'bin·ite** (-it). [<Med. L *rabbinus*] —**rab·bin·is'tic, rab'bin·is'ti·cal, rab'bin·it'ic** (-it'ik) *adj.* —**rab'bin·is'ti·cal·ly** *adv.*

rab·bit (rab'it) *n.* **1** Any of various small burrowing rodents (family *Leporidae*), resembling but smaller than the hare, as the common American cottontail (*Sylvilagus floridanus*). **2** A hare. **3** The pelt of a rabbit or hare. **4** Welsh rabbit. —*v.i.* To hunt rabbits. ◆ Homophone: *rabbet*. [ME *rabette*. Akin to Walloon *robbett*, Flemish *robbe*.] —**rab'bit·er** *n.*

rabbit fever Tularemia.

rab·bit-foot (rab'it·fŏŏt') *n.* **1** A common clover (*Trifolium arvense*) having soft, hairy flower heads supposed to resemble rabbit's paws: also **rabbit's-foot clover**. **2** The left hind foot of a rabbit carried as a good-luck charm.

rabbit hawk The red-tailed hawk (*Buteo borealis*).

rabbit hutch A coop in which domestic rabbits are bred.

rabbit punch In boxing, a short, chopping blow at the base of the skull or back of the neck.

rab·bit·ry (rab'it·rē) *n. pl.* **·ries** A place where rabbits are kept; also, a group of rabbit hutches.

rab·ble[1] (rab'əl) *n.* A rude crowd; mob. —**the rabble** The populace; hoi polloi: used contemptuously. —*adj.* Of or pertaining to, suited to, or characteristic of a rabble; noisy; disorderly. —*v.t.* **·bled, ·bling** To mob. [? <RABBLE[3]]

rab·ble[2] (rab'əl) *n. Metall.* An iron implement, usually bent at one end, for stirring or skimming melted iron in puddling: also **rab'bler.** —*v.t.* **·bled, ·bling** To stir or skim with a rabble. [<F *râble* <L *rutabulum* poker]

rab·ble[3] (rab'əl) *v.t. & v.i.* **·bled, ·bling** *Scot. & Brit. Dial.* To speak or utter in an incoherent or disconnected manner; gabble. [Cf. Du. *rabbelen* speak indistinctly]

rab·ble·ment (rab'əl·mənt) *n.* **1** An uproar; disturbance. **2** A rabble; crowd.

rab·ble-rous·er (rab'əl·rou'zər) *n.* One who tries to incite mobs by arousing prejudices and passions; a demagog.

rab·bo·ni (ra·bō'nē) *n.* My great master: a term of address. [<Hebrew, aug. of RABBI]

rab·do·man·cy (rab'də·man'sē) See RHABDOMANCY.

Rab·e·lais (rab'ə·lā, *Fr.* rà·ble'), **François,** 1494?–1553, French humorist and satirist.

Rab·e·lai·si·an (rab'ə·lā'zē·ən, -zhən) *adj.* Characteristic of or like Rabelais or his works, especially with regard to his boisterous, coarse humor and his extravagance of satire and caricature. —*n.* A student or imitator of Rabelais. —**Rab'e·lai'si·an·ism, Rab'e·la'ism** *n.*

Ra·bi (rä'bē), **Isadore Isaac,** born 1898, U.S. physicist born in Austria.

Ra·bi·a (rə·bē'ə) *n.* Either of two Mohammedan months. See under CALENDAR (Mohammedan). [<Arabic *Rabī*, lit., spring]

rab·id (rab'id) *adj.* **1** Affected with, arising from, or pertaining to rabies; mad. **2** Unreasonably zealous; fanatical; violent. **3** Furious; raging. Also **rab'ic.** [<L *rabidus* <*rabere* be mad. Akin to RAGE.] —**rab'id·ly** *adv.* —**rab'id·ness** *n.*

ra·bies (rā'bēz, -bi·ēz) *n.* An acute infectious disease of animals, especially of dogs, caused by a virus and affecting the central nervous system; hydrophobia: readily transmissible to man by the bite of an affected animal. [<L <*rabere* rave] —**ra'bi·et'ic** (-et'ik) *adj.*

ra·ca (rä'kə, rə·kä') *adj.* Worthless; contemptible. *Matt.* v 22. [<LL <Gk. *rhaka* <Aramaic *rēqā*]

rac·coon (ra·kōōn') *n.* **1** An American nocturnal plantigrade carnivore (genus *Procyon*): the common North American raccoon (*P. lotor*) is grayish-brown, with a black cheek patch, and black-and-white-ringed bushy tail. **2** The fur of this animal. Also spelled *racoon*. [Alter. of Algonquian *arakunem* hand-scratcher]

RACCOON
(Body from 20 to 30 inches long; tail, 10 to 12 inches)

raccoon dog A wild dog (*Nyctereutes procyonoides*) of Japan and northeastern Asia, with long, loose fur, short ears, and a long bushy tail.

race[1] (rās) *n.* **1** One of the major subdivisions of mankind, regarded as having a common origin and exhibiting a relatively constant set of physical traits. On the basis of the more commonly used criteria such as stature, the cephalic index, the nasal index, prognathism, skull capacity, texture of the hair, degree of pilosity, color of the skin, and hair and eye color, mankind has been divided into primary stocks or races, each of which is regarded as including a varying number of ethnic groups. According to some, the primary stocks are: the Caucasoid, the Mongoloid, and the Negroid. A number of races, such as the Australian and Polynesian, are of doubtful classification. **2** Any group of people or any grouping of peoples having, or assumed to have, common characteristics. **3** A nation: the German *race*. **4** A genealogical or family stock; clan: the *race* of MacGregor. **5** Pedigree; lineage: a noble *race*. **6** Any class of beings having characteristics uniting them, or differentiating them from others: the *race* of lawyers. **7** *Biol.* A group of plants or animals, having characteristics clearly differentiating it from other groups within the same species, which breeds true except for minor variations; a variety: a *race* of wheat. **8** A stock, breed, or strain of domestic animals or plants. **9** A quality or aggregate of qualities by which origin is determined; especially, the characteristic flavor or taste of wine. See synonyms under AFFINITY, KIN, PEOPLE, SORT. [<F <Ital. *razza* <L (*gene*)*ratio* lineage, breed]

race[2] (rās) *n.* **1** A contest to determine the relative speed of the contestants. **2** Any contest. **3** Movement or progression; swift movement. **4** Duration of life; course; career. **5** A swift current of water or its channel. **6** A swift current or heavy sea resulting from the meeting of two tides: the Portland *Race*. **7** A sluice or channel by which to conduct water to or from a waterwheel or around a dam. See HEADRACE, MILLRACE, TAILRACE. **8** Any groove or channel along which some part of a machine slides or is guided. **9** Slipstream. —*v.* **raced, rac·ing** *v.i.* **1** To take part in a contest of speed. **2** To move at great or top speed. **3** To move at an accelerated or too great speed, usually because of decreased resistance: said of machinery. —*v.t.* **4** To contend against in a race. **5** To cause to take part in a race. **6** To cause to move at an accelerated or too great speed: to *race* an engine. [<ON *rās*. Akin to OE *rǣs* a rushing.]

race[3] (rās) *n.* A root; specifically, a root of ginger. [<OF *rais* <L *radix* root]

Race (rās), **Cape** The southeasternmost point of Newfoundland.

race-a·bout (rās'ə·bout') *n. Naut.* A sloop-rigged racing boat having a short bowsprit. Compare KNOCK-ABOUT.

race·course (rās'kôrs', -kōrs') *n.* The track over which a horse race, dog race, or the like is run. Also *racetrack*.

race horse A horse bred and trained for contests of speed; a racer.

race knife A tool having a very narrow U-shaped blade, used for tracing or outlining on metal or glass or for scribing on wood.

ra·ceme (rā·sēm', rə-) *n. Bot.* A centripetal or indeterminate flower cluster in which the flowers are arranged singly on distinct, nearly equal pedicels at intervals on an elongated common axis. [<L *racemus* cluster] —**rac·e·mif·er·ous** (ras'ə·mif'ər·əs) *adj.*

ra·ce·mic (rā·sē'mik, -sem'ik, rə-) *adj.* **1** *Bot.* Of, pertaining to, or contained in racemes or in grapes. **2** *Chem.* Indicating or relating to any chemical compound that is optically inactive. Also **rac·e·moid** (ras'ə·moid).

racemic acid *Chem.* A white, crystalline, optically inactive compound, $C_4H_6O_6$, contained with tartaric acid in certain grapes and extracts from tartar: it is separable into dextrorotatory and levorotatory forms.

rac·e·mism (ras'ə·miz'əm, rā·sē'miz·əm) *n. Chem.* The quality or condition of being racemic.

rac·e·mize (ras'ə·mīz, rā·sē'mīz) *v.t.* **·mized, ·miz·ing** *Chem.* To change (an optically active compound) into an optically inactive compound. —**rac'e·mi·za'tion** *n.*

rac·e·mose (ras'ə·mōs) *adj.* Arranged in or as in clusters or racemes: a *racemose* gland. Also **rac'e·mous.** [<L *racemosus*] —**rac'e·mose·ly** *adv.*

race psychology A division of psychology which investigates human traits and behavior in relation to racial factors, actual or assumed.

rac·er (rā'sər) *n.* **1** One who races, or one who contends in a race. **2** Anything having unusually rapid speed, as a race horse, steamer, or yacht; also, an automobile designed for racing. **3** A turntable on which a heavy gun is turned to left or right. **4** One of various colubrine snakes, as the blacksnake.

race riot A violent conflict between groups in the same community, based on differences of color, creed, etc.

race suicide The slow reduction in numbers of a people through voluntary failure on the part of individuals to maintain the birth rate at or above the level of the death rate.

race·track (rās'trak') *n.* A racecourse.

race·way (rās'wā') *n.* **1** A channel for conducting water. **2** A tube for protecting wires, as in a subway. **3** *U.S.* A racecourse for trotting horses.

Ra·chel (rā'chəl; *Fr.* rà·shel', *Pg.* rä·kel', *Sw.* rä'kel) A feminine personal name. Also *Ital.* **Ra·che·le** (rä·kā'lā). [<Hebrew, ewe] —**Rachel** The wife of Jacob; mother of Joseph and Benjamin. *Gen.* xxix 6. —**Rachel** (rà·shel') Stage name of Elisabeth Rachel-Félix, 1821–58, French tragic actress.

ra·chis (rā'kis) *n. pl.* **ra·chi·des** (rā'kə·dēz) or **·chis·es** **1** *Bot.* The axis of an inflorescence; a raceme. **2** *Ornithol.* The shaft of a feather, especially the part filled with pith, which bears the barbs. **3** *Anat.* The spinal column. Sometimes spelled *rhachis*. [<NL <Gk. *rhachis* spine] —**ra·chi·al** (rā'kē·əl) *adj.*

ra·chi·tis (rə·kī'tis) *n. Pathol.* Rickets. [<NL <Gk. *rhachitis* spinal inflammation] —**ra·chit'ic** (-kit'ik)

Rach·ma·ni·nov (räkh·mä'ni·nôf), **Sergei Vassilievich,** 1873–1943, Russian pianist and composer. Also **Rach·ma'ni·noff.**

ra·cial (rā'shəl) *adj.* Pertaining to or characteristic of a race, races, or descent. —**ra'cial·ly** *adv.*

ra·cial·ism (rā'shəl·iz'əm) *n.* **1** The doctrine of the preponderant influence of actual or assumed racial factors in the origin, development, and rank of various human societies; race prejudice. **2** Racism. —**ra'cial·ist** *n.*

Ra·ci·bórz (rä·chē'bōōsh) See RATIBOR.

Ra·cine (rə·sēn') A city on Lake Michigan in SE Wisconsin.

Ra·cine (rà·sēn'), **Jean,** 1639–99, French dramatist.

ra·cism (rā'siz·əm) *n.* An excessive and irrational belief in or advocacy of the superiority of a given group, people, or nation, on racial grounds alone; race hatred. —**ra'cist** *n.*

rack[1] (rak) *n.* **1** An open grating, framework, or the like, in or on which articles may be placed, as a frame to hold dishes, a tier or row of pigeonholes, or a framework to hold fodder for horses, cattle, or sheep. **2** A triangular frame for arranging the balls on a billiard table. **3** A device in an airplane for carrying bombs: also **bomb rack.** **4** *Mech.* A bar or the like having teeth that engage with those of a gearwheel, pinion, or worm gear. **5** A machine for stretching or making tense; especially, an intrument of torture which stretches the limbs of victims. **6** Torture or punishment as by the rack; hence, intense mental or physical suffering. **7** A wrenching or straining, as from a storm. —*v.t.* **1** To place or arrange in or on a rack. **2** To torture on the rack. **3** To cause suffering to; torment. **4** To strain, as with the effort of thinking: to *rack* one's brains. **5** To raise (rents) excessively: see RACK-RENT. ◆ Homophone: *wrack*. [ME *rekke*, prob. <MDu. *rec, recke* <*recken* stretch] —**rack'er** *n.*

rack[2] (rak) *n.* The single-foot. —*v.i.* To proceed or move with this gait. ◆ Homophone: *wrack*. [? Var. of ROCK[2]]

rack[3] (rak) *n.* **1** Thin, flying, or broken clouds. **2** Any floating vapor. —*v.i.* To move rapidly; send, as clouds before the wind. Also spelled *wrack*. [<Scand. Cf. ON *rek* drifting wreckage, *reka* drive, drift.]

rack[4] (rak) *n.* Wrack; wreck; demolition: obsolete except in the phrase "to go to rack and ruin." [Var. of WRACK[2]]
rack[5] *v.t.* To draw off from the lees, as liquor. ◆ Homophone: *wrack*. [<Provencal *arracar* < *raca* refuse of grapes]
rack and pinion *Mech.* A machine movement in which a toothed rack and a pinion mesh together for converting rotary motion into reciprocating motion or vice versa.
rack·et[1] (rak'it) *n.* 1 An implement for striking a ball, as in the game of tennis. It is a nearly elliptical hoop of bent wood, usually strung with catgut, and has a handle. 2 A large wooden sole or shoe to support the weight of a man or horse on swampy ground. 3 A snowshoe. 4 A ratchet: a misnomer. 5 An organ stop. 6 *sing. & pl.* A game resembling court tennis, played in a court with four walls. Often spelled *racquet*. [<F *raquette* <Arabic *rāha* palm of the hand]

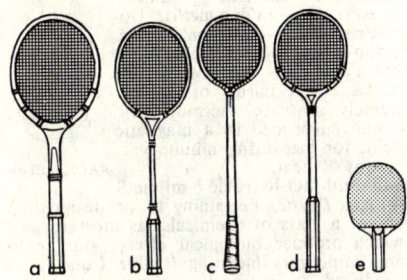

RACK AND PINION

TYPES OF RACKETS
 a. Tennis. c, d. Squash.
 b. Badminton. e. Table tennis.

rack·et[2] (rak'it) *n.* 1 A clattering, vociferous, or confused noise; fuss; commotion. 2 *Colloq.* a A scheme for getting money or other benefits by fraud, intimidation, or other illegitimate means. b Any business or occupation: the retailing *racket*. 3 Social activity or excitement. See synonyms under NOISE, TUMULT. —*v.i.* 1 To make a loud, clattering noise. 2 To indulge in noisy sport or diversion; carouse. [? Metathetic var. of dial. *rattick* make a din, clatter]
rack·et·eer (rak'ə·tir') *n.* 1 One who extorts money from, or seeks to gain control over, a person or organization by intimidation, fraud, violence, or other criminal means; one engaged in a racket. 2 Formerly, a bootlegger or rum-runner. — **rack'et·eer'ing** *n.*
rack·et·y (rak'it·ē) *adj.* Making a racket; noisy.
Rack·ham (rak'əm), **Arthur**, 1867–1939, English artist and illustrator.
rack·le (rak'əl) *adj. Brit. Dial. & Scot.* 1 Hasty or rough in action. 2 Strong; vigorous.
rack railway An inclined railway having a rack or toothed rail (**rack rail**) placed between the regular rails, in the cogs of which open pinions on the driving axle of the locomotive engage: also called *cogway*.
rack-rent (rak'rent') *n.* An exorbitant rent (equal or nearly equal to the full annual value of the property). —*v.t.* To exact rack-rent from or for. [<RACK[1] (stretch) + RENT[1]] — **rack'-rent'er** *n.*
rack·work (rak'wûrk') *n.* A mechanism with a rack, or rack and pinion, as the leading characteristic.
ra·con (rā'kon) *n.* A device for the immediate identification of friendly or hostile aircraft by means of radar signals automatically transmitted in code: adapted also as an aid in navigation. [<RA(DAR) (BEA)CON]
rac·on·teur (rak'on·tûr', *Fr.* rà·kôn'tœr') *n.* A skilled storyteller. [<F *raconter* recount] — **ra'con·teuse'** (-tœz', -tōz'; *Fr.* -tœz') *n. fem.*
ra·coon (ra·kōōn') See RACCOON.
rac·quet (rak'it) See RACKET[1].
rac·y (rā'sē) *adj.* **rac·i·er, rac·i·est** 1 Having a spirited or pungent interest; spicy; piquant: a *racy* style. 2 Having a characteristic flavor assumed to be indicative of origin, as wine; rich, fresh, or fragrant. 3 Suggestive; slightly immodest: a *racy* story. [<RACE[1]] — **rac'i·ly** *adv.* — **rac'i·ness** *n.*
Synonyms: flavorous, forcible, high-flavored, lively, pungent, rich, spicy, spirited. *Racy* applies in the first instance to the pleasing flavor characteristic of certain wines. *Pungent* denotes something sharply stimulating to the organs of taste or smell, as vinegar, ammonia; *piquant* denotes a quality similar in kind to *pungent* but less in degree, alluring and agreeable; *pungent* spices may be deftly compounded into a *piquant* sauce. As applied to literary products, *racy* refers to that which is striking, vigorous, pleasing; *spicy* to that which is stimulating to the mental taste, as spice is to the physical; *piquant* and *pungent* in their figurative use keep very close to their literal sense. *Antonyms:* cold, dull, flat, flavorless, insipid, stale, stupid, tasteless, vapid.
rad[1] (rad) *Scot. adj.* Frightened; afraid. — *v.t.* To fear.
rad[2] (rad) *n. Physics* A unit of absorbed nuclear radiation equivalent to 100 ergs of absorbed energy per gram of absorbing material. Compare REM, REP. [<R(ADIATION) + A(BSORBED) + D(OSE)]
ra·dar (rā'där) *n. Electronics* A locating device which instantaneously detects the presence and indicates the position of aircraft, ships, etc., by measuring the interval between the emission and return of high-frequency radio waves effective under varied conditions. [<RA(DIO) D(ETECTING) A(ND) R(ANGING)]
ra·dar·scope (rā'där·skōp) *n. Electronics* The oscilloscope of a radar set.
Rad·cliffe (rad'klif), **Ann**, 1764–1823, *née* Ward, English novelist.
rad·dle[1] (rad'l) See REDDLE, RUDDLE.
rad·dle[2] (rad'l) *v.t.* **·dled, ·dling** To intertwine or weave together. — *n. obs.* raddle a wattle <AF *reidele*, OF *reddale* stout stick]
ra·deau (ra·dō') *n. French* A raft; float.
Ra·dek (rä'dek), **Karl**, born 1885, U.S.S.R. revolutionist and journalist.
Ra·detz·ky (rä·dets'kē), **Count Joseph Wenzel**, 1766–1858, Austrian field marshal.
Rad·ford (rad'fərd), **Arthur William**, born 1896, U.S. admiral; chairman, joint chiefs of staff 1953–57.
ra·di·ac (rā'dē·ak) *n. Physics* A Geiger counter. [<RA(DIOACTIVITY) D(ETECTION), I(DENTIFICATION) AND C(OMPUTATION)]
ra·di·al (rā'dē·əl) *adj.* 1 Pertaining to, consisting of, or resembling a ray or radius. 2 Extending from a center in the manner of rays. 3 Of or pertaining to the radius or a radiating part. 4 Developing uniformly on all sides. — *n.* A radiating part. — **ra'di·al·ly** *adv.*
radial engine A multicylinder internal-combustion engine having its cylinders arranged like the spokes in a wheel.
ra·di·an (rā'dē·ən) *n. Math.* 1 An arc equal in length to the radius of the circle of which it is a part. 2 The angle subtended by such an arc: 2π radians = $360°$, π radians = $180°$ or 1 radian = $(180/\pi)°$ or $57° 17' 44.80625'' +$. [<RADIUS]
ra·di·ance (rā'dē·əns) *n.* The quality or state of being radiant; brilliant or sparkling luster; brightness; effulgence. Also **ra'di·an·cy, ra'di·ant·ness**.
ra·di·ant (rā'dē·ənt) *adj.* 1 Emitting rays of light or heat. 2 Beaming with light or brightness, kindness, or love: a *radiant* smile. 3 Resembling rays; consisting of or transmitted by radiations: *radiant* heat. See synonyms under BRIGHT. — *n.* 1 A straight line proceeding from and conceived as revolving around a given point. 2 *Astron.* That point in the heavens from the direction of which, during a meteoric shower, the meteors seem to shoot. 3 The luminous point from which light proceeds or is made to radiate. 4 That which radiates. [<L *radians, -antis*, ppr. of *radiare* emit rays < *radius* ray] — **ra'di·ant·ly** *adv.*
radiant energy *Physics* 1 The energy associated with and transmitted by waves emanating from some specified source, as of light, heat, or sound. 2 The energy of radium, atomic disintegration, X-rays, electromagnetic radiation, or the like.
ra·di·ate (rā'dē·āt) *v.* **·at·ed, ·at·ing** *v.i.* 1 To emit rays or radiation; be radiant. 2 To issue forth in rays, as light from the sun. 3 To spread out from a center, as the spokes of a wheel. — *v.t.* 4 To send out or emit in rays. 5 To cause to spread as if from a center; diffuse; disseminate. — *adj.* (-de·it) 1 Divided or separated into rays; having rays; radiating. 2 *Bot.* Bearing rays or ray flowers. 3 *Zool.* Characterized by radial symmetry, as echinoderms and coelenterates. 4 Adorned with rays, as a head on a coin; radiated. — *n.* (-dē·it) 1 An organism having radial symmetry. 2 A ray or raylike projection. [<L *radiatus*, pp. of *radiare* emit rays. See RADIANT.] — **ra'di·a·tive** *adj.*
ra·di·a·tion (rā'dē·ā'shən) *n.* 1 The act of radiating or the state of being radiated. 2 *Physics* a That which is radiated, especially the energy of atoms and molecules undergoing internal changes. b The propagation of radiant energy. c The stages of emission, absorption, and transmission involved in such propagation: distinguished from *conduction*. 3 *Biol.* Adaptive radiation.
radiation pressure *Physics* The force exerted upon an exposed surface by radiant energy, as from light or electromagnetic waves.
radiation sickness *Pathol.* A morbid condition due to the body's absorption of excess radiation and marked by fatigue, nausea, vomiting, internal hemorrhage, and progressive tissue breakdown.
ra·di·a·tor (rā'dē·ā'tər) *n.* 1 That which radiates. 2 A chamber, coil, or flat hollow vessel, through which is passed steam or hot water for warming a building or apartment. 3 In engines, a nest of tubes for cooling water flowing through them. — **ra'di·a·to'ry** (-ə·tôr'ē, -tō'rē) *adj.*
rad·i·cal (rad'i·kəl) *adj.* 1 Of, proceeding from, or pertaining to the root or foundation; essential; fundamental; inherent; basic. 2 Thoroughgoing; unsparing; extreme: a *radical* operation; *radical* measures. 3 *Math.* Pertaining to the root or roots of a number. 4 In philology, belonging or referring to a root or a root syllable; underived. 5 *Bot.* Springing from or belonging or relating to the root: *radical* leaves. 6 *Chem.* Pertaining to a radical. 7 Of or pertaining to political radicals. — *n.* 1 One who carries his theories or convictions to their furthest application; an extremist. 2 In politics, one who advocates wide-spread governmental changes and reforms at the earliest opportunity. 3 The primitive or underived part of a word; a primitive word or syllable; a root; radicle. 4 *Math.* a A quantity of which the root is to be extracted or used in calculation; a radical expression. b The radical sign. 5 *Chem.* A fundamental constituent or part of a compound; specifically, a group of atoms which acts as a unit in a compound and may either pass unchanged through a series of reactions or be replaced as though it were a single atom. ◆ Homophone: *radicle*. [<LL *radicalis* having roots <L *radix, radicis* root] — **rad'i·cal·ness** *n.*
Synonyms (adj.): basic, complete, constitutional, entire, essential, extreme, fundamental, ingrained, inherent, innate, native, natural, organic, original, perfect, positive, primary, primitive, thorough, thoroughgoing, total. The widely divergent senses in which the word *radical* is used, by which it can be at some time interchanged with any word in the above list, are all formed upon the one primary sense of that which is connected with the root (Latin *radix*). A *radical* difference is one that springs from the root, and is thus *constitutional, essential, fundamental, organic, original*; a *radical* change is one that does not stop at the surface, but reaches down to the very root, and is *entire, thorough, total*; since the majority find superficial treatment of any matter the easiest and most comfortable, *radical* measures, which strike at the root of evil or need, are apt to be looked upon as *extreme*. See NATURAL. *Antonyms:* conservative, inadequate, incomplete, moderate, palliative, partial, superficial.

radical expression *Math.* A surd, or an algebraic expression involving a surd.

rad·i·cal·ism (rad′i·kəl·iz′əm) *n.* **1** The state of being radical. **2** Advocacy of thorough-going or extreme measures.

rad·i·cal·ly (rad′ik·lē) *adv.* **1** Completely; thoroughly; fundamentally. **2** With reference to root or origin; originally; primitively.

radical sign *Math.* The symbol √ placed before a quantity to indicate that its root is to be taken: a modification of the letter *r* (Latin *radix* root). A number written above it (called its *index*) shows what root is to be taken; thus $\sqrt[4]{a}$ stands for the fourth root of *a*; when used without a superior number, the symbol means square root of.

rad·i·cand (rad′i·kand′) *n. Math.* The quantity under the radical sign: $x + 1$ is the *radicand* of $\sqrt{x+1}$. [<RADIC(AL) + *-and,* as in *multiplicand*]

rad·i·cel (rad′i·sel) *n.* A rootlet. [<NL *radicella,* dim. of L *radix, radicis* root]

rad·i·cle (rad′i·kəl) *n.* **1** *Bot.* **a** The embryonic root below the cotyledon of a plant. **b** A diminutive root or rootlet. **2** *Anat.* A rootlike part, as the stem of an embryo, the initial fiber of a nerve, the beginning of a vein, etc. **3** *Chem.* A radical. ◆ Homophone: *radical.* [<L *radicula,* dim. of *radix, radicis* root]

ra·di·i (rā′dē·ī) Plural of RADIUS.

ra·di·o (rā′dē·ō) *n. pl.* **·os** **1** The science, art, and process of communicating by means of radiant energy transmitted directly through space in waves. **2** The wireless transmission of radio waves within assigned frequencies, and their reception by devices adapted for reconverting the frequencies into their corresponding original signals. **3** A radio program or broadcast; also, the combined operations for its production. **4** A radio receiving set and its accessories. **5** A radio message or radiogram. **6** The exploitation and development of radio as a commercial enterprise; the radio business and industry. —*adj.* **1** Of, pertaining to, designating, employing, or produced by radiant energy, especially in the form of electromagnetic waves: a *radio* beam. **2** Wireless. —*v.t.* & *v.i.* To transmit (a message, etc.) or communicate with (someone) by radiotelegraphy or radiotelephony. [<RADIO(TELEGRAPHY)]

radio– *combining form* **1** *Anat.* Radial; pertaining to the radius: *radiodigital,* of the fingers on the radial edge of the hand. **2** Radio; produced by or related to radio: *radiogram.* **3** *Chem.* Radioactive; of, produced by, or causing radioactivity: *radioscope, radiothorium.* **4** *Med.* Radiant energy; using radiant energy: *radiotherapy.* [<L *radius* a ray]

ra·di·o·ac·tive (rā′dē·ō·ak′tiv) *adj.* Pertaining to, exhibiting, caused by, or characteristic of radioactivity: a *radioactive* isotope.

radioactive series *Physics* The sequence of disintegration products through which a radioactive element passes before reaching a final stage in one of the stable isotopes of lead. The three principal series are those of uranium, thorium, and actinium.

ra·di·o·ac·tiv·i·ty (rā′dē·ō·ak·tiv′ə·tē) *n. Physics* **1** The propagation of radiant energy. **2** The spontaneous nuclear disintegration of certain elements and isotopes, with the emission of nucleons or of electromagnetic radiation. **3** A particular form of such disintegration: gamma *radioactivity.*

radio announcer A person who introduces a radio program, its sponsor, and participants, identifies the station, reads prepared material, and makes extemporaneous remarks as required.

radio astronomy That branch of astronomy and astrophysics which studies celestial phenomena by the interception and analysis of radio waves emitted by stars and other objects in interstellar space.

ra·di·o·au·tog·ra·phy (rā′dē·ō·ô·tog′rə·fē) *n.* Autoradiography. [<RADIO- + AUTOGRAPH + -Y]

ra·di·o·au·to·gram (rā′dē·ō·ô′tə·gram) *n.* Autoradiograph. Also **ra′di·o·au′to·graph** (-graf, -gräf).

radio beacon A stationary radio transmitter which sends out characteristic signals for the guidance of ships and aircraft.

radio beam **1** A steady flow of radio signals concentrated along a given course or direction. **2** The narrow zone marked out for the guidance of aircraft by the overlapping of recurrent signals transmitted from ground radio stations on either side of an assigned flight course.

ra·di·o·broad·cast (rā′dē·ō·brôd′kast′, -käst′) *v.t.* & *v.i.* **·cast** or **·cast·ed, ·cast·ing** To broadcast by radio. —*n.* Broadcast (def. 1). —**ra′di·o·broad′cast′er** *n.* —**ra′di·o·broad′cast′·ing** *n.*

ra·di·o·car·bon (rā′dē·ō·kär′bən) *n. Physics* The radioactive isotope of carbon of mass 14, with a half–life of about 5,700 years: it is much used in the dating of fossils, artifacts, and certain kinds of geological formations. Also called *carbon 14.*

ra·di·o·car·di·o·gram (rā′dē·ō·kär′dē·ə·gram′) *n.* The graphic record made in radiocardiography.

ra·di·o·car·di·og·ra·phy (rā′dē·ō·kär′dē·og′rə·fē) *n. Med.* A method for studying the blood flow through the heart by recording the passage of injected radioisotopes with the aid of a specially constructed Geiger counter.

radio channel A band of wave frequencies sufficiently wide to permit effective transmission of radio signals of a given type, power, and range.

ra·di·o·chem·is·try (rā′dē·ō·kem′is·trē) *n.* That branch of chemistry dealing with the properties and reactions of radioactive substances, as radium and thorium.

radio circuit A radio system consisting of two stations in direct communication with each other.

radio compass *Aeron.* A direction–finder serving to determine the position of a radio transmitting station.

radio conductor Any material or apparatus that indicates, by some alteration of its conductivity, the presence and strength of electric waves, such as the coherer of a wireless telegraph.

radio control **1** Control by radio signals. **2** *Aeron.* The automatic control of aircraft steering mechanisms by electromagnetic impulses transmitted from a distance.

ra·di·ode (rā′dē·ōd) *n.* **1** A radium container, built to prevent any dangerous leakage of radioactivity. **2** *Med.* An apparatus used in some forms of radiotherapy. [<RADIO- + -ODE[1]]

ra·di·o·dust (rā′dē·ō·dust′) *n.* Radioactive dust particles precipitated from the atmosphere, especially in the fall–out from an atomic or thermonuclear bomb.

ra·di·o·el·e·ment (rā′dē·ō·el′ə·mənt) *n. Physics* **1** An element exhibiting radioactivity. **2** Any of the disintegration products of such elements, as radon, thoron, etc., which are themselves radioactive.

radio fix The position of an aircraft, ship, or radio transmitter, as determined with reference to radio signals from two or more stations, or by similar means.

radio frequency Any wave frequency, or set of frequencies, adapted for the transmission of radio signals. The range is roughly from the upper limit of normal audibility to the lower limit of heat and light waves, or upwards from about 10 kilocycles per second.

ra·di·o·ge·net·ics (rā′dē·ō·jə·net′iks) *n.* The study of genetics in relation to the effects of radioactivity upon the processes of inheritance and the nature of hereditary changes. —**ra′·di·o·ge·net′ic** *adj.*

ra·di·o·ge·nic (rā′dē·ō·jē′nik, -jen′ik) *adj.* Resulting from or developed by radioactivity.

ra·di·o·gram (rā′dē·ō·gram′) *n.* **1** A message sent by wireless telegraphy. **2** A radiographic negative or print.

ra·di·o·graph (rā′dē·ō·graf′, -gräf′) *n.* A negative or picture made by means of radioactivity; an X–ray photograph. —*v.t.* To make a radiograph of. —**ra′di·og′ra·pher** (-og′rə·fər) *n.* —**ra′di·o·graph′ic** or **·i·cal** *adj.* —**ra′di·og′ra·phy** *n.*

ra·di·o·i·so·tope (rā′dē·ō·ī′sə·tōp) *n. Physics* A radioactive isotope, usually one produced artificially from a normally stable element: extensively used in biological and physical research and used in medicine for diagnostic and therapeutic purposes.

ra·di·o·lar·i·an (rā′dē·ō·lâr′ē·ən) *n.* Any member of an order (*Radiolaria,* class *Sarcodina*) of marine protozoans having typically a siliceous skeleton enclosing a perforated membrane. —*adj.* Of or pertaining to the *Radiolaria.* [<NL *Radiolaria,* name of the order <*radiolus,* dim. of L *radius* ray]

ra·di·o·lo·ca·tion (rā′dē·ō·lō·kā′shən) *n.* Radar.

ra·di·ol·o·gy (rā′dē·ol′ə·jē) *n.* That branch of science that relates to radiant energy and its applications, especially in the diagnosis and treatment of disease. [<RADIO- + -LOGY] —**ra·di·o·log·i·cal** (rā′dē·ə·loj′i·kəl) or **ra′di·o·log′ic** *adj.* —**ra′di·ol′o·gist** *n.*

ra·di·o·lu·cent (rā′dē·ō·lo͞o′sənt) *adj.* Permeable to X–rays and other forms of electromagnetic radiation. See RADIOPAQUE.

ra·di·o·lu·mi·nes·cence (rā′dē·ō·lo͞o′mə·nes′·əns) *n.* Luminescence produced by, or resulting from, any form of radiant energy, as X–rays, radioactivity, etc. —**ra′di·o·lu′mi·nes′·cent** *adj.*

ra·di·o·ma·te·ri·al (rā′dē·ō·mə·tir′ē·əl) *n.* Any material that is, or has been made, radioactive.

ra·di·om·e·ter (rā′dē·om′ə·tər) *n.* An instrument for detecting and measuring radiant energy by converting it into mechanical energy, as by the rotation of blackened disks suspended in a vacuum and exposed to sunlight. [<RADIO- + METER[1]] —**ra′di·o·met′ric** (-ō·met′rik) *adj.* —**ra′di·om′e·try** *n.*

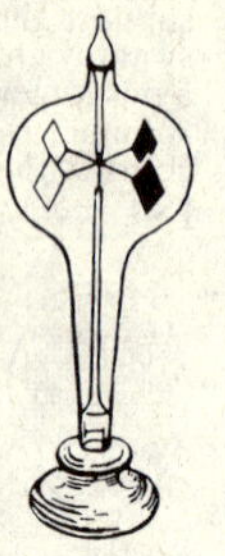
RADIOMETER

ra·di·o·mi·crom·e·ter (rā′dē·ō·mī·krom′ə·tər) *n.* An instrument, consisting primarily of an extremely sensitive thermoelectric couple suspended in a magnetic field, for measuring minute variations of heat.

ra·di·o·mi·met·ic (rā′dē·ō·mi·met′·ik) *adj. Physics* Pertaining to or designating any of a class of chemicals, as mustard gas, which produce biological effects similar to and sometimes indistinguishable from those produced by radioactive substances.

ra·di·o·nu·clide (rā′dē·ō·no͞o′klīd, -nyo͞o′-) *n. Physics* A radioactive nuclide.

ra·di·o·paque (rā′dē·ō·pāk′) *adj.* Impermeable to X–rays or other forms of electromagnetic radiation. [<RADIO- + OPAQUE]

ra·di·o·phone (rā′dē·ō·fōn′) *n.* **1** Any device for the production or transmission of sound by radiant energy. **2** A radiotelephone. —**ra′·di·o·phon′ic** (-fon′ik) *adj.* —**ra′di·oph′o·ny** (-of′ə·nē) *n.*

ra·di·o·pho·tog·ra·phy (rā′dē·ō·fə·tog′rə·fē) *n.* The transmission of a photograph by radio in such a way that each spot on the picture is reproduced by an electric impulse. —**ra′di·o·pho′to·graph** (-fō′tə·graf, -gräf) *n.*

ra·di·o·prax·is (rā′dē·ō·prak′sis) *n.* Radiotherapy.

ra·di·o·scope (rā′dē·ō·skōp′) *n.* An apparatus for detecting radioactivity or X–rays.

ra·di·os·co·py (rā′dē·os′kə·pē) *n.* Examination of opaque bodies with the aid of X–rays or some other form of radiant energy. [<RADIO- + -SCOPY] —**ra′di·o·scop′ic** (-skop′ik) or **·i·cal** *adj.*

ra·di·o·sen·si·tive (rā′dē·ō·sen′sə·tiv) *adj.* **1** Sensitive to X–rays and ultraviolet rays. **2** *Med.* Reducible or destructible by X–rays, as certain tumors.

radio shielding *Aeron.* Metallic covering on the electric wiring and ignition apparatus of an aircraft, intermittently grounded to the frame in order to eliminate disturbances in radio communication.

ra·di·o·sonde (rā′dē·ō·sond′) *n. Meteorol.* A device, attached to a small balloon sent aloft, which measures the pressure, temperature, and humidity of the upper air and radios the data to the ground. Also **ra′di·o·me′te·or·o·graph′** (-mē′tē·ər·ə·graf′, -gräf′). [<RADIO- + F *sonde* sounding]

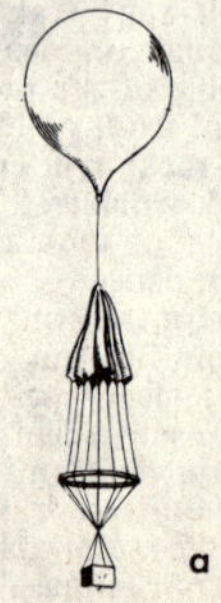
RADIOSONDE
a. Instrument box.

radio spectrum The full range of frequencies pertaining to and associated with radiant energy; specifically, those frequencies employed in radio and television.

radio star Any of a large number of stars which may be identified and studied by means of the characteristic electromagnetic impulses which they emit.

radio station An installation of all the equipment and apparatus necessary for effective radio communication.

ra·di·o·stron·tium (rā'dē·ō·stron'shəm, -tē·əm) *n. Physics* Strontium 90.

ra·di·o·tel·e·gram (rā'dē·ō·tel'ə·gram) *n.* A message sent by radiotelegraphy.

ra·di·o·te·leg·ra·phy (rā'dē·ō·tə·leg'rə·fē) *n.* Telegraphic communication by means of radio waves. **—ra'di·o·tel'e·graph'ic** (-tel'ə·graf'ik) *adj.* **—ra'di·o·tel'e·graph** (-graf, -gräf) *n.*

ra·di·o·tel·e·phone (rā'dē·ō·tel'ə·fōn) *n.* A telephone set that, without the agency of connecting wires, transmits a verbal message to a similar set by means of radio waves. **—ra'di·o·tel'e·phon'ic** (-tel'ə·fon'ik) *adj.* **—ra'di·o·te·leph'o·ny** (-tə·lef'ə·nē) *n.*

radio telescope A sensitive astronomical instrument designed on the principle of a radio receiver, but adapted to intercept and amplify electromagnetic waves in the megacycle range emanating from interstellar space.

ra·di·o·ther·a·py (rā'dē·ō·ther'ə·pē) *n.* The use of X–rays and other forms of radioactivity in the treatment of disease.

ra·di·o·tho·ri·um (rā'dē·ō·thôr'ē·əm, -thō'·rē·əm) *n.* A radioactive product of the thorium series, with a half–life of 1.9 years.

ra·di·o·tox·ic (rā'dē·ō·tok'sik) *adj. Med.* Of or pertaining to the toxic effect of radioactive materials, especially radioisotopes. **—ra'di·o·tox·ic'i·ty** (-tok·sis'ə·tē) *n.*

radio transcription An electrically recorded radio program, speech, musical selection, or the like, intended for subsequent broadcasting.

radio tube A vacuum tube for radio.

radio wave Any of a class of electromagnetic waves propagated at frequencies intermediate between those of audible sound and infrared.

rad·ish (rad'ish) *n.* **1** A tall, branching herb *(Raphanus sativus)* of the mustard family. **2** Its pungent, edible root, commonly eaten raw. [<F *radis* <Ital. *radice* <L *radix, radicis* root. Doublet of RADIX.]

ra·di·um (rā'dē·əm) *n.* A powerfully radioactive metallic element (symbol Ra) chemically related to barium, obtained principally as a disintegration product of uranium but found also in minute quantities in sea water and in certain plants and animals. It has a half–life of about 1,600 years and its atoms undergo spontaneous disintegration associated with a wide range of properties and effects, emitting alpha and beta particles and gamma rays in a succession of stages terminating in radium G, a stable isotope of lead. See ELEMENT. [<NL <L *radius* ray]

radium F Polonium of mass 210.

radium therapy The treatment of skin diseases and of cancer by means of radium.

ra·di·us (rā'dē·əs) *n. pl.* **·di·i** (-dē·ī) **1** A straight line from the center of a circle or sphere to its periphery. **2** *Anat.* The thicker and shorter bone of the forearm, on the same side as the thumb. **3** *Bot.* A ray floret of a composite flower; also, a branch of an umbel. **4** *Zool.* **a** In radiolarians and similar organisms, the imaginary line or plane dividing the body into two theoretically equal parts. **b** A ray or radiating part, as, the barb of a feather. **c** A lateral part of a cirriped shell when overlapping others. **5** *Entomol.* One of the main longitudinal veins of an insect's wings. **6** In a sextant, quadrant, etc., a pivoted arm, mounted so as to move radially, as on a graduated arc or circle. **7** *Mech.* A wheel spoke; a rod or bar which with others extends from a common point. **8** A circular area or boundary measured by the length of its radius. **9** Sphere, scope, or limit, as of activity. **10** A fixed limit of travel beyond which higher fares are charged. [<L, orig., rod, spoke of a wheel, hence radius, ray of light. Doublet of RAY]

radius vector *pl.* **radius vectors** or **ra·di·i vec·to·res** (rā'dē·ī vek·tôr'ēz, -tō'rēz) **1** *Math.* **a** The distance from a fixed origin to any point of a curve. **b** The distance from a point to the pole in the polar coordinate or spherical coordinate system. **2** *Astron.* A line from a center of attraction to a body describing an orbit about it.

ra·dix (rā'diks) *n. pl.* **rad·i·ces** (rad'ə·sēz, rā'·də-) or **ra·dix·es** **1** *Rare* The origin or source. **2** *Math.* A number or symbol used as the basis of a scale of enumeration: 10 is the *radix* of the common system of logarithms. **3** *Bot.* The root of a plant. **4** An original word from which others are derived; radical; root; etymon. [<L, root. Doublet of RADISH.]

Rad·nor·shire (rad'nər·shir) A county of eastern Wales; 471 square miles; county town, Presteigne. Also **Rad'nor.**

Ra·dom (rä'dôm) A city in east central Poland.

ra·dome (rā'dōm) *n. Electronics* A protective housing, radiolucent to radar waves, for the antenna and other equipment of a radar assembly. [<RA(DAR) + DOME]

ra·don (rā'don) *n.* A heavy, gaseous, radioactive element (symbol Rn), an emanation of radium with a half–life of about 4 days: formerly called *niton.* [<RAD(IUM) + -ON, as in *neon*]

rad·u·la (raj'ōō·lə) *n. pl.* **·lae** (-lē) *Zool.* A rasplike organ, the odontophore or lingual ribbon of a mollusk. [<L, scraper <*radere* scrape] **—rad'u·lar** *adj.*

Rae·burn (rā'bərn), **Sir Henry,** 1756–1823, Scottish painter.

Rae·der (rā'dər), **Erich,** born 1876, German admiral in World War II.

Rae·mae·kers (rä'mä·kərz), **Louis,** 1869–1956, Dutch political cartoonist.

Ra·fa·el (rä'fä·el') Spanish form of RAPHAEL. Also *Ital.* **Raf·fa·e·le** (räf'fä·ā'lā) or **Raf·fa·el·lo** (räf'fä·el'lō).

raff (raf) *n.* **1** The rabble; riff–raff. **2** *Scot. & Brit. Dial.* A disorderly collection. [<RIFF-RAFF]

raf·fi·a (raf'ē·ə) *n.* **1** A cultivated palm *(Raphia pedunculata)* of Madagascar, the leafstalks of which furnish fiber for making hats, mats, baskets, etc. **2** Its fiber. Also spelled *raphia.* [<Malagasy *rafia*]

raf·fi·nose (raf'ə·nōs) *n. Chem.* A colorless crystalline carbohydrate, $C_{18}H_{32}O_{16}$, having a mildly sweetish taste, found in cottonseed and in the molasses of the sugar beet: it hydrolyzes into fructose, galactose, and glucose. [<F *raffiner* refine + -OSE[2]]

raff·ish (raf'ish) *adj.* **1** Tawdry; gaudy; flashy. **2** Disreputable. [<RAFF + -ISH[1]]

raf·fle[1] (raf'əl) *n.* A form of lottery in which a number of people buy chances on an object. —*v.* **·fled ·fling** *v.t.* To dispose of by a raffle: often with *off.* —*v.i.* To take part in a raffle. [<OF *rafle* a clean sweep at dice <*rafler* snatch, prob. <Gmc. Cf. G *rafflen* snatch, freq. of *raffen* seize.] **—raf'fler** *n.*

raf·fle[2] (raf'əl) *n.* A jumble of rubbish; tangle: a nautical term. [Prob. <RAFF]

raf·fle·si·a (ra·flē'zhē·ə, -zē·ə) *n.* Any plant of the genus *Rafflesia* (family *Rafflesiaceae*), parasitic on the stems of the Malayan grape, and having huge, stemless, malodorous flowers and no leaves. [<NL, after Sir T. S. *Raffles,* 1781–1826, British governor in Sumatra, who discovered it] **—raf·fle'si·a'·ceous** (-zē·ā'shəs) *adj.*

raft[1] (raft, räft) *n.* A float of logs, planks, etc., fastened together for transportation by water. —*v.t.* **1** To transport on a raft. **2** To form into a raft. —*v.i.* **3** To travel by, be employed on, or manage a raft. [<ON *raptr* log]

raft[2] (raft, räft) *n. Colloq.* A large number or an indiscriminate collection of any kind. [<RAFF]

raft·er (raf'tər, räf'-) *n.* A timber or beam giving form, slope, and support to a roof. [OE *ræfter*]

rafts·man (rafts'mən, räfts'-) *n. pl.* **·men** (-mən) One who manages or works on a raft.

rag[1] (rag) *v.t.* **ragged, rag·ging** *Slang* **1** To tease or irritate. **2** To scold. **3** *Brit.* To play a practical joke on. —*n. Brit.* A ragging. [? <ON *ragna* curse, swear]

rag[2] (rag) *n.* **1** A torn piece of cloth; a fragment or semblance of anything. **2** *pl.* Cotton or linen textile remnants used in the making of rag paper. **3** *pl.* Tattered or shabby clothing; hence, any clothing: a jocular use. **4** A cloth of any kind, or something resembling one or characterized as such: used humorously or in disparagement. **5** In citrus fruits, the axis and carpellary walls. **—glad rags** *Slang* One's best clothes. **—to chew the rag** *Slang* To talk or argue at great length. [<ON *rögg* tuft or strip of fur]

rag[3] (rag) *n.* **1** A roofing slate rough on one side, and measuring 2 x 3 feet. **2** *Brit.* Any hard rock of cellular or coarsely granular texture. [Origin uncertain]

rag[4] (rag) *v.t.* **ragged, rag·ging** To compose or play in ragtime. —*n.* Ragtime.

rag·a·muf·fin (rag'ə·muf'in) *n.* Anyone, especially a boy, wearing very ragged clothes; a vagabond. [after *Ragamoffyn,* demon in a 15th century mystery play <RAG[2] + fanciful ending]

rag–bag (rag'bag') *n.* A bag in which rags or scraps of unused cloth are kept.

rag carpet A carpet made from rags woven together by hand.

rag doll A cloth doll stuffed with rags.

rage (rāj) *n.* **1** Violent anger; wrath; fury. **2** Any great violence or intensity, as of a fever or a storm. **3** Extreme eagerness or emotion; ardent desire; great enthusiasm. **4** Any object eagerly sought after; a fad; fashion: Crossword puzzles are all the *rage.* See synonyms under ANGER, VIOLENCE. —*v.i.* **raged, rag·ing** **1** To speak, act, or move with unrestrained anger; feel or show violent anger. **2** To act or proceed with great violence: The storm *raged* for three days. **3** To spread or prevail uncontrolled, as an epidemic. [<OF <LL *rabia* <L *rabies* madness] **—rag'ing** *adj.* **—rag'ing·ly** *adv.*

rag·ged (rag'id) *adj.* **1** Rent or worn into rags; frayed: a *ragged* coat. **2** Wearing worn, frayed, or shabby garments; ill–dressed. **3** Of rough, broken or uneven character or aspect; harsh; dissonant: *ragged* rocks, *ragged* sounds. **4** Naturally of a rough or shaggy appearance (the original meaning): a *ragged* horse or sheep. See synonyms under ROUGH. **—rag'·ged·ly** *adv.* **—rag'ged·ness** *n.*

ragged edge *Colloq.* The extreme or precarious edge; the verge: the *ragged edge* of starvation; *ragged edge* of insanity. **—on the ragged edge** Dangerously near to losing one's self–control, sanity, etc.

ragged lady Fennelflower.

ragged robin A slender perennial European herb *(Lychnis floscuculi)* having red or pink flowers in panicles; the cuckoo flower.

rag·i (rag'ē, rä'gē) *n.* A cereal grass *(Eleusine coracona)* of the East Indies. Also **rag'ee, rag'·gy.** [<Hind. *rāgī*]

rag·lan (rag'lən) *n.* An overcoat or topcoat, the sleeves of which extend in one piece up to the collar. —*adj.* Denoting a garment with such sleeves. [after Lord Fitzroy *Raglan*]

Rag·lan (rag'lən), **Lord Fitzroy,** 1788–1855, English field marshal.

rag·man (rag'man', -mən) *n. pl.* **·men** (-men', -mən) One who buys and sells old rags and other waste; a ragpicker.

Rag·na·rök (räg'nä·rœk) In Norse mythology, the twilight of the gods, and the doomsday of the world, preceding its regeneration. Also **Rag'na·rok** (-rok). [<ON <*ragna* of the gods (genitive pl. of *regin*) + *rök* judgment]

ra·gout (ra·gōō') *n.* A highly seasoned dish of meat and vegetables stewed; hence, something spicy or piquant. —*v.t.* **ra·gouted** (-gōōd'), **ra·gout·ing** (-gōō'ing) To make into a ragout. [<F <*ragouter* revive the appetite <*re-* anew + *à* (<L *ad*) to + *goût* (<L *gustus*) taste]

rag·pick·er (rag'pik'ər) *n.* One who picks up rags and other junk for a livelihood.

rag rug A rug made of rags.

rag·stone (rag'stōn') *n.* **1** Rag; a rough, sandy, fossiliferous limestone: also **ragg. 2** Stone quarried in thin slabs, as for pavements. Also **ragg'stone'.**

rag–tag (rag'tag') *n.* Ragged people; the rabble. Also **rag–tag and bobtail.**

rag·time (rag'tīm') *n.* **1** A kind of American dance music, developed from about 1890 to 1920, achieving its effects by highly syncopated rhythm in fast time. **2** The rhythm of this dance. [<*ragged time*]

Ra·gu·el (rə·gyōō'el) One of the seven archangels of Hebrew and Christian legend.

Ra·gu·sa (rä·gōō'sä) **1** Italian name for DUBROVNIK. **2** A city in SE Sicily.

rag·weed (rag'wēd') *n.* **1** A coarse, very common, annual or perennial herb (genus *Ambrosia*), especially the common ragweed *(A. artemisiifolia),* which induces hay fever, and the **great ragweed** *(A. trifida),* a tall species

add,āce,câre,pälm; end,ēven; it,īce; odd,ōpen,ôrder; tōōk,pōōl; up,bûrn; ə = a in *above,* e in *sicken,* i in *clarity,* o in *melon,* u in *focus;* yōō = u in *fuse;* oi,oil; ou,pout; ch,check; g,go; ng,ring; th,thin; th,this; zh,vision. Foreign sounds à,œ,ü,kh,ń; and ◆: see page xx. < from; + plus; ? possibly.

with stout hairy stem 5 to 15 feet high: also called *hogweed.* 2 *Brit.* The ragwort.

rag·wort (rag′wûrt′) *n.* Any one of several herbs of the genus *Senecio,* as the European ragwort (*S. jacoboea*), a tall, smooth, cottony plant, with bright-yellow flowers.

rah (rä) *interj.* Hurrah! a cheer used chiefly in college yells. [<HURRAH]

Ra·hab (rā′hab) A harlot of Jericho who sheltered two Israelite spies. *Josh.* ii 1.

Ra·hab (rā′hab) In the Old Testament, a symbolical name for Egypt. *Isaiah* li 9.

Ra·hel (rä′həl) German form of RACHEL.

ra·ia (rä′yə, rī′ə) See RAYAH.

Ra·ia·te·a (rä′yä·tā′ä) The largest of the Society Islands in the Leeward group; 92 square miles.

rai·ble (rā′bəl) *v.t.* & *v.i. Scot.* To gabble.

raid (rād) *n.* 1 A hostile or predatory incursion by a rapidly moving body of troops or an armed vessel; a foray. 2 An attack by military aircraft; an air raid. 3 Any sudden invasion, capture, or irruption, as by the police. 4 An attempt to lower stock prices. See synonyms under INVASION. — *v.t.* To make a raid on. — *v.i.* To participate in a raid. [Scottish var. of ROAD] — **raid′er** *n.*

raiding party A body of troops assigned to make a sudden raid in enemy territory.

rail[1] (rāl) *n.* 1 A bar, usually of wood or iron, resting on supports, as in a fence, at the side of a stairway, or capping the bulwarks of a ship; a horizontal wooden piece between panels, joining the stiles; also, a railing. 2 One of a series of parallel bars, of iron or steel, resting upon cross-ties, forming a support and guide for wheels, as of a railway. 3 A railway track considered as a means of transportation: to ship by *rail.* — **to go by rail** To travel by train. — **to ride (someone) on a rail** To put (a person) astride a rail and carry around or beyond the limits of a community, as a punishment. — *v.t.* To furnish or shut in with rails; fence. [<OF *reille* <L *regula.* Doublet of RULE.]

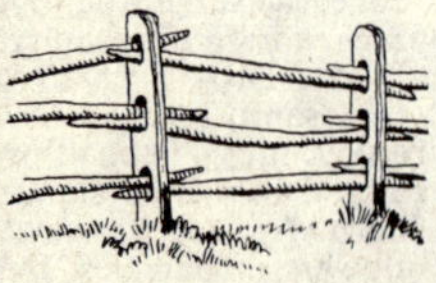

RAIL FENCE

rail[2] (rāl) *n.* 1 Any of numerous marsh-haunting, wading birds (family *Rallidae,* subfamily *Rallinae*) having very short wings, moderately long legs and toes, a short turned-up tail, long compressed bill, and soft, dun-colored plumage; specifically, in North America, the **king rail** (*Rallus elegans*), the **clapper rail** or mud hen (*R. longirostris*), and the sora or **Carolina rail.** They are esteemed as game birds. ◆ Collateral adjective: *ralline.* 2 Any of various other birds of northern Europe, as the corn crake. Also **rail′bird′.** [<OF *raale, ralle,* prob. ult. <L *radere* scratch]

rail[3] (rāl) *v.i.* To use scornful, insolent, or abusive language: with *at* or *against.* — *v.t.* To drive or force by railing. [<F *railler* <Pg. *ralhar* chatter, prob. <L *ragere* shriek. Doublet of RALLY[2].] — **rail′er** *n.* — **rail′ing** *adj.*

rail·head (rāl′hed′) *n.* 1 On an incompleted railroad, the farthest point to which rails have been laid. 2 That point on a railroad from which a military unit draws its supplies, ammunition, etc.

rail·ing (rā′ling) *n.* 1 A series of rails; a balustrade. 2 Rails, or material from which rails are made.

rail·ler·y (rā′lər·ē) *n. pl.* **·ler·ies** Merry jesting or teasing; a merry jest or bantering speech. See synonyms under BANTER, WIT[1]. [<F *raillerie* jesting]

rail·road (rāl′rōd′) *n.* 1 A graded road, having metal rails supported by ties or sleepers, for the passage of rolling stock drawn by locomotives. 2 The system of tracks, stations, rolling stock, etc., used in transportation by rail. 3 The corporation or persons owning or operating such a system. — *v.t.* 1 To transport by railroad. 2 *U.S. Colloq.* To rush or force with great speed or without deliberation: to *railroad* a bill through Congress. 3 *U.S. Slang* To cause to be imprisoned on false charges or without fair trial. — *v.i.* 4 To work on a railroad.

rail·road·er (rāl′rō′dər) *n.* One who works on a railroad.

rail·road·ing (rāl′rō′ding) *n.* The construction, operation, or business of a railroad.

rail-split·ter (rāl′split′ər) *n.* One who splits logs into fence rails. — **the Rail-Splitter** Abraham Lincoln.

rail·way (rāl′wā′) *n.* 1 A railroad: the common British term. 2 A trackway or set of rails, as in a warehouse or factory, for convenience in handling heavy articles, etc.: a parcel *railway* in a store.

railway post office A government post office in a railroad car, sometimes occupying a whole car, sometimes a section of a baggage car.

rai·ment (rā′mənt) *n. Archaic* Wearing apparel; clothing; garb. See synonyms under DRESS. [Aphetic var. of *arrayment* <ARRAY + -MENT]

Rai·mon·do (rī·mōn′dō) Italian form of RAYMOND. Also *Sp.* **Rai·mun·do** (rī·mo͞on′dō).

rain (rān) *n.* 1 The condensed vapor of the atmosphere falling in drops. ◆ Collateral adjective: *hyetal.* 2 The fall of such drops. 3 A fall or shower of anything in the manner of rain, or the substance poured down: a *rain* of bombs. 4 A rainstorm; shower; in the plural, the rainy season in a tropical country; also, a rainy region of the Atlantic Ocean. — *v.i.* 1 To fall from the clouds in drops of water: usually with *it* as the subject. 2 To fall like rain, as tears. 3 To send or pour down rain: said of clouds, God, etc. — *v.t.* 4 To send down like rain; shower. ◆ Homophones: *reign, rein.* [OE *regn*]

rain·band (rān′band′) *n. Astron.* A dark band in the solar spectrum, caused by the presence of water vapor in the atmosphere.

rain·bow (rān′bō′) *n.* 1 An arch of light formed opposite the sun during or after the close of a shower, exhibiting the colors of the spectrum, and caused by refraction, reflection, and dispersion of light in drops of water falling through the air. 2 Hence, any brilliant display of color. [OE *regnboga*]

Rainbow Bridge National Monument A region in southern Utah, site of a natural bridge; 160 acres; established, 1910.

rainbow cactus A cactus (*Echinocereus rigidissimus*) of the SW United States having red and white spines and red flowers.

rain·bow-chas·er (rān′bō′chā′sər) *n.* One who seeks the legendary pot of gold at the foot of the rainbow; a visionary.

rainbow trout See under TROUT.

rain check The stub of a ticket to an outdoor event, as a baseball game, entitling the holder to free admission at a future date if for any reason the event is called off: used figuratively of any postponed invitation.

rain·coat (rān′kōt′) *n.* A cloak or coat intended to be worn in rainy weather.

rain crow The yellow-billed or the black-billed cuckoo (genus *Coccyzus*), so called from the belief among farmers that its cry is a sign of rain.

rain·drop (rān′drop′) *n.* A drop of rain.

rain·fall (rān′fôl′) *n.* 1 A fall of rain. 2 *Meteorol.* The amount of water precipitated in a given region over a stated time, as rain, hail, snow, or the like: measured in inches.

rain gage An instrument for measuring rainfall at a given place or during a given time; a pluviometer. Also **rain gauge.**

Rai·nier (rā·nir′, rā′nir), **Mount** An extinct volcano in the Cascade Range, SW Washington; 14,408 feet; in Mount Rainier National Park.

rain-mak·er (rān′mā′kər) *n.* 1 One reputedly able to cause rain; specifically, among certain American Indians, one who brings rain by incantation. 2 A magician; any wonder-worker.

rain·out (rān′out′) *n. Physics* Precipitation of radioactive water droplets from cloud masses resulting from an underwater nuclear explosion.

rain·proof (rān′pro͞of′) *adj.* Impervious to or shedding rain: said of garments. — *n. Brit.* A raincoat.

rain shadow *Meteorol.* An area of relatively small average rainfall on the leeward side of mountain barriers which serve to break the prevailing rain-bearing winds.

rain·storm (rān′stôrm′) *n.* A storm accompanied by rain.

rain water Water that falls or has fallen directly from the clouds in the form of rain.

rain·y (rā′nē) *adj.* **rain·i·er, rain·i·est** Characterized by, abounding in, or bringing rain. — **rain′i·ly** *adv.* — **rain′i·ness** *n.*

rainy day A time of need; hard times.

Rain·y Lake (rā′nē) A lake in northern Minnesota and SW Ontario; 50 miles long; 350 square miles.

raise (rāz) *v.* **raised, rais·ing** *v.t.* 1 To cause to move upward or to a higher level; lift; elevate. 2 To place erect; set up. 3 To construct or build; erect. 4 To make greater in amount, size, or value: to *raise* the price of corn. 5 To advance or elevate in rank, estimation, etc. 6 To increase the strength, intensity, or degree of. 7 To breed; grow: to *raise* chickens or tomatoes. 8 *U.S.* To rear (children, a family, etc.). 9 To give utterance to; cause to be heard: to *raise* a hue and cry. 10 To cause; occasion, as a smile or laugh. 11 To stir to action or emotion; arouse. 12 To waken; animate or reanimate: to *raise* the dead. 13 To gather together; obtain or collect, as an army, capital, etc. 14 To bring up for consideration, as a question. 15 To cause to swell or become lighter; leaven. 16 To put an end to, as a siege. 17 In poker, to bet more than. 18 *Naut.* To cause to appear above the horizon, as land or a ship, by approaching nearer. 19 *Scot.* To madden; enrage. — *v.i.* 20 *Colloq.* To cough up phlegm. 21 *Dial.* To rise or arise. 22 In poker, to make a raise. — **to raise Cain** (or **the devil, the dickens, a rumpus,** etc.) *Colloq.* To make a great disturbance; stir up confusion. — **to raise steam** To get or produce steam, as in a boiler, for the purpose of starting up a steam engine. — *n.* 1 The act of raising, in any sense; specifically, an increase, as of wages or a bet. 2 *Brit. Dial.* Something raised; an ascent; mound. ◆ Homophone: *raze.* [<ON *reisa* lift, set up. Akin to OE *rǣran* rear.] ◆ In British usage, *rise* is used for an increase in wages. — **rais′er** *n.*

Synonyms (*verb*): aggrandize, elevate, erect, exalt, lift, rear, uplift. See HEIGHTEN, INCREASE, PROMOTE. *Antonyms*: degrade, depress, humble, lower, reduce, sink.

raised (rāzd) *adj.* 1 Elevated in low relief. 2 Made with yeast or leaven.

rai·sin (rā′zən) *n.* A grape of a special sort dried in the sun or in an oven, and used for a dessert or in cookery. [<OF <L *racemus* bunch of grapes]

rais·ing (rā′zing) *n.* 1 The act or process of causing to rise, in any sense. 2 A gathering of persons for the purpose of erecting the frame of a building: also **raising bee.**

Rai·sin River (rā′zən) A river in SE Michigan, flowing 115 miles SE to Lake Erie.

rai·son d'é·tat (re·zôn′ dā·tȧ′) *French* Political motive; literally, reason of state.

rai·son d'ê·tre (re·zôn′ de′tr′) *French* Literally, a reason for being; a reason or excuse for existing.

rai·son·né (re·zô·nā′) *adj. French* Arranged analytically or systematically; logical: a catalog *raisonné.*

Rai·su·li (ra·so͞o′lē), **Ahmed ibn-Muhammed,** 1875?–1925, Berber brigand in Morocco.

raj (räj) *n.* In India, sovereignty; rule. [<Hind. *rāj*]

ra·ja (rä′jə) *n.* A Hindu prince or chief of a tribal state in India; also, a Malay or Javanese ruler: often a mere title of distinction. Also **ra′jah.** [<Hind. *rājā* <Skt. *rājan* king]

Ra·ja·go·pa·la·cha·ria (rä′jə·gō·pä′lä·chä′ryə), **Chakravarti,** born 1879, governor general of India 1948–50.

Raj·ab (ruj′əb) See under CALENDAR (Mohammedan). [<Arabic]

Ra·ja·sthan (rä′jə·stän) A constituent State of NW India, formed by the merger of most of the Rajputana States (1948–50) and the former state of Ajmer (1956); 132,077 square miles; capital, Jaipur. Also **Ra′ja·stan.**

Raj·kot (räj′kōt) The capital city of the former state of Saurashtra, western India, in NW Bombay State after 1956.

Raj·put (räj′po͞ot) *n.* One of a powerful and warlike Hindu caste, said to be a branch of the Kshatriyas, which gives its name to Rajputana. Also **Raj′poot.** [<Hind. *rājpūt* prince <Skt. *rājaputra* < *rājan* a king, ruler + *putra* son]

Raj·pu·ta·na (räj′po͞o·tä′nə) A region in NW India, land of the Rajput princes; 134,959

Rajputana States The former princely states in Rajputana, merged 1948-50, to form Rajasthan, except for four included in Bombay State; 132,559 square miles.

rake[1] (rāk) n. A toothed implement for drawing together loose material, or making a surface loose. — v. **raked, rak·ing** v.t. 1 To scrape or gather together with or as with a rake. 2 To smooth, clean, or prepare with a rake: to *rake a lawn*. 3 To gather by diligent effort; scrape together. 4 To search or examine carefully. 5 To direct heavy gunfire along the length of, as a ship or column of troops; enfilade. — v.i. 6 To use a rake. 7 To scrape or pass roughly or violently: with *across, over*, etc. 8 To make a search. — **to rake in** *Colloq.* To earn or acquire (money, etc.) in large quantities. [OE *raca*] — **rak′er** n.

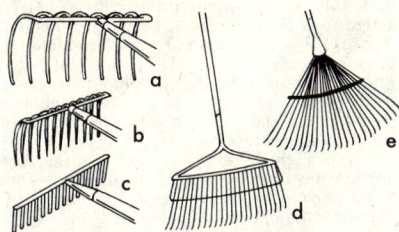

TYPES OF RAKES
a. Refuse. b. Clam. c. Garden.
d. Steel lawn. e. Broom lawn.

rake[2] (rāk) v. **raked, rak·ing** v.i. To lean from the perpendicular, as a ship's masts. — v.t. To cause to lean; incline. — n. Inclination from the perpendicular or horizontal, as of the sustaining surfaces of an airplane, or of the edge of a cutting tool. [Origin uncertain. Cf. G *ragen* project.] — **raked** adj.

rake[3] (rāk) n. A dissolute, lewd person; debauchee. — v. **raked, rak·ing** To play the rake; live a lewd, dissolute life: with *it*. [Short for RAKEHELL]

rake[4] (rāk) v.i. **raked, rak·ing** 1 To hunt with the nose to the ground, thus following by track rather than by wind: said of hunting dogs. 2 To fly after game: said of hawks; also, to fly wide of the game. [OE *racian* go forward, proceed]

rake·hell (rāk′hel′) adj. *Archaic* Recklessly abandoned and dissolute: also **rake′hell′y**. — n. A rake; a profligate debauchee. [ME *rakel* rash, wild; refashioned after RAKE[1] + HELL. Cf. ON *reikal* reckless.]

rake-off (rāk′ôf′, -of′) n. *U. S. Slang* A share, as of profits; commission; rebate, usually illegitimate.

rak·i (rak′ē, rä′kē) n. *Turkish* An aromatic liquor flavored with mastic; mastic brandy; also, a coarse liquor made from grain spirit. Compare ARRACK. Also **rak′ee**. [<Turkish *rāqi* <Arabic *'araq*. Akin to ARRACK.]

rak·ish[1] (rā′kish) adj. 1 *Naut.* Having the masts unusually inclined: usually connoting a suggestion of speed. 2 Dashing; jaunty. [< RAKE[2]; def. 2 infl. by *rakish*[2]] — **rak′ish·ly** adv. — **rak′ish·ness** n.

rak·ish[2] (rā′kish) adj. Like or behaving like a rake; dissolute; profligate. — **rak′ish·ly** adv. — **rak′ish·ness** n.

Rá·kó·czy March (rä′kō·tsē) A national patriotic song of Hungary. [after Francis *Rákóczy*, 1676–1735, Hungarian patriot]

râle (räl) n. *Pathol.* A sound additional to that of normal respiration, heard on auscultation of the chest and indicative of the presence, nature, or stage of a disease. [<F, rattle]

Ra·leigh (rô′lē) n. The capital of North Carolina.

Ra·leigh (rô′lē), **Sir Walter**, 1552–1618, English courtier, colonizer of Roanoke, soldier, and author; beheaded. Also spelled **Ra′legh**.

Ra·lik Chain (rä′lik) The western group of the Marshall Islands, including Kwajalein, Eniwetok, Bikini, Rongelap, Rongerik and others.

ral·len·tan·do (ral′an·tän′dō, *Ital.* räl′len·tän′dō) adv. *Music* Gradually slower. [<Ital., ppr. of *rallentare* slow down <*lento* slow <*lentus*]

ral·li·form (ral′i·fôrm) adj. Pertaining to or like the rails. See RAIL[2]. [<NL *rallus* (<OF *ralle* rail[2]) + -FORM]

ral·line (ral′īn, -in) adj. Of, pertaining to, or belonging to the rail subfamily of birds (*Rallinae*). [<NL *rallus* rail[2]]

ral·ly[1] (ral′ē) n. pl. **·lies** 1 An assembling or reassembling, as of scattered troops. 2 A rapid recovery of a normal condition after exhaustion or depression: such as *rally* from sickness, a *rally* in stocks. 3 A mass meeting to arouse enthusiasm. 4 In tennis, the interchange of several strokes before one side wins the point. — v. **·lied, ·ly·ing** v.t. 1 To bring together and restore to effective discipline: to *rally* fleeing troops. 2 To summon up or revive: to *rally* one's spirits. 3 To bring together for common action. — v.i. 4 To return to effective discipline or action: The enemy *rallied*. 5 To unite for common action. 6 To make a partial or complete return to a normal condition. 7 In tennis, to engage in a rally. See synonyms under ENCOURAGE. [<F *rallier* <*re*- again + *allier* join. See ALLY.] — **ral′li·er** n.

ral·ly[2] (ral′ē) v.t. & v.i. To attack with raillery; joke; tease; banter. See synonyms under RIDICULE. [<F *railler* rail. Doublet of RAIL[3].] — **ral′li·er** n.

Ralph (ralf, *Brit.* rāf) A masculine personal name. Also Lat. **Ra·dul·phus** (ra·dul′fəs). [<Gmc, house wolf]

ram (ram) n. 1 A male sheep. 2 An instrument or device for driving, forcing, or crushing by heavy blows or thrusts; specifically, a battering-ram, the striking weight of a pile driver or steam hammer, or the plunger of a force pump. 3 Formerly, a projection or beak on the bow of a warship, for crushing or cutting into an opposing vessel; also, a warship constructed with such a beak. 4 An instrument for raising water by pressure of condensed air; a hydraulic ram. — v.t. **rammed, ram·ming** 1 To strike with or as with a ram; dash against. 2 To drive or force down or into something. 3 To cram; stuff. [OE] — **ram′mer** n.

Ram (ram) *Astron.* The zodiacal constellation Aries.

Ra·ma (rä′mə) In Hindu mythology, the name of three heroes, especially that of Ramachandra.

Ra·ma·chan·dra (rä′mə·chun′drə) The hero of the *Ramayana*, called the seventh avatar of Vishnu.

Ram·a·dan (ram′ə·dän′) n. The Mohammedan ninth month, the time of the annual fast of thirty days; also, the fast. See under CALENDAR (Mohammedan). Also **Ram′a·dhan′, Ram′a·zan′** (-zän′). [<Arabic *ramaḍan*, lit., the hot month]

Ra·ma·krish·na (rä′mä·krish′nə), 1834–86, religious name of Gadadhar Chatterji, a Hindu mystic and religious teacher, regarded as a divine incarnation by his disciples, who founded the Ramakrishna Mission, a Vedantist Hindu religious order active in the U. S., with headquarters at Beluhr Math, Bengal, near Calcutta, India.

Ra·man (rä′mən), **Sir Chandrasekhara Venkata**, born 1888, Indian physicist.

Ra·man effect (rä′mən) *Physics* The scattering of monochromatic light by a medium, in frequencies both equal to and other than the frequency of the incident light, because of a gain or loss in quanta during transmission. [after Sir Chandrasekhara Venkata *Raman*]

Ra·ma·ya·na (rä·mä′yə·nə) A Hindu epic poem in seven books, of about 400 B.C. Compare MAHABHARATA. [<Skt. *Rāmāyana* <*Rāma* Rama + -*ayana* relating to]

ram·ble (ram′bəl) v.i. **·bled, ·bling** 1 To walk about freely and aimlessly; roam. 2 To write or talk aimlessly or without sequence of ideas. 3 To proceed with turns and twists; meander. — n. 1 The act of rambling; an aimless movement with change of direction; a leisurely stroll. 2 A meandering path; maze. [Origin unknown]

Synonyms (verb): range, roam, rove, stray, stroll, wander. See WANDER.

ram·bler (ram′blər) n. 1 One who or that which rambles. 2 Any of several varieties of roses, as the crimson rambler (*Rosa bar-*

bierana), with climbing stems and huge clusters of small or medium-sized flowers.

Rambler, The A semiweekly publication, 1750–52, published and written for the most part by Dr. Samuel Johnson.

ram·bling (ram′bling) adj. Showing absence of plan or system; aimless; wandering. — **ram′bling·ly** adv.

Ram·bouil·let (ram′boo̅·lā, *Fr.* rän·boo̅·ye′) n. A variety of merino sheep bred in France for meat and wool. [from *Rambouillet*, a town in northern France]

ram·bunc·tious (ram·bungk′shəs) adj. *U.S. Colloq.* Rude and boisterous; rough and uncontrollable. [Prob. <RAM + alter. of BUMPTIOUS]

ram·bu·tan (ram·boo̅′tən) n. 1 The spiny, bright-red, pleasantly acid fruit of an East Indian and Malaysian tree (*Nephelium lappaceum*). 2 The tree that bears it. [<Malay *rambut* hair]

Ra·mée (rə·mā′), **Louise de la** See OUIDA.

ram·e·kin (ram′ə·kin) n. 1 A seasoned dish of bread crumbs baked with eggs and cheese. 2 A dish in which ramekins are baked. 3 Any dish used both for baking and serving. Also **ram′e·quin**. [<F *ramequin*]

ra·men·tum (rə·men′təm) n. pl. **·ta** (-tə) 1 A part of something scraped off; a minute part. 2 *Bot.* A thin, membranous, chaffy scale, formed on the surface of leaves, the stems of ferns, etc.: an outgrowth from the epidermis. [<L, scraping <*radere* scrape] — **ra·men·ta·ceous** (ram′ən·tā′shəs) adj.

Ram·e·ses (ram′ə·sēz) Name of 12 Egyptian monarchs: also spelled **Ramses**.

— **Rameses II**, 1292–25 B.C., built many temples; sometimes said to be the pharaoh who oppressed the Israelites.

ra·met (rā′mit) n. *Bot.* Any individual member of a clon. [<L *ramus* branch]

Ram·gan·ga (räm·gung′gə) A river in northern Uttar Pradesh, India, flowing about 350 miles SW and SE to the Ganges.

ra·mie (ram′ē) n. 1 A shrubby Chinese and East Indian perennial (*Boehmeria nivea*) of the nettle family, with numerous rodlike stems and large heart-shaped leaves. 2 The fine, glossy bast fiber yielded by its stem, used for cordage and certain coarse textile fabrics. Also **ram′ee**. [<Malay *rami*]

ram·i·fi·ca·tion (ram′ə·fə·kā′shən) n. 1 The act or process of ramifying. 2 *Bot.* The arrangement of branches or parts, as on a plant; also, one of the parts. 3 An offshoot or subdivision.

ram·i·form (ram′ə·fôrm) adj. 1 Branch-shaped. 2 Branched. [<L *ramus* branch + -FORM]

ram·i·fy (ram′ə·fī) v.t. & v.i. **·fied, ·fy·ing** To divide or spread out into or as into branches; branch out. [<F *ramifier* <Med. L *ramificare* <L *ramus* branch + *facere* make]

ram·il·lie (ram′ə·lē) n. A type of wig with a plaited tail, worn in 18th century England: named in honor of the British victory at Ramillies. Also **ram′i·lie, ram′i·lies, ram′il·lies**.

Ram·il·lies (ram′ə·lēz, *Fr.* rà·mē·yē′) A village in central Belgium; scene of Marlborough's victory over French forces, 1706. Also **Ra·mil·lies-Of·fus** (rà·mē·yē′ô·fü′).

ram·mish (ram′ish) adj. 1 Like a ram; strongscented. 2 Lustful. Also **ram′my**. — **ram′mish·ness** n.

ram·jet (ram′jet′) n. A type of jet engine which provides continuous jet propulsion on the principle of the athodyd.

Ra·mon (rā′mōn) Spanish form of RAYMOND.

Ra·món y Ca·jal (rä·mōn′ ē kä·häl′), **Santiago**, 1852–1934, Spanish histologist.

ra·mose (rā′mōs, rə·mōs′) adj. 1 Branching. 2 Consisting of or having branches. [<L *ramosus* <*ramus* branch]

ra·mous (rā′məs) adj. 1 Of, pertaining to, or like branches. 2 Ramose. [See RAMOSE]

ramp[1] (ramp) n. 1 An inclined passageway or roadway, as between floors or different levels of a building. 2 In building, a concave part at the top or cap of a baluster, wall, or coping. [<F *rampe* <*ramper* climb]

ramp[2] (ramp) v.i. 1 To rear up on the hind legs and stretch out the forepaws. 2 *Her.* To be in a rampant or threatening position.

rampage 1044 rank

3 To act in a violent or threatening manner; storm; rampage. —*n.* The act of ramping. [<OF *ramper* climb]

ram·page (ram'pāj) *n.* Boisterous agitation or excitement; a dashing about with anger or violence. —*v.i.* (ram·pāj') **·paged, ·pag·ing** 1 To rush or act violently. 2 To storm; rage. [Prob. <RAMP[2]] —**ram·pag'er** *n.*

ram·pa·geous (ram·pā'jəs) *adj.* Violent; boisterous. —**ram·pa'geous·ly** *adv.* —**ram·pa'·geous·ness** *n.*

ram·pan·cy (ram'pən·sē) *n.* The condition or quality of being rampant.

ram·pant (ram'pənt) *adj.* 1 Exceeding all bounds; unrestrained; wild. 2 Widespread; unchecked, as an erroneous belief or superstition. 3 Standing on the hind legs; rearing; leaping: said of a quadruped. 4 *Her.* Standing on the sinister hind leg, with both forelegs elevated, the dexter above the sinister, and the head in profile: said of a beast of prey. 5 *Archit.* Springing from points on an inclined plane. [<OF, ppr. of *ramper* climb] —**ram'pant·ly** *adv.*

RAMPANT

ram·part (ram'pärt, -pərt) *n.* 1 The embankment surrounding a fort, on which the parapet is raised: sometimes including the parapet. 2 A bulwark or defense. —*v.t.* To supply with or as with ramparts; fortify. [<F *rempart* <*remparer* fortify <*re-* again + *emparer* prepare <L *ante* before + *parare* prepare]
Synonyms (noun): barbican, barricade, barrier, breastwork, bulwark, defense, embankment, fence, fortification, guard, mole, mound, outwork, security, wall. See BARRIER, DEFENSE.

ram·pi·on (ram'pē·ən) *n.* 1 A European perennial *(Campanula rapunculus)* cultivated in gardens for its root, which is eaten as a salad. 2 One of various similar plants, as the horned rampion (genus *Phyteuma*), bearing spikes of blue flowers. [<Ital. *ra(m)ponzolo* <L *rapum* turnip]

Ram·pur (räm'po͝or) 1 A former princely state in north central India, included (1949) in Uttar Pradesh State; 894 square miles. 2 A city in north central Uttar Pradesh State, formerly capital of the princely state of Rampur.

ram·rod (ram'rod') *n.* 1 A rod used to drive home the charge of a muzzleloading gun or pistol. 2 A similar rod used for cleaning the barrel of a rifle, etc.

Ram·say (ram'zē), **Allan,** 1686–1758, Scottish poet. —**James Andrew** See DALHOUSIE, MARQUIS OF. —**Sir William,** 1852–1916, Scottish chemist. —**Sir William Mitchell,** 1851–1939, Scottish classicist and geologist.

Ram·ses (ram'sēz) See RAMESES.

Rams·gate (ramz'gāt, *Brit.* -git) A port on the North Sea in eastern Kent, England; a seaside resort.

ram·shack·le (ram'shak'əl) *adj.* About to go to pieces from age and neglect; shaky; unsteady. [Origin uncertain]

ram·son (ram'zən, -sən) *n.* 1 A species of garlic *(Allium ursinum)*; broad-leaved garlic. 2 Its root, used for salads. [OE *hrameson,* pl. of *hramsa*]

ram–stam (ram'stam', räm'stäm') *Brit. Dial.* & *Scot. adj.* Rash; thoughtless; precipitate. —*n.* 1 A hasty and venturesome person. 2 Recklessness. —*adv.* With rashness; heedlessly.

ram·til (ram'til) *n.* An annual herb *(Guizotia abyssinica)* cultivated in Abyssinia and India for its oil-producing seeds, **ramtil seeds.** Also called *Niger seed.* [<Hind.]

ram·u·lose (ram'yə·lōs) *adj. Bot.* Bearing many small branches. [<L *ramulosus* <*ramulus,* dim. of *ramus* branch]

ra·mus (rā'məs) *n. pl.* **·mi** (-mī) 1 A branch. 2 *Biol.* One division of a forked structure, as the branch of a nerve, etc. [<L, branch]

ran (ran) Past tense of RUN.

Ran (rän) In Norse mythology, the wife of Ægir and goddess of the sea.

Ra·na (rä'nä) *n.* Prince: formerly the title of a ruling chief in various parts of India. [<Hind.]

ra·nar·i·um (rə·nâr'ē·əm) *n. pl.* **·nar·i·a** (-nâr'·ē·ə) A place where frogs are raised or kept. [<L *rana* frog]

rance (rans) *n.* A fine hard stone, dull red in color, with blue and white markings; Belgian marble. [<F]

ranch (ranch) *n.* 1 An establishment for rearing or grazing cattle, sheep, horses, etc., in large herds. 2 The buildings, personnel, and lands connected with it. 3 A large farm: a fruit *ranch.* Also **ranche.** —*v.i.* To manage or work on a ranch. [<Sp. *rancho* mess]

ranch·er (ran'chər) *n.* 1 The owner of a ranch. 2 One who works on a ranch; a cowboy.

ran·che·ri·a (ran'chə·rē'ə) *n. SW U.S.* 1 The house or hut of a Mexican ranchero. 2 A cluster of herdsmen's huts. 3 An Indian village. [<Sp. <*rancho* mess]

ran·che·ro (ran·châr'ō) *n. pl.* **·ros** *SW U.S.* A rancher. [<Sp.]

ranch·ing (ran'ching) *n.* 1 The operation of a ranch. 2 Work on a ranch.

ranch·man (ranch'mən) *n. pl.* **·men** (-mən) 1 A herdsman on a ranch. 2 The owner of a ranch; a rancher.

ran·cho (ran'chō, rän'-) *n. pl.* **·chos** *SW U.S.* 1 A hut or group of huts, in which ranchmen lodge. 2 A stock farm; ranch. [<Sp.]

ran·cid (ran'sid) *adj.* Having the peculiar tainted smell of oily substances that have begun to spoil owing to oxidation or hydrolysis; rank; sour. Compare SWEET. [<L *rancidus* <*rancere* be rancid]

ran·cid·i·ty (ran·sid'ə·tē) *n.* 1 The quality or state of being rancid. 2 A rancid smell or taste. Also **ran'cid·ness.**

ran·cor (rang'kər) *n.* Bitter and vindictive enmity; malice; spitefulness. Also *Brit.* **ran'·cour.** See synonyms under ENMITY, HATRED. [<OF <L <*rancere* be rank] —**ran'cor·ous** *adj.* —**ran'cor·ous·ly** *adv.* —**ran'cor·ous·ness** *n.*

rand (rand) *n.* 1 In shoe manufacturing, a strip of leather at the heel of a shoe to which the lifts are attached. 2 *Brit. Dial.* & *Scot.* A river border overgrown with reeds, or the unplowed border round a field; margin; strip. [OE, border, edge]

Rand (rand), **The** See WITWATERSRAND.

ran·dan (ran'dan, ran·dan') *n.* 1 A boat rowed by three persons, the one amidships having two oars and the others one each. 2 This style of rowing. [Origin uncertain]

ran·dem (ran'dəm) *adv.* With three horses harnessed one in front of the other. —*n.* A team or vehicle driven randem. Also **ran'dem-tan'·dem.** [<RANDOM, on analogy with *tandem*]

RANDEM

Ran·dolph (ran'dolf), **John,** 1773–1833, U. S. statesman: known as *Randolph of Roanoke.* —**Peyton,** 1723?–75, American patriot; first president, Continental Congress, 1774–75.

ran·dom (ran'dəm) *n.* 1 Want of definite aim or intention. 2 Something done, made, or chosen without method or purpose. 3 *Printing* A sloping board for holding galleys of type matter intended for making up forms. —**at random** Without definite purpose or aim; haphazardly. —*adj.* 1 Done or chosen without definite aim or deliberate purpose; chance; casual. 2 In statistics, erratic. [<OF *randon* force, violence <*randonner, rander* move rapidly, gallop] —**ran'dom·ly** *adv.*

random sample *Stat.* A limited group of individuals, cases, or observations, so assembled from the total array as to be truly representative of its characteristics, properties, trends, and the like. Also **random selection.**

ran·dy (ran'dē) *Scot. adj.* 1 Disorderly; riotous; also, coarse. 2 Lewd; lustful. —*n.* 1 An impudent beggar. 2 A boisterous, coarse, or loose woman; also, a virago.

ra·nee (rä'nē) See RANI.

rang (rang) Past tense of RING[2].

range (rānj) *n.* 1 The area over which anything moves, operates, or is distributed. 2 *U.S.* An extensive tract of land over which cattle, sheep, etc., roam and graze. 3 *U.S.* Pasturage; grazing ground. 4 *Bot.* & *Zool.* The geographical area throughout which a specific plant or animal exists. 5 The extent or scope of something: the whole *range* of politics. 6 The extent to which any power can be made effective: *range* of vision; *range* of influence. 7 The extent of variation of anything: the temperature *range.* 8 The extent of possible variation in pitch: said of musical instruments or the voice. 9 A line, row, or series, as of mountains. 10 *U.S.* A row of townships, each six miles square, numbered east or west from a base meridian. 11 *Rare* Rank; order. 12 The horizontal distance between a gun and its target. 13 The horizontal distance covered by a projectile. 14 A place for shooting at a mark: a rifle *range.* 15 In archery, the number of ends shot at each given distance: compare ROUND. 16 A large cooking stove for conducting several cooking operations at one time. 17 *Stat.* The inclusive difference between the extreme values in any series of variable data: a *range* of 20 from a value of 0 to a value of 19. —*adj.* Of or pertaining to a range. —*v.* **ranged, rang·ing** *v.t.* 1 To place or arrange in definite order, as in rows or lines. 2 To assign to a class, division, or category; classify; rank. 3 To move about or over (a region, etc.), as in exploration. 4 To put (cattle) to graze on a range. 5 *Mil.* To obtain the range of (a target) by firing alternately above and below it. 6 To place in position; adjust or train, as a telescope or gun. 7 *Naut.* To lay out (the anchor cable) on deck so that the anchor may descend without hindrance. —*v.i.* 8 To move over an area in a thorough, systematic manner, as a dog hunting game. 9 To rove; roam. 10 To occur; extend; be found: said of plants and animals. 11 To extend or proceed: The shot *ranged* to the right. 12 To exhibit variation within specified limits: weights *ranging* from 20 to 50 pounds. 13 To lie in the same direction, line, etc. 14 *Mil.* To be capable of achieving a specified range (def. 12): That old cannon *ranged* about one mile. See synonyms under RAMBLE, WANDER. [<OF <*ranger, rengier* arrange <*renc* row <Gmc. Doublet of RANK[1].]

range finder An instrument with which to determine the distance of an object or target from a given point, as from a gun.

Range·ley Lakes (rānj'lē) A chain of lakes in western Maine.

rang·er (rān'jər) *n.* 1 One who or that which ranges; a rover. 2 One of an armed band, usually mounted, designed to protect large tracts of country. 3 One of a herd of cattle that feeds on a range. 4 *Brit.* A government official in charge of a royal forest or park: formerly a gamekeeper. 5 *U.S.* A warden employed in patrolling forest tracts. —**rang'·er·ship** *n.*

Rang·er (rān'jər) *n.* One of a select group of U. S. soldiers who were trained for raiding action on enemy territory: the equivalent of the English *Commando.*

range rake A T-shaped instrument for obtaining quick angular measurements in correcting deviations in the range of a gun.

Ran·gi·ro·a (rän'gi·rō'ä) Largest of the Tuamotu islands, consisting of 20 islets around a lagoon 45 miles long and 15 miles wide.

Rang·i·tik·ei (räng'gi·tik'ē) A river in southern North Island, New Zealand, flowing 115 miles SW to Cook Strait.

Ran·goon (rang·go͞on') The capital of Burma, a port of Lower Burma on the **Rangoon River,** a marine estuary formed at Rangoon by the junction of two inland rivers, flowing 25 miles SE to the Andaman Sea.

rang·y (rān'jē) *adj.* **rang·i·er, rang·i·est** 1 Disposed to roam, or adapted for roving, as cattle. 2 Having long legs adapted to a long, limber gait. 3 Having long thin arms and legs: said of a person. 4 Affording wide range; roomy. 5 Resembling a mountain range.

ra·ni (rä'nē) *n.* 1 The wife of a raja or prince. 2 A reigning Hindu queen or princess. Also spelled *ranee.* [<Hind.]

Ran·jit Singh (run'jēt sin'hə), 1780–1839, maharajah of the Punjab; founded Sikh empire. Also *Runjeet Singh.*

rank[1] (rangk) *n.* 1 A series of objects ranged in a line or row; a range. 2 Degree of official standing, especially in the army and navy. See table under GRADE. 3 A line of soldiers drawn up side by side in close order: distinguished from *file.* 4 *pl.* An army; also, the mass of soldiery; the order of private soldiers: The colonel rose from the *ranks.* 5 A row of eight squares on a chessboard extending from the left of the player to the right. 6 Relative position in a scale of dignity or of life; degree; grade: the *rank* of baronet; the *rank* of a plant or animal organism. 7 High degree or position; especially, the state of being a member of a titled nobility: a lady of *rank*. 8 Degree of worth or excellence; relative status. See synonyms under CLASS, SORT. — *v.t.* 1 To place or arrange in a rank or ranks. 2 To place in a class, order, etc.; assign to a position or classification. 3 To take precedence of; outrank: Sergeants *rank* corporals. — *v.i.* 4 To hold a specified place or rank: His poetry *ranks* with the best. 5 To have the highest rank or grade. [<OF *ranc, renc* <Gmc. Doublet of RANGE.]

rank[2] (rangk) *adj.* 1 Very vigorous and flourishing in growth as from fertilization or moisture. 2 Strong and disagreeable to the taste or smell. 3 Excessive or immoderate, in unfavorable sense: *rank* injustice. 4 Producing a luxuriant growth; fertile. 5 *Law* Inequitable; excessive. 6 Strong or deep: said of a cut or the adjustment of the tool making a cut. 7 *Obs.* In heat; lustful. [OE *ranc* strong] — **rank′ly** *adv.* — **rank′ness** *n.*

rank and file 1 The common soldiers of an army, including all from the corporals downward. 2 Those who form the bulk of any organization, as distinct from officers or leaders.

Ran·ke (räng′kə), **Leopold von,** 1795–1886, German historian.

rank·er (rangk′ər) *n.* 1 One who has served in the ranks. 2 A commissioned officer who has risen from the ranks.

rank·ing (rangk′ing) *adj.* Superior in rank; taking precedence (over others in the grade): a *ranking* senator, officer, etc.

ran·kle (rang′kəl) *v.* **·kled, ·kling** *v.i.* 1 To cause continued resentment, sense of injury, etc.: The defeat *rankles* in his breast. 2 To become irritated or inflamed; fester. — *v.t.* 3 To irritate; embitter. [<OF *rancler,* alter. of *draoncler* fester <Med. L *dracunculus,* dim. of *draco* dragon]

Ran·noch (ran′əkh), **Loch** A lake in NW Perth, central Scotland; 9 miles long, 1 mile wide.

ran·oid (ran′oid) *adj. Zool.* Of, pertaining, or belonging to a family *(Ranidae)* of frogs, especially as distinguished from the toads. [<L *rana* a frog + -OID]

ran·sack (ran′sak) *v.t.* 1 To search through every part of. 2 To search throughout for plunder; pillage. See synonyms under EXAMINE. [<ON *rannsaka* search a house < *rann* house + *sækja* seek] — **ran′sack·er** *n.*

ran·som (ran′səm) *v.t.* 1 To secure the release of (a person, property, etc.) for a required price, as from captivity or detention. 2 To set free on payment of ransom. 3 To redeem from sin or its consequences. See synonyms under DELIVER. — *n.* 1 The consideration paid for the release of a person or property captured or detained; also, formerly, a heavy fine. 2 Release purchased, as from captivity. [<OF *rançon, raençon* <L *redemptio, -onis* redemption < *redimere* redeem. Doublet of REDEMPTION.] — **ran′som·er** *n.* — **ran′som·less** *adj.*

rant (rant) *v.i.* 1 To speak in loud, violent, or extravagant language; declaim vehemently; rave. 2 *Scot. & Brit. Dial.* To frolic noisily; be uproariously jolly. — *v.t.* 3 To exclaim or utter in a ranting manner. — *n.* 1 Declamatory and bombastic talk. 2 *Scot. & Brit. Dial.* Wild gaiety; a boisterous revel. [<MDu. *ranten* rave] — **rant′ing** *adj.* — **rant′ing·ly** *adv.*

rant·er (ran′tər) *n.* One who rants; a noisy, boisterous speaker or declaimer: applied opprobriously to various religious speakers.

ra·nun·cu·la·ceous (rə·nung′kyə·lā′shəs) *adj. Bot.* Belonging or pertaining to a family *(Ranunculaceae)* of plants, the crowfoot or buttercup family, having acrid, sometimes poisonous, juices, and including larkspur, aconite, peony, and hellebore.

ra·nun·cu·lus (rə·nung′kyə·ləs) *n. pl.* **·lus·es** or **·li** (-lī) Any of a genus *(Ranunculus)* of herbaceous annuals or perennials, the buttercups or crowfoots, typical of the family *Ranunculaceae.* [<L *ranunculus,* a medicinal plant, orig. dim. of *rana* frog]

Ra·oul (rä·o͞ol′) French form of RALPH.

rap[1] (rap) *v.* **rapped, rap·ping** *v.t.* 1 To strike sharply and quickly; hit. 2 To utter in a sharp manner: with *out:* to *rap* out an oath. — *v.i.* 3 To strike sharp, quick blows. — *n.* 1 A sharp blow. 2 A sound caused by or as by knocking; specifically, such a sound ascribed to the agency of spirits. 3 *Slang* A reprimand; blame; also, consequences: to take the *rap.* 4 *Slang* A prison sentence. See synonyms under BLOW[2]. [Imit. Cf. Dan. *rap,* Sw. *rapp.*]

rap[2] (rap) *v.t.* **rapt** or **rapped, rap·ping** 1 *Obs.* To snatch. 2 *Archaic* To seize or transport as with ecstasy; carry away: now current only in the past participle *rapt* (sometimes erroneously spelled *wrapt*). [Back formation <RAPT]

rap[3] (rap) *n.* A counterfeit coin used as a halfpenny in Ireland in the 18th century; hence, anything worthless: I don't care a *rap.* [Origin uncertain. Cf. G. *rappe,* the name of a small coin.]

rap[4] (rap) *n.* A skein of yarn containing 120 yards. [Origin unknown]

ra·pa·cious (rə·pā′shəs) *adj.* 1 Given to plunder or rapine. 2 Extortionate; grasping. 3 Predaceous; subsisting on prey seized alive: said of hawks, etc. See synonyms under AVARICIOUS. [<L *rapax, -acis* < *rapere* seize] — **ra·pa′cious·ly** *adv.*

ra·pac·i·ty (rə·pas′ə·tē) *n.* The quality or character of being rapacious. Also **ra·pa′cious·ness.** [<L *rapacitas, -tatis*]

Ra·pal·lo (rä·päl′lō) A port in NW Italy, at the head of the **Gulf of Rapallo,** an inlet on the Gulf of Genoa.

Ra·pa Nu·i (rä′pä no͞o′ē) The native name for EASTER ISLAND.

rape[1] (rāp) *v.* **raped, rap·ing** *v.t.* 1 To commit rape upon; ravish. 2 To plunder or sack (a city, etc.). 3 *Archaic* To carry off by force. — *v.i.* 4 To commit rape. See synonyms under VIOLATE. — *n.* 1 *Law* The forcible and unlawful carnal knowledge of a woman against her will. 2 A capturing or snatching away by force; abduction. [<AF <L *rapere* seize]

rape[2] (rāp) *n.* An Old World annual *(Brassica napus)* grown as a forage crop for sheep and hogs, and having seeds which yield rape oil. [<L *rapum* turnip]

rape[3] (rāp) *n.* 1 *pl.* In winemaking, refuse stalks and skins of grapes. 2 A filter used in vinegarmaking. [<F *râpe* <Med. L *raspa* < *raspare* grate <Gmc. Cf. OHG *raspon.*]

rape oil A yellowish to brown oil obtained from rapeseed: used as a lubricant and in the manufacture of rubber substitutes, soft soaps, etc. Also *colza oil.*

rape·seed (rāp′sēd′) *n.* 1 The seed of the rape. 2 The plant.

Raph·a·el (raf′ē·əl, rā′fē-; *Fr.* rȧ·fȧ·el′, *Du.* rä′fel, *Ger.* rä′fä·el) A masculine personal name. [<Hebrew, God hath healed]
— **Raphael** One of the seven archangels of Christian legend: also mentioned in Milton's *Paradise Lost.*
— **Raphael,** 1483–1520, Italian painter: full name Rafael or Raffaello Sanzio.

Raph·a·el·esque (raf′ē·əl·esk′) *adj.* Characteristic of, or in the style of Raphael. — **Raph′a·el·ism** *n.* — **Raph′a·el·ite′** *n.*

ra·phe (rā′fē) *n. pl.* **·phae** (-fē) 1 *Anat.* A seamlike appearance often seen in organs, especially at the median line of the body. 2 *Bot.* The fibrovascular cord that connects the hilum of plant ovules with the chalaza. 3 A line or rib connecting the nodules on a diatom valve. Also spelled *rhaphe.* [<NL <Gk. *rhaphē* seam < *rhaptein* stitch together]

ra·phi·a (rā′fē·ə) See RAFFIA.

ra·phide (rā′fid) *n. pl.* **raph·i·des** (raf′ə·dēz) *Bot.* A needle-shaped crystal of oxalate of lime found in many plant cells. Also **ra′phis.** [<Gk. *rhaphis, rhaphidos* needle]

rap·id (rap′id) *adj.* 1 Having great speed, literally or figuratively. 2 Bearing the marks of or characterized by rapidity: a *rapid* style. 3 Done or completed in a short time; advancing speedily to a termination: *rapid* growth. See synonyms under SWIFT[1]. — *n. Usually pl.* A descent in a river less abrupt than a waterfall. [<L *rapidus* < *rapere* seize, rush] — **rap′id·ly** *adv.*

Rap·i·dan River (rap′ə·dan′) A river of northern Virginia, flowing 90 miles east to the Rappahannock; scene of severe fighting in the Civil War.

rap·id-fire (rap′id·fīr′) *adj.* 1 Firing shots rapidly. 2 Designating single-barreled, breech-loading guns larger than small arms, designed for the discharge of projectiles in rapid succession. 3 Characterized by speed: *rapid-fire* repartee. Also **rap′id-fir′ing.**

rapid fire A rate of gunfire lower than that of quick fire.

ra·pid·i·ty (rə·pid′ə·tē) *n.* The quality or state of being rapid; swiftness. Also **rap·id·ness** (rap′id·nis).

rapid transit The local transportation of passengers by means faster than surface vehicles; specifically, elevated or subway passenger transportation.

ra·pi·er (rā′pē·ər, rāp′yər) *n.* 1 In the 16th and 17th centuries, a long, straight, two-edged sword with a large cup hilt, used in dueling, chiefly for thrusting. 2 The French small sword of the 18th century, a shorter straight sword without cutting edge and therefore used for thrusting only. [<F *rapière,* prob. < *raspière* poker, rasper; appar. first used derisively]

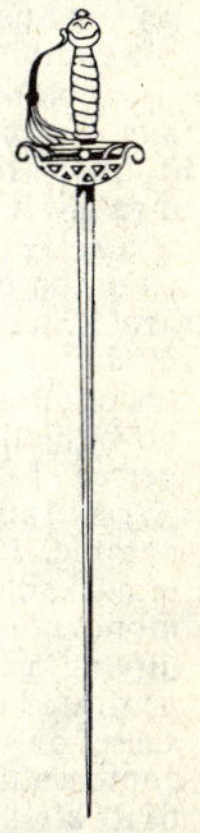
RAPIER

rap·ine (rap′in) *n.* The taking of property by force, as in war; spoliation; pillage. See synonyms under PLUNDER. [<F <L *rapina* < *rapere* seize. Doublet of RAVEN[2], *n.,* RAVINE.]

rap·ist (rā′pist) *n.* One who commits rape.

rap·loch (rap′ləkh) *Scot. & Brit. Dial. adj.* Unkempt; coarse. — *n.* Coarse homespun cloth made of inferior undyed wool.

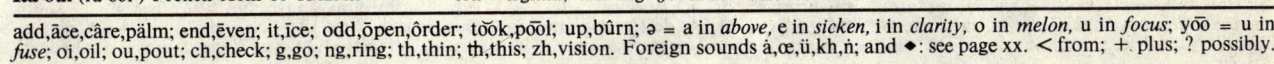

Rap·pa·han·nock (rap′ə·han′ək) A river in northern Virginia, flowing 212 miles SE to Chesapeake Bay.

rap·pa·ree (rap′ə·rē′) *n.* 1 An Irish guerrilla of the 17th century. 2 A freebooter or bandit. [<Irish *rapaire* short pike]

rap·pee (ra·pē′) *n.* A dark, coarse, strong-flavored snuff. [<F *(tabac) râpé* grated (tobacco), pp. of *râper* scrape]

rap·pel (ra·pel′) *v.i.* **·pelled, ·pel·ling** In mountaineering, to descend from a precipitous height by letting oneself down on a rope. — *n.* Descent by means of a rope. [<F]

rap·per (rap′ər) *n.* 1 One who raps. 2 A spiritualist medium. 3 A knocker, as on a door or at the mouth of a mining shaft.

rap·port (ra·pôrt′, -pōrt′; *Fr.* rȧ·pôr′) *n.* Harmony of relation; accordance; sympathetic relation: commonly with *in.* — **en rapport** *French* In close accord. [<F < *rapporter* refer, bring back < *re-* again + *apporter* bring <L *apportare* < *ad-* to + *portare* bring]

rap·proche·ment (rȧ·prôsh·mäṅ′) *n. French* The act of coming or of being brought together; a state of harmony or reconciliation; restoration of cordial relations, as between nations.

rap·scal·lion (rap·skal′yən) *n.* A rogue; scamp; rascal. [<earlier *rascallion* <RASCAL + fanciful ending]

rapt (rapt) *adj.* 1 Carried away with lofty emotion; enraptured; transported. 2 Engrossed; intent; deeply engaged. Sometimes erroneously spelled *wrapt.* [<L *raptus,* pp. of *rapere* seize]

Rap·ti (räp′tē) A river in Nepal and northern India, flowing 400 miles SE to the Gogra river in Uttar Pradesh State.

rap·to·ri·al (rap·tôr′ē·əl, -tō′rē-) *adj.* 1 Seizing and devouring living prey; predatory. 2 *Ornithol.* Having talons adapted for seizing and holding prey: said especially of hawks, vultures, eagles, owls, and other carnivorous birds. [<L *raptor* snatcher < *raptus,* pp. of *rapere* seize]

rap·ture (rap′chər) *n.* 1 The state of being rapt or transported; ecstatic joy; ecstasy. 2 The act of transferring a person from one place to another: Elijah's *rapture* to heaven. 3 An act or expression of excessive delight. 4 *Obs.* A snatching away; violent seizure. — *v.t.* **·tured, ·tur·ing** To enrapture; transport with ecstasy. [<RAPT]

add,āce,câre,pälm; end,ēven; it,īce; odd,ōpen,ôrder; to͝ok,po͞ol; up,bûrn; ə = a in *above,* e in *sicken,* i in *clarity,* o in *melon,* u in *focus;* yo͞o = u in *fuse;* oi,oil; ou,pout; ch,check; g,go; ng,ring; th,thin; th,this; zh,vision. Foreign sounds ȧ,œ,ü,kh,ṅ; and ◆: see page xx. < from; + plus; ? possibly.

Synonyms (noun): bliss, delight, ecstasy, exultation, happiness, joy, rejoicing, transport, triumph. *Rejoicing* is *happiness* or *joy* that finds utterance in word, song, festivity, etc. *Delight* is vivid, overflowing *happiness* of a somewhat transient kind; *ecstasy* is a state of extreme or extravagant *delight*; *rapture* is closely allied to *ecstasy*, but is more serene, exalted, and enduring. *Transport* is the condition of one carried away out of himself by some powerful passion or emotion, whether joyous or the reverse. *Triumph* is such *joy* as results from victory, success, achievement. See ENTHUSIASM, HAPPINESS. *Antonyms*: agony, apathy, dejection, despair, distress, ennui, horror, misery, pain, tedium, torture, woe, wretchedness.

rap·tur·ous (rap′chər·əs) *adj.* Being in a state of, exhibiting, or characterized by rapture. See synonyms under HAPPY. — **rap′tur·ous·ly** *adv.* — **rap′tur·ous·ness** *n.*

Ra·quel (rä·kel′) Spanish form of RACHEL.

ra·ra a·vis (râr′ə ā′vis) *pl.* **ra·rae aves** (râr′ē ā′vēz) *Latin* Literally, a rare bird; any uncommon or peculiar person or thing.

rare[1] (râr) *adj.* **1** Of infrequent occurrence. **2** Highly esteemed because of infrequency or uncommonness; valuable; choice. **3** Rarefied: now chiefly of the atmosphere. **4** *Obs.* Dispersed. [<F <L *rarus* rare]

Synonyms: curious, extraordinary, incomparable, infrequent, odd, peculiar, precious, remarkable, scarce, singular, strange, uncommon, unique, unusual. *Extraordinary*, signifying greatly beyond the ordinary, is a neutral word, capable of a high and good sense or of an invidious, opprobrious, or contemptuous signification. *Unique* is alone of its kind; *rare* is *infrequent* of its kind; great poems are *rare*. To say of a thing that it is *rare* is simply to affirm that it is now seldom found, whether previously common or not; as, a *rare* old book; a *rare* word; to call a thing *scarce* implies that it was at some time more plentiful, as when we say money is *scarce*. A particular coin may be *rare*; *scarce* applies to demand and use, and almost always to concrete things; to speak of virtue, genius, or heroism as *scarce* would be somewhat ludicrous. See CHOICE, EXTRAORDINARY, OBSOLETE, ODD. *Antonyms*: see synonyms for COMMON.

rare[2] (râr) *adj.* Not thoroughly cooked: applied to roasted or broiled meat retaining its redness and juices: in England commonly termed *underdone*. [OE *hrēre* lightly boiled]

rare·bit (râr′bit) *n.* Welsh rabbit. [Alter. of (WELSH) RABBIT]

rare earth *Chem.* Any of the metallic oxides of the rare–earth elements.

rare–earth elements (râr′ûrth′) *Chem.* A group of metallic elements comprising the lanthanide series. Also **rare–earth metals.**

rar·ee show (râr′ē) **1** A show carried or contained in a box; a peepshow. **2** A cheap street show or any street show or spectacle. [Alter. of *rare show*; after the mispronunciation characteristic of the Savoyard promoters of these shows]

rar·e·fac·tion (râr′ə·fak′shən) *n.* The process or act of making rare or less dense. Also **rar′e·fi·ca′tion.** [<L *rarefactus*, pp. of *rarefacere*] — **rar′e·fac′tive** *adj.*

rar·e·fy (râr′ə·fī) *v.* **·fied, ·fy·ing** *v.t.* **1** To make rare, thin, less solid, or less dense; expand by dispersion of the particles. **2** To refine or purify. — *v.i.* **3** To become rare, thin, or less solid. **4** To become more pure. [<F *raréfier* <L *rarefacere* <*rarus* rare + *facere* make] — **rar′e·fi′a·ble** *adj.*

rare·ly (râr′lē) *adv.* **1** Not often; infrequently. **2** With unusual excellence or effect; finely: The breeze blows *rarely*. **3** Exceptionally; extremely; in an unusual degree: She dressed in raiment *rarely* rich.

rare·ness (râr′nis) *n.* The condition or quality of being rare in any sense.

rare·ripe (râr′rīp′) *adj.* Ripening early. — *n.* A fruit that ripens early: applied especially to many varieties of peaches, and to a variety of onion. [OE *hrathe* early, soon + RIPE]

Rar·i·tan River (rar′ə·tən) A river of NE New Jersey, flowing 25 miles south to **Raritan Bay,** a western arm of Lower New York Bay.

rar·i·ty (râr′ə·tē) *n. pl.* **·ties** **1** The quality or state of being rare, uncommon, or infrequent; infrequency. **2** That which is exceptionally valued from scarceness. **3** The state of being rare, thin, or tenuous; tenuity: opposed to *density*. [<L *raritas, -tatis*]

Ra·ro·ton·ga (rä′rō·tông′gə) The largest and southwesternmost of the Cook Islands, capital of the group; 26 square miles.

ras (räs) *n.* In Ethiopia, a prince. [<Arabic *ra's* the head]

ras·cal (ras′kəl) *n.* **1** An unprincipled fellow; a rogue; knave: sometimes used playfully. **2** *Obs.* One of the common herd; a man of low birth or station. — *adj.* Pertaining to the rabble; contemptible; base; mean. [<OF *rascaille* <*rasque* filth, shavings, ult. <L *radere* shave, scrape]

ras·cal·i·ty (ras·kal′ə·tē) *n. pl.* **·ties** **1** The quality of being rascally. **2** A rascally act.

ras·cal·ly (ras′kəl·ē) *adj.* Worthy of a rascal; knavish; base. See synonyms under BAD[1]. — *adv.* After the manner of a rascal.

Ras Da·shan (räs dä·shän′) The highest peak in Ethiopia; 15,157 feet.

rase (rāz) *v.t.* **rased, ras·ing** To raze. [Var. of RAZE]

rash[1] (rash) *adj.* **1** Acting without due caution or regard of consequences; reckless; precipitate. **2** Exhibiting recklessness or precipitancy. **3** *Obs.* Quick; speedy. See synonyms under IMPETUOUS, IMPRUDENT. [ME *rasch*. Akin to Du. & G *rasch* quick.] — **rash′ly** *adv.* — **rash′ness** *n.*

rash[2] (rash) *n.* A superficial eruption of the skin, often localized. [? <F *rache* <OF *rasque*. See RASCAL.]

rash[3] (rash) *n. Scot.* A rush; bulrush.

rash·er[1] (rash′ər) *n.* A thin slice of meat: used especially of bacon. [Prob. <obs. *rash* cut, slash]

rash·er[2] (rash′ər) *n.* A vermilion–colored Californian rockfish (*Sebastichthys miniatus*). [<Sp. *rascacio*, kind of fish]

Rask (räsk), **Rasmus Christian,** 1787–1832, Danish philologist and writer.

Ras·kol·nik (räs·kôl′nik) *n. pl.* **·ni·ki** (-nē·kē) or **·niks** A Russian dissenter; a member of one of the sects that split off from the Orthodox Church in the 17th century. [<Russian *raskolenik* dissenter <*raskole* schism]

Ras·mus·sen (räs′mōōs·ən), **Knud Johan Victor,** 1879–1933, Danish Arctic explorer.

ra·son (rā′son) *n. Meteorol.* A method of obtaining and recording weather information at high altitudes by combining a radiosonde with automatic signal–recording devices and radio direction–finders. [<RA(DIO)SON(DE)]

ra·so·ri·al (rə·sôr′ē·əl, -sō′rē-) *adj.* In the habit of scratching the ground for food, as domestic fowl and other gallinaceous birds. [<NL *Rasores*, lit., scratchers <L *rasum*, pp. of *radere* scrape]

rasp (rasp, räsp) *n.* **1** A filelike tool having coarse pyramidal projections for abrasion. **2** A machine containing a large cylindrical grater. **3** The act or sound of rasping. — *v.t.* **1** To scrape with or as with a rasp. **2** To scrape or rub roughly. **3** To affect unpleasantly; irritate. **4** To utter in a rough voice. — *v.i.* **5** To grate; scrape [<OF *raspe* <*rasper* scrape, prob. <Gmc. Cf. OHG *raspon* grate.] — **rasp′er** *n.*

rasp·ber·ry (raz′ber′ē, -bər·ē, räz′-) *n. pl.* **·ries** **1** The round fruit of certain brambles (genus *Rubus*) of the rose family, composed of drupes clustered around a fleshy receptacle. The typical red raspberry is *R. idaeus*, with many varieties; the native American blackcap is *R. occidentalis*. **2** The plant yielding this fruit. **3** *Slang* A vulgar sound indicating contempt and produced by vibrating the tongue between the lips. [<earlier *rasp* raspberry (? <obs. *raspis* thin, sweet wine <OF (*vin*) *raspé* thin wine <*râpe* RAPE[3]) + BERRY]

rasped (raspt, räspt) *adj.* Rough or roughened, with or as with a coarse file: said of uncut book edges.

rasp·ing (ras′ping, räs′-) *adj.* Making a harsh sound; hence, irritating.

Ras·pu·tin (ras·pyōō′tin, *Russian* räs·pōō′tin), **Grigori,** 1871–1916, Russian monk, favorite of Czar Nicholas II and his wife; assassinated: real name Novikh.

rasp·y (ras′pē, räs′-) *adj.* **rasp·i·er, rasp·i·est** **1** Inclined to rasp; rough; grating. **2** Irritable.

Ras·se·las (ras′ə·ləs) The hero of a philosophical romance of this name by Samuel Johnson.

ra·sure (rā′zhər) *n.* Erasure.

rat (rat) *n.* **1** A destructive and injurious rodent (family *Muridae*) of world–wide distribution, larger and more aggressive than the mouse; especially, the **Norway rat** (*Rattus norvegicus*) and the smaller **roof** or **black rat** (*R. rattus*): both are carriers of the plague bacillus transmitted by the rat flea. **2** Some other mammal like or likened to the rat. **3** *Slang* A cowardly or selfish person who deserts or betrays his associates. **4** A slender cushion of curled hair or the like, worn by women, with the natural hair rolled over it. — *v.i.* **rat·ted, rat·ting** **1** To hunt rats. **2** *Slang* To desert one's party, companions, etc., especially for one's own safety or advantage. **3** *Slang* To inform; act the betrayer. [OE *ræt*]

rat·a·ble (rā′tə·bəl) *adj.* **1** *Brit.* Subject to assessment; legally liable to taxation. **2** Estimated proportionally; pro rata: a *ratable* distribution. **3** That may be rated or valued. Also **rate′a·ble.** — **rat′a·bil′i·ty, rat′a·ble·ness** *n.* — **rat′a·bly** *adv.*

rat·a·fi·a (rat′ə·fē′ə) *n.* **1** A cordial flavored with fruits. **2** A flavoring essence based on the essential oil of bitter almonds. **3** A sweet biscuit. Also **rat′a·fee′** (-fē′). [<F]

Ra·tak (rä′täk) The eastern chain of the Marshall Islands.

ra·tal (rāt′l) *n.* An amount on which rates are assessed. [<RATE + -AL[1]]

ra·tan (ra·tan′) See RATTAN.

rat·a·ny (rat′ə·nē) See RHATANY.

rat·a·plan (rat′ə·plan′) *n.* A rapidly repeated sound, as of the beating of a drum. — *v.t.* & *v.i.* **·planned, ·plan·ning** To sound a rataplan (on). [<F; imit. of drumming]

rat·a·tat·tat (rat′ə·tat′tat′) *n.* A quick, sharp rapping sound, as a knock at a door. [Imit.]

rat·bite fever (rat′bīt′) *Pathol.* An infectious disease caused by the bite of a rat infested with certain bacteria: characterized by local ulcerations, rash, severe muscular pains, and relapsing fever. Also **ratbite disease.**

ratch[1] (rach) *n.* **1** A ratchet or ratchet wheel. **2** A spot on a horse's face. [Short for RATCHET]

ratch[2] (rach) *v.i. Naut.* To sail by the wind on any tack. [Back formation <obs. *raught*, pp. of REACH, on analogy with *caught, catch*]

ratch·et (rach′it) *n.* **1** A mechanism consisting of a notched wheel, the teeth of which engage with a pawl, permitting motion of the wheel in one direction only. **2** The pawl or the wheel thus used. Also **ratchet wheel.** [<F *rochet* spool <Ital. *rochetto* bobbin, dim. of *rocca* distaff <Gmc. Cf. OHG *roccho* spindle.]

rate[1] (rāt) *n.* **1** The measure of a thing by its relation to a standard; proportional or comparative amount or degree: a high *rate* of interest. **2** Degree of value; price: railway *rates*; also, the unit cost of a commodity or service: the *rate* for electricity, gas, water, and the like. **3** Comparative rank or class; condition. **4** The amount of variation of a timepiece; gain or loss in seconds. **5** A ratio for the assessment of property taxes: a *rate* of 40 mills per thousand dollars. **6** *Brit.* A local tax on property. **7** The proportion which a given fact or event bears to the total of relevant cases involved: a death *rate*, marriage *rate*. **8** A fixed allowance or amount. **9** *Obs.* Degree; estimation. See synonyms under TAX. — **at any rate** In any case; under any circumstances; anyhow. — **differential rate** The lower of two rates given usually by two competing railroad lines to one of two places in the same territory in order to make profits even: in England called **preferential rate.** — *v.* **rat·ed, rat·ing** *v.t.* **1** To estimate the value or worth of; appraise. **2** To place in a certain rank or grade. **3** To fix the amount of tax or liability on. **4** To consider; regard: He is *rated* as a great statesman. **5** To fix the rate for the transportation of (goods), as by rail, water, or air. — *v.i.* **6** To have rank, rating, or value. See synonyms under CALCULATE. [<OF <L *rata* (*pars*) reckoned (part), fem. of *ratus*, pp. of *reri* reckon]

rate[2] (rāt) *v.t* & *v.i.* **rat·ed, rat·ing** To reprove with vehemence; rail at; scold. [Origin uncertain. Cf. OF *rater* scold and Sw. *rata* find fault.]

ra·tel (rā′təl, rä′-) *n.* A nocturnal carnivore (genus *Mellivora*) resembling the badger, ashy–gray above and black below, of South and West Africa and India. [<Afrikaans *rateldas* <Du. *raat* honeycomb + *das* badger]

rate·pay·er (rāt′pā′ər) *n. Brit.* One who pays local property taxes or rates.

rat·er[1] (rā′tər) *n.* One who or that which rates or estimates.

rat·er[2] (rā′tər) *n.* One who scolds or berates.

rat–foot dots In Chinese painting, a method of representing pine boughs or branches by brush strokes that resemble the print of a rat's foot: four or five slightly curved strokes radiating from a white center dot.

rath (rath) *adj. Obs.* 1 Unusually early; vehement. 2 Swift; quick; soon. 3 Relating to the forenoon, or to the early part of a period of time. Also **rathe** (rāth). [OE *hrathe* early]

Rat·haus (rät′hous′) *n. German* A government or municipal building; a town hall.

rathe (rāth) *adv. Obs.* Early; betimes; promptly. [OE *hrathe* soon]

Ra·the·nau (rä′tə·nou), **Walther**, 1867–1922, German statesman and industrialist.

rath·er (rath′ər, rä′thər) *adv.* 1 With preference for one of two things or courses; more willingly. 2 With more reason; more wisely; more strictly or accurately. 3 Somewhat; in a greater or less degree; to a certain extent. 4 Very much; exceedingly. 5 *Obs.* Sooner; earlier; more quickly. [OE *hrathor* sooner, compar. of *hrathe* soon, quick]

rath·er·est (rath′ər·ist, rä′thər-) *adv. Brit. Dial.* Most especially; most of all.

raths·kel·ler (rath′skel·ər, räts′kel·ər) *n.* 1 In Germany, the cellar of a city hall, often used as a beer hall or restaurant. 2 Any beer hall or restaurant patterned after the German type, but not necessarily located below the street level. [<G < *rat* town hall + *keller* cellar]

Ra·ti·bor (rä′tē·bôr) A port on the Oder in southern Poland: Polish *Racibórz*.

rat·i·fi·ca·tion (rat′ə·fə·kā′shən) *n.* The act of ratifying, or the state of being ratified.

rat·i·fy (rat′ə·fī) *v.t.* **·fied**, **·fy·ing** To give sanction to, especially official or authoritative sanction; make valid by approving, especially the work of an agent or representative; confirm. [<F *ratifier* <Med. L *ratificare* <L *ratus* fixed, reckoned + *facere* make] — **rat′i·fi′er** *n.*
Synonyms: accept, approve, confirm, corroborate, endorse, establish, justify, sanction, seal, settle, substantiate, validate. See ASSENT, CONFIRM, JUSTIFY. *Antonyms:* abolish, abrogate, annul, cancel, deny, disavow, disown, extinguish, nullify, repeal, rescind, revoke.

rat·ing[1] (rā′ting) *n.* 1 Classification according to a standard; grade; rank. 2 The classification of a vessel. 3 An evaluation of the financial standing of a business firm or an individual. 4 The designation of the operating capacity of a piece of machinery, as expressed in horsepower, kilowatts, etc. 5 Any specialist grade held by an enlisted man or officer: the *rating* of a pilot, gunner, parachutist, etc., in the U.S. Army, or of boatswain's mate in the Navy. 6 *Brit.* An enlisted man in the Royal Navy. See synonyms under TAX.

rat·ing[2] (rā′ting) *n.* A harsh rebuke; scolding. [<RATE[2]]

ra·tio (rā′shō, -shē·ō) *n. pl.* **·tios** 1 Relation of degree, number, etc.; relative amount; proportion; rate: There has always been a *ratio* between demand and supply. 2 The relation between two numbers or two magnitudes of the same kind; especially, the quotient of one magnitude divided by the other, or the factor that, multiplied into one, will produce the other. 3 Formerly, the relation expressed by subtracting one quantity from the other; the difference. 4 *Obs.* A portion; ration. [<L. Doublet of RATION, REASON.]

ra·ti·o·ci·nate (rash′ē·os′ə·nāt) *adj.* Reasoning, as contrasted with *ratiocinate*. [See RATIOCINATE.]

ra·ti·o·ci·nate (rash′ē·os′ə·nāt) *v.i.* **·nat·ed**, **·nat·ing** To make a deduction from premises; reason. — *adj.* Reasoned about. [<L *ratiocinatus*, pp. of *ratiocinari* calculate, deliberate < *ratio* reckoning. See REASON.] — **ra′ti·oc′i·na′tor** *n.*

ra·ti·o·ci·na·tion (rash′ē·os′ə·nā′shən) *n.* The deduction of conclusions from premises; reasoning. See synonyms under REASONING. [<L *ratiocinatio, -onis*]

ra·ti·o·ci·na·tive (rash′ē·os′ə·nā′tiv) *adj.* 1 Of or pertaining to the act or process of reasoning. 2 Given to ratiocination; argumentative. [<L *ratiocinativus*]

ra·tion·ing (rā′shōn·ing) *n.* The reduction or enlargement of a series of aerial photographs so that all are on one scale for use in a mosaic map. [<RATIO + -ING]

ra·tion (rash′ən, rā′shən) *n.* 1 A portion; share. 2 A fixed allowance or portion of food, etc., allotted in time of scarcity. — **emergency ration** Portions of canned beef, hardtack, milk chocolate, etc., for use in the field by soldiers. — *v.t.* 1 To provide with rations; issue rations to, as an army. 2 To give out or allot in rations, as gasoline, rubber, butter, etc. [<F <L *ratio, -onis*. Doublet of RATIO, REASON.] — **ra′tion·ing** *n.*

ra·tion·al (rash′ən·əl) *adj.* 1 Possessing the faculty of reasoning. 2 Conformable to reason; judicious; sensible. 3 Pertaining to reason; attained by reasoning. 4 Pertaining to rationalism. 5 *Math.* **a** Pertaining to a rational number. **b** Denoting an algebraic expression containing variables within radicals, as $\sqrt{x^2 - y^2}$, $\sqrt{4x - 1}$. Compare IRRATIONAL. 6 In Greek and Latin prosody, denoting the measurement of metrical units; capable of being measured in metrical units. — *n.* That which is rational. [<L *rationalis* < *ratio, -onis*] — **ra′tion·al·ly** *adv.* — **ra′tion·al·ness** *n.*
Synonym (adj.): reasonable. A *rational* mind is one that is capable of the ordinary and normal processes of thought; a *reasonable* mood is one at the time susceptible to the influence of reasons. A *rational* man is capable of using his reasoning powers; a *reasonable* man has them habitually in exercise. *Rational* is opposed to *insane*, *reasonable* to *fanatical*, *misguided*, *obstinate*, *unreasonable*, *visionary*. See SAGACIOUS, SANE[1], WISE[1].

ra·tion·ale (rash′ən·al′, -ä′lē, -ā′lē) *n.* 1 A rational exposition of principles. 2 The logical basis of a fact; the reason or reasons collectively. [<L, neut. of *rationalis*]

ra·tion·al·ism (rash′ən·əl·iz′əm) *n.* 1 The formation of opinions by relying upon reason alone, independently of authority or of revelation: opposed to *supernaturalism*. 2 *Philos.* **a** The theory of a priori ideas, that truth and knowledge are attainable through reason rather than through experience: opposed to *empiricism*. **b** The theory that reason itself is a source of knowledge independent of sense perception: opposed to *sensationalism*. — **ra′tion·al·ist** *n.* — **ra′tion·al·is′tic** or **·ti·cal** *adj.* — **ra′tion·al·is′ti·cal·ly** *adv.*

ra·tion·al·i·ty (rash′ən·al′ə·tē) *n. pl.* **·ties** 1 Sanity; reasonableness; naturalness. 2 The cause or reason; rationale. [<LL *rationalitas*]

ra·tion·al·i·za·tion (rash′ən·əl·ə·zā′shən, -ī·zā′shən) *n.* 1 The act or process of rationalizing. 2 *Psychol.* The process of devising acceptable reasons for desires, emotions, acts, beliefs, or opinions which cannot be credibly justified to oneself or to others in terms of their actual motives. 3 *Brit.* The act of bringing an industry into accord with up-to-date methods of organization and operation.

ra·tion·al·ize (rash′ən·əl·īz′) *v.* **·ized**, **·iz·ing** *v.t.* 1 *Psychol.* To explain (one's behavior) on grounds ostensibly rational but not in accord with the actual or unconscious motives. 2 To explain or treat from a rationalistic point of view. 3 To make rational or reasonable; render conformable to reason. 4 *Math.* To remove the radicals containing variables from (an expression or equation); also, to alter the radicals so as to change (the expression) into more workable form: thus, if $\sqrt{x^2 + 2x} = 3$, then, by squaring, $x^2 + 2x = 9$, and $x^2 + 2x - 9 = 0$. — *v.i.* 5 To think in a rational or rationalistic manner. 6 *Psychol.* To rationalize one's behavior. — **ra′tion·al·iz′er** *n.*

rational number See under NUMBER.

Rat·is·bon (rat′is·bon, -iz-) An English name for REGENSBURG.

Rat Islands (rat) A group in the Aleutian Islands, extending 110 miles west of the Andreanof Islands.

rat·ite (rat′īt) *adj.* Designating a division of flightless birds (*Ratitae*), including ostriches, cassowaries, kiwis, emus, etc., which have aborted wings and a breastbone without a keel. — *n.* One of the *Ratitae*. [<L *ratis* raft]

rat·line (rat′lin) *n. Naut.* 1 One of the small ropes fastened across the shrouds of a ship, used as the rounds of a ladder for going aloft or descending. 2 The material so used. See SHROUD[2]. Also **rat′lin** (-lin), **rat′ling** (-ling). [Origin unknown]

RATLINES

ra·toon (ra·tōōn′) *n.* 1 A new shoot from the root of a cropped plant, as from a sugarcane. 2 One of the heart leaves in a tobacco plant. — *v.i.* To sprout from a root planted the previous year. [<Sp. *retoño* <Hind. *ratun*]

rat race *Slang* A frantic, usually fruitless, struggle; a wearisome hustle or strife.

rats·bane (rats′bān′) *n.* Rat poison, as arsenous oxide.

rat–tail (rat′tāl′) *adj.* Resembling a rat's tail in form. Also **rat′–tailed′**.

rat·tan (ra·tan′) *n.* 1 The long, tough, flexible stem of a palm (genera *Calamus* and *Daemonorops*) growing in East India, Africa, and Australia. 2 The palm itself. 3 A cane or switch of rattan. Also spelled *ratan*. [<Malay *rotan*]

rat·teen (ra·tēn′) *n. Obs.* A thick woolen twilled cloth. [<F *ratine*]

rat·ten (rat′n) *v.t. Brit. Slang* To persecute or harass (an employer or employee) because of refusal to join or obey a trade union, as by removing tools or spoiling materials. [<RATTEN[2]] — **rat′ten·er** *n.* — **rat′ten·ing** *n.*

rat·ten[2] (rat′n) *n. Scot. & Brit. Dial.* A rat.

rat·ter (rat′ər) *n.* 1 A dog or cat that catches rats. 2 *Slang* A deserter from his party; traitor.

rat·tish (rat′ish) *adj.* Belonging to or resembling a rat.

rat·tle[1] (rat′l) *v.* **·tled**, **·tling** *v.i.* 1 To make a series of sharp noises in rapid succession, as by striking together: dead limbs *rattling* in the wind. 2 To move or act with such noises. 3 To talk rapidly and foolishly; chatter. — *v.t.* 4 To cause to rattle: to *rattle* pennies in a tin cup. 5 To utter or perform rapidly or noisily. 6 *Colloq.* To confuse; disconcert: Her reaction *rattled* me. See synonyms under SHAKE. — *n.* 1 A series of short, sharp sounds in rapid succession, as from the collision of small, hard objects. 2 A plaything, implement, etc., adapted to produce a rattling noise: a watchman's *rattle*. 3 The series of jointed horny rings in the tail of a rattlesnake, or one of these; also, the noise produced by the vibration of this organ. 4 Rapid and noisy talk; chatter. 5 One who talks fast and foolishly. 6 A râle; the death rattle, caused by the passage of air through mucus. See synonyms under NOISE. [Imit.]

rat·tle[2] (rat′l) *v.t.* **·tled**, **·tling** *Naut.* To fit with ratlines: used in the phrase **to rattle down the rigging**. [<RATLINE]

rat·tle–box (rat′l·boks′) *n.* 1 A toy or the like having a chamber to contain something, as a ball, that will rattle. 2 A low hairy North American annual (genus *Crotalaria*) having seeds which rattle in the inflated pod. 3 The bladder campion.

rat·tle–brain (rat′l·brān′) *n.* A talkative, flighty person; foolish chatterer. Also **rat′tle–head′** (-hed′), **rat′tle–pate** (-pāt′). — **rat′tle–brained′** *adj.*

rat·tler (rat′lər) *n.* 1 One who or that which rattles. 2 A rattlesnake.

rat·tle·snake (rat′l·snāk′) *n.* Any of various venomous, thick-bodied American snakes (genera *Crotalus* and *Sistrurus*, family *Viperidae*) with a tail ending in a series of horny, loosely connected, modified joints, which clash together with a rattling noise when the tail is vibrated.

RATTLESNAKE
(From 2 to 8 feet in length)

rattlesnake flag One of the early flags of the American Revolution, bearing a rattlesnake and the motto "Don't Tread On Me."
rattlesnake plantain A small orchid (*Goodyera pubescens*) of Canada and the eastern United States, with several ovate radical leaves reticulated with white veins and singularly mottled with white and dark green.
rattlesnake root 1 Any of several erect perennial herbs (genus *Prenanthes*) of the composite family, with a thick, tuberous, bitter root; especially, the white lettuce (*P. aspera*) and the slender rattlesnake root (*P. virgata*): considered to be a cure for the bite of a rattlesnake. 2 The root or tuber. 3 Senega.
rattlesnake weed 1 A species of hawkweed (*Hieracium venosum*) of the northern United States. 2 Rattlebox. 3 Button snakeroot (genus *Eryngium*).
rat·tle·trap (rat′l·trap′) *n.* 1 Any rickety, clattering, or worn-out vehicle or article. 2 *Slang* A loquacious or gossipy person. — *adj.* Shaky; dilapidated.
rat·tling (rat′ling) *adj.* 1 Making a clatter. 2 Garrulous; sprightly. 3 *Colloq.* Very; extraordinary; good. — *adv. Colloq.* Extraordinarily; very: a *rattling* good time.
rat·tly (rat′lē) *adj.* 1 Inclined to rattle. 2 Clattering.
rat·ton (rat′n) *n. Scot. & Brit. Dial.* A small rat.
rat·trap (rat′trap′) *n.* 1 A trap for catching rats. 2 A situation from which escape is impossible; any hopeless or fatal predicament.
rat·ty (rat′ē) *adj.* ·ti·er, ·ti·est 1 Ratlike, or abounding in rats. 2 *Slang* Disreputable; shabby.
rau·cle (rô′kəl) *adj. Scot.* Rough; harsh; strong; fearless.
rau·cous (rô′kəs) *adj.* Rough in sound; hoarse; harsh. [< L *raucus*] — **rau′ci·ty** (-sə-tē), **rau′cous·ness** *n.* — **rau′cous·ly** *adv.*
Rau·wol·fi·a (rô·wŏl′fē·ə, -wŏŏl′-) *n.* A genus of tropical trees or shrubs of the dogbane family, several of which contain alkaloids having valuable medicinal properties; especially, *R. serpentina*, an Indian species from which the alkaloid reserpine was first isolated. [after Leonard *Rauwolf*, 17th century German botanist]
rav·age (rav′ij) *v.* ·aged, ·ag·ing *v.t.* To lay waste, as by pillaging or burning; despoil; ruin. — *v.i.* To wreak havoc; be destructive. — *n.* Violent and destructive action, or its result; ruin; desolation. [< F *ravir*. See RAVISH.] — **rav′ag·er** *n.*
rave[1] (rāv) *v.* **raved, rav·ing** *v.i.* 1 To speak wildly or incoherently. 2 To speak with extravagant enthusiasm. 3 To make a wild, roaring sound; rage: The wind *raved* through the trees. — *v.t.* 4 To utter wildly or incoherently. — *n.* The act or state of raving; a frenzy. — *adj. Colloq.* Extravagantly enthusiastic: *rave* reviews. [< OF *raver, rever* < L *rabere* rage]
rave[2] (rāv) *n.* 1 A vertical sidepiece in a wagon body, or in a hand car or sleigh. 2 The wooden or iron piece that fastens the beam to the runners of a logging sled. [Origin unknown]
rav·el (rav′əl) *v.* ·eled or ·elled, ·el·ing or ·el·ling *v.t.* 1 To separate the threads or fibers of; unravel. 2 To make clear or plain; explain: often with *out*. 3 *Archaic* To tangle; confuse. — *v.i.* 4 To become separated thread from thread or fiber from fiber; unravel; fray. 5 *Archaic* To become tangled or confused. — *n.* 1 A broken or rejected thread. 2 A

raveling. [? <MDu. *ravelen* tangle] — **rav′el·er** or **rav′el·ler** *n.*
Ra·vel (ra·vel′), **Maurice Joseph**, 1875-1937, French composer.
rave·lin (rav′lin) *n. Mil.* An outwork with two faces forming a salient angle at the front. [< F < Ital. *ravellino*. Origin uncertain.]
rav·el·ing (rav′əl·ing) *n.* 1 A thread or threads raveled from a fabric. 2 The act of raveling. 3 The process of being raveled. Also **rav′el·ling**.
rav·el·ment (rav′əl·mənt) *n.* A ravel, or the act of raveling; confusion.
ra·ven[1] (rā′vən) *n.* A large, omnivorous, crowlike bird (*Corvus corax*) of North America, Europe, and Asia, having lustrous black plumage, with the feathers of the throat elongated and lanceolate. — *adj.* Black and shining, like the plumage of a raven. [OE *hræfn*]
rav·en[2] (rav′ən) *v.t.* 1 To devour hungrily or greedily. 2 To take by force; ravage. — *v.i.* 3 To search for or take prey or plunder. 4 To eat voraciously; be ravenous. — *n.* The act of plundering; spoliation; pillage. [<OF *raviner* < *ravine* rapine <L *rapina*; a doublet of RAPINE, RAVINE] — **rav′en·er** *n.*
Ra·ven (rā′vən) The southern constellation, *Corvus*. See CONSTELLATION.
Ra·ve·na·la (rav′i·nä′lə) *n.* A genus of palmlike trees of the banana family, having a fan-shaped group of elongated flat leaves arranged around the trunk, especially the traveler's tree (*R. madagascariensis*). [<Malagasy]
rav·en·ing (rav′ən·ing) *adj.* 1 Seeking eagerly for prey; rapacious. 2 Mad; rabid. — *n.* 1 Propensity for prey or booty; rapacity. 2 The prey seized. [ppr. of RAVEN[2]] — **rav′en·ing·ly** *adv.*
Ra·ven·na (rä·ven′nä) A city in north central Italy, 6 miles west of the Adriatic, formerly on it; capital of the Western Roman Empire 402-476.
rav·en·ous (rav′ən·əs) *adj.* 1 Violently voracious or hungry. 2 Extremely eager for gratification. See synonyms under GREEDY. [< OF see RAVEN[2].] — **rav′en·ous·ly** *adv.* — **rav′en·ous·ness** *n.*
Ra·vi (rä′vē) A river in NW India and West Pakistan, flowing SW 474 miles from northern Punjab State, India to the Chenab river above Multan, West Pakistan: ancient *Hydraotes*.
rav·in (rav′in) *n.* 1 The act of plundering or ravaging. 2 That which is obtained by violence or robbery. — *v.t. & v.i.* To raven. [<OF *ravine*. See RAPINE.]
ra·vine (rə·vēn′) *n.* 1 A deep gorge or gully, especially one worn by a stream or flow of water. 2 A long, narrow cleft between heights. See synonyms under VALLEY. [< F. Doublet of RAVEN[2], *n.*, RAPINE.]
rav·ing (rā′ving) *adj.* 1 Furious; delirious; frenzied. 2 *Colloq.* Excessive; extraordinary: a *raving* beauty. — *n.* Furious, incoherent, or irrational utterance. See synonyms under FRENZY. Compare INSANITY.
ra·vi·o·li (rä·vyō′lē, rä′vē·ō′lē, rav′ē-) *n. pl.* Balls of forcemeat, encased in little envelopes of dough and boiled in broth or water: commonly construed in the singular. [<Ital., dim. pl. of dial. *rava* <L *rapa* turnip, beet]
rav·ish (rav′ish) *v.t.* 1 To fill with strong emotion, especially delight; enrapture. 2 To commit a rape upon. 3 To carry off (a woman) by force. 4 To seize and carry off by violence. [<OF *raviss-*, stem of *ravir* carry off <L *rapere* seize. Related to RAPE, RAPTURE.] — **rav′ish·er** *n.* — **rav′ish·ing·ly** *adv.* Synonyms: capitivate, charm, delight, enchant, enrapture, entrance, overjoy, transport. See ABUSE, CHARM[1], POLLUTE, REJOICE. Antonyms: disenchant, disgust, nauseate, repel.
rav·ish·ing (rav′ish·ing) *adj.* Filling with transports of delight; enchanting.
rav·ish·ment (rav′ish·mənt) *n.* The act of ravishing or the state of being ravished; especially, ecstasy; delight. [<OF *ravissement*]
raw (rô) *adj.* 1 Not changed or prepared by cooking, in its natural state; uncooked. 2 Not covered with whole skin; abraded. 3 Bleak; chilling: a *raw* wind. 4 In a natural state; crude; unprepared, as wool, drugs, etc.; also, untempered or without tone, as colors; unrefined; unfinished. 5 Newly done; fresh: *raw* paint, *raw* work. 6 Inexperienced; undisciplined: a *raw* recruit. 7 Unrefined; crude; off-

color: a *raw* joke. 8 Unexposed: said of photographic film. — *n.* 1 A sore or abraded spot; a sensitive point. 2 The state of being raw, untamed, or unpeopled: nature in the raw. [OE *hrēaw*] — **raw′ly** *adv.* — **raw′ness** *n.*
Ra·wal·pin·di (rä′wəl·pen′dē, rôl·pen′dē) A city in the northern Punjab region of West Pakistan, near the border of Jammu and Kashmir state.
raw-boned (rô′bōnd′) *adj.* Having large bones and little flesh; bony; gaunt.
Raw·bones (rô′bōnz′) Death.
raw deal *Slang* Harsh or unfair treatment in a transaction.
raw fibers Textile fibers in their natural state, as silk is in the gum or cotton as it comes from the bale.
raw·hide (rô′hīd′) *n.* 1 A hide dressed without tanning. 2 A whip made of such hide.
raw·ish (rô′ish) *adj.* Somewhat raw.
Raw·lin·son (rô′lin·sən), **George**, 1812-1902, English Orientalist and historian. — **Sir Henry**, 1810-95, English soldier and Assyriologist; brother of the preceding.
raw material Unprocessed material (animal, vegetable, or mineral) needed and used in manufacturing, as contrasted with finished products.
rax (raks) *v.t. & v.i. Scot.* To stretch out; reach.
ray[1] (rā) *n.* 1 A narrow beam of light or other line of propagation of any form of radiant energy; line of radiating force; radiation. 2 A manifestation of intellectual light. 3 One of several lines radiating from an object. 4 *Geom.* A straight line emerging from a center and unlimited in one direction only. 5 A streak or line; a straight row. 6 *Zool.* **a** One of the rods supporting the membrane of a fish's fin. **b** One of the radiating parts of a radiate animal, as a starfish. 7 *Bot.* **a** A raylike flower. **b** One of the pedicels or flower stalks of an umbel. 8 *Physics* A stream of particles spontaneously emitted by a radioactive substance. 9 A trace or minute particle: Not a *ray* of life was present. — *v.i.* 1 To emit rays; shine. 2 To issue forth as rays; radiate. — *v.t.* 3 To send forth as rays. 4 To mark with rays or radiating lines. 5 To irradiate. 6 To treat with or expose to X-rays, etc. [< OF *rai* <L *radius*. Doublet of RADIUS.]
ray[2] (rā) *n.* An elasmobranch fish (order *Selachii*) having a cartilaginous skeleton and a flattened body, with expanded pectoral fins, dorsally placed eyes, and a long caudal appendage; especially, the sting ray or torpedo. [< F *raie* <L *raia*]
Ray (rā), **Cape** A promontory at the SW extremity of Newfoundland at the entrance to the Gulf of Saint Lawrence.
ra·yah (rä′yə, rī′ə) *n.* A non-Moslem inhabitant of Turkey: sometimes spelled *raia*. Also **ra′a**. [<Arabic *ra'iyah* flock, herd]
ray flower *Bot.* Any of the flat marginal flowers of an inflorescence when distinct from the disk, as in the daisy or sunflower. Also **ray floret**.
ray·grass (rā′gras′, -gräs′) *n.* Ryegrass.
Ray·leigh (rā′lē), **Lord**, 1842-1919, John William Strutt, third baron, English physicist.
ray·less (rā′lis) *adj.* 1 Having no light rays. 2 Extremely dark. 3 Having no rays, as certain composite plants. — **ray′less·ly** *adv.* — **ray′less·ness** *n.*
Ray·mond (rā′mənd, *Fr.* rā·môn′) A masculine personal name. Also **Ray′mund**. Also *Lat.* **Ray·mun·dus** (rā·mun′dəs). [<Gmc., wise protection]
ray·on (rā′on) *n.* A lustrous synthetic fiber variously made by chemical means from cellulose or with cellulose as a base, the viscous material being forced through fine spinnerets to produce filaments suitable for textiles and fabrics. [< F, from shine]
raze (rāz) *v.t.* **razed, raz·ing** 1 To level to the ground; tear down; demolish. 2 *Rare* To scrape or shave off. 3 *Obs.* To wound slightly; graze. See synonyms under DEMOLISH. Also spelled *rase*. ◆ Homophone: *raise*. [<F *raser* <L *rasum*, pp. of *radere* scrape]
ra·zee (rā·zē′) *v.t.* To make lower by cutting down, as a ship of war by removing the upper deck or decks; reduce; abridge. — *n.* A vessel that has been reduced by cutting away the upper deck or decks. [<F *rasé*, pp. of *raser* shave, raze]
ra·zor (rā′zər) *n.* A sharp cutting implement

razorback … **1049** … **Reading**

used for shaving off the beard or hair. — **safety razor** A razor provided with a guard or guards for the blade to prevent accidental gashing of the skin. [< OF *rasor* < *raser* scrape]
ra·zor·back (rā′zər·bak′) *n.* 1 A rorqual. 2 A lean-bodied, half-wild hog with long legs, common in the southeastern United States. 3 A hill with a sharp narrow ridge. — **ra′zor·backed′** *adj.*
ra·zor-billed auk (rā′zər·bild′) A small auk (*Alca torda*) of the North Atlantic, having a compressed and deeply furrowed bill. Also **ra′zor·bill′**.
razor clam A clam (genus *Ensis*) having a long, narrow, slightly curved shell resembling a razor. Also **ra′zor-shell′ clam.**
razor grinder One who sharpens or grinds razors or razor blades.

RAZOR-BILLED AUK
(About 16 inches in length)

razor strop A strip, of specially prepared leather, canvas, or other material, upon which the blade of a razor is stroked to give it a fine edge.
razz (raz) *n. Slang* Raspberry (def. 3). — *v.t.* To heckle; deride. [< RASPBERRY]
raz·zi·a (raz′ē·ə) *n.* A foray or armed expedition, as for plunder or conquest, or for the capture of cattle or slaves. [< F < Arabic *ghāziah* < *ghasw* war, battle]
raz·zle-daz·zle (raz′əl·daz′əl) *n. U.S. Slang* Anything bewildering and exciting; dazzling activity or performance. [Varied reduplication of DAZZLE]
re[1] (rā) *n. Music* The second note of any major scale in solmization. [< L *re(sonare)* resound. See GAMUT.]
re[2] (rē) *prep.* Concerning; about; in the matter of: used in business letters: *re* your letter of the 6th instant. [< L, ablative of *res* thing]
Re (rā) See RA.
re- *prefix* 1 Back: *reduce* to lead back, *remit* to send back. 2 Again; anew; again and again: *regenerate*. Re- in this second sense is freely used in Modern English, as in the list of words below. It is hyphenated, in certain cases, to prevent confusion with a similarly spelled word having a different meaning (*retreat* to treat again, *re-treat* to go back), to prevent mispronunciation (*re-argue*, *re-urge*), and also in the coining of new words. Also, before vowels, sometimes *red-*, as in *redeem*. [< L *re-*, *red-* back, again]
reach (rēch) *v.t.* 1 To stretch out or forth, as the hand or foot. 2 To present by means of or as by means of the outstretched hand; deliver; hand over. 3 To extend as far as; touch or grasp, as with the hand: Can you *reach* the top shelf? 4 To arrive at or come to by motion or progress; attain: When do we *reach* Miami? 5 To achieve communication with; gain access to. 6 To amount to; total. 7 To strike or hit, as with a blow or missile. — *v.i.* 8 To stretch the hand, foot, etc., out or forth. 9 To attempt to touch or grasp something: He *reached* for his wallet. 10 To have extent in space, time, amount, or influence: The ladder *reached* to the ceiling. 11 *Naut.* To sail on a tack with the wind on or forward of the beam. — *n.* 1 The act or power of reaching; also, the distance one is able to reach, as with the hand, an instrument, or missile, or by thought, influence, etc.; scope; range. 2 A point, position, or result attained or attainable. 3 An unbroken stretch, as of a stream; a vista or expanse. 4 A pole or bar connecting the rear axle,

truck, or runners of a vehicle with some part at the forward end. 7 *Naut.* The sailing, or the distance sailed, by a vessel on one tack. [OE *rǣcan*. Akin to G *reichen* reach] — **reach′er** *n.*
Synonyms (verb): attain, gain, hit, land, make, strike, touch. To *reach*, in the sense here considered, is to come to by motion or progress. *Attain* is now oftenest used of abstract relations; as, to *attain* success. To *gain* is to *reach* or *attain* a thing eagerly sought; the wearied swimmer *reaches* or *gains* the shore. See ARRIVE, GET, MAKE[1], STRETCH.
reach·less (rēch′lis) *adj.* 1 That cannot be reached; unattainable; lofty. 2 Without a reach (def. 4).
reach-me-down (rēch′mē·doun′) *Slang adj.* Ready-made, as a garment; also, secondhand. — *n.* Usually *pl.* Ready-made or secondhand clothing.
re·act (rē·akt′) *v.i.* 1 To act in response, as to a stimulus. 2 To act in a manner contrary to some preceding act; come into or tend toward a former state or an opposite state. 3 *Physics* To exert an opposite and equal force on an acting or impinging body: said of the body acted upon. 4 *Chem.* To exert mutual action, as substances undergoing chemical change. Compare RE-ACT.
re-act (rē·akt′) *v.t.* To act again.
re·ac·tance (rē·ak′təns) *n. Electr.* In an alternating-current circuit, that component of the resistance that does not oppose the current, but tends to cause a difference of phase between it and the electromotive force: measured in ohms.
re·ac·tion (rē·ak′shən) *n.* 1 Reverse or return action; tendency toward a former or reversed state of things; especially, a trend toward an earlier social, political, or economic policy or condition. 2 *Physiol.* Contrary action or reversed effects following a stimulus; a reflex action. 3 *Psychol.* The partial or total response made to any kind or degree of stimulation. 4 *Physics* In the second law of motion, the equal and opposite force exerted on an agent by the body acted upon. 5 *Chem.* The mutual action of substances subjected to chemical change, or some distinctive result of such action. 6 *Biol.* The effect upon any organism or any of its parts made by the introduction of any foreign substance for diagnostic or therapeutic purposes, or for testing, immunizing, etc. — **re·ac′tive** *adj.*
re·ac·tion·ar·y (rē·ak′shən·er′ē) *adj.* Of, relating to, favoring, or characterized by reaction. — *n. pl.* ·**ar·ies** One who favors political or social reaction; a conservative. Also **re·ac′tion·ist.**
reaction engine An engine which obtains thrust by the expulsion of the hot gases of combustion to the rear; a jet engine.
reaction formation *Psychoanal.* The conscious development of character traits, attitudes, and forms of behavior in contrast with and opposition to original trends of the ego, for which they serve as deceptive and relatively unstable concealment. Compare SUBLIMATION.
reaction time *Physiol.* 1 The time required for a response to a sensory stimulus. 2 The time required for an electric current to act on a muscle.
re·ac·ti·vate (rē·ak′tə·vāt) *v.t.* ·**vat·ed**, ·**vat·ing** To make active or effective again. — **re·ac′ti·va′tion** *n.*
re·ac·tive (rē·ak′tiv) *adj.* 1 Reacting, tending to react, or resulting from reaction. 2 Responsive to a stimulus.
reactive factor *Electr.* In a circuit, the ratio of the reactive volt-amperes to the total volt-amperes.
reactive volt-ampere *Electr.* That component of the volt-amperes in an alternating-current circuit not representing the work done in watts. Also **reactive power.** See VAR.
re·ac·tor (rē·ak′tər) *n.* 1 One who or that which reacts. 2 *Electr.* A device for introducing reactance into a circuit, for starting

motors, controlling current, and the like. 3 *Biol.* An animal or person giving a positive reaction to a specified bacteriological or medical test. 4 *Physics* Any of variously designed assemblies for the initiation and control of nuclear fission, consisting essentially of reserves of fissionable material used as fuel, moderators to check the rate of nuclear reactions, reflectors, and auxiliary structures, equipment, shielding, etc.: also called *pile*.
read (rēd) *v.* **read** (red), **read·ing** (rē′ding) *v.t.* 1 To apprehend the meaning of (a book, writing, etc.) by perceiving the form and relation of the printed or written characters. 2 To utter aloud (something printed or written). 3 To understand the significance of as if by reading: to *read* the sky. 4 To apprehend the meaning of something printed or written in (a foreign language). 5 To make a study of; also, to obtain knowledge of: to *read* law. 6 To discover the true nature of (a person, character, etc.) by observation or scrutiny. 7 To interpret (something read) in a specified manner. 8 To take as the meaning of something read. 9 To have or exhibit as the wording: The passage *reads* "principal," not "principle." 10 To indicate or register: The meter *reads* zero. 11 To bring into a specified condition by reading: I *read* her to sleep. — *v.i.* 12 To apprehend the characters of a book, musical score, etc. 13 To utter aloud the words or contents of a book, etc. 14 To gain information by reading: with *of* or *about*. 15 To learn by means of books; study. 16 To have a specified wording: The contract *reads* as follows. 17 To admit of being read in a specified manner. 18 To give a public reading or recital. — **to read between the lines** To perceive or infer what is not expressed or obvious, as a hidden or true meaning, implication, or motive: He *read* her emotions *between the lines*. — **to read out** To expel from a religious body, political party, etc., by proclamation or concerted action. — **to read up** (or **up on**) To learn by reading. — *adj.* (red) Informed by books or reading; acquainted with books or literature: well *read*. — *n.* (rēd) *Colloq.* A reading; a period spent in reading. [OE *rǣdan* advise, read]
read·a·ble (rē′də·bəl) *adj.* 1 Legible. 2 Easy and pleasant to read. — **read′a·bil′i·ty, read′·a·ble·ness** *n.* — **read′a·bly** *adv.*
Reade (rēd), **Charles,** 1814-84, English novelist.
read·er (rē′dər) *n.* 1 One who reads; specifically, a professional reciter or elocutionist. 2 One who reads and criticizes manuscripts offered to publishers. 3 A proofreader. 4 A layman authorized to read the lesson in church services. 5 A textbook containing matter for exercises in reading. 6 *Brit.* A university or college lecturer.
read·i·ly (red′ə·lē) *adv.* 1 In a ready manner; promptly; easily. 2 Willingly.
read·i·ness (red′ē·nis) *n.* 1 The quality or state of being ready. 2 The quality of being quick or prompt; facility; aptitude. 3 A disposition for prompt compliance; willingness. See synonyms under ABILITY, ADDRESS, DEXTERITY, EASE, INGENUITY.
read·ing (rē′ding) *n.* 1 The act, practice, or art of reading, in any sense of the verb; a public recital; the act of reading formally to a legislative body a bill, etc., proposed for adoption. 2 Literary research; study; scholarship. 3 Matter which is read or is designed to be read. 4 The indication of a graduated instrument, as a thermometer. 5 The form in which any passage or word appears in any copy of a work. 6 An interpretation, as of a riddle, or of any latent and hidden meaning; delineation; rendering. See synonyms under EDUCATION. — *adj.* Pertaining to or suitable for reading.
Read·ing (red′ing), **Marquis of,** 1860-1935, Rufus Daniel Isaacs, British statesman; viceroy of India 1921-26.
Read·ing (red′ing) 1 A county borough and

reabsorb	readdress	readvance	reallege	reanimation	reappear	reapportion
reabsorption	readjourn	reafforest	re-alliance	reannex	reappearance	reapportionment
reaccess	readjournment	reafforestation	re-ally	reannexation	reapply	re-argue
reaccommodate	readopt	reagree	realphabet	reanoint	reappoint	re-argument
reaccuse	readorn	re-alinement	reamputate	reapparel	reappointment	reascend

add, āce, cāre, pälm; end, ēven; it, īce; odd, ōpen, ôrder; tōōk, pōōl; up, bûrn; ə = a in *above*, e in *sicken*, i in *clarity*, o in *melon*, u in *focus*; yōō = u in *fuse*; oi, oil; ou, pout; ch, check; g, go; ng, ring; th, thin; ᵺ, this; zh, vision. Foreign sounds ä, œ, ü, kh, ṅ; and ◆: see page xx. < from; + plus; ? possibly.

county town of Berkshire, England. 2 A city on the Schuylkill River in SE Pennsylvania.
reading desk A desk adapted to hold books, manuscripts, etc., for a speaker or reader, as in church services.
reading room A room provided with periodicals, books, etc., in which the public, or certain classes of readers, may read.
re·ad·just (rē′ə·just′) v.t. & v.i. To adjust again or anew; rearrange. — **re′ad·just′er** n.

READING DESK

re·ad·just·ment (rē′ə·just′mənt) n. 1 The act or process of readjusting, or the state of being readjusted. 2 The reorganization of a company or corporation, usually voluntary.
re·ad·mit (rē′əd·mit′) v.t. ·mit·ted, ·mit·ting To admit again; allow to enter again. — **re′ad·mis′sion, re′ad·mit′tance** n.
read·y (red′ē) adj. read·i·er, read·i·est 1 Prepared for use or action. 2 Prepared in mind; willing. 3 Likely or liable: with to: ready to sink. 4 Quick to act, follow, occur, or appear; prompt. 5 At hand; immediately available; convenient; handy. 6 Designating the standard position in which a rifle is held just before aiming. 7 Quick to understand; alert; quick; facile: a ready wit. 8 Obs. Here; present: used in answering a roll call. See synonyms under ACTIVE, ALERT, GOOD, RIPE[1]. — n. 1 In the manual of arms, the position in which a rifle is held before aiming, the left hand at the balance, the right hand at the small of the stock. 2 Slang Cash: with the. — v.t. read·ied, read·y·ing To make ready; prepare. [OE ræde, geræde]
read·y-made (red′ē·mād′) adj. 1 Not made to order; prepared or kept on hand for general demand: said especially of clothing. 2 Prepared beforehand; not impromptu. 3 Prepared by someone else. 4 Borrowed; lacking in originality; inferior.
ready money Money in hand; cash.
read·y-to-wear (red′ē·tə·wâr′) adj. Ready-made: said of clothing.
read·y-wit·ted (red′ē·wit′id) adj. Quick to apprehend or perceive; alert.
re·af·firm (rē′ə·fûrm′) v.t. To affirm again, as for emphasis. — **re·af·firm′ance, re·af·fir·ma′tion** n.
re·a·gent (rē·ā′jənt) n. 1 One who or that which reacts; a source of reflex action. 2 Chem. Any substance used to ascertain the nature or composition of another by means of their mutual chemical action. 3 Psychol. The subject of an experiment; particularly, one who or that which reacts to a stimulus. [<RE- + AGENT]
re·al[1] (rē′əl, rēl) adj. 1 Having existence or actuality as a thing or state; not imaginary: a real event. 2 Being in accordance with appearance or claim; genuine; not artificial or counterfeit. 3 Representing the true or actual, as opposed to the apparent or ostensible: the real reason. 4 Unaffected; unpretentious: a real person. 5 Philos. Having actual existence, and not merely possible, apparent, or imaginary. 6 Law a Of, pertaining to, or consisting of land and tenements: real property, as contrasted with personal property. b Pertaining to things, as distinguished from persons. — n. That which is real; a real thing. — adv. Colloq. Very; extremely: to be real glad. [<OF <Med. L realis <L res thing] — **re′al·ness** n.
re·al[2] (rē′əl, Sp. rä·äl′) n. 1 pl. **re·als** or **re·a·les** (rä′·äl′ās) A small silver coin of several Spanish countries, including Mexico, and formerly current in the United States, where it was called a bit, and had the value of 12 1/2 cents. 2 pl. **reis** (rās) A former Portuguese

and Brazilian coin; one thousandth of a milreis. [<Sp., lit., royal <L regalis]
real estate Land, including whatever is made part of or attached to it by man or nature, as trees, houses, etc. Also **real property**.
re·al·gar (rē·al′gər) n. A resinous, orange-red arsenic sulfide, As_2S_2, formerly extensively used as a pigment and still employed in pyrotechnics. [<OF <Med. L <Arabic rahj al-ghār powder of the cave. Cf. Sp. rejalgar.]
real image See under IMAGE.
re·al·ism (rē′əl·iz′əm) n. 1 In literature and art, the principle of depicting persons and scenes as they exist, without any attempt at idealization. 2 The tendency to be concerned solely with reality, as opposed to ideals; specifically, the tendency to think and act in the light of actuality, disregarding idealistic motives. 3 Philos. a The doctrine that universals (abstract concepts) have objective existence and are more real than things: opposed to nominalism. Compare CONCEPTUALISM. b The doctrine that things have reality apart from the conscious perception of them: opposed to idealism. — **re·al·is′tic** adj. — **re′al·is′ti·cal·ly** adv.
re·al·ist (rē′əl·ist) n. 1 An adherent of the doctrine of realism in any of its forms, as applied in literature, art, or philosophy. 2 One who is devoted to what is real rather than imaginary or ideal.
re·al·i·ty (rē·al′ə·tē) n. pl. ·ties 1 The fact, state, condition, or quality of being real or genuine. 2 That which is real; an actual person, thing, situation, or event; in the aggregate, the sum of real things; also, the substance that lies back of form and external appearances. 3 That which exists, as contrasted with what is fictitious; that which is objective, not merely an idea. 4 Philos. The absolute; that which is self-existent; the ultimate, as contrasted with phenomena or the apparent. See synonyms under VERACITY. [<Med. L realitas, -tatis <realis real]
reality principle Psychoanal. The adjustment of the ego to meet the requirements of the external world. Compare PLEASURE PRINCIPLE.
re·al·i·za·tion (rē′əl·ə·zā′shən) n. 1 The act of realizing. 2 The state of being realized. 3 A product or instance of realizing. 4 The conversion into fact or action (of plans, ambitions, fears, etc.).
re·al·ize (rē′əl·īz) v. ·ized, ·iz·ing v.t. 1 To understand or appreciate fully. 2 To make real or concrete. 3 To cause to appear real. 4 To obtain as a profit or return. 5 To obtain money in return for: He realized his holdings for a profit. 6 To bring as a profit or return: said of property. — v.i. 7 To sell property for cash. See synonyms under ACCOMPLISH, EFFECT, GAIN[1], KNOW. — **re′al·iz′a·ble** adj. — **re′al·iz′er** n.
re·al·iz·ing (rē′əl·ī′zing) adj. 1 Conceiving of as real; comprehending. 2 Able to visualize vividly. 3 Converting (hopes, plans, etc.) into fact, or (assets) into money.
re·al·ly (rē′ə·lē, rē′lē) adv. In reality; in point of fact; as a matter of fact; actually; indeed; also used without precise meaning, for emphasis.
realm (relm) n. 1 A kingdom. 2 The domain or jurisdiction of any power or influence: the realm of imagination. 3 A primary division of the globe with reference to its fauna; a zoogeographical area larger than a region; also, as used by some authors, a division equivalent to a region. [<OF realme <L regalis royal. See REGAL.]
real number See under NUMBER.
Re·al·po·li·tik (rā·äl′pō·li·tēk′) n. German Literally, practical or realistic politics: a term often used cynically to mean the attainment of political ends by the use or threatened use of armed force.
Real Presence Theol. The actual presence of the body and blood of Christ in the Eucharist.
Re·al·schu·le (rā·äl′shōō′lə) n. (-len (-lən)) German A modern non-classical German secondary school preparing students for commercial or technical occupations that do not require a university education.

re·al·tor (rē′əl·tər, -tôr) n. A realty broker who is affiliated in membership with the National Association of Real Estate Boards, which has for one of its objects the protection of the public against dishonest practices, as of brokers. [<REALTY[1] + -OR]
re·al·ty[1] (rē′əl·tē) n. pl. ·ties Law Real estate or real property in any form. See REAL[1].
re·al·ty[2] (rē′əl·tē) n. Obs. 1 Fealty. 2 Royalty. [<OF realté <L regalis regal]
real wages Wages evaluated in terms of purchasing power, as contrasted with nominal wages, evaluated in money.
ream[1] (rēm) n. 1 Twenty quires of paper; properly, 480 sheets (a **short ream**), but often 500 sheets (a **long ream**) or, in a printer's or **perfect ream**, 516 sheets. 2 pl. Colloq. A prodigious amount of printed, written, or spoken material: reams of footnotes. [<OF reyme <Sp. resma <Arabic rizmah packet <razama pack together]
ream[2] (rēm) v.t. 1 To increase the size of (a hole). 2 To enlarge or taper (a hole) with a rotating cutter or reamer. 3 To turn or roll over the edge of: to ream a cartridge shell. 4 To get rid of (a defect) by reaming. [OE rēman enlarge, make room. Akin to ROOM.]
ream[3] (rēm) n. Scot. & Brit. Dial. Cream; froth; foam. — v.t. Scot. To skim, as cream. — **ream′y** adj.
ream·er (rē′mər) n. 1 One who or that which reams. 2 A finishing tool with a rotating cutting edge for reaming: sometimes spelled rimmer.

REAMERS
a. Adjustable.
b. Square.
c. Center.
d. Rose-shell.
e. Roughening taper.
f. Root reamer.

re·an·i·mate (rē·an′ə·māt) v.t. ·mat·ed, ·mat·ing 1 To bring back to life; resuscitate. 2 To revive; encourage. — **re·an′i·ma′tion** n.
reap (rēp) v.t. 1 To cut and gather (grain); harvest or gather (a fruit or product) with a scythe, reaper, or the like. 2 To cut the growth from or gather the fruit of, as a field. 3 To obtain as the result of action or effort; receive as a return or result. — v.i. 4 To harvest grain, etc. 5 To receive a return or result. See synonyms under GAIN[1]. [OE repan] — **reap′a·ble** adj. — **reap′ing** n.
reap-dole (rēp′dōl′) n. Brit. A gratuity given to reapers after harvesting: a relic of feudalism.
reap·er (rē′pər) n. 1 One who reaps. 2 A machine for harvesting standing grain; a reaping machine.
reaper and binder A reaping machine having a device that binds the grain as it cuts it.
reaping machine A machine for harvesting standing grain. It usually consists of a reciprocating cutter resembling that of a mowing machine, a platform or table on which the cut grain falls, and a dropper which is dropped to deposit the bundles of grain. In addition, it often has a reel for bending the grain toward the cutter, or a raking mechanism for pressing the grain down on the table and sweeping it off in bundles, and a binding mechanism. Also called harvester.
rear[1] (rir) n. 1 The hinder or hindmost part. 2 A place or position at the back of or behind any person or thing. 3 That division of a military force which is last or farthest from the front: opposed to van. — adj. Being in the rear; last; hindmost. [Aphetic form of ARREAR]
rear[2] (rir) v.t. 1 To place upright; raise; elevate. 2 To build; erect. 3 To care for and bring to maturity. 4 To breed or grow. — v.i. 5 To rise upon its hind legs, as a horse. 6 To rise high; tower: The mountain rears above the forest. See synonyms under RAISE. [OE rǣran set upright, causative of rīsan rise. Akin to ON reisa raise.] — **rear′er** n.
rear admiral See under ADMIRAL.

---

reascension	reassertion	reassume	reattempt	rebind	rebury	rechange
reascent	reassign	reattach	reavow	rébloom	recarriage	recharge
reassemblage	reassimilate	reattachment	reawake	reblossom	recarry	recharter
reassemble	reassimilation	reattain	rebaptism	reboil	recelebrate	rechoose
reassert	reassociate	reattainment	rebaptize	rebuild	recelebration	rechristen

rear-end (rir'end') *n.* In automobiles, the after part of the drive train, consisting of the differential gears and rear axles with their housings and the driving wheels. — *adj.* Of or pertaining to the rear-end of an automobile.

rear guard A body of troops to protect the rear of an army. [<AF *reregard*, OF *rereguarde*. Doublet of REARWARD².]

rear-horse (rir'hôrs') *n.* A mantis. [From its habit of rearing when touched]

re-arm (rē-ärm') *v.t. & v.i.* 1 To arm again. 2 To arm with more modern weapons. — **re-arm'a-ment** *n.*

rear-most (rir'mōst') *adj.* Coming or stationed last.

rear-mouse (rir'mous') See REREMOUSE.

re-ar-range (rē'ə-rānj') *v.t. & v.i.* -ranged, -ranging To arrange again or in some new way. — **re'ar-range'ment** *n.*

rear sight The sight of a gun which is nearest the breech.

rear-view mirror (rir'vyōō') In motor vehicles, a mirror so placed in front of the driver that he can see the reflection of the road and vehicles behind. Also **rear'-vi'sion mirror.**

rear-ward¹ (rir'wərd) *adj.* Coming last or toward the rear; hindward. — *adv.* Toward or at the rear; backward. Also **rear'wards.** — *n.* Hindward position; the rear; end.

rear-ward² (rir'wôrd') *n.* *Obs.* A rear guard. [<AF *rerewarde*. Doublet of REAR GUARD.]

rea-son (rē'zən) *n.* 1 That which is thought or alleged as the basis or ground for any opinion, determination, or action; something adduced or adapted to influence the mind in determining or acting; proof; argument; motive; principle. 2 That which explains or accounts for any fact, act, proceeding, or event; loosely, an efficient or final cause, or a condition. 3 The entire mental or rational nature of man, as distinguished from the intelligence of the brute; the mind; in a more limited sense, the purely intellectual faculties. 4 Specifically, the normal exercise of the rational faculties. 5 That which is in conformity to general opinion; common sense: *The anarchist was brought to* reason. 6 A logical ground for thinking; an antecedent; also, the premise or premises of an argument, generally the minor premise. 7 That which is right or befitting; just procedure; a reasonable act or proposition. 8 Intuition. — *v.i.* 1 To think logically; obtain inferences or conclusions from known or presumed facts. 2 To talk or argue logically. — *v.t.* 3 To think out carefully and logically; analyze: with *out.* 4 To influence by means of reason; persuade or dissuade. 5 To argue; debate. See synonyms under ARGUE, DISPUTE. [<OF *raison* <L *ratio, -onis* <*ratus,* pp. of *reri* reckon. Doublet of RATION, RATIO.]
Synonyms (noun): account, aim, argument, cause, consideration, design, end, ground, motive, object, principle, purpose. While the *cause* of any event, act, or fact, as commonly understood, is the power that makes it to be, the *reason* of or for it is the explanation given by the human mind; but *reason* is often used as equivalent to *cause,* especially in the sense of *final cause.* In the statement of any reasoning, the *argument* may be an entire syllogism, or the premises considered together apart from the conclusion, or in logical strictness the middle term only by which the particular conclusion is connected with the general statement. But when the reasoning is not in strict logical form, the middle term following the conclusion is called the *reason;* thus in the statement "All tyrants deserve death; Caesar was a tyrant; therefore Caesar deserved death," "Caesar was a tyrant" would in the strictest sense be called the *argument;* but if we say "Caesar deserved death because he was a tyrant," the latter clause would be termed the *reason.* See CAUSE, INTELLECT, MIND, REASONING, UNDERSTANDING, WISDOM. Compare BECAUSE.

rea-son-a-ble (rē'zən-ə-bəl) *adj.* 1 Conformable to reason; sensible. 2 Having the faculty of reason; rational. 3 Governed by reason in acting or thinking. 4 Moderate, as in price; fair. See synonyms under JUST, LIKELY, PROBABLE, RATIONAL, WISE¹. [<OF *raisonable;* after L *rationabilis*] — **rea'son-a-bil'i-ty, rea'son-a-ble-ness** *n.* — **rea'son-a-bly** *adv.*

rea-soned (rē'zənd) *adj.* Founded upon or characterized by reason; premeditated or studied.

rea-son-er (rē'zən-ər) *n.* One who reasons or argues.

rea-son-ing (rē'zən-ing) *n.* The act or process of the mind by which from propositions known or assumed new propositions are reached; argumentation; also, the reasons, proofs, or arguments employed in such process.
Synonyms: argument, argumentation, debate, ratiocination. *Argumentation* and *debate* always suppose two parties alleging reasons for and against a proposition. *Reasoning* may be the act of one alone, as it is simply the orderly setting forth of reasons, whether for the instruction of inquirers, the confuting of opponents, or the clear establishment of truth for oneself. *Reasoning* may be either deductive or inductive. *Argument* or *argumentation* was formerly used of deductive *reasoning* only. With the rise of the inductive philosophy these words have come to be applied to inductive processes also; but while *reasoning* may be informal or even unconscious, *argument* and *argumentation* strictly imply logical form. Compare INTELLECT, REASON.

rea-son-less (rē'zən-lis) *adj.* Devoid of the faculty of reason; also, not conformable to reason.

re-as-sur-ance (rē'ə-shoor'əns) *n.* 1 The act of reassuring; repeated assurance. 2 Restored confidence. 3 Reinsurance.

re-as-sure (rē'ə-shoor') *v.t.* -sured, -sur-ing 1 To restore to courage or confidence. 2 To assure again. 3 To reinsure. See synonyms under ENCOURAGE. — **re'as-sur'ing** *adj.* — **re'as-sur'ing-ly** *adv.*

Ré-au-mur (rā'ə-myoor', rā'ə-myoor', *Fr.* rā-ō-mür') *adj.* Relating to or designating the thermometric scale devised by de Réaumur, in which the zero point corresponds to the temperature of melting ice, and 80° to the temperature of boiling water. Also **Ré'au-mur'.** *Abbr.* R.

Ré-au-mur (rā'ə-myoor', rā'ə-myoor', *Fr.* rā-ō-mür'), **René Antoine de,** 1683-1757, French physicist and naturalist.

reave (rēv) *v.t.* reaved or reft, reav-ing *Obs.* 1 To carry off as spoil or booty; rape; rob; plunder. 2 To deprive of something; bereave. 3 To tear up or apart; unravel; pull down; strip. [OE *rēafian* rob]

re-bate¹ (rē'bāt, ri-bāt') *v.t.* -bat-ed, -bat-ing 1 To allow as a deduction. 2 To make a deduction from. 3 *Obs.* To blunt, as a sharp edge. — *n.* A deduction from a gross amount; discount: Also **re-bate'ment.** [<OF *rabattre* beat down <*re-* again + *abattre.* See ABATE.] — **re'bat-er** *n.*

re-bate² (rē'bāt, rab'it) See RABBET.

re-ba-to (re-bā'tō) *n. pl.* -tos A collar turned down and falling over the shoulders; worn by both sexes in the 15th and 16th centuries. Also spelled *rabato.* [<MF *rabat* <*rabattre* beat down. See REBATE.]

re-bec (rē'bek) *n.* The earliest form of the violin. Also **re'beck.** [<F alter. of OF *rebebe* <Arabic *rabāb*]

Re-bec-ca (ri-bek'ə, *Ital.* rā-bek'kā) A feminine personal name. Also *Fr.* **Ré-bec-ca** (rā-be-kä'), *Sp.* **Re-be-ca** (rā-bā'kä), *Ger.* **Re-bek-ka** (rā-bek'ə). [<Hebrew, ensnarer]
— **Rebecca** Wife of Isaac; mother of Esau and Jacob. *Gen.* xxiv 15.

re-bel (ri-bel') *v.i.* -belled, -bel-ling 1 To rise in armed resistance against the established government or ruler of one's land. 2 To resist any authority or established usage. 3 To react with violent aversion: usually with *at.*
— **reb-el** (reb'əl) *n.* One who rebels; specifically, one who espoused the American Revolution, or the cause of the South during the Civil War. — *adj.* Rebellious; refractory. [<OF *rebeller* <L *rebellare* make war again <*re-* again + *bellare* make war <*bellum* war. Doublet of REVEL.]

reb-el-dom (reb'əl-dəm) *n.* 1 The domain of rebels; specifically, the Confederate States during the Civil War; also, rebels collectively. 2 Rebellious behavior.

re-bel-lion (ri-bel'yən) *n.* 1 The act of rebelling. 2 Organized resistance to a government or to any lawful authority. See synonyms under REVOLUTION. — **the Rebellion** The American Civil War. [<OF <L *rebellio, -onis*]

re-bel-lious (ri-bel'yəs) *adj.* 1 Being in a state of rebellion; insubordinate. 2 Of or pertaining to a rebel or rebellion. 3 Resisting control; refractory: *rebellious* curls. — **re-bel'lious-ly** *adv.* — **re-bel'lious-ness** *n.*
Synonyms: contumacious, disobedient, insubordinate, intractable, mutinous, refractory, seditious, uncontrollable, ungovernable, unmanageable. *Ungovernable* applies to that which successfully defies authority and power; *unmanageable* to that which resists the utmost exercise of skill or of skill and power combined; *rebellious* to that which is defiant of authority, whether successfully or unsuccessfully; *seditious* to that which partakes of or tends to excite a *rebellious* spirit, *seditious* suggesting more of covert plan, scheming, or conspiracy, *rebellious* more of overt act or open violence. While the *unmanageable* or *ungovernable* defies control, the *rebellious* or *seditious* may be forced to submission. *Insubordinate* applies to the disposition to resist and resent control as such; *mutinous,* to open defiance of authority, especially in the army, navy, or merchant marine. A *contumacious* act or spirit is contemptuous as well as defiant. See RESTIVE, TURBULENT. Compare OBSTINATE, REVOLUTION. *Antonyms:* compliant, controllable, deferential, docile, dutiful, manageable, obedient, submissive, subservient, tractable, yielding.

re-bill (rē-bil') *v.t.* To render another bill; bill again.

re-birth (rē-bûrth', rē'bûrth') *n.* 1 A new birth. 2 A revival or renaissance.

reb-o-ant (reb'ō-ənt) *adj.* Bellowing back; resounding loudly. [<L *reboans, -antis,* ppr. of *reboare* resound <*re-* again + *boare* bellow. Ult. imit.]

re-born (rē-bôrn') *adj.* Born again; having undergone emotional or mental regeneration; renascent.

re-bound (ri-bound') *v.i.* To bound back; recoil. — *v.t.* To cause to rebound. — *n.* (rē'bound', ri-bound') 1 Recoil; elasticity. 2 Something which rebounds or resounds; an echo. 3 Reaction of feeling or emotion after a disappointment: *to fall in love on the rebound.* [<F *rebondir* <*re-* back + *bondir* bound]

re-bo-zo (rā-bō'sō) *n. Spanish* A long scarf of cotton or silk, often embroidered, worn wrapped about the head and shoulders, and sometimes over the face, by women in Spain and Spanish America.

re-broad-cast (rē-brôd'kast', -käst') *v.t.* -cast or -cast-ed, -cast-ing 1 To broadcast (the same program) more than once from the same station. 2 To broadcast (a program received from another station). — *n.* A program so transmitted.

re-buff (ri-buf') *v.t.* 1 To reject or refuse abruptly or rudely. 2 To drive or beat back; repel; repulse. — *n.* 1 A sudden repulse; curt

MEDIEVAL THREE-STRING REBEC

reclasp	recolonize	recommission	recondense	reconquest	reconsecrate	reconstitute	recopy
reclose	recolor	recommittal	reconduct	reconsecration	reconstitution	recross
reclothe	recombine	recompact	reconfirm	reconsolidate	reconvene	recrucify
recoin	recomfort	reconceive	reconjoin	reconsolidation	reconversion	recrystallization
recoinage	recommence	recondensation	reconquer		reconvert	recrystallize

add, āce, cāre, pälm; end, ēven; it, īce; odd, ōpen, ôrder; tōōk, pōōl; up, bûrn; ə = *a* in *above,* e in *sicken,* i in *clarity,* o in *melon,* u in *focus;* yōō = u in *fuse;* oi, oil; ou, pout; ch, check; g, go; ng, ring; th, thin; ŧħ, this; zh, vision. Foreign sounds á, œ, ü, kh, ń; and ♦: see page xx. < from; + plus; ? possibly.

denial. 2 A sudden check; defeat. 3 A beating back. [<MF *rebuffer* <Ital. *ribuffare,* metathetic alter. of *baruffare* <OHG *biroufan* scuffle]
re-buff (rē·buf′) *v.t.* To buff again.
re·buke (ri·byo͞ok′) *v.t.* **·buked, ·buk·ing** 1 To reprove sharply; reprimand. 2 *Obs.* To check or restrain by a command. See synonyms under ADMONISH, BLAME, REPROVE. —*n.* A strong and authoritative expression of disapproval. See synonyms under ANIMADVERSION, REPROOF. [<AF *rebuker,* OF *rebuchier* <*re-* back + *bucher* beat] —**re·buk′a·ble** *adj.*
re·buk·er (ri·byo͞o′kər) *n.* One who rebukes.
re·bus (rē′bəs) *n.* A puzzle representing a word, phrase, or sentence by letters, numerals, pictures, etc., often with pictures of objects whose names have the same sounds as the words represented. [<L, ablative pl. of *res* thing]
re·but (ri·but′) *v.t.* **·but·ted, ·but·ting** 1 *Law* To overthrow by contrary evidence; contradict by countervailing proof; disprove; refute. 2 *Obs.* To push or drive back. [<OF *rebouter* push back <*re-* back + *bouter, boter.* See BUTT[1].]
re·but·ta·ble (ri·but′ə·bəl) *adj.* Capable of being rebutted.
re·but·tal (ri·but′l) *n.* The act of rebutting; refutation.
re·but·ter (ri·but′ər) *n.* 1 One who or that which rebuts. 2 In common-law pleading, a defendant's answer to the plaintiff's surrejoinder.
re·cal·ci·trant (ri·kal′sə·trənt) *adj.* Not complying; obstinate; rebellious; refractory. —*n.* One who is recalcitrant. [<L *recalcitrans, -antis,* ppr. of *recalcitrare* kick back <*re-* back + *calcitrare* kick <*calx, calcis* heel] —**re·cal′ci·trance, re·cal′ci·tran·cy** *n.*
re·cal·ci·trate (ri·kal′sə·trāt) *v.i.* **·trat·ed, ·trat·ing** To refuse compliance or submission; be recalcitrant. [<L *recalcitratus,* pp. of *recalcitrare.* See RECALCITRANT.] —**re·cal′ci·tra′tion** *n.*
re·ca·lesce (rē′kə·les′) *v.i.* **·lesced, ·lesc·ing** To grow hot again; specifically, in physics, to exhibit recalescence. [<L *recalescere* <*re-* again + *calescere* grow warm, inceptive of *calere* be warm]
re·ca·les·cence (rē′kə·les′əns) *n.* 1 A glowing again. 2 *Physics* A phenomenon peculiar to heated iron or steel of glowing more brightly when certain temperatures are reached in the process of gradual cooling from a state of high incandescence. [<L *recalescens,* ppr. of *recalescere.* See RECALESCE.] —**re′ca·les′cent** *adj.*
re·call (ri·kôl′) *v.t.* 1 To call back; order or summon to return. 2 To summon back in awareness or attention. 3 To recollect; remember. 4 To take back; revoke; countermand. 5 *Poetic* To revive; restore. See synonyms under REMEMBER, RENOUNCE. —*n.* (ri·kôl′, rē′kôl′) 1 A calling back or to mind. 2 A signal to call back soldiers, etc., as by a bugle call, the display of a flag, etc. 3 Revocation, as of an order. 4 In certain States, a system whereby public officials may be removed from office by popular vote.
Ré·ca·mier (rā·kà·myā′), **Jeanne Françoise Julie Adélaïde,** 1777–1849, *née* Bernard, French social leader and patroness of literature: commonly known as *Madame Récamier.*
re·cant (ri·kant′) *v.t.* To withdraw formally one's belief in (something previously believed or maintained). —*v.i.* To disavow an opinion or belief previously held. [<L *recantare* <*re-* again + *cantare* sing, freq. of *canere* sing] —**re·can·ta·tion** (rē′kan·tā′shən) *n.* —**re·cant′er** *n.*
Synonyms: abandon, abjure, deny, disavow, discard, disclaim, disown, forswear, recall, renounce, repudiate, retract, revoke. To *recant* is to *deny* formally and publicly some opinion or statement, especially in religion, that one has held or advocated. *Abjure* is etymologically the exact equivalent of the Saxon *forswear,* signifying to put away formally and under oath, as an error, heresy, or evil practice, or a condemned and detested person. A man *recants* his belief, *abjures* or *renounces* his allegiance, *repudiates* another's claim, *renounces* his own, *retracts* a false statement. A person may *deny, disavow, disclaim, disown* what has been truly or falsely imputed to him or supposed to be his. He may *deny* his signature, *disavow* the act of his agent, *disown* his child; he may *repudiate* either a just claim or a base suggestion. Compare ABANDON, RENOUNCE. *Antonyms*: acknowledge, advocate, assert, cherish, claim, defend, hold, maintain, own, proclaim, retain, uphold, vindicate.
re·cap (rē′kap′, rē·kap′) *v.t.* **·capped, ·cap·ping** To reprocess (an automobile tire) by vulcanizing new rubber onto the surface which comes into contact with the road. —*n.* (rē′kap′) A tire which has been so treated. Also *retread.* [RE- + CAP]
re·cap·i·tal·ize (rē·kap′ə·təl·īz′) *v.t.* **·ized, ·iz·ing** To capitalize again or differently. —**re·cap′i·tal·i·za′tion** *n.*
re·ca·pit·u·late (rē′kə·pich′o͞o·lāt) *v.t.* & *v.i.* **·lat·ed, ·lat·ing** To review briefly; sum up. [<LL *recapitulare* <*re-* again + *capitulare.* See CAPITULATE.]
re·ca·pit·u·la·tion (rē′kə·pich′o͞o·lā′shən) *n.* 1 The act of recapitulating; a summing up. 2 *Biol.* The process in which a developing embryo reproduces many of the typical forms of the organisms that precede it in the line of evolution. [<L *recapitulatio, -onis*]
re·ca·pit·u·la·to·ry (rē′kə·pich′o͞o·lə·tôr′ē, -tō′·rē) *adj.* Containing or of the nature of recapitulation. Also **re′ca·pit′u·la′tive** (-lā′tiv).
re·cap·tion (rē·kap′shən) *n. Law* 1 The rearrest of one who has escaped custody. 2 The retaking by peaceable means of one's goods, wife, child, or chattel from one who wrongfully detains them. [<RE- + CAPTION (def. 5)]
re·cap·ture (rē·kap′chər) *v.t.* **·tured, ·tur·ing** 1 To capture again; obtain by recapture. 2 To recall; remember. —*n.* 1 The act of retaking; especially, in war, the forcible recovery of booty or goods. 2 A prize retaken; anything recaptured. 3 The taking by the public of the earnings of a public service corporation over and above a stated profit.
re·cast (rē·kast′, -käst′) *v.t.* **·cast, ·cast·ing** 1 To form anew; cast again. 2 To fashion anew by changing style, arrangement, etc., as a discourse. 3 To calculate anew. —*n.* (rē′kast′, -käst′) Something which has been recast.
re·cede (ri·sēd′) *v.i.* **·ced·ed, ·ced·ing** 1 To move back; withdraw, as flood waters. 2 To withdraw, as from an assertion, position, agreement, etc. 3 To slope backward: a *receding* forehead. 4 To become more distant; incline away. [<L *recedere* <*re-* back + *cedere* go]
re–cede (rē·sēd′) *v.t.* **–ced·ed, –ced·ing** To cede back; grant or yield to a former owner. [<RE- + CEDE]
re·ceipt (ri·sēt′) *n.* 1 The act or state of receiving anything: to be in *receipt* of good news. 2 That which is received: usually in the plural: cash *receipts.* 3 A written acknowledgment of the payment of money, of the delivery of goods, etc. 4 A recipe. —*v.t.* 1 To give a receipt for the payment of. 2 To write acknowledgment of payment on, as a bill. —*v.i.* 3 To give a receipt, as for money paid. [<OF *recete* <L *recepta,* fem. of *receptus,* pp. of *recipere* RECEIVE; refashioned after Latin]
re·ceipt·or (ri·sē′tər) *n. Law* One who gives a receipt; specifically, one who gives a receipt for goods that have been attached.
re·ceiv·a·ble (ri·sē′və·bəl) *adj.* 1 Capable of being received; fit to be received, as legal tender. 2 Maturing for payment: said of a bill.
re·ceiv·a·bles (ri·sē′və·bəlz) *n. pl.* Outstanding accounts listed among the assets of a business.
re·ceive (ri·sēv′) *v.* **·ceived, ·ceiv·ing** *v.t.* 1 To take into one's hand or possession (something given, offered, delivered, etc.); acquire; accept. 2 To gain knowledge or information of: He *received* the news at breakfast. 3 To take from another by hearing or listening: The king *received* his oath of fealty. 4 To bear; support: These columns *receive* the weight of the building. 5 To experience; meet with: to *receive* abuse. 6 To undergo; suffer: He *received* a wound in his arm. 7 To intercept or encounter the force of (a blow, etc.). 8 To contain; hold. 9 To allow entrance to; admit to one's presence; greet. 10 To perceive mentally; understand. 11 To accept as true, proven, authoritative, etc. —*v.i.* 12 To be a recipient; get, obtain, or acquire something from some other person or source. 13 To welcome visitors or callers. 14 To partake of the Eucharist. 15 *Telecom.* To convert incoming radio waves into intelligible sounds or shapes, as a radio or television receiving set. See synonyms under ACCOMMODATE, GET, OBTAIN. [<OF *receivre* <L *recipere* <*re-* back + *capere* take]
re·ceiv·er (ri·sē′vər) *n.* 1 One who receives; a recipient. 2 An official assigned to receive money due. 3 *Law* A person appointed by a court to take into his custody, control, and management the property or funds of another pending judicial action concerning them. 4 One who buys or receives stolen or embezzled goods, knowing them to be stolen. 5 Something which receives; a receptacle. 6 A vessel considered as a receptacle for a gas or fluid, as a jar for receiving and condensing a fluid that has been distilled. 7 A bolthead. 8 *Telecom.* An instrument in an electric circuit serving to receive and reproduce signals transmitted from another part of the circuit: a telephone *receiver.* 9 A radio or television receiving set. 10 The troughlike part of a gun, directly behind the breech, that guides the round into the chamber.
re·ceiv·er·ship (ri·sē′vər·ship) *n.* 1 The office and functions pertaining to a receiver under appointment of a court. 2 The state of being in the hands of a receiver.
receiving set An apparatus for the reception of radio or television signals.
receiving ship A vessel stationed in a harbor to receive and provide for naval recruits, or for men awaiting transfer to new assignments.
re·cense (ri·sens′) *v.t.* **·censed, ·cens·ing** *Obs.* To revise or review, as a book; make a recension of. [<L *recensere.* See RECENSION.]
re·cen·sion (ri·sen′shən) *n.* 1 A critical revision of the text of a book; also, the edition so revised. 2 A review; critique. [<L *recensio, -onis* enumeration <*recensere* examine, survey <*re-* thoroughly + *censere* estimate, value]
re·cent (rē′sənt) *adj.* Pertaining to, or formed, developed, or created in time not long past; modern; fresh; new. See synonyms under FRESH, MODERN, NEW. [<OF <L *recens, -entis*] —**re′cent·ly** *adv.* —**re′cen·cy, re′cent·ness** *n.*
Re·cent (rē′sənt) *adj. Geol.* Pertaining to or designating the present or Holocene geological epoch, succeeding the Pleistocene.
re·cept (rē′sept) *n. Psychol.* A mental image of an external object, formed by the repetition of the same percept, with reinforcing of common characteristics. [<L *receptum,* neut. pp. of *recipere* take back, receive; on analogy with *concept*]
re·cep·ta·cle (ri·sep′tə·kəl) *n.* 1 Anything that serves to contain or hold other things. 2 *Bot.* The base to which the parts of the flower, fruit, or seeds are fixed. 3 An outlet (def. 3). [<OF <L *receptaculum* <*receptare,* freq. of *recipere* RECEIVE] —**re·cep·tac·u·lar** (rē′sep·tak′yə·lər) *adj.*
re·cep·tion (ri·sep′shən) *n.* 1 The act of receiving, or the state of being received; receipt. 2 A formal social entertainment of guests: a wedding *reception*; also, the manner of receiving a person or persons: a warm *reception.* 3 Mental acceptance, as of a proposition. 4 In radio and television, the act or process of receiving or, especially, the quality of reproduction achieved: This radio gives very poor *reception.* [<OF <L *receptio, -onis* <*receptus,* pp. of *recipere* RECEIVE]
reception center A central receiving point; specifically, the point at which newly inducted military personnel are received and examined, and from which they are sent to their assigned units.
re·cep·tion·ist (ri·sep′shən·ist) *n.* A person employed to receive callers, provide information, and the like, at the entrance to an office.
reception room 1 A room for callers in a private house. 2 A waiting room in a hospital, or adjoining a doctor's, dentist's, or

recultivate
recultivation
redecorate
rededicate
rededication
redeliberate
redemonstrate
redeposit
redescend
redescent
redescribe
redetermine
redigest
rediminish
rediscover
rediscovery
redispose
redissolution
redissolve
redistil
redistrainer
redistribute
redistribution
redivide
redo
redraw
redrawer
redrive
re–echo
re–edify
reelaborate
reelect
reelection
re–elevate
reembark

lawyer's office. 3 A large room for a formal reception.

re·cep·tive (ri·sep′tiv) *adj.* Able or inclined to receive, as truths or impressions; able to take in or hold. [<OF *receptif* <Med. L *receptivus* <L *recipere* RECEIVE] — **re·cep′tive·ly** *adv.* — **re·cep·tiv·i·ty** (rē′sep·tiv′ə·tē), **re·cep′tive·ness** *n.*

re·cep·tor (ri·sep′tər) *n.* 1 *Bacteriol.* A combination of atoms in a cell that, by combining with extraneous substances, as drugs or toxins, may be thrown off from the cell and circulate in the blood, thus conferring immunity. 2 *Physiol.* The terminal structure, or free nerve ending, which is specialized to receive various forms of external and internal stimuli, transmitting them to the brain nerve centers: also called *sense organ.* Compare EFFECTOR. [<L, receiver]

re·cess (ri·ses′, rē′ses; *for def. 2, usually* rē′ses) *n.* 1 A depression or indentation in any otherwise continuous line, especially in a wall; niche; alcove. 2 A time of cessation from employment or occupation: The school took a *recess.* 3 *Usually pl.* A quiet and secluded spot; withdrawn or inner place: the *recesses* of the mind. 4 *Anat.* & *Bot.* A depression or cavity. — *v.* (ri·ses′) *v.t.* 1 To place in or as in a recess. 2 To make a recess in, as a wall. — *v.i.* 3 To take a recess. [<L *recessus,* pp. of *recedere.* See RECEDE[1].]

re·ces·sion (ri·sesh′ən) *n.* 1 The act of receding; a withdrawal. 2 The procession of the clergy, choir, etc., as they leave the chancel after a church service. 3 An economic setback in commercial and industrial activity; especially, one occurring as a downward turn during a period of generally rising prosperity; a slight depression. [<L *recessio, -onis* < *recedere.* See RECEDE[1].]

re–ces·sion (rē·sesh′ən) *n.* The act of ceding again; a giving back.

re·ces·sion·al (ri·sesh′ən·əl) *adj.* Of or pertaining to recession. — *n.* A hymn sung as the choir or clergy leave the chancel after service: also **recessional hymn.**

re·ces·sive (ri·ses′iv) *adj.* 1 Having a tendency to recede or go back; receding. 2 Failing to come into expression. 3 *Genetics* Designating that one of a pair of contrasted allelomorphic characters which is suppressed in a hybrid offspring when both are present: opposed to *dominant.* — *n. Genetics* 1 A hybrid which carries and transmits a character suppressed by the corresponding dominant character. 2 The suppressed character itself. — **re·ces′sive·ly** *adv.*

Rech·a·bite (rek′ə·bīt) *n.* 1 One of a Jewish family descended from Jonadab, son of Rechab, who abstained from wine and the planting of vineyards. *Jer.* xxxv 3. 2 Hence, a total abstainer from intoxicants; a teetotaler. 3 A member of a society of teetotalers called the Independent Order of Rechabites, founded in England in 1835 and in the United States in 1842. — **Rech′a·bit′ism** *n.*

ré·chauf·fé (rā·shō·fā′) *n. French* 1 Food warmed over. 2 Reworked or "warmed over" literary work; rehash.

re·cheat (ri·chēt′) *n. Archaic* A strain sounded on a huntsman's horn to recall the hounds from a wrong course or at the end of the hunt; also, the act of sounding this signal. — *v.i. Obs.* To sound the recheat. [<OF *rachater* rally, reassemble]

re·cher·ché (rə·sher·shā′) *adj. French* 1 Much sought after; hence, choice; rare. 2 Far-fetched.

re·cid·i·vist (rə·sid′ə·vist) *n.* 1 Anyone who relapses into a former state or condition. 2 A confirmed criminal; in the United States, one committed to prison for a second term. [<F *récidiviste* <L *recidivus* relapsing < *recidere* < *re-* back + *cadere* fall] — **re·cid′i·vism, rec·i·div·i·ty** (res′ə·div′ə·tē) *n.* — **re·cid′i·vis′tic** *adj.*

re·cid·i·vous (rə·sid′ə·vəs) *adj.* Liable to backslide.

Re·ci·fe (re·sē′fə) A port of NE Brazil; capital of Pernambuco state: also *Pernambuco.*

rec·i·pe (res′ə·pē) *n.* 1 A formula or list of ingredients of a mixture, giving the exact proportions together with proper directions for compounding, cooking, etc. 2 A medical prescription: so called from its opening word: usually abbreviated to ℞. 3 A method prescribed for attaining a desired result. [<L, take, imperative of *recipere.* See RECEIVE.]

re·cip·i·ence (ri·sip′ē·əns) *n.* 1 The process or act of receiving. 2 Receptivity. Also **re·cip′i·en·cy.**

re·cip·i·ent (ri·sip′ē·ənt) *adj.* Receiving or ready to receive; receptive. — *n.* One who or that which receives; one who accepts a gift or favor. [<L *recipiens, -entis,* ppr. of *recipere* RECEIVE]

re·cip·ro·cal (ri·sip′rə·kəl) *adj.* 1 Done or given by each of two to the other; mutual. 2 Mutually interchangeable. 3 Alternating; moving to and fro. 4 So related, as two concepts, that if the first determines the second, then the second determines the first. 5 Expressive of mutual relationship or action: used in connection with certain pronouns and verbs or their meaning. 6 *Math.* Of or pertaining to a fraction the numerator and denominator of which have been reversed. — *n.* 1 That which is reciprocal. 2 *Math.* The quotient obtained by dividing unity by a number or expression, as $\frac{1}{x}$ is the reciprocal of *x*. In a fraction, this reverses the numerator and denominator, as $\frac{3}{2}$ is the reciprocal of $\frac{2}{3}$. See synonyms under MUTUAL. [<L *reciprocus*] — **re·cip′ro·cal′i·ty** (-kal′ə·tē), **re·cip′ro·cal·ness** *n.* — **re·cip′ro·cal·ly** *adv.*

reciprocal pronouns *Gram.* Pronouns or pronominal phrases denoting reciprocal action or relation, as *each other, one another.*

re·cip·ro·cate (ri·sip′rə·kāt) *v.* **·cat·ed, ·cat·ing** *v.t.* 1 To cause to move backward and forward alternately. 2 To give and receive mutually, as favors or gifts; interchange. 3 To give, feel, do, etc., in return; requite, as an emotion. — *v.i.* 4 To move backward and forward. 5 To make a return in kind. 6 To give and receive favors, gifts, etc., mutually. 7 To correspond; be equivalent. See synonyms under REQUITE. [<L *reciprocatus,* pp. of *reciprocare* move to and fro < *reciprocus* returning] — **re·cip′ro·ca′tive** *adj.* — **re·cip′ro·ca′tor** *n.*

reciprocating engine An engine having a piston or pistons which move to and fro: distinguished from *rotary engine.*

re·cip·ro·ca·tion (ri·sip′rə·kā′shən) *n.* The act of reciprocating; a mutual giving and returning; alternation; alternate motion. See synonyms under INTERCOURSE. [<L *reciprocatio, -onis* < *reciprocus* returning]

re·cip·ro·ca·to·ry (ri·sip′rə·kə·tôr′ē, -tō′rē) *adj.* Alternating in direction or movement; reciprocating: opposed to *rotary.*

rec·i·proc·i·ty (res′ə·pros′ə·tē) *n.* 1 Reciprocal obligation, action, or relation. 2 That trade relation or policy between two countries by which each makes concessions favoring the importation of the products of the other. See synonyms under INTERCOURSE. [<F *réciprocité*]

re·ci·sion (ri·sizh′ən) *n.* 1 The act of rescinding. 2 The act of pruning. [<OF <L *recisio, -onis* < *recisum,* pp. of *recidere* cut off < *re-* back + *caedere* cut]

re·cit·al (ri·sīt′l) *n.* 1 A telling over in detail, or that which is thus told; a narration. 2 A public delivery of something previously memorized. 3 A musical program performed by one person, or consisting of works by one person. 4 A detailed statement. See synonyms under HISTORY, REPORT, STORY[1].

rec·i·ta·tion (res′ə·tā′shən) *n.* 1 The act of repeating from memory; the reciting of a lesson, or the meeting of a class for that purpose. 2 That which is allotted for recital or actually recited. [<L *recitatio, -onis* < *recitare.* See RECITE.]

rec·i·ta·tive[1] (res′ə·tā′tiv, ri·sī′tə·tiv) *adj.* Of the nature of a recital as of facts or details; narrative.

rec·i·ta·tive[2] (res′ə·tə·tēv′) *n. Music* Language uttered as in ordinary speech, but in musical tones; that style of singing or a vocal passage so rendered. Also *Italian* **re·ci·ta·ti·vo** (rā′chē·tä·tē′vō). — *adj.* Having the character of a recitative. [<Ital. *recitativo,* ult. <L *recitare*]

re·cite (ri·sīt′) *v.* **·cit·ed, ·cit·ing** *v.t.* 1 To declaim or say from memory, especially formally, as a lesson in class. 2 To tell in particular detail; relate. 3 To enumerate. — *v.i.* 4 To declaim or speak something from memory. 5 To repeat or be examined in a lesson or part of a lesson in class. See synonyms under RELATE. [<F *réciter* <L *recitare* < *re-* again + *citare.* See CITE.] — **re·cit′er** *n.*

re–cite (rē·sīt′) *v.t.* & *v.i.* **–cit·ed, –cit·ing** To cite again. [<RE- + CITE]

reck (rek) *v.t.* & *v.i. Obs.* 1 To have a care or thought (for); heed; mind. 2 To be of concern or interest (to): It *recks* me not. ◆ Homophone: *wreck.* [OE *rēccan*]

reck·less (rek′lis) *adj.* 1 Foolishly heedless of danger; rash. 2 Indifferent; neglectful. See synonyms under IMPROVIDENT, IMPRUDENT, WANTON. [OE *recceléas*] — **reck′less·ly** *adv.* — **reck′less·ness** *n.*

Reck·ling·hau·sen (rek′ling·hou′zən) A city in the Ruhr in NW North Rhine-Westphalia, West Germany.

reck·on (rek′ən) *v.t.* 1 To count; compute; calculate. 2 To look upon as being; regard: They *reckon* him a fool. 3 *Dial.* To suppose or guess; expect. — *v.i.* 4 To make computation; count up. 5 To rely or depend: with *on* or *upon*: to *reckon* on help. See synonyms under CALCULATE. — **to reckon for** To pay for; receive the penalty of. — **to reckon with** 1 To settle accounts with. 2 To take into consideration; bear in mind; consider. — **to reckon without one's host** To reckon a bill without consulting the landlord; hence, to neglect important facts in reaching a conclusion. [OE *recenian* explain. Akin to G *rechnen* count.]

reck·on·er (rek′ən·ər) *n.* 1 One who reckons. 2 A book or device for aiding one to compute: often called **ready reckoner.**

reck·on·ing (rek′ən·ing) *n.* 1 The act of counting; computation; a settlement of accounts. 2 Account; score; bill, as at a hotel. 3 *Naut.* The calculation of a ship's position, especially when made only by log and compass; dead reckoning. 4 An accounting to God.

re·claim (ri·klām′) *v.t.* 1 To bring (swamp, desert, etc.) into a condition to support cultivation or life, as by draining or irrigating. 2 To obtain (a substance) from used or waste products: to *reclaim* rubber. 3 To cause to return from wrong or sinful ways of life; reform. 4 *Obs.* To tame, as a hawk. — *n.* 1 The act of reclaiming or state of being reclaimed; recovery; reformation; that which is reclaimed. 2 A fresh claim. [<OF *reclamer* call back <L *reclamare* < *re-* against + *clamare* cry out] — **re·claim′a·ble** *adj.* — **re·claim′er, re·claim′ant** *n.*

Synonyms (verb): amend, convert, correct, recover, redeem, reform, renew, rescue, restore, subdue, tame. *Antonyms*: corrupt, degrade, deprave, destroy, seduce, vitiate.

re–claim (rē·klām′) *v.t.* To claim again.

rec·la·ma·tion (rek′lə·mā′shən) *n.* 1 The act or process of reclaiming, in any sense. 2 Restoration, as to ownership, cultivation, usefulness, or a moral life. — **Bureau of Reclamation** A branch of the U. S. Department of the Interior which constructs and operates Federal water-power plants and irrigation projects. [<F *réclamation* <L *reclamatio, -onis* a cry of disapproval < *reclamare.* See RECLAIM.]

ré·clame (rā·klàm′) *n. French* 1 Publicity. 2 A striving after publicity.

re·cline (ri·klīn′) *v.* **·clined, ·clin·ing** *v.i.* To assume a recumbent position; lie down or back. — *v.t.* To cause to assume a recumbent position; lay down or back. See synonyms under

reembarkation	**re–emit**	**reencouragement**	**reenjoy**	**reenslave**	**reerect**	**reexamine**
reembody	**reenact**	**reendow**	**reenjoyment**	**reenslavement**	**reestablish**	**reexhibit**
reembrace	**reenaction**	**reengage**	**reenkindle**	**reenstamp**	**reestablishment**	**reexpel**
reemerge	**reenactment**	**reengagement**	**reenlist**	**reenthrone**	**re–evaluate**	**reexperience**
reemergence	**reencourage**	**reengrave**	**reenlistment**	**reenthronement**	**reexamination**	**reexport**

add,āce,câre,pälm; end,ēven; it,īce; odd,ōpen,ôrder; tŏŏk,pōōl; up,bûrn; ə = a in *above,* e in *sicken,* i in *clarity,* o in *melon,* u in *focus;* yōō = u in *fuse;* oi,oil; ou,pout; ch,check; g,go; ng,ring; th,thin; th,this; zh,vision. Foreign sounds à,œ,ü,kh,ṅ; and ◆: see page xx. < from; + plus; ? possibly.

LEAN[1], REST[1]. [<L *reclinare* <*re-* back + *clinare* lean] —**rec·li·na·tion** (rek'lə·nā'shən) *n.* —**re·clin'er** *n.*

rec·luse (rek'lo͞os, ri·klo͞os') *n.* **1** One who lives in retirement or seclusion. **2** One who retires from intercourse with the world, as a religious devotee; specifically, one who lives shut up in a cell and practices exceptional austerities. —**re·cluse** (ri·klo͞os') *adj.* Secluded or retired from the world; solitary. [<OF *reclus* <LL *reclusus*, pp. of L *recludere* shut off <*re-* again + *claudere* close] —**re·clu'sive** *adj.*

re·clu·sion (ri·klo͞o'zhən) *n.* **1** The state of being a recluse; retirement from the world. **2** Rigorous immurement as practiced by certain ascetics in the Middle Ages. **3** Imprisonment; especially, solitary confinement.

re·clu·sive (ri·klo͞o'siv) *adj.* Affording or living in seclusion; recluse.

rec·og·ni·tion (rek'əg·nish'ən) *n.* **1** The act of recognizing; the process of memory that identifies an object, person, etc., as already known or experienced. **2** Acknowledgment of a fact or claim. **3** Friendly notice; salutation; attention: *recognition* of a speaker by the chair. **4** Acknowledgment and acceptance on the part of one government of the independence and validity of another. See synonyms under KNOWLEDGE. [<L *recognitio, -onis* <*recognitus*, pp. of *recognoscere*. See RECOGNIZANCE.] —**re·cog·ni·to·ry** (ri·kog'nə·tôr'ē, -tō'rē), **re·cog'ni·tive** *adj.*

re·cog·ni·zance (ri·kog'nə·zəns, -kon'ə-) *n.* **1** *Law* **a** An acknowledgment or obligation of record, with condition to do some particular act, as to appear and answer, or to keep the peace. **b** A sum of money deposited as surety for fulfilment of such act or obligation, and forfeited by its non-performance. **2** *Obs.* A badge or token to aid in recognition. [<OF *recognoissance* <*reconoissant*, ppr. of *reconoistre* <L *recognoscere* call to mind <*re-* again + *cognoscere* know. See COGNITION.] —**re·cog'ni·zant** *adj.*

rec·og·nize (rek'əg·nīz) *v.t.* **·nized, ·niz·ing** **1** To know again; perceive as identical with someone or something previously known. **2** To identify or know, as by previous experience or knowledge: I *recognize* poor poetry when I see it. **3** To perceive as true; realize: to *recognize* the facts in a case. **4** To acknowledge the independence and validity of, as a newly constituted government. **5** To indicate appreciation or approval of: to *recognize* merit. **6** To approve formally; regard as valid or genuine: to *recognize* a claim. **7** To give (someone) permission to speak, as in a legislative body. **8** To admit the acquaintance of; greet. **9** *Law* To bind by a recognizance. See synonyms under ACKNOWLEDGE, CONFESS, DISCERN. [Back formation <RECOGNIZANCE] —**rec'og·niz'a·ble** *adj.* —**rec'og·niz'a·bly** *adv.* —**rec'og·niz'er** *n.*

re·cog·ni·zee (ri·kog'nə·zē', -kon'ə-) *n.* *Law* One in whose favor a recognizance is made.

re·cog·ni·zor (ri·kog'nə·zər, -kon'ə-) *n.* *Law* One who enters into a recognizance.

re·coil (ri·koil') *v.i.* **1** To start back, as in fear or loathing; shrink: He *recoiled* at the sight. **2** To spring back, as from force of discharge or force of impact. **3** To return to the source; react: with *on* or *upon*: Crime *recoils* upon its perpetrator. **4** To move or draw back; retreat. —*n.* (rē'koil') **1** A backward movement or impulse, as of a gun at the moment of firing; rebound; also, a shrinking. **2** The condition existing as the result of a recoil. [<OF *reculer* <L *re-* again + *culus* buttocks] —**re·coil'er** *n.*

re·coil-op·er·at·ed (rē'koil·op'ə·rā'tid) *adj.* Operated or working by the energy generated in recoil, as certain automatic weapons.

rec·ol·lect (rek'ə·lekt') *v.t.* To call back to the mind; revive in the memory; remember. —*v.i.* To have a recollection of something. See synonyms under REMEMBER. [<L *recollectus*, pp. of *recolligere* gather together again <*re-* again + *colligere*. See COLLECT.]

re-col·lect (rē'kə·lekt') *v.t.* **1** to collect again, as things scattered. **2** To collect or compose (one's thoughts or nerves); compose or recover (oneself). [<RE- + COLLECT]

rec·ol·lect·ed (rek'ə·lek'tid) *adj.* Recalled to mind; remembered.

re-col·lect·ed (rē'kə·lek'tid) *adj.* Calm; composed; collected.

rec·ol·lec·tion (rek'ə·lek'shən) *n.* **1** The act or power of recollecting or remembering; remembrance. **2** Something remembered; a reminiscence; a memory. See synonyms under MEMORY. —**rec'ol·lec'tive** *adj.* —**rec'ol·lec'tive·ly** *adv.*

re-col·lec·tion (rē'kə·lek'shən) *n.* The act of re-collecting, or the state of being re-collected.

re·com·bi·na·tion (rē'kom·bə·nā'shən) *n.* *Genetics* **1** A cross-over. **2** An offspring exhibiting characters caused by such a rearrangement.

rec·om·mend (rek'ə·mend') *v.t.* **1** To commend with favorable representations; praise as desirable, worthy, etc. **2** To make attractive or acceptable: His sagacity *recommends* him. **3** To advise; urge. **4** To give in charge; commend. [<Med. L *recommendare* <*re-* again + *commendare*. See COMMEND.] —**rec'om·mend'er** *n.*

rec·om·men·da·tion (rek'ə·men·dā'shən) *n.* **1** The act of recommending, or that which recommends. **2** A note commending a person to confidence or favor. See synonyms under COUNSEL.

rec·om·mend·a·to·ry (rek'ə·men'də·tôr'ē, -tō'·rē) *adj.* **1** Serving to recommend. **2** Advisory but not imperative, as applied to certain official appointments.

re·com·mit (rē'kə·mit') *v.t.* **·mit·ted, ·mit·ting** **1** To commit again. **2** To refer back to a committee, as a bill.

rec·om·pense (rek'əm·pens) *v.t.* **·pensed, ·pens·ing** **1** To give compensation to; pay or repay; reward; requite. **2** To give compensation for; make up for, as a loss. See synonyms under PAY[1], REQUITE. —*n.* An equivalent for anything given, done, or suffered; payment or repayment; compensation; reward. [<OF *recompenser* <LL *recompensare* <L *re-* again + *compensare*. See COMPENSATE.]

Synonyms (noun): amends, compensation, indemnification, indemnity, remuneration, repayment, requital, retribution, reward, satisfaction. See RESTITUTION, SALARY.

re·com·pose (rē'kəm·pōz') *v.t.* **·posed, ·pos·ing** **1** To restore the composure of; tranquilize. **2** To compose or form anew; rearrange; reconstitute; recombine. —**re·com·po·si·tion** (rē'·kom·pə·zish'ən) *n.*

re·con·cen·tra·do (rā·kōn'sen·trä'dō) *n.* *pl.* **·dos** In Cuba and the Philippine Islands, during and before the Spanish-American War, a dweller in the country who was forced by decree of the Spanish authorities to move within the city limits. [<Sp., pp. of *reconcentrar* move to the center again; so called because the authorities had previously ordered the country population to move within a certain radius of the town]

re·con·cen·trate (rē·kon'sən·trāt) *v.t.* **·at·ed, ·at·ing** To concentrate again; specifically, to treat as reconcentrados. —**re·con'cen·tra'·tion** *n.*

rec·on·cil·a·ble (rek'ən·sī'lə·bəl) *adj.* **1** Capable of being reconciled or of renewing friendship. **2** Capable of being adjusted or harmonized. —**rec'on·cil'a·bil'i·ty, rec'on·cil'a·ble·ness** *n.* —**rec'on·cil'a·bly** *adv.*

rec·on·cile (rek'ən·sīl) *v.t.* **·ciled, ·cil·ing** **1** To bring back to friendship after estrangement; also, to make friendly; win the good will of. **2** To settle or adjust, as a quarrel. **3** To bring to acquiescence, content, or submission: to *reconcile* one to his lot. **4** To make or show to be consistent or congruous; harmonize: often with *to* or *with*: Can he *reconcile* his statement with his conduct? See synonyms under ACCOMMODATE. [<OF *reconciler* <L *reconciliare* <*re-* again + *conciliare* unite. See CONCILIATE.] —**rec'on·cile'ment** *n.* —**rec'on·cil'er** *n.*

rec·on·cil·i·a·tion (rek'ən·sil'ē·ā'shən) *n.* **1** The act of reconciling, or the state of being reconciled; atonement. **2** The effecting or showing of agreement between things; explanation of differences. See synonyms under PROPITIATION. —**rec'on·cil'i·a·to'ry** (-sil'ē·ə·tôr'ē, -tō'rē) *adj.*

rec·on·dite (rek'ən·dīt, ri·kon'dīt) *adj.* **1** Remote from ordinary or easy perception; abstruse; secret. **2** Dealing in abstruse matters; profound. **3** Hidden; not readily observed. See synonyms under MYSTERIOUS, SECRET. [<L *reconditus*, pp. of *recondere* put away, hide <*re-* back + *condere* construct, hide] —**rec'on·dite'ly** *adv.* —**rec'on·dite'ness** *n.*

re·con·nais·sance (ri·kon'ə·səns, -säns) *n.* **1** A reconnoitering; a preliminary examination or survey, as of the territory and resources of a country. **2** The act of obtaining information of military value, especially regarding the position, strength, and movement of enemy forces. Also **re·con'nois·sance.** [<F]

re·con·noi·ter (rē'kə·noi'tər, rek'ə-) *v.t.* To examine by the eye; survey, as for military, engineering, or geological purposes. —*v.i.* To make a reconnaissance. Also **re'con·noi'·tre.** [<OF *reconoistre*. See RECOGNIZANCE.] —**re'con·noi'ter·er, re'con·noi'trer** *n.*

re·con·sid·er (rē'kən·sid'ər) *v.t.* **1** To consider again, especially with a view to a reversal of previous action. **2** In parliamentary usage, to bring before the house for renewed action (a matter previously decided). —*v.i.* **3** To reconsider a matter or decision. —**re'con·sid'er·a'tion** *n.*

re·con·sign (rē'kən·sīn') *v.t.* To consign again; specifically, to consign (goods) to a different place or person while still in transit. —**re'·con·sign'ment** *n.*

re·con·sti·tute (rē·kon'stə·to͞ot, -tyo͞ot) *v.t.* **·tut·ed, ·tut·ing** To constitute again; make over: to *reconstitute* dehydrated fruits and vegetables by adding water. —**re·con'sti·tu'·tion** *n.*

re·con·struct (rē'kən·strukt') *v.t.* To construct again; rebuild.

re·con·struct·ed (rē'kən·struk'tid) *adj.* Rebuilt or made anew: said especially of gems artificially made: a *reconstructed* ruby.

re·con·struc·tion (rē'kən·struk'shən) *n.* **1** The act of reconstructing, or the state of being reconstructed; specifically, the restoration of the seceded States as members of the Union under the **Reconstruction Acts** of March 2 and 23, 1867. **2** The repair of mutilated limbs, as of soldiers, by means of mechanical appliances. —**re'con·struc'tive** *adj.*

Reconstruction Finance Corporation A former (1932–54) branch of the Federal Loan Agency of the U.S. Department of the Interior, authorized to extend financial assistance to agriculture, industry, and commerce, through direct loans to the appropriate financial intermediaries.

Reconstruction period *U.S.* The period following the Civil War during which the seceded Southern States were reorganized in accordance with the Congressional program.

re·con·vey (rē'kən·vā') *v.t.* To convey back to an original owner or place. —**re'con·vey'·ance** *n.*

rec·ord (rek'ərd) *n.* **1** An account in written or other permanent form serving as a memorial or authentic evidence of a fact or event. **2** Something on which such an account is made, as a document or monument. **3** Information on facts or events, preserved and handed down: the heaviest rainfall on *record.* **4** The known career or performance of a person, animal, organization, etc., regarded as a series of things done or achieved: a good *record* in politics. **5** The best listed achievement, as in a competitive sport: to beat the world *record.* **6** *Law* **a** A written account of an act, statement, or transaction made by an officer acting under authority of law, and intended as permanent evidence thereon. **b** An official written account of a judicial or legislative proceeding, including the judgments or enactments and an official copy of all related documents. **7** A cylinder, disk, roll, or other article perforated, indented, or otherwise prepared so as to reproduce sounds. —**off the record** **1** Unofficial or unofficially. **2** Not for quotation or publication, or not from a source to be identified. —*adj.* Surpassing any previously recorded achievement or performance of its kind: a *record* vote. —**re·cord** (ri·kôrd') *v.t.* **1** To write down or otherwise inscribe, as for preserving an

reexportation	**refertilize**	**reflunctuation**	**refortify**	**regerminate**	**regreet**	**rehybridize**
reexpulsion	**reflorescence**	**refold**	**refreeze**	**regermination**	**rehandle**	**rehypothecate**
reface	**reflourish**	**reforge**	**refurbish**	**regild**	**rehearing**	**reillume**
refashion	**reflow**	**reforger**	**refurnish**	**regraft**	**reheel**	**reimplant**
refasten	**reflower**	**refortification**	**regather**	**regrant**	**rehire**	**reimport**

authentic account, evidence, etc. 2 To indicate; register, especially in permanent form, as a cardiograph does. 3 To make a phonograph record of. —*v.i.* 4 To record something. [<OF <*recorder* <L *recordari* call to mind <*re-* again + *cor, cordis* heart, mind]
Synonyms (noun): account, archives, catalog, chronicle, document, enrolment, entry, enumeration, history, inscription, instrument, inventory, memorandum, memorial, muniment, register, roll, scroll. *Record* is a word of wide signification, applying to any writing, mark, or trace that serves as a *memorial* giving enduring attestation of an event or fact; an extended *account, chronicle,* or *history* is a *record*; so, too, may be a brief *inventory* or *memorandum*. A *memorial* is any object, whether a writing, a monument, or other permanent thing that is designed or adapted to keep something in remembrance. A *register* is a formal or official written *record*, especially a series of entries made for preservation or reference; as, a *register* of births and deaths. *Archives*, in the sense here considered, are *documents* or *records*, often legal *records*, preserved in a public or official depository; the word *archives* is also applied to the place where such *documents* are regularly deposited and preserved. *Muniments* are *records* that enable one to defend his title. See CHARACTER, HISTORY, REPORT, STORY[1].

re·cord·er (ri·kôr′dər) *n.* 1 One who records. 2 A magistrate having criminal jurisdiction in a city or borough. 3 A registering apparatus. 4 A fipple flute, having eight holes and any one of four ranges: treble, alto, tenor, or bass. 5 A tape recorder or wire recorder. —**re·cord′er·ship** *n.*

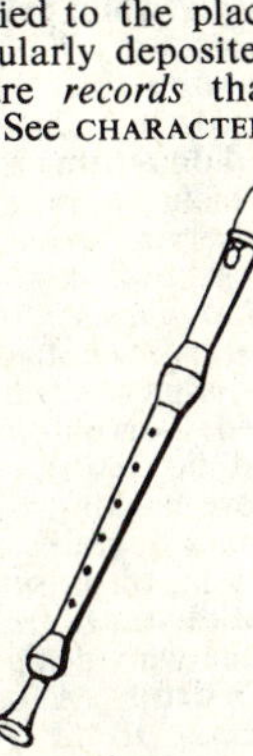

TREBLE RECORDER

re·count (ri·kount′) *v.t.* 1 To relate the particulars of; narrate in detail. 2 To enumerate; recite. See synonyms under RELATE. [<AF, OF *reconter* relate]

re–count (rē·kount′) *v.t.* To count again. —*n.* (rē′kount′, rē·kount′) A repetition of a count; specifically, a second count of votes cast.

re·count·al (ri·koun′təl) *n.* A thing told, or the act of telling; a detailed narration. Also **re·count′ment.**

re·coup (ri·ko͞op′) *v.t.* 1 To recover or obtain an equivalent for; make up, as a loss. 2 To reimburse for a loss; indemnify. 3 *Law* To keep back (something due) in order to make good a counterclaim. —*n.* The act or process of recouping. [<F *recouper* <*re-* again + *couper* cut. See COUP.] —**re·coup′a·ble** *adj.* —**re·coup′ment** *n.*

re·cou·pé (rə·ko͞o′pā′) *adj. Her.* Divided a second time, as an escutcheon. [<F, pp. of *recouper*. See RECOUP.]

re·course (rē′kôrs, -kōrs, ri·kôrs′, -kōrs′) *n.* 1 Resort to or application for help or security in trouble. 2 *Law* The right to exact payment from a party secondarily liable, where the first party liable has failed to pay. 3 A source of help or supply; the person or thing resorted to. 4 *Obs.* Admission; entrance. —**without recourse** A restricted or qualified endorsement of a promissory note or transfer thereof, which signifies that the endorser merely transfers the title to the instrument, but disclaims liability for non-payment. —**to have recourse to** To go to for advice or help. [<OF *recours* <L *recursus* a running back <*recurrere*. See RECUR.]

re·cov·er (ri·kuv′ər) *v.t.* 1 To obtain again after losing; regain, as property, self-control, health, etc. 2 To make up for; retrieve, as a loss. 3 To restore (oneself) to natural balance, health, etc. 4 In sports, to regain (one's normal position of guard, balance, etc.). 5 To reclaim, as land. 6 *Law* **a** To gain in judicial proceedings: to *recover* judgment. **b** To gain or regain by legal process. —*v.i.* 7 To regain health, composure, etc. 8 *Law* To succeed in a lawsuit. 9 In sports, to regain one's balance or position of guard. [<OF *recovrer* <L *recuperare*. See RECUPERATE.] —**re·cov′er·a·ble** *adj.* —**re·cov′er·er** *n.*
Synonyms: cure, heal, reanimate, recruit, recuperate, regain, repossess, restore, resume, retrieve. See RECLAIM. *Antonyms*: die, fail, lapse, sink.

re–cov·er (rē·kuv′ər) *v.t.* To cover again. —**re–cov′er·er** *n.*

re·cov·er·y (ri·kuv′ər·ē) *n. pl.* **·er·ies** 1 The act of recovering. 2 The state of being or having recovered. 3 Restoration from sickness or from a condition of evil. 4 In boating, the forward movement of an oarsman, after having finished one stroke, to take the next. 5 In fencing and sparring, the act of regaining a defensive position after attack. 6 The extraction of valuable substances and materials from original sources, by-products, waste, etc. —**final recovery** *U.S. Law* The final verdict in an action; specifically, the judgment entered upon such a verdict. [<AF *recoverie*]

rec·re·ant (rek′rē·ənt) *adj.* 1 Unfaithful to a cause or pledge; apostate; false. 2 Crying for mercy, as in the old trial by combat; hence, craven; cowardly. See synonyms under PUSILLANIMOUS. —*n.* A cowardly or faithless person; also, a deserter; an apostate. [<OF, ppr. of *recreire* surrender allegiance <Med. L *recredere* <L *re-* back + *credere* believe] —**rec′re·an·cy, rec′re·ance** *n.* —**rec′re·ant·ly** *adv.*

rec·re·ate (rek′rē·āt) *v.* **·at·ed, ·at·ing** *v.t.* To impart fresh vigor to; refresh, especially after toil, by some form of relaxation or entertainment. —*v.i.* To take recreation. See synonyms under ENTERTAIN, RELAX. [<L *recreatus*, pp. of *recreare* create anew <*re-* again + *creare* create] —**rec′re·a′tive** *adj.*

re–cre·ate (rē′krē·āt′) *v.t.* **·at·ed, ·at·ing** To create anew. —**re′–cre·a′tion** *n.*

rec·re·a·tion (rek′rē·ā′shən) *n.* 1 Refreshment of body or mind, but generally of both; diversion; amusement. 2 Any pleasurable exercise or occupation. See synonyms under ENTERTAINMENT, REST[1], SPORT. —**rec′re·a′tion·al** *adj.*

rec·re·ment (rek′rə·mənt) *n.* 1 *Physiol.* A secretion reabsorbed by the body after having performed its function, as gastric juice, saliva, etc. 2 Waste material; dross; scoria; spume. [<F *récrément* <L *recrementum* dross <*re-* back + *cretum*, pp. of *cernere* sift] —**rec′re·men′tal** (-men′təl) *adj.* —**rec′re·men·ti′tial** (-men·tish′əl), **rec′re·men·ti′tious** *adj.*

re·crim·i·nate (ri·krim′ə·nāt) *v.* **·nat·ed, ·nat·ing** *v.t.* To accuse in return. —*v.i.* To repel one accusation by making another in return. [<Med. L *recriminatus*, pp. of *recriminare* <L *re-* again + *criminare*. See CRIMINATE.] —**re·crim′i·na′tive, re·crim·i·na·to·ry** (ri·krim′ə·nə·tôr′ē, -tō′rē) *adj.* —**re·crim′i·na′tor** *n.*

re·crim·i·na·tion (ri·krim′ə·nā′shən) *n.* 1 The act of recriminating. 2 An accusation made in return; a countercharge.

re·cru·desce (rē′kro͞o·des′) *v.i.* **·desced, ·desc·ing** To break out or become active again. [<L *recrudescere* <*re-* again + *crudescere* become harsh, break out <*crudus* raw, harsh]

re·cru·des·cence (rē′kro͞o·des′əns) *n.* 1 A breaking out afresh, as of a disease or wound. 2 A reappearance; return. [<L *recrudescens, -entis* ppr. of *recrudescere*. See RECRUDESCE.] —**re′cru·des′cent** *adj.*

re·cruit (ri·kro͞ot′) *v.t.* 1 To enlist (men) for military or naval service. 2 To muster; raise, as an army, by enlistment. 3 To supply with recruits. 4 To regain or revive (lost health, strength, etc.). 5 *Rare* To replenish. —*v.i.* 6 To enlist new men for military or naval service. 7 To regain lost health or strength. 8 To gain or raise new supplies of anything lost or needed. —*n.* 1 A newly enlisted soldier, sailor, or marine; loosely, any new adherent of a cause, organization, or the like. 2 *Obs.* A new supply of something necessary or useful. [<F *recruter* <*recrute* a recruit <*recrû* grown again, pp. of *recroître* <L *re-* again + *crescere* grow, increase] —**re·cruit′er** *n.* —**re·cruit′ment** *n.*
Synonyms (verb): enlist, reinforce, repair, replenish. See RECOVER. *Antonyms*: decimate, disperse, lose, reduce, scatter.

rec·tal (rek′təl) *adj. Anat.* Relating to, involving, or in the region of the rectum.

rec·tan·gle (rek′tang′gəl) *n.* A right-angled parallelogram. [<F <LL *rectiangulum* <L *rectus* straight + *angulus* angle]

rec·tan·gu·lar (rek·tang′gyə·lər) *adj.* 1 Having one or more right angles. 2 Resembling a rectangle in shape or appearance. —**rec·tan′gu·lar′i·ty** (-lar′ə·tē) *n.* —**rec·tan′gu·lar·ly** *adv.*

rectangular coordinate system See CARTESIAN COORDINATE SYSTEM.

rectangular hyperbola *Math.* A hyperbola with axes of equal length and perpendicular asymptotes.

recti– *combining form* Straight: *rectilinear*. Also, before vowels, **rect–**. [<L *rectus* straight]

rec·ti·fi·ca·tion (rek′tə·fə·kā′shən) *n.* 1 The act or process of rectifying. 2 A setting right of what is wrong. 3 Refining by fractional or renewed distillation. [<F]

rec·ti·fi·er (rek′tə·fī′ər) *n.* 1 One who or that which rectifies. 2 *Electr.* A device used to convert an alternating current into a direct or unidirectional current. 3 A refiner or compounder of spirituous liquors.

rec·ti·fy (rek′tə·fī) *v.t.* **·fied, ·fy·ing** 1 To make right; correct; amend. 2 *Chem.* To refine, as a liquid, by repeated distillations until a desired degree of purity is obtained. 3 *Electr.* To change (an alternating current) into a direct current by reversing the direction of alternate impulses. 4 *Math.* To determine the length of (a curve or arc). 5 To allow for errors or inaccuracies in, as a compass reading. 6 To adjust for accurate calculations: to *rectify* a globe. See synonyms under AMEND. [<OF *rectifier* <LL *rectificare* <L *rectus* right + *facere* make] —**rec′ti·fi′a·ble** *adj.*

rec·ti·graph (rek′tə·graf, -gräf) *n. Optics* A separable part which may be inserted into the tube of an optical instrument in order to reinvert an inverted image. [<RECTI- + -GRAPH] —**rec′ti·graph′ic** *adj.*

rec·ti·lin·e·ar (rek′tə·lin′ē·ər) *adj.* Pertaining to, consisting of, moving in, or bounded by a right line or lines; straight. Also **rec′ti·lin′e·al.** —**rec′ti·lin′e·ar·ly** *adv.*

rec·ti·tude (rek′tə·to͞od, -tyo͞od) *n.* 1 Uprightness in principles and conduct. 2 Freedom from error; correctness of judgment, method, or application; accuracy. 3 *Obs.* Straightness. See synonyms under JUSTICE, VIRTUE. [<F <LL *rectitudo* <L *rectus* right]

rec·to (rek′tō) *n. pl.* **·toes** A right-hand page, as of a book: opposed to *verso* (or *reverso*). [<L *recto (folio)* on the right (page)]

recto– *combining form* Rectal; pertaining to the rectum: *rectocele*, hernia of the rectum. Also, before vowels, **rect–**. [See RECTUM]

rec·tor (rek′tər) *n.* 1 In the Church of England, a priest who has full charge of a parish, and receives the parochial tithes: distinguished from *vicar*. 2 In the Protestant Episcopal Church, a priest in charge of a parish. 3 In the Roman Catholic Church: **a** A priest in charge of a congregation or church, especially one not having parochial status: distinguished from *parish priest*. **b** The head of a seminary or religious house. 4 In certain universities, colleges, and schools, the head or chief officer. [<L *rectus*, pp. of *regere* rule] —**rec′tor·ate** (-it) *n.* —**rec·to·ri·al** (rek·tôr′ē·əl, -tō′rē-) *adj.*

rec·to·ry (rek′tər·ē) *n. pl.* **·ries** 1 A rector's dwelling. 2 In England, a parish domain with its buildings, revenue, etc.

reimportation	**reimpress**	**reincrease**	**reinflame**	**reinoculation**	**reinspection**	**reinter**
reimportune	**reimprison**	**reincur**	**reinform**	**reinscribe**	**reinspire**	**reinterment**
reimpose	**reinaugurate**	**reinduce**	**reinfuse**	**reinsert**	**reinstruct**	**reinterrogate**
reimposition	**reincite**	**reinfect**	**reingratiate**	**reinsertion**	**reintegrate**	**reintrench**
reimpregnate	**reincorporate**	**reinfection**	**reinhabit**	**reinspect**	**reintegration**	**reintroduce**

add,āce,câre,pälm; end,ēven; it,īce; odd,ōpen,ôrder; to͝ok,po͞ol; up,bûrn; ə = a in *above*, e in *sicken*, i in *clarity*, o in *melon*, u in *focus*; yo͞o = u in *fuse*; oi,oil; ou,pout; ch,check; g,go; ng,ring; th,thin; th,this; zh,vision. Foreign sounds à,œ,ü,kh,ṅ; and ◆: see page xx. <from; + plus; ? possibly.

rec·tum (rek'təm) *n. pl.* **·ta** (-tə) *Anat.* The terminal portion of the large intestine, extending from the sigmoid bend of the colon to the anus. [<NL *rectum (intestinum)* straight (intestine)]

rec·tus (rek'təs) *n. pl.* **·ti** (-tī) *Anat.* A straight muscle, as of the eye, the abdomen, the femur, etc. [<NL <L, straight]

rec·u·ba·tion (rek'yə·bā'shən) *n. Obs.* A lying down; specifically, in Roman antiquity, a reclining at table. [<L *recubare* recline]

Re·cu·let (rə·kü·le') The highest peak of the Jura mountains, in eastern France.

re·cum·ben·cy (ri·kum'bən·sē) *n. pl.* **·cies** 1 The state of being recumbent. 2 The act of reclining. 3 A recumbent attitude. Also **re·cum'bence.**

re·cum·bent (ri·kum'bənt) *adj.* 1 Lying down, wholly or partly; reclining; leaning. 2 *Biol.* Tending to rest upon a surface from which they extend: said of certain structures. [<L *recumbens, -entis,* ppr. of *recumbere* < *re-* back + *cumbere* lie, nasalized var. of *cubare* lie down] — **re·cum'bent·ly** *adv.*

re·cu·per·ate (ri·ko͞o'pə·rāt, -kyo͞o'-) *v.* **·at·ed, ·at·ing** *v.i.* 1 To regain health or strength. 2 To recover from loss, as of money. — *v.t.* 3 To obtain again after loss; recover. 4 To restore to vigor and health. See synonyms under RECOVER. [<L *recuperatus,* pp. of *recuperare*]

re·cu·per·a·tion (ri·ko͞o'pə·rā'shən, -kyo͞o'-) *n.* The recovery of lost power or excellence, especially of health or strength.

re·cu·per·a·tive (ri·ko͞o'pə·rā'tiv, -pər·ə·tiv, -kyo͞o'-) *adj.* Tending, assisting, or pertaining to recovery; restorative. Also **re·cu'per·a·to'ry** (-pər·ə·tôr'ē, -tō'rē).

re·cu·per·a·tor (ri·ko͞o'pə·rā'tər, -kyo͞o'-) *n.* 1 One who or that which recuperates. 2 A mechanism, operated by springs or compressed air, for restoring a gun to firing position after the recoil. 3 *Chem.* An apparatus for the recovery of heat from hot gases.

re·cur (ri·kûr') *v.i.* **·curred, ·cur·ring** 1 To happen again or repeatedly, especially at regular intervals: a paroxysm that *recurs.* 2 To come back or return; especially, to return to the mind or in recollection. 3 *Rare* To turn for aid; have recourse. [<L *recurrere* < *re-* back + *currere* run]

re·cur·rence (ri·kûr'əns) *n.* The act or fact of recurring; recourse. Also **re·cur'ren·cy.**

re·cur·rent (ri·kûr'ənt) *adj.* 1 Happening or appearing again or repeatedly; recurring. 2 Running back: said of arteries and nerves. See synonyms under FREQUENT. [<L *recurrens, -entis,* ppr. of *recurrere.* See RECUR.] — **re·cur'rent·ly** *adv.*

recurrent fever Relapsing fever.

recurring decimal A circulating decimal.

re·cur·vant (ri·kûr'vənt) *adj. Her.* Coiled with the head raised to strike: said of a serpent. [<L *recurvans, -antis,* ppr. of *recurvare* bend back]

re·cur·vate (ri·kûr'vit, -vāt) *adj.* Bent back. [<L *recurvatus,* pp. of *recurvare.* See RECURVE.] — **re·cur'va·ture** (-və·chər) *n.*

re·curve (ri·kûrv') *v.t.* & *v.i.* **·curved, ·curv·ing** To curve or bend back or down. [<L *recurvare* < *re-* back + *curvus* curved] — **re·cur·va·tion** (rē'kûr·vā'shən) *n.*

rec·u·sant (rek'yə·zənt, ri·kyo͞o'zənt) *adj.* Persistently refusing to conform; specifically, in English history, refusing to attend services of the Anglican Church. — *n.* One of a recusant character, position, or party; a non-conformist. [<L *recusans, -antis,* ppr. of *recusare.* See RECUSE.] — **rec'u·san·cy** *n.*

rec·u·sa·tion (rek'yə·zā'shən) *n. Law* An exception by which a defendant challenges the judge on grounds of interest or prejudice as to his right to sit. [<L *recusatio, -onis* < *recusatus,* pp. of *recusare.* See RECUSE.]

re·cuse (ri·kyo͞oz') *v.t.* **·cused, ·cus·ing** *Rare* To object to, protest, or challenge (a judge, juror, etc.) as being incompetent, prejudiced, etc. See RECUSATION. [<L *recusare* < *re-* against + *causa* cause, case]

red[1] (red) *adj.* **red·der, red·dest** 1 Of a bright color resembling blood; of the same hue as that color of the spectrum farthest from the violet; also, of a hue approximating red: *red* gold. 2 Ultra-radical in politics; especially, communistic. 3 Pertaining to the pole of a magnet which points to the north. Compare BLUE. — *n.* 1 One of the primary colors, occurring at the opposite end of the spectrum from violet; the color of fresh human blood. 2 Any pigment or dye having or giving this color. 3 An ultra-radical in political views, especially a communist. 4 A red object considered with special reference to its color: the *red* (color) in roulette, the *red* (ball) in billiards. — **in the red** *Colloq.* Operating at a loss; owing money: from the practice of making entries in the debit column of an account book in red ink. — **to see red** To be very angry. [OE *rēad*] — **red'ly** *adv.* — **red'ness** *n.*

red[2] (red) See REDD.

Red (red) *n.* 1 A member of the Communist party of Russia; hence, often, any Russian. 2 A member of the Communist party of any country. 3 Any person who supports or approves of the aims of the Communist party. 4 An ultra-radical; anarchist. [<RED; from the color of their flags and banners]

re·dact (ri·dakt') *v.t.* 1 To prepare, as for publication; edit; revise. 2 To draw up or frame, as a message or edict. [<L *redactus,* pp. of *redigere* reduce to order < *re-* back + *agere* drive] — **re·dac'tor** *n.*

re·dac·tion (ri·dak'shən) *n.* 1 The act of reducing or shaping, as literary matter, into proper form and condition for publication; editing. 2 Literary matter so edited or revised. [<F *rédaction* <LL *redactio, -onis* < *redactus.* See REDACT.]

red algae See RHODOPHYCEAE.

re·dan (ri·dan') *n.* A fortification with two parapets meeting at a salient angle. See also illustration under BASTION. [<F *redan* <OF *redent* < *re-* back + *dent* tooth; from its appearance]

Red Army The army of the U.S.S.R.: now officially the Soviet Army.

red astrachan A large roundish variety of apple, yellow with red stripes or slashes.

red·bay (red'bā') *n.* A tree of the laurel family (*Persea borbonia*) of eastern North America, yielding a bluish-black fruit sometimes called alligator pear.

red·bird (red'bûrd') *n.* 1 The cardinal bird. 2 The scarlet tanager.

red-blood·ed (red'blud'id) *adj.* Having vitality and vigor; hence, manly.

Red Book *Brit.* 1 A book containing a list of all persons in state offices. 2 An official list of the peerage; specifically, a *Royal Kalendar* or *Complete . . . Annual Register* published from 1767 to 1893; also, a similar later publication.

red·breast (red'brest') *n.* 1 A bird having a red breast, as the American or European robin. 2 The long-eared sunfish (*Lepomis auritus*) of the Atlantic coast of the United States: also **red-breasted bream.** 3 An American sandpiper, the knot (*Calidris canutus rufus*): also **red-breasted sandpiper.**

red·bud (red'bud') *n.* The Judas tree.

red·bug (red'bug') *n.* 1 Any of several red insects; especially, the chigger of the southern United States. 2 The cotton stainer (*Dysdercus suturellus*), that stains growing cotton an indelible red.

red·cap (red'kap') *n.* 1 *U.S.* A railroad porter: so called from his red-colored cap. 2 The European goldfinch.

red cedar 1 An American juniper tree (genus *Juniperus*) of the cypress family, having a fine-grained, durable wood of a bright- or dark-red color resembling cedar; especially, the **eastern** (*J. virginiana*) and the **western** (*J. scopulorum*) **red cedar.** 2 The giant arborvitae (*Thuja plicata*) of the western United States, having a light, brittle but durable heartwood: known in the lumber trade and popularly as **western red cedar.** 3 The wood of any of these trees.

red cent A United States copper one-cent piece. — **not worth a red cent** *U.S. Colloq.* Worthless.

red·coat (red'kōt') *n.* 1 A person wearing a red coat. 2 A British soldier of the period when a red coat was part of the uniform worn by the British Army, during the American Revolution and the War of 1812.

red corpuscle An erythrocyte.

red cross 1 The cross of St. George, the emblem of the English. 2 A Greek cross, red on a white ground.

Red Cross Convention See GENEVA CONVENTION.

Red Cross Society A society for the succor of the sick and wounded in war, formed in accordance with the international convention signed at Geneva in 1864, the members wearing a red Geneva cross as a badge of neutrality. These societies are now national organizations, as the **American Red Cross,** and continue their beneficent activities in times of peace, as in fighting disease, etc.

redd (red) *v.t. Dial.* 1 To put in order, as a room; make ready: usually with *up.* 2 To make clear or empty. 3 To adjust, as a quarrel. Also spelled *red.* [OE *hreddan* rescue] — **redd'er** *n.*

red deer 1 The common European and Asian stag (*Cervus elaphus*). 2 The common Virginia white-tailed deer in its rufous summer coat.

red·den (red'n) *v.t.* To make red. — *v.i.* To grow red; flush.

red·den·dum (ri·den'dəm) *n. pl.* **·da** (-də) *Law* A clause in a deed whereby the grantor reserves to himself some new thing, such as rent, out of what he has granted. [<L, neut. of *reddendus,* gerundive of *reddere* give in return. See RENDER.]

red·dish (red'ish) *adj.* Mixed with or somewhat red. — **red'dish·ness** *n.*

red·dle (red'l) *n.* Red ocher or red chalk, used for marking sheep. — *v.t.* **·dled, ·dling** To mark or stain with reddle. Also spelled *raddle.* [Var. of RUDDLE]

red·dle·man (red'l·mən) *n. pl.* **·men** (-mən) One who deals in reddle.

red drum A large drumfish (*Sciaenops ocellatus*) of the Atlantic coast, esteemed as a food fish. Also **red drumfish.**

rede[1] (rēd) *Scot.* or *Obs. v.t.* 1 To advise; counsel. 2 To explain; interpret. — *n.* 1 Advice; counsel. 2 A plan or scheme; decision. 3 A story or narrative; also, interpretation.

rede[2] (rēd) *adj. Scot.* 1 Fierce; impetuous. 2 Drunk.

re·deem (ri·dēm') *v.t.* 1 To regain possession of by paying a price; specifically, to recover, as mortgaged property. 2 To pay off; receive back and satisfy, as a promissory note. 3 To set free; rescue; ransom. 4 *Theol.* To rescue from sin and its penalties. 5 To fulfil, as an oath or promise. 6 To make amends for; compensate for: The play was *redeemed* by its acting. See synonyms under DELIVER, RECLAIM. [<F *rédimer* <L *redimere* < *re-* back + *emere* buy] — **re·deem'a·ble** *adj.*

re·deem·er (ri·dē'mər) *n.* One who redeems. — **The Redeemer** Jesus Christ.

re-de·fec·tor (rē'di·fek'tər) *n.* One who returns to his native country after having previously fled because of real or imagined injustice.

re·de·liv·er (rē'di·liv'ər) *v.t.* 1 To deliver again, as a message or a speech. 2 To give back; return; restore. — **re'de·liv'er·ance, re'de·liv'er·y** *n.*

re·de·mand (rē'di·mand', -mänd') *v.t.* 1 To demand again. 2 To demand or ask the return of.

re·demp·ti·ble (ri·demp'tə·bəl) *adj.* Redeemable. [<L *redemptus* + -IBLE]

re·demp·tion (ri·demp'shən) *n.* 1 The act of redeeming, or the state of being redeemed. 2 The recovery of what is mortgaged or pledged. 3 The payment of a debt or obligation; specifically, the paying of the value of its notes, warrants, etc., by a government. 4 *Theol.* Salvation from sin through the atonement of Christ. [<OF <L *redemptio, -onis* < *redemptus,* pp. of *redimere* redeem. Doublet of RANSOM.]

re·demp·tion·er (ri·demp'shən·ər) *n.* One who redeems himself, as an emigrant by service in payment of passage money.

---

reintroduction
reinundate
reinvent
reinvestigate
reinvestment
reinvigorate
reinvigoration
reinvite
reinvolve
rejoint
rejolt
rejudge
rekindle
relade
reland
relet
reliquidate
reliquidation
relisten
relive
reload
reloan
relocate
relocation
relodge
remake
remast
remasticate
remeasure
remelt
remerge
remigrate
remigration
remix
remodification

re·demp·tive (ri·demp′tiv) *adj.* Serving to redeem, or connected with redemption. Also **re·demp′to·ry** (-tər·ē). [< L *redemptus*, pp. of *redimere*. See REDEEM.]

Re·demp·tor·ist (ri·demp′tər·ist) *n.* A member of a religious order, the Congregation of the Most Holy Redeemer, founded in 1732 by St. Alphonso de Liguori.

re·des·ig·nate (rē·dez′ig·nāt) *v.t.* ·nat·ed, ·nat·ing To designate again.

re·de·vel·op (rē′di·vel′əp) *v.t.* 1 To develop again. 2 *Phot.* To intensify with chemicals and put through a second developing process. —*v.i.* 3 To develop again. —**re′de·vel′op·er** *n.* —**re′de·vel′op·ment** *n.*

red eye 1 *U.S. Colloq.* The danger signal in a railroad semaphore system. 2 *U.S. Slang* Poor-quality whisky. 3 The American rock bass. 4 The red-eyed vireo. 5 The rudd.

red-eyed vireo (red′īd′) See under VIREO.

red-fig·ured (red′fig′yərd) *adj.* Having red figures or markings; specifically, denoting an ancient Greek ceramic ware in which a black glaze was painted over the surface so as to leave the design in the red of the body: a style developed early in the fifth century B.C., and coincident with a great advance in freedom of decorative drawing.

red·fin (red′fin′) *n.* *pl.* **·fins** or **·fin** One of various cyprinoid fishes, especially the common shiner or red dace (*Notropis cornutus*) of eastern North America.

red fir 1 Any of several varieties of fir, as the California red fir (*Abies magnifica*), the largest of the genus. 2 The wood of any of these trees. 3 Douglas fir.

red fire A mixture of easily combustible ingredients, especially strontium salts, that burns with a red light.

red fox The common American fox. See under FOX.

red grouper A large grouper (*Epinephelus morio*) of the southern Atlantic and Gulf coasts.

red gum Strophulus.

red-hand·ed (red′han′did) *adj.* 1 Having hands red with blood, as a murderer caught in the act; hence, having just committed any crime. 2 Caught in the act of doing some particular thing: not always in a bad sense. —**red′-hand′ed·ly** *adv.* —**red′-hand′ed·ness** *n.*

red·head (red′hed′) *n.* 1 A person with red hair. 2 An American duck (*Nyroca americana*), allied to and sometimes mistaken for the canvasback; the pochard. 3 The red-headed woodpecker.

red-head·ed woodpecker (red′hed′id) See under WOODPECKER.

red heat 1 The state of being red-hot. 2 The temperature at which a metal is red-hot.

red hind A serranoid fish (*Epinephelus maculosus*) of the West Indies and southward: one of the groupers. See CABRILLA.

red-hot (red′hot′) *adj.* 1 Heated to redness. 2 New, as if just from the fire: *red-hot* news. 3 Heated; excited: *red-hot* argument. 4 Extreme.

red Indian 1 A North American Indian: also **Red Indian**. 2 The painted cup.

red·in·gote (red′ing·gōt) *n.* An outer coat with long full skirts. [< F *rédingote*, alter. of E *riding coat*]

red·in·te·grate (ri·din′tə·grāt) *v.t.* ·grat·ed, ·grat·ing To restore to a perfect state; make complete; renew. —*adj.* Restored to a whole or perfect state; renewed. [< L *redintegratus*, pp. of *redintegrare* < *red-*, var. of *re-* again + *integrare* to INTEGRATE.]

red·in·te·gra·tion (ri·din′tə·grā′shən) *n.* 1 The act or process of restoration to a whole or sound state. 2 *Psychol.* The act or tendency of the mind to complete again a complex mental state previously experienced, upon the renewal of any part of it.

re·di·rect¹ (rē′di·rekt′) *v.t.* To direct again or anew: to *redirect* a letter. —**re′di·rec′tion** *n.*

re·di·rect² (rē′di·rekt′) *adj.* *Law* Designating the examination of a witness, after cross-examination, by the party who first examined him.

re·dis·count (rē·dis′kount) *n.* 1 A second (or any subsequent) discount on a sum. 2 *Usually pl.* Commercial paper which has been rediscounted. —*v.t.* To discount again.

re·dis·trict (rē·dis′trikt) *v.t.* To district again; especially, to redraw the boundaries of the election districts of.

Red Jacket, 1751–1830, a chief of the Senecas, ally of the United States in the War of 1812: real name *Sagoyewatha*.

red-lat·tice (red′lat′is) *n.* *Archaic* A tavern: named from the red-painted lattice window, the sign of an alehouse.

red lead (led) A lead preparation having a fine red color, used chiefly as a pigment; minium.

red·lead ore (red′led′) Crocoite.

red-let·ter (red′let′ər) *adj.* Happy, fortunate, or memorable: from the use on calendars of red letters to indicate holidays.

red light 1 A traffic signal light meaning stop: opposed to *green light*. 2 Any similar light used to warn of danger or an emergency, to indicate a specified action, location, or result, to insure against interruption, and the like.

red-light district (red′līt′) That part of a city or town in which brothels, sometimes marked by a red light, are numerous.

red lobelia The cardinal flower.

red man An American Indian.

red maple The swamp maple.

Red·mond (red′mənd), **John Edward**, 1851–1918, Irish statesman, leader of the Irish Parliamentary party in the House of Commons.

red oak 1 One of several oaks having a dense, cross-grained wood, as the northern red oak (*Quercus borealis*). 2 The wood of these oaks.

red ocher Ocher.

red·o·lent (red′ə·lənt) *adj.* Full of or diffusing a pleasant fragrance; odorous: often figuratively: *redolent* of the past. [< OF < L *redolens, -entis*, ppr. of *redolere* emit a smell < *red-* thoroughly + *olere* smell] —**red′o·lence**, **red′o·len·cy** *n.* —**red′o·lent·ly** *adv.*

Re·don (rə·dôn′), **Odilon**, 1840–1916, French painter.

Re·don·da (rə·don′də) See ANTIGUA.

red osier 1 A willow (*Salix purpurea*) whose red-tinted twigs are used in making baskets: also called *purple osier*. 2 The red-osier dogwood (*Cornus stolonifera*), with dark-reddish branches and bluish or white fruit.

re·dou·ble (rē·dub′əl) *v.t.* & *v.i.* ·led, ·ling 1 To make or become double. 2 To increase greatly. 3 To echo or re-echo. 4 To fold or turn back. 5 In bridge, to double (an opponent's double).

re·doubt (ri·dout′) *n.* 1 An enclosed fortification, especially a temporary one of any form, employed to defend a pass, a hilltop, etc. 2 An earthwork or simple fortification placed within the main rampart line of a permanent fortification. [< F *redoute* < Ital. *ridotto* < Med. L *reductus*, lit., a refuge, orig. pp. of *reducere* lead back]

re·doubt·a·ble (ri·dou′tə·bəl) *adj.* 1 Inspiring fear; formidable. 2 Deserving respect or deference. Also **re·doubt′ed**. See synonyms under FORMIDABLE. [< F *redoubtable* < *redouter* fear, dread < L *re-* thoroughly + *dubitare* doubt] —**re·doubt′a·ble·ness** *n.* —**re·doubt′a·bly** *adv.*

re·dound (ri·dound′) *v.i.* 1 To have an effect, as by reaction, to the credit, discredit, advantage, etc., of the original agent; return; react; accrue. 2 *Obs.* To surge or flow back. 3 *Obs.* To overflow. —*n.* A return by way of consequence; result; requital. [< F *redonder* < L *redundare* overflow < *red-* back + *undare* surge < *unda* wave]

red·o·wa (red′ə·wə, -və) *n.* Either of two Bohemian dances, one in 3/4 time, resembling a mazurka, the other in 2/4 time, like a polka. [< F < Czech *rejdovák* < *rejdovati* steer, whirl, carouse]

red pepper See PEPPER (def. 3).

red·poll (red′pōl′) *n.* A small finch (genus *Acanthis*) of northern regions, having a reddish crown.

Red Poll One of an English breed of hornless, reddish dairy cattle.

re·draft (rē′draft′, -dräft′) *n.* 1 A second draft or copy. 2 A bill of exchange drawn by the holder of a protested bill on the drawer or endorsers for the reimbursement of the amount of the original bill with costs and charges.

re·dress (ri·dres′) *v.t.* 1 To set right, as a wrong, by compensation or by punishment of the wrongdoer; make reparation for. 2 To make reparation to; compensate: to *redress* the victims of injustice. 3 To remedy; correct. 4 To adjust, as balances. —*n.* (rē′dres, ri·dres′) 1 Satisfaction for wrong done; reparation; amends. 2 A restoration; reformation; correction. [< F *redresser* straighten < *re-* again (< L) + *dresser*. See DRESS.] —**re·dress′er** or **re·dres′sor** *n.*

re-dress (rē·dres′) *v.t.* & *v.i.* To dress again.

Red River 1 A river in Texas, Arkansas, and Louisiana, flowing 1,018 miles east to the Mississippi. 2 A river in the United States and Canada, flowing 545 miles north from NW Minnesota through Manitoba to Lake Winnipeg: also **Red River of the North**. 3 The longest river of Northern Vietnam, flowing 730 miles SE from Yünnan province, China, to the Gulf of Tonkin: Annamese *Song Coi*. Chinese **Yü·an Chiang** (yü·än′ jyäng′) or **Hung Ho** (hŏŏng′ hu′).

red·root (red′rōōt′, -rŏŏt′) *n.* 1 An herb (*Lachnanthes tinctoria*) with sword-shaped, fleshy leaves and fibrous red root, found in swamps along the Atlantic coast of the United States. 2 Any of certain other American plants, as bloodroot, alkanet, green amaranth, bittersweet, etc.

Red Sea An elongated sea between Egypt and Arabia; 1,450 miles long; 170,000 square miles: joined to the Mediterranean by the Suez Canal and connecting with the Indian Ocean by the Gulf of Aden.

red·sear (red′sir′) *v.i.* *Metall.* To break or crack when red-hot, as iron when hammered.

red·shank (red′shangk′) *n.* 1 A Scottish Highlander: so called in allusion to the national costume of Scotland, which leaves the legs bare. 2 A common Old World shore bird (*Totanus totanus*); a tattler.

red shift *Physics* Displacement toward the red end of the spectrum of light from a nebula, star, or other luminous celestial body: caused by an apparent increase in the wavelength of the emitted light. See DOPPLER EFFECT.

red·shirt (red′shûrt′) *n.* A member of Garibaldi's brigade in the struggle for Italian independence.

red·short (red′shôrt′) *adj.* *Metall.* Weak or brittle while red-hot, as iron or steel. [< Sw. *rödskört* < *röd* red + *skör* brittle] —**red′-short′ness** *n.*

red·skin (red′skin′) *n.* A North American Indian. —*adj.* Pertaining to or characteristic of the North American Indians.

red squirrel The chickaree.

red·start (red′stärt′) *n.* 1 A small singing bird (genus *Phoenicura*) allied to the warblers; especially, the common **Old World redstart** (*P. phoenicura*), dark-gray with a black throat, white forehead, and rust-red breast, sides, and tail: also called *brantail*. 2 A small fly-catching warbler (genus *Setophaga*), especially *S. ruticilla*, with bright orange-red patches, common in eastern North America. [< RED¹ + START²]

red stick 1 A stick painted red: war symbol of the Indian chief, Tecumseh. 2 An Indian who carried one of Tecumseh's red sticks. 3 Hence, formerly, any Indian hostile to the United States.

red-tailed buzzard (red′tāld′) See under BUZZARD.

red tape Rigid official procedure involving delay or inaction: from the tying of public

remodify	renerve	reordain	repack	reperuse	replume
remold	renominate	reobtainable	reorder	repaint	rephrase	replunge
remolten	renomination	reoccupation	reordination	repass	replant	repolarization
rename	renumber	reoccupy	reossify	repassage	replantation	repolish
renavigate	renumerate	reoppose	repacify	reperusal	repledge	repopulate

documents with red tape. — **red-tape** (red'-tāp') adj. — **red'-tap'ism** n.

red-top (red'top') n. Any of certain grasses valuable for hay and pasturage; specifically, herd's-grass (*Agrostis alba*). [From the reddish panicle of some varieties]

re-duce (ri-doos', -dyoos') v. **-duced**, **-duc-ing** v.t. 1 To make less in size, amount, number, intensity, etc.; diminish. 2 To bring from a higher to a lower condition; lower; degrade. 3 To bring to submission; subdue; conquer. 4 To bring to a specified condition or state: with *to*: to *reduce* rock to powder; to *reduce* a person to desperation. 5 To thin (paint, etc.) with oil or turpentine. 6 *Math.* To change (an expression) to a more elementary form. 7 *Surg.* To restore (displaced parts) to normal position. 8 *Chem.* a To decrease the positive valence of (an element) by the addition of electrons. b To deprive wholly or partially of oxygen; deoxidize. 9 *Phot.* To diminish the density of (a photographic negative). — v.i. 10 To become less in any way. 11 To decrease one's weight, as by dieting. [< L *reducere* < *re-* back + *ducere* lead] — **re-duc'i-bil'i-ty** n. — **re-duc'i-ble** adj. — **re-duc'i-bly** adv.

— Synonyms: compress, concentrate, condense, consolidate, contract, diminish, solidify, thicken. See ABASE, ABATE, ABBREVIATE, ALLAY, ALLEVIATE, CONQUER, IMPAIR, RELAX, RETRENCH, SCRIMP, SUBDUE, WEAKEN.

re-duc-er (ri-doo'sər, -dyoo'-) n. 1 One who or that which reduces. 2 *Phot.* A chemical solution used for reducing the density of negatives.

reducing agent *Chem.* A substance used to effect a chemical reduction; more specifically, any element which gives up a valence electron to another.

reducing glass A concave lens of considerable diameter used to produce a minified view of drawings, to see how they will appear when they are reduced in size.

reducing valve A valve for maintaining uniform reduced pressure of a fluid, as steam or gas, above or below the valve.

re-duc-tase (ri-duk'tās) n. *Biochem.* Any of a class of enzymes which promote the reduction of compounds to simpler forms. [< REDUCT(ION) + -ASE]

re-duc-ti-o ad ab-sur-dum (ri-duk'shē-ō ad ab-sûr'dəm) *Latin* Literally, reduction to an absurdity; disposal of a proposition by showing that its logical conclusion is absurd; also, proof of a proposition by showing its contradictory to be absurd.

re-duc-tion (ri-duk'shən) n. 1 The act or process of reducing, or its results. 2 *Biol.* The halving of the total number of chromosomes during meiotic cell division. 3 *Chem.* a The process of depriving a compound of oxygen. b The process of decreasing the positive valence of an element by the addition of electrons: distinguished from *oxidation*. 4 *Math.* a One of those formulas by means of which trigonometric functions of angles greater than 90° can be reduced to functions of angles less than 90°. b The process of expressing a fraction in decimal terms. See synonyms under ABBREVIATION. [< F *réduction* < L *reductio*, *-onis* < *reductus*, pp. of *reducere*. See REDUCE.] — **re-duc'tion-al** adj. — **re-duc'tive** adj.

re-dun-dance (ri-dun'dəns) n. 1 The condition or quality of being redundant. 2 That which is redundant. 3 Excess; surplus. See synonyms under CIRCUMLOCUTION, EXCESS.

re-dun-dan-cy (ri-dun'dən-sē) n. pl. **-cies** 1 Redundance. 2 In information theory, deliberate repetition in a message, in whatever medium expressed, in order to lessen the possibility of error.

re-dun-dant (ri-dun'dənt) adj. 1 Being more than is required; constituting an excess. 2 Unnecessarily verbose; tautological. [< L *redundans*, *-antis*, ppr. of *redundare*. See REDOUND.] — **re-dun'dant-ly** adv.

— Synonyms: excessive, exuberant, overflowing, superabundant, superfluous. Antonyms: insufficient, limited, little, scant, scanty, scarce, short, wanting.

re-du-pli-cate (ri-doo'plə-kāt, -dyoo'-) v. **-cat-ed, -cat-ing** v.t. 1 To repeat again and again; redouble; iterate. 2 *Ling.* To affix a reduplication to. — v.i. 3 To undergo reduplication. — adj. (-kit) 1 Repeated again and again; duplicated; doubled. 2 *Bot.* Valvate with the margins reflexed. [< L *reduplicatus*, pp. of *reduplicare* < *re-* again + *duplicare*. See DUPLICATE.]

re-du-pli-ca-tion (ri-doo'plə-kā'shən, -dyoo'-) n. 1 The act of reduplicating, or the state of being reduplicated; a redoubling. 2 A rhetorical figure in which the ending of a sentence, line, or clause is repeated and emphasized at the beginning of the next. 3 *Ling.* a The repetition of an initial element or elements in a word; especially, in the verbs of some Indo-European languages, repetition of some part of the root, usually with vowel modification, serving as a mark of the perfect, as in Greek *bebeka* I have walked, Latin *dedidi* I have given. b The doubling of all or part of a word, often with vowel or consonant change, as in *fiddle-faddle, razzle-dazzle*. c The sound or syllable thus repeated.

re-du-pli-ca-tive (ri-doo'plə-kā'tiv, -dyoo'-) adj. 1 Tending to reduplicate. 2 Of or formed by reduplication. 3 *Bot.* Reduplicate.

red-vein maple (red'vān') The flowering maple.

red-ware (red'wâr') n. A large brown seaweed (*Laminaria digitata*) of the New England coast, sometimes used for food.

red-wat (red'wot') adj. *Scot.* Made wet by something red, as blood.

red-wing (red'wing') n. 1 An American blackbird (*Agelaius phoeniceus*) with bright scarlet patches on the wings of the male: commonly **red-winged blackbird**. 2 An Old World red-winged thrush (*Turdus musicus*), bright reddish-orange on the sides of the body and the under-wing coverts.

red-wood (red'wood') n. 1 An immense California tree (*Sequoia sempervirens*, family Taxodiaceae). See SEQUOIA. 2 Its durable reddish wood. 3 Any one of various other trees yielding a reddish wood, or the wood itself, which yields a red dye.

red-wud (red'wud') adj. *Scot.* Raging mad; furious; insane.

red-yel-low (red'yel'ō) n. One of the range of colors situated between the red and yellow portions of the visible spectrum, sharing the hue of each but identical with neither.

ree¹ (rē) adj. *Scot.* Wild; tipsy; delirious.

ree² (rē) See REEVE³.

reed (rēd) n. 1 The slender, frequently jointed stem of certain tall grasses growing in wet places, or the grasses themselves. 2 A thin, elastic plate of reed, wood, or metal nearly closing an opening, as in a pipe: used in reed organs, the reed pipes of pipe organs, and instruments of the bassoon and clarinet order, to produce a musical tone either by itself or when reinforced by the vibration of air in a pipe. 3 A musical pipe made of the hollow stem of a plant; a shepherd's pipe. 4 *Archit.* A semicylindrical ornamental molding or bead. 5 That part of a loom that drives the filling against the woven fabric, consisting of two horizontal parallel bars near together and connected by numerous thin parallel slips. See illustration under LOOM. 6 An arrow. 7 An ancient Hebrew measure of length; six cubits. 8 The abomasum. — v.t. 1 To fashion into or decorate with reeds. 2 To thatch with reeds. [OE *hrēod*]

Reed (rēd), **John**, 1887-1920, U.S. journalist and poet. — **Walter**, 1851-1902, U.S. army surgeon; demonstrated the transmission of yellow fever by mosquitos.

reed-bird (rēd'bûrd') n. The bobolink: so called chiefly in the southern United States.

reed-buck (rēd'buk') n. An antelope (*Redunca arundineum*) of southern Africa that frequents reedy places; the reitbok.

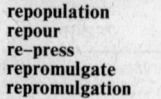
REDWOOD (From 200 to 240 feet high)

reed bunting The European black-headed bunting (genus *Emberiza*), with a white collar, common in marshy places.

reed-ing (rē'ding) n. 1 Beading or semicylindrical moldings collectively. 2 Ornamentation by such moldings. 3 A molding of this kind: the reverse of *fluting*. 4 The knurling on the edge of a coin, as distinguished from *milling*.

reed-ling (rēd'ling) n. The European bearded titmouse (*Panurus biarmicus*), common in reedy places. The male has a black tuft of feathers on each side of the chin.

reed-mace (rēd'mās') n. A cat-tail; any plant of the genus *Typha*, especially *T. latifolia* and *T. angustifolia*.

reed organ A keyboard musical instrument sounding by means of free reeds.

reed pipe An organ pipe having a reed whose vibrations set in motion the air column: distinguished from *flue pipe*.

reed-stop (rēd'stop') n. An organ stop controlling a set of reed pipes.

re-ed-u-cate (rē-ej'oo-kāt) v.t. **-cat-ed, -cat-ing** 1 To educate again. 2 To rehabilitate, as a criminal, by education. — **re'-ed-u-ca'-tion** n.

reed warbler A bird (genus *Acrocephalus*) with moderately rounded tail, found in most parts of the Old World.

reed-y (rē'dē) adj. **reed-i-er, reed-i-est** 1 Full of reeds. 2 Like a reed. 3 Having a thin, sharp tone, like a reed instrument. — **reed'i-ness** n.

reef¹ (rēf) n. 1 A ridge of sand or rocks, or especially of coral, at or near the surface of the water. 2 A lode, vein, or ledge. 3 A shoal. [< ON *rif* rib, reef] — **reef'y** adj.

reef² (rēf) *Naut.* n. 1 The part of a sail that is folded and secured or untied and let out in regulating its size on the mast. 2 The tuck taken in a sail when reefed. — v.t. 1 To reduce (a sail) by folding a part and tying it round, and usually fastening it to, a yard or boom. 2 To shorten or lower, as a topmast by taking part of it in. [ME *riff*, prob. < ON *rif* rib]

reef-band (rēf'band') n. *Naut.* A strip of canvas used to give additional strength to sails along the lines where the reef-points are attached.

reef-er¹ (rē'fər) n. 1 One who reefs. 2 A short double-breasted coat or jacket of heavy material.

reef-er² (rē'fər) n. *U.S. Slang* A marihuana cigarette. [? from its resemblance to the reef of a sail]

reef knot A square knot. See illustration under KNOT.

reef-point (rēf'point') n. *Naut.* One of a series of short lines attached by their centers to the eyelets of a reef-band, and used to fasten the sail in reefing.

reek (rēk) v.i. 1 To give off smoke, vapor, etc. 2 To give off a strong, offensive smell. 3 To be pervaded with anything offensive. — v.t. 4 To expose to smoke or its action. 5 To give off; emit. — n. *Scot.* Smoke; vapor; steam. — Homophone: *wreak*. [OE *rēc*] — **reek'er** n.

reek-y (rē'kē) adj. **reek-i-er, reek-i-est** Having been smoked; smoky; soiled by or emitting smoke. Also **reek'ie**.

reel¹ (rēl) n. 1 A rotatory device or frame for winding rope, cord, photographic film, or other flexible substance. 2 In cinematography, the film wound on one reel: used as a unit of length, usually from 1,000 to 2,000 feet. 3 A wooden spool for wire, thread, etc. 4 Material, such as thread, paper, and the like, when wound on a reel. — v.t. 1 To wind on a reel or bobbin, as a line. 2 To draw in by reeling a line: with *in*: to *reel* a fish in. 3 To say, do, etc., easily and fluently: with *off*. [OE *hrēol*] — **reel'a-ble** adj.

reel² (rēl) v.i. 1 To stagger, sway, or lurch, as when giddy or drunk. 2 To whirl round and round. 3 To have a sensation of giddiness or whirling: My head *reels*. 4 To waver or fall back, as attacking troops. — v.t. 5 To cause to reel. See synonyms under SHAKE. — n. 1 A

repopulation	re-prove	requicken	rerestitution	reseek	resettle	resolder
repour	reprune	reread	rerise	reseize	resettlement	re-solve
re-press	repurchase	re-record	resail	reseizure	reshape	resow
repromulgate	repurge	re-refer	resale	resend	resharpen	restipulate
repromulgation	repurify	rereign	resalute	re-serve	resmooth	restipulation

staggering motion; giddiness. 2 A lively Scottish dance, or its music; also, the Virginia reel. [<REEL[1]] **— reel'er** *n.*

re·em·pha·size (rē·em'fə·sīz) *v.t.* **·sized, ·siz·ing** To stress or emphasize again.

re·en·force (rē'en·fôrs', -fōrs'), **re·en·force·ment** (rē'en·fôrs'mənt, -fōrs'-), etc. See REINFORCE, etc.

re·en·ter (rē·en'tər) *v.t.* & *v.i.* To enter again. **— re·en'trance** *n.*

re·en·ter·ing (rē·en'tər·ing) *adj.* 1 Entering again. 2 Extending inward, as an angle.

reentering angle An angle which is turned inward, as in a figure or structure.

re·en·trant (rē·en'trənt) *adj.* Reentering; extending inward. *—n.* 1 One who or that which reenters. 2 A reentering angle, as in a fortification wall.

re·en·try (rē·en'trē) *n.* 1 The act of entering again. 2 *Law* The act of resuming possession of lands or tenements. 3 In whist and bridge, a card by which a player gains or can gain the lead. 4 In astronautics, the controlled return of a rocket or artificial satellite to the surface of the earth: compare PLANETFALL.

reest[1] (rēst) *v.t.* & *v.i. Scot.* & *Brit. Dial.* To check; balk. **— reest'y** *adj.*

reest[2] (rēst) *v.t.* & *v.i. Scot.* To dry or cure, as by smoking.

reeve[1] (rēv) *v.t.* **reeved** or **rove** (*for pp. also* **rov·en), reev·ing** *Naut.* 1 To pass, as a rope or rod, through a hole, block, or aperture. 2 To fasten in such manner. 3 To pass a rope, etc., through (a block or pulley). [<Du. *reven* reef a sail]

reeve[2] (rēv) *n.* In medieval England, a high administrative officer formerly holding authority over landed areas; bailiff; overseer; steward. [OE *gerēfa* steward]

reeve[3] (rēv) *n.* The female of the ruff: also called *ree.* [Origin unknown]

re·ex·change (rē'iks·chānj') *v.t.* **·changed, ·chang·ing** To exchange again. *—n.* 1 A second or renewed exchange. 2 The sum that the holder of a bill of exchange may demand of the drawer or indorser as indemnity for the loss incurred by its dishonor in a foreign country, where it was payable.

re·fect (ri·fekt') *v.t. Obs.* To refresh after weariness or hunger; restore; repair. [<L *refectus* refreshed. See REFECTION.]

re·fec·tion (ri·fek'shən) *n.* 1 Refreshment by food; a light meal. 2 In civil law, repair of property. 3 *Med.* Spontaneous recovery, as from an ailment or the effects of a vitamin deficiency. [<OF <L *refectio, -onis* <*refectus,* pp. of *reficere* remake, refresh <*re-* again + *facere* make] **— re·fec'tion·er** *n.* **— re·fec'tive** *adj.*

re·fec·to·ry (ri·fek'tər·ē) *n. pl.* **·ries** A room for eating; usually, in a religious house or college, a hall set apart for meals. [<Med. L *refectorium* <L *refectus.* See REFECTION.]

re·fer (ri·fûr') *v.* **·ferred, ·fer·ring** *v.t.* 1 To direct or send for information or other purpose: I *refer* you to another department. 2 To hand over or submit for consideration, settlement, etc.: They *referred* the bill to a special committee. 3 To attribute the cause or source of; assign; relate: He *refers* his success to unceasing application. 4 To assign or attribute to a group, class, period, etc. *—v.i.* 5 To make reference; allude. 6 To turn, as for information, help, or authority; have recourse: to *refer* to the dictionary. See synonyms under ALLUDE, ATTRIBUTE. [<OF *referer* <L *referre* <*re-* back + *ferre* bear, carry] **— ref·er·a·ble** (ref'ər·ə·bəl), **re·fer'ra·ble** or **re·fer'ri·ble** *adj.* **— re·fer'ral** *n.* **— re·fer'rer** *n.*

ref·e·ree (ref'ə·rē') *n.* 1 A person to whom a thing is referred. 2 In certain games, as football, an official who has general control of the game. 3 *Law* A person to whom a case is sent by order of court for investigation and report; an arbitrator. See synonyms under JUDGE. *—v.t.* & *v.i.* To judge as a referee.

ref·er·ence (ref'ər·əns, ref'rəns) *n.* 1 The act of referring. 2 An incidental allusion or direction of the attention: *reference* to a recent event. 3 A note or other indication in a book, referring to some other book or passage: compare CROSS-REFERENCE. 4 One who or that which is or may be referred to. 5 The state of being referred or related: used in the phrases *with* or *in reference to.* 6 *Law* The act or process of submitting a matter to a referee; also, the proceedings of and before a referee. 7 The person or persons to whom one seeking employment may refer for recommendation; also, a written statement or testimonial, as of character or dependability. **— ref'er·enc·er** *n.*

ref·er·end (ref'ə·rend) *n.* The instrument, vehicle, or means by which an act of reference is made. [<REFERENDUM]

ref·er·en·dum (ref'ə·ren'dəm) *n. pl.* **·dums** or **·da** (-də) 1 The submission, by a diplomatic representative to his government, of a proposition not covered by his original instructions. 2 The submission of a proposed public measure or law, which has been passed upon by a legislature or convention, to a vote of the people for ratification or rejection. See INITIATIVE. [<L, gerund of *referre.* See REFER.]

ref·er·ent (ref'ər·ənt) *n.* The particular object, concept, class, event, or the like to which reference is made in any verbal statement or its symbolic equivalent. [<L *referens, -entis,* ppr. of *referre.* See REFER.]

re·fill (rē·fil') *v.t.* To fill again. *—n.* (rē'fil') Any commodity packaged to fit and fill a container originally containing that commodity: a *refill* for a lipstick case.

re·fine (ri·fīn') *v.* **·fined, ·fin·ing** *v.t.* 1 To make fine or pure; free from impurities or extraneous matter. 2 To make polished or cultured; free from coarseness or vulgarity. *—v.i.* 3 To become fine or pure. 4 To become more polished or cultured. 5 To make fine distinctions; use subtlety in thought or speech. See synonyms under CHASTEN, PURIFY. [<RE- + FINE[1], *v.*] **— re·fin'er** *n.*

re·fined (ri·fīnd') *adj.* 1 Characterized by refinement or polish. 2 Free from impurity; purified; clarified. 3 Exceedingly precise or exact; subtle: *refined* tortures. See synonyms under FINE.

re·fine·ment (ri·fīn'mənt) *n.* 1 Fineness of thought, taste, language, etc.; freedom from coarseness or vulgarity; delicacy; culture. 2 The act, effect, or process of refining; purification. 3 A nice distinction; subtlety; also, fastidiousness. [<REFINE]

Synonyms: civilization, cultivation, culture. *Civilization* applies to nations, denoting the sum of those civil, social, economic, and political attainments by which a community is removed from barbarism; a people may be civilized while still far from *refinement* or *culture,* but *civilization* is susceptible of various degrees and of continued progress. *Refinement* applies either to nations or individuals, denoting the removal of what is coarse and rude, and a corresponding attainment of what is delicate, elegant, and beautiful. *Culture* in the fullest sense, as distinct from *cultivation,* denotes that degree of *refinement* and development which results from continued *cultivation* through successive generations; a man's faculties may be brought to a high degree of *cultivation* in some specialty, while he himself remains uncultured even to the extent of coarseness and rudeness. See HUMANITY. *Antonyms*: barbarism, boorishness, brutality, clownishness, coarseness, grossness, rudeness, rusticity, vulgarity.

re·fin·er·y (ri·fī'nər·ē) *n. pl.* **·er·ies** A place where some crude material, as sugar or petroleum, is purified.

re·fit (rē·fit') *v.t.* & *v.i.* **·fit·ted, ·fit·ting** To make or be made fit or ready again; return to serviceable condition, as by making repairs, replacing equipment, etc. *—n.* The repair of damages or wear, especially of a ship.

re·flate (rē·flāt') *v.t.* **·flat·ed, ·flat·ing** To inflate again. [<RE- + (IN)FLATE]

re·flect (ri·flekt') *v.t.* 1 To turn or throw back, as rays of light, heat, or sound. 2 To give back an image of; mirror. 3 To cause to rebound or return; cast: He *reflects* credit on his teacher. 4 *Obs.* To bend or fold back. *—v.i.* 5 To send back rays, as of light or heat. 6 To return in rays: The light *reflects* into my eyes. 7 To give back an image; also, to be mirrored. 8 To think carefully; ponder. 9 To bring blame, discredit, etc.: with *on* or *upon.* See synonyms under CONSIDER, DELIBERATE, MUSE. [<OF *reflecter* <L *reflectere* <*re-* back + *flectere* bend]

re·flec·tance (ri·flek'təns) *n. Physics* The ratio of the radiant or luminous flux reflected from a given surface to the total light falling upon it.

reflecting telescope See under TELESCOPE.

re·flec·tion (ri·flek'shən) *n.* 1 The act of reflecting, or the state of being reflected. 2 *Physics* The throwing off or back (from a surface) of impinging light, heat, sound, or any form of radiant energy. 3 The result of reflecting; reflected rays or an image thrown by reflection. 4 Consideration of or meditation upon past knowledge or experience; thought: *Reflection* increases wisdom; also, its result: a wise *reflection.* 5 The casting of blame; censure. 6 *Anat.* The folding of a part upon itself; a fold, as in a membrane. 7 Reflex action, as of the nerves. Also spelled *reflexion.* [<OF *reflexion* <L *reflexio, -onis*] **— re·flec'tion·al** or **re·flex'ion·al** *adj.*

Synonyms: cogitation, consideration, contemplation, deliberation, meditation, musing, rumination, study, thinking, thought. See ANIMADVERSION, THOUGHT[1]. *Antonyms*: carelessness, heedlessness, imprudence, inconsiderateness, negligence, thoughtlessness.

re·flec·tive (ri·flek'tiv) *adj.* 1 Given to reflection or thought; meditative: a *reflective* person. 2 Used in or capable of consideration or reflection. 3 Having the quality of throwing back light, heat, sound, etc. **— re·flec'tive·ly** *adv.* **— re·flec'tive·ness** *n.*

re·flec·tiv·i·ty (rē'flek·tiv'ə·tē) *n.* 1 The state or quality of being reflective. 2 *Physics* That portion of light or other forms of radiant energy which is reflected by a surface exposed to uniform radiation.

re·flec·tor (ri·flek'tər) *n.* 1 That which reflects. 2 A polished surface, of glass or metal (usually concave), for reflecting light, heat, or sound, and also pictures or slides in a particular direction. 3 A telescope which transmits an image from a reflecting surface to the eyepiece. 4 *Physics* A substance placed around the core of a nuclear reactor for the purpose of reducing neutron leakage and maintaining the level of the chain reaction: sometimes called a *tamper.* 5 *Telecom.* The rear portion of an antenna, serving to increase its directional characteristics.

re·flet (rə·flē') *n.* 1 Iridescence of surface; especially, the metallic glaze on pottery. 2 Pottery having metallic or iridescent luster. [<F, reflection]

re·flex (rē'fleks) *adj.* 1 Turned or thrown backward; reflected, as light. 2 *Physiol.* Of, pertaining to, or produced by a reflex. 3 Turned back upon itself or in the direction whence it came: *reflex* motion. 4 Bent back; reflexed. 5 *Telecom.* Designating a radio receiving circuit in which a single vacuum tube serves for the simultaneous amplification of two different frequencies. *—n.* 1 Reflection, or an image produced by reflection, as from a mirror or like surface. 2 An image or copy; also, an adaptation from another language or dialect, as of a word. 3 Light reflected from an illuminated surface to a shady one. 4 *Physiol.* An involuntary movement or action produced by the transmission of an afferent impulse to a nerve center and its reflection thence as an efferent impulse, as in winking when the eye is threatened: also **reflex action.** *—v.t.* (ri·fleks') To bend back; turn back or reflect. [<L *reflexus* reflected, pp. of *reflectere.* See REFLECT.]

restrengthen	**resubjection**	**resurprise**	**retrim**	**re-utter**	**revegetate**	**revindication**
restrike	**resummon**	**resurvey**	**re-urge**	**revaluation**	**revictual**	**reweigh**
restrive	**resummons**	**retraverse**	**re-use**	**revalue**	**revictualment**	**rewin**
resubject	**resupply**	**retrial**	**re-utilize**	**revarnish**	**revindicate**	**rework**

add,āce,câre,pälm; end, ēven; it,īce; odd,ōpen,ôrder; tŏŏk,pōōl; up,bûrn; ə = a in *above,* e in *sicken,* i in *clarity,* o in *melon,* u in *focus;* yōō = u in *fuse;* oi,oil; ou,pout; ch,check; g,go; ng,ring; th,thin; th,this; zh,vision. Foreign sounds à,œ,ü,kh,ṅ; and ◆: see page xx. < from; + plus; ? possibly.

reflex angle See under ANGLE.
reflex arc *Physiol.* The entire path covered by a nerve impulse from the point of origin in the receptors to the nerve center, and thence outwards to the effectors.
re·flex·ive (ri·flek′siv) *adj.* 1 Reflex. 2 *Gram.* Reflected upon or referring to itself or its subject: in the sentence "He dresses himself," "dresses" is a *reflexive* verb, "himself" is a *reflexive* pronoun. —*n.* A reflexive verb or pronoun. —**re·flex′ive·ly** *adv.* —**re·flex′ive·ness, re·flex·iv·i·ty** (rē′flek·siv′ə·tē) *n.*
re·flight (rē′flīt′) *n. Aeron.* A subsequent flight made over a given area to obtain supplementary photographs or to obtain other necessary details of information.
ref·lu·ent (ref′lōō·ənt) *adj.* Flowing back; ebbing, as the tide. [<L *refluens, -entis,* ppr. of *refluere* flow back < *re-* back + *fluere* flow] —**ref′lu·ence, ref′lu·en·cy** *n.*
re·flux (rē′fluks′) *n.* A flowing back; ebb; return: the flux and *reflux* of fortune. [<L *refluxus,* pp. of *refluere.* See REFLUENT.]
re·for·est (rē·fôr′ist, -for′-) *v.t.* & *v.i.* To replant (an area) with trees. —**re′for·es·ta′tion** *n.*
re·form (ri·fôrm′) *v.t.* 1 To make better by removing abuses, altering, etc.; restore to a better condition. 2 To make better morally; persuade or educate from a sinful to a moral life. 3 To put an end to; stop (an abuse, malpractice, etc.). —*v.i.* 4 To give up sin or error; become better. See synonyms under AMEND, RECLAIM. —*n.* An act or result of reformation; change for the better, especially in administration; correction of evils or abuses; abandonment of vicious habits. [<OF *reformer* <L *reformare* < *re-* again + *formare* form] —**re·form′a·tive** *adj.* —**re·form′er, re·form′ist** *n.*
Re·form (ri·fôrm′) *adj.* Designating a movement in Judaism which arose in Germany in the 19th century: it stresses the historical continuity of the Jewish community and the moral and ethical values of Judaism, reinterprets traditional nationalistic and Messianic teachings, and follows revised or modified forms of many older ceremonial or ritual practices.
re–form (rē·fôrm′) *v.t.* & *v.i.* To form again. [<RE- + FORM] —**re′–for·ma′tion** *n.*
ref·or·ma·tion (ref′ər·mā′shən) *n.* 1 The act of reforming. 2 The state of being reformed. 3 Moral or religious restoration or revival.
Ref·or·ma·tion (ref′ər·mā′shən) *n.* The religious revolution of the 16th century in Europe which began as a movement to reform Catholicism and ended with the establishment of Protestantism in many parts of northern and western Europe.
re·form·a·to·ry (ri·fôr′mə·tôr′ē, -tō′rē) *adj.* Having a tendency or aiming to produce reformation. —*n. pl.* **·ries** An institution for the reformation and instruction of juvenile offenders.
Reform Bill The electoral reform bill passed by the British Parliament in 1832 for the correction and extension of the suffrage.
re·formed (ri·fôrmd′) *adj.* Restored to a better state; corrected or amended; delivered from vicious habits.
Re·formed (ri·fôrmd′) *adj.* Designating those Protestant churches which separated from the Lutherans in the 16th century on questions of doctrine; specifically, those churches which follow the teachings of Calvin and Zwingli. See CALVINISM, ZWINGLIAN.
reform school A reformatory.
re·fract (ri·frakt′) *v.t.* 1 To deflect (a ray) by refraction. 2 *Optics* To determine the degree of refraction of (an eye or lens). [<L *refractus,* pp. of *refringere* turn aside < *re-* back + *frangere* break]
refracting telescope See under TELESCOPE.
re·frac·tion (ri·frak′shən) *n. Physics* The change of direction of a ray, as of light or heat, in oblique passage from one medium to another of different density, or in traversing a medium whose density is not uniform.

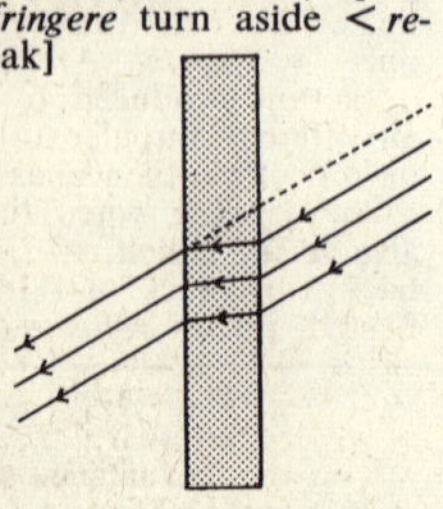

LIGHT REFRACTION

—**double refraction** The property possessed by certain types of crystals of breaking up a beam of light into two differently refracted and polarized rays. —**re·frac′tive** *adv.* —**re·frac′tive·ness, re·frac·tiv·i·ty** (rē′frak·tiv′ə·tē) *n.* —**re·frac′tor** *n.*
re·frac·tom·e·ter (rē′frak·tom′ə·tər) *n.* Any instrument for measuring indices of refraction. [<REFRACT + -(O)METER]
re·frac·to·ry (ri·frak′tər·ē) *adj.* 1 Not amenable to control; disobedient; unmanageable; obstinate. 2 Resisting ordinary methods of reduction: said of an ore. See synonyms under OBSTINATE, REBELLIOUS, RESTIVE, TURBULENT. —*n. pl.* **·ries** 1 A refractory or obstinate person or thing. 2 Any of various materials highly resistant to the action of great heat, as fireclay, graphite, magnesite, etc. [<L *refractarius*] —**re·frac′to·ri·ly** *adv.* —**re·frac′to·ri·ness** *n.*
ref·ra·ga·ble (ref′rə·gə·bəl) *adj.* Capable of being refuted. [<Med. L *refragabilis* <L *refragari* oppose]
re·frain[1] (ri·frān′) *v.i.* To keep oneself back; abstain from action; forbear. —*v.t.* To restrain; curb. [<OF *refrener* <L *refrenare* curb < *re-* back + *frenum* a bridle] —**re·frain′er** *n.*
Synonyms: abstain, forbear, restrain. See CEASE, KEEP. *Antonyms*: begin, continue, persevere, persist.
re·frain[2] (ri·frān′) *n.* 1 A phrase or strain repeated at intervals, generally regular, in a poem or a song; the burden. It generally recurs at the end of a stanza or strophe, and is common in old ballads and in Provençal poetry. 2 Any saying that is repeated over and over. [<OF < *refraindre* check, repeat <L *refringere* break off. See REFRACT.]
re·fran·gi·ble (ri·fran′jə·bəl) *adj.* Capable of being refracted, as light. [<RE- + L *frangere* break + -IBLE] —**re·fran′gi·bil′i·ty, re·fran′gi·ble·ness** *n.*
re·fresh (ri·fresh′) *v.t.* 1 To make (a person) fresh or vigorous again, as by food or rest; reinvigorate; revive. 2 To make fresh, clean, cool, etc. 3 To stimulate, as the memory. 4 To renew or replenish with or as with new supplies. —*v.i.* 5 To become fresh again; revive. 6 To take refreshment. 7 To lay in provisions. [<OF *refreschier* < *re-* again (<L) + *fres* fresh. See FRESH.]
re·fresh·er (ri·fresh′ər) *n.* 1 One who or that which refreshes. 2 A refresher course. —*adj.* Designating something that reacquaints one with the material of subjects previously studied and forgotten: a *refresher* course.
re·fresh·ing (ri·fresh′ing) *adj.* Serving to refresh: often used sarcastically: *refreshing* impudence. See synonyms under DELIGHTFUL. —**re·fresh′ing·ly** *adv.*
re·fresh·ment (ri·fresh′mənt) *n.* 1 The act of refreshing, or the state of being refreshed; restoration of vigor or liveliness. 2 That which refreshes, as food or drink. 3 *pl.* Food, or food and drink, served as a light meal.
re·frig·er·ant (ri·frij′ər·ənt) *adj.* Cooling or freezing; allaying heat or fever. —*n.* 1 Any medicine or material, as ice, which reduces abnormal heat of the body. 2 A substance used for obtaining and maintaining a low temperature, as carbon dioxide, ammonia, or methyl chloride; a freezing mixture; a freezing agent. [<L *refrigerans, -antis,* ppr. of *refrigerare.* See REFRIGERATE.]
re·frig·er·ate (ri·frij′ə·rāt) *v.t.* **·at·ed, ·at·ing** 1 To keep or cause to become cold; cool. 2 To freeze or chill (foodstuffs) for preservative purposes. [<L *refrigeratus,* pp. of *refrigerare* < *re-* thoroughly + *frigerare* cool < *frigus, frigoris* cold] —**re·frig′er·a′tion** *n.* —**re·frig′er·a′tive** *adj.* & *n.*
re·frig·er·a·tor (ri·frij′ə·rā′tər) *n.* 1 That which makes or keeps cold. 2 A box, cabinet, room, railroad car, etc., equipped with apparatus for preserving the freshness of perishable foods, etc., by means of ice or other refrigerant.
re·frig·er·a·to·ry (ri·frij′ər·ə·tôr′ē, -tō′rē) *adj.* Reducing heat. —*n. pl.* **·ries** That which cools or refrigerates. [<L *refrigeratorius*]
re·frin·gen·cy (ri·frin′jən·sē) *n.* Power to refract. Also **re·frin′gence.** [< obs. *refringe* <L *refringere.* See REFRACT.] —**re·frin′gent** *adj.*
reft (reft) Past tense and past participle of REAVE.
re·fu·el (rē·fyōō′əl, -fyōōl′) *v.* **·eled** or **·elled, ·el·ing** or **·el·ling** *v.t.* To replenish with fuel. —*v.i.* To take on a fresh supply of fuel.
ref·uge (ref′yōōj) *n.* 1 Shelter or protection, as from danger or distress. 2 One who or that which shelters or protects. 3 A safe place; asylum. 4 *Brit.* A raised or enclosed safety area for the use of pedestrians at busy street crossings. —*v.t.* & *v.i. Obs.* To give or take refuge. [<OF <L *refugium* < *refugere* retreat < *re-* back + *fugere* flee]
Synonyms (noun): asylum, cover, covert, harbor, hiding-place, protection, retreat, sanctuary, stronghold. See SHELTER.
ref·u·gee (ref′yōō·jē′) *n.* 1 One who flees to a refuge. 2 One who flees from invasion, persecution, or political danger. [<F *réfugié,* pp. of *réfugier* <L *refugere.* See REFUGE.]
re·ful·gence (ri·ful′jəns) *n.* Splendor; brilliant radiance. Also **re·ful′gen·cy.**
re·ful·gent (ri·ful′jənt) *adj.* Shining with a bright light; brilliant; splendid. See synonyms under BRIGHT. [<L *refulgens, -entis,* ppr. of *refulgere* reflect light < *re-* back + *fulgere* shine] —**re·ful′gent·ly** *adv.*
re·fund (ri·fund′) *v.t.* 1 To give or pay back (money, etc.). 2 *Obs.* To pour back. —*v.i.* 3 To make repayment. —*n.* (rē′fund) A repayment; refunding; also, the amount repaid. [<L *refundere* pour back < *re-* back + *fundere* pour out, discharge] —**re·fund′er** *n.* —**re·fund′ment** *n.*
re–fund (rē·fund′) *v.t.* To fund anew; replace (an old loan) by issuing new securities.
re·fus·al (ri·fyōō′zəl) *n.* 1 The act of refusing; denial of what is asked. 2 The privilege of accepting or rejecting; an option.
re·fuse[1] (ri·fyōōz′) *v.* **·fused, ·fus·ing** *v.t.* 1 To decline to do, permit, take, or yield. 2 *Mil.* To turn back (the wing of a line of troops), so that it stands at an angle with the main body. 3 To decline to jump over: said of a horse at a ditch, hedge, etc. 4 *Obs.* To disown; renounce; resign. —*v.i.* 5 To decline to do, permit, take, or yield something. [<OF *refuser* <L *refusus,* pp. of *refundere.* See REFUND.] —**re·fus′er** *n.*
ref·use[2] (ref′yōōs) *adj.* Rejected as worthless. —*n.* Anything worthless; rubbish. See synonyms under WASTE. [<OF *refus,* pp. of *refuser.* See REFUSE[1].]
re–fuse (rē·fyōōz′) *v.t.* & *v.i.* **–fused, –fus·ing** To fuse again.
re–fu·sion (rē·fyōō′zhən) *n. Med.* The temporary withdrawing of blood from circulation, as for exposing it to air or other treatment. Compare TRANSFUSION.
ref·u·ta·tion (ref′yōō·tā′shən) *n.* The act of refuting or proving the falsity or error in a statement, proposition, or argument; evidence applied to overthrow an erroneous statement or position. Also **re·fu·tal** (ri·fyōōt′l). [<L *refutatio, -onis* < *refutare* stop, repel]
re·fute (ri·fyōōt′) *v.t.* **·fut·ed, ·fut·ing** 1 To prove the incorrectness or falsity of (a statement). 2 To prove (a person) to be in error; confute. [<L *refutare*] —**re·fut′a·bil′i·ty** *n.* —**re·fut′a·ble** *adj.* —**re·fut′a·bly** *adv.* —**re·fut′er** *n.*
Synonyms: confound, confute, disprove. To *refute* and to *confute* are to answer so as to admit of no reply. *Refute* applies either to arguments and opinions or to accusations; *confute* is not applied to accusations and charges, but to overwhelming arguments or opinions that confound; a person is *confuted* when his arguments are *refuted.*
re·gain (ri·gān′) *v.t.* 1 To get possession of again, as something lost; gain anew. 2 To reach again; get back to: He *regained* the street. See synonyms under RECOVER. [<MF *regaigner*] —**re·gain′er** *n.*
re·gal (rē′gəl) *adj.* Belonging to or fit for a king; royal; also, stately. See synonyms under IMPERIAL, KINGLY. [<OF <L *regalis* < *rex, regis* king. Doublet of ROYAL.] —**re′gal·ly** *adv.*
re·gale (ri·gāl′) *v.* **·galed, ·gal·ing** *v.t.* 1 To give unusual pleasure to; delight: He *regaled* us with stories. 2 To entertain royally or sumptuously; feast. —*v.i.* 3 To feast. —*n. Obs.* 1 A sumptuous feast. 2 Refreshment. 3 A choice dish. [<F *régaler*; ult. origin uncertain] —**re·gale′ment** *n.*
re·ga·li·a (ri·gā′lē·ə, -gāl′yə) *n. pl.* 1 The insignia and emblems of royalty, as the crown, scepter, verge, vestments, etc. 2 The distinctive symbols, insignia, etc., of any society, order, or rank; hence, fine clothes; fancy trappings. 3 In old English law, royal rights; the six prerogatives of sovereignty: the powers of judicature, life and death, war and peace,

taxation, minting money, and taking masterless goods, as waifs, strays, etc. [< L, neut. pl. of *regalis* kingly < *rex, regis* king]

re·gal·i·ty (ri·gal′ə·tē) *n. pl.* **·ties** 1 Sovereign jurisdiction; royalty. 2 A territorial jurisdiction conferred by the crown on a subject. 3 A country subject to royal authority; a kingdom. [<OF *regalité*]

Re·gan (rē′gən) In Shakespeare's *King Lear*, the second daughter of Lear. See LEAR.

re·gard (ri·gärd′) *v.t.* 1 To look at or observe closely or attentively. 2 To look on or think of in a certain or specified manner; consider: I *regard* him as a friend. 3 To take into account; consider. 4 To have relation or pertinence to; concern. 5 *Obs.* To care for.. — *v.i.* 6 To pay attention. 7 To gaze or look. See synonyms under ESTEEM, LOOK, PERTAIN. — *n.* 1 Observant attention or notice; heed; consideration. 2 Common estimation or repute, especially good repute: a man of *regard*. 3 Reference; relation. 4 A look or aspect; view. 5 *Usually pl.* Respect; affection: My kindest *regards* to your family. 6 Motive. [<OF *regarder* look at < *re-* again + *garder* guard, heed. Doublet of REWARD.]
— *Synonyms (noun):* esteem, favor, respect. *Regard* is more personal and less distant than *esteem*, and adds a special kindliness; *respect* is a more distant word than *esteem*. *Respect* may be wholly on one side, while *regard* is more often mutual; *respect* in the fullest sense is given to what is lofty, worthy, and honorable, or to a person of such qualities; we may pay an external *respect* to one of lofty station, regardless of personal qualities, showing *respect* for the office. See ATTACHMENT, ESTEEM, FAVOR, FRIENDSHIP, LOVE. *Antonyms:* abhorrence, antipathy, aversion, contempt, dislike, hatred, loathing, repugnance.

re·gard·ant (ri·gär′dənt) *adj. Her.* Looking backward. Compare GARDANT. [<F, ppr. of *regarder* look at]

re·gard·ful (ri·gärd′fəl) *adj.* 1 Having or showing regard; heedful. 2 Respectful. — **re·gard′ful·ly** *adv.* — **re·gard′ful·ness** *n.*

re·gard·ing (ri·gär′ding) *prep.* In reference to; with regard to.

re·gard·less (ri·gärd′lis) *adj.* Having no regard or consideration; heedless; negligent. See synonyms under INATTENTIVE. — *adv. Colloq.* In spite of everything.

re·gat·ta (ri·gat′ə, -gä′tə) *n.* A boat race, or a series of such races. [< Ital. *regatar* strive]

re·ge·late (rē′jə·lāt) *v.i.* **·lat·ed**, **·lat·ing** To unite by regelation. [<RE- + L *gelatus*, pp. of *gelare* freeze]

re·ge·la·tion (rē′jə·lā′shən) *n.* The refreezing of melting ice by reducing the pressure to which it is subjected, thus raising the freezing point.

re·gen·cy (rē′jən·sē) *n. pl.* **·cies** 1 The government or office of a regent or body of regents; vicarious government. 2 The period during which a regent or body of regents governs. 3 A body of regents. 4 The district under the rule of a regent. Also **re′gent·ship**. — **the Regency** 1 In English history, the years 1811-20. 2 In French history, the years 1715-1723.

re·gen·er·a·cy (ri·jen′ər·ə·sē) *n.* The state of being regenerate.

re·gen·er·ate (ri·jen′ə·rāt) *v.* **·at·ed**, **·at·ing** *v.t.* 1 To cause complete moral and spiritual reformation or regeneration in. 2 To produce or form anew; re-create; reproduce. 3 To make use of (heat or other energy that might otherwise be wasted) by means of various devices. 4 *Biol.* To grow or form by regeneration. 5 *Electronics* To raise the amplification of (a vacuum tube) by transferring to the input circuit some of the power of the output circuit. — *v.i.* 6 To form anew; be reproduced. 7 To become spiritually regenerate. 8 To effect regeneration. — *adj.* (ri·jen′ər·it) 1 Having new life; restored. 2 Spiritually renewed; regenerated. [< L *regeneratus*, pp. of *regenerare* generate again < *re-* again + *generare*. See GENERATE.]

re·gen·er·a·tion (ri·jen′ə·rā′shən) *n.* 1 The act of regenerating, or the state of being regenerated. 2 The impartation of spiritual life by divine grace. 3 *Biol.* **a** The reproduction of a lost part or organ, as in lizards. **b** The renewal or reproduction of cells, tissues, etc., in the ordinary vital processes: the *regeneration* of the ectodermic layers. 4 The process by which, in various devices, heat or other forms of energy are saved and re-utilized. 5 *Electronics* The amplification of radio signal strength by returning part of the output of a vacuum tube to the grid: an effect of feedback. [<OF] — **re·gen·er·a·tive** (ri·jen′ə·rā′tiv, -ər·ə·tiv) *adj.* — **re·gen′er·a′tive·ly** *adv.*

re·gen·er·a·tor (ri·jen′ə·rā′tər) *n.* 1 One who or that which regenerates. 2 A device in a furnace, gas burner, or similar apparatus, by which the waste heat of escaping gases is used to heat the gas and air just entering. 3 A furnace containing such a device.

Re·gens·burg (rā′gənz·bŏŏrkh) A city of eastern Bavaria, West Germany, a port on the Danube at its northernmost point: English *Ratisbon*.

re·gent (rē′jənt) *n.* 1 One who rules in the name and place of the sovereign. 2 Any ruler or governor; one who governs. 3 A resident master who takes part in the government of a university or college. 4 One of various officers having charge of the higher education, as of a state. — *adj.* 1 Exercising authority in another's place. 2 Governing; ruling. [<OF <L *regens, -entis*, ppr. of *regere* rule]

Reg·gio di Ca·la·bri·a (red′jō dē kä·lä′brē·ä) A port in southern Italy on the Strait of Messina. Also **Reggio Calabria**. Latin **Rhe·gi·um** (rē′jē·əm).

Reg·gio nell'E·mi·lia (red′jō nel·ā·mē′lyä) A city in north central Italy. Also **Reggio Emilia**.

reg·i·cide (rej′ə·sīd) *n.* 1 The killing of a king or sovereign. 2 The killer of a king or sovereign. [<L *rex, regis* king + -CIDE] — **reg′i·ci′dal** *adj.*

Re·gil·lus (ri·jil′əs) In ancient geography, a small lake near Rome; scene of the victory of the Romans over the Latins in 496 B.C.

re·gime (ri·zhēm′) *n.* 1 System of government or administration. 2 Prevalent mode in social matters; social system. 3 Regimen (def. 1). Also **ré·gime** (rā·zhēm′). [<F *régime* < L *regimen*. Doublet of REGIMEN.]

reg·i·men (rej′ə·mən) *n.* 1 A systematized course of living, as to food, clothing, etc. 2 Government; control. 3 *Gram.* The influence of one word in determining the form of another connected with it; grammatical government. See synonyms under FOOD. [<L *regimen* < *regere* rule. Doublet of REGIME.]

reg·i·ment (rej′ə·mənt) *n.* 1 A body of soldiers constituting the unit of infantry, cavalry, artillery, etc., commanded by a colonel. 2 *Obs.* Government over a people or country. — *v.t.* 1 To form into a regiment or regiments; organize. 2 To assign to a regiment. 3 To form into well-defined or specific units or groups; systematize. 4 To make uniform at the expense of individual differences: Certain types of education *regiment* children. [<OF <LL *regimentum* <L *regere* rule] — **reg′i·men′tal** *adj.*

reg·i·men·tals (rej′ə·men′təlz) *n. pl. Military* uniform; the uniform worn by the men and officers of a regiment.

reg·i·men·ta·tion (rej′ə·men·tā′shən) *n.* 1 The act of regimenting; formation into or as into a regiment. 2 Organization into disciplined, uniform groups.

Re·gin (rā′gin) In Norse mythology, a dwarf, foster father of Sigurd, by whom he was slain. Also **Re′ginn**. See FAFNIR, SIGURD.

re·gi·na (ri·jī′nə) *n. Latin* Queen.

Re·gi·na (ri·jī′nə) The capital of Saskatchewan province, Canada.

re·gi·nal ((ri·jī′nəl) *adj.* Pertaining to a queen; queenly; also, supporting or favoring a queen. [<Med. L *reginalis*]

Reg·i·nald (rej′ə·nəld) A masculine personal name. Also *Lat.* **Reg·i·nal·dus** (rej′ə·nal′dəs), **Rey·nal·dus** (rā·nal′dəs). [<Gmc., wise power]

re·gion (rē′jən) *n.* 1 A portion of territory or space; a country or district; also, realm; specifically, one of the strata into which the air or the sea is divided by imaginary boundaries. 2 A zoogeographical division of the earth's surface: the Australian *region*. 3 A portion of the body, arbitrarily circumscribed for anatomical and medical purposes: the abdominal *region*. See synonyms under LAND. [<AF *regiun*, OF *regium* <L *regio, -onis* < *regere* rule]

re·gion·al (rē′jən·əl) *adj.* 1 Of or pertaining to a particular region; sectional; local: *regional* planning. 2 Of or pertaining to an entire region or section, especially a geographic one: *regional* features. — **re′gion·al·ly** *adv.*

ré·gis·seur (rā·zhē·sœr′) *n. French* Director; manager.

reg·is·ter (rej′is·tər) *n.* 1 An official record, the book containing it, or an entry therein; roll; list; schedule; a registry. 2 A registrar. 3 That which registers; a registering apparatus, as for recording velocity, pressure, etc. 4 A device for regulating the admission of heated air to a room. 5 A machine or apparatus which automatically records cash intake; a cash register. 6 *Music* **a** The range or compass of a voice or musical instrument. **b** A class or series of tones of a particular quality or belonging to a particular portion of the compass of a voice or of some instruments. The normal and natural register of the voice is the chest, or thick, register; a middle and an upper register are also recognized, the latter being also termed a head, or thin, register. 7 *Phot.* Relation of position between the sensitive plate or film and the focusing screen. 8 *Printing* **a** Exact correspondence of the lines and margins on the opposite sides of a printed sheet. **b** Correct relation of the colors in color printing. See synonyms under HISTORY, RECORD. — *v.t.* 1 To enter in or as in a register; enrol; specifically, to record formally, as a document, securities, etc. 2 To indicate on a scale. 3 To express or indicate: His face *registered* his disapproval. 4 To effect the exact correspondence of (parts), as the two sides of a printed sheet, the separate plates or films of a color print, etc. 5 To cause (mail) to be recorded, on payment of a fee, when deposited with the postal system, so as to insure delivery. — *v.i.* 6 To enter one's name in a register, poll, etc. 7 To have effect; make an impression. 8 *Printing* To be in register. See synonyms under ENROL. [<OF *registre* <Med. L *registrum* <L *regesta* records, neut. pl. of *regestus*, pp. of *regerere* record < *re-* back + *gerere* carry] — **reg·is·tra·ble** (rej′is·trə·bəl) *adj.*

reg·is·tered (rej′is·tərd) *adj.* 1 Recorded, as a birth, a voter, an animal's pedigree, etc. 2 Having a required or official certificate, as a nurse.

registered mail First-class mail, specially entered and recorded at a higher fee, to insure safe delivery.

registered nurse A graduate nurse licensed to practice by the appropriate State authority and entitled to add R.N. after her name.

reg·is·trant (rej′is·trənt) *n.* One who registers, as a voter; especially, one who registers a trademark or patent. [<F]

reg·is·trar (rej′is·trär, rej′is·trär′) *n.* The authorized keeper of a register or of records; especially, a college or university officer who records the enrolment of students, their grades, etc. [<Med. L *registrarius*]

reg·is·tra·tion (rej′is·trā′shən) *n.* 1 The act of entering in a registry; also, an entry in a registry. 2 The registering of voters; also, the number of voters registered. 3 Enrolment in a school, college, or university. 4 The combination of stops used in playing a composition on the organ. [<Med. L *registratio, -onis*]

reg·is·try (rej′is·trē) *n. pl.* **·tries** 1 Registration. 2 A register, or the place where it is kept. 3 The condition of being registered: a certificate of *registry*.

re·gi·us (rē′jē·əs) *adj. Latin* Royal: a designation of certain English university professorships founded by the crown, or of their incumbents, and also of certain Scottish professors appointed by the crown.

reg·let (reg′lit) *n.* 1 A flat, narrow molding. 2 *Printing* A thin wooden strip used for making space between lines of type, as in posters; also, the strips collectively or the material of which they are made. [<OF, dim. of *regle* <L *regula*. See RULE.]

reg·ma (reg′mə) *n.* *pl.* **·ma·ta** (-mə·tə) *Bot.* A capsular fruit made up of two or more carpels, each of which dehisces at maturity. [<NL <Gk. *rhēgma* fracture < *rhēgnynai* break]

reg·nal (reg′nəl) *adj.* Of or pertaining to a reign, a king, or a kingdom. [<LL *regnalis* <L *regnum* reign]

reg·nant (reg′nənt) *adj.* Reigning in one's own right; hence, dominant; commanding. [<L *regnans, -antis,* ppr. of *regnare* < *regnum* reign]

reg·nant pop·u·li (reg′nənt pop′yŏŏ·lī) *Latin* The people rule: motto of Arkansas.

Re·gnauld (rə·nyō′) French form of REGINALD. Also **Re·gnault′**.

Ré·gnier (rā·nyā′), **Henri de,** 1864–1936, French author.

re·gorge (ri·gôrj′) *v.* **·gorged, ·gorg·ing** *v.t.* To vomit up; disgorge. —*v.i.* To gush or flow back. [<F *regorger* < *re-* again + *gorger* gorge < *gorge* throat <L *gurges* whirlpool. Related to REGURGITATE.]

re·grade (ri·grād′) *v.t.* **·grad·ed, ·grad·ing** To grade again.

re·grate (ri·grāt′) *v.t.* **·grat·ed, ·grat·ing** **1** To buy up, as provisions, for the purpose of selling at a higher price in or near the same market. **2** To retail, as provisions. [<OF *regrater*; ult. origin uncertain]

re·gress (rē′gres) *n.* **1** Passage back; return; also, the power or right of passing back or returning. **2** Retrogression. —*v.i.* (ri·gres′) **1** To go back; move backward; return. **2** *Astron.* To move in a direction opposite to that of the general motion of the heavenly bodies, as the moon's nodes. **3** *Stat.* To return to the mean value of a series of observations. [<L *regressus,* pp. of *regredi* go back < *re-* back + *gradi* walk] —**re·gres′sor** *n.*

re·gres·sion (ri·gresh′ən) *n.* **1** The act of moving back or returning. **2** *Astron.* Motion in a direction opposite to that of the general motion of the heavenly bodies. **3** *Psychoanal.* A retreat of the libido to earlier levels of development or to infantile tendencies belonging to a period preceding the obstacles which prevented their normal fulfilment. **4** *Stat.* The return to a mean or average value. **5** *Med.* The subsidence of a disease or of its symptoms.

re·gres·sive (ri·gres′iv) *adj.* **1** Passing back; returning. **2** Retroactive. **3** Retrogressive. —**re·gres′sive·ly** *adv.*

re·gret (ri·gret′) *v.t.* **·gret·ted, ·gret·ting** **1** To look back upon with a feeling of distress or loss. **2** To feel sorrow or grief concerning. See synonyms under MOURN. —*n.* **1** Distress of mind in recalling some past event; a wish that something had or had not happened. **2** Remorseful sorrow; compunction. **3** An expression of sorrow or disappointment. **4** *pl.* A polite declination in response to an invitation. See synonyms under GRIEF, REPENTANCE. [<OF *regreter*; ult. origin uncertain] —**re·gret′ta·ble** *adj.* —**re·gret′ta·bly** *adv.* —**re·gret′ter** *n.*

re·gret·ful (ri·gret′fəl) *adj.* Feeling, expressive of, or full of regret. —**re·gret′ful·ly** *adv.* —**re·gret′ful·ness** *n.*

reg·u·la (reg′yə·lə) *n.* *pl.* **·lae** (-lē) *Archit.* A fillet, especially one of a series in a Doric architrave, placed under the taenia and bearing six guttae on the under side. [<L, ruler < *regere* rule, lead straight]

reg·u·lar (reg′yə·lər) *adj.* **1** Made according to rule; symmetrical; normal. **2** Acting according to rule; recurring without fail; methodical; orderly: *regular* habits. **3** Constituted, appointed, or conducted in the proper manner; duly authorized: a *regular* meeting, a *regular* practitioner. **4** *Gram.* Undergoing the inflection that is normal or most common to the class of words to which it belongs; following the rule; not exceptional. **5** *Bot.* Having all the parts or organs of the same kind uniform in structure or shape and size: said mainly of flowers. **6** *Zool.* Conforming to an established type; exhibiting radial or bilateral symmetry. **7** *Music* Following strict and classical rules of composition: a *regular* movement. **8** *Eccl.* Bound by a religious rule; pertaining or belonging to a religious order: the *regular* clergy. **9** *Mil.* Belonging to the standing army; permanent. **10** In politics, adhering loyally to a party organization or platform; also, nominated by the official party organization: said of a candidate. **11** *Geom.* Having equal sides and angles. **12** Controlled or governed by one law or operation throughout: a *regular* equation. **13** *Colloq.* Thorough; unmitigated; absolute. **14** *U.S.* Designating that component of a branch of the armed services which consists of persons in continuous service on active duty in both peace and war: the **Regular Army, Regular Navy, Regular Air Force.** See synonyms under CONTINUAL, GRADUAL, HABITUAL, NORMAL, SOBER, USUAL. —*n.* **1** A soldier belonging to a standing army as opposed to a volunteer, draftee, or member of a reserve unit. **2** *Colloq.* One regularly employed or engaged; also, a habitual customer. **3** *Eccl.* A member of a religious or monastic order. **4** A person loyal to a certain political party; a stand-patter. [<L *regularis* < *regula* rule] —**reg′u·lar·ness** *n.*

reg·u·lar·i·ty (reg′yə·lar′ə·tē) *n.* *pl.* **·ties** The state, quality, or character of being regular: *regularity* of form or in occurrence. See synonyms under SYMMETRY, SYSTEM.

reg·u·lar·ize (reg′yə·lə·rīz′) *v.t.* **·ized, ·iz·ing** To make regular. —**reg′u·lar·i·za′tion** *n.*

reg·u·lar·ly (reg′yə·lər·lē) *adv.* In a regular manner; according to the usual method or order.

reg·u·late (reg′yə·lāt) *v.t.* **·lat·ed, ·lat·ing** **1** To direct, manage, or control according to certain rules, principles, etc. **2** To adjust according to a standard, degree, etc.: to *regulate* currency. **3** To adjust to accurate operation: to *regulate* a watch. **4** To put in order; set right. [<LL *regulatus,* pp. of *regulare* rule <L *regula* a rule < *regere* rule, lead straight] —**reg′u·la′tive** *adj.*

Synonyms: adjust, arrange, conduct, direct, dispose, govern, guide, manage, methodize, order, rule, systematize. See SET, SETTLE. *Antonyms*: confuse, derange, disorder, displace, distract, disturb, unsettle.

reg·u·la·tion (reg′yə·lā′shən) *n.* **1** The act of regulating, or the state of being regulated. **2** A rule prescribed for conduct: army *regulations*: also used adjectively. See synonyms under LAW[1], RULE.

reg·u·la·tor (reg′yə·lā′tər) *n.* **1** One who or that which regulates. **2** A clock used as a standard; also, an index arm for regulating the rate of a watch. **3** *Mech.* A contrivance for governing or equalizing motion or flow; the governor of a steam engine; a damper or other device for regulating a draft; a throttle valve. **4** A register (def. 4). **5** A thermostat. **6** *Electr.* A device for keeping at constant strength the current produced by a dynamo. —**reg′u·la′tor·ship** *n.*

Reg·u·la·tor (reg′yə·lā′tər) *n.* **1** A member of any of several bands or committees organized in North Carolina (1768–71) to resist official extortion, and in South Carolina (1767–69) to exterminate horse thieves. **2** One belonging to a volunteer band or committee, which, in the absence of lawful authority, took it upon itself to preserve order and punish crime, but which often deteriorated into lawless bands of violent men.

reg·u·la·to·ry (reg′yə·lə·tôr′ē, -tō′rē) *adj.* Tending or serving to regulate: *regulatory* measures. Also **reg′u·la′tive.**

reg·u·lus (reg′yə·ləs) *n.* *pl.* **·li** (-lī) *Metall.* **1** The metallic mass that sinks to the bottom of the vessel in which slag is being treated. **2** An intermediate product obtained in smelting ores of copper, lead, silver, and nickel. [<L, lit., kinglet, dim of *rex, regis* king] —**reg′u·line** (-lin, -līn) *adj.*

Reg·u·lus (reg′yə·ləs) A white star, Alpha in the constellation Leo; magnitude, 1.34: sometimes called *Cor Leonis.* [<L]

Reg·u·lus (reg′yə·ləs), **Marcus Attilius** Roman general; put to death by the Carthaginians about 250 B.C.

re·gur·gi·tate (ri·gûr′jə·tāt) *v.* **·tat·ed, ·tat·ing** *v.i.* To rush, pour, or surge back; vomit. —*v.t.* To cause to surge back, as partially digested food; vomit. [<LL *regurgitatus,* pp. of *regurgitare* < *re-* back + *gurgitare* flood, engulf <L *gurges, gurgites* whirlpool] —**re·gur′gi·tant** *adj.*

re·gur·gi·ta·tion (ri·gûr′jə·tā′shən) *n.* **1** The act of rushing back or reswallowing. **2** *Physiol.* The backward rush of blood into the heart, due to defective valves.

re·ha·bil·i·tate (rē′hə·bil′ə·tāt) *v.t.* **·tat·ed, ·tat·ing** **1** To restore to a former state, capacity, privilege, rank, etc.; reinstate. **2** To make one capable of becoming a useful member of society again: to *rehabilitate* a crippled soldier. [<Med. L *rehabilitatus,* pp. of *rehabilitare* < *re-* back + *habilitare.* See HABILITATE.] —**re′ha·bil′i·ta′tion** *n.*

re·hash (rē·hash′) *v.t.* To work into a new form; go over again. —*n.* (rē′hash′) Something hashed over, or made or served up from something used before, as old matter issued under a new name.

re·hears·al (ri·hûr′səl) *n.* **1** The act of rehearsing, as a play. **2** The act of reciting or telling over again.

re·hearse (ri·hûrs′) *v.* **·hearsed, ·hears·ing** *v.t.* **1** To perform privately in preparation for public performance, as a play or song. **2** To cause to perform or recite by way of preparation; instruct by rehearsal. **3** To say over again; repeat aloud; recite. **4** To give an account of; relate. **5** To enumerate. —*v.i.* **6** To rehearse a play, song, dance, etc. See synonyms under RELATE. [<OF *reherser* harrow over, repeat < *re-* again + *herser* harrow < *herse.* See HEARSE.] —**re·hears′er** *n.*

re·heat (rē·hēt′) *v.t.* To heat again or anew.

re·heat·er (rē·hē′tər) *n.* **1** An apparatus for reheating a substance which has cooled or partly cooled during some process. **2** A device for reheating exhaust steam in transition in a compound steam engine.

Re·ho·bo·am (rē′ō·bō′əm) Son and successor of Solomon; king of Judah after the revolt of the ten tribes. II *Chron.* ix 31.

rei (rā) Erroneous English form for Portuguese *real.* See MILREIS, REAL[2].

Reich (rīkh) Germany or its government. —**First Reich** The Holy Roman Empire from its establishment in the ninth century to its collapse in 1806. —**Second Reich** The German Empire, 1871–1919, or the Weimar Republic, 1919–1933, or both German governments in the period 1871–1933. —**Third Reich** The Nazi state under Adolf Hitler, 1933–45. [<G, realm]

Reich (rīkh), **Wilhelm,** 1897–1957, U. S. psychotherapist and natural scientist, born in Germany.

Reich·en·berg (rīkh′ən·berkh) The German name for LIBEREC.

Reichs·bank (rīkhs′bängk) *n.* The state or national bank of Germany, founded in 1876. [<G]

Reichs·land (rīkhs′länt) **1** From 1806 to 1871, all German crown lands. **2** From 1871 to 1918, Alsace-Lorraine.

reichs·mark (rīkhs′märk) See MARK[2] (def. 1). [<G]

reichs·pfen·nig (rīkhs′pfen′ikh) See PFENNIG. [<G]

Reichs·rat (rīkhs′rät) *n.* **1** The former parliament of the Austrian Empire, excluding Hungary. **2** The Council of the Reich under the Weimar Republic. Also **Reichs′rath.** [<G, lit., council of the empire]

Reichs·tag (rīkhs′tākh) *n.* The former legislative assembly of Germany. [<G, lit., day of the empire. Cf. DIET for analogous development.]

Reid (rēd), **Whitelaw,** 1837–1912, U. S. journalist and diplomat.

reif (rēf) *n. Scot.* Robbery; plunder.

re·i·fy (rē′ə·fī) *v.t.* **·fied, ·fy·ing** To make real or concrete; materialize: to *reify* an idea. [<L *res, rei* thing + -FY] —**re′i·fi·ca′tion** *n.* —**re′i·fi′er** *n.*

reign (rān) *n.* **1** The possession or exercise of supreme political power; sovereignty; dominion. **2** The time or duration of a sovereign's rule. —*v.i.* **1** To hold and exercise sovereign power; be the head of a monarchy. **2** To hold sway; be predominant; prevail: Winter *reigns.* See synonyms under GOVERN. ◆ Homophones: *rain, rein.* [<F *règne* <L *regnum* rule]

Reign of Terror The period of the French Revolution from May, 1793, to August, 1794, during which Louis XVI, Marie Antoinette, and thousands of other persons were guillotined, and confiscation, violence, and terror reigned under the revolutionary leaders.

re·im·burse (rē′im·bûrs′) *v.t.* **·bursed, ·burs·ing** **1** To pay back (a person) an equivalent for what has been spent or lost; recompense; indemnify. **2** To pay back; refund. [< RE- + obs. *imburse* <LL *imbursare* <L *in-* in

+ *bursa* purse] — **re′im·burse′ment** *n.* — **re′im·burs′er** *n.*

re·im·plan·ta·tion (rē′im·plan·tā′shən) *n. Surg.* The act of restoring in place a bone, or part of a bone, removed in an operation.

re·im·pres·sion (rē′im·presh′ən) *n.* **1** A new or second impression of anything. **2** A reprint of a book without editorial change.

Reims (rēmz, *Fr.* raṅs) A city in NE France; its cathedral, former coronation place of the French kings, was greatly damaged by German bombardment in 1870 and 1914: also *Rheims.*

rein (rān) *n.* **1** *Usually pl.* A strap attached to the bit to control a horse or other draft animal. **2** Any means of restraint or control; government. — *v.t.* **1** To guide, check, or halt with or as with reins. **2** To furnish with reins. — *v.i.* **3** To check or halt a horse by means of reins: with *in* or *up.* **4** To obey the reins. See synonyms under REPRESS. ◆ Homophones: *rain, reign.* [<AF *redne,* OF *resne* <L *retinere.* See RETAIN.]

Rei·nach (re·nȧk′), **Salomon,** 1858–1932, French archeologist.

re·in·car·nate (rē′in·kär′nāt) *v.t.* **·nat·ed, ·nat·ing** To cause to undergo reincarnation.

re·in·car·na·tion (rē′in·kär·nā′shən) *n.* A rebirth of the soul in successive bodies; specifically, in Vedic religions, the becoming of an avatar again: one of the series in the transmigrations of souls. — **re′in·car·na′tion·ist** *n.*

rein·deer (rān′dir′) *n. pl.* **·deer** A deer (genus *Rangifer*) of northern regions, having branched antlers in both sexes: long domesticated for its milk, hide, and flesh, and used as a draft and pack animal. [<ON *hreindȳri* < *hreinn* reindeer + *dȳr* deer]

Reindeer Lake A lake in northern Saskatchewan and Manitoba provinces, Canada; 2,444 square miles.

reindeer moss A gray, branched lichen (*Cladonia rangiferina*) found as far as the extreme limits of arctic vegetation, and furnishing food for reindeer and sometimes man.

re in·fec·ta (rē in·fek′tə) *Latin* The business being unfinished.

re·in·force (rē′in·fôrs′, -fōrs′) *v.t.* **·forced, ·forc·ing 1** To give new force or strength to. **2** To increase the military or naval strength of by providing with more troops or ships. **3** To add some strengthening part or material to; thicken; strengthen; support. See synonyms under RECRUIT. — *n.* That which strengthens or reinforces, as the part of a cannon near the breech that is cast thicker than the rest. Also spelled *reenforce.* [<RE- + *inforce,* var. of ENFORCE]

reinforced concrete Concrete containing metal bars, rods, or netting disposed through the mass in such a way as to increase its tensile strength and durability; ferroconcrete.

re·in·force·ment (rē′in·fôrs′mənt, -fōrs′-) *n.* **1** The act of reinforcing. **2** Increase of force; a fresh body of troops or additional vessels: often in the plural. See synonyms under INCREASE. Also spelled *reenforcement.*

Rein·hardt (rīn′härt), **Max,** 1873–1943, Austrian theatrical director and producer active in Germany and the United States.

Rein·hold (rīn′hōld; *Dan.* rīn′hōlth, *Ger.* rīn′·hōlt, *Sw.* rīn′hōld) See REGINALD. Also *Ger.* **Rei·nald** (rī′nält), *Du.* **Rei·nold** (rī′nōlt).

reins (rānz) *n. pl. Archaic* **1** The kidneys. **2** The region near the kidneys. **3** The affections and passions, formerly thought to have their seat in the loins. [<OF <L *renes,* pl. of *ren*]

re·in·stall (rē′in·stôl′) *v.t.* To install again. — **re·in·stal·la·tion** (rē′in·stə·lā′shən) *n.* — **re′in·stal′ment** or **re′in·stall′ment** *n.*

re·in·state (rē′in·stāt′) *v.t.* **·stat·ed, ·stat·ing** To restore to a former state, position, etc. — **re′in·state′ment** *n.*

re·in·sure (rē′in·shŏŏr′) *v.t.* **·sured, ·sur·ing 1** To protect (the risk on a policy already issued) by obtaining insurance from a second insurer: said of a first insurer. **2** To insure anew. — **re′in·sur′ance** *n.* — **re′in·sur′er** *n.*

re·in·vest (rē′in·vest′) *v.t.* To invest (money) again; especially, to invest earnings from previous investments. — **re′in·vest′ment** *n.*

Rein·wald (rīn′vält) A German form of REGINALD.

reis (rēs) Plural of REAL[2] (def. 2). See MILREIS.

reise (rēs) *n. Scot.* A twig; brush; brushwood.

re·is·sue (rē·ish′ōō) *n.* **1** A second or subsequent issue, as of a publication changed only in form or price. **2** A second printing of postage stamps from the same plates. — *v.t.* **·sued, ·su·ing** To issue again.

reit·bok (rēt′bok) *n.* The reedbuck. [<Du. *rietbok*]

re·it·er·ate (rē·it′ə·rāt) *v.t.* **·at·ed, ·at·ing** To say or do again and again; repeat. [<L *reiteratus,* pp. of *reiterare* < *re-* again + *iterare.* See ITERATE.] — **re·it′er·a′tion** *n.*

re·it·er·a·tive (rē·it′ə·rā′tiv) *adj.* Characterized by reiteration. — *n.* **1** A word or syllable repeated, usually with some slight change, so as to make a reduplicated word; also, the word so formed, as *tittle-tattle.* **2** A word expressing repeated action. — **re·it′er·a′tive·ly** *adv.*

Ré·jane (rā·zhän′), **Gabrielle Charlotte,** 1857–1920, French actress and comedienne: real name *Réju.*

re·ject (ri·jekt′) *v.t.* **1** To refuse to accept, recognize, believe, etc. **2** To refuse to grant; deny, as a petition. **3** To refuse (a person) recognition, acceptance, etc. **4** To expel, as from the mouth; vomit. **5** To cast away as worthless; discard. — *n.* (rē′jekt) A person or thing that has been discarded or rejected. [<L *rejectus,* pp. of *reicere* fling back < *re-* back + *jacere* throw] — **re·ject′er** or **re·jec′tor** *n.*

re·jec·ta·men·ta (ri·jek′tə·men′tə) *n. pl.* Things thrown away; especially, things rejected from a living organism; excrement. [<NL <L *rejectare,* freq. of *reicere* fling back]

re·jec·tion (ri·jek′shən) *n.* **1** The act of rejecting. **2** That which is rejected.

re·joice (ri·jois′) *v.* **·joiced, ·joic·ing** *v.i.* To feel joyful; be glad. — *v.t.* To fill with joy; gladden. [<OF *rejoiss-, resjoiss-,* stem of *resjoir* enjoy < *re-* again (<L) + *esjoir* <L *ex-* thoroughly + *gaudere* be joyous < *gaudium* joy] — **re·joic′er** *n.*

Synonyms: cheer, delight, enjoy, enrapture, exhilarate, exult, gladden, gratify, joy, please, ravish, triumph. Compare HAPPINESS, HAPPY. *Antonyms*: afflict, agonize, bewail, grieve, lament, mourn, pain, regret, sadden, sorrow.

re·joic·ing (ri·joi′sing) *adj.* Pertaining to or characterized by joyfulness. See synonyms under HAPPY. — *n.* The feeling or expression of joy. See synonyms under HAPPINESS, LAUGHTER, RAPTURE.

re·join[1] (ri·join′) *v.t.* **1** To say in reply; answer. — *v.i.* **2** To answer. **3** *Law* To make answer to the plaintiff's replication. [<F *rejoindre* < *re-* again (<L) + *joindre.* See JOIN.]

re·join[2] (rē·join′) *v.t.* **1** To come again into company with. **2** To join together again; reunite. — *v.i.* **3** To come together again. [<RE- + JOIN]

re·join·der (ri·join′dər) *n.* **1** An answer to a reply; also, any reply or retort. **2** *Law* The answer filed by a defendant to a plaintiff's replication. See synonyms under ANSWER. [< F *rejoindre* answer, reply]

re·ju·ve·nate (ri·jōō′və·nāt) *v.t.* **·nat·ed, ·nat·ing 1** To make young; give new vigor or youthfulness to. **2** *Geog.* To restore (a mature or old river) to its youthful condition by the development of lakes, as by obstruction through mountain growth or elevation. Also **re·ju′ve·nize.** [<RE- again + L *juvenis* young + -ATE[1]] — **re·ju′ve·na′tion** *n.*

re·ju·ve·nes·cence (ri·jōō′və·nes′əns) *n.* **1** A renewal of youth; the state of being or growing young again. **2** *Biol.* The transformation of the entire protoplasm of a vegetative cell into a primordial cell, which subsequently invests itself with a new cell wall, and forms the starting point of the life of a new individual. [<L *rejuvenescens,* ppr. of *rejuvenescere* renew youth < *re-* again + *juvenescere* grow young < *juvenis* young] — **re·ju′ve·nes′cent** *adj.*

re·lapse (ri·laps′) *v.i.* **·lapsed, ·laps·ing 1** To lapse back, as into disease after partial recovery. **2** To return to bad habits or sin; backslide. — *n.* (*also* rē′laps) A relapsing; lapse into a former evil state. [<L *relapsus,* pp. of *relabi* slide back < *re-* back + *labi* slide] — **re·laps′er** *n.*

relapsing fever *Pathol.* An acute infectious disease occurring in several forms and due to certain spirochetes transmitted by lice and ticks. It is characterized by febrile paroxysms recurring every five or seven days. Also called *recurrent fever.*

re·late (ri·lāt′) *v.* **·lat·ed, ·lat·ing** *v.t.* **1** To tell the events or the particulars of; narrate. **2** To bring into connection or relation. — *v.i.* **3** To have relation: with *to.* **4** To have reference: with *to.* [<F *relater* <L *relatus,* pp. to *referre.* See REFER.] — **re·lat′er** *n.*

Synonyms: describe, detail, narrate, recite, recount, rehearse, report, state, tell. See PERTAIN. *Antonyms*: deny, hide, suppress, withhold.

re·lat·ed (ri·lā′tid) *adj.* **1** Standing in relation; connected. **2** Of common ancestry; connected by blood or marriage; akin. **3** Narrated. **4** Belonging to the same harmonic or melodic series. — **re·lat′ed·ness** *n.*

re·la·tion (ri·lā′shən) *n.* **1** The fact or condition of being related or connected, or that by which things are connected, either objectively or in the mind; interdependence; connection. **2** The act of relating or narrating; also, that which is related or told. **3** Connection by blood or marriage; kinship. **4** A person connected by blood or marriage; a kinsman: now mostly supplanted by *relative.* **5** *Law* **a** The statement of the grounds of a complaint or grievance by a relator. **b** The reaching back and taking effect of an act or judicial decree at a date anterior to its actual occurrence: Assignment in bankruptcy operates by *relation* back to the date of filing the petition. **6** Reference; regard; allusion: chiefly in the phrase, **in relation to. 7** The position of one person with respect to another: the *relation* of ruler to subject. **8** *pl.* Conditions in general which bring an individual in touch with his fellows; also, the various ways in which one country may come into contact with another politically and commercially. See synonyms under ANALOGY, KINDRED, KINSMAN, REPORT, STORY[1]. [<F <L *relatio, -onis* < *relatus,* pp. to *referre.* See REFER.]

re·la·tion·al (ri·lā′shən·əl) *adj.* **1** Pertaining to or expressing relation: said especially of certain parts of speech. **2** Having relation or kinship.

re·la·tion·ship (ri·lā′shən·ship) *n.* The state of being related; connection. See synonyms under AFFINITY, KIN.

rel·a·tive (rel′ə·tiv) *adj.* **1** Having connection; pertinent: an inquiry *relative* to one's health. **2** Resulting from or depending upon relation; comparative: a *relative* truth. **3** Intelligible only in relation to each other: the *relative* terms "father" and "son." **4** Referring to, relating to, or qualifying an antecedent term: a *relative* pronoun. **5** Having the same key signature, as major and minor keys and scales. — *n.* **1** One who is related; a kinsman. **2** A relative word or term; especially, a relative pronoun. See synonyms under KINDRED, KINSMAN. [<F *relatif* <LL *relativus* <L *relatus*] — **rel′a·tive·ly** *adv.* — **rel′a·tive·ness** *n.*

relative pronoun See under PRONOUN.

rel·a·tiv·ism (rel′ə·tiv·iz′əm) *n. Philos.* The theory that truths are relative and may vary according to the individual, the group, the place, or the time. — **rel′a·tiv·ist** *n.* — **rel′a·tiv·is′tic** *adj.*

rel·a·tiv·i·ty (rel′ə·tiv′ə·tē) *n.* **1** The quality or condition of being relative; relativeness. **2** *Philos.* Existence only as an object of, or in relation to, a thinking mind; phenomenality: sometimes called the doctrine of the relativity of existence. **3** A condition of dependence or of close relation, as of the solar system on the sun. **4** *Physics* The principle of the interdependence of matter, energy, space, and time, as mathematically formulated by A. Einstein. The **special theory of relativity** states that the velocity of light is independent of the motion of its source and that motion itself is a meaningless concept except as between two physical systems or material bodies moving relatively to each other. The **general theory of relativity** extends these principles to the law of gravitation and the motions of the heavenly bodies.

relativity of knowledge *Philos.* The theory that knowledge of what things really are is impossible, since knowledge itself is dependent upon the mind's purely subjective forms of relating its objects.

re·la·tor (ri·lā′tər) *n.* 1 One who relates; a relater. 2 *Law* One who institutes a special proceeding by relation or by information: the *relator* in the writ of quo warranto. [<L]

re·lax (ri·laks′) *v.t.* 1 To make lax or loose; make less tight or firm. 2 To make less stringent or severe, as discipline. 3 To abate; slacken, as efforts. 4 To relieve from strain or effort: to *relax* the eyes. —*v.i.* 5 To become lax or loose; loosen. 6 To become less stringent or severe. 7 To rest; engage in relaxation. 8 To unbend; become less formal. [<L *relaxare* <*re-* again + *laxare* loosen <*laxus* loose. Doublet of RELEASE.] —**re·lax′a·ble** *adj.* —**re·lax′er** *n.*

Synonyms: abate, divert, ease, loose, loosen, mitigate, recreate, reduce, relieve, remit, slacken, unbend. Compare WEAKEN. *Antonyms*: bind, confine, contract, strain, stretch, tighten.

re·lax·a·tion (rē′lak·sā′shən) *n.* 1 The act of relaxing, or the state of being relaxed. 2 Indulgence in diversion, or the diversion indulged in; entertainment. [<L *relaxatio, -onis*] —**re·lax·a·tive** (ri·lak′sə·tiv) *adj.* & *n.*

re·lay (rē′lā, ri·lā′) *n.* 1 A fresh set, as of men, horses, or dogs, to replace or relieve a tired set. 2 A supply of anything kept in store for anticipated use or need. 3 A relay race, or one of its laps or legs. 4 *Electr.* A device which utilizes variations in the condition or strength of a current in a circuit to effect the operation of similar devices in the same or another circuit: a telegraph *relay.* —*v.t.* 1 To send onward by or as by relays. 2 To provide with relays. 3 *Electr.* To operate or retransmit by means of a relay. [<F *relais* <Ital. *rilascio* <*rilasciare, rilassare* leave behind, release <L *relaxare* loosen again. See RELAX.]

re–lay (rē·lā′) *v.t.* **-laid, -lay·ing** To lay again.

relay race A race between two or more teams of runners, each of whom runs a set part of the course and is relieved by a teammate.

re·lease (ri·lēs′) *v.t.* **·leased, ·leas·ing** 1 To set free; liberate; deliver from worry, pain, obligation, etc. 2 To free from something that holds, binds, etc. 3 To permit the circulation, sale, performance, etc., of, as a motion picture, phonograph record, or news item. —*n.* 1 The act of releasing or setting free, or the state of being released; liberation from restraint of any kind. 2 A deliverance or final relief, as from anything grievous or oppressive. 3 A discharge from responsibility or penalty, as from a debt. 4 *Law* An instrument of conveyance by which one of two persons having a mutual interest in lands surrenders and relinquishes all his interest and estate to the other; quitclaim. 5 A motion picture, phonograph record, news item, or the like ready for distribution or circulation. 6 Exhaust of motive fluid in a steam engine; also, the point at which such exhaust begins. 7 *Mech.* Any catch or device to hold and release a mechanism, weights, etc. [<OF *relaisser* let free <L *relaxare*. Doublet of RELAX.] —**re·leas′er** *n.*

Synonyms (verb): deliver, discharge, disengage, emancipate, exempt, extricate, free, liberate, loose, unbind, unfasten, unloose, untie. See ABSOLVE. *Antonyms*: bind, capture, catch, confine, constrain, fetter, hold, imprison, keep, restrain, shackle.

re–lease (rē·lēs′) *v.t.* **-leased, -leas·ing** To lease again.

rel·e·gate (rel′ə·gāt) *v.t.* **·gat·ed, ·gat·ing** 1 To send off or consign, as to an obscure position or place. 2 To assign, as to a particular class or sphere. 3 To refer (a matter) to someone for decision. 4 To banish; exile. See synonyms under COMMIT. [<L *relegatus,* pp. of *relegare* send away <*re-* away, back + *legare* send] —**rel′e·ga′tion** *n.*

re·lent (ri·lent′) *v.i.* To soften in temper; become more gentle or compassionate. —*v.t. Obs.* To cause to relent. [<OF *ralentir* <L *re-lentescere* grow soft <*re-* again + *lentus* soft]

re·lent·less (ri·lent′lis) *adj.* 1 Indifferent to the pain of others; not relenting; pitiless. 2 Unremitting; continuous. See synonyms under AUSTERE, IMPLACABLE. —**re·lent′less·ly** *adv.* —**re·lent′less·ness** *n.*

rel·e·vant (rel′ə·vənt) *adj.* 1 Fitting or suiting given requirements; pertinent; applicable: commonly with *to.* 2 *Ling.* Designating those features of a phoneme which function to distinguish it from other phonemes in a language, as place of articulation in English consonants. [<Med. L *relevans, -antis,* ppr. of *relevare* bear upon <L, raise up. See RELIEVE.] —**rel′e·vance, rel′e·van·cy** *n.* —**rel′e·vant·ly** *adv.*

re·li·a·ble (ri·lī′ə·bəl) *adj.* 1 That may be relied upon; worthy of confidence; trustworthy. 2 *Stat.* Exhibiting a reasonable consistency in results obtained, as in a group of repeated tests: distinguished from *valid.* [<RELY + -ABLE] —**re·li′a·bil′i·ty, re·li′a·ble·ness** *n.* —**re·li′a·bly** *adv.*

Synonyms: trustworthy, trusty. *Trusty* and *trustworthy* refer to inherent qualities of a high order, *trustworthy* being especially applied to persons, and denoting moral integrity and truthfulness; we speak of a *trusty* sword, a *trustworthy* man. *Reliable* is inferior in meaning, denoting merely the possession of such qualities as are needed for safe reliance; as, a *reliable* pledge, *reliable* information. A man is said to be *reliable* with reference not only to moral qualities, but to judgment, knowledge, skill, habit, or perhaps pecuniary ability. A *reliable* messenger is one who may be depended on to do his errand correctly and promptly; a *trusty* or *trustworthy* messenger is one who may be admitted to knowledge of the views and purposes of those who employ him.

re·li·ance (ri·lī′əns) *n.* 1 The act of relying or the condition of being reliant; confidence; trust; dependence. 2 That upon which one relies; a ground of confidence. See synonyms under BELIEF, FAITH. [<RELY + -ANCE]

re·li·ant (ri·lī′ənt) *adj.* Confident; manifesting reliance, especially upon oneself. [<RELY + -ANT] —**re·li′ant·ly** *adv.*

rel·ic (rel′ik) *n.* 1 Some remaining portion or fragment of that which has vanished or is destroyed: a *relic* of barbarism. 2 Something cherished in memory of one deceased; an object of sacred reverence or of affection; a keepsake or memento. 3 The body or part of the body of a saint, or an object connected with a saint or his tomb; a sacred memento. 4 *pl. Obs.* A corpse; remains. Also spelled *relique.* [<F *relique* <L *reliquiae* remains, leavings <*relinquere* leave. See RELINQUISH.]

rel·ict (rel′ikt) *n.* 1 A widow; rarely, a widower. 2 *Biol.* A plant or animal species persisting in a given area as a survival from an earlier period or type. —*adj.* (ri·likt′) *Geol.* Left by gradual erosion; residual. [<L *relicta* widow, fem. of *relictus,* pp. of *relinquere* leave behind. See RELINQUISH.]

re·lief (ri·lēf′) *n.* 1 The act of relieving, or the state of being relieved; removal in whole or in part of any evil, hardship, or trial; alleviation; comfort. 2 That which relieves. 3 Charitable aid, given in the form of money or food to the needy. 4 The release, as of a sentinel or guard, from his post or duty, and the substitution of some other person or persons; also, the person or persons so substituted. 5 In architecture and sculpture, the projection of a figure, ornament, etc., from a surface; also, any such figure: opposed to *round.* Sculptural relief is of three principal kinds: *alto-relievo, bas-relief,* and *mezzo-relievo.* Extremely low relief is called *stiacciato.* 6 In painting, the apparent projection of forms and masses from the plane or ground of a picture given by the arrangement of the lines, colors, or gradations of color; hence, sharpness of outline caused by contrast. 7 The elevation of any person, deed, object, or characteristic above or beyond a common or ordinary plane. 8 In feudal law, a tribute of a fee paid to the lord by the vassal-heir of a deceased tenant for the right of assuming the lapsed tenancy. 9 *Geog.* **a** The unevenness of land surface, as caused by mountains, hills, etc. **b** The parts of a map which portray the configuration of the district represented; contour lines. [<OF <*relever.* See RELIEVE.]

re·li·er (ri·lī′ər) *n.* One who or that which relies. See RELY.

re·lieve (ri·lēv′) *v.t.* **·lieved, ·liev·ing** 1 To free wholly or partly from pain, embarrassment, etc. 2 To lessen or alleviate, as pain or anxiety. 3 To give aid or assistance to: to *relieve* a besieged city. 4 To free from obligation, injustice, etc. 5 To release from duty, as a sentinel, by providing or serving as a substitute. 6 To make less monotonous, harsh, or unpleasant; vary. 7 To bring into relief or prominence; display by contrast. See synonyms under ALLAY, ALLEVIATE, RELAX. [<OF *relever* give assistance to, succor <L *relevare* lift up <*re-* again + *levare* lift, raise <*levis* light] —**re·liev′a·ble** *adj.* —**re·liev′er** *n.*

re·lie·vo (ri·lē′vō) *n. pl.* **·vos** Relief (defs. 5 and 6). [<Ital. <*rilevare* emphasize, elevate <L *relevare.* See RELIEVE.]

re·li·gieuse (rə·lē·zhyœz′) *n. pl.* **·gieuses** (-zhyœz′) *French* A nun.

re·li·gieux (rə·lē·zhyœ′) *n. pl.* **·gieux** (-zhyœ′) *French* A man under monastic vows; a monk.

re·lig·ion (ri·lij′ən) *n.* 1 A belief binding the spiritual nature of man to a supernatural being, as involving a feeling of dependence and responsibility, together with the feelings and practices which naturally flow from such a belief. 2 Any system of faith and worship: the Christian *religion.* 3 An essential part or a practical test of the spiritual life. See *James* i 27. 4 An object of conscientious devotion or scrupulous care: His work is a *religion* to him. 5 *Obs.* Religious practice or belief. [<OF <L *religio, -onis*]

Synonyms: devotion, faith, godliness, holiness, pietism, piety, worship. *Piety* is primarily filial duty, and hence, in its purest sense, a loving obedience and service to God as the heavenly Father; *pietism* often denotes a mystical, sometimes an affected *piety*; *religion* is the reverent acknowledgment of a divine being. *Religion* includes *worship* whether it be external and formal, or the reverence of the human spirit for the divine, seeking outward expression. *Devotion,* which in its fullest sense is self-consecration, is often used to denote an act of *worship,* especially prayer or adoration; as, He is engaged in his *devotions. Godliness* is a character and spirit like that of God. *Holiness* is the highest, sinless perfection of any spirit, whether divine or human, and often used for purity or for consecration. *Faith,* strictly a firm reliance on the truth of religious doctrines, is often used as a comprehensive word for a whole system of *religion* considered as the object of *faith*; as, the Christian *faith,* the Buddhist *faith. Antonyms*: atheism, blasphemy, godlessness, impiety, infidelity, irreligion, profanity, sacrilege, unbelief, ungodliness.

re·lig·ion·ism (ri·lij′ən·iz′əm) *n.* The practice of or adherence to religion: used derogatorily to imply affectation and insincerity. —**re·lig′ion·ist** *n.*

re·lig·i·os·i·ty (ri·lij′ē·os′ə·tē) *n.* Religiousness; also, pious sentimentality. [<LL *religiositas, -tatis*]

re·lig·ious (ri·lij′əs) *adj.* 1 Feeling and manifesting religion; devout; pious. 2 Of or pertaining to religion; teaching or setting forth religion: a *religious* teacher. 3 Having thorough and genuine fidelity; strict in performance; conscientious: a *religious* loyalty. 4 Belonging to the monastic life; bound by monastic vows; following or devoted to a life of religion and devotion. —*n. pl.* **·ious** A person or people devoted to a life of piety and devotion; a monk or nun. [<OF *religious* <L *religiosus*] —**re·lig′ious·ly** *adv.* —**re·lig′ious·ness** *n.*

re·lin·quish (ri·ling′kwish) *v.t.* 1 To give up; abandon; surrender. 2 To cease to demand; renounce: to *relinquish* a claim. 3 To let go (a hold or something held). See synonyms under ABANDON, SURRENDER. [<OF *relinquiss-,* stem of *relinquir* <L *relinquere* <*re-* back, from + *linquere* leave] —**re·lin′quish·er** *n.* —**re·lin′quish·ment** *n.*

rel·i·quar·y (rel′ə·kwer′ē) *n. pl.* **·quar·ies** A casket, coffer, shrine, or other repository for relics. [<F *reliquaire* <L *reliquiae* remains. See RELIC.]

rel·ique (rel′ik, ri·lēk′) See RELIC.

re·liq·ui·ae (ri·lik′wi·ē) *n. pl. Latin* Fossil organisms; relics; organic remains.

rel·ish (rel′ish) *n.* 1 Appetite; appreciation; liking: a *relish* for excitement. 2 The flavor, especially when agreeable, in food and drink; figuratively, the quality in anything that lends spice or zest: Danger gives *relish* to adventure. 3 A slight savory dish served to stimulate appetite; also, something taken with food to lend it flavor or zest; a condiment. 4 An admixture or a small but important characteristic; flavoring: no *relish* of nature in his poetry. — *v.t.* **1** To like the taste or savor of; enjoy: to *relish* a dinner or a joke. **2** To give pleasant flavor to. — *v.i.* **3** To have an agreeable flavor; afford gratification. See synonyms under LIKE. [ME *reles* <OF *reles,* var. of *relais* remainder <*relaisser* leave behind. See RELEASE.] — **rel'ish·a·ble** *adj.*

Synonyms (noun): appetite, appreciation, fondness, gusto, inclination, partiality, predilection, taste, zest. See APPETITE, SAVOR. *Antonyms*: antipathy, aversion, disgust, dislike, distaste, loathing, repugnance.

re·lo·cate (rē·lō'kāt) *v.t.* & *v.i.* **·cat·ed, ·cat·ing** To locate again or anew.

re·lu·cent (ri·lōō'sənt) *adj.* Shining back; reflecting light; gleaming. [<L *relucens, -entis,* ppr. of *relucere* <*re-* back + *lucere* shine. See LUCENT.]

re·luct (ri·lukt') *v.i.* **1** To show reluctance; hesitate. **2** To rebel; make opposition. [<L *reluctari.* See RELUCTANT.]

re·luc·tance (ri·luk'təns) *n.* **1** The state of being reluctant; unwillingness. **2** *Electr.* Capacity for opposing magnetic induction: the reciprocal of *permeance.* **3** *Obs.* Resistance; opposition. Also **re·luc'tan·cy.** [<RELUCTANT]

re·luc·tant (ri·luk'tənt) *adj.* **1** Disinclined to yield to some requirement; unwilling. **2** Marked by unwillingness or rendered unwillingly. **3** *Obs.* Struggling; offering opposition. [<L *reluctans, -antis,* ppr. of *reluctari* fight back <*re-* back + *luctari* fight] — **re·luc'·tant·ly** *adv.*

Synonyms: averse, backward, disinclined, indisposed, loath, opposed, slow, unwilling. *Reluctant* signifies struggling against what one is urged or impelled to do, or is actually doing; *averse* signifies turned away as with dislike or repugnance; *loath* signifies having a repugnance, disgust, or loathing for, but the adjective *loath* is not so strong as the verb *loathe.* A man may be *slow* or *backward* in entering upon that to which he is by no means *averse.* A man is *loath* to believe evil of his friend, *reluctant* to speak of it, absolutely *unwilling* to use it to his injury. A legislator may be *opposed* to a certain measure, while not *averse* to what it aims to accomplish. Compare ANTIPATHY. *Antonyms*: desirous, disposed, eager, favorable, inclined, willing.

rel·uc·tiv·i·ty (rel'ək·tiv'ə·tē) *n. Electr.* The specific electrical reluctance, or the resistance to magnetization of a given substance per unit of length or cross-section: the reciprocal of *permeability.*

re·lume (ri·lōōm') *v.t.* **·lumed, ·lum·ing** **1** To light again; rekindle. **2** To illuminate again. Also **re·lu·mine** (ri·lōō'min). [<RE- + (IL)LUME]

re·ly (ri·lī') *v.i.* **·lied, ·ly·ing** To place trust or confidence: with *on* or *upon.* See synonyms under LEAN[1]. [<OF *relier* bind (together), adhere to <L *religare* <*re-* again + *ligare* bind]

rem (rem) *n. Physics* That dose of absorbed ionizing radiation which has the same biological effect as one roentgen of high-voltage X-ray radiation. [<R(OENTGEN) + E(QUIVALENT) + M(AN)]

Re·ma·gen (rā'mä·gən) A town on the Rhine in northern Rhineland-Palatinate, West Germany: site of the Ludendorff bridge, which collapsed in March, 1945, after being crossed by Allied troops.

re·main (ri·mān') *v.i.* **1** To stay or be left behind after the removal, departure, or destruction of other persons or things. **2** To continue in one place, condition, or character: He *remained* in office. **3** To be left as something to be done, dealt with, etc.: It *remains* to be proved. **4** To endure or last; abide. See synonyms under ABIDE, PERSIST, STAND. [<OF *remaindre* <L *remanere* <*re-* back + *manere* stay, remain]

re·main·der (ri·mān'dər) *n.* **1** That which remains; something left after a subtraction, expenditure, or passing over of a part; a residue; remnant. **2** *Math.* **a** That which is left after the subtraction of one quantity from another. **b** In division, the excess of the dividend over the product of the divisor by the integral part of the quotient. **3** *Law* An estate in expectancy, but not in actual possession and enjoyment; that remnant or residue of interest which, on the creation of a particular prior estate, is by the same instrument limited to another to be enjoyed on the termination of that estate. **4** In philately, an obsolete issue of stamps, demonetized by the government and sold at a large discount, generally to dealers. **5** A copy or part of an edition of a book remaining with a publisher after sales have ceased. — *adj.* Left over; remaining. — *v.t.* To sell as a remainder (def. 5). [<AF <OF *remaindre* REMAIN]

re·mains (ri·mānz') *n. pl.* **1** That which is left after a part has been removed or destroyed; remnants. **2** The body of a deceased person; a corpse. **3** Writings of an author published after his death. **4** Survivals of the past, as fossils, monuments, etc.: the *remains* of ancient Troy. See synonyms under BODY, TRACE[1].

re·man (rē·man') *v.t.* **·manned, ·man·ning** **1** To furnish with a fresh complement of men. **2** To instil courage or manliness into.

re·mand (ri·mand', -mänd') *v.t.* **1** To order or send back: to *remand* a soldier to his post. **2** *Law* **a** To recommit to custody, as an accused person after a preliminary examination. **b** To send back to a lower court, as a case improperly brought before the court so ordering. — *n.* **1** Recommittal, as of an accused person to custody; also, the recommitted person. **2** A judicial order of recommittal. [<OF *remander* <LL *remandare* <L *re-* back + *mandare* order] — **re·mand'ment** *n.*

rem·a·nence (rem'ə·nəns) *n.* **1** The state or quality of remaining; permanence; also, the remainder. **2** *Electr.* That part of magnetic induction remaining in a material after the removal of an applied magnetomotive force. [<L *remanens, -entis,* ppr. of *remanere* remain] — **rem'a·nent** *adj.*

re·mark (ri·märk') *n.* **1** A comment or saying, oral or written; a casual observation; also, conversational speech in general: I enjoyed his *remarks.* **2** The act of observing or noticing; observation; notice. **3** Remarque. — *v.t.* **1** To say or write by way of comment. **2** To take particular notice of. **3** *Obs.* To mark; distinguish. — *v.i.* **4** To make remarks: with *on* or *upon.* [<F *remarque* observation <*remarquer* notice <*re-* again + *marquer* mark. See MARK.] — **re·mark'er** *n.*

Synonyms (noun): annotation, comment, note, observation, utterance. A *comment* is an explanatory or critical *remark,* as upon some passage in a literary work or some act or speech in common life. A *note* is something to call attention, hence a brief written statement; in correspondence, a *note* is briefer than a letter. *Annotations* are especially brief *notes,* commonly marginal, and closely following the text. *Comments, observations,* or *remarks* may be oral or written, *comments* being oftenest written, and *remarks* oftenest oral. An *observation* is properly the result of fixed attention and reflection; a *remark* may be the suggestion of the instant. *Remarks* are more informal than a speech.

re·mark·a·ble (ri·mär'kə·bəl) *adj.* Worthy of special notice; hence, extraordinary; unusual; conspicuous; distinguished. See synonyms under EMINENT, RARE, EXTRAORDINARY. — **re·mark'a·ble·ness** *n.* — **re·mark'a·bly** *adv.*

re·marque (ri·märk') *n.* **1** A small engraved picture or other distinguishing mark on an engraved plate, appearing on the engraved surface or in the margin, to indicate a stage in its progress before completion. **2** A print from an engraved or etched plate bearing such a mark. [<F]

Re·marque (rə·märk'), **Erich Maria,** born 1897, U.S. novelist born in Germany: real name *Erich Paul Kramer.*

re·mar·ry (rē·mar'ē) *v.t.* & *v.i.* **·ried, ·ry·ing** To marry again. — **re·mar'riage** (-mar'ij) *n.*

Rem·brandt (rem'brant, *Du.* rem'bränt), 1606–1669, Dutch painter and etcher: full name *Rembrandt Harmenszoon van Rijn* or *van Ryn.*

re·me·di·a·ble (ri·mē'dē·ə·bəl) *adj.* Capable of being cured or remedied. [<F *remédiable*] — **re·me'di·a·bly** *adv.*

re·me·di·al (ri·mē'dē·əl) *adj.* Of the nature of or adapted to be used as a remedy: *remedial* measures. [<L *remedialis*] — **re·me'di·al·ly** *adv.*

rem·e·di·less (rem'ə·dē·lis) *adj.* Without remedy; incurable; irreparable.

rem·e·dy (rem'ə·dē) *v.t.* **·died, ·dy·ing** **1** To cure or heal, as by medicinal treatment. **2** To make right; repair; correct. **3** To overcome or remove (an evil or defect). — *n. pl.* **·dies** **1** That which cures or affords relief to bodily disease or ailment; a medicine; also, remedial treatment. **2** A means of counteracting or removing evil; relief. **3** *Law* A legal mode for enforcing a right or redressing or preventing a wrong. **4** Tolerance (def. 5). [<AF <L *remedium* <*re-* thoroughly + *mederi* heal, restore]

re·mem·ber (ri·mem'bər) *v.t.* **1** To bring back or present again to the mind or memory; recall; recollect. **2** To keep in mind carefully, as for a purpose. **3** To bear in mind with affection, respect, awe, etc. **4** To bear in mind as worthy of a reward, gift, etc.: She *remembered* me in her will. **5** To reward; tip: *Remember* the steward. **6** *Obs.* To remind. — *v.i.* **7** To have or use one's memory. — **to remember (one) to** To inform a person of the regard of: *Remember* me *to* your wife. [<OF *remembrer* <LL *rememorari* <L *re-* again + *memorare* bring to mind <*memor* mindful] — **re·mem'ber·er** *n.*

Synonyms: recall, recollect, retain. Compare synonyms for MEMORY. *Antonyms*: forget, overlook.

re·mem·brance (ri·mem'brəns) *n.* **1** The act or power of remembering; the state of being remembered; memory. **2** The period within which one can remember. **3** That which is remembered; a reminiscence. **4** A memento; keepsake; also, a token or message of friendship: often in the plural. **5** Mindful regard. See synonyms under MEMORY.

re·mem·branc·er (ri·mem'brən·sər) *n.* **1** One who or that which causes one to remember; a reminder. **2** One of the recording officers of the Exchequer in England, as the **King's** or **Queen's remembrancer,** responsible for collecting debts due to the sovereign: since 1873, an officer of the Supreme Court.

re·mex (rē'meks) *n. pl.* **rem·i·ges** (rem'ə·jēz) *Ornithol.* One of the large quill feathers of a bird's wing: usually in the plural. [<L, oarsman <*remus* oar] — **re·mig·i·al** (ri·mij'ē·əl) *adj.*

re·mind (ri·mīnd') *v.t.* To bring to (someone's) mind; cause to remember. See synonyms under ADMONISH. [<RE- + MIND] — **re·mind'er** *n.*

re·mind·ful (ri·mīnd'fəl) *adj.* **1** Tending to remind; serving as a reminder: said of things. **2** Mindful: said of persons.

Rem·ing·ton (rem'ing·tən), **Frederic,** 1861–1909, U.S. painter and sculptor. — **Philo,** 1816–89, U.S. inventor and gunsmith.

rem·i·nisce (rem'ə·nis') *v.i.* **·nisced, ·nisc·ing** To recall incidents or events of the past; indulge in reminiscences. [Back formation <REMINISCENT]

rem·i·nis·cence (rem'ə·nis'əns) *n.* **1** The recalling to mind of past incidents and events; also, the narration of past experiences. **2** The act or power of reproducing past cognitions in consciousness. **3** An expression, fact, or feature serving as a reminder of something else. See synonyms under MEMORY. [<F]

rem·i·nis·cent (rem'ə·nis'ənt) *adj.* **1** Of the nature of or possessing reminiscence; also, recalling or dwelling upon the past; remembering. **2** Inducing a reminiscence of a person or thing; suggestive. [<L *reminiscens, -entis,* ppr. of *reminisci* recollect <*re-* again + *memini* remember] — **rem'i·nis'cent·ly** *adv.*

re·mise (ri·mīz') *Law v.t.* **·mised, ·mis·ing** To give; surrender; release; relinquish: used in conveyancing. — *n.* The act of remising. [<F, fem. of *remis,* pp. of *remettre* <L *remittere* send back. See REMIT.]

re·miss (ri·mis') *adj.* Slack or careless in matters requiring attention; dilatory; negligent; hence, lacking in earnestness or energy. See synonyms under INATTENTIVE. [<L *remissus,* pp. of *remittere* send back, slacken. See REMIT.] — **re·miss'ness** *n.*

re·mis·si·ble (ri·mis'ə·bəl) *adj.* Capable of being remitted or pardoned, as sins. [<F *rémissible*] — **re·mis'si·bil'i·ty** *n.*

re·mis·sion (ri·mish'ən) *n.* **1** The act of remitting, or the state of being remitted; specifically, discharge from penalty; pardon; deliverance, as from a debt or obligation.

2 Abatement, as of a fine erroneously imposed. 3 Relaxation, as from work or study. 4 Temporary abatement of a disease or of pain. 5 The act of sending a remittance. [<OF <L *remissio, -onis*]

re·mit (ri·mit′) *v.* **·mit·ted, ·mit·ting** *v.t.* 1 To send, as money in payment for goods; transmit. 2 To refrain from exacting or inflicting, as a penalty. 3 To pardon; forgive, as a sin or crime. 4 To abate; relax, as vigilance. 5 To restore; replace. 6 To put off; postpone. 7 To refer or submit for judgment, settlement, etc., as to one in authority. 8 *Law* To refer (a legal proceeding) to a lower court for further consideration. 9 *Rare* To send back, as to prison. 10 *Obs.* To resign; renounce. 11 *Obs.* To free; release. —*v.i.* 12 To send money, as in payment. 13 To diminish; abate. —*n.* The act of remitting; specifically, the sending of a legal cause from one tribunal to another. [<L *remittere* send back <*re-* back + *mittere* send] —**re·mit′ta·ble** *adj.* —**re·mit′ter** or **re·mit′tor** *n.*

re·mit·tal (ri·mit′l) *n.* Remission.

re·mit·tance (ri·mit′ns) *n.* The act of transmitting money or credit; also, that which is remitted, as money.

remittance man A ne'er-do-well living outside his home country on money transmitted at regular intervals by friends or relatives: originally applied to British persons living in the colonies or in the western United States.

re·mit·tent (ri·mit′nt) *adj.* 1 Having remissions. 2 Having partial, irregular, or temporary diminutions of energy or action: a *remittent* fever or geyser. —*n.* A remittent fever. [<L *remittens, -entis,* ppr. of *remittere.* See REMIT.]

remittent fever *Pathol.* A form of malaria in which the fever fluctuates daily but does not entirely disappear.

rem·nant (rem′nənt) *n.* 1 That which remains of anything; specifically, the piece of cloth, silk, etc., left over after the last cutting. 2 A remaining trace or survival of anything, suggestive of former condition, use, or belief. 3 A small piece or quantity. 4 A small remaining number of people. See synonyms under TRACE[1]. —*adj.* Remaining. [<OF *remenant,* ppr. of *remaindre.* See REMAIN.]

re·mod·el (rē·mod′l) *v.t.* **·eled** or **·elled, ·el·ing** or **·el·ling** 1 To model again. 2 To make over or anew.

re·mon·e·tize (ri·mon′ə·tīz) *v.t.* **·tized, ·tiz·ing** To reinstate, especially silver, as lawful money. [<RE- again + L *moneta* money + -IZE] —**re·mon′e·ti·za′tion** *n.*

re·mon·strance (ri·mon′strəns) *n.* 1 The act of remonstrating; protest; expostulation. 2 Expostulatory counsel or reproof. [<OF]

Re·mon·strance (ri·mon′strəns) *n.* The document formulating the five points of Arminian dissent from strict Calvinism, presented to the states of Holland and Friesland in 1610 and condemned by the synod of Dort in 1619. —**the Grand Remonstrance** A document presented by Parliament to King Charles I of England, Nov. 22, 1641, protesting against his misgovernment. —**Re·mon′strant** *n.*

re·mon·strant (ri·mon′strənt) *adj.* Having the character or tendency of a remonstrance; expostulatory. —*n.* One who presents or signs a remonstrance. [<Med. L *remonstrans, -antis,* ppr. of *remonstrare.* See REMONSTRATE.]

re·mon·strate (ri·mon′strāt) *v.* **·strat·ed, ·strat·ing** *v.t.* 1 To say or plead in protest or opposition. 2 *Obs.* To point out; demonstrate. —*v.i.* 3 To urge strong reasons against any course or action; protest; object. [<Med. L *remonstratus,* pp. of *remonstrare* demonstrate <L *re-* again + *monstrare* show] —**re·mon·stra·tion** (rē′mon·strā′shən, rem′ən-) *n.* —**re·mon′stra·tive** (-strə·tiv) *adj.* —**re·mon′stra·tor** (strā·tər) *n.*

re·mon·ta (rā·mōn′tä) *n. SW U.S.* A group of saddle horses. [<Sp.]

re·mon·tant (ri·mon′tənt) *adj. Bot.* Ascending again: said of roses that bloom more than once in a season. —*n.* A remontant rose. [<F, ppr. of *remonter.* See REMOUNT.]

rem·on·toir (rem′ən·twär′, *Fr.* rə·môn·twär′) *n. Mech.* An apparatus that utilizes force from the train of a clock to give new impulse to the escape wheel at certain intervals, usually once in 30 seconds. [<F]

rem·o·ra (rem′ər·ə) *n.* 1 Any of a genus (*Remora*) of fish (family *Echeneididae*) having on its head an oval suctorial disk by means of which it attaches itself to sharks, other fishes, or floating objects, thus being carried great distances. 2 Any delay or impediment. [<L, hindrance <*re-* back + *mora* delay]

re·morse (ri·môrs′) *n.* 1 The keen or hopeless anguish caused by a sense of guilt; compunction; distressing self-reproach. 2 *Obs.* Compassion; pity. See synonyms under REPENTANCE. [<OF *remors* <LL *remorsus* a biting back <L *remordere* keep biting <*re-* again + *mordere* bite] —**re·morse′ful** *adj.* —**re·morse′ful·ly** *adv.* —**re·morse′ful·ness** *n.*

re·morse·less (ri·môrs′lis) *adj.* Having no compassion; pitiless; cruel. —**re·morse′less·ly** *adv.* —**re·morse′less·ness** *n.*

re·mote (ri·mōt′) *adj.* 1 Located far from a specified place or some place regarded as a point of reference: *remote* regions. 2 Removed far from present time; distant in time: the *remote* future. 3 Having slight relation or connection; separated; foreign; distant in relation: a *remote* cause, *remote* kinship. 4 Not obvious; inconsiderable; slight: a *remote* likeness or analogy. 5 Abstracted; absent-minded; hence, aloof. —*n.* A television or radio broadcast made from a mobile camera or microphone operated at a distance from the station, and sent to the transmitter by cable or through relay towers. See synonyms under ALIEN. [<L *remotus,* pp. of *removere* remove <*re-* again + *movere* move] —**re·mote′ly** *adv.* —**re·mote′ness** *n.*

remote control Control from a distance, as of a machine, apparatus, aircraft, guided missile, etc., by electrical or radio circuits.

re·mo·tion (ri·mō′shən) *n.* 1 The act of removing; removal. 2 *Obs.* Departure. [<OF]

ré·mou·lade (rā′mə·läd′, *Fr.* rā·mōō·làd′) *n.* A sharp sauce made of hard-boiled egg yolks, oil, vinegar, and seasoning. [<F <Ital. *remolata,* lit., vigorously stirred]

re·mount (rē·mount′) *v.t.* & *v.i.* To mount again or anew. —*n.* (rē′mount′) 1 A new setting or framing. 2 A fresh riding horse. [<OF *remonter*]

re·mov·a·ble (ri·mōō′və·bəl) *adj.* Capable of being removed; movable; also, capable of being displaced, dismissed, or obliterated: *removable* walls, officials, or stains. —**re·mov′a·bil′i·ty, re·mov′a·ble·ness** *n.* —**re·mov′a·bly** *adv.*

re·mov·al (ri·mōō′vəl) *n.* 1 The act of removing or the state of being removed. 2 Dismissal, as from office. 3 Changing of place, especially of habitation.

re·move (ri·mōōv′) *v.* **·moved, ·mov·ing** *v.t.* 1 To take or move away or from one place to another. 2 To take off; doff, as a hat. 3 To get rid of; do away with: to *remove* abuses. 4 To kill; assassinate. 5 To displace or dismiss, as from office. 6 To take out; extract: with *from.* —*v.i.* 7 To change one's place of residence or business; move. 8 *Poetic* To go away; depart. See synonyms under ABOLISH, ABSTRACT, ALLEVIATE, CANCEL, CARRY, CONVEY, DISPLACE, EXTERMINATE, SEPARATE. —*n.* 1 A removal; a move; the act of removing, as one's business or belongings. 2 The space moved over in changing an object from one position to another; hence, a degree of difference; step; interval: He is only one *remove* from a fool. 3 *Brit.* A dish or course at dinner removed to give place to another. 4 *Obs.* A period of absence. [<OF *remouvoir* <L *removere* <*re-* again + *movere* move] —**re·mov′er** *n.*

re·moved (ri·mōōvd′) *adj.* 1 Separated, as by intervening space, time, or relationship, or by difference in kind: a cousin twice *removed.* 2 Taken away; transferred. —**re·mov·ed·ness** (ri·mōō′vid·nis) *n.*

Rem·scheid (rem′shīt) An industrial city in central North Rhine-Westphalia, West Germany.

Rem·sen (rem′sən), **Ira,** 1846–1927, U.S. chemist and educator.

re·mu·da (rā·mōō′dä) *n. SW U.S.* The extra mounts or saddle horses of each cowboy herded together, usually a herd of 90 to 100 geldings for an outfit of eight to ten men: called a *saddle band* in the Northwest. [<Sp., lit., exchange <*remudar* replace]

re·mu·ner·ate (ri·myōō′nə·rāt) *v.t.* **·at·ed, ·at·ing** To make just or adequate return to or for; compensate; pay or pay for; reward. See synonyms under PAY, REQUITE. [<L *remuneratus,* pp. of *remunerari* <*re-* again + *munus, muneris* gift] —**re·mu′ner·a·bil′i·ty** *n.* —**re·mu′ner·a·ble** *adj.*

re·mu·ner·a·tion (ri·myōō′nə·rā′shən) *n.* 1 The act or fact of remunerating. 2 That which remunerates; pay; compensation; recompense. See synonyms under RECOMPENSE, RESTITUTION, SALARY.

re·mu·ner·a·tive (ri·myōō′nə·rā′tiv, -nər·ə·tiv) *adj.* 1 Profitable; lucrative. 2 Serving to pay or remunerate: *remunerative* justice. See synonyms under PROFITABLE. —**re·mu′ner·a′tive·ly** *adv.* —**re·mu′ner·a′tive·ness** *n.*

Re·mus (rē′məs) In Roman mythology, the twin brother of Romulus, by whom he was killed.

Remus (rē′məs), **Uncle** See UNCLE REMUS.

ren- Var. of RENI-.

ren·ais·sance (ren′ə·säns′, -zäns′, ri·nā′səns; *Fr.* rə·ne·säṅs′) *n.* A new birth; resurrection; renascence. [<F <*renaître* be reborn <*re-* again + L *natus,* pp. of *nasci* be born]

Ren·ais·sance (ren′ə·säns′, -zäns′, ri·nā′səns; *Fr.* rə·ne·säṅs′) *n.* 1 The revival of letters and art in Europe, marking the transition from medieval to modern history: it began in Italy in the 14th century and gradually spread to other countries. 2 The period of this revival, from the 14th to the 16th century; also, the style of art, literature, etc., marked by a classical influence, that was developed in and characteristic of this period. Also *Renascence.* —*adj.* Pertaining to or characteristic of the Renaissance.

Renaissance architecture A style of building and decoration that followed the medieval, originating in Italy in the 15th century, and based on the classic Roman style.

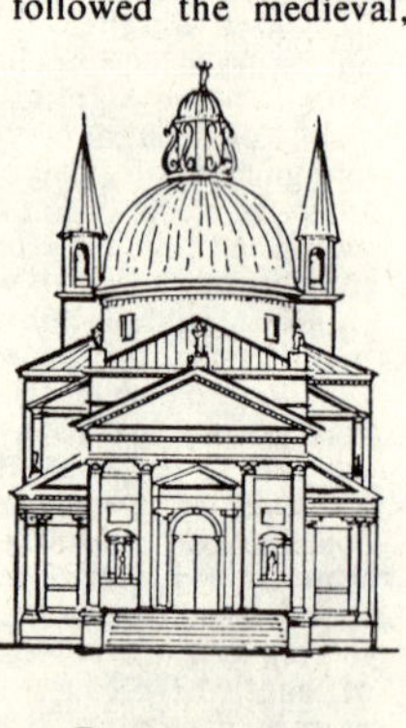

RENAISSANCE ARCHITECTURE
Church of the Redentore, Venice, 1578–80.

re·nal (rē′nəl) *adj. Med.* Of, pertaining to, affecting, or situated near the kidneys. [<F *rénal* <L *renalis* <*renes* kidneys]

renal capsule or **gland** The suprarenal gland.

Re·nan (rə·näṅ′), **Joseph Ernest,** 1823–1892, French historian, philologist, and critic.

Ren·ard (ren′ərd) See REYNARD.

re·nas·cence (ri·nas′əns) *n.* Rebirth; new birth or life; a renaissance; a revival. [<L *renascens, -entis,* ppr. of *renasci* <*re-* again + *nasci* be born] —**re·nas′cent** *adj.*

Re·nas·cence (ri·nas′əns) *n.* The Renaissance.

Re·naud (rə·nō′) French form of REGINALD.

ren·con·tre (ren·kon′tər, *Fr.* räṅ·kôṅ′tr′) *n. French* A rencounter.

ren·coun·ter (ren·koun′tər) *n.* 1 *Obs.* A sudden hostile collision, as with an enemy. 2 An unexpected encounter, as of travelers. 3 A contest or debate. —*v.t.* & *v.i. Obs.* To meet unexpectedly or by surprise. [<F *rencontrer.* See RE- and ENCOUNTER.]

rend (rend) *v.* **rent** or **rend·ed, rend·ing** *v.t.* 1 To tear apart forcibly; split; break. 2 To pull or remove forcibly: with *away, from, off,* etc. 3 To pass through (the air) violently and noisily. 4 To distress (the heart, etc.), as with grief or despair. —*v.i.* 5 To split; part. [OE *rendan* tear, cut down] —**rend′er** *n.*

Synonyms: break, burst, cleave, lacerate, mangle, rip, rive, rupture, sever, slit, sunder, tear. *Rend* and *tear* are applied usually to the sundering of textile substances, *tear* being the milder, *rend* the stronger word. To *rip,* as applied to articles made by sewing or stitching, is to divide along the line of a seam by cutting or breaking the stitches. *Rive* is a woodworkers' word for parting wood in the way of the grain without a clean cut, as by splitting. To *lacerate* is to *tear* roughly the flesh or animal tissue, as by the teeth of a wild beast. *Mangle* is a stronger word than *lacerate*; *lacerate* is more superficial, *mangle* more complete. To *burst* or *rupture* is to tear or rend by force from within, *burst* denoting the greater violence; as, to *burst* a gun; to

rupture a blood vessel. Compare BREAK. *Antonyms*: heal, join, mend, reunite, secure, stitch, unite, weld.

ren·der (ren′dər) *v.t.* **1** To give, present, or submit for action, approval, payment, etc. **2** To provide or furnish; give: to *render* aid to the poor. **3** To give as due: to *render* obedience. **4** To perform; do: to *render* great service. **5** To give or state formally: to *render* judgment. **6** To give by way of requital or retribution: to *render* double for one's sins. **7** To represent or depict, as in music or painting. **8** To cause to be or become: to *render* a ship seaworthy. **9** To express in another language; translate. **10** To melt and clarify, as lard. **11** To give back; return: often with *back*. **12** To surrender; give up: to *render* a fortress. See synonyms under INTERPRET. — *n.* **1** A payment, specifically of rent, made to a superior. **2** A coat of plaster applied without intervening lathing. [<F *rendre* <L *reddere* give back < *re-* back + *dare* give] — **ren′der·a·ble** *adj.* — **ren′der·er** *n.*

ren·dez·vous (rän′dā·vōō, -də-; *Fr.* räṅ·de·vōō′) *n. pl.* **·vous** (-vōōz, *Fr.* -vōō′) **1** An appointed place of meeting. **2** A meeting or an appointment to meet. **3** A base for naval ships or for military units. **4** *Obs.* A resort; refuge. — *v.t.* & *v.i.* **·voused** (-vōōd), **·vous·ing** (-vōō′ing) To assemble or cause to assemble at a certain place or time. [<F *rendez-vous*, lit., betake yourself < *se rendre* betake oneself]

ren·di·tion (ren·dish′ən) *n.* **1** A translation; the interpretation of a text. **2** Artistic, dramatic, or musical interpretation; also, the performance or execution of a dramatic or musical composition. **3** A surrendering, especially of a person. **4** The act of rendering, or the amount rendered. [< obs. F < *rendre* render]

Ren·do·va (ren·dō′və) An island in the New Georgia group of the Solomon Islands; of volcanic origin; 75 square miles.

Re·né (rə·nā′, *Fr.* rə·nā′) A masculine personal name. [<F, reborn] — **Re·née** (rə·nā′) *fem.*

ren·e·gade (ren′ə·gād) *n.* **1** An apostate. **2** A traitor; deserter. Also **ren′e·ga′do** (-gā′dō). — *adj.* Traitorous. [<Sp. *renegado*, pp. of *renegar* deny <Med. L *renegare* <L *re-* again and again + *negare* deny]

re·nege (ri·nig′, -neg′, -nēg′) *v.i.* **·neged, ·neg·ing** **1** In card games, to fail to follow suit when able to do so. See REVOKE. **2** *Colloq.* To fail to fulfil a promise. **3** *Obs.* To renounce; deny. Also **re·nig′.** [<Med. L *renegare.* See RENEGADE.] — **re·neg′er** *n.*

re·new (ri·nōō′, -nyōō′) *v.t.* **1** To make new or as if new again; restore to a former or sound condition. **2** To begin again; resume: to *renew* an argument. **3** To repeat: to *renew* an oath of loyalty. **4** To acquire again; regain (vigor, strength, etc.). **5** To cause to continue in effect; extend: to *renew* a subscription. **6** To revive; reestablish. **7** To replenish or replace, as provisions. — *v.i.* **8** To become new again. **9** To begin or commence again. See synonyms under RECLAIM. [<RE- again + NEW] — **re·new′a·ble** *adj.*

re·new·al (ri·nōō′əl -nyōō′-) *n.* The act of renewing, or the state of being renewed.

re·newed (ri·nōōd′, -nyōōd′) *adj.* Made new; restored; revived; repeated. See synonyms under FRESH. — **re·new·ed·ly** (ri·nōō′id·lē, -nyōō′-) *adv.*

Renewed Church of the United Brethren See MORAVIAN.

Ren·frew (ren′frōō) A county in SW Scotland; 240 square miles; county town, Renfrew. Also **Ren′frew·shire** (-shir).

Re·ni (rā′nē), **Guido,** 1575–1642, Italian painter.

reni- *combining form* Kidney; of or related to the kidneys: *reniform*: also, before vowels, *ren-*. Also **reno-.** [<L *ren, renis* a kidney]

ren·i·form (ren′ə·fôrm, rē′nə-) *adj.* Kidney-shaped. [<RENI- + -FORM]

ren·in (ren′in) *n. Biochem.* A protein substance secreted by an ischemic kidney or blood vessel and supposed to be responsible for a rise in blood pressure. [<L *ren* kidney]

re·ni·tent (ri·nī′tənt, ren′ə·tənt) *adj.* Offering resistance to any influence or force; continuously reluctant; recalcitrant; specifically, presenting elastic resistance to pressure. [<L *renitens, -entis,* ppr. of *reniti* resist < *re-* back + *niti* struggle] — **re·ni′tence, re·ni′ten·cy** *n.*

Rennes (ren) A city in NW central France; the intellectual center of Brittany.

ren·net (ren′it) *n.* **1** The dried stomach of certain young hoofed animals, especially the mucous membrane lining the fourth stomach of a suckling calf or sheep, which is capable of curdling milk. **2** Anything used to curdle milk. **3** An aqueous or vinous infusion of animal rennet. **4** Rennin. [Alter. of ME *rennels* <OE *rinnan* run together, coagulate]

ren·nin (ren′in) *n. Biochem.* An enzyme present in rennet; the milk-curdling ferment: also called *chymosin.* [<RENN(ET) + -IN]

Re·no (rē′nō) A city in western Nevada.

Re·noir (rə·nwàr′), **Pierre Auguste,** 1841–1919, French Impressionist painter.

re·nounce (ri·nouns′) *v.* **·nounced, ·nounc·ing** *v.t.* **1** To give up, especially by formal statement. **2** To disown; repudiate. **3** In card games, to indicate inability to follow (a suit led) by playing a card of another suit. — *v.i.* **4** In card games, to renounce the suit led. [<F *renoncer* <L *renuntiare* protest against, announce < *re-* back, against + *nuntiare* report < *nuntius* messenger] — **re·nounce′ment** *n.* — **re·nounc′er** *n.*

Synonyms: abandon, abjure, deny, disavow, discard, disclaim, disown, forswear, recall, recant, refuse, reject, repudiate, retract, revoke. *Abjure, discard, forswear, recall, recant, renounce, retract,* and *revoke,* like *abandon,* imply some previous connection. *Renounce* is to declare against and give up formally and definitively; as, to *renounce* the pomps and vanities of the world. *Retract* is to take back something that one has said as not true or as what one is not ready to maintain; as, to *retract* a charge or accusation; one *recants* his own opinions or beliefs. *Repudiate* is to put away with emphatic and determined repulsion; as, to *repudiate* a debt. To *deny* is to affirm to be not true or not binding; as, to *deny* a statement or relationship; or to refuse to grant, as a request or petition. To *discard* is to cast away as useless or worthless; thus, one *discards* a worn garment. *Revoke,* etymologically the equivalent of the English *recall,* is to take back something given or granted; as, to *revoke* a command, a will, or a grant; *recall* may be used in the exact sense of *revoke,* but is often applied to persons, as *revoke* is not; we *recall* a messenger and *revoke* an order. Compare ABANDON, ABDICATE, ABJURE, RECANT. *Antonyms*: acknowledge, advocate, assert, avow, cherish, claim, defend, hold, maintain, own, proclaim, retain, uphold, vindicate.

ren·o·vate (ren′ə·vāt) *v.t.* **·vat·ed, ·vat·ing** **1** To make as good as new; repair. **2** To renew; refresh; reinvigorate. — *adj.* Renovated. [<L *renovatus,* pp. of *renovare* < *re-* again + *novare* make new < *novus* new] — **ren′o·va′tion** *n.* — **ren′o·va′tor** *n.*

re·nown (ri·noun′) *n.* **1** Exalted reputation; celebrity; the state of being widely known for great achievements or merits; fame. **2** *Obs.* Rumor; report. See synonyms under FAME. — *v.t. Obs.* To spread the fame of; render famous. [<OF *renon* < *renomer* name again, make famous <L *re-* again + *nominare* name < *nomen* a name]

re·nowned (ri·nound′) *adj.* Having renown; famous. See synonyms under ILLUSTRIOUS.

rens·se·laer·ite (ren′sə·lə·rīt′, ren′sə·lâr′īt) *n.* A light-colored variety of talc of such waxlike consistency that it may be worked on a lathe. [after Stephen Van *Rensselaer,* 1764–1839, U. S. soldier and politician]

rent[1] (rent) *n.* **1** Compensation made in any form by a tenant to a landlord or owner for the use of land, buildings, etc.; especially, such compensation paid in money at regular or specified intervals. **2** Similar payment for the use of any property, movable or fixed. **3** *Econ.* **a** Income derived by the owner from the use of his land or property. **b** The return afforded by cultivated land in excess of the costs, as of labor or materials. **c** That which is yielded by land in excess of the yield of the poorest land cultivated under equal conditions: also called **economic rent. d** Hence, a return derived from a similar advantage, as in a monopoly of natural resources. **4** *Obs.* **a** Landed or other property affording revenue. **b** Income or revenue. — **for rent** Available for use or occupancy by the paying of rent. — *v.t.* **1** To obtain the temporary possession and use of for a compensation, usually made at fixed intervals. **2** To grant the temporary possession and use of for a rent. — *v.i.* **3** To be let for rent. [<OF *rente* <LL *rendita,* L *reddita* what is given back or paid, fem. of pp. of *reddere.* See RENDER.] — **rent′a·ble** *adj.*

rent[2] (rent) Alternative past tense and past participle of REND. — *n.* **1** A hole or slit made by rending or tearing; tear; rip; fissure. **2** A schism; violent separation; split. See synonyms under BREACH, HOLE. [<REND]

rent·al (ren′təl) *n.* **1** The revenue derived from rented property. **2** A schedule of rents. — *adj.* Of or pertaining to rent. [<AF]

rente (räṅt) *n. French* **1** *pl.* The bonds and other securities representing the government indebtedness of France; also, the sums paid as interest on this indebtedness: also **rentes sur l'État** (räṅt sür lā·tà′). **2** Income or revenue in general; annuity.

rent·er (ren′tər) *n.* One who rents; specifically, one who rents an estate or tenement; a tenant.

ren·tier (räṅ·tyā′) *n. French* One who owns, or derives a fixed income from, invested capital or lands.

re·nun·ci·a·tion (ri·nun′sē·ā′shən, -shē-) *n.* **1** The act of renouncing or disclaiming; repudiation. **2** A declaration, statement, or formula in which something is renounced. [<L *renunciatio, -onis* a proclamation] — **re·nun′ci·a′tive** *adj.* — **re·nun·ci·a·to·ry** (ri·nun′sē·ə·tôr′ē, -tō′rē-, -shē-) *adj.*

re·o·pen (rē·ō′pən) *v.t.* & *v.i.* **1** To open again. **2** To begin again; resume.

re·or·gan·i·za·tion (rē′ôr·gən·ə·zā′shən, -ī·zā′-) *n.* **1** The act of reorganizing, or the condition of being reorganized. **2** The legal reconstruction of a corporation, usually after or to avert a failure.

re·or·gan·ize (rē·ôr′gən·īz) *v.t.* & *v.i.* **·ized, ·iz·ing** To organize anew. — **re·or′gan·iz′er** *n.*

re·o·ri·ent (rē·ôr′ē·ənt, -ō′rē-) *adj. Rare* Rising again. [See ORIENT]

rep[1] (rep) *n.* A silk, cotton, rayon, or wool fabric having a distinctive crosswise rib: also spelled *repp.* [<F *reps* <E *ribs*]

rep[2] (rep) *n. Slang* Reputation.

rep[3] (rep) *n. Slang* A representative.

rep[4] (rep) *n. Physics* **1** A unit of absorbed nuclear radiation equivalent to the release of from 83 to 97 ergs per gram of absorbing material. **2** The rad. [<R(OENTGEN) + E(QUIVALENT) + P(HYSICAL)]

re·pair[1] (ri·pâr′) *v.t.* **1** To restore to sound or good condition after damage, injury, decay, etc.; mend. **2** To make amends for (an injury); remedy. **3** To make up, as a loss; compensate for. See synonyms under AMEND, RECRUIT. — *n.* **1** Restoration, as after decay, waste, injury, etc.; reparation. **2** Condition after use or after repairing: in good *repair.* [<OF *reparer* <L *reparare* < *re-* again + *parare* prepare, make ready] — **re·pair′er** *n.*

re·pair[2] (ri·pâr′) *v.i.* **1** To betake oneself; go: to *repair* to the garden. **2** To return. — *n.* **1** The act of repairing, or the place to which one repairs; a haunt. **2** *Scot.* A concourse of people to a certain spot. [<OF *repairer* <LL *repatriare* < *re-* again + *patria* native land]

re·pair·man (ri·pâr′man′, -mən) *n. pl.* **·men** (-men′, -mən) A man whose work is to make repairs.

re·pand (ri·pand′) *adj. Bot.* Having a wavy or uneven outline: said of leaves. [<L *repandus* bent back < *re-* back + *pandus,* pp. of *pandare* bend]

rep·a·ra·ble (rep′ər·ə·bəl) *adj.* Capable of repair or reparation. Also **re·pair·a·ble** (ri·pâr′ə·bəl). [<F *réparable* <L *reparabilis*] — **rep′a·ra·bil′i·ty** *n.* — **rep′a·ra·bly** *adv.*

rep·a·ra·tion (rep′ə·rā′shən) *n.* **1** The act of making amends; atonement; amends; indemnity; also, that which is done by way of amends or satisfaction. **2** The act of repairing or the state of being repaired. **3** *pl.* Repairs; specifically, indemnities paid by defeated countries for acts of war. See synonyms under RESTITUTION. [<L *reparatio,*

-onis a renewal] **—re·par·a·tive** (ri·par′ə·tiv) *adj.*

rep·ar·tee (rep′är·tē′, -ər-) *n.* 1 A witty or quick reply; a sharp rejoinder. 2 Conversation characterized by such replies. 3 Skill or quickness in such wit. See synonyms under ANSWER. [<F *repartie*, pp. of *repartir* depart again, reply < *re-* again + *partir* depart]

re·par·ti·tion (rē′pär·tish′ən) *n.* 1 Distribution; allotment. 2 Redistribution.

re·past (ri·past′, -päst′) *n.* 1 Food taken at a meal; hence, a meal. 2 Food in general; also, mealtime. [<OF *repas* <Med. L *repastum*, orig. pp. of LL *repascere* feed again <L *re-* again + *pascere* feed]

re·pa·ten·cy (ri·pāt′n·sē, -pat′n-) *n.* The reopening of a part or vessel that had been closed. [<RE- + L *patentia*, neut. pl. of *patens, patentis*, ppr. of *patere* be open]

re·pa·tri·ate (rē·pā′trē·āt) *v.t.* & *v.i.* **·at·ed, ·at·ing** To send back or return to his own country, as a soldier interned in a neutral territory; restore to citizenship. **—** *n.* (rē·pā′trē·it) A person who has been repatriated. [<LL *repatriatus*, pp. of *repatriare* <L *re-* again + *patria* native land] **—re·pa′tri·a′tion** *n.*

re·pay (ri·pā′) *v.* **·paid, ·pay·ing** *v.t.* 1 To pay back; refund. 2 To pay back or refund something to. 3 To make compensation or retaliation for; give a reward or inflict a penalty for. **—** *v.i.* 4 To make repayment or requital. See synonyms under REQUITE. [<OF *repaier*] **—re·pay′a·ble** *adj.* **—re·pay′ment** *n.*

re·peal (ri·pēl′) *v.t.* 1 To rescind, as a law; revoke. 2 *Obs.* To summon back, as from exile. See synonyms under ABOLISH, ANNUL, CANCEL. **—** *n.* 1 The act of repealing; revocation; rescission. 2 *Obs.* Recall, as from exile. [<OF *rapeler* recall < *re-* again + *appeler*. See APPEAL.] **—re·peal′a·ble** *adj.* **—re·peal′·er** *n.*

re·peat (ri·pēt′) *v.t.* 1 To say again; reiterate: to *repeat* a question. 2 To recite from memory. 3 To say (what another has just said). 4 To tell, as a secret, to another. 5 To do, make, or experience again. **—** *v.i.* 6 *U.S.* To vote more than once at the same election: an offense punishable by law. **—** *n.* 1 The act of repeating; a repetition. 2 *Music* **a** A sign consisting of dots placed in the spaces at the left hand of a bar, to indicate that the preceding passage is to be repeated. **b** A repeated passage, song, refrain, etc. 3 Anything repeated, as a new supply of goods, or a renewed order for such supply. [<OF *repeter* <L *repetere* do or say again < *re-* again + *petere* seek]

re·peat·ed (ri·pē′tid) *adj.* Occurring or spoken again and again; reiterated. See synonyms under FREQUENT. **—re·peat′ed·ly** *adv.*

re·peat·er (ri·pē′tər) *n.* 1 One who or that which repeats. 2 A timepiece, especially a watch, which will strike again the hour last struck when a spring is pressed. 3 A repeating firearm. 4 An instrument for automatically retransmitting electromagnetic signals: a telegraph *repeater*. 5 *U.S.* One who votes, or attempts to vote, more than once at the same election. 6 One who has been repeatedly imprisoned for criminal offenses.

repeating decimal *Math.* 1 A decimal fraction in which one figure is repeated indefinitely. 2 A circulating decimal.

repeating firearm A gun, rifle, or pistol arranged to deliver several shots without reloading.

re·pêch·age (rə·pesh·äzh′) *n. French* Consolation race; a second heat to afford another chance to those running second best in preliminary heats.

re·pel (ri·pel′) *v.* **·pelled, ·pel·ling** *v.t.* 1 To force or drive back; repulse. 2 To reject; refuse, as a suggestion. 3 To cause to feel distaste or aversion: His manner *repels* me. 4 To refuse to mix with or adhere to: Mercury *repels* iron. 5 To push or keep away, especially with invisible force: Like magnetic poles *repel* each other: opposed to *attract*. **—** *v.i.* 6 To act so as to drive something back or away. 7 To cause distaste or aversion. [<L *repellere* < *re-* back + *pellere* drive] **—re·pel′ler** *n.*

Synonyms: check, oppose, repulse, resist. *Repulse* is stronger and more conclusive than *repel*; one may be *repelled* by the very aspect of the person whose favor he seeks, but is not *repulsed* except by a direct refusal of his suit. See DRIVE. *Antonyms*: accept, admit, encourage, entertain, favor, grant, welcome.

re·pel·lent (ri·pel′ənt) *adj.* 1 Serving, tending, or having power to repel. 2 Waterproof. 3 Repugnant. **—** *n.* 1 A waterproof cloth. 2 A remedial application that tends to repel fluids from a swollen part. 3 A chemical compound intended to be distasteful to insects and other vermin and to keep them at a distance. **—re·pel′len·cy, re·pel′lence** *n.*

re·pent[1] (ri·pent′) *v.i.* 1 To feel remorse or regret, as for something done or undone; be contrite. 2 To change one's mind concerning a past action because of disappointment, failure, etc.: with *of*: He *repented* of his generosity to the old man. 3 *Theol.* To feel such sorrow for one's sins as to reform. **—** *v.t.* 4 To feel remorse or regret for (an action, sin, etc.). 5 To change one's mind concerning (a past action): He *repented* his decision. [<OF *repentir* <L *re-* again + *poenitere* cause to repent < *poena* punishment] **—re·pent′er** *n.*

re·pent[2] (rē′pənt) *adj.* 1 *Bot.* Lying flat and rooting, as certain plants; procumbent. 2 *Zool.* Reptant. [<L *repens, repentis*, ppr. of *repere* creep]

re·pen·tance (ri·pen′təns) *n.* A turning with sorrow from a past course or action; loosely, regret or contrition; also, the condition of being penitent.

Synonyms: compunction, contrition, penitence, regret, remorse, sorrow. *Regret* is *sorrow* for any painful or annoying matter. One is moved with *penitence* for wrongdoing. To speak of *regret* for a fault of our own marks it as slighter than one for which we should express *penitence*. *Repentance* is *sorrow* for sin with self–condemnation, and complete turning from the sin. *Compunction* is a momentary sting of conscience, in view either of a past or of a contemplated act. *Contrition* is a subduing *sorrow* for sin, as against the divine holiness and love. *Remorse* is, as its derivation indicates, a biting or gnawing back of guilt upon the heart. *Antonyms*: approval, comfort, complacency, content, hardness, impenitence, obduracy, obstinacy, recusancy, stubbornness.

re·pen·tant (ri·pen′tənt) *adj.* Showing, experiencing, or characterized by repentance. [<OF] **—re·pen′tant·ly** *adv.*

re·peo·ple (rē·pē′pəl) *v.t.* **·pled, ·pling** 1 To people anew. 2 To provide again with animals; restock.

re·per·cus·sion (rē′pər·kush′ən) *n.* 1 The act of driving or throwing back, or the state of being driven back; repulse; also, echo; reverberation. 2 A stroke or blow given in return; recoil after impact; hence, the indirect result of something; aftereffect: the *repercussions* of the peace treaty. 3 *Music* The repetition of the subject and answer in a fugue, or the frequent reiteration in a composition of a tone, note, or chord; also, a tone or chord often repeated. 4 *Med.* The motion produced on a fetus by the process of ballottement. [<L *repercussio, -onis* < *repercussus*, pp. of *repercutere* rebound < *re-* again + *percutere* strike. See PERCUSS.]

re·per·cus·sive (rē′pər·kus′iv) *adj.* Causing, of the nature of, or produced by repercussion; reverberated.

rep·er·toire (rep′ər·twär, -twôr) *n.* A list of songs, plays, operas, or the like, that a person or company is prepared to perform; also, such pieces collectively. [<F <LL *repertorium*. See REPERTORY.]

rep·er·to·ry (rep′ər·tôr′ē, -tō′rē) *n. pl.* **·ries** 1 A place where things are gathered together, or the things so gathered; a repository; collection. 2 Repertoire. [<LL *repertorium* inventory <L < *repertus*, pp. of *reperire* find, discover < *re-* again + *parire* produce]

rep·e·tend (rep′ə·tend, rep′ə·tend′) *n.* 1 *Math.* That part of a circulating decimal which is repeated indefinitely: often indicated by dots over the first and last digits of the group repeated. 2 Something repeated or to be repeated, as a refrain. [<L *repentendus* to be repeated, gerundive of *repetere*. See REPEAT.]

rep·e·ti·tion (rep′ə·tish′ən) *n.* 1 The act of repeating; the doing, making, or saying of something again; recital from memory. 2 *Music* The singing or playing of the same note, chord, or passage over again. 3 That which is repeated; a copy. [<F *répetition*]

rep·e·ti·tious (rep′ə·tish′əs) *adj.* Characterized by or containing useless or tedious repetition. **—rep′e·ti′tious·ly** *adv.* **—rep′e·ti′tious·ness** *n.*

re·pet·i·tive (ri·pet′ə·tiv) *adj.* Marked by repetition; recurrent.

re·pine (ri·pīn′) *v.i.* **·pined, ·pin·ing** To be discontented or fretful; complain; murmur. See synonyms under COMPLAIN. [<RE- + PINE[2]] **—re·pin′er** *n.* **—re·pin′ing** *n.*

re·place (ri·plās′) *v.t.* **·placed, ·plac·ing** 1 To put back in place. 2 To take or fill the place of; supersede. 3 To refund; repay. **—re·place′a·ble** *adj.* **—re·plac′er** *n.*

re·place·ment (ri·plās′mənt) *n.* 1 The act of replacing; also, that which takes the place of anything discarded or worn out. 2 *Mineral.* The formation of a new crystal face which obliterates an edge or angle. 3 A soldier available for assignment to fill a vacancy or a quota. 4 The act of putting a thing back in place. 5 *Chem.* A substitution. 6 A substitute.

re·plead·er (ri·plē′dər) *n. Law* 1 An order of court directing the parties to file new pleadings in order to present a better issue for trial. 2 The right of pleading again. [<RE- + obs. *pleader* a pleading in court]

re·plen·ish (ri·plen′ish) *v.t.* 1 To fill again, as something that has been wholly or partially emptied. 2 To bring back to fullness or completeness, as diminished supplies. 3 To repeople. See synonyms under RECRUIT. [<OF *repleniss-*, stem of *replenir* < *re-* again + L *plenus* full] **—re·plen′ish·er** *n.* **—re·plen′·ish·ment** *n.*

re·plete (ri·plēt′) *adj.* 1 Full to the uttermost. 2 Gorged with food or drink; sated. 3 Abundantly supplied or stocked; abounding. [<OF *replet* <L *repletus*, pp. of *replere* fill again < *re-* again + *plere* fill]

re·ple·tion (ri·plē′shən) *n.* 1 The state of complete or excessive fullness; surfeit. 2 The satisfaction of a want or desire. 3 *Med.* Plethora.

re·plev·in (ri·plev′in) *Law n.* 1 An action to regain possession of personal property unlawfully retained, on giving security to try the title and respond to the judgment; recovery of property by such action. 2 The judicial writ or process by which such proceedings are instituted. **—** *v.t.* To replevy. [<AF *replevine* <OF *replevir* warrant, pledge < *re-* back + *plevir* pledge <Gmc.]

re·plev·y (ri·plev′ē) *Law v.t.* **·plev·ied, ·plev·y·ing** 1 To recover possession of (chattels) by proceedings in replevin. 2 To admit to bail or give bail for. **—** *n.* Replevin. [<OF *replevir*. See REPLEVIN.] **—re·plev′i·a·ble, re·plev′is·a·ble** *adj.*

rep·li·ca (rep′lə·kə) *n.* 1 A duplicate, as of a picture, executed by the original artist. 2 Any close copy or reproduction. See synonyms under DUPLICATE, MODEL. [<Ital. <L *replicare* reply, answer to. See REPLY.]

rep·li·cate (rep′lə·kit) *adj.* Folded backward, as the upper part of a leaf on the lower, or the wing of an insect. Also **rep′li·cat′ed** (-kā′·tid). **—** *v.t.* (-kāt) **·cat·ed, ·cat·ing** 1 To fold over. 2 To make a replica of. 3 To answer; reply. [<L *replicatus*, pp. of *replicare* answer. See REPLY.]

rep·li·ca·tion (rep′lə·kā′shən) *n.* 1 A reply. 2 *Law* A plaintiff's reply to a defendant's plea or answer. 3 A repetition or copy. 4 A methodical or systematic doubling over of a surface. 5 The sending back again of sound; reverberation; echo. [<OF] **—rep′li·ca′tive** *adj.*

re·ply (ri·plī′) *v.* **·plied, ·ply·ing** *v.i.* 1 To give an answer, orally or in writing. 2 To respond by some act, gesture, etc.: He *replied* with a blow. 3 To echo. 4 *Law* To file a pleading in answer to the statement of the defense. **—** *v.t.* 5 To say in answer: often with a clause as object: She *replied* that she would do it. **—** *n. pl.* **·plies** Something said, written, or done by way of answer; a response; rejoinder. See synonyms under ANSWER. [<OF *replier* bend back <L *replicare* fold back, answer to, make a reply < *re-* back + *plicare* fold] **—re·pli′er** *n.*

ré·pon·dez s'il vous plaît (rā·pôn·dā′ sēl vōō ple′) *French* Reply, if you please: used on formal invitations: abbr. *R.S.V.P.*

re·port (ri·pôrt′, -pōrt′) *v.t.* 1 To make or give an account of, especially formally: to *report* the minutes of a meeting, or an event for a newspaper. 2 To relate, as information obtained by investigation: Please *report* your findings. **3** To bear back or repeat to another, as an answer. **4** To complain about, especially to a superior: I'll *report* you to the manager. **5** To state the result of consideration concerning: The committee *reported* the bill. —*v.i.* **6** To make a report. **7** To act as a reporter. **8** To present oneself, as for duty. See synonyms under ANNOUNCE. —*n.* **1** That which is reported; an announcement, statement, or account; the formal statement of the result of an investigation: a medical *report.* **2** Common talk; rumor; hence, fame, reputation, or character: good *report; reports* grossly untrue. **3** A record with more or less detail of the transactions of a deliberative body. **4** An account of any occurrence prepared for publication through the press. **5** *Law Usually pl.* A published narration (usually official) of a case or series of cases judicially considered: the Supreme Court *reports.* **6** An explosive sound: the *report* of a gun. [<OF *reporter* carry back <L *reportare* < *re-* back + *portare* carry] —**re·port'a·ble** *adj.*

Synonyms (noun): account, description, narration, narrative, recital, record, rehearsal, relation, rumor, statement, story, tale. *Account,* primarily a commercial summary, carries a similar meaning in the derived sense; an *account* of an occurrence is circumstantial, adequate, complete, and unembellished; we speak of a clear, a full, or a partial *account*; a glowing *account* is still supposed to be circumstantially as well as substantially correct. A *statement* is definite, confined to essentials and properly to matters within the personal knowledge of the one who states them. A *narrative* is a somewhat extended and embellished *account* of events in order of time, ordinarily with a view to please or entertain. A *description* gives especial scope to the pictorial element. A *report* is supposed or intended to bring back the past, and may be concise and formal or highly descriptive and dramatic. Compare ALLEGORY, ANECDOTE, HISTORY, NEWS, RECORD.

report card *U.S.* A periodic statement of a pupil's scholastic record, which is presented to the parents or guardian.

re·port·ed·ly (ri·pôr'tid·lē, -pōr'-) *adv.* According to report.

re·port·er (ri·pôr'tər, -pōr'-) *n.* **1** A bearer of news; specifically, one employed by a newspaper to gather and report news for publication. **2** One who edits reports of important cases in court for official publication. [<OF *reporteur*] —**rep·or·to·ri·al** (rep'ər·tôr'ē·əl, -tō'rē-) *adj.*

re·pose[1] (ri·pōz') *n.* **1** The act of taking rest, or the state of being at rest; especially, rest in a recumbent posture. **2** Freedom from excitement or anxiety; composure; hence, ease of manner; graceful and dignified calmness. **3** That which conduces to rest or calm. See synonyms under REST. —*v.* **·posed, ·pos·ing** *v.t.* **1** To lay or place in a position of rest: to *repose* oneself on a bed. —*v.i.* **2** To lie at rest. **3** To rely; depend: with *on, upon,* or *in.* See synonyms under REST. [<F *reposer* <LL *repausare* < *re-* again + *pausare* pause] —**re·pos'al** *n.* —**re·pos'er** *n.*

re·pose[2] (ri·pōz') *v.t.* **·posed, ·pos·ing** **1** To place, as confidence or hope: with *in.* **2** *Rare* To deposit. [<L *repositus,* pp. of *reponere* put back, on analogy with *depose, oppose,* etc.] —**re·pos'al** *n.*

re·pose·ful (ri·pōz'fəl) *adj.* Full of repose; restful.

re·pos·it (ri·poz'it) *v.t.* To put in some secure and proper place; deposit. [<L *repositus.* See REPOSE.] —**re·po·si·tion** (rē'pə·zish'ən, rep'ə-) *n.*

re·pos·i·to·ry (ri·poz'ə·tôr'ē, -tō'rē) *n. pl.* **·ries** **1** A place in which goods are or may be stored; a depository. **2** A person to whom a secret is entrusted. **3** A building used as a place of exhibition and sale. **4** A burial vault. **5** A sepulcher (def. 2). [<L *repositorium* < *repositus.* See REPOSE.]

re·pos·sess (rē'pə·zes') *v.t.* **1** To have possession of again; regain possession of. **2** To give back possession or ownership to. **3** *Scot.* To reinstate: with *in.* See synonyms under RECOVER. —**re'pos·ses'sion** (-zesh'ən) *n.*

re·pous·sé (rə·pōō·sā') *adj.* Formed in relief, as a design in metal, or adorned with such designs. [<F, lit., thrust back <L *repulsus.* See REPULSE.]

repp (rep) See REP[1].

Rep·plier (rep'lir), **Agnes,** 1855–1950, U.S. essayist.

rep·re·hend (rep'ri·hend') *v.t.* To criticize sharply; find fault with; blame. See synonyms under BLAME, REPROVE. [<L *reprehendere* < *re-* back + *prehendere* hold]

rep·re·hen·si·ble (rep'ri·hen'sə·bəl) *adj.* Deserving blame or censure. —**rep're·hen'si·bil'·i·ty, rep're·hen'si·ble·ness** *n.* —**rep're·hen'si·bly** *adv.*

rep·re·hen·sion (rep'ri·hen'shən) *n.* A finding fault; expression of blame; rebuke. See synonyms under ANIMADVERSION, REPROOF. —**rep'·re·hen'sive** *adj.* —**rep're·hen'sive·ly** *adv.*

rep·re·hen·so·ry (rep'ri·hen'sər·ē) *adj.* Censorious.

rep·re·sent (rep'ri·zent') *v.t.* **1** To serve as the symbol, expression, or designation of; symbolize: The letters of the alphabet *represent* the sounds of speech. **2** To express or symbolize in this manner: to *represent* royal power with a scepter. **3** To set forth a likeness or image of; depict; portray, as in painting or sculpture. **4 a** To produce on the stage, as an opera. **b** To act the part of; impersonate, as a character in a play. **5** To serve as or be the delegate, agent, etc., of: He *represents* the State of Maine. **6** To describe as being of a specified character or condition: They *represented* him as a genius. **7** To set forth in words; state; explain: He *represented* the circumstances of his case. **8** To bring before the mind; present clearly. **9** To serve as an example, specimen, type, etc., of; typify: His use of words *represents* an outmoded school of writing. See synonyms under IMITATE. [<OF *representer* <L *repraesentare* < *re-* again + *praesentare.* See PRESENT[2].] —**rep're·sent'a·ble** *adj.* —**rep're·sent'a·bil'i·ty** *n.*

re-present (rē'pri·zent') *v.t.* To present again. —**re'-pre'sen·ta'tion** *n.*

rep·re·sen·ta·tion (rep'ri·zen·tā'shən) *n.* **1** The act of representing, or the state of being represented. **2** That which represents; a likeness; model; picture; statue; statement; description; also, a dramatic performance. **3** The right of acting authoritatively for others, especially in a legislative body; also, the system of electing delegates to act for a constituency. **4** Representatives collectively. **5** The stage or process of mental conservation that consists in the presenting to itself by the mind of objects previously known. **6** *Law* The authorized acting for or in the stead of another in regard to that other's affairs. **7** A setting forth by statement or account; specifically, an argument against some object or proposal. See synonyms under IMAGE, MODEL, PICTURE. [<OF]

rep·re·sen·ta·tive (rep'ri·zen'tə·tiv) *adj.* **1** Typifying or typical of a group or class. **2** Acting, having the power or authority to act, or qualified to act, as an agent. **3** Made up of representatives. **4** Based on or pertaining to the political principle of representation. **5** Presenting, portraying, or representing, or capable of so doing. **6** Having to do with cognition of a memory image: distinguished from *presentative.* —*n.* **1** One who or that which is fit to stand as a type; a typical instance. **2** One who is a qualified agent of any kind. **3** A member of a deliberative or legislative body chosen by vote of the people; specifically, in the United States, a member of the lower house of Congress or of a State legislature. See synonyms under DELEGATE. —**rep're·sen'ta·tive·ly** *adv.* —**rep'·re·sen'ta·tive·ness** *n.*

re·press (ri·pres') *v.t.* **1** To keep under restraint or control; curb. **2** To put down; quell, as a rebellion. **3** *Psychoanal.* To effect the repression of, as fears, impulses, etc. [<L *repressus,* pp. of *reprimere* < *re-* back + *premere* press] —**re·press'er** or **re·pres'sor** *n.* —**re·press'i·ble** *adj.*

Synonyms: bridle, chasten, check, crush, curb, overcome, overpower, quiet, rein, restrain, stay, still, subdue, suppress. See LIMIT, RESTRAIN, SUBDUE. *Antonyms*: agitate, animate, arouse, awaken, encourage, excite, incite, inspirit, instigate, kindle, provoke, rouse, stimulate.

re·pressed (ri·prest') *adj.* Suppressed.

re·pres·sion (ri·presh'ən) *n.* **1** The act of repressing, or the condition of being repressed. **2** That which holds in check; a restraint. **3** *Psychoanal.* The exclusion from consciousness of painful, unpleasant, or unacceptable psychic material, as memories, desires, and impulses, which are thus compelled to manifest themselves through the unconcious.

re·pres·sive (ri·pres'iv) *adj.* **1** Tending to repress. **2** Capable of repressing. —**re·pres'·sive·ly** *adv.* —**re·pres'sive·ness** *n.*

re·prieve (ri·prēv') *v.t.* **·prieved, ·priev·ing** **1** To suspend temporarily the execution of a sentence upon. **2** To relieve for a time from suffering, danger, or trouble. **3** To postpone or delay, as a danger. —*n.* **1** The temporary suspension of a sentence, or the instrument officially ordering such a suspension. **2** Temporary relief or cessation of pain or ill; respite. See synonyms under RESPITE. [< earlier *repry* <F *repris,* pp. of *reprendre* take back; infl. in form by ME *repreven* <OF *reprover* reprove]

rep·ri·mand (rep'rə·mand, -mänd) *v.t.* To reprove sharply or formally. See synonyms under ADMONISH, REPROVE. —*n.* Severe reproof or formal censure, public or private. See synonyms under REPROOF. [<F *réprimande* reproof <L *reprimenda,* fem. of *reprimendus* to be repressed, gerundive of *reprimere.* See REPRESS.]

re·print (rē'print') *n.* An edition of a printed work that is a verbatim copy of the original; specifically, a copy of matter already printed, as in another country. —*v.t.* (rē·print') To print a new edition or copy of; print anew or again. —**re·print'er** *n.*

re·pri·sal (ri·prī'zəl) *n.* **1** Forcible seizure of anything from an enemy by way of retaliation or indemnity. **2** Anything taken from an enemy as indemnification or in retaliation; also, any act or infliction by way of retaliation; specifically, the infliction of suffering or death on a prisoner of war in retaliation for acts of inhumanity inflicted by him. **3** Any act of retaliation. **4** *Obs.* A prize seized or gained. [<OF *reprisaille* < *repris,* pp. of *reprendre* take back <L *reprehendere.* See REPREHEND.]

re·prise (ri·prīz' *for def. 1*; rə·prēz', -prīz' *for def. 2*) *n.* **1** *pl. Brit. Law* Deductions and payments (as for annuities) out of lands: a manor's yearly value over and above *reprises.* **2** *Music* A repeated phrase; specifically, the repetition of or return to the subject after an intermediate movement. [<OF fem. of *repris.* See REPRISAL.]

re·proach (ri·prōch') *v.t.* **1** To charge with or blame for something wrong; rebuke; censure; upbraid. **2** To bring discredit and disgrace upon; to disgrace. See synonyms under ABUSE, BLAME, REPROVE, REVILE. —*n.* **1** The act of reproaching, or the words of one who reproaches; censure; reproof; rebuke. **2** A cause of blame or disgrace; hence, disgrace or discredit. See synonyms under BLEMISH, REPROOF, SCANDAL. [<F *reprocher.* Origin uncertain.] —**re·proach'a·ble** *adj.* —**re·proach'a·ble·ness** *n.* —**re·proach'a·bly** *adv.* —**re·proach'er** *n.* —**re·proach'less** *adj.*

re·proach·ful (ri·prōch'fəl) *adj.* **1** Containing or full of reproach; expressing reproach. **2** *Obs.* Reproachable. —**re·proach'ful·ly** *adv.* —**re·proach'ful·ness** *n.*

rep·ro·bate (rep'rə·bāt) *adj.* **1** Abandoned in sin; lost to all sense of duty; utterly depraved; profligate. **2** Abandoned to punishment; condemned. **3** *Obs.* Not enduring proof or trial; inferior or base. —*n.* One lost to all sense of duty or decency; one abandoned to depravity or doom. —*v.t.* **·bat·ed, ·bat·ing** **1** To disapprove of heartily; condemn. **2** *Theol.* To abandon, condemn, or foreordain to damnation. See synonyms under BLAME, CONDEMN. [<LL *reprobatus,* pp. of *reprobare.* See REPROVE.]

rep·ro·ba·tion (rep'rə·bā'shən) *n.* **1** The act of reprobating, or the condition of being reprobated; censure. **2** *Theol.* Rejection or condemnation by God's purpose. See synonyms under OATH.

rep·ro·ba·tive (rep′rə·bā′tiv) *adj.* Of, pertaining to, or expressing reprobation. — **rep′ro·ba′tive·ly** *adv.*

re·proc·ess (rē·pros′es) *v.t.* To process again.

reprocessed wool Wool fibers previously woven or knitted but never used by a consumer, unraveled and spun and rewoven into fabric.

re·pro·duce (rē′prə·do͞os′, -dyo͞os′) *v.* **·duced, ·duc·ing** *v.t.* **1** To make a copy, image, or reproduction of. **2** *Biol.* **a** To give rise to (offspring) by sexual or asexual generation. **b** To replace (a lost part or organ) by regeneration. **3** To cause the reproduction of (plant life, etc.). **4** To produce again; bring forward or exhibit anew. **5** To bring into existence again; recreate; revive. **6** To recall to the mind; visualize again; re-create mentally. — *v.i.* **7** To produce offspring. **8** To undergo copying, reproduction, etc. — **re′pro·duc′i·ble** *adj.*

re·pro·duc·er (rē′prə·do͞o′sər, -dyo͞o′-) *n.* **1** One who or that which reproduces. **2** A diaphragm used for the reproduction of sounds in a phonograph, etc.

re·pro·duc·tion (rē′prə·duk′shən) *n.* **1** The act or power of reproducing. **2** *Biol.* The process by which an animal or plant gives rise to another of its kind; generation. **3** *Psychol.* The process of the memory by which objects that have previously been known are brought back into consciousness. **4** That which is reproduced, as a revival in drama or a copy in art. See synonyms under DUPLICATE.

re·pro·duc·tive (rē′prə·duk′tiv) *adj.* Pertaining to, employed in, or tending to reproduction. — **re′pro·duc′tive·ly** *adv.* — **re′pro·duc′tive·ness** *n.*

re·proof (ri·pro͞of′) *n.* **1** The act of reproving; rebuke; blame; censure. **2** *Obs.* Ignominy; reproach. Also **re·prov·al** (ri·pro͞o′vəl). [<OF *reprove* < *reprover.* See REPROVE.]

Synonyms: admonition, animadversion, blame, censure, check, chiding, comment, condemnation, criticism, denunciation, disapproval, objurgation, rebuke, reflection, reprehension, reprimand, reproach, reproval, upbraiding. *Blame, censure,* and *disapproval* may either be felt or uttered; *comment, criticism, rebuke, reflection, reprehension,* and *reproof* are always expressed. The same is true of *admonition* and *animadversion. Comment* and *criticism* may be favorable as well as censorious; they imply no superiority or authority on the part of him who utters them; nor do *reflection* or *reprehension,* which are simply turning the mind back upon what is disapproved. *Reprehension* is supposed to be calm and just, and with good intent; *reflection* is often from mere ill feeling, and is likely to be more personal and less impartial than *reprehension. Rebuke,* literally a stopping of the mouth, is administered to a forward or hasty person; *reproof* is administered to one intentionally or deliberately wrong; both words imply authority in the reprover, and direct expression of *disapproval* to the face of the person *rebuked* or *reproved. Reprimand* is official *censure* formally administered by a superior to one under his command. *Rebuke* may be given at the outset, or in the midst of an action; *reflection, reprehension, reproof,* always follow the act; *admonition* is anticipatory, and meant to be preventive. *Check* is allied to *rebuke,* and given before or during action; *chiding* is nearer to *reproof,* but with more personal bitterness and less authority. Compare CONDEMN, REPROVE. *Antonyms*: applause, approbation, approval, commendation, encomium, eulogy, panegyric, praise.

re·prove (ri·pro͞ov′) *v.t.* **·proved, ·prov·ing** **1** To censure, as for a fault; rebuke. **2** To express disapproval of (an act). **3** *Obs.* To convince; convict. [<OF *reprover* <LL *reprobare* < *re-* again + *probare* test < *probus* upright] — **re·prov′a·ble** *adj.* — **re·prov′er** *n.* — **re·prov′ing·ly** *adv.*

Synonyms: admonish, blame, censure, chasten, check, chide, condemn, rebuke, reprehend, reprimand, reproach, upbraid. To *censure* is to pronounce an adverse judgment that may or may not be expressed to the person *censured*; to *rebuke* is to *reprove* sharply, and often abruptly; to *blame* is a familiar word signifying to pass *censure* upon, make answerable, as for a fault. To *reproach* is to *censure* openly and vehemently, and with intense personal feeling as of grief or anger; as, to *reproach* one for ingratitude; *reproach* knows no distinction of rank or character; a subject may *reproach* a king or a criminal a judge. Compare REPROOF. See ADMONISH, BLAME, CONDEMN. *Antonyms*: see synonyms for PRAISE.

rep·tant (rep′tənt) *adj. Zool.* Creeping; crawling: also *repent.* [<L *reptans, -antis,* ppr. of *reptare,* intens. of *repere* creep]

rep·tile (rep′til, -tīl) *n.* **1** A cold-blooded, air-breathing vertebrate, especially one with scales, as a lizard, snake, or crocodile; a reptilian; any member of the class *Reptilia.* **2** A groveling, abject person; one morally base or odious. — *adj.* **1** Crawling on the belly; creeping; reptant. **2** Groveling morally; sly and base; treacherous; venomous. **3** Of, pertaining to, or resembling a reptile. [<LL, neut. sing. of *reptilis* crawling < *reptus,* pp. of *repere* creep]

rep·til·i·an (rep·til′ē·ən) *adj.* **1** Of or pertaining to a class *(Reptilia)* of cold-blooded, air-breathing vertebrates, the reptiles, having fully ossified skeletons and bodies usually covered with horny plates or scales. In addition to the limbless snakes, the class includes crocodiles, alligators, lizards, and turtles. **2** Malicious; base; mean. — *n.* One of the *Reptilia*; any reptile.

re·pub·lic (ri·pub′lik) *n.* **1** A state in which the sovereignty resides in the people or a certain portion of the people, and the legislative and administrative powers are lodged in officers elected by and representing the people; a representative democracy: applied to almost every form of government except kingdoms, empires, and dictatorships. **2** A community of persons working freely in or devoted to the same cause: the *republic* of letters. — **The Republic** **1** The United States. **2** Plato's dialog on government. [<F *république* <L *respublica* commonwealth < *res* thing + *publica,* fem. of *publicus* public]

re·pub·li·can (ri·pub′li·kən) *adj.* Pertaining to, of the nature of, or suitable for a republic; agreeable to the nature of a republic; also, of or pertaining to any party supporting republican government. — *n.* One who advocates or upholds a republican form of government or belongs to a party upholding republican government; one who believes in equality and liberty.

Re·pub·li·can (ri·pub′li·kən) *adj.* Pertaining to or belonging to the Republican party of the United States, or to any political group which calls itself by this name: the *Republican* parties of Spain or France. — *n.* A member of the Republican party. — **black Republican** Formerly, a member of the Republican party: derisively so called in allusion to their opposition to Negro slavery.

Republican calendar See under CALENDAR.

re·pub·li·can·ism (ri·pub′li·kən·iz′əm) *n.* **1** The theory or principles of republican government. **2** A liking for republican principles.

Re·pub·li·can·ism (ri·pub′li·kən·iz′əm) *n.* The policy and principles of the Republican party of the United States.

re·pub·li·can·ize (ri·pub′li·kən·īz′) *v.t.* **·ized, ·iz·ing** To make republican in spirit or character. — **re·pub′li·can·i·za′tion** *n.*

Republican party **1** One of the two major political parties of the United States, founded in 1854 in opposition to the extension of slavery. **2** The political party founded by Thomas Jefferson in 1792: full name, *Democratic-Republican party.* One of its several factions became, in 1828, the present Democratic party. **3** One of various political parties of foreign countries, devoted to the overthrow of monarchy or the establishment or extension of democratic ideals.

Republican River A river in Colorado, Nebraska, and Kansas, flowing 445 miles east to the Kansas River.

re·pub·li·ca·tion (rē′pub·lə·kā′shən) *n.* The act of republishing, or that which is republished.

re·pub·lish (rē·pub′lish) *v.t.* **1** To publish again. **2** *Law* To revive, as a canceled will, by executing anew. — **re·pub′lish·er** *n.*

re·pu·di·ate (ri·pyo͞o′dē·āt) *v.t.* **·at·ed, ·at·ing** **1** To refuse to accept as valid, true, or authorized; reject; condemn. **2** To refuse to acknowledge or pay. **3** To cast off; disown, as a son. **4** *Obs.* To divorce; put away (a wife). See synonyms under ABANDON, RECANT, RENOUNCE. [<L *repudiatus,* pp. of *repudiare* divorce < *repudium* divorce, separation, ? < *re-* back + *pudere* feel shame] — **re·pu′di·a′tive** *adj.* — **re·pu′di·a′tor** *n.*

re·pu·di·a·tion (ri·pyo͞o′dē·ā′shən) *n.* **1** The act of repudiating. **2** The state of being repudiated. **3** The rejection of the whole or a part of a contract, debt, or obligation, as by a government.

re·pugn (ri·pyo͞on′) *v.t.* & *v.i. Obs.* To oppose; resist. [<OF *repugner* <L *repugnare* < *re-* back + *pugnare* fight]

re·pug·nance (ri·pug′nəns) *n.* **1** A feeling of aversion and resistance. **2** *Logic* The relation of contradictories; inconsistency. **3** *Obs.* Opposition. Also **re·pug′nan·cy.** See synonyms under ANTIPATHY, HATRED.

re·pug·nant (ri·pug′nənt) *adj.* **1** Offensive to taste or feeling; exciting aversion or repulsion. **2** Being inconsistent or opposed; antagonistic. **3** *Law* Contrary to or in conflict with something else in the same or in another document or statute. **4** Hostile; rebellious; resisting. See synonyms under INCONGRUOUS, INIMICAL. [<OF <L *repugnans, -antis,* ppr. of *repugnare.* See REPUGN.]

re·pulse (ri·puls′) *v.t.* **·pulsed, ·puls·ing** **1** To drive back; repel, as an attacking force. **2** To repel by coldness, discourtesy, etc.; reject; rebuff. See synonyms under DRIVE, REPEL. — *n.* **1** The act of repulsing, or the state of being repulsed. **2** Rejection; refusal. [<L *repulsus,* pp. of *repellere.* See REPEL.] — **re·puls′er** *n.*

re·pul·sion (ri·pul′shən) *n.* **1** The act of repelling or repulsing, or the state of being repelled or repulsed. **2** Aversion; repugnance. **3** *Physics* The mutual action of two bodies which tends to drive them apart: opposed to *attraction.*

re·pul·sive (ri·pul′siv) *adj.* **1** Exciting such feelings, as of dislike, disgust, or horror, that one is repelled; grossly offensive; causing aversion. **2** Such as to forbid approach or familiarity; forbidding. **3** Acting by repulsion: *repulsive* forces. — **re·pul′sive·ly** *adv.* — **re·pul′sive·ness** *n.*

rep·u·ta·ble (rep′yə·tə·bəl) *adj.* **1** Having a good reputation; estimable; honorable. **2** Consistent with honorable standing; complying with the usage of the best writers and speakers. — **rep′u·ta·bil′i·ty** *n.* — **rep′u·ta·bly** *adv.*

rep·u·ta·tion (rep′yə·tā′shən) *n.* **1** The general estimation in which a person or thing is held by others, especially by a community; repute, either good or bad. **2** The state of being in high regard or esteem; good repute: to ruin one's *reputation.* **3** A particular credit or character ascribed to a person or thing: usually with *for*: a *reputation* for honesty. See synonyms under CHARACTER, FAME. [<L *reputatio, -onis* < *reputatus,* pp. of *reputare* be reputed. See REPUTE.]

re·pute (ri·pyo͞ot′) *v.t.* **·put·ed, ·put·ing** To regard or consider to be as specified; esteem: usually in the passive: They are *reputed* to be an intelligent people. — *n.* **1** Reputation, good or bad. **2** Public opinion; general report. [<L *reputare* reckon, be reputed < *re-* again + *putare* think, count]

re·put·ed (ri·pyo͞o′tid) *adj.* Generally thought or supposed; having a specified reputation. — **re·put′ed·ly** *adv.*

re·quest (ri·kwest′) *v.t.* **1** To express a desire for, especially politely; ask for; solicit. **2** To address a request to; ask: to *request* a person to do one a favor. See synonyms under ASK, DEMAND, PRAY. — *n.* **1** The act of requesting; entreaty; petition. **2** That which is asked for. **3** The state of being so esteemed as to be in demand; demand: in *request.* See synonyms under PETITION, PRAYER. — *adj.* Having been asked for; in response to a request: a *request* program. [<OF *requeste* <Med. L *requisita,* orig. fem. of L *requisitus,* pp. of *requirere* seek again. See REQUIRE.]

re·qui·em (rē′kwē·əm, rek′wē-) *n.* **1** Any musical hymn, composition, or service for the dead. **2** *Often cap. Eccl.* In the Roman Catholic Church, a solemn mass sung for the repose of the souls of the dead, the **Requiem mass.** **3** *Often cap.* A musical setting for such a mass; also a similar piece of music using different words. [<L *Requiem (aeternam dona eis, Domine)* rest (eternal give unto them, O Lord), the opening words of the introit of this mass]

req·ui·es·cat (rek′wē·es′kat) *n.* A prayer for the repose of a departed soul: the first word of the Latin petition **requiescat in pa·ce** (in pā′sē), may he rest in peace. Abbr. *R.I.P.* [<L]

re·quire (ri·kwīr′) *v.* **·quired, ·quir·ing** *v.t.* **1** To have need of; find necessary. **2** To demand authoritatively; insist upon: to *require* absolute silence. **3** To command; order: He *requires* us to be punctual. —*v.i.* **4** To make demand or request. See synonyms under ASK, DEMAND, DICTATE, MAKE. [<L *requirere* seek again, be in want of < *re-* again + *quaerere* ask, seek] —**re·quir′a·ble** *adj.* —**re·quir′er** *n.*

re·quire·ment (ri·kwīr′mənt) *n.* **1** That which is required; a requisite. **2** The act of requiring, or that which requires; a demand. See synonyms under NECESSITY, ORDER.

req·ui·site (rek′wə·zit) *adj.* Required by the nature of things or by circumstances; indispensable. See synonyms under NECESSARY. —*n.* That which cannot be dispensed with; a necessity; requirement. See synonyms under NECESSITY. [<L *requisitus,* pp. of *requirere.* See REQUEST.] —**req′ui·site·ly** *adv.* —**req′ui·site·ness** *n.*

req·ui·si·tion (rek′wə·zish′ən) *n.* **1** A formal request, summons, or demand, as by a government. **2** A necessity or requirement. **3** The state of being required. **4** A demand for the surrender of a fugitive from justice made by the governing official of one state or country upon another. —*v.t.* To make a requisition for or upon; demand or take upon requisition. [<L *requisitio, -onis* < *requisitus,* pp. of *requirere.* See REQUIRE.]

re·qui·tal (ri·kwīt′l) *n.* **1** The act of requiting. **2** That which requites; adequate return for good or ill; in the favorable sense, reward or compensation; in the unfavorable sense, retaliation. See synonyms under RECOMPENSE, REVENGE. [<REQUITE]

re·quite (ri·kwīt′) *v.t.* **·quit·ed, ·quit·ing** **1** To make equivalent return for, as kindness, service, or injury; make up for. **2** To make return to; compensate or repay in kind: Does she *requite* me for my love? **3** To give or do in return. [<RE- + *quite,* obs. var. of QUIT] —**re·quit′a·ble** *adj.* —**re·quit′er** *n.*

Synonyms: avenge, compensate, pay, punish, quit, reciprocate, recompense, remunerate, repay, retaliate, return, revenge, reward, satisfy. *Requite* is often used in the more general sense of *recompense* or *repay,* but always with the suggestion, at least, of the original idea of full equivalent. To *repay* or to *retaliate,* to *punish* or to *reward,* may be to make some return very inadequate to the benefit or injury received or the right or wrong done; but to *requite* is to make such return as to *quit* oneself of all obligation of favor or hostility, of punishment or reward. See PAY. *Antonyms*: absolve, acquit, excuse, forget, forgive, neglect, overlook, pardon, slight.

re–ra·di·a·tion (rē′rā′dē·ā′shən) *n.* **1** *Telecom.* The emission of one or more radio frequencies from the antenna of a radio receiver through improper oscillation of the tubes, with resulting confusion of signals. **2** *Physics* Secondary emission.

rere·dos (rir′dos) *n.* **1** An ornamental screen behind an altar. **2** The back of an open fire hearth; a fireback. **3** In old armor, a backplate. [<AF *areredos* <OF *arere* at the back (<L *ad* to + *retro* behind) + *dos* back <L *dorsum*]

rere·mouse (rir′mous) *n.* *pl.* **·mice** (-mīs) *Brit. Dial.* A bat: also spelled *rearmouse.* [OE *hreremūs*]

re·run (rē′run′) *n.* **1** A running over again or a second time. **2** The presenting of a motion picture after its original presentation. —*v.t.* (rē·run′) **·ran, ·run·ning** To run again.

Re·sa·ca (ri·sä′kə) A town of NW Georgia; scene of a Civil War battle, 1864.

Re·sa·ca de la Pal·ma (rā·sä′kä dā lä päl′mä) A locality in southern Texas north of Brownsville; scene of an American victory over the Mexicans, 1846.

res ad·ju·di·ca·ta (rēz a·jōō′də·kā′tə) See RES JUDICATA.

re·scind (ri·sind′) *v.t.* To make void, as an act; abrogate; repeal: to *rescind* a resolution. See synonyms under ANNUL, CANCEL. [<L *rescindere* < *re-* back + *scindere* cut] —**re·scind′a·ble** *adj.* —**re·scind′er** *n.*

re·scis·si·ble (ri·sis′ə·bəl) *adj.* Capable of being rescinded.

re·scis·sion (ri·sizh′ən) *n.* The act of rescinding or abrogating.

re·scis·so·ry (ri·sis′ər·ē, -siz′-) *adj.* Having power to rescind; rescinding; revoking. [<LL *rescissorius* <L *rescissus,* pp. of *rescindere.* See RESCIND.]

re·script (rē′skript) *n.* **1** In ancient Rome, an imperial decree, consisting of the emperor's answer to questions on matters of state or law. **2** Any decree, edict, order, or formal announcement, especially one made by a monarch or ruler. **3** A formal, written reply by the pope to a petition or question of morality or canon law submitted to him. **4** A facsimile; counterpart; something written over again. [<L *rescriptum* edict, orig. neut. pp. of *rescribere* write back (in reply) < *re-* again + *scribere* write]

res·cue (res′kyōō) *v.t.* **·cued, ·cu·ing** **1** To save or free from danger, captivity, evil, etc.; deliver. **2** *Law* To take or remove forcibly from the custody of the law. See synonyms under DELIVER, RECLAIM. —*n.* The act of rescuing; deliverance. [<OF *rescourre* <Med. L *rescutere* <L *re-* again + *excutere* shake off < *ex-* off, out + *quatere* shake. Related to QUASH.] —**res′cu·a·ble** *adj.* —**res′cu·er** *n.*

rescue grass A high bromegrass (*Bromus catharticus*), cultivated for hay. [Origin uncertain; perhaps confused with FESCUE]

re·search (ri·sûrch′, rē′sûrch) *n.* **1** Diligent, protracted investigation; studious inquiry. **2** A systematic investigation of some phenomenon or series of phenomena by the experimental method. See synonyms under INQUIRY. —*v.i.* To make research; investigate. [<F *recherche*] —**re·search′er** *n.*

re–search (rē·sûrch′) *v.t.* & *v.i.* To search again or anew.

re·seat (rē·sēt′) *v.t.* **1** To seat again. **2** To put a new seat or seats in or on.

ré·seau (rā·zō′) *n.* *pl.* **·seaux** (-zō′) **1** In textile work, a laceground composed of regular meshes; netground. **2** *Astron.* The small lines forming squares cut upon a glass plate: used in mapping out the heavens by photography. **3** A network. **4** *Meteorol.* A group of weather stations operating in the same territory or under common direction. **5** *Phot.* A sensitive filter screen for use in making color films. [<F, dim. of OF *roix* net <L *rete*]

re·sect (ri·sekt′) *v.t.* *Surg.* To cut or pare off: distinguished from *excise.* [<L *resectus,* pp. of *resecare* < *re-* back + *secare* cut, amputate]

re·sec·tion (ri·sek′shən) *n.* **1** A cutting or paring off. **2** *Surg.* The operation of cutting out part of a bone, organ, etc. **3** The determination of a position with reference to points of known location, whether on the ground or on a map or chart. [<L *resectio, -onis*]

re·se·da (ri·sē′də) *n.* **1** An herb of the mignonette family (genus *Reseda*). **2** A light or grayish green. [<L, prob. < *resedare* assuage; because once thought to be a sedative]

res·e·da·ceous (res′ə·dā′shəs) *adj.* *Bot.* Designating a family (*Resedaceae*) of annual or perennial herbs with alternate simple leaves and terminal spikes of small unsymmetrical flowers; the mignonette family. [<RESEDA + -ACEOUS]

re·sell (rē·sel′) *v.t.* **·sold, ·sell·ing** To sell anew or again. —**re·sell′er** *n.*

re·sem·blance (ri·zem′bləns) *n.* **1** The quality of similarity in nature, form, etc.; relative identity; likeness. **2** That which resembles; a semblance or likeness of a person or thing. **3** *Obs.* A characteristic quality or attribute. **4** *Obs.* Probability or likelihood. See synonyms under ANALOGY, APPROXIMATION, PICTURE. [<AF]

re·sem·ble (ri·zem′bəl) *v.t.* **·bled, ·bling** **1** To be similar to in appearance, quality, or character. **2** *Obs.* To compare; liken. See synonyms under IMITATE. [<OF *resembler* < *re-* again and again + *sembler* seem <L *simulare.* See SIMULATE.] —**re·sem′bler** *n.*

re·sent (ri·zent′) *v.t.* To feel or show resentment at; be indignant at, as an injury or insult. [<F *ressentir* feel the effects < *re-* again + *sentir* feel <L *sentire*]

re·sent·ful (ri·zent′fəl) *adj.* Disposed to resent; full of or characterized by resentment. See synonyms under MALICIOUS. —**re·sent′ful·ly** *adv.* —**re·sent′ful·ness** *n.*

re·sent·ment (ri·zent′mənt) *n.* Anger and ill will in view of real or fancied wrong or injury. See synonyms under ANGER, HATRED, OFFENSE, PIQUE.

re·ser·pine (ri·sûr′pēn, -pin, res′ər-) *n.* An ataractic drug prepared from alkaloids found in certain species of *Rauwolfia,* especially *R. serpentina.*

res·er·va·tion (rez′ər·vā′shən) *n.* **1** The act of reserving. **2** That which is reserved, kept back, or withheld. **3** The unexpressed qualification of a statement, promise, etc., that would, if uttered, so affect or alter its meaning for the person addressed as to vitiate its truth: also **mental reservation.** **4** Hence, any limitation. **5** A tract of government land reserved for the use and occupancy of an Indian tribe or for some other special purpose, as the preservation of forests, wild birds, etc. See synonyms under RESERVE. [<OF <LL *reservatio, -onis*]

re·serve (ri·zûrv′) *v.t.* **·served, ·serv·ing** **1** To hold back or set aside for special or future use; store up. **2** To keep as one's own; retain: He *reserves* that privilege for himself. **3** To arrange for ahead of time; have set aside for one's use: I *reserved* two tickets on the train. **4** To set aside (a portion of the consecrated elements of the Eucharist) for communion of the sick. See synonyms under RETAIN. —*n.* **1** That which is reserved; something stored up for future use, as in a reservoir; something set apart for a particular purpose; specifically, a reservation of land. **2** In banking, the amount of funds reserved from investment, in order promptly to meet regular or emergent demands. **3** The act of reserving; reservation. **4** The state of being reserved; silence as to one's feelings, opinions, or affairs; reticence; also, absence of exaggeration. **5** A fighting force held back from action to meet possible emergencies or demands. **6** That component of the armed forces of a nation composed of civilians trained for military service or assignment and subject to call to active duty in emergencies or under particular circumstances; specifically, *U.S.,* the **Army Reserve, Air Force Reserve, Naval Reserve, Marine Corps Reserve,** and **Coast Guard Reserve.** —*adj.* Held in reserve; constituting a reserve: a *reserve* supply of money. [<OF *reserver* <L *reservare* keep back < *re-* back + *servare* keep] —**re·serv′a·ble** *adj.* —**re·serv′er** *n.*

Synonyms (noun): backwardness, coldness, constraint, coyness, haughtiness, limitation, modesty, pride, reservation, reservedness, restraint, reticence, shyness, taciturnity. *Reserve* is the holding oneself aloof from others, or holding back one's feelings from expression, or one's affairs from communication to others; *reserve* may spring from *coldness* or *pride,* but is not identical with either and may arise from timidity or policy. See MODESTY. Compare TACITURN. *Antonyms*: abandon, forwardness, frankness, freedom, indiscretion, loquaciousness, pertness, presumption.

reserve bank A member of the Federal Reserve System.

re·served (ri·zûrvd′) *adj.* **1** Showing or characterized by reserve of manner; distant; undemonstrative. **2** Retained; kept back. See synonyms under HAUGHTY, TACITURN. —**re·serv·ed·ly** (ri·zûr′vid·lē) *adv.* —**re·serv′ed·ness** *n.*

re·serv·ist (ri·zûr′vist) *n.* A member of the military reserve.

res·er·voir (rez′ər·vwôr, -vwär, -vôr) *n.* **1** A receptacle where some material, especially of a liquid or gas, may be kept in store. **2** A basin, either natural or artificial, for collecting and containing a supply of water, as for use in a city or for water power. **3** An attachment to a stove, machine, or instrument, for containing a fluid to be used in its operation: the *reservoir* of a lamp. **4** An

extra supply; a store of anything. [<F *réservoir*]
re·set (rē·set′) *v.t.* **·set, ·set·ting** To set again. —*n.* (rē′set′) The act of resetting, or that which is reset; specifically, a resetting of type. —**re·set′ter** *n.*
res ges·ta (rēz jes′tə) *pl.* **res ges·tae** (jes′tē) *Latin* 1 Anything done; a transaction. 2 *Usually pl.* All the essential circumstances attending a transaction.
resh (resh) *n.* The twentieth Hebrew letter. See ALPHABET. [<Hebrew *rēsh,* lit., the head]
re·ship (rē·ship′) *v.* **·shipped, ·ship·ping** *v.t.* 1 To ship again. 2 To transfer (oneself) to another vessel. —*v.i.* 3 To go on a vessel again. 4 To sign for another voyage as a crew member or a passenger.
re·ship·ment (rē·ship′mənt) *n.* 1 The act of reshipping. 2 The thing reshipped.
Resht (resht) A city in northern Iran, near the Caspian Sea.
re·side (ri·zīd′) *v.i.* **·sid·ed, ·sid·ing** 1 To dwell for a considerable time; make one's home; live. 2 To exist as an attribute or quality: with *in.* 3 To be vested: with *in.* See synonyms under ABIDE. [<F *résider* <L *residere* sit back, abide <*re-* back + *sedere* sit] —**re·sid′er** *n.*
res·i·dence (rez′ə·dəns) *n.* 1 The place or the house where one resides. 2 The act of residing. 3 Inherence in a thing, as of an attribute in a subject. 4 The fact of being officially present; the statutory presence of an incumbent in a benefice, as a bishop in his diocese: especially in the phrase **in residence:** the canon *in residence.* 5 The seat or place of power or government. 6 The length of time one resides in a place. See synonyms under HOME, HOUSE. [<OF <LL *residentia*]
res·i·den·cy (rez′ə·dən·sē) *n. pl.* **·cies** 1 Residence. 2 In the East Indies, the official abode of the representative of the governor general, as at a native court. 3 Formerly, a government division of the Dutch East Indies.
res·i·dent (rez′ə·dənt) *n.* 1 One who resides or dwells in a place. 2 A diplomatic representative residing at a foreign court or seat of government; specifically, a **minister resident,** a diplomatic agent of the third rank, accredited by the sovereign or head of one country to the sovereign or head of another country; also, an agent in a protectorate. —*adj.* 1 Having a residence; residing. 2 Abiding in a place in connection with one's official work: a *resident* physician. 3 Inherent: Pungency is *resident* in pepper. 4 Not migratory: said of certain birds. [<OF]
res·i·den·tial (rez′ə·den′shəl) *adj.* 1 Pertaining to, fitted for, or resulting from residence; having residence. 2 Used by residents.
res·i·den·ti·ar·y (rez′ə·den′shē·er′ē, -shər·ē) *adj.* 1 Having or maintaining a residence, especially an official residence. 2 Pertaining to residence. —*n. pl.* **·ar·ies** A resident.
re·sid·u·al (ri·zij′o͞o·əl) *adj.* 1 Pertaining to or having the nature of a residue or remainder. 2 Left over as a residue. —*n.* 1 That which is left over from a total mass, magnitude, or quantity which has been acted upon in any specified way; a remainder or remnant. 2 *Stat.* a The difference between observed results and those obtained by computation according to formula. b The difference between the value of a given observation and the mean of a series to which it belongs.
re·sid·u·ar·y (ri·zij′o͞o·er′ē) *adj.* Of or pertaining to a residuum or remainder; residual.
res·i·due (rez′ə·do͞o, -dyo͞o) *n.* 1 A remainder or surplus after a part has been separated or otherwise treated. 2 *Chem.* a Insoluble matter left after filtration or separation from a liquid. b An atom or radical separated from a molecule of a substance. c A residuum. 3 *Law* That portion of an estate which remains after all charges, debts, and particular bequests have been satisfied. [<OF *residu* <L *residuum,* neut. of *residuus* remaining <*residere.* See RESIDE.]
re·sid·u·um (ri·zij′o͞o·əm) *n. pl.* **·u·a** (-o͞o·ə) 1 That which remains after any process of subtraction; a residue. 2 *Chem.* A residual product: the *residuum* from the distillation of coal tar. 3 Residue (def. 3). [<L]
re·sign (ri·zīn′) *v.t.* 1 To give up, as a position, office, or trust. 2 To relinquish (a privilege, claim, etc.). 3 To give over (oneself, one's mind, etc.), as to fate or domination. —*v.i.* 4 To resign a position, etc. See synonyms under ABANDON. [<OF *resigner* <L *resignare* sign back, transfer, cancel <*re-* back + *signare* sign] —**re·sign′er** *n.*
re-sign (rē·sīn′) *v.t.* To sign again.
res·ig·na·tion (rez′ig·nā′shən) *n.* 1 The act of resigning, as a position, office, or trust, or the formal document declaring such act. 2 The quality of being submissive: unresisting acquiescence. See synonyms under PATIENCE, SUBMISSION. [<F *résignation*]
re·signed (ri·zīnd′) *adj.* Characterized by resignation; submissive. —**re·sign·ed·ly** (ri·zī′nid·lē) *adv.* —**re·sign′ed·ness** *n.*
re·sile (ri·zīl′) *v.i.* **·siled, ·sil·ing** 1 To spring back; recoil. 2 To resume original shape or position after being stretched or compressed. 3 *Brit.* To depart from a line of conduct without changing ultimate intent. [<MF *resiler* <L *resilire* rebound <*re-* back + *salire* leap]
re·sil·ience (ri·zil′yəns) *n.* 1 The act or power of springing back to a former position or shape; elasticity. 2 *Physics* The quantity of work given back by a body that is compressed to a certain limit and then allowed freely to recover its former size or shape. Also **re·sil′ien·cy.**
re·sil·ient (ri·zil′yənt) *adj.* 1 Springing back to a former shape or position. 2 Capable of recoiling from pressure or shock unchanged or undamaged. 3 Elastic; buoyant. [<L *resiliens, -entis,* ppr. of *resilire.* See RESILE.] —**re·sil′ient·ly** *adv.*
res·in (rez′in) *n.* 1 An amorphous organic substance exuded from plants, especially from fir or pine trees, yellowish or dark in color and usually translucent or transparent: it is soluble in alcohol and ether, and is a non-conductor of electricity. 2 Any of various substances made by chemical synthesis, especially those used in the making of plastics. 3 The resinous precipitate obtained from a vegetable tincture by treatment with water: used in pharmacy. 4 Rosin. —*v.t.* To apply resin to. [<OF *resine* <L *resina* <Gk. *rhētinē*] —**res·i·na·ceous** (rez′ə·nā′shəs) *adj.*
res·in·ate (rez′ən·āt) *v.t.* **·at·ed, ·at·ing** To infuse or impregnate with resin.
res·in·if·er·ous (rez′ən·if′ər·əs) *adj.* Producing resin.
res·in·og·ra·phy (rez′ən·og′rə·fē) *n.* The microscopic study of the etched or polished surfaces of synthetic resins in order to identify the pigments, fillers, or other substances composing them. —**res′in·og′ra·pher** *n.*
res·in·oid (rez′ən·oid) *adj.* Resembling resin. —*n.* 1 A substance either wholly or partially of a resinous nature. 2 Any of a class of thermosetting synthetic resins.
res·i·nous (rez′ə·nəs) *adj.* 1 Of the nature of resins, or containing more or less resin as an ingredient. 2 Obtained from resin: *resinous* electricity; electronegative.
res·in·y (rez′ən·ē) *adj.* Resinous.
re·sist (ri·zist′) *v.t.* 1 To strive against; act counter to for the purpose of stopping, preventing, defeating, etc. 2 To be proof against; withstand; defeat. —*v.i.* 3 To offer opposition. See synonyms under DRIVE, HINDER, OPPOSE, REPEL. —*n.* Any substance used to prevent the entrance or action of another substance, as a coating applied to a surface to protect it from an acid, or a material applied to some portions of a fabric to protect it from the action of a dye. [<OF *resister* <L *resistere* cause to stand back, withstand <*re-* back + *sistere,* causative of *stare* stand] —**re·sist′er** *n.*
re·sis·tance (ri·zis′təns) *n.* 1 The act of resisting. 2 Any force tending to hinder motion. 3 *Electr.* a The opposition offered by a body to the passage through it of an electric current: expressed in ohms: the reciprocal of *conductance.* b Impedance. 4 *Psychol.* The force tending to prevent the return to consciousness of unpleasant incidents and experiences. 5 The underground and guerrilla movement opposing an occupying power. See synonyms under DEFENSE. [<F *résistance*]
re·sis·tant (ri·zis′tənt) *adj.* Offering or tending to produce resistance; resisting. —*n.* One who or that which resists. [<F *résistant*]
Re·sis·ten·cia (rā′sēs·ten′syä) The capital of Chaco province, northern Argentina.
re·sist·i·ble (ri·zis′tə·bəl) *adj.* Capable of being resisted. —**re·sist′i·bil′i·ty** *n.* —**re·sist′i·bly** *adv.*
re·sis·tive (ri·zis′tiv) *adj.* Having or exercising the power of resistance. —**re·sis′tive·ly** *adv.*
re·sis·tiv·i·ty (rē′zis·tiv′ə·tē) *n.* 1 The capacity to resist, or the degree of that capacity. 2 *Electr.* Specific resistance to the electric or magnetic force of a substance as tested in a cube measuring one centimeter: the reciprocal of *conductivity.*
re·sist·less (ri·zist′lis) *adj.* 1 Irresistible. 2 Offering no resistance; powerless. —**re·sist′less·ly** *adv.* —**re·sist′less·ness** *n.*
re·sis·tor (ri·zis′tər) *n. Electr.* A device, as a coil of wire, for introducing resistance into an electrical circuit.
res ju·di·ca·ta (rēz jo͞o′də·kā′tə) *Latin* Literally, a matter decided; an issue or point of law that has been previously decided by a court of authoritative or competent jurisdiction and which when pleaded is conclusive of the matter in controversy. Also *res adjudicata.*
re·sole (rē·sōl′) *v.t.* **·soled, ·sol·ing** To sole again.
res·o·lu·ble (rez′ə·lo͞o·bəl, ri·zol′yə·bəl) *adj.* Capable of being resolved; soluble. [<LL *resolubilis*] —**res′o·lu·bil′i·ty, res′o·lu·ble·ness** *n.*
res·o·lute (rez′ə·lo͞ot) *adj.* Having a fixed purpose; determined; constant; steady; also, bold; unflinching. See synonyms under FIRM, INFLEXIBLE, OBSTINATE. [<L *resolutus,* pp. of *resolvere.* See RESOLVE.] —**res′o·lute·ly** *adv.* —**res′o·lute·ness** *n.*
res·o·lu·tion (rez′ə·lo͞o′shən) *n.* 1 The act of resolving or of reducing to a simpler form. 2 The state of being resolute; active fortitude; resoluteness. 3 The making of a resolve; also, the purpose or course resolved upon; a resolve; determination. 4 Chemical, mechanical, or mental analysis; separation of anything into component parts. 5 A proposition offered to or adopted by an assembly. 6 *Law* A judgment or decision of a court. 7 *Med.* The termination of an abnormal condition. 8 *Music* a The replacement of a dissonant tone or chord by a higher or lower one so that a consonance, or, sometimes, another dissonance occurs. b The tone or chord replacing the original dissonant tone or chord. See synonyms under COURAGE, DETERMINATION, FORTITUDE, PURPOSE, PERSEVERANCE, WILL. —**concurrent resolution** A resolution adopted by both of the houses of Congress and having the force of law without the signature of the President. —**joint resolution** A resolution which, when passed by both houses of Congress and approved by the President, has the force of law. [<L *resolutio, -onis* <*resolutus.* See RESOLUTE.] —**res′o·lu′tion·er, res′o·lu′tion·ist** *n.*
re·solv·a·ble (ri·zol′və·bəl) *adj.* Capable of being resolved, analyzed, or solved. —**re·solv′a·bil′i·ty, re·solv′a·ble·ness** *n.*
re·solve (ri·zolv′) *v.* **·solved, ·solv·ing** *v.t.* 1 To decide; determine (to do something). 2 To cause to decide or determine. 3 To separate or break down into constituent parts; analyze. 4 To make clear; explain or solve, as a problem. 5 To explain away; remove (doubts, etc.). 6 To state or decide by vote, as in a legislative assembly. 7 To transform; convert: He *resolves* his anger into pride. 8 *Music* To change, as a chord, from dissonance to consonance; cause to undergo resolution. 9 *Chem.* To separate (a racemic compound) into its optically active components. 10 *Optics* To make distinguishable the structure or parts of. 11 *Med.* To cause to disperse or be absorbed without the formation of pus. 12 *Obs.* To melt; dissolve. 13 *Obs.* To inform. —*v.i.* 14 To make up one's mind; arrive at a decision: with *on* or *upon.* 15 To become separated into constituent parts. 16 *Music* To undergo resolution. —*n.* 1 Fixity of purpose; resolution. 2 A fixed determination; a resolution. 3 The action of a deliberative body expressing formally its intention or purpose. See synonyms under DETERMINATION, PURPOSE. [<L *resolvere* loosen again, relax <*re-* again + *solvere* loosen] —**re·solv′er** *n.*
re·solved (ri·zolvd′) *adj.* Fixed or set in purpose; determined; also, having formed a resolve. See synonyms under OBSTINATE. —**re·solv·ed·ly** (ri·zol′vid·lē) *adv.*
re·solv·ent (ri·zol′vənt) *adj.* Having the power to cause the dissolution or resolution of a thing into its elements; solvent. —*n.* 1 That

which has the power of resolving or dissolving; a solvent. 2 *Med.* A preparation which has the property of reducing or dispersing a swelling. [<L *resolvens, -entis,* ppr. of *resolvere.* See RESOLVE.]

res·o·nance (rez′ə·nəns) *n.* 1 The state or quality of being resonant; resonant sound. 2 *Physics* **a** The phenomenon exhibited by any vibratory system responding with large amplitude to a series of imposed vibrations of equal, or nearly equal, frequency. **b** That property of a molecule by virtue of which it assumes an electronic structure intermediate between two other theoretically possible structures. **c** The prolongation and amplification of sound by reverberation within a cavity. 3 *Electr.* The condition of an electric circuit in which maximum flow of current is obtained by impressing an electromotive force of given frequency. [<L *resonantia* echo]

res·o·nant (rez′ə·nənt) *adj.* 1 Sending back or having the quality of sending back or prolonging sound. 2 Resounding; specifically, having resonance. [<L *resonans, -antis,* ppr. of *resonare* resound, echo < *re-* back, again + *sonare* sound] — **res′o·nant·ly** *adv.*

res·o·nate (rez′ə·nāt) *v.i.* **·nat·ed, ·nat·ing** 1 To have or produce resonance. 2 To manifest sympathetic vibration, as a resonator. [<L *resonatus,* pp. of *resonare.* See RESONANT.]

res·o·na·tor (rez′ə·nā′tər) *n.* 1 That which resounds. 2 *Electronics* Any device used to exhibit or utilize the effects of resonance. 3 *Physics* A set or cluster of electrons which absorbs electromagnetic waves of certain frequencies. [<NL]

re·sorb (ri·sôrb′) *v.t.* To reabsorb. [<L *resorbere* drink in again, suck back < *re-* back, again + *sorbere* drink in, suck up] — **re·sorp·tion** (ri·sôrp′shən) *n.*

re·sor·cin·ol (ri·zôr′sin·ōl, -ol) *n. Chem.* A colorless crystalline compound, $C_6H_6O_2$, of peculiar odor and sweetish taste, used as an antiseptic and in the treatment of skin eruptions. Also **re·sor′cin.** [<RES(IN) + ORCINOL] — **re·sor′cin·al** *adj.*

re·sort (ri·zôrt′) *v.i.* 1 To go frequently or habitually; repair. 2 To have recourse; apply or betake oneself for relief or aid: with *to.* — *n.* 1 The act of frequenting a place. 2 A place resorted to or frequented to regain health, or for amusement or entertainment. 3 The use of something as a means; a recourse; refuge. [<OF *resortir* < *re-* again + *sortir* go out] — **re·sort′er** *n.*

re·sound (ri·zound′) *v.i.* 1 To be filled with sound; echo; reverberate. 2 To make a loud, prolonged, or echoing sound. 3 To ring; echo: said of sounds. 4 *Poetic* To be famed or extolled. — *v.t.* 5 To give back (a sound, etc.); re-echo. 6 *Poetic* To celebrate; extol. 7 *Rare* To utter or repeat loudly. See synonyms under ROAR. [<OF *resoner* <L *resonare.* See RESONANT.]

re-sound (rē·sound′) *v.t.* & *v.i.* To sound again.

re·source (ri·sôrs′, -sōrs′, rē′sôrs, -sōrs) *n.* 1 That which is resorted to for aid or support; resort. 2 *pl.* Available means or property; a supply that can be drawn on; any natural advantages or products: natural *resources.* 3 Capacity for finding or adapting means; power of achievement. 4 Fertility in expedients; resourcefulness; skill or ingenuity in meeting any situation. See synonyms under ALTERNATIVE, PROPERTY. [<OF *ressource* < *resourdre* rise again < *re-* (<L *re-*) back + *sourdre* <L *surgere* rise, surge]

re·source·ful (ri·sôrs′fəl, -sōrs′-) *adj.* 1 Fertile in resources or expedients. 2 Full of resources. — **re·source′ful·ly** *adv.* — **re·source′·ful·ness** *n.*

re·spect (ri·spekt′) *v.t.* 1 To have deferential regard for; esteem. 2 To treat with propriety or consideration. 3 To regard as inviolable; avoid intruding upon. 4 To have relation or reference to; concern. See synonyms under ADMIRE, DEFER, VENERATE. — *n.* 1 A just regard for and appreciation of worth; honor and esteem: I have great *respect* for the man. 2 Demeanor or deportment indicating deference; courteous regard: to have *respect* for one's elders. 3 *pl.* Expressions of consideration or esteem; compliments: to pay one's *respects.* 4 Conformity to duty or obligation; compliance or observance: *respect* for the law. 5 The condition of being honored or respected: He is held in *respect* by his colleagues. 6 A specific aspect or feature; detail: In what *respect* is he wanting? 7 Reference or relation: usually with *to*: with *respect* to profits. 8 Undue inclination or bias of mind: to have *respect* of persons. 9 *Obs.* Consideration. [<L *respectare* < *respectus,* pp. of *respicere* look back, consider < *re-* again + *specere* look]

re·spect·a·bil·i·ty (ri·spek′tə·bil′ə·tē) *n. pl.* **·ties** 1 The characteristic or quality of being respectable; fair social standing; good repute. 2 The respectable people of a community, collectively. 3 *pl.* Certain conventions and other features of conduct presumed to be signs of gentility, social position, morality, etc. Also **re·spect′a·ble·ness.**

re·spect·a·ble (ri·spek′tə·bəl) *adj.* 1 Deserving of respect; being of good name or repute; also, respected. 2 Being of moderate excellence; fairly good; considerable in number, quantity, size, quality, etc.; average. 3 Having a good appearance; presentable. 4 Conventionally correct or socially acceptable in conduct; of decent character. — **re·spect′a·bly** *adv.*

re·spect·er (ri·spek′tər) *n.* One who respects, usually one who respects persons: often in the phrase **respecter of persons,** one who shows favoritism or is influenced in his opinions or actions by others.

re·spect·ful (ri·spekt′fəl) *adj.* Marked by or manifesting respect; deferential. — **re·spect′·ful·ly** *adv.* — **re·spect′ful·ness** *n.*

re·spect·ing (ri·spek′ting) *prep.* In relation to; regarding.

re·spec·tive (ri·spek′tiv) *adj.* 1 Pertaining or relating severally to each of those under consideration; several; particular. 2 *Obs.* Characterized by partiality. 3 *Obs.* Attentive.

re·spec·tive·ly (ri·spek′tiv·lē) *adv.* As singly or severally considered; singly in the order designated: The first, second, and third seats belong to John, James, and William *respectively.*

re·spell (rē·spel′) *v.t.* To spell again, especially in a system whereby pronunciation is indicated. — **re·spell′ing** *n.*

Re·spi·ghi (rā·spē′gē), **Ottorino,** 1879–1936, Italian composer.

re·spir·a·ble (ri·spīr′ə·bəl, res′pər·ə·bəl) *adj.* 1 Capable of being respired or breathed; fit for respiration. 2 Able to breathe or respire. [<F]

res·pi·ra·tion (res′pə·rā′shən) *n.* 1 The act of inhaling air into the lungs and expelling it; breathing. 2 The process by which a plant or animal takes in oxygen from the air and gives off carbon dioxide and other products of oxidation in the tissues. [<L *respiratio, -onis*]

res·pi·ra·tor (res′pə·rā′tər) *n.* 1 A screen, as of fine gauze, worn over the mouth or nose, as a protection against dust, etc. 2 A device worn over the nose and mouth for the inhalation of medicated vapors, or to warm or sift the air for lung patients. 3 A gas mask. 4 An apparatus for artificial respiration, as a Pulmotor. [<L *respiratus,* pp. of *respirare.* See RESPIRE.]

respirator cabinet An iron lung.

re·spir·a·to·ry (ri·spīr′ə·tôr′ē, -tō′rē, res′pər·ə-) *adj.* Of, pertaining to, employed in, or caused by respiration.

re·spire (ri·spīr′) *v.* **·spired, ·spir·ing** *v.i.* 1 To inhale and exhale air; breathe. 2 To breathe again; recover vitality, hope, ambition, courage, etc. — *v.t.* 3 To inhale and exhale; breathe. 4 *Rare* To breathe or give forth; exhale. [<F *respirer* <L *respirare* < *re-* again + *spirare* breathe]

res·pite (res′pit) *n.* 1 Postponement; delay. 2 Temporary intermission of labor or effort; an interval of rest. 3 *Law* Temporary suspension of the execution of a sentence for a capital offense; reprieve. — *v.t.* **·pit·ed, ·pit·ing** 1 To relieve by a pause or rest. 2 To grant delay in the execution of (a penalty, sentence, etc.). 3 To put off or postpone. [<OF *respit* <Med. L *respectus* delay <L, consideration, regard < *respicere.* See RESPECT.]

Synonyms (noun): delay, forbearance, interval, pause, postponement, reprieve, rest, stay. *Antonyms*: accomplishment, completion, consummation, effect, execution, operation, performance.

re·splen·dence (ri·splen′dəns) *n.* The state or quality of being resplendent; brilliant luster; splendor. Also **re·splen′den·cy.**

re·splen·dent (ri·splen′dənt) *adj.* Shining with brilliant luster; vividly bright; splendid; gorgeous. See synonyms under BRIGHT. [<L *resplendens, -entis,* ppr. of *resplendere* glitter < *re-* again and again + *splendere* shine] — **re·splen′dent·ly** *adv.*

re·spond (ri·spond′) *v.i.* 1 To give an answer; reply. 2 To act in reply or return. 3 *Law* To be liable or answerable. — *v.t.* 4 To say in answer; reply. — *n. Archit.* A pilaster, semi column, or similar feature placed against a wall, to receive an arch. [<L *respondere* give back in return < *re-* back + *spondere* pledge, promise] — **re·spond′er** *n.*

re·spon·dence (ri·spon′dəns) *n.* 1 The character or condition of being respondent. 2 The act of responding. 3 Agreement. Also **re·spon′den·cy.**

re·spon·dent (ri·spon′dənt) *adj.* 1 Giving response, or given as a response; answering; responsive. 2 *Law* Occupying the position of defendant. 3 *Obs.* Correspondent. — *n.* 1 One who responds or answers. 2 *Law* The party called upon to answer an appeal or petition; a defendant; especially, the defendant in a suit in equity, admiralty, or divorce. [<L *respondens, -entis,* ppr. of *respondere.* See RESPOND.]

re·sponse (ri·spons′) *n.* 1 The act of responding, or that which is responded; words or acts evoked by the words or acts of another or others; an answer; reply. 2 *Eccl.* A portion of a liturgy or church service said or sung by the congregation or choir in reply to the officiating priest; also, an anthem sung or said during or after a reading. 3 *Biol.* The action of an organism or a part, or the cessation of action, resulting from a stimulus or influence; a reaction. [<OF <L *responsum,* neut. of pp. of *respondere.* See RESPOND.]

Synonyms: answer, rejoinder, repartee, reply, retort. A *rejoinder* is strictly an *answer* to a *reply,* while often used in the general sense of *answer,* but always with the implication of something more or less controversial or opposed, yet lacking the conclusiveness implied in *answer.* A *response* is accordant or harmonious, designed or adapted to carry on the thought of the words that called it forth, or to meet the wish of him who seeks it; as, The appeal for aid met a prompt and hearty *response. Repartee* is a prompt, witty, and commonly good-natured *answer* to some argument or attack; a *retort* may also be witty, but is severe and may be even savage in its intensity. See ANSWER.

re·spon·ser (ri·spon′sər) *n. Electronics* The receiving element connected with the transponder of an interrogator assembly. Also **re·spon′sor.**

re·spon·si·bil·i·ty (ri·spon′sə·bil′ə·tē) *n. pl.* **·ties** 1 The state of being responsible or accountable. 2 That for which one is answerable; a duty or trust. 3 Ability to meet obligations or to act without superior authority or guidance. See synonyms under DUTY. Also **re·spon′si·ble·ness.**

re·spon·si·ble (ri·spon′sə·bəl) *adj.* 1 Answerable legally or morally for the discharge of a duty, trust, or debt. 2 Having capacity to perceive the distinctions of right and wrong; having ethical discrimination. 3 Able to meet legitimate claims; having sufficient property or means for the payment of debts. 4 Involving accountability or obligation. 5 Denoting the status of a cabinet or ministry with respect to the legislative body to which it is answerable. [< obs. F *responsible* <L *responsus,* pp. of *respondere.* See RESPOND.] — **re·spon′si·bly** *adv.*

re·spon·sion (ri·spon′shən) *n.* 1 *Rare* A response; reply. 2 *pl.* At Oxford University, the first of the three examinations to be passed

add,āce,câre,pälm; end,ēven; it,īce; odd,ōpen,ôrder; to͝ok,po͞ol; up,bûrn; ə = a in *above,* e in *sicken,* i in *clarity,* o in *melon,* u in *focus*; yo͞o = u in *fuse*; oi,oil; ou,pout; ch,check; g,go; ng,ring; th,thin; th,this; zh,vision. Foreign sounds à,œ,ü,kh,ṅ; and ◆: see page xx. < from; + plus; ? possibly.

by a candidate for a B.A. degree. [<L *responsio, -onis*]

re·spon·sive (ri·spon′siv) *adj.* 1 Inclined or ready to respond; being or reacting in accord, sympathy, or harmony; responding. 2 Constituting, or of the nature of, response or reply. 3 Characterized by or containing responses. 4 *Obs.* Correspondent. —**re·spon′sive·ly** *adv.* —**re·spon′sive·ness** *n.*

re·spon·so·ry (ri·spon′sər·ē) *adj. Obs.* Of or pertaining to response; containing answer; responsive. —*n. pl.* **·ries** *Eccl.* 1 A response sung between readings. 2 A response of the people or congregation to the officiating priest or clergyman. [<Med. L *responsorium*]

res pub·li·ca (rēz pub′li·kə) *pl.* **res pub·li·cae** (pub′li·sē) *Latin* 1 The commonwealth. 2 *pl.* Things that belong to the state.

rest[1] (rest) *v.i.* 1 To cease working, exerting oneself, etc., so as to refresh oneself. 2 To cease from effort or activity for a time. 3 To seek or obtain ease or refreshment by lying down, sleeping, etc. 4 To sleep. 5 To be at peace; be tranquil. 6 To lie in death; be dead. 7 To remain unchanged: And there the matter *rests.* 8 To be supported; stand, lean, lie, or sit: with *against, on,* or *upon.* 9 To be founded or based: with *on* or *upon.* 10 To rely; depend: with *on* or *upon*: Our hopes *rest* on you. 11 To be placed as a burden or responsibility: with *on* or *upon.* 12 To be or lie in a specified place: The blame *rests* with me. 13 To be directed; remain on something, as the gaze or eyes. 14 *Law* To cease presenting evidence in a case. 15 *Agric.* To lie fallow. —*v.t.* 16 To give rest to; refresh by rest. 17 To put, lay, lean, etc., as for support or rest. 18 To found; base. 19 To direct (the gaze, eyes, etc.). 20 *Law* To cease presenting evidence in (a case). —*n.* 1 The act or state of resting; cessation from labor, exertion, action, or motion of any kind; repose; quiet. 2 Freedom from disturbance or disquiet; peace; tranquillity. 3 Sleep; also, death. 4 That on which anything rests; a support; base; basis; foundation; specifically, in billiards and pool, a support for a cue; a bridge. 5 A place of repose or quiet; a stopping place; abode. 6 *Music* **a** A pause, or an interval of silence. **b** A character indicating such pause: an eighth *rest.* 7 In prosody, a pause in a verse; a caesura. 8 *Obs.* Restored or renewed strength. 9 *Mil.* A command given troops, allowing them to relax. ◆ Homophone: *wrest.* [OE *restan*] —**rest′er** *n.*

Synonyms: (*verb*): abide, acquiesce, cease, desist, halt, hold, lean, lie, pause, recline, repose, sleep, slumber, stand, stay, stop, unbend. See ABIDE, LEAN[1]. *Antonyms*: contend, fight, labor, strive, struggle, toil, wake, watch, work.

Synonyms (*noun*): calm, calmness, cessation, ease, pause, peace, peacefulness, quiescence, quiet, quietness, quietude, recreation, repose, sleep, slumber, stay, stillness, stop, tranquillity. *Ease* denotes freedom from cause of disturbance, whether external or internal. *Quiet* denotes freedom from agitation, or especially from annoying sounds. *Rest* is a *cessation* of activity, especially of wearying or painful activity. *Recreation* is some pleasing activity of certain organs or faculties that affords *rest* to other parts of our nature that have become weary. *Repose* is a laying down, primarily of the body, and figuratively, a relaxing freedom from toil or strain of mind. *Sleep* is the perfection of *repose,* the most complete *rest*; *slumber* is a light and ordinarily pleasant form of *sleep.* See REMAINDER, RESPITE. *Antonyms*: agitation, commotion, disquiet, disturbance, excitement, motion, movement, restlessness, stir, strain, toil, tumult, unrest, work.

rest[2] (rest) *n.* 1 That which remains or is left over; a remainder. 2 Those remaining or not enumerated; the others: in this sense a collective noun taking a plural verb. 3 A balance, as of resources. —*v.i.* 1 To be and remain; continue; stay: *Rest* content. 2 *Obs.* To be left: Nothing *rests* but hope. —*v.t.* 3 *Obs.* To cause to remain: God *rest* you well. ◆ Homophone: *wrest.* [<OF *reste* < *rester* remain <L *restare* stop, stand < *re-* again + *stare* stand]

rest[3] (rest) *n.* A support for a lance attached to medieval armor: an aphetic form of *arrest.* ◆ Homophone: *wrest.*

re·state (rē·stāt′) *v.t.* **·stat·ed, ·stat·ing** To state again or anew. —**re·state′ment** *n.*

res·tau·rant (res′tər·ənt, -tə·ränt) *n.* A place where refreshments or meals are provided; a public dining-room. [<F, lit., restoring, ppr. of *restaurer* <OF *restorer.* See RESTORE.]

res·tau·ra·teur (res′tər·ə·tûr′, *Fr.* res·tō·rà·tœr′) *n.* The proprietor or keeper of a restaurant. [<F]

rest-balk (rest′bôk′) *n. Agric.* An unplowed ridge between furrows.

rest cure A treatment, as of nervous disorders, prescribing seclusion and quiet, generous diet, massage, etc.

rest·ful (rest′fəl) *adj.* 1 Full of or giving rest; affording freedom from disturbance, work, or trouble. 2 Being at rest or in repose; quiet. —**rest′ful·ly** *adv.* —**rest′ful·ness** *n.*

rest·har·row (rest′har′ō) *n.* A low European undershrub (*Ononis hircina*) of the bean family, with pink and white flowers. [Aphetic form of ARREST + HARROW; from the resistance offered by its tough roots]

res·ti·form (res′tə·fôrm) *adj. Anat.* Ropelike; twisted, as the **restiform bodies,** ropelike bundles of nerve fibers of the medulla oblongata that pass upward to the cerebellum. [<L *restis* cord + -FORM]

rest·ing (res′ting) *adj.* 1 At rest; reposing; also, dead. 2 Dormant.

resting spore *Bot.* A spore that germinates only after a lapse of a number of weeks or months, or at the end of the winter season.

res·ti·tu·tion (res′tə·tōō′shən, -tyōō′-) *n.* 1 The act of restoring something that has been taken away or lost. 2 The act of making good or rendering an equivalent for injury or loss; indemnification. 3 Restoration to, return to, or recovery of a former position or condition. 4 *Physics* The property of elastic bodies by which they tend to recover their shape after compression. 5 Establishment of the true nature or position of objects distorted in an aerial photograph. [<OF <L *restitutio, -onis* < *restitutus,* pp. of *restituere* restore, set up again < *re-* again + *statuere* set up]

Synonyms: amends, compensation, indemnification, indemnity, recompense, remuneration, reparation, repayment, restoration, return. *Antonyms*: cheat, cheating, defrauding, embezzlement, extortion, fraud, plunder, robbery, stealing, theft.

res·tive (res′tiv) *adj.* 1 Impatient of control; unruly. 2 Restless; fidgety; also, stubborn; balky. [<F *restif* < *rester* remain, balk <L *restare.* See REST[2].] —**res′tive·ly** *adv.* —**res′tive·ness** *n.*

Synonyms: fidgety, fractious, fretful, frisky, impatient, intractable, mutinous, rebellious, refractory, restless, skittish, unruly, vicious. The disposition to offer active resistance to control by any means whatever is what is commonly indicated by *restive.* A horse may be made *restless* by flies or by martial music, but with no refractoriness; the *restive* animal impatiently resists or struggles to break from control, as by bolting, flinging his rider, or otherwise. With this the metaphorical use of the word agrees, which is always in the sense of such terms as *impatient, intractable, rebellious,* and the like; a people *restive* under despotism are not disposed to "rest" under it, but to resist it and fling it off. *Antonyms*: docile, gentle, manageable, obedient, peaceable, quiet, submissive, tractable, yielding.

rest·less (rest′lis) *adj.* 1 Having no rest; never quiet; unresting: the *restless* waves. 2 Unable or disinclined to rest. 3 Uneasy; constantly seeking change. 4 Discontented. 5 Devoid of or destructive to rest or repose; obtaining no rest or sleep; sleepless. See synonyms under ACTIVE. —**rest′less·ly** *adv.* —**rest′less·ness** *n.*

re·stock (rē·stok′) *v.t.* To stock again or anew.

res·to·ra·tion (res′tə·rā′shən) *n.* 1 The act of restoring a person or thing to a former place or condition. 2 The state of being restored; rehabilitation; renewal. 3 The bringing back of a building or work of art as nearly as may be to its original state; also, the restored building or object. 4 *Paleontol.* The reconstruction of the skeleton of a fossil animal. 5 *Theol.* The doctrine that all men will eventually be restored to a sinless state and divine favor. See UNIVERSALISM. —**the Restoration** 1 The return of Charles II to the English throne in 1660, after the overthrow of the Cromwellian Protectorate; also, the following period until 1685. 2 The return of the Bourbons to power in 1814 under Louis XVIII; also, the period following the return. 3 The return of the Jews to Palestine after the Babylonian captivity. [<OF *restauration* <LL *restauratio, -onis* <L *restauratus,* pp. of *restaurare.* See RESTORE.]

re·sto·ra·tive (ri·stôr′ə·tiv, -stō′rə-) *adj.* 1 Tending or able to restore. 2 Pertaining to restoration. —*n.* That which restores; specifically, something to restore consciousness after a fainting fit.

re·store (ri·stôr′, -stōr′) *v.t.* **·stored, ·stor·ing** 1 To bring into existence or effect again: to *restore* peace. 2 To bring back to a former or original condition, appearance, etc.: to *restore* a great painting. 3 To put back in a former place or position; reinstate, as a deposed monarch. 4 To bring back to health and vigor. 5 To give back (something lost or taken away); return. See synonyms under RECLAIM, RECOVER. [<OF *restorer* <L *restaurare* < *re-* again + *-staurare* make firm, as in *instaurare* repair] —**re·stor′er** *n.*

re-store (rē·stôr′, -stōr′) *v.t.* **-stored, -stor·ing** To store again or anew.

re·strain (ri·strān′) *v.t.* 1 To hold back from acting, proceeding, or advancing; keep in check; repress. 2 To deprive of freedom or liberty, as by placing in a prison or asylum. 3 To restrict or limit. [<OF *restraindre, restreindre* <L *restringere* < *re-* back + *stringere* draw tight] —**re·strain′a·ble** *adj.* —**re·strain′ed·ly** *adv.*

Synonyms: abridge, bridle, check, circumscribe, confine, constrain, curb, hinder, hold, keep, repress, restrict, suppress. *Constrain* is positive; *restrain* is negative; one is *constrained* to an action; he is *restrained* from an action. *Constrain* refers almost exclusively to moral force, *restrain* frequently to physical force, as when we speak of putting one under restraint. To *restrain* an action is to hold it partially or wholly in check, thus controlling it even in performance; to *restrict* an action is to fix a limit or boundary which it may not pass, but within which it is free. To *repress,* literally to press back, is to hold in check, and perhaps only temporarily, that which is still very active; it is a feebler word than *restrain*; to *suppress* is finally and effectually to put down; *suppress* is a much stronger word than *restrain*; as, to *suppress* a rebellion. See ARREST, BIND, GOVERN, KEEP, LIMIT, REFRAIN, REPRESS, TEMPER. *Antonyms*: aid, animate, arouse, emancipate, encourage, excite, free, impel, incite, release.

re·strain·er (ri·strā′nər) *n.* 1 One who or that which restrains. 2 *Phot.* A chemical agent used to retard the action of the developer.

re·straint (ri·strānt′) *n.* 1 The act of restraining. 2 The state of being restrained; abridgment of liberty; confinement. 3 That which restrains; a restriction. 4 Self-repression; constraint. See synonyms under BARRIER, RESERVE. [<OF *restrainte,* noun use of pp. of *restraindre.* See RESTRAIN.]

re·strict (ri·strikt′) *v.t.* To hold or keep within limits or bounds; confine. See synonyms under BIND, CIRCUMSCRIBE, LIMIT, RESTRAIN. [<L *restrictus,* pp. of *restringere.* See RESTRAIN.]

re·strict·ed (ri·strik′tid) *adj.* 1 Limited; confined. 2 Not for general consumption, use, or service: *restricted* traffic or supplies. 3 Denoting specified defense information the unauthorized publication or dissemination of which is prohibited by law. —**re·strict′ed·ly** *adv.*

re·stric·tion (ri·strik′shən) *n.* 1 The act of restricting, or the state of being restricted; limitation. 2 That which restricts; a restraint. 3 Reservation; self-repression. See synonyms under BARRIER.

re·stric·tive (ri·strik′tiv) *adj.* 1 Serving, tending, or operating to restrict. 2 *Gram. & Logic* Limiting in thought, expression, or application. —**re·stric′tive·ly** *adv.*

rest-room (rest′rōōm′, -rŏŏm′) *n.* A room in a public building, as a railroad station, theater, office building, etc., provided with means and conveniences for the rest and comfort of patrons or employees; also, a toilet in a public building.

re·sult (ri·zult′) *n.* 1 The outcome of an action, course, process, or agency; consequence;

effect; conclusion. 2 *Math.* A quantity or value ascertained by calculation. 3 The final determination of a deliberative assembly. See synonyms under CONSEQUENCE, END, EVENT, HARVEST, OPERATION, PRODUCT. — *v.i.* 1 To be a result or outcome; be a physical or logical consequent; follow: with *from.* 2 To have an issue; terminate; end: with *in.* [<Med. L *resultare* <L, spring back, freq. of *resilire* rebound. See RESILE.]

re·sul·tant (ri·zul′tənt) *adj.* Arising or following as a result. — *n.* 1 That which results; a consequence. 2 *Physics* A force, velocity, etc., resulting from the action of two or more quantities of the same kind. [<L *resultans, -antis,* ppr. of *resultare.* See RESULT.]

re·sume (ri·zōōm′) *v.* **·sumed, ·sum·ing** *v.t.* 1 To begin again; take up again after cessation or interruption. 2 To take or occupy again: *Resume* your places. 3 To take for oneself again: to *resume* a title. — *v.i.* 4 To continue after cessation or interruption. See synonyms under RECOVER. [<F *résumer* <L *resumere* take up again, take back < *re-* again + *sumere* take, seize] — **re·sum′a·ble** *adj.* — **re·sum′er** *n.*

rés·u·mé (rez′ōō·mā′, rez′ōō·mā) *n.* A summary, as of one's employment record. [<F]

re·sump·tion (ri·zump′shən) *n.* The act of resuming. [<L *resumptio, -onis* < *resumptus,* pp. of *resumere.* See RESUME.]

re·su·pi·nate (ri·sōō′pə·nāt) *adj. Bot.* Having the appearance of being upside down; inverted; reversed: said of the flowers of orchids. [<L *resupinatus,* pp. of *resupinare* bend back < *resupinus.* See RESUPINE.] — **re·su′pi·na′·tion** *n.*

re·su·pine (rē′sōō·pīn′) *adj.* Lying on the back; supine. [<L *resupinus* < *re-* again + *supinus* on the back]

re·sur·face (rē·sûr′fis) *v.t.* **·faced, ·fac·ing** To provide with a new surface.

re·sur·gam (ri·sûr′gam) *Latin* I shall rise again.

re·surge (ri·sûrj′) *v.i.* **·surged, ·surg·ing** 1 To rise again; be resurrected. 2 To surge or sweep back again, as the tide. [<L *resurgere* < *re-* again + *surgere* rise]

re·sur·gence (ri·sûr′jəns) *n.* A rising again.

re·sur·gent (ri·sûr′jənt) *adj.* 1 Rising again, as from the grave. 2 Surging back or again. [<L *resurgens, -entis,* ppr. of *resurgere*]

res·ur·rect (rez′ə·rekt′) *v.t.* 1 To bring back to life; raise from the dead. 2 To bring back into use or to notice. — *v.i.* 3 To rise again from the dead. [Back formation <RESURRECTION]

res·ur·rec·tion (rez′ə·rek′shən) *n.* 1 A rising again from the dead. 2 The state of those who have risen from the dead. 3 Any revival or renewal, as of a practice or custom, after disuse, decay, etc.; restoration; rebirth. 4 In Christian Science, spiritualization of thought; a new and higher idea of immortality, or spiritual existence; material belief yielding to spiritual understanding. — **the Resurrection** *Theol.* 1 The rising of Christ from the dead. 2 The rising again of all the dead at the day of final judgment. [<L *resurrectio, -onis* < *resurrectus,* pp. of *resurgere.* See RESURGE.] — **res′ur·rec′tion·al** *adj.*

res·ur·rec·tion·ar·y (rez′ə·rek′shən·er′ē) *adj.* 1 Of or pertaining to resurrection. 2 Of or pertaining to the exhuming of dead bodies.

res·ur·rec·tion·ist (rez′ə·rek′shən·ist) *n.* 1 One who steals bodies from the grave; a body-snatcher. 2 One who brings to light anything buried in obscurity. 3 A believer in the rising again of the dead. — **res′ur·rec′tion·ism** *n.*

resurrection plant The rose of Jericho.

re·sus·ci·tate (ri·sus′ə·tāt) *v.t.* & *v.i.* **·tat·ed, ·tat·ing** To bring or come back to life; revive from unconsciousness or apparent death. [<L *resuscitatus,* pp. of *resuscitare* < *re-* again + *suscitare* revive < *sub-* under + *citare* call, rouse. See CITE.] — **re·sus′ci·ta′tive** *adj.* — **re·sus′ci·ta′tor** *n.*

re·sus·ci·ta·tion (ri·sus′ə·tā′shən) *n.* The act of resuscitating, or the state of being resuscitated; revivification; reanimation.

Resz·ke (resh′ke), **Édouard de,** 1856–1917, Polish basso. — **Jean de,** 1853–1925, Polish tenor; brother of the preceding.

ret (ret) *v.t.* **ret·ted, ret·ting** To steep or soak, as flax, to facilitate the separation of the fibers: also *rot.* [ME *reten,* ? <MDu. *reten* soak]

re·ta·ble (ri·tā′bəl) *n.* 1 A shelf or ledge raised above the back of an altar to support ornaments, lights, etc. 2 A panel containing a picture or bas-relief of subjects from sacred history. [<F <OF *rere-table* <Med. L *retrotabulum* <L *retro-* behind + *tabula* plank]

re·tail (rē′tāl) *n.* The selling of goods in small quantities: opposed to *wholesale.* — *adj.* Of, pertaining to, or concerned in the sale of goods in small quantities or parcels. — *v.t.* 1 To sell in small quantities; sell directly to the ultimate consumer. 2 (ri·tāl′) To repeat, as gossip. — *v.i.* 3 To be sold at retail. [<OF, cutting < *retailler* cut up < *re-* again + *tailler* cut <LL *taliare* split]

re·tail·er (rē′tā·lər) *n.* One who sells in small quantities to the consumer.

re·tain (ri·tān′) *v.t.* 1 To keep or continue to keep in one's possession; hold. 2 To maintain in use, practice, etc.: to *retain* one's standards. 3 To keep in a fixed condition or place. 4 To keep in mind; remember. 5 To hire, as a servant; also, to engage (an attorney or other representative) by paying a retainer. [<OF *retenir* <L *retinere* < *re-* back + *tenere* hold] — **re·tain′a·ble** *adj.*

Synonyms: detain, employ, engage, hire, hold, keep, maintain, preserve, reserve, secure, withhold. See KEEP, REMEMBER. *Antonyms:* abandon, cede, discard, discharge, dismiss, eject, relinquish, renounce, resign, surrender.

re·tain·er[1] (ri·tā′nər) *n.* 1 One retained in the service of a person of rank or position. 2 One who retains or keeps. 3 *Mech.* A device for holding the parts of ball or roller bearings in place.

re·tain·er[2] (ri·tā′nər) *n.* 1 The fee paid, or the agreement made, to employ an attorney to serve in a suit; a retaining fee. 2 A similar fee paid to anyone to retain his services. [<OF *retenir* hold back, in a noun use]

retaining wall A wall to prevent the material of an embankment or cut from sliding; sometimes, a revetment.

re·take (rē·tāk′) *v.t.* **·took, ·tak·en, ·tak·ing** 1 To take back; receive again. 2 To recapture. 3 To photograph again. — *n.* (rē′·tāk′) A motion-picture scene or sequence photographed again.

re·tal·i·ate (ri·tal′ē·āt) *v.* **·at·ed, ·at·ing** *v.i.* To return like for like; especially, to repay evil with evil. — *v.t.* To repay (an injury, wrong, etc.) in kind; revenge. See synonyms under AVENGE. [<L *retaliatus,* pp. of *retaliare* < *re-* back + *talio* punishment in kind < *talis* such] — **re·tal′i·a′tive** *adj.*

re·tal·i·a·tion (ri·tal′ē·ā′shən) *n.* The act of retaliating; reprisal; requital. See synonyms under REVENGE.

re·tal·i·a·to·ry (ri·tal′ē·ə·tôr′ē, -tō′rē) *adj.* Pertaining to, containing, or of the nature of retaliation.

re·tard (ri·tärd′) *v.t.* To cause to move or proceed slowly; hinder the advance or course of; impede; delay. — *v.i.* To be delayed. See synonyms under HINDER, OBSTRUCT. — *n.* Delay; retardation. [<F *retarder* <L *retardare* < *re-* back + *tardare* make slow < *tardus* slow] — **re·tard′a·tive** *adj.* — **re·tard′er** *n.*

re·tard·ant (ri·tär′dənt) *n.* Something that retards. — *adj.* Tending to hinder.

re·tar·da·tion (rē′tär·dā′shən) *n.* 1 The act of retarding. 2 The state of being retarded. 3 A lessening of velocity, gain, or progress; a delaying. 4 The amount of delay or hindrance effected. 5 That which retards; a hindrance. 6 Slowness. 7 *Music* A gradual slackening of the time. [<L *retardatio, -onis*]

retch (rech) *v.i.* To make an effort to vomit; strain; heave. ◆ Homophone: *wretch.* [OE *hrǣcan* bring up (blood or phlegm)]

re·te (rē′tē) *n. pl.* **·ti·a** (-shē·ə, -tē·ə) A plexiform arrangement, as of vessels or nerves; network. [<L, net]

re·tell (rē·tel′) *v.t.* **·told, ·tell·ing** To count or relate again.

re·tem (rē′təm) *n.* A desert shrub (genus *Retama*) of Arabia and Syria, with small white flowers: the Old Testament juniper. [<Arabic *ratam,* pl. of *ratamah*]

ret·ene (ret′ēn, rē′tēn) *n. Chem.* A colorless crystalline compound, $C_{18}H_{18}$, contained in resinous pine wood and fir wood, also in fossil pine stems found in beds of peat and lignite. [<Gk. *rhētinē* resin]

re·tent (ri·tent′) *n.* That which is retained. [<L *rententus,* pp. of *retinere.* See RETAIN.]

re·ten·tion (ri·ten′shən) *n.* 1 The act of retaining. 2 The ability to remember; memory. 3 The keeping up or maintenance, as of a custom, practice, opinion, or intention. 4 *Med.* A holding within the body of materials normally excreted, as urine, etc. [<OF <L *retentio, -onis*]

re·ten·tive (ri·ten′tiv) *adj.* Having the power or tendency to retain; retaining: a *retentive* memory.

re·ten·tive·ness (ri·ten′tiv·nis) *n.* 1 The capacity of holding or retaining. 2 *Psychol.* The preservative function of memory.

re·ten·tiv·i·ty (rē′ten·tiv′ə·tē) *n.* 1 Retentiveness. 2 *Physics* The capacity of a material to retain magnetism after the withdrawal of the magnetizing force.

Re·thondes (rə·tônd′) A village 5 miles east of Compiègne, eastern France; armistice to suspend hostilities of World War I signed here, Nov. 11, 1918; during World War II, armistice to suspend hostilities between Germany and France signed here, June 22, 1940.

re·ti·ar·i·us (rē′shē·âr′ē·əs) *n. pl.* **·ar·i·i** (-âr′ē·ī) One of a class of ancient Roman gladiators, armed with a net to enmesh their adversaries, and a trident and dagger to dispatch them. See illustration under GLADIATOR. [<L *rete* net]

re·ti·ar·y (rē′shē·er′ē) *adj.* 1 Of or pertaining to the making of nets or netlike structures. 2 Furnished with a net. 3 Skilful in entangling (something). [<L *rete, retis* net + -ARY[1]]

ret·i·cence (ret′ə·səns) *n.* The quality, act, or an instance of being reserved in speech; reserve; taciturnity. Also **ret′i·cen·cy.** See synonyms under RESERVE. [<L *reticentia,* orig. neut. pl. of *reticens.* See RETICENT.]

ret·i·cent (ret′ə·sənt) *adj.* Habitually silent or reserved in utterance. See synonyms under TACITURN. [<L *reticens, -entis,* ppr. of *reticere* remain silent < *re-* again + *tacere* be silent] — **ret′i·cent·ly** *adv.*

ret·i·cle (ret′i·kəl) *n. Optics* The network of fine threads or lines of reference in the focal plane of a telescope or other optical instrument, serving to determine the position of an observed object: also spelled *reticule.* [<L *reticulum.* Doublet of RETICULUM.]

re·tic·u·lar (ri·tik′yə·lər) *adj.* 1 Like a network; reticulate; intricate. 2 *Anat.* Of or pertaining to a reticulum. Also **re·tic′u·lar′y.** [<NL *reticularis* <L *reticulum.* See RETICULUM.]

re·tic·u·late (ri·tik′yə·lāt) *v.* **·lat·ed, ·lat·ing** *v.t.* 1 To make a network of. 2 To cover with or as with lines of network. — *v.i.* 3 To form a network. — *adj.* (-lit, -lāt) Having the form or appearance of a network; having lines or veins crossing, as in leaves: also **re·tic′u·lat′ed.** [<L *reticulatus* < *reticulum.* See RETICULUM.]

re·tic·u·la·tion (ri·tik′yə·lā′shən) *n.* Any formation that is reticulated; a network. [< RETICULATE]

ret·i·cule (ret′ə·kyōōl) *n.* 1 A small bag formerly used by women for carrying personal articles, sewing materials, etc. 2 *Optics* A reticle. [<F *réticule*]

re·tic·u·lum (ri·tik′yə·ləm) *n. pl.* **·la** (-lə) 1 *Anat.* A protoplasmic network of cells or cellular tissue. 2 *Zool.* The honeycomb bag or second stomach of a ruminant, with the lining membrane raised into folds forming hexagonal cells. [<L, dim. of *rete* net. Doublet of RETICLE.]

Re·tic·u·lum (ri·tik′yə·ləm) A southern constellation, the Net. See CONSTELLATION. [<NL]

re·ti·form (rē′tə·fôrm, ret′ə-) *adj.* Arranged like a network; reticulate. [<F *rétiforme* <L *rete* net + *forma* shape]

ret·i·na (ret′ə·nə, ret′nə) *n. pl.* **·nas** or **·nae** (-nē) *Anat.* The inner membrane at the back of the eyeball, containing the light-sensitive rods and cones which receive the optical image. See illustration under EYE. [<LL <L *rete* net] — **ret′i·nal** *adj.*

ret·in·ene (ret′ən·ēn) *n. Biochem.* The yellow

pigment found in visual yellow and associated also in the production of vitamin A. [<RETINA]

ret·i·nite (ret′ə·nīt) *n.* A hard, brittle, vitreous resin obtained from lignite. [<Gk. *rhētinē* + -ITE[1]]

ret·i·ni·tis (ret′ə·nī′tis) *n. Pathol.* Inflammation of the retina.

ret·i·nol (ret′ə·nōl, -nol) *n.* A yellowish liquid hydrocarbon obtained by the distillation of various resins, and used as a solvent, especially in pharmacy. Also called *rosin oil.* [< Gk. *rhētinē* resin + -OL[2]]

ret·i·nos·co·py (ret′ə·nos′kə·pē) *n.* Skiascopy. [<L *retino* (<RETINA) + -SCOPY] —**ret′i·no·scop′ic** (-nō·skop′ik) *adj.*

ret·i·nue (ret′ə·nōō, -nyōō) *n.* The body of retainers attending a person of rank; an escort; cortège. [<F *retenue*, fem. of *retenu*, pp. of *retenir.* See RETAIN.]

re·tire (ri·tīr′) *v.* **·tired, ·tir·ing** *v.i.* **1** To go away or withdraw, as for privacy, shelter, or rest. **2** To go to bed. **3** To withdraw oneself from business, public life, or active service. **4** To fall back; retreat, as troops under attack. **5** To move back; recede or appear to recede. —*v.t.* **6** To remove from active service, as an officer of the army or navy. **7** To pay off and withdraw from circulation: to *retire* bonds. **8** To withdraw (troops, etc.) from action. **9** In baseball, etc., to keep (a batter or runner) from reaching base or scoring by putting him out, or to remove (a side) from an opportunity of scoring. [<F *retirer* < *re-* back + *tirer* draw]

re·tired (ri·tīrd′) *adj.* **1** Withdrawn from public view; existing or passed in seclusion; solitary; secluded: a *retired* life. **2** Withdrawn from active service, business, office, or public life: a *retired* sea captain. **3** Due or received by a person withdrawn from active service: *retired* pay. See synonyms under SECRET.

retired list One of the lists of officers or enlisted men voluntarily or involuntarily retired from an active status in one of the armed services of the United States, either on account of disability, age, or years of service, with or without retired pay, and with or without continuing legal inactive status as part of the armed services.

re·tire·ment (ri·tīr′mənt) *n.* **1** The act of retiring, or the state of being retired; withdrawal; seclusion. **2** A secluded place; a retreat.

Synonyms: loneliness, privacy, seclusion, solitude. In *retirement* one withdraws from association he has had with others; in *seclusion* one shuts himself off from the society of all except intimate friends or attendants; in *solitude* no other person is present. As private denotes what concerns ourselves individually, *privacy* denotes freedom from the presence or observation of those not concerned or whom we do not wish to have concerned in our affairs; *privacy* is more temporary than *seclusion*; we speak of a moment's *privacy.* There may be *loneliness* without *solitude*, as amid an unsympathizing crowd, and *solitude* without *loneliness*, as when one is glad to be alone. See SECLUSION, SOLITUDE. *Antonyms*: association, companionship, company, fellowship, society.

re·tir·ing (ri·tīr′ing) *adj.* **1** Shy; modest; reserved; unobtrusive. **2** Pertaining to retirement: a *retiring* pension. See synonyms under MODEST.

re·tort[1] (ri·tôrt′) *v.t.* **1** To direct (a word or deed) back upon the originator. **2** To reply to, as an accusation or argument, by a similar one. —*v.i.* **3** To make answer, especially sharply. —*n.* A retaliatory speech; a turning back of an accusation or insult upon the one who makes it; a keen rejoinder or caustic repartee; also, the act of making such reply: to be quick at *retort.* See synonyms under ANSWER. [<L *retortus*, pp. of *retorquere* < *re-* back + *torquere* twist] —**re·tort′er** *n.*

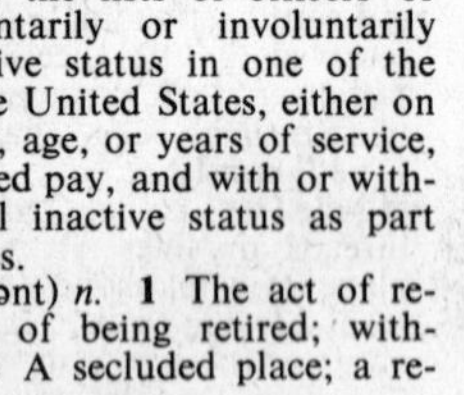

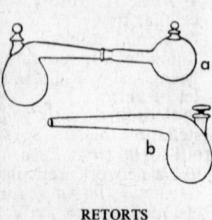

RETORTS
a. Retort with receiver.
b. Common retort.

re·tort[2] (ri·tôrt′) *n.* **1** *Chem.* A vessel with a bent tube, for the heating of substances, or for distillation. **2** *Metall.* A vessel in which ore may be heated for the removal of its metal content. [<L *retortus* bent back. See RETORT[1].]

re·tor·tion (ri·tôr′shən) *n.* **1** The act of retorting. **2** A bending, turning, or twisting back. **3** Retaliation; in international law, the infliction by one nation upon the subjects of another of the same ill treatment that its own citizens have received from the latter government. Also **re·tor′sion.** [<Med. L *retortio, -onis*]

re·touch (rē·tuch′) *v.t.* **1** To add new touches to; modify; revise. **2** *Phot.* To change, or improve, as a print, by a hand process in which a hard, sharp pencil or fine brush is used. —*n.* (*also* rē′tuch′) An additional touch, as to a picture, model, or other work of art, previously regarded as finished. [<F *retoucher*] —**re·touch′er** *n.*

re·trace (ri·trās′) *v.t.* **·traced, ·trac·ing** **1** To go back over; follow backward, as a path. **2** To trace the whole story of, from the beginning. **3** To go back over with the eyes or mind. [<F *retracer*] —**re·trace′a·ble** *adj.*

re–trace (rē·trās′) *v.t.* **–traced, –trac·ing** To trace again, as an engraving, drawing, or map.

re·tract (ri·trakt′) *v.t.* & *v.i.* **1** To take back (an assertion, accusation, admission, etc.); make a disavowal (of); recant. **2** To draw back or in, as the claws of a cat. See synonyms under RECANT, RENOUNCE. [<F *rétracter* <L *retractare* draw back < *re-* back + *tractare* draw violently, freq. of *trahere* draw] —**re·tract′a·ble** or **·i·ble** *adj.* —**re·trac·ta·tion** (rē′·trak·tā′shən) *n.*

re·trac·tile (ri·trak′til) *adj. Zool.* Capable of being drawn back or in, as a cat's claws or the head of a tortoise. [<F *rétractile*] —**re·trac·til·i·ty** (rē′trak·til′ə·tē) *n.*

re·trac·tion (ri·trak′shən) *n.* **1** The act of retracting or drawing something back or in. **2** The state of being retracted. **3** The act of withdrawing or recalling something said or avowed; recantation; revocation.

re·trac·tive (ri·trak′tiv) *adj.* Having the power or tendency to retract; retracting.

re·trac·tor (ri·trak′tər) *n.* **1** One who or that which retracts. **2** *Surg.* An instrument used to hold apart the edges of a wound.

re·tral (rē′trəl) *adj.* Situated at the back; posterior. [<L *retro* backward + -AL[1]]

re·tread (rē′tred′) *n.* A new outer covering of a pneumatic tire, to replace a worn or damaged one. —*v.t.* (rē·tred′) **·tread·ed, ·tread·ing** To fit or furnish (an automobile tire) with a new tread. Also *recap.*

re–tread (rē·tred′) *v.t.* **–trod, –trod·den, –tread·ing** To tread again.

re·treat (ri·trēt′) *v.i.* **1** To go back or backward; withdraw; retire. **2** To curve or slope backward. —*v.t.* **3** In chess, to move (a piece) back. —*n.* **1** The act of retreating, as from contest or danger. **2** The retirement of a naval or land force from a position of danger or from an enemy; also, a signal for retreating, made by trumpet or drum. **3** In the army or navy, a signal, as by bugle, for the lowering of the flag at sunset. **4** Retirement; seclusion; solitude. **5** A place of retirement, quiet, or security; a refuge; shelter; haunt. **6** Religious retirement; also, the time spent in religious retirement. **7** An establishment for the mentally ill, for alcoholics, etc. See synonyms under REFUGE, SECLUSION, SHELTER. Compare RETIREMENT. [<F *retraite*, orig. fem. of pp. of *retraire* draw back <L *retrahere* < *re-* again + *trahere* draw]

re·trench (ri·trench′) *v.t.* **1** To cut down or reduce; curtail (expenditures). **2** To cut off or away; remove; omit. —*v.i.* **3** To make retrenchments; economize. [<MF *retrencher* < *re-* back + *trencher* cut. See TRENCH.]

Synonyms: abridge, clip, curtail, cut, decrease, diminish, economize, lessen, reduce. *Antonyms*: elongate, expand, extend, lavish, lengthen, prolong, protract, squander, waste.

re·trench·ment (ri·trench′mənt) *n.* **1** The act of retrenching. **2** Reduction, as of expenses, for the sake of economy. **3** An interior breastwork or rampart from which the enemy can be resisted should the outer line be taken.

ret·ri·bu·tion (ret′rə·byōō′shən) *n.* **1** The act of requiting; impartial infliction of punishment. **2** That which is done or given in requital. **3** A reward or (especially) a punishment. See synonyms under RECOMPENSE, REVENGE. [<OF <L *retributio, -onis* < *retributus*, pp. of *retribuere* pay back < *re-* back + *tribuere* pay]

re·trib·u·tive (ri·trib′yə·tiv) *adj.* Tending to reward or punish. Also **re·trib′u·to′ry** (-tôr′ē, -tō′rē).

re·triev·al (ri·trē′vəl) *n.* **1** The act of retrieving. **2** Restoration from loss, damage, or failure.

re·trieve (ri·trēv′) *v.* **·trieved, ·triev·ing** *v.t.* **1** To get back; regain. **2** To restore; revive, as flagging spirits. **3** To make up for; remedy the consequences of. **4** To call to mind; remember. **5** To find and bring in (wounded or dead game): said of dogs. —*v.i.* **6** To retrieve game. See synonyms under RECOVER. —*n.* The act of retrieving; retrieval; recovery. [ME *retreve* <OF *retroev-*, stressed stem of *retrouver* find again < *re-* again + *trouver* find] —**re·triev′a·bil′i·ty** *n.* —**re·triev′a·ble** *adj.* —**re·triev′a·bly** *adv.*

re·triev·er (ri·trē′vər) *n.* **1** A sporting dog variously bred and specifically trained to retrieve game. **2** A person who retrieves.

retro– *prefix* **1** Back; backward: *retroflex, retrograde.* **2** Chiefly in scientific terms, behind: *retrolental.* [<L *retro-* < *retro* back, backward]

ret·ro·act (ret′rō·akt′, rē′trō-) *v.i.* **1** To act reciprocally or in return; react. **2** *Law* To affect past acts, obligations, or penalties. [Back formation <RETROACTIVE] —**ret′ro·ac′·tion** *n.*

ret·ro·ac·tive (ret′rō·ak′tiv, rē′trō-) *adj.* Having or designed to have a retrospective effect or reversed action; in effect also during a specified prior period. —**ret′ro·ac′tive·ly** *adv.* —**ret′ro·ac·tiv′i·ty** *n.*

retroactive law A law legalizing past proceedings; a retrospective law.

ret·ro·cede (ret′rō·sēd′) *v.* **·ced·ed, ·ced·ing** *v.t.* To cede, grant, or give back. —*v.i.* To go back; recede. [<L *retrocedere* < *retro-* back + *cedere* go]

ret·ro·ces·sion (ret′rō·sesh′ən) *n.* **1** The act of retroceding or giving back. **2** *Law* The conveyance of an estate to a former owner. [<LL *retrocessio, -onis*]

ret·ro·choir (ret′rə·kwīr) *n.* That part of a church interior which is east of or beyond the altar. [<RETRO- + CHOIR, modeled on Med. L *retrochorus*]

ret·ro·flex (ret′rə·fleks) *adj.* **1** Bent or turned backward; reflexed. **2** *Phonet.* Cacuminal. Also **ret′ro·flexed.** [<LL *retroflexus*, pp. of *retroflectere* <L *retro-* back + *flectere* bend]

ret·ro·flex·ion (ret′rə·flek′shən) *n.* **1** A bending or being bent backward. **2** *Anat.* A position or condition of the uterus in which its body is bent back at an angle with the cervix. Also **ret′ro·flec′tion.**

ret·ro·grade (ret′rə·grād) *v.* **·grad·ed, ·grad·ing** *v.i.* **1** To move or appear to move backward; recede. **2** To grow worse; decline; degenerate. **3** *Astron.* To have a retrograde motion. —*v.t.* **4** To cause to move backward; reverse. —*adj.* **1** Going, moving, or tending backward; contrary; reversed. **2** Declining to or toward a worse state or character. **3** *Astron.* Apparently moving from east to west relatively to the fixed stars. **4** Reversed; inverted. **5** *Obs.* Opposed; contrary. —*n.* A retrograde movement; decline. [<L *retrogradus*] —**ret′·ro·gra·da′tion** (-grā·dā′shən) *n.*

ret·ro·gress (ret′rə·gres) *v.i.* To go back to an earlier or worse condition. [<L *retrogressus*, pp. of *retrogradi* < *retro-* backward + *gradi* walk]

ret·ro·gres·sion (ret′rə·gresh′ən) *n.* **1** A retreat; degeneration; motion in a reverse direction. **2** A moving toward a lower plane. **3** *Biol.* Descent to or toward a less complex or less perfect structure.

ret·ro·gres·sive (ret′rə·gres′iv) *adj.* **1** Retrograde. **2** Deteriorating; degenerating. **3** *Biol.* Descending from a higher to a less complex organization.

ret·ro·len·tal (ret′rō·len′təl) *adj.* Behind the lens of the eye. [<RETRO- + L *lens, lentis.* See LENS.]

retrolental fi·bro·pla·sia (fī′brō·plā′zhə, -zhē·ə) *Pathol.* The persistence or growth of embryonic vascular tissue behind the lens of the eye.

ret·ro–rock·et (ret′rō·rok′it) *n.* An auxiliary jet engine whose thrust acts to lessen

the velocity of fall of a rocket or spaceship to the surface of the earth or other celestial body.

re·trorse (ri·trôrs′) *adj.* Turned, bent, or directed backward. [<L *retrorsus,* contraction of *retroversus* < *retro-* backward + *versus,* pp. of *vertere* turn] — **re·trorse′ly** *adv.*

ret·ro·spect (ret′rə·spekt) *v.i. Rare* **1** To think about the past. **2** To look or refer back. — *v.t.* **3** *Rare* To consider or think about in retrospect. — *n.* A looking back on things past; view or contemplation of something past. See synonyms under MEMORY. [<L *retrospectus,* pp. of *retrospicere* reexamine, look back < *retro-* back + *specere* look]

ret·ro·spec·tion (ret′rə·spek′shən) *n.* A calling to remembrance; a looking back upon or recollection of the past.

ret·ro·spec·tive (ret′rə·spek′tiv) *adj.* **1** Looking back on the past; of, pertaining to, or referring to the past. **2** Retroactive: said of some legislation. **3** Characterized by retrospection. — **ret′ro·spec′tive·ly** *adv.*

re·trous·sage (rə·troo̅·säzh′) *n.* In etching, a process of wiping a soft cloth across the ink-filled incisions of an etched plate before printing to produce an effect of softness or richness. [<F < *retrousser* turn up]

ret·rous·sé (ret′roo̅·sā′, *Fr.* rə·troo̅·sā′) *adj.* Turned up at the end: said of noses. [<F, pp. of *retrousser* turn up, tuck up < *re-* back + *trousser* fasten together]

ret·ro·ver·sion (ret′rə·vûr′zhən, -shən) *n.* **1** A tipping or bending backward. **2** The state of being turned backward. **3** The act of looking or turning back.

ret·ro·vert (ret′rə·vûrt) *v.t.* To turn back. [<LL *retrovertere* <L *retro-* back + *vertere* turn]

re·turn (ri·tûrn′) *v.i.* **1** To come or go back, as to or toward a former place or condition. **2** To come back or revert in thought or speech. **3** To revert to a former owner. **4** To answer; respond. — *v.t.* **5** To bring, carry, send, or put back; restore; replace. **6** To give in return for something: to *return* ingratitude for kindness. **7** To repay or requite, especially with an equivalent: to *return* a compliment. **8** To yield or produce, as a profit or interest. **9** To send back; reflect, as light or sound. **10** To render (a verdict, etc.). **11** To submit, as a report or writ, to one in authority. **12** To report or announce officially. **13** To replace (a weapon, etc.) in its holder. **14** In card games, to lead (a suit previously led by one's partner). — *n.* **1** The act, process, state, or result of coming back or returning; also, that which is returned; resumption; restoration or replacement; repayment or requital; response; answer; retort; reappearance or recurrence. **2** That which accrues, as from investments, labor, or use; profit. **3** A coming back, reappearance, or recurrence, as of a periodical event or season. **4** A report, list, etc.; especially, a formal or official report, or, in the plural, a set of tabulated statistics: election *returns.* **5** *Archit.* **a** A continuation of a dripstone, hood molding, etc., to form a termination having a different direction from the main part. **b** A part or face of a building at an angle with the main part of the façade. **6** The sending back by a sheriff of a writ to the court from which it was issued; also, a sheriff's report on such writ. **7** *Law* A brief statement, usually endorsed on a writ by the officer to whom it was issued, of what has been done under it; also, the filing of the writ thus endorsed in the office of the clerk or the tribunal whence it was issued. **8** In card games, a returned lead. **9** Any volley, stroke, or thrust received from an opponent; specifically, in a game, the sending of an object, as a tennis ball, from one player to another from whom he has received it. See synonyms under HARVEST, INCREASE, PRODUCT, PROFIT, RESTITUTION. — *adj.* Of or pertaining to a return; given, taken, or done in return; returning: a *return* visit; a *return ticket.* [<OF *returner*] — **re·turn′er** *n.*

re–turn (rē·tûrn′) *v.t.* & *v.i.* To turn again; fold over or back.

re·turn·a·ble (ri·tûr′nə·bəl) *adj.* **1** Capable of being or suitable to be returned. **2** Due and required: said of a judicial writ in reference to the time when and the place where it is to be returned by the officer to whom it is directed.

re·tuse (ri·to̅o̅s′, -tyo̅o̅s′) *adj. Bot.* Having a rounded end or apex in which there is a slight depression, indentation, or notch: said of leaves. [<L *retusus,* pp. of *retundere* beat back < *re-* back + *tundere* beat]

Retz (rets), **Cardinal de,** 1614–79, Jean François Paul de Gondi, French ecclesiastic and author.

Reu·ben (roo̅′bin) A masculine personal name. [<Hebrew, behold, a son]
— **Reuben** The eldest son of Jacob. *Gen.* xxix 32.

Reuch·lin (roikh′lēn, roikh·lēn′), **Johann,** 1455–1522, German humanist and Hebraist. — **Reuch·lin′i·an** *adj.* — **Reuch′lin·ism** *n.*

re·un·ion (rē·yo̅o̅n′yən) *n.* **1** The act of reuniting; renewed harmony. **2** A social gathering of persons who have been separated: a family *reunion.*

Ré·un·ion (rē·yo̅o̅n′yən, *Fr.* rā·ü·nyôṅ′) A French island of the Mascarene group, east of Madagascar; 970 square miles; capital, Saint-Denis: formerly *Bourbon Island.*

re·un·ion·ism (rē·yo̅o̅n′yən·iz′əm) *n.* The principle of renewed union as a policy; specifically, advocacy of reunion of the various Christian churches. — **re·un′ion·ist** *n.* — **re·un′ion·is′tic** *adj.*

re·u·nite (rē′yo̅o̅·nīt′) *v.t.* & *v.i.* **·nit·ed, ·nit·ing** To unite, cohere, or combine again after separation. — **re′u·nit′er** *n.*

Reu·ter·dahl (roi′tər·däl), **Henry,** 1871–1935, U. S. marine and naval painter born in Sweden.

Reu·ters (roi′tərz) *n.* A British organization for collecting news and distributing it to member newspapers. Also **Reuter's News Agency.** [after Baron Paul Julius von *Reuter,* 1816–99, English founder of the agency]

Reu·ther (roo̅′thər), **Walter Philip,** born 1907, U.S. labor leader.

rev (rev) *n.* A revolution, as of a motor or machine part. — *v.t.* & *v.i.* **revved, rev·ving** To alter the speed of (a motor): with *up* or *down.*

Re·val (rā′väl) The German name for TALLINN.

rev·a·len·ta (rev′ə·len′tə) *n. Brit.* Meal made from ground lentils, prepared as food for invalids. [<NL, alter. of earlier *ervalenta* <L *ervum* vetch + *lens, lentis* lentil]

re·vamp (rē·vamp′) *v.t.* **1** To vamp (a boot or shoe) anew. **2** To patch up; make over. — *n.* A thing which is revamped. [<RE- + VAMP]

re·veal (ri·vēl′) *v.t.* **1** To make known; disclose; divulge. **2** To make visible; expose to view; exhibit; show. See synonyms under ANNOUNCE, INFORM, PUBLISH. — *n. Archit.* The vertical side of an aperture or opening in a wall; especially, the portion of the side of a door or window between the line where the window frame or door frame stops and the outer edge of the opening. [<OF *reveler* <L *revelare* unveil < *re-* back + *velum* veil] — **re·veal′a·ble** *adj.* — **re·veal′er** *n.*

re·veal·ment (ri·vēl′mənt) *n.* A revelation; act of revealing; disclosure.

rev·eil·le (rev′i·lē) *n.* **1** A morning signal by drum or bugle, notifying soldiers or sailors to rise. **2** The hour at which this signal is sounded. [<F *reveillez-vous,* imperative of *se reveiller* wake up < *re-* (<L *re-*) again + L *vigilare* watch. See VIGIL.]

rev·el (rev′əl) *v.i.* **·eled** or **·elled, ·el·ing** or **·el·ling** **1** To take delight; indulge freely: with *in*: He *revels* in his freedom. **2** To make merry; engage in boisterous festivities. — *n.* **1** Merry-making; carousing; noisy festivity. **2** An occasion of boisterous festivity; a celebration. [<OF *reveler* make an uproar <L *rebellare.* Doublet of REBEL.] — **rev′el·er** or **rev′el·ler** *n.*
Synonyms (noun): carnival, carousal, carouse, feast, festivity, jollification, merrymaking, revelry, rout.

Re·vel (re′vel·y′) The Russian name for TALLINN.

rev·e·la·tion (rev′ə·lā′shən) *n.* **1** The act or process of revealing, or the state of being revealed. **2** That which is or has been revealed. **3** *Theol.* **a** The act of revealing or communicating divine truth, especially by divine agency or supernatural means. **b** That which has been so revealed, as concerning God in his relations to man. **c** That which is revealed in the Bible; loosely, the Bible itself. [<OF <LL *revelatio, -onis* <L *revelatus,* pp. of *revelare.* See REVEAL.]

Rev·e·la·tion (rev′ə·lā′shən) The Apocalypse, or Book of Revelation: in full, **The Revelation of Saint John the Divine;** the last book of the Bible.

rev·e·la·tion·ist (rev′ə·lā′shən·ist) *n.* One who holds that God has made a supernatural revelation of himself and his will.

rev·e·la·tor (rev′ə·lā′tər) *n.* A revealer. [<LL]

rev·el·ry (rev′əl·rē) *n. pl.* **·ries** Noisy or boisterous merriment.

rev·e·nant (rev′ə·nənt) *n.* **1** One who or that which returns. **2** A ghost; an apparition. [<F, ppr. of *revenir* come back < *re-* back + *venir* come]

re·venge (ri·venj′) *v.* **·venged, ·veng·ing** *v.t.* **1** To inflict punishment, injury, or loss in return for; to take vengeance for; avenge. **2** To take or seek vengeance in behalf of. — *v.i.* **3** *Obs.* To take vengeance. — *n.* **1** The act of returning injury for injury; the infliction of injury or punishment in the spirit of personal vindictiveness; retaliation. **2** A mode or means of avenging oneself or others. **3** The desire for vengeance. [<OF *revenger* < *re-* (<L *re-*) again + *venger* take vengeance <L *vindicare.* See VINDICATE.] — **re·veng′er** *n.*
Synonyms (noun): avenging, requital, retaliation, retribution, vengeance. *Retaliation* and *revenge* are personal and often bitter. *Retaliation* may be partial; *revenge* is meant to be complete and may be excessive. *Vengeance,* which once meant an indignant vindication of justice, now signifies the most furious and unsparing *revenge. Revenge* emphasizes more the personal injury in return for which it is inflicted. A *requital* is an even return, such as to quit one of obligation for what has been received, and may be good or bad. *Avenging* and *retribution* give a solemn sense of exact justice, *avenging* being more personal in its infliction, and *retribution* the impersonal visitation of the doom of righteous law. See HATRED. *Antonyms*: compassion, excuse, forgiveness, grace, mercy, pardon, pity.

re·venge·ful (ri·venj′fəl) *adj.* Vindictive; disposed to, or full of, revenge. — **re·venge′ful·ly** *adv.* — **re·venge′ful·ness** *n.*

re·ve·nons à nos mou·tons (rəv·nôṅ′ zà nō mo̅o̅·tôṅ′) *French* Let us return to our sheep; that is, to our subject.

rev·e·nue (rev′ə·nyo̅o̅, -no̅o̅) *n.* **1** Total current income of a government, except duties on imports: also **internal revenue.** **2** Income from any form of property. **3** The department of government or civil service which collects the national funds: in the United States, the **Internal Revenue Service** of the Department of the Treasury. **4** A source or an item of income. [<F, fem. of *revenu,* pp. of *revenir* return]

revenue cutter An armed vessel in the government revenue service used to enforce customs regulations and prevent smuggling.

Rev·e·nue–Cut·ter Service (rev′ə·nyo̅o̅·kut′ər, -no̅o̅-) Originally, the Revenue-Marine Service created by Congress in 1790: merged into the United States Coast Guard in 1915.

re·ver·ber·ant (ri·vûr′bər·ənt) *adj.* Resounding. [<L *reverberans, -antis,* ppr. of *reverberare.* See REVERBERATE.]

re·ver·ber·ate (ri·vûr′bə·rāt) *v.* **·at·ed, ·at·ing** *v.i.* **1** To resound or re-echo. **2** To be reflected or repelled. **3** To bend back, as flames in a reverberatory furnace. **4** To rebound or recoil. — *v.t.* **5** To echo back (a sound); re-echo. **6** To reflect. **7** To cause to bend back, as flames in a reverberatory furnace; deflect. **8** To expose to heat in a reverberatory furnace. See synonyms under ROAR. [<L *reverberatus,* pp. of *reverberare* strike back, cause to rebound < *re-* back + *verberare* beat]

re·ver·ber·a·tion (ri·vûr′bə·rā′shən) *n.* **1** The act or process of reverberating. **2** That which constitutes reverberating. **3** The rebound or reflection of light, heat, or sound waves. — **re·ver′ber·a′tive** *adj.*

re·ver·ber·a·tor (ri·vûr′bə·rā′tər) *n.* **1** One who or that which causes reverberation. **2** A reflecting lamp, or a reverberatory furnace.

re·ver·ber·a·to·ry (ri·vûr′bər·ə·tôr′ē, -tō′rē)

adj. Producing or intended to produce reverberation; reverberative. — *n.* *pl.* **·ries** A reverberatory furnace.

reverberatory furnace A furnace having a vaulted ceiling that deflects the flame and heat toward the hearth or the upper surface of the substance to be treated.

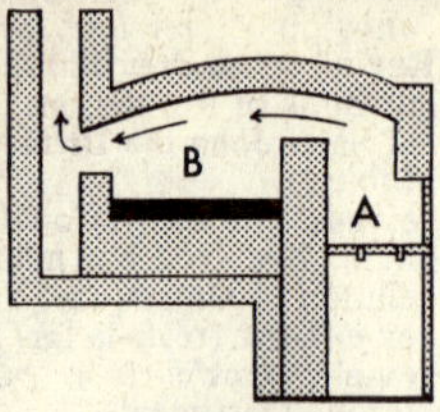

REVERBERATORY FURNACE
A. Flames and gases.
B. Bed of molten iron.

re·vere (ri·vir′) *v.t.* **·vered, ·ver·ing** To regard with veneration; reverence; venerate. See synonyms under ADMIRE, DEFER, VENERATE, WORSHIP. [<L *revereri* feel awe of < *re-* again and again + *vereri* fear] — **re·ver′er** *n.*

Re·vere (ri·vir′), **Paul,** 1735–1818, American silversmith, famous for his midnight ride from Charlestown to Lexington, Mass., the night of April 17–18, 1775, to warn the colonists of the approach of British troops.

rev·er·ence (rev′ər·əns) *n.* **1** A feeling of profound respect often mingled with awe and affection; veneration. **2** An act of respect; an obeisance. **3** The quality or character that commands respect. **4** A reverend person: used as a respectful appellation or title, especially applied to a clergyman. — *v.t.* **·enced, ·enc·ing** To regard with reverence. See synonyms under VENERATE. [<OF <L *reverentia*]

Synonyms (noun): adoration, awe, homage, honor, veneration, worship. See VENERATION. *Antonyms*: contumely, derision, dishonor, insult, irreverence, mockery, outrage, ridicule, scoff, scoffing.

rev·er·end (rev′ər·ənd) *adj.* **1** Worthy of reverence. **2** Being a clergyman; of or pertaining to the clergy or the clerical office. — *n. Colloq.* A clergyman; minister. [<L *reverendus*, gerundive of *revereri*. See REVERE.]

rev·er·ent (rev′ər·ənt) *adj.* **1** Impressed with or feeling reverence. **2** Expressing reverence. [<L *reverens, -entis*] — **rev′er·ent·ly** *adv.*

rev·er·en·tial (rev′ə·ren′shəl) *adj.* Proceeding from or expressing reverence. — **rev′er·en′·tial·ly** *adv.*

rev·er·ie (rev′ər·ē) *n.* *pl.* **·er·ies 1** Abstracted musing; dreaming. **2** A product of such musing in written or musical composition. Also **rev′er·y**. See synonyms under DREAM, THOUGHT. [<F *rêverie* < *rêver* dream, rave, ? <L *rabere* rage]

re·vers (rə·vir′, -vâr′) *n.* *pl.* **·vers** (-virz′, -vârz′) **1** A part of a garment folded over to show the inside, as the lapel of a coat. **2** Material used to cover such a part. [<OF. See REVERSE.]

re·ver·sal (ri·vûr′səl) *n.* **1** The act of reversing. **2** *Physics* The change of a dark to a bright spectral line, or vice versa. **3** *Law* An annulling or setting aside: the *reversal* of a decree.

re·verse (ri·vûrs′) *adj.* **1** Turned backward; contrary or opposite in direction, character, order, etc. **2** On the other side; backward; inverted. **3** Causing backward motion: the *reverse* gear of an automobile. — *n.* **1** That which is directly opposite or contrary: The *reverse* of what you say is true. **2** The back, rear, or secondary side or surface, as distinguished from the front or principal side. **3** A reversing; change to an opposite position, direction, or state; reversal: a *reverse* of a gun or gun carriage. **4** A change or alteration for the worse; a check or partial defeat; misfortune. **5** *Mech.* A reversing gear or movement. See synonyms under MISFORTUNE. — *v.* **·versed, ·vers·ing** *v.t.* **1** To turn upside down or inside out; invert or overturn. **2** To turn in an opposite direction. **3** To transpose; exchange. **4** To change into something different or opposite; alter: to *reverse* policy. **5** To set aside; annul: to *reverse* a decree. **6** *Mech.* To cause to have an opposite motion or effect: *Reverse* engines! — *v.i.* **7** To move or turn in the opposite direction, as in dancing. **8** To reverse its action: said of engines, etc. See synonyms under ABOLISH. [<OF *revers* <L *reversus*, pp. of *revertere*. See REVERT.] — **re·vers′er** *n.*

reverse fault *Geol.* A thrust fault.

re·verse·ly (ri·vûrs′lē) *adv.* In a reverse or contrary manner.

re·vers·i·ble (ri·vûr′sə·bəl) *adj.* **1** Capable of being reversed in either direction or position. **2** Capable of going either forward or backward, as a chemical reaction or physiological process. **3** Capable of being used or worn inside out or backward: a *reversible* coat. **4** Having the finish on both sides, as a fabric. — *n.* A reversible coat. — **re·vers′i·bil′i·ty, re·vers′i·ble·ness** *n.* — **re·vers′i·bly** *adv.*

re·ver·sion (ri·vûr′zhən, -shən) *n.* **1** A return to or toward some former state or condition. **2** The act of reversing or the state of being reversed. **3** A return, as to a former practice or belief. **4** *Biol.* **a** The recurrence or reappearance in an individual of characteristics which had not been evident for two or more generations; atavism. **b** An example of such recurrence. **5** *Law* **a** The return of an estate to the grantor or his heirs after the expiration of the grant. **b** The estate so returning. **c** The right of succession to an estate. **6** *Obs.* Remainder. [<OF <L *reversio, -onis.* See REVERT.]

re·ver·sion·al (ri·vûr′zhən·əl, -shən-) *adj.* Reversionary.

re·ver·sion·ar·y (ri·vûr′zhən·er′ē, -shən-) *adj.* Of, pertaining to, characterized by, or involving reversion.

re·ver·sion·er (ri·vûr′zhən·ər, -shən-) *n. Law* One entitled to an estate in reversion.

re·ver·so (ri·vûr′sō) *n.* *pl.* **·sos** A left-hand page: opposed to *recto.* [<Ital. *riverso* reverse]

re·vert (ri·vûrt′) *v.i.* **1** To go or turn back to a former place, condition, attitude, topic, etc. **2** *Biol.* To return to or show characteristics of an earlier, primitive type. **3** *Law* To return to the former owner or to his heirs. — *n.* **1** One who is reconverted to a former faith. **2** That which reverts. [<OF *revertir* <L *revertere* turn back < *re-* back + *vertere* turn] — **re·vert′i·ble** *adj.* — **re·ver′tive** *adj.*

re·vest (rē·vest′) *v.t.* **1** To vest again, as with rank, authority, or ownership; reinvest. **2** To vest again, as office or powers. — *v.i.* **3** To take effect again, as a title reverting to a former owner. [<OF *revestir* <LL *revestire* reclothe <L *re-* again + *vestire* clothe < *vestis* a garment]

re·vet (ri·vet′) *v.t.* **·vet·ted, ·vet·ting** To face, as an embankment, with masonry. [<F *revêtir* clothe <L *revestire.* See REVEST.]

re·vet·ment (ri·vet′mənt) *n.* A facing, sheathing, or retaining wall, as of masonry, for protecting earthworks, river banks, etc. [<F *revêtement*]

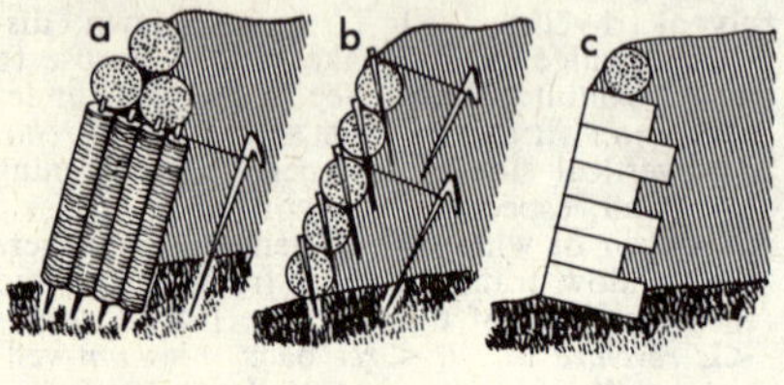

REVETMENTS
a. Built of gabions supporting fascines.
b. Built of fascines only.
c. Built of timbers or concrete.

re·view (ri·vyōō′) *v.t.* **1** To go over or examine again; look at or study again. **2** To look back upon, as in memory; think of retrospectively. **3** To go over, as a manuscript, so as to correct defects. **4** To make an inspection of, especially formally. **5** To write or make a critical review of, as a new book. **6** *Law* To examine (something done or adjudged by a lower court) so as to determine its legality or correctness. — *v.i.* **7** To write a review or reviews, as for a magazine. [<RE- + VIEW; modeled on F *revoir* look at again] — *n.* **1** A second, repeated, or new view, examination, consideration, or study of something; a retrospective survey. **2** A lesson studied or recited again. **3** Critical study or examination. **4** An article or essay containing a critical examination, discussion, or notice of some work; a criticism; critique. **5** A periodical devoted to essays in criticism and on general subjects. **6** A formal or official inspection or view, as of troops. **7** *Law* A judicial revision by a superior court of the order or decree of a subordinate court. **8** A revision, as of a work by its author; examination with a view to correction or improvement. [<MF *reveue* <pp. of *revoir* <L *revidere* < *re-* again + *videre* see]

re·view·al (ri·vyōō′əl) *n.* A review; the act of reviewing.

re·view·er (ri·vyōō′ər) *n.* A critic or examiner; an essayist in critical periodicals; a book reviewer.

re·vile (ri·vīl′) *v.* **·viled, ·vil·ing** *v.t.* To assail with abusive or contemptuous language; vilify; abuse. — *v.i.* To use abusive or contemptuous language. [<OF *reviler* treat as vile < *re-* + *vil* vile] — **re·vile′ment** *n.* — **re·vil′er** *n.* — **re·vil′ing·ly** *adv.*

Synonyms: abuse, asperse, calumniate, defame, malign, reproach, slander, traduce, upbraid, vilify. See ABUSE, ASPERSE. *Antonyms*: see synonyms for PRAISE.

Re·vil·la·gi·ge·do Island (ri·vil′ə·gi·gē′dō) An island in the Alexander Archipelago, SE Alaska; 1,120 square miles.

re·vis·al (ri·vī′zəl) *n.* Revision; the act of revising.

re·vise (ri·vīz′) *v.t.* **·vised, ·vis·ing 1** To read or read over so as to correct errors, suggest or make changes, etc.: to *revise* a manuscript or the proofs of a book. **2** To change; alter: He has *revised* his opinions. — *n.* **1** The act or result of revising or reviewing; a revision. **2** A corrected proof after revision. [<F *reviser* <L *revisere* look back, see again < *re-* again + *visum,* pp. of *videre* see] — **re·vis′er** or **re·vi′sor** *n.*

Revised Version A translation of the Bible into English, made by two bodies, one of English and one of American scholars, in the years 1870–84.

re·vi·sion (ri·vizh′ən) *n.* The act or result of revising; a revised version or edition. — **re·vi′·sion·al, re·vi′sion·ar′y** *adj.*

re·vi·sion·ism (ri·vizh′ən·iz′əm) *n.* The advocacy of revision.

re·vi·sion·ist (ri·vizh′ən·ist) *n.* **1** One who advocates revision. **2** A reviser.

re·vis·it (rē·viz′it) *v.t.* To visit again. — *n.* A return visit. — **rē′vis·i·ta′tion** *n.*

re·vi·so·ry (ri·vī′zər·ē) *adj.* Effecting, or capable of effecting, revision; revising: *revisory* powers.

re·vi·tal·ize (rē·vī′təl·īz) *v.t.* **·ized, ·iz·ing** To restore vitality to; bring back to life; revive. — **re·vi′tal·i·za′tion** *n.*

re·vi·val (ri·vī′vəl) *n.* **1** The act of reviving, or the state of being revived; specifically, a recovery, as from depression. **2** A restoration or resuscitation after neglect, oblivion, or obscurity: the *revival* of letters. **3** A renewal of special interest in and attention to religious services and duties and the subject of personal salvation; a religious awakening. **4** A series of emotional and sensational evangelical meetings.

re·vi·val·ism (ri·vī′vəl·iz′əm) *n.* **1** The spirit and methods of religious revivals or revivalists, or that promote revivals. **2** A tendency to restore former conditions or principles.

re·vi·val·ist (ri·vī′vəl·ist) *n.* A preacher or leader in a religious revival movement.

revival of learning or **literature** See RENAISSANCE.

re·vive (ri·vīv′) *v.* **·vived, ·viv·ing** *v.t.* **1** To bring to life again after real or apparent death; restore to consciousness. **2** To give new vigor, health, etc., to. **3** To bring back into use or currency. **4** To make effective or operative again. **5** To renew in the mind or memory; refresh; reawaken. **6** To produce again, as an old play. — *v.i.* **7** To come back to life again; return to consciousness. **8** To assume new vigor, health, etc. **9** To come back into use or currency. **10** To become effective or operative again. [<F *revivre* <L *revivere* < *re-* again + *vivere* live] — **re·viv′er** *n.*

re·viv·i·fy (ri·viv′ə·fī) *v.t.* **·fied, ·fy·ing** To give new life or spirit to; revive. [<F *revivifier* <L *revivificare* < *re-* again + *vivificare* vivify < *vivus* alive + *facere* make] — **re·viv′i·fi·ca′·tion** *n.*

rev·i·vis·cence (rev′ə·vis′əns) *n.* A renewal of life or of vital activities and vigor; a return to life; restoration; revival. Also **rev′i·vis′·cen·cy.** [<L *reviviscens, -entis,* ppr. of *reviviscere* < *re-* again + *viviscere* come to life, freq. of *vivere* live] — **rev′i·vis′cent** *adj.*

rev·o·ca·ble (rev′ə·kə·bəl) *adj.* Capable of being revoked. [<F *révocable*] — **rev′o·ca·bil′i·ty** *n.* — **rev′o·ca·bly** *adv.*

rev·o·ca·tion (rev′ə·kā′shən) *n.* **1** The act of

re·voice revoking, or the state of being revoked; repeal; reversal. 2 *Law* The annulment or cancellation of an instrument, act, or promise by or in behalf of the party who made it. 3 *Obs.* A summoning back or recalling. [<OF *revocacion*] —**rev·o·ca·to·ry** (rev′ə·kə·tôr′ē, -tō′rē) *adj.*

re·voice (rē·vois′) *v.t.* **·voiced, ·voic·ing 1** To restore or give the proper quality of tone to: to *revoice* an organ pipe. **2** To voice again or in return; echo.

re·voke (ri·vōk′) *v.* **·voked, ·vok·ing** *v.t.* **1** To annul or make void by recalling; cancel; rescind. **2** *Obs.* To call or summon back; recall. —*v.i.* **3** In card games, to fail to follow suit when possible and when required by the rules. See synonyms under ABOLISH, ANNUL, CANCEL, RECANT, RENOUNCE. —*n.* **1** An annulling or cancellation. **2** In card games, neglect to follow suit; a renege. [<OF *revoquer* <L *revocare* <*re-* back + *vocare* call] —**re·vok′er** *n.*

re·volt (ri·vōlt′) *n.* **1** A throwing off of allegiance and subjection; an uprising against authority; a rebellion or mutiny; insurrection. **2** An act of protest, refusal, revulsion, or disgust. See synonyms under REVOLUTION. —*v.i.* **1** To rise in rebellion against constituted authority; renounce allegiance; mutiny; rebel. **2** To turn away in disgust or abhorrence; be shocked or repelled: with *against, at,* or *from.* —*v.t.* **3** To cause to feel disgust or revulsion; repel. [<F *révolte* <*révolter* <Ital. *rivoltare* <L *revolutus,* pp. of *revolvere.* See REVOLVE.] —**re·volt′er** *n.*

re·volt·ing (ri·vōl′ting) *adj.* Abhorrent; loathsome; nauseating. —**re·volt′ing·ly** *adv.*

rev·o·lute (rev′ə·loot) *adj. Bot.* Rolled backward or downward from the margins upon the under surface, as certain leaves. [<L *revolutus.* See REVOLT.]

rev·o·lu·tion (rev′ə·loo′shən) *n.* **1** The act or state of revolving. **2** A motion in a closed curve around a center, or a complete or apparent circuit made by a body in such a course: used generally in this sense in distinction from *rotation.* **3** Rotation about an axis; especially, a complete rotation so that every part of the moving body returns to the position from which it started. **4** *Mech.* Any winding or turning about an axis, as in a spiral or other bend, so as to come to a point corresponding to the starting point. **5** A group, round, or cycle of successive events or changes; a cycle; also, the period of space or time occupied by a cycle or by the accomplishment of a circuit. **6** The overthrow and replacement of a government or political system by those governed. **7** An extensive or drastic change in a condition, method, idea, etc.: a *revolution* in industry. [<OF *revolucion* <LL *revolutio, -onis* <L *revolutus,* pp. of *revolvere.* See REVOLVE.] —*Synonyms:* anarchy, confusion, disintegration, disorder, insubordination, insurrection, lawlessness, mutiny, rebellion, revolt, riot, sedition, tumult. The essential idea of *revolution,* in definition 6, is a change in the form of government or constitution, or a change of rulers, otherwise than as provided by existing laws of succession, election, etc.; while such change is apt to involve armed hostilities, these make no necessary part of a *revolution,* which may be accomplished without a battle. *Anarchy* refers to the condition of a state when government is superseded or destroyed by factions. A *revolt* is an uprising against existing authority without the comprehensive views of change in the form or administration of government that are involved in *revolution.* See CHANGE. Compare ANARCHY, REBELLION, REVOLT. *Antonyms:* authority, command, control, domination, dominion, empire, government, law, loyalty, obedience, order, rule, sovereignty, submission, supremacy.
—**American Revolution** The war for independence carried on by the thirteen American colonies against Great Britain, 1775–83. Also *Revolutionary War.* See table under WAR.
—**Chinese Revolution** The events in China during the years 1911–12, inspired by Sun Yat-sen, which overthrew the Dowager Empress and the Manchu Empire, and resulted in the establishment of a republic.
—**English Revolution** The course of events in England in 1642–89 that brought about the execution of Charles I, the rise of the Commonwealth, the dethronement of James II, and the establishment of a constitutional government under William of Orange and Mary: called in England **The Revolution,** sometimes with reference to the events of 1688.
—**French Revolution** The revolution which began in 1789, overthrew the French monarchy, and culminated in the Empire of Napoleon I.
—**Russian Revolution** The conflict (1917–22), beginning in a Petrograd uprising on March 12, 1917, that resulted in a provisional moderate government and the abdication of Nicholas II. On November 6 (October 24, Old Style), the Bolsheviks under Lenin overthrew this government (the *October Revolution*), and after resisting counter-revolution and libertarian revolution until December, 1922, united the soviet states in the Union of Soviet Socialist Republics under Communist (Bolshevik) control.

rev·o·lu·tion·ar·y (rev′ə·loo′shən·er′ē) *adj.* **1** Pertaining to or of the nature of revolution, especially political; causing or tending to produce revolution. **2** Rotating; revolving. —*n. pl.* **·ar·ies** A revolutionist.

Revolutionary calendar See CALENDAR (Republican).

Revolutionary War See AMERICAN REVOLUTION and REVOLUTION.

rev·o·lu·tion·ist (rev′ə·loo′shən·ist) *n.* One who takes part in a revolution.

rev·o·lu·tion·ize (rev′ə·loo′shən·īz) *v.t.* **·ized, ·iz·ing** To effect a radical or entire change in the character, government, or affairs of: to *revolutionize* a country.

re·volve (ri·volv′) *v.* **·volved, ·volv·ing** *v.i.* **1** To move in an orbit about a center; move in a circle. **2** To rotate. **3** To move in cycles; recur periodically. —*v.t.* **4** To cause to move in a circle or orbit. **5** To cause to rotate. **6** To turn over mentally; consider; ponder. [<L *revolvere* <*re-* back + *volvere* roll, turn] —**re·volv′a·ble** *adj.* —**re·volv′ing·ly** *adv.*
—*Synonyms:* roll, rotate, turn. Any round body *rolls* which continuously touches with successive portions of its surface successive portions of another surface; a wagon wheel *rolls* along the ground. To *rotate* is said of a body that has a circular motion about its own center or axis; to *revolve* is said of a body that moves about a center outside of itself. A *revolving* body may also either *rotate* or *roll* at the same time; the earth *revolves* around the sun, and *rotates* on its own axis. Any object that is in contact with or connected with a *rolling* body is often said to *roll;* as, The car *rolls* smoothly along the track. Objects whose motion approximates or suggests a rotary motion along a supporting surface are also said to *roll;* as, Ocean waves *roll* in upon the shore. *Antonyms:* bind, chafe, grind, slide, slip, stick.

re·volv·er (ri·vol′vər) *n.* **1** One who or that which revolves. **2** A type of pistol with a revolving cylinder in the breech chambered to hold several cartridges so that it may be fired in succession without reloading.

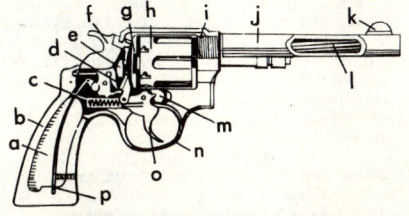

NOMENCLATURE OF THE REVOLVER

a. Stock.	f. Hammer.	l. Rifling.
b. Frame.	g. Extractor.	m. Cylinder stop.
c. Trigger spring	h. Cylinder.	n. Trigger guard.
	i. Barrel pin.	
d. Sear.	j. Barrel.	o. Trigger.
e. Bolt.	k. Front sight.	p. Mainspring.

revolving door A door rotating like a turnstile about a central post and consisting of three or four adjustable leaves so encased in a doorway as to exclude drafts of air.

revolving fund A fund set up to finance loans or operations which yield returns that are placed in the fund for re-use.

revolving stage A circular stage divided in sections, each set for a different scene: by revolving the stage, scenes may be rapidly changed.

re·vue (ri·vyoo′) *n.* A kind of musical comedy, without plot or dramatic sequence, characterized by songs and dances, and by a series of skits which lampoon or burlesque contemporary people and events. [<F. See REVIEW.]

re·vul·sion (ri·vul′shən) *n.* **1** A sudden change of feeling, conduct, or conditions; a strong reaction of any kind. **2** The drawing back or away from something; violent movement, or recoil. **3** *Med.* A turning or diverting of any disease from one part of the body to another, as by counterirritation. [<OF <L *revulsio, -onis* < *revulsus,* pp. of *revellere* pluck away <*re-* back + *vellere* pluck, pull] —**re·vul′sive** *adj.*

Re·wa (rē′wə) **1** A trading city in northern Madhya Pradesh State, India; capital of the former state of Vindhya Pradesh, 1948–56; before 1948, capital of the princely state of Rewa. **2** A former princely state of central India; merged with the former state of Vindhya Pradesh, 1948, which merged with Madhya Pradesh State, 1956; 12,830 square miles.

re·ward (ri·wôrd′) *n.* **1** Something given or done in return; especially, a gift, prize, or recompense for merit, service, or achievement; also, punishment or retribution for evil. **2** Money offered for information, for the return of lost goods, the apprehension of criminals, etc. **3** Merited results; just deserts: He has gone to his *reward.* See synonyms under RECOMPENSE, SUBSIDY. —*v.t.* To give a reward to or for; requite; be a reward for; recompense. See synonyms under PAY, REQUITE. [<AF *rewarder,* OF *regarder* look at. Doublet of REGARD.] —**re·ward′er** *n.*

re·wind (rē·wīnd′) *v.t.* **·wound, ·wind·ing** To wind or coil anew.

re·wire (rē·wīr′) *v.t.* **·wired, ·wir·ing** To wire again, as a house or a machine.

re·word (rē·wûrd′) *v.t.* **1** To say again in other words; express differently. **2** To utter or say again in the same words; repeat.

re·write (rē·rīt′) *v.t.* **·wrote, ·writ·ten, ·writ·ing 1** To write over again. **2** In American journalism, to put into publishable form (a story submitted by a reporter). —*n.* (rē′rīt′) A news item sent in by a reporter and rewritten for publication.

Rex (reks) A masculine personal name. [<L, king]

Rey·kja·vik (rā′kyä·vēk′) The capital of Iceland, a port on the SW coast.

Rey·mont (rā′mônt), **Władysław Stanisław,** 1867–1925, Polish novelist.

Rey·nal·do (rā·näl′thō) Spanish form of REGINALD.

Reyn·ard (ren′ərd, rā′nərd) *n.* The fox, especially as the personification of cunning. [<MDu. <OF *Renard* <OHG *Reginhard,* name of the protagonist in *Reynard the Fox,* the medieval beast epic]

Rey·naud (rā·nō′), **Paul,** born 1878, French statesman; premier, 1940.

Reyn·old (ren′əld) See REGINALD.

Reyn·olds (ren′əldz), **Sir Joshua,** 1723–92, English painter.

Re·za·i·yeh (ri·zä′ē·yä′) See RIZAHIYEH.

rhab·do·man·cy (rab′də·man′sē) *n.* Divination; the discovery of springs, precious metals, etc., by means of a divining rod: also spelled *rabdomancy.* [<LL *rhabdomantia* <Gk. *rhabdomanteia* < *rhabdos* a rod + *manteia* divination] —**rhab′do·man′tist** *n.*

rha·chis (rā′kis) See RACHIS.

Rhad·a·man·thus (rad′ə·man′thəs) In Greek mythology, a son of Zeus and Europa who was noted for justice during his lifetime, and in the afterworld was made a judge, together with Minos and Aeacus. Also **Rhad′a·man′thys.** —**Rhad′a·man′thine** (-thin) *adj.*

Rhae·ti·a (rē′shē·ə) An ancient Roman province, including part of modern Tirol and the Grisons, and later extended to the Danube. Also **Rhæ′ti·a.** —**Rhae·tian** (rē′shən) *adj. & n.*

Rhaetian Alps A division of the central Alps

on the Italo-Swiss and Swiss-Austrian borders, within the boundaries of ancient Rhaetia; highest peak, 13,300 feet.
Rhae·tic (rē′tik) adj. 1 Geol. Of or pertaining to a group of rock strata representing the upper division of the Triassic system in England and western Europe. 2 Of or pertaining to the Rhaetian Alps. Also **Rhe′tic**. [< L *Rhaeticus*]
Rhae·to-Ro·man·ic (rē′tō-rō-man′ik) adj. Of or pertaining to the peoples of SE Switzerland, northern Italy, and Tirol, or to their Romance dialects known as Ladin, Romansch, and Friulian. — n. These dialects as a group.
-rhage, -rhagia, -rhagy See -RRHAGIA.
rham·na·ceous (ram-nā′shəs) adj. Bot. Of, pertaining to, or designating a family (*Rhamnaceae*) of spiny shrubs and small trees, the buckthorn family, having simple leaves and regular flowers in cymes. [< Gk. *rhamnos*, a kind of prickly shrub]
rha·phe (rā′fē) See RAPHE.
-raphy See -RRHAPHY.
rhap·so·dist (rap′sə-dist) n. 1 Among the ancient Greeks, a wandering minstrel who recited epic poems, either his own or another's; especially, one who declaimed the Homeric poems. 2 One who expresses himself with exaggeration of sentiment in speech or writing.
rhap·so·dize (rap′sə-dīz) v.t. & v.i. **·dized, ·diz·ing** To express or recite rhapsodically.
rhap·so·dy (rap′sə-dē) n. pl. **·dies** 1 A series of disconnected and often extravagant sentences, extracts, or utterances, gathered or composed under excitement; rapt or rapturous utterance. 2 In ancient Greece, an epic poem, or a part of such a poem, especially from the *Odyssey* or *Iliad*, recited by a rhapsodist; also, the recitation itself. 3 *Music* An instrumental composition of irregular form, often suggesting the qualities of improvisation. 4 A miscellaneous collection; a medley. [< F *rapsodie* < L *rhapsodia* < Gk. *rhapsōidia* < *rhapsōidos* rhapsodist < *rhaptein* stitch together + *ōidē* song] — **rhap·sod·ic** (rap·sod′ik) or **·i·cal** adj. — **rhap·sod′i·cal·ly** adv.
rhat·a·ny (rat′ə-nē) n. 1 Either of two perennial, shrubby South American plants of the pea family (genus *Krameria*), the **Peruvian rhatany** (*K. triandra*) or the **Brazilian rhatany** (*K. argentea*), whose dried roots are used in medicine. 2 The roots of these plants, or medicinal substances prepared from them. Also spelled *ratany*. [< NL < Sp. *ratania* < Quechua]
rhe·a (rē′ə) n. A ratite bird (genus *Rhea*) of the plains of South America, smaller than true ostriches, and having three toes: also called *ostrich*. [< NL]
Rhe·a (rē′ə) In Greek mythology, the daughter of Uranus and Gaea, wife of her brother Kronos, and mother of Zeus, Poseidon, Hades, Hera, Demeter, and Hestia: identified with the Phrygian *Cybele* and the Roman *Ops*: also called *Mother of the Gods*. See KRONOS.
-rhea See -RRHEA.
Rhea Sylvia In Roman legend, a vestal, the mother by Mars of Romulus and Remus.
Rhee (rē), **Syngman**, born 1875, Korean statesman; president 1948–1960.
Rheims (rēmz, *Fr.* raṅs) See REIMS.
rhe·in (rē′in) n. *Chem.* A yellow crystalline acid, $C_{15}H_8O_6$, obtained from senna leaves and Chinese rhubarb: sometimes used as a purgative. [< Gk. *rhēon* rhubarb + -IN]
Rhein (rīn) The German name for RHINE.
Rhein·fall (rīn′fäl) The German name for SCHAFFHAUSEN FALLS.
Rhein·gold (rīn′gōld, *Ger.* rīn′gōlt) 1 In Wagner's *Der Ring des Nibelungen,* the gold snatched from the Rhine by Alberich, from which he made the magical ring. 2 The title of the first of the tetralogy of music dramas by Wagner forming *Der Ring des Nibelungen.* Also spelled *Rhinegold*. [< G]
Rhein·land (rīn′länt) The German name for RHINELAND.
Rhein·pfalz (rīn′pfälts) The German name for RHINE PALATINATE.
rhe·mat·ic (ri-mat′ik) adj. 1 Relating to or derived from a verb. 2 Pertaining to the formation of words. [< Gk. *rhēma* word, verb]
Rhen·ish (ren′ish) adj. Pertaining to the river Rhine, or to the adjacent regions. — n. Rhine wine. [< L *Rhenus* Rhine]

Rhenish Hesse An administrative division of Rhineland-Palatinate, West Germany; 517 square miles.
Rhenish Prussia See RHINE PROVINCE.
rhe·ni·um (rē′nē-əm) n. A heavy, lustrous, metallic chemical element (symbol Re) of the manganese group. See ELEMENT. [< NL < L *Rhenus* Rhine]
Rhe·nus (rē′nəs) Ancient name for the RHINE.
rheo- *combining form* Current or flow, as of water or electricity: *rheostat*. [< Gk. *rheos* a current]
rhe·o·base (rē′ə-bās) n. *Physiol.* The minimum voltage of an electric current required to stimulate a nerve or muscle. Compare CHRONAXY.
rhe·ol·o·gy (rē-ol′ə-jē) n. The study of the properties and behavior of flowing substances; the science of flow. [< RHEO- + -LOGY] — **rhe·ol′o·gist** n.
rhe·om·e·ter (rē-om′ə-tər) n. A device for indicating the force or velocity of blood circulation. [< RHEO- + -METER]
rhe·o·scope (rē′ə-skōp) n. A galvanoscope. — **rhe′o·scop′ic** (-skop′ik) adj.
rhe·o·stat (rē′ə-stat) n. *Electr.* A device for regulating current-strength of electricity, as by resistance coils. [< RHEO- + Gk. *statos* standing] — **rhe′o·stat′ic** adj.

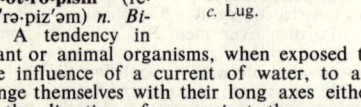

rhe·o·tax·is (rē′ə-tak′sis) n. *Biol.* The response of an organism to the influence of a current of water, especially of water. — **rhe′o·tac·tic** (-tak′tik) adj.
rhe·ot·ro·pism (rē-ot′rə-piz′əm) n. *Biol.* A tendency in plant or animal organisms, when exposed to the influence of a current of water, to arrange themselves with their long axes either in the direction of or against the current. — **rhe·o·trop·ic** (rē′ə-trop′ik) adj.
rhe·sus (rē′səs) n. A macaque (*Macaca mulatta*) with a moderate tail, common throughout India. [< NL < Gk. *Rhēsos* Rhesus; arbitrarily assigned]
Rhe·sus (rē′səs) In the *Iliad,* a king of Thrace and ally of the Trojans, killed by Odysseus the night of his arrival before Troy.
Rhesus factor (rē′səs) See Rh FACTOR.

RHESUS
(From 12 to 18 inches long; tail, 6 to 8 inches)

rhe·tor (rē′tər) n. 1 Formerly, one who taught rhetoric. 2 An orator. [< L < Gk. *rhētōr*]
rhet·o·ric (ret′ə-rik) n. 1 The art of discourse; skill in the use of language. 2 The power of pleasing or persuading. 3 A textbook treating of discourse; especially, written discourse. 4 Affected and exaggerated display in the use of language. 5 Prose, as opposed to verse. [< F *rhétorique* < L *rhetorica* < Gk. *rhētorikē* (*technē*) rhetorical (art)]
rhe·tor·i·cal (ri-tôr′i-kəl, -tor′-) adj. 1 Pertaining to rhetoric; oratorical; declamatory. 2 Designed for showy oratorical effect. — **rhe·tor′i·cal·ly** adv. — **rhe·tor′i·cal·ness** n.
rhetorical question A question put only for oratorical or literary effect, the answer being implied in the question.
rhetorical stress The emphasis required by the meaning of a line or the lines in a poem: opposed to *metrical stress*.
rhet·o·ri·cian (ret′ə-rish′ən) n. 1 A master or teacher of rhetoric. 2 An orator; one who writes or speaks eloquently. [< F *rhétoricien*]
rheum (rōōm) n. 1 *Pathol.* Catarrhal discharge from the nose and eyes; hence, a cold. 2 *Med.* Any thin watery flux, as tears or saliva. [< OF *reume* < L *rheuma* < Gk. *rheuma* a flow < *rheein* flow] — **rheum′y** adj.
rheu·mat·ic (rōō-mat′ik) adj. Pertaining to, causing, or affected with rheumatism. — n.

1 One affected with or liable to rheumatism. 2 *pl. Colloq.* Rheumatic pains. [< OF *reumatique* < L *rheumaticus* < Gk. *rheumatikos* < *rheuma*. See RHEUM.]
rheumatic fever *Pathol.* A severe, probably infectious disease chiefly affecting children and young adults, characterized by painful inflammation around the joints, typically intermittent fever, and inflammation of the pericardium and valves of the heart.
rheu·ma·tism (rōō′mə-tiz′əm) n. *Pathol.* 1 A variable, shifting, painful inflammation and stiffness of the muscles, joints, or other structures. 2 Rheumatic fever. 3 Rheumatoid arthritis. [< L *rheumatismus* rheum < Gk. *rheumatismos* < *rheuma* rheum]
rheu·ma·toid (rōō′mə-toid) adj. *Pathol.* 1 Resembling rheumatism or rheumatic symptoms: *rheumatoid* arthritis. 2 Afflicted with rheumatism. Also **rheu′ma·toi′dal** (-toid′l). — **rheu′ma·toi′dal·ly** adv.
rheumatoid arthritis *Pathol.* A persisting inflammatory disease of the joints, marked by atrophy, rarefaction of the bones, and deformities.
Rheydt (rīt) A city in western North Rhine-Westphalia, West Germany; the twin city of München-Gladbach.
Rh factor *Biochem.* Any of a group of genetically transmitted agglutinogens present in the blood of most individuals (Rh positive) and which may cause hemolytic reactions under certain conditions, as during pregnancy or following transfusions with blood lacking this factor (Rh negative). Also called *Rhesus factor*.
rhig·o·lene (rig′ə-lēn) n. *Chem.* A colorless, volatile, inflammable liquid distillate of petroleum: used in medicine as a local freezing anesthetic for minor operations. [< Gk. *rhigos* frost + L *oleum* oil]
rhin- Var. of RHINO-.
rhi·nal (rī′nəl) adj. Of or pertaining to the nose; nasal. [< RHIN- + -AL]
Rhine (rīn) The principal river of west central Europe, flowing 810 miles north from SE Switzerland, through Germany and Netherlands, to the North Sea, forming part of the SW boundary of Germany, and dividing, in the Netherlands, into the *Waal,* the *Lek,* the *Oude Rijn,* and the *Ijssel*: ancient *Rhenus,* German *Rhein,* Dutch *Rijn.* French *Rhin* (raṅ).
Rhine (rīn), **Joseph Banks,** born 1895, U.S. psychologist.
Rhine·gold (rīn′gōld) The hoard of the Nibelungs, secreted in the Rhine. Compare RHEINGOLD.
Rhine·land (rīn′land′) 1 That part of Germany west of the Rhine. 2 The Rhine Province. German *Rheinland.*
Rhine·land-Pa·lat·i·nate (rīn′land′pə-lat′ə-nāt) A state of West Germany; 7,654 square miles; capital, Mainz. German *Rhein·land-Pfalz* (rīn′länt-pfälts′).
rhi·nen·ceph·a·lon (rī′nen·sef′ə-lon) n. pl. **·la** (-lə) *Anat.* That portion of the brain which forms the olfactory lobe, consisting of the olfactory tubercle, tract, and bulb, which give origin to the sense of smell. [< RHIN- + ENCEPHALON] — **rhi·nen·ce·phal·ic** (rī′nen·sə·fal′ik) adj.
Rhine Palatinate See PALATINATE, THE. German *Rheinpfalz.*
Rhine Province A former Prussian province in western Germany, included since 1945 in Rhineland-Palatinate; 9,451 square miles; former capital, Coblenz: also *Rhenish Prussia.*
rhine·stone (rīn′stōn′) n. A highly refractive, colorless glass or paste, used as an imitation gemstone. [Trans. of F *caillou du Rhin;* orig. made at Strasbourg]
Rhine wine Wine made from grapes grown in the neighborhood of the Rhine; specifically, the white, still wines of this region, noted for their delicate bouquet; hock.
rhi·ni·tis (rī·nī′tis) n. *Pathol.* Inflammation of the mucous membranes of the nose; nasal catarrh. [< RHIN- + -ITIS]
rhi·no (rī′nō) n. pl. **·nos** A rhinoceros.
rhino- *combining form* Nose; nasal: *rhinoplasty.* Also, before vowels, *rhin-.* [< Gk. *rhis, rhinos* nose]
rhi·noc·er·os (rī·nos′ər·əs) n. pl. **·ros·es** or **·ros** A large, herbivorous, odd-toed mammal (family *Rhinocerotidae*) of Africa and Asia, with one or two keratin-fiber horns on the

snout, a very thick hide, and the upper lip protruded and prehensile. [<LL <Gk. *rhinokerōs* < *rhis, rhinos* nose + *keras* horn]

RHINOCEROS
a. African: About 5 feet at the shoulder; to 3000 pounds.
b. Indian: About 5 1/2 feet at the shoulder; to 4000 pounds.

rhi·nol·o·gy (rī·nol′ə·jē) *n.* The branch of medicine that relates to the nose and its diseases. [<RHINO- + -LOGY] — **rhi·nol′o·gist** *n.*
rhi·no·plas·ty (rī′nō·plas′tē) *n.* Plastic surgery of the nose. — **rhi′no·plas′tic** *adj.*
rhi·no·scope (rī′nə·skōp) *n.* An instrument for inspecting the nasal cavities.
rhi·nos·co·py (rī·nos′kə·pē) *n.* Inspection of the nasal passages.
rhizo- *combining form* Root; pertaining to a root or to roots: *rhizogenic*. Also, before vowels, **rhiz-**. [<Gk. *rhiza* a root]
rhi·zo·bi·um (rī·zō′bē·əm) *n. pl.* **·bi·a** (-bē·ə) *Bacteriol.* One of a genus (*Rhizobium*) of rod-shaped, nitrogen-fixing bacteria causing nodules on the roots of leguminous plants. [<NL <RHIZO- + Gk. *bios* life]
rhi·zo·car·pous (rī′zō·kär′pəs) *adj. Bot.* Having annual stems and foliage growing from perennial roots: said of perennial plants. Also **rhi′zo·car′pic.**
rhi·zo·ceph·a·lous (rī′zō·sef′ə·ləs) *adj. Zool.* Naming or pertaining to a suborder (*Rhizocephala*) of parasitic cirripeds, without antennae or feet, which attach themselves to crabs by a short peduncle from which rootlike processes branch out.
rhi·zo·gen·ic (rī′zō·jen′ik) *adj. Bot.* Root-producing: said of the layer of mother cells at the periphery of the central cylinder of a root that gives rise to rootlets. Also **rhi·zog·e·nous** (rī·zoj′ə·nəs).
rhi·zoid (rī′zoid) *adj.* Rootlike; similar to or resembling a root. — *n. Bot.* A delicate filiform or hairlike organ developed on all kinds of thalli, moss stems, etc.: the analog of the roots of flowering plants, serving for absorption and attachment. — **rhi·zoi·dal** (rī·zoid′l) *adj.*
rhi·zome (rī′zōm) *n. Bot.* A procumbent or subterranean rootlike stem, producing roots from its lower surface and leaves or shoots from its upper surface; a rootstock. Also **rhi·zo·ma** (rī·zō′mə). [<NL *rhizoma* < Gk. *rhizōma* mass of roots, ult. < *rhiza* root] — **rhi·zom·a·tous** (rī·zom′ə·təs, -zō′mə-) *adj.*

RHIZOME
The bearded iris.

rhi·zo·morph (rī′zō·môrf) *n. Bot.* One of the rootlike parts of the mycelium, composed of many united hyphal strands, by which certain fungi attach themselves to and penetrate the higher plants.
rhi·zo·mor·phous (rī′zō·môr′fəs) *adj. Bot.* Branching after the manner of rootlets: said of mycelia.
rhi·zoph·a·gous (rī·zof′ə·gəs) *adj.* Feeding on roots. [<RHIZO- + -PHAGOUS]
rhi·zo·pod (rī′zə·pod) *n.* Any member of a subclass (*Rhizopoda*) of protozoans with pseudopodia for locomotion and the ingestion of food. — **rhi·zop·o·dan** (rī·zop′ə·dən) *adj. & n.* — **rhi·zop′o·dous** *adj.*
rhi·zot·o·my (rī·zot′ə·mē) *n. Surg.* The division of the roots of the spinal nerves, for the relief of pain or spastic paralysis. [<RHIZO- + -TOMY]
rho (rō) *n.* The seventeenth letter and twelfth consonant in the Greek alphabet (Ρ,ρ): equivalent to the English *r* aspirated. As a numeral it denotes 100. [<Gk. *rhō*]
Rho·da (rō′də) A feminine personal name. [<Gk., rose]

— **Rho·da** A damsel in the house of Mary, the mother of John. *Acts* xii 13.
rho·da·mine (rō′də·mēn, -min) *n. Chem.* Any of various red or pink dyestuffs obtained by condensing an amino derivative of phenol with phthalic anhydride. The solution shows green fluorescence. Also **rho′da·min** (-min). [<Gk. *rhodon* rose + AMINE]
Rhode Island A southern New England State of the United States; 1,214 square miles; capital, Providence; entered the Union May 29, 1790, one of the original thirteen States: officially **The State of Rhode Island and Providence Plantations;** the smallest State in the Union; nickname *Little Rhody*: abbr. *R.I.* — **Rhode Islander**
Rhode Island Red An American breed of domestic fowls, reddish and black in color, with yellow naked legs and small single comb.
Rhodes (rōdz) 1 The largest island of the Dodecanese group; 545 square miles. 2 Its chief city, capital of the Dodecanese Islands. Italian *Rodi.* Greek **Ró·dhos** (rô′thôs). See COLOSSUS OF RHODES.
Rhodes (rōdz), **Cecil (John)**, 1853–1902, British South African financier and statesman. — **James Ford**, 1848–1927, U.S. industrialist and historian.
Rho·de·sia (rō·dē′zhə, -zhē·ə) A region in south central Africa, north of Transvaal province; divided by the Zambesi River into **Southern Rhodesia**, a British self-governing territory; 150,333 square miles, capital, Salisbury; and **Northern Rhodesia**, a British protectorate, 288,130 square miles, capital, Lusaka; and associated (1953) with Nyasaland in the **Federation of Rhodesia and Nyasaland.** — **Rho·de′sian** *adj. & n.*
Rho·de·sian man (rō·dē′zhən) An African forerunner (*Homo Rhodesiensis*) of Neanderthal man, represented by the massive upper jaw and cranium of a skull discovered in 1921 at Broken Hill, Rhodesia.
Rhodes scholarships Any of a number of scholarships, tenable at Oxford University, provided for in the will of Cecil Rhodes, for selected scholars from the United States and the British dominions and colonies.
Rho·di·an (rō′dē·ən) *adj.* Of or pertaining to the island of Rhodes or to the Knights of Rhodes. — *n.* A Knight of Rhodes; also, a native of that island.
rho·dic (rō′dik) *adj. Chem.* Of, pertaining to, or derived from rhodium: *rhodic sulfate*.
rho·di·um (rō′dē·əm) *n.* A whitish-gray metallic element (symbol Rh) of the platinum group, whose salts are for the most part rose-colored: used for plating silver and in alloys, especially with steel. See ELEMENT. [<NL <Gk. *rhodon* rose; from the color of its salts]
rho·do·chro·site (rō′də·krō′sīt) *n.* A vitreous rose-red or variously colored rhombohedral manganese carbonate, MnCO₃. [<G *rhodochrosit* <Gk. *rhodochrōs* rose-colored + *rhodon* rose + *chrōs* color]
rho·do·den·dron (rō′də·den′drən) *n.* Any of a genus (*Rhododendron*) of showy evergreen shrubs or small trees of the heath family, with profuse clusters of beautiful flowers, found growing wild in mountainous regions; especially, the **great rhododendron** (*R. macrophyllum*), the State flower of Washington, and the **rosebay rhododendron** (*R. maximum*), the State flower of West Virginia. [<L <Gk. < *rhodon* rose + *dendron* tree]
rho·do·lite (rō′də·līt) *n.* A pale rose-colored garnet, used as a gem. [<Gk. *rhodon* rose + -LITE]
rho·do·nite (rō′də·nīt) *n.* A vitreous, red or pink manganese silicate, MnSiO₃, crystallizing in the triclinic system, and often used as an ornamental stone. [<Gk. *rhodon* rose]
Rhod·o·pe Mountains (rod′ə·pē) A mountain chain of the Balkan Peninsula, dividing Bulgaria from Thrace and Macedonia; highest peak, 9,591 feet.
Rho·do·phy·ce·ae (rō′də·fī′sē·ē) *n. pl.* A major division or group of algae characterized by a red, purple, or reddish-brown color and found chiefly in seas of the temperate zone; the red algae. [<NL <Gk. *rhodon* rose + *phykos* seaweed]
rho·dop·sin (rō·dop′sin) *n. Biochem.* The rose-colored component of visual purple, breaking down into retinene on exposure to light. [<Gk. *rhodon* rose + *opsis* appearance]
rho·do·ra (rō·dôr′ə, -dō′rə) *n.* A handsome shrub (*Rhododendron canadense*), from 1 to 3 feet high, with terminal clusters of pale-purple flowers preceding the leaves. It is found in cool bogs, from Pennsylvania to Canada. [<L *rhodora* meadowsweet]
-rhoea See -RRHEA.
rhomb (rom, romb) *n.* A rhombus. [<F *rhombe*. See RHOMBUS.]
rhom·ben·ceph·a·lon (rom′ben·sef′ə·lon) *n. Anat.* The parts of the cerebrospinal axis that develop from the posterior cerebral vesicle; medulla oblongata and cerebellum taken together. [<NL]
rhom·bic (rom′bik) *adj.* 1 Pertaining to or having the shape of a rhombus. 2 Orthorhombic. Also **rhom′bi·cal.**
rhom·bo·he·dral (rom′bə·hē′drəl) *adj. Geom.* Pertaining to a rhombohedron.
rhombohedral system In the classification of some authors, the trigonal division of the hexagonal crystal system.
rhom·bo·he·dron (rom′bə·hē′drən) *n. pl.* **·drons** or **·dra** (drə) *Geom.* A prismatic form included within six equal rhombic faces.

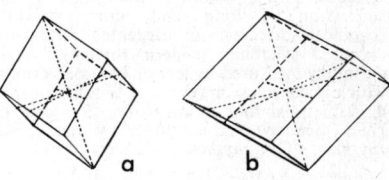

RHOMBOHEDRONS
a. Acute. b. Obtuse.

rhom·boid (rom′boid) *n. Geom.* 1 A parallelogram having opposite sides and opposite angles equal but no right angle. 2 A solid bounded by such parallelograms. — *adj.* 1 Having the character or shape of a rhomboid. 2 Having a shape approaching that of a rhombus, as one of two muscles attached to the shoulder blades. [<F *rhomboïde*] — **rhom·boi·dal** (rom·boid′l) *adj.*
rhom·bus (rom′bəs) *n. pl.* **·bus·es** or **·bi** (-bī) *Geom.* 1 An equilateral parallelogram having the angles usually, but not necessarily, oblique; A square may be considered as a special case of the *rhombus*. 2 A rhombohedron. [<L <Gk. *rhombos* spinning top, rhomb]
rhon·chus (rong′kəs) *n. pl.* **·chi** (-kī) *Pathol.* A rattling or whistling sound in respiration, especially when it resembles snoring; a râle. [<L <Gk. *rhonchos*] — **rhon′chal, rhon′chi·al** *adj.*
Rhond·da (ron′də) An urban district of Glamorganshire, SE Wales: also *Ystradyfodwg*.
Rhond·da (ron′də), **Viscount**, 1856–1918, David Alfred Thomas, British industrialist and administrator.
Rhône (rōn) A river in Switzerland and SE France, flowing 504 miles to the Mediterranean including 45 miles through Lake Geneva above Geneva; enters the Gulf of the Lion below Arles. Also **Rhone**.
rhu·barb (rōō′bärb) *n.* 1 A stout, coarse, perennial herb (genus *Rheum*) of the buckwheat family, having large leaves and small clusters of flowers on tall fleshy stalks; especially, the common rhubarb or pie plant (*R. rhaponticum*), whose acid leafstalks are used in cooking. 2 The dried roots of the medicinal rhubarb (*R. officinale* and *R. palmatum*), used as a cathartic and bitter tonic. 3 *U.S. Slang* A heated argument; scuffle or quarrel. [<OF *reubarbe* <LL *rhabarbarum* <Gk. *Rha* Volga river, Volga plant, rhubarb + *barbaron* foreign; so called because orig. imported from Russia]
rhumb (rum, rumb) *n. Naut.* 1 One of the 32 points of the mariners' compass, separated by arcs of 11° 15′. 2 One of these arcs or divisions. [<OF *rumb*]
rhum·ba (rum′bə) See RUMBA.
rhumb line A line or course along the surface of a sphere crossing successive meridians at the same angle; a loxodromic curve.
Rhus (rus) *n.* A large genus of trees or shrubs

of the cashew family, including the true sumacs. Poison ivy and poison oak, also often included, are now placed in the genus *Toxicodendron.* [<NL <L <Gk. *rhous* sumac]

rhyme (rīm), **rhym·er** (rī′mər), **rhyme·ster** (rīm′stər), etc. See RIME, etc.

rhyn·cho·ce·pha·li·an (ring′kō·sə·fā′lē·ən) *adj.* Pertaining to or designating a nearly extinct order of lizardlike reptiles *(Rhynchocephalia)*, represented by only one genus *(Sphenodon)*, the tuatara of New Zealand. — *n.* One of the *Rhynchocephalia.* [<NL *Rhynchocephalia*, name of the order <Gk. *rhynchos* snout + *kephalē* head]

rhy·o·lite (rī′ə·līt) *n.* A highly acidic, variously colored volcanic rock. [<Gk. *rhyax* stream + -LITE]

rhythm (rith′əm) *n.* **1** Movement characterized by regular measured or harmonious recurrence of stress, beat, sound, accent, or motion: the *rhythm* of the pulse, the *rhythm* of moving oars. **2** The musical property dependent on the regular succession of accents or tone–impulses; accent–movement or accent–structure; also, a system or kind of accentuation as determined by the make–up of the accentual divisions. **3** In poetry, the cadenced flow of sound as determined by the succession of long and short syllables (**classical rhythm**), or accented and unaccented syllables (**modern rhythm**). When definitely measured by feet or bars or periods, which make lines or verses, it becomes *meter.* **4** A metrical foot or measure. **5** Verse or rime. See synonyms under METER. [<F *rhythme* <Gk. *rhythmos* < *rheein* flow]

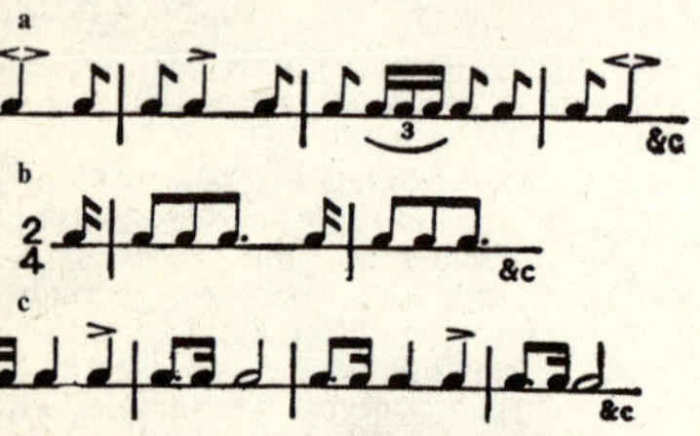

CHARACTERISTIC DANCE RHYTHMS
a. Cracovienne. *b.* Polka. *c.* Mazurka.

rhyth·mic (rith′mik) *adj.* Relating to or characterized by rhythm: contrasted with *harmonic.* Also **rhyth′mi·cal.** — **rhyth′mi·cal·ly** *adv.*

rhyth·mics (rith′miks) *n.* The science of rhythm.

rhyth·mist (rith′mist) *n.* A master of rhythmical composition; also, one versed in rhythmics.

ri·al (rī′al) *n.* The monetary unit of Iran; a silver coin, twenty of which equal one pahlavi. [<OF *rial, real* royal]

ri·al·to (rē·al′tō) *n. pl.* **·tos** A market or place of exchange. [from *Rialto* <*Rivo Alto* ancient name of the island on which Venice was founded about 800 <Ital. *rivo* channel (<L *rivus* brook) + *alto* deep <L *altus*]

Ri·al·to (rē·al′tō, *Ital.* rē·äl′tō) **1** An island comprising the ancient business quarter of Venice. **2** A bridge over the Grand Canal connecting the old Rialto with the island of San Marco at Venice, Italy: short for **Ponte del Rialto.** **3** In New York City, the theater district.

ri·ant (rī′ənt) *adj.* Laughing. [<F, laughing, ppr of *rire* laugh] — **ri′ant·ly** *adv.*

ri·a·ta (rē·ä′tə) *n.* A lasso; lariat. [<Sp. *reata* < *reatar* tie again <L *re-* again + *aptare* fit]

Ri·au Archipelago (rē′ou) See RIOUW ARCHIPELAGO.

rib (rib) *n.* **1** *Anat.* One of the series of bony rods attached to the spine of most vertebrates, and nearly encircling the thoracic cavity. In man there are twelve ribs on each side, forming the walls of the thorax, of which the first seven (**true** or **sternal ribs**) are attached to the sternum, the last five (**false** or **asternal ribs**) being either attached by their edges to the rib above, as in the upper three, or free distally (**floating ribs**), as in the lower two. ◆ Collateral adjective: *costal.* **2** Something likened to the rib of an animal; a ridge, strip, or band. **3** A curved side timber bending away from the keel in a boat or ship, or a curved timber or support in a vault. **4** A raised wale or stripe in cloth or knit goods, as stockings. **5** *Aeron.* An element in the construction of an airplane wing, usually extending fore and aft and crossing the wing spars, to hold the fabric of the wing in shape. **6** *Bot.* A vein or nerve of a leaf, especially the middle one; any ridge on a plant. **7** Cut of meat including one or more ribs. **8** A wife: in jocular allusion to *Gen.* ii 22. **9** *Slang* A practical joke. — *v.t.* **ribbed, rib·bing 1** To make with ridges: to *rib* a piece of knitting. **2** To strengthen or protect by or enclose within ribs. **3** *Slang* To make fun of; tease. [OE *ribb*]

rib·ald (rib′əld) *adj.* Pertaining to or indulging in coarse or offensive language or vulgar jokes; coarsely jocular. — *n.* One who uses coarse or abusive language. [<OF *ribauld* <Gmc. Cf. MHG *riben* copulate, MDu. *ribe* whore.]

rib·ald·ry (rib′əl·drē) *n.* Coarse or ribald language. [<OF *ribauderie*]

rib·and (rib′ənd) *n. Archaic* A decorative ribbon. [Earlier form of RIBBON]

Ri·bault (rē·bō′), **Jean,** 1520–65, French Protestant explorer and colonizer in Florida. Also **Ri·baut′.**

rib·band (rib′band′, rib′ənd, -ən) *n. Naut.* A lengthwise strip following a vessel's curves and bolted to its ribs, to hold them in place until they receive the planking or plating. Also **rib′-band′.** [<RIB + BAND[1]]

Rib·ben·trop (rib′ən·trôp), **Joachim von,** 1893–1946, German Nazi diplomat; executed.

rib·bing (rib′ing) *n.* An arrangement or collection of ribs, as in ribbed cloth, etc.

rib·ble-rab·ble (rib′əl·rab′əl) *n.* **1** Meaningless talk; gabble. **2** Rabble; crowd; mob. [Reduplication of RABBLE]

rib·bon (rib′ən) *n.* **1** A narrow strip of fine fabric, usually silk or satin, having two selvages, and commonly less than eight inches wide, made in a variety of weaves: used as trimming. **2** Something shaped like or suggesting a ribbon, as a watch spring, or a painted stripe on the side of a vessel. **3** A narrow strip; a shred: torn to *ribbons.* **4** An ink–bearing strip of cloth in a typewriter. **5** A ribband. **6** *pl. Colloq.* Driving reins. **7 a** A colored strip of cloth worn to signify membership in an order, the award of a prize, etc. **b** A similar strip of cloth worn on the left breast of a military or naval uniform to indicate campaigns served in, medals won, etc. **8** A ticker tape. — *v.t.* To ornament with ribbons; also, to form or tear into ribbons. — *adj.* **1** Made of or like ribbon. **2** Having parallel bands or streaks, as certain minerals: *ribbon* jasper. **3** Of a standard to receive a prize in a competitive show: a *ribbon* hog. [<OF *riban;* origin unknown]

rib·bon·fish (rib′ən·fish′) *n. pl.* **·fish** or **·fish·es** A long marine fish with a compressed, ribbonlike body, as an oarfish or dealfish.

Rib·bon·man (rib′ən·man′) *n. pl.* **·men** (-men′) A member of the **Ribbon Society,** a secret association established in 1808 in Ireland, in opposition to the Orangemen and to oppose the eviction of tenant farmers: so called from the badge of its members, a green ribbon. — **Rib′bon·ism** *n.*

ribbon snake The American garter snake.

Ri·bei·rão Prê·to (rē·bā·roun′ prā′tōō) A city in north central São Paulo state, SE Brazil.

Ri·be·ra (rē·vā′rä), **José,** 1588–1656, Spanish painter: sometimes called "Lo Spagnoletto."

ri·bo·fla·vin (rī′bō·flā′vin) *n. Biochem.* A member of the vitamin B complex, vitamin B_2, an orange–yellow, crystalline compound, $C_{17}H_{20}N_4O_6$, found in milk, green leafy vegetables, egg yolk, and meats, and also made synthetically: formerly called *lactoflavin, vitamin G.* [<RIBO(SE) + FLAVIN]

ri·bon·ic acid (rī·bon′ik) *Chem.* An acid, $C_5H_{10}O_6$, produced by the oxidation of ribose.

ri·bose (rī′bōs) *n. Chem.* A sugar, $C_5H_{10}O_5$, derived from pentose and occurring in certain nucleic acids. [<RIB(ONIC ACID) + -OSE[2]]

rib·wort (rib′wûrt′) *n.* The English plantain *(Plantago lanceolata)*, or a related species. See PLANTAIN.

-ric *combining form* Realm or jurisdiction of: *bishopric.* [OE *rīce* kingdom, realm]

Ri·car·do (*Ital.* rē·kär′dō, *Pg.* rē·kär′thōō, *Sp.* -thō) Italian, Portuguese, and Spanish form of RICHARD. Also *Ital.* **Ric·car·do** (rēk·kär′dō), *Lat.* **Ri·car·dus** (rē·kär′dəs).

Ri·car·do (ri·kär′dō), **David,** 1772–1823, English political economist. — **Ri·car′di·an** *adj.* & *n.*

Ric·cio (rēt′chō), **David** See RIZZIO.

rice (rīs) *n.* **1** An annual cereal grass *(Oryza sativa)*, widely cultivated on wet land in warm climates. **2** The edible grain or seeds of this plant. [<F *riz* <L *oryza* <Gk. *oryza*]

Rice (rīs), **Elmer,** born 1892, U. S. dramatist. — **Grantland,** 1888–1954, U. S. journalist.

rice·bird (rīs′bûrd′) *n.* **1** Any bird frequenting rice fields; especially, in the southern United States, the bobolink: also **rice bunting.** **2** The Java sparrow.

rice·braid (rīs′brād′) *n.* Braid made to resemble rice grains strung together lengthwise.

rice paper 1 Paper made from rice straw. **2** A delicate vegetable paper made from the pith of a Chinese shrub, the **rice–paper plant** *(Tetrapanax papyriferus)*, pared into thin rolls and flattened into sheets.

ric·er (rī′sər) *n.* A kitchen utensil consisting of a perforated container through which potatoes and other vegetables are pressed, emerging in small particles resembling grains of rice.

rice weevil A small brown weevil *(Sitophilus* or *Calandra oryza)* destructive to growing rice and the stored grain. For illustration see INSECTS (injurious).

rich (rich) *adj.* **1** Having large possessions, as of money, goods, or lands; wealthy; opulent. **2** Composed of rare or precious materials; valuable; costly: *rich* fabrics. **3** Having in a high degree qualities pleasing to the senses; luscious to the taste: often implying an unwholesome excess of butter, fats, flavoring, etc. **4** Full, satisfying, and pleasing, as a tone, voice, color, or perfume. **5** Luxuriant; abundant: *rich* hair; *rich* crops. **6** Yielding abundant returns; fruitful. **7** Abundantly supplied: often with *in* or *with.* **8** Abounding in desirable qualities; of full strength, as blood. **9** *Colloq.* Exceedingly humorous; amusing or ridiculous: a *rich* joke. See synonyms under FERTILE, RACY. [OE *rīce;* infl. in form by OF *riche* <Gmc.] — **rich′ly** *adv.* — **rich′ness** *n.*

Rich·ard (rich′ərd; *Fr.* rē·shàr′, *Ger.* rē′khärt) A masculine personal name. Also *Lat.* **Ri·char·dus** (rē·kär′dəs), *Du.* **Ri·chart** (rē′shärt). [<Gmc., strong ruler]

— **Richard I,** 1157–99, king of England 1189–1199; went on Third Crusade: called "Coeur de Lion" or "the Lion–Hearted."

— **Richard II,** 1367–1400, king of England 1377–99, deposed by Henry IV.

— **Richard III,** 1452–85, king of England 1483–85; usurped throne; killed at Bosworth.

Richard Roe See JOHN DOE.

Rich·ards (rich′ərdz), **Theodore William,** 1868–1928, U. S. chemist.

Rich·ard·son (rich′ərd·sən), **Henry Handel** Pseudonym of Henrietta Richardson, 1878?–1946, Australian novelist. — **Henry Hobson,** 1838–1886, U. S. architect. — **Owen Willans,** born 1879, English physicist. — **Samuel,** 1689?–1761, English novelist.

Ri·che·lieu (rē·shə·lyœ′), **Duc de,** 1585–1642, Armand Jean Duplessis, French cardinal and statesman; prime minister of Louis XIII.

Ri·che·lieu River (rē·shə·lyœ′, rish′ə·lōō) A river of southern Quebec, flowing 75 miles north from Lake Champlain to the Saint Lawrence.

rich·es (rich′iz) *n. pl.* [In Middle English, this was a singular noun and spelled *richess* or *richesse;* now, from its form, used in the plural] **1** Abundant possessions; wealth. **2** Hence, abundance of whatever is precious. See synonyms under WEALTH. [<F *richesse* < *riche* <Gmc.]

Ri·chet (rē·she′), **Charles Robert,** 1850–1935, French physiologist.

Rich·mond (rich′mənd) **1** The capital of Virginia, a port on the James River: capital of the Confederacy, 1861–65. **2** A borough on the Thames in northern Surrey, England. **3** A borough of New York City coextensive with Staten Island.

richt (rikht) *n. Scot.* Right.

Rich·ter (rikh′tər), **Johann Paul Friedrich,** 1763–1825, German author and humorist: pseudonym *Jean Paul.*

Richt·ho·fen (rikht′hō·fən), **Baron Manfred von,** 1892–1918, German aviator in World War I; killed in action.

rich·weed (rich′wēd′) *n.* **1** An herb *(Pilea pumila)* of the nettle family growing in wet, cool places: also called *clearweed.* **2** A strong-scented herb *(Collinsonia canadensis)* of the mint family: also called *horse balm.* **3** Ragweed. **4** White snakeroot.

ri·cin (rī′sin, ris′in) *n. Chem.* A very toxic protein isolated from the castor bean in the form of a white powder: it agglutinates red blood corpuscles. [<L *ricinus* castor bean]

ric·in·o·le·ic (ris′in·ō·lē′ik) *adj.* Of, pertaining to, or derived from the castor bean.

ricinoleic acid *Chem.* An unsaturated fatty acid, $C_{18}H_{34}O_3$, present in castor oil and hardening in a thick, yellow, crystalline or viscid mass.

ric·in·o·le·in (ris′in·ō′lē·in) *n. Chem.* The glycerol ester derivative of ricinoleic acid, preponderant in castor oil. [<L *ricinus* castor bean + *oleum* oil + -IN]

rick (rik) *n.* **1** A stack, as of hay, having the top rounded and thatched to protect the interior from rain. **2** A haycock in the field. —*v.t.* To pile in ricks. [OE *hrēac*]

Rick·en·back·er (rik′ən·bak′ər), **Edward Vernon,** born 1890, U. S. aviation executive; military aviator in World War I.

rick·ets (rik′its) *n. Pathol.* A disease of early childhood, chiefly due to a deficiency of calcium salts as provided by vitamin D, characterized by softening of the bones and consequent deformity; rachitis. [Origin uncertain]

rick·ett·si·a (rik·et′sē·ə) *n. pl.* **·si·ae** (-si·ē) Any of a genus *(Rickettsia)* of micro-organisms typically parasitic in the bodies of certain ticks and lice, but transmissible to other animals and to man; especially, *R. prowazeki,* the causative agent of typhus. [after Howard T. *Ricketts,* 1871–1910, U. S. pathologist]

rick·ett·si·al (rik·et′sē·əl) *adj.* Pertaining to or designating any of the various infective diseases caused by micro-organisms of *Rickettsia* or related genera, as typhus, Rocky Mountain spotted fever, or trench fever.

rick·et·y (rik′it·ē) *adj.* **1** Ready to fall; tottering. **2** Affected with rickets. —**rick′et·i·ly** *adv.* —**rick′et·i·ness** *n.*

rick·ey (rik′ē) *n.* A cooling drink of which spirits, lime juice, and carbonated water are the chief ingredients. [Origin uncertain]

rick·le (rik′əl) *n. Scot.* **1** A heap or bundle. **2** A small rick of grain or hay; a stook.

rick-rack[1] (rik′rak′) *n.* Flat braid in zigzag form, made of cotton, rayon, silk, or wool; also, the openwork trimming made with this serpentine braid. [Reduplication of RACK[1]]

rick-rack[2] (rik′rak′) *n.* The sound made by the rhythmic motion of oars in oarlocks. [Imit.]

rick·shaw (rik′shô) *n.* A jinriksha. Also **rick′·sha.** [Short for JINRIKSHA]

ric·o·chet (rik′ə·shā′, -shet′) *v.i.* **·cheted** (-shād′) or **·chet·ted** (-shet′id), **·chet·ing** (-shā′ing) or **·chet·ting** (-shet′ing) To glance from a surface, as a projectile over the water; make a series of skips or bounds. —*n.* **1** A bounding, as of a projectile over a surface. **2** The method of firing by which a projectile is made to rebound. **3** A projectile so rebounding. [<OF; origin uncertain]

ric·tus (rik′təs) *n.* **1** The expanse of the open mouth; a gaping. **2** A fissure or cleft. [<L, open, gaping mouth <*ringi* open the mouth wide] —**ric′tal** *adj.*

rid[1] (rid) *v.t.* **rid** or **rid·ded, rid·ding** **1** To free, as from a burden or annoyance; clear: usually with *of*: to *rid* a house of vermin. **2** *Obs.* To rescue; deliver. **3** *Obs.* To drive away; expel; banish. —*adj.* Free; clear; quit: with *of*: We are well *rid* of him. [Fusion of OE *geryddan* clear (land) + ON *rythja* clear (land) of trees]

rid[2] (rid) Obsolete past tense and past participle of RIDE.

rid·a·ble (rī′də·bəl) *adj.* That may be ridden on, through, or over, as an animal or a road.

rid·dance (rid′ns) *n.* A ridding of something undesirable, or the state of being rid.

rid·den (rid′n) Past participle of RIDE.

rid·dle[1] (rid′l) *v.t.* **·dled, ·dling** **1** To perforate in numerous places, as with shot. **2** To sift through a coarse sieve. **3** To damage, injure, refute, etc., as if by perforating: to *riddle* a theory. [< *n.*] —*n.* **1** A coarse sieve, such as one used in a foundry or in washing for gold. **2** A board set with pins, used for straightening wire. [OE *hriddel* sieve] —**rid′·dler** *n.*

rid·dle[2] (rid′l) *n.* **1** A puzzling question or conundrum; anything ambiguous or puzzling. **2** Any mysterious object or person. —*v.* **·dled, ·dling** *v.t.* To solve; explain. —*v.i.* To utter or solve riddles; speak in riddles. [OE *rǣdels* < stem of *rǣdan* interpret, solve]

Synonyms (noun): conundrum, enigma, paradox, problem, puzzle. *Conundrum* signifies some question or statement in which some hidden and fanciful resemblance is involved, the answer often depending upon a pun; an *enigma* is a dark saying; a *paradox* is a true statement or fact that appears absurd or contradictory. The *riddle* is not so petty as the *conundrum*; it is an ambiguous or paradoxical statement with a hidden meaning to be guessed by the mental acuteness of the one to whom it is proposed; a *problem* may require simply study and scholarship, as a *problem* in mathematics; a *puzzle* may be in something other than a verbal statement, as a dissected map or any perplexing mechanical contrivance. Both *enigma* and *puzzle* may be applied to any matter difficult of answer or solution, *enigma* conveying an idea of greater dignity, *puzzle* applying to something more commonplace and mechanical. *Antonyms*: answer, axiom, explanation, proposition, solution.

ride (rīd) *v.* **rode** (*Obs.* **rid**), **rid·den** (*Obs.* **rid**), **rid·ing** *v.i.* **1** To sit on and be borne along by a horse or other animal, especially while guiding or controlling its motion. **2** To be borne along as if on horseback. **3** To travel or be carried on or in a vehicle or other conveyance. **4** To be supported in moving: The wheel *rides* on the shaft. **5** To move; be borne; float: The ship *rides* on the waves. **6** To support and carry a rider in a specified manner: This car *rides* easily. **7** To seem to float in space, as a star. **8** *Naut.* To lie at anchor, as a ship. **9** To overlap or overlie, as broken bones. **10** To work or move upward out of place: with *up*: His sleeve has *ridden* up. **11** *Slang* To continue unchanged: Let it *ride.* —*v.t.* **12** To sit on and control the motion of (a horse, bicycle, etc.). **13** To move or be borne or supported upon: The glider *rides* air currents. **14** To overlap or overlie. **15** To travel or traverse (an area, etc.) on horseback, in an automobile, etc. **16** To control imperiously or oppressively: usually in the past participle: a king-*ridden* people. **17** To accomplish by riding: to *ride* a race. **18** To cause to ride. **19** To place (someone) astride something and carry him, especially as a punishment: They *rode* him out of town on a rail. **20** *Naut.* To keep at anchor. **21** *Colloq.* To tease or harass by ridicule or petty criticisms; tyrannize. See synonyms under DRIVE. —**to ride out** To survive; endure successfully. —*n.* **1** An excursion by any means of conveyance, as on horseback, by car, etc. **2** A road intended for riding. [OE *rīdan*]

ri·dent (rī′dənt) *adj.* Laughing; smiling; grinning. [<L *ridens, -entis,* ppr. of *ridere* laugh, smile]

rid·er (rī′dər) *n.* **1** One who or that which rides; a horseman; a bicyclist; specifically, one who breaks in horses. **2** Any device that rides upon or weighs down something else, actually or figuratively. **3** A separate piece of writing or print added to a document, record, or the like. **4** An addition or proposed addition to a legislative bill, adding to or modifying its original purport. **5** A metallic weight for use astride the graduated beam of a delicate balance. **6** The top rail of a rail fence.

rid·er·less (rī′dər·lis) *adj.* Without a rider, as a horse.

ridge (rij) *n.* **1** An elevation or protuberance long in proportion to its width and height and generally having sloping sides; a raised strip; especially, a lengthened elevation of land; a long hill, or range of hills. **2** That part of a roof where the rafters meet the ridge pole. **3** A slight elevation of earth in a garden or field thrown up by the plow, hoe, or other implement. **4** The back or backbone of an animal, especially of a whale. **5** *Meteorol.* A relatively narrow band of high pressure between two cyclone areas, as shown on a weather map. —*v.* **ridged, ridg·ing** *v.t.* **1** To mark with ridges. **2** To form into ridges. —*v.i.* **3** To form ridges. [OE *hrycg* spine, ridge]

ridge pole A horizontal timber at the ridge of a roof, to which the upper ends of the rafters are nailed. Also **ridge beam, ridge piece, ridge plate.**

Ridg·way (rij′wā), **Matthew Bunker,** born 1895, U. S. general; chief of staff 1953–55.

ridg·y (rij′ē) *adj.* Having ridges; raised in a ridge; ridged.

rid·i·cule (rid′ə·kyōōl) *n.* **1** Language calculated to make a person or thing the object of contemptuous humorous disparagement; also, looks or acts expressing amused contempt; derision; mockery. **2** An object of mocking merriment; butt. **3** *Obs.* Ridiculousness. —*v.t.* **·culed, ·cul·ing** To make fun of; hold up as a laughingstock; deride. [<OF <L *ridiculum* a jest, joke, orig. neut. of *ridiculus* comical < *ridere* laugh] —**rid′i·cul′er** *n.*

Synonym (noun): derision. *Ridicule* may be merely sportive or thoughtless; *derision* is always hostile or malicious. See BANTER.

Synonyms (verb): banter, chaff, deride, flout, jeer, lampoon, mock, quiz, rally, satirize, scoff, scout, taunt. *Antonyms*: applaud, celebrate, compliment, eulogize, extol, honor, praise.

ri·dic·u·lous (ri·dik′yə·ləs) *adj.* Exciting or calculated to excite ridicule; absurdly comical; unworthy of consideration. [<L *ridiculus*] —**ri·dic′u·lous·ly** *adv.* —**ri·dic′u·lous·ness** *n.*

Synonyms: absurd, comical, droll, farcical, funny, grotesque, laughable, ludicrous, preposterous, risible, silly, trifling, trivial. See ABSURD, QUEER. *Antonyms*: clever, commendable, grave, imposing, judicious, majestic, sensible, venerable, wise.

rid·ing[1] (rī′ding) *n.* The act of one who rides; a ride. —*adj.* **1** To be ridden on or in; suitable for riding: a *riding* horse. **2** To be used while riding: *riding* boots. **3** For use while at anchor: a *riding* light.

rid·ing[2] (rī′ding) *n.* **1** One of the three administrative divisions of Yorkshire, England: North Riding, East Riding, and West Riding. **2** Any similar administrative division, as in Canada, New Zealand, etc. [OE *thrithing* the third part (of a county); the initial *th* having been lost through the influence of the final *t* or *th* of *East, West,* and *North.* Related to THIRD.]

riding habit Apparel worn by horseback riders, especially that designed for women, consisting usually of a jacket and breeches or jodhpurs.

riding horse A horse used for riding.

riding school An establishment where the art of riding on horseback is taught.

Rid·ley (rid′lē), **Nicholas,** 1500?–55, Anglican bishop, reformer, and martyr.

ri·dot·to (ri·dot′ō) *n. pl.* **·tos** A public musical and dancing entertainment much in vogue in England in the 18th century. [<Ital., a festival, redoubt. See REDOUBT.]

Rid·path (rid′path, -päth), **John Clark,** 1840–1900, U. S. historian.

Rie·mann (rē′män), **Georg Friedrich Bernhard,** 1826–66, German mathematician.

Ri·en·zi (rē·en′zē), **Cola di,** 1313–54, Italian popular orator and leader. Also **Ri·en′zo.**

Rie·sen·ge·bir·ge (rē′zən·gə·bir′gə) The highest range of the Sudetes, in Lower Silesia and northern Bohemia; highest point, 5,259 feet. *Czech* **Kr·ko·no·še** (kûr′kô·nô·she′), *Polish* **Kar·ko·no·sze** (kär′kô·nô′she).

Riet (ryet) A river in Orange Free State and Cape of Good Hope Province of the Union of South Africa, flowing 250 miles NW to the Vaal.

Rif (rif) The mountain range of NW Africa, bordering the northern coast of Morocco; highest point, 8,060 feet. Also **Riff.**

ri·fa·ci·men·to (rē·fä′chē·men′tō) *n. pl.* **·ti** (-tē) *Italian* A remaking; recasting: said of literary or musical adaptations.

rife (rīf) *adj.* **1** Great in number or quantity; plentiful; abundant; prevalent; current. **2** Containing in abundance: followed by *with.* [OE *rīfe* abundant]

riff (rif) *n.* In jazz music, a melodic phrase or

motif, played repeatedly as background or used as the main theme. [Prob. back formation of RIFFLE]
Riff (rif) *n.* One of a Berber tribe inhabiting the mountainous region of northern Morocco. — **Rif′fi·an** *adj.* & *n.*
rif·fle[1] (rif′əl) *n.* 1 *U.S.* A shoal or rocky obstruction lying beneath the surface of a river or other stream. 2 A stretch of shallow, choppy water caused by such a shoal; a rapid. 3 A way of shuffling cards. — *v.t.* & *v.i.* **·fled, ·fling** 1 To cause or form a rapid. 2 To shuffle (cards) by bending up adjacent corners of two halves of the pack, and permitting the cards to slip together as they are released. 3 To thumb through (the pages of a book). [? Blend of RIPPLE and RUFFLE]
rif·fle[2] (rif′əl) *n. Mining* 1 A groove or indentation set in the bottom of an inclined trough or sluice, for arresting gold contained in sands or gravels. 2 A cross slat or cleat rising above the bottom of such a sluice and adapted for catching gold: also **riffle bar, riffle block.** [Cf. LG *riffel* furrow]
rif·fler (rif′lər) *n.* 1 A file with curved working surfaces at one or both ends and a smooth center serving as a handle: used in sculpture, woodcarving, diemaking, etc. 2 A workman in any of these fields who handles such a tool. [<RIFFLE[2]]
riff-raff (rif′raf′) *n.* 1 The populace; rabble. 2 Miscellaneous rubbish. [<OF *rif et raf* every bit]
ri·fle[1] (rī′fəl) *n.* 1 A firearm, of any size, having grooves, now always spiral, on the surface of the bore for imparting rotation to the projectile and increasing the accuracy of the weapon. 2 One of these grooves. 3 Such a weapon fired from the shoulder, as distinguished from pistols, a carbine, or artillery, and provided with a device for attaching a bayonet. 4 *pl.* A body of soldiers

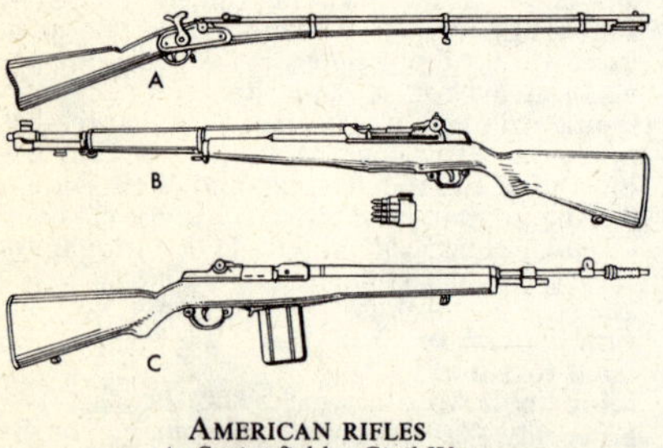

AMERICAN RIFLES
A. Springfield—Civil War.
B. Garand—World War II.
C. M-14, Automatic—1958.

equipped with rifles. 5 An emery-coated stick for whetting scythes. — **magazine rifle** A rifle with a chamber containing extra cartridges which are brought one by one into position for firing; a semi-automatic or repeating rifle. — *v.t.* **·fled, ·fling** To cut a spirally grooved bore in (a firearm, etc.). [Cf. G *reifeln* flute, LG *rifeln* furrow, F *rifler* scratch; *n.*, short for *rifled gun*]
ri·fle[2] (rī′fəl) *v.t.* **·fled, ·fling** 1 To search through and rob, as a safe. 2 To search and rob (a person). 3 To seize and take away by force. [<OF *rifler* scratch, plunder <Gmc.]
rifle grenade A grenade designed to be discharged from a rifle by means of a launching device.
ri·fle·man (rī′fəl·mən) *n. pl.* **·men** (-mən) One armed or skilled with the rifle.
rifle pit A trench, the earth from which is thrown up in front, as a protection for riflemen.
ri·fler (rī′flər) *n.* A robber.
rifle salute A salute in the position of right shoulder arms or order arms, with the left hand carried smartly to the rifle, palm down and fingers together.
ri·fling (rī′fling) *n.* 1 The operation of forming the grooves in a rifle. 2 The grooves of a rifle collectively: shallow or deep *rifling.* [<RIFLE[1]]
rift[1] (rift) *n.* An opening made by riving or splitting; a cleft; fissure. — *v.t.* & *v.i.* To rive; burst open; split. [<Scand. Cf. Dan. *rift* cleft, ON *ript* < *ripta* break. Akin to RIVE.]
rift[2] (rift) *n.* 1 A shallow place in a stream; fording place. 2 The wash up the beach after a wave has broken. [? Alter. of *riff,* obs. var. of REEF[1]]
Rift Valley See GREAT RIFT VALLEY.
rig[1] (rig) *v.t.* **rigged, rig·ging** 1 To fit out; equip. 2 *Naut.* **a** To fit, as a ship, with rigging. **b** To fit (sails, stays, etc.) to masts, yards, etc. 3 *Colloq.* To dress; clothe, especially in finery. 4 To make or construct hurriedly or by makeshifts: often with *up:* to *rig* up a door from old boards. — *n.* 1 *Naut.* The arrangement of sails, rigging, spars, etc., on a vessel. 2 *Colloq.* A style of dress; costume. 3 *U.S. Colloq.* A turnout for driving; a horse or horses and vehicle. 4 An outfit; apparatus or tackle: an oil-well *rig.* 5 Fishing tackle. [<Scand. Cf. ON *rigga* wrap around, Norw. *rigga* bind.]
rig[2] (rig) *v.t.* **rigged, rig·ging** To control fraudulently; manipulate: to *rig* an election. — **to rig the market** To manipulate the exchange market by raising or lowering prices without regard to the value of the security or commodity traded in, in order to derive a profit. — *n.* 1 A practical joke; a trick; jest. 2 A tumult; frolic. [Origin uncertain]
rig[3] (rig) *n. Scot.* & *Brit. Dial.* 1 A ridge or strip of ground. 2 The back of an animal. 3 A path; way. [Var. of RIDGE]
Ri·ga (rē′gə) The capital of Latvia, a port on the **Gulf of Riga,** an arm of the Baltic Sea between Estonia and Latvia.
rig·a·doon (rig′ə·do͞on′) *n.* 1 A gay, quick dance for two, originating probably in Provence. 2 The music for such a dance. [<F *rigodon* a dance]
Ri·gel (rī′jəl, -gəl) A star, Beta in the constellation of Orion; magnitude, 0.34. See STAR. [<Arabic *rijl* foot]
rig·ger (rig′ər) *n.* 1 One who rigs. 2 One who fits the rigging of ships. 3 One who assembles and alines the major parts of an aircraft.
rig·ging[1] (rig′ing) *n.* 1 *Naut.* The entire cordage system of a vessel. 2 Tackle used in logging.
rig·ging[2] (rig′ing) *n. Scot.* 1 The back or top of anything. 2 A ridge of a house; roof.
Riggs's disease (rig′ziz) Pyorrhea alveolaris. [after J. M. *Riggs,* 1810–85, U. S. dentist]
Ri·ghi (rē′gē) See RIGI.
right (rīt) *adj.* 1 Done in accordance with or conformable to moral law or to some standard of rightness; equitable; just; righteous. 2 Conformable to truth or fact; correct; true; accurate; not mistaken. 3 Conformable to a standard of propriety or to the conditions of the case; proper; fit; suitable. 4 Most desirable or preferable; also, fortunate. 5 Pertaining to that side of the body which is toward the south when one faces the sunrise: opposed to *left.* 6 Holding one direction, as a line; straight; direct. 7 Properly placed, disposed, or adjusted; well-regulated; orderly; correctly done. 8 Sound in mind or body; healthy; well. 9 *Geom.* Formed with reference to a line or plane perpendicular to another line or plane: a *right* angle. See ANGLE. 10 Designed to be worn outward or placed toward an observer in use: the *right* side of cloth. 11 *Law* Rightful; legal. 12 *Obs.* Real or genuine in character; not spurious. — *adv.* 1 In accordance with justice or moral principle. 2 According to the fact or truth; correctly. 3 In a straight line; directly. 4 Very: used dialectically or in some titles: a *right* good time, *Right* Reverend. 5 Suitably; properly. 6 Precisely; just; also, immediately. 7 Without delay or evasion. 8 Toward the right. 9 Completely or quite: The house burned *right* to the ground. — *n.* 1 That which is right; moral rightness: opposed to *wrong;* also, justice. 2 A just and proper claim or title to anything, or that which may be claimed on just, moral, legal, or customary grounds: often in the plural. 3 *Law* A claim or title to, or interest in, anything whatsoever that is enforceable by law. 4 The right hand, side, or direction. 5 Anything adapted for right-hand use or position. 6 In some legislative bodies, as those of continental Europe, the party occupying seats on the right side of the presiding officer, usually the traditionalist party; the Conservative party: opposed to the *Left* or Liberals: also **the Right.** 7 The outside or front side of a thing: opposed to *reverse.* 8 In boxing, a blow delivered with the right hand. 9 A stockholder's privilege to purchase new stock in a corporation at a special price, usually at par. — **natural rights** Rights with which mankind is supposedly endowed by nature, such as the right to life, liberty, security, and the pursuit of happiness. — *v.t.* 1 To restore to an upright or normal position. 2 To put in order; set right. 3 To make correct or in accord with facts. 4 To make reparation for; redress or avenge: to *right* a wrong. 5 To make reparation to (a person); do justice to. — *v.i.* 6 To regain an upright or normal position. ◆ Homophones: *rite, wright, write.* [OE *riht*] — **right′er** *n.*

Synonyms (adj.): correct, direct, equitable, fair, good, honest, just, lawful, perpendicular, rightful, straight, true, unswerving, upright. See CORRECT, INNOCENT, JUST, MORAL, PRECISE, VIRTUOUS. *Antonyms:* bad, crooked, evil, false, improper, incorrect, indirect, iniquitous, unfair, unjust, unrighteous, wrong.

Synonyms (noun): advantage, claim, exemption, franchise, immunity, liberty, license, prerogative, privilege. In the sense of that which one may rightly claim, a *right* may be either general or special, natural or artificial. "Life, liberty, and the pursuit of happiness" are the natural and inalienable *rights* of all men; *rights* of property, inheritance, etc., are individual and special, and often artificial, as the *right* of inheritance by primogeniture. A *privilege* is always special, exceptional, and artificial. It is something peculiar to one or some, as distinguished from others. A *privilege* may be of doing or avoiding; in the latter case it is an *exemption* or *immunity;* as, a *privilege* of hunting or fishing; *exemption* from military service; *immunity* from arrest. A *franchise* is a specific *right* or *privilege* granted by the government or established as such by governmental authority; as, the elective *franchise,* a railroad *franchise.* A *prerogative* is an official *right* or *privilege,* especially one inherent in the royal or sovereign power; in a wider sense it is an exclusive and peculiar *privilege* which one possesses by reason of being what he is; as reason is the *prerogative* of man; kings and nobles have often claimed *prerogatives* and *privileges* opposed to the inherent *rights* of the people. See DUTY, JUSTICE, PROPERTY.

right-a·bout (rīt′ə·bout′) *n.* 1 The opposite direction. 2 A turning in or to the opposite direction, physically or mentally.
right angle See under ANGLE.
right-an·gled (rīt′ang′gəld) *adj.* Forming or containing a right angle or angles: a *right-angled* triangle.
right ascension *Astron.* The angular distance of a celestial body from the vernal equinox, measured eastward along the celestial equator in hours, minutes, and seconds from 0 hours to 24 hours.
right away At once; immediately.
right·eous (rī′chəs) *adj.* Conforming in disposition and conduct to a standard of right and justice; upright; virtuous; blameless; morally right; equitable; right-thinking. See synonyms under GOOD, INNOCENT, JUST, MORAL, VIRTUOUS. [OE *rihtwis* < *riht* right + *wis* wise] — **right′eous·ly** *adv.*
right·eous·ness (rī′chəs·nis) *n.* 1 The quality or character of being righteous; uprightness; rectitude. 2 A righteous act or quality. 3 Rightfulness; justice. See synonyms under DUTY, JUSTICE, VIRTUE.
right face In military drill, a 90-degree turn to the right, using the ball of the left foot and the heel of the right.
right·ful (rīt′fəl) *adj.* 1 Characterized by or conformed to a right or just claim according to established laws or usage; also, owned or held by just claim: *rightful* heritage. 2 Consonant with moral right or with justice and truth. 3 Proper. 4 Upright; just. See synonyms under JUST, RIGHT. [OE *rihtful*] — **right′·ful·ly** *adv.* — **right′ful·ness** *n.*
right-hand (rīt′hand′) *adj.* 1 Of, pertaining to, or situated on the right side; dextral. 2 Chiefly depended on: He was my *right-hand* man.
right-hand·ed (rīt′han′did) *adj.* 1 Using the right hand habitually or more easily than the left. 2 Done with the right hand. 3 Turning or moving from left to right, as the hands of a clock. 4 Adapted for use by the right hand, as a tool. 5 In conchology, having the spirals rising from left to right. — **right′-hand′ed·ness** *n.*
right-hand rope Plain-laid rope.

right·ism (rī'tiz·əm) *n.* The advocacy of conservative or reactionary policies. — **right'ist** *n.* & *adj.*

right·ly (rīt'lē) *adv.* 1 Correctly. 2 Honestly; uprightly. 3 Properly; aptly.

right–mind·ed (rīt'mīn'did) *adj.* Having approved feelings or opinions.

right·ness (rīt'nis) *n.* 1 The quality or condition of being right. 2 Moral rectitude. 3 Correctness. 4 Straightness. See synonyms under VIRTUE.

right·o (rī'tō) *interj. Brit. Colloq.* An exclamation of satisfaction or assent.

right off Right away.

right of search In international law, the right of a belligerent vessel in time of war to verify the nationality of a vessel and to ascertain, if neutral, whether it carries contraband goods. Also *right of visit and search.*

right of way 1 *Law* The right, general or special, of a person to pass over the land of another; also, the path or piece of land over which passage is made. 2 The strip of land, acquired by easement, condemnation, or purchase, over which a railroad lays its tracks, or that land on which a public highway is built; also, the strip of land above which a high-tension power line is built. 3 The legal or customary precedence which allows one vehicle or vessel to cross in front of another.

right shoulder arms The position in which the rifle is held at an angle of 45 degrees on the right shoulder, barrel uppermost.

right smart *Colloq.* 1 A large quantity; a great number: He chopped a *right smart* of wood. 2 Large: a *right smart* pile of potatoes. 3 Rather well: I feel *right smart.*

right triangle A plane triangle containing one right angle.

right whale A whale, especially *Balaena mysticetus* of circumpolar seas, having a large head with long, narrow, highly elastic whalebone plates in its mouth, for straining food: it yields more oil than any other species. [Prob. orig. so called because advantageous to pursue]

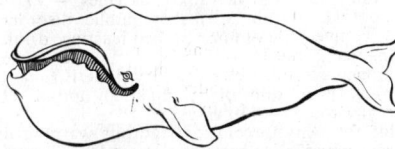

ATLANTIC, OR SOUTHERN, RIGHT WHALE
(From 50 to 60 feet in length;
the pigmy right whale to 20 feet)

right wing 1 A political party or group advocating moderate or conservative policies. 2 That part of any group advocating conservative policies. Also **Right Wing**. — **right'-wing'** *adj.* — **right'-wing'er** *n.*

Ri·gi (rē'gē) A mountain between Lakes Lucerne and Zug, central Switzerland; highest peak, 5,908 feet: also **Righi.**

rig·id (rij'id) *adj.* 1 Resisting change of form; stiff. 2 Rigorous; inflexible; severe. 3 Strict; exact, as reasoning. 4 *Aeron.* Designating a type of airship whose gas compartments are enclosed within a rigid structure. See synonyms under AUSTERE, HARD, INFLEXIBLE, PRECISE, SEVERE. [<L *rigidus* < *rigere* be stiff] — **rig'id·ly** *adv.* — **rig'id·ness** *n.*

ri·gid·i·ty (ri·jid'ə·tē) *n.* 1 The character of being rigid; inflexibility. 2 The property of bodies by which they resist a change in shape: opposed to *ductility.*

Rig·il Ken·tau·rus (rij'il ken·tôr'əs) A star of .06 magnitude in the constellation Centaurus. See STAR.

rig·ma·role (rig'mə·rōl) *n.* A succession of confused or nonsensical statements; incoherent talk or writing; nonsense. [Alter. of *ragman (roll)* document (with pendant seals), catalog, ME *rageman* document; origin unknown]

rig·ol (rig'əl) *n. Obs.* A ring; circle; hence, a crown. [<F *rigole* groove]

ri·go·let·to (rē'gō·let'ō) *n. Italian* A round dance.

rig·or (rig'ər) *n.* 1 The condition of being stiff or rigid. 2 Stiffness of opinion or temper; harshness. 3 Exactness without allowance or indulgence; inflexibility; strictness; severity. 4 Inclemency, as of the weather; hardship. 5 A severe, harsh, or cruel act. 6 *Med.* **a** A violent chill from cold or nervous shock. **b** The trembling observed in the chill preceding a fever. 7 *Biol.* A rigid state in an organism or in any of its parts, caused by adverse or unfavorable conditions. Also *Brit.* **rig'our.** [<L < *rigere* be stiff]

rig·or·ism (rig'ə·riz'əm) *n.* Stiffness in opinion or conduct; severity in style or living, etc.; strictness; austerity. Also *Brit.* **rig'our·ism.** — **rig'or·ist** *n.* — **rig'or·is'tic** *adj.*

rig·or mor·tis (rig'ər môr'tis, rī'gər) The muscular rigidity that ensues within a few hours after death. [<L, stiffness of death]

rig·or·ous (rig'ər·əs) *adj.* 1 Marked by or acting with rigor; uncompromising; severe. 2 Logically accurate; exact; strict. 3 Inclement; severe; bitter; causing hardship: a *rigorous* climate. See synonyms under AUSTERE, SEVERE. [<OF *rigoureux*] — **rig'or·ous·ly** *adv.* — **rig'or·ous·ness** *n.*

Rigs·dag (rigz'däg) *n.* The two chambers that form the Danish parliament: the Landsting and Folketing. [<Dan. < *rige* kingdom + *dag* day. See REICHSTAG.]

rigs·da·ler (rigz'dä'lər), **rijks·daal·der** (rēks'däl'dər) See RIX-DOLLAR.

Rig–Ve·da (rig·vā'də, -vē'-) The oldest collection of hymns and verses in Hindu sacred literature; supposed date, 2000 B.C. See VEDA. [<Skt. *Rigveda* < *ric* praise, hymn + *veda* knowledge]

rig·wid·die (rig·wid'ē) *adj. Scot.* Bony; sapless; scrawny. Also **rig·wood'ie** (-wŏŏd'ē).

Riis (rēs), **Jacob August,** 1849–1914, U. S. journalist and sociologist born in Denmark.

Ri·je·ka (rē·ye'kä) A port of NW Croatia, Yugoslavia, on the Adriatic SW of Zagreb: 40 miles, across Istria, from Trieste: Italian *Fiume.*

Rijn (rīn) The Dutch name for the RHINE.

Rijs·wijk (rīs'wīk) The Dutch name for RYSWICK.

Riks·mål (rēks'môl) *n.* One of the two official forms of Norwegian, based on literary Danish: also called *Dano–Norwegian.* Danish **Rigs·mål** (rēks'môl). Compare LANDSMÅL. [<Norw., speech of the kingdom]

rile (rīl) *v.t. Colloq.* or *Dial.* 1 To vex; irritate. 2 To roil; make muddy. [Var. of ROIL]

ri·ley (rī'lē) *adj.* 1 Roiled; muddy. 2 Ill-tempered; also, irritated.

Ri·ley (rī'lē), **James Whitcomb,** 1849?–1916, U. S. poet.

ri·lie·vo (rē·lye'vō) *n.* Relief (defs. 5 and 6). [<Ital.]

Ril·ke (ril'kə), **Rainer Maria,** 1875–1926, German poet born in Prague.

rill (ril) *n.* 1 A small stream; rivulet. 2 A long, narrow, and generally straight valley on the face of the moon: also **rille**. See synonyms under STREAM. [Cf. Du. *ril,* G *rille*]

rill·et (ril'it) *n.* A little rill (def. 1).

rim (rim) *n.* 1 The edge of an object, usually of a circular object; a margin; border. 2 The peripheral part of a wheel, connected to the hub by spokes. 3 On an automobile wheel, the detachable band over which the tire is fitted. 4 The frame of a pair of spectacles surrounding the lenses. See synonyms under BANK. — *v.t.* **rimmed, rim·ming** 1 To provide with a rim; border. 2 In sports, to roll around the edge of (the basket, cup, etc.) without falling in: The ball *rimmed* the cup. [OE *rima*]

Rim·baud (raṅ·bō'), **Arthur,** 1854–91, French poet.

rime[1] (rīm) *n.* [The spelling *rhyme,* introduced in the 17th century through association with *rhythm,* is etymologically unjustified.] 1 A correspondence of sounds in two or more words, especially at the ends of lines of poetry. See also NEAR RIME, INTERNAL RIME, TERMINAL RIME. 2 A verse, line, etc., corresponding in terminal sound with another. 3 A word corresponding in sound with another. 4 Poetry; verse; also, a tale in verse. See synonyms under POETRY. — *v.* **rimed, rim·ing** *v.i.* 1 To make rimes or verses; compose poetry. 2 To correspond in sound or in terminal sounds. — *v.t.* 3 To put or write in rime or verse. 4 To use as a rime. 5 To cause to correspond in sound. Also spelled *rhyme.* [Prob. fusion of OF *rime* <Gmc. + OE *rīm* a number] — **rime'less** *adj.*

rime[2] (rīm) *n.* 1 Hoarfrost. 2 *Meteorol.* A rough or feathery coating of ice deposited by fog on terrestrial objects. — *v.t.* & *v.i.* **rimed, rim·ing** To cover with or congeal into rime. [OE *hrīm* frost]

rim·er (rī'mər) *n.* One who makes riming verse, especially inferior verse: also spelled *rhymer.*

rime riche (rēm rēsh') *French* In prosody, rime involving words identical in sound but of different meaning. Also **rich rime.**

rime royal A stanza of seven lines in iambic pentameter, rimed *ababbcc:* first used in Chaucer's *Complaint unto Pity.*

rime scheme The pattern of rimes in a stanza or poem, usually represented by letters: A standard *rime scheme* is *abab.*

rime·ster (rīm'stər) *n.* One who makes rimes; a mere versifier; a maker of inferior verses: also spelled *rhymester.*

Rim·i·ni (rim'i·nē, *Ital.* rē'mē·nē) A port in north central Italy on the Adriatic: ancient *Ariminum.*

ri·mose (rī'mōs, rī·mōs') *adj.* Full of fissures or cracks; chinky. Also **ri'mous.** [<L *rimosus* < *rima* chink] — **ri'mose·ly** *adv.* — **ri·mos'i·ty** (rī·mos'ə·tē) *n.*

rim·ple (rim'pəl) *n.* A fold or wrinkle. — *v.t.* & *v.i.* **·pled, ·pling** To wrinkle; rumple. [OE *hrympel*]

Rim·sky–Kor·sa·kov (rim'skē·kôr'sə·kôf, *Russian* rēm'skē·kor·sä·kôf'), **Nicholas Andreievich,** 1844–1908, Russian composer.

rim·y (rī'mē) *adj.* 1 White with rime. 2 Cold; frosty.

rin (rin) *v.t.* & *v.i. Scot.* 1 To run. 2 To melt.

Ri·nal·do (rē·näl'dō) Italian form of REGINALD.

rind[1] (rīnd) *n.* The skin or outer coat that may be peeled or taken off, as of flesh, fruit, or trees. [OE *rinde* bark, crust]

rind[2] (rind, rīnd) See RYND.

rin·der·pest (rin'dər·pest) *n.* An infectious disease of cattle and sometimes of sheep, characterized by inflammation of the mucous membranes of the intestines; cattle plague: formerly known as *murrain.* [<G < *rinder* cattle + *pest* plague]

Rine·hart (rīn'härt), **Mary Roberts,** 1876–1958, U. S. fiction writer.

rin·for·zan·do (rēn'fôr·tsän'dō) *adj.* Reinforcing or increasing the power and emphasis: a musical direction. [<Ital., ppr. of *rinforzare* reinforce]

ring[1] (ring) *n.* 1 Any circular object having an opening of nearly its own diameter. 2 A circular band of precious metal, worn on a finger. 3 Any metal or wooden band used for holding or carrying something: a napkin *ring*; also, a hoop. 4 A group of persons or things in a circle. 5 A combination of persons, often for corrupt or mercenary cooperation, as in business or politics; a clique. 6 A place where the bark has been cut away around a branch or tree trunk. 7 One of a series of concentric layers of wood in an exogenous stem, formed by annual growth: also **annual ring**. 8 An area or arena, as that in which boxers fight; hence, prize fighting in general; a circular racecourse or track, as of a circus or horse show. 9 The field of competition or rivalry: He tossed his hat into the *ring*. 10 The area set apart for bookmakers and other betters at a racetrack. 11 *Chem.* An arrangement of atoms in a closed chain: the benzene *ring.* 12 The space between two concentric circles. — *v.* **ringed, ring·ing** *v.t.* 1 To surround with a ring; encircle. 2 To form into a ring or rings. 3 To provide or decorate with a ring or rings. 4 To cut a ring of bark from (a branch or tree); girdle. 5 To put a ring in the nose of (a pig, bull, etc.). 6 To hem in (cattle, etc.) by riding in a circle around them. 7 In certain games, to cast a ring over (a peg or pin). — *v.i.* 8 To form a ring or rings. 9 To move or fly in rings or spirals; circle. ◆ Homophone: *wring.* [OE]

ring[2] (ring) *v.* **rang, rung, ring·ing** *v.i.* 1 To give forth a resonant, sonorous sound, as a bell when struck. 2 To sound loudly or be filled with sound or resonance; reverberate; resound. 3 To cause a bell or bells to sound, as in summoning a servant. 4 To have or

suggest a sound expressive of a specified quality: His story *rings* true. 5 To have a continued sensation of ringing or buzzing: My ears *ring*. — *v.t.* 6 To cause to ring, as a bell. 7 To produce, as a sound, by or as by ringing. 8 To announce or proclaim by ringing: to *ring* the hour. 9 To summon, escort, usher, etc., in this manner: with *in* or *out*: to *ring* out the old year. 10 To strike (coins, etc.) on something so as to test their quality by the sound produced. 11 To call on the telephone: often with *up*. — **to ring the changes** See under CHANGE. — *n.* 1 The sound produced by a bell or other vibrating, sonorous body; the act of sounding a bell; also, a telephone call. 2 Any reverberating sound, as of acclamation. 3 A sound that is characteristic or indicative: His words have the *ring* of truth. 4 A set, chime, or peal of bells. ◆ Homophone: *wring*. [OE *hringan*]
ring-billed (ring'bild') *adj.* Having a ring of color around the beak: said of certain birds.
ring-billed gull The common gull (*Larus delawarensis*) having a black ring around the bill.
ring bolt A bolt having a ring through an eye in its head.
ring-bone (ring'bōn') *n.* A bony enlargement or excrescence on the pastern bones of a horse, usually causing lameness.
ring dove 1 The cushat; also called *wood pigeon*. 2 One of several other pigeons related to the common, or to the collared, turtle dove (*Streptopelia risoria*) of southeastern Europe.
ringed (ringd) *adj.* 1 Having a wedding ring; hence, lawfully married. 2 Encircled by raised or depressed lines or bands, as the stems or roots of some plants. 3 Encircled by a ring or rings of color; composed of rings.
rin-gent (rin'jənt) *adj. Biol.* Gaping, as a two-lipped corolla in which the lips are widely separated, or as the valves of certain bivalves. [< L *ringens, -entis,* ppr. of *ringi* gape]
ring-er[1] (ring'ər) *n.* 1 One who or that which rings (a bell or chime). 2 *Slang* An athlete who illegally enters a contest by concealing facts which would disqualify him, as his age, professional status, etc. 3 *Slang* A person who bears a marked resemblance to another: You are a *ringer* for Jones.
ring-er[2] (ring'ər) *n.* 1 One who or that which rings or encircles. 2 A quoit or horseshoe that falls around one of the posts.
Ring-er's solution (ring'ərz) *Chem.* A physiologically balanced solution of the chlorides of sodium, potassium, and calcium, used to keep organs alive outside the body. [after Sidney *Ringer*, 1835-1910, English physiologist]
ring finger The third finger of the left hand, on which the marriage ring is worn.
ring-hals (ring'hals) *n.* The spitting snake. [< G, lit., ring-neck]
ring-head (ring'hed') *n.* An instrument for stretching woolen cloth.
ring-lead-er (ring'lē'dər) *n.* A leader or organizer of any undertaking, especially of a party or mob in an unlawful undertaking, such as a riot.
ring-let (ring'lit) *n.* 1 A long, spiral lock of hair; a curl. 2 A small ring.
ring-mas-ter (ring'mas'tər, -mäs'-) *n.* One who has charge of a circus ring and of the performances in it.
ring-neck (ring'nek') *n.* 1 The ring snake. 2 The ring plover. 3 The ring-necked duck.
ring-necked (ring'nekt') *adj.* Having a ring of color around the neck: said of certain birds and animals.
ring-necked duck A North American scaup (*Nyroca* or *Marila collaris*), blackish with a chestnut collar about the neck: also called *marsh bluebill*. Also spelled *ringneck*.
Ring of the Ni-be-lung (nē'bə-lŏŏng) In German legend, the ring which Alberich made from the Rheingold. In his tetralogy of music dramas, *Das Rheingold, Die Walküre, Siegfried,* and *Die Götterdämmerung*, which collectively bear this title, Richard Wagner traces the story of the ring. Also **German Der Ring des Ni-be-lung-en** (der ring des nē'bə-lŏŏng'ən).
ring plover Any of certain small plovers (genus *Charadrius*) marked with a single black breast-encircling band; especially *C. semipalmatus* and the smaller piping plover (*C. melodus*) of eastern North America; also, a

European plover (*C. hiaticula*); also *ringneck*.
ring-shake (ring'shāk') *n.* A cupshake.
ring-side (ring'sīd') *n.* The space or seats immediately surrounding a ring, as at a prize fight.
ring snake 1 A small, harmless, grayish-green snake of North America (*Diadophis punctatus*) having a bright yellow ring around the neck: also *ringneck*. 2 The hoop snake.
ring-ster (ring'stər) *n. U.S. Colloq.* A member of a political ring.
ring-streaked (ring'strēkt') *adj.* Streaked with encircling rings, as an animal. Also *Archaic* **ring'straked'** (-strākt').
ring-tailed (ring'tāld') *adj. Slang* Very extraordinary or superior; stupendous.
ring-tailed roarer In U.S. folklore, a person of extraordinary size, strength, or athletic prowess; a bragging, swaggering fellow.
ring-time (ring'tīm') *n. Obs.* The time of marriage or betrothal.
ring-worm (ring'wûrm') *n. Pathol.* One of several contagious skin diseases affecting both man and domestic animals, caused by certain fungi, and marked by the localized appearance of discolored, scaly patches on the skin and by disorders of the scalp.
rink (ringk) *n.* 1 A smooth, artificial surface of ice, usually covered, used for ice-skating. 2 A smooth floor, similarly enclosed, used for roller-skating. 3 A building containing a surface smoothed and prepared for ice-skating or roller-skating. 4 An area on a field of ice marked off for the game of curling. 5 The part of a bowling green occupied by one side. 6 In bowling, quoits, and curling, the players on one side. [< dial. E (Scottish), prob. <OF *renc* row, rank]
rinse (rins) *v.t.* **rinsed, rins-ing** 1 To remove soap from by putting through clear water. 2 To wash lightly, as by dipping in water or by running water over or into. 3 To remove (dirt, etc.) by this process. See synonyms under CLEANSE. — *n.* The act of rinsing. [<OF *rincer, reincer,* ? ult. <L *recens* recent, fresh] — **rins'er** *n.*
rins-ing (rin'sing) *n.* 1 A rinse. 2 The liquid in which anything is rinsed. 3 That which is removed by rinsing.
Rí-o Bran-co (rē'ōō vrаńg'kŏō) 1 Capital of Acre territory, western Brazil. 2 A federal territory of northern Brazil; 89,035 square miles; capital, Boa Vista. 3 A river in Rio Branco territory, northern Brazil, flowing 350 miles south from the Uraricoera to the Río Negro.
Río Bra-vo (rē'ō brä'vō) The Mexican name for the RIO GRANDE. Also **Río Bravo del Norte** (thel nôr'te).
Rí-o da Dú-vi-da (rē'ōō thə thŏō'vē-thə) A former Portuguese name for the ROOSEVELT RIVER.
Rí-o de Ja-nei-ro (rē'ō də nâr'ō, zhə-nâr'ō; *Pg.* rē'ōō thə zhə-nā'rōō) The capital of Brazil (until the transfer of the federal capital, in April, 1960, to the new city of **Bra-sí-li-a** (brə-zē'lē-ə), located in a federal district, on the central plateau of Goiás state, about 120 miles NE of Goiânia, the state capital), a port on Guanabara Bay or **Rio de Janeiro Bay**: also **Río**. An inhabitant of the city is known as a *Carioca*. 2 A state in SE Brazil; 16,439 square miles; capital, Niterói.
Rí-o de la Pla-ta (rē'ō thä lä plä'tä) See PLATA, RÍO DE LA.
Rí-o de O-ro (rē'ō thä ō'rō) 1 The undefined area of Spanish interest on the NW coast of Africa SW of Morocco. 2 A zone of Spanish Sahara; 73,362 square miles; capital, Villa Cisneros.
Rí-o Gal-le-gos (rē'ō gä-yä'gōs) The capital of Santa Cruz national territory, southern Argentina.
Rí-o Grande (rē'ō grand') 1 A river flowing 1,800 miles from the Rocky Mountains in SW Colorado to the Gulf of Mexico and forming the boundary between Texas and Mexico: Mexican *Río Bravo*. 2 See RIO GRANDE DO SUL (def. 2).

Rí-o Gran-de do Nor-te (rē'ōō grаńn'də thŏō nôr'tə) A maritime state in NE Brazil; 20,482 square miles; capital, Natal.
Rí-o Gran-de do Sul (rē'ōō grаńn'də thŏō sŏŏl') 1 The southernmost state of Brazil; 109,037 square miles; capital, Pôrto Alegre. 2 A port in Rio Grande do Sul state; formerly *São Pedro de Rio Grande do Sul*: also *Rio Grande*.
Rio-ja (ryō'hä) See LA RIOJA.
Rí-o Mu-ni (rē'ō mŏŏ'nē) The mainland district of Spanish Guinea, including the islands of Annobon, Corisco, and Great and Little Elobey; mainland area, 10,040 square miles; chief town, Bata: also *Continental Guinea*.
Rí-o Ne-gro (rē'ō nā'grō) 1 A river in the southern Argentine Republic, flowing 400 miles east and SE to the Atlantic. 2 A river in NW Brazil, flowing SE about 1,400 miles to the Amazon. 3 A national territory of south central Argentina; 78,363 square miles; capital, Viedma. 4 A river in central Uruguay, rising in southern Brazil, and flowing 500 miles SW to the Uruguay river.
Rí-on Strait (rē-ôn') A strait of the Ionian Sea, joining the Gulf of Patras to the Gulf of Corinth: 1 mile wide: formerly *Strait of Lepanto*.
Rí-o Pie-dras (rē'ō pyä'thräs) A city of northern Puerto Rico, the second largest of the commonwealth.
Rí-o Roo-se-velt (rē'ō rō'zə-velt) A Spanish name for the ROOSEVELT RIVER.
ri-ot (rī'ət) *n.* 1 A disturbance consisting of wild and turbulent conduct of a large number of persons, as a mob; uproar; tumult. 2 *Law* Specifically, a tumultuous disturbance of the public peace by three or more assembled persons who, in the execution of some private object, do an act, lawful or unlawful, in a manner calculated to terrorize the people. 3 A state of confusion; a jumble: The garden was a *riot* of color. 4 Boisterous festivity; revelry. 5 *U.S. Slang* An uproariously amusing person, thing, or performance. See synonyms under REVOLUTION, TUMULT. — **to run riot** 1 To act or move wildly and without restraint. 2 To grow rankly, as vines. — *v.i.* 1 To take part in a riot or public disorder. 2 To live a life of unrestrained feasting, drinking, etc.; revel. — *v.t.* 3 To spend (time, money, etc.) in riot or revelry. [<OF *riote* < *rioter*, prob. dim. of *ruir* make an uproar <L *rugire* roar] — **ri'ot-er** *n.*
riot act Any forceful or vigorous warning or reprimand. — **to read the riot act to** To reprimand bluntly and severely.
Riot Act English statute of George I (1715) for preventing tumultuous and riotous assemblages. The act provides that if any twelve persons or more are unlawfully assembled and disturbing the peace, any sheriff, under-sheriff, justice of the peace, or mayor may, by proclamation, command them to disperse, and that if they refuse to obey they are all guilty of felony.
Rí-o Té-o-do-ro (rē'ō tā'ō-thō'rō) A Spanish name for the ROOSEVELT RIVER.
riot gun A short-barreled shotgun for use on guard duty or against rioters.
ri-ot-ous (rī'ət-əs) *adj.* 1 Pertaining to riot; engaged in riot or tumultuous disorder; tumultuous. 2 Indulging in revelry; also, profligate: more *riotous* spending. See synonyms under NOISY, TURBULENT. — **ri'ot-ous-ly** *adv.* — **ri'ot-ous-ness** *n.*
Ri-ouw Archipelago (rē'ou) An Indonesian island group south of Singapore; 2,279 square miles; comprising, with other islands, a province of Indonesia; 12,503 square miles; capital, Tandjungpinang. Also *Riau Archipelago*.
rip[1] (rip) *v.* **ripped, rip-ping** *v.t.* 1 To tear or cut apart roughly or violently; slash. 2 To tear or cut from something else in a rough or violent manner: with *off*, *away*, *out*, etc. 3 To saw or split (wood) in the direction of the grain. — *v.i.* 4 To be torn or cut apart; split. 5 *Colloq.* To utter with vehemence: with *out*. 6 *Colloq.* To rush headlong. See synonyms under REND. — *n.* 1 A place torn or ripped open, especially along a seam; a tear. 2 A ripsaw. [ME *rippen*, prob. <LG. Cf. Frisian *rippe*, Flemish *rippen*.]
rip[2] (rip) *n.* 1 A ripple; a rapid in a river. 2 A riptide. [? <RIP[1]]
rip[3] (rip) *n. Colloq.* 1 A dissipated or worthless person. 2 A worn-out, worthless animal or object. [? Var. of *rep*, short for REPROBATE]

rip[4] (rip) *n. Scot.* A handful of unthreshed grain or of hay.

ri·par·i·an (ri·pâr′ē·ən, rī-) *adj.* **1** Pertaining to the bank of a river: *riparian* rights. **2** Growing naturally in the sides or banks of watercourses, ponds, etc. [<L *riparius* < *ripa* bank of a river]

ri·par·i·ous (ri·pâr′ē·əs, rī-) *adj.* Growing or living along the banks of streams, as an animal or a plant.

rip·cord (rip′kôrd′) *n. Aeron.* **1** The cord, together with the handle and fastening pins, which, when pulled, releases the canopy of a parachute from its pack. **2** A cord attached to the rip panel of a balloon, which, when pulled, frees the panel from the envelope.

ripe[1] (rīp) *adj.* **1** Grown to maturity and fit for food, as fruit or grain. **2** Brought by keeping and care to a condition for use, as wine. **3** Fully developed; matured. **4** In full readiness to do or try; prepared; ready: The men are *ripe* for mutiny. **5** Fit; opportune: The times are *ripe* for war. **6** Resembling ripe fruit; rosy; luscious. **7** *Surg.* Ready for an operation of removal or opening, as an appendix or an abscess. [OE *rīpe* ready for reaping] **—ripe′ly** *adv.* **—ripe′ness** *n.*

Synonyms: complete, consummate, finished, fit, mature, matured, mellow, perfect, perfected, ready, seasoned. *Antonyms*: budding, callow, crude, green, immature, imperfect, sour, undeveloped.

ripe[2] (rīp) *v.t.* **riped, rip·ing** *Scot. & Brit. Dial.* **1** To cleanse. **2** To examine thoroughly. **3** To search.

rip·en (rī′pən) *v.t. & v.i.* To make or become ripe; mature. **—rip′en·er** *n.*

Rip·ley (rip′lē), **William Zebina,** 1867–1941, U.S. economist.

Rip·on (rip′ən) A municipal borough of West Riding, Yorkshire, England.

Rip·on Falls (rip′ən) A waterfall in SE Uganda on the Victoria Nile just below Lake Victoria; about 16 feet high; 900 feet wide.

ri·post (ri·pōst′) *n.* **1** A return thrust, as in fencing. **2** A quick, clever reply; repartee. **—***v.i.* **1** To make a ripost. **2** To reply quickly. Also **ri·poste′.** [<F *riposte* <Ital. *risposta,* properly fem. of pp. of *rispondere* <L *respondere.* See RESPOND.]

rip panel *Aeron.* A segment of the fabric in a balloon or nonrigid airship that may be ripped open quickly to permit emergency deflation.

rip·per (rip′ər) *n.* **1** One who or that which rips. **2** A tool for ripping, as a ripsaw. **3** A double-ripper. **4** *Brit. Slang* A thorough-going or efficient person or thing; something or someone very good.

rip·ping (rip′ing) *Brit. Slang adj.* Splendid; excellent. **—***adv.* Very; extraordinarily: a *ripping* good time.

rip·ple[1] (rip′əl) *v.* **·pled, ·pling** *v.i.* **1** To become slightly agitated on the surface, as water running over a rough, pebbly surface or blown on by a light breeze; form small waves or undulations. **2** To flow with small waves or undulations on the surface. **3** To make a sound like water flowing in small waves. **—***v.t.* **4** To cause to form ripples. **—***n.* **1** One of the wavelets on the surface of water; a ruffle, or slight curling wave. **2** Any sound like that made by rippling. **3** Any appearance like a wavelet. See synonyms under WAVE. [Origin uncertain] **—rip′pler** *n.* **—rip′pling** *adj.* **—rip′pling·ly** *adv.*

rip·ple[2] (rip′əl) *n.* A toothed tool, especially a comblike instrument for cleaning flax fiber or broomcorn. **—***v.t.* **·pled, ·pling** To cleanse, as flax or hemp, by removing the seeds and capsules from the stalk. [<Gmc. Cf. Frisian *ripelje.*]

ripple current *Electr.* The alternating-current component of a pulsating current when this component is small relative to the direct-current component.

rip·plet (rip′lit) *n.* A small ripple.

rip·ply (rip′lē) *adj.* Marked by or sounding like ripples.

rip-rap (rip′rap′) *n.* **1** Broken stones loosely thrown together for a foundation, as in deep water or on a soft bottom, or for a sustaining wall, as along a river bank; also, the stones used, or the foundation so made. **2** *pl.* Artificial islands in Chesapeake Bay. **—***v.t.* **-rapped, -rap·ping** To make a rip-rap in or upon; strengthen with rip-raps.

rip-roaring (rip′rôr′ing, -rōr′-) *adj. U.S. Slang* **1** Excellent; superior; exciting: a *rip-roaring* time. **2** Lively; full of vigor.

rip-roar·i·ous (rip·rôr′ē·əs, -rōr′-) *adj. U.S. Slang* Uproarious; boisterous; violent. **—rip-roar′i·ous·ly** *adv.*

rip·saw (rip′sô′) *n.* A coarse-toothed saw used for cutting wood in the direction of the grain.

rip·snort·er (rip′snôr′tər) *n.* **1** Any person or thing excessively noisy, violent, or striking. **2** A violent windstorm.

rip·tide (rip′tīd′) *n.* Water agitated and made dangerous for swimmers by conflicting tides or currents. Also called *rip, tiderip.*

Rip·u·ar·i·an (rip′yōō·âr′ē·ən) *adj.* Designating or pertaining to a branch of the Frankish people that dwelt on both sides of the Rhine, near Cologne, in the fourth century. **—***n.* A Ripuarian Frank. [<L *ripuarius* < *ripa* bank]

Rip Van Win·kle (rip van wing′kəl) In Washington Irving's tale by that name in *The Sketch Book,* a Dutch villager, who, while out hunting in the Catskills, falls asleep for twenty years, and awakes to find his world changed and himself forgotten.

rise (rīz) *v.* **rose, ris·en, ris·ing** *v.i.* **1** To move upward; go from a lower to a higher position. **2** To slope gradually upward: The ground *rises* here. **3** To have height or elevation; extend upward: The city *rises* above the plain. **4** To gain elevation in rank, status, fortune, or reputation. **5** To swell up: Dough *rises.* **6** To become greater in force, intensity, height, etc. **7** To become greater in amount, value, etc. **8** To become erect after lying down, sitting, etc.; stand up. **9** To get out of bed. **10** To return to life. **11** To revolt; rebel: The people *rose* against the tyrant. **12** To adjourn: The House passed the bill before *rising.* **13** To appear above the horizon: said of heavenly bodies. **14** To come to the surface, as a fish after a lure. **15** To have origin; begin: The river *rises* in the mountains. **16** To become perceptible to the mind or senses: The scene *rose* in his mind. **17** To occur; happen. **18** To be able to cope with an emergency, danger, etc.: Will he *rise* to the occasion? **—***v.t.* **19** To cause to rise. **20** *Naut.* To cause, as a ship, to appear above the horizon by drawing nearer to it. **—to rise above** To prove superior to; show oneself indifferent to. **—***n.* **1** The act of rising; ascent. **2** Degree of ascent; elevation; also, an ascending course. **3** The act of beginning to be or appear, as from a source: the *rise* of a stream. **4** An elevated place; rising ground; a small hill. **5** The act of appearing above the horizon. **6** Increase or advance, as in price. **7** Advance, as in rank, prosperity, or importance; also, elevation morally, mentally, or spiritually. **8** The spring or height of an arch above the impost level. **9** The height of a stair step. **10** Ascent in the diatonic scale; also, increase in volume of tone; a swell. **11** The ascent of a fish to food or bait; also, the flying up of a game bird. **12** *Colloq.* An emotional reaction; a response or retort. **13** *Brit.* An increase in salary. See synonyms under BEGINNING. [OE *rīsan*]

Synonyms (verb): arise, ascend, flow, spring. Compare RAISE.

ris·en (riz′ən) Past participle of RISE.

ris·er (rī′zər) *n.* **1** One who rises or gets up, as from bed: He is an early *riser.* **2** The vertical part of a step or stair. **3** *Metall.* An aperture in the top of a mold to permit the escape of gases and allow for the pouring in of new molten metal to fill cavities.

ris·i·bil·i·ty (riz′ə·bil′ə·tē) *n. pl.* **·ties** **1** A tendency to laughter. **2** *pl.* Impulses to laughter; appreciation of what seems ridiculous: also **ris′i·bles.**

ris·i·ble (riz′ə·bəl) *adj.* **1** Having the power of laughing. **2** Of a nature to excite laughter. **3** Pertaining to laughter. See synonyms under RIDICULOUS. [<F <LL *risibilis* <L *risus,* pp. of *ridere* laugh] **—ris′i·bly** *adv.*

ris·ing (rī′zing) *adj.* **1** Increasing in wealth, power, or distinction. **2** Ascending: the *rising* moon; also, sloping upward: a *rising* hill. **3** Advancing to adult years or to a state of vigor and activity; growing: the *rising* generation. **—***n.* **1** The act of one who or that which rises. **2** That which rises above the surrounding surface; specifically, a tumor; wen. **3** An insurrection or revolt; an uprising. **4** Yeast or leaven used to make dough rise; also, the quantity of dough prepared at once. **—***prep. Dial.* **1** Approaching; going on: He's six years old, *rising* seven. **2** More than; upwards of: a crop *rising* 5,000 bushels.

risk (risk) *n.* **1** A chance of encountering harm or loss; hazard; danger. **2** In insurance, hazard of loss, as of a ship or cargo, or of goods or other property; also, degree of exposure to loss or injury. **3** An obligation or contract of insurance on the part of the insurer: to take a *risk* on a cargo. **4** An applicant for an insurance policy considered with regard to the advisability of placing insurance upon him. See synonyms under DANGER, HAZARD. **—***v.t.* **1** To expose to a chance of injury or loss; hazard. **2** To incur the risk of. [<F *risque* <Ital. *rischio* < *risicare* dare, ult. <Gk. *rhiza* cliff, root] **—risk′er** *n.*

risk·y (ris′kē) *adj.* **risk·i·er, risk·i·est** Attended with risk; hazardous; dangerous. See synonyms under PRECARIOUS.

Ri·sor·gi·men·to (rē·sôr′jē·men′tō) *n.* The movement for the liberation and unification of Italy in the 19th century. [<Ital., resurgence]

ri·sot·to (rē·sôt′tō) *n.* Rice cooked in broth and served with meat, cheese, and various condiments. [<Ital. < *riso* rice]

ris·qué (ris·kā′, *Fr.* rēs·kā′) *adj.* Bordering on or suggestive of impropriety; bold; daring; off-color: a *risqué* story or play. [<F]

Riss (ris) See GLACIAL EPOCH. [from *Riss,* name of a German stream]

ris·sole (ris′ōl, *Fr.* rē·sôl′) *n.* In cookery, a sausagelike roll consisting of minced meat or fish, enclosed in a thin puff paste and fried. [<F, <OF *ruissolle, rousole* <LL *russeola,* fem. of L *russeolus* reddish < *russus* red]

ris·so·lé (rē·sô·lā′) *adj. French* Browned by frying.

Rist (rēst), **Charles,** 1874–1955, French economist.

ri·sus (rī′səs) *n.* A grin or laugh, especially the **risus sar·do·ni·cus** (sär·don′i·kəs), the twisted, grinning expression caused by spasm of the facial muscles, as in tetanus. [<L, a grimace < *ridere* laugh]

ri·tar·dan·do (rē′tär·dän′dō) *adj. Music* Slackening the speed gradually; retarding. [<Ital., gerund of *ritardare* delay]

rite (rīt) *n.* **1** A solemn or religious ceremony performed in an established or prescribed manner, or the words or acts constituting or accompanying it. **2** Any formal practice or custom. See synonyms under FORM, SACRAMENT. ◆ Homophones: *right, wright, write.* [<L *ritus*]

Rit·ter (rit′ər) *n. German* A knight; one of the lowest of the noble orders in Austria and Germany.

rit·u·al (rich′ōō·əl) *n.* A prescribed form or method for the performance of a religious or solemn ceremony; any body of rites or ceremonies; also, a book setting forth such a system of rites or observances. See synonyms under FORM. **—***adj.* Of, pertaining to, or consisting of a rite or rites. [<OF <L *ritualis* < *ritus* rite] **—rit′u·al·ly** *adv.*

rit·u·al·ism (rich′ōō·əl·iz′əm) *n.* **1** A system of conducting public worship according to prescribed or established forms. **2** Strenuous insistence upon ritual.

rit·u·al·ist (rich′ōō·əl·ist) *n.* One who practices or advocates ritualism. **—***adj.* Ritualistic.

rit·u·al·is·tic (rich′ōō·əl·is′tik) *adj.* **1** Of or pertaining to ritual or ritualism. **2** Advocating or adhering to ritualism. **—rit′u·al·is′ti·cal·ly** *adv.*

rit·u·al·ly (rich′ōō·əl·ē) *adv.* According to ritual or to a certain ritual.

ritz·y (rit′sē) *adj. U.S. Slang* Smart; elegant; classy. [after César *Ritz,* 1850–1918, Swiss hotelier who founded world-famous hotels bearing his name in London, Paris, and New York]

riv·age (riv′ij) *n. Archaic* A shore; coast; bank. [<OF < *rive* <L *ripa* shore]

ri·val (rī′vəl) *n.* **1** One who strives to equal

or excel another, or is in pursuit of the same object as another; a competitor. 2 One equaling or nearly equaling another, in any respect. 3 *Obs.* An associate, or companion in office. See synonyms under ENEMY. — *v.* **·valed** or **·valled, ·val·ing** or **·val·ling** *v.t.* 1 To strive to equal or excel; compete with. 2 To be the equal of or a match for. — *v.i.* 3 *Archaic* To be a competitor. — *adj.* Standing in competition or emulation; having opposing claims to the same object; competing. [<F <L *rivalis*]

ri·val·ry (rī′vəl·rē) *n. pl.* **·ries** 1 The act of rivaling. 2 The state of being a rival or rivals; competition. See synonyms under AMBITION, COMPETITION, EMULATION.

rive (rīv) *v.* **rived, rived** or **riv·en, riv·ing** *v.t.* 1 To split asunder by force; cleave. 2 To break (the heart, etc.). — *v.i.* 3 To become split. See synonyms under BREAK, REND. [<ON *rifa* tear, rend] — **riv·er** (rī′vər) *n.*

rived (rīvd) Alternative past participle of RIVE. — *adj.* Split instead of sawed.

riv·en (riv′ən) Alternative past participle of RIVE. — *adj.* Rent, burst, or torn asunder; split; cleaved.

riv·er (riv′ər) *n.* 1 A large, natural stream of water, usually fed by converging tributaries along its course and discharging into a larger body of water, as into the ocean, a lake, or another stream. ◆ Collateral adjective: *fluvial.* 2 A large stream of any kind; copious flow. See synonyms under STREAM. — **to sell down the river** 1 Formerly, to sell (a Negro slave) into unsparing and rigorous servitude: from the severe conditions on the lower Mississippi cane and cotton plantations. 2 Hence, to betray the trust of; deceive. — **to send up the river** To send to the penitentiary: from the fact that Sing Sing is up the Hudson from New York. [<OF *rivière* <LL *riparia* <L *riparius.* See RIPARIAN.]

Ri·ve·ra (rē·vä′rä), **Diego,** 1886–1957, Mexican painter.

Ri·ve·ra y Or·ba·ne·ja (rē·vä′rä ē ôr′vä·nā′·hä), **Miguel Primo de,** 1870–1930, Spanish general; chief of state 1923–30.

river basin *Geog.* An extensive area of land drained by a river and its branches.

river bottom Low-lying alluvial land along a river.

river fever Tsutsugamushi disease.

riv·er·head (riv′ər·hed′) *n.* The source of a river.

river horse A hippopotamus.

River Indians Formerly, those Indians collectively who lived on the upper Connecticut River, as distinguished from the coastal tribes.

riv·er·ine (riv′ə·rīn, -ər·in) *adj.* Pertaining to or like a river; riparian.

River of Doubt A former name for the ROOSEVELT RIVER.

Riv·ers (riv′ərz), **William Halse,** 1864–1922, English physiologist and anthropologist.

riv·er·side (riv′ər·sīd′) *n.* The space alongside of or adjacent to a river.

riv·er·weed (riv′ər·wēd′) *n.* A small aquatic plant (*Podostemon ceratophyllum*) resembling a seaweed, found in the eastern and southern United States.

Rives (rēvz), **Amélie,** 1863–1945, Princess Troubetzkoy, U. S. novelist.

riv·et (riv′it) *n.* A short, soft metal bolt, having a head on one end, used to join objects, as metal plates, by passing the shank through holes and forming a new head by flattening out the headless end. — *v.t.* 1 To fasten with or as with a rivet. 2 To batter the headless end of (a bolt, etc.) so as to make fast. 3 To fasten firmly. 4 To engross or attract (the eyes, attention, etc.). [<OF <*river* clench] — **riv′et·er** *n.*

Riv·i·er·a (riv′ē·âr′ə, *Ital.* rē·vyâ′rä) The coastal strip between the southernmost Alpine ranges and the Mediterranean, extending from Hyères, France, about 230 miles to La Spezia, Italy.

ri·vière (rē·vyâr′) *n. French* A necklace of diamonds or other gems, usually in several strings.

riv·u·let (riv′yə·lit) *n.* A small stream or brook; streamlet. See synonyms under STREAM. [<Ital. *rivoletto,* dim. of *rivolo* <L *rivulus,* dim. of *rivus* brook]

rix-dol·lar (riks′dol′ər) *n.* 1 Any one of several small silver coins formerly current in the Scandinavian countries and the Netherlands: also called *rigsdaler, rijksdaalder.* 2 A former British silver coin of Ceylon, Cape Colony, etc. [<Du. *rijksdaler* dollar of the realm]

Ri·yadh (rē·yäd′) The capital of Nejd and (with Mecca) of Saudi Arabia. Also **Ri·yad′.**

Ri·za·i·yeh (rē·zä′ē·yā′) 1 A city of NE Iran: also *Rezaiyeh.* 2 See URMIA.

Ri·zal (rē·säl′), **José,** 1861–96, Filipino patriot and author; shot for alleged conspiracy against Spain.

Rizal Day A holiday observed on December 30 in the Philippine Islands in memory of José Rizal.

Ri·za Shah Pah·la·vi (ri·zä′ shä′ pä′lə·vē), 1877–1944, shah of Iran 1925–41; abdicated.

riz·zer (riz′ər) *v.t. Scot.* To parch or dry in the sun. Also **riz′zar.**

Riz·zi·o (rēt′tsyō), **David,** 1533?–66, Italian musician; secretary of Mary Queen of Scots; assassinated: also *Riccio.*

ro (rō) *n. Archit.* In Japanese houses, a firepan set into the floor and used in connection with formal tea ceremonies.

Ro (rō) *n.* An artificial, international language based on the classification of ideas and dispensing with existing words and roots. [Coined by Rev. E. P. Foster of Ohio, who devised it in 1906]

roach[1] (rōch) *n.* 1 A European fresh-water fish (*Rutilus rutilus*) of the carp family, with a greenish back. 2 One of certain other related cyprinoid fishes, as the American fresh-water sunfish. [<OF *roche*]

roach[2] (rōch) *n.* A cockroach. [See COCKROACH]

roach[3] (rōch) *v.t.* To clip or trim, as the mane of an animal. [Origin unknown] — **roached** *adj.*

road (rōd) *n.* 1 An open way for public passage, especially from one city, town, or village to another; a highway: distinguished from a *street.* 2 Any way of advancing or progressing; any course followed in a journey; a path. 3 A roadstead: commonly in the plural: Hampton *Roads.* 4 *U.S.* A railroad. — **on the road** 1 On tour: said of circuses, theatrical companies, etc. 2 Traveling, as a canvasser or salesman. 3 Living the life of a tramp or hobo. [OE *rād* a ride, a riding <*rīdan* ride. Related to RIDE.]

Synonyms: course, highway, lane, passage, path, pathway, route, street, thoroughfare, track, turnpike, way. See WAY.

road·a·gent (rōd′ā′jənt) *n.* A highway robber; highwayman, especially on stage routes of the western United States.

road·bed (rōd′bed′) *n.* 1 The graded foundation of gravel, etc., on which the ties, rails, etc., of a railroad are laid. 2 The graded foundation or surface of a road.

road·block (rōd′blok′) *n.* 1 An obstruction in a road. 2 Any arrangement of men and materials for blocking passage, as of enemy troops along a course of advance or retreat.

road hog An automobilist or other driver who keeps his vehicle in or near the middle of a road, making it difficult for other drivers to pass.

road·house (rōd′hous′) *n.* A restaurant, dance hall, or similar establishment located at the side of the road in a rural area.

road metal Broken stone or the like, used for making or repairing roads.

road·run·ner (rōd′·run′ər) *n.* A long-tailed ground cuckoo (genus *Geococcyx*), especially *G. californianus,* inhabiting open regions of southwestern North America, and running with great swiftness: also called *chaparral cock* or *hen.*

ROADRUNNER
(Length about 22 inches over-all)

road·stead (rōd′sted) *n. Naut.* A sheltered place of anchorage offshore, but less sheltered than a harbor. [<ROAD + STEAD (def. 4)]

road·ster (rōd′stər) *n.* 1 A light, open automobile, usually single-seated and having a luggage compartment or a rumble seat in the rear. 2 A horse adapted for use on the road, as in light driving; also, a buggy or light carriage. 3 One who journeys a great deal on roads.

road·way (rōd′wā′) *n.* A road; specifically, that part over which vehicles pass.

Ro·ald (rō′äl) A Norwegian masculine personal name. [<Norw., lit., famous power <Gmc.]

roam (rōm) *v.i.* To move about purposelessly from place to place; wander; rove. — *v.t.* To wander over; range: to *roam* the fields. See synonyms under RAMBLE, WANDER. — *n.* The act of roaming; a ramble. [ME *romen*; origin unknown] — **roam′er** *n.*

roan (rōn) *adj.* 1 Of a color consisting of bay, sorrel, or chestnut, thickly interspersed with gray or white, as a horse. 2 Made of roan leather. — *n.* 1 A roan color. 2 An animal of a roan color. 3 A soft sheepskin leather, tanned to a roan color and used in bookbinding: also **roan leather.** [<OF <Sp. *roano,* ? ult. <L *ravidus* grayish <*ravus* grayish-yellow]

Ro·a·noke (rō′ə·nōk) A city in western Virginia.

Roanoke Island An island off the eastern coast of North Carolina north of Cape Hatteras; settlements attempted by Raleigh in 1585 and 1587 failed; 12 miles long, 3 miles wide. See CROATAN.

Roanoke River A river in Virginia and North Carolina, flowing 410 miles to the head of Albemarle Sound.

roar (rôr, rōr) *v.i.* 1 To utter a deep, prolonged cry, as of rage or distress. 2 To make a loud noise or din, as the sea or a cannon. 3 To laugh loudly. 4 To move, proceed, or act noisily. 5 To make a labored, rasping sound in breathing, as a horse. — *v.t.* 6 To utter or express by roaring: The crowd *roared* its disapproval. — *n.* 1 A full, deep, resonant cry, as of a beast; a similar cry of a human being, as in pain, grief, or anger. 2 Any loud, prolonged sound, as of wind or waves, or a confused mingling of sounds suggesting the cry of wild beasts. See synonyms under NOISE. [OE *rārian*]

Synonyms (verb): bawl, bellow, boom, bray, shout, shriek, yell. See CALL.

roar·er (rôr′ər, rō′rər) *n.* 1 One who or that which roars. 2 An oil gusher. 3 *U.S. Slang* A person who excels.

roar·ing (rôr′ing, rō′ring) *adj.* 1 Emitting or uttering roars; bellowing. 2 *Archaic* Characterized by riotous merriment; boisterous. 3 *Colloq.* Very prosperous or brisk: a *roaring* business. — *n.* 1 A loud, deep, continued sound, as of some animals, or of the waves. 2 A disease among horses, characterized by labored, rasping breathing.

roast (rōst) *v.t.* 1 To cook by subjecting to the action of heat, as in an oven. 2 Originally, to cook before an open fire, or by placing in hot ashes, embers, etc. 3 To heat excessively, or to an extreme degree. 4 To dry and parch under the action of heat: to *roast* coffee. 5 *Metall.* To heat (ores) with access of air, but without fusing, for the purpose of driving off or volatilizing impurities, or for oxidizing them. 6 *Colloq.* To banter or ridicule severely. — *v.i.* 7 To roast food in an oven, etc. 8 To be cooked or prepared by this method. 9 To be uncomfortably hot. — *n.* 1 Something roasted; a piece of meat that is adapted or prepared for roasting, or that is roasted. 2 The act of roasting. — *adj.* Roasted. [<OF *rostir* <OHG *rosten* <*rost* a gridiron, a roast]

roast·er (rōs′tər) *n.* 1 A person who roasts. 2 A pan for roasting. 3 Something suitable for roasting, especially a pig.

rob (rob) *v.* **robbed, rob·bing** *v.t.* 1 To seize and carry off the property of by unlawful violence or threat of violence; commit robbery upon. 2 To deprive (a person) of something belonging or due; defraud. 3 To plunder; rifle, as a house. 4 To steal. — *v.i.* 5 To commit robbery. See synonyms under STEAL. [<OF *rober* <OHG *roubon.* Akin to REAVE, ROBE.]

Rob (rob) Diminutive of ROBERT.

rob·a·lo (rob′ə·lō, rō′bə-) *n. pl.* **·los** or **·lo** Any of a family (*Centropomidae*) of perchlike fishes of tropical American seas, especially *Centropomus undecimalis,* a large and esteemed food fish; a sergeant fish. [<Sp. *róbalo* <Catalan *elobarro,* ult. <L *lupus* a wolf]

rob·and (rob′ənd) *n. Naut.* A piece of spun yarn for fastening the head of a sail to a spar: sometimes called *rope band.* Also **rob′bin.**

[Earlier *raband*, ult. <ON *rābenda* bend a sail on a yard < *ra* a yard for a sail + *benda* bend, bind]

rob·ber (rob′ər) *n.* A plunderer, as a burglar or highwayman.
Synonyms: bandit, brigand, buccaneer, burglar, depredator, footpad, freebooter, highwayman, marauder, pillager, pirate, plunderer, thief. A *robber* seeks to obtain the property of others by force or intimidation; a *thief* by stealth and secrecy.

robber fly The assassin fly.

rob·ber·y (rob′ər-ē) *n. pl.* **·ber·ies** The act of robbing; the taking away of the property of another unlawfully, by force or fear. See synonyms under PLUNDER.

Rob·bia (rôb′byä), **del·la** (del′lä) A family of Italian sculptors and workers in glazed terra cotta; especially **Luca**, 1400-82; his nephew, **Andrea**, 1435-1525; and grandnephew, **Giovanni**, 1469-1529.

robe (rōb) *n.* 1 A long, loose, flowing garment, worn over other dress; a gown. 2 *pl.* Such a garment worn as a badge of office or rank. 3 Any kind of costume; dress; figuratively, anything that covers in the manner of a robe. 4 A blanket or covering, as for use in a carriage or automobile: lap *robe*. 5 The dressed skin of an animal, formerly especially of the American bison, used as a garment or blanket. — *v.* **robed**, **rob·ing** *v.t.* To put a robe upon; clothe; dress. — *v.i.* To put on robes. [< OF, orig. booty <OHG *roub* spoils, robbery. Akin to ROB.]

robe de chambre (rôb′ də shän′br′) *French* A dressing gown. Also **robe′-de-cham′bre**.

robe de nuit (rôb′ də nwē′) *French* A nightgown.

Rob·ert (rob′ərt; *Du., Ger., Sw.* rō′bert, *Fr.* rō·bâr′) A masculine personal name. Also *Ital., Pg., Sp.* **Ro·ber·to** (rō·ber′tō), *Lat.* **Ro·ber·tus** (rə·bûr′təs). [<Gmc., bright fame] — **Robert I**, died 1035, duke of Normandy 1028-35; father of William the Conqueror; called "Robert the Devil." — **Robert II**, 1054?-1134, duke of Normandy 1087-1134; son of William the Conqueror; invaded England, defeated by his brother Henry I. — **Robert the Bruce** See BRUCE.

Ro·ber·ta (rə·bûr′tə) A feminine personal name. [Fem. of ROBERT]

Rob·erts (rob′ərts), **Frederick Sleigh**, 1832-1914, Earl Roberts of Kandahar, Pretoria, and Waterford, British field marshal; known as **Bobs**. — **Kenneth**, 1885-1957, U. S. novelist.

Rob·ert·son (rob′ərt·sən), **William**, 1721-93, Scottish historian. — **Sir William Robert**, 1860-1933, English field marshal; chief of British general staff 1915-18.

Ro·ber·val (rô·ber·val′), **Gilles Personne de**, 1602-75, French mathematician.

Robe·son (rōb′sən), **Paul**, born 1898, U. S. Negro singer and actor.

Robes·pierre (rōbz′pir, *Fr.* rô·bəs·pyâr′), **Maximilien François Marie Isidore de**, 1758-1794, French revolutionist; guillotined.

rob·in (rob′in) *n.* 1 A large North American thrush (*Turdus migratorius*) with black head and tail, grayish wings and sides, and reddish-brown breast and underparts. 2 A small European bird (*Erithacus rubecula*) of the thrush family, especially common in Great Britain, with the forehead, cheeks, and breast yellowish-red. [<OF *Robin*, dim. of ROBERT]

Rob·in (rob′in) Diminutive of ROBERT.

Rob·in Good·fel·low (rob′in good′fel′ō) 1 In English folklore, a merry and mischievous sprite: originally identified with Puck, but later believed to work his mischief around houses. Compare PUCK. 2 Any fairy or elf.

Robin Hood A legendary medieval hero of England, bold, chivalrous, courteous, and generous, an outlaw of great skill in archery, who robbed the rich to relieve the poor, especially in Sherwood Forest in Nottinghamshire, England. Compare ALLAN-A-DALE, FRIAR TUCK.

robin redbreast The European or American robin.

rob·in's-egg blue (rob′inz·eg′) A light greenish hue, the color of the egg shell of the American robin.

Rob·in·son (rob′in·sən), **Edwin Arlington**, 1869-1935, U. S. poet. — **James Harvey**, 1863-1936, U. S. historian. — **Lennox**, born 1886, Irish playwright and theater manager. — **Sir Robert**, born 1886, English biochemist.

Rob·in·son Cru·soe (rob′in·sən krōō′sō) In Defoe's *Robinson Crusoe* (1719), the hero, a sailor shipwrecked on a tropical island, where, by ingenious devices, he maintained himself until rescued. See FRIDAY; SELKIRK, ALEXANDER.

ro·ble (rō′blä) *n.* One of various trees of the oak family, especially the Californian white oak (*Quercus lobata*). [<Sp. <L *robur*, a hard variety of oak]

rob·o·rant (rob′ər·ənt) *adj.* Restoring strength; strengthening. — *n.* Any strengthening medicine; a tonic. [<L *roborans*, *-antis*, ppr. of *roborare* strengthen <*robur*, *-oris*. See ROBUST.]

ro·bot (rō′bət, rob′ət) *n.* 1 An automaton; a manufactured, mechanical person that performs all hard work. 2 One who works mechanically and heartlessly. [after a creation introduced by Karel Čapek, Bohemian playwright, in his *Rossom's Universal Robots* (R. U. R.) in 1921; ult. <Czech *robota* work, compulsory service < *robotiti* drudge]

robot bomb See under BOMB.

robot pilot An automatic pilot.

Rob Roy (rob roi) Nickname of Robert Macgregor, 1671?-1734, a Highland outlaw; hero and title of one of Scott's novels.

Rob·son (rob′sən), **Mount** The highest peak in the Canadian Rockies, in eastern British Columbia, near Alberta; 12,972 feet.

ro·bust (rō·bust′, rō′bust) *adj.* 1 Possessing or characterized by great strength or endurance; rugged; healthy. 2 Requiring strength. 3 Violent; rude. See synonyms under FIRM, POWERFUL, STRONG. [<L *robustus* < *robur*, *roboris*, a hard variety of oak, strength] — **ro·bust′ly** *adv.* — **ro·bust′ness** *n.*

ro·bus·tious (rō·bus′chəs) *adj. Archaic* Of a robust character; also, rough: now often used humorously. — **ro·bus′tious·ly** *adv.* — **ro·bus′tious·ness** *n.*

roc (rok) *n.* In Arabian and Persian legend, an enormous and powerful bird of prey. [<Arabic *rokh*, *rukhkh* <Persian *rukh*]

Ro·ca (rō′kä), **Cape** A cape near Lisbon in Portugal; westernmost point of continental Europe. *Portuguese* **Ca·bo da Ro·ca** (kä′vōō thə rō′kə).

roc·am·bole (rok′əm·bōl) *n.* A European perennial (*Allium scorodoprasum*), allied to the leek, with bulbs or cloves resembling those of garlic. [<F <G *rokenbolle* rye bulb]

Ro·cham·beau (rō·shän·bō′), **Comte de**, 1725-1807, Jean Baptiste Donatien de Vimeure, French marshal; commanded French allies in the American Revolution.

Roch·dale (roch′dāl) A county borough in SE Lancashire, England, where, in 1844, the first cooperative stores were established.

Roche·fort (rōsh·fôr′) A port on the Charente in western France, 10 miles above the Bay of Biscay. Also **Roche·fort′-sur-mer′** (-sür·mâr′).

Ro·chelle (rō·shel′), **La** See LA ROCHELLE.

Rochelle powder (rō·shel′) Seidlitz powder.

Rochelle salt Potassium sodium tartrate, $KNaC_4H_4O_6 \cdot 4H_2O$, a white crystalline salt used as a cathartic. [from LA ROCHELLE]

roches mou·ton·nées (rôsh′ mōō·tô·nā′, rôsh′) Rounded knobs of rock ground down and smoothed by glacial action: so called because smooth and rounded like a sheep's back: also called *sheepbacks*. [<F, sheep-shaped rocks]

Roch·es·ter (roch′es·tər, -is-) 1 A municipal borough of northern Kent, England. 2 A city of western New York, near Lake Ontario.

Roch·es·ter (roch′es·tər, -is-), **Earl of**, 1648?-1680, John Wilmot, English courtier and poet.

roch·et (roch′it) *n.* A ceremonial garment similar to a surplice, but with closer sleeves or without sleeves: worn by bishops and other high churchmen. [<OF, dim. of *roc* a cloak <Gmc. Cf. G *rock* coat.]

Ro·ci·nan·te (rō′si·nän′tā) The raw-boned steed of Don Quixote; hence, any ill-looking riding horse: also spelled *Rosinante*. [<Sp. *rocín* nag]

rock[1] (rok) *n.* 1 Any large mass of stone or stony matter; a boulder; also, a stone small enough to throw; stony fragments; a cliff. 2 A firm or immovable support; refuge; defense. 3 That on which one may be wrecked, as a reef; some source of ruin or injury. 4 *Geol.* The consolidated material forming the crust of the earth; any mass of mineral matter forming an essential part of the earth's crust. 5 The rockfish, or striped bass. 6 The rock dove. 7 A hard confection, of varied flavors. 8 Any of several very hard objects, as ice, rock candy, rock salt, etc.; also, a kind of cooky. 9 *U.S. Slang* A dollar; in the plural, money. — **on the rocks** *U.S. Slang* 1 Ruined; also, destitute; bankrupt. 2 Served with ice cubes but without soda or water: said of whisky or other spirituous beverage. — *adj.* Made or composed of rock; hard; stony: a *rock* wall. [<OF *roque*, *roke*, ult. origin uncertain]

rock[2] (rok) *v.i.* 1 To move backward and forward or from side to side; sway. 2 To sway, reel, or stagger, as from a blow; shake. 3 *Mining* To be washed in a cradle, as ores. — *v.t.* 4 To move backward and forward or from side to side, especially so as to soothe or put to sleep. 5 To cause to sway or reel: The earthquake *rocked* the houses. 6 *Mining* To wash (ores) in a cradle. 7 In mezzotint engraving, to prepare (a plate) by roughing its surface with a rocker (def. 8). See synonyms under SHAKE. — *n.* The act of rocking; a rocking motion. [OE *roccian*]

rock[3] (rok) *n. Archaic* A distaff. [<MDu. *rocke*]

Rock, the Gibraltar.

rock·a·by (rok′ə·bī) *interj.* Go to sleep: from a nursery song intended to lull a child to slumber. — *n.* A lullaby. Also **rock′a·bye**, **rock′-a-bye**.

rock·a·hom·i·ny (rok′ə·hom′ə·nē) *n.* Indian corn parched and pounded; hominy. [<N. Am. Ind. (Algonquian) < *roc* corn + *oham* grind + termination *-min*]

rock·air (rok′âr′) *n.* A rocket launched from an aircraft, usually equipped with instruments for the investigation and recording of conditions in the upper atmosphere. Compare ROCKOON.

rock-and-roll (rok′ən·rōl′) *adj.* Describing a form of popular music, derived from hillbilly styles, achieving its effect by repetition of simple melodic elements, strongly marked rhythms, and exaggerated vocal mannerisms. — *n.* Rock-and-roll music. Also **rock 'n' roll**.

rock·a·way (rok′ə·wā) *n.* A four-wheeled, two-seated pleasure carriage with standing top. [from *Rockaway*, town in New Jersey]

rock bass A fresh-water food fish (*Ambloplites rupestris*) common in eastern North America.

rock bottom 1 The very bottom; the lowest possible level: Prices have hit *rock bottom*. 2 The basis or foundation of any issue. — **rock′-bot′tom** *adj.*

rock-bound (rok′bound′) *adj.* Encircled by or bordered with rocks.

rock candy Sugar candied in hard, clear crystals.

rock cork A variety of asbestos. Also called *rock leather*.

rock crystal Colorless transparent quartz.

rock dove The European wild pigeon (*Columba livia*), the parent of domestic varieties.

Rock·e·fel·ler (rok′ə·fel′ər) *n.* Name of a family of American capitalists and philanthropists, including **John Davison**, 1839-1937, his son, **John Davison, Jr.**, 1874-1960, and the latter's sons, **John Davison, III**, born 1906, **Nelson Aldrich**, born 1908, **Laurance S.**, born 1910, **Winthrop**, born 1912, and **David**, born 1915.

rock·er[1] (rok′ər) *n.* 1 One who or that which rocks, in any sense. 2 One of the curved pieces on which a rocking chair or a cradle rocks. 3 A rocking chair. 4 A rock shaft. 5 A rocking-horse. 6 *Mining* A cradle. 7 An ice skate having a curved runner. 8 A small steel plate with a serrated edge for preparing a copper plate for a mezzotint.

rock·er[2] (rok′ər) *n.* The rock dove.

rocker arm *Mech.* An arm on a rock shaft, as in the valve mechanism of a steam engine.

rocker cam A cam on a rock shaft.

rocker shaft A rock shaft.

rock·er·y (rok′ər·ē) *n. pl.* **·er·ies** 1 Rockwork. 2 A rock garden.

rock·et¹ (rok′it) *n.* 1 A firework, projectile, missile, or other device, usually cylindrical in form, that is propelled by the reaction of escaping gases produced during flight. 2 A type of vehicle operated by rocket propulsion and designed for space travel. —*v.i.* 1 To move like a rocket. 2 To fly straight up into the air, as a bird when alarmed. [< Ital. *rocchetta* spool, dim. of *rocca* distaff < OHG *roccho*; from its resemblance to a distaff]

rock·et² (rok′it) *n.* 1 Any of several ornamental Old World herbs (genus *Hesperis*), especially the common garden **dame rocket** (*H. matronalis*), or damewort. 2 An annual (*Eruca sativa*) used in southern Europe as a salad. [< F *roquette*, ult. < L *eruca* colewort]

rocket bomb See under BOMB.

rock·et·eer (rok′ə·tir′) *n.* One who designs or launches rockets; a student of rocket flight.

rocket gun A gun having the barrel open at both ends and used for the discharge of rocket projectiles. Compare BAZOOKA.

rocket launcher See under LAUNCHER.

rocket projector A device for aiming and discharging rockets.

rock·et·ry (rok′it·rē) *n.* The science, art, and technology of rocket flight, including all aspects from fundamental research to design, engineering, construction, and operation.

rock·et·sonde (rok′it·sond′) *n. Meteorol.* A radiosonde adapted for use on high-altitude rockets.

Rock fever Undulant fever. [from ROCK (OF GIBRALTAR)]

rock·fish (rok′fish′) *n. pl.* **·fish** or **·fish·es** 1 A fish living about rocks. 2 Any of several food fishes (*Sebastodes* and related genera) of the west coast of North America. The black, orange, red, and spotted rockfish, as well as other species, are familiar in California markets. 3 One of various other fishes, as the striped bass, or the killifish.

rock flour Finely pulverized rock produced by the grinding action of glacier ice: also called *glacier meal*.

rock garden A garden with flowers and plants growing in rocky ground or among rocks arranged to imitate this.

rocking chair A chair having the legs set on rockers.

Rock·ing·ham (rok′ing·əm), **Marquis de,** 1730–1782, Charles Watson-Wentworth, English statesman.

rock·ing-horse (rok′ing-hôrs′) *n.* A toy horse mounted on rockers, large enough to be ridden by a child.

rocking stone A stone, often very large, so poised as to rock under little pressure.

rock leather Rock cork.

rock lobster The spiny lobster.

rock maple The sugar maple.

rock milk Agaric mineral.

Rock·ne (rok′nē), **Knute Kenneth,** 1888–1931, U. S. football coach born in Norway.

rock oil Petroleum.

rock·oon (rok·ōōn′) *n.* A small rocket equipped with various meteorological recording devices and attached to a balloon from which it is released at altitudes determined chiefly by its weight. Compare ROCKAIR. [< ROCK(ET) + (BALL)OON]

rock rabbit A hyrax.

rock·rose (rok′rōz′) *n.* One of several plants (genera *Cistus, Helianthemum,* and *Crocanthemum*) having flowers resembling the wild rose.

rock salt Halite.

rock shaft A shaft made to rock on its bearings; particularly, such a shaft used for operating a slide valve in an engine: also called *rocker, rocker shaft.*

rock·weed (rok′wēd′) *n.* Any one of various coarse seaweeds (genera *Fucus* and *Sargassum*) growing on rocks.

rock wool Mineral wool.

rock·work (rok′wûrk′) *n.* 1 A mound or wall of stones set with mortar and arranged to imitate a rocky surface. 2 An artificial grotto.

rock·y¹ (rok′ē) *adj.* **rock·i·er, rock·i·est** 1 Consisting of, abounding in, or resembling rocks. 2 Tough; unfeeling; hard; also, disreputable.

rock·y² (rok′ē) *adj.* **rock·i·er, rock·i·est** *Colloq.* Shaky or dizzy, as if rocking; unsteady in the head, as from past intoxication. — **rock′i·ness** *n.*

Rocky Mountain goat A conspicuous, typically white antelope (*Oreamnos americanus*) found in the mountains of NW North America.

Rocky Mountain National Park A mountainous region in northern Colorado; 395.5 square miles; established, 1915.

ROCKY MOUNTAIN GOAT
(About 40 inches high at the shoulder)

Rocky Mountains The major mountain system of western North America, extending from the Arctic to Mexico; highest peak, Mount Elbert, 14,431 feet. Also **Rock′ies.**

Rocky Mountain sheep The bighorn.

Rocky Mountain spotted fever *Pathol.* An acute infectious rickettsial disease caused by a micro-organism (*Dermacentroxenus rickettsii*) transmitted by the bite of certain ticks (genus *Dermacentor*): it is marked by fever, chills, headache, and diffuse pains, and is endemic in Rocky Mountain and Pacific coast States.

ro·co·co (rə·kō′kō) *n.* 1 A style of decoration and architecture, developed from the baroque and distinguished by profuse, elaborate, and often delicately executed ornament in imitation of rockwork, shells, foliage, and scrolls massed together: prevalent during the 17th and 18th centuries. 2 Anything regarded as florid, fantastic, or odd in literature. — *adj.* 1 Having, or built in, the style of rococo. 2 Overelaborate; florid. [< F, fanciful alter. of *rocaille* shellwork < *roc* rock]

rod (rod) *n.* 1 A shoot or twig of any woody plant; a straight, slim piece of wood or other material, used as an instrument of punishment, a badge of office, etc.; hence, with the definite article, discipline; correction. 2 A scepter; hence, dominion; power. 3 A bar, commonly of metal, forming part of a machine; a connecting rod. 4 A light pole used to suspend and manipulate a fishing line. 5 A measure of length, equal to 5.5 yards or 16.5 feet, or 5.02 meters; also, in England, a **cubic rod,** a unit of volume equal to 1,000 cubic feet. 6 A measuring rule. 7 One of the rodlike bodies of the retina sensitive to faint light. 8 A particular line of family descent. 9 *U.S. Slang* A pistol. 10 A lightning rod. 11 The drawbar of a freight train. — **to ride the rods** *U.S. Slang* To steal a ride by getting on the drawbars of a freight train. See synonyms under STICK. [OE *rod.* Related to ROOD.]

rode (rōd) Past tense of RIDE.

ro·dent (rōd′nt) *n.* A gnawing mammal (order *Rodentia*) having in each jaw two (rarely four) incisors, growing continually from persistent pulps, and no canine teeth, as a squirrel, beaver, or rat. — *adj.* 1 Gnawing. 2 Pertaining to the rodents. [< L *rodens, -entis,* ppr. of *rodere* gnaw] — **ro·den′tial** (rō·den′shəl) *adj.*

rodent ulcer *Pathol.* A malignant ulcer that progressively destroys soft tissues and bones, especially of the face. Also called *noli-me-tangere.*

ro·de·o (rō′dē·ō, rō·dā′ō) *n. pl.* **·de·os** 1 The driving of cattle together to be branded, counted, inspected, etc.; a roundup. 2 An enclosure in a stock farm, in which cattle are collected to be counted and branded. 3 A public spectacle in which the more exciting features of a roundup are presented, as the riding of broncos, branding, lariat-throwing, etc. [< Sp. < *rodear* go around < *rueda* wheel < L *rota*]

Rod·er·ick (rod′ər·ik) *n.* A masculine personal name. Also **Rod′er·ic,** *Lat.* **Ro·der′i·cus** (rō′·də·rī′kəs), *Ger.* **Ro·de·rich** (rō′də·riKH), *Fr.* **Ro·drigue** (rô·drēg′), *Ital., Sp.* **Ro·dri·go** (*Ital.*

rō·drē′gō, *Sp.* -thrē′-). [< Gmc., famous king] — **Roderick,** died 711, last king of the Visigoths.

Rod·gers (roj′ərz), **Richard,** born 1902, U. S. composer.

Ro·di (rō′dē) Italian name for RHODES.

Ro·din (rō·daṅ′), **Auguste,** 1840–1917, French sculptor.

rod·man (rod′mən) *n. pl.* **·men** (-mən) One who uses or carries a surveyor's leveling rod. Also **rods′man.**

Rod·ney (rod′nē), **George Brydges,** 1719?–1792, Baron Rodney, English admiral.

Ro·dó (rō·thō′), **José Enrique,** 1872–1917, Uruguayan essayist.

Ro·dol·fo (*Ital.* rō·dōl′fō, *Sp.* rō·thōl′fō) Italian and Spanish form of RUDOLPH. Also *Fr.* **Ro·dolphe** (rō·dôlf′), *Ital.* **Ro·dol·pho** (rō·dōl′fō), **Ro·dol·phus** (*Du.* rō·dol′fōōs, *Lat.* rō·dol′fəs).

rod·o·mon·tade (rod′ə·mon·tād′, -täd′) *n.* Vainglorious boasting; bluster. — *adj.* Bragging. — *v.i.* **·tad·ed, ·tad·ing** To boast; bluster; brag. [< F < Ital. *rodomontata* < *Rodomonte,* name of a boastful Saracen king in Ariosto's *Orlando Furioso*]

roe¹ (rō) *n.* 1 The spawn or eggs of female fish. 2 The milt of male fish. 3 The eggs of crustaceans. ♦ Homophone: *row.* [Var. of dial. *roan,* appar. < ON *hrogn*]

roe² (rō) *n.* 1 A small, graceful deer (genus *Capreolus*) of Europe and western Asia, with slender antlers rising vertically from the head. Also **roe deer.** 2 Improperly, the doe of the red deer. ♦ Homophone: *row.* [OE *rā*]

Roeb·ling (rōb′ling), **John Augustus,** 1806–1869, U. S. engineer born in Germany; built the Niagara and Cincinnati suspension bridges. — **Washington Augustus,** 1837–1926, son of preceding; built Brooklyn Bridge, completed in 1883.

roe·buck (rō′buk′) *n.* A roe, especially the male.

Roe·mer (rœ′mər), **Olaus,** 1644–1710, Danish astronomer.

roent·gen (rent′gən, runt′-; *Ger.* rœnt′gən) *n.* The international unit of X-ray intensity; the quantity of radiation which, with full use of secondary electrons and without loss to the walls of the chamber, produces in 1 cubic centimeter of air at normal temperature and pressure 1 electrostatic unit of electricity of either sign: also spelled *röntgen.* [after Wilhelm Konrad *Roentgen*]

Roent·gen (rent′gən, runt′-; *Ger.* rœnt′gən), **Wilhelm Konrad,** 1845–1923, German physicist; discoverer of Roentgen rays, better known as X-rays.

roentgen equivalent man See REM.

roentgen equivalent physical See REP.

roent·gen·ize (rent′gən·īz, runt′-) *v.t.* **·ized, ·iz·ing** To subject or expose to the action of X-rays. — **roent′gen·i·za′tion** *n.*

roentgeno- *combining form* X-rays; using, produced by, or producing X-rays: *roentgenogram.* Also, before vowels, **roentgen-.** [< ROENTGEN]

roent·gen·o·gram (rent′gən·ō·gram′, runt′-) *n.* An X-ray photograph, especially one taken for medical or therapeutic purposes; a skiagraph.

roent·gen·og·ra·phy (rent′gən·og′rə·fē, runt′-) *n. Med.* Photography by means of X-rays; radiography.

roent·gen·ol·o·gy (rent′gən·ol′ə·jē, runt′-) *n.* The science which treats of the properties, action, and effects of X-rays. — **roent·gen·ol′o·gist** *n.*

roent·gen·o·paque (rent′gən·ō·pāk′, runt′-) *adj.* Impervious to X-rays.

roent·gen·o·ther·a·py (rent′gən·ō·ther′ə·pē′, runt′-) *n. Med.* Treatment of disease by means of X-rays.

Roentgen rays X-rays.

Roer·ich (rœr′iKH), **Nicholas Konstantin,** 1874–1947, Russian painter.

ro·ga·tion (rō·gā′shən) *n.* 1 In ancient Rome, the submission of a proposed law by the executive (consul or tribune) to the people, requesting its adoption; also, a law submitted in this manner and accepted. 2 Litany; supplication. [< L *rogatio, -onis* < *rogatus,* pp. of *rogare* ask]

Rogation days *Eccl.* The three days immediately preceding Ascension Day, observed as days of special supplication by litanies, processions, etc.

ro·ga·to·ry (rō′gə·tôr′ē, -tō′rē) *adj.* 1 Commissioned to gather information. 2 Officially requesting another court to ascertain and report certain facts: letters *rogatory.*

Rog·er (roj′ər) *interj.* 1 Message received: a code signal used in radiotelephone communication. 2 *U.S. Colloq.* All right; O.K. [from *Roger,* personal name]

Roger (roj′ər, *Fr.* rô·zhā′) A masculine personal name. Also *Lat.* **Ro·ger·us** (rō·jir′əs), *Ital.* **Ro·ge·ro** (rō·jā′rō), **Ro·ge·rio** (*Pg.* rō·zhā′ryōō, *Sp.* rō·hā′ryō). [<Gmc., spear of fame]

Rog·ers (roj′ərz), **Will,** 1879–1935, U. S. actor and humorist: full name *William Penn Adair Rogers.*

Ro·get (rô·zhā′), **Peter Mark,** 1779–1869, English physician and philologist; compiled *Roget's Thesaurus of English Words and Phrases,* 1852.

rogue (rōg) *n.* 1 A dishonest and unprincipled person; trickster; rascal. 2 One who is innocently mischievous or playful: sometimes said familiarly and endearingly. 3 An idle, sturdy beggar; a roving vagrant. 4 *Biol.* A variation from a standard. 5 A fierce and dangerous elephant separated from the herd: in this sense also used adjectively: a *rogue* elephant. — *v.* **rogued, ro·guing** *v.t.* 1 To practice roguery upon; defraud. 2 To eliminate (inferior individuals) from a plot of plants undergoing selection. — *v.i.* 3 To live or act like a rogue. [Origin uncertain]

Rogue River A river in SW Oregon, flowing 200 miles SW to the Pacific.

ro·guer·y (rō′gər·ē) *n. pl.* **·guer·ies** 1 Knavery, cheating, or dishonesty, or an instance of it. 2 Playful mischievousness.

rogues' gallery A collection of photographs of criminals taken to aid the police in their future identification.

rogues' march Music played in derision of a person when he is expelled or driven away in disgrace, as from a military body or community.

ro·guish (rō′gish) *adj.* 1 Playfully mischievous. 2 Knavish; dishonest. — **ro′guish·ly** *adv.* — **ro′guish·ness** *n.*

Ro·han (rô·än′), **de** A feudal family of France; especially, its descendants, **Henri,** 1579–1638, duke and Huguenot leader; and **Louis René Édouard,** 1734–1803, grand almoner and cardinal.

Ro·hil·khand (rō′hil·kund′) A division of north central Uttar Pradesh State, India; 11,759 square miles; capital, Bareilly.

roil (roil) *v.t.* 1 To make muddy, as a liquid, by stirring up sediment. 2 To irritate or anger. Also spelled *rile.* [<F *rouiller* rust, make muddy <OF *rouil* mud, rust]

roil·y (roi′lē) *adj.* 1 Full of sediment; stirred up; turbid. 2 Irritated; vexed.

roist·er (rois′tər) *v.i.* 1 To act in a blustery manner; swagger. 2 To engage in revelry; riot. [<earlier *roister* loud bully <OF *ruistre* <L *rusticus.* See RUSTIC.] — **roist′er·er** *n.* — **roist′er·ing** *adj.*

rok·e·lay (rok′ə·lā) See ROQUELAURE.

Ro·kos·sov·sky (ro·kə·sôf′skē), **Konstantin,** born 1893?, U.S.S.R. marshal.

Ro·land (rō′lənd, *Dan.* rō′län, *Fr.* rô·läṅ′, *Ger.* rō′länt) A masculine personal name. Also *Du.* **Roe·land** (rōō′länt), *Ital., Sp.* **Ro·lan·do** (rō·län′dō), *Pg.* **Ro·lan·do** (rō·läṅ′dōō), **Rol·dão** (rōl·douṅ′), *Lat.* **Ro·lan·dus** (rō·lan′dəs). [< Gmc., fame of the land]

— **Roland** Hero of the Anglo-Norman epic *Chanson de Roland* and of many other stories of the Charlemagne cycle. According to legend he was the nephew of Charlemagne, and a bulwark of Christianity against the Saracens, dying in battle at Roncesvalles in 788. He is known as *Orlando* in Italian romances concerning Charlemagne. — **a Roland for an Oliver** Action taken in retaliation, or by way of matching something said or done by another; a tit for tat: in allusion to an indecisive battle between Roland and Oliver, his companion-in-arms.

role (rōl) *n.* A part or character taken by an actor; any assumed character or function. Also **rôle.** ◆ Homophone: *roll.* [<F]

Rolfe (rolf), **John,** 1585–1622, English colonist in Virginia; husband of Pocahontas.

roll (rōl) *v.i.* 1 To move forward upon a surface by turning round and round, as the wheel of a vehicle. 2 To move or be moved on wheels: The cart *rolled* down the hill. 3 To rotate wholly or partially: Her eyes *rolled* with pleasure. 4 To assume the shape of a ball or cylinder by turning over and over upon itself. 5 To move or appear to move in undulations or swells, as waves or plains. 6 To sway or move from side to side, as a ship: to pitch and *roll.* 7 To walk with a swaying motion; swagger; also, to stagger. 8 To make a sound as of heavy, rolling wheels; rumble: Thunder *rolled* across the sky. 9 To become spread or flat because of pressure applied by a roller, etc.: The metal *rolls* easily. 10 To perform a periodic revolution, as the sun. 11 To move ahead; progress. — *v.t.* 12 To cause to move along a surface by turning round and round, as a ball, log, etc. 13 To move, push forward, etc., on wheels or rollers. 14 To impel or cause to move onward with a steady, surging motion: The ocean *rolls* its waves upon the shore. 15 To rotate, as the eyes. 16 To impart a swaying motion to. 17 To spread or make flat by means of a roller. 18 To wrap round and round upon itself. 19 To cause to assume the shape of a ball or cylinder by means of rotation and pressure: to *roll* a cigarette. 20 To wrap or envelop in or as in a covering. 21 To utter with a trilling sound: to *roll* one's r's. 22 To emit in a full and swelling manner, as musical sounds. 23 To beat a roll upon, as a drum. 24 To cast (dice) in the game of craps. 25 *Printing* To apply ink to (a form) by means of a roller or rollers. See synonyms under REVOLVE. — **to roll back** To cause (prices or wages) to return to a previous, lower level, as by government order. — **to roll in** 1 To arrive. 2 To gather. 3 *Colloq.* To luxuriate; wallow. — **to roll out** 1 To unroll. 2 *Colloq.* To leave. 3 To flatten by means of rollers. — **to roll up** 1 To assume or cause to assume the shape of a ball or cylinder by turning over and over upon itself. 2 To accumulate; amass: to *roll up* large profits. — *n.* 1 Anything rolled up in cylindrical form: a *roll* of parchment. 2 Hence, an official writing, especially a list of names or a register. 3 *U.S. Slang* A wad of paper money; also, money in general. 4 A long strip, as of ribbon or carpet, rolled upon itself or upon a core: sometimes of an agreed length used as a measure of quantity. 5 Any food rolled up in preparation for use, as bread by rolling up pieces of dough, meat for roasting, or a pudding or cake formed in a similar way: a jelly *roll.* 6 A roller; particularly, a cylinder in fixed bearings used as a roller. 7 A reverberation, as of thunder. 8 A trill. 9 The rapid beating of a drum to make its sound continuous. 10 A rolling gait or movement; also, motion from side to side, as of a ship in a seaway. 11 *Aeron.* A single turn of an airplane about its long axis without change in the direction of flight: also called **barrel roll;** when performed quickly, called a **snap roll.** 12 A strip of leather or other material fitted with pockets to hold tools or toilet articles, etc., around which it is rolled and fastened. See synonyms under RECORD. ◆ Homophone: *role.* [<OF *roller* <L *rotula* < *rota* wheel]

Rol·land (rô·läṅ′), **Romain,** 1868–1944, French novelist and dramatist.

roll·a·way (rōl′ə·wā′) *adj.* Mounted on rollers for easy movement into storage: a *rollaway* bed.

roll-back (rōl′bak′) *n.* A return, by government order, to a previous, lower price or wage level.

roll call 1 The act of calling over a roll or list of the names of a number of persons, as soldiers or workmen, to ascertain which are present. 2 The time of or signal for calling the roll.

roll·er (rō′lər) *n.* 1 One who or that which rolls anything. 2 Any cylindrical device that rolls. 3 The wheel of a caster or roller skate. 4 A rod for carrying a curtain, towel, map, or the like. 5 A heavy cylinder for rolling, smoothing, or crushing something: a steam *roller.* 6 *Printing* A cylindrical device, often of hard rubber, to spread the ink on a form before impressing on paper. 7 *Surg.* A long rolled bandage to be wrapped around a limb or the like. 8 One of a series of long, swelling waves which break on a coast, especially after a storm. 9 *Ornithol.* **a** An Old World bird of crowlike form with gaudy colors, remarkable for its irregular rolling or tumbling flight, especially the common roller (*Coracias garrula*) found in Europe. **b** A tumbler pigeon.

roller bearing A bearing employing steel rollers to lessen friction between the parts of a mechanism.

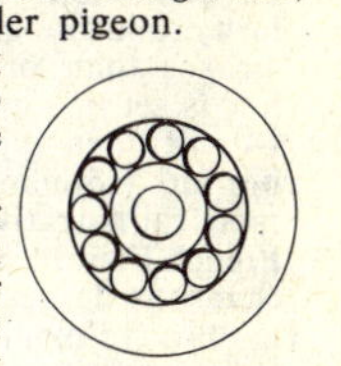

ROLLER BEARING

roller coaster A circular switchback railway with many steep inclines, over which small cars are run: common at amusement parks.

roller derby *U.S.* A race between two teams on roller skates: a player scores points for his team by overtaking opposing players after skating completely around the track within a given time limit.

roll·er-skate (rō′lər·skāt′) *v.i.* **-skat·ed, -skat·ing** To go on roller skates.

roller skate A skate having rollers or wheels instead of a runner.

roller towel An endless towel for use on a roller.

rol·lick (rol′ik) *v.i.* To move in a careless, frolicsome manner; act carelessly and jovially. [Blend of ROMP and FROLIC]

rol·lick·ing (rol′ik·ing) *adj.* 1 Moving in a careless or swaggering manner; jovial. 2 Expressive of a careless, frolicsome spirit: *rollicking* behavior. Also **rol′lick·some** (-səm), **rol′lick·y.**

roll·ing (rō′ling) *adj.* 1 Having a succession of sloping elevations and depressions; undulating: *rolling* prairies. 2 Turned back or down as if over a roll: a *rolling* collar. 3 Of or pertaining to rolling; used in rolling. 4 Moving on or as if on wheels; rotating. 5 Surging in puffs or billows, as smoke, clouds, etc. 6 Recurring; elapsing: said of time. 7 Swaying from side to side: a *rolling* gait. — *n.* The act of a person or thing that rolls, or of one who uses a rolling tool.

rolling barrage *Mil.* An artillery barrage in which the range is steadily increased so that the shells fall just ahead of advancing ground troops.

rolling hitch A hitch with one or more intermediate turns between the first and last hitch. See illustration under HITCH.

rolling kitchen *Mil.* A field kitchen equipped to move with troops.

rolling mill An establishment in which metal is rolled into sheets, bars, etc.

rolling pin A roller, usually of wood, with a handle at each end, for rolling out dough, etc.

rolling stock The wheeled transportation equipment of a railroad.

rolling stone 1 A stone worn smooth by friction and wear. 2 A person of restless, unsettled habits and occupation.

Rol·lo (rol′ō) A masculine personal name. [See RUDOLPH]

— **Rollo,** 860?–932?, Norwegian Viking leader; first duke of Normandy: also *Hrolf.*

roll-top (rōl′top′) *adj.* Having a cover which slides back out of the way: a *roll-top* desk.

roll·way (rōl′wā′) *n.* An inclined way, natural or artificial, down which logs may be rolled or shot; chute.

Röl·vaag (rœl′väg), **Ole Edvart,** 1876–1931, U. S. educator and novelist born in Norway.

ro·ly-po·ly (rō′lē·pō′lē) *adj.* Short and fat; pudgy; dumpy. — *n.* 1 *Brit.* A pudding made of a sheet of pastry dough spread with fruit, preserves, etc., rolled up and cooked. 2 A pudgy person. [Reduplication of ROLL]

Ro·ma·gna (rō·mä′nyä) A region and former province of the Papal States in north central Italy on the Adriatic.

Ro·ma·ic (rō·mā′ik) *adj.* Pertaining to or characteristic of the language or people of modern Greece. — *n.* Modern Greek, especially the popular spoken form. [<LL *Romaicus* <Gk. *Rhōmaikos* Roman <*Rhōmē* Rome]

ro·maine (rō·mān′) *n.* A variety of lettuce (*Lactuca sativa longifolia*) characterized by long, crisp leaves. [<F, fem. of *romain* Roman]

Ro·mains (rô·maṅ′), **Jules** Pseudonym of Louis Farigoule, born 1885, French novelist.

roman[1] (rō′mən) *adj. Printing* Designating or pertaining to a common style of type or letter, characterized chiefly by serifs, perpendicularity, and the greater thickness of its upright strokes than of its horizontal strokes: This line is set in roman: distinguished from *italic*. — *n.* Roman type. Also **Ro′man.**

ro·man[2] (rō·mäṅ′) *n. French* **1** A type of metrical narrative, especially common in Old French literature, developed from the ancient chansons de geste. **2** A romantic novel.

Ro·man (rō′mən) *adj.* **1** Of, pertaining to, or characteristic of Rome or its people. **2** Belonging to or connected with the Church of Rome or its head; Roman Catholic. **3** Somewhat aquiline: a *Roman* nose. — *n.* **1** A native, resident, or citizen of modern Rome or a citizen of ancient Rome. **2** A Roman Catholic. **3** *pl.* The Epistle to the Romans. — **Epistle to the Romans** One of the books of the New Testament; a letter from the apostle Paul to the Christians at Rome. [<OF *romain* <L *Romanus* <*Roma* Rome]

ro·man à clef (rô·mäṅ′ ä kle′) *French* A novel in which actual persons and places appear under fictitious names; literally, a novel with a key.

Roman alphabet The Latin alphabet.

Roman architecture A style of architecture

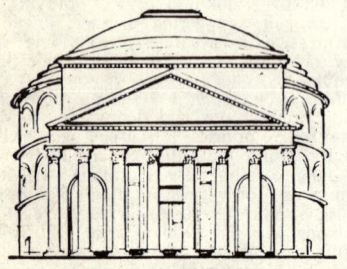

ROMAN ARCHITECTURE
Pantheon, Rome, A.D. 123

which is characterized by the size, massiveness, and boldness of its round arches and vaults, by the somewhat lavish adoption of Greek embellishments, and by excellent stonemasonry and brickmasonry of every kind.

Roman calendar See under CALENDAR.

Roman candle A firework consisting of a tube filled with a composition which discharges colored balls and sparks of fire.

Roman Catholic A member of the Roman Catholic Church.

Roman Catholic Church The church in communion with the pope, whom it recognizes as its supreme head on earth: an official designation. Also called the *Catholic Church*.

ro·mance (rō·mans′, rō′mans) *n.* **1** Adventurous, heroic, or picturesque character or nature; strange and fascinating appeal: the *romance* of faraway places. **2** A disposition to delight in the mysterious or adventurous: a child of *romance*. **3** A love affair. **4** A long narrative from medieval legend, presenting chivalrous ideals and aristocratic society and usually involving heroes in strange adventures and affairs of love. **5** Any long fictitious narrative embodying scenes and events remote from common life and filled with extravagant adventures and often long digressions. **6** The class of literature consisting of romances (defs. 4 and 5). **7** An extravagant or fanciful falsehood. **8** *Music* A simple rhythmic melody, often sentimental, suggestive of a love song. See synonyms under DREAM, FICTION. — *v.* (rō·mans′) **·manced**, **·manc·ing** *v.i.* **1** To tell romances. **2** To think or act in a romantic manner. **3** *Colloq.* To make love. — *v.t.* **4** *Colloq.* To make love to; woo. [<OF *romans* a story written in French <L *Romanice* in Roman style <*Romanus* Roman]

Ro·mance (rō·mans′) *adj.* Pertaining or belonging to one or more, or all, of the languages which have developed from the vulgar Latin speech, and which exist now as French, Italian, Spanish, Portuguese, Catalan, Provençal, Rhaeto-Romanic, and Rumanian. — *n.*

One, or all collectively, of the Romance languages.

ro·manc·er (rō·man′sər) *n.* **1** A writer of romances. **2** One who indulges in extravagant fictions or fancies.

Ro·man de la Rose (rô·mäṅ′ də là rôz′) An allegorical Old French verse romance, begun by Guillaume de Lorris about the middle of the 13th century, and completed in satirical tone by Jean de Meung toward the end of the century: source of Chaucer's *Romaunt of the Rose.*

ROMAN EMPIRE
At Its Greatest Extent A.D. 117

Roman Empire The empire of ancient Rome, established by Augustus in 27 B.C. and continuing until the reign of Theodosius in A.D. 395, when it was divided into the Eastern Roman Empire and the Western Roman Empire.

ro·man·esque (rō′mən·esk′) *adj.* Romantic; fabulous; fanciful. [<F <Ital. *romanesco* <Med. L *romaniscus* <L *romanus* Roman]

Ro·man·esque (rō′mən·esk′) *adj.* **1** Pertaining to or designating the Romanesque style of architecture. **2** Pertaining to or characterized by the Romance languages, especially Provençal. — *n.* **1** Romanesque architecture. **2** The vernacular of Languedoc and other provinces in southern France.

Romanesque architecture The prevailing style, developed from Roman principles, of Western architecture from the 5th to the 12th centuries, embracing the Saxon, Norman, Lombard, etc., characterized by the round arch and general massiveness. It reached its

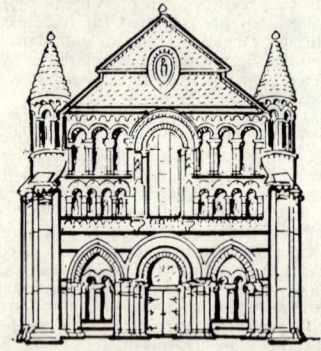

ROMANESQUE ARCHITECTURE
Notre Dame la Grande, Poitiers, France,
A.D. 11th Century.

best form in France in the 11th and 12th centuries.

Roman holiday **1** A day of gladiatorial and other contests in ancient Rome. **2** Enjoyment or profit whereby others suffer.

Ro·mâ·ni·a (rō·mäṅ′ya, *Rumanian* rô·mœ′nyä) The Rumanian name for RUMANIA.

Ro·man·ic (rō·man′ik) *adj.* Roman; also, Romance.

Ro·man·ism (rō′mən·iz′əm) *n.* The dogmas, forms, etc., of the Roman Catholic Church: a term used chiefly in disparagement. — **Ro′man·ist** *adj. & n.*

Ro·man·ize (rō′mən·īz) *v.t. & v.i.* **·ized**, **·iz·ing** **1** To make or become Roman or Roman Catholic. **2** To write or speak in a Latinized style. — **Ro′man·i·za′tion** *n.*

Roman mile See MILE.

Roman nose A nose that is somewhat aquiline.

Roman numerals The letters used by the ancient Romans as symbols in arithmetical notation. See NUMERAL.

Ro·ma·nov (rō′mə·nôf, *Russian* rô·mä′nôf) A Russian dynasty, 1613–1917, founded by Czar Michael, 1596–1645. Also **Ro′ma·noff.**

Roman punch An ice consisting of the white of eggs beaten with rum and lemon juice.

Ro·mansch (rō·mansh′, -mänsh′) *n.* **1** A Rhaeto-Romanic dialect spoken in the Grisons canton, Switzerland. **2** The Rhaeto-Romanic dialects as a group. Also **Ro·mansh′.** [<L *Romanicus* <*Roma* Rome]

ro·man·tic (rō·man′tik) *adj.* **1** Characterized or influenced by romance or the extravagantly ideal; imaginative; marvelous; fanciful: a *romantic* tale. **2** Given to feelings or thoughts of romance; dreamy: a *romantic* girl. **3** Characterized by or conducive to love or amorousness. **4** Visionary; impractical: a *romantic* scheme. **5** Strangely wild or picturesque: *romantic* scenery. **6** Of, pertaining to, or characteristic of a style of art and literature tending toward free expression of subjective feeling, impressive picturesqueness, imagination, sensuousness, etc.: opposed to *classic* or *classical.* **7** Of or pertaining to romanticism in art and literature in the 19th century. — *n.* **1** An adherent of romanticism; a romanticist. **2** A romantic person. **3** A romantic trait, idea, etc. [<F *romantique* <*romant, roman* romance, novel] — **ro·man′ti·cal·ly** *adv.*

Synonyms (*adj.*): airy, chimerical, dreamy, extravagant, fanciful, fantastic, fictitious, ideal, imaginative, picturesque, poetic, sentimental, visionary, wild. *Antonyms:* exact, historical, literal, precise, truthful, unadorned, unimaginative, unvarnished.

ro·man·ti·cism (rō·man′tə·siz′əm) *n.* **1** The quality or characteristic of being romantic. **2** In art, music, and literature, a romantic style as opposed to the classical. **3** In the late 18th century and the 19th, a social and esthetic movement, beginning as a reaction to neo-classicism, that sought to free the individual from unpleasant realities by appealing to his aspirations for wonder and mystery. It emphasized a love for strange beauty, for the past and the far-away, and for the wild, irregular, or grotesque in nature, and found creative expression in spontaneity, lyricism, reverie, sentimentalism, mysticism, and individualism. — **ro·man′ti·cist** *n.*

ro·man·ti·cize (rō·man′tə·sīz) *v.t.* **·cized**, **·ciz·ing** To regard or interpret in a romantic manner.

Romantic Movement See ROMANTICISM (def. 3).

Rom·a·ny (rom′ə·nē) *adj.* Of or pertaining to the Gipsies or their language. — *n.* **1** A Gipsy. **2** The Indic language of the Gipsies, containing elements of the language of each country in which they live: also called *Gipsy*. Also **Rom′ma·ny.** [<Romany *romani* <*rom* man]

ro·maunt (rō·mänt′, -mônt′) *n.* A romance, usually in verse. [<OF *romant*, var. of *romans*. See ROMANCE.]

Rom·berg (rom′bûrg), **Sigmund**, 1887–1951, U. S. composer born in Hungary.

Rom·blon (rôm·blôn′) A Philippine province, comprising a group of Visayan Islands including **Romblon Island** (32 square miles); 512 square miles; capital, Romblon.

Rome (rōm) **1** A city on the Tiber river, capital of Italy and the site of Vatican City, center of the Roman Catholic Church; formerly the capital of the Roman republic, the Roman Empire, and the States of the Church. *Italian* and *Latin* **Ro·ma** (rō′mä). **2** The Roman Catholic Church. **3** Roman Catholicism. **4** A city in central New York.

Ro·me·o (rō′mē·ō) In Shakespeare's tragedy *Romeo and Juliet*, the hero of the play, son of Montague, in love with Juliet, daughter of Capulet who is the enemy of the house of the Montagues.

Rom·ford (rum′fərd, rom′-) A municipal borough in SW Essex, England.

Rom·ish (rō′mish) *adj.* Pertaining to the Roman Catholic Church: an invidious usage.

Rom·mel (rum′əl, *Ger.* rôm′əl), **Erwin**, 1891–1944, German field marshal in World War II.

Rom·ney (rom′nē, rum′-), **George**, 1734–1802, English painter.

romp (romp) *v.i.* **1** To play boisterously. **2** To win easily. — *n.* **1** One, especially a girl, who

romps. 2 Noisy, exciting frolic or play. [Var. of RAMP².]
romp·er (rom′pər) *n.* 1 One who romps. 2 *pl.* A combination of waist and bloomers, as worn by young children at play.
romp·ing (rom′ping) *n.* Boisterous playing. — **romp′ing·ly** *adv.*
romp·ish (rom′pish) *adj.* Inclined toward boisterousness in play. — **romp′ish·ly** *adv.* — **romp′ish·ness** *n.*
Rom·u·lus (rom′yə·ləs) In Roman mythology, a son of Mars and founder of Rome, later deified as *Quirinus:* abandoned in the Tiber with his twin brother Remus, the infant Romulus was reared by a she-wolf, later killing his brother to become the first ruler of Rome.
Ron·ald (ron′əld, *Norw.* rō′näl′) A masculine personal name. [See REGINALD.]
Ron·ces·val·les (ron′sə·valz, *Sp.* rôn′thes·vä′lyäs) A village in the Pyrenees, northern Spain; nearby **Roncesvalles Pass** was the scene of Roland's death and the defeat of Charlemagne's rear guard, 788. *French* **Ronce·vaux** (rôns·vō′).
ron·deau (ron′dō, ron·dō′) *n.* A poem of French origin, consisting of thirteen lines with only two rimes: the opening words of the first line are added, as an unrimed refrain, after the eighth and thirteenth lines. [<F *rondel* < *rond* round]
ron·del (ron′dəl, -del) *n.* A form of French verse consisting of 13 or 14 lines, in two stanzas of four and one of five or six lines, the first two lines being repeated, as a refrain, in the seventh and eighth lines, and again in the thirteenth and fourteenth. The names *rondeau* and *rondel* are often used interchangeably in English. [<F. See RONDEAU.]
ron·de·let (ron′də·let) *n.* A brief French verse form with a refrain, which generally consists of two or more words of the first line. [<OF, dim. of *rondel*. See RONDEAU.]
ron·do (ron′dō, ron·dō′) *n.* 1 *Music* A composition or movement having a main theme and several contrasting episodes, the former being repeated in its original key after each subordinate theme. 2 The musical setting of a rondeau. [<Ital., round]
Ron·dô·nia (rôn·dô′nyä) A federal territory of western Brazil; 1,381,877 square miles; capital, Porto Velho; formerly *Guaporé.*
ron·dure (ron′jər) *n.* Anything circular or spherical; a curve or swell. [<F *rondeur* roundness]
Ron·ge·lap (rông′ə·lap) An atoll in the Ralik chain of the Marshall Islands; 35 miles long; 3 square miles. Also **Rong′e·lab.**
Ron·ge·rik (rông′ə·rik) An atoll in the Ralik chain of the Marshall Islands; about 30 miles in circumference.
ron·ion (run′yən) *n. Obs.* A mangy or scabby animal or person. Also **ron′yon.** [<F *rogne* scab]
Rön·ne (rœn′ə) A Danish city, chief port of Bornholm island.
ron·quil (ron′kil) *n.* A deep-water fish (family Bathymasteridae) of the North Pacific. [<Sp. *ronquillo,* dim. of *ronco* hoarse <L *raucus* hoarse]
Ron·sard (rôn·sàr′), **Pierre de,** 1524–85, French poet.
Rönt·gen (rent′gən, runt′-; *Ger.* rœnt′gən) See ROENTGEN.
rood (rōōd) *n.* 1 A cross or crucifix; specifically, a crucifix or a representation of the Crucifixion over the altar screen of a church. 2 A square land measure, **square rood,** equivalent to one fourth of a statute acre, or 40 square rods. 3 A linear measure varying locally between six and eight yards. ♦ Homophone: *rude.* [OE *rōd* rod, measure of land, cross. Related to ROD.]
rood beam A beam over the entrance to a choir for supporting a cross or crucifix.
Roo·de·poort-Ma·rais·burg (rō′də·pôrt·mar′is·bûrg, -pôrt-, rō′-) A town of southern Transvaal province, Union of South Africa.
rood screen An enriched screen, usually surmounted by a rood, separating the choir presbytery from the nave.
roof (rōōf, rōōf) *n.* 1 The exterior upper covering of a building. 2 Any top covering, as of a car or oven. 3 A house; home. 4 The most elevated part of anything; top; summit. — *v.t.* To cover with or as with a roof. [OE *hrōf*]

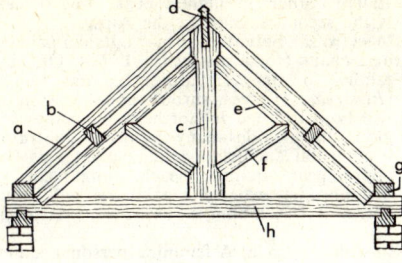

ROOF CONSTRUCTION–KINGPOST TYPE
a. Common rafters.   e. Principal rafters.
b. Purlin.   f. Struts.
c. Kingpost.   g. Pole plate.
d. Ridge pole.   h. Tie beams.

roof·age (rōō′fij, rōōf′ij) *n.* The material forming a roof; roofing.
roof·er (rōō′fər, rōōf′ər) *n.* One who makes or repairs roofs.
roof garden A garden on the roof of a building; especially, a space on a roof used for public entertainments, restaurants, etc.
roof·ing (rōō′fing, rōōf′ing) *n.* 1 Roofs collectively. 2 Material for roofs. 3 Shelter. 4 The act of covering with a roof.
roof·less (rōōf′lis, rōōf′-) *adj.* 1 Having no roof. 2 Destitute of shelter; homeless.
roof of the mouth The hard palate.
roof·tree (rōōf′trē, rōōf′-) *n.* 1 The ridge pole of a roof. 2 The roof. 3 A home or dwelling.
rook¹ (rōōk) *n.* 1 An Old World corvine bird with the feathers of the face lost in the adult state; especially, the common *Corvus frugilegus,* noted for its gregariousness. 2 A sharper; cheat; trickster. — *v.t.* & *v.i.* To cheat; defraud. [OE *hrōc*]
rook² (rōōk) *n.* One of a pair of castle-shaped chessmen which can move any number of unoccupied squares parallel to the sides of the board; a castle. [<OF *roc* <Persian *rukh*; orig. meaning unknown]
rook·er·y (rōōk′ər·ē) *n. pl.* **·er·ies** 1 A colony or breeding place of rooks. 2 A breeding place of sea birds, seals, etc. 3 A rambling building; an old tenement densely populated.
rook·ie (rōōk′ē) *n. Slang* 1 A raw recruit in the army, police, or any other service. 2 A novice in professional baseball. [Prob. alter. of RECRUIT]
rook·y (rōōk′ē) *adj.* 1 Pertaining to rooks and their habits. 2 Gregarious. 3 Abounding in rooks.
room (rōōm, rōōm) *n.* 1 Extent of space considered with regard to its sufficiency for some implied or specified purpose; free or open space. 2 A space for occupancy or use enclosed on all sides, as in a building; an apartment; chamber. 3 Suitable or warrantable occasion; opportunity: *room* for doubt. See synonyms under PLACE. — *v.i.* To occupy a room; lodge. [OE *rūm* space]
room·er (rōō′mər, rōōm′ər) *n.* A lodger; especially, one who rents a room and eats elsewhere.
room·ette (rōō·met′, rōōm·et′) *n.* A compartment with a single bed in some railroad sleeping-cars.
room·ful (rōōm′fōōl′, rōōm′-) *n.* 1 As many or as much as a room will hold. 2 A number of persons present in a room considered collectively.
rooming house A house for roomers; lodging house.
room·mate (rōōm′māt′, rōōm′-) *n.* One who occupies a room with another or others.
room·y (rōō′mē, rōōm′ē) *adj.* **room·i·er, room·i·est** Having abundant room; spacious. — **room′i·ly** *adv.* — **room′i·ness** *n.*
roon¹ (rōōn) *adj. Scot.* Round.
roon² (rōōn) *n. Scot.* A shred; border; strip of cloth.
roop (rōōp) *n. Brit. Dial.* 1 An outcry; call. 2 Hoarseness. [OE *hrōp* clamor]
roor·back (rōōr′bak) *n. U.S.* A fictitious report especially circulated for political purposes. [after *Roorback,* purported author of a (non-existent) book of travel, which was cited as authority for certain defamatory charges made against President Polk in the 1844 campaign]
roose (rōōz, rœz) *v.t. Scot.* To praise. — *n.* Praise. — **roos′er** *n.*
Roo·se·velt (rō′zə·velt, rōz′velt, -vəlt), **(Anna) Eleanor,** born 1884, *née* Roosevelt, U. S. lecturer, writer, and humanitarian; wife of F. D. Roosevelt. — **Franklin Delano,** 1882–1945, president of the United States 1933–45; reelected to fourth consecutive term 1944. — **Theodore,** 1858–1919, president of the United States 1901–09.
Roosevelt Dam A dam in the Salt River, central Arizona; 280 feet high; 1,125 feet long; completed 1911.
Roosevelt River A river in western Brazil, flowing 400 miles north to the Aripuaña; formerly *River of Doubt* (Portuguese *Rio da Dúvida*): Spanish *Río Roosevelt, Río Teodoro.*
roost (rōōst) *n.* 1 A perch upon which fowls rest at night; also, any place where birds resort to spend the night. 2 Any temporary resting place. — *v.i.* 1 To sit or perch upon a roost. 2 To come to rest; settle. [OE *hrōst*]
roost·er (rōōs′tər) *n.* The male of the chicken; cock. [<ROOST + -ER.]
roost·it (rōōs′tit) *adj. Scot.* 1 Rusty. 2 Dry; parched.
root¹ (rōōt, rōōt) *n.* 1 The underground portion or descending axis of a plant, which absorbs moisture, obtains or stores nourishment, and provides support. It differs from the stem in that it branches irregularly and lacks joints or leaves. 2 Loosely, any underground growth, as a tuber or bulb. 3 One of certain other growths serving for attachment, support, etc., as in the ivy or mistletoe. 4 That from which anything derives origin, growth, or life and vigor: Money is the *root* of evil; Industry is the *root* of prosperity. 5 An antecedent; ancestor. 6 Some rootlike part of an organ or structure: the *root* of a tooth or nerve. 7 *Ling.* A morpheme serving as the common center or basic constituent element of a related group of words, as *know* in *unknown, knowledge, knowable,* and *knowingly.* A root to which affixes or other morphemes may be added directly is equivalent to a stem. 8 *Math.* A quantity that, taken a specified number of times as a factor, will give another quantity called its *power:* 2 is the fourth *root* of 16. The number of times the root is thus taken as a factor is called its *index,* and roots are named from the indices, the words **square root** and **cube root** being often used for *second* and *third root.* 9 A tone on which a chord is built up. — *v.i.* 1 To put forth roots and begin to grow; take root. 2 To be or become firmly fixed or established. — *v.t.* 3 To fix or implant by or as by roots. 4 To pull, dig, or tear up by or as by the roots; extirpate; eradicate: with *up* or *out.* [OE *rōt* <ON *rōt*]
root² (rōōt, rōōt) *v.t.* 1 To turn up or dig with the snout or nose, as swine. — *v.i.* 2 To turn up the earth with the snout. 3 To search for something; rummage. 4 To work hard; toil. [OE *wrōtan* root up < *wrōt* snout]
root³ (rōōt, rōōt) *v.i. U.S. Colloq.* To cheer for or encourage a contestant: with *for:* He *rooted* for Harvard. [Prob. var. of ROUT³]
Root (rōōt, rōōt), **Elihu,** 1845–1937, U. S. lawyer and statesman.
root beer A beverage made with yeast and the extracts of several roots.
root climber Any plant that climbs by means of adventitious roots developed from stems.
root·er¹ (rōō′tər, rōōt′ər) *n.* One who or that which takes root.
root·er² (rōō′tər, rōōt′ər) *n.* One who or that which roots, as a swine, or tears up as by rooting; a destroyer; eradicator.
root·er³ (rōō′tər, rōōt′ər) *n. U.S. Colloq.* One who gives encouragement, as by applauding.
root hair *Bot.* Hairlike outgrowths of plant roots, having an absorbent and protective function.
root·less (rōōt′lis, rōōt′-) *adj.* Without roots.
root·let (rōōt′lit, rōōt′-) *n.* A small root.
root sheath The tough membrane covering the root portion of a hair. See illustration under HAIR.
root·stalk (rōōt′stôk′, rōōt′-) *n. Bot.* An underground rootlike stem; a rhizome.

root·stock (root'stok', root'-) n. 1 A rhizome. 2 Original source; origin.
root·y (roo'tē, roo'tē) adj. **root·i·er**, **root·i·est** 1 Full of or consisting of roots. 2 Resembling roots. — **root'i·ness** n.
rope (rōp) n. 1 A construction of twisted fibers, as of hemp, cotton, flax, etc., so intertwined in several strands as to form a thick cord. 2 A collection of things plaited or united in a line. 3 A slimy or glutinous filament or thread. 4 A cord or halter used in hanging; hence, execution or death by strangling or hanging. 5 A lasso. — **to give (one) plenty of rope** To allow (a person) to pursue unchecked a course that will end in disaster. — **to know the ropes** To be familiar with all the conditions in any sphere of activity; hence, to be sophisticated in the ways of the world. — v. **roped**, **rop·ing** v.t. 1 To tie or fasten with or as with rope. 2 To enclose, border, or divide with a rope: usually with *off*: He *roped off* the arena. 3 To catch with a lasso. 4 *Colloq.* To deceive; take in: with *in*. — v.i. 5 To become drawn out or extended into a filament or thread. [OE *rāp*]
rope band Roband.
rope-dancer (rōp'dan'sər, -dän'-) n. One who performs on the tightrope. — **rope'-danc'ing** n.
rope ferry A set of ropes overhanging a stream or defile, over which supplies and equipment may be pulled by a towline.
rope ladder A ladder made of ropes or with rope sides and wooden or other rounds.
rop·er·y (rō'pər·ē) n. 1 A ropewalk. 2 *Archaic* Roguery.
rope's end 1 A short piece of rope used for flogging: a nautical term. 2 A hangman's noose.
rope·walk (rōp'wôk) n. A long alley formerly used for the spinning of rope yarn: now in general superseded by some structure using improved machinery.
rope·walk·er (rōp'wô'kər) n. One who performs on the tightrope.
rop·y (rō'pē) adj. **rop·i·er**, **rop·i·est** 1 That may be drawn into threads, as a glutinous substance; stringy. 2 Resembling ropes or cordage. — **rop'i·ly** adv. — **rop'i·ness** n.
roque (rōk) n. A form of croquet requiring more skill than the ordinary game. [Aphetic alter. of CROQUET]
Roque·fort (rōk'fərt, *Fr.* rôk·fôr') A village in south central France. Also **Roquefort-sur-Soul·zon** (-sür-sool·zôn').
Roquefort cheese A strong cheese with a blue mold (*Penicillium roqueforti*) made from ewe's and goat's milk, originally at Roquefort, France.
roqu·e·laure (rōk'ə·lôr, -lōr, -lor) n. A form of short cloak worn by men in the 18th century: also spelled *rokelay*. [after Duc de Roquelaure, 1656-1738, French nobleman]
ro·quet (rō·kā') v.t. & v.i. **·quet·ed** (-kād'), **·quet·ing** (-kā'ing) In croquet, to strike (another player's ball). — n. The act of roqueting. [See ROQUE]
Ro·rai·ma (rō·rī'mä) Mount A peak at the junction of the Brazil-Venezuela-British Guiana boundaries; 9,219 feet.
Ro·rer (rôr'ər, rōr'-) **Sarah Tyson**, 1849-1937, née Heston, U. S. home economist and writer.
ror·qual (rôr'kwəl) n. Any of a genus (*Balaenoptera*) of whales of the Atlantic and Pacific oceans; especially, *B. physalis* of the North Atlantic: also called *finback*, *finback whale*. [< F < Norw. *röyrkval*]

RORQUAL
(About 60 feet in length)

Ror·schach test (rôr'shäk, -shäkh, rôr'-) *Psychol.* A test in which personality characteristics are made accessible to analysis by the subject's interpretation of the nature and meaning of a series of standard inkblot patterns. [after Hermann Rorschach, 1884-1922, Swiss psychiatrist]
Ro·sa (rō'zə) A feminine personal name. See ROSE.
Ro·sa (rō'zä), **Monte** The highest mountain group of the Pennine Alps on the Swiss-Italian border; its highest peak, 15,216 feet, is the second highest in the Alps.
Ro·sa (rō'zä), **Salvator**, 1615-73, Italian painter.
ro·sa·ceous (rō·zā'shəs) adj. 1 *Bot.* Of, pertaining to, or designating the rose family (*Rosaceae*) of trees, shrubs, and herbs: it is widely distributed in northern temperate regions, and includes many important genera of ornamental and fruit-yielding plants, as the apple, pear, quince, peach, plum, cherry, hawthorn, strawberry, blackberry, and raspberry. 2 Resembling a rose; rosy. [< L *rosaceus*]
Ros·a·lie (rō'zə·lē) A feminine personal name. Also **Ro·sa·li·a** (rō·zā'lē·ə). [< L, little rose]
Ros·a·lind (roz'ə·lind) A feminine personal name. [< L, pretty rose]
— **Rosalind** In Shakespeare's *As You Like It*, the heroine, daughter of the banished duke, who assumes male attire.
Ros·a·mond (roz'ə·mənd) A feminine personal name. Also *Fr.* **Rose-monde** (rōz·mônd'), **Ro·sa·mun·da** (*Lat.* rō'zə·mun'də, *Sp.* rō'sä·moon'dä). [< L, rose of the world]
ro·san·i·line (rō·zan'ə·lin, -lēn) n. *Chem.* 1 A colorless, crystalline organic compound, $C_{20}H_{21}ON_3$, having basic properties, obtained from aniline by treatment with reagents, as arsenic acid, nitric acid, and stannic chloride. It forms reddish salts, used as dyestuffs. 2 Some salt of this base used as a dyestuff, as fuchsin or rosein. [< ROSE[1] + ANILINE]
Ro·sa·rio (rō·sä'ryō) A port on the Paraná in Santa Fé province, Argentina.
ro·sa·ry (rō'zə·rē) n. pl. **·ries** 1 *Eccl.* a A series of prayers, consisting in its common form (**Dominican rosary**) of fifteen decades, each containing ten Aves preceded by a paternoster and followed by the Gloria Patri, and each related to a mystery or event in the life of Christ or the Virgin Mary which is contemplated during its recitation. b A string of beads for keeping count of the prayers thus recited. 2 A garden or bed of roses. 3 A chaplet or garland, as of roses. 4 A collection of literary selections. [< LL *rosarium* a rose garden < L *rosa* a rose]
Ros·ci·us (rosh'əs), **Gallus Quintus**, died 62? B.C., Roman comic actor. — **Ros'cian** adj.
Ros·coe (ros'kō), **Sir Henry Enfield**, 1833-1915, English chemist.
Ros·com·mon (ros·kom'ən) A county of eastern Connacht province, Ireland; 951 square miles; county town, Roscommon.
rose[1] (rōz) n. 1 A hardy, erect or climbing shrub (genus *Rosa*) grown in many varieties, with rodlike, prickly stems. In cultivation the stamens are transformed into petals and the flowers become double. It is the national flower of England and the State flower of New York, North Dakota, and Iowa. 2 The flower, having 5, or rarely 4, sepals. 3 Any one of various other plants or flowers having some real or fancied likeness to the true rose. 4 A light pinkish red, like the color of many roses. 5 An ornamental knot, as of ribbon or lace; a rosette. 6 A perforated cap, plate, or nozzle at the end of a pipe, for throwing water in a fine spray. 7 A compass rose. 8 A form in which gems, especially diamonds, are often cut, characterized by a flat base with a hemispherical upper surface covered with small facets; also, a diamond so cut. 9 Erysipelas. — **golden rose** A rose of wrought gold, blessed by the pope and presented, usually to a Roman Catholic sovereign, as a distinguished honor. — **under the rose** In secret. See SUB ROSA. — v.t. **rosed**, **ros·ing** To cause to blush; redden; flush. [OE < *rosa* < Gk. *rhodon*]
rose[2] (rōz) Past tense of RISE.
Rose (rōz) A feminine personal name. Also **Ro·sa** (rō'zə); *Fr.* rō·zä', *Ger.* rō'zä, *Ital.* rō'zä, *Sp.* rō'sä). [< L, rose]
rose acacia A locust tree (*Robinia hispida*) of the SE United States, bearing racemes of large rose or pale-purple flowers.
ro·se·ate (rō'zē·it, -āt) adj. 1 Of a rose color. 2 Rosy; rose-colored; hence, optimistic. [< L *roseus*] — **ro'se·ate·ly** adv.
roseate spoonbill A tropical American wading bird (*Ajaia ajaja*) having a bare head and throat and pink plumage.
rose-bay (rōz'bā') n. 1 Any rhododendron, especially *Rhododendron maximum*. 2 The oleander. 3 The willow herb.
rose beetle 1 The goldsmith beetle. 2 The rose chafer.
Rose·ber·y (rōz'bər·ē), **Earl of**, 1847-1929, Archibald Philip Primrose, English statesman and author.
rose-breast·ed grosbeak (rōz'bres'tid) See under GROSBEAK.
rose·bud (rōz'bud') n. 1 The bud of a rose. 2 A young girl; a debutante.
rose bush A rose-bearing shrub or vine.
rose campion 1 Any species of *Lychnis*, especially *L. coronaria*. 2 The corncockle.
rose chafer A hairy, fawn-colored beetle (*Macrodactylus subspinosus*) injurious to roses: also called *rose beetle*. Also **rose bug**. For illustration see INSECTS (injurious).
rose cold *Pathol.* A variety of hay fever, assumed to be caused by rose pollen. Also **rose fever**.
rose-col·ored (rōz'kul'ərd) adj. Pink or crimson, as a rose. — **to see through rose-colored glasses** To see things in an unduly favorable light; to look too much or only on the bright side. Compare COULEUR DE ROSE.
Rose·crans (rōz'kranz), **William Starke**, 1819-1898, U. S. general.
rose-cross (rōz'krôs', -kros') n. The symbol of the Rosicrucians, a rose and cross combined in some form.
rose·fish (rōz'fish') n. pl. **·fish** or **·fish·es** An orange-red scorpaenoid food fish (*Sebastes marinus*) of the North Atlantic.
rose geranium A cultivated geranium (*Pelargonium capitatum*) with rose-scented leaves and dense clusters of rose-purple flowers, grown extensively in South Africa.
Rose Island Easternmost island in the American Samoa group.
rose·mal·low (rōz'mal'ō) n. 1 The hibiscus: also called *mallow rose*. 2 The hollyhock.
rose·mar·y (rōz'mâr'ē) n. pl. **·mar·ies** An evergreen, fragrant shrub (*Rosmarinus officinalis*) of the mint family of southern Europe and western Asia, with usually blue flowers: cultivated for its stimulating and refreshing perfume, for an oil obtained from it, and for use in cookery. [Alter. of L *rosmarinus* < *ros* dew + *marinus* marine; infl. by *rose*, *Mary*]
rose moss A garden variety of portulaca (*Portulaca grandiflora*).
Ro·sen·wald (rō'zən·wôld), **Julius**, 1862-1932, U. S. businessman and philanthropist.
rose of Jericho A small annual (*Anastatica hierochuntica*) growing in desert places from Syria to Algeria, which rolls up when dry and expands again when moist: also called *resurrection plant*, *Jericho rose*.
rose of Sharon 1 In Scripture (*Canticles* ii 1), an unknown flower, perhaps the autumn crocus or the narcissus. 2 The shrub-althea. 3 A species (*Hypericum calycinum*) of St. Johnswort with large, yellow flowers.
ro·se·o·la (rō·zē'ə·lə) n. *Pathol.* A rose-colored rash appearing on the skin. Also **rose rash**. [< NL, dim. of L *roseus* rosy]
rose quartz A translucent to semitransparent variety of quartz, pink or rose in color and often asteriated: used for ornament, as a gemstone, etc.
ros·et (roz'it) n. *Scot.* Resin.
Ro·set·ta (rō·zet'ə) 1 The western branch of the Nile in its delta, Lower Egypt. Ancient **Bol·bi·ti·ne** (bol'bə·tī'nē). 2 A town on the Rosetta.
Ro·set·ta stone (rō·zet'ə) A tablet of basalt containing an inscription in two forms of Egyptian hieroglyphics (demotic and hieratic) and in Greek, found near Rosetta, Egypt, in 1799. It supplied Champollion with the key to the ancient inscriptions of Egypt.
ro·sette (rō·zet') n. 1 An ornament or badge having some resemblance to a rose; specifically, a painted or sculptured architectural ornament with parts circularly arranged. 2 A ribbon badge worn in the lapel buttonhole of civilian clothes to indicate possession of a certain military decoration. 3 A ribbon decoration shaped like a full-blown

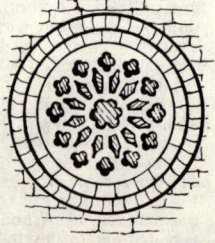
ROSETTE

or double rose and made of gathered or pleated silk, lace, etc. 4 A flowerlike cluster or combination of leaves, organs, parts, or markings, arranged in circles, as in certain plants. [<F, little rose]
rose·wa·ter (rōz'wô'tər, -wot'ər) n. A fragrant toilet and pharmaceutical water made variously by the distillation of rose petals or rose oil with water. — adj. 1 Made with or resembling rosewater. 2 Extremely or affectedly delicate or sentimental: *rosewater* philosophy.
rose window A circular window filled with tracery, called, when this takes the form of spokes, a *wheel window*.
rose·wood (rōz'wŏŏd') n. 1 A hard, close-grained, dark-colored, fragrant wood yielded by different Brazilian trees of the genus *Dalbergia*, etc., especially that produced by *D. nigra*, the most highly prized for cabinet work. Some species are said to be rose-scented when fresh. 2 Any of various other woods in some way resembling the true rose-woods. 3 Any tree yielding such a wood.
Rosh Ha·sha·na (rosh hə·shä'nə, rōsh) The Jewish New Year, celebrated on Tisri 1st and 2nd (September–early October). Also **Rosh Ha·sho'nah** (-shō'-). [<Hebrew *rōsh* head of + *hash-shānāh* the year]
Ro·si·cru·cian (rō'zə·krōō'shən, roz'ə-) n. One who is a member of an international fraternity, said to have originated in Egypt, and devoted to the practical application of an occult philosophy to human relationship. See ILLUMINATI. — adj. Of or pertaining to this society, its members, or its doctrines. [<L *rosae crucis* roses of the cross; said to be the trans. of the name of Christian *Rosenkranz*, 1387–1484, a German to whom the founding of this order has been attributed] — **Ro'si·cru'cian·ism** n.
ros·in (roz'in) n. 1 Resin. 2 The hard, amber-colored resin forming the residue after the distillation of oil of turpentine from crude turpentine; colophony. — v.t. 1 To apply rosin to. [Var. of RESIN] — **ros'in·y** adj.
Ros·i·nan·te (roz'ə·nan'tē) See ROCINANTE.
rosin oil Retinol.
ros·in·weed (roz'in·wēd') n. 1 A coarse perennial herb (genus *Silphium*) of the composite family, with copious resinous juice, growing in the central and western United States. 2 The compass plant.
ro·so·lio (rō·zō'lyō) n. A cordial made from raisins and brandy in the Mediterranean countries. [<Ital. <Med. L *ros solis* <L *ros* dew + *solis* of the sun) sundew, from which it was once extracted]
Ross (rôs), **Betsy**, 1752–1836, American patriot; made first American flag, 1777. — **Sir James Clark**, 1800–62, English Arctic explorer; discovered the north magnetic pole in 1831. — **Sir John**, 1777–1856, Scottish Arctic explorer; uncle of the preceding. — **Sir Ronald**, 1857–1932, English physician; first investigator of malaria-bearing mosquitoes.
Ross and Crom·ar·ty (krom'ər·tē, krum'-) A maritime county of NW Scotland; 3,089 square miles; county town, Dingwall.
Ross Dependency An uninhabited, ice-covered region of the Antarctic Zone under the jurisdiction of New Zealand; 175,000 square miles.
Ros·set·ti (rō·set'ē, -zet'ē), **Christina Georgina**, 1830–94, English poet; sister of Dante Gabriel. — **Dante Gabriel**, 1828–82, English painter and poet.
Ros·si (rôs'sē), **Bruno**, born 1905, Italian physicist.
Ros·si·ni (rôs·sē'nē), **Gioacchino Antonio**, 1792–1868, Italian composer.
Ross Island 1 An island off the NE tip of Palmer Peninsula, Antarctica; 39 nautical miles long, 31 nautical miles wide. 2 A volcanic island in the western Ross Sea off Victoria Land, Antarctica, on which Mount Erebus is located; 43 nautical miles long, 45 nautical miles wide.
Ros·si·ya (rôs·sē'yä) The Russian name for RUSSIA.
Ross Sea An inlet of the South Pacific in Antarctica south of New Zealand.
Ross Shelf Ice The extensive area of shelf ice in Antarctica, occupying the southern part of the Ross Sea; about 400 miles wide on its seaward side. Also called **Ross Barrier**.
Ros·tand (rôs·tän'), **Edmond**, 1869?–1918, French dramatist and poet.
ros·tel·late (ros'tə·lāt, -lit) adj. Having a small beak or rostellum. [<NL *rostellatus*]
ros·tel·lum (ros·tel'əm) n. pl. **-tel·la** (-tel'ə) 1 *Bot.* A small, beaklike structure developed from the stigma of an orchid. 2 *Zool.* The hooked scolex of a tapeworm. [<L, dim. of *rostrum* beak]
ros·ter (ros'tər) n. 1 A list of officers and men enrolled for duty; also, a list of active military organizations. 2 Any list of names. [<Du. *rooster* list]
Ros·tock (ros'tok, Ger. rôs'tôk) 1 A port on the Baltic in Rostock district, northern East Germany. 2 A district of East Germany, formerly part of the former state of Mecklenburg; 2,722 square miles.
Ros·tov (ros'tof) A city in southern Russian S.F.S.R., on the Don near its mouth on the Sea of Azov. Also **Ros·tov'-on-Don'** (-dôn').
Ros·tov·tzeff (ro·stôf'tsəf), **Michael Ivanovich**, 1870–1952, U.S. historian and archeologist born in Russia.
ros·tral (ros'trəl) adj. 1 Of or pertaining to a rostrum. 2 Having a rostrum, or beaklike process; beaked: often used in combination, as in *curvirostral*, having a crossed or curved-down beak. Also **ros'trate** (-trāt). [<LL *rostralis*]
ros·trum (ros'trəm) n. pl. **-trums** or **-tra** (-trə) 1 A pulpit or platform. 2 pl. **ros·tra** The orators' platform in the Roman forum: embellished with the beaks of the Latin ships captured 338 B.C. 3 A beak or snout; a beaklike process or part. 4 One of various beaklike parts, as the beak or prow of an ancient war galley. [<L *rostrum* beak]

ROSTRUM *(def. 4)*

ros·y (rō'zē) adj. **ros·i·er**, **ros·i·est** 1 Like a rose; rose-red; blooming; blushing. 2 Figuratively, bright, pleasing, or flattering. 3 Made of or ornamented with roses. 4 Auguring success; favorable: *rosy* predictions. 5 Optimistic. — **ros'i·ly** adv. — **ros'i·ness** n.
rot (rot) v. **rot·ted**, **rot·ting** v.i. 1 To undergo decomposition; decompose; decay. 2 To fall or pass by decaying: with *away*, *off*, etc. 3 To become morally rotten. — v.t. 1 To cause to decompose; decay. 5 To ret. See synonyms under DECAY, PUTREFY. — n. 1 That which is rotten, or the process of rotting. 2 A wasting disease, as of the lungs. 3 A parasitic disease affecting sheep and other domestic animals. 4 A form of decay in plants, caused by fungi and bacteria. 5 *Colloq.* Trashy and nonsensical opinions or expressions; twaddle; bosh. — *interj.* Nonsense; bosh. [OE *rotian*]
ro·ta (rō'tə) n. 1 A roll of names, giving order of duty; a roster. 2 A routine. 3 A wheel. [<L, wheel]
Ro·ta (rō'tə) In the Roman Catholic Church, an ecclesiastical court composed of ten prelates or auditors, subject only to papal authority, and serving as a court of final appeal: also known as *Sacra Romana Rota*.
Ro·ta (rō'tä) An island in the southern Marianas group; 33 square miles.
Ro·tar·i·an (rō·târ'ē·ən) n. A member of a Rotary Club. — adj. Of or pertaining to the organization of Rotary Clubs or to their members. — **Ro·tar'i·an·ism** n.
ro·ta·ry (rō'tər·ē) adj. 1 Turning around its axis, like a wheel, or so constructed as to turn thus. 2 Having some part that so turns: a *rotary* press. [<LL *rotarius* < *rota* wheel]
Rotary Club A club belonging to an international association of clubs, Rotary International, whose aim is to improve civic service, and whose motto is "Service".
rotary engine 1 An engine in which rotary motion is directly produced without reciprocating parts, as in a steam turbine: distinguished from *reciprocating engine*. 2 In internal-combustion engines, a radial engine revolving about a fixed crankshaft.
rotary harrow *Agric.* A harrow with many spikes set along the rim of a wheel which turns on a horizontal axis as it is pulled along the ground.
rotary motor See under MOTOR.
rotary plow *Agric.* A set of plowshares arranged on the rim of a rotating, power-driven shaft.
rotary press A printing press using curved type plates which revolve against the paper.
ro·tate (rō'tāt) v.t. & v.i. **·tat·ed**, **·tat·ing** 1 To turn or cause to turn on or as on its axis. 2 To alternate in a definite order or succession. See synonyms under REVOLVE. — adj. 1 Wheel-shaped; circular, as the corollas of certain flowers. 2 Forming a circle around a part, as spines or hairs. [<L *rotatus*, pp. of *rotare* turn <*rota*] — **ro'tat·a·ble** adj.
ro·tat·ed (rō'tā·tid) adj. 1 Turned around. 2 Rotate.
ro·ta·tion (rō·tā'shən) n. 1 The act or state of rotating; rotary motion. 2 Change by alternation; order of succession, variation, or sequence: *rotation* of crops or office. 3 The period represented by the age of a forest, or a part of a forest, as the time when it is cut, or intended to be cut. — **ro·ta'tion·al** adj.
ro·ta·tive (rō'tə·tiv) adj. Pertaining to or causing rotation; turning.
ro·ta·tor (rō'tā·tər) n. 1 One who or that which rotates or causes rotation. 2 pl. **ro·ta·to·res** (rō'tə·tôr'ēz, -tō'rēz) *Anat.* A muscle that rolls or rotates a part upon its axis. [<L]
Ro·ta·to·ri·a (rō'tə·tôr'ē·ə, -tō'rē·ə) See ROTIFER.
ro·ta·to·ry (rō'tə·tôr'ē, -tō'rē) adj. 1 Having, pertaining to, or producing rotation. 2 Following in succession. 3 Alternating or recurring.
rotche (roch) n. A bird, the dovekie. Also **rotch**. [Var. of *rotge* <Du. *rotje* petrel]
rote[1] (rōt) n. 1 Mechanical routine. 2 Repetition of words as a means of learning them, with slight attention to the sense. — **by rote** Mechanically; without intelligent attention: to learn *by rote*. [Var. of ROUTE]
rote[2] (rōt) n. Rare The roar of the surf. [Cf. ON *rōt* breaking of waves]
ro·te·none (rō'tə·nōn) n. *Chem.* A white crystalline substance, $C_{23}H_{22}O_6$, the effective principle in insecticides and fish poisons, obtained from the roots of various plants, especially an Amazonian tree (genus *Lonchocarpus*) and the Indian derris. [Origin unknown]
rot·gut whisky (rot'gut') *U.S.* An inferior raw whisky. Also **rot'gut'**.
Roth·er·ham (roth'ər·əm) A county borough in the West Riding, southern Yorkshire, England.
Roth·schild (rôth'chīld, *Ger.* rōt'shilt) A family of European bankers, of whom the first, **Meyer Amschel**, 1743–1812, established a bank in Frankfort on the Main. His sons opened branches: **James**, 1792–1868, at Paris; **Karl**, 1788–1855, at Naples; **Nathan Meyer**, 1777–1836, at London; **Salomon**, 1774–1855, at Vienna.
ro·ti·fer (rō'tə·fər) n. One of a division (*Rotifera*) of many-celled, microscopic, aquatic organisms usually found in stagnant fresh water, having rings of cilia which in motion resemble revolving wheels; a wheel animalcule. Some authorities place the *Rotifera* in a separate class or phylum, *Rotatoria*. [<NL <L *rota* wheel + *ferre* bear] — **ro·tif·er·al** (rō·tif'ər·əl), **ro·tif'er·ous** adj.
ro·ti·form (rō'tə·fôrm) adj. Shaped like a wheel; rotate. [<L *rota* wheel + -FORM]
ro·tis·se·rie (rō·tis'ə·rē') n. 1 A restaurant where patrons select uncooked food and have it roasted and served. 2 A shop where food is roasted and sold. 3 A rotating device for roasting meat, etc. [<F <*rôtir* roast]
rotl (rot'l) n. pl. **ar·tal** (är'täl) A weight used in Moslem countries, varying in different localities between one and five pounds. [<Arabic *raṭl*]
ro·to·chute (rō'tə·shōōt) n. *Aeron.* A long, darklike, high-altitude parachute equipped with a propeller having rapidly rotating

blades for breaking the speed of descent. [<L *rota* wheel + CHUTE]

ro·to·graph (rō′tə·graf, -gräf) *n.* One of a series of photographs printed from a developed roll of sensitized paper that bears the images. [<L *rota* wheel + -GRAPH]

ro·to·gra·vure (rō′tə·grə·vyŏŏr′, -grāv′yər) *n.* **1** A picture engraved on a cylindrical printing surface and run through a rotary press that prints both sides of the paper at the same time. **2** The process of making such pictures. [<L *rota* wheel + GRAVURE]

ro·tor (rō′tər) *n.* **1** *Electr.* The portion of an alternating-current motor which revolves. **2** A revolving part of a machine, as the wheel or wheels of a turbine. Compare STATOR. **3** *Aeron.* The horizontally rotating unit of a helicopter. [Contraction of ROTATOR]

rotor ship A vessel propelled by rotors operated by wind power but fitted with auxiliary power for propulsion when the wind fails.

rot·ten (rot′n) *adj.* **1** Decomposed by natural process; putrid. **2** Unsound; liable to break. **3** Untrustworthy; treacherous; also, venal; corrupt. **4** Afflicted with the rot, as sheep. **5** *Colloq.* Worthless. [<ON *rotinn*] —**rot′·ten·ly** *adv.* —**rot′ten·ness** *n.*

Synonyms: carious, corrupt, decayed, deceitful, decomposed, defective, fetid, offensive, putrefied, putrescent, putrid, tainted, treacherous, unsound. See BAD. *Antonyms*: complete, fresh, healthful, healthy, perfect, pure, sound, sweet, untainted, wholesome.

rotten borough **1** Any English borough prior to 1832 having few voters, yet entitled to send a member to Parliament. **2** Any election district or political unit no longer having sufficient population to justify the representation alloted to it.

rot·ter (rot′ər) *n. Slang* A scoundrel; a worthless scamp; any objectionable person.

rot·ten·stone (rot′n·stōn′) *n.* A soft, friable rock, consisting largely of siliceous particles, used for polishing: also called *tripoli.*

Rot·ter·dam (rot′ər·dam) The largest port of the Netherlands, in the western part.

Ro·tu·ma (rō·to͞o′mə) An island dependency of Fiji; 18 square miles.

ro·tund (rō·tund′) *adj.* **1** Rounded out; spherical; plump. **2** Full-toned, as a voice or utterance; in style, using sonorous words. **3** Complete; entire. **4** Circular, or nearly so; orbicular. See synonyms under ROUND. [<L *rotundus* <*rota* wheel. Doublet of ROUND.] —**ro·tund′ly** *adv.* —**ro·tund′ness** *n.*

ro·tun·da (rō·tun′də) *n.* A circular building or an interior hall, surmounted with a dome. [<Ital. *rotonda* <L *rotunda*, fem. of *rotundus.* See ROTUND.]

ro·tun·di·ty (rō·tun′də·tē) *n.* **1** The condition of being rotund; sphericity. **2** A protuberance.

ro·ture (rō·tür′) *n. French* **1** A plebeian condition or rank. **2** In French-Canadian law, a tenure of feudal lands by a constituted rent, without feudal duties and charges.

ro·tu·rier (rō·tü·ryā′) *n. French* **1** A person without rank; plebeian or peasant. **2** In French-Canadian law, one who holds lands by the tenure of roture.

Rou·ault (ro͞o·ō′), **Georges,** 1871–1958, French painter.

Rou·baix (ro͞o·be′) A city in northern France, NE of Lille; produces most of France's woolen textiles.

rou·ble (ro͞o′bəl) See RUBLE.

rouche (ro͞osh) See RUCHE.

rou·é (ro͞o·ā′) *n.* A sensualist; debauchee. [<F, jaded, orig. pp. of *rouer* break on the wheel, beat severely < *roue* wheel <L *rota*; from the appearance of a debauchee]

Rou·en (ro͞o·äṅ′) A city on the Seine in Normandy, northern France; noted for its cathedral; scene of the burning of Joan of Arc.

rouge (ro͞ozh) *n.* **1** Any cosmetic used for coloring the cheeks or lips pink or red. **2** A ferric oxide used in polishing metals and glass. —*v.* **rouged, roug·ing** *v.t.* To color, as the face, with rouge. —*v.i.* To apply rouge. [<F, red <L *rubeus* ruby]

rouge et noir (ro͞ozh′ e nwär′) A gambling game played with cards on a table having four diamond-shaped figures, two red and two black. See TRENTE-ET-QUARANTE. [<F, red and black]

Rou·get de l'Isle (ro͞o·zhe′ də lēl′), **Claude Joseph,** 1760–1836, French poet; author and composer of the *Marseillaise.*

rough (ruf) *adj.* **1** Having an uneven surface; having small inequalities on the surface; not smooth or polished: *rough* stone. **2** Coarse in texture; shaggy; also, disordered or ragged; shabby: said of dress or appearance: a *rough* suit, a *rough* shock of hair. **3** Having the surface broken; uneven: a *rough* country. **4** Characterized by rude or violent action: *rough* sports. **5** *Naut.* Boisterous or tempestuous; stormy: a *rough* passage. **6** Characterized by harshness of spirit; brutal. **7** Lacking the finish and polish bestowed by art or culture; unpolished; crude. **8** Done or made hastily and without attention to details; approximate. **9** *Phonet.* Uttered with an aspiration; aspirated: a *rough* breathing. **10** Harsh to the ear; grating; inharmonious: *rough* sounds. —*n.* **1** A low, rude, and violent fellow; a ruffian; a rowdy. **2** A crude, incomplete, or unpolished object, material, or condition. **3** Any part of a golf course on which tall grass, bushes, etc., grow. **4** A spike for insertion in a horseshoe, to prevent slipping. —*v.t.* **1** To make rough; roughen. **2** To treat roughly; specifically, in football, to treat (a player) with needless and intentional violence. **3** To make, cut, or sketch roughly: with *in* or *out*: to *rough* in the details of a plan. —*v.i.* **4** To become rough. **5** To behave roughly. —**to rough it** To live under rough, hard, or impoverished conditions; also, to camp out or travel in a rough manner; rusticate. —*adv.* In a rude manner; roughly. ◆ Homophone: *ruff.* [OE *rūh*] —**rough′ly** *adv.* —**rough′ness** *n.*

Synonyms (adj.) coarse, craggy, harsh, jagged, ragged, rude, rugged, shaggy, uneven, unfinished, unhewn, unpolished. See AWKWARD, BLUFF. *Antonyms*: bland, even, glossy, level, plain, polished, sleek, smooth.

rough·age (ruf′ij) *n.* **1** Any coarse or tough substance. **2** Food material containing a high percentage of indigestible constituents, as cellulose.

rough-and-read·y (ruf′ən·red′ē) *adj.* **1** Characterized by or acting with rude but effective promptness. **2** Unpolished but good enough.

rough-and-tum·ble (ruf′ən·tum′bəl) *adj.* **1** Disregarding all rules: said of a certain kind of fighting. **2** Scrambling; disorderly. —*n.* **1** A fight disregarding procedure according to rule, or in which anything goes; also, a scuffle. **2** Rough or adventurous existence.

rough breathing See under BREATHING.

rough-cast (ruf′kast′, -käst′) *v.t.* **-cast, -cast·ing** **1** To shape or prepare in a preliminary or incomplete form. **2** To roughen the surface of (pottery) before firing. **3** To coat, as a wall, with coarse plaster, and cover with thin mortar by dashing it on. —*n.* **1** Very coarse plaster for the outside of buildings. **2** A rude model; the form of a thing in its first rough stage. —**rough′-cast′er** *n.*

rough-draft (ruf′draft′, -dräft′) *v.t.* To make a rough or unfinished draft of; design or sketch hastily, as a plan or discourse.

rough-draw (ruf′drô′) *v.t.* **-drew, -drawn, -draw·ing** To sketch hastily or crudely.

rough-dry (ruf′drī′) *v.t.* **-dried, -dry·ing** To dry without ironing, as washed clothes.

rough·en (ruf′ən) *v.t.* & *v.i.* To make or become rough.

rough·er (ruf′ər) *n.* One who makes things in the rough.

rough-hew (ruf′hyo͞o′) *v.t.* **-hewed, -hewed** or **-hewn, -hew·ing** **1** To hew or shape roughly or irregularly or without smoothing. **2** To make crudely; rough-cast.

rough-house (ruf′hous′) *Slang n.* A noisy, boisterous or violent game or disturbance; rough play, especially within a room or house. —*v.* **-housed, -hous·ing** *v.i.* To make a disturbance; engage in horseplay or violence. —*v.t.* To handle or treat roughly but without hostile intent.

rough·ie (ruf′ē) *n. Scot.* **1** Brushwood or heath for fuel; a dead and fallen branch. **2** A kind of torch. **3** A clogged wick.

rough-leg·ged hawk (ruf′leg′id, -legd′) A large hawk *(Buteo lagopus sanctijohannis)* of Alaska and Canada, having the legs feathered all the way to the toes.

rough·neck (ruf′nek′) *n. U.S. Slang* A rowdy.

rough-rid·er (ruf′rī′dər) *n. U.S.* **1** One skilled in breaking broncos or performing dangerous feats in horsemanship. **2** A western cowboy.

Rough Riders The 1st U.S. Volunteer Cavalry in the Spanish-American War of 1898, mainly organized and subsequently commanded by Theodore Roosevelt.

rough-shod (ruf′shod′) *adj.* Shod with rough shoes to prevent slipping, as a horse. —**to ride rough-shod (over)** To act overbearingly; domineer without consideration.

rou·lade (ro͞o·läd′) *n.* **1** In singing, a run of short notes on one syllable; also, a roll or flourish, as on a drum. **2** A slice of meat rolled around a filling and cooked. [<F < *rouler* roll]

rou·leau (ro͞o·lō′) *n. pl.* **·leaux** (-lōz) or **·leaus** **1** A roll of coins in paper. **2** *Usually pl.* In millinery, a roll or fold of ribbon used for piping. [<F, dim. of *rôle* roll]

Rou·lers (ro͞o·lâr′) A town in western Belgium south of Bruges; scene of several battles of World War I, 1914, 1917. *Flemish* **Roe·se·la·re** (ro͞o′sə·lä′rə).

rou·lette (ro͞o·let′) *n.* **1** A game played at a table divided into spaces numbered and colored red and black, and having in the center a rotating disk on which a ball is rolled until it drops into one of 37 correspondingly numbered and colored spaces, a player winning if he has staked his money on that space or its color or on a combination including it. **2** An engraver's disk of tempered steel, as for tracing points on a copperplate; also, a draftsman's wheel for making dotted lines. **3** In philately, a series of incisions, made in any of several shapes, without removal of paper. Compare PERFORATION. —*v.t.* **·let·ted, ·let·ting** To use or produce a roulette upon. [<F, dim. of *rouelle*, dim. of *roue* wheel <L *rota*]

Roum (ro͝om) See RUM.

Rou·ma·ni·a (ro͞o·mā′nē·ə, -mān′yə), **Rou·ma·ni·an** (ro͞o·mā′nē·ən, -mān′yən) See RUMANIA, etc.

Rou·me·li·a (ro͞o·mē′lē·ə) See RUMELIA.

rounce (rouns) *n.* A game of cards with a full pack for two to nine persons, in which each player seeks to efface the score of 15 with which he starts, each trick taken subtracting one from it. [<Du. *rondse*]

round[1] (round) *adj.* **1** Having such a contour that a section in some direction will be circular or approximately so; circular, spherical, or cylindrical. **2** Having a curved contour or surface; not angular or flat; convex or concave. **3** Liberal; ample; large: a good *round* fee. **4** Easy and free, as in motion; brisk: a *round* pace. **5** Of full cadence; well-balanced; full-toned: a *round* sentence or tone. **6** Made without reserve; bold; outspoken: a *round* assertion. **7** Open; just; honorable. **8** Formed or moving in rotation or a circle: a *round* dance. **9** Returning to the point of departure, usually by the same means of transportation: a *round* trip. **10** Passing through the same or a like series of mutations: the *round* year. **11** Free from fractions; also, not exact in the small denominations; especially, evenly divisible by 10: *round* numbers. **12** Semicircular: a *round* arch; also, characterized by the round arch: the *round* style. **13** *Phonet.* Labialized; rounded. —*n.* **1** Something round, as a globe, ring, or cylinder, a rung of a ladder, a crossbar connecting the legs of a chair, a portion of the thigh of a beef, etc. **2** A circular course or range; circuit; beat: often in the plural; also, revolving motion or one revolution. **3** A series of recurrent movements; a routine; a completed succession or order: the daily *round* of life. **4** One of a series of concerted actions performed in succession by a number of persons: a *round* of toasts or applause. **5** One of the divisions of a boxing match; a bout. **6** In archery, the total number of arrows shot; the sum of all arrows in two or three ranges. **7** A short melody taken up at intervals by several voices; a rondo, roundel, or roundelay. **8** A firing by a company or squad in which each soldier fires once; volley. **9** A single charge of ammunition. **10** A round dance. **11** The state of being carved out on all sides: opposed to *relief.* **12** The state or condition of being circular; roundness. **13** A thick slice from a haunch: a *round* of beef. —**to go the rounds** **1** To take the usual walk of inspection. **2** To pass from mouth to mouth or person to person of a certain group. —*v.t.* **1** To make

round. 2 To bring to completion; perfect: usually with *off* or *out.* 3 To free of angularity; fill out to fullness of form. 4 *Phonet.* To utter (a vowel) with the lips in a rounded position; labialize. 5 To travel or go around; make a circuit of. 6 *Archaic* To encircle; surround. — *v.i.* 7 To become round. 8 To come to completeness or perfection. 9 To fill out; become plump. 10 To make a circuit; travel a circular course. 11 To turn around. — **to round off** *Math.* To reduce the number of decimal places to which a number is carried in a calculation: usually, a final figure less than 5 is eliminated and a final figure of 5 or greater increases the preceding figure to its next highest value, as, 2.1414, rounded off, becomes 2.141; 3.14159 becomes 3.1416. — **to round up** 1 To collect (cattle, etc.) in a herd, as for driving to market. 2 *Colloq.* To gather together; assemble. — *adv.* 1 On all sides; in such a manner as to encircle: A crowd gathered *round.* 2 With a circular or rotating motion: The wheel turns *round.* 3 Through a circle or circuit; more or less completely from person to person or to point to point: provisions enough to go *round.* 4 In circumference: a log 3 feet *round.* 5 From one view or position to another; hither and yon; to and fro. 6 In the vicinity: to hang *round.* See AROUND. — *prep.* 1 Enclosing; encircling: a belt *round* his waist. 2 On every side of, or from every side toward; surrounding. 3 Toward every side from; about. He peered *round* him. [< OF *roonde,* fem. of *roond* < L *rotundus.* Doublet of ROTUND.] — **round′ness** *n.*

Synonyms (adj.): circular, curved, curvilinear, cylindrical, globose, globular, orbed, orbicular, plump, rotund, spherical, spheroidal. See BLUNT. *Antonyms:* angular, conical, cubical, flat, polygonal, quadrangular, quadrilateral, rectangular, square, triangular.

Round may appear as a combining form in hyphemes or solidemes:

round-arched	round-hoofed
round-armed	round-horned
round-backed	round-leaved
round-barreled	round-limbed
round-bellied	round-lobed
round-billed	round-mouthed
round-bodied	round-nosed
round-boned	round-pointed
round-bottomed	round-ribbed
round-bowled	round-rooted
round-celled	round-sided
round-cornered	round-skirted
round-crested	round-spun
round-eared	round-stalked
round-edged	round-tailed
round-faced	round-toed
round-fenced	round-topped
round-footed	round-trussed
round-furrowed	round-visaged
round-handed	round-winged
round-headed	round-wombed

round² (round) *v.t.* & *v.i. Obs.* To whisper (to). [OE *rūnian*]
round-a·bout (round′ə·bout′) *adj.* 1 Circuitous; indirect. 2 Covering the whole field; ample. 3 Encircling. — *n.* 1 An outer garment reaching to the waist; a jacket. 2 *Brit.* A merry-go-round.
round-about chair A corner chair.
round clam A quahaug.
round dance 1 A country dance in which the dancers form a circle. 2 A dance with a revolving motion, as a waltz or polka, performed by two persons.
round·ed (roun′did) *adj.* 1 Round or spherical. 2 *Phonet.* Labialized.
roun·del (roun′dəl) *n.* 1 A roundelay. 2 In prosody, a modification of the rondeau, introduced by Swinburne, written in three stanzas of three lines each, with a refrain after the first and third. Compare RONDEL. 3 *Archit.* A semicircular recess, small round window, etc. [< OF *rondel* a roundelay]
roun·de·lay (roun′də·lā) *n.* 1 A simple melody. 2 A musical setting of a poem with a recurrent refrain. 3 A dance performed in a circle. [< OF *rondelet,* dim. of *rondel* < *rond* round]
round·er (roun′dər) *n.* 1 *U.S. Slang* A dissolute person who makes the rounds of public resorts at night, or who is often arrested for misdemeanors. 2 A tool for rounding. 3 *pl.* An old English game of ball somewhat resembling baseball: construed as singular.
round·hand (round′hand′) *n.* A style of handwriting in which the tendency is to make all letters round, full, and distinct.
Round·head (round′hed′) *n.* A member of the Parliamentary party in England in the civil war of 1642–49: so called in contempt by the Royalists, from their close-cropped hair.
round·house (round′hous′) *n.* 1 A cabin on the after part of the quarter-deck of a vessel. 2 A round building with a turntable in the center for housing and switching locomotives. 3 *Obs.* A lockup. 4 A round trip in pinochle.
round·ing (roun′ding) *adj.* 1 Pertaining to or denoting something, as a tool, used in or for rounding. 2 Becoming round; also, somewhat round.
round·ish (roun′dish) *adj.* Somewhat round. — **round′ish·ness** *n.*
round·let (round′lit) *n.* A little circle. [< F *rondelet*]
round·ly (round′lē) *adv.* 1 In a round manner or form; circularly; spherically. 2 Severely; vigorously: to be *roundly* denounced. 3 Frankly; bluntly. 4 Thoroughly; completely.
round-nose (round′nōz′) *adj.* Designating a kind of pliers whose gripping surfaces meet in a round, tapering point. See illustration under PLIERS.
round ringing A method of change-ringing a set of chimes in sequence from the bell of the highest note to that of the lowest, and then repeating this sequence while the earlier overtones are still vibrating, thus producing an effect like a round.
round robin 1 A number of signatures, as to a petition, written in a circle so as to avoid giving prominence to any single name; also, a paper so signed. 2 The cigar fish. 3 A tournament, as in tennis or chess, in which each player meets every other player.
rounds (roundz) *n. pl.* The position of a set of chiming bells when struck in a descending scale from highest to lowest.
round-shoul·dered (round′shōl′dərd) *adj.* Having the back rounded or the shoulders stooping.
rounds·man (roundz′mən) *n. pl.* ·**men** (-mən) A police officer having charge of a group of patrolmen.
round table Any meeting place for conference or discussion; also, any discussion group. — **round-ta·ble** (round′tā′bəl) *adj.*
Round Table The table of King Arthur, made exactly circular so as to avoid any question of precedence among his knights; also, collectively, King Arthur and the body of knights having places there.
round-the-clock (round′thə·klok′) *adj.* Through all twenty-four hours of the day.
round tower 1 Any cylindrical tower; especially, a slender, tapering tower of circular plan, with a conical cap. 2 In Ireland, a detached campanile built as a watchtower to guard church treasures, etc., against viking raids.
round trip 1 A trip to a place and back again; a return trip. 2 In pinochle, a meld of four kings and four queens: also called *roundhouse.* — **round′-trip′** *adj.*
round-up (round′up′) *n.* 1 The bringing together of cattle scattered over a range for inspection, branding, or selection for sale. 2 The cowboys, horses, etc., employed in this work. 3 *U.S. Colloq.* A bringing together of several persons: a *roundup* of pickpockets by the police.
round·worm (round′wûrm′) *n.* A nematode worm, especially one parasitic in the human intestines.
roup¹ (rōōp) *n.* An infectious respiratory and catarrhal disease affecting poultry. [Origin uncertain]
roup² (roup, rōōp) *Scot. n.* An auction. — *v.t.* To auction. [OE *hrōpan*]
roup·it (rōō′pit, rou′pit) *adj. Scot.* Roupy.
Rou·phi·a (rōō·fē′ə) The former name for the ALPHEUS RIVER.
roup·y (rōō′pē) *adj.* 1 Pertaining to, like, or affected with roup. 2 *Scot.* Hoarse.

Rous (rous), **Francis Peyton,** born 1879, U.S. pathologist.
rouse¹ (rouz) *v.* **roused, rous·ing** *v.t.* 1 To cause to awaken from slumber, repose, unconsciousness, etc. 2 To excite to vigorous thought or action; stir up. 3 To startle or drive (game) from cover. — *v.i.* 4 To awaken from sleep or unconsciousness. 5 To become active. 6 To start from cover: said of game. See synonyms under PIQUE, SPUR, STIR. — *n.* 1 The act of rousing; an awakening to or signal for action. 2 *Brit.* Reveille. [Origin unknown] — **rous′er** *n.*
rouse² (rouz) *n.* 1 *Archaic* A full draft of liquor; a bumper. 2 Noisy mirth; a drinking bout; carousal. [Aphetic form of CAROUSE]
rouse³ (rouz) *v.t.* & *v.i.* **roused, rous·ing** *Naut.* To pull together and with vigor. [? < ROUSE¹]
rouse·ment (rouz′mənt) *n.* A stirring up of interest or enthusiasm; especially, a widespread religious awakening or excitement.
rous·ing (rou′zing) *adj.* 1 Able to rouse or excite: a *rousing* speech. 2 Lively; active; vigorous: a *rousing* trade. 3 *Colloq.* Outrageous; astonishing: a *rousing* lie.
Rous·seau (rōō·sō′), **Henri,** 1844–1910, French painter: called "Le Douanier." — **Jean Jacques,** 1712–78, French philosopher and author. — **Pierre Étienne Théodore,** 1812–67, French painter.
Rous·sil·lon (rōō·sē·yôn′) A region and former province of southern France on the Spanish border; former capital, Perpignan.
roust (roust) *v.t.* & *v.i. Colloq.* To arouse and drive (a person or thing); stir up: usually with *out.* [Blend of ROUSE and ROUT]
roust·a·bout (rous′tə·bout′) *n.* 1 A laborer on river craft or on the waterfront; a deck hand. 2 One employed for casual work, especially, a transient laborer. 3 A man of all work on a cattle ranch or in a cow camp.
rout¹ (rout) *n.* 1 A disorderly and overwhelming defeat or flight. 2 A boisterous and disorderly assemblage; the rabble. 3 An entourage; a retinue. 4 *Law* A disturbance of the peace by three or more persons with riotous intent. 5 *Archaic* A large and festive evening social gathering. 6 *Archaic* Any assembly; a throng. See synonyms under REVEL. — *v.t.* To defeat disastrously; put to flight. See synonyms under CONQUER. [< OF *route* < L *rupta,* fem. of *ruptus,* pp. of *rumpere* break]
rout² (rout) *v.i.* 1 To root, as swine. 2 To search; rummage. — *v.t.* 3 To dig or turn up with the snout. 4 To disclose to view; turn up as if with the snout: with *out.* 5 To hollow, gouge, or scrape, as with a scoop. 6 To drive or force out. [Var. of ROOT²]
rout³ (rout, rōōt) *v.i. Scot.* or *Obs.* To make a loud noise; snore. — *n. Obs.* 1 Snoring. 2 A roaring noise; uproar. [OE *hrūtan*]
route (rōōt, rout) *n.* 1 A course, road, or way taken in passing from one point to another by any person or moving object. 2 The specific course over which mail is sent; also, the territory covered by a newsboy. See synonyms under ROAD, WAY. — *v.t.* **rout·ed, rout·ing** To dispatch or send by a certain way, as passengers, goods, etc. [< OF < L *rupta (via)* broken (road), fem. of *ruptus,* pp. of *rumpere* break] — **rout′er** *n.*
route column Close marching order for troops.
route formation An open formation of military aircraft prior or subsequent to action.
route march A troop march with discipline reduced to permit singing, talking, etc. Also **route step.**
route of march In a military march order, the designation of the way to be taken and the location of headquarters for each evening.
rout·er (rou′tər) *n.* 1 One who scoops or routs. 2 A tool for routing. 3 A plane devised for working a molding around a circular sash. [< ROUT²]
routh (rōōth, routh) *Scot. adj.* Abundant. — *n.* Plenty; abundance. Also spelled *rowth.*
rou·tine (rōō·tēn′) *n.* 1 A detailed method of procedure, regularly followed; prescribed course of action: an official *routine.* 2 Habitual methods of action induced by circumstances. See synonyms under HABIT. — *adj.* Customary; habitual; everyday. [< F < *route* way, road]

rou·tin·ism (ro͞o·tē′niz·əm) *n.* Adherence to routine or routine methods in general. — **rou·tin′ist** *n.*

roux (ro͞o) *n. French* Melted butter mixed with browned flour for thickening soups, gravies, etc.

rove[1] (rōv) *v.* **roved, rov·ing** *v.i.* To wander from place to place; go or move without any definite destination. — *v.t.* To roam over, through, or about. See synonyms under RAMBLE, WANDER. — *n.* The act of roving or roaming; a ramble. [<Du. *rooven* rob]

rove[2] (rōv) *v.t.* **roved, rov·ing 1** To join and elongate, as a number of slivers from a carding machine, by passing between one or more pairs of rollers. **2** To pass through an eye. **3** To draw into thread; ravel out. **4** To reduce the diameter of with a hooked, flat tool: to *rove* a grindstone. — *n.* **1** A slightly twisted wool, cotton, flax, jute, or silk sliver. **2** A metal ring or washer for use in clinching a nail in boatbuilding. [Origin uncertain]

rove[3] (rōv) Past participle of REEVE[1].

rove beetle Any of a family (*Staphylinidae*) of beetles having elongated bodies with very short elytra: most species are scavengers. For illustration see INSECTS (beneficial).

rov·er[1] (rō′vər) *n.* **1** One who roves; a wanderer. **2** A pirate, or pirate vessel. **3** A croquet ball that has been sent through all the arches and has only to strike the final stake to go out. [<MDu., a robber. Akin to ROBBER.]

rov·er[2] (rō′vər) *n.* In archery, any object, usually distant, chosen as a mark. Also **roving mark.** [Origin unknown]

Ro·vu·ma (rō·vo͞o′mə) The Portuguese name for the RUVUMA.

row[1] (rō) *n.* An arrangement or series of persons or things in a continued line; a rank; file; specifically, a line of houses on a street, or the street: Park *Row*; also, a line of plants, trees, etc., in a field or garden. — **a long row to hoe** A hard task or undertaking. — **at the end of one's row** Exhausted; also, having used up one's resources. — *v.t.* To arrange in a row: with *up.* ◆ Homophone: *roe.* [OE *rāw, rǣw* line]

row[2] (rō) *v.i.* **1** To use oars, sweeps, etc., in propelling a boat. — *v.t.* **2** To propel across the surface of the water with oars, as a boat. **3** To transport by rowing. **4** To be propelled by (a specific number of oars): said of boats. **5** To make use of (oars or rowers), especially in a race. **6** To row against in a race. — *n.* A trip in a rowboat; also, a turn at the oars, or the distance covered. ◆ Homophone: *roe.* [OE *rōwan*]

row[3] (rou) *n.* A noisy disturbance or quarrel; dispute; brawl; hence, any disturbance. — *v.t.* & *v.i.* To engage in a row or brawl. [Prob. back formation <ROUSE[2] (taken as a pl.)]

row[4] (rō) *Scot. n.* A roll, as of wool. — *v.t.* & *v.i.* To roll.

row·an (rō′ən, rou′-) *n.* **1** The European mountain ash (*Sorbus aucuparia*). **2** The related American mountain ash (*S. americana*). **3** The fruit of these trees. [<Scand. Cf. Norw. *raun, roun,* ON *reynir.*]

row·an·ber·ry (rō′ən·ber′ē) *n. pl.* **·ries** The fruit of the rowan.

row·boat (rō′bōt′) *n.* A boat propelled by oars.

row·dy (rou′dē) *n. pl.* **·dies** One inclined to create disturbances or engage in rows; a rough, quarrelsome person. — *adj.* **·di·er, ·di·est** Rough and loud; disorderly. [Origin uncertain] — **row′dy·ish** *adj.* — **row′dy·ism, row′di·ness** *n.*

Rowe (rō), **Nicholas,** 1674–1718, English poet and dramatist; poet laureate 1715.

row·el (rou′əl) *n.* **1** A spiked or toothed wheel, as on a spur. **2** The spur so furnished. **3** A hair or silk thread passed through a horse's skin, to facilitate the discharge of pus. — *v.t.* **·eled** or **·elled, ·el·ing** or **·el·ling 1** To prick with a rowel; spur. **2** To attach or apply a rowel to. [<OF *roele* <LL *rotella* little wheel <L *rota* wheel]

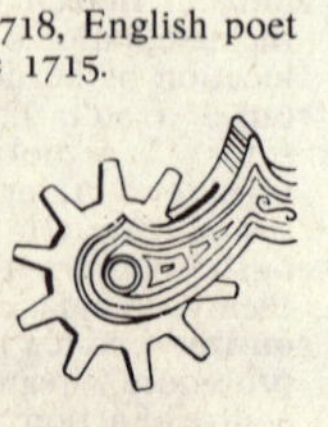

ROWEL ON SPUR

row·en (rou′ən) *n.* A second growth of grass or hay; aftermath. [<OF *regain*]

Row·lands (rō′ləndz), **John** See STANLEY, HENRY.

Row·land·son (rō′lənd·sən), **Thomas,** 1756–1827, English artist and caricaturist.

row·lock (rō′lok′) *n. Brit.* A device in which an oar plays and which serves as a point for applying its power to a boat: also called *oarlock.* [Alter. of OARLOCK; infl. by *row*[2]]

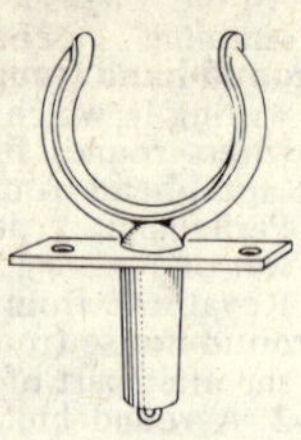

ROWLOCK

rowth (ro͞oth, routh) See ROUTH.

Rox·an·a (rok·san′ə) A feminine personal name. Also **Rox·y** (rok′sē), *Fr.* **Rox·ane** (rôk·sȧn′). [<Persian, dawn of day]
— **Roxane** In Rostand's *Cyrano de Bergerac,* the heroine.

Ro·xas y A·cu·ña (rō′häs ē ä·ko͞o′nyä), **Manuel,** 1892–1948, Philippine statesman; president of the Philippines 1946–48.

Rox·burgh·shire (roks′bûr·ə·shir′) An inland county in southern Scotland; 666 square miles; capital, Jedburgh. Also **Rox′burgh** (-bûr·ə).

roy·al (roi′əl) *adj.* **1** Pertaining to a monarch; kingly. **2** Under the patronage or authority of a king, or connected with a monarchical form of government: the *Royal* Society; a *royal* governor. **3** Like a king; princely; regal. **4** Of superior quality or size: *royal* octavo. **5** Surpassingly pleasant or fine: We had a *royal* time. See synonyms under IMPERIAL, KINGLY. — *n.* **1** A size of paper, 19 × 24 for writing, 20 × 25 for printing. **2** *Naut.* A sail next above the topgallant, used in a light breeze. [<F <L *regalis* kingly < *rex* king. Doublet of REGAL.] — **roy′al·ly** *adv.*

Royal Academy A society established in 1768 by George III of England for the advancement of painting, sculpture, and design: in full, *Royal Academy of Arts.*

Royal Air Force The air force of Great Britain.

Royal Australian Air Force The air force of Australia.

royal blue 1 Originally, the color of smalt, or cobalt blue; also, Prussian blue. **2** A more modern, brilliant blue; a reddish blue.

Royal Canadian Air Force The air force of Canada.

royal fern An attractive, deep-rooted fern (*Osmunda regalis*) of Asia, Africa, and America, having branched stems with oval or elliptical leaflets.

royal flush See under FLUSH.

Royal Gorge A canyon of the Arkansas River in south central Colorado, extending 10 miles; over 1,000 feet deep.

roy·al·ism (roi′əl·iz′əm) *n.* Adherence to the principles or cause of royalty.

roy·al·ist (roi′əl·ist) *n* A supporter of a royal dynasty. — *adj.* Supporting a royal house; pertaining to royalists: also **roy′al·is′tic.**

Roy·al·ist (roi′əl·ist) *n.* **1** In English history, a Cavalier or adherent of King Charles I, as against the Parliament, in the middle of the 17th century. **2** In French history, a supporter of the Bourbon or Orléans claims to the throne since 1793. **3** In the American Revolution, a supporter of the king; Loyalist; Tory.

roy·al·mast (roi′əl·mast′, -mäst′) *n. Naut.* The section of a mast next above the topgallant mast.

Royal Navy The naval forces of Great Britain.

royal purple 1 A very deep violet color verging toward blue. **2** Originally, a rich crimson: a very old color term.

Royal Society A society founded about 1660 in London and chartered in 1662, concerned with the advancement of science.

royal tine The tine of an antler projecting away from or above the bez tine: also called *trez tine.* For illustration see ANTLER.

royal touch The touch of a reigning monarch once believed to cure scrofula (king's evil).

roy·al·ty (roi′əl·tē) *n. pl.* **·ties 1** Royal rank, birth, or lineage; kingly nature or quality; kingliness; regal authority; sovereignty. **2** A royal personage; royal persons collectively. **3** A share of proceeds paid to a proprietor, author, or inventor, by those doing business under some right belonging to him. **4** A tax or seigniorage paid to the crown on the produce of royal mines, or on gold and silver coinage. **5** A royal possession or domain; hence, domain or province in general. [<OF *roialté*]

Royce (rois), **Josiah,** 1855–1916, U. S. philosopher and psychologist.

Ro·za·mond (rō′zə·mont) Dutch form of ROSAMOND.

-rrhagia *combining form Pathol.* A morbid or violent discharge or flow; an eruption: *metrorrhagia*: also spelled *-rhagia.* Also **-rrhage, -rrhagy.** Corresponding adjectives are formed in **-rrhagic.** [<Gk. < *rrhag-,* root of *rrhēgnynai* burst]

-rrhaphy *combining form* A sewing together; a suture: *neurorrhaphy,* the suturing of a nerve. Also spelled *-rhaphy.* [<Gk. *rrhaphē* a seam]

-rrhea *combining form Pathol.* An abnormal or excessive flow or discharge: *diarrhea*: also spelled *-rhea, -rhoea.* Also **-rrhoea.** [<Gk. *-rrhoia* < *rheein* flow]

Ru·an·da-U·run·di (ro͞o·än′dä·o͞o·ro͞on′dē) A United Nations Trust Territory in east central Africa, administered by Belgium; 20,900 square miles; capital, Usumbura; formerly in German East Africa.

rub (rub) *v.* **rubbed, rub·bing** *v.t.* **1** To move or pass over the surface of with pressure and friction. **2** To cause (something) to move or pass with friction; scrape; grate. **3** To cause to become frayed, worn, or sore from friction: This collar *rubs* my neck. **4** To clean, shine, burnish, etc., by means of pressure and friction, or by means of a substance applied thus. **5** To apply or spread with pressure and friction: to *rub* polish on a table. **6** To force by rubbing: with *in* or *into*: to *rub* oil into wood. **7** To remove or erase by friction: with *off* or *out.* — *v.i.* **8** To move along a surface with friction; scrape. **9** To exert pressure and friction. **10** To become frayed, worn, or sore from friction; chafe. **11** To undergo rubbing or removal by rubbing: with *off, out,* etc. — **to rub it in** *Slang* To harp on someone's errors, faults, etc. — **to rub out** *Slang* To kill. — **to rub the wrong way** *Slang* To irritate; annoy. See synonyms under WEAR. Compare FRICTION. — *n.* **1** A subjection to frictional pressure; rubbing: Give it a *rub.* **2** That which renders progress difficult; a hindrance or a doubt: There's the *rub.* **3** Something that rubs or is rough to the feelings; a sarcasm: a *rub* in debate. **4** A roughness or unevenness of surface, quality, or character. [ME *rubben,* prob. <LG. Cf. G *reiben.*]

rub-a-dub (rub′ə·dub′) *n.* The sound of a drum when beaten; hence, any clatter. [Imit.]

ru·bái·yát (ro͞o′bī·yät, -bē-) *n. pl.* **1** In Persian poetry, four-lined stanzas; quatrains. **2** Hence, **Rubáiyát,** a poem by Omar Khayyám and an English translation of it by Edward FitzGerald. [<Arabic *rubā′īyāt,* pl. of *rubā′īyah* quatrain, fem. of *rubā′ī* fourfold < *rubā* four]

Rub al Kha·li (ro͞ob′ äl khä′lē) The desert region of southern Arabia; 250,000 square miles: English *Empty Quarter*: also *Ar Rimal.*

ru·basse (ro͞o·bas′, -bäs′) *n.* A crystalline variety of quartz stained a ruby red by spangles of hematite. Also **ru·bace′.** [<F *rubace* < *rubi.* See RUBY.]

ru·ba·to (ro͞o·bä′tō) *adj. Music* Literally, robbed; noting the lengthening of one note at the expense of another. — *n. pl.* **·tos** A rubato modification. [<Ital.]

rub·ber[1] (rub′ər) *n.* **1** A tenacious, elastic material obtained by coagulating the milky latex of certain tropical plants, especially the tree *Hevea brasiliensis.* When purified, the crude rubber or caoutchouc is a white polymerized isoprene; it is insoluble in water or alcohol, and for commercial use is mixed with various vulcanizing agents, fillers, and pigments, then heated and molded into the desired form. **2** Anything used for rubbing, erasing, polishing, etc. **3** One who or that which rubs. **4** An article made of rubber, as an elastic band or an overshoe. — *adj.* Made of rubber. [<RUB] — **rub′ber·y** *adj.*

rub·ber[2] (rub′ər) *n.* In bridge, whist, and other card games, a series of two or three games played by the same partners against the same adversaries, terminated when one side has won two games; also, the odd game which breaks a tie between the players. [Origin unknown]

rub·ber·ize (rub′ər·īz) *v.t.* **·ized, ·iz·ing** To coat, impregnate, or cover, as silk, with a preparation of rubber.

rub·ber·neck (rub′ər·nek′) *n. U.S. Slang* One who cranes his neck in order to see something;

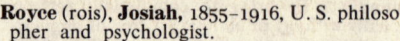

rubber plant 1 Any of several plants yielding rubber. 2 An East Indian tree of the mulberry family (*Ficus elastica*) having large, glossy, leathery leaves: much cultivated as an ornamental house plant.

rub·ber-stamp (rub′ər-stamp′) *v.t.* 1 To endorse, initial, or approve with the mark made by a rubber stamping device. 2 *Colloq.* To pass or approve as a matter of course or routine.

rub·bish (rub′ish) *n.* Waste refuse, or broken matter; trash. [Origin unknown]

rub·bish·y (rub′ish-ē) *adj.* Worthless; without value.

rub·ble (rub′əl) *n.* 1 Rough, irregular pieces of broken stone. 2 The debris to which buildings of brick, stone, etc., have been reduced by a violent action, such as earthquake or bombing. 3 (also roo′bəl) a In quarrying, the weathered or friable surface layer of rock. b Rough pieces of stone for use in construction, especially in residences. 4 Water-worn stones. 5 Rubblework. [Origin uncertain. Prob. related to RUBBISH.]

rub·ble·work (rub′əl-wûrk′) *n.* Masonry composed of irregular or broken stone, or fragments of stone mingled with cement or clay.

rub-down (rub′doun′) *n.* A massage.

rube (roob) *n. Slang* A countryman; farmer; rustic. [Abbreviation of REUBEN]

ru·be·fa·cient (roo′bə-fā′shənt) *adj.* Causing redness, as of the skin. — *n.* A medicament for producing irritation of the skin. [< L *rubefaciens, -entis* < *rubefacere* redden < *rubeus* red + *facere* make] — **ru′be·fa′cience** *n.* — **ru′be·fac′tion** (-fak′shən) *n.*

ru·bel·la (roo-bel′ə) *n. Pathol.* A contagious eruptive fever intermediate between scarlatina and measles: also called *German measles*. [< NL, neut. pl. of L *rubellus* reddish, dim. of *ruber* red]

ru·bel·lite (roo′bə-līt) *n.* A red, usually transparent, tourmaline: used as a gem. [< L *rubellus*. See RUBELLA.]

Ru·bens (roo′bənz, *Flemish* rü′bəns), **Peter Paul**, 1577-1646, Flemish painter.

ru·be·o·la (roo-bē′ə-lə) *n. Pathol.* 1 Measles. 2 Rubella. [< NL, neut. pl. dim. of L *rubeus* red] — **ru·be′o·lar** *adj.*

Ru·ber·to (roo-ber′tō) Italian and Spanish form of RUPERT.

ru·bes·cent (roo-bes′ənt) *adj.* Becoming red; reddening. [< L *rubescens, -entis*, ppr. of *rubescere* grow red, inceptive of *rubere* < *rubeus* red] — **ru·bes′cence** *n.*

ru·bi·a·ceous (roo′bē-ā′shəs) *adj. Bot.* Belonging or pertaining to a large, chiefly tropical family (*Rubiaceae*) of trees, shrubs, and herbs, the madder family of the order *Rubiales*, with simple opposite or whorled leaves and perfect, often dimorphous, flowers, including plants yielding coffee, quinine, and ipecac. [< L *rubia* madder]

Ru·bi·con (roo′bi·kon) A river in north central Italy, flowing 15 miles NE to the Adriatic; modern *Fiumicino*; it formed the boundary separating Caesar's province of Gaul from Italy, and by crossing it under arms he committed himself to a civil war with the Roman government then controlled by Pompey; hence, **to cross the Rubicon**, to be committed definitely to some course of action; make an irrevocable move.

ru·bi·cund (roo′bə-kənd) *adj.* Red, or inclined to redness; rosy. [< L *rubicundus* red] — **ru′bi·cun′di·ty** *n.*

ru·bid·i·um (roo-bid′ē-əm) *n.* A soft, rare, silvery-white metallic element (symbol Rb) resembling potassium, discovered by Bunsen and Kirchoff with the aid of the spectroscope. See ELEMENT. [< NL < L *rubidus* red]

ru·big·i·nous (roo-bij′ə-nəs) *adj.* Having a rusty or brownish-red color: *rubiginous* plants. Also **ru·big′i·nose** (-nōs). [< LL *rubiginosus* < L *rubigo, rubiginis* rust]

ru·bi·go (roo-bī′gō, -bē′-) *n.* Red iron oxide, used as a polishing powder and pigment. [< L, rust]

Ru·bin·stein (roo′bin·stīn), **Anton Gregor**, 1829-94, Russian pianist and composer. — **Artur**, born 1886, U.S. pianist born in Poland.

ru·bi·ous (roo′bē-əs) *adj.* Red; ruby-colored. [< RUBY]

ru·ble (roo′bəl) *n.* The Russian monetary unit containing 100 kopecks, and equivalent to one tenth of a chervonets: also spelled *rouble*. [< Russian *rubl*′]

ru·bric (roo′brik) *n.* 1 That exceptional part of an early manuscript or a book that appears in red, or in some distinctive type: once used to indicate initial letters, caption words, headings, etc. 2 The heading or title of a statute or of a section in a code of law, formerly written in red. 3 *Eccl.* A direction or rule printed in devotional or liturgical office, as in a prayer book, missal, or breviary; also, such rules collectively. 4 A division, group, or category. 5 The color red. 6 *Obs.* Red ochre or chalk; reddle. 7 Any direction or rule of conduct. 8 A distinguishing flourish or mark after a person's signature. — *adj.* 1 Red or reddish. 2 Written or printed in red. — *v.t.* **·bricked, ·brick·ing** *Rare* To rubricate. [< OF *rubrique* < L *rubrica* red earth < *ruber* red] — **ru′bri·cal** *adj.* — **ru′bri·cal·ly** *adv.*

ru·bri·cate (roo′brə-kāt) *v.t.* **·cat·ed, ·cat·ing** 1 To mark or tint with red; illuminate with red, as a book. 2 To furnish with a rubric or rubrics; arrange in permanent form. — *adj.* Marked, written, or printed in red. [< L *rubricatus*, pp. of *rubricare* redden < *rubrica*. See RUBRIC.] — **ru′bri·ca′tion** *n.* — **ru′bri·ca′tor** *n.*

ru·bri·cian (roo-brish′ən) *n.* One versed in the knowledge of, or punctiliously adhering to, rubric or rubrics.

ru·by (roo′bē) *n. pl.* **·bies** 1 A translucent gemstone of a deep-red color, usually a variety of corundum, as the Oriental ruby, but sometimes a variety of spinel, as the **almandine, balas**, and **spinel rubies**. 2 A rich red color like that of a ruby. 3 Something like a ruby in color, as red wine or a carbuncle. 4 Something made of a ruby; especially, in watchmaking, a bearing or roller made of a ruby or similar material. 5 In England, a size of type (5 1/2 points): equivalent to the American *agate*. — *adj.* Pertaining to or like a ruby; being of a rich crimson; ruby lips. — *v.t.* **·bied, ·by·ing** To redden; tint with the color of a ruby. [< OF *rubi* < L *rubeus* red]

ru·by-crowned kinglet (roo′bē-kround′) See under KINGLET.

ru·by-throat (roo′bē-thrōt′) *n.* The hummingbird (*Archilochus colubris*) of eastern North America, having in the male a gorget of brilliant metallic red. Also **ruby-throated hummingbird.**

ru·cer·vine (roo-sûr′vēn, -vin) *adj.* 1 Of or pertaining to a genus (*Rucervus*) of large deer native in southeastern Asia. 2 Denoting the antlers characteristic of this genus. For illustration see ANTLER. [< NL *Rucervus*, name of the genus < *Malay rusa* deer + CERVINE]

ruche (roosh) *n.* A quilted or ruffled strip of fine fabric, worn about the neck or wrists of a woman's costume: also spelled *rouche*. [< F, beehive <Med. L *rusca* tree bark, ? <Celtic]

ruch·ing (roo′shing) *n.* Material for ruches; ruches collectively.

ruck[1] (ruk) *n.* The common herd or run; a crowd; also, trash; rubbish. [< Scand. Cf. Norw. *ruka* heap, crowd.]

ruck[2] (ruk) *v.t. & v.i.* 1 To wrinkle, rumple, crease, etc. 2 To annoy; ruffle: usually with *up.* — *n.* A wrinkle, crease, or ridge, as in cloth or paper; a wrinkled place. [< ON *hrukka* wrinkle]

ruck·sack (ruk′sak′, rook′-) *n.* A canvas knapsack. [< G, lit., back sack]

ruck·us (ruk′əs) *n. U.S. Slang* An uproar; commotion; rumpus. [Prob. blend of RUMPUS and RUCTION]

ruc·ta·tion (ruk-tā′shən) *n.* Eructation.

ruc·tion (ruk′shən) *n. Colloq.* A riotous outbreak; quarrel; uproar. [Prob. alter. of INSURRECTION]

ruc·tious (ruk′shəs) *adj. Slang* Difficult; quarrelsome. [< RUCTION]

rud·beck·i·a (rud-bek′ē-ə) *n.* Any of a genus (*Rudbeckia*) of North American herbs of the composite family, the coneflowers, with alternate simple or compound leaves and showy yellow heads; especially, the black-eyed Susan.

[after Olaus *Rudbeck*, 1630-1702, Swedish botanist]

rudd (rud) *n.* A European fresh-water fish (*Scardinius erythrophthalmus*), olive-brown with red fins: also called *red eye*. [OE *rudu* red color]

rud·der (rud′ər) *n.* 1 *Naut.* A broad, flat device hinged vertically at the stern of a vessel to direct its course. 2 Anything that guides or directs a course. 3 *Aeron.* A hinged or pivoted surface, used to control the position of an aircraft about its vertical axis. [OE *rōthor*, scull] — **rud′der·less** *adj.*

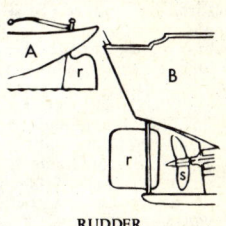

RUDDER
A. Sailboat B. Motorboat.
r. Rudder.
s. Screw.

rudder bar *Aeron.* A foot-operated rod by which the pilot controls the rudder of an airplane.

rudder fish Any of various fishes that follow vessels, as the pilot fish, etc.

rudder stock The vertical shaft to which the rudder of a ship or boat is attached, having at its upper portion a yoke (the **rudder crosshead**) or tiller by which it may be turned. Also **rudder post.**

rud·dle (rud′l) *n.* A variety of red ocherous iron ore; reddle. — *v.t.* **·dled, ·dling** To color or stain with red ocher. Also spelled *raddle*. [OE *rudu* red color]

rud·dle·man (rud′l-mən) *n. pl.* **·men** (-mən) A reddleman.

rud·dock (rud′ək) *n.* The European robin. [OE *rudduc* robin < *rudu* red color]

rud·dy (rud′ē) *adj.* **·di·er, ·di·est** 1 Tinged with red. 2 Having a healthy glow; rosy: a *ruddy* complexion. See synonyms under FRESH. [OE *rudig*] — **rud′di·ly** *adv.* — **rud′di·ness** *n.*

ruddy duck A small North American duck (*Erismatura jamaicensis rubida*) having the tail feathers stiffened with narrow webs. The adult male is bright chestnut-reddish above. Also called *paddywhack*.

rude (rood) *adj.* **rud·er, rud·est** 1 Rough or abrupt; severe or tempestuous; offensively blunt or uncivil; impudent. 2 Characterized by lack of polish or refinement; uncultivated; uncouth. 3 Unskilfully made or done; lacking in skill or training; crude; rough: *rude* workmanship. 4 Characterized by robust vigor; strong: *rude* health. 5 Barbarous; savage. 6 Humble; lowly; rustic. See synonyms under BARBAROUS, BLUFF, IMPUDENT, ROUGH, RUSTIC, VULGAR. ♦ Homophone: rood. [< OF < L *rudis* rough] — **rude′ly** *adv.*

rude·ness (rood′nis) *n.* 1 The state or quality of being rude. 2 A rude action. See synonyms under IMPUDENCE.

rudes·by (roodz′bē) *n. Archaic* An ill-bred boor.

Rü·di·ger (rü′di·ger) German form of ROGER.

ru·di·ment (roo′də-mənt) *n.* 1 A first principle, step, stage, or condition. 2 That which is as yet undeveloped or only partially developed. 3 *Biol.* a Something in a first, embryonic, incomplete, or early stage that may develop by growth; a germ. b A part, organ, or other structure that has become aborted or stunted and will always be undeveloped; a vestige; vestigial part. [< F < L *rudimentum* first attempt < *rudis* rough]

ru·di·men·ta·ry (roo′də-men′tər-ē) *adj.* 1 Pertaining to or of the nature of a rudiment: *rudimentary* knowledge. 2 Being or remaining in an imperfectly developed state; germinal; undeveloped; abortive. Also **ru′di·men′tal.** — **ru′di·men·ta·ri·ly** *adv.* — **ru′di·men·ta·ri·ness** *n.*

Ru·dolf (roo′dolf; *Du.* rü′dolf, *Ger.* roo′dôlf, *Sw.* roo′dôlf) A masculine personal name. Also **Ru′dolph.** [<Gmc., wolf of fame] — **Rudolf I**, 1218-91, Holy Roman Emperor 1273-91; founded the Hapsburg dynasty. — **Rudolf II**, 1552-1612, Holy Roman Emperor 1576-1612; persecuted Protestants. — **Rudolf of Hapsburg**, 1858-89, crown prince of Austria; son of Francis Joseph; committed suicide.

Rudolf, Lake A lake in NW Kenya, extending into Ethiopia on the northwest; 3,500 square miles.

rue[1] (rōō) *v.* **rued, ru·ing** *v.t.* To feel sorrow or remorse for; regret extremely. — *v.i.* To feel sorrow or remorse; be regretful. See synonyms under MOURN. — *n.* **1** Sorrowful remembrance; regret. **2** *Scot.* Repentance. [OE *hrēowan* be sorry] — **ru′er** *n.*

rue[2] (rōō) *n.* **1** A small, bushy herb (*Ruta graveolens*) with bitter, acrid leaves, formerly much used in medicine for stimulating effects: formerly also an emblem of bitterness or grief. **2** An infusion made from this plant; hence, any bitter draft. [<F <L *ruta* <Gk. *rhytē*]

rue anemone A delicate little American woodland perennial (*Anemonella thalictroides*), having white flowers in the spring.

Rue de la Paix (rü də lä pe′) A street in Paris, France, famous for its fashionable shops.

rue·ful (rōō′fəl) *adj.* **1** Feeling or causing sorrow, regret, or pity; deplorable; sorrowful. **2** Expressing sorrow or pity. — **rue′ful·ly** *adv.* — **rue′ful·ness** *n.*

ru·fes·cent (rōō·fes′ənt) *adj.* Inclining to reddishness; somewhat reddish or rufous. [<L *rufescens, -entis,* ppr. of *rufescere* redden < *rufus* red] — **ru·fes′cence** *n.*

ruff[1] (ruf) *n.* **1** A pleated, round, heavily starched collar popular in the 16th century. **2** Ruffle[1] (def. 1). **3** A natural collar of projecting feathers or hair around the neck of a bird or mammal. **4** An Old World sandpiper (*Philomachus pugnax*) of which the male in the breeding season has an erectile frill of elongated feathers about the neck. The female is called a *reeve.* — *v.i.* To become ruffled; stand out like a ruff. ◆ Homophone: *rough.* [Short for RUFFLE[1]]

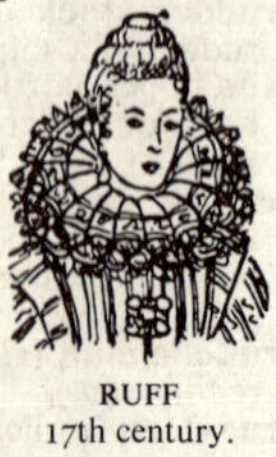

RUFF
17th century.

ruff[2] (ruf) *n.* **1** The playing of a trump upon another suit when one has no cards of that suit. **2** An old game, the predecessor of whist. — *v.t.* & *v.i.* To trump when unable to follow suit. ◆ Homophone: *rough.* [<OF *roffle, rouffle, ronfle,* aphetic alter. of *triomphe* triumph. Cf. Ital. *ronfa* a game at cards < *trionfo* triumph. Related to TRUMP[1].]

ruff[3] (ruf) *n.* A small perchlike fish (*Acerina cernua*) of European fresh waters. Also **ruffe.** ◆ Homophone: *rough.* [<ROUGH]

ruff[4] (ruf) See RUFFLE[3].

ruffed (ruft) *adj.* Having a ruff, ruffle, or frill; ruffled.

ruffed grouse A North American grouse (*Bonasa umbellus*): called *partridge* in the northern and *pheasant* in the southern United States.

ruf·fi·an (ruf′ē·ən, ruf′yən) *n.* A lawless, brutal, cruel fellow; a rough; one ready for or given to riotous, cruel, or murderous deeds. — *adj.* Lawlessly or recklessly brutal or cruel. [<F *rufian* <Ital. *rufiano* pimp, ? <OHG *ruf* dirty] — **ruf′fi·an·ism** *n.* — **ruf′fi·an·ly** *adj.*

ruf·fle[1] (ruf′əl) *n.* **1** A pleated strip; frill, as for trim or ornament; also, anything resembling this: also *ruff.* **2** A temporary discomposure. **3** A ripple. — *v.* **·fled, ·fling** *v.t.* **1** To disturb or destroy the smoothness or regularity of: The wind *ruffles* the lake. **2** To draw into folds or ruffles; gather. **3** To furnish with ruffles. **4** To erect (the feathers) in a ruff, as a bird when frightened. **5** To disturb or irritate; upset. **6 a** To riffle (the pages of a book). **b** To shuffle (cards). — *v.i.* **7** To be or become rumpled or disordered. **8** To become disturbed or irritated. [ME *ruffelen.* Cf. LG *ruffelen* rumple, ON *hrufla* scratch.]

ruf·fle[2] (ruf′əl) *n.* A low, continuous beat of a drum, not as loud as a roll: also *ruff.* — *v.t.* **·fled, ·fling** To beat a ruffle upon, as a drum. [<earlier *ruff*; imit.]

ruf·fle[3] (ruf′əl) *v.i.* **·fled, ·fling** To act in a rough or turbulent manner; swagger; bluster. [? Special use of RUFFLE[2]] — **ruf′fler** *n.*

ru·fous (rōō′fəs) *adj.* Dull-red; rust-colored. [<L *rufus* red]

Ru·fus (rōō′fəs) A masculine personal name. [<L, red]

rug[1] (rug) *n.* **1** A heavy textile fabric, made in one piece, to cover a portion of a floor. **2** A covering made from the skins of animals dressed with the hair or wool on. **3** A heavy coverlet or lap robe. [<Scand. Cf. Norw. *rugga* coarse coverlet, *skinnrugga* skin rug, ON *rögg* long, rough fleece.]

rug[2] (rug) *v.t.* **rugged, rug·ging** *Scot.* & *Brit. Dial.* To tug or tear roughly.

ru·ga (rōō′gə) *n. pl.* **·gae** (-jē) A fold, wrinkle, or crease. [<L]

ru·gate (rōō′gāt, -git) *adj.* Covered with or having rugae; corrugated; wrinkled. [<L *rugatus,* pp. of *rugare* wrinkle < *ruga* a wrinkle]

Rug·bei·an (rug·bē′ən) *adj.* Of or pertaining to Rugby, England, or to the school there located. — *n.* A native or inhabitant of Rugby; a pupil at Rugby school.

Rug·by (rug′bē) A municipal borough of eastern Warwickshire, England; seat of a boys' school founded in 1567.

Rugby football A form of football played between two teams of fifteen men each, in which the ball is propelled toward the opponents' goal by kicking or carrying, but in which no player of the side in possession of the ball may be ahead of the ball while it is in play.

Rü·gen (rü′gən) The largest German island in the Baltic, NE of Stralsund; 358 square miles.

rug·ged (rug′id) *adj.* **1** Having a surface full of abrupt inequalities; broken into irregular points or crags; steep and rocky; rough; uneven. **2** Shaggy; unkempt; disordered; ragged. **3** Rough in temper, character, or action; harsh; stern. **4** Having strongly marked features; wrinkled; frowning; furrowed. **5** Lacking culture or refinement; rude. **6** Rough to the ear; grating . **7** Robust; sturdy. **8** Tempestuous; stormy. See synonyms under FIRM, ROUGH. [<Scand. Cf. Sw. *rugga* roughen. Prob. related to RUG[1].] — **rug′ged·ly** *adv.* — **rug′ged·ness** *n.*

Rug·gie·ro (rōōd·jä′rō) Italian form of ROGER.

ru·gose (rōō′gōs) *adj.* **1** Covered with or full of rugae or wrinkles; corrugate; rugate. **2** *Bot.* Having a rough or wrinkled surface, as some strongly veined leaves. Also **ru′gous.** [<L *rugosus* < *ruga* wrinkle] — **ru·gos·i·ty** (rōō·gos′ə·tē) *n.*

Ruhr (rōōr) **1** A river of western Germany, flowing 142 miles west to the Rhine. **2** The region south of which the Ruhr river flows, noted as an industrial and coal-mining district; about 2000 square miles; included in North Rhine-Westphalia, West Germany.

ru·in (rōō′in) *n.* **1** Total destruction of value or usefulness; in morals, the loss of character, chastity, or honor; seduction; corruption. **2** That which remains of something demolished, destroyed, or decayed: often in the plural. **3** A condition of desolation or degradation. **4** That which causes destruction, downfall, decay, or injury: Gambling was his *ruin.* **5** The act of falling down; collapse. — *v.t.* **1** To bring to ruin; destroy; demolish. **2** To bring to bankruptcy or poverty. **3** To deprive of chastity; seduce. — *v.i.* **4** To fall into ruin. See synonyms under ABUSE, DEMOLISH. [<OF *ruine* <L *ruina* < *ruere* fall] — **ru′in·a·ble** *adj.* — **ru′in·er** *n.*

Synonyms (noun): collapse, decay, defeat, desolation, destruction, discomfiture, downfall, fall, overthrow, perdition, subversion, undoing, wreck. See ADVERSITY, MISFORTUNE. *Antonyms*: conservation, preservation, prosperity, recovery, regeneration, reparation, success.

ru·in·ate (rōō′in·āt) *v.t.* **·at·ed, ·at·ing** *Rare* To ruin. — *adj.* Ruined. [<Med. L *ruinatus,* pp. of *ruinare* ruin < *ruina*]

ru·in·a·tion (rōō′in·ā′shən) *n.* **1** The act of ruining. **2** The state of being ruined. **3** Something that ruins.

ru·ined (rōō′ind) *adj.* **1** Destroyed; ravaged; in ruins. **2** Bankrupt.

ru·in·ous (rōō′in·əs) *adj.* **1** Causing or tending to ruin. **2** Falling to ruin; decayed; dilapidated; ruined. See synonyms under PERNICIOUS. [<OF *ruineux* <L *ruinosus*] — **ru′in·ous·ly** *adv.* — **ru′in·ous·ness** *n.*

Ruis·dael (rois′däl, *Du.* rœis′däl) See RUYSDAEL.

Ru·iz (rōō·ēth′, -ēs′) A Spanish masculine personal name.

rule (rōōl) *n.* **1** Controlling power, or its possession and exercise; government; dominion; authority. **2** A method or principle of action; common or regular course of procedure, or customary standard or form: I make early rising my *rule.* **3** An authoritative direction or enactment; a concise direction respecting the doing or method of doing something, as one of the regulations of a legislative or deliberative body for the government of its own proceedings, or a regulation to be observed in playing a given game. **4** A regulation for the conduct of religious services or for the government of life; specifically, the body of directions laid down by or for a religious order: the *rule* of St. Francis. **5** A prescribed form, method, or set of instructions for solving a given class of mathematical problems. **6** An established usage or law, fixing the form or use of words or the construction of sentences: a *rule* for forming the plural. **7** What belongs to the ordinary course of events or condition of things: In some communities illiteracy is the *rule.* **8** Regular or proper method; propriety, as of conduct; regularity. **9** *Law* A formal regulation prescribed by authority touching a certain matter: a *rule* of court; also, a judicial decision on some motion or special application: a *rule* to show cause. A **rule of court** is an order made by a court, and is either *general,* as for regulating the practice of the court, or *special,* as an order sending a case before a referee. **10** A straight-edged instrument for use in measuring, or as a guide in drawing lines; a ruler, usually marked in inches, feet, etc. **11** *Printing* A strip of type-high metal for handling type or for printing a rule or line. **12** A ruled line. — **as a rule** Ordinarily; usually. — *v.* **ruled, rul·ing** *v.t.* **1** To have authority or control over; govern. **2** To influence greatly; dominate: Greed has *ruled* his life. **3** To decide or determine judicially or authoritatively. **4** To restrain; keep in check: *Rule* your temper. **5** To mark with straight, parallel lines. **6** To make (such a line) with or as with a ruler. — *v.i.* **7** To have authority or control; be in command. **8** To maintain a standard of rates: Prices *ruled* high. **9** To form and express a decision: The judge *ruled* on that point. See synonyms under GOVERN, REGULATE. [<OF *reule* <L *regula* ruler, rule < *regere* lead straight, direct. Doublet of RAIL[1].] — **rul′a·ble** *adj.*

Synonyms (noun): canon, formula, guide, maxim, method, order, regulation, standard. See HABIT, LAW, STICK, SYSTEM.

ruled surface *Math.* A surface capable of being generated by a straight line, as a hyperboloid of one sheet: the various positions of the straight line are called **rulings of a ruled surface.**

rule of three *Math.* A rule for finding any term of a proportion, the three others being given.

rule of thumb **1** Measurement by the thumb. **2** Roughly practical rather than scientifically accurate measure.

rul·er (rōō′lər) *n.* **1** One who rules or governs, as a sovereign. **2** A straight-edged strip for guiding a marking implement; a rule; a ruling machine. **3** One who rules lines, as with a ruling machine. See synonyms under CHIEF.

rul·ing (rōō′ling) *adj.* Exercising dominion; controlling; predominant. — *n.* **1** The act of one who rules or governs. **2** A decision, as of a judge or presiding officer. **3** The act of making ruled lines, or the lines so made.

rum[1] (rum) *n.* **1** An alcoholic liquor distilled from fermented molasses or cane juice. **2** Any alcoholic liquor. [Short for obs. *rumbullion* rum, alter. of *Rambouillet,* town in France]

rum[2] (rum) *adj. Brit. Slang* Queer; strange; peculiar. [? <Romany *rom* man]

Rum (rōōm) The Arabic name for the BYZANTINE EMPIRE: also *Roum.*

Ru·ma·ni·a (rōō·mā′nē·ə, -mān′yə), **People's Republic of** A state in SE Europe; 91,671 square miles; capital, Bucharest: also *Romania, Roumania*: Rumanian *România.*

Ru·ma·ni·an (rōō·mā′nē·ən, -mān′yən) *adj.* Of Rumania, its people, or their language. — *n.* **1** A native or inhabitant of Rumania. **2** The Romance language of the Rumanians. Also *Romanian, Roumanian.*

rum·ba (rum′bə, *Sp.* rōōm′bä) *n.* **1** A frenzied dance formerly performed by Cuban Negroes. **2** A modern dance based on this. Also spelled *rhumba.* [<Sp.]

rum·ble (rum′bəl) *v.* **·bled, ·bling** *v.i.* **1** To make a low, heavy, rolling sound, as thunder.

2 To move or proceed with such a sound. —*v.t.* 3 To cause to make a low, heavy, rolling sound. 4 To utter with such a sound. 5 To subject to the action of a tumbling box. —*n.* 1 A continuous low, heavy, rolling sound; a muffled roar. 2 A tumbling box: also **rum′bler.** 3 A seat or baggage compartment in the rear of a carriage. 4 A folding seat in the back of a coupé or roadster: in full, **rumble seat.** 5 *U.S. Slang* A fight involving a group, usually deliberately provoked. [ME *romblen* <MDu. *rommelen*] —**rum′bler** *n.* —**rum′bling·ly** *adv.*

Ru·me·li·a (rōō·mē′lē·ə) The possessions of the former Ottoman Empire in the Balkan Peninsula, including Macedonia, Thrace, and Albania: also *Roumelia.*

ru·men (rōō′men) *n. pl.* **ru·mi·na** (rōō′mə·nə) 1 The first stomach of a ruminant. 2 The cud of a ruminant. [<L, throat]

Rum·ford (rum′fərd), **Count,** 1753–1814, Benjamin Thompson, physicist, born in America, active in Germany, England, and France.

ru·mi·nant (rōō′mə·nənt) *n.* One of a division *(Ruminantia)* of even-toed ungulates, as a deer, antelope, sheep, goat, or cow, that has a stomach with four complete cavities: the rumen, the reticulum, the manyplies (omasum, psalterium), and the reed or abomasum, the food entering the first being returned to the mouth, rechewed and swallowed, and digested in the other compartments. —*adj.* 1 Chewing the cud. 2 Of or pertaining to the *Ruminantia.* 3 Meditative or contemplative; thoughtful; drowsily quiet. [<L *ruminans, -antis,* ppr. of *ruminare* ruminate <*rumen* gullet]

ru·mi·nate (rōō′mə·nāt) *v.t.* & *v.i.* **·nat·ed, ·nat·ing** 1 To chew (food previously swallowed and regurgitated) over again; chew (the cud). 2 To meditate or reflect (upon); ponder. See synonyms under MUSE. —*adj.* Perforated or mottled, as the albumen of a betelnut or nutmeg: also **ru′mi·nat′ed.** [<L *ruminatus,* pp. of *ruminare.* See RUMINANT.] —**ru′mi·nat′ing·ly** *adv.* —**ru′mi·na′tive** *adj.* —**ru′mi·na′tive·ly** *adv.* —**ru′mi·na′tor** *n.*

ru·mi·na·tion (rōō′mə·nā′shən) *n.* 1 The act, process, or characteristic of chewing the cud. 2 The act of ruminating mentally. 3 The regurgitation of imperfectly digested food. See synonyms under REFLECTION.

Ruml (rum′əl, rŏŏm′əl), **Beardsley,** 1894–1960, U.S. businessman and financier.

rum·mage (rum′ij) *v.* **·maged, ·mag·ing** *v.t.* 1 To search through (a place, box, etc.) by turning over and disarranging the contents; ransack. 2 To find or bring out by searching: with *out* or *up.* —*v.i.* 3 To make a thorough search. —*n.* 1 Any act of rummaging; especially, disarranging things by searching thoroughly. 2 An upheaval or stirring up; bustle. 3 *Obs.* Room in a ship for stowing cargo; also, the arrangement or stowing of the cargo. 4 A rummage sale. [<obs. F *arrumage* place or act of stowage <*arrumer* stow away, ? <*rum* ship's hold <OE *rūm* room] —**rum′mag·er** *n.*

rummage sale 1 A sale of all sorts of second-hand objects gathered up from benevolent givers, to obtain money for some charitable object. 2 A sale of unclaimed articles, or a clearing-out sale prior to restocking.

rum·mer (rum′ər) *n.* A glass or cup for drinking; specifically, a tall stemless glass; also, its contents. [<Du. *roemer* <*roemen* praise; from its use in drinking toasts]

rum·my[1] (rum′ē) *n.* A card game in which each player in turn draws a card from the talon or the discard pile beside it, and discards another card, the object being to get rid of one's hand in sequences of three cards or more of the same suit. [Origin unknown]

rum·my[2] (rum′ē) *n. pl.* **·mies** *Slang* A drunkard. —*adj.* 1 Of or pertaining to rum: a *rummy* flavor. 2 Affected by rum; befuddled; drunk.

ru·mor (rōō′mər) *n.* 1 Popular report; common gossip; also, reputation. 2 A story circulating without known foundation or authority; an unverified report passing from person to person. 3 *Obs.* A confused sound; confusion; murmur. —*v.t.* To tell or spread as a rumor; report abroad. Also *Brit.* **ru′mour.** [<OF <L, noise]

rump (rump) *n.* 1 The hinder parts or buttocks. 2 The fag-end of anything; an inferior remnant. 3 A legislative group having only a remnant of its former membership and therefore lacking authority because unrepresentative. 4 The piece of beef between aitchbone and loin. [<Scand. Cf. Dan. *rumpl,* ON *rumpr.*]

Rum·pel·stilts·kin (rum′pəl·stilt′skin, *Ger.* rŏŏm′pəl·shtilts′kin) In German folklore, a dwarf who saves the life of a girl who has married a king, by spinning for her the fabulous quantity of flax her mother has boasted that she can spin. In return for this, the dwarf demands her first child. When it is born the distressed mother begs to be released from her promise. The dwarf consents if she can guess his name within three days. She does so at the crucial moment and Rumpelstiltskin's power is broken. Also **Rum·pel·stiltz·chen** (rŏŏm′pəl·shtilts′khən).

rum·ple (rum′pəl) *v.t.* & *v.i.* **·pled, ·pling** To form into creases or folds; wrinkle; ruffle. —*n.* 1 An irregular fold; a rumpled fabric. 2 The condition of being rumpled. [<MDu. *rumpelen*]

Rump Parliament See LONG PARLIAMENT under PARLIAMENT.

rum·pus (rum′pəs) *n. Colloq.* A row; wrangle; to-do. [Origin uncertain]

rumpus room A room for games, informal gatherings, etc.

rum-run·ner (rum′run′ər) *n.* One who illicitly transports or smuggles alcoholic liquors across a border; also, a vessel employed in illegal liquor traffic.

run (run) *v.* **ran, run, run·ning** *v.i.* 1 To move by rapid steps, faster than walking, in such a manner that both feet are off the ground for a portion of each step. 2 To move rapidly; go swiftly. 3 To flee; take flight. 4 To make a brief or rapid journey: We *ran* over to Staten Island last night. 5 To make regular trips; ply: This steamer *runs* between New York and Liverpool. 6 **a** To take part in a race. **b** To be a candidate or contestant: to *run* for dogcatcher. 7 To finish a race in a specified position: I *ran* a poor last. 8 To move or pass easily: The rope *runs* through the block. 9 To pass continuously and rapidly; elapse: The hours *run* by. 10 To proceed in direction or extent: This road *runs* north. 11 To move in or as in a stream; flow. 12 To become liquid and flow, as wax; also, to spread or mingle confusedly, as colors when wet. 13 To move or pass inadvertently: The ship *ran* aground. 14 To pass into a specified condition: to *run* to seed. 15 To come undone; unravel, as a fabric. 16 To give forth a discharge or flow; suppurate. 17 To leak. 18 To continue or proceed without restraint: The conversation *ran* on and on. 19 To be in operation; be operative; work: Will the engine *run*? 20 To continue in existence or effect; extend in time: Genius *runs* in her family. 21 To be reported or expressed: The story *runs* as follows. 22 To migrate, as salmon from the sea to spawn. 23 To occur or return to the mind: An idea *ran* through his head. 24 To occur with specified variation of size, quality, etc.: The corn is *running* small this year. 25 To be performed or repeated in continuous succession: The play *ran* for forty nights. 26 To make a rapid succession of demands for payment, as on a bank. 27 To continue unexpired or unpaid, as a debt; become payable. —*v.t.* 28 To run or proceed along, as a route or path. 29 To make one's way over, through, or past: to *run* rapids. 30 To perform or accomplish by or as by running: to *run* a race or an errand. 31 To compete against in or as in a race. 32 To enter (a horse) for a race. 33 To present and support as a candidate. 34 To hunt or chase, as game. 35 To bring to a specified condition by or as by running: to *run* oneself out of breath. 36 To drive or force: with *out of, off, into, through,* etc. 37 To cause (a vessel) to move rapidly or freely: They *ran* the ship into port. 38 To move (the eye, hand, etc.) quickly or lightly: He *ran* his hand over the table. 39 To cause to move, slide, etc., as into a specified position: to *run* up a flag. 40 To cause to go or ply: to *run* a train between New York and Washington. 41 To transport or convey in a vessel or vehicle. 42 To smuggle. 43 To cause to flow: to *run* water into a pot. 44 To give forth a flow of; emit: Her eyes *ran* tears. 45 To mold, as from melted metal; found. 46 To sew (cloth) in a continuous line, usually by taking a number of stitches with the needle at a time. 47 To maintain or control the motion or operation of. 48 To direct or control; manage; oversee. 49 To allow to continue or mount up, as a bill. 50 In games, to make (a number of points, strokes, etc.) successively. 51 To publish in a magazine or newspaper: to *run* an ad. 52 To mark, set down, or trace, as a boundary line. 53 To suffer from (a fever, etc.). —**to run across** To meet by chance. —**to run down** 1 To pursue and overtake, as a fugitive. 2 To strike down while moving. 3 To exhaust, damage, lessen in worth, vigor, etc., as by abuse or overwork. 4 To speak of disparagingly; decry. —**to run in** 1 To insert; include. 2 *Printing* To print without a paragraph or break. 3 *Slang* To arrest and place in confinement. —**to run into** 1 To meet by chance. 2 To collide with. —**to run off** 1 To produce on a typewriter, printing press, etc. 2 To decide (a tied race, game, etc.) by the outcome of another, subsequent race, game, etc. 3 To flee or escape; elope. —**to run out** To come to an end; be exhausted, as supplies. —**to run out of** To exhaust one's supply of. —**to run over** 1 To ride or drive over; run down. 2 To overflow. 3 To go over or examine hastily or quickly; rehearse. —**to run through** 1 To spend wastefully; squander. 2 To stab or pierce. 3 To run over (def. 3). —**to run up** To produce; make hurriedly, as on a sewing machine. —*n.* 1 The act, or an act, of running or going rapidly. 2 A running pace: to break into a *run.* 3 Flow; movement; sweep: the *run* of the tide. 4 A distance covered by running. 5 A journey or passage, especially between two points, made by a vessel, train, etc.: the *run* from New York to Albany. 6 A rapid journey or excursion, marked by a brief stay at the destination: to take a *run* into town. 7 A swift stream or brook. 8 A migration of fish, especially to up-river spawning grounds. 9 A grazing or feeding ground for animals or fowl; a range: a sheep *run.* 10 The regular trail or path of certain animals: an elephant *run.* 11 The bower of a bowerbird. 12 The privilege of free use or access: to have the *run* of the place. 13 A runway. 14 *Music* A rapid succession of tones; a roulade. 15 A series or succession. 16 A sequence of three or more playing cards in consecutive order. 17 A trend or tendency: the general *run* of the market. 18 The direction or course (of something): the *run* of the grain of wood. 19 A continuous length (of something): a *run* of pipe. 20 A continuous spell (of some condition): a *run* of luck. 21 A surge of demands made upon a bank or treasury to meet its obligations. 22 Any great sustained demand. 23 A period of continuous performance, occurrence, popularity, etc.: a play with a long *run.* 24 Class or type: the general *run* of readers. 25 A period of operation of a machine or device: an experimental *run.* 26 The output during such a period. 27 A period during which a liquid is allowed to run. 28 The amount of liquid allowed to flow at one time. 29 A measure of yarn (about 1,600 yards). 30 A narrow, lengthwise ravel, as in a sheer stocking. 31 An approach to a target made by a bombing plane. 32 In baseball, a complete circuit of the bases from home plate and back before three outs are made, thus scoring a point. 33 In cricket, an act in which both batsmen successfully run to opposite popping creases, thereby scoring a point. 34 A hunt, especially on horseback; a chase. 35 *Naut.* The after part of a ship's bottom where it narrows off from the floor timbers to the sternpost. 36 *Mining* A vein of ore or rock. See synonyms under STREAM. —**dry run** Any practice test; specifically, an approach to a target made by a bombing plane, without dropping bombs. —**in the long run** As the ultimate outcome of any train of circumstances. — *adj.* **1** Made liquid; melted. **2** Made by a process of melting and casting or molding: *run* metal; *run* butter. **3** Extracted or drained: *run* honey. **4** Smuggled; contraband: *run* liquor. [OE *rinnan* flow]

run–a·bout (run′ə·bout′) *n.* **1** A light, handy, open automobile for ready service. **2** A light, open wagon. **3** A small motorboat.

run·a·gate (run′ə·gāt) *n. Archaic* **1** A deserter; renegade. **2** A vagabond; homeless wanderer. [Alter. of RENEGADE; infl. by *run*, dial. *agate* on the way]

run–a·round (run′ə·round′) *n.* **1** *Slang* Artful deception; evasion. **2** Run–round. **3** *Printing* Type set narrower than the body of the text, as around illustrations.

run·a·way (run′ə·wā′) *adj.* **1** Escaping or escaped from restraint or control; fugitive. **2** Brought about by running away: a *runaway* marriage. **3** Easily won: said of a horse race; hence, decisive; one–sided. — *n.* **1** One who or that which runs away or flees; a fugitive or deserter; also, a horse of which the driver has lost control. **2** An act of running away: said especially of a horse or a team: to be injured in a *runaway*.

run·ci·ble spoon (run′sə·bəl) A fork having three broad tines, of which one has a sharp edge. [<RUNC(INATE) + -IBLE]

run·ci·nate (run′sə·nāt, -nit) *adj. Bot.* Sawtoothed, with the incisions or teeth inclined backward: said of leaves. [<L *runcinatus*, pp. of *runcinare* plane off < *runcina* plane, saw]

run·dle (run′dəl) *n.* **1** A rung, as of a ladder. **2** Something that rotates about an axis, as the drum of a capstan. [Var. of ROUNDEL]

rund·let[1] (rund′lit) *n.* A small barrel, or the measure of wine it contains, about 18 wine gallons. Also *runlet*. [<OF *rondelet*. See ROUNDELAY.]

rund·let[2] (rund′lit) See RUNLET[1].

run–down (run′doun′) *adj.* **1** Debilitated; physically weak; tired out. **2** Dilapidated; shabby. **3** Stopped because not wound: said of a timepiece. — *n.* (run′doun′). **1** A summary; resumé. **2** In baseball, a play in which a base runner is put out when trapped between two bases.

Rund·stedt (ro͝ont′shtet), **Karl Rudolf Gerd von,** 1875–1953, German field marshal in World War II.

rune (ro͞on) *n.* **1** A character of the primitive runic alphabet. **2** A Finnish poem or one of its cantos. **3** *pl.* Old Norse lore expressed, or considered as if expressed, in runes; hence, early rimes or poetry in general. **4** Any obscure or mystic song, poem, verse, or saying; a mystery. [<ON *rūn* mystery, rune] — **ru′nic** *adj.*

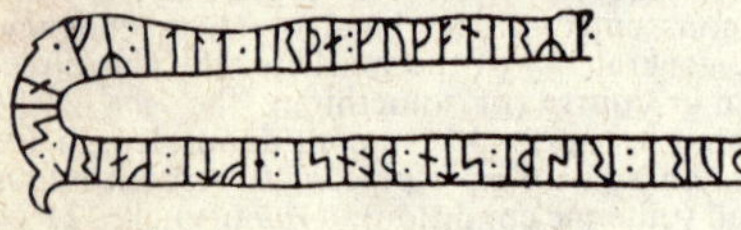

RUNES
Tomb inscription, Sweden, eleventh century.

Ru·ne·berg (ro͞o′nə·ber′y′), **Johann,** 1804–77, Finnish national poet; wrote mainly in Swedish.

rung[1] (rung) *n.* **1** A round crosspiece of a ladder or chair; a round; also, a spoke of a wheel. **2** *Naut.* **a** One of the handles on the rim of a ship's tiller. **b** A floor timber of a ship. **3** *Scot.* & *Brit. Dial.* A heavy club or staff; cudgel. [OE *hrung* crossbar]

rung[2] (rung) Past participle of RING[2].

ru·nic alphabet (ro͞o′nik) An old Germanic alphabet, probably originating in both the Latin and Greek, consisting originally of 24 characters, or runes, later reduced to 16 in Scandinavian writings. The earliest inscriptions in this alphabet are of the second or third century A.D. In England it was still in occasional use at the end of the Old English period, and was finally completely replaced by the Roman alphabet through the spread of Christian writings. Also called *futhark*.

runic staff A clog almanac.

run–in (run′in′) *n.* **1** A quarrel; bicker. **2** *Printing* Inserted or added matter. — *adj.* (run′in′) *Printing* That is inserted or added.

Run·jeet Singh (run′jēt sin′hə) See RANJIT SINGH.

run·kle (rung′kəl) *n.*, *v.t. Scot.* Wrinkle.

run·let[1] (run′lit) *n.* A little stream; rivulet; a runnel. Also *rundlet*. See synonyms under STREAM.

run·let[2] (run′lit) See RUNDLET[1].

run·nel (run′əl) *n.* A streamlet; brooklet; rivulet. [OE *rynel* < *rinnan* run]

run·ner (run′ər) *n.* **1** One who or that which runs; especially, one who runs a race; also, a fugitive or deserter. **2** One who operates or manages anything; especially, the driver of a locomotive. **3** One who runs errands or goes about on any kind of business; a messenger, as for a bank; specifically, one who drums up or solicits patronage or business, as for a hotel. **4** That part on which an object runs or slides: the *runner* of a skate. **5** *Mech.* A device to assist sliding motion. **6** A slender fish (*Elegatis bipinnulatus*) of warm seas with single dorsal and anal pinnules; also, the jurel of the Atlantic coast of America. **7** A cursorial bird; the water rail. **8** *Bot.* **a** A slender, procumbent stem disposed to root at the end and nodes, as in the strawberry; also, sometimes, the plant itself. **b** Any of various twining plants: the scarlet *runner*. **9** A smuggler. **10** A blacksnake. **11** A long, narrow rug or carpeting, used in hallways, etc. **12** A narrow strip of cloth, usually of fine quality, used on tables, dressers, etc.

run·ner–up (run′ər·up′) *n.* A contestant or team finishing in second place.

run·ning (run′ing) *adj.* **1** Such as runs: said specifically of horses inclined or trained to a running gait rather than to pacing or trotting. **2** Following one another without intermission; successive: used with words expressing periods of time: He talked for three hours *running*. **3** Continuous; repeated: said of a design: a *running* ornament, a *running* molding, etc. **4** Kept up continuously; also, passing; cursory: *running* comments, a *running* glance. **5** Characterized by easy flowing curves; cursive: a *running* hand. **6** Discharging, as pus from a sore. — *n.* **1** The act or movement of one who or that which runs: a horse trained for fast *running*. **2** That which runs or flows; the amount or quantity that runs. **3** A discharge, as from a sore. **4** Ability or power to run. **5** Competition; race; rivalry: He's out of the *running*. **6** Climbing; sending out runners, as certain plants.

running board A footboard on the side of a locomotive, street car, automobile, etc.

running expenses Daily expenses.

running fits Fright disease.

running gear **1** *Mech.* **a** The wheels and axles of any vehicle and their immediate attachments, as distinguished from the body, frame, etc., which they support. **b** Those parts of a mechanism or construction that have partially independent motion: the *running gear* of a watch. **2** *Naut.* The movable ropes and wires on a boat or ship by which sails, etc., are raised, lowered, and trimmed.

running hand Writing done with a continuous easy motion without lifting the pen from the paper and usually having the letters slanted forward.

running knot A knot made so as to slip along a noose and tighten when pulled upon.

running lights The sidelights of a vessel.

running mate **1** A horse that is teammate for another; also, a horse entered to set the pace for another entered to run in a horse race. **2** The candidate for the lesser of two offices closely linked by constitutional provisions, as the vice–presidency with the presidency.

running title *Printing* A title or headline repeated at the head of succeeding pages throughout a book or chapter. Also **running head.**

Run·ny·mede (run′i·mēd) A meadow in Surrey, England, on the Thames west of London, where King John is said to have met his barons to sign the Magna Carta in 1215.

run–off (run′ôf′, -of′) *n.* **1** That part of the rainfall in a particular area which is not absorbed directly by the soil but is drained off in rills or streams. **2** A special contest held to break a tie.

run–of–the–mill (run′əv·thə·mil′) See MILL–RUN.

run–on (run′on′, -ôn′) *n. Printing* Appended or added matter.

run–on line Enjambement.

run–out (run′out′) *n.* That portion of a motion–picture film immediately following the last frame of the picture itself.

run–round (run′round′) *n. Pathol.* A circumscribed inflammation of the skin, as on the fingers or toes; a felon: also called *run-around*.

runt (runt) *n.* **1** An unusually small, weak, or stunted animal; also, the smallest and weakest of a litter. **2** A dwarf. **3** *Scot.* An old ox or cow; a withered old man or hag. **4** *Scot.* & *Brit. Dial.* A stump of a tree or shrub; also, the stem or stalk of a plant. [Origin uncertain] — **runt′i·ness** *n.* — **runt′y** *adj.*

run·way (run′wā′) *n.* **1** A way or path over which something runs. **2** The channel or bed of a stream, or the path over which animals pass to and from their places of feeding or watering. **3** In lumbering, an incline down which logs are slid; a chute. **4** Any track specially laid for wheeled vehicles. Also *run*. **5** *Aeron.* An artificial landing strip for airplanes.

Run·yon (run′yən), **Damon,** 1884–1946, U.S. journalist and writer.

ru·pee (ro͞o·pē′) *n.* The standard monetary unit of British India: it contains 16 annas. [< Hind. *rupiya* <Skt. *rūpya* silver]

Ru·pert (ro͞o′pərt; *Fr.* rü·pâr′, *Ger.* ro͞o′pert) A masculine personal name: variant of ROBERT. Also *Ger.* **Ru·precht** (ro͞o′prekht). [See ROBERT]

— **Rupert, Prince,** 1619–82, Royalist general in English Civil War; born in Bavaria.

ru·pi·ah (ro͞o·pē′ä) *n.* The principal Indonesian currency unit.

rup·ture (rup′chər) *n.* **1** The act of breaking apart or the state of being broken apart. **2** Hernia. **3** Breach of peace and concord between individuals or nations. — *v.t.* & *v.i.* **·tured, ·tur·ing** **1** To break apart; separate into parts. **2** To affect with or suffer a rupture. See synonyms under BREAK, REND. [<F <L *ruptus*, pp. of *rumpere* break] — **rup′tur·a·ble** *adj.*

Synonyms (noun): blast, breach, break, burst, disruption, fracture. See BREACH, QUARREL[1].

ru·ral (ro͝or′əl) *adj.* **1** Pertaining to the country as distinguished from the city or the town; rustic. **2** Pertaining to farming or agriculture. See synonyms under RUSTIC. [<F <L *ruralis* < *rus, ruris* country] — **ru′ral·ism** *n.* — **ru′ral·ist** *n.* — **ru′ral·ly** *adv.* — **ru′ral·ness** *n.*

rural dean Dean (def. 2).

rural free delivery A government service of house–to–house free mail delivery by carrier in rural districts, as distinguished from the general delivery service: in addresses abbreviated *R.F.D.* Often shortened to *R.D.*

ru·ral·i·ty (ro͞o·ral′ə·tē) *n.* **1** Ruralness. **2** A rural peculiarity. **3** A place in the country.

ru·ral·ize (ro͝or′əl·īz) *v.* **·ized, ·iz·ing** *v.t.* To make rural. — *v.i.* To go into or live in the country; rusticate. — **ru′ral·i·za′tion** *n.*

Ru·rik (ro͞o′rik) Russian form of RODERICK.

— **Rurik,** died 879, Scandinavian conqueror who founded the Russian monarchy, the **House of Rurik,** which lasted from 862 to 1598.

ruse (ro͞oz) *n.* An action intended to mislead or deceive; a stratagem; trick. See synonyms under ARTIFICE, PRETENSE. [<F <*ruser* dodge, detour, drive back. Related to RUSH[1].]

Ru·se (ro͞o′se) A city on the Danube in NE Bulgaria.

ruse de guerre (rüz də gâr′) *French* A stratagem of war.

rush[1] (rush) *v.i.* **1** To move or go swiftly or with violence. **2** To make an attack; charge: with *on* or *upon*. **3** To proceed recklessly or rashly; plunge: with *in* or *into*. — *v.t.* **4** To drive or push with haste or violence; hurry. **5** To do or perform hastily or hurriedly: to *rush* one's work. **6** **a** To make a sudden assault upon. **b** To capture by such an assault. **7** *Slang* To seek the favor of with assiduous attentions. **8** In football, to move (the ball) toward the goal of the other team. See synonyms under HUSTLE. — *n.* **1** The act of rushing; a sudden turbulent movement, drive, or onset. **2** A sudden pressing demand; a run: a *rush* on foreign bonds. **3** A sudden exigency; urgent pressure: a *rush* of business. **4** A sudden flocking of people to a new region, especially to an area rumored to be rich in a precious mineral: a gold *rush*. **5** *U.S.* A general contest or scrimmage between students from different classes, as between sophomores and freshmen. 6 In football: **a** An attempt to take the ball through the opposing linemen and toward the goal. **b** Formerly, a player in the rush line: a center *rush.* 7 In motion pictures, the first film prints of a scene or series of scenes, before editing or selection. See synonyms under CAREER. —*adj.* 1 Requiring urgency or haste: a *rush* order. 2 Characterized by much traffic, business, etc.: the *rush* hours. 3 Denoting a time or function set aside for fraternity or sorority members to meet new students to consider them for membership: *rush* week; a *rush* smoker. [<AF *russher* push, var. of *russer,* OF *ruser, reuser* push back, dodge <LL *recusare* push back. See RECUSANT.]

rush[2] (rush) *n.* 1 Any one of various grasslike, usually aquatic herbs (family *Juncaceae*). The common or **soft rush** *(Juncus effusus)* grows in marshy ground and has soft and pliant, cylindrical, leafless stems: used for mats, seats of chairs, etc. 2 A thing of little or no value. 3 A rushlight. [OE *rysc*]

Rush (rush), **Benjamin,** 1745–1813, American physician, signer for Pennsylvania of the Declaration of Independence.

rush·er (rush′ər) *n.* 1 One who rushes. 2 In football, a lineman.

rush–hold·er (rush′hōl′dər) *n.* A candlestick with a clip for supporting a rushlight.

rush hour A time when traffic or business is at its height. —**rush–hour** (rush′our′) *adj.*

rush·ing (rush′ing) *n. U.S.* The series of activities in which fraternity and sorority members meet and evaluate new college students wishing to be pledged.

rush·light (rush′līt′) *n.* A candle made by dipping a rush in tallow. Also **rush candle.**

rush line In football, the linemen collectively: a term no longer in general use.

Rush·more (rush′môr), **Mount** A mountain in the Black Hills of western South Dakota, on the side of which are carved gigantic faces of Washington, Jefferson, Lincoln, Theodore Roosevelt, and Wilson: in **Mount Rushmore National Memorial**; 1,220 acres; established, 1929.

rush·y (rush′ē) *adj.* **rush·i·er, rush·i·est** Abounding in or made of rushes.

ru·sine (rōō′sin, -sīn) *adj.* Of, pertaining to, or designating a genus *(Rusa)* of deer native in the East Indies. Compare SAMBUR. [<Malay *rūsa* deer]

rusine antler An antler having a simple brow tine and a simple fork at the tip of the beam.

rus in ur·be (rus′ in ûr′bē) *Latin* The country in the city.

rusk (rusk) *n.* 1 A light, sweetened bread or biscuit. 2 Bread or cake that has been crisped and browned in an oven, then often pounded fine to be eaten with milk. [<Sp. *rosca,* twisted loaf of bread]

Rus·kin (rus′kin), **John,** 1819–1900, English art critic and author.

Russ (rus) *adj.* & *n.* Russian.

Rus·sell (rus′əl), **Bertrand Arthur William,** born 1872, third Earl Russell, English mathematician and philosopher. —**Countess Elizabeth Mary,** 1866–1941, *née* Beauchamp, English novelist: pen name *Elizabeth.* —**George William** See Æ. —**Lord John,** 1792–1878, first Earl Russell, English statesman. —**Lillian,** 1861–1922, U.S. soprano: original name Helen Louise Leonard.

rus·set (rus′it) *n.* 1 A color formed by combining orange and purple; popularly, any reddish or yellowish brown. 2 Russet cloth, clothing, etc.; hence, any coarse homespun cloth or garment; a country dress. 3 Russet leather. 4 A winter apple of greenish color, mottled with brown. —*adj.* 1 Of a reddish- or yellowish-brown color. 2 Made of russet cloth; hence, coarse; homespun; rustic. 3 Finished, but not blacked: said of leather: *russet* shoes. [<OF *rousset,* dim. of *rous* <L *russus* reddish] —**rus′set·y** *adj.*

Rus·sia (rush′ə) 1 Before 1917, an empire of eastern Europe and northern Asia; capital, Saint Petersburg (Petrograd). 2 The Union of Soviet Socialist Republics. 3 The Russian Soviet Federated Socialist Republic: Russian *Rossiya.*

Rus·sian (rush′ən) *adj.* Pertaining to Russia, its people, or their language. —*n.* 1 An inhabitant of Russia; especially, one of any of the Slavic peoples of the U.S.S.R., including the **Great Russians** of the central and northwestern region, the Ukrainians (or **Little Russians**) of the Ukrainian S.S.R. and eastern Poland, which group includes also the Cossacks and Ruthenians, and the **White Russians** of the west, all speaking Indo-European Balto-Slavic languages, such as Russian, Polish, Lithuanian, and Lettish; also, one of any of the peoples of Russia speaking any of the Uralic languages, especially the Finno-Ugric branch; also, one of any of the native peoples of the Caucasus speaking languages unrelated to these others, as Circassian, Georgian, and the Lesghian group. 2 The language of Russia, belonging to the East Slavic branch of the Balto-Slavic languages, having a separate alphabet including several characters not found in other alphabets. Its subdivisions are **Great Russian,** the principal subdivision and standard literary language in northern and central Russia, Ukrainian or Ruthenian (or **Little Russian**), and **White Russian,** the literary language of western Russia.

Russian Church A division of the Greek Church, independent since 1589, and governed by the Holy Synod. See GREEK CHURCH.

Russian dressing Mayonnaise dressing to which chili sauce, pimientos, and chopped pickles have been added.

Rus·sian·ize (rush′ən·īz) *v.t.* **·ized, ·iz·ing** To make Russian.

Russian leather A smooth, well-tanned, high-grade leather of calfskin or light cattle hide, dressed with birch oil and having a characteristic odor.

Russian Revolution See under REVOLUTION.

Russian Soviet Federated Socialist Republic The largest of the constituent republics of the U.S.S.R., occupying 76 per cent of the U.S.S.R. and extending across northern Asia and eastern Europe; 6,590,564 square miles; capital, Moscow: also *(Soviet) Russia*: abbr. *R.S.F.S.R.*

Russian Turkestan See under TURKESTAN.

Russian wolfhound The borzoi.

Russo– *combining form* Russia; pertaining to the Russians: *Russophobia.* [<RUSSIA]

Rus·so·phile (rus′ə·fil, -fīl) *n.* One who favors Russia, or its principles, policy, or methods.

Rus·so·pho·bi·a (rus′ə·fō′bē·ə) *n.* Fear of the policy or influence of Russia. —**Rus′so·phobe** *n.*

rust (rust) *n.* 1 The reddish or yellow coating caused on iron and steel by oxidation, as by the action of air and moisture, consisting of ferric hydroxide, $Fe(OH)_3$, and ferric oxide, Fe_2O_3. 2 A film of oxide formed on any metal by corrosion. 3 Any of the parasitic fungi of the order *Uredinales,* living on the tissues of higher plants. 4 The diseases caused by such fungi; incorrectly, any one of several diseases not caused by these fungi. 5 Any coating or accretion formed by a corrosive or degenerative process: *rust* on salted meat. 6 A condition, affection, or tendency that destroys or weakens energy or active qualities: the *rust* of idleness. 7 Any of several shades of reddish brown, somewhat like the color of rust, but containing more orange. —*v.t.* & *v.i.* 1 To become or cause to become rusty; undergo or cause to undergo oxidation. 2 To contract or cause to contract rust. 3 To become or cause to become weakened or impaired because of inactivity or disuse: to allow one's powers to *rust.* 4 To make or become rust-colored. [OE]

rus·tic (rus′tik) *adj.* 1 Rural; hence, plain; homely: *rustic* garments. 2 Uncultured; rude; awkward: *rustic* manners. 3 Unaffected; artless: *rustic* simplicity. 4 Pertaining to any irregular style of work or decoration appropriate to the country or to work in natural, unpolished wood. —*n.* 1 One who lives in the country; a country person of simple manners or character; also, a coarse or clownish person. 2 Rusticwork. 3 Country dialect. [<F *rustique* <L *rusticus* <*rus* country] —**rus′ti·cal·ly** *adv.*

Synonyms (adj.): agricultural, artless, awkward, boorish, bucolic, clownish, coarse, countrified, country, hoydenish, inelegant, outlandish, pastoral, plain, rude, rural, sylvan, uncouth, unpolished, unsophisticated, untaught, verdant. *Rural* refers especially to scenes or objects in the country, considered as the work of nature; *rustic* refers to their effect upon man or to their condition as affected by human agency; as, a *rural* scene; a *rustic* party; a *rustic* lass. We speak, however, of the *rural* population, *rural* simplicity, etc. *Rural* has always a favorable sense; *rustic* often an unfavorable one, as denoting lack of culture and refinement; thus, *rustic* politeness expresses that which is well-meant, but awkward. *Rustic* is, however, often used of a studied simplicity, an artistic rudeness, which is pleasing and perhaps beautiful; as, a *rustic* cottage. *Pastoral* refers to the care of flocks and to the shepherd's life with the pleasing associations suggested by the old poetic ideal of that life; as, *pastoral* poetry. *Bucolic* is kindred to *pastoral,* but is a less elevated term, and sometimes slightly contemptuous. *Antonyms*: accomplished, cultured, elegant, polished, polite, refined, urban, urbane.

rus·ti·cate (rus′tə·kāt) *v.* **·cat·ed, ·cat·ing** *v.i.* 1 To go to the country. 2 To stay or live in the country. —*v.t.* 3 To send or banish to the country. 4 *Brit.* To suspend (a student) and send away temporarily, as from a college. 5 To make rustic. 6 To construct (masonry) with rusticwork. [<L *rusticatus,* pp. of *rusticari* rusticate <*rusticus.* See RUSTIC.] —**rus′ti·ca′tion** *n.* —**rus′ti·ca′tor** *n.*

rus·tic·i·ty (rus·tis′ə·tē) *n. pl.* **·ties** 1 Rustic condition, characters, or manners; simplicity; homeliness; awkwardness. 2 A rustic trait or peculiarity. [<L *rusticitas, -tatis*]

rus·tic·work (rus′tik·wûrk′) *n.* 1 Ashlar masonry, or a method of making it, with rough surfaces, and often with deeply sunk grooves at the joints, to make them conspicuous. 2 Woodwork made of the natural limbs and roots of trees, fancifully arranged.

rus·tle[1] (rus′əl) *v.t.* & *v.i.* **·tled, ·tling** To fall, move, or cause to move with a quick succession of small, light, rubbing sounds, as dry leaves or sheets of paper. —*n.* A rustling sound. [OE *hrūxlian* make a noise. Cf. OE *gehyrstan* murmur.] —**rus′tler** *n.* —**rus′tling** *adj.* —**rus′tling·ly** *adv.*

rus·tle[2] (rus′əl) *v.t.* & *v.i.* **·tled, ·tling** 1 *Colloq.* To act with or obtain by energetic or vigorous action. 2 *U.S. Colloq.* To steal (cattle, etc.). [Blend of RUSH and HUSTLE]

rus·tler (rus′lər) *n. U.S.* 1 *Slang* Any person who is active, pushing, and bustling in any enterprise. Compare HUSTLER. 2 *Colloq.* **a** A cowboy or ranchman. **b** A cook on a ranch. **c** A cattle or horse thief.

rust·y[1] (rus′tē) *adj.* **rust·i·er, rust·i·est** 1 Covered or affected with rust. 2 Consisting of or produced by rust. 3 Having the appearance of rust; having a reddish or yellowish discoloration, as from decomposition: said especially of salted fish or meat that has become rancid. 4 Impaired by inaction or want of exercise; also, lacking nimbleness; stiff. 5 **a** Weakened through neglect of use: My Latin is *rusty.* **b** Having lost skill for want of practice: *rusty* in math. 6 *Biol.* Appearing as if covered with rust; brownish-red. See synonyms under TRITE. [OE *rustig* <*rust* rust] —**rust′i·ly** *adv.* —**rust′i·ness** *n.*

rust·y[2] (rus′tē) *adj. Brit. Dial.* Restive; stubborn; obstinate.

rut[1] (rut) *n.* 1 A sunken track worn by a wheel, as in a road; hence, a groove forming a path for anything. 2 A settled habit or course of procedure; routine. —*v.t.* **rut·ted, rut·ting** To wear or make a rut or ruts in. [? Var. of ROUTE]

rut[2] (rut) *n.* 1 The sexual excitement of various animals, especially deer; estrus; also, the period during which it lasts. 2 A roaring or uproar; especially, the noise made by a rutting stag. —*v.* **rut·ted, rut·ting** *v.i.* To be in rut. —*v.t. Rare* To unite with in copulation; cover. [<F <L *rugitus* a roaring, tumult <*rugire* roar] —**rut′ting** *adj.*

ru·ta·ba·ga (rōō′tə·bā′gə) *n.* 1 A cultivated plant *(Brassica napobrassica)* allied to the common turnip. 2 Its edible, yellowish root. Also *Swedish turnip.* [<dial. Sw. *rotabagge*]

ru·ta·ceous (rōō·tā′shəs) *adj. Bot.* Of or pertaining to the rue family (*Rutaceae*) of shrubs, trees, and, rarely, herbs, including the lemon, lime, citron, etc. [<L *ruta* <Gk. *rhytē* rue]
Rut·ger (rut′gər, rōōt′gər) Dutch form of ROGER.
ruth (rōōth) *n.* Sorrow; compassion; pity; also, grief; misery; repentance; regret. [ME *reuthe, reowthe* <OE *hrēow* sad]
Ruth (rōōth, *Fr.* rüt) A feminine personal name. [<Hebrew, companion]
— **Ruth** A woman of Moab, daughter-in-law of the Israelite Naomi; she left her own people and went to Bethlehem, where she married Boaz, thus becoming an ancestress of David. Her story is told in the Old Testament book of this name.
Ru·the·ni·a (rōō·thē′nē·ə) A region of western Ukrainian S.S.R.; formerly a province of Czechoslovakia; annexed by Hungary, 1939; ceded to U.S.S.R., 1945; since 1945, the **Transcarpathian Oblast** of the Ukrainian S.S.R.; 5,000 square miles; capital, Uzhgorod: also *Carpatho-Ukraine*.
Ru·the·ni·an (rōō·thē′nē·ən) *n.* 1 One of a group of Ukrainians living in Ruthenia and eastern Czechoslovakia, formerly in Austria. 2 The East Slavic language of the Ukrainians; Ukrainian. See RUSSIAN. — *adj.* Pertaining to the Ruthenians or their language.
ru·then·ic (rōō·then′ik) *adj. Chem.* Of, pertaining to, or derived from ruthenium, especially when combined in its higher valence.
ru·the·ni·ous (rōō·thē′nē·əs) *adj. Chem.* Of, pertaining to, or derived from ruthenium, especially when combined in its lower valence.
ru·the·ni·um (rōō·thē′nē·əm) *n.* A gray, brittle, rare metallic element (symbol Ru) of the platinum group. See ELEMENT. [<NL, after *Ruthenia*]
ruth·er·ford (ruth′ər·fərd) *n.* A unit of radioactivity larger than the curie: equal to that quantity of a radioisotope which decays at the rate of a million disintegrations per second. [after Sir Ernest *Rutherford*]
Ruth·er·ford (ruth′ər·fərd), **Sir Ernest,** 1871–1937, English physicist. — **Joseph,** 1869–1942, U.S. leader of Jehovah's Witnesses.
ruth·ful (rōōth′fəl) *adj. Archaic* 1 Full of sorrow or pity; sorrowful; merciful. 2 Causing sorrow. — **ruth′ful·ly** *adv.* — **ruth′ful·ness** *n.*
ruth·less (rōōth′lis) *adj.* Having no compassion; unrestrained by pity; merciless: the *ruthless* cruelty of the barbaric Hun. — **ruth′less·ly** *adv.* — **ruth′less·ness** *n.*
ru·ti·lant (rōō′tə·lənt) *adj.* Of a shining red color; glittering. [<L *rutilans, -antis,* ppr. of *rutilare* glow red <*rutilus.* See RUTILE.]
ru·ti·lat·ed (rōō′tə·lā′tid) *adj.* Enclosing rutile needles: *rutilated* quartz.
ru·tile (rōō′til, -tēl, -tīl) *n.* An adamantine, reddish-brown, transparent to opaque titanium dioxide, TiO₂, usually containing a small quantity of iron. [<F, shining <L *rutilus* red]
Rut·land (rut′lənd) 1 A county of eastern England; 152 square miles: county town Oakham. Also **Rut′land·shire** (-shir). 2 A city in central Vermont, important as a center of the marble-cutting industry.
Rut·ledge (rut′lij), **Ann,** 1816–35, fiancée of Abraham Lincoln. — **Edward,** 1749–1800, American jurist; signer for South Carolina of Declaration of Independence. — **John,** 1739–1800, jurist; a framer of the U.S. Constitution; brother of the preceding.
rut·tish (rut′ish) *adj.* Disposed to rut; lustful; libidinous.
rut·ty (rut′ē) *adj.* Full of ruts. — **rut′ti·ness** *n.*
Ru·vu·ma (rōō·vōō′mə) A river in eastern Africa, flowing 450 miles north and east to the Indian Ocean, and forming the Tanganyika-Mozambique boundary: Portuguese *Rovuma*.
Ru·wen·zo·ri (rōō′wən·zôr′ē, -zō′rē) A mountain group in east central Africa between Albert and Edward lakes, on the boundary between the Belgian Congo and Uganda: identified with the *Mountains of the Moon* of ancient writers; highest peak, 16,795 feet.
Ru·y (*Sp.* rōō·ē′, *Pg.* rōō′ē) Spanish and Portuguese form of RODERICK.
Ruys·dael (rois′däl, *Du.* rœis′däl), **Jacob van,** 1625?–82, Dutch painter: also spelled *Ruisdael*.
Ruy·ter (roi′tər), **Michel Adriaanszoon de,** 1607–76, Dutch admiral.

Ru·žič·ka (rōō′zhich·kä), **Leopold,** born 1887, Yugoslav organic chemist.
-ry Var. of -ERY.
Rya·zan (rē·ə·zän′, *Russian* ryä·zän′) A city near the Oka river in the central European Russian S.F.S.R.
Ry·binsk (ri′binsk) See SHCHERBAKOV.
Ry·binsk Reservoir (ri′binsk) The largest artificial lake of the U.S.S.R. in north central European Russian S.F.S.R. on the upper Volga; 1,800 square miles: also **Rybinsk Sea.**
Ry·der (ri′dər), **Albert Pinkham,** 1847–1917, U.S. painter.
rye¹ (rī) *n.* 1 The grain or seeds of a hardy cereal grass (*Secale cereale*) closely allied to wheat. 2 The plant. 3 Whisky distilled from rye. ♦ Homophone: *wry.* [OE *ryge* rye]
rye² (rī) *n.* In Gipsy dialect, a gentleman. ♦ Homophone: *wry.* [<Romany *rei, rae,* prob. <Skt. *rājan* a king]
Rye (rī) A municipal borough of east Sussex, England; an important Channel port before the sea receded in the early 19th century.
rye·grass (rī′gras′, -gräs′) *n.* Common darnel: sometimes called *raygrass*.
ryke (rīk, rēk) *v.i. Scot.* To reach.
rynd (rind, rīnd) *n.* An iron fitting supporting an upper millstone, having a central hollow bearing which rests upon the upper pointed end of the mill spindle: also spelled *rind*. [Prob. <M Du. *rijn*]
Ryo·jun (ryô·jōn) The Japanese name for PORT ARTHUR.
ry·ot (rī′ət) *n.* In India, a tenant; tiller of the soil; peasant. [<Hind. *raiyat* <Arabic *ra'iyah*]
Rys·kind (ris′kind), **Morris,** born 1895, U.S. playwright.
Rys·wick (riz′wik) A village in southern Netherlands, near The Hague; site of the signing of a treaty by France, Germany, the Netherlands, England, and Spain, 1697: Dutch *Rijswijk*.
Ryu·kyu Islands (ryōō·kyōō) An archipelago between Kyushu and Taiwan; 1,803 square miles; chief island, Okinawa; Japanese possessions, administered by the U.S. after 1945; by 1958 the islands north of Okinawa had been returned to Japan. Also *Nansei Islands*.

S

s, S (es) *n. pl.* **s's, S's** or **ss, Ss** or **ess·es** (es′iz) 1 The nineteenth letter of the English alphabet, from Phoenician *shin,* through Hebrew *shin,* Greek *sigma,* Roman S. 2 The sound of the letter *s,* usually a voiceless sibilant. See ALPHABET. — *symbol* 1 *Chem.* Sulfur (symbol S). 2 Anything shaped like an S.
-s¹ A variant of *-es¹,* inflectional ending of the plurals of nouns, attached to nouns not ending in a sibilant or an affricate: *books, words, cars.* It is pronounced (s) after a voiceless consonant, and (z) after a voiced consonant or a vowel.
-s² An inflectional ending used to form the third person singular present indicative of verbs not ending in a sibilant, affricate, or vowel: *reads, walks, sings.* Compare —ES².
-s³ *suffix* On; of; a; at: often used in adverbs without appreciable force: *nights, Mondays, always, towards.* [OE *-es,* genitive ending]
-'s¹ An inflectional ending used to form the possessive of singular nouns and of plural nouns not ending in *-s:* a *man's* world, *women's* fashions. In plurals ending in *-s* (or *-es*) a simple apostrophe is used as a sign of the possessive: a *girls'* school, the *churches'* steeples, the *Joneses'* claim to the inheritance.
-'s² Contraction of: 1 Is: *He's* here. 2 Has: *She's* left. 3 Us: *Let's* go.
Saa·di (sä′dē), **Muslih-ud-Din,** 1184?–1291?, Persian poet. Also spelled *Sadi*.
Saa·le (zä′lə) 1 A river in central East Germany, flowing 265 miles north to the Elbe. Also **Sax·o·ni·an Saale** (sak·sō′nē·ən). 2 A river in northern Bavaria, West Germany, flowing 84 miles west, south and SW from the East German border to the Main. Also **Fran·co·ni·an Saale** (frang·kō′nē·ən).
Saa·mi (sä′mē) *n.* Lapp.
Saar (zär) A river in NE France and western Germany, flowing 152 miles north from the Vosges Mountains to the Moselle.
Saar (zär), **The** A state and industrial region in the Saar valley, SW West Germany; 989 square miles; capital, Saarbrücken. French *Sarre,* German **Saar·land** (zär′länt). Also **Saar Basin, Saar Territory.**
Saar·brück·en (zär′brük·ən) The capital of The Saar.
Saa·re (sä′re) The largest Estonian island in the Baltic, at the mouth of the Gulf of Riga; 1,046 square miles: Swedish *Ösel:* also *Sarema*. Also **Saa·re·maa** (sä′re·mä). *Russian* **E·zel** (ā′zel).
Saa·ri·nen (sä′ri·nen), **Eero,** born 1910, U.S. architect. — **Eliel,** 1873–1950, U.S. architect born in Finland; father of the preceding.
Saa·ve·dra La·mas (sä·vä′thrä lä′mäs), **Carlos,** born 1878, Argentine lawyer and statesman.
sab (sab) *n., v.t.* & *v.i. Scot.* Sob.
sa·ba (sä′bä′) *n.* A fine Philippine fabric made from fibers of a plant resembling the banana. [<Tagalog]
Sa·ba (sä′bä) 1 The Arabic name for SHEBA. 2 An island in the eastern group of the Netherlands West Indies; 5 square miles.
sab·a·dil·la (sab′ə·dil′ə) *n.* 1 The acrid seeds of a Mexican and Central American bulbous plant (*Schoenocaulon officinale*), used as a source of veratrine, and formerly as an anthelmintic. 2 The plant. Also spelled *cebadilla, cevadilla.* [<Sp. *cebadilla,* dim. of *cebada* barley]
Sa·ba·ism (sä′bə·iz′əm) *n.* Star worship. [<

Hebrew *tsābhā* (heavenly) host, army + -ISM] — **Sa′ba·ist** *n.*
Sab·a·oth (sab′ē·oth, sə·bā′ōth) *n. pl.* Armies; hosts: chiefly in the phrase *the Lord of Sabaoth. Rom.* ix 29, *James* v 4. [<L *Sabaōth* <Hebrew *tsebāōth,* pl. of *tsābhā* host, army]
Sa·bar·ma·ti (sä′bər·mu′tē) A river in southern Rajasthan and northern Bombay, India, flowing 250 miles south to the Gulf of Cambay.
Sa·ba·tier (sà·bà·tyā′), **Paul,** 1854–1941, French chemist.
sab·bat (sab′ət) *n.* The witches' sabbath. Also **Sab′bat.** [<OF. See SABBATH.]
sab·ba·tar·i·an (sab′ə·târ′ē·ən) *adj.* Pertaining to the Sabbath or its strict observance. — *n.* 1 A Christian who observes Sunday with strict propriety. 2 A Christian who observes the seventh day as the Sabbath: opposed to *dominical.* [<L *sabbatarius* <*sabbatum* SABBATH] — **Sab′ba·tar′i·an·ism** *n.*
Sab·bath (sab′əth) *n.* 1 The seventh day of the week, appointed in the decalog as a day of rest to be observed by the Jews; now, Saturday. 2 The first day of the week as observed by Christians; Sunday. 3 The institution or observance of a day of rest; a time of rest, peace, or quiet. 4 The sabbatical year of the Jews. *Lev.* xxv 4. [Fusion of OE *sabat* and OF *sabbat, sabat,* both <L *sabbatum* <Gk. *sabbaton* <Hebrew *shabbāth* <*shābath* rest] — **Sab·bat′ic** or **-i·cal** *adj.* **Sab·bat′i·cal·ly** *adv.*
— *Synonym:* Sunday. *Sabbath* carries a more direct reference to the Mosaic economy, with a suggestion of sacred rest that is not in the name *Sunday,* given by the heathen to the first day of the week. Compare FIRST DAY.

Sabbath school A Sunday school.
sab·bat·i·cal (sə·bat′i·kəl) *adj.* Of the nature of the Sabbath as a day of rest; offering rest at regular intervals. Also **sab·bat′ic**. — *n.* A sabbatical year. [< *sabbatic* < F *sabbatique* < Gk. *sabbatikos* < *sabbaton* SABBATH]
sabbatical year 1 In the ancient Jewish economy, every seventh year, in which the people were required to refrain from tillage. 2 A year's vacation awarded to teachers in some American educational institutions every seven years.
sa·be (sä′bē) *SW U.S. v.i.* To understand; know. — *n.* Understanding; knowledge. [< Sp. *saber* know]
Sa·be·an (sə·bē′ən) *adj.* Of or pertaining to ancient Sheba, its people, or their language. — *n.* 1 One of an ancient people of the kingdom of Sheba in SW Arabia in the first millennium B.C.: noted for their commerce and their wealth. 2 The Southwest Semitic language of these people. Also **Sa·bae′an**. [< L *Sabaeus* < Gk. *Sabaios* < *Saba* Sheba < Arabic *Saba*′ < Hebrew *Shebā*]
Sa·bel·li·an (sə·bel′ē·ən) *n.* A branch of the Italic subfamily of Indo-European languages, including the ancient Aequian, Marsian, Sabine, and Volscian. [< L *Sabellus* a Sabine]
sa·ber (sā′bər) *n.* A heavy one-edged cavalry sword, with a thick-backed blade, often curved. — *v.t.* **·bered** or **·bred**, **·ber·ing** or **·bring** To strike, wound, kill, or arm with a saber. Also spelled *sabre*. [< F *sabre*, *sable* < MHG *sabel*, prob. <Slavic]
sa·ber–toothed (sā′bər·tootht′) *adj.* Having very long, curved, upper canine teeth, likened to sabers.
saber–toothed tiger *Paleontol.* A large, ferocious, extinct carnivore (subfamily *Machaerodontinae*), characterized by very large, trenchant, upper canine teeth; especially, *Smilodon californicus*, common in the western hemisphere until its extinction in the Pleistocene.

SABER–TOOTHED TIGER
(About 3 feet high at the shoulders)

Sa·bi (sä′bē) A river in SE Africa, flowing 400 miles SE to the Indian Ocean: Portuguese *Save*.
Sa·bi·an (sā′bē·ən) *n.* One of an ancient religious sect dwelling in Mesopotamia and described in the Koran as monotheistic: identified by some with the Mandeans. — *adj.* Pertaining to the Sabians or to their religious worship. [< Arabic *sabi'ah* < Aramaic *tsebhaʹ* immerse, baptize] — **Saʹbi·an·ism** *n.*
sab·i·cu (sab′i·kōō′) Horseflesh (def. 3).
sa·bin (sā′bin) *n. Physics* A unit of sound absorption, equivalent to one square foot of a completely absorbing substance. [after W. C. W. *Sabine*, 1868–1919, U. S. physicist]
sab·ine (sab′in) See SAVIN.
Sa·bine (sab′in) *n.* 1 One of an ancient central Italian people, conquered and absorbed by Rome in 290 B.C., whose daughters the early Romans married by force. 2 The language of these people, belonging to the Sabellian branch of the Italic languages. — *adj.* Of or pertaining to the Sabines. [< L *Sabinus*]
Sabine River (sə·bēn′) A river in eastern Texas and western Louisiana, flowing 578 miles SE, passing through **Sabine Lake** (17 miles long) and entering the Gulf of Mexico through **Sabine Pass** (7 miles long).
Sa·bir (sa·bēr′) *n.* The lingua franca of the Mediterranean ports, largely a mixture of French, Italian, and Spanish. [< *Se ti sabir*, a jargon phrase in Molière's *Bourgeois Gentilhomme* < Provençal *sabir* knowledge, science]
sa·ble (sā′bəl) *n.* 1 A carnivore (*Martes zibellina*), of northern Asia and Europe, related to the marten. ◆ Collateral adjective: *zibeline*. 2 The dressed fur of a sable, specifically, of the Asian sable. 3 *pl.* Garments made wholly or partly of this fur. 4 The color black; hence, mourning or a mourning garment. 5 *Her.* Black: represented, when uncolored, by a network of lines crossing each other at right angles. — **Alaska sable** A trade name for natural or dyed skunk. — *adj.* 1 Black, especially as the color of mourning. 2 Made of or having the color of sable fur; dark-brown. See synonyms under DARK. [< OF *sable*, *saible* < Med. L *sabelum* <Slavic]
Sable, Cape 1 The southernmost point of the United States, at the SW tip of Florida. 2 The southernmost extremity of Nova Scotia, on an islet just south of **Cape Sable Island** (7 miles long, 3 miles wide).
sable antelope A large, black, African antelope (*Hippotragus niger*) having annular curved horns.
sa·ble·fish (sā′bəl·fish′) *n. pl.* **·fish** or **·fish·es** The coalfish (def. 2).
Sable Island An island SW of Nova Scotia; 30 miles by 2 miles; opposite Cape Sable, Nova Scotia.
sa·bot (sab′ō, *Fr.* sȧ·bō′) *n.* 1 A wooden shoe, as of a French peasant. 2 A shoe having a wooden sole but flexible shank. See GETA. 3 A disk formerly attached to a projectile to cause it to maintain its position in the bore of a firearm or to take the rifling of the gun. [< OF *sabot*, alter. of *savate* an old shoe, ult. <Arabic *sabbāt* a sandal; infl. in form by *bot* a boot]
sab·o·tage (sab′ə·täzh, *Fr.* sȧ·bô·tazh′) *n.* An act of malicious damage; deliberately poor workmanship intended to cause damage, obstruction of plans, aims, etc., as in secret resistance to an enemy: sometimes resorted to by workmen to secure compliance with demands. — *v.t. & v.i.* **·taged**, **·tag·ing** To engage in, damage, or destroy by sabotage. [< F *saboter* work badly, damage < *sabot* a sabot; with ref. to damage done to machinery with sabots]
sab·o·teur (sab·ə·tûr′, *Fr.* sȧ·bô·tœr′) *n.* One who engages in sabotage. [< F]
sa·bre (sā′bər) See SABER.
sa·bre·tache (sā′bər·tash, sab′ər-) *n.* A leather pocket hung from the sword belt of a mounted man. [< F <G *säbeltasche* < *säbel* a saber + *tasche* a pocket]
sab·u·lous (sab′yə·ləs) *adj.* Gritty, like sand; arenaceous. Also **sab′u·lose** (-lōs). [< L *sabulosus* < *sabulum* sand] — **sab′u·los′i·ty** (-los′ə·tē) *n.*
sac (sak) *n. Biol.* A membranous pouch; a cavity or receptacle: the ink *sac* of a squid. [< F < L *saccus*. Doublet of SACK.]
Sac (sak, sōk) See SAUK.
sac·a·ton (sak′ə·tōn′) *n.* A coarse perennial grass (*Sporobolus wrightii*) of the United States and Mexico, yielding hay. [< Sp. *zacatón* < *zacate*, *sacate* <Nahuatl *çacatl* a kind of grass]
sac·cate (sak′it, -āt) *adj.* 1 Sac-shaped. 2 Having a sac, bag, or pouch. [< Med. L *saccatus* < L *saccus* a sack]
sac·cha·rate (sak′ə·rāt) *n. Chem.* 1 A salt of saccharic acid. 2 A compound of a sugar with a metallic oxide.
sac·char·ic (sak·ar′ik) *adj. Chem.* 1 Of, pertaining to, or derived from sugar or a sweetish substance. 2 Designating a dibasic acid, $C_6H_{10}O_8$, obtained by the oxidation of glucose and other sugars: it occurs in three optically different forms.
sac·cha·ride (sak′ə·rīd, -rid) *n. Chem.* Any of a class of carbohydrates containing sugar, as a *monosaccharide*, *polysaccharide*, etc.
sac·char·i·fy (sə·kar′ə·fī, sak′ər·ə·fī) *v.t.* **·fied**, **·fy·ing** To convert, as starches, into sugar; impregnate with sugar. [< SACCHAR(O)- + (I)FY] — **sac·char′i·fi·ca′tion** *n.*
sac·cha·rim·e·ter (sak′ə·rim′ə·tər) *n.* A polariscope for detecting the strength or concentration of sugar in a solution. [< F *saccharimètre* <Gk. *sakchari* sugar + *metron* measure] — **sac′cha·rim′e·try** *n.*
sac·cha·rin (sak′ər·in) *n. Chem.* A white crystalline compound, $C_7H_5O_3NS$, derived from toluene. It is 300 to 500 times sweeter than cane sugar, and is used as a sweetening agent, especially by diabetics. Also spelled *saccharine*. [< Med. L *saccharum* sugar (< *saccharon* <Gk. *sakcharon*, ult. <Skt. *sharkarā* grit, gravel, sugar) + -IN]
sac·cha·rine (sak′ər·in, -ēn) *adj.* 1 Of, pertaining to, or of the nature of sugar; sweet. 2 Ingratiatingly or cloyingly sweet. — *n.* Saccharin. [< SACCHAR(O)- + -INE[1]] — **sac′cha·rine·ly** *adv.* — **sac·cha·rin′i·ty** *n.*
sac·cha·rize (sak′ə·rīz) *v.t.* **·rized**, **·riz·ing** *Chem.* To convert into sugar; ferment. — **sac′·cha·ri·za′tion** *n.*
saccharo- *combining form* Sugar; of or pertaining to sugar: *saccharometer*. Also, before vowels, **sacchar-**. [< Gk. *sakcharon* sugar]
sac·cha·roid (sak′ə·roid) *adj.* 1 Resembling sugar. 2 *Geol.* Having crystalline granular structure: *saccharoid* marble. Also **sac′cha·roi′dal** (-roid′l).
sac·cha·rom·e·ter (sak′ə·rom′ə·tər) *n.* A hydrometer for determining the concentration of sugar in saccharine solutions.
sac·cha·ro·my·ce·tous (sak′ə·rō·mī·sē′təs, -mī′sə-) *adj.* Of or pertaining to a genus (*Saccharomyces*) of fungi, the yeast family, commonly unicellular, but sometimes developing a septate mycelium. Several produce endogenous spores, while most of them cause alchoholic fermentation with evolution of carbon dioxide. [< NL <Gk. *sakcharon* sugar + *mykēs*, -*ētos* a mushroom, fungus]
sac·cha·rose (sak′ə·rōs) *n.* Sucrose.
Sac·co (sak′ō, *Ital.* säk′kō), **Nicola**, 1891–1927, philosophical anarchist in the United States who with *Bartolomeo Vanzetti* was convicted of murder in connection with a payroll robbery, and executed. Their trial aroused international protest because it was thought to have been influenced by political considerations.
sac·cu·late (sak′yə·lāt) *adj.* Formed into a series of saclike expansions; dilated and constricted alternately. Also **sac′cu·lat′ed**. [< SACCULUS + -ATE[1]]
sac·cule (sak′yōōl) *n.* 1 A little sac. 2 *Anat.* Part of the membranous labyrinth of the ear. [< L *sacculus* a sacculus]
sac·cu·lus (sak′yə·ləs) *n. pl.* **·li** (-lī) A small sac or pouch; a saccule. [< L, dim. of *saccus* a sack]
sac·er·do·tal (sas′ər·dōt′l) *adj.* 1 Pertaining to a priest or priesthood; priestly. 2 Believing in the divine authority of the priesthood. [< OF <L *sacerdotalis* < *sacerdos*, *-dotis* a priest < *sacer* holy + *do-*, stem of *dare* give] — **sac′er·do′tal·ly** *adv.*
sac·er·do·tal·ism (sas′ər·dōt′l·iz′əm) *n.* 1 The character and methods of the priesthood; priestcraft. 2 Zeal for priestly things.
sa·chem (sā′chəm) *n.* 1 A North American Indian hereditary chief. 2 Any chief; the head of a political party; specifically, one of the leaders of the Tammany Society in New York. See synonyms under CHIEF. [< Algonquian (Narraganset). Akin to SAGAMORE.]
sa·chet (sa·shā′, *esp. Brit.* sash′ā) *n.* A small ornamental bag for perfumed powder. [< OF, dim. of *sac* < L *saccus* a sack]
Sachs (zäks), **Hans**, 1494–1576, German shoemaker, poet, and playwright; also, the hero of Wagner's *Die Meistersinger*.
Sach·sen (zäkh′sən) The German name for SAXONY.
sack[1] (sak) *n.* 1 A bag for holding bulky articles. 2 A measure or weight of varying amount. 3 A loose garment with sleeves worn by women: also spelled *sacque*. 4 *Slang* Dismissal: especially in the phrases **to get the sack, to give (someone) the sack.** 5 In baseball slang, a base. 6 *Slang* A bed; mattress. — **to be left holding the sack** *Slang* To be left to take the consequences of a bad situation. — *v.t.* 1 To put into a sack or sacks. 2 To dismiss, as a servant. [OE *sacc* < L *saccus* <Gk. *sakkos* <Hebrew *saq* sackcloth, a grain sack. Doublet of SAC.] *Sack* may appear as a combining form in hyphemes or solidemes, or as the first element in two-word phrases, with the meaning of the noun (def. 1):

sack baler	sack-formed
sack baling	sackmaker
sack carrier	sackmaking
sack checker	sackman
sack cleaner	sack mender
sack cutter	sack repairer
sack emptier	sack-shaped
sack examiner	sack sorter

sack[2] (sak) *v.t.* To plunder or pillage (a town or city) after capturing. — *n.* 1 The pillaging of a captured town or city. 2 Loot; booty obtained by pillage. [<MF *sac* <Ital. *sacco,* orig. plunder <Med. L *saccare* pillage <L *saccus* a sack; from the use of sacks in carrying off plunder] — **sack'er** *n.*

sack[3] (sak) *n.* Light-colored Spanish dry wine; also, any strong white wine from southern Europe. [Earlier *(wyne)seck* <F *(vin) sec* a dry (wine) <L *siccus* dry]

sack·but (sak'but) *n.* 1 A primitive instrument resembling the trombone. 2 In the Bible, a stringed instrument. [<MF *saquebute,* orig. a hooked lance for horseback fighting <OF *saquer* pull + *bouter* push]

sack·cloth (sak'klôth', -kloth') *n.* 1 A coarse cloth used for making sacks. 2 Coarse cloth or haircloth worn in penance.

sack coat A man's short, loose-fitting coat with no waist seam, for informal wear.

sack·ful (sak'fo͝ol') *n. pl.* **·fuls** Enough to fill a sack.

sack·ing (sak'ing) *n.* A coarse cloth made of hemp or flax and used for sacks; bagging.

sack posset A posset formerly brewed with sack.

sack race A race in which each contestant has a sack tied over both feet.

sack suit A man's suit having a sack coat.

Sack·ville (sak'vil), **Thomas,** 1536–1608, first Earl of Dorset and Baron Buckhurst, English poet and diplomat.

Sack·ville-West (sak'vil·west'), **V(ictoria Mary),** born 1892, English novelist and poet.

sacque (sak) See SACK[1] (def. 3).

sa·cral[1] (sā'krəl) *adj.* Of, pertaining to, or situated near the sacrum. — *n.* A sacral vertebra or nerve. [<NL *sacralis* <*sacrum* SACRUM]

sa·cral[2] (sā'krəl) *adj.* Pertaining to sacred rites. [<L *sacrum* a rite, orig. neut. sing. of *sacer* holy]

sac·ra·ment (sak'rə·mənt) *n.* 1 *Eccl.* A rite ordained by Christ or by the church as an outward and visible sign of an inward and spiritual grace: in the Greek Church, also called *mystery.* Traditionally they are seven in number (baptism, the Eucharist, confirmation, matrimony, orders, penance, and unction) in the Greek, Roman Catholic, and some other churches; since the Reformation only two of these (baptism and the Eucharist) are recognized by most Protestant churches. 2 *Often cap. Eccl.* **a** The Eucharist; the Lord's Supper. **b** The consecrated bread and wine of the Eucharist: often with *the.* See BLESSED SACRAMENT. 3 Any sign or token of a solemn covenant or pledge. 4 Any thing considered to have a secret or mysterious meaning. [<OF *sacrement* <LL *sacramentum* a mystery <L, an oath, pledge <*sacrare.* See SACRED.]

Synonyms: ceremony, communion, Eucharist, observance, ordinance, rite, service, solemnity. A *ceremony* is a form expressing reverence, or respect; as, religious *ceremonies,* the *ceremonies* of a coronation or of a wedding. An *observance* has more than a formal obligation, approaching a religious sacredness; a religious *observance* viewed as established by authority is called an *ordinance*; viewed as an established custom, it is a *rite.* Any religious act, especially a public act, viewed as a means of serving God is called a *service. Sacrament* and *ordinance* in the religious sense are often used interchangeably; the *ordinance* derives its sacredness from the authority that ordained it, while the *sacrament* possesses a sacredness due to something in itself, even when viewed simply as a memorial. The Lord's Supper is the Scriptural name for the *observance* commemorating the death of Christ; the word *communion* is once applied to it (I *Cor.* x 16). *Eucharist,* called *The Sacrament,* describes the Lord's Supper as a thanksgiving *service.*

sac·ra·men·tal (sak'rə·men'təl) *n.* 1 One of certain rites, such as the use of holy water, oil, or salt, employed as adjuncts to sacraments, or regarded as analogous to a sacrament. 2 *pl.* The objects, words, or ceremonies used in administering a sacrament. — *adj.* 1 Of or pertaining to a sacrament; constituting or composing a sacrament; having the influence or efficacy of a sacrament. 2 Consecrated, as by sacred vows: the *sacramental* host of God's elect. — **sac'ra·men'tal·ism** *n.* — **sac'ra·men'tal·ist** *n.* — **sac'ra·men'tal·ly** *adv.*

sac·ra·men·tar·i·an (sak'rə·men·târ'ē·ən) *n.* One who regards the sacraments as channels of divine grace. Also **sac'ra·men'ta·rist, sac'·ra·ment·er.** — *adj.* Of or pertaining to a sacrament or sacraments, or to sacramentarians. — **sac'ra·men·tar'i·an·ism** *n.*

Sac·ra·men·tar·i·an (sak'rə·men·târ'ē·ən) *n.* One who regards the sacraments as simply symbols or signs: the name given to Calvinists and followers of Zwingli.

sac·ra·men·ta·ry (sak'rə·men'tər·ē) *n. pl.* **·ries** Any of several books containing the ritual for mass, the sacraments, and various other rites: now used only as a basis for modern rituals.

Sac·ra·men·to (sak'rə·men'tō) The capital of California, on the **Sacramento River,** the largest river in the State, flowing 382 miles south from Central Valley to Suisun Bay.

sa·crar·i·um (sə·krâr'ē·əm) *n. pl.* **·i·a** (-ē·ə) 1 Any sacred or secluded place or shrine of the ancient Romans where venerated things were deposited. 2 The sanctuary of a church. 3 In the Roman Catholic Church, a piscina. [<L <*sacer* holy, sacred]

sa·cred (sā'krid) *adj.* 1 Set apart or dedicated to religious use; hallowed: a *sacred* edifice. 2 Pertaining or related to deity, religion, or hallowed places or things. 3 Consecrated by love or reverence; dedicated to a person or purpose. 4 Entitled to reverence or respect; not to be profaned; inviolable. 5 *Rare* Set apart for evil; accursed. See synonyms under HOLY. [Orig. pp. of obs. *sacre* consecrate <OF *sacrer* <L *sacrare* <*sacer* holy] — **sa'·cred·ly** *adv.* — **sa'cred·ness** *n.*

Sacred College See COLLEGE OF CARDINALS.

sac·ri·fice (sak'rə·fīs) *n.* 1 The act of making an offering to a deity, in worship or atonement. 2 That which is sacrificed; a victim. 3 A giving up of some cherished or desired object. 4 Loss incurred or suffered without return; destruction, as of life. 5 A reduction of price that leaves little or no profit or involves loss. 6 In baseball, a sacrifice hit. — *v.* **·ficed, ·fic·ing** *v.t.* 1 To make an offering or sacrifice of, as to a god or deity in propitiation, supplication, etc. 2 To give up, yield, permit injury to, or relinquish (something valued) for the sake of something else, as a person, thing, or idea. 3 To sell at a reduced price; part with at a loss. 4 In baseball, to advance (one or more runners) by means of a sacrifice hit. — *v.i.* 5 To make a sacrifice. 6 To make a sacrifice hit. See synonyms under SURRENDER. [<OF <L *sacrificium* <*sacra* rites, orig. neut. pl. of *sacer* holy + *facere* make] — **sac'ri·fic'er** *n.* — **sac'·ri·fic'ing·ly** *adv.*

sacrifice hit In baseball, a hit by which the batter is retired but by which a base runner is advanced another base, the batter not being charged with a time at bat: when batted into the air also called **sacrifice fly.**

sac·ri·fi·cial (sak'rə·fish'əl) *adj.* Pertaining to, performing, or of the nature of a sacrifice. — **sac'ri·fi'cial·ly** *adv.*

sac·ri·lege (sak'rə·lij) *n.* The act of violating or profaning anything sacred, including sacramental vows. [<OF <L *sacrilegium* <*sacrilegus* a temple robber <*sacer* holy + *legere* gather] — **sac'ri·le'gist** (-lē'jist) *n.*

sac·ri·le·gious (sak'rə·lij'əs, -lē'jəs) *adj.* 1 Having committed, or being ready to commit, sacrilege; impious. 2 Of the nature of sacrilege. See synonyms under PROFANE. — **sac'ri·le'gious·ly** *adv.* — **sac'ri·le'gious·ness** *n.*

sa·cring bell (sā'kring) A small bell rung at the elevation during mass; the tolling of the church bell at this time; the Sanctus bell. [<*sacring,* ppr. of obs. *sacre* consecrate + BELL]

sa·crist (sā'krist) *n.* 1 A sacristan. 2 A person who takes charge of choir books and copy music; also, a sexton. [<OF *sacrist* <L *sacrista* <*sacra* holy (objects), neut. pl. of *sacer*]

sac·ris·tan (sak'ris·tən) *n.* An officer having charge of the sacristy of a church or religious house and its contents, and of the proper arrangement of all objects needed for divine service. The sacristan of a cathedral is commonly in orders. Compare SEXTON. [<Med. L *sacristanus* <L *sacrista.* Doublet of SEXTON.]

sac·ris·ty (sak'ris·tē) *n. pl.* **·ties** A room in a religious house for the sacred vessels and vestments; a vestry. [<F *sacristie* <Med. L *sacristia* <L *sacrista* a sacrist]

sacro– *combining form Med.* Near, or related to the sacrum: *sacrosciatic.* [<L *(os) sacrum* the sacral (bone)]

sac·ro·il·i·ac (sak'rō·il'ē·ak) *adj. Anat.* Pertaining to the sacrum and the ilium and to the joints or ligaments connecting them. [<SACRO- + ILIAC]

sac·ro·sanct (sak'rō·sangkt) *adj.* Peculiarly and exceedingly sacred; inviolable; preeminent for sanctity: sometimes used ironically. [<L *sacrosanctus* <*sacro,* ablative of *sacrum* a rite (<*sacer* holy) + *sanctus,* pp. of *sancire* make holy, inviolable] — **sac'ro·sanc'ti·ty** *n.*

sa·cro·sci·at·ic (sā'krō·sī·at'ik) *adj. Anat.* Of or pertaining to the sacrum and the ischium: the *sacrosciatic* ligaments, connecting the sacrum and the hip bone.

sa·crum (sā'krəm) *n. pl.* **·cra** (-krə) *Anat.* A composite bone formed by the union of the five vertebrae between the lumbar and caudal regions, constituting the dorsal part of the pelvis. [<NL <L *(os) sacrum* sacred (bone); from its being offered in sacrifices]

Sa·cy (sà·sē'), **Baron de,** 1758–1838, Antoine Isaac Silvestre, French Oriental scholar.

sad (sad) *adj.* **sad·der, sad·dest** 1 Sorrowful or depressed in spirits; expressing, or having the external appearance of grief or sorrow; unhappy; mournful; gloomy. 2 Causing sorrow or pity; distressing; unfortunate. 3 *Dial.* Heavy; soggy: said of food. 4 *Colloq.* Vexatious, mischievous, or bad: often humorously or as a mild intensive: That boy is a *sad* tease. 5 Dark-hued; somber. [OE *sæd,* orig. sated] — **sad'ly** *adv.* — **sad'ness** *n.*

Synonyms: afflicted, dejected, depressed, desolate, despondent, disconsolate, dismal, distressed, doleful, downcast, dreary, dull, gloomy, grave, heavy, lugubrious, melancholy, miserable, mournful, sober, somber, sorrowful, sorry, unhappy, woebegone, woeful. *Sad, melancholy, unhappy,* and many similar words may be used either of the personal experience of grief, sorrow, mental depression, etc., or of that which causes grief or pain; a person is *sad* on account of a *sad* event. See synonyms under BAD. *Antonyms:* see synonyms for HAPPY.

sad·den (sad'n) *v.t.* & *v.i.* To make or become sad or unhappy.

sad·dle (sad'l) *n.* 1 A seat or pad for a rider, as on the back of a horse or on a bicycle. 2 A padded cushion for a horse's back, as part of a harness or to support a pack, etc. For illustration see HARNESS. 3 The two hindquarters of a carcass, as of mutton, veal, or venison; also, the undivided loins of such a carcass. 4 Some part like or likened to a saddle, as the lower part of the back of a fowl. See illustration under FOWL. 5 *Geog.* A depression across the summit of a ridge; a pass. 6 *Meteorol.* A low-pressure area between two anticyclones; a col. 7 Something resembling a saddle in form or position, as a bearing for a car axle. — **in the saddle** In control. — *v.* **·dled, ·dling** *v.t.* 1 To put a saddle on: to *saddle* a horse. 2 To load, as with a burden. 3 To place as a burden or responsibility: with *upon.* — *v.i.* 4 To get into a saddle. [OE *sadol*]

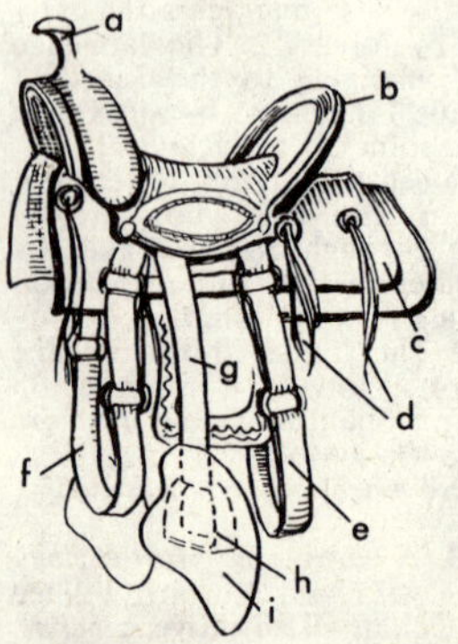

NOMENCLATURE—AMERICAN STOCK SADDLE
a. Pommel or horn.
b. Cantle.
c. Saddle.
d. Saddle strings.
e. Back cinch.
f. Front cinch.
g. Stirrup strap or leather.
h. Stirrup.
i. Tapadera or stirrup hood.

saddle bags A pair of pouches connected by a strap or band and slung over an animal's back or attached to a saddle.

saddle band A remuda.

sad·dle·bow (sad'l·bō') *n.* The arched front upper part of a saddletree.

sad·dle·cloth (sad'l·klôth', -kloth') *n.* A cloth

under and attached to a saddle, or one under the saddle of a harness.
saddle horse A horse used with or trained for the saddle.
sad·dler (sad′lər) *n.* 1 A maker of saddles, harness, etc. 2 A saddle horse.
saddle roof A roof consisting of two gables and one ridge.
sad·dler·y (sad′lər·ē) *n. pl.* **·dler·ies** 1 Saddles, harness, and fittings, collectively. 2 A saddler's shop. 3 The business of a saddler.
saddle shoe A white sport shoe with a dark band of leather across the instep.
saddle soap A softening and preserving soap for leather, containing pure white soap, usually Castile, and neat's-foot oil.
sad·dle·tree (sad′l·trē′) *n.* 1 The frame of a saddle. 2 The tuliptree: so called from its saddle-shaped leaf.
Sad·du·cee (saj′ŏŏ·sē, sad′yŏŏ·sē) *n.* A member of a strict Jewish school that arose in the second century B.C., and later became skeptical and traditionalistic, adhering only to the Mosaic law. [<LL *Sadducaeus* <Gk. *Saddoukaios* <Hebrew *tsaddūgī*, appar. after *Tsaddūg* Zadok (*Ezek.* xl 46)] — **Sad′du·ce′an, Sad′du·cae′an** *adj.* — **Sad′du·cee′ism** *n.*
sa·de (sä·dā′) *n.* The eighteenth Hebrew letter: also spelled *tsade.* Also **sa·dhe′**. See ALPHABET.
Sade (sȧd), **Comte Donatien de,** 1740–1814, French writer and libertine: known as *Marquis de Sade.*
Sa·di (sä·dē′) See SAADI.
sad–i·ron (sad′ī′ərn) *n.* A flatiron for smoothing clothes: distinguished from a *box-iron.* [<SAD, in obs. sense "heavy" + IRON]
sad·ism (sā′diz·əm, sad′iz·əm) *n. Psychiatry* 1 Sexual gratification obtained through the infliction of pain upon others. 2 A morbid delight in being cruel. [after Comte Donatien de *Sade*; with reference to the various sexual aberrations described in his writings] — **sad·ist** (sā′dist, sad′ist) *n.* & *adj.* — **sa·dis·tic** (sə·dis′tik, sā-) *adj.* — **sa·dis′ti·cal·ly** *adv.*
Sa·do·va (sä′dô·vä) A town in southern Bohemia, Czechoslovakia; scene of the culminating defeat of the Austrians by Prussian forces, 1866. *German* **Sa′do·wa.**
sae (sā) *adv. Scot.* So.
sa·fa·ri (sə·fä′rē) *n.* 1 An expedition or journey, often on foot, as for hunting. 2 The caravan and animals employed in this; also, a day's march: also spelled *suffari.* [<Swahili <Arabic *safara* travel]
safe (sāf) *adj.* 1 Free or freed from danger or evil. 2 Having escaped injury or damage; unharmed. 3 Not hazardous; not involving risk or loss; also, conferring safety; of persons, trusty; prudent. 4 Not likely to disappoint; free from doubt or error: It is *safe* to promise. 5 Not likely to cause or do harm or injury. 6 In politics, adhering to party principles; to be depended on to support certain interests: said of a candidate; also, sure to vote for a certain candidate: said of a district. 7 In baseball, having reached base without being retired: He was ruled *safe* at second. See synonyms under SECURE. — *n.* 1 A strong iron-and-steel receptacle, usually fireproof, for protecting valuables, as money or jewels. 2 Any place of safe storage, as a room, tank, refrigerator, or box, for preserving perishable articles, as meat or fish. [<OF *sauf* <L *salvus* whole, healthy] — **safe′ly** *adv.* — **safe′ness** *n.*
safe–blow·ing (sāf′blō′ing) *n.* The act of using explosives to open a safe to be robbed. — **safe′-blow′er** *n.*
safe–break·er (sāf′brā′kər) *n.* A safe-cracker.
safe–con·duct (sāf′kon′dukt) *n. Law* 1 An official document assuring protection on a journey or voyage, as in time of war; a passport. 2 The act of conducting in safety. — *v.t.* (sāf′kən·dukt′) 1 To convoy in safety. 2 To provide with a safe-conduct.
safe–crack·er (sāf′krak′ər) *n.* One who breaks into safes to rob them.
safe–crack·ing (sāf′krak′ing) *n.* The breaking open of safes for robbery.
safe deposit A room, vault, or other fireproof storage place for valuables.
safe–de·pos·it box (sāf′di·poz′it) A box, safe, drawer, or other fireproof receptacle for valuable jewelry, papers, etc., generally in a bank.
safe·guard (sāf′gärd′) *n.* 1 One who or that which guards or keeps in safety, as an escort, guard, or safe-conduct. 2 A mechanical device designed to prevent accident or injury. See synonyms under DEFENSE. — *v.t.* To defend; protect; guard.
safe–hand (sāf′hand′) *n.* 1 A safe method of transmitting official secret papers. 2 A trustworthy courier.
safe hit In baseball, a fair hit by which the batter reaches first base.
safe·keep·ing (sāf′kē′ping) *n.* The act or state of keeping or being kept in safety; protection.
safe·ty (sāf′tē) *n. pl.* **·ties** 1 The state or condition of freedom from danger or risk. 2 Freedom from injury. 3 A device or catch designed as a safeguard, as in a firearm. 4 In football, the act or play of touching the ball to the ground behind the player's own goal line when the impetus which sent the ball over the goal line was given to it by one of his own side: also **safe′ty-touch′down′** (-tuch′doun′). 5 In baseball, a safe hit.
safety belt 1 An extensible strap encircling the user and a permanently fixed object so that the user may move freely but be safe from falling or slipping: used by linemen, window cleaners, etc. 2 *Aeron.* A strap fixed to the seat of an aircraft by which the passenger is secured against sudden shocks or turning movements. 3 A life belt.
safety bicycle A bicycle having equal or nearly equal wheels, operated by pedal cranks communicating with the driving wheel.
safety glass See under GLASS.
Safety Islands A group of three islands off the coast of French Guiana, including Devil's Island. *French* **Îles du Sa·lut** (ēl dü sȧ·lü′).
safety lamp 1 A miner's lamp having the flame surrounded by fine wire gauze, which prevents the ignition of explosive gases: called a *davy* from its inventor, Sir Humphry Davy. 2 A specially protected incandescent electric lamp.
safety lever *Mech.* 1 A device for controlling the movement of machine parts. 2 A similar contrivance for preventing the accidental discharge of a grenade, automatic pistol, etc.: also **safety catch.**
safety match A match that will ignite only when struck upon a chemically prepared surface.
safety pin 1 A pin whose point springs into place within a protecting sheath. 2 A pin which prevents the premature detonation of a hand grenade.
safety razor See under RAZOR.
safety valve 1 *Mech.* A valve in a steam boiler, etc., for automatically relieving excessive pressure. 2 Any outlet for pent-up energy or emotion.
saf·flow·er (saf′lou′ər) *n.* 1 A thistlelike herb (*Carthamus tinctorius*) about 2 feet high, with spiny heads of orange-red flowers. 2 The dried flower heads of this plant pressed into small cakes for export: also **safflower cake.** 3 The reddish dyestuff obtained from the dried flowers. [<Du. *saffloer* <OF *saffleur, safour* <Ital. *saffiore*; infl. in form by SAFFRON and FLOWER]
saf·fron (saf′rən) *n.* 1 An autumn-flowering species of crocus (*Crocus sativus*). 2 The dried orange-colored stigmas of this plant used for coloring confectionery, varnishes, etc., and in parts of the Old World as a flavoring and coloring ingredient in cookery. 3 A deep yellow orange: also **saffron yellow.** — *adj.* Of the orange color of saffron. [<OF *safran* <Sp. *azafran* <Arabic *az-za'farān* the saffron]
Sa·fi (sä′fē) A port in Morocco, SW of Casablanca. Formerly **Saf′fi.**
saf·ra·nine (saf′rə·nēn, -nin) *n. Chem.* 1 Any of a class of basic compounds considered as symmetrical diamine derivatives of the azo group bases. Their salts form important dyes. 2 Any of various mixtures of safranine salts used in dyeing. See SAFFRON. Also **saf′ra·nin** (-nin). [<F *safran* SAFFRON + -INE[2]]
saf·role (saf′rōl) *n. Chem.* A poisonous liquid, $C_{10}H_{10}O_2$, forming a large portion of sassafras oil: contained also in other oils, as oil of camphor: used in medicine and perfumery. Also **saf′rol.** [<F *safran* SAFFRON + -OLE[1]]
saft (saft, säft) *adj. Scot.* 1 Soft. 2 Wet: said of the weather.
sag (sag) *v.* **sagged, sag·ging** *v.i.* 1 To bend or sink downward from weight or pressure, especially in the middle. 2 To hang unevenly. 3 To lose firmness or determination; weaken, as from exhaustion, age, etc. 4 To decline, as in price or value. 5 *Naut.* To drift. — *v.t.* 6 To cause to sag. — *n.* 1 A sagging, or its extent or degree; a sagging place or part, as of a roof. 2 *Naut.* A sidewise drift, as of a vessel. 3 A depressed or sunken place in flat land; a marsh. [ME *saggen*, ? <MDu. *zakken* subside, ? <dial. ON (nautical) *sakka* plummet]
sa·ga (sä′gə) *n.* 1 A medieval Scandinavian (specifically, Icelandic) prose narrative of conventionalized form dealing with legendary or historical exploits, usually of a single hero or a single family. 2 A story, sometimes poetic, having the saga form or manner, often chronicling the history of a family, as Galsworthy's *Forsyte Saga.* [<ON, history, narrative. Akin to SAW[2].]
sa·ga·cious (sə·gā′shəs) *adj.* 1 Ready and apt to apprehend and to decide on a course. 2 Characterized by discernment, shrewdness, and wisdom. 3 Quick of scent, as a hound. [<L *sagax, sagacis* wise, foreseeing] — **sa·ga′cious·ly** *adv.* — **sa·ga′cious·ness** *n.*
Synonyms: able, acute, apt, clear-sighted, discerning, intelligent, judicious, keen, keen-sighted, keen-witted, perspicacious, quick-witted, rational, sage, sensible, sharp, sharp-witted, shrewd, wise. *Sagacious* refers to a power of tracing the hidden or recondite by slight indications, as by instinct or intuition; with reference to inferior animals it is often applied to special keenness of sense-perception as of a hound in following a trail. In human affairs *sagacious* refers to a power of ready, far-reaching, and accurate inference from observed facts, perhaps in themselves very slight, that seems like a special sense; or to a similar readiness to foresee the results of any action, a kind of prophetic common sense, especially upon human motives or conduct. *Sagacious* is a broader word than *shrewd*, and not capable of the invidious sense which the latter often bears; on the other hand, *sagacious* is less lofty than *wise* in its full sense, and more limited to practical matters. See ACUTE, ASTUTE, KNOWING, POLITIC, WISE[1]. *Antonyms*: absurd, dull, foolish, futile, ignorant, irrational, obtuse, senseless, silly, simple, sottish, stupid, unintelligent.
sa·gac·i·ty (sə·gas′ə·tē) *n.* The quality of being sagacious; discernment and judgment; shrewdness. See synonyms under ACUMEN, WISDOM. [<MF *sagacité* <L *sagax* sagacious]
sa·ga·man (sä′gə·man′, -mən) *n.* The author, singer, or narrator of a saga; a Scandinavian poet or bard. [Trans. of ON *sögumadhr* < *sögu*, genitive of *saga* a saga + *madhr* a man]
Sa·ga·mi Sea (sä·gä·mē) An inlet of the Philippine Sea in central Honshu island, Japan.
sag·a·more (sag′ə·môr, -mōr) *n.* A tribal or lesser chief among the Algonquian Indians of North America, usually inferior to *sachem.* [<Algonquian (Penobscot) *sagamo.* Akin to SACHEM.]
sag·a·nash (sag′ə·nash) *n.* A white man: an Algonquian Indian term. [<Algonquian *sagannash*]
sage[1] (sāj) *n.* A venerable man of recognized experience, prudence, and foresight; a profoundly wise counselor or philosopher. — *adj.* 1 Characterized by or proceeding from calm, far-seeing wisdom and prudence. 2 Befitting a sage; profound; learned; also, grave; serious; shrewd. See synonyms under SAGACIOUS, WISE[1]. [<OF *saige, savie*, ult. <L *sapiens, -entis* wise, ppr. of *sapere* be wise. Doublet of SAPIENT.] — **sage′ly** *adv.* — **sage′ness** *n.*
sage[2] (sāj) *n.* 1 A plant of the mint family (genus *Salvia*), especially the common garden sage (*S. officinalis*), a stiff, shrubby perennial with gray-green leaves and purple, blue, or white flowers: used for flavoring meats, etc. 2 A light, greenish-gray color, like the color of sage leaves. 3 Any other plant of the genus *Salvia*, as the scarlet sage (*S. splendens*). 4 The Jerusalem sage (genus *Phlomis*), also of the mint family. 5 The sagebrush. [<OF

add, āce, câre, pälm; end, ēven; it, īce; odd, ōpen, ôrder; tŏŏk, pōōl; up, bûrn; ə = a in *above*, e in *sicken*, i in *clarity*, o in *melon*, u in *focus*; yōō = u in *fuse*; oi, oil; ou, pout; ch, check; g, go; ng, ring; th, thin; th, this; zh, vision. Foreign sounds à, œ, ü, kh, ṅ; and ◆: see page xx. < from; + plus; ? possibly.

sauce < L *salvia*, ? < *salvus* safe; with ref. to its reputed healing powers]
Sage (sāj), **Russell**, 1816-1906, U.S. financier and philanthropist.
sage·brush (sāj′brush′) *n*. An aromatic, bitter, typically perennial herb or small shrub (genus *Artemisia*) of the composite family, widely distributed on the alkali plains of the western United States; especially, *A. tridentata*, the State flower of Nevada. The Old World species are called *wormwood*.
Sagebrush State Nickname of NEVADA.
sage cock See COCK OF THE PLAINS.
sage hen 1 The female of the sage cock. See GROUSE. **2** *Usually cap.* A native of Nevada: a nickname.
sage sparrow A small, pale-gray, fringilline bird of the western United States (*Amphispiza nevadensis*).
sag·gar (sag′ər) *n*. **1** A vessel of baked fireproof clay in which are fired delicate pieces of pottery that would be injured by direct exposure to the heat. **2** Clay used for making saggars. Also spelled *seggar*. Also **sag′gard** (-ərd). —*v.t.* To place or treat in a saggar, as pottery. Also **sag′ger**. [Contraction of SAFEGUARD]
Sa·ghal·ien (sə-gäl′yən) A former spelling of SAKHALIN.
Sag·i·naw (sag′ə-nô) A city in east central Michigan, a port of entry on the **Saginaw River**, which flows about 22 miles into Saginaw Bay just below Bay City.
Saginaw Bay A SW arm of Lake Huron, extending 60 miles into eastern Michigan.
Sa·git·ta (sə-jit′ə) *Astron.* A northern constellation between Aquila and Cygnus; the Arrow. See CONSTELLATION. [< L, lit., an arrow]
sag·it·tal (saj′ə-təl) *adj.* **1** Pertaining to or resembling an arrow or arrowhead. **2** *Anat.* **a** Straight: the *sagittal* suture between the two parietal bones of the skull. **b** Of or pertaining to the longitudinal plane dividing an animal into right and left halves. [< L *sagitta* an arrow] —**sag′it·tal·ly** *adv.*
Sag·it·ta·ri·us (saj′ə-târ′ē-əs) **1** A zodiacal constellation, pictured as a centaur shooting an arrow; the Archer. See CONSTELLATION. **2** The ninth sign of the zodiac, with the symbol ♐. [< L, lit., an archer < *sagitta* an arrow]
sag·it·tate (saj′ə-tāt) *adj. Bot.* Shaped like an arrowhead, as certain leaves. Also **sag′it·tat′ed**, **sa·git·ti·form** (sə-jit′ə-fôrm). [< L *sagitta* an arrow]
sa·go (sā′gō) *n.* Any of several varieties of East Indian palm (genus *Metroxylon*). **2** The dried, powdered pith of this palm, used as a thickening agent in puddings, etc. [< Malay *sāgū*]
Sa·guache (sə·wach′) See SAWATCH MOUNTAINS.
sa·gua·ro (sə-gwä′rō, -wä′-) *n. pl.* **·ros** The giant cactus of the SW United States (*Cereus giganteus*): its blossom is the State flower of Arizona. Also **sa·hua′ro** (-wä′-). [< Sp. < Piman]
Sag·ue·nay River (sag′ə-nā′) A river in southern Quebec province, Canada, flowing 110 miles east from Lake St. John to the St. Lawrence River; total length, including sections above and traversing Lake St. John, about 475 miles.
Sa·guia el Ham·ra (sä′gyä el häm′rä) A territory of Spanish West Africa; 32,047 square miles. Also **Se·kia el Hamra** (se′kyä).
sa·gum (sā′gəm) *n. pl.* **·ga** (-gə) The ancient Roman soldiers' military cloak; as the toga was of peace. [< L, ? ult. < Celtic]
Sa·gun·to (sä-gōōn′tō) An ancient town in eastern Spain near the Gulf of Valencia; destroyed by Hannibal, 219 B.C.: formerly *Murviedro*. Ancient **Sa·gun·tum** (sə-gun′təm).
sag·y (sāj′ē) *adj.* Flavored or seasoned with or like sage.
Sa·ha·ra (sə-har′ə, -hâr′ə, -hä′rə) The world's largest desert area, extending from the Atlantic to the Red Sea in northern Africa; about 3,000,000 square miles. Also **Sahara Desert**.
Sa·ha·ran·pur (sə-hä′rən-pōōr′) A city in northern Uttar Pradesh State, India.
Sa·hib (sä′ib) *n.* Master; lord; Mr.; sir: used in India by natives in speaking of or addressing Europeans, also by Hindus and Moslems for people of rank: *Raja Sahib*. Also **Sa′heb**.

[< Urdu *sāhib* < Arabic *ṣāḥib*, lit., a friend]
saice (sīs) See SYCE.
said¹ (sed) Past tense and past participle of SAY¹. —*adj. Law* Previously mentioned; aforesaid.
sa·id² (sä′yid, sī′id) See SAYID.
Sa·i·da (sä′ē-dä) A port in SW Lebanon: ancient *Sidon*: also *Sayida*.
Sai·du (sī′dōō) A market town in northern West Pakistan, NE of Peshawar; the capital of the former princely state of Swat.
sai·ga (sī′gə) *n.* An antelope (*Saiga tatarica*) of the Siberian steppes, resembling a sheep. [< Russian *saiga*]
Sai·gon (sī-gon′, *Fr.* sà-ē-gôn′) A city in SW South Vietnam, capital of the Republic of Vietnam.
sail (sāl) *n.* **1** *Naut.* A piece of canvas, etc., attached to the mast of a vessel, to secure its propulsion by the wind: variously shaped and rigged: fore-and-aft or square *sails*. **2** Sails collectively: full *sail*. **3** A sailing vessel or craft: plural same as singular: 30 *sail* in sight. **4** A trip or passage in a sailing vessel, or in any watercraft. **5** Anything resembling a sail in form or use, as the broad part of the arm of a windmill or a bird's wing. —**to make sail**

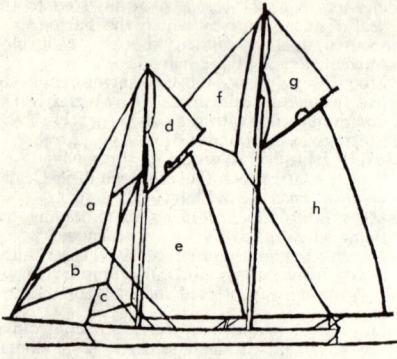

SAILS OF A CLUB TOPSAIL SCHOONER
a. Jib topsail. e. Foresail.
b. Flying jib. f. Maintopmast staysail.
c. Jib. g. Main club topsail.
d. Fore club topsail. h. Mainsail.

1 To unfurl a sail or sails. **2** To set out on a voyage. —**to set sail** To begin a voyage; get under way. —**under sail** Sailing; with sails spread and driven by the wind. —*v.i.* **1** To move across the surface of water by the action of wind or, by extension, steam. **2** To travel over water in a ship or boat. **3** To begin a voyage; set sail. **4** To manage a sailing craft: Can you *sail*? **5** To move, glide, or float in the air; soar. **6** To move along in a stately or dignified manner: She *sailed* by haughtily. **7** *Colloq.* To pass rapidly. **8** *Colloq.* To proceed boldly into action: with *in.* —*v.t.* **9** To move or travel across the surface of (a body of water) in a ship or boat. **10** To navigate (a ship, etc.). —**to sail into 1** To begin with energy. **2** To attack violently. ♦ Homophone: *sale*. [OE *segl*] —**sail′a·ble** *adj.*
sail·boat (sāl′bōt′) *n.* A small boat propelled by a sail or sails.
sail·cloth (sāl′klôth′, -kloth′) *n.* A very strong, firmly woven, cotton canvas suitable for sails: also called *duck*.
sail·er (sā′lər) *n.* A vessel that sails; a ship having a specified sailing power: a fast *sailer*.
sail·fish (sāl′fish′) *n. pl.* **·fish** or **·fish·es 1** Any of a genus (*Istiophorus*) of marine fishes allied to the swordfish, having a large or conspicuous dorsal fin likened to a sail. **2** The basking shark.

SAILFISH (Up to 6 feet in length)

sail·ing (sā′ling) *n.* **1** The setting forth on or prosecution of a voyage: the *sailing* of a vessel. **2** The art and method of determining the direction and distance sailed by a ship at sea, the point reached, and the course to be taken; navigation; seamanship.

sailing orders Instructions given to a ship's captain, covering all details of a voyage.
sail-loft (sāl′lôft′, -loft′) *n.* A room where sails are cut out and sewed.
sail·or (sā′lər) *n.* **1** A seaman; mariner. **2** A sailor hat. —**sail′or·ly** *adj.*
  Synonyms: mariner, seafarer, seaman. In nautical language *sailors* and *seamen* are exclusive of officers, but in literary use all whose vocation is navigation are figuratively termed *sailors* or *seamen. Mariner* is one who navigates or assists in navigating a ship; in the United States statutes *mariner* denotes any person, from captain to cook, who serves in any capacity on a ship. *Antonym:* landsman.
sailor hat A low-crowned, flat-topped straw hat with a brim, worn by both sexes. Also *sailor*.
sail·or's-choice (sā′lərz-chois′) *n.* **1** The hogfish. **2** The pinfish. **3** A West Indian grunt (*Haemulon parra*) or related fish.
sail·plane (sāl′plān′) *n. Aeron.* A glider having a high degree of maneuverability.
Sai·maa (sī′mä) A lake system of SE Finland near the U.S.S.R. border; 1,699 square miles.
sain (sān) *v.t. Scot.* or *Archaic* To sign or bless with the sign of the cross to preserve against malign influence: also spelled *sane*. [OE *segnian* < L *signare* sign, make a sign of the cross < *signum* a sign]
sain·foin (sān′foin) *n.* An Old World perennial, cloverlike herb (*Onobrychis viciaefolia*) of the bean family, with variegated flowers, cultivated for forage: also called *esparcet*. Also **saint′foin**. [< F < *sain* wholesome (< L *sanus* healthy) + *foin* hay]
saint (sānt) *n.* **1** A holy, godly, or sanctified person; in the New Testament, any Christian believer. *Eph.* i 1. **2** Such a person who has died and been canonized by certain churches, as the Roman Catholic. **3** Any one of the blessed in heaven. **4** An angel. **5** A very patient, unselfish person. —*v.t.* To canonize; venerate as a saint. —*adj.* Holy; canonized: as a title, often abbreviated to *St.* [< OF *seint, saint* < L *sanctus*, orig. holy, consecrated, pp. of *sancire* make sacred]
Saint For entries not found under *Saint*, see under ST.
Saint (sānt) *n.* A member of one of the religious bodies known as **Saints**: Latter-day *Saint*.
Saint-Bar·thé·le·my (san′bàr·tā·lə·mē′) An island dependency of Guadeloupe, 125 miles NW of it, in the Leeward Islands; 9 1/2 square miles.
Saint Bernard A working dog of great size and strength, originally bred in Switzerland, characterized by a massive head, and a thick, white coat combined with red or brindle; used to rescue travelers by the hospice at Great St. Bernard Pass in the Swiss Alps.
Saint-Cloud (san·klōō′) A town in northern France, 5 miles west of Paris; former residence of French monarchs.

SAINT BERNARD
(About 28 inches high at the shoulder)

Saint-Cyr (san·sēr′), **Marquis Laurent de Gouvion**, 1764-1830, French marshal.
Saint-Cyr-l'É·cole (san·sēr·lä·kôl′) A town near Versailles in north central France; site of a national military academy.
Saint-De·nis (san·də·nē′) **1** A northern suburb of Paris; burial place of many French kings. **2** The capital of Réunion island.
Sainte-Beuve (sant·bœv′), **Charles Augustin**, 1804-69, French poet, critic, and historian.
saint·ed (sān′tid) *adj.* **1** Canonized. **2** Of holy character; saintly.
Saintes (sant), **Les** See LES SAINTES.
Saint-É·tienne (san·tā·tyen′) A city in SE France, SW of Lyon; a key industrial center.
Saint-Ex·u·pé·ry (san·teg·zü·pā·rē′), **Antoine de**, 1900-44, French writer and aviator.
Saint-Gau·dens (sānt·gô′dənz), **Augustus**, 1848-1907, U.S. sculptor born in Ireland.
Saint-Ger·main (san·zher·maṅ′) A NW suburb of Paris; scene of the signing of the peace treaty between France and Austria,

sainthood / 1109 / **sale**

1919. Also **Saint-Germain-en-Laye** (-än·lā′).
saint·hood (sānt′hood) *n.* **1** The character or condition of a saint. **2** Saints collectively.
Saint-Just (san·zhüst′), **Louis Antoine Léon de,** 1767-94, French revolutionist; one of the triumvirate of the Reign of Terror.
Saint-Lô (san·lō′) A town of NW France; partially destroyed during the Normandy campaign of World War II.
Saint-Lou·is (san·lōō·ē′) **1** A city at the mouth of the Senegal, capital of Senegal and of Mauritania. **2** A town on the SW coast of Réunion island.
saint·ly (sānt′lē) *adj.* Like a saint; godly; pious; holy. See synonyms under HOLY. — **saint′li·ness** *n.*
Saint-Ma·lo (san·ma·lō′) A port in NW France on the **Gulf of Saint-Malo,** an inlet of the English Channel between Normandy and Brittany.
Saint-Mi·hiel (san·mē·yel′) A town on the Meuse in NE France, scene of an American victory of World War I, 1918.
Saint-Na·zaire (san·na·zâr′) A port in central western France, at the mouth of the Loire.
Saint-O·mer (san·tō·mâr′) A town in northern France, 22 miles SE of Calais.
Saint-Ouen (san·twän′) A northern suburb of Paris, on the Seine.
Saint·paul·i·a (sānt·pô′lē·ə) *n.* The African violet, much cultivated as an ornamental house plant. [<NL, after Baron Walter von *Saint Paul,* German botanist, its discoverer]
Saint-Pierre (san·pyâr′), **Jacques Henri Bernardin de,** 1737-1814, French author.
Saint-Quen·tin (san·kan·tan′) A city in NE France, on the Somme.
Saint-Quentin Canal A canal in NE France, connecting the Oise with the Somme and the Scheldt; 58 miles long.
Saint-Saëns (san·säns′), **(Charles) Camille,** 1835-1921, French composer.
Saints·bur·y (sānts′ber·ē), **George Edward Bateman,** 1845-1933, English literary critic and historian.
saint·ship (sānt′ship) *n.* Sainthood.
Saint-Si·mon (san·sē·môn′), **Comte de,** 1760-1825, Claude Henri, founder of French socialism. — **Duc de,** 1675-1755, Louis de Rouvroy, French historian, memoirist, and diplomat.
Saint-Si·mon·ism (sānt·sī′mən·iz′əm) *n.* The socialistic principles of the Comte de Saint-Simon, advocating the state ownership of all property and the distribution of earnings based on the amount and quality of the work done by each laborer: also called *Simonianism.*
Saint-Vaast-la-Hogue (san·väst′lä·ôg′) A port of the NE Cotentin Peninsula, NW France; site of a French naval defeat by English and Dutch forces, 1692: also *La Hogue.*
Sai·pan (sī·pän′, -pan′, sī′pan) The largest island of the Marianas group; 47 square miles; capital, Garapan; captured from Japan by United States forces in World War II, 1944; after 1947, a district of the Trust Territory of the Pacific Islands, administered by the U. S. under the United Nations.
sair[1] (sâr) *v.t. & v.i. Scot.* To serve.
sair[2] (sâr) *adj. Scot.* Sore; sorrowful; heavy, great. — **sair′ly** *adv.*
sair·y (sâr′ē) *adj. Scot.* Sorry; sorrowful; wretched; poor. Also **sair′ie.**
Sa·is (sā′is) An ancient city of Lower Egypt on the Rosetta branch of the Nile. — **Sa·ite** (sā′īt) *n.* — **Sa·it·ic** (sā·it′ik) *adj.*
saith (seth) *Archaic* Present indicative third person singular of SAY[1].
sai·yid (sī′id, sā′yid) See SAYID.
sa·jou (sə·jōō′) *n.* Sapajou.
Sa·ki (sä′kē) A city on Osaka Bay, southern Honshu, Japan; industrial center; once a port.
Sa·kart·ve·lo (sä·kärt′ve·lō) The Georgian name for GEORGIA, U.S.S.R.
Sa·kar·ya (sä·kär′yä) A river in west central Turkey in Asia, flowing 490 miles to the Black Sea.
sake[1] (sāk) *n.* **1** Purpose of obtaining or accomplishing: preceded by *for* and followed by *of:* to open the window for the *sake* of air. **2** Interest, regard, or affectionate or reverent consideration, felt for any person or thing; account; well-being; advantage: commonly with *for* and a possessive: for your *sake,* for the *sake* of your children. [OE *saccu* (legal) case]

sa·ke[2] (sä′kē) *n.* A fermented liquor made from rice; by extension, in Japan, any spirituous liquor. Also **sa′ki.** [<Japanese]
Sa·kel (zä′kəl), **Manfred,** 1906-57, Austrian psychiatrist; originator of insulin shock therapy.
sa·ker (sā′kər) *n.* A falcon; specifically, the Old World *Falco cherrug* or *sacer,* and the American prairie falcon *(F. mexicanus).* [< OF *sacre* <Sp. *sacro* <Arabic *şaqr* falcon]
Sa·kha·lin (sä′hä·lēn′, sak′ə·lēn) An island of Siberian Russian S.F.S.R. off its eastern coast; 29,700 square miles; the portion south of latitude 50° (13,930 square miles), known as Japanese *Sakhalin* or *Karafuto,* was ceded to Japan by the Treaty of Portsmouth; reoccupied by Russia after World War II. Formerly *Saghalien.*
Sa·ki (sä′kē) Pseudonym of Hector Hugh Munro, 1870-1916, English writer.
Sa·ki·shi·ma Islands (sä·kē·shē′mä) A southern group of the Ryukyu Islands; 343 square miles.
Sak·ka·ra (sə·kä′rə) A village of Upper Egypt; site of excavations of many ancient ruins: also *Saqqara.*
Sak·ti (säk′tē, sak′-; *Sanskrit* shuk′tē) See SHAKTI.
Sa·kun·ta·la (sə·koon′tə·lä, shə-) The heroine of a famous Sanskrit play of this name by Kalidasa. Also **Sakuntala.**
sal (sal) *n.* Salt. [<L]
sa·laam (sə·läm′) *n.* An oriental salutation or obeisance resembling prostration, the palm of the right hand being held to the forehead; also, a respectful or ceremonious verbal greeting. — *v.t. & v.i.* To greet with or make a salaam. [<Arabic *salām,* orig. peace, in *(as)salām ('alaikum)* peace (be upon you), a salutation]
sal·a·ble (sā′lə·bəl) *adj.* Such as can be sold; marketable: also spelled *saleable.* See synonyms under VENAL[1]. — **sal′a·bil′i·ty,** **sal′a·ble·ness** *n.* — **sal′a·bly** *adv.*
sa·la·cious (sə·lā′shəs) *adj.* Lustful; lecherous. [<L *salax, salacis* <*salire* leap] — **sa·la′cious·ly** *adv.* — **sa·la′cious·ness, sa·lac′i·ty** (-las′ə·tē) *n.*
sal·ad (sal′əd) *n.* **1** A dish of green herbs or vegetables, usually uncooked and served with a dressing, sometimes mixed with chopped cold meat, fish, etc. **2** The course consisting of such a dish. [<OF *salade* <Provençal *salada* <L *salata,* pp. of *salare* salt < *sal* salt]
sa·la·dang (sə·lä′däng) *n.* The East Indian ox *(Bos gaurus):* also called *gaur.* Also spelled *seladang.* [Var. of *seladang* <Malay *sĕladañ*]
salad days Days of youth, freshness, and inexperience.
salad dressing A savory sauce used on salads, as mayonnaise, or a mixture of salt, oil, and vinegar, etc.
Sal·a·din (sal′ə·din), 1137?-93, sultan of Egypt and Syria, 1174-93; defended Acre against Crusaders.
Sa·la·do (sä·lä′thō), **Río 1** A river of north central Argentina, flowing 1,250 miles SE from the Andes to the Paraná: also *Salado del Nor·te* (thel nôr′tä). **2** A river of central Argentina, flowing 750 miles south from the Andes to the salt marshes near the Colorado: in its upper courses known as the **Des·a·gua·de·ro** (des′ä·gwä·thä′rō). **3** A river in Buenos Aires province, eastern Argentina, flowing 400 miles SE from the border of Santa Fe province to the Río de la Plata.
Sa·la·jar (sä·lä′yär) An Indonesian island off SW Celebes; 259 square miles. Also **Sa·la′yar.**
Sal·a·man·ca (sal′ə·mang′kə, *Sp.* sä′lä·mäng′kä) A city in west central Spain; scene of a victory of Wellington against the French, 1812.
sal·a·man·der (sal′ə·man′dər) *n.* **1** Any of an order *(Caudata)* of tailed, lizardlike amphibians having a smooth, moist, scaleless skin and usually two pairs of limbs, as the American tiger salamander *(Ambystoma tigrinum):* once popularly believed able to live in fire. **2** One of

SPOTTED SALAMANDER
(From 6 to 7 inches long over-all)

the genii fabled to live in fire; an elemental fire spirit in Paracelsus' theory of elementals; hence, a creature fabled to live in fire. **3** Any person or thing that can stand great heat. **4** A large poker or other implement used around or in fire, or when red-hot. **5** A mass of hardened metal or slag remaining in the hearth of a furnace after the fires are drawn: also called *shadrach.* [<OF *salamandre* <L *salamandra* <Gk.] — **sal′a·man′drine** (-drin) *adj.*
Sa·lam·bri·a (sä′läm·brē′ä, *Greek* sä′läm·bre′ä) The former name for the PENEUS.
sa·la·mi (sə·lä′mē) *n.* A salted, spiced sausage, originally Italian. [< Ital., pl., preserved meat, salt pork, ult. <L *salare* salt < *sal* salt]
Sal·a·mis (sal′ə·mis) **1** An ancient ruined city of eastern Cyprus. **2** An island in the Saronic Gulf of the Aegean, off the coast of Attica, Greece; 39 square miles; the **Bay of Salamis,** nearly bisecting the island, was the scene of the Greek victory over the Persian fleet, 480 B.C.
sal ammoniac A white, soluble ammonium chloride. [<L *sal Ammoniacum,* lit., salt of Ammon; so called because orig. made from camel's dung near the shrine of Jupiter *Ammon* in Libya]
sal·a·ried (sal′ər·ēd) *adj.* **1** In receipt of a salary. **2** Yielding a salary.
sal·a·ry (sal′ər·ē) *n. pl.* **·ries** A periodic allowance as compensation for official or professional services. — *v.t.* **·ried, ·ry·ing** To pay or allot a salary to. [<AF *salarie* <L *salarium* money paid Roman soldiers for their salt, orig. neut. of *salarius* of salt < *sal* salt]
Synonyms (*noun*): allowance, compensation, earnings, fee, hire, honorarium, pay, payment, recompense, remuneration, requital, stipend, wages. An *allowance* is a stipulated amount furnished at regular intervals, as of food or of money. *Compensation* signifies a return for service done. *Remuneration* is applied to matters of great amount or importance. *Recompense* has a still wider meaning; there are services for which affection and gratitude are the sole and sufficient *recompense*; *earnings, fees, hire, pay, salary,* and *wages* are forms of *compensation* and may be included in *compensation, remuneration,* or *recompense. Pay* is commercial, and signifies an exact pecuniary equivalent for a thing or service, except when the contrary is expressly stated, as when we speak of high *pay* or poor *pay.* A *wage* is what a worker receives, and is usually estimated on an hourly or daily rate. *Earnings* is often equivalent to *wages,* but may be used with reference to the real value of work done or service rendered, and even applied to inanimate things; as, the *earnings* of capital. *Hire* is distinctly mercenary or menial. *Salary* is for professional, literary, executive, or clerical work, and is usually estimated on a weekly, monthly, or annual rate. A *fee* is given for a single service or privilege, and is sometimes a gratuity. Compare REQUITE.
Sa·la·zar (sä′lə·zär′), **Antonio de Oliveira,** born 1889, Portuguese statesman; prime minister 1932-.
sale (sāl) *n.* **1** The act of selling; the exchange or transfer of property for money or its equivalent. **2** An auction or selling-off at bargain prices. **3** Opportunity of selling; demand by purchasers; market: Stocks find no *sale.* — **for sale** (or **on sale**) Offered or ready for sale. ♦ Homophone: sail. [OE *sala,* prob. <ON]
Synonyms: bargain, barter, change, deal, exchange, trade. A *bargain* is strictly an agreement or contract to buy and sell; (see CONTRACT) but the word is often used to denote the entire transaction and also the thing sold or purchased. *Change* and *exchange* are words of wider signification, applying only incidentally to the transfer of property or value; a *change* secures something different in any way or by any means; an *exchange* secures something as an equivalent or return, but not necessarily as payment for what is given. *Barter* is the *exchange* of one commodity, generally a portable one, for another. *Trade* in the broad sense may apply to vast businesses (as the book *trade*), but as denoting

a single transaction is used chiefly in regard to things of moderate value, when it becomes nearly synonymous with *barter*. *Sale* is commonly limited to the transfer of property for money, or for something estimated at a money value or considered as equivalent to so much money. A *deal* in the political sense is a *bargain*, substitution, or transfer for the benefit of certain persons or parties against all others; as, The nomination was the result of a *deal*; in business it may have a similar meaning, but it frequently signifies simply a *sale* or *exchange*, a dealing.

Sa·lé (sä·lā′) A port of NW Morocco NE of Rabat. *Arabic* **Sla** (slä).

sale·a·ble (sā′lə·bəl) See SALABLE.

Sa·lem (sā′ləm) 1 A port of entry in NE Massachusetts; the original settlement of the Massachusetts Bay Colony. 2 The capital of Oregon, on the Willamette River in the NW part of the State. 3 An Old Testament name for JERUSALEM. 4 A city in west central Madras State, India.

sal·ep (sal′ep) *n.* A farinaceous meal obtained from the dry tubers of various orchids, used as food and formerly as medicine. [<F <Turkish *sālep* <Arabic *tha'leb, sa'leb*, prob. contraction of *khasyu'th-tha'lab* orchis, lit., fox's testicles]

sal·e·ra·tus (sal′ə·rā′təs) *n.* Sodium (or formerly potassium) bicarbonate, for use in cookery; baking soda. [<NL *sal aëratus* aerated salt <L *sal* salt + *aër* air, gas; so called because it produces carbon dioxide]

Sa·ler·no (sä·ler′nō) A port in SW Italy on the **Gulf of Salerno**, an arm of the Tyrrhenian Sea; scene of fierce fighting in World War II between Germans and Allied landing forces, 1943. Ancient **Sa·ler·num** (sə·lûr′nəm).

sales·clerk (sālz′klûrk′) *n.* A clerk who sells goods in a store.

sales·girl (sālz′gûrl′) *n.* A woman or girl hired to sell merchandise, especially in a store.

Sa·le·sian (sə·lē′shən) *n.* A member of an order of priests and nuns founded in Italy by Don Bosco for the rescue and education of poor and neglected children and named for St. Francis de Sales, patron of the order. —*adj.* Pertaining to the spirit of St. Francis de Sales or to his works.

sales·la·dy (sālz′lā′dē) *n. pl.* **·dies** *Colloq.* A woman or girl hired to sell merchandise, especially in a store.

sales·man (sālz′mən) *n. pl.* **·men** (-mən) A man hired to sell goods, stock, etc., in a store or by canvassing.

sales·man·ship (sālz′mən·ship) *n.* 1 The work or profession of a salesman. 2 Ability or skill in selling.

sales·peo·ple (sālz′pē′pəl) *n. pl.* Salespersons.

sales·per·son (sālz′pûr′sən) *n.* A person hired to sell merchandise, especially in a store.

sales resistance An attitude or state of mind in an individual or in the buying public that resists buying certain goods because of something antipathetic in the salesman, the advertising, or in the product.

sales·room (sālz′rōōm′, -rŏŏm′) *n.* A room where merchandise is displayed for sale.

sales tax A tax on money received from sales of goods.

sales·wom·an (sālz′wŏŏm′ən) *n. pl.* **·wom·en** (-wim′in) A woman or girl hired to sell merchandise, especially in a store.

Sal·ford (sôl′fərd) A county borough in SE Lancashire, England.

Sa·li·an (sā′lē·ən) *adj.* Of or pertaining to the **Sal·i·i** (sal′ē·ī), a tribe of Franks who, in the fourth century A.D., settled on both sides of the lower Rhine, near the Zuyder Zee. —*n.* One of the Salii. Also **Salian Frank.** [<LL *Salii*]

sal·ic (sal′ik) *adj. Geol.* Belonging to a group of igneous rocks composed chiefly of silica and alumina, as the feldspars, quartz, etc. [<S(ILICA) + AL(UMINUM) + -IC]

Sal·ic (sal′ik) *adj.* 1 Characterizing a law, the **Salic Law**, derived from Germanic sources in the fifth century, and providing that males only could inherit lands: later applied to the succession to the French and Spanish thrones. 2 Pertaining to the Salian Franks. Also spelled *Salique.* [<MF *salique* <Med. L *salicus* <LL *Salii* the Salii]

sal·i·ca·ceous (sal′ə·kā′shəs) *adj. Bot.* Of or pertaining to a family *(Salicaceae)* of shrubs and trees forming the order *Salicales*, having alternate undivided leaves and dioecious flowers; the willow family. It includes the willows and the poplars. [<NL *salicaceus* <L *salix, -icis* a willow]

sal·i·cin (sal′ə·sin) *n. Chem.* A white, crystalline, bitter glycoside, $C_{13}H_{18}O_7$, contained in the bark of certain willows and poplars, and also made synthetically: used in medicine for rheumatism and as an antiperiodic. Also **sal′i·cine** (-sēn, -sin). [<F *salicine* <L *salix, -icis* a willow + F *-ine* -INE[2]]

sal·i·cyl·ate (sal′ə·sil′āt, sə·lis′ə·lāt) *n. Chem.* A salt or ester of salicylic acid.

sal·i·cyl·ic (sal′ə·sil′ik) *adj.* Of, pertaining to, or derived from certain willows. [<F *salicyle* (<L *salix, -icis* a willow + F *-yle* -YL) + -IC]

salicylic acid *Chem.* A white crystalline compound, $C_7H_6O_3$, occurring naturally in many plants and also made synthetically from phenol. It is an antiseptic and is used sparingly in preserving foods, and, in the form of its salts, for treating rheumatism.

sa·li·ence (sā′lē·əns) *n.* 1 The condition of being salient or, figuratively, noteworthy. 2 A protruding feature or detail. 3 That which arrests attention because of its importance. Also **sa′li·en·cy.**

sa·li·ent (sā′lē·ənt) *adj.* 1 Standing out prominently; striking; conspicuous: a *salient* feature. 2 Extending beyond the general line; projecting. 3 Leaping; springing. —*n.* An angle pointing outwards, as of a fortification (see illustration under BASTION); projecting line or lines of trenches; a sharp curve in a military line protruding toward the enemy. [<L *saliens, -entis*, ppr. of *salire* leap] —**sa′li·ent·ly** *adv.* —**sa′li·ent·ness** *n.*

sa·li·en·ti·an (sā′lē·en′shē·ən) *n.* Any of an order *(Salientia)* of amphibians characterized by broad, stocky bodies and the absence of tails, and having hind legs adapted for leaping, including the frogs and toads. —*adj.* Belonging or pertaining to the *Salientia*. [<NL <L *saliens*. See SALIENT.]

sa·lif·er·ous (sə·lif′ər·əs) *adj.* Containing a considerable proportion of salt in beds or as brine: *saliferous* strata. [<L *sal, salis* salt + -FEROUS]

sal·i·fy (sal′ə·fī) *v.t.* **·fied, ·fy·ing** 1 To combine or impregnate with a salt. 2 To form into a salt, as with an acid. [<F *salifier* <NL *salificare* <L *sal, salis* salt + *facere* make] —**sal′i·fi′a·ble** *adj.* —**sal′i·fi·ca′tion** *n.*

sa·lim·e·ter (sə·lim′ə·tər) *n.* A salinometer. [<L *sal, salis* salt + -METER]

sa·li·na (sə·lī′nə) *n.* 1 A pool, pond, or marsh containing salt water diked in from the sea; also, a salt spring; a saltlick. 2 A saltworks or salt mine. [<Sp. <L *salinae (fodinae)* salt (pits) < *sal, salis* salt]

Sa·li·na Cruz (sä·lē′nä krōōs′) A port of Oaxaca, southern Mexico, on the Pacific coast.

sa·line (sā′līn) *adj.* Constituting, consisting of, or characteristic of salt; containing salt; salty. —*n.* 1 A metallic salt, especially a salt of one of the alkalis or of magnesium. 2 A salt solution used in the investigation of biological and physiological processes, and also in medicine, as for an injection. 3 A natural deposit of common or other soluble salt; salina. [<F *salin* <LL (assumed) *salinus* <L *sal, salis* salt]

sa·lin·i·ty (sə·lin′ə·tē) *n.* 1 The state or degree of being salt or saline. 2 The quantity of solid material dissolved in one kilogram of water: expressed in parts per thousand. Compare CHLORINITY.

sa·lin·i·za·tion (sā′lin·ə·zā′shən, -ī·zā′-) *n.* 1 The accumulation of salt. 2 The process by which a soil acquires various kinds of salts, as sodium chloride, calcium sulfate, or the like.

sal·i·nom·e·ter (sal′ə·nom′ə·tər) *n.* A hydrometer graduated to show the percentage of salt in a solution and to measure the density of sea water. [< *salino-* (<SALINE) + -METER] —**sal′i·nom′e·try** *n.*

Sa·lique (sə·lēk′, sal′ik, sā′lik) See SALIC.

Salis·bur·y (sôlz′ber·ē, -brē) 1 A municipal borough, county town of Wiltshire, England, on the Avon River; noted for its 13th century cathedral: also *New Sarum.* 2 The capital of Southern Rhodesia.

Salis·bur·y (sôlz′ber·ē, -brē), **Marquis of,** 1830–1903, Robert Gascoigne Cecil, English statesman.

Salisbury Plain An undulating chalk-plateau in southern Wiltshire, England; 300 square miles; site of Stonehenge.

Salisbury steak Hamburger (def. 2).

Sa·lish (sā′lish) *n.* 1 A North American Indian of Salishan stock: commonly called *Flathead.* 2 Any of the languages of the Salishan Indians.

Sa·lish·an (sā′lish·ən, sal′ish-) *adj.* Of or pertaining to a linguistic stock of North American Indians, formerly inhabiting Oregon, Washington, British Columbia, and Montana. —*n.* Any of the Salishan languages. Also **Sa′lish.**

sa·li·va (sə·lī′və) *n. Physiol.* The slightly alkaline fluid secreted by the glands of the mouth; spittle. It contains a specific amylase called ptyalin, which converts starch into maltose and is therefore considered a promoter of digestion. [<L] —**sal·i·var·y** (sal′ə·ver′ē) *adj.*

sal·i·vate (sal′ə·vāt) *v.* **·vat·ed, ·vat·ing** *v.i.* To secrete saliva. —*v.t.* To produce salivation in. [<L *salivatus*, pp. of L *salivare* < *saliva* saliva]

sal·i·va·tion (sal′ə·vā′shən) *n.* An abnormally increased flow of saliva, especially when due to the effect of drugs, as mercury.

Salk (sôk, sôlk), **Jonas Edward,** born 1914, U.S. bacteriologist; developed vaccine for poliomyelitis.

sall (sal) *v.i. Dial.* Shall. [Var. of SHALL]

salle à man·ger (sål à mäṅ·zhā′) *French* Dining-room.

sal·len·ders (sal′ən·dərz) *n. pl.* An eczematic inflammation about the hock joint of a horse. Compare MALANDERS. [<F *solandre*; ult. origin uncertain]

sal·let (sal′it) *n.* A hemispherical helmet of the 15th century. [<OF *salade* <Ital. *celata* <L *caelata (cassis)* an engraved (helmet), orig. pp. fem. of *caelare* engrave]

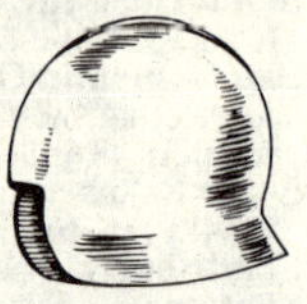

SALLET
Of Italian archer, 15th century.

sal·low[1] (sal′ō) *adj.* Of an unhealthy yellowish color: said chiefly of the human skin. [OE *salo*] —**sal′low·ish** *adj.* —**sal′low·ly** *adv.* —**sal′low·ness** *n.*

sal·low[2] (sal′ō) *n.* 1 A European willow with less flexible shoots than the osiers, especially the goat willow *(Salix caprea)*, sometimes called the **great sallow**, and the **gray sallow** *(S. caprea cinerea).* 2 An osier; a willow shoot. [OE *sealh*]

sal·low·y (sal′ō·ē) *adj.* Fringed with or abounding in sallows.

Sal·lust (sal′əst), 86–35 B.C., Roman historian: full name *Gaius Sallustius Crispus.*

sal·ly (sal′ē) *v.i.* **·lied, ·ly·ing** 1 To rush out suddenly. 2 To set out energetically. 3 To go out, as from a room or building. —*n. pl.* **·lies** 1 A rushing forth, as of besieged troops against besiegers; sortie. 2 A going forth, as on a walk or excursion. 3 A sudden overflow of spirits; a witticism or bantering remark. [<OF *saillie*, orig. pp. fem. of *saillir* <L *salire* leap]

Sal·ly (sal′ē) A diminutive of SARAH; also, a feminine personal name.

sally lunn (lun) A raised and sweetened teacake resembling a muffin. [after *Sally Lunn*, pastry cook, of Bath, England, in the 18th century]

Sal·ma·cis (sal′mə·sis) In Greek mythology, a nymph of a fountain in ancient Caria, the waters of which were supposed to render effeminate all who drank of them. See HERMAPHRODITUS.

sal·ma·gun·di (sal′mə·gun′dē) *n.* 1 A dish of chopped meat, anchovies, eggs, onions, etc., mixed and seasoned. 2 Hence, any medley or miscellany; a potpourri. [<F *salmigondis*, prob. <Ital. *salami conditi* pickled meats < *salame* preserved meat, sausage + *conditi*, pp. of *condire* flavor <L, preserve, pickle]

Sal·ma·gun·di (sal′mə·gun′dē) A series of humorous and satirical papers published periodically in 1807–08 by Washington Irving and others.

sal·mi (sal′mē) *n.* A spiced dish of birds or game roasted, minced, and stewed in wine; a ragout. Also **sal·mis** (sal′mē, *Fr.* sål·mē′). [<F, prob. contraction of *salmigondis* SALMAGUNDI]

salm·on (sam′ən) *n.* 1 A clupeid fish (family *Salmonidae*, genus *Salmo*), especially *S. salar* of the North Atlantic, brownish above, silvery on the sides, with black spots. The salmon ascends to the headwaters of rivers to spawn, and surmounts obstructions, as waterfalls of considerable height. It is a highly prized game and food fish, and has delicate reddish-orange flesh. 2 One of other salmonoid fishes, especially the quinnat, ascending rivers flowing to the North Pacific. 3 A color of a reddish- or pinkish-orange tint: also **salm′on-pink′**. — *adj.* Having the color salmon. [<AF *samoun, saumoun, salmun* <L *salmo, -onis*]

salm·on·ber·ry (sam′ən·ber′ē) *n. pl.* **·ries** 1 A hardy raspberry (*Rubus spectabilis*) of the Pacific coast. 2 The cloudberry. 3 A raspberry (*R. parviflorus*) of the United States, having a white blossom.

Sal·mo·nel·la (sal′mō·nel′ə) *n.* A genus of aerobic, motile, rodlike bacteria capable of fermenting certain carbohydrates with the formation of acid and gas; especially, *S. paratyphi*, which causes a form of paratyphoid in man. [<NL, after Daniel Elmer *Salmon*, U.S. pathologist, 1850–1914]

Sal·mo·ne·us (sal·mō′nē·əs) In Greek mythology, a son of Aeolus and king of Elis who was destroyed by thunderbolts for claiming to be the equal of Zeus.

sal·mo·noid (sal′mə·noid) *adj.* Resembling a salmon; belonging to the salmon family. — *n.* A salmonoid fish.

Salmon River A river in central Idaho, flowing 425 miles NE to the Snake River.

salmon trout 1 The European sea trout (*Salmo trutta*). 2 Certain other salmonoid fish, as the namaycush or the steelhead. See TROUT.

sal·ol (sal′ōl, -ol) *n. Chem.* A colorless crystalline compound, $C_{13}H_{10}O_3$, derived from salicylic acid: used in medicine as a substitute for salicylic acid and as an antineuralgic, antirheumatic, antipyretic, and an internal antiseptic. [<SAL(ICYLIC ACID) + -OL[1]]

Sa·lo·me (sə·lō′mē, *Ger.* zä′lō·mā) A feminine personal name. Also *Fr.* **Sa·lo·mé** (sà·lô·mā′). [<Hebrew, peace]

— **Salome** The daughter of Herodias, who asked from Herod the head of John the Baptist on a silver charger in return for her dancing. *Matt.* xiv 8.

Sal·o·mon (sal′ə·mən; *Fr.* sà·lô·môn′, *Hungarian* shol′o·mōn, *Polish* sä·lō′môn, *Sp.* sä′lō·mōn′) See SOLOMON. Also *Du., Ger.* **Sa·lo·mo** (sä′lō·mō), *Ital.* **Sa·lo·mo·ne** (sä′lō·mō′nā), *Pg.* **Sa·lo·mão** (sä′lō·moun′).

Sal·o·mon (sal′ō·mən), **Haym,** 1740?–85, American patriot born in Poland; advanced large loans to American treasury during the Revolutionary War.

sa·lon (sə·lon′, *Fr.* sà·lôn′) *n.* 1 A room in which guests are received; a drawing-room. 2 The periodic gathering or reception of noted persons, under the auspices of some distinguished woman, especially in Paris in the 17th and 18th centuries. 3 A hall or gallery used for exhibiting works of art. 4 An exhibition of works of art. 5 An establishment devoted to some specific purpose: a beauty *salon*. [<F <Ital. *salone*, aug. of *sala* a room, hall <OHG *sal*]

Sa·lon (sə·lon′, *Fr.* sà·lôn′) *n.* An annual exhibition of works by living artists at the Grand Palais des Champs Élysées in Paris: so called because formerly held in the Salon Carré of the Louvre. Since 1891 there have been two rival Salons in Paris, the Salon of the Champs Élysées, or **Old Salon,** and the Salon of the Champ de Mars, or **New Salon,** which both exhibit in the Grand Palais.

Sa·lo·ni·ka (sä′lō·nē′kä) A port of NE Greece, in Macedonia on the **Gulf of Salonika,** an arm of the Aegean between Thessaly and Macedonia: ancient *Therma*: Greek *Thessalonike*. Also **Sa′lo·ni′ki** (-kē), **Sa′lo·ni′ca.**

sa·loon (sə·lōōn′) *n.* 1 *U.S.* A place where alcoholic drinks are sold; a bar. 2 *Brit.* In a public house, a section of the bar set aside for patrons of a higher social status than those in the public bar. 3 A large apartment or room for assemblies, public entertainment, exhibitions, etc. 4 The main cabin of a passenger ship, used by the passengers in general. 5 A salon (def. 1). 6 *Brit.* A sedan (def. 1). [<F *salon* a salon]

sa·loon-keep·er (sə·lōōn′kē′pər) *n.* One who keeps a saloon; a liquor dealer.

sa·loop (sə·lōōp′) *n. Brit.* An infusion of sassafras chips, salep, or similiar aromatic herbs, formerly used largely as a beverage, as a cure for rheumatism, etc.; sassafras tea. [Var. of SALEP]

Sal·op (sal′əp) See SHROPSHIRE. — **Sa·lo·pi·an** (sə·lō′pē·ən) *adj.* & *n.*

sal·pa (sal′pə) *n.* Any of a genus (*Salpa*) of free-swimming, transparent, cylindrical tunicates (class *Thaliacea*) found in warm seas: they exhibit a solitary asexual stage and a colonial sexual one. Also **sal′pi·an, sal′pid.** [<NL <L <Gk. *salpē*, a kind of sea fish] — **sal′pi·form** *adj.*

Sal·pi·glos·sis (sal′pə·glos′is) *n.* A small genus of South American solanaceous, downy herbs having entire leaves and handsome variegated flowers. [<NL <Gk. *salpinx, -ingos* a trumpet + *glōssa* tongue]

sal·pin·gec·to·my (sal′pin·jek′tə·mē) *n. Surg.* The excision of a Fallopian tube; sterilization of women. [<NL *salpinx, salpingos* a Fallopian tube (<Gk., a trumpet) + -ECTOMY]

sal·pinx (sal′pingks) *n. pl.* **sal·pin·ges** (sal·pin′·jēz) *Anat.* A tube in man and other mammals, especially the Eustachian or Fallopian tube. [<NL <Gk., a trumpet]

sal·si·fy (sal′sə·fē) *n.* An Old World plant (*Tragopogon porrifolius*) of the composite family, with a white, edible root: from its flavor called *oyster plant, vegetable oyster*. [<F *salsifis*, prob. <Ital. *sassefrica*; ult. origin unknown]

sal·sil·la (sal·sil′ə) *n.* Any of several tropical American plants of the amaryllis family (genus *Bomarea*), yielding edible tubers resembling those of the Jerusalem artichoke. [<Sp., dim. of *salsa* a sauce]

sal soda Sodium carbonate; washing soda. See SODA.

salt (sôlt) *n.* 1 Sodium chloride, NaCl, a widely distributed compound, used by men from time immemorial as a seasoning and preservative: a necessary ingredient of food for most mammals. It is obtained by evaporation or freezing of the water of the ocean, of saline lakes and springs or wells, and by mining in beds of rock salt. ◆ Collateral adjective: *saline*. 2 *Chem.* Any compound produced when all or part of the hydrogen of an acid is replaced by an electropositive radical or a metal. Salts are usually formed by treating a metal with an acid or by the interaction of a base and an acid. Usually the salts derived from acids whose names end in *-ic* take the suffix *-ate*, and those ending in *-ous* take *-ite*. 3 *pl.* A salt used as a laxative or cathartic; also, smelling salts. 4 Piquant humor; dry wit; repartee: from the phrase *Attic salt*. 5 That which preserves, corrects, or purifies: the *salt* of criticism; seasoning. 6 A sailor: an old *salt*. 7 A saltcellar. — **below the salt** In inferior, subordinate, or servile position. — **to take with a grain of salt** To allow for exaggeration; have doubts about. — *adj.* 1 Flavored with salt; salty; briny: opposed to *sweet*. 2 Cured or preserved with salt. 3 Containing, or growing or living in or near, salt water. 4 *Obs.* Salacious; licentious; gross. — *v.t.* 1 To season with salt. 2 To preserve or cure with salt. 3 To furnish with salt: to *salt* cattle. 4 To season as if with salt; add zest or piquancy to. 5 To add something to so as fraudulently to increase the value: to *salt* a mine with gold. — **to salt away** 1 To pack in salt for preserving. 2 *Colloq.* To store up; save. — **to salt out** To separate (coal-tar colors) by adding salt to solutions containing them. [OE *sealt*. Akin to SAL.] — **salt′ness** *n.*

Sal·ta (säl′tä) A province of NW Argentina; 59,743 square miles; capital, Salta.

sal·tant (sal′tənt) *adj.* Leaping; jumping; saltatory. [<L *saltans, -antis*, ppr. of *saltare* dance, freq. of *salire* leap]

sal·ta·rel·lo (sal′tə·rel′ō, *Ital.* säl′tä·rel′lō) *n. pl.* **·rel·li** (-rel′ē, *Ital.* -rel′lē) 1 A quick Italian dance, diversified by skips. 2 Music for such a dance. [<Ital., lit., a firecracker <*saltare* dance, leap <L. See SALTANT.]

sal·ta·tion (sal·tā′shən) *n.* 1 A leaping or leap, as in a dance. 2 A throbbing or palpitation, as of a blood vessel. 3 *Biol.* Mutation. [<L *saltatio, -onis* <*saltatus*, pp. of *saltare*. See SALTANT.]

sal·ta·to·ri·al (sal′tə·tôr′ē·əl, -tō′rē-) *adj.* 1 Built or adapted for leaping. 2 *Zool.* Adapted for or characterized by leaping. [<SALTATORY]

sal·ta·to·ry (sal′tə·tôr′ē, -tō′rē) *adj.* 1 Of or pertaining to leaping or dancing. 2 Moving by leaps; fitted for leaping; specifically, moving the feet synchronously, as certain birds. [<L *saltatorius* <*saltator* a leaper <*saltare*. See SALTANT.]

salt cake Crude sodium sulfate, especially as obtained by the action of sulfuric acid on sodium chloride.

salt·cel·lar (sôlt′sel′ər) *n.* A small receptacle for salt; a saltshaker.

salt·ed (sôl′tid) *adj.* 1 Treated with or as with salt for any purpose; hence, preserved. 2 Immune from infectious disease by reason of previous attack: a term used in South Africa. 3 *Colloq.* Experienced or expert in some occupation.

salt·er (sôl′tər) *n.* 1 One who applies salt to cure fish, meat, etc. 2 One who manufactures or deals in salt. [OE *sealtere*]

salt·ern (sôl′tərn) *n.* A place or building where salt is manufactured. [OE *sealtærn*]

salt grass Any of certain grasses found growing on salt marshes or on alkaline western plains, as *Distichlis spicata* or some species of *Spartina*.

salt hay Hay made from salt grass.

salt-horse (sôlt′hôrs′) *n.* Salted beef; corned beef: a sailor's term. Also **salt′-junk′** (-jungk′).

sal·ti·grade (sal′tə·grād) *adj.* Adapted for leaping: said especially of certain insects, as grasshoppers. [<NL *Saltigradae*, group name of saltigrade spiders <L *saltus* a leap + *gradi* step]

Sal·til·lo (säl·tē′yō) The capital of Coahuila state, NE Mexico.

sal·tire (sal′tir) *n. Her.* An ordinary formed by a bend and a bend sinister crossing as in St. Andrew's cross: also spelled *sautoir*. Also **sal′tier.** [<OF *sauteoir* a stirrup cord <Med. L *saltatorium* <L *saltatorius* SALTATORY]

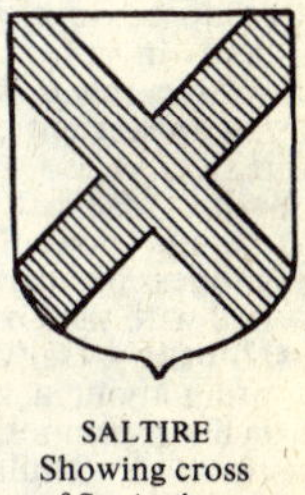

SALTIRE
Showing cross of St. Andrew.

salt·ish (sôl′tish) *adj.* Somewhat salty.

Salt Lake See GREAT SALT LAKE.

Salt Lake City The capital and largest city of Utah, SE of Great Salt Lake; center of Mormonism.

salt·lick (sôlt′lik′) *n.* A place to which animals resort to lick salt from superficial deposits; a salt spring or dried salt pond.

salt marsh Low coastal land frequently overflowed by the tide, usually covered with coarse grass. Also **salt meadow.**

salt of the earth The fundamentally fine people of the world; those who add value to mankind. *Matt.* v 13.

Sal·ton Sink (sôl′tən) A depression in southern California; lowest part, 280 feet below sea level; in 1905 and 1906 became a shallow, saline lake, the **Salton Sea,** by an overflow of the Colorado River; originally about 450 square miles, reduced to about 300 square miles by evaporation.

salt·pan (sôlt′pan′) *n.* 1 A vessel in which salt is made by evaporating saline water. 2 A pond or basin from which salt is obtained by natural evaporation.

salt·pe·ter (sôlt′pē′tər) *n.* Niter: so called colloquially and in commerce. — **Chile saltpeter** Mineral sodium nitrate occurring in beds in a desert region near the boundary of Chile and Peru, but chiefly in Chile. Also **salt′pe′tre.** [<OF *saltpetre* <Med. L *sal petrae*, lit., salt of rock <L *sal* salt + *petra* a rock <Gk.]

salt rheum *Pathol.* One of various skin eruptions, as eczema.

salt-ris·ing (sôlt′rī′zing) *n.* Salted batter used as leaven, or bread made from it.

Salt River 1 A river in central Arizona, flowing about 200 miles west to the Gila River. 2 A river in NE Missouri, flowing 200 miles SE to the Mississippi. 3 A river in north central Kentucky, flowing 125 miles NW to the Ohio. — **to row** (or **be rowed**) **up**

Salt River *Colloq.* To suffer political defeat.
salt·shak·er (sôlt′shā′kər) *n.* A container with small apertures for sprinkling table salt.
salt spring A flow of salt water from the earth.
salt–wa·ter (sôlt′wô′tər, -wot′ər) *adj.* Of, composed of, or living in salt water.
salt well A well from which brine is obtained.
salt·works (sôlt′wûrks′) *n. pl.* **·works** An establishment where salt is made on a commercial scale: in England the form **saltwork** is preferred in describing a single factory.
salt·wort (sôlt′wûrt′) *n.* **1** Any of various maritime plants (genus *Salsola*), especially the common saltwort (*S. kali*), used in making soda ash. **2** Any of various glassworts, as the dwarf glasswort (*Salicornia bigelovi*) of the New England coast. [Prob. trans. of Du. *zoutkruid*]
salt·y (sôl′tē) *adj.* **salt·i·er, salt·i·est** **1** Tasting somewhat like salt; of or containing salt. **2** Reminiscent of the sea; smelling of the sea. **3** Piquant; sharp; pungent, as literature, speech, etc. — **salt′i·ly** *adv.* — **salt′i·ness** *n.*
sa·lu·bri·ous (sə·lo͞o′brē·əs) *adj.* Conducive to health; healthful; wholesome. See synonyms under HEALTHY. [<L *salubris* < *salus* health] — **sa·lu′bri·ous·ly** *adv.* — **sa·lu′bri·ty, sa·lu′bri·ous·ness** *n.*
Sa·lu·da River (sə·lo͞o′də) A river in west central South Carolina, flowing 200 miles SE to the Congaree River.
sa·lu·ki (sə·lo͞o′kē) *n.* A very old breed of hound, having feathered ears, tail, and legs, and a greyhoundlike body; the "dog" of the Bible, known as the Royal Dog of Egypt: introduced into England in 1840. [< Arabic *salūqi* <*Salūq* an ancient Arabian city]

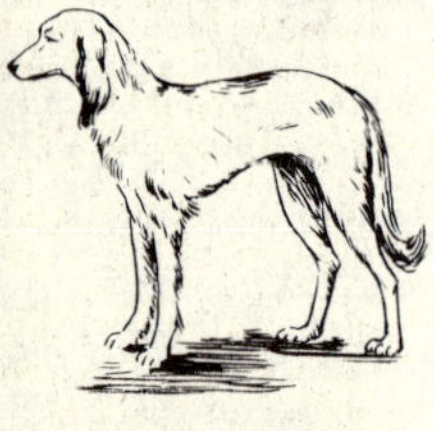

SALUKI
(From 23 to 28 inches high at the shoulder)

Sa·lus (sā′ləs) In Roman mythology, goddess of health and prosperity: later identified with the Greek *Hygeia.* [<L, health]
sal·u·tar·y (sal′yə·ter′ē) *adj.* **1** Calculated to bring about a sound condition by correcting evil or promoting good; corrective; beneficial. **2** Salubrious; wholesome; healthful. See synonyms under HEALTHY, USEFUL. [<F *salutaire* <L *salutaris* < *salus, salutis* health] — **sal′u·tar′i·ly** *adv.* — **sal′u·tar′i·ness** *n.*
sal·u·ta·tion (sal′yə·tā′shən) *n.* **1** The act of saluting. **2** Any form of greeting. **3** The opening words of a letter, as *Dear Sir.* [<OF *salutacion* <L *salutatio, -onis* < *salutatus,* pp. of *salutare* SALUTE]
sa·lu·ta·to·ri·an (sə·lo͞o′tə·tôr′ē·ən, -tō′rē-) *n. U.S.* In colleges and schools, the graduating student, usually the second (sometimes the first) honor man, who delivers the salutatory at commencement. [<SALUTATORY]
sa·lu·ta·to·ry (sə·lo͞o′tə·tôr′ē, -tō′rē) *n. pl.* **·ries** An opening oration, as at a college commencement. —*adj.* Pertaining to or consisting in greeting or welcome; specifically, relating to a salutatory address. [<L *salutatorius* pertaining to salutation < *salutare* SALUTE]
sa·lute (sə·lo͞ot′) *n.* **1** A greeting by display of military, naval, or other official honors, as by presenting arms, firing cannon, etc. **2** The act of or attitude assumed in giving a military salute. **3** A gesture of greeting, compliment, respect, or the like, as a bow, kiss, etc. —*v.* **·lut·ed, ·lut·ing** *v.t.* **1** To greet with an expression or sign of welcome, respect, etc.; welcome. **2** To honor in some prescribed way, as by raising the hand to the cap, presenting arms, firing cannon, etc. —*v.i.* **3** To make a salute. See synonyms under ADDRESS. [<L *salutare* < *salus, salutis* health] — **sa·lut′er** *n.*
sal·va·ble (sal′və·bəl) *adj.* Capable of being saved or salvaged. [<LL *salvare* SAVE] — **sal′·va·bil′i·ty** *n.*
Sal·va·dor (sal′və·dôr′, *Sp.* säl′vä·thôr′) **1** El Salvador. **2** The capital of Bahia state, Brazil: formerly *Bahia, São Salvador.*
Sal·va·do·ri·an (sal′və·dôr′ē·ən, -dō′rē-) *adj.* Relating to El Salvador or its people. —*n.* A native or inhabitant of El Salvador. Also **Sal′·va·do′ran.** [<(*El*) *Salvador* <Sp., the Saviour <LL *salvator* < *salvare* save]
sal·vage (sal′vij) *v.t.* **·vaged, ·vag·ing** To save, as a ship or its cargo, from wreck, capture, etc.; salve. —*n.* **1** The saving of a ship, cargo, etc., from loss; hence, any act of saving property. **2** The compensation allowed to persons by whose voluntary exertions a vessel, her cargo, or the lives of those belonging to her are saved from danger or loss: termed legally **civil salvage,** as distinguished from **military salvage,** which consists in the liberation of property from the enemy in time of war. **3** That which is saved from a wrecked or abandoned vessel or from or after a fire; hence, anything saved from destruction. [< OF < *salver* SAVE] — **sal′vag·er** *n.*
sal·va·gee (sal′və·jē′) *n.* In maritime law, a person in whose favor or behalf salvage has been effected.
Sal·va·ges (səl·vä′zhēsh) See SELVAGENS.
Sal·var·san (sal′vər·san) *n.* Proprietary name for a brand of arsphenamine. [<G <LL *salvare* save + G *arsen* arsenic]
sal·va·tion (sal·vā′shən) *n.* **1** The process or state of being saved; preservation from impending evil. **2** *Theol.* Deliverance from sin and penalty, realized in a future state; redemption. **3** Any means of deliverance from danger, evil, or ruin. [<OF *sauvacion* <LL *salvatio, -onis* < *salvatus,* pp. of *salvare* SAVE]
Salvation Army A religious and charitable organization on semimilitary lines, founded by William Booth in England in 1865 as the Christian Mission, which took the title of Salvation Army in 1878.
Sal·va·tion·ist (sal·vā′shən·ist) *n.* A member of the Salvation Army.
salve[1] (sav, säv) *n.* **1** A thick, adhesive ointment for local ailments. **2** Anything that heals, soothes, or mollifies; hence, praise or flattery. —*v.t.* **salved, salv·ing** **1** To dress with salve or ointment. **2** To soothe; appease, as conscience, pride, etc. [OE *sealf*]
salve[2] (salv) *v.t.* **salved, salv·ing** To save from loss; salvage. [Back formation <SALVAGE]
sal·ve[3] (sal′vē) *interj.* Hail: literally, be well. [<L, imperative of *salvere* be well]
Sal·ve·mi·ni (säl·vā′mē·nē), **Gaetano,** 1873–1957, Italian historian active in the United States.
sal·ver (sal′vər) *n.* A tray, as of silver. [<OF *salve* <Sp. *salva,* orig. the foretasting of food, as for a king < *salvar* taste, save <LL *salvare* SAVE]
Sal·ve Re·gi·na (sal′vē ri·jī′nə) *Eccl.* **1** A hymn to the Virgin Mary, contained in the Roman Catholic breviary. **2** A translation of this. [<L, Hail, O queen; the opening words]
sal·vi·a (sal′vē·ə) *n.* Any of a genus (*Salvia*) of ornamental plants of the mint family; the sage. [<NL <L, SAGE[2]]
Sal·vi·ni (säl·vē′nē), **Tommaso,** 1829–1916, Italian actor.
sal·vo[1] (sal′vō) *n. pl.* **·vos** or **·voes** **1** A simultaneous discharge of artillery, or of two or more bombs from an aircraft. **2** A salute given by firing all the guns, as at the funeral of an officer; hence, any salute or simultaneous discharge: *salvos* of applause, a *salvo* of rockets. **3** The concentrated fire of many pieces, as in a naval engagement. **4** A successive and specified number of discharges of guns, from right to left, or left to right, at prescribed intervals. [Orig. *salva* <Ital., a salute <L *salve* SALVE[3]]
sal·vo[2] (sal′vō) *n. pl.* **·vos** **1** A saving clause; proviso. **2** An evasion, reservation, or bad excuse. **3** An expedient. [<L *salvo* (*jure*) (right) being reserved, ablative of *salvus* uninjured, safe]
sal vo·lat·i·le (sal vō·lat′ə·lē) Ammonium carbonate. Compare HARTSHORN. [<NL, volatile salt <L]
sal·vor (sal′vər) *n.* One who or a ship which saves or helps to save vessels or property from loss at sea; a salvager. Also **salv′er.** [<SALVE[2] + -OR]
Sal·ween (sal′wēn′) A river of eastern Tibet, SW China, and eastern Burma, flowing about 1,750 miles east, SE, and south to the Andaman Sea at Moulmein. Also **Sal′win′.**
Salz·burg (zälts′bo͝orkh) A city in west central Austria; the birthplace of Mozart.
Sa·ma·ni (sä·mä′nē) *n. pl.* A Persian dynasty ruling from A.D. 874–1005, celebrated because of its encouragement of literature and the arts.
Sa·mar (sä′mär) One of the Visayan Islands, third largest of the Philippines; 5,050 square miles.
sam·a·ra (sam′ər·ə, sə·mâr′ə) *n. Bot.* A one-seeded indehiscent fruit, as of the elm, ash, or maple, provided with a membrane or wing; a key or key fruit. [<NL <L, elm seed]
Sa·ma·ra (sä·mä′rä) The former name for KUIBYSHEV.
Sa·ma·rang (sə·mä′räng) See SEMARANG.
Sa·mar·i·a (sə·mâr′ē·ə) **1** In the Bible, a city of Palestine, capital of the northern kingdom of Israel, or the hill on which it was built, on the site of modern Sebastye in western Jordan. **2** In the Bible, the territory occupied by the kingdom of Israel, or, later, a restricted portion of central Palestine west of the Jordan occupied by the Samaritans.
Sa·mar·i·tan (sə·mar′ə·tən) *n.* **1** One of the people of Samaria, a mixed population. II *Kings* xvii. **2** The Northwest Semitic language of this people. — **Good Samaritan** A humane, compassionate person who helps one in trouble: from the parable in *Luke* x 30–37. —*adj.* Of or pertaining to Samaria. [<LL *Samaritanus* <Gk. *Samareitēs* < *Samareia* Samaria]
sa·mar·i·um (sə·mâr′ē·əm) *n.* A hard, brittle, yellowish-gray metallic element (symbol Sm) of the lanthanide series. See ELEMENT. [<NL <SAMAR(SKITE); so called because first found in the spectrum of samarskite]
Sam·ar·kand (sam′ər·kand′, *Russian* sä′mär·känt′) The second largest city and former capital of the Uzbek S.S.R., in the extreme eastern part; the ancient capital of Tamerlane's empire and the site of his tomb: ancient *Maracanda.* Also **Sam′ar·cand′.**
sa·mar·skite (sə·mär′skīt) *n.* An orthorhombic, vitreous, black mineral, source of several elements, as samarium, etc. [<G *samarskit,* after Col. Samarski, 19th c. Russian mine officer]
sam·ba (sam′bə, säm′bä) *n.* A dance of Brazilian origin in two-four time. —*v.i.* To dance the samba. [<Pg. <a native African name]
sam·bo (sam′bō) *n. pl.* **·bos** A half-breed of mixed Negro and Indian or Negro and mulatto blood. [<Sp. *zambo* a Negro, a mulatto, a monkey, prob. <Bantu *nzambu* a monkey]
Sam·bre (säN′br′) A river in northern France and SW Belgium, flowing 100 miles NE to the Meuse.
Sam Browne belt (sam′ broun′) A military belt, with one or two light shoulder straps running diagonally across the chest from right to left: designed by General Sir Samuel J. Browne, 1824–1901, of the British army, to carry the pistol when on horseback, and the sword when dismounted: later used by the armies of the United States and other countries.
sam·bu·ca (sam·byo͞o′kə) *n.* In ancient music, a sharp-toned, triangular, stringed instrument resembling a harp: of Asian origin. Also **sam·buke** (sam′byo͞ok). [<L <Gk. *sambykē,* prob. <Aramaic *sabbĕkhā*]
sam·bur (sam′bər, sam′-) *n.* A rusine deer, especially *Cervus aristotelis,* of hilly districts in India, Burma, and China. Also **sam′bar.** [<Hind. *sābar* <Skt. *shambara*]
same (sām) *adj.* **1** Having individual or specific identity or quality; identical; equal: with *the.* **2** Similar in kind or quality. **3** Aforesaid; identical: said of a person or thing just mentioned or held in mind. **4** Equal in degree of preference; indifferent. **5** Unchanged; monotonous. See synonyms under IDENTICAL, SYNONYMOUS. — **all the same** **1** Nevertheless. **2** Equally significant; equally acceptable or unacceptable. — **just the same** **1** Nevertheless. **2** Exactly identical or corresponding; unchanged. —*pron.* The identical person, thing, event, etc. —*adv.* In like manner; equally: with *the.* [ME <ON *samr, sami.* Akin to OE *same* equally.]
sa·mek (sä′mek) *n.* The fifteenth Hebrew letter. Also **sa′mech, sa′mekh.** See ALPHABET.
same·ness (sām′nis) *n.* **1** Lack of change or variety; dull monotony. **2** Close similarity. **3** Identity.
Sam Hill (sam′ hil′) *U.S. Slang* Hell: a euphemism.
sam·iel (sam′yel) *n.* The simoom. [<Turkish *samyel* < *sam* poison (<Arabic *samm*) + *yel* wind]

sam·i·sen (sam′i-sen) *n.* A Japanese guitarlike instrument with three strings, played with a plectrum. [<Japanese <Chinese *san hsien* three strings]

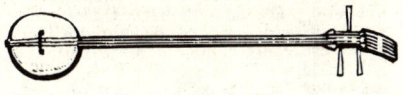
SAMISEN

sa·mite (sā′mīt, sam′īt) *n.* A rich medieval fabric of silk, often interwoven with gold or silver. [<OF *samit* <Med. L *samitum*, var. of *examitum* <Med. Gk. *hexamiton* <*hexamitos*, woven with six threads <Gk. *hex* six + *mitos* a thread]

sam·let (sam′lit) *n.* A young salmon; a parr. [Contracted dim. of SALMON; infl. by earlier SALMONET]

Sam·nite (sam′nīt) *adj.* Of or pertaining to ancient Samnium, its people, or their language. — *n.* **1** One of the people of ancient Samnium, descended from the Sabines. **2** The Italic language of these people: also called *Oscan.* [<L *Samnis, Samnitis*]

Sam·ni·um (sam′nē-əm) An ancient country of central Italy, on the Adriatic, conquered by the Romans by 290 B.C.

Sa·mo·a (sə-mō′ə) An island group in the SW Pacific; 1,209 square miles; formerly *Navigators' Islands*; divided by the 171st meridian into: (1) **American** (or **Eastern**) **Samoa**, an unincorporated territory of the United States, comprising Tutuila, Rose, Swains, and Manua; 76 square miles; capital, Pago Pago, on Tutuila. (2) The United Nations Trust Territory of **Western Samoa**, administered by New Zealand and comprising Savaii, Upolu, and several smaller islands; 1,133 square miles; capital, Apia, on Upolu.

Sa·mo·an (sə-mō′ən) *adj.* Of or pertaining to Samoa, to its aboriginal Polynesian inhabitants, or to their language. — *n.* **1** A native of the Samoan islands. **2** The Polynesian language of the Samoans.

Sam·o·gi·ti·a (sam′ō-jish′ē-ə) A historical region of western Lithuania.

Sa·mos (sā′mos, *Greek* sä′môs) A Greek island in the Aegean, north of the Dodecanese; 194 square miles. — **Sa·mi·an** (sā′mē-ən) *adj.* & *n.*

Sam·o·thrace (sam′ə-thrās) A Greek island in the NE Aegean; 71 square miles. *Greek* **Sa·mo·thra·ki** (sä′mô-thrä′kē). — **Sam′o·thra′cian** (-thrā′shən) *adj.* & *n.*

sam·o·var (sam′ə-vär, sam′ə-vär′) *n.* A metal urn containing a tube for charcoal for heating water, as for making tea. [<Russian, lit., self-boiler <*samo-* self + *varit* boil]

Sam·o·yed (sam′ə-yed′) *n.* **1** One of a Mongoloid people inhabiting the Arctic coasts of Siberia. **2** A large dog characterized by a thick white coat of long hair, originally bred by the Samoyeds as a sled dog and for herding reindeer. — *adj.* Samoyedic. Also **Sam′o·yede** (-yed′). [<Russian, lit., self-eater, i.e., a cannibal]

Sam·o·yed·ic (sam′ə-yed′ik) *adj.* Of or pertaining to the Samoyeds or their language. — *n.* A subfamily of the Uralic languages, including the language of the Samoyeds.

samp (samp) *n.* Coarse, hulled Indian corn; also, a porridge made of it. [<Algonquian (Narraganset) *nasaump* softened with water]

sam·pan (sam′pan) *n.* A small flat-bottomed boat or skiff used along rivers and coasts of China and Japan. [<Chinese *san-pan* <*san* three + *pan* board]

sam·phire (sam′fīr) *n.* **1** A European herb (*Crithmum maritimum*) of the parsley family, having fleshy leaves (formerly used in pickles). **2** A species (*Salicornia europaea*) of glasswort. [Earlier *sampere* <F (*l'herbe de*) *Saint Pierre* (the herb of) Saint Peter; ? infl. in form by CAMPHIRE]

SAMPAN
Shown with typical lateen rig.

sam·ple (sam′pəl) *n.* A portion, part, or piece taken or shown as a representative of the whole. — *v.t.* **·pled, ·pling** To test or examine by means of a portion or sample. [ME, aphetic var. of *asample* <OF *essample* EXAMPLE]
Synonyms (*noun*): case, example, exemplification, illustration, instance, specimen. A *sample* is a portion taken at random out of a quantity supposed to be homogeneous, so that the qualities found in the *sample* may reasonably be expected to be found in the whole; as, a *sample* of sugar, a *sample* of cloth. A *specimen* is one unit of a series, or a fragment of a mass, all of which is supposed to possess the same essential qualities; as, a *specimen* of coinage, or of quartz. No other unit or portion may be exactly like the *specimen*, while all the rest is supposed to be exactly like the *sample*. An *instance* is a *sample* or *specimen* of action. See EXAMPLE.

sam·pler[1] (sam′plər) *n.* **1** One who tests by samples; one who exhibits samples. **2** A device for removing a portion of a substance for testing.

sam·pler[2] (sam′plər) *n.* A piece of needlework, as a sample, designed to show a beginner's skill. [Aphetic var. of OF *essamplaire* <LL *examplarium* <L *exemplum* EXAMPLE]

sam·pling (sam′pling) *n.* **1** A small part of something or a number of items from a group selected for examination or analysis in order to estimate the quality or nature of the whole. **2** The act or process of making this selection.

Samp·son (samp′sən), **William Thomas**, 1840–1902, U. S. rear admiral.

sam·sa·ra (səm-sä′rə) *n.* **1** In Buddhism, the course of mundane existence; the endless cycle of birth, death and rebirth; the wheel of causation. **2** Transmigration; metempsychosis. [<Skt. *saṃsāra*, lit., a passage through a succession of states]

sam·shu (sam′shōō′) *n.* Alcoholic liquor resembling arrack, distilled in China from fermented rice or millet; loosely, any kind of spirits. [<Pidgin English, ? <dial. Chinese (Cantonese) *sam shiu* <Chinese *san shao* thrice distilled]

Sam·son (sam′sən, *Fr.* sän-sôn′) A masculine personal name. Also **Samp′son**. Also *Pg.* **San·são** (sän-soun′), *Sp.* **San·són** (sän-sōn′). [<Hebrew, the sun]
— **Samson** A Hebrew judge of great physical strength, betrayed to the Philistines by Delilah. *Judges* xiii 24.

Sam·u·el (sam′yōō·əl; *Ger.* zä′mōō·el, *Fr.* sà·mwel′, *Sp.* sä·mōō·el′) A masculine personal name. Also *Hungarian* **Sá·mu·el** (shä′mōō·el), *Ital.* **Sa·mue·le** (sä·mwä′lā). [<Hebrew, name of God]
— **Samuel** A Hebrew judge and prophet, I *Sam.* i 20; also, either of two historical books of the Old Testament.

Sam·u·rai (sam′ōō-rī) *n. pl.* **·rai** Japanese Under the Japanese feudal system, a member of the soldier class of the lower nobility, acting as a military retainer of the daimios; also, the class itself.

San (sän) A river in SE Poland, flowing 247 miles NW from the Carpathians to the Vistula.

Sa·naa (sä-nä′) The capital of Yemen, Arabia. Also **Sa·na′**.

San An·to·ni·o (san an-tō′nē-ō) A city in south central Texas, the third largest in the State; a port of entry with a free port zone on the **San Antonio River**, which, rising here, flows 195 miles SE to join the Guadalupe River near its mouth on San Antonio Bay.

San Antonio Bay An inlet of the Gulf of Mexico, extending 19 miles into southern Texas.

san·a·tive (san′ə-tiv) *adj.* Healing; sanatory; health-giving. [<OF, fem. of *sanatif* <Med. L *sanativus* <L *sanatus*, pp. of *sanare* heal]

san·a·to·ri·um (san′ə-tôr′ē-əm, -tō′rē-) *n. pl.* **·to·ri·ums** or **·to·ri·a** (-tôr′ē-ə, -tō′rē-ə) **1** A health retreat, especially one in the mountains. **2** An institution for treatment of disease by curative waters or climate, or for the care of invalids. [<NL <LL *sanatorius* SANATORY]

san·a·to·ry (san′ə-tôr′ē, -tō′rē) *adj.* Promotive of health; curative. [<LL *sanatorius* <L *sanatus*, pp. of *sanare heal*]

san·be·ni·to (san′bə-nē′tō) *n. pl.* **·tos** A black garment worn by a condemned heretic or a yellow cloak worn by a penitent under the Inquisition. [<Sp. *sambenito* <*San Benito* Saint Benedict; so called from its resemblance to a Benedictine's cloak.]

San Ber·nar·di·no Mountains (san bûr′nər-dē′nō) A range in SE California south of the Mojave Desert; highest peak, 11,485 feet.

San Bernardino Pass A pass in the Lepontine Alps, SE Switzerland; 6,770 feet.

San Blas (sän bläs′), **Gulf of** An inlet of the Caribbean on the north coast of Panama, east of the Panama Canal.

San·cho Pan·za (sang′kō pan′zə, *Sp.* sän′chō pän′thä) In Cervantes' *Don Quixote*, a credulous peasant who acts as squire to the Don.

San Cris·to·bal (san kris·tō′bəl) One of the British Solomon Islands; 80 miles long, 25 miles wide. Also **San Cris·to′val** (-tō′vəl).

San Cris·tó·bal Island (san kris·tō′bəl) The chief island of the Galápagos group; 195 square miles: also *Chatham Island*.

sanc·ti·fied (sangk′tə-fīd) *adj.* Made holy; freed from sin; consecrated; also, sanctimonious.

sanc·ti·fy (sangk′tə-fī) *v.t.* **·fied, ·fy·ing** **1** To set apart as holy or for holy purposes; consecrate. **2** To free of sin; purify or make holy. **3** To give religious sanction to; render sacred or inviolable, as a vow. **4** To render productive of or conducive to holiness or spiritual blessing. [<OF *saintifier*, *sanctifier* <LL *sanctificare* <L *sanctus* holy + *facere* make]
— **sanc′ti·fi·ca′tion** *n.* — **sanc′ti·fi′er** *n.*

sanc·ti·mo·ni·ous (sangk′tə-mō′nē-əs) *adj.* **1** Making an ostentatious display or a hypocritical pretense of sanctity. **2** *Obs.* Holy. — **sanc′ti·mo′ni·ous·ly** *adv.* — **sanc′ti·mo′ni·ous·ness** *n.*

sanc·ti·mo·ny (sangk′tə-mō′nē) *n.* Assumed or outward sanctity; a show of holiness or devoutness; exaggerated gravity or solemnity. See synonyms under HYPOCRISY, SANCTITY. [<OF *sanctimonie* <L *sanctimonia* holiness <*sanctus* holy]

sanc·tion (sangk′shən) *v.t.* **1** To approve authoritatively; confirm; ratify. **2** To countenance; allow. See synonyms under ABET, ALLOW, CONFIRM, RATIFY. — *n.* **1** Final and authoritative confirmation; justification or ratification. **2** A formal decree. **3** A provision for securing conformity to law, as by the enactment of rewards or penalties or both; a reward or penalty. **4** *pl.* In international law, a coercive measure adopted, usually by several nations at the same time, to force a nation which is violating international law to desist or yield to adjudication, by withholding loans, limiting trade relations, or by military force and blockade. **5** In ethics, that which makes virtue morally obligatory, or which furnishes a motive for man to seek it. [<MF <L *sanctio, -onis* ordaining something inviolable, a decree <*sanctus*, pp. of *sancire* make sacred, decree]

sanc·ti·ty (sangk′tə-tē) *n. pl.* **·ties** **1** The state of being sanctified; holiness. **2** Sacredness; solemnity. [<OF *sainteté* <L *sanctitas, -tatis* <*sanctus* holy]
Synonyms: holiness, sanctimoniousness, sanctimony. As referring to character, *sanctity* is *holiness*, while *sanctimoniousness*, or *sanctimony* is the pretense or affectation of *holiness*. Compare synonyms for HOLY.

sanc·tu·ar·y (sangk′chōō-er′ē) *n. pl.* **·ar·ies** **1** A holy or sacred place; especially, a building or space, as a church, mosque, temple, or structure devoted to the worship of any deity. **2** The most sacred part of a place in a sacred structure; especially, the part of a church where the principal altar is situated; in Scripture, the holy of holies of the Jewish tabernacle and temple; also, the adytum of an ancient Greek or Roman temple. **3** A place of refuge; asylum; hence, immunity. See synonyms under REFUGE, SHELTER. [<OF *saintuarie* <LL *sanctuarium* <L *sanctus* holy]

sanc·tum (sangk′təm) *n. pl.* **·tums** or **·ta** (-tə) **1** A sacred place. **2** A private room where one is not to be disturbed. [<L, neut. of *sanctus* holy]

sanc·tum sanc·to·rum (sangk′təm sangk·tôr′əm, -tō′rəm) **1** The holy of holies. **2** A place

Sanctus of great privacy: often used humorously. [< L *sanctum*, neut. nominative sing. + *sanctorum*, neut. genitive pl. of *sanctus* holy]

Sanc·tus (sangk'təs) *n. Eccl.* 1 An ascription of praise to God, occurring at the end of the Preface in many eucharistic liturgies. 2 A musical setting for this. [< L *sanctus* holy, its thrice repeated opening word]

Sanctus bell *Eccl.* In the celebration of the Eucharist, a bell rung at the singing of the Sanctus, the elevation of the Host, etc.: also called *mass bell*, *sacring bell*.

San·cy (sän·sē') *Puy de* The highest peak of the Massif Central, France; 6,817 feet.

sand (sand) *n.* 1 A hard, granular, comminuted rock material finer than gravel and coarser than dust. 2 *pl.* Sandy wastes; stretches of sandy beach. 3 *pl.* Sand grains or particles, as those of the hourglass; hence, moments of time or life. 4 *Slang* Strength of character; endurance; grit; courage. 5 A reddish-yellow color. — *v.t.* 1 To sprinkle or cover with sand. 2 To smooth or abrade with sand or sandpaper. 3 To mix sand with: to *sand* sugar. 4 To fill with sand, as a harbor by the action of currents. [OE]

Sand (sand, *Fr.* sänd), **George** Pseudonym of Amandine Aurore Lucie Dudevant, 1803–76, née Dupin, French novelist.

San·da·kan (sän·dä'kän) A port of British North Borneo, its chief town and former capital.

san·dal[1] (san'dəl) *n.* 1 A foot covering, consisting usually of a sole only, held to the foot by thongs. 2 A light slipper. 3 An overshoe of rubber, cut very low. 4 A strap or latchet for fastening a low shoe on the foot. 5 Sendal. [< L *sandalium* < Gk. *sandalion*, dim. of *sambalon*, *sandalon*] — **san'daled** *adj.*

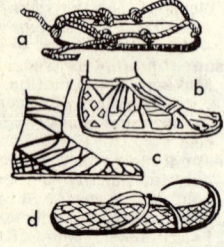
SANDALS
a. Japanese. c. Greek.
b. Roman. d. Egyptian.

san·dal[2] *n.* Sandalwood.

sandal tree A Burmese evergreen tree (*Sandoricum koetjape*), extensively cultivated in the tropics. Its fruit is an applelike edible berry.

san·dal·wood (san'dl·wood') *n.* 1 The fine-grained, dense, fragrant wood of any of several East Indian trees (genus *Santalum*). 2 The similar wood of other trees, as the East Indian **red sandalwood** (*Pterocarpus santalinus*); also called **sanderswood**. [< obs. *sandal* sandalwood (< Med. L *sandalum*, ult. < Skt. *śandana*) + WOOD]

Sandalwood Island A former name for SUMBA.

san·da·rac (san'də·rak) *n.* A pale-yellow aromatic gum resin that exudes in drops from the sandarac tree: used as a lacquer and as an incense. See GUM[1]. Also **san'da·rach**. [< L *sandaraca* < Gk. *sandarakē* < an Oriental source]

sandarac tree A medium-sized North African tree (*Tetraclinis articulata*), yielding sandarac gum and a hard, dark-colored, fragrant wood susceptible of a high polish and used in ornamental work. Also **sandarach tree**.

sand·bag (sand'bag') *n.* 1 A bag filled with or intended for holding sand: used for building fortifications, for ballast, etc. 2 A long, narrow bag filled with sand and used as a club or weapon. — *v.t.* **·bagged**, **·bag·ging** 1 To fill or surround with sandbags. 2 To strike or attack with or as with a sandbag. — **sand'bag'ger** *n.*

sand·bar (sand'bär') *n.* A ridge of silt or sand in rivers, along beaches, etc., formed by the action of currents or tides.

sand bird Any of various birds frequenting the seashore, as a snipe or sandpiper.

sand·blast (sand'blast', -bläst') *n.* 1 An apparatus for propelling a jet of sand, as for engraving patterns on glass. 2 The jet of sand. 3 A sandstorm. — *v.t.* To clean or engrave by means of a sandblast.

sand·blind (sand'blīnd') *adj.* Partially blind; having the vision affected by appearance of moving specks, etc. — **sand'blind'ness** *n.*

sand·box (sand'boks') *n.* 1 A box with a perforated top, formerly used for sanding freshly written paper to avoid blotting. 2 A reservoir on a locomotive filled with sand to be poured on the rail treads in front of the forward drivers to prevent slipping. 3 A box of sand for children to play in. 4 The sandbox tree.

sandbox tree A tropical American tree (*Hura crepitans*), often cultivated for its curious woody capsules which burst with a loud report when ripe.

sand·bur (sand'bûr') *n.* 1 A pernicious weed (*Solanum rostratum*) of the great plains of the western United States, having prickly foliage. 2 An ambrosiaceous weed (*Franseria acanthicarpa*) common in western North America. Also **sand'burr'**.

Sand·burg (sand'bûrg, san'-), **Carl**, born 1878, U.S. poet.

sand-cast (sand'kast', -käst') *v.t.* **-cast**, **-casting** To make (a casting) by pouring metal into a mold of sand.

sand-crack (sand'krak') *n.* A crack running down from the coronet of a horse's hoof and apt to cause lameness if neglected. See QUARTER-CRACK.

sand dab See under DAB[1].

sand dollar Any small, flat sea urchin (genus *Echinarachnius*) having a circular shell, found on sandy bottoms from New Jersey to Labrador and on the Pacific coast.

sand·ed (san'did) *adj.* 1 Filled, covered, or clogged with sand. 2 Of a sandy color; minutely speckled.

sand eel One of a family (*Ammodytidae*) of fishes with elongate bodies. Also **sand lance** or **sand launce**.

sand·er (san'dər) *n.* 1 One who or that which sands, as a locomotive sandbox. 2 A sandpapering machine.

san·der·ling (san'dər·ling) *n.* A small sandpiper (*Crocethia alba*) of arctic breeding habits, the adult gray and white in winter but having a rusty breast in summer. [< SAND + OE *yrthling* a kind of small bird, a ploughman]

san·ders·wood (san'dərz·wood') *n.* Sandalwood (def. 2). Also **san'ders**.

sand flea The chigoe. 2 A beach flea.

sand fly Any of various minute hairy flies (family *Psychodidae*) found near the seashore and in damp places: some of the genus *Phlebotomus* are carriers of the tropical disease leishmaniasis.

sand grouse An Old World bird (family *Pteroclidae*) of pigeonlike form, with long pointed wings and short feathered legs, inhabiting sandy tracts.

san·dhi (san'dē, sän'-) *n. Ling.* 1 A phonetic environment in which a word undergoes assimilative change from its absolute form under the influence of neighboring words: "Did you" becomes (dij'ōō) in *sandhi*. 2 The assimilative changes occurring in combined sounds in consecutive speech: "Has" becomes (s) by *sandhi* in the sentence "Jack's done that." [< Skt. *samdhi* a placing together]

sand-hill·er (sand'hil'ər) *n.* A poor white inhabitant of the sand-hill districts of Georgia and South Carolina; a cracker.

sand·hog (sand'hôg', -hog') *n.* One who works under air pressure, as in caisson-sinking, tunnel-building, etc.: also called *ground hog*.

sand hopper A flea (def. 2).

Sand·hurst (sand'hûrst) A village in Berkshire, England; seat of the Royal Military College.

San Di·e·go (san dē·ā'gō) A port and U.S. naval base in SW California, on **San Diego Bay**, a landlocked natural harbor separated from the Pacific Ocean by overlapping peninsulas.

sand lily A low-growing herb (*Leucocrinum montanum*) of the lily family, with fragrant white flowers, native in western and Pacific States: also *star lily*.

sand-lot (sand'lot') *adj.* Of or in a vacant lot in or near an urban area: applied to games played in such lots: *sand-lot* baseball.

sand·man (sand'man') *n.* In nursery lore, a mythical person supposed to make children sleepy by casting sand in their eyes.

sand martin The bank swallow. See under SWALLOW[2].

sand painting An indigenous Amerindian art form practiced especially by the Navaho. Pigments of finely ground sand in five colors are trickled on a ground base of neutral-colored sand to give highly symbolic representations (usually the gods, a rainbow, lightning, etc.). Each painting, whether three or

twenty feet in diameter, has to be started at dawn and finished by sunset.

sand·pa·per (sand'pā'pər) *n.* Stout paper coated with sand for smoothing or polishing. — *v.t.* To rub or polish with sandpaper.

sand pine The smooth-barked pine (*Pinus clausa*) of sandy areas of the southern United States, especially common to the Gulf coast of Florida.

sand·pi·per (sand'pī'pər) *n.* Any of certain small wading birds (family *Scolopacidae*), mostly frequenting seashores in flocks. The two best known are the **common sandpiper** (*Actitis hypoleuca*) of Europe, and the **spotted sandpiper** (*A. macularia*) of North America. — **least sandpiper** A tiny, common, American marsh and shore bird (*Pisobia minutilla*): also **sand'peep'** (-pēp').

SANDPIPER
(From 7 to 9 inches long)

San·dring·ham (san'dring·əm) A royal estate and parish of NW Norfolk, England.

San·dro·cot·tus (san'drō·kot'əs) See CHANDRAGUPTA I.

sand·stone (sand'stōn') *n.* A rock consisting chiefly of quartz sand cemented with silica.

sand·storm (sand'stôrm') *n.* A high wind by which sand or dust is carried along.

San·dus·ky (san·dus'kē) A port of entry on Lake Erie in northern Ohio.

sand verbena A trailing annual or perennial plant (genus *Abronia*) with vivid red, yellow, or white flowers, native in deserts of the western United States.

sand viper 1 The hog-nosed snake. 2 The horned viper.

sand·wich (sand'wich, san'-) *n.* Two thin slices of bread, having between them meat, cheese, etc.; hence, any combination of alternating dissimilar things pressed together. — *v.t.* To place between two layers or objects; insert between dissimilar things. [after John Montagu, fourth Earl of *Sandwich*, 1718–92, who is said to have originated it in order to eat without leaving the gaming table]

Sand·wich (sand'wich) A municipal borough in Kent, England, near Dover; the most ancient of the Cinque Ports.

Sandwich Islands A former name for the HAWAIIAN ISLANDS.

sand·wich·man (sand'wich·man', -mən, san'-) *n. pl.* **-men** (-men', -mən) A man carrying advertising boards slung in front and behind.

sand·wort (sand'wûrt') *n.* Any of a genus (*Arenaria*) of low, usually tufted herbs, with opposite sessile leaves and small white flowers.

sand·y (san'dē) *adj.* **sand·i·er**, **sand·i·est** 1 Consisting of or characterized by sand; containing, covered with, or full of sand. 2 Yellowish-red: a *sandy* beard. — **sand'i·ness** *n.*

Sandy Hook A peninsula, 6 miles long, extending north from eastern New Jersey, at the entrance to New York Bay.

sane[1] (sān) *adj.* 1 Mentally sound; not deranged. 2 Proceeding from a sound mind. [< L *sanus* whole, healthy] — **sane'ly** *adv.* — **sane'ness** *n.*
Synonyms: healthy, lucid, rational, sober, sound, underanged, unperverted. See SOBER.

sane[2] (sān) See SAIN.

San·ford (san'fərd), **Mount** The highest peak of the Wrangell Mountains in southern Alaska; 16,208 feet.

San·for·ize (san'fə·rīz) *v.t.* **·ized**, **·iz·ing** To treat (cloth) by a special mechanical process so as to prevent more than slight shrinkage: a trade name. [after *Sanford* L. Cluett, U.S. inventor of the process, born 1874] — **San'for·ized** *adj.* — **San'for·iz'ing** *adj.* & *n.*

San Fran·cis·co (san'frən·sis'kō) The second largest city of California, a port on **San Francisco Bay**, a landlocked inlet of the Pacific Ocean in western Cali-

San' Fran·cis'can n. & adj.
San Francisco Peaks Three peaks of an extinct eroded volcano in northern Arizona; highest peak, also highest in State, 12,655 feet.
sang[1] (sang) Past tense of SING.
sang[2] (sang) n. Scot. Song.
San·gal·lo (sän-gäl'lō) **Giuliano da,** 1445-1516, Italian architect and sculptor.
San·ga·mon River (sang'gə-mən) A river in central Illinois, flowing 250 miles west to the Illinois River.
sang·ar (sang'ər, sung'ər) n. A shelter, breastwork, or rifle pit for two or three men. Also **sang'er, san'ga.** [< Hind. *sangar*]
san·ga·ree (sang·gə·rē') n. A tropical drink of wine or brandy and water, spiced and sweetened. [< Sp. *sangría*, lit., bleeding < *sangre* blood < L *sanguis*]
San·gay (säng·gī') An active volcano in east central Ecuador; 17,454 feet.
sang de bœuf (sän də bœf') French Oxblood.
Sang·er (sang'ər), **Margaret,** born 1883, U. S. advocate of birth-control education.
sang-froid (sän·frwä') n. Calmness amid trying circumstances; coolness; composure. [< F, lit., cold blood]
San·gha (sung'gə) n. 1 The assembly; one of the three jewels of the Buddhist triad; the union of the generative power of Buddha with the productive power of the female Dharma. 2 Any order or community of Buddhist monks. 3 The total body of Buddhist monks everywhere. 4 A community of Jain monks. [< Skt. *samgha* close contact, an assemblage < *samhan* strike together, unite closely]
San·gihe Islands (säng'ir) An Indonesian island group between the Celebes Sea and the Molucca Sea; total, 314 square miles. Also **San·gi Islands** (säng'ē).
San·gre·al (sang'grē·əl) n. The Holy Grail. Also **San·graal** (sang-gräl'). [< OF *Saint Graal* < *saint* holy < L *sanctus*) + *graal* GRAIL]
San·gre de Cris·to Mountains (säng'grā dā krēs'tō) A mountain range in southern Colorado, the southernmost range of the Rocky Mountains; highest point, 14,363 feet.
san·guic·o·lous (sang·gwik'ə·ləs) adj. Inhabiting the blood, as a parasite. [< L *sanguis* blood + *colere* inhabit]
san·guif·er·ous (sang·gwif'ər·əs) adj. Conducting blood, as the organs of circulation. [< L *sanguis* blood + -FEROUS]
san·gui·nar·i·a (sang·gwə·nâr'ē·ə) n. The bloodroot, or its medicinal preparation which is emetic. [< NL < L (*herba*) *sanguinaria*, fem. of *sanguinarius* SANGUINARY]
san·gui·nar·y (sang'gwə·ner'ē) adj. 1 Attended with bloodshed. 2 Prone to shed blood; bloodthirsty. 3 Consisting of blood. [< L *sanguinarius* < *sanguis*, -*inis* blood] — **san'gui·nar'i·ly** adv. — **san'gui·nar'i·ness** n.
Synonyms: bloodthirsty, bloody, cruel, inhuman, murderous, sanguine, savage. *Sanguinary* applies either to the act of shedding blood or to the spirit that delights in bloodshed; *bloody* applies more directly to the actual staining with blood; we may say either a *sanguinary* or a *bloody* battle, but a *bloody* (not a *sanguinary*) field. *Sanguine* is sometimes used in poetic or elevated style in the sense of *bloody*; as, a *sanguine* stain. See BLOODY.
san·guine (sang'gwin) adj. 1 Of buoyant disposition; hopeful; confident; originally, having a temperament supposed to be due to active blood. 2 Having the color of blood; of, like, or full of blood. 3 *Obs.* Bloodthirsty; sanguinary. [< OF *sanguin* < L *sanguineus* < *sanguis, sanguinis* blood] — **san'guine·ly** adv. — **san'guine·ness** n.
Synonyms: animated, ardent, buoyant, confident, enthusiastic, hopeful. *Sanguine,* from the same root as *sanguinary,* came to denote full-blooded or plethoric, hence, *ardent, confident, hopeful,* because these qualities were supposed to be associated with fullness of blood. For the rare use of *sanguine* in direct literal sense, see synonyms under SANGUINARY.
san·guin·e·ous (sang·gwin'ē·əs) adj. 1 Pertaining to, consisting of, or forming blood. 2 Full-blooded; sanguine; hence, hopeful. 3 Of the color of blood. [< L *sanguineus* SANGUINE]

san·guin·o·lent (sang·gwin'ə·lənt) adj. Tinged or mixed with blood; bloody. [< OF < L *sanguinolentus* < *sanguis, sanguinis* blood]
San·he·drin (san'hi·drin, san'i-) n. 1 In ancient times, the supreme council and highest court of the Jewish nation: also **Great Sanhedrin.** 2 Figuratively, any council or assembly. Also spelled *Synedrion, Synedrium.* Also **San'he·drim** (-drim). [< Hebrew *sanhedrin* < Gk. *synedrion,* lit., a sitting together < *syn-* together + *hedra* a seat]
san·i·cle (san'i·kəl) n. Any of a genus (*Sanicula*) of smooth perennial herbs of the carrot family, reputed to have medicinal roots. [< OF < Med. L *sanicula,* prob. dim. < L *sanus* healthy; with ref. to its reputed healing powers]
sa·ni·es (sā'ni·ēz) n. *Pathol.* A serous, greenish, blood-tinged fluid discharged from ulcers. [< NL < L]
San Il·de·fon·so (sän ēl'thä·fôn'sō) A town in central Spain, 38 miles NW of Madrid; site of royal palace; scene of the signing of a treaty by Spain, France, and England, 1796: also **La Granja.**
sa·ni·ous (sā'nē·əs) adj. 1 Of or like sanies; watery and blood-tinged. 2 Producing or discharging sanies.
san·i·tar·i·an (san'ə·târ'ē·ən) n. A person skilled in matters relating to sanitation and public health.
san·i·tar·i·um (san'ə·târ'ē·əm) n. pl. ·tar·i·ums or ·tar·i·a (-târ'ē·ə) A sanatorium. [< NL < L *sanitas* health]
san·i·tar·y (san'ə·ter'ē) adj. 1 Relating to the preservation of health. 2 Cleanly; disease-preventing. See synonyms under HEALTHY. — n. pl. ·tar·ies A public watercloset or urinal. [< F *sanitaire* < L *sanitas* health < *sanus* healthy] — **san'i·tar'i·ly** adv.
sanitary belt A belt, usually made of elastic, that has tabs to which a sanitary napkin may be attached.
sanitary cordon See CORDON SANITAIRE.
sanitary napkin An absorbent pad used by women during menstruation.
san·i·tate (san'ə·tāt) v.t. ·tat·ed, ·tat·ing To apply sanitary measures to. Also **san'i·tize.** [Back formation < SANITATION]
san·i·ta·tion (san'ə·tā'shən) n. The practical application of sanitary science; the removal or neutralization of elements injurious to health. [< SANIT(ARY) + -ATION]
san·i·ty (san'ə·tē) n. 1 The state of being sane or sound; soundness of mind; mental health. 2 Sane moderation or reasonableness. [< MF *sanité* < L *sanitas* health < *sanus* healthy]
Synonyms: rationality, reasonableness, soundness. *Antonyms:* insanity.
San Ja·cin·to (san' jə·sin'tō) A locality in eastern Texas, scene of a Texan victory against Mexico, 1836; at the mouth of the **San Jacinto River,** which flows 115 miles south to Galveston Bay.
san·jak (sän'jäk') n. Formerly, an administrative subdivision of a Turkish province or vilayet. [< Turkish *sanjāg,* lit., a banner]
San Joa·quin River (san' wô·kēn', wä·kēn') A river of south central California, flowing 317 miles through Central Valley to the Sacramento River just above its mouth.
San Jo·sé (sän hō·zā') 1 The capital of Costa Rica. 2 (san' hō·zā') A city on San Francisco Bay in western California.
San José scale A scale insect (*Quadraspidiotus perniciosus*) destructive to various fruit trees: so called because it first appeared in the United States at San José, California. For illustration see INSECTS (injurious).
San Juan (sän hwän') 1 A port, capital of Puerto Rico. 2 A province of west central Argentina; 33,249 square miles; capital, San Juan.
San Juan de la Cruz (thä lä krōōth') See JOHN OF THE CROSS, SAINT.
San Juan de los Mor·ros (thä lōs môr'rōs) The capital of Guárico state, north central Venezuela.
San Juan Hill A hill near Santiago, east Cuba; captured by United States troops in the Spanish-American War.
San Juan Islands An American island group lying between SE Vancouver Island and the mainland of NW Washington at the northern end of Puget Sound.
San Juan Mountains A range of the Rocky Mountains in SW Colorado; highest peak, 14,306 feet.
sank (sangk) Past tense of SINK.
San·key (sang'kē), **Ira David,** 1840-1908, U. S. evangelist and hymn writer.
San·khya (säng'kyə) n. The oldest system of Indian philosophy, professing unqualified dualism: founded by Kapila, fabled son of Brahma. [< Skt. *Sāmkhya* < *samkhyā* enumeration; with ref. to its enumeration of twenty-four material principles (*tattva*) and one independent immaterial principle]
Sankt Mo·ritz (zängt mō'rits) The German name for ST. MORITZ.
San Lu·is (sän lōō·ēs') A province in west central Argentina; 29,625 square miles; capital, San Luis.
San Luis Po·to·sí (pō'tō·sē') A state in central Mexico; 24,415 square miles; capital, San Luis Potosí.
San Ma·ri·no (mä·rē'nō) An independent republic in eastern Italy near the coast of the Adriatic Sea; 23 square miles; capital, San Marino.
San Mar·tín (mär·tēn'), **José de,** 1778-1850, South American general and statesman.
San Mi·guel Gulf (mē·gel') The eastern part of the Gulf of Panama, adjacent to eastern Panama.
san·nup (san'up) n. A married male American Indian; the husband of a squaw. Also **san'nop.** [< Algonquian (Narraganset) *sannop*]
San Pa·blo Bay (san' päb'lō) The northern part of San Francisco Bay, California.
San Re·mo (sän rā'mō) A port and resort on the Gulf of Genoa in NW Italy.
sans (sanz, *Fr.* sän) prep. Without. [< OF *sens, sanz,* alter. of L *absentia* absence, infl. by *sine* without]
San Sal·va·dor (san' sal'və·dôr, *Sp.* sän säl'vä·thôr') The capital of El Salvador.
San Salvador Island An island in the central Bahamas, the first landing place of Columbus in the New World, 1492: also called **Watling Island.**
san·sar (sän'sər) n. A sarsar.
sans cé·ré·mo·nie (sän sā·rā·mô·nē') French Without ceremony; informal.
sans-cu·lotte (sanz'kyōō·lot', *Fr.* sän·kü·lôt') n. 1 A name originally applied by the aristocrats as a term of contempt for those who started the revolution of 1789; later it became a popular name for one of a revolutionary mob; a Jacobin. 2 Any revolutionary republican or radical. 3 Any ragged or strangely dressed person. [< F, lit., without knee breeches] — **sans'cu·lot'tic** adj. — **sans'·cu·lot'tism** n.
sans-cu·lot·tides (sanz'kyōō·lot'idz, *Fr.* sän'·kü·lô·tēd') See CALENDAR (Republican).
sans doute (sän dōōt') French Without doubt; unquestionably.
San Se·bas·tián (san sə·bas'chən, *Sp.* sän sä·väs·tyän') A port on the Bay of Biscay, in northern Spain.
San·sei (sän·sā) n. pl. ·sei An American citizen of Japanese descent whose grandparents settled in the United States; a third-generation Japanese American. [< Japanese, third generation]
san·se·vi·e·ri·a (san'sə·vi·ir'ē·ə) n. Any of a genus (*Sansevieria*) of erect perennial herbs of the lily family, native in Africa but sometimes grown as an ornamental plant. [< NL, after the Prince of *Sanseviero,* 1710-71, a Neapolitan savant]
San·skrit (san'skrit) n. The ancient and classical language of the Hindus of India, belonging to the Indic branch of the Indo-Iranian subfamily of Indo-European languages. It includes specifically **Vedic Sanskrit,** the language of the Vedas, and the later **classical Sanskrit** of India's great religious, philosophical, and poetic literature, still used for sacred or learned writings, and distinguished from the vernacular Prakrit. Also **San'scrit.** Abbr. *Skt.* [< Skt. *samskrita* well-formed < *sam-* together + *kṛ* make, do] — **San'skrit·ist** n.
San·skrit·ic (san-skrit'ik) adj. 1 Of, pertaining to, or written in the ancient and sacred language of India. 2 Designating a group

Sansovino

of some 30 to 40 ancient and modern languages and dialects of India, embracing Sanskrit, Prakrit, Pali, Assamese, Bengali, Eastern and Western Hindi, Punjabi, Singhalese, Romany or Gipsy, etc.

San·so·vi·no (sän'sō·vē'nō), 1486?–1570, Italian sculptor and architect: original name *Jacopo Tatti.*

sans pa·reil (sän pá·rā'y) *French* Without equal.

sans peur et sans re·proche (sän pœr ā sän rə·prōsh') *French* Without fear and without reproach.

sans ser·if (sanz ser'if) *Printing* A style of type without serifs.

sans–sou·ci (sän·sōō·sē') *adj. French* Careless; free and easy.

San Ste·fa·no (sän stä'fä·nō) A village on the Sea of Marmara west of Istanbul, in Turkey in Europe; scene of the signing of a Russo-Turkish treaty, 1878: Turkish *Yesilköy.*

San·ta An·a (sän'tä ä'nä) A city in NW El Salvador, the second largest; a coffee center and rail junction.

San·ta An·na (sän'tä ä'nä), **Antonio Lopez de,** 1795–1876, Mexican general: president and dictator of Mexico; massacred the surviving defenders of the Alamo, Mar. 6, 1836; defeated by the U.S. Army in 1847. Also **San'ta A'na.**

San·ta Bar·ba·ra Islands (sän'tə bär'bər·ə) A chain of small islands, extending 150 miles along the coast of southern California.

San·ta Cat·a·li·na (sän'tə kat'ə·lē'nə) One of the Santa Barbara Islands, 24 miles south of Los Angeles, California; 22 miles long: also *Catalina Island.*

San·ta Ca·ta·ri·na (sän'tə kä'tə·rē'nə) A maritime state in southern Brazil; 36,592 square miles; capital, Florianópolis.

San·ta Cla·ra (sän'tä klä'rä) A city in central Cuba.

San·ta Claus (sän'tə klôz') In nursery folklore, a friend of children who brings presents at Christmas time: usually represented as a fat, jolly old man. The patron saint of children, figuring in the nursery lore of many countries and identified with *St. Nicholas.* [< dial. Du. *Sante Klaus* Saint Nicholas]

San·ta Cruz (sän'tä krōōz', *Sp.* sän'tä krōōth') 1 St. Croix. 2 A national territory in southern Patagonia, Argentina; 77,822 square miles; capital, Río Gallegos.

Santa Cruz de Ten·er·ife (də ten'ə·rif', *Sp.* thä tä'nä·rē'fā) 1 One of the two provinces in the Canary Islands, comprising Tenerife, La Palma, Gomera, and Hierro; 1,238 square miles. 2 A port of NE Tenerife, capital of this province and of Tenerife island.

Santa Cruz Islands An island group north of the New Hebrides, comprising part of the British Solomon Islands protectorate; total, 370 square miles; scene of U. S. naval victory over Japanese, 1942.

San·ta Fe (sän'tä fā') 1 The capital of New Mexico, in the northern part of the State. 2 (sän'tä fā') A province of NE central Argentina; 51,341 square miles; capital, Santa Fe.

Santa Fe trail The trade route, important from 1821 to 1880, between Independence, Missouri, and Santa Fe, New Mexico.

San·ta Is·a·bel (sän'tä ē'sä·bel) 1 A city on Fernando Pó island, capital of Spanish Guinea. 2 (sän'tə iz'ə·bel) One of the British Solomon Islands; 1,800 square miles: also *Ysabel.*

san·ta·la·ceous (san'tə·lā'shəs) *adj. Bot.* Of or pertaining to a family (*Santalaceae*) of apetalous shrubs, herbs, and some trees; the sandalwood family. [< NL <*Santalum,* genus name < Med. L]

san·tal·ic (san·tal'ik) *adj.* Of, pertaining to, or derived from sandalwood, as **santalic acid,** a red crystalline coloring matter, $C_{15}H_{14}O_5$. [< NL *santal(um)* sandalwood (< Med. L) + -IC]

San·ta Ma·ri·a (sän'tə mə·rē'ə, *Sp.* sän'tä mä·rē'ä) One of the three ships of Columbus on his maiden voyage to America.

San·ta Ma·ri·a (sän'tä mä·rē'ä) 1 An island in the SE Azores; 37 square miles. 2 An active volcano in SW Guatemala; 12,362 feet.

San·ta Mau·ra (sän'tä mou'rä) The Italian name for LEVKAS.

San·tan·der (sän'tän·der') A port of northern Spain on the Bay of Biscay.

San·ta·rém (sän'tə·rän') The largest second

1116

city of Pará state, Brazil, on the Tapajós at its influx into the Amazon.

San·ta·ya·na (sän'tä·yä'nä), **George,** 1863–1952, U. S. philosopher and author born in Spain.

San·tee River (san·tē') A river in east central South Carolina, flowing 143 miles SE from the junction of the Congaree and Wateree rivers, over **Santee Dam** (45 feet high, 7.8 miles long; completed 1941), to the Atlantic.

San·ti·a·go (sän·tē·ä'gō) 1 The capital of Chile. Also *Santiago de Chi·le* (thā chē'lā). 2 Santiago de Compostela. 3 Santiago de los Caballeros.

Santiago de Com·po·ste·la (thā kom'pō·stā'lä) A city and chief pilgrimage center of NW Spain: also *Santiago.*

Santiago de Cu·ba (thā kōō'bä) 1 The second largest city of Cuba and capital of Oriente province, on the southern coast. 2 The former name for ORIENTE province, Cuba.

Santiago del Es·te·ro (thel es·tā'rō) A province of northern Argentina; 22,208 square miles; capital, Santiago del Estero.

Santiago de los Ca·bal·le·ros (thā lōs kä'vä·yā'rōs) A city in northern Dominican Republic, the second largest city of the Republic: also *Santiago.*

San·to Do·min·go (sän'tō dō·ming'gō) A former name for the DOMINICAN REPUBLIC and for its capital, CIUDAD TRUJILLO.

san·ton·i·ca (san·ton'i·kə) *n.* 1 An Old World plant of the composite family, especially the European wormwood (*Artemisia maritima*). 2 The unexpanded flower heads of this plant, used as a vermifuge. [< NL < L (*herba*) *Santonica* a kind of wormwood, fem. sing. of *Santonicus* of the Santoni < *Santoni* the Santoni, a people of Aquitania]

san·to·nin (san'tə·nin) *n. Chem.* A colorless crystalline poisonous compound, $C_{15}H_{18}O_3$, contained in santonica: used in medicine as a vermifuge. Also **san'to·nine** (-nēn, -nin). [< F *santonine* < NL *santon(ica)* + F *-ine* -INE²]

San·to·rin (sän'tō·rēn') A former name for THERA.

San·tos (sän'tōōs) A port and the second largest city in São Paulo state, SE Brazil.

San·tos–Du·mont (sän'tōōz·dü·môn'), **Alberto,** 1873–1932, Brazilian airship pioneer active in France.

São Fran·cis·co (souñ frän·sēs'kōō) A river of eastern Brazil, flowing 1,800 miles to the Atlantic; the third largest drainage basin of Brazil; developed for hydroelectric power.

São Jor·ge (souñ zhôr'zhə) An island in the central Azores; 92 square miles.

São Lu·ís (souñ lōō·ēs') A port of NE Brazil, capital of Maranhão state. Formerly **São Luís do Ma·ra·nhão** (thōō mä'rə·nyouñ'). Also **São Luiz.**

São Ma·nuel (souñ mə·nwel') A river of northern Mato Grosso, Brazil, flowing 700 miles NW to the Tapajós.

São Mi·guel (souñ mē·gel') An island in the eastern Azores; 288 square miles; chief town, Ponta Delgada.

Saône (sōn) A river in eastern France, flowing 268 miles SW to the Rhône at Lyon.

São Pau·lo (souñ pou'lōō) A maritime state in SE Brazil; 95,428 square miles; capital, São Paulo.

São Paulo de Lo·an·da (thə lō·än'də) A former name for LUANDA.

São Pe·dro de Ri·o Gran·de do Sul (souñm pā'thrōō thə rē'ōō grañn'də thōō sōōl') A former name for the port of RIO GRANDE DO SUL.

Saor·stat Éir·eann (sâr'stät âr'ən, *Gaelic* â'rôn) Gaelic name for IRISH FREE STATE.

São Sal·va·dor (souñ säl'və·thôr') The former name for Salvador, Brazil.

São Tia·go (souñ tyä'gōō) Largest of the Cape Verde Islands; 383 square miles; capital, Praia. Also **São Thia'go.**

Sao To·mé e Prín·ci·pe (souñ tô·me' e preñ'sē·pə) A Portuguese province in the Bight of Biafra, comprising the islands of **São Tomé** (also **São Thomé,** English *St. Thomas*); 320 square miles; and **Principe** (English *Prince Island*); 52 square miles.

sap¹ (sap) *n.* 1 The aqueous juices of plants, which contain and transport the materials necessary to vegetable growth. 2 Any vital fluid; vitality. 3 Sapwood. 4 *Slang* A foolish, stupid, or ineffectual person; a saphead. [OE *sæp*]

saponite

sap² (sap) *v.* **sapped, sap·ping** *v.t.* 1 To weaken or destroy gradually and insidiously; enervate; exhaust. 2 To approach or undermine (an enemy fortification) by digging a sap or saps. — *v.i.* 3 To dig a sap or saps; undermine an enemy fortification. See synonyms under WEAKEN. — *n.* A deep, narrow trench or tunnel dug so as to approach or undermine a fortification. [< MF *saper, sapper* < *sappe* a spade < Ital. *zappe* < *zappa* a goat; with ref. to resemblance of the handle to a goat's horns]

sap·a·jou (sap'ə·jōō, *Fr.* sá·pá·zhōō') *n.* A South American monkey, the capuchin: often seen in captivity. Also called *sajou.* [< F < Tupian]

SAPAJOU
(Head and body about 1 1/2 feet long)

sa·pan·wood (sə·pan'wood) *n.* 1 The brownish–red dyewood obtained from a medium–sized East Indian tree (*Caesalpinia sappan*) of the bean family. 2 The tree. Also spelled *sappanwood:* also called *brazil.* [Trans. of Du. *sapanhout* < Malay *sapang* sapanwood + Du. *hout* wood]

Sa·phar (sä·fär') See under CALENDAR (Mohammedan).

sap·head (sap'hed') *n. Slang* A soft–headed person; simpleton. [<SAP¹ (def. 4) + HEAD] — **sap'head'ed** *adj.*

sa·phe·na (sə·fē'nə) *n. pl.* **-nae** (-nē) *Anat.* One of the two large superficial veins of the leg. [< Med. L, a vein in the leg < Arabic *sāfīn*] — **sa·phe'nous** *adj.*

sap·id (sap'id) *adj.* Affecting the sense of taste; savory; agreeable. [< L *sapidus* < *sapere* taste] — **sa·pid'i·ty, sap'id·ness** *n.*

sa·pi·ence (sā'pē·əns) *n.* Wisdom; learning: often ironical. Also **sa'pi·en·cy.** [<OF < L *sapientia* wisdom < *sapiens, -entis* SAPIENT]

sa·pi·ent (sā'pē·ənt) *adj.* Wise; sagacious: often ironical. See synonyms under WISE¹. [< L *sapiens, -entis,* ppr. of *sapere* know, taste] — **sa'pi·ent·ly** *adv.*

sa·pi·en·tial (sā'pē·en'shəl) *adj.* Of, marked by, or expounding books of wisdom; especially, the *sapiential* books of the Bible, as Proverbs. — **sa'pi·en'tial·ly** *adv.*

sap·in·da·ceous (sap'in·dā'shəs) *adj. Bot.* Of or pertaining to a family (*Sapindaceae*) of mostly tropical trees, shrubs, and vines, the soapberry family, including some genera with edible fruit, as the litchi tree. [<NL <*Sapindus,* genus name < L *sapo* soap + *Indicus* Indian]

sap·less (sap'lis) *adj.* 1 Destitute of sap; withered. 2 Wanting vitality, spirit, or vivacity; insipid; dull.

sap·ling (sap'ling) *n.* 1 A young tree. 2 A youth. [Dim. of SAP¹]

sap·o·dil·la (sap'ə·dil'ə) *n.* 1 A large evergreen tree (*Achras zapota*) of the West Indies and Central America. 2 Its luscious apple–shaped fruit, the **sapodilla plum,** for which it is cultivated. Often called *mamey, marmalade tree.* Also **sa·po·ta** (sə·pō'tə), **sap'a·dil'lo, sap'o·dil'lo.** [<Sp. *zapotille,* dim of *zapota* < Nahuatl *zapotl, sapotl*]

sap·o·na·ceous (sap'ə·nā'shəs) *adj.* Of the nature of soap; soapy. [<NL *saponaceus* <L *sapo, saponis* soap]

sa·pon·i·fi·ca·tion (sə·pon'ə·fə·kā'shən) *n.* 1 The process or result of making soap. 2 *Chem.* a A decomposition in which an ester is changed into an acid and an alcohol. b The conversion of certain acid derivatives, as nitrates, acid amides, etc., into the corresponding acids.

sa·pon·i·fy (sə·pon'ə·fī) *v.t.* **·fied, ·fy·ing** To convert (a fat or oil) into soap by the action of an alkali. [<F *saponifier* <NL *saponificare* <L *sapo, saponis* soap + *facere* make] — **sa·pon'i·fi'a·ble** *adj.* — **sa·pon'i·fi'er** *n.*

sap·o·nin (sap'ə·nin) *n. Biochem.* One of several nearly white amorphous glycosides contained in various plants and characterized by their ability to form emulsions and produce soapy lathers. Also **sap'o·nine** (-nēn, -nin). [<F *saponine* <L *sapo, saponis* soap + *-ine* -INE²]

sap·o·nite (sap'ə·nīt) *n.* A soft, hydrous silicate of magnesium and aluminum, found as an

sa·por (sā′pər, -pôr) *n.* That quality of a substance affecting the sense of taste; flavor; taste. Also *Brit.* **sa′pour**. [<L *taste, know*] — **sap·o·rif·ic** (sap′ə·rif′ik), **sap′o·rous** *adj.*

sap·o·ta·ceous (sap′ə·tā′shəs) *adj. Bot.* Of or pertaining to a family *(Sapotaceae)* of trees and shrubs yielding a milky juice of considerable economic importance, and also some edible fruits, as the sapodilla family. [<NL <*sapota* <Sp. *zapote* SAPODILLA]

sap·pan·wood (sə·pan′wōōd′) See SAPANWOOD.

sap·per (sap′ər) *n.* 1 One who or that which saps. 2 A soldier employed in making trenches, tunnels, and underground fortifications. [<SAP² + -ER]

Sap·phic (saf′ik) *adj.* 1 Pertaining to or in the manner of Sappho. 2 Denoting a meter or verse form used by Sappho, especially a stanza of three Sapphics followed by an Adonic. — *n.* A line of trochaic pentameter with a dactyl in the third foot: much used by Sappho. [<F *sapphique*, *saphique* <L *Sapphicus* <Gk. *Sapphikos* <*Sapphō* Sappho]

Sap·phi·ra (sə·fī′rə) Wife of Ananias. *Acts* v.

sap·phire (saf′īr) *n.* 1 Any one of the hard, transparent, colored varieties of corundum which when cut are used as gems: usually and specifically, the blue variety. 2 Deep pure blue. — **star sapphire** A sapphire cut en cabochon, showing six rays on the dome. [<OF *sapir* <L *sapphirus*, *sapp(h)ir* <Gk. *sappheiros*, a gemstone <Semitic, ? ult. <Skt. *sanipriya* dear to the planet Saturn]

sap·phi·rine (saf′ər·in, -ə·rēn) *adj.* Consisting of or like sapphire. — *n.* 1 A vitreous pale blue or green silicate of aluminum and magnesium, crystallizing in the monoclinic system. 2 Sapphire quartz. 3 A blue variety of spinel.

Sap·pho (saf′ō) Greek poetess of Lesbos; lived about 600 B.C.

Sap·po·ro (säp′pō·rō) The capital of Hokkaido island, Japan.

sap·py (sap′ē) *adj.* **·pi·er**, **·pi·est** 1 Full of sap; juicy. 2 *Slang* Immature; silly. 3 Vital; pithy. — **sap′pi·ly** *adv.* — **sap′pi·ness** *n.*

sa·pre·mi·a (sə·prē′mē·ə) *n. Pathol.* Blood poisoning by the products of putrefaction. Also **sa·prae′mi·a**. [<NL <Gk. *sapros* putrid + *haima* blood] — **sa·pre′mic** *adj.*

sapro- *combining form* 1 Decomposition or putrefaction: *saprogenic*. 2 Saprophytic: *saproplankton*. [<Gk. *sapros* rotten]

sap·ro·gen·ic (sap′rə·jen′ik) *adj.* 1 Productive of putrefaction. 2 Developing in or living upon putrefying matter. Also **sa·prog·e·nous** (sə·proj′ə·nəs).

sap·ro·lite (sap′rə·līt) *n. Geol.* Thoroughly decomposed, earthy rock, lying in its original place. [<SAPRO- + -LITE] — **sap′ro·lit′ic** (-lit′ik) *adj.*

sa·proph·a·gous (sə·prof′ə·gəs) *adj.* Feeding on decaying substances. [<SAPRO- + -PHAGOUS]

sap·ro·phyte (sap′rə·fīt) *n.* An organism that lives on dead or decaying organic matter, as certain fungous or other plants, various bacteria, etc. [<SAPRO- + -PHYTE] — **sap′ro·phyt′ic** (-fit′ik) *adj.*

sap·ro·plank·ton (sap′rə·plangk′tən) *n.* Plankton found on the surface of stagnant water. [<SAPRO- + PLANKTON]

sap·sa·go (sap′sə·gō) *n.* A hard green Swiss cheese flavored with melilot, used chiefly in cooking. [Alter. of G *schabzieger* < *schaben* shave, scrape + *zieger* whey]

sap·suck·er (sap′suk′ər) *n.* Any small black-and-white woodpecker (genus *Sphyrapicus*), especially the yellow-bellied sapsucker (*S. varius*), which damages orchard trees by exposing and devouring the sapwood.

YELLOW-BELLIED SAPSUCKER (About 8 1/2 inches long)

sap sugar Maple sugar.

sap·wood (sap′wŏŏd′) *n. Bot.* The new wood next the bark of an exogenous tree; alburnum. See illustration under EXOGEN.

Saq·qa·ra (sə·kä′rə) See SAKKARA.

sar·a·band (sar′ə·band) *n.* A stately Spanish dance in triple time, of the 17th and 18th centuries; also, the music for or in the rhythm of this dance, often used as one of the movements of the classical suite. Also **sar′a·bande**. [<F *sarabande* <Sp. *zarabanda*, ult. <Persian *sarband* a kind of dance and song]

Sa·ra·bat (sä′rä·bät′) Former name of the GEDIZ.

Sar·a·cen (sar′ə·sən) *n.* 1 Originally, a nomad Arab of the Syrian-Arabian desert, who harassed the frontiers of the Roman Empire. 2 A Moslem enemy of the Crusaders. 3 Any Arab. 4 *Obs.* A heathen; pagan. [Fusion of OE *Saracene* and OF *Sarazin*, *Saracin*, both <LL *Saracenus* <LGk. *Sarakēnos*, ? <Arabic] — **Sar′a·cen′ic** (-sen′ik) or **·i·cal** *adj.*

Sar·a·gos·sa (sar′ə·gos′ə) A city of NE Spain, on the Ebro; former capital of Aragon: Spanish *Zaragoza*.

Sar·ah (sâr′ə) A feminine personal name. Also **Sar·a** (sâr′ə; *Fr.* sá·rá′, *Ital., Sp.* sä′rä, *Ger.* zä′rä). [<Hebrew *sārāh* a princess] — **Sarah** The wife of Abraham. *Gen.* xvii 15.

Sa·ra·je·vo (sä′rä·yā′vō) A city in central Yugoslavia; the former capital of Bosnia where Archduke Francis Ferdinand was assassinated, June 28, 1914: also *Serajevo*.

sa·ran (sə·ran′) *n.* Any of a class of synthetic fibers and textile materials obtained by the chemical treatment of petroleum and natural brines. [Coined by Dow Chemical Co.]

Sar·a·nac Lakes (sar′ə·nak) Three lakes in NE New York, in the Adirondack Mountains, Upper, Middle, and Lower Saranac, linked by the **Saranac River,** which flows 50 miles NE to Lake Champlain at Plattsburg.

Sa·ransk (sä·ränsk′) The capital of Mordvinian Autonomous S.S.R.

Sar·a·to·ga (sar′ə·tō′gə) A former name for SCHUYLERVILLE.

Saratoga Springs A resort city in eastern New York, noted for horse-racing and mineral waters.

Saratoga trunk A very large traveling trunk used formerly by ladies. [after *Saratoga Springs*]

Sa·ra·tov (sä·rä′tôf) A port on the Volga in SE European Russian S.F.S.R.; a major industrial and natural-gas-producing center.

Sa·ra·wak (sə·rä′wäk) A British crown colony in NW Borneo; 47,071 square miles; capital, Kuching. — **Sa·ra·wak·ese** (sə·rä′wäk·ēz′, -ēs′) *adj. & n.*

sar·casm (sär′kaz·əm) *n.* 1 A keenly ironical or scornful utterance; contemptuous and taunting language. 2 The use of biting gibes or cutting rebukes. See synonyms under BANTER. [<LL *sarcasmus* <Gk. *sarkasmos* < *sarkazein* tear flesh, speak bitterly < *sarx, sarkos* flesh]

sar·cas·tic (sär·kas′tik) *adj.* 1 Characterized by or of the nature of sarcasm. 2 Taunting. Also **sar·cas′ti·cal**. — **sar·cas′ti·cal·ly** *adv.*

sarce·net (särs′nit) See SARSENET.

Sar·ci·na (sär′si·nə) *n.* A genus of parasitic, usually Gram-positive bacteria, which divide to form clusters of individuals: many are saprophytic. See illustration under BACTERIA. [<NL <L, a bundle < *sarcire* patch, mend]

sarco- *combining form* Flesh; of or related to flesh: *sarcogenic*. Also, before vowels, **sarc-**. [<Gk. *sarx*, *sarkos* flesh]

sar·co·carp (sär′kō·kärp) *n. Bot.* The succulent part of a drupaceous fruit, as the fleshy edible part of a plum or peach. [<F *sarcocarpe* <Gk. *sarx*, *sarkos* flesh + *karpos* a fruit]

Sar·co·di·na (sär′kō·dī′nə) *n. pl.* A class of marine and fresh-water protozoa which move by means of pseudopodia, including both naked forms, as the *Amoebae*, and those with protective shell covering, as the *Foraminifera*. [<NL <Gk. *sarkōdēs* fleshy < *sarx*, *sarkos* flesh]

sar·co·gen·ic (sär′kō·jen′ik) *adj.* Flesh-producing. Also **sar·cog·e·nous** (sär·koj′ə·nəs).

sar·co·lem·ma (sär′kō·lem′ə) *n. Anat.* The elastic membrane that invests striated muscular fibers. [<NL <Gk. *sarx*, *sarkos* flesh + *lemma* a husk]

sar·co·ma (sär·kō′mə) *n. pl.* **·ma·ta** (-mə·tə) *Pathol.* A tumor, or group of tumors, often malignant, composed of embryonal lymphoid or connective tissue in which the cell elements predominate. [<NL <Gk. *sarkōma* < *sarkein* become fleshy < *sarx*, *sarkos* flesh] — **sar·co′ma·toid**, **sar·co′ma·tous** (-kō′mə·təs, -kom′ə-) *adj.*

sar·co·ma·to·sis (sär·kō′mə·tō′sis) *n. Pathol.* The formation of sarcomatous growths in the body. [<NL Gk. *sarkōma*, *-ōmatos* SARCOMA + -*osis* -OSIS]

sar·coph·a·gus (sär·kof′ə·gəs) *n. pl.* **·gi** (-jī) 1 A stone coffin or tomb; hence, a large ornamental coffin of marble or stone placed in a crypt or exposed to view. 2 A kind of limestone, used by the Greeks for coffins and said to reduce flesh to dust. [<L <Gk. *sarkophagos*, orig. *(adj.)*, flesh-eating < *sarx*, *sarkos* flesh + *phagein* eat]

sar·co·plasm (sär′kō·plaz′əm) *n. Anat.* The substance resembling hyaloplasm that lies between the columns of a striated muscle fiber.

sar·cous (sär′kəs) *adj.* Of, pertaining to, or composed of flesh or muscle. [<Gk. *sarx*, *sarkos* flesh]

sard (särd) *n.* The deep brownish-red variety of chalcedony, translucently blood-red: used as a gem. Also called *sardine*, *sardius*. [<OF *sarde* <L *sarda* <Gk. *sardios*. See SARDIUS.]

Sar·da·na·pa·lus (sär′də·nə·pā′ləs) Greek form of ASHURBANIPAL.

sar·dine¹ (sär·dēn′) *n.* 1 A small fish preserved in oil as a delicacy, especially the California pilchard (*Sardinia coerulea*). 2 The young of the herring or some like fish similarly prepared. [<OF <Ital. *sardina* <L <Gk. *sardēnē* < *sarda* a kind of fish, prob. <*Sardo* Sardinia]

sar·dine² (sär′dīn) *n.* See SARD.

Sar·din·i·a (sär·din′ē·ə) 1 An Italian island in the Mediterranean, west of Italy; 9,196 square miles; forming, with its neighboring islands, an autonomous region of Italy; 9,298 square miles; capital, Cagliari. *Italian* **Sar·de·gna** (sär·dā′nyä). 2 A former kingdom (1720–1860) of northern Italy, including the island of Sardinia with Savoy and Piedmont. — **Sar·din′i·an** *adj. & n.*

Sar·dis (sär′dis) An ancient city of Asia Minor, capital of Lydia; destroyed by Tamerlane. Also **Sar′des**.

sar·di·us (sär′dē·əs) *n.* 1 A sard. 2 A stone in the breastplate of the Hebrew high priest. *Ex.* xxviii 17. [<LL <Gk. *sardios*, *sardion* <*Sardeis* Sardis]

sar·don·ic (sär·don′ik) *adj.* Scornful or derisive; sneering; mocking; cynical. [<F *sardonique* <L *sardonius* <Gk. *sardonios* < *sardanios* bitter, scornful; infl. in form by *Sardō* Sardinia, because thought to be < *sardanē*, a bitter plant of Sardinia causing fatal, laughterlike convulsions] — **sar·don′i·cal·ly** *adv.* — **sar·don′i·cism** *n.*

sar·do·nyx (sär′dō·niks) *n.* A variety of onyx, consisting of alternate layers of light-colored chalcedony and reddish carnelian. [<L <Gk., appar. <*sardios* sardius + *onyx* onyx]

Sar·dou (sár·dōō′), **Victorien**, 1831–1908, French dramatist.

Sa·re·ma (sä′re·mä) See SAARE.

Sarg (särg), **Tony**, 1882–1942, U. S. artist born in Germany; maker of marionettes: full name *Anthony Frederick Sarg*.

sar·gas·so (sär·gas′ō) *n.* Any of a large genus (*Sargassum*) of brown algae found growing in tropical seas; the gulfweed. Also **sar·gas′sum**. [<Pg. *sargaço* < *sarga* a kind of grape]

Sargasso Sea A part of the North Atlantic, extending from the West Indies to the Azores, known for its relatively still water and its large amounts of floating seaweed.

Sar·gent (sär′jənt), **John Singer,** 1856–1925, U. S. painter.

Sar·gon II (sär′gon), died 705 B.C., king of Assyria 722–705 B.C.

sa·ri (sä′rē) *n.* A long piece of cotton or silk cloth, constituting the principal garment of Hindu women: worn round the waist, one end falling to the feet, and the other crossed over the bosom, shoulder, and sometimes over the head. Also **sa′ree**. [<Hind. *sarī*, *sarhī* <Skt. *śāṭī*]

sark (särk) *n. Scot.* A shirt or chemise; hence, a shroud. [OE *serc*]

Sark (särk) The smallest of the Channel Islands; 2 square miles. *French* **Sercq** (serk).

sark·it (sär′kit) *adj. Scot.* Provided with shirts.
Sar·ma·ti·a (sär-mā′shē-ə, -shə) An ancient name for a region of NE Europe, in Poland and U.S.S.R. between the Vistula and the Volga. — **Sar·ma′tian** *adj. & n.* — **Sar·mat′ic** (-mat′ik) *adj.*
sar·men·tose (sär-men′tōs) *adj. Bot.* Having or producing sarmenta; having runners. Also **sar·men·ta·ceous** (sär′mən-tā′shəs), **sar·men′tous.** [< L *sarmentosus* full of twigs < *sarmentum.* See SARMENTUM.]
sar·men·tum (sär-men′təm) *n. pl.* ·**ta** (-tə) *Bot.* The slender runner of a plant, as in a vine. Also **sar′ment.** [< NL < L, a twig lopped off < *sarpere* prune (trees)]
Sar·mien·to (sär-myen′tō), **Domingo**, 1811–88, Argentine educator, journalist, and statesman.
sa·rong (sə-rông′) *n.* 1 A skirtlike garment of colored silk or cloth worn by both sexes in the Malay Archipelago, etc. 2 The material used for this garment. [< Malay *sārung,* prob. < Skt. *sāranga* variegated]
Sa·ron·ic Gulf (sə-ron′ik) An inlet of the Aegean in central Greece, separating Attica from Peloponnesus; 50 miles long, 30 miles wide: also **Gulf of Aegina.**
Sa·ros (sā′rōs), **Gulf of** An arm of the Aegean in Turkey in Europe north of Gallipoli Peninsula; 37 miles long, 22 miles wide.
Sa·roy·an (sə-roi′ən), **William**, born 1908, U.S. novelist and playwright.
Sar·pe·don (sär-pē′dən) In Greek mythology: 1 A son of Zeus and Europa who was allowed to live for three generations. 2 A Lycian prince and warrior killed by Patroclus in the Trojan War.
sar·ra·ce·ni·a (sar′ə-sē′nē-ə) *n.* Any of a genus (*Sarracenia*) of North American insectivorous marsh plants, having trumpetlike or pitcher-shaped leaves by which insects are entrapped and then digested by the plants; a pitcherplant. [< NL, orig. *Sarracena,* after Dr. D. Sarrazin, 17th–18th c. physician of Quebec who sent a specimen to the botanist Tournefort in 1700] — **sar′ra·ce′ni·a′ceous** (-sē′nē-ā′shəs) *adj.*
Sarre (sâr) The French name for THE SAAR.
sar·rus·o·phone (sa-rus′ə-fōn) *n.* A musical instrument resembling a bassoon but with a metal tube. [after *Sarrus,* 19th c. French bandmaster, its inventor + -(O)PHONE]
sar·sa·pa·ril·la (sas′pə-ril′ə, sär′sə-pə-ril′ə) *n.* 1 The dried roots of certain tropical American climbing plants (genus *Smilax*). 2 A medicinal preparation or a beverage made from them. 3 Any one of various plants, so called from some resemblance to true sarsaparilla, as the wild sarsaparilla (*Aralia nudicaulis*). [< Sp. *zarzaparilla* < *zarza* a bramble + *parilla,* dim. of *parra* a vine]
sar·sar (sär′sər) *n.* A cold, whistling wind of Moslem lands: also spelled *sansar.* [< Arabic *ṣarṣar* a cold wind]
sarse·net (särs′nit) *n.* A fine, thin silk, used for linings: also spelled *sarcenet.* [< AF *sarzinet,* dim. of ME *sarzin* a Saracen; prob. infl. by OF *drap sarrasinois,* lit., Saracen cloth < Med. L *pannus saracenicus*]
Sar·to (sär′tō), **Andrea del**, 1487–1531, Florentine painter.
Sar·ton (sär′tən), **George Alfred**, 1884–1956, U.S. historian of science born in Belgium.
sar·tor (sär′tər) *n.* A tailor: a humorous or literary term. [< L, a patcher, mender < *sartus,* pp. of *sarcire* mend]
sar·to·ri·al (sär-tôr′ē-əl, -tō′rē-) *adj.* 1 Pertaining to a tailor or his work; also, pertaining to men's clothes: *sartorial* perfections. 2 *Anat.* Relating to the sartorius. — **sar·to′ri·al·ly** *adv.*
sar·to·ri·us (sär-tôr′ē-əs, -tō′rē-) *n. Anat.* A long, narrow muscle of the thigh that aids in flexing the knee; the longest muscle in the human body: so called from its use in crossing the legs, as in the manner in which tailors traditionally sat down to work. [< NL < L *sartor* a tailor]
Sar·tre (sär′tr′), **Jean Paul**, born 1905, French philosopher, novelist, and dramatist.
Sar·um (sâr′əm), **New** See SALISBURY.
Sa·se·bo (sä′se-bō) A port on NW Kyushu island, Japan.
Sa·se·no (sä′se-nô) An Albanian island in the Strait of Otranto, at the entrance to the Bay of Valona; 2 square miles. Albanian **Sa·zan** (sä′zän).
sash[1] (sash) *n.* An ornamental band or scarf, worn as a girdle, or around the waist or over the shoulder, often as part of a uniform or as a badge of distinction: [Orig. *shash* < Arabic *shāsh* muslin, turban]
sash[2] (sash) *n.* A frame, as of a window, in which glass is set. — *v.t.* To furnish with a sash. [Alter. of CHASSIS, taken as a pl.]
sa·shay (sa-shā′) See CHASSÉ[1].
sa·sin (sā′sin) *n.* The common black buck. [< Nepalese]
sa·sine (sā′sin, -sīn) *n. Scot.* The act of giving legal possession of feudal property; also, the instrument granting such possession.
Sas·katch·e·wan (sas-kach′ə-wən) A province of west central Canada; 251,700 square miles; capital, Regina: abbr. *Sask.*
Saskatchewan River A river of west central Canada, flowing 340 miles east to Lake Winnipeg from the confluence of the **North Saskatchewan**, flowing 760 miles east, and the **South Saskatchewan**, flowing 550 miles NE.
sas·ka·toon (sas′kə-tōōn′) *n.* A small tree (*Amelanchier alnifolia*) of the rose family, with thick leaves and a globular purple fruit; a shadbush. [< Algonquian (Cree) *misāskwatomin* < *misāskwat* the shadbush + *min* a fruit, a berry]
Sas·ka·toon (sas′kə-tōōn′) *n.* A city in south central Saskatchewan province, Canada, on the South Saskatchewan River.
sass (sas) *Colloq. n.* Impudence; back talk. — *v.t.* To talk to impudently or disrespectfully. [Dial. alter. of SAUCE]
sas·sa·by (sas′ə-bē) *n. pl.* ·**bies** A large dark-red South African antelope (genus *Damaliscus*), with almost black back and face. [< Bantu *tsessébe, tsessábi*]
sas·sa·fras (sas′ə-fras) *n.* 1 A tree (genus *Sassafras*) of the laurel family. 2 The bark of the roots, yielding an aromatic stimulant and an essential oil used in cosmetics. [< Sp. *sasafrás,* prob. < N. Am. Ind. name; infl. in form by Sp. *sassifragia* < L *saxifraga* saxifrage]
Sas·sa·nid sas′ə-nid) *n. pl.* **Sas·sa·nids** or **Sas·san·i·dae** (sa-san′ə-dē) A member of the last national dynasty of ancient Persia (226–651). — *adj.* Of or pertaining to the Sassanids. Also **Sas·sa·ni·an** (sa-sā′nē-ən), **Sas·sa·nide.** [< Med. L *Sassanidae,* pl. < *Sassan* Sasan, grandfather of Ardashir I, the first Sassanian king]
Sas·se·nach (sas′ə-nakh) *n. Scot. & Irish* A person of Saxon blood; an Englishman; a Protestant. [< Irish *sasanach,* Scottish Gaelic *Sasunnach* < Gaelic *Sasunn* a Saxon]
Sas·soon (sa-sōōn′), **Siegfried**, born 1886, English poet and author.
sas·sy[1] (sas′ē) *adj.* ·**si·er**, ·**si·est** *U. S. Dial.* Saucy; impertinent.
sas·sy[2] (sas′ē) *n.* A West African tree (*Erythrophleum guineense*) with poisonous bark and juice. Also **sas′sy·wood**′ (-wood′). [< native W. African name, < E *saucy*]
sas·tru·ga (sas-trōō′gə) See ZASTRUGA.
sat (sat) Past tense of SIT.
Sa·tan (sā′tən) In the Bible, the great adversary of God and tempter of mankind; the Devil: identified with *Lucifer* who, in Semitic mythology, led a revolt against God, was defeated by the archangel Michael, and cast into hell as punishment for his pride. *Luke* iv 5–8; *Rev.* xii 7–9. [< Hebrew *sātān* an enemy < *sātan* oppose, plot against]
sa·tang (sä-täng′) *n. pl.* **sa·tang** A bronze coin and money of account in Thailand; one one-hundredth of a baht. [< Siamese *satāņ*]
sa·tan·ic (sā-tan′ik) *adj.* Devilish; infernal; wicked. Also **sa·tan′i·cal.** See synonyms under INFERNAL. — **sa·tan′i·cal·ly** *adv.*
Sa·tan·ism (sā′tən-iz′əm) *n.* Satan-worship; specifically, a cult addicted to profane mockeries of the holy rites of Christian worship. — **Sa′tan·ist** *n.*
sat·a·ra (sat′ər-ə, sə-tä′rə) *n.* A lustrous ribbed woolen fabric. [from *Satara,* a town about 100 miles from Bombay, India]
satch·el (sach′əl) *n.* A small handbag. [< OF *sachel* < L *sacellus,* dim. of *saccus* a sack]
sate[1] (sāt) *v.t.* **sat·ed, sat·ing** To satisfy the appetite of; satiate. See synonyms under SATISFY. [Appar. alter. of obs. *sade* sate, OE *sadian;* refashioned after L *sat, satis* enough]
sate[2] (sāt) Archaic past tense of SIT.
sa·teen (sa-tēn′) *n.* A cotton fabric woven so as to give it a satin surface: usually mercerized cotton. [Alter. of SATIN; infl. in form by VELVETEEN]
sat·el·lite (sat′ə-līt) *n.* 1 *Astron.* A smaller body attending upon and revolving round a larger one; a moon. 2 One who attends upon a person in power. 3 Any obsequious attendant. 4 A small nation politically, economically, or militarily dependent on a great power. 5 A town or community whose activities are largely determined by those of a neighboring metropolis. 6 An airfield, base, or installation dependent upon a larger one. 7 A man-made object launched from and revolving around the earth: compare SPUTNIK. [< F < L *satelles, satellitis* an attendant, a guard]

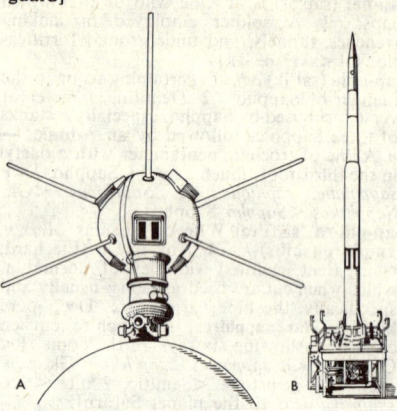

TYPE OF AMERICAN SATELLITE
A. Satellite. *B.* Rocket, which carries satellite into orbit, in position for launching.

sa·tem (sä′təm, sā′-) *n.* The eastern division of the Indo-European family of languages, including the Indo-Iranian, Armenian, Albanian, and Balto-Slavic subfamilies, in which proto-Indo-European palatal (k) is typically sibilated, as in the Avestan word *satem* "hundred." Compare CENTUM.
Sa·than (sā′tən), **Sath·a·nas** (sath′ə-nas) See SATAN.
sa·ti·a·ble (sā′shē-ə-bəl, -shə-bəl) *adj.* Capable of being satiated. — **sa′ti·a·bil′i·ty, sa′ti·a·bleness** *n.* — **sa′ti·a·bly** *adv.*
sa·ti·ate (sā′shē-āt) *v.t.* ·**at·ed**, ·**at·ing** 1 To satisfy the appetite or desire of; gratify. 2 To fill or gratify beyond natural desire; glut; surfeit. See synonyms under SATISFY. — *adj.* Filled with satiety; satiated. [< L *satiatus,* pp. of *satiare* fill < *satis* enough] — **sa′ti·a′tion** *n.*
Sa·tie (sà·tē′), **Erik Alfred Leslie**, 1866–1925, French composer.
sa·ti·e·ty (sə-tī′ə-tē) *n. pl.* ·**ties** Repletion; surfeit. [< F *satieté* < L *satietas, -tatis* < *satis* enough]
sat·in (sat′ən) *n.* A silk, cotton, rayon, or acetate fabric of thick texture, with glossy face and dull back. — *adj.* Of or similar to satin; glossy; smooth. [< OF < Med. L *satinus, setinus,* ult. < L *seta* silk]
sat·i·net (sat′ə·net′) *n.* 1 A strong fabric with cotton warp and woolen filling. 2 A thin satin. Also **sat′i·nette′.** [< F, dim. of *satin*]
sat·in-flow·er (sat′ən-flou′ər) *n.* The garden flower honesty, so called from the satiny luster of its silvery silicles. Also **sat′in·pod**′ (-pod′).
satin spar A silky fibrous mineral, a variety either of calcite, aragonite, or gypsum.
sat·in·wood (sat′ən-wood′) *n.* 1 The satinlike wood of an East Indian tree (*Chloroxylon swietenia*) of the mahogany family. 2 The tree. 3 A West Indian tree (*Zanthoxylum flavum*) of the rue family, having a fine-textured, golden-yellow wood much used in fine cabinetwork.
sat·in·y (sat′ən-ē) *adj.* Resembling or characteristic of satin; glossy.
sat·ire (sat′īr) *n.* 1 The use of sarcasm, irony, or keen wit in denouncing abuses or follies; ridicule. 2 A written composition in which vice, folly, or incapacity is held up to ridicule. See synonyms under BANTER. [< MF < L *satira, satura* a satire, earlier, a discursive verse composition on a number of subjects, orig. a medley < (*lanx*) *satura* a fruit salad, lit., a full (dish), fem. of *satur* full]

sa·tir·ic (sə·tir′ik) *adj.* Of, pertaining to, or resembling satire, especially literary satire: *satiric* verse.

sa·tir·i·cal (sə·tir′i·kəl) *adj.* **1** Given to or characterized by satire: a *satirical* writer. **2** Severely sarcastic; biting; caustic: a *satirical* laugh. **3** Satiric. **—sa·tir′i·cal·ly** *adv.* **—sa·tir′i·cal·ness** *n.*

sat·i·rist (sat′ə·rist) *n.* A writer of satire; a satirical person.

sat·i·rize (sat′ə·rīz) *v.t.* **·rized, ·riz·ing** To subject to or criticize in satire. See synonyms under RIDICULE. **—sat′i·riz′er** *n.*

sat·is·fac·tion (sat′is·fak′shən) *n.* **1** The act of satisfying, or the state of being satisfied; complete gratification. **2** The making of amends, reparation, or payment; extinguishment of a claim or obligation by payment, performance, restitution, or the rendering of an equivalent. **3** That which satisfies; atonement; compensation. [<OF *satisfactiun* <L *satisfactio, -onis* < *satisfactus,* pp. of *satisfacere* SATISFY]
Synonyms: comfort, complacence, content, contentment, enjoyment, gratification. See COMFORT, HAPPINESS, PROPITIATION, RECOMPENSE. *Antonyms*: annoyance, discontent, dislike, displeasure, dissatisfaction, disturbance, pain, sorrow, trouble, vexation.

sat·is·fac·tion–piece (sat′is·fak′shən·pēs′) *n.* A formal acknowledgment given by one who has received satisfaction of a mortgage or judgment, to authorize the entry of such satisfaction on the record.

sat·is·fac·to·ry (sat′is·fak′tər·ē) *adj.* **1** Giving satisfaction; answering fully all desires, expectations, or requirements; sufficient. **2** Making satisfaction; atoning or expiatory. See synonyms under ADEQUATE, COMFORTABLE. **—sat′is·fac′to·ri·ly** *adv.* **—sat′is·fac′to·ri·ness** *n.*

sat·is·fy (sat′is·fī) *v.* **·fied, ·fy·ing** *v.t.* **1** To supply fully with what is desired, expected, or needed; cause to have enough; gratify; content. **2** To free from doubt or anxiety; assure; convince. **3** To give what is due to. **4** To pay or discharge (a debt, obligation, etc.). **5** To answer sufficiently or convincingly, as a question or objection. **6** To fulfil the conditions or requirements of, as an equation. **7** To make reparation for; expiate. *—v.i.* **8** To give satisfaction. [<OF *satisfier* <L *satisfacere* < *satis* enough + *facere* do] **—sat′is·fi′er** *n.* **—sat′is·fy′ing** *adj.* **—sat′is·fy′ing·ly** *adv.*
Synonyms: cloy, content, fill, glut, sate, satiate, suffice, surfeit. To *satisfy* is to furnish enough to meet physical, mental, or spiritual desire. To *sate* or *satiate* is to gratify desire so fully as to extinguish it for a time. To *cloy* or *surfeit* is to gratify to the point of revulsion or disgust. *Glut* is a strong word applied to the utmost satisfaction of vehement appetites and passions; as, to *glut* a vengeful spirit with slaughter; we speak of *glutting* the market with a supply so excessive as to extinguish the demand. Much less than is needed to *satisfy* may *suffice* a frugal or abstemious person; less than a sufficiency may *content* one of a patient and submissive spirit. See INDULGE, PAY[1], REQUITE. *Antonyms*: check, deny, disappoint, refuse, restrain, restrict, starve, stint, tantalize.

sa·to·ri (sä·tō·rē) *n.* In Japanese Buddhism, enlightenment; especially, the abrupt or "sudden" enlightenment of Zen Buddhism. [<Japanese, lit., comprehension, perception]

Sat·pu·ra Range (sät·po͞o′rə) A line of hills in northern India, forming the northern edge of the Deccan Plateau; highest point, 4,429 feet.

sa·trap (sā′trap, sat′rap) *n.* **1** A governor of a province in ancient Persia. **2** Any petty ruler under a despot. **3** A subordinate ruler or governor. [<L *satrapes* <Gk. *satrapēs* <OPersian *shathraparan,* lit., a protector of a province]

sa·trap·y (sā′trə·pē, sat′rə·pē) *n. pl.* **·trap·ies** The territory or the jurisdiction of a satrap. Also **sa·trap·ate** (sā′trə·pit, sat′rə-).

Sa·tsu·ma (sä·tso͞o·mä) A former province of southern Kyushu island, Japan.

Satsuma ware A kind of Japanese pottery, originally made at Satsuma.

Sa·tu–Ma·re (sä′to͞o·mä′rä) A city in NW Rumania near the Hungarian border.

sat·u·rant (sach′ər·ənt) *adj.* Saturating. *—n.* A substance that fully neutralizes another. [<L *saturans, -antis,* ppr. of *saturare* SATURATE]

sat·u·rate (sach′ə·rāt) *v.t.* **·rat·ed, ·rat·ing** **1** To soak or imbue thoroughly; fill or impregnate to the utmost capacity for absorbing or retaining. **2** *Chem.* To utilize fully the combining powers of the atoms in (a molecule). *—adj.* **1** Filled to repletion; saturated. **2** Very intense; deep: said of colors. [<L *saturatus,* pp. of *saturare* fill up < *satur* full] **—sat·u·ra·ble** (sach′ər·ə·bəl) *adj.* **—sat′u·ra′tor** or **sat′u·rat′er** *n.*

sat·u·rat·ed (sach′ə·rā′tid) *adj.* **1** Completely satisfied; replete; incapable of holding more of a substance or material: *saturated* vapor; a *saturated* solution. **2** *Chem.* Designating an organic compound having no free valences and without double or triple bonds, as paraffin, methane, and other *saturated* hydrocarbons. **3** Designating a pure color or hue, as in the spectrum; exhibiting high saturation. **4** *Geol.* Designating rocks or minerals with a maximum content of silica.

sat·u·ra·tion (sach′ə·rā′shən) *n.* **1** The act of saturating, or the state of being saturated; full impregnation. **2** The impregnation of one substance with another till no more can be received. Saturation may be by solution or by chemical combination. **3** *Meteorol.* The filling of the atmosphere with any vapor to the point of condensation. **4** The maximum magnetization of which a body is capable. **5** The degree of vividness or purity of chromatic color, as indicated by its freedom from admixture with white.

Sat·ur·day (sat′ər·dē, -dā) *n.* The seventh or last day of the week; the day of the Jewish Sabbath. [OE *Sæterdæg, Sæternesdæg,* trans. of L *Saturni dies* Saturn's day]

Sat·urn (sat′ərn) **1** The planet next beyond Jupiter and next to Jupiter in size, remarkable for its 9 satellites and its flat, luminous, encircling rings. In astrology it was regarded as a melancholy planet. See PLANET. **2** In Roman mythology, the god of agriculture: identified with the Greek *Kronos* [OE *Sætern, Saturnus* <L *Saturnus,*? < *satus,* pp. of *serere* sow]

sat·ur·na·li·a (sat′ər·nā′lē·ə) *n.* Any season or period of general license or revelry: generally construed as singular: a *saturnalia* of crime. [<L. See SATURNALIA.]

Sat·ur·na·li·a (sat′ər·nā′lē·ə) *n. pl.* The feast of Saturn held at Rome in mid–December, celebrating the winter solstice, and marked by wild reveling and licentious abandon. [<L, orig. neut. pl. of *Saturnalis* of Saturn < *Saturnus* Saturn] **—Sat′ur·na′li·an** *adj.*

Sa·tur·ni·an (sə·tûr′nē·ən) *adj.* Of or pertaining to the god, or to the planet, Saturn, especially to a fabled golden age in the reign of Saturn, marked by simplicity, virtue, and happiness.

sa·tur·ni·id (sə·tûr′nē·id) *n.* Any of a family *(Saturniidae)* of large, hairy, brightly–colored moths widely distributed in most temperate regions. Many of them produce cocoons useful in the production of silk. *—adj.* Of or pertaining to the *Saturniidae.* [<NL <*Saturnia,* genus name <L *Saturnius* of Saturn < *Saturnus* Saturn]

sat·ur·nine (sat′ər·nīn) *adj.* **1** Having a grave, gloomy, or morose disposition or character; heavy; dull. **2** In old chemistry, pertaining to lead. **3** *Pathol.* Pertaining to or produced by lead. [<OF *saturnin* of Saturn, of lead, heavy <Med. L *Saturnus* lead, Saturn <L, Saturn]

Sat·ur·nine (sat′ər·nīn) *adj.* **1** Of or pertaining to the planet Saturn. **2** Born or being under the influence of the planet Saturn; hence, gloomy; heavy.

sat·urn·ism (sat′ərn·iz′əm) *n.* Lead poisoning. [<Med. L *Saturnus.* See SATURNINE.]

Sat·ya·gra·ha (sut′yə·gru′hə) *n.* **1** A movement characterized by non–violent resistance and non–cooperation, adopted in India, 1919, by the followers of M. K. Gandhi in protest against certain civil and religious abuses. **2** The non–violent force characterizing this movement, defined as an active love for one's opponents and a radical insistence on truth. [<Hind., truth–force, lit., a grasping for truth <Skt. *satya* truth + *graha* a grasping]

sat·yr (sat′ər, sā′tər) *n.* **1** In Greek mythology, a woodland deity in human form, having pointed ears, pug nose, short tail and budding horns, and of wanton nature. **2** A very lascivious man. **3** Any butterfly of the family *Agapetidae,* commonly brown and gray with eyelike spots. [<L *satyrus* <Gk. *satyros*] **—sa·tyr·ic** (sə·tir′ik) or **·i·cal** *adj.*

SATYR

sat·y·ri·a·sis (sat′ə·rī′ə·sis) *n.* *Psychiatry* A morbid lasciviousness in males. [<NL <Gk. *satyriaein* suffer from satyriasis < *satyros* a satyr]

sauce (sôs) *n.* **1** An appetizing dressing or liquid relish for food; loosely, any appetizing garnish of a meal; formerly, any condiment, as salt, pepper. **2** A dish of fruit pulp stewed and sweetened: cranberry *sauce.* **3** *Colloq.* Table vegetables, as roots or greens: also **garden sauce.** **4** *Colloq.* Pert or impudent language. *—v.t.* **sauced, sauc·ing** **1** To flavor with sauce; season. **2** To give zest or piquancy to. **3** *Colloq.* To be saucy to. [<OF <LL *salsa,* orig. fem. of L *salsus* salted, pp. of *salire* salt < *sal* salt]

sauce·box (sôs′boks′) *n. Colloq.* A saucy person: said generally of a child.

sauce·pan (sôs′pan′) *n.* A metal or enamel pan with projecting handle, for cooking food.

sau·cer (sô′sər) *n.* **1** A small dish for holding a cup. **2** Any small, round, shallow vessel of similar shape. [<OF *saussier* < *sauce* sauce]

sau·cy (sô′sē) *adj.* **·ci·er, ·ci·est** **1** Disrespectful to superiors; impudent. **2** Piquant; sprightly; amusing. See synonyms under IMPUDENT. **—sau′ci·ly** *adv.* **—sau′ci·ness** *n.*

Sa·ud (sä·o͞od′), **King,** born 1902, king of Saudi Arabia 1953–; son of Ibn Saud: full name *Ibn Abdul Aziz al Faisal al Saud.*

Sa·u·di Arabia (sä·o͞o′dē) A kingdom (1932) in the northern and central part of Arabia; 927,000 square miles; dual capitals, Mecca and Riyadh.

sauer·bra·ten (sour′brät′n, *Ger.* zou′ər·brä′·tən) *n.* Beef marinated in vinegar before being braised. [<G < *sauer* sour + *braten* roast]

sauer·kraut (sour′krout′) *n.* Shredded and salted cabbage fermented in its own juice: also spelled *sourcrout.* [<G < *sauer* sour + *kraut* cabbage, vegetable, a plant]

sau·ger (sô′gər) *n.* A percoid fish, the smaller American pike perch *(Cynoperca canadensis),* resembling the walleye. [<N. Am. Ind.]

saugh (sôkh) *n. Scot.* The sallow; the willow. **—saugh′y** *adj.*

saught (sôkht) *n. Scot.* Agreement; peace; ease; rest.

Sauk (sôk) *n.* One of a tribe of North American Indians of Algonquian stock, formerly occupying Michigan, later Wisconsin and the Mississippi valley: now on reservations in Oklahoma, Iowa, and Kansas. Also spelled *Sac.*

saul (sôl) *n. Scot.* Soul; mettle.

Saul (sôl) A masculine personal name. [<Hebrew, asked (of God)]
—Saul The first king of Israel. I *Sam.* ix 2.
—Saul The Hebrew name of the Apostle Paul. *Acts* xiii 9. Also **Saul of Tarsus.**

sau·lie (sô′lē) *n. Scot.* A paid mourner.

Sault Sainte Ma·rie (so͞o′ sānt′ mə·rē′) **1** A city in northern Michigan, on St. Marys River. **2** A city opposite it in south central Ontario. Also **Sault Ste. Marie.**

Sault Sainte Marie Canals Three ship canals at the rapids in the St. Marys River, connecting Lake Superior with Lake Huron: also, *Colloq., Soo Canals.*

sau·mont (sô′mənt) *n. Scot.* The salmon: also spelled *sawmunt.*

saunt (sänt, sônt) *n. Scot.* Saint.

saun·ter (sôn′tər) *v.i.* To walk in a leisurely or lounging way; stroll. See synonyms under LINGER. *—n.* **1** A slow, aimless manner of walking. **2** An idle stroll. [ME *santren* muse, meditate; ult. origin unknown]

Sau·rash·tra (sou·räsh′trə) A former constituent State of western India, comprising most of Kathiawar peninsula and including 222

former states; merged into Bombay State, 1956; 21,062 square miles; former capital, Rajkot.
sau·rel (sôr′əl) *n.* A horse mackerel (genus *Trachurus*), especially *T. trachurus* and *T. symmetricus* of America and Europe. [<F <Gk. *sauros* a horse mackerel]
sau·rian (sôr′ē·ən) *n.* One of a suborder *(Sauria)* of reptiles, the lizards: formerly including also crocodiles, dinosaurians, pterodactyls, and other fossil forms. —*adj.* Pertaining to the *Sauria.* [<NL <Gk. *sauros* a lizard]
sau·ris·chi·an (sô·ris′kē·ən) *adj. Paleontol.* Of, pertaining to, or belonging to an order *(Saurischia)* of reptilelike dinosaurs that flourished through most of the Mesozoic era. —*n.* A member of this order. [<NL <Gk. *sauros* a lizard + *ischion* a hip]
sauro– *combining form* Lizard: *sauropod.* Also, before vowels, **saur–.** [<Gk. *sauros* a lizard]
sau·ro·pod (sôr′ə·pod) *n. Paleontol.* One of a suborder *(Sauropoda)* of amphibious four-footed dinosaurs of the Triassic, Jurassic, and Cretaceous periods. —*adj.* Of or pertaining to the *Sauropoda.* [<NL <Gk. *sauros* a lizard + *pous, podos* a foot] —**sau·rop·o·dous** (sô·rop′ə·dəs) *adj.*
–saurus *combining form Zool.* Lizard: used to form genus names: *Brontosaurus, Plesiosaurus.* Corresponding class names end in **–sauria,** family names in **–sauridae,** and individual names in **–saur** or **–saurid.** [<Gk. *sauros* a lizard]
sau·ry (sôr′ē) *n. pl.* **·ries** An edible fish *(Scomberesox saurus)* of the Atlantic, having the jaws developed into a slim beak. It travels in predatory shoals. Also **saury pike.** [<NL *saurus* <Gk. *sauros* a lizard]
sau·sage (sô′sij) *n.* **1** Finely chopped and highly seasoned meat, commonly stuffed into the cleaned and prepared entrails of some animal or artificial casings. **2** *Aeron.* A type of airship or captive observation balloon, shaped like a sausage. [<AF *saussiche* <LL *salsicia,* ult. <L *salsus.* See SAUCE.]
saus·su·rite (sô·sŏŏr′īt, sôs′yə·rīt) *n.* A tough, compact, impure form of labradorite. [after Prof. H. B. de *Saussure,* 1740–99, Swiss geologist] —**saus·su·rit·ic** (sôs′yə·rit′ik) *adj.*
saut (sät, sôt) *adj.* & *n. Scot.* Salt.
sau·té (sō·tā′, sô–) *adj.* Fried quickly with little grease. —*v.t.* **·téed, ·té·ing** To fry quickly in a little fat. [<F, pp. of *sauter* leap]
sau·terne (sō·tûrn′, sô–; *Fr.* sō·tern′) *n.* A sweet, white French wine; often, in America, any white wine, dry or sweet. Also **sau·ternes′.** [from *Sauternes,* district in SW France]
sau·toir (sō·twär′) *n. Her.* A saltire. Also **sau·toire′.** —**en sautoir** Worn saltirewise, or diagonally about the body, as a ribbon. [<F. See SALTIRE.]
sauve qui peut (sōv kē pœ′) *French* A stampede; rout; literally, save himself who can.
Sa·va (sä′vä) A river of northern Yugoslavia, flowing about 583 miles east to the Danube near Belgrade; the longest river entirely in Yugoslavia. *French* **Save** (sàv), *German* **Sau** (sou).
sav·age (sav′ij) *adj.* **1** Of a wild and untamed nature; not domesticated; hence, ferocious; fierce. **2** Living in or belonging to the most primitive and rude condition of human life and society; uncivilized; uncultivated: *savage* tribes. **3** Enraged; cruel; furious: said of man or beast. **4** *Obs.* Remote from human abode; belonging to the wilderness: a *savage* trail. See synonyms under BARBAROUS, BITTER, FIERCE, GRIM, SANGUINARY. —*n.* **1** A primitive or uncivilized human being. **2** A brutal, fierce, and cruel person; a barbarian. [<OF *salvage, sauvage* <L *silvaticus, salvaticus* < *silva* a wood] —**sav′age·ly** *adv.*
Sav·age (sav′ij), **Arthur William,** 1857–1938, U. S. inventor; manufacturer of rifles, etc. —**Richard,** 1697?–1743, English poet.
Savage Island See NIUE.
sav·age·ry (sav′ij·rē) *n. pl.* **·ries** **1** The state of being savage: also **sav′age·ness.** **2** Cruelty in disposition or action; a cruel or savage act. **3** Savages collectively: also **sav′age·dom.** Also **sav′ag·ism.**
Savage's Station A battlefield near Richmond, Virginia; scene of an unsuccessful Confederate attack (1862) during the Civil War.
Sa·vai·i (sä·vī′ē) The largest island in Western Samoa; 700 square miles.
sa·van·na (sə·van′ə) *n.* **1** A tract of level land covered with low vegetation; a treeless plain. **2** Any large area of tropical or subtropical grassland, covered in part with trees and spiny shrubs. Also **sa·van′nah.** [Earlier *zavana* <Sp. <Cariban]
Sa·van·nah (sə·van′ə) A port in eastern Georgia, at the mouth of the **Savannah River,** which flows 314 miles SE to the Atlantic and forms the boundary between Georgia and South Carolina.
sa·vant (sə·vänt′, sav′ənt; *Fr.* sà·vän′) *n.* A man of exceptional learning. See synonyms under SCHOLAR. [<F, orig. ppr. of *savoir* know <L *sapere* be wise]
Sa·vart (sà·vàr′), **Felix,** 1791–1841, French physician and physicist.
save[1] (sāv) *v.* **saved, sav·ing** *v.t.* **1** To preserve or rescue from danger, harm, etc. **2** To keep from being spent, expended, or lost; avoid the loss or waste of. **3** To set aside for future use; accumulate: often with *up.* **4** To treat carefully so as to avoid fatigue, harm, etc.: to *save* one's eyes. **5** To avoid the need or trouble of; prevent by timely action: A stitch in time *saves* nine. **6** *Theol.* To deliver from spiritual death or the consequences of sin; redeem. —*v.i.* **7** To avoid waste; be economical. **8** To preserve something from danger, harm, etc. **9** To admit of preservation, as food. See synonyms under DELIVER, PRESERVE, SCRIMP. [<OF *salver, sauver* <LL *salvare* save <L *salvus* safe] —**sav′a·ble** or **save′a·ble** *adj.* —**sav′a·ble·ness** *n.* —**sav′er** *n.*
save[2] (sāv) *prep.* Except; but. —*conj.* **1** Except; but. **2** *Archaic* Unless. See synonyms under BUT[1]. [<OF *sauf* being excepted, orig. safe <L *salvus*]
Sa·ve (sä′və) The Portuguese name for the SABI.
save–all (sāv′ôl′) *n.* **1** A contrivance for preventing waste; anything that saves fragments. **2** A child's savings bank. **3** An overall or pinafore.
saved (sāvd) *adj.* **1** Delivered from punishment after death. **2** Converted to religion. **3** Not spent or lost; amassed.
sav·e·loy (sav′ə·loi) *n.* A kind of highly seasoned, dried sausage made of salted pork. [Alter. of F *cervelas* <Ital. *cervellata* < *cervello* the brain <L *cerebellum.* See CEREBELLUM.]
Sav·ile (sav′il), **Sir Henry,** 1549–1622, English classical and biblical scholar.
Savile Row A street in London famous for fashionable men's tailor shops; hence, sartorially magnificent.
sav·in (sav′in) *n.* **1** A bushy shrub or small tree *(Juniperus sabina)* of the cypress family. **2** The young shoots of this plant, yielding an acrid volatile oil used in medicine. **3** The red cedar *(Juniperus virginiana).* Also called *sabine.* [OE *safine* <OF *savine* <L *(herba) Sabina* the Sabine (herb), fem. of *Sabinus*]
sav·ing (sā′ving) *adj.* **1** That saves; preserving, as from destruction. **2** Redeeming; delivering. **3** Avoiding needless waste or expense; economical; frugal. **4** Incurring no loss, if not gainful: a *saving* investment. **5** Holding in reserve; making an exception; qualifying: a *saving* clause. —*n.* **1** Preservation from loss or danger. **2** Avoidance of waste; economy. **3** The result of this; reduction in cost: a *saving* of 16 percent. **4** That which is saved; especially, in the plural, sums of money not expended. **5** *Law* Reservation; exception. See synonyms under FRUGALITY. —*prep.* **1** With the exception of; save. **2** With due respect for: *saving* your presence. —*conj.* Save. —**sav′ing·ly** *adv.* —**sav′ing·ness** *n.*
savings account An account drawing interest at a savings bank.
savings bank **1** An institution for receiving and investing savings and paying interest on deposits. **2** A container with a slot for coins.
sav·ior (sāv′yər) *n.* One who saves. Also *Brit.* **sav′iour.** [<OF *saveour* <LL *salvator, -oris* <L *salvare* SAVE]
Sav·iour (sāv′yər) *n.* He who saves men from death and sin: a title sometimes applied directly to God, but chiefly to Jesus Christ, as the Redeemer: usually with *the.* Also **Sav′ior.**
sa·voir faire (sà·vwàr fâr′) *French* Ability to see and to do the right thing; readiness in proper and gracious actions and speech; tact; literally, to know how to act.
sa·voir vi·vre (sà·vwàr vē′vr′) *French* Good breeding; good social manners; literally, to know how to live.
Sa·vo·na (sä·vō′nä) A port on the Gulf of Genoa in NW Italy.
Sav·o·na·ro·la (sav′ə·nə·rō′lə, *Ital.* sä′vō·nä·rō′lä), **Girolamo,** 1452–98, Italian monk; reformer; burned at the stake for heresy.
sa·vor (sā′vər) *n.* **1** That quality of a thing that affects the taste and smell, or both; flavor; odor. **2** Specific or characteristic quality or approach to a quality; flavor. **3** Relish; zest: The conversation had *savor.* **4** *Archaic* Character; reputation. —*v.i.* **1** To have savor; taste or smell: with *of.* **2** To have a specified savor or character: with *of.* —*v.t.* **3** To give flavor to; season. **4** To taste or enjoy with pleasure; relish. **5** To have the savor or character of. Also *Brit.* **sa′vour.** [<OF *savour* <L *sapor* taste < *sapere* taste, know] —**sa′vor·er** *n.* —**sa′vor·ous** *adj.*
Synonyms (noun): flavor, fragrance, odor, relish, scent, smell, taste. See SMELL.
sa·vor·less (sā′vər·lis) *adj.* Tasteless; insipid.
sa·vor·y[1] (sā′vər·ē) *adj.* **1** Of an agreeable taste and odor; appetizing. **2** Piquant to the taste. **3** In good repute. See synonyms under DELICIOUS. —*n. Brit.* A small, hot serving of food eaten at the end or beginning of a dinner. Also *Brit.* **sa′vour·y.** [<OF *savouré,* pp. of OF *savourer* taste < *savour* SAVOR] —**sa′vor·i·ly** *adv.* —**sa′vor·i·ness** *n.*
sa·vor·y[2] (sā′vər·ē) *n.* A hardy, annual, aromatic culinary herb of the mint family *(Satureia hortensis)* used for seasoning. Also **summer savory.** [<OF *savoreie,* alter. of L *satureia;* infl. in form by OF *savour* savor]
sa·voy (sə·voi′) *n.* A variety of cabbage with wrinkled leaves and a compact head. [<F *(chou de) Savoie* (cabbage of) Savoy]
Sa·voy (sə·voi′) A region and former duchy of the kingdom of Sardinia, between Italy and France; ceded to France in 1860. *French* **Sa·voie** (sà·vwà′).
Sa·voy (sə·voi′), **House of** A family of French nobles, reigning in Italy from 1861–1946. Its members were descended from Humbert I, Count of Savoy (11th century).
Sa·voy·ard (sə·voi′ərd, *Fr.* sà·vwà·yàr′) *n.* **1** A native or inhabitant of Savoy, France. **2** An actor or actress in the Gilbert and Sullivan operas of which most were originally produced at the Savoy Theatre in London. **3** An admirer or producer of these operas. —*adj.* **1** Of or pertaining to Savoy, France. **2** Of the Savoy Theatre, London. [<F < *Savoie* Savoy]
Sa·vu Sea (sä′vōō) That part of the Indian Ocean bounded by the islands of Flores, Sumba, and Timor.
sav·vy (sav′ē) *Slang v.i.* **·vied, ·vy·ing** To understand; comprehend. —*n.* Understanding; good sense. [Alter. of Sp. ¿ *Sabe (usted)*? Do (you) know? < *saber* know <L *sapere* know, taste]
saw[1] (sô) *n.* **1** A cutting instrument with pointed teeth arranged continuously along the edge of the blade: used to cut or divide wood, bone, metal, etc. See illustrations under BUCKSAW, FRET SAW, HACKSAW. **2** A machine for operating a saw or gang of saws. **3** Any tool or instrument without teeth used like a saw, as a steel disk for cutting armor plate, etc. —**circular saw** A disk having saw teeth in or on its periphery, and mounted on an arbor, with which it is rotated, usually at high speed. —*v.* **sawed, sawed** or **sawn, saw·ing** *v.t.* **1** To cut or divide with a saw. **2** To shape or fashion with a saw. **3** To cut or slice (the air, etc.) as if using a saw: The speaker *saws* the air. **4** To cause to move with a to-and-fro motion like that of a saw. —*v.i.* **5** To use a saw. **6** To cut: said of a saw. **7** To be cut with a saw: This wood *saws* easily. [OE *sagu, saga*] —**saw′er** *n.*
saw[2] (sô) *n.* A proverbial or familiar saying; old maxim. See synonyms under ADAGE. [OE *sagu.* Akin to SAGA.]
saw[3] (sô) Past tense of SEE[1].
saw[4] (sô) *v.t. Scot.* To sow.
Sa·watch Range (sə·wach′) A range of the Rocky Mountains in central Colorado; highest peak, 14,431 feet: also *Saguache.*
saw·bill (sô′bil′) *n.* A motmot.
saw·bones (sô′bōnz′) *n. Slang* A surgeon.

saw·buck (sô′buk′) *n.* **1** A rack or frame consisting of two X-shaped ends joined by a connecting bar or bars, for holding sticks of wood while they are being sawed. Compare SAWHORSE. **2** *U.S. Slang* A ten-dollar bill: so called from the resemblance of X, Roman numeral ten, to the ends of a sawbuck. [Trans. of Du. *zaagbok*]

SAWBUCK AND SAWHORSE
a. Sawbuck. *b.* Bucksaw. *c.* Sawhorse.

saw·dust (sô′dust′) *n.* Small particles of wood cut or torn out by sawing.
sawed-off (sôd′ôf′, -of′) **1** *adj.* Having one end sawed off. **2** Short; not of average height or length: a *sawed-off* shotgun.
saw·fish (sô′fish′) *n. pl.* **·fish** or **·fish·es** A sharklike, tropical ray (genus *Pristis*) having an elongated body and the snout prolonged into a flat blade with teeth on each edge.
saw·fly (sô′flī′) *n. pl.* **·flies** A hymenopterous insect (family *Tenthredinidae*) having in the female a sawlike ovipositor for piercing plants, soft wood, etc., in which to lay eggs.
saw·grass (sô′gras′, -gräs′) *n.* A sedge (genus *Mariscus*) with saw-toothed leaves, growing in marshes along the Atlantic coast from North Carolina to Florida and westward.
saw·horse (sô′hôrs′) *n.* **1** A frame consisting of a long wooden bar or plank supported by four extended legs: used by carpenters. Compare SAWBUCK. **2** A packsaddle.
saw log A log of suitable size for sawing.
saw·mill (sô′mil′) *n.* **1** An establishment for sawing logs with power-driven machinery. **2** A large sawing machine.
saw·munt (sô′mənt) See SAUMONT.
sawn (sôn) Alternative past participle of SAW[1].
saw palmetto Either of two palmettos (*Serenoa repens* and *Paurotis wrighti*) of the southern United States and the West Indies.
saw·pit (sô′pit′) *n.* A pit over which a timber is laid to be sawed by two sawyers, one of whom stands in the pit and the other above.
saw set An instrument to give set to, or bend slightly outward, the teeth of a saw.
saw-toothed (sô′tōōtht′) *adj.* Serrate; having teeth or toothlike processes similar to those of a saw.
saw·yer (sô′yər) *n.* **1** One who saws logs; specifically, a lumberman who fells trees by sawing, or one who works in a sawmill: also spelled *sawer.* **2** Any beetle of the genus *Monochamus* having wood-boring larvae. [Alter. of SAWER]
sax[1] (saks) *n.* **1** A chopping tool for trimming edges of roofing slates: also called *slate ax.* **2** A long knife. **3** A short, broad sword. [OE *seax* a knife]
sax[2] (saks) *n. Colloq.* A saxophone. [Short for SAXOPHONE]
sax·a·tile (sak′sə·til) *adj.* **1** Pertaining to rocks. **2** Saxicoline. [<L *saxatilis* <*saxum* a rock]
Saxe (sȧks) The French name for SAXONY.
Saxe (sȧks), **Comte de,** 1696–1750, Hermann Maurice, French marshal.
Saxe-Al·ten·burg (saks′äl′tən·bûrg) A former duchy in central Germany.
Saxe-Co·burg (saks′kō′bûrg) A former duchy in central Germany; united in 1826 with **Saxe-Go·tha** (-gō′thə), another duchy, to form the duchy of **Saxe-Coburg-Gotha,** which was divided between Thuringia and Bavaria in 1918.
Saxe-Mei·ning·en (saks′mī′ning·ən) A former duchy in central Germany.
Saxe-Wei·mar (saks′vī′mär) A former grand duchy in central Germany; became the duchy of **Saxe-Weimar-Ei·se·nach** (ī′zə·näkh) in 1741.
sax·horn (saks′hôrn′) *n.* A brass wind instrument having a long winding tube and cup-shaped mouthpiece: much used in military bands. [after Charles Joseph *Sax,* 1791–1865, Belgian instrument maker + HORN]
sax·ic·o·line (sak·sik′ə·lēn, -lin) *adj. Ecol.* Living or growing among rocks. Also **sax·ic′o·lous.** [<NL *saxicola* <L *saxum* rock + *colere* inhabit]
sax·i·fra·ga·ceous (sak′sə·frə·gā′shəs) *adj. Bot.* Of or pertaining to a widely distributed family (*Saxifragaceae*) of herbs, shrubs, and trees, including gooseberries and witch hazel. [<NL <L *saxifraga.* See SAXIFRAGE.]
sax·i·frage (sak′sə·frij) *n.* **1** Any plant of the genus *Saxifraga,* growing in rocky places. **2** Any of various related plants. Also called *stonebreak.* [<OF <L (*herba*) *saxifraga,* lit., stone-breaking (herb)]
Sax·o Gram·mat·i·cus (sak′sō grə·mat′i·kəs), 1150?–1220?, Danish historian.
Sax·on (sak′sən) *n.* **1** A member of a Germanic tribal group living in the southern part of what is now Schleswig-Holstein in the early centuries of the Christian era. **2** A member of any of the offshoots of this group, as those who, with the Angles and Jutes, invaded England in the fifth and sixth centuries A.D. **3** An Anglo-Saxon. **4** An inhabitant of Saxony. **5** The modern High German dialect of Saxony. **6** A Teuton. **—Old Saxon** The dialect of Low German current in the valley of the lower Elbe in the early Middle Ages. *—adj.* **1** Of or pertaining to the Saxons, or to their language. **2** Germanic; Anglo-Saxon; also, English: said of words, phrases, etc. [<F <L *Saxo, Saxonis* <WGmc.]
Sax·on·ism (sak′sən·iz′əm) *n.* A word, phrase, etc., of English, specifically Anglo-Saxon, origin.
Sax·on·ist (sak′sən·ist) *n.* An authority on pre-Norman England or the Saxon language, especially Old Saxon.
Sax·o·ny (sak′sə·nē) *n. pl.* **·nies** **1** A fabric made from wool raised in Saxony, central Germany. **2** A variety of fine yarn. **3** A glossy woolen cloth.
Sax·o·ny (sak′sə·nē) **1** A former duchy, electorate, and kingdom of central Germany. **2** A former Prussian province of central Germany, constituted in 1816, largely from the territories of the kingdom of Saxony; 9,753 square miles; capital, Magdeburg. **3** A former state of east central Germany, 1918–45; 5,789 square miles; capital, Dresden. **4** A former state of SE East Germany, 1949–52; 6,561 square miles; capital, Dresden. French *Saxe,* German *Sachsen.*
Sax·o·ny-An·halt (sak′sə·nē·än′hält) A former state (1949) of east central East Germany; 9,515 square miles; capital, Halle.
sax·o·phone (sak′sə·fōn) *n.* A brass wind instrument with about 20 finger keys, tonally like, but more powerful than, a clarinet. [after Antoine Joseph *Sax* (called Adolphe), 1814–94, Belgian instrument maker, who invented it about 1840 + -PHONE] **—sax′o·phon′ist** *n.*

SAXOPHONE

sax·tu·ba (saks′tōō′bə, -tyōō′-) *n.* A large saxhorn. [<SAX(HORN) + TUBA]
say[1] (sā) *v.* **said, say·ing** *v.t.* **1** To pronounce or utter; speak. **2** To declare or express in words; tell; state. **3** To state positively or as an opinion: *Say* which you prefer. **4** To recite; repeat: to *say* one's prayers. **5** To report; allege. **6** To assume as possibly true or as a hypothesis: He is worth, *say,* a million. *—v.i.* **7** To make a statement; speak. **—that is to say** In other words. *—n.* **1** What one has said or has to say; testimony; word: Let him have his *say.* **2** *Colloq.* Right or turn to speak or choose: Now it is my *say.* **3** Authority: to have the *say.* *—interj. U.S. Colloq.* A hail or an introductory exclamation to command attention: also *Brit.* **I say!** Compare LISTEN. [OE *secgan*] **—say′er** *n.*
Synonyms (*verb*): allege, assert, speak.
say[2] (sā) *n.* **1** A fine, thin serge used in the 16th century, sometimes partly of silk, later entirely of wool. [<OF *saie* <L *saga,* pl. of *sagum* a military cloak]
Say (sā), **Thomas,** 1787–1834, U.S. zoologist.
Sa·yan Mountains (sä·yän′) A mountain system on the Siberia-Mongolia border.
Say·ers (sā′ərz, sârz), **Dorothy L(eigh),** 1893–1957, English author.
say·id (sī′id, sä′yid) *n.* Lord: a title applied to men who claimed to be descendants of Mohammed through his elder grandson, Husain: also spelled *said, saiyid.* Also **say′yid.** [<Arabic *sayyid*]
Sa·yi·da (sä′yə·dä) See SAIDA.
say·ing (sā′ing) *n.* An utterance; also, a maxim. See synonyms under ADAGE.
says (sez) Third person singular, present indicative of SAY.
say-so (sā′sō′) *n. Colloq.* **1** An unsupported assertion or decision. **2** Right or power to make decisions: He has the *say-so.*
'sblood (zblud) *interj. Archaic* God's blood: an imprecation. [Short for *God's blood*]
S-brack·et (es′brak′it) *n.* In mechanical construction, a bracket or other piece in the shape of the letter S: also called *S-piece.*
scab (skab) *n.* **1** A crust formed on the surface of a wound or sore. **2** A contagious disease among sheep, resembling mange; scabies. **3** Any of certain plant diseases of bacterial or fungous origin, in which there is a roughened or warty exterior. **4** *Slang* A mean, paltry fellow. **5** A workman who does not belong to or will not join or act with a labor union; one who takes the place of a striker; a strikebreaker. *—v.i.* **scabbed, scab·bing** **1** To form or become covered with a scab. **2** To take the place of a striker; act as a scab. [Fusion of ON *skabbr* (assumed) and OE *sceabb*; infl. in meaning by L *scabies.* See SCABIES.] **—scabbed** *adj.* **—scab′bi·ly** *adv.* **—scab′bi·ness** *n.* **—scab′by** *adj.*
scab·bard (skab′ərd) *n.* A sheath for a weapon, as for a bayonet or a sword. *—v.t.* To sheathe in or furnish with a scabbard. [<OF *escalberc,* prob. <OHG *scar* a sword + *bergan* hide, protect]
scabbard fish **1** The cutlas fish (*Trichiurus lepturus*) having a long, eel-like body, found in the warm coastal waters of the United States and West Indies. **2** Any long, slender, silvery fish of the genus *Lepidopus* of European coasts.
scab·ble (skab′əl) *v.t.* **·bled, ·bling** In stoneworking, to dress or shape roughly. [Earlier *scapple* <OF *escapeler* dress timber]
scab·bling (skab′ling) *n.* A stone chip or fragment.
sca·bi·es (skā′bi·ēz, -bēz) *n.* The itch; especially, a contagious skin disease of sheep caused by any of certain itch mites, as *Psoroptes communis.* [<L, roughness, an itch <*scabere* scratch, scrape. Akin to SHAVE.] **—sca·bi·et·ic** (skā′bē·et′ik) *adj.*
sca·bi·ous[1] (skā′bē·əs) *adj.* **1** Pertaining to scabies. **2** Having scabs. [<L *scabiosus* <*scabies.* See SCABIES.]
sca·bi·ous[2] (skā′bē·əs) *n.* Any of a genus (*Scabiosa*) of herbs of the teasel family, with involucrate heads of variously colored flowers, as the sweet scabious (*S. atropurpurea*). Also **sca′bi·o′sa** (-ō′sə). [<NL <Med. L (*herba*) *scabiosa* fem. sing. of *scabiosus* SCABIOUS[1]]
sca·brous (skā′brəs) *adj.* **1** Roughened with minute points; rugged; scurfy. **2** Knotty; difficult to handle tactfully. [<LL *scabrosus* <*scabere* scratch] **—sca′brous·ly** *adv.* **—sca′·brous·ness** *n.*
scac·cog·ra·phy (ska·kog′rə·fē) *n.* The literature pertaining to the science and art of chess. [<Ital. *scacchi* chess, pl. of *scacco* a square on a chessboard + -(O)GRAPHY] **—scac·chic** (skak′ik) *adj.* **—scac·cog′ra·pher** *n.*
scad (skad) *n.* A saurel. [? Var. of SHAD]
scads (skadz) *n. pl. Colloq.* A large amount or quantity. [? Var. of dial. E *scald* a large amount, great number]
Sca·fell Pike (skô·fel′) A mountain in Cumberland, England, the highest peak in England; 3,210 feet.
scaff (skaf, skäf) *n. Scot. & U.S. Dial.* Food; provisions. *—v.t. & v.i.* **1** To beg for (food). **2** *U.S. Slang* To eat. Also spelled *scauff.* [Prob. <G and Du. *schaffen* provide (food)]
scaf·fold (skaf′əld, -ōld) *n.* **1** A temporary elevated structure for the support of workmen, materials, etc., as in building. **2** A raised wooden framework used for drying hay, tobacco, fish, etc. **3** A platform for the execution of criminals. **4** A stage, as for

add,āce,câre,pälm; end,ēven; it,īce; odd,ōpen,ôrder; tōōk,pōōl; up,bûrn; ə = a in *above,* e in *sicken,* i in *clarity,* o in *melon,* u in *focus*; yōō = u in *fuse*; oi,oil; ou,pout; ch,check; g,go; ng,ring; th,thin; th,this; zh,vision. Foreign sounds à,œ,ü,kh,ṅ; and ◆: see page xx. < from; + plus; ? possibly.

scaffolding 1122 **scan**

exhibition purposes. 5 A raised wooden frame formerly used by certain North American Indians for the disposal of their dead. —v.t. 1 To furnish or support with a scaffold. 2 To place on a scaffold. [< OF (e)schaffaut, escadafaut. Related to CATAFALQUE.]

scaf·fold·ing (skaf′əl-ding) n. A scaffold, or system of scaffolds, or the materials for constructing them. Also **scaff·fold·age**.

scaff-raff (skaf′raf′, skäf′räf′) n. Scot. The rabble.

sca·gl·ia (skäl′yə) n. An Italian calcareous rock, corresponding to the chalk of England. [<Ital., a scale, a chip of marble <Med. L scalia <Gmc.]

sca·gli·o·la (skal·yō′lə) n. Hard, polished plasterwork imitating marble, granite, or other stone: made of powdered gypsum and glue, colored in various ways. [<Ital. scagliuola, dim. of scaglia SCAGLIA.]

scaith (skāth) n. Scot. Scathe; damage; mar. —**scaith′less** adj.

sca·lade (skə-lād′) n. An escalade. Also **sca·la·do** (skə-lā′dō). [<Ital. scalada <scalare scale <scala a ladder <L. See SCALE².]

scal·age (skā′lij) n. 1 A percentage by which something is scaled down to allow for shrinkage. 2 The amount of lumber estimated to be in a log or logs being scaled. [<SCALE² + -AGE]

sca·lar (skā′lər) adj. Completely definable by a single number or by a point on a scale: said of a quantity having magnitude but no direction, as a volume or mass: distinguished from vector. —n. Math. A pure number, especially one representing only a magnitude. [<L scalaris of a ladder <scala a ladder. See SCALE².]

sca·la·re (skə-lā′rē, -lä′rä) n. 1 A deep-bodied cichlid fish of South American rivers (genus Pterophyllum), noted for its striking coloration and popular as an aquarium fish: also called angelfish. 2 A related fish of the Amazon, the **blue scalare** (Symphysodon discus), with a brownish-green, disk-shaped body. [<NL <L scalaris of a ladder; so called because marked with dark crossbars]

sca·lar·i·form (skə-lar′ə-fôrm) adj. Biol. Ladderlike: said of cells or vessels. [<NL scalariformis <L scalaris of a ladder + forma form]

scal·a·wag (skal′ə-wag) n. 1 Colloq. A worthless fellow; scamp. 2 Specifically, during the Reconstruction period (1865-70), a native Southern white Republican, as distinguished from a carpetbagger: an opprobrious epithet used by Southern Democrats. Also spelled scallawag, scallywag. [? from Scalloway, Shetland Islands; orig. with ref. to undersized ponies or cattle grown there]

scald¹ (skôld) v.t. 1 To burn with or as with hot liquid or steam. 2 To cleanse or treat with boiling water. 3 To heat (a liquid) to a point just short of boiling. 4 To cook in a liquid which is just short of the boiling point. —v.i. 5 To be or become scalded. —n. 1 A burn or injury to the skin by a hot fluid, as steam or water. 2 An act of scalding. 3 A destructive parasitic disease of cranberries. 4 A discoloration of plant tissue due to improper conditions of growth, bad storage, etc.: apple scald. [<AF escalder <LL excaldare wash with hot water <ex- very + calidus hot]

scald² (skôld, skäld) n. An ancient Scandinavian bard, minstrel, or reciter of eulogies: also spelled skald. [<ON skald] —**scal·dic** (skôl′dik, skäl′-) adj.

scald³ (skäld, skôld) v.t. & v.i. Scot. To scold.

scald⁴ (skôld) n. Pathol. Favus.

scald-head (skôld′hed′) n. Pathol. Favus. [< scald, var. of scalled (<SCALL) + HEAD]

scale¹ (skāl) n. 1 One of the thin, flat, horny, membranous or bony outgrowths of the skin of various vertebrates, as most fishes, usually overlapping and forming a nearly complete investment. 2 A scab. 3 A scale insect. 4 Bot. A rudimentary or metamorphosed leaf, as of a pine cone. 5 Metall. The coating of oxide that forms on heated iron, etc.: also, an incrustation, as on the inside of boilers. 6 Any hard, thin, scalelike formation, as a flake, husk, shell, pod, or exfoliation. —v. **scaled, scal·ing** v.t. 1 To strip or clear of scale or scales. 2 To form scales on; cover with scales. 3 To take off in layers or scales; pare off. 4 To throw (a thin, flat object) so that its edge cuts the air or so that it skips along the surface of water. —v.i. 5 To come off in layers or scales; peel. 6 To shed scales. 7 To become incrusted with scales. [<OF escale a husk <Gmc.; infl. in meaning by OF escaille a fish's scale, an oyster's shell <Med. L scalia <Gmc.]

scale² (skāl) n. 1 A piece of metal, wood, or glass bearing accurately spaced lines or graduations for use in measurement, or the series of marks so used. 2 Any system of designating units of measurement or in which a fixed proportion is used in determining quantities: a scale of 1 inch to the mile. 3 Math. A system of notation in which the successive places determine the value of figures, as the decimal system. 4 Any progressive or graded series; a graduation: the social scale. 5 Music All the tones or notes of a key in regular ascending or descending order, in an octave or

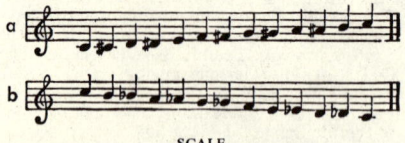

SCALE
a. Ascending. b. Descending.

more. 6 Phot. The range of light values which may be reproduced by a photographic paper. 7 An escalade. 8 A succession of steps; ladder; stairs: the original meaning. —**major scale** Music A scale having semitones between the 3-4 and 7-8 notes. —**minor scale** Music A scale having semitones between 2-3, 5-6, 7-8 notes (the harmonic form); or between 2-3, 7-8 ascending, 6-5, 3-2 descending (the melodic form). —v. **scaled, scal·ing** v.t. 1 To climb to the top of; go up by or as by means of a ladder. 2 To make according to a scale. 3 To regulate or adjust according to a scale or ratio: with up, down, etc. 4 To measure (logs) or estimate the amount of timber in (standing timber). —v.i. 5 To climb; ascend. 6 To rise, as in steps or stages: Mountains scaling to the skies. [<Ital. scala a ladder <L <scandere climb] —**scal′a·ble** adj. —**scal′er** n.

scale³ (skāl) n. 1 The bowl, scoop, or platform of a weighing instrument or balance. 2 The balance itself; hence, figuratively, the scale or scales of Justice. 3 Usually pl. Any form of weighing machine. —**to turn the scales** To determine; decide. —v. **scaled, scal·ing** v.t. 1 To weigh in scales. 2 To amount to in weight. —v.i. 3 To be weighed in scales. [<ON skál a bowl, in pl. a weighing balance. Akin to SHALE, SHELL.]

scale board 1 A very thin, veneerlike piece of board, as for the back of a picture. 2 Printing A narrow strip of wood used in justifying a line of type. [<SCALE¹ + BOARD]

scale insect One of numerous small, hemipterous, plant-feeding insects (family Coccidae) which as adults are degenerate, sedentary, and covered with a scalelike, waxy protective shield.

scale moss Any plant belonging to the class Hepaticae; any of the liverworts: so called because of their regularly arranged scalelike leaves.

sca·lene (skā′lēn, skā-lēn′) adj. 1 Geom. a Having no two sides equal: said of a triangle. b Having the axis inclined to the base: said of a cone or cylinder. 2 Anat. Designating one of several deeply placed muscles attached to the cervical vertebrae and first and second ribs and acting to flex or bend the neck. Also **sca·le·nous** (skā-lē′nəs). [<LL scalenus <Gk. skalēnos uneven]

sca·le·nus (skə-lē′nəs) n. A scalene muscle. [<NL (musculus) scalenus <L, SCALENE (def. 2)]

Scales (skālz) A sign of the zodiac, called also Libra or The Balance.

Scal·i·ger (skal′ə-jər), **Joseph Justus**, 1540-1609, French philologist. —**Julius Caesar**, 1484-1558, Italian author, critic, and Latin scholar; father of the preceding.

scall (skôl) n. Pathol. 1 A cutaneous eruption of small pustular vesicles: often epidemic among children. 2 Any scabby or scaly eruption. Also called scald. [<ON skalle a bald head]

scal·la·wag (skal′ə-wag), **scal·ly·wag** (skal′ē-wag) See SCALAWAG.

scal·lion (skal′yən) n. 1 The shallot. 2 Any bulbous onion with a long thick neck, as a leek. [<AF scalun, OF eschalogne, ult. <L (caepa) Ascalonia (onion) of Ashkelon <Ascalon Ashkelon, a Palestinian seaport]

scal·lop (skal′əp, skol′-) n. 1 A bivalve (genus Pecten) having a nearly circular shell with radiating ribs and wavy edge. 2 Its adductor muscle, which as a rule is edible and very succulent. 3 Its shell, formerly worn as a pilgrim's badge. 4 A dish or pan (originally a scallop shell) in which oysters are cooked or served. 5 One of a series of semicircular curves along an edge, as for ornament. —v.t. 1 To shape the edge of with scallops; ornament with scallops. 2 To bake (food) in a casserole with a liquid or sauce, often topped with bread crumbs. Also spelled escallop, scollop. [<OF escalope <Gmc. Akin to SCALE¹.] —**scal′lop·er** n.

SCALLOP SHELL

scalp (skalp) n. 1 The skin of the top and back of the human skull, usually covered with hair; also, a portion of this, cut or torn away as a war trophy among certain North American Indians, particularly of the St. Lawrence region. 2 A similar piece taken from the head of a wild animal as an evidence that it has been killed for the collection of a bounty. 3 A political victory or defeat. 4 A denuded or bare summit, as of a hill or cliff. 5 On the stock exchange, a small profit taken by a speculator. —v.t. 1 To cut or tear the scalp from. 2 Colloq. To buy (tickets) and sell again at prices exceeding the established rate. 3 Colloq. To buy and sell quickly in order to make a small profit. —v.i. 4 Colloq. To scalp bonds, tickets, etc. [ME, prob. <Scand. Cf. ON skálpr a sheath.] —**scalp′er** n.

scalp dance A ceremonial victory dance of certain North American Indians, in which the women of the tribe display the trophies and perform the dances, accompanied by the singing of the warriors.

scal·pel (skal′pəl) n. A small pointed knife with a very sharp, thin blade, used in dissections and in surgery. [<L scalpellum, dim. of scalprum a knife <scalpere cut]

scalp lock A long lock of hair left on the crown of the head by certain North American Indians, often braided and interwoven with feathers or fur: a challenge to an enemy.

scal·y (skā′lē) adj. **scal·i·er, scal·i·est** 1 Having a covering of scales; hence, also, exfoliated; scurfy. 2 Of the nature of scales; squamous. 3 Incrusted, as a boiler. 4 Slang Mean; dishonorable. [<SCALE¹ + -Y¹] —**scal′i·ness** n.

scaly ant-eater A pangolin.

Sca·man·der (skə-man′dər) Ancient name for the MENDERES (def. 2).

scam·ble (skam′bəl) v. **·bled, ·bling** Brit. Dial. v.t. 1 To scatter (something) to a crowd. 2 To gather confusedly. —v.i. 3 To scramble. 4 To stumble along. [Cf. SCRAMBLE and SHAMBLE¹]

scam·mo·ny (skam′ə-nē) n. 1 A climbing plant (Convolvulus scammonia) of the morning-glory family, native in Asia Minor, with tuberous roots containing a milky juice. 2 The dried resin of scammony roots, used as a strong cathartic. [<L scammonia <Gk. skammōnia]

scamp¹ (skamp) n. A confirmed rogue; good-for-nothing fellow; rascal. [<obs. scamp, v., roam, contraction of SCAMPER] —**scamp′ish, scamp′y** adj.

scamp² (skamp) v.t. To perform (work) carelessly or dishonestly. [Orig. dial. E, ?<ON skemma shorten <skammr short. Akin to SCANT, SKIMP.] —**scamp′er** n.

scam·per (skam′pər) v.i. To run quickly or hastily, as from danger; hurry away. —n. A hurried flight. [?<obs. Du. schampen run away <AF escamper, OF eschamper decamp, run off hurriedly, escape, ult. <L ex out from + campus a plain, battlefield] —**scamp′er·er** n.

scan (skan) v. **scanned, scan·ning** v.t. 1 To examine in detail; scrutinize closely. 2 To pass

scan·dal (skan'dəl) *n.* 1 The heedless or malicious repetition of evil reports; aspersion of character. 2 Reproach caused by outrageous or improper conduct. 3 A discreditable circumstance, event, or action; cause of reproach. 4 Injury to reputation, or general comment causing it. 5 *Law* Malicious defamation by word of mouth. 6 One whose conduct disgraces. [<AF *escandle* <L *scandalum* a cause of stumbling <Gk. *skandalon* a snare; refashioned after MF *scandale* <L *scandalum.* Doublet of SLANDER.] – *Synonyms:* aspersion, backbiting, calumny, defamation, detraction, obloquy, odium, reproach, slander. *Scandal* may be odious truth; *slander* is certain falsehood. *Antonyms:* applause, celebrity, credit, eulogy, fame, glory, honor, renown, reputation, repute.
scan·dal·i·za·tion (skan'dəl·ə·zā'shən, -ī·zā'-) *n.* 1 The act of offending moral feelings. 2 That which scandalizes; a scandal.
scan·dal·ize (skan'dəl·īz) *v.t.* **·ized, ·iz·ing** To shock the moral feelings of, as by improper, frivolous, or offensive conduct; outrage. – **scan'dal·iz'er** *n.*
scan·dal·ous (skan'dəl·əs) *adj.* 1 Causing, or tending to cause, scandal; being a scandal; opprobrious; disgraceful; shocking to the sense of truth, decency, or propriety. 2 Consisting of evil or malicious reports; tending to injure reputation. 3 *Law* Libelous; irrelevant. See synonyms under FLAGRANT, INFAMOUS. – **scan'dal·ous·ly** *adv.* – **scan'dal·ous·ness** *n.*
scan·dent (skan'dənt) *adj.* Climbing, or aiding to climb, as a plant. [<L *scandens, -entis,* ppr. of *scandere* climb]
Scan·der·beg (skan'dər·beg), 1403–68, Albanian chief and national hero: real name George Castriota.
scan·di·a (skan'dē·ə) *n. Chem.* Scandium oxide, Sc$_2$O$_3$, a colorless, amorphous powder soluble in acids. [<NL <*scandium* SCANDIUM]
Scan·di·an (skan'dē·ən) *adj.* 1 Relating to Scandia, the Scandinavian Peninsula. 2 Scandinavian. [<L *Scandia*]
scan·dic (skan'dik) *adj. Chem.* Pertaining to or derived from scandium, especially in its higher valence.
Scan·di·na·vi·a (skan'də·nā'vē·ə) The region of NW Europe occupied by Sweden, Norway, and Denmark; 315,156 square miles; Finland, Iceland, and the Faroe Islands are often included: total area, 485,539 square miles. Ancient **Scan·di·a** (skan'dē·ə).
Scan·di·na·vi·an (skan'də·nā'vē·ən) *adj.* Of or pertaining to Scandinavia, its people, or their languages. – *n.* 1 A native or inhabitant of Scandinavia. 2 The North Germanic group of languages. See under GERMANIC. – **Old Scandinavian** Old Norse. See under NORSE. [<L *Scandinavia,* var. of *Scadinavia* <Gmc.]
Scandinavian Peninsula The peninsula of NW Europe containing Norway and Sweden; 298,550 square miles.
scan·di·um (skan'dē·əm) *n.* A metallic element (symbol Sc) of the lanthanide series, found in certain Swedish yttrium minerals. See ELEMENT. [<NL <L *Scandia* Scandinavia]
scan·ning (skan'ing) *n.* 1 Scansion. 2 The process by which the electron beam of a television transmitting unit passes rapidly over every point of the image on the photosensitive screen.
scan·sion (skan'shən) *n.* The act or art of scanning verse to show its metrical parts. Compare METER^2 (def. 2). [<F <LL *scansio, -onis* <L *scandere.* See SCAN.]
scan·so·ri·al (skan·sôr'ē·əl, -sō'rē-) *adj. Zool.* Pertaining to or adapted for climbing. Also

scan·so'ri·ous. [<L *scansorius* <*scansus,* pp. of *scandere* climb]
scant (skant) *adj.* 1 Scarcely enough; meager in measure or quantity. 2 Being just short of the measure specified; of limited extent: often with the indefinite article even with a plural noun: a *scant* half-hour, a *scant* five yards. 3 Insufficiently supplied with something: with *of:* We were *scant* of breath. See synonyms under SCANTY. – *v.t.* 1 To restrict or limit in supply; stint. 2 To treat briefly or inadequately. See synonyms under SCRIMP. – *adv. Dial.* Scarcely; barely; not quite. [<ON *skamt,* neut. of *skammr* short] – **scant'ly** *adv.* – **scant'ness** *n.*
scant·ling (skant'ling) *n.* 1 A timber of moderate cross-section, used for studding, etc. 2 Such timbers collectively. 3 The dimensions of a timber in breadth and depth, but not in length; also, the dimensions of a stone in length, breadth, and thickness. 4 A small quantity or part; a sample. [Alter. of obs. *scantillon* <OF *eschantillon* specimen, corner-piece, chip; ? infl. in meaning by SCANT]
scant·y (skan'tē) *adj.* **scant·i·er, scant·i·est** 1 Limited in extent; small; close; cramped. 2 Restricted in quantity or number; scarcely sufficient. 3 Sparing. [<SCANT] – **scant'i·ly** *adv.* – **scant'i·ness** *n.* – *Synonyms:* deficient, insufficient, limited, narrow, niggardly, parsimonious, poor, scant, scarce, scrimped, scrimping, scrimpy, short, sparing, sparse. *Antonyms:* see synonyms under AMPLE.
Sca·pa Flow (skä'pə flō', skap'ə) A sea basin and British naval base in the Orkney Islands, northern Scotland; 50 square miles: the Germans scuttled part of their own fleet here, June 21, 1919.
scape^1 (skāp) *n.* 1 *Bot.* A long, naked peduncle rising from a depressed stem, as in the dandelion. 2 *Biol.* A stemlike part, as of an insect antenna, or the shaft of a feather. 3 *Archit.* The shaft of a column, or the apophyge of a shaft. [<L *scapus* <dial. Gk. (Doric) *scapos.* Akin to SCEPTER.]
scape^2 (skāp) *n.* A scene, as of land, sea, clouds, or the like. [Back formation <LANDSCAPE]
scape^3 (skāp) *Archaic v.t. & v.i.* To escape: generally written 'scape. – *n.* 1 An escape or means of escape. 2 A fault; an escapade. [Aphetic var. of ESCAPE]
scape·goat (skāp'gōt') *n.* 1 The goat upon whose head Aaron symbolically laid the sins of the people on the day of atonement, after which it was led away into the wilderness. *Lev.* xvi. 2 Any animal or person on whom the bad luck or sins of an individual or group are symbolically placed, and which is then turned loose: a world-wide folk custom of great antiquity. 3 Any person bearing blame for others. [<SCAPE^3, *n.* + GOAT]
scape·grace (skāp'grās') *n.* A mischievous or incorrigible person. [<SCAPE^3 + GRACE (def. 4)]
scape wheel An escape wheel.
scaph·oid (skaf'oid) *adj.* Boat-shaped. – *n. Anat.* A proximal bone of the wrist on the radial side; the navicular; also, a bone of the tarsus. [<NL *scaphoides* <Gk. *skaphoeidēs* <*skaphē* a boat + *eidos* form]
scapi- combining form A stalk or stem; a shaft: *scapiform,* resembling a scape. Also, before vowels, **scap-**. [<L *scapus* a stalk]
scap·o·lite (skap'ə·līt) *n.* Any of a definite group of tetragonal silicates, chiefly of aluminum, calcium, and sodium: also called *wernerite.* [<G *skapolith* <Gk. *skapos* a rod + *lithos* stone]
sca·pose (skā'pōs) *adj.* 1 Bearing a scape. 2 Like a scape. [<SCAPE^1 + -OSE]
scap·u·la (skap'yə·lə) *n. pl.* **·lae** (-lē) *Anat.* The shoulder blade; the superior or proximal element of the pectoral girdle in the skeleton of vertebrates. [<LL, shoulder <L *scapulae* shoulder blades]
scap·u·lar (skap'yə·lər) *adj.* 1 An outer garment worn by members of certain religious orders and consisting of two strips of cloth hanging down front and back and joined across the shoulders; formerly, a monastic working dress. 2 Two small rectangular pieces of

woolen cloth connected by strings: worn as a badge of membership by certain religious orders. 3 *Surg.* A bandage passing over the shoulder blade. 4 *Ornithol.* The shoulder feathers of a bird lying along the sides of the back. – *adj.* Of or pertaining to the scapula or scapulars. Also **scap'u·lar'y** (-ler'ē). [< Med. L *scapulare* <LL *scapula.* See SCAPULA.]
scar^1 (skär) *n.* 1 The mark left on the skin after the healing of a wound or sore; a cicatrix. 2 Any mark resulting from past injury: often applied figuratively to the effects on a character of crimes or sorrows. 3 The mark left on or made by an organ, as by leaves after separation from a stem or branch. 4 An indentation or mark made by use, motion, or contact. – *v.t.* & *v.i.* **scarred, scar·ring** To mark or become marked with a scar. [<OF *escare* <LL *eschara* a scab <Gk.] – **scar'less** *adj.*
scar^2 (skär) *n.* 1 A bare rock standing alone. 2 A cliff or rocky place on the side of a hill or mountain. Also, *Scot.* **scaur.** [<ON *sker*]
scar·ab (skar'əb) *n.* 1 A scarabaeid beetle, especially the large, black, dung beetle (*Scarabaeus sacer*), held sacred by the ancient Egyptians as the symbol of resurrection and fertility. 2 A gem representing this beetle and inscribed with symbols, used in ancient Egypt as an amulet. Also **scar·a·bee** (skar'ə·bē). [<MF *scarabée* <L *scarabaeus*]

SCARAB

scar·a·bae·id (skar'ə·bē'id) *adj.* Pertaining to a large family (*Scarabaeidae*) of beetles, including cockchafers, etc. – *n.* A member of the Scarabaeidae family of beetles. Also **scar'a·bae'an, scar'a·bae'oid.** [<NL <L *scarabaeus* scarab]
scar·a·bae·us (skar'ə·bē'əs) *n. pl.* **·bae·i** (-bē'ī) 1 The scarab beetle. 2 A gem or conventionalized design resembling it. Also **scar'a·be'us.** [<L]
scar·a·boid (skar'ə·boid) *adj.* Resembling or of the nature of a scarab or scarabaeid. – *n.* A scarab or scarabaeid.
scar·a·mouch (skar'ə·mouch, -moosh) *n.* A boastful, cowardly character; a swaggering buffoon: so called from a character in old Italian comedy. [<F *Scaramouche* <Ital. *Scaramuccia,* lit., a skirmish <Gmc.]
Scar·bor·ough (skär'bûr·ə) A municipal borough and resort town in North Riding, eastern Yorkshire, England, on the North Sea.
scarce (skârs) *adj.* 1 Rarely met with; infrequent. 2 Not plentiful; scant; insufficient. 3 Characterized or attended by insufficiency or want. See synonyms under RARE^1, SCANTY. – **to make oneself scarce** *Colloq.* To go away, or stay away. [<AF *scars, escars,* OF *eschars* scanty, insufficient, ult. <L *excerptus.* See EXCERPT.] – **scarce'ness** *n.*
scarce·ly (skârs'lē) *adv.* 1 Only just; barely. 2 Not quite; hardly.
scarce·ment (skârs'mənt) *n.* 1 *Archit.* A plain flat ledge or set-off in a wall. 2 *Mining* A ledge or projection left in the side of a shaft, embankment, or wall, as for a ladder. [Appar. <obs. *scarce, v.,* lessen (<SCARCE) + -MENT]
scar·ci·ty (skâr'sə·tē) *n. pl.* **·ties** Scantiness; insufficiency; lack of necessities; dearth. See synonyms under WANT.
scare^1 (skâr) *v.* **scared, scar·ing** *v.t.* 1 To strike with sudden fear; frighten. 2 To drive or force by frightening: with *off* or *away:* to *scare* away an intruder. – *v.i.* 3 To take fright; become scared. See synonyms under FRIGHTEN. – **to scare up** *Colloq.* To get together hurriedly; discover; produce: to *scare* up votes, food, a group of people, etc. – *n.* 1 Sudden fright, especially from slight or imaginary cause; terror. 2 A panic (def. 2). [<ON *skirra* frighten <*skiarr* shy] – **scar'er** *n.* – **scar'ing·ly** *adv.*
scare^2 (skâr) *n.* In golf, the smaller end of a club head at the part where, formerly, it was spliced to the handle. [<dial. E (Scottish), a joint <ON *skor*]
scare·crow (skâr'krō) *n.* 1 Any effigy set up to scare crows and other birds from growing crops. 2 A cause of false alarm. 3 A wretched-looking person.

Synonyms: bogy, bugbear, fright, goblin, hobgoblin.

scare·head (skâr′hed′) *n.* An exceptionally large newspaper headline in very bold type for news of sensational interest.

scare·mon·ger (skâr′mung′gər, -mong′-) *n.* One who spreads an alarming rumor; an alarmist.

scarf[1] (skärf) *n. pl.* **scarfs** 1 In carpentry, a lapped joint made as by notching two timbers at the ends, and bolting them together so as to form one continuous piece without increased thickness: also **scarf joint.** 2 The notched end of either of the timbers so cut. 3 A cut or incision in the blubber of a whale. —*v.t.* 1 To unite with a scarf joint. 2 To cut a scarf in. [? <ON *skarfr* a notch in a timber]

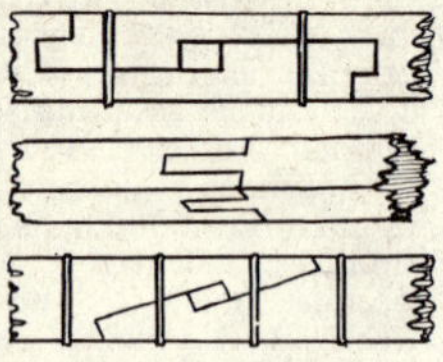
TYPES OF SCARF JOINTS

scarf[2] (skärf) *n. pl.* **scarfs** or **scarves** (skärvz) 1 A long and wide band, especially when worn about the head and neck; also, any sash. 2 A necktie or cravat. 3 A runner for a bureau or dresser. 4 An official sash, denoting rank. 5 A tippet or neckpiece. —*v.t.* 1 To cover or decorate as with a scarf. 2 To use as a scarf; wrap loosely around one. [<AF *escarpe*, OF *escharpe*, ? <*escreppe* a scrip[2]]

scarf[3] (skärf) See SCART[3].

scarf·pin (skärf′pin′) *n.* An ornamental pin worn on a tie or scarf.

scarf·skin (skärf′skin′) *n.* The epidermis; cuticle. [<SCARF[2] + SKIN]

scar·i·fi·ca·tor (skar′ə·fə·kā′tər) *n.* A surgical instrument, consisting of several lancet points, and used for making incisions in the skin to draw blood.

scar·i·fi·er (skar′ə·fī′ər) *n.* One who or that which scarifies.

scar·i·fy (skar′ə·fī) *v.t.* **·fied, ·fy·ing** 1 To scratch or make slight incisions in, as the skin in surgery. 2 To criticize severely; make cutting comments on. 3 *Agric.* To stir the surface of, as soil. 4 To prune. [<MF *scarifier* <LL *scarificare* <L *scarifare* <Gk. *skariphasthai* scratch an outline, sketch <*skariphos* a stylus] —**scar′i·fi·ca′tion** *n.*

scar·i·ous (skâr′ē·əs) *adj.* 1 *Bot.* Thin, dry, membranaceous, and not green: said of plants. 2 Scaly. Also **scar′i·ose** (-ōs). [<F *scarieux* <NL *scariosus* <L *scaria* a thorny shrub]

scar·la·ti·na (skär′lə·tē′nə) *n.* Scarlet fever; popularly, a mild form of scarlet fever. [<NL <Ital. *scarlattina*, fem. dim. of *scarlatto* <Med. L *scarlatum* SCARLET]

scar·la·ti·noid (skär′lə·tē′noid, skär·lat′ə·noid) *adj.* Resembling scarlet fever. —*n. Pathol.* One of a group of erythemas closely resembling scarlet fever.

Scar·lat·ti (skär·lät′tē), **Alessandro,** 1659–1725, Italian composer. —**Domenico,** 1683–1757, Italian composer; son of preceding.

scar·let (skär′lit) *n.* 1 A brilliant red, inclining to orange. 2 A bright-red dye formerly obtained from the kermes or cochineal insect. 3 Any one of several coal-tar colors, varying from yellow to brown and used for dyeing. 4 Cloth or clothing of a scarlet color. —*adj.* 1 Brilliant-red, inclining to orange. 2 Clothed in scarlet. [<OF *escarlate* <Med. L *scarlatum*, prob. <Arabic *siqillāt* <Persian *saqalāt* a rich, scarlet cloth]

scarlet fever *Pathol.* An acute infectious fever caused by certain strains of hemolytic streptococci and characterized by a diffused scarlet rash followed by scaling off of the skin. Also *scarlatina.*

scarlet letter A scarlet "A," a badge of shame, which women convicted of adultery were once compelled to wear.

scarlet runner A tall climbing bean *(Phaseolus coccineus)* of tropical America, with vivid red flowers and long seed pods, now widely cultivated as a vegetable; a string bean.

Scarlet Woman The woman of *Rev.* xvii 4–6: an abusive epithet, first applied to pagan Rome, latterly to the Roman Catholic Church.

scarp (skärp) *n.* Any steep slope; an abrupt declivity; an escarpment. Compare illustration under BASTION. —*v.t.* To cut to a steep slope. [<AF *escarpe* <Ital. *scarpa*]

Scar·pan·to (skär′pän·tō) The Italian name for KARPATHOS.

Scarpe (skärp) A river of northern France, flowing 60 miles NE to the Scheldt.

scar·pet·ti (skär·pet′tē) *n. pl.* Rope-soled shoes used by mountain climbers. [<Ital. *scarpetto* a light shoe, dim. of *scarpa* a shoe]

Scar·ron (skä·rôṅ′), **Paul,** 1610–60, French poet and dramatist; first husband of Madame de Maintenon.

scart[1] (skärt) *Scot. n.* 1 A scratch; slight wound. 2 A pen or pencil mark. 3 A scrap or small irregular portion. —*v.t.* & *v.i.* To mark slightly; scratch or scrape. [ME *scratte*; of obscure origin]

scart[2] (skärt) *n. Scot.* A puny- or scrawny-looking person; also, a miserly person.

scart[3] (skärt) *n. Scot.* A cormorant. Also called *scarf.* [<ON *skarfr*]

scarves (skärvz) Alternative plural of SCARF[2].

scar·y (skâr′ē) *adj.* **scar·i·er, scar·i·est** *Colloq.* 1 Easily scared. 2 Somewhat frightened; anxious; timid. 3 Giving cause for alarm.

scat[1] (skat) *v.i.* **scat·ted, scat·ting** *Colloq.* To go away; depart: usually in the imperative. [? <*ss*, imit. of a hiss + CAT]

scat[2] (skat) *n. Slang* A type of jazz singing in which meaningless syllables are improvised on the melodic line. [Prob. <SCAT[1]]

scathe (skāth) *v.t.* **scathed, scath·ing** 1 To criticize severely. 2 To injure severely; harm; blast. —*n.* Severe injury; harm; loss. Also **scath** (skath). Also, *Scot.*, *skaith.* [<ON *skatha* <*skathi* harm] —**scathe′ful** *adj.*

scathe·less (skāth′lis) *adj.* Free from harm. Also **scath′less** (skath′-).

scath·ing (skā′thing) *adj.* Damaging by scorching or blasting; withering: now usually figuratively: a *scathing* rebuke. —*n.* Harm; injury. —**scath′ing·ly** *adv.*

scato- *combining form* Dung; excrement: *scatology.* Also, before vowels, **scat-.** [<Gk. *skōr, skatos* dung]

scat·o·log·ic (skat′ə·loj′ik) *adj.* Of or pertaining to scatology; obscene. Also **scat′o·log′i·cal.**

sca·tol·o·gy (skə·tol′ə·jē) *n.* 1 The study of excrement, considered as a branch of paleontology, medicine, and psychiatry. 2 Preoccupation with filth or obscenity, as in literature. Also **scat·o·lo·gi·a** (skat′ə·lō′jē·ə). [<SCATO- + -LOGY] —**sca·tol′o·gist** *n.*

scat·o·man·cy (skat′ō·man′sē) *n.* In folklore, divination, or determination of disease, by means of feces.

scat·ter (skat′ər) *v.t.* 1 To throw about in various places; sprinkle; strew, as seed. 2 To separate and drive away in different directions; disperse; rout. 3 *Physics* To reflect (heat or light) irregularly. —*v.i.* 4 To separate and go in different directions; disperse; dissipate. See synonyms under SPREAD, SQUANDER. [ME *scateren* squander. ? Akin to SHATTER.] —**scat′ter·er** *n.*

scat·ter·brain (skat′ər·brān′) *n.* A person without concentration of mind; a heedless person. —**scat′ter-brained′** *adj.*

scat·ter·good (skat′ər·gŏŏd′) *n.* 1 One who wastes that which is good; a spendthrift. 2 One who or that which distributes charities.

scat·ter·ing (skat′ər·ing) *n.* 1 Dispersion. 2 *Physics* The spreading out of a stream of particles, as alpha rays, upon striking a film of gaseous or solid matter. 3 The dispersion, over an area, of votes for candidates. —*adj.* Placed at intervals or at a distance apart. —**scat′ter·ing·ly** *adv.*

scat·ter·ling (skat′ər·ling) *n.* A person without fixed home or connections; a vagrant. [<SCATTER + -LING[1]]

sca·tu·ri·ent (skə·chŏŏr′ē·ənt) *adj. Obs.* Gushing forth, as a fountain. [<L *scaturiens, -entis*, ppr. of *scaturire* <*scatere* flow out]

scaud (skôd) *v.t. Scot.* To scald.

scauff (skôf) See SCAFF.

scaup[1] (skôp) *n.* A sea duck (genus *Nyroca*) of northern regions, related to the canvasback, having the head and neck black in the male; especially, the common American bay duck *(N. marila).* Also **scaup duck.** [Short for *scaup duck* <SCAUP[2] + DUCK]

scaup[2] (skôp) *n. Obs. & Scot.* 1 The scalp; skull. 2 A mussel bed. [Var. of SCALP]

scaur (skär, skôr) See SCAR[2].

scav·enge (skav′inj) *v.* **·enged, ·eng·ing** *v.t.* 1 To remove filth, rubbish, and refuse from, as streets. 2 To remove exhaust gases from (the cylinder of an internal-combustion engine). 3 *Metall.* To remove impurities from (a metal or alloy). —*v.i.* 4 To act as a scavenger. 5 To search for food. [Back formation <SCAVENGER]

scav·en·ger (skav′in·jər) *n.* 1 A street-cleaner. 2 An animal that feeds on carrion, as the buzzard. [ME *scavager* <AF *scawager* <*scawage* inspection <*escauwer* inspect <Flemish *scauwen* see]

sce·nar·i·o (si·nâr′ē·ō, -nä′rē·ō) *n. pl.* **·nar·i·os** 1 The plot of a dramatic work, or a skeleton libretto. 2 The written plot and arrangement of incidents of a motion picture. [<Ital. <LL *scenarius* of stage scenes <L *scena.* See SCENE.]

sce·nar·ist (si·nâr′ist, -nä′rist) *n.* One who writes scenarios.

scend (send) *Naut. v.i.* To heave upward, as a vessel on a wave. —*n.* The upward angular displacement of a vessel: correlative of *pitch.* Also spelled *send.* —**pitch and scend** The longitudinal rocking of a vessel. [Var. of SEND[2]; infl. in form by *ascend*]

scene (sēn) *n.* 1 A locality and all connected with it, as presented to view; a landscape. 2 The place in which the action of a drama is supposed to occur; setting or locality. 3 The place and surroundings of any event, real or imagined, as in literature or art. 4 A division of an act of a play; one comprehensive event in a play. 5 Any incident or episode that may serve as the subject of a description. 6 The painted canvas screen or screens for the background for a play. 7 Any striking exhibition or display; especially, a display of passion or excited feeling. —**behind the scenes** 1 Out of sight of a theater audience; backstage. 2 Privately; in secret. [<OF <L *scena, scaena* <Gk. *skēnē* a tent, a stage]

Synonyms: action, display, event, exhibition, incident, landscape, place, prospect, situation, view. See QUARREL[1], SPECTACLE.

scen·er·y (sē′nər·ē) *n. pl.* **·er·ies** Natural or theatrical scenes collectively. [<Ital. *scenario.* See SCENARIO.]

sce·nic (sē′nik, sen′ik) *adj.* 1 Artistic in grouping and effect. 2 Picturesque. 3 Relating to stage scenery. Also **sce′ni·cal.** —**sce′ni·cal·ly** *adv.*

sce·nog·ra·phy (sē·nog′rə·fē) *n.* The art of making drawings in perspective. [<F *scénographie* <L *scaenographia* <Gk. *skēnographia* <*skēnē* a scene, tent + *graphein* write] —**scen·o·graph·ic** (sen′ə·graf′ik, sē′nə-), **scen′o·graph′i·cal** *adj.*

scent (sent) *n.* 1 An odor, pleasant or unpleasant. 2 The effluvium by which an animal can be tracked. 3 A clue aiding investigation. 4 Scraps of paper, in the game of hare and hounds, dropped by the hares in their flight to enable the hounds to follow them. 5 A fluid essence containing extracts from flowers or other fragrant bodies; perfume. 6 The sense of smell. —*v.t.* 1 To perceive by the sense of smell. 2 To form a suspicion of. 3 To cause to be fragrant; perfume. —*v.i.* 4 To hunt by the sense of smell: said of hounds. ◆ Homophone: *cent.* [<OF *sentir* discern by the senses, feel <L *sentire*] —**scent′less** *adj.*

Synonyms (noun): savor, smell.

scep·ter (sep′tər) *n.* 1 A staff or wand carried as the badge of command or sovereignty. 2 Hence, kingly office or power. —*v.t.* 1 To confer the scepter on; invest with royal power. 2 To furnish with or as with a scepter or scepters. Also **scep′tre** (-tər). [<OF *ceptre, sceptre* <L *sceptrum* <Gk. *skēptron* a staff <*skēptesthai* prop oneself, lean on]

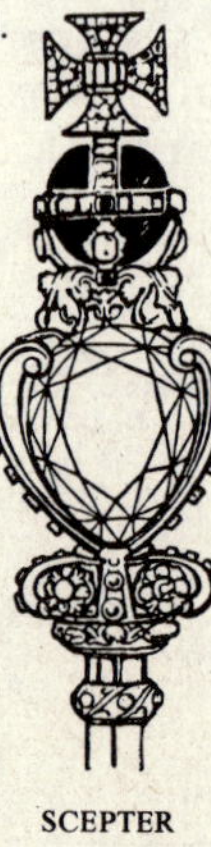
SCEPTER

scep·tic (skep′tik), **scep·ti·cal, scep·ti·cism,** etc. See SKEPTIC, etc.

Schac·a·bac (shak′ə·bak) In the *Arabian Nights*, the beggar invited to the Barmecide feast.

Schacht (shäkht), **(Horace Greeley) Hjalmar,** born 1877, German financier and economic expert.

Schaff·hau·sen (shäf′hou′zən) A town in northern Switzerland, on the Rhine near **Schaffhausen Falls** (German *Rheinfall*), a group of several cataracts falling a total of 100 feet, now harnessed for hydroelectric power.

schanz (skhäns) *n.* A breastwork of earth and stones. —*v.t.* To protect with a schanz. [< Du. *schans*]

schap·pe (shä′pə) *n.* A fabric woven from spun silk. —*v.t.* **schapped, schap·ping** To ferment (silk) so as to remove its gum coating. [<dial. G (Swiss), a waste, impurity]

Scharn·horst (shärn′hôrst), **Gerhard von,** 1755–1813, Prussian general and military writer.

schat·chen (shät′khən) *n.* One who arranges marriages for a fee; a marriage-broker: chiefly among Russian Jews. Also **schad′chan.** [<Yiddish, a marriage broker <Hebrew *shadhkhān*]

Schaum·burg-Lip·pe (shoum′bŏŏrkh·lip′ə) A former state of NW Germany, comprised in Lower Saxony, northern West Germany, 1945; 131 square miles; former capital, Bückeburg.

sched·ule (skej′ōōl, *Brit.* shed′yōōl) *n.* **1** A written or printed statement, usually in tabular form, specifying the details of some matter, and often annexed to statutes, petitions, and other documents. **2** A list; catalog; an inventory. **3** A timetable; also, a detailed and timed plan for any procedure. **4** A program. —*v.t.* **·uled, ·ul·ing 1** To place in or on a schedule. **2** To make a schedule of. **3** *Colloq.* To appoint or plan for a specified time or date: He *scheduled* his appearance for five o'clock. [Alter. of ME *sedule* <OF *cedule* <LL *scedula,* dim. of L *scida, scheda* a leaf of paper <Gk. *schidē* a wood splinter < *schizein* split; infl. in form by Med. L *schedula*]

Schee·le (shā′lə), **Karl Wilhelm,** 1742–86, Swedish chemist; discovered chlorine in 1774, oxygen in 1777, and other elements.

schee·lite (shē′līt) *n.* A vitreous, variously colored, tetragonal calcium tungstate. [after K. W. *Scheele,* who discovered tungstic acid]

Schef·fel (shef′əl), **Joseph Victor von,** 1826–1886, German poet.

schef·fer·ite (shef′ə·rīt) *n.* A brown manganese pyroxene often containing iron. [after H. T. *Scheffer,* 1710–59, Swedish chemist]

Sche·her·e·za·de (shə·her′ə·zä′də, -zäd′) In the *Arabian Nights,* the bride of a sultan who had sworn to kill each of his wives after her wedding night. Scheherezade tricked the sultan into sparing her life by telling him an exciting story each night, not revealing the ending until the following day. Also **Sche·her′a·za′de.**

scheik (shēk) See SHEIK.

Scheldt (skelt) A river in northern France, Belgium, and the Netherlands, flowing 270 miles NE to North Sea: French *Escaut. Flemish & Dutch* **Schel·de** (skhel′də).

Schel·ling (shel′ing), **Friedrich Wilhelm Joseph von,** 1775–1854, German philospher. —**Schel·lin·gi·an** (she·lin′jē·ən) *adj.*

sche·ma (skē′mə) *n. pl.* **sche·ma·ta** (skē′mə·tə) **1** A scheme, synopsis, or summary. **2** A diagrammatic representation of certain relations in some system of knowledge. **3** Any figure drawn in outline; formerly, a geometric diagram. [<L. See SCHEME.]

sche·ma·tism (skē′mə·tiz′əm) *n.* **1** A particular form or disposition of anything. **2** Orderly arrangement of parts, as in a philosophic system, the classification of knowledge, etc.; design.

sche·ma·tize (skē′mə·tīz) *v.t.* **·tized, ·tiz·ing** To form into or arrange according to a scheme or schema. [<Gk. *schēmatizein* < *schēma, -atos* a form] —**sche′ma·ti·za′·tion** *n.*

scheme (skēm) *n.* **1** A plan of something to be done; a plot or device for the accomplishment of an object. **2** A combination of various things according to a general plan or design; systematic arrangement. **3** A formal plan or arrangement, or a statement of such a plan; also, a table or schedule. **4** An outline drawing or sketch; diagram. **5** In astrology, a plan representing the aspects of the heavenly bodies at any given time. See synonyms under DESIGN, HYPOTHESIS, PLAN, PROJECT. —*v.* **schemed, schem·ing** *v.t.* **1** To make a scheme for; devise; plan. **2** To plan or plot in an underhand manner. —*v.i.* **3** To make schemes; plan or plot; connive. [<L *schema* a shape, figure of speech <Gk. *schēma, -atos* a form, plan] —**sche·mat·ic** (skē·mat′ik) or **·i·cal** *adj.* —**sche·mat′i·cal·ly** *adv.* —**schem′er** *n.* —**schem′ing** *adj.*

Sche·nec·ta·dy (skə·nek′tə·dē) An industrial city on the Mohawk River in eastern New York.

schenk beer (shengk) Beer fermented in 4 to 6 weeks and brewed for immediate use in the winter. [<G *schenkbier* < *schenken* pour out + *bier* beer; so called with ref. to its being put on schenk (draft) as soon as it is made, to keep it from turning sour]

scher·zan·do (sker·tsän′dō) *adv. Music* In a sportive or playful manner. [<Ital., ppr. of *scherzare* play < *scherzo.* See SCHERZO.]

scher·zo (sker′tsō) *n. pl.* **·zos** or **·zi** (-tsē) *Music* A sportive or lightsome movement, usually following a slow movement, especially in a symphony or sonata. [<Ital., a jest <G *scherz*]

Sche·ven·ing·en (skhā′vən·ing′ən) A resort town of western Netherlands just NW of The Hague on the North Sea; scene of a British naval victory over the Dutch, 1653.

Schia·pa·rel·li (skyä′pä·rel′lē), **Giovanni,** 1835–1910, Italian astronomer.

Schick (shik), **Béla,** born 1877, Austrian pediatrician, born in Hungary and active in the United States.

Schick test A test to determine the susceptibility of a person to diphtheria by injecting a diluted diphtheria toxin: a positive reaction gives a reddening of the skin. [after Dr. Béla *Schick,* who devised it]

Schie·dam (skhē·däm′) A town of western Netherlands, just west of Rotterdam.

schil·ler (shil′ər) *n. Mineral.* A bronzelike luster or iridescence due to the reflection of particles dispersed in certain minerals. [<G, a play of colors < *schillern* change color]

Schil·ler (shil′ər), **Johann Christoph Friedrich von,** 1759–1805, German poet and dramatist.

schil·ler·ize (shil′ə·rīz) *v.t.* **·ized, ·iz·ing** To impart schiller to. —**schil′ler·i·za′tion** *n.*

schil·ling (shil′ing) *n.* The monetary unit of Austria since 1924; a former North German silver coin. [<G]

Schip·hol (skhip′hôl) A village and international airport in western Netherlands, 6 miles SW of Amsterdam.

schip·per·ke (skip′ər·kē) *n.* A Belgian breed of dog used as a watchdog and sometimes for hunting. It is usually tailless, has a thick-set body, foxlike head, and a rather thick, short, black coat. Formerly called *Spits* or *Spitske.* [< dial. Du., a little boatman, dim. of Du. *schipper;* so called because orig. used as watchdogs on boats]

schism (siz′əm) *n.* **1** A division of a church into factions. **2** The offense of causing division in a church or a religious community. **3** An ecclesiastical body separated from a larger or older body, as from an established church. **4** The act of dividing, or the state of being divided; division. See synonyms under SECT. [<OF *cisme, scisme* <LL *schisma* <Gk., a split < *schizein* split]

schis·mat·ic (siz·mat′ik) *adj.* Relating to, having the character of, implying, or promoting schism: also **schis·mat′i·cal.** —*n.* One who makes or participates in an ecclesiastical schism: a term of opprobrium. [<OF *cismatique, scismatique* <LL *schismaticus* <L Gk. *schismatikos* <Gk. *schisma, -atos.* See SCHISM.] —**schis·mat′i·cal·ly** *adv.* —**schis·mat′i·cal·ness** *n.*

schist (shist) *n. Geol.* Any rock that readily splits or cleaves; specifically, a rock that has had a parallel or foliated structure secondarily developed in it: also spelled *shist.* [<F *schiste* <L *schistos* readily split <Gk. < *schizein* split] —**schist′ous, schist·ose** (shis′·tōs) *adj.*

schis·ta·ceous (shis·tā′shəs) *adj.* Bluish-gray; of a light slaty color. [<SCHIST + -ACEOUS]

schisto- *combining form* Split: *schistosome.* Also, before vowels, **schist-.** [<Gk. *schistos* split]

schis·to·some (shis′tə·sōm) *n.* Any of a genus *(Schistosoma)* of trematode worms, including certain species parasitic in the blood of man, as the blood fluke *(S. haematobium)* common in Africa. [<NL <Gk. *schistos* split (<*schizein* split) + *sōma* a body]

schis·to·so·mi·a·sis (shis′tə·sō·mī′ə·sis) *n. Pathol.* A wasting disease caused by infestation with worms of the genus *Schistosoma,* endemic in Egypt and other parts of Africa: also called *bilharziasis.* [<NL < *Schistosoma* a schistosome]

schizo- *combining form* Split; divided: *schizophrenia.* Also, before vowels, **schiz-.** [<Gk. *schizein* split]

schiz·o·carp (skiz′ə·kärp) *n. Bot.* A split fruit; a pericarp splitting at maturity into two or more one-seeded indehiscent portions. Compare illustration under FRUIT. —**schiz′o·car′·pous, schiz′o·car′pic** *adj.*

schiz·o·gen·e·sis (skiz′ō·jen′ə·sis) *n. Biol.* Reproduction by fission.

schi·zog·o·ny (ski·zog′ə·nē) *n. Biol.* Reproduction by multiple fission, as in certain protozoa. [<NL *schizogonia* <Gk. *schizein* split + *genesthai* become]

schiz·oid (skiz′oid) *Psychiatry adj.* Of, pertaining to, or afflicted with schizophrenia. —*n.* One affected with schizophrenia. [< SCHIZ(OPHRENIA) + -OID]

schiz·o·my·cete (skiz′ō·mī·sēt′) *n.* One of the *Schizomycetes;* a bacterium. [<NL <Gk. *schizein* split + *mykēs, -ētos* a mushroom] —**schiz′o·my·ce′tous** *adj.*

Schiz·o·my·ce·tes (skiz′ō·mī·sē′tēz) *n. pl.* A class of widely distributed, minute, unicellular plants reproducing by fission and allied to the fungi: it comprises the bacteria and, in the Bergey classification, includes the following orders:

Eubacteriales	Simple undifferentiated forms
Actinomycetales	Moldlike bacteria
Chlamydobacteriales	Algalike iron bacteria
Caulobacteriales	Aquatic, gum-secreting bacteria
Thiobacteriales	Sulfur bacteria
Myxobacteriales	Slime-mold bacteria
Spirochaetales	Protozoanlike bacteria

schiz·o·my·co·sis (skiz′ō·mī·kō′sis) *n. Pathol.* A morbid condition or disease due to the presence of schizomycetes. [<NL]

schiz·ont (skiz′ont, skī′zont) *n.* The mature trophozoite of a sporozoan, as the malaria parasite, from which new cells, or merozoites, are liberated into the blood by schizogony. [<Gk. *schizōn, -ontos,* ppr. of *schizein* split]

schiz·o·phre·ni·a (skiz′ō·frē′nē·ə) *n. Psychiatry* A mental derangement characterized by the presence of conflicting impulses, emotions, and ideas, and resulting in a disintegration of personality resembling, but more inclusive than, that found in dementia precox. [<NL <Gk. *schizein* split + *phrēn* mind] —**schiz′·o·phren′ic** (-fren′ik) *adj.* & *n.*

schiz·o·phyte (skiz′ə·fīt) *n.* One of a division or phylum *(Schizophyta)* of unicellular or simple multicellular plants which reproduce by fission or by asexual spores: it includes the bacteria and the blue-green algae. —*adj.* Of or pertaining to the *Schizophyta*: also **schiz′o·phyt′ic** (-fit′ik). [<NL <Gk. *schizein* split + *phyton* a plant]

schiz·o·pod (skiz′ə·pod) *n.* Any of a former order *(Schizopoda)* of crustaceans having a soft carapace and resembling the shrimp: now included in the subclass *Malacostraca.* [<NL <Gk. *schizopous, -podos* having parted toes < *schizein* split + *pous, podos* a foot]

schiz·o·thy·mi·a (skiz′ō·thī′mē·ə) *n. Psychiatry* A schizophrenic condition marked by introversion and a withdrawing from the world, but milder than schizophrenia. [<NL <Gk. *schizein* split + *thymos* spirit] —**schiz′o·thyme** *n.* —**schiz′o·thy′mic** *adj.*

Schle·gel (shlā′gəl), **August Wilhelm von,** 1767–1845, German philologist, poet, and literary critic. —**Friedrich von,** 1772–1829, German philosopher and critic: brother of the preceding.

Schlei·er·ma·cher (shlī′ər·mäkh′ər), **Friedrich Ernst Daniel,** 1768–1834, German theologian and philosopher.

schle·miel (shlə·mēl′) *n. Slang* An inept, easily duped person; a bungler; dolt. Also **schle·mihl′**. [<Yiddish, an unlucky person <Hebrew *Shelumiēl*, a personal name]

Schle·si·en (shlā′zē·ən) The German name for SILESIA.

Schles·wig (shläs′vikh) **1** A city in NW Germany, former capital of Schleswig-Holstein. **2** The southern part of Jutland Peninsula, divided between Germany and Denmark: Danish *Sleswig*.

Schles·wig-Hol·stein (shläs′vikh·hōl′shtīn) A state of NW Germany (NE West Germany); 6,052 square miles; capital, Kiel.

Schley (slī), **Winfield Scott**, 1839–1911, U.S. rear admiral.

Schlie·mann (shlē′män), **Heinrich**, 1822–90, German archeologist.

schlie·re (shlē′rə) *n. pl.* **·ren** (-rən) *Geol.* In an igneous rock, an irregular, commonly not sharply bounded, portion differing in composition or texture from the general mass of the rock. [<G, lit., a streak] —**schlie′ric** *adj.*

schlie·ren (shlē′rən) *n. Physics* Any disturbance in the light path of an interferometer which alters the density of the air and thus changes the interference pattern of the light waves. [<G, pl. of *schliere*, lit., a streak]

Schmal·kal·den (shmäl′käl′dən) A town in the former state of Thuringia, central Germany (SW East Germany).

Schmal·kal·dic League (shmäl·käl′dik) A league formed in Schmalkalden by German Protestants in 1530 against Emperor Charles V.

schmaltz (shmälts) *n. Slang* **1** Anything which is overly sentimental, as in music or literature. **2** Extreme sentimentalism. [<Yiddish <G *schmalz*, lit., melted fat] —**schmaltz′y** *adj.*

Schna·bel (shnä′bəl), **Artur**, 1882–1951, U.S. pianist and composer born in Austria.

schnap·per (shnap′ər, snap′-) *n.* A sparoid fish (*Pagrosomus auratus*), reddish with blue bars or spots. It is one of the most important food fishes of Australia, Tasmania, and New Zealand. [Alter. of SNAPPER; infl. in form by G *schnapper* a schnapper]

schnapps (shnäps, shnaps) *n.* Holland gin; Hollands; loosely, any ardent spirits. Also **schnaps**. [<G, a dram, a nip <Du. *snaps*, lit., a gulp, mouthful]

schnau·zer (shnou′zər) *n.* A small, active terrier originally developed in Germany, having a wiry black or pepper-and-salt coat. —**miniature schnauzer** A toy terrier bred from the standard schnauzer and a pinscher. [<G, lit., a growler < *schnauzen* growl, snarl]

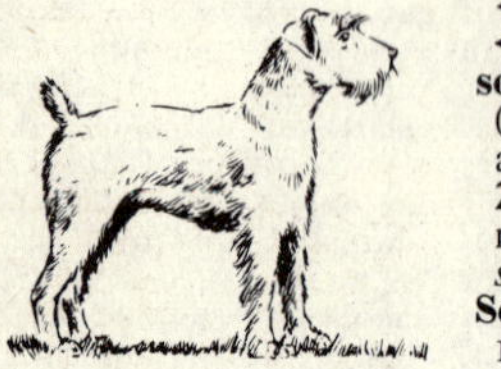

STANDARD SCHNAUZER
(From 18 to 20 inches high at the shoulder)

Schnitz·ler (shnits′lər), **Arthur**, 1862–1931, Austrian playwright and novelist.

schnor·kel (shnôr′kəl) *n.* **1** An apparatus for the ventilation of a submerged submarine, consisting of retractable tubes for the intake of fresh air and the removal of the toxic gases of respiration and power-plant operation. **2** A snorkel. [<G *schnörkel* spiral]

schnor·rer (shnôr′ər) *n.* A professional or habitual beggar. [<Yiddish <G *schnurrer* < slang *schnurren* go begging, orig. whirr, purr; with ref. to musical instruments carried by beggars]

schnoz·zle (shnoz′əl) *n. Slang* Nose. [<Yiddish <G *schnauze*. Akin to SNOUT.]

Scho·field (skō′fēld), **John McAllister**, 1831–1906, U.S. general.

schol·ar (skol′ər) *n.* **1** A person eminent for learning. **2** The holder of a scholarship. **3** One who learns under a teacher; a pupil. —**Rhodes scholar** A male student selected from a college or university of the United States or of any British dominion or colony, to receive one of the scholarships established by Cecil Rhodes for attendance at Oxford University, England. [Prob. fusion of OE *scolere* and OF *escoler*, both <LL *scholaris* <L *schola*. See SCHOOL¹.]

Synonyms: disciple, learner, pupil, savant, student. Historically the primary sense of a *scholar* is one who is being schooled; thence the word passes to designate one who is apt in school work, and finally one who is thoroughly schooled, master of what the schools can teach, an erudite or accomplished person: when used without qualification, the word is generally understood in this sense; as, He is manifestly a *scholar*. *Pupil* signifies one under the close personal supervision or instruction of a teacher or tutor. *Antonyms*: dunce, fool, idiot, idler, ignoramus.

schol·arch (skol′ärk) *n.* **1** In Greek antiquity, the head of a school of philosophy in Athens. **2** The head of any school. [<Gk. *scholarchē* < *scholē* a school + *archein* rule]

schol·ar·ly (skol′ər·lē) *adj.* Like a scholar; learned; erudite. —*adv.* After the manner of a scholar.

schol·ar·ship (skol′ər·ship) *n.* **1** Learning; erudition. **2** Maintenance or a stipend for a student awarded by an educational institution: also, the position of such a student. See synonyms under KNOWLEDGE, LEARNING.

scho·las·tic (skō·las′tik, skə-) *adj.* **1** Pertaining to or characteristic of scholars, education, or schools. **2** Pertaining to or characteristic of the medieval schoolmen. **3** Precise; pedantic. **4** Pertaining to the theological grade of students of the Jesuit order. Also **scho·las′ti·cal**. —*n.* **1** A student; pupil. **2** *Often cap.* A schoolman; an advocate of scholasticism. **3** A pedant. [<L *scholasticus* <Gk. *scholastikos* < *scholazein* be at leisure, devote leisure to study < *scholē*. See SCHOOL¹.]

scho·las·ti·cate (skō·las′tə·kāt, skə-) *n.* A general house of higher studies for Jesuit scholastics. [<NL *scholasticatus* <L *scholasticus* SCHOLASTIC]

scho·las·ti·cism (skō·las′tə·siz′əm, skə-) *n.* **1** *Often cap.* The systematized Christian logic, philosophy, and theology of medieval scholars from the 10th to the 15th centuries, based on Aristotle's *Logic* and *Metaphysics* and the writings of the early Christian fathers. See HUMANISM. **2** Any system of teaching which insists on traditional doctrines and forms. **3** A similar teaching of the present day given especially in seminaries of the Roman Catholic Church.

scho·li·ast (skō′lē·ast) *n.* A commentator; especially, an ancient grammarian or annotator of classical texts. See SCHOLIUM. [<L *scholiasta* <Gk. *scholiastēs* < *scholion* a commentary < *scholē* a school] —**scho′li·as′tic** *adj.*

scho·li·um (skō′lē·əm) *n. pl.* **·li·ums** or **·li·a** (-lē·ə) **1** An explanatory marginal note, as on a classical text by an ancient grammarian. **2** An interpolated note accompanying a mathematical proof. [<LL *scholium* <Gk. *scholion*. See SCHOLIAST.]

Schön·berg (shœn′berkh), **Arnold**, 1874–1951, Austrian composer and conductor active in the United States.

school¹ (skōōl) *n.* **1** An educational institution. **2** The place in which formal instruction is given; a schoolhouse or schoolroom. **3** A period or session of an educational institution; a course of study at a school: *School* begins tomorrow. **4** The pupils in an educational institution. **5** A subdivision of a university devoted to a special branch of higher education: a *school* of education, medicine, etc. **6** The prescribed drill, duties, instruction, and training of any branch of the army or navy: gunnery *school*, aviation *school*; also, the manual of such instruction. **7** A body of disciples of a teacher or system; a sect, etc.; also, the system, methods, or opinions characteristic of those thus associated: the Scottish *school* of philosophy, a painting of the Flemish *school*. **8** A general style of life, manners, etc. **9** In medieval times, specifically, a seminary of logic, metaphysics, and theology; in the plural, the seats of the scholastic philosophy. **10** Any sphere or means of instruction: the *school* of example. —*v.t.* **1** To instruct in a school; train; educate. **2** To subject to rule or discipline. [OE *scōl* <L *schola* <Gk. *scholē* leisure or that which is done during leisure time, a school]

—**common school** One of the free public elementary schools in the United States.

—**consolidated school** A school, usually rural, consisting of several elementary schools and sometimes a high school, merged into one organization, for pupils from outlying districts.

—**continuation school** Comprehensively, a school for the further education of persons already employed; specifically, a school for employed boys and girls below the legal age for leaving school, attended a few hours a week on the employers' time.

—**dame school** An early form of school kept by a woman who drilled young children in their ABC's and the beginnings of reading: forerunner of the modern private primary school.

—**elementary school** A school giving a course of education of from six to nine years, pupils usually entering at about six years of age.

—**finishing school** A school that prepares girls for entrance into society.

—**grammar school** **1** In graded public schools, the grades between primary and high school, grades one to eight inclusive. **2** Popularly, an elementary school. **3** *Brit.* A secondary school, often preparatory for college; originally for the teaching of Latin and Greek, but now offering broader curriculums.

—**high school** The highest division of the common schools, typically comprising grades 9, 10, 11, and 12: often preparatory for college or the vocations. —**junior high school** A modern division of the common schools, consisting of the 7th, 8th, and 9th grades, sometimes of only the 8th and 9th grades: also **intermediate school**. —**senior high school** A corresponding division of the public schools, consisting usually of grades 10, 11, and 12.

—**industrial school** **1** An institute for the practical development of aptitudes or skill for application in industry. **2** A school for the care and training or reformation of neglected children.

—**parochial school** A school, usually elementary, supported by the parish of a church, especially by a Roman Catholic church.

—**primary school** A school for the teaching of the youngest pupils; the first grades of common schools beyond kindergarten.

—**private school** A school maintained under private or corporate management, usually for profit: in the United States, contradistinguished from *public school*.

—**public school** **1** A school maintained by public funds for the free education of the children of the community, usually covering elementary and secondary grades. **2** In England, a private or endowed school not run for profit; specifically, the exclusive endowed schools preparing students for the universities, as Eton, Harrow, etc.: so called because it serves the country at large, and not merely one community.

—**secondary school** A high school or preparatory school intermediate between the grammar school and college.

—**trade school** A vocational school designed to give a knowledge of processes and a skill of hand adequate for work in a specific trade.

—**vocational school** In general, a school training in the practical application of knowledge to business, the professions, or the technical arts and crafts.

school² (skōōl) *n.* A large number of fish, whales, etc., swimming together; shoal. —*v.i.* To swim together in a school. [<Du., a crowd, school of fishes. Akin to SHOAL².]

school board A legal board or committee having oversight of public schools.

school·book (skōōl′bŏŏk′) *n.* A book for use in school; textbook.

school·boy (skōōl′boi′) *n.* A boy attending school.

school·fel·low (skōōl′fel′ō) *n.* A schoolmate.

school·girl (skōōl′gûrl′) *n.* A girl attending school.

school·house (skōōl′hous′) *n.* A building in which a school is conducted.

school·ing (skōō′ling) *n.* **1** Instruction given at school; also, any preparatory training or discipline. **2** Price paid for instructing pupils. **3** The training of horses and riders. See synonyms under EDUCATION, NURTURE.

school·man (skōōl′mən) *n. pl.* **·men** (-mən) One of the theologians of the Middle Ages; a scholastic.

school·marm (skōōl′märm′) *n. Colloq.* A woman schoolteacher, especially one considered to be prudish, spinsterish, or strict. Also **school′ma′am′** (-mam′).

school·mas·ter (skōōl′mas′tər, -mäs′-) *n.* **1** A man who teaches school. **2** One who or that which instructs or disciplines in any way:

Necessity was his *schoolmaster.* **3** A Caribbean fish *(Neomaenis apoda)* of the snapper family. See synonyms under MASTER.
school·mate (skool′māt′) *n.* A fellow pupil; a schoolfellow.
school·mis·tress (skool′mis′tris) *n.* A woman who teaches school.
school·room (skool′room′, -room′) *n.* A room in which instruction is given.
school ship A vessel in which boys and young men are trained in seamanship, navigation, etc.
school·teach·er (skool′tē′chər) *n.* One who gives instruction in a school below the college level.
school·yard (skool′yärd′) *n.* The grounds about a school used for play.
schoon·er (skoo′nər) *n.* **1** A fore-and-aft rigged vessel having originally two masts, but now often three or more. **2** A prairie schooner. **3** A large beer glass, holding usually about a pint or more. [Appar. coined in New England < dial. *scoon* skim on water, prob. <Scand.]

SCHOONER

schoon·er-yacht (skoo′nər-yot′) *n.* A yacht rigged like a schooner.
Scho·pen·hau·er (shō′pən·hou′ər), **Arthur**, 1788–1860, German philosopher. — **Scho′pen·hau·er·i·an** *adj.*
Scho·pen·hau·er·ism (shō′pən·hou′ə·riz′əm) *n.* The philosophy (pessimistic determinism) of Arthur Schopenhauer, who taught that egoism, manifested in the "will to live," must be overcome; that the world is evil and should not be perpetuated; and that God, free will, and the immortality of the soul are illusions.
schorl (shôrl) *n.* Tourmaline, especially the black variety: also spelled *shorl.* [<G *schörl*]
schor·la·ceous (shôr·lā′shəs) *adj.* Containing black tourmaline, as granite. [<SCHORL + -ACEOUS]
schot·tische (shot′ish) *n.* A dance in 2/4 time similar to the polka, but somewhat slower; also, the music for such a dance. [<G *(der) schottische (tanz)* (the) Scottish (dance)]
Schou·ten Islands (skhou′tən, -ən) An island group off NW New Guinea, comprising part of Netherlands New Guinea; total, 1,231 square miles.
schrik (skhrik) *n. Afrikaans* Panic or sudden fright.
Schrö·ding·er (shrœ′ding·ər), **Erwin**, 1887–1961, Austrian physicist.
Schu·bert (shoo′bərt, *Ger.* shoo′bert), **Franz Peter**, 1797–1828, Austrian composer.
schuit (skoit) *n.* A Dutch vessel, sloop-rigged, used in rivers and canals. Also **schuyt.** [<Du. *schuit, schuyt* <MDu. *schute*]
schule[1] (skül, skœl) *n. Scot.* A school.
schule[2] (shool) *n. Scot.* A shovel.
Schu·man (shü·màn′), **Robert**, born 1886, French political leader born in Luxembourg.
Schu·mann (shoo′män), **Robert**, 1810–56, German composer.
Schu·mann–Heink (shoo′män·hīngk′), **Ernestine**, 1861–1936, U.S. dramatic contralto singer born in Austria.
Schur·man (shûr′mən), **Jacob Gould**, 1854–1942, U.S. philosopher and diplomat born in Canada; president of Princeton University 1892–1920.
Schurz (shoorts), **Carl**, 1829–1906, U.S. statesman, journalist, and general, born in Germany.
Schusch·nigg (shoosh′nik), **Kurt von**, born 1897, Austrian statesman.
schuss (shoos) *v.i.* To ski down a steep slope at high speed. — *n.* A straight, steep ski course, or the act of skiing this course. [<G, lit., a shot]
Schutz·staf·fel (shoots′shtä′fel) *n. pl.* **·feln** (-fəln) *German* Hitler's personal bodyguard, known as the Black Shirts; later, the chief section, the Elite Guard, of the Nazi militia, used to maintain order in Germany and occupied countries. Abbr. *SS.*
Schuy·ler (skī′lər), **Philip John**, 1733–1804, American statesman and soldier.

Schuy·ler·ville (skī′lər·vil) A town in eastern New York near Saratoga Springs; scene of Battle of Saratoga and General Burgoyne's surrender, October 17, 1777, in the American Revolution. See also *Saratoga.*
Schuyl·kill River (skool′kil, skoo′kəl) A river in SE Pennsylvania, flowing 150 miles SE to the Delaware River.
schwa (shwä, shvä) *n.* **1** *Phonet.* A weak or obscure, central vowel sound occurring in most of the unstressed syllables in English speech. The sound, regardless of spelling, is that of the *a* in *alone,* the *o* in *lemon,* or the *u* in *circus:* written ə. **2** In Hebrew, the obscure vowel sound: written : and often transliterated by *e.* [<G <Hebrew *shewa*]
Schwab (shwäb), **Charles M.** 1862–1939, U.S. industrialist.
Schwa·ben (shvä′ben) The German name for SWABIA.
Schwann (shvän), **Theodor**, 1810–82, German physiologist.
schwan·pan (shwän′pän′) See SWANPAN.
Schwarz·wald (shvärts′vält) The German name for the BLACK FOREST.
Schwein·furt (shvīn′foort) A city in NW Bavaria, West Germany.
Schweit·zer (shvī′tsər), **Albert**, born 1875, Alsatian clergyman, physician, missionary, philosopher, and musicologist.
Schweitzer's reagent *Chem.* An aqueous solution of cupric hydroxide precipitated in ammonium hydroxide: used as a solvent for cellulose, especially in the cuprammonium process. Also called *cuprammonia.* [after Mathias E. *Schweitzer,* 1818–1860, German chemist]
Schweiz (shvīts) The German name for SWITZERLAND.
Schwei·zer·kä·se (shvī′tsər·kā′zə) *n. German* Swiss cheese.
Schwe·rin (shve·rēn′) The capital of the former state of Mecklenburg, northern East Germany.
Schwyz (shvēts) A city of east central Switzerland near Lake Lucerne.
sci·ae·noid (sī·ē′noid) *n.* Any of a family *(Sciaenidae)* of spiny-finned, carnivorous fishes (order *Percomorphi*), as croakers, drums, weakfish, etc. They are mostly marine or estuarine, and in all the air bladder is large and complicated, enabling them to make grunting or drumming noises. — *adj.* Of or pertaining to the *Sciaenidae.* Also **sci·ae′nid** (-ē′nid). [<NL <L *sciaena* <Gk. *skiaina,* a kind of fish]
sci·a·ma·chy (sī·am′ə·kē) *n. pl.* **·chies** An imaginary fight; struggle with a shadow or with an imaginary foe; useless combat: also spelled *sciomachy.* [<Gk. *skiamachia* < *skia* a shadow + *machein* fight]
sci·at·ic (sī·at′ik) *adj.* Pertaining to or affecting the hip or its nerves; ischial. — *n.* A sciatic nerve or part. [<MF *sciatique* <Med. L *sciaticus,* alter. of L *ischiadicus* <Gk. *ischiadikos* < *ischion* hip, hip joint]
sci·at·i·ca (sī·at′i·kə) *n. Pathol.* **1** Neuralgia, affecting the sciatic nerve traversing the hip and thigh. **2** Any painful affection of these or adjoining parts. [<Med. L *sciatica (passio)* (the) sciatic (disease), fem. of *sciaticus* SCIATIC]
sci·ence (sī′əns) *n.* **1** Knowledge as of facts, phenomena, laws, and proximate causes, gained and verified by exact observation, organized experiment, and correct thinking; also, the sum of universal knowledge. **2** An exact and systematic statement or classification of knowledge concerning some subject or group of subjects. **3** Any department of knowledge in which the results of investigation have been systematized in the form of hypotheses and general laws subject to verification. **4** Expertness, skill, or proficiency resulting from knowledge. **5** Any one of the seven liberal arts (grammar, rhetoric, logic, arithmetic, music, geometry, astronomy): an ancient use. [<OF <L *scientia* < *sciens, -entis,* ppr. of *scire* know]
Synonyms: knowledge, art, learning, scholarship. *Knowledge* may be a medley of facts which gain real value only when coordinated and systematized by the man of *science. Art* relates to something to be done or produced by skill, *science* to something to be known.

Creative *art* seeking beauty for its own sake is closely akin to fundamental *science* seeking *knowledge* for its own sake. See ART[1], KNOWLEDGE.
sci·ence-fic·tion (sī′əns·fik′shən) *n.* Fiction employing scientific ideas or devices, usually projections or extrapolations of present-day science, as elements of plot or background; loosely, including, in addition, fantasy, weird and occult tales, utopian romances, and other stories of the future.
sci·en·tial (sī·en′shəl) *adj.* Of, characterized by, or producing knowledge or science; also, skilful; scientific; knowing; capable.
sci·en·tif·ic (sī′ən·tif′ik) *adj.* **1** Of, pertaining to, discovered by, derived from, or used in science; of the nature of science. **2** Agreeing with the rules, principles, or methods of science; accurate; systematic; exact. **3** Versed in science or a science; eminently learned or skilful. Also **sci·en·tif′i·cal.** [<F *scientifique* <LL *scientificus* < *scientia* knowledge + *facere* make; orig. trans. of Gk. *epistēmonikos* pertaining to knowledge, science] — **sci′en·tif′i·cal·ly** *adv.*
sci·en·tism (sī′ən·tiz′əm) *n.* **1** Adherence to or belief in the aims and methods of scientists. **2** *Sociol.* The application of quantitative methods and mathematical analysis in the interpretation of sociological data, especially in the fields of education, economics, and psychology.
sci·en·tist (sī′ən·tist) *n.* One versed in science or devoted to scientific study or investigation.
Sci·en·tist (sī′ən·tist) *n.* A Christian Scientist.
scil·i·cet (sil′ə·set) *adv.* Namely; to wit; that is to say: introducing a word to be supplied, or an explanation: generally abbreviated *scil., sc.,* or *ss.* [<L, contraction of *scire licet* it is permitted to know]
Scil·la (sil′ə) A town at the NE end of the Strait of Messina, southern Italy, on a small promontory supposed to be the site of the cave of the legendary Scylla.
Scil·ly Islands (sil′ē) A group of 140 islands off the SW coast of Cornwall, SW England; 6.3 square miles.
scim·i·tar (sim′ə·tər) *n.* **1** An Oriental sword or saber of extreme curve. **2** A billhook of somewhat similar form. Also **scim′e·tar, scim′i·ter:** formerly variously spelled with *si-, ci-,* etc. [<MF *cimeterre;* infl. in form by Ital. *scimitarra;* both ? <Persian *shamshīr*]

SCIMITAR

scin·coid (sing′koid) *n.* One of a family *(Scincidae)* of lizardlike viviparous reptiles (order *Squamata*) with typically smooth scales; a skink. — *adj.* Of or pertaining to the *Scincidae.* Also **scin·coi·di·an** (sing·koi′dē·ən). [<NL *scincoides* <L *scincus* skink]
scin·til·la (sin·til′ə) *n.* A spark; hence, a trace; iota: usually of something abstract; a *scintilla* of truth. See synonyms under PARTICLE. [<L]
scin·til·lant (sin′tə·lənt) *adj.* Emitting sparks; scintillating.
scin·til·late (sin′tə·lāt) *v.* **·lat·ed, ·lat·ing** *v.i.* **1** To give off sparks. **2** To sparkle; glitter. **3** To twinkle, as a star. — *v.t.* **4** To give off as a spark or sparks. See synonyms under SHINE. [<L *scintillatus,* pp. of *scintillare* scintillate < *scintilla* a spark] — **scin′til·lat′ing** *adj.*
— **scin′til·lat′ing·ly** *adv.*
scin·til·la·tion (sin′tə·lā′shən) *n.* **1** The act or state of scintillating; a sparkling, tremulous flashing or twinkling. **2** A spark or sparkle. **3** The twinkling of the stars. See synonyms under LIGHT.
sci·o·lism (sī′ə·liz′əm) *n.* Charlatanism; pretentious superficial knowledge. [<LL *sciolus* a smatterer, dim. of L *scius* knowing < *scire* know]
sci·o·list (sī′ə·list) *n.* One who has a smattering of knowledge, especially a pretender to scientific attainment. — **sci′o·lis′tic, sci′o·lous** *adj.*
sci·o·ma·chy (sī·om′ə·kē) *n.* See SCIAMACHY.
sci·on (sī′ən) *n.* **1** *Bot.* A cion. **2** A child or descendant. [<OF *cion,* prob. blend of *scier* saw and L *sectio, -onis* a cutting, both <L *secare* cut]

sci·oph·i·lous (sī·of′ə·ləs) *adj. Ecol.* Shade-loving; able to live in shade; thriving in the shade. [<Gk. *skia* shade + -PHILOUS]

sci·o·phyte (sī′ə·fīt) *n.* A plant growing or adapted to live in the shade. [<Gk. *skia* shade + -PHYTE] —**sci′o·phyt′ic** (-fit′ik) *adj.*

sci·os·o·phy (sī·os′ə·fē) *n.* Any system of thought founded on beliefs which are at variance with contemporary scientific knowledge and resistant to the procedures of scientific method. [<Gk. *skia* a shadow + -SOPHY] —**sci·os′o·phist** *n.*

Sci·o·to River (sī·ō′tə, -tō) A river in central and southern Ohio, flowing 237 miles south to the Ohio River.

Scip·i·o (sip′ē·ō) Name of a great Roman family, especially including **Publius Cornelius Scipio Africanus Major**, 237–183 B.C., general; defeated Hannibal at Zama 202 B.C.: called "The Elder"; and **Publius Cornelius Scipio Aemilianus Africanus Minor**, 185–129 B.C., general; burned Carthage: called "The Younger."

sci·re fa·ci·as (sī′rē fā′shē·əs) *Law Latin* A writ (or the proceeding under it) commanding the party against whom it is issued to show cause why the plaintiff should not have advantage of or execution on a judicial record, or why a nonjudicial record should not be repealed or annulled; literally, that you cause to know: abbr. *sci. fa.*, or *s. f.*

scir·rhus (skir′əs, sir′-) *n. pl.* **scir·rhi** (skir′ī) or **scir·rhus·es** *Pathol.* A hard tumor; specifically, a hard cancerous tumor. [<NL <L *scirros* <Gk. *skirrhos* a tumor < *skiros* hard] —**scir·rhos·i·ty** (ski·ros′ə·tē, si-) *n.* —**scir′·rhous, scir′rhoid** *adj.*

scis·sile (sis′il) *adj.* Capable of being cut or split easily and evenly. [<L *scissilis* < *scissus*, pp. of *scindere* cut]

scis·sion (sizh′ən, sish′-) *n.* The act of cutting or splitting, or the state of being cut; hence, any division. [<OF <LL *scissio, -onis* < *scissus*, pp. of *scindere* cut]

scis·sor (siz′ər) *v.t.* & *v.i.* To cut with scissors.

scis·sor·er (siz′ər·ər) *n.* One using scissors; hence, a compiler.

scis·sors (siz′ərz) *n. pl.* & *sing.* **1** A cutting implement with handles and a pair of blades pivoted face to face: sometimes a **pair of scissors.** **2** In wrestling, a hold secured by clasping the legs about the body or head of the opponent. **3** Gymnastic feats in which the movement of the legs suggests the opening and closing of scissors. [<OF *cisoires* <LL *cisoria*, pl. of *cisorium* a cutting instrument < *caedere* cut; infl. in form by L *scissor* one who cuts < *scindere* cut]

scissors kick In swimming, a kick performed usually with the side stroke, in which both legs are thrust apart, the upper leg bent at the knee while the lower is kept straight, then brought sharply together.

scis·sor·tail (siz′ər·tāl′) *n.* A flycatcher *(Muscivora forficata)* of the SW United States and Mexico having a scissorlike tail.

scis·sure (sizh′ər, sish′-) *n.* **1** A lengthwise cut; fissure. **2** Any division, rupture, or schism. [<MF <L *scissura* < *scissus*. See SCISSION.]

sci·u·rine (sī′yōō·rīn, -rin) *adj.* Belonging or pertaining to a family *(Sciuridae)* of rodents, including squirrels, chipmunks, woodchucks, marmots, etc. —*n.* One of the *Sciuridae*. [< L *sciurus* a squirrel <Gk. *skiouros* < *skia* a shadow + *oura* a tail + -INE¹]

sci·u·roid (sī·yōōr′oid) *adj.* **1** Of or pertaining to the *Sciuridae*. **2** *Bot.* Resembling a squirrel's tail, as the tufted spikes of certain cereal grasses. [<NL <L *sciurus* a squirrel + Gk. *eidos* a form]

sclaff (sklaf) *v.i.* **1** In golf, to strike the ground with the club before hitting the ball. —*v.t.* **2** In golf: **a** To strike (the ball) or make (a stroke) in this manner. **b** To drag (the club) thus. —*n.* **1** A slight slap or blow; the noise so made. **2** A light shoe; a slipper. **3** The golf stroke made by sclaffing. [< dial. E (Scottish) *sclaf* slap, shuffle; imit.]

Sclav (skläv), **Sclav·ic** (sklä′vik), etc. Obsolete forms of SLAV, etc.

scle·ra (sklir′ə) *n. Anat.* The hard, firm, fibrous outer coat of the eye, continuous with the cornea; the white of the eye. Also **scle·rot·i·ca** (sklə·rot′i·kə). See illustration under EYE. [<NL <Gk. *sklēros* hard]

scle·ren·chy·ma (sklə·reng′kə·mə) *n. Bot.* The tough, stony, thick-walled tissue composing the hard parts of plants. [<NL <Gk. *sklēros* hard + *enchyma* an infusion] —**scle·ren·chym·a·tous** (sklir′eng·kim′ə·təs) *adj.*

scle·ri·a·sis (sklə·rī′ə·sis) *n. Pathol.* Any morbid hardening or induration of parts. [<NL <Gk. *sklēriasis* < *sklēria* hardness < *sklēros* hard]

scle·rite (sklir′īt) *n.* **1** *Zool.* **a** One of the definite hard pieces of the integument of an arthropod. **b** A hard element in the integument of a polyp. **2** A spicule. [<Gk. *sklēros* hard + -ITE¹] —**scle·rit·ic** (sklə·rit′ik) *adj.*

scle·ri·tis (sklə·rī′tis) *n. Pathol.* Rheumatic ophthalmia; inflammation of the sclera of the eye. Also **scle·ro·ti·tis** (sklir′ō·tī′tis, skler′-). [<NL < *sclera* the white of the eye] —**scle′·ro·tit′ic** (-tit′ik) *adj.*

sclero– *combining form* Hardness; hard: *scleroderma*. Also, before vowels, **scler–**. [<Gk. *sklēros* hard]

scle·ro·der·ma (sklir′ō·dûr′mə, skler′-) *n. Pathol.* Hardening of the skin. [<NL <Gk. *sklēros* hard + *derma* skin]

scle·ro·der·ma·tous (sklir′ō·dûr′mə·təs, skler′-) *adj. Zool.* Provided with a horny or bony covering, as an armadillo. [<Gk. *sklēros* hard + *derma, -atos* skin]

scle·roid (sklir′oid) *adj. Biol.* Hard; sclerous; hard in texture, as the shells of nuts, etc. [<Gk. *sklēroeidēs* < *sklēros* hard + *eidos* form]

scle·ro·ma (sklə·rō′mə) *n. Pathol.* Hardening of the cellular tissue; sclerosis; scleroderma. [<NL <Gk. *sklērōma* < *sklēroein* harden < *sklēros* hard]

scle·rom·e·ter (sklə·rom′ə·tər) *n.* An instrument for determining the degree of hardness of a mineral.

scle·rosed (sklə·rōst′) *adj.* Affected with sclerosis; grown abnormally hard. [<SCLEROS(IS) + -ED³]

scle·ro·sis (sklə·rō′sis) *n.* **1** *Pathol.* The morbid thickening and hardening of a tissue; especially, the hardening of the coats of the arteries. **2** *Bot.* The hardening of a plant cell wall by the formation of lignin in it. [<Med. L *sclirosis* <Gk. *sklērōsis* < *sklēroein* harden < *sklēros* hard] —**scle·ro′sal** *adj.*

scle·rot·ic (sklə·rot′ik) *adj.* **1** Dense; hard, as the white of the eye. **2** *Pathol.* Pertaining to or affected with sclerosis. [<NL *scleroticus* <Gk. *sklērotēs* hardness < *sklēroein*. See SCLEROMA.]

scle·ro·ti·um (sklə·rō′shē·əm) *n. pl.* **·ti·a** (-shē·ə) *Bot.* A compact horny mass of mycelium, found in certain higher fungi; especially, in the myxomycetes, a plasmodium, or part of a plasmodium, dry and hard, which for some time remains dormant. [<NL <Gk. *sklēros* hard] —**scle·ro′ti·oid** (-shē·oid), **scle·ro′tial** (-shəl) *adj.*

scle·rot·o·my (sklə·rot′ə·mē) *n. Surg.* Incision of the sclera. [<SCLER(A) + -(O)TOMY]

scle·rous (sklir′əs) *adj.* Hard or indurated; bony. [<SCLER(O)- + -OUS]

scob (skob) *n.* A defect in fabric caused by failure of the warp to interlace in the weaving. [? <Irish and Scottish Gaelic *sgolb* a splinter]

scoff (skôf, skof) *v.i.* To speak with contempt or derision; jeer: often with *at*. —*v.t.* To deride; mock. —*n.* An expression or an object of contempt or derision. See synonyms under SNEER. [ME *scof*, prob. <Scand. Cf. Dan. *skof* a jest, mockery.] —**scoff′er** *n.* —**scoff′·ing·ly** *adv.*

Synonyms (verb): deride, flout, gibe, jeer, mock, sneer, taunt. See RIDICULE; SCORN. *Antonyms*: see synonyms for PRAISE.

scoff·law (skôf′lô′, skof′-) *n.* One who scoffs at the law; especially, a habitual or deliberate violator of traffic, safety, or public-health regulations.

scog·ger (skog′ər) *n. Brit. Dial.* A heavy woolen garment worn for protection as a gaiter, or as a sleeve. [Cf. *cocker* a boot, quiver, OE *cocer*]

scold (skōld) *v.t.* To find fault with harshly. —*v.i.* To find fault harshly or continuously. —*n.* One who scolds, especially a virago: also **scold′er.** [Appar. <ON *skāld* a poet, satirist] —**scold′ing** *adj.* & *n.* —**scold′ing·ly** *adv.*

scol·e·cite (skol′ə·sīt, skō′lə-) *n.* A vitreous or silky, colorless, hydrous silicate of calcium and aluminum; a zeolite, isomorphous with natrolite. [<Gk. *skōlēx, -ēkos* a worm; so called because it sometimes curls up when heated]

sco·lex (skō′leks) *n. pl.* **sco·le·ces** (skō·lē′sēz) or **scol·i·ces** (skol′ə·sēz, skō′lə-) *Zool.* The knoblike head of a tapeworm, equipped with a circular disk of hooks and a group of two or four suckers. [<NL <Gk. *skōlēx* a worm]

sco·li·o·sis (skō′lē·ō′sis, skol′ē-) *n. Pathol.* A lateral curvature of the spine. Also **sco′li·o′·ma.** [<NL <Gk. *skoliōsis* < *skolios* curved] —**sco′li·ot′ic** (-ot′ik) *adj.*

scol·lop (skol′əp), etc. See SCALLOP, etc.

scol·o·pen·drid (skol′ə·pen′drid) *n.* One of a family of chilopods *(Scolopendridae)* including the typical centipedes. [<NL <L *scolopendra* <Gk. *skolopendra* a milliped] —**scol′o·pen′·drine** (-drīn, -drin) *adj.*

scom·broid (skom′broid) *adj.* Of or pertaining to a family *(Scombridae)* of acanthopterygian fishes, including mackerels, tunnies, and related genera. —*n.* One of the *Scombridae*. [<NL <L *scomber* a mackerel <Gk. *skombros*]

sconce¹ (skons) *n.* **1** A small earthwork or fort. **2** A protective shelter, covering, or screen. [<Du. *schanz* a fortress, wicker basket; infl. in form by SCONCE²]

sconce² (skons) *n.* An ornamental wall bracket for holding a candle or other light. [<OF *esconse* a dark lantern, hiding place <Med. L *sconsa*, short for L *absconsa*, pp. fem. of *abscondere* hide]

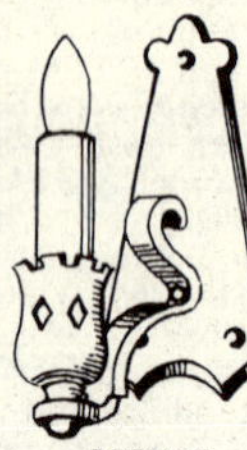
SCONCE

sconce³ (skons) *n. Colloq.* **1** The head or skull. **2** Brains; wit. [? Special use of SCONCE¹]

sconce⁴ (skons) *Brit. n.* A light fine or penalty. —*v.t.* **sconced, sconc·ing** To fine; mulct. [? <SCONCE³]

scone (skōn, skon) *n. Scot.* Originally, a thin oatmeal cake, baked on a griddle; hence, a teacake or soda biscuit.

Scone (skōōn, skōn) A village in SE Perthshire, Scotland; coronation place of Scottish kings, 1153 to 1488. —**the Stone of Scone** The stone on which early Scottish kings were crowned: brought to England by Edward I and placed under the seat of the Coronation Chair in Westminster Abbey.

scon·ner (skon′ər) *Scot. v.i.* To feel loathing or disgust. —*n.* Loathing; abhorrence. Also spelled *scunner*.

scoop (skōōp) *n.* **1** A shovel-like instrument or large shovel with high sides for scooping. **2** A small shovel-like implement or ladle used by grocers, druggists, etc. **3** An implement for bailing, as water from a boat. **4** A spoon-shaped instrument for using in a cavity: a surgeons' *scoop*. **5** An act of scooping; a movement in a curved line convex downward. **6** The amount scooped at once: a *scoop* of water. **7** *Colloq.* A large gain, especially in speculation: He made a big *scoop* on that deal. **8** A bowl-shaped cavity; hollow excavation. **9** In newspaper slang, a news story obtained and published ahead of rival papers. —*v.t.* **1** To take or dip out with or as with a scoop. **2** To hollow out, as with a scoop; excavate. **3** To empty with a scoop. **4** *Colloq.* To heap up or gather in as if in scoopfuls; amass. **5** In newspaper slang, to obtain and publish a news story before (a rival). [Fusion of MDu. *schope* a vessel for bailing out water, and *schoppe* a shovel] —**scoop′er** *n.* —**scoop′·ful′** *n.*

scoot (skōōt) *v.i. Colloq.* To go quickly; dart off. —*n.* The act of scooting; a darting off hurriedly. [Prob. <Scand.; cf. ON *skiōta* shoot. Akin to SHOOT.]

scoot·er (skōō′tər) *n.* **1** A child's vehicle consisting of a board mounted on two tandem wheels and steered by a long handle attached to the front axle: the rider stands with one foot on the board, using the other foot to push. **2** A similar vehicle powered by an internal-combustion motor and provided with a driver's seat: also *motor scooter*. **3** A sailboat so constructed that it may be sailed in water and on ice. **4** A small plow with a single shovel used for opening the soil: also *scooter plow*.

scop (skop) *n. Obs.* A bard, minstrel, or poet. [OE]

Sco·pas (skō′pəs) Greek sculptor of the fourth century B.C.

scope (skōp) *n.* **1** A range of view or action; outlook. **2** Room for the exercise of faculties or function; extent; capacity for achievement. **3** End in view; aim; purpose. **4** Length or sweep, as of a cable. **5** The range of a missile. [<Ital. *scopo* <L *scopus* <Gk. *skopos* a watcher <*skopeein* look at]

-scope *combining form* An instrument for viewing, observing, or indicating: *microscope.* [<Gk. *skopos* a watcher <*skopeein* watch]

Scopes (skōps), **John T(homas),** born 1901, U. S. educator; prosecuted for teaching evolution in Tennessee.

sco·po·drom·ic (skō′pə·drom′ik) *adj.* Pursuing a course in the line of sight; homing: said of guided missiles. [<Gk. *skopos* a watcher + *dromos* a running]

sco·pol·a·mine (skō·pol′ə·mēn, -min, skō′pə·lam′ēn, -in) *n. Chem.* An alkaloid, $C_{17}H_{21}O_4N$, extracted from the dried rhizomes of certain solanaceous plants, as *Scopolia carniolica*: its salts are used in medicine as a mydriatic, hypnotic, and sedative: also called *hyoscine.* [<G *scopolamin* <NL *Scopolia,* genus name of plants from which it is obtained, after G. A. *Scopoli,* 1723–88, Italian naturalist]

sco·po·line (skō′pə·lēn, -lin) *n. Chem.* A crystalline compound, $C_8H_{13}O_2N$, derived from scopolamine: also called *oscine.* [<SCO-POL(AMINE) + -INE[2]]

sco·po·phil·i·a (skō′pə·fil′ē·ə) *n. Psychiatry* Pleasure, especially of a sexual nature, derived from the act of observing, contemplating, or looking at something. Also **scop′to·phil′i·a** (skop′tə-). [<Gk. *skopos* a watcher + -PHILIA]

sco·po·pho·bi·a (skō′pə·fō′bē·ə) *n. Psychiatry* A morbid fear of being looked at. [<Gk. *skopos* a watcher + -PHOBIA]

scop·u·late (skop′yə·lit, -lāt) *adj.* Broom-shaped. [<L *scopulae* a little broom, pl. of *scopula* a broom twig, dim. of *scopa* a twig, a broom]

Sco·pus (skō′pəs), **Mount** A peak in central Palestine, NE of Jerusalem; 2,736 feet.

-scopy *combining form* Observation; viewing: *microscopy.* [<Gk. *-skopia* <*skopeein* watch]

scor·bu·tic (skôr·byōō′tik) *adj.* Relating to, like, or affected with scurvy: also **scor·bu′ti·cal.** —*n.* A person affected with scurvy. [<NL *scorbuticus* <Med. L *scorbutus* SCORBUTUS] —**scor·bu′ti·cal·ly** *adv.*

scor·bu·tus (skôr·byōō′təs) *n. Pathol.* Scurvy. [<NL <Med. L, appar. <MDu. *scheurbot, scheurbuik* <*scheuren* break, lacerate + *bot, buik* belly]

scorch (skôrch) *v.t.* **1** To change the color, taste, etc., of by slight burning; char the surface of. **2** To wither or shrivel by heat. **3** To affect painfully, as if by heat; criticize severely. —*v.i.* **4** To become scorched. **5** *Colloq.* To go at high speed. —*n.* **1** A superficial burn. **2** A mark caused by heat, as a slight burn. [Prob. related to ME *skorken* <ON *skorpna* dry up, shrivel; infl. in form by OF *escorchier* flay <L *excorticare* <*ex-* off + *cortex, -icis* bark] —**scorch′ing** *adj.* —**scorch′·ing·ly** *adv.*

scorched-earth policy (skôrcht′ûrth′) The policy of destroying all crops, industrial equipment, dwellings, etc., before an advancing enemy so as to leave nothing for his use or aid.

scorch·er (skôr′chər) *n.* **1** Something that scorches or is hot enough to scorch: Today was a *scorcher.* **2** Something severe or caustic, as criticism. **3** One who or that which moves or may move at great speed.

scor·da·to (skôr·dä′tō) *adj. Music* Out of tune; altered in tuning; made discordant. [<Ital., pp. of *scordare* be out of tune, short for *discordare* <L. See DISCORD.]

scor·da·tu·ra (skôr′dä·tōō′rä) *n. Music* An intentional changing of the normal tuning of a stringed instrument: resorted to for effect. [<Ital. *scordato* SCORDATO]

score (skôr, skōr) *n.* **1 a** A notch, cut, groove, mark, or line. **b** A notch or line used in keeping a tally or score; hence, an account or reckoning kept by notches or marks. **2** Any record, especially of indebtedness; debt; bill: to run up a *score* at a grocery. **3** Something charged or laid up against one; grudge; difference: to pay off old *scores*; an account; a credit; motive. **4** The record of the winning points, counts, or runs in competitions and games; also, the whole number of such points made by a player or side or in the game. **5** *Music* The collective notes in which a composition is written, when placed on two or more connected staffs one above another. **6** The number twenty, originally indicated by a special notch on a tally; twenty units or things: in the plural often indicating indefinitely large numbers. **7** *Psychol.* A quantitative value assigned to an individual or group response to a test or series of tests, as of intelligence or performance. —*v.* **scored, scor·ing** *v.t.* **1** To mark with notches, cuts, or lines. **2** To mark with cuts or lines for the purpose of keeping a tally or record. **3** To obliterate or cross out by means of a line drawn through: with *out.* **4** To make or gain, as points, runs, etc. **5** To count for a score of, as in games: A touchdown *scores* six points. **6** To rate or grade, as an examination paper; evaluate. **7** *Music* **a** To orchestrate. **b** To arrange or adapt for an instrument. **8** *U.S.* To criticize severely; scourge. **9** In cooking, to make superficial cuts in (meat, etc.). —*v.i.* **10** To make points, runs, etc., as in a game. **11** To keep score. **12** To make notches, cuts, etc. **13** To win an advantage; achieve a success. [OE *scoru* <ON *skor* a notch, tally] —**scor′er** *n.*

score-keep·er (skôr′kē′pər, skōr′-) *n.* One who keeps score.

Scores·by Sound (skôrz′bē, skōrz′-) An inlet and fjord system extending 200 miles into eastern Greenland from the Greenland Sea.

sco·ri·a (skôr′ē·ə, skō′rē·ə) *n. pl.* **·ri·ae** (-i·ē) Fragmentary lava; slag; refuse of ores or metals. [<L <Gk. *skōria* refuse <*skōr* dung] —**sco′ri·a′ceous** (-ā′shəs) *adj.*

sco·ri·form (skôr′ə·fôrm, skō′rə-) *adj.* Resembling scoria; in the form of dross.

sco·ri·fy (skôr′ə·fī, skō′rə-) *v.t.* **·fied, ·fy·ing** *Metall.* **1** To separate, as gold or silver, from an ore by smelting with lead, borax, etc. **2** To reduce to scoria or dross. [<SCORI(A) + -FY] —**sco′ri·fi·ca′tion** *n.*

scorn (skôrn) *n.* **1** Disdain; a feeling entertained toward someone or something as so inferior as to be unworthy of attention. **2** The expression of such a feeling; derision. **3** An object of supreme contempt. —*v.t.* **1** To hold in or treat with contempt; despise. **2** To reject with scorn; disdain; spurn. —*v.i.* **3** *Obs.* To mock; jeer. [<OF *escarn* <*escarnir* <Gmc.] —**scorn′er** *n.* —**scorn′ful** *adj.* —**scorn′ful·ly** *adv.* —**scorn′ful·ness** *n.*

Synonyms (noun): contempt, contumely, derision, despite, disdain, dishonor, mockery, scoff, scoffing, sneer, sneering, taunt. *Antonyms*: admiration, approbation, approval, attention, consideration, courtesy, deference, esteem, honor, regard, respect, reverence.

Synonyms (verb): abhor, contemn, despise, detest, disdain, spurn. *Antonyms*: see synonyms for CHERISH.

scor·pae·noid (skôr·pē′noid) *adj.* Belonging to a family (*Scorpaenidae*) of spiny-finned marine fishes. —*n.* A scorpaenoid fish: also **scor·pae′nid** (-nid). [<NL <L *scorpaena* a kind of fish + Gk. *eidos* form]

Scor·pi·o (skôr′pē·ō) **1** *Astron.* The Scorpion, a zodiacal constellation between Libra and Sagittarius, containing the brilliant red star Antares. See CONSTELLATION. **2** The eighth sign of the zodiac. Also **Scor′pi·on** (-ən), **Scor′·pi·us** (-əs). [<L, a scorpion]

scor·pi·oid (skôr′pē·oid) *adj.* **1** Scorpionlike. **2** Rolled or curled like the tail of a scorpion: specifically said of a terminal unilateral inflorescence, as in the borage family of plants. [<Gk. *skorpioeidēs* <*skorpios* a scorpion + *eidos* form]

scor·pi·on (skôr′pē·ən) *n.* **1** One of an order (*Scorpionida*) of rapacious arachnids with elongated, lobsterlike bodies and segmented tails which bear a poisonous sting: they are chiefly tropical but occur as far north as Canada.

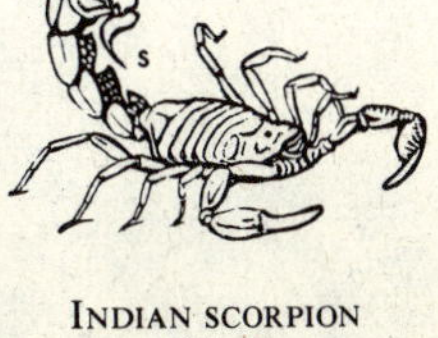

INDIAN SCORPION
s. Stinger.

2 The harmless pine lizard (genus *Sceloporus*) of the southern United States. **3** An instrument of chastisement; a whip or scourge. I *Kings* xii 11. **4** An ancient ballistic engine. [<OF <L *scorpio, -onis* <Gk. *skorpios*]

scorpion fly A mecopterous insect (genus *Panorpa*) living on the banks of shaded streams and in moist woods: in the male, the end of the abdomen is upcurved like a scorpion's sting. See illustration under INSECTS (beneficial).

Scorpion's Heart *Astron.* The star Antares in the constellation Scorpio.

scot (skot) *n.* An assessment; tax; a contribution, reckoning, or fine. [Fusion of ON *skot,* OF *escot*; ? infl. by OE *sceot, scot* payment]

Scot (skot) *n.* **1** A native of Scotland; a Scotsman; formerly, a Gaelic Highlander. **2** One of a Gaelic people who migrated in the fifth century to northwestern Britain from Ireland. [OE *Scottas,* pl., the Irish <LL *Scotus, Scoti* <OIrish *Scuit*]

scot and lot An assessment in Great Britain formerly laid on all of a parish or borough, according to their ability to pay; also, figuratively, obligations of every kind.

scotch[1] (skoch) *v.t.* **1** To cut; scratch. **2** To wound so as to maim or cripple. **3** To put down; crush or suppress. **4** To dress, as stone, with a pick. —*n.* **1** A superficial cut; a scratch; a notch. **2** A line traced on the ground, as for hopscotch. [Origin uncertain]

scotch[2] (skoch) *v.t.* To block, as a wheel or log, with a chock or wedge to prevent moving or slipping. —*n.* A block put behind or under something, as a wheel, to prevent rolling or sliding. [Origin unknown]

Scotch (skoch) *n.* **1** The people of Scotland collectively: with *the.* **2** One or all of the dialects spoken by the people of Scotland. **3** Scotch whisky. —*adj.* Of or pertaining to Scotland, its inhabitants, or their language; Scottish; Scots.

◆ **Scotch, Scots, Scottish** Of these three proper adjectives, the form *Scotch* developed in the dialects of the Midland and southern England, and is accepted even in Scotland as applying to *Scotch* plaid, *Scotch* terriers, *Scotch* whisky, etc.; in Scotland and in northern England, however, the forms *Scots* and *Scottish* (earlier *Scottis*) prevailed, and are preferred as applying to the people, culture, and institutions of Scotland: *Scots* or *Scottish* English, the *Scottish* church. This distinction is now widely accepted.

Scotch broom See under BROOM.

Scotch elm The wych-elm.

Scotch grain Heavy, durable, chrome-tanned leather with pebbled grain, usually of cowhide.

Scotch·man (skoch′mən) *n. pl.* **·men** (-mən) A Scot; Scotsman.

Scotch stone See AYR STONE.

Scotch tape A rolled strip of transparent cellulose tape having an adhesive on one side: a trade name.

Scotch terrier See under TERRIER.

Scotch whisky Whisky having rather a smoky flavor and made (originally in Scotland) from malted barley.

Scotch woodcock Eggs cooked and served on toast or crackers spread with anchovy paste.

sco·ter (skō′tər) *n.* A sea duck (genera *Oidemia* and *Melanitta*) of northern regions, having the bill gibbous or swollen at the base, especially the **American scoter** (*O. americana*) also called *coot,* or **scoter duck.** [<dial. E *scote,* var. of SCOOT]

scot-free (skot′frē′) *adj.* Free from scot; untaxed; unharmed.

sco·ti·a (skō′shē·ə, -shə) *n. Archit.* A concave molding common in the bases of classical columns. [<L <Gk. *skotia* darkness <*skotos*; so called from the darkness in its concavity]

Sco·tia (skō′shə) The Medieval Latin name for SCOTLAND.

Sco·tism (skō′tiz·əm) *n.* The scholastic system and metaphysical doctrines of the Scottish philosopher John Duns Scotus (13th century): a kind of formalism. —**Sco′tist** *n.* —**Sco·tis′·tic** *adj.*

Scot·land (skot′lənd) A political division and the northern part of Great Britain; a separate kingdom until its legislative union with England, 1707; 30,405 square miles; capital, Edinburgh.

Scotland Yard 1 The former headquarters of the London metropolitan police, situated in Great Scotland Yard, a short street in central London; removed to **New Scotland Yard,** on the Thames Embankment, in 1890. 2 The police force at headquarters; specifically, the detective bureau of the London police.

scoto– *combining form* Darkness: *scotophobia.* Also, before vowels, **scot–.** [<Gk. *skotos* darkness]

sco·to·ma (skə·tō'mə) *n. pl.* **·ma·ta** (-mə·tə) *Pathol.* A defect in the field of vision; a blind or dark spot. [<LL <Gk. *skotōma* dizziness < *skotoein* darken < *skotos* darkness]

scot·o·phil·i·a (skot'ə·fil'ē·ə) *n.* A love of darkness. [<SCOTO- + -PHILIA]

scot·o·pho·bi·a (skot'ə·fō'bē·ə) *n.* A morbid fear of darkness: also called *nyctophobia.* [<SCOTO- + -PHOBIA]

sco·to·pi·a (skə·tō'pē·ə) *n. Physiol.* Adaptation of the eye for night vision. [<NL <Gk. *skotos* darkness + *ōps, ōpos* an eye] **—sco·top'ic** (-top'ik) *adj.*

Scots (skots) *adj.* Scottish. **—***n.* The Scottish dialect of English. [Earlier *Scottis,* var. of SCOTTISH]

Scots·man (skots'mən) *n. pl.* **·men** (-mən) A Scot: the preferred term.

Scott (skot), **Dred,** 1795?–1858, U.S. Negro, central figure in Supreme Court decision. **—Sir George Gilbert,** 1811–78, English architect. **—Robert Falcon,** 1868–1912, English Antarctic explorer; reached South Pole, Jan. 17, 1912; perished on return journey. **—Sir Walter,** 1771–1832, Scottish novelist and poet. **—Winfield,** 1786–1866, U.S. general in the War of 1812 and Mexican and Civil Wars.

Scot·ti·cism (skot'ə·siz'əm) *n.* A form of expression or an idiom peculiar to the Scottish people.

Scot·tish (skot'ish) *adj.* Pertaining to or characteristic of Scotland, its inhabitants, or their language. **—***n.* 1 The dialect of English spoken in Scotland, especially in the Lowlands; Scots. 2 The people of Scotland collectively: with *the.* [OE *Scottisc* < *Scotta* a Scot]

Scottish Gaelic The Goidelic language of the Scottish Highlands.

scoun·drel (skoun'drəl) *n.* A mean, thoroughgoing rascal; a rogue; villain. **—***adj.* Scoundrelly. [Prob. dim. <AF *escoundre,* OF *escondre* abscond <L *ex-* off + *condere* hide]

scoun·drel·dom (skoun'drəl·dəm) *n.* Scoundrels collectively; scoundrelism.

scoun·drel·ism (skoun'drəl·iz'əm) *n.* The conduct or characteristics of scoundrels; rascality.

scoun·drel·ly (skoun'drəl·ē) *adj.* 1 Having the character of a scoundrel. 2 Pertaining to or characteristic of a scoundrel; rascally.

scour[1] (skour) *v.t.* 1 To clean or brighten by thorough washing and rubbing, as with sand or steel wool. 2 To remove dirt, etc., from; clean: to *scour* wool. 3 To remove by or as by rubbing away. 4 To clear by means of a strong current of water; flush. 5 To purge the bowels of. 6 To clean (wheat) before milling. **—***v.i.* 7 To rub something vigorously so as to clean or brighten it. 8 To become bright or clean by rubbing. See synonyms under CLEANSE. **—***n.* 1 The act of scouring. 2 A place scoured, as by running water. 3 A cleanser used in cleaning wool. 4 *Usually pl.* A watery diarrhea in cattle. [Prob. <MDu. *schuren* <OF *escurer,* ult. <L *ex-* out + *curare* take care of < *cura* care]

scour[2] (skour) *v.t.* 1 To range over or through, as in making a search. 2 To move or run swiftly over or along. **—***v.i.* 3 To range about, as in making a search. 4 To move or run swiftly. [ME *scoure.* Cf. ON *skura* rush, run.]

scour·er[1] (skour'ər) *n.* 1 One who or that which cleanses, removes stains, etc. 2 A cathartic. 3 A grain scourer. [<SCOUR[1]]

scour·er[2] (skour'ər) *n.* One who prowls about the streets by night; a vagabond. [<SCOUR[2]]

scourge (skûrj) *n.* 1 A whip for inflicting suffering or punishment. 2 Any instrumentality or means for causing suffering or death; hence, severe punishment; also, a cause of suffering. **—***v.t.* **scourged, scourg·ing** 1 To whip severely; lash; flog. 2 To punish severely; chastise; afflict. See synonyms under BEAT. [<AF *escorge* <LL *excoriare* flay <L *ex-* off + *corium* a hide] **—scourg'er** *n.*

Scourge of God Attila, king of the Huns.

scour·ing rush (skour'ing) Any species of horsetail, formerly much used for polishing wood and metal; scrub grass.

scour·ings (skour'ingz) *n. pl.* The residue after scouring: said especially of grain.

scouse (skous) *n.* A sailor's dish of sea biscuit and vegetables with or without meat; a hasty pudding of corn and rye meal. [Short for LOBSCOUSE]

scout[1] (skout) *n.* 1 One who or that which is engaged in scouting; specifically, a person sent out to observe and get information, as of the position or strength of an enemy in war. 2 The act of scouting. 3 At Oxford University, an undergraduate's manservant. 4 In cricket, a fielder: applied chiefly to one who fields at a distance in practice. 5 A boy scout; a girl scout. See synonyms under SPY. **—***v.t.* To observe or spy upon for the purpose of gaining information; reconnoiter, as an enemy position. **—***v.i.* To go or act as a scout. **—to scout around** To go in search. [<OF *escoute* a listener, listening < *escouter* listen <L *auscultare*] **—scout'er** *n.*

scout[2] (skout) *v.t.* & *v.i.* To reject with disdain; mock; jeer. See synonyms under RIDICULE. [<Scand. Cf. ON *skūta* a taunt.]

scout car A lightly armored motor car for reconnaissance work.

scouth (sko͞oth) *n. Scot.* Room for movement; scope.

scouth·er (skō'thər, sko͞o'-) *v.t. Scot.* To toast over a gridiron; scorch; singe. Also **scowd·er** (sko͞o'dər).

scout·mas·ter (skout'mas'tər, -mäs'-) *n.* The leader of a troop of Boy Scouts.

scow (skou) *n.* A large boat with a flat bottom and square ends: chiefly used as a lighter. [<Du. *schouw* a boat propelled by a pole <MDu. *schoude*]

scowl (skoul) *n.* 1 A lowering of the brows, as in anger, strong disapproval, or sullenness. 2 Gloomy aspect. **—***v.i.* 1 To lower and contract the brows in anger, sullenness, or disapproval. 2 To look threatening; lower. **—***v.t.* 3 To affect or express by scowling. [ME *skoul,* prob. <Scand. Cf. Dan. *skule.*] **—scowl'er** *n.* **—scowl'ing·ly** *adv.*

scrab·ble (skrab'əl) *v.* **·bled, ·bling** *v.i.* 1 To scratch, scrape, or paw, as with the hands. 2 To make irregular or meaningless marks; scribble. 3 To struggle or strive. **—***v.t.* 4 To make meaningless marks on; scribble on. 5 To gather hurriedly; scrape together. **—***n.* 1 The act of scrabbling; a moving on hands and feet or knees. 2 A scrambling effort. 3 A sparse growth: a *scrabble* of underbrush. [<Du. *schrabbelen,* freq. of *schrabben* scratch]

scrag (skrag) *v.t.* **scragged, scrag·ging** *Colloq.* To use roughly; wring the neck of; specifically, to kill by hanging; garrote. [< *n.*] **—***n.* 1 Something thin or lean and rough; a lean or bony piece or end of meat, especially from the neck. 2 *Slang* The neck. 3 A lean, bony person or animal. [Prob. <Scand. Cf. Norw. *skragg* a lean, feeble person.]

scrag·gly (skrag'lē) *adj.* **·gli·er, ·gli·est** Unkempt; shaggy; irregular; jagged. [Prob. <SCRAGG(Y) + -LY]

scrag·gy (skrag'ē) *adj.* **·gi·er, ·gi·est** 1 Rough. 2 Lean; scrawny; bony. [<SCRAG + -Y[1]] **—scrag'gi·ly** *adv.* **—scrag'gi·ness** *n.*

scraich (skrākh) *Scot. v.i.* To scream harshly; screech, as a fowl. **—***n.* A shrill cry; scream; screech. Also **scraigh.**

scram (skram) *v.i.* **scrammed, scram·ming** *U.S. Slang* To go away; leave quickly. [Prob. short for SCRAMBLE]

scram·ble (skram'bəl) *v.* **·bled, ·bling** *v.i.* 1 To move by clambering or crawling on hands and feet. 2 To struggle with others in a disorderly manner; scuffle; also, to strive for something in such a manner. **—***v.t.* 3 To mix together haphazardly or confusedly. 4 To gather or collect hurriedly or confusedly. 5 To cook (eggs) with the yolks and whites stirred together, usually with milk and butter. 6 *Telecom.* To invert or otherwise alter the frequency spectrum of (radio or wireless messages) so as to insure secrecy. **—***n.* The act of scrambling; a disorderly performance or struggle. [Prob. nasalized var. of SCRABBLE]

scrambled eggs Eggs prepared by stirring together the whites and yolks while cooking, usually with milk and butter.

scram·bler (skram'blər) *n.* 1 One who or that which scrambles. 2 *Telecom.* A device for altering the frequencies of radio and wireless signals during transmission.

scran·nel (skran'əl) *Archaic adj.* Thin; lean; reedy; slight; also, harsh. **—***n.* A lean person. [Prob. <Scand. Cf. Norw. *skrann* lean.]

Scran·ton (skran'tən) A city in NE Pennsylvania; an anthracite and manufacturing center.

scrap[1] (skrap) *n.* 1 A small piece cut or broken from something; fragment. 2 A brief extract. 3 *pl.* Pieces of crisp fat tissue after the oil has been expressed by cooking. 4 Old or refuse metal. See synonyms under PARTICLE. **—***v.t.* **scrapped, scrap·ping** 1 To break up into scrap; make scrap of. 2 To discard; throw away. **—***adj.* Having the form of scraps; discarded after use: *scrap* metal. [< ON *skrap* scrapings, scraps < *skrappa* scrape. Akin to SCRAPE.]

scrap[2] (skrap) *v.i.* **scrapped, scrap·ping** To fight; quarrel. **—***n.* A scrimmage; slight disagreement; scuffle; squabble. [<SCRAPE, *n.* (def. 2)]

scrap·book (skrap'bo͝ok') *n.* 1 A blank book in which to paste pictures, cuttings from periodicals, etc. 2 A personal notebook.

scrape (skrāp) *v.* **scraped, scrap·ing** *v.t.* 1 To rub, as with something rough or sharp, so as to abrade or to remove an outer layer or adherent matter. 2 To remove thus: with *off, away,* etc. 3 To rub (a rough or sharp object) across a surface. 4 To rub roughly across or against (a surface). 5 To dig or form by scratching or scraping. 6 To gather or accumulate with effort or difficulty: usually with *up* or *together.* **—***v.i.* 7 To scrape something. 8 To rub with a grating noise. 9 To emit or produce a grating noise. 10 To draw the foot backward along the ground in bowing: to bow and *scrape.* 11 To manage or get along with difficulty. 12 To be very or overly economical. **—to scrape acquaintance** To make acquaintance without an introduction. **—***n.* 1 The act or effect of scraping; also, the noise made by scraping. 2 A difficult situation; predicament. 3 A scraping or drawing back of the foot in bowing. [Prob. fusion of OE *scrapian* and ON *skrapa* scrape, erase]

scrap·er (skrā'pər) *n.* 1 Any instrument used for scraping. 2 A horse-drawn or motor-driven apparatus having a large metal scoop or scoops, for scraping up, transporting, and dumping dirt: a road *scraper*; a road leveller. 3 One who or that which scrapes. 4 A miser. 5 An unskilful player on the violin.

scrap·ing (skrā'ping) *n.* 1 The act of someone or something that scrapes. 2 The sound so produced. 3 Something scraped off or together.

scrap iron Old pieces of iron suitable for reworking.

scrap·ple (skrap'əl) *n.* A mixture of meal or flour boiled with scraps of pork, seasoned, and allowed to set: usually cooked by frying. [Dim. of SCRAP[1]]

scrap·py[1] (skrap'ē) *adj.* **·pi·er, ·pi·est** Composed of scraps; disconnected; fragmentary. [<SCRAP[1] + -Y[1]] **—scrap'pi·ly** *adv.* **—scrap'pi·ness** *n.*

scrap·py[2] (skrap'ē) *adj.* **·pi·er, ·pi·est** Pugnacious; given to picking fights. [<SCRAP[2] + -Y[1]] **—scrap'pi·ly** *adv.* **—scrap'pi·ness** *n.*

scratch (skrach) *v.t.* 1 To tear or mark the surface of with something sharp or rough. 2 To scrape or dig with something sharp or rough, as the claws or nails. 3 To scrape lightly with the nails, etc., as to relieve itching. 4 To rub with a grating sound; scrape. 5 To write or draw awkwardly or hurriedly. 6 To erase or cancel by or as by scratches or marks. 7 To erase or cancel the name of (a candidate) from a political ticket, while supporting the rest of the ticket; also, to bolt (a ticket or party) in this way. 8 To withdraw (an entry) from a race, game, etc. **—***v.i.* 9 To use the nails or claws, as in fighting or digging. 10 To scrape the skin, etc., lightly, as to relieve itching. 11 To make a grating noise. 12 To manage or get along with difficulty. 13 To withdraw from a game, race, etc. 14 In billiards and pool, to make

a scratch. — *n.* **1** A mark or incision made on a surface by scratching; a shallow mark, groove, furrow, or channel. **2** A slight flesh wound or cut. **3** The line from which contestants start, as in racing: to start from *scratch*. **4** The contestant who competes against an allowance: also **scratch-man. 5** In pugilism, a line across a prize ring at which fighters formerly began each round. **6** A disease of horses, consisting of dry scabs or chaps on the heel: also **scratch′es. 7** In billiards, a chance shot; also, a fluke; in billiards and pool, a shot resulting in a penalty; specifically, a shot in which the cue ball goes into a pocket, leaves the table, or fails to hit an object ball. — **from scratch** From the beginning; from nothing. — **up to scratch** *Colloq.* Meeting the standard or requirement in courage, stamina, or performance; in proper or fit condition: He was never *up to scratch* in writing. — *adj.* **1** Done by chance; haphazard. **2** In sports, without handicap or allowance. **3** Made as, or used for, a first try: a *scratch* pad. **4** Chosen at random or by chance: a *scratch* team. [Prob. blend of ME *scratte* scratch (prob. <Scand.; cf. Sw. *kratta* rake) and *cracchen* scratch <MDu. *cratsen*] — **scratch′er** *n.*
scratch test *Med.* A test to determine the substances to which a person is allergic by rubbing allergens in small scratches made in his skin.
scratch·y (skrach′ē) *adj.* **scratch·i·er, scratch·i·est 1** Characterized by scratches. **2** Making a scratching noise. **3** Straggling; shaggy; rough. — **scratch′i·ly** *adv.* — **scratch′i·ness** *n.*
scrawl (skrôl) *v.t. & v.i.* To write hastily or illegibly; scribble. — *n.* Irregular or careless writing. [? < dial. E, var. of CRAWL; ? infl. in meaning by *scribble, scroll*, etc.] — **scrawl′er** *n.*
scrawl·y (skrô′lē) *adj.* **scrawl·i·er, scrawl·i·est** Consisting of or characterized by ill-formed or irregular characters.
scraw·ny (skrô′nē) *adj.* **·ni·er, ·ni·est** Lean and bony; skinny; thin. [< dial. E *scranny*, var. of SCRANNEL] — **scraw′ni·ness** *n.*
screak (skrēk) *v.i.* To creak; screech. — *n.* A screech; also, a creak. [<ON *skrækja*; prob. imit.]
scream (skrēm) *v.i.* **1** To utter a prolonged, piercing cry, as of pain, terror, or surprise. **2** To make a prolonged, piercing sound. **3** To laugh loudly or immoderately. **4** To use heated, hysterical language. **5** To have an effect as of screaming: This color *screams* in contrast to green. — *v.t.* **6** To utter with a scream. — *n.* A loud, shrill, prolonged cry or sound, generally denoting fear or pain. See synonyms under CALL, ROAR. [ME *scraemen,* ? <ON *skraema* scare]
scream·er (skrē′mər) *n.* **1** One who or that which screams. **2** A South American bird, related to the ducks (family *Anhimidae* or *Palamedidae* including the **horned screamer** (*Anhima* or *Palamedea cornuta*) and the **crested screamers** (genus *Chauna*). **3** *U.S. Slang* Something calculated to call forth screams of admiration, astonishment, or the like; hence, a person of great size, strength, or skill. **4** *U.S. Slang* A sensational headline in a newspaper.
scream·ing (skrē′ming) *adj.* **1** Uttering or emitting screams. **2** Provocative of screams or of laughter: a *screaming* farce. **3** Like a scream.
scree (skrē) *n.* Debris of stones and rock fragments at the foot of a cliff or steep, rocky face: usually a sloping mass. See TALUS. [Back formation < *screes,* earlier *screethes* <ON *skridha* a landslide]
screech (skrēch) *n.* A shrill, harsh cry; shriek. — *v.t.* To utter with or as with a screech. — *v.i.* To make a prolonged, harsh, piercing sound; shriek. [Var. of obs. *scritch,* prob. < Scand. <ON *skrækja,* prob. ult. imit.] — **screech′er** *n.* — **screech′y** *adj.*
screech owl 1 Any of various small owls (genus *Otus*) common from Canada to Brazil; especially, the small, gray *O. asio* of the eastern United States. **2** The English barn owl.
screed (skrēd) *n.* **1** A prolonged tirade; harangue. **2** A wooden strip or a strip of mortar laid on a wall at intervals, to gage the thickness of the plastering. **3** A long torn strip or shred; hence, any detached strip or fragment: the original meaning, now chiefly Scottish. **4** *Scot.* A tearing; rent; tear; also, a drinking spree. — *v.t.* **1** To rend or tear into shreds. **2** *Scot.* To repeat glibly. [Var. of SHRED]

screen (skrēn) *n.* **1** That which separates or cuts off, shelters or protects, as a light partition. **2** A sieve or riddle, for sifting. **3** A smooth surface, as a canvas or curtain, on which motion pictures, etc., may be shown. **4** A motion picture or motion pictures collectively. **5** A plate of glass bearing very finely ruled lines, placed between the object and the camera in photographing for reproduction by the half-tone process. **6 a** *Mil.* A detachment of troops sent to deceive an enemy as to the movement of the main force. **b** *Nav.* A formation of ships arranged for the protection of heavier vessels from enemy submarines, etc. **7** *Physics* Any of various devices for confining the action of a physical agency or instrument to a definite area: a magnetic *screen*. **8** *Psychoanal.* A person who stands for someone else or others having some common characteristic, as in a dream: a form of concealment. — *v.t.* **1** To shield from observation or annoyance with or as with a screen. **2** To pass through a screen or sieve; sift. **3** To show or exhibit on a screen, as a motion picture. **4** *Psychol.* To separate from a group (those individuals showing indications of, or tendencies toward, mental or physical incapacity for specified activities): often with *out*. See synonyms under HIDE, PALLIATE, SHELTER. [Prob. <OF *escren, escrin,* prob. < OHG *skirm*] — **screen′a·ble** *adj.* — **screen′er** *n.*
screen·ing (skrē′ning) *n.* **1** *Physics* The effect of those electrons nearer the nucleus of an atom upon the nuclear attraction for the outer electrons. **2** *Psychol.* The process of weeding out from a group all individuals who do not conform to certain standards, as of loyalty, intelligence, ability, performance, and the like.
screen·ings (skrē′ningz) *n. pl.* The waste of anything passed through a sieve, as coal or defective grains; siftings.
screw (skroo) *n.* **1** A device resembling a nail but having a slotted head and a tapering or cylindrical spiral for driving into wood with a screwdriver, or for insertion into a corresponding grooved part: called **male** or **external screw. 2** A cylindrical socket with a

TYPES OF SCREWS
a. Lagscrew.   f. Shoulder screw.
b. Wood screw. g, h. Thumbscrews.
c. Saw screw. i. Collar screw.
d. Cap screw. j. Slotted screw.
e. Skein screw.

spiral groove: called **female** or **internal screw. 3** Anything having the form of a screw. **4** A screw propeller. **5** A turn of or as of a screw. **6** Pressure; force. **7** *Brit. Slang* Salary; pay. **8** *Slang* A prison guard. **9** A haggler over prices; a crafty bargainer. **10** *Brit.* A worthless horse. **11** *Brit.* A small packet of tobacco. — **to have a screw loose** *Slang* To be mentally deranged, eccentric, etc. — **to put the screws on** (or **to**) *Slang* To exert pressure or force upon. — *v.t.* **1** To tighten, fasten, attach, etc., by or as by a screw or screws. **2** To turn or twist. **3** To force as if by the pressure of a screw; urge: to *screw* one's courage to the sticking point. **4** To twist out of shape; contort, as one's features. **5** To practice oppression or extortion on; defraud. **6** To obtain by extortion. — *v.i.* **7** To turn or admit of being turned as a screw. **8** To be attached or become detached by means of a screw or screws: with *on, off,* etc. **9** To have turns like those of a screw. **10** To practice oppression or extortion. [Appar. <OF *escroue* a nut, female screw, ? <L *scrofa* sow; infl. in OF by L *scrobis* vulva] — **screw′er** *n.*
screw·ball (skroo′bôl′) *n.* **1** In baseball, a pitch thrown with a wrist motion opposite to that used for the out-curve, and breaking sharply and often unpredictably. **2** *U.S. Slang* An unconventional or erratic person.
screw·bean (skroo′bēn′) *n.* **1** The seed of the spirally twisted pod of a species of mesquite (*Strombocarpa odorata*). **2** The tree bearing this seed.
screw·driv·er (skroo′drī′vər) *n.* A tool for turning screws.
screwed (skrood) *adj.* **1** Having screw threads. **2** *Brit. Slang* Intoxicated.
screw jack 1 A hoisting or lifting jack operated by a screw; jackscrew. **2** A dental implement for regulating the position of the teeth.
screw log A patent log.
screw·pile (skroo′pīl′) *n.* A pile having a strong metal base with a screw thread to ensure firm penetration of hard ground or bedrock. See illustration under LIGHTHOUSE.

SCREW JACK

screw·pine (skroo′pīn′) *n.* Any of a tropical genus (*Pandanus*) of plants having a screwlike arrangement of the clustered leaves and aerial roots.
screw propeller A mechanism consisting of a revolving shaft with radiating blades set at an angle to produce a spiral action: used in propelling ships, etc.
screw thread 1 The projecting spiral ridge of uniform pitch on the outer or inner surface of a cylinder or cone, as of a screw or nut. **2** A complete revolution of any point in this ridge.
screw·worm fly (skroo′wûrm′) A shiny, blue-green blowfly (genus *Cochliomyia,* family *Calliphoridae*) about twice the size of the common housefly, whose larvae breed in living flesh; especially, *C. americana,* the destructive cattle pest of the southern and western United States.
screw·y (skroo′ē) *adj.* **screw·i·er, screw·i·est** *Slang* Extremely irrational; crazy.
Scria·bin (skryä′bēn), **Alexander,** 1872–1915, Russian composer.
scrib·ble (skrib′əl) *v.* **·bled, ·bling** *v.t.* **1** To write hastily and carelessly. **2** To cover with careless or illegible writing or marks. — *v.i.* **3** To write in a careless or hasty manner. **4** To make illegible or meaningless marks. — *n.* **1** Hasty, careless writing. **2** Meaningless lines and marks; any scrawl. [<Med. L *scribillare,* freq. of L *scribere* write]
scrib·bler (skrib′lər) *n.* **1** One who scribbles. **2** A writer of no reputation; a petty or inferior author.
scribe (skrīb) *n.* **1** One who writes or copies manuscripts. **2** A clerk, public writer, or amanuensis. **3** An author, penman, or journalist: used humorously. **4** An ancient Jewish teacher, interpreter, or writer of the Mosaic law. **5** A pointed instrument for marking wood, bricks, etc. — *v.* **scribed, scrib·ing** *v.t.* **1** To mark or scratch with a pointed instrument. **2** To write, inscribe, or engrave. **3** In carpentry, to mark and fit closely. — *v.i.* **4** *Rare* To write; work as a scribe. [<L *scriba* < *scribere* write] — **scrib′al** *adj.*
Scribe (skrēb), **Augustin Eugène,** 1791–1861, French dramatist.
scrib·er (skrī′bər) *n.* **1** One who or that which scribes. **2** Any sharp-pointed tool used in scribing.
scrieve (skrēv) *v.i.* *Scot.* To glide swiftly along. [Prob. <ON *skrefa* stride]
scrim (skrim) *n.* **1** A lightweight, open-mesh, coarse cotton fabric, usually white or écru, used for draperies, etc. **2** In the theater, a

similar fabric, often painted, used as a transparency, to support artificial foliage, etc. [Origin unknown]

scrim·mage (skrim′ij) *n.* **1** A rough-and-tumble contest; fracas; formerly, a skirmish. **2** In American football, a mass play from the line of scrimmage after the ball has been placed on the ground and snapped back, the play ending when the ball is dead. **3** In Rugby football, a scrummage. —**line of scrimmage** In football, the hypothetic line, parallel to the goal lines, on which the ball rests and along which the opposing linemen take position at the start of play. —*v.t.* & *v.i.* **·maged, ·mag·ing** To engage in a scrimmage. Also spelled *scrummage.* [Alter. of *scrimish,* var. of SKIRMISH]

scrimp (skrimp) *v.i.* **1** To be very or overly economical; be niggardly. —*v.t.* **2** To be overly sparing toward; skimp. **3** To cut too small, narrow, etc. —*adj.* Scanty; short: also **scrimp′y.** See synonyms under SCANTY. —*n.* A miser; niggard. [?. Related to OE *scrimman* shrink, shrivel] —**scrimp′i·ness** *n.*
Synonyms (verb): contract, curtail, economize, limit, pinch, reduce, save, scant, shorten, straiten. *Antonyms*: dissipate, lavish, squander, waste.

scrimp·it (skrim′pit) *adj. Scot.* Niggardly; scanty.

scrim·shaw (skrim′shô) *v.t.* & *v.i.* To ornament (ivory, whale's teeth, etc.) by cutting or carving: a sailor's term. —*n.* A neat example of mechanical work; especially, a scrimshawed article, ornamented with fanciful carving. [?< the surname *Scrimshaw*]

scrip[1] (skrip) *n.* **1** A scrap of paper, especially one containing writing. **2** A writing; a certificate, schedule, or written list. **3** A piece of paper money less than a dollar formerly issued in the United States: also called *shinplaster.* [<SCRIPT, prob. infl. in form by SCRAP]

scrip[2] (skrip) *n.* A provisional document (or documents collectively) certifying that the holder is entitled to receive something else, as shares of stock or land. [Short for obs. *subscription receipt*]

scrip[3] (skrip) *n.* A wallet or small bag. [Prob. fusion of ON *skreppa* a bag and OF *escrepe,* in phrase *escrepe et bordon* wallet and staff]

scrip dividend A distribution of surplus to stockholders in the form of scrip or promises to pay the amount of the dividend at a certain time.

Scripps (skrips) Name of a family of U.S. newspaper publishers, including **James Edmund,** 1835–1906, born in England; his half-brother, **Edward Wyllis,** 1854–1926; and **Robert Paine,** 1895–1938, son of Edward Wyllis.

scrip·sit (skrip′sit) *Latin* He (or she) wrote (it): used after an author's name on manuscripts, etc.

script (skript) *n.* **1** Writing of the ordinary cursive form. **2** Type, or printed or engraved matter, in imitation of handwriting. **3** *Law* A writing, especially an original; in English practice, a will; codicil. **4** A piece of writing; a manuscript or typescript; especially, a prepared copy, often containing suggestions, for the use of actors in a theatrical, radio, or television performance. —*v.t.* & *v.i. U.S. Colloq.* To prepare a script for (a radio, television, or theatrical performance). [<OF *escript* <L *scriptum,* pp. neut. of *scribere* write]

This line is in script.

scrip·to·ri·um (skrip·tôr′ē·əm, -tō′rē-) *n. pl.* **·ri·ums** or **·ri·a** (-ē·ə) The writing-room of a monastery, where records, annals, and manuscripts were written, copied, or illuminated. [<Med. L <L *scriptus,* pp. of *scribere* write]

scrip·tur·al (skrip′chər·əl) *adj.* Relating to writing; written. —**scrip′tur·al·ly** *adv.* —**scrip′tur·al·ness** *n.*

Scrip·tur·al (skrip′chər·əl) *adj.* Pertaining to, contained in, quoted from, or warranted by the Bible; Biblical. —**Scrip′tur·al·ly** *adv.* —**Scrip′tur·al·ness** *n.*

Scrip·tur·al·ism (skrip′chər·əl·iz′əm) *n.* The quality or character of being Scriptural; also, strict or literal adherence to the Scriptures. —**Scrip′tur·al·ist** *n.*

scrip·ture (skrip′chər) *n.* **1** The sacred writings of any people. **2** Originally, anything written, as a document, book, or inscription, or its contents; a writing. [<OF *escripture* <L *scriptura* <*scriptus,* pp. of *scribere* write]

Scrip·ture (skrip′chər) *n.* **1** The books of the Old and New Testaments, including often the Apocrypha; specifically, the Bible: usually plural. **2** A text or passage from the Bible.

script·writ·er (skript′rī′tər) *n.* A writer who prepares copy for the use of a radio or television actor or announcer.

scrive (skrīv) *v.t.* **scrived, scriv·ing** **1** To engrave. **2** *Obs.* To write; scribe. [<OF *escrivre* write <L *scribere*]

scri·vel·lo (skri·vel′ō) *n. pl.* **·loes** or **·los** An elephant's tusk. [<Pg. *escrevelho,* ? var. of *escaravelho* a pin, peg]

scriv·en·er (skriv′ən·ər, skriv′nər) *n.* **1** One who prepares deeds, contracts, and other writings; a clerk or scribe. **2** Formerly, a money-lender. [<obs. *scrivein* <OF *escrivain* <Ital. *scrivano* <L *scribere* write]

scro·bic·u·late (skrō·bik′yə·lit, -lāt) *adj. Biol.* Marked with many small depressions; furrowed or pitted. Also **scro·bic′u·lat′ed** (-lā′·tid). [<L *scrobiculus,* dim. of *scrobis* a trench]

scrod (skrod) *n.* A young codfish, especially when split and prepared for broiling. [? <MDu. *schrode* a piece cut off. Akin to SHRED.]

scrof·u·la (skrof′yə·lə) *n. Pathol.* A tuberculous condition of the lymphatic glands, characterized by enlargement, suppurating abscesses, and cheeselike degeneration; the king's evil. [Orig. pl. <LL *scrofulae,* dim. pl. <*scrofa* a breeding sow; so called because sows were supposed to be subject to the disease]

scrof·u·lous (skrof′yə·ləs) *adj.* **1** Pertaining to, affected with, or of the nature of scrofula. **2** Like scrofula; hence, morally diseased. —**scrof′u·lous·ly** *adv.* —**scrof′u·lous·ness** *n.*

scrog·gy (skrog′ē) *adj. Scot.* & *Brit. Dial.* Stunted; dwarfed; shriveled; also, abounding with brushwood. [Prob. <Scand. Cf. Dan. *skrog* a lean carcass.]

scroll (skrōl) *n.* **1** A roll of parchment, paper, or the like, especially one containing or intended for writing; also, the writing on such a roll; specifically, an outline; draft. **2** Anything resembling or suggestive of a parchment roll; specifically, a convoluted ornament or an ornamental space or tablet on sculptured work. **3** The curved head of a violin or similar instrument. **4** *Her.* A ribbon bearing a motto. See synonyms under RECORD. [Earlier *scrowle,* alter. of obs. *scrow* <AF *escrowe* a scroll; prob. infl. in form by ME *rowle* a roll]

scroll saw A narrow-bladed saw, or a sawing machine bearing such a blade, for doing curved or irregular work.

scroll·work (skrōl′wûrk′) *n.* Ornamental work of scroll-like pattern; particularly, fanciful designs cut from thin material by means of scroll saws.

Scrooge (skrōōj), **Ebenezer** In Dickens's *Christmas Carol,* a miser whose hard nature is transformed by the revelations of human joy and sorrow given to him by three spirits that visit him on Christmas Eve.

scroop (skrōōp) *v.i.* To give forth a harsh, scraping sound or cry; creak; grate. —*n.* A harsh grating or crunching sound; harsh cry. [Imit.; infl. by SCRAPE]

scroph·u·lar·i·a·ceous (skrof′yə·lâr′ē·ā′shəs) *adj.* Of or pertaining to a family *(Scrophulariaceae)* of herbs, shrubs, and a few trees, the figwort family, including the veronica, snapdragon and digitalis. [<NL <*Scrophularia,* type genus <Med. L *scrophula* SCROFULA; so called from its supposed power to cure scrofula]

scro·tum (skrō′təm) *n. pl.* **·ta** (-tə) *Anat.* The pouch that contains the testes. [<L] —**scro′·tal** *adj.*

scrouge (skrōōj, skrouj) *v.t. Brit. Dial.* To squeeze or grind down; crowd; press. [Earlier *scruze,* blend of SCREW and SQUEEZE]

scrounge (skrounj) *v.t.* & *v.i.* **scrounged, scroung·ing** *Slang* **1** To hunt about and take (something); pilfer. **2** To mooch; sponge; beg. [? <dial. E *scrunge* steal, var. of SCROUGE] —**scroung′er** *n.*

scroung·y (skroun′jē) *adj.* **scroung·i·er, scroung·i·est** *Slang* **1** Given to scrounging. **2** Unkempt; unclean; grubby.

scrub[1] (skrub) *v.* **scrubbed, scrub·bing** *v.t.* **1** To rub vigorously, as with the hand or a brush, in washing. **2** To remove (dirt, etc.) by such action. **3** To cleanse (a gas). —*v.i.* **4** To rub something vigorously in washing. See synonyms under CLEANSE. —*n.* The act of scrubbing. [? <Scand. Cf. Dan. *skrubbe.*]

scrub[2] (skrub) *n.* **1** A stunted tree. **2** A thicket or tract of stunted trees or shrubs. **3** A domestic animal of inferior or impure breed. **4** A poor, insignificant person. **5** In sports, a player not on the varsity or regular team. **6** A game of baseball contrived hastily by a few players. —*adj.* **1** Undersized or stunted-looking; inferior. **2** Consisting of or participated in by untrained players or scrubs: *scrub* team; *scrub* game. [Dial. var. of SHRUB[1]]

scrub·ber (skrub′ər) *n.* **1** One who or that which scrubs. **2** Any apparatus that removes undesired material through the medium of a liquid by washing.

scrub·by (skrub′ē) *adj.* **·bier, ·bi·est** **1** Of stunted growth. **2** Covered with or consisting of scrub or underbrush. [<SCRUB[2]] —**scrub′·bi·ness** *n.*

scrub grass The scouring rush.

scrub·land (skrub′land′) *n.* Land covered with scrub.

scrub oak Any of various dwarf oaks of the United States, as *Quercus ilicifolia* and *Q. prinoides,* common in New England; especially, the turkey oak or *Q. laevis* of the sandy barrens of the South.

scrub pine Any of several American dwarf pines; especially, the common Jersey pine *(Pinus virginiana)* and the shore pine of California, a variety of lodgepole pine *(P. contorta).*

scrub typhus *Pathol.* Tsutsugamushi disease.

scruff (skruf) *n.* The nape or outer back part of the neck. [Earlier *scuff* (<ON *skopt* hair); infl. in form by *scruff,* var. of SCURF]

scrum (skrum) *n. Brit. Colloq.* Scrummage: an abbreviated form.

scrum·mage (skrum′ij) *v.t.* & *v.i.* **·maged, ·mag·ing** To scrimmage. —*n.* **1** Scrimmage. **2** In Rugby football, a formation, around the ball, of the opposing sets of forwards, each of which endeavors by superior weight or compactness to dislodge the opponent, secure and break away with the ball, or kick it out. [Var. of SCRIMMAGE] —**scrum′mag·er** *n.*

scrump·tious (skrump′shəs) *adj. Slang* **1** Elegant or stylish; fine; delightful; splendid. **2** Fastidious; overly particular; nice. [<dial. E, mean, stingy, ult. <SCRIMP; prob. infl. in meaning by SUMPTUOUS]

scrunch (skrunch) *v.t.* & *v.i.* To crush; squeeze; crunch. —*n.* A crunch. [Imit. alter. of CRUNCH]

scru·ple (skrōō′pəl) *n.* **1** Doubt or uncertainty regarding a question of moral right or duty; reluctance arising from conscientious disapproval. **2** An apothecaries' weight of twenty grains, or 1.295 grams (symbol: ℈). **3** A minute quantity. **4** An ancient Roman coin. —*v.t.* & *v.i.* **·pled, ·pling** To have scruples (about); hesitate (doing) from considerations of right or expediency. [<OF *scrupule* <L *scrupulus,* dim. of *scrupus* a sharp stone]

scru·pu·lous (skrōō′pyə·ləs) *adj.* **1** Cautious in action for fear of doing wrong; nicely conscientious. **2** Resulting from the exercise of scruples; exact; careful. See synonyms under PRECISE, SQUEAMISH. [<L *scrupulosus* <*scrupulus* a scruple] —**scru′pu·lous·ly** *adv.* —**scru′pu·lous·ness, scru′pu·los′i·ty** (-los′·ə·tē) *n.*

scru·ti·nize (skrōō′tə·nīz) *v.t.* **·nized, ·niz·ing** To observe carefully; examine in detail. See synonyms under EXAMINE. —**scru′ti·niz′er** *n.* —**scru′ti·niz′ing·ly** *adv.*

scru·ti·ny (skrōō′tə·nē) *n. pl.* **·nies** **1** The act of scrutinizing; close investigation. **2** A method of electing the pope by secret ballot. **3** An official examination of votes after an election. See synonyms under INQUIRY. [<LL *scrutinium* <L *scrutari* examine, appar. <*scruta* trash, rags; with ref. to a careful search, including even trash and rags]

scry·ing (skrī′ing) *n. Archaic* Crystal-gazing. [Aphetic var. of DESCRY]

scud (skud) *v.i.* **scud·ded, scud·ding** **1** To move, run, or fly swiftly. **2** *Naut.* To run rapidly before the wind; especially, to run before a gale with little or no sail set. —*n.* **1** The act of scudding or moving swiftly. **2** Light clouds driven rapidly before the wind; a misty rain. **3** *Brit. Slang* A swift runner. **4** *Scot.* A slap with the open hand. **5** *pl. Scot.* Foaming beer or ale. [Prob. <Scand. (cf. Norw. *skudda* push); ? infl. in meaning by *scut,* in earlier sense of "a hare"]

Scud·der (skud′ər), **Horace,** 1838–1902, U.S. author.

Scu·dé·ry (skü·dā·rē′), **Madeleine de,** 1607–1701, French novelist.

scu·do (skōō′dō) *n. pl.* **scu·di** (skōō′dē) A former Italian and Sicilian silver or gold coin. [<Ital. <L *scutum* a shield]

scuff (skuf) *v.i.* **1** To walk with a dragging movement of the feet; shuffle. —*v.t.* **2** To scrape (the floor, ground, etc.) with the feet. **3** To make the surface of rough by rubbing or scraping. —*n.* The act of scuffing; also, the noise so made. [Prob. <ON *skūfa* shove]

scuf·fle[1] (skuf′əl) *v.i.* **·fled, ·fling 1** To struggle roughly or confusedly. **2** To drag one's feet; shuffle. —*n.* A disorderly struggle carried on by grappling, pulling, pushing, or the like; confused fracas. [Prob. freq. of SCUFF] —**scuf′fler** *n.*

scuf·fle[2] (skuf′əl) *n.* A form of hoe used by pushing in the manner of a spade. Also **scuffle hoe.** See illustration under HOE. [<Du. *schoffel* a weeding hoe]

scul·dud·der·y (skul·dud′ər·ē) *n. Scot.* Obscenity.

scull[1] (skul) *n.* **1** A long oar worked from side to side over the stern of a boat. **2** A light, short-handled spoon oar, used in pairs by one person. **3** A small boat for sculling. —*v.t. & v.i.* To propel (a boat) by a scull or sculls. ◆ Homophone: *skull.* [ME *sculle, skulle*; origin unknown] —**scull′er** *n.*

scull[2] (skul) *n. Scot.* A large, shallow wicker basket. ◆ Homophone: *skull.*

scul·ler·y (skul′ər·ē) *n. pl.* **·ler·ies** A room where kitchen utensils are kept and cleaned; a back kitchen. [<OF *escuelerie* care of dishes <*escuelle* a dish <L *scutella* a tray]

scul·lion (skul′yən) *n.* **1** A servant who washes and scours dishes, pots, and kettles. **2** A low wretch. [<OF *escouillon* a mop <*escouve* a broom <L *scopae* a bundle of twigs, pl. of *scopa* a twig]

sculp (skulp) *v.t. & v.i. Colloq.* To sculpture. [Short for SCULPTURE]

scul·pin (skul′pin) *n.* **1** One of several broad-mouthed fishes (family *Cottidae*), of inferior food value, with large, spiny head. The **daddy sculpin** (*Acanthocothus scorpius*) is a common North Atlantic species of which the North American form is a variety. **2** A fish (*Scorpaena guttata*) having a large head and spiny fins, found in southern California. **3** *Brit.* A contemptible fellow; mischief-maker. [Prob. alter. of F *escorpene* <L *scorpaena,* a scorpionlike fish <Gk. *skorpaina* <*skorpios* a scorpion]

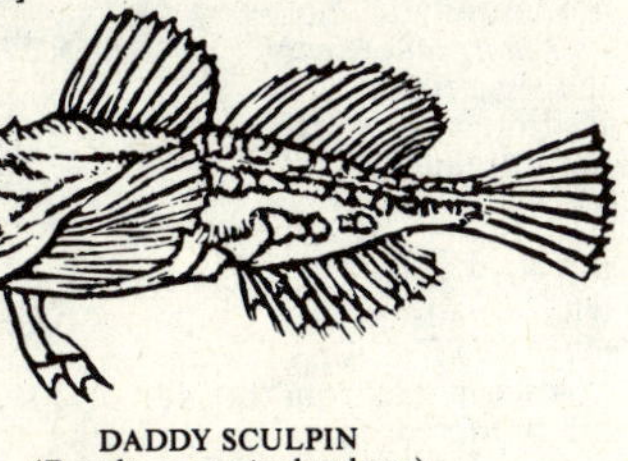

DADDY SCULPIN
(Rarely over 4 inches long)

sculp·sit (skulp′sit) *Latin* He (or she) sculptured it: used on a piece of statuary or sculpture following the name of the person who executed it.

sculp·tor (skulp′tər) *n.* One who designs sculpture by carving wood, modeling plastics, or chiseling stone. [<L <*sculpere* sculpture] —**sculp′tress** (-tris) *n. fem.*

sculp·ture (skulp′chər) *n.* **1** The art of fashioning figures of stone, wood, clay, or bronze. **2** Figures or groups carved, cut, hewn, cast, or modeled in wood, stone, clay, or metal. **3** Raised or incised lines, or markings, as upon a shell. —*v.t.* **·tured, ·tur·ing 1** To fashion, as statuary, by modeling, carving, or casting. **2** To represent or portray in sculpture. **3** To embellish with sculpture. **4** To change, as the face of a valley or canyon, by erosion and deposition. [<L *sculptura* <*sculptus,* pp. of *sculpere* carve in stone <*scalpere* cut] —**sculp′tur·al** *adj.*

sculp·tur·esque (skulp′chə·resk′) *adj.* Resembling sculpture; coldly, calmly, or grandly beautiful; statuesque; well-proportioned; majestic. —**sculp′tur·esque′ly** *adv.* —**sculp′tur·esque′ness** *n.*

scum (skum) *n.* **1** Impure or extraneous matter that rises to the surface of boiling or fermenting liquids. **2** Minute vegetation on stagnant water. **3** Scoria or dross of molten metals; also, foam; froth. **4** Figuratively, vile element; refuse. See synonyms under WASTE. —*v.* **scummed, scum·ming** *v.t.* To take scum from; skim. —*v.i.* To become covered with or form scum. [<MDu. *schuum*] —**scum′mer** *n.* —**scum′my** *adj.*

scum·ble (skum′bəl) *v.t.* **·bled, ·bling** In drawing and painting, to soften the outlines or blend the colors of by rubbing, as with comparatively dry or opaque color. —*n.* **1** The softening or blending of colors so produced. **2** The material used in scumbling. [Freq. of SCUM]

scun·ner (skun′ər) See SCONNER.

Scun·thorpe (skun′thôrp) A municipal borough of NW Lincolnshire, England.

scup (skup) *n.* **1** A valuable sparoid food fish (*Stenotomus chrysops*) of the eastern coast of the United States; the porgy: also **scup·paug** (skup′ôg, skə·pôg′). **2** A related species (*S. aculeatus*) found southward from Cape Hatteras and on the Gulf Coast to Texas. [<Algonquian (Narraganset) *mishcup* thick-scaled <*mishe* large + *cuppi* a scale]

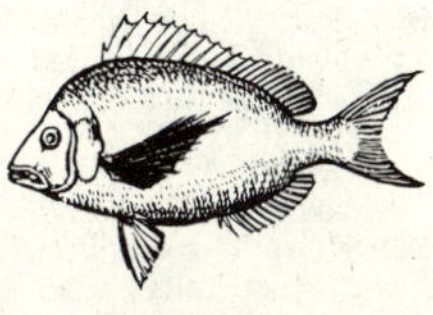

COMMON SCUP
(About 12 inches long)

scup·per[1] (skup′ər) *n. Naut.* A hole or gutter bordering a ship's deck, to let water run off. [Prob. short for *scupper hole* <OF *escope* a bailing scoop <Gmc. Akin to SCOOP.]

scup·per[2] (skup′ər) *v.t. Brit.* To put in great difficulty or danger; surprise or surprise and annihilate. [? <SCUPPER[1]]

scup·per·nong (skup′ər·nông, -nong) *n.* **1** A variety of muscadine grape cultivated in the southern United States. **2** A sweet, straw-colored wine made from this grape. [from the *Scuppernong* River in Tyrrell County, N.C.]

scurf (skûrf) *n.* **1** Loose scarfskin thrown off in minute scales, as in dandruff. **2** Any extraneous scaly matter adhering to a surface. **3** Worthless or impure coating or covering. [OE, alter. of *sceorf*; prob. infl. in form by Scand. Cf. Dan. *skurv.*] —**scurf′i·ness** *n.* —**scurf′y** *adj.*

scur·ril·i·ty (skə·ril′ə·tē) *n. pl.* **·ties** Coarse, vulgar abuse; a scurrilous remark. **2** The quality of being obscenely jocular. [<MF *scurrilité* <L *scurrilitas* <*scurrilis* SCURRILOUS]

scur·ri·lous (skûr′ə·ləs) *adj.* **1** Grossly offensive or vulgar; opprobrious. **2** Expressed with or given to low buffoonery. Also **scur·rile** (skûr′·il), **scur′ril.** [Earlier *scurrile* <L, neut. of *scurrilis* buffoon-like <*scurra* a buffoon] —**scur′ri·lous·ly** *adv.* —**scur′ri·lous·ness** *n.*

scur·ry (skûr′ē) *v.i.* **·ried, ·ry·ing** To move or go hurriedly; scamper. —*n. pl.* **·ries 1** The act or sound of scurrying; a precipitate movement. **2** A flurry, as of snow; whirl. **3** A short, fast run or race on horseback. [Short for HURRY-SCURRY; ? infl. by SCOUR[2]]

S-curve (es′kûrv′) *n.* A curve shaped like an *S.*

scur·vy (skûr′vē) *adj.* **·vi·er, ·vi·est 1** Meanly low or contemptible; base. **2** *Obs.* Afflicted with scurvy; also, scabby. See synonyms under BAD[1], BASE[2]. —*n. Pathol.* A disease characterized by livid spots under the skin, swollen and bleeding gums, and great prostration: caused by lack of vitamin C in the diet. [<SCURF] —**scur′vi·ly** *adv.* —**scur′vi·ness** *n.*

scurvy grass A biennial herb (*Cochlearia officinalis*) highly prized by Arctic explorers as a remedy for scurvy.

scut (skut) *n.* **1** A short or docked tail. **2** *Slang* A contemptible person. —*v.t. Obs.* To dock (an animal's tail). —*adj.* Short. [ME, a tail, a hare, prob. <Scand. Cf. Icelandic *skott* a fox's tail.]

scu·tage (skyōō′tij) *n.* A tax exacted from feudal knights instead of personal military service for their lands. [<Med. L *scutagium* <L *scutum* a shield]

Scu·ta·ri (skōō′tä·rē) **1** Üsküdar. **2** The largest city of northern Albania, at the SE end of **Lake Scutari,** a lake in SW Yugoslavia and NW Albania; 205 square miles. *Albanian* **Shko·dër** (shkô′dər), **Shko·dra** (shkô′drə). Ancient **Sco·dra** (skō′drə).

scu·tate (skyōō′tāt) *adj. Biol.* **1** Covered with horny, shieldlike plates or large scales. **2** Shaped like a shield. Also *scutellate.* See PELTATE. [<L *scutatus* provided with a shield <*scutum* a shield]

scutch (skuch) *v.t.* **1** To dress (textile fiber) by beating. **2** To separate the woody parts from the valuable fiber of (flax, etc.) by beating. —*n.* An implement for scutching hemp and flax. [Prob. <OF *escousser* shake, ? <Scand. Cf. Norw. *skoka* a scutch.] —**scutch′er** *n.*

scutch·eon (skuch′ən) *n.* **1** An escutcheon or anything shaped like it. **2** A metal plate or shield; a name plate or the like. [Aphetic var. of ESCUTCHEON]

scute (skyōōt) *n.* **1** *Zool.* A thin plate or scale, as a scale on a reptile. **2** Scutellum. [<L *scutum* a shield]

scu·tel·late (skyōō·tel′it, skyōō′tə·lāt) *adj. Zool.* **1** Platterlike; shield-shaped. **2** Covered with transverse scales; scutate. Also **scu′tel·lat′ed** (-lā′tid). [<NL *scutellatus* <L *scutella* a platter, dim. of *scutra* a tray; infl. in meaning by L *scutum* a shield]

scu·tel·la·tion (skyōō′tə·lā′shən) *n. Ornithol.* The presence or the arrangement of the scales on a bird's tarsus and toes.

scu·tel·lum (skyōō·tel′əm) *n. pl.* **·la** (-ə) **1** *Bot.* A small shieldlike organ or part, as in the cotyledon of a plant. **2** *Ornithol.* A scale on the foot of a bird. [<NL, dim. of L *scutum* a shield] —**scu·tel′lar** *adj.*

scu·ti·form (skyōō′tə·fôrm) *adj.* Shield-shaped. [<NL *scutiformis* <L *scutum* a shield + *forma* form]

scut·ter (skut′ər) *v.i.* To scurry; scuttle. —*n.* A hasty running. [SCUTT(LE)[3] + -ER[4]]

scut·tle[1] (skut′l) *n.* **1** A small opening or hatchway with movable lid or cover, especially in the roof or wall of a house, or in the deck or side of a ship. **2** The lid closing such an opening. **3** A sea cock in the bottom of a ship. —*v.t.* **·tled, ·tling** To sink (a ship) by making holes in the bottom or by opening the sea cocks. [<MF *escoutille* a hatchway <Sp. *escotilla,* prob. <Gmc.]

scut·tle[2] (skut′l) *n.* **1** A metal vessel or hod for coal. **2** Rarely, a vessel or pail for other purposes. [OE *scutel* a dish, platter <L *scutella*]

scut·tle[3] (skut′l) *v.i.* **·tled, ·tling** To run in haste; scurry. —*n.* A hurried run or departure. [? Var. of *scuddle,* freq. of SCUD; prob. infl. in form by dial. E *scut* a hare, a short tail; with ref. to the rapid movement of the hare]

scut·tle·butt (skut′l·but) *n.* **1** A drinking fountain aboard ship; formerly, a cask containing the day's drinking water. **2** *U.S. Slang* Rumor; gossip. [Orig. *scuttled butt* a lidded cask for drinking water]

scut·tler (skut′lər) *n.* The striped lizard of the southern United States.

scu·tum (skyōō′təm) *n. pl.* **·ta** (-tə) **1** The large oval or rectangular shield of the Roman legionaries. **2** *Zool.* Some platelike piece or part in a turtle, fish, etc.; a large scale. [<L]

Scu·tum (skyōō′təm) The Shield, a zodiacal constellation. See CONSTELLATION. [<L]

Scyl·la (sil′ə) In Greek mythology, a six-headed sea monster who dwelt in a cave on the Italian coast opposite the whirlpool Charybdis. See SCILLA. —**between Scylla and Charybdis** Between two dangers, where one cannot be avoided without incurring equally great peril from the other.

scypho- *combining form* Cup; vessel: also,

scyphozoan before vowels, **scyph-**. Also **scyphi-**, as in *scyphiform*, cup-shaped. [< L *scyphus* and Gk. *scyphos* a cup]

scy·pho·zo·an (sī′fə·zō′ən) *n.* Any of a class (*Scyphozoa*) of coelenterates including the sea anemones, corals, and jellyfish. — *adj.* Of or resembling the Scyphozoa. [< NL < Gk. *skyphos* a cup + *zōon* an animal]

Scy·ros (sī′rəs) The Latin name for SKYROS.

scythe (sīth) *n.* **1** A long curved blade for mowing, reaping, etc., fixed at an angle to a long bent handle or snath. **2** The implement so formed. **3** A curved blade attached to the axles or wheels of some ancient war chariots. — *v.t.* **scythed, scyth·ing** To cut or mow as with a scythe. [OE *sīthe*]

Scyth·i·a (sith′ē·ə) An ancient region of southern Europe, generally considered as lying north of the Black Sea.

Scyth·i·an (sith′ē·ən) *n.* **1** One of an ancient nomadic and fiercely savage people dwelling along the north shore of the Black Sea and extending as far east as the Aral Sea: last known in history about 100 B.C. **2** The Iranian language of the Scythians. — *adj.* Of or pertaining to the Scythians, their land, or their language. [< L *Scythia* < Gk. *Skythia* < *Skythēs* a Scythian]

'sdeath (zdeth) *interj. Archaic* God's death: an imprecation.

Sdot Yam (sdōt yäm) A settlement in NW Israel, on the site of ancient Caesarea.

sea (sē) *n.* **1** The great body of salt water covering the larger portion of the earth's surface; the ocean. **2** A large or considerable body of oceanic water partly or almost entirely enclosed by land: the Adriatic *Sea.* **3** A large inland body of water, salt or fresh: the Dead *Sea* or the *Sea* of Galilee. **4** The swell of the ocean; the course, flow, or set of the waves. **5** Anything that resembles or suggests the sea, as something vast, boundless, or wide-spread. — **at sea 1** On the ocean. **2** At a loss what to do or think; bewildered. — **to follow the sea** To follow the occupation of a sailor. — **the high seas** The unenclosed expanse of the ocean; also, that part of the ocean beyond a country's territorial waters. — **the seven seas** All the oceans of the world: the North and the South Atlantic, the North and the South Pacific, the Indian, the Arctic, and the Antarctic oceans. ◆ Homophone: *see*. [OE *sǣ*]

sea anchor A drag anchor; a heavy float or canvas bag or sail serving to hold a ship's head to the wind in order to ride out a gale or reduce drifting.

sea anemone A soft-bodied marine coelenterate (class *Anthozoa*, order *Actinaria*), that attaches itself to rocks, etc., suggesting a flower by its coloring and outspread tentacles.

sea bag A cylindrical canvas bag, fastened with a drawstring, in which sailors stow their clothes.

sea bass 1 A dusky-brown or black seranoid food fish (*Centropristes striatus*) common from Cape Cod to Florida: also called *blackfish*. **2** A related fish of California waters (*Stereolepis gigas*). Also **black sea bass. 3** The white sea bass of California (*Cynoscion nobilis*). **4** The related shortfin sea bass (*C. parvipinnis*).

Sea·bee (sē′bē) *n.* A member of the Construction Battalions of the U. S. Navy, which build base facilities, airfields, etc. [< C(*onstruction*) B(*attalion*)]

sea bird Any web-footed bird frequenting the oceans or their coasts, as albatrosses, gulls, gannets, petrels, frigate birds, shearwaters, etc.

sea biscuit Hardtack.

sea·board (sē′bôrd′, -bōrd′) *adj.* Bordering on the sea. — *n.* The seashore or seacoast; also, the land or region bordering the sea. [< SEA + *board* a border, OE *bord*]

Sea·borg (sē′bôrg), **Glenn Theodore**, born 1912, U.S. physical chemist.

SEA ANEMONE
a. Tentacles contracted.
b. Tentacles extended.

sea bread An unsalted hard biscuit used at sea; hardtack.

sea bream Any of several Old World sparoid food fishes; specifically, a common migratory species (*Pagellus centrodontus*).

sea breeze A cool breeze blowing from the ocean toward the land.

sea butterfly A pteropod.

sea calf The common harbor seal (*Phoca vitulina*) of the North Atlantic.

sea captain The captain of a seagoing vessel.

sea·coast (sē′kōst′) *n.* The seashore; seaboard.

sea cock A cock or valve controlling connection with the water through a vessel's hull.

sea coconut The very large and heavy bilobate fruit of a palm (*Lodoicea maldivica*) native to islands of the Indian Ocean, weighing 40 or 50 pounds and containing four nuts 18 inches long: also called *double coconut*.

sea cow 1 Any aquatic herbivorous mammal of the order *Sirenia*, sometimes attaining a length of about 25 feet; especially, the manatee or the dugong. **2** The walrus. **3** The hippopotamus.

sea craft 1 Skill in navigation. **2** Seagoing vessels.

sea cucumber A large holothurian (genera *Cucumaria* and *Thyone*) found on both coasts of the Atlantic: named from the form it commonly assumes.

sea devil 1 A devilfish. **2** An angelfish.

sea dog 1 The harbor seal or the California sea lion. **2** The piked or spiny dogfish. **3** A sailor with long experience at sea. **4** A fog dog.

sea drake 1 The male of the eider duck. **2** A cormorant.

sea drift Anything cast up by the sea; flotsam, especially vegetable or animal matter.

sea·drome (sē′drōm′) *n. Aeron.* An airport established at sea for the accommodation and servicing of aircraft making overseas flights. [< SEA + -DROME]

sea duck Any duck that frequents salt water, belonging to the subfamily *Nyrocinae*; especially, the American eider duck (*Somateria mollissima dresseri*), ranging from Labrador to Maine and as far westward as the Great Lakes. See DUCK.

sea eagle 1 An eagle, related to the bald eagle, which lives principally on fish; especially, Steller's sea eagle (*Thalassoaëtus pelagicus*), found on the islands off Alaska. **2** The osprey.

sea fan A coral (*Gorgonia flabellum*) of Florida and the West Indies, with fanlike branches.

sea·far·er (sē′fâr′ər) *n.* A seaman; a mariner. See synonyms under SAILOR. [< SEA + FARER]

sea·far·ing (sē′fâr′ing) *adj.* Following the sea as a calling. — *n.* Traveling over the ocean.

sea fight A conflict between vessels on the high seas.

sea fire The phosphorescence of sea water.

sea floor The bottom of the sea.

sea flower A sea anemone or related anthozoan.

sea foam 1 Foam of the ocean. **2** Meerschaum. **3** A fluffy candy made of spun sugar.

sea food Edible fish, shellfish, etc.

sea fowl A sea bird or sea birds collectively.

sea fox The thresher shark.

sea front Land that borders on the sea; buildings, etc., that face the sea.

sea gage 1 The depth to which a vessel sinks in the water; the draft of a vessel. **2** A sounding instrument showing the depth of water by the pressure on a column of fluid. Also **sea gauge.**

sea-girt (sē′gûrt′) *adj.* Surrounded by waters of the sea or ocean. [< SEA + GIRT²]

sea·go·ing (sē′gō′ing) *adj.* **1** Adapted for use on the ocean. **2** Skilful in navigation; seafaring.

sea grape A tropical American tree (*Coccolobis uvifera*) of the buckwheat family, with glossy, red-veined leaves, white flowers, and clusters of a purple fruit resembling grapes.

sea green A deep bluish green, like the color of sea water.

sea gull Any gull or large tern.

sea hog A porpoise.

sea holly A European coarse herb (*Eryngium maritimum*) of the carrot family.

sea horse 1 A teleost fish, usually 3 inches long, found in warm seas and allied to the pipefish; especially, *Hippocampus guttatus*, having a head resembling that of a horse. **2** A hippopotamus. **3** A walrus. **4** A fabulous animal, half horse and half fish, driven by Neptune. **5** A large white-crested wave.

Sea Island cotton A valuable long-staple variety of cotton formerly grown on the Sea Islands, now also cultivated elsewhere.

Sea Islands A chain of small islands off the coasts of South Carolina, Georgia, and northern Florida.

sea kale A hardy perennial herb (*Crambe maritima*) of the mustard family, cultivated for its edible young shoots.

sea king 1 A viking as a maritime leader; Norse pirate king of the Middle Ages. **2** Neptune.

SEA HORSE
(From 2 to 12 inches in length)

seal¹ (sēl) *n.* **1** An instrument or device used for making an impression upon some tenacious substance, as wax or a wafer; also, the impression made. **2** The wax, wafer, or similar token affixed to a document as a proof of authenticity; also, an impression, scroll, or mark on the paper. **3** A substance employed to secure a letter, door, lid, wrapper, joint, etc., firmly. **4** Anything that confirms or ratifies; a pledge; authentication. **5** Any instrumentality that keeps something close, secret, or unknown. **6** The fluid filling the trap of a drainage pipe and preventing the upward flow of gas. **7** An ornamental stamp for packages, etc. — *v.t.* **1** To affix a seal to, as to prove authenticity or prevent tampering. **2** To stamp or otherwise impress a seal upon in order to attest to weight, fineness, quality, etc. **3** To fasten or close with or as with a seal: to *seal* a letter; to *seal* a glass jar. **4** To grant or assign under seal. **5** To establish or settle finally; determine. **6** In Mormon usage, to solemnize forever, as a marriage or the adoption of a child. **7** To sign with the cross; also, to baptize or confirm. **8** To secure, set, or fill up, as with plaster. **9** To supply with a device or trap for preventing a return flow of gas or air. ◆ Homophone: *ceil.* [< OF *seel* < L *sigillum* a small picture, seal, dim. of *signum* a sign] — **seal′a·ble** *adj.*

seal² (sēl) *n.* **1** An aquatic carnivorous mammal (order or suborder *Pinnipedia*) mostly of high latitudes, of which some species, as the **fur seal**, yield valuable fur; any member of *Pinnipedia* except the walrus. Seals feed mostly on fish, and frequent seacoast rocks, ice floes, etc. In the breeding season they congregate on seacoasts, wild islands, etc. ◆ Collateral adjective: *phocine.* **2** The fur of a fur seal; sealskin. **3** Leather made from the hide of a seal. **4** Any fur prepared so as to look like sealskin. — *v.i.* To hunt seals. ◆ Homophone: *ceil.* [OE *seolh*]

SEAL
(Species vary from 7 to 12 feet long)

seal·ant (sē′lənt) *n.* Any substance which secures the contents of a container against contamination, evaporation, spoilage, or leakage.

sea lavender Any of a genus (*Limonium*) of mostly Old World maritime herbs bearing lavender-colored flowers.

sea lawyer A sailor given to criticizing and querying at every opportunity; a captious or argumentative person.

sea leather The skins of sharks, porpoises, and dogfishes prepared for use as leather.

sealed orders Orders given in a sealed envelope, with instructions to open at a given time or place under specified conditions; specifically, such orders given to the master of a ship before sailing.

sea legs The ability to walk aboard ship without losing one's balance.

seal·er[1] (sē′lər) *n.* 1 A person or thing that seals. 2 An officer who attests and certifies weights, materials, etc. [<SEAL[1]]
seal·er[2] (sē′lər) *n.* A person or ship employed in hunting seals. [<SEAL[2]]
seal·er·y (sē′lər·ē) *n. pl.* **·er·ies** 1 The business of hunting seals. 2 A place where seals are regularly hunted.
sea lettuce A green seaweed (genus *Ulva*) often used for food.
sea level The level continuous with that of the surface of the ocean at mean tide, between high and low water: used in reckoning altitudes.
sea lily A crinoid; a stalked marine invertebrate resembling a flower.
sealing wax A mixture of shellac and resin with turpentine and pigment that is fluid when heated but becomes solid as it cools: used for sealing papers and bottles.
sea lion One of various large, eared seals (family *Otariidae*), especially the California sea lion *(Zalophus californianus)*.
seal ring A signet ring; a finger ring containing an engraved stone.
seal·skin (sēl′skin′) *n.* 1 The under fur of the fur seal when prepared for use by removing the long hairs and dyeing dark-brown or black. 2 A coat or other article made of this fur.
sea lungwort An attractive American herb *(Mertensia maritima)* of the borage family, with white, long-stalked flowers, common to northern coasts.
Sea·ly·ham terrier (sē′lē·ham, -əm) See under TERRIER.
seam[1] (sēm) *n.* 1 A visible line of junction between parts, especially the edges of two pieces of cloth sewn together. 2 A crack; fissure; rent. 3 A ridge made in joining two pieces or left by a mold upon a casting. 4 A scar or cicatrix; also, a wrinkle. 5 A thin layer or stratum of rock. 6 A suture. — *v.t.* 1 To unite by means of a seam. 2 To mark with a cut, furrow, wrinkle, etc. 3 In knitting, to give the appearance of a seam to; purl. — *v.i.* 4 To crack open; become fissured. 5 In knitting, to form seams. ◆ Homophone: *seem.* [OE *sēam*] — **seam′er** *n.*
seam[2] (sēm) *n. Obs.* Any kind of grease; hence, fatness. ◆ Homophone: *seem.* [<OF *saim,* ult. <L *sagina* a fattening]
sea maiden *Poetic* A sea nymph or a mermaid. Also **sea maid.**
sea·man (sē′mən) *n. pl.* **·men** (-mən) 1 An enlisted man in the Navy or in the Coast Guard, graded according to his rank. 2 One skilled in the work of a ship and the ways of the sea; mariner; sailor. — **sea′man·like′** (-līk′) *adj.* — **sea′man·ly** *adj.* & *adv.*
sea·man·ship (sē′mən·ship) *n.* The skill and ability of a seaman in the operation and handling of a boat or ship.
sea·mark (sē′märk′) *n.* Any landmark that serves as a guide in navigation; a beacon; lighthouse.
Seam·as (shā′məs) Irish form of JAMES. Also **Seam′us.**
sea mew A gull, especially the European mew *(Larus canus).* [<SEA + MEW[3]]
sea mile See under MILE.
sea milkwort See under MILKWORT.
seam·less (sēm′lis) *adj.* Having no seam.
sea monster 1 Any huge, terrifying, or strange marine creature, as a devilfish or octopus. 2 A fabulous or mythical man-eating monster of the sea.
sea·mount (sē′mount′) *n.* Any of a widely distributed group of orogenic formations which rise to various heights from the ocean floor and serve as indicators of geologic processes; a submarine mountain.
sea mouse One of a family *(Aphroditidae)* of annelids with iridescent hairlike setae.
seam·ster (sēm′stər) *n.* A person employed in sewing. [OE *seamestre*]
seam·stress (sēm′stris) *n. fem.* A woman skilled in needlework, especially one whose occupation is sewing. Also spelled *sempstress.* [<OE *seamestre* a seamster + -ESS]
seam·y (sē′mē) *adj.* **seam·i·er, seam·i·est** 1 Full of seams, as the wrong side of a garment. 2 Showing the worst aspect: the *seamy* side. — **seam′i·ness** *n.*
Seán (shôn, shän) Irish form of JOHN.
Sean·ad Eir·eann (san′ad âr′ən) The Senate, or upper house, of the Irish Free State legislature. [<Irish *seanad* a senate + *Eireann* of Ireland]
sé·ance (sā′äns, *Fr.* sā·äns′) *n.* 1 A session or sitting. 2 A meeting of persons seeking spiritualistic manifestations. [<F <OF *seoir* sit <L *sedere*]
sean·na·chie (shan′ə·kē) *n. Scot.* A bard who preserved and repeated the traditions of the Scottish Highland tribes.
sea onion A bulbous herb *(Urginea maritima)* of the Old World, the source of squill.
sea otter A nearly extinct otter *(Enhydra lutris)* of the rocky shores of the North Pacific, about four feet long, and feeding principally on shellfish. The deep, rich fur, silvery-gray brown superficially, liver-brown beneath, is extremely valuable.
sea palm See under KELP.
sea pen A polyp (genus *Pennatula*) having a rodlike base with the polyps borne on lateral pinnae, giving the appearance of a feather.
sea·plane (sē′plān′) *n.* An airplane designed to rise from and descend upon the water.
sea·port (sē′pôrt′, -pōrt′) *n.* 1 A harbor or port on a coast accessible to seagoing ships. 2 A town located on such a harbor.
sea potato A brown alga (genus *Leathesia*) having a rounded, tuberous appearance.
sea power 1 A nation of great naval importance. 2 The naval strength of a nation.
sea purse *Zool.* The rectangular capsule enclosing the eggs or embryo of certain sharks, skates, and rays.

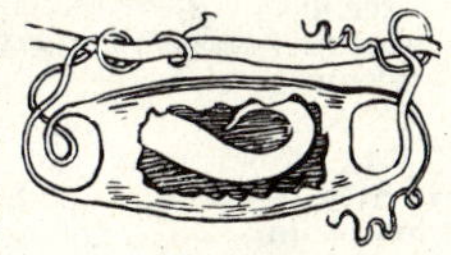

SEA PURSE

sea·quake (sē′kwāk′) *n.* An agitation of the sea from a submarine earthquake; a seismic disturbance under the sea.
sear[1] (sir) *v.t.* 1 To wither; dry up. 2 To burn the surface of; scorch. 3 To burn or cauterize, as with a hot iron; brand. 4 To make callous; harden. — *v.i.* 5 To become withered; dry up. — *adj.* Dried or blasted; withered. — *n.* A scar or brand. Also spelled *sere.* ◆ Homophones: *cere, sere.* [OE *sēarian* wither <*sēar* dry]
sear[2] (sir) *n.* The pawl in a gunlock, which holds the hammer at half or full cock. ◆ Homophones: *cere, sere.* [<OF *serre* a grasp <*serrer* close, press <LL *serrare* bolt, bar <L *sera* a lock; infl. in LL by L *serrare* saw]
sea raven 1 A deep-water sculpin. 2 The cormorant.
search (sûrch) *n.* 1 The act of seeking or looking diligently. 2 Investigation; inquiry. 3 A critical examination or scrutiny. 4 *Law* Right of search. — *v.t.* 1 To look through or explore thoroughly in order to find something; go over or through in making a search. 2 To subject (a person) to a search, as for concealed weapons, etc. 3 To examine with close attention; probe. 4 To penetrate or pierce: The wind *searches* my clothes. 5 To learn by examination or investigation: with *out.* — *v.i.* 6 To make a search. See synonyms under EXAMINE, HUNT. [<OF *cercher* <L *circare* go round, explore <*circus* a ring] — **search′a·ble** *adj.* — **search′er** *n.*
search·ing (sûr′ching) *adj.* 1 Investigating minutely. 2 Keenly penetrating. — **search′·ing·ly** *adv.* — **search′ing·ness** *n.*
search·light (sûrch′līt′) *n.* An apparatus containing a reflector, and so mounted that a beam of intensely brilliant light may be thrown in various directions for search or signaling; the beam of light from this apparatus.
search warrant A warrant directing an officer to search a house or other specified place for things alleged to be unlawfully concealed there.
sea risk Danger or hazard at sea; specifically, in marine insurance, a peril of the sea.
sea robin One of various gurnards, especially the American brown-finned species *(Prionotus strigatus).*
sea room Sufficient offing or space for a vessel to be maneuvered.
sea·scape (sē′skāp′) *n.* 1 An ocean view, especially when picturesque. 2 A picture presenting a marine view. [<SEA + (LAND)SCAPE]
sea-scout·ing (sē′skou′ting) *n.* Training in seamanship and water activities given to older boy scouts, called **sea scouts.**
sea serpent A snakelike animal, of monstrous size, believed by many to inhabit the ocean in very limited numbers.
sea·shell (sē′shel′) *n.* The shell of a marine mollusk.
sea·shore (sē′shôr′, -shōr′) *n.* Land adjacent to or bordering on the ocean; the ground between high- and low-water marks.
sea·sick (sē′sik′) *adj.* Suffering from seasickness.
sea·sick·ness (sē′sik′nis) *n.* Nausea, dizziness, and prostration caused by the motion of a vessel.
sea·side (sē′sīd′) *n.* The seashore, especially as a place of resort; also, the side abutting or facing the sea.
sea snake 1 A venomous fish-eating snake (subfamily *Hydrophinae*) of tropical seas, especially of the Indian Ocean. 2 A sea serpent.
sea·son (sē′zən) *n.* 1 A division of the year as determined by the earth's position with respect to the sun, and as marked by the temperature, moisture, vegetation, etc. The ancient Greeks had three seasons, spring, summer, and winter (mentioned by Homer and Hesiod); autumn appears first in Alcman: these four seasons are still used. 2 A period of time. 3 Any of the periods into which the Christian year is divided. 4 A period of special activity: usually with the definite article: the opera or hunting *season.* 5 A fit or suitable time. 6 That which imparts relish; seasoning. See synonyms under OPPORTUNITY, TIME. — **in season** 1 In condition and obtainable for use: Clams are *in season* during the summer. 2 In good or sufficient time; opportunely. 3 To be killed or taken by permission of the law. 4 Ready to mate or breed: said of animals. — *v.t.* 1 To increase the flavor or zest of (food), as by adding spices, etc. 2 To add zest or piquancy to. 3 To render more suitable for use, especially by drying or hardening, as timber. 4 To make accustomed or inured; harden: to *season* troops by strict discipline. 5 To mitigate or soften; moderate. — *v.i.* 6 To become seasoned. [<OF *seson* <LL *satio, -onis* sowing time <L, a sowing <*satus,* pp. of *serere* sow] — **sea′son·er** *n.*
sea·son·a·ble (sē′zən·ə·bəl) *adj.* 1 Being in keeping with the season. 2 Done at the proper time. See synonyms under CONVENIENT. — **sea′son·a·ble·ness** *n.* — **sea′son·a·bly** *adv.*
sea·son·al (sē′zən·əl) *adj.* Characteristic of, or occurring at, a certain season. — **sea′son·al·ly** *adv.*
sea·son·er (sē′zən·ər) *n.* 1 One who or that which seasons or gives added relish; a seasoning. 2 *U.S.* One engaged to serve for the season on a fishing vessel.
sea·son·ing (sē′zən·ing) *n.* 1 The act or process by which something, as lumber, is rendered fit for use. 2 Something added to food to give relish; especially, a condiment; hence, figuratively, something added to increase enjoyment or to relieve monotony. 3 The gradual process of acclimation to a new country or climate.
season ticket A ticket or pass entitling the holder to daily trips on a train for a certain period or to admission to a series of entertainments.
sea squirt An ascidian.
seat (sēt) *n.* 1 That on which one sits; a chair, bench, or stool. 2 That part of a thing upon which one rests in sitting, or upon which an object or another part rests. 3 That part of the person which sustains the weight of the body in sitting, or the corresponding portion of a garment. 4 The place where anything is situated, settled, or established: the *seat* of pain, the *seat* of a government; a site. 5 A place of abode; an estate or mansion, especially a country estate. 6 The privilege or right of membership in a legislative body, stock exchange, or the like. 7 The manner of sitting, as on horseback. 8 A surface or part upon which the base of anything rests. 9 A position in a legislature or an office. — *v.t.* 1 To place on a seat or seats; cause to sit down. 2 To have seats for; furnish

add,āce,câre,pälm; end,ēven; it,īce; odd,ōpen,ôrder; to͝ok,po͞ol; up,bûrn; ə = a in *above,* e in *sicken,* i in *clarity,* o in *melon,* u in *focus;* yo͞o = u in *fuse;* oi,oil; ou,pout; ch,check; g,go; ng,ring; th,thin; th,this; zh,vision. Foreign sounds à,œ,ü,kh,ṅ; and ◆: see page xx. < from; + plus; ? possibly.

with seats: The theater *seats* only 299 people. 3 To put a seat on or in; renew or repair the seat of. 4 To locate, settle, or center: usually in the passive: The French government is *seated* in Paris. 5 To fix or set firmly or in place. [<ON *sæti*. Akin to SIT.]

sea tangle A large brown seaweed (genus *Laminaria*) of the temperate zones.

seat·ing (sē′ting) *n.* 1 The act of providing with seats. 2 Fabric for upholstering seats. 3 A fitted support or base; a seat.

SEATO (sē′tō) Southeast Asia Treaty Organization.

seat of government 1 Any city (usually the capital) of a state or nation where the administrative offices of the government are located. 2 A town where a county court sits; a county seat.

sea trout 1 A trout that descends to the sea after spawning. 2 A weakfish.

seat·stone (sēt′stōn′) *n.* Underclay.

Se·at·tle (sē·at′l) A port on Puget Sound in west central Washington.

sea urchin An echinoderm (class *Echinoidea*) having a soft rounded body covered with a variously shaped shell bearing numerous movable spines.

sea wall 1 A wall or an embankment for preventing the encroachments of the sea or for breaking the force of the waves. 2 A ridge of stones, etc., washed up by the sea. — **sea-walled** (sē′wôld′) *adj.*

sea walnut Any of various ctenophores having an ovate body somewhat resembling a walnut, especially of the genus *Pleurobrachia.*

sea·wan (sē′wən) *n.* An oblong bead made from shell; hence, wampum: used by the Algonquian Indians of North America: also spelled *sewan.* Also **sea′want** (-wənt). [<Algonquian (Narraganset) *seawohn* scattered, i.e., unstrung (shell beads)]

sea·ward (sē′wərd) *adj.* 1 Going toward the sea. 2 Blowing, as wind, from the sea. — *adv.* In the direction of the sea: also **sea′wards.**

sea·ware (sē′wâr′) *n.* Seaweed; especially, coarse seaweed thrown up on the beach: used for manure and other purposes. [OE *sǣwār* < *sǣ* sea + *wār* alga]

sea·way (sē′wā′) *n.* 1 A way or lane over the sea. 2 An inland waterway that receives ocean shipping. 3 The headway made by a ship. 4 A rough sea: usually in *in a seaway.*

sea·weed (sē′wēd′) *n.* Any of a widely distributed class *(Algae)* of plants growing in the sea, including the kelps, rockweeds, dulse, sea lettuce, etc.

sea·wor·thy (sē′wûr′thē) *adj.* In fit condition for a voyage: said of a vessel. See synonyms under STAUNCH. — **sea′wor′thi·ness** *n.*

sea wrack Seaweed, especially a kelp or other large species.

se·ba·ceous (si·bā′shəs) *adj. Physiol.* 1 Pertaining to, appearing like, or secreting fat. 2 Designating the compound, saclike glands in the corium of the skin. [<NL *sebaceus* <L, a tallow candle < *sebum* tallow]

se·bac·ic (si·bas′ik, -bā′sik) *adj.* 1 Of or derived from fat. 2 *Chem.* Designating a white crystalline acid, $C_{10}H_{18}O_4$, contained in various oils, from which it is obtained by distillation. [<SEBAC(EOUS) + -IC]

Se·bas·tian (si·bas′chən, *Ger.* zä·bäs′tē·än) A masculine personal name. Also *Du., Sw.* **Se·bas·ti·aan** (sä·bäs′tē·än), *Fr.* **Sé·bas·tien** (sä·bäs·tyaṅ′), *Ital.* **Se·bas·tia·no** (sä′bäs·tyä′nō), *Lat.* **Se·bas·ti·a·nus** (si·bas′tē·ā′nəs), *Pg.* **Se·bas·tião** (sä′bäs·tyouṅ′). [<Gk., venerable]

Se·bas·to·pol (si·bas′tə·pōl) A former spelling of SEVASTOPOL.

Se·bas·tye (sa·bäs′tē·yə) A town in western Jordan, on the site of ancient Samaria.

Se·bat (shi·bät′) See SHEBAT.

Seb·ha (seb′hə) The capital of Fezzan, Libya.

sebi- *combining form* Fat; fatty matter: *sebiferous*: also, before vowels, **seb-.** Also **sebo-.** [<L *sebum* tallow]

se·bif·er·ous (si·bif′ər·əs) *adj.* Secreting or producing fat or fatty matter; sebaceous: *sebiferous* glands; *sebiferous* plants. Also **se·bip·a·rous** (si·bip′ər·əs). [<SEBI- + -FEROUS]

seb·or·rhe·a (seb′ə·rē′ə) *n. Pathol.* A morbid increase of secretion from the sebaceous glands: also called *steatorrhea.* Also **seb′or·rhoe′a.** [<L *sebum* tallow + -RRHEA]

se·bum (sē′bəm) *n. Physiol.* A fatty matter secreted by the sebaceous glands. [<L, tallow]

sec (sek) *adj. French* Dry: said of wines. Also *Italian* **sec·co** (sek′kō).

se·cant (sē′kənt, -kant) *adj.* Cutting, especially into two parts; intersecting. — *n.* 1 *Geom.* A straight line intersecting a given curve. 2 *Trig.* **a** A line drawn from the center of a circle through one extremity of an arc to the tangent drawn from the other extremity of the same arc. **b** The ratio of this line to the radius of the circle: the reciprocal of the cosine. [<L *secans, -antis,* ppr. of *secare* cut]

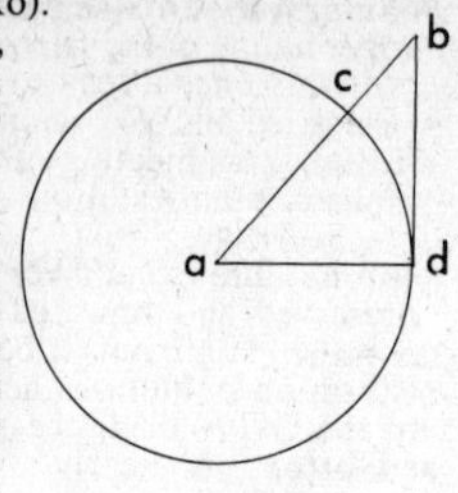

SECANT
Ratio of *ab* to *ad* is the secant of angle *a*. *ab* is the secant of arc *cd.*

sec·co painting (sek′ō) Painting done on dry plaster, as opposed to fresco painting on wet plaster. [<Ital., dry <L *siccus*]

se·cede (si·sēd′) *v.i.* **·ced·ed, ·ced·ing** To withdraw formally from a union, fellowship, or association, especially from a political or religious organization. [<L *secedere* withdraw < *se-* apart + *cedere* go] — **se·ced′er** *n.*

se·cern (si·sûrn′) *v.t.* 1 To separate; also, to distinguish. 2 *Physiol.* To secrete: said of a gland or follicle. [<L *secernere* < *se-* apart + *cernere* separate] — **se·cern′ent** *adj.* — **se·cern′ment** *n.*

se·cesh (si·sesh′) *n. U.S. Slang* A secessionist during the American Civil War; also, secessionists collectively. — *adj.* Belonging to, supporting, or sympathetic toward the Southern Confederacy. [Short for SECESSIONIST]

se·ces·sion (si·sesh′ən) *n.* 1 The act of seceding; withdrawal from fellowship, especially from political or religious association. 2 *Usually cap. U.S.* The withdrawal of the Southern States from the Union in 1860–61. [<L *secessio, -onis* < *secedere* SECEDE] — **se·ces′sion·al** *adj.*

se·ces·sion·ism (si·sesh′ən·iz′əm) *n. U.S.* The principles and doctrines of those who favored the withdrawal of the Southern States from the Union. — **se·ces′sion·ist** *adj.* & *n.*

seck (sek) *adj.* Barren; profitless; unenforceable by distress: said of rent. [<F *sec* <L *siccus* dry]

Seck·el (sek′əl, sik′əl) *n.* A variety of small, sweet pear. Also called *sickle pear.* [after the Pennsylvania farmer who introduced it]

se·clude (si·klōōd′) *v.t.* **·clud·ed, ·clud·ing** 1 To remove and keep apart from company or society of others; isolate. 2 To screen or shut off, as from view: usually in the past participle. [<L *secludere* < *se-* apart + *claudere* shut]

se·clud·ed (si·klōō′did) *adj.* 1 Separated; withdrawn; living apart from others. 2 Protected or screened. — **se·clud′ed·ly** *adv.* — **se·clud′ed·ness** *n.*

se·clu·sion (si·klōō′zhən) *n.* 1 The act of secluding, or the state or condition of being secluded; solitude; retirement. 2 A secluded place. [<Med. L *seclusio, -onis* <L *seclusus,* pp. of *secludere* SECLUDE]

Synonyms: privacy, retirement, retreat, secrecy, separation, solitude. See RETIREMENT, SOLITUDE. *Antonyms*: crowd, multitude, numbers, publicity, society, throng, world.

se·clu·sive (si·klōō′siv) *adj.* Having a tendency to seclusion. — **se·clu′sive·ly** *adv.* — **se·clu′sive·ness** *n.*

sec·ond[1] (sek′ənd) *n.* 1 A unit of time, 1/60 of a minute. 2 *Geom.* A unit of angular measure, 1/60 of a minute of arc. Symbol: ″. 3 In the duodecimal notation, 1/12 of an inch or prime. [<OF *seconde* <Med. L *seconda (minuta),* lit., second (minute), i.e., the result of the second operation of sexagesimal division, fem. of L *secundus* SECOND[2]]

sec·ond[2] (sek′ənd) *adj.* 1 Next in order, authority, responsibility, etc., after the first: the ordinal of *two.* 2 Ranking next to or below the first or best; of inferior quality or value; secondary; subordinate. 3 Identical in character with another or preceding one; another; other. 4 *Music* Lower in pitch, or rendering a lower part than the principal one. — *n.* 1 The one next after the first in position, rank, importance, or quality. 2 An attendant who supports or aids another, as in a duel. 3 *pl.* Articles of merchandise of imperfect manufacture, of second grade, or of inferior quality. 4 *Music* **a** The interval between any note and the next above or below in the diatonic scale. **b** A note separated by this interval from any other. **c** Two notes at this interval written or sounded together. **d** The resulting dissonance. **e** A second or subordinate part, instrument, or voice. 5 In parliamentary law, an utterance whereby a motion is seconded: Do I hear a *second*? — **major second** *Music* A second between whose tones is a difference of pitch of a step. — *v.t.* 1 To act as a supporter or assistant of; promote; stimulate; encourage. 2 In deliberative bodies, to support formally, as a motion, resolution, etc., as a prerequisite to discussion or adoption. See synonyms under AID, HELP. — *adv.* In the second order, place, or rank: also, in formal discourse, **sec′ond·ly.** [<OF <L *secundus* following < *sequi* follow]

Second Advent The expected second coming of Christ, to judge the world. Also **Second Coming.** — **Second Adventist**

sec·on·dar·y (sek′ən·der′ē) *adj.* 1 Of second rank, grade, or influence; subordinate; auxiliary; subsequent; resultant. 2 Depending on what is primary or original. 3 *Ornithol.* Of or pertaining to the secondaries of a bird's wings. 4 *Electr.* Of, pertaining to, or noting an induced current or its circuit, especially in an induction coil. 5 *Chem.* Formed by replacement of atoms or radicals in the molecules of certain organic compounds: a *secondary* alcohol. 6 *Geol.* Subsequent in origin; involving some chemical or physical change of the original mineral: contrasted with *primary.* 7 Pertaining to instruction in a secondary school. — *n. pl.* **·dar·ies** 1 One who acts in a secondary or subordinate capacity; an assistant; a deputy or delegate. 2 Anything of secondary size, position, or importance. 3 A secondary planet; a satellite. 4 *Ornithol.* One of the feathers that grow on the second joint or forearm of a bird's wing. See illustrations under BIRD, FOWL. 5 One of the hind wings of an insect. — **sec′on·dar′i·ly** *adv.*

Sec·on·dar·y (sek′ən·der′ē) *n. Geol.* 1 The Mesozoic era. 2 The rocks formed in this era. — *adj.* Belonging to or occurring in the Mesozoic era.

secondary education High school or preparatory school education; schooling beyond the elementary or primary, and below the college, level.

secondary electron *Physics* An electron emitted from a surface by the direct impact of electrons or ions, as produced by an X-ray machine.

secondary emission *Physics* The emission of secondary electrons from a substance exposed to direct radiation, as by X-rays, etc. Also **secondary radiation.**

secondary school See under SCHOOL.

second base In baseball, the second base reached by the runner, situated between first and third base. See illustration under BASEBALL.

second childhood A time or condition of foolishness or dotage; senility.

sec·ond-class (sek′ənd·klas′, -kläs′) *adj.* 1 Ranking next below the first or best; inferior; mediocre. 2 Of, pertaining to, or belonging to a class next below the first: *second-class* mail, *second-class* standing, *second-class* ticket, etc. — *adv.* By second-class ticket or by using second-class conveniences: to travel *second-class.*

second class A class of mail including all periodical printed matter.

se·conde (si·kond′, *Fr.* sə·gôṅd′) *n.* The second position in fencing. [<F, fem. of *second* <OF, SECOND]

sec·ond·er (sek′ən·dər) *n.* One who seconds, supports, or approves what is attempted, moved, or proposed by another.

second fiddle 1 The part played by the second violins in an orchestral composition. 2 Any secondary status; a substitute. — **to be** (or **play**) **second fiddle** To be of secondary importance in an undertaking or in the affections of another.

sec·ond-hand (sek′ənd·hand′) *adj.* 1 Having been previously owned, worn, or used by another; not new. 2 Received from another;

not direct from the original source: *second-hand* information. 3 Employed in handling or dealing in merchandise that is not new. 4 Of inferior grade; being a poor imitation: a *second-hand* statesman. — *n.* That which is second-hand or a poor imitation.

second hand The hand that marks the seconds on a clock or a watch.

sec·on·dine (sek′ən·dīn, -din) See SECUNDINE.

second mortgage A mortgage given next after and subordinate to a first mortgage.

second nature A disposition or character that is acquired and not innate; deep-seated habits that have become fixed.

se·con·do (sā·kôn′dō) *n. pl.* **·di** (-dē) *Italian* The second part in concerted music, especially in a pianoforte duet; also, the performer of this part.

sec·ond-rate (sek′ənd·rāt′) *adj.* Second in quality, size, rank, importance, etc.; second-class. — *n.* That which is mediocre or of inferior value: also **sec′ond-rat′er.**

second sight 1 The faculty or power of seeing the invisible. 2 The power of prophecy; intuition; clairvoyance. — **sec′ond-sight′ed** *adj.*

second sound *Physics* The peculiar vibratory motion, resembling that of sound waves, associated with the rapid transfer of heat by helium atoms cooled to within two degrees of absolute zero.

Second World War See WORLD WAR II in table under WAR.

sec·par (sek′pär) *n. Astron.* Parsec. [< *sec(ond of) par(allax)*]

se·cre·cy (sē′krə·sē) *n. pl.* **·cies** 1 The condition or quality of being secret or hidden; concealment. 2 The character of being secretive; secretiveness. 3 Privacy; retirement; solitude. Also **se′cret·ness.** See synonyms under SECLUSION. [Earlier *secretee* < obs. *secre* <OF *secré* secret; refashioned after *primacy, lunacy,* etc.]

se·cret (sē′krit) *adj.* 1 Kept separate or hidden from view or knowledge, or from all persons except the individuals concerned; not immediately apparent; unseen; occult. 2 Affording privacy; secluded. 3 Good at keeping secrets; close-mouthed. 4 Unrevealed or unavowed as such: a *secret* partner. 5 *U.S.* Designating defense information classified second to top-secret material with regard to required security and protection. Compare TOP-SECRET, CONFIDENTIAL (def. 4). — *n.* 1 Something not to be told. 2 A thing undiscovered or unknown. 3 An underlying reason; that which, when known, explains; key. 4 A secret contrivance. 5 Secrecy. — **in secret** In privacy; in a hidden place. [<OF *secré, secret* <L *secretus,* orig. pp. of *secernere* < *se-* apart + *cernere* separate] — **se′cret·ly** *adv.*

Synonyms (adj.): clandestine, concealed, covered, covert, furtive, hid, hidden, latent, mysterious, obscure, occult, private, recondite, retired, unknown, unrevealed, unseen, veiled. See MYSTERIOUS. *Antonyms*: aboveboard, apparent, clear, evident, manifest, obvious, plain, transparent, unconcealed, undisguised.

se·cret·age (sē′krə·tij) *n.* A process of preparing or dressing furs by means of mercury or some of its salts, in order to facilitate felting and matting; carroting. Also **se′cret·ing.** [<F *sécréter* conceal; because it was at first a secret process]

sec·re·tar·i·at (sek′rə·târ′ē·it, -at) *n.* 1 A secretary's position. 2 The place where a secretary transacts his business and preserves his official records. 3 The entire staff of secretaries in an office; especially, the department headed by a governmental secretary. Also **sec′re·tar′i·ate.** [<F *secrétariat* <Med. L *secretariatus* the office of secretary < *secretarius* SECRETARY]

Sec·re·tar·i·at (sek′rə·târ′ē·it, -at) *n.* The administrative organ of the former League of Nations and of the present United Nations, consisting of the Secretary General, his officials, and secretaries.

sec·re·tar·y (sek′rə·ter′ē) *n. pl.* **·tar·ies** 1 A person employed to deal with correspondence, keep records, and handle clerical business for a person, business, committee, or organization. 2 An executive officer presiding over and managing a department of government. 3 A writing desk with a bookcase or cabinet with pigeonholes on top. — **under-secretary** In a government department, the official who ranks next below the secretary. [<Med. L *secretarius* <L *secretum* a secret, neut. of *secretus* SECRET] — **sec′re·tar′i·al** (-târ′ē·əl) *adj.*

secretary bird A South African bird (genus *Sagittarius*), having long legs and a crested head: so named from the resemblance of its crest to quill pens stuck behind the ear. It preys on serpents.

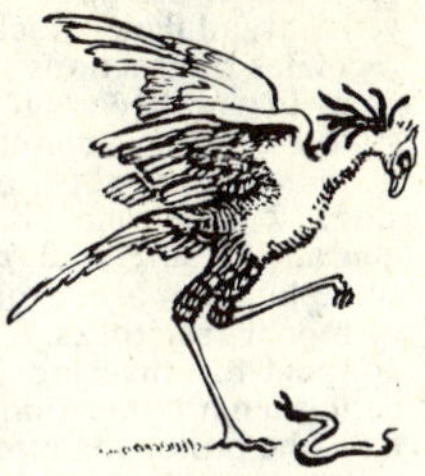

SECRETARY BIRD (About 4 feet high)

secretary general *pl.* **secretaries general** A chief secretary; an assistant to a governor general. — **sec′re·tar′y-gen′er·al·cy** *n.*

sec·re·tar·y·ship (sek′rə·ter′ē·ship) *n.* The work or position of a secretary.

sec·re·tar·y-treas·ur·er (sek′rə·ter′ē·trezh′ər·ər) *n.* 1 A person who performs the combined duties of secretary and treasurer; especially, an official in an organization. 2 In Canada, a town or city clerk.

se·crete (si·krēt′) *v.t* **·cret·ed, ·cret·ing** 1 To remove or keep from observation; conceal; hide. 2 *Biol.* To separate or elaborate from blood or sap. [Alter. of obs. *secret, v.* conceal; refashioned after L *secretus* SECRET] — **se·cre′tor** *n.*

Synonym: conceal. *Secrete* is a stronger word than *conceal,* and is used chiefly of such material objects as may be separated from the person, or from their ordinary surroundings, and put in unlooked-for places; a man *conceals* a scar on his face, but does not *secrete* it; a thief *secretes* stolen goods; an officer may also be said to *secrete* himself to watch the thief. See HIDE.

se·cre·tin (si·krē′tin) *n. Biochem.* A hormone found in the lining of the intestinal wall and stimulating the flow of pancreatic juice. [<SECRET(ION) + -IN]

se·cre·tion (si·krē′shən) *n.* 1 *Biol.* The process by which materials are separated from blood or sap and elaborated into new substances: the *secretion* of milk, gastric juice, or urine. Secretion in animals is generally performed by glandular epithelial cells. Compare EXCRETION. 2 The substance secreted, as saliva or milk. 3 The act of concealing. 4 A deposit of mineral matter in successive coatings, filling cavities, and fissures.

se·cre·tive (si·krē′tiv) *adj.* 1 (*also* sē′krə·tiv) Inclined to secrecy; reticent. 2 Producing or causing secretion. — **se·cre′tive·ly** *adv.* — **se·cre′tive·ness** *n.*

se·cre·to·ry (si·krē′tər·ē) *adj.* Pertaining to secretion. — *n. pl.* **·ries** A secreting vessel or gland.

secret service 1 Investigation conducted secretly for a government. 2 The secret or espionage work of various government agencies in time of war.

Secret Service A section of the Department of the Treasury concerned with the suppression of counterfeiting, the protection of the president of the United States, etc.

secret society A society or association that uses secret signs, oaths, rites, or symbols.

sect (sekt) *n.* 1 A body of persons distinguished by peculiarities of faith and practice from other bodies adhering to the same general system; specifically, the adherents collectively of a particular creed or confession; a denomination, or an organized body of dissenters from an established or older form of faith. 2 Adherents of a particular philosophical system or teacher. 3 Any number of persons united in opinion or interest, as in the state or in society; a party or faction; an order. 4 A cutting in horticulture. [<OF *secte* <L *secta* a following, a faction < *sequi* follow. Doublet of SET.]

Synonyms: church, communion, denomination, heresy, heterodoxy, party, schism, school. *Heresy* or *heterodoxy* is a departure from the established doctrine; a *schism* is a division of the *church* either on matters of faith or practice; *schism* is applied also to non-religious organizations. A *sect* or *denomination* is an organized body of believers distinct in doctrine or practice, or in both, from others: *sect* is an opprobrious and *denomination* an honorable term for the same body. Within a *denomination* there may be *schools* differing on minor matters, or *parties* favoring or opposing certain persons or measures, without breach of essential and organic unity. *Church* is often used as synonymous with *denomination*; as, the Presbyterian *Church. Communion* designates those who share a common faith with reference to their spiritual unity.

-sect *combining form* Cut; divided (in a specified manner or number of parts): *vivisect.* Also **-sected,** as in *bisected.* [<L *sectus,* pp. of *secare* cut]

sec·tar·i·an (sek·târ′ē·ən) *adj.* Pertaining to a sect; bigoted. — *n.* A member of a sect, especially if bigoted.

sec·tar·i·an·ism (sek·târ′ē·ən·iz′əm) *n.* Sectarian character or tendency; excessive devotion to or zeal for a particular sect.

sec·tar·i·an·ize (sek·târ′ē·ən·īz′) *v.t.* **·ized, ·iz·ing** To make sectarian.

sec·ta·ry (sek′tər·ē) *n. pl.* **·ries** 1 A sectarian: mostly used opprobriously. 2 A dissenter from an established church; a nonconformist. 3 *Obs.* A religious sect. Also **sec′ta·rist.** [< MF *sectaire* <Med. L *sectarius* <L *secta* a sect]

sec·tile (sek′til) *adj.* Admitting of being cut or severed smoothly. [<F <L, neut. of *sectilis* < *sectus,* pp. of *secare* cut] — **sec·til·i·ty** (sek·til′ə·tē) *n.*

sec·tion (sek′shən) *n.* 1 A separate part or division; a portion of a book, treatise, or writing; a subdivision of a chapter; also, a division of law. 2 A distinct part of a country, community, etc. 3 *U.S.* An area of public land one mile square, containing 640 acres and constituting 1/36 of a township. 4 A portion of a railway company's tracks under the care of a particular set of men. 5 In a sleeping-car, a space containing two berths. 6 A tactical unit of the U.S. Army, smaller than a platoon and larger than a squad. 7 A division of an animal group, of indeterminate rank. 8 A representation, picture, or drawing of a building, machine, geological formation, etc., as if cut by an intersecting plane; also, the thing so cut or viewed. 9 A very thin slice of anything, especially for microscopic examination. 10 The character §, indicating a subdivision: used also as a reference mark. 11 The act of cutting; division by cutting, as in surgical operations. 12 The figure formed by the intersection of a plane or other surface with a solid. In mechanical drawing the following sections are distinguished: **lengthwise** or **longitudinal section,** usually representing objects as cut lengthwise through the center; **cross-section** or **transverse section,** cut crosswise; **horizontal section,** cut horizontally, and usually through the center; **oblique section,** cut at various angles. See synonyms under PART. — **frozen section** A cutting, slice, or sliced surface of a frozen part: much employed in anatomy. — *v.t.* 1 To cut or divide into sections. 2 To shade (a drawing) so as to designate a section or sections. [<MF <L *sectio, -onis* < *sectus,* pp. of *secare* cut]

-section *combining form* The act or process of cutting or dividing: *vivisection.* [<L *sectio, -onis* a cutting < *secare* cut]

sec·tion·al (sek′shən·əl) *adj.* 1 Pertaining to a section, as of a country; local; characteristic of the people of a certain section or area: a *sectional* dialect. 2 Dividing or alienating one section from another: *sectional* problems. 3 Made up of sections. — **sec′tion·al·ly** *adv.*

sectional feeling Intense consciousness of the differences between the interests of one section of a country and those of another.

sec·tion·al·ism (sek′shən·əl·iz′əm) *n.* Regard for a particular section of the country rather than the whole; sectional feeling. — **sec′tion·al·ist** *n.*

sec·tion·al·ize (sek′shən·əl·īz′) *v.t.* **·ized, ·iz·ing** 1 To make sectional. 2 To divide into sections. — **sec′tion·al·i·za′tion** *n.*

add,āce,câre,pälm; end,ēven; it,īce; odd,ōpen,ôrder; to͝ok,po͞ol; up,bûrn; ə = a in *above,* e in *sicken,* i in *clarity,* o in *melon,* u in *focus*; yo͞o = u in *fuse*; oi,oil; ou,pout; ch,check; g,go; ng,ring; th,thin; th,this; zh,vision. Foreign sounds à,œ,ü,kh,ṅ; and ◆: see page xx. < from; + plus; ? possibly.

section gang A work crew assigned to a certain section of a railroad.

sec·tor (sek′tər) *n.* 1 *Geom.* A part of a circle bounded by two radii and the arc subtended by them. 2 A mathematical instrument consisting of two arms marked with various scales and hinged together at one end. 3 *Mil.* A part of a front in contact with the enemy. —*v.t.* To divide into sectors. [<LL <L, a cutter < *sectus,* pp. of *secare* cut]

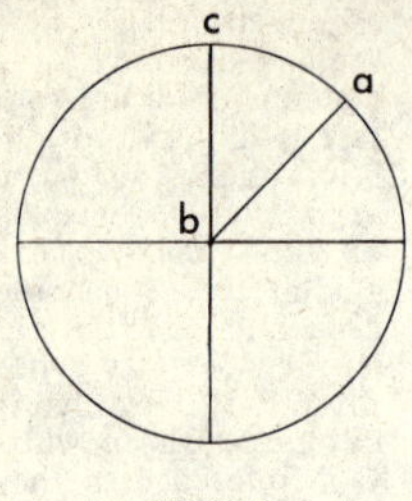

SECTOR
abc is a sector of the circle.

sec·to·ri·al (sek·tôr′ē·əl, -tō′rē-) *adj.* 1 Of or pertaining to a sector. 2 *Zool.* Adapted for cutting; carnassial.

sec·u·lar (sek′yə·lər) *adj.* 1 Of or pertaining to this world or the present life; temporal; worldly: contrasted with *religious* or *spiritual.* 2 Not under the control of the church; civil; not ecclesiastical. 3 Not concerned with religion; not sacred: *secular* art. 4 Not bound by monastic vows: opposed to *regular*: the *secular* clergy. 5 Occurring or observed but once in an age or century. 6 Lasting for ages. See synonyms under PROFANE. —*n.* 1 One in holy orders who is not bound by monastic vows. 2 A layman. [<OF *seculer* <LL *saecularis* <L, belonging to an age < *saeculum* a generation, an age]

sec·u·lar·ism (sek′yə·lə·riz′əm) *n.* Regard for worldly as opposed to spiritual matters; specifically, the belief of secularists.

sec·u·lar·ist (sek′yə·lə·rist) *n.* 1 A person who bases morality on the well-being of mankind in this world without any consideration of religious systems and forms of worship. 2 One who believes that religion should not be introduced into public education or the management of public affairs. —**sec′u·lar·is′tic** *adj.*

sec·u·lar·i·ty (sek′yə·lar′ə·tē) *n.* 1 Secularism; worldliness. 2 Any practice or interest belonging exclusively to the present life.

sec·u·lar·ize (sek′yə·lə·rīz′) *v.t.* **·ized, ·iz·ing** 1 To make secular; convert from sacred to secular uses. 2 To make worldly. 3 To change from a monastic or regular to a secular, as a monk. —**sec′u·lar·i·za′tion** *n.*

se·cund (sē′kund, sek′und) *adj. Bot.* Having the parts or organs arranged on one side only, as certain flowers; unilateral. [<L *secundus* following. See SECOND².]

Se·cun·der·a·bad (si·kun′dər·ä·bäd′) A northern suburb of Hyderabad, Andhra Pradesh, India, where conduction of malaria by mosquitoes was discovered by Sir Ronald Ross, 1898.

sec·un·dine (sek′ən·dīn, -din) *n.* 1 *Bot.* The inner, first-developed coat or integument of an ovule. 2 That which remains in the womb to be expelled after childbirth: usually in the plural. Also spelled *secondine.* [<LL *secundinae,* pl., the afterbirth <L *secundus* following. See SECOND².]

se·cun·dum na·tu·ram (si·kun′dəm nə·tyōōr′əm) *Latin* According to nature.

se·cun·dum u·sum (si·kun′dəm yōō′səm) *Latin* According to usage or ritual.

se·cure (si·kyōōr′) *adj.* 1 Guarded against or not likely to be exposed to danger; safe. 2 Free from fear, apprehension, etc. 3 Confident; careless. 4 Assured; certain; sure: followed by *of,* sometimes by an infinitive. 5 So strong or well made as to render loss, escape, or failure impossible. —*v.* **·cured, ·cur·ing** *v.t.* 1 To make secure; protect. 2 To make firm, tight, or fast; fasten. 3 To make sure or certain; insure; guarantee. 4 To obtain possession of; get. —*v.i.* 5 To be or become secure; take precautions. See synonyms under ARREST, BIND, CATCH, GET, OBTAIN, PRESERVE, PURCHASE, RETAIN. [<L *securus* < *se-* without + *cura* care. Doublet of SURE.] —**se·cur′a·ble** *adj.* —**se·cure′ly** *adv.* —**se·cure′ment** *n.* —**se·cure′ness** *n.* —**se·cur′er** *n.*

Synonyms (adj.): assured, careless, certain, confident, defended, guarded, impregnable, insured, protected, safe, sure, unassailable, undisturbed, unmolested, unsuspecting, untroubled. See FIRM. *Antonyms*: dangerous, dubious, exposed, hazardous, imperiled, insecure, perilous, risky.

Securities and Exchange Commission An agency of the U.S. government which supervises the registration of security issues, prevents fraudulent stock manipulations, and regulates transactions in securities.

se·cur·i·ty (si·kyōōr′ə·tē) *n. pl.* **·ties** 1 The state of being secure; specifically, freedom from danger, risk, care, poverty, or apprehension. 2 One who or that which secures or guarantees; surety. 3 *pl.* Written promises or something deposited or pledged for payment of money, as stocks, bonds, etc. 4 Methods adopted for insuring freedom or secrecy of action, communications, etc., as in wartime; also, the protection afforded by such methods.

Synonyms: bail, collateral, earnest, gage, pledge, surety. The first four words agree in denoting something given or deposited as an assurance of something to be given, paid, or done. An *earnest* is a portion delivered in advance, as when part of the purchase money is paid, "to bind the bargain." A *pledge* or *security* may be wholly different in kind from that to be given or paid; it may greatly exceed it in value, and may be of real or personal property; a *pledge* (as here considered) is always of personal property or chattels. Every pawnshop contains unredeemed *pledges*; land, merchandise, bonds, etc., are frequently offered and accepted as *security. Collateral* is property, as stocks, bonds, etc., actually deposited as *security,* often termed *collateral security.* A person may become *security* or *surety* for another's payment of a debt, appearance in court, etc.; in the latter case, he is said to become *bail* for that person; the person accused gives *bail* for himself. *Gage* survives only as a literary word, chiefly in certain phrases; as, "the *gage* of battle."

security battalions Military forces, as under an enemy-controlled puppet government, to protect rural areas.

Security Council A permanent organ of the United Nations charged with the maintenance of international peace and security and consisting of five permanent members (China, France, the U.S.S.R., the United Kingdom, and the United States) and six elected members, three of whom are replaced each year.

se·dan (si·dan′) *n.* 1 A closed automobile having one compartment for passengers and driver. 2 A closed chair, for one passenger, carried by two or more men by means of poles at the sides: also **sedan chair.** [? <Ital. *sedere* sit <L]

Se·dan (si·dan′, *Fr.* sə·dän′) A city in NE France on the Meuse; scene of the decisive French defeat in the Franco-Prussian War, 1870.

se·date (si·dāt′) *adj.* Characterized by habitual composure; staid. [<L *sedatus,* pp. of *sedare* make calm, settle < *sedere* sit] —**se·date′ly** *adv.* —**se·date′ness** *n.*

Synonyms: calm, contemplative, demure, grave, quiet, serene, serious, sober, solemn, staid, still, thoughtful, tranquil, undisturbed, unruffled. See CALM, SERIOUS, THOUGHTFUL. *Antonyms*: agitated, disturbed, excited, flighty, flurried, frolicsome, gay, lively, mad, merry, wild.

se·da·tion (si·dā′shən) *n. Med.* The act or process of reducing distress, irritation, excitement, etc., particularly by administering sedatives; also, the amount of sedative administered.

sed·a·tive (sed′ə·tiv) *adj.* 1 Having a soothing tendency. 2 *Med.* Allaying irritation; assuaging pain. —*n.* Any means, as a medicine, of allaying irritation or pain.

sed·en·tar·y (sed′ən·ter′ē) *adj.* 1 Sitting much of the time; accustomed to sit much or to work in a sitting posture; hence, settled in one place, as certain tribes; sluggish; inactive. 2 Characterized by sitting. 3 Resulting from much or long sitting. 4 *Zool.* Remaining in one place; attached or fixed to an object; sessile. [<L *sedentarius* < *sedens, -entis,* ppr. of *sedere* sit] —**sed′en·tar′i·ly** *adv.* —**sed′en·tar′i·ness** *n.*

sedge (sej) *n.* 1 A grasslike cyperaceous herb (genus *Carex*) with flowers densely clustered in spikes: widely distributed in marshy places. 2 Any coarse, rushlike or flaglike herb growing in a wet place. [OE *secg*] —**sedged** *adj.* —**sedg′y** *adj.*

Sedge·moor (sej′mōōr) A tract in Somersetshire, England; scene of the victory of James II over the Duke of Monmouth, 1685.

Sedg·wick (sej′wik), **Anne Douglas,** 1873–1935, U.S. novelist.

se·di·le (si·dī′lē) *n. pl.* **·dil·i·a** (-dil′ē·ə) A seat (usually one of three) near the altar in the chancel of a church, for officiating clergy: usually in the plural. Also **se·dil′i·um.** [<L, a seat < *sedere* sit]

sed·i·ment (sed′ə·mənt) *n.* 1 Matter that settles to the bottom of a liquid; settlings; dregs; lees. 2 *Geol.* Fragmentary material deposited by water or air. See synonyms under WASTE. [<MF *sédiment* <L *sedimentum* a settling < *sedere* sit, settle]

sed·i·men·ta·ry (sed′ə·men′tər·ē) *adj.* 1 Pertaining to or having the character of sediment. 2 *Geol.* Designating rocks, as shale and sandstone, composed of fragments of other rocks deposited after transportation from their sources, and including also rocks formed by precipitation, as gypsum, or by calcareous secretions of animals, as certain limestones. Also **sed′i·men′tal.**

sed·i·men·ta·tion (sed′ə·men·tā′shən) *n.* 1 The accumulation or deposition of sediment. 2 The depositing of an insoluble material.

se·di·tion (si·dish′ən) *n.* 1 Language or conduct directed against public order and the tranquillity of the state. 2 The incitement of such disorder, tending toward treason, but lacking an overt act. 3 Dissension; revolt. See synonyms under REVOLUTION. [<OF <L *seditio, -onis* < *sed-* aside + *itio, -onis* a going < *ire* go]

se·di·tion·ar·y (si·dish′ən·er′ē) *adj.* Seditious. —*n. pl.* **·ar·ies** One who promotes sedition: also **se·di′tion·ist.**

se·di·tious (si·dish′əs) *adj.* 1 Pertaining to, promotive of, or having the character of sedition. 2 Inclined to, taking part in, or guilty of sedition. See synonyms under REBELLIOUS, TURBULENT. [OF *seditieux* <L *seditiosus* < *seditio, -onis* SEDITION] —**se·di′tious·ly** *adv.* —**se·di′tious·ness** *n.*

Se·dl·ča·ny (sed′l·chä′nē) A village in southern Bohemia, Czechoslovakia. *German* **Sed·litz** (zed′lits). Also *Seidlitz.*

se·duce (si·dōōs′, -dyōōs′) *v.t.* **·duced, ·duc·ing** 1 To lead astray; entice into wrong, disloyalty, etc.; tempt. 2 To induce, as a woman, to surrender chastity; debauch. See synonyms under ALLURE. [<L *seducere* lead apart < *se-* apart + *ducere* lead] —**se·duc′er** *n.* —**se·duc′i·ble** or **se·duce′a·ble** *adj.*

se·duc·tion (si·duk′shən) *n.* 1 The act of seducing. 2 Something which seduces; an enticement. Also **se·duce′ment.** [<MF *séduction* <L *seductio, -onis* < *seductus,* pp. of *seducere.* See SEDUCE.]

se·duc·tive (si·duk′tiv) *adj.* Tending to seduce; enticing. —**se·duc′tive·ly** *adv.* —**se·duc′tive·ness** *n.*

se·duc·tress (si·duk′tris) *n.* A female seducer.

se·du·li·ty (si·dōō′lə·tē, -dyōō′-) *n.* The state or character of being sedulous.

sed·u·lous (sej′ōō·ləs) *adj.* Constant in application or attention; persevering in effort; assiduous. See synonyms under INDUSTRIOUS. [<L *sedulus* careful, appar. < *sedulo* sincerely < *se dolo* without guile] —**sed′u·lous·ly** *adv.* —**sed′u·lous·ness** *n.*

se·dum (sē′dəm) *n.* Any of a large genus (*Sedum*) of chiefly perennial smooth plants, the stonecrops, having very thick leaves and cymose flowers. [<L, house leek]

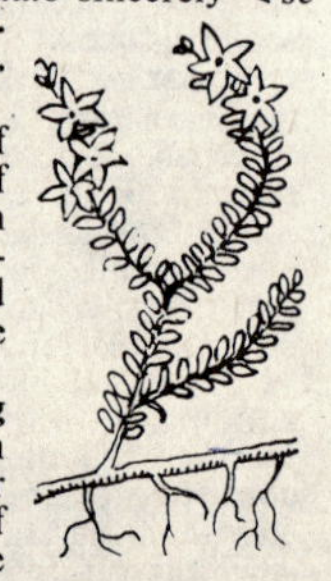
SEDUM

see¹ (sē) *v.* **saw, seen, see·ing** *v.t.* 1 To perceive with the eyes; gain knowledge or awareness of by means of one's vision. 2 To perceive with the mind; understand; comprehend. 3 To find out or ascertain; inquire about: *See* who is at the door. 4 To have experience or knowledge of; undergo: We have *seen* more peaceful times. 5 To encounter; chance to meet: I *saw* your husband today. 6 To have a meeting or interview with; visit or receive as a guest, visitor, etc.: The doctor will *see* you now. **7** To attend as a spectator; view. **8** To accompany; escort. **9** To take care; be sure: with a clause as object: *See* that you do it! **10** In poker, to accept (a bet) or equal the bet of (a player) by betting an equal sum. — *v.i.* **11** To have or exercise the power of sight. **12** To find out; inquire: I will go and *see.* **13** To understand; comprehend. **14** To think; consider. **15** To take care; be attentive: *See* to your work. **16** To gain certain knowledge, as by awaiting an outcome: We will *see* if you are right or wrong. — **to see about** **1** To inquire into the facts, causes, etc., of. **2** To take care of; attend to. — **to see through** **1** To penetrate, as a disguise or deception. **2** To aid or protect, as throughout a period of difficulty or danger. See synonyms under LOOK. ◆ Homophone: *sea.* [OE *sēon*]

see[2] (sē) *n.* **1** The local seat from which a bishop, an archbishop, or the pope exercises jurisdiction; espiscopal or papal jurisdiction, authority, or rank; a bishop's or pope's office. **2** *Obs.* A seat, especially of dignity or power. — **Holy See** The pope's jurisdiction, court, or office; erected as an independent state, Feb. 11, 1929: also **See of Rome.** ◆ Homophone: *sea.* [<OF *se, sie, sed* <L *sedes* a seat]

See (sē), **Thomas Jefferson,** born 1866, U.S. astronomer and mathematician.

see·catch (sē'kach') *n. pl.* **·catch·ie** An adult male fur seal. [<Russian *sekach,* prob. < Aleut]

seed (sēd) *n.* **1** The ovule from which a plant may be reproduced; the fertilized ovule containing an embryo. **2** That from which anything springs; source. **3** Offspring; children. **4** The male fertilizing element; semen; milt. **5** Any small seedlike fruit; also, any part of a plant from which it may be propagated, as bulbs, tubers, etc. **6** A young oyster fit for transplanting. **7** Race; generation; birth. **8** The seed-bearing stage; hence, overripeness. **9** *U.S. Dial.* An animal or animals used for breeding. — *v.t.* **1** To sow with seed. **2** To sow (seed). **3** To remove the seeds from: to *seed* raisins. **4** In sports: **a** To arrange (the drawing for positions in a tournament, etc.) so that the more skilled competitors meet only in the later events. **b** To rank (a skilled competitor) thus. — *v.i.* **5** To sow seed. **6** To grow to maturity and produce or shed seed. — **to go to seed** **1** To develop and shed seed. **2** To become shabby, useless, etc.; deteriorate. ◆ Homophone: *cede.* [OE *sǣd*] — **seed'less** *adj.*

seed bud *Bot.* The germ within a seed; also, the ovule.

seed cake **1** A sweet cake containing aromatic seeds, as caraway. **2** Cottonseed-oil cake.

seed·case (sēd'kās') *n. Bot.* A seed vessel; pericarp.

seed capsule *Bot.* A testa (def. 1).

seed coat *Bot.* The integument of a seed, usually the outer one or testa.

seed coral Small pieces of coral used in jewelry and ornaments.

seed corn Corn or grain of high quality, especially maize, used or intended for seed.

seed crystal A crystallon.

seed·er (sē'dər) *n.* **1** One who or that which sows seed, as a machine. **2** A device for removing seeds from fruit.

seed leaf *Bot.* A cotyledon.

seed·ling (sēd'ling) *n.* **1** *Bot.* A plant grown from seed, as distinguished from one propagated by grafting. **2** A very small or young tree or plant.

seed oyster A young oyster, especially one transplanted to another bed: also *oyster seed.*

seed pearl A small pearl, especially one used for ornamenting bags, etc., or in embroidery.

seed plant A plant which bears seeds; spermatophyte.

seeds·man (sēdz'mən) *n. pl.* **·men** (-mən) **1** A dealer in seeds. **2** A sower. Also **seed'man.**

seed·time (sēd'tīm') *n.* The proper time for sowing seed.

seed vessel *Bot.* The part of a plant that contains the seeds; pericarp.

seed·y (sē'dē) *adj.* **seed·i·er, seed·i·est** **1** Abounding with seeds; going to seed. **2** Poor and ragged; shabby. **3** Feeling or looking wretched. — **seed'i·ly** *adv.* — **seed'i·ness** *n.*

See·ger (sē'gər), **Alan,** 1888–1916, U.S. poet.

see·ing (sē'ing) *n.* The act of seeing; vision; sight. — *conj.* Taking into consideration; since; in view of the fact.

Seeing Eye A philanthropic organization located near Morristown, New Jersey, that trains and supplies dogs (**Seeing Eye dogs**) as guides and companions to the blind.

seek (sēk) *v.* **sought, seek·ing** *v.t.* **1** To go in search of; look for. **2** To strive for; try to get or obtain: to *seek* glory. **3** To endeavor or try: with an infinitive as object: He *seeks* to mislead me. **4** To ask or inquire for; request: to *seek* information. **5** To go to; betake oneself to: to *seek* a warmer climate. **6** *Obs.* or *Dial.* To search or explore. — *v.i.* **7** To make a search or inquiry. [OE *sēcan*] — **seek'er** *n.*

Seekt (zākt), **Hans von,** 1866–1936, German general in World War I.

See·land (zā'länt) The German name for ZEALAND.

see·ly (sē'lē) *adj. Obs.* Weak; wretched; feeble. [OE *gesǣlig* punctual, happy, innocent < *sǣl* time, due time, happiness]

seem (sēm) *v.i.* **1** To give the impression of being; appear. **2** To appear to oneself: a form of reflexive use: I *seem* to hear strange voices. Compare MESEEMS. **3** To appear to exist: There *seems* no reason for hesitating. **4** To be evident or apparent: It *seems* to be raining. [<ON *sǣma* honor, conform to] — **seem'er** *n.*

seem·ing (sē'ming) *adj.* Having the appearance of reality; apparent: often implying non-reality. See synonyms under APPARENT. — *n.* Appearance; semblance; especially, false show. See synonyms under PRETENSE. — **seem'ing·ly** *adv.* — **seem'ing·ness** *n.*

seem·ly (sēm'lē) *adj.* **·li·er, ·li·est** Befitting the proprieties; becoming; proper; decorous; suited to the occasion. See synonyms under BECOMING. — *adv.* Becomingly; decently; appropriately. [<ON *sǣmiligr* honorable, becoming < *sǣmr* fitting] — **seem'li·ness** *n.*

seen (sēn) Past participle of SEE.

seep (sēp) *v.i.* To soak through pores or small interstices; percolate; ooze. — *n.* A small spring; a place out of which water, oil, or other liquid oozes. [OE *sipian* soak]

seep·age (sē'pij) *n.* **1** The oozing or percolation of fluid. **2** The fluid or moisture that oozes.

seer[1] (sē'ər *for def. 1;* sir *for defs. 2 and 3*) *n.* **1** One who sees. **2** One who foretells events; a prophet. **3** One believed to have second sight. [<SEE[1] + -ER] — **seer'ess** *n. fem.*

seer[2] (sir) *n.* **1** A weight used in different parts of India, and having varying local values: also spelled *ser.* **2** A measure of capacity: used chiefly in Bombay and Ceylon. [<Hind. *ser*]

seer·suck·er (sir'suk'ər) *n.* **1** A thin linen or linen and silk fabric, usually striped in colors, with crinkled surface. **2** A similar lightweight cotton or rayon crinkled fabric made by having some of the warp threads slack and others tight. [<Hind. *shirshaker* <Persian *shīr o shakkar,* lit., milk and sugar]

see-saw (sē'sô') *n.* **1** A sport in which persons sit or stand on opposite ends of a balanced plank and make it move up and down. **2** A plank or board balanced for this sport. **3** Any up-and-down or to-and-fro movement. **4** A crossruff. — *v.t.* & *v.i.* To move or cause to move on or as if on a see-saw. — *adj.* Moving to and fro; vacillating. [Reduplication of SAW[1] <*See saw sack a downe,* a sawyer's jingle]

seethe (sēth) *v.* **seethed** (*Obs.* **sod**), **seethed** (*Obs.* **sod·den, sod**), **seeth·ing** *v.i.* **1** To boil. **2** To foam or bubble as if boiling. **3** To be agitated or excited, as by rage. — *v.t.* **4** To soak in liquid; steep. **5** *Archaic* To boil. — *n.* The act of seething; turmoil. [OE *sēothan*]

se·gar (si·gär') See CIGAR.

Se·ges·ta (si·jes'tə) An ancient city of NW Sicily.

seg·gar (seg'ər) See SAGGAR.

seg·ment (seg'mənt) *n.* **1** A part cut off or divided from the other parts of anything; a section. **2** *Geom.* **a** A part of a figure cut off by a line or plane; especially, the part of a circle included within a chord and its arc. **b** A finite part of a divided line. **3** *Zool.* One of the serial divisions of an animal; somite; metamere; also, the portion of a limb between two joints. See synonyms under PART. — *v.t.* & *v.i.* To divide into segments. [<L *segmentum* < *secare* cut] — **seg·men·tal** (seg·men'təl) *adj.* — **seg·men'tal·ly** *adv.* — **seg·men·tar·y** (seg'mən·ter'ē) *adj.*

seg·men·ta·tion (seg'mən·tā'shən) *n.* **1** The act of cutting or dividing into segments. **2** The state of being so divided. **3** The cleavage of a cell into parts.

segmentation cavity *Biol.* The cavity formed by segmentation of a fertilized ovum; blastocele.

se·gno (sā'nyō) *n. pl.* **·gni** (-nyē) *Music* A sign; specifically, the musical sign *:S:* or 𝄋, indicating the beginning or end of a repeat. [< Ital. <L *signum*]

se·go (sē'gō) *n. pl.* **·gos** **1** A perennial herb (*Calochortus nuttalli*) of the lily family, having white flowers lined with purple: it is the State flower of Utah. **2** Its edible bulb. Also **sego lily.** [<Shoshonean (Ute) *sīgo*]

Se·go·via (sā·gō'vyä) A city of Old Castile, central Spain; remarkable for its architecture, and a Roman aqueduct still supplying the city with water.

Se·go·via (sā·gō'vyä), **Andrés,** born 1894, Spanish classical guitarist.

seg·re·gate (seg'rə·gāt) *v.* **·gat·ed, ·gat·ing** *v.t.* **1** To place apart from others or the rest; isolate. — *v.i.* **2** To separate from a mass and gather about nuclei or along lines of fracture, as in crystallization or solidification. **3** To undergo segregation. — *adj.* **1** Separated or set apart from others; select. **2** Simple; solitary; not compound. [<L *segregatus,* pp. of *segregare* separate < *se-* apart + *grex, gregis* a flock] — **seg're·ga'tive** *adj.* — **seg're·ga'tor** *n.*

seg·re·ga·tion (seg'rə·gā'shən) *n.* **1** The act or process of segregating. **2** *Biol.* The separation and distribution of inherited characters in the offspring of crossbred parents. **3** The provision for separate facilities, as in housing, schools, and transportation, for whites and non-whites, especially Negroes.

se·gui·dil·la (sā'gē·dē'lyä) *n. Spanish* **1** A lively Spanish dance, in triple time, for two dancers. **2** The music of such a dance, or its movement, based on a stanza of four to seven short lines, partly assonant. **3** *pl.* An air to which the dancers sing a group of these stanzas.

sei·cen·to (sā·chen'tō) *n.* The 17th century, in reference to Italian art and literature. [<Ital., short for *mil seicento* one thousand six hundred]

seiche (sāsh) *n.* An occasional oscillation of water above and below the mean level of lakes or landlocked seas, lasting from a few minutes to an hour or more. [< dial. F (Swiss), ? ult. <L *siccus* dry]

Seid·litz (zīd'lits) A German name for SEDLČANY: also *Sedlitz.*

Seid·litz powder (sed'lits) An aperient powder consisting of two separate parts: tartaric acid and sodium bicarbonate mixed with Rochelle salt: a mild cathartic used by dissolving separately, mixing the solutions, and drinking while effervescing: also called *Rochelle* powder. [from *Seidlitz;* so called because of its aperient property, similar to that of the water from the spring there]

seign·ior (sēn'yər) *n.* **1** A lord; in southern Europe, equivalent to English *sir.* **2** A lord or feudal lord. Also **sei·gneur** (sēn·yûr'). [< AF *segnour,* OF *seignor* <L *senior* older] — **sei·gnio·ri·al** (sēn·yôr'ē·əl, -yō'rē-) *adj.*

seign·ior·age (sēn'yər·ij) *n.* **1** Something charged or claimed as a prerogative. **2** A charge made by a government for coining bullion; also, the difference between the cost of bullion and the face value of coin made from it. **3** A royalty. Compare BRASSAGE.

seign·ior·y (sēn'yər·ē) *n.* **1** The territory or jurisdiction of a seignior; a manor; lordship. **2** Right or priority belonging to feudal superiority.

Seim (sām) A river in SW European U.S.S.R., rising in SW Russian S.F.S.R. and flowing 435 miles west to the Desna in central northern Ukrainian S.S.R., above Chernigov: also *Seym.*

seine (sān) *n.* Any long fishnet, having floats at the top edge and weights at the bottom, and

add,āce,câre,pälm; end, ēven; it,īce; odd,ōpen,ôrder; to͝ok,po͞ol; up,bûrn; ə = a in *above,* e in *sicken,* i in *clarity,* o in *melon,* u in *focus;* yo͞o = u in *fuse;* oi,oil; ou,pout; ch,check; g,go; ng,ring; th,thin; th,this; zh,vision. Foreign sounds à,œ,ü,kh,ṅ; and ◆: see page xx. < from; + plus; ? possibly.

hauled by its ends to close around a body of fish. — *v.t.* & *v.i.* **seined, sein·ing** To fish or catch with a seine. [OE *segne* <L *sagena* <Gk. *sagēnē* a fishing net]

Seine (sān, *Fr.* sen) A river of NE France, flowing 482 miles NW to the English Channel between Le Havre and Honfleur.

Seine, Bay of the A bay of the Normandy coast, NW France, indented by the estuary of the Seine; 65 miles wide; 25 miles long.

seise (sēz), **sei·sin** (sē′zin) See SEIZE, SEIZIN.

seism (sī′zəm, -səm) *n.* An earthquake. [<Gk. *seismos.* See SEISMIC.]

seis·mic (sīz′mik, sīs′-) *adj.* Pertaining to, characteristic of, or produced by earthquakes. Also **seis′mal, seis′mi·cal.** [<Gk. *seismos* an earthquake < *seiein* shake]

seis·mism (sīz′miz·əm, sīs′-) *n.* The process or phenomena involved in earth movements.

seismo– *combining form* Earthquake: *seismograph.* Also, before vowels, **seism–.** [<Gk. *seismos* an earthquake]

seis·mo·gram (sīz′mə·gram, sīs′-) *n.* The record of an earthquake or earth tremor made by a seismograph.

seis·mo·graph (sīz′mə·graf, -gräf, sīs′-) *n.* An instrument for automatically recording the intensity, direction, and duration of an earthquake shock. —**seis′mo·graph′ic** *adj.* —**seis·mog·ra·pher** (sīz·mog′rə·fər, sīs-) *n.*

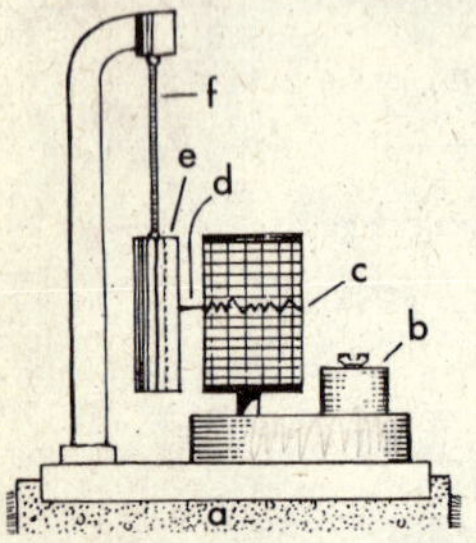

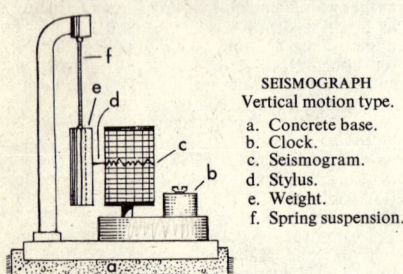

SEISMOGRAPH
Vertical motion type.
a. Concrete base.
b. Clock.
c. Seismogram.
d. Stylus.
e. Weight.
f. Spring suspension.

seis·mog·ra·phy (sīz·mog′rə·fē, sīs-) *n.* The study or description of earthquakes. [<SEISMO- + -GRAPHY]

seis·mol·o·gy (sīz·mol′ə·jē, sīs-) *n.* The science of earthquake phenomena. [<SEISMO- + -LOGY] —**seis·mo·log·ic** (sīz′mə·loj′ik, sīs′-) or **·i·cal** *adj.* —**seis′mo·log′i·cal·ly** *adv.* — **seis·mol′o·gist** *n.*

seis·mom·e·ter (sīz·mom′ə·tər, sīs-) *n.* A seismograph. —**seis·mo·met·ric** (sīz′mō·met′rik, sīs′-) or **·ri·cal** *adj.*

seis·mom·e·try (sīz·mom′ə·trē, sīs-) *n.* The scientific recording of facts regarding earthquake phenomena.

seis·mo·scope (sīz′mə·skōp, sīs′-) *n.* A simple form of seismograph; a device for indicating the time and occurrence of earthquake waves without measuring them. —**seis′mo·scop′ic** (-skop′ik) *adj.*

Seis·tan (sās·tän′) A region and inland lake depression of eastern Iran and SW Afghanistan: also *Sistan.*

seize (sēz) *v.* **seized, seiz·ing** *v.t.* **1** To take hold of suddenly and forcibly; clutch; grasp. **2** To grasp mentally; comprehend; understand. **3** To take possession of by authority or right. **4** To take possession of by or as by force: The usurper *seized* the throne. **5** To take prisoner; capture; arrest. **6** To act upon with sudden and powerful effect; attack; strike: Terror *seized* the attackers and they fled. **7** To take advantage of immediately, as an opportunity. **8** *Law* To put into legal possession: usually spelled *seise.* **9** *Naut.* To fasten or bind by turns of cord, line, or small rope; lash. —*v.i.* **10** To take a sudden or forcible hold. See synonyms under ARREST, CATCH, GRASP. [<OF *saisir, seisir* <Med. L *(ad propriam) sacire* take (into one's own possession), prob. <Gmc.] —**seiz·a·ble** *adj.*

seiz·er (sē′zər) *n.* **1** One who seizes in any sense. **2** *Law* One who takes livery of seizin: also **seiz′or, seis′or.**

sei·zin (sē′zin) *n. Law* **1** The possession of land under a claim of a freehold. **2** That which is possessed; property. **3** The act of taking possession. —**livery of seizin** The delivery of corporeal possession of lands and tenements of freehold. Also spelled *seisin.* [<OF *saisine* < *saisir.* See SEIZE.]

seiz·ing (sē′zing) *n.* **1** The act of grasping or taking forcible possession. **2** The process of fastening or binding together with turns of cord. **3** A small cord used in making such fastenings, and the fastening itself.

sei·zure (sē′zhər) *n.* **1** The act of seizing. **2** A sudden or violent attack, as of epilepsy or neuralgia; fit; spell.

se·jant (sē′jənt) *adj. Her.* Sitting with the fore limbs erect, as a lion. Also **se′jeant.** [<AF *sejant,* OF *seant,* ppr. of AF *seier,* OF *seoir* sit <L *sedere*]

Se·ja·nus (si·jā′nəs), **Lucius Aelius,** died A.D. 31, Roman favorite of Tiberius; executed.

Sejm (sām) *n. Polish* An assembly or diet having legislative power; specifically, the former Constituent Assembly of the Polish Republic.

sel (sel) *n. Scot.* Self. —**a body's sel** Oneself alone.

Se·la·chi·i (si·lā′kē·ī) *n. pl.* An order or subclass of elasmobranch fishes, including the sharks, skates, dogfishes, and rays, with their immediately related fossil allies. [<NL <Gk. *selachos* a shark] —**se·la′chi·an** *adj.* & *n.* — **sel·a·choid** (sel′ə·koid) *adj.* & *n.*

se·la·dang (sə·lä′däng) See SALADANG.

sel·a·gi·nel·la (sel′ə·ji·nel′ə) *n.* One of a widely distributed genus *(Selaginella)* of flowerless branching herbs with scalelike leaves. [<NL, dim. of L *selago, -inis,* a plant like the savin]

se·lah (sē′lə) *n.* A word of unknown meaning in the Psalms and Habakkuk, usually considered as a direction to readers or musicians. [<Hebrew *selāh*]

se·lam·lik (si·läm′lik) *n.* The men's quarters in a Turkish house, where guests are received; formerly, the official visit of the Turkish sultan to a mosque on a Friday. [<Turkish *selāmliq* <Arabic *salām* health, peace]

Se·lan·gor (se·läng′gôr, -gōr) A state in the Federation of Malaya, on the Strait of Malaya; 3,160 square miles; capital, Kuala Lumpur.

Sel·den (sel′dən), **John,** 1584–1654, English jurist and antiquary.

sel·dom (sel′dəm) *adv.* At widely separated intervals, as of time or space; infrequently. [OE *seldum, seldan,* dative pl. of *seld-* rare, strange]

se·lect (si·lekt′) *v.t.* To take in preference to another or others; pick out; choose. —*v.i.* To make a choice; choose. See synonyms under ALLOT, CHOOSE. —*adj.* **1** Chosen in preference to others; taken as being most fit or desirable; choice. **2** Exclusive. **3** Very particular in selecting. See synonyms under CHOICE, EXCELLENT. [<L *selectus,* pp. of *seligere* < *se-* apart + *legere* choose] —**se·lect′ness** *n.* —**se·lec′tor** *n.*

se·lec·tee (si·lek′tē′) *n.* One selected; specifically, a person called up for military service under selective service.

se·lec·tion (si·lek′shən) *n.* **1** The act of selecting; choice. **2** Anything selected; a collection made with care. **3** *Biol.* The process, natural or artificial, by which certain organisms, or any of their characteristics, are favored in the struggle for perpetuation and survival.

se·lec·tive (si·lek′tiv) *adj.* **1** Pertaining to selection; tending to select. **2** Having or characterized by good selectivity.

selective service Compulsory military service according to specified conditions of age, fitness, etc. —**se·lec′tive–ser′vice** *adj.*

selective transmission *Mech.* A transmission for motor vehicles effected by a single lever which directly changes the gear from one speed to another.

se·lec·tiv·i·ty (si·lek′tiv′ə·tē) *n.* **1** The state or condition of being selective. **2** *Telecom.* That characteristic of a radio receiver by which certain frequencies can be received to the exclusion of others.

se·lect·man (si·lekt′mən) *n. pl.* **·men** (-mən) One of a board of town officers, elected annually in New England, except in Rhode Island, to exercise executive authority in local affairs.

selector box A watertight metal box which contains the mechanisms controlling a set of submarine mines and operated electrically from a shore station.

sel·e·nate (sel′ə·nāt) *n. Chem.* A salt of selenic acid. [<SELEN(IC) + -ATE[3]]

Se·le·ne (si·lē′nē) In Greek mythology, goddess of the moon: identified with the Roman *Luna.* Also **Se·le′na** (-nə). [<Gk. *Selēnē,* lit., the moon]

Se·len·ga (se′leng·gä′) A river in the Mongolian People's Republic, flowing 897 miles NE to Lake Baikal.

se·len·ic (si·len′ik, -lē′nik) *adj. Chem.* Of, pertaining to, or derived from selenium, especially in its higher valence. [<SELEN(IUM) + -IC]

selenic acid *Chem.* A transparent, colorless liquid, H_2SeO_4, obtained variously, as by decomposing a selenate with hydrogen sulfide.

se·le·ni·ous (si·lē′nē·əs) *adj. Chem.* Of, pertaining to, or derived from selenium, especially in its lower valence, as the colorless, crystalline **selenious acid,** H_2SeO_3.

sel·e·nite[1] (sel′ə·nīt) *n.* A pearly, usually transparent variety of gypsum. [<L *selenites* <Gk. *selēnitēs (lithos),* lit., moonstone < *selēnē* the moon; so called because it was thought to wax and wane with the moon]

sel·e·nite[2] (sel′ə·nīt) *n.* A salt of selenious acid. [<SELEN(IUM) + -ITE[2]]

sel·e·nite[3] (sel′ə·nīt) *n. Often cap.* An imaginary inhabitant of the moon. [<Gk. *selēnitēs* < *selēnē* the moon]

se·le·ni·um (si·lē′nē·əm) *n.* A gray, crystalline, non-metallic element of the sulfur group (symbol Se) varying greatly in electrical resistance under the influence of light. See ELEMENT. [<NL <Gk. *selēnē* the moon]

selenium cell A photoelectric cell in which plates of metallic selenium respond in accordance with the action of light upon them.

seleno– *combining form* Moon; pertaining to the moon; lunar: *selenography.* Also, before vowels, **selen–.** [<Gk. *selēnē* the moon]

sel·e·nog·ra·phy (sel′ə·nog′rə·fē) *n.* The science or study of the moon's surface. [<SELENO- + -GRAPHY] —**sel′e·nog′ra·pher** or **·phist** *n.* —**sel′e·no·graph′ic** (-nō·graf′ik) or **·i·cal** *adj.*

sel·e·nol·o·gy (sel′ə·nol′ə·jē) *n.* The science that treats of the movements and astronomical relations of the moon. [<SELENO- + -LOGY] —**se·le·no·log·i·cal** (si·lē′nō·loj′i·kəl) *adj.* —**sel′e·nol′o·gist** *n.*

Se·leu·cia (si·lo͞o′shə) **1** An ancient city on the NE Mediterranean, in extreme southern Turkey near Syria; formerly the port of Antioch. Also **Seleucia Pi·e·ri·a** (pī·ir′ē·ə). **2** An ancient city of Mesopotamia, on the Tigris, the site of which is 20 miles SE of Baghdad, Iraq. **3** An ancient city of Cilicia, SW of Tarsus, the site of modern Silifke, Turkey. Also **Seleucia Tra·che·o·tis** (trā′kē·ō′tis).

Se·leu·cid (si·lo͞o′sid) *adj.* Pertaining to the Seleucids: also **Se·leu′ci·dan, Se·leu·cid·i·an** (sē′lo͞o·sid′ē·ən). —*n.* One of the Seleucids. [<L *Seleucides* <Gk. *Seleukidēs,* a descendant of Seleucus <*Seleukos* Seleucus]

Se·leu·cids (si·lo͞o′sids) *n. pl.* The members of the dynasty that ruled Syria from 312 B.C. till the Roman conquest, 64 B.C.: named from Seleucus. Also **Se·leu′ci·dae** (-dē).

Se·leu·cus (si·lo͞o′kəs) Name of six kings of the Seleucid dynasty, especially **Seleucus Ni·ca·tor** (nī·kā′tər), 358?–280 B.C., Macedonian general under Alexander the Great, and founder of the dynasty.

self (self) *adj.* **1** Same; identical: obsolete except in the compound *selfsame.* **2** Pure; unmixed: applied especially to colors. —*n. pl.* **selves 1** An individual known or considered as the subject of his own consciousness; anything considered as having a distinct personality. **2** Personal interest or advantage. **3** Any thing, class, or attribute that, abstractly considered, maintains a distinct and characteristic individuality or identity. [OE]

Self may appear as a combining form with various meanings in solidemes and hyphemes, as shown in the list beginning at the foot of this page.

1 Of the self (the object of the root word); as in:

self–abandonment	**self–abhorrence**	**self–administer**	**self–adornment**	**self–advertise**	**self–aggrandizement**	**self–applause**
self–abasement	**self–accusation**	**self–admiration**	**self–adulation**	**self–advertisement**	**self–analysis**	**self–appreciation**
self–abasing	**self–adaptive**	**self–admission**	**self–advancement**	**self–affliction**	**self–annihilation**	**self–approbation**

self-ab·ne·ga·tion (self′ab′ni·gā′shən) *n.* The complete putting aside of self and claims of self for the sake of some person or object; self-sacrifice.
Synonyms: self-control, self-denial, self-devotion, self-renunciation, self-sacrifice. *Self-control* is holding oneself within due limits in pleasures and duties, as in all things else; *self-denial*, the giving up of pleasures for the sake of duty. *Self-renunciation* surrenders conscious rights; *self-abnegation* forgets that there is anything to surrender. A mother will care for a sick child with complete *self-abnegation*, but without a thought of *self-denial*. *Self-devotion* is whole-hearted consecration of self to a person or cause with readiness for any needed sacrifice. *Self-sacrifice* is the strongest term of all, and contemplates the gift of self as actually made. *Antonyms:* self-gratification, self-indulgence, self-will.
self-a·buse (self′ə·byōōs′) *n.* 1 The disparagement of one's own person or powers. 2 Masturbation.
self-ad·dressed (self′ə·drest′) *adj.* Addressed to and by oneself.
self-as·sured (self′ə·shoord′) *adj.* Confident in one's own abilities; self-reliant. — **self′-as·sur′ance** *n.*
self-col·ored (self′kul′ərd) *adj.* 1 Having the natural color. 2 Of but one color or tint. Also *Brit.* **self′-col′oured.**
self-com·mand (self′kə·mand′, -mänd′) *n.* The state of having all the faculties and powers fully and effectively at command: more positive and less repressive than *self-control.*
self-com·posed (self′kəm·pōzd′) *adj.* Calm; controlling one's emotions.
self-con·ceit (self′kən·sēt′) *n.* An unduly high opinion of oneself or of one's own abilities, acquirements, etc.; self-esteem; vanity; egotism. See synonyms under EGOTISM, PRIDE. — **self′-con·ceit′ed** *adj.*
self-con·fi·dence (self′kon′fə·dəns) *n.* Confidence in oneself or in one's own unaided powers, judgment, etc. See synonyms under ASSURANCE, EGOTISM. — **self′-con′fi·dent** *adj.* — **self′-con′fi·dent·ly** *adv.*
self-con·scious (self′kon′shəs) *adj.* 1 Unduly conscious that one is observed by others, or manifesting such consciousness; embarrassed by inability to forget oneself; ill at ease. 2 Conscious of one's existence. — **self′-con′scious·ly** *adv.* — **self′-con′scious·ness** *n.*
self-con·tained (self′kən·tānd′) *adj.* 1 Keeping one's thoughts and feelings to oneself; uncommunicative; impassive. 2 Exercising self-control. 3 Complete and independent; bearing its own motor, as a machine; mounted on its own boiler, as a steam engine.
self-con·tra·dic·tion (self′kon′trə·dik′shən) *n.* 1 The contradicting of oneself or itself. 2 That which contradicts itself. — **self′-con′tra·dic′to·ry** *adj.*
self-con·trol (self′kən·trōl′) *n.* The act, power, or habit of having one's faculties or energies under control of the will. Compare SELF-COMMAND.
self-de·fense (self′di·fens′) *n.* Defense of oneself, one's property, or one's reputation. Also **self′-de·fence′.** — **self′-de·fen′sive** *adj.*
self-de·ni·al (self′di·nī′əl) *n.* The act or power of denying oneself gratification; passive self-sacrifice. See synonyms under ABSTINENCE, SELF-ABNEGATION. — **self′-de·ny′ing** *adj.* — **self′-de·ny′ing·ly** *adv.*
self-de·ter·mi·na·tion (self′di·tûr′mə·nā′shən) *n.* 1 The principle of free will; decision by oneself without extraneous force or influence. 2 Decision by the people of a country or section as to its future political status. — **self′-de·ter′min·ing** *adj.* & *n.*
self-de·vo·tion (self′di·vō′shən) *n.* The devoting of oneself, with one's claims, wishes, or interests, to the service of a person or a cause. See synonyms under SELF-ABNEGATION. — **self′-de·vo′tion·al** *adj.*
self-driv·en (self′driv′ən) *adj.* Driven by itself; automotive.
self-ed·u·cat·ed (self′ej′ōō·kā′tid) *adj.* 1 Educated through one's own efforts without the aid of instructors. 2 Educated at one's own expense. — **self′-ed′u·ca′tion** *n.*
self-es·teem (self′es·tēm′) *n.* A good opinion of oneself; an overestimate of oneself. See synonyms under EGOTISM, PRIDE.
self-ev·i·dent (self′ev′ə·dənt) *adj.* Carrying its evidence or proof in itself; requiring no proof of its truth. — **self′-ev′i·dence** *n.* — **self′-ev′i·dent·ly** *adv.*
self-ex·e·cut·ing (self′ek′sə·kyōō′ting) *adj.* Containing provisions for securing its own execution independent of legislation: said of a law, etc.
self-ex·ist·ence (self′ig·zis′təns) *n.* Inherent, underived, independent existence: an attribute of God. — **self′-ex·ist′ent** *adj.*
self-ex·pres·sion (self′ik·spresh′ən) *n.* Expression of one's own temperament or emotions, as in art.
self-feed·er (self′fē′dər) *n.* A machine, boiler, or other mechanical device that feeds itself automatically. — **self′-feed′ing** *adj.*
self-fer·til·i·za·tion (self′fûr′təl·ə·zā′shən, -ī·zā′shən) *n. Biol.* Fertilization of an ovum by semen from the same animal or of a plant ovule by its own pollen.
self-gov·ern·ment (self′guv′ərn·mənt, -ər·mənt) *n.* 1 Self-control. 2 Government of a country or region by its own people; especially, government of a colony by the inhabitants rather than by the mother country. — **self′-gov′ern·ing, self′-gov′erned** *adj.*
self-hard·en·ing (self′här′də·ning) *adj. Metall.* Pertaining to or designating certain steels which will harden properly without the need for quenching.
self·heal (self′hēl′) *n.* 1 A weedy, perennial herb (genus *Prunella*) with violet or purple flowers, formerly reputed to cure disease, especially the common selfheal of North America (*P. vulgaris*). 2 One of various similar plants, as the sanicle.
self·hood (self′hood) *n.* 1 The state of being an individual, or that which constitutes such a state; personality. 2 Selfishness.
self-i·den·ti·ty (self′ī·den′tə·tē) *n.* 1 The identity of a thing with itself. 2 *Psychol.* That state of consciousness by or through which the self recognizes itself as one and the same.
self-im·por·tance (self′im·pôr′təns) *n.* Pompous self-conceit. — **self′-im·por′tant** *adj.*
self-in·duced (self′in·dōōst′, -dyōōst′) *adj. Electr.* Characterizing an electromotive force induced in a circuit because of variations of the current in that circuit.
self-in·duc·tion (self′in·duk′shən) *n. Electr.* The production of an induced or extra current in a circuit by the variation of the current in that circuit, especially when it is started or stopped. — **self′-in·duc′tive** *adj.*
self-in·sur·ance (self′in·shoor′əns) *n.* That proportion of the insurance risk which the insured assumes himself by the premium payments he makes.
self-in·ter·est (self′in′tər·ist, -in′trist) *n.* Personal interest or advantage, or the pursuit of it; selfishness. — **self′-in′ter·est·ed** *adj.*
self·ish (sel′fish) *adj.* 1 Caring chiefly for self or for one's own interests or comfort; influenced by personal motives to the disregard of the welfare or wishes of others. 2 Proceeding from or characterized by undue love of self. See synonyms under GREEDY. — **self′ish·ly** *adv.*
self·ish·ness (sel′fish·nis) *n.* The quality of being selfish; undue regard for one's own interest, regardless of others.
Synonym: self-love. *Self-love* is a due care for one's own happiness and well-being, which is perfectly compatible with justice, generosity, or benevolence toward others; *selfishness* is an undue or exclusive care for one's own

self-approval	self-correction	self-disposal	self-humbling	self-lashing	self-persuasion	self-scrutinizing
self-asserting	self-corruption	self-disquieting	self-humiliation	self-laudatory	self-pitiful	self-scrutiny
self-assertion	self-creation	self-dissolution	self-hypnosis	self-limitation	self-pity	self-searching
self-assertive	self-criticism	self-distrust	self-hypnotism	self-limited	self-pitying	self-serve
self-awareness	self-cure	self-doubt	self-hypnotize	self-limiting	self-pleasing	self-slaughter
self-bedizenment	self-damnation	self-easing	self-idolatry	self-loss	self-praise	self-soothing
self-betrayal	self-debasement	self-enriching	self-idolizing	self-loving	self-praising	self-study
self-blame	self-deceit	self-estimate	self-ignorance	self-maceration	self-preparation	self-subjection
self-castigation	self-deceiving	self-evacuation	self-ignorant	self-maintenance	self-presentation	self-subordination
self-chastisement	self-dedication	self-exalting	self-imitation	self-martyrdom	self-preserving	self-support
self-cognizance	self-defeating	self-examination	self-immolation	self-mastery	self-projection	self-supporting
self-commendation	self-deflation	self-exculpation	self-immurement	self-mistrust	self-protecting	self-suppression
self-committal	self-degradation	self-excuse	self-impairment	self-mortification	self-protection	self-surrender
self-comparison	self-deifying	self-expansion	self-improvement	self-murder	self-punishment	self-suspicious
self-comprehending	self-dejection	self-expatriation	self-indignation	self-murderer	self-raising	self-taxation
self-condemnation	self-delation	self-exploiting	self-indulgence	self-mutilation	self-realization	self-teacher
self-condemning	self-delusion	self-exposure	self-indulgent	self-neglect	self-recollection	self-terminating
self-conditioning	self-depreciation	self-extermination	self-indulgently	self-neglectful	self-reconstruction	self-tolerant
self-confinement	self-depreciative	self-fearing	self-indulging	self-nourishment	self-reduction	self-torment
self-confounding	self-destroying	self-flatterer	self-inspection	self-objectification	self-regulation	self-torture
self-congratulatory	self-destruction	self-flattering	self-instruction	self-observation	self-representation	self-treatment
self-conquest	self-destructive	self-flattery	self-insurer	self-offense	self-repressing	self-trust
self-conservative	self-direction	self-folding	self-integration	self-opinion	self-repression	self-trusting
self-conserving	self-disapproval	self-formation	self-intensifying	self-painter	self-reproach	self-undoing
self-consideration	self-discipline	self-glorification	self-interrogation	self-paying	self-reproachful	self-upbraiding
self-consoling	self-disclosure	self-gratification	self-introduction	self-perceiving	self-restriction	self-usurp
self-consuming	self-discovery	self-guidance	self-judgment	self-perceptive	self-revealing	self-valuing
self-contempt	self-disgrace	self-harming	self-justification	self-perfecting	self-revelation	self-vaunting
self-contradicting	self-disparagement	self-help	self-justifying	self-perfection	self-ruin	self-vindication
self-conviction	self-display	self-helpful	self-knowledge	self-perpetuation	self-satirist	self-worship

2 By oneself or itself; by one's own effort (the agent of the root word); as in:

self-abandoned	self-balanced	self-caused	self-conducted	self-corrupted	self-deprived	self-divided
self-appointed	self-beguiled	self-chosen	self-confuted	self-declared	self-destroyed	self-doomed
self-approved	self-betrayed	self-commissioned	self-constituted	self-defended	self-determined	self-elaborated
self-authorized	self-blinded	self-condemned	self-convicted	self-deluded	self-devised	self-elected

comfort or pleasure, regardless of the happiness, and often of the rights, of others. *Self-love* is necessary to high endeavor, and even to self-preservation; *selfishness* limits endeavor to a narrow circle of intensely personal aims. *Antonyms:* See synonyms under BENEVOLENCE.

self·less (self'lis) *adj.* Regardless of self; unselfish.

self·liq·ui·dat·ing (self'lik'wə·dā'ting) *adj.* Designating a business transaction in which goods in great demand are converted into cash over a short period.

self·load·ing (self'lō'ding) *adj.* Automatically reloading: said of a gun using the energy of recoil to eject and reload.

self·love (self'luv') *n.* Love of oneself; the desire or tendency that leads one to seek to promote his own well-being. See synonym under SELFISHNESS.

self·made (self'mād') *adj.* 1 Having attained honor, wealth, etc., by one's own efforts. 2 Made by oneself.

self·per·cep·tion (self'pər·sep'shən) *n.* Perception of one's own existence or mental states; introspection.

self·pol·li·na·tion (self'pol'ə·nā'shən) *n. Bot.* The transfer of pollen from stamens to pistils of the same flower.

self·pos·ses·sion (self'pə·zesh'ən) *n.* 1 The full possession or control of one's powers or faculties; freedom from perturbation, perplexity, or excitement. 2 Presence of mind; self-command. — **self'-pos·sessed'** *adj.*

self·pres·er·va·tion (self'prez'ər·vā'shən) *n.* 1 The protection of oneself from destruction. 2 The urge to protect oneself regarded as an instinct.

self·prof·it (self'prof'it) *n.* Self-interest.

self·pro·nounc·ing (self'prə·noun'sing) *adj.* Having marks of pronunciation and stress applied to a word without phonetic alteration of the spelling.

self·re·li·ance (self'ri·lī'əns) *n.* Reliance on one's own abilities, resources, or judgment. See synonyms under ASSURANCE. — **self'-re·li'ant** *adj.*

self·re·nun·ci·a·tion (self'ri·nun'sē·ā'shən) *n.* Renunciation of one's own rights, privileges, or claims. — **self'-re·nun'ci·a·to'ry** (-sē·ə·tôr'ē, -tō'rē) *adj.*

self·re·spect (self'ri·spekt') *n.* Such regard for one's own character as will restrain one from unworthy action; rational self-esteem. See synonyms under PRIDE. — **self'-re·spect'ing** *adj.*

self·re·straint (self'ri·strānt') *n.* Restraint, as of the passions, by the force of one's own will; self-control.

self·right·eous (self'rī'chəs) *adj.* Righteous in one's own estimation; pharisaic. — **self'-right'eous·ly** *adv.* — **self'-right'eous·ness** *n.*

self·ris·ing (self'rī'zing) *adj.* 1 That rises of itself. 2 Having the leaven already added by the millers, as some flours.

self·sac·ri·fice (self'sak'rə·fis) *n.* The sacrifice or subordination of one's self or one's personal welfare or wishes, for the sake of duty or for others' good. See synonyms under SELF-ABNEGATION. — **self'-sac'ri·fic'ing** *adj.*

self·same (self'sām') *adj.* Exactly the same; identical. See synonyms under IDENTICAL. — **self'-same'ness** *n.*

self·sat·is·fac·tion (self'sat'is·fak'shən) *n.* Satisfaction with one's own actions and characteristics; conceit; self-complacency. — **self'-sat'is·fied** *adj.* — **self'-sat'is·fy'ing** *adj.*

self·seek·ing (self'sē'king) *adj.* Given to the exclusive pursuit of one's own interests or gain. — *n.* Self-aggrandizement; selfishness. — **self'-seek'er** *n.*

self·ser·vice (self'sûr'vis) *adj.* Designating a particular type of café, restaurant, or store where patrons serve themselves.

self·start·er (self'stär'tər) *n.* 1 An internal-combustion engine, with automatic or semiautomatic starting mechanism; also, such mechanism. 2 *Slang* One who requires no outside stimulus to start or accomplish work.

self·styled (self'stīld') *adj.* Characterized (as such) by oneself: a *self-styled* gentleman.

self·suf·fi·cient (self'sə·fish'ənt) *adj.* 1 Able to support or maintain oneself without aid or cooperation from others. 2 Having overweening confidence in oneself. Also **self'-suf·fic'ing** (-sə·fī'sing). — **self'-suf·fi'cien·cy** *n.*

self·will (self'wil') *n.* Pertinacious adherence to one's own will or wish, especially with disregard of the wishes of others; obstinacy. — **self'-willed'** *adj.*

self·wind·ing (self'wīn'ding) *adj.* Having a magnetic, electrical, or other attachment which automatically winds a clock or other mechanism at certain times.

self·wrong (self'rông', -rong') *n.* Injury done to one's self.

Sel·juk (sel·jōōk') *n.* A member of one of several Turkish dynasties which reigned over a large part of central and western Asia from the 11th to the 13th centuries. — *adj.* Pertaining to a Seljuk. Also **Sel·ju·ki·an** (sel·jōō'kē·ən). [<Turkish *seljūq*, after *Seljūq*, a Turkish chieftain, reputed ancestor of the Seljuk dynasties]

Sel·kirk (sel'kûrk) A county of SE Scotland; 267 square miles; county burgh, Selkirk. Also **Sel'kirk·shire** (-shir).

Sel·kirk (sel'kûrk), **Alexander**, 1676–1721, Scottish sailor who was marooned on Juan Fernandez Island, Pacific Ocean, for four years. His adventures are said to have suggested Defoe's *Robinson Crusoe*.

Selkirk Mountains A range of the Rocky Mountains in SE British Columbia.

sell[1] (sel) *v.* **sold, sell·ing** *v.t.* 1 To transfer (property) to another for a consideration; dispose of by sale. 2 To deal in; offer for sale. 3 To deliver, surrender, or betray for a price or reward: to *sell* one's honor. 4 *Colloq.* To cause to accept or approve something: They *sold* him on the scheme. 5 *Colloq.* To cause the acceptance or approval of. 6 *Slang* To deceive; cheat. — *v.i.* 7 To transfer ownership for a consideration; engage in selling. 8 To be on sale; be sold. See synonyms under CONVEY. — *n.* 1 *Slang* A trick; joke; swindle. 2 On the stock exchange, a stock that ought to be sold. ◆ Homophone: *cell.* [OE *sellan* give]

sell[2] (sel) *n.* 1 An elevated seat; an honorable place; also, any seat. 2 A saddle. ◆ Homophone: *cell.* [<OF *selle* <L *sella* a seat, ult. <*sedere* sit]

sell·er (sel'ər) *n.* 1 One who sells. 2 Something with a measure of salability: This book is a good *seller.*

sell·ing-plat·er (sel'ing·plā'tər) *n.* A horse that runs in a selling race.

selling race *Brit.* A horse race in which the entrants may be claimed for a set price, and the winning horse must be offered at auction. Compare CLAIMING RACE. Also **sell'er, sell'ing·er.**

sell-out (sel'out') *n.* 1 An act of selling out. 2 *Colloq.* A performance for which all seats have been sold. 3 *Slang* A betrayal through a secret bargain or agreement.

Selt·zer (selt'sər) *n.* An effervescing mineral water. Also **Seltzer water, Sel·ters** (sel'tərz). [Alter. of G *Selterser*, from *Nieder Selters*, a village in SW Prussia, its place of origin]

sel·vage (sel'vij) *n.* 1 The edge of a woven fabric so finished that it will not ravel. 2 An edge. 3 The edge plate of a lock having an opening for a bolt. Also **sel'vedge.** [<SELF + EDGE, trans. of MDu. *selfegghe*]

Sel·va·gens (sel·vä'zhēnsh) A group of uninhabited islets in Madeira: also *Salvages*.

selves (selvz) Plural of SELF.

se·man·tic (si·man'tik) *adj.* 1 Of or pertaining to meaning. 2 Of or relating to semantics. [<Gk. *sēmantikos* <*sēmainein* signify]

se·man·ti·cist (si·man'tə·sist) *n.* A specialist in semantics.

se·man·tics (si·man'tiks) *n. pl.* (construed as singular) 1 *Ling.* The study of the meanings of speech forms, especially of the development and changes in meaning of words and word groups. 2 *Logic* The relation between signs or symbols and what they signify or

---

self-employed	self-honored	self-instructed	self-maimed	self-pampered	self-proclaimed	self-schooled	
self-exhibited	self-idolized	self-invited	self-matured	self-performed	self-professed	self-sown	
self-explained	self-illumined	self-irrecoverable	self-misused	self-perpetuated	self-punished	self-subdued	
self-exposed	self-improvable	self-judged	self-mortified	self-perplexed	self-renounced	self-supported	
self-extolled	self-incurred	self-justified	self-named	self-planted	self-repressed	self-sustained	
self-furnished	self-inflicted	self-kindled	self-offered	self-pollinated	self-restrained	self-taught	
self-hidden	self-initiated	self-limited	self-paid	self-posed	self-revealed	self-tempted	

3 To, toward, in, for, on, or with oneself; as in:

self-absorbed	self-centered	self-contented	self-dissatisfied	self-injurious	self-preference	self-repellent
self-absorption	self-comment	self-delight	self-elation	self-injury	self-preoccupation	self-repose
self-aid	self-communing	self-dependence	self-enamored	self-kindness	self-prescribed	self-reproof
self-aim	self-compassion	self-dependent	self-enclosed	self-liking	self-pride	self-repulsive
self-amusement	self-compensation	self-desire	self-exultation	self-loathing	self-procured	self-resentment
self-angry	self-complacence	self-despair	self-focusing	self-oblivious	self-produced	self-resigned
self-application	self-complacency	self-directed	self-gain	self-occupied	self-profit	self-respectful
self-applied	self-complacent	self-direction	self-helpfulness	self-panegyrical	self-purifying	self-responsibility
self-assumed	self-concentration	self-disdain	self-helpless	self-penetration	self-reflection	self-rigorous
self-assuming	self-conflict	self-disgust	self-hope	self-permission	self-regard	self-sent
self-benefit	self-consistency	self-dislike	self-imposture	self-pictured	self-relation	self-tenderness
self-care	self-consistent	self-dissatisfaction	self-infliction	self-pleased	self-relying	self-vexation

4 From oneself or itself; from one's own nature or power; as in:

self-apparent	self-derived	self-explaining	self-initiative	self-moving	self-refuting	self-rewarding
self-arising	self-desirable	self-explanatory	self-intelligible	self-operative	self-renewing	self-sprung
self-born	self-developing	self-forbidden	self-interpretative	self-originating	self-resourceful	self-stability
self-coherence	self-distinguishing	self-fruition	self-issuing	self-perfect	self-resplendent	self-stimulated
self-complete	self-effort	self-healing	self-luminous	self-poise	self-restoring	self-sustaining
self-defining	self-evolving	self-inclusive	self-manifestation	self-poised	self-reward	self-warranting

5 Independent; as in:

		self-agency	self-authority			
self-credit	self-dominance	self-entity	self-existence	self-ownership	self-rule	self-sovereignty

6 *Technol.* Automatic or automatically; as in:

self-acting	self-binder	self-cleaning	self-feed	self-lubricating	self-propelled	self-registering
self-adapting	self-burning	self-closing	self-filling	self-lubrication	self-propelling	self-regulated
self-adjustable	self-changing	self-cocking	self-inking	self-oiling	self-propulsion	self-regulating
self-adjusting	self-charging	self-cooled	self-lighting	self-primer	self-raker	self-righting
self-alining	self-checking	self-emptying	self-locking	self-priming	self-recording	self-setting

denote: also called *semasiology, semiotics.* Compare GENERAL SEMANTICS. **3** Loosely, verbal trickery, especially by adulteration or shift of meaning within a word; amphibology.

sem·a·phore (sem'ə-fôr, -fōr) *n.* An apparatus for making signals, as with movable arms, disks, flags, or lanterns; a signal telegraph. [<F *sémaphore* <Gk. *sēma* a sign + *pherein* carry] — **sem'a·phor'ic** (-fôr'ik, -for'ik) or **·i·cal** *adj.*

Se·ma·rang (sə-mä'räng) A port of northern Java: also *Samarang.*

SEMAPHORE
a. Clear. b. Approach. c. Stop.

se·ma·si·ol·o·gy (sĭ-mā'sē-ol'ə-jē, -zē-) *n.* Semantics (def. 2). [<Gk. *sēmasia* the signification of a word <*sēma* sign + -LOGY] — **se·ma·si·o·log·i·cal** (si-mā'sē-ə-loj'i-kəl, -zē-) *adj.*

se·mat·ic (si-mat'ik) *adj.* Of the nature of a sign; significant; warning; in animal coloration, serving to distinguish as a means of recognition or warning. [<Gk. *sēma, -atos* a sign]

sem·bla·ble (sem'blə-bəl) *adj.* **1** Resembling; similar. **2** Apparent; not real. — *n.* A thing resembling another thing. Also **sem'bla·tive.** [<OF <*sembler.* See SEMBLANCE.] — **sem'bla·bly** *adv.*

sem·blance (sem'bləns) *n.* **1** A mere show without reality; pretense. **2** Outward appearance; look; aspect. **3** A pictorial representation; likeness; resemblance; similarity; image. See synonyms under ANALOGY, IMAGE, PRETENSE. [<OF <*sembler* seem <L *simulare, similare* simulate <*similis* like]

sem·ble (sem'bəl) *v.i.* **·bled, ·bling** It seems; it would seem: used only impersonally in law, and generally in abbreviated form, **sem.** or **semb.** [<F, it seems <*sembler.* See SEMBLANCE.]

se·mé (sə-mā', *Fr.* se·mā') *adj. Her.* Strewn or scattered over with small bearings, as fleurs-de-lis; powdered. [<OF, pp. of *semer* sow <L *seminare* <*semen* a seed]

se·mei·ol·o·gy (sē'mī·ol'ə-jē, sē'mē-), **se·mei·ot·ics** (sē'mī·ot'iks, sē'mē-), etc. See SEMIOLOGY, SEMIOTICS, etc.

Sem·e·le (sem'ə-lē) In Greek mythology, the mother of Dionysus by Zeus: she was destroyed by lightning when she asked to see Zeus as he appeared to the gods.

se·meme (sē'mēm) *n. Ling.* The meaning of a morpheme. [<Gk. *sēma* a sign; on analogy with *phoneme*]

se·men (sē'mən) *n.* **1** The impregnating fluid of male animals. **2** Seed. [<L <*serere* sow]

se·mes·ter (si·mes'tər) *n.* A college half-year; hence, a period of instruction, usually lasting 17 or 18 weeks. [<G <L (*cursus*) *semestris* (a period) of six months <*sex* six + *mensis* a month] — **se·mes'tral** *adj.*

semi- *prefix* **1** Half; partly; not fully: *semiautomatic, semicivilized.* **2** Exactly half: *semicircle.* **3** Occurring twice (in the period specified): *semiweekly.* [<L]

Semi-, meaning not fully, partially, or partial, is found in solidemes and hyphemes, as in the list beginning at the foot of this page.

sem·i·an·nu·al (sem'ē-an'yōō-əl) *adj.* Issued or occurring twice a year; half-yearly. — *n.* A publication issued twice a year. — **sem'i·an'nu·al·ly** *adv.*

sem·i·a·quat·ic (sem'ē-ə-kwat'ik, -kwot'ik) *adj. Biol.* Adapted for living or growing near water, as certain types of plants and animals.

sem·i·au·to·mat·ic (sem'ē-ô'tə-mat'ik) *adj.* Only partly automatic: said especially of guns which are self-loading but not self-firing.

sem·i·breve (sem'ē-brēv') *n. Music* A note equal to half a breve; a whole note.

sem·i·cell (sem'ē-sel') *n. Biol.* Half of a complete cell, usually joined to the other half by an isthmus, as in certain green algae. Compare DESMID.

sem·i·cen·ten·ni·al (sem'ē-sen-ten'ē-əl) *adj.* Occurring or celebrated at the end of fifty years from some event. — *n.* The fiftieth anniversary of an event, or its celebration.

sem·i·cir·cle (sem'ē-sûr'kəl) *n.* **1** A half-circle; an arc or a segment of 180°. **2** Anything formed or arranged in a half-circle. — **sem'i·cir'cu·lar** *adj.*

semicircular canal *Anat.* One of the three tubular structures in the inner ear of most vertebrates, which together serve as the organ of balance. See illustration under EAR.

sem·i·cir·cum·fer·ence (sem'ē-sər-kum'fər-əns, -frəns) *n.* One half of a circumference.

sem·i·civ·i·lized (sem'ē-siv'ə-līzd) *adj.* Half or partly civilized.

sem·i·co·lon (sem'ē-kō'lən) *n.* A mark (;) of punctuation, indicating a greater degree of separation than the comma.

sem·i·con·duc·tor (sem'ē-kən-duk'tər) *n. Physics* **1** One of a class of crystalline solids, as germanium, silicon, and lead sulfide, which are electrical conductors at ordinary temperatures: used in the manufacture of transistors. **2** Any substance or material having an electrical conductivity intermediate between metals and dielectrics.

sem·i·con·scious (sem'ē-kon'shəs) *adj.* Partly conscious; half-conscious.

sem·i·de·tached (sem'ē-di-tacht') *adj.* Joined to another on one side only: said of two houses built side by side with one common wall.

sem·i·di·am·e·ter (sem'ē-dī-am'ə-tər) *n.* A radius; half of a diameter.

sem·i·di·ur·nal (sem'ē-dī-ûr'nəl) *adj.* **1** Pertaining to or continuing during a half-day; occurring or accomplished in a half-day, or once each half-day. **2** Designating either half of the arc described by a heavenly body during its rising or setting. [<SEMI- + DIURNAL]

sem·i·dome (sem'ē-dōm') *n. Archit.* A roof structure resembling a portion, approximately half, of a dome divided vertically.

sem·i·el·lip·ti·cal (sem'ē-i-lip'ti·kəl) *adj.* Having the form of half of an ellipse that has been divided along either diameter.

SEMIDOME

sem·i·fi·nal (sem'ē-fī'nəl) *n.* **1** A competition which precedes the final in a list of sporting events. **2** One of two competitions in a tournament, the winners of each meeting in the final. — *adj.* Next before the final. — **sem'i·fi'nal·ist** *n.*

sem·i·flu·id (sem'ē-flōō'id) *adj.* Fluid, but thick and viscous. — *n.* A thick, viscous fluid. — **sem'i·flu·id'ic** (-flōō-id'ik) *adj.*

sem·i·liq·uid (sem'ē-lik'wid) *adj.* Half liquid. — *n.* A partly liquid substance.

sem·i·lu·nar (sem'ē-lōō'nər) *adj.* Resembling or shaped like a half-moon; crescentic. Also **sem'i·lu'nate** (-lōō'nāt).

semilunar bone *Anat.* The middle bone in the upper row of wrist bones.

semilunar valve *Anat.* One of the crescent-shaped pockets at the entrances to the aorta and to the pulmonary artery respectively: their function is to prevent the backward flow of blood.

sem·i·mo·bile (sem'ē-mō'bēl) *adj.* Partly mobile: said especially of military units not fully equipped with motor vehicles.

sem·i·month·ly (sem'ē-munth'lē) *adj.* Taking place twice a month. — *n. pl.* **·lies** A publication issued twice a month. — *adv.* At half-monthly intervals.

sem·i·mute (sem'ē-myōōt') *adj.* Having imperfectly developed or partially lost speech.

sem·i·nal (sem'ə-nəl) *adj.* **1** Pertaining to or containing seeds, germs, or primal elements. **2** Having productive power; germinal; propagative. **3** Not developed; embryonic; rudimentary. [<OF <L *seminalis* <*semen, seminis* semen, a seed] — **sem'i·nal·ly** *adv.*

sem·i·nar (sem'ə-när) *n.* **1** A group of advanced students at a college or university, meeting regularly and informally with a professor for discussion of research problems. **2** The course thus conducted. [<G <L *seminarium.* See SEMINARY.]

sem·i·nar·y (sem'ə-ner'ē) *n. pl.* **·nar·ies 1** A special school, as of theology; also, a school of higher education. **2** A seminar. **3** The place where anything is nurtured. **4** A seminary priest. — *adj.* **1** Seminal. **2** Pertaining to a seminary. [<MF *séminaire* <L *seminarium* a seed plot, orig. neut. of *seminarius* seminal <*semen, seminis* a seed, semen]

sem·i·na·tion (sem'ə-nā'shən) *n.* **1** The act of sowing or spreading; dispersion of seeds. **2** Propagation. [<L *seminatio, -onis* <*semen, seminis* a seed, semen]

sem·i·nif·er·ous (sem'ə-nif'ər-əs) *adj.* **1** Carrying or producing semen. **2** Seed-bearing. [<L *semen, seminis* a seed, semen + *ferre* bear]

sem·i·niv·o·rous (sem'ə-niv'ər-əs) *adj.* Feeding on seeds. [<L *semen, seminis* a seed, semen + -VOROUS]

Sem·i·nole (sem'ə-nōl) *n.* One of a Florida tribe of North American Indians of Muskhogean linguistic stock, an offshoot of the Creeks: now chiefly in Oklahoma, a remnant remaining in Florida. [<Muskhogean (Creek) *Simanóle,* lit., a separatist, a runaway]

sem·i·of·fi·cial (sem'ē-ō-fish'əl) *adj.* Having official authority or sanction; official to a certain extent. — **sem'i·of·fi'cial·ly** *adv.*

se·mi·ol·o·gy (sē'mē-ol'ə-jē, sē'mī-) *n.* **1** The science that relates to sign language. **2** *Med.* Symptomatology. **3** The use of signs in signaling. Also spelled *semeiology.* [<Gk. *sēmeion,* dim. of *sēma* a mark + -LOGY]

sem·i·o·paque (sem'ē-ō-pāk') *adj.* Half-opaque; translucent but not transparent.

se·mi·ot·ic (sē'mē-ot'ik, sē'mī-) *adj.* **1** Of or pertaining to semantics (def. 2). **2** *Med.* Relating to symptomatology. Also spelled *semeiotic.* Also **se'mi·ot'i·cal.** [<Gk. *sēmeiōtikos* <*sēmeion.* See SEMIOLOGY.]

se·mi·ot·ics (sē'mē-ot'iks, sē'mī-) *n. pl.* (construed as singular) **1** Semantics (def. 2). **2** *Med.* Symptomatology. Also spelled *semeiotics.* [<Gk. *sēmeiōtikos.* See SEMIOTIC.]

sem·i·o·vip·a·rous (sem'ē-ō-vip'ər-əs) *adj.* Giving birth to imperfectly developed offspring, as a marsupial.

Se·mi·pa·la·tinsk (sye-mē-pə-lä'tyinsk) A city of eastern Kazakh S.S.R., on the Irtysh.

sem·i·pal·mate (sem'ē-pal'māt, -mit) *adj. Ornithol.* Having the toes connected by webs for less than half their length, as many shore birds. Also **sem'i·pal'mat·ed.**

semipalmated plover A common plover (*Charadrius semipalmatus*) of the Atlantic coast, which breeds only in the Arctic.

sem·i·par·a·sit·ic (sem'ē-par'ə-sit'ik) *adj. Biol.*

semiaccomplishment	semiarchitectural	semibleached	semiclosure	semiconversion	semidiaphanous	semifailure
semiacquaintance	semiarid	semiblind	semicoagulated	semicooperative	semidigested	semifatalistic
semiaffectionate	semiatheist	semiblunt	semicollapsible	semicured	semidirect	semifeudalism
semiagricultural	semiattached	semiboiled	semicolonial	semicylindrical	semidomesticated	semifictional
semialcoholic	semi-autonomous	semibourgeois	semicomplete	semidangerous	semidry	semifinished
semiallegiance	semi-autonomy	semichannel	semiconceal	semidarkness	seni-Empire	semifit
semianarchist	semibald	semichaotic	semiconfident	semideaf	semienclosed	semifitting
semiangular	semibarbarian	semichivalrous	semiconfinement	semidelirious	semierect	semifixed
semianimal	semibarbarism	semi-Christian	semiconformist	semidenatured	semiremitical	semiflexed
semianimated	semibarbarous	semiclerical	semiconnection	semidependent	semiexposed	semifluctuating
semiarborescent	semibarren	semiclosed	semiconservative	semidestructive	semiextinction	semiforeign

add, āce, câre, päIm; end, ēven; it, īce; odd, ōpen, ôrder; tōōk, pōōl; up, bûrn; ə = a in *above,* e in *sicken,* i in *clarity,* o in *melon,* u in *focus;* yōō = u in *fuse;* oi, oil; ou, pout; ch, check; g, go; ng, ring; th, thin; th, this; zh, vision. Foreign sounds á, œ, ü, kh, ṅ; and ◆: see page xx. < from; + plus; ? possibly.

Semi-Pelagian Partly parasitic: said especially of certain bacteria and of chlorophyll-bearing plants, as the mistletoe.

Sem·i-Pe·la·gi·an (sem′ē-pə-lā′jē-ən) *n.* One of a theological party in the fifth century which held a middle ground between the predestination doctrine of Augustine and the free-will doctrine of Pelagius. [<SEMI- + PELAGIAN]

sem·i·per·me·a·ble (sem′ē-pûr′mē-ə-bəl) *adj.* Partially permeable: said especially of osmotic membranes that separate a solvent from the dissolved substance.

sem·i·por·ce·lain (sem′ē-pôr′sə-lin, -pôr′-, -pôrs′lin, -pōrs′-) *n.* 1 A grade of porcelain having little or no translucency. 2 Earthenware resembling porcelain.

sem·i·post·al (sem′ē-pōs′təl) *adj.* Designating a postage stamp or series of stamps sold by postal authorities for more than the franking value, the additional proceeds usually going to a philanthropic purpose. — *n.* A semipostal stamp.

sem·i·pre·cious (sem′ē-presh′əs) *adj.* Precious, but not sufficiently so to be used as gems: *semiprecious* stones.

sem·i·qua·ver (sem′ē-kwā′vər) *n. Music* A note one sixteenth the value of a semibreve or whole note.

Se·mir·a·mis (si-mir′ə-mis) In Assyrian legend, the wife of Ninus and founder of Babylon, known for her beauty and wisdom.

sem·i·rig·id (sem′ē-rij′id) *adj. Aeron.* Partly rigid, as an airship in which an exterior stiffener supports the load. — *n.* A semirigid airship.

sem·i·round (sem′ē-round′) *adj.* Having one side round and the other flat. — *n.* A semiround object.

sem·i·skilled (sem′ē-skild′) *adj.* Partly skilled, but not enough to perform specialized work.

sem·i·sol·id (sem′ē-sol′id) *adj.* Partly solid; so viscous as to be nearly solid.

Sem·ite (sem′it, sē′mit) *n.* 1 A person believed to be or considered as a descendant of Shem. 2 One of a people of Caucasian stock, now represented by the Jews and Arabs, but originally including the ancient Babylonians, Assyrians, Arameans, Phoenicians, etc. Also *Shemite.* [<NL *Semita* <LL *Sem* Shem <Gk. *Sēm* <Hebrew *shēm*]

Se·mit·ic (sə-mit′ik) *adj.* Of or pertaining to the Semites, or to any of their languages. — *n.* A subfamily of the Hamito-Semitic family of languages, divided into three groups — **East Semitic** (Akkadian), **Northwest Semitic** (Phoenician, ancient and modern Hebrew, Aramaic, etc.), and **Southwest Semitic** (Arabic, Ethiopic, Amharic, etc.).

Se·mit·ics (sə-mit′iks) *n.* The scientific study of the history, language, and literature of the Semitic peoples.

Sem·i·tism (sem′ə-tiz′əm) *n.* 1 A Semitic word or idiom. 2 Semitic practices, opinions, or customs collectively. 3 Any political or economic policy favoring the Jews.

sem·i·tone (sem′ē-tōn′) *n. Music* An interval approximately equal to half a major tone on the scale: the smallest interval in most European music. — **sem′i·ton′ic** (-ton′ik) *adj.*

sem·i·trail·er (sem′ē-trā′lər) *n.* A trailer having wheels only at the rear, the front end resting on the towing vehicle.

sem·i·trans·lu·cent (sem′ē-trans·lōō′sənt, -tranz-) *adj.* Half or partly translucent.

SEMITRAILER

sem·i·trans·par·ent (sem′ē-trans-pâr′ənt) *adj.* Half or partly transparent.

sem·i·trop·i·cal (sem′ē-trop′i·kəl) *adj.* Nearly tropical.

sem·i·vit·ri·fied (sem′ē-vit′rə-fīd) *adj.* Half vitrified; partially made into glass.

sem·i·vow·el (sem′ē-vou′əl) *n. Phonet.* A vowel-like sound used as a consonant, as (w), (y), and (r): also called *glide.* — **sem′i·vo′cal** (-vō′kəl) *adj.*

sem·i·week·ly (sem′ē-wēk′lē) *adj.* Issued or occurring twice a week. — *n. pl.* **·lies** A publication issued twice a week. — *adv.* At half-weekly intervals.

Sem·lin (zem-lēn′) The German name for ZEMUN.

Sem·mel·weiss (sem′əl-vīs), **Ignaz Philipp**, 1816–65, Austrian obstetrician; pioneer in prevention of puerperal fever.

Semmes (semz), **Raphael,** 1809–77, American Confederate naval officer.

sem·o·li·na (sem′ə-lē′nə) *n.* The gritty or grain-like portions of wheat retained in the bolting machine after the fine flour has been passed through. [Alter. of Ital. *semolino,* dim. of *semola* bran <L *simila* fine flour]

Sem·pach (zem′päkh) A town in central Switzerland; scene of a Swiss victory over the Austrians, 1386.

sem·per fi·de·lis (sem′pər fi·dē′lis, fi·dā′lis) *Latin* Always faithful: motto of the U. S. Marine Corps.

sem·per pa·ra·tus (sem′pər pə·rā′təs) *Latin* Always prepared: motto of the U. S. Coast Guard.

sem·per·vi·rent (sem′pər·vī′rənt) *adj.* Evergreen. [<L *semper* always + *virens, -entis,* ppr. of *virere* be green]

sem·pi·ter·nal (sem′pə·tûr′nəl) *adj.* Enduring or existing to all eternity; everlasting. See synonyms under IMMORTAL, PERPETUAL. [<OF *sempiternel* <LL *sempiternalis* <L *sempiternus* everlasting <*semper* always] — **sem′pi·ter′ni·ty** *n.*

sem·pli·ce (sem′plē-chä) *adj. Music* Simple; unaffected: a direction to performers. [<Ital. <L *simplex, simplicis*]

sem·pre (sem′prä) *adv. Music* Always; throughout the passage or composition: *sempre* legato, piano, etc. [<Ital. <L *semper*]

semp·stress (semp′stris, sem′-) See SEAMSTRESS.

sen (sen) *n.* Japanese A Japanese copper or bronze coin, equal to 1/100 of a yen.

sen′ (sen) *v.t. & v.i., n. Scot.* Send[1].

sen·a·ry (sen′ər-ē) *adj.* Of or pertaining to six; containing six units. [<L *senarius* <*seni* six each <*sex* six]

sen·ate (sen′it) *n.* 1 The governing body of some universities and institutions of learning. 2 An advisory body of members of the faculty and representative students in a school or college. 3 A body of distinguished or venerable men; council; legislative body. [<OF *senat* <L *senatus,* lit., a council of old men <*senex, senis* old]

Sen·ate (sen′it) *n.* 1 The upper branch of national or state legislative bodies of the United States, and of France and other governments; especially, the **United States Senate,** composed of two Senators elected by popular vote from each State. 2 In ancient Rome, the state council, whose originally very extensive powers were curtailed under the empire: limited to 100 patricians under the kings, it consisted, under the republic, of 300 patricians, plebeians, and high officials; under Augustus, there were 600 senators.

sen·a·tor (sen′ə-tər) *n.* A member of a senate. [<OF *senateur* <L *senator* <*senex, senis* an old man, old] — **sen′a·tor·ship′** (-ship′) *n.*

sen·a·to·ri·al (sen′ə-tôr′ē-əl, -tō′rē-) *adj.* 1 Pertaining to or befitting a senator or senate. 2 Entitled to elect a senator, as a district. — **sen′a·to′ri·al·ly** *adv.*

se·na·tus con·sul·tum (sə-nā′təs kən-sul′təm) *Latin* A decree of the ancient Roman Senate, pronounced upon some matter of law or public policy: originally only advisory and finally authoritative as laws. Also **se·na′tus con·sult′.**

send[1] (send) *v.* **sent, send·ing** *v.t.* 1 To cause or direct to go; dispatch, as a **messenger.** 2 To cause to be conveyed to another place; transmit; forward: to *send* a letter. 3 To cause to issue; emit or discharge, as heat, light, smoke, etc.: with *forth, out, etc.* 4 To throw or drive by force; impel. 5 To cause to come, happen, etc.; grant: God *send* us peace. 6 To bring into a specified state or condition; drive: The decision *sent* him into bankruptcy. 7 To transmit, as a current or electromagnetic impulses. — *v.i.* 8 To dispatch an agent, messenger, or message. — **to send for** To summon by a message or messenger. — **to send in one's papers** To resign. — *n.* A messenger. [OE *sendan*] — **send′er** *n.*
— *Synonyms (verb):* cast, delegate, depute, discharge, dispatch, dismiss, emit, fling, forward, hurl, impel, lance, launch, project, propel, sling, throw, transmit. *Send* in its most common use involves personal efficiency without personal presence; according to the adage, "If you want your business done, go; if not, *send*"; one *sends* a letter or a bullet, a messenger or a message. To *dispatch* is to *send* hastily or very promptly, ordinarily with a destination in view; to *dismiss* is to *send* away from oneself without reference to a destination; as, to *dismiss* a clerk, an application, or an annoying subject. To *discharge* is to *send* away so as to relieve a person or thing of a load; we *discharge* a gun or *discharge* the contents; as applied to persons, *discharge* is a harsher term than *dismiss.* To *emit* is to *send* forth from within, with no reference to a destination; as, The sun *emits* light and heat. *Transmit,* from the Latin, is a dignified term, often less vigorous than the Saxon *send,* but preferable at times in literary or scientific use; as, to *transmit* a charge of electricity. *Transmit* fixes the attention more on the intervening agency, as *send* does upon the points of departure and destination. *Antonyms:* bring, carry, convey, get, give, hand, hold, keep, receive, retain.

send[2] (send) *Naut. n.* 1 The flow or impulse of the waves. 2 Scend. — *v.i.* 1 To move by the force of waves. 2 To scend. [<SEND[1]; prob. infl. in meaning by ASCEND]

Sen·dai (sen·dī′) A city on NE Honshu island, Japan.

sen·dal (sen′dəl) *n.* 1 A light, thin, silken fabric much used for dresses, etc., in the Middle Ages. 2 An article made of it. Also spelled *sandal.* [<OF *cendal, sendal,* ult. <Gk. *sidōn* fine linen]

send-off (send′ôf′, -of′) *n.* 1 The act of sending off; a start. 2 A farewell dinner or other celebration or demonstration at parting. 3 Encouragement, as in starting a career.

Sen·e·ca (sen′ə-kə) *n.* One of a tribe of North American Indians of Iroquoian stock formerly inhabiting western New York, the largest tribe of the confederation known as the Five Nations: still numerous in New York and Ontario. [<Du. *Sennacaas* the Five Nations <Algonquian (Mohegan) *A′sinnika,* trans. of Iroquoian *Oneñiute,* short for *oneñiute′ roñ non* Oneida, lit., people of the standing rock]

Sen·e·ca (sen′ə-kə), **Lucius Annaeus,** 3? B.C.–A.D. 65, Roman Stoic philosopher, statesman, and tragic dramatist.

Sen·e·ca Lake (sen′ə-kə) One of the Finger Lakes in west central New York, extending 35 miles north and south; 67 square miles.

sen·e·ga (sen′ə-gə) *n.* 1 The dried root of an herb *(Polygala senega)* of the milkwort family, used as a stimulating expectorant, as in treating bronchitis. 2 The plant itself. Also **senega**

semifriable	semihostile	semiliberal	semiopened	semipolitician	semi-Romanized	semistagnation
semifrontier	semihumanitarian	semilined	semiorganized	semiprivate	semiroyal	semistarvation
semifunctional	semihumorous	semilucent	semiovoid	semiprofessional	semirustic	semistarved
semigala	semi-idle	semimilitary	semipagan	semipublic	semisacred	semisuccess
semigenuflection	semi-idleness	semimonastic	semipanic	semiraw	semisatiric	semitailored
semi-Gnostic	semi-incandescent	semimonopoly	semiparallel	semirebellion	semiscientific	semitrained
semi-Gothic	semi-independence	semimystical	semiparalysis	semireligious	semisecrecy	semitruth
semigranulate	semi-intoxicated	seminationalization	semipastoral	semiresolute	semiseriousness	semivirtue
semihard	semi-intoxication	seminecessary	semipeace	semirespectability	semisocial	semivital
semihigh	semi-invalid	seminervous	semiperfect	semirespectable	semisocialism	semivoluntary
semihistorical	semileafless	semioblivious	semipermanent	semiretirement	semisoft	semiwarfare
semihobo	semilegendary	semiobscurity	semiperspicuous	semiriddle	semispontaneity	semiwild

root. [<NL, alter. of SENECA; so called because thought, by the Seneca Indians, to be good for snakebites]

Sen·e·gal (sen′ə·gôl′) A river of western Africa, flowing about 1,000 miles NW from SW Guinea to the Atlantic at Saint-Louis, Senegal, forming the border between Senegal and Mauritania.

Sen·e·gal (sen′ə·gôl′), **Republic of** An independent republic of the French Community in west Africa; 76,124 square miles; capital, Dakar: formerly a French overseas territory. — **Sen′e·ga·lese′** (-gə·lēz′, -lēs′) *adj.* & *n.*

Sen·e·gam·bi·a (sen′ə·gam′bē·ə) The former name for the territory between the Senegal and Gambia rivers, in west Africa.

se·nes·cent (si·nes′ənt) *adj.* 1 Growing old. 2 Characteristic of old age. [<L *senescens, -entis*, ppr. of *senescere* grow old <*senex* old] — **se·nes′cence** *n.*

sen·e·schal (sen′ə·shəl) *n.* 1 An official in the household of a medieval prince or noble who had charge of feasts, etc.; a steward or major-domo. 2 A magistrate or governor. 3 *Brit.* A cathedral official. [<OF <Gmc. Cf. OHG *siniskalk* old servant.]

se·nile (sē′nīl, -nil) *adj.* 1 Pertaining to, proceeding from, or characteristic of old age. 2 Infirm; weak; doting. 3 *Geog.* Almost worn away to base level: a *senile* continent. [<L *senilis*< *senex* old] — **se′nile·ly** *adv.*

senile dementia *Psychiatry* The progressive deterioration of cerebral functions and mental faculties associated with old age. Also **senile psychosis.**

se·nil·i·ty (si·nil′ə·tē) *n.* Mental and physical infirmity due to old age; old age accompanied by infirmity.

sen·ior (sēn′yər) *adj.* 1 Older in years; elder; specifically, after personal names (usually in the abbreviated form *Sr.*), to denote the elder of two related persons of the same name, especially a father and his son. 2 Older in office; more advanced in service; superior in rank or dignity. 3 Pertaining to the closing year of a high school or college course. — *n.* 1 One older in years or office, or more advanced in rank or dignity than another. 2 Hence, any elderly person. 3 A member of a senior class. 4 A graduate or one of the older fellows of an English college. [<L, compar. of *senex, senis* old]

sen·ior·i·ty (sēn·yôr′ə·tē, -yor′-) *n.* *pl.* **·ties** 1 The state of being older in years or in office; priority of age, service, or rank. 2 An assembly of seniors or, in England, senior fellows of a college.

Sen·lac (sen′lak) A hill, near Hastings, in Sussex, England; scene of the battle of Hastings, 1066.

sen·na (sen′ə) *n.* 1 The dried leaflets of any one of several leguminous plants (genus *Cassia*), used medicinally for their purgative properties; especially, the Old World species *C. acutifolia* and *C. angustifolia*. 2 Any one of the plants yielding true senna or a similar product. [<NL *senna, sena* <Arabic *sanā*]

Sen·nach·er·ib (si·nak′ər·ib), died 681 B.C.; king of Assyria 705–681 B.C.; invaded Palestine.

Sen·nar (sen·när′) An ancient city in the Sudan between the White Nile and the Blue Nile, capital of a large native kingdom, 15th to 19th centuries. Also **Sen·naar′.**

sen·net (sen′it) *n.* A signal of exit or entrance sounded on a horn: chiefly as a stage direction in Elizabethan plays. [<OF *senet, sinet, signet.* Double of SIGNET.]

sen·night (sen′īt, -it) *n. Archaic* A week. Also **se′n′night, sev′en–night′.** [OE *seofan nihta* <*seofan* seven + *nihta*, pl. of *niht* a night]

sen·nit (sen′it) *n.* 1 Plaited cordage, of from 3 to 9 strands, used for gaskets on ships. 2 Plaited grass or straw for hatmaking. [Earlier *sinnet*, ? <SEVEN + KNIT]

se·no·pi·a (si·nō′pē·ə) *n.* An apparent restoration of normal vision in formerly myopic people who have become hypermetropic in old age. Also called *gerontopia.* [<NL <L *senex* old + Gk. *ōps, ōpos* an eye]

se·ñor (sā·nyôr′) *n.* *pl.* **·ño·res** (-nyō′rās) *Spanish* A Spanish title of courtesy; a gentleman; Mr.; sir: used before a name, like *Mr.*, or alone, like *Sir.*

se·ño·ra (sā·nyō′rä) *n. Spanish* A Spanish lady; Mrs.; madam.

se·ño·ri·ta (sā′nyō·rē′tä) *n. Spanish* A young, unmarried Spanish lady; miss.

sen·sate (sen′sāt) *adj.* Perceived or appreciated by the senses: *sensate* matters: also **sen′sat·ed.** — *v.t.* **·sat·ed, ·sat·ing** To perceive by the senses. [<LL *sensatus* gifted with sense <L *sensus* sense]

sen·sa·tion (sen·sā′shən) *n.* 1 *Physiol.* **a** That aspect of consciousness resulting from the stimulation of a nerve process beginning at any point in the body and passing through the brain, especially by those stimuli affecting any of the sense organs, as hearing, taste, touch, smell, and sight. **b** The capacity to respond to such stimulation. 2 That which produces interest or excitement; an excited condition: to cause a *sensation.* 3 A condition of mind resulting from inherent feeling; emotion. [<Med. L *sensatio, -onis* <LL *sensatus.* See SENSATE.]

Synonyms: emotion, feeling, perception, sense. *Sensation* is the mind's consciousness due to bodily response to stimuli, as heat or sound; *perception* is the cognition of some external object which causes the *sensation.* While *sensations* are connected with the body, *emotions* add the reactions of the mind. *Feeling* is a term popularly denoting what is felt, whether through the body or by the mind alone, and includes both *sensation* and *emotion.* A *sense* is an organ or faculty of *sensation* or of *perception.* See FEELING.

sen·sa·tion·al (sen·sā′shən·əl) *adj.* 1 Pertaining to emotional excitement. 2 Pertaining to physical sensation. 3 Causing excitement; startling. 4 Causing unnatural emotional excitement; melodramatic; trashy: a *sensational* story.

sen·sa·tion·al·ism (sen·sā′shən·əl·iz′əm) *n.* 1 *Philos.* The theory that all knowledge originates in sensation, or is composed of transformed sense elements, that all consciousness is modified sensation, and all mental phenomena have a sensory basis: a branch of modern empiricism, often combined with *associationism.* 2 The use of melodramatic methods in writing or speaking. 3 The theory that feeling is the only criterion of good. — **sen·sa′tion·al·ist** *n.* — **sen·sa′tion·al·is′tic** *adj.*

sense (sens) *n.* 1 The faculty of sensation; sense perception. 2 Any of certain agencies by or through which an individual receives impressions of the external world; popularly, one of the five senses. 3 *Physiol.* Any receptor, or group of receptors, specialized to receive and transmit stimuli, either external, as of sight, taste, smell, etc., or internal, as of hunger, thirst, sex, equilibrium, muscular and visceral movements, etc. 4 Rational perception accompanied by feeling; realization; discriminating cognition: a *sense* of wrong. 5 Normal power of mind or understanding; sound or natural judgment: The fellow has no *sense*; often in the plural: She is coming to her *senses.* 6 Signification; import; meaning. 7 Opinion, view, or judgment of the majority: The *sense* of the meeting was manifest. 8 That which commends itself to the understanding as being in accordance with reason and good judgment: to talk *sense.* 9 Capacity to perceive or appreciate: a *sense* of color. 10 *Geom.* One of two opposite directions in which a magnitude may be described or generated. 11 Direction; trend. See synonyms under FEELING, MIND. — **the five senses** The Aristotelian division of senses into sight, hearing, smell, taste, and touch: now collectively known as the **special senses.** — **sixth sense** 1 Capacity for perception beyond the normal range of the senses; extrasensory perception. 2 Intuitive or premonitory knowledge, especially as affecting or affected by the senses. 3 Cenesthesia. — *v.t.* **sensed, sens·ing** 1 To become aware of through the senses. 2 *Colloq.* To comprehend; understand. [<MF *sens* <L *sensus* perception <*sentire* feel]

sense datum *Psychol.* That which is experienced as a result of the stimulation of a sense organ.

sense·less (sens′lis) *adj.* 1 Deprived of consciousness; unconscious. 2 Incapable of feeling or perception; insensate. 3 Devoid of sense; foolish; stupid. — **sense′less·ly** *adv.* — **sense′less·ness** *n.*

sense organ *Physiol.* A structure specialized to receive sense impressions, as the eye, nose, ear, etc.; a receptor (def. 2).

sense perception Immediate knowledge of things through the senses, as distinguished from mediate or inferred knowledge.

sense stress See SENTENCE STRESS.

sen·si·bil·i·ty (sen′sə·bil′ə·tē) *n.* *pl.* **·ties** 1 The capability of sensation; power to perceive or feel. 2 The capacity of sensation and rational emotion, as distinguished from intellect and will. 3 Susceptibility or sensitiveness to outside influences or mental impressions; sometimes, abnormal sensitiveness: often in the plural. 4 Appreciation accompanying mental apprehension; discerning judgment. 5 Delicacy or sensitiveness of an instrument. 6 Responsiveness to pathos or to artistic or esthetic values. 7 *Archaic* Sentimentality.

Synonyms: sensitiveness, sensibility, susceptibility. In popular use *sensibility* denotes sometimes capacity of feeling of any kind; as, *sensibility* to heat or cold; sometimes, a peculiar readiness to be the subject of feeling, especially of the higher feelings; as the *sensibility* of the artist or the poet. *Sensitiveness* denotes an especial delicacy of *sensibility*, ready to be excited by the slightest cause, as displayed, for instance, in the sensitive plant. *Susceptibility* is rather a capacity to receive, to contain feeling, so that a person of great *susceptibility* is capable of being readily and deeply moved; *sensitiveness* is more superficial, *susceptibility* more pervading. In physics, the *sensitiveness* of a magnetic needle is the ease with which it may be deflected, as by another magnet; its *susceptibility* is the degree to which it can be magnetized by a given magnetic force or the amount of magnetism it will hold. A person of great *sensitiveness* is quickly and keenly affected by any external influence, as by music, pathos, or ridicule, while a person of great *susceptibility* is not only touched, but moved to his utmost soul. See FEELING. *Antonyms*: coldness, deadness, hardness, insensibility, numbness, unconsciousness.

sen·si·ble (sen′sə·bəl) *adj.* 1 Possessed of good mental perception; exhibiting sound sense and judgment; discreet; judicious. 2 Capable of physical sensation; sensitive: *sensible* to pain. 3 Perceptible or appreciable through the senses: *sensible* heat. 4 Emotionally or mentally sensitive. 5 Having a perception or cognition; fully aware; persuaded. 6 Great enough to be perceived; appreciable. 7 *Obs.* Sensitive to minute changes. See synonyms under CONSCIOUS, EXPEDIENT, INTELLIGENT, PHYSICAL, SAGACIOUS. — *n.* 1 A substance capable of being felt or observed. 2 A sentient being. 3 *Music* The leading note; the seventh of a scale: also **sensible note** (or **tone**). [<OF <L *sensibilis* <*sensus*, pp. of *sentire* feel, perceive] — **sen′si·ble·ness** *n.* — **sen′si·bly** *adv.*

sen·si·tive (sen′sə·tiv) *adj.* 1 Easily affected by outside operations or influences; excitable or impressible; touchy; easily offended. 2 *Chem.* & *Phot.* Reacting readily to the proper agents or forces: paper *sensitive* to light. 3 Pertaining to or depending on the senses or sensation: *sensitive* motions. 4 Closing or moving when touched or irritated, as certain plants. 5 Liable to fluctuation. 6 *Obs.* Wise; sensible. 7 Capable of indicating minute changes or differences; delicate. See synonyms under FINE, MOBILE. [<OF *sensitif* <Med. L *sensitivus* <L *sensus.* See SENSIBLE.] — **sen′si·tive·ly** *adv.* — **sen′si·tive·ness** *n.*

sensitive plant A shrubby tropical herb (*Mimosa pudica*), whose leaves close at a touch: often cultivated in hothouses.

sen·si·tiv·i·ty (sen′sə·tiv′ə·tē) *n.* 1 The state or degree of being sensitive; sensitiveness. 2 *Physiol.* The degree of acuteness with which sensations are discriminated; irritability, as of organs: distinguished from *sensibility*, in which the mental side is more prominent. 3 The degree of responsiveness to an electric current or to radio waves. 4 *Phot.* Sensitiveness to light.

sen·si·tize (sen′sə·tīz) *v.t.* **·tized, ·tiz·ing** 1 To render sensitive. 2 *Phot.* To make sensitive to light, as a plate or film. 3 *Med.* To make susceptible or hypersensitive to the action of a drug by repeated injections. [<SENSIT(IVE) + -IZE] —**sen′si·ti·za′tion** *n.* —**sen′si·tiz′er** *n.*

sen·si·tom·e·ter (sen′sə·tom′ə·tər) *n.* An apparatus by which the sensitiveness to light of a photographic film or body tissue may be tested or measured. [<SENSIT(IVE) + -(O)-METER]

sen·sor (sen′sər) *adj.* Sensory: applied to nerves and nerve organs. [Short for SENSORY]

sen·so·ri·mo·tor (sen′sə·ri·mō′tər) *adj. Physiol.* Of or pertaining to muscular and nervous responses induced by sensory stimuli. Compare IDEOMOTOR. [<SENSORY + MOTOR]

sen·so·ri·um (sen·sôr′ē·əm, -sō′rē-) *n. pl.* **·ri·a** (-ē·ə) 1 *Anat.* The nervous system, including the cerebrum, as the collective organ of sensation. 2 *Biol.* The entire sensory apparatus of an organism. [<LL <L *sensus.* See SENSIBLE.]

sen·so·ry (sen′sər·ē) *adj.* 1 Pertaining to the sensorium or to sensation. 2 Conveying or producing sense impulses. Also **sen·so·ri·al** (sen·sôr′ē·əl, -sō′rē-). [<LL *sensorium* SENSORIUM]

sen·su·al (sen′shōō·əl) *adj.* 1 Unduly indulgent to the appetites or sexual pleasure; exhibiting a predominance of the animal nature; lewd. 2 Pertaining to the body or the physical senses; also, fleshly; carnal: opposed to *spiritual.* 3 Pertaining to sensualism: usually opprobrious. See synonyms under BRUTISH. [<MF *sensuel* <LL *sensualis* <L *sensus* SENSE] —**sen′su·al·ly** *adv.*

sen·su·al·ism (sen′shōō·əl·iz′əm) *n.* 1 Sensuality. 2 *Philos.* A debased sensationalism. 3 A system of ethics predicating the pleasures of sense to be the highest good. 4 Emphasis on the sensuous elements of beauty, rather than the ideal. —**sen′su·al·ist** *n.* —**sen′su·al·is′tic** *adj.*

sen·su·al·i·ty (sen′shōō·al′ə·tē) *n.* 1 The state of being sensual, or sensual acts collectively. 2 Sensual or animal indulgence. Also **sen′su·al·ness.**

sen·su·al·ize (sen′shōō·əl·īz′) *v.t.* **·ized, ·iz·ing** To make sensual. Also *Brit.* **sen′su·al·ise′.** —**sen′su·al·i·za′tion** *n.*

sen·su·ous (sen′shōō·əs) *adj.* 1 Pertaining or appealing to or derived from the senses: used in a higher and purer sense than *sensual.* 2 Keenly appreciative of and aroused by beauty, refinement, or luxury. 3 Resembling imagery that appeals to the senses: a *sensuous* portrayal. [<L *sensus* SENSE + -OUS] —**sen′su·ous·ly** *adv.* —**sen′su·ous·ness** *n.*

sent (sent) Past tense and past participle of SEND.

sen·tence (sen′təns) *n.* 1 *Gram.* A word or a related group of words expressing a complete thought, whether a statement of fact (declarative), a question (interrogative), a command (imperative), or an exclamation (exclamatory). Declarative and interrogative sentences usually contain a subject (that which is spoken of) and a predicate (what is said about the subject), but either or both of these elements may be missing in an utterance that, nevertheless, conveys full meaning, as in "Where is John?" "At home." or "Look!" —**simple sentence** A sentence consisting of one independent clause, as *The dog barked.* Its subject and predicate may be simple (having one substantive or one verb) or compound (having two or more substantives or verbs), and there may be modifying words and phrases. —**compound sentence** A sentence consisting of more than one independent clause, as *The sun shone and the birds sang.* —**complex sentence** A sentence consisting of a principal clause and one or more subordinate clauses, as *After I have read it, I shall give the book to you.* 2 *Law* A final judgment; penalty pronounced upon a person convicted. 3 A determination; opinion, especially as expressed formally. 4 An instructive saying; a maxim. 5 *Music* A complete idea or period, usually consisting of several phrases, as the half of a four-line hymn tune or song. —*v.t.* **·tenced, ·tenc·ing** To pass sentence upon; condemn to punishment. See synonyms under CONDEMN. [<OF <L *sententia* an opinion <*sentire* feel, be of opinion] —**sen·ten·tial** (sen·ten′shəl) *adj.*

sen·tenc·er (sen′tən·sər) *n.* One who pronounces sentence; a judge.

sentence stress The variation in emphasis given to successive words in a sentence to stress the meaning: also called *sense stress.* Also **sentence accent.**

sen·ten·tious (sen·ten′shəs) *adj.* 1 Abounding in or giving terse expression to thought; axiomatic; sometimes, opprobriously, pompously formal, or moralizing. 2 Habitually using terse, laconic, or aphoristic language. See synonyms under TERSE. [<L *sententiosus* <*sententia* a maxim. See SENTENCE.] —**sen·ten′tious·ly** *adv.* —**sen·ten′tious·ness, sen·ten·ti·os·i·ty** (sen·ten′shē·os′ə·tē) *n.*

sen·ti·ence (sen′shē·əns, -shəns) *n.* 1 The state of being sentient. 2 Capacity for sensation or sense perception. 3 Consciousness. 4 Sensation regarded as immediate experience and so distinguished from thought or perception. Also **sen′ti·en·cy.**

sen·ti·ent (sen′shē·ənt, -shənt) *adj.* Possessing powers of sense or sense perception; having or actually experiencing sensation or feeling: opposed to *inanimate* and *vegetal.* —*n.* One capable of sensation or perception; loosely, the mind, as the seat of consciousness. [<L *sentiens, -entis,* ppr. of *sentire* feel] —**sen′ti·ent·ly** *adv.*

sen·ti·ment (sen′tə·mənt) *n.* 1 Noble, tender, or artistic feeling, or susceptibility to such feeling; sensibility; also, its verbal expression. 2 A mental attitude or response to a person, object, or idea conditioned entirely by feeling instead of reason; loosely, an exaggerated emotional reaction. 3 Idealistic, personal, or esthetic reaction as distinguished from intellectual or practical. 4 An opinion or judgment; thought as distinguished from its expression: often in the plural. 5 An expressive thought or idea dressed in appropriate language, as a toast aptly uttered. See synonyms under FEELING, IDEA. [<OF *sentement* <Med. L *sentimentum* <L *sentire* feel]

sen·ti·men·tal (sen′tə·men′təl) *adj.* 1 Characterized by sentiment or intellectual emotion; involving or exciting tender emotions or aspirations. 2 Experiencing, displaying, or given to sentiment, often in an extravagant or mawkish manner: a *sentimental* person. See synonyms under ROMANTIC. —**sen′ti·men′tal·ly** *adv.*

sen·ti·men·tal·ism (sen′tə·men′təl·iz′əm) *n.* The state of being sentimental, or its manifestation; tendency to be emotional. Also **sen′ti·men·tal′i·ty** (-men·tal′ə·tē). —**sen′ti·men′tal·ist** *n.*

sen·ti·men·tal·ize (sen′tə·men′təl·īz) *v.* **·ized, ·iz·ing** *v.t.* 1 To affect with sentiment. 2 To cherish sentimentally. —*v.i.* 3 To behave sentimentally. Also *Brit.* **sen′ti·men′tal·ise.**

sen·ti·nel (sen′tə·nəl) *n.* A sentry; hence, any watcher or guard. —*v.t.* **·neled** or **·nelled, ·nel·ing** or **·nel·ling** 1 To watch over as a sentinel. 2 To protect or furnish with sentinels. 3 To station or appoint as a sentinel. [<OF *sentinelle* <Ital. *sentinella* <LL *sentinare* avoid danger <*sentire* perceive]

sen·try (sen′trē) *n. pl.* **·tries** 1 A soldier placed on guard to see that only authorized persons pass his post and to give warning of approaching danger; a sentinel. 2 The watch or guard kept by a sentry. [? Short for obs. *centrenel,* var. of SENTINEL]

sentry box A small shelter or cabin to protect a sentry from the weather.

Se·nus·si (se·nōō′sē) *n. pl.* A belligerent Moslem religious sect once influential in northern Africa and Arabia, founded by Sidi Mohammed ben Ali ben Es Senussi about 1842. Also **Se·nu′si, Se·nus′sites.** —**Se·nus′si·an** *adj.*

Se·oul (sā·ōōl′, sōl; *Korean* syœ·ōōl) The capital of the Republic of Korea (South Korea): also *Kyongsong*: Japanese *Keijo.*

se·pal (sē′pəl) *n. Bot.* One of the individual leaves of a calyx. [<F *sépale* <NL *sepalum* <L *sep(aratus)* separate + *(pet)alum* a petal] —**sep·a·line** (sep′ə·lin, -līn), **sep′a·lous** *adj.*

sep·a·ra·ble (sep′ər·ə·bəl, sep′rə-) *adj.* Capable of being separated or divided. [<L *separabilis* <*separare* separate] —**sep′a·ra·bil′i·ty, sep′a·ra·ble·ness** *n.* —**sep′a·ra·bly** *adv.*

sep·a·rate (sep′ə·rāt) *v.* **·rat·ed, ·rat·ing** *v.t.* 1 To set asunder; disunite or disjoin; sever. 2 To occupy a position between; serve to keep apart: The Hudson River *separates* New York from New Jersey. 3 To divide into components, parts, etc. 4 To isolate or obtain from a compound, mixture, etc.: to *separate* the wheat from the chaff. 5 To consider separately; distinguish between. 6 *Law* To part by separation. —*v.i.* 7 To become divided or disconnected; draw apart. 8 To part company; withdraw from association or combination. —*adj.* (sep′ər·it, sep′rit) 1 Existing or considered apart from others; distinct; individual: *separate* rooms. 2 Disembodied; disunited from the body. 3 Separated; disjoined. See synonyms under PARTICULAR. [<L *separatus,* pp. of *separare* <*se-* apart + *parare* prepare] —**sep′a·rate·ly** *adv.* —**sep′a·rate·ness** *n.*

Synonyms (verb): alienate, detach, disconnect, disengage, disjoin, dissever, disunite, divide, part, remove, sever, split, sunder, withdraw. *Antonyms*: see synonyms for MIX.

sep·a·ra·tion (sep′ə·rā′shən) *n.* 1 The act or process of separating; division. 2 The state of being disconnected or apart. 3 A dividing line. 4 *Law* Relinquishment of cohabitation between husband and wife by mutual consent: distinguished from *divorce.* See synonyms under SECLUSION.

separation center A central army or navy point that handles all types of discharges and releases of personnel except medical discharges.

sep·a·ra·tist (sep′ər·ə·tist, sep′rə-) *n.* One who advocates or upholds separation; specifically, a seceder; dissenter. Also **sep′a·ra′tion·ist.** —**sep′a·ra·tism** *n.*

sep·a·ra·tive (sep′ə·rā′tiv, -rə·tiv) *adj.* Tending to or inducing separation; useful in separating. Also **sep·a·ra·to·ry** (sep′ər·ə·tôr′ē, -tō′rē, sep′rə-).

sep·a·ra·tor (sep′ə·rā′tər) *n.* 1 Any device, implement, or apparatus for dividing or separating things into their component parts. 2 A machine for separating the chaff from grain. 3 A centrifugal mechanism for separating cream from milk. 4 One who separates.

sep·a·ra·trix (sep′ə·rā′triks) *n.* A separating point or line; decimal point. [<LL *(linea) separatrix* the separating or dividing (line) <*separatus.* See SEPARATE.]

sep·a·ra·tum (sep′ə·rā′təm) *n. pl.* **·ta** (-tə) A paper published separately from a series to which it belongs; a reprint of an article previously published as a part of a report. [<L, neut. of *separatus.* See SEPARATE.]

Se·phar·dim (si·fär′dim) *n. pl.* The Spanish and Portuguese Jews or their descendants: distinguished from the *Ashkenazim.* Also **Se·phar′a·dim** (-ə·dim). [<Hebrew *sephārādhīm* <*Sephāradh,* a country mentioned in *Ob.* iii 20, identified by the rabbis with Spain, but prob. orig. in Asia Minor] —**Se·phar′dic, Se·phar′a·dic** *adj.*

se·pi·a (sē′pē·ə) *n.* 1 A reddish-brown pigment prepared from the ink of the cuttlefish; the color of this pigment. 2 A picture done in this pigment. 3 The ink of the cuttlefish. 4 Any of a genus *(Sepia)* of decapod mollusks having an internal shell, especially the common Atlantic cuttlefish *(S. officinalis).* —*adj.* Executed in or colored like sepia; dark-brown with a tinge of red. [<L <Gk. *sēpia* a cuttlefish]

Se·pik (sā′pik) A river in NE New Guinea, in the Australian Territory of New Guinea, rising near the border of Netherlands New Guinea and flowing 700 miles NE to the Bismarck Sea.

se·pi·o·lite (sē′pē·ə·līt′) *n.* Meerschaum. [<G *sepiolith* <NL *sepium* cuttlebone (<Gk. *sēpion,* dim. of *sēpia* a cuttlefish) + Gk. *lithos* a stone]

se·poy (sē′poi) *n.* A native Indian soldier outfitted and trained in European style; especially, one employed in the former British Indian Army. [<Pg. *sipae* <Urdu *sipāhī* a soldier <Persian <*sipāh* an army]

Sepoy Mutiny The Indian Mutiny. Also **Sepoy Rebellion.**

sep·pu·ku (sep·pōō·kōō) *n. Japanese* Hara-kiri.

sep·sis (sep′sis) *n. Pathol.* 1 Poisonous putrefaction. 2 Infection of the blood by putrescent material containing pathogenic microorganisms. [<NL <Gk. *sēpsis* <*sēpein* make putrid]

sept (sept) *n.* A division of a tribe ruled by a hereditary chief, especially in ancient and medieval Ireland; any similar social unit or group descended from a common ancestor. [Prob. <OF *septe*, var. of *secte* SECT] — **sep'tal** *adj.*

sept-[1] Var. of SEPTI-[1].

sept-[2] Var. of SEPTI-[2].

sep·ta (sep'tə) Plural of SEPTUM.

sep·tan·gle (sep'tang'gəl) *n.* A heptagon. [<LL *septangulus* <*septem* seven + *angulus* an angle] — **sep·tan·gu·lar** (sep·tang'gyə·lər) *adj.*

sep·tar·i·um (sep·târ'ē·əm) *n. pl.* **·tar·i·a** (-târ'ē·ə) *Geol.* A rock nodule or concretion, usually several feet in diameter and roughly spherical, having a compact crust and an internal mass broken up by angular radiating or intersecting cracks usually filled with a foreign mineral: also called *turtlestone.* [<NL <L *septum* an enclosure, wall] — **sep·tar'i·an** *adj.*

sep·tate (sep'tāt) *adj.* Divided by or provided with a partition or partitions; having a septum or septa. [<NL *septatus* <LL, surrounded <L *septum* an enclosure, wall]

sep·tec·to·my (sep·tek'tə·mē) *n. Surg.* Excision of a part of the nasal septum. [<SEPT-[2] + -ECTOMY]

Sep·tem·ber (sep·tem'bər) The ninth month of the year, containing 30 days; the seventh month in the old Roman calendar. — **massacre of September** The massacre in Paris in September, 1792, when 10,000 persons were put to death in prison by order of Danton: also **September massacre** or **massacres.** [<L *septem* seven]

Sep·tem·brist (sep·tem'brist) *n.* A member of the Parisian mob that massacred political prisoners in the massacre of September 2 to 6, 1792; hence, a cruel and bloodthirsty person; a butcher; murderer.

sep·te·mi·a (sep·tē'mē·ə) *n.* Septicemia. Also **sep·tae'mi·a.** [<NL <Gk. *sēptos* putrid + *haima* blood]

sep·tem·vir (sep·tem'vər) *n. pl.* **·virs** or **·vi·ri** (-vi·rī) One of seven men in Roman history associated in some office, authority, or work. [<L <*septem* seven + *vir* a man]

sep·te·nar·y (sep'tə·ner'ē) *adj.* 1 Consisting of, pertaining to, or being seven. 2 Septennial. 3 Septuple. — *n. pl.* **·nar·ies** 1 The number seven; heptad. 2 A group of seven things of any kind; anything that has some definite relation to the number seven. 3 A verse containing seven feet. Also **sep'te·nar'i·us** (sep'tə·nâr'ē·əs). [<L *septenarius* <*septeni* seven each <*septem* seven]

sep·te·nate (sep'tə·nāt) *adj.* Having seven parts, or the parts in sevens. [<L *septeni* seven each (<*septem* seven) + -ATE[1]]

sep·ten·nate (sep·ten'āt) *n.* A period of seven years; a term of office or the like lasting seven years. [<F *septennat* <L *septennis* (<*septem* seven + *annus* a year)]

sep·ten·ni·al (sep·ten'ē·əl) *adj.* 1 Recurring every seven years. 2 Continuing or capable of lasting seven years. [<L *septennium* a period of seven years <*septem* seven + *annus* a year] — **sep·ten'ni·al·ly** *adv.*

Sep·ten·tri·o (sep·ten'trē·ō) The constellation Ursa Major; the Dipper. See CONSTELLATION. Also **Sep·ten'tri·on.** [<L, sing. of *septentriones*, orig. *septem triones* the seven stars of the Big Dipper <*septem* seven + *triones*, pl. of *trio* a plow ox]

sep·ten·tri·on (sep·ten'trē·on) *adj.* Of, pertaining to, or coming from the north; boreal. — *n.* The north; northern regions. [<L *septentrionalis* <*septentrio* SEPTENTRIO] — **sep·ten'tri·o·nal** (-trē·ə·nəl) *adj.*

sep·tet (sep·tet') *n.* 1 A group of seven singers, players, or other persons, things, or parts. 2 *Music* A composition for seven voices or instruments. Also **sep·tette'.** [<G <L *septem* seven]

septi-[1] *combining form* Seven: *septilateral.* Also, before vowels, *sept-.* [<L *septem* seven]

septi-[2] *combining form* 1 A partition; fence: *septicidal.* 2 *Med.* The nasal septum: *septectomy.* Also, before vowels, *sept-.* Also **septo-.** [<L *septum* an enclosure, wall]

sep·tic (sep'tik) *adj.* 1 Of, pertaining to, or caused by sepsis. 2 Productive of putrefaction; putrid. Also **sep'ti·cal.** — *n.* Any substance that produces or promotes putrefaction. [<LL *septicus* <Gk. *sēptikos* <*sēpein* putrefy]

sep·ti·ce·mi·a (sep'tə·sē'mē·ə) *n. Pathol.* A morbid condition of the blood due to infection by pathogenic micro-organisms; blood poisoning: also called *septemia.* Also **sep'ti·cae'mi·a.** [<NL <Gk. *sēptikos* putrefactive + *haima* blood] — **sep'ti·ce'mic** (-sē'mik) *adj.*

sep·ti·ci·dal (sep'tə·sīd'l) *adj. Bot.* Dividing at the partitions: said of the dehiscence of a plant capsule that resolves itself at maturity into its component carpels by splitting through the septa. Also **sep'ti·cide.** [<SEPTI-[2] + L *caedere* cut] — **sep'ti·ci'dal·ly** *adv.*

sep·tic·i·ty (sep·tis'ə·tē) *n.* The quality of being septic; sepsis.

septic tank A tank in which sewage is allowed to remain until purified by the action of anaerobic bacteria.

sep·tif·ra·gal (sep·tif'rə·gəl) *adj. Bot.* Breaking away from the partitions: said of a form of dehiscence in plants. [<SEPTI-[2] + L *frangere* break]

sep·ti·lat·er·al (sep'tə·lat'ər·əl) *adj.* Seven-sided. [<SETPI-[1] + LATERAL]

sep·til·lion (sep·til'yən) *n.* A cardinal number: in the French system and in the United States, 1 followed by 24 ciphers; in the English system, 1 followed by 42 ciphers. — *adj.* Numbering a septillion. [<F *septillion* <L *sept(em)* + F *(m)illion* a million] — **sep·til'·lionth** *adj. & n.*

sep·time (sep'tēm) *n.* The seventh position of a swordsman in fencing. [<L *septimus* seventh <*septem* seven]

sep·tu·a·ge·nar·i·an (sep'cho͞o·ə·jə·nâr'ē·ən, sep'to͞o-) *n.* A person 70 years old, or between 70 and 80. [<L *septuagenarius* <*septuaginta* seventy]

sep·tu·ag·e·nar·y (sep'cho͞o·aj'ə·ner'ē, sep'to͞o-) *adj.* 1 Containing or consisting of 70. 2 Pertaining to a septuagenarian.

sep·tu·a·ges·i·ma (sep'cho͞o·ə·jes'ə·mə, sep'·to͞o-) *n.* A period of 70 days. [<L *septuagesima (dies)* the seventieth (day), fem. of *septuagesimus* seventieth <*septuaginta* seventy] — **sep'tu·a·ges'i·mal** *adj.*

Sep·tu·a·ges·i·ma (sep'cho͞o·ə·jes'ə·mə, sep'·to͞o-) *n.* The third Sunday before Lent. Also **Septuagesima Sunday.** [<L, seventieth, on analogy with *Quadragesima, Quinquagesima*]

Sep·tu·a·gint (sep'cho͞o·ə·jint', sep'to͞o-) *n.* An old Greek version of the Old Testament Scriptures, made in Alexandria between 280 and 130 B.C. It is the version used by the Greek Church. [<L *septuaginta* seventy; from a tradition that it was produced for Ptolemy II in 70 days by a group of 72 scholars] — **Sep'·tu·a·gin'tal** *adj.*

sep·tum (sep'təm) *n. pl.* **·ta** (-tə) *Biol.* 1 A dividing wall between two cavities: the nasal *septum.* 2 A partition, as in coral or in a spore. [<L *sepere* enclose <*sepes* a hedge] — **sep'tal** *adj.*

sep·tu·ple (sep'to͞o·pəl, -tyo͞o-, sep·to͞o'-, -tyo͞o'-) *adj.* 1 Consisting of seven; sevenfold. 2 Multiplied by seven; seven times repeated. — *v.t. & v.i.* **·pled, ·pling** To multiply by seven; make or become septuple. — *n.* A number or sum seven times as great as another. [<L *septuplus* <*septem* seven]

sep·tu·pli·cate (sep·to͞o'plə·kit, -tyo͞o'-) *adj.* 1 Sevenfold. 2 Raised to the seventh power. — *v.t.* (-kāt) **·cat·ed, ·cat·ing** To multiply by seven; septuple. — *n.* (-kit) One of seven like things. [<L *septuplus*, on analogy with *duplicate, triplicate,* etc.] — **sep·tu'pli·cate·ly** *adv.* — **sep·tu'pli·ca'tion** *n.*

sep·ul·cher (sep'əl·kər) *n.* A burial place, especially one found or made in a rock or solidly built of stone; tomb; vault. 2 A receptacle for relics, especially in an altar slab; a box or urn in a chapel to receive the Holy Sacrament: also called the *repository.* — **the Holy Sepulcher** The rock-hewn tomb in which the body of Jesus was buried. — *v.t.* **·chered** or **·chred, ·cher·ing** or **·chring** To place in a grave; entomb; bury. Also **sep'ul·chre** (-kər). [<OF *sepulcre* <L *sepulcrum* a burial place, tomb <*sepultus*, pp. of *sepelire* bury]

se·pul·chral (si·pul'krəl) *adj.* 1 Pertaining to a sepulcher. 2 Suggestive of burial or the grave; dismal in color or aspect, or unnaturally low or hollow in tone; gloomy: a *sepulchral* color, a *sepulchral* voice. — **se·pul'chral·ly** *adv.*

sep·ul·ture (sep'əl·chər) *n.* 1 The act of entombing; burial. 2 A sepulcher. [<OF <L *sepultura* burial <*sepultus.* See SEPULCHER.]

se·qua·cious (si·kwā'shəs) *adj.* 1 Disposed to follow; following; attendant. 2 Logically consecutive. 3 Ductile; pliable. [<L *sequax, -acis* following, pursuing <*sequi* attend, follow] — **se·qua'cious·ly** *adv.* — **se·quac·i·ty** (si·kwas'ə·tē) *n.*

se·quel (sē'kwəl) *n.* 1 Something which follows and serves as a continuation; a development from what went before. 2 A narrative discourse which, though entire in itself, develops from a preceding one. 3 A consequence; upshot; result. [<OF *sequelle* <L *sequela* <*sequi* follow]

se·que·la (si·kwē'lə) *n. pl.* **·lae** (-lē) 1 One who or that which follows. 2 *Pathol.* A morbid condition resulting from a preceding disease. [<L, a sequel]

se·quence (sē'kwəns) *n.* 1 The process or fact of following in space, time, or thought; succession or order: also **se'quen·cy.** 2 Order of succession; arrangement. 3 A number of things following one another, considered collectively; a series. 4 An effect or consequence. 5 *Music* A regular succession of similar melodic phrases at different pitches. 6 *Eccl.* In the Eucharistic liturgy, a prose or hymn sung immediately after the gradual and before the gospel. 7 In card games, a set of three or more cards next each other in value; in poker, a straight. 8 A section of motion-picture film presenting a single episode, without time lapses or interruptions. 9 *Math.* An ordered succession of quantities, as $2x$, $4x^2$, $8x^3$, $16x^4$ . . . $2^n x^n$, a finite sequence, and $x_1, x_2, x_3, \ldots x_n, \ldots$, or x_n, an infinite sequence. See synonyms under TIME. [<MF *sequence* <L *sequentia* <*sequens, -entis,* ppr. of *sequi* follow]

sequence of tenses See under TENSE[2].

se·quent (sē'kwənt) *n.* That which follows; a consequence; result. — *adj.* 1 Following in the order of time; succeeding. 2 Consequent; resultant. [<OF <L *sequens.* See SEQUENCE.]

se·quen·tial (si·kwen'shəl) *adj.* 1 Characterized by or forming a sequence, as of parts. 2 Sequent. — **se·quen·ti·al·i·ty** (si·kwen'shē·al'ə·tē) *n.* — **se·quen'tial·ly** *adv.*

se·ques·ter (si·kwes'tər) *v.t.* 1 To place apart; separate; segregate. 2 To seclude; withdraw: often used reflexively. 3 *Law* To take (property) into custody until a controversy, claim, etc., is settled or satisfied. 4 In international law, to confiscate and control (enemy property) by preemption. [<OF *sequestrer* <LL *sequestrare* remove, lay aside <*sequester* a trustee] — **se·ques'tra·ble** *adj.*

se·ques·tered (si·kwes'tərd) *adj.* Retired; secluded.

se·ques·trant (si·kwes'trənt) *n.* 1 That which sets apart, divides, or sequesters. 2 *Chem.* A substance which forms a colorless mixture with certain metals precipitated in a solution.

se·ques·trate (si·kwes'trāt) *v.t.* **·trat·ed, ·trat·ing** 1 To seize, especially for the use of the government; confiscate. 2 To take possession of for a time, with a view to the just settlement of the claims of creditors. 3 To seclude; sequester. [<LL *sequestratus,* pp. of *sequestrare.* See SEQUESTER.] — **se·ques·tra·tion** (sē'·kwes·trā'shən, sek'wəs-) *n.* — **se·ques·tra·tor** (sē'kwes·trā'tər, si·kwes'trā·tər) *n.*

se·ques·trum (si·kwes'trəm) *n. pl.* **·tra** (-trə) *Pathol.* A piece of dead bone remaining in its place, but separated from the living bone. [<NL <L, something separated, orig. neut. of *sequester* standing apart]

se·quin (sē'kwin) *n.* 1 An obsolete gold coin of the Venetian republic later introduced into Turkey: also spelled *zecchino.* 2 A spangle or coinlike ornament sewn on clothing. [<F <Ital. *zecchino* <*zecca* the mint <Arabic *sikka* a coining-die]

se·quoi·a (si·kwoi'ə) *n.* One of a genus *(Sequoia)* of gigantic trees (family *Taxodiaceae*) of the western United States, including only two species, the redwood *(S. sempervirens)* and the mammoth or "big" tree *(S. gigantea* or *Sequoiadendron giganteum),* both natives of California. [<NL, after *Sikwayi,* 1770?–1843,

a half-breed Cherokee Indian who invented the Cherokee alphabet]

Sequoia National Park A government reservation in east central California including Mount Whitney and containing many giant sequoias and redwoods; 602 square miles; established, 1890.

ser (sir) See SEER².

ser– Var. of SERO–.

se·ra (sir′ə) Plural of SERUM.

sé·rac (sā·räk′) *n. Geol.* One of the largest angular blocks or tower-shaped forms into which glacier ice breaks in passing down steep inclines. [< dial. F (Swiss), a cheese put up in cubic form; from its resemblance to the shape of this cheese]

ser·a·file (ser′ə·fīl) See SERREFILE.

se·ra·glio (si·ral′yō, -räl′-) *n.* **1** A harem. **2** Loosely, a place of debauchery. **3** The old palace of the sultans at Constantinople with its mosques, official buildings, and gardens. **4** Hence, any residence of a sultan. Also **se·rail** (se·räl′). [<Ital. *serraglio* an enclosure, ult. <LL *serrare,* var. of L *serare* lock up < *sera* lock; used to render Turkish *serai* a palace, lodging, because of similarity of sound]

se·ra·i (se·rä′ē) *n.* **1** In the Orient, an inn or caravansary. **2** A Turkish palace. [<Turkish <Persian *sarāī*]

Se·ra·je·vo (se′rä·yā′vō) See SARAJEVO.

se·ra·pe (se·rä′pē) *n.* A shawl or blanketlike outer garment worn in Latin America, especially in Mexico: also spelled *zarape.* [<Sp.]

ser·aph (ser′əf) *n. pl.* **ser·aphs** or **ser·a·phim** (ser′ə·fim) A celestial being; an angel of the highest order. *Is.* vi 2–6. [Back formation from *Seraphim,* pl. <LL <Hebrew *serāphīm,* ? ult. < *sāraph* burn] **— se·raph·ic** (si·raf′ik), **se·raph′i·cal** *adj.* **— se·raph′i·cal·ly** *adv.*

Seraphic Doctor Saint Bonaventura: so called because of the religious purity and fervor of his life.

ser·a·phim (ser′ə·fim) *n.* **1** Plural of SERAPH: also **ser′a·phin** (-fin). **2** *pl.* **·phims** A seraph, as in *Isaiah* vi 2, 6: an erroneous usage.

ser·a·phine (ser′ə·fēn) *n.* A coarse-toned musical instrument, a kind of harmonium, played with a keyboard. Also **ser·a·phi·na** (ser′ə·fē′·nə). [<SERAPH + -INE¹]

Se·ra·pis (si·rā′pis) In Egyptian mythology, a god of the lower world in the form of the dead Apis. **— Se·ra′pic** *adj.*

ser·as·kier (ser′əs·kir′, si·ras′kir) *n.* A Turkish minister of war, or commander in chief. Also **se·ras·ker** (si·ras′kər), **ser·as·quier** (ser′əs·kir′). [<Turkish <Persian *ser asker* head of an army < *ser* head + Arabic *'asker* an army]

Ser·bi·a (sûr′bē·ə) A constituent republic of eastern Yugoslavia; 34,107 square miles; capital, Belgrade; formerly an independent kingdom: formerly *Servia. Serbo-Croatian* **Sr·bi·ja** (sûr′bē·yä). **— Serb, Ser′bi·an** *adj.* & *n.*

Ser·bo-Cro·a·tian (sûr′bō·krō·ā′shən) *n.* **1** The South Slavic language of Yugoslavia, including all the old languages and dialects of Serbia, Montenegro, Bosnia, Herzegovina, Croatia, Slavonia, and Dalmatia. **2** A native of Yugoslavia; any person whose native tongue is Serbo-Croatian. Also spelled *Servo-Croatian.* Also **Ser′bo-Cro′at.** **—** *adj.* Of or pertaining to the people of Yugoslavia, or to their language.

Ser·bo·ni·an Bog or **Lake** (sûr·bō′nē·ən) A large marshy tract once existing near the Red Sea littoral of Lower Egypt, in which Herodotus said whole armies were engulfed; hence, a strait; difficulty; complication: also spelled *Sirbonian.*

Serbs, Croats, and Slovenes, Kingdom of the See YUGOSLAVIA.

ser·dab (sûr′dab, sûr·däb′) *n.* A secret cell within the masonry of an ancient Egyptian tomb, in which images of the deceased were deposited. [<Arabic *serdāb* a cellar <Persian, an icehouse, a grotto]

sere¹ (sir) See SEAR¹.

sere² (sir) *n. Ecol.* The series of changes found in a given plant formation from the initial to the ultimate stage. ◆ Homophones: *cere, sear.* [Back formation <SERIES] **— ser′al** *adj.*

se·rein (sə·ran′) *n. Meteorol.* A fine rain that falls sometimes from an apparently clear sky, especially in the tropics after sunset. [<F]

ser·e·nade (ser′ə·nād′) *n.* **1** An evening song, usually that of a lover beneath his lady's window; also, by extension, music performed in honor of some person in front of his residence in the open air at night. **2** The music for such a song. **—** *v.t.* & *v.i.* **·nad·ed, ·nad·ing** To entertain with a serenade. [<F *sérénade* <Ital. *serenata* < *sereno* serene, open air <L *serenus* clear, serene; infl. in meaning by L *sera (hora)* the evening (hour), fem. of *serus* late] **— ser′·e·nad′er** *n.*

ser·e·na·ta (ser′ə·nä′tə) *n. Music* **1** A dramatic cantata on any imaginative or simple subject not sacred, intended to be performed in the open air. **2** An instrumental work or orchestral suite usually beginning with a march and including a minuet. [<Ital. See SERENADE.]

ser·en·dip·i·ty (ser′ən·dip′ə·tē) *n.* The faculty of happening upon or making fortunate discoveries when not in search of them. [Coined by Horace Walpole (1754), in *The Three Princes of Serendip* (Ceylon), the heroes of which make such discoveries]

se·rene (si·rēn′) *adj.* **1** Clear, or fair and calm; having its brightness undimmed: a *serene* sky. **2** Marked by peaceful repose; tranquil; unruffled; placid: a *serene* spirit. **3** Of exalted rank: chiefly in the titles of certain continental European princes: His *Serene* Highness. See synonyms under SEDATE. **—** *n. Rare* or *Poetic* **1** Clearness, or a serene or clear region. **2** Calmness; placidity. [<L *serenus*] **— se·rene′ly** *adv.* **— se·ren·i·ty** (si·ren′ə·tē), **se·rene′ness** *n.*

Se·reth (zā′ret) See SIRET.

serf (sûrf) *n.* **1** A person who is attached to the estate on which he lives; loosely, a peasant. **2** Figuratively, one in servile subjection. ◆ Homophone: *surf.* [<OF <L *servus* a slave] **— serf′dom, serf′age, serf′hood** *n.*

serge (sûrj) *n.* **1** A strong twilled fabric made of wool yarns and characterized by a diagonal rib on both sides of the cloth. **2** In the Middle Ages, a coarse woolen cloth. **3** A rayon lining fabric. ◆ Homophone: *surge.* [<OF *sarge, serge* <L *serica (lana)* (wool) of the Seres <*Seres* the Seres, an eastern Asian people]

ser·geant (sär′jənt) *n.* **1** A non-commissioned military officer ranking next above a corporal. See the table under GRADE. **2** In the United States, a police officer of rank next below a captain (sometimes lieutenant); in England, one next below an inspector. **3** *Brit.* Formerly, one who held land of the king by tenure of military service, or a squire or gentleman of less than knightly rank; one of the household officials of a sovereign. **4** A sergeant at arms. **5** A sergeant at law. **6** A constable or bailiff. **7** The sergeant fish. Also *serjeant.* **— color sergeant** A sergeant who carries the regimental or national colors or standard. **— lance sergeant** A corporal acting as sergeant. **— mess sergeant** A non-commissioned officer who plans meals, issues rations, and superintends the company mess under the mess officer. [<OF *sergent, serjant* <L *serviens,* ppr. of *servire* serve] **— ser′gean·cy, ser′geant·cy, ser′geant·ship** *n.*

sergeant at arms **1** An executive officer in a legislative body who enforces order; especially, *Brit.,* the attendant on the lord chancellor or on the speaker of the House of Commons. **2** The title of certain court or city officials who have ceremonial duties.

sergeant at law Formerly, a barrister of high order or rank taking social but not professional precedence of king's counsel.

sergeant fish **1** A large, dusky fish *(Rachycentron canadum)* of warm seas, with a broad black band suggesting a chevron on the sides. **2** The robalo.

sergeant major **1** In the U.S. Army, the principal enlisted assistant to the adjutant of a battalion or higher unit. **2** The highest non-commissioned officer in the U.S. Marine Corps.

ser·geant·y (sär′jən·tē) *n. Brit.* Formerly, a tenure of lands on condition of rendering some personal or menial service directly to the king or to some nobleman; also, the service rendered.

Ser·gi·pe (sər·zhē′pə) A state in eastern Brazil on the Atlantic, 8,502 square miles; capital, Aracajú.

se·ri·al (sir′ē·əl) *adj.* **1** Of the nature of a series. **2** Published in a series at regular intervals. **3** Successive; arranged in rows or ranks: also **se·ri·ate** (sir′ē·it, -āt). **—** *n.* **1** A novel or other story regularly presented in successive instalments, as in a magazine, on radio or television, or in motion pictures. **2** *Brit.* A periodical. **3** A subdivision of a military unit organized for transport or for marching. ◆ Homophone: *cereal.* [<NL *serialis* <L *series* a row, order] **— se′ri·al·ly, se′ri·ate·ly** *adv.*

serial comma A comma placed before an *and* which joins the last two in a series of three or more substantives, adjectives, phrasal modifiers, or adverbs. ◆ Opinion is evenly divided as to whether the serial comma is needed. The best argument for its use, which is observed in this dictionary, is that it makes the meaning unmistakable and the relation between the various modifiers unmistakably clear: The motion was opposed by Lords Arundel, Salisbury, Somerset, and Say and Sele. If the last comma (the serial comma) were omitted here, there could be confusion as to the name of the last peer, which is *Say and Sele.*

se·ri·al·ize (sir′ē·əl·īz′) *v.t.* **·ized, ·iz·ing** To arrange or publish in serial form. **— se′ri·al·i·za′tion** *n.*

serial number A number assigned to a person, object, item of merchandise, etc., as a means of identification.

serial symmetry **1** The symmetry of serial parts. **2** Metamerism. Also **serial homology.**

se·ri·a·tim (sir′ē·ā′tim, ser′ē-) *adv.* One after another; in connected order; serially. [< Med. L <L *series,* on analogy with *gradatim*]

se·ri·a·tion (sir′ē·ā′shən) *n.* The arrangement of unorganized material or data in an orderly series.

se·ri·ceous (si·rish′əs) *adj.* **1** Lustrous like silk; silky. **2** *Bot.* Having fine, soft, appressed hairs, as the leaves of certain plants. [<L *sericeus* < *sericum* silk, orig. neut. of *sericus* silken, belonging to the Seres. See SERGE.]

ser·i·cin (ser′ə·sin) *n. Biochem.* A viscous substance formed on the surface of raw silk fiber and usually removed by boiling in soapy water. [<L *sericus* silken + -IN]

ser·i·cul·ture (sir′ə·kul′chər) *n.* The raising and care of silkworms for the production of raw silk. [Contraction of F *sériciculture* <L *sericum* silk + *cultura* a raising, culture] **— ser′i·cul′tur·al** *adj.* **— ser′i·cul′tur·ist** *n.*

ser·i·e·ma (ser′i·ē′mə, -ā′mə) *n.* **1** A long-legged crested bird *(Cariama cristata)* of the plains of Brazil and Paraguay. **2** The smaller species, Burmeister's cariama *(Chunga burmeisteri)* of Argentina. [<NL *seriema, cariama* <Tupian *siriema, sariama* crested]

se·ries (sir′ēz) *n. pl.* **se·ries** **1** An arrangement of one thing after another; a connected succession of persons, books, objects, observations, etc., on the basis of like relationships. **2** *Math.* An ordered, finite or infinite arrangement of expressions, each a function of another, the sum of which is indicated, as $x_1 + x_2 + x_3 \ldots + x_n + \ldots$, or $\sum_{i=1}^{\infty} x_i$ for infinite series, and $x_1 + x_2 + x_3 \ldots + x_n$, or $\sum_{i=1}^{n} x_n$ for finite series. **3** *Chem.* A group of compounds or elements resembling one another more or less in their chemical characters and crystalline forms, or differing from each other by a constant difference of certain factors. **4** *Electr.* An arrangement of sources or utilizers of electricity, as batteries or lamps, in which the positive electrode of one is connected with the negative electrode of another. **5** *Gram.* A group of successive coordinate elements of a sentence. [<L <*serere* join, weave together]

series motor A motor whose field and armature windings are in series.

series winding *Electr.* The winding of a dynamo or an electric motor in such a way that the field-magnet coil is a part of the armature and exterior circuit. Compare SHUNT-WOUND. **— se·ries-wound** (sir′ēz·wound′) *adj.*

ser·if (ser′if) *n. Printing* A hairline; a light line or stroke crossing or projecting from the end of a main line or stroke in a letter: also spelled *ceriph.* [<Du. *schreef* a stroke, line < *schrijve* write <L *scribere*]

ser·i·graph (ser′ə·graf, -gräf) *n.* **1** An artist's color print made by serigraphy. **2** A method for testing the tensile strength and elasticity of textile fabrics, paper, leather, rubber, etc., under specified conditions. [<L *sericum* silk + GRAPH]

se·rig·ra·phy (si·rig′rə·fē) *n.* An adaptation of the silk-screen process in which hand-made color prints are made on any desired surface by the use of stencils painted upon or cemented to the screen, one stencil to each color, the finished print being in all details the work of an individual artist in distinction from those commercially reproduced by silk-screen printing. — **se·rig′ra·pher** *n.* — **ser·i·graph·ic** (ser′ə·graf′ik) *adj.*

ser·in (ser′in) *n.* A small greenish finch (*Serinus canarius*), related to and closely resembling the wild canary, but smaller. [<F; ult. origin unknown]

ser·ine (ser′ēn, -in) *n. Biochem.* A white crystalline amino acid, $C_3H_7NO_3$, obtained as a dissociation product of various proteins. Also **ser′in.** [<L *sericus* silken + -INE[2]; so called because originally obtained from dissociation of sericin]

se·rin·ga (si·ring′gə) *n.* Any of several Brazilian trees (genus *Hevea*) yielding rubber. [<Pg. <L *syringa* SYRINGA]

Se·rin·ga·pa·tam (sə·ring′gə·pə·tam′) A town in southern Mysore, India; former seat of the sultans of Mysore, 7 miles NE of Mysore city.

se·ri·o-com·ic (sir′ē·ō·kom′ik) *adj.* Mingling mirth and seriousness, or the comic with an appearance of gravity. Also **se′ri·o-com′i·cal.** [<*serio-* partly serious (<SERIOUS) + COMIC]

se·ri·ous (sir′ē·əs) *adj.* **1** Grave and earnest in quality, feeling, or disposition; thoughtful; sober. **2** Said, planned, or done with full practical intent; not jesting or making a false pretense; being or done in earnest. **3** Of grave importance; weighty; attended with considerable danger or loss: a *serious* matter, a *serious* accident. **4** Particularly attentive to religion. [<MF *sérieux* <LL *seriosus* <L *serius*] — **se′ri·ous·ly** *adv.* — **se′ri·ous·ness** *n.*

Synonyms: dangerous, demure, earnest, grave, great, important, momentous, sedate, sober, solemn. A *serious* person is *sedate, sober, solemn*; a *serious* purpose is *earnest*; a *serious* illness is *dangerous*; a *serious* business is *important*, and may be *momentous*. See BAD, GOOD, IMPORTANT, SEDATE. *Antonyms*: careless, gay, insignificant, jocose, jolly, light, slight, thoughtless, trifling, trivial, volatile.

ser·jeant (sär′jənt) See SERGEANT.

ser·mon (sûr′mən) *n.* **1** A discourse based on a passage or text of the Bible, delivered as part of a church service; hence, any discourse intended for the pulpit. **2** Any discourse of a serious kind; an exhortation to duty or a formal reproof. See synonyms under SPEECH. [<AF *sermun*, OF *sermon* <L *sermo, -onis* talk]

ser·mon·et (sûr′mən·et′) *n.* A brief sermon. Also **ser′mon·ette′.** [Dim. of SERMON]

ser·mon·ic (sər·mon′ik) *adj.* Pertaining to or of the nature of a sermon or sermonizing; didactic. Also **ser·mon′i·cal.**

ser·mon·ize (sûr′mən·īz) *v.t.* & *v.i.* **·ized, ·iz·ing** To compose or deliver sermons (to); address or discourse at length in a didactic manner. — **ser′mon·iz′er** *n.*

Sermon on the Mount The discourse of Jesus found recorded in *Matt.* v, vi, vii: properly distinguished from the **Sermon on the Plain,** *Luke* vi 20–49.

sero- *combining form* Connected with or related to serum: *serology.* Also, before vowels, *ser-.* [<L *serum* whey]

se·rol·o·gy (si·rol′ə·jē) *n.* The science of serums and their actions: also called *orrhology.* [SERO- + -LOGY] — **se·ro·log·i·cal** (sir′ə·loj′i·kəl) *adj.*

se·roon (si·rōōn′) *n.* A bale of goods, as Spanish dates, figs, etc., packed in an animal's hide: also spelled *ceroon.* [<Sp. *serón* a hamper, crate <*sera* a large basket]

se·ros·i·ty (si·ros′ə·tē) *n.* **1** The condition of being serous or watery. **2** A watery or serous secretion. Also **se′rous·ness.** [<F *sérosité* <NL *serositas* <*serosus* SEROUS]

se·ro·ther·a·py (sir′ō·ther′ə·pē) *n. Med.* The treatment of disease by injecting into the veins serum from immunized animals.

se·rot·i·nous (si·rot′ə·nəs) *adj.* Produced, blossoming, or developing relatively late in the season: used also figuratively. Also **se·rot′i·nal, ser·o·tine** (ser′ə·tin, -tīn). [<L *serotinus* <*serus* late]

ser·o·to·nin (ser′ə·tō′nin) *n. Biochem.* A crystalline protein found in the serum of clotted blood and in various animals and plants: it is associated with a wide range of physiological processes, especially in the brain and blood vessels. [<SERO- + TON- + -IN]

se·rous (sir′əs) *adj.* Pertaining to, producing, or resembling serum. [<F *séreux* <L *serosus* <*serum* serum, whey]

serous fluid Any of the thin watery fluids secreted by the serous membranes.

serous membrane *Anat.* A tissue of endothelial cells lining the large cavities of the body, as the peritoneum and the pleura.

ser·ow (ser′ō) *n.* Any of a genus (*Capricornis*) of antelopes ranging from the Himalayas to Japan; especially, the large goat antelope (*C. bubalinus*). [<Tibetan]

Ser·pens (sûr′penz) *Astron.* An equatorial constellation, the Serpent, between Corona Borealis and Libra. See CONSTELLATION. [<L, a serpent]

ser·pent (sûr′pənt) *n.* **1** A scaly, limbless reptile; a snake, especially when of large size. **2** Anything of serpentine form or appearance, as a certain kind of twisting firework. **3** An obsolete musical wind instrument, bent several times in serpentine form. **4** An insinuating and treacherous person. **5** Satan. [<OF <L *serpens, -entis* a serpent, creeping thing, orig. ppr. of *serpere* creep]

serpent fence A worm fence. See under FENCE.

ser·pen·tine (sûr′pən·tēn, -tīn) *adj.* **1** Pertaining to or like a serpent; zigzag or sinuous; crawling sinuously. **2** Subtle; cunning. — *n.* A massive or fibrous, often mottled green or yellow, hydrous magnesium silicate, the fibrous varieties of which are important sources of asbestos, the massive as architecturally decorative stones. [<OF *serpentin* <L *serpentinus* <*serpens* SERPENT]

Ser·pen·to (sûr′pən·tō) *n. pl.* **·to** The Shoshone Indians: a tribe belonging to the Shoshoneans.

ser·pi·go (sər·pī′gō) *n. Pathol.* An eruption on the skin; spreading ringworm. [<Med. L <L *serpere* creep] — **ser·pig·i·nous** (sər·pij′ə·nəs) *adj.*

Ser·ra (ser′rä), **Junípero,** 1713–84, Spanish Franciscan missionary in California.

ser·ra·noid (ser′ə·noid) *adj.* Of or pertaining to the *Serranidae*, a family of fishes including the sea bass, striped bass, and their allies. — *n.* A serranoid fish. [<NL <*Serranus*, genus name (<L *serra* a saw) + Gk. *eidos* form]

ser·rate (ser′āt, -it) *adj.* Toothed or notched like a saw, as the margins of certain leaves. Also **ser′rat·ed.** [<L *serratus* <*serra* a saw]

ser·ra·tion (se·rā′shən) *n.* **1** The state of being edged as with saw teeth. **2** *Biol.* One of the projections of a serrate formation, or a series of such projections. Also **ser·ra·ture** (ser′ə·chər). [<NL *serratio, -onis* <L *serratus*, pp. of *serrare* saw <*serra* a saw]

ser·re·file (ser′ə·fīl) *n.* **1** One of the non-commisioned officers drawn up in line in the rear of a troop or squadron. **2** *pl.* The line of serrefiles. Also spelled *serafile.* [<F <*serrer* tighten (<LL *serrare* lock <L *sera* a lock) + *file* a file[1]]

ser·ried (ser′ēd) *adj.* Compacted in rows or ranks, as soldiers in company formation. [Pp. of obs. *serry* press close together in ranks <MF *serré*, pp. of *serrer.* See SERREFILE.]

ser·ri·form (ser′ə·fôrm) *adj.* Formed like a saw; saw-toothed. [<L *serra* a saw + -FORM]

ser·ru·late (ser′ə·lit, -lāt, ser′yə-) *adj.* Diminutively serrate; serrate with small, fine teeth. Also **ser′ru·lat′ed** (-lā′tid). [<L *serrula*, dim. of *serra* a saw + -ATE[1]]

ser·ru·la·tion (ser′ə·lā′shən, ser′yə-) *n.* **1** The state of being or becoming serrulate; a fine notching. **2** One of the teeth of a serrulate margin.

Ser·to·ri·us (sər·tôr′ē·əs, -tō′rē-), **Quintus,** 121?–72 B.C., Roman general; assassinated.

ser·tu·lar·i·an (sûr′chōō·lâr′ē·ən) *n. Zool.* One of a genus (*Sertularia*) of branching colonial hydroids common between tide lines. [<NL <L *sertula*, dim. of *serta* a garland]

se·rum (sir′əm) *n. pl.* **se·rums** or **se·ra** (sir′ə) **1** The more fluid constituent of blood, lymph, milk, and similar animal liquids. **2** The serum of the blood of an animal which has been subjected to the process of immunization; any antitoxic blood serum or lymph. **3** Whey; serum of milk. **4** Any similar secretion. [<L, whey, watery fluid]

serum sickness Illness caused by inoculation of serum.

serum therapy Serotherapy.

ser·val (sûr′vəl) *n.* An African wildcat (*Felis serval*), yellow with black spots and having a ringed tail and long legs. [<F <Pg. *lobo cerval* <*lobo* a wolf (<L *lupus*) + *cerval* a stag <L *cervus*]

ser·vant (sûr′vənt) *n.* **1** A person employed to work for another; especially, in law, one employed to render service and assistance in some trade or vocation; an employee. **2** A person hired to assist in domestic matters, sometimes living within the employer's house; hired help. **3** A slave or bondman. **4** A government official. [<OF, orig. ppr. of *servir* SERVE]

serve (sûrv) *v.* **served, serv·ing** *v.t.* **1** To work for, especially as a servant; be in the service of. **2** To be of service to; wait on. **3** To promote the interests of; aid; help: to *serve* one's country. **4** To obey and give homage to: to *serve* God. **5** To satisfy the requirements of; suffice for. **6** To perform the duties connected with, as a public office. **7** To go through (a period of enlistment, term of punishment, etc.). **8** To furnish or provide, as with a regular supply. **9** To offer or bring food or drink to (a guest, etc.); wait on at table. **10** To bring and place on the table or distribute among guests, as food or drink. **11** To operate or handle; tend: to *serve* a cannon. **12** To copulate with: said of male animals. **13** In tennis, etc., to put (the ball) in play by hitting it to one's opponent. **14** *Law* **a** To deliver (a summons or writ) to a person. **b** To deliver a summons or writ to. **15** *Naut.* To wrap (a rope, stay, etc.), as with marlin or spun yarn, so as to strengthen or protect. — *v.i.* **16** To work as or perform the functions of a servant; wait at table. **17** To perform the duties of any employment, office, etc. **18** To go through a term of service, as in the army or navy. **19** To be suitable or usable, as for a purpose; perform a function. **20** To be favorable, as weather. **21** In tennis, etc., to put the ball in play. — *n.* **1** In tennis, etc., the delivering of the ball by striking it toward an opponent. **2** The turn of the server. [<OF *servir* <L *servire* <*servus* a slave]

Synonyms: advance, aid, assist, attend, benefit, help, minister, obey, promote, subserve, succor, suffice. See ACCOMMODATE. *Antonyms*: command, control, desert, disobey, hinder, obstruct, oppose, retard, thwart, withstand.

serv·er (sûr′vər) *n.* **1** One who serves; especially, an attendant aiding a priest at low mass. **2** That which is used in serving, as a tray. **3** The male of any domestic animal used for breeding. **4** The player who serves the ball in games.

Ser·ve·tus (sər·vē′təs), **Michael,** 1511–53, Spanish physician and theologian, burned at the stake in Geneva for heresy. Also *Sp.* **Miguel Ser·ve·to** (ser·vā′tō). — **Ser·ve′tian** (-shən) *adj.* & *n.*

Ser·vi·a (sûr′vē·ə) A former name for SERBIA.

ser·vice (sûr′vis) *n.* **1** Assistance or benefit afforded another: to render a *service*; to be of *service.* **2** A useful result or product of labor which is not a tangible commodity: in the plural, often contrasted with *goods.* **3** The manner in which one is waited upon or served: The *service* in this restaurant is only fair. **4** A system of labor and material aids used to accomplish some regular work or accommodation for the public: telephone *service*, train *service*, postal *service.* **5** A division of public employment devoted to a particular function: the diplomatic *service.* **6** Employment as a public servant in government: to enter public *service.* **7** A public duty or function: jury *service.* **8** Any branch of the armed forces: to enter the *service.* **9** Military duty or assignment: to volunteer for foreign *service.*

10 Devotion to God, as demonstrated by obedience and good works. **11** A formal and public exercise of worship: to attend Sunday *services.* **12** A ritual prescribed for a particular ministration or observance: a burial *service*; a marriage *service.* **13** The music for a liturgical office or rite. **14** The state or position of a servant, especially a domestic servant. **15** A set of tableware for a specific purpose: a tea *service.* **16** Installation, maintenance, and repair of an article provided a buyer by a seller. **17** *Law* **a** The legal communication of a writ or process to a designated person. **b** Duty or work rendered by one person for another. **c** A duty rendered by a feudal tenant as recompense to his lord. **18** In tennis and similar games, the act or manner of serving a ball. **19** *Naut.* The protective cordage wrapped around a rope. **20** In animal husbandry, the copulation or covering of a female. — *adj.* **1** Pertaining to or for service. **2** Used by, or for the use of, servants or tradespeople: a *service* entrance. **3** Of, pertaining to, or belonging to a military service: a *service* flag. **4** Worn during active military service: distinguished from *dress*: a *service* cap, hat, or uniform. — *v.t.* **·viced, ·vic·ing** **1** To maintain or repair: to *service* a car or radio. **2** To supply service to. [<OF *servise* <L *servitium* <*servus* a slave]

Synonyms (noun): advantage, avail, benefit, good, purpose, serviceableness, use, utility. See PROFIT, SACRAMENT, UTILITY.

Ser·vice (sûr′vis), **Robert William,** born 1874, Canadian writer.

ser·vice·a·ble (sûr′vis·ə·bəl) *adj.* **1** That can be made of service; beneficial; such as serves or can serve a useful purpose. **2** Capable of rendering long service; durable. **3** *Obs.* Obliging; attentive. See synonyms under GOOD, USEFUL. — **ser′vice·a·ble·ness, ser′vice·a·bil′i·ty** *n.* — **ser′vice·a·bly** *adv.*

ser·vice·ber·ry (sûr′vis·ber′ē) *n. pl.* **·ries** The Juneberry; the shadberry. [<*service,* the service tree + BERRY]

service book A book containing the offices, or forms of service, of any church that uses liturgical forms.

service cap A uniform cap of cotton, wool, or felt, with visor, authorized to be worn by all officers, warrant officers, and flight officers when not in formation with troops: also worn, when issued, by enlisted men.

service club An organization to promote community welfare and further the interests of its members.

service coat A single-breasted jacket worn with the U.S. Army service uniform: also called a *blouse.*

service hat A hat worn in the U. S. Army when full-dress uniform is not worn: its shape and material vary according to requirements of climate. Formerly, a khaki-colored felt hat with broad flat brim, and crown dented in four places.

ser·vice·man (sûr′vis·man′) *n. pl.* **·men** (-men′) A member of one of the armed forces. — **ser′vice·wom′an** *n. fem.*

service man A man who performs services of maintenance, supply, repair, etc.

service ribbon A distinctively colored ribbon worn on the U. S. service uniform to indicate the wearer's right to the corresponding campaign medal or decoration.

service station **1** A place for supplying automobiles with gasoline, oil, water, etc. **2** A place where adjustments and repairs can be made and parts obtained for electrical or mechanical devices.

service stripe A stripe worn on the sleeve of a uniform to denote years of service or employment, as in an army or on a police force: also *hash mark.*

service tree **1** An Old World tree *(Sorbus domestica)* with odd-pinnate leaves, panicled cream-colored flowers, and small edible fruit. **2** The American mountain ash *(S. americana).* **3** The Juneberry. [Orig. *serves,* pl. of obs. *serve,* OE *syrfe* <L *sorbus*]

service uniform The regulation uniform to be worn during routine or active service in the army or navy.

ser·vi·ette (sûr′vē·et′, -vyet′) *n.* A table napkin. [<MF, prob. <*servir* SERVE]

ser·vile (sûr′vil) *adj.* **1** Having the spirit of a slave; slavish; abject: a *servile* flatterer. **2** Pertaining to or appropriate for slaves or servants: a *servile* insurrection, *servile* employment. **3** Being of a subject class; existing in a condition of servitude. **4** Obedient; subject: with *to*: *servile* to applause. **5** *Ling.* Not belonging to the original root; serving only to modify the construction or pronunciation of a word. **6** Designating tenures of land in England subject to conditions distinguished from those of freehold, as labor instead of rent. See synonyms under BASE, OBSEQUIOUS. — *n.* **1** A slave, or one of slavish spirit; menial. **2** *Ling.* A letter, syllable, or sound used only to modify a word, and not part of its radical form. [<L *servilis* <*servus* a slave] — **ser′vile·ly** *adv.* — **ser·vil′i·ty, ser′vile·ness** *n.*

ser·vi·tor (sûr′və·tər) *n.* **1** One who waits upon and serves another; an attendant; follower; servant. **2** Formerly, an undergraduate at Oxford University partly supported by a college grant and partly earning his living by service. [<OF <LL <L *servire* SERVE] — **ser′vi·tor·ship′** *n.*

ser·vi·tude (sûr′və·to͞od, -tyo͞od) *n.* **1** The condition of a slave; slavery; bondage; now, especially, enforced service as a punishment for crime: penal *servitude.* **2** A state of subjection to any claim, demand, or control: *servitude* to vice. **3** The condition or duties of a servant; menial service. **4** The subjection of a person to a person or to a thing, or of a thing to a person or thing. **5** *Law* An easement; a right that one man may have to use the land of another for a special purpose. See synonyms under BONDAGE. [<MF <L *servitudo* <*servus* a slave]

servo- *combining form* In technical use, auxiliary: *servomechanism.* [<L *servus* a slave]

Ser·vo-Cro·a·tian (sûr′vō·krō·ā′shən) See SERBO-CROATIAN.

ser·vo·mech·a·nism (sûr′vō·mek′ə·niz′əm) *n.* Any of various relay devices which can be actuated by a comparatively weak force in the automatic control of a complex machine, instrument, operation, or process, as artillery fire, the course of an airplane or ship, etc. [<SERVO- + MECHANISM]

ser·vo·mo·tor (sûr′vō·mō′tər) *n.* An electric motor connected with and supplying power for a servomechanism.

ses·a·me (ses′ə·mē) *n.* An East Indian herb *(Sesamum indicum),* containing seeds which are used as food and as a source of the pale yellow **sesame oil,** used as an emollient. — **open sesame** A charm to secure admission, originally to the robbers' cave in the story of Ali Baba and the Forty Thieves in the *Arabian Nights.* [<F *sésame* <L *sesamum, sesama* <Gk. *sēsamon, sēsamē,* prob. <an Oriental source]

ses·a·moid (ses′ə·moid) *adj. Anat.* **1** Having the shape of a sesame seed; obovate; nodular: said specifically of certain bones, cartilages, and nodules. **2** Pertaining to a sesamoid. — *n.* A sesamoid bone or cartilage, as the kneecap. [<L *sesamoides* <Gk. *sēsamoeidēs* <*sēsamon* sesame + *eidos* form]

sesqui- *prefix* **1** One and a half; one-half more; one and a half times: *sesquicentennial.* **2** *Chem.* Indicating the presence of three atoms of one element and two of another in a compound, as chromium *sesquioxide,* Cr_2O_3. [<L *sesqui-* one-half more <*semis* half + *que* and]

ses·qui·cen·ten·ni·al (ses′kwi·sen·ten′ē·əl) *adj.* Of or pertaining to a century and a half. — *n.* A 150th anniversary, or its celebration.

ses·qui·pe·da·li·an (ses′kwi·pi·dā′lē·ən) *adj.* **1** Measuring a foot and a half. **2** Long and ponderous, as polysyllabic words. Also **ses·quip·e·dal** (ses·kwip′ə·dəl, ses′kwi·pēd′l). — *n.* A very long word. [<L *sesquipedalis* <*sesqui-* more by a half + *pes, pedis* a foot]

ses·qui·plane (ses′kwi·plān′) *n. Aeron.* A type of biplane one wing of which has half or less than half the area of the other.

ses·sile (ses′il) *adj.* **1** *Bot.* Attached by its base, without a stalk, as a leaf. **2** *Zool.* Fixed; sedentary; firmly or permanently attached. [<L *sessilis* sitting down, stunted <*sessus,* pp. of *sedere* sit] — **ses·sil′i·ty** *n.*

ses·sion (sesh′ən) *n.* **1** The sitting together of a legislative assembly, court, etc., for the transaction of business. **2** A single meeting or series of meetings of an assembly, court, or other body, for conducting business. **3** The governing body of a Presbyterian Church congregation. **4** In some educational institutions, a term. **5** *Law* The term for which a court or legislative body sits continuously for the transaction of business. **6** *pl.* The sitting of a certain court: the quarter-*sessions*; and in the United States, the Court of *Sessions,* a court of criminal jurisdiction; any one of certain courts, especially in England: general *sessions,* petty *sessions.* **7** *Obs.* The act of sitting, or the state of one who is seated. ◆ Homophone: *cession.* [<F <L *sessio, -onis* <*sessus,* pp. of *sedere* sit] — **ses′sion·al** *adj.* — **ses′sion·al·ly** *adv.*

sess·pool (ses′po͞ol′) See CESSPOOL.

ses·terce (ses′tûrs) *n. pl.* **·ter·ces** (-tûr′sēz) A coin of ancient Rome equal to 1/4 denarius: originally of silver, later of bronze. Also **ses·ter′ti·us** (-shē·əs). [<L *sestertius (nummus)* (a coin) that is two and a half <*semis* half + *tertius* third; so called because worth two and a half asses]

ses·ter·ti·um (ses·tûr′shē·əm) *n. pl.* **·ti·a** (-shē·ə) An ancient Roman money of account equivalent to 1,000 sesterces. [<L, short for *(mille) sestertium* (a thousand) sesterces, gen. pl. of *sestertius* a sesterce]

ses·tet (ses·tet′) *n.* **1** The last six lines of a sonnet; any six-line stanza. **2** *Music* A sextet. [<Ital. *sestetto* <*sesto* sixth (<L *sextus*) + *-etto,* dim. suffix]

ses·ti·na (ses·tē′nə) *n.* A verse form consisting of six stanzas of six, generally unrimed, lines each and a three-line envoy: the end words of the first stanza are progressively changed in order in the remaining five, and appear medially and terminally in the envoy. Also **ses′tine** (-tin). [<Ital. <*sesto* sixth <L *sextus*]

Ses·tos (ses′tos) A ruined town on the Dardanelles in Turkey in Europe. See ABYDOS.

set[1] (set) *v.* **set, set·ting** *v.t.* **1** To put in a certain place or position; place. **2** To put into a fixed or immovable position, condition, or state: to *set* brick; to *set* one's jaw. **3** To bring to a specified condition or state: *Set* your mind at ease; to *set* a boat adrift. **4** To restore to proper position for healing, as a broken bone. **5** To place in readiness for operation or use: to *set* a trap. **6** To adjust according to a standard: to *set* a clock. **7** To adjust (an instrument, dial, etc.) to a particular calibration or position. **8** To place knives, forks, etc., on (a table) in preparing for a meal. **9** To bend the teeth of (a saw) to either side alternately. **10** To appoint or establish; prescribe: to *set* a time or limit. **11** To fix or establish a time for: We *set* our departure for noon. **12** To assign for performance, completion, etc.; allot: to *set* a task. **13** To assign to some specific duty or function; appoint; station: to *set* a guard. **14** To cause to sit. **15** To present or perform so as to be copied or emulated: to *set* the pace; to *set* a bad example. **16** To give a specified direction to; direct: He *set* his course for the Azores. **17** To put in place so as to catch the wind: to *set* the jib. **18** To place in a mounting or frame, as a gem. **19** To stud or adorn with gems: to *set* a crown with rubies. **20** To place (a hen) on eggs to hatch them. **21** To place (eggs) under a fowl or in an incubator for hatching. **22** To place (a price or value): with *by* or *on*: to *set* a price on an outlaw's head. **23** To point (game): said of hunting dogs. **24** *Printing* **a** To arrange (type) for printing; compose. **b** To put into type, as a sentence, manuscript, etc. **25** *Music* To arrange (music) for words or write (words) to accompany music. **26** To describe (a scene) as taking place: to *set* the scene in Monaco. **27** In the theater, to arrange (a stage) so as to depict a scene. **28** In some games, as bridge, to defeat. — *v.i.* **29** To go or pass below the horizon, as the sun. **30** To wane; decline. **31** To sit on eggs, as fowl. **32** To become hard or firm; solidify; congeal. **33** To begin a journey; start: with *forth, out, off,* etc. **34** To have a specified direction; tend. **35** To hang or fit, as clothes. **36** To point game: said of hunting dogs. **37** *Bot.* To begin development or growth, as a rudimentary fruit. — **to set about** To start doing; begin. — **to set against** **1** To balance; compare. **2** To make unfriendly to; prejudice against. — **to set aside** **1** To place apart or to one side. **2** To reject; dismiss. **3** To declare null and void. — **to set back** To reverse; hinder. — **to set down** **1** To place on a surface. 2 To write or print; record. 3 To judge or consider. 4 To attribute; ascribe. — **to set forth** To state or declare; express. — **to set in** 1 To begin. 2 To blow or flow toward shore, as wind or tide. — **to set off** 1 To put apart by itself. 2 To serve as a contrast or foil for; enhance. 3 To cause to explode. — **to set on** To incite or instigate; urge. — **to set out** 1 To present to view; display; exhibit. 2 To lay out or plan (a garden, etc.). 3 To establish the limits or boundaries of, as a town. 4 To plant. — **to set to** 1 To start; begin. 2 To start fighting. — **to set up** 1 To place in an upright position. 2 To raise. 3 To place in power, authority, etc. 4 **a** To construct or build. **b** To put together; assemble. **c** To found; establish. 5 To provide with the means to start a new business. 6 To cause to be heard: to *set up* a cry. 7 To propose or put forward (a theory, etc.). 8 To cause. 9 *Colloq.* **a** To pay for the drinks, etc., of; treat. **b** To pay for (drinks, etc.). 10 *Colloq.* To encourage; exhilarate. — *adj.* 1 Established by authority or agreement; prescribed; appointed: a *set* time; a *set* method. 2 Customary; conventional: a *set* phrase. 3 Deliberately and systematically conceived; formal: a *set* speech. 4 Fixed and motionless; rigid. 5 Fixed in opinion or disposition; obstinate. 6 Formed; built; made: with a qualifying adverb: deep-*set* eyes; a low-*set* man. 7 Ready; prepared: to get *set*. — *n.* 1 The act or condition of setting. 2 Permanent change of form, as by chemical action, cooling, pressure, strain, etc. 3 The arrangement, tilt, or hang of a garment, hat, sail, etc. 4 Carriage or bearing: the *set* of his shoulders. 5 The sinking of a heavenly body below the horizon. 6 The direction of a current or wind. 7 A young plant ready for setting out; a cutting, slip, or seedling. 8 *Mech.* The spread in opposite directions given to the alternate teeth of certain saws. 9 *Psychol.* A temporary condition assumed by an organism preparing for a particular response or activity. 10 In tennis, a group of games completed when one side wins six games, or in the event of a score tied at five games, the group of games terminated when one side wins two more games consecutively. [OE *settan* cause to sit. Akin to SIT.]

Synonyms (verb): adapt, adjust, appoint, arrange, assign, determine, dispose, establish, fix, locate, place, plant, post, prescribe, put, regulate, settle, station. See ALLOT, PLANT, PREPARE, PUT, RAISE. *Antonyms*: detach, disestablish, disturb, eradicate, loosen, overthrow, remove, transfer, unsettle, uproot.

set² (set) *n.* 1 A number of persons regarded as associated through status, common interests, etc.: a new *set* of customers. 2 A social group having some exclusive character; coterie; clique: the fast *set*. 3 A number of things belonging together and customarily used together: a *set* of instruments; a *set* of teeth; a *set* of dishes. 4 A number of specific things so grouped as to form a whole: a *set* of lyrics; a *set* of motives; a *set* of features. 5 A group of volumes issued together and related by common authorship or subject. 6 The number of couples needed for a square dance or country dance. 7 The group of movements that compose a square dance. 8 In motion pictures, the complete assembly of properties, structures, etc., required in a scene. 9 Radio or television receiving equipment assembled for use. See synonyms under CLASS, FLOCK. [<OF *sette* <L *secta* a sect; infl. by SET¹. Doublet of SECT.]

Set (set) In Egyptian mythology, the animal-headed god of darkness, night, and evil; opponent and slayer of Osiris. Also *Seth*.

SET

se·ta (sē′tə) *n. pl.* **·tae** (-tē) *Biol.* 1 A bristle, or slender, bristle-like part or process of an organism. 2 A slender spine or prickle. 3 A coarse, rigid hair. [<L]

set-a·side (set′ə·sīd′) *n.* An amount or quantity of something put in reserve for future use.

se·ta·ceous (si·tā′shəs) *adj.* 1 Bristly; more or less covered with bristles. 2 Of the nature or form of setae. Also **se′tal** (sēt′l). [<NL *setaceus* <L *seta* a bristle]

set·back (set′bak′) *n.* 1 A check; forced return to a point already passed; a reverse in fortune or plan. 2 A countercurrent; eddy. 3 *Archit.* In mammoth buildings, the stepping of sections in such a way that, while the first section is erected on the street line, the remaining sections are erected in step formation, so as to permit of better light and ventilation in the street below.

Sète (set) A port on the Mediterranean in southern France SW of Montpellier: formerly *Cette*.

Set·e·bos (set′ə·bos) A supposed deity of the Patagonians: alluded to in Shakespeare's *The Tempest* as the power worshiped by Sycorax, the witch, mother of Caliban.

Seth (seth) A masculine personal name. [< Hebrew, appointed]
— **Seth** The third son of Adam. *Gen.* v 3.
— **Seth** The Egyptian god Set.

set-ham·mer (set′ham′ər) *n.* A hammer the head of which may be easily removed from the handle. See illustration under HAMMER.

seti- *combining form* A bristle: *setiferous*. Also, before vowels, **set-**. [<L *seta* a bristle]

se·tif·er·ous (si·tif′ər·əs) *adj.* Bearing setae; bristly. Also **se·tig·er·ous** (si·tij′ər·əs). [< SETI- + -FEROUS]

se·ti·form (sē′tə·fôrm) *adj.* Having the form of a seta; setaceous.

set-off (set′ôf′, -of′) *n.* 1 An offset or counterpoise. 2 A decorative contrast or setting. 3 A counterclaim or the discharge of a debt by a counterclaim. 4 *Archit.* A ledge; offset.

se·ton (sē′tən) *n. Surg.* A bristle, or a few threads, passed through a fold of the skin and left there to produce an issue for relief of subjacent parts. [<Med. L *seto, -onis,* appar. <*seta* silk <L, a bristle]

Se·ton (sē′tən), **Ernest Thompson,** 1860–1946, U.S. naturalist and writer born in England.

se·tose (sē′tōs) *adj.* Setaceous; bristly. Also **se′tous.** [<L *setosus* <*seta* a bristle]

set·screw (set′skrōō′) *n.* A screw used as a clamp; especially, one having a cup instead of a point: used to screw through one part and slightly into another to bind the parts tightly. See illustration under SCREW.

set·tee¹ (se·tē′) *n.* A long wooden seat with a high back; a kind of sofa. [<SET¹ + *-ee*, dim. suffix; infl. in meaning by SEAT]

set·tee² (se·tē′) *n.* A Mediterranean vessel with long prow, single deck, two or three masts, and lateen sails. [<Ital. *saettia*, prob. <*saetta* an arrow]

set·ter (set′ər) *n.* 1 One who or that which sets. 2 One of a breed of medium-sized, silky-coated, lithe bird dogs of great intelligence, originally trained to indicate the presence of game birds by crouching, now by standing rigid — **English setter** A setter, white, or white marked with black, tan, yellow, or orange, trained since the 16th century to find and point game. The most famous British strains, the Laveracks and the Llewellins, are popular in field trials and bench shows. — **Gordon setter** A setter having a black coat marked with tan, chestnut, or red, probably crossbred with black-and-tan spaniels: named for the original breeder, the Duke of Gordon, and used especially for cover shooting. — **Irish** (or **red**) **setter** A handsome, useful, and companionable golden-chestnut or red setter, probably a setter-spaniel-pointer combination: extensively bred in America and very popular.

ENGLISH SETTER
(About 25 inches high at the shoulder)

set·ting (set′ing) *n.* 1 The act of anything that sets. 2 An insertion. 3 That in which something is set; a frame; environment. 4 The act of indicating game like a setter. 5 A number of eggs placed together for hatching. 6 The music adapted to a song or poem. 7 The scene or background of a play or narrative. 8 The apparent sinking of the sun, etc., below the horizon.

set·ting-out (set′ing·out′) *n. Colloq.* The trousseau and household equipment given to a bride or newly married couple by the parents.

set·tle (set′l) *v.* **·tled, ·tling** *v.t.* 1 To put in order; set to rights; to *settle* affairs. 2 To put firmly in place; establish or fix permanently or as if permanently: He *settled* himself on the couch. 3 To free of agitation or disturbance; calm; quiet: to *settle* one's nerves. 4 To cause (sediment or dregs) to sink to the bottom. 5 To cause to subside or come to rest; make firm or compact: to *settle* dust or ashes. 6 To make clear or transparent, as by causing sediment or dregs to sink. 7 *Colloq.* To make quiet or orderly: One blow *settled* him. 8 To decide or determine finally, as an argument or difference. 9 To pay, as a debt; satisfy, as a claim. 10 To establish residents or residence in (a country, town, etc.). 11 To establish as residents. 12 To establish in a permanent occupation, home, etc. 13 To decide (a suit at law) by agreement between the litigants. 14 *Law* To make over or assign (property) by legal act: with *on* or *upon*. — *v.i.* 15 To come to rest, as after moving about or flying. 16 To sink gradually; subside. 17 To sink or come to rest, as dust or sediment. 18 To become more firm or compact. 19 To become clear or transparent, as by the sinking of sediment. 20 To take up residence; establish one's abode or home. 21 To come to a decision; determine; resolve: with *on, upon,* or *with*. 22 To pay a bill, etc. — **to settle down** 1 To start living a regular, orderly life, especially after a period of wandering or irresponsibility. 2 To apply steady effort or attention. — *n.* 1 A long seat or bench, generally of wood, with a high back, originally to direct the draft up the chimney and to provide a warm nook: often with arms and sometimes having a chest from seat to floor. 2 A wide step; platform. 3 *Obs.* A ledge. [OE *setlan* <*setl* a seat]

Synonyms (verb): adjust, allay, arrange, calm, compose, decide, determine, establish, finish, fix, pay, quiet, regulate. Compare CONFIRM, PAY¹, RATIFY, REQUITE, SET. *Antonyms*: agitate, confuse, derange, disarrange, discompose, disorder, disturb, fluster, flutter, mix, muss, scatter.

set·tle·ment (set′l·mənt) *n.* 1 The act of settling, or state of being settled; specifically, an adjustment of affairs by public authority. 2 Colonization. 3 Subsidence of a structure, or its effect. 4 An area of country newly occupied by those who intend to live and labor there; a colonized region, village, or town. 5 *Brit.* A regular or settled place of living; one's dwelling place. 6 An accounting; adjustment; liquidation in regard to amounts. 7 The conveyance of property in such form as to provide for some future object, especially the support of members of the settler's family; also, the property so settled. 8 A religious community. 9 *pl.* A collection or series of frontier dwellings and clearings: distinguished from wild, unsettled territory. 10 Formerly, Negro quarters on a southern plantation. 11 A welfare institution established in a congested part of a city, having a resident staff of workers to conduct educational and recreational activities for the community: also **settlement house.**

set·tler (set′lər) *n.* 1 One who settles; especially, one who establishes himself in a colony or new country; a colonist. 2 One who or that which settles or decides something.

set·tling (set′ling) *n.* 1 The act of settling or sinking. 2 *pl.* Dregs; sediment.

set-to (set′tōō′) *n.* A bout at fighting, fencing, arguing, or any other mode of contest. [< *set to*; see under SET]

Se·tú·bal (sə·tōō′bəl) A port of south central Portugal, on **Setúbal Bay,** an inlet of the Atlantic off the west coast of Portugal. Formerly **St. Yves** (sänt ēvz).

set·u·lose (sech′ōō·lōs) *adj.* Clothed or covered with setae or bristles. [<NL, dim. of L *seta* a bristle]

set-up (set′up′) *n.* 1 Physique; physical build; make-up. 2 Carriage of the body; bearing.

3 *U.S. Slang* A system or scheme of organization or construction; the salient elements of a situation; circumstances. 4 *U.S. Slang* A contest or match arranged to result in an easy victory; a contest in which the strength of the contestants is so unequal that the result is easily foreseen; also, the weaker of two such contestants. 5 *U.S. Colloq.* Ice, soda water, etc., provided to a customer who has brought his own liquor.

Seu·rat (sœ·rä′), **Georges,** 1859–91, French painter.

Se·van (se·vän′), **Lake** The largest lake in Transcaucasia, in central Armenian U.S.S.R.; 546 square miles. *Turkish* **Gök·cha** (gœk′chä).

Se·vas·ti·an (se·väs′tē·än′) Russian form of SEBASTIAN.

Se·vas·to·pol (si·vas′tə·pōl, sev′əs·tō′pəl, *Russian* si′väs·tô′pəl) A port and naval base in the SW Crimea, Russian S.F.S.R.: formerly *Sebastopol.*

sev·en (sev′ən) *adj.* Being one more than six. — *n.* 1 The sum of one and six. 2 The symbols (7, vii, VII) representing that number. 3 A playing card with seven spots. 4 Something composed of seven units. [OE *seofon*]

Seven against Thebes In Greek legend, the seven heroes (Adrastus, Amphiaraus, Capaneus, Hippomedon, Parthenopaeus, Polynices, and Tydeus) who unsuccessfully marched on Thebes to restore Polynices to the throne which had been usurped by his brother Eteocles: all were killed save Adrastus. See ANTIGONE, EPIGONI.

Seven Cities of Ci·bo·la (sē′bō·lä) Seven Zuñi towns sought by the early Spanish explorers as centers of hidden gold and fabulous wealth: found in what is now New Mexico by Coronado about 1540.

Seven Deadly Sins Pride, Lechery, Envy, Anger, Covetousness, Gluttony, and Sloth as personified in medieval literature: also known as *cardinal sins.*

sev·en·fold (sev′ən·fōld′) *adj.* 1 Seven times as many or as great. 2 Made up of seven; septuple. 3 Folded seven times. — *adv.* In sevenfold manner or degree.

Seven Hills of Rome The group of seven hills on and around which the city of Rome was built: the Palatine, Caelian, Esquiline, Capitoline, Quirinal, Viminal, and Aventine.

seven lively arts All forms of popular entertainment; originally, motion pictures, vaudeville, popular music, popular dancing, writing in the vernacular, musical shows, comic strips. [from *The Seven Lively Arts* (1924) by Gilbert Seldes]

Seven Pines See FAIR OAKS.

Seven Seas See under SEA.

sev·en·teen (sev′ən·tēn′) *adj.* Being seven more than ten. — *n.* The sum of ten and seven, or the symbols (17, xvii, XVII) representing this number. [OE *seofontȳne*]

sev·en·teenth (sev′ən·tēnth′) *adj.* 1 Seventh in order after the tenth. 2 Being one of seventeen equal parts. — *n.* 1 One of seventeen equal parts of anything. 2 A seventeenth object or unit. [OE *seofontēotha* <*seofontȳne* seventeen]

sev·en·teen–year locust (sev′ən·tēn′yir′) A dark-bodied, wedge-shaped cicada *(Magicicada septemdecim)* native to the eastern United States: the northern variety has an underground nymphal stage of 17 years, and the southern variety of 13 years.

sev·enth (sev′ənth) *adj.* 1 Next in order after the sixth. 2 Being one of seven equal parts. — *n.* 1 One of seven equal parts; the quotient of a unit divided by seven. 2 A seventh object or unit. 3 *Music* **a** The interval between any note and the seventh note above it on the diatonic scale, counting the starting point as one. **b** A note separated by this interval from any other, considered with reference to that other; specifically, the seventh above the keynote. **c** Two notes at this interval written or sounded together. **d** The resulting dissonance. See INTERVAL (def. 5). — *adv.* In the seventh order, place, or rank: also, in formal discourse, **sev′enth·ly.** [ME *seventhe* <SEVEN + -TH, replacing OE *seofande* and *seofotha*]

sev·enth–day (sev′ənth·dā′) *adj.* 1 Pertaining to the seventh day of the week. 2 Advocating the observance of the seventh day as the Sabbath: a *Seventh-Day* Adventist.

seventh day 1 Saturday; the seventh day of the week; the Sabbath of the Jews and of some other religious groups. 2 Saturday, in the speech of the Society of Friends.

Seventh–Day Adventist See under ADVENTIST.

seventh heaven 1 The highest abode or condition of happiness. 2 The highest heaven according to various ancient systems of astronomy or in certain theologies.

sev·en·ti·eth (sev′ən·tē·ith) *adj.* 1 Tenth in order after the sixtieth. 2 Being one of seventy equal parts. — *n.* 1 One of seventy equal parts; the quotient of a unit divided by seventy. 2 A seventieth object or unit. [ME *seventithe* <SEVENTY + -TH]

sev·en·ty (sev′ən·tē) *adj.* Being ten more than sixty, or seven times ten. — *n. pl.* **·ties** The sum of ten and sixty, or the symbols (70, lxx, LXX) representing this number. [OE *(hund-) seofontig*] — **sev′en·ty·fold′** *adj.* & *adv.*

Sev·en·ty–Six (sev′ən·tē·siks′) Short for 1776: the year of the Declaration of Independence.

sev·en–up (sev′ən·up′) *n.* A game of cards: said to be originally so called from the number of points required to win: also called *all fours, old sledge.*

Seven Wonders of the World The seven works of man considered the most remarkable in the ancient world: generally considered to be the Egyptian pyramids, the hanging gardens of Babylon, the temple of Diana at Ephesus, the statue of Zeus by Phidias at Olympia, the Mausoleum at Halicarnassus, the Colossus of Rhodes, and the Pharos or lighthouse of Alexandria.

sev·er (sev′ər) *v.t.* 1 To put or keep apart; separate. 2 To cut or break into two or more parts. 3 To break off; dissolve, as a relationship or tie. — *v.i.* 4 To come or break apart or into pieces. 5 To go away or apart; separate. See synonyms under BREAK, CUT, REND, SEPARATE. [<AF *severer,* OF *sevrer* <L *separare* SEPARATE]

sev·er·a·ble (sev′ər·ə·bəl) *adj.* 1 Capable of being severed. 2 *Law* That can be severed from something to which it is attached or of which it forms part: said of a contract consisting of several obligations, when non-fulfilment of one obligation does not invalidate the contract.

sev·er·al (sev′ər·əl, sev′rəl) *adj.* 1 Being of an indefinite number, more than two, yet not large; divers. 2 Considered individually; pertaining to an individual; single; separate. 3 *Law* Pertaining individually and separately to each tenant or party to a bond: opposed to *joint:* a joint and *several* note. 4 Individually different; various or diverse. [<AF <Med. L *separalis* <L *separ* separate, distinct]

sev·er·al·ly (sev′ər·əl·ē, sev′rəl·ē) *adv.* 1 Individually; separately. 2 Respectively.

sev·er·al·ty (sev′ər·əl·tē, sev′rəl-) *n. pl.* **·ties** 1 *Law* The holding of land in one's own right without participation; a sole tenancy. 2 The character of being several or distinct.

sev·er·ance (sev′ər·əns, sev′rəns) *n.* 1 The act of severing, or the condition of being severed. 2 Separation; partition.

se·vere (si·vir′) *adj.* 1 Trying to one's powers or endurance; hard to bear. 2 Rigorous in the treatment of others; unsparing; harsh; merciless. 3 Conforming to rigid rules; marked by pure and simple excellence; accurate. 4 Serious and austere in disposition or manner; grave; sedate; austerely plain. 5 Causing sharp pain or anguish; extreme: a *severe* pain. [<MF *sévère* <L *severus*] — **se·vere′ly** *adv.*

Synonyms: austere, rigid, rigorous, stern, stiff, unrelenting. That is *severe* which is devoid of all softness, mildness, indulgence, or levity, or (in literature and art) devoid of unnecessary ornament, amplification, or embellishment of any kind; as, a *severe* style; as said of anything painful, *severe* signifies such as heavily taxes endurance or power to resist; as, a *severe* pain, fever, or winter. *Rigid* signifies primarily *stiff,* resisting any effort to change its shape, its will, or course of conduct. *Rigorous* is nearly akin to *rigid,* but is a stronger word, having reference to action or active qualities: a *rigid* rule may be *rigorously* enforced. *Strict* signifies bound or stretched tight, tense, strenuously exact. *Stern* unites harshness and authority with strictness or severity; *stern,* as said even of inanimate objects, suggests something authoritative or forbidding. *Austere* signifies severely simple or temperate, *strict* in self-restraint or discipline, and similarly *unrelenting* toward others. See ARDUOUS, AUSTERE, BAD, DIFFICULT, HARD, IMPLACABLE, MOROSE, VIOLENT.

se·ver·i·ty (si·ver′ə·tē) *n. pl.* **·ties** 1 The quality of being severe. 2 Harshness or cruelty of disposition or treatment; power of paining or distressing. 3 Extreme strictness in character or rigor in operation; exactness. 4 Seriousness; austerity. 5 Strict conformity to truth or law. See synonyms under ACRIMONY, VIOLENCE.

Sev·ern (sev′ərn) A river in northern Wales and western England, flowing 210 miles NE, SE, south, and SW from near Aberystwith to the Bristol Channel.

Severn River A river in NW Ontario, Canada, flowing 610 miles NE to Hudson Bay.

Se·ver·na·ya Zem·lya (sā′vir·nə·yə zim·lyä′) An archipelago in the Arctic Ocean in Krasnoyarsk territory, Russian S.F.S.R.; total, 14,300 square miles.

Se·ver·sky (si·ver′skē), **Alexander de,** born 1894, U. S. airplane designer and manufacturer born in Russia.

Se·ve·rus (si·vir′əs), **Lucius Septimius,** 146–211, Roman emperor; rebuilt Hadrian's Wall across northern England.

Sé·vi·gné (sā·vē·nyā′), **Madame de,** 1626–96, Marie de Rabutin-Chantal, French writer.

Se·ville (sə·vil′) A city of SW Spain; the leading city of Andalusia. *Spanish* **Se·vil·la** (sā·vēl′lyä).

Sè·vres (se′vr′) *n.* A fine porcelain originally made at Sèvres, France. Also **Sèvres ware.**

Sè·vres (se′vr′) A city of north central France on the Seine just SW of Paris.

sew (sō) *v.* **sewed, sewed** or **sewn, sew·ing** *v.t.* 1 To make, mend, or fasten with needle and thread. 2 To affect by sewing: often with *up.* — *v.i.* 3 To work with needle and thread. [OE *siwan, siowian*]

sew·age (sōō′ij) *n.* 1 The waste matter from domestic, commercial, and industrial establishments carried off in sewers. 2 Loosely, sewerage. [<SEW(ER) + -AGE]

Sew·all (sōō′əl), **Samuel,** 1652–1730, Massachusetts jurist and diarist.

se·wan (sē′wən) See SEAWAN.

Sew·ard (sōō′ərd), **William Henry,** 1801–72, U. S. statesman; secretary of state 1861–69.

Seward Peninsula (sōō′ərd) The westernmost part of Alaska, extending 210 miles west to Cape Prince of Wales; 90–140 miles wide.

se·wel·lel (si·wel′el) *n.* A brown, burrowing, nocturnal, vegetarian rodent *(Aplodontia rufa)* of the Pacific coast north of California; the mountain beaver. [<Chinook *shewallal* dual, a blanket of two sewellel skins sewn together (mistaken by Lewis and Clark for the animal's name) < *ogwoolal* a sewellel]

sew·er[1] (sōō′ər) *n.* 1 A conduit, usually laid underground, to carry off drainage and excrement. ◆ Collateral adjective: *cloacal.* 2 Any large public drain. [<OF *seuwiere* a channel from a fish pond, ult. <L *ex-* off + *aqua* water]

sew·er[2] (sōō′ər) *n.* Formerly, in England, an attendant who supervised the serving of meals and seating of guests. Also **sew′ar.** [<AF *asseour,* OF *asseoir* cause to sit <L *assidere* <*ad-* to + *sedere* sit]

sew·er[3] (sō′ər) *n.* One who or that which sews.

sew·er·age (sōō′ər·ij) *n.* 1 A system of sewers. 2 Systematic draining by sewers. 3 Sewage.

sew·ing (sō′ing) *n.* 1 The act, business, or occupation of one who sews. 2 That which is sewed; material on which one is at work with needle and thread; needlework.

sewing bee A social gathering of the women and girls of a community to sew for some charitable purpose.

sewing circle 1 A group of women, usually organized within a church or other welfare organization, meeting periodically to sew for some charitable purpose. 2 A meeting of such a group. Also **sewing society.**

sewing machine A machine for stitching or sewing cloth, leather, etc.

sewing silk Finely twisted silk thread used for sewing.

sewn (sōn) Alternative past participle of SEW.

sex (seks) *n.* 1 Either of two divisions, male and female, by which organisms are distinguished with reference to the reproductive functions. 2 Males or females collectively. 3 The character of being male or female. 4 The

sex- activity or phenomena of life concerned with sexual desire or reproduction. 5 *Colloq.* Sexual gratification. — **the fair sex** Women. [< OF *sexe* < L *sexus*, prob. orig. division]
sex- *combining form* Six: *sexpartite*. Also spelled *sexi-*. [< L *sex* six]
sex·a·ge·nar·i·an (sek′sə·jə·nâr′ē·ən) *n.* A person between sixty and seventy years of age. — *adj.* 1 Sixty years old, or between sixty and seventy. 2 Of or pertaining to a sexagenarian. [< SEXAGENARY]
sex·ag·e·nar·y (seks·aj′ə·ner′ē) *adj.* 1 Of or pertaining to the number sixty; progressing by sixties. 2 Sixty years old, or between sixty and seventy. — *n. pl.* **·nar·ies** A sexagenarian. [< L *sexagenarius* < *sexageni* sixty each < *sexaginta* sixty]
Sex·a·ges·i·ma (sek′sə·jes′ə·mə) *n.* The second Sunday before Lent. Also **Sexagesima Sunday**. [< L *sexagesima (dies)* sixtieth (day). See SEXAGESIMAL.]
sex·a·ges·i·mal (seks′ə·jes′ə·məl) *adj.* Pertaining to or founded on the number sixty. [< Med. L *sexagesimalis* < L *sexagesimus* sixtieth < *sexaginta* sixty]
sex·an·gle (seks′ang′gəl) *n.* A six-angled figure; a hexagon. [< L *sexangulus* < *sex* six + *angulus* an angle] — **sex·an′gu·lar** (-ang′gyə·lər), **sex·an′gled** *adj.* — **sex·an′gu·lar·ly** *adv.*
sex appeal A physical quality or charm which attracts sexual interest.
sex cell A gamete; a sperm or ovum.
sex·cen·te·nar·y (seks·sen′tə·ner′ē, seks′sen·ten′ər·ē) *adj.* Pertaining to or consisting of six hundred, especially six hundred years. — *n. pl.* **·nar·ies** 1 A period of six hundred years or a collection of six hundred units. 2 A six-hundredth anniversary. [< L *sexcenteni* six hundred each < *sexcenti* six hundred < *sex* six + *centum* a hundred]
sex chromosome *Biol.* A chromosome whose presence in the reproductive cells of certain plants and animals is associated with the determination of the sex of their offspring. In mammals the ovum carries two X-chromosomes and sperm an X- and a Y-chromosome; females are produced by a paired XX in the fertilized ovum, males by a paired XY.
sex·en·ni·al (seks·en′ē·əl) *adj.* Happening once every six years, or lasting six years. — *n.* A sixth anniversary. [< L *sexennis, sexennium* < *sex* six + *annus* a year] — **sex·en′ni·al·ly** *adv.*
sex·fid (seks′fid) *adj. Bot.* Six-cleft, as a calyx. Also **sex′i·fid**. [< SEX- + -FID]
sex gland A gonad; either of the testes or ovaries.
sex hygiene The division of hygiene which has to do with sexual conduct as related to the health of the individual and of the community.
sexi- Var. of SEX-.
sex·less (seks′lis) *adj.* Having no sex; neuter. — **sex′less·ly** *adv.* — **sex′less·ness** *n.*
sex linkage *Biol.* That type of inheritance which is associated with the transmission of genes attached to the sex chromosomes. — **sex-linked** (seks′lingkt′) *adj.*
sex·ol·o·gy (seks·ol′ə·jē) *n.* The study of human sexual behavior. [< SEX + -(O)LOGY] — **sex·o·log·ic** (sek′sə·loj′ik) or **·i·cal** *adj.* — **sex·ol′o·gist** *n.*
sex·par·tite (seks·pär′tīt) *adj.* Divided into or made up of six parts, as a groined arch or other structure. [< NL *sexpartitus* < L *sexus* six + *partitus* PARTITE]
sex ratio The ratio of males to females in a given population, usually expressed as the number of males per 100 females.
sex reversal The transformation of an individual from one sex into the other, especially as induced by changes in or abnormalities of the gonads.
sext (sekst) *n.* 1 One of the canonical hours; the office for the sixth hour or noon. 2 The sixth book of the decretals, added by Pope Boniface VIII. [< LL *sexta* < L *sexta (hora)* the sixth (hour), fem. of *sextus* sixth < *sex* six]
sex·tan (seks′tən) *adj.* Occurring or returning at intervals of six days. — *n.* A fever returning at intervals of six days. ◆ Homophone: *sexton*. [< NL *sextana (febris)* sextan (fever) < L *sextus*. See SEXT.]
Sex·tans (seks′tənz) *n.* An equatorial constellation between Leo and Hydra; the Sextant. See CONSTELLATION. [< L, a sextant]

sex·tant (seks′tənt) *n.* 1 An instrument for measuring angular distance between two objects, as between a heavenly body and the horizon, by a double reflection from two mirrors: used especially in determining latitude at sea. 2 The sixth part of a circle; an arc of 60 degrees. [< L *sextans, -antis* the sixth part < *sextus* sixth]

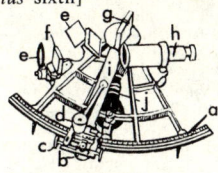

SEXTANT
a. Scale.   d. Reading lens.   g. Index glass.
b. Clamp screw.   e. Glass shades.   h. Telescope.
c. Tangent screw.   f. Horizon glass.   i. Movable arm.
  j. Handle.

sex·tar·i·us (seks·târ′ē·əs) *n. pl.* **·tar·i·i** (-târ′ē·ī) An ancient Roman measure of capacity. See CONGIUS. [< L, a sixth part < *sextus* sixth < *sex* six]
sex·tet (seks·tet′) *n.* 1 A band of six singers or players; also, a musical composition for six parts. 2 Any collection of six persons or things. Also **sex·tette′**. [Alter. of SESTET; refashioned after L *sex* six]
sex·tile (seks′til) *adj.* Indicated or measured by a distance of 60 degrees. — *n.* 1 *Astron.* The aspect of two planets at a distance of 60 degrees from each other. 2 *Stat.* One of the divisions of a frequency distribution containing exactly one sixth of the total number of cases or observations included. [< L *sextilis (mensis)* the sixth (month), i.e., August < *sextus* sixth]
sex·til·lion (seks·til′yən) *n.* A cardinal number: in the French system and in the United States, 1 followed by 21 ciphers; in the English system, 1 followed by 36 ciphers. [< F < L *sex* six + F (m)*illion* a million]
sex·to·dec·i·mo (seks′tō·des′ə·mō) *n.* Sixteenmo. [< L *sextusdecimus* sixteenth < *sextus* sixth + *decimus* tenth]
sex·ton (seks′tən) *n.* 1 A janitor of a church having charge also of ringing the bell, overseeing burials, etc.; also, formerly, a gravedigger. 2 Any of certain carrion beetles (genus *Necrophorus*) that bury small dead animals by excavating the ground beneath them: also called *burying beetle*. The larvae feed on the maggots in the rotting flesh. ◆ Homophone: *sextan*. [< AF *segerstaine*, OF *secrestein* < Med. L *sacristanus*. Doublet of SACRISTAN.] — **sex′ton·ship** *n.*
sex·tu·ple (seks′tōō·pəl, -tyōō-, seks·tōō′-, -tyōō′-) *adj.* 1 Sixfold. 2 Multiplied by six; six times repeated. 3 *Music* Having six beats to the measure. — *v.t.* **·pled, ·pling** To make sextuple; multiply by six. — *n.* A number or sum six times as great as another. [< L *sextus* sixth < *sex* six, formed on analogy with *quadruple, quintuple*, etc.] — **sex′tu·ply** *adv.*
sex·tu·plet (seks′tōō·plit, -tyōō-, seks·tōō′-, -tyōō′-) *n.* 1 A set of six similar things. 2 One of six offspring produced at a single birth. [< SEXTUPLE on analogy with *triplet*]
sex·tu·pli·cate (seks·tōō′plə·kit, -tyōō′-) *adj.* 1 Sixfold. 2 Raised to the sixth power. — *v.t.* (-kāt) **·cat·ed, ·cat·ing** To multiply by six; sextuple. — *n.* One of six like things. [< Med. L *sextuplicatus*, pp. of *sextuplicare* < *sextuplex* multiplied by six < *sex* six] — **sex·tu′pli·cate·ly** *adv.* — **sex·tu′pli·ca′tion** *n.*
sex·u·al (sek′shōō·əl) *adj.* 1 Of, pertaining to, peculiar to, characteristic of, or affecting sex, the sexes, or the organs or functions of sex. 2 Characterized by or having sex: opposed to *asexual*. [< LL *sexualis* < L *sexus* sex] — **sex′u·al·ly** *adv.*
sexual affinity 1 The affinity or attraction exhibited by an individual of one sex for a particular member of the opposite sex. 2 A relationship between members of different species enabling them to have sexual intercourse.
sex·u·al·i·ty (sek′shōō·al′ə·tē) *n.* 1 The state of having, or of being distinguished by, sex. 2 Preoccupation with sex. 3 Possession of sexual power.
sexual selection In the theory of evolution, a phase of natural selection whereby characters, as bright colors, or fine song, considered as especially attractive to the opposite sex, have a tendency to become perpetuated or enhanced.
sex·y (sek′sē) *adj.* **sex·i·er, sex·i·est** *Slang* 1 Provocative of sexual desire: a *sexy* dress; a *sexy* woman. 2 Concerned in large or excessive degree with sex: a *sexy* novel.
Sey·chelles (sā·shel′, -shelz′) An island group in the western Indian Ocean, comprising a British colony; 156 square miles; capital, Victoria, on Mahé.
Sey·han (sā·hän′) A river in south central Turkey in Asia, flowing 320 miles SW to the Mediterranean.
Seym (sām) See SEIM.
Sey·mour (sē′môr, -mōr), **Jane**, 1509?-37, third wife of Henry VIII of England; mother of Edward VI.
Sfax (sfäks) A port in central eastern Tunisia, on the north shore of the Gulf of Gabès; the second largest city of Tunisia.
sfer·ics (sfer′iks) *n. Meteorol.* 1 A cathode-ray tube connected with a directional antenna, used for the detection and plotting of electrical discharges in the atmosphere up to distances of several thousand miles. 2 *pl.* Atmospherics. [Short for ATMOSPHERICS]
Sfor·za (sfôr′tsä) A Milanese ducal family which flourished in the 15th century. — **Count Carlo**, 1873-1952, Italian anti-Fascist leader.
sfor·zan·do (sfôr·tsän′dō) *adj. Music* Accented more forcibly than the rhythm requires; especially, sounded, as a note or chord, with sudden explosive force: also spelled *forzando*. Also **sfor·za′to** (-tsä′tō). [< Ital., forcing < *sforzare* force]
's Gra·ven·ha·ge (skhrä′vən·hä′khə) The Dutch name for THE HAGUE.
Shaa·ban (shä·bän′) See CALENDAR (Mohammedan).
shab·by (shab′ē) *adj.* **·bi·er, ·bi·est** 1 Threadbare; ragged; soiled or defaced, as from hard use. 2 Characterized by worn or defaced garments. 3 Mean; paltry. See synonyms under BAD[1], BASE[2]. [OE *sceabb* a scab < -Y[1]] — **shab′bi·ly** *adv.* — **shab′bi·ness** *n.*
Sha·bu·oth (shä·vōō′ōth, shə·vōō′əs) *n. pl.* The Jewish festival of Pentecost or Feast of Weeks. [< Hebrew *shebuôth*, lit., weeks]
shach·le (shäkh′əl) *Scot. v.t.* To pull or wrench out of shape, as by excessive use. — *n.* Anything misshapen by or as by excessive use.
shack[1] (shak) *n.* 1 A rude cabin, as of logs. [? < dial. Sp. (Mexican) *jacal* a wooden hut < Nahuatl *xacalli*; prob. infl. by RAMSHACKLE]
shack[2] (shak) *n.* 1 Fallen acorns or nuts of any kind; mast. 2 Any bait picked up at sea, as dead sea birds, refuse fish, etc.: distinguished from bait regularly carried or newly caught: also **shack bait**. 3 A catch of miscellaneous, unsorted fish. [< *shack, v.*, dial. var. of SHAKE]
shack[3] (shak) *n.* A slow trot: also **shack gait**. — *v.t.* To go at a slow trot. [Short for *shack-rag*, var. of *shake-rag* a vagabond, a worthless horse < SHAKE + RAG[2]]
shack[4] (shak) *v.t.* To go after; retrieve. [Origin unknown]
shack·le (shak′əl) *n.*

SHACKLES

1 A ring, clasp, or braceletlike fastening for encircling and fettering a limb; fetter; gyve. 2 Impediment or restraint. 3 One of various forms of fastenings, as the bow of a padlock, a clevis, or a link for coupling railway cars. See synonyms under FETTER. — *v.t.* **·led, ·ling** 1 To restrain or confine with shackles; fetter. 2 To keep or restrain from free action or speech. 3 To connect or fasten with a shackle. See synonyms under BIND. [OE *sceacul*] — **shack′ler** *n.*
shackle bolt 1 A bolt having on its end a shackle or clevis, or a bolt that is passed through the eyes of a shackle. 2 The

of a padlock, chain, etc. 3 *Her.* Shackle and padlock: used as a bearing.

Shack·le·ton (shak′əl·tən), **Sir Ernest Henry,** 1874–1922, English Antarctic explorer.

shack·o (shak′ō) See SHAKO.

shad (shad) *n. pl.* **shad** A deep-bodied food fish (genus *Alosa*) related to the herring, especially the common or American shad *(Alosa sapidissima)* of the Atlantic coast, which is highly esteemed as food. [OE *sceadd*]

shad–bel·lied (shad′bel′ēd) *adj.* **1** Cutaway: said of a coat. **2** Lean and lank: said of persons.

shad·ber·ry (shad′ber′ē) *n. pl.* **·ries** The shadbush, or its fruit. [<SHAD(BUSH) + BERRY]

shad·bush (shad′bŏŏsh′) *n.* **1** The Juneberry. **2** Other smaller and shrublike related forms of the same genus, as *Amelanchier alnifolia* of the northern and western United States. Also **shad′blow′** (-blō′). [<SHAD + BUSH[1]; so called because it flowers when the shad appear in U. S. rivers]

shad·dock (shad′ək) *n.* **1** The large, pale-yellow fruit of a tropical tree (genus *Citrus*), varying in size from the smaller grapefruit or pomelo of the United States to the pompelmous, which may be 8 inches in diameter. **2** The tree. [after Capt. *Shaddock,* commander of an East India ship, who brought the seed to the West Indies from the East Indies in 1696]

shade (shād) *v.* **shad·ed, shad·ing** *v.t.* **1** To screen from light by intercepting its rays; put in shade. **2** To make dim with or as with shade; darken; overcast. **3** To screen or protect with or as with a shade. **4** To cause to change, pass, blend, or soften, by gradations. **5 a** To represent (degrees of shade, colors, etc.) by gradations of light or dark lines or shading. **b** To represent varying shades, colors, etc., in (a picture or painting) thus. **6** To make slightly lower, as a price. —*v.i.* **7** To change or vary by degrees. [<*n.*] —*n.* **1** Relative obscurity from interception of the rays of light: distinguished from *shadow*; hence, gloom; darkness; obscurity; the state of being outshone. **2** A shady place; secluded retreat. **3** Something that serves to intercept or screen from light; hence, a screen that shuts off light, heat, air, dust, etc. **4** A gradation of color; also, slight degree; minute difference. **5** The unilluminated part of a picture, drawing, or engraving: opposed to *light.* **6** A disembodied spirit; ghost; something unreal. See synonyms under SPECTER. —**the shades** The abode of departed spirits; Hades. [OE *sceadu*] —**shade′less** *adj.*

shade grass Pachysandra.

shad·fly (shad′flī′) *n. pl.* **·flies** Any of several flies that appear when the shad are running; especially, a mayfly.

shad·ing (shā′ding) *n.* **1** Protection against light or heat. **2** The lines, dots, etc., by which degrees of darkness, color, or depth are represented in a picture or painting. **3** A slight difference or variation.

sha·doof (shä·dōōf′) *n.* A water-raising device, operating on the principle of a well sweep: used in the Orient for irrigation, etc., as on the Nile. Also **sha·duf′.** [Arabic *shādūf*]

shad·ow (shad′ō) *n.* **1** A comparative darkness within an illuminated area caused by the interception of light by an opaque body. **2** The dark figure or image thus produced on a surface and representing the approximate shape of the intercepting body: the *shadow* of a man. **3** The shaded or dark portion of a picture. **4** A mirrored image: to see one's *shadow* in a pool. **5** A delusive image or semblance; anything unreal or unsubstantial. **6** A phantom; ghost; shade. **7** A faint representation or indication; a symbol: the *shadow* of things to come. **8** A remnant; vestige: *shadows* of his former glory. **9** An insignificant trace or portion: not a *shadow* of evidence. **10** *Archaic* Shelter; protection. **11** Gloom; a saddening influence. **12** An inseparable companion. **13** One who trails or follows another, as a detective or spy. See synonyms under IMAGE. —*v.t.* **1** To cast a shadow upon; overspread with shadow; shade. **2** To darken or cloud; make gloomy. **3** To represent or foreshow dimly or vaguely: with *forth* or *out.* **4** To follow closely or secretly; spy on. **5** To shade in painting, drawing, etc. **6** *Archaic* To screen; shelter. [OE *sceadwe,* genitive and dative of *sceadu* a shade] —**shad′ow·er** *n.*

shad·ow·box (shad′ō·boks′) *v.i.* To spar with an imaginary opponent as a form of exercise. —**shad′ow·box′ing** *n.*

shad·ow·graph (shad′ō·graf, -gräf) *n.* **1** A pictorial image formed by casting a shadow, usually of the hands, upon a lighted surface or screen. **2** A drama produced by a series of these images: also **shadow play.** **3** A radiograph.

shad·ow·less (shad′ō·lis) *adj.* Destitute of shadow; also, supernatural; weird.

shadow test Skiascopy.

shad·ow·y (shad′ō·ē) *adj.* **1** Full of or affording shadow; dark; shady: a *shadowy* grove. **2** Like shadows in indistinctness; vague; dim. **3** Unsubstantial or illusory; unreal; ghostly. **4** Symbolic. See synonyms under DARK, IMAGINARY, VAIN. —**shad′ow·i·ness** *n.*

sha·drach (shā′drak, shad′rak) See SALAMANDER (def. 5).

Sha·drach (shā′drak, shad′rak) A Jewish captive in Babylon, who, with Meshach and Abednego, was cast into a fiery furnace by Nebuchadnezzar, but came out unscathed. *Dan.* i 7; iii 1–30.

Shad·well (shad′wel), **Thomas,** 1642?–92, English dramatist and poet.

shad·y (shā′dē) *adj.* **shad·i·er, shad·i·est** **1** Full of shade; casting a shade. **2** Shaded or sheltered. **3** Morally questionable; dubious; suspicious. **4** Quiet; hidden. See synonyms under DARK. —**to keep shady** **1** To stay in hiding; keep out of the way. **2** To hide and protect (another). —**on the shady side of** Older than; past the age of. —**shad′i·ly** *adv.* —**shad′i·ness** *n.*

SHAEF (shāf) In World War II, Supreme Headquarters, Allied Expeditionary Forces.

shaft[1] (shaft, shäft) *n.* **1** The long narrow rod of an arrow, spear, lance, harpoon, etc. **2** An arrow. **3** Anything resembling a missile in appearance or effect: *shafts* of ridicule. **4** A beam or streak of light. **5** A long handle, as of a hammer, ax, etc. **6** *Mech.* A long and usually cylindrical bar, especially if rotating and transmitting motive power. **7** *Archit.* **a** The portion of a column between capital and base. **b** A slender column. **8** An obelisk or memorial column. **9** The vertical part of a cross. **10** The stem of a feather. **11** *Anat.* **a** A long slender portion, as the diaphysis of a bone. **b** The portion of a hair from the root to the end. **12** On a loom, one of the long laths at the ends of the heddles. **13** One of the thills of a one-horse vehicle. [OE *sceaft*]

shaft[2] (shaft, shäft) *n.* **1** A narrow, vertical or inclined, excavation connected with a mine; also, a passage for light or air. **2** The tunnel of a blast furnace. **3** An opening through the floors of a building, as for an elevator. [<LG *schacht* rod, shaft; infl. by SHAFT[1]]

Shaftes·bur·y (shafts′bər·ē, shäfts′-), **Earl of,** 1621–83, Anthony Ashley Cooper, English statesman; lord chancellor 1672–73.

shaft·ing (shaf′ting, shäf′-) *n.* **1** A system of shafts or rods, as in pulleys or gearwheels, for communicating power. **2** Material from which to make shafts.

shag[1] (shag) *n.* **1** A rough coat or mass, as of hair. **2** A wild growth, as of weeds. **3** A long nap on cloth. **4** Cloth having a rough or long nap; formerly, a silk or worsted cloth having a velvet nap. **5** A cormorant. **6** A coarse, strong tobacco: also **shag tobacco.** —*v.* **shagged, shag·ging** *v.t.* **1** To make shaggy or hairy; roughen. **2** In baseball, to catch (flies) in practice. —*v.i.* **3** To become shaggy or rough. —*adj.* Shaggy: also **shag·ged** (shag′id). [OE *sceacga* rough hair, wool]

shag[2] (shag) *n.* A dance of the late 1930's, consisting of hopping quickly on alternate feet. —*v.i.* **shagged, shag·ging** To dance the shag.

shag·bark (shag′bärk′) *n.* **1** The white hickory *(Carya ovata),* which yields high-grade hickory nuts. **2** Its wood. Also called *shellbark.*

shag·gy (shag′ē) *adj.* **·gi·er, ·gi·est** **1** Having, consisting of, or resembling rough hair or wool; rugged; rough. **2** Covered with any rough, tangled growth; fuzzy; scrubby. **3** Unkempt; unpolished: said of manners. See synonyms under ROUGH. —**shag′gi·ly** *adv.* —**shag′gi·ness** *n.*

sha·green (shə·grēn′) *n.* **1** The rough skin of various sharks and rays: used for polishing. **2** A rough-grained Russian or Oriental leather or parchment, usually dyed green, or a pressed leather made in imitation of it. **3** Chagrin. [<F *chagrin* <Turkish *sāghrī* horse's hide]

shah (shä) *n.* An eastern king or ruler, especially of Persia. [<Persian *shāh,* short for *pādshāh.* See PADISHAH.]

Sha·hap·ti·an (shä·hap′tē·ən) *n.* A linguistic stock of North American Indians of which the Nez Percés were the chief tribe, formerly occupying the upper Columbia River valley: now on reservations in Oregon.

Shah·ja·han·pur (shä′jə·hän′pŏŏr) A city in Uttar Pradesh State, northern India.

Shah Je·han (shä jə·hän′), 1592–1666, Mogul emperor, 1627–58, of Delhi, celebrated for his peacock throne and as the builder of the Taj Mahal. Also **Shah Ja·han′.**

shaik (shīk) See SHEIK.

shaird (shârd) *n. Scot.* **1** A shard. **2** A fragment.

shairn (shârn) *n. Scot.* Dung of cattle. —**shairn′y** *adj.*

Shairp (shârp, shärp), **John Campbell,** 1819–1885, Scottish educator and critic.

shai·tan (shī·tän′) *n.* **1** In Moslem countries, the devil. **2** Any evil spirit; an evilly disposed person. **3** In India, a duststorm. Also spelled *sheitan.* [<Arabic *shaiṭān* <Hebrew *sāṭān.* See SATAN.]

shake (shāk) *v.* **shook, shak·en, shak·ing** *v.t.* **1** To cause to move to and fro or up and down with short, rapid movements. **2** To affect in a specified manner by or as by vigorous action: with *off, out, from,* etc.: to *shake* out a sail; to *shake* off a tackler. **3** To cause to tremble or quiver; jolt; vibrate: The blows *shook* the door. **4** To cause to stagger or totter. **5** To weaken or disturb; unsettle: I could not *shake* his determination. **6** To agitate or rouse; stir: often with *up.* **7** *Slang* To get rid of or away from. **8** *Music* To trill. **9** In dice games, to mix (the dice) before casting. —*v.i.* **10** To move to and fro or up and down in short, rapid movements. **11** To be affected in a specified way by vigorous action: with *off, out, from,* etc. **12** To tremble or quiver, as from cold or fear. **13** To become unsteady; totter. **14** *Music* To trill a note, etc. —**to shake down** **1** To cause to fall by shaking; bring down. **2** To cause to settle; make compact. **3** *Slang* To extort money from. —**to shake hands** To clasp hands as a form of greeting, agreement, etc. —*n.* **1** A shaking; concussion; agitation; vibration; shock; jolt. **2** The state of being shaken. **3** *pl. Colloq.* The chill or ague of intermittent fever. **4** A rough, unshaved shingle used to cover barns and shanties. **5** A frost or wind crack in timber; also, a tight fissure in rock. **6** An earthquake. **7** *Slang* An instant; a jiffy. **8** *Music* A trill. **9** *Colloq.* A bargain. —**to give (someone) the shake** To get rid of (someone). [OE *scacan*]

Synonyms (verb): agitate, brandish, flap, fluctuate, flutter, jar, joggle, jolt, jounce, oscillate, quake, quaver, quiver, rattle, reel, rock, shiver, shudder, sway, swing, thrill, totter, tremble, vibrate, wave, waver. A thing is *shaken* which is subjected to short and abruptly checked movements as forward and backward, up and down, from side to side, etc. A thing *rocks* that is held up from below; it *swings* if suspended from above, as a pendulum, or pivoted at the side, as a crane or a bridge draw; to *oscillate* is to *swing* with a smooth and regular returning motion; a *vibrating* motion may be tremulous or *jarring.* The pendulum of a clock may be said to *swing* or *oscillate*; a steel bridge *vibrates* under the passage of a heavy train; the term *vibrate* is also applied to molecular movements. *Jolting* is a lifting from and letting down suddenly upon an unyielding surface; a *jarring* motion is abruptly and very rapidly repeated through an exceedingly limited space; the *jolting* of the carriage *jars* the windows. *Rattling* refers directly to the sound produced by *shaking.* To *joggle* is to *shake* slightly; as, A passing touch *joggles* the desk on which one is writing. To *agitate* in its literal use is nearly the same as to *shake,* but we speak of the sea as *agitated* when we could not say it is *shaken*; in the metaphorical use *agitate* is more transitory and superficial, *shake* more fundamental and

shake-down enduring; a person's feelings are *agitated* by distressing news; his courage, his faith, his credit, or his testimony is *shaken*. Compare FLUCTUATE, QUAKE, SWAY, TREMBLE.

shake-down (shāk′doun′) *n.* 1 A bed of straw shaken down; hence, any makeshift bed. 2 *U.S. Slang* A swindle; a share of graft; extortion money. 3 A noisy, energetic dance common among Negroes of the southern United States. — *adj.* For the purpose of adjusting mechanical parts or habituating people: a *shake-down* cruise.

shak-er (shā′kər) *n.* 1 One who or that which shakes; a container for shaking something: a *saltshaker*, cocktail *shaker*, etc. 2 One who shivers or shakes; a totterer.

Shak-er (shā′kər) *n.* One of a sect practicing celibacy and communal living, introduced in America in 1774 under the leadership of Mother Ann Lee, at Lebanon, New York: so called from their characteristic bodily movements during religious meetings. Their official name is *The United Society of Believers in Christ's Second Appearing.* — **Shak′er·ism** *n.*

Shake-speare (shāk′spir), **William**, 1564–1616, English poet and dramatist. Also **Shake′spere**, **Shak′speare**, **Shak′spere**.

Shake-spear-i-an (shāk-spir′ē-ən) *adj.* Of, pertaining to, or characteristic of Shakespeare, his work, or his style. — *n.* A specialist on Shakespeare or his writings. Also **Shake·spear′e·an**.

Shake-spear-i-an-ism (shāk-spir′ē-ən-iz′əm) *n.* 1 An expression peculiar to Shakespeare. 2 Shakespearian style.

Shakespearian sonnet See under SONNET.

shake-up (shāk′up′) *n.* A change of personnel or organization, as in a government administration, a business office, etc.

Shakh-ty (shäkh′tē) A city in southern European Russian S.F.S.R.

shaking palsy *Pathol.* A chronic disorder of the central nervous system, characterized by alternations of muscular rigidity and tremor and peculiar gait: also called *paralysis agitans*.

shak-o (shak′ō) *n. pl.* **·os** A kind of high, stiff military headdress, having a peak and an upright plume: originally of fur: also spelled *shacko*. [<F *schako* <Hungarian *csákó*]

Shak-ti (shuk′tē) *n.* The female energy of the Hindu god, Siva: worshiped under various forms; Devi: also spelled *Sakti*. [<Skt. *śakti* power] — **Shak′tism** *n.*

Sha-kun-ta-la (shə-kŏŏn′tə-lə) See SAKUNTALA.

shak-y (shā′kē) *adj.* **shak·i·er**, **shak·i·est** 1 Habitually shaking or tremulous; tottering; weak; unsound. 2 Of doubtful credit or solvency; embarrassed. — **shak′i·ly** *adv.* — **shak′i·ness** *n.*

shale[1] (shāl) *n.* A fissile argillaceous rock resembling slate, with fragile, uneven laminae. [<G *schale* shale] — **shal′y** *adj.*

shale[2] (shāl) *n.* Shell or husk. [OE *scealu*] — **shaled** *adj.*

shale oil Petroleum obtained by the distillation of bituminous shales.

shall (shal) A defective verb having a past tense **should**, an archaic present second person singular, (thou) **shalt**, past (thou) **shouldst** or **shouldest**, and no other inflected forms. It is now used only as an auxiliary followed by the infinitive without *to*, or elliptically with the infinitive unexpressed. Its function is to indicate, now chiefly in formal discourse: 1 In the first person, simple futurity, with a matter-of-fact attitude toward the action or state projected: We *shall* take only the usual precautions. (But see usage note below.) 2 In the second and third persons, futurity combined with a mood or feeling of: **a** Determination: They *shall* not pass. **b** Promise: You *shall* have whatever you need. **c** Threat: You *shall* pay for this. **d** Command: No one *shall* twice be put in jeopardy. **e** Inevitability: When earthly time *shall* end, will life survive? 3 In all persons, indefinite future time in conditional statements: If and when you or we or the divers *shall* locate the treasure, it will (or, in legal use, the mandatory *shall*) be shared out according to the agreement. 4 In all persons, futurity involving ideal certainty, in clauses following expressions of anxiety, demand, or desire: They are anxious, indeed insist, that you or I or both of us *shall* go, rather than any outsider. [OE *sceal* I am obliged, 1st person sing. of *sceolan*]
◆ **shall vs. will** The traditional view on the use of *shall* and *will* is that to indicate simple futurity *shall* is used in the first person, *will* in the second and third: their roles are reversed to express determination, promise, threat, command, inevitability, etc.; while in questions, the choice between them depends on which one is expected in the answer. These statements hold fairly well for legal usage, but they are too arbitrary to describe accurately the facts of current American usage in speech and writing, except at the most stilted formal level. *Shall* and *will* have had a tendency gradually to exchange roles once each century since 1500, and during the present century it has been the turn of *will* to make its way into the lead. In the important task of indicating simple future time in the first person, *will* has largely replaced *shall*, aided in doing so by the leveling effect of the contraction '*ll*: I'll (= I will or I shall) be free at ten. *Shall*, thus displaced, takes on one role assigned by traditional formula to *will*, and is used in the first person to express determination plus inevitability, as in General MacArthur's "I *shall* return" and Winston Churchill's "... and win we *shall*." If *will* in the first person is to express determination according to the formula, it must be stressed or qualified in some way: I *will* too go out and play. In questions in the first person, *shall* is still commonly used to express the simple future, but it is also found as the hortatory *shall*, either humorously formal: *Shall* we (= Let's) dance? or politely threatening: *Shall* we (= Let's) do it my way for once. Again, *will* has won out over *shall* when it comes to giving routine or polite, as distinct from peremptory, commands: You *will* proceed to Hill 90 and occupy it. The peremptory *shall* in the second person is now usually replaced, except in legal usage, by *will* with *have to*: You *will* have to (= *shall*) go whether you want to or not. It is hazardous to try to sum up the present position of these two forms, but with the few exceptions noted, in American usage *will* now usually indicates the simple future in all persons, while *shall* expresses the future complicated by some feeling about it.

shal·loon (sha-lōōn′) *n.* A light, woven woolen fabric used for linings. [<F *chalon*, from *Châlons*-sur-Marne, France]

shal·lop (shal′əp) *n.* An open boat propelled by oars or sails. [<F *chaloupe* <Du. *sloep*. See SLOOP.]

shal·lot (shə-lot′) *n.* 1 An onionlike culinary vegetable (*Allium ascalonicum*) allied to garlic but having milder bulbs which are used in seasoning and for pickles. 2 A small onion. Also spelled *eschalot*. [<OF *eschalotte*, alter. of *eschaloigne*. See SCALLION.]

shal·low (shal′ō) *n.* 1 Having the bottom not far below the surface or top; lacking depth; shoal. 2 Lacking intellectual depth; not wise or profound; superficial. — *n.* A shallow place in a body of water; shoal. — *v.t.* & *v.i.* To make or become shallow. [ME *schalowe*. Prob. related to SHOAL[1].] — **shal′low·ly** *adv.* — **shal′low·ness** *n.*

shalt (shalt) Archaic or poetic second person singular, present tense of SHALL: used with *thou*.

shal·war (shul′wär) *n.* Oriental trousers or pajamas. [<Persian *shalwār*]

sham (sham) *v.* **shammed**, **sham·ming** *v.t.* 1 To assume or present the appearance of; counterfeit; feign. 2 To represent oneself as; pretend to be. 3 *Obs.* To delude; deceive. — *v.i.* 4 To make false pretenses; feign something. See synonyms under COUNTERFEIT, PRETEND. — *adj.* False; pretended; counterfeit; mock. See synonyms under FACTITIOUS. — *n.* 1 A pretense; imposture; deception. 2 One who affects or simulates a certain character; an impostor: also **sham′mer**. 3 A deceptive imitation; simulation; counterfeit. 4 A bordered strip simulating the edge of a sheet on a made-up bed; an embroidered covering simulating a pillow cover. See synonyms under HYPOCRISY. [Prob. dial. var. of SHAME]

sha·man (shä′mən, shā′-, sham′ən) *n.* 1 A priest of Shamanism; a magician. 2 Among certain northwestern North American Indians, a tribal medicine man or wizard. — *adj.* Of or pertaining to a shaman: also **sha·man·ic** (shə-man′ik). [<Russian <Tungusic *samán* <Skt. *samana* ascetic]

Sha·man·ism (shä′mən-iz′əm, shā′-, sham′ən-) *n.* A primitive religion of NE Asia and Europe holding that gods, demons, ancestral spirits, etc., work for the good or ill of mankind through the sole medium of its priests, the shamans. Certain Indians of the American Northwest have similar beliefs and practices. — **Sha′man·is′tic** *adj. & n.*

Sha·mash (shä′mäsh) In Assyro-Babylonian religion, the sun god regarded as the deity controlling crops and personifying righteousness.

sham·ble (sham′bəl) *v.i.* **·bled**, **·bling** To walk with shuffling or unsteady gait. — *n.* A shambling walk; shuffling gait. [Origin uncertain]

sham·bles (sham′bəlz) *n. pl.* (generally construed as singular) 1 A place where butchers kill animals; slaughterhouse. 2 Any place of carnage or execution: The trench was a *shambles*. 3 A place marked by great destruction or disorder. 4 *Brit. Dial.* A meat market; in the singular, a table or stall in such a market. [OE *scamel* a bench, stool <L *scamellum*, dim. of *scamnum* bench, stool]

shame (shām) *n.* 1 A painful sense of guilt or degradation caused by consciousness of guilt or of anything degrading, unworthy, or immodest. 2 The restraining sense of pride, decency, or modesty. 3 That which brings reproach; a disgrace. 4 A state of ignominy; sensitiveness or susceptibility to humiliation. See synonyms under ABOMINATION, CHAGRIN. — **to put to shame** 1 To disgrace; make ashamed. 2 To surpass or eclipse. — *v.t.* **shamed**, **sham·ing** 1 To make ashamed; cause to feel shame. 2 To bring shame upon; disgrace. 3 To impel by a sense of shame: with *into* or *out of*. See synonyms under ABASH. [OE *scamu*]

shame·faced (shām′fāst′) *adj.* Easily abashed; showing shame or bashfulness in one's face; modest; bashful. [Alter. of ME *shamefast*, OE *scamfæst* abashed] — **shame·fac·ed·ly** (shām′fā′sid·lē, shām′fāst′lē) *adv.* — **shame′fac′ed·ness** *n.*

shame·ful (shām′fəl) *adj.* 1 Deserving or bringing shame or disgrace; disgraceful; scandalous. 2 Exciting shame; indecent. See synonyms under FLAGRANT. — **shame′ful·ly** *adv.* — **shame′ful·ness** *n.*

shame·less (shām′lis) *adj.* 1 Impudent; brazen; immodest. 2 Done without shame, indicating a want of pride or decency. See synonyms under INFAMOUS, IMMODEST, IMPUDENT. — **shame′less·ly** *adv.* — **shame′less·ness** *n.*

sham·my (sham′ē), **sham·ois** (sham′ē) See CHAMOIS (def. 2).

Sha·mo (shä′mō) A Chinese name for the GOBI DESERT.

sham·poo (sham-pōō′) *v.t.* 1 To lather, rub, and wash (the hair and scalp) thoroughly. 2 To cleanse by rubbing. — *n.* The act or process of shampooing, or a preparation used for it. [<Hind. *champnā* press] — **sham·poo′er** *n.*

sham·rock (sham′rok) *n.* Any one of several trifoliate plants, accepted as the national emblem of Ireland, especially the wood sorrel (*Oxalis acetosella*), the white clover (*Trifolium repens*), and the black medic (*Medicago lupulina*). [<Irish *seamróg*, dim. of *seamar* trefoil]

SHAMROCK
a. White clover.
b. Wood sorrel.

Shan (shan, shän) *n.* 1 One of a group of Mongoloid tribes of southern China, Assam, Burma, and Thailand. 2 The Thai language

shand (shand, shänd) *Scot. adj.* Worthless; mean, paltry. — *n.* Spurious coin.

shan·dry·dan (shan'drə-dan) *n. Irish* 1 A two-wheeled Irish cart, or hooded chaise: also **shan'da·ra·dan', shan'der·y·dan'**. 2 An old-fashioned or rickety vehicle. [Origin unknown]

shan·dy·gaff (shan'dē-gaf) *n.* An alcoholic drink composed of two liquids mixed, at least one being effervescent: usually ale or beer and ginger beer. [Origin unknown]

shang·hai (shang'hī, shang-hī') *v.t.* **-haied, -hai·ing** 1 To drug or render unconscious and kidnap for service aboard a ship. 2 To cause to do something by force or deception. [from *Shanghai*]

Shang·hai (shang'hī) *n.* One of a former large breed of domestic fowls, with long legs and feathered shanks, said to have originated in Shanghai, China.

Shang·hai (shang'hī', *Chinese* shäng'hī') A port of eastern China, the largest city on the Asian continent.

Shan·gri-la (shang'grǐ-lä') *n.* 1 Any imaginary hidden utopia or paradise. 2 The reported taking-off place of the United States Army bombers that raided Tokyo April 18, 1942: a term used by Franklin Roosevelt. 3 Any secret base for air force military operations. [From the locale of James Hilton's novel *Lost Horizon*]

Shan·hai·kwan (shän'hī'gwän') A city in NE Hopeh province, China, on the Gulf of Liaotung on the Manchurian boundary; the easternmost end of the Great Wall. Formerly called **Lin·yü** (lin'yoo').

shank (shangk) *n.* 1 The leg proper; that part of the lower limb between the knee and the ankle. 2 A cut of meat from the leg of an animal; the shin. 3 The tarsus of a bird. See the illustration under FOWL. 4 Something resembling a leg. 5 The part of a tool connecting the handle with the working part, as the stem of a drill. 6 The projecting piece or loop by which some forms of buttons are attached. 7 The stem of an anchor. 8 The stem of a key between the bow and the bit. 9 The straight part of a hook. 10 The narrow part of a spoon handle. 11 A continuation of the tang of a tool or instrument. 12 *Printing* The body of a type. 13 The narrow part of a shoe sole in front of the heel. See illustration under SHOE. 14 *Bot.* A pedicel. 15 *Colloq.* The remainder or last part of a thing: the *shank* of the evening. — *v.i.* 1 *Bot.* To decay or fall off the stem because of disease. 2 *Scot.* To travel on foot. [OE *sceanca*]

Shan·ka·ra (shung'kə-rə), flourished about 800 A.D., Hindu religious reformer, teacher and writer; foremost exponent of Vedanta philosophy; considered an incarnation of Shiva. Also **Shan·ka·ra·char·ya** (shung'kə-rä-chär'yə).

shanks' mare One's own legs as a means of conveyance.

shan·na (shan'na, shän'nä) *Scot.* Shall not.

Shan·non (shan'ən) An international airport in southern County Clare, Ireland, on the Shannon River west of Limerick.

Shannon River (shan'ən) The chief river of Ireland, in the west central part, flowing 224 miles south and west to the Atlantic.

Shan·si (shän'sē') A province of NE China; 50,000 square miles; capital, Taiyüan.

Shan State (shän, shan) A constituent unit of Burma in east Upper Burma, consisting of **Northern Shan State** and **Southern Shan State**; 61,090 square miles; capital, Lashio.

sha'nt (shant, shänt) Shall not: a contraction. Also **shan't**.

shan·tung (shan'tung, shan-tung') *n.* A silk fabric similar to pongee and having the same rough, nubby surface: originally made in China of wild silk, now often made of rayon combined with cotton. [from SHANTUNG]

Shan·tung (shan'tung', *Chinese* shän'dŏŏng') A province of NE China, extending as a peninsula in the eastern part into the Yellow Sea; 55,000 square miles; capital, Tsinan.

shan·ty¹ (shan'tē) *n. pl.* **·ties** A hastily built shack or cabin; a ramshackle or rickety dwelling. See synonyms under HOUSE, HUT. [<F (Canadian) *chantier* lumberer's shack]

shan·ty² (shan'tē) See CHANTEY.

shan·ty·man (shan'tē-man') *n. pl.* **·men** (-mən) One who lives in a shanty; specifically, a woodcutter or lumberman.

shan·ty·town (shan'tē-toun') *n.* 1 That section of a city or town comprised of ramshackle or hastily constructed shacks. 2 The inhabitants collectively of such a section: All *shantytown* turned out for the parade.

Shao·hing (shou'shing') A city in northern Chekiang province, China. Also **Shao'-hsing'**.

shape (shāp) *n.* 1 Outward form or construction; configuration; contour. 2 A developed expression or definite formulation; realization or application; embodiment; cast: to put an idea into *shape*. 3 A being, image, or appearance considered with reference to its form, generally incorporeal; ghost; phantom. 4 The character or form in which a thing appears; guise; aspect. 5 Something that gives or determines form; a pattern or mold; in millinery, a stiff frame. 6 The lines of a per-

SHARKS
A. Great white shark (to 40 feet). B. Blue shark (to 15 feet). C. Hammerhead shark (to 15 feet).

son's body; figure. 7 Manner of execution. 8 Condition as regarding fitness. 9 A blancmange, jelly, etc., cooled and shaped in a mold. — **to take shape** To have or assume a definite form. — *v.* **shaped, shaped** (*Rare* **shap·en**), **shap·ing** *v.t.* 1 To give shape to; mold; form. 2 To adjust or adapt; modify. 3 To devise; prepare. 4 To give direction or character to: to *shape* one's course of action. 5 To put into or express in words. 6 *Obs.* To appoint; ordain. — *v.i.* 7 To take shape; develop; form: often with *up* or *into*. 8 *Rare* To become adapted; conform. 9 *Rare* To happen; come about. See synonyms under MAKE. [OE *gesceap* creation] — **shap'er** *n.*

shaped (shāpt) *adj.* 1 Formed. 2 Resembling in shape: used in compounds, as in *leaf-shaped, club-shaped, key-shaped.*

shaped charge An explosive charge so placed in a shell or projectile as to deliver most of its force directly through the nose of the shell instead of scattering it at random: developed especially for anti-tank guns.

shape·less (shāp'lis) *adj.* Having no definite shape; lacking symmetry; formless. — **shape'-less·ly** *adv.* — **shape'less·ness** *n.*

shape·ly (shāp'lē) *adj.* **·li·er, ·li·est** Having a pleasing shape; well-formed; graceful. — **shape'li·ness** *n.*

shape-up (shāp'up') *n.* The selection of a work crew by an employer representative, a labor union deputy, or other agent, who chooses from among a number of men assembled for a work shift: a common practice in hiring longshoremen and workers in other industries in which the relationship between an employee and a specific employer is by the day or otherwise casual.

Shap·ley (shap'lē), **Harlow**, born 1885, U.S. astronomer.

shard (shärd) *n.* 1 A broken piece of a brittle substance, as of an earthen vessel; a potsherd; a fragment: also spelled *sherd*. 2 *Zool.* A hard, thin shell, or a wing cover, of an insect. [OE *sceard*. Related to SHEAR.]

share (shâr) *n.* 1 A portion; allotted or equitable part. 2 Specifically, one of the equal parts into which the capital stock of a company or corporation is divided. 3 An equitable part of something enjoyed or suffered in common. 4 A plowshare: also spelled *shear*. 5 A blade of a cultivator, seeder, etc. See synonyms under PART. — *v.* **shared, shar·ing** *v.t.* 1 To divide and give out in shares or portions; apportion. 2 To enjoy or endure in common; participate in. 3 To have a part; participate: with *in*. See synonyms under APPORTION. [OE *scearu* < *sceran* shear. Related to SHEAR.] — **shar'er** *n.*

share-crop·per (shâr'krop'ər) *n.* A tenant farmer who pays a share of his crop as rent for his land.

share·hold·er (shâr'hōl'dər) *n.* An owner of a share or shares of a company's stock; a stockholder.

shares·man (shârz'mən) *n. pl.* **·men** (-mən) A member of a cooperative fishing crew who shares in the risks and profits of the cruise or season. Also **share'man**.

Sha·ri (shä'rē) A river of central French Equatorial Africa, forming the principal tributary of Lake Chad and flowing 500 miles NW: French *Chari*.

shark¹ (shärk) *n.* One of a group of voracious elasmobranch fishes (order *Selachii*), mostly marine, of medium to large size, having a cartilaginous skeleton, lateral gill slits, and dun-colored bodies covered with placoid scales. Most species do not molest man; the great white shark (*Carcharodon carcharias*) is the man-eater frequenting warm seas. — *v.i.* To fish for sharks. [Origin uncertain]

shark² (shärk) *n.* 1 A bold and dishonest person; a rapacious swindler. 2 *Slang* A person of exceptional skill or ability in some special line. Also **shark'er**. — *v.t. Archaic* To obtain by unscrupulous or deceitful means. — *v.i.* To live by trickery or deceit. [Prob. <G *schurke* scoundrel]

shark·skin (shärk'skin') *n.* 1 The skin of a shark. 2 A summer fabric with a smooth, almost shiny surface, made of acetate rayon and used for sports clothes; originally, a weave of woolen yarns of two colors: so called from its resemblance to sharkskin leather.

sharn (shärn) *n. Scot.* Cow dung. — **sharn'y** *adj.*

Shar·on (shar'ən), **Plain of** A part of the coastal plain of western Israel, extending 50 miles between the Hills of Ephraim and the Mediterranean.

sharp (shärp) *adj.* 1 Having a keen edge or an acute point; capable of cutting or piercing. 2 Coming to an acute angle; not obtuse; angular; abrupt: a *sharp* peak. 3 Keen of perception or discernment; also, shrewd in bargaining; artful; overreaching: *sharp* practice. 4 Ardent; quick; eager; keen, as the appetite; impetuous or fiery, as a combat or debate; vigilant or attentive. 5 Affecting the mind or senses, as if by cutting or piercing; afflicting; poignant; painful; harsh; censorious; acrimonious; rigorous; stern; sarcastic; bitter. 6 Shrill. 7 Pinching; cutting; as cold. 8 Having an acid or pungent taste. 9 Distinct, as an outline; not blurred or hazy; well-defined. 10 *Music* Being above the proper or indicated pitch; specifically, being a half-step higher than the indicated note; sharped. 11 Hard and rough; gritty, as sand. 12 *Phonet.* Surd; voiceless: opposed to *flat*: said of consonants. — *adv.* 1 In a sharp manner; sharply. 2 Promptly; exactly; on the instant: at 4 o'clock *sharp*. 3 *Music* Above the proper pitch. — *n.* 1 *Music* A character (♯) used on a natural degree of the staff to make it represent a pitch a half-step higher; the tone so indicated; on the pianoforte, the next higher key; one of the black keys: a loose use in the phrase *sharps and flats*. 2 A sewing needle of long, slender shape. 3 A cheating rogue; sharper: a *cardsharp*. 4 *Obs.* A cheating sword; rapier. — *v.t. Music* To raise in pitch, as by a half-step. — *v.i. Music* To sing, play, or sound above the right pitch. [OE *scearp*] — **sharp'ly** *adv.* — **sharp'ness** *n.*

Synonyms (adj.): acute, cutting, keen, penetrating, piercing, pointed. See ACID, ACUTE, ASTUTE, BITTER, CLEVER, FINE, KNOWING, SAGACIOUS, STEEP, VIOLENT. *Antonyms*: blunt, dull, dulled, edgeless, flat, obtuse, pointless, round, rounded.

Sharp (shärp), **William**, 1856?-1905, Scottish poet and novelist: pseudonym, *Fiona McLeod*.

sharp·en (shär'pən) *v.t. & v.i.* To make or become sharp. — **sharp'en·er** *n.*

sharp·er (shär'pər) *n.* A swindler; cheat.

sharp-eyed (shärp′īd′) *adj.* 1 Having acute eyesight. 2 Keenly observant; alert.
sharp·ie (shär′pē) *n.* A long, sharp, flat-bottomed sailboat having a centerboard and one or two masts, each having a triangular sail: originally used in the oyster and scallop fisheries. [<SHARP; in allusion to its outline]

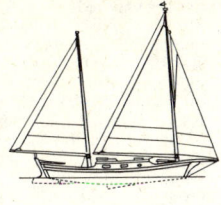
SHARPIE

Sharps·burg (shärps′bûrg) A town in NW Maryland; site of the battle of Antietam, 1862, in the Civil War.
sharp-set (shärp′set′) *adj.* 1 Set at a sharp angle; prepared like a saw for cutting. 2 Keen; eager; fierce. 3 Ravenous; hungry; thin and hungry-looking.
sharp-shinned (shärp′shind′) *adj.* Having slender shanks, somewhat angular in front: specifically said of the North American **sharp-shinned hawk** (*Accipiter velox*).
sharp·shoot·er (shärp′shoo̅′tər) *n.* 1 A skilled marksman, especially in the use of the rifle. 2 The second grade of skill in small-arms shooting, ranking next above *marksman* and below *expert*; also, a soldier having this grade. — **sharp′shoot′ing** *n.*
sharp-sight·ed (shärp′sī′tid) *adj.* Having keen vision. — **sharp′-sight′ed·ness** *n.*
sharp-tongued (shärp′tungd′) *adj.* Bitter or caustic in speech.
sharp-wit·ted (shärp′wit′id) *adj.* Acute; intelligent; discerning. See synonyms under INTELLIGENT, SAGACIOUS. — **sharp′-wit′ted·ness** *n.*
Sha·si (shä′sē′) A city on the Yangtze in south central Hupeh province, China.
Shas·ta (shas′tə), **Mount** A volcanic cone in the Cascade Range of northern California; 14,161 feet.
Shasta daisy A cultivated variety of a short-lived perennial (*Chrysanthemum maximum*) having large, white-rayed flowers.
Shas·tan (shas′tən) *adj. & n.* Comanchean.
Shatt-el-Ar·ab (shat′al·ar′əb) A river in SE Iraq, formed by the Tigris and Euphrates which unite 40 miles NW of Basra and flow 120 miles SE, forming the Iraq-Iran boundary from below Basra to the Persian Gulf.
shat·ter (shat′ər) *v.t.* 1 To break into pieces suddenly, as by a blow. 2 To break the health or tone of, as the body or mind; disorder; damage. 3 *Obs.* To scatter. — *v.i.* 4 To break into pieces. See synonyms under BREAK. — *n. Obs.* 1 A shattered fragment; a splinter: a tree rent into *shatters*. 2 A shattered or disordered condition; a cornet or horn. [ME *schateren*. ? Akin to SCATTER.]
shat·ter·proof glass (shat′ər·proof′) See under GLASS.
shauch·le (shäkh′l, shä′khəl, shô′-) *Scot. v.t.* To put out of shape; distort, as a shoe. — *v.i.* To shuffle; shamble. Also **shaugh′le.**
shaul (shôl) See SHOAL[1].
shave (shāv) *v.* **shaved, shaved** or **shav·en, shav·ing** *v.i.* 1 To cut hair or beard close to the skin with a razor. — *v.t.* 2 To remove hair or beard from (the face, head, etc.) with a razor. 3 To cut (hair or beard) close to the skin with a razor: often with *off*. 4 To trim closely as if with a razor: to *shave* a lawn. 5 To cut thin slices from, as in preparing the surface; pare; plane. 6 To cut into thin slices: to *shave* ice. 7 To touch or scrape in passing; graze; come close to. 8 *U.S.* To buy (commercial paper) at a greater reduction than the bank discount. — *n.* 1 The act or operation of cutting off the beard with a razor. 2 A knife or blade, mounted between two handles, as for shaving wood: also **draw shave, spoke shave.** 3 A shaving; thin slice. 4 An extra or exorbitant discount paid for cashing a note or draft, as a premium given for an extension of time. 5 *Colloq.* The act of rushing by or barely grazing something; hence, a narrow escape: a close *shave*. 6 One who drives hard bargains. [OE *scafan* shave]
shave·ling (shāv′ling) *n.* 1 One who is shaven; opprobriously, a monk or priest. 2 A youth.
shav·en (shā′vən) Alternative past participle of SHAVE. — *adj.* 1 Shaved; also, tonsured. 2 Trimmed closely.
shav·er (shā′vər) *n.* 1 One who shaves; specifically, a barber. 2 A plunderer; cheat; sharper. 3 *Colloq.* A lad.
shave·tail (shāv′tāl′) *n. U.S. Slang* 1 A second lieutenant, especially one recently commissioned. 2 An untrained or intractable mule. 3 A tenderfoot. [Formerly in allusion to young, unbroken army mules with their tails bobbed]
Sha·vi·an (shā′vē·ən) *n.* An admirer of George Bernard Shaw, his books, or his theories. — *adj.* Of, pertaining to, or like George Bernard Shaw, or his style and methods.
shav·ie (shā′vē) *n. Scot.* A deceptive trick.
shav·ing (shā′ving) *n.* 1 The act of one who shaves; that which shaves. 2 A thin paring shaved from anything, as a board.
shaw[1] (shô) *v.t. Scot.* To show.
shaw[2] (shô) *n. Brit.* 1 A thicket; copse: also **shaugh.** 2 The leaves and tops of vegetables: usually in the plural. [OE *scaga* copse]
Shaw (shô), **George Bernard,** 1856-1950, Irish dramatist, critic, and novelist. — **Henry Wheeler,** 1818-85, U.S. humorist: pseudonym *Josh Billings*. — **Thomas Edward** See LAWRENCE, THOMAS EDWARD.
Sha·wan·gunk Mountains (shong′gum, -gungk) A range of the Appalachians in SE New York; about 45 miles long; highest point, 2,289 feet.
shawl (shôl) *n.* A wrap, as a square cloth, or large broad scarf, worn over the upper part of the body. [<Persian *shāl*]
shawm (shôm) *n.* An ancient, double-reed instrument; inaccurately, a cornet or horn. [<OF *chalemie* pipe <LL *calamellus*, dim. of L *calamus* reed]
Shaw·nee (shô·nē′) *n.* One of a warlike tribe of North American Indians of Algonquian stock, formerly living in Tennessee and South Carolina; now in Oklahoma. [<Algonquian (Shawnee) *Shawunogi* southerners < *shawun* south]
Shaw·wal (shô·wäl′) See under CALENDAR (Mohammedan).
shay (shā) *n.* A chaise: a back formation due to mistaking *chaise* for a plural.
Shays (shāz), **Daniel,** 1747?-1825, a captain in the American Revolution and leader of **Shays' Rebellion,** 1786-87, a popular insurrection in western Massachusetts, caused by economic distress of that time. — **Shays′ite** *n.*
Shcher·ba·kov (shchir·bä·kôf′) A city on the Volga in north central European Russian S.F.S.R.: formerly *Rybinsk*.
she (shē) *pron.* 1 The female person or being previously mentioned or understood, in the nominative case. 2 That woman or female; any woman: *She* who listens learns. — *n.* A female person or being: This puppy is a *she*. [OE *sēo, sio,* fem. of *sē* the, replacing *hēo* she]
she- combining form Female; feminine: in hyphenated compounds: a *she*-lion.
shea (shē) *n.* A large tree (*Butyrospermum parkis*) growing only in western tropical Africa and yielding **shea butter,** used for food, illumination, and making soap. [<Mundingo *si, se*]
sheaf[1] (shēf) *n. pl.* **sheaves** (shēvz) 1 A quantity of the stalks of cut grain or the like, bound together; a bundle of straw. 2 Any collection of things, as papers, held together by a band or tie. 3 The quiverful of arrows carried by an archer, usually 24. — *v.t.* To bind in a sheaf; sheave. [OE *scēaf*]
sheaf[2] See SHEAVE[2].
sheal[1] (shēl) *n. Scot. & Brit. Dial.* A shealing.
sheal[2] (shēl) *n. Brit.* A pod or shell. [Var. of SHELL]
sheal·ing (shē′ling) *n.* 1 *Brit. Dial.* A hut or cabin for the use of shepherds or sportsmen in the hills, for fishermen at the shore, etc.: also spelled *shieling.* 2 *Scot.* A shed for sheltering sheep at night in the hills. Also called *sheal.*
shealing hill *Scot.* A hill upon which grain is winnowed by the wind: also spelled *sheeling hill.*
shear (shir) *n.* 1 A two-bladed cutting instrument: obsolete except in the plural. See SHEARS. 2 *Physics* A deformation of a solid body, equivalent to a sliding over each other of adjacent laminar elements, with a progressive relative displacement: also **shearing stress.** 3 The act or result of shearing. 4 A plowshare. 5 *Naut.* Sweep; sheer. — *v.* **sheared** (*Archaic* **shore**), **sheared** or **shorn, shear·ing** *v.t.* 1 To cut the hair, fleece, etc., from. 2 To remove by cutting or clipping: to *shear* wool. 3 To deprive; strip, as of power or wealth. 4 To cut or clip with shears or other sharp instrument: to *shear* a cable. 5 *Dial.* To reap, as grain, with a sickle. — *v.i.* 6 To use shears or other sharp instrument. 7 To slide or break from a shear (def. 2). 8 To proceed by or as by cutting a way: with *through*. 9 *Dial.* To reap with a sickle. See synonyms under CUT. ◆ Homophone: *sheer*. [OE *sċeāra* scissors <*sċeran* shear. Akin to SHARD, SHARE.] — **shear′er** *n.*
shear·ling (shir′ling) *n.* 1 The fleece from the second shearing of a sheep. 2 The sheep from which one fleece has been cut.
shears (shirz) *n. pl.* 1 Any large cutting or clipping instrument worked by the crossing of cutting edges. 2 The ways or guides, as of a lathe. 3 An apparatus for hoisting and moving heavy objects, consisting of two or more spars with lower ends spread out and upper ends jointed to receive the tackle: also **shear legs:** sometimes spelled *sheers.* 4 The side frames of a steam fire engine. [See SHEAR]
shear-wa·ter (shir′wô′tər, -wot′ər) *n.* One of several sea birds (genus *Puffinus*) related to the petrels and albatrosses, found in most seas: so called because they skim close to the water.
sheat-fish (shēt′fish′) *n. pl.* **-fish** or **-fish·es** A catfish (*Silurus glanis*) of the fresh waters of central and eastern Europe. It is the largest fresh-water fish in Europe. [OE *sċēota* trout + FISH]
sheath (shēth) *n.* 1 An envelope or case, as for a sword; scabbard. 2 *Bot.* A case enclosing a part of an organ, as the lower part of the leaves in grasses. 3 *Zool.* Any covering in animals that resembles a sheath. 4 *Entomol.* An elytron of a beetle. 5 *Agric.* A bar connecting the beam and sole in a plow. [OE *scǣth*] — **sheath′less** *adj.*
sheath-bill (shēth′bil′) *n.* Any of a small number of species of sea birds of the family *Chionididae*, natives of the Antarctic islands. They are pure white in plumage and have a horny sheath at the base of the bill.
sheathe (shēth) *v.t.* **sheathed, sheath·ing** 1 To put into a sheath. 2 To plunge (a sword, etc.) into flesh, as if into a sheath. 3 To incase or protect with a covering, as the hull of a ship with metal. 4 To draw in, as claws. [<SHEATH]
sheath·ing (shē′thing) *n.* 1 A casing, as of a building, or the protective covering of a ship's hull; that which sheathes; also, the material used. 2 The act of one who sheathes. 3 *Archit.* The covering or waterproof material on outside walls or roof.
sheath knife A large case knife carried in a sheath attached to a belt, worn by sailors and riggers.
sheave[1] (shēv) *v.t.* **sheaved, sheav·ing** To gather into sheaves; collect. [<SHEAF]
sheave[2] (shēv) *n.* 1 A grooved pulley wheel; also, a pulley wheel and its block. 2 An eccentric or its disk. 3 *Scot.* A slice or cut. Also spelled *sheaf, sheeve.* [Var. of SHIVE[1]]

SHEAVE

sheaves (shēvz) Plural of SHEAF.
She·ba (shē′bə) The Old Testament name for a region of the SW Arabian peninsula, corresponding to modern Yemen: Arabic *Saba*.
She·ba (shē′bə), **Queen of** A queen, called *Balkis* in the Koran, who visited Solomon to test his wisdom. I Kings x 1-13.
she·bang (shi·bang′) *n. U.S. Slang* 1 A building, vehicle, saloon, theater, etc. 2 Any matter of present concern; thing; contrivance; outfit: tired of the whole *shebang*. [Var. of SHEBEEN]
She·bat (shi·bät′) See under CALENDAR (Hebrew). Also spelled *Sebat*.
she·been (shi·bēn′) *n. Irish & Scot.* A groggery; specifically, a place where liquors are sold without a license; hence, weak ale or beer. [<Irish *sibín* little mug]
She·be·li (shi·bä′lē), **Web·be** (web′a) See SHIBELI, WEBBE.

She·chem (shē′kem) The ancient name for NABLUS.

shed¹ (shed) v. **shed, shed·ding** v.t. 1 To pour forth in drops; emit, as tears or blood. 2 To cause to pour forth. 3 To send forth or abroad; diffuse; radiate, as light. 4 To throw off without allowing to penetrate, as rain; repel. 5 To cast off by natural process, as hair, skin, a shell, etc. —v.i. 6 To cast off or lose hair, skin, etc., by natural process. 7 To fall or drop, as leaves or seed. —**to shed blood** To kill. —n. 1 That which sheds, as a sloping surface or watershed. 2 The act of shedding: *bloodshed*. 3 A separation or division; parting: applied technically to the opening in the warp through which the shuttle is thrown in weaving, and in parts of Great Britain to the parting of the hair. See illustration under LOOM. 4 The slope of a hill. [OE *scēadan* separate, part]

shed² (shed) n. 1 A small low building, often with front or sides open; also, a lean-to: a wagon *shed*. 2 *Brit.* A storehouse; barn. 3 A temporary covering; cabin. 4 A hangar. See synonyms under HUT. [Var. of SHADE]

she'd (shēd) 1 She had. 2 She would.

shed·der (shed′ər) n. 1 One who sheds. 2 An animal that sheds or has lately shed its skin, as a crab.

she-dev·il (shē′dev′əl) n. 1 A bad-tempered and spiteful woman. 2 A female demon.

shee (shē) See SID.

sheel·ing hill (shē′ling) See SHEALING HILL.

sheen (shēn) n. 1 A glistening brightness, as if from reflection. 2 Bright, shining attire. See synonyms under LIGHT. —*adj.* Shining; radiant; beautiful. —v.i. To shine; gleam; glisten. [OE *scēne* beautiful; infl. in meaning by SHINE. Akin to G *schön* beautiful.] —**sheen′y** *adj.*

sheep (shēp) n. pl. **sheep** 1 A medium-sized, domesticated ruminant of the genus *Ovis* (family *Bovidae*), highly prized for its flesh, wool, and skin. ◆ Collateral adjective: *ovine*. 2 Leather made from the skin of the sheep, as for bookbinding: also *sheepskin*. 3 Someone with the supposed temperament of a sheep; hence, a meek, bashful, or timid person. [OE *scēap*]

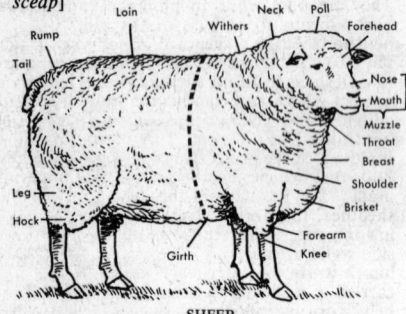

SHEEP
Nomenclature of anatomical parts.

sheep·backs (shēp′baks′) n. pl. Roches moutonnées.

sheep·ber·ry (shēp′ber′ē) n. pl. **·ries** 1 One of the black, oval, edible drupes of the sweet viburnum (*Viburnum lentago*). 2 The tree itself.

sheep·cote (shēp′kōt′) n. A small enclosure for the protection of sheep; a sheepfold. Also **sheep′cot′** (-kot′).

sheep dip Any of several liquid disinfectants which contain creosote, nicotine, cresol, arsenic, etc., used for dipping sheep.

sheep dog 1 A dog trained to guard and control sheep; shepherd's dog: often a collie, but also a rough-coated, heavy, short-tailed dog much used by drovers in England. 2 Figuratively, a chaperon. —**old English sheep dog** A bob-tailed dog of undetermined origin, used as a sporting dog, and, in Great Britain, to herd flocks: characterized by a strong, muscular, thick-set body, covered with a very thick gray, grizzle, or blue-gray shaggy coat.

sheep·fold (shēp′fōld′) n. A place where sheep are enclosed at night; a pen for sheep.

sheep·herd·er (shēp′hûr′dər) n. A herder of sheep. —**sheep′herd′ing** n.

sheep·ish (shē′pish) adj. Foolish, as a sheep; awkwardly diffident; abashed. —**sheep′ish·ly** adv. —**sheep′ish·ness** n.

sheep laurel Lambkill.

sheep ranch A ranch and range where sheep are bred and raised. Also *Brit.* **sheep′walk′**, *Austral.* **sheep run.**

sheep's eyes Bashful, sidelong, or amorous glances.

sheeps·head (shēps′hed′) n. 1 A common deep-bodied sparoid food fish (*Archosargus probatocephalus*) of the Atlantic coast of the United States. 2 The Great Lakes drumfish, also found in the Mississippi region. 3 The dollarfish. 4 A foolish or silly person.

sheep·shear·ing (shēp′shir′ing) n. 1 The act of shearing sheep. 2 The shearing season; an occasion at which sheep are shorn, and the feast or celebration given at the occasion. —**sheep′shear′er** n.

sheep·skin (shēp′skin′) n. 1 The skin of a sheep, tanned or untanned, or anything made from it, as parchment. 2 A document written on parchment; hence, a diploma.

sheep sorrel An herb (*Rumex acetosella*) of the buckwheat family, widely distributed in dry places, and having leaves of an acrid taste.

sheer¹ (shir) v.i. To swerve from a course; turn aside. —v.t. To cause to swerve or deviate. —n. 1 *Naut.* **a** The rise, or the amount of rise from a level, of the lengthwise lines of a vessel's hull. **b** A position of a vessel that enables it to swing clear of a single anchor. 2 A swerving or curving course. ◆ Homophone: *shear*. [<SHEAR]

sheer² (shir) adj. 1 Having no modifying conditions; unmitigated; absolute; downright; utter: *sheer* folly; *sheer* nonsense. 2 Exceedingly thin and fine: said of fabrics. 3 Perpendicular; steep; ascending vertically: a *sheer* precipice. 4 Pure; pellucid. 5 *Obs.* Bright; shining. See synonyms under PURE, STEEP. —n. Any very thin fabric used for clothes. —adv. Entirely; perpendicularly: also **sheer′ly.** ◆ Homophone: *shear*. [ME *schere.* Cf. ON *skærr* clear, bright and OE *scīr* bright, shining.] —**sheer′ness** n.

Sheer·ness (shir′nes′) An urban district and port on the Isle of Sheppey at the mouth of the Medway, in the Thames estuary, northern Kent, England.

sheers (shirz) See SHEARS (def. 3).

sheet (shēt) n. 1 A very thin and broad piece of any substance; that which is or can be spread, as upon a surface, or can be laid in broad folds; anything having a considerable expanse with very little thickness. 2 A large rectangular piece of linen or cotton cloth, used in making up a bed. 3 A piece of paper, especially one of a regular size; hence, a newspaper, or a leaf of a book. 4 A piece of metal or other substance hammered, rolled, fused, or cut very thin: a *sheet* of glass. 5 A broad, flat surface; superficial expanse: a *sheet* of water; a *sheet* of flame. 6 *Naut.* **a** A rope or chain from a lower corner of a sail to extend it or move it. **b** *pl.* In an open boat, the space at the bow and stern not occupied by the thwarts. The former is termed the **fore sheets** and the latter the **stern sheets.** 7 A sail: a literary use. 8 *Geol.* **a** An originally horizontal or moderately inclined layer of igneous rock of small thickness as compared with its lateral extent. **b** Any superficial deposit, as of gravel left by a glacier, or of soil or ice. 9 The large, unseparated block of stamps printed by one impression of a plate. —**three sheets in the wind** *Slang* Tipsy; drunk. —v.t. 1 To stretch by hauling on a sheet: used only in the expression **to sheet home,** to stretch the clews of a sail to the extremities of the next lower yard. 2 To cover with or wrap in a sheet. 3 To furnish with sheets. —v.i. 4 To extend in a particular direction: said of the sheets of a sail. [OE *scēte* linen cloth]

sheet anchor 1 One of two anchors for use only in emergency; formerly, the main anchor. 2 A sure dependence on occasion of danger or emergency.

sheet bend *Naut.* A knot used to join two ropes' ends, made by passing one end through a loop of the other rope, carrying it around the loop, and slipping it under its own running part.

sheet·ing (shē′ting) n. 1 The act of sheeting, in any sense. 2 Cotton, muslin, linen, or cotton percale, for making bleached, unbleached, or colored sheets for beds.

sheet lightning Lightning appearing in sheetlike form as a momentary and broadly diffused radiance in the sky, caused by the reflection of a distant lightning flash.

sheet metal Metal rolled and pressed into sheets.

sheet music Music printed on separate sheets of paper.

sheeve (shēv) See SHEAVE².

Shef·field (shef′ēld) A city and county borough in West Riding, southern Yorkshire, England.

sheik (shēk, *Brit.* shāk) n. 1 A Moslem high priest, a venerable man; the chief or head of an Arab tribe or family: often used as a title of respect. 2 A man who fascinates women; a lady-killer: from *The Sheik,* a novel (1921) by Edith M. Hull. Also spelled *scheik, shaik, sheyk.* [<Arabic *sheikh, shaykh,* lit., an elder, chief < *shakha* grow old] —**sheik·dom** (shēk′dəm) n. The land ruled by a sheik. Also **sheikh′dom.**

Sheik ul Is·lam (shēk′ ool is-läm′) Formerly, the head of the hierarchy in Turkey; the grand mufti.

shei·tan (shī-tän′) See SHAITAN.

shek·el (shek′əl) n. 1 An Assyrian, Babylonian, and, later, Hebrew unit of weight and money; a coin having this weight. 2 *pl. Slang* Money; riches. [<Hebrew *sheqel* < *shāqal* weigh]

She·ki·nah (shi-kī′nə) n. A cloud of glory which accompanied the tabernacle of the Jews, especially when over the mercy seat: a symbol and manifestation of the divine presence. [<Hebrew *shekhinah,* lit., dwelling place < *shākhan* dwell]

Shel·don (shel′dən), **Charles Monroe,** 1857–1946, U.S. clergyman and author.

shel·drake (shel′drāk′) n. 1 A large Old World duck of either of the genera *Tadorna* or *Casarca,* as the common sheldrake (*T. tadorna*), or the ruddy sheldrake (*C. rutila*) of southeastern Europe and North Africa. 2 A merganser, especially the red-breasted merganser or salt-water sheldrake (*Mergus serrator*). 3 The canvasback duck. [<dial. E *sheld* piebald, dappled + DRAKE]

shelf (shelf) n. pl. **shelves** (shelvz) 1 A board or slab set horizontally into or against a wall to support articles, as books; one of the boards in a bookcase or closet; the contents of a shelf. 2 Any flat projecting ledge, as of rock. 3 A steep-sided bank or shallow place in a body of water; a reef; shoal. 4 The stratum of bedrock met in sinking a shaft. —**on the shelf** No longer in use; discarded. [<LG *schelf* set of shelves]

Shel·i·kof Strait (shel′i-kôf) A channel between Alaska and Kodiak Island, connecting the North Pacific with the Gulf of Alaska, 130 miles long, 30 miles wide.

shell (shel) n. 1 A hard structure incasing an animal, as a mollusk, or an egg or fruit. 2 A mollusk; shellfish: much used in composition. 3 A hollow structure or vessel, generally thin and weak; also, a framework with its interior removed or destroyed, or one to be filled out or built upon. 4 A very light, long, and narrow racing rowboat. 5 A hollow metallic projectile filled with an explosive or chemical; especially, an artillery projectile filled with high explosive: used against materiel and fortifications and distinguished from shrapnel used against personnel. 6 The plates, etc., constituting the framework of a steam boiler or the like. 7 A metallic or paper cartridge case for breechloading small arms (see illustration under CARTRIDGE); also, any paper case used to contain the explosives of fireworks, such as torpedos. 8 *Physics* One of the orbits in which the electrons of an atom revolve. 9 A shape or outline that merely simulates a reality; hollow form; external semblance. 10 The external ear; auricle. 11 The lyre: originally a stringed tortoise shell. 12 A reserved or impersonal attitude: to come out of one's *shell*. —v.t. 1 To divest of or remove from a shell; strip from the husk, pod, or shell. 2 To separate from the cob, as Indian corn. 3 To bombard with shells, as a fort. 4 To cover with shells. —v.i. 5 To shed or become freed from the shell or pod. 6 To fall off, as a shell or scale. —**to shell out** *Colloq.* To hand over, as money. [OE *scell* shell] —**shell′er** n. —**shell′less** adj. —**shell′y** adj.

she'll (shēl) She will.

shel·lac (shə-lak′) n. 1 A purified lac obtained as plates or cakes and extensively used in varnish, sealing wax, insulators, etc. 2 A solution, orange or white, of flake shellac

dissolved in methylated spirit: used for coating floors, woodwork, etc. —*v.t.* **·lacked, ·lack·ing 1** To cover or varnish with shellac. **2** *Slang* **a** To belabor; beat. **b** To defeat utterly. Also **shel·lack′.** [<SHELL + LAC[1], trans. of F *laque en écailles* lac in fine sheets]
shel·lack·ing (shə·lak′ing) *n. Slang* **1** A beating; assault. **2** A thorough defeat.
shellac varnish Any of several varnishes containing dissolved shellac and giving a thin, hard, sometimes glossy, coat.
shell·back (shel′bak′) *n.* A veteran sailor; an old salt; especially, one who has crossed the equator. [Prob. with reference to the shell of the sea turtle]
shell·bark (shel′bärk′) *n.* The shagbark or one of its nuts.
shell bean Any of various beans cultivated for their edible mature seeds.
Shel·ley (shel′ē), **Mary Wollstonecraft,** 1797–1851, *née* Godwin, English novelist; wife of the following. —**Percy Bysshe,** 1792–1822, English poet.
shell·fire (shel′fīr′) *n.* The firing of artillery shells.
shell·fish (shel′fish′) *n. pl.* **·fish** or **·fish·es** Any aquatic animal having a shell, as a mollusk.
shell game 1 A swindling game in which the victim bets on the location of a pea covered by one of three nutshells; thimblerig. **2** Any game in which the victim cannot win.
shell·heap (shel′hēp′) *n.* A kitchen midden. Also **shell′mound′** (-mound′).
shell hole A hole made by an exploding shell; specifically, a craterlike depression in the ground. Also **shell crater.**
shell jacket A snugly fitted jacket, short at the back, worn in place of the tuxedo in tropical countries.
shell pink Any of several shades of light, pure pink, like the color in certain seashells.
shell·proof (shel′proof′) *adj.* Built to resist the destructive effect of projectiles and bombs.
shell shock Combat fatigue. —**shell-shocked** (shel′shokt′) *adj.*
shel·ter (shel′tər) *v.t.* To provide protection or shelter for; shield, as from danger or inclement weather. —*v.i.* To take shelter. —*n.* **1** That which covers or shields from exposure or danger; a place of safety. **2** The state of being sheltered or protected. **3** A cover from the weather, as a box for meteorological instruments, etc. **4** One who protects; a guardian. [Appar. alter. of ME *scheltrum* <OE *scēld-truma,* a body of men armed with shields, phalanx, protection] —**shel′ter·er** *n.* —**shel′ter·less** *adj.*
Synonyms (verb): cover, defend, guard, harbor, protect, screen, shield, ward. To *cover* generally means to extend completely over something; a vessel is *covered* with a lid; the head is *covered* with hair. To *shelter* is to *cover* so as to *protect* from injury or annoyance; as, The roof *shelters* from the storm. To *defend* implies the actual, *protect* implies the possible use of force or resisting power; *guard* implies sustained vigilance with readiness for conflict. *Protect* is more complete than *guard* or *defend*; an object may be faithfully *guarded* or bravely *defended* in vain, but that which is *protected* is secure. See CHERISH. Compare synonyms for DEFENSE. *Antonyms*: betray, expel, expose, refuse, reject, surrender.
Synonyms (noun): asylum, cover, covert, defense, harbor, haven, protection, refuge, retreat, sanctuary, shield. See DEFENSE. *Antonyms*: assault, attack, danger, exposure, onslaught, peril.
shelter belt Natural or artificial forest maintained as a protection from wind or snow.
shelter tent A tent large enough to accommodate two men: divided into two sections, each of which, called a **shelter half,** is carried as part of a soldier's field equipment. Also called *pup tent.*
shelt·ie (shel′tē) *n. Scot.* A Shetland pony. Also **shelt′y.**
shelve (shelv) *v.* **shelved, shelv·ing** *v.t.* **1** To place on a shelf. **2** To postpone indefinitely; put aside. **3** To retire. **4** To provide or fit with shelves. —*v.i.* **5** To incline gradually; slope. [<SHELF] —**shelv′y** *adj.*
shelves (shelvz) Plural of SHELF.
shelv·ing (shel′ving) *n.* **1** Shelves collectively. **2** Material for the construction of shelves. **3** The act of putting away on shelves; hence, putting aside; dismissing. **4** A slight inclining.
Shem (shem) The eldest son of Noah. *Gen.* v 32.
Shem·ite (shem′īt) See SEMITE.
Shen·an·do·ah (shen′ən·dō′ə) A river in Virginia and West Virginia, flowing 55 miles to the Potomac at Harper's Ferry. The **Shenandoah Valley,** part of the Great Appalachian Valley, was the scene of many battles during the Civil War.
Shenandoah National Park A region in the Blue Ridge Mountains of NW Virginia; 302 square miles; established 1935.
she·nan·i·gan (shi·nan′ə·gən) *n. Colloq.* Trickery; foolery; nonsense; also, treacherous action or a treacherous act. [Prob. <Irish *sionnach* fox]
Sheng·king (sheng′jing′) A former name for LIAONING.
Shen·si (shen′sē′) A province of NW central China; 75,000 square miles; capital, Sian.
Shen·stone (shen′stən, -stōn), **William,** 1714–1763, English poet.
Shen·yang (shun′yäng′) A city of NE China, capital of Liaoning province; the capital city of the former Manchuria region; a major metal-fabricating center: formerly *Mukden, Fengtien.*
she·ol (shē′ōl) *n.* Hell. [<Hebrew *she'ōl* cave < *shā'al* dig]
She·ol (shē′ōl) In the Old Testament, a place under the earth where the departed spirits were believed to go.
Shep·ard (shep′ərd), **Alan B., Jr.,** born 1923, U. S. naval officer; first American astronaut to make a rocket flight into space and first astronaut to control the space vehicle himself, May 5, 1961.
shep·herd (shep′ərd) *n.* **1** A keeper or herder of sheep. **2** Figuratively, a pastor, leader, or guide. —*v.t.* To watch and tend as a shepherd; guard; protect. [OE *scēaphyrde*] —**shep′herd·ess** *n. fem.*
shep·herd·clock (shep′ərd·klok′) *n.* Goatbeard. Also **shep′herd's-clock′.**
shepherd dog A sheep dog, as the Scotch collie or the German shepherd dog.
shepherd kings See HYKSOS.
shep·herd's-nee·dle (shep′ərdz·nēd′l) *n.* Venus's-comb.
shep·herd's-purse (shep′ərdz·pûrs′) *n.* A common herbaceous weed *(Capsella bursa-pastoris)* bearing small white flowers and notched triangular pods (whence its name).
Shep·pey (shep′ē), **Isle of** An island in the Thames estuary, northern Kent, England; 36 square miles.
sher·ard·ize (sher′ər·dīz) *v.t.* **·ized, ·iz·ing** *Metall.* To give a coating of zinc to (steel or iron) by packing in zinc dust, placing in a furnace, and subjecting to a heat sufficient to cause the zinc vapor to soak in. [after *Sherard* Cowper-Coles, died 1936, British inventor] —**sher′ard·iz′ing** *n.* & *adj.*
Sher·a·ton (sher′ə·tən) *adj.* Denoting the graceful, straight-lined style of English furniture developed by Thomas Sheraton.
Sher·a·ton (sher′ə·tən), **Thomas,** 1751–1806, English furniture maker and designer.
sher·bet (shûr′bit) *n.* **1** A flavored water ice. **2** An Oriental drink, made of fruit juice sweetened and diluted with water and sometimes cooled with snow. [<Turkish *sharbat* <Arabic *sharbah* a drink < *shariba* drink. Doublet of SIRUP.]
Sher·brooke (shûr′brook) A city in southern Quebec province, Canada.
sherd (shûrd) *n.* A fragment of pottery: often in composition: *potsherd*: also spelled *shard.* [Var. of SHARD]
Sher·i·dan (sher′ə·dən), **Philip Henry,** 1831–1888, U. S. general in the Civil War. —**Richard Brinsley,** 1751–1816, English dramatist and politician.
she·rif (she·rēf′) *n.* **1** A member of a princely Moslem family which claims descent from Mohammed through his daughter Fatima. **2** The chief magistrate of Mecca: also **grand sherif. 3** An Arab chief. Also **she·reef′.** [<Arabic *sharīf* noble]
sher·iff (sher′if) *n.* The chief administrative officer of a county, who executes the mandates of courts, etc. In the United States, the sheriff is elected by the legislature or by direct vote of the citizens and must be of age, a citizen of the country, and reside in the county he represents. [OE *scīr-gerēfa* shire reeve] —**sher′iff·dom** *n.*
Sher·iff·muir (sher′if·myoor) A locality in southern Perthshire, Scotland; scene of a battle between the Scottish Jacobites and the English, 1715.
sher·lock (shûr′lok) *n. Slang* A detective. [after *Sherlock* Holmes]
Sher·lock Holmes (shûr′lok hōmz′) A fictitious English detective, the central character of numerous stories by Arthur Conan Doyle.
Sher·man (shûr′mən), **John,** 1823–1900, U. S. statesman. —**Roger,** 1721–93, American statesman; signer of Declaration of Independence. —**William Tecumseh,** 1820–91, U. S. general in the Civil War; led march from Atlanta to the sea, 1864.
she·root (shə·root′) See CHEROOT.
Sher·ra·moor (sher′ə·moor) *n. Scot.* & *Brit. Dial.* The Scottish rebellion of 1715, so called because the Jacobites were stopped in their advance at Sheriffmuir; hence, any turmoil or tumult. Also **Sher′ry·moor.**
Sher·riff (sher′if), **Robert Cedric,** born 1896, English writer.
Sher·ring·ton (sher′ing·tən), **Sir Charles Scott,** 1861–1952, English physiologist.
sher·ry (sher′ē) *n. pl.* **·ries** The fortified wines of Jerez (formerly Xerez), Spain, or a wine made in imitation of these, as in California. [from *Xerez,* Spain]
sherry cobbler A mixed beverage of sherry, lemon, sugar, water, and ice.
s' Her·to·gen·bosch (ser′tō·khən·bôs′) The capital of North Brabant province in south central Netherlands: French *Bois-le-Duc.*
Sher·wood (shûr′wood), **Robert Emmet,** 1896–1955, U. S. playwright.
Sherwood Forest An ancient forest, chiefly in Nottinghamshire, central England; celebrated as the home of Robin Hood and his men.
she's (shēz) **1** She is. **2** She has.
Shet·land Islands (shet′lənd) A Scottish island group NE of the Orkney Islands, comprising a county of northern Scotland (**Shetland**: also *Zetland*); 551 square miles; several hundred islands, 24 inhabited; capital, Lerwick, on Mainland, the largest island.
Shetland pony A small, hardy, shaggy breed of pony originally bred on the Shetland Islands.
Shetland wool Thin, very loosely twisted yarn from the wool of Shetland sheep; also, the wool.
sheuch (shūkh) *n. Scot.* A ditch or open drain. Also **sheugh.**
sheuk (shœk) *Scot.* Past tense and past participle of SHAKE.
shew (shō) Older spelling of SHOW.
shew·bread (shō′bred′) *n.* Unleavened bread formerly displayed in the Jewish temple: also spelled *showbread.*
shew·er (shō′ər) See SHOWER[2].
she-wolf (shē′woolf′) *n. pl.* **-wolves** (-woolvz′) A female wolf.
Shey·enne River (shī·en′) A river in North Dakota, flowing 325 miles east and south to the Red River of the North.
Shi·ah (shē′ə) *n.* **1** One of the two great sects (Sunni and Shiah) of Islam, consisting of followers of Ali, the cousin and son-in-law of Mohammed, who maintain that Ali was the first Imam and true Successor to the Prophet. **2** An adherent of Shiah: also called *Shiite*: also **Shie·ite** (shē′īt). [<Arabic *shī'i* a follower, sect]
shib·bo·leth (shib′ə·leth) *n.* A test word or pet phrase of a party; a watchword: from the Hebrew word *shibboleth,* given by Jephthah (*Judges* xii 4–6) as a test to distinguish his own men from the Ephraimites, who used the pronunciation *sibboleth.* [<Hebrew *shibbōleth* ear of corn]
Shi·be·li (shi·bā′lē), **Web·be** (web′ā) A river in Ethiopia and Somalia, flowing NE, then SE and south 1,200 miles to a swamp 25 miles inland from the Indian Ocean: also *Shebeli.* Also **Webi Shebeli, Webi Shibeli.**
Shi·de·ha·ra (shē·de·hä·rä), **Baron Kijuro,**

add,āce,câre,pälm; end,ēven; it,īce; odd,ōpen,ôrder; to͝ok,po͞ol; up,bûrn; ə = a in *above,* e in *sicken,* i in *clarity,* o in *melon,* u in *focus*; yo͞o = u in *fuse*; oi,oil; ou,pout; ch,check; g,go; ng,ring; th,thin; t͟h,this; zh,vision. Foreign sounds à,œ,ü,kh,ṅ; and ◆: see page xx. < from; + plus; ? possibly.

1872–1951, Japanese diplomat and statesman.
shied (shīd) Past tense and past participle of SHY.
shield (shēld) *n.* **1** A broad piece of defensive armor, commonly carried on the left arm; a large buckler. **2** Something that protects or defends; a defender; shelter. **3** Any device for covering or protecting something. **4** *Mil.* A screen of steel attached to a gun to protect the men who are serving it. **5** *Mining* A framework or screen of wood or iron protecting the workers: pushed forward as the work advances. **6** *Her.* The escutcheon upon which emblems of heraldry are depicted. **7** *Zool.* A platelike protective part, as the carapace of a crustacean. **8** A policeman's badge. See synonyms under DEFENSE, SHELTER. —*v.t.* **1** To protect from danger as with a shield; defend; guard. **2** *Archaic* To avert; forbid. —*v.i.* **3** To act as a shield or safeguard. See synonyms under SHELTER. [OE *sceld*] —**shield′er** *n.* —**shield′-bear′er** (-bâr′ər) *n.* —**shield′-shaped′** (-shāpt′) *adj.*

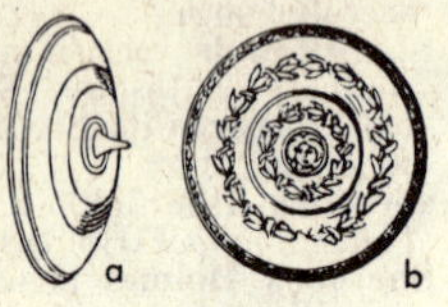

SHIELDS
a. Anglo-Saxon.
b. Greek.

shield bone *Anat.* The scapula or shoulder bone.
shield-fern (shēld′fûrn′) *n.* A fern (genus *Dryopteris*), so called from its shield-shaped sporangia.
Shield of David See MOGEN DAVID.
shiel·ing (shē′ling) *n.* A shepherd's or sportsman's hut: also spelled *shealing.* [Var. of SHEALING]
shi·er[1] (shī′ər) *n.* A horse in the habit of shying: also spelled *shyer.*
shi·er[2] (shī′ər), **shi·est** (shī′ist) Comparative and superlative of SHY.
shift (shift) *v.t.* **1** To change or move from one position, place, etc., to another. **2** To change for another or others of the same class. **3** To change (gears) from one arrangement to another. **4** *Ling.* To alter phonetically as part of a systematic change. —*v.i.* **5** To change position, place, etc. **6** To try varied expedients; do the best one can; manage. **7** To evade; equivocate. **8** To shift gears: The car *shifts* automatically. —*n.* **1** The act of shifting. **2** A recourse or contrivance adopted in the absence of direct means: We'll make *shift* to get along; hence, a dodge; artifice; trick; evasion. **3** *Archaic* or *Dial.* An undergarment; chemise. **4** A change of clothes. **5** A change of place, direction, or form: a *shift* in the wind; transfer, as of a burden. **6** A change of the position of the hand when playing on the fingerboard of an instrument of the viol class. **7** A relay of workers; also, the working time of each group. **8** *Physics* Any of various displacements of spectral lines caused by velocity of the light source, gravitational effect, etc. Compare EINSTEIN SHIFT, DOPPLER EFFECT. **9** *Geol.* The relative displacement of areas on opposite sides of a rock fault and outside of the zone of dislocation. **10** *Ling.* **a** A patterned phonetic or phonemic change, as the consonant *shift* described in Grimm's Law. **b** Functional shift. See synonyms under CHANGE, CONVEY. [OE *sciftan* divide] —**shift′er** *n.*
shift·less (shift′lis) *adj.* **1** Unable or unwilling to shift for oneself; inefficient or lazy. **2** Inefficiently done; showing lack of energy or resource. See synonyms under IMPROVIDENT. —**shift′less·ly** *adv.* —**shift′less·ness** *n.*
shift·y (shif′tē) *adj.* **shift·i·er, shift·i·est** **1** Full of expedients; alert; capable. **2** Artful; tricky; fickle. —**shift′i·ly** *adv.* —**shift′i·ness** *n.*
Shi·ge·mi·tsu (shē·ge·mē·tsōō), **Mamoru,** born 1887, Japanese diplomat.
Shi Huang Ti (shir′ hwäng′ tē′), 259–210 B.C., Chinese emperor.
Shi·ism (shē′iz·əm) *n.* The doctrine held by the Shiah or Persian branch of Moslems, showing traces of the earlier Persian faith. See SHIAH.
Shi·ite (shē′īt) *n.* A Shiah. —**Shi·it′ic** (-it′ik) *adj.*
shi·kar (shi·kär′) *Anglo-Indian v.t.* To hunt. —*n.* Hunting; sport; the chase. [<Urdu <Persian]
shi·ka·ree (shi·kä′rē) *n.* A hunter or sportsman; especially, a native attendant and guide in the chase. Also **shi·kar′ree, shi·ka′ri.** [<Urdu *shikari*]
Shi·kar·pur (shi·kär′pŏŏr) A city in NE Sind, West Pakistan.
Shi·ko·ku (shē·kō·kōō) An island of SW Japan, east of Kyushu; 7,248 square miles.
Shil·ka (shil′kə) A river in SE Siberian Russian S.F.S.R., flowing 345 miles NE to the Amur.
shill[1] (shil) *adj. Scot.* Shrill.
shill[2] (shil) *n. Slang* The assistant of a sidewalk peddler or gambler who makes a purchase or bet to encourage onlookers to buy or bet; a capper. [Origin unknown]
shil·le·lagh (shi·lā′lə, -lē) *n.* In Ireland, a stout cudgel made of oak or blackthorn. See synonyms under STICK. Also **shil·la′lah, shil·lea′lah, shil·le′lah.** [from *Shillelagh,* a town in Ireland famed for its oaks]
shil·ling (shil′ing) *n.* **1** A current silver coin of Great Britain, first issued in 1504; twelvepence. Compare SOLIDUS (def. 2). **2** A former denomination of money in the United States varying in value from 12 1/2 to 16 2/3 cents. —**King's shilling** An English shilling formerly handed to a recruit on his joining the British military service: considered as binding as the signing of a contract: also **Queen's shilling.** [OE *scilling*]

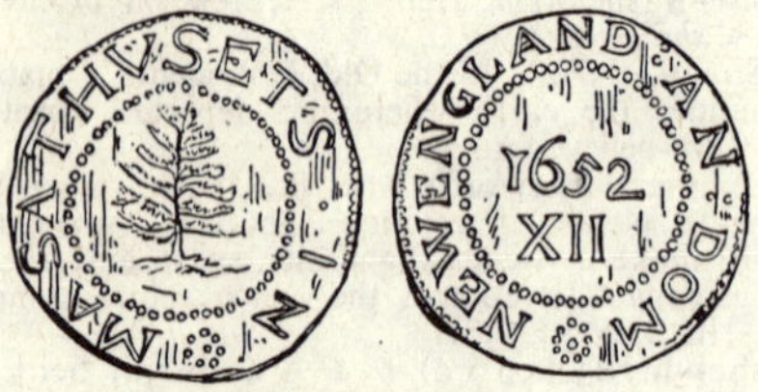

PINE-TREE SHILLING
Issued by Massachusetts in 1652 (actual size).

shilling side Formerly, the west side of lower Broadway, New York, where the shops carried a cheaper grade of merchandise than on the other (the dollar) side.
Shil·long (shi·lông′) The capital of Khasi and Jaintia Hills district, India.
shil·ly-shal·ly (shil′ē·shal′ē) *v.i.* **·lied, ·ly·ing** **1** To act with indecision; be irresolute; vacillate. **2** To trifle. —*adj.* Weak; hesitating. —*n.* Weak or foolish vacillation; irresolution; any trifling. —*adv.* In an irresolute manner. [Dissimilated reduplication of *shall I?*] —**shil′ly-shal′li·er** *n.*
Shi·loh (shī′lō) **1** An ancient Israelite sanctuary in central Palestine, NW of the Dead Sea. **2** A national military park in SW Tennessee; 6 square miles; scene of a Confederate defeat in the Civil War, 1862; established 1894.
shil·pit (shil′pit) *adj. Scot.* **1** Watery and insipid; weak: *shilpit* drink. **2** Sickly; puny: a *shilpit* girl.
shi·ly (shī′lē) See SHYLY.
shim (shim) *n.* In machinery, stoneworking, and railroading, a piece of metal or other material used to fill out space, as where joints are worn loose, or between something and its support. —*v.t.* **shimmed, shim·ming** To wedge up or fill out to a proper position or level by inserting a shim. [Origin uncertain]
Shi·mi·zu (shē·mē·zōō) A port on central Honshu island, Japan.
shim·mer (shim′ər) *v.i.* To shine faintly; give off or emit a tremulous light; glimmer. —*n.* A tremulous shining or gleaming; glimmer; gleam. See synonyms under LIGHT. [OE *scimerian,* prob. freq. of *scīnan* shine] —**shim′·mer·y** *adj.*
shim·my (shim′ē) *n. pl.* **·mies** *U.S.* **1** *Colloq.* A chemise. **2** A jazz dance accompanied by shaking movements: also **shimmy shake. 3** Unusual vibration, as in automobile wheels. —*v.i.* **·mied, ·my·ing** **1** To vibrate or wobble. **2** To dance the shimmy. [Alter. of CHEMISE]
Shim·o·no·se·ki (shim′ə·nō·sä′kē, *Japanese* shē·mō·nō·sä·kē) A port of SW Honshu island, Japan, on **Shimonoseki Strait,** a narrow channel between Honshu and Kyushu, connecting the Sea of Japan with the Inland Sea.
shin[1] (shin) *n.* **1** The front part of the leg below the knee; also, the shin bone. **2** The lower foreleg: a *shin* of beef. —*v.t.* & *v.i.* **shinned, shin·ning** **1** To climb (a pole) by gripping with the hands or arms and the shins or legs: usually with *up.* **2** To kick (someone) on the shins. [OE *scinu*]
shin[2] (shēn) *n.* The twenty-first Hebrew letter. See ALPHABET.
Shi·nar (shī′när) An ancient country along the lower Tigris and Euphrates. *Gen.* x 10.
shin bone The tibia.
shin·dig (shin′dig) *n. U.S. Slang* A dance or noisy party. [? < *a dig on the shin*]
shin·dy (shin′dē) *n. pl.* **·dies** **1** *Slang* A riotous quarrel; row; also, a dance or shindig. **2** The game of shinny. [Var. of SHINNY]
shine (shīn) *v.i.* **shone** or (*esp. for def.* **5**) **shined, shin·ing** **1** To emit light; beam; glow. **2** To gleam, as by reflected light. **3** To excel or be conspicuous in splendor, beauty, or intellectual brilliance; be preeminent. —*v.t.* **4** To cause to shine. **5** To brighten by rubbing or polishing. —**to shine up to** *Slang* To try to please. —*n.* **1** The state or quality of being bright or shining; radiance; luster; sheen. **2** Fair weather; sunshine. **3** *U.S. Colloq.* A liking or fancy. **4** *U.S. Colloq.* A smart trick or prank. **5** A gloss or polish on shoes. See synonyms under LIGHT. —**to take a shine to** *U.S. Colloq.* To become fond of. [OE *scīnan*]

Synonyms (verb): beam, coruscate, glare, gleam, glisten, glitter, glow, scintillate, sparkle.
shin·er (shī′nər) *n.* **1** One who or that which shines or causes to shine. **2** A bright or gold coin. **3** One of various silvery cyprinoid freshwater fishes (genus *Notropis*) common in North America. **4** A bristletail. **5** *Slang* A black eye from a blow.
shin·gle[1] (shing′gəl) *n.* **1** A thin, tapering piece of wood or other material, usually about 18 inches long and 4 or more inches wide, used in courses to cover roofs. **2** A small sign board, as a shingle or a brass plate, bearing the name of a doctor, a lawyer, etc., and placed outside his office. **3** A short haircut. —*v.t.* **·gled, ·gling** **1** To cover (a roof, building, etc.) with or as with shingles. **2** To cut (the hair) short all over the head. [Alter. of ME *schindle* <L *scindula,* var. of *scandula* a shingle] —**shin′gler** *n.*
shin·gle[2] (shing′gəl) *n.* **1** Rounded, waterworn detritus, coarser than gravel, found on the seashore. **2** A place strewn with shingle, as a beach. [Cf. Norw. *singl* coarse gravel] —**shin′gly** *adj.*
shin·gle[3] (shing′gəl) *v.t.* **·gled, ·gling** *Metall.* To drive out impurities from (puddled iron) by heavy blows or pressure. [Origin unknown]
shingle oak The jack oak.
shin·gles (shing′gəlz) *n. Pathol.* A skin disease, most commonly due to an infection, but also to nervous trouble, accompanied by neuralgia, with eruptions sometimes extending half round the body like a girdle: also called *herpes zoster.* [Alter. of Med. L *cingulus* <L *cingulum* girdle < *cingere* gird]
shin·ing (shī′ning) *adj.* **1** Emitting or reflecting a continuous light; gleaming; luminous. **2** Of unusual brilliance or excellence; conspicuous. —**shin′ing·ly** *adv.*
shin·leaf (shin′lēf′) *n.* A low perennial herb (*Pyrola elliptica*), with rounded evergreen root leaves, common in the woods of the northern United States. [From the use of its leaves for shinplasters]
shin·ny[1] (shin′ē) *n.* A game resembling hockey, or one of the sticks or clubs used by the players. Also **shin′ney.** [< *shin ye,* a cry used in the game]
shin·ny[2] (shin′ē) *v.i.* **·nied, ·ny·ing** *U.S. Colloq.* To climb using one's shins: usually with *up.*
shin·plas·ter (shin′plas′tər, -pläs′-) *n.* **1** *U.S.* **a** Fractional currency issued by other than the constituted authorities. See FRACTIONAL CURRENCY. **b** Any scrip or paper money issued by private enterprises. **2** A plaster for a sore shin.
shin·ti·yan (shin′tē·yan) *n. pl.* Wide loose trousers worn by Moslem women. [<Arabic < Turkish *chintiyan*]
Shin·to (shin′tō) *n.* The primitive religion of Japan, consisting chiefly in ancestor worship, nature worship, and the worship of many ethnic divinities, from the chief of whom the Emperor is thought to be descended, and thus himself a god: as **State Shinto,** it was the state religion of Japan, 1868–1945, and in that

shiny period incorporated many nationalistic and militaristic elements, later minimized. Also **Shin′to·ism**. [< Japanese *shin*, way of the gods < Chinese *shin* god + *tao* way or law] — **Shin′-to·ist** *n.*
shin·y (shī′nē) *adj.* **shin·i·er, shin·i·est** 1 Glistening; glossy; polished. 2 Bright; clear.
ship (ship) *n.* 1 Any vessel suitable for deep-water navigation: a *steamship*, sailing *ship*. 2 A large seagoing sailing vessel with at least three masts, carrying square-rigged sails on all three. 3 An airship or airplane. 4 Figuratively, fortune: when my *ship* comes in. — **capital ship** Any vessel of war of the first rank, as a battleship, battle cruiser, or aircraft carrier. — *v.* **shipped, ship·ping** *v.t.* 1 To transport by ship or other mode of conveyance. 2 To send by any established mode of transportation, as by rail. 3 To hire and receive for service on board a vessel, as sailors. 4 *Naut.* To receive over the side, as in rough weather: to *ship* a wave. 5 *Colloq.* To get rid of. 6 To set or fit in a prepared place on a boat or vessel, as a mast, or a rudder; also, to draw (oars) inside a boat from rowlocks. — *v.i.* 7 To go on board ship; embark. 8 To undergo shipment: Raspberries do not *ship* well. 9 To enlist as a seaman. [OE *scip*]

FULL–RIGGED SHIP
With double topsails and staysails.
a. Flying jib. *b.* Jib. *c.* Foretopmast staysail. *d.* Foresail. *e.* Mainsail. *f.* Crossjacksail. *g.* Spanker. *h.* Maintopmast staysail. *i.* Mizzenmast staysail. *j.* Lower foretopsail. *k.* Lower maintopsail. *l.* Lower mizzentopsail. *m.* Upper foretopsail. *n.* Upper maintopsail. *o.* Upper mizzentopsail. *p.* Foretopgallant sail. *q.* Maintopgallant sail. *r.* Mizzentopgallant sail. *s.* Fore royal. *t.* Main royal. *u.* Mizzen royal. *v.* Main skysail. *w.* Maintopgallant staysail. *x.* Mizzentopgallant staysail. *y.* Main royal staysail.

-ship *suffix of nouns* 1 The state, condition, or quality of: *friendship*. 2 Office, rank, or dignity of: *kingship*. 3 The art or skill of: *marksmanship*. [OE *-scipe*]
ship biscuit Hardtack; sea biscuit.
ship·board (ship′bôrd′, -bōrd′) *n.* The side or deck of a ship; hence, a vessel: only in phrase **on shipboard**.
ship broker A mercantile agent who buys and sells ships, cargoes, etc. Also **ship·bro·ker** (ship′brō′kər). — **ship′bro′ker·age** *n.*
ship·build·er (ship′bil′dər) *n.* One who designs, superintends, contracts for, or works at the building of vessels. — **ship′build′ing** *n.*
ship canal A waterway or canal deep enough for seagoing vessels.
ship carpenter 1 A carpenter who builds or repairs vessels; a shipwright. 2 A carpenter attached to a vessel.
ship chandler One who deals in cordage, canvas, and other furniture of vessels.
ship fever Prison fever.
Ship·ka Pass (ship′kä) A pass in the central Balkan Mountains, central Bulgaria; elevation, 4,166 feet; scene of a defeat of the Turks by the Bulgarians, 1877.
ship·load (ship′lōd′) *n.* The quantity that a ship carries or can carry; a cargo.
ship·man (ship′mən) *n. pl.* **·men** (-mən) A sailor; mariner.
ship·mas·ter (ship′mas′tər, -mäs′-) *n.* The captain or master of a merchant ship.
ship·mate (ship′māt′) *n.* A fellow sailor.
ship·ment (ship′mənt) *n.* The act of shipping, or that which is shipped; a consignment.
ship money An impost levied by the sovereign on English maritime towns and counties, for providing and arming a fleet for the protec-

tion of the coast: originated about 1007 and declared illegal by Parliament in 1640.
ship of the line Formerly, a man-of-war large enough to take a position in a line of battle.
ship owner A person owning a ship, ships, or shares in them. — **ship owning**
ship·pa·ble (ship′ə·bəl) *adj.* That can be shipped or transported.
ship·pen (ship′ən) *n. Scot.* A cow shed; barn. Also **ship′pon**.
ship·per (ship′ər) *n.* 1 One who or that which ships. 2 Any appliance for shifting some part of a machine, as in a loom. 3 A skipper; mariner.
ship·ping (ship′ing) *n.* 1 Ships collectively; the body of vessels belonging to a country or port; also, tonnage. 2 The act of shipping, in any sense. 3 *Obs.* A voyage.
shipping ton A freight ton. See under TON (def. 3).
ship–rigged (ship′rigd′) *adj. Naut.* Rigged as a ship; square-rigged. See illustration under SHIP.
ship·shape (ship′shāp′) *adj.* Well arranged; trim; orderly; neat. — *adv.* In a seamanlike manner; neatly.
ship's papers The documents required by international law to be carried by a ship, as bills of lading, bill of health, invoices, log-book, proofs of ownership; also, certificate of registry, crew-list, clearance, license, and shipping articles. Compare MANIFEST.
ship's time *Naut.* The time as shown by the deck clock: usually local mean time at whatever meridian a vessel happens to be.
ship·way (ship′wā′) *n.* 1 The ways on which a ship is built or examined. 2 A ship canal.
ship·worm (ship′wûrm′) *n.* One of a family (*Teredinidae*) of marine bivalves, resembling worms, especially *Teredo navalis*, which burrows into the timbers of ships, piers, wharfs, etc.: also called *borer*.
ship·wreck (ship′rek′) *n.* 1 The partial or total destruction of a ship at sea. 2 Utter or practical destruction; ruin. 3 Scattered remnants, as of a wrecked ship; wreckage. — *v.t.* 1 To wreck, as a vessel. 2 To bring to disaster; ruin; destroy.
ship·wright (ship′rīt′) *n.* A ship carpenter or builder; one who works on the wooden parts of ships.
ship·yard (ship′yärd′) *n.* An enclosure where ships are built or repaired.
shipyard eye *Pathol.* Kerato-conjunctivitis.
Shi·raz (shē·räz′) A city in SW Iran.
shire (shīr) *n.* 1 A territorial division of Great Britain; a county. 2 A county in America: used only in compounds and proper names borrowed from England. [OE *scīr*]
Shi·ré (shē′rā) A river of southern Nyasaland and central Mozambique, SE Africa, flowing 250 miles south from Lake Nyasa to the Zambezi. Portuguese **Chi·re** (shē′rə).
shire horse One of a breed of large draft horses originating in the shires or midland counties of England. Also **Shire**.
shire town The capital of a county; county seat; county town.
shirk (shûrk) *v.t.* 1 To avoid the doing of; evade doing (something that should be done). 2 *Obs.* To obtain by trickery. 3 To avoid work or evade obligation. [< n.] — *n.* One who shirks: also **shirk′er**. [Prob. <G *schürke* rascal. Akin to SHARK²]
Shir·ley (shûr′lē) **James**, 1596–1666, English dramatist.
Shir·pu·la (shir·pōō′lə) See LAGASH.
shirr (shûr) *v.t.* 1 To gather on parallel gathering threads. 2 To bake with crumbs in a buttered dish, as eggs. — *n.* 1 A fulling or gathering by threads. 2 A rubber thread woven into a fabric to make it elastic. [Origin unknown]
shirt (shûrt) *n.* 1 A loose garment for the upper part of the body, usually having collar and cuffs and a front closing. 2 A closely fitting undergarment for the upper part of the body. 3 The inner lining of a blast furnace. — **to keep one's shirt on** *Slang* To remain calm; keep one's temper. — **to lose one's shirt** *Slang* To lose everything. [OE *scyrte* shirt, short garment. Akin to SKIRT.] — **shirt′less** *adj.*

shock absorber
shirt·ing (shûr′ting) *n.* Closely woven material of cotton, linen, silk, etc., used for making shirts, blouses, dresses, etc.
shirt–waist (shûrt′wāst′) *n.* A tailored, sleeved blouse or shirt: usually worn tucked in under skirt or trousers.
Shir·wa (shir′wä) See CHILWA.
shist (shist) See SCHIST.
shit·tim–wood (shit′im·wood′) *n.* In the Bible, the wood of a species of acacia (the **shit′tah** or **shittah tree**) used in making the furniture of the Jewish tabernacle. Also **shit′tim**. [< Hebrew *shittīm*, pl. of *shittāh*]
shiv (shiv) *n. Slang* In the criminal underworld, a knife or razor: often spelled *chevy, chiv*. Also **shive, shiv′y**. [<Romany *chiv* goad]
Shi·va (shē′və) See SIVA.
shiv·a·ree (shiv′ə·rē′) *n. U.S.* A charivari, especially in the sense of the burlesque serenade of newly-weds. [Alter. of CHARIVARI]
shive¹ (shīv) *n. Brit. Dial.* 1 A short flat cork; a thin wooden bung. 2 *Brit.* A slice cut off, as of bread. [Cf. ON *skifa* slice]
shive² (shīv) *n.* A thin fragment; shiver; a woody fragment separated from flax by breaking. [Back formation <SHIVER²]
shiv·er¹ (shiv′ər) *v.i.* To tremble, as with cold or fear; shake; vibrate; quiver. — *v.t. Naut.* To cause to flutter in the wind, as a sail. See synonyms under QUAKE. — *n.* The act of shivering; a shaking or quivering from any cause. [? Blend of SHAKE and QUIVER]
shiv·er² (shiv′ər) *v.t. & v.i.* To break suddenly into fragments; shatter. See synonyms under BREAK, SHAKE. — *n.* A splinter; sliver. [ME *schivere*; origin uncertain]
shiv·er·y¹ (shiv′ər·ē) *adj.* Chilly; tremulous.
shiv·er·y² (shiv′ər·ē) *adj.* Easily shivered; brittle.
Shi·zu·o·ka (shē·zōō·ō·kä) A city on central Honshu island, Japan.
Sho·a (shō′ə) A province and former kingdom of central Ethiopia; 30,400 square miles; capital, Addis Ababa.
shoal¹ (shōl) *n.* 1 A shallow place in any body of water. 2 A sandbank or bar, especially one seen at low water. Compare BANK and REEF. — *v.i.* 1 To become shallow. — *v.t.* 2 To make shallow. 3 To sail into a lesser depth of (water), as shown by soundings: The ship *shoaled* her water off Cape Hatteras. — *adj.* Of little depth; shallow. Also, *Scot., shaul.* [OE *sceald* shallow]
shoal² (shōl) *n.* An assemblage or multitude; throng, as of fish. — *v.i.* 1 To throng in shoals or multitudes. 2 To school: said of fish. [OE *scolu* shoal of fish. Akin to SCHOOL².]
shoal duck The American eider duck: so called from Isles of Shoals, off Portsmouth, New Hampshire.
shoal·y (shō′lē) *adj.* Abounding in shoals. — **shoal′i·ness** *n.*
shoat (shōt) *n.* 1 A young hog. 2 A worthless fellow. Also spelled *shote*. [Cf. West Flemish *schote* young pig]
shock¹ (shok) *n.* 1 A violent collision or concussion; impact; blow. 2 A sudden and violent sensation, as if causing one to shake or tremble; a stroke: a *shock* of paralysis. 3 A sudden agitation of the mind; startling emotion. 4 *Pathol.* Prostration of bodily functions, as from sudden injury. 5 The passage of a strong electric current through the body, or the phenomena it produces: characterized by involuntary muscular contractions. See synonyms under BLOW, COLLISION. — *v.t.* 1 To shake by sudden collision; jar; give a shock to. 2 To disturb the emotions or mind of; horrify; disgust. 3 To encounter with hostile intent; meet with sudden encounter. — *v.i.* 4 *Archaic* To come into violent contact; collide. [<F *choc* <*choquer* <Gmc. Cf. MDu. *schokken* collide.]
shock² (shok) *n.* A number of sheaves of grain, stalks of maize, or the like, stacked for drying upright in a field. — *v.t. & v.i.* To gather (grain) into a shock or shocks. [ME *schokke* <Gmc. Cf. MLG *schok*.] — **shock′er** *n.*
shock³ (shok) *adj.* Shaggy; bushy. — *n.* 1 A coarse, tangled mass of hair. 2 A dog with a woolly coat. [? Var. of SHAG]
shock absorber *Mech.* 1 A device designed to absorb the energy of sudden impacts or of abrupt changes in velocity, as the springs of

an automobile, or an airplane landing gear. 2 A type of damper which absorbs motion, as of a part or mechanism, by hydraulic action, friction, etc.

shock action *Mil.* A sudden, violent attack by mobile and massed military units, as tanks, artillery, infantry with bayonets fixed, etc.

shock·er (shok′ər) *n.* 1 One who or that which shocks or startles. 2 *Brit. Colloq.* A sensational novel.

shock excitation *Electr.* The excitation of an oscillatory circuit by an impulse of different frequency, as in radio: also called *impulse excitation.*

shock·head (shok′hed′) *adj.* Having thick, bushy hair. Also **shock′-head′ed.**

shock·ing (shok′ing) *adj.* Causing a mental shock; striking as with horror or disgust; repugnant; distressing. See synonyms under AWFUL, FLAGRANT, FRIGHTFUL. — **shock′ing·ly** *adv.* — **shock′ing·ness** *n.*

shock tactics *Mil.* The use of a preponderating mass of picked troops in an attack in which hand-to-hand encounter is relied upon more than gunfire.

shock therapy *Med.* The treatment of certain nervous and mental disorders by the subcutaneous injection of drugs, as Metrazol, insulin, camphor, etc., or by electrical shocks.

shock troops *Mil.* Seasoned or picked men selected to lead an attack.

shock wave *Physics* A wave (of air, sound, etc.) having a pattern of flow which changes abruptly, with corresponding changes in temperature, pressure, and density: characteristic of bodies moving at or above the speed of sound.

shod (shod) Past tense and alternative past participle of SHOE.

shod·dy (shod′ē) *n. pl.* **·dies** 1 Reclaimed wool obtained by shredding discarded woolens or worsteds: longer fiber than mungo and better quality. 2 Fiber or cloth manufactured of inferior material or of shredded woolen rags. 3 Vulgar assumption or display; pretension; sham. 4 Refuse; waste. — *adj.* **·di·er, ·di·est** 1 Made of or containing shoddy. 2 Sham; inferior. [Origin uncertain] — **shod′di·ly** *adv.* — **shod′di·ness** *n.*

shoe (sho͞o) *n. pl.* **shoes** (*Obs.* **shoon**) 1 An outer covering, usually of leather, for the human foot, usually distinguished from a *boot* by not reaching above the ankle. 2 Something resembling a shoe in position or use. 3 A rim or plate of iron to protect the hoof of an animal from wear or injury. 4 A strip of iron, steel, or other hard material fitted under a sleigh or sledge runner to receive friction. 5 A drag of iron or wood placed under the wheel of a vehicle to retard its motion in going downhill; also, the part of a brake that presses upon the wheel. 6 An iron socket or ferrule for protecting the point of a wooden pile, or the end of a handspike, pole, or staff. 7 The tread or outer covering of a pneumatic tire, as for an automobile. 8 The part of a bridge on which the superstructure rests. 9 The sliding contact plate on an electric car. — *v.t.* **shod, shod** or **shod·den, shoe·ing** 1 To furnish with shoes or the like. 2 To furnish with a guard of metal, wood, etc., for protection, as against wear. [OE *scōh*]

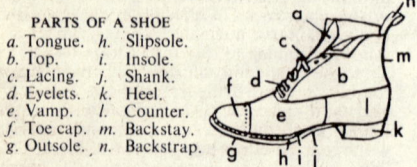

PARTS OF A SHOE

a. Tongue. *h.* Slipsole.
b. Top. *i.* Insole.
c. Lacing. *j.* Shank.
d. Eyelets. *k.* Heel.
e. Vamp. *l.* Counter.
f. Toe cap. *m.* Backstay.
g. Outsole. *n.* Backstrap.

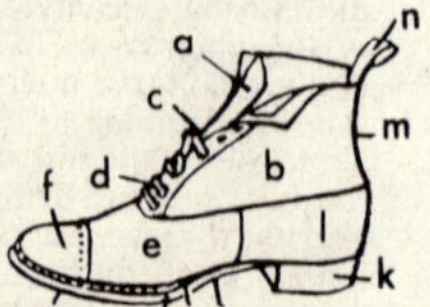

shoe·bill (sho͞o′bil′) *n.* A heron (*Balaeniceps rex*) of central Africa, with a huge vaulted and hooked bill.

shoe·black (sho͞o′blak′) *n.* A bootblack.

shoe findings Shoemakers' tools and supplies, with the exception of leather.

shoe·horn (sho͞o′hôrn′) *n.* A smooth curved implement of horn or other material shaped to aid in putting on a shoe.

shoe·mak·er (sho͞o′mā′kər) *n.* 1 One who makes shoes, boots, etc. 2 A cobbler. — **shoe′mak′ing** *n.*

sho·er (sho͞o′ər) *n.* One who supplies or fits on shoes; specifically, a blacksmith.

shoe·shine (sho͞o′shīn′) *n.* 1 The waxing and polishing of a pair of shoes. 2 The polished appearance thus given to the shoes.

shoe string A lace, cord, or ribbon for tying a shoe. Also **shoe lace.**

shoe·tree (sho͞o′trē′) *n.* A wooden or metal form for inserting in boots and shoes to preserve their shape or to stretch them: also called *boot-tree.*

sho·far (shō′fär) *n.* A ram's horn used in Jewish ritual, sounded on solemn occasions and in war. It is still blown on the Jewish New Year and on the Day of Atonement: also spelled *shophar.* [<Hebrew *shōphār*]

sho·gun (shō′gun, -go͞on) *n.* The hereditary commander in chief of the Japanese army until 1868: known to foreigners as the *tycoon.* [<Japanese <Chinese *chiang-chün* leader of an army] — **sho′gun·ate** (-it, -āt) *n.*

Sho·la·pur (shō′lə·po͝or) A town in central southern Bombay State, India.

shone (shōn, shon) Past tense and past participle of SHINE.

shoo (sho͞o) *interj.* Begone! be off! away!: used in driving away fowls. — *v.t.* To drive away by crying "shoo." — *v.i.* To cry "shoo." [Imit.]

shoo·fly (sho͞o′flī) *U.S. n.* 1 A shuffling dance; also, the music for it. 2 An enclosed child's rocker with sides representing horses, swans, etc. 3 A kind of pie with a sirupy filling made with molasses and brown sugar: also **shoofly pie.** 4 In railroading, a temporary track circumventing an obstructed regular line.

shook[1] (sho͝ok) Past tense of SHAKE.

shook[2] (sho͝ok) *n.* 1 A collection of barrel staves, shaped, chamfered, and arranged for assembling, conveniently bundled for transportation. 2 A set of boards in order for nailing together into a packing box, and conveniently bundled for transportation. 3 A shock of sheaves. [? Var. of SHOCK[2]]

shool (sho͞ol) *n. Scot.* A shovel.

shoon (sho͞on) Obsolete plural of SHOE.

shoot (sho͞ot) *v.* **shot, shoot·ing** *v.t.* 1 To hit, wound, or kill with a missile discharged from a weapon. 2 To discharge (a missile) from a bow, rifle, etc. 3 To discharge (a weapon): often with *off*: to *shoot* a cannon. 4 To take the altitude of with a sextant, etc.: to *shoot* the sun. 5 To send forth as if from a weapon, as questions, glances, etc. 6 To pass over or through swiftly: to *shoot* rapids. 7 To go over (an area) in hunting game. 8 To emit, as rays of light. 9 To photograph; film. 10 To cause to stick out or protrude; extend. 11 To put forth in growth; send forth (buds, leaves, etc.). 12 To push into or out of the fastening, as the bolt of a door. 13 To propel, discharge, or dump, as down a chute or from a container. 14 To variegate, as with streaks of color: usually in the past participle: The morning clouds were *shot* with silver. 15 In games: **a** To score (a goal, point, etc.) by kicking or otherwise forcing the ball, etc., to the objective. **b** To play (golf, craps, pool, etc.). **c** To propel (a marble) from between the thumb and forefinger; play (marbles). **d** To cast (the dice). — *v.i.* 16 To discharge a missile from a bow, firearm, etc.: Don't *shoot*! 17 To go off; discharge. 18 To move swiftly; dart. 19 To hunt game. 20 To jut out; extend or project. 21 To put forth buds, leaves, etc.; germinate; sprout. 22 To take a photograph. 23 To start the cameras, as in motion pictures. 24 In games, to make a play by propelling the ball, puck, etc., in a certain manner. — **to shoot at** (or **for**) *Colloq.* To strive for; attempt to attain or obtain. — **to shoot down** To bring to earth by shooting. — **to shoot off one's mouth** *Slang* To talk too freely or too much. — **to shoot up** 1 To move or grow upward quickly. 2 To strike with several or many shots. 3 *SW U.S.* To ride through (a town, etc.) shooting recklessly in all directions. — *n.* 1 A young branch or sucker of a plant; offshoot. 2 A narrow passage in a stream; a rapid. 3 An inclined passage down which anything may be shot; a chute. 4 The act of shooting; a shot. 5 A shooting match, hunting party, etc. 6 The thrust of an arch. 7 An antler or horn just pushing up. 8 Shooting distance; range. 9 A rapid thrusting movement. [OE *scēotan*]

shoot·ing (sho͞o′ting) *n.* The act of one who or that which shoots.

shooting affair An argument or difference to be settled with firearms; a pistol fight.

shooting box A small house in a game district, furnishing accommodation for sportsmen. Also **shooting lodge.**

shooting gallery A place where one can go for target practice.

shooting iron *U.S. Slang* A firearm.

shooting star 1 A meteor. 2 Any of certain small perennial herbs (genus *Dodecatheon*); especially, the American cowslip (*D. meadia*) with oblong leaves and clusters of cyclamenlike flowers.

shop (shop) *n.* 1 A place for the sale of goods at retail: in the United States commonly called a *store.* 2 A place for making or repairing any article, or the carrying on of any artisan craft: a blacksmith's *shop*, car *shops.* 3 One's own craft or business as a subject of conversation: to talk *shop.* — *v.i.* **shopped, shop·ping** To visit shops or stores to purchase or look at goods. [OE *sceoppa* booth]

shop·boy (shop′boi′), **shop·girl** (-gûrl′) *n.* A boy or a girl who works in a shop.

sho·phar (shō′fär) See SHOFAR.

shop·keep·er (shop′kē′pər) *n.* One who keeps a shop or store; a tradesman.

shop·lift·er (shop′lif′tər) *n.* One who steals goods exposed for sale in a shop. — **shop′·lift′ing** *n.*

shop·per (shop′ər) *n.* One who purchases or inspects goods in shops. — **shop′ping** *n.*

shop talk Conversation limited to one's job or profession.

shop·walk·er (shop′wô′kər) *n. Brit.* A floorwalker; a person who walks about a shop to supervise employees and help customers.

shop·worn (shop′wôrn′, -wōrn′) *adj.* Soiled or otherwise deteriorated from having been handled or on display in a shop.

sho·ran (shôr′an, shō′ran) *n.* A high-precision electronic navigation system which transmits pulses, usually from an aircraft or ship, to ground stations at distances determined by the elapsed time between emission and return of the pulses. [<SHO(RT) RA(NGE) N(AVIGATION)]

shore[1] (shôr, shōr) *n.* 1 The coast or land adjacent to an ocean, sea, lake, or large river. ◆ Collateral adjective: *littoral.* 2 *Law* The ground between the ordinary high-water mark and low-water mark. See synonyms under BANK[1], LAND, MARGIN. — **in shore** Near or toward the shore. — *v.t.* **shored, shor·ing** 1 To set on shore. 2 To surround as with a shore. [ME *schore*; origin uncertain]

shore[2] (shôr, shōr) *v.t.* **shored, shor·ing** To prop, as a wall, by a vertical or sloping timber: usually with *up.* — *n.* A beam set endwise as a prop, as against the side of a building, a ship on the stocks, etc., especially as a temporary support. See illustration under DRYDOCK. [Cf. Du. *schoor* prop, ON *skordha* stay]

shore[3] (shôr, shōr) Archaic past tense of SHEAR.

Shore (shôr, shōr), **Jane,** 1445?–1527, favorite of Edward IV of England.

shore bird Any of various birds (suborder *Charadrii*) which frequent beaches and also the shores of inland waters, including the snipe, sandpiper, and plover.

shore·less (shôr′lis, shōr′-) *adj.* Having no shore; boundless.

shore·line (shôr′līn′, shōr′-) *n.* The line or contour of a shore.

shore patrol A detail of the U. S. Navy, Coast Guard, or Marine Corps assigned to police duties ashore.

shore·ward (shôr′wərd, shōr′-) *adj.* & *adv.* Toward the shore. Also **shore′wards.**

shor·ing (shôr′ing, shō′ring) *n.* 1 The operation of propping, as with shores. 2 Shores, collectively.

shorl (shôrl) See SCHORL.

shorn (shôrn, shōrn) Alternative past participle of SHEAR.

short (shôrt) *adj.* 1 Having little linear extension; not long; of little extent; of no great distance. 2 Being below the average stature; not tall. 3 Having little extension in time; of limited duration; brief. 4 Abrupt in manner or spirit; curt; petulant; cross. 5 Not reaching or attaining a requirement, result, or mark; deficient; inadequate; scant: often with *of.* 6 In finance or commerce, not having in possession when selling, but having to procure in time to deliver as contracted; not being in

possession of the seller, as stocks or shares; of or pertaining to short stocks or commodities: *short* sales. **7** Not comprehensive or retentive; at fault; in error; narrow: said of persons or their faculties: *short* memory. **8** Breaking easily; friable; crisp. **9** *Phonet.* **a** Relatively brief in pronunciation: said of vowels. **b** Designating a set of vowel sounds which contrast with the "long" vowels. See LONG[1] (def. 9). **10** In classical prosody, requiring a relatively short time to pronounce: said of syllables containing a short vowel (epsilon, omicron, etc.) not followed by two consonants or a double consonant. **11** In English prosody, unaccented. **12** Less than: with *of.* **13** Concise; compressed. See synonyms under LITTLE, SCANTY, TERSE, TRANSIENT. **—** *n.* **1** The compressed substance or pith of a matter. **2** Anything that is short; a short syllable or vowel. **3** A deficiency, as in a payment. **4** A short contract or sale; one who has sold short; a bear. **5** *pl.* Bran mixed with coarse meal or flour. **6** *pl.* Trousers with legs extending part way to the knees: worn by both men and women. **7** *pl.* A man's undergarment covering the loins and often a portion of the legs. **8** In baseball slang, a shortstop. **9** *pl.* Clippings, scraps, etc., left over in the manufacture of different products and used to make an inferior quality of the product. **10** *Electr.* A short circuit. **11** A motion picture of relatively short duration as compared with the feature attraction on a program. **— for short** For brevity: Edward was called Ed *for short.* **— in short** In a word; briefly. **— the short and the long** The whole; the entire sum and substance. **—** *adv.* In a short manner or method, in any sense of the adjective: to stop *short,* to turn *short,* to sell *short.* **—** *v.t.* & *v.i.* To short-circuit. [OE *sceort* short] **— short'·ish** *adj.* **— short'ness** *n.*

Short (shôrt), **Walter Campbell,** 1880–1949, U. S. general.

short account **1** The account of a person who sells short on the stock market. **2** The open short sales as a whole.

short·age (shôr'tij) *n.* The amount by which anything is short; deficiency.

short·bread (shôrt'bred') *n.* A rich, dry cake or cooky made with shortening.

short·cake (shôrt'kāk') *n.* **1** A cake made short and crisp with butter or other shortening. **2** Cake or biscuit served with fruit usually between layers: strawberry *shortcake.*

short·change (shôrt'chānj') *v.t.* **·changed, ·chang·ing** To give less change than is due to; hence, to cheat or swindle. **— short'chang'er** *n.*

short–cir·cuit (shôrt'sûr'kit) *v.t.* & *v.i.* To make a short circuit (in).

short circuit *Electr.* **1** A path of low resistance established between any two points in an electric circuit, thus shortening the distance traveled by the current. **2** Any defect in an electric circuit or apparatus which may result in a dangerous or wasteful leakage of current.

short·com·ing (shôrt'kum'ing) *n.* **1** Failure; remissness; delinquency. **2** A falling off; shortage, as of a crop.

short commons A scanty supply of food; a meager ration.

short covering The buying of stocks or securities to close out a short sale.

short–cut (shôrt'kut') *v.t.* & *v.i.* To take a short cut (in).

short cut **1** A byway or path between two places shorter than the regular road. **2** A means or method that saves distance or time.

short–eared owl (shôrt'ird') See under OWL.

short·en (shôr'tən) *v.t.* **1** To make short or shorter; curtail. **2** To reduce; diminish; lessen. **3** To furl or reef (a sail) so that less canvas is exposed to the wind. **4** To make brittle or crisp, as pastry, by adding shortening. **—** *v.i.* **5** To become short or shorter. See synonyms under ABBREVIATE, SCRIMP. **— short'en·er** *n.*

short·en·ing (shôr'tən·ing) *n.* **1** A fat, such as lard or butter, used to make pastry crisp. **2** An abbreviation. **3** The act of one who shortens.

short·hand (shôrt'hand') *n.* Any system of rapid writing, as stenography or phonography. **—** *adj.* **1** Written in shorthand. **2** Using shorthand.

short–hand·ed (shôrt'han'did) *adj.* Not having a sufficient or the usual number of assistants, workmen, or hands.

short·head (shôrt'hed') *n.* A brachycephalic individual. **— short'–head'ed** *adj.*

short·horn (shôrt'hôrn') *n.* One of a breed of cattle with short horns, originally from northern England.

shor·ti·a (shôr'tē·ə) *n.* Any of a genus *(Shortia)* of perennial evergreen herbs with bell-shaped, nodding, white flowers. [after C. W. *Short,* 1794–1863, U. S. botanist]

Short·land Islands (shôrt'lənd) A group in the British Solomon Islands SE of Bougainville; total, 200 square miles.

short–lived (shôrt'līvd', -livd') *adj.* Living or lasting but a short time.

short·ly (shôrt'lē) *adv.* **1** At the expiration of a short time; quickly; soon. **2** In few words; briefly. **3** Curtly; abruptly.

short sale A sale for future delivery of goods or stocks not in possession at time of sale.

short shrift **1** A short time in which to confess before dying. **2** Little or no mercy or delay in dealing with a person or disposing of a matter.

short–sight·ed (shôrt'sī'tid) *adj.* **1** Unable to see clearly at a distance; myopic; nearsighted. **2** Lacking foresight. **3** Resulting from or characterized by lack of foresight. See synonyms under IMPRUDENT. **— short'–sight'ed·ly** *adv.* **— short'–sight'ed·ness** *n.*

short–spo·ken (shôrt'spō'kən) *adj.* Characterized by shortness or curtness of speech or manner; abrupt in address; gruff.

short–sta·ple (shôrt'stā'pəl) *adj.* Having a short fiber: in the United States, said of cotton fibers less than 1 1/8 inches long.

short·stop (shôrt'stop') *n.* In baseball, an infielder stationed between second and third bases.

short story A narrative prose story presenting a central theme or impression, usually subordinated to a single mood or characterization: shorter than a novel or novelette, usually under 10,000 words.

short–tem·pered (shôrt'tem'pərd) *adj.* Easily aroused to anger.

short–term (shôrt'tûrm') *adj.* Payable a short time after issue: said of securities.

short ton See under TON.

short wave A radio wave having a length of about 100 meters or less, corresponding to a frequency ranging upwards from about 3000 kilocycles. **— short'–wave'** *adj.*

short–wind·ed (shôrt'win'did) *adj.* Affected with difficulty of breathing; becoming easily out of breath.

Sho·sho·ne (shō·shō'nē) *n.* **1** One of a large and important tribe of North American Indians of northern Shoshonean stock of the Uto-Aztecan family, formerly occupying western Wyoming, central and southern Idaho, northeastern Nevada, and western Utah. **2** The Shoshonean language of this tribe. Also **Sho·sho'ni.**

Sho·sho·ne·an (shō·shō'nē·ən, shō'shə·nē'ən) *n.* The largest branch of the Uto-Aztecan linguistic family of North American Indians, including the Comanche, Paiute, Ute, and Shoshone plateau tribes, and the Hopi Indians. **—** *adj.* Of or pertaining to this linguistic branch. Also **Sho·sho'ni·an.**

Shoshone Cavern A national monument on the Shoshone River SW of Cody, Wyoming; 212.4 acres; established 1909.

Shoshone Falls A cascade in the Snake River, southern Idaho; over 200 feet high.

Shoshone River A river in NW Wyoming, flowing 100 miles NE to the Bighorn River.

Shos·ta·ko·vich (shos'tə·kō'vich), **Dmitri,** born 1906, Russian composer.

shot[1] (shot) *n.* **1** *pl.* **shot** A solid missile, as a ball of iron, or a bullet or pellet of lead, to be discharged from a firearm; also, such spherules or pellets collectively. See illustration under CARTRIDGE. **2** The act of shooting; any stroke, hit, or blow. **3** One who shoots; a marksman. **4** The distance traversed or that can be traversed by a projectile; reach; range. **5** A blast, as in mining. **6** A stroke, especially in certain games, as in billiards. **7** A conjecture; guess. **8** An attempted performance. **9** A metal sphere which a competitor puts, pushes, or slings, in a distance contest. **10** *Slang* A hypodermic injection of a drug. **11** *Slang* A drink of liquor. **12** An action or scene recorded on motion-picture film. **13** A picture taken with a camera; a photograph or a snapshot. **14** *Naut.* A unit of chain length: in the United States, 15 fathoms; in Great Britain, 12 1/2 fathoms. **15** *Obs.* Any projectile. **—** *v.t.* **shot·ted, shot·ting** **1** To load or weight with shot. **2** To clean, as bottles, by partially filling with shot and shaking. **—** *adj.* **1** Of changeable color, as when warp and weft are of different colors. **2** *Slang* More or less intoxicated. **3** *Colloq.* Completely done for; ruined. [OE *scot*]

shot[2] (shot) Past tense and past participle of SHOOT.

shot[3] (shot) *n.* A reckoning or charge, or a share of such a reckoning; scot. [Var. of SCOT]

shote (shōt) See SHOAT.

shot effect *Electronics* The background noise resembling the patter of small shot, developed in a vacuum tube by the fluctuating emission of electrons from the heated filament. Also **shot noise.**

shot·gun (shot'gun') *n.* A light, smoothbore gun, either single- or double-barreled, adapted for the discharge of shot at short range. **—** *adj.* **1** Having a clear passageway straight through: a *shotgun* house. **2** Coerced with, or as with, a shotgun: a *shotgun* wedding.

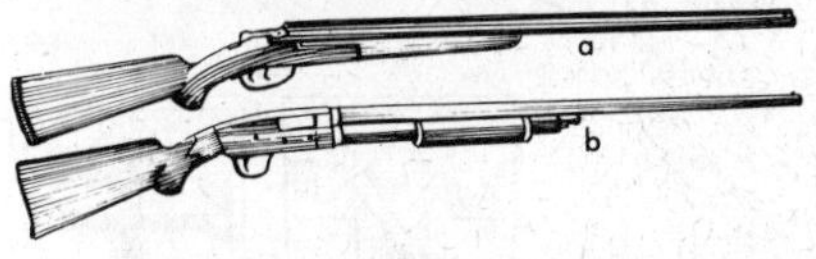

SHOTGUNS
a. Double-barrel hammerless shotgun.
b. Repeating shotgun.

shot peening A method for improving the mechanical properties of steel parts by bombarding the surfaces with metallic shot delivered under pressure or by centrifugal action.

shot–put (shot'po͝ot') *n.* **1** An athletic contest in which a shot is thrown, or put, for distance. **2** A single put of the shot. **— shot'–put'ter** *n.*

shot·ten (shot'n) *adj.* Having spawned: said of a fish, especially a herring. [Obs. pp. of SHOOT]

should (sho͝od) Past tense of SHALL, but rarely a true past, rather chiefly used as a modal auxiliary which, while conveying varying shades of present and future time, expresses a wide range of subtly discriminated feelings and attitudes: **1** Obligation or propriety in varying degrees, but milder than *ought*: You *should* write that letter; *Should* we tell him the truth about his condition? His father thought that he *should* go; You *should* really taste that cake! **2** Condition: **a** Simple contingency, but involving less probability than *shall* or the present with future sense: If I *should* die before I wake . . . If I *should* go, he would go too. **b** Assumption: *Should* (=*Assuming that*) the space platform prove practicable, as seems almost certain, a trip to the moon will be easy. **3** Surprise at an unexpected event in the past: When I reached the station, whom *should* I run into but the detective! **4** Expectation: I *should* be at home by noon. ("I said that I *should* be home by noon" implies expectation, whereas "I said that I *would* be home by noon" implies intention.) **5** *U.S. Colloq.* Irony, in positive statement with negative force: He'll be fined heavily, but with all his money he *should* (=*need not*) worry! **6** Hesitation or deprecatory modesty, in the first person: I *should* hardly think so; We *should* like to have you come to dinner, if you are free and have nothing better to do. (Ordinarily, in American usage, but not in British, *would* is used in the first person, as well as in the second and third, before *like, prefer,* etc.: We *would,* or We'*d,* like to have you come to visit us.) See usage note under WOULD. [OE *scolde,* pt. of *sculan* owe]

shoul·der (shōl'dər) *n.* **1** The part of the trunk between the neck and the free portion of the arm or forelimb; also, the joint connecting the arm or forelimb with the body. **2** Anything which supports, bears up, or

projects like a shoulder. 3 The forequarter of various animals. 4 An enlargement, projection, or offset, as for keeping something in place, or preventing movement past the projection. 5 *Printing* The top of the shank of a type when extending above or below the face of the letter. 6 Either edge of a road or highway. 7 The angle of a bastion included between a face and the adjacent flank: also **shoulder angle.** — **shoulder to shoulder 1** Side by side and close together. 2 With united effort; in cooperation. — **straight from the shoulder** *Colloq.* Candidly; straightforwardly. — *v.t.* 1 To assume as something to be borne; sustain; bear. 2 To push with or as with the shoulder or shoulders. 3 To fashion with a shoulder or abutment; make a shoulder on. — *v.i.* 4 To push with the shoulder or shoulders. — **to shoulder arms** To rest a rifle against the shoulder, holding the butt with the hand on the same side, the arm being held bent and close to the side. [OE *sculder* shoulder]
shoulder blade The scapula.
shoulder loop 1 A strap worn on or over the shoulder to support an article of dress. 2 A strap of cloth marked with insignia of rank, worn by army and navy officers. Also **shoulder strap.**
shoulder patch A cloth insignia worn on the upper part of the sleeve of a uniform to indicate the branch or unit to which the wearer belongs.

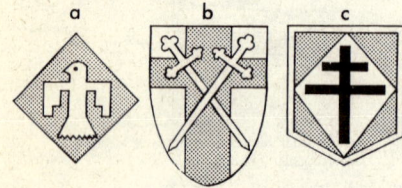

SHOULDER PATCHES
a. United States — 46th Division.
b. Great Britain — Army of Liberation.
c. France — Fighting French Commandos.

shoulder screw A screw having a shoulder, as for limiting the depth to which it may be sunk. See illustration under SCREW.
shoulder weapon Any small-arm weapon designed to be held against the shoulder in firing, as a rifle, carbine, etc.
should·na (shood′nə) *Scot.* Should not.
should·n't (shood′nt) Should not.
shout (shout) *n.* A sudden and loud outcry, such as a call or command, but also expressing emotion, as of joy, exultation, courage, or derision; a loud burst of voice or voices. — *v.t.* To utter with a shout; say or express loudly. — *v.t.* To utter a shout; cry out loudly. See synonyms under CALL, ROAR. [Origin unknown]
shout·er (shou′tər) *n.* 1 One who utters cries of religious exultation during a religious meeting. 2 A vociferous supporter of a political candidate.
shouth·er (shoo′thər) *n. Scot.* The shoulder.
shove (shuv) *v.t. & v.i.* **shoved, shov·ing 1** To push, as along a surface: to *shove* a boat with a pole. 2 To press forcibly (against); jostle. See synonyms under PUSH. — **to shove off 1** To push along or away, as a boat. 2 *Colloq.* To depart. — *n.* 1 The act of pushing or shoving; strong push. 2 The woody center of flax. 3 *Can.* A forward movement of ice in a river. [OE *scūfan*] — **shov′er** *n.*
shov·el (shuv′əl) *n.* 1 A flattened scoop with a handle, as for digging, lifting earth, rock, etc. 2 *Colloq.* A shovel hat. — *v.* ·**eled, ·el·ing** or **·elled, ·el·ling** *v.t.* 1 To take up and move or gather with a shovel. 2 To toss hastily or in large quantities as if with a shovel. 3 To clear or clean with a shovel, as a path. — *v.i.* 4 To work with a shovel. [OE *scofl*]
shov·el·board (shuv′əl-bôrd′, -bōrd′) *n.* Shuffleboard.
shov·el·er (shuv′əl·ər, shuv′lər) *n.* 1 One who or that which shovels. 2 A large river duck (genus *Spatula*) with spatulate bill broadening roundly toward the end; especially, the **common shoveler** (*S. clypeata*) of the northern hemisphere; also *shovelbill.* Also **shov′el·ler.**
shovel hat A hat with broad brim turned up at the sides and projecting in front.

shov·el·head (shuv′əl·hed′) *n.* 1 A shark (*Sphyrna tiburo*) resembling the hammerhead, about 5 feet long. 2 The paddlefish. 3 The shovelnose (def. 1).
shov·el·nose (shuv′əl·nōz′) *n.* 1 A sturgeon (*Scaphirhynchus platyrhynchus*), common in the Mississippi valley, having a broad, depressed, shovel-shaped snout. 2 Any of several varieties of shark with a shovel-like nose; especially, the cow shark (*Hexanchus corinus*), found on the Pacific coast of the United States.
shov·el·nosed (shuv′əl·nōzd′) *adj.* Having a broad, flattened snout or beak.
show (shō) *v.* **showed, shown** or **showed, show·ing** *v.t.* 1 To cause or permit to be seen; present to view; exhibit; manifest; display. 2 To give in a marked or open manner; confer; bestow: to *show* favor. 3 To cause or allow (something) to be understood or known; explain; reveal; tell. 4 To cause (someone) to understand or see; explain something to; convince; teach. 5 *Law* To advance an allegation; plead: to *show* cause. 6 To make evident by logical process; prove; demonstrate. 7 To guide; lead; introduce, as into a room or building: with *in* or *up*: to *show* a caller in. 8 To indicate: The thermometer *shows* the temperature. 9 To enter in a show or exhibition. — *v.i.* 10 To become visible or known; be manifested or displayed. 11 To appear; seem. 12 To make one's or its appearance; be present. 13 *Colloq.* To give a theatrical performance; appear: to *show* in Newark. 14 *Colloq.* In racing, to be the third (horse, dog, etc.) to finish in a race. — **to show off 1** To exhibit proudly or ostentatiously. 2 To make an ostentatious display of oneself, or of one's accomplishments. — **to show up 1** To expose or be exposed, as faults. 2 To be evident or prominent. 3 To attend; arrive; make an appearance. 4 *Colloq.* To be better than. — *n.* 1 That which is shown; a public spectacle; a theatrical performance, circus, or motion picture; exhibition. 2 The act of showing; specifically, display; parade. 3 Pretense; semblance. 4 That which shows; an indication; promise; specifically, a sign of precious metal in a mine: a *show* of ore. 5 *Colloq.* An opportunity or chance. 6 *U.S. Colloq.* The third place in a race. — **the whole show** The center of interest or notice. [OE *scēawian*]
show·bill (shō′bil′) *n.* A poster announcing a play or show.
show biz *U.S. Slang* Show business.
show·boat (shō′bōt′) *n.* A boat, such as the old stern-wheelers on the Mississippi, on which a traveling troupe gives a theatrical performance.
show·bread (shō′bred′) See SHEWBREAD.
show business The entertainment arts, especially the theater, motion pictures, television, etc., collectively considered as an industry.
show·case (shō′kās′) *n.* A glass case for exhibiting and protecting articles for sale.
show·down (shō′doun′) *n.* 1 In poker, the play in which the hands are laid on the table face up. 2 Any action or any disclosure of facts, plans, etc., that brings an issue to a head.
show·er[1] (shou′ər) *n.* 1 A fall of rain, hail, or sleet, especially heavy rain of short duration within a local area. 2 A copious fall, as of tears, sparks, or other small objects. 3 A shower bath. 4 A variety of fireworks for simulating a shower of stars. 5 A party for the bestowal of gifts, as to a bride; also, the gifts. — *v.t.* 1 To sprinkle or wet with or as with showers. 2 To discharge in a shower; pour out. 3 To bestow with liberality. — *v.i.* 4 To fall as in a shower. 5 To take a shower bath. [OE *scūr*] — **show′er·y** *adj.*
show·er[2] (shō′ər) *n.* One who shows.
shower bath A bath in which water is sprayed on the body from an overhead, perforated nozzle.
show·folk (shō′fōk′) *n. pl.* Persons engaged in the entertainment business.
show-how (shō′hou′) *n. U.S. Colloq.* The teaching which imparts know-how: used especially in connection with the export of U.S. technological and agricultural aid to the backward areas of the world. [First in print in State Department publication on Point 4, Nov. 1949.]
show·ing (shō′ing) *n.* 1 Show; display; ex-

quality. 2 Presentation; statement, as of a subject.
show·man (shō′mən) *n. pl.* ·**men** (-mən) 1 One who exhibits or owns a show. 2 One who is skilled in presenting shows, etc. — **show′man·ship** *n.*
Show Me State Nickname of MISSOURI.
shown (shōn) Past participle of SHOW.
show-off (shō′ôf′, -of′) *n. Colloq.* One who makes a pretentious display of himself; a swaggerer.
show·piece (shō′pēs′) *n.* 1 A prized object considered worthy of special exhibit. 2 An object on display.
show place A place exhibited for its beauty, historic interest, etc.
show ring A circular enclosure at a fair, cattle show, or other exhibition, where animals are shown to compete for prizes, or for sale.
show·y (shō′ē) *adj.* **show·i·er, show·i·est** 1 Making a great display; gaudy; gay; splendid. 2 Given to display; ostentatious. — **show′i·ly** *adv.* — **show′i·ness** *n.*
shrank (shrangk) Past tense of SHRINK.
shrap·nel (shrap′nəl) *n. pl.* ·**nel** *Mil.* 1 A field artillery projectile for use against personnel, containing a quantity of metal balls and a time fuze and base charge which expel the balls in mid-air. 2 Shell fragments. [after Henry Shrapnel, 1761–1842, British artillery officer]

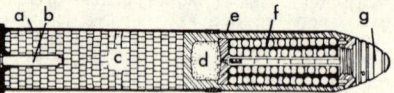

SHRAPNEL SHELL
a. Brass casing.   *e.* Steel shell body.
b. Percussion primer.   *f.* Shrapnel balls.
c. Smokeless powder.   *g.* Time fuze.
d. Black powder.

shred (shred) *n.* 1 A small irregular strip torn or cut off. 2 A bit; fragment; particle. See synonyms under PARTICLE. — *v.t.* **shred·ded** or **shred, shred·ding** 1 To tear or cut into shreds, as fibrous material. 2 *Brit. Dial.* To lop off; trim. [OE *scrēade* cutting]
shred·der (shred′ər) *n.* 1 One who or that which shreds. 2 A machine for cutting up corn or cane stalks, or for shredding wheat.
Shreve·port (shrēv′pôrt, -pōrt) A city on the Red River in NW Louisiana.
shrew (shroo) *n.* 1 Any of numerous diminutive, mouse-like, insectivorous mammals (family *Soricidae*) having a long pointed snout and soft fur, as the long-tailed shrew (*Sorex personatus*); also *shrew′mouse.*

SHREW
(Species vary from 1 1/2 to 6 inches in body length)

◆ Collateral adjective: *soricine.* 2 A woman of vexatious, scolding, or nagging disposition. — *v.t. Obs.* To berate; curse. [OE *scrēawa*]
shrewd (shrood) *adj.* 1 Having keen insight; sharp; sagacious. 2 Artful; sly. 3 *Obs.* Keen or sharp; biting. 4 *Obs.* Shrewish; also, vexatious, vicious; dangerous. See synonyms under ACUTE, ASTUTE, INTELLIGENT, KNOWING, POLITIC, SAGACIOUS. [ME *shrewed,* pp. of *schrewen* curse < *shrew* malicious person] — **shrewd′ly** *adv.* — **shrewd′ness** *n.*
shrew·ish (shroo′ish) *adj.* Like a shrew; ill-tempered. — **shrew′ish·ly** *adv.* — **shrew′ish·ness** *n.*
Shrews·bur·y (shrooz′ber·ē, -bər·ē) A municipal borough and county town of Shropshire, England.
shriek (shrēk) *n.* A sharp shrill outcry or scream. — *v.i.* To utter a shriek. — *v.t.* To utter with or as a shriek. See synonyms under CALL, ROAR. [<ON *skrækja*] — **shriek′er** *n.*
shriev·al·ty (shrē′vəl·tē) *n. pl.* ·**ties** The office, term, or jurisdiction of a sheriff. — **shriev′al** *adj.*
shrieve (shrēv) *n. Obs.* A sheriff. [Contraction of SHERIFF]
shrift (shrift) *n.* The act of shriving; confession; absolution. [OE *scrift*]
shrike (shrīk) *n.* Any of numerous birds (family *Laniidae*) with hooked bill, short wings, and long tail; especially, the **loggerhead shrike** (*Lanius ludovicianus*) of the southern Atlantic coast. [OE *scrīc* thrush]

shrill (shril) *adj.* **1** Having a high and piercing quality; sharp and piercing, as a sound. **2** Emitting a sharp, piercing sound. **3** *Poetic* Sharp to other senses than that of hearing; keen. — *v.t.* To cause to utter a shrill sound. — *v.i.* To make a shrill sound. — *adv.* Shrilly. [<Gmc. Cf. LG *schrell* having a sharp tone.] — **shrill′ness** *n.*

shrill·y (shril′ē) *adj. Poetic* Shrill, or somewhat shrill. — *adv.* (shril′lē) In a shrill manner.

shrimp (shrimp) *n.* **1** Any of numerous diminutive, long-tailed, principally marine crustaceans (genus *Crago*), especially the common edible shrimp (*C. vulgaris*) of the northern hemisphere. **2** *Slang* A small, wizened, or shrunken person. [Akin to OE *scrimman* shrink]

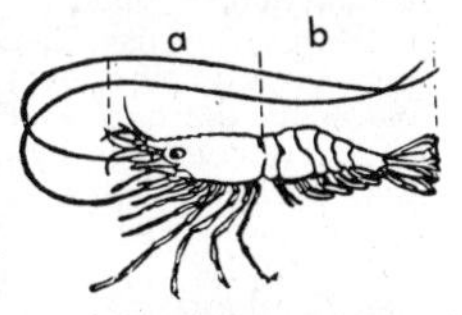

SHRIMP
a. Cephalothorax.
b. Abdomen.

shrine (shrīn) *n.* **1** A receptacle for sacred relics. **2** A place, as a tomb or a chapel, sacred to some holy personage, or considered as sanctified by the remains or presence of such. **3** A thing or spot made sacred by historic or other association. — *v.t.* **shrined, shrin·ing** To enshrine. [OE *scrīn* <L *scrinium* case, chest]

Shrine (shrīn) *n.* A secret fraternal order said to have been founded in Mecca, A.D. 646, and established in the United States in 1872: officially called *Ancient Arabic Order of Nobles of the Mystic Shrine.*

Shrin·er (shrī′nər) *n.* A member of the Shrine.

shrink (shringk) *v.* **shrank** or **shrunk, shrunk** or *less commonly* **shrunk·en, shrink·ing** *v.i.* **1** To draw together; contract, as from heat, cold, etc. **2** To diminish; become less or smaller. **3** To draw back, as from disgust, horror, or timidity; withdraw; recoil: with *from.* **4** To flinch; wince. — *v.t.* **5** To cause to shrink, contract, or draw together. See synonyms under WITHER. — *n.* The act of shrinking; contraction. [OE *scrincan*] — **shrink′a·ble** *adj.* — **shrink′er** *n.* — **shrink′ing** *adj.* — **shrink′ing·ly** *adv.*

shrink·age (shringk′ij) *n.* **1** Contraction, as of metal by cooling, or wood by drying. **2** The amount lost by contraction, depreciation, etc. **3** Decrease in value; depreciation.

shrive (shrīv) *v.* **shrove** or **shrived, shriv·en** or **shrived, shriv·ing** *v.t.* **1** To receive the confession of and give absolution to. **2** To obtain absolution for (oneself) by confessing one's sins and doing penance. — *v.i.* **3** To make confession. **4** To hear confession. [OE *scrīfan,* ult. <L *scribere* write, prescribe] — **shriv′er** *n.*

shriv·el (shriv′əl) *v.t.* & *v.i.* **·eled** or **·elled, ·el·ing** or **·el·ling** **1** To contract into wrinkles; shrink and wrinkle: often with *up.* **2** To make or become impotent; wither. [Origin uncertain. Cf. Sw. *skryvla.*]

shriv·en (shriv′ən) Alternative past participle of SHRIVE.

shroff (shrof) *n.* **1** In China and Japan, an expert detector of counterfeit money or base coin. **2** In India, a money-changer. [<Hind. *sarrāf* <Arabic]

Shrop·shire (shrop′shir, -shər) *n.* A breed of black-faced hornless sheep, noted for heavy fleece and superior mutton, originating in Shropshire.

Shrop·shire (shrop′shir, -shər) A county in western England on the border of Wales; 1,347 square miles; county town, Shrewsbury: also *Salop.*

shroud[1] (shroud) *n.* **1** A dress or garment for the dead; winding sheet. **2** Something that envelops or conceals like a garment. — *v.t.* **1** To dress for the grave; clothe in a shroud. **2** To envelop, as with a garment. **3** *Archaic* To shelter. — *v.i.* **4** *Obs.* To take shelter; go under cover; also, to gather together, as beasts, for warmth. See synonyms under MASK[1]. [OE *scrūd* a garment] — **shroud′less** *adj.*

shroud[2] (shroud) *n.* **1** *Naut.* **a** One of a set of ropes fitted in pairs and constituting part of the standing rigging of a vessel; specifically, one of the ropes, often of wire, stretched from a masthead to the sides or rims of a top, serving as means of ascent and as a lateral strengthening stays to the masts. **b** One of a pair or set of stay ropes or chains to give lateral support to a topmast, bowsprit, etc. **2** A guy, as a support for a smokestack: usually in the plural. **3** One of the supporting ropes attached to the edges of a parachute canopy. [<SHROUD[1]]

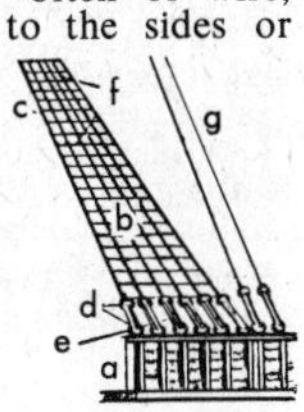

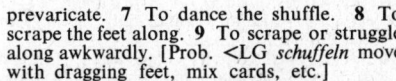

SHROUDS
a. Chain plates.
b. Shrouds.
c. Swifter.
d. Deadeyes.
e. Lanyards.
f. Ratlines.
g. Topmast backstays.

shroud-laid (shroud′lād′) *adj.* Made of four strands twisted around a core: said of rope.

shrove (shrōv) Alternative past tense of SHRIVE.

Shrove·tide (shrōv′tīd′) *n.* The three days immediately preceding Ash Wednesday (**Shrove Sunday, Shrove Monday, Shrove Tuesday**), on which confession is made in preparation for Lent. Compare QUINQUAGESIMA. [ME *schroftide* < stem of SHRIVE + TIDE[1], *n.* (def. 4)]

shrub[1] (shrub) *n.* A woody perennial plant of low stature, characterized by persistent stems and branches springing from the base. ◆ In popular language a shrub is a *bush.* [OE *scrybb* brushwood]

shrub[2] (shrub) *n.* A beverage of sweetened fruit juice, sometimes with spirits. [<Arabic *sharab.* See SHERBET.]

shrub-al·the·a (shrub′al·thē′ə) *n.* A hardy shrub (*Hybiscus syriacus*) of the mallow family: also called *rose of Sharon.*

shrub·ber·y (shrub′ər·ē) *n. pl.* **·ber·ies** **1** Shrubs collectively. **2** A shrubby place; a collection of shrubs, as in a garden.

shrub·by (shrub′ē) *adj.* **·bi·er, ·bi·est** **1** Containing many shrubs; covered with shrubs. **2** Of or pertaining to, or like a shrub or shrubs; stunted. — **shrub′bi·ness** *n.*

shrug (shrug) *v.t.* & *v.i.* **shrugged, shrug·ging** To draw up (the shoulders), as in displeasure, doubt, surprise, etc. — *n.* The act of shrugging the shoulders. [Origin uncertain]

shrunk (shrungk) Alternative past tense and past participle of SHRINK.

shrunk·en (shrungk′ən) Alternative past participle of SHRINK. — *adj.* Contracted and atrophied.

shuck (shuk) *n.* **1** A husk, shell, or pod, as of maize or peas; the outer covering of nuts. **2** A shell of an oyster or a clam. **3** *U.S. Colloq.* Something of little or no value: usually plural: not worth *shucks.* — *v.t.* **1** To remove the shucks of or from; remove the husk or shell from (corn, oysters, etc.). **2** *Colloq.* To take off or cast off, as clothes, or any outer covering. [? Metathetic alter. of HUSK]

shuck·er (shuk′ər) *n.* One who or that which shucks.

shuck·ing (shuk′ing) *n. U.S. Colloq.* **1** A husking bee. **2** The removing of shucks, especially from corn.

shucks (shuks) *interj. U.S. Colloq.* A mild ejaculation expressing annoyance, disgust, etc.

shud·der (shud′ər) *v.i.* To tremble or shake, as from fright or cold; shiver; quake. — *n.* The act of shuddering; convulsive shiver, as from horror or fear; tremor. See synonyms under QUAKE, SHAKE. [Prob. freq. of OE *scūdan* move, shake] — **shud′der·ing** *adj.* — **shud′der·ing·ly** *adv.*

shuf·fle (shuf′əl) *n.* **1** A mixing or changing of the order of things, as of cards in a pack before each deal. **2** A hesitating, evasive, or tricky course; prevarication; artifice. **3** A scraping of the feet, as in walking; a slow, dragging gait. **4** A dance, or the step used in it, where the dancer pushes his foot along the floor at each step. — *v.* **·fled, ·fling** *v.t.* **1** To shift this way and that; mix; confuse; disorder; especially, to change the order of by mixing, as cards in a pack. **2** To move (the feet) along the ground or floor with a dragging gait. **3** To change from one place to another. **4** To make up or remove fraudulently or hastily; also, to put aside carelessly: with *up, off,* or *out.* — *v.i.* **5** To change position; shift ground. **6** To resort to indirect methods; prevaricate. **7** To dance the shuffle. **8** To scrape the feet along. **9** To scrape or struggle along awkwardly. [Prob. <LG *schuffeln* move with dragging feet, mix cards, etc.]

shuf·fle·board (shuf′əl·bôrd′, -bōrd′) *n.* **1** A game in which wooden or composition disks are slid by means of a pronged cue along a smooth surface toward numbered spaces. **2** The board or surface on which the game is played. Also spelled *shovelboard.*

shuf·fler (shuf′lər) *n.* **1** One who shuffles. **2** The scaup duck. **3** The coot.

shuf·fling (shuf′ling) *adj.* **1** Marked by awkward or clumsy movements. **2** Evading the truth; prevaricating.

Shu·fu (shōō′fōō′) The Chinese name for KASHGAR.

Shu·lam·ite (shōō′ləm·īt) The chief female character in the Song of Solomon. *Cant.* vi 13.

shun (shun) *v.t.* **shunned, shun·ning** **1** To keep clear of; avoid; refrain from. **2** *Obs.* To escape; evade. **3** *Obs.* To abhor. See synonyms under ABHOR, ESCAPE. [OE *scunian*] — **shun′ner** *n.*

shun·pike (shun′pīk′) *n. U.S.* A byway or side road used to avoid tollgates.

shunt (shunt) *n.* **1** A turning aside; the act of using a switch or shunt. **2** A railroad switch. **3** *Electr.* A conductor joining two points in a circuit and serving to divert part of the current. The proportion of the current diverted is regulated by the resistance of the shunt employed. — *v.t.* **1** To turn aside. **2** In railroading, to switch, as a train or car, from one track to another. **3** *Electr.* To distribute by means of shunts. **4** To evade by turning away from; put off on someone else, as a task. — *v.i.* **5** To move to one side. **6** *Electr.* To be diverted by a shunt: said of current. **7** To shift or transfer one's views or course. [Origin uncertain] — **shunt′er** *n.*

shunt-wound (shunt′wound′) *adj. Electr.* Designating a type of direct-current motor in which the armature circuit and field circuit are connected in parallel: distinguished from *series-wound.*

shure (shür) *Scot.* Past tense of SHEAR.

Shu·shan (shōō′shän) Old Testament name for SUSA.

shut (shut) *v.* **shut, shut·ting** *v.t.* **1** To bring into such position as to close an opening or aperture; close, as a door, lid, or valve. **2** To close (an opening, aperture, etc.) so as to prevent ingress or egress. **3** To close and fasten securely, as with a latch or lock. **4** To forbid entrance into or exit from. **5** To keep from entering or leaving; confine or exclude; bar: with *in, out, from,* etc. **6** To close, fold, or bring together, as extended, expanded, or unfolded parts: to *shut* an umbrella. **7** To hide from view; obscure. — *v.i.* **8** To be or become closed or in a closed position. — **to shut down** **1** To cease from operating, as a factory or mine; close up; stop work. **2** To lower; come down close: The fog *shut down.* **3** *Colloq.* To suppress: with *on.* — **to shut one's eyes to** To ignore. — **to shut out** In sports, to keep (an opponent) from scoring during the course of a game. — **to shut up** **1** *Colloq.* To stop talking or cause to stop talking. **2** *Colloq.* To become exhausted and stop running, as a horse in a race. **3** To close all the entrances to, as a house. **4** To imprison; confine. — *adj.* **1** Made fast or closed. **2** Not sonorous; dull: said of sound. **3** *Phonet.* **a** Formed by closing the oral and nasal passages completely, preparatory to uttering certain sounds: said of certain consonants, as *t, p, k, b,* and *d.* **b** Cut off sharply by succeeding consonants: said of vowels, as *i* in *pit* and *o* in *top.* **4** *Dial.* Freed, as from something disagreeable; rid: with *of.* — *n.* **1** The act of shutting; also, the time of shutting, closing, or ending: the *shut* of day. **2** The place of shutting or closing together; specifically, the junction or line of junction between welded pieces of metal. [OE *scyttan*]

Synonyms (verb): bar, beleaguer, block, blockade, close, confine, enclose, exclude, imprison, intercept, preclude, prohibit, seal, stop. *Antonyms*: expand, liberate, open, unbar, unclose, undo, unfasten.

shut·down (shut′doun′) *n.* The closing of or ceasing of work in a mine, mill, factory, or other industrial plant.

Shute (sho͞ot), **Nevil,** 1899–1960, English aeronautical engineer and writer: full name Nevil Shute Norway.

shut–eye (shut′ī′) *n. Slang* Sleep.

shut–in (shut′in′) *n.* An invalid who has to stay at home. —*adj.* **1** Obliged to stay at home. **2** Inclined to avoid people.

shut·off (shut′ôf′, -of′) *n. Mech.* A device for shutting something off.

shut·out (shut′out′) *n.* **1** A shutting out; especially, a lock-out. **2** In sports, a game in which one side is prevented from scoring; also, the action or the play that prevents scoring.

shut·ter (shut′ər) *n.* **1** One who or that which shuts. **2** That which shuts out or excludes; specifically, a cover, usually hinged, for closing an opening. **3** A hinged screen or cover for a window. **4** *Phot.* Any of various mechanisms for momentarily admitting light through a camera lens to the film or plate. —*v.t.* To furnish, close, or divide off with shutters.

shut·tle (shut′l) *n.* **1** A device used in weaving to carry the weft to and fro between the warp threads. **2** A similar rotating or other device in a sewing machine or the like. **3** A transport system operating between two nearby points. —*v.t.* & *v.i.* **·tled, ·tling** To move to and fro, like a shuttle. —*adj.* Pertaining to or designating any contrivance, action, etc., intended to operate back and forth between two points: *shuttle* bombing. [OE *scytel* missile; so called because shot to and fro in weaving]

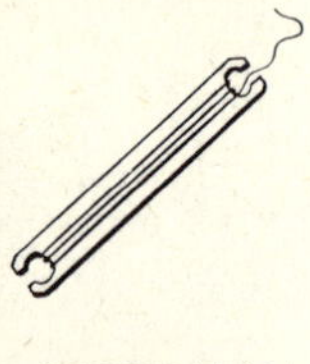

SHUTTLE (*def. 1*)

shuttle armature An H-armature.

shut·tle·cock (shut′l·kok′) *n.* A rounded piece of cork, with a crown of feathers, used in the game of badminton and of battledore and shuttlecock; the game itself. —*v.t.* To send or knock back and forth like a shuttlecock. [<SHUTTLE + COCK[1]]

shwan·pan (shwän′pän′) See SWANPAN.

shy[1] (shī) *v.i.* **shied, shy·ing** **1** To start suddenly aside, as in fear: said of a horse. **2** To draw back, as from doubt or caution: with *off* or *away.* [<*adj.*] —*adj.* **shy·er, shy·est,** or **shi·er, shi·est** **1** Easily frightened or startled, timorous. **2** Bashful; reserved; coy. **3** Circumspect, as from motives of caution; watchful; wary: with *of.* **4** Not easy to perceive, seize, or secure; elusive: a *shy* expression. **5** Not prolific: said of plants, trees, or, rarely, birds. **6** *Colloq.* Having a less amount of money than is called for or required: originally a term used in poker: to be *shy* a dollar in the pool. **7** Short; lacking: often with *on.* —*n.* A starting aside, as in fear. [OE *scēoh* timid. Akin to ESCHEW.] —**shy′ly** *adv.* —**shy′ness** *n.*

shy[2] (shī) *v.t.* & *v.i.* **shied, shy·ing** To throw with a swift, sidelong motion. —*n. pl.* **shies** **1** A careless throw; fling, hence, a verbal fling; a sneer. **2** A trial; experiment. [Origin unknown]

shy·er (shī′ər) *n.* **1** One who shies. **2** A shying horse. Also spelled *shier.*

Shy·lock (shī′lok) In Shakespeare's *Merchant of Venice,* a revengeful usurer who endeavors to exact a pound of flesh from Antonio's body as a forfeit for non-payment of a debt; hence, any relentless creditor.

shy·ster (shīs′tər) *n.* **1** Anyone who conducts his business in an unscrupulous or tricky manner. **2** A lawyer who practices in an unprofessional manner, preys on petty criminals, etc. [? <SHY[1], in slang sense of "disreputable" + -STER]

si[1] (sē) See TI[1].

si[2] (sē) *adv.* Italian, Portuguese, and sometimes French, for "yes". [<L *sic* thus]

si·al (sī′al) *n. Geol.* A rock formation rich in silica and alumina which underlies sedimentary rock in continental land masses. —**si·al′·ic** *adj.* [<SI(LICA) + AL(UMINA)]

si·a·lid (sī′ə·lid) *n.* Any member of a family of insects (*Sialidae,* order *Megaloptera*), with enlarged or elongated thorax, including the hellgrammite and related genera. —*adj.* Of or pertaining to the *Sialidae.* Also **si·al·i·dan** (sī·al′ə·dən). [<Gk. *sialis,* kind of bird]

Si·al·kot (sē·äl′kōt) A city in the NE part of the former province of Punjab, NE West Pakistan.

sialo– *combining form* Saliva; pertaining to saliva: *sialogog.* Also, before vowels, **sial–.** [<Gk. *sialon* saliva]

si·al·o·gog (sī·al′ə·gog) *n.* Any agent exciting a flow of saliva. Also **si·al′a·gogue, si·al′o·gogue.** [<SIAL(O)- + -AGOG] —**si·a·lo·gog·ic** (sī′ə·lō·goj′ik) *adj.* & *n.*

si·a·loid (sī′ə·loid) *adj.* Like or resembling saliva.

Si·am (sī·am′) See THAILAND.

Siam, Gulf of An arm of the South China Sea, separating the Malay Peninsula from Indochina; 300 to 350 miles wide, 450 miles long.

si·a·mang (sē′ə·mang) *n.* A large black gibbon (genus *Symphalangus*) found in Sumatra. [<Malay *siaman* < *iaman* black]

Si·a·mese (sī′ə·mēz′, -mēs′) *adj.* **1** Pertaining to Thailand (Siam), its people, or their language. **2** Closely connected; twin. —*n.* **1** A native or the natives of Siam, belonging to the Thai stock. **2** The Thai language of these people.

Siamese cat A breed of short-haired cat native in Siam, now extensively bred in the United States, typically fawn-colored or pale cream, with dark-tipped ears, tail, feet, and dark mask, a wedge-shaped head, and bright- or deep-blue, gently slanting eyes.

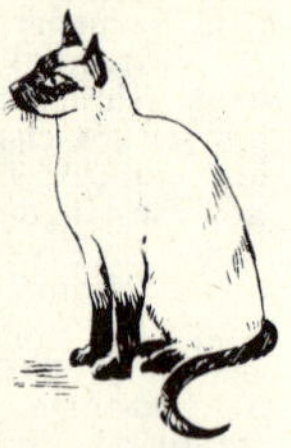

SIAMESE CAT
(About 11 inches at the shoulder)

Siamese twins **1** Originally, the two Chinese males, Eng and Chang, 1811–74, born in Siam, whose bodies were joined by a fleshy band from the navel to the xiphoid cartilage. **2** Any twins joined together at birth.

Si·an (sē′än′, shē′-) **1** The capital of Shensi province, NW China: formerly *Singan.* **2** A city in northern Lianoing province, south central Manchuria, China.

Siang (syäng, shyäng) **1** A river in Hunan province, China, flowing 715 miles NE to Tungting Lake. **2** See YÜ RIVER.

Siang·tan (syäng′tän′, shyäng′-) A city in eastern Hunan province, China.

sib (sib) *n.* **1** A blood-relation; kinsman. **2** Kinsmen collectively; relatives. —*adj.* **1** Related by blood; akin. **2** Related; similar. Also **sibb.** [OE *sibb*]

Sib·bo·leth (sib′ə·leth) See SHIBBOLETH.

Si·be·li·us (si·bā′lē·əs, -bāl′yəs; *Finnish* si·bā′·lyo͞os), **Jean,** 1865–1957, Finnish composer.

Si·be·ri·a (sī·bir′ē·ə) A region of the U.S.S.R. in Asia extending from the Ural Mountains and the Caspian Sea to the Pacific Ocean in northern Asia and roughly corresponding to the Asiatic part of Russian S.F.S.R.; 5,000,000 square miles. *Russian* **Si·bir** (si·bēr′). —**Si·be′ri·an** *adj.* & *n.*

Siberian husky A breed of working dog of medium size with a strong, closely knit body, head resembling that of a fox, brush tail, and thick, soft outer coat.

Siberian Sea, East A section of the Arctic Ocean north of NE Siberia, east of the New Siberian Islands and west of Wrangell Island (def. 2): Russian *Vostochno-Sibirskoye More.*

sib·i·lant (sib′ə·lənt) *adj.* **1** Hissing. **2** *Phonet.* Describing those consonants which are uttered with a hissing sound, as (s), (z), (sh), and (zh). —*n. Phonet.* A sibilant consonant. [<L *sibilans, -antis,* ppr. of *sibilare* hiss] —**sib′i·lance, sib′i·lan·cy** *n.* —**sib′i·lant·ly** *adv.*

sib·i·late (sib′ə·lāt) *v.t.* **·lat·ed, ·lat·ing** To give a hissing sound to, as in pronouncing the letter *s.* [<L *sibilatus,* pp. of *sibilare* hiss] —**sib′i·la′tion** *n.*

Si·biu (si·byo͞o′) A city in central Rumania: German *Hermannstadt.*

sib·ling (sib′ling) *n.* A blood-relation; a relative: used in eugenics, psychology, and anthropology to denote brothers and sisters. [OE, a relative]

Si·bu·yan Sea (si·bo͞o′yän) A part of the Pacific in the central Philippines, bounded by Mindoro, Luzon, Masbate, and Panay.

sib·yl (sib′əl) *n.* **1** In ancient Greece and Rome, any of several women who prophesied under the supposed inspiration of some deity, chiefly of Apollo, and delivered their oracles in a frenzied state. **2** A fortune-teller; sorceress. [<L *sibylla* <Gk.]

Sib·yl (sib′əl) A feminine personal name. Also *Du.* **Si·byl·la** (sē·bil′ə), *Fr.* **Si·bylle** (sē·bēl′), *Ger.* **Si·byl·le** (sē·bē′lə), *Lat.* **Si·byl·la** (si·bil′ə). [<L, soothsayer]

sib·yl·line (sib′əl·īn, -ēn, -in) *adj.* **1** Pertaining to or characteristic of the sibyls; uttered or composed by sibyls; hence, prophetic; oracular; occult. **2** Exorbitant; excessive. Also **si·byl·ic** (si·bil′ik), **si·byl′lic.**

Sibylline Books A collection of nine books which were reputed to set forth the destiny of Rome. The last three were brought from the Cumaean sibyl by Tarquin the Proud and placed in the temple of Jupiter Capitolinus, and were consulted by the senate on momentous occasions.

sic[1] (sik) *adv.* So; thus: sometimes inserted in brackets after something quoted, to indicate that the quotation is literal, and that, in the opinion of the one making the insertion, what immediately precedes is questionable or incorrect. [<L]

sic[2] (sik) *adj. Scot.* Such. Also **sic′can.**

sic[3] (sik) See SICK[2].

Si·ca·ni·an (si·kā′nē·ən) *adj.* Sicilian.

sic·ca·tive (sik′ə·tiv) *adj.* Causing to dry; drying. —*n.* That which has a drying effect; a drying agent or medicine. [<LL *siccativus* <L *siccatus,* pp. of *siccare* dry < *siccus* dry]

sice (sīs) See SYCE.

Sic·el (sis′əl) *n.* **1** A member of an ancient people of Sicily. **2** The Indo-European language of the Sicels, possibly related to Ligurian or Latin. —*adj.* Of or pertaining to the Sicels or their language.

sicht (sikht) *n.* & *v. Scot.* Sight.

Si·cil·i·an Vespers (si·sil′ē·ən, -sil′yən) A general massacre of the French in Sicily (1282) by Sicilians rising against the French rule of Charles of Anjou: so called because the toll that called to Vespers on Easter Monday was the signal for attack.

Sic·i·lies (sis′ə·lēz), **The Two** See TWO SICILIES, THE.

Sic·i·ly (sis′ə·lē) The largest island in the Mediterranean, just SW of Italy (9,831 square miles); comprising with some small neighboring islands an autonomous region of Italy; 9,926 square miles; capital, Palermo: ancient *Trinacria. Italian* **Si·ci·lia** (sē·chē′lyä). —**Si·cil′i·an** *adj.* & *n.*

sick[1] (sik) *adj.* **1** Affected with disease; ill; ailing. **2** Of or used by ill persons: often used in combination: *sickroom.* **3** Affected by nausea; nauseated; desiring to vomit. **4** Expressive or suggestive of nausea; sickly: a *sick* laugh. **5** Impaired or unsound from any cause; weakened; out of condition. **6** Pallid; wan: said of colors. **7** Depressed and longing because of some unattained desire; languishing: *sick* for the sea. **8** Disinclined by reason of satiety or disgust; surfeited: with *of:* *sick* of music. **9** Exhausted, as soil; unable to produce a profitable yield; also, diseased. —*n.* Sick people collectively: with *the.* [OE *sēoc*]

sick[2] (sik) *v.t.* **1** To seek or attack: used in the imperative to order a dog to attack. **2** To urge to attack: I'll *sick* the dog on you. Also spelled *sic.* [Var. of SEEK]

sick·bay (sik′bā′) *n.* That part of a ship or of a naval base set aside for the care of the sick, including operating room, dispensary, and hospital.

sick·bed (sik′bed′) *n.* The bed upon which a sick person lies.

sick call *Mil.* **1** The daily period for reporting to the medical officer all non-hospitalized sick or injured military personnel. **2** The call or signal which announces it.

sick·en (sik′ən) *v.t.* & *v.i.* To make or become sick or disgusted. —**sick′en·er** *n.*

sick·en·ing (sik′ən·ing) *adj.* Disgusting; nauseating. —**sick′en·ing·ly** *adv.*

sick·er[1] (sik′ər) *adj.* More sick.

sick·er[2] (sik′ər) *adj. Scot.* & *Brit. Dial.* Safe; sure; also, cautious. —*adv.* Surely; securely. Also spelled *siker.* [OE *sicor* <L *securus* safe]

sick headache Headache accompanied by nausea and stomach disorders; migraine.

sick·ish (sik′ish) *adj.* **1** Somewhat sick. **2** Slightly nauseating: a sweet, *sickish* odor. See synonyms under SQUEAMISH. —**sick′ish·ly** *adv.* —**sick′ish·ness** *n.*

sick·le (sik′əl) *n.* A reaping implement with a

Sickle long, curved blade mounted on a short handle. — v.t. **·led, ·ling** To cut with a sickle, as grass, hay, etc. [OE *sicel* <L *secula* <*secare* cut]

Sickle A sickle-shaped group of stars in the constellation Leo.

sick·le·bill (sik'əl·bil') n. Any of several birds having a strongly curved bill, as a hummingbird or the long-billed curlew (*Numenius americanus*).

sickle feather One of the long curved feathers in the tail of the domestic cock. See illustration under FOWL.

sickle pear A seckel.

sick·list (sik'list') n. A list of those incapacitated by illness, especially in an army or navy.

sick·ly (sik'lē) adj. **·li·er, ·li·est** 1 Habitually indisposed; ailing; unhealthy: a *sickly* child. 2 Marked by the prevalence of sickness: a *sickly* summer. 3 Nauseating; disgusting; also, mawkish; sickening. 4 Pertaining to or characteristic of the sick or sickness: a *sickly* appearance. 5 Weak- or sick-looking; faint: a *sickly* moon. — adv. In a sick manner; poorly. — v.t. **·lied, ·ly·ing** To make sickly or sickish, as in color or complexion. — **sick'li·ly** adv. — **sick'li·ness** n.

Synonyms (adj.): ailing, diseased, faint, feeble, frail, ill, infirm, invalided, languid, unhealthy, unwell, weak.

sick man of Europe Turkey or the Turkish Empire, as having chronic financial and political troubles and apparently nearing dissolution: an epithet first applied by Czar Nicholas in 1853.

sick·ness (sik'nis) n. 1 Illness; the state of being sick. 2 A particular form of disease. 3 Specifically, nausea. 4 Any disordered and weakened state: the soul's *sickness*. See synonyms under DISEASE, ILLNESS.

sick·room (sik'room', -room') n. A room for the sick.

sic·like (sik'lik') Scot. adj. Similar; such; suchlike. — adv. Similarly.

sic pas·sim (sik pas'im) *Latin* Thus everywhere (as throughout a book).

sic sem·per ty·ran·nis (sik sem'pər ti·ran'is) *Latin* Thus ever to tyrants: motto of Virginia.

sic tran·sit glo·ri·a mun·di (sik tran'sit glō'rē·ə mun'dī) *Latin* Thus passes away the glory of the world.

Sic·y·on (sish'ē·on) An ancient city NW of Corinth in Peloponnesus, southern Greece: Greek *Sikyon*.

sid (shē) n. 1 The fairy folk of Ireland; originally, the Tuatha De Danann. 2 A tumulus or mound believed to be the abode of the fairies; a fairy hill. Also *sidhe*. [<Irish *sid*, *side* fairy hill, fairy folk]

Sid·dons (sid'nz), **Sarah**, 1755-1831, *née* Kemble, English tragic actress.

sid·dur (sid'oor) n. The Jewish prayer book, containing the year's prayers for weekdays, Sabbaths, fast days, and holy days. [< Hebrew *siddūr* arrangement]

side¹ (sīd) n. 1 Any one of the bounding lines of a surface or of the bounding surfaces of a solid object: often limited to a particular bounding line or surface, as distinguished from top, or bottom: the *side* of a box, house, or mountain. 2 A lateral part of a surface or object. 3 One of two or more contrasted surfaces, parts, or places: *inside* and *outside*. 4 Any distinct party or body of competitors or partisans; a faction. 5 An opinion, aspect, or point of view considered with respect to its opposite: my *side* of the question. 6 Family connection, especially by descent through one parent: my grandfather on my father's *side*. 7 The lateral half of a slaughtered animal or of a tanned skin or hide. 8 Either half of the human body as divided by the median plane. The space beside someone. 10 A page of written or printed paper. 11 *Naut.* The part of a ship's hull from stem to stern above the waterline. 12 In billiards, a lateral spin given to the cue ball; english. 13 A bounding line of a geometrical figure. 14 *Brit. Slang* Superciliousness of manner; pretentiousness. — **off side** See OFFSIDE. — adj. 1 Situated at or on one side; lateral: a *side* window. 2 Being or viewed as if from one side; oblique: a *side* glance; incidental: a *side* issue. — v.t. **sid·ed, sid·ing** 1 To provide with sides, as a building. 2 To cut into sides, as a carcass. 3 To thrust aside. — **to side with** To range oneself on the side of; take the part of. [OE]

side² (sīd) adj. 1 *Scot. & Brit. Dial.* Relatively long or wide; large: said of garments. 2 *Scot.* Far; distant.

side arms Weapons worn at the side, as swords, pistols, bayonets, etc.

side·bands (sīd'bandz') n. pl. *Telecom.* The bands of frequencies on either side of the carrier wave within which fall the frequencies produced by modulation.

side·board (sīd'bôrd', -bōrd') n. 1 A piece of dining-room furniture for holding tableware. 2 A board at the side of something, as on sailing canoes.

side·burns (sīd'bûrnz') n. pl. 1 Whiskers grown on the cheeks; burnsides. 2 The hair growing on the sides of a man's face below the hairline: usually worn with the rest of the beard shaved off. [Alter. of BURNSIDES]

side-by-side (sīd'bī'sīd') adj. Beside or next to each other; together.

side·car (sīd'kär') n. 1 A small, one-wheeled passenger car attached to the side of a motorcycle. 2 A cocktail containing equal parts of lemon juice, brandy, and curaçao or Cointreau. 3 A jaunting car.

side chain *Chem.* A group of atoms, specifically an alkyl group, attached to a carbon atom of a ring compound.

side dish A portion of food subordinate to the main dish or dishes of a course; also, the small dish in which it is served.

side-dress (sīd'dres') v.t. *Agric.* To apply fertilizer along only one side of (a row of growing plants). — **side'-dress'ing** n.

side effect *Med.* A secondary, often injurious effect resulting from a drug or other form of therapy whose action is not restricted to the condition for which it was administered.

side·kick (sīd'kik') n. *U.S. Slang* A close friend; buddy.

side·light (sīd'līt') n. 1 A side window. 2 A light coming from the side; hence, incidental illustration or information. 3 *Naut.* One of the colored lights (red on the port side, green on the starboard) displayed on the sides of ships at night; a running-light; also, a night light in the gangway of a war vessel.

side·line (sīd'līn') n. 1 An auxiliary line of goods sold by a store or a commercial traveler. 2 Any additional or secondary work differing from one's main job. 3 A track or road, especially of a railroad, branching off from the main line. 4 A line used to hobble a horse by connecting the fore and hind feet of the same side: also **side'-hob'ble** (-hob'əl). 5 One of the lines bounding the two sides of a football field, tennis court, or the like; also, the area just outside these lines: often in the plural. 6 The point of view of an outsider or non-participant.

side·ling (sīd'ling) adj. Having a slanting or oblique position or motion; indirect. — adv. Sidewise; obliquely; indirectly.

side·long (sīd'lông', -long') adj. Inclining or tending to one side; lateral. — adv. 1 In a lateral or oblique direction. 2 Steeply inclined.

side mark A marker on a river bank giving a river-boat pilot his bearings.

side meat A side of salt pork or bacon, or a piece of it.

side·piece (sīd'pēs') n. 1 A piece at or forming the side of anything. 2 The jamb or check in any finished aperture in a wall, as of a doorway.

si·de·re·al (sī·dir'ē·əl) adj. 1 Pertaining or relative to stars; constituted of or containing stars. 2 Measured by means of the stars: said of periods of time. [<L *sidereus* <*sidus*, *sideris* star] — **si·de're·al·ly** adv.

sidereal time See under TIME.

sid·er·ite (sid'ə·rīt) n. 1 A vitreous, native ferrous carbonate, FeCO₃; spathic iron ore: also called *chalybite*. 2 An indigo-blue variety of quartz. 3 An iron meteorite. [<L *siderites* <Gk. *siderītēs* of iron <*sideros* iron] — **sid'er·it'ic** (-rit'ik) adj.

sidero-¹ *combining form* Iron, of or pertaining to iron: *siderolite*. Also, before vowels, **sider-**. [<Gk. *sidēros* iron]

sidero-² *combining form* Star; stellar: *siderostat*. Also, before vowels, **sider-**. [<L *sidus, sideris* a star]

sid·er·o·lite (sid'ər·ə·līt') n. 1 A spongy meteoric iron containing embedded grains of certain minerals, as chrysolite. 2 A meteorite. [<SIDERO-¹ + -LITE]

sid·er·o·scope (sid'ər·ə·skōp') n. A magnetic device for detecting the presence of iron or steel particles in the eyes.

sid·er·o·sis (sid'ə·rō'sis) n. *Pathol.* 1 Abnormal deposit of iron in the tissues of the body, and especially of the lungs. 2 Any lung disease caused by the inhalation of metallic dust; pneumonoconiosis.

sid·er·o·stat (sid'ər·ə·stat') n. *Astron.* A mirror turning by clock motion so as to reflect the light of a star in an invariable direction into a fixed telescope or other astronomical instrument. [<SIDERO-² + Gk. *statos* standing] — **sid'er·o·stat'ic** adj.

side·sad·dle (sīd'sad'l) n. A woman's saddle having but one stirrup and a cushioned horn on the same side, about which the right knee fits.

side show 1 A small show incidental to a larger or more important one; especially, one connected with a circus but charging an extra entrance fee; also, a minor exhibit at a fair. 2 Any subordinate issue or attraction.

side-slip (sīd'slip') v.i. **-slipped, -slip·ping** To slip or skid sideways. — n. 1 A lateral skid, as of an automobile. 2 A downward, sidewise slipping of an airplane along the lateral axis: executed to lose altitude without a gain in forward speed.

side-split·ting (sīd'split'ing) adj. Having a tendency as if to split the sides with laughter; mirth-provoking.

side-step (sīd'step') v. **-stepped, -step·ping** v.i. To step to one side; avoid responsibility. — v.t. To avoid, as an issue, or postpone, as a decision; evade. — n. 1 A step or a movement to one side, as of a pugilist. 2 A step on the side of a thing for ascending and descending. — **side'-step'per** n.

side stroke In swimming, a stroke made while lying on the side, the arms being thrust forward alternately, the upper arm above the water, the lower arm below the water: performed with a scissors kick.

side·swipe (sīd'swīp') n. A sweeping blow along the side. — v.t. & v.i. **·swiped, ·swip·ing** To strike or collide with such a blow.

side·track (sīd'trak') v.t. & v.i. 1 To move to a siding, as a railroad train. 2 To divert or depart from the main issue or subject; distract or be distracted. — n. A railroad siding; also, a branch line.

side·walk (sīd'wôk') n. A path or pavement at the side of the street for the use of pedestrians.

side·ward (sīd'wərd) adj. Directed or moving toward or from the side; lateral. — adv. Toward or from the side; laterally: also **side'wards**.

side·ways (sīd'wāz') adv. 1 From one side. 2 So as to incline toward the side, or with the side forward: Hold it *sideways*. 3 Toward one side; askance; obliquely; indirectly. — adj. Moving to or from one side: a *sideways* glance. Also **side'way', side'wise'**.

side wheel A wheel at the side; specifically, one of two paddle wheels on either side of a steamboat. — **side'-wheel'** adj. — **side'-wheel'er** n.

side-wind·er (sīd'wīn'dər) n. 1 One of several small rattlesnakes found in the American Southwest, and particularly the horned rattler (*Crotalus cerastes*): so called because of its characteristic lateral motion. 2 A heavy, swinging, sideways blow with the fist.

SIDE WHEEL

Si·di If·ni (sē'dē ēf'nē) The capital of Ifni.

sid·ing (sī'ding) n. 1 A railway track by the side of the main track. 2 The boarding that covers the side of a wooden house or is prepared for that purpose: often in the plural. 3 The act of dressing timbers to correct

sidle 1168 sign

breadths, as in shipbuilding, or the timbers themselves. 4 The act of taking sides, as in a controversy.

si·dle (sīd′l) *v.i.* **·dled, ·dling** To move sideways, especially in a cautious or stealthy manner. —*n.* A sideways step or movement. [<obs. *sidling* sidelong] —**si′dler** *n.*

Sid·ney (sid′nē) A masculine personal name. Also *Sydney*. [<F, St. Denis]

Sid·ney (sid′nē), **Sir Philip,** 1554–86, English soldier, courtier, poet, and writer: also spelled *Sydney*.

Si·don (sīd′n) The capital of ancient Phoenicia, on the site of modern *Saida*. —**Si·do·ni·an** (sī·dō′nē·ən) *adj.* & *n.*

Sid·ra (sid′rə), **Gulf of** An inlet of the Mediterranean in the coast of Libya; 275 miles wide: ancient *Syrtis Major*.

Sie·ben·ge·bir·ge (zē′bən·gə·bir′gə) A range of hills along the Rhine south of Bonn in West Germany; highest point, 1,509 feet.

siè·cle (sye′kl′) *n. French* Century; age; period.

Sie·dl·ce (she′dəl·tse) A city in eastern Poland; formerly, capital of a political subdivision of Russian Poland.

Sieg·bahn (sēg′bän), **Karl Manne Georg,** born 1886, Swedish physicist.

siege (sēj) *n.* **1** The besieging of a town or fortified place; beleaguerment. ◆ Collateral adjective: *obsidional*. **2** A steady attempt to win something; also, the protracted period spent in the effort: He laid *siege* to her heart. **3** The time during which one undergoes a protracted illness or difficulty. **4** *Obs.* A seat; chair; throne. **5** *Obs.* Rank, station. —*v.t.* **sieged, sieg·ing** To besiege. [<OF <L *sedes* seat <*sedere* sit; infl. in meaning by L *obsidium* siege]

Siege Perilous A seat at King Arthur's Round Table, fatal to all occupants save Sir Galahad, the knight destined to find the Holy Grail.

Sieg·fried (sēg′frēd, *Ger.* zēkh′frēt) The hero of the *Nibelungenlied* and several other Germanic legends. [<G, peace of victory]

Siegfried Line See LIMES.

Sieg Heil (zēkh′ hīl′) *German* Hail to victory: a Nazi salute.

Sie·mens (sē′mənz, *Ger.* zē′məns), **Ernst Werner von,** 1816–92, German electrical engineer, inventor, and manufacturer. —**Sir William,** 1823–83, Karl Wilhelm Siemens, German engineer who settled in England; invented the electrodynamometer; brother of the preceding.

Si·en·a (sē·en′ə, *Ital.* syä′nä) A city in Tuscany, central Italy. —**Si·en·ese** (sē′ən·ēz′, -ēs′) *adj.* & *n.*

si·en·ite (sī′ən·īt) See SYENITE.

Sien·kie·wicz (shen·kyä′vich), **Henryk,** 1846–1916, Polish novelist.

si·en·na (sē·en′ə) *n.* **1** A brownish orange-yellow natural clay colored with oxides of iron and manganese: used as a pigment. **2** Orange-yellow, the color of this pigment. [<Ital. *(terra di) Siena* (earth of) Siena]

sier·o·zem (sir′ə·zem) *n.* A grayish-brown soil that merges gradually into a calcareous or hardpan layer: formed usually in a temperate to cool climate. [<Russian, gray earth]

si·er·ra (sē·er′ə) *n.* **1** A mountain range or chain, especially one having a jagged or serrated outline: a term occurring in the names of ranges in Spain and former Spanish colonies. **2** Any of several large mackerel-like fishes, as the cero. [<Sp. <L *serra* saw]

Si·er·ra de Cór·do·ba (sē·er′rä thā kôr′thō·vä) A mountain range of central Argentina; highest point, 9,450 feet.

Si·er·ra de Gre·dos (sē·er′rä thā grā′thōs) A mountain range in central Spain, 25 miles west of Madrid; highest point, 8,504 feet.

Si·er·ra de Gua·dar·ra·ma (sē·er′rä thā gwä′·thär·rä′mä) A mountain range in central Spain, NW of Madrid; highest point, 7,972 feet.

Si·er·ra Le·o·ne (sē·er′rä lā·ō′nā) **1** An independent state on the west coast of Africa; 27,925 square miles, mostly on the **Sierra Leone Peninsula,** extending 25 miles into the Atlantic; capital, Freetown: formerly a British dependency. **2** An estuary in western Sierra Leone, flowing 25 miles past Freetown to the Atlantic.

Si·er·ra Ma·dre (sē·er′rä mä′thrä) A Mexican mountain chain bordering the central plateau on the east and west and divided into the **Sierra Madre del Sur** in the south, the **Sierra Madre Occidental** in the west, and the **Sierra Madre Oriental** in the east; highest point, 18,700 feet.

Si·er·ra Mo·re·na (sē·er′rä mō·rā′nä) A mountain range in southern Spain; highest point, 4,340 feet.

Si·er·ra Ne·vad·a (sē·er′ə nə·vad′ə, -vä′də; *Sp.* sē·er′rä nā·vä′thä) **1** A mountain range of eastern California, extending 400 miles north and south; highest point, 14,495 feet. **2** A mountain range in southern Spain; highest peak, 11,411 feet.

si·es·ta (sē·es′tə) *n.* A midday or afternoon nap. [<Sp. <L *sexta (hora)* sixth (hour), noon <*sex* six]

sieur (syœr) *n.* Sir; master: a former French title of respect. [<F <L *senior* older]

sieve (siv) *n.* **1** A utensil or apparatus for sifting, consisting of a frame provided with a bottom of mesh wire. **2** A garrulous person. —*v.t.* & *v.i.* **sieved, siev·ing** To sift. [OE *sife* sieve]

sieve cell *Bot.* A thin-walled, elongated cell having perforations, **sieve pores,** and sieve plates that permit communication between contiguous cells, forming sieve tubes.

sieve plate *Bot.* One of the thickened terminal sections of a sieve cell.

Sie·vers (sē′vərz, *Ger.* zē′fərs), **Georg Eduard,** 1850–1932, German philologist.

sieve tissue *Bot.* Phloem tissue containing or made up of vascular bundles of sieve cells.

sieve tube *Bot.* An arrangement of sieve cells in plants by means of which conduction is accomplished.

Sie·yès (syā·yes′), **Emmanuel,** 1748–1836, French revolutionist: called "Abbé Sieyès."

si·fak·a (si·fak′ə) *n.* Any of a genus *(Propithecus)* of lemuroid primates related to the indris and characterized by long tails, black skin, short arms, and powerful hind limbs: native in Madagascar: also called *propitheque*. Also **si·fac** (sē′fak). [<Malagasy]

sif·fle (sif′əl) *v.t.* & *v.i.* **·fled, ·fling** To whistle; hiss. —*n.* A sibilant râle. [<F *siffler* <L *sibilare* hiss]

sift (sift) *v.t.* **1** To pass through a sieve in order to separate the fine parts from the coarse. **2** To scatter by or as by a sieve. **3** To examine carefully. **4** To separate as if with a sieve; distinguish: to *sift* fact from fiction. —*v.i.* **5** To use a sieve; sift something. **6** To fall or pass through or as through a sieve: The light *sifts* through the trees. [OE *siftan* sift] —**sift′er** *n.*

sift·ings (sif′tingz) *n. pl.* Something removed or separated by a sieve.

sigh (sī) *v.i.* **1** To draw in and exhale a deep, audible breath, as in expressing sorrow, weariness, pain, etc. **2** To make a sound suggestive of a sigh, as the wind. **3** To yearn; long. —*v.t.* **4** To express with a sigh. **5** To lament with sighs. —*n.* The act or sound of or as of sighing. [Back formation <ME *sighte,* pt. of *siken* <OE *sīcan* sigh]

sight (sīt) *n.* **1** The faculty, act, or fact of seeing; vision. **2** That which is seen; a view; spectacle; show; as used absolutely, something remarkable and strange. **3** *pl.* Things worth seeing: the *sights* of the town. **4** The range or scope of vision; limit of eyesight. **5** A point of view; estimation. **6** Insight; opportunity for investigation or study. **7** A device to assist aim, as on a gun, leveling instrument, etc. **8** An aim or observation taken with a telescope or other sighting instrument. **9** A view; glimpse. **10** The part of a drawing or painting within the marginal lines or the frame. **11** *Colloq.* A great quantity or number: a *sight* of people. —**at** (or **on**) **sight 1** As soon as seen: to read or shoot *at sight*. **2** On presentation for payment: said of drafts, bills, and notes. —**battle sight** The position of the rear sight on a rifle in which the leaf is laid down. —**bore sight** An auxiliary sighting device with parts attached to the muzzle and breech of a gun, used to secure alinement of the axis of the bore with the axis of the gun sight. —**leaf sight** A rear sight for small arms, containing a movable peep sight and hinged to permit raising and lowering. —**peep sight** A sight attached to the breech end of a firearm, and provided with a small hole in the center for close aiming. —*v.t.* **1** To perceive with the eyes; see: to *sight* a whale. **2** To take a sight of; observe; look at through a telescope or similar instrument. **3** To furnish with sights, or adjust the sights of, as a gun. **4** To give the proper aim or elevation to, as a gun; take aim with. **5** *Colloq.* To bring to notice; present, as a bill to its drawee. —*v.i.* **6** To take aim. **7** To make an observation or sight. ◆ Homophones: *cite, site*. [OE *gesiht*]

sight draft A draft or bill payable on presentation. Also **sight bill.**

sight-hole (sīt′hōl′) *n.* A hole to look through; peephole.

sight·less (sīt′lis) *adj.* **1** Without the power of sight; blind. **2** Invisible. —**sight′less·ly** *adv.* —**sight′less·ness** *n.*

sight·ly (sīt′lē) *adj.* **·li·er, ·li·est 1** Pleasant to the view; comely. **2** Affording a grand view. —**sight′li·ness** *n.*

sight·see·ing (sīt′sē′ing) *n.* The visiting of objects of interest. —**sight′se′er** *n.*

sight unseen Without examining: To exchange stamps *sight unseen*.

sig·il (sij′il) *n.* A seal or signature; also, a mark or sign supposed to exercise occult power. [<L *sigillum* seal] —**sig′il·lary** (-ə·ler′ē) *adj.*

Sig·is·mund (sij′əs·mənd, sig′-; *Ger.* zē′gis·mo͞ont) A masculine personal name. Also *Fr.* **Si·gis·mond** (sē·zhēs·môn′), *Ital.* **Si·gis·mon·do** (sē′jēs·mōn′dō), **Si·gis·mun·do** (*Pg.* sē′zhēs·mo͞on′do͞o, *Sp.* sē′hēs·mo͞on′dō), *Du.* **Si·gis·mun·dus** (sē′gēs·mun′dus). [<Gmc., protecting conqueror]

—**Sigismund,** 1368–1437, king of Hungary 1387–1437, Holy Roman Emperor 1411–37.

sig·ma (sig′mə) *n.* **1** The 18th letter in the Greek alphabet, written Σ (capital), σ (small initial), or ς (small final): corresponding to English *s* in *so*. As a numeral it denotes 200. **2** *Math.* The symbol signifying that the sum is to be taken of a series or sequence following. **3** Something shaped like a sigma. [<Gk. *sigma* the letter *s*]

sig·mate (sig′māt) *adj.* Having the shape or form of S or of sigma.

sig·ma·tism (sig′mə·tiz′əm) *n.* Excessive use of *s*-sounds in speech or writing.

sig·moid (sig′moid) *adj.* **1** Shaped like the Greek capital letter sigma (Σ), or like the letter S. **2** Pertaining to the sigmoid flexure. Also **sig·moi·dal** (sig·moid′l). [<Gk. *sigmoeidēs*]

sigmoid flexure *Anat.* A fold in the colon just above the rectum.

sign (sīn) *n.* **1** A motion or action indicating thought, desire, or command; a pantomimic gesture. **2** A board, plate, or representation of any sort, generally bearing an inscription and used to indicate a place of business or resort. **3** An arbitrary mark used to express meaning, rank, condition, value, etc. **4** Any evidence of a recent presence, as tracks, droppings, etc.; a vestige; trace. **5** A mark used in place of a signature by persons unable to write. **6** *Music* Any mark used in musical notation, as a flat or sharp. **7** *Math.* A conventional mark to indicate an operation or relation, as one of the symbols +, −, ×, ÷, indicating the four fundamental operations of addition, subtraction, multiplication, and division. **8** Any indicative or significant object or event; a symbol; token. **9** In the Bible, a miraculous deed as a proof of divine commission or supernatural power; a miracle. **10** *Astron.* One of the twelve equal divisions of the zodiac, named from the constellations that formerly occupied them. See ZODIAC. **11** In hunting, a trace left by an animal; spoor. **12** *Med.* A symptom of disease that is apparent to someone other than the patient. **13** *Eccl.* The sign of the Cross: used in service books and before signatures of bishops. —*v.t.* **1** To write one's signature or initials on. **2** *Law* To acknowledge an instrument by affixing a mark or seal to. **3** To indicate or represent by a sign; stand for. **4** To mark or consecrate with a sign, especially with a cross. **5** To engage by obtaining the signature of to a contract: to *sign* a baseball player; also, to hire (oneself) out for work: often with *on*. **6** To dispose of by signature: with *off* or *away*. **7** To express or indicate with a sign. —*v.i.* **8** To make signs or signals. **9** To write one's signature or initials. —**to sign off** *Telecom.* To announce the close of a program from a broadcasting station and stop transmission. —**to sign up** To enlist, as in a branch of military service.

◆ Homophone: *sine.* [<OF *signe* <L *signum*] **—sign′er** *n.*

Synonyms (noun): emblem, indication, manifestation, mark, note, omen, pattern, presage, prognostic, signal, symbol, symptom, token, type. A *sign* is any distinctive *mark* by which a thing may be recognized or its presence known, and may be intentional or accidental, natural or artificial, suggestive, descriptive or wholly arbitrary. While a *sign* may be involuntary, and even unconscious, a *signal* is always voluntary; a ship may show *signs* of distress to the casual observer, but *signals* of distress are a distinct appeal for aid. A *symptom* is a vital phenomenon resulting from a diseased condition; in medical language a *sign* is an *indication* of any physical condition, whether morbid or healthy; thus, a hot skin and rapid pulse are *symptoms* of pneumonia; dulness of some portion of the lungs under percussion is one of the physical *signs.* See CHARACTERISTIC, EMBLEM, LETTER, MARK[1], TRACE[1].

sig·nal (sig′nəl) *n.* 1 A sign or means of communication agreed upon or understood, and used to convey information or command, as at a distance. 2 *Telecom.* A radio wave or electric current which transmits intelligence, whether direct or in code. 3 An event that incites to action or movement. 4 In some card games, a lead or play that conveys certain information to one's partner. See synonyms under SIGN. *—adj.* 1 Distinguished by some special sign or characteristic; notable; conspicuous. 2 Used to signal: a *signal* fire. See synonyms under EMINENT, EXTRAORDINARY. *—v.* **·naled** or **·nalled, ·nal·ing** or **·nal·ling** *v.t.* 1 To make signals to; inform or notify by signals. 2 To communicate by signals. *—v.i.* 3 To make a signal or signals. [<F <L *signalis* < *signum* sign] **—sig′nal·er** or **sig′nal·ler** *n.*

Signal Corps A branch of the U. S. Army; a body of officers and enlisted men in charge of signaling apparatus and the transmitting of intelligence by telegraph, telephone, radio, visual signs, etc.

signal fire A fire used as a signal; a beacon fire.

signal generator An electromagnetic oscillator used to supply currents of known frequencies through a specified range in testing the performance of a radio receiver.

sig·nal·ize (sig′nəl·īz) *v.t.* **·ized, ·iz·ing** 1 To render noteworthy. 2 To point out with care.

sig·nal·ly (sig′nəl·ē) *adv.* In a signal manner; eminently.

sig·nal·man (sig′nəl·mən) *n. pl.* **·men** (-mən) 1 One who makes or interprets signals; a signaler. 2 One who operates a railroad signal.

sig·nal·ment (sig′nəl·mənt) *n.* 1 The act of signaling. 2 Description of a person for identification by peculiar or characteristic marks, as in the case of a criminal. [<F *signalement*]

signal smoke A smoke from a fire used to signal to a distance, by a system either of puffs, spirals, or clouds.

signal tower 1 Any tower from which signals are displayed. 2 A small railroad tower from which semaphore or block-system signals are controlled.

sig·na·to·ry (sig′nə·tôr′ē, -tō′rē) *adj.* Bound by the terms of a signed document; having signed: *signatory* powers. *—n.* One who has signed or is bound by a document; specifically, a nation so bound. [<L *signatorius* < *signatus,* pp. of *signare* sign < *signum* a sign]

sig·na·ture (sig′nə·chər) *n.* 1 The name of a person, or something representing his name, written, stamped, or inscribed by himself or by deputy, as a sign of agreement or acknowledgment. 2 *Printing* **a** A distinguishing mark, letter, or number on the first page of each form or sheet of a book, as a guide to the binder. **b** The form or sheet on which this mark is placed. **c** One of the fractional parts of a book; a folded printed sheet, usually comprising 16 pages. 3 *Music* A symbol or group of symbols at the beginning of a staff, indicating time or key. See KEY SIGNATURE, TIME SIGNATURE. 4 In radio, the musical number or sound effect that introduces or closes a given program. 5 *Zool.* A color mark resembling a letter. 6 *Med.* The part of a physician's or pharmacist's prescription that indicates how the medicine is to be taken: usually preceded by *S.* or *Sig.* [<F <Med. L *signatura* <L *signatus.* See SIGNATORY.]

sign·board (sīn′bôrd′, -bōrd′) *n.* A board on which a sign, direction, or advertisement is displayed.

sig·net (sig′nit) *n.* 1 A seal; especially, in England, one of the seals of the sovereign, used in sealing his private letters and bills of grants, etc. 2 An impression made by or as if by a seal. *—v.t.* To mark or make official with a signet or seal. ◆ Homophone: *cygnet.* [<F, dim. of *signe* sign <L *signum*]

sig·nif·i·cance (sig·nif′ə·kəns) *n.* 1 The character or state of being significant; expressiveness. 2 That which is signified or intended to be expressed; meaning. 3 Importance; consequence: opposed to *insignificance.* Also **sig·nif′i·can·cy.**

sig·nif·i·cant (sig·nif′ə·kənt) *adj.* 1 Having or expressing a meaning; bearing or embodying a meaning. 2 Betokening or standing as a sign for something; having some covert meaning; significative: His manner was *significant.* 3 Important, as pointing out something weighty; momentous: opposed to *insignificant.* 4 *Math.* Having value or the determining or influential value: the *significant* figures in a number. See synonyms under IMPORTANT. *—n.* Something bearing a meaning; specifically, a token or letter. [<L *significans, -antis,* ppr. of *significare* make a sign, mean < *signum* sign + *facere* do, make] **—sig·nif′i·cant·ly** *adv.*

sig·ni·fi·ca·tion (sig′nə·fə·kā′shən) *n.* 1 That which is signified; meaning; sense; import. 2 The act of signifying; communication. [<OF *significacion* <L *significatio, -onis.* See SIGNIFICANT.]

sig·nif·i·ca·tive (sig·nif′ə·kā′tiv) *adj.* 1 Representing, as a sign; symbolical. 2 Conveying, or tending to convey, a meaning; significant.

sig·ni·fy (sig′nə·fī) *v.* **·fied, ·fy·ing** *v.t.* 1 To make known by signs or words; express; communicate; announce; declare. 2 Hence, to betoken in any way; mean; import. 3 To amount to; mean: What does his opinion *signify*? 4 To denote (medical use) by signature or markings. *—v.i.* 5 To have some meaning or importance; matter. See synonyms under ALLUDE, IMPORT. **—sig′ni·fi′er** *n.*

sign language 1 Dactylology. 2 A system of communication by means of signs, largely manual; specifically, the system used by the Plains Indians to communicate with tribes speaking other languages.

sign manual *pl.* **signs manual** 1 The personal signature of the British sovereign written at the top of state papers. 2 A sign made with the hand; also, any code consisting of manual signs.

si·gnor (sēn′yôr) *n.* 1 An Anglicized form of the Italian title **signore,** used in respectful address to a gentleman: in society equivalent to the English *sir* when no name follows, to *Mr.* with a name, and to the French *monsieur.* 2 A lord or gentleman; especially, an Italian of rank, official position, or social distinction. Also **si′gnior.** [<Ital. *signore* <L *senior* senior]

si·gno·ra (sē·nyō′rä) *n. Italian* Madam; Mrs.: a title of respectful address.

Si·gno·rel·li (sē′nyō·rel′lē), **Luca,** 1442?–1524?, Italian painter.

si·gno·ri·na (sē′nyō·rē′nä) *n. Italian* The equivalent to *miss*: diminutive of Italian *signora.*

si·gno·ri·no (sē′nyō·rē′nō) *n. Italian* A title of respectful address to a young man: diminutive of Italian *signore,* sir.

si·gno·ry (sēn′yə·rē) See SEIGNIORY.

sign·post (sīn′pōst′) *n.* A post bearing a sign; sometimes, a guideboard.

Sigs·bee (sigz′bē), **Charles Dwight,** 1845–1923, U. S. admiral.

Sig·urd (sig′ərd, *Ger.* zē′gŏŏrt) In the *Volsunga Saga,* the hero who slays Fafnir. He corresponds to Siegfried, the hero of the *Nibelungenlied.*

Si·kang (sē′kang′, *Chinese* shē′käng′) A former province of SW China bordering on Tibet, incorporated, 1955, in Szechwan province; 204,194 square miles; former capital, Yaan.

sike (sīk, sik) *n. Scot.* & *Brit. Dial.* 1 A gutter; rill. 2 A marshy bottom with a stream flowing through it: also spelled *syke.* [OE *sīc* streamlet]

sik·er (sik′ər) See SICKER[2].

Sikh (sēk) *n.* One of a religious and military sect founded by Guru Nanak (1469–1538) in India early in the 16th century. *—adj.* Of or pertaining to the Sikhs. [<Hind., lit., disciple]

Sikh·ism (sēk′iz·əm) *n.* The creed and practices of the Sikhs: it is a pantheistic system, combining the teachings of the Persian Sufis with those of Hinduism, rejecting caste, and enjoining purity of life and toleration.

Si·kho·te–A·lin Range (sē′khō·tā·ä·lēn′) A mountain range along the Sea of Japan in Russian S.F.S.R.; highest point, 5,200 feet.

Si Kiang (sē′ kyäng′, shē′ jyäng′) The Chinese name for the WEST RIVER.

Sik·kim (sik′im) A protectorate of India in the eastern Himalayas, south central Asia; 2,818 square miles; capital, Gangtok.

Si·kor·sky (si·kôr′skē), **Igor,** born 1889, U. S. aeronautical engineer born in Russia: inventor of the helicopter.

Sik·y·on (sik′ē·on) The Greek name for SICYON.

si·lage (sī′lij) *n.* Ensilage. [<ENSILAGE]

sil·ane (sil′ān) *n. Chem.* 1 A compound of silicon and hydrogen, SiH_4; silicon hydride or monosilane. 2 Any of a series of similar compounds, named according to the number of silicon atoms present in the molecule. [< SIL(ICON) + -ANE[1]]

Si·las (sī′ləs) A masculine personal name. [See SILVANUS]

sil·a·zane (sil′ə·zān) *n. Chem.* Any of a class of nitrogen-containing silicon compounds having the general formula $H_3Si(NHSiH_2)_nNHSiH_3$. [<SIL(ICON) + AZ(OTE) + -ANE[1]]

sile (sīl) *Brit. Dial. v.t.* 1 To strain; skim. 2 To glide or pass through. *—v.i.* 3 To sink; subside. 4 To boil gently; simmer. *—n.* 1 A strainer. 2 Filth; sediment. [Cf. Sw. & Norw. *sila* strain]

si·le·na·ceous (sī′lə·nā′shəs) *adj.* Caryophyllaceous. [<NL *Silene,* a genus of plants <L *Silenus* Silenus + -ACEOUS]

si·lence (sī′ləns) *n.* 1 The state or quality of being silent; abstinence from speech or noise; taciturnity. 2 Absence of sound or noise; stillness. 3 Absence of note; failure to mention; oblivion; secrecy. 4 *Music* A rest. *—v.t.* **·lenced, ·lenc·ing** 1 To render silent; take away the authority to speak or the power of reply from. 2 To stop the motion or activity of; put to rest; quiet. 3 To force (guns, etc.) to cease firing, as by return fire, bombing, or the like. *—interj.* Be silent. [<F <L *silentium* < *silere* be silent]

si·lenc·er (sī′lən·sər) *n.* 1 A tubular device attached to the muzzle of a firearm rendering the discharge noiseless. 2 A muffler (def. 2). 3 A device to prevent the buzzing of telegraph or telephone wires.

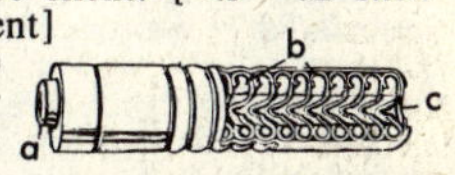

MAXIM SILENCER
a. Socket for attaching to gun.
b. Vortex chamber for gases.
c. Passage groove for bullet.

si·lent (sī′lənt) *adj.* 1 Not making any sound or noise; noiseless; still; also, unspoken; unuttered: *silent* grief. 2 Not speaking, or not given to speech; mute; taciturn. 3 Making no mention or allusion; passing by without notice or record. 4 Free from activity, motion, or disturbance; calm; quiet: a *silent* retreat. 5 Interested financially in a business, but having no authority to act: a *silent* partner. 6 Written, but not pronounced: said of a letter, as the *b* in *debt.* [<F <L *silens, -entis,* ppr. of *silere* be silent] **—si′lent·ly** *adv.* **—si′lent·ness** *n.*

silent butler A small receptacle with a handle and hinged lid, used for collecting refuse from ashtrays, etc.

si·len·ti·ar·y (sī·len′shē·er′ē) *n. pl.* **·ar·ies** 1 One appointed to keep silence and order in court. 2 A Byzantine official sworn not to divulge secrets of state; a privy councilor.

add,āce,câre,pälm; end,ēven; it,īce; odd,ōpen,ôrder; tŏŏk,pōōl; up,bûrn; ə = a in *above,* e in *sicken,* i in *clarity,* o in *melon,* u in *focus*; yōō = u in *fuse*; oi,oil; ou,pout; ch,check; g,go; ng,ring; th,thin; th,this; zh,vision. Foreign sounds à,œ,ü,kh,ṅ; and ◆: see page xx. < from; + plus; ? possibly.

3 An observer of silence because of religious beliefs. [<LL *silentiarius*]

silent partner See under PARTNER.

silent system A system of prison discipline imposing silence on all prisoners.

si·le·nus (sī·lē′nəs) *n.* *pl.* **·ni** (-nī) In Greek mythology, any woodland deity resembling a satyr.

Si·le·nus (sī·lē′nəs) In Greek mythology, the foster father and teacher of Bacchus and leader of the satyrs: traditionally represented as a fat, drunken old man with pointed ears and goat's legs, riding on an ass. [<L <Gk. *Seilēnos* Silenus]

si·le·sia (si·lē′shə, sī-) *n.* **1** A glazed linen cloth first made in Prussian Silesia. **2** A thin, twilled cotton fabric for linings.

Si·le·sia (si·lē′shə, sī-) A region of east central Europe divided between north central Czechoslovakia and SW Poland; formerly a province of Prussia and a crownland of Austria; total area, about 20,000 square miles: German *Schlesien,* Polish *Śląsk,* Czech *Slezsko.* — **Si·le′sian** *adj.* & *n.*

si·lex (sī′leks) *n.* Silica. [<L, flint]

sil·hou·ette (sil′ōō·et′) *n.* **1** A profile drawing or portrait having its outline filled in with uniform color, commonly black: often cut out, as from cardboard. **2** The figure or likeness cast by a shadow; the outline of a solid figure. — *v.t.* **·et·ted, ·et·ting** To cause to appear in silhouette; make a silhouette profile of. [after Étienne de *Silhouette,* 1709–1767, French minister of finance; in mockery of the petty economies for which he was notorious]

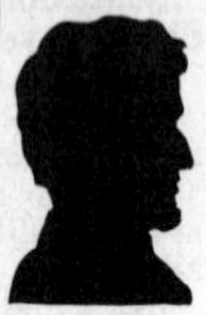

SILHOUETTE
Abraham Lincoln

silic– Var. of SILICO–.

sil·i·ca (sil′i·kə) *n.* A white or colorless, extremely hard, crystalline silicon dioxide, SiO_2, the principal constituent of quartz and sand. [<NL <L *silex, silicis* flint]

silica gel A highly adsorbent colloidal silica, used for deodorizing and cleaning air, purifying blast–furnace gases, etc.

sil·i·cane (sil′i·kān) *n. Chem.* Silane.

sil·i·cate (sil′i·kit) *n. Chem.* A salt of silicic acid. The silicates are mineralogically of great importance, and make up a large part of the earth's crust.

si·li·ceous (si·lish′əs) *adj.* **1** Pertaining to, resembling, or containing silica. **2** Growing or living on siliciferous soil. Also **si·li′cious.** [<L *siliceus*]

si·lic·ic (si·lis′ik) *adj.* Pertaining to, derived from, or consisting of silica or silicon. [< SILIC- + -IC]

silicic acid *Chem.* Any of several gelatinous and easily decomposed compounds of silica and water; especially, orthosilicic acid, H_4SiO_4, associated in the formation of many metallic silicates.

sil·i·cide (sil′ə·sīd) *n. Chem.* A binary compound of silicon with a metal, such as iron, cobalt, nickel, chromium, copper, or magnesium.

sil·i·cif·er·ous (sil′ə·sif′ər·əs) *adj.* Containing or producing silica; united partially with silica. [<SILIC- + -(I)FEROUS]

si·lic·i·fied wood (si·lis′ə·fīd) Wood that has been replaced by silica crystallizing out from solution so as to become a mass of quartz of the original form and structure of the wood; petrified wood. See PETRIFIED FOREST.

si·lic·i·fy (si·lis′ə·fī) *v.* **·fied, ·fy·ing** *v.t.* To convert into silica, as wood. — *v.i.* To become silica, or become impregnated with it. [< SILIC- + -(I)FY] — **si·lic′i·fi·ca′tion** *n.*

sil·i·cle (sil′i·kəl) *n.* A very short, flat silique. [<L *silicula,* dim. of *siliqua* pod]

silico– *combining form* Silicon; of, related to, or containing silicon. Also, before vowels, *silic–,* as in *silicosis.* [<L *silex, silicis* flint]

sil·i·con (sil′ə·kən) *n.* A widely–distributed nonmetallic element (symbol Si) prepared as a dull–brown amorphous powder, as shining metallic scales resembling graphite, or as a steel–gray crystalline mass, by heating silica with carbon in an electric furnace. See ELEMENT. Also **si·li·ci·um** (si·lish′ē·əm, -lis′-). [<L *silex, silicis* flint]

sil·i·cone (sil′ə·kōn) *n. Chem.* Any of various organosilicon compounds containing a silicon–carbon bond: their great physical, chemical, and electrical stability adapts them for many industrial uses as lubricants, greases, polishes, insulating resins, waterproofing materials, and for the making of a special type of synthetic rubber. [<SILICON]

sil·i·co·sis (sil′ə·kō′sis) *n. Pathol.* A pulmonary disease caused by the inhalation of finely powdered silica or quartz.

si·lic·u·lose (si·lik′yə·lōs), **si·lic·u·lous** (-ləs) *adj.* Siliquose. [<SILIQUOSE]

Si·lif·ke (si·lif·ke′) A town in central southern Turkey near the Mediterranean: ancient *Seleucia Trachea.*

si·lique (si·lēk′, sil′ik) *n. Bot.* A narrow, dry, two–valved pod or fruit characteristic of plants of the mustard family. For illustration see under FRUIT. Also **sil·i·qua** (sil′ə·kwə). [<F <L *siliqua* pod]

sil·i·quose (sil′ə·kwōs) *adj.* Silique–bearing; pertaining to or resembling a silique. Also **sil′i·quous** (-kwəs). [<NL *siliquosus* <L *siliqua* pod]

Si·lis·tri·a (si·lis′trē·ə) A city on the Danube, in NE Bulgaria. Also **Si·li·stra** (sē·lē′strä). Ancient **Du·ros·to·rum** (dōō·ros′tə·rəm, dyōō-).

silk (silk) *n.* **1** The creamy–white or yellowish, very fine natural fiber produced by various insects, especially by the larvae of silkworms, to form their cocoons. **2** A similar thread spun by other insects or arachnids. **3** Cloth, thread, or garments made of silk. **4** Anything resembling or suggestive of silk, as the fine soft styles of an ear of corn. — **to hit the silk** *Slang* To descend from an aircraft by parachute. — *adj.* Consisting of silk; silken; silky. — *v.t.* To clothe or cover with silk: grand ladies plumed and *silked.* — *v.i.* To produce the portion of the flower called silk: said of corn. [OE *seoloc,* ult. <L *sericus* silken, lit., pertaining to the Seres (Chinese), from whom silk was bought. Related to SERGE.]

silk·a·line (sil′kə·lēn′) *n.* A soft and thin mercerized cotton fabric resembling silk. Also **silk′a·lene′.**

silk cotton The silky seed covering of various species of a genus *(Bombax)* of tropical American trees, and of the West Indian god tree *(Ceiba pentandra)* or corkwood *(Ochroma pyramidale).* Its principal use is for stuffing cushions, packing, etc.

silk–cot·ton tree (silk′kot′n) Any tree producing silk cotton.

silk·en (sil′kən) *adj.* **1** Made of silk. **2** Like silk; glossy; delicate; smooth. **3** Dressed in silk; hence, luxurious.

silk hat A high cylindrical hat covered with fine silk plush: worn by men and used as a dress hat.

silk·man (silk′mən) *n.* *pl.* **·men** (-mən) A dealer in or a manufacturer of silk; also, an operative in a silk factory.

silk paper A granite paper made with occasional silk fibers in the pulp.

silk–screen print (silk′skrēn′) A reproduction made by the silk–screen process.

silk–screen process A printing process which forces ink through the meshes of a silk screen on which the desired pattern or design has been imposed.

silk–stock·ing (silk′stok′ing) *adj.* Wearing silk stockings; hence, wealthy; luxurious. — *n.* **1** One who wears silk stockings; a member of the wealthy class. **2** A supporter of a branch of the Whig party in the United States in the early 19th century.

silk vine A deciduous shrub of the milkweed family *(Periploca graeca)* growing in the neighborhood of the Black Sea: its bark yields periplocin. Also called *wolf's–bane.*

silk·weed (silk′wēd′) *n.* Milkweed.

silk·worm (silk′wûrm′) *n.* The larva of a moth that produces a dense silken cocoon, especially the common silkworm *(Bombyx mori),* from whose cocoon commercial silk is made.

silk·y (sil′kē) *adj.* **silk·i·er, silk·i·est** **1** Like silk in any way; soft; lustrous. **2** Made of or consisting of silk; silken. **3** Long, fine, and appressed, as hairs, or covered with such hairs, as leaves. **4** Gentle or insinuating in manner: usually in a bad sense, implying insincerity. — **silk′i·ly** *adv.* — **silk′i·ness** *n.*

silky oak See under LACEWOOD.

sill (sil) *n.* **1** A horizontal member forming the foundation, or part of the foundation, of a structure of any kind, as at the bottom of a casing in a building; especially, a door sill or a window sill. **2** A timber in the frame of the floor of a railroad car: end *sill;* side *sill.* **3** *Geol.* A relatively thin stratum of igneous rock intruded between level or gently inclined beds of other rock. [OE *syll*]

sil·la·bub (sil′ə·bub) *n.* **1** A dish made by combining milk or cream with wine or cider, and thus forming a soft curd, which is then flavored. It may be whipped into a froth, or made solid by boiling after adding water and gelatin. **2** Figuratively, something frothy, as flowery language. Also spelled *syllabub.* [Alter. of obs. *sillibouk* <SILLY + OE *būc* belly]

Sil·lan·pää (sil′län·pa), **Frans Eemil,** born 1888, Finnish writer.

sil·ler (sil′ər) *adj.* & *n. Scot.* Silver; money.

Sil·li·man (sil′i·mən), **Benjamin,** 1779–1864, U. S. chemist and geologist.

sil·ly (sil′ē) *adj.* **·li·er, ·li·est** **1** Destitute of ordinary good sense; simple; foolish; imbecile; fatuous; sometimes, senile. **2** Characterized by or resulting from foolishness or imbecility; stupid: *silly* talk. **3** *Rare* Simple; plain; rustic. **4** *Colloq.* Stunned; dazed, as by a blow. **5** *Obs.* or *Brit. Dial.* Frail; feeble; weak; helpless. **6** *Scot.* Mentally or physically incapable; idiotic; imbecilic. **7** *Obs.* Scanty; meager. See synonyms under CHILDISH, RIDICULOUS. — *n.* *pl.* **·lies** *Colloq.* A silly person. [OE *gesælig* happy] — **sil′li·ly** *adv.* — **sil′li·ness** *n.*

si·lo (sī′lō) *n.* *pl.* **·los** A structure, usually of wood or concrete, as a cylindrical pit or a tower, in which fodder, grain, or other food is stored green to be fermented and used as feed for cattle, etc. See ENSILAGE. — *v.t.* **·loed, ·lo·ing** To put or preserve in a silo; turn into ensilage. [<Sp. <L *sirus* <Gk. *siros* pit for corn]

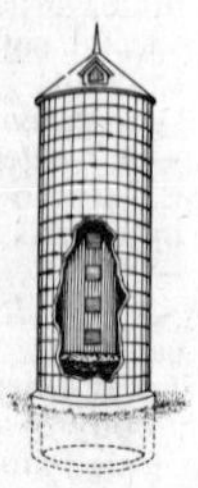

SILO
Stave type, showing interior construction and pit.

Si·lo·am (si·lō′əm, sī-) A spring and pool outside Jerusalem. *John* ix 7.

sil·ox·ane (sil′ok′sān) *n. Chem.* Any of a class of oxygen–containing silicon compounds of the general formula $H_3Si(OSiH_2)_nOSiH_3$. [<SIL(ICON) + OX(YGEN) + -ANE[1]]

silt (silt) *n.* **1** An earthy sediment consisting of extremely fine particles of rock and soil suspended in and carried by water. **2** A deposit of such sediment, as at the mouth of a river. — *v.i.* **1** To become filled or choked with silt: usually with *up.* **2** To ooze; drift. — *v.t.* **3** To fill or choke with silt or mud: usually with *up.* [ME *sylte.* Cf. Dan. *sylt* salt marsh, Norw. *sylta* coast–land washed by the sea.] — **silt′y** *adj.*

sil·ta·tion (sil·tā′shən) *n.* The process of depositing silt.

sil·thi·ane (sil′thē·ān) *n. Chem.* Any of a class of sulfur–containing silicon compounds having the general formula $H_3Si(SSiH_2)_nSSiH_3$. [<SIL(ICON) + THI- + -ANE[1]]

Sil·u·res (sil′yə·rēz) *n. pl.* The pre–Celtic inhabitants of ancient Britain, occupying what is now SE Wales, described by Tacitus as of Iberian origin. [<L]

Si·lu·ri·an (si·lŏŏr′ē·ən, sī-) *adj.* **1** *Geol.* Of or pertaining to the period or rock system of the Paleozoic era following the Ordovician and preceding the Devonian, sometimes called the era of invertebrates: so called because first identified in southern Wales, the home of the ancient Silures. **2** Of or pertaining to the Silures. — *n.* **1** The Silurian period or system. **2** Originally, the period between the Cambrian and the Devonian.

si·lu·rid (si·lŏŏr′id, sī-) *n.* Any one of a large family of fishes *(Siluridae),* the catfishes, including many fresh–water food fishes of the United States. — *adj.* Of or pertaining to the *Siluridae.* Also **si·lu′roid.** [<NL *Siluridae,* name of the family <L *silurus* a river fish <Gk. *silouros*]

sil·va (sil′və), **sil·van** (sil′vən), etc. See SYLVA, etc.

Sil·va·nus (sil·vā′nəs) In Roman mythology, a god of woods and farming: also *Sylvanus.* [<L < *silva* forest]

Sil·va·nus (sil·vā′nəs, *Du., Ger.* sĕl·vä′nōōs) A masculine personal name. Also *Fr.* **Sil·vain** (sĕl·vaṅ′) or **Sil·vie** (sĕl·vē′), *Ital., Sp.* **Sil·va·no** (sĕl·vä′nō) or **Sil·vio** (sĕl′vyō). [<L, of the forest]

sil·ver (sil′vər) *n.* **1** A white, ductile, and very

malleable metallic element (symbol Ag), crystallizing in the isometric system and possessing a high electric conductivity: one of the precious metals. It is found native, as well as in combination. See ELEMENT. 2 Silver regarded as a commodity or as a standard of currency. 3 Silver coin considered as money; hence, ready cash or change; money in general. 4 Articles for domestic use, as tableware, made of silver; silver plate; silverware. 5 A luster or color resembling that of silver; also, the color of silver. 6 *Phot.* Silver nitrate or one of the other salts of silver, used for sensitizing paper. —*adj.* 1 Made of or coated with silver. 2 Resembling silver; having a silvery lustre. 3 Having the soft, clear tones of a silver bell; hence, enticing; persuasive; eloquent. 4 Relating to, connected with, or producing silver. 5 Designating a 25th wedding anniversary. 6 White; hoary: said of the hair or beard. 7 Favoring the use of silver as a monetary standard. —*v.t.* 1 To coat or plate with silver. 2 To coat with some substance having a resemblance to silver; specifically, to coat with amalgam of tin and mercury, as a mirror. 3 To make silverlike; cause to glitter like silver. 4 To coat, as photographic paper, with a film of a silver salt. —*v.i.* 5 To become silver or white, as with age; to become silverlike. [OE *siolfor*] —**sil'ver·er** *n.*

silver age 1 In Latin literature, the age, following the Augustan age, of which Martial and Tacitus are representatives. 2 In classical mythology, the age of Jupiter's rule, succeeding that of Kronos or Saturn and preceding the brazen age.

sil·ver·bell (sil'vər·bel') *n.* A small tree (*Halesia carolina*) of the southern United States, with showy white flowers: sometimes called *snowdrop tree.* Also **silverbell tree.**

sil·ver·ber·ry (sil'vər·ber'ē) *n.* *pl.* **·ber·ries** A shrub (*Elaeagnus commutata*) of the northwestern United States, with silvery foliage, flowers, and edible fruit.

silver bromide *Chem.* A photosensitive compound, $AgBr$, of silver salts and a bromide: used in photography.

silver certificate Paper currency issued by the United States treasury and validated by silver currency or bullion.

silver chloride *Chem.* A white, curdy precipitate, $AgCl$, made by treating silver salts with chloride solutions: used in photography for developing and printing.

sil·ver·fish (sil'vər·fish') *n.* *pl.* **·fish** or **·fish·es** 1 A silvery-white variety of the goldfish. 2 The tarpon. 3 The silversides. 4 Any of numerous primitive, flat-bodied, wingless insects (genus *Lepisma*, order *Thysanura*) having three bristlelike tails and feeding on starchy matter: often called *bristletail.* 5 Any of several similar insects, as the firebrat.

silver fox 1 The red fox (*Vulpes fulva*) of the United States and Canada, in that color phase when the pelage is black with interspersed silver-tipped hairs. 2 The fur.

silver glance Silver sulfide, Ag_2S; argentite.

silver gray A slightly bluish gray, the color of silver.

sil·ver·ing (sil'vər·ing) *n.* 1 A plating or covering of silver, or an imitation of it, as applied to any surface. 2 The art or process of coating surfaces with or as with silver. 3 Sensitization of photographic paper with a silver salt.

sil·ver·ling (sil'vər·ling) *n.* 1 An old Hebrew or Persian silver coin. 2 A tarpon.

sil·ver·ly (sil'vər·lē) *adv.* In the manner of silver; brightly; with sweet tone.

silver maple White maple.

sil·vern (sil'vərn) *adj.* *Archaic* or *Poetic* Made of or like silver.

silver nitrate *Chem.* A white, crystalline, poisonous compound, $AgNO_3$, obtained by treating silver with nitric acid. It is widely used in industry and photography, and in medicine as an astringent, antiseptic, etc.

silver plate 1 Table utensils made of silver. 2 *U.S.* Plated silverware: a trade term.

silver point 1 A drawing implement consisting of a slender silver rod pointed at one end, or of silver wire held in an etching-needle holder. 2 The process of drawing with such an implement. 3 A drawing made with a silver point on paper coated with a white pigment, as Chinese white, characterized by delicacy of line, and often by a tarnish which is highly esteemed. 4 *Physics* The melting point of silver at normal atmospheric pressure, 960.5° C.: one of the basic points in the international temperature scale.

silver poplar The white poplar.

sil·ver·sides (sil'vər·sīdz') *n.* 1 Any of certain small fishes (family *Atherinidae*) related to the mullets and blennies, having a silver band along each side of the body; especially, the **common silversides** (*Menidia notata*) of the coast of the eastern United States. 2 Any small cyprinoid; a fresh-water minnow. Also **sil'ver·side'.**

sil·ver·smith (sil'vər·smith') *n.* A worker in silver; a maker of silverware.

silver standard A monetary standard or system based on silver.

Silver Star A U. S. military decoration in the form of a bronze star inset with a small raised silver star, awarded for gallantry in action: first issued in 1932, and ranking next in honor to the Distinguished Service Cross.

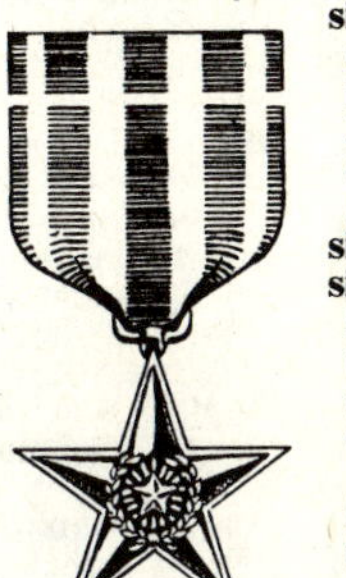

SILVER STAR

Silver State Nickname of Nevada: so called from its native silver ores.

sil·ver-tongued (sil'vər·tungd') *adj.* Persuasive; eloquent.

sil·ver·ware (sil'vər·wâr') *n.* Articles made of silver; silver plate; especially, tableware.

sil·ver·weed (sil'vər·wēd') *n.* A perennial herb (*Potentilla anserina*) of the rose family, growing on shores and meadows, with pinnate silvery leaves and large yellow flowers.

sil·ver·y (sil'vər·ē) *adj.* 1 Containing or adorned with silver. 2 Resembling silver, as in luster, hue, or sound: a *silvery* laugh. —**sil'ver·i·ness** *n.*

Sil·ves·ter (sil·ves'tər) A masculine personal name. Also *Fr.* **Sil·ves·tre** (sēl·ves'tr'). *Ital.* **Sil·ves·tro** (sēl·ves'trō). [<L, forest dweller]

sil·vi·cul·ture (sil'vi·kul'chər) *n.* The art of producing and tending a forest and forest trees. See FORESTRY. [<L *silva* forest + CULTURE] —**sil'vi·cul'tur·al** *adj.* —**sil'vi·cul'tur·al·ly** *adv.* —**sil'vi·cul'tur·ist** *n.*

s'il vous plaît (sēl vōō ple') *French* If you please; please.

si·ma (sī'mə) *n. Geol.* An igneous rock rich in silica and magnesium underlying sial formations in continental land masses. [<SI(LICA) + MA(GNESIUM)]

si·mar (si·mär') *n.* A light, flowing robe for women. [<F *simarre* <Ital. *cimarra* <Arabic *sammūr* sable]

sim·a·ru·ba (sim'ə·rōō'bə) *n.* Any of a genus (*Simaruba*) of tropical American trees of the quassia family, having diclinous flowers and drupaceous fruits. *S. amara* yields a bark used in pharmacy. Also **sim'a·rou'ba.** [<NL <native Carib name] —**sim'a·ru·ba'ceous** or **·rou·ba'ceous** *adj.*

Sim·bor (sim·bôr') A continental territory of Diu district, Portuguese India, on the southern coast of the Kathiawar peninsula at the mouth of the Gulf of Cambay.

Sim·chath To·rah (sim'khäs tō'rə) Literally, the rejoicing over the law; a Jewish holiday which falls on the 23rd of Tishri and closes the feast of Sukkoth. See CALENDAR (Hebrew), JEWISH HOLIDAYS. Also **Sim'hath To'rah.**

Sim·coe (sim'kō), **Lake** A lake in southern Ontario, Canada; 280 square miles.

Sim·e·on (sim'ē·ən, *Ger.* zē'mā·ōn) A masculine personal name. Also *Fr.* **Si·mé·on** (sē·mā·ôn'), *Pg.* **Si·ma·ão** (sē'mä·ouṅ') or **Si·mão** (sē·mouṅ'), *Sp.* **Si·ma·on** (sē'mä·ōn'). See also SIMON. [<Hebrew, obedient]

—**Simeon** The second son of Jacob, or the tribe descended from him. *Gen.* xxix 33.

—**Simeon** A Biblical personage. See NUNC DIMITTIS.

—**Simeon Sty·li·tes** (stī·lī'tēz), 390?–459, Syrian ascetic and stylite.

Sim·fer·o·pol (sim'fər·ō'pəl, *Russian* sēm'fi·rô'pəl) The capital of the Crimea, SW Russian S.F.S.R., in the south central part of the peninsula.

sim·i·an (sim'ē·ən) *adj.* Like or pertaining to the apes and monkeys. —*n.* An ape or monkey. [<L *simia* ape]

sim·i·lar (sim'ə·lər) *adj.* 1 Bearing resemblance to one another or to something else; like, but not completely identical. 2 Of like characteristics, nature, or degree; of the same scope, order, or purpose. 3 *Music* Having motion in the same direction; ascending or descending together, as two parts. 4 *Geom.* Shaped alike: said of two figures, each of which may become congruous with the other by altering all its linear dimensions in one and the same ratio, its angles remaining unchanged. See synonyms under ALIKE, SYNONYMOUS. [<F *similaire* <L *similis* like]

similar fraction See under FRACTION.

sim·i·lar·i·ty (sim'ə·lar'ə·tē) *n.* *pl.* **·ties** 1 The quality or state of being similar. 2 The point in which the objects compared are similar. 3 *pl.* Things that coincide with or resemble each other. See synonyms under ANALOGY, APPROXIMATION.

sim·i·lar·ly (sim'ə·lər·lē) *adv.* Likewise.

sim·i·le (sim'ə·lē) *n.* A rhetorical figure expressing comparison or likeness, by the use of such terms as *like, as, so,* etc.: distinguished from *metaphor* and *comparison* proper. [<L, neut. of *similis* similar]

Synonyms: comparison, figure, illustration, image, imagery, likeness, metaphor, similitude, symbol. The *simile* carries its note of *comparison* on the surface, in the words *as, like, such as,* or similar expressions; the *metaphor* is given directly without any note of *comparison.* "God is *like* a rock" is a *simile*; "God *is* a rock" is a *metaphor.* In order that a *comparison* may become a *simile*, objects of different classes must be compared, bringing in some imaginative element. To say, "The Hudson is like the Rhine" is not *simile,* but direct and literal *comparison*; but to say, "The Hudson flows like the march of time" is to lift the river out of its class and associate it with a great elemental conception, and thus to transform the *comparison* into a *simile. Similitude* is broader in meaning than *simile* or *metaphor,* and may include direct and literal *comparison.* Compare ALLEGORY, ANALOGY, EMBLEM.

si·mil·i·a si·mil·i·bus cu·ran·tur (si·mil'ē·ə si·mil'ə·bəs kōō·ran'tər) *Latin* Like (ailments) are cured by like, i.e., by remedies that produce the effects of the disease itself: the principle of homeopathy.

si·mil·i·tude (si·mil'ə·tōōd, -tyōōd) *n.* 1 Similarity. 2 One who or that which is similar. 3 A rhetorical figure involving comparison or likeness; loosely, a metaphor or a simile. 4 *Geom.* The relation of identity between two figures irrespective of magnitude. See synonyms under ANALOGY, IMAGE, PICTURE, SIMILE. [<L *similitudo* <*similis* like]

sim·i·ous (sim'ē·əs) *adj.* Simian. Also **sim'i·oid.** [See SIMIAN]

sim·i·tar (sim'ə·tər), **sim·i·ter** See SCIMITAR.

Sim·la (sim'lə) A city of far northern central India; capital of Himachal Pradesh. Formerly the summer capital of India under British rule.

sim·lin (sim'lin) *n.* A cymlin. [Var. of SIMNEL]

sim·mer[1] (sim'ər) *v.i.* 1 To boil gently or with a singing sound; be or stay at or just below the boiling point. 2 To be on the point of breaking forth, as with rage. —*v.t.* 3 To keep at or just below the boiling point. —*n.* The state or process of simmering; figuratively, a busy pondering over something, or a state of repressed emotion. [<obs. *simper* boil; origin unknown; prob. imit.]

sim·mer[2] (sim'ər) *n. Scot.* Summer.

Simms (simz), **William Gilmore,** 1806–70, U. S. novelist and poet.

sim·nel (sim'nəl) *n. Brit.* 1 A brittle cake or bread made of fine flour. 2 A rich cake for mid-Lent Sunday, Easter, or Christmas. [<OF *simenel* <LL *siminellus* <L *simila* fine wheat flour, prob. <Gk. *semidalis,* ult. <Babylonian *samidu* fine flour]

si·mo·le·on (si·mō'lē·ən) *n. Slang* A dollar. [Origin unknown]

Si·mon (sī'mən; *Fr.* sē·môn'; *Ger.* zē'mōn; *Sp.,*

Sw. sē·mōn′; *Hungarian* shē′mōn) A masculine personal name. Also *Ital.* **Si·mo·ne** (sē·mō′nā). See also SIMEON. [<Hebrew, obedient]
— **Simon Ma·gus** (mā′gəs) A Samarian magician of the first century, founder of the Simonians, a sect competing with Gentile Christianity.
— **Simon Peter** See PETER.
— **Simon Ze·lo·tes** (zē·lō′tēz) One of the twelve apostles: also called "Simon the Canaanite."

Si·mon (sī′mən) *n.* A credulous unsophisticated person: from the Mother Goose nursery rime *Simple Simon.*

Si·mon (sī′mən), **Sir John Allsebrook,** 1873–1954, first viscount Simon, English lawyer and statesman.

si·mo·ni·ac (si·mō′nē·ak) *n.* One who carries on or is guilty of simony. [<Med. L *simoniacus.* See SIMONY.] — **sim·o·ni·a·cal** (sim′ə·nī′ə·kəl) *adj.* — **sim′o·ni′a·cal·ly** *adv.*

Si·mo·ni·an (si·mō′nē·ən, sī-) *adj.* Of or pertaining to Simon Magus or his sect. — *n.* One of an early sect that held Simon Magus to be the Messiah.

Si·mo·ni·an·ism (si·mō′nē·ən·iz′əm, sī-) *n.* Saint-Simonism.

Si·mon·i·des of Ceos (sī·mon′ə·dēz) A Greek lyric poet of the sixth and early fifth centuries B.C.

Si·mon·iz (sī′mən·īz) *n.* A compound devised for cleaning and waxing lacquered or enameled surfaces: a trade name. — *v.t.* **·ized, ·iz·ing** To clean and polish with Simoniz.

Simon Le·gree (li·grē′) In Harriet Beecher Stowe's *Uncle Tom's Cabin,* the cruel overseer; hence, any brutal master.

si·mon-pure (sī′mən·pyŏŏr′) *adj.* Real; genuine; authentic. [after a character in a 17th c. comedy, who is impersonated by a rival; the rival is discomfited when the real Simon Pure appears]

si·mo·ny (sī′mə·nē, sim′ə-) *n.* Traffic in sacred things; the purchase or sale of ecclesiastical preferment. [<Med. L *simonia* <*Simon (Magus),* who offered Peter money for the gift of the Holy Spirit]

si·moom (si·mōōm′, sī-) *n.* A hot, dry, dust-laden, exhausting wind of the desert, as in Africa and Arabia: also spelled *samoun.* Also **si·moon′** (-mōōn′). [<Arabic *samūm* < *samma* poison]

simp (simp) *n. U.S. Slang* A simpleton.

sim·per (sim′pər) *v.i.* To smile in a silly, self-conscious manner; smirk. — *v.t.* To say with a simper. — *n.* A silly, self-conscious smile; smirk. [Prob. <Scand. Cf. Sw. and Norw. *semper* affected, coy.] — **sim′per·er** *n.* — **sim′per·ing·ly** *adv.*

sim·ple (sim′pəl) *adj.* **·pler, ·plest 1** Consisting of one thing; single; uncombined; unmingled. **2** Not complex or complicated; easy. **3** Without embellishment; plain; unadorned. **4** Free from affectation; sincere; artless; unsophisticated; also, of humble rank or condition; lowly. **5** Of weak intellect; silly; feeble-minded. **6** Not worth much consideration; insignificant; trifling; ordinary. **7** Without luxury; frugal. **8** Having nothing added; mere: the *simple* truth. **9** *Chem.* That cannot be or has not been decomposed; elementary; also, unmixed. **10** *Bot.* Not subdivided: a *simple* leaf; entire; not divided. **11** *Music* **a** Single. **b** Without overtones. **c** Not developed or elaborated: *simple* harmony. — *n.* **1** That which is simple; an unartificial, uncomplex, or natural thing; an element. **2** A medicinal plant, or the medicine extracted from it: from the former supposition that each single herb was or provided a specific for some disease. **3** A simpleton; a stupid or ignorant person; also, a person of humble position or birth. **4** *Eccl.* A feast of lowest rank which is merely commemorated at the canonical hours. **5** *pl. Colloq.* Foolishness; insanity: He suffers from the *simples.* [<OF <L *simplex, simplus*]
Synonyms (adj.): chaste, modest, natural, neat, plain, quiet, unadorned, unaffected, unembellished, unpretentious, unstudied, unvarnished. See CANDID, PURE. *Antonyms*: affected, artful, artificial, complex, complicated, elaborate, intricate, involved, ostentatious, pretentious, showy.

simple fraction See under FRACTION.

simple fruit *Bot.* A fruit consisting of a single enlarged and matured ovary, as the date, cherry, peach, apple, and quince.

sim·ple-heart·ed (sim′pəl·här′tid) *adj.* **1** Tender-hearted. **2** Ingenuous in disposition; open; sincere.

simple honors In bridge, three honors of the trump suit held by a player and his partner.

simple interest Interest computed on the original principal alone.

simple machine 1 Any one of certain elementary mechanical contrivances, as the lever, the wedge, the inclined plane, the screw, the wheel and axle, and the pulley. **2** A hand tool having no parts, as a hammer or chisel, or two parts working in simple combination, as shears.

sim·ple-mind·ed (sim′pəl·mīn′did) *adj.* **1** Artless or unsophisticated in character. **2** Defective in intellect; mentally imbecile. — **sim′ple-mind′ed·ly** *adv.* — **sim′ple-mind′ed·ness** *n.*

sim·pler (sim′plər) *n.* A collector or dispenser of herbs or medicinal remedies extracted from them; herbalist.

simple sentence See under SENTENCE.

Simple Simon A simpleton: from a character in an old English nursery rime of this name.

simple sugar A monosaccharide.

sim·ple·ton (sim′pəl·tən) *n.* A weak-minded or silly person.

sim·plex (sim′pleks) *adj.* **1** Simple. **2** Noting a form of telegraphy in which only one message is sent over a wire at a time. [<L, simple]

simplici- *combining form* Simple. Also, before vowels, **simplic-.** [<L *simplex, simplicis* simple]

sim·pli·ci·den·tate (sim′plə·si·den′tāt) *adj.* Pertaining or belonging to a suborder of rodents (*Simplicidentata*) with a single pair of upper incisors, which includes mice, squirrels, porcupines, and all others with the exception of hares and pikas.

sim·plic·i·dent (sim·plis′ə·dənt) *adj.* Simplicidentate. — *n.* A simplicidentate rodent.

sim·plic·i·ty (sim·plis′ə·tē) *n. pl.* **·ties 1** The state of being simple; freedom from admixture, ornament, formality, ostentation, subtlety, or difficulty; sincerity; unaffectedness. **2** Deficiency of intelligence or good sense, or an instance of it. See synonyms under INNOCENCE. Also **sim′ple·ness.** [<L *simplicitas, -tatis*]

sim·pli·fy (sim′plə·fī) *v.t.* **·fied, ·fy·ing** To make more simple or less complex. [<F *simplifier* <Med. L *simplificare* <L *simplex* simple + *facere* make] — **sim′pli·fi·ca′tion** *n.* — **sim′pli·fi′er** *n.*

Sim·plon Pass (sim′plon, *Fr.* saṅ·plôṅ′) A pass over the Alps in SW Switzerland; elevation, 6,592 feet; traversed by a road built (1800–06) by Napoleon near the **Simplon Tunnel** (1906), 12 1/4 miles, the longest in the world.

sim·ply (sim′plē) *adv.* **1** In a simple manner; intelligibly; without ostentation or extravagance; without subtlety or affectation; unassumingly. **2** Merely. **3** Without sense or discretion; foolishly. **4** Really; absolutely: *simply* charming: often used ironically.

Sims (simz), **William Sowden,** 1858–1936, U. S. admiral born in Canada.

Sim·son (sim′sôn) Swedish form of SAMSON.

sim·u·la·cre (sim′yə·lā′kər) *n. Obs.* An image.

sim·u·la·crum (sim′yə·lā′krəm) *n. pl.* **·cra** (-krə) **1** That which is made in the likeness of a being or thing; an image. **2** An imaginary, visionary, or shadowy semblance. **3** A sham. [<L, image < *simulare.* See SIMULATE.]

sim·u·lant (sim′yə·lənt) *adj.* Simulating. — *n.* One who or that which simulates. [<L *simulans, -antis,* ppr. of *simulare.* See SIMULATE.]

sim·u·lar (sim′yə·lər) *n.* One who simulates; a pretender. — *adj.* **1** Given to simulation; pretending. **2** Counterfeit.

sim·u·late (sim′yə·lāt) *v.t.* **·lat·ed, ·lat·ing 1** To assume or have the appearance or form of, without the reality; counterfeit; imitate. **2** To make a pretense of. See synonyms under IMITATE, PRETEND. — *adj.* (-lāt, -lit) Simulated; pretended. [<L *simulatus,* pp. of *simulare* imitate < *similis* like] — **sim′u·la′tor** *n.*

sim·u·la·tion (sim′yə·lā′shən) *n.* The act of simulating; counterfeit; sham. See synonyms under PRETENSE. — **sim′u·la′tive, sim′u·la·to′ry** (-lə·tôr′ē, -tō′rē) *adj.* — **sim′u·la′tive·ly** *adv.*

si·mul·cast (sī′məl·kast′, -käst′) *v.t.* **·cast, ·cast·ing** To broadcast by radio and television simultaneously. — *n.* A broadcast transmitted by radio and television simultaneously. [< SIMUL(TANEOUS) + (BROAD)CAST]

si·mul·ta·ne·ous (sī′məl·tā′nē·əs, sim′əl-) *adj.* Occurring, done, or existing at the same time. [<LL *simultaneus* <L *simul* at the same time] — **si′mul·ta′ne·ous·ly** *adv.* — **si′mul·ta′ne·ous·ness, si′mul·ta·ne′i·ty** (-tə·nē′ə·tē) *n.*

simultaneous equations *Math.* A series of algebraic equations such that each will satisfy the conditions of two or more variables, as, $x + y = 7$, and $2x + 3y = 19$, where $x = 2$, $y = 5$.

si·murg (si·mŏŏrg′) *n.* In Persian mythology, an immense bird possessing great knowledge, who has witnessed the destruction of the world three times; perhaps, the roc. Also **si·murgh′.** [<Persian *simurgh*]

sin[1] (sin) *n.* **1** A lack of conformity to, or a transgression, especially when deliberate, of a law, precept, or principle regarded as having divine authority. **2** The state or condition of having thus transgressed; wickedness. **3** A particular instance of such transgression. **4** Any fault or error; an offense against a standard: a literary *sin.* — *v.* **sinned, sin·ning** *v.i.* **1** To commit sin, transgress, neglect, or disregard the divine law or any requirement of right, duty, or propriety; do wrong. — *v.t.* **2** To commit or do wrongfully: to *sin* a great sin. **3** To effect, consume, drive, etc., by sin. [OE *synn*]
Synonyms (noun): crime, criminality, delinquency, depravity, evil, guilt, ill-doing, immorality, iniquity, misdeed, offense, transgression, ungodliness, unrighteousness, vice, viciousness, wickedness, wrong, wrong-doing. *Sin,* in religious teaching, is any lack of holiness, any defect of moral purity and truth, whether in heart or life, whether of commission or of omission. *Transgression,* as its etymology indicates, is the stepping over a specific enactment, whether of God or man, ordinarily by overt act, but in the broadest sense in volition or desire. *Sin* may be either act or state; *transgression* is always an act, mental or physical. *Crime* is often used for a flagrant violation of right, but in the technical sense denotes specific violation of human law. *Depravity* denotes no act, but a perverted moral condition from which any act of *sin* may proceed. *Immorality* denotes outward violation of the moral law. Compare OFFENSE. *Antonyms*: decorum, godliness, goodness, holiness, integrity, morality, purity, right, righteousness, sinlessness, uprightness, virtue. Compare synonyms for VIRTUE.

sin[2] (sin) *adv., prep., & conj. Scot. & Brit. Dial.* Since.

Si·nai (sī′nī, -nē·ī′) A peninsula between the Mediterranean and the Red Sea, constituting the easternmost part of Egypt NW of Arabia. Also **Sinaitic Peninsula.** — **Si·na·ic** (sī·nā′ik), **Si·na·it·ic** (sī′nā·it′ik) *adj.*

Si·nai (sī′nī, -nē·ī′), **Mount** The mountain where Moses received the law from God: generally identified with a mountain in the southern part of Sinai. *Ex.* xix.

sin·al·bin (sin·al′bin) *n. Chem.* A white, bitter, crystalline alkaloid, $C_{30}H_{42}O_{15}N_2S_2$, found in the seeds of the white mustard. [<L *sinapi* mustard + *albus* white]

Si·na·lo·a (sē′nä·lō′ä) A state in western Mexico, on the Gulf of California; 22,580 square miles; capital, Culiacán.

Sin·an·thro·pus (sin·an′thrə·pəs, sī′nan·thrō′·pəs) *n. Paleontol.* A large-brained, well-developed hominid primate identified from extensive fossil remains discovered between 1927 and 1939 in the Pleistocene deposits of a cave near Peking, China. Also called *Peking man.* [<NL <Gk. *Sinai* Chinese + *anthropos* man]

sin·a·pine (sin′ə·pēn, -pin) *n. Chem.* A bitter, unstable alkaloid, $C_{16}H_{25}O_6N$, contained in the seed of the black mustard. Also **sin′a·pin** (-pin). [<L *sinapi* mustard + -INE[2]]

sin·a·pism (sin′ə·piz′əm) *n.* A mustard plaster. [<L *sinapismus* <Gk. *sinapismos* < *sinapi* mustard]

Sin·ar·quist (sin′är·kwist) *n.* A member of an armed fascist group (*Unión Nacional Sinarquista*), formed about 1937, pledged to destroy liberalism and democracy in Mexico and to establish an authoritarian clerical state. [< Sp. *sinarquista* < *sinarquismo* < *sin-* (<L *sine*) without + *anarquismo* anarchism] — **Sin′ar·quism** *n.* — **Sin′ar·quis′tic** *adj.*

Sin·bad (sin′bad) See SINDBAD THE SAILOR.

since (sins) *adv.* **1** From a past time, mentioned or referred to, up to the present. **2** At some time between a certain past time or event and the present: He was willing at first, but has *since* refused. **3** In time before the present; ago; before now. — *prep.* **1** During or within the time after or later than: Things have changed *since* you left. **2** Continuously throughout the time after: He has been working *since* noon. — *conj.* **1** During or within the time after which. **2** Continuously from the time when: She has been ill *since* she arrived. **3** Because of or following upon the fact that; inasmuch as. See synonyms under BECAUSE. [ME *sithens* < OE *siththan* afterwards + -*s* (adverbial termination)]

sin·cere (sin·sir′) *adj.* **1** Being in reality as it is in appearance; real; genuine: *sincere* regret. **2** Intending precisely what one says or what one appears to intend; free from hypocrisy; honest in one's action or profession: a *sincere* friend. **3** *Obs.* Being without admixture; free; pure. **4** *Obs.* Blameless. **5** *Obs.* Sound; whole. See synonyms under CANDID, HONEST. [< L *sincerus* uncorrupted < *sin-* without + stem of *caries* decay] — **sin·cere′ly** *adv.*

sin·cer·i·ty (sin·ser′ə·tē) *n.* The state or quality of being sincere; honesty of purpose or character; freedom from hypocrisy, deceit, or simulation. See synonyms under INNOCENCE. Also **sin·cere′ness.**

sin·ci·put (sin′si·put) *n. Anat.* The top of the head, especially the anterior portion. Compare OCCIPUT. [< L < *semi-* half + *caput* head] — **sin·cip·i·tal** (sin·sip′ə·təl) *adj.*

Sin·clair (sin·klâr′), **Upton,** born 1878, U.S. author and socialist.

sind (sind) *Scot. v.t.* To rinse; wash down (food) with drink; quench. — *n.* A slight washing; a drink with or after food.

Sind (sind) A former province of West Pakistan, on the Arabian Sea; incorporated into West Pakistan province, 1955; 47,569 square miles; former capital, Hyderabad; total area of Sind and Khairpur, 56,447 square miles, divided into the Commissioners' Divisions of Hyderabad and Khairpur, 1955.

Sind·bad the Sailor (sind′bad) In the *Arabian Nights*, a traveling merchant of Baghdad, who relates the marvelous adventures that befell him on his seven voyages. Also spelled *Sinbad*.

sin·dry (sin′drē) *Scot. adj.* Sundry; several. — *adv.* Asunder.

sine[1] (sīn) *n. Trig.* **1** A function of an angle in a right triangle expressible as the ratio of the side opposite the angle to the hypotenuse. **2** A function of any acute angle expressible, when plotted in Cartesian coordinates, as the ratio of the ordinate to the distance from the point where the ordinate crosses one leg of the angle to the origin. Abbreviated *sin*, as, *sin A*. See TRIGONOMETRIC FUNCTION. — **versed sine** One of the trigonometric functions, equal to one minus the cosine: also *versine*. ♦ Homophone: *sign*. [< L *sinus* bend (trans. of Arabic *jayb* bosom of a garment, sine). Doublet of SINUS.]

SINE.
AB. Arc.
AO. Radius.
BC. Perpendicular.
BC/AO is sine of the arc AB.

si·ne[2] (sī′nē) *prep. Latin* Without.

sin eater One who takes upon himself the sins of a dead person by eating food placed upon the breast of the dead: an ancient Celtic custom closely related to the scapegoat theory.

si·ne·cure (sī′nə·kyŏor, sin′ə-) *n.* **1** An office having emoluments but few or no duties. **2** A benefice without cure of souls. [< L *sine* without + *cura* care] — **si′ne·cur·ism** *n.* — **si′ne·cur·ist** *n.*

sine curve *Math.* The plane curve of the equation $y = \sin x$. The curve has a period of $x = 2\pi$ (radians) along the abscissa and a limit along the ordinate of $y = \pm 1$.

si·ne di·e (sī′nē dī′ē) *Latin* Without a day; indefinitely: an adjournment *sine die* (that is, without setting a day for reassembling).

si·ne mo·ra (sī′nē mō′rə) *Latin* Without delay.

si·ne pro·le (sī′nē prō′lē) *Latin* Without offspring: used in genealogical tables: abbr. *s.p.*

si·ne qua non (sī′nē kwä non′) *Latin* That which is indispensable; an essential: literally, without which not.

sin·ew (sin′yōō) *n.* **1** A tendon or other fibrous cord. **2** Strength, or that which supplies strength. **3** *Obs.* A nerve. — *v.t.* To strengthen or knit together, as with sinews; supply with sinews. [OE *sinu, seonu*]

sin·ew·less (sin′yōō·lis) *adj.* **1** Without sinews. **2** Without strength or vigor.

sinews of war Money as a means of carrying on war.

sin·ew·y (sin′yōō·ē) *adj.* **1** Characteristic or consisting of a sinew or nerve. **2** Well braced with sinews; strong; brawny. See synonyms under STRONG.

sin·fo·ni·a (sin·fō′nē·ə, *Ital.* sēn′fō·nē′ä) *n. Italian* **1** A symphony. **2** The overture, in operas of early date.

sin·ful (sin′fəl) *adj.* Consisting in, suggestive of, or tainted with sin. [OE *synfull*] — **sin′ful·ly** *adv.* — **sin′ful·ness** *n.*
Synonyms (adj.): bad, criminal, depraved, evil, faulty, flagitious, immoral, iniquitous, nefarious, unholy, unrighteous, unworthy, vicious, vile, villainous, wicked, wrong. See BAD[1], CRIMINAL, IMMORAL. Compare synonyms for SIN[1]. *Antonyms:* godly, good, holy, immaculate, incorrupt, incorruptible, innocent, just, right, righteous, sinless, spotless, stainless, undefiled, unfallen, unperverted, unstained, unsullied, untainted, upright, virtuous, worthy.

sing (sing) *v.* **sang** or (*now less commonly*) **sung**, **sung**, **sing·ing** *v.i.* **1** To utter words or sounds with musical inflections of the voice. **2** To perform vocal compositions professionally or in a specified manner: She *sings* well. **3** To utter melodious sounds, as a bird. **4** To make a continuous, melodious sound suggestive of singing, as a teakettle, the wind, etc. **5** To buzz or hum; ring: My ears are *singing*. **6** To be suitable for singing. **7** To relate something in verse; hence, to compose poetry. **8** *Slang* To confess the details of a crime, and so implicate others. — *v.t.* **9** To perform (a song, etc.) vocally. **10** To chant; intone. **11** To bring to a specified condition by singing: *Sing* me to sleep. **12** To accompany or escort with songs. **13** To acclaim or relate in or as in song: Generations *sing* his deeds. — *n.* **1** The humming sound made by a bullet in flight. **2** *Colloq.* A social gathering at which songs are sung: a community *sing*. [OE *singan*] — **sing′a·ble** *adj.*
Synonyms (verb): carol, chant, chirp, chirrup, hum, warble. To *sing* is primarily and ordinarily to utter a succession of articulate musical sounds with the human voice. The word has come to include any succession of musical sounds; we say the bird or the rivulet *sings*, or the teakettle or the cricket *sings*. To *chant* is to *sing* in solemn and somewhat uniform cadence; *chant* is ordinarily applied to non-metrical religious compositions. To *carol* is to *sing* joyously, and to *warble* is to *sing* with trills or quavers, usually also with the idea of joy. *Carol* and *warble* are especially applied to the *singing* of birds. To *chirp* is to utter a brief musical sound, perhaps often repeated in the same way, as by certain small birds, insects, etc. To *chirrup* is to utter a somewhat similar sound; the word is often used of a brief sharp sound uttered as a signal to animate or rouse a horse or other animal. To *hum* is to utter murmuring sounds with somewhat monotonous musical cadence, usually with closed lips; we speak also of the *hum* of machinery, etc.

Si·ngan (sē′ngän′) A former name for SIAN (def. 1).

Sin·ga·pore (sing′gə·pôr, -pōr, sing′ə-) **1** An island (224 square miles) off the southern end of the Malay Peninsula, comprising with adjacent islands and Christmas Island the British Crown Colony of Singapore; part of the Straits Settlements, 1826–1946; 286 square miles. **2** Its capital, a port on **Singapore Strait**, a channel between Singapore island and the Malay Peninsula, connecting the South China Sea and the Strait of Malacca; 10 miles wide.

Sin·ga·ra·dja (sing′gə·rä′jə) The capital of Bali, and of Nusa Tenggara province, Indonesia, near the north coast of Bali; site of many Hindu temples. Also **Sin′ga·ra′ja.**

singe (sinj) *v.t.* **singed**, **singe·ing** **1** To burn slightly or superficially; discolor by burning; scorch: to *singe* the nap of cloth. **2** To remove bristles or feathers from by passing through flame. **3** To burn the ends of (hair, etc.). See synonyms under BURN[1]. — *n.* **1** The act of singeing, especially as performed by a barber. **2** A heat that singes. **3** An injury or risk, as if from or of singeing. [OE *sengan* scorch, hiss, causative of *singan* sing; from the singing sound produced]

sing·er[1] (sing′ər) *n.* One who sings, especially as a profession; also, a poet. See synonyms under POET.

sing·er[2] (sing′ər) *n.* One who or that which singes.

Sing·er (sing′ər), **Charles,** born 1876, English historian of science. — **Isaac Merrit,** 1811–75, U.S. inventor; first manufacturer of the sewing machine.

Sin·gha·lese (sing′gə·lēz′, -lēs′) *adj.* Of or pertaining to Ceylon, to the people constituting the majority of the inhabitants of Ceylon, or to their language. — *n.* **1** One of the Singhalese people. **2** The language of the Singhalese, belonging to the Indic branch of the Indo-Iranian languages, but containing many Dravidian words: the official language of Ceylon since 1956. Also spelled *Sinhalese*. [< Skt. *Siṅhala* Ceylon]

sin·gle (sing′gəl) *adj.* **1** Consisting of one only; separate; individual. **2** Having no companion or assistant; alone. **3** Unmarried; also, pertaining to the unmarried state. **4** Of or pertaining to one alone; hence, uncommon; singular; unique. **5** Consisting of only one part; simple; uncompounded. **6** In good condition; sound; also, upright; sincere. **7** Designed for use by only one person: a *single* bed. **8** Designed for use with one thing of which there might be more: a *single* harness (for one horse). **9** *Bot.* Solitary, as a flower when it is the only one on a stem: opposed to *clustered*; in popular usage, having only one row of petals: opposed to *double*. **10** *Obs.* Of medium strength; mild; not double or strong: said of malt liquors. **11** Simplex. See synonyms under PARTICULAR, SOLITARY. — *n.* **1** That which or one who is single; a unit; individual. **2** In baseball, a hit by which the batter reaches first base. **3** A hotel room for one person. **4** A golf match between two players only: opposed to *foursome*. **5** In cricket, a hit which scores one run. **6** In falconry, a talon. — *v.* **gled**, **gling** *v.t.* **1** To choose or select (one) from others: usually with *out*. — *v.i.* **2** To go with the single-foot gait, as a horse. **3** In baseball, to make a single. [< OF < L *singulus*] — **sin′gle·ness** *n.* — **sin′gly** *adv.*

sin·gle-act·ing (sing′gəl·ak′ting) *adj.* Doing effective work in only one direction, as a motor having a reciprocating motion.

sin·gle-ac·tion (sing′gəl·ak′shən) *adj.* Designating a type of firearm of which the trigger must be cocked by one action and released by another.

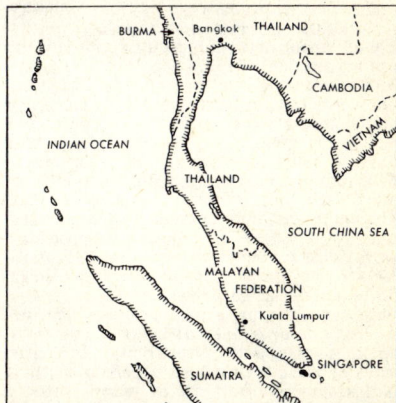

sin·gle-breast·ed (sing′gəl·bres′tid) *adj.* Having only one thickness of cloth over the breast; fastening in front with a single row of buttons, loops, or like means of engagement: said of a coat, waistcoat, etc.

sin·gle-cross (sing′gəl·krôs′, -kros′) *n. Genetics*

The first generation of a cross between two inbred lines.

single entry A method of bookkeeping in which the daybook and ledger are the essential books, transactions being carried to a single account only. — **sin′gle-en′try** *adj.*

sin·gle-foot (sing′gəl·fŏŏt′) *n.* The gait of a horse in which the footfall sequence is right hind, right fore, left hind, left fore, the support of the body being alternately upon one foot and two feet: sometimes called *amble* or *rack.* — *v.i.* To go at this gait.

sin·gle-hand·ed (sing′gəl·han′did) *adj.* 1 Without assistance; unaided. 2 Having but one hand. 3 Capable of being used with a single hand. 4 Having only one workman. Also **sin′gle-hand′.** — **sin′gle-hand′ed·ly** *adv.*

sin·gle-heart·ed (sing′gəl·här′tid) *adj.* Of sincere and frank disposition. — **sin′gle-heart′·ed·ly** *adv.*

sin·gle-mind·ed (sing′gəl·mīn′did) *adj.* 1 Having but one purpose or end in view. 2 Free from duplicity; ingenuous; sincere. — **sin′gle-mind′ed·ly** *adv.* — **sin′gle-mind′ed·ness** *n.*

sin·gle-phase (sing′gəl·fāz′) *adj. Electr.* Applied to the current generated by a two-pole alternating dynamoelectric machine.

sin·gles (sing′gəlz) *n. pl.* In lawn tennis, table tennis, etc., a match with only one person on each side: opposed to *doubles.* — *adj.* Having but one player on a side: a *singles* match.

sin·gle·stick (sing′gəl·stik′) *n.* 1 A cudgel; specifically, a basket-hilted stick used in fencing. 2 The art of fencing with singlesticks; also, a bout with cudgels.

sin·gle-stick·er (sing′gəl·stik′ər) *n. Colloq.* A one-masted sailboat; a sloop.

sin·gle-sur·faced (sing′gəl·sûr′fist) *adj.* Surfaced, covered, or finished on one side only: applied especially to fabric-covered aircraft sections.

sin·glet (sing′glit) *n.* 1 A woolen or cotton undershirt or jersey. 2 An unlined waistcoat.

sin·gle·ton (sing′gəl·tən) *n.* 1 A single card of a suit in the hand of a player at the deal. 2 Any single thing, distinct from a pair.

sin·gle·tree (sing′gəl·trē′) *n.* A swingletree.

sin·gly (sing′glē) *adv.* 1 Without companions or associates; alone; unaided, as an individual. 2 One by one; one at a time. 3 *Obs.* Uprightly; honestly.

Sing Sing (sing′ sing′) 1 A State prison near Ossining, New York. 2 The former name of OSSINING.

sing·song (sing′sông′, -song′) *n.* 1 Monotonous cadence in speaking or reading. 2 Inferior verse; doggerel. — *adj.* 1 Monotonous; droning. 2 Rising and falling in pitch, as speech.

sing·spiel (sing′spēl, *Ger.* zing′shpēl) *n.* 1 A dramatic representation in which dialog and song alternate. 2 Opera in which music is subordinated to words, especially in dramatic movement. [<G, lit., sing-play]

sin·gu·lar (sing′gyə·lər) *adj.* 1 Extraordinary; remarkable; uncommon: her *singular* beauty. 2 Odd; unconventional; peculiar; not customary or usual: to be *singular* in one's dress. 3 Representing the only one of its type; unique: a *singular* instance. 4 *Gram.* Of or designating a word form which denotes one person or thing, or a class considered as a unit, as *man, dog, he*; not dual or plural. 5 *Logic* Embodying something specific or individual; not general: a *singular* idea. See synonyms under EXTRAORDINARY, QUEER, RARE[1]. — *n. Gram.* The singular number, or a word form having this number. [<OF *singuler* <L *singularis* <*singuli* single] — **sin′gu·lar·ly** *adv.* — **sin′gu·lar·ness** *n.*

sin·gu·lar·i·ty (sing′gyə·lar′ə·tē) *n. pl.* **·ties** 1 The state or quality of being singular; uncommonness; oddity; eccentricity. 2 A character or quality by which a person or thing is distinguished from all or many others; a peculiarity. 3 Something or someone of uncommon or remarkable character.

sin·gu·lar·ize (sing′gyə·lə·rīz′) *v.t.* **·ized, ·iz·ing** To make singular; convert into the singular number.

Sin·ha·lese (sin′hə·lēz′, -lēs′) See SINGHALESE.

Sin·i·cism (sin′ə·siz′əm) *n.* Something peculiar to the Chinese, as their manners or customs. [<LL *Sinae* the Chinese]

sin·i·grin (sin′ə·grin) *n. Chem.* A white crystalline glycoside, $C_{10}H_{16}KNO_9S_2$, found principally in the seeds of the black mustard. [<NL *sinapis negra* black mustard]

Si·ning (shē′ning′) The capital of Tsinghai province, NW China.

sin·is·ter (sin′is·tər) *adj.* 1 Morally wrong; malevolent; evil; bad; perverse: *sinister* purposes; a *sinister* expression on the face. 2 Boding, tending toward, or attended with disaster; unlucky; inauspicious: from a superstition that omens seen on the left boded ill. 3 Situated on the left side or hand: opposed to *right* or *right-hand.* 4 *Her.* Of a shield, left as regards the wearer; hence, right as regards the observer: opposed to *dexter.* Compare illustration under ESCUTCHEON. [<F *sinistre* <L *sinister* left] — **sin′is·ter·ly** *adv.* — **sin′is·ter·ness** *n.*

sin·is·trad (sin′is·trad) *adv.* Toward the left aspect of the body: opposed to *dextrad.* [<L *sinister* left]

sin·is·tral (sin′is·trəl) *adj.* Of, pertaining to, or turned toward the left side or left hand. [<OF] — **sin′is·tral·ly** *adv.*

sin·is·trorse (sin′is·trôrs, sin′is·trôrs′) *adj.* 1 Sinistral. 2 Twined or twining from right to left, as the hop. Compare DEXTRORSE. [<L *sinistrorsus,* contraction of *sinistroversus* turned toward the left side <*sinister* left + *versum* turned, pp. of *vertere*] — **sin′is·tror′sal** *adj.* — **sin′is·tror′sal·ly** *adv.*

sin·is·trous (sin′is·trəs) *adj.* 1 Of, pertaining to, or directed toward the left; sinistral. 2 Sinister; unpropitious; ill-omened. [<L *sinister* left] — **sin′is·trous·ly** *adv.*

sink (singk) *v.* **sank** or **sunk, sunk** (*Obs.* **sunk·en), sink·ing** *v.i.* 1 To go beneath the surface or to the bottom, as of water or snow. 2 To descend to a lower level; go down, especially slowly or by degrees: The flames are *sinking.* 3 To descend toward or below the horizon, as the sun. 4 To incline downward; slope, as land. 5 To pass into a specified state: to *sink* into sleep or a coma. 6 To fail, as from ill-health or lack of strength; approach death: He's *sinking* fast. 7 To become less in force, volume, or degree: His voice *sank* to a whisper. 8 To become less in value, price, etc. 9 To decline in moral level, prestige, wealth, etc.: to *sink* into vice. 10 To penetrate a softer body: The oil *sank* into the wood. 11 To be impressed or fixed, as in the heart or mind: with *in*: I think that lesson will *sink* in. — *v.t.* 12 To cause to go beneath the surface or to the bottom. 13 To cause to fall or drop; lower: He *sank* his head upon his breast. 14 To force or drive into place: to *sink* a fence post. 15 To make (a mine shaft, well, etc.) by digging or excavating. 16 To reduce in force, volume, or degree. 17 To debase or degrade, as one's character or honor. 18 To suppress or hide; also, to omit. 19 To defeat; ruin. 20 To invest. 21 To invest and subsequently lose: I *sank* a million in that deal. — *n.* 1 A box-shaped, basinlike, porcelain or metal receptacle with a drainpipe and usually with a water supply; a cesspool or the like. 2 A place where corruption and vice gather or are rampant. 3 A natural pool, marsh, or basin in which a river terminates by evaporation or percolation. [OE *sincan*] — **sink′a·ble** *adj.*

sink boat A boat sunk to the rim in water: used in duck shooting: also called *surface boat.*

sink·er (singk′ər) *n.* 1 One who or that which sinks, or causes to sink: a die-*sinker.* 2 A weight for sinking a fishing or sounding line. 3 *U.S. Slang* A doughnut.

sink·hole (singk′hōl′) *n.* A natural cavity, especially a drainage cavity, as a hole worn by water through a rock along a joint or fracture.

Sin·kiang-Ui·gur Autonomous Region (shin′·jyäng′wē′gŏŏr′) The westernmost division of China, formerly, as **Sinkiang,** her largest province, comprising all of northwestern China between Mongolia and Tibet; 700,000 square miles; capital, Urumchi (Tihwa): also *Chinese Turkestan.*

sinking fund A fund so instituted and invested that its gradual accumulations will wipe out a debt at maturity.

sin·less (sin′lis) *adj.* Having no sin; guiltless; innocent. See synonyms under INNOCENT, PERFECT. — **sin′less·ly** *adv.* — **sin′less·ness** *n.*

sin·ner (sin′ər) *n.* 1 One who has sinned. 2 An irreligious person.

Sinn Fein (shin fān) Literally, we ourselves; an Irish society aiming at both independence and the cultural development of the Irish people. It originated about 1905 and in 1916 became active politically, advocating republicanism and causing a revolt in the spring of that year. — **Sinn Fein′er** — **Sinn Fein′ism**

Sino- *combining form* Chinese; of or pertaining to the Chinese people, language, etc. See CHINO-. [<LL *Sinae* the Chinese]

sin offering An offering made in atonement for sin.

Sin·o·log (sin′ə·lôg, -log, sī′nə-) *n.* One who studies or is versed in Sinology. Also **Sin′o·logue.** [<SINO- + Gk. *logos* discourse]

Si·nol·o·gy (sī·nol′ə·jē, si-) *n.* The systematic study or investigation of the Chinese people, language, literature, history, and characteristics. — **Sin·o·log·i·cal** (sin′ə·loj′i·kəl, sī′nə-) *adj.* — **Si·nol′o·gist** *n.*

Si·non (sī′nən) In the *Iliad,* the Greek who induced the Trojans to drag the wooden horse into Troy.

Si·no·phile (sī′nə·fīl, sin′ə-) *n.* An admirer of the Chinese. — *adj.* Having admiration for the Chinese or things Chinese.

Si·no·phobe (sī′nə·fōb, sin′ə-) One antipathetic toward the Chinese. — *adj.* Having hostility toward the Chinese or Chinese customs.

Si·no-Ti·bet·an (sī′nō·ti·bet′n) *n.* A family of tone languages spoken over a wide area in central and SE Asia, including the languages of China, Burma, Tibet, and Indochina: characterized by monosyllabic, uninflected wordforms which indicate their syntactic roles by position only. Also called *Indochinese.*

Sin·siang (shin′shyäng′) A city on the Wei in northern Honan province, east central China; from 1949 to 1952, capital of the former province of Pingyuan, NE central China.

sin·syne (sin′sīn) *adv. Scot.* Since; ago.

sin·ter (sin′tər) *n.* 1 Calcareous or siliceous material deposited by springs. 2 That which is produced by sintering. — *v.t.* & *v.i. Metall.* To bring about the cohesion of (metal particles) by the combined action of heat and pressure. [<G, dross of iron]

Sint Eu·sta·ti·us (sint ŏŏ·stä′tē·ŏŏs) See under ST. EUSTATIUS.

Sint Maar·ten (sint mär′tən) See under ST. MARTIN.

Sin·tra (sēn′trə) The Portuguese spelling of CINTRA.

sin·u·ate (sin′yŏŏ·it, -āt) *adj.* 1 Winding in and out, as a margin; tortuous; sinuous; wavy. 2 *Bot.* Having a sinus, or sinuses: a *sinuate* leaf. Also **sin′u·at′ed.** — *v.i.* (sin′yŏŏ·āt) **·at·ed, ·at·ing** To curve in and out; turn; wind. [<L *sinuatus,* pp. of *sinuare* turn, wind <*sinus* curve] — **sin′u·ate·ly** *adv.* — **sin′u·a′tion** *n.*

Sin·ui·ju (shin·ē·jŏŏ) A city of NW North Korea, on the Yalu; heavily bombed in the Korean war, 1950.

sin·u·os·i·ty (sin′yŏŏ·os′ə·tē) *n.* 1 Sinuous quality. 2 A winding; deflection.

sin·u·ous (sin′yŏŏ·əs) *adj.* 1 Characterized by bends or folds; winding; undulating. 2 *Bot.* Sinuate. 3 Devious; erring. [<L *sinuosus* <*sinus* bend] — **sin′u·ous·ly** *adv.* — **sin′u·ous·ness** *n.*

si·nus (sī′nəs) *n.* 1 A recess formed by a bending or folding; an opening or cavity. 2 *Anat.* **a** An air cavity in one of the cranial bones communicating with the nostrils: the frontal *sinus.* **b** A channel or receptacle for venous blood; also, a dilated part of a blood vessel. 3 *Pathol.* Any narrow opening leading to an abscess. 4 *Bot.* A recess or rounded curve between two projecting lobes or teeth of a leaf. [<L. Doublet of SINE.]

si·nu·si·tis (sī′nə·sī′tis) *n. Pathol.* Inflammation of a sinus. Also **sin·u·i·tis** (sin′yŏŏ·ī′tis). [<SINUS + -ITIS]

si·nu·sot·o·my (sī′nə·sot′ə·mē) *n. Surg.* Incision of a sinus.

-sion Var of -TION.

Si·on (sī′ən) See ZION.

Sion (syôṅ) The capital of Valais canton, SW Switzerland: German *Sitten.*

Siou·an (sŏŏ′ən) *n.* A large linguistic stock of North American Indians, formerly ranging from the west banks of the Mississippi to the Rocky Mountains, and comprising the languages of the Dakota or Sioux tribes proper, those of a group including the Omaha, Osage, etc., those of the Iowa, Missouri, and Oto, also Winnebago, Mandan, Crow, Catawba, and several others now extinct. —*adj.* Of or pertaining to this linguistic stock.

Sioux (so͞o) *n. pl.* **Sioux** One of a group of North American Indian tribes of Siouan linguistic stock formerly occupying the Dakotas and parts of Minnesota and Nebraska. They were Plains Indians and called themselves Dakota.

Sioux City A city on the Missouri River in western Iowa.

Sioux Falls A city in SE South Dakota.

sip (sip) *v.* **sipped, sip·ping** *v.t.* **1** To imbibe in small quantities. **2** To drink from by sips. **3** To take in; absorb. —*v.i.* **4** To drink in sips. —*n.* **1** A very small draft; a mere taste. **2** The act of sipping. [OE *sypian* drink in]

sipe (sīp) *v.i. Scot.* To seep.

si·phon (sī′fən) *n.* **1** A tube having a bend used for transferring liquids from a higher to a lower level over an intervening elevation by making use of atmospheric pressure. **2** A siphon bottle. **3** *Zool.* A tubular structure in certain aquatic animals, as the squid, for drawing in or expelling liquids. —*v.t.* To draw off by or cause to pass through or as through a siphon. —*v.i.* To pass through a siphon. Also spelled *syphon.* [<F <L *sipho, -onis* <Gk. *siphōn*] —**si′phon·al** *adj.*

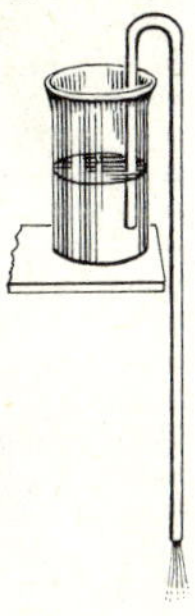

SIPHON

si·phon·age (sī′fən·ij) *n.* The use or action of a siphon.

si·pho·nap·ter·ous (sī′fə·nap′tər·əs) *adj.* Of or pertaining to the order of insects (*Siphonaptera*) including the fleas: small, flattened, wingless, bloodsucking insects with great jumping ability. [<NL *Siphonaptera,* name of the order <SIPHON + A-⁴ + -PTEROUS]

siphon bottle A bottle containing aerated or carbonated water, which is expelled through a bent tube in the neck of the bottle by the pressure of the gas.

si·pho·no·phore (sī′fə·nə·fôr′, -fōr′, sī·fon′ə-) *n.* A marine organism (order *Siphonophora*) with free-swimming pelagic colonies arising by budding, as the Portuguese man-of-war. [<NL *Siphonophora,* name of the order <Gk. *siphōnophoros* tube-carrying <*siphōn* tube + *pherein* bear]

si·pho·no·stele (sī′fə·nə·stēl′, -stē′lē) *n. Bot.* The hollow tubular stem of certain plants, as ferns. [<Gk. *siphōn* tube + STELE]

sip·pet (sip′it) *n.* **1** A triangular or finger-shaped piece of toasted or fried bread used to garnish a dish of hash or minced meat; a crouton. **2** Any eatable, especially bread, cut into small pieces and soaked in some liquid: frequently used in the plural. **3** Hence, any very small quantity. [Blend of SIP and SOP in a dim. form]

Si·quei·ros (sē·kā′rōs), **David Alfaro,** born 1898, Mexican painter and muralist.

sir (sûr) *n.* **1** The conventional term of respectful address to men: used absolutely, and not followed by a proper name. **2** A title given to persons of rank or to officials: *sir* herald, *sir* clerk. **3** An influential or important person. **4** *Archaic* A title of respect for a priest. [<SIRE]

Sir (sûr) *n.* A title of baronets and knights, used before the Christian name or the full name.

Sir·bo·ni·an (sər·bō′nē·ən) See SERBONIAN.

sir·dar (sər·där′) *n.* **1** In India and Oriental countries, a chief or lord. **2** In Egypt, the commander in chief of the army. **3** In India, a head servant; a leader of palanquin-bearers; also, a body-servant or valet: also **sir·dar′-bear′er.** [<Hind. *sardār* leader <*sar* head + *dār* holding] —**sir·dar′ship** *n.*

sire (sīr) *n.* **1** A father; begetter: used also in composition: *grandsire.* The feminine correlative is *dame.* **2** The male parent of a mammal, the female parent (of lower animals) being usually termed the *dam.* **3** A form of address to a superior: now used only in addressing a king or other sovereign. **4** *Obs.* A master; lord; also, a gentleman. —*v.t.* **sired, sir·ing** To beget; procreate: now used chiefly of domestic animals. [<OF <L *senior* older. See SENIOR.]

si·ren (sī′rən) *n.* **1** One of two, or, in later Greek legend, three nymphs, living on an island, who lured sailors to destruction by their sweet singing. They are represented as birds with women's heads, later as women with birds' feet and wings. Odysseus escaped them by sealing his companions' ears with wax and having himself bound to his ship's mast. **2** Hence, a fascinating, dangerous woman; also, a sweet singer. **3** An eel-like amphibian (genus *Siren*) having well developed gills and lacking hind legs; a mud eel. **4** An apparatus having a device with a perforated rotating disk or disks through which sharp puffs of steam or compressed air are permitted to escape in such rapid succession as to produce a continued musical note or a loud whistle: used in acoustical investigations and as a warning signal. —*adj.* Of or pertaining to a siren; hence, alluring; bewitching; dangerously fascinating. Also spelled *syren.* [<L <Gk. *seirēn*]

si·re·ni·an (sī·rē′nē·ən) *n.* One of an order (*Sirenia*) of aquatic mammals, including the manatee, dugong, etc., of somewhat fishlike form with the lower jaw as in ordinary mammals, and having mostly molariform teeth for a herbivorous diet. —*adj.* Of or pertaining to the *Sirenia.* [<NL *Sirenia,* name of the order <L *siren.* See SIREN.]

Si·ret (sē·ret′) A river in eastern Rumania, flowing 270 miles SE to the Danube: also *Sereth.*

si·ri·a·sis (si·rī′ə·sis) *n. Pathol.* Sunstroke or thermic fever. [<Gk. *seiriasis* <*seirios* hot, scorching]

Sir·i·us (sir′ē·əs) The Dog Star, Alpha in the constellation of Canis Major; magnitude, −1.6. [<Gk. *seirios* hot, scorching]

sir·loin (sûr′loin) *n.* A loin of beef, especially the upper portion: also spelled *surloin.* [Alter. (after *Sir,* from a legend that the cut was knighted for its excellence) of obs. *surloyn* <OF *surlonge* <*sur-* over, above + *longe* loin <L *lumbus*]

si·roc·co (si·rok′ō) *n. pl.* **·cos** **1** A hot, dry, and dusty southerly wind blowing from the African coast to Italy, Sicily, and Spain. **2** A warm, sultry wind blowing from a warm region toward a center of low barometric pressure. **3** A southeast wind: the popular Italian name. [<Ital. *scirocco* <Arabic *sharq* the east, the rising sun <*sharaqa* rise]

Si·ros (sī′ros, *Greek* sē′rôs) See SYROS.

sir·rah (sir′ə) *n. Archaic* Fellow; sir: used in contempt or annoyance. [Var. of *sir,* after Provençal *sira*]

Sir·rah (sir′ə) See ALPHERATZ.

sir-rev·er·ence (sûr·rev′ər·əns) *interj. Obs.* Save your reverence; begging your pardon: used as an apology before any unbecoming expression. [Misspelling of *sa' reverence,* contraction of *save (your) reverence,* erroneous trans. of L *salva reverentia* with due regard for decency]

sir·up (sir′əp) *n.* A thick, sweet liquid, as the boiled juice of fruits, sugarcane, etc.: also spelled *syrup.* [<OF *sirop* <Turkish *sharbat.* Doublet of SHERBET.] —**sir′up·y** *adj.*

sir·vente (sēr·vänt′) *n.* A lyric in Provençal troubadour literature, characterized by great formal elaboration and a satirical treatment of themes of a political and courtly nature. [<F <Provençal <L *servens, -entis* ppr. of *servire* serve]

sis (sis) *n. Colloq.* Sister.

si·sal (sī′səl, sis′əl, sē′səl) *n.* **1** The strong, tough fiber obtained from the leaves of a West Indian agave (*Agave sisalina*). **2** Henequen. Also **sisal grass, sisal hemp.** [from *Sisal,* town in Yucatán, Mexico]

sis·co·wet (sis′kə·wet) *n.* **1** The namaycush. **2** The cisco. Also **sis′ka·wet, sis′ki·wit.** [<F (Canadian) *ciscoette* <Algonquian (Ojibwa) *pemitewiskawet* oily-fleshed creature]

Sis·er·a (sis′ər·ə) A Canaanite chieftain defeated by the Israelites; murdered by Jael. *Judges* iv 2; v 20, 26.

sis·er·ar·y (sis′ə·rer′ē) *n. Brit. Dial.* **1** An effective proceeding. **2** A writ of certiorari. —**with a siserary** With a vengeance; like a thunderclap. [Alter. of CERTIORARI]

sis·kin (sis′kin) *n.* A finch (genus *Spinus*) related to the goldfinch, as the **European siskin** (*S. spinus*) olive-green and yellow barred with black, or the North American **pine siskin** (*S. pinus*). [<MDu. *cijsken* <LG *zieske* <Polish *czyżik,* dim. of *czyż* finch]

Sis·ley (sēs·lē′), **Alfred,** 1830–99, French painter of English descent.

Sis·mon·di (sēs·môṅ·dē′), **Jean Charles Léonard Simonde de,** 1773–1842, Swiss historian and economist.

siss (sis) *v.i.* To hiss; sizzle. —*n.* A hissing or sizzling sound. [Imit.]

sis·si·fied (sis′i·fīd) *adj. U.S. Colloq.* Like a sissy; effeminate.

sis·sy (sis′ē) *n. pl.* **·sies** *U.S. Colloq.* An effeminate man or boy; a milksop; a weakling, male or female. [<SIS] —**sis′sy·ish** *adj.*

Si·stan (se·stän′) See SEISTAN.

sis·ter (sis′tər) *n.* **1** A female person or animal having the same parent or parents as another person or animal. Daughters of the same parents are **full** or **whole sisters,** called in law **sisters german.** Those having only one parent in common are **half-sisters.** **2** A woman or girl allied to another or others by some association: *sisters* in spirit: also used figuratively: Astronomy and astrology are *sisters.* **3** *Eccl.* A member of a sisterhood; a nun. **4** A head nurse in the ward of a hospital; also, popularly, any nurse. —**the three** (or **Fatal**) **Sisters** The Fates. —*adj.* Bearing the relationship of a sister or one suggestive of sisterhood; sisterly. [<ON *systir*] —**sis′ter·ly** *adj.*

sis·ter·hood (sis′tər·ho͝od) *n.* **1** A body of sisters united by some bond of fellowship or sympathy. **2** *Eccl.* **a** A community of women bound by monastic vows. **b** An association of women set apart for works of mercy and faith, sometimes bound by a revocable vow. **3** The sisterly relationship.

sis·ter-in-law (sis′tər·in·lô′) *n. pl.* **sis·ters-in-law** A sister by marriage: a sister of one's husband, a sister of one's wife, a brother's wife, or, loosely, a brother-in-law's wife.

Sis·tine (sis′tēn, -tin) *adj.* Belonging or relating to one of the five popes named Sixtus (Italian *Sisto*), particularly to Sixtus IV and Sixtus V. [<Ital. *Sistino* <*Sisto* <L *sextus* sixth]

Sistine Chapel The principal chapel in the Vatican Palace at Rome, constructed by Sixtus IV, and afterward decorated with frescos by Michelangelo and others.

Sistine choir Formerly, a select choir of thirty-two cultivated voices attached to the court of the pope.

Sistine Madonna The Madonna painted by Raphael for the Church of St. Sixtus (Italian *di San Sisto*) in 1515.

sis·troid (sis′troid) *adj. Geom.* Included by the convex sides of two intersecting curves: said of an angle and opposed to *cissoid.* [<SISTR(UM) + -OID]

sis·trum (sis′trəm) *n. pl.* **·tra** (-trə) or **·trums** A musical rattle used in the worship of Isis in ancient Egypt. [<L <Gk. *seistron* <*seiein* shake]

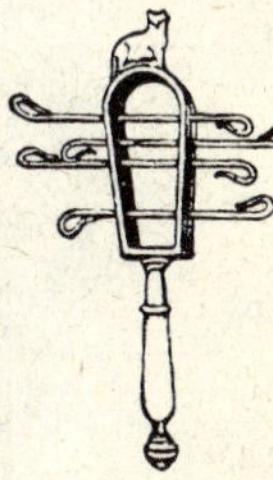

SISTRUM

Sis·y·phe·an (sis′ə·fē′ən) *adj.* **1** Of or pertaining to Sisyphus. **2** Difficult and interminable: a *Sisyphean* task.

Sis·y·phus (sis′ə·fəs) In Greek mythology, a crafty, greedy king of Corinth, condemned in Hades forever to roll uphill a huge stone that always rolled down again.

sit (sit) *v.* **sat** (*Archaic* **sate**), **sat, sit·ting** *v.i.* **1** To rest, as upon a chair, with the body bent at the hips, and the spine nearly vertical; rest upon the haunches; take or occupy a seat. **2** To perch or roost, as a bird; brood; also, to cover eggs so as to give warmth for hatching. **3** To be or remain in a seated or settled position. **4** To remain passive or inactive, or in a position of idleness or rest. **5** To assume an attitude of readiness; take a position for a special purpose; pose, as for a portrait. **6** To meet in assembly for deliberation or business; hold a session. **7** To occupy or be entitled to a seat in a deliberative body. **8** To have or exercise judicial authority. **9** To fit or be adjusted; suit: That dress *sits* well. **10** To be suffered or borne, as a burden. **11** To be situated or located; be in some position or direction: The wind *sits* in the east. —*v.t.*

add, āce, câre, pälm; end, ēven; it, īce; odd, ōpen, ôrder; to͝ok, po͞ol; up, bûrn; ə = a in *above,* e in *sicken,* i in *clarity,* o in *melon,* u in *focus;* yo͞o = u in *fuse;* oi, oil; ou, pout; ch, check; g, go; ng, ring; th, thin; *th*, this; zh, vision. Foreign sounds à, œ, ü, kh, ṅ; and ◆: see page xx. < from; + plus; ? possibly.

12 To have or keep a seat or a good seat upon: to *sit* a horse. 13 To seat (oneself): *Sit* yourself down. —**to sit in (on)** To join: to *sit in on* a game of cards, or a business deal. —**to sit out** 1 To sit quietly till the end of: to *sit out* an entertainment. 2 To sit aside during: They *sat out* a dance. 3 To stay longer than. —**to sit tight** *Colloq.* To wait quietly for the next move on the part of somebody else. [OE *sittan*]

si·tar (si·tär′) *n.* A modern East Indian stringed instrument, resembling a guitar, with four steel and three brass strings, plucked with a plectrum. [<Hind. *sitār*]

sit-down strike (sit′doun′) A strike during which strikers refuse to leave the factory or other place of employment until agreement is reached. Also **sit′-down′**.

site (sīt) *n.* 1 Situation; local position. 2 A plot of ground set apart for some specific use. 3 The degree of inclination from the horizontal of a line joining the target and the muzzle of a gun: also **angle of site.** See synonyms under PLACE. ◆ Homophones: *cite, sight.* [<F <L *situs* position]

sit·fast (sit′fast′, -fäst′) *n.* An induration on a horse arising from uneven pressure of a saddle or load and tending to ulceration.

sith (sith) *adv., prep., & conj. Archaic* Since. Also **sith·ence** (sith′əns). [OE *siththan* after]

Sit·ka (sit′kə) A town and naval base on Baranof Island in SE Alaska.

sito- *combining form* Food; related to food: *sitotropism.* [<Gk. *sitos* food]

si·tol·o·gy (sī·tol′ə·jē) *n.* The science of foods, diet, and nutrition. —**si·to·log·ic** (sī′tə·loj′ik) or **·i·cal** *adj.*

si·to·ma·ni·a (sī′tō·mā′nē·ə, -mān′yə) *n.* An abnormal craving for food.

si·to·pho·bi·a (sī′tō·fō′bē·ə) *n.* A morbid fear of food.

si·tos·ter·ol (sī·tos′tə·rōl, -rol) *n. Biochem.* Any of a group of sterols found in higher plants and related to cholesterol, especially *a-sitosterol,* $C_{29}H_{50}O$, from wheat embryos. [<SITO- + STEROL]

si·to·ther·a·py (sī′tō·ther′ə·pē) *n.* Medical treatment by means of food and dieting.

si·to·tox·in (sī′tō·tok′sin) *n.* Any poison evolved in vegetable foods, especially in cereals, by the action of micro-organisms. [<SITO- + TOXIN]

si·to·tox·ism (sī′tō·tok′siz′əm) *n.* Food poisoning, especially that due to infected vegetables and cereals.

si·to·tro·pism (sī′tō·trō′piz·əm) *n.* The automatic response of an organism to the positive or negative influence of food. [<SITO- + TROPISM]

Si·tsang (sē′tsäng′) The Chinese name for TIBET.

Sit·tang (sit′täng′) A river in south central Burma, flowing 350 miles south to the Andaman Sea at the head of the Gulf of Martaban.

Sit·ten (zit′n) The German name for SION.

sit·ter (sit′ər) *n.* 1 One who sits. 2 A person hired to care for children while the parents are out: also *baby-sitter.* 3 A person sitting as a model. 4 A setting hen.

Sit·ter (sit′ər), **Willem de,** 1872–1934, Dutch astronomer.

sit·ting (sit′ing) *adj.* Being in the position of a sitter; also, used for sitting: a *sitting*-room. —*n.* 1 The act or position of one who sits; hence, a seat; also, the place of or the right to a seat, as in a church. 2 A single period of uninterrupted application, as for the painting of a portrait. 3 A session or term. 4 An incubation; period of hatching; also, the number of eggs on which a bird sits at one incubation.

Sitting Bull, 1834?–90, Sioux Indian chief; defeated Custer at the battle of Little Big Horn, 1876.

sitting duck 1 A duck resting on water, and therefore an easy target for a hunter. 2 *Colloq.* Any easy target.

sit·ting-room (sit′ing·ro͞om′, -ro͝om′) *n.* A parlor; living-room.

sit·u·ate (sich′o͞o·āt) *v.t.* **·at·ed, ·at·ing** 1 To fix a site for. 2 To place in a certain position or under certain conditions or circumstances; locate. —*adj.* (sich′o͞o·it, -āt) *Archaic* Situated. [<Med. L *situatus,* pp. of *situare* place <*situs* a place]

sit·u·at·ed (sich′o͞o·ā′tid) *adj.* 1 Having a fixed place or location; placed. 2 Placed in (usually specified) circumstances or conditions: He is *well* situated.

sit·u·a·tion (sich′o͞o·ā′shən) *n.* 1 The place in which something is situated; relative local position; locality. 2 Condition as modified or determined by surroundings; status. 3 A salaried post of employment, usually subordinate. 4 A combination of circumstances; complication; specifically, in the drama, a conjuncture, climax, or crisis. See synonyms under CIRCUMSTANCE, PLACE, SCENE. —**sit′u·a′tion·al** *adj.*

si·tus (sī′təs) *n.* 1 Site; situation; place. 2 A fitting or natural position, as of a part of a plant. [<L]

Sit·well (sit′wel) Name of an English literary family, including **Edith,** born 1887, and her brothers, **Osbert,** born 1892, and **Sacheverell,** born 1897, all three poets, essayists, biographers, and critics.

sitz bath (zits) 1 A small bathtub in which one bathes in a sitting posture. 2 A bath taken in such a tub. [<G *sitzbad*]

Si·va (sē′və, shē′-) The Hindu god of destruction and reproduction: forming with Brahma and Vishnu the Hindu trinity: also called *Shiva.* [<Hind. *Shiva* <Skt. *śivás* propitious]

Si·va·ism (sē′və·iz′əm, shē′-) *n.* The worship of Siva. —**Si′va·ist** *n.* —**Si′va·is′tic** *adj.*

Si·va·ji (sē·vä′jē), 1627–80, founder of the Mahratta power in India.

Si·van (sē·vän′) A Jewish month. See under CALENDAR (Hebrew). Also **Si·wan′.**

Siv·a·pi·the·cus (siv′ə·pi·thē′kəs) *n. Paleontol.* An extinct ape related to Dryopithecus. [<NL <SIVA + Gk. *pithēkos* ape]

Si·vas (sē·väs′) A city in central Turkey in Asia.

si·ver (sī′vər) *n. Scot.* An open drain; sewer.

Si·wa·lik Range (si·wä′lik) The southernmost range of the Himalayas in south central Asia; highest peak about 5,000 feet.

Si·wash (sī′wosh) *n.* 1 An Indian of the northern Pacific coast. 2 The lingua franca used between the Siwashes and white traders. [<Chinook jargon <F *sauvage* savage]

six (siks) *n.* The cardinal number following five and preceding seven, or any of the symbols (6, vi, VI) used to represent it; also, anything made up of six units or members, as a playing card with six pips. —*adj.* Being one more than five; twice three. [OE] —**six′fold′** *adj. & adv.*

six bits Seventy-five cents.

Six Nations The Iroquois confederation known as the Five Nations, plus the Tuscarora, who joined them in the 18th century. Also **Six Allied Nations.**

six·pence (siks′pəns) *n.* A British silver coin of the value of six English pennies; the half of a shilling.

six·pen·ny (siks′pen′ē, -pən·ē) *adj.* 1 Worth, valued at, or sold for sixpence; hence, paltry; trashy. 2 Denoting a size of nails. See -PENNY.

six·score (siks′skôr′, -skōr′) *adj.* One hundred and twenty.

six-shoot·er (siks′sho͞o′tər) *n. Colloq.* A revolver that will fire six shots without reloading. —**six′-shoot′ing** *adj.*

sixte (sikst) *n.* In fencing, a parry in which the hand is opposite the right breast and the foil is carried to the right. [<F <L *sextus* sixth]

six·teen (siks′tēn′) *n.* The cardinal number following fifteen and preceding seventeen, or any of the symbols (16, xvi, XVI) representing it. —*adj.* Being one more than fifteen; four times four. [OE *sixtēne*]

six·teen·mo (siks·tēn′mō) *n. pl.* **·mos** 1 The page size of a book or pamphlet made up of printer's sheets folded 16 leaves to the sheet, the pages being usually 4 1/2 × 6 7/8 inches. 2 A book having pages of this size. Often written *16mo.* Also called *sextodecimo.* —*adj.* Consisting of pages of this size.

six·teenth (siks′tēnth′) *adj.* 1 Sixth in order after the tenth: the ordinal of *sixteen.* 2 Being one of sixteen equal parts. —*n.* 1 One of sixteen equal parts of anything; the quotient of a unit divided by sixteen. 2 *Music* A sixteenth note.

sixteenth note *Music* A note of one sixteenth of the value of a whole note; semiquaver.

sixteenth section The section of a township, numbered 16, reserved for school purposes; a school section.

sixth (siksth) *adj.* 1 Next in order after the fifth. 2 Being one of six equal parts. —*n.* 1 One of six equal parts. 2 *Music* **a** The interval between any note and the sixth note above or below it on the diatonic scale. **b** A note separated by this interval from any other, considered with reference to that other. **c** The sixth above the keynote. **d** Two notes at this interval written or sounded together, or the resulting consonance. —**chord of the sixth** *Music* A chord consisting of a tone with its minor third and its sixth: also **sixth chord.** —*adv.* In the sixth order, place, or rank: also, in formal discourse, **sixth′ly.** [OE *sixta*; refashioned to conform to *fourth*]

sixth sense Intuitive perception supposedly not employing the five senses.

six·ti·eth (siks′tē·ith) *adj.* 1 Tenth in order after fiftieth: the ordinal of *sixty.* 2 Being one of sixty equal parts. —*n.* One of sixty equal parts of anything; the quotient of a unit divided by sixty.

Six·tine (siks′tēn, -tin) See SISTINE.

Six·tus (siks′təs) Appellation of five popes including:

—**Sixtus IV,** 1414–84, pope 1471–84, real name Francesco della Rovere; built the Sistine Chapel.

—**Sixtus V,** 1521–90, pope 1585–90, real name Felice Peretti; built the Lateran Palace.

six·ty (siks′tē) *n. pl.* **·ties** The cardinal number following fifty-nine and preceding sixty-one, or any of the symbols (60, lx, LX) representing it. —*adj.* Being one more than fifty-nine; ten times six. [OE *sixtig*]

siz·a·ble (sī′zə·bəl) *adj.* Of comparatively large or convenient size. Also **size′a·ble.** —**siz′a·ble·ness** *n.* —**siz′a·bly** *adv.*

siz·ar (sī′zər) *n.* At Cambridge University, England, and Trinity College, Dublin, a student allowed free commons, etc.: formerly required to perform menial services. [<SIZE[1] (def. 4)] —**siz′ar·ship** *n.*

size[1] (sīz) *n.* 1 Measurement or extent of a thing as compared with some standard; comparative magnitude or bulk: when unqualified, implying relative largeness. 2 One of a series of graded measures, or the magnitude between two such limits, as of hats, shoes, etc. 3 A standard of measurement; specified quantity. 4 At Cambridge University, an allotted quantity of provisions; ration. 5 Mental caliber; importance; character. 6 *Colloq.* State of affairs; true situation: That's the *size* of it. 7 Measure or amount. See synonyms under MAGNITUDE. —*v.t.* **sized, siz·ing** 1 To estimate the size of. 2 To distribute or classify according to size. 3 To cut or otherwise shape (an article) to the required size. —**to size up** *Colloq.* 1 To form an estimate, judgment, or opinion of. 2 To meet specifications. [<F *assise.* See ASSIZE.]

size[2] (sīz) *n.* 1 A solution of gelatinous material, usually glue, casein, wax, or clay, used to finish fabrics. 2 A gelatinous substance used to glaze paper, or applied to walls before papering, etc. 3 A viscous preparation used as in fixing gilding. Also *sizing.* —*v.t.* **sized, siz·ing** 1 To treat with size or any size-like substance. 2 To make plastic, as clay. [<OItal. *sisa* painter's glue, aphetic var. of *assisa,* orig., pp. of *assidere* make sit down <L *assidere.* See ASSIZE.]

sized (sīzd) *adj.* 1 Having graded dimensions or a definite size: chiefly in composition: large-*sized.* 2 Arranged according to size.

siz·ing (sī′zing) *n.* 1 Size[2], *n.* 2 The process of adding size to a fabric, yarn, etc., to give it additional strength, stiffness, smoothness, weight, etc.

siz·y (sī′zē) *adj.* Glutinous. [<SIZE[2]]

sizz (siz) *v.i.* To make a hissing sound; sizzle. [Imit.]

siz·zle (siz′əl) *v.i.* **·zled, ·zling** To burn or scorch with or as with a hissing sound; emit a hissing sound under the action of heat. —*n.* A hissing sound as from frying or effervescence. [Freq. of SIZZ]

siz·zler (siz′lər) *n. Colloq.* Anything extremely hot, especially a summer day. [<SIZZLE]

siz·zling (siz′ling) *adj.* 1 Extremely hot. 2 That sizzles: a *sizzling* steak.

Sjael·land (shel′län) The Danish name for ZEELAND.

sjam·bok (sham′bok) *n.* A short, heavy whip of rhinoceros hide. [<Afrikaans <Du. <Malay *chamboq* <Persian *chābuq* whip]

Ska·gen (skä′gən), **Cape** The northernmost point of Jutland, Denmark, at the junction of the Skagerrak and the Kattegat: also *The Skaw.*

Skag·er·rak (skag′ə·rak, *Norw.* skäg′ûr·räk) An arm of the North Sea between Jutland and Norway; 150 miles long, 80 to 90 miles wide.

Skag·way (skag′wā) A city of SE Alaska near Chilkoot Pass: gateway to the Yukon and Klondike gold fields in the 1890's.

skail (skāl) *v.t. Brit. Dial.* **1** To scatter; spill. **2** To separate; disperse, as the members of an assembly. [Cf. ON *skilja* divide, part]

skaith (skāth) *v.t.* & *n. Scot.* Scathe; damage.

skald (skôld, skäld) See SCALD[2].

skat (skät) *n.* **1** A three-handed game played with 32 cards. Any of several varieties of the game can be chosen by the highest bidder who, alone, must oppose the other players. **2** In the game of skat, two cards dealt face down and taken into his hand by the successful bidder or otherwise treated according to rule. [<G, orig. *skart* <Ital. *scartare* discard]

skate[1] (skāt) *n.* **1** A keel-shaped metal runner attached to a plate or frame, with suitable clamps or straps for fastening it to the sole of a boot or shoe, enabling the wearer to glide rapidly over ice; also, such a runner affixed to a shoe or boot. **2** A similar contrivance with wheels instead of a runner, for use on a floor or other smooth surface; a roller skate. **3** An ice-boat runner. —*v.i.* **skat·ed, skat·ing** To glide or move over ice or some other smooth surface, on or as on skates. [< earlier *skates* <Du. *schaats* <OF *escache* stilt <Gmc.]

skate[2] (skāt) *n.* Any of several flat-bodied rays (genus *Raia*) with enlarged pectoral fins, ventral gill slits, and a pointed snout; especially, the **barn-door skate** *(R. laevis)* of eastern North America, or the common European **gray skate** *(R. batis).* [<ON *skata*]

skate[3] (skāt) *n. Slang* **1** A miserable, contemptible person. **2** An old, worn-out horse. [Origin uncertain]

skat·er (skā′tər) *n.* **1** One who skates. **2** One of various insects with long legs that run over the surface of the water, as if skating: also called *water strider.*

skat·ole (skat′ōl) *n. Biochem.* A white crystalline compound, C_9H_9N, contained in the feces and urine, and formed in the alimentary canal by the decomposition of proteins. Also **skat′·ol.** [<Gk. *skōr, skatos* dung + -OLE]

Skaw (skô), **The** See SKAGEN, CAPE.

skean (shkēn, skēn) *n.* An early Irish double-edged dagger or short sword. Also **skeen.** [<Irish *sgian* knife]

Skeat (skēt), **Walter William,** 1835–1912, English lexicographer and philologist.

ske·dad·dle (ski·dad′l) *Colloq. v.i.* **·dled, ·dling** To flee in haste; run away; scamper. —*n.* The act of running away; hasty flight. [< dial. E, spill, scatter. Cf. Gk. *skedannynai* scatter.]

skee (skē) See SKI.

Skee·na River (skē′nə) A river in western British Columbia, Canada, flowing 360 miles south and SW to the Pacific.

skeet[1] (skēt) *n.* A variety of trapshooting in which a succession of saucer-shaped targets hurled in such a way as to resemble the flight of quail are fired at from various angles by the shooter. [Origin unknown]

skeet[2] (skēt) *n.* A long-handled scoop or dipper, used for wetting sails and decks. [Origin unknown]

skeet gun A short-barreled shotgun.

skeg (skeg) *n. Naut.* The after part of a vessel's keel, or a projection on or continuation of it, as for supporting the lower end of the rudder of a screw steamer. [<Du. *schegge,* prob. <Scand. Cf. ON *skegg* beard.]

skeigh (skekh, skēkh) *Scot. adj.* **1** Shy; skittish; mettlesome: said of a horse. **2** Coy; disdainful; proud: said especially of women. —*adv.* Proudly.

skein (skān) *n.* **1** A fixed quantity of yarn, thread, silk, wool, etc., wound to a certain length and then doubled and knotted. **2** A measure of length, 360 feet, or 109.73 meters. [<OF *escaigne,* prob. <Celtic. Cf. Irish *sgainne.*]

skein screw A screw with a broad shallow thread. See illustration under SCREW.

skel·dock (skel′dok) *n. Scot.* Wild mustard *(Brassica kaber).* Also **skel′lock.**

skel·e·tal (skel′ə·təl) *adj.* Of, pertaining to, forming, or like a skeleton.

skel·e·ton (skel′ə·tən) *n.* **1** The framework of an animal body, composed of bone and cartilage. The skeletal structure either surrounds and shields the vital organs, as the *exoskeleton* of a turtle, or is embedded within the body, as the *endoskeleton* of man and the vertebrates. **2** Any open framework constituting the main supporting parts of a structure: the *skeleton* of a house. **3** A mere sketch or outline of anything, especially of some literary production: the *skeleton* of an address. **4** A person or animal very thin by nature or loss of flesh; also, a band or troop whose numbers have been greatly thinned out. See synonyms under SKETCH. —**skeleton in the closet** A secret source of shame or discredit. See FAMILY SKELETON. —*adj.* Consisting merely of a framework or outline; resembling a skeleton in use or appearance; meager; emaciated. [<NL <Gk. *skeleton (sōma)* dried (body), mummy < *skeletos* dried up]

skeleton construction A construction in which the main support is an internal framework of steel, to which the outer walls are affixed, their weight being carried, story by story, by the framework.

skel·e·ton·ize (skel′ə·tən·īz′) *v.t.* **·ized, ·iz·ing** **1** To reduce to a skeleton or framework by removing soft tissues or parts; make a skeleton of. **2** To reduce greatly in size or numbers. **3** To draft in outline.

skeleton key A slender false key designed to avoid the wards of a lock, for use as a master key.

skel·lum (skel′əm) *n. Scot.* A scamp.

skelp[1] (skelp) *Brit. Dial.* & *Scot. v.t.* **1** To kick severely. **2** To slap with the hand; spank. **3** To cause to move rapidly. —*n.* A glancing blow with the open hand; slap.

skelp[2] (skelp) *n.* A strip of iron or steel; especially, one from which tubes are made. —*v.t.* To beat out into a skelp, as iron. [Origin unknown]

Skel·ton (skel′tən), **John,** 1460?–1529, English poet, scholar, and clergyman.

skep (skep) *n.* **1** A beehive, especially one made of straw. **2** A receptacle of wickerwork or wood, especially for grain; a basket. [<ON *skeppa* basket]

skep·tic (skep′tik) *n.* **1** One who questions the fundamental doctrines of religion, especially of the Christian religion. **2** One who refuses concurrence in generally accepted conclusions in science, philosophy, etc. **3** An adherent of any philosophical school of skepticism; especially, an adherent of the **Skeptic school** in ancient Greece, of which the Pyrrhonists, with their doctrine of the relativity of knowledge, were the first systematic exponents. **4** One who doubts any particular statement. —*adj.* Skeptical. Also, *Brit., sceptic.* [<F *sceptique* <L *scepticus* <Gk. *skeptikos* reflective < *skeptesthai* consider]

Synonyms (noun): agnostic, atheist, deist, disbeliever, freethinker, infidel, unbeliever. The *skeptic* doubts divine revelation; the *disbeliever* and the *unbeliever* reject it, the *disbeliever* with more of intellectual dissent, the *unbeliever* (in the common acceptation) with indifference or with opposition of heart as well as of intellect. *Infidel* is an opprobrious term that is commonly applied to any decided opponent of an accepted religion. The *atheist* denies that there is a God; the *deist* admits the existence of God, but denies that the Christian Scriptures are a revelation from him; the *agnostic* denies either that we do know or that we can know whether there is a God. *Antonyms*: believer, Christian.

skep·ti·cal (skep′ti·kəl) *adj.* Doubting; questioning; of or pertaining to a skeptic or skepticism. Also, *Brit., sceptical.* —**skep′ti·cal·ly** *adv.* —**skep′ti·cal·ness** *n.*

skep·ti·cism (skep′tə·siz′əm) *n.* **1** A doubting or incredulous state of mind. **2** *Philos.* The doctrine that absolute knowledge is unattainable and that judgments must be continually questioned and doubted in order to attain approximate or relative certainty: opposed to *dogmatism.* Also, *Brit., scepticism.*

sker·ry (sker′ē) *n. pl.* **·ries** *Scot.* An insulated rock or reef.

sketch (skech) *n.* **1** An incomplete but suggestive delineation or presentation of anything, whether graphic or literary; an outline. **2** An artist's preliminary study, graphic or plastic, of a work of art intended for elaboration. **3** A literary or dramatic composition, short, discursive, and of slight construction. **4** A short scene, play, or musical act, especially in vaudeville. —*v.t.* To make a sketch or sketches of; outline. —*v.i.* To make a sketch or sketches. [<Du. *schets* <Ital. *schizzo* <L *schedium* improvisation <Gk. *schedios*] —**sketch′a·ble** *adj.* —**sketch′er** *n.*

Synonyms (noun): brief, delineation, draft, drawing, outline, picture, plan, skeleton. An *outline* gives only the bounding or determining lines of a figure or a scene; a *sketch* may give lines, shading and color, but is hasty and incomplete. The lines of a *sketch* are seldom so full and continuous as those of an *outline. Draft* and *plan* apply especially to mechanical drawing, of which *outline, sketch,* and *drawing* are also used; a *plan* is strictly a view from above, as of a building or machine, giving the lines of a horizontal section, originally at the level of the ground, now in a wider sense at any height; as, a *plan* of the cellar; a *plan* of the attic. A *design* is such a preliminary *sketch* as indicates the object to be accomplished or the result to be attained, and is understood to be original. One may make a *drawing* of any well-known mechanism, or a *drawing* from another man's *design*; but if he says "The *design* is mine," he claims it as his own invention or composition. In written composition an *outline* gives simply the main divisions, and is often called a *skeleton*; a somewhat fuller suggestion of illustration, treatment, and style is given in a *sketch.* A lawyer's *brief* is a succinct statement of the main facts in a case, and of the main heads of his argument on points of law, with reference to authorities. See PICTURE.

sketch·book (skech′bŏŏk′) *n.* **1** A blank book used for sketching. **2** A printed volume of literary sketches. Also **sketch book.**

sketch·y (skech′ē) *adj.* **sketch·i·er, sketch·i·est** Like or in the form of a sketch; roughly suggested without detail; hence, incomplete; superficial; slight. —**sketch′i·ly** *adv.* —**sketch′·i·ness** *n.*

skew (skyōō) *v.i.* **1** To take an oblique direction; move or turn aside; swerve. **2** To look obliquely or askance; squint. —*v.t.* **3** To put askew; give an oblique position or direction to. **4** To shape or form in an oblique manner; distort. —*adj.* **1** Placed or turned obliquely; twisted to one side; askew; hence, perverted in use or meaning. **2** *Stat.* Having some elements on opposite sides of a median line reversed or unbalanced; distorted: a *skew* curve. —*n.* **1** A deviation from symmetry or straightness; distortion. **2** A sidelong glance; squint. **3** A slanting coping, as at the corner of a gable. [<AF *eskiuer,* OF *eschiuver* shun <Gmc. Related to ESCHEW.] —**skew′ly** *adv.*

skew arch *Archit.* An arch whose axis is in a vertical plane making other than right angles with its abutments.

skew·back (skyōō′·bak′) *n. Archit.* **1** An abutment with inclined face receiving the thrust of a segmented arch. See also illustration under ARCH. **2** A cap or other casting, on the end of a truss, to receive the pull of a tie rod.

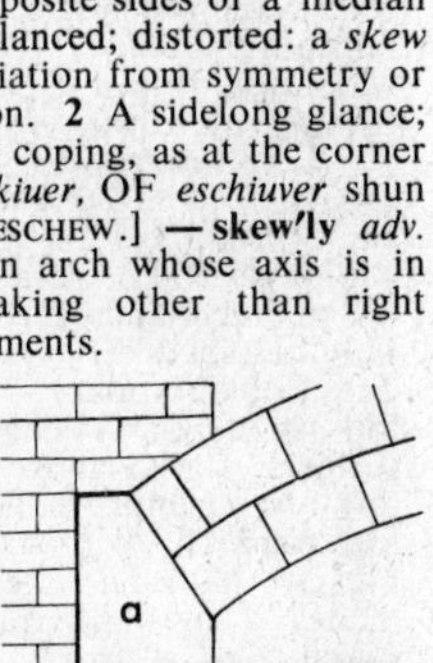

SKEWBACK
a. Skewback *(def. 1).*

skew·bald (skyōō′·bôld′) *adj.* Piebald, especially when the spots are white and some other color than black. [ME *skewed* piebald]

skew·er (skyōō′ər) *n.* **1** A long pin of wood or metal, used chiefly for fastening meat to keep it in shape while roasting. **2** Any of various articles of similar shape or use. —*v.t.* To run through or fasten with or as with a skewer. [Var. of SKIVER]

skew-gee (skyōō′jē) *adj. U.S. Colloq.* Crooked; off center; hence, mentally confused; uncertain.
skew·ness (skyōō′nis) *n.* **1** The state of being unsymmetrical or distorted. **2** *Stat.* The deviation of a frequency distribution curve from a symmetrical form.

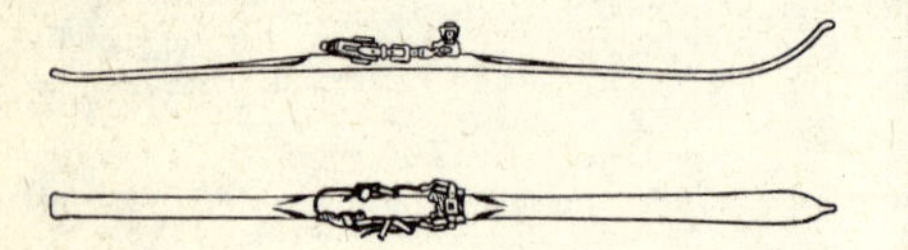

SKI
Side and top view.

ski (skē, *Norw.* shē) *n. pl.* **skis** or **ski** One of a pair of wooden runners, about 7 feet long and 3 1/2 inches wide, attached to the feet and used in gliding over snow or ice. —*v.i.* **skied, ski·ing 1** To glide or travel on skis. **2** To engage in the sport of gliding over snow-covered inclines on skis. Also spelled *skee.* [<Norw. <ON *skidh* snowshoe] —**ski′er** *n.*
ski·a·graph (skī′ə·graf, -gräf) *n.* Roentgenogram. [<Gk. *skia* shadow + -GRAPH]
Ski·ap·o·des (skī·ap′ə·dēz) *n. pl.* In Greek mythology, an ancient people of Libya who had feet so enormous that they used them for sunshades. [<Gk. *skia* shadow + *podes*, pl. of *pous* foot]
ski·a·scope (skī′ə·skōp) *n.* An instrument for examining the refractive power of the eye by the response of the retina to lights and shadows. [<Gk. *skia* shadow + -SCOPE]
ski·as·co·py (skī·as′kə·pē) *n.* Examination of the eye by the skiascope: also called *retinoscopy, shadow test.*
skid (skid) *n.* **1** One of a pair of timbers used to support a heavy tilting or rolling object, as a cask, boat, or cannon; also, a log used as a track in sliding heavy articles about, or forming an inclined plane to ease their descent. **2** In lumbering, one of several logs used to make a track on which other logs are slid or piled; also, one of the cross-logs of a skid road. **3** A shoe or drag on a wagon wheel. **4** *Naut.* A fender hung over a vessel's side to protect it from rubbing and scraping: usually in the plural. **5** *Aeron.* A runner in an airplane's landing gear. **6** The act of skidding; a side-slip. **7** A small frame or platform upon which merchandise is stacked to be moved about or temporarily stored. —**on the skids** *Slang* Rapidly declining in prestige or power. —*v.* **skid·ded, skid·ding** *v.i.* **1** To slide instead of revolving, as a wheel which does not rotate though the vehicle is in motion. **2** To slip sideways through inability to grip the road: said of wheels, and, by extension, of vehicles. **3** *Aeron.* To slide sideways away from the center of curvature when turning, by reason of insufficient banking. —*v.t.* **4** To furnish with skids; put, drag, or haul on skids. **5** To brake or hold back with a skid. [? <ON *skidh* piece of wood]
Skid·daw (skid′ô, ski·dô′) A mountain in central Cumberland, England; 3,054 feet.
skid·doo (ski·dōō′) *interj. Slang* Go away; get out. [<SKEDADDLE]
skid fin *Aeron.* A lengthwise vertical surface formerly placed above the upper wing of an airplane to improve lateral stability.
skid road 1 A road or track along which logs are hauled to the skidway. **2** A road made of logs laid transversely and spaced about five feet apart. **3** Skid row.
skid row *Slang* An urban section inhabited by vagrants and derelicts and consisting mainly of cheap bars, flophouses, etc.
skid·way (skid′wā′) *n.* A structure made of two logs or skids, about 10 feet apart and laid alongside a log road, on which logs are piled before loading.
skied[1] (skīd) Past tense and past participle of SKY.
skied[2] (skēd) Past tense and past participle of SKI.
skiff (skif) *n.* A light rowboat; formerly, a small sailing vessel. —**St. Lawrence skiff** A small boat, carrying centerboard and spritsail, light enough to be rowed with ease. [<F *esquif* <Ital. *schifo* <OHG *scif* ship, boat]
ski·ing (skē′ing) *n.* The act or sport of gliding on skis.
ski·jor·ing (skē·jôr′ing, -jō′ring) *n.* The sport of traveling over ice or snow on skis, towed by a horse or motor vehicle. [<Norw. *skikjöring* <*ski* ski + *kjöring* driving]
ski jump 1 A jump or leap made by a person wearing skis. **2** A course prepared for making such jumps.
skil·ful (skil′fəl) *adj.* **1** Having skill; clever; dexterous; able. **2** Showing or requiring skill. Also **skill′ful.** —**skil′ful·ly** *adv.* —**skil′ful·ness** *n.*
Synonyms: adept, adroit, apt, deft, dexterous, expert, handy, happy, proficient, skilled, trained. One is *adept* in that for which he has a natural gift, improved by practice; he is *expert* in that of which training, experience, and study have given him a thorough mastery; he is *dexterous* in that which he can do effectively with or without training, especially in work of the hand or bodily activities. A *skilled* workman is one who has thoroughly learned his trade, but he may be naturally quite dull: a *skilful* workman has some natural brightness, ability, and power of adaptation, in addition to his acquired knowledge and dexterity. See CLEVER, GOOD. *Antonyms*: awkward, bulky, clumsy, helpless, inexpert, maladroit, unskilled, untaught, untrained.
skill[1] (skil) *n.* **1** The familiar knowledge of any science, art, or handicraft, as shown by dexterity in execution or performance, or in its application to practical purposes; technical ability. **2** A specific art or trade; also, a gift; accomplishment. **3** *Obs.* Intellect; understanding. See synonyms under ABILITY, DEXTERITY, INGENUITY, WISDOM. [<ON *skil* knowledge] —**skill′-less** *adj.*
skill[2] (skil) *v.i. Obs.* To matter; make a difference: usually used impersonally and with a negative: It *skills* not what I do. [<ON *skilja* separate]
skilled (skild) *adj.* Possessing or requiring skill; expert; proficient. See synonyms under SKILFUL.
skil·let (skil′it) *n.* **1** A frying pan. **2** A small kettle or stew pan, often with a bail and short legs. [? <OF *escuellete*, dim. of *escuelle* porringer <L *scutella*, dim. of *scutra* dish]
skil·ling (skil′ing) *n.* A former small copper coin of the Scandinavian countries. [<Dan. and Sw.]
skill·y (skil′ē) *adj. Scot.* & *Brit. Dial.* Skilful.
skim (skim) *v.* **skimmed, skim·ming** *v.t.* **1** To remove floating matter from the surface of, as with a ladle: to *skim* milk. **2** To remove thus: to *skim* cream. **3** To cover with a thin film, as of ice. **4** To move lightly and quickly across or over. **5** To cause to pass swiftly and lightly, as a coin across a pond. **6** To read or glance over hastily or superficially. —*v.i.* **7** To move quickly and lightly across or near a surface; glide. **8** To make a hasty and superficial perusal; glance: with *over* or *through.* **9** To become covered with a thin film. —*n.* **1** The act of skimming. **2** That which is skimmed off; scum. **3** Something from which floating matter has been removed, as skim milk. **4** A thin scum of ice. —*adj.* Skimmed: *skim* milk. [Var. of SCUM]
skim·ble-scam·ble (skim′bəl·skam′bəl) *adj.* Incoherent; rambling. —*n.* Meaningless talk; nonsense. Also **skim′ble-skam′ble.** [Prob. reduplication of dial. *scamble* ramble, struggle on the ground (said of horses)]
skim·mer (skim′ər) *n.* **1** A flat ladle or other utensil for skimming. **2** One who or that which skims. **3** A ternlike bird (genus *Rhynchops*) having the lower mandible compressed, that skims up the small fishes from near the surface of the water. *R. nigra* is the **black skimmer.**

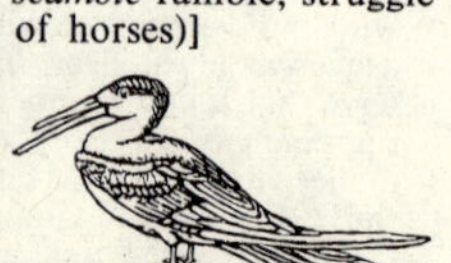

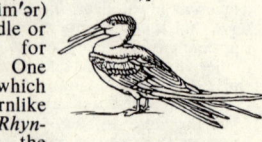

SKIMMER
(Length 16 to 20 inches; wingspread 42 to 50 inches)

skim-milk (skim′milk′) *adj.* Weak; inferior.
skim milk Milk from which the cream has been removed: often used to designate a type of inferiority.
skim·ming (skim′ing) *n.* **1** The act of one who or that which skims. **2** That which is skimmed off: usually in the plural.
skimp (skimp) *v.t.* & *v.i.* To scrimp or scamp. —*adj.* Scant; meager. [Prob. <ON *skemma* shorten; infl. in meaning by SCRIMP]
skimp·y (skim′pē) *adj.* **skimp·i·er, skimp·i·est 1** Carelessly done. **2** Scanty. **3** Niggardly. —**skimp′i·ly** *adv.* —**skimp′i·ness** *n.*
skin (skin) *n.* **1** The membranous external investment of an animal; the integument. ◆ Collateral adjective: *dermal.* **2** The pelt of a small animal, removed from its body, whether raw or dressed, as distinguished from the *hide* of a large animal. **3** A vessel for holding liquids, made of the skin of an animal: a wine-*skin.* **4** An outside layer, coat, or covering resembling skin, as the epidermis of a plant, fruit, etc.; rind; of pearls, the outermost layer of nacreous matter. **5** Planking or plating of a vessel. **6** A membrane resembling the integument. **7** *Slang* A mean person; skinflint; also, a sharper; blackleg. **8** One's life or physical existence: to save one's *skin.* —**by the skin of one's teeth** Very closely or narrowly; barely. —**under one's skin** Provoking; beneath the surface of control (of irritation, excitation, emotion, etc.). —*v.* **skinned, skin·ning** *v.t.* **1** To remove the skin of; flay; peel. **2** To cover with or as with skin. **3** To remove as if taking off skin: to *skin* a dollar from a roll of bills. **4** *Slang* To cheat or swindle. —*v.i.* **5** To become covered with skin; cicatrize. **6** *Slang* To make off hastily; run away: usually with *off.* [<ON *skinn*]
skin-bound (skin′bound′) *adj.* Affected with a rigid contraction of the skin and hardening of the connective tissue.
skin-deep (skin′dēp′) *adj.* Superficial. —*adv.* Superficially.
skin diving Underwater exploration in which the swimmer is equipped with a self-contained breathing apparatus, goggles, foot fins, rubber garments, etc. —**skin diver**
skin effect *Electr.* An increase of current density on the surface of an alternating-current conductor, giving an increase in resistance: especially marked at high frequencies, as in radio.
skin·flint (skin′flint) *n.* A miser; one who drives a hard bargain.
skin friction 1 *Physics* The component of a fluid force tangential to a given point on a surface. **2** *Aeron.* Resistance of air particles due to friction while in contact with the moving surfaces of an airplane: also **skin drag.**
skin·game (skin′gām′) *n.* **1** A gambling game at cards in which the players have no chance of winning against the house or the bank. **2** Any swindle.
skink[1] (skingk) *n.* One of a group of lizards (family *Scincidae*) with short limbs and a conical tail; especially, the **blue-tailed skink** *(Eumeces skiltonianus)* of the United States. [<L *scincos* <Gk. *skinkos*, kind of lizard]

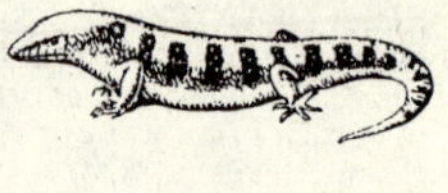

SKINK
(Up to 8 inches long)

skink[2] (skingk) *v.t. Brit. Dial.* **1** To draw or pour out. **2** To fill with liquor. [<MDu. *schenken*]
skink·er (skingk′ər) *n. Brit. Dial.* A bartender; also, an innkeeper. [<SKINK[2]]
skink·ing (skingk′ing) *adj. Scot.* Thin; sloppy.
skin·less (skin′lis) *adj.* Without skin.
skin·ner (skin′ər) *n.* **1** One who skins; a flayer of animals. **2** *U. S. Slang* A cheat; swindler. **3** A dealer in skins. **4** *U. S. Slang* A mule driver.
Skin·ner (skin′ər), **Cornelia Otis,** born 1901, U. S. actress and author; daughter of the following. —**Otis,** 1858–1942, U. S. actor.
skin·ny (skin′ē) *adj.* **·ni·er, ·ni·est 1** Wanting flesh; lean. **2** Consisting of or like skin. See synonyms under MEAGER. —**skin′ni·ly** *adv.* —**skin′ni·ness** *n.*
skin-tight (skin′tīt′) *adj.* Fitting tightly to the skin, as a garment.
skip (skip) *v.* **skipped, skip·ping** *v.i.* **1** To move with light springing steps; caper; leap lightly. **2** To be deflected from a surface; ricochet; skim. **3** To pass from one point to another without noticing what lies between. **4** *Colloq.* To leave or depart hurriedly; flee. **5** To be advanced in school beyond the next grade in order —*v.t.* **6** To leap lightly over. **7** To cause to ricochet. **8** To pass over or by without notice. **9** *Colloq.* To leave (a place) hurriedly. —*n.* **1** A light bound or spring; especially, a hop alternating between steps in walking. **2** A passing over without notice;

skip distance That area within which signals from a radio transmitter are not received: it is between the farthest point reached by the ground wave and the nearest point at which the reflected sky wave strikes the earth.

skip floor A type of apartment house in which there is one public corridor for alternate floors, each dwelling unit occupying two floors connected by internal stairways.

skip·jack (skip'jak) *n.* 1 Any of various fishes that skip along the surface of the water, as the bonito. 2 Any snapping or click beetle (family *Elateridae*). 3 A Chesapeake Bay sailing vessel with a centerboard and one mast: used in dredging oysters.

skip·per[1] (skip'ər) *n.* 1 One who or that which skips. 2 The saury. 3 A butterfly of the family *Hesperiidae*: so named from its flight. 4 A cheese maggot.

skip·per[2] (skip'ər) *n.* The master or captain of a small vessel; hence, one in charge of any craft. [<Du. *schipper* <*schip* ship]

skip·pet (skip'it) *n.* A round flat box for containing and protecting the large heavy seal formerly tied to a document. [Dim. of SKEP]

skirl (skûrl, skirl) *Scot. v.i.* To shriek shrilly, as a bagpipe. — *v.t.* To play the bagpipe. — *adj.* Shrill. — *n.* A shrill cry; a squall of wind with rain or snow. [Metathetic var. of ME *scrille* <Scand. Cf. Norw. *skrylla*.]

skir·mish (skûr'mish) *v.i.* To fight in a preliminary or desultory way. — *n.* 1 A light engagement, as between small parties; desultory fighting between two armies on a skirmish line. 2 Figuratively, any light movement or operation evasive of the main contention or business. See synonyms under BATTLE. [<OF *eskermiss-*, stem of *eskermir* fence, fight <Gmc. Cf. OHG *skirman* defend <*skirm* shield. Related to SCRIMMAGE.] — **skir'mish·er** *n.*

skirmish line A line of infantry spread out in extended order for attack.

skirr (skûr) *v.t.* 1 To scour. 2 To skim over. — *v.i.* 3 To move rapidly. — *n.* A whirring sound. [Imit.]

skir·ret (skir'it) *n.* An Old World herb (*Sium sisarum*) formerly much cultivated in Europe for its white tubers, which are cooked and served like salsify. [ME *skirwhit*, prob. OF *eschervis* <Arabic *karawya*. Cf. CARAWAY.]

skirt (skûrt) *n.* 1 That part of a dress, gown, or robe that hangs from the waist downward. 2 A separate garment hanging from the waist and covering the lower portion of the body. 3 *Slang* A girl; woman. 4 The margin, border, or outer edge of anything. 5 *pl.* The border, fringe, or edge of a particular area, path, geographical feature, etc.: on the *skirts* of the town, forest, highway, etc. 6 One of the flaps or loose, hanging parts of a saddle: also **saddle skirt.** 7 *Naut.* The leech of a sail. 8 The diaphragm or midriff of a butchered animal. See synonyms under MARGIN. — *v.t.* 1 To lie along or form the edge of; to border. 2 To surround or border: with *with.* 3 To pass around or about, usually to avoid crossing: to *skirt* the town. — *v.i.* 4 To be or pass along the edge or border: to *skirt* along the coast. [ON *skyrt* shirt. Akin to SHIRT.]

ski-run·ner (skē'run'ər) *n.* One who travels on skis.

skit (skit) *n.* 1 A short literary article, theatrical sketch, etc., usually humorous or satirical. 2 A bantering jest. [<Scand.; cf. ON *skjōta* shoot. Prob. akin to SHOOT.]

skite (skīt) *Scot. n.* 1 A quick, sharp slap. 2 A quick, heavy, splashing shower; dash, as of rain. 3 A trick. 4 A squire. — *v.i.* 1 To squirt. 2 To glide away quickly; scoot.

ski troops Soldiers equipped and trained for action on skis.

skit·ter (skit'ər) *v.i.* 1 To glide or skim along, touching ground or water at intervals. 2 To fish by the method known as skittering. [Freq. of SKIT]

skit·ter·ing (skit'ər·ing) *n.* A style of fishing with a hook twitched along the water.

skit·tish (skit'ish) *adj.* 1 Easily frightened, as a horse; hence, shy; timid. 2 Capricious; uncertain; unreliable. 3 Tricky; deceitful. See synonyms under RESTIVE. [<dial. E *skit* caper (said of horses)] — **skit'tish·ly** *adv.* — **skit'tish·ness** *n.*

skit·tle (skit'l) *n.* 1 *pl.* A game of ninepins, in which a flattened ball or thick rounded disk is thrown to knock down the pins. 2 One of the pins used in this game: also **skittle–pin.** — **beer and skittles** Carefree existence, consisting of drink and play; unruffled enjoyment: usually with a negative: Life is not all *beer and skittles.* [Prob. <Dan. *skyttel* child's earthen ball]

skive[1] (skīv) *v.t.* **skived, skiv·ing** To shave or pare the surface of, as leather. [<ON *skīfa* slice]

skive[2] (skīv) *n.* A gem-cutter's diamond wheel. — *v.t.* **skived, skiv·ing** To grind off, as the surface of a gem. [<Du. *schÿf*]

skiv·er (skī'vər) *n.* 1 Leather split with a knife: used for bookbinding. 2 One who skives. 3 A knife or machine used in skiving.

skiv·vies (skiv'ēz) *n. pl. Slang* Men's underwear. [Origin uncertain]

sklent (sklent) *Scot. v.i.* 1 To move in a slanting manner. 2 To glance hostilely; squint. 3 To tell a lie. — *n.* 1 A slant. 2 A lie. — *adj.* Slanting.

skoal (skōl) *interj.* Hail: a toast or salutation in Scandinavian use. — *n.* The act of saluting or toasting with the word "skoal!" [<Scand. Cf. Dan. *skaal* bowl, toast, ON *skāl* bowl.]

Ško·da (shkô'dä), **Emil von,** 1839–1900, Czech engineer and industrialist.

skook·um (skōōk'əm) *adj. U.S. Slang* Strong; powerful. [<N. Am. Ind. *skukum* powerful, evil spirit]

Skop·lje (skôp'lye) A city in SE Yugoslavia: the economic, cultural, and Islamic religious center of modern Macedonia: Turkish *Usküb.*

skreigh (skrēkh, skräkh) *v. & n. Scot.* Shriek; screech. Also **skreegh.**

Skry·mer (skrī'mər) See UTGARD-LOKI.

sku·a (skyoō'ə) *n.* A gull-like bird; a jaeger. Also **skua gull.** [<Faroese *skūgver* <ON *skūfr*]

Skuld (skoōld) In Norse mythology, one of the Norns.

skul·dug·ger·y (skul-dug'ər-ē) *n. U.S.* Trickery; underhandedness. [Var. of dial. *sculduddery*; origin uncertain]

skulk (skulk) *v.i.* 1 To move about furtively or slily; lie close or keep hidden; lurk. 2 To shirk; evade work or responsibility. — *n.* 1 One who skulks. 2 A troop of foxes. [<Scand. Cf. Dan. *skulke.*] — **skulk'er** *n.*

HUMAN SKULL

a.	Parietal bone.	i.	Inferior maxillary.
b.	Squamosal suture.	j.	Superior maxillary.
c.	Temporal bone.	k.	Malar bone.
d.	Occipital bone.	l.	Nasal bone.
e.	Opening of ear.	m.	Zygomatic bone.
f.	Mastoid process.	n.	Sphenoid bone.
g.	Styloid process.	o.	Frontal bone.
h.	Zygomatic arch.	p.	Coronal suture.

skull (skul) *n.* 1 The bony framework of the head of a vertebrate animal; the cranium. 2 The head considered as the seat of the brain; the mind. ◆ Homophone: *scull.* [<Scand. Cf. dial. Norw. *skul* shell.]

skull and crossbones A representation of the human skull over two crossed thigh bones: used as a symbol of death, as a warning label on poison, and as an emblem of piracy.

skull·cap (skul'kap') *n.* 1 The sinciput. 2 Any plant of the genus *Scutellaria*, especially *S. galericulata*, of wet shady places, with large blue flowers.

skull cap 1 A cap closely fitting the skull. 2 A light cap without brim or peak, for indoor wear.

skunk (skungk) *n.* 1 A nocturnal, burrowing carnivore of North America (family *Mustelidae*), usually black with a white stripe running from the nape of the neck to a large, bushy tail: under the tail are perineal glands that secrete a liquid of very offensive odor ejected at will. The common striped skunk (*Mephitis mephitis*) of the United States is about the size of a cat, and there are spotted varieties (genus *Spilogale*). 2 *Colloq.* A low, contemptible person. — *v.t.* 3 *Slang* To defeat, as in a contest, so thoroughly as to keep from scoring. [<Algonquian *seganku*]

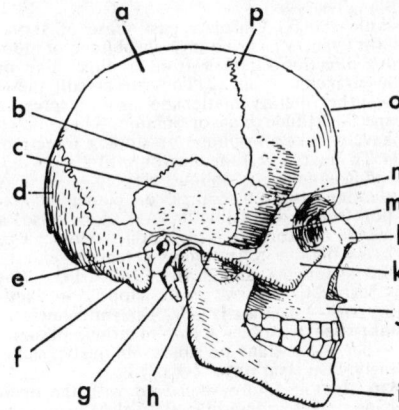

SKUNK (Up to 16 inches in length; tail: 7 inches)

skunk cabbage 1 A stemless perennial herb (*Symplocarpus foetidus*) of the United States, producing in the early spring a horn-shaped, brownish-purple spathe which encloses the oval spadix and emits a strong odor, especially when crushed or bruised: also called *swamp cabbage.* 2 A somewhat similar plant (*Lysichitum americanum*) of western North America. Also **skunk'weed** (-wēd').

sky (skī) *n. pl.* **skies** 1 The blue vault, or part of it, that seems to bend over the earth; the firmament. 2 The upper atmosphere; especially, the region of the clouds. 3 The celestial regions or powers; heaven. 4 Climate; weather. 5 *Obs.* A cloud. — *v.t.* **skied, sky·ing** *Colloq.* 1 In games, to bat or throw (a ball) high into the air. 2 To hang or put (a picture) in a high place or near the ceiling, where it cannot be easily seen. [<ON *sky* cloud]

sky-blue (skī'bloō') *adj.* Of the color of the sky; azure.

sky blue A blue like the color of the sky.

Skye (skī), **Isle of** Largest of the Inner Hebrides Islands; 670 square miles.

Skye terrier See under TERRIER.

sky·ey (skī'ē) *adj.* Pertaining to or resembling the sky; heavenly.

sky-high (skī'hī') *adj. & adv.* Extremely high.

sky·lark (skī'lärk') *n.* A lark (*Alauda arvensis*) that utters a sweet song as it flies. — *v.i.* To indulge in hilarious or boisterous frolic. — **sky'lark'er** *n.* — **sky'lark'ing** *n.*

sky·light (skī'līt') *n.* A window facing skyward.

sky line 1 The line of the visible horizon. 2 The sharp outline of an object, or of a group of buildings, seen against the sky; silhouette.

sky·man (skī'mən) *n. pl.* **·men** (-mən) *Colloq.* An aviator.

sky pilot *Slang* A clergyman; also, a chaplain.

skyre (skīr) *Scot. v.i.* To appear gaudy; glisten. — *n.* Anything brightly colored.

sky·rock·et (skī'rok'it) *n.* A rocket that is shot high into the air, where it explodes, often with brilliant pyrotechnic effect. — *v.i.* To rise or cause to rise or ascend steeply, like a skyrocket: used figuratively of wages, prices, etc.

Sky·ros (skē'ros) The largest island of the Northern Sporades group, in the Aegean east of Euboea; 80 square miles. Latin *Scyros.*

sky·sail (skī'səl, -sāl') *n. Naut.* A light sail above the royal in a square-rigged vessel.

sky·scrap·er (skī'skrā'pər) *n.* A very high building.

sky·ward (skī'wərd) *adv.* Toward the sky: also **sky'wards.** — *adj.* Moving or directed toward the sky.

sky wave A radio wave projected into the upper atmosphere by a transmitter and reflected back to earth from the Kennelly-Heaviside layer.

sky·writ·ing (skī'rī'ting) *n.* The forming of

words in the air by an aviator, by releasing a jet of vapor from the tail of an airplane. —**sky'writ'er** *n.*

slab[1] (slab) *n.* **1** The outside cut made from a log in sawing it into boards, planks, etc., often bearing the bark on one side. **2** A flat plate, piece, mass, or slice, as of metal, stone, chocolate, or the like. **3** *U.S. Slang* In baseball, the pitcher's plate. —*v.t.* **slabbed, slab·bing 1** To saw slabs from, as a log; to square by removing the slabs. **2** To cover with, or form of or into slabs. [ME; origin uncertain]

slab[2] (slab) *n.* Slime; viscous mud; mire. —*adj. Archaic* Slimy; viscous. [<ON *slabb* mud]

slab·ber (slab'ər) *v.* & *n.* Slobber. [Prob. <LG. Cf. Du. *slabberen.*]

slab·by (slab'ē) *adj.* **·bi·er, ·bi·est** *Archaic* **1** Thick; viscous. **2** Sloppy; wet. [<SLAB[2]]

slab·sid·ed (slab'sī'did) *adj. U.S. Colloq.* **1** Having flat sides. **2** Lanky; gawky; ungainly.

slack[1] (slak) *adj.* **1** Hanging or extended loosely. **2** Loose or careless in performance; remiss; tardy; slovenly; slow; also, weak; loose: a *slack* mouth. **3** Lacking activity; not brisk or pressing: a *slack* season. **4** Listless; limp: a *slack* grip. **5** Flowing sluggishly, as water between the ebb and flow of the tide; also, blowing slowly, as a wind. **6** Incomplete; underdone; unfinished. See synonyms under SLOW. —*v.t.* **1** To slacken. **2** To slake, as lime. —*v.i.* **3** To be or become slack. —*n.* **1** The part of anything, as a rope, that is slack or loose; also, a slack condition; looseness. **2** A period of inactivity; a slack season. **3** An extent of water where there is no current. **4** *pl.* Loose-fitting trousers worn by both men and women as part of a casual sports costume; also, cotton or wool trousers in a military uniform. —*adv.* In a slack manner; slackly. [OE *slæc*] —**slack'ly** *adv.* —**slack'·ness** *n.*

slack[2] (slak) *n.* Small coal; screenings. [Cf. LG *slacke*]

slack[3] (slak) *n. Scot.* & *Brit. Dial.* **1** A dry hollow or gully. **2** A bog. **3** A common. **4** A natural slope of ground. [<ON *slakki* dip, depression]

slack–baked (slak'bākt') *adj.* Not thoroughly cooked; underdone.

slack·en (slak'ən) *v.i.* **1** To become slack, as business; diminish; retard. **2** To become less tense or tight; loosen. **3** To become slow or less intense. —*v.t.* **4** To be or become negligent of or remiss in; to avoid, as duty, especially a military duty; shirk. **5** To make slack.

slack·er (slak'ər) *n.* One who shirks his duties or avoids military service in wartime; shirker.

slae (slē) *n. Scot.* The sloe or blackthorn.

slag (slag) *n.* **1** *Metall.* **a** The fused residue separated in the reduction of ores; metallic dross. **b** A basic iron silicate that floats on the surface of molten iron. **2** Volcanic scoria. —*v.t.* & *v.i.* **slagged, slag·ging** To form into slag. [<MLG *slagge*] —**slag'gy** *adj.*

slag wool Mineral wool.

slain (slān) Past participle of SLAY.

slake (slāk) *v.* **slaked, slak·ing** *v.t.* **1** To render inoperative or harmless, especially by satisfying, as an appetite. **2** To lessen the force of in any way; quench; appease; assuage: to *slake* thirst or flames. **3** To mix with water or moist air, so that a chemical combination shall ensue. **4** To disintegrate and hydrate, as lime. **5** To make loose, slow, or less tense. —*v.i.* **6** To become disintegrated and hydrated: said of lime. **7** To slacken; become loose, slow, or less tense. **8** To ease up on one's efforts; slow down. —*n.* The act or period of slackening; an abatement. [OE *slacian* retard <*slæc* SLACK[1]]

sla·lom (slä'ləm, slā'-) *n.* In skiing, a race over a downhill, serpentine course laid out between posts and marked with flags, victory going to the skier who makes the best speed with the best grace and form. —*v.i.* To ski in or as in a slalom. [<Norw.]

slam (slam) *v.* **slammed, slam·ming** *v.t.* **1** To shut with violence and a loud noise; pull or push to loudly: to *slam* a door. **2** To put, dash, throw, or bring with violence and a loud noise; bang: to *slam* a book down. **3** *Slang* To strike with the fist. **4** *Colloq.* To take to task; criticize severely. —*v.i.* **5** To be shut, enter a place, etc., with force and noise. —*n.* **1** A closing or striking with a bang; the act or noise of slamming. **2** *Colloq.* Severe criticism. **3** A card game of the 16th century, resembling ruff. **4** In bridge, the winning of more than eleven tricks: **grand slam** is the winning of all 13 tricks; **little slam** is the winning of 12 tricks. [<Scand. Cf. dial. Norw. *slamra* slam.]

slam–bang (slam'bang') *adv.* Violently; noisily; also, recklessly. —*v.i.* To move with noise and violence.

slan·der (slan'dər) *n.* A false tale or report, or such tales or reports collectively, uttered with malice and designed or tending to injure the reputation of another; calumny; also, the utterance of such tales or reports; defamation. See synonyms under SCANDAL. —*v.t.* To injure by maliciously uttering a false report; defame; calumniate. —*v.i.* To utter slander. See synonyms under ABUSE, ASPERSE, REVILE. [<AF *esclaundre,* OF *esclandre,* ult. <L *scandalum.* Doublet of SCANDAL.] —**slan'der·er** *n.*

slan·der·ous (slan'dər·əs) *adj.* **1** Uttering slander; guilty of slander. **2** Containing slander; calumnious. —**slan'der·ous·ly** *adv.* —**slan'der·ous·ness** *n.*

slang (slang) *n.* **1** A type of popular language comprised of words and phrases of a vigorous, colorful, or facetious nature, which are invented as needed or derive from the unconventional use of the standard vocabulary. The vocabulary of slang, although usually ephemeral, may achieve wide colloquial currency, and, in the evolution of language, many words originally slang have been adopted by good writers and speakers, and ultimately taken their place as accepted English. **2** The special vocabulary of a certain class, group, or profession: college *slang.* **3** Originally, the argot or jargon of thieves and vagrants. —*v.t.* To abuse or address with slang; also, to scold. —*v.i.* To use slang. [Origin uncertain]

Synonyms: argot, cant, jargon, lingo. The language of the underworld is *argot,* stressing its secrecy; *cant* often signifies the vocabulary of a special occupational group; *jargon* emphasizes unintelligibility and cacophonous sound. *Cant,* originally the beggar's whine, then the preacher's drone, acquired, in later usage, the more common meaning of sanctimonious moralizing. *Jargon* has as its commonest sense barbarous-sounding gabble. *Lingo* commonly designates foreign-sounding speech, or a language with which we are unfamiliar.

slang·y (slang'ē) *adj.* **slang·i·er, slang·i·est 1** Of the nature of or containing slang. **2** Using or given to using slang. —**slang'i·ly** *adv.* —**slang'i·ness** *n.*

slank (slangk) Obsolete past tense of SLINK.

slant (slant) *v.t.* **1** To give an oblique or sloping direction to; turn from a direct line or level; incline; lean. **2** To write or edit (news or other literary matter) so as to express a special attitude, bias, or opinion. —*v.i.* **3** To have or take an oblique or sloping direction. **4** To have a certain bias or attitude. —*adj.* Lying at an angle; oblique; sloping —*n.* **1** A slanting direction, course, or plane; inclination from a direct line or level; slope; also, a mental or moral bent, opinion, attitude, etc. **2** An oblique reflection; a sarcastic remark. See synonyms under TIP[1]. [<earlier *slent* <Scand. Cf. Norw. *slenta* slope.] —**slant'·ing** *adj.* —**slant'ing·ly** *adv.* —**slant'ing·ness** *n.*

slant·wise (slant'wīz') *adj.* Slanting; oblique. —*adv.* At a slant or slope; obliquely; slantingly: also **slant'ways'** (-wāz').

slap (slap) *n.* A blow delivered with the open hand or with something flat; also, an insult; slur. —*v.* **slapped, slap·ping** *v.t.* **1** To hit or strike with the open hand or with something flat; also, to rebuff; insult. **2** To put or place violently or carelessly. —*v.i.* **3** To strike or beat as if with slaps: The waves *slapped* against the dock. —*adv.* **1** Suddenly and forcibly; abruptly. **2** *Colloq.* Directly; straight: *slap* into his face. [<LG *slapp*] —**slap'per** *n.*

slap·bang (slap'bang') *adj.* & *adv. Slang* Slapdash.

slap–dash (slap'dash') *adj.* Done or acting in a dashing or reckless way; impetuous; careless. —*n.* **1** Offhand or careless work, or thoughtless conduct. **2** Rough casting, or rough plastering. —*adv.* In a dashing or heedless manner.

slap–hap·py (slap'hap'ē) *adj. Slang* Giddy and weak-minded because of concussion of the brain; punch-drunk.

slap·jack (slap'jak) *n.* **1** A griddlecake; flapjack. **2** A children's game of cards.

slap·stick (slap'stik') *n.* **1** A flexible, double paddle formerly used in farces and pantomimes to make a loud report when an actor was struck with it. **2** The use of this apparatus, or the type of rough comedy in which it is used. —*adj.* Using or suggestive of the slapstick: *slapstick* comedy.

slash (slash) *v.t.* **1** To cut by striking violently and without attempt at accuracy; cut with long sweeping strokes; strike violently with or as with an edged instrument; slit; gash. **2** To strike with long sweeping blows of a whip; lash; scourge. **3** To make long gashes, cuts, or slits in; specifically, to slit, as a garment, so as to expose ornamental material or lining in or under the slits. **4** To criticize severely; censure harshly. **5** To cut down wastefully, as timber in a forest. **6** To reduce sharply, as salaries. —*v.i.* **7** To make a long sweeping stroke or several such strokes with or as with something sharp; cut. —*n.* **1** The act or result of slashing; a sweeping, random cut with a cutting weapon or whip; a slit or gash; specifically, an ornamental slit or cut in a garment showing some other material in or through the slit. **2** An opening or gap made in a forest. **3** The loose tops and branches of trees left in a forest after logging or a high wind. **4** A swampy thicket; low-lying boggy land: usually in the plural. **5** *Printing* A virgula. [? <OF *esclachier* break] —**slash'er** *n.*

slash·ing (slash'ing) *adj.* **1** Striking or cutting at random: a *slashing* warrior or critic. **2** *Colloq.* Of uncommonly high degree; very large; very fine, swift, etc.; exceptionally brilliant. —*n.* **1** A slash; the act of slashing; especially, the wasteful destruction of timber. **2** A region where timber trees have been cut down. **3** A mass of felled trees heaped for burning. —**slash'ing·ly** *adv.* —**slash'ing·ness** *n.*

slash pine 1 A pine (*Pinus caribae*) growing in the slashes along the southeastern coast of the United States; also, its wood. **2** The loblolly pine.

Śląsk (shlônsk) The Polish name for SILESIA.

slat[1] (slat) *n.* **1** One of a number of thin, flat, narrow strips of wood used to support the springs or mattress of a bed. **2** Any thin, narrow strip of wood or metal; a lath. **3** *Aeron.* A movable auxiliary airfoil attached to the leading edge of an airplane wing. —*v.t.* **slat·ted, slat·ting** To provide or make with slats. [<OF *esclat* splinter, chip]

slat[2] (slat) *v.* **slat·ted, slat·ting** *v.t. Dial.* **1** To throw or dash violently; fling carelessly. **2** To beat; slap. —*v.i.* **3** To flap, as sails against yards. —*n. Dial.* A sudden, sharp blow. [? <ON *sletta* slap]

slate[1] (slāt) *n.* **1** Any rock that splits readily into thin and even laminae; specifically, an argillaceous, fine-grained rock that so splits; also, an artificial material made in imitation of it. **2** A piece, slab, or plate of slate used for roofing, writing upon, etc. **3** A record of one's past performance or behavior: a clean *slate.* **4** A list of political candidates made up before their nomination or election; any prearranged list. **5** A dull bluish-gray color resembling that of slate: also **slate gray.** —*adj.* **1** Made of slate: a *slate* roof. **2** Slate-colored. —*v.t.* **slat·ed, slat·ing 1** To roof with slate. **2** To put on a political slate or a list of any sort; hence, to register or designate as if by writing on a slate: He is *slated* for promotion. **3** To remove hair from (hides) with a slater. [<OF *esclate,* fem. of *esclat* a chip, splinter] —**slat'y** *adj.*

slate[2] (slāt) *v.t.* **slat·ed, slat·ing 1** To censure, criticize, or review severely; berate. **2** To punish severely. [OE *slætan* bait]

slate ax See SAX[1] (def. 1).

slate pencil A pencil made of soft slate: used for writing on a slate.

slat·er[1] (slā'tər) *n.* **1** A person whose trade is to lay slates. **2** A slate-edged implement for removing hair from hides. **3** A terrestrial isopod crustacean, as the common pill bug or wood louse.

slat·er[2] (slā'tər) *n.* One who censures severely; a caustic critic.

slath·er (slath'ər) *Colloq.* or *Dial. v.t.* To daub thickly; spend or use profusely; lavish. —*n.* **1** A thick layer or spread. **2** *pl.* A lot; very

slating — sleet

much: *slathers* of fun. [Var. of dial. E *slither slip*]

slat·ing (slā′ting) *n.* 1 The act or occupation of laying slates. 2 Slates or slate collectively. 3 A liquid for giving a slatelike surface to blackboards, etc.

slat·tern (slat′ərn) *n.* An untidy or slovenly woman. — *adj.* Untidy; slovenly. [< dial. E *slatter* slop, spill] — **slat′tern·li·ness** *n.* — **slat′tern·ly** *adj.* & *adv.*

slaugh·ter (slô′tər) *n.* 1 The act of killing; specifically, the butchering of cattle and other animals for market. 2 Wanton or savage killing, especially of human beings; massacre; carnage. 3 *Slang* A sweeping or ruinous reduction in prices. See synonyms under MASSACRE. — *v.t.* 1 To kill for the market; butcher. 2 To kill wantonly or savagely, especially in large numbers. 3 *Slang* To reduce greatly the price of; sell at a low figure. See synonyms under KILL¹. [< ON *slātr* buteher's meat. Akin to SLAY.] — **slaugh′ter·er** *n.* — **slaugh′ter·ous** *adj.* — **slaugh′ter·ous·ly** *adv.*

slaugh·ter·house (slô′tər·hous′) *n.* A place where animals are butchered; a scene of carnage.

Slav (släv, slav) *n.* A member of any of the Slavic-speaking peoples of northern or eastern Europe, the northern group comprising the Russians, Poles, Czechs, Moravians, Sorbs or Wends, Slovaks, etc.; the southeastern group comprising the Bulgarians, Serbians, Croats, and Slovenes. Also, *Obs. Sclav.* [< G *Sklave* < Med. L *Sclavus* < LGk. *Sklabos* < Slavic]

slave (slāv) *n.* 1 One whose person is held as property; a person in slavery; a bondsman; serf. 2 *Law* A person over whose life, liberty, and property someone has absolute control. 3 A person in mental or moral subjection to a habit, vice, or influence: a *slave* of tobacco. 4 One who labors like a slave; a drudge. 5 A person of slavish disposition; an abject creature. — *v.* **slaved, slav·ing** *v.i.* To work like a slave; toil; drudge. — *v.t. Rare* To enslave. [< F *esclave* < Med. L *slavus, sclavus,* orig. a Slav; because many Slavs were conquered and enslaved]

slave ant An ant enslaved by members of another species.

slave auction An auction at which slaves are sold.

Slave Coast The coastal region of western Africa extending westward from the mouths of the Niger along the Bight of Benin to Ghana: named for its former trade in slaves.

slave-driv·er (slāv′drī′vər) *n.* 1 A person hired for or charged with the overseeing of slaves at work. 2 Any severe or exacting employer.

slave·hold·er (slāv′hōl′dər) *n.* An owner of slaves. — **slave′hold′ing** *adj.* & *n.*

slav·er¹ (slav′ər) *v.t.* To dribble saliva over. — *v.i.* To dribble saliva; drool. — *n.* Saliva issuing or dribbling from the mouth. [Prob. < ON *slafra*] — **slav′er·er** *n.*

slav·er² (slā′vər) *n.* 1 A person or a vessel engaged in the slave trade. 2 One who procures white slaves.

Slave River A river in NE Alberta and the southern region of Mackenzie district, Northwest Territories, Canada, flowing 258 miles NW to Great Slave Lake: also *Great Slave River.*

slav·er·y (slā′vər·ē, slāv′rē) *n.* 1 Involuntary servitude; specifically, the legalized social institution in which humans are held as property or chattels; complete subjection of one person to another. 2 Mental, moral, or spiritual bondage. 3 Slavish toil; drudgery. See synonyms under BONDAGE.

Slave State Any of the United States in which slavery was not prohibited by statute before the Civil War: Alabama, Arkansas, Delaware, Florida, Georgia, Kentucky, Louisiana, Maryland, Mississippi, Missouri, North Carolina, South Carolina, Tennessee, Texas, Virginia.

slave trade The business of dealing in slaves; specifically, the bringing of Negro slaves to America for sale. — **slave′trad′er** *n.*

slav·ey (slā′vē, slav′ē) *n. Brit.* A household servant; drudge; usually, a maidservant.

Slav·ic (slä′vik, slav′ik) *adj.* Of or pertaining to the Slavs or their language. — *n.* A branch of the Balto-Slavic subfamily of the Indo-European language family, consisting of three groups — **East Slavic** (Russian or Great Russian, Ukrainian or Ruthenian or Little Russian, White Russian), **West Slavic** (Czechoslovakian, Sorbian or Wendish, Polish), **South Slavic** (Church Slavonic, Bulgarian, Serbo-Croatian, Slovenian). Also *Slavonic.*

slav·ish (slā′vish) *adj.* 1 Pertaining to or befitting a slave; servile; base. 2 Extremely hard or laborious. 3 Enslaved. See synonyms under BASE², OBSEQUIOUS. — **slav′ish·ly** *adv.* — **slav′ish·ness** *n.*

Slav·ism (slä′viz·əm, slav′iz·əm) *n.* The characteristics or aims of the Slavs, collectively.

Slavo- *combining form* Slavic; of or pertaining to the Slavs: *Slavophobe.* [< SLAV]

slav·oc·ra·cy (slāv·ok′rə·sē) *n. pl.* **·cies** Slaveholders or slaveholding interests as a political power, especially for the maintenance of slavery. [< SLAV(E) + -(O)CRACY] — **slav·o·crat** (slā′və·krat) *n.* — **slav′o·crat′ic** *adj.*

Sla·vo·ni·a (slə·vō′nē·ə) A region of Croatia, northern Yugoslavia, between the Sava and Drava rivers. — **Sla·vo′ni·an** *adj.* & *n.*

Sla·von·ic (slə·von′ik) *adj.* & *n.* Slavic. — **Church Slavonic** A member of the South Slavic group of the Balto-Slavic languages: now in use only as the language of the Slavic Greek Orthodox Church and certain Roman Catholic dioceses. See GLAGOL.

Slav·o·phile (slāv′ə·fīl, -fil, slav′-) *n.* An admirer of the Slavs or their ideas, institutions, art, etc. Also **Slav′o·phil** (-fil). [< SLAVO- + -PHILE] — **Sla·voph·i·lism** (slə·vof′ə·liz′əm) *n.*

Slav·o·phobe (slāv′ə·fōb, slav′-) *n.* One who is fearful of the Slavs or Slavic influence.

slaw¹ (slô) *n.* Cabbage sliced, shredded, or chopped, and served, usually raw, as a salad. [< Du. *sla,* short for *salade* salad]

slaw² (slô) *adj. Scot.* Slow.

slay (slā) *v.t.* **slew, slain, slay·ing** 1 To kill, especially by violence; put to death; destroy by, or as by, killing. 2 *Obs.* To smite; strike. See synonyms under KILL¹. ◆ Homophones: *sleigh, sley.* [OE *slēan*] — **slay′er** *n.*

sleave (slēv) *v.t.* **sleaved, sleav·ing** To separate, as a mass of threads; disentangle. — *n.* Something tangled, matted, knotted, or unspun, as silk or thread. ◆ Homophone: *sleeve.* [OE *slēfan* divide]

sleave silk Raw untwisted silk; floss.

slea·zy (slē′zē, slā′-) *adj.* **·zi·er, ·zi·est** Lacking firmness of texture or substance. [Origin uncertain] — **slea′zi·ly** *adv.* — **slea′zi·ness** *n.*

sled (sled) *n.* 1 A vehicle on runners, designed for carrying people or loads over snow and ice; a sledge. 2 A small, light frame mounted on runners, used especially by children for sliding on snow and ice. — *v.* **sled·ded, sled·ding** *v.t.* To convey on a sled. — *v.i.* To ride on or use a sled. [< MLG *sledde*]

sled·der (sled′ər) *n.* 1 One who rides on or hauls with a sled. 2 An animal that draws a sled.

sled·ding (sled′ing) *n.* 1 Condition of roads admitting of the use of sleds: usually with a qualifying word: fine *sledding.* 2 The act of using a sled; use of sleds in hauling, traveling, etc. 3 State or circumstances of progress, work, etc.: We have had hard *sledding.*

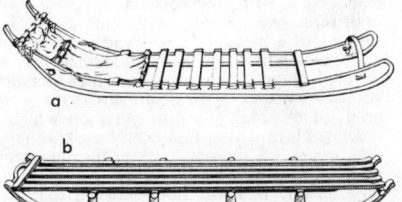

SLEDGES
a. Peary North Pole Expedition.
b. Byrd Antarctic Expedition.

sledge¹ (slej) *n.* A vehicle mounted on low runners for moving loads; especially, one designed to be drawn over snow and ice by dogs, horses, or reindeer, or one designed to be drawn on the ground by draft animals; also, a sled. — *v.t.* & *v.i.* **sledged, sledg·ing** To travel or convey on a sledge. [< MDu. *sleedse*]

sledge² (slej) *n.* A heavy hammer wielded with one or both hands, for blacksmiths' use, or for breaking stone, coal, etc.: also **sledge′ham′mer** (-ham′ər). — *v.t.* **sledged, sledg·ing** To hammer, break, or strike with a sledge. [OE *slecg*]

slee (slē) *adj. Scot.* Sly; dexterous.

sleek (slēk) *adj.* 1 Smooth and glossy; polished. 2 Smooth-spoken; flattering; unctuous; insinuating. See synonyms under SMOOTH. — *v.t.* 1 To make smooth, even, or glossy; polish. 2 To soothe; mollify; also, to make less disagreeable or offensive. Also, *U.S., slick.* [Var. of SLICK] — **sleek′ly** *adv.* — **sleek′ness** *n.* — **sleek′y** *adj.*

sleek·it (slēk′it) *adj. Scot.* 1 Sleek. 2 Deceitful.

sleep (slēp) *n.* 1 A state or period of complete or partial unconsciousness, normal and periodic in man and the higher animals. In animals it is sometimes much prolonged, as in hibernation. 2 A period of slumber. 3 Any condition of inactivity, torpor, or rest; specifically, the sleep of the grave; death. 4 Nyctitropism. See synonyms under REST¹. — *v.* **slept, sleep·ing** *v.i.* 1 To be or fall asleep; slumber. 2 To be in a state resembling sleep; to be dormant, inactive or quiet, or to rest in death. 3 To be in a benumbed state from retarded circulation of the blood: My foot *sleeps.* 4 To spin with such velocity as to be without apparent motion, as a top. 5 To undergo nyctitropism. 6 *Bot.* To assume a different position at night, as petals. — *v.t.* 7 To rest or repose in: with a cognate object: to *sleep* the sleep of the dead. 8 To provide with sleeping quarters; lodge: The hotel can *sleep* a hundred guests. See synonyms under REST¹. — **to sleep away** (or **off** or **out**) To pass or get rid of by sleep or as by sleep: to *sleep off* a hang-over. — **to sleep on** To postpone a decision upon. [OE *slēp*]

Sleep may appear as a combining form in hyphemes; as in:

sleep-bringing	sleep-inducing
sleep-compelling	sleep-inviting
sleep-dispelling	sleep-loving
sleep-disturber	sleep-producing
sleep-disturbing	sleep-provoking
sleep-filled	sleep-resisting

sleep·er (slē′pər) *n.* 1 One who sleeps; figuratively, a dead person. 2 A railroad sleeping-car. 3 A hibernating animal. 4 In football, a member of the backfield or an end stationed far out at either side before the ball is put in motion. 5 A heavy beam resting on or in the ground, as a support for a roadway, rails, etc.; a like support of iron or stone; also, a timber on or near the ground for the lower joists of a building. 6 A deadman. 7 *U.S. Colloq.* A play, motion picture, or book which achieves unexpected and striking success.

sleep·ing-car (slē′ping·kär′) *n.* A passenger railroad car with accommodations for sleeping.

sleeping partner See under PARTNER.

sleeping pill *Med.* A sedative; especially, one of the barbiturates taken to relieve acute or persistent insomnia.

sleeping sickness *Pathol.* 1 The terminal stage of a form of trypanosomiasis prevalent in tropical Africa: it is caused by the presence in the cerebrospinal fluid of certain trypanosomes usually transmitted by the bite of the tsetse fly, and is marked by progressive lethargy, recurrent fever and headaches, terminating in somnolence and death. 2 Epidemic encephalitis lethargica.

sleep·less (slēp′lis) *adj.* Unable to sleep; wakeful; restless; unquiet. See synonyms under VIGILANT. — **sleep′less·ly** *adv.* — **sleep′less·ness** *n.*

sleep·walk·er (slēp′wô′kər) *n.* A somnambulist. — **sleep′walk′ing** *n.*

sleep·y (slē′pē) *adj.* **sleep·i·er, sleep·i·est** 1 Inclined to sleep. 2 Drowsy; sluggish; dull; heavy. 3 Conducive to sleep. — **sleep′i·ly** *adv.* — **sleep′i·ness** *n.*

sleep·y-head (slē′pē·hed′) *n.* A sleepy person. — **sleep′y-head′ed** *adj.*

sleet (slēt) *n.* 1 A mixture of snow or hail and rain. 2 A drizzle or shower of partly frozen rain, or rain that freezes as it falls. 3 A thin coating of ice, as on rails, wires, roads, etc. — *v.i.* To pour or shed sleet. [Akin to MLG *slote* hail] — **sleet'y** *adj.*

sleeve (slēv) *n.* 1 The part of a garment that serves especially as a covering for the arm. 2 *Mech.* **a** A tube surrounding something, as a shaft, for protection or to permit motion of itself or of the shaft. **b** A short pipe receiving the ends of two other pipes or rods; a sleeve coupling or sleeve valve. 3 *Electr.* The cylindrical contacting part of a telephone-circuit plug. — **up one's sleeve** Hidden but at hand. — *v.t.* **sleeved, sleev·ing** To furnish with a sleeve or sleeves. ◆ Homophone: *sleave.* [OE *slēfe*]

sleeve coupling *Mech.* A short tube for connecting shafts or pipes.

sleeve·less (slēv'lis) *adj.* 1 Having no sleeves. 2 *Archaic* Unprofitable; fruitless; futile: a *sleeveless* errand.

sleeve valve *Mech.* A valve consisting of a hollow slotted sleeve in the cylinder of an internal-combustion engine, operating with the piston to allow for intake or exhaust of gases.

sleigh (slā) *n.* A light vehicle with runners for use on snow and ice, adapted especially for pleasure use or travel, as distinguished from hauling. Compare SLED, SLEDGE[1]. — *v.i.* To ride or travel in a sleigh. ◆ Homophones: *slay, sley.* [<Du. *slee,* contraction of *slede* sledge] — **sleigh'er** *n.*

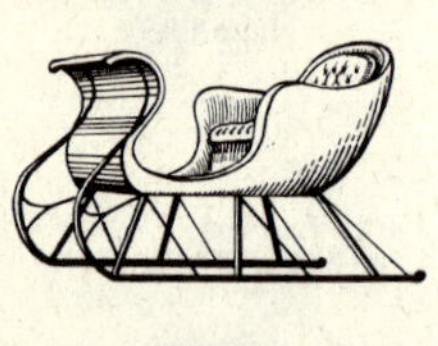
SLEIGH

sleigh·ing (slā'ing) *n.* 1 The act of riding in a sleigh. 2 The condition of the snow or ice that admits of using a sleigh.

sleight (slīt) *n.* 1 The quality of being skilful in manipulation; mechanical expertness; skill; dexterity. 2 A juggler's trick so deftly done that the manner of performance escapes observation; feat of legerdemain. 3 Craft; cunning. ◆ Homophone: *slight.* [<ON *slǣgdh* slyness]

sleight of hand 1 Skill in performing tricks in juggling. 2 The art or practice, or an instance, of legerdemain.

slen·dang (slen'däng) *n.* A scarf or shawl worn over the shoulders by women in the Philippines. [<Malay *sĕlendan*]

slen·der (slen'dər) *adj.* 1 Having a small diameter or circumference, in proportion to the length or height; slim; thin. 2 Having little strength or vigor; feeble; frail; delicate. 3 Having slight basis or foundation; of little validity. 4 Small or inadequate; moderate; insignificant: a *slender* income or diet. 5 Meagerly or insufficiently supplied: a *slender* table. 6 Thin in sound or quality; lacking volume. 7 *Phonet.* Denoting vowels which are pronounced with a narrow opening above the tongue, as (ē); close; narrow; opposed to *broad.* See synonyms under FINE[1], LITTLE, MINUTE[2]. [ME *slendre,* prob. <OF *esclendre*] — **slen'der·ly** *adv.* — **slen'der·ness** *n.*

slen·der·ize (slen'də·rīz) *v.t.* & *v.i.* **·ized, ·iz·ing** To make or become slender.

slept (slept) Past tense and past participle of SLEEP.

Sle·svig (sles'vikh) The Danish name for SCHLESWIG.

sleuth (slōōth) *n.* 1 *U.S. Colloq.* A detective. 2 A sleuthhound. 3 *Obs.* The track of a man or beast, as followed by the scent. — *v.t.* To follow, as a detective. — *v.i.* To play the detective. [<ON *slōdh* track, trail. Doublet of SLOT[2].]

sleuth·hound (slōōth'hound') *n.* A bloodhound.

slew[1] (slōō) Past tense of SLAY[1].

slew[2] (slōō) See SLOUGH[2].

slew[3] (slōō) *n. U.S. Colloq.* A large number, crowd, or amount; a lot: also spelled *slue.* [Cf. Irish *sluagh* a large crowd]

slew[4] (slōō) See SLUE[1].

sley (slā) *n.* 1 The reed guiding the warp threads of a loom. 2 In knitting machines, a groove, slot, or bar for directing the action of a part. — *v.t.* To separate and arrange the threads of (yarn) in a reed for weaving. ◆ Homophones: *slay, sleigh.* [OE *slege*]

Slez·sko (sles'kô) The Czech name for SILESIA.

slice (slīs) *n.* 1 A piece; especially, a thin, broad piece cut off from a larger body. 2 One of various tools or devices, used for slicing or resembling a slice in broadness and thinness; specifically, a broad knife used for serving fish, or a broad flat knife used by printers to remove ink; also, a druggist's spatula. 3 In golf, a blow delivered crosswise from right to left, causing the ball to curve to the right. — *v.* **sliced, slic·ing** *v.t.* 1 To cut or remove from a larger piece: often with *off.* 2 To cut into broad, thin pieces; divide; apportion. 3 To sunder, as with a sharp knife; split. 4 To clear out with a slice bar. 5 In golf, to hit (the ball) with a slice. — *v.i.* 6 In golf, to slice a ball. See synonyms under CUT. [<OF *esclice* <*esclicer* <OHG *slizan* slit] — **slic'er** *n.*

slice bar A thin, wide iron tool for cleaning clinkers from the grate bars of a furnace.

slick (slik) *adj.* 1 Smooth; slippery; sleek. 2 Flattering; obsequious; smooth-tongued; plausible. 3 *Colloq.* Dexterously done; cleverly said; specious; tricky. 4 Smart; clever: said of people. 5 Healthy; plump: said of animals. 6 Smooth; oily, as the surface of water. 7 Glazed, as paper; also, printed on glazed paper: *slick* magazines. 8 *Slang* Agreeable; excellent: a *slick* time. — *n.* 1 A smooth place on a surface of water, as from oil or the presence of fish; also, a sleek place in the fur or hair of an animal. 2 A broad chisel for paring or slicking: also **slick chisel.** 3 *pl. U.S.* Magazines printed on glazed paper: distinguished from *pulps.* — *adv. Slang* In a slick or smooth manner; deftly; quickly. — *v.t.* 1 To make smooth, trim, glossy, or oily. 2 *Colloq.* To trim up; make presentable: often with *up.* [ME *slike* <OE *slician* make smooth]

slick·en·sides (slik'ən·sīdz) *n. pl. Geol.* Polished and scratched or striated rock surfaces, exhibited on the opposed faces of veins or faults where they have moved one upon another. [<dial. E *slicken* slick + SIDE] — **slick'en·sid'ed** *adj.* — **slick'en·sid'ing** *n.*

slick·er (slik'ər) *n.* 1 An implement for dressing leather, having a wooden handle. 2 *U.S.* A waterproof overcoat of oilskin. 3 *Colloq.* A cheat; clever person.

slid (slid) Past tense and past participle of SLIDE.

slid·den (slid'n) Alternative past participle of SLIDE.

slid·der·y (slid'ər·ē) *adj. Scot.* Slippery; sly; deceitful; mutable.

slide (slīd) *v.* **slid, slid** or **slid·den, slid·ing** *v.i.* 1 To pass along over a surface with a smooth, slipping movement: to *slide* on ice. 2 To slip off, as scales in shedding. 3 To move or pass imperceptibly, smoothly, deftly, or easily; pass gradually or imperceptibly: The years *slide* away swiftly. 4 To move, pass, or proceed by sufferance merely; also, to take care of oneself or itself; go by default or without heed: with *let*: to let the matter *slide.* 5 *Music* To glide from tone to tone without breaking the sound. 6 To make a moral slip; err; sin. 7 To slip; lose one's equilibrium or foothold. 8 In baseball, to throw oneself along the ground toward a base, in order to avoid being tagged by the baseman. — *v.t.* 9 To cause to slide, as over a surface. 10 To move, put, enter, etc., with quietness or dexterity: with *in* or *into.* — *n.* 1 An act of sliding. 2 The slipping of a mass of earth, snow, etc., from a higher to a lower level; an avalanche. 3 An inclined plane or channel on which persons, goods, logs, etc., slide downward to a lower level. 4 A small plate of glass on which a specimen is mounted and examined through a microscope. 5 A small plate of transparent material bearing a single image for projection on a screen. 6 *Phot.* In a camera, that part of a plate holder which covers and uncovers the negative. 7 *Music* **a** A series of short musical notes leading smoothly to a principal note: a type of ornamentation. **b** A portamento. **c** In a trumpet or trombone, a U-shaped portion of the tubing which is pushed in and out to vary the pitch. 8 *Mech.* **a** A sliding part. **b** A groove, rail, etc., on which something slides. [OE *slīdan*]

slide fastener A fastening device for use on fabrics, dress goods, etc., having two rows of interlocking teeth or scoops which may be closed or separated by a sliding element: often called *zipper.*

slide·knot (slīd'not') *n.* A slipknot, particularly one made of two half-hitches on a fishing line.

Sli·dell (slī·del'), **John,** 1793–1871, American lawyer; Confederate agent to France in 1861.

Slide Mountain The highest peak of the Catskill Mountains, in SE New York, 18 miles west of Kingston; 4,204 feet.

slid·er (slī'dər) *n.* 1 One who or that which slides. 2 A ring or strap through which something slides.

slide rule A device consisting of a rigid ruler with a central sliding piece, both ruler and slide being graduated in a similar logarithmic scale to permit of rapid calculations.

slide valve *Mech.* 1 A sliding piece in the cylinder of a steam engine, regulated to move back and forth over the ports and connect them alternately with the boiler and the exhaust passage, thus imparting reciprocating motion to the piston. 2 A valve that slides on its seat.

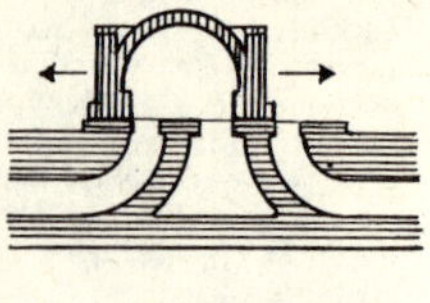
SLIDE VALVE
Arrows show reciprocal action.

sliding scale 1 A schedule affecting imports, prices, or wages, varying under conditions of consumption, demand, or market price of some article. 2 Any graduated scale, as in a clinometer or slide rule, designed to move against a fixed scale in order to facilitate rapid and accurate measurements and computations.

slight (slīt) *adj.* 1 Of small importance; small in quantity, intensity, or degree; inconsiderable. 2 Slender; frail; delicate; flimsy. 3 Of weak intellect or character. 4 *Scot.* Smooth; slippery; unscrupulous. See synonyms under FINE[1], FRAGILE, INSIGNIFICANT, LITTLE, SMALL. — *v.t.* 1 To manifest intentional neglect of or disregard for; snub; omit due courtesy toward or respect for: to *slight* a friend. 2 To omit due care in the doing or performance of; do imperfectly or thoughtlessly; shirk. 3 To treat as trivial or insignificant. — *n.* An act or omission involving failure in courtesy or respect toward another; any contemptuous or neglectful action. ◆ Homophone: *sleight.* [ME. Akin to ON *sléttr* smooth.] — **slight'ness** *n.*

Synonyms (noun): disregard, neglect, scorn. *Disregard* is chiefly a matter of intellectual estimate; *slight* is a matter of outward action; *neglect* may be of thought or act. *Disregard* of a thing is setting it aside as not worthy of regard. *Neglect* of a person or thing may be the result of ignorance, thoughtlessness, or preoccupation with other things; a *slight* is an intentional omission of kindness, courtesy, or attention. *Scorn* expresses mingled contempt and bitterness. See NEGLECT. *Antonyms*: esteem, honor, regard, respect, reverence.

slight·ing (slī'ting) *adj.* Conveying, containing, or characterized by a slight: a *slighting* remark. — **slight'ing·ly** *adv.*

slight·ly (slīt'lē) *adv.* In a slight manner; inconsiderably; partially; carelessly.

Sli·go (slī'gō) 1 A county of NW Ireland in Connacht province; 694 square miles. 2 Its county town, a port on **Sligo Bay,** an inlet of the Atlantic extending 7 miles into County Sligo; 10 miles wide.

sli·ly (slī'lē) *adv.* In a sly manner: also spelled *slyly.*

slim (slim) *adj.* **slim·mer, slim·mest** 1 Small in thickness in proportion to height or length, as a human figure or a tree. 2 Having little logical strength; weak. 3 Constructed unsubstantially; flimsy. 4 Lacking robustness; frail. 5 Insufficient; narrow; meager: a *slim* attendance; a *slim* chance. 6 *Brit. Dial.* Sly; crafty; worthless; bad. — *v.t.* & *v.i.* **slimmed, slim·ming** To make or become thin or thinner. [<Du. *slim* bad] — **slim'ly** *adv.* — **slim'ness** *n.*

slime (slīm) *n.* 1 Any soft, sticky, or dirty thing; hence, any offensive quality or thing. 2 Soft, moist, adhesive mud or earth; muck. 3 A mucous exudation from the bodies of certain animals, as fishes and snails, and certain plants. 4 Bitumen; asphalt. 5 *Usually pl.* A mudlike substance formed of ore in an almost impalpable powder, mixed with water. — *v.* **slimed, slim·ing** *v.t.* 1 To smear or cover with or as with slime. 2 To remove slime from, as fishes. — *v.i.* 3 To become covered with or as with slime. [OE *slīm*]

slime flux A watery or viscous flow from the injured bark of various deciduous trees, sometimes providing a medium for parasitic growths.
slime mold A fungus belonging to the class *Myxomycetes.* Also **slime fungus.**
slim·sy (slim′zē) *adj. Colloq.* **1** Utterly limp, as from fatigue or illness. **2** Lacking in stiffness or texture, as limp fabric; flimsy. Also **slimp·sy** (slimp′sē). [Blend of SLIM and FLIMSY]
slim·y (slī′mē) *adj.* **slim·i·er, slim·i·est 1** Covered or bedaubed with slime. **2** Containing slime. **3** Slimelike; foul. **— slim′i·ly** *adv.* **— slim′i·ness** *n.*
sling[1] (sling) *n.* **1** A strap or pocket with a string attached to each end, for hurling a stone or other missile by centrifugal force. **2** One of various ropes, straps, chains, or the like, for suspending or hoisting something, for holding up an injured limb, lifting and supporting an animal, in case of lameness or other need, carrying a rifle, etc. **3** *Naut.* A rope or chain by which a lower yard or a gaff is suspended; also, in the plural, the middle portion of a yard. **4** The act of slinging; a sudden throw; cast; fling. [< *v.*] **—** *v.* **slung, sling·ing** *v.t.* **1** To fling from or as from a sling; hurl. **2** To place or hang up in or as in a sling; move or hoist, as by a rope or tackle. **—** *v.i.* **3** To move at an easy gait. See synonyms under SEND[1]. [<ON *slyngva* hurl] **— sling′er** *n.*
sling[2] (sling) *n. U.S.* A drink of brandy, whisky, or gin, with sugar and nutmeg, lemon juice, and hot or cold water. **—** *v.i. U.S. Colloq.* To drink slings; take an alcoholic drink. [Cf. G *schlingen* swallow]
sling·shot (sling′shot′) *n.* A weapon or toy consisting of a forked stick with an elastic strap attached to the prongs for catapulting small missiles.
slink (slingk) *v.* **slunk** (*Obs.* **slank**), **slunk, slink·ing** *v.i.* To creep or steal along furtively or stealthily, as in fear. **—** *v.t.* To give birth to prematurely; miscarry: said of animals, especially cows. **—** *adj.* Produced prematurely, as a calf; too immature to be eaten. **—** *n.* An animal, especially a calf, prematurely born; also, its flesh, too immature for proper food. [OE *slincan* creep] **— slink′ing·ly** *adv.*
slink·y (slingk′ē) *adj.* **slink·i·er, slink·i·est 1** Sneaking; stealthy. **2** *Slang* Sinuous or feline in movement or form.
slip[1] (slip) *v.* **slipped** or **slipt, slip·ping** *v.t.* **1** To cause to move smoothly and easily; cause to glide or slide. **2** To put on or off easily, as a ring or a loose garment. **3** To convey slily or secretly. **4** To free oneself or itself from, as a fetter or bridle. **5** To let loose; unleash, as hounds. **6** To release from its fastening and let run out, as a cable. **7** To give birth to prematurely; slink; cast: said of animals. **8** To dislocate, as a bone. **9** To escape or pass unobserved: It *slipped* my mind. **10** To overlook; omit negligently: to *slip* an opportunity. **—** *v.i.* **11** To slide so as to cause harm or inconvenience; lose one's footing; become misplaced by failing to hold. **12** To fall into an error or fault; err. **13** To escape, as a ship. **14** To move smoothly and easily; slide; glide. **15** To get free of restraint; be unleashed. **16** To go or come stealthily or unnoticed: often with *off, away,* or *from.* **— to let slip** To say without intending to. **—** *n.* **1** An act of slipping; a sudden slide. **2** A lapse or error in speech, writing, or conduct; a slight mistake. **3** *U.S.* A narrow space between two wharves. **4** An artificial pier sloping down to the water, serving as a landing place. **5** An inclined plane leading down to the water, on which vessels are repaired or constructed. **6** A woman's undergarment. **7** A pillowcase. **8** A leash containing a device which permits quick release of the dog. **9** In cricket, a position on the off side a few yards behind the wicket; also, the player who stands at this position. **10** *Naut.* **a** The difference between the speed of a screw propeller and that of the ship. **b** The velocity of the back current generated by a propeller. **11** *Physics* The difference between the advance made by a propeller moving in a fluid and the advance it would make if moving in a solid substance. **12** *Mech.* **a** The relative motion of two surfaces which are meant to be immovable with respect to each other, as a belt on a pulley. **b** Allowance made for slipping or play, as between connected members of a mechanism; slippage. **13** *Geol.* A small dislocation of rock strata. **— to give (someone) the slip** To elude (someone). [<MLG *slippen*]
slip[2] (slip) *n.* **1** A cutting from a plant for planting or grafting; a cion. **2** A small, slender person, especially a youthful one. **3** A small piece of something, as of paper or cloth, rather long relative to its width; a strip. **4** A small piece of paper for jotting down memoranda, a record, etc. **5** *U.S.* A narrow pew in a church. **—** *v.t.* **slipped, slip·ping** To cut off for planting; make a slip or slips of. [<MDu. *slippe* < *slippen* cut]
slip[3] (slip) *n.* Liquid potter's clay, used for decorating and coating rough surfaces. [OE *slype, slypa*]
slip·cov·er (slip′kuv′ər) *n.* **1** A fitted cloth cover for a chair, couch, sofa, or other piece of furniture, that can be readily removed. **2** A paper or cloth jacket for a book.
slip·knot (slip′not′) *n.* **1** A knot so formed, by having part of the material drawn through in a bow, as to be readily untied: also called *bowknot.* **2** A running knot.
slip–on (slip′on′, -ôn′) *n.* A garment which can be easily donned or taken off. **—** *adj.* Denoting such a garment: a *slip–on* blouse.
slip–o·ver (slip′ō′vər) *adj.* Designating a garment easily donned by drawing over the head: a *slip–over* shirt. **—** *n.* A garment of this type.
slip·page (slip′ij) *n.* **1** The amount by which or distance through which anything slips, as a screw propeller. **2** The difference between actual and calculated speed, due to slipping. **3** The act of slipping; slip.
slip·per[1] (slip′ər) *n.* **1** A low, light shoe, chiefly for indoor wear, into or out of which the foot is easily slipped. **2** One who or that which slips. [<SLIP[1] + -ER[2]]
slip·per[2] (slip′ər) *adj. Archaic* Slippery; deceitful; unreliable; forgetful; voluble. [OE *slipor*]
slip·pered (slip′ərd) *adj.* Wearing slippers.
slip·per·wort (slip′ər·wûrt′) *n.* Calceolaria.
slip·per·y (slip′ər·ē) *adj.* **·per·i·er, ·per·i·est 1** Having a surface so smooth that bodies slip or slide easily on it. **2** That evades one's grasp; tricky; elusive. **3** Unreliable; undependable; tricky. **4** *Obs.* Wanton. [<SLIPPER[2] + -Y[3]] **— slip′per·i·ly** *adv.* **— slip′per·i·ness** *n.*
slippery elm 1 A tree (*Ulmus fulva*) of eastern North America. **2** Its hard wood. **3** Its mucilaginous inner bark, used in medicine as a nutritious demulcent.
slip–ring (slip′ring′) *n. Electr.* One of two or more metal rings of an electric machine serving, through contact with stationary brushes, to deliver or transmit a current.
slip–sheet (slip′shēt′) *Printing n.* A blank piece of paper interleaved between newly printed press sheets to prevent offset. **—** *v.t.* To insert slip–sheets in.
slip·shod (slip′shod′) *adj.* Wearing shoes or slippers down at the heels; hence, slovenly. [<SLIP[1] + SHOE]
slip·slop (slip′slop′) *n. Colloq.* **1** Sloppy victuals; any weak drink; slop. **2** A blunder, as in speaking.
slip·stream (slip′strēm′) *n. Aeron.* The stream of air driven backwards by the propeller of an aircraft: also called *race.*
slip–up (slip′up′) *n. Colloq.* A mistake; error.
slip·way (slip′wā′) *n.* A slip (def. 5).
slit (slit) *n.* A cut that is relatively straight and long; also, a long, narrow opening. [< *v.*] **—** *v.t.* **slit, slit·ting 1** To make a long incision in; slash. **2** To cut lengthwise into strips. See synonyms under REND. [ME *slitten* cut] **— slit′ter** *n.*
slith·er (slith′ər) *v.i.* **1** To slide; slip, as on a loose surface. **2** To glide, as a snake. **—** *v.t.* **3** To cause to slither. [Var. of dial. E *slidder* <OE *slidrian,* freq. of *slīdan* slide] **— slith′er·y** *adj.*
slit trench A narrow, shallow trench, similar to a foxhole.
sliv·er (sliv′ər) *n.* **1** A slender piece, as of wood, cut or torn off lengthwise; a splinter. **2** Corded textile fibers drawn into a fleecy strand. **3** A piece cut longitudinally from the side of a fish: used as bait; also, a filet. **—** *v.t.* & *v.i.* To cut or split into long thin pieces; splinter. [< dial. E *slive* cleave] **— sliv′·er·er** *n.*
Sliv·no (slēv′nō) A city in east central Bulgaria. Also **Sliv·en** (slē′vən).
sli·vo·vitz (slē′vô·vēts) *n.* A white, dry plum brandy drunk especially in central European countries. [<Serbo–Croatian < *sliva* a plum]
Sloan (slōn), **John,** 1871–1951, U.S. painter.
slob (slob) *n.* **1** Mud; mire. **2** Slush; mushy snow. **3** *Slang* A stupid, careless, or unclean person. [<Irish *slab,* prob. <SLAB[2]]
slob·ber (slob′ər) *v.t.* **1** To wet and foul with liquids oozing from the mouth. **2** To shed or spill, as liquid food, in eating. **—** *v.i.* **3** To drivel; slaver. **4** To talk or act gushingly. **—** *n.* **1** Liquid spilled as from the mouth; slaver. **2** Gushing, sentimental talk. [Var. of SLABBER] **— slob′ber·er** *n.* **— slob′ber·y** *adj.*
slock (slok) *v.t. Brit. Dial.* To slake; drench; extinguish (a fire). Also **slock′en.**
sloe (slō) *n.* **1** A small, plumlike, astringent fruit. **2** The shrub (*Prunus spinosa*) that bears it; the blackthorn. **3** The blackhaw. **4** The wild yellow plum (*Prunus americana*); also, the Allegheny plum (*P. alleghaniensis*). ◆ Homophone: *slow.* [OE *slā*]
sloe–eyed (slō′īd′) *adj.* Having eyes dark as sloes.
sloe gin A cordial with a gin base; flavored with sloes.
slog (slog) *v.t.* & *v.i.* **slogged, slog·ging 1** To slug, as a pugilist. **2** To plod (one's way), as through deep mud. **—** *n.* A heavy blow. [Var. of SLUG[3]] **— slog′ger** *n.*
slo·gan (slō′gən) *n.* **1** A battle or rallying cry: originally of the Highland clans. **2** A catchword or motto adopted by a manufacturer, political party, or the like. [<Scottish Gaelic *sluagh* army + *gairm* yell]
sloid (sloid), **slojd** See SLOYD.
sloop (slo͞op) *n. Naut.* A single–masted, fore–and–aft rigged sailing vessel having a fixed bowsprit and carrying at least one jib: now used principally as a racing vessel. [<Du. *sloep*]

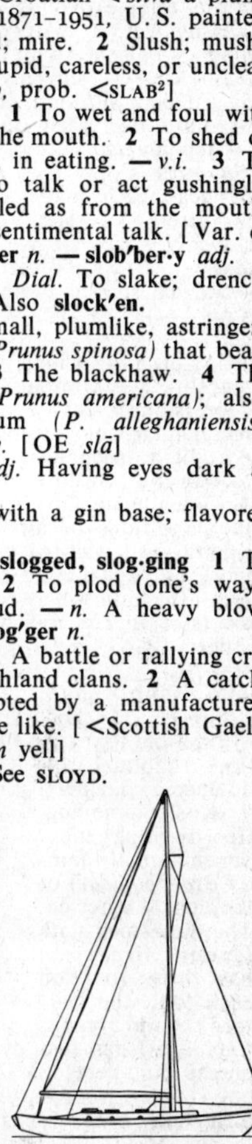

SLOOP

sloop of war In old navies, a vessel rigged either as ship, brig, or schooner, and mounting between 18 and 32 guns; later, any war vessel larger than a gunboat and carrying guns on one deck only.
slop[1] (slop) *v.* **slopped, slop·ping** *v.i.* **1** To splash or spill. **2** To walk or move through slush. **—** *v.t.* **3** To cause (a liquid) to spill or splash. **4** *U.S.* To feed (a domestic animal) with slops. **— to slop over 1** To overflow and splash. **2** To do or say more than is necessary, because of excess zeal, sentimentality, etc. **—** *n.* **1** Slush; watery mud. **2** A dash or puddle of liquid that has been slopped. **3** An unappetizing liquid or watery food. **4** Refuse liquid. **5** *pl.* Waste food or swill, as from a kitchen, used to feed cattle, pigs, etc. **6** *pl.* Distiller's mash which has been deprived of its alcohol. [ME *sloppe*]
slop[2] (slop) **1** *Obs.* A loose outer garment; a smock; in the plural, wide baggy breeches. **2** *pl.* Articles of clothing and other merchandise sold to sailors on shipboard. [ME *sloppe*]
slope (slōp) *v.* **sloped, slop·ing** *v.i.* **1** To be inclined from the level or the vertical; slant. **2** To move on an inclined path; go obliquely. **3** *Colloq.* To leave suddenly; run off. **—** *v.t.* **4** To cause to slope. See synonyms under TIP[1]. [< *adj.*] **—** *n.* **1** Any slanting surface or line; a declivity or acclivity; an inclined plane: the Atlantic *slope* of North America. **2** The degree of inclination of a line or surface from the plane of the horizon. **3** *Math.* **a** The tangent of the positive angle of less than 180° made between the *x*–axis and a tangent to a curve traced in the Cartesian coordinate system; also, the derivative of such a curve at a given point. **b** The tangent of the positive

angle of less than 180° made between the *x*-axis and a straight line traced in the Cartesian coordinate system. —*adj.* Slanting; oblique. [Aphetic var. of *aslope*, OE *āslopen*, ppr. of *āslūpan* slip away] —**slop′er** *n.* —**slop′·ing** *adj.* —**slop′ing·ly** *adv.* —**slop′ing·ness** *n.*

slop·o·ver (slop′ō′vər) *n.* A breakover.

slop·py (slop′ē) *adj.* **·pi·er, ·pi·est 1** Slushy; splashy; wet. **2** Watery or pulpy: *sloppy* pudding. **3** Splashed with liquid or slops. **4** *Colloq.* Messy; slovenly; extremely untidy. **5** *Colloq.* Slipshod; careless. **6** *Colloq.* Maudlin; overly sentimental. —**slop′pi·ly** *adv.* —**slop′pi·ness** *n.*

slop·work (slop′wûrk′) *n.* **1** The manufacture of cheap ready-made clothing; also, the clothing itself. **2** Hence, any inferior, cheap, slovenly work.

slosh (slosh) *v.t.* To throw about, as a liquid. —*v.i.* To splash; flounder: to *slosh* through a pool. —*n.* Slush. [Var. of SLUSH] —**slosh′y** *adj.*

slot[1] (slot) *n.* **1** A long narrow groove or opening; slit. **2** A comparatively long and narrow depression or cavity, particularly one that is rectangular, cut to receive some corresponding part in a mechanism. **3** The opening to receive the coin in a slot machine. **4** *Aeron.* An opening in an airplane wing to improve the conditions of airflow at high angles of flight. —*v.t.* **slot·ted, slot·ting 1** To adjust in a slot. **2** To cut a slot in; groove. [<OF *esclot* the hollow between the breasts]

slot[2] (slot) *n.* The trail of an animal, especially a deer. [<AF *esclot* <ON *slōdh*. Doublet of SLEUTH.]

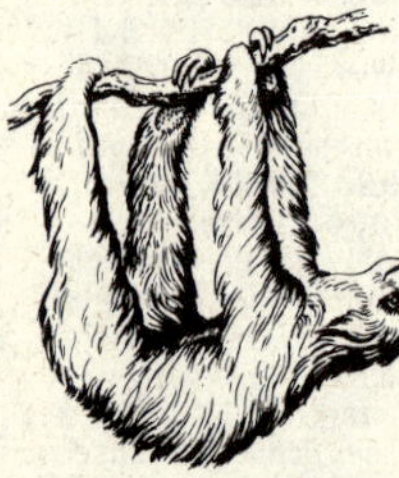

THREE-TOED SLOTH (Head and body about 21 inches long)

sloth (slōth, slôth, sloth) *n.* **1** Disinclination to exertion; habitual indolence; laziness. **2** A slow-moving, tree-dwelling edentate mammal (family *Bradypodidae*) of tropical America. The **three-toed sloth** (genus *Bradypus*) has three toes on each foot; the **two-toed sloth** (genus *Choloepus*) has two on the front and three on the hind feet. **3** A related fossil edentate (family *Megatheriidae*). [<SLOW + -TH[1]]

sloth bear A black bear of India and Ceylon (genus *Melursus*), feeding mainly on honey and fruit.

sloth·ful (slōth′fəl, slôth′-, sloth′-) *adj.* Sluggish; lazy; indolent. See synonyms under IDLE. —**sloth′ful·ly** *adv.* —**sloth′ful·ness** *n.*

slot machine A vending machine or gambling machine having a slot in which a coin is dropped to cause operation.

slouch (slouch) *v.i.* **1** To have a downcast or drooping gait, look, or posture. **2** To hang or droop in a careless manner, as a hat. —*n.* **1** A hanging down awkwardly or carelessly; movement or appearance caused by depression or drooping. **2** An awkward, heavy, or incompetent person. [Origin uncertain] —**slouch′y** *adj.* —**slouch′i·ly** *adv.* —**slouch′·i·ness** *n.*

slough[1] (slou *for defs. 1 and 4*; slōō *for defs. 2 and 3*) *n.* **1** A place of deep mud or mire; bog. **2** A depression in a prairie, often dry but sometimes deeply miry, forming part of the natural drainage system. **3** A stagnant swamp, backwater, bayou, inlet, or pond in which water backs up: also spelled *slew, slue.* **4** A state of moral depravity or of despair. [OE *slōh*] —**slough′y** *adj.*

slough[2] (sluf) *n.* **1** Dead tissue separated and thrown off from the living parts, as in gangrene; also, a scab. **2** The skin of a serpent that has been or is about to be shed; cast. —*v.t.* **1** To cast off, as dead from living tissue; shed. **2** To discard; shed, as a habit or a growth; get rid of as useless or needless. —*v.i.* **3** To be cast off. **4** To cast off a slough or tissue; form a scab. [ME *slouh*] —**slough′y** *adj.*

Slough (slou) A municipal borough of SE Buckingham, England.

Slo·vak (slō′vak, slō·vak′) *n.* **1** One of a Slavic people of NW Hungary and parts of Moravia, who united with the Czechs to form the Czechoslovak republic in 1918: now in central Czechoslovakia. **2** The dialect of Czechoslovakian spoken by the Slovaks. —*adj.* Of or pertaining to the Slovaks or to their language. Also **Slo·vak·i·an** (slō·vak′ē·ən, -vä′kē·ən). [<Czech *slovák* a Slav]

Slo·vak·i·a (slō·vak′ē·ə, -vä′kē·ə) The eastern geographical region and a former province (1920–39 and 1945–49) of Czechoslovakia; an independent state, under German protection, 1939–45; divided into six administrative regions, 1949; 18,897 square miles; capital, Bratislava. *Czech* **Slo·ven·sko** (slō′ven·skō).

slov·en (sluv′ən) *n.* One who is careless of dress or of cleanliness; one habitually untidy. [Cf. Flemish *sloef* dirty] —**slov′en·li·ness** *n.* —**slov′en·ly** *adj.* & *adv.*

Slo·vene (slō′vēn, slō·vēn′) *n.* One of a group of southern Slavs now living in NW Yugoslavia. —*adj.* Of or pertaining to the Slovenes or to their language. [<G *Slowene*]

Slo·ve·ni·a (slō·vē′nē·ə) A constituent republic of NW Yugoslavia; 7,717 square miles; capital, Ljubljana: formerly part of Austria.

Slo·ve·ni·an (slō·vē′nē·ən) *adj.* Of or pertaining to Slovenia, its people, or their language. —*n.* The South Slavic language of the Slovenes.

slow (slō) *adj.* **1** Having relatively small velocity; not quick in motion, performance, or occurrence; not advancing or growing rapidly. **2** Behind the standard time: said of a timepiece. **3** Taking sufficient time; not precipitate or hasty: *slow* to anger. **4** Dull or tardy in comprehending; mentally sluggish: a *slow* student. **5** Lacking promptness, spirit, or liveliness; also, colloquially, dull or tedious in character. **6** Denoting a condition of a racetrack that retards the horses' speed, but in less degree than a muddy or heavy track: a *slow* track. —*v.t.* **1** To make slow or slower; cause to go at a slower pace; slacken in speed: often with *up* or *down.* **2** To retard; delay. —*v.i.* **3** To go or become slow or slower: often with *up* or *down.* —*adv.* In a slow or cautious manner or speed. ◆ Homophone: *sloe.* [OE *slāw*] —**slow′ly** *adv.* —**slow′ness** *n.*

Synonyms (adj.): deliberate, dilatory, drowsy, dull, gradual, inactive, inert, lingering, moderate, slack, sluggish, tardy. *Tardy* is applied to that which is behind the proper or desired time, especially in doing work or arriving at a place; *slow* applies to that which is a relatively long time in passing from one point to another, or in beginning or executing something. A person is *deliberate* who takes a noticeably long time to consider and decide before acting, or who acts or speaks as if he were deliberating at every point; a person is *dilatory* who lays aside, or puts off as long as possible, necessary or required action. *Gradual* signifies advancing by steps, and refers to *slow* but regular and sure progression. *Slack* refers to action that seems to indicate a lack of tension, as of muscle or of will, *sluggish* to action that seems as if reluctant to advance. See GRADUAL, HEAVY, RELUCTANT, TEDIOUS. *Antonyms*: see synonyms for IMPETUOUS, NIMBLE.

slow-down (slō′doun′) *n.* A slackening of pace.

slow match A slowly burning fuse used in firing explosives.

slow-mo·tion (slō′mō′shən) *adj.* Pertaining to or designating a motion picture filmed at greater than standard speed so that the action appears slow in normal projection.

slow·poke (slō′pōk′) *n. Slang* A person who works or moves at an exceedingly slow pace; a laggard.

slow·worm (slō′wûrm′) *n.* A blindworm.

sloyd (sloid) *n.* A system of elementary manual training originating in Sweden, having exercises graduated from the simplest use of tools to the most complete joinery: also spelled *sloid, slojd.* [<Sw. *slöjd* skill]

slub (slub) *v.t.* **slubbed, slub·bing** To twist (slivers of wool) slightly in preparation for spinning. —*n.* **1** A slightly twisted roll of cotton, wool, or silk. **2** A thick, uneven lump in yarn. [Origin unknown]

sludge (sluj) *n.* **1** Soft, water-soaked mud; mire. **2** A slush of snow or broken or half-formed ice. **3** Muddy or pasty refuse of various kinds, as that produced by the action of a rock drill and in the purification of sewage. **4** The sediment in a water tank or boiler. [<earlier *slutch.* ? Related to SLUSH.] —**sludg′y** *adj.*

slue[1] (slōō) *v.* **slued, slu·ing** *v.t.* **1** To cause to move sidewise, as if some portion were pivoted; swing, slide, or skid to the side. **2** To cause to twist or turn in its seat or fastenings: said of a boom or mast. —*v.i.* **3** To move sidewise. —*n.* The act of sluing around sidewise; a skidding or pivoting about; also, the position of a body that has slued. Also spelled *slew.* [Origin unknown]

slue[2] (slōō) See SLEW[3].

slue[3] (slōō) See SLOUGH[1] (def. 3).

slug[1] (slug) *n.* **1** A bullet or shot of irregular or oblong shape, especially as used in old muskets. **2** *Printing* **a** A strip of type metal, thicker than a lead and less than type-high, for spacing matter, etc. **b** A metal strip bearing a type-high number, abbreviated title, or the like, used as a compositor's mark. **3** A slungshot, or its metal weight. **4** Any small chunk of metal; especially, one used as a coin in automatic machines, as dial telephones. **5** *Physics* A unit of mass; the mass of a body which, when acted upon by a force of one pound, acquires an acceleration of one foot per second per second: it has the value of about 32.174 pounds or 14.59 kilograms, and is also called *geepound.* —*v.* **slugged, slug·ging** *v.i.* To take shape to fit the grooves of a rifle, as a bullet. —*v.t.* To load with slugs. [Origin uncertain. ? Akin to SLAG.]

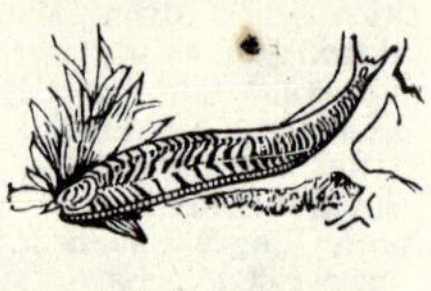

SLUG

slug[2] (slug) *n.* **1** A gastropod (order *Pulmonata*) related to the snail, having an elongated body and a rudimentary shell concealed in the mantle, especially of the genus *Limax.* **2** The larva of a sawfly or other insect resembling a gastropod. **3** A sluggard. [ME *slugge* a sluggard, ? < Scand. Cf. dial. Norw. *slugg* a large, heavy object.]

slug[3] (slug) *Colloq. n.* **1** A heavy blow, as with the fist or a baseball bat. **2** A drink of undiluted liquor. —*v.t.* **slugged, slug·ging** To strike heavily or brutally, or without science, as with the fist or a baseball bat. [Origin uncertain] —**slug′ger** *n.*

slug·a·bed (slug′ə·bed′) *n.* One who lounges late in bed, because of laziness.

slug·gard (slug′ərd) *n.* A person habitually lazy or idle; a drone. —*adj.* Lazy; sluggish. [< SLUG[2] + -ARD]

slug·gish (slug′ish) *adj.* **1** Having little motion or power of motion; slow; inactive; torpid. **2** Habitually idle and lazy. **3** Not active; slow; stagnant: a *sluggish* season. See synonyms under HEAVY, IDLE, SLOW, TEDIOUS. —**slug′·gish·ly** *adv.* —**slug′gish·ness** *n.*

sluice (slōōs) *n.* **1** Any artificial channel for conducting water, or the stream so conducted; specifically, a body of water controlled by a floodgate. **2** A floodgate. **3** A flume. **4** *Mining* A board trough having at the bottom baffles holding quicksilver to separate gold from placer dirt carried through the trough by a current of water. **5** That through which anything issues or flows. —*v.* **sluiced, sluic·ing** *v.t.* **1** To wet or drench, water or irrigate by or as by means of a sluice. **2** To wash in or by a sluice. **3** To draw out or conduct by or through a sluice. **4** To send (logs) down a sluiceway. —*v.i.* **5** To flow out or issue from a sluice. [<OF *escluse* <L *exclusa,* pp. fem. of *excludere* shut out]

sluice·gate (slōōs′gāt′) *n.* The gate of a sluice; a watergate or floodgate.

sluice·way (slōōs′wā′) *n.* An artificial channel for the passage of water; a sluice, as in mining; flume.

sluit (slōōt) *n. Afrikaans* A narrow, natural or artificial channel, through which water flows; a gully.

slum[1] (slum) *n.* A squalid, dirty, overcrowded street or section of a city, marked by the poverty and poor living conditions of its inhabitants. —*v.i.* **slummed, slum·ming** To visit slums, as for reasons of curiosity or philanthropy. [< slang E, a room; ult. origin unknown] —**slum′mer** *n.* —**slum′ming** *n.*

slum[2] (slum) *n. Slang* Slumgullion.

slum·ber (slum′bər) *v.i.* **1** To sleep, especially lightly or quietly. **2** To be inactive; stagnate. —*v.t.* **3** To spend or pass in sleeping. See synonyms under REST. —*n.* Sleep; formerly, light sleep; more recently, complete, quiet sleep. [OE *slumerian* < *slūma*] — **slum′ber·er** *n.* — **slum′ber·ing·ly** *adv.* — **slum′ber·less** *adj.*

slum·ber·ous (slum′bər·əs) *adj.* Inviting to, being in, suggesting, or resembling slumber; soporific; drowsy; sleepy. Also **slum·brous** (slum′brəs). — **slum′ber·ous·ly** *adv.* — **slum′·ber·ous·ness** *n.*

slum·ber·y (slum′bər·ē) *adj.* Slumberous; somnolent.

slum·gul·lion (slum·gul′yən) *n.* **1** *Slang* **a** A stew made principally of meat and vegetables. **b** A weak beverage. **2** A servant, especially one who performs menial chores. **3** Refuse drainage from blubber; also, fish offal. **4** A reddish, muddy deposit in mine sluiceways. [< slang E; origin uncertain]

slum·gum (slum′gum′) *n.* The residue of propolis, cocoons, etc., after beeswax is extracted from honeycombs.

slump (slump) *v.i.* **1** To break through a crust, as of snow or ice, and sink; sink, as a foot, into any soft material. **2** To slide with perceptible motion down a declivity: said of loose earth or rock. **3** To fall or fail suddenly, as in value or quality. **4** To stand, walk, or proceed with a stooping posture; slouch: He *slumps* badly. — *n.* **1** The act of slumping; a collapsing fall. **2** A collapse or failure; also, a sudden fall of prices: a *slump* in stocks. **3** A decline, as of interest, excitement, etc. [Prob. imit.]

slung (slung) Past tense and past participle of SLING.

slung·shot (slung′shot′) *n.* A weight attached to a thong or cord, used as a weapon.

slunk (slungk) Past tense and past participle of SLINK. — *n.* The body of a stillborn animal, especially of a calf when cut away from the mother's womb. See SLINK.

slur (slûr) *v.t.* **slurred, slur·ring** **1** To slight; disparage; depreciate. **2** To pass over lightly or hurriedly; suppress; conceal: to *slur* a fact. **3** To pronounce, as a syllable, hurriedly and indistinctly. **4** *Music* **a** To sing or play as indicated by the slur. **b** To mark with a slur. **5** To smear; soil; contaminate. — *n.* **1** A disparaging remark or insinuation; also, the occasion for it, or the resulting state; a stigma. **2** *Music* **a** A curved line (‿ or ⁀) indicating that tones so tied are to be sung to the same syllable or performed without a break between them. **b** The legato effect indicated or produced by this mark. **3** A blur. **4** A slurred pronunciation. [< dial. E, orig. fluid mud]

slurp (slûrp) *v.t.* & *v.i.* *Slang* To sip noisily. [Imit.]

slur·ry (slûr′ē) *n.* *pl.* **·ries** **1** Any one of several watery mixtures used to make repairs in furnace linings, to neutralize poisonous chemicals, etc. **2** A mixture used in the manufacture of Portland cement. **3** The slushy matter which results from grinding. [< dial. E *slur* fluid mud]

slush (slush) *n.* **1** Soft, sloppy material, as melting snow or soft mud. **2** Greasy material used for lubrication, etc. **3** The greasy refuse of cooking, especially from a ship's galley: used on shipboard for lubricating the masts. **4** A mixture of lime with white lead or tallow, for coating bright iron or steel parts of machinery to keep them from rusting. **5** Emotional talk or writing; gush; drivel. — *v.t.* **1** To cover or daub with slush, as for lubrication. **2** To fill (spaces in masonry) with mortar: usually with *up.* **3** To wash by throwing water upon, as a deck. [Origin unknown] — **slush′y** *adj.*

slush fund *U.S.* **1** Formerly, on naval vessels, money obtained from the sale of garbage and used to buy small luxuries. **2** Money collected or spent for corrupt purposes, as bribery, lobbying, propaganda, etc.

slut (slut) *n.* **1** A female dog; bitch. **2** A slatternly woman. **3** A drudge. **4** A woman of loose character; hussy. [Origin uncertain]

slut·tish (slut′ish) *adj.* Slatternly; dirty. — **slut′·tish·ly** *adv.* — **slut′tish·ness** *n.*

sly (slī) *adj.* **sli·er** or **sly·er, sli·est** or **sly·est** **1** Artfully dexterous in doing things secretly; cunning in evading notice or detection. **2** Playfully clever; roguish; mischievous. **3** Meanly or stealthily clever; crafty. **4** Done with or marked by artful secrecy: a *sly* trick. **5** Skilful; possessed of practical ability; wise. See synonyms under INSIDIOUS. — **on the sly** In a stealthy way; with concealment. [<ON *slœgr*] — **sly′ness** *n.*

sly·boots (slī′bōōts′) *n.* *Colloq.* A roguish, cunning, and sly person or animal.

sly·ly (slī′lē) See SLILY.

smack[1] (smak) *n.* **1** A quick, sharp sound, as of the lips when separated rapidly; a noisy kiss. **2** A sounding blow or slap. **3** The sound of a snapping whip. — *v.t.* & *v.i.* To give or make a smack, as in tasting, kissing, striking, etc.; slap. [Cf. MDu. *smack* a blow]

smack[2] (smak) *v.i.* **1** To have a taste or flavor, especially as tested by smacking: usually with *of.* **2** To have, keep, or disclose a slight suggestion: with *of.* — *n.* **1** A suggestive tincture, taste, or flavor. **2** A mere taste; smattering. [OE *smæc* taste]

smack[3] (smak) *n.* A small, decked or half-decked vessel of various rig used chiefly for fishing; especially, one having a well for fish in its hold. [<Du. *smak, smacke*]

smack·ing (smak′ing) *adj.* Making a sharp sound; hence, brisk; lively: a *smacking* breeze. — *n.* A quick, sharp sound; smack.

smaik (smāk) *n.* *Scot.* A rascal; petty rogue; mean, contemptible, or silly fellow.

Smal·kal·dic League (smôl·kôl′dik) See SCHMALKALDIC LEAGUE.

small (smôl) *adj.* **1** Comparatively less than another or than a standard; diminutive; little. **2** Being of slight moment, weight, or importance. **3** Lacking in moral or mental breadth; narrow; ignoble; mean; paltry. **4** Lacking in the qualities of greatness; not largely gifted. **5** Acting or transacting business in a limited way. **6** Weak in characteristic properties; mildly alcoholic: said of liquors: *small* beer. **7** Having little body or volume; slender; fine; soft, as a voice. **8** Of low degree; obscure. **9** Lacking in power or strength. — *adv.* **1** In a low or faint tone: to sing *small.* **2** Into small pieces. **3** In a small way; trivially; also, timidly; to talk *small.* — *n.* **1** A small or slender part: the *small* of the back. **2** A small thing or quantity. [OE *smæl*] — **small′ness** *n.*

Synonyms (*adj.*): diminutive, fine, little, mean, microscopic, minute, narrow, petty, puny, tiny. See FINE[1], INSIGNIFICANT, LITTLE.

small·age (smô′lij) *n.* Celery, especially in the wild state. [<SMALL + F *ache* wild celery <L *apium* parsley]

small arms Arms that may be carried on the person, as a rifle, automatic pistol, or revolver.

small beer **1** Insipid or weak beer. **2** *Brit.* An insignificant person or thing.

small calorie See under CALORIE.

small capital A capital letter cut slightly larger than the lower-case letters of a specified type size. Abbr. *s.c., s. cap., small cap., sm. cap.*

THIS LINE IS IN CAPITAL LETTERS
THIS LINE IS IN SMALL CAPITAL LETTERS
this line is in lower-case letters

small circle The circumference formed by a plane cutting a sphere but not passing through its center.

small-clothes (smôl′klōthz′, -klōz′) *n. pl.* **1** A man's nether garments. **2** Close-fitting knee breeches worn by men in the 18th century. Also *smalls.*

small craft **1** Small vessels collectively. **2** Small things or persons generally.

small-fry (smôl′frī′) *n. pl.* **1** Small, young fish. **2** Young children. **3** Small or insignificant people or things.

small hours The early hours of the morning.

small·ish (smô′lish) *adj.* Somewhat small.

small-minded (smôl′mīn′did) *adj.* **1** Having a petty mind; interested in trivialities. **2** Narrow; intolerant; ungenerous.

small·mouth (smôl′mouth′) *n.* An American black bass (*Micropterus dolomieu*).

small potatoes *U.S. Colloq.* Unimportant, insignificant persons or things.

small·pox (smôl′poks′) *n.* *Pathol.* An acute, infectious, highly contagious disease caused by a filtrable virus and characterized by high inflammatory fever, followed by an eruption of deep-seated pustules; variola.

smalls (smôlz) *n. pl.* **1** Small-clothes. **2** *Brit. Colloq.* The first examination after matriculation; responsions: used at Oxford to denote the first university examination counting toward a degree. [<SMALL]

small stores Small, miscellaneous items, as tobacco, soap, thread, etc., stocked by a ship's store to be sold to the crew.

small sword **1** A light sword used on dress occasions. **2** The straight sword of modern fencing, introduced about 1700.

small talk Unimportant or trivial conversation.

small-time (smôl′tīm′) *adj.* *U.S. Slang* Petty; unimportant.

smalt (smôlt) *n.* A deep-blue glass colored with cobalt oxide: used when pulverized for painting, etc. [<F <Ital. *smalto* <Gmc.]

smalt·ite (smôl′tīt) *n.* A tin-white to steel-gray cobalt arsenide, crystallizing in the isometric system. Also **smalt·ine** (smôl′tin, -tēn). [<SMALT]

smal·to (smäl′tō) *n.* *pl.* **·ti** (-tē) *Italian* Colored glass, or a piece of it, employed in mosaics.

smar·agd (smar′agd) *n.* *Obs.* A green precious stone, as the beryl or the emerald. Also **smar′·agde.** [<L *smaragdus* <Gk. *smaragdos.* Doublet of EMERALD.] — **sma·rag·dine** (smə·rag′dēn, -din) *adj.*

sma·rag·dite (smə·rag′dīt) *n.* A thin, foliated, light grass-green variety of amphibole.

smarm (smärm) *Brit. Colloq.* *v.t.* To smear or plaster (the hair) with oil. — *v.i.* To behave in a servilely flattering manner; toady. [Var. of dial. E *smalm* smear, plaster]

smarm·y (smär′mē) *adj.* *Brit. Colloq.* Unctuously flattering; oily; toadying.

smart (smärt) *v.i.* **1** To experience a stinging sensation, generally superficial, either bodily or mental. **2** To cause a stinging sensation. **3** To experience remorse. **4** To have one's feelings hurt. **5** To pay a severe penalty. — *v.t.* **6** To cause to smart. — *adj.* **1** Quick in thought or action; bright; acute; clever. **2** Impertinently witty: often used contemptuously. **3** Vigorous; emphatic; severe; brisk. **4** Causing a smarting sensation; stinging; pungent. **5** Keen or sharp, as at trade; shrewd. **6** In active health; well. **7** *Colloq.* Superior, as in speed, strength, or skill. **8** *Colloq.* Large; considerable: a *smart* crop of wheat. **9** Sprucely dressed; showy. **10** Belonging to the stylish classes; fashionable: a *smart* set. **11** Making a creditable showing: a *smart* regiment. See synonyms under CLEVER. — *n.* **1** An acute stinging sensation, as from a scratch or an irritant. **2** Any distress; poignant mental suffering. **3** *Dial.* A degree, number, or amount: with ***right***: a right *smart* of people. [OE *smeortan*] — **smart′ly** *adv.* — **smart′ness** *n.*

smart al·eck (al′ik) *Colloq.* A cocky, offensively conceited person. — **smart-al·eck·y** (smärt′al′ik·ē) *adj.*

smart·en (smär′tən) *v.t.* To improve in appearance; make smart, as oneself or one's habitation: with *up.*

smart money **1** *Law* Damages awarded against a defendant because of great aggravation attending the wrong committed. **2** Money paid for a release from an engagement or from a painful situation. **3** Money paid by an employer to a workman injured in his service. **4** *Brit.* Money allowed to soldiers or sailors for injuries received in the service. **5** *Colloq.* Money bet by gamblers who supposedly know the result of a contest beforehand.

smart set Fashionable society.

smart·weed (smärt′wēd′) *n.* Any of several species of widely distributed herbs (genus *Polygonum*) having jointed stems, long, grass-like leaves, and inconspicuous, greenish flowers, especially the **common smartweed** or water pepper. Also called *knotweed.*

smart·y (smär′tē) *n.* *pl.* **smart·ies** *Slang* A person affecting to be smart or witty; a smart aleck.

smash (smash) *v.t.* **1** To break in many pieces suddenly, as by a blow, pressure, or collision. **2** To flatten; crush: to *smash* a hat. **3** To dash or fling violently so as to crush or break in pieces. **4** To strike with a sudden blow. **5** To make bankrupt. **6** To destroy, as a theory. **7** In tennis, to strike (the ball) with a hard, swift, overhand stroke. — *v.i.* **8** To go bankrupt; fail, as a business, etc. **9** To move or be moved with force; come into violent contact so as to crush or be crushed; collide; dash: The boats *smashed* together. See synonyms under BREAK. — **to go to smash** *Colloq.* To be ruined; fail. — *n.* 1 An act or instance of smashing, or the state of being smashed: often compounded with *up*: a *smash-up* on a railroad. 2 Any disaster or sudden break-up of any kind: a *smash* in business. 3 A beverage of spirituous liquors, usually brandy, with mint, water, sugar, and ice. 4 In tennis, a strong overhand shot. 5 *Colloq.* Something acclaimed by the public: The film is a box-office *smash.* [Prob. imit.] — **smash′er** *n.*

smash·ing (smash′ing) *adj. Colloq.* Extremely impressive; overwhelmingly good: a *smashing* success.

smash-up (smash′up′) *n.* A smash; a disastrous collision.

smat·ter (smat′ər) *v.t.* To talk of, dabble in, study, or use superficially. — *n.* A slight knowledge. [ME *smateren,* ? <Scand. Cf. Sw. *smattra* patter.] — **smat′ter·er** *n.*

smat·ter·ing (smat′ər·ing) *n.* A superficial degree or kind of anything, especially of knowledge. — **smat′ter·ing·ly** *adv.*

smear (smir) *v.t.* 1 To spread, rub, or cover with grease, paint, dirt, etc.; bedaub. 2 To spread or apply in a thick layer or coating: to *smear* grease on an axle. 3 To sully the reputation of; defame; slander. 4 *U.S. Slang* To defeat utterly; overwhelm or stop. 5 To cover with smear (def. 3). 6 *Obs.* To anoint. — *v.i.* 7 To be or become smeared. — *n.* 1 A soiled spot; stain. 2 A small quantity of material, as blood, sputum, etc., placed on a microscope slide or bacterial culture for analysis. 3 A volatile flux for glazing ware. 4 *Obs.* Ointment; grease. 5 A slanderous attack; defamation. 6 *Slang* Anything to spread on bread, as butter, jam, etc. [OE *smerian* < *smeoru* grease]

smear·case (smir′kās′) *n. U.S.* Cottage cheese. [<G *schmierkäse*]

smear·y (smir′ē) *adj.* **smear·i·er, smear·i·est** Greasy, viscous, or staining; also, smeared. — **smear′i·ness** *n.*

smeath (smēth) *n.* A smew. Also **smee** (smē).

smed·dum (smed′əm) *n. Scot.* & *Brit. Dial.* 1 Fine ore particles that have passed through a wire sieve; fine coal slack. 2 Powder; especially, ground malt. 3 Vigor of mind; sense.

smeek (smēk) *v.* & *n. Scot.* Smoke. — **smeek′y** *adj.*

smell (smel) *v.* **smelled** or **smelt, smell·ing** *v.t.* 1 To perceive by means of the nose and its olfactory nerves. 2 To perceive the odor or perfume of; scent. 3 To test by odor or smell. 4 To discover, detect, or seek to know, as if by smelling: often with *out.* — *v.i.* 5 To emit an odor or perfume; give off a particular odor: frequently with *of*; also, to give indications of, as if by odor: to *smell* of treason. 6 To be malodorous. 7 To use the sense of smell. 8 To pry; investigate: with *about.* — *n.* 1 That special sense by means of which odors are perceived. 2 The sensation excited through the olfactory nerves. 3 That which is directly perceived by this sense; an odor; perfume. 4 A faint suggestion; hint; trace. 5 An act of smelling. [ME *smellen*]

Synonyms (noun): aroma, bouquet, fragrance, odor, perfume, savor, scent, stench, stink. *Smell* is the generic word including all the rest. *Aroma, fragrance,* and *perfume* are ordinarily pleasing; *odor, savor,* and *scent* may be so. *Odor* is nearly synonymous with *smell,* but is susceptible of more delicate use; as, the *odor* of incense. An *aroma* is a delicate and spicy *odor,* as of fine coffee; *bouquet* is said chiefly of the delicate *odor* of certain wines. We speak of the *fragrance* or *perfume* of flowers, but *fragrance* is more delicate; a *perfume* may be so strong and rich as to be repulsive by excess. There is a tendency to restrict the application of *perfume* to the artificial preparations called collectively "perfumery." *Scent* is chiefly used for the characteristic *odor* of an animal by which it is tracked or avoided by other animals; the word is also applied to any *odor,* natural or artificial, especially when faintly diffused through the air; as, the *scent* of mignonette or of new-mown hay. *Savor* is chiefly said of the appetizing *odor* evolved from articles of food by the processes of cooking. Any *smell* that is at once foul, strong, and pervasive may be called a *stench.* See SAVOR.

smell·er (smel′ər) *n.* 1 One who smells (anything). 2 *Slang* The nose. 3 A feeler, as of an animal.

smell·feast (smel′fēst′) *n. Archaic* 1 A person who frequents good tables. 2 A greedy sponger; parasite.

smelling salts Pungent or aromatic salts, or mixtures of such, often scented, used as stimulants by smelling; specifically, a preparation of ammonium carbonate.

smell·y (smel′ē) *adj.* **smell·i·er, smell·i·est** Having an unpleasant smell; malodorous.

smelt¹ (smelt) *v.t. Metall.* 1 To reduce (ores) by fusion in a furnace. 2 To obtain (a metal) from the ore by a process including fusion. — *v.i.* 3 To melt or fuse, as a metal. [<MDu. *smelten* melt]

smelt² (smelt) *n. pl.* **smelts** or **smelt** Any of certain small silvery food fishes (genus *Osmerus* or a related genus), of the northern Atlantic and Pacific. [OE]

smelt³ (smelt) Alternative past tense and past participle of SMELL.

smelt·er (smel′tər) *n.* 1 One engaged in smelting ore. 2 An establishment for smelting: also **smelt′er·y.**

Sme·ta·na (sme′tä·nä), **Bedřich,** 1824–84, Czech composer.

Smeth·wick (smeth′wik) A county borough of southern Staffordshire, England.

smew (smyōō) *n.* A small merganser (*Mergus albellus*) of northern parts of the Old World. The male is black and white, with a white crest. Also called *smeath, smee.* [Origin unknown]

smid·die (smid′ē) *n. Scot.* A smithy. Also **smid′dy.**

smidg·en (smij′ən) *n. U.S. Colloq.* A tiny bit or part; mite; trifle.

Smig·ly-Rydz (shmēg′wē-rēts), **Edward,** 1886–1943?, Polish marshal; ruled Poland 1935–39.

smi·la·ca·ceous (smī′lə·kā′shəs) *adj. Bot.* Of or pertaining to a family (*Smilacaceae*) of herbs or woody-stemmed vines, having dioecious flowers and globular fruits. [<NL <L *smilax.* See SMILAX.]

smi·la·cin (smī′lə·sin) *n.* Parillin.

smi·lax (smī′laks) *n.* 1 Any of a large, widely scattered genus (*Smilax*) of shrubby or herbaceous plants having net-veined leaves, dioecious flowers in umbels, and globular fruit, especially *S. aristolochiifolia,* a source of sarsaparilla: also called *catbrier, greenbrier.* 2 A delicate twining plant (*Asparagus asparagoides*) of the lily family, from South Africa, with greenish flowers: cultivated in greenhouses and extensively used for bouquets, etc. [<L <Gk. *smilax* yew]

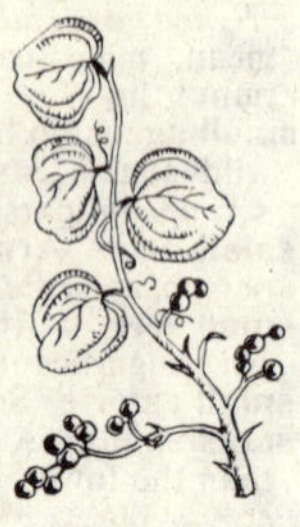

SMILAX
Greenbrier.

smile (smīl) *n.* 1 A pleased or amused expression of the face, characterized by lateral upward extension of the lips. 2 A pleasant aspect: the *smile* of spring. 3 Propitious or favorable disposition; favor; blessing: the *smile* of fortune. — *v.* **smiled, smil·ing** *v.i.* 1 To give a smile; wear a cheerful aspect. 2 To show approval or favor: often with *upon.* — *v.t.* 3 To express by means of a smile; effect as by a smile. [ME *smilen,* prob. <LG] — **smil′er** *n.* — **smil′ing** *adj.* — **smil′·ing·ly** *adv.* — **smil′ing·ness** *n.*

Smiles (smīlz), **Samuel,** 1812–1904, Scottish writer and physician.

smirch (smûrch) *v.t.* 1 To soil, as by contact with grime; smear. 2 Figuratively, to defame; degrade: to *smirch* a reputation. — *n.* The act of smirching, or the state of being smirched; a smutch; smear; a moral stain or defect. See synonyms under BLEMISH. [ME *smorchen,* appar. <OF *esmorcher* hurt, torment]

smirk (smûrk) *v.i.* To smile in a silly, self-complacent, or affected manner. — *n.* An affected or artificial smile. Also *Obs.* **smerk.** [OE *smercian*] — **smirk′ing·ly** *adv.*

smit (smit) *Brit. Dial. v.t.* 1 To destroy. 2 To infect. — *n.* 1 A smutch; spot. 2 Infection.

smite (smīt) *v.* **smote** (*Obs.* **smit**), **smit·ten** or **smit** or **smote, smit·ing** *v.t.* 1 To strike (something). 2 To strike a blow with (something); cause to strike. 3 To cut, sever, or break by a blow: usually with *off* or *out.* 4 To strike with disaster; afflict; destroy by a catastrophe. 5 To affect powerfully with sudden feeling; in the passive, to affect with love. 6 To cause to feel regret or remorse: His conscience *smote* him. 7 To affect as if by a blow; come upon suddenly: The thought *smote* him. 8 To kill by a sudden blow. — *v.i.* 9 To come with sudden force; also, to knock against something: His knees *smote* together. See synonyms under BEAT. [OE *smītan*] — **smit′er** *n.*

smith (smith) *n.* 1 One who shapes metals by hammering: often used in combination: *goldsmith, tinsmith.* 2 A blacksmith. [OE]

Smith (smith), **Adam,** 1723–90, Scottish economist. — **Alfred Emanuel,** 1873–1944, U.S. politician. — **Edmund Kirby,** 1824–93, American Confederate general and educator. — **Francis Hopkinson,** 1838–1915, U.S. civil engineer, artist, and author. — **Goldwin,** 1823–1910, English historian. — **Captain John,** 1579–1631, English adventurer; president of Virginia Colony 1608. — **Joseph,** 1805–44, founder and first prophet of the Mormon Church; assassinated. — **Sydney,** 1771–1845, English clergyman and author. — **Theobald,** 1859–1934, U.S. pathologist. — **Walter Bedell,** born 1895, U.S. Army officer and diplomat. — **William,** 1769–1839, English geologist. — **Sir William,** 1813–93, English classical scholar. — **William Robertson,** 1846–94, Scottish theologian and author.

smith·er·eens (smith′ə·rēnz′) *n. pl. Colloq.* Fragments produced as by a blow. Also **smith′·ers.** [Cf. dial. E (Irish) *smidirin* a fragment]

smith·er·y (smith′ər·ē) *n. pl.* **·er·ies** 1 The art or trade of a smith. 2 A smith's shop; a smithy.

Smith·son (smith′sən), **James,** 1765–1829, English chemist: known in his youth as James Lewis Macie.

Smith·so·ni·an Institution (smith·sō′nē·ən) An institution founded in 1846 at Washington, D.C., from funds left by James Smithson "for the increase and diffusion of knowledge among men."

smith·son·ite (smith′sən·īt) *n.* 1 A vitreous zinc carbonate, $ZnCO_3$. 2 *Brit.* Hemimorphite. [after James *Smithson*]

Smith Sound (smith) A sea passage between Ellesmere Island and NW Greenland; 45 miles wide.

smith·y (smith′ē, smith′ē) *n. pl.* **smith·ies** A blacksmith's shop; a forge.

smit·ten (smit′n) Alternative past participle of SMITE. — *adj.* 1 Struck with sudden force; gravely afflicted. 2 Having the affections suddenly attracted.

smock (smok) *n.* 1 A loose outer garment of light material worn like a coat to protect one's clothes. 2 In colonial times, a woman's undergarment; chemise. — *v.t.* 1 To furnish with or clothe in a smock. 2 To shirr (def. 1). See SMOCKING. [OE *smoc*]

smock-frock (smok′frok′) *n.* A loose-fitting outer garment or jacket worn by laborers.

smock·ing (smok′ing) *n.* Shirred work; decorative stitching holding fullness in regular patterns.

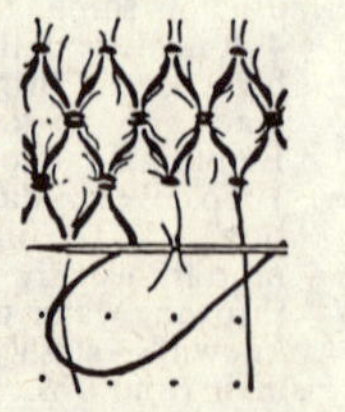

SMOCKING

smog (smog) *n. Colloq.* A combination of smoke and fog, especially as seen in and about heavy industry and manufacturing areas. [Blend of SM(OKE) + (F)OG]

smok·a·ble (smō′kə·bəl) *adj.* Capable of being smoked. — *n.* Something to be smoked, as a cigarette, cigar, etc.: usually in the plural.

smoke (smōk) *n.* 1 The volatilized products of the combustion of an organic compound, as coal, wood, etc., charged with fine particles of carbon or soot; less properly, fumes, steam, etc. 2 *Chem.* A colloid system of solid particles in a gas. See COLLOID SYSTEM. 3 Anything transient and unsubstantial; a useless or ephemeral result. 4 The act of smoking a pipe, cigar, etc. 5 A period of time during which one smokes tobacco. 6 *Colloq.* A cigarette, cigar, or pipeful of tobacco. 7 A chemical-warfare agent producing a smoke-like cloud; also, a smudge. 8 A column of smoke, used as a signal by the North American Indians. 9 *U.S. Slang* In baseball, speed in a pitch. 10 A cheap drink, usually of wood

alcohol. — v. **smoked, smok·ing** v.i. 1 To emit or give out smoke: The embers *smoke*; also, to emit smoke excessively or in an undesired direction, as a stove or lamp. 2 To raise dust in rapid riding or driving; hence, to travel rapidly; speed. — v.t. 3 To inhale and exhale the smoke of (tobacco, opium, etc.); also, to use, as a pipe, for this purpose. 4 To treat or affect with smoke; treat by the application of smoke; cure; medicate; fumigate; tinge; flavor with smoke. 5 To apply smoke to in order to drive away or expel: to *smoke* bees; hence, to force out of hiding: usually with *out*: to *smoke* out a criminal. 6 To get the scent of; hence, to suspect. 7 To change the color of (glass, etc.) by darkening with smoke. [OE *smoca*]
smoke candle A small, portable, short-burning smokepot.
smoke helmet A protective headdress, such as worn by soldiers, firemen, etc. for protection against poisonous gas fumes and smoke.
smoke·house (smōk'hous') n. 1 A building or close room in which meat, fish, hides, etc., are cured by the action of smoke. 2 A building in which anything is disinfected by the use of smoke.
smoke·jack (smōk'jak') n. A mechanism by which to turn a roasting spit, operated by the ascending combustion gases in a chimney.
smoke·jump·er (smōk'jum'pər) n. A fireman trained and equipped to fight forest fires when parachuted to the affected area by aircraft.
smoke·less (smōk'lis) adj. Having or emitting little or no smoke: *smokeless* powder.
smoke·pot (smōk'pot') n. A small container for generating a dense cloud of smoke.
smok·er (smō'kər) n. 1 One who or that which smokes; one who smokes tobacco habitually. 2 A firebox for blowing smoke upon bees to quiet them. 3 A smoking car. 4 A social gathering of men. 5 A smoking jacket.
smoke screen A dense cloud of smoke emitted to screen an attack or bombardment by land or sea, or to cover a retreat. Also **smoke blanket**.
smoke·stack (smōk'stak') n. 1 An upright pipe, usually of sheet or plate iron, through which combustion gases from a boiler furnace are discharged into the air. 2 The funnel of a steamboat or locomotive, or the tall chimney of a factory, etc.
smoke·tree (smōk'trē') n. 1 An ornamental Old World shrub or tree (*Cotinus coggygria*) with long feathery stalks resembling smoke or mist. 2 A related American species (*C. americanus*).
smoking jacket A short coat worn instead of a regular suit coat as a lounging jacket.
smok·y (smō'kē) adj. **smok·i·er, smok·i·est** 1 Giving forth smoke. 2 Mixed with or containing smoke: *smoky* air. 3 Liable to be filled with smoke, as a house. 4 Emitting smoke improperly and unpleasantly, as from bad draft. 5 Discolored with smoke. 6 Smoke-colored; dark-gray. 7 Covered with mist: said of certain mountains. — **smok'i·ly** adv. — **smok'i·ness** n.
Smoky Hill River A river in Colorado and Kansas, flowing 560 miles east, SE, north, and NE, joining the Republican River to form the Kansas River.
Smoky Mountains See GREAT SMOKY MOUNTAINS.
smol·der (smōl'dər) v.i. 1 To burn and smoke in a smothered way, showing little smoke and no flame. 2 Figuratively, to exist in a latent state; to manifest suppressed feeling: His wrath was *smoldering*. — n. Smother; smoke. Also **smoulder**. [Dissimilated var. of ME *smorther*. See SMOTHER.]
Smo·lensk (smô·lyensk') A city on the Dnieper in western Russian S.F.S.R.
Smol·lett (smol'it), **Tobias George**, 1721–71, English novelist and physician.
smolt (smōlt) n. A young salmon on its first descent from the river to the sea. [? Related to SMELT[2]]
smooch (smōōch) n. A smear; a smutch. — v.t. To smear; smudge. — v.i. *Slang* To neck. [Cf. SMUTCH]
smoor (smōōr) *Scot.* v.t. To smother. — n. Stifling smoke or atmosphere.

smooth (smōōth) adj. 1 Having a surface without irregularities; not rough; continuously even. 2 Having no impediments or obstructions; easy; free from shocks or jolts. 3 Calm and unruffled; bland; pleasant; mild. 4 Flowing melodiously: opposed to *rugged*: a *smooth* style. 5 Suave, as in speech; flattering: often implying deceit. 6 *Phonet.* Sounded without the aspirate: opposed to *rough*: a *smooth* breathing. 7 Free from hair; beardless. 8 Having no acidulous or astringent taste or quality: said of liquors. 9 Without lumps; having the elements perfectly blended: a *smooth* mayonnaise. 10 Offering no resistance to a body sliding along its surface; without friction. 11 Having the high points removed by wear, as the surface of a tire. — adv. Calmly; evenly. — v.t. 1 To make smooth or even on the surface. 2 To make easy or less difficult: to *smooth* one's path. 3 To free from obstructions. 4 To remove (an obstruction): often with *away*: to *smooth* away a mound. 5 To render less harsh or softer and more flowing: to *smooth* one's verses. 6 To soften the worst features of; palliate; extenuate: usually with *over*. 7 To make calm; mollify: to *smooth* one's feelings. — v.i. 8 To become smooth. Also **smooth'en**. — **to smooth (someone's) ruffled feathers** To mollify. — n. 1 The smooth portion or surface of anything: the *smooth* of the neck. 2 The act of smoothing. [OE *smōth*] — **smooth'er** n. — **smooth'ly** adv. — **smooth'ness** n.
Smooth may appear as a combining form in hyphemes; as in:

smooth-ankled	smooth-headed
smooth-barked	smooth-limbed
smooth-billed	smooth-necked
smooth-bodied	smooth-paced
smooth-browed	smooth-polished
smooth-cheeked	smooth-riding
smooth-combed	smooth-rimed
smooth-cut	smooth-sculptured
smooth-edged	smooth-sided
smooth-filed	smooth-skinned
smooth-fibered	smooth-sliding
smooth-flowing	smooth-speaking
smooth-fronted	smooth-stalked
smooth-gliding	smooth-stemmed
smooth-going	smooth-surfaced
smooth-grained	smooth-tempered
smooth-haired	smooth-voiced
smooth-handed	smooth-woven

Synonyms (adj.): even, flat, glossy, level, plain, plane, polished, sleek, undisturbed, unruffled. An *even* surface is free from any considerable irregularities, as knobs, or splinters, or abrupt changes of direction or curvature; a *smooth* surface is one that the hand may be passed over without friction or in which the eye discerns no noticeable break or flaw. That which is *polished* is brought to a very high degree of smoothness, so as to be not only frictionless to touch but lustrous to the eye. A board is sawed to an *even* surface, planed till it is *smooth*, and sandpapered till it is *polished*. A thing may be *smooth* or *polished* and yet very uneven, as a warped piece of veneering. See BLAND, BLUNT, CALM, FINE[1], LEVEL, PACIFIC. Antonyms: see synonyms for ROUGH.
smooth-bore (smōōth'bôr', -bōr') n. A firearm, as a shotgun, with an unrifled bore. Also **smooth bore**.
smooth breathing See under BREATHING.
smooth-faced (smōōth'fāst') adj. 1 Beardless. 2 Of smooth surface, as a wall, etc. 3 Bland or mild in expression, especially with deceitful intent.
smooth-shod (smōōth'shod') adj. Shod without sharp projections on the shoes, as a horse.
smör·gås·bord (smôr'gəs·bôrd', *Sw.* smœr'gös·bôrd) n. Scandinavian hors d'oeuvres. Also spelled **smor'gas·bord**. [<Sw.]
smor·zan·do (zmôr·tsän'dō) adj. *Music* Fading away; growing softer. [<Ital.]
smote (smōt) Past tense of SMITE.
smoth·er (smuth'ər) v.t. 1 To prevent the respiration of, as by filling or covering the mouth and nostrils; also, to kill by such means; suffocate; stifle. 2 To cover, or cause to smolder, as a fire. 3 Figuratively, to hide or suppress: to *smother* a scandal. 4 In cooking, to enclose and cook in a covered dish or under a close mass of some other substance. 5 To daub; smear. — v.i. 6 To be covered without vent or air, as a fire. 7 To be hidden or suppressed, as wrath. — n. 1 That which smothers, as stifling vapor or dust. 2 The state of being smothered; suppression; also, a smoldering fire. 3 A surging of foam or water; a welter. [Earlier *smorther*. Related to OE *smorian* suffocate] — **smoth'er·y** adj.
smouch (smōōch, smouch) v.t. *Dial.* To lift; pilfer.
smoul·der (smōl'dər) See SMOLDER.
smout·y (smōō'tē) adj. *Scot.* Smutty; obscene. Also **smout'ie, smoot'ie**.
smudge (smuj) v. **smudged, smudg·ing** v.t. 1 To smear; soil. 2 To protect (from frost, insects, etc.) by a heavy, smoky pall. — v.i. 3 To cause a smudge. 4 To be smudged. — n. 1 A soiling, as of dry dirt or soot; smear; stain. 2 A smoky fire or its smoke for driving away insects, preventing frost, etc. 3 Paint-pot scrapings and cleanings. [Var. of SMUTCH] — **smudg'i·ly** adv. — **smudg'i·ness** n. — **smudg'y** adj.
smug (smug) adj. **smug·ger, smug·gest** 1 Characterized by a smoothly self-satisfied or extremely complacent air. 2 Trim; neat; spruce, especially with suggestions of respectability or self-satisfaction. [Cf. LG *smuk* neat] — **smug'ly** adv. — **smug'ness** n.
smug-faced (smug'fāst') adj. Having a prim, self-satisfied face or expression.
smug·gle (smug'əl) v. **·gled, ·gling** v.t. 1 To take (merchandise) into or out of a country without payment of lawful duties. 2 To bring in or introduce illicitly or clandestinely. — v.i. 3 To engage in or practice smuggling. [<LG *smuggeln*]
smug·gler (smug'lər) n. 1 One who smuggles. 2 A vessel used in smuggling.
smug·gling (smug'ling) n. The offense or practice of fraudulently and illegally importing or exporting merchandise without payment of lawful duties.
smut (smut) n. 1 The blackening made by soot, smoke, etc. 2 Obscenity; obscene language. 3 Any of various fungus diseases of plants, in which the affected parts change into a dusty black powder. 4 The parasitic fungus (order *Ustilaginales*) causing such a disease. — v. **smut·ted, smut·ting** v.t. 1 To blacken or stain, as with soot or smoke. 2 To affect with smut, as growing grain. 3 To remove the smut from (grain). 4 Figuratively, to pollute; defame. — v.i. 5 To give off smut. 6 To be or become stained. 7 To be affected with smut, as growing grain. [<LG *schmutt* dirt]
smutch (smuch) n. & n. Soil; smear; smudge. [Cf. MHG *smutzen* smear] — **smutch'y** adj.
Smuts (smuts), **Jan Christiaan**, 1870–1950, South African statesman and general.
smut·ty (smut'ē) adj. **·ti·er, ·ti·est** 1 Soiled with smut; black; stained. 2 Affected by smut: *smutty* corn. 3 Obscene; coarse; indecent. — **smut'ti·ly** adv. — **smut'ti·ness** n.
Smyr·na (smûr'nə) A port of western Turkey in Asia, on the **Gulf of Smyrna**, an inlet of the Aegean 35 miles long; Turkish *Izmir*. — **Smyr'ni·ot** (-nē·ot), **Smyr'ni·ote** (-nē·ōt) adj. & n.
smyte·rie (smī'trē, smīt'rē) n. *Scot.* A numerous collection of small things. Also **smy'trie**.
Smyth (smith), **Henry De Wolf**, born 1898, U.S. physicist.
snack (snak) n. 1 A sip or bite. 2 A slight, hurried meal. 3 A share of something. [Orig. a verb <MDu. *snacken* bite, snap]
snaf·fle (snaf'əl) n. A horse's bit without a curb, jointed in the middle. Also **snaf'fle·bit'** (-bit'). — v.t. **·fled, ·fling** To control with a snaffle. [<Du. *snavel* muzzle]
sna·fu (sna·fōō') *Slang* adj. In a state of utter confusion; chaotic. — v.t. **·fued, ·fu·ing** To put into a confused or chaotic condition. — n. Anything which is confused or chaotic. [From the initial letters of the words "Situation normal, all fouled up"]
snag (snag) n. 1 A jagged or stumpy knot or protuberance, especially the stumpy base of a branch left in pruning. 2 The root or remnant of a tooth remaining in the jaw; also, a projecting tooth. 3 A branch or point of a deer's antler. 4 The trunk of a tree fixed in

the bottom of a river, bayou, etc., by which boats are sometimes pierced. 5 Hence, any unsuspected or hidden obstacle or difficulty. — *v.* **snagged, snag·ging** *v.t.* 1 To injure, destroy, or impede by or as by a snag. 2 To clear of snags. 3 *Colloq.* To block; impede. — *v.i.* 4 To run upon a snag: said especially of river craft. [Prob. <Scand. Cf. dial. Norw. *snag* sharp point, projection.] —**snagged** *adj.*

snag boat A vessel equipped with machinery for removing snags from river beds.

snag chamber A watertight room or compartment in the bow of a river steamboat to keep the boat from sinking if snagged.

snag·gle-tooth (snag′əl·to͞oth′) *n.* A tooth that is broken, projecting, or conspicuously out of alinement with the others. —**snag′gle-toothed′** (-to͞otht′, -to͞othd′) *adj.*

snag·gy (snag′ē) *adj.* **·gi·er, ·gi·est** 1 Full of snags, as a river. 2 Full of knots or stubs, as a tree or swamp. 3 Like a snag.

snail (snāl) *n.* 1 Any of numerous slow-moving gastropod mollusks of largely terrestrial habits, with a spiral shell; especially, the common garden snail (*Helix aspersa*) and the edible snail (*H. pomatia*). 2 A slow or lazy person. [OE *snægl*]

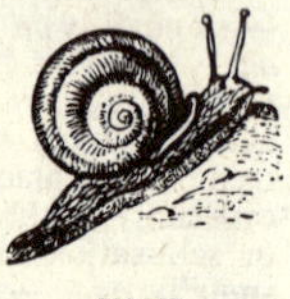

SNAIL

snail hawk The small, bluish-gray everglade kite (*Rostrhamus sociabilis plumbeus*) ranging from Florida to Mexico, that feeds on snails.

snail pace A very slow gait or advance movement. Also **snail's pace.** —**snail′-paced′** *adj.*

snake (snāk) *n.* 1 An ophidian reptile (suborder *Serpentes*), having a greatly elongated, scaly body, no limbs, and a specialized swallowing apparatus. The bite of most snakes is non-venomous, but some have much enlarged fangs, connected with venom glands from which a deadly poison flows into the punctures they make. ◆ Collateral adjective: *anguine.* 2 A treacherous or insinuating person. 3 A flexible, resilient wire used to clean clogged drains, etc. — *v.* **snaked, snak·ing** *v.t. Colloq.* To drag by seizing an end or limb and pulling forcibly or quickly; haul along the ground, as a log. — *v.i.* To wind or move like a snake. [OE *snaca*]

Snake (snāk) *n.* A member of any of various Shoshonean tribes of North American Indians, but especially the Walpapi and Yahuskin of eastern Oregon.

snake·bird (snāk′bûrd′) *n.* One of several birds (genus *Anhinga*), with very long slender neck, frequenting southern swamps and feeding upon fish; the water turkey; the darter.

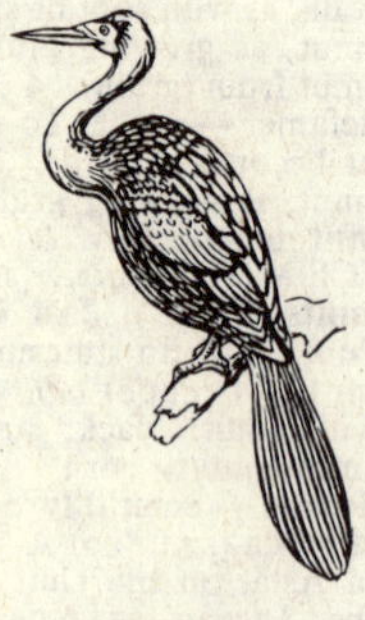

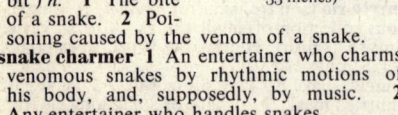

SNAKEBIRD
(Average length about 33 inches)

snake·bite (snāk′bīt′) *n.* 1 The bite of a snake. 2 Poisoning caused by the venom of a snake.

snake charmer 1 An entertainer who charms venomous snakes by rhythmic motions of his body, and, supposedly, by music. 2 Any entertainer who handles snakes.

snake dance A ceremonial dance of the Hopi Indians of Arizona in which live rattlesnakes are carried in the mouths of the dancers. The Hopis believe the snakes to be influential with the rain gods.

snake fence A worm fence; a stake-and-rider. See under FENCE.

snake hawk The swallow-tailed kite (*Elanoides forficatus*) of North and South America.

snake·head (snāk′hed′) *n.* The turtlehead.

Snake Mountains A range in eastern Nevada; highest point, 12,049 feet.

snake·mouth (snāk′mouth′) *n.* A terrestrial orchid (*Pogonia ophioglossoides*) native in eastern North America, with fragrant rose-pink flowers.

Snake River The principal tributary of the Columbia River, flowing 1,038 miles from NW Wyoming through southern Idaho, then northwards to form part of the boundary between Oregon and Idaho, to SE Washington.

snake·root (snāk′ro͞ot′, -ro͝ot′) *n.* 1 One of various plants having roots reputed to be effective against snakebite; especially, the bugbane; the **Seneca snakeroot** (*Polygala senega*), growing east of the Mississippi; the **Virginia snakeroot** (*Aristolochia serpentaria*), with purplish-brown flowers and fibrous roots; and the **white snakeroot** (*Eupatorium rugosum*) of Europe and the United States. 2 The root of any of these plants.

snake·skin (snāk′skin′) *n.* The skin of a snake.

snake·stone (snāk′stōn′) *n.* 1 An ammonite: so called because it resembles a fossil coiled snake. 2 Any porous or absorbent stone-like material popularly believed to be a specific for snakebite.

snake·weed (snāk′wēd′) *n.* The bistort.

snak·y (snā′kē) *adj.* **snak·i·er, snak·i·est** 1 Of or like a snake; serpentine; winding. 2 Insinuating; cunning; treacherous. 3 Full of snakes. —**snak′i·ly** *adv.* —**snak′i·ness** *n.*

snap (snap) *v.* **snapped, snap·ping** *v.i.* 1 To make a sharp, quick sound, as of percussion. 2 To break suddenly with a cracking noise; part with a snap. 3 To fly off or give way quickly, as when tension is suddenly relaxed. 4 To make the jaws come suddenly together in an effort to bite: often with *up* or *at.* 5 To seize or snatch suddenly: often with *up* or *at.* 6 To speak sharply, harshly, or irritably: often with *at.* 7 To emit, or seem to emit, a spark or flash of light: said of the eyes. 8 To close, fasten, etc., with a click or snapping sound, as a lock. 9 To move or act with sudden, neat gestures: He *snapped* to attention. — *v.t.* 10 To seize suddenly or eagerly, with or as with the teeth; snatch: often with *up.* 11 To sever with a snapping sound. 12 To utter, address, or interrupt harshly, abruptly, or irritably: often with *out.* 13 To cause to make a sharp, quick sound. 14 To close, fasten, etc., with a snapping sound. 15 To strike, press, etc., with a snap: to *snap* a whip. 16 To cause to move suddenly, neatly, etc. 17 To photograph instantaneously with a camera. 18 In football, to put in play: said of the ball when sent to a back by the center. — **to snap out of it** *Colloq.* 1 To recover. 2 To change one's attitude. — *n.* 1 The act of snapping, or a sharp, quick sound produced by it: the *snap* of a whip. 2 A sudden breaking of anything, or the sound so produced. 3 Any catch, fastener, or other device that closes or springs into place with a snapping sound. 4 A sudden seizing or effort to seize with or as with the teeth; a sharp shutting, as of the jaws or of a trap. 5 A quick blow of the thumb sprung from the finger or of the finger from the thumb. 6 The sudden release of the tension of a spring or elastic cord. 7 A small, thin, crisp cake, usually containing ginger; a gingersnap. 8 Brisk energy; vigor; vim; zip. 9 A brief spell; a sudden turn: said chiefly of cold weather. 10 A hasty meal; snack. 11 Any task or duty easy to perform: often in the phrase **a soft snap.** 12 A bit: It is not worth a *snap.* 13 The instantaneous taking of a photograph; also, the photograph so taken; a snapshot. 14 A stringbean. — *adj.* 1 Made or done suddenly and without consideration; offhand. 2 Contrived to take unawares and at an advantage: a *snap* policy. 3 Fastening with a snap. — *adv.* With a snap; quickly. [<MDu. *snappen* bite at]

snap·back (snap′bak′) *n.* 1 The center rusher in football. 2 The act of snapping back.

snap bean A wax bean.

snap·drag·on (snap′drag′ən) *n.* 1 A plant (genus *Antirrhinum*) of the figwort family, especially the **large-flowered snapdragon** (*A. majus*) having solitary axillary flowers, likened to dragons' heads. 2 A game in which raisins, etc., are snatched from burning brandy; also, the articles so snatched; flapdragon.

snap·per (snap′ər) *n.* 1 One who or that which snaps, as a cracker. 2 A large food fish (genus *Lutianus*) of the Gulf coast, as the **red snapper** (*L. blackfordii*). 3 One of various other fishes, as the bluefish, rosefish, etc. 4 A snapping turtle.

snapping beetle An elaterid beetle which by a quick, snapping movement of its body is able to right itself when on its back; especially, the eyed elater (*Alaus oculatus*) of eastern North America: also called *click beetle, skipjack.* For illustration see under INSECTS (injurious).

snapping turtle 1 A large voracious turtle of North America, especially *Chelydra serpentina,* much used as food. 2 The alligator turtle (*Macrochelys temminickii*), a related species.

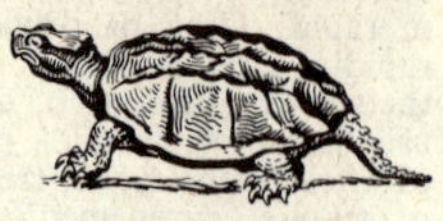

SNAPPING TURTLE
(Up to 2 feet in length; weight to 100 pounds or more)

snap·pish (snap′ish) *adj.* 1 Apt to speak crossly or tartly. 2 Disposed to snap, as a dog. See synonyms under FRETFUL. —**snap′pish·ly** *adv.* —**snap′pish·ness** *n.*

snap·py (snap′ē) *adj.* **·pi·er, ·pi·est** 1 *Colloq.* Brisk, vivid, and energetic; vivacious. 2 Smart or stylish in appearance. 3 Snappish. —**snap′pi·ly** *adv.* —**snap′pi·ness** *n.*

snap·shot (snap′shot′) *n.* A photograph taken with a small camera without timing.

snap shot A shot made without aim.

snap·weed (snap′wēd′) *n.* Any plant of the genus *Impatiens*; a touch-me-not.

snare[1] (snâr) *n.* 1 A device, as a noose, for catching birds or other animals; a gin; trap. 2 Anything by which one is brought into trouble or caused to sin; an allurement; wile. 3 *Surg.* A loop of wire used to remove tumors and other growths from the body. — *v.t.* **snared, snar·ing** 1 To catch with a snare; ensnare; entrap. 2 To capture by trickery; entice; inveigle. [<ON *snara.* Akin to SNARE[2].] —**snar′er** *n.*

snare[2] (snâr) *n.* 1 A cord to produce a rattling on a drumhead. 2 A snare drum. [<MDu., a string. Akin to SNARE[1].]

snare drum A small drum to be beaten on one head and having snares or strings of catgut stretched across the other.

snarl[1] (snärl) *n.* A sharp, harsh, angry growl; harsh or quarrelsome utterance. — *v.i.* 1 To growl harshly, as a dog. 2 To speak angrily and resentfully. — *v.t.* 3 To utter or express with a snarl. [Freq. of obs. *snar* growl] —**snarl′er** *n.* —**snarl′ing·ly** *adv.* —**snarl′y** *adj.*

snarl[2] (snärl) *n.* 1 A tangle, as of hair or yarn. 2 Any complication, perplexity, or entanglement. 3 *Colloq.* A wrangle; quarrel. 4 A knot or gnarl in wood. — *v.i.* 1 To get into a snarl or tangle; become entangled. — *v.t.* 2 To put into a snarl or tangle. 3 To confuse; entangle mentally; embarrass; make entanglements in. 4 To emboss or flute (thin metalware). [<SNARE[1]] —**snarl′er** *n.* —**snarl′y** *adj.*

snarling iron A curved tool for snarling hollow metalware, etc. Also **snarling tool.**

snash (snash, snäsh) *n. Scot.* Impertinent, abusive, or sneering language.

snatch (snach) *v.t.* 1 To seize or lay hold of suddenly, hastily, or eagerly. 2 To take or remove suddenly. 3 To take or obtain as the opportunity arises: to *snatch* a few hours of sleep. 4 *Slang* To kidnap. — *v.i.* 5 To attempt to seize swiftly and suddenly: with *at.* 6 To accept with great eagerness: with *at.* — *n.* 1 An act of snatching; a hasty grab or grasp: usually with *at.* 2 A brief period: a *snatch* of rest. 3 A small amount; fragment: *snatches* of a conversation; *snatches* of melody. 4 *Slang* A kidnaping. [ME *snacchen.* ? Related to SNACK.] —**snatch′er** *n.*

snatch block *Naut.* A single block having an opening in one cheek to receive a rope, and usually having a swivel hook.

snatch·y (snach′ē) *adj.* Interrupted; spasmodic.

snath (snath) *n.* The long curved handle of a scythe. Also **snathe** (snāth). [Var. of dial. E *snead,* OE *snǣd*]

snaw (snô, snä) *v. & n. Scot.* Snow.

sneak (snēk) *v.i.* 1 To move or go in a stealthy manner. 2 To act with covert cowardice or servility. — *v.t.* 3 To put, give, transfer, move, etc., secretly or stealthily. 4 *Colloq.* To pilfer. — *n.* 1 One who sneaks; a mean, cowardly fellow. 2 *pl. Colloq.* Sneakers. 3 A stealthy movement. — *adj.* Stealthy; covert: a *sneak* attack. [Akin to OE *snīcan* creep]

sneak boat A small, shallow boat used for duck hunting. Also **sneak-box** (snēk′boks′).

sneak·er (snē′kər) *n.* 1 One who sneaks; a sneak. 2 *pl. U.S. Colloq.* Rubber-soled canvas shoes.

sneak·ing (snē′king) *adj.* 1 Cringing; meanly secret and underhand. 2 Secretly entertained

or cherished; unavowed: a *sneaking* suspicion. See synonyms under BASE[2]. —**sneak′ing·ly** *adv.*

sneak thief One who steals small miscellaneous articles, without violence, by sneaking in through unfastened doors or windows.

sneak·y (snē′kē) *adj.* **sneak·i·er, sneak·i·est** Like a sneak; sneaking. —**sneak′i·ly** *adv.* —**sneak′i·ness** *n.*

snecked (snekt) *adj.* **1** Twisted to one side of the vertical plane of the shank, as the point of a fish hook. **2** Built of rubblework. [? Related to SNATCH]

sned (sned) *v.t. Scot.* To cut; trim; lop off.

sneer (snir) *n.* **1** A grimace of contempt or derision made by slightly raising the upper lip and nostrils. **2** A mean or contemptuous insinuation; a fling. —*v.i.* **1** To make or show a sneer. **2** To express derision or contempt in speech, writing, etc. —*v.t.* **3** To utter with a sneer or in a sneering manner. See synonyms under SCOFF. [ME *sneren*; ult. origin uncertain] —**sneer′er** *n.* —**sneer′ing** *adj.* —**sneer′·ing·ly** *adv.*

Synonyms (noun): fling, gibe, jeer, scoff, taunt. A *sneer* may be simply a contemptuous facial contortion or some brief satirical utterance that throws a contemptuous sidelight on what it attacks without attempting to prove or disprove. The *jeer* and *gibe* are uttered; the *gibe* is bitter, and often sly or covert; the *jeer* is rude and open. A *scoff* may be in act or word, and is commonly directed against that which claims honor, reverence, or worship. A *fling* is careless and commonly pettish; a *taunt* is intentionally insulting and provoking; the *sneer* is supercilious; the *taunt* is defiant. See SCORN.

sneesh (snēsh) *n. Scot.* Snuff. Also **sneesh′ing.**

sneeze (snēz) *v.i.* **sneezed, sneez·ing** *v.i.* To drive air forcibly and audibly out of the mouth and nose by a spasmodic involuntary action. —*v.t.* To utter with or as with a sneeze: often with *out.* —**not to be sneezed at** *Colloq.* Of a character entitling to consideration. —*n.* An act of sneezing: also **sneez′ing.** [Misreading of ME *fnese,* OE *fnēosan* sneeze] —**sneez′er** *n.* —**sneez′y** *adj.*

sneeze gas A sternutator.

sneeze·weed (snēz′wēd′) *n.* Any plant of a genus *(Helenium)* of the composite family, especially *H. autumnale*: from the effect of the powdered leaves and flowers when snuffed up: also called *bitterweed.*

sneeze·wort (snēz′wûrt′) *n.* **1** A perennial Eurasian plant *(Achillea ptarmica)* resembling the yarrow. Its powdered dry leaves produce sneezing. **2** Sneezeweed.

snell[1] (snel) *n.* A short line of gut, horsehair, etc., bearing a fish hook, to be attached to a longer line. [Origin unknown]

snell[2] (snel) *adj.* **1** *Scot.* Sharp; keen; piercing. **2** Austere; severe. **3** Nimble; quick.

Snel·len test (snel′ən) *Med.* A test for determining visual acuity performed by reading a standard set of graded letters at a specified distance. [after Herman *Snellen,* 1834–1908, Dutch opthalmologist, who devised it]

snick (snik) *n.* **1** A small cut; nick; snip. **2** A knot in thread or the like. **3** In cricket, a glancing hit. —*v.t.* **1** To cut a nick in. **2** To hit (a ball) a glancing blow. —**to snick and snee** To thrust and cut. [Back formation < *snick or snee.* See SNICKERSNEE.]

snick·er (snik′ər) *n.* A half-suppressed or smothered laugh. —*v.i.* To utter a snicker; laugh slyly and foolishly with audible catches of the voice; giggle. —*v.t.* To utter or express with a snicker. Also *snigger.* [Imit.]

snick·er·snee (snik′ər·snē′) *n.* **1** A fight with knives. **2** A knife suitable for thrusting and cutting. Also **snick and snee, snick′-a-snee, snick or snee.** [Alter. of earlier *snick or snee* thrust or cut, ult. <Du. *steken* thrust + *snijen* cut]

snide (snīd) *adj.* Malicious or derogatory; nasty. —*n.* A snide person. [Origin unknown]

sniff (snif) *v.i.* **1** To breathe through the nose in short, quick, audible inhalations. **2** To express contempt, etc., by sniffing: often with *at.* **3** To inhale a scent in sniffs. —*v.t.* **4** To breathe in through the nose; inhale. **5** To smell or attempt to smell with sniffs: to *sniff* smoke. **6** To perceive as if by sniffs: to *sniff* peril. **7** To express (contempt) by sniffs. —*n.* **1** An act or the sound of sniffing. **2** Perception by or as by sniffing; that which is inhaled by sniffing. [Appar. back formation <SNIVEL]

snif·fle (snif′əl) *v.i.* **·fled, ·fling** **1** To snuffle. **2** To snivel or whimper; whine; sniff. —*n.* A snuffle. [Freq. of SNIFF]

snif·fy (snif′ē) *adj.* **·fi·er, ·fi·est** *Colloq.* Disposed to sniff or be disdainful or scornful.

snif·ter (snif′tər) *n.* **1** A liquor glass, pear-shaped, with a small opening to concentrate the aroma. **2** *U.S. Slang* A small drink of liquor, usually a dram. [< *snift,* var. of SNIFF]

snig·ger (snig′ər) *n.* A snicker, especially a derisive snicker. —*v.i.* To snicker, especially in derision. [Var. of SNICKER] —**snig′ger·er** *n.*

snig·gle (snig′əl) *v.t.* **·gled, ·gling** *Brit.* **1** To fish for or catch, as eels, by thrusting the bait into their hiding places. **2** To entrap, as in a net; ensnare. [< dial. E *snig* eel]

snip (snip) *v.* **snipped, snip·ping** *v.t.* To clip, remove, or cut with a short, light stroke or strokes of scissors or shears: often with *off.* —*v.i.* To cut with small, quick strokes. —*n.* **1** An act of snipping. **2** A small piece snipped off. **3** *U.S. Colloq.* A small or insignificant person or thing. **4** *pl.* Small shears for cutting metal. [<Du. *snippen*]

snipe (snīp) *n. pl.* **snipe** or **snipes** **1** A shore bird (genus *Capella*), allied to the woodcock and much esteemed as a game bird; especially, the common **European** or **whole snipe** *(C. gallinago),* and the common **American** or **Wilson's snipe** *(C. delicata).* **2** One of other snipelike birds, as the **lesser snipe** or **jack snipe** of Europe *(Limnocryptes minimus).* **3** *U.S. Slang* A cigarette or cigar butt. —*v.i.* **sniped, snip·ing** **1** To hunt or shoot snipe. **2** To shoot at or pick off individual enemies from cover or ambush. **3** *U.S. Slang* To hunt for cigarette or cigar butts. [<ON *snipa*]

snip·er (snī′pər) *n.* One who shoots an enemy from cover; a sharpshooter.

snip·er·scope (snī′pər·skōp′) *n.* An electronic optical device which may be mounted on a carbine or rifle in order to permit accurate night-firing by means of infrared rays focused on a fluorescent screen.

snip·pet (snip′it) *n.* **1** A small piece snipped off. **2** A small portion or share.

snip·pet·y (snip′it·ē) *adj.* **1** Arrogant; brusk; snippy. **2** Trivial, as if composed of little pieces snipped off; small; trifling. —**snip′pet·i·ness** *n.*

snip·py (snip′ē) *adj.* **·pi·er, ·pi·est** *Colloq.* **1** Supercilious; pert; impertinent. **2** Fragmentary.

snitch (snich) *Slang v.t.* To grab quickly; steal. —*v.i.* To inform; peach: usually with *on.* [? Var. of SNATCH]

snits (snits) *n. pl.* Slices of dried fruit, especially of dried apples. [<Pennsylvania Dutch *schnitz* sections of apple]

sniv·el (sniv′əl) *v.i.* **·eled** or **·elled, ·el·ing** or **·el·ling** To cry in a snuffling manner; run at the nose; snuffle; make affectedly tearful professions. —*n.* **1** Discharge from the nose. **2** The act of sniveling. [OE (assumed) *snyflan* < *snyflung* mucus from the nose] —**sniv′el·er** or **sniv′el·ler** *n.* —**sniv′el·ing** or **sniv′el·ling** *adj.* & *n.*

snob (snob) *n.* **1** One who makes birth, wealth, or education the sole criterion of worth. **2** One who is cringing to superiors and overbearing with inferiors in position. **3** Any pretender to gentility. **4** *Obs. Brit.* A scab; rat: said of a workingman. [Origin uncertain] —**snob′·ber·y** *n.*

snob·bish (snob′ish) *adj.* Pertaining to, characteristic of, or befitting a snob or snobs. —**snob′bish·ly** *adv.* —**snob′bish·ness** *n.*

snod (snod) *Scot. v.t.* To make trim or neat; prune; tidy. —*adj.* Neat; also, sly; demure.

snood (snōōd) *n.* **1** A small meshlike cap or bag attached to the back of a hat, worn by women to keep the hair in place. **2** *Scot.* A fillet formerly worn about the hair by an unmarried woman in Scotland as an emblem of virginity. —*v.t.* To bind with a snood, as hair. [OE *snōd*]

snook (snōōk, snōōk) *v.i.* **1** *Scot.* To sniff. **2** To lurk. —*n.* **1** A smell; sniff; a bite. **2** *Slang* An informer.

snook·er (snōōk′ər) *n.* A pool game played with fifteen red object-balls (one point each) and six variously colored object-balls (2 to 7 points). The player pocketing a red ball may try for any varicolored ball. When all red balls have been pocketed, the varicolored balls must be played in order. Also **snooker pool.** [Origin uncertain]

snool (snōōl) *Scot. v.i.* **1** To snivel. **2** To yield submissively. —*n.* One who is meanly subservient.

snoop (snōōp) *Colloq. v.i.* To look or pry into things with which one has no business; thrust one's nose into things. —*n.* One who snoops: also **snoop′er.** [<Du. *snoepen* eat goodies on the sly] —**snoop′y** *adj.*

snoop·er·scope (snōō′pər·skōp) *n.* An optical device operating on the same principle as the sniperscope but designed for carrying in the hand or on the head.

snoot (snōōt) *n. Colloq.* A person's nose or face; also, a wry face; a grimace. [Var. of SNOUT]

snoot·y (snōō′tē) *adj.* **snoot·i·er, snoot·i·est** *U.S. Colloq.* Conceited or supercilious.

snooze (snōōz) *Colloq. v.i.* **snoozed, snooz·ing** To sleep lightly; doze. —*n.* A short and light sleep. [Origin uncertain]

Sno·qual·mie Falls (snō·kwol′mē) A waterfall of 270 feet in the **Snoqualmie River,** a river flowing 45 miles west and NW in central Washington from the Cascade Range east of Seattle.

snore (snôr, snōr) *v.i.* **snored, snor·ing** To breathe in sleep through the nose and open mouth, with a hoarse rough noise and rattling vibrations of the soft palate. —*n.* An act or the noise of snoring. [Imit.] —**snor′er** *n.*

snor·kel (snôr′kəl) *n.* An adaptation of the schnorkel for underwater breathing in shallow depths, as in skin diving. [<SCHNORKEL]

Snor·ri Stur·lu·son (snôr′ē stûr′lə·sən, stōōr′-), 1179–1241, Icelandic historian and poet.

snort (snôrt) *v.i.* **1** To force the air violently and noisily through the nostrils, as spirited horses. **2** To express indignation, ridicule, etc., by a snort. **3** *Colloq.* To laugh with a boisterous outburst. —*v.t.* **4** To utter or express by snorting. **5** To expel by or as by a snort. —*n.* **1** The act or sound of snorting. **2** *Slang* A small drink. [ME *snorten.* ? Related to SNORE.] —**snort′er** *n.*

snot (snot) *n.* **1** Mucus from or in the nose: a vulgar usage. **2** *Slang* A low or mean fellow. [OE *gesnot*]

snot·ty (snot′ē) *adj.* **·ti·er, ·ti·est** **1** Dirtied with snot: a vulgar usage. **2** *Slang* Contemptible; mean; paltry. **3** *Slang* Impudent; proudly conceited; saucy.

snout (snout) *n.* **1** The forward projecting part of a beast's head, especially of a swine's; proboscis; muzzle. **2** Some similar anterior prolongation of the head of an animal, as the rostrum of a gastropod or that of a weevil. **3** Something resembling a hog's snout, such as the nozzle of a hose, a pipe, or the like; a blunt projection, as of rock; or, contemptuously, a person's nose. —*v.t.* To provide with a snout or nozzle. [ME *snūte.* Related to OE *snȳtan* blow the nose.]

snout beetle The curculio.

snow[1] (snō) *n.* **1** Precipitation taking the form of minute ice crystals formed from an aqueous vapor in the air when the temperature is below 32° F., and usually falling in irregular masses or flakes. ◆ Collateral adjective: *nival.* **2** A similar aggregation that resembles snow in being white or composed of flakes: a *snow* of blossoms. **3** A fall of snow; snowstorm. **4** A winter. **5** *Slang* Cocaine. **6** The pattern of snowlike drops appearing on a television screen as a result of weakened signals in a receiver. —*v.i.* **1** To fall as snow: usually used impersonally: It is *snowing.* —*v.t.* **2** To scatter or cause to fall as or like snow. **3** To cover, enclose, or

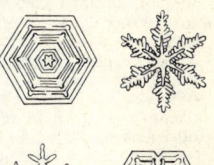

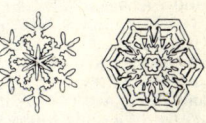

SNOW CRYSTALS

obstruct with snow: with *in, over, under,* or *up.* [OE *snāw*]

Snow may appear as a combining form in hyphemes or solidemes, or as the first element in two-word phrases; as in:

snowbank	**snow–driven**
snow–beaten	**snow field**
snow–blast	**snow–haired**
snow–blown	**snowland**
snow–bright	**snowless**
snow–clad	**snowlike**
snow cloud	**snow–lined**
snow–cold	**snow peak**
snow–colored	**snowscape**
snow–covered	**snow–tipped**
snow–crested	**snow–topped**

snow[2] (snō) *n. Naut.* A two–masted square–rigged vessel characterized by having a trysail mast close behind the mainmast. [<MDu. *snauw* snout]

snow apple The Fameuse.

snow·ball (snō′bôl′) *n.* **1** A small round mass of snow compressed to be thrown, as in sport. **2** The guelder–rose *(Viburnum opulus)*: so called from its ball–shaped clusters of white flowers: also **snowball bush** or **tree.** — *v.i.* **1** To throw snowballs. **2** To gain in size, importance, etc., as a snowball that rolls over snow. — *v.t.* **3** To throw snowballs at.

snow·bell (snō′bel′) *n.* Any of a genus *(Styrax)* of trees and shrubs of warm regions, bearing showy white flowers in racemes; especially, *S. americana* of the SE United States.

snow·ber·ry (snō′ber′ē) *n. pl.* **·ries 1** A bushy American shrub *(Symphoricarpos albus)* having a loose, leafy cluster of snow–white berries. **2** A West Indian shrub *(Chiococca alba)* of the madder family; it produces the cainca root and is often cultivated in greenhouses for its white berries: also called *milkberry.*

snow·bird (snō′bûrd′) *n.* **1** A small finch (genus *Junco*) of northern North America, commonly seen in flocks during the winter. **2** The snow bunting. **3** *Slang* A cocaine or heroin addict.

snow blindness An impairment of vision, caused by exposure of the eye to the glare of snow. — **snow′–blind′** *adj.*

snow·blink (snō′blingk′) *n.* The dazzling scintillation of light reflected from a field of ice or snow.

snow–bound (snō′bound′) *adj.* Hemmed in or forced to remain in a place by heavy snow; snowed in.

snow bridge A natural arch formation of snow bridging a crevasse.

snow·broth (snō′brôth′, -broth′) *n.* **1** Melted snow or snow and water mixed. **2** Any very cold liquid.

snow bunting A bird (genus *Plectrophenax*), especially *P. nivalis* of northern regions, the male of which in the breeding season is snow–white with black markings. Also called *snowbird, snowflake.*

snow·bush (snō′bo͝osh′) *n.* A California shrub of the genus *Ceanothus,* as *C. cordulatus,* that bears numerous small white flowers.

snow·cap (snō′kap′) *n.* A crest of snow, as on a mountain peak. — **snow′–capped′** *adj.*

snow cover The blanket of snow, variable in thickness and duration, which covers the ground over a given area, affecting ground temperatures and vegetation.

Snow·den (snōd′n), **Philip,** 1864–1937, first viscount, English statesman and economist.

Snow·don (snōd′n) A mountain in Caernarvonshire, the highest point in Wales; 3,560 feet. *Welsh* **Er·y·ri** (er′i·rē).

snow·drift (snō′drift′) *n.* A pile of snow heaped up by the wind.

snow·drop (snō′drop′) *n.* **1** A low, European, early–blooming bulbous plant *(Galanthus nivalis)* bearing a single, white, drooping flower. **2** The common anemone.

snowdrop tree The silverbell.

snow–eat·er (snō′ē′tər) *n.* The chinook.

snow·fall (snō′fôl′) *n.* **1** A fall of snow. **2** The amount of snow that falls in a given period.

snow fence Portable fencing consisting of thin, closely placed pickets, used to prevent the drifting of snow over roads, fields, etc.

snow·flake (snō′flāk′) *n.* **1** One of the small feathery masses in which snow falls. **2** The snow bunting. **3** Any of certain plants (genus *Leucojum*) allied to and resembling the snowdrop; especially, the **spring snowflake** *(L. vernum),* the **summer snowflake** *(L. aestivum),* and the **autumn snowflake** *(L. autumnale).*

snow goose Any of certain North American geese (genus *Chen*) which breed in the Arctic, snow–white with black primary feathers; specifically, *C. hyperborea atlantica* of the North Atlantic and *C. hyperborea hyperborea* of the West.

snow leopard The ounce.

snow lily An attractive spring–blooming herb *(Erythronium grandiflorum)* of the lily family native in the Rocky Mountains.

snow line 1 The limit of perpetual snow on the sides of mountains, varying in position with the latitude, the season, and the climate. **2** The extreme distance north and south of the equator within which snow never falls. Also **snow limit.**

Snow Mountains The mountain ranges of central Netherlands New Guinea. *Dutch* **Snee·uw Ge·berg·te** (snā′o͞o khə·berkh′tə).

snow pellets Snowlike particles sometimes precipitated from a cloud during showers.

snow plant A handsome, blood–red saprophytic herb *(Sarcodes sanguinea)* found in the rich humus of mountain forests in southern California, frequently covered with snow in its blooming season.

snow·plow (snō′plou′) *n.* Any large, plowlike device for turning fallen snow aside from a road or railroad, or for the removal of snow from such surfaces. Also **snow′plough′.**

snow pudding A pudding containing gelatin, sugar, and white of egg whipped into a snowlike foam.

snow·shed (snō′shed′) *n.* A timber structure, as one built over portions of a railway, as a protection from snowslides.

snow·shoe (snō′sho͞o′) *n.* A device, usually a network of sinew in a wooden frame, to be fastened on the foot by a strap across the toes, as a support in walking over snow; also, a ski. — *v.i.* **·shoed, ·shoe·ing** To walk on snowshoes. — **snow′shoed′** (-sho͞od′) *adj.* — **snow′sho′er** (-sho͞o′ər) *n.*

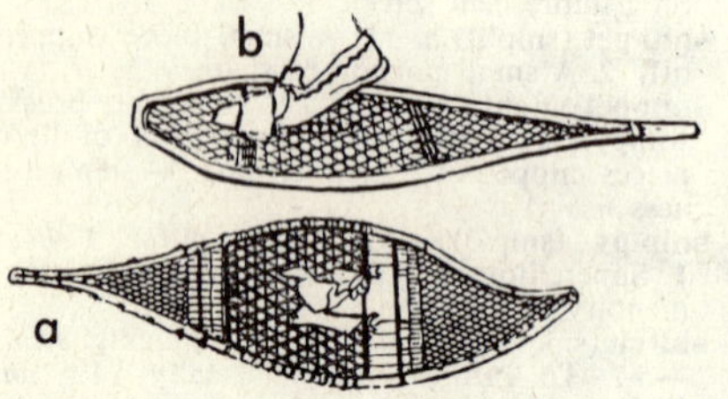

SNOWSHOES
a. Sioux Indian. *b.* Iroquois Indian.

snow·slide (snō′slīd′) *n.* An avalanche of snow.

snow squall A flurry of wind and snow.

snow·storm (snō′stôrm′) *n.* A storm with a heavy fall of snow.

snow·suit (snō′so͞ot′) *n.* A heavy outer garment for winter wear, either in one piece or consisting of ankle–length, tight–fitting pants and a short, snug jacket.

snow–white (snō′hwīt′) *adj.* White as snow.

snow·y (snō′ē) *adj.* **snow·i·er, snow·i·est 1** Abounding in or full of snow. **2** Snow–white; hence, pure; unblemished; spotless: *snowy* linen. — **snow′i·ly** *adv.* — **snow′i·ness** *n.*

snowy heron The common small white egret *(Egretta thula)* of the southern United States and northern South America.

snowy owl See under OWL.

snub (snub) *v.t.* **snubbed, snub·bing 1** To treat with contempt or disdain; slight. **2** To rebuke or check with a sharp or cutting remark. **3** To stop or check, as a rope in running out, by taking a turn about a post, etc.; also, to make fast (a boat, etc.) thus. **4** *Obs.* To clip; stunt; nip. — *adj.* Short; pug: said of the nose. — *n.* **1** An act of snubbing; a deliberate and intentional slight. **2** A sudden checking, as of a running rope or cable. **3** A snub nose. [<ON *snubba* snub] — **snub′ber** *n.*

snub–nosed (snub′nōzd′) *adj.* Having a pug or snub nose.

snuff[1] (snuf) *v.t.* **1** To draw in (air, etc.) through the nose. **2** To catch the scent of; smell; sniff; also, to examine by smelling. — *v.i.* **3** To snort; sniff. **4** To inhale air in disdain or anger. — *n.* **1** An act of snuffing; sniff; also, perception by smelling. **2** Resentment expressed by sniffing. [<MDu. *snuffen*]

snuff[2] (snuf) *n.* The charred portion of a wick. — *v.t.* **1** To crop the snuff from (a wick). **2** To put out or extinguish: with *out.* [Cf. G *schnuppe* snuff of a candle]

snuff[3] (snuf) *n.* **1** Pulverized tobacco to be inhaled into the nostrils. **2** The quantity of it taken at one time. **3** Any medicinal powder to be drawn into the nostrils. — **up to snuff** *Colloq.* **1** Meeting the usual standard, as in quality, health, etc. **2** Not easily deceived; sharp–witted. — *v.i.* To take or use snuff. [<Du. *snuf,* appar. short for *snuiftabak,* lit., tobacco to be inhaled]

snuff·box (snuf′boks′) *n.* A small box for carrying snuff about the person.

snuff color The color of snuff; yellowish brown. — **snuff′–col′ored** *adj.*

snuf·fer[1] (snuf′ər) *n.* **1** One who or that which snuffs. **2** A porpoise.

snuf·fer[2] (snuf′ər) *n.* **1** One who or that which snuffs (a candle). **2** *pl.* A scissorlike instrument for removing the snuff from a candle: also called **pair of snuffers.**

snuf·fle (snuf′əl) *v.* **·fled, ·fling** *v.i.* **1** To breathe through the nose noisily and with difficulty, as when it is obstructed by mucus. **2** To breathe noisily, as a dog following a scent. **3** To talk through the nose; snivel. — *v.t.* **4** To utter in a nasal tone. — *n.* **1** An act of snuffling, or the sound made by it. **2** *pl.* Nasal catarrh. **3** An affected nasal or emotional voice or twang; hence, cant. [Freq. of SNUFF[1]] — **snuf′fler** *n.* — **snuf′fly** *adj.*

snuff·y (snuf′ē) *adj.* **snuf·fi·er, snuf·fi·est 1** Pertaining to or like snuff. **2** Soiled with or smelling of snuff; hence, offensive; unattractive. — **snuf′fi·ly** *adv.* — **snuf′fi·ness** *adj.*

snug (snug) *adj.* **snug·ger, snug·gest 1** Closely and comfortably sheltered, covered, or situated. **2** Close or compact; having room enough, but not too much; comfortable; cozy; also, having everything closely secured; trim: said of a ship. **3** Fitting closely but comfortably, as a garment. See synonyms under COMFORTABLE. — *v.* **snugged, snug·ging** *v.t.* To make snug. — *v.i.* To snuggle; move close. — **to snug down** To make a vessel ready for a storm by reducing sail, etc. [Prob. <LG. Cf. Du. *snugger* clean, smooth.] — **snug′ly** *adv.* — **snug′ness** *n.*

snug·ger·y (snug′ər·ē) *n. pl.* **·ger·ies** A cozy and comfortable place or room.

snug·gle (snug′əl) *v.t.* & *v.i.* **·gled, ·gling** To lie or draw close; nestle; cuddle: often with *up* or *together.* [Freq. of SNUG, *v.*]

so[1] (sō) *adv.* **1** To this or that or such a degree; to this or that extent; in the same degree, quantity, or proportion: either used alone, the degree being implied or understood: Why *so* long; or followed by or preceded by a dependent expression introduced by *as, that,* or *but.* **2** In this, that, or such a manner; in the same or a like or corresponding manner; in the manner mentioned: often following a clause beginning with *as,* or preceding one beginning with *that.* **3** Just as said, directed, suggested, or implied; also, according to fact: referring to a preceding (sometimes following) statement or suggestion. **4** To an extreme degree; extremely; very. **5** The fact being thus: used as an expletive. **6** About as many or as much as stated; thereabouts: I shall stay a day or *so.* **7** At all events; in any case; at all: now only in the compounds *whosoever, whichsoever,* etc. **8** According to the truth of what is sworn to or averred: said in oaths or asseverations: *So* help me God. **9** Indeed! elliptical for *Is it so*? **10** To such an extent: used elliptically for *so much*: I love him *so*! **11** Too: used in emphatic denial: You can *so*! **12** Indicative of surprise or disapproval: *So* there you are. **13** So as to follow immediately; then; therefore. **14** Let it be that way; very well. — *conj.* **1** With the purpose that: often with *that*: They left the hotel early *so* (that) they would not encounter him. **2** In such a way that; as a consequence of which: He consented, *so* they went away. **3** *Obs.* As. — *interj.* **1** Stay still! **2** Is that so! **3** In nautical parlance, steady! [OE *swā*]

so[2] (sō) *n. Music* The fifth of the syllables used in singing the scale: also *sol.* [See GAMUT]

soak (sōk) *v.t.* **1** To place in liquid till thoroughly saturated; steep. **2** To wet thoroughly;

drench: The rain *soaks* the earth. 3 To take in through or as through pores or interstices; suck up; absorb: with *in* or *up.* 4 *Colloq.* To drink, especially to excess. 5 *U.S. Slang* **a** To charge exorbitantly. **b** To pawn. **c** To strike hard; beat. —*v.i.* 6 To remain or be placed in liquid till saturated. 7 To penetrate; pass: with *in* or *into.* 8 *U.S. Slang* To drink to excess. —*n.* 1 The process or act of soaking, or state of being soaked. 2 Liquid in which something is soaked. 3 *Slang* A hard drinker; a drinking spree. ◆ Homophone: *soke.* [OE *socian.* Akin to SUCK.]

soak·age (sō′kij) *n.* 1 The process of soaking, or the state of being soaked. 2 The quantity of liquid that soaks in or through, or seeps out.

soak·er (sō′kər) *n.* 1 One who or that which soaks. 2 *Slang* A habitual drunkard.

soak·ers (sō′kerz) *n. pl.* Short pants of absorbent material, usually wool, worn by babies over diapers.

soak·y (sō′kē) *adj.* **soak·i·er, soak·i·est** Covered or filled with moisture; steeped; soggy.

so–and–so (sō′ən·sō′) *n.* 1 An unnamed or undetermined person or thing. 2 *Colloq.* A euphemism for many offensive epithets.

soap (sōp) *n.* 1 A cleansing agent consisting of sodium or potassium salts of fatty acids, made by decomposing the glyceryl esters of fats and oils with alkalies; a detergent. *Hard* soaps are made by the use of soda, while the potash soaps are *soft.* 2 A metallic salt of one of the fatty acids. 3 *U.S. Slang* Money used for sinister purposes; hence, any means of obtaining an end. —*v.t.* To rub with soap; treat with soap. [OE *sāpe*]

soap·bark (sōp′bärk′) *n.* 1 The bark of the quillai. 2 The bark of a tropical American shrub (genus *Pithecellobium*), used as a substitute for soap. Also **soapbark tree.**

soap·ber·ry (sōp′ber′ē) *n. pl.* **·ries** 1 The fruit of any one of several trees or shrubs (genus *Sapindus*) of the family *Sapindaceae.* 2 Any one of the trees producing it, especially *S. saponaria,* of tropical America and southern Florida, the pulp of whose fruit is used in washing textile fabrics.

soap·box (sōp′boks′) *n.* 1 A box or crate for soap. 2 Any box or crate used as a platform by street orators. —**soapbox oratory** Impromptu or crude oratory, marked by vigor rather than logic. Also **soap box.**

soap·box·er (sōp′bok′sər) *n. Colloq.* A loud and ranting speaker; a street–corner orator; a tubthumper.

soap bubble 1 An inflated bubble of soapsuds, forming a hollow globule. 2 Anything attractive but unsubstantial.

soap opera A daytime television or radio drama presented serially and usually dealing with domestic themes of a highly emotional character: so called in reference to the soap commercials often presented on such programs.

soap plant Any of several plants whose bulbs are used for soap, especially a lilywort (*Chlorogalum pomeridianum*) of California.

soap·stone (sōp′stōn′) *n.* Steatite: so called from its soapy feel.

soap·suds (sōp′sudz′) *n. pl.* Soapy water, especially when worked into a foam.

soap·wort (sōp′wûrt′) *n.* A perennial herb (*Saponaria officinalis*) of the pink family having clusters of pink or whitish, often double, flowers: so called because its juice forms a lather with water. Also called *bouncing Bet.*

soap·y (sō′pē) *adj.* **soap·i·er, soap·i·est** 1 Resembling, containing, or consisting of soap. 2 Smeared with soap. 3 *Slang* Flattering.

soar (sôr, sōr) *v.i.* 1 To float aloft through the air on wings, as a bird. 2 To sail through the air without perceptibly moving the wings, as a hawk or vulture. 3 To glide without losing altitude, as an airplane. 4 To rise above any usual level: Prices will *soar* if the ceilings are removed. See synonyms under FLY[1]. —*n.* An act of soaring; a range of upward flight. [<F *essorer* <L *ex* out + *aura* breeze, air] —**soar′er** *n.*

sob (sob) *n.* A convulsive, audible inhalation of air under the impulse of painful or hysterical emotion, and usually accompanied with tears; the act or the sound of sobbing; also, any similar sound, as of the wind. Also **sob′·bing.** —*v.* **sobbed, sob·bing** *v.i.* 1 To weep with audible, convulsive catches of the breath. 2 To make a sound like a sob, as the wind. —*v.t.* 3 To utter with sobs. 4 To bring to a specified condition by sobbing: to *sob* oneself to sleep. [Imit.]

So·bat (sō′bat) A river in SE Sudan, flowing 205 miles NW from the Ethiopian border to the White Nile.

so·be·it (sō·bē′it) *n.* An amen. —*conj.* If so; if only; provided: originally **so be it.**

so·ber (sō′bər) *adj.* 1 Possessing properly controlled faculties; even–tempered; well–balanced; temperate in action or thought. 2 Grave; sedate; realizing the importance and seriousness of life. 3 Not under the influence of an intoxicant; not drunk. 4 Moderate in or abstinent from the use of intoxicating drink. 5 Of subdued or modest color. —*v.t.* & *v.i.* To make or become sober. [<OF *sobre* <L *sobrius*] —**so′ber·ly** *adv.* —**so′ber·ness** *n.*

Synonyms (*adj.*): abstemious, abstinent, calm, collected, cool, dispassionate, moderate, quiet, regular, sane, staid, steady, temperate, unimpassioned, unintoxicated. See SAD, SANE[1], SEDATE, SERIOUS. *Antonyms*: agitated, crazy, drunk, drunken, ecstatic, excited, extravagant, extreme, frantic, furious, immoderate, impassioned, intemperate, intoxicated, passionate, unreasonable.

So·bies·ki (sô·byes′kē), **John** See JOHN III OF POLAND.

So·bran·je (sô·brän′yə) *n.* The national assembly or legislature of Bulgaria. Also **So·bran·i·ye** (sô·brän′ē·yə), **So·bran′ye.**

so·bri·e·ty (sō·brī′ə·tē) *n. pl.* **·ties** 1 The state of being sober. 2 Moderateness in temper or conduct; sedateness; seriousness; temperance. See synonyms under ABSTINENCE. [<L *sobrietas, -tatis* <*sobrius* sober]

so·bri·quet (sō′bri·kā) *n.* A fanciful or humorous appellation; a nickname: also spelled *soubriquet.* [<F; ult. origin unknown]

sob sister *U.S. Slang* A woman journalist who writes mawkishly sentimental news stories.

sob story *Slang* A sad personal narrative told to elicit pity or sympathy.

soc·age (sok′ij) *n.* The feudal tenure of land by certain determinate services other than knight–service; hence, later, tenure by any fixed service other than military. [<*soc,* var. of SOKE] —**soc′ag·er** *n.*

so–called (sō′kôld′) *adj.* Called as stated; generally styled thus: usually implying a doubtful or improper form.

soc·cer (sok′ər) *n.* A form of football in which the ball is propelled toward the opponents' goal by kicking or by striking with the body or head, the goalkeepers being the only players allowed to touch the ball with the hands or forearms: officially called *association football.* [Alter. of ASSOCIATION]

So·che (sō′che′) The Chinese name for YARKAND.

So·chi (sô′chē) A port on the Black Sea in southern Krasnodar territory, European Russian S.F.S.R.

so·cia·bil·i·ty (sō′shə·bil′ə·tē) *n.* The quality or character of being sociable. Also **so′cia·ble·ness.**

so·cia·ble (sō′shə·bəl) *adj.* 1 Inclined to seek company; social. 2 Agreeable in company; companionable; genial. 3 Characterized by or affording occasion for agreeable conversation and friendliness. See synonyms under AMICABLE, FRIENDLY. —*n.* 1 An informal social gathering: also *social.* 2 A four–wheeled open carriage with facing seats. [<F <L *sociabilis* <*socius* friend] —**so′cia·bly** *adv.*

so·cial (sō′shəl) *adj.* 1 Of or pertaining to society or its organization; relating to persons as living in society or to the public as an aggregate body: *social* life, *social* questions. 2 Disposed to hold friendly intercourse with others; sociable. 3 Constituted to live in society; having developed or fulfilled tendencies to organize in society as a race or people: *social* beings. 4 Of or pertaining to public welfare: *social* insurance. 5 Pertaining to or characteristic of persons of fashion: *social* register. 6 Living in communities: *social* ants or bees; aggregate; compound; colonial. 7 Grouping compactly, as individual plants; partly or wholly covering a large area of land: said of plant species. 8 Venereal: *social* disease: a euphemism. 9 Pertaining to or between allies or confederates, as the wars waged by Rome in 90–89 B.C., and by Athens in 357–355 B.C. against their allies. See synonyms under FRIENDLY, GOOD. —*n.* A sociable. [<L *socialis* <*socius* ally]

social climber A person who attempts to become friendly with prominent or wealthy people.

social contract *Philos.* The supposed original agreement by which individuals were united in political associations for their mutual protection, the surrender of their individual sovereignty having been made not through force but by mutual consent: a theory of Hobbes, Locke, Rousseau, etc.

Social Democrat 1 A member of the Social Democratic party of Germany, founded by Bebel and Liebknecht in 1869 and based on Marxian principles. In 1875 it was merged with Lassalle's General German Workmen's Association. Under the Republic (1918–33) it advocated the principles embedded in the Weimar Constitution. 2 A member of a similar party in other countries. —**so′cial–dem′o·crat′ic** *adj.*

social evil Prostitution.

social insurance Government insurance designed to protect wage earners against unemployment, illness, accident, or the like, and not always requiring the payment of a premium on the part of the insured.

so·cial·ism (sō′shəl·iz′əm) *n.* Public collective ownership or control of the basic means of production, distribution, and exchange, with the avowed aim of operating for use rather than for profit, and of assuring to each member of society an equitable share of goods, services, and welfare benefits: as a system of social and economic organization planned, attempted, or achieved through various methods—in **Utopian** or **Christian Socialism,** through cooperative communal groups holding all things in common (approximating the philosophic anarchism of Thoreau, Tolstoy, and Kropotkin, and the communalism and commensalism of the early and undivided church); in **Guild Socialism,** through organization of producer groups and the professions in syndicalist guilds to be represented in a federal legislative body; in **Fabian** or **British Labour Party Socialism,** through parliamentary democracy using gradualist evolutionary processes; in **Marxist–Leninist State Socialism,** through revolution, expropriation, and dictatorship of the so–called proletariat, in short, Communism. Compare MIXED ECONOMY. —**creeping socialism** Anything considered as a gradual or piecemeal encroachment upon the system of private property and free enterprise through state action: used as an epithet.

so·cial·ist (sō′shəl·ist) *n.* An advocate of socialism. —*adj.* Socialistic.

so·cial·is·tic (sō′shəl·is′tik) *adj.* Of, pertaining to, advocating, like, or practicing socialism.

Socialist Labor party A U. S. political party, originally formed in 1877 as the **Socialistic Labor party** and renamed in 1891.

Socialist party A U. S. political party formed in 1901 by the combination of a dissident group from the Socialist Labor party with the **Social Democratic party** (established in 1898 by E. V. Debs and others). After the **Social Democratic Federation,** a dissident group formed in 1936, rejoined in 1957, the name **Socialist party–Social Democratic Federation** was officially adopted.

so·cial·ite (sō′shəl·īt) *n.* A person prominent in fashionable society.

so·ci·al·i·ty (sō′shē·al′ə·tē) *n. pl.* **·ties** 1 The state or character of being social; sociability. 2 A social custom or action. 3 The instinct or tendency which is the basis of social organization.

so·cial·ize (sō′shəl·īz) *v.* **·ized, ·iz·ing** *v.t.* 1 To convert from an anti–social to a social attitude; make friendly, cooperative, or sociable. 2 To arouse to an interest in humanity. 3 To convert or adapt to social uses or needs. 4 To put under group control; especially, to

add,āce,câre,pälm; end,ēven; it,īce; odd,ōpen,ôrder; to͝ok,po͞ol; up,bûrn; ə = a in *above,* e in *sicken,* i in *clarity,* o in *melon,* u in *focus*; yo͞o = u in *fuse*; oi,oil; ou,pout; ch,check; g,go; ng,ring; th,thin; th,this; zh,vision. Foreign sounds à,œ,ü,kh,ṅ; and ◆: see page xx. < from; + plus; ? possibly.

regulate according to socialistic principles. —*v.i.* 5 To take part in social activities. Also *Brit.* **so′cial·ise.** —**so′cial·i·za′tion** *n.*
socialized medicine A system proposing to supply the public with medical care at nominal cost, by regulating services and fees, by government subsidies to physicians and medical projects, or by cooperative projects.
social register A directory of persons prominent in fashionable society.
social science 1 The body of knowledge that relates to man as a member of society, or of any component part of society, as the state, family, or any systematized human institution. 2 Any field of knowledge dealing with human society, as economics, history, sociology, education, politics, ethics, etc.
social security 1 Any public system which provides welfare services for members of the community in need. 2 *U.S.* A Federal program of old-age and unemployment insurance, public assistance to the blind, disabled, and dependent, and maternal and child welfare services, administered by the **Social Security Administration.**
social service Activity intended to advance human welfare. —**so′cial-ser′vice** *adj.*
social settlement An institution or settlement, usually in the poor quarters of a large city, devoted to the aid and instruction of the poor.
social work Any clinical, social, or recreational service for improving community welfare, as through health clinics, recreational facilities, aid to the poor and the aged, etc. —**social worker**
so·ci·e·ty (sə·sī′ə·tē) *n. pl.* **·ties** 1 The system of community life, in which individuals, ordinarily in a territorial establishment, form a continuous and regulatory association for their mutual benefit and protection. 2 The body of persons composing such a community. 3 A number of persons in a community regarded as forming a class having certain common interests, status, etc.: high *society.* 4 The fashionable or cultured portion of a community, considered as constituting a class. 5 A body of persons associated for a common purpose or object; an association: a medical *society.* 6 *U.S.* In some States, an incorporated religious congregation. 7 A club or fraternity. 8 Association based on friendship or intimacy; companionship; company: to enjoy the *society* of workingmen. 9 *Ecol.* A group of plants or animals living together under the same physiographic conditions and influences and characterized by a principal species. See synonyms under ASSOCIATION. [<OF *societe* <L *societas, -tatis* < *socius* a friend]
Society Islands A part of French Oceania, comprising an island group south of the Tuamotu Islands; 690 square miles; capital, Papeete, on Tahiti; divided into two clusters: the *Windward Islands,* including Tahiti, Moorea, and other adjacent islands; and the *Leeward Islands,* including the major island of Raiatea.
Society of Friends A Christian religious group, founded in England by George Fox in the middle of the seventeenth century, characterized by their doctrine of "waiting upon the Spirit" for direct guidance and their repudiation of ritual, formal sacraments, oaths, and violence: commonly known as *Quakers.*
Society of Jesus See JESUIT.
So·cin·i·an (sō·sin′ē·ən) *adj.* Pertaining to either or both of the Italian theologians named Socinus or to their religious teachings, as the denial of the Trinity, of the natural depravity of man, of vicarious atonement, and of the efficacy of sacraments. —*n.* A believer in the Socinian theory. —**So·cin′·i·an·ism** *n.*
So·ci·nus (sō·sī′nəs) Latinized name of two Italian theologians and reformers: **Faustus,** 1539–1604, and **Laelius,** 1525–62, uncle of the preceding. Real name *Sozzini.*
socio– *combining form* 1 Society; social: *sociology.* 2 Sociology; sociological: *sociobiology.* [<F <L *socius* a companion]
so·ci·o·bi·ol·o·gy (sō′sē·ō·bī·ol′ə·jē, -shē·ō-) *n.* The study of sociology from the viewpoint and in terms of the biological and physical sciences; biophysics as applied to sociology. —**so′ci·o·bi′o·log′i·cal** (-bī′ə·loj′i·kəl) *adj.*
so·ci·o·ge·net·ic (sō′sē·ō·jə·net′ik, -shē·ō-) *adj.* Of or pertaining to the origin, development, and preservation of human society in any of its aspects.
so·ci·o·log·i·cal (sō′sē·ə·loj′i·kəl, -shē·ə-) *adj.* 1 Of or concerned with human social relations or conditions. 2 Of or pertaining to sociology. Also **so′ci·o·log′ic.** —**so′ci·o·log′i·cal·ly** *adv.*
so·ci·ol·o·gy (sō′sē·ol′ə·jē, sō′shē-) *n.* The science that treats of the origin and evolution of human society and social phenomena, the progress of civilization, and the laws controlling human institutions and functions. —**so′ci·ol′o·gist** *n.*
so·ci·om·e·try (sō′sē·om′ə·trē, sō′shē-) *n.* The study of the interrelationships of individuals within a community or social group, especially as expressed by attitudes of acceptance or rejection. —**so′ci·o·met′ric** (-ə·met′rik) *adj.*
sock[1] (sok) *n.* 1 A short stocking. 2 The light shoe worn by comic actors in the Greek and Roman drama; hence, comedy. Compare BUSKIN. [OE *socc* <L *soccus* slipper]
sock[2] (sok) *Slang v.t.* To strike or hit, especially with the fist; to punch. —*n.* A hard blow. [Origin unknown]
sock·dol·a·ger (sok·dol′ə·jər) *n. Slang* 1 That which gives the finishing stroke, or is decisive, especially in a dispute; a decisive blow, a conclusive reply or argument, or the like. 2 Something of great size; a rouser. Also **sock·dol′o·ger.** [Alter. of DOXOLOGY]
sock·et (sok′it) *n.* 1 *Mech.* A cavity or an opening specially adapted to receive and hold some corresponding piece or fixture: the *socket* for an electric-light bulb. 2 *Anat.* A cavity or hollowed depression for the reception of an organ or part. —*v.t.* To furnish with, hold by, or put into a socket. [<AF *soket,* dim. of OF *soc* a plowshare <Celtic]
sock·eye (sok′ī′) *n.* The red salmon of the Pacific coast (*Oncorhynchus nerka*), highly valued as a food fish. [Alter. of Salishan *sukkegh*]
so·cle (sō′kəl) *n. Archit.* 1 A plain, square block, higher than a plinth, supporting a statue or other work of art. 2 A base supporting a wall or a range of ornaments. [<F <Ital. *zoccolo* a pedestal, shoe <L *socculus,* dim. of *soccus* a sock]
soc·man (sok′mən) *n. pl.* **·men** (-mən) In old English law, one who holds land in socage: often spelled *sokeman.* Also **sock′man.**
So·co·tra (sō·kō′trə) An island, administered by the Eastern Aden Protectorate, in the Indian Ocean near the entrance to the Gulf of Aden; 1,400 square miles; capital, Tamarida: also *Sokotra, Soqotra.*
Soc·ra·tes (sok′rə·tēz), 469?–399 B.C., Athenian philosopher; the chief character in the dialogs of Plato: accused of impiety and innovation, he was imprisoned, condemned to death, and forced to drink an infusion of hemlock.
So·crat·ic (sō·krat′ik) *adj.* Pertaining to or characteristic of Socrates: also **So·crat′i·cal.** —*n.* A disciple of Socrates. —**So·crat′i·cal·ly** *adv.* —**Soc·ra·tism** (sok′rə·tiz′əm) *n.* —**Soc′·ra·tist** *n.*
Socratic irony A pretense of ignorance, though one may be wise, in order to expose the errors in an opponent's reasoning.
Socratic method The dialectic method of instruction by questions and answers, as adopted by Socrates in his disputations, leading either to a foreseen conclusion or to admissions damaging to an opponent.
sod[1] (sod) *n.* 1 Grassy surface soil held together by the matted roots of grass and weeds; sward; also, a piece of such soil; a turf or divot. 2 Grassy ground; lawn; the earth or soil. —*v.t.* **sod·ded, sod·ding** To cover with sod. [<MDu. *sode* piece of turf]
sod[2] (sod) Obsolete past tense of SEETHE.
so·da (sō′də) *n.* 1 Any of several white alkaline compounds widely used in medicine, industry, and the arts, especially sodium bicarbonate (baking soda), sodium carbonate, sodium hydroxide, and sodium oxide. 2 Soda alum. 3 Soda salts. 4 Soda water; also, a soft drink containing carbonated water, flavoring, and, sometimes, ice-cream. 5 In faro, the first card to appear face up in the dealing box before the start of play. [<Med. L <Ital. *soda* (*cenere*) solid (ash) <L *solidus*]
soda alum *Chem.* A double salt of sodium sulfate and aluminum.
soda ash Crude sodium carbonate.
soda biscuit 1 A biscuit leavened with sodium bicarbonate. 2 A soda cracker.
soda cracker A thin, crisp cracker made with yeast-leavened dough containing soda.
soda fountain 1 An apparatus from which soda water is drawn, usually containing receptacles for sirups, ice, and ice-cream. 2 A counter at which soft drinks and ice-cream are dispensed.
soda jerk *U.S. Slang* A clerk who serves at a soda fountain.
soda lime A mixture made from sodium hydroxide and calcium oxide, which, if heated with a nitrogenous compound, yields ammonia, from which the nitrogen may be extracted.
so·da·lite (sō′də·līt) *n.* A vitreous, translucent silicate of sodium and aluminum, with some chlorine. [<SODA + -LITE]
so·dal·i·ty (sō·dal′ə·tē) *n. pl.* **·ties** A brotherhood or fraternity; especially, a brotherhood for devotional or charitable purposes. [<L *sodalitas, -tatis* < *sodalis* companion]
so·dar (sō′där) *n.* A device for obtaining information on local weather conditions by projecting sound waves directly overhead and analyzing the echoes as recorded on an oscilloscope. [<SO(UND) D(ETECTING) A(ND) R(ANGING)]
soda water 1 An effervescent drink consisting of water strongly charged under pressure with purified carbon dioxide gas, formerly generated from sodium bicarbonate: often flavored with a fruit sirup. 2 Alkaline water as found in natural reservoirs or springs.
sod·den (sod′n) *adj.* 1 Soaked with moisture: *sodden* ground. 2 Doughy; soggy, as bread, biscuits, etc. 3 Flabby and pale; flaccid, especially from dissipation: said of persons or their features. 4 Dull; dreary: a *sodden* life. —*v.t.* & *v.i.* To make or become sodden. [ME *sothen,* orig. pp. of SEETHE] —**sod′den·ly** *adv.* —**sod′den·ness** *n.*
Sod·dy (sod′ē), **Frederick,** 1877–1956, English chemist and physicist.
Sö·der·blom (sœ′dər·blŏŏm), **Nathan,** 1866–1931, Swedish theologist.
sod house A dwelling built of sod or turf walls, often having a wooden roof: used by early settlers on the prairies.
so·di·um (sō′dē·əm) *n.* A silver-white, highly reactive, alkaline, metallic element (symbol Na), very similar to potassium. It is soft, malleable, and lighter than water, which it decomposes with the liberation of hydrogen; many of its compounds are of the highest importance. See ELEMENT. [<NL <Med. L *soda* SODA]
Sodium Am·y·tal (am′i·tôl, -tal) Proprietary name of a white, hygroscopic powder, $C_{11}H_{17}O_3N_2Na$, used in medicine as a sedative and hypnotic.
sodium benzoate *Chem.* A white, odorless, amorphous, granular or crystalline powder, $NaC_7H_5O_2$, used as an antipyretic, antirheumatic, antiseptic, as a food preservative, and to disguise taste, as of food of poor quality.
sodium bicarbonate *Chem.* A white crystalline compound, $NaHCO_3$, of alkaline taste, used in medicine and cookery; baking soda.
sodium borate *Chem.* Any sodium salt of boric acid; specifically, borax.
sodium carbonate *Chem.* A strongly alkaline compound, Na_2CO_3: in crystalline hydrated form known as washing soda, $Na_2CO_3{\cdot}10H_2O$. It is widely used in the manufacture of glass, soap, paper, etc., and in medicine and photography.
sodium chlorate *Chem.* A white crystalline compound, $NaClO_3$, used as a mordant, insecticide, weed-killer, and as an oxidizing agent.
sodium chloride Common salt, NaCl.
sodium cyanide *Chem.* A white, extremely poisonous salt of hydrocyanic acid, NaCN: used in electroplating and casehardening of metals, as a fumigant and chemical reagent.
sodium dichromate *Chem.* A red crystalline compound, $Na_2Cr_2O_7.2H_2O$, used as a dye and in making inks.
sodium hydroxide *Chem.* A white, caustic, fusible compound, NaOH: used in various solutions in chemistry, metallurgy, as a bleaching agent, etc.; caustic soda.
sodium hypochlorite *Chem.* An oxidizing and

bleaching compound, NaOCl, used also as a decontaminating agent for war gases.
sodium hyposulfite 1 Sodium thiosulfate. 2 A colorless crystalline salt, $Na_2S_2O_4$.
sodium nitrate A white compound, $NaNO_3$, used in the manufacture of nitric acid and as a manure. It occurs abundantly in nature.
sodium oxide A gray, highly reactive compound, Na_2O; sodium monoxide.
sodium perborate *Chem.* A colorless crystalline compound, $NaBO_3 \cdot 4H_2O$, used as a bleaching agent and disinfectant.
sodium peroxide *Chem.* A yellowish solid, Na_2O_2, used in combination with other chemicals as a bleaching agent. When treated with water it yields oxygen, and hydrogen peroxide. It is a powerful oxidizing agent.
sodium phosphate *Chem.* A sodium salt of phosphoric acid, Na_2HPO_4, crystallizing in the presence of water: the tribasic form is used as a laxative and also as a fixing agent in textile coloring.
sodium propionate *Chem.* A colorless, crystalline, water-soluble compound, $NaC_3H_5O_2$, used in medicine as a fungicide and to retard bacterial and mold growth in foods.
sodium silicate A material used in making artificial stone and in various industrial processes: known also as *soluble glass* or *waterglass.*
sodium sulfate *Chem.* A compound, Na_2SO_4, made by the action of sulfuric acid on common salt or on Chile saltpeter. It is used in glassmaking, and in one form (Glauber's salt) is important in medicine.
sodium sulfide *Chem.* A bleaching and decontaminating agent, Na_2S, especially effective against mustard gas.
sodium sulfite *Chem.* A decontaminating agent, Na_2SO_3, effective against chlorpicrin.
sodium thiosulfate *Chem.* A crystalline salt, $Na_2S_2O_3$, used industrially and medicinally and in photography as a fixing agent: also called *hypo.*
Sod·om (sod′əm) In the Bible, a city on the Dead Sea, destroyed with Gomorrah because of the wickedness of its people. *Gen.* xiii 10.
sod·om·ite (sod′əm·īt) *n.* One guilty of sodomy.
Sod·om·ite (sod′əm·īt) *n.* One of the people of Sodom.
sod·om·y (sod′əm·ē) *n.* Carnal copulation between male persons or with beasts. [<OF *sodomie* <LL *Sodoma* Sodom, to whose people this practice was imputed]
Soem·ba (sōōm′bä) The Dutch name for SUMBA.
Soem·ba·wa (sōōm·bä′wä) The Dutch name for SUMBAWA.
Soe·ra·ba·ja (sōō′rä·bä′yä) The Dutch name for SURABAYA.
so·ev·er (sō·ev′ər) *adv.* To or in some conceivable degree: used in generalizing and emphasizing what follows: a word often added to *who, which, what, where, when, how,* etc., to form the compounds *whosoever,* etc., giving them specific force. Often used separately: *how* great *soever* he might be.
so·fa (sō′fə) *n.* A wide seat, upholstered and having a back and raised ends. [<F <Arabic *ṣoffah* a part of a floor raised to form a seat]
sofa bed A sofa which may be opened up to form a large bed.
so·far (sō′fär) *n.* A system for locating stranded ships or aircraft by means of underwater sound waves, set up by depth charges released by the survivors and detected by hydrophones operated from ground stations. [<SO(UND) F(IXING) A(ND) R(ANGING)]
sof·fit (sof′it) *n. Archit.* The under side of a staircase, entablature, lintel, archway, or cornice. [<F *soffite* <Ital. *soffita* <L *suffixus.* Doublet of SUFFIX.]
So·fi·a (sō·fē′ä) German, Italian, Russian, Spanish, and Swedish form of SOPHIA.
So·fi·a (sō′fē·ə, sō·fē′ə) The capital of Bulgaria, in the western part. *Bulgarian* **So·fi·ya** (sô′fē·yä).
soft (sôft, soft) *adj.* 1 Being or composed of a substance whose shape is changed easily by pressure, without fracture; impressible; pliable, ductile, or malleable; easily worked: *soft* wood: opposed to *hard.* 2 Smooth and delicate to the touch: *soft* skin. 3 Gentle in its effect upon the ear; not loud or harsh. 4 Mild in any mode of physical action; gentle; bland: a *soft* breeze; a *soft* ripple. 5 Of subdued coloring or delicate shading; not glaring or abrupt: *soft* tints; *soft* outline. 6 Gentle; conciliatory; expressing mildness or sympathy; courteous: *soft* words. 7 Giving or enjoying rest; placid: *soft* sleep. 8 Easily or too easily touched in feeling; tender; sympathetic: a *soft* heart. 9 Incapable of bearing hardship; susceptible; tender; delicate: *soft* muscles. 10 Of yielding character; weak; effeminate. 11 *Colloq.* Of weak intellect; also, yielding to emotion; maudlin. 12 Free from mineral salts which prevent the detergent action of water and soap: said of water. 13 Bituminous, as opposed to anthracite: said of coal. 14 Describing *c* and *g* when articulated fricatively as in *cent* and *gibe*: opposed to *hard*; also, voiced and weakly articulated; also, palatalized, as certain consonants in the Slavic languages. 15 *Colloq.* Easy: a *soft* job. 16 *Scot.* & *Brit. Dial.* Characterized by moisture or thawing: said of the weather. See synonyms under BLAND, SUPPLE. —*n.* 1 That which is soft; softness; a soft part or material. 2 *Colloq.* One who is soft or foolish; a softy. —*adv.* 1 Softly. 2 Quietly; gently. —*interj. Archaic* Proceed softly; be quiet or slow. [OE *sōfte*] —**soft′ly** *adv.* & *interj.* —**soft′ness** *n.*
sof·ta (sof′tə) *n. Turkish* A student at a Moslem mosque; especially, a student of the Moslem theology.
soft·ball (sôft′bôl′, soft′-) *n.* A variation of baseball, requiring a smaller diamond, a larger, softer ball, ten players on a team, and seven innings for play.
soft-boiled (sôft′boild′) *adj.* 1 Boiled, as an egg, to an extent of incomplete coagulation of the albumen. 2 *Colloq.* Mild in disposition; lenient.
soft clam The common long clam (*Mya arenaria*) of the north Atlantic coast.
soft coal Bituminous coal.
soft drink A nonalcoholic beverage, as sweetened soda water, ginger ale, etc.
sof·ten (sôf′ən, sof′-) *v.t.* & *v.i.* To make or become soft or softer. See synonyms under ALLAY, ALLEVIATE, CHASTEN, TEMPER. —**sof′ten·er** *n.*
softening of the brain 1 *Pathol.* Degeneration of the brain tissue, especially as resulting from paresis; encephalomalacia. 2 *Colloq.* Dementia.
soft-finned (sôft′find′, soft′-) *adj. Zool.* Having fins whose membrane is supported on flexible or jointed rays: opposed to *spiny-finned.*
soft focus *Phot.* A slightly blurred effect obtained by an imperfect focusing of the lens upon a scene or object.
soft·head (sôft′hed′, soft′-) *n.* A foolish or simple person. —**soft′-head′ed** *adj.*
soft·heart·ed (sôft′här′tid, soft′-) *adj.* Tenderhearted; merciful. —**soft′heart′ed·ly** *adv.* —**soft′heart′ed·ness** *n.*
soft-ped·al (sôft′ped′l, soft′-) *v.t.* **·aled** or **·alled, ·al·ing** or **·al·ling** 1 To mute the tone of by depressing the soft pedal. 2 *Colloq.* To render less emphatic; moderate; tone down.
soft pedal A pedal which mutes the tone, as in a piano.
soft sell *U.S. Colloq.* The use of subtle, unobtrusive, non-insistent methods of salesmanship.
soft-shell (sôft′shel′, soft′-) *adj.* 1 Having a soft shell, as certain clams, or a crab or lobster after shedding its shell: also **soft′-shelled′.** 2 *U.S. Colloq.* Somewhat moderate in opinion or doctrine; somewhat liberal; not hidebound: a *soft-shell* Baptist. —*n.* A crab which has lately shed its shell: also **soft-shelled crab.**
soft-shelled turtle Any member of a family (*Trionychidae*) of turtles having a long snout and a soft, leathery shell, especially *Trionyx* (or *Amyda*) *spinifera,* common from the Gulf States to the St. Lawrence River.
soft-soap (sôft′sōp′, soft′-) *v.t.* To flatter. —**soft′-soap′er** *n.*
soft soap 1 Fluid or semifluid soap. 2 *Colloq.* Flattery; blarney.
soft spot 1 A tendency toward impracticality. 2 Sympathy; tenderness; also, sentimentality.
soft·wood (sôft′wŏŏd′, soft′-) *n.* 1 A coniferous tree or its wood. 2 Any soft wood, or any tree with soft wood.
soft·y (sôf′tē, sof′-) *n. pl.* **soft·ies** *Colloq.* 1 An extremely sentimental person. 2 A weak or effeminate man or boy; one not inured to hardship; a sissy.
Sog·di·an (sog′dē·ən) *n.* 1 One of an ancient Iranian people inhabiting Sogdiana. 2 Their extinct Iranian language, known from fragmentary manuscripts of the eighth or ninth century A.D. discovered in Chinese Turkestan.
Sog·di·a·na (sog′dē·ā′nə) An ancient region of central Asia comprising part of the Persian Empire; capital, Samarkand: also *Transoxiana.*
sog·gy (sog′ē) *adj.* **·gi·er, ·gi·est** 1 Saturated with water or moisture; wet and heavy; soaked. 2 Heavy: said of pastry. 3 Soft; boggy: said of land. 4 Dull; logy: said of a person or an animal. Also **sog·ged** (sog′id). [<dial. E *sog* a swamp, bog <Scand. Cf. dial. Norw. *soggjast* get wet.] —**sog′gi·ly** *adv.* —**sog′gi·ness** *n.*
Sog·ne Fjord (sông′nə) The longest and deepest fjord in Norway, in the western part; 112 miles long; an inlet of the North Sea.
So·ho (sō·hō′, sō′hō) The foreign quarter of London, noted for its restaurants.
soi-di·sant (swà·dē·zäṅ′) *adj. French* Self-styled; pretended: usually implying false pretense.
soi·gné (swà·nyā′) *adj. French* Cared for; well-groomed.
soil[1] (soil) *n.* 1 Finely divided rock mixed with decayed vegetable or animal matter, constituting that portion of the surface of the earth in which plants grow. 2 The ground in general; native land; country. 3 A mixture of lampblack, glue, and water used in plumbing. See synonyms under LAND. [<OF *soile, sueil* <L *solium* a seat, mistaken for *solum* the ground]
soil[2] (soil) *v.t.* 1 To make dirty; smudge. 2 To disgrace; defile. 3 *Obs.* To manure. —*v.i.* 4 To become dirty. See synonyms under BLEMISH, DEFILE[1], POLLUTE, STAIN. —*n.* 1 That which soils; foul matter; a foul spot; hence, a taint. 2 Manure: confused in use with *soil*[1]. 3 A slough or marshy place in which a hunted boar takes refuge; hence, water or a wet place resorted to by other game. [<OF *soillier,* ult. <L *suculus,* dim. of *sus* a pig]
soil[3] (soil) *v.t.* 1 To feed and fatten, as stalled cattle, with freshly cut, green food. 2 To purge with green food. [? <OF *saoler, saouler* fill <L *satullare* <*satullus,* dim. of *satur* sated]
soil·age (soi′lij) *n.* Green crops for feeding animals.
soil·ure (soil′yər) *n.* Soiling, or the condition of being soiled.
soi·rée (swä·rā′, *Fr.* swà·rā′) *n.* A party or reception given in the evening. Also **soi·ree′.** [<F <*soir* evening]
Sois·sons (swà·sôṅ′) A city in NE France on the Aisne.
so·ja (sō′jə, sō′yə) *n.* The soybean. [<NL <Du. *soya* the soybean]
so·journ (sō′jûrn, sō·jûrn′) *v.i.* To stay or dwell temporarily; abide for a time. See synonyms under ABIDE. —*n.* (sō′jûrn) A temporary residence or stay, as of one in a foreign land. [<OF *sojorner, sojourner,* ult. <L *sub-* under + *diurnus* daily] —**so′journ·er** *n.*
soke (sōk) *n.* 1 In feudal law, a franchise, privilege, or liberty; jurisdiction; a privilege to administer justice within a certain territory, as a manor. 2 The district within which such privilege was exercised. ◆ Homophone: *soak.* [<Med. L *soca* <OE *sōcn* jurisdiction]
soke·man (sōk′mən) See SOCMAN.
Soke of Peterborough An administrative county in NE Northamptonshire, England; 83 square miles.
So·kol (sō′kôl) *n.* 1 A Czech patriotic organization of gymnasts started (1862) as a democratic fraternal body to develop strength, litheness, alertness, and fearlessness. 2 A member of this organization. [<Czechoslovakian *sokol* falcon]
So·ko·to (sō′kō·tō) A province of northern

add,āce,câre,pälm; end,ēven; it,īce; odd,ōpen,ôrder; tŏŏk,pōōl; up,bûrn; ə = a in *above,* e in *sicken,* i in *clarity,* o in *melon,* u in *focus*; yōō = u in *fuse*; oi,oil; ou,pout; ch,check; g,go; ng,ring; th,thin; th,this; zh,vision. Foreign sounds à,œ,ü,kh,ṅ; and ◆: see page xx. < from; + plus; ? possibly.

Nigeria; 39,965 square miles; capital, Sokoto.

So·ko·tra (sō·kō′trə) See SOCOTRA.

sol[1] (sōl) *n. Music* The fifth note of the diatonic scale. [See GAMUT]

sol[2] (sol) *n.* A former French silver or copper coin, equivalent to 12 deniers. [<OF <LL *solidus,* a gold coin <L, solid]

sol[3] (sōl) *n. pl.* **so·les** (sō′lās) A Peruvian monetary unit, equivalent to 1/10 libra. [<Sp., sun]

sol[4] (sol, sōl) *n.* A colloidal suspension in a liquid. [<(HYDRO)SOL]

sol[5] (sol) *n.* In alchemy, gold. [<L, sun]

Sol (sol) **1** The sun. **2** In Roman mythology, the god of the sun. [<L]

so·la (sō′lə) See SOLUS.

sol·ace (sol′is) *v.t.* **·aced, ·ac·ing 1** To comfort or cheer in trouble, grief, or calamity; console. **2** To alleviate, as grief; soothe; assuage; mitigate. —*n.* Comfort in grief, trouble, or calamity; also, that which supplies such comfort or alleviation: also **sol′ace·ment.** [<OF *solacier, solasier* < *solas* comfort <L *solacium*] —**sol′ac·er** *n.*

so·lan (sō′lən) *n.* The gannet, a bird related to the pelicans. Also **so·land** (sō′lənd, -lən), **solan goose.** [<ON *súla* the gannet]

sol·a·na·ceous (sol′ə·nā′shəs) *adj. Bot.* Pertaining or belonging to a widely distributed family *(Solanaceae)* of frequently narcotic poisonous plants, the nightshade family, having colorless juice and alternate simple leaves. The family includes belladonna, tobacco, eggplant, and potato. [<NL <L *solanum* nightshade]

so·lan·der (sə·lan′dər) *n.* A hinged case or box, usually in the form of a book, adapted to hold a variety of objects, as jewelry, cigarettes, writing materials, pamphlets, maps, rare books and the like. [after Daniel C. *Solander,* 1736–82, English inventor born in Sweden]

so·la·no (sō·lä′nō) *n.* A hot, violent, southeasterly wind of the Mediterranean. [<Sp. <L *sol* sun]

so·la·num (sə·lā′nəm) *n.* Any of a genus *(Solanum)* of herbs and shrubs, the nightshades, typifying the family *Solanaceae,* especially *S. tuberosum,* the common potato. [<NL <L, nightshade]

so·lar (sō′lər) *adj.* **1** Pertaining to, proceeding from, or connected with the sun. **2** Affected, determined, or measured by the sun. **3** Operated by the action of the sun's rays: a *solar* engine. [<L *solaris* < *sol* sun]

solar constant The amount of solar energy falling on one square centimeter of the earth's surface at normal incidence, having a mean value of 1.92 small calories per minute.

so·lar·im·e·ter (sō′lə·rim′ə·tər) *n.* An instrument for determining the intensity of solar radiation.

so·lar·ism (sō′lə·riz′əm) *n.* The belief that folk tales symbolizing solar phenomena presuppose a sun-cult or sun-worship: now considered unsubstantiated.

so·lar·i·um (sō·lâr′ē·əm) *n. pl.* **·i·a** (-ē·ə) A room or enclosed porch exposed to the sun's rays, as in a sanatorium. [<L]

so·lar·i·za·tion (sō′lər·ə·zā′shən, -ī·zā′-) *n.* **1** Exposure to the sun's rays. **2** *Phot.* Injury to a sensitized film resulting from overexposure to strong light, or from overprinting.

so·lar·ize (sō′lə·rīz) *v.* **·ized, ·iz·ing** *v.t.* **1** To affect or injure by the action of the sun's rays. **2** *Phot.* To overexpose. —*v.i.* **3** *Phot.* To be overexposed.

solar month A twelfth of a solar year; the time during which the sun is passing through one of the signs of the zodiac.

solar myth A primitive etiological story explaining symbolically some natural phenomenon of the sun, such as its splendor contrasted with the pallor of the moon, its course, etc.; also, a folk tale arising among an agricultural people explaining or symbolizing the power or influence of the sun.

solar plexus 1 *Anat.* The large network of the sympathetic nervous system, found behind the stomach, and containing important ganglia serving the abdominal viscera. **2** *Colloq.* The pit of the stomach.

solar system The sun and the heavenly bodies that revolve about it.

solar time See under TIME.

solar year See under YEAR.

sol·ate (sol′āt) *v.i.* **·at·ed, ·at·ing** *Chem.* To change from a gel to a sol. [<SOL[4] + -ATE[1]] —**so·la·tion** (sə·lā′shən) *n.*

so·la·ti·um (sə·lā′shē·əm) *n. pl.* **·ti·a** (-shē·ə) **1** Compensation; solace. **2** *Law* Compensation for injury to the feelings as distinguished from pecuniary loss or physical suffering. [<L, var. of *solacium* solace]

sold (sōld) Past tense and past participle of SELL.

sol·dan (sol′dən) *n. Archaic* A ruler or sovereign of a Moslem country, especially Egypt: also spelled *suldan, soudan.* [<OF *soudan* <Arabic *sulṭān* king, sovereign]

sol·der (sod′ər) *n.* **1** A fusible metal or alloy used for joining metallic surfaces or margins: applied in a melted state, either as a **hard solder,** melting only at a red heat, or as a **soft solder,** melting below a red heat. **2** Anything that unites or cements. —*v.t.* **1** To unite or repair with solder. **2** To join together; bind. —*v.i.* **3** To work with solder. **4** To be united by or as by solder. [<OF *soldure* < *souder* make hard <L *solidare* < *solidus* firm, hard] —**sol′der·er** *n.*

sol·dier (sōl′jər) *n.* **1** A person serving in an army. **2** A private in an army, as distinguished from a commissioned officer. **3** A brave, skilful, or experienced warrior. **4** One who serves loyally in any cause. **5** *Colloq.* One who makes a show of working but does little; a shirker; malingerer. **6** *Entomol.* **a** An asexual form (neuter or worker) of a termite or white ant, in which the head and jaws are largely developed, and whose office is to defend the community. **b** A similar neuter of certain true ants. See synonyms under ARMY. —*v.i.* **1** To be a soldier; perform military service. **2** To make a show of working; shirk; malinger. [<OF < *soude* pay, wages <LL *solidus.* See SOL[2].]

sol·dier·ly (sōl′jər·lē) *adj.* Like a true soldier; brave; martial. See synonyms under WARLIKE.

soldier of fortune A military adventurer; a soldier who serves where fortune summons him.

Soldier's Medal A decoration in the form of a bronze octagon on which is displayed an eagle standing on fasces between two groups of stars: awarded to any member of the U.S. Army, or of a military organization connected with it, for heroism not involving actual conflict with the enemy.

sol·dier·y (sōl′jər·ē) *n. pl.* **·dier·ies 1** Soldiers collectively. **2** Military service.

sol·do (sol′dō, *Ital.* sôl′dō) *n. pl.* **·di** (-dē) A small Italian copper coin worth, generally, one twentieth of a lira. [<Ital. <LL *solidus,* a gold coin <L, solid]

sole[1] (sōl) *n.* **1** The bottom surface of the foot. ◆ Collateral adjectives: *plantar, volar*[2]. **2** The bottom surface of a shoe, boot, etc. **3** The lower part of a thing, or the part on which it rests when standing; especially, the bottom part of a plowshare. **4** The bottom part of the head of a golf club. —*v.t.* **soled, sol·ing 1** To furnish with a sole; resole, as a shoe. **2** In golf, to allow (the clubhead) to rest flat on the ground, just behind the ball. ◆ Homophone: *soul.* [<OF <Med. L *sola,* var. of L *solea* a sandal]

sole[2] (sōl) *n.* **1** Any of several flatfishes allied to the flounders, having a small mouth and small eyes set close together on one side of the head; especially, the common **European sole** *(Solea solea),* highly esteemed as food, and the **American sole** (genus *Achirus*), common on the Atlantic coast of the United States. **2** One of various flounders, as *Psettichthys melanostictus,* a food fish of the Pacific coast of the United States. ◆ Homophone: *soul.* [<OF <L *solea*]

sole[3] (sōl) *adj.* **1** Being alone or the only one; existing or acting without another; only; individual. **2** *Law* **a** Unmarried; single: feme *sole* (an unmarried woman). **b** Having exclusive rights; absolute: opposed to *joint:* a *sole* tenant. **3** *Archaic* Solitary. See synonyms under SOLITARY. ◆ Homophone: *soul.* [<OF *sol* <L *solus* alone]

sol·e·cism (sol′ə·siz′əm) *n.* **1** A violation of grammatical rules or of the approved idiomatic usage of language. **2** Any impropriety or incongruity. [<L *soloecismus* <Gk. *soloikismos* < *soloikos* speaking incorrectly < *Soloi,* a Cilician town whose people spoke a substandard Attic dialect] —**sol′e·cist** *n.* —**sol′e·cis′tic** or **·ti·cal** *adj.*

sol·e·cize (sol′ə·sīz) *v.i.* **·cized, ·ciz·ing** *Rare* To use solecisms. Also *Brit.* **sol′e·cise.**

sole·ly (sōl′lē) *adv.* **1** By oneself or itself alone; singly. **2** Completely; entirely. **3** Without exception; exclusively.

sol·emn (sol′əm) *adj.* **1** Characterized by majesty, mystery, or power; exciting grave or serious thought; impressive; awe-inspiring. **2** Characterized by ceremonial observances; religious; sacred. **3** Marked by gravity; serious; earnest; also, affectedly serious. **4** *Law* Done in due form of law; executed formally: a *solemn* protest. **5** *Obs.* Of great reputation, dignity, or importance. **6** *Obs.* Somber; sober: said of color. See synonyms under AWFUL, SEDATE, SERIOUS. [<OF *solemne* <L *solemnis*] —**sol′em·ness, sol′emn·ness** *n.* —**sol′emn·ly** *adv.*

so·lem·ni·ty (sə·lem′nə·tē) *n. pl.* **·ties 1** The state or quality of being solemn; solemn feeling; gravity; reverence. **2** A rite expressive of religious reverence; also, any ceremonious observance. **3** A thing of a solemn or serious nature. **4** Mock seriousness; affected gravity. **5** *Law* A formality to be seriously observed and requisite to the validity or legality of an act. See synonyms under SACRAMENT.

sol·em·nize (sol′əm·nīz) *v.t.* **·nized, ·niz·ing 1** To perform as a ceremony or solemn rite, or according to legal or ritual forms: to *solemnize* a marriage. **2** To dignify as with a ceremony; celebrate. **3** To make solemn, grave, or serious. Also *Brit.* **sol′em·nise.** See synonyms under CELEBRATE. —**sol′em·ni·za′tion** *n.* —**sol′em·niz′er** *n.*

Solemn League and Covenant See under COVENANT.

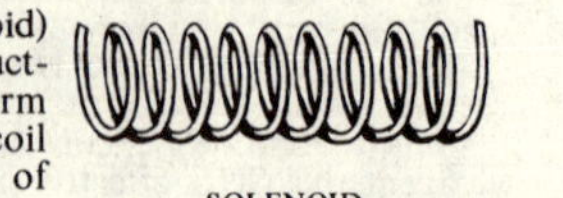

SOLENOID

so·le·noid (sō′lə·noid) *n. Electr.* A conducting wire in the form of a cylindrical coil or helix, capable of setting up a magnetic field by the passage through it of an electric current. [<Gk. *sōlēn* a channel + -OID] —**so′le·noi′dal** *adj.* —**so′le·noi′dal·ly** *adv.*

So·lent (sō′lənt), **The** A strait between the Isle of Wight and Southampton, England; 3/4 to 5 miles wide, 15 miles long.

sol·er·et (sol′ə·ret) See SOLLERET.

sole trader See FEME-SOLE TRADER.

So·leure (sô·lœr′) The French name for SOLOTHURN.

sol–fa (sōl′fä′) *Music v.t.* & *v.i.* **–faed, –fa·ing** To sing syllables instead of words to (notes); sing solfeggi. —*n.* **1** Syllables collectively used in solmization; the act of singing them. **2** Rarely, a scale. —**tonic sol–fa** See TONIC. [<Ital. *solfa* the gamut. See GAMUT.] —**sol′–fa′ist** *n.*

sol·fa·ta·ra (sōl′fä·tä′rä) *n. Geol.* An area or phase of volcanic action characterized by the escape of steam, various gases, and sublimates. [<Ital., a dormant crater near Naples, Italy < *solfo* sulfur] —**sol′fa·ta′ric** *adj.*

sol·feg·gio (sōl·fej′ō) *n. pl.* **·feg·gi** (-fej′ē) or **·feg·gios** *Music* **1** A singing exercise of runs, broken chords, etc., sung either to different syllables or all to the same syllable or vowel. **2** Solmization. [<Ital. < *solfa.* See SOL–FA.]

sol·fe·ri·no (sol′fe·rē′nō) *n.* **1** A bright purplish red. **2** Fuchsin. [from *Solferino;* named in honor of a battle fought there in 1859]

Sol·fe·ri·no (sōl′fā·rē′nō) A village in northern Italy; scene of a French and Sardinian victory over Austria, 1859.

so·lic·it (sə·lis′it) *v.t.* **1** To ask for earnestly; seek to obtain by persuasion or entreaty. **2** To beg or entreat (a person) persistently. **3** To influence to action; tempt; especially, to entice (one) to an unlawful or immoral act. —*v.i.* **4** To make petition or solicitation. See synonyms under ASK, PLEAD. [<OF *solliciter* <L *sollicitare* agitate]

so·lic·i·ta·tion (sə·lis′ə·tā′shən) *n.* **1** Importunity; the act of soliciting. **2** An attempt to entice.

so·lic·i·tor (sə·lis′ə·tər) *n.* **1** A person who does any kind of soliciting; especially, one who solicits gifts of money or subscriptions to magazines. **2** The legal advisor to certain branches of the public service. **3** In England, a lawyer who may advise clients or who prepares cases for presentation in court, but who may appear as an advocate in the lower courts only. See BARRISTER. Also **so·lic′i·ter.** —**so·lic′i·tor·ship′** *n.*

Solicitor General *pl.* **Solicitors General** 1 In the United States, an officer who ranks after the Attorney General, and, in the absence of the latter, acts in his place. 2 The principal law officer in some of the States, corresponding to the Attorney General in others. 3 In England, a law officer of the Crown, ranking next after the Attorney General.

so·lic·i·tous (sə·lis′ə·təs) *adj.* 1 Full of anxiety or concern, as for the attainment of something. 2 Full of eager desire; willing. See synonyms under URGENT. —**so·lic′i·tous·ly** *adv.* —**so·lic′i·tous·ness** *n.*

so·lic·i·tude (sə·lis′ə·to͞od, -tyo͞od) *n.* 1 The state of being solicitous; uneasiness of mind. 2 That which makes one solicitous. See synonyms under ANXIETY, CARE.

sol·id (sol′id) *adj.* 1 Having its constituent particles so firmly coherent as to resist stress; compact, firm, and unyielding: opposed to *fluid.* 2 Substantial; firm and stable. 3 Filling the whole of the space occupied by its apparent form; completely filled; not hollow. 4 Having no aperture or crevice; compact. 5 Manifesting strength and firmness; not weak or sickly; sound. 6 Characterized by reality; substantial or satisfactory. 7 Exhibiting united and unbroken characteristics, opinions, etc.; being or acting in unison; unanimous: the *solid* vote; This county is *solid* for the Democratic party; also, blindly or unreasonably partisan. 8 Financially sound or safe. 9 *U.S. Colloq.* Certain and safe in approval and support: They were *solid* with the boss. 10 Having or relating to the three dimensions of length, breadth, and thickness. 11 Written without a hyphen: said of a compound word. See SOLIDEME. 12 Cubic in shape: a *solid* yard. 13 Unadulterated; unalloyed: *solid* gold. 14 Carrying weight or conviction: a *solid* argument. 15 Serious; reliable; exhibiting sound judgment: a *solid* citizen. 16 Continuous; unbroken: a *solid* hour. 17 *Printing* Having no leads or slugs between the lines; not open. See synonyms under FIRM[1], HARD, IMPENETRABLE. —*n.* 1 A mass of matter of which the shape cannot be changed permanently and greatly without fracture. 2 A magnitude that has length, breadth, and thickness, as a cone, cube, pyramid, prism, or sphere. [<F *solide* <L *solidus*] —**sol′id·ly** *adv.* —**sol′id·ness** *n.*

sol·i·da·go (sol′ə·dā′gō) *n. pl.* **·gos** Any of a large North American genus *(Solidago)* of perennial plants of the composite family; a goldenrod: the State flower of Alabama, Kentucky, and Nebraska. [<NL <L *solidare* strengthen; with ref. to its alleged curative powers]

solid angle See under ANGLE.

sol·i·dar·i·ty (sol′ə·dar′ə·tē) *n. pl.* **·ties** Coherence and oneness in nature, relations, or interests, as of a race, class, etc.

sol·i·dar·y (sol′ə·der′ē) *adj.* United in nature or interests.

sol·i·deme (sol′ə·dēm) *n.* A solid compound word. Compare HYPHEME. [<SOLID + *-eme,* as in *phoneme*]

solid geometry That part of geometry which includes all three dimensions of space in its reasoning.

so·lid·i·fy (sə·lid′ə·fī) *v.t.* & *v.i.* **·fied, ·fy·ing** 1 To make or become solid, hard, firm, or compact, as water crystallizing into ice. 2 To bring or come together in unity. —**so·lid′i·fi·ca′tion** *n.*

so·lid·i·ty (sə·lid′ə·tē) *n. pl.* **·ties** 1 The quality or state of being solid; the property of occupying space; extension in the three dimensions of space; incompressibility. 2 Mental, moral, or financial soundness; substantial or reliable character or quality; firm standing; stability. 3 *Aeron.* The ratio of the total blade area of a rotor or propeller to the area of the disk swept by the blades. 4 *Geom.* Cubic contents; volume.

Solid South The Southern States of the United States, regarded as a political unit because of their support of the Democratic party.

solid state physics That branch of physics which deals with the physical properties of solids, especially as exhibited by atoms and molecules when in the solid state. It includes the study of crystal structure, elasticity, and friction, semiconductors and plastics, defects in materials, thermal properties, and a wide range of electrical and magnetic phenomena.

sol·i·dus (sol′ə·dəs) *n. pl.* **·di** (-dī) 1 A gold coin of the Byzantine Empire: first issued under Constantine, it remained the standard unit of currency during the Middle Ages, when it was called a *bezant.* 2 A medieval coin, equal to 12 denarii: often called *shilling.* 3 The sign (/) used to divide shillings from pence: 10/6 (10*s.* 6*d.*), being originally the long *ſ* written for shilling: sometimes also used instead of a horizontal line to express fractions: 3/4. See VIRGULE. [<LL]

sol·i·fid·i·an (sol′ə·fid′ē·ən) *n.* One who maintains that faith alone, without works, is the one requisite to salvation. —*adj.* Maintaining that faith alone is necessary to insure salvation; also, pertaining to such belief. [<L *solus* alone + *fides* faith]

so·lil·o·quize (sə·lil′ə·kwīz) *v.i.* **·quized, ·quiz·ing** To discourse to oneself; utter a soliloquy. Also *Brit.* **so·lil′o·quise.**

so·lil·o·quy (sə·lil′ə·kwē) *n. pl.* **·quies** A talking to oneself, regardless of the presence or absence of others; a monolog. [<LL *soliloquium* <L *solus* alone + *loqui* talk]

So·li·mões (sô′lē·moiṅzh′) The upper reaches of the Amazon river, extending from the Peruvian border to the Río Negro.

So·ling·en (zō′ling·ən) A city in North Rhine–Westphalia, West Germany.

sol·i·on (sol′ī′ən) *n. Physics* A small electrochemical cell so constructed that the movement of ions in solution serves to indicate minute changes in temperature, pressure, sound or light waves, acceleration, and other external conditions: used as an electronic control device. [< *ion(s in) sol(ution)*]

sol·ip·sism (sol′ip·siz′əm) *n.* The theory or belief that only knowledge of the self is possible, and that, for each individual, the self itself is the only thing really existent, and therefore that reality is subjective. [<L *solus* alone + *ipse* self] —**sol′ip·sist** *n.*

sol·i·taire (sol′ə·târ′) *n.* 1 A diamond or other gem set alone. 2 One of many games, especially of cards, played by one person. 3 A bird *(Pezophaps solitarius)* somewhat resembling the dodo but more slender and graceful: formerly a native of Réunion but now extinct. [<F <L *solitarius* solitary]

sol·i·tar·y (sol′ə·ter′ē) *adj.* 1 Living, being, or going alone. 2 Made, done, or passed alone: a *solitary* life. 3 Unfrequented by human beings; secluded; lonely; desolate. 4 Lonesome; lonely. 5 Single; one; sole: Not a *solitary* soul was there. —*n. pl.* **·tar·ies** A hermit; recluse; one who lives alone. [<L *solitarius* < *solus* alone] —**sol′i·tar′i·ly** *adv.* —**sol′i·tar′i·ness** *n.*

Synonyms (adj.): alone, companionless, deserted, lone, lonely, lonesome, only, single, sole, unaccompanied, unattended. *Antonyms*: manifold, many, multiplied, multitudinous, myriad, numerous.

sol·i·ter·ra·ne·ous (sol′ə·tə·rā′nē·əs) *adj.* Pertaining to the joint influence of solar and terrestrial forces, especially in relation to meteorological phenomena. [<L *sol, solis* the sun + *terra* the earth]

sol·i·tude (sol′ə·to͞od, -tyo͞od) *n.* 1 Loneliness; seclusion. 2 A deserted or lonely place; hence, a desert. [<OF <L *solitudo* < *solus* alone]

Synonyms: isolation, loneliness, privacy, retirement. See RETIREMENT, SECLUSION.

sol·ler·et (sol′ə·ret′) *n.* In medieval armor, a mounted warrior's steel shoe or one of its overlapping splints: also spelled *soleret.* [<OF, dim. of *soller, soler* a shoe <L *solea* sole of the foot]

sol·mi·zate (sol′mə·zāt) *v.i.* **·zat·ed, ·zat·ing** To sing by syllables; sol-fa.

sol·mi·za·tion (sol′mə·zā′shən) *n. Music* The use of syllables as names for the notes or tones of the scale. The syllables now commonly used are *do, re, mi, fa, sol, la, ti.* [<SOL + MI]

so·lo (sō′lō) *n. pl.* **·los** or **·li** (-lē) 1 A musical composition or passage for a single voice or instrument, with or without accompaniment. 2 Any of several card games, especially one in which the player who bids to take the highest number of tricks plays alone against the others. 3 Any performance accomplished alone or without assistance. —*adj.* 1 Composed or written for, or executed by, a single voice or instrument; performed as a solo. 2 Done by a single person alone: a *solo* flight. —*v.i.* **·loed, ·lo·ing** To fly an airplane alone, especially for the first time. [<Ital. <L *solus* alone]

So·lo (sō′lō) 1 A city in central Java: formerly *Surakarta.* 2 The longest river of Java, flowing 335 miles north, east, and NE from south central Java to the Java Sea, opposite Madura.

so·lo·ist (sō′lō·ist) *n.* One who performs a solo.

So·lo man (sō′lō) *Paleontol.* A species of early man *(Homo soloensis)* identified from a group of skulls found near the Solo river at Ngandong, Java, in 1931: it is thought to be an evolutionary advance over Pithecanthropus. Also called *Ngandong man.*

Sol·o·mon (sol′ə·mən) A masculine personal name. Also (diminutive) **Sol.** See also SALOMON. [<Hebrew, peaceful]

—**Solomon** King of Israel during the tenth century B.C.; noted for his wisdom and magnificence; a son of David.

Solomon Islands An archipelago in the SW Pacific east of New Guinea; about 16,500 square miles; the SE islands, including Guadalcanal, Santa Isabel, San Cristobal, Choiseul, New Georgia, and Malaita, comprise a British protectorate; total, 11,500 square miles; capital, Honiara, on Guadalcanal; the NW islands, including Bougainville, Buka and adjacent islands (total 4,320 square miles) are part of the Territory of Papua and New Guinea, administered by Australia.

Sol·o·mon's-seal (sol′ə·mənz·sēl′) *n.* Any one of several rather large perennial herbs of the lily family (genus *Polygonatum*), having tubular, six-toothed flowers and rootstocks marked at intervals by circular scars.

Solomon's seal A six-pointed star. See MOGEN DAVID.

So·lon (sō′lən), 638?–558? B.C., Athenian lawgiver; hence, **solon,** any wise lawmaker. —**So·lo·ni·an** (sə·lō′nē·ən) *adj.*

so long *Colloq.* Good-by.

So·lor Islands (sô·lôr′) An island group east of Flores, part of the Lesser Sunda Islands, Nusa Tenggara, Indonesia; total, 785 square miles.

So·lo·thurn (zō′lō·to͞orn) 1 A canton of NW Switzerland; 305 square miles. 2 The capital of Solothurn canton, NW Switzerland, on the Aar river: French *Soleure.*

sol·pu·gid (sol·pyo͞o′jid) *n.* A predatory, spider-like arachnid (order *Solpugida*) of warm climates that hides by day. [<L *solpuga, solipuga,* a kind of venomous ant or spider]

sol·stice (sol′stis) *n.* 1 *Astron.* The time of year when the sun is at its greatest distance from the celestial equator, either north or south, and seems to pause before returning on its course; either the **summer solstice,** about June 22 in the northern hemisphere, or the **winter solstice,** about December 22. 2 A culminating or high point; epoch; limit. [<F <L *solstitium* < *sol* sun + *sistere* cause to stand] —**sol·sti·tial** (sol·stish′əl) *adj.*

sol·u·bil·i·ty (sol′yə·bil′ə·tē) *n. pl.* **·ties** The state of being soluble, or the capability of being dissolved. Also **sol′u·ble·ness.**

sol·u·bil·ize (sol′yə·bəl·īz′) *v.t.* **·ized, ·iz·ing** *Chem.* To make soluble; specifically, to disperse (normally insoluble oils and fats) by the action of detergents and certain protein molecules. —**sol′u·bil·i·za′tion** *n.*

sol·u·ble (sol′yə·bəl) *adj.* 1 Capable of being uniformly dissolved in a liquid: Sugar is *soluble* in water. 2 Susceptible of being solved or explained. [<OF <L *solubilis* < *solvere* solve, dissolve] —**sol′u·bly** *adv.*

soluble cotton Nitrocellulose which is soluble in acetone, amyl acetate, ethanol, and certain other solvents: used in making nail polish and similar lacquers.

soluble glass Sodium silicate.

so·lum (sō′ləm) *n.* That part of a soil profile above the parent material in which the processes of soil formation take place; the soil proper. [<L *solum* ground]

so·lus (sō′ləs) *adj. Latin* Alone: used in stage directions. — **so·la** (sō′lə) *adj. fem.*

sol·ute (sol′yōōt, sō′lōōt) *n.* The substance dissolved in a solution as distinguished from the solvent.

so·lu·tion (sə·lōō′shən) *n.* **1** A homogeneous mixture formed by dissolving one or more substances, whether solid, liquid, or gaseous, in another substance, usually a liquid but sometimes a solid or a gas. **2** Any homogeneous mixture of which the solute is uniformly dispersed through the solvent, and whose composition may undergo continuous variation within certain limits. **3** The act or process by which such a mixture is made. **4** The act or process of explaining, settling, or disposing, as of a difficulty, problem, or doubt. **5** *Law* Payment or satisfaction of a claim or debt. **6** *Med.* The crisis of a disease; termination of a disease with critical signs. **7** *Math.* The answer to a problem; also, the method of finding the answer. **8** Separation; disruption: the *solution* of continuity. [<OF <L *solutio, -onis* < *solutus,* pp. of *solvere* dissolve]

solution pressure *Chem.* The pressure caused by the tendency of atoms or molecules to dissolve. In the case of metals it produces the current in a primary battery.

sol·u·tive (sol′yə·tiv) *adj.* **1** Loosening; laxative. **2** Soluble.

So·lu·tre·an (sə·lōō′trē·ən) *adj. Anthropol.* Pertaining to or characteristic of an Upper Paleolithic culture preceding the Magdalenian in western Europe: it is typified by a skilled technique in the making of bladed flint implements and by marked improvements in polychrome cave painting. Also **So·lu′tri·an.** [after *Solutré,* a village in central France, where remains were discovered]

solv·a·ble (sol′və·bəl) *adj.* **1** That may be solved. **2** That may be dissolved. — **solv′a·bil′i·ty, solv′a·ble·ness** *n.*

sol·va·tion (sol·vā′shən) *n. Chem.* A loose combination sometimes formed by the solute and solvent of a solution, as copper sulfate crystallizing from water.

Sol·vay process (sol′vā) A process of making soda by treating a concentrated solution of common salt with ammonia and carbon dioxide, yielding sodium bicarbonate, which is converted into soda by heat, carbon dioxide and water being expelled. [after Ernst *Solvay,* 1838–1922, Belgian chemist]

solve (solv) *v.t.* **solved, solv·ing** To arrive at or work out the correct explanation or solution of; find the answer to; resolve. [<L *solvere* solve, loosen] — **solv′er** *n.*

Synonyms: clear, decipher, do, elucidate, explain, guess, interpret, resolve, understand, unfold. *Antonyms*: confound, confuse, perplex.

sol·ven·cy (sol′vən·sē) *n.* The condition of being solvent.

sol·vent (sol′vənt) *adj.* **1** Having means sufficient to pay all debts; having more assets than liabilities. **2** Having the power of dissolving. — *n.* **1** That which solves. **2** A substance, generally a liquid, capable of dissolving other substances; that in which another substance is dissolved. **3** A medicine used for dissolving morbid concretions or obstructions in or upon some organ. [<L *solvens, -entis,* ppr. of *solvere* solve, loosen]

sol·vol·y·sis (sol·vol′ə·sis) *n. Chem.* Any of various double-decomposition reactions similar to hydrolysis, as the reaction of mercuric chloride with liquid ammonia to form a basic salt. [<L *solvere* loosen + Gk. *lysis* a loosening] — **sol·vo·lyt·ic** (sol′və·lit′ik) *adj.*

Sol·way Firth (sol′wā) An inlet of the Irish Sea on the boundary between England and Scotland; extends 40 miles inland; 20 miles wide at its mouth.

Sol·y·man (sol′i·mən) See SULEIMAN.

so·ma (sō′mə) *n. pl.* **·ma·ta** (-mə·tə) *Biol.* The body of any organism, excluding the germ, or germ plasm. [<Gk. *sōma* body]

-soma See -SOME².

So·ma·li (sō·mä′lē) *n.* **1** A member of one of certain Hamitic tribes of Somaliland. **2** The Hamitic language of the Somalis. Also **So·mal** (sō·mäl′).

So·ma·lia (sō·mä′lyə) An independent republic in eastern Africa comprising the former United Nations Trust Territory of Somalia, administered by Italy, and British Somaliland; about 270,000 square miles; capital, Mogadishu. — **So·ma′li** *adj.* & *n.*

So·ma·li·land (sō·mä′lē·land), **French** A French overseas territory in eastern Africa; 8,494 square miles; capital, Djibouti.

so·ma·scope (sō′mə·skōp) *n.* A photographic device combining the principles of sonar, radar, and television to facilitate the detection of cancer and other diseased conditions in the body. — *v.t.* **·scope, ·scop·ing** To apply the somascope to (a patient). [<Gk. *sōma* body + -SCOPE] — **so·mas·co·py** (sō·mas′kə·pē) *n.*

so·mas·the·ni·a (sō′məs·thē′nē·ə) *n. Pathol.* General debility, accompanied by poor appetite, insomnia, and chronic exhaustion. [<Gk. *sōma* body + ASTHENIA] — **so′mas·then′ic** (-then′ik) *adj.*

so·ma·tal·gi·a (sō′mə·tal′jē·ə) *n. Pathol.* Organic pain; pain due to purely physical causes: distinguished from *psychalgia.* [<SOMAT(O)- + Gk. *algos* a pain]

so·mat·ic (sō·mat′ik) *adj. Biol.* **1** Of or relating to the body, as opposed to the spirit; physical; corporeal. **2** Of or pertaining to the framework or walls of a body, as distinguished from the viscera; parietal. **3** Pertaining to those elements or processes of an organism which are concerned with the maintenance of the individual as distinguished from the reproduction of the species: *somatic* cells. [<Gk. *sōmatikos* < *sōma* body]

somatic cell *Biol.* A cell that assists with maintenance of the body rather than reproduction of the species: distinguished from *germ cell.*

so·mat·ics (sō·mat′iks) *n. pl.* (*construed as singular*) Somatology.

so·ma·tism (sō′mə·tiz′əm) *n.* Materialism. — **so′ma·tist** *n.*

somato- *combining form* Body; of, pertaining to, or denoting the body: *somatology.* Also, before vowels, **somat-**. [<Gk. *sōma, sōmatos* the body]

so·ma·to·gen·ic (sō′mə·tō·jen′ik) *adj. Biol.* Originating in the soma or body cells of an organism: said of variations due to the direct influence of environment: *somatogenic* or acquired characters. Also **so′ma·to·ge·net′ic** (-jə·net′ik). — **so′ma·to·gen′e·sis** *n.*

so·ma·tol·o·gy (sō′mə·tol′ə·jē) *n.* **1** The science of organic bodies, especially of the human body: embracing anatomy and physiology. **2** The branch of anthropology that treats of the physical nature of man. [< SOMATO- + -LOGY] — **so′ma·to·log′ic** (-tō·loj′ik), **so′ma·to·log′i·cal** *adj.* — **so′ma·to·log′i·cal·ly** *adv.* — **so′ma·tol′o·gist** *n.*

so·ma·to·pleure (sō′mə·tō·plōōr′) *n. Biol.* In the embryonic development of vertebrates, the outer of the two layers into which the mesoblast divides, together with its investing epiblast. [<SOMATO- + Gk. *pleura* side]

som·ber (som′bər) *adj.* **1** Partially deprived of light or brightness; dusky; murky; gloomy. **2** Somewhat melancholy; producing or denoting gloomy feelings; depressing. Also **som′bre,** *Obs.* **som′brous.** See synonyms under DARK, SAD. [<F *sombre*; ult. origin uncertain] — **som′ber·ly** *adv.* — **som′ber·ness** *n.*

som·bre·ro (som·brâr′ō) *n. pl.* **·ros** A broad-brimmed hat, usually of felt, much worn in Spain, Latin America, and the southwestern United States: humorously called a *ten-gallon hat.* [<Sp. < *sombra* shade]

Som·bre·ro (sôm·brâr′ō) See ST. CHRISTOPHER, NEVIS, AND ANGUILLA.

some (sum) *adj.* **1** Of indeterminate quantity, number, or amount. **2** Limited in degree or amount; moderate. **3** Conceived or thought of, but not definitely known: *some* person. **4** *Logic* Part (more than one) but not all of a class; a quantifier. **5** *U.S. Slang* Of considerable account; worthy of notice; extraordinary: That was *some* cake. — *pron.* **1** A certain undetermined quantity or part; a portion. **2** Certain particular ones not definitely known or not specifically designated. — *adv.* **1** *U.S. Colloq.* In an approximate degree; as nearly as may be estimated; about: *Some* eighty people were present. **2** *Slang* Somewhat. ♦ Homophone: *sum.* [OE *sum* some]

-some¹ *suffix of adjectives* Characterized by, or tending to be (what is indicated by the main element): *blithesome, frolicsome, darksome.* [OE *-sum* like, resembling]

-some² *suffix of nouns* A body: *chromosome, merosome.* Also spelled *-soma.* [<Gk. *sōma* a body]

-some³ *suffix of nouns* A group consisting of (a specified number): *twosome, foursome.* [<SOME]

some·bod·y (sum′bod′ē, -bəd·ē) *pron.* A person unknown or unnamed: *Somebody* loves me. — *n. pl.* **·bod·ies** A person of consequence or importance: She thinks herself a *somebody.*

some·day (sum′dā′) *adv.* At some future time.

some·deal (sum′dēl′) *adv. Archaic* Somewhat.

some·how (sum′hou′) *adv.* In some way or in some manner not explained.

some·one (sum′wun′, -wən) *pron.* Some person; somebody. — *n.* A somebody.

som·er·sault (sum′ər·sôlt) *n.* A leap in which a person turns heels over head and lights on his feet. — *v.i.* To perform a somersault. Also spelled *summersault, summerset.* Also **som′er·set.** [<OF *sombresault,* alter. of *sobresault,* ult. <L *supra* above + *saltus* a leap]

Som·er·set (sum′ər·set) A maritime county in SW England; 1,613 square miles; county town, Taunton. Also **Som′er·set·shire′** (-shir′).

Som·er·vell (sum′ər·vel), **Brehon,** 1892–1955, U. S. general.

So·mes (sô·mesh′) A river in northern Rumania and NE Hungary, flowing 145 miles NW to the Tisza river; including either headstream, 250 miles.

so·mes·the·sis (sō′məs·thē′sis) *n.* A diffuse, generalized awareness of the body and of bodily sensation. Also **so′mes·the′si·a** (-thē′zhē·ə, -zhə). [<Gk. *sōma* body + *esthesis*] — **so′mes·thet′ic** (-thet′ik) *adj.*

some·thing (sum′thing) *n.* **1** A particular thing indefinitely conceived or stated. **2** Some portion or quantity. **3** A person or thing of importance. — *adv.* Somewhat: archaic except in special phrases, as **something like.**

some·time (sum′tīm′) *adv.* **1** At some future time not precisely stated; eventually. **2** At some indeterminate time or occasion. — *adj.* Former; quondam: a *sometime* student at Oxford.

some·times (sum′tīmz′) *adv.* **1** At times; occasionally. **2** *Obs.* Formerly; once.

some·way (sum′wā′) *adv.* In some way or other; somehow. Also **some way, some′ways′.**

some·what (sum′hwot′, -hwət) *n.* **1** An uncertain quantity or degree; something. **2** An individual or thing of consequence. — *adv.* In some degree.

some·when (sum′hwen′) *adv.* At some time.

some·where (sum′hwâr′) *adv.* **1** In, at, or to some place unspecified or unknown. **2** In one place or another. **3** In or to some existent place: opposed to *nowhere.* **4** Approximately. — *n.* An unspecified or unknown place.

some·wheres (sum′hwârz′) *adv. Chiefly Dial.* Somewhere: not considered an acceptable form in standard English.

some·whith·er (sum′hwith′ər) *adv. Archaic* To some indefinite or unknown place; somewhere.

some·why (sum′hwī′) *adv.* For some reason.

some·wise (sum′wīz′) *adv.* In some way or other: obsolete except in the phrase, **in somewise.**

so·mite (sō′mīt) *n. Zool.* A serial segment of the body of an animal, especially of an annelid or arthropod. [<Gk. *sōma* body + -ITE¹] — **so·mi·tal** (sō′mə·təl), **so·mit·ic** (sō·mit′ik) *adj.*

Somme (sôm) A river in northern France, flowing 150 miles west to the English Channel; scene of battles in World War I (1916, 1918), and in World War II (1940, 1944).

som·me·lier (sô·me·lyā′) *n. French* A wine steward.

som·nam·bu·late (som·nam′byə·lāt) *v.* **·lat·ed, ·lat·ing** *v.i.* To walk or wander about while asleep. — *v.t.* To walk over or through while asleep. [<L *somnus* sleep + AMBULATE]

som·nam·bu·lism (som·nam′byə·liz′əm) *n.* The act or state of walking during sleep. Also **som·nam′bu·la′tion.** — **som·nam′bu·lant** (-lənt) *adj.* — **som·nam′bu·list** *n.* — **som·nam′bu·lis′tic** *adj.*

somni- *combining form* Sleep; of or pertaining to sleep: *somnifacient.* [<L *somnus* sleep]

som·ni·fa·cient (som′nə·fā′shənt) *adj.* Promoting sleep; hypnotic. — *n.* A drug which induces sleep. [<SOMNI- + -FACIENT]

som·nif·er·ous (som·nif′ər·əs) *adj.* Tending to produce sleep; soporiferous; narcotic. Also **som·nif′ic.** [<SOMNI- + -FEROUS]

som·nil·o·quy (som·nil′ə·kwē) *n.* **1** The act of talking when asleep, especially in mesmeric sleep. 2 The words so spoken. [<SOMNI- + L *loqui* speak] —**som·nil′o·quist** *n.*

som·no·lence (som′nə·ləns) *n.* Oppressive drowsiness or inclination to sleep. Also **som′·no·len·cy.**

som·no·lent (som′nə·lənt) *adj.* 1 Inclined to sleep; drowsy. 2 Tending to induce drowsiness. [<F <L *somnolentus* <*somnus* sleep] —**som′no·lent·ly** *adv.*

Som·nus (som′nəs) In Roman mythology, the god of sleep: identified with the Greek *Hypnos.*

son (sun) *n.* 1 A male child considered with reference to either parent or to both parents. 2 Any male descendant. 3 One who occupies the place of a son, as by adoption, marriage, or regard. 4 A person regarded as a native of a particular country or place. 5 A male person who is characterized or influenced by some quality or thing or by a being representing some quality or character: a *son* of liberty; *sons* of Belial. Homophone: *sun.* [OE *sunu*]

Son (sun) Jesus Christ; the second person of the Trinity.

so·nance (sō′nəns) *n.* 1 A sound, as of music; also, a tune or air. 2 The state or quality of being sonant.

so·nant (sō′nənt) *adj.* 1 Sounding; resonant. 2 *Phonet.* Voiced: opposed to *surd, voiceless.* —*n. Phonet.* 1 A voiced speech sound. 2 A syllabic sound; in the Indo-European languages, a sonorant. [<L *sonans, -antis,* ppr. of *sonare* resound]

so·nar (sō′när) *n.* 1 A method of using underwater sound waves, at either audible or ultrasonic frequencies, for sounding, navigating, range finding, detection of submerged objects, communication, etc. 2 The equipment for accomplishing the transmission or reception of underwater sound waves. —*adj.* Of, or pertaining to, the equipment, personnel, or methods employed in underwater acoustic signaling. [<SO(UND) NA(VIGATION AND) R(ANGING)]

so·na·ta (sə·nä′tä) *n. Music* A composition for one or two instruments, written in three or four movements, each of which is distinct from the others in tempo and mood but akin to them in style and key. [<Ital. <*sonare* sound]

sonata form *Music* The outline upon which the construction of a movement, specifically the first, of a sonata, quartet, symphony, etc., is based. A movement written in sonata form falls into three sections, called *exposition,* or statement of themes, *development,* and *recapitulation.*

so·na·ti·na (son′ə·tē′nə) *n. pl.* **·ti·ne** (-tē′nā) *Music* A short or easy sonata. [<Ital., dim. of *sonata* SONATA]

son·der (zon′dər) *n. Naut.* A class of small yachts, of which the sum of the water-line length, extreme beam, and extreme draft must not be greater than thirty-two feet. Also **son′·der·class′** (-klas′, -kläs′). [Short for G *sonderklasse* <*sonder* particular + *klasse* class]

sone (sōn) *n. Physics* A unit of loudness, equivalent to a simple tone having a frequency of 1,000 cycles per second at 40 decibels above the threshold of hearing. [<L *sonus* sound]

song (sông, song) *n.* 1 The rendering of vocal music; more widely, any melodious utterance, as of a bird. 2 A musical composition for the voice or for several voices. 3 A short poem whether intended to be sung or not; a lyric or ballad. 4 Poetry; verse. 5 A mere trifle: to sell something for a *song.* [OE] —**song′less** *adj.*

Synonyms: air, anthem, ballad, canticle, carol, chant, descant, ditty, hymn, lay, lyric, melody, poem, poesy, poetry, psalm, sonnet, strain. Compare synonyms for SING.

song and dance 1 A short theatrical act consisting of a song and dance, often having no connection with the rest of the program; especially, a vaudeville act. 2 *Colloq.* Any highly interesting or entertaining statement of no pertinence to the subject under consideration; a rigmarole.

song·bird (sông′bûrd′, song′-) *n.* A bird that utters a musical call; an oscine bird.

Song Bo (sông′ bō′) The Annamese name for the BLACK RIVER, North Vietnam.

Song Coi (sông′ koi′) The Annamese name for the RED RIVER, North Vietnam.

song·ful (sông′fəl, song′-) *adj.* Full of song or melody.

Song of Solomon A Hebrew dramatic love poem in the Old Testament, attributed to Solomon; Canticles. Also **Song of Songs.**

song sparrow A common sparrow (*Melospiza melodia*) of the eastern United States, noted for its song.

song·ster (sông′stər, song′-) *n.* 1 A person or bird given to singing; one skilled in song; a poet. 2 A book of favorite songs.

song·stress (sông′stris, song′-) *n.* A female songster.

son·ic (son′ik) *adj.* 1 Of, pertaining to, determined or affected by sound: *sonic* vibrations. 2 Having a speed approaching that of sound. [<L *sonicus* <*sonus* sound]

sonic barrier *Aeron.* The transonic barrier.

so·nif·er·ous (sō·nif′ər·əs) *adj.* Producing or conducting sound. [<L *sonus* sound + -FEROUS]

son-in-law (sun′in·lô′) *n. pl.* **sons-in-law** The husband of one's daughter.

son·net (son′it) *n.* 1 A poem of fourteen decasyllabic or (rarely) octosyllabic lines, originally composed of an octave and a sestet, properly expressing two successive phases of a single thought or sentiment. In the **Petrarchan,** or **Italian, sonnet** the rime scheme for the octave is *abbaabba,* followed by two or three other rimes in the sestet, with a slight change in thought after the octave. In the **Elizabethan,** or **Shakespearean, sonnet** the rime scheme is *ababcdcdefefgg.* 2 A short poem; an amatory lyric. See synonyms under SONG. —*v.t.* To celebrate in sonnets. —*v.i.* To compose sonnets. [<F <Ital. *sonnetto* <Provençal *sonet,* dim. of *son* a sound <L *sonus*]

son·net·eer (son′ə·tir′) *n.* A composer of sonnets. —*v.i.* To compose sonnets.

son·ny (sun′ē) *n. pl.* **·nies** *Colloq.* Youngster: a familiar form of address.

son of liberty Originally, one who fought against the British in the American Revolution, or a member of any of various organizations formed to oppose the Stamp Act.

So·no·ra (sō·nō′rä, -nō′rä) 1 A state in NW Mexico; 70,465 square miles; capital, Hermosillo. 2 A river in Sonora state, flowing 250 miles south and SW to the Gulf of California.

so·no·rant (sə·nôr′ənt, -nō′rənt) *n. Phonet.* A voiced consonant of relatively high resonance, as (l), (r), (m), and (n), capable of constituting a syllable.

so·nor·i·ty (sə·nôr′ə·tē, -nor′-) *n.* Sonorous quality or state; resonance. Also **so·no′rous·ness.**

so·no·rous (sə·nôr′əs, -nō′rəs) *adj.* 1 Productive or capable of sound vibrations; sounding. 2 Loud and full-sounding; resonant. 3 *Phonet.* Sonant. [<L *sonorus* <*sonare* resound] —**so·no′rous·ly** *adv.*

son·ship (sun′ship) *n.* The state or relation of being a son.

Sons of Liberty The various patriotic societies, or members thereof, organized to oppose British rule in the American colonies.

son·sy (son′sē) *adj. Scot. & Brit. Dial.* Having sweet, engaging looks; happy; jolly; well-conditioned. Also **son′sie.**

Soo Canals (sōō) A colloquial name for the SAULT SAINTE MARIE CANALS.

soo·chong (sōō′chong′, -jong′) See SOUCHONG.

Soo·chow (sōō′chou′, *Chinese* sōō′jō′) See SÜCHOW.

soom (sōōm) *v. & n. Scot.* Swim.

soon (sōōn) *adv.* 1 At a future or subsequent time not long distant; shortly. 2 Without delay; in a speedy manner; also, with ease; readily. 3 With willingness or readiness: usually with *would as, had as,* etc. 4 In good season; early. 5 *Obs.* At once; immediately. [OE *sōna* immediately]

soon·er (sōō′nər) *n. U.S. Slang* 1 A person who goes before the appointed time to take up free public land, and thus obtains one of the most desirable sites. 2 One who makes an unfair and premature start.

Soon·er (sōō′nər) *n. U.S.* A nickname for a native of Oklahoma.

Sooner State Nickname for OKLAHOMA.

Soong (sōōng) Name of a distinguished Chinese family influential in politics, whose members include **Tse-ven,** born 1891, banker; **Ai-ling,** born 1888, the wife of H. H. Kung; **Ching-ling,** born 1890, the widow of Sun Yat-sen, and **Mei-ling,** born 1898, the wife of Chiang Kai-shek.

soop (sōōp) *v.t. & v.i. Scot.* To sweep.

soor (sōōr) *adj. Scot.* Sour.

soot (sŏŏt, sōōt) *n.* A black substance, essentially carbon from the combustion of wood or coal, as deposited on the inside of chimneys and other surfaces in contact with smoke. —*v.t.* To soil or cover with soot. [OE *sōt*]

sooth (sōōth) *Archaic adj.* 1 True; real. 2 Soothing; smooth. —*n.* Truth. Also spelled *soth.* [OE *sōth*] —**sooth′ly** *adv.*

soothe (sōōth) *v.* **soothed, sooth·ing** *v.t.* 1 To restore to a quiet or normal state; calm. 2 To mitigate, soften, or relieve, as pain or grief. 3 *Obs.* To yield assent to; agree with. —*v.i.* 4 To afford relief; have a calming or relieving effect. See synonyms under ALLAY, TEMPER, TRANQUILIZE. [OE *sōthian* verify <*sōth* truth] —**sooth′er** *n.*

sooth·fast (sōōth′fast′, -fäst′) *adj. Archaic* 1 Truthful; also, steadfast; loyal. 2 Real; true. —**sooth′fast′ly** *adv.* —**sooth′fast′ness** *n.*

sooth·ing (sōō′thing) *adj.* Calming; quieting, as a sedative; pacifying. —**sooth′ing·ly** *adv.*

sooth·say (sōōth′sā′) *v.i.* **·said, ·say·ing** To announce the future, as a soothsayer. —**sooth′·say′ing** *n.*

sooth·say·er (sōōth′sā′ər) *n.* 1 One who claims to have supernatural insight and to be able to foretell events. 2 *Obs.* A truthful person: the original meaning.

soot·y (sŏŏt′ē, sōō′tē) *adj.* **soot·i·er, soot·i·est** 1 Blackened or stained by soot. 2 Producing or consisting of soot. 3 Black like soot. —**soot′i·ly** *adv.* —**soot′i·ness** *n.*

sooty grouse A blue grouse (*Dendrogapus fuliginosus*).

sop (sop) *v.* **sopped, sop·ping** *v.t.* 1 To dip or soak in a liquid. 2 To drench. 3 To take up by absorption: often with *up.* —*v.i.* 4 To be absorbed; soak in. 5 To be or become saturated or drenched. —*n.* 1 Anything softened in liquid, as bread. 2 Anything given to pacify, as a bribe. 3 Any soggy mass. [OE *sopp*] —**sop′py** *adj.*

So·phi·a (sō·fī′ə, -fē′ə) A feminine personal name. Also **So·phie** (sō′fē; *Dan., Du.* sō·fē′ə, *Fr.* sō·fē′). [<Gk., wise]

soph·ism (sof′iz·əm) *n.* 1 A false argument intentionally used to deceive. 2 The doctrine or method of the sophists. See synonyms under SOPHISTRY. [<L *sophisma* <Gk., ult. <*sophos* wise]

soph·ist (sof′ist) *n.* 1 A philosopher; a learned man; a thinker. 2 One who argues cleverly but fallaciously or unnecessarily minutely. —*adj.* Pertaining to the art or method of sophists, or to sophistry. [<L *sophista* <Gk. *sophistēs,* ult. <*sophos* wise]

Soph·ist (sof′ist) *n.* 1 A member of a certain school of early Greek philosophy, preceding the Socratic school. 2 One of the later Greek teachers of philosophy and rhetoric, who acquired great skill in subtle disputation under logical forms.

soph·is·ter (sof′is·tər) *n.* 1 A student in one of the later years of a course in some English universities. At Cambridge, students of the second year are called **junior sophisters** or **junior sophs,** and the third-year men **senior sophisters.** 2 A sophist or a Sophist.

so·phis·tic (sə·fis′tik) *adj.* Pertaining to a Sophist, sophists, or sophistry. —*n.* The art or method of the Sophists. Also **so·phis′ti·cal.** —**so·phis′ti·cal·ly** *adv.* —**so·phis′ti·cal·ness** *n.*

so·phis·ti·cate (sə·fis′tə·kāt) *v.* **·cat·ed, ·cat·ing** *v.t.* 1 To make less simple or ingenuous in mind or manner; render worldly-wise or artificial. 2 To mislead or corrupt (a person). 3 To adulterate. 4 To falsify (a text, statement, etc.) by unauthorized or deceptive alterations. —*v.i.* 5 To indulge in sophistry; be sophistic. —*n.* (-kit, -kāt) A sophisticated person. [<Med. L *sophisticatus,* pp. of *sophisticare* <*sophisticus* sophistic] —**so·phis′ti·ca′·tor** *n.*

so·phis·ti·cat·ed (sə·fis′tə·kā′tid) *adj.* 1 Worldly-wise; deprived of natural simplicity; disillusioned. 2 Pretentiously wise; possessing superficial information. 3 Of a kind that appeals to the worldly-wise.

so·phis·ti·ca·tion (sə·fis′tə·kā′shən) *n.* 1 The act of sophisticating; a quibble; a misrepresentation in reasoning or argument. 2 The state of being sophisticated. 3 Adulteration or falsification.

soph·is·try (sof′is·trē) *n. pl.* **·tries** 1 Subtly fallacious reasoning or disputation. 2 The art or methods of the Greek Sophists.
Synonyms: casuistry, chicanery, evasion, fallacy, hair-splitting, paralogism, prevarication, quibbling, sophism, subterfuge, trickery.

Soph·o·cles (sof′ə·klēz), 495?–406 B.C., Athenian tragic poet. — **Soph′o·cle′an** *adj.*

soph·o·more (sof′ə·môr, -mōr) *n.* In American high schools, colleges, and universities having a four-year course, a second-year student. [Earlier *sophomer* a dialectician < *sophom*, var. of SOPHISM (def. 1), because they studied dialectics; later infl. in meaning by Gk. *sophos* wise + *mōros* a fool]

soph·o·mor·ic (sof′ə·môr′ik, -mor′-) *adj.* Of, pertaining to, or like a sophomore; hence, marked by a shallow assumption of learning or by empty grandiloquence; immature; callow. Also **soph′o·mor′i·cal.** — **soph′o·mor′i·cal·ly** *adv.*

So·phro·ni·a (sə·frō′nē·ə) A feminine personal name. [<Gk., prudent]

So·phy (sō′fē) *n. Obs.* A title formerly given to kings of Persia: also **So′phi.** [<Persian *Safawi*, a Persian royal family]

So·phy (sō′fē) A diminutive of SOPHIA.

-sophy *combining form* Knowledge pertaining to a (specified) field: *theosophy.* [<Gk. *sophia* wisdom]

so·por (sō′pər) *n.* Deep lethargic sleep. [<L]

so·po·rif·er·ous (sō′pə·rif′ər·əs, sop′ə-) *adj.* Bringing sleep. — **so′po·rif′er·ous·ly** *adv.* — **so′po·rif′er·ous·ness** *n.*

so·po·rif·ic (sō′pə·rif′ik, sop′ə-) *adj.* 1 Causing or tending to cause sleep. 2 Drowsy; sleepy; characterized by lethargy. — *n.* A medicine that produces sleep.

sop·ping (sop′ing) *adj.* Wet through; drenched; soaking.

sop·py (sop′ē) *adj.* **·pi·er, ·pi·est** Saturated and softened with moisture; soft and sloppy; very wet.

so·pran·o (sə·pran′ō, -prä′nō) *n. pl.* **so·pran·os** or **so·pra·ni** (sə·prä′nē) 1 A woman's or boy's voice of the highest range, usually extending from middle C upward about two octaves. 2 The music intended for such a voice; the treble. 3 A person having a treble or high-range voice, or singing such a part. — *adj.* Of or pertaining to a soprano voice or part. [<Ital. < *sopra* above <L *supra*. Related to SOVEREIGN.]

So·pron (shō′prōn) A city in western Hungary near the Austrian border: German *Ödenburg.*

So·qo·tra (sō·kō′trə) See SOCOTRA.

so·ra (sôr′ə, sō′rə) *n.* A small grayish-brown North American rail (*Porzana carolina*), esteemed as food. Also **sora rail.** [? <N. Am. Ind.]

So·ra·ta (sō·rä′tä) See ILLAMPU.

sorb (sôrb) *n.* 1 The service tree, or the rowan. 2 The fruit of either of these. [<F *sorbe* <L *sorbus* service tree]

Sorb (sôrb) *n.* A Wend.

sorb apple The fruit of the service tree.

sor·be·fa·cient (sôr′bə·fā′shənt) *adj.* Conducive to absorption; absorptive. — *n.* A medicine that promotes absorption. [<L *sorbere* absorb + -FACIENT]

Sorb·i·an (sôr′bē·ən) *adj.* Of or pertaining to the Sorbs or Wends or to their language. — *n.* 1 A Sorb or Wend. 2 The West Slavic language of the Sorbs; Wendish.

sor·bite (sôr′bīt) *n. Metall.* A mixture of ferrite and cementite forming an important constituent of tempered steels. [after H. C. *Sorby*, 1826–1908, British metallurgist]

sor·bi·tol (sôr′bə·tōl, -tol) *n. Chem.* A white, sweetish, crystalline alcohol, $C_6H_{14}O_6$, found in mountain-ash berries, cherries, apples, pears, and some other fruits: it is used as a plasticizer and moistening agent and in the manufacture of ascorbic acid. [<SORB + -IT(E)[1] + -OL[1]]

Sor·bon·ist (sôr′bən·ist) *n.* A doctor of the Sorbonne; a student at the Sorbonne.

Sor·bonne (sôr·bôn′) 1 A former theological college founded in Paris by Robert de Sorbon in 1255–59. 2 The seat of the faculties of literature and science of the University of Paris.

sor·bose (sôr′bōs) *n.* A non-fermentable monosaccharide, $C_6H_{12}O_6$, obtained from sorbitol by bacterial action and important in the synthesis of vitamin C. [<SORB + -OSE[2]]

sor·cer·er (sôr′sər·ər) *n.* A wizard; conjurer; magician. — **sor′cer·ess** *n. fem.*

sor·cer·y (sôr′sər·ē) *n. pl.* **·cer·ies** 1 Pretended employment of supernatural agencies; magic; witchcraft. 2 Any remarkable or inexplicable means of accomplishment; witchery. [<OF *sorcerie* < *sorcier* <L *sors* fate] — **sor′cer·ous** *adj.* — **sor′cer·ous·ly** *adv.*
Synonyms: divination, enchantment, incantation, magic, necromancy, spell, voodoo, witchcraft.

sor·del·li·na (sôr′del·lē′nä) *n. Italian* A kind of small bagpipe.

Sor·del·lo (sôr·del′ō, *Ital.* sôr·del′lō), 1180?–1255?, Provençal troubadour.

sor·did (sôr′did) *adj.* 1 Of, pertaining to, or actuated by a low desire for gain; mercenary. 2 Of degraded character; vile; base; squalid. 3 Of a dull, dirty, or muddy hue. 4 Foul: the old sense. See synonyms under AVARICIOUS, BASE[2]. [<L *sordidus* squalid] — **sor′did·ly** *adv.* — **sor′did·ness** *n.*

sor·di·no (sôr·dē′nō) *n. pl.* **·ni** (-nē) *Music* A mute. [<Ital.]

sore (sôr, sōr) *n.* 1 A place on an animal body where the skin or flesh is bruised, broken, or inflamed; an ulcer or diseased spot. 2 A painful memory; distressing evil; trouble; grief; controversy. — *adj.* **sor·er, sor·est** 1 Morbidly tender; having a sore or sores. 2 Pained or distressed in mind; aggrieved; touchy. 3 Arousing painful feelings; irritating; distressing. 4 Causing extreme distress; severe; also, very great; extreme: He was in *sore* need of money. 5 *Colloq.* Offended; aggrieved; angry. — *adv. Archaic* Sorely. [OE *sār*] — **sore′ness** *n.*

so·re·di·um (sə·rē′dē·əm) *n. pl.* **·di·a** (-dē·ə) *Bot.* A scalelike structure of algal cells in a lichen, enveloped in a network of hyphae and capable of independent vegetative growth. Also **so·rede** (sō′rēd). [<NL <Gk. *sōros* a heap]

sore·head (sôr′hed′, sōr′-) *U.S. Slang n.* A disgruntled or offended person. — *adj.* Dissatisfied; discontented.

sor·el (sôr′əl, sor-) See SORREL[2].

sore·ly (sôr′lē, sōr′-) *adv.* 1 Grievously; distressingly. 2 Greatly; in high degree: His aid was *sorely* needed.

sor·ghum (sôr′gəm) *n.* 1 A stout canelike tropical grass (genus *Sorghum*) cultivated for its saccharine juice and as fodder, especially any of the varieties of *Sorghum vulgare.* 2 Molasses prepared from the sweet juices of the plant. [<NL <Ital. *sorgo*, ult. <L *Syricus* of Syria, where originally grown]

sor·go (sôr′gō) *n. pl.* **·gos** *Spanish* Any variety of sorghum cultivated for its sweet juices and for forage. Also **sor′gho.**

so·ri (sôr′ī, sō′rī) Plural of SORUS.

sor·i·cine (sôr′ə·sīn, -sin, sor′-) *adj.* Pertaining or belonging to a subfamily (*Soricinae*) typical of a family (*Soricidae*) of small, mouselike mammals, the shrews, widely distributed in the northern hemisphere; shrewlike. [<NL <L *sorex, soricis* a shrew]

so·ri·tes (sō·rī′tēs, sō-) *n. Logic* A form of compound syllogism made up of successive coordinate members: Bucephalus is a horse; a horse is a quadruped; a quadruped is an animal; therefore Bucephalus is an animal. [<L <Gk. *sōreitēs* < *sōros* a heap] — **so·rit′i·cal** (-rit′i·kəl) *adj.*

sorn (sôrn) *v.i. Scot.* To force oneself on others for food and lodging. — **sorn′er** *n.*

So·ro·ca·ba (sō′rōō·kä′və) A city in SE São Paulo state, Brazil.

so·ro·che (sō·rō′chā) *n. Spanish* Mountain sickness; puna.

So·rol·la y Bas·ti·da (sō·rō′lyä ē väs·tē′thä), **Joaquín**, 1863–1923, Spanish painter.

so·ror·ate (sôr′ə·rāt, sō′rə·rāt) *n. Anthropol.* The marriage of a man with the sister or sisters of his wife, or with other close female relatives. Compare LEVIRATE. [<L *soror* a sister]

so·ror·i·cide (sə·rôr′ə·sīd) *n.* 1 The killing of a sister. 2 One who kills a sister. [<LL *sororicidium* < *soror* a sister + *caedere* kill; def. 2 <L *sororicida*]

so·ror·i·ty (sə·rôr′ə·tē, -ror′-) *n. pl.* **·ties** A sisterhood; specifically, a women's national or local association having chapters in a secondary school, college, or university. [<Med. L *sororitas, -tatis*, <L *soror* a sister]

so·ro·sis (sə·rō′sis) *n. Bot.* A type of multiple fruit consisting of a fleshy mass formed by the merging of many flowers, as in the mulberry. [<NL <Gk. *sōros* a heap]

So·ro·sis (sə·rō′sis) *n.* A women's club, the first to be organized (1868) in America; hence, **sorosis**, any women's club or society.

sorp·tion (sôrp′shən) *n.* Any process by which one substance takes up and holds the molecules of another substance, as by absorption or adsorption. [<NL *sorptio, -onis* <L *sorbere*]

sor·rel[1] (sôr′əl, sor′-) *n.* 1 Any of several low perennial herbs (genus *Rumex*) with acid leaves, especially the common sorrel (*R. acetosa*). 2 The wood sorrel (genera *Oxalis* or *Xanthoxalis*). [<F *surele* < *sur* <OHG, sour]

sor·rel[2] (sôr′əl, sor′-) *n.* 1 A reddish- or yellowish-brown color. 2 An animal of this color. 3 A buck of the third year. Also spelled *sorel.* [<OF *sorel* < *sor*, a hawk with red plumage]

sorrel tree An American tree (*Oxydendrum arboreum*) of the heath family, with drooping clusters of white flowers and sour evergreen leaves.

Sor·ren·to (sôr·ren′tō) A port on the Bay of Naples, SW Italy.

sor·row (sor′ō, sôr′ō) *n.* 1 Pain or distress of mind because of loss, injury, or misfortune, the commission of sin, or sympathy with suffering; grief. 2 An event that causes pain or distress of mind; affliction; a trial; misfortune; woe. 3 The expression of grief; lamentation; mourning. See synonyms under GRIEF, MISFORTUNE, REPENTANCE. — *v.i.* To feel sorrow; grieve; lament; be sad. See synonyms under MOURN. [OE *sorg* care] — **sor′row·er** *n.*

sor·row·ful (sor′ə·fəl, sôr′-) *adj.* Sad; unhappy; mournful. See synonyms under BAD[1], PITIFUL, SAD. — **sor′row·ful·ly** *adv.* — **sor′row·ful·ness** *n.*

sor·ry (sor′ē, sôr′ē) *adj.* **·ri·er, ·ri·est** 1 Grieved or pained; affected by sorrow from any cause. 2 Causing sorrow; melancholy; dismal. 3 Pitiable or worthless; poor; paltry. 4 Painful; grievous. See synonyms under BAD, SAD. [OE *sārig* < *sār* sore] — **sor′ri·ly** *adv.* — **sor′ri·ness** *n.*

sort (sôrt) *n.* 1 Any number or collection of persons or things characterized by the same or similar qualities; a kind; species; class; set. 2 Form of being or acting; character; nature; quality; also, manner; way; style. 3 *Printing* A character or type considered as a portion of a font: usually in the plural. 4 *Obs.* Social rank, especially high rank. 5 *Obs.* A lot; destiny. — **of sorts** Originally, of various or different kinds; now, of a poor or unsatisfactory kind: used disparagingly: an actor *of sorts.* — **sort of** Somewhat. — *v.t.* 1 To arrange or separate into grades, kinds, or sizes; classify; assort. — *v.i.* 2 To agree; be suitable; correspond. 3 To associate; consort. [<OF *sorte* <L *sors, sortis* lot, condition] — **sort′a·ble** *adj.* — **sort′a·bly** *adv.* — **sort′er** *n.*
Synonyms (noun): character, condition, degree, denomination, description, kind, nature, order, race, rank, style.

sor·tie (sôr′tē) *n. Mil.* 1 A sally of troops from a besieged place to attack the besiegers. 2 A single trip of an aircraft on an assigned military or naval mission. [<F < *sortir* go forth]

sor·ti·lege (sôr′tə·lij) *n.* The act or practice of drawing lots; divination by lot; also, sorcery. [<OF *sortilege* <LL *sortilegus* a diviner <L *sors, sortis* a lot + *legere* pick, choose]

so·rus (sôr′əs, sō′rəs) *n. pl.* **so·ri** (sôr′ī, sō′rī) *Bot.* In ferns and fernlike plants, a cluster of spore cases (sporangia); a fruit dot. [<NL <Gk. *sōros* a heap]

S O S The code signal of distress adopted by the Radiotelegraphic Convention in 1912, and used by airplanes, ships, etc.; hence, any call for assistance.

So·sno·wiec (sô·snô′vyets) A city in SW Poland, important as an industrial center.

so-so (sō′sō′) *adj.* Passable; neither very good nor very bad; mediocre. — *adv.* Indifferently; tolerably.

sos·te·nu·to (sôs′te·nōō′tō) *Music adj.* Sustained or continuous in tone; prolonged or held. — *n.* A sostenuto passage or movement.

Also **sos'ti·nen'to** (-tē·nen'tō), **sos'te·nen'do** (-te·nen'dō). [<Ital.]

sot (sot) *n.* A habitual drunkard. [OE <OF <LL *sottus* a drunkard]

so·te·ri·ol·o·gy (sə·tir'ē·ol'ə·jē) *n.* The branch of theology that treats of salvation by Jesus Christ. [<Gk. *sōtērios* <*sōtēr* savior + -LOGY] — **so·te'ri·o·log'ic** (-ə·loj'ik), **so·te'ri·o·log'i·cal** *adj.*

soth (sōth) See SOOTH.

Soth·ern (suth'ərn), **Edward Hugh,** 1859–1933, U. S. actor.

So·thic cycle (sō'thik, soth'ik) A period of about 1,460 years, based on an ordinary year of 365 days; or 1,461 years, based on a Sothic year of the Egyptians. See under YEAR. Also **Sothic period.**

So·this (sō'this) Sirius, the Dog Star. [<Gk. <Egyptian] — **So'thic** *adj.*

So·tho (sō'thō) *n.* A Bantu language of southern Africa.

so·tol (sō'tōl) *n.* Any one of a genus (*Dasylirion*) of yuccalike plants found in the SW United States. [<Mexican Sp. <Nahuatl *tzotolli*]

sot·ted (sot'id) *adj.* Drunk or infatuated; besotted.

sot·tish (sot'ish) *adj.* Having the manner or character of a sot; stupefied with drink; hence, stupid; doltish. — **sot'tish·ly** *adv.* — **sot'tish·ness** *n.*

sot·to vo·ce (sot'ō vō'chē, *Ital.* sôt'tō vō'chā) Softly; in an undertone; privately; under the breath. [<Ital., under the (normal) voice]

sou (sōō) *n.* A former French coin of varying value; now, colloquially, something trivial or negligible. [<F <LL *solidus,* a gold coin]

sou·a·ri (sōō·ä'rē) *n.* Any of several tropical American trees (genus *Caryocar*), yielding a durable timber known as **souari wood,** and edible nuts called **souari nuts** or butternuts; especially, *C. nuciferum.* [<F *saouari* < native name]

sou·bise (sōō·bēz') *n. French* A sauce of onions, butter, and white sauce: also **soubise sauce.**

sou·brette (sōō·bret') *n.* **1** An actress in light comedy; originally, a pert, intriguing lady's maid. **2** A frivolous or coquettish maidservant. [<F <Provençal *soubreto* <*soubret* shy, coy] — **sou·bret'tish** *adj.*

sou·bri·quet (sōō'bri·kā) See SOBRIQUET.

sou·car (sou·kär') *n. Anglo-Indian* A native banker: also spelled *sowcar.*

sou·chong (sōō'chong', -shong') *n.* A variety of black tea, made from the youngest leaves of the earliest pickings, or the infusion made from it: also spelled *soochong.* [<F <Chinese *siao* small + *chung* plant]

sou·dan (sou'dən) See SOLDAN.

Sou·dan (sōō·dän') The French name for SUDAN.

souf·fle (sōō'fəl) *n.* A low whispering or blowing sound or murmur heard on auscultation: the respiratory *souffle.* [<F <*souffler* blow]

souf·flé (sōō·flā') *adj.* Made light and frothy, and fixed in that condition by heat: also **souf·fléed'** (-flād'). — *n.* A light, baked dish made fluffy with beaten egg whites combined with the yolks, and often with cheese, mushrooms, or other ingredients. [<F, orig. pp. of *souffler* blow <L *sufflare* <*sub-* under + *flare* blow]

Sou·fri·ère (sōō·frē·âr') **1** La Soufriere. **2** Either of two dormant volcanoes, one on Guadaloupe, French West Indies, the other on St. Lucia, The West Indies.

sough (suf, sou) *v.i.* To make a sighing sound, as the wind. — *n.* A deep, murmuring sound, as of wind through trees. — **to keep a calm sough** *Scot.* To be silent. [OE *swōgan* sound, roar, rustle]

sought (sôt) Past tense and past participle of SEEK.

soul (sōl) *n.* **1** The rational, emotional, and volitional faculties in man, conceived of as forming an entity distinct from, and often existing independently of, his body. **2** *Theol.* **a** The divine principle of life in man. **b** The moral or spiritual part of man as related to God, considered as surviving death and liable to joy or misery in a future state. **3** The emotional faculty of man as distinguished from his intellect: He puts his *soul* into his acting. **4** Fervor; emotional force; heartiness; vitality; nobleness: His music lacks *soul.* **5** The animating principle of a thing; an essential or vital element: Justice is the *soul* of law. **6** The leading figure or inspirer of a cause, movement, party, etc.: Lee was the *soul* of the Confederacy. **7** A person considered as the embodiment of a quality or attribute: He is the *soul* of generosity. **8** A living person; a human being: Every *soul* trembled at the sight. **9** The disembodied spirit of one who has died; a ghost. **10** In Christian Science, Spirit; Deity. ◆ Homophone: *sole.* [OE *sāwol*] — **souled** *adj.*

Synonyms: mind, spirit. The *soul* includes the intellect, sensibilities, and will; beyond what is expressed by the word *mind,* the *soul* denotes especially the moral, the immortal nature; we say of a dead body, the *soul* (not the *mind*) has fled. *Spirit* is used especially in contradistinction from matter; it may in many cases be substituted for *soul,* but *soul* has commonly a fuller and more determinate meaning; we can conceive of *spirits* as having no moral nature; the fairies, elves, and brownies of mythology might be termed *spirits,* but not *souls.* In the figurative sense, *spirit* denotes animation, excitability, perhaps impatience; as, a lad of *spirit*; He sang with *spirit*; He replied with *spirit. Soul* denotes energy and depth of feeling, as when we speak of *soulful* eyes; or it may denote the very life of anything; as, the *soul* of harmony. Compare MIND.

soul·ful (sōl'fəl) *adj.* Full of that which appeals to or satisfies the higher feelings; emotional; spiritual. — **soul'ful·ly** *adv.* — **soul'ful·ness** *n.*

soul·less (sōl'lis) *adj.* **1** Having no soul. **2** Heartless; unemotional. **3** Devoid of activity or expression. — **soul'less·ly** *adv.* — **soul'less·ness** *n.*

Soult (sōōlt), **Nicolas Jean de Dieu,** 1769–1851, Duke of Dalmatia; marshal of France under Napoleon I.

sou mar·qué (sōō mär·kā') *French* **1** An 18th century copper coin of France. **2** Hence, a trifle; something of little value. Also **sou mar·kee'** (-kē'), **sou mar·quee'** (-kē').

sound[1] (sound) *n.* **1** The sensation of hearing, produced by stimulation of the auditory centers of the brain by vibratory waves propagated through the atmosphere or other elastic medium. **2** The vibrations that produce sound waves, having for the normal human ear frequencies from about 20 to 20,000 cycles per second. **3** Noise of any specified quality: the *sound* of bugles; any tone, voice, or note. **4** Significance; implication: The story has a sinister *sound.* **5** Sounding or hearing distance; earshot: We were within the *sound* of battle. **6** Mere noise without significance: full of *sound* and fury. **7** *Obs.* Rumor. — *v.i.* **1** To give forth a sound or sounds. **2** To give a specified impression; seem: The story *sounds* true. — *v.t.* **3** To cause to give forth sound. **4** To give a signal or order for or announcement of: to *sound* retreat; to *sound* the hour. **5** To utter audibly; pronounce. **6** To make known or celebrated: to *sound* a hero's fame. **7** To test or examine by sound; auscultate. — **to sound in tort** To act as or have the nature of a tort. [<OF *son* <L *sonus*]

Synonyms (noun): noise, note, tone. *Sound* is the most comprehensive word, applying to anything that is audible. *Tone* is sound considered from the point of view of quality or pitch, or as expressive of some feeling; *noise* is *sound* considered without reference to musical quality or as distinctly unmusical or discordant. *Noise* and *sound* scarcely differ, and we say almost indifferently "I heard a *sound*" or "I heard a *noise.*" We speak of a fine, musical, or pleasing *sound,* but never thus of a *noise.* In music, *tone* may denote a musical *sound* or the interval between two such *sounds,* but in the most careful usage the latter is now distinguished as the "interval." *Note* in music strictly denotes the character representing a *sound,* but in loose popular usage it denotes the *sound* also, and becomes practically equivalent to *tone.* Aside from its musical use, *tone* is also applied to that quality of the human voice by which feeling is expressed; as, He spoke in a cheery *tone.* See NOISE.

sound[2] (sound) *adj.* **1** Having all the organs or faculties complete and in normal action and relation; healthy. **2** Free from injury, flaw, mutilation, defect, or decay: *sound* timber. **3** Founded in truth; right; substantial; valid; legal. **4** Correct in views or processes of thought. **5** Solvent. **6** Profound, as rest; deep; unbroken; also, resting profoundly. **7** Complete and effectual; thorough. **8** Solid; stable; firm; safe; hence, trustworthy. **9** Based on good judgment. See synonyms under HEALTHY, SANE[1], STAUNCH, WISE[1]. [OE *gesund*] — **sound'ly** *adv.* — **sound'·ness** *n.*

sound[3] (sound) *n.* **1** A long and narrow body of water, more extensive than a strait, connecting larger bodies. **2** The air bladder of a fish. [Fusion of OE *sund* sea, a swimming and ON *sund* a strait, swimming]

sound[4] (sound) *v.t.* **1** To test the depth of (water, etc.), especially by means of a lead weight at the end of a line. **2** To measure (depth) thus. **3** To explore or examine (the bottom of the sea, etc.) by means of a sounding lead adapted for bringing up adhering particles. **4** To discover or try to discover the views and attitudes of (a person) by means of conversation and round-about questions: usually with *out.* **5** To try to ascertain or determine (beliefs, attitudes, etc.) in such a manner. **6** *Surg.* To search or examine, as with a sound. — *v.i.* **7** To measure depth, as with a sounding lead. **8** To dive down suddenly and deeply, as a whale when harpooned. **9** To make investigation; inquire. — *n. Surg.* An instrument for exploring a cavity; a probe. [<OF *sonder,* ? <L *sub-* under + *unda* a wave] — **sound'a·ble** *adj.*

Sound (sound), **The** See ÖRESUND.

sound barrier *Aeron.* The transonic barrier.

sound·board (sound'bôrd', -bōrd') *n.* A thin board, as in a piano or violin, forming the upper plate of a resonant box: also called *belly.*

sound·box (sound'boks') *n.* That part of a phonograph which by means of a sensitive diaphragm relays to the surrounding air the acoustic vibrations transmitted to it by the stylus in the record groove.

sound effects In motion pictures, radio, etc., the incidental and often mechanically produced sounds, as of rain, hoofbeats, fire, etc., required to heighten the illusion of reality.

sound·er (soun'dər) *n.* **1** One who or that which sounds or gives a sound. **2** An apparatus for taking soundings, as at sea. **3** A probe. **4** A telegraphic device for converting electromagnetic code impulses into sound, thus enabling messages to be interpreted.

sound·ing[1] (soun'ding) *adj.* **1** Giving forth a full sound; sonorous. **2** Having much sound with little significance; noisy and empty. — **sound'ing·ly** *adv.*

sound·ing[2] (soun'ding) *n.* **1** The act of one who or that which sounds, in any sense. **2** Measurement of the depth of water. **3** *pl.* The depth of water as sounded; also, water of such depth that the bottom may be reached by sounding.

sounding board **1** A structure or suspended dome over a pulpit or speaker's platform to amplify and clarify the speaker's voice. **2** Any device that gives force to an opinion or speech.

sounding lead The lead or other weight used on a sounding line; a plummet.

sounding line A weighted line marked at fathom intervals with pieces of leather, cloth, etc., used for determining the depth of water.

sound·less (sound'lis) *adj.* Having or making no sound; silent. — **sound'less·ly** *adv.* — **sound'·less·ness** *n.*

sound locator An apparatus for locating the position of aircraft by means of the sound waves which they emit.

sound picture A motion picture with a sound track.

sound·proof (sound'prōōf') *adj.* Resistant to the penetration or spread of sound. — *v.t.* To make soundproof.

sound ranging A method of locating the point of origin of a sound by checking time intervals as recorded from microphones of known position.

sound spectrograph See under SPECTROGRAPH.

add,āce,câre,pälm; end,ēven; it,īce; odd,ōpen,ôrder; tŏŏk,pōōl; up,bûrn; ə = a in *above,* e in *sicken,* i in *clarity,* o in *melon,* u in *focus*; yōō = u in *fuse*; oi,oil; ou,pout; ch,check; g,go; ng,ring; th,thin; th,this; zh,vision. Foreign sounds à,œ,ü,kh,ṅ; and ◆: see page xx. < from; + plus; ? possibly.

sound track That portion along the edge of a motion-picture film which carries the sound record.

soup (sōōp) *n.* **1** Liquid food made by boiling meat, vegetables, etc., in water: distinguished from *broth*, which is usually strained. **2** *Phot.* A developer. **3** *U.S. Slang* Nitroglycerin. — **in the soup** *U.S. Slang* In difficulties; in a quandary. — **to soup up** *U.S. Slang* To supercharge or otherwise modify (an automobile) for high speed. [< F *soupe* < Gmc.]

soup·çon (sōōp-sôn′) *n.* French Literally, a suspicion; hence, a minute quantity; a taste.

soup kitchen A place where soup is served to the needy either free or at very low cost.

sou·ple (sōō′pəl) *adj.* *Scot.* **1** Supple. **2** Swift.

soup·spoon (sōōp′spōōn′) *n.* A spoon used in eating soup.

soup·y (sōō′pē) *adj.* **soup·i·er, soup·i·est** Like soup in appearance or consistency.

sour (sour) *adj.* **1** Sharp to the taste; acid; tart, like vinegar: designating one of the four fundamental taste sensations. **2** Having an acid or rancid taste as the result of fermentation; also, pertaining to fermentation. **3** Having a rancid, acid smell or vapor; dank. **4** Misanthropic and crabbed; cross; morose: a *sour* person, a *sour* smile. **5** Cold and wet; unpleasant: *sour* weather. **6** Acid; harsh to crops: said of land. **7** Containing sulfur compounds: said of gasoline. — *v.t.* & *v.i.* To become or make sour. — *n.* **1** Something sour or distasteful. **2** An acid solution used in bleaching or in curing skins. **3** A treatment with such a solution. **4** A sour acid beverage: a whisky *sour*. [OE *sūr*] — **sour′ly** *adv.* — **sour′ness** *n.*

source (sôrs, sōrs) *n.* **1** That from which any act, movement, or effect proceeds; an originator; creator; origin. **2** A place where something is found or whence it is taken or derived. **3** The spring or fountain from which a stream of water proceeds; a fountainhead; fountain. **4** A person, writing, or agency from which information is obtained. **5** The initiator of a payment, dividend, etc. [< OF, orig. pp. of *sourdre* rise < L *surgere*]
— *Synonyms:* beginning, fountain, fountainhead, origin, spring. See BEGINNING, CAUSE. *Antonyms:* close, completion, conclusion, consequence, end, event, expiration, result, sequel, termination.

sour·crout (sour′krout′) See SAUERKRAUT.

sour·dine (sōōr-dēn′) *n.* **1** A mute, especially a trumpet mute. **2** A stop on the harmonium producing a soft effect. **3** An obsolete, soft-toned musical instrument. **4** In telegraphy, a silencer. [< F < Ital. *sordino* < *sordo* < L *surdus* deaf]

sour·dough (sour′dō′) *n.* **1** *Dial.* Fermented dough for use as leaven in making bread. **2** *U.S. & Can. Slang* A pioneer or prospector; especially, an Alaskan or Canadian prospector who carries fermented dough for use in making bread.

sour gourd **1** One of a genus (*Adansonia*) of trees with huge trunk, having a woody gourdlike capsule; especially, the Australian tree (*A. gregorii*). **2** The acid fruit of this tree. **3** The Madagascar baobab.

sour grapes That which a person affects to despise, because it is beyond his attainment: in allusion to the fable of the fox and the grapes.

sour·gum (sour′gum′) *n.* Any of several species of trees of the genus *Nyssa*, especially the blackgum tree and the tupelo.

Sou·ris (sōōr′is) A river in North Dakota and Canada, flowing 435 miles SE and north from SE Saskatchewan, through northern North Dakota, to the Assiniboine river in southern Manitoba.

sour·puss (sour′pōōs′) *n. Slang* A person with a sullen, peevish expression or character.

sour·sop (sour′sop′) *n.* **1** A tree (*Annona muricata*) of tropical America. **2** The pulpy, somewhat acid fruit of this tree.

Sou·sa (sōō′zə), **John Philip**, 1854–1932, U.S. bandmaster and composer.

sou·sa·phone (sōō′zə-fōn, -sə-) *n.* A brass wind instrument, resembling a tuba, but circular and with flaring bell frontward: used chiefly in military bands. [after John P. *Sousa*]

souse¹ (sous) *v.t.* & *v.i.* **soused, sous·ing** **1** To dip or steep in a liquid. **2** To pickle. **3** *Slang* To make or get drunk. — *n.* **1** Pickled meats; especially, the feet and ears of swine, pickled or soused in brine; formerly, any salt pickle. **2** A plunge in water. **3** Brine. **4** *Slang* A drunkard; sot. [< OF *sous* < OHG *sulza* brine]

souse² (sous) *Archaic v.* **soused, sous·ing** *v.t.* To pounce upon. — *v.i.* To swoop suddenly, as a hawk: with *on* or *upon*. [< *n.*] — *n.* A swoop, as of a hawk on its prey; a downright blow or stroke. — *adv.* Suddenly; with a plunge or swoop headlong; all over. [Var. of SOURCE, in earlier sense of "a rising"]

Sousse (sōōs) A port of eastern Tunisia: ancient *Hadrumetum*; formerly *Susa*.

sou·tache (sōō-täsh′) *n. French* **1** A very narrow, flat, decorative braid in a herringbone effect. **2** Mohair or silk rounded braid.

sou·tane (sōō-tän′) *n.* A Roman Catholic priest's cassock. [< F < Ital. *sottana* < *sotto* under < L *subtus*]

sou·ter (sōō′tər) *n. Scot.* A shoemaker; cobbler: also spelled *sowter*. Also **sou′tar**.

south (south) *n.* **1** That one of the four cardinal points of the compass which is directly opposite to north, and at the right hand of an observer who faces the sunrise. **2** The direction in which the point lies. **3** A region lying in this direction. **4** A south wind. — *adj.* **1** Situated in a southern direction relatively to the observer or to any given place or point. **2** Facing toward the south. **3** Belonging to or proceeding from the south; southern. — *v.i.* **1** To turn southward. **2** *Astron.* To cross the meridian. — *adv.* **1** Toward or at the south. **2** From the south. [OE *sūth*]

South, the **1** The portion of the United States lying south of the Mason-Dixon line, and east and south of the western and northern borders of Missouri. **2** The Confederacy.

South Africa, Union of An independent republic (since 1961), consisting of the provinces of Cape of Good Hope, Natal, Transvaal, and Orange Free State; 472,359 square miles; seat of government, Pretoria; seat of legislature, Cape Town.

South African **1** Pertaining to South Africa, especially to the Union of South Africa. **2** A native of the Union of South Africa; an Afrikander.

South African Dutch Afrikaans.

South African Republic A former state of South Africa, coextensive with the Transvaal.

South African War See BOER WAR in table under WAR.

South America The southern continent of the western hemisphere; about 6,900,000 square miles. — **South American**

South·amp·ton (south-hamp′tən, sou-thamp′-) **1** An administrative county and the mainland part of Hampshire, southern England; 1,503 square miles; county town, Winchester. **2** A port and county borough of southern Hampshire, England, the major European terminal of most transatlantic shipping lines, at the head of **Southampton Water**, an inlet of The Solent in Southampton county, 6 miles long.

South Australia A state of southern Australia; 380,070 square miles; capital, Adelaide.

South Bass Island An island in Lake Erie, 15 miles NW of Sandusky, Ohio; site of Put-in-Bay; 3 1/2 miles long, 1 1/2 miles wide.

South Bend A city in northern Indiana.

south·bound (south′bound′) *adj.* Going southward. Also **south′bound′**.

south by east One point east of south on the mariner's compass. See COMPASS CARD.

south by west One point west of south on the mariner's compass. See COMPASS CARD.

South Carolina A SE State of the United States, on the Atlantic; 31,055 square miles; capital, Columbia; entered the Union May 23, 1788; one of the thirteen original States; nickname *Palmetto State*: abbr. *S.C.* — **South Carolinian**

South China Sea The arm of the Pacific between the SE Asian mainland and the Malay Archipelago.

South Dakota A State in the north central United States; 77,047 square miles; capital, Pierre; entered the Union Nov. 2, 1889; nickname *Coyote State*: abbr. *S. Dak., S.D.* — **South Dakotan**

South·down (south′doun′) *n.* One of an English breed of hornless sheep with dusky-brown legs and faces: originally bred in the South Downs, England.

South Downs A range of hills in southern England, chiefly in Sussex, terminating at Beachy Head.

south·east (south′ēst′, *in nautical usage* sou′-ēst′) *n.* That point on the mariner's compass midway between south and east; any region lying toward that point on the horizon. — *adj.* Of, pertaining to, toward, or from the southeast. — *adv.* Toward or from the southeast. — **south′east′ern** *adj.* — **south′east′ern·most** *adj.* — **south′east′ward** *adj.* & *adv.* — **south′east′ward·ly, south′east′wards** *adv.*

southeast by east One point east of southeast on the mariner's compass. See COMPASS CARD.

southeast by south One point south of southeast on the mariner's compass. See COMPASS CARD.

south·east·er (south′ēs′tər, *in nautical usage* sou′ēs′tər) *n.* A gale from the southeast.

south·east·er·ly (south′ēs′tər-lē, *in nautical usage* sou′ēs′tər-lē) *adj.* & *adv.* **1** Toward the southeast. **2** From the southeast: said of wind.

South·end-on-Sea (south′end′on-sē′) A resort and county borough in SE Essex, England.

south·er (sou′thər) *n.* A gale from the south.

sou·ther² (sō′thər) *n. Brit. Dial.* Solder: also spelled *sowther*.

south·er·ly (suth′ər-lē) *adj.* **1** Situated in or tending toward the south. **2** Proceeding from the south. — *adv.* Toward or from the south. — **south′er·li·ness** *n.*

south·ern (suth′ərn) *adj.* **1** Pertaining to the south or a place relatively in the south. **2** Proceeding from the south, as a wind [OE *sutherne*] — **south′ern·er** *n.* — **south′ern·ly** *adv.* — **south′ern·most** *adj.*

South·ern (suth′ərn) *adj.* Of or pertaining to the South.

Southern Alps A mountain range in west central South Island, New Zealand; highest peak, 12,349 feet.

Southern Cross A southern constellation having four bright stars in the form of a cross. See CONSTELLATION.

Southern Crown A southern constellation near Sagittarius; Corona Australis.

South·ern·er (suth′ərn-ər) *n.* A native of the South.

Southern Kar·roo (ka-rōō′) A plateau region in SW Cape of Good Hope Province, Union of South Africa: also *Little Karroo*.

southern lights The aurora australis.

Southern Pines A resort town in central North Carolina.

Southern Protectorate of Morocco A Spanish protectorate on the NW coast of Africa, part of Spanish West Africa; 10,039 square miles.

Southern Rhodesia See under RHODESIA.

Southern Territories **1** A division of southern Algeria; 769,827 square miles. **2** A former French military territory of southern Tunisia; about 17,800 square miles. *French* **Ter·ri·toires du Sud** (ter-ē-twär′ dü süd′).

south·ern·wood (suth′ərn-wood′) *n.* A European plant (*Artemisia abrotanum*) allied to wormwood.

South·ey (suth′ē), **Robert**, 1774–1843, English poet; poet laureate 1813–43.

South Georgia A barren island between the Falkland Islands and Cape Horn, considered part of the Falkland Island Dependencies in the South Atlantic; 1,600 square miles.

South Holland A province of western Netherlands; 1,085 square miles; capital, The Hague. Dutch **Zuid·hol·land** (zoit′hô′länt).
south·ing (sou′thing) n. 1 The difference of latitude measured toward the south between any position and the last one determined. 2 Astron. **a** The passage across the meridian, in its diurnal motion, of a celestial object that culminates south of the zenith. **b** The attainment of this position, or the time at which it is reached. 3 Deviation or progression toward the south.
South Island The largest island of the New Zealand group; 58,093 square miles.
South Jutland The southern part of the Jutland peninsula, northern Germany and southern Denmark, coextensive with the former duchy of Schleswig.
South Korea See under KOREA.
south·land (south′land′) n. A land or region situated to the south. — **south′land′er** n.
South Mountain The northernmost part of the Blue Ridge, in western Pennsylvania, Maryland, and northern Virginia; highest peak, 2,145 feet; scene of a Union victory in the Civil War, Sept. 14, 1862.
South Orkney Islands An island group in the South Atlantic NE of the Palmer Peninsula; 240 square miles; claimed by Great Britain and by Argentina; administered by Great Britain as a dependency of the Falkland Islands.
South Os·se·tian Autonomous Region (o-sē′shən) An administrative division of northern Georgian S.S.R.; 1,428 square miles; capital, Staliniri.
south·paw (south′pô′) Slang n. 1 In baseball, a left-handed pitcher. 2 Any left-handed person or player. — adj. Left-handed.
South Platte River A river in Colorado and Nebraska, flowing 450 miles east and NE to join the North Platte, forming the Platte River.
South Pole The southern extremity of the earth's axis; the 90th degree of south latitude, from which all terrestrial directions are north.
South·port (south′pôrt, -pōrt) A county borough on the Irish Sea in SW Lancashire, England, north of Liverpool.
south·ron (suth′rən, south′-) adj. Southern. — n. A person who lives in the south. [Alter. of dial. E southren, var. of SOUTHERN; infl. in form by Saxon, Briton, etc.]
South·ron (suth′rən, south′-) n. Scot. An Englishman or native of southern Britain: formerly used by the Scots as a term of derision: also **Suthron**.
South Sandwich Islands An island group in the South Atlantic, included in the Falkland Island Dependencies; 130 square miles.
South Sea Islands The islands of the South Pacific Ocean.
South Seas 1 The South Pacific Ocean. 2 The seas of the world south of the Equator.
South Shetland Islands An archipelago in the South Atlantic, between South America and Antarctica; claimed by Great Britain, Argentina, and Chile; administered by Great Britain as a dependency of the Falkland Islands; 1,800 square miles.
South Shields A port and county borough of NE Durham, England, at the mouth of the Tyne on the North Sea.
south–south·east (south′south′ēst′, in nautical usage sou′sou′ēst′) n. That point on the mariner's compass midway between south and southeast. — adj. & adv. Midway between south and southeast. See COMPASS CARD.
south–south·west (south′south′west′, in nautical usage sou′sou′west′) n. That point on the mariner's compass midway between south and southwest. — adj. & adv. Midway between south and southwest. See COMPASS CARD.
South Vietnam 1 The Republic of Vietnam, comprising the former French colony of Cochin China and the southern part of the former Empire of Annam; 65,709 square miles; capital, Saigon. 2 Cochin China alone, now the southern part of the Republic of Vietnam; 24,750 square miles; capital, Saigon.
south·ward (south′wərd, in nautical usage suth′ərd) adj. Situated in or toward the south. — adv. In a southerly direction: also **south′ward·ly**, **south′wards**. — n. The direction of south; also, a region to the south.
South·wark (suth′ərk) A borough of London, England, on the south bank of the Thames.
South·well (south′wel, suth′əl) A town in central Nottinghamshire, England; known for its Norman cathedral.
south·west (south′west′, in nautical usage sou′west′) n. 1 That point on the mariner's compass midway between south and west; any region lying toward that point on the horizon. — adj. Of, pertaining to, facing, or toward the southwest; blowing from the southwest. — adv. Toward or from the southwest. — **south′west′ern**, **south′west′ern·most** adj. — **south′west′ward** adj. & adv. — **south′west′ward·ly**, **south′west′wards** adv.
Southwest, the The SW part of the United States: generally including Oklahoma, Texas, New Mexico, Arizona, and southern California.
South–West Africa A self-governing mandated territory on the Atlantic north of and administered by the Union of South Africa; 317,887 square miles; capital, Windhoek: formerly German Southwest Africa.
southwest by south One point south of southwest on the mariner's compass. See COMPASS CARD.
southwest by west One point west of southwest on the mariner's compass. See COMPASS CARD.
south·west·er (south′wes′tər, in nautical usage sou′wes′tər) n. 1 A wind, gale, or storm from the southwest. 2 A waterproof hat of oilskin, canvas, etc., with a broad brim behind to protect the neck: worn in stormy weather. Also **sou′'west′er**.
south·west·er·ly (south′wes′tər·lē, in nautical usage sou′wes′tər·lē) adj. & adv. 1 Toward the southwest. 2 From the southwest: said of wind.
Southwest Semitic See under SEMITIC.
sou·ve·nir (sōō′və·nir′, sōō′və·nir) n. A token of remembrance; memento. [<F, remember <L subvenire come to mind]
sov·er·eign (sov′rin, suv′-) n. 1 One who possesses supreme authority, especially a person or a determinate body of persons in whom the supreme power of the state is vested; a monarch. 2 An English gold coin equivalent to one pound sterling or twenty shillings, first issued by Henry VII. 3 A former gold coin of Austria. See synonyms under MASTER. — adj. 1 Exercising or possessing supreme jurisdiction or power; royal. 2 Free, independent, and in no way limited by external authority or influence: a sovereign state. 3 Possessing supreme excellence or greatness; preeminent; paramount. 4 Superior in efficacy; potent: a sovereign remedy. See synonyms under IMPERIAL, PREDOMINANT. Also, Poetic, sovran. [<OF soverain, ult. <L super above. Related to SOPRANO.] — **sov′er·eign·ly** adv.
sov·er·eign·ty (sov′rin·tē, suv′-) n. pl. ·ties 1 The state of being sovereign; supreme authority. 2 The ultimate, supreme power in a state. 3 A sovereign state. 4 The status or dominion of a sovereign. Also, Poetic, sovranty. — **popular sovereignty** The theory that the right to legislate and choose a government belongs to the body of the people.
So·vetsk (so-vyetsk′) A city on the Neman River in western European Russian S.F.S.R.: formerly Tilsit.
So·vet·ska·ya Ga·van (so-vyet′skə·yə gä′vən·y) A port and naval base on the Sea of Japan, Russian S.F.S.R.
so·vi·et (sō′vē·et, sō′vē·et′) n. 1 A council; especially, a soldiers', peasants', and workmen's council in Russia. 2 A local governing body consisting of representatives of the people elected to the **village soviet** or **town soviet** which sends delegates to the higher councils or soviet congresses. 3 Any of various similar socialistic bodies. [<Russian sovyet a council] — **so′vi·et·dom** n.
So·vi·et (sō′vē·et, sō′vē·et′) adj. Of or pertaining to the Soviet Union.
Soviet Central Asia The constituent Soviet republics of Russian Turkestan: the Kazakh, Kirghiz, Tadzhik, Turkmen, and Uzbek Soviet Socialist Republics; 1,541,530 square miles: also Central Asia.

soviet congress The administrative body of each constituent republic of the Union of Soviet Socialist Republics, empowered to adopt its own constitution based on the Union constitution: autonomous republics are governed by executive committees elected by the local Congress of Soviets.
Soviet Far East The easternmost part of Siberia, together with Sakhalin and the Kurile Islands, in Russian S.F.S.R.; 1,204,700 square miles.
so·vi·et·ism (sō′vē·ə·tiz′əm) n. The policies and principles of, or goverment by peoples' councils or congresses, especially as practiced in Soviet Russia. — **so′vi·et·ist** n.
so·vi·et·ize (sō′vē·ə·tiz′) v.t. ·ized, ·iz·ing To bring under a soviet form of government. — **so′vi·et·i·za′tion** n.
Soviet Russia 1 The Union of Soviet Socialist Republics. 2 The Russian Soviet Federated Socialist Republic.
Soviet Union The Union of Soviet Socialist Republics.
sov·ran (sov′rən, suv′-), **sov·ran·ty** (sov′rən·tē, suv′-), etc. See SOVEREIGN, etc.
sow¹ (sou) v. **sowed**, **sown** or **sowed**, **sow·ing** v.t. 1 To scatter (seed) over land for growth. 2 To scatter seed over (land). 3 To spread abroad; disseminate; implant: to sow the seeds of distrust. 4 To cover or sprinkle. — v.i. 5 To scatter seed. See synonyms under PLANT. [OE sāwan] — **sow′er** n. — **sow′ing** n.
sow² (sou) n. 1 A female hog. 2 Metall. **a** The connection between pieces of pig iron before breaking up. **b** The conduit to the pig bed for molten metal. [OE sū, sugu]
so·war (sō·wär′, -wôr′) n. Anglo-Indian Formerly, in the British–Indian army, a native trooper of the irregular cavalry; a mounted orderly.
sow–bel·ly (sou′bel′ē) n. U.S. Colloq. Salt pork.
sow·bread (sou′bred) n. The cyclamen.
sow bug A small, terrestrial, isopodous crustacean (family Oniscidae) found under logs and stones; a wood louse.
sow·car (sou·kär′) See SOUCAR.
sow·ens (sō′ənz) n. Scot. 1 Sour flummery or porridge made from the refuse of the oatmeal mill. 2 A weaver's paste for stiffening yarn. Also **sow′ans**, **sow′ins**.
sowth (sōōth) v.t. & v.i. Scot. To hum or whistle (an air) softly.
sow·ther (sou′thər) See SOUTHER².
sow thistle Any of a genus (Sonchus) of spiny plants, especially the common sow thistle (S. oleraceus), a coarse annual weed.
soy (soi) n. 1 A small, erect herb (Glycine soja) of the bean family, growing in India and China and cultivated for forage. 2 Its edible bean, a source of oil, flour, and other products: also **soy·a** (soi′ə), **soy′bean′**. 3 A sauce prepared in China and Japan from soybeans that have been fermented and steeped in brine: also **soy sauce**. [<Japanese soy, shoy, short for shōyū soy]
so·zin (sō′zin) n. Biochem. Any protein normally contained in the body of an animal and forming a natural protection against germs. Also **so′zine**. [<Gk. sōzein save + -IN]
spa (spä) n. Any locality frequented for its mineral springs; a mineral spring. [from SPA]
Spa (spä) A resort town in eastern Belgium.
Spaak (späk), **Paul Henri**, born 1899, Belgian and international statesman.
Spaatz (späts), **Carl**, born 1891, U.S. general.
space (spās) n. 1 An interval between points or objects; a limited portion of extension; distance; area. 2 The abstract possibility of extension; that which is characterized by illimitable dimension; continuous boundless extension in all directions. 3 An interval of time; period; hence, a little while. 4 An occasion or opportunity. 5 Printing A piece of type metal, less than type-high, used for spacing between lines; specifically, one less than one en in width. 6 One of the degrees of a musical staff. 7 One of the intervals during the transmission of a telegraph message when the key is open or not in contact. 8 Reserved accommodations, as on a train or airplane. 9 Math. A system of continuous, unlimited, corresponding points in a series; an ordered set of infinite numbers. 10 Outer space. See synonyms under PLACE. — v.t.

spaced, spac·ing 1 To separate by spaces. 2 To divide into spaces. [<OF *espace* <L *spatium*] —**space′less** *adj.* —**spac′er** *n.*

Space, meaning for, pertaining to, or concerned with travel in outer space, may appear as a combining form in solidemes or as the first element in two-word phrases; as in:

space age	**space lawyer**	**space sociology**
space crew	**spaceman**	**space station**
space fiction	**space patrol**	**space suit**
space flight	**space rocket**	**space vehicle**

space band *Printing* In Linotype operation, an adjustable wedge-shaped metal strip used for spacing.

space-borne (spās′bôrn′, -bōrn′) *adj.* Carried in or transported through space: a *space-borne* observatory; *space-borne* troops.

space-cab·in simulator (spās′kab′in) A chamber built to resemble the cabin of a spaceship, having instruments, mechanisms, and control equipment for testing human or animal reactions under physiological conditions simulating those to be expected in actual space travel.

space charge *Physics* 1 An electric charge uniformly distributed through a given space. 2 A grouping of electrons around the filament of a vacuum tube, imparting a negative charge which inhibits the free emission of other electrons.

space·craft (spās′kraft′, -kräft′) *n.* 1 A spaceship or other manned vehicle for navigating in outer space. 2 Spacemanship.

space ebullism *Pathol.* The vaporizing or bubbling out of body fluids under the abnormal conditions of temperature and pressure experienced at altitudes in excess of about 63,000 feet.

space lattice *Physics* The characteristic arrangement of the atoms or structural units in a crystal, such that corresponding units are separated by constant intervals along any straight line drawn through their centers.

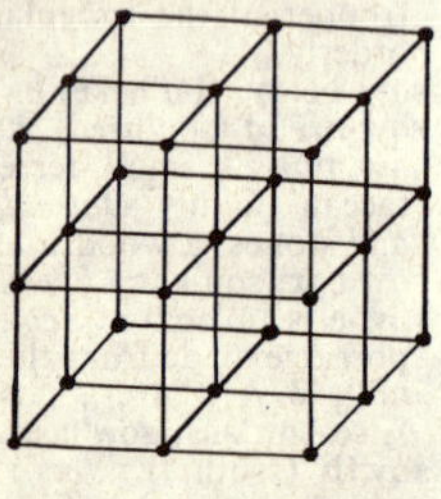

SPACE LATTICE

space·man·ship (spās′mən·ship) *n.* The science and art of space travel, especially as regards the design, construction, fueling, launching, and operation of vehicles and missiles equipped for flight beyond the earth's atmosphere.

space medicine The branch of aviation medicine which deals with the biological, physiological, and psychological aspects of travel in outer space.

space·ship (spās′ship′) *n.* Any of various vehicles designed for the transport of men and materials through outer space, especially between and among the planets.

space time A four-dimensional continuum within which may be precisely located any magnitude having both extension and duration: it consists of three spatial coordinates and one coordinate of time.

space-time continuum (spās′tīm′) See FOURTH DIMENSION.

space travel Travel in regions above the earth's atmosphere or beyond its gravitational field, whether within or outside of the solar system.

spa·cial (spā′shəl) See SPATIAL.

spac·ing (spā′sing) *n.* 1 The arrangement of spaces. 2 A space or spaces, as in a line of print.

spa·cious (spā′shəs) *adj.* 1 Of indefinite or vast extent. 2 Affording ample room; capacious: in literal or figurative sense. See synonyms under LARGE. —**spa′cious·ly** *adv.* —**spa′cious·ness** *n.*

Spack·le (spak′əl) *n.* A dry powder which, in the form of a paste, is used for filling cracks, holes, and other surface defects before painting and decorating: a trade name. —*v.t.* **·led, ·ling** To apply Spackle to (a crack or surface defect).

spade[1] (spād) *n.* 1 An implement used for digging in the ground, ditching, cutting turf, etc., heavier than a shovel and having a flatter blade. 2 A tool or implement resembling a spade; specifically, a large chisel-like implement for flensing whales. 3 A heavy piece of metal at the end of a gun-carriage trail which helps to keep the carriage in position when the gun recoils. —**to call a spade a spade** To call a thing by its right name; speak the plain, uncompromising truth. —*v.t.* **spad·ed, spad·ing** To dig or cut with a spade. [OE *spadu*] —**spade′ful** *n.* —**spad′er** *n.*

spade[2] (spād) *n.* 1 A figure, resembling a heart with a triangular handle, on a playing card. 2 A card so marked. 3 The suit of cards so marked: usually in the plural. [<Sp. *espada* a sword <L *spatha* <Gk. *spathē*]

spade·fish (spād′fish′) *n. pl.* **·fish** or **·fish·es** 1 A spiny-finned food fish (*Chaetodipterus faber*) of the Atlantic coast from Massachusetts to the West Indies. 2 The paddlefish.

spade·work (spād′wûrk′) *n.* 1 Work done with a spade. 2 Any preliminary work necessary to get a project under way.

spa·di·ceous (spā·dish′əs) *adj.* 1 Of or like a spadix. 2 Of a clear brown or bay color.

spa·dix (spā′diks) *n. pl.* **spa·di·ces** (spā·dī′sēz) *Bot.* A spike or head of flowers with a fleshy axis, usually enclosed within a spathe. [<Gk. *spadix* <*spaein* break]

spae (spā) *v.t. Scot.* To foretell; divine. —**spae′·man** (-mən), **spae′wife′** (-wīf′) *n.*

spa·ghet·ti (spə·get′ē) *n.* 1 A cordlike food paste, in size between macaroni and vermicelli. 2 Insulated cloth tubing through which wire is passed, as in a radio circuit. [<Ital., pl. dim. of *spago* a small cord]

spa·gyr·ic (spə·jir′ik) *adj.* Alchemical. Also **spa·gir′ic, spa·gyr′i·cal.** [<NL *spagyricus*; prob. coined by Paracelsus]

spa·hi (spä′hē) *n.* 1 Formerly, a Turkish corps of irregular cavalry; a member of such a corps. 2 One of a native Algerian cavalry corps in the French service. Also **spa′hee.** [<Turkish *sipāhi* <Persian *sipāh* an army. Cf. SEPOY.]

spail (spāl) *n. Scot.* A lath; chip or shaving. Also **spale.**

Spain (spān) A nominal monarchy in SW Europe; continental Spain alone, 189,626 square miles; including the Balearic and Canary Islands, 194,368 square miles; capital, Madrid: Spanish *España.*

spairge (spârj) *v.t. Scot.* To besprinkle; sparge.

spake (spāk) Archaic past tense of SPEAK.

Spa·la·to (spä′lä·tō) The Italian name for SPLIT.

spall (spôl) *v.t.* To break up; chip; prepare for sorting, as ore. —*v.i.* To chip at the edges, as a stone under pressure. —*n.* A chip, splinter, or flake, as from a stone. [ME *spalle.* ? Related to MLG *spalden* split.]

Spal·lan·za·ni (späl′län·dzä′nē), **Lazzaro,** 1729–99, Italian naturalist; disproved theory of spontaneous generation.

spal·la·tion (spa·lā′shən) *n.* 1 The act or process of reducing to fragments. 2 *Physics* The splitting of an atomic nucleus into numerous parts instead of the two or three characteristic of ordinary fission. [<SPALL + -ATION]

spal·peen (spal·pēn′, spal′pēn) *n.* A wandering harvester; hence, a good-for-nothing. [<Irish *spailpīn* laborer]

span[1] (span) *v.t.* **spanned, span·ning** 1 To measure, especially with the hand with the thumb and little finger extended. 2 To encircle or grasp with the hand, as in measuring. 3 To stretch across; extend over or from side to side of: This road *spans* the continent. 4 To provide with something that stretches across or extends over. [<*n.*] —*n.* 1 The extreme space over which the hand can be expanded: 9 inches, or 22.86 centimeters. 2 Any small interval or distance, in space or in time. 3 *Archit.* The space or distance between the supports of an arch, abutments of a bridge, etc. 4 That which spans. 5 *Aeron.* The maximum lateral distance from tip to tip of airplane wings. [OE *spann*]

span[2] (span) *v.* **spanned, span·ning** *v.t.* To bind; make fast; fetter. —*v.i.* To match in color and size: said of horses. —*n.* 1 A rope or chain used as a fastening on a ship. 2 A pair of matched horses or oxen. 3 In South Africa, a team of oxen or bullocks, of two or more yokes. [<MDu. *spannen* fasten, join, draw together]

span[3] (span) Archaic past tense of SPIN.

Span·dau (shpän′dou) A district of western Berlin, Germany: formerly an independent city.

span·drel (span′drəl) *n. Archit.* 1 The triangular space between the outer curve of an arch and the rectangular figure formed by the moldings or framework surrounding it. 2 The space between the shoulders of two adjoining arches. See illustration under ARCH. Also **span′dril.** [Dim. of AF *spaundre*, prob. <OF *espandre* expand]

spa·ne·mi·a (spə·nē′mē·ə) *n. Pathol.* Poverty of blood; anemia. Also **spa·nae′mi·a.** [<NL <Gk. *spanos* lacking + *haima* blood] —**spa·ne′mic** (-nē′mik, -nem′ik) *adj.*

spang (spang) *v.t. Brit. Dial.* To throw or bang down. —*v.i. Brit. Dial.* To spring. —*adv. U.S. Colloq.* Abruptly; straight: He ran *spang* into the wall.

span·gle (spang′gəl) *n.* 1 A small bit of brilliant tin or other metal foil, or other substances, used for decoration in dress, as in theatrical costume. 2 Any small sparkling object. —*v.* **·gled, ·gling** *v.t.* To adorn with or as with spangles; cause to glitter. —*v.i.* To sparkle as spangles; glitter. [Dim. of MDu. *spang* a clasp, brooch] —**span′gly** *adj.*

Span·iard (span′yərd) *n.* 1 A native or citizen of Spain. 2 A prickly bush of New Zealand (*Aciphylla colensoi*).

span·iel (span′yəl) *n.* 1 A small or medium-sized dog having large pendulous ears and long silky hair: used especially for hunting small game in the fields, retrieving water birds, etc. 2 One who follows like a dog; an obsequious follower. [<OF *espaignol* Spanish (dog)]

—**Blenheim spaniel** A variety of English toy spaniel having a white coat with rich chestnut or ruby-red markings: originally bred at Blenheim, England, from cocker spaniels sent to the Duke of Marlborough from China: formerly used for woodcock shooting, now usually a pet.

—**Clumber spaniel** A small, stout-bodied, short-legged spaniel having a straight silky coat usually white with lemon markings: named from Clumber, the estate of the second Duke of Newcastle, where they were bred.

—**cocker spaniel** The smallest of the sporting spaniels, of solid or various coloring, characterized by sturdy body and rather short legs: an excellent retriever, especially in thick covers and swamps, and named for its special skill in woodcock hunting.

—**English springer spaniel** A spaniel of moderate size, strongly built, with a long and broad skull, deep chest, and wavy or flat coat; usually liver and white or tan, black and white or tan, etc.: named for its characteristic method of flushing game.

—**field spaniel** A black or varicolored spaniel used for hunting small game, having a long, low body, short legs, and a larger, stronger appearance than the cocker.

—**Irish water spaniel** A sporting dog of a breed developed in Ireland, having a curly, waterproof, liver-colored coat: used especially as a duck retriever.

—**Japanese spaniel** A breed of toy dog, thought to have originated in China, squarely built with small-boned feathered legs, a proportionately large head, and a long, straight coat which may be black and white or red and white.

—**King Charles spaniel** An English toy spaniel, originating in the Far East at an unknown date, having a long, silky, black-and-tan coat, feathery ears and feet, rounded head, and a short, turned-up nose.

—**Sussex spaniel** A field spaniel bred in Sussex county, England, somewhat slow in speed, having a very keen nose, massive muscular body, heavy head, rather large ears, long back, short large-boned legs, and a thick coat of a rich, golden-liver color.

—**Welsh springer spaniel** A dark, rich red-and-white sporting spaniel of uncertain origin, found chiefly in Wales and the west of England: larger than the cocker, and an excellent watchdog.

Span·ish (span′ish) *adj.* Pertaining to Spain, the Spaniards, or their language. —*n.* 1 The Romance language of Spain, Spanish America, and the Philippine Islands. 2 The inhabitants of Spain collectively: with *the.*

Spanish America The parts of the western hemisphere in which Spanish is the common language: Mexico, the countries of Central America, except British Honduras, the countries of South America except Brazil and

Spanish-American

the Guianas, and most of the Caribbean islands: also *Hispanic America.*
Span·ish-A·mer·i·can (span'ish-ə-mer'ə-kən) *adj.* 1 Pertaining to the parts of America where Spanish is the vernacular tongue or is in common use. 2 Designating or pertaining to the war between the United States and Spain, 1898. — *n.* One of Spanish blood living in America, especially Central or South America; a citizen of a Spanish-American country.
Spanish-American War See table under WAR.
Spanish Armada See under ARMADA.
Spanish cedar Cedar (def. 4).
Spanish dagger Any of various species of yucca, with sword-shaped leaves. Also **Spanish bayonet.**
Spanish fly A bright-green blister beetle (*Lytta vesicatoria*) of the Mediterranean region, used in the preparation of the drug cantharidin. See illustration under INSECTS (beneficial).

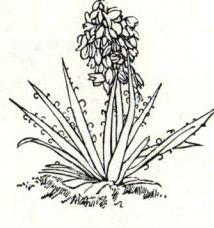
SPANISH DAGGER

Spanish Guinea A Spanish colony in western Africa, divided into the two districts of Fernando Pó and Río Muni (continental Guinea, together with Annobon, Corisco, Great Elobey, and Little Elobey); total area, 10,853 square miles; capital, Santa Isabel, on Fernando Pó.
Spanish mackerel 1 The chub. 2 The cero.
Spanish Main 1 The mainland of Spanish America, especially the coastal region of northern South America between Panama and the mouth of the Orinoco. 2 That part of the Caribbean comprising the course of Spanish merchantmen formerly sailing between the eastern and western hemispheres.
Spanish Morocco Formerly, a Spanish protectorate on the coastal strip of Morocco north of French Morocco; incorporated in Morocco, 1956; 10,808 square miles; former capital, Tetuán.
Spanish moss A long, pendent, epiphytic plant (*Tillandsia* or *Dendropogon usneoides*) that grows upon trees of the southern United States near the seacoast: not a true parasite: sometimes called *long moss, Florida moss.* Also **Spanish beard.**
Spanish needles 1 A smooth annual plant (*Bidens bipinnata*) of the composite family, with bipinnate leaves and spiny achenia. 2 The barbed, prickly fruit of this plant.
Spanish onion A large, fleshy variety of onion, usually mild flavored.
Spanish paprika A cultivated variety of paprika, widely used as a condiment.
Spanish Peaks Two peaks in the Sangre de Cristo Mountains of southern Colorado; 13,623 feet and 12,683 feet.
Spanish Sahara A subdivision of Spanish West Africa, comprising Río de Oro and Saguia el Hamra; 105,409 square miles.
Spanish West Africa The Spanish possessions in NW Africa: Spanish Sahara (Río de Oro and Saguia el Hamra), Ifni, and the Southern Protectorate of Morocco; 116,189 square miles.
spank (spangk) *v.t.* To slap or strike, especially on the buttocks with the open hand as a punishment. — *v.i.* To move briskly. — *n.* A smack on the buttocks; a spanking. [Imit.]
spank·er (spangk'ər) *n.* 1 One who or that which spanks. 2 *Naut.* A fore-and-aft sail extended by a boom and a gaff from the mizzenmast of a ship or bark. 3 Any person or thing uncommonly large or fine. 4 One who or that which proceeds rapidly.
spank·ing (spangk'ing) *adj.* 1 Moving or blowing rapidly; swift; dashing; lively; strong. 2 *Brit. Colloq.* Uncommonly large or fine. — *n.* A series of slaps on the buttocks; the act of administering such punishment.
span·less (span'lis) *adj.* That cannot be spanned.
span·ner (span'ər) *n.* 1 One who or that which spans. 2 *Brit.* A hand-tool used to turn nuts, bolts, etc.: a form of wrench. 3 A measuring worm. [def. 2 <G]
span-new (span'noo', -nyoo') *adj. Dial.* Really or freshly new. [<ON *spān-nȳr* < *spānn* chip + *nȳr* new]
span worm A measuring worm.
spar[1] (spär) *n.* 1 *Naut.* A round timber for extending a sail, as a mast, yard, or boom. 2 A similar heavy, round timber forming part of a derrick, crane, etc., or used for various other purposes. 3 *Aeron.* That part of an airplane wing which carries the ribs. 4 *Naut.* A spar buoy: see under BUOY. — *v.t.* **sparred, spar·ring** 1 To furnish with spars. 2 *Archaic* To fasten, as with a bolt. [<ON *sparri* a beam]
spar[2] (spär) *v.i.* **sparred, spar·ring** 1 To box, especially with care and adroitness. 2 To bandy words; wrangle. 3 To fight, as cocks, by striking with spurs. — *n.* The act or practice of boxing, as by pugilists; a boxing match: also **spar'ring.** [<OF *esparer* <Ital. *sparare* kick <L *parare* prepare]
spar[3] (spär) *n.* A vitreous, crystalline, easily cleavable, lustrous mineral. [<MDu. Akin to OE *spær* gypsum.]
Spar (spär) *n.* A member of the women's reserve of the United States Coast Guard: also SPAR. [<L *s(emper) par(atus)* always ready, the motto of the U. S. Coast Guard]
spar·a·ble (spar'ə-bəl) *n.* A species of small headless nail used by shoemakers in soling boots. [Alter. of *sparrow bill*; so called from resemblance in shape]
spar buoy *Naut.* See under BUOY.
spar deck *Naut.* The light upper deck of a vessel extending from bow to stern: including the quarter-deck and the forecastle; the deck on which extra spars are stowed.
spare (spâr) *v.* **spared, spar·ing** *v.t.* 1 To refrain from injuring, molesting, or killing; treat with mercy or lenience. 2 To free or relieve (someone) from (pain, expense, etc.): *Spare* us the sight. 3 To use frugally; refrain from using or exercising: *Spare* the rod and spoil the child. 4 To dispense or dispense with; do without: Can you *spare* a dime? — *v.i.* 5 To be frugal; live or act economically. 6 To be lenient or forgiving; show mercy. — *adj.* **spar·er, spar·est** 1 That can be spared or used at will; disposable; available. 2 Held in reserve; additional; extra; surplus. 3 Having little flesh; thin; lean. 4 Not lavish or abundant; scanty. 5 Economical; chary; stingy; parsimonious. See synonyms under MEAGER. — *n.* 1 That which has been saved or stored away; something unused. 2 A duplicate; an item kept as a substitute in case the original breaks down, as an automobile tire or a mechanical part. 3 In bowling, the act of overturning all the pins with the first two balls; also, the score thus made. [OE *sparian*] — **spare'ly** *adv.* — **spare'ness** *n.* — **spar'er** *n.*
spare·rib (spâr'rib') *n.* A piece of meat, especially pork, consisting of ribs somewhat closely trimmed.
sparge (spärj) *v.t.* & *v.i.* **sparged, sparg·ing** To scatter; sprinkle; shower. — *n.* A sprinkling. [<OF *espargier* <L *spargere* sprinkle]
sparg·er (spär'jər) *n.* 1 A sprinkler or sprinkling apparatus. 2 In brewing, a hot-water sprinkler for use in a mashing tub.
spar·ing (spâr'ing) *adj.* 1 Scanty; slight. 2 Frugal; stingy. 3 Merciful; forbearing. See synonyms under SCANTY. — *n.* The act of one who spares; frugality; parsimony. See synonyms under FRUGALITY. — **spar'ing·ly** *adv.* — **spar'ing·ness** *n.*
spark[1] (spärk) *n.* 1 An incandescent particle thrown off from a red-hot or burning body or struck from a flint. 2 Any glistening or brilliant point or transient luminous particle. 3 Anything that kindles or animates. 4 *Electr.* **a** The luminous effect of a disruptive electric discharge, or the discharge itself. **b** A small transient arc or an incandescent particle thrown off from such an arc. 5 A small diamond, or bit of diamond used as in cutting glass. 6 A small trace or indication. — *v.i.* 1 To give off sparks; sparkle; scintillate. 2 In an internal-combustion engine, to have the electric ignition operating. — *v.t.* 3 To bring

Spartacist

into action or being; activate or cause: The shooting *sparked* a revolution. [OE *spearca*]
spark[2] (spärk) *n.* 1 A man fond of gallantry. 2 A lover; suitor; gallant. — *v.t.* & *v.i.* To play the spark (to); woo; court. [Special use of SPARK[1]]
spark arrester 1 A sievelike device for catching sparks, as on a locomotive. 2 *Electr.* An apparatus to prevent injurious sparking at the opening of a circuit made and broken frequently.
spark coil *Electr.* An induction coil used with an internal-combustion engine, wireless telegraph equipment, etc., to secure sparking.
spark·er (spärk'ər) *n.* 1 One who or that which sparks. 2 An electrical spark arrester.
spark gap *Electr.* 1 An arrangement of two electrodes between which a disruptive electric charge may pass. 2 The space so covered.
spark generator *Electr.* Any device capable of generating a sufficiently high voltage to discharge across a spark gap.
spark·ish (spärk'ish) *adj.* 1 Jaunty; sprightly; airy; gay. 2 Showy; fine; well-dressed.
spark killer *Electr.* A device, usually a condenser, or condenser and resistance in series, for reducing harmful sparking at frequently interrupted points in a circuit. Also **spark suppressor.**
spar·kle (spär'kəl) *v.i.* **·kled, ·kling** 1 To give off flashes of light; scintillate; glitter. 2 To emit sparks. 3 To effervesce. 4 To be brilliant or vivacious: His words *sparkle* with wit. See synonyms under SHINE. — *n.* 1 A spark; gleam. See synonyms under LIGHT[1]. [Freq. of SPARK[1]]
spar·kler (spär'klər) *n.* 1 Something that sparkles. 2 A sparkling gem. 3 A thin, rod-like firework that emits sparks. 4 A person who shines with spirit or vivacity.
spar·kling (spär'kling) *adj.* Giving out sparks or flashes; glittering; figuratively, brilliant; vivacious. — **spar'kling·ly** *adv.* — **spar'kling·ness** *n.*
spark plug A device for igniting the charge in an internal-combustion engine by means of an electric current.
Sparks (spärks), **Jared,** 1789–1866, U. S. editor and historian.
spark transmitter *Telecom.* A radio transmitter which obtains its alternating current from the discharge of a condenser across a spark gap.
spar·ling (spär'ling) *n.* 1 A smelt, parr, or other young fish. 2 A young herring. [<OF *esperlinge* <Gmc.]
spar·oid (spâr'oid, spär'-) *adj.* Of or pertaining to a family (*Sparidae*) of spiny-finned marine fishes allied to the grunts and including the porgy, sheepshead, etc. — *n.* A sparoid fish; the sea bream. [<L *sparus* <Gk. *sparos* gilthead + -OID]
spar·row (spar'ō) *n.* 1 Any of various small, plainly colored, passerine birds (family *Fringillidae*) related to the finches, grosbeaks, and buntings; especially, the European **house sparrow** (*Passer domesticus*), known in the United States as the **English sparrow.** 2 Some other singing bird like or likened to the house sparrow, as the song sparrow. [OE *spearwa*]
sparrow bill A sparable.
spar·row-grass (spar'ō-gras', -gräs') *n. Dial.* Asparagus: a corruption. Also **spar'ry-grass'** (spar'ē-).
sparrow hawk 1 A small falconine bird that preys on sparrows, as the kestrel, or the **eastern sparrow hawk** (*Falco sparverius*). 2 A small European hawk (*Accipiter nisus*) that preys on other birds.
spar·ry (spär'ē) *adj.* **·ri·er, ·ri·est** Of, abounding in, or like spar.
sparse (spärs) *adj.* Scattered at considerable distances apart; thinly diffused; not dense. [<L *sparsus,* pp. of *spargere* sprinkle, scatter] — **sparse'ly** *adv.* — **sparse'ness, spar·si·ty** (spär'sə-tē) *n.*
Spar·ta (spär'tə) An ancient city in the Peloponnesus, southern Greece; capital of ancient Laconia: also *Lacedaemon.*
Spar·ta·cist (spär'tə-sist) *n.* A member of a party of extreme socialists in Germany (1918–19): name derived from the pseudonym *Spartacus* used by their leader, Karl Liebknecht. Also **Spar'ta·cide** (-sīd).

---

add, āce, câre, pälm; end, ēven; it, īce; odd, ōpen, ôrder; tōōk, pōōl; up, bûrn; ə = a in *above,* e in *sicken,* i in *clarity,* o in *melon,* u in *focus;* yōō = u in *fuse;* oi, oil; ou, pout; ch, check; g, go; ng, ring; th, thin; ᵺ, this; zh, vision. Foreign sounds à, œ, ü, kh, ṅ; and ♦: see page xx. < from; + plus; ? possibly.

Spar·ta·cus (spär′tə·kəs) Thracian leader of slaves in the gladiatorial war against Rome 73–71 B.C.

Spar·tan (spär′tən) *adj.* Pertaining to Sparta or the Spartans; heroically brave and enduring. —*n.* A native or citizen of Sparta; hence, one of exceptional valor and fortitude. —**Spar′tan·ism** *n.*

spar·te·ine (spär′ti·ēn, -tē·in) *n. Chem.* A colorless, oily, poisonous alkaloid, $C_{15}H_{26}N_2$, contained in the common broom: it resembles digitalis in action. [<Gk. *spartos* broom + -INE²]

spasm (spaz′əm) *n.* **1** Any sudden or convulsive action or effort, as of the body, mind, or nature, especially such a one as is abnormal or temporary. **2** *Pathol.* Any involuntary convulsive contraction of muscles: when manifested by alternate contractions and relaxations it is a **clonic spasm**; when persistent and steady, it is a **tonic spasm.** [<L *spasma, spasmus* <Gk. *spasmos* < *spaein* draw, pull]

spas·mod·ic (spaz·mod′ik) *adj.* **1** Of the nature of a spasm; convulsive. **2** Violent, or impulsive and transitory. Also **spas·mod′i·cal.** —**spas·mod′i·cal·ly** *adv.*

spas·mol·y·sis (spaz·mol′ə·sis) *n. Med.* The checking or relief of spasms. —**spas·mo·lyt·ic** (spaz′mə·lit′ik) *adj.*

spas·mo·phil·i·a (spaz′mə·fil′ē·ə) *n. Pathol.* A constitutional tendency to spasms and convulsions. —**spas′mo·phil′ic** *adj.*

spas·tic (spas′tik) *adj.* Of, pertaining to, or characterized by spasms; spasmodic; tetanic: *spastic* hemiplegia. —*n.* A person afflicted with spastic seizures. [<L *spasticus* <Gk. *spastikos* < *spaein* draw, pull] —**spas′ti·cal·ly** *adv.*

spat¹ (spat) Past tense and past participle of SPIT¹.

spat² (spat) *n.* **1** Spawn of shellfish; specifically, spawn of the oyster. **2** A young oyster, or young oysters collectively. —*v.i.* **spat·ted, spat·ting** To spawn, as oysters. [? Related to SPIT¹]

spat³ (spat) *n.* **1** A slight blow; slap. **2** A splash, as of rain; spatter. **3** A petty dispute. —*v.* **spat·ted, spat·ting** *v.i.* **1** To strike with a slight sound; slap. **2** To engage in a petty quarrel. —*v.t.* **3** To slap. [Prob. imit.]

spat⁴ (spat) *n.* A short gaiter worn over a shoe and fastened underneath with a strap: usually in the plural. [Short for SPATTERDASH]

spate (spāt) *n.* **1** A freshet; overflow. **2** A sudden, violent rainstorm; also, a waterspout. **3** A sudden or vigorous outpouring, as of words, feeling, etc. Also **spait.** [Origin uncertain]

spa·tha·ceous (spə·thā′shəs) *adj. Bot.* Bearing or of the nature of a spathe. Also **spa·thal** (spā′thəl).

spathe (spāth) *n. Bot.* A large bract or pair of bracts sheathing a flower cluster, as a spadix. [<L *spatha* <Gk. *spathē* broadsword] —**spa·those** (spā′thōs, spath′ōs) *adj.*

spath·ic (spath′ik) *adj. Mineral.* Of, pertaining to, or resembling spar. Also **spath·ose** (spath′ōs). [<G *spath* spar]

spa·tial (spā′shəl) *adj.* Pertaining to space; involving or having the nature of space. Also *spacial.* [<L *spatium* space] —**spa·ti·al·i·ty** (spā′shē·al′ə·tē) *n.* —**spa′tial·ly** *adv.*

spa·ti·o·tem·po·ral (spā′shē·ō·tem′pər·əl) *adj.* Of or pertaining to both space and time.

spat·ter (spat′ər) *v.t.* **1** To scatter in drops or splashes, as mud or paint. **2** To splash with such drops; bespatter. **3** To defame. —*v.i.* **4** To throw off drops or splashes; sputter. **5** To fall in a shower, as raindrops. —*n.* **1** The act of spattering, or the matter spattered; a splash. **2** A pattering noise, as of falling rain. [OE *spat-*, stem of *spatlian* spit out + -ER⁴]

spat·ter·dash (spat′ər·dash′) *n.* A legging reaching to the knee, worn as a protection from mud, especially when riding: used chiefly in the plural. —**spat′ter·dashed′** *adj.*

spat·ter·dock (spat′ər·dok′) *n.* The yellow pondlily (*Nuphar advena).*

spat·u·la (spach′ŏŏ·lə) *n.* **1** A knifelike instrument with a flat, flexible blade, used to spread plaster, cake icing, etc. **2** *Med.* An instrument used to press the tongue down or aside, as in examinations. [<L, dim. of *spatha.* See SPATHE.] —**spat′u·lar** *adj.*

spat·u·late (spach′ŏŏ·lit, -lāt) *adj.* **1** Shaped like a spatula. **2** *Bot.* Oblong, with an attenuated base, as many leaves.

spav·in (spav′in) *n.* A disease of the hock joint of horses, occurring either as an infusion of lymph within the joint (**blood spavin** or **bog spavin**) or as a bony deposit stiffening the joint (**bone spavin**). Also *Scot.* **spa·vie** (spā′vē, spav′ē). [<OF *espavain, esparvain*; ult. origin uncertain] —**spav′ined,** *Scot.* **spa·viet** (spā′vit, spav′it) *adj.*

spawn (spôn) *n.* **1** *Zool.* The eggs of fishes, amphibians, mollusks, etc., especially in masses. **2** Derisively, the offspring of any animal; also, outcome or results; product; yield. **3** The spat of the oyster. **4** Very small fish; fry. **5** *Bot.* The mycelium of mushrooms or other fungi. [< *v.*] —*v.i.* **1** To produce spawn; deposit eggs or roe. **2** To come forth as or like spawn. —*v.t.* **3** To produce (spawn). **4** To give rise to; originate. **5** To bring forth abundantly or in great quantity. **6** To plant with spawn or mycelium. [<AF *espaundere,* OF *espendre* <L *expandere.* Doublet of EXPAND.]

spay (spā) *v.t.* To remove the ovaries from (a female animal). [<AF *espeier,* OF *espeer* cut with a sword < *espee* a sword <L *spatha*]

speak (spēk) *v.* **spoke** (*Archaic* **spake**), **spo·ken** (*Archaic* **spoke**), **speak·ing** *v.i.* **1** To employ the vocal organs in ordinary speech; utter words. **2** To express or convey ideas, opinions, etc., in or as in speech: to *speak* about a matter; Actions *speak* louder than words. **3** To make a speech; deliver an address. **4** To converse. **5** To make a sound; also, to bark, as a dog. —*v.t.* **6** To express or make known in or as in speech. **7** To utter in speech: to *speak* words of love. **8** To use or be capable of using (a language) in conversation. **9** To speak to. **10** *Naut.* To hail and exchange communications with (a vessel) at sea. —**to speak daggers** To express hatred. —**to speak for 1** To speak in behalf of; represent officially. **2** To lay claim to; bespeak; engage. [OE *specan, spreccan*]

Synonyms: announce, articulate, converse, declaim, declare, deliver, dictate, enunciate, express, pronounce, say, talk, tell, utter. See TALK.

speak·a·ble (spē′kə·bəl) *adj.* That may, or may properly, be spoken.

speak–eas·y (spēk′ē′zē) *n. pl.* **–eas·ies** A saloon where liquor is sold contrary to law.

speak·er (spē′kər) *n.* **1** One who speaks; an orator. **2** The presiding officer in any one of various legislative bodies. **3** A volume of oratorical selections, for declamation. **4** A loudspeaker. —**speak′er·ship** *n.*

speak·ing (spē′king) *adj.* **1** Having the power of effective speech; uttering speech. **2** Expressive; vivid; telling; lifelike. —*n.* **1** The act of utterance; vocal expression. **2** Oratory; public declamation. —**speak′ing·ly** *adv.*

spean (spēn) *v.t. Brit. Dial.* To wean. [<MDu. *spene* a teat]

spear (spir) *n.* **1** A weapon consisting of a pointed head on a long shaft. **2** A similar instrument, barbed and usually forked, as for use in spearing fish; a fishgig. **3** A spearman. **4** A leaf or slender stalk, as of grass: sometimes called a *spire.* —*v.t.* **1** To pierce or capture with a spear. —*v.i.* **2** To pierce as a spear does. **3** To send forth spears or spires, as a plant. [OE *spere* spear] —**spear′er** *n.*

spear·fish (spir′fish′) *n. pl.* **·fish** or **·fish·es** A powerful marine fish (genus *Tetrapturus*) with a long, sword–shaped snout, related to the swordfish.

spear·head (spir′hed′) *n.* **1** The point of a spear or lance. **2** The military units which lead in a massed attack on enemy positions. —*v.t.* To be in the lead of (an attack, etc.).

spear·man (spir′mən) *n. pl.* **·men** (-mən) A man armed with a spear. Also **spears′man.**

spear·mint (spir′mint′) *n.* An aromatic herb (*Mentha spicata)* similar to peppermint.

spear side *Archaic* The male branch of a family: opposed to the female or *distaff* or *spindle side.* Also **spear half.**

spear·wort (spir′wûrt′) *n.* Any of several species of crowfoot having lance–shaped or linear leaves, especially, the **lesser spearwort** (*Ranunculus flammula).*

spe·cial (spesh′əl) *adj.* **1** Having some peculiar or distinguishing characteristic or characteristics; out of the ordinary; uncommon; particular. **2** Designed for or assigned to a specific purpose; limited or specific in range, aim, or purpose. **3** Of or pertaining to, constituting, or designating a species; specific; distinguishing; differential. **4** Unique; singular; exceptional. **5** Extra or additional, as a dividend. **6** Intimate; esteemed; beloved: a *special* favorite. See synonyms under PARTICULAR. —*n.* **1** A person or thing made, detailed for, or appropriated to a specific service or occasion, as a train, a newspaper edition, etc. **2** A featured dish or course in a restaurant or cafeteria. [<OF *especial* <L *specialis* < *species* kind, species] —**spe′cial·ly** *adv.*

special delivery *U.S.* Mail delivery by special courier in advance of regular delivery: a postal service obtained for an additional fee.

spe·cial·ism (spesh′əl·iz′əm) *n.* The confining of oneself to a particular line of study or work; a special field of work.

spe·cial·ist (spesh′əl·ist) *n.* A person devoted to some one line of study, occupation, or professional work. —**spe′cial·is′tic** *adj.*

spe·ci·al·i·ty (spesh′ē·al′ə·tē) *n. pl.* **·ties 1** A specific or individual characteristic; peculiarity. **2** Specialty (defs. 3, 4, 5). ◆ In British usage, this form is preferred instead of *specialty.*

spe·cial·i·za·tion (spesh′əl·ə·zā′shən, -ī·zā′-) *n.* **1** The act or process of specializing; also, the state of being or becoming specialized. **2** *Biol.* The development of an organ or part for a special function; differentiation.

spe·cial·ize (spesh′əl·īz) *v.* **·ized, ·iz·ing** *v.i.* **1** To concentrate on one particular activity or subject; engage in a specialty. **2** *Biol.* To take on a special form or forms by specialization or adaptation. —*v.t.* **3** To adapt for some special use or purpose; endow with a particular character. **4** *Biol.* To develop by specialization or adaptation, as an organ or part. **5** To endorse, as a check, to a particular payee. **6** To mention specifically; particularize. Also *Brit.* **spe′cial·ise.**

special pleading 1 *Law* **a** A pleading made with reference to some new or particular matter instead of the general issue. **b** The allegation of new or special matter in reply to the opposing party's averments, rather than an offer of a direct denial. **2** A presentation of the favorable aspects of an argument while avoiding or suppressing the unfavorable.

spe·cial·ty (spesh′əl·tē) *n. pl.* **·ties 1** The state of being special or of having peculiar characteristics. **2** An individual characteristic; peculiarity; distinguishing mark. **3** An occupation or study limited to one particular line. **4** An article dealt in exclusively or chiefly, or a manufactured product of peculiar character. **5** *Law* A sealed contract; deed. —**specialty of the house** A featured dish or course in a restaurant.

spe·ci·a·tion (spē′shē·ā′shən) *n. Biol.* The formation of a species by the action of evolutionary processes upon plant and animal organisms.

spe·cie (spē′shē) *n.* Coined money; coin. See synonyms under MONEY. —**in specie 1** In coin. **2** *Law* In kind; in the shape mentioned; in sort. [<L *(in) specie* (in) kind]

spe·cies (spē′shēz, -shiz; *Lat.* spē′shi·ēz) *n. pl.* **·cies 1** *Biol.* A category of animals or plants subordinate to a genus but above a breed, race, strain, or variety. The species name follows immediately after the name of the genus to which it belongs, and with it forms the scientific name of the individual plant or animal, as *Oreamnos americanus,* the Rocky Mountain goat. **2** A group of individuals or objects agreeing in some common attribute or attributes and designated by a common name. **3** A mental image considered as having the likeness of some object in nature. **4** A kind; sort; variety; form. **5** *Eccl.* **a** The visible form of bread or of wine retained by the eucharistic elements after consecration. **b** The consecrated elements of the Eucharist. **6** *Obs.* Specie; coin. [<L, form, kind. Doublet of SPICE.]

spec·i·fi·a·ble (spes′ə·fī′ə·bəl) *adj.* Such as can be specified.

spe·cif·ic (spi·sif′ik) *adj.* **1** Distinctly and plainly set forth; definite or determinate; particular; explicit. **2** Of, pertaining to, or distinguishing a species: a *specific* name of an animal. **3** Peculiar; special. **4** Having some distinct medicinal or pathological property;

specifically — **speech**

distinguishable or determinate: a *specific* medicine, a *specific* germ. 5 Having or designating a particular property, composition, ratio, or quantity serving to identify a given substance or phenomenon in relation to some arbitrary but constant standard of comparison: *specific* heat, *specific* volume, etc. 6 Denoting a customs duty chargeable upon imported merchandise by quantity, weight, or number, without regard to value: contrasted with ad valorem duty. Also *Rare* **spe·cif′i·cal**. — *n.* Anything specific or adapted to effect a specific result, as a medicine specially indicated to cure or prevent some particular disease. [<L *specificus* < *species* kind, class + *facere* make] — **spec·i·fic·i·ty** (spes′ə·fis′ə·tē) *n.*

spe·cif·i·cal·ly (spi·sif′ik·lē) *adv.* 1 In a specific manner; explicitly; particularly; definitely. 2 As to or in respect to species: *specifically* distinct. 3 In a particular sense or case.

spec·i·fi·ca·tion (spes′ə·fə·kā′shən) *n.* 1 The act of specifying. 2 A definite and complete statement, as in a contract; also, one detail in such a statement. 3 In patent law, the detailed statement of an inventor's scheme, setting forth the nature of the invention and the precise method of constructing and applying it. 4 A specific description of certain dimensions, types of material, etc., to be used in a construction or engineering project; also, any item in this description.

specific gravity *Physics* The ratio of the mass of a body to that of an equal volume of some standard substance, water in the case of solids and liquids, and air or hydrogen in the case of gases.

specific heat *Physics* The amount of heat required to raise the temperature of a given quantity of a substance one degree.

specific impulse The thrust in pounds produced in one second by the burning with its oxidizer of one pound of a specified fuel, as in a rocket motor or jet engine.

spec·i·fy (spes′ə·fī) *v.t.* **·fied**, **·fy·ing** 1 To mention specifically; state in full and explicit terms. 2 To embody in a specification. [<OF *specifier* <L *species* kind, species + *facere* make]

spec·i·men (spes′ə·mən) *n.* 1 One of a class of persons or things regarded as representative of the class; an example; sample. 2 *Biol.* A plant or an animal, entire or in part, prepared and kept as an example to illustrate a species or variety. 3 A sample for urinalysis. 4 *Colloq.* A person of pronounced or curious type; a character; a case: What a *specimen!* See synonyms under EXAMPLE, SAMPLE. [<L *specere* look at]

spe·ci·os·i·ty (spē′shē·os′ə·tē) *n. pl.* **·ties** 1 One who or that which is specious; a thing that appears just and plausible at first view but actually is not. 2 *Obs.* The state of being beautiful.

spe·cious (spē′shəs) *adj.* 1 Apparently good or right, but without merit; plausible: *specious* reasoning. 2 Pleasing or attractive in appearance, but deceptive; fair–seeming: a *specious* promise. 3 Beguiling, but lacking in sincerity: a *specious* hypocrite. 4 *Archaic* Showy; pleasing to the view. See synonyms under OSTENSIBLE. [<L *speciosus* fair] — **spe′cious·ly** *adv.* — **spe′cious·ness** *n.*

speck (spek) *n.* 1 A small spot; a little stain or discoloration. 2 Any very small thing; a particle. See synonyms under BLEMISH. — *v.t.* To mark with spots or specks; speckle. [OE *specca*]

speck·le (spek′əl) *v.t.* **·led**, **·ling** To mark with specks or speckles. — *n.* A diminutive spot; speck.

speck·led (spek′əld) *adj.* 1 Dotted with specks or spots. 2 Of motley appearance or mixed character.

specs (speks) *n. pl. Colloq.* Spectacles. Also **specks**.

spec·ta·cle (spek′tə·kəl) *n.* 1 That which is exhibited to public view; a grand display; pageant; parade; show. 2 An unwelcome or deplorable exhibition; a painful sight. 3 *pl.* A pair of eyeglasses, with hinged bows to secure them before the eyes: used to correct defects in vision, or to protect the eyes from glare. 4 *pl.* A marking on animals resembling a pair of spectacles. — **compound spectacles** *Optics* 1 Spectacles having supplementary colored glasses hinged to them for use when desired. 2 Supplementary lenses of greater power, similarly hinged. 3 Bifocals; trifocals. [<F <L *spectaculum* < *spectare*, freq. of *specere* see]

Synonyms: display, exhibition, pageant, parade, scene, show, sight. See SIGHT.

spec·ta·cled (spek′tə·kəld) *adj.* 1 Wearing spectacles. 2 Having markings resembling a pair of spectacles: the *spectacled* cobra.

spec·tac·u·lar (spek·tak′yə·lər) *adj.* Characterized by grand scenic display; exciting wonder by dramatic or unusual display. — *n.* 1 An imposing exhibition. 2 In television, a lavish dramatic or musical production of 90 minutes duration, especially designed for reproduction in color. 3 An elaborate, illuminated sign. — **spec·tac′u·lar·ly** *adv.* — **spec·tac′u·lar′i·ty** (-lar′ə·tē) *n.*

spec·ta·tor (spek′tā·tər, spek·tā′-) *n.* 1 One who beholds; an eyewitness; an onlooker. 2 One who is present at and views a show, game, spectacle, etc. [<L < *spectare* look at]

Synonyms: beholder, bystander, onlooker, observer, witness.

Spec·ta·tor (spek′tā·tər, spek·tā′-), **The** An English periodical, conducted by Joseph Addison and Richard Steele from March, 1711, to Dec., 1712; revived by Addison, June–Dec., 1714.

spec·ter (spek′tər) *n.* A phantom of the dead or of a disembodied spirit; especially, one of a grisly or horrible nature; ghost; apparition. Also **spec′tre**. [<F *spectre* <L *spectrum* vision]

Synonyms: apparition, phantom, ghost, shade, spirit.

specter of the Brocken See BROCKEN.

spec·tra (spek′trə) Plural of SPECTRUM.

spec·tral (spek′trəl) *adj.* 1 Pertaining to a specter; ghostly. 2 Pertaining to the spectrum or to spectra. See synonyms under GHASTLY. — **spec′tral·ly** *adv.* — **spec·tral·i·ty** (spek·tral′ə·tē) *n.*

spectro– combining form 1 Radiant energy, as exhibited in the spectrum: *spectroscope*. 2 Spectroscope; spectroscopic: *spectrobolometer*. [<SPECTRUM]

spec·tro·bo·lom·e·ter (spek′trō·bō·lom′ə·tər) *n.* A bolometer combined with a spectroscope for measuring the heat of different parts of the spectrum.

spec·tro·chem·is·try (spek′trō·kem′is·trē) *n.* The study of chemical phenomena and properties by means of spectrum analysis. — **spec′tro·chem′i·cal** *adj.*

spec·tro·gram (spek′trə·gram) *n.* The record of a spectrograph.

spec·tro·graph (spek′trə·graf, -gräf) *n. Physics* 1 An apparatus for photographing a spectrum or for forming a representation of the spectrum in any way. 2 A photograph of a spectrum. — **sound spectrograph** An electronic instrument designed to record the frequencies of speech sounds as measured in cycles per second, and the amplitude at any given frequency: used in acoustic phonetics.

spec·tro·he·li·o·gram (spek′trō·hē′lē·ə·gram′) *n.* A photograph made by the spectroheliograph.

spec·tro·he·li·o·graph (spek′trō·hē′lē·ə·graf′, -gräf′) *n.* An instrument for photographing the sun with its prominences by means of monochromatic light.

spec·trom·e·ter (spek·trom′ə·tər) *n.* 1 An instrument by means of which the angular deviation of a ray of light produced by a prism or by a refraction grating can be determined, or a wavelength of a ray of light can be accurately measured. 2 A spectroscope provided with such an instrument. — **spec·tro·met·ric** (spek′trō·met′rik) *adj.*

spec·tro·pho·tom·e·ter (spek′trō·fō·tom′ə·tər) *n.* An instrument for determining the relative intensity of two spectra or of the corresponding bands of color in two spectra.

spec·tro·ra·di·om·e·ter (spek′trō·rā′dē·om′ə·tər) *n.* A form of spectrometer for determining the distribution of the intensity of any type of radiation, especially in the infrared region of the spectrum. — **spec′tro·ra′di·om′e·try** *n.*

spec·tro·scope (spek′trə·skōp) *n.* An optical instrument for forming and analyzing spectra emitted by bodies or substances. — **spec′tro·scop′ic** (-skop′ik) or **·i·cal** *adj.* — **spec′tro·scop′i·cal·ly** *adv.*

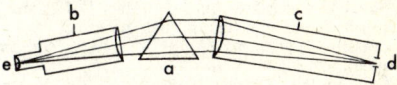

PRINCIPLE OF SIMPLE SPECTROSCOPE
a. Prism. *b.* Telescope for viewing prism through eyepiece *(e)*. *c.* Collimator with slit *(d)*.

spec·tros·co·py (spek·tros′kə·pē) *n.* 1 The branch of physical science treating of the phenomena observed with the spectroscope. 2 The art of using the spectroscope. — **spec·tros′co·pist** *n.*

spec·trum (spek′trəm) *n. pl.* **·tra** (-trə) 1 The continuously varying band of color observed when a beam of white light is passed through a prism which separates each component of the light according to frequencies ranging from low for red to high for violet: also **visible spectrum, chromatic spectrum.** 2 An image formed by radiant energy directed through a spectroscope and brought to a focus in which each wavelength corresponds to a specific band or line in a progressive series characteristic of the emitting source. 3 An after–image. [<L, a vision]

spectrum analysis The investigation or qualitative analysis of bodies or substances by means of their spectra; spectroscopy.

spec·u·la (spek′yə·lə) Plural of SPECULUM.

spec·u·lar (spek′yə·lər) *adj.* 1 Pertaining to or assisted by a speculum or a mirror; reflecting. 2 *Obs.* Affording a view; aiding vision. [<L *specularis* < *speculum* mirror]

specular iron A lustrous, crystalline variety of hematite.

specular pig iron Spiegeleisen.

spec·u·late (spek′yə·lāt) *v.i.* **·lat·ed**, **·lat·ing** 1 To form conjectures regarding anything without experiment; theorize; conjecture. 2 To make an investment involving a risk, but with hope of gain. [<L *speculatus*, pp. of *speculari* look at, examine < *specere* see]

spec·u·la·tion (spek′yə·lā′shən) *n.* 1 The act of theorizing or conjecturing; speculating. 2 A theory or conjecture. 3 A conclusion reached by or based upon conjecture. 4 An investment involving risk with hope of large profit. 5 The act of engaging in risky business transactions that offer a possibility of large profit. 6 *Archaic* Vision; observation; intuition. See synonyms under HYPOTHESIS, THOUGHT[1].

spec·u·la·tive (spek′yə·lā′tiv, -lə·tiv) *adj.* 1 Of, pertaining to, engaged in, or given to speculation: opposed to *experimental*. 2 Strictly theoretical or purely scientific: opposed to *practical*. 3 Engaging in or involving financial speculation. 4 *Archaic* Pertaining to vision or observation; affording a good view. 5 *Archaic* Prying; observing. — **spec′u·la′tive·ly** *adv.* — **spec′u·la′tive·ness** *n.*

spec·u·la·tor (spek′yə·lā′tər) *n.* One who speculates, in any sense. — **spec′u·la·to′ry** (-lə·tôr′ē, -tō′rē) *adj.*

spec·u·lum (spek′yə·ləm) *n. pl.* **·la** (-lə) or **·lums** 1 A mirror of polished metal or of glass coated with a metal film used for telescope reflectors and other optical instruments. 2 *Med.* An instrument that dilates a passage of the body for examination. 3 *Ornithol.* A specially colored, typically iridescent area on the wings of certain birds, as ducks. [<L, a mirror < *specere* see]

sped (sped) Alternative past tense and past participle of SPEED.

Spee (shpā), **Count Maximilian von,** 1861–1914. German rear admiral in World War I.

speech (spēch) *n.* 1 The faculty of expressing thought and emotion by spoken words; the power of speaking. 2 The act of speaking, involving the production of meaningful combinations of distinctive speech sounds. 3 That which is spoken; conversation; talk; a saying or remark. 4 A public address; a discourse. 5 A characteristic manner of speaking: His *speech* is loud and unpleasant. **6** A particular language, idiom, or dialect: American *speech*. **7** Any audible or visible method of communication, including cries, gestures, and sign language. **8** The study of oral communication, including the physiology of articulation, the nature of speech sounds, and the techniques of effective expression. [OE *spec, sprec* < *specan, sprecan* speak]

Synonyms: address, discourse, discussion, disquisition, dissertation, eloquence, harangue, oration, oratory, sermon. *Speech* is the general word for utterance of thought in language. A *speech* is the simplest mode of delivering one's sentiments; an *oration* is an elaborate and prepared *speech*; a *harangue* is a vehement appeal to passion, or a *speech* that has something disputatious and combative in it. A *discourse* is a set *speech* on a definite subject intended to convey instruction. See LANGUAGE. *Antonyms*: hush, silence, stillness.

speech clinic A place where speech disorders are corrected by training and re-education.

speech community All the speakers of a given language or dialect in both contiguous and geographically distributed areas.

speech defect The manifestation or end product of a speech disorder.

speech disorder Disorganization or impairment of speech caused either by physical defect or by mental disorder, such as aphasia, stuttering, etc.

speech·i·fy (spē'chə·fī) *v.i.* **·fied, ·fy·ing** To make speeches: often used derisively. — **speech'i·fi'er** *n.*

speech·less (spēch'lis) *adj.* **1** Unable to speak or temporarily deprived of speech because of physical weakness or strong emotion, etc.: *speechless* with rage. **2** Mute; dumb. **3** Silent; reticent. **4** *Archaic* Unspoken in words: the *speechless* message in her eyes. **5** Unaccompanied by speech: *speechless* joy. **6** *Archaic* Inexpressible. — **speech'less·ly** *adv.* — **speech'less·ness** *n.*

speech·mak·er (spēch'mā'kər) *n.* One who delivers a speech or speeches. — **speech'mak'·ing** *n.*

speech sound An articulation which functions in oral communication.

speed (spēd) *n.* **1** The act or state of moving or progressing swiftly; rapidity of motion; celerity; swiftness. **2** *Physics* **a** Rate of motion, especially as considered without reference to direction: a scalar quantity distinguished from *velocity*. **b** Rate of performance, as shown by the ratio of work done to time spent. **3** *Mech.* A transmission gear in a motor vehicle. **4** *Phot.* In a camera lens, the minimum time required for an effective exposure under given conditions, expressed as the ratio of focal length to effective aperture. **5** *Archaic* Good luck; success; prosperity. — *v.* **sped** or **speed·ed, speed·ing** *v.i.* **1** To move or go with speed. **2** *Obs.* To prosper. **3** *Obs.* To fare in a specified manner. — *v.t.* **4** To promote the forward progress of; cause to move or go with speed. **5** To promote the success of. **6** To wish Godspeed to: *Speed* the parting guest. — **to speed up** To accelerate in speed or action. See synonyms under FLY[1]. — *adj.* Having, pertaining to, characterized by, regulating, or indicating speed: used chiefly in compounds:

speed-cone	**speed-lathe**	**speed-test**
speed-gage	**speed-pulley**	**speed-trap**
speed-gear	**speed-recorder**	

[OE *spēd* power]

speed·boat (spēd'bōt') *n.* A motorboat capable of high speed.

speed·er (spē'dər) *n.* Someone or something that speeds; specifically, a motorist who drives at a speed exceeding a safe or legally specified limit.

speed indicator **1** An instrument showing the rotation speed of a machine or part of a machine. **2** A speedometer.

speed·ing (spē'ding) *adj.* Moving with speed. — *n.* Travel at high speed; especially, by motor vehicles, travel at an unsafe or reckless speed or above a specified speed limit.

speed limit A legally set maximum speed at which vehicles may travel on certain stretches of roads or through specified districts.

speed·om·e·ter (spi·dom'ə·tər) *n.* A device for indicating the speed of a vehicle or the distance traveled.

speed·ster (spēd'stər) *n.* **1** A speeder. **2** An automobile, usually having two seats, designed for speed.

speed-up (spēd'up') *n.* An acceleration in work, output, movement, etc.

speed·way (spēd'wā') *n.* A specially reserved or prepared road for vehicles traveling at high speed.

speed·well (spēd'wel) *n.* One of various low herbs (genus *Veronica*) of the figwort family, bearing blue or white flowers, especially the common speedwell (*V. arvensis*) and the **germander speedwell** (*V. chamaedrys*), with bright-blue flowers: also called *birdseye*.

speed·y (spē'dē) *adj.* **speed·i·er, speed·i·est** **1** Characterized by speed. **2** Without delay. See synonyms under NIMBLE, SWIFT[1]. — **speed'·i·ly** *adv.* — **speed'i·ness** *n.*

speed·y-cut (spē'dē·kut') *n.* An injury on the side of the knee or carpus of a horse caused by a blow from the shoe of the foot of the opposite leg when trotting or moving at any other rapid gait.

speel (spēl) *v.t.* & *v.i.* *Scot.* To climb.

speer (spir) *v.t.* & *v.i.* *Scot.* To inquire; ask: also spelled *spier*. — **to speer at** To question.

speer·ing (spir'ing) *n.* *Scot.* Inquiry; news; information.

speiss (spīs) *n.* An impure mixture consisting of the arsenides of certain metals, as copper, iron, and nickel, that concentrate in smelting certain ores. Also *Ger.* **spei·se** (shpī'zə). [<G *speise* amalgam]

spe·le·an (spi·lē'ən) *adj.* **1** Dwelling in a cave or caves. **2** Of or pertaining to a cave or caverns. Also **spe·lae'an.** [<L *spelaeum* <Gk. *spēlaion* a cave]

spe·le·ol·o·gy (spē'lē·ol'ə·jē) *n.* **1** The scientific study of caves in their physical, geological, and biological aspects. **2** The exploration of caves as a sport or profession. [<L *spelaeum* a cave + -LOGY] — **spe'le·o·log'i·cal** (-ə·loj'i·kəl) *adj.* — **spe'le·ol'o·gist** *n.*

spell[1] (spel) *v.* **spelled** or **spelt, spell·ing** *v.t.* **1** To pronounce or write the letters of (a word); especially, to do so correctly. **2** To form or be the letters of: C-a-t *spells* cat; hence, to compose; make up. **3** To read with difficulty; hence, to puzzle out and learn: sometimes with *over* or *out*. **4** To signify; mean: Extravagance *spells* disaster. — *v.i.* **5** To form words out of letters, especially correctly. [<OF *espeler* <Gmc. Akin to SPELL[2].]

spell[2] (spel) *n.* A formula used as a charm; incantation; charm; hence, fascination. — *v.t.* **spelled, spell·ing** To cast a spell upon; fascinate; bewitch. [OE, story, statement. Akin to SPELL[1].]

spell[3] (spel) *n.* **1** A period of time, usually of short length. **2** *Colloq.* A continuous period characterized by a certain type of weather. **3** *Colloq.* A short distance. **4** *Colloq.* A fit of illness, debility, etc. **5** A turn of duty in relief of another. **6** A period of work or employment. **7** *Austral.* A period of relaxation; rest. — *v.t.* **1** To relieve temporarily from some work or duty. **2** *Austral.* To give a rest to, as a horse. — *v.i.* **3** To take a rest. [OE *gespelia* a substitute, one who spells another]

spell·bind (spel'bīnd') *v.t.* **·bound, ·bind·ing** To bind or enthral, as if by a spell.

spell·bind·er (spel'bīn'dər) *n.* One who casts a spell over others; specifically, a political orator.

spell·bound (spel'bound') *adj.* Bound as by a spell; fascinated.

spell·er (spel'ər) *n.* **1** One who spells. **2** A spelling book.

spell·ing (spel'ing) *n.* **1** The act of one who spells. **2** The art of correct spelling; orthography. **3** The way in which a word is spelled.

spelling bee A gathering at which contestants engage in spelling words, those who spell wrongly usually being retired until only one remains.

spelling book A book of exercises for training students to spell.

Spell·man (spel'mən), **Francis Joseph,** born 1889, U. S. cardinal; archibishop of New York.

spelt[1] (spelt) Alternative past tense and past participle of SPELL[1].

spelt[2] (spelt) *n.* A species of wheat (*Triticum spelta*) or any of its winter or spring varieties. [OE]

spel·ter (spel'tər) *n.* Zinc: a commercial term. — **brazing spelter** See BRAZING SOLDER. [Var. of PEWTER]

spe·lun·ker (spē·lung'kər) *n.* An enthusiast in the exploration and study of caves; a speleologist. [<L *spelunca* a cave]

Spe·mann (shpā'män), **Hans,** 1869–1941, German zoologist.

spence (spens) *n.* *Brit. Dial.* **1** A pantry or larder. **2** The parlor of a cottage. [<OF *despense* <L *dispendere* DISPENSE]

spen·cer[1] (spen'sər) *n.* A trysail.

spen·cer[2] (spen'sər) *n.* **1** A man's short jacket of the early 19th century. **2** A similar outer garment for women, usually tight-fitting and often knitted or fur-trimmed. [after 2nd Earl *Spencer*, 1758–1834, English nobleman]

Spen·cer (spen'sər), **Herbert,** 1820–1903, English philosopher.

Spen·cer Gulf (spen'sər) An inlet of the Indian Ocean in South Australia; 200 miles long; 80 miles wide.

Spen·ce·ri·an (spen·sir'ē·ən) *adj.* **1** Pertaining to Herbert Spencer, or to his doctrine. **2** Pertaining to a system of freehand penmanship devised by P. R. Spencer about 1855. — *n.* A follower of Herbert Spencer and his system.

Spen·cer·ism (spen'sə·riz'əm) *n.* The doctrine of Herbert Spencer that the universe has evolved through mechanical forces from relative simplicity to relative complexity; synthetic philosophy. Also **Spen·ce·ri·an·ism** (spen·sir'ē·ən·iz'əm).

spend (spend) *v.* **spent, spend·ing** *v.t.* **1** To pay out or disburse (money). **2** To expend by degrees; use up. **3** To apply or devote, as thought or effort, to some activity, purpose, etc. **4** To pass: to *spend* one's life in jail. **5** To lose: now chiefly in the nautical phrase **to spend a mast.** **6** To emit, as a milt or spawn. — *v.i.* **7** To pay out or disburse money, etc. **8** *Obs.* To be wasted or exhausted. See synonyms under SQUANDER. [OE *aspendan* <L *expendere* EXPEND] — **spend'er** *n.*

Spen·der (spen'dər), **Stephen,** born 1909, English poet and critic.

spend·thrift (spend'thrift') *n.* One who is wastefully lavish of money: also **spend'er.** — *adj.* Excessively lavish; wasteful; prodigal.

Speng·ler (speng'glər, *Ger.* shpeng'lər), **Oswald,** 1880–1936, German philosopher of history.

Spen·ser (spen'sər), **Edmund,** 1552–99, English poet.

Spen·se·ri·an (spen·sir'ē·ən) *adj.* Of or pertaining to Edmund Spenser or to his style.

Spenserian sonnet See under SONNET.

Spenserian stanza A nine-line stanza consisting of eight lines of ten syllables and one of twelve syllables and riming *ababbcbcc*: used by Edmund Spenser in *The Faerie Queene*.

spent (spent) Past tense and past participle of SPEND. — *adj.* **1** Worn out or exhausted. **2** Deprived of force: a *spent* bullet or cannon ball.

Sper·lon·ga (sper·lông'gä) A village and port of south central Italy; site of a cave containing numerous ancient statues.

sperm[1] (spûrm) *n.* **1** The male fertilizing fluid; semen. **2** A male reproductive cell; spermatozoon. [<Gk. *sperma* a seed < *speirein* sow]

sperm[2] (spûrm) *n.* **1** A sperm whale. **2** Spermaceti. **3** Sperm oil. [Short for SPERMACETI]

-sperm *combining form Bot.* A seed (of a specified kind): *gymnosperm*. [<Gk. *sperma, spermatos* a seed]

sper·ma·ce·ti (spûr'mə·sē'tē, -set'ē) *n.* A white, waxy substance separated from the oil contained in the head of the sperm whale: used for making candles, ointments, etc. [<F <L *sperma ceti* seed of a whale]

sper·ma·ry (spûr'mər·ē) *n. pl* **·ries** The sperm-generating gland of the male; testis.

sper·ma·the·ca (spûr'mə·thē'kə) *n.* *pl.* **·cae** (-sē) *Zool.* A receptacle for receiving and retaining spermatozoa in the females of many invertebrates, as insects, worms, and mollusks. [<L *sperma* a seed + THECA] — **sper'·ma·the'cal** (-thē'kəl) *adj.*

sper·mat·ic (spûr·mat'ik) *adj.* Of or pertaining to sperm or a spermary.

spermatic cord *Anat.* The cord, made up of the spermatic duct and its accompanying vessels and nerves, that passes from the testis through the inguinal canal into the abdominal cavity.

spermatic fluid *Physiol.* Semen.

spermatic sac *Anat.* The scrotum.
sper·ma·tid (spûr′mə-tid) *n. Biol.* A cell resulting from the division of the secondary spermatocytes, and developing into a spermatozoon.
sper·ma·ti·um (spûr-mā′shē-əm) *n. pl.* **·ti·a** (-shē-ə) *Bot.* 1 A minute spore in certain lichens and fungi: formerly regarded as a nonmotile male gamete. 2 A non-motile gamete which, in the red algae, unites with the carpogonium. [< NL < Gk. *spermation*, dim. of *sperma* a seed]
spermato- *combining form* 1 Seed; pertaining to seeds: *spermatophyte*. 2 Spermatozoa; of or related to spermatozoa: *spermatophore*. Also spelled *spermo-*. Also, before vowels, **spermat-**. [< Gk. *sperma, spermatos* a seed]
sper·ma·to·cyte (spûr′mə-tə-sīt′) *n. Biol.* A primary cell from which spermatozoa are developed through primary and secondary divisions, resulting in the spermatids.
sper·ma·to·gen·e·sis (spûr′mə-tə-jen′ə-sis) *n. Biol.* The development of spermatozoa. — **sper·ma·to·ge·net·ic** (spûr′mə-tō-jə-net′ik) *adj.*
sper·ma·to·go·ni·um (spûr′mə-tə-gō′nē-əm) *n. pl.* **·ni·a** (-nē-ə) *Biol.* One of the cells of the seminal tubules that produce the spermatocytes. — **sper′ma·to·go′ni·al** *adj.*
sper·ma·toid (spûr′mə-toid) *adj.* Resembling sperm.
sper·ma·to·phore (spûr′mə-tə-fôr′, -fōr′) *n. Zool.* A capsule or case containing spermatozoa, as in many mollusks, worms, and other invertebrates. — **sper′ma·toph′o·ral** (-tof′ər-əl) *adj.*
sper·ma·to·phyte (spûr′mə-tə-fīt′) *n.* Any plant of a phylum or division (*Spermatophyta*) of the most highly developed plants; a flowering and seed-bearing plant. — **sper′ma·to·phyt′ic** (-fit′ik) *adj.*
sper·ma·tor·rhe·a (spûr′mə-tə-rē′ə) *n. Pathol.* Excessive or frequent seminal discharge without sexual excitement. Also **sper′ma·tor·rhoe′a.**
sper·ma·to·zo·id (spûr′mə-tə-zō′id) *adj.* Resembling a spermatozoon. — *n.* 1 A spermatozoon. 2 *Bot.* A motile male germ cell in plants. Also **sper′ma·to·zo′oid** (-zō′oid).
sper·ma·to·zo·on (spûr′mə-tə-zō′on) *n. pl.* **·zo·a** (-zō′ə) *Biol.* The male fertilizing element of an animal, usually in the form of a nucleated cell with a long flagellate process or tail by which it swims actively about. [< SPERMATO- + Gk. *zōion* an animal] — **sper′ma·to·zo′al, sper′ma·to·zo′ic** *adj.*
sper·mic (spûr′mik) *adj.* Of or pertaining to sperm or semen; spermatic.
sper·mine (spûr′mēn, -min) *n. Biochem.* A colorless, crystalline, strongly basic compound, $C_{10}H_{26}N_4$, salts of which are contained in semen and other animal tissues. [< SPERM + -INE²]
sperm·ism (spûr′miz·əm) *n.* The old theory that the spermatozoon alone is responsible for the development of the future animal.
sperm·ist (spûr′mist) *n.* A believer in spermism.
spermo- See SPERMATO-.
sper·mo·go·ni·um (spûr′mə-gō′nē-əm) *n. pl.* **·ni·a** (-nē-ə) *Bot.* In fungi, a cup- or flask-shaped receptacle, bearing a great number of spermatia.
sperm oil Oil obtained from the head and blubber cavities of the sperm whale.
sperm·o·phile (spûr′mə-fīl, -fil) *n.* A squirrel-like burrowing rodent (*Citellus* and related genera), as the striped gopher or the suslik.
-spermous *combining form* Having (a specified number or kind of) seeds; -seeded: *polyspermous*. Also **-spermal, -spermic.** [< -SPERM + -OUS]

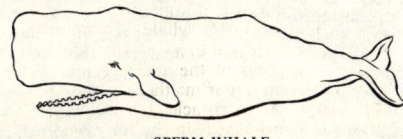

SPERM WHALE
(Up to 80 feet in length)

sperm whale A large, toothed whale (*Physeter catodon*) of warm seas, having a huge truncate head containing a reservoir of sperm oil; the cachalot.
Sper·ry (sper′ē), **Elmer Ambrose,** 1860–1930. U. S. engineer and inventor.
sper·ry·lite (sper′i-līt) *n.* A metallic tin-white platinum arsenide, $PtAs_2$, crystallizing in the isometric system. [after F. L. *Sperry*, Canadian mineralogist + -LITE]
spet (spet) *Obs. v.t. & v.i.* To spit. — *n.* Spittle. [OE *spǣtan*]
spetch·es (spech′iz) *n. pl.* The offal of skins, hides, etc., used for making glue. [Origin uncertain]
spew (spyōō) *v.t. & v.i.* To vomit; throw up. — *n.* That which is spewed; vomit. [OE *spīwan*]
Spey·er (shpī′ər) A city in SE Rhineland-Palatinate, West Germany; former capital of Rhine Palatinate: English *Spires*.
Spe·zia (spe′tsyä) La Spezia, a port on the Gulf of Spezia, an inlet of the Gulf of Genoa in NW Italy.
sphac·e·late (sfas′ə-lāt) *v.i.* **·lat·ed, ·lat·ing** *Pathol.* To become gangrenous; decay; die. [< Gk. *sphakelos* gangrene] — **sphac′e·la′tion** *n.*
sphag·num (sfag′nəm) *n.* Any of a genus (*Sphagnum*) of whitish-gray mosses constituting the family *Sphagnaceae*, the bog or peat mosses: used as packing and in surgical dressings. [< Gk. *sphagnos*, kind of moss] — **sphag′nous** *adj.*
sphal·er·ite (sfal′ər-īt) *n.* A resinous to adamantine native zinc sulfide, ZnS, crystallizing in the isometric system; zinc blende. [< Gk. *sphaleros* deceptive + -ITE¹]
sphene (sfēn) *n.* An adamantine, variously colored silicate of calcium and titanium, crystallizing in the monoclinic system: also called *titanite*. [< F *sphène* < Gk. *sphēn* wedge]
sphen·ic (sfē′nik) *adj.* Wedge-shaped.
sphenic number See under NUMBER.
spheno- *combining form* 1 Wedge-shaped: *sphenogram*. 2 *Med.* Pertaining to the sphenoid bone. Also, before vowels, **sphen-**. [< Gk. *sphēn, sphēnos* a wedge]
sphe·no·don (sfē′nə-don) *n.* A lizardlike reptile (*Sphenodon punctatum*), the sole surviving representative of the order *Rhynchocephalia*; the hatteria or tuatara of New Zealand. [< NL < Gk. *sphēn, sphēnos* a wedge + *odous, odontos* a tooth]
sphe·no·gram (sfē′nə-gram) *n.* A cuneiform character or symbol.
sphe·noid (sfē′noid) *n.* 1 *Mineral.* In the tetragonal and orthorhombic crystal systems, a hemihedral form enclosed by four faces, each of which cuts all three axes. 2 The sphenoid bone. — *adj.* Wedge-shaped: the *sphenoid* bone. [< SPHEN(O)- + -OID] — **sphe·noi′dal** (sfi-noid′l) *adj.*
sphenoid bone *Anat.* An irregular, compound bone situated at the base of the skull.
sphe·ra·di·an (sfi-rā′dē-ən) *n.* A steradian.
spher·al (sfir′əl) *adj.* 1 Shaped like a sphere; spherical; rounded; symmetrical. 2 Of or pertaining to a sphere. 3 Belonging to or relating to the celestial sphere; harmonious.
sphere (sfir) *n.* 1 The surface described by a semicircle making one complete rotation on its diameter as a fixed axis; a globular figure enclosed by a surface, every point of which is equidistant from a point within called the center. 2 An approximately globular body; a globe; ball; orb. 3 One of the heavenly bodies; a planet, sun, or star. 4 The apparent outer dome of the heavens on which the heavenly bodies appear to lie. 5 In old astronomy, one of the concentric and transparent globes believed to revolve about the earth and carry the various heavenly bodies, their movement supposedly producing mysteriously beautiful music. 6 Compass or field of activity, endeavor, influence, etc.; range; scope; province. 7 Social rank or position. — *v.t.* **sphered, spher·ing** 1 To place in or as in a sphere; encircle; encompass. 2 To set among the celestial spheres. 3 To make spherical. [< OF *espere* < L *sphaera* < Gk. *sphaira* a ball]
-sphere *combining form* 1 Denoting an enveloping spherical mass: *hydrosphere, atmosphere*. 2 A sphere-shaped body: *oosphere*. 3 Denoting a spherical form: *planisphere*. [< Gk. *sphaira* a ball, sphere]
sphere of influence A country or region, usually backward politically or economically undeveloped, in which a state or states claim and are allowed exclusive rights to colonize, exploit natural and economic resources, or eventually annex.
spher·ic (sfer′ik) *adj.* Pertaining to a sphere or spheres; spherical.
spher·i·cal (sfer′i-kəl) *adj.* 1 Shaped like a sphere; globular. 2 Pertaining to a sphere or spheres. 3 Pertaining to the heavenly bodies; celestial. See synonyms under ORBICULATE, ROUND¹. — **spher′i·cal·ly** *adv.* — **spher′i·cal·ness** *n.*
spherical aberration See under ABERRATION.
spherical angle See under ANGLE.
spherical coordinate system *Math.* A three-dimensional system for indicating the shape of a solid by means of a sphere with the pole at the center. A point is located in terms of its distance along its radius vector from the pole, and in terms of two angles—the colatitude, or angle the radius vector forms with the vertical or polar axis of the sphere; and the longitude, or angle the radius vector makes with a fixed, vertical plane or initial meridian axis.
spherical sailing Navigation in which calculations are based upon a consideration of the spherical or spheroidal shape of the earth: distinguished from *plane sailing*.
spherical triangle *Math.* A spherical polygon the three sides of which are arcs of great circles of a sphere.
spherical trigonometry *Math.* The study of spherical triangles.
sphe·ric·i·ty (sfi-ris′ə-tē) *n. pl.* **·ties** The state of being a sphere; spherical form; roundness.
spher·ics (sfer′iks) *n.* 1 The geometry and trigonometry of figures on the surface of a sphere. 2 Atmospherics.
sphe·roid (sfir′oid) *n. Geom.* A body having nearly the form of a sphere; an ellipsoid. — **sphe·roi′dal** (sfi-roid′l), **sphe·roi′dic** or **·di·cal** *adj.* — **sphe·roi′dal·ly** *adv.*
sphe·roi·dic·i·ty (sfir′oi-dis′ə-tē) *n.* The state or character of being a spheroid. Also **sphe·roi·di·ty** (sfi-roi′də-tē).
sphe·rom·e·ter (sfi-rom′ə-tər) *n.* An instrument for measuring curvature or radii of spherical and other curved surfaces. [< SPHERE + -(O)METER]
spher·ule (sfer′ōōl) *n.* A small or minute sphere; globule. — **spher·u·lar** (sfer′ōō-lər) *adj.*
spher·u·lite (sfer′ōō-līt) *n.* A radiating spherical group of minute acicular crystals common in acidic glassy rocks. [< SPHERULE + -ITE¹] — **spher′u·lit′ic** (-lit′ik) *adj.*
spher·y (sfir′ē) *adj. Poetic* 1 Like a sphere or star. 2 Of or relating to the celestial spheres.
sphinc·ter (sfingk′tər) *n. Anat.* A muscle that surrounds an opening or tube and serves to close it. [< LL < Gk. *sphinktēr* < *sphingein* close] — **sphinc′ter·al** *adj.*
sphinx (sfingks) *n. pl.* **sphinx·es** or **sphin·ges** (sfin′jēz) 1 In Egyptian mythology, a wingless monster with a lion's body and the head of a man (*androsphinx*), or simply *sphinx*), or of a ram (*criosphinx*), or of a hawk (*hieracosphinx*); also, any monumental representation of such a creature. 2 In Greek mythology, a winged monster with a woman's head and breasts and a lion's body, that destroyed those unable to guess her riddle. See OEDIPUS. 3 A mysterious or enigmatical person. 4 A large, stout-bodied, swift-flying moth. — the **Sphinx** The colossal androsphinx at Gizeh, having the body of a couchant lion, representing Harmachis, the Egyptian god of the morning, and dating to the IV dynasty. [< Gk. *sphinx* < *sphingein* close, strangle]
sphinx moth A hawk moth.
sphra·gis·tics (sfrə-jis′tiks) *n.* The study of signet rings or engraved seals, including their

SPHINX (def. 2)

authenticity, age, history, etc. [<Gk. *sphragistikos* of sealing <*sphragis* seal] **—sphra·gis′tic** *adj.*

sphyg·mic (sfig′mik) *adj. Physiol.* Pertaining to the pulse; pulsatory. [<Gk. *sphygmikos* < *sphygmos* pulse]

sphygmo- *combining form* Pulse; of or related to the pulse: *sphygmogram.* Also, before vowels, **sphygm-.** [<Gk. *sphygmos* pulse]

sphyg·mo·gram (sfig′mə·gram) *n.* A series of connected curves traced by a sphygmograph.

sphyg·mo·graph (sfig′mə·graf, -gräf) *n.* An instrument that, when applied over the heart or an artery, notes and records the character of the pulse and its rate, force, and variations: also called *pulsimeter.* **—sphyg′mo·graph′ic** *adj.* **—sphyg·mog·ra·phy** (sfig·mog′rə·fē) *n.*

sphyg·moid (sfig′moid) *adj. Physiol.* Pulselike.

sphyg·mo·ma·nom·e·ter (sfig′mō·mə·nom′ə·tər) *n.* An instrument for measuring the pressure of the blood in the arteries. Also **sphyg·mom′e·ter** (-mom′ə·tər).

sphyg·mo·scope (sfig′mə·skōp) *n.* An apparatus designed to make the pulse beat visible and to exhibit the varying pressures of the blood in the arteries during circulation.

sphyg·mus (sfig′məs) *n.* The pulse. [<NL <L <Gk. *sphygmos* pulse]

spi·ca (spī′kə) *n. pl.* **·cae** (-sē) **1** An ear of grain; a spike. **2** *Surg.* A bandage having a reversed spiral form, somewhat resembling an ear of wheat. [<L, spike, ear of grain]

Spi·ca (spī′kə) *Astron.* A spectroscopic binary star, Alpha in the constellation Virgo; magnitude, 1.21.

spi·cate (spī′kāt) *adj.* **1** *Bot.* Arranged in spikes: said of flowers. **2** *Ornithol.* Having a spur, as the legs of some birds. Also **spi′cat·ed.** [<L *spicatus* <*spica* spike]

spic·ca·to (spēk·kä′tō) *adj.* & *adv. Music* Detached; not legato. [<Ital., pp. of *spiccare* detach]

spice (spīs) *n.* **1** An aromatic, pungent vegetable substance, as cinnamon, cloves, etc., used to flavor food and beverages. **2** Such substances collectively. **3** That which gives zest or adds interest. **4** An aromatic odor; an agreeable perfume. **5** *Obs.* Sort; kind; species: the original meaning; also, a specimen. **—*v.t.*** **spiced, spic·ing** To season with spice; hence, to add zest or piquancy to. [< OF *espice* <L *species.* Doublet of SPECIES.] **—spic′er** *n.*

spice·ber·ry (spīs′ber′ē) *n. pl.* **·ries 1** A small tree *(Eugenia rhombea)* found in the West Indies and Florida. **2** The black or orange fruit of this tree. **3** The wintergreen or checkerberry.

spice·bush (spīs′bo͝osh′) *n.* An aromatic American shrub *(Lindera benzoin)* of the laurel family, the leaves of which have been used for tea, and the drupes, when powdered, for allspice. Also **spice′wood′** (-wo͝od′).

Spice Islands A former name for the MOLUCCA ISLANDS.

spic·er·y (spī′sər·ē) *n. pl.* **·er·ies 1** Spices collectively. **2** *Obs.* A place where spices are kept. **3** Spicy property or character; also, that which has spiciness.

spick (spik) *n. U.S. Slang* A Spanish-speaking person: an offensive term. Also **spic, spig** (spig).

spick-and-span (spik′ən·span′) *adj.* **1** Neat and clean. **2** Perfectly new, or looking as if new. [Prob. <*spick,* var. of SPIKE[1] + SPAN-NEW]

spic·ule (spik′yo͞ol) *n.* **1** A small, slender, sharp-pointed body; a spikelet. **2** *Zool.* One of the small, needlelike, calcareous growths supporting the soft tissues of certain invertebrates, as sponges, radiolarians, etc. Also **spic·u·la** (spik′yə·lə). [<L *spiculum,* dim. of *spicum* point, spike] **—spic′u·lar, spic′u·late** (-lāt, -lit) *adj.*

spic·u·lum (spik′yə·ləm) *n. pl.* **·la** (-lə) **1** A spicule. **2** *Zool.* Any small, dartlike organ, as the spines of a sea urchin. [<L]

spic·y (spī′sē) *adj.* **spic·i·er, spic·i·est 1** Containing, flavored, or fragrant with spices. **2** Producing spices. **3** Highly flavored; pungent; having zest; hence, somewhat improper; risqué. See synonyms under RACY. **—spic′i·ly** *adv.* **—spic′i·ness** *n.*

spi·der (spī′dər) *n.* **1** Any one of a large number of wingless arachnids (order *Araneae*) having an unsegmented abdomen and capable of spinning silk in the construction of webs for the capture of prey such as flies or other insects. **2** A long-handled iron frying pan, often having legs. **3** A portable electric switching apparatus for use in motion-picture studios. **4** A three-legged iron stool for the support of pots and pans over a fire; a trivet. **5** An apparatus for pulverizing the ground during cultivation. **6** *Electr.* The central part of an armature core. **7** *Naut.* **a** An iron hoop around the mast of a ship for the attachment of shrouds. **b** A magnifying glass for a ship's compass. **8** Any of several vehicles of different types having unusually light frames. [OE *spithra* <*spinnan* spin]

spider crab Any of a genus *(Libinia)* of decapod crustaceans with long legs, retractable eyes, and spiny growths on the carapace, especially *L. emarginata,* common on the Atlantic coast of North America.

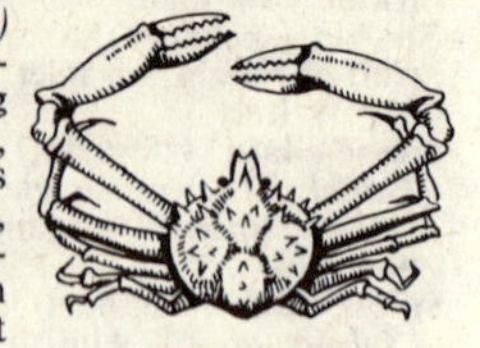

SPIDER CRAB
(Up to 10 inches in breadth)

spi·der·flow·er (spī′dər·flou′ər) *n.* A cleome.

spider monkey An arboreal American monkey (genus *Ateles*) of slender form, with very long limbs, thumbs absent or vestigial, and a long prehensile tail: range from Mexico to Paraguay.

spider phaeton A type of carriage of light construction, having a covered seat in front, and a rear seat for a footman or attendant.

spi·der·wort (spī′dər·wûrt′) *n.* **1** Any species of a genus *(Tradescantia)* of plants, especially *T. virginiana,* an American perennial with deep-blue, three-petaled flowers in umbels. **2** Any plant of the same family *(Commelinaceae).*

spi·der·y (spī′dər·ē) *adj.* Spiderlike.

S-piece (es′pēs′) *n.* An S-bracket.

spied (spīd) Past tense and past participle of SPY.

spie·gel·ei·sen (spē′gəl·ī′zən) *n.* A white, very hard and brittle cast iron containing manganese, largely used in the manufacture of steel: when more than about 20 percent of manganese is in the alloy it is called *ferromanganese.* Also **spie′gel, spiegel iron.** [<G <*spiegel* a mirror + *eisen* iron]

spiel (spēl) *U.S. Slang v.i.* To talk; orate. **—*n.*** A speech, especially a long speech. [<G, a game, play <*spielen* play]

spi·er[1] (spī′ər) *n. Brit. Dial.* A spy; scout.

spier[2] (spir) See SPEER.

spif·fy (spif′ē) *adj. Slang* Smartly dressed; spruce. [<dial. E *spiff* a dandy]

spi·ge·li·a (spī·jē′lē·ə) *n.* Pinkroot: used as a vermifuge. [<NL, after Adrian van den *Spiegel,* 1578–1625, Flemish anatomist]

spig·ot (spig′ət) *n.* **1** A plug or faucet for the bunghole of a cask. **2** A turning plug fitting into a faucet, or the faucet itself. [ME *spigote.* Prob. akin to SPIKE[1].]

spike[1] (spīk) *n.* **1** A stout piece of metal, like a large nail, but thicker in proportion. **2** A projecting, pointed piece of metal, or any similar object, as in the soles of shoes to keep the wearer from slipping. **3** A very high heel on a woman's shoe, narrow at the bottom. **4** A steel pin for plugging cannon vents. **5** A straight, unbranched antler, as of a young deer. **6** A young mackerel. **—*v.t.*** **spiked, spik·ing 1** To fasten with spikes. **2** To set or provide with spikes. **3** To block the vent of (a cannon) with a spike, rendering it useless. **4** To block; put a stop to. **5** To pierce with or impale on a spike. **6** In baseball, to injure (another player) with the spikes on one's shoes. **7** *Colloq.* To add spirituous liquor to. [ME <Scand. Cf. ON *spikr* a nail.]

spike[2] (spīk) *n.* **1** An ear of corn, barley, wheat, or other grain. **2** *Bot.* A flower cluster in which there are numerous flowers arranged closely on an elongated common axis. [<L *spica* ear of grain]

spike lavender See under LAVENDER.

spike·let (spīk′lit) *n. Bot.* A small spike bearing few flowers and forming the compound inflorescence of cereal grasses and sedges.

spike·nard (spīk′nərd, -närd) *n.* **1** An ancient fragrant and costly ointment prepared mainly from a plant of the same name. **2** A perennial East Indian herb *(Nardostachys jatamansi)* of the valerian family. **3** An American herb *(Aralia racemosa)* of the ginseng family. [<L *spica* spike + *nardus* nard]

spik·er (spī′kər) *n.* One who or that which spikes; specifically, a workman who drives spikes in railroad ties.

spik·y (spī′kē) *adj.* **spik·i·er, spik·i·est 1** Resembling a spike; pointed. **2** Having spikes.

spile[1] (spīl) *n.* **1** A large timber driven into the ground to serve as a foundation; a pile. **2** A wooden pin or plug used as a vent in a cask; a spigot. **3** A spout driven into a sugar-maple tree to lead the sap to a bucket. **—*v.t.*** **spiled, spil·ing 1** To pierce for and provide with a spigot. **2** To drive spiles into. [<MDu., skewer, splinter]

spile[2] (spīl) *v.t.* & *v.i. Dial.* To spoil.

spil·i·kin (spil′i·kin) *n.* **1** A jackstraw; one of the thin straws used in playing jackstraws. **2** *pl.* The game of jackstraws. Also **spil′li·kin.** [Dim. of SPILL[2]]

spil·ing (spī′ling) *n.* Spiles collectively; piling.

spill[1] (spil) *v.* **spilled** or **spilt, spill·ing** *v.t.* **1** To allow or cause to fall or run out or over, as a liquid or a powder. **2** To shed, as blood. **3** *Naut.* To empty (a sail) of wind. **4** *Colloq.* To cause to fall, as from a horse. **5** *Colloq.* To divulge; make known, as a secret. **—*v.i.*** **6** To fall or run out or over: said of liquids, etc. **—to spill the beans** *Colloq.* To divulge, especially a secret. **—*n.*** **1** *Colloq.* A fall to the ground, as from a horse or vehicle; tumble. **2** *Colloq.* A downpour, as of rain. **3** A crack, seam, or other defect in iron or steel castings, forgings, etc. [OE *spillan* destroy] **—spill′·age** *n.* **—spill′er** *n.*

spill[2] (spil) *n.* **1** A slip of wood, or rolled strip of paper, used for lighting lamps, etc.; a lamplighter. **2** A slender peg, pin, or bar of wood or metal; especially, a slender plug for stopping a hole in a cask; a spile. [Var. of SPILE[1]]

spill·way (spil′wā′) *n.* **1** A passageway in or about a dam to release the water in a reservoir. **2** The paved upper surface of a dam over which surplus water escapes.

SPILLWAY

spil·o·site (spil′ə·sīt) *n.* A greenish schistous rock spotted with chlorite, produced by the shearing of a basic amygdaloid. [<Gk. *spilos* spot + -ITE[1]]

spilt (spilt) Alternative past tense and past participle of SPILL[1].

spilth (spilth) *n.* That which is spilled or poured out profusely; effusion; excess of supply.

spin (spin) *v.* **spun** (*Archaic* **span**), **spun, spin·ning** *v.t.* **1** To draw out and twist into threads; also, to draw out and twist fiber into (threads, yarn, etc.). **2** To make or produce as if by spinning. **3** To form (a net, etc.) from filaments of a viscous substance extruded from the body: said of spiders, silkworms, etc. **4** To tell, as a story or yarn. **5** To protract; prolong, as a period of time by delays or a story by additional details: with *out.* **6** To cause to whirl rapidly: to *spin* a top. **—*v.i.*** **7** To make thread or yarn. **8** To extrude filaments of a viscous substance from the body: said of spiders, etc. **9** To whirl rapidly; rotate. **10** To seem to be whirling, as from dizziness: My head is *spinning.* **11** To move rapidly. **12** To fish with a spoon bait or swivel. **—*n.*** **1** An act or instance of spinning; a rapid whirling. **2** Any rapid movement or action. **3** *Aeron.* The downward spiral motion of an airplane about a vertical axis, with its longitudinal axis steeply inclined. **4** *Physics* The angular momentum of an atomic particle or nuclide. **—flat spin** The descent of a stalled airplane while rotating about its vertical axis. [OE *spinnan* spin]

spi·na·ceous (spi·nā′shəs) *adj.* Of, relating to, or resembling spinach or plants allied to it.

spin·ach (spin′ich, -ij) *n.* **1** An edible garden pot herb *(Spinacia oleracea)* of the goosefoot family. **2** Its fleshy, edible leaves. Also **spin′·age.** [<OF *espinage* <LL *spinacia* <Arabic *isbānah*; infl. in form by L *spina* a thorn]

spi·nal (spī′nəl) *adj.* **1** Pertaining to the backbone; vertebral. **2** Pertaining to a spine, spines, or spinous processes. **3** Dependent upon or functioning with a spinal cord, as the vertebrates.

spinal canal *Anat.* The tubular cavity on the dorsal side of the spinal column, in which the spinal cord and its membranes are lodged.

spinal column *Anat.* The series of articulated vertebrae which, with their associated structures, enclose the spinal cord and provide dorsal support for the ribs; the backbone or spine.

spinal cord *Anat.* That portion of the central nervous system enclosed by the spinal column. It is composed of an inner region of gray matter and an outer, larger region of white matter, the whole divided into the cervical, thoracic, lumbar, sacral, and coccygeal areas.

spi·nate (spī′nāt) *adj.* Spinelike, or bearing spines or thorns. Also **spi′nat·ed**.

spin·dle (spin′dəl) *n.* **1** A rod having a slit or catch in the top and a whorl of wood or metal at its lower end, formerly used in hand spinning, and on which was wound the thread from the distaff. **2** The slender rod in a spinning wheel by the rotation of which the thread is twisted and wound on a spool or bobbin on the same rod; also, a small rod or pin bearing the bobbin of a spinning machine or a shuttle. **3** *Mech.* A rotating rod, pin, axis, arbor, or shaft, especially when small and bearing something that rotates: the *spindle* of a lathe. **4** The pin on which rotates a fusee in a watch, or the fusee itself. **5** The tapering end of a vehicle axle that enters the hub. **6** A small shaft passing through the lock of a door and bearing the knobs or handles. **7** *Biol.* A spindle-shaped structure of elongated achromatic fibers formed during the mitosis of a cell. **8** A measure of length for cotton or linen yarn, varying according to the number of hanks or cuts: generally 18 hanks, or 15,120 yards. **9** *Naut.* An iron pile or pipe, surmounted by a lantern or other conspicuous object, placed on a rock or shoal for the guidance of seamen. **10** A hydrometer. **11** A needlelike rod mounted on a weighted base, for impaling bills, checks, etc. — *v.* **·dled, ·dling** *v.i.* **1** To grow into a long, slender stalk or body; become extremely long and slender. — *v.t.* **2** To form into or as into a spindle. **3** To provide with a spindle. [OE *spinel* < *spinnan* spin]

spin·dle-leg·ged (spin′dəl·leg′id, -legd′) *adj.* Having long, slender legs. Also **spin′dle-shanked**′ (-shangkt′).

spin·dle-legs (spin′dəl·legz′) *n.* **1** Long, slender legs. **2** *Colloq.* A person having long, slender legs. Also **spin′dle-shanks**′ (-shangks′).

spindle side The female or distaff side of a family.

spindle tree A European shrub or low-spreading tree (*Euonymus europaeus*), so called from the use of its compact wood in making spindles, slender pins, skewers, etc.

spin·dling (spind′ling) *adj.* Long and thin; disproportionately slender. — *n.* A spindling person or plant shoot.

spin·dly (spind′lē) *adj.* Of a slender, lanky growth or form, suggesting weakness.

spin·drift (spin′drift) *n.* Blown spray or scud: also called *spoondrift*. [Alter. of *spoondrift* < *spoon*, var. of SPUME + DRIFT]

spine (spīn) *n.* **1** The spinal column of a vertebrate; backbone. **2** *Zool.* Any of various hard, pointed outgrowths on the bodies of certain animals, as the porcupine and starfish; a spicule; the fin ray of a fish. **3** *Bot.* A stiff, short-pointed woody process on the stems of certain plants, as the honey locust; thorn. **4** The back of a bound book. **5** A projecting eminence or ridge. **6** Any slender, thornlike process, as of a vertebra or nerve. **7** The central ridge on the underside of a horse's hoof. [< OF *espine* < L *spina* spine, thorn]

spi·nel (spi·nel′, spin′əl) *n.* A hard isometric mineral of various colors and composition, the red variety of which is used as a gem under the name of **ruby spinel**. Compare BALAS. [< F *spinelle* < Ital. *spinella*, dim. of L *spina* spine]

spine·less (spīn′lis) *adj.* **1** Having no spine or backbone; invertebrate. **2** Lacking spines. **3** Having a very flexible backbone; limp. **4** Figuratively, lacking decision of character or steadfastness. — **spine′less·ness** *n.*

spi·nes·cent (spī·nes′ənt) *adj.* **1** *Bot.* Bearing spines; spinous; terminating in a spine. **2** *Zool.* Tending to become spinous, as certain animals during the period of racial decline. — **spi·nes′cence** *n.*

spin·et (spin′it) *n.* A small keyboard musical instrument of the harpsichord class in use from the 16th century to the 18th. Each note had one string, sounded by plucking with quills or leather plectra attached to jacks. [Perhaps after G. *Spinetti*, 16th century Venetian inventor]

Spin·garn (spin′gärn), **Joel Elias**, 1875–1939, U. S. poet and critic.

spini- *combining form* A spine; thorn: *spiniferous*. [< L *spina* a thorn]

spi·nif·er·ous (spī·nif′ər·əs) *adj.* Bearing or producing spines. Also **spi·nig′er·ous** (-nij′ər·əs).

spin·i·fex (spin′i·feks, spī′ni-) *n.* An Australian grass (genus *Spinifex*) with sharp-pointed leaves.

spin·na·ker (spin′ə·kər) *n. Naut.* A large jib-shaped sail sometimes carried on the mainmast of a racing vessel, opposite the mainsail, and used when sailing before the wind. The foot slides on a spar called the **spinnaker boom**. [? < *spinx*, a mispronunciation of *Sphinx*, the name of the first vessel to carry this kind of sail]

spin·ner (spin′ər) *n.* **1** One who or that which spins, as a spider or a machine. **2** In angling, a whirling spoon bait. **3** *Aeron.* A streamlined fairing fitted over the boss of an airplane propeller and revolving with it. See illustration under AIRPLANE. **4** A play in football wherein the ball carrier spins around to conceal the direction of the play from his opponents.

spin·ner·et (spin′ər·et) *n.* **1** An organ, as of spiders and silkworms, for spinning silk. **2** A metal plate pierced with holes through which filaments of plastic material are forced, as in the making of rayon fibers.

spin·ner·y (spin′ər·ē) *n. pl.* **·ner·ies** A spinning mill.

spin·ney (spin′ē) *n.* A small wood or thicket. Also **spin′ny**. [< OF *espinei* < LL *spinetum* < L *spina* a thorn]

spin·ning (spin′ing) *n.* **1** The action of, or activities involved in, converting fibers into thread or yarn. **2** The product of spinning. — *adj.* **1** That spins, in any sense. **2** Of, belonging to, or used in the process of spinning.

spinning gland A gland that secretes silk or a silky substance, as in silkworms.

spinning house A house of correction for prostitutes: so called because the inmates were formerly forced to spin yarn, etc.

spinning jenny A framed mechanism for spinning more than one strand of yarn at a time: also called *jenny*.

spinning mill A mill or factory devoted to spinning.

spinning mule A kind of spinning jenny.

spinning wheel A household implement formerly used for spinning yarn or thread, consisting of a rotating spindle operated by a treadle and flywheel.

SPINNING WHEEL

spin-off (spin′ôf′, -of′) *n.* Action of a corporation in divesting itself, tax free, of a segment or division of its operations by transfer to a new, independently owned and managed company, the stockholders of the original corporation receiving the new shares pro rata.

spi·nose (spī′nōs) *adj.* Bearing, armed with, or having many spines. [< L *spinosus* < *spina* a thorn] — **spi′nose·ly** *adv.*

spi·nos·i·ty (spī·nos′ə·tē) *n. pl.* **·ties** **1** The state of being spinous or spinose. **2** A spinous part or thing.

spi·nous (spī′nəs) *adj.* **1** Spinelike; prickly. **2** Spinose.

Spi·no·za (spi·nō′zə), **Baruch**, 1632–77, Dutch philosopher and pantheist: also known as Benedict Spinoza.

Spi·no·zism (spi·nō′ziz·əm) *n. Philos.* A system of absolute monism developed from Cartesianism by Spinoza. This system regards the entire universe as one infinite and universal substance, namely, God, of which extension and mind are but attributes, and of which specific things, ideas, and states of mind are modes. — **Spi·no′zist** *n.* — **Spin·o·zis·tic** (spin′-ō·zis′tik) *adj.*

spin·ster (spin′stər) *n.* **1** An unmarried woman, especially when no longer young; an old maid. **2** *Law* In England, a woman who has never married: a legal title. **3** A woman who spins; a spinner. [ME < SPIN + -STER] — **spin′ster·hood** *n.* — **spin′ster·ish** *adj.*

spin·thar·i·scope (spin·thar′i·skōp) *n.* A device for showing the radioactivity of a substance by the scintillations of the alpha rays emitted from a minute particle of the substance and thrown against a fluorescent screen. [< Gk. *spintharis* spark + -SCOPE] — **spin·thar′i·scop′ic** (-skop′ik) *adj.*

spi·nule (spī′nyōōl, spin′yōōl) *n.* A small spine; spicule. Also **spin·u·la** (spin′yə·lə). [< L *spinula*, dim. of *spina* spine]

spin·u·les·cent (spin′yə·les′ənt, spī′nyə-) *adj.* Furnished with or producing spinules; somewhat spiny.

spin·u·lose (spin′yə·lōs, spī′nyə·lōs) *adj.* Having spinules. Also **spin′u·lous** (-ləs).

spin·y (spī′nē) *adj.* **spin·i·er, spin·i·est** **1** Having spines; thorny. **2** Difficult; perplexing. — **spin′i·ness** *n.*

spiny ant-eater The echidna.

spin·y-finned (spī′nē-find′) *adj.* Characterized by fins bearing one or more sharp, unsegmented rays, as the perch, mackerel, and bass. Also **spine′-finned**′.

spiny lobster One of various large-bodied marine crustaceans (genus *Palinurus*) with spiny shells but lacking claws; especially, the California spiny lobster (*P. interruptus*), valued as a sea food. Also called *crayfish*.

Spi·on Kop (spē′ən kop) A hill in NW Natal province, Union of South Africa; scene of a battle in the Boer War, 1900.

spir- Var. of SPIRO-.

spir·a·cle (spir′ə·kəl, spī′rə-) *n.* **1** *Zool.* **a** An aperture or orifice for the passage of air or water in the respiration of terrestrial arthropods, as the grasshopper and locust. **b** A breathing hole, as the blowhole or nostril of a cetacean. **2** A minute cone formed on a stream of lava by escaping gases. **3** Any opening to admit or expel air; an airhole. [< OF < L *spiraculum* airhole < *spirare* breathe]

spi·rae·a (spī·rē′ə) *n.* Any of a genus (*Spiraea*) of ornamental shrubs of the rose family, having alternate simple or pinnate leaves and small, white or pink flowers; especially, an American variety, the meadowsweet. Also **spi·re′a**. [< L, meadowsweet <Gk. *speiraia* < *speira* coil]

spi·ral (spī′rəl) *adj.* **1** Winding about and constantly receding from a center. **2** Winding and advancing; helical. **3** Winding and rising in a spire, as some springs. — *n.* **1** *Geom.* Any plane curve formed by a point that moves around a fixed center and continually increases its distance from it. **2** A curve winding like a screw thread. **3** Something wound as a spiral, as woolen puttees which are wound around the legs, or having a spiral shape, as a spiral spring or a whorled shell. **4** *Aeron.* A flight of an airplane in a spiral path. **5** In football, the motion of a ball rotating on its long axis. — *v.t. & v.i.* **·raled** or **·ralled, ·ral·ing** or **·ral·ling** To take or cause to take a spiral form or course. [<Med. L *spiralis* < L *spira* SPIRE²] — **spi′ral·ly** *adv.*

spiral nebula *Astron.* An extragalactic system of celestial bodies exhibiting a spiral configuration, known to be composed of aggregates of stars resembling the Milky Way, as the *spiral nebula* in Andromeda; an island universe.

spiral of Archimedes *Math.* The polar curve traced by a point starting at the pole and moving along its radius vector at a constant velocity while the radius vector moves at a constant angular velocity.

spi·rant (spī′rənt) *n.* & *adj.* *Phonet.* Fricative.

spire[1] (spīr) *n.* **1** The tapering or pyramidal roof of a tower; a pinnacle; also, loosely, a steeple. **2** A slender stalk or blade. **3** The summit or tapering end of anything; a sharp point. — *v.* **spired, spir·ing** *v.t.* **1** To furnish with a spire or spires. — *v.i.* **2** To shoot or point up in or as in a spire. **3** To put forth a spire or spires; sprout. [OE *spīr* a stalk, stem] — **spired** *adj.*

SPIRE

spire[2] (spīr) *n.* **1** A spiral or a single turn of one; whorl; twist. **2** The portion of a spiral formed by a single revolution about the central point. **3** *Zool.* The convoluted portion of a spiral shell. [<F <L *spira* <Gk. *speira* coil] — **spired** *adj.*

spi·reme (spī′rēm) *n.* *Biol.* **1** The stage in the division of a cell during which the chromatin appears like a skein of filaments. **2** One of these filaments. Also **spi′rem** (-rem). [<Gk. *speirēma* a coil]

Spires (spīrz) The English name for SPEYER.

spi·rif·er·ous (spī·rif′ər·əs) *adj.* **1** Bearing spiral appendages. **2** Having a spire, as a univalve. Also **spi·rig′er·ous** (-rij′ər·əs). [<L *spira* coil + -(I)FEROUS]

spir·il·lo·sis (spir′ə·lō′sis) *n.* **1** *Pathol.* Any disease caused by the presence of spirilla in the body. **2** A disease of domestic fowls caused by a spirochete transmitted by a tick. [<SPIRILLUM + -OSIS]

spi·ril·lum (spī·ril′əm) *n.* *pl.* **·ril·la** (-ril′ə) Any of a genus *(Spirillum)* of flagellate bacteria with cells in spirally twisted and rigid filaments. See illustration under BACTERIUM. [< NL, dim. of L *spira* a coil]

spir·it (spir′it) *n.* **1** The principle of life and energy in man and animals, at one time regarded as being composed of an especially refined substance, such as breath or warm air, separable from the body, mysterious in nature, and ascribable to a divine origin. **2** An entity conceived of as that part of a human being that is incorporeal and invisible and is characterized by intelligence, personality, self-consciousness, and will; the mind: opposed to *body.* **3** The substance or universal aspect of reality, regarded as independent of and opposed to matter. **4** In the Bible, the creative, animating power or divine influence of God. *Joel* ii 28. **5** A rational, supernatural being without a material body, as an angel, demon, elf, fairy, etc.; specifically, such a being with a certain character or a particular abode or area of activity: an evil *spirit.* **6** A disembodied soul regarded as manifested to the senses, often as visible or having some kind of immaterial body: a ghost; specter: Hamlet saw his father's *spirit.* **7** A person regarded with reference to any peculiar activity, characteristic, or temper: a leading *spirit* in the community. **8** *Usually pl.* A state of mind; mood; temper: Success raised his *spirits.* **9** Vivacity or energy; ardor; dash; fire: an attack made with *spirit.* **10** Ardent loyalty or devotion: school *spirit.* **11** True intent or meaning as opposed to outward, formal signification: to keep the *spirit* of the law. Compare LETTER (def. 5). **12** The emotional or affective faculty of man; the heart: Great poetry stirs the *spirit.* **13** The characteristic temper or disposition of a period or of a movement: the *spirit* of the Reformation. **14** *pl.* A strong alcoholic liquor or liquid obtained by distillation. **15** *Usually pl. Chem.* **a** The essence or distilled extract of a substance: *spirits* of turpentine. **b** Ethanol. **16** *Often pl.* In pharmacy, a solution of a volatile principle in alcohol; a tincture; essence: *spirits* of ammonia. **17** In dyeing, a solution of a tin salt in acid. **18** In alchemy, one of four substances, mercury, sal ammoniac, sulfur, and arsenic (or orpiment). **19** In medieval physiology, one of the three degrees of spirit inherent in the human body: **natural spirit,** located in the liver and underlying the processes of nutrition, growth, and reproduction; **vital spirit,** located in the heart, which circulated heat and life through the body; **animal spirit,** located in the brain, which guided reason and conveyed the powers of motion and sensation to and through the nerves. **20** *Obs.* Breathed air; breeze; wind. **21** *Obs.* The breath; life. See synonyms under CHARACTER, COURAGE, MIND, SPECTER. — *v.t.* **1** To carry off secretly or mysteriously, as if by the agency of a spirit: with *away, off,* etc. **2** To infuse with spirit or animation; inspirit; encourage: often with *up.* — *adj.* **1** Of or pertaining to ghosts or the belief in the existence of departed souls; spiritualistic. **2** Operated by the burning of alcohol: a *spirit* lamp. [<OF *espirit* <L *spiritus* breathing <*spirare* breathe. Doublet of SPRITE.]

Spir·it (spir′it) *n.* In Christian theology, the Holy Spirit.

spir·it·ed (spir′it·id) *adj.* Full of spirit; animated: used in various compound adjectives: high-*spirited,* mean-*spirited.* See synonyms under RACY. — **spir′it·ed·ly** *adv.* — **spir′it·ed·ness** *n.*

spir·it·ing (spir′it·ing) *n.* Movement as of a spirit; hence, something dexterously done; the work or ministering of a spirit; inspiration; encouragement.

spir·it·ism (spir′i·tiz′əm) *n.* **1** Loosely, spiritualism. **2** *Rare* Animism; the theory that inanimate objects possess spirits. — **spir′it·ist** *n.* — **spir′it·is′tic** *adj.*

spirit lamp A lamp that burns alcohol: used in laboratory work, etc.

spir·it·less (spir′it·lis) *adj.* Lacking in enthusiasm, energy, or courage; lacking in the sense of well-being. — **spir′it·less·ly** *adv.* — **spir′it·less·ness** *n.*

spirit level An instrument for adjusting any deviation from the horizontal or perpendicular by reference to the position of a bubble of air in a tube of alcohol or other liquid.

spi·ri·to·so (spir′i·tō′sō, *Ital.* spē′rē·tō′sō) *Music adj.* Spirited; animated. — *adv.* With spirit. [<Ital. <*spirito* spirit]

spir·it·ous (spir′i·təs) *adj.* **1** Like spirits; refined. **2** Spirituous. **3** Spirited; ardent.

spirit rapping The professed communication with the spirits of departed persons by raps, as on a table; also, the rapping believed to be made by spirits.

spirits of hartshorn See under AMMONIA.

spirits of turpentine Oil of turpentine.

spirits of wine Rectified alcohol.

spir·i·tu·al (spir′i·chōō·əl) *adj.* **1** Of or pertaining to spirit, as distinguished from matter; having the nature of spirit; consisting of spirit; incorporeal. **2** Pertaining to or affecting the immaterial nature or soul of man. **3** Of or pertaining to God, his Spirit, or his law, or to the soul as acted upon by the Holy Spirit; holy; pure; not carnal. **4** Sacred or religious; not lay or temporal; ecclesiastical: *spiritual* authorities: contrasted with *secular.* **5** Marked or characterized by the highest qualities of the human mind; intellectualized. — *n.* **1** Anything pertaining to spirit or to sacred matters. **2** A religious folk song originating among the Negroes of the southern United States, typified by colorful rhythm and emotion: sometimes in narrative or ballad form; also, any song composed in imitation of a Negro spiritual. — **the Spirituals** See under FRATICELLI. [<L *spiritualis* <*spiritus* spirit] — **spir′i·tu·al·ly** *adv.* — **spir′i·tu·al·ness** *n.*

spiritual incest *Eccl.* Sexual intercourse between persons spiritually related, as between godparent and godchild.

spir·i·tu·al·ism (spir′i·chōō·əl·iz′əm) *n.* **1** The belief that the spirits of the dead in various ways communicate with and manifest their presence to the living, usually through the agency of a person called a medium; also, the doctrines and practices of those so believing. **2** The doctrine that there are beings not cognizable by the senses or characterized by the properties of matter, and that are therefore spiritual, as distinguished from material: opposed to *materialism.* **3** The doctrine that man is an immortal spirit and as such may know, love, or worship God. **4** A non-materialistic philosophy, a form of idealism which identifies ultimate reality as one universal conscious mind. **5** The state or character of being spiritual. — **spir′i·tu·al·ist** *n.* — **spir′i·tu·al·is′tic** *adj.*

spir·i·tu·al·i·ty (spir′i·chōō·al′ə·tē) *n.* *pl.* **·ties** **1** The state of being spiritual. **2** That which belongs to the church or to an ecclesiastic: opposed to *temporality.*

spir·i·tu·al·ize (spir′i·chōō·əl·īz′) *v.t.* **·ized, ·iz·ing** **1** To make spiritual; free of grossness or materialism: to *spiritualize* the thoughts. **2** To imbue with spirit; animate. **3** To treat as having a spiritual meaning or sense. Also *Brit.* **spir′i·tu·al·ise′.** — **spir′i·tu·al·i·za′tion** *n.* — **spir′i·tu·al·iz′er** *n.*

spir·i·tu·al·ty (spir′i·chōō·əl·te) *n.* *pl.* **·ties** Ecclesiastical bodies collectively; the clergy.

spiritual wife Among the Mormons, a woman who has been married for eternity in accordance with the doctrine of the Mormon gospel; hence, **spiritual wifeism, spiritual wifehood.**

spir·i·tu·el (spir′i·chōō·el′, *Fr.* spē·rē·tü·el′) *adj.* Characterized by esprit, or wit, and by the higher and finer qualities of the mind generally. [<F] — **spir′i·tu·elle′** *adj. fem.*

spir·i·tu·ous (spir′i·chōō·əs) *adj.* **1** Containing alcohol. **2** Intoxicating; distilled. **3** *Obs.* Spiritlike; ethereal. **4** *Rare* Lively. — **spir′i·tu·ous·ness** *n.*

spir·i·tus (spir′i·təs) *n.* *pl.* **·tus** **1** A breathing or an aspirate. In Greek grammar the rough breathing is called **spiritus asper,** the smooth **spiritus lenis.** See BREATHING. **2** Any liquid product of distillation; especially, alcoholic liquor. See SPIRIT. [<L, a breathing]

spirit writing Visible but automatic writing done without the conscious will of the writer, and believed to be a manifestation of spirit guidance; pneumatography.

spiro-[1] *combining form* Breath; respiration: *spirograph.* Also, before vowels, *spir-.* [<L *spirare* breathe]

spiro-[2] *combining form* Spiral; coiled: *spirochete.* Also, before vowels, *spir-.* [<Gk. *speira* a coil]

spi·ro·chete (spī′rə·kēt) *n.* **1** Any of a genus *(Spirochaeta)* of typically saprophytic bacteria commonly found in water and sewage, and characterized by spiral flexible filaments with apparently rotary movements. See illustration under BACTERIUM. **2** Any of various other similar micro-organisms of the order *Spirochaetales,* including those which cause syphilis and relapsing fever. Also **spi′ro·chaete.** [<Gk. *speira* coil + *chaitē* bristle] — **spi′ro·che′tal** *adj.*

spi·ro·che·to·sis (spī′rə·kē·tō′sis) *n.* **1** *Pathol.* Infection by spirochetes. **2** An infectious septicemia in chickens caused by a spirochete *(Borrelia anserina).*

spi·ro·graph (spī′rə·graf, -gräf) *n.* An instrument for recording the breathing movement. [<SPIRO-[1] + -GRAPH] — **spi′ro·graph′ic** *adj.* — **spi·rog·raph·y** (spī·rog′rə·fē) *n.*

spi·ro·gy·ra (spī′rə·jī′rə) *n.* Any of a genus *(Spirogyra)* of bright-green, fresh-water algae forming dense masses or beds of growth in slow-running or stagnant water, and characterized by having the chlorophyll bands winding spirally to the right. [<SPIRO-[2] + Gk. *gyros* ring, coil]

spi·roid (spī′roid) *adj.* Resembling a spiral.

spi·rom·e·ter (spī·rom′ə·tər) *n.* An instrument for measuring the capacity of the lungs. [<SPIRO-[1] + -METER] — **spi·ro·met·ric** (spī′rə·met′rik) *adj.* — **spi·rom′e·try** *n.*

spirt (spûrt) See SPURT.

spir·u·la (spir′yə·lə, spir′ōō-) *n.* *pl.* **·lae** (-lē) Any of a genus *(Spirula)* of cephalopods with an internal spiral chambered shell having whorls detached and in the same plane. [<NL <Gk. *speira* coil]

spir·y[1] (spīr′ē) *adj.* **1** Pertaining to or having the form of a spire. **2** Abounding in spires, as a city.

spir·y[2] (spīr′ē) *adj.* Having the form of a spiral; coiled; whorled.

spis·sat·ed (spis′ā·tid) *adj.* Thickened. [<L *spissatus,* pp. of *spissare* thicken]

spit[1] (spit) *v.* **spat** or **spit, spit·ting** *v.t.* **1** To eject (saliva, blood, etc.) from the mouth. **2** To throw off, eject, or utter with violence. **3** To light, as a fuse. — *v.i.* **4** To eject saliva from the mouth. **5** To make a noise like that made in ejecting saliva. **6** To fall in scattered drops or flakes, as rain or snow. — *n.* **1** Spittle; saliva. **2** An act of spitting or expectorating. **3** A frothy, spitlike secretion of the spittle insect; also, a spittle insect. **4** A light, scattered fall or short, driving flurry of snow or rain. **5** *Colloq.* Exact image; likeness; counterpart: He's the *spit* of John. [OE *spittan*] — **spit′ter** *n.*

spit[2] (spit) *n.* **1** A pointed rod on which meat is turned and roasted before a fire. **2** A

spital

point of low land, or a long, narrow shoal, extending from a shore into the water. — *v.t.* **spit·ted**, **spit·ting** To transfix or impale with or as with a spit. [OE *spitu* spit]

spit·al (spit′l) *n. Obs.* A hospital. Also **spital house, spit′tle, spittle house**.

spit·ball (spit′bôl′) *n.* 1 Paper chewed in the mouth and shaped into a ball for use as a missile. 2 In baseball, a pitched ball wet with saliva, and rotating deceptively in its course: no longer permitted by the rules. — **spit′ball′er** *n.*

spitch·cock (spich′kok) *v.t.* To split and broil, as a bird or fish. — *n.* An eel split and broiled. [Origin unknown]

spite (spīt) *n.* 1 Malicious bitterness prompting to vexatious acts; mean hatred; grudge. 2 That which is done in spite. 3 *Archaic* Trouble; bad luck: a Shakespearean usage. See synonyms under ENMITY, HATRED. — **in spite of** (or spite of) Formerly, in contempt of; now, notwithstanding. — *v.t.* **spit·ed, spit·ing** 1 To show one's spite toward; vex maliciously; thwart. 2 *Obs.* To fill with spite; offend; vex. [Short for DESPITE]

spite fence *U.S.* A fence or wall put up to spite a neighbor, usually of such a nature as to detract from the desirability or value of the adjoining property: now illegal.

spite·ful (spīt′fəl) *adj.* 1 Filled with spite. 2 Prompted by spite. See synonyms under MALICIOUS. — **spite′ful·ly** *adv.* — **spite′ful·ness** *n.*

spit·fire (spit′fīr′) *n.* A quick-tempered person who is given to saying spiteful things.

Spit·head (spit′hed) A roadstead between the Isle of Wight and southern England, at Portsmouth.

Spits·ber·gen (spits′bûr·gən) A group of Norwegian islands in the Arctic Ocean, east of Greenland and north of Norway, and comprising most of Svalbard; 23,658 square miles. Also **Spitz′ber·gen**.

Spits (spits), **Spits·ke** (spits′kē) See SCHIPPERKE.

Spit·te·ler (shpit′ə·lər), **Carl**, 1845–1924, Swiss writer; pseudonym *Felix Tandem*.

spit·ter (spit′ər) *n.* 1 A young deer whose antlers have emerged, but have not branched. 2 One who cooks meat on a spit.

spit·ting image (spit′ing) *Colloq.* An exact likeness or counterpart. Also **spit and image**.

spitting snake A venomous snake of South Africa (*Sepedon haemachates*) related to the cobras, that is able to eject its poison for some distance; the ringhals.

spit·tle (spit′l) *n.* 1 The fluid secreted by the mouth; saliva; spit. 2 The salivalike matter in which the larvae of spittle insects live. [OE *spātl*; infl. in form by *spit*[1]]

spittle insect A froghopper.

spit·toon (spi·tōōn′) *n.* A receptacle for spit; a cuspidor.

spitz (spits) *n.* One of a breed of small dogs with a tapering muzzle; a Pomeranian. Also **spitz dog**. [<G, short for *spitzhund* < *spitz* pointed + *hund* dog]

Spitz·en·burg (spit′sən·bûrg) *n.* A variety of apple, yellow and red in color, prized for its delicate flavor. [Prob. <Du. '*spits* a point + *berg* a hill; so called because it is pointed in shape and was discovered on an upstate New York hillside]

spiv (spiv) *n.* 1 *Brit. Colloq.* A flashy chiseler, sharper, or one who lives by his wits. 2 In Scotland Yard usage, a low and common thief.

spiv·er·y (spiv′ər·ē) *n. Brit. Colloq.* The obtaining of a livelihood by the least possible personal effort, relying on government subsidy, sinecures, or private income.

splanch·nic (splangk′nik) *adj. Anat.* Pertaining to or supplying the viscera: a *splanchnic* nerve. [<Gk. *splanchnikos* < *splanchnon* entrail]

splanchno– combining form *Anat. & Med.* The viscera; of or related to the viscera. Also, before vowels, **splanchn–**. [<Gk. *splanchnon* entrail]

splanch·nol·o·gy (splangk·nol′ə·jē) *n.* The anatomy and physiology of the viscera. [< SPLANCHNO– + -LOGY]

splash (splash) *v.t.* 1 To dash or spatter (a liquid, etc.) about. 2 To spatter, wet, or soil with a liquid dashed about. 3 To make by throwing splashes: to *splash* one's way. 4 To decorate with splashed ornament. 5 To make a splash or splashes. 6 To move, fall, or strike with a splash or splashes. — *n.* 1 The act or noise of splashing. 2 The result of splashing; a spot made by a liquid or color splashed on. 3 In logging, a head of water released suddenly from a splash dam to drive a body of logs. [Var. of PLASH[1]]

splash·board (splash′bôrd′, -bōrd′) *n.* 1 Any of various devices to protect against splashes, especially a dashboard for a vehicle. 2 A board for closing the spillway or sluice of a dam. Also **splash′wing** (-wing′).

splash dam A dam used in logging to back up water to be used in a log drive: sometimes portable.

splash·er (splash′ər) *n.* 1 One who or that which splashes. 2 A piece of oilcloth, toweling, rug, or other device to protect a surface against splashing, as at the back of a washstand or over a wheel.

splash·y (splash′ē) *adj.* 1 Slushy; wet. 2 Marked by or as by splashes; blotchy. 3 *Colloq.* Sensational; showy: They made a *splashy* appearance.

splat (splat) *n.* A thin, broad piece of wood, as that forming the middle of a chair back. [Origin uncertain]

splat·ter (splat′ər) *v.t. & v.i.* To spatter or splash. — *n.* A spatter; splash. [Blend of SPLASH and SPATTER]

splay (splā) *adj.* Spread out; displayed; broad; clumsy; clumsily formed: a *splay* mouth. — *n. Archit.* A slanted surface or beveled edge, as of the sides of a doorway or window, or of a joist. — *v.t.* To make with a splay; bevel or chamfer away a corner or angle of, as a window opening. 2 To open to sight; spread; cut open; display. 3 In farriery, to dislocate. — *v.i.* To spread out; open. 5 To slant; slope. [Aphetic var. of DISPLAY]

splay·foot (splā′fo͝ot′) *n.* 1 Abnormal flatness and turning outward of the feet. 2 A foot so deformed. — **splay′–foot′ed** *adj.*

spleen (splēn) *n. Anat.* 1 A highly vascular, flattened, ductless organ found near the stomach of most vertebrates, which effects certain modifications in the blood. ◆ Collateral adjective: *lienal*. 2 This organ regarded as the seat of various emotions. 3 Ill temper; spitefulness: to vent one's *spleen*. 4 *Archaic* Lowness of spirits; melancholy; hypochondria. 5 *Obs.* Mode or state of mind; also, caprice; a fit of pique. 6 *Obs.* Violent mirth. [<L *splen* <Gk. *splēn*] — **spleen′ish** *adj.* — **spleen′y** *adj.*

spleen·ful (splēn′fəl) *adj.* Affected with spleen; peevish; ill-tempered. — **spleen′ful·ly** *adv.*

spleen·wort (splēn′wûrt′) *n.* Any of a genus (*Asplenium*) of hardy and cultivated ferns with simple or compound fronds: so called from the use formerly made of some species in disorders of the spleen.

splen·dent (splen′dənt) *adj.* 1 Shining; lustrous. 2 Illustrious. [<L *splendens, -entis,* ppr. of *splendere* shine]

splen·did (splen′did) *adj.* 1 Magnificent; imposing. 2 Inspiring to the imagination; glorious; illustrious. 3 Giving out or reflecting brilliant light; shining. 4 *Colloq.* Very good; excellent: a *splendid* offer. See synonyms under FINE[1], BRIGHT. [<L *splendidus* < *splendere* shine] — **splen′did·ly** *adv.* — **splen′did·ness** *n.*

splen·di·de men·dax (splen′di·dē men′daks) *Latin* Splendidly false; nobly untruthful.

splen·dif·er·ous (splen·dif′ər·əs) *adj. Colloq.* Exhibiting great splendor; very magnificent: a facetious usage. [<SPLEND(OR) + -(I)FEROUS]

splen·dor (splen′dər) *n.* 1 Exceeding brilliance from emitted or reflected light. 2 Magnificence. 3 Conspicuous greatness of achievement; preeminence. Also *Brit.* **splen′dour**. — **splen′dor·ous, splen′drous** *adj.*

sple·net·ic (spli·net′ik) *adj.* 1 Pertaining to the spleen. 2 Fretfully spiteful; peevish. See synonyms under MOROSE. Also **sple·net′i·cal, splen·i·tive** (splen′ə·tiv). — *n.* 1 One suffering from disease of the spleen. 2 A peevish person. — **sple·net′i·cal·ly** *adv.*

splen·ic (splen′ik, splē′nik) *adj.* Of, in, or pertaining to the spleen.

splinter

sple·ni·tis (spli·nī′tis) *n. Pathol.* Inflammation of the spleen.

sple·ni·um (splē′nē·əm) *n. pl.* **·ni·a** (-nē·ə) 1 *Surg.* A compress or bandage. 2 *Anat.* The rounded posterior end of the corpus callosum. [<NL <Gk. *splēnion* a bandage] — **sple′ni·al** *adj.*

sple·ni·us (splē′nē·əs) *n. pl.* **·ni·i** (-nē·ī) *Anat.* A large, thick muscle of the back of the neck, extending in two parts from the skull to the vertebral spines in the cervical and upper thoracic region. [<NL <Gk. *splēnion* a bandage] — **sple′ni·al** *adj.*

spleno– combining form *Anat. & Med.* The spleen; of or related to the spleen. Also, before vowels, **splen–**, as in *splenitis*. [<Gk. *splēn, splēnos* the spleen]

spleu·chan (sploo′khən) *n. Scot. & Irish* A small bag or wallet to hold tobacco, etc.: sometimes used as a purse. [<Irish *spliúcán* a leather pouch]

splice (splīs) *v.t.* **spliced, splic·ing** 1 To unite, as two ropes or parts of a rope, so as to form one continuous piece, by intertwining the strands. 2 To connect, as timbers, by beveling, scarfing, or overlapping at the ends. 3 *Slang* To join in marriage: usually in the passive. — **to splice the main brace** To serve or take a glass of grog: chiefly jocular. — *n.* 1 A union at the ends of joined parts, especially of ropes, made by intertwining the strands. 2 The place at which two parts are spliced. [<MDu. *splissen*]

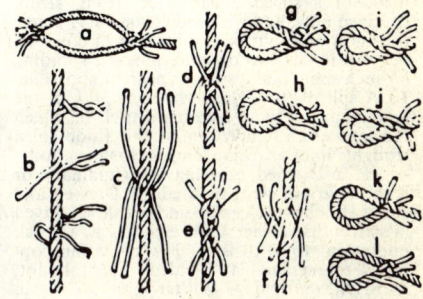

SPLICES
a. Cut splice.
b–c. Long splices.
d–f. Short splices.
g–l. Eye splices.

splic·er (splī′sər) *n.* 1 One who makes splices. 2 An implement by which a splice is made in a rope; a fid for splicing.

spline (splīn) *n.* 1 *Mech.* A metal key permanently set into a slot in one of two connected rotating mechanical parts, as a shaft and a pulley, and engaging with a similar slot cut in the other, thus permitting both parts to have relative lengthwise motion, but not to rotate upon each other: also called *feather, feather key*. 2 A long, flexible strip of wood or hard rubber, used by mechanical draftsmen to lay down ship lines, railway curves, or similar work. 3 A thin strip or tongue of wood or metal used in matching grooved planks, making partitions, filling air spaces, etc. — *v.t.* **splined, splin·ing** 1 To make a slot or groove in for a spline. 2 To fit with a spline. [? Related to SPLINT] — **splined** *adj.*

splint (splint) *n.* 1 A thin, flat piece split off; a splinter. 2 A thin, flexible strip of split wood used for basket-making, chair bottoms, etc. 3 In plate armor, one of the flexibly adjusted overlapping laminae. 4 *Surg.* An appliance, as of wood or metal, used for keeping a fractured limb or other injured part in a fixed position. 5 A splint bone. 6 An osseous tumor on the splint bone of a horse, due to inflammation of the periosteum; also, a bony callosity resulting from disease of the splint bones. — *v.t.* To confine, support, or brace, as a fractured limb, with or as with splints. [<MDu. *splinte*]

splint armor Armor made of overlapping metal plates.

splint bone 1 One of the small rudimentary bones of the metacarpus or metatarsus of the horse and related animals. Compare illustration under HORSE. 2 The fibula.

splin·ter (splin′tər) *n.* A thin, sharp piece of wood, glass, metal, etc., split or torn off

---

add, āce, câre, pälm; end, ēven; it, īce; odd, ōpen, ôrder; tōōk, pōōl; up, bûrn; ə = a in *above*, e in *sicken*, i in *clarity*, o in *melon*, u in *focus*; yōō = u in *fuse*; oi, oil; pout; ch, check; g, go; ng, ring; th, thin; th, this; zh, vision. Foreign sounds å, œ, ü, kh, ñ; and ◆: see page xx. < from; + plus; ? possibly.

lengthwise; a sliver. — *v.t.* & *v.i.* To split into thin sharp pieces or fragments; shatter; shiver. [<MDu.] — **splin′ter·y** *adj.*

splin·ter-proof (splin′tər·prōōf′) *adj.* Resistant to the penetration of splinters: said especially of shelters affording protection from machine-gun fire and shell fragments.

split (split) *v.* **split, split·ting** *v.t.* **1** To separate into parts by force, especially into two approximately equal parts. **2** To break or divide lengthwise or along the grain; rive; separate into layers. **3** To divide into groups or factions; disrupt, as a political party. **4** To divide and distribute by portions or shares. — *v.i.* **5** To break apart; divide lengthwise or along the grain. **6** To become divided or disunited through disagreement, etc. **7** To share something with others. — **to split hairs** To make fine distinctions; be unnecessarily precise or subtle. — **to split off 1** To break off by splitting. **2** To separate by or as by splitting. — **to split the difference** To divide equally a sum in dispute. — **to split up 1** To separate into parts and distribute. **2** To cease association; separate. — *n.* **1** The act or result of splitting; a longitudinal fissure; cleft; rent. **2** Separation of an aggregate body into factions; rupture; schism: a *split* in the church. **3** A sliver; splinter. **4** A share or portion, as of loot or booty. **5** A six-ounce bottle of an alcoholic beverage or of mineral water. See BABY. **6** A split osier, used in certain phases of basketweaving. **7** A confection made of a sliced banana, ice-cream, sirup, chopped nuts, and whipped cream. **8** A single thickness of a split skin or hide. **9** In bowling, the position of two or more pins left standing on such spots that a spare is nearly impossible. **10** A split ballot: There were 47 *splits* in the ballot box. **11** An acrobatic trick in which the legs are extended upon the floor in a straight line at right angles to the body. — *adj.* **1** Divided, especially longitudinally or with the grain; cleft; fissured. **2** Dressed and cured after being cleaned: said of fish. **3** Given in sixteenths, rather than eighths, as a stock quotation: 10 1/16 is a *split* quotation: opposed to *regular.* **4** Divided: a *split* ballot. [<MDu. *splitten*] — **split′ter** *n.*

Split (splēt) The chief Dalmatian city of southern Croatia, Yugoslavia; a major port on the Adriatic: Italian *Spalato.*

split infinitive See under INFINITIVE.

split-lev·el (split′lev′əl) *adj.* Designating a type of dwelling in which the floors of adjoining parts are at different levels, connected by short flights of stairs, permitting a compact arrangement of living and service rooms.

split product *Chem.* Any product of a decomposition, as of a protein into amino acids, or the like.

split saw A ripsaw.

split ticket 1 A ballot on which the voter has distributed his vote among candidates of different parties. **2** A ballot containing names of candidates of more than one party or party faction. Compare STRAIGHT TICKET.

split·ting (split′ing) *adj.* Acute or extreme in kind or degree: a *splitting* pain.

splore (splôr, splōr) *n. Scot.* A noisy frolic; carouse.

splotch (sploch) *n.* A discolored spot, as of ink, etc.; a daub; splash; spot. — *v.t.* To soil or mark with a splotch or splotches. [Cf. OE *splot* spot] — **splotch′y** *adj.*

splurge (splûrj) *Colloq. n.* **1** An ostentatious display. **2** An extravagant expenditure. — *v.i.* **splurged, splurg·ing 1** To show off; be ostentatious. **2** To spend money lavishly or wastefully. [Imit.] — **splurg′y** *adj.*

splut·ter (splut′ər) *v.i.* **1** To make a series of slight, explosive sounds, or throw off small particles, as meat frying in fat. **2** To speak hastily, confusedly, or incoherently, as from surprise or indignation. — *v.t.* **3** To utter excitedly or confusedly; sputter. **4** To spatter or bespatter. — *n.* A noise as of spluttering; bustle; confused stir. [Blend of SPLASH and SPUTTER] — **splut′ter·er** *n.*

spode (spōd) *n.* A fine porcelain or pottery made at the works founded by Josiah Spode, in Staffordshire, England.

Spode (spōd), **Josiah,** 1754–1827, English potter.

spod·u·mene (spoj′ōō·mēn) *n.* A vitreous, transparent to translucent lithium-aluminum silicate, belonging to the pyroxene group and crystallizing in the monoclinic system. [<Gk. *spodoumenos,* ppr. of *spodoesthai* be burned to ashes < *spodos* ashes]

spoil (spoil) *v.* **spoiled** or **spoilt, spoil·ing** *v.t.* **1** To impair or destroy the value, usefulness, or beauty of; injure: to *spoil* a book. **2** To weaken or impair the character or personality of, especially by overindulgence: Spare the rod and *spoil* the child. **3** *Obs.* To take property from by force; despoil. **4** *Obs.* To seize by force. — *v.i.* **5** To lose normal or useful qualities; specifically, to become tainted or decayed, as food. **6** *Obs.* To plunder; rob. See synonyms under CORRUPT, DECAY, DEFILE[1], INDULGE, PAMPER. — **to be spoiling for** To long for; crave: He is *spoiling for* a fight. — *n.* **1** Plunder seized by violence; booty; loot. **2** *pl.* The emoluments of public office as the objects of political contests and rewards of political service. **3** The act of pillaging; spoliation. **4** An object to be forcibly seized and taken away. **5** *Obs.* Ruin; destruction. **6** Material removed in digging trenches or excavations. **7** *Obs.* Damage; waste. See synonyms under PLUNDER. [<OF *espoillier* <L *spoliare* < *spolium* booty]

spoil·age (spoi′lij) *n.* **1** Spoiled material collectively. **2** Something that is or has been spoiled. **3** The process of spoiling. **4** The state of being spoiled.

spoil·er (spoi′lər) *n.* **1** One who takes spoil; a robber; despoiler. **2** One who or that which causes to spoil; a corrupter.

spoil-five (spoil′fīv′) *n.* A card game played by 2 to 10 persons, in which each player tries to take the pool or to spoil it, taking being accomplished by winning, spoiling by preventing the other players from winning, three out of five possible tricks in a deal.

spoils·man (spoilz′mən) *n. pl.* **·men** (-mən) One who advocates the spoils system or works for a political party for spoils.

spoil-sport (spoil′spôrt′, -spōrt′) *n.* A person whose actions or attitudes spoil the pleasures of others.

spoils system The theory, or the practice of a political party after a victorious campaign, of making public offices the rewards of partisan services.

Spo·kane (spō·kan′) The second largest city in Washington, located on a falls in the **Spokane River,** a river which flows 100 miles west from western Idaho to the Columbia River in eastern Washington.

spoke[1] (spōk) *n.* **1** One of the members of a wheel which serve to support the rim (or felly) by connecting it to the hub. **2** One of the radial handles of a ship's steering wheel. **3** A stick or bar for insertion in a wheel to prevent it from turning, as in descending a hill. **4** A rung of a ladder. — **to put a spoke in (someone's) wheel** To hinder or prevent (someone's) action. — *v.t.* **spoked, spok·ing 1** To provide with spokes. **2** To fasten (a wheel) with a stick or spoke to prevent its turning. [OE *spāca*]

spoke[2] (spōk) Past tense and archaic past participle of SPEAK.

spo·ken (spō′kən) Past participle of SPEAK. — *adj.* **1** Uttered orally, as opposed to written. **2** Speaking or having a specified kind of speech: smooth-*spoken.*

spoke·shave (spōk′shāv′) *n.* A wheelwright's tool having a blade set between two handles, used with a drawing motion in rounding and smoothing wooden surfaces.

spokes·man (spōks′mən) *n. pl.* **·men** (-mən) One who speaks in the name and behalf of another or others. — **spokes′wom′an** (-wŏŏm′·ən) *n. fem.*

Spo·le·to (spō·lā′tō) A town of central Italy, site of extensive Roman ruins and several medieval churches, including an 11th century cathedral. Ancient **Spo·le·ti·um** (spō·lē′shē·əm).

spo·li·a·tion (spō′lē·ā′shən) *n.* **1** The act of despoiling; specifically, the plundering of neutral commerce by a belligerent. **2** *Law* Destruction; mutilation; alteration; specifically, the erasure, alteration, mutilation, or destruction of a paper to prevent its being used as evidence. **3** In English canon law, the taking of the fruits of a benefice under a pretended but illegal title, or a writ or suit brought on such grounds. **4** *Law* The destruction of a ship's papers so as to conceal its nationality, the character of its trade, cargo, etc. [<L *spoliatio, -onis* < *spoliare* despoil] — **spo′li·a′tor** *n.*

spo·li·a·tive (spō′lē·ā′tiv) *adj.* Tending to abstract from or lessen; in medicine, resulting in a considerable loss of blood.

spon·da·ic (spon·dā′ik) *adj.* **1** Pertaining to or of the nature of a spondee; composed of spondees. **2** Having a spondee in a position where another kind of metrical foot is usual. Also **spon·da′i·cal.** [<L *spondaicus* <Gk. *spondeiakos* < *spondē.* See SPONDEE.]

spon·dee (spon′dē) *n.* A metrical foot consisting of two long syllables or, in English verse, of two accented syllables. [<F *spondée* <Gk. *spondeios (pous)* libation (meter) < *spondē* a libation; because used in the solemn chants accompanying a libation]

spon·du·lics (spon·dōō′liks) *n. U.S. Slang* Cash money. Also **spon·du′licks, spon·du′lix.**

spon·dy·li·tis (spon′də·lī′tis) *n. Pathol.* Pott's disease.

spondylo– *combining form Anat.* & *Med.* A vertebra; of or pertaining to vertebrae. Also, before vowels, **spondyl–.** [<Gk. *spondylos* a vertebra]

sponge (spunj) *n.* **1** Any of a phylum *(Porifera)* of fixed, usually marine organisms characterized by a highly porous body without specialized internal organs. **2** The skeleton or network of elastic fibers that remains after the removal of the living matter from certain sponges and that readily absorbs liquids: used as an absorbent, for bathing, etc. **3** Some spongelike implement or substance that serves as an absorbent, as a swabbing implement for cleaning a cannon bore after discharge. **4** Leavened dough, or dough in the process of leavening and before kneading. **5** A porous, spongelike form assumed by finely divided metals, as iron and platinum. **6** *Surg.* An absorbent pad, as of sterilized gauze, used in operations, etc., to absorb blood or other fluid matter. **7** One who consumes or absorbs a great deal, as of food or drink. **8** *Colloq.* A person who lives at the expense of another or others; a parasite. — **to throw** (or **toss**) **up** (or **in**) **the sponge** *Colloq.* To yield; give up; abandon the struggle. — *v.* **sponged, spong·ing** *v.t.* **1** To wipe, wet, or clean with a sponge. **2** To wipe out; expunge; erase. **3** To absorb; suck in, as a sponge does. **4** *Colloq.* To get by mean device or at another's expense. — *v.i.* **5** To be absorbent. **6** To gather or fish for sponges. **7** *Colloq.* To live at the expense of others. See synonyms under CLEANSE. [OE <L *spongia,* ult. <Gk. *spongos.* Akin to FUNGUS.]

sponge cake A cake of sugar, eggs, and flour, containing no shortening and beaten very light.

spong·er (spun′jər) *n.* **1** One who or that which sponges in any sense. **2** A person or vessel that gathers sponges. **3** A human parasite.

spon·gi·form (spun′jə·fôrm, spon′-) *adj.* Resembling a sponge in form or structure.

spon·gin (spun′jin) *n. Biochem.* A protein from the skeletal tissue of sponges and corals.

spon·gi·o·blast (spon′jē·ə·blast′) *n. Biol.* **1** An epithelial cell of the embryonic neural tube which becomes transformed into a cell of the tissue lining the central cavities of the brain and spinal cord. **2** A spongoblast. [<Gk. *spongia,* var. of *spongos* a sponge + -BLAST]

spon·go·blast (spon′gō·blast) *n. Biol.* An ameboid cell in the mesenchyme of sponges by which spongin is secreted. Also **spon·gin·blast** (spun′jin·blast, spon′-). [<Gk. *spongos* a sponge + -BLAST]

spong·y (spun′jē) *adj.* **spong·i·er, spong·i·est 1** Having the nature or character of a sponge; elastic, compressible, and porous. **2** Having the quality of imbibing fluids; absorptive. **3** Existing in a condition of fine division and loose coherence. **4** *Obs.* Wet; soaked. Also **spon·gi·ose** (spun′jē·ōs). — **spong′i·ness** *n.*

spon·sal (spon′səl) *adj.* Relating to marriage or to a spouse. [<L *sponsus,* pp. of *spondere* promise]

spon·sion (spon′shən) *n.* **1** The act of becoming surety or sponsor for another. **2** In international law, an undertaking on behalf of his state by a public officer not specifically empowered to enter into it.

spon·son (spon′sən) *n.* **1** A curved projection from the hull of a vessel or seaplane, to give greater stability or increase the surface area. **2** A similar protuberance on a ship or

sponsor

tank, for storage purposes or for the training of a gun. 3 An air tank built into the side of a canoe, to improve stability and prevent sinking. [Appar. alter. of EXPANSION]
spon·sor (spon′sər) *n.* 1 One who makes himself responsible for a statement by, or the debt or duty of, another; a surety. 2 One who makes the required professions and promises for an infant at baptism and becomes responsible for its religious training; a godfather or godmother. 3 A business firm or enterprise that assumes all the costs of a radio or television program which advertises its product or service. — *v.t.* To act as sponsor for; answer or vouch for. — **spon′so·ri·al** (spon-sôr′ē-əl, -sō′rē-) *adj.* — **spon′sor·ship** *n.*
spon·ta·ne·i·ty (spon′tə-nē′ə-tē) *n. pl.* **·ties** 1 Spontaneous quality. 2 The tendency to action or behavior independent of external forces, conditions, or influences.
spon·ta·ne·ous (spon-tā′nē-əs) *adj.* 1 Arising from inherent qualities or tendencies without external efficient cause; done or acting from one's own impulse, prompting, or desire. 2 Not having material causation outside itself. 3 Generated or produced without human labor; wild or sporadic; indigenous. 4 *Biol.* Apparently arising independently of external stimulus, influence, or conditions. [<LL *spontaneus* <L *sponte* of free will] — **spon·ta′ne·ous·ly** *adv.* — **spon·ta′ne·ous·ness** *n.*
— *Synonyms*: automatic, instinctive, involuntary, unbidden, voluntary, willing. That is *spontaneous* which is freely done, with no external compulsion and, in human actions, without special premeditation or distinct determination of the will; that is *voluntary* which is freely done with distinct act of will; that is *involuntary* which is independent of the will, and perhaps in opposition to it; a *willing* act is not only in accordance with will, but with desire. Thus *voluntary* and *involuntary*, which are antonyms of each other, are both partial synonyms of *spontaneous*. An infant's smile in answer to that of its mother is *spontaneous*; the smile of a pouting child wheedled into good humor is *involuntary*. In physiology the action of the heart and lungs is *involuntary* action; the growth of the hair and nails is *spontaneous*; the action of swallowing is *voluntary* up to a certain point, beyond which it becomes *involuntary* or *automatic*.
spontaneous combustion The oxidation of a substance with such rapidity as to engender heat sufficient to ignite it, as masses of oiled rags, finely powdered ores, coal, and certain metals.
spontaneous generation *Biol.* Abiogenesis.
spon·toon (spon-tōōn′) *n.* A half-pike usually armed with a hook, carried by infantry officers in the 18th century: also spelled *espontoon*. [<F *sponton* <Ital. *spontone* pike <*puntone* a point]
spoof (spōōf) *Colloq. v.t. & v.i.* To deceive or hoax; joke. — *n.* Deception; humbug; hoax. [after a nonsensical game invented by Arthur Roberts, 1852–1933, English comedian]
spook (spōōk) *n.* A ghost; an apparition; specter. — *v.t. Colloq.* To haunt (a person or place). [<Du.] — **spook′ish** *adj.*
spook·nik (spōōk′nik) *n. U.S. Slang* An unidentified flying object reported to be a flying saucer or the like, but officially interpreted either as a result of faulty observation or as an outright hallucination. [<SPOOK + (SPUT)NIK]
spook·y (spōō′kē) *adj.* **spook·i·er, spook·i·est** 1 Like a ghost; ghostly. 2 Haunted. 3 Suggesting the presence or agency of ghosts; eerie. — **spook′i·ly** *adv.* — **spook′i·ness** *n.*
spool (spōōl) *n.* 1 A small cylinder, commonly of wood and with a flange at each end and an axial bore, upon which thread or yarn is or may be wound. 2 The quantity of thread held by a spool; also, the spool and the thread upon it. 3 Anything resembling a spool in shape or purpose. — *v.t.* To wind on a spool. [<MLG *spole*]
spoon (spōōn) *n.* 1 A utensil having a shallow, generally ovoid bowl and a handle, used in preparing, serving, or eating food. 2 Something resembling a spoon or its bowl. 3 A metallic lure attached to a fishing line: also **spoon bait, trolling spoon.** 4 A concave overhanging extension on a torpedo tube to keep the launched torpedo in a straight course. 5 A wooden golf club with lofted face and comparatively short, stiff shaft, used by some players for approaching. — *v.t.* 1 To lift up or out with a spoon. 2 To hollow out like the bowl of a spoon. 3 In certain games, to play or hit (the ball) with little force up into the air; in croquet, to shove or scoop (the ball) with the mallet. — *v.i.* 4 To fish with a spoon. 5 In certain games, to spoon the ball. 6 *Colloq.* To make love, especially openly and demonstratively. [OE *spōn* sliver, chip]

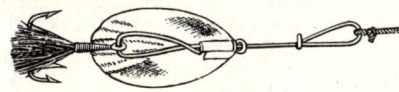
SPOON BAIT

spoon·bill (spōōn′bil′) *n.* 1 A wading bird (genera *Platalea* or *Ajaia*) related to the ibises, having the bill broad and flattened. 2 The shoveler (def. 2). 3 The paddlefish. — **spoon′-billed′** *adj.*
spoon·bread (spōōn′bred′) *n.* A quick bread made of cornmeal, eggs, milk, and shortening, baked soft enough to be served with a spoon: also called *batter bread*.
spoon·drift (spōōn′drift) *n.* Spindrift.
spoon·er·ism (spōō′nə-riz′əm) *n.* The unintentional transposition of sounds or of parts of words in speaking, as in "half-*warmed fish*" for "half-*formed wish*." [after William A. Spooner, 1844–1930, of New College, Oxford, who was renowned for such slips of the tongue]
spoon-fed (spōōn′fed′) *adj.* Fed with a spoon; hence, pampered.
spoon·ful (spōōn′fōōl′) *n. pl.* **·fuls** As much as a spoon will hold; especially, a teaspoonful.
spoon hook A fish hook with a bright, revolving, spoon-shaped piece of metal attached.
spoon·y (spōō′nē) *Colloq. adj.* **spoon·i·er, spoon·i·est** Sentimental or silly, as in lovemaking; soft. — *n. pl.* **spoon·ies** A foolish, demonstrative lover; sentimental simpleton. Also **spoon′ey.**
spoor (spōōr) *n.* 1 A track; trail. 2 Footprint or other trace of a wild animal. — *v.t. & v.i.* To track by or follow a spoor. [<Du.]
Spor·a·des (spôr′ə-dēz, *Greek* spô-rä′thes) 1 Loosely, all the Greek islands in the Aegean, exclusive of the Cyclades. 2 Anciently, the islands of the SE Aegean, off western Asia Minor, including the Dodecanese, Icaria, Samos, and, in some usages, Chios and Lesbos. 3 Strictly, the **Northern Sporades,** a group of Greek islands in the western Aegean, off the coasts of Euboea and Thessaly, the chief of which is Skyros: Greek *Voriaí Sporádes.*
spo·rad·ic (spə-rad′ik, spō-) *adj.* 1 Occurring here and there; occasional. 2 Separate; isolated. 3 Not widely diffused; neither epidemic nor endemic: said of disease. Also **spo·rad′i·cal.** [<Med. L *sporadicus* <Gk. *sporadikos* <*sporas* scattered] — **spo·rad′i·cal·ly** *adv.* — **spo·rad′i·cal·ness** *n.*
spor·a·do·sid·er·ite (spôr′ə-dō-sid′ər-īt, spō-) *n.* A meteorite consisting of a typically crystalline mass with disseminated particles of nickeliron alloy. [<SPORADIC + SIDERITE]
spo·ran·gi·o·spore (spô-ran′jē-ə-spôr′, spō-ran′jē-ə-spōr′) *n. Bot.* A spore produced within a sporangium.
spo·ran·gi·um (spô-ran′jē-əm, spō-) *n. pl.* **·gi·a** (-jē-ə) *Bot.* A sac in which asexual spores are produced endogenously, as in certain algae and fungi. Also called *spore case.* [<SPOR(O)- + Gk. *angeion* a vessel] — **spo·ran′gi·al** *adj.*
spore (spôr, spōr) *n.* 1 *Bot.* The reproductive body in flowerless plants, analogous to the seeds of flowering plants, but containing no embryo. They are free, usually single-celled and highly resistant bodies, produced externally or in some closed sac or cavity, and are capable of developing at once or after a time into an independent organism or individual. 2 A minute body that develops into a new individual; any minute organism; a germ. — *v.i.* **spored, spor·ing** To develop spores: said of plants. [<Gk. *spora* seed, sowing] — **spo·ra·ceous** (spô-rā′shəs, spō-) *adj.*
spore case A sporangium.
spore fruit *Bot.* An ascocarp; any plant structure producing spores.
spo·rif·er·ous (spô-rif′ər-əs, spō-) *adj.* Bearing spores.
sporo- *combining form* Seed; spore: *sporophyte.* Also, before vowels, **spor-.** [<Gk. *spora* a seed]
spo·ro·carp (spôr′ə-kärp, spō′rə-) *n. Bot.* 1 A many-celled form of fruit produced from a fertilized archicarp in certain of the lower cryptogams, especially red algae and ascomycetous fungi: also called *cystocarp.* 2 The sporogonium in mosses.
spo·ro·cyst (spôr′ə-sist, spō′rə-) *n. Zool.* 1 An asexual form of a trematode worm that develops directly from the embryo and in which mouth and intestinal tract are wanting. 2 An encysted organism, especially a protozoan, that gives rise to spores.
spo·ro·cyte (spôr′ə-sīt, spō′rə-) *n. Biol.* The mother cell from which spores are produced.
spo·ro·gen·e·sis (spôr′ə-jen′ə-sis, spō′rə-) *n. Biol.* 1 Reproduction by spores. 2 Sporogony. — **spo·rog·e·nous** (spə-roj′ə-nəs, spō-) *adj.*
spo·ro·go·ni·um (spôr′ə-gō′nē-əm, spō′rə-) *n. pl.* **·ni·a** (-nē-ə) *Bot.* An elongated stalk having upon its summit a capsule in which the asexual spores of liverworts and mosses are produced.
spo·rog·o·ny (spə-rog′ə-nē, spō-) *n. Biol.* Spore formation; specifically, in sporozoans, the development of spores from a mature zygote. [<SPORO- + -GONY]
spo·ro·phore (spôr′ə-fôr, spō′rə-fōr) *n.* 1 A spore-bearer or seed-bearer. 2 *Bot.* In fungi, a branch from the thallus which bears the spores.
spo·ro·phyll (spôr′ə-fil, spō′rə-) *n. Bot.* The leaf, or modified leaf, which bears the sporangia. Also **spo′ro·phyl.**
spo·ro·phyte (spôr′ə-fīt, spō′rə-) *n. Bot.* The spore-bearing individual or generation in certain plants which reproduce by alternation of generations.
spo·ro·tri·cho·sis (spôr′ə-tri-kō′sis, spō′rə-) *n. Pathol.* A chronic disease caused by a fungus (genus *Sporotrichum*) and marked by the formation of ulcerated lesions in the lymph nodes or subcutaneous tissue. [<NL *Sporotrichum,* genus of fungi + -OSIS]
-sporous *combining form* Having (a specified number or kind of) spores: *homosporous.* [<SPOR(O)- + -OUS]
spo·ro·zo·an (spôr′ə-zō′ən, spō′rə-) *adj.* Designating or belonging to a class (*Sporozoa*) of parasitic protozoans developing by asexual and sexual stages and reproducing by sporulation, as the malaria parasite. — *n.* One of the class *Sporozoa.* [<SPORO- + Gk. *zōion* animal]
spo·ro·zo·ite (spôr′ə-zō′īt, spō′rə-) *n. Zool.* An aggregation of protoplasm of a sporozoan zygote, segmented off as a minute sickle-shaped germ: the initial phase of the malaria parasite in its host.
spor·ran (spor′ən) *n.* A skin pouch, generally with the fur on, worn in front of the kilt by Highlanders. [<Scottish Gaelic *sporan* <LL *bursa* purse]
sport (spôrt, spōrt) *n.* 1 That which amuses in general; diversion; pastime. 2 A particular game or play pursued for diversion, especially an outdoor or athletic game, as baseball, football, track, tennis, swimming, etc. 3 A spirit of jesting or raillery. 4 That with which one sports; a toy; plaything. 5 Mockery; an object of derision: to make *sport* of someone; also, a laughingstock; butt. 6 *Biol.* An animal or plant, or one of its parts, that exhibits sudden and spontaneous variation from the normal type; a mutation. 7 *Bot.* A bud variation. 8 *Colloq.* One whose interest in sport lies chiefly in gambling; a gamester or gambler. 9 *Colloq.* One who lives a fast, gay, or flashy life. 10 A person characterized by his observance of the rules of fair play, or by his ability to get along with others: a good *sport.* 11 *Archaic* Amorous fondling; wanton dalliance. — *v.i.* 1 To amuse oneself; play; frolic.

2 To participate in games. 3 To make sport or jest; trifle. 4 *Bot.* **a** To vary suddenly or spontaneously from the normal type; mutate. **b** To display bud variation. 5 *Archaic & Dial.* To make love in a sportive or trifling manner. — *v.t.* 6 *Colloq.* To display or wear ostentatiously; show off. 7 *Obs.* To amuse; divert. See synonyms under FRISK. — *adj.* Of, pertaining to, or fitted for sports; also, appropriate for informal outdoor wear: a *sport* coat: also **sports.** [Aphetic var. of DISPORT] — **sport′er** *n.* — **sport′ful** *adj.* — **sport′ful·ly** *adv.* — **sport′ful·ness** *n.*

Synonyms (noun): amusement, diversion, entertainment, frolic, fun, gaiety, gambol, game, jollity, joviality, merriment, merrymaking, mirth, pastime, play, playfulness, pleasantry, pleasure, prank, recreation. See ENTERTAINMENT, FROLIC. Compare RIDICULE.

sport·ing (spôr′ting, spōr′-) *adj.* 1 Pertaining to, engaged in, or used in connection with athletic games or field sports. 2 Characterized by the spirit of sportsmanship; conforming to the codes or standards of sportsmanship. 3 Interested in or associated with sports for gambling or betting: a *sporting* man. — **sport′ing·ly** *adv.*

sporting chance *Colloq.* A chance involving the risk of loss.

spor·tive (spôr′tiv, spōr′-) *adj.* 1 Relating to or fond of sport or play; frolicsome. 2 Interested in, active in, or related to sports. 3 *Obs.* Wanton or amorous. See synonyms under HUMOROUS, JOCOSE, MERRY, VIVACIOUS, WANTON. — **spor′tive·ly** *adv.* — **spor′tive·ness** *n.*

sports car A low, rakish automobile, usually seating two persons, and built for high speed and maneuverability.

sports·cast·er (spôrts′kas′tər, -käs′-, spōrts′-) *n. U.S.* One who broadcasts sports events, news, and comment.

sport shirt A shirt for informal wear, often cut square at the bottom so as to be worn inside or outside slacks. Also **sports shirt.**

sports·man (spôrts′mən, spōrts′-) *n. pl.* **·men** (-mən) 1 One who pursues field sports, especially hunting and fishing. 2 A professional gambler; also, one who bets on horse races. 3 One who abides by a code of fair play in games or in daily practice.

sports·man·like (spôrts′mən·līk′, spōrts′-) *adj.* Pertaining to sportsmen; honorable; generous; conforming to the rules of sportsmanship. Also **sports′man·ly.**

sports·man·ship (spôrts′mən·ship, spōrts′-) *n.* 1 The art or practice of field sports. 2 Honorable or sportsmanlike conduct.

sports·wear (spôrts′wâr′, spōrts′-) *n.* Clothes made for informal and outdoor activities.

sports·wom·an (spôrts′wŏŏm′ən, spōrts′-) *n. pl.* **·wom·en** (-wim′in) A woman who participates in sports.

sport·y (spôr′tē, spōr′-) *adj.* **sport·i·er, sport·i·est** *Colloq.* Relating to or characteristic of a sport; hence, gay, loud, or dissipated. — **sport′i·ly** *adv.* — **sport′i·ness** *n.*

spor·u·late (spôr′yə·lāt, spor′-) *v.i.* **·lat·ed, ·lat·ing** To form spores.

spor·u·la·tion (spôr′yə·lā′shən, spor′-) *n. Biol.* The act or condition of spore formation, especially by multiple cell division after encystment.

spor·ule (spôr′yōōl, spor′-) *n.* A spore; sometimes, a little spore. [Dim. of SPORE]

spot (spot) *n.* 1 A particular place of small extent; a definite locality. 2 Any small portion of a surface differing as in color from the rest; blot. 3 A stain or blemish on character; a fault; a reproach. 4 A congenital birthmark. 5 A food fish (*Leiostomus xanthurus*) of the Atlantic coast of the United States, marked with a spot above each pectoral fin; the oldwife. 6 One of the figures or pips with which a playing card is marked; also, a card having (a certain number of) such marks: the five *spot* of clubs. 7 *Slang* A currency note having a specified value: a ten *spot.* 8 *Chiefly Brit.* A portion or bit: a *spot* of tea. 9 *Slang* Position or situation: He was in a good *spot.* 10 *U.S. Slang* A spotlight. See synonyms under BLEMISH, PLACE. — **in a spot** *Slang* In a difficult or embarrassing situation; in trouble. — **in spots** Now and then, in some respects: He is bright *in spots.* — **to go to the spot** To satisfy a definite need or craving. — **to hit the spot** *Slang* To gratify an appetite or need. — **on the spot** 1 At once; immediately. 2 At the very place. 3 *Slang* **a** In danger of death. **b** Accountable or in danger of being held accountable for some action. — *v.* **spot·ted, spot·ting** *v.t.* 1 To mark or soil with spots. 2 To decorate with spots; dot. 3 To place on a designated spot; locate; station. 4 *Colloq.* To recognize or detect; see. 5 *Colloq.* To yield (an advantage or handicap) to (someone): We *spotted* them five points. — *v.i.* 6 To become marked or soiled with spots. 7 To make a stain or discoloration. 8 *Mil.* To observe the effect of gunfire to obtain data for improving its accuracy. See synonyms under STAIN. — *adj.* 1 Being on the place or spot. 2 Paid or prepared for payment on delivery; also, ready for instant delivery following sale. [ME <LG. Cf. MDu. *spotte* a spot.] — **spot′ta·ble** *adj.*

spot brake Disk brake.

spot cash Immediate payment on actual delivery.

spot·less (spot′lis) *adj.* Free from spot, stain, or impurity. See synonyms under INNOCENT, PERFECT, PURE. — **spot′less·ly** *adv.* — **spot′less·ness** *n.*

spot·light (spot′līt′) *n.* 1 A circle of powerful light thrown on the stage to bring an actor or actors into clearer view. 2 The apparatus that produces such a light. 3 A pivoted automobile lamp for illuminating objects beyond the range of the fixed lights. 4 Notoriety; publicity.

Spot·syl·va·ni·a (spot′sil·vā′nē·ə) A village in NE Virginia; scene of a 13-day battle in the Civil War, May, 1864. Formerly called **Spotsylvania Courthouse.**

spot·ted (spot′id) *adj.* 1 Discolored in spots; stained; soiled. 2 Characterized or marked by spots. 3 Blazed: said of trees, trails, boundary lines, etc.

spotted adder The house snake.

spotted crake A small European rail (*Porzana porzana*), allied to the American sora.

spotted cranebill See under CRANEBILL.

spotted fever *Pathol.* 1 Meningitis. 2 Typhus. 3 Rocky Mountain spotted fever.

spotted sandpiper See under SANDPIPER.

spot·ter (spot′ər) *n.* 1 One who or that which spots. 2 *Colloq.* A private detective. 3 An observation balloon. 4 A device on a railroad car that marks irregularities along the track. 5 In drycleaning, one who removes spots.

spot·ty (spot′ē) *adj.* **·ti·er, ·ti·est** 1 Having many spots. 2 Occurring in spots; lacking uniformity; unevenly distributed. — **spot′ti·ly** *adv.* — **spot′ti·ness** *n.*

spous·al (spou′zəl) *adj.* Pertaining to marriage. See synonyms under MATRIMONIAL. — *n.* Marriage; espousal.

spouse (spouz, spous) *n.* A partner in marriage; one's husband or wife. — *v.t.* **spoused, spous·ing** *Obs.* To wed; marry; espouse. [<OF *espous, espouse* <L *sponsus,* pp. of *spondere* promise, betroth]

spout (spout) *v.i.* 1 To pour out copiously and forcibly, as a liquid under pressure. 2 To discharge a fluid either continuously or in jets. 3 *Colloq.* To speak or orate pompously; declaim. — *v.t.* 4 To cause to pour or shoot forth. 5 To utter grandiloquently or pompously. 6 *Brit. Slang* To pawn or pledge. — *n.* 1 A tube, trough, etc., for the discharge of a liquid. 2 A continuous stream of fluid. 3 A shoot or lift; specifically, the shoot or lift in a pawnbroker's shop. 4 *Brit. Slang* A pawnbroker's shop. [ME *spoute*; origin uncertain] — **spout′er** *n.*

sprag (sprag) *n.* A billet of wood used to prevent a vehicle from slipping backward, or in mining as a prop to support coal when undermined. [Origin uncertain]

Sprague's pipit (sprāgz) The Missouri skylark.

sprain (sprān) *n.* 1 A violent straining or twisting of the ligaments surrounding a joint. 2 The condition due to such strain. [< *v.*] — *v.t.* To cause a sprain in; wrench the muscles of (a joint). [<OF *espreindre* squeeze <L *exprimere.* See EXPRESS.]

sprang (sprang) Alternative past tense of SPRING.

sprat (sprat) *n.* 1 A herringlike fish (*Clupea sprattus*) found in shoals on the Atlantic coast of Europe. 2 The young of the herring. [OE *sprott*]

sprat·tle (sprat′l) *Scot. v.i.* To struggle or scramble. — *n.* A struggle; scramble.

sprawl (sprôl) *v.i.* 1 To sit or lie with the limbs stretched out ungracefully. 2 To be stretched out ungracefully, as the limbs. 3 To move with awkward motions of the limbs. 4 To spread out in a straggling manner, as handwriting, vines, etc. — *v.t.* 5 To cause to spread or extend awkwardly or irregularly. — *n.* The act or position of sprawling; an awkward recumbent posture or movement. [OE *spreawlian* move convulsively] — **sprawl′er** *n.*

spray[1] (sprā) *n.* 1 Water or other liquid dispersed in fine particles. 2 An instrument for discharging small particles of liquid; an atomizer. [< *v.*] — *v.t.* 1 To disperse (a liquid) in fine particles. 2 To apply spray to, as with an atomizer. — *v.i.* 3 To send forth or scatter spray. 4 To go forth as spray. [Akin to MDu. *sprayen* sprinkle] — **spray′er** *n.*

spray[2] (sprā) *n.* 1 A small branch bearing dependent branchlets or flowers. 2 Any ornament, pattern, etc., resembling a collection of twigs or flowers. [ME; origin uncertain]

spread (spred) *v.* **spread, spread·ing** *v.t.* 1 To open or unfold to full width, extent, etc., as wings, sail, a map, etc. 2 To distribute over a surface, especially in a thin layer; scatter or smear. 3 To cover with a layer of something: to *spread* toast with marmalade. 4 To force apart or farther apart: The heavy train has *spread* the rails. 5 To extend over a period of time; prolong: He *spread* the payments over a six-month period. 6 To make more widely known, active, etc.; promulgate or diffuse: to *spread* a rumor; to *spread* contagion. 7 To set (a table, etc.), as for a meal. 8 To arrange or place on a table, etc., as a meal or feast. 9 To set forth or record in full. — *v.i.* 10 To be extended or expanded; increase in size, width, etc. 11 To be distributed or dispersed, as over a surface or area; scatter. 12 To become more widely known, active, etc. 13 To be forced farther apart; separate. — *n.* 1 The act of spreading: the *spread* of the gospel. 2 An open extent or expanse. 3 The limit or extent of expansion of some designated object, as of sail or a bird's wings. 4 *Aeron.* The maximum distance from tip to tip of an airplane wing. 5 A cloth or covering for a bed, table, or the like. 6 *Colloq.* An informal feast or banquet; also, a table with a meal set out on it. 7 Anything used to spread on bread: a cheese *spread.* 8 Two pages of a magazine or newspaper facing each other and covered by related material; also, print spread across two or more columns or on facing pages for advertising or display. 9 In finance and commerce, a straddle. 10 Diffusion; dispersion. — *adj.* Having a broad surface; expanded; outstretched. [OE *sprǣdan*]

Synonyms (verb): circulate, diffuse, disperse, disseminate, distribute, divulge, expand, extend, promulgate, propagate, scatter. See PUBLISH, STRETCH. *Antonyms*: check, confine, condense, contract, restrain.

spread-ea·gle (spred′ē′gəl) *adj.* Resembling a spread eagle; hence, extravagant; bombastic: applied especially to patriotic American oratory. — *v.* **-ea·gled, -ea·gling** *v.t.* To lash to the mast or shrouds in spread-eagle position as a punishment: a former practice. — *v.i.* To deliver an oration in bombastic, patriotic style. — **spread′-ea′gle·ism** *n.*

spread eagle 1 The figure of an eagle with extended wings: used as an emblem of the United States. 2 Any position or movement resembling this, as a figure in skating. 3 Extravagant speech; especially, American bombastic, patriotic oratory.

spread·er (spred′ər) *n.* 1 One who or that which spreads, as a small knife for spreading butter. 2 A bar of wood, metal, etc., to keep stays or wires apart, etc. 3 *Agric.* An implement for spreading hay, manure, or the like.

spreagh (sprākh) *n. Scot.* Property, particularly cattle, taken as plunder; booty; prey; a foray. Also **spreagh′er·y.**

spreck·le (sprek′əl) *n. & v. Scot. & Brit. Dial.* Speckle.

spree (sprē) *n.* 1 A drinking spell; drunken carousal. 2 A gay frolic. See synonyms under FROLIC. Compare SPORT. [Origin uncertain]

Spree (sprā, shprā) A river of eastern East Germany, flowing 250 miles north to the Havel River.

sprig (sprig) *n.* 1 A shoot or sprout of a tree or plant; an ornament in this form. 2 An offshoot from an ancestral stock; a young man. 3 One of various small, pointed implements.

4 A brad without a head. 5 A small, wedge-shaped piece of metal used to hold glass in a window sash. — *v.t.* **sprigged, sprig·ging** 1 To ornament with a design of sprigs. 2 To form (twigs or plants) into sprays. 3 To fasten with sprigs or brads. 4 To pluck sprigs from. [ME *sprigge*; origin uncertain] — **sprig′ger** *n.*

sprig·gy (sprig′ē) *adj.* **·gi·er, ·gi·est** Abounding in sprigs or small branches.

spright (sprīt) See SPRITE.

spright·ly (sprīt′lē) *adj.* **·li·er, ·li·est** Full of animation and spirits; vivacious; lively. — *adv.* Spiritedly; briskly; gaily. — **spright′li·ness** *n.*
Synonyms: airy, animated, brisk, bustling, cheerful, lively, nimble, spry, vivacious. The *sprightly* display a cheerful, pleasing lightness and quickness, spiritlike; *lively* has a similar meaning, as abounding in cheerful life. The *brisk* and *bustling* are full of stir, the former generally to some purpose. The *spry* are quick within a narrow range, according to the common proverb, "*spry* as a cricket." See ACTIVE, AIRY, CHEERFUL, HAPPY, NIMBLE, VIVACIOUS, VIVID.

spring (spring) *v.* **sprang** or **sprung, sprung, spring·ing** *v.i.* 1 To move or rise suddenly and rapidly; leap; dart: He *sprang* across the creek; The cat *sprang* into the air. 2 To move suddenly as by elastic reaction; snap: The jaws of the heavy trap *sprang* shut. 3 To move as if with a leap: An angry retort *sprang* to his lips. 4 To rise up suddenly, as birds from cover. 5 To work or snap out of place, as a mechanical part. 6 To become warped or bent, as boards. 7 To explode: said of a mine. 8 To rise above surrounding objects. 9 To come into being: New towns have *sprung* up. 10 To originate; proceed, as from a source. 11 To develop; grow, as a plant. 12 To be descended: He *springs* from good stock. 13 *Poetic* To begin to appear, as light or dawn. — *v.t.* 14 To cause to spring or leap. 15 To cause to act, close, open, etc., unexpectedly or suddenly, as by elastic reaction: to *spring* a trap. 16 To cause to happen, become known, or appear suddenly: to *spring* a surprise. 17 To leap over; vault. 18 To start (game) from cover; flush. 19 To explode (a mine). 20 To warp or bend; split. 21 To cause to snap or work out of place. 22 To force into place, as a beam or bar. 23 To suffer (a leak). 24 *Slang* To obtain the release of (a person) from prison or custody. See synonyms under LEAP, RISE. — *n.* 1 *Mech.* An elastic body or contrivance that yields under stress, and returns to its normal form when the stress is removed. 2 Elastic quality or energy. 3 The act of flying back from a position of tension; recoil. 4 An energy or power; a cause of action; impelling motive. 5 The act of leaping up or forward suddenly; a jump; bound. 6 The season in which vegetation starts anew; in the north temperate zone, the three months of March, April, and May; in the astronomical year, the period from the vernal equinox to the summer solstice. 7 A flow or fountain, as of water; hence, any source or origin of continued supply; a flow of curative water. 8 A crack or break, as of a plank, beam, or spar, or a thing sprung or warped. 9 *Archit.* The commencement of curvature in an arch. 10 A hinge. See illustration under HINGE. 11 *Scot.* A quick, lively tune. See synonyms under BEGINNING, CAUSE, SOURCE. — *adj.* 1 Pertaining to the season of spring. 2 Resilient; acting like or having a spring. 3 Hung on springs. [OE *springan*]

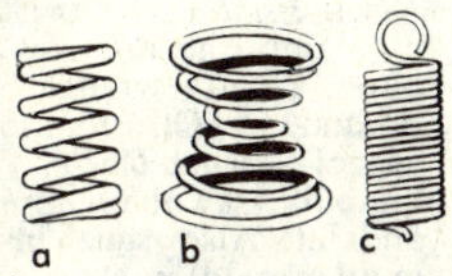

SPRING
a. Compression coil.
b. Double spiral.
c. Extension coil.

spring·al¹ (spring′əl) *n.* An engine like the ballista, used in medieval warfare: also *espringal.* Also **spring′ald** (-əld). [<AF *springalde,* OF *espringale* < *espringuer* spring < Gmc.]

spring·al² (spring′əl) *n. Scot.* A youth. Also **spring′all, spring′ald** (-əld).

spring·beau·ty (spring′byo͞o′tē) *n. pl.* **·ties** One of a genus (*Claytonia*) of perennial wild flowers of the purslane family; especially, *C. virginica* of the eastern United States, with pink-tinged white flowers. See CLAYTONIA.

spring·board (spring′bôrd′, -bōrd′) *n.* 1 An elastic board used to aid in leaping; a springy board secured at one end, used to give impetus to a dive into the water below. Also *diving board.* 2 A short board inserted by one end in a notch in a tree, on which a workman stands when felling large trees.

spring·bok (spring′bok′) *n.* A small South African gazelle (*Antidorcas marsupialis*) noted for its ability to leap high in the air. Also **spring′·buck′** (-buk′). [< Afrikaans]

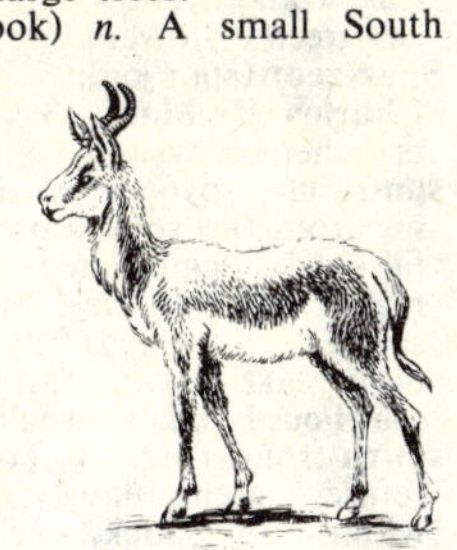

SPRINGBOK
(About 2 feet high at the shoulder)

spring chicken 1 A young chicken, 10 weeks to 10 months old, especially tender for cooking: so called because usually hatched in the spring. 2 *Colloq.* A young, immature, or unsophisticated person.

springe (sprinj) *n.* A snare or noose, arranged with a spring to catch small game. [ME *sprenge.* Related to SPRING.]

spring·er (spring′ər) *n.* 1 One who or that which springs. 2 *Archit.* The bottom stone of an arch, lying upon the impost (see illustration under ARCH); the lowest stone in the coping of a gable; a rib in a groined roof or vault. 3 A spaniel valuable for flushing birds. See under SPANIEL. 4 The springbok. 5 The grampus. 6 A spring chicken.

spring fever The listlessness and restlessness that overtakes a person with the first warm days of spring.

Spring·field (spring′fēld) 1 The capital of Illinois. 2 A city in southern Massachusetts; site of a U. S. arsenal. 3 A city in SW Missouri. 4 A city in SW Ohio.

Springfield rifle A magazine-fed, bolt-action, .30-caliber U. S. Army rifle. Also **Springfield.** [from the U. S. arsenal at *Springfield,* Mass.]

spring·halt (spring′hôlt′) *n.* A stringhalt.

spring·head (spring′hed′) *n.* A fountainhead; source.

spring hinge A hinge the leaves of which are connected with a spring to insure automatic closing.

spring·house (spring′hous′) *n.* A small building constructed over a spring, and used for keeping milk, meats, etc., cool.

spring·ing (spring′ing) *n.* 1 The act of one who or that which springs. 2 *Archit.* A springer: also **springing line.**

spring·let (spring′lit) *n.* A small spring; streamlet or rill.

Springs (springz) A city of southern Transvaal province, Union of South Africa.

spring·tail (spring′tāl′) *n.* Any of certain very small wingless insects (order *Collembola*) having a tail comprised of two united parts, which bends beneath it and enables it to jump.

spring tide 1 A high tide occurring under the combined attraction of sun and new or full moon. 2 Any great wave of feeling, etc.

spring·time (spring′tīm′) *n.* The season of spring. Also **spring′tide′** (-tīd′).

spring water Water found in or obtained from a spring.

spring·y (spring′ē) *adj.* **spring·i·er, spring·i·est** 1 Elastic. 2 Spongy; wet. — **spring′i·ly** *adv.* — **spring′i·ness** *n.*

sprin·kle (spring′kəl) *v.* **·kled, ·kling** *v.t.* 1 To scatter in drops or small particles. 2 To besprinkle; specifically, to apply drops of water to, as a form of baptism: opposed to *immerse.* — *v.i.* 3 To fall or rain in scattered drops. — *n.* A falling in drops or particles, or that which so falls; a sprinkling; hence, a small quantity. [ME *sprenkelen.* Akin to LG *sprinkeln* scatter.]

sprin·kler (spring′klər) *n.* 1 A nozzle or other device for spraying water on lawns, built either as a portable apparatus or as a unit in a stationary network fed by underground pipes. 2 An outlet in a sprinkler system.

sprinkler system An arrangement of pipes distributed through a building, with outlets suitably placed for sprinkling water or other extinguishing fluid to put out fire: often with automatic temperature control.

sprin·kling (spring′kling) *n.* 1 That which is sprinkled. 2 A small number or quantity. 3 A mottling. 4 The act of scattering drops of liquid.

sprint (sprint) *n.* A short race run at top speed. [< *v.*] — *v.i.* To run fast, as in a sprint. [ME *sprenten* <Scand. Cf. ON *spretta* run.] — **sprint′er** *n.*

sprit¹ (sprit) *n. Naut.* 1 A small spar reaching diagonally from a mast to the peak of a fore-and-aft sail. 2 *Brit.* A pole used for propelling a boat. 3 A bowsprit. [OE *sprēot* pole]

sprit² (sprit) *n. Scot.* A rush or rushlike plant. — **sprit′tie** *adj.*

sprite (sprīt) *n.* 1 A fairy, elf, or goblin. 2 A disembodied spirit; a ghost. Also spelled *spright.* [<OF *esprit* <L *spiritus.* Doublet of SPIRIT.]

sprit·sail (sprit′səl, sprit′sāl′) *n. Naut.* A sail extended by a sprit.

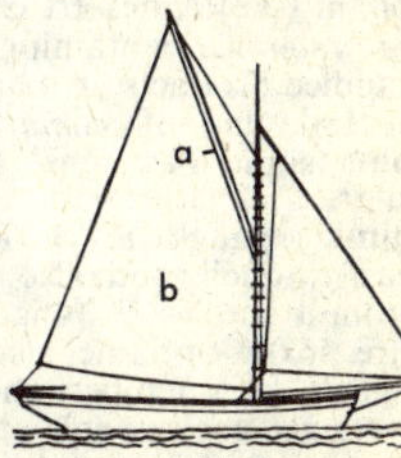

SPRITSAIL
a. Sprit. *b.* Spritsail.

sprock·et (sprok′it) *n. Mech.* 1 A projection, as on the periphery of a wheel, for engaging with the links of a chain. 2 A wheel bearing such projections: also **sprocket wheel.** [Origin uncertain]

sprout (sprout) *v.i.* 1 To put forth shoots; begin to grow; germinate. 2 To develop or grow rapidly. — *v.t.* 3 To cause to sprout. 4 To remove shoots from. — *n.* 1 A new shoot or bud on a plant; hence, something like or suggestive of a sprout; a scion. 2 *pl.* Brussels sprouts. — **a course of sprouts** A period of training. [OE *sprūtan*]

sprout·ling (sprout′ling) *n.* A little sprout.

spruce¹ (spro͞os) *n.* 1 Any of a genus (*Picea*) of evergreen trees of the pine family, having a sharp-pointed pyramidal crown, needle-shaped leaves, and pendulous cones; especially, the ornamental **Norway spruce** (*P. abies*), and the **Engelmann spruce** (*P. engelmanni*) of the Pacific coast. 2 The wood of any of these trees. 3 Any of certain other coniferous trees, as the Douglas fir. [Earlier *pruce* Prussian <*Pruce* Prussia <Med. L *Prussia*; so called because first known as a product of Prussia]

spruce² (spro͞os) *adj.* 1 Having a smart, trim appearance. 2 Fastidious. See synonyms under NEAT¹. — *v.* **spruced, spruc·ing** *v.t.* To make spruce; dress or arrange neatly: often with *up.* — *v.i.* To make oneself spruce: usually with *up.* [Special use of SPRUCE³] — **spruce′ly** *adv.* — **spruce′ness** *n.*

spruce³ (spro͞os) *n.* A kind of superior Prussian leather. Also **spruce leather.** [See SPRUCE¹]

spruce beer A slightly fermented beverage made by boiling leaves and twigs of spruce with sugar or molasses.

sprue¹ (spro͞o) *n.* 1 In founding, a channel connecting with the gate through which the melted metal is poured into the mold; also, dross. 2 A pouring hole in a mold; gate. [Origin uncertain]

sprue² (spro͞o) *n. Pathol.* 1 A disease of tropical regions marked by anemia, emaciation, and gastrointestinal disturbances; psilosis. 2 Thrush. [<Du *spruw*]

sprug (sprug) *n. Scot. & Brit. Dial.* The common sparrow.

sprung (sprung) Past participle and alternative past tense of SPRING.

sprung rhythm In prosody, a rhythm involving feet of varying number of syllables but of equal time length, the stress usually falling on the first syllable: a term coined by Gerard Manley Hopkins.

sprung weight In automobiles, the weight supported by the suspension system: opposed to *unsprung weight.*

add,āce,câre,pälm; end,ēven; it,īce; odd,ōpen,ôrder; to͝ok,po͞ol; up,bûrn; ə = a in *above,* e in *sicken,* i in *clarity,* o in *melon,* u in *focus*; yo͞o = u in *fuse*; oi,oil; ou,pout; ch,check; g,go; ng,ring; th,thin; th,this; zh,vision. Foreign sounds à,œ,ü,kh,ṅ; and ◆: see page xx. < from; + plus; ? possibly.

spry (sprī) *adj.* **spri·er** or **spry·er, spri·est** or **spry·est** Quick and active; agile. See synonyms under ACTIVE, SPRIGHTLY. [< dial. E *sprey* <Scand. Cf. Sw. *sprygg* active.] —**spry'ly** *adv.* —**spry'ness** *n.*

spud (spud) *n.* **1** A spadelike tool with narrow blade or prongs for removing the roots of weeds by digging or cutting. **2** *Colloq.* A potato. —*v.t.* **spud·ded, spud·ding** To remove, as weeds, with a spud. [ME *spudde* <Scand. Cf. Dan. *spyd* a spear.]

spud·der (spud'ər) *n.* A tool for removing bark from trees; also, one who uses such an implement.

spul·yie (spül'yē) *Scot. n.* The act of despoiling; spoil; booty. Also **spul'yie·ment.** —*v.t.* & *v.i.* To plunder. Also **spuil'zie, spul'zie** (-yē).

spume (spyōōm) *n.* Froth, as on an agitated or effervescing liquid; foam; scum. —*v.i.* **spumed, spum·ing** To foam; froth. [<F <L *spuma* foam] —**spu'mous** *adj.* —**spum'y** *adj.*

spu·mes·cent (spyōō·mes'ənt) *adj.* Resembling or producing froth or foam; spumy. —**spu·mes'cence** *n.*

spu·mo·ne (spə·mō'nē, *Ital.* spōō·mō'nā) *n. pl.* **·ni** (-nē) A dessert or mousse of ice-cream or water ice containing fruit, nuts, or other candied products, in a base of whipped cream. [<Ital., aug. of *spuma* froth <L *spuma*]

spun (spun) Past tense and past participle of SPIN.

spunk (spungk) *n.* **1** Dry wood that burns easily; touchwood; also, a kind of tinder made from a species of fungus; punk. **2** A small fire, spark, or flame; also, a match. **3** *Colloq.* Quick, fiery temper; mettle; pluck; courage. —**to get one's spunk up** To become defiant or angry; also, to take heart; show courage. —*v.i.* To take fire; flare up; kindle. [<Irish *sponnc* tinder <L *spongia* sponge]

spunk·ie (spungk'ē) *Scot. n.* **1** The ignis fatuus. **2** A small flame. **3** Liquor; whisky. —*adj.* Spunky.

spunk·y (spungk'ē) *adj.* **spunk·i·er, spunk·i·est** *Colloq.* Spirited; courageous; also, touchy. —**spunk'i·ly** *adv.* —**spunk'i·ness** *n.*

spun rayon Yarns or fabrics made from short rayon fibers instead of from one long filament.

spun silk **1** Short fibers of silk from cocoons which the worms have pierced, and which cannot be reeled. **2** Yarn or cloth made from these fibers.

spun yarn *Naut.* A two- to four-stranded, left-handed line made from loosely twisted rope yarn: used for seizings, etc.

spur (spûr) *n.* **1** A pricking or goading instrument worn on a horseman's heel, and bearing a sharp point or a series of points on a rotating wheel. **2** Anything that incites or urges; instigation; incentive. **3** A part or attachment projecting like or suggestive of a spur, as a crag or mountain peak, a steel gaff fastened to a gamecock's leg, the ergot of rye, etc. **4** A stiff, sharp spine, as on the legs of some insects and the wings of some birds; especially, the spine on the tarsus of the domestic cock. See illustration under FOWL. **5** *Archit.* A buttress or other offset from a wall; also, a claw or the like projecting upon the plinth at the four angles of the base of a column. **6** In carpentry, a brace reinforcing a rafter or post; a strut. **7** *Bot.* A tubular expansion of a foliaceous part, usually some part of the flower, as in the columbine and larkspur. **8** A branch of a lode, railroad, etc. —**on the spur of the moment** Hastily; prompted by an impulse. —*v.* **spurred, spur·ring** *v.t.* **1** To prick or urge with or as with spurs. **2** To furnish with spurs. **3** To injure or gash with the spur, as a gamecock. —*v.i.* **4** To spur one's horse. **5** To hasten; hurry. [OE *spura*] —**spur'rer** *n.*

Synonyms (verb): goad, impel, incite, instigate, provoke, rouse, stimulate, sting, stir, urge. *Antonyms*: check, deter, discourage, dissuade, hold, moderate, rein, restrain.

spur·gall (spûr'gôl') *n.* A galled place on a horse's side, caused by the spur. —*v.t.* To injure or gall with a spur.

spurge (spûrj) *n.* **1** Any of several shrubs (genus *Euphorbia*) having fertile flowers with 3-lobed ovaries on long pedicels and yielding a milky juice of bitter taste. **2** One of various related plants of the spurge family (*Euphorbiaceae*). [<OF *espurge* < *espurgier* purge <L *expurgare* < *ex-* out + *purgare* cleanse]

spur gear *Mech.* **1** A spur wheel. **2** Spur gearing.

spur gearing *Mech.* Gearing composed of spur wheels.

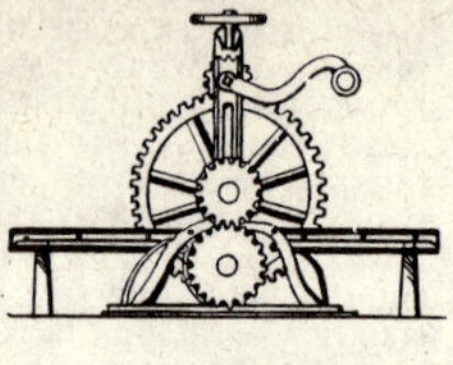

SPUR GEARING

spurge laurel An evergreen shrub of Europe and Asia (*Daphne laureola*), with oblanceolate leaves and yellowish-green flowers.

Spur·geon (spûr'jən), **Charles Haddon,** 1834–92, English Baptist preacher and writer.

spu·ri·ous (spyŏŏr'ē·əs) *adj.* **1** Not proceeding from the source pretended; not genuine; false. **2** Illegitimate. **3** Apparent, but not real; resembling in appearance but not in structure: a *spurious* fruit. See synonyms under COUNTERFEIT, FACTITIOUS. [<L *spurius*] —**spu'ri·ous·ly** *adv.* —**spu'ri·ous·ness** *n.*

spurn (spûrn) *v.t.* **1** To reject with disdain; refuse contemptuously; scorn. **2** To strike with the foot; kick. —*v.i.* **3** To reject something with disdain. See synonyms under SCORN. —*n.* The act of spurning; also, a kick. [OE *spurnan* kick, reject] —**spurn'er** *n.*

spurred (spûrd) *adj.* Wearing or having spurs; having sharp spikes, claws, or shoots.

spur·ri·er (spûr'ē·ər) *n.* A maker of spurs.

spur·ry (spûr'ē) *n. pl.* **·ries** Any of several low annual herbs (genus *Spergula*); especially, the **corn spurry** (*S. arvensis*), which is a common weed. Also **spur'rey.** [<Du. *spurrie*]

spurt (spûrt) *n.* **1** A sudden gush of liquid. **2** Any sudden outbreak, as of anger. **3** An extraordinary effort of brief duration; a sudden rise in activity or price. **4** A brief period. —*v.i.* **1** To come out in a jet; gush forth. **2** To make a sudden and extreme effort. —*v.t.* **3** To force out in a jet; squirt. Also spelled *spirt.* [Var. of earlier *spirt,* metathetic var. of *sprit* <OE *spryttan* come forth]

spur·tle (spûr'təl) *n. Scot.* A stirring stick for porridge.

spur track A short side track connecting with the main track of a railroad. Also **spur.**

spur wheel A toothed wheel having external radial teeth on the periphery; a spur gear.

sput·nik (spŏŏt'nik, sput'-) *n.* A Russian artificial earth satellite: the first to be recorded in world history, called Sputnik I, containing various scientific instruments, was launched October 4, 1957, to an initial height of 560 miles, orbiting at a mean velocity of 18,000 miles per hour. [<Russian, a satellite; lit., that which travels with something else]

sput·ter (sput'ər) *v.i.* **1** To throw off solid or fluid particles in a series of slight explosions. **2** To emit particles of saliva from the mouth, as when speaking excitedly. **3** To speak rapidly or confusedly. —*v.t.* **4** To throw off or emit in small particles. **5** To utter in a confused or excited manner. —*n.* **1** The act or sound of sputtering; especially, excited talk; jabbering. **2** That which is thrown out in sputtering. **3** Trouble; fuss. [Freq. of SPOUT, *v.*] —**sput'ter·er** *n.*

spu·tum (spyōō'təm) *n. pl.* **·ta** (-tə) Saliva; spittle; expectorated matter. [<L < *spuere* spit]

Spuy·ten Duy·vil Creek (spīt'n dī'vəl) A narrow stream in New York City, connecting the Hudson and Harlem rivers and separating Manhattan Island from the mainland on the north; used as a ship canal.

spy (spī) *n. pl.* **spies** **1** One who enters an enemy's military lines covertly to get information; a secret agent. **2** One who watches others secretly: often used contemptuously. **3** A peep; glance; hence, an eye. **4** The act of watching secretly. —*v.* **spied, spy·ing** *v.t.* **1** To keep watch closely or secretly; act as a spy. **2** To make careful examination; pry: with *into.* —*v.t.* **3** To observe stealthily and with hostile intent: usually with *out.* **4** To catch sight of; see; espy. **5** To discover by careful or secret investigation: with *out.* **6** To examine or scrutinize carefully. [<OF *espie* < *espier* espy <Gmc.]

Synonyms (noun): emissary, scout. The *scout* and the *spy* are both employed to obtain information of the numbers, movements, etc., of an enemy. The *scout* lurks on the outskirts of the hostile army with such concealment as the case admits of, but without disguise; a *spy* enters in disguise within the enemy's lines. A *scout,* if captured, has the rights of a prisoner of war; a *spy* is held to have forfeited all rights, and is liable, in case of capture, to capital punishment. Soldiers not in disguise or military aviators are not considered *spies,* even while passing through or over hostile territory. An *emissary* is rather political than military, sent to influence opponents secretly rather than to bring information concerning them.

spy·glass (spī'glas', -gläs') *n.* A small field glass or telescope.

Spy·ri (shpē'rē), **Johanna,** 1827–1901, *née* Heusser, Swiss author of *Heidi.*

squab (skwob) *n.* **1** A young pigeon, especially when an unfledged nestling. **2** A fat, short person. **3** A soft, stuffed cushion; sofa; ottoman. —*adj.* **1** Fat and short; low and bulky; squat. **2** Unfledged or but half-fledged; half-grown, as a pigeon, or figuratively, any fowl. [< dial. E <Scand. Cf. dial. Norw. *skvabb* a soft, wet mass.]

squab·ble (skwob'əl) *v.* **·bled, ·bling** *v.i.* To engage in a petty wrangle or scuffle; quarrel. —*v.t. Printing* To twist (composed type) so as to mix the lines. —*n.* The act of squabbling; a petty wrangle. See synonyms under QUARREL[1]. [Cf. dial. Sw. *skvabbel* dispute, argue] —**squab'bler** *n.*

squab·by (skwob'ē) *adj.* **·bi·er, ·bi·est** Short and fat. Also **squab'bish.**

squad (skwod) *n.* **1** A small group of persons organized for the performance of a specific function; a small detachment of troops or police; specifically, the smallest tactical unit in the infantry of the U. S. Army. **2** Hence, a team: a football *squad.* —*v.t.* **squad·ded, squad·ding** **1** To form into a squad or squads. **2** To assign to a squad. [<F *escouade* <OF *esquadre* a square <Ital. *squadra* <L *quattuor* four]

squad car An automobile used by police for patrolling, and equipped with radiotelephone for communicating with headquarters.

squad·ron (skwod'rən) *n.* **1** An assemblage of war vessels smaller than a fleet; one of the divisions of a fleet. **2** A division of a cavalry regiment. **3** The basic unit of the United States Air Force, usually consisting of two or more flights operating as a unit. **4** Any regularly arranged or organized body, as of men. —*v.t.* To arrange in a squadron or squadrons. [<Ital. *squadrone,* aug. of *squadra* SQUAD]

squail (skwāl) *n.* A disk used in the game of squails.

squails (skwālz) *n. pl.* A game played with small wooden disks on a table, the object being to approach as nearly as possible to a mark at the center of the board, by snapping the disks from the edge. [Origin uncertain]

squal·id (skwol'id) *adj.* Having a foul, mean, or poverty-stricken appearance; dirty, neglected, and wretched. See synonyms under BASE[2]. [<L *squalidus* < *squalere* be foul] —**squal'id·ly** *adv.* —**squal'id·ness, squa·lid·i·ty** (skwo·lid'ə·tē) *n.*

squall[1] (skwôl) *n.* A loud, screaming outcry. —*v.i.* To cry loudly; scream; bawl. [Cf. ON *skvala* shout, bawl] —**squall'er** *n.*

squall[2] (skwôl) *n.* A sudden, violent burst of wind, often accompanied by rain or snow. —*v.i.* To blow a squall; be squally. [Cf. Sw. *skval-regn* a sudden rainstorm]

squall cloud A grayish cloud rolling beneath an approaching thunderstorm.

squall line *Meteorol.* A cold front characterized along its edge by a sharp change of wind and the occasional formation of line squalls.

squall·y (skwô'lē) *adj.* **squall·i·er, squall·i·est** **1** Stormy; blustering. **2** *Colloq.* Threatening a squall or trouble of any kind.

squal·or (skwol'ər) *n.* The state of being squalid, or the filth of thriftless poverty. [<L < *squalere* be foul]

squa·lus (skwā'ləs) *n.* Any of a genus (*Squalus*) of cartilaginous fishes (class *Chondrichthyes*), including the spiny dogfish or shark (*S. acanthias*) common in shore waters of the Atlantic. [<L, large marine fish]

squa·ma (skwā'mə) *n. pl.* **·mae** (-mē) A thin, scalelike structure; a scale. [<L] —**squa'·mate** (-māt) *adj.*

Squa·ma·ta (skwə·mā'tə) *n. pl.* An order of reptiles, including lizards, chameleons, and serpents. [<NL <L *squama* a scale]

squa·ma·tion (skwə-mā′shən) *n.* **1** The state of being scaly. **2** The arrangement of epidermal scales.

squa·mo·sal (skwə-mō′səl) *adj.* **1** Like a scale; squamous. **2** *Anat.* Relating to the squamous portion of the temporal bone or the analogous bone in lower animals. — *n.* The squamosal bone.

squa·mous (skwā′məs) *adj.* **1** Covered with scales; scaly; scalelike. **2** *Anat.* Designating the vertical plate of the temporal bone. Also **squa′mose** (-mōs). [< L *squamosus* < *squama* a scale] — **squa′mous·ly** *adv.* — **squa′mous·ness** *n.*

squam·u·lose (skwam′yə-lōs, skwā′myə-) *adj. Bot.* Provided with small bracts or scales, as a plant; minutely squamate.

squan·der (skwon′dər) *v.t.* **1** To spend (money, time, etc.) wastefully; lavish profusely; dissipate. **2** *Obs.* To scatter. — *n.* Prodigality; the act of squandering. [Cf. dial. E *squander* scatter] — **squan′der·er** *n.* — **squan′der·ing·ly** *adv.*
Synonyms (verb): dissipate, expend, lavish, scatter, spend, waste. *Antonyms:* economize, hoard, hold, husband, preserve, reserve, save.

squan·tum[1] (skwon′təm) *n.* Among North American Indians, especially the Narragansets, a spirit or god; an evil spirit.

squan·tum[2] (skwon′təm) *n.* In New England, a picnic or shore dinner; a chowder party; hence, any merrymaking or frolic. [from *Squantum*, Mass., after *Tisquantum*, a Massachuset chief]

square (skwâr) *n.* **1** A parallelogram having four equal sides and four right angles. **2** Any object, part, or surface that is square or nearly so, as a pane of glass, or one of the spots on a checkerboard. **3** An instrument by which to measure or lay out right angles, consisting usually of two legs or branches at right angles to each other, in L-shape or T-shape (in the latter case called a *T-square*).* **4** An open area in a city or village, left between streets at their intersection or formed by their expansion. **5** A town or city block; also, the distance between one street and the next. **6** *Math.* The product of a number or quantity multiplied by itself. **7** Formerly, a body of troops formed in a four-sided array. **8** *Obs.* A standard or pattern; rule. **9** *Slang* A person not conversant with developments in the popular arts, especially the latest fashions in jazz, slang, etc. — **on the square 1** At right angles. **2** On equal terms. **3** *Colloq.* In a fair and honest manner. **4** In Freemasonry, in good standing: said of members. — **out of square 1** Not at right angles; obliquely. **2** Incorrectly; askew; out of order. — *adj.* **1** Having four equal sides and four right angles; loosely, approaching a square in form. **2** Formed with or characterized by a right angle; rectangular. **3** Adapted to forming squares or computing in squares: a *square* measure. **4** Direct; fair; just; equitable; honest. **5** Having debit and credit balanced; even; settled. **6** Absolute; complete; unequivocal. **7** Having a broad, stocky frame; hence, strong, sturdy. **8** *Colloq.* Solid; full; satisfying: a *square* meal. **9** *Naut.* At right angles to the mast and keel: said of the yards of a square-rigged ship. **10** *Math.* Raised to the second power; squared: 10 *square* equals 100. **11** Steady: said of a horse's gait. **12** *Mech.* Having the cylinder bore equal, or nearly equal to the piston stroke: said of engines. See synonyms under JUST[1]. — *v.* **squared, squar·ing** *v.t.* **1** To make square; form with four equal sides and four right angles. **2** To shape or adjust so as to form a right angle, or a right angle with something else. **3** To mark with or divide into squares. **4** To test for the purpose of adjusting to a straight line, right angle, or plane surface. **5** To bring to a position suggestive of a right angle: *Square* your shoulders. **6** To make satisfactory settlement of: to *square* accounts. **7** To make (the score of a game or contest) equal. **8** To cause to conform; adapt; reconcile: to *square* one's opinions to the times. **9** *Math.* **a** To multiply (a number or quantity) by itself. **b** To determine the contents of in square measure. **c** To find the square equivalent of: to *square* a circle. **10** *Slang* To bribe: to *square* a jockey. — *v.i.* **11** To be at right angles. **12** To conform; agree; harmonize. **13** In golf, to make the scores equal. **14** *Obs.* To squabble; quarrel. — **to square away 1** *Naut.* To set (the yards) at right angles to the keel. **2** To square up. — **to square off** To assume a position for attack or defense; prepare to fight. — **to square up** To adjust satisfactorily. — *adv.* **1** So as to be square, or at right angles. **2** Honestly; fairly. **3** Directly; firmly. [< OF *esquire, esquarre*, ult. < L *quattuor* four] — **square·ness** *n.*

square bracket Bracket (def. 4).

square dance Any dance, as a quadrille, in which the couples form sets in squares.

squared circle *Colloq.* A boxing ring; the prize ring. Also **squared ring.**

square deal *Colloq.* **1** In card games, an honest deal. **2** Hence, fair or just treatment.

square·head (skwâr′hed′) *n. U.S. Slang* **1** A Scandinavian. **2** A German.

square knot A common knot, formed of two overhand knots: also called *reef knot.* See illustration under KNOT.

square league An old Spanish land measure equal to 4,438 acres.

square·ly (skwâr′lē) *adv.* **1** In a direct or straight manner; He looked her *squarely* in the eyes. **2** Honestly; fairly. **3** *U.S.* Plainly; unequivocally. **4** In a square form: *squarely* built. **5** At right angles (to a line or plane).

square meal *Colloq.* A full and substantial meal.

square measure A unit or system of units for measuring areas, as in the following table of principal customary standards. See also METRIC SYSTEM.

144 square inches (sq. in.; in²)	=	1 square foot (sq. ft.; ft²)
9 square feet	=	1 square yard (sq. yd.; yd²)
30.25 square yards	=	1 square rod (sq. rd.; rd²)
160 square rods	=	1 acre (A.)
640 acres	=	1 square mile (sq. mi.)

square number See under NUMBER.

square piano See under PIANO[1].

squar·er (skwâr′ər) *n.* **1** One who squares. **2** *Archaic* A brawler.

square-rigged (skwâr′rigd′) *adj. Naut.* Having the principal sails extended by horizontal yards; ship-rigged: distinguished from *fore-and-aft-rigged.* Compare illustrations under BARK, BRIG, SHIP.

square-rig·ger (skwâr′rig′ər) *n.* A square-rigged ship.

square root *Math.* A number or quantity that, multiplied by itself, produces the given number or quantity: 4 is the *square root* of 16; a second root. See under CUBE[1], ROOT[1].

square sail *Naut.* A quadrilateral sail usually rigged on a yard set at right angles to the mast.

square shooter *Colloq.* An upright person; one who acts honestly and justly.

square-toed (skwâr′tōd′) *adj.* Having the toes square, as the shoes worn by the Puritans; hence, exact; punctilious.

square-toes (skwâr′tōz′) *n.* An old-fashioned, exact person.

squaring a log Sawing a log so as to give it four equal sides.

squaring the circle Quadrature of the circle.

squar·rose (skwar′ōs, skwo·rōs′) *adj.* **1** *Biol.* Rough with projecting scalelike processes. **2** *Bot.* Crowded and rigid: *squarrose* leaves. Also **squar·rous** (skwar′əs). [< L *squarrosus* scurfy]

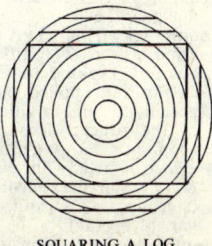

SQUARING A LOG

squash[1] (skwosh) *v.t.* **1** To beat or press into a pulp or soft mass; crush. **2** To quell or suppress. — *v.i.* **3** To be smashed or squashed. **4** To make a splashing or sucking sound. — *n.* **1** A soft or overripe object; also, a crushed mass. **2** The sudden fall of a heavy, soft, or bursting body; also, the sound made by such a fall. **3** The sucking, squelching sound made by walking through ooze or mud. **4** Either of two games played on an indoor court with rackets and a ball. In one (**squash rackets**) a slow rubber ball is used; in the other (**squash tennis**), a livelier, smaller ball. **5** A beverage of which one ingredient is a fruit juice: lemon *squash.* — *adv.* With a squelching, oozy sound. [< OF *esquasser*, ult. < L *ex-* thoroughly + *quassare* crush] — **squash′er** *n.*

squash[2] (skwosh) *n.* **1** The edible fruit of various trailing annuals (genus *Cucurbita*) of the gourd family. **2** The plant that bears it. [< Algonquian. Cf. Massachuset *askootasquash*, lit., eaten raw.]

squash bug A large, brownish-black, evil-smelling North American hemipterous insect (*Anasa tristis*) which is destructive to squash vines.

squash·y (skwosh′ē) *adj.* **squash·i·er, squash·i·est** Soft and moist; easily squashed. — **squash′i·ly** *adv.* — **squash′i·ness** *n.*

squat (skwot) *v.* **squat·ted** or **squat, squat·ting** *v.i.* **1** To sit on the heels or hams, or with the legs near the body. **2** To crouch or cower down, as to avoid being seen. **3** To settle on a piece of land without title or payment. **4** To settle on government land in accordance with certain government regulations that will eventually give title. — *v.t.* **5** To cause (oneself) to squat. — *adj.* **1** Short and thick; squatty. **2** Being in a squatting position. — *n.* A squatting attitude or position. [< OF *esquatir* < *es-* thoroughly (< L *ex-*) + *quatir* press down < L *coactus,* pp. of *cogere* force < *co-* together + *agere* drive]

squat tag A game of tag in which the players cannot be tagged while squatting.

squat·ter (skwot′ər) *n.* **1** One who or that which squats; specifically, one who settles on land without permission or right, as on public or unimproved land. **2** In the United States and Australia, one who settles on government land subject to regulations with a view to obtaining title.

squatter sovereignty 1 The political theory that the people or settlers of a Territory had the right to make their own laws, specifically whether or not slavery should be permitted; popular sovereignty. **2** The right of settlers to the lands they have settled.

squat·ty (skwot′ē) *adj.* Disproportionately short and thick.

squaw (skwô) *n.* **1** An American Indian woman or wife. **2** *Colloq.* Any woman or girl. [< Algonquian, woman]

squaw·bush (skwô′bŏŏsh′) *n.* **1** Any shrub of the genus *Cornus*; especially, the red-osier dogwood. **2** The cranberry tree.

squaw·fish (skwô′fish′) *n. pl.* **·fish** or **·fish·es 1** A cyprinoid fish (genus *Ptychocheilus*) found in the rivers of the northern Pacific coast. **2** A surf fish.

squawk (skwôk) *v.i.* **1** To utter a shrill, harsh cry, as a parrot. **2** *Slang* To utter loud complaints or protests. — *n.* **1** The harsh cry of certain birds; also, the act of squawking. **2** *Slang* A loud protest or complaint. **3** The black-crowned night heron (*Nycticorax nycticorax*). [Prob. imit.] — **squawk′er** *n.*

squaw man 1 Among the American Indians, a man who lives and works among the women. **2** A white man married to an Indian woman and in possession of tribal rights on that account.

squaw·root (skwô′rōōt′, -rōōt′) *n.* **1** A yellowish-brown leafless North American herb (*Conopholis americana*) parasitic on roots. **2** Of certain other plants, as the blue cohosh.

squaw vine The partridgeberry.

squeak (skwēk) *n.* **1** A thin, sharp, penetrating sound. **2** *Colloq.* A narrow margin; the least amount; a hairbreadth: in the phrase **narrow** (or **a near**) **squeak.** — *v.i.* **1** To make a squeak. **2** *Colloq.* To let out information; squeal. — *v.t.* **3** To utter or effect with a squeak. 4 To cause to squeak. [ME *squeke*, prob. <Scand. Cf. Sw. *sqväka* croak.] — **squeak′er** *n.*

squeak·y (skwē′kē) *adj.* **squeak·i·er, squeak·i·est** Making a squeaking noise. — **squeak′i·ly** *adv.* — **squeak′i·ness** *n.*

squeal (skwēl) *v.i.* 1 To utter a sharp, shrill, somewhat prolonged cry. 2 *Slang* To turn informer; betray an accomplice or a plot. — *v.t.* 3 To utter with a squeal. — *n.* A shrill, prolonged cry, as of a pig. [Imit.] — **squeal′er** *n.*

squeam·ish (skwē′mish) *adj.* 1 Easily disgusted or shocked; unduly scrupulous. 2 Easily nauseated; affected with nausea. [< earlier *squeamous* <AF *escoymous*; ult. origin unknown] — **squeam′ish·ly** *adv.* — **squeam′ish·ness** *n.*

Synonyms: affected, dainty, difficult, fastidious, finical, foolish, hypercritical, overnice, oversensitive, particular, prudish, qualmish, scrupulous, sickish.

squee·gee (skwē′jē) *n.* 1 A wooden implement having a stout straight-edged strip of rubber or leather inserted in its blade, used for removing water from wet decks or floors, window panes, etc. 2 *Phot.* A smaller similar implement, made in the same way or in the form of a roller, used for pressing a film closer to its mount, or for squeezing the moisture from a print. — *v.t.* 1 To smooth down, as a photographic film, with a squeegee. 2 To cleanse with a squeegee. Also spelled *squilgee*, *squillagee*. [< *squeege*, var. of SQUEEZE]

squeeze (skwēz) *v.* **squeezed, squeez·ing** *v.t.* 1 To press hard upon; compress. 2 To extract something from by pressure: to *squeeze* oranges. 3 To draw forth by pressure; express: to *squeeze* juice from apples. 4 To force or push; cram. 5 To oppress, as with burdensome taxes. 6 To exert pressure upon (someone) to act as one desires, as by blackmailing. 7 To take a squeeze (def. 4) of. — *v.i.* 8 To apply pressure. 9 To force one's way; push: with *in*, *through*, etc. 10 To be pressed; yield to pressure: These lemons *squeeze* well. See synonyms under JAM[1]. — **to squeeze out** To force out of business, or ruin financially, by unscrupulous methods. — *n.* 1 The act or process of squeezing; pressure. 2 A firm grasp of someone's hand; a hearty handclasp; also, an embrace; hug. 3 Something, as juice, extracted or expressed. 4 A facsimile, as of a coin or inscription, produced by pressing some soft substance upon it. 5 *Colloq.* A crowded social gathering. 6 *Colloq.* Pressure exerted for the extortion of money or favors; also, financial pressure. [? <OF *es-* thoroughly (<L *ex-*) + ME *queisen*, OE *cwēsan* crush] — **squeez′a·ble** *adj.*

squeeze play In baseball, a play in which the batter bunts the ball so that a man on third base may score by starting while the pitcher is about to deliver the ball.

squeez·er (skwē′zər) *n.* One who or that which squeezes; especially, a mechanical device for applying pressure on fruit.

squeez·ing (skwē′zing) *n.* 1 The act or process of squeezing. 2 *Often pl.* That which is squeezed out. 3 A crowding together; press or crowd.

squelch (skwelch) *v.t.* 1 To crush; squash. 2 *Colloq.* To subdue utterly; silence, as with a crushing reply. — *v.i.* 3 To make a splashing or sucking noise, as when walking in deep mud. 4 To walk with such a sound. — *n.* 1 A noise made when walking in wet boots. 2 A heavy fall or blow. 3 *Colloq.* A squelcher. [Prob. imit.]

squelch·er (skwel′chər) *n.* 1 One who or that which squelches. 2 *Colloq.* A silencing retort; crushing reply.

sque·teague (skwi·tēg′) *n.* A weakfish. [<Algonquian (Narraganset) *pesukwiteaug* they make glue]

squib (skwib) *n.* 1 A roll or case filled with gunpowder, to be thrown or rolled swiftly, finally exploding like a rocket. 2 A tubular case filled with gunpowder and connected with an electric circuit, used for firing a charge in a blasthole, igniting a smokepot, or the like. 3 A broken firecracker that burns with a spitting sound. 4 A short speech or writing in a witty or satirical vein; a mild lampoon. 5 An undistinguished or petty person. — *v.* **squibbed, squib·bing** *v.i.* 1 To write or use squibs. 2 To fire a squib. 3 To explode or sound like a squib. 4 To move quickly or restlessly. — *v.t.* 5 To attack with squibs; lampoon. 6 To fire or use as a squib. [Origin unknown]

squid (skwid) *n.* 1 One of various ten-armed cephalopods (genera *Loligo* and *Ommastrephes*) with a slender conical body, ink sac, and broad caudal fins; especially, the common squid (*L. pealei*) of the Atlantic coast. 2 Fish bait prepared from squid; also, an artificial fish bait, often made in imitation of a squid. — *v.i. Aeron.* To assume a narrow, squidlike shape, as a parachute under excess wind or air pressure. [Origin uncertain]

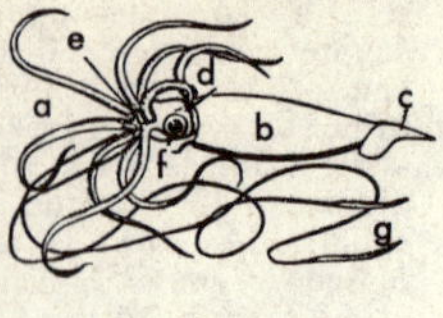

SQUID
a. Arm. e. Mouth.
b. Body. f. Siphon.
c. Fluke. g. Tentacles.
d. Eye.

squig·gle (skwig′əl) *Colloq.* *n.* A meaningless scrawl. — *v.i.* To wriggle. [Blend of SQUIRM and WRIGGLE]

squil·gee (skwil′jē, skwil·jē′) See SQUEEGEE.

squill[1] (skwil) *n.* 1 A bulbous plant (*Urginea maritima*) of the lily family, growing in the Mediterranean region; the sea onion. 2 Its bulb, dried and sliced, the white variety having diuretic and expectorant properties, and the red variety yielding a rat poison. 3 Any plant of the genus *Scilla*, the more common ones usually called by some other name, as the common English bluebell or wild hyacinth. [<L *squilla* <Gk. *skilla* sea onion]

squill[2] (skwil) *n.* Any of a genus (*Squilla*) of burrowing crustaceans having the form and appearance of a mantis: sometimes called *mantis shrimp*. Also **squil·la** (skwil′ə). [<L *squilla* shrimp]

squil·la·gee (skwil′ə·jē) See SQUEEGEE.

squinch (skwinch) *n.* *Archit.* A small stone arch or series of arches, or of projecting courses, across an interior angle of a square tower, to support an oblique side of an octagonal spire or lantern. [Alter. of obs. *scunch*, abbreviation of *scuncheon* <OF *escoinson*]

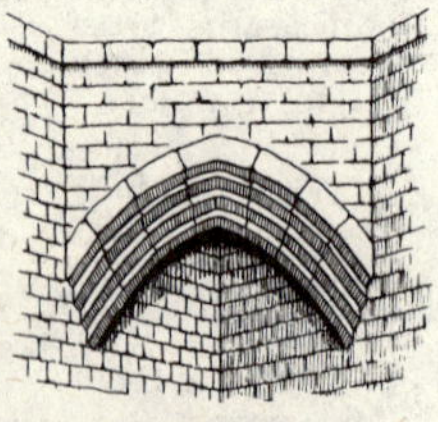
SQUINCH
Salisbury Cathedral, England.

squin·ny (skwin′ē) *v.i.* & *n.* *Obs.* Squint. Also **squin′y.** [Var. of SQUINT]

squint (skwint) *v.i.* 1 To look with half-closed eyes, as into bright light. 2 To look with a side glance; look askance. 3 To be cross-eyed. 4 To incline or tend: with *toward*, etc. — *v.t.* 5 To hold (the eyes) half shut, as in glaring light. 6 To cause to squint. — *adj.* 1 Having the optic axes not coincident; affected with strabismus: said of the eyes. 2 Looking obliquely or askance; indirect. — *n.* 1 *Pathol.* An affection of the eyes in which their axes are differently directed; strabismus. 2 The act or habit of squinting. 3 Hence, an indirect leaning, tendency, or drift. 4 A hagioscope. [Origin uncertain] — **squint′er** *n.*

squint-eye (skwint′ī′) *n.* Strabismus, or one afflicted with it.

squint-eyed (skwint′īd′) *adj.* 1 Affected with strabismus; cross-eyed. 2 Looking sidewise; aiming in two directions. 3 Apt to see awry; malignant; evil.

squire (skwīr) *n.* 1 A knight's attendant; an armorbearer. 2 A title of dignity, office, or courtesy ranking in England below that of *knight*, and applied in the United States especially to rural or village lawyers and justices of the peace; also, in England, a landed proprietor. 3 A gentleman who acts as the escort of a lady in public; a gallant. — *v.t.* & *v.i.* **squired, squir·ing** To attend or serve (someone) as a squire or escort. [Aphetic var. of ESQUIRE]

squire·ar·chy (skwīr′är·kē) *n.* *pl.* **·chies** 1 English country gentlemen collectively; also, any body of squires. 2 Government by squires. Also **squir′ar·chy.**

squire·ling (skwīr′ling) *n.* A petty squire. Also **squire′let** (-lit).

squirm (skwûrm) *v.i.* 1 To bend and twist the body; wriggle; writhe. 2 To show signs of pain or distress. — *n.* A squirming motion; a wriggle. [Origin uncertain] — **squirm′er** *n.* — **squirm′y** *adj.*

squir·rel (skwûr′əl, *Brit.* skwir′əl) *n.* 1 Any of various slender rodents (family *Sciuridae*) with a very long bushy tail, living mainly in trees and feeding chiefly on nuts, but occasionally on eggs and small birds. The **red squirrel** (*Sciurus hudsonicus*), the **gray squirrel** (*S. carolinensis*), and the **fox squirrel** (*S. niger*) are North American types. ◆ Collateral adjective: *sciurine*. 2 One of various sciuroid rodents, as the **rock squirrel** (*Otospermophilus grammurus*) of the western United States. 3 The fur of a squirrel. 4 In Australia, a phalanger. [<OF *esquireul* <LL *scurellus*, dim. of L *sciurus* <Gk. *skiouros* < *skia* shadow + *oura* tail]

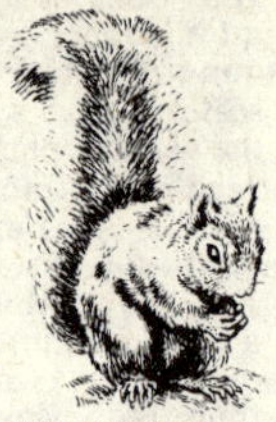
GRAY SQUIRREL
(Body to 10 inches; tail to 8 inches)

squirrel corn A smooth and delicate plant (*Dicentra canadensis*) of the northern United States, having white or cream-colored flowers with the spurs rounded and yellow tubers resembling grains of corn.

squirrel monkey A marmoset.

squirt (skwûrt) *v.i.* 1 To come forth in a thin stream or jet; spurt out. 2 To eject water, etc., thus. — *v.t.* 3 To eject (water or other liquid) forcibly and in a jet. 4 To wet or bespatter with a squirt or squirts. — *n.* 1 The act of squirting or spurting; also, a jet of liquid squirted forth. 2 A syringe or squirt gun. 3 *Colloq.* A conceited, brainless fellow. [Cf. LG *swirtjen*] — **squirt′er** *n.*

squirt gun An instrument, as a syringe or child's toy, shaped like a gun and used for squirting.

squirting cucumber The fruit of a procumbent branching herb (*Ecballium elaterium*) of the gourd family, which, when ripe, ejects its seeds and juice.

squish (skwish) *v.t.* & *v.i.* *Colloq.* To squash. — *n.* A squashing sound. [Var. of SQUASH[1]] — **squish′y** *adj.*

Sri·nag·ar (srē·nug′ər) The capital of Jammu and Kashmir State, NW India; site of extensive eighth century Buddhist ruins.

St. For entries not found under *St.*, see under SAINT.

-st See -EST[2].

stab (stab) *v.* **stabbed, stab·bing** *v.t.* 1 To pierce with a pointed weapon; wound, as with a dagger. 2 To thrust (a dagger, etc.), as into a body. 3 To penetrate; pierce. — *v.i.* 4 To thrust or lunge with a knife, sword, etc. 5 To inflict a wound thus. See synonyms under PIERCE. — *n.* A thrust made with any pointed weapon. [? <Irish *stob* push, thrust, fix a stake < *stob* a stake] — **stab′ber** *n.*

Sta·bat Ma·ter (stä′bät mä′tər, stā′bat mā′tər) *Latin* A 13th century hymn commemorating the agony of Mary at the crucifixion of Christ and so called from its opening words: literally, the mother was standing.

sta·bile (stā′bil, stab′il) *adj.* 1 Not kept in motion. 2 *Med.* **a** Not affected by moderate heat. **b** Denoting a form of electrotherapy in which one of the electrodes is kept stationary on a part. Compare LABILE. — *n.* An amorphous piece of stationary sculpture. Compare MOBILE. [<L *stabilis*. See STABLE[1].]

sta·bil·i·ty (stə·bil′ə·tē) *n.* *pl.* **·ties** 1 The condition of being stable; steadiness. 2 The quality or character of being steady or constant; steadfastness of purpose or resolution. 3 *Physics* The state of being in stable equilibrium, or the degree of such equilibrium as measured by the force with which a body tends to maintain its condition of rest or steady motion. 4 *Aeron.* The ability of an aircraft to resume equilibrium when disturbed. 5 A vow to continue in the same profession and order, taken by some Benedictine monks. 6 *Obs.* Rigidity: opposed to *fluidity*. [<L *stabilitas*, *-tatis* < *stabilis*. See STABLE[1].]

sta·bi·lize (stā′bə·līz) *v.t.* **·lized, ·liz·ing** 1 To make firm or stable. 2 To keep steady; keep from fluctuating, as money or currency: to *stabilize* prices. 3 *Aeron.* To secure or maintain the equilibrium of (an aircraft) by means of fixed surfaces, gyroscopes, etc. [<L

stabilis steady + -IZE] — **sta'bi·li·za'tion** n.
sta·bi·liz·er (stā'bə·lī'zər) n. 1 Aeron. An automatic balancing device; especially, one which steadies the flight of an airplane. See illustration under AIRPLANE. 2 Chem. A substance which increases the stability of another substance or compound, especially one which reduces the spontaneous combustion of an explosive.
sta·ble[1] (stā'bəl) adj. 1 Standing firmly in place; not easily moved, shaken, or overthrown; fixed. 2 Marked by fixity of purpose; steadfast; inflexible. 3 Having durability or permanence; abiding. 4 Chem. Not easily decomposed: said of compounds. 5 Physics Resisting forces which tend to cause or distort motion. See synonyms under FIRM, PERMANENT. [<F <L stabilis <stare stand] — **sta'bly** adv. — **sta'ble·ness** n.
sta·ble[2] (stā'bəl) n. 1 A building set apart for lodging and feeding horses or cattle. 2 Specifically, race horses belonging to a particular stable; also, the owner and personnel of a particular stable collectively. — v.t. & v.i. **·bled, ·bling** To put or lodge in a stable. [<OF estable <L stabulum <stare stand]
sta·ble·boy (stā'bəl·boi') n. A boy employed in a stable.
sta·ble·man (stā'bəl·man', -mən) n. pl. **·men** (-men', -mən) One who works in a stable; a hostler; groom.
sta·bling (stā'bling) n. 1 The act of one who stables. 2 Room or accommodation for a stable.
stab·lish (stab'lish) v.t. Archaic To establish. [Aphetic var. of ESTABLISH]
stac·ca·to (stə·kä'tō) adj. 1 Music Played, or to be played, in an abrupt, disconnected manner: opposed to legato. 2 Marked by abrupt, sharp emphasis: a staccato style of speaking. [<Ital., pp. of staccare detach]
stack (stak) n. 1 A large, orderly pile of unthreshed grain, hay, or straw, usually conical. 2 Any systematic pile or heap, as a pile of poker chips purchased or won by a player. 3 A group of rifles (usually three) set upright and supporting one another. 4 A case composed of several rows of bookshelves one above the other. 5 pl. That part of a library where most of the books are shelved. 6 A vertical main smoke flue, especially of a furnace or boiler; a chimney; smokestack; also, a collection of such chimneys or flues. 7 Brit. A measure of fuel (coal or wood), equal to 108 cubic feet or 4 cubic yards. 8 Colloq. A great amount; plenty. — v.t. To gather or place in a pile; pile up in a stack. — **to stack the cards** 1 To arrange cards in the pack in a manner favorable to the dealer. 2 To have an advantage secured beforehand. [<ON stakkr]
stack arms The command to place several rifles close together and upright on the ground in a slanting position, three rifles linked at their swivels providing support for the others.
stack·er[1] (stak'ər) n. Agric. An attachment or apparatus for depositing straw from a threshing machine on a wagon or on a stack.
stack·er[2] (stak'ər) v. & n. Scot. Stagger. Also **stach·er** (stakh'ər).
stac·te (stak'tē) n. One of the spices, of uncertain composition, anciently used by the Jews in preparing incense. Ex. xxx 34. [<L, oil of myrrh <Gk. stakté <stazein drip]
stac·tom·e·ter (stak·tom'ə·tər) n. A tube having a minute orifice in one end, for measuring a liquid in drops: also called stalagmometer. [<Gk. staktos trickling + -METER]
stad·dle (stad'l) n. Brit. Dial. 1 Anything that serves as a foundation or support; a prop; staff; crutch. 2 Agric. A raised platform or frame, or an arrangement of short posts, for a stack of hay or straw, to keep it dry and free from vermin. [OE stathol base]
stad·hold·er (stad'hōl·dər) n. Formerly, a viceroy or governor of a province or town of the Netherlands; specifically, the chief magistrate of the Netherlands, a hereditary office in the family of the princes of Orange. Also **stadt'·hold·er** (stat'-). [<Du. stadhouder lieutenant <stad place + houder holder]
sta·di·a (stā'dē·ə) n. 1 A temporary surveying station. 2 A form of sighting instrument for measuring distances used in connection with a vertical graduated rod (**stadia rod**). 3 The rod alone or the method of using it. 4 A graduated stick held at arm's length as a simple aid in measuring short distances. [Prob. <F stade a stage, a measure of length <L stadium. See STADIUM.]
sta·di·um (stā'dē·əm) n. pl. **·di·a** (-dē·ə), for def. 2 **·di·ums** 1 In ancient Greece, a course for footraces, with banked seats for spectators, as at Olympia and Athens, where games were held. 2 A similar modern structure in which athletic games are played: the stadium at Harvard. 3 An ancient Greek measure of length, equaling 606.75 feet. 4 A degree of progress or development. 5 Med. A given stage or period in the course of a disease. [<L <Gk. stadion, a measure of length]
Staël (stäl), **Madame de,** 1766–1817, Baronne de Staël-Holstein, née Anne Louise Germaine Necker, French writer; famous for her salon.
staff[1] (staf, stäf) n. pl. **staves** (stāvz) or **staffs** for defs. 1–3, **staffs** for defs. 4–6. 1 A stick or piece of wood carried for some special purpose, as an aid in walking or climbing, or as a cudgel or weapon, or as an emblem of authority. 2 A shaft or pole that forms a support or handle: the staff of a spear; a flagstaff. 3 A stick used in measuring or testing, as a surveyors' leveling rod. 4 Mil. **a** A body of officers not having command but attached in an executive or advisory capacity to an army or navy unit as assistants to the officer in command. The central body is known as the **general staff**. **b** The personnel of a military establishment, as the officers in charge of construction, ordnance, repairs, equipment, provisions, medicine and surgery, the paymasters, and engineers. 5 A body of persons associated in carrying out some special enterprise under the supervision of a manager or chief: the editorial staff of a newspaper. 6 Music The combined lines and spaces used to represent the pitches of tones. The staff has always five long horizontal lines and the accompanying long spaces, but is enlarged as the occasion may require, by short lines above or below and the short spaces they bring. See synonyms under STICK. — v.t. To provide (an office, etc.) with a staff. [OE stæf stick]
staff[2] (staf, stäf) n. A composition of plaster, fiber, etc., for temporary buildings, statues, etc. [Prob. <G staffieren fill, decorate]
Staf·fa (staf'ə) An islet of the Inner Hebrides group, NW Argyll, Scotland, on which is Fingal's Cave.
staf·fel·ite (staf'əl·īt) n. A variety of apatite believed to be the result of carbonated waters acting on phosphorite. [from Staffel, town in Germany where it was found]
staff officer An officer serving on a staff.
Staf·ford (staf'ərd) A county in west central England; 1,153 square miles; county town, Stafford. Also **Staf'ford·shire** (-shir). Shortened form **Staffs.**
stag (stag) n. 1 The male of the red deer (Cervus elaphus), especially the matured male. 2 The male of other large deer, as the caribou, and of certain other animals. 3 A castrated bull or boar. 4 Scot. A colt: also spelled staig. 5 U.S. Slang A man, especially when not in the company of women. 6 U.S. Slang A social gathering for men only. — adj. U.S. Slang Of or for men only: a stag party. — v.i. **stagged, stag·ging** Slang 1 Brit. To turn informer; squeal. 2 U.S. To attend a social affair unaccompanied by a woman. [OE stagga]

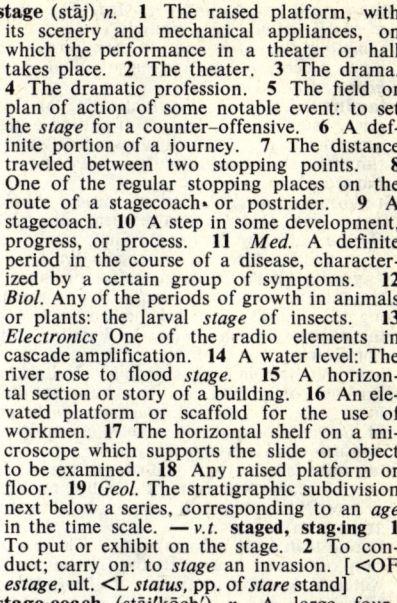
STAG HEAD
Showing antlers.

stag beetle A large, lamellicorn beetle (family Lucanidae), the male of which has the jaws enormously developed and branched like the antlers of a stag; specifically, the European Lucanus cervus and the American L. dama: also called pinchbug. They are injurious to trees.
stage (stāj) n. 1 The raised platform, with its scenery and mechanical appliances, on which the performance in a theater or hall takes place. 2 The theater. 3 The drama. 4 The dramatic profession. 5 The field or plan of action of some notable event: to set the stage for a counter-offensive. 6 A definite portion of a journey. 7 The distance traveled between two stopping points. 8 One of the regular stopping places on the route of a stagecoach or postrider. 9 A stagecoach. 10 A step in some development, progress, or process. 11 Med. A definite period in the course of a disease, characterized by a certain group of symptoms. 12 Biol. Any of the periods of growth in animals or plants: the larval stage of insects. 13 Electronics One of the radio elements in cascade amplification. 14 A water level: The river rose to flood stage. 15 A horizontal section or story of a building. 16 An elevated platform or scaffold for the use of workmen. 17 The horizontal shelf on a microscope which supports the slide or object to be examined. 18 Any raised platform or floor. 19 Geol. The stratigraphic subdivision next below a series, corresponding to an age in the time scale. — v.t. **staged, stag·ing** 1 To put or exhibit on the stage. 2 To conduct; carry on: to stage an invasion. [<OF estage, ult. <L status, pp. of stare stand]
stage·coach (stāj'kōch') n. A large four-wheeled vehicle having a regular route from town to town: formerly the principal mode of conveyance for passengers and mail.
stage·craft (stāj'kraft', -kräft') n. Skill in writing or staging plays.
stage door A door to a theater used by actors and stagehands which leads to the stage or behind the scenes.
stage-door Johnny (stāj'dôr') U.S. Slang A man who frequents stage doors seeking the companionship of actresses.
stage·hand (stāj'hand') n. A worker in a theater who handles scenery and props, operates lights, etc.
stage manager One who superintends the stage during the production of a play.
stag·er (stā'jər) n. 1 One who has had long experience at anything; an old hand: often **old stager**. 2 Archaic An actor.
stage-set·ting (stāj'set'ing) n. 1 The scene or background of a stage presentation. 2 The act of arranging scenery: often used figuratively.
stage-struck (stāj'struk') adj. Possessed of the idea of becoming an actor or an actress; enamored of theatrical life.
stage whisper A loud whisper, such as one uttered on the stage for the benefit of the audience.
stag·y (stā'jē) See STAGY.
stag·gard (stag'ərd) n. The male of the red deer in his fourth year. Also **stag'gart** (-ərt). [<STAG + -ARD]
stag·ger (stag'ər) v.i. 1 To move unsteadily; totter; reel. 2 To begin to give way; become less confident or resolute; waver; hesitate. — v.t. 3 To cause to stagger. 4 To affect strongly; overwhelm, as with surprise or grief. 5 To place in alternating rows or groups. 6 To arrange so as to prevent congestion or confusion, as by distributing: to stagger lunch hours. 7 Aeron. To adjust (two surfaces, as the wings of a biplane) so that the edge of one extends beyond the other. — n. 1 The act of staggering; a reeling motion. See STAGGERS. 2 Aeron. The amount of advance of the leading edge of one wing of a biplane over that of the other. [<obs. stacker <ON stakra] — **stag'ger·er** n. — **stag'ger·ing·ly** adv.
stag·ger·bush (stag'ər·boosh') n. A shrub (Lyonia mariana), 2 to 4 feet high, with white or pale-red flowers, common to Tennessee and the North Atlantic seaboard: poisonous to stock.
stag·gers (stag'ərz) n. pl. (construed as singular) 1 Any of various diseases of domestic animals, as horses, characterized by vertigo, staggering, and sudden falling, due to disorder of the brain and spinal cord: also **blind staggers**. 2 A giddy or reeling sensation.
stag·gy (stag'ē) n. Scot. A colt. Also **stag'gie**.

stag·hound (stag′hound′) *n.* One of a breed of nearly extinct, large hounds, somewhat resembling the foxhound, formerly used for hunting deer, wolves, etc.: also called *buckhound, deerhound.*

stag·ing (stā′jing) *n.* **1** A scaffolding or temporary platform. **2** The act of putting a play upon the stage. **3** The business of driving or running stagecoaches; also, traveling by stagecoach.

Sta·gi·ra (stə·jī′rə) A city of ancient Macedonia, on the Chalcidice peninsula, NE Greece, near the Strymonic Gulf; birthplace of Aristotle. Also **Sta·gi·rus** (stə·jī′rəs).

Stag·i·rite (staj′ə·rīt) *n.* A native of Stagira; specifically, Aristotle. — *adj.* Of or pertaining to Stagira; also, Aristotelian.

stag·nant (stag′nənt) *adj.* **1** Standing still; not flowing: said of water, as in a pool; hence, foul from long standing. **2** Lacking briskness or activity, as life or business; dull; inert; sluggish. [<F <L *stagnans, -antis,* ppr. of *stagnare* stagnate <*stagnum* a pool] — **stag′nan·cy** *n.* — **stag′nant·ly** *adv.*

stag·nate (stag′nāt) *v.i.* **·nat·ed, ·nat·ing** **1** To be or become stagnant. **2** To become dull or inert. [<L *stagnatus,* pp. of *stagnare.* See STAGNANT.]

stag·na·tion (stag·nā′shən) *n.* **1** The condition of being stagnant: the *stagnation* of water; *stagnation* in trade. **2** *Physiol.* Accumulation and retardation of a circulating fluid in the body.

St. Agnes' Eve The evening of January 20th, when, by old superstition, a girl might have prevision of her future husband. Also **St. Agnes's Eve.**

stag·y (stā′jē) *adj.* **stag·i·er, stag·i·est** Having a theatrical manner; of or suited to the stage. Also spelled *stagey.* — **stag′i·ly** *adv.* — **stag′i·ness** *n.*

staid (stād) *adj.* Fixed; steady and sober; sedate. See synonyms under SOBER, SEDATE. [Orig. pt. and pp. of STAY[1]] — **staid′ly** *adv.* — **staid′ness** *n.*

staig (stag) See STAG (def. 4).

stain (stān) *n.* **1** A discoloration from foreign matter; a spot; smirch; blot. **2** The act of discoloring, or the state of being discolored. **3** A dye or thin pigment used in staining. **4** A chemical reagent for coloring microscopic specimens. **5** A moral taint; tarnish. [<*v.*] — *v.t.* **1** To make a stain upon; discolor; soil. **2** To color by the use of a dye or stain. **3** To bring a moral stain upon; blemish. **4** To impregnate, as a microscopic specimen, with a substance whose reaction colors some part without affecting others, thus rendering form or structure visible. — *v.i.* **5** To take or impart a stain. [Aphetic var. of DISTAIN] — **stain′a·ble** *adj.* — **stain′er** *n.* — **stain′less** *adj.* — **stain′less·ly** *adv.*

Synonyms (verb): blot, color, discolor, disgrace, dishonor, dye, soil, spot, sully, tarnish, tinge, tint. To *color* is to impart a color desired or undesired, temporary or permanent, or, in the intransitive use, to assume a color in any way. To *dye* is to impart a color intentionally and with a view to permanence, and especially so as to pervade the substance or fiber of that to which it is applied. To *stain* is primarily to *discolor,* to impart a color undesired and perhaps unintended, and which may or may not be permanent. *Stain* is, however, used of giving an intended and perhaps pleasing color to wood, glass, etc., by an application of coloring matter which enters the substance a little below the surface, in distinction from painting, in which coloring matter is spread upon the surface; *dyeing* is generally said of wool, yarn, cloth, or similar materials which are dipped into the *coloring* liquid. To *tinge* is to *color* slightly. It may be used of giving a slight flavor, or a slight admixture of one ingredient or quality with another that is more pronounced. See BLEMISH, DEFILE[1], POLLUTE. Compare FOUL.

stained glass See under GLASS.

stainless steel A steel alloy made resistant to corrosion and atmospheric influences by the addition of from 10 to 30 percent chromium, and other ingredients.

stair (stâr) *n.* **1** A step, or one of a series of steps, for mounting or descending from one level to another. **2** A series of steps: usually in the plural. ◆ Homophone: *stare.* [OE *stǣger*]

stair·case (stâr′kās′) *n.* A flight or set of stairs, usually from one floor to another, complete with the supports, balusters, etc.

stair·head (stâr′hed′) *n.* The top of a staircase.

stair·way (stâr′wā′) *n.* A flight of stairs; staircase.

stair·well (stâr′wel′) *n.* A vertical shaft enclosing a staircase.

stake (stāk) *n.* **1** A stick or post, as of wood sharpened for driving into the ground: used as a boundary mark, sign of ownership, to support the rails of a fence, etc. **2** A post to which a person is bound to be burned alive; hence, death by burning at the stake. **3** An upright, set in a socket at the edge of the floor of a car or wagon, to confine loose material. **4** Something wagered or risked, as the money bet on a race. **5** A prize in a contest: sometimes in the plural. **6** An interest in an enterprise; contingent gain or loss. **7** A grubstake. — **at stake** In hazard or jeopardy; in question: My whole future was *at stake.* — **to pull up stakes** To wind up one's business in a place and move on; move out. — *v.t.* **staked, stak·ing** **1** To fasten or support by means of a stake; tether to a stake. **2** To mark the boundaries of with stakes: often with *off* or *out.* **3** *Colloq.* To put at hazard; wager; risk. **4** *Colloq.* To grubstake; also, to supply with working capital; finance. ◆ Homophone: *steak.* [OE *staca*]

stake-and-rid·er (stāk′ən·rī′dər) *n.* A split rail fence having the ends of the rails or riders laid at an angle across each other and supported by pairs of stakes: also called *snake fence.* — **staked′-and-rid′ered** *adj.*

Staked Plain See LLANO ESTACADO.

Sta·kha·no·vism (stä·khä′no·viz′əm) *n.* The efficiency system of Stakhanovite competition and awards. [after Aleksei G. *Stakhanov,* a Russian miner who originated it in 1935]

Sta·kha·no·vite (stä·khä′no·vīt) *n.* In the U.S.S.R., a worker awarded special privileges and bonuses for having displayed marked efficiency and initiative.

sta·lac·ti·form (stə·lak′tə·fôrm) *adj.* Resembling or having the form of a stalactite.

sta·lac·tite (stə·lak′tīt) *n.* **1** An elongated, downward-hanging form in which certain minerals, especially calcium carbonate, are sometimes deposited by slow dripping, as in a cave. **2** Any similar formation. **3** A downward-projecting ornament of a vaulted surface. [<NL *stalactites* < Gk. *stalaktitos* dripping <*stalassein* trickle, drip] — **stal·ac·tit·ic** (stal′ək·tit′ik) or **·i·cal** *adj.*

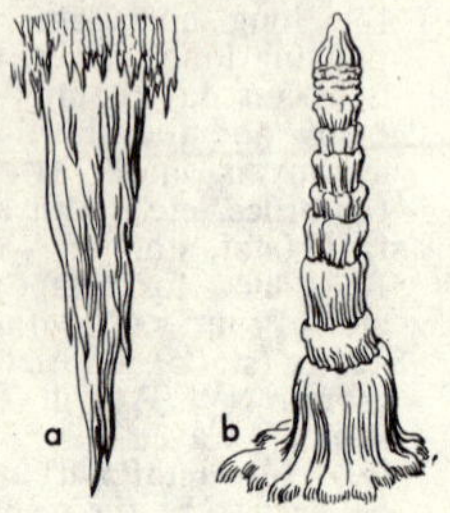

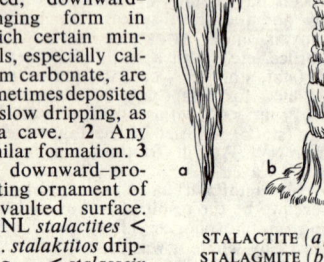

STALACTITE *(a)*
STALAGMITE *(b)*

sta·lag (stal′ag, *Ger.* shtä′läkh) *n.* A German prison camp for captured enlisted men. [<G, contraction of *stammlager* <*stamm* base + *lager* camp]

sta·lag·mite (stə·lag′mīt) *n.* **1** An incrustation, usually cylindrical, or conical, on the floor of a cavern: the counterpart of a stalactite, often fusing with it into the stalactite column. **2** Any similar formation. [<NL *stalagmites* <Gk. *stalagmos* a dripping <*stalassein* drip] — **stal·ag·mit·ic** (stal′əg·mit′ik) or **·i·cal** *adj.* — **stal′ag·mit′i·cal·ly** *adv.*

stal·ag·mom·e·ter (stal′əg·mom′ə·tər) *n.* A stactometer.

St. Al·bans (sânt′ ôl′bənz) A city in southern Hertfordshire, England: scene of a Yorkist victory in the Wars of the Roses, 1455; site of one of the oldest inhabited houses in England: Roman *Verulamium.*

stale[1] (stāl) *adj.* **1** Having lost freshness; slightly changed or deteriorated by standing, as air, vapid wine or beer, old bread, etc. **2** Lacking in interest from age or familiarity; worn out; trite: a *stale* joke. **3** In poor condition from prolonged activity, as from overstudy or, in athletics, from overtraining: especially in the phrase **gone stale.** **4** Inactive; dull: said of a stock market after a period of overactivity. **5** *Law* In courts of equity, impaired in legal force, due to long neglect in pressing or asserting a claim or to a change in the condition or situation of the parties. See synonyms under TRITE. — *v.i.* **staled, stal·ing** To become stale or trite. [Origin uncertain] — **stale′ly** *adv.* — **stale′ness** *n.*

stale[2] (stāl) *n.* The urine of cattle or horses. — *v.i.* **staled, stal·ing** To urinate: said of horses and cattle. [Prob. <MLG *stal* horse urine]

stale[3] (stāl) *n. Obs.* **1** A prostitute; drab. **2** A snare; trap. **3** Concealment; stealth; also, theft. [? <AF *estale, estal* a decoy]

stale·mate (stāl′māt′) *n.* **1** In chess, a position in which a player, not in check, can make no move without putting his king in check. The result is a drawn game. **2** Hence, any tie or deadlock. — *v.t.* **·mat·ed, ·mat·ing** **1** To put into a condition of stalemate. **2** To bring to a standstill. [<AF *estale* a fixed position + MATE[2]]

Sta·lin (stä′lin, -lēn) **1** See VARNA. **2** A city in Transylvania, central Rumania: Rumanian *Brașov,* German *Kronstadt.* **3** See STALINO.

Sta·lin (stä′lin, -lēn), **Joseph,** 1879–1953, U.S.S.R. statesman: real name *Iosif Dzhugashvili.*

Sta·li·na·bad (stä′lyi·nä·bät′) The capital of Tadzhik S.S.R., in the eastern part: formerly *Dyushambe.*

Sta·lin·grad (stä′lin·grad, *Russian* stä′lyin·grät′) A city on the lower Volga in SE European Russian S.F.S.R.; scene of a battle and ultimate Russian victory over German forces in World War II, Sept., 1942, to Jan., 1943: formerly *Tsaritsyn.*

Sta·li·ni·ri (stä′lyi·nyē′rē) The capital of the South Ossetian Autonomous Region, north central Georgian S.S.R. Formerly **Tskhin·va·li** (tskhin′vä·lē).

Sta·lin·ism (stä′lin·iz′əm) *n.* The doctrines or practices of Stalin; especially, communism involving a rigid implementation of government policy, through coercion, intimidation, and ruthless suppression of opposition, and characterized by ardent patriotism focused upon the Soviet Union and its leader. — **Sta′lin·ist** *n.*

Sta·li·no (stä′lyi·no) A city in the Donbas in SE Ukrainian S.S.R.: formerly *Yuzovka:* also *Stalin.*

Stalin Peak **1** The highest peak of the Carpathians and of Czechoslovakia; 8,737 feet: formerly *Gerlachovka.* **2** The highest point in the U.S.S.R., in SE Tadzhik S.S.R.; 24,590 feet: formerly *Garmo Peak.* **3** The highest peak of the Rhodope Mountains and of Bulgaria, in SW Bulgaria: formerly *Mus Allah.*

Sta·linsk (stä′lyinsk) A city on the Tom river in SW Asiatic Russian S.F.S.R.: formerly *Novo Kuznetsk.*

stalk[1] (stôk) *n.* The stem or axis of a plant, especially when herbaceous. **2** Any support on which an organ is borne, as a pedicel. **3** A supporting part or stem: the jointed *stalk* of a sea lily, the *stalk* of a quill. **4** Any stem or main axis, as of a goblet. [ME *stalke,* dim. of OE *stæla* stem of a plant] — **stalked** *adj.* — **stalk′less** *adj.*

stalk[2] (stôk) *v.i.* **1** To approach game, etc., stealthily. **2** To walk in a stiff, dignified manner: also used figuratively: Murder *stalked* through the streets. **3** *Obs.* To go stealthily; creep. — *v.t.* **4** To approach (game, etc.) stealthily. **5** To pace through: Famine *stalked* the countryside. — *n.* **1** The act of stalking game. **2** A stately step or walk. [OE *bestealcian* move stealthily] — **stalk′er** *n.*

stalk·ing-horse (stô′king·hôrs′) *n.* **1** A horse behind which a hunter conceals himself in stalking game. **2** Anything serving to conceal one's intention.

stalk·y (stô′kē) *adj.* **stalk·i·er, stalk·i·est** **1** Long and slender, like a stalk. **2** Consisting of stalks.

stall (stôl) *n.* **1** A compartment in which a horse or bovine animal is confined and fed. **2** A small booth or compartment in a street, market, etc., for the sale or display of small articles. **3** A partially enclosed seat, as in the orchestra of a theater or the choir of a cathedral. **4** A working compartment in a coal mine. **5** A space set aside for the parking of an automobile. **6** A sheath or covering for a finger or thumb; a cot. **7** *Aeron.* The condition of an airplane which has lost the relative speed necessary for control; the act of stalling. **8** *Colloq.* An evasion or argument made to postpone action or decision. — *v.t.* **1** To place or keep in a stall. **2** To keep in a stall for fattening, as cattle. 3 To bring to a standstill; stop the progress or motion of, especially unintentionally. 4 To cause to stick fast in mud, snow, etc. —*v.i.* 5 To come to a standstill; stop, especially unintentionally. 6 To stick fast in mud, snow, etc. 7 *Colloq.* To make delays; be evasive: to *stall* for time. 8 To live or be kept in a stall. 9 *Aeron.* To go into a stall. [OE *steall*]

stall-feed (stôl′fēd′) *v.t.* **-fed, -feed·ing** To feed (cattle) in a stall or stable; fatten. —**stall′-fed′** *adj.*

stal·lion (stal′yən) *n.* An uncastrated male horse. [<OF *estalon* <OHG *stal* stable]

stal·wart (stôl′wərt) *adj.* 1 Strong and brawny; robust. 2 Resolute: determined; unwavering. 3 Brave; courageous. —*n.* 1 An uncompromising partisan, as in politics. 2 *U.S.* A member of a conservative faction of the Republican party (1874–85) which opposed civil service reform and liberal policies toward the South. [Var. of STALWORTH] —**stal′wart·ly** *adv.* —**stal′wart·ness** *n.*

stal·worth (stôl′wərth) *adj. Obs.* Stalwart. [OE *stælwierthe* serviceable < *stæl* place + *wierthe* worth]

stam·bou·line (stam′bə·lēn′) *n.* A coat for formal occasions, worn by officials in Turkey. [from *Stamboul,* var. of *Stambul*]

Stam·bul (stäm·bōōl′) 1 Istanbul. 2 The old part of the city of Istanbul. Also **Stam·boul′.**

sta·men (stā′mən) *n. pl.* **sta·mens,** *Rare* **stam·i·na** (stam′ə·nə) *Bot.* The pollen-bearing floral organ of a flower, standing inside the floral envelopes and consisting of two parts: the *filament,* or support, and the *anther,* or pollen sac. [<L, warp, thread < *stare* stand]

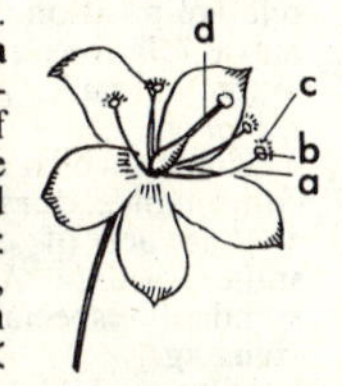

STAMEN (*a, b, c*)
a. Filament.
b. Anther.
c. Pollen.
d. Pistil.

Stam·ford (stam′fərd) 1 A municipal borough in SW Lincolnshire, England; a 14th century center of learning, at that time comparable to Oxford. 2 A city on Long Island Sound in SW Connecticut.

stam·i·na (stam′ə·nə) *n.* 1 Supporting vitality; strength; vigor; physical or moral capacity to endure or withstand hardship or difficulty. 2 The supporting part of a body. [<L, pl. of *stamen* warp, thread. See STAMEN.]

stam·i·nal (stam′ə·nəl) *adj.* 1 Of or pertaining to a stamen. 2 Relating to or furnishing stamina or lasting strength and vigor; essential.

stam·i·nate (stam′ə·nit, -nāt) *adj. Bot.* 1 Having stamens but no pistils, as certain flowers. 2 Having stamens.

stamini- *combining form Bot.* Stamen; of or pertaining to stamens: *staminiferous,* bearing stamens. Also, before vowels, **stamin-.** [<L *stamen, -inis* a fiber, thread]

stam·i·no·di·um (stam′ə·nō′dē·əm) *n. pl.* **·di·a** (-dē·ə) *Bot.* An abortive or sterile stamen, or an organ resembling one. Also **stam′i·node** (-nōd). [<NL <L *stamen, -inis* a stamen + Gk. *eidos* form]

stam·i·no·dy (stam′ə·nō′dē) *n. Bot.* The conversion of other parts of a flower, such as bracts, sepals, petals, or pistils, into stamens.

stam·mel (stam′əl) *n.* A linsey-woolsey of a dull-scarlet color; also, a dull scarlet. —*adj.* Of or pertaining to stammel or its color; dull-red. [<OF *estamel* < *estamine* <L *stamen* thread]

stam·mer (stam′ər) *v.t.* & *v.i.* To speak or utter with a halting articulation, commonly with nervous repetitions or prolongations of a sound or syllable, and involuntary pauses: to *stammer* an apology. —*n.* A halting, defective utterance. [OE *stamerian*] —**stam′·mer·er** *n.*

Synonym: stutter. *Stammer* and *stutter* are virtually interchangeable in general use. Frequently, however, *stammer* is associated with nervousness, excitement, or embarrassment, while *stutter* is reserved by the speech therapists for a particular speech disorder of obscure origin.

stamp (stamp) *v.t.* 1 To strike heavily with the sole of the foot. 2 To bring down (the foot) heavily and noisily. 3 To affect in a specified manner by or as by stamping with the foot: to *stamp* a fire out; to *stamp* out opposition. 4 To make marks or figures upon by means of a die, stamp, etc. 5 To imprint or impress with a die, stamp, etc. 6 To fix or imprint permanently: The deed was *stamped* on his memory. 7 To assign a specified quality to; characterize; brand: to *stamp* a story false. 8 To affix an official seal, stamp, etc., to. 9 To crush, break, or pulverize, as ore. —*v.i.* 10 To strike the foot heavily on the ground. 11 To walk with heavy, resounding steps. See synonyms under IMPRESS[1], INSCRIBE. —*n.* 1 A characteristic mark made by stamping; a device or design impressed upon any object, as by a die. 2 An implement or machine for stamping; specifically, a die having a pattern as for coinworking; any instrument for impressing a mark, design, or copy upon any object or surface: a hand *stamp.* 3 The weight or block as in an ore mill, which by its impact crushes the ore; by extension, the stamping mill itself. 4 A cutting tool for making articles of outline corresponding to the cutting edges: operated by pressure or by blows. 5 Any characteristic mark, as a label or imprint; a brand. 6 Hence, figuratively, characteristic quality or form; kind; sort: I dislike men of his *stamp.* 7 The act of stamping. 8 A printed device prepared and sold by a government, for attachment, as to a letter (**postage stamp**), commodity (**revenue stamp**), etc., as proof that the tax or fee has been paid; also, a trading stamp. See synonyms under MARK[1]. —**trading stamp** A stamp of fixed value given by a tradesman to a purchaser and exchangeable, in quantities, for goods selected from a premium list. [ME *stampen.* Akin to OE *stempan* pound.]

Stamp Act An act of the British Parliament, passed in March, 1765, and repealed in March, 1766, which required the American colonists to affix to various legal and commercial papers, as well as to pamphlets, newspapers, vellum, parchment, or paper, a government stamp varying in price from a halfpenny up to £10.

stam·pede (stam·pēd′) *n.* 1 A sudden starting and rushing off through panic: said primarily of a herd of cattle, horses, etc. 2 Any sudden, impulsive, tumultuous running movement of a crowd, as of a mob. 3 A movement or rush of people toward a certain region or object, as a gold rush or for homestead sites. 4 The sudden unplanned movement to support a certain candidate at a political convention, as from common impulse. —*v.* **·ped·ed, ·ped·ing** *v.t.* To cause a stampede or panic in. —*v.i.* To rush or flee in a stampede. [<Am. Sp. *estampida* crash < *estampar* stamp] —**stam·ped′er** *n.*

stamp·er (stam′pər) *n.* 1 One who stamps, in any sense. 2 One who cancels stamps, as in a post office. 3 Any tool or machine for stamping.

stamping ground 1 A place where horses or other animals gather in numbers. 2 A favorite resort; a habitual gathering place.

stamp mill A machine for pulverizing rock for the purpose of extracting the ore it contains.

stance (stans) *n.* 1 Mode of standing; posture. 2 In golf, the position of a player's feet, with reference to the ball and to each other, when making a stroke. 3 *Scot.* A position; a station; hence, a site; foundation. [<OF *estance* <L *stans, stantis,* ppr. of *stare* stand]

stanch (stanch, stänch) *v.t.* 1 To stop or check the flow of (blood, etc.). 2 To stop the flow of blood from (a wound). 3 *Obs.* To quench; quell; put an end to. Also spelled *staunch.* —*adj.* & *n.* Staunch. [<OF *estanchier* halt, bring to a stop, make stand, ult. <L *stare* stand] —**stanch′er** *n.*

◆ The spelling *stanch* is preferred for the verb in both England and the United States, and *staunch* for the adjective. Many writers use one or the other spelling for both.

stan·chion (stan′shən) *n.* 1 An upright bar forming a principal support. 2 A vertical bar or pair of bars used to confine cattle in a stall. —*v.t.* 1 To provide with stanchions. 2 To support or confine with stanchions. [<OF *estanchon* < *estance* situation, position. See STANCE.]

stand (stand) *v.* **stood, stand·ing** *v.i.* 1 To assume or maintain an erect position on the feet: distinguished from *sit, lie, kneel,* etc. 2 To be in a vertical position; be erect. 3 To measure a specified height when standing: He *stands* six feet. 4 To assume a specified position: to *stand* aside. 5 To be situated; have position or location; lie. 6 To have or be in a specified state, condition, or relation: We *stand* ready to fight; He *stood* in fear of his life. 7 To assume an attitude for defense or offense: *Stand* and fight! 8 To be or remain firm or resolute, as in determination. 9 To be consistent; accord; agree. 10 To remain unimpaired, unchanged, or valid: My decision still *stands.* 11 To collect and remain; also, to be stagnant, as water. 12 To be of a specified rank or class: He *stands* third. 13 To stop or pause; halt. 14 To scruple; hesitate. 15 *Naut.* To take a direction; steer: The brig *stood* into the wind. 16 To point, as a hunting dog. 17 *Brit.* To be a candidate, as for election. —*v.t.* 18 To place upright; set in an erect position. 19 To put up with; endure; tolerate. 20 To be subjected to; undergo: He must *stand* trial. 21 To withstand; resist. 22 *Colloq.* To pay for; bear the expense of: to *stand* a treat. —**to stand a chance** (or **show**) To have a chance or likelihood, as of success. —**to stand by** 1 To stay near and be ready to help or operate. 2 To help; support. 3 To abide by; make good; adhere to. 4 To remain passive and watch, as when help is needed. 5 *Telecom.* To wait, as for the continuance of an interrupted transmission. —**to stand clear** To remain at a safe distance. —**to stand down** *Law* To leave the witness stand. —**to stand for** 1 To represent; symbolize. 2 To put up with; tolerate. —**to stand from under** To move from beneath, as something about to fall. —**to stand in** *Colloq.* To cost. —**to stand in for** To act as a substitute for. —**to stand off** *Colloq.* 1 To keep at a distance. 2 To fail to agree or comply. —**to stand on** 1 To be based on or grounded in; rest. 2 To insist on or demand observance of: to *stand on* ceremony. 3 *Naut.* To keep on the same tack or course. —**to stand on one's own (two) feet** (or **legs**) To be independent; manage one's own affairs. —**to stand out** 1 To stick out; project or protrude. 2 To be prominent; appear in relief or contrast. 3 To refuse to consent or agree; remain in opposition. —**to stand over** 1 To remain near and watch, as a subordinate. 2 To be postponed. —**to stand pat** 1 In poker, to play one's hand as dealt, without drawing new cards. 2 To resist change. See STAND-PATTER. —**to stand to reason** To conform to reason. —**to stand up** 1 To stand erect. 2 To withstand wear, criticism, analysis, etc. 3 *Slang* To fail, usually intentionally, to keep an appointment with. —**to stand up for** To side with; take the part of. —**to stand up to** To confront courageously; face. —**to stand up with** To be best man or bridesmaid for. —*n.* 1 A structure upon which persons or things may stand, or on which articles may be kept or displayed. 2 A small table on which things may be placed conveniently. 3 A rack or other piece of furniture on which hats may be hung, or canes, umbrellas, etc., supported: a hall *stand.* 4 A stall, counter, or the like, where merchandise is displayed: a *bookstand.* 5 A structure upon which persons may sit or stand, as a platform, or a series of raised seats: a *bandstand,* a judges' *stand;* also, a small platform in court from which a witness testifies. 6 Any place where or in which something stands; position; place; specifically, the place of one's customary occupation; an assigned or chosen location. 7 The act of standing, especially of standing firmly: to make a *stand* against the enemy. 8 Cessation from motion or progress; a standstill. 9 A complete set; outfit: chiefly in the phrase **stand of arms.** 10 A growth on the field, as of corn or grass. 11 A tree grown from seed; also, a young tree left when others are cut down.

12 The growing trees in a forest or in part of a forest. **13** In the theater, a stop made while on tour to give a performance; also, the place: a one-night *stand.* **14** *Obs.* A troop; force. **15** A curved metal bar attached to the base of a force pump and serving as a fulcrum for the brake which moves the piston up and down. **16** In prosody, an epode: so called because it was originally sung while the chorus stood still. [OE *standan*] **—stand′er** *n.*
Synonyms (verb): abide, continue, endure, halt, pause, remain, stay, stop. See REST[1]. *Antonyms*: decline, droop, drop, fail, faint, falter, flee, fly, sink, succumb, yield.

stan·dard (stan′dərd) *n.* **1** A flag, ensign, or banner, used as a distinctive emblem of a government, body of men, or special cause: the *standard* of freedom or revolt. **2** A figure or an image adopted as the emblem of a nation. **3** A long, narrow flag carried by mounted and motorized units of the U. S. Army. **4** Any established measure of extent, quantity, quality, or value. **5** Any type, model, or example for comparison; a criterion of excellence; test: a *standard* of conduct or taste. **6** In coinage, the established proportion by weight of fine metal and alloy. **7** An upright timber, post, pole, or beam, especially as a support. **8** *Bot.* **a** Any tree, shrub, bush, or herb not dwarfed by grafting, and growing on a vigorous upright stem without support of a wall or trellis. **b** The vexillum (def. 2). **9** A heavy or stationary article of furniture. See synonyms under EXAMPLE, IDEAL, RULE. **—National Bureau of Standards** A branch of the U. S. Department of Commerce that maintains scientific, technical, and industrial standards, and acts as a research and testing agency for the government. **—***adj.* **1** Having the accuracy or authority of a standard; serving as a gage, test, or model; hence, of recognized excellence or authority: a *standard* book or author. **2** Designating or belonging to the form of a language which, through its use in a region of economic and cultural importance, has gained acceptance and social prestige among all the speakers of the language. ◆ In this dictionary, words and meanings not considered to be at the level of the standard language are appropriately labeled colloquial, slang, dialectal, or illiterate. [<OF *estandard* banner <Gmc.]

stan·dard–bear·er (stan′dərd·bâr′ər) *n.* **1** An officer or soldier of a regiment or other military body who carries the flag or ensign. **2** Hence, one who leads, as a candidate; specifically, a presidential nominee.

stan·dard·bred (stan′dərd·bred′) *n.* A breed of horse notable for its trotters and pacers: descendent from the thoroughbred stallion *Messenger,* imported from England in 1788.

stan·dard–bred (stan′dərd·bred′) *adj.* Bred so as to be of a required strain, quality, or pedigree, as poultry, horses, etc.

standard candle *Physics* A unit of luminous intensity equal to one lumen.

standard cell *Electr.* A voltaic cell which serves as a standard of electromotive force.

standard deviation *Stat.* The square root of the arithmetic average of the squares of all the deviations from the mean value of a series of observations.

standard dollar See under DOLLAR.

standard gage **1** A gage for determining whether tools, etc., are of a recognized standard size. **2** A railroad track width of 56 1/2 inches, considered as standard. **3** A railroad having such a gage, or a locomotive or car made to run on this gage. **—stan′dard–gage′** (-gāj′) *adj.*

stan·dard·ize (stan′dər·dīz) *v.t.* **·ized, ·iz·ing** To make to or regulate by a standard. **—stan′·dard·i·za′tion** *n.* **—stan′dard·iz′er** *n.*

standard lamp **1** Any of several standard lighting units used in photometric determinations. **2** In the United States, the pentane-burning lamp, equal to 10 international candles, or the Hefner lamp, burning amyl acetate, equal to 0.9 international candle.

standard time Civil time as reckoned from a certain meridian officially established as standard over a large area. Reckoning from the meridian of Greenwich, each time zone, comprising a sector of 15 degrees of longitude, is considered to represent a time interval of one hour, although in practice these zones are adjusted to meet various geographic and other regional conditions, as along the International Date Line. See table above. In the United States the four standard time zones are the Eastern, Central, Mountain, and Pacific, using respectively the mean local time of the 75th, 90th, 105th, and 120th meridians west of Greenwich, and being 5, 6, 7, and 8 hours slower or earlier than Greenwich time. Canada, in addition to the above, has a fifth zone, the Atlantic (or Provincial), based on the local time of the 60th meridian, which is 4 hours earlier than Greenwich time. See also TIME ZONE.

STANDARD TIME IN PRINCIPAL CITIES

Referred to noon in Washington, D.C. (*Time Zone +5*) and in Greenwich, England (*Time Zone 0*). Times have been calculated to the even hour, disregarding a few deviations (up to 30 minutes) resulting from local time-zone adjustments. All hours are for the same day except as indicated by (*) when the time is for the *following* day. Compare the table under TIME ZONE for explanation of the plus and minus factor in column 4.

(1) *City*	(2) *When* NOON *Washington Time*	(3) *When* NOON *Greenwich Time*	(4) *Time Zone Number*
Alexandria	7 P.M.	2 P.M.	–2
Amsterdam	5 P.M.	NOON	0
Athens	7 P.M.	2 P.M.	–2
Auckland	*4 A.M.	11 P.M.	–11
Baghdad	8 P.M.	3 P.M.	–3
Bangkok	MIDNIGHT	7 P.M.	–7
Belfast	5 P.M.	NOON	0
Berlin	6 P.M.	1 P.M.	–1
Bogotá	NOON	7 A.M.	+5
Bombay	10 P.M.	5 P.M.	–5
Boston	NOON	7 A.M.	+5
Brussels	5 P.M.	NOON	0
Bucharest	7 P.M.	2 P.M.	–2
Budapest	6 P.M.	1 P.M.	–1
Buenos Aires	1 P.M.	8 A.M.	+4
Cairo	7 P.M.	2 P.M.	–2
Calcutta	11 P.M.	6 P.M.	–6
Cape Town	6 P.M.	7 P.M.	–7
Caracas	1 P.M.	8 A.M.	+4
Chicago	11 A.M.	6 A.M.	+6
Copenhagen	6 P.M.	1 P.M.	–1
Delhi	10 P.M.	5 P.M.	–5
Denver	10 A.M.	5 A.M.	+7
Detroit	11 A.M.	6 A.M.	+6
Dublin	5 P.M.	NOON	0
Geneva	6 P.M.	1 P.M.	–1
Greenwich	5 P.M.	NOON	0
Halifax	1 P.M.	8 A.M.	+4
Havana	1 P.M.	8 A.M.	+4
Hong Kong	*1 A.M.	.8 P.M.	–8
Istanbul	7 P.M.	2 P.M.	–2
Jakarta	MIDNIGHT	7 P.M.	–7
Johannesburg	7 P.M.	2 P.M.	–2
LeHavre	5 P.M.	NOON	0
Leningrad	7 P.M.	2 P.M.	–2
Lima	NOON	7 A.M.	+5
Lisbon	5 P.M.	NOON	0
Liverpool	5 P.M.	NOON	0
London	5 P.M.	NOON	0
Madrid	6 P.M.	1 P.M.	–1
Manila	*1 A.M.	8 P.M.	–8
Melbourne	*3 A.M.	10 P.M.	–10
Mexico City	11 A.M.	6 A.M.	+6
Montreal	NOON	7 A.M.	+5
Moscow	7 P.M.	2 P.M.	–2
New York	NOON	7 A.M.	+5
Oslo	6 P.M.	1 P.M.	–1
Ottawa	NOON	7 A.M.	+5
Paris	5 P.M.	NOON	0
Peking	*1 A.M.	8 P.M.	–8
Philadelphia	NOON	7 A.M.	+5
Quebec	NOON	7 A.M.	+5
Río de Janeiro	2 P.M.	9 A.M.	+3
Rome	6 P.M.	1 P.M.	–1
St. Louis	11 A.M.	6 A.M.	+6
San Francisco	9 A.M.	4 A.M.	+8
Shanghai	*1 A.M.	8 P.M.	–8
Singapore	MIDNIGHT	7 P.M.	–7
Stockholm	6 P.M.	1 P.M.	–1
Sydney	*3 A.M.	10 P.M.	–10
Teheran	8 P.M.	3 P.M.	–3
Tokyo	*2 A.M.	9 P.M.	–9
Toronto	NOON	7 A.M.	+5
Vancouver	9 A.M.	4 A.M.	+8
Vienna	6 P.M.	1 P.M.	–1
Vladivostok	*2 A.M.	9 P.M.	–9
Warsaw	6 P.M.	1 P.M.	–1
Washington	NOON	7 A.M.	+5
Winnipeg	11 A.M.	6 A.M.	+6
Yokohama	*2 A.M.	9 P.M.	–9
Zurich	6 P.M.	1 P.M.	–1

standard wavelength *Physics* The wavelength of the red cadmium line observed in dry air at 15 degrees Celsius and 760 millimeters of mercury: equal to 6438.4696 angstrom units.

stand–by (stand′bī′) *n. pl.* **–bys** Any person or thing that can be relied on in time of stress or emergency.

stand·ee (stan·dē′) *n. Colloq.* A person who must stand for lack of chairs or seats, as at a theater or on a train.

stand–fast (stand′fast′, -fäst′) *n.* That which stands firm and strong; a solid or settled position. **—***adj.* Firm; settled.

stand–in (stand′in′) *n.* **1** A position of influence or favor; a pull. **2** A person who relieves a motion-picture player from tedious waiting intervals and substitutes for him in hazardous actions.

stand·ing (stan′ding) *adj.* **1** Remaining erect; not prostrated or cut down, as grain. **2** Continuing for regular or permanent use; remaining the same indefinitely; not special or temporary: a *standing* rule, a *standing* army. **3** Stagnant; not flowing: *standing* water. **4** Begun while standing: distinguished from *running*: a *standing* high jump. **5** Established; permanent: the *standing* church. **—***n.* **1** Place; relative position, as in social, commercial, or moral relations; repute; grade; especially, high grade or rank; good reputation: a man of *standing.* **2** A place to stand in; station. **3** Time in which something stands or goes on; continuance; duration: a feud of long *standing.* **4** The act of one who stands; erectness; stance. **—***adv.* At or to a sudden stop or standstill, especially in the phrase **to bring up standing.**

standing order **1** A military order always in force and not subject to change or modification. **2** In parliamentary procedure, a general regulation governing the manner in which the business of a body shall be conducted: in force from session to session until rescinded or voided.

standing rigging *Naut.* The heavy ropes or cables which support the masts and fixed spars of a ship.

standing room Place in which to stand, as in a building, theater, etc., where the seats are all occupied.

stand·ish (stan′dish) *n.* A receptacle for pens and ink. [<STAND + DISH]

Stan·dish (stan′dish), **Miles,** 1584?–1656, English soldier and emigrant in the "Mayflower"; military leader of the Pilgrims; subject of Longfellow's *The Courtship of Miles Standish.*

stand–off (stand′ôf′, -of′) *n.* **1** A draw or tie, as in a game. **2** A counterbalancing or neutralization. **3** A feeling or state of indifference or coldness; aloofness. **4** A postponement. **—stand′–off′ish** *adj.*

stand oil Linseed or other oil heated for several hours at a temperature of about 300° C. to thicken it and remove coagulated impurities: used in varnishes and paints.

stand–out (stand′out′) *n.* **1** Someone or something that is outstanding, excellent, etc. **2** *Colloq.* One who stubbornly refuses to agree, consent, or cooperate.

stand–pat (stand′pat′) *adj.* Characterized by or pertaining to the policy of opposition to change; conservative.

stand–pat·ter (stand′pat′ər) *n.* In U. S. politics, one who adheres obstinately to a policy or party; specifically, a politician who advocates maintaining the existing tariff schedules; a conservative. **—stand′–pat′tism** *n.*

stand·pipe (stand′pīp′) *n.* A vertical pipe, as at a reservoir, into which the water is pumped to give it a head; a water tower.

stand·point (stand′point′) *n.* A position from which things are viewed or judged; point of view; basal principle.

St. An·drews (sānt an′drōōz) A port and burgh in eastern Fifeshire, Scotland; the rules of golf are established at a famous course nearby, founded 1754.

St. Andrew's cross The oblique cross; also, a saltire. See under CROSS.

stand·still (stand′stil′) *n.* A pause; cessation of motion or action; halt; rest. **—***adj.* In a state of rest or inactivity; standing still.

stand–up (stand′up′) *adj.* **1** Having an erect

Stanford revision

position: a *stand-up* collar. 2 Done, consumed, etc., while standing.
Stan·ford revision (stan'fərd) *Psychol.* A modification of the Binet-Simon scale designed to cover a wider range of mental age and to provide a constant number of tests, usually six, for each year group. [from *Stanford* University, Calif., where it was originated]
stang[1] (stang) *Scot. & Brit. Dial. v.t.* To sting. — *v.i.* To throb with pain. — *n.* A sting; throbbing pain.
stang[2] (stang) Obsolete past tense of STING.
stan·hope (stan'hōp) *n.* A light, open, one-seated carriage. [after Fitzroy *Stanhope*, 1787-1864, English clergyman, for whom it was first made]
stan·iel (stan'yəl) *n.* The kestrel. Also **stan·nel** (stan'əl). [OE *stāngella* < *stān* stone + *gellan* scream]
sta·nine (stā'nīn) *n. Psychol.* A composite weighted score of aptitudes and performance based on a scale of nine: it is a form of frequency distribution about a median of 5, 1 being lowest and 9 highest: originally developed to test air crews in the U.S. Army Air Forces. [<STA(NDARD SCALE OF) NINE]
Sta·ni·slav (stä'ni-slȧf) A city in western Ukrainian S.S.R., formerly, 1919-45, in Poland. German **Stan·is·lau** (stän'is.lou), Polish **Sta·ni·sła·wow** (stä·nē·swä'voof).
Stan·is·lav·sky (stä'ni-slȧf'skē), **Constantin**, 1863-1938, Russian actor, director, and producer: real name *Konstantin Sergeyevich Alekseyev.*
stank[1] (stangk) Past tense of STINK.
stank[2] (stangk) *n. Brit. Dial.* A pool; reservoir; pond; ditch; dam. [<OF *estanc* <L *stagnum* < *stagnare* stagnate. See STAGNANT.]
Stan·ley (stan'lē) A masculine personal name. [OE, stone lea]
Stan·ley (stan'lē), **Arthur Penrhyn**, 1815-81, English author and divine. — **Sir Henry Morton,** 1841-1904, British journalist and explorer active in the United States; sent out to find David Livingstone in Africa; original name *John Rowlands.* — **Wendell Meredith,** born 1904, U.S. biochemist.
Stanley, Mount See RUWENZORI.
Stanley Falls The seven cataracts of the upper Congo river, near the equator; total fall, 200 feet.
Stanley Pool A lakelike expansion of the Congo river on the border between SW Belgian Congo and SE French Equatorial Africa; 320 square miles.
stan·na·ry (stan'ər.ē) *n. pl.* **·ries** A tin mine or region of tin mines. [<Med. L *stannaria* <L *stannum* tin]
stan·nic (stan'ik) *adj. Chem.* Of, pertaining to, or containing tin, especially in its higher valence. [<L *stannum* tin]
stannic acid *Chem.* Any of three compounds derived from stannic chloride by the action of alkalis.
stannic chloride *Chem.* A thin, colorless liquid, SnCl₄, made by exposing metallic tin to the action of chlorine: used as a mordant in dyeing.
stannic oxide *Chem.* A white, amorphous, pulverulent compound, SnO₂, found native or formed by heating the lower (stannous) oxide in air: extensively used as a polishing agent called *putty powder.*
stannic sulfide *Chem.* A yellow compound, SnS₂, precipitated from a solution of a stannic salt by hydrogen sulfide: used in bronzing.
stan·nif·er·ous (stə·nif'ər·əs) *adj.* Yielding or containing tin. [<L *stannum* tin + -FEROUS]
stan·nite (stan'īt) *n.* A granular, metallic, steel-gray to iron-black mineral containing tin, copper, iron, sulfur, and sometimes zinc; tin pyrites. [<L *stannum* tin + -ITE[1]]
stan·nous (stan'əs) *adj. Chem.* Of, pertaining to, or containing tin, especially in its lower valence. [<L *stannum* tin + -OUS]
stan·num (stan'əm) *n. Tin.* [<L]
Sta·no·voi Range (stä'nō·voi') A mountain chain in SE Siberian Russian S.F.S.R.; highest point, 8,143 feet; forms the watershed between the Lena and Amur river basins. Also **Sta·no·vy Range** (stä·nō'vē).
St. Anthony's fire *Pathol.* Erysipelas.
St. Anthony's nut An earthnut (*Conopodium denudatum*), fed to pigs: so called because St. Anthony was once a swineherd: also called *pignut, groundnut.*

Stan·ton (stan'tən), **Edwin McMasters,** 1814-1869, U.S. statesman; secretary of war 1862-1868. — **Elizabeth Cady,** 1815-1902, U.S. woman-suffrage leader.
stan·za (stan'zə) *n.* A certain number of lines of verse grouped in a definite scheme of meter and sequence; a metrical division of a poem: often incorrectly called a *verse.* [<Ital., room, stanza <L *stans, stantis* standing. See STANCE.] — **stan·za·ic** (stan.zā'ik) *adj.*
sta·pe·li·a (stə·pē'lē·ə) *n.* Any of a genus (*Stapelia*) of fleshy African plants of the milkweed family, having leafless toothed stems and showy, starlike, ill-smelling, purple or yellowish flowers sometimes a foot in diameter; a carrion flower. [<NL, after J. B. van *Stapel,* died 1636, Dutch botanist]
sta·pes (stā'pēz) *n. Anat.* The innermost ossicle of the middle ear of mammals. See illustration under EAR. [<LL *stapes* a stirrup] — **sta·pe·di·al** (stə·pē'dē·əl) *adj.*
staphylo- combining form 1 *Anat.* The uvula: *staphyloplasty.* 2 *Med.* Staphylococcic. Also, before vowels, **staphyl-.** [<Gk. *staphylē* bunch of grapes]
staph·y·lo·coc·cus (staf'ə·lō·kok'əs) *n. pl.* **·coc·ci** (-kok'sī) Any of a genus (*Staphylococcus*) of typically parasitic bacteria occurring singly, in pairs, or in irregular clusters; especially, *S. aureus,* an infective agent in boils, furuncles, and suppurating wounds. See illustration under BACTERIUM. [<NL <Gk. *staphylos* bunch of grapes + *kokkos* a berry] — **staph'y·lo·coc'cic** (-kok'sik) *adj.*
staph·y·lo·plas·ty (staf'ə·lō·plas'tē) *n.* Reparative surgery of the soft palate and uvula. — **staph'y·lo·plas'tic** *adj.*
staph·y·lor·rha·phy (staf'ə·lôr'ə·fē, -lor'-) *n. Surg.* The operation of uniting a cleft palate. Also **staph'y·lor'a·phy.** [<STAPHYLO- + -RHAPHY]
sta·ple[1] (stā'pəl) *n.* 1 A principal commodity or production of a country or region; a well-established article of commerce. 2 A chief element or main constituent of something. 3 The carded or combed fiber of cotton, wool, or flax. 4 Raw material. 5 A commercial emporium; mart. 6 Hence, a source of supply; storehouse. — *adj.* 1 Regularly and constantly produced or sold; hence, main; chief. 2 Commercially established; having regular commercial channels. 3 Marketable. — *v.t.* **·pled, ·pling** To sort or classify according to length, as wool fiber. [<OF *estaple* market, support <Gmc.]
sta·ple[2] (stā'pəl) *n.* A U-shaped piece of metal with pointed ends, or a loop of thin wire, driven into wood, fabrics, paper, etc., to serve as a fastening. — *v.t.* **·pled, ·pling** To fix or fasten by a staple or staples. [OE *stapol* post, prop]
sta·pler[1] (stā'plər) *n.* 1 A sorter of wool according to its staple. 2 A merchant who participated in one of the monopolies formerly granted by royal authority.
sta·pler[2] (stā'plər) *n.* A wire-stitching machine that binds pamphlets, books, etc.
star (stär) *n.* 1 *Astron.* One of a class of self-luminous celestial bodies, exclusive of comets, meteors, and nebulae, but including the sun. The stars are classified according to their relative brightness in what are known as magnitudes, the first being the brightest and the sixth the faintest visible to the naked eye. The table below gives the names of the principal navigational stars and their apparent magnitudes, with the constellation in which each may be found. ◆ Collateral adjectives: *astral, sidereal, stellar.* 2 Loosely, any heavenly body; a planet. 3 A conventional figure having five or more radiating points: used as an emblem or device, as on the shoulder strap of a general. 4 An asterisk (*). 5 A white spot on the forehead of a horse or bovine animal. 6 An actor or actress who plays the leading part; hence, anyone who shines prominently in a calling or profession: a literary *star.* 7 A heavenly body considered as influencing one's fate; hence, fortune; destiny. — **binary star** A pair of stars revolving about a common center. Three types have been noted: *eclipsing,* in which the members successively eclipse each other; *spectroscopic,* in which the members are distinguishable only by shifts in their spectral lines; and *visual,* in which the members may be distinguished through the telescope. — **dark star** An invisible star, non-shining or dimly shining; known only through relation to visible stars, as during eclipsing action. — **double star** Two stars so near to each other as to be almost indistinguishable except through a telescope. — **dwarf star** Any of a class of stars which have reached their greatest temperature and are in the phase of contraction, with luminosity passing from bluish to orange. — **giant star** Any of a class of stars of great mass and high luminosity which are passing through the early stages of their evolution. — **variable star** Any of several groups of stars whose apparent magnitude varies at different times. The cause may be external, as with binary stars, one of which regularly eclipses the other; or internal, as with true variable stars whose periodic fluctuations in light are caused by internal changes. See CEPHEID VARIABLE, NOVA. — *v.* **starred, star·ring** *v.t.* 1 To set or adorn with spangles or stars. 2 To mark with an asterisk. 3 To transform into a star. 4 To present as a star in a play or motion picture. — *v.i.* 5 To shine brightly as a star; be prominent or brilliant. 6 To play the leading part; be the star. — *adj.* 1 Of or pertaining to a star or stars. 2 Prominent; brilliant: a *star* football player. [OE *steorra*]

TABLE OF PRINCIPAL STARS

Star	Constellation	Magnitude
Achernar	Eridanus	0.60
Acrux	Crucis	1.05
Aldebaran	Taurus	1.06
Alpheratz	Andromeda	2.15
Altair	Aquila	0.89
Antares	Scorpio	1.22
Arcturus	Boötes	0.24
Betelgeuse	Orion	1.20
Canopus	Argo	−0.86
Capella	Auriga	0.21
Deneb	Cygnus	1.33
Fomalhaut	Piscis Austrinus	1.29
Peacock	Pavo	2.12
Polaris	Ursa Minor	2.12
Pollux	Gemini	1.21
Procyon	Canis Minor	0.48
Regulus	Leo	1.34
Rigel	Orion	0.34
Rigil Kentaurus	Centaurus	0.06
Sirius	Canis Major	−1.58
Spica	Virgo	1.21
Vega	Lyra	0.14

star apple 1 The edible fruit of a West Indian tree (*Chrysophyllum cainito*), resembling an apple in size and appearance, and having ten cells and as many seeds disposed stellately around its center. 2 The tree itself.
star·board (stär'bərd) *Naut. n.* The right-hand side of a vessel as one looks from stern to bow: opposed to *larboard, port.* — *adj.* Of or pertaining to the right of the observer on a vessel when facing the bow. — *adv.* Toward the starboard side. — *v.t.* To put, move, or turn (the helm) to the starboard side. [OE *steorbord* steering side]
star boarder The senior boarder in a boarding house, or one who pays more than the others, considered as entitled to special privileges.
starch (stärch) *n.* 1 *Biochem.* A white, odorless, tasteless, amorphous, powdery carbohydrate (C₆H₁₀O₅)ₙ, insoluble in cold water, alcohol, and other liquids, found in the seeds, pith, or tubers of most plants. Starch is an exceedingly important component of vegetable foods, reacting with certain digestive enzymes to produce maltose and dextrin; it is also used in the commercial production of glucose, for stiffening linen, and for many industrial purposes. 2 Stiffness or formality; a stiff or formal manner. 3 *U.S. Slang* Energy; vigor. — *v.t.* 1 To apply starch to; stiffen with or as with starch. [ME *sterche* <OE *stercan* stiffen < *stearc* stiff. Related to STARK.]
Star Chamber A former English court which met in secret and dispensed justice without jury, and which was noted for its arbitrary

and inquisitorial proceedings: abolished by Parliament in 1641; hence, any arbitrary or secret tribunal. [Prob. because it met in Westminster Palace in a chamber whose ceiling was decorated with stars]
starch gum Dextrin.
starch sugar Dextrose.
starch·y (stär′chē) *adj.* **starch·i·er, starch·i·est** 1 Stiffened with starch; stiff; figuratively, prim; formal; precise: also **starched.** 2 Formed of or combined with starch; farinaceous. —**starch′i·ly** *adv.* —**starch′i·ness** *n.*
star cluster *Astron.* Any of numerous groupings of stars associated in the same region of space, as the Pleiades and Coma Berenices: they are classified as open or galactic, and globular.
star·dom (stär′dəm) *n.* The status of a movie or theatrical star.
star drift *Astron.* A common proper motion of stars in the same region of the heavens: noticed in close groups of stars and in pairs of widely separated stars.
stare (stâr) *v.* **stared, star·ing** *v.i.* 1 To gaze fixedly, usually with the eyes open wide, as from admiration, fear, or insolence. 2 To be conspicuously or unduly apparent; glare. 3 To stand on end, as hair. —*v.t.* 4 To stare at. 5 To affect in a specified manner by a stare: to *stare* a person into silence. See synonyms under LOOK. —*n.* A steady, fixed gaze with wide-open eyes. ◆ Homophone: *stair.* [OE *starian*] —**star′er** *n.*
sta·re de·ci·sis (stā′rē di·sī′sis) *Law Latin* The doctrine that precedents are law and should be followed; literally, to stand by decisions.
star facet One of eight triangular facets adjoining the table in the crown of a brilliant-cut gem. For illustration see DIAMOND.
star·fish (stär′fish′) *n. pl.* **·fish** or **·fish·es** Any of various radially symmetrical echinoderms (class *Asteroidea*), commonly with a star-shaped body having five or more arms. Starfish feed mainly on mollusks, including oysters.

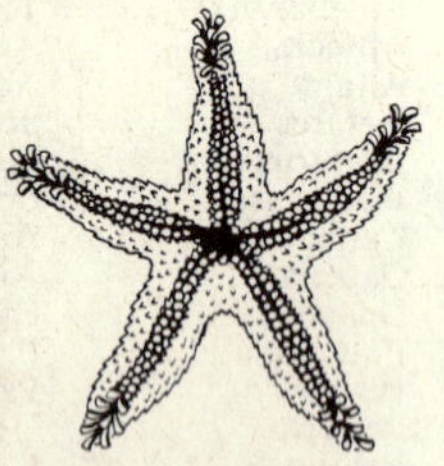
STARFISH
Ventral side, showing tube feet.

star·flow·er (stär′flou′ər) *n.* 1 Any of various plants with conventionally star-shaped flowers; especially, a low perennial *(Trientalis borealis)* with one or more white star-shaped flowers. 2 A starwort. 3 A star of Bethlehem.
star-gaze (stär′gāz′) *v.i.* **-gazed, -gaz·ing** 1 To gaze at or study the stars. 2 To daydream.
star-gaz·er (stär′gā′zər) *n.* 1 One who gazes at or studies the stars; especially, an astrologer or astronomer. 2 A marine carnivorous fish with eyes small and near the front of the top of the head, as *Uranoscopus scaber* of the Mediterranean, and *Astroscopus anoplus* of the Atlantic coast of the United States.
star-gaz·ing (stär′gā′zing) *adj.* Given to watching the stars. —*n.* 1 The act or practice of watching or studying the stars. 2 An absent-minded state; abstraction.
star·grass (stär′gras′, -gräs′) *n.* Any of various grasslike plants (genus *Hypoxis*) of the amaryllis family, with starlike flowers.
star·ing (stâr′ing) *adj.* 1 Gazing with a stare. 2 Standing out prominently, as from high relief or gaudy color; glaring. —**star′ing·ly** *adv.*
stark (stärk) *adj.* 1 Stiff or rigid, as in death. 2 *Obs.* Stubborn; inflexible. 3 Severe; tempestuous, as weather; strict or grim, as a person; also, deserted or barren, as a landscape. 4 *Obs.* Strong and powerful. 5 Without ornamentation; blunt; complete; utter; downright: *stark* misery. 6 Naked: short for *stark naked.* —*adv.* 1 In a stark manner. 2 Completely; utterly: *stark* mad. [OE *stearc* stiff. Related to STARCH.] —**stark′ly** *adv.*
Stark (shtärk), **Johannes,** 1874–1957, German physicist.
Stark (stärk), **John,** 1728–1822, American Revolutionary general.
stark naked Entirely without clothing. [Alter. of ME *stert-naked* <OE *steort* tail + *nacod* naked; infl. in form by STARK]
star·less (stär′lis) *adj.* Being without stars or starlight.
star·let (stär′lit) *n.* 1 A small star. 2 *Colloq.* A young movie actress aspiring to stardom.
star·light (stär′līt) *n.* The light given by a star or stars. —*adj.* Lighted by or only by the stars: also **star′lit′** (-lit′).
star·like (stär′līk′) *adj.* Like a star; bright; luminous; shining.
star lily The sand lily.
star·ling[1] (stär′ling) *n.* Any of several Old World passerine birds (genus *Sturnus*). The common starling *(S. vulgaris)* is brown glossed with black, with metallic purple and green reflections and a buff tip to each feather. It is often caged. [OE *stærling* < *stær* starling]
star·ling[2] (stär′ling) *n.* 1 An enclosure of close piling, as around a pier of a bridge for protection. 2 One of the piles of such an enclosure. [OE *statholung* foundation]
Starn·ber·ger·see (shtärn′ber′gər·zā′) A lake SW of Munich, in southern Bavaria, West Germany; 22 square miles: also *Würmsee.*
star·nose (stär′nōz′) *n.* A North American mole *(Condylura cristata)* having a radiate arrangement of fleshy processes around the end of the nose. Also **star-nosed mole.**
star of Bethlehem 1 The large star by which the three Magi were guided to the manger in Bethlehem where the child Jesus lay. 2 An Old World plant *(Ornithogalum umbellatum)* of the lily family, having white stellate flowers striped with green on the outside: naturalized in the eastern United States.
star of David The six-pointed star used as a symbol by the Hebrews; the mogen David.

STAR OF DAVID

star of Jerusalem Goatbeard.
starred (stärd) *adj.* 1 Spangled with stars; marked with stars or a star; specifically, marked with an asterisk. 2 Affected by astral influence: chiefly in composition: ill-*starred.* 3 Presented or advertised as the star of a play or motion picture; featured.
star·ry (stär′ē) *adj.* **·ri·er, ·ri·est** 1 Set with stars or starlike spots or points; abounding in stars. 2 Lighted by the stars. 3 Shining as or like the stars. 4 Star-shaped. 5 Of, pertaining to, proceeding from, or connected with stars. 6 Consisting of stars; stellar. —**star′ri·ness** *n.*
star·ry-eyed (stär′ē·īd′) *adj.* Given to fanciful wishes or yearnings.
Stars and Bars The first flag authorized by the Congress of the Southern Confederacy, consisting of a field of three bars, red, white, and red, and a blue canton with a circle of white stars, one for each State of the Confederacy.
Stars and Stripes The flag of the United States of America, a field of thirteen horizontal stripes, alternate red and white, and blue union with as many white stars as States: with the definite article.
star sapphire See under SAPPHIRE.
star·shell (stär′shel′) *n.* An artillery shell that explodes in mid-air with a shower of bright light: used for illuminating objectives, signaling, etc.
star shower A meteoric shower.
star-span·gled (stär′spang′gəld) *adj.* Spangled with stars or starlike spots or points: said especially of the United States flag.
Star-Spangled Banner, The 1 The flag of the United States. 2 A poem written by Francis Scott Key in 1814 during the bombardment by the British of Fort McHenry, Md., and adopted by Congress in 1931 as the national anthem of the United States. The music to which it is sung is that of an old English drinking song, *To Anacreon in Heaven.*
start[1] (stärt) *v.i.* 1 To make an involuntary, startled movement, as from fear or surprise. 2 To move suddenly, as with a spring, leap, or bound; jump. 3 To make a beginning or start; set out. 4 To begin; commence: The play *starts* at eight o'clock. 5 To protrude; seem to bulge: His eyes *started* from his head. 6 To be displaced or dislocated; become loose, warped, etc.: The rivets have *started.* —*v.t.* 7 To set in motion: to *start* an engine; to *start* a rumor. 8 To begin; commence: to *start* a lecture. 9 To set up; establish. 10 To introduce (a subject) or propound (a question). 11 To displace or dislocate; loosen, warp, etc.: The collision *started* the ship's seams. 12 To rouse from cover; cause to take flight; flush, as game. 13 To draw the contents from; tap, as a cask. 14 *Archaic* To startle. See synonyms under INSTITUTE. —**to start in** To begin. —**to start off** To begin a journey; set out. —**to start out** 1 To start off. 2 To make a beginning or start. —**to start up** 1 To rise or appear suddenly. 2 To begin or cause to begin operation, as an engine. —**to start with** In the first place; to begin with. —*n.* 1 A quick, startled movement or feeling; a sudden quickening of sense, pulse, or nerve at something unexpected. 2 A setting out or going forth; beginning. 3 A temporary or spasmodic action or attempt; a brief, intermittent effort: by fits and *starts.* 4 *Archaic* A sudden impulse or effusion; burst; sally: *starts* of wit. 5 Advantage or distance in advance at the outset; lead: I had a *start* of five miles in the race. 6 Impetus at the beginning of motion or, figuratively, of a course of action: to get a *start* in business. 7 A loosened place or condition; crack: a *start* in a ship's planking. See synonyms under BEGINNING. [ME *sterten* start, leap, fusion of ON *sterta* overturn and OE *styrtan* start, jump]
start[2] (stärt) *n.* 1 The sharp point of an antler. 2 A tail-like piece. 3 The tail of a bird or animal: the original sense, now obsolete except in compounds: a *redstart.* [OE *steort* tail]
start·er (stär′tər) *n.* 1 One who or that which starts; specifically, one who sees to it that buses, trolleys, etc., leave on schedule. 2 A self-starter. 3 A competitor at the start of a race. 4 A person who gives the signal for the start of a race.
star thistle An Old World weed *(Centaurea calcitrapa)* with spiny heads of tubular flowers, naturalized in the United States; also, another species *(C. solstitialis)* with yellow flowers.
star·tle (stär′təl) *v.* **·tled, ·tling** *v.t.* To arouse or excite suddenly; cause to start involuntarily; alarm. —*v.i.* To be aroused or excited suddenly; take alarm. —*n.* A sudden fright or shock; a scare. [OE *steartlian* kick, struggle] —**star′tler** *n.*
star·tling (stärt′ling) *adj.* Rousing sudden surprise, alarm, or the like. —**star′tling·ly** *adv.*
star·va·tion (stär·vā′shən) *n.* 1 The act of starving. 2 The state of being starved.
starve (stärv) *v.* **starved, starv·ing** *v.i.* 1 To die or perish from lack of food. 2 To suffer from extreme hunger. 3 To suffer from lack or need: to *starve* for friendship. 4 *Dial.* To die of cold. 5 *Obs.* To die. —*v.t.* 6 To cause to die of hunger; deprive of food. 7 To bring to a specified condition by starving: to *starve* an enemy into surrender. [OE *steorfan* die] —**starv′er** *n.*
starve·ling (stärv′ling) *n.* A person or animal that is starving, starved, or emaciated. —*adj.* 1 Starving; emaciated; hungry. 2 Failing to meet needs; inadequate: a *starveling* religion. See synonyms under MEAGER.
star·wort (stär′wûrt′) *n.* A stitchwort.
stase (stās) *n. Ecol.* A deposit of fossil plants which has not moved from its original position, often occurring as a series of layers of related species. [<Gk. *stasis.* See STASIS.]
stash (stash) *v.t. Slang* To hide or conceal (money or valuables), for storage and safekeeping: often with *away.* [? Blend of STORE + CACHE]
sta·sis (stā′sis, stas′is) *n. Pathol.* 1 Stoppage of the blood in its circulation, especially in the small vessels and capillaries: caused by abnormal resistance of the capillary walls, rather than by any lessening of the heart's action. 2 Retarded movement of the intestinal contents due to obstruction or muscular malfunction. [<NL <Gk., a standing < *histanai* stand]
Stass·furt (shtäs′fŏŏrt) A town in the former state of Saxony-Anhalt, eastern East Germany.
stat- Var. of STATO-.
-stat *combining form* A device which stops or makes constant: *thermostat, rheostat.* [<Gk. *-statēs* causing to stand < *histanai* stand]
state (stāt) *n.* 1 Mode of existence as determined by circumstances, external or internal; nature; condition; situation. 2 Frame of mind; mood: a *state* of anxiety. 3 Mode or style of living; station; especially, grand and ceremonious style; pomp; formality. 4 A sovereign political community organized under a distinct government recognized and

conformed to by the people as supreme, and having jurisdiction over a given territory; a nation. 5 One of a number of political communities or bodies politic united to form one sovereign state; specifically, one of the United States: in this sense usually written **State.** 6 *pl.* The legislative bodies of a nation; estates. 7 Authority of government; the territorial, political, and governmental entity comprising a state or nation. 8 *Obs.* A person of rank; a noble. 9 *Obs.* An estate; order; class of persons. See synonyms under PEOPLE. — **Department of State** An executive department of the U.S. government (established in 1789), headed by the Secretary of State, which supervises the conduct of foreign affairs, directs the activities of all diplomatic and consular representatives, protects national interests abroad, and assists in the formulation of policies in relation to international problems. Also **State Department.** — **to lie in state** To be placed on public view, with ceremony and honors, before burial. — *adj.* 1 Of or pertaining to the state, nation, or government: *state* papers. 2 Intended for use on occasions of ceremony. — *v.t.* **stat·ed, stat·ing** 1 To set forth explicitly in speech or writing; assert; declare. 2 To fix; determine; settle. 3 *Law* To make known specifically; declare as a matter of fact. See synonyms under AFFIRM, ALLEGE, ASSERT, RELATE. [Aphetic var. of OF *estat* <L *status* condition, state < *stare* stand; defs. 4, 5, and 7 directly <L, as in *status rei publicae* the state of the republic. Doublet of STATUS.] — **sta·tal** (stā′·tal) *adj.*

state·craft (stāt′kraft′, -kräft′) *n.* The art of conducting affairs of state.

stat·ed (stā′tid) *adj.* Established; regular; fixed. See synonyms under HABITUAL. — **stat′ed·ly** *adv.*

State flower A flower or plant adopted by popular consent or official designation as the floral emblem of one of the United States.

State·hood (stāt′ho͝od) *n.* The condition or status of one of the United States as opposed to that of a Territory.

State House A building used for sessions of a State legislature and for other public purposes; a State capitol.

state·less (stāt′lis) *adj.* 1 Without nationality: a *stateless* person. 2 Without a state or community of states: a *stateless* society.

state·ly (stāt′lē) *adj.* **·li·er, ·li·est** Dignified; lofty. See synonyms under AWFUL, GRAND, HAUGHTY, SUBLIME. — *adv.* Loftily: also **state′·li·ly.** — **state′li·ness** *n.*

state·ment (stāt′mənt) *n.* 1 A summary of facts; narration; the act of stating. 2 That which is stated. 3 *Law* A formal narration of facts filed as the foundation for judicial proceeding; a pleading. 4 A summary of the assets and liabilities of a bank or firm, showing the balance due. 5 A report sent, usually at monthly intervals, to a debtor of a business firm or to a depositor in a bank. See synonyms under REPORT.

Stat·en Island (stat′n) An island SW of the mouth of the Hudson River, at the entrance to New York Harbor, coextensive with the borough of Richmond, New York City; 57 square miles.

State prison A prison built and controlled by a State, usually for felons.

stat·er[1] (stā′tər) *n.* One who makes a statement.

sta·ter[2] (stā′tər) *n.* Any of several standard coins of the ancient Greek city-states, made variously of gold, silver, and electrum, and differing widely in value. [<Gk. *statēr*]

State rights 1 The rights and powers not delegated to the United States by the Constitution, nor prohibited by it to the States: reserved by the Constitution to the respective States, or to the people of the States, under the Tenth Amendment. 2 That construction of the Constitution which makes these rights and powers as large as possible. 3 The doctrine that the States, being sovereign, have the right to judge and nullify an act of the Federal government. See NULLIFICATION. Also **States' rights.**

state·room (stāt′ro͞om′, -ro͝om′) *n.* 1 A small private room having sleeping accommodations on a passenger boat. 2 A private sleeping compartment on a railroad car.

State's attorney *U.S.* A lawyer appointed by a State to represent it in court.

state's evidence 1 One who confesses himself guilty of a crime and testifies as a witness against his accomplices. 2 Evidence produced by the State in criminal prosecutions. Also, in Great Britain, Canada, Australia, etc., *king's* or *queen's evidence.*

States General A general as opposed to a provincial legislature: the name of the legislative body of the Netherlands and that of France before the Revolution.

state·side (stāt′sīd′) *adj.* Of or in the continental United States. — *adv.* In or to the continental United States.

states·man (stāts′mən) *n. pl.* **·men** (-mən) One skilled in the science of government; a political leader of distinguished ability; also, one engaged in government matters, or influential in state affairs or policy. — **states′man·like′, states′man·ly** *adj.* — **states′man·ship** *n.*

state socialism A political theory advocating government ownership of utilities and industries for the purpose of equalizing income.

States of the Church A part of central Italy which, before the unification of Italy in 1870, was under the sovereignty of the pope: also *Papal States.*

States' Rights party A political party founded during May, 1948, in Jackson, Miss., by southern Democrats who were opposed to the civil rights program of the regular Democratic party. It nominated Gov. James Strom Thurmond of South Carolina as its candidate for president. Popularly called *Dixiecrats.*

states·wom·an (stāts′wo͝om′ən) *n. pl.* **·wom·en** (-wim′in) A woman engaged or skilled in the conduct of government affairs.

state-wide (stāt′wīd′) *adj.* Throughout a state.

stat·ic (stat′ik) *adj.* 1 Pertaining to bodies at rest or forces in equilibrium: opposed to *dynamic.* 2 *Physics* Acting as weight, but not moving: *static* pressure. 3 *Electr.* Pertaining to electricity at rest, or to stationary electric charges. 4 At rest; quiescent; dormant; not active. 5 Of or pertaining to non-active elements. 6 In art, simply posed; monumental. 7 Treating of fixed or stable conditions rather than of fluctuations of sales: said of capital or goods. Also **stat′i·cal.** — *n. Electr.* A condition in which electromagnetic waves produced by atmospheric disturbances affect a radio receiving set, interfering with normal reception. [<Gk. *statikos* causing to stand < *histanai* stand] — **stat′i·cal·ly** *adv.*

stat·ics (stat′iks) *n. pl. (construed as singular)* The science of bodies at rest and of the relations required to produce equilibrium. Compare DYNAMICS.

static tube *Aeron.* A small closed tube with openings around the side, facing into the wind on an airplane and designed to measure the static pressure of the air.

sta·tion (stā′shən) *n.* 1 A place where a person or thing usually stands or is; an assigned location. 2 The headquarters of some official person or body of men: a police *station.* 3 An established building or place serving as a starting point, stage, stopping place, or post; specifically, a building for the accommodation of passengers or freight, as on a railroad or bus line; terminal; depot. 4 Social condition; rank; standing. 5 *Mil.* A military post; the place to which an individual, unit, or ship is assigned for duty. 6 The administrative offices, studios, and technical installations of a radio broadcasting unit operating on its assigned frequency. 7 *Mining* A recess in a shaft or passage of a mine. 8 *Austral.* A cattle or sheep run with its appertaining buildings and grounds. 9 In surveying, a point around or from which measurements of angles or distances are made; also, the distance adopted for the standard length. 10 *Eccl.* **a** A stopping place, as a church, shrine, etc., for a solemn religious procession, at which certain prayers are said. **b** A Station of the Cross. — *v.t.* To assign to a station; set in position. [<F <L *statio, -onis* < *status,* pp. of *stare* stand]

Synonym (noun): depot. Properly, a train stops at a *station* to take on and discharge passengers or freight. Freight is kept in a *depot,* which is a storage room or a storehouse. However, the *station* and the *depot* were so often located in one building in the early days of railroads that the word *depot,* which formerly was thought to be more elegant but now has less dignity than *station,* came to be used for both. See PLACE.

sta·tion·ar·y (stā′shən·er′ē) *adj.* 1 Remaining in one place. 2 Fixed: opposed to *portable.* 3 Exhibiting no change of character or condition. — *n. pl.* **·ar·ies** One who or that which is stationary; especially, a member of a stationary military force. ◆ Homophone: *stationery.*

sta·tion·er (stā′shən·ər) *n.* 1 A dealer in stationery and kindred wares. 2 *Obs.* A bookseller; publisher. [<Med. L *stationarius* stationary, having a fixed location (for business)]

Stationers' Company A guild incorporated in London in 1577 comprising printers, bookbinders, booksellers, etc., which until 1911 in England exercised a copyright monopoly requiring all publications to be registered at its office, **Stationers' Hall.**

sta·tion·er·y (stā′shən·er′ē) *n.* Writing materials in general; paper, pens, pencils, ink, notebooks, etc. — *adj.* Dealing in or pertaining to stationery. ◆ Homophone: *stationary.*

station house A police station.

sta·tion·mas·ter (stā′shən·mas′tər, -mäs′-) *n.* The person having charge of a bus or railroad station.

Stations of the Cross The fourteen images or pictures ranged in a church or on church property, which form in series the representation of the successive scenes of the Passion of Christ, and before which devotions are performed.

station wagon An automotive vehicle with one or more rows of removable or folding seats located behind the front seat and with a hinged tailgate for admitting luggage, or the like.

stat·ism (stā′tiz·əm) *n.* 1 A theory of government which holds that the returns from group or individual enterprise are vested in the state, as in communism. 2 Loosely, adherence to state sovereignty, as in a republic. 3 *Obs.* Statecraft.

stat·ist (stā′tist) *n.* 1 An adherent of statism. 2 A statistician. 3 *Obs.* A statesman; politician.

sta·tis·tic (stə·tis′tik) *adj.* Statistical. — *n.* 1 Any element entering into a statistical statement or array, as the mean, the standard deviation, number of cases, etc. 2 Statistics. [<G *statistik* <Med. L *statisticus* statesmanlike, ult. <L *status.* See STATE.]

stat·is·ti·cian (stat′is·tish′ən) *n.* One skilled in collecting and tabulating statistical data.

sta·tis·tics (stə·tis′tiks) *n.* 1 Quantitative data, collectively, pertaining to any subject or group, especially when systematically gathered and collated; specifically, such data relating to a large body of people: *statistics* of population: construed as plural. 2 The science that deals with the collection, tabulation, and systematic classification of quantitative data, especially with reference to frequency distribution and as a basis for inference and induction respecting probable future trends: construed as singular. — **sta·tis′ti·cal** *adj.* — **sta·tis′ti·cal·ly** *adv.*

Sta·tius (stā′shəs), **Publius Papinius,** A.D. 45?–96?, Roman poet.

stato- *combining form* Position: *statoscope.* Also, before vowels, *stat-.* [<Gk. *statos* standing, fixed < *histanai* stand]

stat·o·blast (stat′ə·blast) *n. Zool.* One of the chitinous internal buds developed in fresh-water sponges and on the funiculus of fresh-water polyzoans.

stat·o·cyst (stat′ə·sist) *n. Anat.* One of the sacs in the labyrinth of the internal ear, provided with sensitive hairs and otoliths which are believed to aid in maintaining body equilibrium.

stat·o·lith (stat′ə·lith) *n.* 1 *Bot.* A starch grain or other minute particle in a plant cell, believed to influence the response of a plant organ to the action of gravity. 2 *Anat.* An otolith.

sta·tor (stā′tər) *n.* The stationary portion of a dynamo, turbine, or other power generator.

add,āce,câre,pälm; end,ēven; it,īce; odd,ōpen,ôrder; to͝ok,po͞ol; up,bûrn; ə = a in *above,* e in *sicken,* i in *clarity,* o in *melon,* u in *focus;* yo͞o = u in *fuse;* oi,oil; ou,pout; ch,check; g,go; ng,ring; th,thin; th,this; zh,vision. Foreign sounds à,œ,ü,kh,ṅ; and ◆: see page xx. < from; + plus; ? possibly.

Compare ROTOR. [<NL <L, a supporter < *status*. See STATE.]

stat·o·scope (stat′ə·skōp) *n.* **1** *Meteorol.* A very sensitive form of aneroid barometer having a large reservoir of air, for indicating minute fluctuations in pressure. **2** *Aeron.* A device which indicates small variations in air pressure: used to show changes in altitude of an aircraft.

stat·u·ar·y (stach′o͞o·er′ē) *n. pl.* **·ar·ies** **1** Statues collectively. **2** One who makes statues; a sculptor. **3** The art of making statues. —*adj.* Of or suitable for statues. [<L *statuaria* < *statua* statue. See STATUE.]

stat·ue (stach′o͞o) *n.* A representation of a human or animal figure in marble, bronze, etc., especially when nearly life-size or larger, and preserving the proportions in all directions: distinguished from *painting* or *relief.* See synonyms under IMAGE. —*v.t.* **·ued, ·u·ing** To make a statue of. [<F <L *statua* < *status,* pp. of *stare* stand]

Statue of Liberty National Monument The site of Bartholdi's giant bronze statue *Liberty Enlightening the World* (presented to the U. S. by France, unveiled 1886) on Liberty Island in Upper New York Bay; 10 acres; established 1924. The statue is over 150 feet high and depicts a crowned woman holding aloft a burning torch.

stat·u·esque (stach′o͞o·esk′) *adj.* Resembling a statue, as in grace, pose, or dignity. [<STATUE + -ESQUE] —**stat′u·esque′ly** *adv.* —**stat′u·esque′ness** *n.*

stat·u·ette (stach′o͞o·et′) *n.* A statue not more than half life-size. [<F, dim. of *statue*]

stat·ure (stach′ər) *n.* **1** The natural height of an animal body, especially of a human body. **2** The height of anything, especially of a tree. **3** Development; growth: used figuratively: moral *stature.* [<OF <L *statura* < *status.* See STATE.]

sta·tus (stā′təs, stat′əs) *n.* **1** State, condition, or relation. **2** Relative position or rank. [<L. Doublet of STATE.]

sta·tus quo (stā′təs kwō, stat′əs) The condition or state in which (a person or thing is or has been): often used with the definite article: to maintain the *status quo.* Also **status in quo.** [<L]

sta·tus quo an·te bel·lum (stā′təs kwō an′tē bel′əm) *Latin* The state of affairs existing before the war.

stat·u·ta·ble (stach′o͞o·tə·bəl) *adj.* Statutory; agreeing or conforming with statute. —**stat′u·ta·bly** *adv.*

stat·ute (stach′o͞ot) *n.* **1** *Law* A legislative enactment duly sanctioned and authenticated by constitutional rule; act of Parliament, Congress, etc.; also, any authoritatively declared rule, ordinance, decree, or law. **2** The act of a corporation or its founder, intended as a permanent rule or law: the *statutes* of a university. See synonyms under LAW[1]. —*adj.* Consisting of or regulated by statute. [<F *statut* <LL *statutum,* neut. of L *statutus,* pp. of *statuere* set, found, constitute]

statute law The law as set forth in statutes.

statute mile See under MILE.

statute of limitations A statute which imposes time limits upon the right of action in certain cases, as by obliging a creditor to demand payment of a debt within a specified time.

stat·u·to·ry (stach′ə·tôr′ē, -tō′rē) *adj.* Pertaining to a statute; created by or dependent upon legislative enactment.

St. Augustine (sānt ô′gəs·tēn) A city on the Atlantic in NE Florida; oldest permanent town in the United States; founded by Spain in 1565.

staum·rel (stôm′rəl) *Scot. adj.* Half-witted. —*n.* A half-wit.

staunch (stônch, stänch) *adj.* **1** Firm in principle; constant; faithful; loyal; trustworthy: a *staunch* friend. **2** Stout; sound; tight; seaworthy: a *staunch* ship; having firm constitution or construction; strong and vigorous; hearty. —*v.t.* To stanch. —*n. Brit. Dial.* A floodgate; weir; dam. Also spelled *stanch.* [<OF *estanche* watertight, reliable < *estanchier* make stand. See STANCH.] —**staunch′ly** *adv.* —**staunch′ness** *n.*

Synonyms (adj.): firm, seaworthy, sound, stout, strong, taut, tight, trim, trustworthy, trusty. See FAITHFUL. *Antonyms*: crazy, leaky, rotten, unseaworthy, untrustworthy.

stau·ro·lite (stôr′ə·līt) *n.* A brown to brownish-black native silicate of iron and aluminum, found in prismatic crystals and sometimes used as a gem. [<Gk. *stauros* cross + -LITE; from the crosslike twin crystals] —**stau′ro·lit′ic** (-lit′ik) *adj.*

stau·ro·scope (stôr′ə·skōp) *n. Optics* An instrument used to determine the directions of the planes of vibration of polarized light in crystals. [<Gk. *stauros* cross + -SCOPE] —**stau′ro·scop′ic** (-skop′ik) *adj.*

Sta·vang·er (stä·väng′ər) A port in SW Norway; a fishing and industrial center; site of an 11th century cathedral.

stave (stāv) *n.* **1** A curved strip of wood, forming a part of the sides of a barrel, tub, or the like; hence, any narrow strip of material used for a like purpose: iron *staves.* **2** A straight board forming part of a curb, as about a well. **3** *Music* A staff. **4** A stanza; verse. **5** A rod, cudgel, or staff. **6** A rung of a rack or ladder. —*v.* **staved** or **stove, stav·ing** *v.t.* **1** To break in the staves or strakes of (a cask or a boat); crush the shell or surface of; smash. **2** To make (a hole) by crushing or collision. **3** To furnish with staves. **4** To ward off, as with a staff; keep at a distance: usually with *off*: to *stave* off hunger. —*v.i.* **5** To be broken in, as a vessel's hull. [Back formation < *staves,* pl. of STAFF]

staves (stāvz) **1** Alternative plural of STAFF. **2** Plural of STAVE.

staves·a·cre (stāvz′ā′kər) *n.* **1** A tall larkspur (*Delphinium staphisagria*) of southern Europe. **2** Its seeds, yielding a poisonous alkaloid formerly used as a purgative and antispasmodic. [<OF *stafisagre* <Med. L *staphis agria* <Gk. *staphis* raisin + *agrios* wild]

Stav·ro·pol (stav·rô′pəl) A city on the Volga in southern European Russian S.F.S.R.: formerly (1935–43) *Voroshilovsk.*

staw (stô) *Scot.* Past tense of STEAL.

stay[1] (stā) *v.i.* **1** To cease motion; stop; halt. **2** To continue in a specified place, condition, or state: to *stay* indoors; to *stay* healthy. **3** To remain temporarily as a guest, resident, etc.: Where are you *staying*? **4** To pause; wait; tarry. **5** *Colloq.* To have endurance; stand up; last. **6** *Colloq.* To keep pace with a competitor, as in a race. **7** In poker, to remain in a hand by meeting an ante, bet, or raise. **8** *Archaic* To cease. **9** *Archaic* To stand firm. —*v.t.* **10** To bring to a stop; halt; check. **11** To hinder; delay. **12** To put off; postpone. **13** To satisfy the demands of temporarily; quiet; appease: to *stay* the pangs of hunger. **14** To remain for the duration of: I will *stay* the night. **15** To remain till or beyond the end of: with *out*: to *stay* out one's welcome. **16** *Archaic* To quell, as strife. **17** *Obs.* To wait for. See synonyms under ABIDE, HINDER[1], OBSTRUCT, PERSIST, REPRESS, REST[1], STAND. —*n.* **1** The act or time of staying; continuance in a place; sojourn; visit. **2** That which checks or stops; specifically, a suspension of judicial proceedings. **3** Staying power; endurance; persistence. **4** A state of rest; standstill. See synonyms under RESPITE, REST. [<AF *estaier,* OF *ester* <L *stare* stand] —**stay′er** *n.*

stay[2] (stā) *v.t.* **1** To be a support to; prop or hold up. **2** To support mentally; comfort; sustain. **3** To cause to depend or rely, as for support: with *on* or *upon.* —*n.* **1** Anything which props or supports; a prop, buttress, or the like. **2** *pl.* A corset. [<OF *estayer*]

stay[3] (stā) *Naut. n.* **1** A large, strong rope, often of wire, used to support, steady, or fasten a mast or spar. **2** Any rope supporting a mast or funnel; a guy rope. —**in stays** In the act of turning about on another tack. —*v.t.* **1** To support with a stay or stays, as a mast. **2** To put (a vessel) on the opposite tack. —*v.i.* **3** To tack: said of vessels. [OE *stæg*]

stay-at-home (stā′at·hōm′) *adj.* Given to remaining at home; not in the habit of traveling. —*n.* A person accustomed to staying home.

staying power The ability to endure.

stay·sail (stā′səl, -sāl′) *n. Naut.* A sail, usually triangular, extended on a stay.

St. Ber·nard (sānt bər·närd′, *Fr.* saṅ ber·nàr′) Either of two passes in the Alps: (1) **Great St. Bernard,** between Switzerland and Italy, east of Mont Blanc; elevation 8,120 feet; **St. Bernard hospice,** founded in the 11th century for the rescue of snowbound travelers, is at its summit. (2) **Little St. Bernard,** between France and Italy, south of Mont Blanc; elevation, 7,180 feet; site of a 10th century hospice for travelers.

St. Chris·to·pher (kris′tə·fər), **Ne·vis** (nē′vis, nev′is) **and An·guil·la** (ang·gwil′ə) A British colony, and a federating unit of The West Indies (federation), formerly a presidency of the British Leeward Islands, comprising the islands of **St. Christopher** (commonly *St. Kitts*); 68 square miles; *Nevis*; 50 square miles; and *Anguilla*; 34 square miles; capital, Basseterre, on St. Kitts.

St. Clair (klâr), **Lake** A lake between southern Ontario, Canada, and SE Michigan; 460 square miles; connected with Lake Huron by the **St. Clair River,** a river which flows 40 miles south, forming part of the boundary between Ontario, Canada, and Michigan.

St. Croix (sānt kroi′) The largest of the Virgin Islands of the United States; 82 square miles; capital, Christiansted: also *Santa Cruz.*

St. Croix River **1** A river forming the boundary between NW Wisconsin and eastern Minnesota and flowing 164 miles SW and south, through **Lake St. Croix** (a natural widening of the river extending 24 miles south) to the Mississippi River. **2** A river forming the boundary between Maine and New Brunswick and flowing 75 miles south and east to Passamaquoddy Bay.

stead (sted) *n.* **1** Place of another person or thing: preceded by *in*: Serfdom came *in* the *stead* of slavery. Compare INSTEAD. **2** Place or attitude of support; use; avail; service: in the phrase **to stand one in stead** or **in good stead.** **3** A steading or farm: used chiefly in compounds: *homestead, Hempstead.* **4** *Archaic* Position; condition; place, in general. —*v.t.* *Archaic* To be of advantage to; help; benefit; support. [OE *stede* place]

stead·fast (sted′fast′, -fäst′, -fəst) *adj.* **1** Firmly fixed in faith or devotion to duty; constant; unchanging. **2** Directed fixedly at one point or to one end, as the gaze or purpose; steady. Also spelled *stedfast.* See synonyms under FIRM, INFLEXIBLE, PERMANENT. [OE *stedefæst*] —**stead′fast′ly** *adv.* —**stead′fast′ness** *n.*

stead·ing (sted′ing) *n. Brit. Dial.* A farmhouse, sheds, and offices; a farmstead.

stead·y (sted′ē) *adj.* **stead·i·er, stead·i·est** **1** Stable in position; firmly supported; fixed. **2** Moving or acting with uniform regularity; constant; unfaltering: a *steady* light; hence, not readily disturbed or upset: *steady* nerves. **3** Free from intemperance and dissipation; industrious, sober, and reliable: *steady* habits. **4** Constant in mind or conduct; not wavering; steadfast; also, regular: a *steady* customer. **5** Uninterrupted; continuous: a *steady* flow of conversation. **6** *Naut.* Having the direction of the ship's head unchanged. See synonyms under FIRM, SOBER. —*v.t.* & *v.i.* **stead·ied, stead·y·ing** To make or become steady. —*interj.* **1** *Naut.* Keep her steady: an order to a helmsman to keep the ship's head pointed in the same direction. **2** Not so fast; keep calm: an order enjoining self-control or composure. —*n. Slang* A sweetheart or steady companion. [<STEAD + -Y[3]] —**stead′i·er** *n.* —**stead′i·ly** *adv.* —**stead′i·ness** *n.*

steak (stāk) *n.* **1** A slice of meat, as of beef, usually broiled or fried; specifically, beefsteak. **2** Meat, chopped for cooking like a steak: hamburger *steak.* ◆ Homophone: *stake.* [<ON *steik*]

steal (stēl) *v.* **stole, sto·len, steal·ing** *v.t.* **1** To take from another without right, authority, or permission, and usually in a secret manner. **2** To take or obtain in a surreptitious, artful, or subtle manner: He has *stolen* the hearts of the people. **3** To move, place, or convey stealthily: with *away, from, in, into,* etc. **4** In baseball, to reach (a base) without the aid of a hit or error. —*v.i.* **5** To commit theft; be a thief. **6** To move secretly or furtively. —*n.* **1** The act of stealing or that which is stolen; a theft. **2** In baseball, the act of stealing a base. **3** Any financial transaction or other deal that benefits no one but the originators. ◆ Homophones: *steel, stele.* [OE *stelan*] —**steal′er** *n.* —**steal′ing** *n.*

Synonyms (verb): abstract, embezzle, extort, filch, pilfer, pillage, plunder, purloin, rob, swindle. To *steal* is, in law, to commit simple *larceny*; but the word may be applied to any

stealage

furtive, covert, or surreptitious taking of anything, whether material or immaterial. To *pilfer* is to *steal* petty articles. *Filch* especially emphasizes the secrecy and slyness of the act, and is ordinarily applied to things of little value, but may apply to the most precious, as in Shakespeare, "he that *filches* from me my good name." To *purloin* is etymologically to carry far away, and is commonly applied to the dishonest removal of articles of value or importance. To *rob* is, in law, to take feloniously from the person by force or fear, as in highway robbery; it is also applied to the felonious taking of articles of value from places as well as persons generally with suggestion of force and violence. To *abstract* is to take secretly and feloniously from among other things belonging to another. To *embezzle* is to appropriate fraudulently to oneself funds received and held in trust. To *swindle* is to cheat grossly, commonly by false pretenses, but is not a recognized legal offense under that name; one form of *swindling*, "obtaining money by false pretenses," is an indictable offense, but much *swindling* may be carried on under the forms of law. To *plunder* is to take property from an enemy in time of war, and is not a crime at law. See ABSTRACT. *Antonyms*: refund, repay, restore, return, surrender.

steal·age (stē′lij) *n*. Losses suffered from stealing.

stealth (stelth) *n*. 1 The quality or habit of acting secretly; a concealed manner of acting; a secret or clandestine act, movement, or proceeding. 2 *Obs.* Theft or the thing stolen. [ME *stelthe, stalthe* <OE *stelan* steal]

stealth·y (stel′thē) *adj*. **stealth·i·er, stealth·i·est** Moving or acting secretly or slily; done or characterized by stealth; furtive. — **stealth′i·ly** *adv*. — **stealth′i·ness** *n*.

steam (stēm) *n*. 1 Water in the form of vapor. 2 The gas or vapor into which water is changed by boiling. 3 The visible mist into which aqueous vapor is condensed by cooling. 4 Any kind of vaporous exhalation. 5 Energy, force, or power derived from water vapor under pressure, as in cooking, heating, etc. 6 *Colloq.* Vigor; force; speed. — **to let off steam** *Colloq.* To give expression to pent-up emotions or opinions. — *v.i.* 1 To give off or emit steam or vapor. 2 To rise or pass off as steam. 3 To become covered with condensed water vapor: often with *up*. 4 To generate steam. 5 To move or travel by the agency of steam. — *v.t.* 6 To treat with steam, as in softening, cooking, cleaning, etc. — *adj*. 1 Of, driven, or operated by steam: a *steam* gage, *steam* shovel. 2 Containing or conveying steam: a *steam* boiler. 3 Treated by steam. [OE *stēam*]

steam·boat (stēm′bōt′) *n*. A boat or vessel propelled by steam.

steam boiler A closed vessel used in generating steam.

steam chest The box or chest through which steam is delivered from a boiler to an engine cylinder. Also **steam box.**

steam engine An engine that derives its motive force from the action of steam, usually by pressure against a piston sliding within a closed cylinder.

steam·er (stē′mər) *n*. Something propelled or worked by steam, as a steamship. 2 A vessel in which something is steamed, as for cooking, washing, etc.

steamer trunk A trunk small enough to fit under a berth in a ship's cabin.

steam·fit·ter (stēm′fit′ər) *n*. A man who sets up or repairs steampipes and their fittings. — **steam′fit′ting** *n*.

steam point *Physics* The boiling point of water at standard atmospheric pressure; 100° C.: one of the fixed points of the international temperature scale.

steam roller 1 A road-rolling machine driven by steam. 2 Any force that ruthlessly overcomes opposition. — **steam′roll′er** *adj*.

steam·ship (stēm′ship′) *n*. A large vessel used for ocean traffic and propelled, usually, by one or more screws operated by steam; a steamer.

steam shovel A steam-operated shovel for digging and excavation.

steam table A long table with openings in which containers of food are placed to be kept warm by hot water or steam circulating beneath them.

steam-tight (stēm′tīt′) *adj*. Preventing the escape of steam.

steam turbine A turbine operated by steam power.

steam·y (stē′mē) *adj*. **steam·i·er, steam·i·est** Consisting of, like, or full of steam; misty. — **steam′i·ly** *adv*. — **steam′i·ness** *n*.

Ste. Anne de Beau·pré (sànt an′ də bō·prā′) A village and shrine on the north bank of the St. Lawrence River, 21 miles NE of Quebec, Canada.

ste·ap·sin (stē·ap′sin) *n. Biochem.* A lipase contained in pancreatic juice. [<STEA(RIN) + (PE)PSIN]

ste·a·rate (stē′ə·rāt) *n. Chem.* A salt or ester of stearic acid.

ste·ar·ic (stē·ar′ik, stir′ik) *adj. Chem.* 1 Of, pertaining to, or derived from stearin. 2 Designating a white fatty acid, $C_{17}H_{35}COOH$, contained in the more solid animal fats and in many vegetable oils. [<F *stéarique* <Gk. *stear* suet]

ste·a·rin (stē′ə·rin, stir′in) *n. Chem.* 1 A white, crystalline ester of glycerol and stearic acid, $C_3H_5(C_{18}H_{35}O_2)_3$, obtained from various animal and vegetable fats: more correctly called *glyceryl stearate* or *tristearin.* 2 Stearic acid, especially as combined with palmitic acid for making candles, etc. 3 Fat in solid form. Also **ste·a·rine** (stē′ə·rin, -rēn). [<Gk. *stear* suet]

ste·a·rop·tene (stē′ə·rop′tēn) *n. Chem.* A solid crystalline compound that separates from a volatile oil on standing or exposure to cold. Compare ELAEOPTENE. [<STEAR(IC) + (ELAE)OPTENE]

ste·ar·rhe·a (stē′ə·rē′ə) *n. Pathol.* Steatorrhea. [<Gk. *stear* suet + -RRHEA]

ste·a·tite (stē′ə·tīt) *n*. Massive talc; soapstone: found in extensive beds and quarried for hearths, sink linings, coarse utensils, etc. See TALC. [<L *steatitis* <Gk. *stear, steatos* suet, tallow] — **ste·a·tit′ic** (-tit′ik) *adj*.

ste·a·to·py·gi·a (stē′ə·tō·pī′jē·ə, -pij′ē·ə) *n*. Abnormal growth of fat on the buttocks: noted especially in women among the Bushmen and Hottentots of Africa. Also **ste′a·to·py′ga** (-pī′gə). [<NL <Gk. *stear, steatos* suet, fat + *pygē* buttock] — **ste′a·to·py′gous** (-pī′gəs) *adj*.

ste·a·tor·rhe·a (stē′ə·tə·rē′ə) *n. Pathol.* 1 Seborrhea. 2 Excess fat in the stools. Also **ste′a·tor·rhoe′a.** [<Gk. *stear, steatos* suet + -RRHEA]

stech (stekh) *v.t. & v.i. Scot.* To cram; stuff.

sted·fast (sted′fast′, -fäst, -fəst) See STEADFAST.

steed (stēd) *n*. A horse; especially, a spirited war horse: now chiefly a literary use. [OE *stēda* studhorse]

Steed (stēd), **Henry Wickham**, 1871–1956, English journalist.

steek (stēk) *Scot. v.t. & v.i.* To shut; close. — *n*. A stitch.

steel (stēl) *n*. 1 A tough alloy of iron containing carbon in variable amounts up to about 2.0 percent, malleable under proper conditions, and greatly hardened by sudden cooling. Commercial grades are classified, on the basis of carbon content, as: **mild** or **soft steel,** with up to about 0.30 percent of carbon; **medium steel,** 0.30 to 0.60 percent of carbon, and **high** or **hard steel,** containing more than 0.60 percent of carbon. The addition of other components gives a large range of alloys having special properties, as **chrome steel, nickel steel,** etc. 2 Something made of steel, as an implement or weapon; a sword; a knife sharpener. 3 Hardness of character; steel-like nature or quality. 4 A strip or band of steel, as for stiffening a corset. 5 The quotation for shares in a steel company. — *adj*. Made or composed of steel; also, resembling steel; hence, hard; obdurate; adamant; unyielding. — *v.t.* 1 To cover with steel; plate, edge, point, or face with steel. 2 To make hard or strong like steel; make unfeeling or unyielding; harden: to *steel* one's heart against misery. ♦ Homophones: *steal, stele.* [OE *stēl*]

steel-blue (stēl′blōō′) *adj*. Having a color similar to the bluish tinge of certain steels.

steel blue A steel-blue color.

steel-die printing (stēl′dī′) Intaglio printing.

Steele (stēl), **Sir Richard,** 1672–1729, English dramatist and essayist.

steel engraving 1 The art and process of engraving on a steel plate. 2 The impression made from such a plate.

steel gray Any of several shades of dark, dull gray, like the color of finished steel.

steel·head (stēl′hed′) *n*. 1 A species of migratory trout (*Salmo gairdneri*), found from California to Alaska. 2 The black spotted trout (*S. purpuratus*) of the western United States, especially in its adult marine stage.

steel·ing (stē′ling) *n*. 1 The coating of an engraved copper plate with a protective film of iron by electrolysis to increase its durability. 2 Casehardening.

steel·mak·er (stēl′mā′kər) *n*. A maker of steel; especially, the operator or owner of a steel mill.

steel wool Steel fibers matted together for use as an abrasive or in cleaning, polishing, and finishing utensils and the like.

steel·work (stēl′wûrk′) *n*. 1 Any article or construction of steel. 2 *pl.* A shop or factory where steel is made or fabricated. — **steel′work′ing** *n*.

steel·work·er (stēl′wûr′kər) *n*. One who works in a steel mill.

steel·y (stē′lē) *adj*. **steel·i·er, steel·i·est** Made of, resembling, or containing steel; suggesting steel; figuratively, having a steel-like hardness: a *steely* obduracy; a *steely* gaze. — **steel′i·ness** *n*.

steel·yard (stēl′yärd′, -yərd) *n*. A simple device for weighing, consisting of a scaled beam, counterpoise, and hooks, the article to be weighed being hung at the short end, and the counterpoise moved along the long arm. Also **steel′yards.** [from *Steelyard,* formerly, the London headquarters for Hanseatic traders; a mistranslation of MLG *stalhof* a court where samples of goods are displayed]

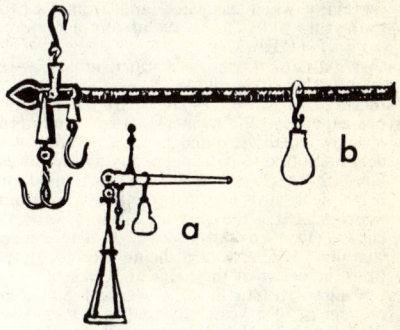

STEELYARDS
a. Pompeian. *b.* Modern.

Steen (stān), **Jan,** 1626–79, Dutch painter.

steen·bok (stān′bok, stēn′-) See STEINBOK.

steep[1] (stēp) *adj*. 1 Making a large angle with the plane of the horizon; precipitous. 2 *Colloq.* Exorbitant; excessive; high, as a price. — *n*. A cliff; hill; precipice; a precipitous place. [OE *stēap*] — **steep′ly** *adv*. — **steep′ness** *n*.

— *Synonyms (adj.):* abrupt, high, precipitous, sharp, sheer. *High* is used of simple elevation; *steep* is said only of an incline where the vertical measurement is sufficiently great in proportion to the horizontal to make it difficult of ascent. *Steep* is relative; an ascent of 100 feet to the mile on a railway is a *steep* grade; a rise of 500 feet to the mile makes a *steep* wagon road; a roof is *steep* when it makes with the horizontal line an angle of more than 45°. A *sharp* ascent or descent is one that makes a sudden, decided angle with the plane from which it starts; a *sheer* ascent or descent is perpendicular, or nearly so; *precipitous* applies to that which is of the nature of a precipice, and is used especially of a descent; *abrupt* is as if broken sharply off, and applies to either acclivity or declivity. See HIGH. *Antonyms*: easy, gentle, gradual, level, low, slight.

steep[2] (stēp) *v.t.* 1 To soak in a liquid, as for

softening, cleansing, etc. 2 To saturate; imbue thoroughly: *steeped* in crime. — *v.i.* 3 To undergo soaking in a liquid. — *n.* 1 The process of steeping, or the state of being steeped. 2 A liquid or bath in which anything is or is to be steeped; especially, a fertilizing liquid for seeds. [ME *stepen*, ? <Scand. Cf. ON *steypa* pour.] — **steep'er** *n.*

steep·en (stē'pən) *v.t.* & *v.i.* To make or become steep or steeper.

stee·ple (stē'pəl) *n.* A lofty structure rising above the tower of a church; a spire. [OE *stēpel, stȳpel*]

stee·ple·bush (stē'pəl·bŏŏsh') *n.* An erect shrub (*Spirae tomentosa*) of the rose family, with dense terminal clusters of rose-colored flowers; the hardhack.

stee·ple·chase (stē'pəl·chās') *n.* 1 A race on horseback across country, in which obstacles are to be leaped: originating from a race to see which of several riders could first reach a distant church steeple. 2 A race over a course artificially prepared, as with hedges, rails, and water jumps. 3 Any cross-country run. — **stee'ple·chas'ing** *n.*

stee·ple·chas·er (stē'pəl·chā'sər) *n.* A person who takes part in a steeplechase; also, a horse used in or trained for steeplechasing.

stee·ple·jack (stē'pəl·jak') *n.* A man whose occupation is to climb steeples and other tall structures to inspect or make repairs. [<STEEPLE + obs. *jack* workman]

steer[1] (stir) *v.t.* 1 To direct the course of (a vessel or vehicle) by means of a rudder, steering wheel, or other device. 2 To follow (a course). 3 To direct; guide; control. — *v.i.* 4 To direct the course of a vessel, vehicle, etc. 5 To undergo guiding or steering: The car *steers* easily. 6 To follow a course: to *steer* for land. — **to steer clear of** To avoid; keep away from. — *n. U.S. Slang* A tip; piece of advice. [OE *stēoran*] — **steer'a·ble** *adj.* — **steer'er** *n.*

steer[2] (stir) *n.* 1 A young male of the ox kind, especially when castrated and from two to four years old. 2 An ox of any age raised for beef. [OE *stēor*]

steer[3] (stir) *Scot. v.t.* To disturb; molest. — *n.* A disturbance; a nudge.

steer·age (stir'ij) *n.* 1 That part of an ocean passenger vessel, formerly near the stern, but now usually situated in the forward lower decks, allotted to passengers paying the lowest fares. 2 In a war vessel, the portion of the berth deck just forward of the wardroom, appropriated as the quarters of junior officers, clerks, etc. See GUNROOM. 3 The act of steering. 4 The state of being steered; direction; the effect of the helm on a vessel.

steer·age·way (stir'ij·wā') *n. Naut.* 1 Sufficient movement of a vessel to enable it to answer the helm. 2 The lowest speed at which a vessel can be accurately steered.

steering committee A committee in a legislature or other assemblage that arranges or directs the course of the business that is to be considered.

steering gear *Mech.* Any arrangement of parts for converting action on the steering wheel into corresponding motion of the rudder of a ship, or, on an automotive vehicle, of the steering axle and its connected members.

steering wheel *Mech.* 1 A vertical wheel with handles along the rim, by which motion is communicated to the rudder of a ship by the wheel ropes or other connections. 2 A hand wheel for guiding an automobile or other heavy vehicle.

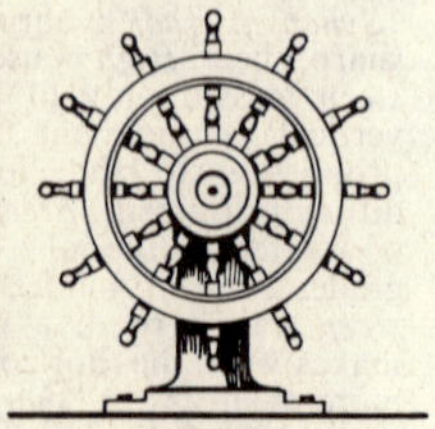

SHIP'S STEERING WHEEL

steers·man (stirz'mən) *n. pl.* **·men** (-mən) One who steers a boat; a helmsman.

steer·y (stir'ē) *Scot. n.* A stir; bustle. — *adj.* Busy; bustling.

steeve[1] (stēv) *n.* A derrick or a spar with a block at one end used in stowing cargo. [<*v.*] — *v.t.* **steeved, steev·ing** 1 To stow, as cargo in the hold of a vessel, by using a steeve or a jackscrew. 2 *Scot.* To pack; cram. [<F *estiver* <L *stipare* compress]

steeve[2] (stēv) *Naut. n.* The angular elevation of a bowsprit from the horizontal: also **steev'ing.** — *v.t.* & *v.i.* **steeved, steev·ing** To set or be set upward at an angle with the horizon. [? <OF *estive* tail of a plough <L *stiva*]

Stef·ans·son (stef'ən·sən), **Vilhjalmur,** born 1879, U.S. Arctic explorer, born in Canada of Icelandic parentage.

Stef·fens (stef'ənz), **(Joseph) Lincoln,** 1866–1936, U.S. journalist.

steg·o·my·ia (steg'ə·mī'ə) *n.* Any of a former genus (*Stegomyia*) of mosquitos; especially, the yellow-fever mosquito (*S. fasciata* or *S. calopus*), which is now called *Aëdes aegypti.* [<NL <Gk. *stegos* a roof + *myia* fly]

steg·o·sau·rus (steg'ə·sôr'əs) *n. pl.* **·sau·ri** (-sôr'ī) *Paleontol.* Any of a genus (*Stegosaurus*) of herbivorous armored dinosaurs of great size which flourished in the western United States during the Upper Jurassic and Lower Cretaceous periods. [<NL <Gk. *stegos* roof + -SAURUS]

Stei·er·mark (shtī'ər·märk) The German name for STYRIA.

stein (stīn) *n.* A beer mug, holding usually a pint; also, the quantity of beer it contains. [<G]

Stein (shtīn), **Baron vom und zum,** 1757–1831, Heinrich Friedrich Karl, Prussian statesman.

Stein (stīn), **Gertrude,** 1874–1946, U.S. writer, resident in France.

Stein·am·ang·er (shtīn'äm·äng'ər) See SZOMBATHELY.

Stein·beck (stīn'bek), **John Ernst,** born 1902, U.S. novelist.

stein·bok (stīn'bok) *n.* A small fawn-colored African antelope (*Raphicerus campestris*): also spelled *steenbok.* Also **stein'buck'** (-buk'). [<Du. *steenbok* <*steen* stone + *bok* buck]

STEINBOK
(About 20 inches shoulder height)

Stein·metz (stīn'mets), **Charles Proteus,** 1865–1923, U.S. electrical engineer born in Germany.

St. E·li·as (sānt i·lī'əs), **Mount** A peak (18,008 feet) in the **St. Elias Mountains** in SW Yukon and SE Alaska; filled by the world's most extensive glacier system apart from the polar ice caps; highest point, 19,850 feet.

ste·le[1] (stē'lē) *n. pl.* **·lae** (-lē) or **·les** (-lēz) An upright sculptured slab or tablet of stone, either sepulchral or intended for public use, as for laws, decrees, treaties, milestones, etc. Also **ste·la** (stē'lə). [<L *stela* <Gk. *stēlē*] — **ste'lar, ste'lene** (-lēn) *adj.*

stele[2] (stēl) *n. Bot.* An axial cylinder of vascular tissue in plants, sometimes more than one. ◆ Homophones: *steal, steel.* [<STELE[1]] — **ste'lic** *adj.*

Stel·la (stel'ə) A feminine personal name. [<L, star]

stel·lar (stel'ər) *adj.* 1 Of or pertaining to the stars; astral. 2 Of or pertaining to an actor or actress who plays a principal role, or to other persons prominent in the arts. [<LL *stellaris* <L *stella* star]

stel·lar·a·tor (stel'ə·rā'tər) *n. Physics* A device for the study of controlled thermonuclear reactions, consisting essentially of a series of magnetizing coils surrounding a hollow glass tube in which ionized gases may be briefly heated to temperatures of several million degrees. [<STELLAR + -ATOR; from the great temperatures developed]

stel·late (stel'it, -āt) *adj.* Star-shaped or starlike; radiating. See illustration under FROST. Also **stel·lat·ed** (stel'ā·tid). [<L *stellatus*, pp. of *stellare* cover with stars <*stella* star] — **stel'late·ly** *adv.*

stel·lif·er·ous (ste·lif'ər·əs) *adj.* Abounding with stars. [<L *stella* star + -(I)FEROUS]

stel·li·form (stel'ə·fôrm) *adj.* Star-shaped. [<NL *stelliformis* <L *stella* star + *forma* form]

Stell·ite (stel'īt) *n.* A class of hard cobalt alloys containing varying amounts of tungsten and chromium: used chiefly in the manufacture of cutting tools: a trade name.

stel·lu·lar (stel'yə·lər) *adj.* Bespangled with fine stars; shaped like or resembling little stars. [<LL *stellula* little star]

St. El·mo's fire or **light** (sānt el'mōz) A luminous charge of atmospheric electricity sometimes appearing on the masts and yardarms of ships, on church steeples, etc.: also called *corposant.*

stem[1] (stem) *n.* 1 The ascending axis or stalk of a plant, as distinguished from the descending axis or *root*; the main body or stalk of a tree, shrub, or other plant, rising above the ground or other rooting place. 2 The relatively slender growth supporting the fruit, flower, or leaf of a plant; a stalk, peduncle, pedicel, or petiole. ◆ Collateral adjective: *cauline.* 3 A bunch of bananas. 4 The main line of descendants from a particular ancestor. 5 An ethnic line; race. 6 The long, slender, usually cylindrical portion of an instrument: a pipe *stem.* 7 The slender upright support of a goblet, wineglass, vase, etc. 8 A shaft, as of a hair or feather. 9 In a watch, the small projecting rod used for winding the mainspring. 10 In some locks, the central circular part about which the key turns. 11 *Printing* The upright stroke of a type face or letter. 12 *Music* The line attached to the head of a written musical note. 13 *Ling.* The element common to all the members of a given inflection or related groups of words. A stem often consists of more than one morpheme, as the Latin stem *luci-* "light" in *lucifer* "light-bearer" is composed of the root *luc-* plus the thematic vowel *-i-*. 14 *Electr.* The air-sealed, tubular glass section at the base of an incandescent lamp, serving to lead the filaments into the evacuated bulb. See illustration under INCANDESCENT. — *v.* **stemmed, stem·ming** *v.t.* 1 To remove the stems of or from. 2 To supply with stems. — *v.i.* 3 To be descended or derived: to *stem* from John Alden. [OE *stemm, stemn, stæfn* stem of a tree, prow of a ship] — **stem'less** *adj.*

stem[2] (stem) *n. Naut.* 1 A nearly upright timber or metal piece uniting the two sides of a vessel at the fore-end. 2 The bow or prow of a vessel. — **from stem to stern** From end to end; hence, thoroughly. — *v.* **stemmed, stem·ming** *v.t.* 1 To resist or make progress against, as a current: said of a vessel. 2 To stand firm or make progress against (any opposing force): to *stem* the tide of public opinion. 3 To strike with the stem (of a vessel). [<STEM[1], in obs. sense "a tree trunk"]

stem[3] (stem) *v.t.* **stemmed, stem·ming** 1 To stop, hold back, or dam up, as a current; stanch. 2 To make tight, as a joint; to plug. [<ON *stemma* stop]

stem·mer (stem'ər) *n.* 1 One who stems. 2 In tobacco manufacture, one who takes out the main stem from the tobacco plant in making strips. 3 A device for stemming fruits, as grapes.

stem·son (stem'sən) *n. Naut.* A curved supporting timber bolted to the stem and keelson of a vessel near the bow. [<STEM[2] + (KEEL)SON]

stem turn In skiing, a turn made by placing the points of the skis nearly together and the ends wide apart, then placing the weight on the outside ski.

stem·ware (stem'wâr') *n.* Drinking vessels with stems, as goblets, taken collectively.

stem-wind·er (stem'wīn'dər) *n.* 1 A watch wound by turning the crown of the stem. 2 *U.S. Slang* A very superior person or thing.

stem-wind·ing (stem'wīn'ding) *adj.* Wound by turning a knob on an outside stem connected with inside mechanism.

stench (stench) *n.* A foul or offensive odor; stink. See synonyms under SMELL. [OE *stenc*]

sten·cil (sten'səl) *n.* 1 A thin sheet or plate in which a pattern is cut by means of spaces or dots, through which applied paint or ink penetrates to a surface beneath. 2 A decoration or the like produced by stenciling. — *v.t.* **·ciled** or **·cilled, ·cil·ing** or **·cil·ling** To mark with a stencil. [Prob. ME *stansel* decorate with many colors <OF *estenceler*, ult. <L *scintilla* a spark] — **sten'cil·er** or **sten'cil·ler** *n.*

Sten·dhal (stän·dàl') Pen name of *Marie Henri Beyle*, 1783–1842, French novelist.

steno- *combining form* Tight; narrow; contracted: *stenography.* Also, before vowels, **sten-**. [<Gk. *stenos* narrow]

sten·o·graph (sten'ə·graf, -gräf) *n.* 1 A character or writing in shorthand. 2 A keyboard machine for printing in shorthand.

ste·nog·ra·pher (stə·nog′rə·fər) *n.* One who writes stenography or is skilled in shorthand; especially, a writer of phonography. Also **ste·nog′ra·phist.**

ste·nog·ra·phy (stə·nog′rə·fē) *n.* **1** The art of writing by the use of contractions or arbitrary symbols; shorthand. **2** Loosely, phonography. — **sten·o·graph·ic** (sten′ə·graf′ik) or **·i·cal** *adj.* — **sten′o·graph′i·cal·ly** *adv.*

sten·o·morph (sten′ə·môrf) *n. Ecol.* A plant form that is abnormally undersized because of a cramped habitat. — **sten′o·mor′phic** *adj.*

sten·o·phyl·lous (sten′ō·fil′əs) *adj. Bot.* Characterized by narrow leaves, as certain plants.

ste·no·sis (sti·nō′sis) *n. Pathol.* Narrowing of a duct or canal in the body. [<NL <Gk. *stenōsis* < *stenos* narrow]

sten·o·ther·mal (sten′ō·thûr′məl) *adj. Ecol.* Adapted to a limited range of temperature variations: said especially of certain plants. — **sten′o·ther′my** *n.*

sten·o·trop·ic (sten′ə·trop′ik) *adj. Ecol.* Having a narrow range of adaptability to environmental changes: said of plant and animal species.

sten·o·type (sten′ə·tīp) *n.* A letter or combination of letters representing a word or phrase, especially in shorthand.

Sten·o·type (sten′ə·tīp) *n.* A keyboard-operated machine used in stenotypy: a trade name.

sten·o·typ·y (sten′ə·tī′pē) *n.* A system of shorthand representing, by ordinary letters or type, shortened forms of words or phrases.

stent[1] (stent) *Scot. v.t.* To assess for taxation; rate; tax. — *n.* A tax, levy, or due.

stent[2] (stent) *n., v.t.* & *v.i. Brit. Dial.* Stint.

stent[3] (stent) *adj. Scot.* Drawn tight; taut.

sten·tor (sten′tôr) *n.* **1** One who possesses an uncommonly strong, loud voice. **2** Any of a genus *(Stentor)* of fresh-water protozoans (class *Ciliata*) having contractile trumpet-shaped bodies capable of attachment by their lower ends. [after *Stentor*]

Sten·tor (sten′tôr) In the *Iliad*, a herald famous for his loud voice.

sten·to·ri·an (sten·tôr′ē·ən, -tō′rē-) *adj.* Extremely loud.

step (step) *n.* **1** An act of progressive motion that requires one of the supporting limbs of the body to be thrust in the direction of the movement, and to reassume its function of support; a pace. **2** The distance passed over in making such a motion; in military quick-time marching, 30 inches. **3** Any short distance; a space easily traversed. **4** That upon which the foot rests in ascending or descending, as a stair or ladder rung. **5** A single action or proceeding regarded as leading to something: a *step* toward emancipation. **6** An advance or promotion that forms one of a series, especially in military usage; grade; degree. **7** The manner of stepping; walk; gait; also, the sound of a footfall. **8** A footprint; track. **9** *pl.* Progression by walking; walk. **10** A combination of foot movements in dancing, forming a pattern that may be repeated, varied, or elaborated: the tango *step.* **11** An interval measuring a difference of musical pitch, corresponding to a degree of the scale or staff. **12** A socket, supporting framework, pocket, or the like: the *step* of a mast. **13** A steplike projection or part, as of the bit of a key. **14** *Mech.* The radial distance between the face of one pulley and that of another stepped on the same shaft. **15** A break in the contour of a float or hull, as of a seaplane, designed to lessen resistance and improve control. **16** A stage in cascade amplification. — **in step** In agreement or synchronism when marching, dancing, etc.; walking evenly with another by taking corresponding steps. — **out of step** Not in step. — **to take steps** To adopt measures, as to attain an end. — *v.* **stepped, step·ping** *v.i.* **1** To move forward or backward by taking a step or steps. **2** To go by foot; walk a short distance: to *step* across the street. **3** To move with measured, dignified, or graceful steps. **4** To move or act quickly or briskly: The old man was *stepping* down the road. **5** To pass into a situation, circumstance, etc., as if in a single step: He *stepped* into a fortune. — *v.t.* **6** To take (a pace, stride, etc.). **7** To perform the steps of: to *step* a quadrille. **8** To place or move (the foot) in taking a step. **9** To measure by taking steps: often with *off:* to *step* off five yards. **10** To cut or arrange in steps. **11** *Naut.* To place the lower end of (a mast) in its step. — **to step down 1** To decrease gradually, or by steps or degrees. **2** To resign from an office or position; abdicate. — **to step in** To begin to take part; intervene. — **to step on** (or **upon**) **1** To put the foot down on; tread upon. **2** To put the foot on so as to activate, as a brake or treadle. **3** *Colloq.* To reprove or subdue. — **to step on it** To hurry; hasten. — **to step out 1** To go outside, especially for a short while. **2** *Colloq.* To go out for fun or entertainment. **3** To step down (def. 2). **4** To walk vigorously and with long strides. — **to step up** To increase; raise. ◆ Homophone: *steppe.* [OE *stæpe*]

step– *combining form* Related through the previous marriage of a parent or spouse, but not by blood: *stepchild.* [OE *steop-* < stem of *astypan, astepan* bereave, orphan]

step·broth·er (step′bruth′ər) *n.* The son of one's step-parent by a former marriage.

step·child (step′chīld′) *n.* The child of one's husband or wife by a former marriage.

step·dame (step′dām′) *n. Archaic* A stepmother.

step·daugh·ter (step′dô′tər) *n.* A female stepchild.

step-down (step′doun′) *adj.* **1** That decreases gradually. **2** *Electr.* Converting a small current of high voltage into a large one of low voltage: said of the usual form of transformer: opposed to *step-up.* **3** Designating a ratio-reducing gear.

step·fa·ther (step′fä′thər) *n.* The husband of one's mother other than one's own father.

Steph·a·nie (stef′ə·nē) A feminine personal name: feminine of STEPHEN. Also **Steph′a·na,** *Fr.* **Sté·pha·nie** (stā·fà·nē′).

Ste·phen (stē′vən) A masculine personal name: often spelled *Steven.* Also *Sw.* **Ste·fan** (stā′fän), *Ital.* **Ste·fa·no** (stā·fä′nō), *Russian* **Ste·pan** (stā·pän′), *Dan., Ger.* **Ste·phan** (shtā′fän). [<Gk., crown]
— **Stephen** The first Christian martyr. *Acts* vii 60.
— **Stephen I,** 975?–1038, first king of Hungary: known as *St. Stephen.*
— **Stephen of Blois,** 1097?–1154, king of England 1135–54.

Ste·phen (stē′vən), **Sir Leslie,** 1832–1904, English biographer and critic.

Ste·phens (stē′vənz), **Alexander Hamilton,** 1812–83, U. S. statesman; vice president of the Confederate States. — **James,** 1882–1950, Irish poet and novelist.

Ste·phen·son (stē′vən·sən), **George,** 1781–1848, English engineer; invented the locomotive. — **Robert,** 1803–59, English engineer; son of the preceding; invented the tubular bridge.

step-in (step′in′) *n.* **1** An undergarment like short drawers, without actual legs: also **step′-ins′.** **2** A pumplike shoe. — *adj.* Put on, as undergarments or shoes, by being stepped into.

step·lad·der (step′lad′ər) *n.* A set of portable steps with, usually, a hinged frame at the back, which may be extended to support the steps in an upright position.

step·moth·er (step′muth′ər) *n.* The wife of one's father, other than one's own mother.

Step·ney (step′nē) A metropolitan borough of eastern London, including the districts of Whitechapel and Limehouse; site of the Tower of London.

step-par·ent (step′pâr′ənt) *n.* A stepfather or stepmother.

steppe (step) *n.* A vast plain devoid of forest; specifically, one of the extensive plains in Russia and Siberia. ◆ Homophone: *step.* [<Russian *step′*]

step·per (step′ər) *n.* **1** One who or that which steps: The horse is a high *stepper.* **2** *Slang* A dancer.

Steppes (steps), **The** See KIRGHIZ STEPPE.

step·ping·stone (step′ing·stōn′) *n.* **1** A stone affording a footrest, as for crossing a stream, etc. **2** That by which one advances or rises: *steppingstones* to fortune.

step·re·la·tion (step′ri·lā′shən) *n.* A person related through the remarriage of a parent or spouse and not by blood. — **step′re·la′tion·ship** *n.*

step·sis·ter (step′sis′tər) *n.* The daughter of one's step-parent by a former marriage.

step·son (step′sun′) *n.* A male stepchild.

step-up (step′up′) *adj.* **1** Increasing by stages: a *step-up* transformer: opposed to *step-down.* **2** Designating a ratio-increasing gear.

step·wise (step′wīz′) *adv.* In the manner of steps; step by step.

-ster *suffix of nouns* **1** One who makes or is occupied with: often with pejorative force: *songster, prankster.* **2** One who belongs or is related to: *gangster.* **3** One who is: *youngster.* [OE *-estre,* feminine suffix]

ste·ra·di·an (sti·rā′dē·ən) *n.* The unit of measurement for solid angles; that solid angle which, on a sphere, encloses a surface equivalent to the square of the radius: also called *spheradian.* [<Gk. *stereos* solid + RADIAN]

ster·co·ra·ceous (stûr′kə·rā′shəs) *adj.* Consisting of or pertaining to excrement or dung: *stercoraceous* vomiting. Also **ster·co·rous** (stûr′kər·əs). [<STERCOR(I)- + -ACEOUS]

stercori– *combining form* Dung; excrement: *stercoricolous:* also, before vowels, **stercor–.** Also **sterco–.** [<L *stercus, stercoris* dung]

ster·co·ric·o·lous (stûr′kə·rik′ə·ləs) *n.* Living in manure, as some insects. [<STERCORI- + -COLOUS]

ster·cu·li·a·ceous (stûr·kyōō′lē·ā′shəs) *adj. Bot.* Designating or belonging to a family *(Sterculiaceae)* of chiefly tropical herbs, shrubs, and trees, including the cacao and the colanut tree, having a variously shaped inflorescence of regular perfect flowers. [<NL <L *Sterculius,* the deity of manuring < *stercus* dung]

stere (stir) *n.* A measure of capacity in the metric system, equal to one cubic meter. See METRIC SYSTEM. [<F *stère* <Gk. *stereos* solid]

stereo– *combining form* Solid; firm; hard: *stereoscope.* Also, before vowels, **stere–.** [<Gk. *stereos* hard]

ster·e·o·bate (ster′ē·ə·bāt′, stir′-) *n. Archit.* A substructure, continuous base, or solid platform without columns, as distinguished from a *stylobate,* which has them. [<STEREO- + Gk. *batēs* that which steps] — **ster′e·o·bat′ic** (-bat′ik) *adj.*

ster·e·o·chem·is·try (ster′ē·ō·kem′is·trē, stir′-) *n.* The branch of chemistry that treats of the spatial arrangement of atoms and molecules.

ster·e·o·chro·my (ster′ē·ō·krō′mē, stir′-) *n.* The art or process of painting with pigments mixed with waterglass. [<STEREO- + Gk. *chrōma* color] — **ster′e·o·chro′mic** *adj.* — **ster′e·o·chro′mi·cal·ly** *adv.*

ster·e·o·com·pa·ra·graph (ster′ē·ō·kom′pər·ə·graf′, -gräf′, stir′-) *n.* A mapmaking device which utilizes data provided by stereoscopic photographs. [<STEREO- + COMPARE + -GRAPH]

ster·e·og·no·sis (ster′ē·og·nō′sis, stir′-) *n.* Perception of shape, solidity, and weight, especially by the sense of touch. [<STEREO- + Gk. *gnōsis* knowing] — **ster′e·og·nos′tic** (-nos′tik) *adj.*

ster·e·o·gram (ster′ē·ə·gram′, stir′-) *n.* **1** A picture or diagram giving the impression of a solid in relief, or two pictures of an object combined so as to produce the effect of a solid, as in a stereoscopic picture. **2** A stereograph.

ster·e·o·graph (ster′ē·ə·graf′, -gräf′, stir′-) *n.* **1** A photograph or pair of photographs representing objects so that they appear solid; a stereoscopic photograph. **2** An instrument for making projections of solid objects.

ster·e·og·ra·phy (ster′ē·og′rə·fē, stir′-) *n.* **1** The art of representing solids on a plane by means of lines; perspective. **2** The branch of geometry that treats of solids and of the construction of regularly bounded solids. — **ster′e·o·graph′ic** (-ə·graf′ik) or **·i·cal** *adj.* — **ster′e·o·graph′i·cal·ly** *adv.*

ster·e·o·i·som·er·ism (ster′ē·ō·ī·som′ə·riz′əm, stir′-) *n. Chem.* An isomerism which depends on the spatial arrangement of the atoms or groups in an organic compound. — **ster′e·o·i′so·mer′ic** (-ī′sō·mer′ik), **ster′e·o·mer′ic** *adj.*

ster·e·ome (ster′ē·ōm, stir′-) *n. Bot.* The solid supporting elements of the fibrovascular tissues of plants. [<Gk. *stereōme* solid body < *stereos* solid]

ster·e·om·e·try (ster′ē·om′ə·trē, stir′-) *n.* The art of measuring the volume and other spatial elements of solids. [<STEREO- + -METRY] —**ster′e·o·met′ric** (-ō·met′rik) or **·ri·cal** *adj.* —**ster′e·o·met′ri·cal·ly** *adv.*

ster·e·o·phone (ster′ē·ə·fōn′, stir′-) *n.* Any sound-transmitting system equipped with stereophonic devices.

ster·e·o·phon·ic (ster′ē·ə·fon′ik, stir′-) *adj.* **1** Pertaining to, designed for, or characterized by the perception of sound by both ears; binaural. **2** Denoting a system of sound transmission in which two or more microphones or loudspeakers are so placed as to give the effect of hearing with both ears simultaneously, as in wide-screen motion pictures and certain types of radio receivers. —**ster′e·o·phon′i·cal·ly** *adv.*

ster·e·o·phon·ics (ster′ē·ə·fon′iks, stir′-) *n. pl.* (*construed as singular*) The branch of acoustics which investigates the stereophonic reproduction of sound and develops its practical applications.

ster·e·o·phon·ism (ster′ē·ə·fō′niz·əm, stir′-) *n.* The condition of being stereophonic; binaural hearing.

ster·e·oph·o·ny (ster′ē·of′ə·nē, stir′-) *n.* The art and techniques of designing, producing, and applying stereophonic devices for the recording and transmission of sound.

ster·e·op·sis (ster′ē·op′sis, stir′-) *n.* Vision characterized by stereoscopy; stereoscopic vision. [<STERE(O)- + -OPSIS]

ster·e·op·ti·con (ster′ē·op′ti·kon, stir′-) *n.* A double magic lantern arranged to combine two images of the same object or scene, or used to bring one image after another on the screen by the alternate use of the lanterns; a projection lantern. [<STEREO- + Gk. *optikos* of sight]

ster·e·o·scope (ster′ē·ə·skōp, stir′-) *n.* An instrument for blending into one image two pictures of an object from slightly different points of view, so as to produce upon the eye the impression of relief and solidity. [<STEREO- + -SCOPE] —**ster′e·o·scop′ic** (-skop′ik) or **·i·cal** *adj.* —**ster′e·o·scop′i·cal·ly** *adv.*

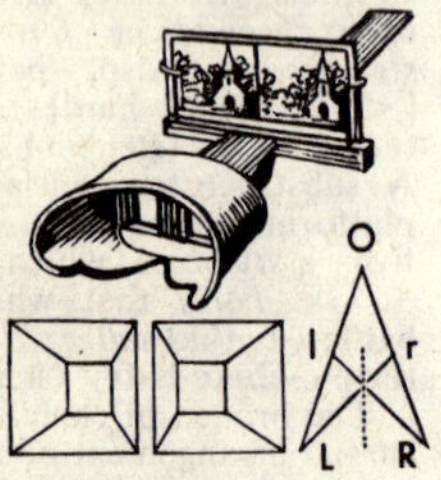

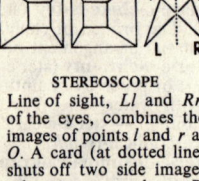

STEREOSCOPE
Line of sight, *Ll* and *Rr*, of the eyes, combines the images of points *l* and *r* at *O*. A card (at dotted line) shuts off two side images otherwise seen along *Rl* at *l* and *Lr* at *r*.

ster·e·os·co·py (ster′ē·os′kə·pē, stir′-) *n.* **1** The art of making or using stereoscopes and stereoscopic slides. **2** The viewing of objects as in three dimensions. —**ster′e·os′co·pism** *n.* —**ster′e·os′co·pist** *n.*

ster·e·o·ski·ag·ra·phy (ster′ē·ō·skī·ag′rə·fē, stir′-) *n.* Stereoscopic photography by means of X-rays. [<STEREO- + SKIAGRAPHY]

ster·e·ot·ro·pism (ster′ē·ot′rə·piz′əm, stir′-) *n.* Involuntary response of an organism to contact with a foreign body. Also **ster′e·o·tax′is** (-ō·tak′sis) —**ster′e·o·trop′ic** (-trop′ik) *adj.*

ster·e·o·type (ster′ē·ə·tīp′, stir′-) *n.* **1** A plate taken in type metal from a matrix, as of paper, reproducing the surface from which the matrix was made. **2** Stereotypy. **3** Anything made or processed in this way. **4** A conventional or hackneyed expression, custom, or mode of thought. —*v.t.* **·typed, ·typ·ing 1** To make a stereotype of. **2** To fix firmly or unalterably.

ster·e·o·typed (ster′ē·ə·tīpt′, stir′-) *adj.* Formalized as if produced from a stereotype; hackneyed; without originality.

ster·e·o·typ·er (ster′ē·ə·tī′pər, stir′-) *n.* **1** One who makes stereotype plates. **2** A stereotype-making machine for making embossed plates from which printing for the blind is done. Also **ster′e·o·typ′ist.** —**ster′e·o·typ′ic** (-tip′ik), **ster′e·o·typ′i·cal** *adj.*

ster·e·o·typ·y (ster′ē·ə·tī′pē, stir′-) *n.* The art or act of making stereotypes. Also **ster′e·o·typ′er·y** (-tī′pər·ē).

ster·e·o·vi·sion (ster′ē·ō·vizh′ən, stir′-) *n.* Three-dimensional vision.

ster·ic (ster′ik, stir′-) *adj. Chem.* Denoting relative position in space: said of the component atoms in a molecule. Also **ster′i·cal.** [<Gk. *stereos* solid]

ster·il·ant (ster′əl·ənt) *n.* **1** That which makes sterile or induces sterility. **2** *Agric.* Any of various chemical compounds whose use as weed-killers renders the soil infertile for one or more growing seasons.

ster·ile (ster′əl) *adj.* **1** Having no reproductive power; barren. **2** *Bot.* Producing no pistil or no spores; incapable of germinating, as certain plants. **3** Lacking productiveness or fertility; hence, useless; being without result: *sterile* soil. **4** Containing no pathogenic bacteria or other micro-organisms; aseptic: a *sterile* fluid. **5** Destitute of attractiveness or suggestiveness: said especially of literary work: *sterile* verse. [<L *sterilis* barren] —**ster′ile·ly** *adv.* —**ste·ril·i·ty** (stə·ril′ə·tē), **ster′ile·ness** *n.*

ster·il·i·za·tion (ster′əl·ə·zā′shən, -ī·zā′-) *n.* **1** The act or process of making sterile. **2** The condition of being sterile. **3** The deliberate procedure of destroying reproductive power by surgical means.

ster·il·ize (ster′əl·īz) *v.t.* **·ized, ·iz·ing 1** To deprive of productive or reproductive power, especially by surgical operation on the Fallopian tubes or on the vas deferens. **2** To destroy bacteria in; free from germs. **3** To make barren; exhaust the productiveness of. **4** To make powerless. —**ster′il·iz′er** *n.*

ster·let (stûr′lit) *n.* A small sturgeon (*Acipenser ruthenus*) found in the Black, Caspian, and Azov seas, and in rivers of Russia, yielding superior caviar and isinglass. [<Russian *sterlyad′*]

ster·ling (stûr′ling) *n.* **1** The official standard of fineness for British coins: for silver (**sterling silver**), 0.925 until 1920, 0.500 since then; for gold, 0.91666 or 11/12. **2** Sterling silver, 0.925 fine, as used in manufacturing articles, as tableware, etc.; also, an article or articles made of it. **3** A former silver penny of England and Scotland, in circulation as early as the 12th century. —*adj.* **1** Made of or payable in sterling: pounds *sterling.* **2** Made of sterling silver. **3** Having accepted worth; genuine; hence, valuable; esteemed: *sterling* qualities. See synonyms under GOOD. [Prob. OE *steorra* star + -LING; because a star was stamped on some of the coins]

stern[1] (stûrn) *adj.* **1** Proceeding from or marked by severity or harshness; unyielding: a *stern* command. **2** Having an austere disposition; strict; severe: a *stern* judge. **3** Inspiring fear; repelling. **4** Resolute; stout: a *stern* resolve. See synonyms under AUSTERE, GRIM, HARD, SEVERE. [OE *styrne*] —**stern′ly** *adv.* —**stern′ness** *n.*

stern[2] (stûrn) *n.* **1** *Naut.* The aft part of a ship, boat, etc. **2** The buttocks or tail part of an animal: now chiefly humorous. **3** The hindmost part of any object. —*adj.* Situated at or belonging to the stern. [<ON *stjoren* steering, rudder <*styra* steer]

Stern (stûrn), **Otto,** born 1888, U. S. physicist born in Germany.

ster·nal (stûr′nəl) *adj.* Pertaining to the breastbone or sternum.

stern chase *Naut.* A chase in which the pursuing vessel follows in the other's course.

stern chaser A cannon mounted in the stern to fire at a pursuing ship.

Sterne (stûrn), **Laurence,** 1713–68, English clergyman and novelist.

stern·fore·most (stûrn′fôr′mōst′, -məst, -fōr′-) *adv.* Hind side foremost; moving with the stern in advance; backward; hence, awkwardly.

stern·most (stûrn′mōst′, -məst) *adj.* Farthest to the rear or stern.

sterno- *combining form Anat.* & *Med.* The sternum: *sternotomy,* cutting through the sternum. Also, before vowels, **stern-.** [<L *sternum* breast]

stern·post (stûrn′pōst′) *n. Naut.* The main vertical post of the stern frame of a vessel, to which the rudder is attached.

stern·sheets (stûrn′shēts′) *n. Naut.* The inside stern portion of a boat; the space in a boat abaft the thwarts.

stern·son (stûrn′sən) *n. Naut.* An inner sternpost attached to the center keelson, to strengthen the stern frame. Also **stern′knee′** (-nē′), **stern′son-knee′.** [<STERN + (KEEL)-SON]

ster·num (stûr′nəm) *n. pl.* **·na** (-nə) or **·nums 1** *Anat.* The breastbone which forms the ventral support of the ribs in most vertebrates. **2** *Zool.* The ventral portion of a somite in an arthropod, as an insect or crustacean. [<L <Gk. *sternon* breast]

ster·nu·ta·tion (stûr′nyə·tā′shən) *n.* **1** The act of sneezing. **2** A sneeze or the noise produced by it. [<L *sternutatio, -onis* <*sternutare,* freq. of *sternuere* sneeze]

ster·nu·ta·tor (stûr′nyə·tā′tər) *n.* One of a class of chemical-warfare agents having a strongly irritant effect upon the nasal and respiratory passages, with resulting physical exhaustion; a sneeze gas.

ster·nu·ta·to·ry (stər·nyōō′tə·tôr′ē, -tō′rē, -nōō′-) *adj.* Causing or tending to cause sneezing: also **ster·nu′ta·tive** (-tə·tiv). —*n. pl.* **·ries** Any substance tending to cause sneezing, as snuff.

stern·ward (stûrn′wərd) *adj.* & *adv.* Toward the stern; astern. Also **stern′wards.**

stern·way (stûrn′wā′) *n. Naut.* Backward or sternforemost movement of a vessel: opposed to *headway.*

stern-wheel·er (stûrn′hwē′lər) *n.* A steamboat of small draft propelled by one large paddle wheel at the stern.

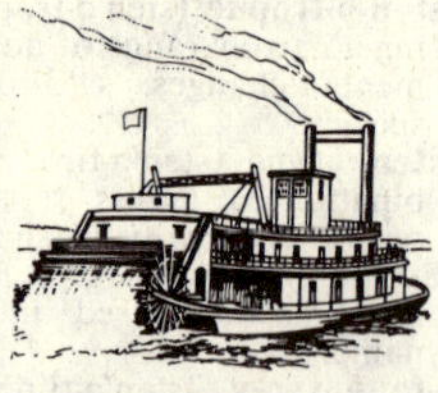
STERN-WHEELER

ster·oid (ster′oid) *n. Biochem.* Any of a sizable group of organic compounds widely distributed in nature, including the sterols, the bile acids, and the sex hormones. [<STER(OL) + -OID]

ster·ol (ster′ōl, -ol) *n. Biochem.* Any of a class of complex, chiefly unsaturated, solid alcohols widely distributed in plant and animal tissue, as cholesterol. [Contraction of CHOLESTEROL]

Ster·o·pe (ster′ə·pē) One of the Pleiades: also called *Asterope.*

ster·tor (stûr′tər) *n.* A deep snore or snoring. [<NL <L *stertere* snore]

ster·tor·ous (stûr′tər·əs) *adj.* Characterized by snoring; accompanied by a snoring sound: *stertorous* breathing. —**ster′tor·ous·ly** *adv.* —**ster′tor·ous·ness** *n.*

ster·ule (ster′ōōl, -yōōl) *n.* A small glass container holding a sterile solution. Compare AMPOULE. [<STER(ILE) + -ULE]

stet (stet) Let it stand: a direction used in proofreading to indicate that a word, letter, etc., marked for omission or correction is to remain. —*v.t.* **stet·ted, stet·ting** To cancel a former correction or omission of by marking with the word *stet.* Compare DELE. [<L, 3rd person sing. subjunctive of *stare* stand, stay]

stetho- *combining form* The breast or chest; pectoral: *stethoscope.* Also, before vowels, **steth-.** [<Gk. *stēthos* breast]

ste·thom·e·ter (ste·thom′ə·tər) *n.* An instrument to measure the expansion of the chest in breathing. [<STETHO- + -METER]

steth·o·scope (steth′ə·skōp) *n. Med.* An apparatus for auscultation, of various forms, sizes, and materials, adapted for conveying the sounds of the body to the examiner's ear or ears. —**steth′o·scop′ic** (-skop′ik), **steth′o·scop′i·cal** *adj.* —**steth′o·scop′i·cal·ly** *adv.* —**ste·thos·co·py** (ste·thos′kə·pē) *n.*

Stet·son (stet′sən) *n.* A hat; especially, one of felt with high crown and wide brim: a trade name. [after John Batterson *Stetson,* 1830–1906, U. S. hatmaker]

Stet·tin (stet′in, *Ger.* shte·tēn′) A port on the Oder in NW Poland, formerly the capital of Pomerania. *Polish* **Szcze·cin** (shche·tsēn′).

Stet·tin·i·us (stə·tin′ē·əs, -tin′yəs), **Edward Riley,** 1900–49, U. S. industrialist and statesman; secretary of state December 1944–June 1945.

Steu·ben (stōō′bən, *Ger.* shtoi′bən), **Baron Friedrich Wilhelm von,** 1730–94, Prussian general; served in American Revolutionary War.

St. Eu·sta·tius (sānt yōō·stā′shəs, -shē·əs) An island in the eastern group of the Netherlands West Indies; 8 square miles. *Dutch* **Sint Eu·sta·ti·us** (sint ōō·stä′tē·ōōs).

Steve (stēv) Familiar shortening of STEPHEN. Also **Ste′vie.**

ste·ve·dore (stē′və·dôr, -dōr) *n.* One whose business is stowing or unloading the holds of vessels. —*v.t.* & *v.i.* **·dored, ·dor·ing** To load

or unload (a vessel or vessels). [<Sp. *estivador* <*estivar* stow <L *stipare* compress, stuff]

stevedore knot A knot used by stevedores to prevent unreeving.

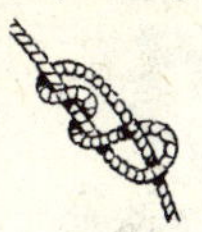

STEVEDORE KNOT

Ste·ven (stē′vən) See STEPHEN.

Ste·vens (stē′vənz), **Thaddeus,** 1792–1868, U. S. statesman; abolitionist. — **Wallace,** 1879–1955, U. S. poet and businessman.

Ste·ven·son (stē′vən·sən), **Adlai Ewing,** born 1900, U. S. lawyer and political leader. — **Robert Louis,** 1850–94, Scottish novelist and essayist.

ste·vi·o·side (stē′vē·ə·sīd) *n. Chem.* A glycoside extracted from the dried leaves of a small South American shrub *(Stevia rebaudiana)* and having a sweetness 300 times that of cane sugar. [<STEVI(A), the genus of the shrub + (GLYC)OSIDE]

stew (stōō, styōō) *v.t. & v.i.* **1** To boil slowly and gently; seethe; keep or be at the simmering point. **2** *Colloq.* To worry. — *n.* **1** Stewed food, especially a preparation of meat or fish cooked by stewing. **2** *Colloq.* Mental agitation; worry. **3** *pl. Archaic* A brothel. **4** *Obs.* A room heated for bathing or drying purposes. [<OF *estuver*, prob. ult. <L *ex-* out + Gk. *typhos* steam, vapor]

stew·ard (stōō′ərd, styōō′-) *n.* **1** A person entrusted with the management of estates or affairs not his own; an administrator. **2** A person put in charge of the domestic affairs of an establishment. **3** On shipboard, a petty officer in charge of the service of provisions, or a man who waits on table and takes care of passengers' rooms. **4** *Brit.* A fiscal officer in certain ancient guilds. [OE *stīweard* <*stī* hall, sty + *weard* ward, keeper] — **stew′ard·ess** *n. fem.* — **stew′ard·ship** *n.*

Stew·art (stōō′ərt, styōō′-), **Dugald,** 1753–1828, Scottish philosopher.

Stewart Island An island of New Zealand south of South Island; 670 square miles.

stewed (stōōd, styōōd) *adj.* **1** Cooked by stewing. **2** *Slang* Drunk.

stew pan A cooking vessel used for stewing.

stey (stā) *adj. Scot.* **1** Steep. **2** Haughty; lofty.

St. Fran·cis River (sānt fran′sis) **1** A river in southern Quebec, Canada, flowing 150 miles SW to the St. Lawrence. **2** A river in SW Missouri and NE Arkansas, flowing 470 miles south to the Mississippi.

St. Gall (saṅ gȧl′) **1** A canton in NE Switzerland; 778 square miles. **2** Its capital, site of a seventh century Benedictine abbey. *German* **Sankt Gal·len** (zängt gäl′ən). Also **Saint Gal·len** (sānt gal′in).

St. George's (sānt jôr′jiz) The capital of Grenada, The West Indies (federation); former administrative capital of the Windward Islands. Also **St. George.**

St. George's Channel A strait between SE Ireland and Wales, connecting the Irish Sea with the Atlantic Ocean; 100 miles long, 50 to 95 miles wide.

St. George's cross The Greek cross, used on the British flag. See under CROSS.

St. Gott·hard (sānt got′ərd, *Fr.* saṅ gô·tȧr′) A mountain group in the Lepontine Alps, south central Switzerland; highest peak 10,483 feet; site of the **St. Gotthard Pass,** at 6,929 feet, and of the **St. Gotthard tunnel,** extending 9 1/4 miles at an elevation of 3,786 feet.

St. He·le·na (sānt hə·lē′nə) An island in the South Atlantic, 1,200 miles west of Africa, to which Napoleon was exiled from 1815–21; 47 square miles; comprising a British crown colony with the dependencies of Ascension Island and the Tristan da Cunha group; 133 square miles; capital, Jamestown, on St. Helena.

St. Hel·ens (sānt hel′ənz) A county borough in SW Lancashire, England.

St. Hel·ier (sānt hel′yər) Capital of Jersey, Channel Islands.

sthe·ni·a (sthē′nē·ə, sthi·nī′ə) *n.* Unusual energy or vigor; excited force: opposed to *asthenia.* [<NL <Gk. *sthenos* strength]

sthen·ic (sthen′ik) *adj.* **1** Exhibiting activity or energy, especially in morbid states. **2** Having power to inspire or animate; indicating vigor. [<Gk. *sthenos* strength]

Sthe·no (sthē′nō, sthen′ō) One of the Gorgons.

stiac·cia·to (styät·chä′tō) *n.* Sculpture or a piece of sculpture in lower relief than bas-relief, as the very low relief used on coins. — *adj.* Of or pertaining to this kind of sculpture; in very low relief. [<Ital., crushed, flattened, pp. of *stiacciare*]

stib·ble (stib′əl) *n. Scot.* Stubble. Also **stib′bul.**

stib·bler (stib′lər) *n. Scot.* **1** A gleaner. **2** A minister without a ministerial charge.

stib·ine (stib′ēn, -in) *n. Chem.* A colorless poisonous gas, SbH_3, resembling arsine, formed by decomposing antimony or any of its compounds in the presence of hydrogen. [<STIB(IUM) + -INE²]

stib·i·um (stib′ē·əm) *n.* Antimony. [<L <Gk. *stibi*] — **stib′i·al** *adj.*

stib·nite (stib′nīt) *n.* A metallic steel-gray antimony sulfide, Sb_2S_3, crystallizing in the orthorhombic system: the most important ore of antimony. [<STIB(I)N(E) + -ITE²]

stich (stik) *n.* **1** A line of the Bible. **2** A line of poetry; a verse: used often in composition: *hemistich.* [<Gk. *stichos* row]

stich·ic (stik′ik) *adj.* **1** Relating to or consisting of stichs. **2** Metrically the same throughout: said of verses.

sti·chom·e·try (sti·kom′ə·trē) *n.* **1** The measurement of the text of a manuscript by lines of measured length into which it is divided; also, the appendix stating the number of lines. **2** The practice of writing prose in line lengths corresponding to the sense of the phrasal cadence. [<Gk. *stichos* line + -METRY] — **stich·o·met·ric** (stik′ə·met′rik) or **·ri·cal** *adj.*

sti·chom·y·thy (sti·kom′ə·thē) *n.* The arrangement of a dialog in alternate lines of verse: characteristic of ancient Greek drama, poetry, and disputation: also spelled *stychomythia.* [<Gk. *stichos* line + *mythos* speech] — **stich·o·myth·ic** (stik′ə·mith′ik) *adj.*

-stichous *combining form* Having (a specified number of) rows: *tristichous.* [<Gk. *stichos* a row, line]

stich·wort (stich′wûrt′) See STITCHWORT.

stick (stik) *n.* **1** A piece of wood that is long, compared with its cross-section; a stiff shoot or branch cut from a tree or bush and used as a rod, wand, staff, club, etc.; also, sometimes one much bigger: a *stick* of timber. **2** *Brit.* A cane. **3** Anything resembling a stick in form: a *stick* of candy or dynamite. **4** *Printing* **a** A composing stick. **b** As much type as a composing stick will hold: about two inches in depth. **c** Copy which will fill this space in a newspaper column: also **stick′ful′.** **5** A piece of wood of any size, cut for fuel, lumber, or timber. **6** *Aeron.* The control lever of an airplane which operates the elevators and ailerons. **7** A poke, stab, or thrust with a stick or pointed instrument. **8** *Archaic* A difficulty or obstacle; hesitation; stop. **9** The state of being stuck together; adhesion. **10** In sports, a baseball bat, hockey stick, racing hurdle, etc. **11** A timber tree. **12** *Colloq.* A stiff, inert, or dull person. **13** *Slang* Any alcoholic ingredient in an otherwise non-alcoholic drink. **14** A revolver or rifle. **15** *Colloq.* The mast of a ship. **16** *Mil.* A group of bombs released consecutively in a straight line crossing the target area. **17** A stalk, as of asparagus. **18** *Colloq.* A conductor's baton. — **the sticks 1** A timber forest. **2** *Colloq.* The backwoods; an obscure rural district. — *v.* **stuck** or (*for defs.* 15, 16) **sticked, stick·ing** *v.t.* **1** To pierce, stab, or penetrate with a pin, knife, or other pointed object. **2** To kill or wound by piercing; stab. **3** To thrust or force, as a sword or pin, into or through something else. **4** To force the end of (a nail, etc.) into something so as to be fixed in place: to *stick* a nail in a wall. **5** To fasten in place with or as with pins, nails, etc.: to *stick* a ribbon on a dress. **6** To cover with objects piercing the surface: a paper *stuck* with pins. **7** To fix on a pointed object; impale; transfix. **8** To put or thrust: He *stuck* his hand into his pocket. **9** To fasten to a surface by or as by an adhesive substance. **10** To bring to a standstill; obstruct; halt: usually in the passive: We were *stuck* in Rome. **11** *Colloq.* To smear with something sticky. **12** *Colloq.* To baffle; puzzle. **13** *Slang* To impose upon; cheat. **14** *Slang* To force great expense, an unpleasant task, responsibility, etc., upon. **15** To provide with sticks or brush on which to grow, as a vine. **16** *Printing* To set or compose (type). — *v.i.* **17** To be or become fixed in place by being thrust in: The pins are *sticking* in the cushion. **18** To become or remain attached by or as by adhesion; adhere; cling. **19** To come to a standstill; become blocked or obstructed; stop; halt. **20** To be baffled or disconcerted. **21** To hesitate; scruple: with *at* or *to.* **22** To persist; persevere, as in a task or undertaking: with *at* or *to.* **23** To remain firm or resolute; be faithful, as to an ideal or bargain. **24** To be extended; protrude: with *from, out, through, up,* etc. — **to be stuck on** *Colloq.* To be enamored of. — **to stick around** *Slang* To remain near or near at hand. — **to stick by** To remain faithful to; be loyal to. — **to stick it out** To persevere to the end. — **to stick up** *Slang* To stop and rob. — **to stick up for** *Colloq.* To take the part of; support; defend. [OE *sticca*]

Synonyms (noun): bat, baton, birch, bludgeon, cane, club, cudgel, ferule, joist, partisan, pole, rod, rule, ruler, shillelagh, staff, stock, switch, timber, truncheon, wand.

stick·er (stik′ər) *n.* **1** One who holds tenaciously to anything. **2** One who or that which fastens with or as with paste. **3** A paster. **4** *Colloq.* Anything that confuses or silences a person; a puzzle. **5** A prickly stem, thorn, or bur.

stick·er·ei (stik′ə·rī′, *Ger.* shtik′ə·rī′) *n.* Braid of even weave, having embroidered, scalloped, or notched edges. [<G, embroidery]

sticking plaster An adhesive material for covering slight cuts, etc.; a court plaster.

stick insect An orthopterous insect (family *Phasmidae*), typically wingless and characterized by a long, sticklike body, as the green or pinkish *Timema* of the Pacific coast.

stick-in-the-mud (stik′in·thə·mud′) *n. Colloq.* A person too sluggish or lacking in initiative to take any progressive action.

stick·it (stik′it) *adj. Scot.* Stuck; unsuccessful; having failed in or given up something.

stickit minister *Scot.* A probationer who fails to qualify for a license, or a licentiate without pastoral charge.

stick·le¹ (stik′əl) *v.i.* **·led, ·ling 1** To contend about trifling matters. **2** To insist or hesitate for petty reasons. [ME *stightlen* set in order, freq. of OE *stihtan* arrange, dispose]

stick·le² (stik′əl) *n.* A prickle; spine: obsolete except in compounds. [OE *sticel* sting]

stick·le·back (stik′əl·bak′) *n.* A small fresh- or salt-water fish (genera *Gasterosteus* and *Eucalia*) of northern regions, having sharp dorsal spines. The male builds nests for the reception of the eggs laid by the female.

stick·ler (stik′lər) *n.* **1** One who contends over trifles. **2** *Obs.* A referee.

stick·pin (stik′pin′) *n.* An ornamental pin for a necktie.

stick·seed (stik′sēd′) *n.* Any of a genus *(Lappula)* of coarse weeds, whose prickly seeds stick in clothing, the wool of sheep, etc.

stick·tight (stik′tīt′) *n.* A coarse herb (genus *Bidens*) of the composite family with prickly achenes; a bur marigold.

stick-to-it-ive (stik·tōō′it·iv) *adj. Colloq.* Persevering; dogged; pertinacious. — **stick-to′-it·ive·ly** *adv.* — **stick-to′-it·ive·ness** *n.*

stick-up (stik′up′) *n. Slang* **1** A robbery or hold-up. **2** A robber who intimidates his victims with a weapon, compelling them to hold their hands in the air.

stick·weed (stik′wēd′) *n.* Ragweed.

stick·y (stik′ē) *adj.* **stick·i·er, stick·i·est 1** Adhering to a surface; adhesive. **2** Warm and humid. See synonyms under ADHESIVE. — **stick′i·ly** *adv.* — **stick′i·ness** *n.*

Stieg·litz (stēg′lits), **Alfred,** 1864–1946, U. S. photographer and art patron.

stiff (stif) *adj.* **1** Resisting the action of a bending force; not flaccid, limp, pliant, or flexible; rigid. **2** Not easily moved; acting with difficulty or friction. **3** Not natural, graceful, or easy; constrained and awkward; formal. **4** Not liquid or fluid; thick; viscous. **5** Taut; tightly drawn. **6** Having a strong, steady movement: a *stiff* breeze. **7** Firm in resistance; obstinate; stubborn. **8** Difficult to achieve, understand, or accept; harsh; severe: a *stiff* penalty. **9** High; dear: a *stiff* price. **10** Firm in prices; strong and steady: a *stiff* market. **11** *Naut.* Heeling over but little, while carrying much sail; not crank: a *stiff* ship. **12** *Scot.* & *Brit. Dial.* Lusty; strong; sturdy. **13** Dense; not porous, as soil. **14** Strong; potent: a *stiff* drink. **15** Difficult; arduous: a *stiff* climb. **16** *Obs.* Formidable; serious: said of news. See synonyms under INFLEXIBLE, SEVERE. — *n. Slang* **1** A corpse. **2** An awkward or unresponsive person; especially, a bore. **3** A man; fellow: *working* stiff; also, a roughneck. **4** A hobo. **5** An accomplice in dishonest dealings; also, a prospective victim. [OE *stīf*] — **stiff′ly** *adv.* — **stiff′ness** *n.*

stiff·en (stif′ən) *v.t.* & *v.i.* To make or become stiff or stiffer.

stiff·en·er (stif′ən·ər) *n.* One who or that which stiffens. — **bow stiffener** *Aeron.* A rigid structural member to reinforce the bow of a dirigible or other airship: also **nose stiffener.**

stiff–necked (stif′nekt′) *adj.* Not yielding; stubborn; incorrigible; obstinate.

sti·fle[1] (stī′fəl) *v.* **·fled, ·fling** *v.t.* **1** To kill by stopping respiration; suffocate; choke. **2** To keep back; suppress or repress, as sobs. — *v.i.* **3** To die of suffocation. **4** To experience difficulty in breathing, as in a stuffy room. [<ON *stīfla* stop up, choke] — **sti′fler** *n.* — **sti′fling** *adj.* — **sti′fling·ly** *adv.*

sti·fle[2] (stī′fəl) *n.* **1** The stifle joint. **2** Any abnormal condition of the stifle joint or stifle bone. [Origin unknown]

stifle bone The patella or kneepan of a horse, situated at the stifle joint, formerly thought of as stopping or damming up the joint.

sti·fled (stī′fəld) *adj.* Having some disease of the stifle joint; affected with stifle.

stifle joint The joint in the upper leg of a horse or a dog. See illustration under DOG, HORSE.

stig·ma (stig′mə) *n. pl.* **stig·ma·ta** (stig′mə·tə, stig·mä′tə) or *(for defs.* 1–3, *usually)* **stig·mas 1** A mark of infamy, or token of disgrace; blemish; a blot on one's good name. **2** Formerly, a brand made with a branding iron on slaves and criminals. **3** *Bot.* That part of a pistil which receives the pollen. **4** *Biol.* **a** A mark or spot, as on the wings of certain insects. **b** An aperture, as the gill slit of a tunicate. **5** A small mark or scar; a birthmark. **6** *Pathol.* A small red or bleeding spot on the skin caused by nervous tension or by capillary congestion. **7** *pl.* The wounds that Christ received during the Passion and Crucifixion; also, marks on the body corresponding to these wounds: said to be miraculously impressed on certain persons as a token of divine favor. **8** One of the characteristic signs or marks of a disease. See synonyms under BLEMISH. [<L, mark, brand <Gk., pointed end, mark < *stizein* prick, brand]

stig·mas·ter·ol (stig·mas′tər·ōl, -ol) *n. Biochem.* A sterol, $C_{29}H_{47}OH$, obtained chiefly from the calabar bean, and in lesser amounts from soybean oil. [<STIGMA + STEROL]

stig·mat·ic (stig·mat′ik) *adj.* **1** Of, pertaining to, or marked with a stigma or stigmata. **2** Infamous; ignominious or vicious; hence, deformed. **3** Anastigmatic. Also **stig·mat′i·cal** — *n.* One marked with or bearing a stigma or stigmata.

stig·ma·tism (stig′mə·tiz′əm) *n.* **1** The state of being affected with stigmas. **2** *Optics* The quality or condition of a lens or of the cornea of the eye through which rays of light are accurately focused.

stig·ma·tist (stig′mə·tist) *n.* One bearing miraculous stigmata.

stig·ma·tize (stig′mə·tīz) *v.t.* **·tized, ·tiz·ing 1** To characterize or brand as ignominious. **2** To mark with a stigma. **3** To cause stigmata to appear on. Also *Brit.* **stig′ma·tise.** [<Med. L *stigmatizare* <Gk. *stigmatizein* mark < *stigma* pointed end, mark] — **stig′ma·ti·za′tion** *n.* — **stig′ma·tiz′er** *n.*

Sti·kine River (sti·kēn′) A river in NW British Columbia and SE Alaska flowing 335 miles SW to the Pacific from the **Stikine Mountains,** a range in northern British Columbia; highest point 8,200 feet.

stilb (stilb) *n.* A unit of illumination, equal to one candle per square centimeter. [<Gk. *stilbein* glitter]

stil·bene (stil′bēn) *n. Chem.* A crystalline unsaturated hydrocarbon, $C_{14}H_{12}$, used in making dyestuffs. [<Gk. *stilbein* glitter + -ENE]

stil·bes·trol (stil′bəs·trōl, -trol) *n. Chem.* A synthetic sex hormone, $C_{18}H_{20}O_2$, similar in action to but more potent than the naturally occurring estrogens. [<STILB(ENE) + ESTR(ONE) + -OL[1]]

stil·bite (stil′bīt) *n.* A vitreous native hydrous silicate of aluminum, calcium, and sodium crystallizing in the monoclinic system. [<Gk. *stilbein* glitter + -ITE[1]]

stile[1] (stīl) *n.* A step, or series of steps, on each side of a fence or wall to aid in surmounting it; loosely, a turnstile. ◆ Homophone: *style.* [OE *stigel* < *stigan* climb]

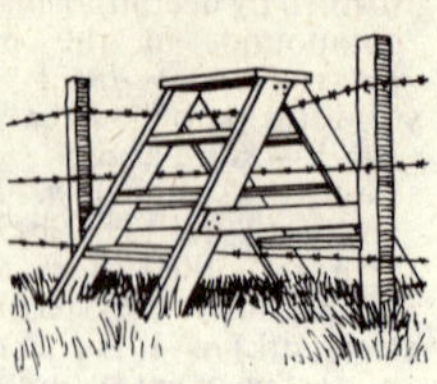
STILE
Over wire fence.

stile[2] (stīl) *n.* One of the vertical side-pieces in a door or a window sash. ◆ Homophone: *style.* [< Du. *stijl* doorpost]

sti·let·to (sti·let′ō) *n. pl.* **·tos** or **·toes 1** A small dagger with a slender blade. **2** A small, sharp-pointed instrument, as of bone, for puncturing eyelets. — *v.t.* To pierce with a stiletto; stab. Also **sti·let′, sti·lette′.** [<Ital., dim. of *stilo* dagger <L *stilus.* See STYLE[1].]

Stil·i·cho (stil′ə·kō), **Flavius,** 359?–408, Roman general and statesman.

still[1] (stil) *adj.* **1** Being without movement; motionless. **2** Free from disturbance or agitation; peaceful; tranquil. **3** Making no sound; silent. **4** Low in sound; hushed. **5** Subdued; soft. **6** Dead; inanimate. **7** Having no effervescence: opposed to *sparkling*: said of wines. **8** *Phot.* Showing no movement. See synonyms under CALM, PACIFIC, SEDATE. — *n.* **1** Absence of sound or noise; stillness; calm. **2** A still-life picture. **3** *Phot.* A still photograph; especially, one taken with a still camera on a motion-picture set, for advertising, promotion, etc. **4** A still alarm. — *adv.* **1** Now as previously; up to this or that time; yet: He is *still* here. **2** After or in spite of something; all the same; nevertheless. **3** In increasing degree; even more; even yet: *still* more. **4** *Poetic & Dial.* Always; constantly. See synonyms under BUT[1], NOTWITHSTANDING, YET. — *conj.* Nevertheless. — *v.t.* **1** To cause to be still or calm. **2** To silence or hush. **3** To quiet or allay, as fears. — *v.i.* **4** To become still. See synonyms under ALLAY, REPRESS, TRANQUILIZE. [OE *stille*] — **still′ness** *n.*

still[2] (stil) *n.* **1** An apparatus in which a substance is vaporized by heat, and the vapor then liquefied in a condenser and collected: used especially for distilling liquors. **2** A distillery: also **still house.** — *v.t.* & *v.i.* To distil. [<L *stillare* drip < *stilla* a drop]

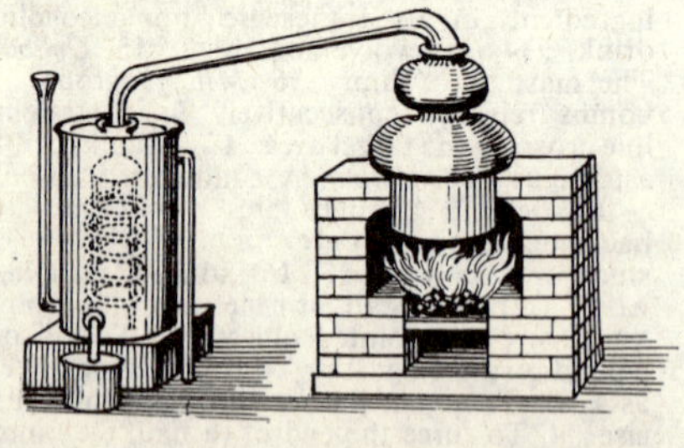
STILL

Still (stil), **Andrew Taylor,** 1828–1917, U.S. physician; founder of osteopathy.

still alarm A fire alarm given by telephone or other call without sounding the regular signal apparatus.

still·birth (stil′bûrth′) *n.* The bringing forth or birth of a dead child.

still·born (stil′bôrn′) *adj.* Dead at birth.

still–hunt (stil′hunt′) *v.t.* & *v.i.* To hunt (game) stealthily; stalk. — *n.* **1** The hunting of game by stealth. **2** The cautious, guarded pursuit of anything; specifically, secret or underhand methods in politics.

stil·li·form (stil′ə·fôrm) *adj.* Drop-shaped. [<NL *stilliformis* <L *stilla* drop + *forma* shape]

still–life (stil′līf′) *n.* **1** In painting, the representation of fruit, flowers, lifeless animals, and inanimate objects. **2** A picture of such a subject.

Still·son wrench (stil′sən) A wrench closely resembling a monkey wrench, but with one serrated jaw capable of slight angular movement about the other, so that the grip is increased by pressure on the handle: a trade name.

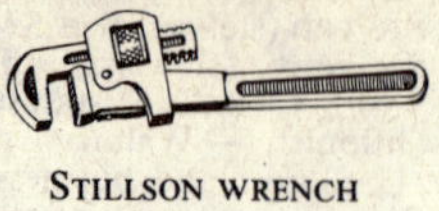
STILLSON WRENCH

Still·wa·ter (stil′wô′tər, -wot′ər) A village near Saratoga Springs, eastern New York; scene of several battles of the Revolutionary War, 1777.

still·y (stil′ē) *adj.* Still; silent, calm. — *adv.* (stil′lē) Calmly; quietly; without noise.

stilt (stilt) *n.* **1** One of a pair of slender poles made with a projection to support the foot above the ground in walking. **2** A tall post or pillar used as a support for a dock or building. **3** Any of several long-legged, three-toed birds (genera *Himantopus* and *Cladorhynchus*) related to the avocet, inhabiting ponds and fresh- and salt-water marshes. The American stilt *(H. mexicanus)* is mostly white with back, wings, crown, and nape a greenish black. The Old World stilt *(H. candidus)* is white except for wings and back. **4** *Scot.* A crutch. — *v.t.* To raise on stilts. — *v.i. Scot.* To hobble on crutches. [ME *stilte,* ? <LG. Cf. MLG *stelte.*]

stilt·ed (stil′tid) *adj.* Artificially elevated in manner; bombastic; inflated. — **stilt′ed·ly** *adv.* — **stilt′ed·ness** *n.*

stilted arch *Archit.* An arch whose curve springs from a level some distance above that of the impost.

Stil·ton cheese (stil′tən) A rich cheese permeated when ripe with a blue-green mold: originally made at Stilton, England. Also **Stil′ton.**

stilt–walk·er (stilt′wô′kər) *n.* One who walks or runs on stilts.

Stil·well (stil′wel), **Joseph Warren,** 1883–1946, U.S. general.

stime (stīm) *n. Scot.* A particle of light; a glimpse.

stim·part (stim′pərt) *n. Scot.* The fourth of a peck.

Stim·son (stim′sən), **Henry Lewis,** 1867–1950, U.S. statesman; secretary of war 1911–13, 1940–45; secretary of state 1929–33.

stim·u·lant (stim′yə·lənt) *n.* **1** Anything that quickens or promotes the activity of some physiological process, as a drug or alcoholic beverage. **2** An intoxicant. — *adj.* Acting as a stimulant; serving to stimulate. [<L *stimulans, -antis,* ppr. of *stimulare.* See STIMULATE.]

stim·u·late (stim′yə·lāt) *v.* **·lat·ed, ·lat·ing** *v.t.* **1** To rouse to activity or to quickened action by some agency or motive; spur. **2** To arouse, or to increase action in, by applying some form of stimulus: to *stimulate* the skin. **3** To affect by intoxicants. — *v.i.* **4** To act as a stimulant. See synonyms under ENCOURAGE, PIQUE[1], SPUR, STIR[1]. [<L *stimulatus,* pp. of *stimulare* prick, goad < *stimulus* a goad] — **stim′u·lat′er, stim′u·la′tor** *n.* — **stim′u·la′tion** *n.*

stim·u·la·tive (stim′yə·lā′tiv) *adj.* Having the power or tendency to stimulate. — *n.* A stimulus.

stim·u·lus (stim′yə·ləs) *n. pl.* **·li** (-lī) **1** Anything that rouses the mind or spirits; an incentive; a stimulant; a sting; a spur, or goad. **2** *Physiol.* **a** Any agent or form of excitation which influences the activity of an organism as a whole or in any of its parts. **b** That which initiates an impulse, as in a nerve or muscle, or produces an altered state of consciousness, as by arousing new or stronger sensations. [<L]

sti·my (stī′mē) See STYMIE.

sting (sting) *v.* **stung** (*Obs.* **stang**), **stung, sting·ing** *v.t.* **1** To pierce or prick painfully, as with a sharp, poisonous organ: The bee *stung* me. **2** To cause to suffer sharp, smarting pain from or as from a sting: The blow *stung* his cheek. **3** To cause to suffer mentally; pain: His heart was *stung* with remorse. **4** To stimulate or rouse as if with a sting; goad; spur. **5** *Slang* To impose upon; get the better of; also, to overcharge. — *v.i.* **6** To have or use a sting, as a bee. **7** To suffer or cause a sharp, smarting pain. **8** To suffer or cause mental distress; pain. See synonyms under INCENSE¹, PIQUE¹, SPUR. —n. 1 Zool. A sharp offensive or defensive organ, as of a bee or wasp, capable of inflicting a painful and especially a poisonous wound. 2 The act of stinging; the wound made by a sting, or the pain caused by it. 3 Any sharp, smarting sensation; stinging quality: the *sting* of remorse. 4 A keen stimulus; spur; goad. 5 *Bot.* One of the sharp-pointed hairs of a nettle; a stinging hair. 6 The point of an epigram. [OE *stingan*] — **sting′ing·ly** *adv.*

sting–and–ling (sting′ən·ling′) *adv. Scot.* As a whole; forcibly.

sting·a·ree (sting′ə·rē, sting′ə·rē′) *n.* A sting ray. [Alter. of STING RAY]

stinge (stinj) *v.i.* **stinged, stinge·ing** To act in a miserly, stingy way. [Back formation <STINGY]

sting·er (sting′ər) *n.* 1 One who or that which stings. 2 A plant or animal that stings. 3 An insect's sting. 4 A cocktail made of brandy and white crème de menthe.

stinging hair *Bot.* One of the hairs of a nettle, charged at the base with an irritating fluid which is injected beneath the skin when touched.

stin·go (sting′gō) *n. Brit. Slang* 1 A strong ale or beer. 2 Zest; vim. [<STING; from the sharpness of the taste]

sting ray One of the various flat-bodied selachian fishes (*Dasyatis* and related genera) with broad pectoral fins and a whiplike tail having one or more stinging spines which are capable of inflicting severe, often poisoned, wounds. Also called *stingaree*.

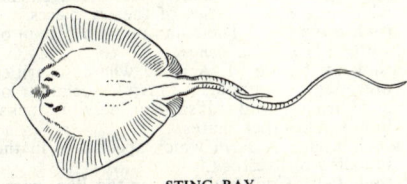

STING RAY
(Body about 20 inches in length; the stinger, 8 to 15 inches)

stin·gy¹ (stin′jē) *adj.* **·gi·er, ·gi·est** 1 Extremely penurious or selfish; miserly. 2 Scanty, as from penurious giving. See synonyms under AVARICIOUS. [<dial. E *stinge* a sting] — **stin′gi·ly** *adv.* — **stin′gi·ness** *n.*

sting·y² (sting′ē) *adj. Colloq.* Stinging; piercing. — **sting′i·ly** *adv.*

stink (stingk) *n.* A strong, foul odor; stench. See synonyms under SMELL. —v. **stank** or **stunk, stunk, stink·ing** *v.i.* 1 To give forth a foul odor. 2 To be extremely offensive or hateful. —v.t. 3 To cause to stink. — **to stink out** To drive from a den, hideaway, etc., by a foul or suffocating odor. [OE *stincan* smell] — **stink′ing** *adj.* — **stink′ing·ly** *adv.*

stink·ard (stingk′ərd) *n.* 1 A mean, detestable fellow. 2 The dogfish.

stink·ball (stingk′bôl′) *n.* A jar containing a mixture of various compounds, as gunpowder, asafetida, etc., formerly used for throwing from one warship to another when at close quarters: also called *stinkpot.* Also **stink′bomb′** (-bom′).

stink·bug (stingk′bug′) *n.* Any of a family (*Pentatomidae*) of hemipterous insects, including mostly rather large, broad, flattened bugs which emit a sickening, sweetish odor when disturbed.

stink·er (stingk′ər) *n.* 1 One who or that which stinks, as a stinkball. 2 The fulmar or other petrel that feeds on carrion. 3 *Slang* An unpleasant, disgusting, or irritating person.

stink·horn (stingk′hôrn′) *n.* Any of an order (*Phallales*) of basidiomycetous, ill-smelling fungi, especially the carrion fungus (*Ithyphallus impudicus*).

stinking hellebore Helleboraster.

stinking smut The bunt disease of wheat.

stink·pot (stingk′pot′) *n.* 1 A stinkball. 2 The musk turtle.

stink·stone (stingk′stōn′) *n.* Any kind of rock that gives off a fetid odor under percussion, as certain limestones.

stink·weed (stingk′wēd′) *n.* The jimsonweed or stramonium.

stink·wood (stingk′wŏod′) *n.* 1 Any of various trees having wood of a disagreeable odor. 2 The wood.

Stin·nes (shtin′əs), **Hugo,** 1870–1924, German industrialist.

stint (stint) *v.t.* 1 To limit, as in amount or share; be stingy with: Don't *stint* yourself. 2 *Archaic* To stop. —v.i. 3 To be frugal or sparing. 4 *Archaic* To stop. —n. 1 A fixed amount, as of work; a task to be performed within a specified time; allowance. 2 A bound; restriction. 3 A small sandpiper. 4 *Obs.* A cessation. See synonyms under TASK, TOIL¹. [ME *stynten* cause to stop <OE *styntan* stupefy <*stunt* stupid] — **stint′er** *n.* — **stint′ing** *adj.* — **stint′ing·ly** *adv.*

stipe (stīp) *n.* 1 *Zool.* A stalk or support. 2 *Bot.* **a** A stalklike support of a gynoecium or carpel. **b** The petiole or support of a fern's frond. **c** The stem supporting the cap of a mushroom or similar fungus. See illustration under MUSHROOM. [<F <L *stipes* branch]

sti·pel (stī′pəl) *n. Bot.* A secondary or small stipule standing at the base of a leaflet. [<NL *stipella*, dim. of *stipes* a branch] — **sti·pel·late** (stī·pel′it, stī′pəl·it, -āt) *adj.*

sti·pend (stī′pend) *n.* 1 An allowance or salary; a fixed payment for services, especially a salary that affords a bare livelihood. 2 *Scot.* A clergyman's salary. 3 *Eccl.* In the Roman Catholic Church, an offering given to a priest for saying a mass with a special intention. See synonyms under SALARY. [<L *stipendium* tax, tribute <*stips* coin, payment in coin + *pendere* weigh, pay out]

sti·pen·di·ar·y (stī·pen′dē·er′ē) *adj.* 1 Receiving a stipend. 2 Paying tribute; owing feudal service; performing services for a fixed payment. —n. pl. **·ar·ies** 1 One who receives a stipend, as a clergyman. 2 A person owing feudal service. 3 A province paying a special tribute to a Roman emperor, instead of a tax. [<L *stipendiarius* <*stipendium* STIPEND]

sti·pes (stī′pēz) *n.* 1 A stipe. 2 *Entomol.* The subbasal, central, and usually the largest part of an insect's maxilla. [<L] — **sti′pi·form** (-pə·fôrm), **stip′i·ti·form** (stip′ə·tə·fôrm′) *adj.*

stip·i·tate (stip′ə·tāt) *adj.* Having or borne on a stipe; stalked. [<NL *stipitatus* <L *stipes* stock]

stip·ple (stip′əl) *v.t.* **·pled, ·pling** To draw, paint, or engrave with dots or short touches instead of lines, so as to produce a shaded effect. —n. In painting, etching, etc., a method of representing light and shade by employing dots instead of lines, or the effect thus produced: also **stip′pling.** [<Du. *stippelen* <*stippen* speckle <*stip* dot] — **stip′pler** *n.*

stip·u·lar (stip′yə·lər) *adj. Bot.* 1 Growing on stipules. 2 Of, resembling, or pertaining to stalks or stems.

stip·u·late¹ (stip′yə·lāt) *v.* **·lat·ed, ·lat·ing** *v.t.* 1 To specify as the terms of an agreement, contract, etc. 2 To specify as a requirement or condition for agreement. 3 To promise; guarantee. —v.i. 4 To demand something as a requirement or condition: with *for.* 5 To make an agreement. [<L *stipulatus*, pp. of *stipulari* bargain] — **stip′u·la′tor** *n.*

stip·u·late² (stip′yə·lit, -lāt) *adj.* Furnished with stipules. Also **stip′u·lat′ed** (-lā′tid).

stip·u·la·tion (stip′yə·lā′shən) *n.* 1 The act of stipulating, or the condition of being stipulated. 2 An agreement or contract. See synonyms under CONTRACT. — **stip′u·la·to′ry** (-lə·tôr′ē, -tō′rē) *adj.*

stip·ule (stip′yōol) *n. Bot.* One of a pair of leaflike appendages at the base of the petiole of certain leaves. [<L *stipula* stalk]

stir¹ (stûr) *v.* **stirred, stir·ring** *v.t.* 1 To agitate so as to alter the relative position of the particles or components of, as soup with a spoon. 2 To cause to move, especially slightly or irregularly; disturb: The tide *stirred* the boat. 3 To move vigorously; bestir: *Stir* yourself! 4 To rouse, as from sleep, indifference, or inactivity; stimulate. 5 To incite; provoke: often with *up.* 6 To affect strongly; move with emotion. —v.i. 7 To move, especially slightly: The log wouldn't *stir.* 8 To be active; move about: They heard him *stirring* in his room. 9 To take place; happen. 10 To undergo stirring: This molasses *stirs* easily.

—n. 1 The act of stirring, or state of being stirred; activity. 2 Public interest; excitement; to-do; commotion. 3 A poke; nudge. [OE *styrian*] — **stir′rer** *n.*

Synonyms (verb): agitate, animate, arouse, awake, awaken, excite, incite, instigate, move, prompt, provoke, rouse, stimulate, wake. See ACTUATE, INFLUENCE, SPUR. *Antonyms:* see synonyms for ALLAY, ALLEVIATE.

stir² (stûr) *n. Slang* A jail; prison. [Origin uncertain]

stir·a·bout (stûr′ə·bout′) *n. Brit.* A porridge made of oatmeal or cornmeal stirred in boiling milk or water; a hasty pudding.

stirk (stûrk) *n.* 1 A yearling ox or cow. 2 *Scot.* A stupid fellow. [OE *stirc* calf <*stēor* steer]

Stir·ling (stûr′ling) A county in central Scotland; 451 square miles; county town, Stirling. Also **Stir′ling·shire** (-shir).

stir·pi·cul·ture (stûr′pi·kul′chər) *n.* The breeding of special races or strains of animals and plants. [<L *stirps, stirpis* stem, stock + CULTURE] — **stir′pi·cul′tur·al** *adj.* — **stir′pi·cul′tur·ist** *n.*

stirps (stûrps) *n. pl.* **stir·pes** (stûr′pēz) 1 Race; family. 2 A stock as regards lineage: a source of property-descent: Descent per *stirpes* (as a family) is distinguished from descent per capita (as an individual). 3 *Biol.* The number of organic units existing in and determining the development of a fertilized ovum. [<L]

stir·ring (stûr′ing) *adj.* 1 Stimulating; inspiring. 2 Full of activity or stir; lively. See synonyms under VIVID. — **stir′ring·ly** *adv.*

stir·rup (stûr′əp, stir′-) *n.* 1 A loop, as an inverted U-shaped piece of metal or wood with flat footpiece, suspended from a saddle to support the rider's foot in and after mounting. 2 A loop or metal strap, as for supporting a beam. 3 *Naut.* A rope on a ship depending from a yard and having at its end an eye or thimble to carry a footrope. [OE *stigrāp* mounting rope]

stirrup bone *Anat.* The stapes.

stir·rup·cup (stûr′əp·kup′, stir′-) *n.* A cup of liquor, as that taken by a mounted horseman on departing; hence, a farewell drink.

stirrup leather The strap by which the stirrup iron is hung from the saddle. Also **stirrup strap.**

stitch¹ (stich) *n.* 1 A single passage of a threaded needle or other implement through fabric and back again, as in sewing or embroidery, or, in surgery, through skin or flesh. 2 A single turn of thread or yarn around a needle or other implement, as in knitting or crocheting; also, the link or loop resulting from such a turn. 3 Any peculiar or individual arrangement of a thread or threads used in sewing, embroidery, or crocheting: a chain *stitch.* 4 A sharp sudden pain, especially in the back or side. 5 A ridge between two furrows. 6 *Colloq.* A garment: I haven't a *stitch* to wear. — **to be in stitches** *Colloq.* To laugh uproariously; be overcome with laughter. —v.t. 1 To join together with stitches. 2 To ornament with stitches. —v.i. 3 To make stitches; sew. [OE *stice* prick, stab]

stitch² (stich) *n. Brit. Dial.* 1 A space passed over; a span of time; distance. 2 A fragment. [OE *stycce* piece]

stitch·er (stich′ər) *n.* One who or that which stitches; especially, a machine for that purpose, as in bookbinding.

stitch·wort (stich′wûrt′) *n.* Any of various plants (genus *Stellaria*), especially the common chickweed: also called *starwort, stichwort.* [OE *sticwyrt* <*stice* prick + *wyrt* plant]

stith·y (stith′ē, stith′ē) *n. pl.* **stith·ies** 1 A smithy or forge. 2 An anvil. —v.t. **stith·ied, stith·y·ing** *Archaic* To forge on an anvil. [<ON *stedhi*]

stive (stīv) *v.t. Obs.* To stow closely; cram; stifle. [<OF *estiver* <L *stīpare* crowd]

sti·ver (stī′vər) *n.* 1 A small Dutch coin, 1/20 of a guilder. 2 Anything of little value. [<Du. *stuiver*]

St. James's Palace (sānt jām′ziz) The Tudor palace in Pall Mall, London, residence of the British sovereigns from Henry VIII to the accession of Victoria: the British royal court is still called the **Court of Saint James's.**

St. John (sānt jon′) **1** One of the Virgin Islands of the United States; 19 square miles. **2** A port on the Bay of Fundy, in southern New Brunswick, Canada. **3** St. John's, Leeward Islands.
St. John (sānt jon′, sin′jən), **Henry** See BOLINGBROKE.
St. John, Lake A lake in south central Quebec, Canada; 375 square miles.
St. John River A river flowing 400 miles NE and east through northern Maine and western New Brunswick to the Bay of Fundy, forming part of the boundary between Maine and New Brunswick.
St. John's (jonz) **1** The capital and largest city of Newfoundland, a port on the SE coast. **2** The capital of Antigua, The West Indies (federation), and former administrative capital of the Leeward Islands.
St. John's bread See CAROB.
St. Johns River A river in NE Florida, flowing 285 miles north and east to the Atlantic.
St. Johns·wort (sānt jonz′wûrt′) Any of a genus (*Hypericum*) of hardy perennial shrubs and herbs, with deep-yellow flowers: found in dry fields. Also **St.-John's-wort.**
St. Jo·seph (sānt jō′zif) A city in NW Missouri on the Missouri River.
St. Kitts (sānt kits′) See ST. CHRISTOPHER, NEVIS, AND ANGUILLA.
St. Lau·rent (saṅ lô·räṅ′), **Louis,** born 1882, Canadian prime minister 1948–57.
St. Law·rence Island (sānt lôr′əns, lor′-) An island of western Alaska in the Bering Sea; 90 miles long, 8 to 22 miles wide.
St. Lawrence River A river of SE Canada, the outlet of the Great Lakes system, flowing 744 miles NE from the NE end of Lake Ontario to the **Gulf of St. Lawrence,** an inlet of the North Atlantic between Newfoundland and eastern Canada; together with the Great Lakes and the St. Marys River it forms a waterway about 2,350 miles long, from the western end of Lake Superior to the Atlantic.
St. Lou·is (sānt lo͞o′is, lo͞o′ē) A city in eastern Missouri, on the Mississippi River below the influx of the Missouri; a major center of transportation, industry, and commerce.
St. Lu·ci·a (sānt lo͞o′shē·ə, lo͞o·sē′ə, lo͞o′shə) A British colony in the Windward Islands, a federating unit of The West Indies (federation); 233 square miles; capital, Castries.
St. Mar·tin (sānt mär′tin) An island in the NW Leeward Islands; the southern part, *Dutch* **Sint Maar·ten** (sint mär′tən), 13 square miles, in the Netherlands Antilles; the northern part, *French* **Saint-Mar·tin** (saṅ·mär·taṅ′), 20 square miles, a dependency of Guadeloupe; total 33 square miles.
St. Mar·y·le·bone (sānt mâr′ē·lə·bōn′) See MARYLEBONE.
St. Mar·ys River (sānt mâr′ēz) **1** A river flowing 63 miles SE from Lake Superior to Lake Huron and forming the boundary between northern Michigan and Ontario. **2** A river in SE Georgia and NE Florida, flowing 175 miles south, east, and north from the Okefinokee Swamp to the Atlantic, and forming part of the Georgia-Florida border.
St. Mau·rice River (sānt môr′is, mor′is; *Fr.* saṅ mô·rēs′) A river in Quebec province, Canada, flowing 325 miles SE and south to the St. Lawrence.
St. Mo·ritz (sānt môr′its, mō′rits) A resort town in SE Switzerland; elevation, 6,080 feet: German *Sankt Moritz. French* **Saint-Mo·ritz** (saṅ·mô·rēts′).
sto·a (stō′ə) *n. pl.* **sto·ae** (stō′ē) or **sto·as** In Greek architecture, a covered colonnade, portico, cloister, or promenade. [<Gk., porch]

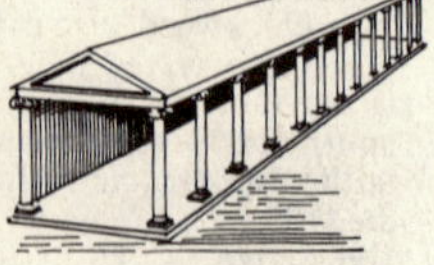

STOA

stoat¹ (stōt) *n.* The ermine, especially in its summer coat, red-brown above, yellow below. [ME *stote*; origin uncertain]
stoat² (stōt) *v.t.* To sew with an invisible stitch that passes only half-way through the cloth. [Origin unknown] **— stoat′ing** *n.*
stob (stob) *Dial. n.* A stake or post, usually short; also, the stump of a tree. *— v.t.* To stab. [Var. of STUB]
stoc·ca·do (stə·kä′dō, -kä′-) *n. Archaic* A stabbing or thrusting movement with a rapier. Also **stoc·ca′ta** (-tə). [<Ital. *stoccata* < *stocco* rapier]
sto·chas·tic (stō·kas′tik) *adj.* **1** Of, pertaining to, characterized by, or skilled in conjecture; conjectural. **2** *Physics* Subject to the laws of probability; not predictable within a given time limit or spatial framework, as the disintegration of a single radioactive element: the *stochastic* phenomena of microphysics. **3** Denoting the process of selecting, from among a group of theoretically possible alternatives, those elements or factors whose combination will most closely approximate a desired result: a *stochastic* model. [<Gk. *stochastikos* < *stochazesthai* guess at < *stochos* mark, aim]
stock (stok) *n.* **1** The trunk or main stem of a tree or other plant, as distinguished from a branch or root. **2** A line of familial descent. **3** The original progenitor of a family line. **4** An ethnic group; race. **5** *Ling.* A family of languages. **6** A related group or family of plants or animals. **7** *Bot.* **a** A rhizome. **b** A stem upon which a graft is made. **8** *Zool.* A zooid which reproduces by generation. **9** Livestock. **10** A quantity of something acquired or kept for future use: to lay in a *stock* of provisions. **11** The merchandise or goods which a trader or merchant has on hand. **12** In card games and dominoes, the part of the pack or group of dominoes that is left on the table and drawn from. **13** The broth from boiled meat or fish used in preparing soups, etc. **14** Raw material: paper *stock.* **15** *pl.* A timber frame with holes for confining the ankles and often the wrists, formerly used in punishing petty offenders. **16** *pl.* The timber frame on which a vessel rests during construction. **17** *pl.* A frame for confining an animal for shoeing or veterinary treatment. **18** *Naut.* An anchor crossbar. **19** The wooden block suspending a bell. **20** In firearms: **a** The rear wooden portion of a rifle, musket, or shotgun, to which the barrel and mechanisms are secured. **b** The arm on rapid-fire guns connecting the shoulder piece to the slide. **c** The handle of a pistol or similar firearm. **d** That member of a gun carriage which usually bears the prolonge and trails along the ground. **21** The handle of certain instruments, as of a whip or fishing rod. **22** A theatrical stock company. **23** The collection of dramas produced by a theatrical stock company. **24** A broad stiffened band, formerly worn as a cravat. **25** *Geol.* The rounded mass of plutonic rock rising above ground level: also called *boss.* **26** *Mech.* An adjustable wrench used for grasping and turning thread-cutting dies. **27** An ornamental garden plant, as the gilliflower, or common stock (*Mathiola incana*). **28** In finance: **a** The capital or fund raised by a corporation through the sale of shares, which entitle the holder to interest or dividends and to part ownership of the corporation. The stockholder may not claim repayment of the principal, though he may sell his shares to other investors at the current market value. **b** The proportional part of this capital credited to an individual stockholder and represented by the number of shares he owns. **c** A certificate showing ownership of a specific number of shares. **— common stock** The stock of a corporation which entitles the holder to dividends, or a share in the profits, only after all other obligations have been met and dividends have been rendered to the owners of preferred stock. Direction of a corporation is usually vested in the owners of common stock. **— debenture stock** *Brit.* A debenture of a corporation or public body issued in the form of stock, the certificates of which are usually transferable but not redeemable and entitle the holder to a perpetual annuity. **— no-par stock** Stock issued without a face value on the certificate and sold at whatever price it will command on the market. **— preferred stock** The stock of a corporation which gives the holder prior claim to dividends up to a certain amount. **— to take stock 1** To take an inventory. **2** To make a careful estimate or appraisal. **— to take stock in** To have trust or belief in; give credence to. *— v.t.* **1** To furnish with stock; supply with cattle, as a farm, or with merchandise, as a store. **2** To keep for sale: to *stock* black ink. **3** To put aside for future use. **4** To provide with a handle or stock. **5** *Obs.* To put (a person) in the stocks for punishment. *— v.i.* **6** To lay in supplies or stock: often with *up.* **7** To send out new shoots; sprout. *— adj.* **1** Kept continually ready or constantly brought forth, like old goods: a *stock* joke. **2** Kept on hand: a *stock* size. **3** Banal; commonplace: a *stock* phrase. **4** Used for breeding purposes: a *stock* mare. **5** Employed in handling or caring for the stock: a *stock* clerk. *— adv.* Motionlessly; like a stump or block of wood: used in combination: *stockstill.* [OE *stocc*]
Synonyms (noun): accumulation, capital, fund, hoard, material, provision, store, supply. See STICK.
stock·ade (sto·kād′) *n.* **1** A line of stout posts, stakes, etc., set upright in the earth to form a fence or barrier; also, the area thus enclosed. **2** Specifically, a strong, high barrier of upright posts, stakes, etc., formerly used by American settlers as a defense against Indians. **3** A breakwater of piling, as for protecting a pier. *— v.t.* **·ad·ed, ·ad·ing** To surround or fortify with a stockade. [<OF *estocade, estacade* < *estaque* a stake <Gmc.]
stock·breed·er (stok′brē′dər) *n.* One who breeds and raises livestock.
stock·breed·ing (stok′brē′ding) *n.* The breeding and raising of livestock.
stock·bro·ker (stok′brō′kər) *n.* One who buys and sells stocks or securities for others. **— stock′bro′ker·age, stock′bro′king** *n.*
stock car 1 An automobile, as one selected at random, typifying the regular factory stock. **2** Such an automobile, usually a sedan, modified for racing.
stock company 1 An incorporated company that issues stock. **2** A more or less permanent dramatic company under one management, which presents a series of theater pieces.
stock dove (duv) The common wild pigeon of Europe (*Columba oenas*).
stock exchange 1 A place where securities are bought and sold. **2** An association of stockbrokers who transact business in stocks, bonds, and other shares.
stock farm A farm which specializes in the breeding of livestock.
stock fish Cod, haddock, or the like, cured by splitting and drying in the air, without salt.
stock·hold·er (stok′hōl′dər) *n.* One who holds certificates of ownership in a company or corporation.
Stock·holm (stok′hōm, *Sw.* stôk′hôlm) The capital of Sweden, a port on the east coast, on the Baltic Sea; called "the Venice of the North" because of its waterways.
stock·i·net (stok′i·net′) *n.* **1** An elastic knitted fabric used chiefly for undergarments. **2** A style of knitting in which the rows are alternately knitted and purled: also **stockinet stitch.** Also **stock′i·nette′.** [Alter. of *stockinget* <STOCKING + -ET]
stock·ing (stok′ing) *n.* **1** A close-fitting woven or knitted covering for the foot and lower leg. **2** Something resembling such a covering. [<STOCK, in obs. sense of "a stocking" + -ING³] **— stock′ing·less** *adj.*
stock in trade 1 The goods which a storekeeper has for sale. **2** Resources, either material or spiritual.
stock·ish (stok′ish) *adj.* Like a stock or block of wood; stupid.
stock·job·ber (stok′job′ər) *n.* A dealer or speculator in stocks in his own interest; also, a stockbroker. **— stock′job′ber·y, stock′job′·bing** *n.*
stock·man (stok′mən) *n. pl.* **·men** (-mən) **1** A man having charge of stock, as on a ranch. **2** One who raises or owns livestock; a cattleman.
stock market 1 A stock exchange. **2** The business transacted in such a place: The *stock market* was active. **3** The rise and fall of prices of securities.
stock·pile (stok′pīl′) *n.* A storage pile of materials or supplies. Also **stock pile.** *— v.t. & v.i.* **·piled, ·pil·ing** To accumulate a supply or stockpile (of).
Stock·port (stok′pôrt, -pōrt) A county borough of NE Cheshire, England.
stock·pot (stok′pot′) *n.* A pot for preparing and keeping soup stock.
stock-rais·ing (stok′rā′zing) *n.* Breeding and raising of various types of livestock. **— stock′-rais′er** *n.*
stock·room (stok′ro͞om′, -ro͝om′) *n.* A room where reserve stocks of goods are stored.

stock·still (stok′stil′) *adj.* Still as a stock or post; motionless.

Stock·ton (stok′tən), **Frank,** 1834–1902, U.S. author; full name Francis Richard Stockton. — **Richard,** 1730–81, American statesman; signer of the Declaration of Independence.

Stock·ton-on-Tees (stok′tən·on·tēz′) A port and industrial municipal borough in SE Durham, England.

stock·work (stok′wûrk′) *n. Geol.* An irregular mass of rock interlaced by a network of small ore-bearing veins.

stock·y (stok′ē) *adj.* **stock·i·er, stock·i·est** Short and stout; thick-set. — **stock′i·ly** *adv.* — **stock′·i·ness** *n.*

stock·yard (stok′yärd′) *n.* A large yard with pens, stables, etc., where cattle are kept ready for shipping, slaughter, etc.

stodge (stoj) *v.* **stodged, stodg·ing** *v.t.* To render dull and heavy by stuffing with food. — *v.i.* To become muddy or marshy. [<dial. E *stodge* fill to distention]

stodg·y (stoj′ē) *adj.* **stodg·i·er, stodg·i·est** 1 Distended; crammed full; bulky; lumpy. 2 Stupid; dull; heavy. 3 Indigestible; satiating. 4 Sticky; muddy. 5 Thick-set; clumsy and stiff. — **stodg′i·ly** *adv.* — **stodg′i·ness** *n.*

sto·gy (stō′gē) *n. pl.* **·gies** 1 A stout, coarse boot or shoe. 2 A long, slender, inexpensive cigar: also **sto′gie.** [Earlier *stoga* <(CONE)-STOGA (WAGON), because their drivers wore heavy boots and smoked coarse cigars]

sto·ic (stō′ik) *n.* A person apparently unaffected by pleasure or pain. — *adj.* Indifferent to pleasure or pain; impassive; uncomplaining. Also **sto′i·cal.** — **sto′i·cal·ly** *adv.* — **sto′i·cal·ness** *n.*

Sto·ic (stō′ik) *n.* A member of a school of Greek philosophy founded by Zeno about 308 B.C., holding the pantheistic beliefs that the world is a manifestation of a divine mind, that there is no reality but matter, even the human soul being sublimated substance doomed to dissolution, that wisdom lies in being superior to passion, joy, grief, etc., and in unperturbed submission to the divine will. — *adj.* Of or pertaining to the Stoics or Stoicism. [<L *Stoicus* <Gk. *Stoikos* <*Stoa (Poikilē)* (Painted) Porch, the colonnade at Athens where Zeno taught]

stoi·chi·ol·o·gy (stoi′kē·ol′ə·jē) *n.* The science of the elements or constituent processes in the physiology of animal tissues. Also **stoe′·chi·ol′o·gy, stoi′chei·ol′o·gy.** [<Gk. *stoicheion* element + -LOGY] — **stoi′chi·o·log′i·cal** (-ə·loj′·i·kəl) *adj.*

stoi·chi·om·e·try (stoi′kē·om′ə·trē) *n.* The branch of chemistry that treats of the combining proportions of elements or compounds involved in reactions, and the methods of calculating them. Also **stoe′chi·om′e·try, stoi′·chei·om′e·try.** [<Gk. *stoicheion* element + -METRY] — **stoi′chi·o·met′ric** (-ə·met′rik) or **·ri·cal** *adj.*

sto·i·cism (stō′ə·siz′əm) *n.* Indifference to pleasure or pain; stoicalness. See synonyms under APATHY.

Sto·i·cism (stō′ə·siz′əm) *n.* The doctrines of the Stoics.

stoit (stōt, stoit) *v.i. Scot. & Irish* 1 To walk in a reeling, stumbling manner: also **stoit′er, stoit′ur.** 2 To rebound; bounce. 3 To leap from the water: said of certain fish.

stoke[1] (stōk) *v.t. & v.i.* **stoked, stok·ing** To supply (a furnace) with fuel; stir up or tend (a fire or furnace). [Back formation <STOKER]

stoke[2] (stōk) *n. Physics* A unit of kinematic viscosity, equivalent to 1 poise in a fluid having a density of 1 gram per cubic centimeter referred to a specified temperature. [after Sir George G. *Stokes,* 1819–1903, English mathematician and physicist]

stoke·hold (stōk′hōld′) *n. Naut.* 1 The furnace room of a steamer. 2 The space in front of the furnaces from which they are stoked.

stoke·hole (stōk′hōl′) *n.* 1 The space about the mouth of a furnace; the fireroom. 2 The mouth of a furnace. 3 A stokehold.

Stoke-on-Trent (stōk′on·trent′) A county borough in NW Stafford, England. Also **Stoke′-up·on′-Trent′.**

Stoke Po·ges (pō′jis) A village in SE Buckingham, England; generally regarded as the scene of Gray's *Elegy.*

stok·er (stō′kər) *n.* 1 One who or that which supplies fuel to a furnace, especially of a steam boiler, as in a ship or locomotive; a fireman on a locomotive, ship, etc. 2 A device for feeding coal to a furnace. [<Du. <*stoken* stir a fire <*stok* stick]

Stokes mortar A light, muzzleloading mortar for high-angle, short-range fire. [after Sir Frederick W. S. *Stokes,* 1860–1927, English inventor]

Sto·kow·ski (stə·kôf′skē, -kou′skē), **Leopold,** born 1882, U. S. orchestra conductor born in England.

stole[1] (stōl) *n.* 1 *Eccl.* A long, narrow band, usually of decorated silk or linen, worn about the shoulders by priests and bishops, and over the left shoulder only by deacons, when officiating; loosely, any ecclesiastical vestment. 2 A fur, scarf, or garment resembling a stole, worn by women. 3 In ancient Rome, a long outer garment worn by matrons. [OE <L *stola* a robe <Gk. *stolē* a garment] — **stoled** *adj.*

stole[2] (stōl) Past tense of STEAL.

sto·len (stō′lən) Past participle of STEAL.

stol·id (stol′id) *adj.* Having or expressing no power of feeling or perceiving; impassible; dull. See synonyms under BRUTISH, HEAVY. [<L *stolidus* dull] — **sto·lid·i·ty** (stə·lid′ə·tē), **stol′id·ness** *n.* — **stol′id·ly** *adv.*

sto·lon (stō′lon) *n.* 1 *Bot.* **a** A trailing branch that is capable of taking root. **b** A runner or rootstock by which grasses may propagate. 2 *Zool.* A prolongation of the body of various animals, as corals. [<NL <L *stolo, stolonis*]

sto·ma (stō′mə) *n. pl.* **sto·ma·ta** (stō′mə·tə, stom′ə·tə) 1 A minute orifice; pore. 2 *Biol.* An aperture in the walls of blood vessels or in serous membranes, or in the epidermis of leaves, young stems, etc. [<Gk. *stoma* mouth]

-stoma See -STOME.

stom·ach (stum′ək) *n.* 1 The pouchlike, highly vascular dilation of the alimentary canal, situated in most vertebrates between the esophagus and the small intestine, and serving as one of the principal organs of digestion. ◆ Collateral adjective: *gastric.* 2 Any digestive cavity, as of an invertebrate. 3 The abdomen; belly: an anatomically incorrect use. 4 Desire for food; appetite; hence, any desire or inclination. 5 Temper; spirit. 6 *Obs.* Pride; haughtiness. — *v.t.* 1 To accept without apparent opposition; to put up with; endure. 2 To take into and retain in the stomach; digest. 3 *Obs.* To resent. [<OF *estomac* <L *stomachus* <Gk. *stomachos* gullet, stomach <*stoma* a mouth]

stomach ache Pain in the stomach, as from indigestion or inflammation.

stom·ach·er (stum′ək·ər) *n.* A former ornamental article of dress for the breast and stomach.

sto·mach·ic (stō·mak′ik) *adj.* 1 Pertaining to the stomach. 2 Strengthening the activity of the stomach. Also **stom·ach·al** (stum′ək·əl), **sto·mach′i·cal.** — *n.* Any medicine strengthening or stimulating the stomach.

stomach tooth *Dent.* A lower canine tooth of the first dentition: so called because its emergence is frequently accompanied by digestive disturbances.

stomach worm Any of various nematode worms which are parasitic in the stomachs of man and animals, especially the sheep stomach worm *(Haemonchus contortus).*

stom·ach·y (stum′ək·ē) *adj.* 1 Having a paunch. 2 *Brit. Dial.* Spirited; haughty; proud; also, choleric; resentful.

sto·ma·ta (stō′mə·tə, stom′ə·tə) Plural of STOMA.

sto·ma·tal (stō′mə·təl, stom′ə-) *adj.* Of or pertaining to stomata.

sto·mat·ic (stō·mat′ik) *adj.* 1 Of or pertaining to the mouth. 2 Of, pertaining to, or like a stoma.

sto·ma·tif·er·ous (stō′mə·tif′ər·əs, stom′ə-) *adj.* Bearing stomata. [<STOMAT(O)- + -(I)FEROUS]

sto·ma·ti·tis (stō′mə·tī′tis, stom′ə-) *n. Pathol.* Inflammation of the mouth.

stomato- *combining form* The mouth; of or pertaining to the mouth: *stomatoplasty.* Also, before vowels, **stomat-.** [<Gk. *stoma, stomatos* the mouth]

sto·ma·tol·o·gy (stō′mə·tol′ə·jē, stom′ə-) *n.* The science treating of the mouth and of its diseases.

sto·ma·to·plas·ty (stō′mə·tə·plas′tē, stom′ə-) *n.* Plastic surgery of the mouth; specifically, the operation of forming a mouth when from any cause the mouth has been contracted or closed.

sto·ma·to·pod (stō′mə·tə·pod′, stom′ə-) *n.* Any of an order *(Stomatopoda)* of crustaceans having abdominal gills and legs near the mouth, including the squills. — **sto′ma·top′o·dous** (-top′ə·dəs) *adj.*

sto·ma·tous (stō′mə·təs, stom′ə-) *adj.* Having a stoma or stomata.

-stome *combining form* Mouth; mouthlike opening: *peristome.* Also spelled *-stoma.* [<Gk. *stoma* the mouth]

sto·mo·de·um (stō′mə·dē′əm, stom′ə-) *n. pl.* **·de·a** (-dē′ə) *Biol.* The invagination of the ectoderm, or outer layer of the embryo, that forms the mouth. Also **sto′mo·dae′um.** [<NL <Gk. *stoma* mouth + *hodaios* on the way <*hodos* way] — **sto′mo·de′al** or **·dae′al** *adj.*

-stomous *combining form* Having a (specified kind of) mouth: *microstomous.* Also **-stomatous.** [<Gk. *stoma, stomatos* the mouth]

stomp (stomp) *Dial. v.t. & v.i.* To stamp; tread heavily (upon). — *n.* A dance involving a heavy and lively step. [Var. of STAMP]

-stomy *combining form Surg.* An operation to form an artificial opening for or into (a specified organ or part): *colostomy, ileostomy.* [<Gk. *stoma* the mouth]

stone (stōn) *n.* 1 A small piece of rock, as a cobble or pebble. 2 Rock, or a piece of rock hewn or shaped; a milestone; a gravestone; hard, concreted mineral or earthy matter. 3 A precious stone; gem. 4 Anything resembling a stone in shape or hardness: a *hailstone.* 5 *Pathol.* A stony concretion in the bladder, or a disease characterized by such concretions. 6 *Bot.* The hard covering of the kernel in a fruit. 7 (*pl.* **stone**) *Brit.* A measure of weight, avoirdupois, usually 14 pounds. 8 A testicle: usually in the plural. 9 *Printing* An imposing table for type, whether made of stone or metal. — *adj.* 1 Made of stone: a *stone* ax. 2 Made of coarse hard earthenware: a *stone* bottle. 3 Characterized by the use of stone implements: the *Stone* Age. — *v.t.* **stoned, ston·ing** 1 To hurl stones at; pelt or kill with stones. 2 To remove the stones or pits from. 3 To furnish or line, as a well, with stone. 4 To castrate; geld, as a hog. 5 *Obs.* To make hard or unyielding, as the heart. [OE *stān*] — **ston′er** *n.*

Stone (stōn), **Harlan Fiske,** 1872–1946, U. S. educator; Supreme Court justice 1941–46. — **Lucy,** 1818–93, U.S. suffragist: wife of *Henry Broun Blackwell.*

Stone Age The earliest known period of the cultural evolution of mankind, marked by the creation and use of stone implements and weapons, preceding the Bronze Age, and subdivided into the Eolithic, Paleolithic, and Neolithic periods.

stone-blind (stōn′blīnd′) *adj.* Blind as a stone; totally blind.

stone·boat (stōn′bōt′) *n. U.S.* A runnerless plank sled used for transporting rocks or similar heavy objects or, when weighted, dragged across a field to break clods of earth, etc.; also, a platform swung under the axles of a wagon.

stone·break (stōn′brāk′) *n.* Saxifrage.

stone-broke (stōn′brōk′) *adj. Colloq.* Without any money; having no funds. Also **ston′y-broke′.**

stone·chat (stōn′chat′) *n.* A small thrushlike European bird (genus *Saxicola*) with upper parts black and breast dark-reddish. [<STONE + CHAT[1] (def. 2); from its cry suggesting the knocking together of pebbles]

stone coal Hard or anthracite coal.

stone color Bluish gray. — **stone′-col′ored** *adj.*

stone·crop (stōn′krop′) *n.* A low spreading mosslike herb *(Sedum acre)* with small fleshy leaves and yellow flowers.

stone·cut·ter (stōn′kut′ər) *n.* One who or that which cuts stone; specifically, a machine for facing stone. — **stone′cut′ting** *n.*

stone-deaf (stōn′def′) *adj.* Completely deaf.

stone fly A plecopteran.

stone fruit A fruit having a stone; a drupe.

add,āce,câre,pälm; end,ēven; it,īce; odd,ōpen,ôrder; to͝ok,po͞ol; up,bûrn; ə = a in *above,* e in *sicken,* i in *clarity,* o in *melon,* u in *focus;* yo͞o = u in *fuse;* oi,oil; ou,pout; ch,check; g,go; ng,ring; th,thin; th,this; zh,vision. Foreign sounds à,œ,ü,kh,ṅ; and ◆: see page xx. < from; + plus; ? possibly.

Stone·henge (stōn'henj) A prehistoric megalithic structure on Salisbury Plain, SE Wiltshire, England. It consists primarily of circles of dressed stones, some with lintels, the main structure dating probably from 1500 B.C.

STONEHENGE

stone lily A fossil sea lily or other crinoid.
stone marten 1 An Old World marten (*Martes foina*). 2 Its fur, marked in white on throat and breast.
stone·ma·son (stōn'mā'sən) *n*. One whose occupation or trade is to prepare and lay stones in building. — **stōn'ma'son·ry** *n*.
stone·mint (stōn'mint') *n*. Dittany (def. 1).
Stone Mountain A granite dome (1,686 feet) of NW central Georgia; a Confederate monument is carved on one side.
stone parsley An Old World herb of the parsley family, especially a British perennial (*Sison amomum*) with cream-colored flowers and aromatic seeds.
stone roller 1 A cyprinoid fish (*Campostoma anomalum*) of North America. 2 A North American sucker (*Catastomus nigricans*).
stone's-cast (stōnz'kast', -käst') *n*. The distance a stone may be cast by hand. Also **stone's-throw'** (-thrō').
Stones River A river of central Tennessee, flowing 39 miles NW to the Cumberland River; scene of a Union victory in the Civil War, 1862–63.
stone-still (stōn'stil') *adj*. Perfectly motionless.
stone·wall (stōn'wôl') *v.i.* 1 In cricket, to play on the defensive so as to secure a draw. 2 *Austral*. To oppose by a policy of obstruction; filibuster: a political term.
stone wall A wall built of stone; especially, a fence built of stones.
Stone·wall Jack·son (stōn'wôl' jak'sən) See JACKSON, THOMAS JONATHAN.
stone·ware (stōn'wâr') *n*. A variety of very hard, glazed pottery, made from siliceous clay or clay mixed with flint or sand.
stone·work (stōn'wûrk') *n*. 1 Work concerned with cutting or setting stone; work made of stone. 2 *pl*. A place where stone is shaped or stoneware is made. — **stone'work'er** *n*.
stone·wort (stōn'wûrt') *n*. Any of a genus (*Chara*) of green algae growing submerged in fresh or brackish waters and often incrusted with deposits of calcium carbonate.
ston·ish (ston'ish) *v.t. Obs*. To astonish. [Aphetic var. of ASTONISH]
ston·y (stō'nē) *adj*. **ston·i·er, ston·i·est** 1 Abounding in stone. 2 Made or consisting of stone. 3 Hard as stone; hence, unfeeling or inflexible. 4 Converting into stone; petrifying; cold and stiff. 5 *Slang* Stone-broke; having no money. — **ston'i·ly** *adv*. — **ston'i·ness** *n*.
stony coral A coral having a calcareous skeleton.
ston·y-heart·ed (stō'nē-här'tid) *adj*. Hardhearted; unfeeling; pitiless.
Stony Point A village in SE New York; scene of an American victory in the Revolutionary War, July, 1779.
stood[1] (stood) Past tense and past participle of STAND.
stood[2] (stood) *Illit. & Dial*. Stayed: He should have *stood* in bed.
stooge (stōōj) *Colloq. n*. 1 An actor placed in the audience to heckle a comedian on the stage. 2 An actor who feeds lines to the principal comedian, acts as a foil for his jokes, etc. 3 Anyone who acts as or is the tool or dupe of another. — *v.i.* **stooged, stoog·ing** To act as a stooge: usually with *for*. [Origin unknown]
stook (stook, stook) *n*. A collection of sheaves set together in the field; a shock of corn. — *v.t.* To set up in stooks. [Cf. MLG *stuke* a bundle] — **stook'er** *n*.
stool (stool) *n*. 1 A backless and armless seat intended for one person. 2 A low bench or portable support for the feet or for the knees in kneeling. 3 A seat used in defecating; a privy. 4 The matter evacuated from the bowels. 5 *Bot*. **a** A plant from which young plants are produced, as from runners. **b** A stump or root of any kind from which suckers or sprouts shoot up. **c** The shoots from such a root or stump. 6 A decoy, as a bird or likeness of one. — *v.i.* 1 To send up shoots or suckers. 2 To decoy wild fowl with a stool or stools. 3 To void feces. 4 *U.S. Slang* To be a stool pigeon; inform. [OE *stōl*]
stool pigeon 1 A living or artificial pigeon attached to a stool or perch to decoy others. 2 Any decoy, as a person employed to decoy others into a gambling house, etc. 3 *U.S. Slang* An informer or spy, especially for the police.
stoop[1] (stoop) *v.i.* 1 To bend or lean the body forward and down; bow; crouch. 2 To stand or walk with the upper part of the body habitually bent forward; slouch. 3 To bend; lean; sink: said of trees, cliffs, etc. 4 To lower or degrade oneself; condescend; deign. 5 To pounce or swoop, as a hawk on prey. 6 *Obs*. To submit; yield. — *v.t.* 7 To bend (one's head, shoulders, etc.) forward. 8 *Obs*. To humble or subdue. See synonyms under BEND[1]. — *n*. 1 An act of stooping; a downward and forward bending of the body; also, a habitual forward inclination of the body and shoulders. 2 A decline from dignity or superiority. 3 A swoop, as of a bird of prey. [OE *stūpian*]
stoop[2] (stoop) *n. U.S.* 1 Originally, a platform at the door of a house approached by steps and having seats. 2 A small porch or platform at the entrance to a house. [< Du. *stoep*]
stoop[3] (stoop) *n. Brit. Dial*. A post set in the ground; a pillar. [< ON *stolpi*]
stoop[4] (stoop) See STOUP.
stop (stop) *v*. **stopped** or (*chiefly Poetic*) **stopt, stop·ping** *v.t.* 1 To bring (something in motion) to a halt; arrest the progress of: to *stop* an automobile. 2 To prevent the doing or completion of: to *stop* a revolution. 3 To prevent (a person) from doing something; restrain. 4 To keep back, withhold, or cut off, as wages or supplies. 5 To cease doing; desist from; discontinue: *Stop* that! 6 To intercept in transit, as a letter. 7 To block up, obstruct, or clog (a passage, road, etc.): often with *up*. 8 To fill in, cover over, or otherwise close, as a hole, cavity, etc. 9 To close (a bottle, barrel, etc.) with a cork, plug, or other stopper. 10 To stanch (a wound, etc.). 11 To order a bank not to pay or honor: to *stop* a check. 12 To defeat; also, to kill. 13 *Music* To press down (a string) on the fingerboard, or to close (a finger hole) in order to vary pitch. 14 To punctuate. 15 In boxing, etc., to parry. — *v.i.* 16 To come to a halt; cease progress or motion. 17 To cease doing something; pause or desist. 18 To come to an end. See synonyms under ABIDE, ARREST, CEASE, END, HINDER[1], OBSTRUCT, REST[1], SHUT, STAND, SUSPEND. — **to stop off** To cease traveling temporarily before reaching one's destination. — **to stop over** *Colloq*. 1 To stay at a place temporarily. 2 To interrupt a journey; make a stopover. — *n*. 1 The act of stopping, or the state of being stopped; a halt; pause; cessation; end. 2 That which stops or limits the range or time of a movement: a camera *stop*; an obstruction or obstacle; a hindrance. 3 *Music* The pressing down of a string or the closing of an aperture on a musical instrument, to change the pitch of the tone emitted; a key, lever, or handle for stopping a string or an aperture; a fret for a guitar. 4 *Music* In an organ, a set of pipes or reeds producing tones of the same quality, and arranged in regular musical progression. 5 **a** *Brit*. A punctuation mark; a period. **b** In cables, etc., a period. 6 In joinery, a block, pin, or the like to check sliding motion, as of a drawer. 7 *Naut*. A small line for lashing or fastening anything temporarily on a ship. 8 *Phonet*. **a** Complete blockage of the breath stream (implosion), as with the lips or tongue, followed by a sudden release (explosion). **b** A consonant so produced; a plosive: opposed to *continuant*. The stops in English are the bilabials (p) and (b), the alveolars (t) and (d), and the velars (k) and (g); the nasals (m) and (n) may also be included in this category. 9 In dogs, the short incline between the forepart of the skull and the face. See illustration under DOG. 10 *pl*. A card game in which certain cards, called **stop cards**, terminate play when they appear: a variety of *newmarket*. See synonyms under REST[1]. [OE *-stoppian*, as in *forstoppian* stop up]
stop·cock (stop'kok') *n*. A faucet or short pipe having a valve for stopping or regulating the passage of liquid, gas, etc.
stope (stōp) *Mining n*. An excavation from which the ore is removed, either above or below a level, in a series of steps. — *v.t. & v.i.* **stoped, stop·ing** To excavate in stopes. [Appar. related to STEP]
stop·gap (stop'gap') *n*. That which stops a gap; also, an expedient.
stop key A key so made that when inserted in one side of a lock no key may be used on the other.
stop knob The knob by which a set of organ pipes is opened.
stop light 1 A red light on a traffic sign, directing a motorist or pedestrian to stop. 2 A red light on the rear of a motor vehicle which shines upon application of the brakes.
stop-loss (stop'lôs', -los') *adj*. Intended to prevent further loss, in a brokerage account, from falling prices on financial markets.
stop net A small net joined to a seine, to increase its length or prevent the escape of fish.
stop order An order to an agent or stockbroker to buy or sell a stock at the market only when it reaches a specified price.
stop·o·ver (stop'ō'vər) *adj*. Giving permission to stop over, as a railway ticket. — *n*. A stopover check, the act of stopping over, or permission to stop over, as from one train to a later train on the same railroad. Also **stop'-off'**.
stop·page (stop'ij) *n*. 1 The act of stopping or the state of being stopped. 2 A deduction from pay to repay something.
stop payment An order to a bank to refuse payment on a certain check.
stop·per (stop'ər) *n*. 1 One who or that which stops up or closes. 2 A plug or cork which checks movement or action of any kind. — *v.t.* To secure or close with a stopper.
stop-pit (stop'it) *Scot*. Past participle of STOP.
stop·ple (stop'əl) *n*. A stopper, plug, cork, or bung. — *v.t.* **·pled, ·pling** To close with or as with a stopple. [ME *stoppel*, prob. < *stoppen* stop]
stop sign A sign in a traffic system, instructing a pedestrian or vehicle to stop.
stopt (stopt) Past tense and past participle of STOP: chiefly a poetic form of *stopped*.
stop thrust In fencing, a slight thrust designed to frustrate the attack of an opponent.
stop·watch (stop'woch') *n*. A watch which has a hand indicating fractions of a second and which may be stopped or started by the pressure of a spring: used for timing races, etc.
stop·way (stop'wā') *n*. An extension of an airfield runway to permit safe landing in the event of engine failure during take-off.
stor·age (stôr'ij, stō'rij) *n*. 1 The depositing of articles in a warehouse for safekeeping. 2 Space for storing goods. 3 A charge for storing. 4 The preparing of a storage battery for the generation of electrical energy.

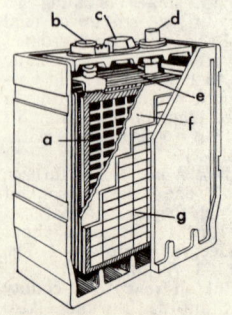

STORAGE BATTERY
a. Positive plate.
b. Positive terminal.
c. Vent cap or plug.
d. Negative terminal.
e. Electrolyte space.
f. Separator.
g. Negative plate.

storage battery A connected group of two or more electrolytic cells for the generation of electric energy by the passage of a current which, on being reversed in direction, serves to recharge the cells for another period of use.
sto·rax (stôr'aks, stō'raks) *n*. 1 A fragrant

store balsam obtained from the wood and inner bark of either of two trees (*Liquidambar orientalis*, or *L. styraciflua*) of Asia Minor: used in medicine and as a perfume. 2 A gum resin obtained from certain trees of a family (*Styracaceae*), especially *Styrax officinalis*. [< L < Gk. *styrax*]

store (stôr, stōr) *v.t.* **stored, stor·ing 1** To put away for future use; to accumulate. **2** To furnish or supply; provide. **3** To place in a warehouse or other place of deposit for safe-keeping. — *n.* **1** That which is stored or laid up against future need; hence, a large amount at hand. **2** *pl.* Supplies, as of ammunition, arms, or clothing; necessary articles, especially of food. **3** A place where commodities are stored; warehouse. **4** *U.S.* A place where merchandise of any kind is kept for sale; a shop. See synonyms under HEAP, STOCK. — **department store** A large retail establishment selling various types of merchandise and service, and organized by departments. — **in store** Set apart for the future; forthcoming; impending. — **to set store by** To value or esteem; regard. [Aphetic var. of earlier *astore* < OF *estorer* erect, equip, store < L *instaurare* restore, erect]

store·house (stôr'hous', stōr'-) *n.* A building in which goods are stored; a warehouse; depository.

store·keep·er (stôr'kē'pər, stōr'-) *n.* **1** A person who keeps a retail store or shop; a shopkeeper. **2** One who has charge of receiving and distributing stores; especially, one in charge of naval or military stores.

store·room (stôr'rōōm', stōr'-) *n.* A room in which things are stored, as supplies.

sto·rey (stôr'ē, stōr'ē) See STORY[2].

sto·ried[1] (stôr'ēd, stōr'-) *adj.* Having or consisting of stories, as a building: usually in compounds: a six-*storied* house. Also **stor'eyed.**

sto·ried[2] (stôr'ēd, stōr'ēd) *adj.* **1** Having a notable history. **2** Related in a story. **3** Ornamented with designs representing scenes from history or story.

sto·ri·ette (stôr'ē·et', stō'rē-) *n.* A short story or tale.

stork (stôrk) *n.* A wading bird with a long neck and long legs (family *Ciconiidae*), related to the herons and ibises, especially the Old World **migratory** or **white stork** (*Ciconia ciconia*), which often nests on buildings. [OE *storc*]

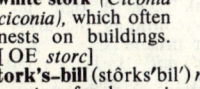

WHITE STORK
(About 20 inches tall)

stork's-bill (stôrks'bil') *n.* **1** Heronbill. **2** Any species of pelargonium.

storm (stôrm) *n.* **1** A disturbance of the atmosphere, generally a great whirling motion of the air, accompanied by rain, snow, etc. **2** In the Beaufort scale, a wind force of the 11th degree. **3** Figuratively, a furious flight or shower of objects, especially of missiles. **4** A violent outburst, as of passion or excitement: a *storm* of applause or rage. **5** *Mil.* A violent and rapid assault on a fortified place. **6** A violent commotion, as in politics, society, or domestic life. — *v.i.* **1** To blow with violence; rain, snow, hail, etc., heavily: used impersonally. It *stormed* all day. **2** To be very angry; rage. **3** To move or rush with violence or rage: He *stormed* about the room. — *v.t.* **4** *Mil.* To take or try to take by storm. [OE] — **Synonyms** (noun): agitation, disturbance, tempest. A *storm* is properly a *disturbance* of the atmosphere, with or without rain, snow, hail, or thunder and lightning. Thus we have *rainstorm, snowstorm,* etc., and by extension, *magnetic storm,* etc. A *tempest* is a *storm* of extreme violence, always attended with some precipitation, as of rain, from the atmosphere. In the moral and figurative use *tempest* commonly implies greater intensity. We speak of *agitation* of feeling, *disturbance* of mind, a *storm* of passion, a *tempest* of rage. See WIND. **Antonyms:** calm, hush, peace, serenity, stillness, tranquillity.

Storm may appear as a combining form in hyphemes or solidemes, or as the first element in two-word phrases:

storm area	storm god	storm-rocked
storm-beaten	storm goddess	storm shutter
storm blast	storm gust	storm-swept
storm-boding	storm jacket	stormtight
stormbound	storm lane	storm-tossed
storm-bringer	stormlike	storm-washed
storm cloud	storm path	stormwind
storm coat	storm-rent	storm-worn

Storm (shtôrm), **Theodor Woldsen**, 1817–88, German poet and novelist.

storm-belt (stôrm'belt') *n.* A strip of territory along which storms most frequently move.

storm cellar A cyclone cellar.

storm center 1 *Meteorol.* The center or place of lowest pressure and comparative calm in a cyclonic storm. **2** The central point of a heated argument; the focus of any trouble or turmoil.

storm door A strong outer door for added protection during storms and inclement weather.

Stor·month (stôr'mənth), **James**, 1825–82, Scottish lexicographer.

storm petrel Any of certain petrels of the North Atlantic; especially, *Hydrobates pelagicus,* thought to portend storm. Also **stormy petrel.**

storm-proof (stôrm'prōōf') *adj.* Capable of keeping out storms.

storm trooper A member of the Nazi party militia unit, the *Sturmabteilung.*

storm warning A signal, as a flag or light, used to warn mariners of coming storm. Also **storm signal.**

storm window An extra window outside the ordinary one as a protection against storms or for greater insulation against cold.

storm·y (stôr'mē) *adj.* **storm·i·er, storm·i·est 1** Characterized by storms; boisterous; also, turbulent; violent: a *stormy* life. **2** Accompanying storms; also, passionate. See synonyms under BLEAK[1]. [OE *stormig*] — **storm'i·ly** *adv.* — **storm'i·ness** *n.*

Stor·thing (stôr'ting', stōr'-) *n.* The Norwegian parliament. Also **Stor'ting'.** [< Norw. < *stor* great + *thing* meeting]

sto·ry[1] (stôr'ē, stō'rē) *n. pl.* **·ries 1** A narrative or recital of an event, or a series of events, whether real or fictitious. **2** A narrative, usually of fictitious events, intended to entertain a reader or hearer; a short tale. **3** An account or allegation of the facts relating to a particular person, thing, or incident: He tells a more plausible *story* of the conflict. **4** A news article in a newspaper or magazine. **5** The material for a news article. **6** An anecdote. **7** *Colloq.* A lie; falsehood. **8** The series of events in a novel, play, etc. **9** Celebrated or romantic legend or history: to live on in *story*. — *v.t.* **·ried, ·ry·ing 1** To relate as a story. **2** To adorn with designs representing scenes from history, legend, etc. [< OF *estoire* < L *historia*. Doublet of HISTORY.] — **Synonyms** (noun): allegory, anecdote, incident, narrative, recital, record, relation, tale. *Tale* is nearly synonymous with *story,* but is somewhat archaic; it is used for an imaginative, legendary, or fictitious *recital,* especially if of ancient date; as, a fairy *tale*; also, for an idle or malicious report; as, Do not tell *tales.* See FICTION, HISTORY, REPORT.

sto·ry[2] (stôr'ē, stō'rē) *n. pl.* **·ries** A division in a building comprising the space between two successive floors; a floor; habitable rooms on the same level; also, a horizontal architectural division of a building: also spelled *storey.* [Special use of STORY[1]; ? from earlier sense of "a tier of painted windows or sculptures that narrated an event"]

Sto·ry (stôr'ē, stō'rē), **Joseph**, 1779–1845, U.S. jurist. — **William Wetmore**, 1819–95, U.S. sculptor; son of preceding.

story board 1 A bulletin board in a newspaper office on which are posted reportorial assignments to specific stories. **2** The set of original drawings illustrating each stage in the sequence of a motion picture, television program, animated cartoon, etc.

sto·ry·tell·er (stôr'ē·tel'ər, stō'rē-) *n.* **1** One who relates stories or anecdotes. **2** *Colloq.* A prevaricator; liar; fibber. — **sto'ry·tell'ing** *n. & adj.*

stoss (stos, Ger. shtōs) *adj. Geol.* Facing the direction whence a glacier moves. [< G *stoss* a thrust, push]

sto·tin·ka (stō·tiŋ'kä) *n. pl.* **·ki** (-kē) A small copper coin of Bulgaria; one one-hundredth of a lev. [< Bulgarian]

stound (stound) *n. Obs.* **1** A short time. **2** A sharp pain; pang; heavy blow. — *v.i. Scot.* To ache; hurt. [OE *stund*]

stoup (stōōp) *n.* **1** *Eccl.* A basin for holy water at the entrance of a church. **2** *Scot.* A pail; bucket; flagon; cup; also, its contents. **3** A measure for liquids: a pint *stoup.* Also spelled *stoop, stowp.* [< ON *staup* bucket]

stour[1] (stour) *n. Obs.* **1** A battle; conflict. **2** Dust in motion; chaff. Also **stoure.** [< OF *estour* tumult] — **stour'ie, stour'y** *adj.*

stour[2] (stōōr) *Scot. adj.* **1** Sturdy; also, harsh; rough; surly. **2** Grievous; painful. — *n.* Pressure of circumstances.

stout (stout) *adj.* **1** Strong or firm of structure or material; sound; tough. **2** Determined; resolute. **3** Fat; bulky; thick-set. **4** Strong in effects or active qualities; substantial; solid. **5** Having muscular strength; robust. **6** Proud; stubborn. See synonyms under CORPULENT, STAUNCH, STRONG. — *n.* **1** A stout person. **2** A dress or suit made for a stout person.

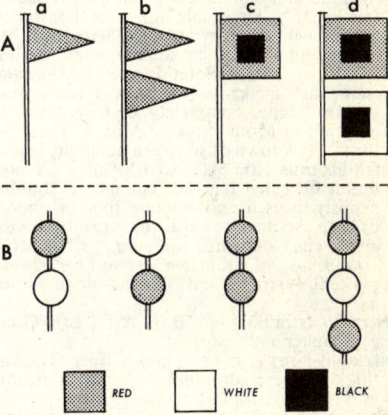

STORM WARNINGS
A. Daylight signals. B. Night signals.
a. Small-craft warning. b. Gale. c. Whole gale.
d. Hurricane.

3 A strong, very dark porter or ale: also **brown stout.** [< OF *estout* bold, strong < Gmc. Cf. MDu. *stolt* bold.] — **stout'ly** *adv.* — **stout'ness** *n.*

stout-heart·ed (stout'här'tid) *adj.* Brave; courageous. — **stout'heart'ed·ly** *adv.* — **stout'heart'ed·ness** *n.*

Sto·va·ine (stō'və·ēn, stō·vā'in) *n.* Proprietary name for a local anesthetic, $C_{14}H_{21}NO_2 \cdot HCl$, used especially intraspinally: invented by E. F. A. Fourneau, born 1872, French chemist.

stove[1] (stōv) *n.* **1** An apparatus, usually of metal, in which fuel is consumed for heating or cooking. **2** A drying room or box used in some factories. **3** An artificially heated greenhouse. **4** A pottery kiln. [OE *stofa* a heated room]

stove[2] (stōv) Alternative past tense and past participle of STAVE.

stove·pipe (stōv'pīp') *n.* **1** A pipe, usually of thin sheet iron, for conducting the smoke and gases of combustion from a stove to a chimney flue. **2** *U.S. Colloq.* A tall silk hat: also **stovepipe hat.**

stov·er (stō'vər) *n.* Fodder or feed for cattle; cornstalks. [< OF *estover.* See ESTOVERS.]

stow[1] (stō) *v.t.* **1** To place or arrange compactly; pack. **2** To fill by packing. **3** To have room for; hold: said of a room, receptacle, etc. **4** *Slang* To stop; cease. **5** *Obs.* To furnish lodging for. — **to stow away 1** To put in a place of safekeeping, hiding, etc. **2** To be a stowaway. [OE *stōwian* < *stōw* a place]

stow[2] (stō) *v.t. Scot. & Brit. Dial.* To lop or cut off; crop.

stow[3] (stō) *n. Scot. & Brit. Dial.* A stump or shoot of a tree; also, a slice; cut.

Stow (stō), **John,** 1525?–1605, English historian and antiquary.

stow·age (stō′ij) *n.* **1** The act or manner of stowing, or the state of being packed away. **2** Space for stowing goods. **3** Charge for stowing goods. **4** The goods stowed.

stow·a·way (stō′ə·wā′) *n.* One who conceals himself, as on a vessel, to obtain free passage.

Stowe (stō), **Harriet Beecher,** 1811–96, U.S. author; wrote *Uncle Tom's Cabin.*

stown·lins (stoun′linz) *adv. Scot.* Secretly; by stealth.

stowp (stōōp) See STOUP.

St. Pat·rick's Day (sānt pat′riks) March 17. See under PATRICK.

St. Paul (sānt pôl′) The capital of Minnesota, in the SE part of the State, on the Mississippi River: one of the Twin Cities.

St. Pe·ter (sānt pē′tər), **Lake** An expansion of the St. Lawrence River in southern Quebec, Canada; 130 square miles.

St. Pe·ters·burg (sānt pē′tərz·bûrg) **1** The capital of the former Russian Empire; renamed *Petrograd* in 1914 and *Leningrad* in 1924. **2** A city on Tampa Bay, western Florida.

St. Pierre (sānt pyâr′, *Fr.* saṅ pyâr′) An island group off Newfoundland; 10 square miles; with the adjacent island group of **Mi·que·lon** (mē·kə·lôṅ′), 83 square miles, it constitutes the French territory of **St. Pierre and Miquelon;** total area, 93 square miles; capital, St. Pierre. *French* **Saint-Pierre-et-Mi·que·lon** (saṅ·pyâr′ā·mē·kə·lôṅ′). **2** A former town in Martinique, completely destroyed by the eruption of Mont Pelée on May 8, 1902. See PELÉE. **3** A town of southern Réunion Island.

stra·bis·mus (strə·biz′məs) *n. Pathol.* A condition in which the eyes cannot be simultaneously focused on the same spot: when one or both eyes turn inward, the patient is *cross-eyed*: when outward, *walleyed.* [<NL <Gk. *strabismos* <*strabizein* squint <*strabos* twisted] —**stra·bis′mal, stra·bis′mic** or **·mi·cal** *adj.*

Stra·bo (strā′bō), 63? B.C.–A.D. 24?, Greek geographer and historian.

stra·bot·o·my (strə·bot′ə·mē) *n. Surg.* The cutting of the eyeball muscles to correct strabismus. [<Gk. *strabos* oblique + -TOMY]

Stra·chey (strā′chē), **(Evelyn) John,** born 1901, English politician and writer. —**(Giles) Lytton,** 1880–1932, English author and biographer.

strad·dle (strad′l) *v.* **·dled, ·dling** *v.i.* **1** To stand, walk, or sit with the legs spread apart. **2** To stand wide apart: said of the legs. **3** *Colloq.* To appear to favor both sides of an issue; refuse to commit oneself. —*v.t.* **4** To stand, walk, or sit with the legs on either side of. **5** To spread (the legs) wide apart. **6** *Colloq.* To appear to favor both sides of (an issue). **7** *Mil.* To fire shots both beyond and in front of (a target) so as to determine the range. —*n.* **1** A going, standing, or sitting with legs wide apart; the space between the feet or legs of one who straddles. **2** A noncommittal or vacillating position in any issue. **3** A stock transaction in which the holder obtains the privilege of either delivering or calling for a stock at a fixed price. **4** A long position in some stocks while being short in others. **5** *Mil.* Successive range settings that have bracketed the target. [Freq. of STRIDE] —**strad′dler** *n.* —**strad′dling·ly** *adv.*

Stra·di·va·ri (strä′dē·vä′rē), **Antonio,** 1644–1737, violin-maker of Cremona, Italy. Also **Strad·i·var·i·us** (strad′i·vâr′ē·əs).

Strad·i·var·i·us (strad′i·vâr′ē·əs) *n.* One of the famous violins produced by the fine workmanship of Antonio Stradivari.

strafe (strāf, sträf) *v.t.* **strafed, straf·ing 1** To attack (troops, emplacements, etc.) with machine-gun fire from low-flying airplanes. **2** To bombard or shell heavily. **3** *Slang* To punish. —*n.* A heavy bombardment. [<G *strafen* punish] —**straf′er** *n.*

Straf·ford (straf′ərd), **Earl of,** 1593–1641, Thomas Wentworth, English statesman; beheaded.

strag·gle (strag′əl) *v.i.* **·gled, ·gling 1** To wander from the road, main body, etc.; stray. **2** To wander aimlessly about; ramble. **3** To occur at irregular intervals. [? Freq. of obs. *strake* move, go about] —**strag′gler** *n.*

strag·gly (strag′lē) *adj.* **·li·er, ·li·est** Scattered or spread out irregularly.

straight (strāt) *adj.* **1** Extending uniformly in the same direction without curve or bend. **2** Free from kinks; not curly, as hair. **3** Not stooped or inclined; erect, as in posture. **4** Not deviating from truth, fairness, or honesty; accurate; honest; upright; reliable; also, candid. **5** Free from obstruction; uninterrupted; unbroken. **6** Correctly kept, ordered, or arranged. **7** Sold without discount for number or quantity taken. **8** *Colloq.* Adhering without reservation or exception to a particular party or policy; representing the regular or older organization; accepting the whole, as of a plan, party, or policy: a *straight* ticket. **9** In poker, consisting of five cards forming a sequence: a *straight* flush. **10** Having nothing added; unmixed; undiluted: *straight* whisky. —*n.* **1** A straight part or piece. **2** The part of a racecourse between the winning post and the last turn. **3** In poker, a numerical sequence of five cards not of the same suit, or a hand containing this. **4** A straight line. —*adv.* **1** In a straight line or a direct course. **2** Closely in line; correspondingly. **3** At once; straightway. ◆ Homophone: *strait.* [ME *stregt* <OE *streht,* pp. of *streccan* stretch] —**straight′ly** *adv.* —**straight′·ness** *n.*

straight angle See under ANGLE.

straight-arm (strāt′ärm′) *v.t.* In football, to ward off (an opposing tackler) with the outstretched arm.

straight-a·way (strāt′ə·wā′) *adj.* Having no curve or turn; straightforward. —*n.* A straight course or track. —*adv.* At once; straightway.

straight·edge (strāt′ej′) *n.* A bar of wood or metal having one edge true to a straight line: used for ruling, etc. —**straight′-edged′** *adj.*

straight·en (strāt′n) *v.t.* **1** To make straight. **2** To lay out (a corpse). —*v.i.* **3** To become straight. —**to straighten out** To restore order to; set right; rectify. —**to straighten up 1** To free from disorder; make neat; tidy. **2** To stand in erect posture. **3** To reform; become honorable after being dishonest. —**straight′·en·er** *n.*

straight face A sober, expressionless, or unsmiling face. —**straight-faced** (strāt′fāst′) *adj.*

straight flush See under FLUSH.

straight·for·ward (strāt′fôr′wərd) *adj.* Proceeding in a straight course or direct manner; frank. See synonyms under CANDID, CLEAR, HONEST, JUST[1], PLAIN[1]. —**straight′for′ward·ly** *adv.* —**straight′for′ward·ness** *n.*

straight·for·wards (strāt′fôr′wərdz) *adv.* In a straight course or direct manner; straightforwardly.

straight-line (strāt′līn′) *adj. Mech.* Designating a linkage or similar apparatus intended to copy or generate motion in a straight or nearly straight line.

straight man *U.S. Colloq.* An entertainer who acts as a foil for a comedian.

straight-out (strāt′out′) *adj.* **1** Showing the true sentiments or feelings; unreserved; also, shown without reserve. **2** Real; genuine. **3** Adhering to a political party or to party principles; uncompromising. **4** Straight: said of a political ticket.

straight ticket 1 A political party ballot or ticket that presents the regular party candidates without addition or change. **2** A ballot cast for all the candidates of one party. Compare SPLIT TICKET.

straight·way (strāt′wā′) *adv.* Immediately; straight-away.

straik[1] (strāk) *v.t. Scot.* To stroke.

straik[2] (strāk) *Scot.* Past participle of STRIKE.

strain[1] (strān) *v.t.* **1** To pull or draw tight; stretch. **2** To exert to the utmost. **3** To injure by overexertion; sprain; also, to wrench or twist. **4** To deform in structure or shape as a result of pressure or stress. **5** To stretch beyond the true intent, proper limit, etc.: to *strain* a point. **6** To embrace tightly; hug. **7** To pass through a strainer (def. 2). **8** To remove by filtration. **9** *Mech.* To alter in size or shape by applying external force. **10** *Obs.* To force; constrain. —*v.i.* **11** To make violent efforts; strive. **12** To be or become wrenched or twisted. **13** To filter, trickle, or percolate. See synonyms under STRETCH. —**to strain at 1** To push or pull with violent efforts. **2** To strive for. **3** To scruple or balk at accepting. —*n.* **1** An act of straining or the state of being strained; a violent effort or exertion. **2** The injury due to excessive tension or effort. **3** *Physics* Change of shape or size of a body, especially of a solid, produced by the action of a stress; deformation, temporary or permanent; thrust; force. [<OF *estrein-,* stem of *estreindre* <L *stringere* bind tight]

strain[2] (strān) *n.* **1** Line of descent, or the individuals, collectively, in that line; race; stock. **2** Inborn or hereditary disposition; natural tendency; trace; an element or admixture: to have a heroic *strain* in one's character. **3** *Biol.* A special line of individuals belonging to a certain race or species and maintained at a high standard of perfection by selection: said of animals or plants. **4** *Rare* Distinguishing nature or quality; kind; sort. **5** A section, in hymn tunes, divided off by a double bar; a melody; tune; air. **6** A distinctive portion of a poem; also, a composition in verse. **7** Prevailing tone, style, or manner; mood. See synonyms under SONG, TUNE. [? Var. of ME *strene,* OE *strēon* offspring]

strain·er (strā′nər) *n.* **1** One who or that which strains. **2** A utensil or device, containing meshes or porous parts, through which liquids are passed to separate them from coarse particles. **3** A device used for tightening, strengthening, or stretching.

straining arch Any arch erected to exert a corrective strain or to resist a destructive strain in a building.

straining beam A tie beam receiving a lengthwise pulling stress, and connecting the rafters of a roof with the tops of the queenposts. Also **straining piece.**

strait (strāt) *adj.* **1** Of small transverse dimensions; narrow. **2** *Archaic* Restricted as to space or room; close; tight. **3** Destitute, as of money; needy. **4** *Archaic* Strict; rigorous. **5** *Obs.* Difficult; hard-pressed. —*n.* **1** A narrow passage of water connecting two larger bodies of water. **2** Any narrow pass or passage. **3** A position of perplexity or distress; necessity: frequently plural. **4** *Obs.* An isthmus. ◆ Homophone: *straight.* [<OF *estreit* <L *strictus,* pp. of *stringere* bind tight. Doublet of STRICT.] —**strait′ly** *adv.* —**strait′·ness** *n.*

strait·en (strāt′n) *v.t.* **1** To make strait or narrow; contract; restrict. **2** To embarrass, as in finances; also, to distress; hamper. Also **strait.** See synonyms under SCRIMP.

strait·ened (strāt′nd) *adj.* **1** Contracted; narrowed. **2** Suffering privation or hardship, especially from pecuniary difficulties.

strait·jack·et (strāt′jak′it) *n.* A tight jacket of strong canvas, for confining the arms of dangerous lunatics or prisoners.

strait-laced (strāt′lāst′) *adj.* **1** Tightly laced, as stays; encased in tight corsets. **2** Strict, especially in morals or manners: a *strait-laced* Puritan.

Straits (strāts), **The** The Bosporus and the Dardanelles considered as a single passage from the Mediterranean to the Black Sea.

Straits Settlements A former British crown colony comprising Singapore, Penang, Malacca, and Labuan: dissolved, 1946.

strake (strāk) *n. Naut.* A breadth of planking or a line of plating on a vessel's hull from stem to stern: also spelled *streak.* [Appar. akin to STRETCH; infl. in meaning by STREAK]

Stral·sund (shträl′zōōnt) A port on the Baltic Sea in northern East Germany, in the former state of Mecklenburg.

stra·min·e·ous (strə·min′ē·əs) *adj.* **1** Straw-colored. **2** Strawlike; chaffy. [<L *stramineus* <*stramen* straw]

stra·mo·ni·um (strə·mō′nē·əm) *n.* **1** The jimsonweed. **2** A drug prepared from the dried leaves and flowering tops of this plant, used as a sedative, especially in asthma. Also **stram·o·ny** (stram′ə·nē). [<NL <Med. L *stramonia,* ? ult. <Tatar *turman,* a medicine for horses]

strand[1] (strand) *n.* A shore or beach; especially, that portion of an ocean shore between high and low tides. See synonyms under BANK[1]. —*v.t.* & *v.i.* **1** To drive or run aground. **2** To leave or be left in straits or difficulties: usually in the passive. [OE *strand*]

strand[2] (strand) *n.* **1** One of the principal twists or members of a rope. **2** A fiber, hair, or the like. **3** Wires twisted into a cable.

4 Anything plaited or twisted. —*v.t.* 1 To break a strand of (a rope). 2 To make by twisting strands. [? <OF *estran* <Gmc.]

strand line A line marking the boundary between the shore and the ocean, especially a line higher than the present one.

strang (strang) *adj. Scot.* Strong.

strange (strānj) *adj.* 1 Previously unknown, unseen, or unheard of; unfamiliar. 2 Not according to the ordinary way; unaccountable; remarkable. 3 Pertaining to another or others; of a different class, character, or kind. 4 Foreign; alien. 5 Distant in manner; reserved; shy. 6 Inexperienced; unskilled; unaccustomed. See synonyms under ALIEN, EXTRAORDINARY, ODD, QUEER, RARE[1]. —*adv.* Strangely. [<OF *estrange* <L *extraneus* foreign <*extra* on the outside. Doublet of EXTRANEOUS.] —**strange′ly** *adv.* —**strange′·ness** *n.*

stran·ger (strān′jər) *n.* 1 One who is not an acquaintance. 2 An unfamiliar visitor; guest. 3 A foreigner. 4 One unversed in or unacquainted or unfamiliar with something specified: with *to.* 5 *Law* Any person who is neither a party to a transaction nor privy to it. See synonyms under ALIEN. [<OF *estrangier* <*estrange.* See STRANGE.]

stran·gle (strang′gəl) *v.* **·gled, ·gling** *v.t.* 1 To choke to death; throttle; suffocate; stifle. 2 To repress; suppress. —*v.i.* 3 To suffer or die from strangulation. [<F *estrangler* <L *strangulare* <Gk. *strangalaein* <*strangalē* a halter <*strangos* twisted] —**stran′gler** *n.*

strangle hold 1 In wrestling, a hold which chokes one's opponent: usually forbidden. 2 Any influence or power that chokes freedom or progress.

stran·gles (strang′gəlz) *n. pl.* An infectious bacterial disease of the horse characterized by fever and inflammation of the respiratory mucous membrane.

stran·gu·late (strang′gyə·lāt) *v.t.* **·lat·ed, ·lat·ing** 1 To strangle. 2 *Pathol.* To compress, contract, or obstruct, especially so as to cut off circulation of the blood or flow of fluid. —*adj.* Strangulated. [<L *strangulatus,* pp. of *strangulare.* See STRANGLE.]

stran·gu·lat·ed (strang′gyə·lā′tid) *adj. Pathol.* Characterized by strangulation.

strangulated hernia *Pathol.* A form of hernia in which the protruded organ or part is so tightly constricted as to cut off normal circulation of the blood, with possible necrosis and mortification.

stran·gu·la·tion (strang′gyə·lā′shən) *n.* 1 The act of strangling or the state of being strangled. 2 *Pathol.* The state of being strangulated; constriction of a part, as of the intestine in strangulated hernia, to cut off circulation.

stran·gu·ry (strang′gyə·rē) *n. Pathol.* Difficult and painful urination. [<L *stranguria* <Gk. *strangouria* <*stranx, strangos* a drop + *ouron* urine]

strap (strap) *n.* 1 A long, narrow, and flexible strip of leather or the like, usually having a buckle or other fastener, for binding about objects. 2 A razor strop. 3 A shoulder strap. 4 Something made of, resembling, or used as a strap. 5 A thin metal band or plate. —*v.t.* **strapped, strap·ping** 1 To fasten or bind with a strap. 2 To beat with a strap. 3 To sharpen or strop. 4 *Scot.* To hang. 5 To embarrass financially. [Var. of STROP] —**strap′less** *adj.*

strap hinge A hinge having long leaves, designed for attaching to the flat surfaces of a door and jamb. See illustration under HINGE.

strap·pa·do (strə·pā′dō, -pä′dō) *n. pl.* **·does** 1 A former punishment in which one was drawn up by a rope attached usually to the wrists, and let fall to the length of the rope; also, the machine used. 2 Erroneously, a beating with a strap. [<Ital. *strappata* a pulling, orig. fem. pp. of *strappare* pull]

strap·pan (strap′ən) *adj. Scot.* Tall and handsome; strapping.

strap·per (strap′ər) *n.* 1 One who uses a strap or straps. 2 One who bolts the straps to rails. 3 *Colloq.* A strong, tall person. 4 One who grooms horses.

strap·ping (strap′ing) *adj. Colloq.* Large and muscular; robust.

Stras·bourg (stras′bûrg, sträz′-; *Fr.* stråz·bōōr′) A city in NE France, the chief city of Alsace. *German* **Strass·burg** (shträs′bōōrkh).

strass (stras) *n.* A lead glass of great brilliance used in the manufacture of gems; paste. [after Josef *Strasser,* 18th century German jeweler]

strasse (stras) *n.* Refuse of silk left in making skeins. [<F *strasse* <Ital. *straccio* rag, something torn <*stracciare* tear, lacerate]

stra·ta (strā′tə, strat′ə) Plural of STRATUM.

strat·a·gem (strat′ə·jəm) *n.* 1 A maneuver designed to deceive or outwit an enemy in war. 2 A deception; any device for obtaining advantage. See synonyms under ARTIFICE. [<F *stratagème* <L *strategema* <Gk. *stratēgēma* piece of generalship <*stratēgos* a general <*stratos* army + *agein* lead]

stra·tal (strāt′l) *adj.* Pertaining to, derived from, characteristic of, or caused by a stratum or strata.

stra·te·gic (strə·tē′jik) *adj.* Of or pertaining to strategy; characterized by, used in, or having relation to strategy. Also **stra·te′gi·cal, strat·e·get·ic** (strat′ə·jet′ik) or **·i·cal.** —**stra·te′gi·cal·ly, strat′e·get′i·cal·ly** *adv.*

strategic material Any of several, chiefly raw, materials essential to national defense and industry, especially those that are wholly lacking or in insufficient supply within a nation's boundaries and have to be obtained from sources outside the country: the stockpiling of *strategic materials.*

stra·te·gics (strə·tē′jiks) *n. pl.* (*construed as singular*) The art or science of strategy; generalship.

strat·e·gist (strat′ə·jist) *n.* One versed in strategy, or skilled in managing affairs.

strat·e·gy (strat′ə·jē) *n. pl.* **·gies** 1 The science and art of conducting a military campaign by the combination and employment of means on a broad scale for gaining advantage in war; generalship: distinguished from *tactics.* 2 The use of stratagem or artifice, as in business, politics, etc. 3 Skill in management. [<F *stratégie* <Gk. *stratēgia* <*stratēgos* general. See STRATAGEM.]

Strat·ford-on-A·von (strat′fərd·on·ā′von) A town on the Avon river in SW Warwickshire, England; birthplace and place of burial of Shakespeare.

strath (strath) *n. Scot.* A wide, open valley; a river course.

Strath·clyde and Cum·bri·a (strath′klīd; kum′·brē·ə) An early medieval British kingdom comprising territory now in southern Scotland and northern England.

Strath·co·na and Mount Royal (strath·kō′·nə), **Lord,** 1820–1914, Donald Alexander Smith, Canadian railroad builder and administrator born in Scotland.

Strath·more (strath′môr) A plain extending 100 miles across Scotland, south of the Grampians, between Dumbarton and the North Sea coast of Kincardine.

strati- *combining form* A stratum; of or pertaining to a stratum or to strata: *stratiform.* Also, before vowels, **strat-.** [<L *stratum* a covering]

stra·tic·u·late (strə·tik′yə·lit, -lāt) *adj. Geol.* Arranged in thin layers or strata: said of sedimentary rocks and certain minerals, as the agate. [<NL *straticulum,* dim. of L *stratum* a layer + -ATE[1]] —**stra·tic′u·la′tion** *n.*

strat·i·form (strat′ə·fôrm) *adj.* 1 *Geol.* Having the form of or constituting a stratum. 2 *Anat.* Denoting a fibrous cartilage enclosed in a channel in a bone as a support for tendons. 3 *Meteorol.* Resembling a stratus. [<STRATI- + -FORM]

strat·i·fy (strat′ə·fī) *v.* **·fied, ·fy·ing** *v.t.* 1 To form or arrange in strata. 2 To preserve (seeds) by spreading in alternating layers of earth and sand. —*v.i.* 3 To form in strata. 4 To be formed in strata. [<F *stratifier* <Med. L *stratificare* <L *stratum* layer + *facere* make] —**strat′i·fi·ca′tion** *n.*

stra·tig·ra·phy (strə·tig′rə·fē) *n.* 1 The order and relative position of the strata of the earth's crust. 2 The study or description of such strata; stratigraphic geology. [<STRATI- + -GRAPHY] —**strat·i·graph·ic** (strat′·ə·graf′ik) or **·i·cal** *adj.* —**strat′i·graph′i·cal·ly** *adv.*

stra·toc·ra·cy (strə·tok′rə·sē) *n. pl.* **·cies** Government by the military. [<Gk. *stratos* army + -CRACY] —**strat·o·crat·ic** (strat′ə·krat′ik) *adj.*

Strat·o·cruis·er (strat′ō·krōō′zər) *n.* A large airliner designed to transport passengers or freight at stratospheric altitudes: a trade name.

stra·to·cu·mu·lus (strā′tō·kyōō′myə·ləs) *n. pl.* **·li** (-lī) *Meteorol.* Large rolls or globular masses of cloud, gray to dark in color, disposed in waves, groups, or bands, and often covering the whole sky: also called *cumulo-stratus.* See table under CLOUD. [<*strato-* (<STRATUS) + CUMULUS]

Strat·o·li·ner (strat′ō·lī′nər) *n.* A passenger-carrying airplane designed to operate in altitudes of about six miles: a trade name.

strat·o·pause (strat′ə·pôz) *n. Meteorol.* The zone of transition between the stratosphere and the ionosphere.

strat·o·sphere (strat′ə·sfir, strā′tə-) *n. Meteorol.* The portion of the atmosphere lying above the troposphere and beginning at a height of about six miles. In it the systematic fall of temperature with increasing altitude, characteristic of the region below it, ceases, often giving place to a more or less uniform temperature. —**strat′o·spher′ic** (-sfer′ik) or **·i·cal** *adj.*

stra·tum (strā′təm, strat′əm) *n. pl.* **·ta** (-tə) or **·tums** 1 A natural or artificial layer, bed, or thickness. 2 *Geol.* A more or less homogeneous layer of rock, often in two or more beds, and serving to identify a geological group, system, or series. 3 *Biol.* A sheet or layer of tissue. 4 Something corresponding to a stratum of the earth: a low *stratum* of society. [<L, orig. neut. of *stratus,* pp. of *sternere* spread]

stra·tus (strā′təs, strat′əs) *n. pl.* **·ti** (-tī) *Meteorol.* A cloud of foglike appearance, low-lying and arranged in a uniform layer. See table under CLOUD. [<L, orig. pp. of *sternere* spread]

straught (strôkht) *v.t. Scot.* To straighten; stretch. —*adj.* Straight. Also **straucht.**

Straus (strous, *Ger.* shtrous), **Oscar,** 1870–1954, Austrian composer.

Strauss (strous, *Ger.* shtrous), **David Friedrich,** 1808–74, German rationalistic theologian.—**Johann,** 1804–49, Austrian composer of dance music. —**Johann,** 1825–99, Austrian composer; son of the preceding. —**Richard,** 1864–1949, German composer.

stra·vage (strə·vāg′) *v.i.* & *n. Irish* & *Scot.* Stroll; ramble. Also **stra·vaig′, stra·vague′.** —**stra·vaig′er** *n.*

Stra·vin·sky (strə·vin′skē, *Russian* strä·vēn′·skē), **Igor Fëdorovich,** born 1882, Russian composer active in Europe and the United States.

straw (strô) *n.* 1 A dry or ripened stalk. 2 Stems or stalks of grain, collectively, after the grain has been thrashed out. 3 A mere trifle or slight indication. 4 A slender tube, originally a wheat straw, now made of paper, glass, etc., used to suck up a beverage. —**the last straw** The final test of patience or endurance; the culminating element in any state of circumstances. —*adj.* 1 Like or of straw; of straw color. 2 Of no value; worthless; sham. 3 Made of straw. [OE *strēaw* straw] —**straw′y** *adj.*

straw·ber·ry (strô′ber′ē, -bər-ē) *n. pl.* **·ries** 1 The edible fruit of any plant of the genus *Fragaria,* technically neither a fruit nor a berry, but an enlarged fleshy achene receptacle. 2 The plant that bears this fruit, a stemless perennial of the rose family, with radical trifoliolate leaves, usually white flowers on scapes, and slender runners by which it propagates: also **strawberry vine.** [OE *strēaw* straw + BERRY]

strawberry bass The calico bass.

strawberry blond A person having reddish-blond hair; a red-headed person.

strawberry bush 1 An upright or straggling shrub (*Euonymus americanus*) of the United States and Canada, with rough, warty, depressed crimson pods and scarlet aril. 2 The wahoo or burningbush.

strawberry festival A sociable gathering, church bazaar, etc., at which strawberries are served.

strawberry shrub A shrub (genus *Calycanthus*), named for the strawberrylike fragrance of its purple or dark-red flowers.
strawberry tomato The ground cherry.
strawberry tree A small evergreen tree (*Arbutus unedo*) of southern Europe, having racemose white flowers and edible fruit resembling strawberries.
straw·board (strô′bôrd′, -bōrd′) *n.* Coarse board, made of straw, used for paper boxes and book covers.
straw boss *U.S. Colloq.* In construction work, logging, etc., an under-foreman.
straw color A pale-yellow color, as of clean ripe straw. Also *Brit.* **straw colour.** — **straw′-col′ored** (-kul′ərd) *adj.*
straw man 1 A figure of a man made of straw. 2 A puppet. 3 A fraudulent surety or a perjured witness.
straw ride A nighttime pleasure ride taken by a group in a hayrack or large wagon half full of straw or hay; a hay ride.
straw vote A vote taken at a chance gathering to test the strength of opposing candidates; an unofficial test vote.
straw wine A sweet wine made from grapes dried or partly dried in the sun on straw.
straw worm The larva of a caddis fly.
stray (strā) *v.i.* 1 To wander from the proper course, an area, group, etc.; straggle; roam. 2 To wander about; rove. 3 To deviate from right or goodness; go astray. See synonyms under RAMBLE, WANDER. —*adj.* 1 Having strayed; straying. 2 Irregular; occasional; casual; unrelated. —*n.* 1 A domestic animal that has strayed; an estray. 2 A person who is lost or wanders aimlessly. 3 The act of straying or wandering. 4 *pl. Electronics* Electromagnetic waves, affecting a radio receiver, produced by atmospheric electric discharges and electrical storms. [<OF *estraier* wander about, ult. <L *extra vagare* wander outside] — **stray′er** *n.*
streak (strēk) *n.* 1 A long, narrow, somewhat irregularly shaped mark, line, or stripe: a *streak* of lightning. 2 A not very marked characteristic; a vein; trace; dash: a *streak* of meanness; also, a transient mood; whim. 3 *Mineral.* The color of the line of powder left when a mineral is rubbed on an unglazed porcelain plate known as a **streak plate.** 4 A strake. 5 A layer or strip: meat with a *streak* of fat and a *streak* of lean. 6 *Bacteriol.* The application of an inoculum in a thin stripe or line, as across the surface of a culture. —*v.t.* 1 To mark with a streak; form streaks in or on; stripe. —*v.i.* 2 To form a streak or streaks. 3 To move, run, or travel at great speed. [OE *strica.* Akin to STRIKE.] — **streaked** (strēkt) *adj.*
streak·y (strē′kē) *adj.* 1 Marked with or occurring in streaks; streaked. 2 Of variable quality or character; not uniform. — **streak′i·ly** *adv.* — **streak′i·ness** *n.*
stream (strēm) *n.* 1 A current or flow of water or other fluid. 2 Anything continuously flowing, moving, or passing, as people. 3 A continuous course or advance; drift; current. 4 Anything issuing out or flowing from a source; a ray. — **on stream** In full commercial production, as an oil refinery, chemical plant, etc. —*v.i.* 1 To pour forth or issue in a stream. 2 To pour forth a stream: eyes *streaming* with tears. 3 To move in continuous succession; proceed uninterruptedly, as a crowd from a hall. 4 To float with a waving movement, as a flag. 5 To move with a trail of light, as a meteor. 6 In mining or dyeing, to wash in running water. [OE *strēam*] — **stream′y** *adj.*
Synonyms (noun): brook, channel, course, creek, current, drift, eddy, flow, flume, flux, race, rill, river, rivulet, run, runlet, runnel, streamlet, tide, watercourse.
stream·er (strē′mər) *n.* 1 An object that streams forth, or hangs extended. 2 A flag, pennant, or ensign; a long, narrow flag or standard. 3 A stream or shaft of light, such as shoots up from the horizon into or across the sky in certain forms of the aurora borealis. 4 A newspaper headline that runs across the whole page.
stream·let (strēm′lit) *n.* A rivulet.
stream·line (strēm′līn′) *n.* 1 The course of a fluid relative to a solid body past which it is moving, especially a course free of turbulence or eddies. 2 Any shape or contour designed to lessen air resistance. —*adj.* 1 Designating an uninterrupted flow or drift. 2 Denoting a form, body, or the like so constructed as to permit an uninterrupted flow of fluid around it: a *streamline* flow, a *streamline* shape, a *streamline* body for a motor car. —*v.t.* **·lined, ·lin·ing** To design with a streamline shape.
stream·lin·er (strēm′lī′nər) *n.* A fast, streamlined train.
stream of consciousness *Psychol.* The uninterrupted series of individual conscious states moving continuously as though in a stream. Also **stream of thought.**
stream-of-con·scious·ness technique (strēm′·əv·kon′shəs·nis) A method of writing fiction in which an author objectifies the inward thoughts, feelings, and sometimes sensations of the characters to supplement or replace dialog and narrated action.
streek (strēk) *Scot. v.t.* 1 To stretch or extend; hence, to lay out, as a corpse. 2 To stretch forth; stretch. —*n.* Extent; progress.
street (strēt) *n.* 1 A public way, with buildings on one or both sides, in a city, town, or village. 2 The highway on which buildings front; also, the roadway for vehicles, between sidewalks. 3 *Colloq.* The people living, habitually gathering, or doing business in a street. See synonyms under ROAD, WAY. [OE *strǣt* <LL *strata (via)* paved (road)]
Street may appear as a combining form in hyphemes or as the first element in two-word phrases, with the following meanings:
1 Of or pertaining to a street or streets:

street-cleaner	**street-sprinkler**
street-cleaning	**street-sprinkling**
street directory	**street-sweeper**
street layer	**street-sweeping**
street name	**street-widening**

2 In the streets:

street beggar	**street music**	**street-pacing**
street-bred	**street musician**	**street peddler**
street fight	**street noise**	**street singer**

3 On or abutting a street:

street corner	**street entrance**	**street gate**
street door	**street floor**	**street lamp**

street Arab A homeless or outcast child who lives in the streets; a gamin.
street car A passenger car that runs on rails laid on the surface of the streets.
street-walk·er (strēt′wô′kər) *n.* A prostitute who solicits in the streets. — **street′-walk′ing** *n.* & *adj.*
strength (strength) *n.* 1 The quality or property of being strong; power; muscular force; physical vitality. 2 The capacity of material bodies to sustain the application of force without yielding or breaking; solidity; tenacity; toughness. 3 Power in general; operative energy; ability to do or bear. 4 Binding force or validity, as of a law. 5 Vigor or force of style. 6 Available numerical force in a military unit or other organization. 7 Degree of intensity; vehemence: *strength* of passion. 8 The degree in which a thing possesses its distinctive properties or essential elements; concentration. 9 Potency, as of a drug, chemical, or liquor. 10 Rising prices; firmness of prices. 11 One regarded as an embodiment of sustaining or protecting power; in archaic or poetic use, a fortress. See synonyms under POWER, PROWESS. [OE *strengthu* <*strang* strong]
strength·en (streng′thən) *v.t.* 1 To make strong. 2 To encourage; hearten; animate. —*v.i.* 3 To become or grow strong or stronger. See synonyms under CONFIRM. — **strength′·en·er** *n.*
stren·u·ous (stren′yōō·əs) *adj.* 1 Eagerly pressing or urgent; earnest. 2 Necessitating or marked by strong effort or exertion. [<L *strenuus.* Akin to Gk. *strēnēs* strong.] — **stren′u·ous·ly** *adv.* — **stren′u·os′i·ty** (-os′ə·tē), **stren′u·ous·ness** *n.*
streph·o·sym·bo·li·a (stref′ō·sim·bō′lē·ə) *n.* 1 *Pathol.* A defect of vision in which objects are seen in reverse, as in a mirror. 2 *Psychol.* A condition marked by an inability to differentiate between certain oppositely oriented letters, as *b, d; p, q,* resulting in difficulty in learning to read. [<NL <Gk. *strephein* twist + *symbolon* sign, symbol]
strep·i·tous (strep′i·təs) *adj.* Noisy; boisterous. Also **strep′i·tant.** [<L *strepitus* din, noise]
strep·to·coc·cus (strep′tə·kok′əs) *n. pl.* **·coc·ci** (-kok′sī) Any of a genus (*Streptococcus*) of Gram-positive, typically non-motile ovoid or spherical bacteria, grouped in long chains, and dividing in one plane, including highly pathogenic species causing many diseases, as pneumonia, erysipelas, etc. See illustration under BACTERIUM. [<NL <Gk. *streptos* twisted + COCCUS] — **strep′to·coc′cal** (-kok′əl), **strep′to·coc′cic** (-kok′sik) *adj.*
strep·to·my·cin (strep′tō·mī′sin) *n.* A potent antibiotic isolated from a moldlike organism (*Streptomyces griseus*), effective against certain pathogenic bacteria. [<Gk. *streptos* twisted + *mykēs* fungus]
strep·to·thri·cin (strep′tō·thrī′sin, -thris′in) *n.* A bactericidal substance isolated from a soil fungus (*Streptomyces lavendulae*): used therapeutically in certain intestinal infections. [<NL *Streptothrix,* former genus name <Gk. *streptos* twisted + *thrix* hair + -IN]
Stre·se·mann (shtrā′zə·män), **Gustav,** 1878–1929, German statesman.
stress (stres) *n.* 1 Special weight, importance, or significance. 2 *Physics* Force exerted between contiguous portions of a body or bodies and generally expressed in pounds per square inch; strain; tension. 3 *Mech.* A force or system of forces which tends to produce deformation in a body on which it acts. 4 Influence exerted forcibly; pressure; compulsion. 5 In pronunciation and oral reading, the relative force with which a sound, syllable, or word is uttered. See also METRICAL STRESS, RHETORICAL STRESS. —*v.t.* 1 To subject to mechanical stress, as a timber. 2 To put stress or emphasis on; accent, as a syllable. 3 To put into straits or difficulties; distress. [<OF *estrece* <*estrecier* constrain <L *strictus,* pp. of *stringere* draw tight] — **stress′ful** *adj.* — **stress′less** *adj.*
-stress *suffix of nouns* Feminine form of -STER: *songstress.* [<-STER + -ESS]
stretch (strech) *v.t.* 1 To extend or draw out, as to full length or width. 2 To extend or draw out forcibly, especially beyond normal or proper limits: The weight has *stretched* the cable; to *stretch* the truth. 3 To cause to reach, as from one place to another or over an area; extend: They *stretched* telegraph wires across the continent. 4 To put forth, hold out, or extend (the hand, an object, etc.): often with *out*: to *stretch* out the hands in appeal. 5 To draw tight; tighten. 6 To strain or exert to the utmost: to *stretch* every nerve. 7 *Slang* To fell with a blow. —*v.i.* 8 To reach or extend over an area or from one place to another: The road *stretches* on and on. 9 To become extended, especially beyond normal or proper limits. 10 To extend one's body or limbs, as in relaxing or reaching for something. 11 To lie down and extend one's limbs to full length: usually with *out.* —*n.* 1 An act of stretching, or the state of being stretched; tension. 2 Extent or reach of that which stretches; scope; especially, an overstrain. 3 A continuous extent of space or time. 4 In racing, that part of the track which, being straight, admits of the greatest speed being made; the straight-away. 5 Direction. 6 *Slang* A term of imprisonment. [OE *streccan* stretch] — **stretch′a·ble** *adj.* — **stretch′i·ness** *n.* — **stretch′y** *adj.*
Synonyms (verb): elongate, exaggerate, expand, extend, lengthen, reach, spread, strain, tighten. See PERVERT. *Antonyms*: loosen, relax, slacken.
stretch·er (strech′ər) *n.* 1 One who or that which stretches; any device for stretching, as a device for loosening the fit of gloves, shoes, etc., a frame for drying curtains, sweaters, etc., in shape. 2 A frame, as of stretched canvas, for carrying the wounded or dead; a litter. 3 In masonry, a brick or stone lying lengthwise of a course. 4 A tie beam in the frame of a building.
stretch·er-bear·er (strech′ər·bâr′ər) *n.* One who carries one end of a stretcher or litter. Also **stretch′er·man′** (-man′).
stretch-out (strech′out′) *n.* 1 A system of industrial operation in which employees are required to perform more work per unit of time worked, as by tending additional machines, usually without proportionate increase in pay. 2 A slow-down practiced by employees so as to make the work last longer: see CA' CANNY.
stret·to (stret′tō) *n. pl.* **·ti** (-tē) or **·tos** *Music*

1 A portion of a fugue, near the close, in which the answer crowds closely on the subject. 2 In an oratorio or operatic piece, the portion at the close accelerated in time to produce a climax: also **stret′ta** (-tä). [<Ital., lit., drawn tight <L *strictus*. See STRESS.]

strew (strōō) *v.t.* **strewed, strewed** or **strewn, strew·ing** 1 To spread about loosely or at random; scatter; sprinkle. 2 To cover with something scattered or sprinkled. 3 To be scattered over (a surface). [OE *strēawian*]

stri·a (strī′ə) *n. pl.* **stri·ae** (strī′ē) 1 A narrow streak, stripe, or band of distinctive color, structure, or texture, often parallel with others. 2 *Geol.* A small groove, channel, or ridge on a rock surface, due to the action of glacier ice. [<L, a groove]

stri·ate (strī′āt) *adj.* 1 Having fine linear markings; grooved. 2 Constituting a stria or striae. Also **stri′at·ed.** —*v.t.* **·at·ed, ·at·ing** To mark with striae. [<L *striatus*, pp. of *striare* groove <*stria* a groove]

stri·a·tion (strī·ā′shən) *n.* 1 The act of striating, or the state of being striated. 2 A striate form or appearance. 3 One of a series of parallel striae, as in a muscle or mineral.

stri·a·ture (strī′ə·chər) *n.* 1 The manner in which striae are disposed or arranged; striation. 2 A stria.

strick (strik) *n. Brit. Dial.* 1 A bunch of fibers, as flax, hackled or ready for hackling. 2 A bundle of silk fibers prepared for the second combing. [Prob. <STRICKEN (def. 4)]

strick·en (strik′ən) *adj.* 1 Wounded, especially by a missile: a *stricken* hare. 2 Struck down; afflicted, as by calamity or disease: *stricken* with polio. Compare STRIKE *v.* 3 Advanced or far gone, as in age: *stricken* in years. 4 Having the contents leveled off even with the top of a container. [OE *stricen*, pp. of *strican* strike]

strick·le (strik′əl) *n.* 1 A straightedge used for striking off an even measure of grain. 2 A template or curved piece of wood used in smoothing a sand or loam mold to form a core. 3 A straightedge, to which emery is applied, for sharpening rotary knives. —*v.t.* **·led, ·ling** To shape or smooth with a strickle. [OE *stricel*]

strict (strikt) *adj.* 1 Observing or enforcing rules exactly; also, containing exact or severe rules or provisions; exacting. 2 Strenuously enjoined and maintained; rigidly observed. 3 Exactly defined, distinguished, or applied; not indefinite or loose. 4 Stretched tight; not lax; tense. 5 Close, narrow, and upright; straight: said of the panicles of certain plants. See synonyms under AUSTERE, PRECISE. [<L *strictus*, pp. of *stringere* draw tight. Doublet of STRAIT.] —**strict′ly** *adv.* —**strict′ness** *n.*

stric·tion (strik′shən) *n.* Constriction. [<L *strictio, -onis* <*strictus*. See STRICT.]

stric·ture (strik′chər) *n.* 1 Severe criticism. 2 *Pathol.* A morbid contraction of some duct or channel of the body. 3 *Obs.* Strictness. [<L *strictura* <*strictus* strict]

strid·den (strid′n) Past participle of STRIDE.

stri·dle (strid′l) *v.t.* & *v.i. Brit. Dial.* To straddle. Also **strid′dul.** [Freq. of STRIDE; infl. in meaning by STRADDLE]

stride (strīd) *n.* 1 A long and sweeping or measured step; also, the space passed over by such a step. 2 In animal locomotion, an act of progressive motion, completed when all the feet are returned to the same relative positions they occupied at the beginning of the movement. 3 A stage of progress. —**to hit one's stride** To attain one's normal speed. —**to make rapid strides** To make quick progress. —**to take (something) in one's stride** To do (something) without undue effort as part of one's normal activity. —*v.* **strode, strid·den, strid·ing** *v.i.* 1 To walk with long steps, as from haste or pride. 2 *Archaic* To straddle. —*v.t.* 3 To walk through, along, etc., with long steps. 4 To pass over with a single stride. 5 To straddle; bestride. [OE *strīdan* stride] —**strid′er** *n.*

stri·dent (strīd′nt) *adj.* Giving a loud and harsh sound; shrill; grating. [<L *stridens, -entis*, ppr. of *stridere* creak] —**stri′dence, stri′den·cy** *n.* —**stri′dent·ly** *adv.*

stri·dor (strī′dər) *n.* 1 A harsh, shrill, creaking, screechy, or grating noise. 2 *Pathol.* A harsh grating noise, particularly one heard in laryngeal obstruction. [<L]

strid·u·late (strij′ōō·lāt) *v.i.* **·lat·ed, ·lat·ing** To make a shrill, creaking noise, as a locust, cicada, or the like. [<NL *stridulatus*, pp. of *stridulare* <*stridulus* rattling <*stridere* rattle, rasp] —**strid′u·la′tion** *n.* —**strid′u·la·to′ry** (-lə·tôr′ē, -tō′rē), **strid′u·lous** *adj.* —**strid′u·lous·ly** *adv.* —**strid′u·lous·ness** *n.*

strife (strīf) *n.* 1 Angry contention; fighting. 2 Any contest for advantage or superiority; rivalry. 3 The act of striving; strenuous endeavor. See synonyms under BATTLE, FEUD[1], QUARREL[1]. [<OF *estrif* <*estriver*. See STRIVE.]

strig·il (strij′əl) *n.* 1 In ancient Greece and Rome, a scraper, as of metal, bone, or ivory, used for scraping the skin, as at the bath. 2 *Archit.* One of a group of wavy flutings carved on flat or curved surfaces, as in Roman architecture. [<L *strigilis* scraper]

strig·i·la·tion (strij′ə·lā′shən) *n.* 1 The application of a strigil to the skin. 2 The friction thus caused.

strig·il·lose (strij′ə·lōs) *adj.* Diminutively or minutely strigose. [<NL *strigilla*, dim. of *striga* a furrow]

stri·gose (strī′gōs, stri·gōs′) *adj.* 1 *Bot.* Rough with short, sharp, appressed stiff hairs or bristles, as a leaf; hispid. 2 *Zool.* Marked with stripes or striae. [<NL *strigosus* <L *striga* a furrow]

strike (strīk) *v.* **struck, struck** (*chiefly Archaic* **strick·en**), **strik·ing** *v.t.* 1 To come into violent contact with; hit; crash into: The car *struck* the wall. 2 To hit with a blow; deal a blow to; smite: It *struck* him in the face. 3 To deal (a blow, etc.). 4 To cause to hit forcibly: He *struck* his hand on the table. 5 To attack; assault: We *struck* the enemy on his left flank. 6 To remove, separate, or take off by or as by a blow: with *off, from*, etc.: *Strike* it from the record. 7 **a** To ignite (a match, etc.). **b** To produce (a light, etc.) thus. 8 To form by stamping, printing, etc.; impress; coin. 9 To announce; sound: The clock *struck* two. 10 To fall upon; reach; catch: A sound of crying *struck* his ear. 11 To arrive at; come upon: to *strike* a trail. 12 To discover; find: to *strike* oil. 13 To affect suddenly or in a specified manner: He was *struck* speechless. 14 To come to the mind of; occur to: An idea *strikes* me. 15 To impress in a specified manner; seem to: He *strikes* me as an honest man. 16 To attract the attention of; impress: The dress *struck* her fancy. 17 To assume; take up: to *strike* an attitude. 18 To cause to enter or penetrate deeply or suddenly: to *strike* dismay into one's heart. 19 To lower or haul down; take or let down, as a sail, or a flag in token of surrender. 20 To cease working at in order to compel compliance to a demand, etc. 21 In the theater, to dismantle (a set or scene). 22 To make level (a measure of grain, etc.); strickle. 23 To make and confirm, as a bargain. 24 To harpoon (a whale). 25 To hook (a fish that has taken the lure) by a sharp pull on the line. 26 To arrive at by reckoning: to *strike* a balance. —*v.i.* 27 To come into violent contact; crash; hit. 28 To deal or aim a blow or blows. 29 To make an assault or attack. 30 To sound from a blow or blows. 31 To be indicated by the sound of blows or strokes: Noon has just *struck*. 32 To ignite. 33 To run aground, as on a reef or shoal: The ship *struck* and heeled over. 34 To lower a flag in token of surrender or in salute. 35 To come suddenly or unexpectedly; chance: with *on* or *upon*: to *strike* upon an unknown path. 36 To take a course; start and proceed: to *strike* for home. 37 To move quickly; dart. 38 To cease work in order to enforce demands, etc. 39 To snatch at or swallow the lure: said of fish. —**to strike camp** To take down the tents of a camp. —**to strike down** 1 To fell with a blow. 2 To affect disastrously; incapacitate completely. —**to strike dumb** To astonish; amaze. —**to strike hands** To clasp hands, especially in confirming a bargain. —**to strike home** 1 To deal an effective blow. 2 To have telling effect. —**to strike it rich** 1 To find a valuable vein or pocket of ore. 2 To come into wealth or good fortune. —**to strike off** 1 To remove or take off by or as by a blow or stroke. 2 To cross out or erase by or as by a stroke of the pen. 3 To deduct. —**to strike out** 1 To strike off (def. 2). 2 To aim a blow or blows. 3 To originate; devise; contrive. 4 To begin; start. 5 In baseball: **a** To put out (the batter) by pitching three strikes. **b** To be put out because of taking three strikes. —**to strike up** 1 To begin to play, sing, or sound, as a band or musical instrument. 2 To start up; begin, as a friendship. —*n.* 1 An act of striking or hitting; a blow. 2 In baseball, an unsuccessful attempt by the batter to hit the ball; a pitched ball that passes over home plate above the level of the batter's knees and below that of his shoulders; a foul bunt; any foul tip held by the catcher; any ball hit foul except when there have been two strikes. 3 In bowling, the knocking down by a player of all the pins with the first bowl in any frame. 4 The quitting of work by a body of workers to enforce some demand. 5 A new or unexpected discovery, as of oil or ore. 6 Any unexpected or complete success. 7 A straight-edged implement for leveling something, as grain in a measure; strickle. 8 *Geol.* The direction, referred to the meridian, of a horizontal line in a given structural plane, or of the intersection of the structural plane with a horizontal surface. 9 In coining, the quantity of coin or the number of medals made or struck at one time. 10 Full measure; hence, excellence. 11 The act of attempting to obtain money or some valuable thing, as by simple request, or by the introduction of a bill in a legislative body for the purpose of being bought off. 12 The sudden rise and taking of the bait by a fish; a bite. —**general strike** Concerted cessation of work on the part of the employees of all or nearly all industries, including public utilities, in a certain town, region, or nation. —**sit-down strike** See SIT-DOWN. [OE *strīcan* stroke, move. Akin to STREAK.]

strike·break·er (strīk′brā′kər) *n.* 1 One who takes the place of a workman on strike. 2 A person who supplies workmen to take the place of strikers. —**strike′break′ing** *n.*

strike fault *Geol.* A fault lying parallel with the strike of the rocks through which it cuts.

strike figure A percussion figure.

strike-out (strīk′out′) *n.* An instance of striking out, especially in baseball.

strik·er (strī′kər) *n.* 1 One who or that which strikes. 2 In certain torpedoes, a plunger which strikes the priming cap and ignites the charge. 3 An employee who is on strike. 4 One whose business is to strike something in a mechanical occupation, as in forging. 5 *U.S. Colloq.* One who makes a blackmailing strike in politics. 6 In the U. S. Navy, an apprentice in training for a specific technical rating: a radioman *striker.* 7 Formerly, the engineer's apprentice on a river steamboat. 8 In the U. S. Army, a soldier assigned to run errands and do odd jobs for an officer.

strik·ing (strī′king) *adj.* Notable; impressive. See synonyms under EXTRAORDINARY. —**strik′ing·ly** *adv.* —**strik′ing·ness** *n.*

Strind·berg (strind′bûrg, *Sw.* strēn′ber·y′), **John August,** 1849–1912, Swedish dramatist and novelist.

string (string) *n.* 1 A slender line, thinner than a cord and thicker than a thread, used for tying or lacing; twine; also, a slender strip, as of cloth or leather; the cord of a bow; prepared wire or catgut for musical instruments. 2 A stringlike organ or formation; a fibrous vegetable formation; an animal nerve or tendon. 3 A thin cord upon which anything is strung; a row or series of things connected by a small cord: a *string* of pearls. 4 A connected series or succession as of things, acts, or events: sometimes implying unusual length: a *string* of carriages; a *string* of lies. 5 *U.S. Colloq.* A drove or small collection of stock, especially of saddle horses. 6 *pl.* Stringed instruments, especially those of an orchestra; those who play on these. 7 In billiards, the score; the buttons, strung on a wire, by which the score is kept; the string line; the act of stringing. 8 *Archit.* **a** A string-course, as of bricks. **b** A stout inclined plank, notched and set edgewise as a support for the steps of a wooden stairway; a ramp or sidepiece of

add,āce,câre,pälm; end,ēven; it,īce; odd,ōpen,ôrder; tŏŏk,pōōl; up,bûrn; ə = a in *above*, e in *sicken*, i in *clarity*, o in *melon*, u in *focus*; yōō = u in *fuse*; oi,oil; ou,pout; ch,check; g,go; ng,ring; th,thin; th,this; zh,vision. Foreign sounds à,œ,ü,kh,ṅ; and ◆: see page xx. < from; + plus; ? possibly.

solid-built stairs. 9 In sports, a group of contestants ranked as to skill. 10 The conditions, limitations, or restrictions attached to any proposition, gift, or donation, whereby the terms may not be binding, or whereby the donor retains some control. —**to pull strings** To manipulate or influence others, secretly or underhandedly, to gain some advantage. —*v.* **strung, string·ing** *v.t.* 1 To thread, as beads, on or as on a string. 2 To fit with a string or strings, as a guitar. 3 To bind, fasten, or adorn with a string or strings. 4 To tighten the strings of (a musical instrument). 5 To brace; strengthen. 6 To make tense or nervous. 7 To arrange or extend like a string. 8 To remove the strings from (vegetables). 9 *Colloq.* To hang: usually with *up.* 10 *Slang* To fool or deceive; hoax: often with *along.* —*v.i.* 11 To extend, stretch, or proceed in a line or series. 12 To form into strings. 13 In billiards, to drive the cue ball from within the string against the farther cushion and back. —**to string along (with)** To follow with trust or confidence. [OE *streng* string]

string bass The double bass.

string bean 1 Any of several varieties of beans (genus *Phaseolus*) cultivated for their edible pods, especially *P. vulgaris.* 2 The pod itself: usually in the plural.

string·board (string′bôrd′, -bōrd′) *n. Archit.* A board serving as a stringpiece in which the ends of steps of a staircase are set.

string–course (string′kôrs′, -kōrs′) *n. Archit.* A horizontal molding or ornamental course, usually projecting along the face of a building.

stringed (stringd) *adj.* 1 Furnished with strings. 2 Produced from stringed instruments. 3 Tied with string.

strin·gen·do (strin·jen′dō) *adj. Music* Hastening the tempo as toward a climax; accelerando. [<Ital., ppr. of *stringere* draw tight <L]

strin·gent (strin′jənt) *adj.* 1 Keeping one closely to strict requirements; rigid; severe, as regulations. 2 Hampered by obstructing conditions or scarcity of money; close or tight: The money market is very *stringent.* 3 Convincing; forcible. [<L *stringens, -entis,* ppr. of *stringere* draw tight] —**strin′gen·cy, strin′gent·ness** *n.* —**strin′gent·ly** *adv.*

string·er (string′ər) *n.* 1 A heavy timber, generally horizontal, supporting other members of a structure, and usually running in the direction of the greatest length of the collection of supported members. 2 Any horizontal framing timber, as a tie beam; a stringpiece. 3 A lengthwise timber on which rails are laid, as distinguished from a *cross–tie* or *sleeper.* 4 One who makes, sells, or applies bowstrings. 5 One who strings; especially, one who puts the strings in a piano.

string·halt (string′hôlt′) *n.* A convulsive movement of the hind legs of a horse: also called *springhalt.* —**string′halt′ed, string′halt′y** *adj.*

string line In billiards, a line passing across the table through the cue spot: used in stringing for lead.

string·piece (string′pēs′) *n.* A heavy supporting timber, horizontal or inclined, forming the margin or edge of a framework, as of a floor or staircase; a stringer, or stringboard.

string·y (string′ē) *adj.* **string·i·er, string·i·est** 1 Containing fibrous strings. 2 Forming in strings, as thick glue; ropy. 3 Having tough sinews; wiry. —**string′i·ly** *adv.* —**string′i·ness** *n.*

strip[1] (strip) *n.* 1 A narrow piece, comparatively long, as of cloth, wood, etc. 2 A number of stamps attached in a row. 3 A narrow piece of land; a minor civil division in Maine. 4 An act of destruction or spoliation. 5 A comic strip. —**Cherokee strip** A strip of land formerly leased to the Cherokee Indians but now part of Oklahoma. —*v.t.* **stripped, strip·ping** To cut or tear into strips. [? <MLG *strippe* a strap]

strip[2] (strip) *v.* **stripped** (*Rare* **stript**), **strip·ping** *v.t.* 1 To pull the covering, clothing, etc., from; denude; lay bare. 2 To pull off (the covering or clothing). 3 To rob or plunder; spoil. 4 To make bare or empty. 5 To remove; take away. 6 To deprive of something; divest: He was *stripped* of his rank. 7 To separate the leaves of (tobacco) from the stalks. 8 To milk (a cow) dry by a downward stroke and compression of the thumb and forefinger. 9 *Mech.* To damage or break the teeth, thread, etc., of (a gear, bolt, or the like). —*v.i.* 10 To remove one's clothing; undress. 11 *Mech.* To suffer breaking or jamming of the teeth or thread. [ME <OE *-strȳpan,* as in *bestrȳpan* despoil, plunder]

stripe[1] (strīp) *n.* 1 A line, band, or long strip of material of different color or finish from the adjacent surface. 2 Distinctive quality or character; kind; sort; also, a certain kind of religious or political belief or opinion: a man of Democratic *stripe.* 3 Striped cloth. 4 *pl.* Prison uniform. 5 A piece of material or braid on the sleeve of a uniform to indicate rank, etc.; a chevron; a service or wound stripe. —*v.t.* **striped, strip·ing** To mark with a stripe or stripes. [<MDu.]

stripe[2] (strīp) *n.* 1 A blow struck with a whip or rod, as in flogging. 2 A weal or welt on the skin caused by such a blow. See synonyms under BLOW[2]. [Prob. <LG. Cf. Du. *strippen* whip.]

striped (strīpt, strī′pid) *adj.* Having stripes; marked with stripes.

striped bass See under BASS[1] (def. 1).

striped snake A garter snake.

striped squirrel A chipmunk.

strip·er (strī′pər) *n. U.S. Colloq.* A person who wears stripes on his or her sleeves: a one–*striper* or ensign in the Navy.

strip·ling (strip′ling) *n.* A mere youth; a lad. [<STRIP[1] + -LING]

strip·per (strip′ər) *n.* 1 One who or that which strips. 2 *U.S. Slang* A strip–teaser.

strip·ping (strip′ing) *n.* 1 The act or process of one who or that which strips. 2 *pl.* The milk drawn from a cow by stripping.

strip–tease (strip′tēz′) *n.* In burlesque, a gradual disrobing by a female performer. —**strip′–teas′er** *n.*

strip·y (strī′pē) *adj.* **strip·i·er, strip·i·est** Being in or suggesting stripes or streaks; having or marked with stripes.

Stritch (strich), **Samuel Alphonsus,** 1887–1958, U. S. cardinal.

strive (strīv) *v.i.* **strove, striv·en** (striv′ən) or **strived, striv·ing** 1 To make earnest effort. 2 To engage in strife; contend; fight. 3 To vie; emulate. See synonyms under CONTEND, ENDEAVOR, STRUGGLE. [<OF *estriver,* prob. <Gmc.] —**striv′er** *n.*

stroan (strōn) *Scot. v.i.* To urinate. —*n.* The act of urinating; a streamlet. Also spelled *strone.*

strob·ic (strō′bik) *adj.* 1 Resembling a top. 2 Seeming to spin: said of concentric circles that appear to spin when moved. [<Gk. *strobos* whirling]

stro·bi·la (strō·bī′lə) *n. pl.* **·lae** (-lē) *Zool.* 1 A stage in the life cycle of a jellyfish characterized by a series of annular plates each of which separates as a new organism. 2 The segmented body of a tapeworm. [<NL <Gk. *strobilē* plug of lint shaped like a fir cone <*strobilos* fir cone, anything twisted <*strobos* twisted, ult. <*strephein* twist]

strob·i·la·ceous (strob′ə·lā′shəs) *adj.* 1 Resembling or relating to a strobile or cone. 2 Producing strobiles.

strob·i·late (strob′ə·lāt) *v.i.* **·lat·ed, ·lat·ing** To divide metamerically; undergo strobilation.

strob·i·la·tion (strob′ə·lā′shən) *n. Zool.* Asexual reproduction by division, as in jellyfish and tapeworms; metameric division.

strob·ile (strob′il) *n. Bot.* 1 A multiple fruit consisting of an oblong, oval, or conical mass of dry imbricated scales, as in the pines, spruces, firs, etc.; a cone. 2 A cone–shaped mass of sporophylls producing spore cases, as in the horsetails, clubmosses, etc. Also **strob′il.** [See STROBILA]

strob·o·scope (strob′ə·skōp) *n.* An instrument for observing or studying periodic motion by rendering the moving body visible only at certain points of its path. [<Gk. *strobos* twisting + -SCOPE] —**strob′o·scop′ic** (-skop′ik) or **·i·cal** *adj.* —**strob·os·co·py** (strob·os′kə·pē) *n.*

strob·o·tron (strob′ə·tron) *n.* A low–pressure electron tube filled with a rare gas or a mixture of rare gases and used in stroboscopic photography, or to transmit supersonic impulses by means of its periodic discharges.

strode (strōd) Past tense of STRIDE.

stroke (strōk) *n.* 1 The act or movement of striking; a knock; an impact. 2 One of a series of recurring movements, as of oars, a piston, etc.; also, the rate, extent, or manner of such movement. 3 Stroke oar. 4 A single movement, as of the hand, arm, or some instrument, by which something is made or done. 5 A single movement of some instrument, as of a pen or pencil. 6 A blow or any ill effect caused as if by a blow: a *stroke* of misfortune, a *sunstroke.* 7 *Pathol.* An attack of paralysis or apoplexy. 8 A blow or the sound of a blow of a striking mechanism, as of a clock. 9 A sudden or brilliant mental act; feat; coup: a great *stroke* of diplomacy, a *stroke* of wit. 10 A pulsation, as of the heart. 11 A mark or dash of a pen or tool. 12 A light caressing movement; a stroking. See synonyms under BLOW[2], MISFORTUNE. —*v.t.* **stroked, strok·ing** 1 To pass the hand over gently or caressingly, or with light pressure. 2 To set the pace for (a rowboat or its crew); act as stroke for. 3 To sound (time), as a gong or clock. [ME *strok, strak* <OE *strācian* strike]

stroke oar 1 The aftmost oar of a boat, whose movement sets the rate of rowing. 2 The person who rows with this oar: also **stroke–oars·man** (strōk′ôrz′mən, -ōrz′-), **strokes·man** (strōks′mən). 3 The position occupied by such an oarsman.

stroll (strōl) *v.i.* 1 To walk in a leisurely or idle manner; saunter. 2 To go from place to place. —*n.* An idle or leisurely walk; a wandering. See synonyms under RAMBLE. [Origin uncertain]

stroll·er (strō′lər) *n.* 1 One who strolls; especially, a strolling showman or player. 2 A tramp. 3 A light baby carriage.

stro·ma (strō′mə) *n. pl.* **·ma·ta** (-mə·tə) 1 *Physiol.* The ground substance or connective tissue that forms the framework of an organ or cell. 2 *Bot.* In fungi, the union of mycelial threads into a dense crust on or in which the sporophores are borne. [<Gk. *strōma* bed] —**stro·mat·ic** (strō·mat′ik) *adj.*

Strom·bo·li (strôm′bō·lē) The northernmost of the Lipari Islands in the Tyrrhenian Sea; 5 square miles; site of an active volcano, 3,040 feet.

stro·mey·er·ite (strō′mī′ər·īt) *n.* A metallic, lustrous, steel–gray native sulfide of copper and silver, crystallizing in the orthorhombic system. [after F. *Stromeyer,* 1786–1835, German chemist]

strone (strōn) See STROAN.

strong (strông, strong) *adj.* 1 Physically or bodily powerful; muscular; vigorous. 2 Healthy; robust: a *strong* constitution. 3 Morally powerful; firm; resolute; courageous. 4 Mentally powerful or vigorous. 5 Especially competent or able (in a certain subject or field): *strong* in mathematics. 6 Abundantly or richly supplied (with something): *strong* in trumps; *strong* in literary interest. 7 Solidly made or constituted; not easily destroyed, injured, or strained: *strong* walls, paper, etc. 8 Powerful as a rival or combatant: a *strong* team, army, etc. 9 Easy to defend; difficult to capture: a *strong* hill position. 10 In numerical force: an army 20,000 *strong.* 11 Well able to exert influence, authority, etc.: a *strong* government. 12 Financially sound: a *strong* bank. 13 Powerful in effect: *strong* poison, medicine, etc. 14 Concentrated; not diluted or weak: *strong* coffee. 15 Containing much alcohol: a *strong* drink. 16 Powerful in flavor or odor; also, rank; unpleasant: a *strong* breath. 17 Intense in degree or quality; not mild: a *strong* pulse; *strong* light, heat, etc. 18 Loud and firm: a *strong* voice. 19 Firm; tenacious: a *strong* grip; a *strong* opinion. 20 Deeply earnest; fervid: a *strong* desire. 21 Cogent; convincing: *strong* evidence. 22 Distinct; marked; definite: a *strong* resemblance. 23 Extreme; high–handed: *strong* measures. 24 Emphatic; not moderate: *strong* language. 25 Moving with great force: said of a wind, stream, or tide; specifically, *Meteorol.,* designating a breeze (No. 6) or a gale (No. 9) on the Beaufort scale. 26 Characterized by steady or rising prices: a *strong* market. 27 *Phonet.* Stressed; accented, as a syllable. 28 *Gram.* In Germanic languages: **a** Of verbs, indicating changes in tense by means of ablaut vowel alteration in the stem, rather than by the addition of inflectional endings; as, English *drink, drank, drunk*; *write, wrote, written*; German *singen, sang, gesungen*: also called *irregular.*

b Of nouns and adjectives (in German and Old English), showing distinctive declensional endings for case, number, and gender. For example, in German, a descriptive adjective is used in the strong form when not preceded by a limiting word *(guter Mann)* or when preceded by one having no distinctive case and gender inflection *(mein guter Mann)*. Compare WEAK (def. 12). — *adv.* Strongly: usually employed in combination: *strong*-talking. Many self-explaining compound adjectives have *strong* as the first element: *strong*-armed, *strong*-smelling, etc. [OE] — **strong′ly** *adv.*

Synonyms (adj.): cohesive, compact, hardy, robust, sinewy, stalwart, stout, stubborn, sturdy, tenacious, vigorous. See FIRM, HEALTHY, POWERFUL, STAUNCH. *Antonyms*: brittle, debilitated, delicate, feeble, fragile, frail, frangible, perishable, tender, weak.

strong-arm (strông′ärm′, strong′-) *Colloq. adj.* Violent; having physical power. — *v.t.* To use physical force upon; assault.

strong·bark (strông′bärk′, strong′-) *n.* A small tree *(Bourreria ovata)*, often a mere shrub, native to the West Indies and Florida. The wood is brown, hard, and strong, but is used only for fuel; the berries are edible.

strong·box (strông′boks′, strong′-) *n.* A strongly built chest or safe for keeping valuables.

strong conjugation The conjugation of an irregular or strong verb. See STRONG (def. 28).

strong drink Alcoholic liquors.

strong·hold (strông′hōld′, strong′-) *n.* A place that nature or man has made strongly defensible; hence, a refuge. See synonyms under FORTIFICATION, REFUGE.

strong-mind·ed (strông′mīn′did, strong′-) *adj.* Having a determined, vigorous mind. — **strong′-mind′ed·ly** *adv.* — **strong′-mind′ed·ness** *n.*

strong-willed (strông′wild′, strong′-) *adj.* Having a strong will; decided; often, obstinate.

stron·gyle (stron′jil) *n.* Any of an order *(Strongyloidea)* of parasitic nematode worms, many of them very injurious to man and certain animals; especially, the hookworm and the gapeworm. Also **stron′gyl.** [<Gk. *strongylos* round]

stron·gy·lo·sis (stron′ji·lō′sis) *n. Pathol.* Disease due to intestinal worms (genus *Strongylus*) of the family *Strongylidae.* [<NL]

stron·ti·a (stron′shē·ə) *n. Chem.* **1** A grayish-white, infusible strontium monoxide, SrO. **2** Strontium hydroxide, $Sr(OH)_2$. [<NL <STRONTIUM]

stron·ti·an (stron′shē·ən) *n.* **1** Strontia. **2** Strontium.

stron·ti·an·ite (stron′shē·ən·īt′) *n.* A vitreous, native strontium carbonate, $SrCO_3$, occurring in various forms and colors.

stron·ti·um (stron′shē·əm, -shəm, -tē·əm) *n.* A hard yellowish metallic element (symbol Sr) of the alkaline earth group, known chiefly through its salts, which burn with a red flame and are used largely in pyrotechnics, but also in medicine and ceramics. See ELEMENT. [< NL, from *Strontian*, Argyll, Scotland, where first discovered] — **stron′tic** (-tik) *adj.*

strontium 90 *Physics* A radioactive isotope of strontium chemically resembling calcium and with a half-life of about 28 years: as a component in the fall-out of a thermonuclear bomb it contaminates the soil and is a radiation hazard through progressive concentration in the bones of men and animals. Also called *radiostrontium.*

strop (strop) *n.* **1** A strip of leather or canvas on which to sharpen a razor; also, a rectangular implement with strops on it. **2** A strap. — *v.t.* **stropped, strop·ping** To sharpen on a strop. [OE *stropp* <L *struppus* <Gk. *strophos* a band, cord]

stro·phan·thin (strō·fan′thin) *n.* A bitter, poisonous, crystalline glycoside contained in certain varieties of a tropical plant (genus *Strophanthus*) and resembling digitalis in its action on the heart. [<NL *Strophanthus*, a genus name (<Gk. *strophos* cord + *anthos* flower) + -IN]

stro·phe (strō′fē) *n.* **1** In ancient Greek poetry, the verses sung by the chorus in a play while moving from right to left. **2** In classical prosody, the lines of an ode comprising a stanza and alternating with the antistrophe. **3** The first of two alternating metrical systems in a poem. [<Gk. *strophē* a turning, twist < *strephein* turn] — **stroph·ic** (strof′ik, strō′fik) or **·i·cal** *adj.*

stroph·i·ole (strof′ē·ōl, strō′fē-) *n. Bot.* An aril-like appendage attached to the base of certain seeds. [<L *strophiolum*, dim. of *strophium* a band <Gk. *strophos* cord < *strephein* twist] — **stroph′i·o·late′** (-lāt′), **stroph′i·o·lat′ed** *adj.*

stroph·u·lus (strof′yə·ləs) *n. Pathol.* Any of various types of miliaria common in children: also called *tooth rash, red gum.* [<NL, dim. of Gk. *strophos* a cord. See STROPHE.]

stross·ers (stros′ərz) *n. pl. Obs.* Trousers.

stroud (stroud) *n.* A coarse, heavy, woolen material used for blankets; also, a blanket made of this material, formerly used for trading with North American Indians. [from *Stroud*, England]

strove (strōv) Past tense of STRIVE.

strow (strō) *v.t. Obs.* To strew.

struck (struk) Past tense and past participle of STRIKE.

struck jury A jury specially selected by a process in which each party strikes twelve names from a list of forty-eight eligible persons, and the remaining twenty-four are summoned as the panel from which the jury of twelve men is drawn.

struck measure A measure, as of meal, smoothed down: opposed to *heaped measure.*

struck mine, plant, etc. A mine, manufacturing plant, etc., in which work has been stopped by a strike.

struc·tur·al (struk′chər·əl) *adj.* **1** Of, pertaining to, possessing, characterized, or caused by structure. **2** *Geol.* Having a form, position, or character determined by the preexistent structure of the earth's crust; tectonic. **3** *Biol.* Morphological. **4** *Chem.* Pertaining to or denoting the spatial arrangements of atoms in a molecule: a *structural* formula. **5** Used in or essential to construction. — **struc′tur·al·ly** *adv.*

structural iron **1** Shapes of iron used in constructing buildings, bridges, etc. **2** Iron cast in shapes for this purpose.

structural shape Any particular shape or form of a structural material, especially of iron or steel, designed for a particular function and place in a structure. Commercially the term refers to the standard sizes, shapes, etc.

structural steel **1** Steel prepared after the manner of structural iron for use in building. **2** Rolled steel peculiarly adapted for use in construction: of considerable toughness and strength.

struc·ture (struk′chər) *n.* **1** That which is constructed; a combination of related parts, as a building or machine. **2** *Biol.* The arrangement and functional union of parts, tissues, and organs of a plant or animal. **3** *Geol.* **a** The spatial arrangement of rock strata in a larger formation. **b** The gross physical characteristics of a rock. **4** *Chem.* The disposition of atoms within a molecule or of molecules in a compound. **5** The manner of construction or organization: the social *structure* of a primitive society. **6** *Archaic* The act of constructing. — *v.t.* **·tured, ·tur·ing** **1** To form into an organized structure; build. **2** To conceive as a structural whole; ideate: He *structured* the plan before proposing it. See synonyms under FRAME. [<F <L *structura* < *structus*, pp. of *struere* build]

stru·del (strood′l, *Ger.* shtrōō′dəl) *n.* A kind of pastry made of a thin sheet of dough, spread with fruit or cheese, nuts, etc., rolled, and baked. [<G, lit., eddy]

strug·gle (strug′əl) *n.* A violent effort or series of efforts; a labored contest; sometimes, a war; battle. See synonyms under ENDEAVOR. [< *v.*] — *v.* **·gled, ·gling** *v.i.* **1** To contend with an adversary in physical combat; fight. **2** To put forth violent efforts; strive: to *struggle* against odds. **3** To make one's way by violent efforts: to *struggle* through mud. — *v.t.* **4** To accomplish with a struggle. [ME *strogelen*; origin unknown] — **strug′gler** *n.* — **strug′gling·ly** *adv.*

Synonyms (verb): battle, contend, contest, endeavor, fight, labor, strain, strive, toil, try, vie, wrestle, writhe.

Struld·brug (struld′brug) *n.* One of a class of immortal human beings, described in Swift's *Gulliver's Travels* (Voyage to Laputa), who became senile after the age of eighty, gradually losing all power of communication and becoming wards of the state.

strum (strum) *v.t.* & *v.i.* **strummed, strum·ming** To play (on a stringed instrument) without expression; thrum. — *n.* The act of strumming. [Prob. imit.] — **strum′mer** *n.*

stru·ma (strōō′mə) *n. pl.* **·mae** (-mē) **1** *Pathol.* **a** Scrofula. **b** Goiter. **2** *Bot.* A wenlike cushion or swelling of or on an organ, as at the base of the capsule in certain mosses. [<L < *struere* build] — **stru·mat·ic** (strōō·mat′ik), **stru′mose** (-mōs), **stru′mous** (-məs) *adj.*

Stru·ma (strōō′mä) A river in SW Bulgaria and NE Greece, flowing 215 miles SE to the Aegean: Greek *Strymon.*

strum·pet (strum′pit) *n.* A whore; harlot. [? Ult. <OF *strupe* concubinage <L *stuprum* dishonor]

strung (strung) Past tense and past participle of STRING.

strunt (strunt, strōōnt) *Scot. v.i.* **1** To strut. **2** To be sullen. — *n.* **1** A sullen mood; umbrage. **2** Spirituous liquor; whisky and water; toddy. — **to take the strunt** To be or become sulky.

strut (strut) *n.* **1** A proud or pompous step or walk. **2** A compression member in a framework, keeping two others from approaching nearer together, as the vertical members of the wing truss of a biplane. **3** An instrument used in adjusting the plaits of a ruff. — *v.* **strut·ted, strut·ting** *v.i.* To walk pompously, conceitedly, and affectedly. — *v.t.* To brace or support, as a framing or structure, by compression pieces, as struts or posts. [OE *strūtian* be rigid, stand stiffly] — **strut′ter** *n.* — **strut′ting** *adj.* — **strut′ting·ly** *adv.*

Stru·thi·on·i·dae (strōō′thē·on′i·dē) *n. pl.* A family of large, terrestrial, swift-running ratite birds; the ostriches, especially *Struthio camelus* of Africa and Arabia. [<NL <L *struthio* <Gk. *strouthiōn* ostrich]

stru·thi·ous (strōō′thē·əs) *adj.* **1** Like an ostrich. **2** Pertaining to the *Struthionidae.* [<L *struthio* an ostrich]

strych·nine (strik′nin, -nēn, -nīn) *n.* A white, crystalline, bitter, extremely poisonous alkaloid, $C_{21}H_{22}N_2O_2$, contained in various plants (genus *Strychnos*) of the logania family, especially *S. nuxvomica.* Its salts are used in medicine, chiefly as a neural stimulant; a large dose produces tetanic spasms. Also **strych′ni·a** (-nē·ə), **strych′nin** (-nin). [<F <L *strychnos* <Gk., nightshade]

strych·nin·ism (strik′nin·iz′əm) *n. Pathol.* The morbid condition resulting from the excessive or improper use of strychnine.

Stry·mon (strī′mən) The Greek name for the STRUMA.

Stry·mon·ic Gulf (strī·mon′ik) **1** An inlet of the northern Aegean in Greek Macedonia; 14 miles wide, 17 miles long. **2** The entire section of the Aegean between Thasos and Akti: also *Gulf of Orfani.*

St. Si·mons Island (sānt sī′mənz) One of the Sea Islands off the coast of SE Georgia; about 13 miles long, 3 to 7 miles wide; site of a national monument and of a decisive battle between England and Spain, 1742.

St. Swith·in's Day (sānt swith′ənz) July 15th, the day that commemorates St. Swithin, a former patron saint of Winchester Cathedral, England: rain occurring on this day is said to foretell wet weather for the following 40 days. Also **St. Swith′un's Day.**

St. Thom·as (sānt tom′əs) **1** An island of the Virgin Islands of the United States; 28 square miles; capital, Charlotte Amalie. **2** The English name for SÃO TOMÉ.

Stu·art (stōō′ərt, styōō′-) Name of the royal family of Scotland, 1371–1603, and of Great Britain, 1603–1714: also spelled *Stewart.* — **Charles Edward,** 1720–88, English prince, grandson of James II: called "Bonnie Prince Charlie" and the "Young Pretender." — **James Edward,** 1688–1766, English prince, son of James II: called the "Old Pretender."

add,āce,câre,pälm; end,ēven; it,īce; odd,ōpen,ôrder; tŏŏk,pōōl; up,bûrn; ə = a in *above*, e in *sicken*, i in *clarity*, o in *melon*, u in *focus*; yōō = u in *fuse*; oi,oil; ou,pout; ch,check; g,go; ng,ring; th,thin; th,this; zh,vision. Foreign sounds à,œ,ü,kh,ṅ; and ♦: see page xx. < from; + plus; ? possibly.

Stuart (stōō′ərt, styōō′-), **Gilbert Charles,** 1755–1828, U.S. portrait painter. — **James Ewell Brown,** 1833–64, American Confederate cavalry general: nickname *Jeb.*

stub (stub) *n.* **1** The part of a tree trunk, bush, etc., that remains when the main part is cut down. **2** Any short projecting part or piece; a remnant, as of a pencil, candle, cigarette, cigar, or broken tooth. **3** In a checkbook or the like, one of the inner ends upon which a memorandum is entered, and which remains when the check is detached; also, the detachable coupon of a theater or other ticket. **4** Anything blunt, short, or stumpy, as a worn horseshoe nail or a stub pen. **5** *Obs.* A log; block; blockhead. **6** The title of a row in a statistical table; also, the first or reading column in such a table. — *v.t.* **stubbed, stub·bing 1** To strike, as the toe, against a low obstruction or projection. **2** To grub up, as roots; root out. **3** To clear or remove the stubs or roots from. — *adj.* Thick-set; stocky. [OE *stubb*]

stub·bed (stub′id, stubd) *adj.* **1** Made into or resembling a stub. **2** Full of stubs. **3** Sturdy; blunt in manner; stout and rough; rugged. — **stub′bed·ness** *n.*

stub·ble (stub′əl) *n.* **1** The stubs of grain stalks, sugarcane, etc., covering a field after the crop has been cut. **2** The field itself. **3** Any surface or growth resembling stubble, as short bristly hair or beard. [<OF *stuble,* ult. <L *stipula* stalk] — **stub′bled** *adj.* — **stub′·bly** *adj.*

stub·born (stub′ərn) *adj.* **1** Inflexible in opinion or intention; unreasonably obstinate. **2** Not easily handled, bent, or overcome; intractable: *stubborn* facts. **3** Characterized by perseverance or persistence: *stubborn* fighting. See synonyms under HARD, INFLEXIBLE, OBSTINATE, PERVERSE, STRONG. [Prob. OE *stubb* a stump] — **stub′born·ly** *adv.* — **stub′·born·ness** *n.*

Stubbs (stubz), **William,** 1825–1901, English bishop and historian.

stub·by (stub′ē) *adj.* **·bi·er, ·bi·est 1** Short, stiff, and bristling: a *stubby* beard. **2** Short and thick; like a stub: a *stubby* pencil. **3** Having many stubs. — **stub′bi·ly** *adv.* — **stub′bi·ness** *n.*

stub nail 1 A short thick nail. **2** An old horseshoe nail.

stub pen A very blunt-pointed pen for writing.

stuc·co (stuk′ō) *n. pl.* **·coes** or **·cos 1** A fine plaster for walls or their relief ornaments, usually of Portland cement, sand, and a small amount of lime. **2** Any plaster or cement used for the external coating of buildings. — *adj.* Stucco-coated. — *v.t.* **·coed, ·co·ing** To apply stucco to; decorate with stucco. [<Ital. <Gmc. Akin to OHG *stucchi* crust.] — **stuc′·co·er** *n.*

stuck (stuk) Past tense and past participle of STICK.

stuck-up (stuk′up′) *adj. Colloq.* Conceited; very vain; supercilious and arrogant; snobbish. — **stuck′-up′ness** *n.*

stud[1] (stud) *n.* **1** A short intermediate post, as in a building frame; a post to which laths are nailed; a scantling. **2** A knob, round-headed nail, or small protuberant ornament, as an ornamental button in a shirt front. **3** A crosspiece in a link, as in a chain cable. **4** A small pin such as is used in a watch. **5** Stud poker. — *v.t.* **stud·ded, stud·ding 1** To set thickly with small points, projections, or knobs. **2** To be scattered or strewn over: Daisies *stud* the meadows. **3** To support or stiffen by means of studs or upright props. [OE *studu* post]

stud[2] (stud) *n.* **1** A collection of horses and mares for breeding. **2** The place where they are kept. **3** A collection of horses for riding, hunting, or racing. **4** A stallion: also *studhorse.* — *adj.* **1** Of or pertaining to a stud. **2** Kept for breeding: a *stud* mare. [OE *stōd*]

stud book A record of the pedigree of a stud, or of thoroughbred racing stock collectively.

stud·die (stud′ē) *n. Brit. Dial.* An anvil; a stithy. [Prob. var. of STITHY]

stud·ding (stud′ing) *n.* **1** Studs or joists collectively, or material from which to make them. **2** The height of a room from floor to ceiling.

stud·ding·sail (stun′səl, stud′ing·sāl′) *n. Naut.* A light auxiliary sail set out beyond one of the principal sails by extensible booms during a following wind.

student (stōōd′nt, styōōd′nt) *n.* **1** A person engaged in a course of study; especially, one in a secondary school, college or university. **2** One who closely examines or investigates; one devoted to study. See synonyms under SCHOLAR. [<OF *estudiant* <L *studens, -entis,* ppr. of *studere* be eager, apply oneself, study]

student lamp A reading lamp easily adjustable for direction or distance of light rays.

student nurse One who is in training in a hospital school of nursing.

stu·dent·ship (stōōd′nt·ship, styōōd′nt-) *n.* **1** A scholarship. **2** The condition of being a student.

stud·fish (stud′fish′) *n. pl.* **·fish** or **·fish·es** Any of several minnows (genus *Fundulus*) having the sides studded with orange or brown spots, as *Fundulus stellifer* of the Alabama River.

stud·horse (stud′hôrs′) *n.* A stallion kept for breeding. Also **stud horse.**

stud·ied (stud′ēd) *adj.* **1** Deliberately and intentionally designed or undertaken; planned; premeditated: a *studied* insult. **2** Acquired or prepared by study. **3** *Rare* Learned; versed. — **stud′ied·ly** *adv.* — **stud′ied·ness** *n.*

stu·di·o (stōō′dē·ō, styōō′-) *n. pl.* **·di·os 1** The workroom of an artist, photographer, etc. **2** A place where motion pictures are filmed. **3** A room or rooms where radio or television programs are broadcast or recorded. [<Ital. <L *studium* zeal < *studere* apply oneself, be diligent]

studio couch A backless couch with a bed frame underneath which may be drawn out and made level with the couch to form a double bed or twin beds.

stu·di·ous (stōō′dē·əs, styōō′-) *adj.* **1** Given to study; devoting oneself to the acquisition of knowledge. **2** Earnest in the use of means; assiduous: *studious* to please. **3** Done with deliberation; studied: *studious* politeness. **4** Favorable to study; for study: *studious* halls. [<L *studiosus* < *studium* zeal. See STUDIO.] — **stu′di·ous·ly** *adv.* — **stu′di·ous·ness** *n.*

stud poker A game of poker in which the cards of the first round are dealt face down and the rest face up, betting opening on the second round.

stud·work (stud′wûrk′) *n.* **1** Walls of brick-work between studs. **2** Studded leather armor. **3** Anything set or supported with studs.

stud·y (stud′ē) *v.* **stud·ied, stud·y·ing** *v.t.* **1** To apply the mind in acquiring a knowledge of: to *study* physics. **2** To examine; search into: to *study* a problem. **3** To look at attentively; scrutinize: to *study* one's reflection in a mirror. **4** To endeavor to memorize, as a part in a play. **5** To give thought and attention to, as something to be done or devised: often with *out.* — *v.i.* **6** To apply the mind in acquiring knowledge. **7** To follow a regular course of instruction; be a student. **8** To muse; meditate. See synonyms under CONSIDER, EXAMINE, MUSE. — **to study up on** To acquire more complete information concerning, as by investigation. — *n. pl.* **stud·ies 1** The act of studying; the process of acquiring information; application of the mind to books, to art or science, etc. **2** A particular instance or form of mental work. **3** Something to be studied; a branch or department of knowledge. **4** A specific product of studious application. **5** In art, a first sketch; a student's art exercise. **6** A carefully elaborated literary treatment of a subject. **7** A room devoted to study, reading, etc. **8** A studious state of mind; profound thought; absent-mindedness: in a brown *study.* See BROWN STUDY. **9** Earnest endeavor; thoughtful care or its object: Our *study* is to please you. **10** *Music* A composition designed to aid development in technical facility; an étude. See synonyms under EDUCATION, INQUIRY, LEARNING, REFLECTION, TASK, THOUGHT[1]. [<OF *estudier* < *estudie* a study <L *studium* zeal. See STUDIO.] — **stud′·i·a·ble** *adj.*

study hall In a school, a large room equipped and reserved for study.

stuff (stuf) *v.t.* **1** To fill completely; pack; cram full. **2** To fill (an opening, etc.) with something forced in; plug. **3** To obstruct or stop up; choke. **4** To fill or expand with padding, as a cushion. **5** To fill (a fowl, roast, etc.) with stuffing. **6** In taxidermy, to fill the skin of (a bird, animal, etc.) with a material preparatory to mounting. **7** To fill too full; overload; distend. **8** To fill or cram with food: He *stuffed* himself with oysters. **9** To fill with knowledge, ideas, or attitudes, especially unsystematically: His head is *stuffed* with prejudices. **10** To force or cram, as into a small space. **11** To fill the pores of (a skin or pelt) with a preservative of oil and tallow. — *v.i.* **12** To eat to excess; gluttonize. — **to stuff a ballot box** To put fraudulent votes into a ballot box. [< *n.*] — *n.* **1** The material out of which something may be shaped or made; hence, raw or unwrought material. **2** Figuratively, the fundamental element of anything, material or spiritual. **3** Possessions generally, especially household goods. **4** A worthless collection of things; rubbish; hence, worthless ideas: often used as an interjection: *Stuff* and nonsense! **5** Woven material, especially of wool. **6** Any textile fabric. **7** Any one of various substances, mixtures, or compounds prepared for use, as paper pulp; in leathermaking, dubbing or stuffing. **8** A medicinal mixture or potion. **9** *Scot.* Luggage; belongings; corn; grain. **10** *Slang* Money; means. **11** In journalism, copy ready for the printer or engraver. [<OF *estoffe,* prob. <L *stuppa* tow] — **stuff′er** *n.*

stuffed shirt *Colloq.* A pretentious person; especially, a pompous boob.

stuff·ing (stuf′ing) *n.* **1** The material with which anything is stuffed. **2** A mixture, as of bread or cracker crumbs with meat and seasoning, used in stuffing fowls, etc., for cooking. **3** The process of stuffing anything.

stuff·ing box *Mech.* A device consisting of a chamber affording passage and lengthwise or rotary motion of a piece, as of a piston rod or shaft, while preventing leakage about the moving part by using packing material to fill the free space.

stuffing nut The nut which encloses a stuffing box.

stuff·y (stuf′ē) *adj.* **stuff·i·er, stuff·i·est 1** Badly ventilated. **2** Impeding respiration. **3** *U.S. Colloq.* Angry; sulky. **4** Old-fashioned; stodgy; stiffly precise; strait-laced. — **stuff′·i·ly** *adv.* — **stuff′i·ness** *n.*

Stu·ka (stōō′kə, *Ger.* shtōō′kä) *n.* A German dive bomber: contraction of *Sturzkampfflugzeug.*

stull (stul) *n. Mining* A cross-timbering or platform in an excavation, especially in a stope to support workmen or to protect workers from falling stones. [Prob. <G *stollen* post, prop]

Stülp·na·gel (shtülp′nä·gəl), **Otto von,** 1880–1948, German general.

stul·ti·fy (stul′tə·fī) *v.t.* **·fied, ·fy·ing 1** To cause to appear absurd; give an appearance of foolishness to. **2** To bring to naught; nullify. **3** *Law* To allege to be of unsound mind. [<LL *stultificare* make foolish <L *stultus* foolish + *facere* make] — **stul′ti·fi·ca′tion** *n.* — **stul′ti·fi′er** *n.*

stum (stum) *n.* **1** Unfermented or partly fermented grape juice. **2** Wine revived, as by adding must, to produce increased fermentation; must. — *v.t.* **stummed, stum·ming 1** To stop fermentation in by some admixture. **2** To revive (wine), as by adding must, so as to increase fermentation. [<Du. *stom* must, lit., silent]

stum·ble (stum′bəl) *v.* **·bled, ·bling** *v.i.* **1** To miss one's step in walking or running; trip. **2** To walk or proceed unsteadily or in a blundering manner. **3** To happen upon something by chance: with *across, on, upon,* etc. **4** To fall into sin or error. — *v.t.* **5** To cause to stumble. — *n.* The act of stumbling; hence, a blunder; false step. [Cf. Norw. *stumla* stumble in the dark] — **stum′bler** *n.* — **stum′bling** *adj.* — **stum′bling·ly** *adv.*

stum·bling-block (stum′bling·blok′) *n.* Any obstacle or hindrance; something that may cause one to err: now only figurative.

stump (stump) *n.* **1** That portion of the trunk of a tree left standing when the tree is felled. **2** The part of anything, as of a limb, that remains when the main part has been removed; a stumplike part; a stub. **3** *pl. Colloq.* The legs: chiefly in the phrase **to stir one's stumps. 4** A place or platform where a stump speech is made; hence, any place or platform from which speeches are made; also, political haranguing. 5 *Colloq.* A challenge; a dare. 6 In cricket, any one of the three posts (the **off stump,** the **middle stump,** and the **leg stump**) forming the wicket. 7 A pencil-like soft leather or rubber bar, with conical ends, used to soften drawings of crayon or charcoal or to apply powdered pigments. 8 A short, thick-set person or animal. 9 A heavy step; a clump. —**to be up a stump** To be in trouble or in a dilemma. —**to take the stump** To electioneer in a political campaign. —*adj.* 1 Being or resembling a stump; stumpy. 2 Of or pertaining to political oratory or campaigning: a *stump* speaker, *stump* speech. —*v.t.* 1 To reduce to a stump; truncate; lop. 2 To remove stumps from (land). 3 To canvass (a district) by making political speeches: The candidate *stumped* the State. 4 *Colloq.* To challenge, as to a contest; dare; defy. 5 *Colloq.* To bring to a halt by real or fancied obstacles; nonplus; baffle. 6 To strike against an obstacle; stub, as one's toe. 7 To shade (a drawing) by rubbing with a stump (def. 7). —*v.i.* 8 To go about on or as on stumps; hence, to walk heavily, noisily, and stiffly; hobble. [<MLG]

stump·age (stum′pij) *n.* 1 Standing timber considered with reference to its value for cutting; also, its price. 2 A tax on lumber cut, rated by the amount cut and the price.

stump·er (stum′pər) *n.* 1 One who or that which stumps. 2 A political speaker. 3 Any problem, situation, etc., beyond one's powers of decision.

stump·y (stum′pē) *adj.* **stump·i·er, stump·i·est** 1 Full of stumps. 2 Like a stump; short and thick. —**stump′i·ness** *n.*

stun (stun) *v.t.* **stunned, stun·ning** 1 To render unconscious or incapable of action by a blow, fall, etc. 2 To astonish; astound. 3 To daze or overwhelm by loud or explosive noise. —*n.* A stupefying blow, shock, or concussion; also, the condition of being stunned. [<OF *estoner* resound, stun <L *ex-* thoroughly + *tonare* thunder, crash]

Stun·dist (shtōōn′dist) *n.* A member of a Russian body of Christians originating among peasants about 1860. As fundamentalists, the Stundists rejected forms and ceremonies, took only the Bible as their guide, and emphasized brotherly love and the need to labor: they were subjected to persecution. [<Russian *shtundist′* <G *stunde* hour, lesson] —**Stun′dism** *n.*

stung (stung) Past tense and past participle of STING.

stunk (stungk) Past participle and alternative past tense of STINK.

stun·ner (stun′ər) *n.* 1 One who or that which stuns. 2 *Slang* A person or thing of extraordinary or surprising qualities, such as beauty.

stun·ning (stun′ing) *adj.* 1 Rendering unconscious. 2 *Colloq.* Surprising; impressive; wonderful; beautiful. —**stun′ning·ly** *adv.*

stun·sail (stun′səl) *n. Naut.* A studdingsail. Also **stun′s'le.** [Contraction of STUDDINGSAIL]

stunt[1] (stunt) *v.t.* To check the natural development of; dwarf; cramp. —*n.* 1 A check in growth, progress, or development. 2 A stunted animal or thing. [OE *stunt* dull, foolish; prob. infl. in meaning by ON *stuttr* short] —**stunt′ed** *adj.* —**stunt′ed·ness** *n.*

stunt[2] (stunt) *U.S. Colloq. n.* 1 A sensational feat, as of bodily skill. 2 Any remarkable feat, enterprise, or undertaking. —*v.i.* To perform a stunt or stunts. —*v.t.* To perform stunts with (an airplane, etc.). [Prob. <G *stunde* lesson; orig. college slang]

stunt man In motion pictures, a man employed to perform dangerous actions, such as falling, jumping, etc., often as a temporary substitute for an actor.

stu·pa (stōō′pə) *n.* Tope[4]. [<Skt., heap]

stupe (stōōp, styōōp) *n. Med.* A compress or medicated cloth to be applied to a wound. [<L *stupa, stuppa* tow]

stu·pe·fa·cient (stōō′pə·fā′shənt, styōō′-) *adj.* Having power to stupefy; stupefying: also **stu′pe·fac′tive.** (-fak′tiv). —*n.* Anything that stupefies, as a narcotic. [<L *stupefaciens, -entis,* ppr. of *stupefacere* stun. See STUPEFY.]

stu·pe·fac·tion (stōō′pə·fak′shən, styōō′-) *n.* The act of stupefying or state of being stupefied; stupor. See synonyms under STUPIDITY.

stu·pe·fy (stōō′pə·fī, styōō′-) *v.t.* **·fied, ·fy·ing** 1 To dull the senses or faculties of; stun. 2 To amaze; astound. [<F *stupéfier* <L *stupefacere* stun <*stupere* be stunned + *facere* make] —**stu′pe·fied** *adj.* —**stu′pe·fi′er** *n.*

stu·pen·dous (stōō·pen′dəs, styōō-) *adj.* Of prodigious size, bulk, or degree; characterized by any highly impressive feature: a *stupendous* structure, a *stupendous* error. See synonyms under IMMENSE. [<L *stupendus* amazed, orig. gerundive of *stupere* be benumbed, stunned] —**stu·pen′dous·ly** *adv.* —**stu·pen′dous·ness** *n.*

stu·pid (stōō′pid, styōō′-) *adj.* 1 Very slow of apprehension or understanding; dull-witted; sluggish. 2 Affected with stupor; stupefied: *stupid* from drink. 3 Marked by, or resulting from, lack of understanding, reason, or wit; senseless; doltish: *stupid* acts. See synonyms under ABSURD, BRUTISH, FLAT[1], HEAVY. [<L *stupidus* struck dumb <*stupere* be stunned] —**stu′pid·ly** *adv.* —**stu′pid·ness** *n.*

stu·pid·i·ty (stōō·pid′ə·tē, styōō-) *n.* The state, quality, or character of being stupid; great mental dulness. [<L *stupiditas, -tatis* <*stupidus.* See STUPID.]

Synonyms: apathy, dulness, insensibility, obtuseness, slowness, sluggishness, stupefaction, stupor. *Stupidity* is sometimes loosely used for temporary *dulness* or partial *stupor,* but chiefly for innate and chronic *dulness* and *sluggishness* of mental action, *obtuseness* of apprehension, etc. *Apathy* may be temporary, and be dispelled by appeal to the feelings or by the presentation of an adequate motive, but *stupidity* is inveterate and often incurable. Compare APATHY, IDIOCY, STUPOR. *Antonyms*: acuteness, alertness, animation, brilliancy, cleverness, intelligence, keenness, quickness, readiness, sagacity, sense, sensibility.

stu·por (stōō′pər, styōō′-) *n.* 1 A condition of the body in which the senses and faculties are suspended or greatly dulled, as by drugs or intoxicants. 2 Extreme intellectual or moral dulness; gross stupidity. [<L <*stupere* be stunned] —**stu′por·ous** *adj.*

Synonyms: apathy, asphyxia, coma, fainting, insensibility, lethargy, swoon, swooning, syncope, unconsciousness. The *apathy* of disease is a mental state of morbid indifference; *lethargy* is a morbid tendency to heavy and continued sleep, from which the patient may perhaps be momentarily aroused. *Coma* is a deep, abnormal sleep, from which the patient cannot be aroused, or is aroused only with difficulty, a state of profound *insensibility* perhaps with full pulse and deep, stertorous breathing, and is due to brain-oppression. *Syncope* or *swooning* is a sudden loss of sensation and of power of motion, with suspension of pulse and of respiration, and is due to failure of heart action, as from sudden nervous shock or intense mental emotion. *Insensibility* is a general term denoting loss of feeling from any cause, as from cold, intoxication, or injury. *Stupor* is especially profound and confirmed *insensibility,* properly comatose. *Asphyxia* is a special form of *syncope* resulting from partial or total suspension of respiration, as in strangulation or drowning. See STUPIDITY.

stupp (stup, *Ger.* shtōōp) *n.* A deposit of finely divided metallic mercury, as in the condensers of mercury smelters. Also **stup.** [<G]

stur·dy[1] (stûr′dē) *adj.* **·di·er, ·di·est** 1 Possessing rugged health and strength; hardy; enduring; vigorous; lusty: *sturdy* health, *sturdy* blows. 2 Firm and unyielding; resolute: a *sturdy* defense. See synonyms under POWERFUL, STRONG. [<OF *estourdi* dazed, reckless <*estourdir* stun, amaze <LL *exturdire* deafen; ult. origin uncertain] —**stur′di·ly** *adv.* —**stur′di·ness** *n.*

stur·dy[2] (stûr′dē) *n.* A disease of sheep; gid. [Special use of STURDY[1]] —**stur′died** *adj.*

stur·geon (stûr′jən) *n.* A large ganoid fish of northern regions (family *Acipenseridae*), with coarse, edible flesh, especially *Acipenser sturio,* the common sturgeon of both coasts of the Atlantic, which ascends rivers. Sturgeons are the principal source of isinglass and caviar. [<AF *sturgeon,* OF *sturgiun* <Med. L *sturio, -onis* <OHG *sturjo*]

Stur·gis (stûr′jis), **Russell,** 1836–1909, U.S. architect and writer.

stur·ine (stûr′ēn, -in) *n.* A bactericidal protamine from the sperm of sturgeons. [< STUR(GEON) + (PROTAM)INE]

Stur·lu·son (stûr′lə·sən, stōōr′-), **Snorri** See SNORRI STURLUSON.

Sturm·ab·teil·ung (shtōōrm′äp·tī′lōōng) *n. pl.* **·teil·ung·en** *German* Literally, storm detachment; a political militia of the Nazi party, organized to keep order at Nazi mass meetings. After 1934, as *Brown Shirts,* the organization became a national army of political soldiers in charge of pre- and post-military indoctrination.

Sturm und Drang (shtōōrm′ ōōnt dräng′) *German* Storm and stress: used to designate the late 18th century period of German literary romanticism.

sturt (stûrt) *Brit. Dial. v.t.* 1 To annoy; vex; trouble. 2 To startle. —*v.i.* 3 To start with fear; be frightened. —*n.* 1 Vexation. 2 Strife; wrath. 3 Unrest. [Prob. var. of START[1]]

sturt·in (stûr′tin) *adj. Scot.* Frightened; overwhelmed.

stut·ter (stut′ər) *v.t.* & *v.i.* To utter or speak with spasmodic repetition, blocking, and prolongation of sounds and syllables, especially those in initial position in a word. —*n.* The act or habit of stuttering. See synonyms under STAMMER. [Freq. of ME *stutten* stutter] —**stut′ter·er** *n.* —**stut′ter·ing** *adj.* & *n.* —**stut′ter·ing·ly** *adv.*

Stutt·gart (stut′gärt, *Ger.* shtōōt′gärt) A city in SW West Germany, capital of Baden-Württemberg.

Stuy·ve·sant (stī′və·sənt), **Peter,** 1592–1672, last Dutch governor of New Amsterdam 1647–64.

St. Val·en·tine's Day (sānt val′in·tīnz) See under VALENTINE.

St. Vin·cent (sānt vin′sənt) A British colony of the Windward Islands; 150 square miles including dependencies in the Grenadines; capital, Kingstown; a component unit of The West Indies (federation).

St. Vincent, Cape The SW extremity of Portugal and of continental Europe. *Portuguese* **Ca·bo de São Vi·cen·te** (kä′bōō thə souṅ vē·sāṅn′tə).

St. Vi·tus's dance (sānt vī′təs·iz) *Pathol.* Chorea. Also **St. Vitus dance.**

sty[1] (stī) *n. pl.* **sties** 1 A pen for swine. 2 Any filthy habitation or place of bestiality or debauchery. —*v.t.* & *v.i.* **stied, sty·ing** To keep or live in a sty or hovel. [OE *stī, stig*]

sty[2] (stī) *n. pl.* **sties** *Pathol.* A small, inflamed swelling of a sebaceous gland on the edge of the eyelid. Also **stye.** [<obs. *styanye* <OE *stīgend,* ppr. of *stīgan* rise + *ye* eye]

stych·o·myth·i·a (stik′ə·mith′ē·ə) See STICHOMYTHY.

Styg·i·an (stij′ē·ən) *adj.* 1 Pertaining to the river Styx; hence, infernal; dark and gloomy. 2 Inviolable, like the oath, "By the Styx." [<L *Stygius* <Gk. *Stygios* <*Styx* the Styx, prob. <*stygein* hate]

style (stīl) *n.* 1 Manner of expressing thought, in writing or speaking; distinctive or characteristic form of expression: a florid *style*; the *style* of Mark Twain. 2 A good or suitable mode of expression: His writing lacks *style.* 3 A particular form of composition, construction, or appearance, as in art, music, etc.: the Gothic *style*; the American *style* of automobile. 4 The manner in which some action or work is performed: The horse ran in fine *style.* 5 A good or exemplary manner of performing: a team with *style.* 6 A mode of conduct or behavior; a way of living: to

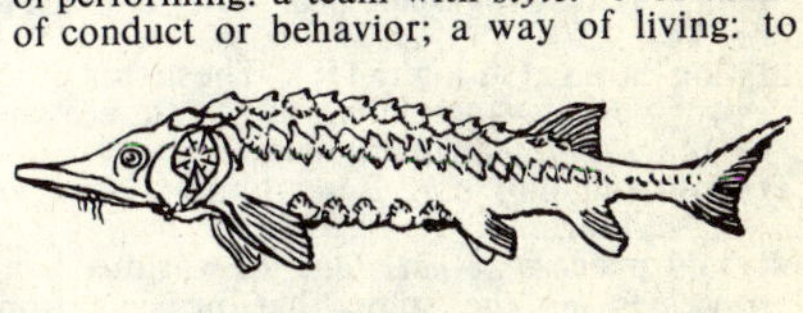

STURGEON
(Length up to 10 feet)

live in makeshift *style.* 7 A fashionable manner or appearance: to live in *style.* 8 A particular fashion in clothing. 9 A particular type or fashion suitable for or agreeable to a person: That coat is not my *style.* 10 *Printing* The

stylet

conventions of typography, design, etc., observed in a given printing office. **11** The legal or official title or appellation of a person, organization, etc. **12** A stylus (in any sense). **13** The gnomon of a sundial. **14** *Surg.* A slender probe with a blunt point: also called *stylet.* **15** *Bot.* The prolongation of a carpel or ovary, bearing the stigma. **16** *Zool.* A stylet. **17** A system of arranging the length of the calendar years so as to average that of the true solar year: called **New Style,** when following the arrangement made by Pope Gregory XIII (Gregorian calendar) and used in nearly all Christian countries; and **Old Style** when following the Julian calendar. England adopted the New Style by act of Parliament in 1752. Since 1900 New Style has been 13 days later than Old Style. See synonyms under AIR[1], CUSTOM, DICTION, MANNER, NAME. — *v.* **styled, styl·ing** *v.t.* **1** To name; give a title to. **2** To make consistent in typography, spelling, punctuation, etc., as copy to be printed; stylize. — *v.i.* **3** In ornamentation, to use a style or stylus. ◆ Homophone: *stile.* [<OF <L *stilus, stylus* writing instrument] — **sty′lar, sty′li·form** *adj.* — **styl′er** *n.*

sty·let (stī′lit) *n.* **1** Any slender pointed instrument, as a poniard or stiletto. **2** *Surg.* A style. **3** *Zool.* Any pointed, bristlelike process or appendage. [<F <Ital. *stiletto.* See STILETTO.]

sty·li·form (stī′lə-fôrm) *adj.* Resembling or shaped like a stylus. [<NL *styliformis* <L *stylus* a stylus + *forma* a form]

styl·ish (stī′lish) *adj.* Having style; especially, very fashionable. — **styl′ish·ly** *adv.* — **styl′ish·ness** *n.*

styl·ist (stī′list) *n.* **1** One who is a master of literary or rhetorical style. **2** An adviser concerning style in clothes, interior decoration, etc.

sty·lis·tic (stī-lis′tik) *adj.* Pertaining to style, especially literary style. — *n.* Stylistics. — **sty·lis′ti·cal·ly** *adv.*

sty·lis·tics (stī-lis′tiks) *n. pl.* (construed as singular) The art or study of literary expression.

sty·lite (stī′līt) *n.* One of a class of early religious ascetics who lived most of the time on the tops of pillars, without shelter. The practice was originated by Simeon Stylites in A.D. 420. [<Gk. *stylitēs* < *stylos* column]

styl·ize (stī′līz) *v.t.* **ized, ·iz·ing** To conform to a distinctive mode or style; conventionalize. Also *Brit.* **styl′ise.** — **styl′i·za′tion** *n.* — **styl′iz·er** *n.*

stylo- *combining form* **1** A pillar: *stylobate.* **2** *Bot.* & *Zool.* A style; of or related to a style: *stylopodium.* **3** *Anat.* Denoting relationship to a styloid process. Also, before vowels, **styl-.** [<Gk. *stylos* a column, pillar]

sty·lo·bate (stī′lə-bāt) *n. Archit.* A continuous base for two or more columns, in contradistinction to a pedestal, which is a base for only one column or object. Compare STEREOBATE. [<L *stylobates* <Gk. *stylobatēs* <*stylos* pillar + *-batēs* a treader < *bainein* walk, step]

STYLOBATE *(a)*

sty·lo·graph (stī′lə-graf, -gräf) *n.* A fountain pen from which ink is fed to a conical writing point. Also **stylographic pen.** — **sty′lo·graph′ic** or **·i·cal** *adj.*

sty·log·ra·phy (stī-log′rə-fē) *n.* The art or process of writing, engraving, etc., with a stylus or other pointed instrument.

sty·loid (stī′loid) *adj.* Resembling a style or peg; styliform.

styloid process *Anat.* One of various bony processes, as the spine that projects from the base of the temporal bone; a projection on the head of the fibula; the pointed lower extremity of either the radius or the ulna; the proximal end of the third metacarpal bone.

sty·lo·lite (stī′lə-līt) *n. Geol.* A small columnar body of the same composition as the surrounding rock. — **sty′lo·lit′ic** (-lit′ik) *adj.*

sty·lo·po·di·um (stī′lə-pō′dē-əm) *n. pl.* **·di·a** (-dē-ə) *Bot.* The fleshy disk that bears the style in umbelliferous flowers. [<NL]

sty·lus (stī′ləs) *n.* **1** An ancient writing instrument, having one end pointed for writing on wax tablets and the other end blunt for erasure. **2** A pointed instrument for marking or engraving, as on carbons, stencils, etc. **3** The needle of a phonograph or of a sound-recording instrument. [<L]

sty·mie (stī′mē) *n.* A condition obtaining in golf when an opponent's ball lies in, the line of the player's putt on the green, the balls being more than six inches apart. — *v.t.* **·mied, ·my·ing** To block (an opponent) by or as by a stymie. Also spelled *stimy.* [Origin uncertain]

Stym·pha·lus (stim-fā′ləs) A district of NE Arcadia, central Peloponnesus, Greece.

styp·sis (stip′sis) *n.* The application or action of a styptic. [<LL <Gk., a contraction < *styphein* contract]

styp·tic (stip′tik) *adj.* **1** Causing contraction of living tissues, as blood vessels. **2** Preventing hemorrhage; astringent: a *styptic* pencil. Also **styp′ti·cal.** — *n.* A substance or agent that, applied locally, arrests bleeding. [<L *stypticus* <Gk. *styptikos* < *stypsis* a contraction. See STYPSIS.] — **styp·tic·i·ty** (stip-tis′ə-tē) *n.*

Styr (stir) A river in western Ukrainian S.S.R., flowing 280 miles north to the Pripet River.

Sty·ra·ca·ce·ae (stī′rə-kā′si-ē) *n. pl.* An order of gamopetalous trees or shrubs yielding resins and gums, the storax family, having alternate simple leaves and usually white racemed flowers with a corolla of 4 to 8 united petals. They are found in all parts of the world. [<NL <L *styrax* storax] — **sty′ra·ca′ceous** (-shəs) *adj.*

sty·rene (stī′rēn, stir′ēn) *n. Chem.* A colorless aromatic hydrocarbon, C_8H_8, contained in liquid storax, from which it may be derived by distillation: also obtainable from cinnamic acid by treatment with lime. [<L *styrax* storax + -ENE]

Sty·ri·a (stir′ē-ə) A province and former duchy in central and SE Austria; 6,324 square miles; capital, Graz: German *Steiermark.*

stythe (stīth) *n.* Chokedamp. [OE *stith* harsh]

Styx (stiks) In Greek mythology, the river of hate, one of the five rivers surrounding Hades.

su·a·ble (soo′ə-bəl) *adj.* Legally subject to civil process; able to be sued. — **su′a·bil′i·ty** *n.*

Sua·kin (swä′kən) A port on the Red Sea in NE Sudan.

sua·sion (swā′zhən) *n.* The act of persuading; persuasion: archaic except in the phrase **moral suasion.** [<OF <L *suasio, -onis* < *suadere* persuade] — **sua·sive** (swā′siv), **sua·so·ry** (swā′sər-ē) *adj.*

suave (swäv, swāv) *adj.* Smooth and pleasant in manner; bland; gracious. [<F <L *suavis* sweet] — **suave′ly** *adv.* — **suave′ness** *n.*

suav·i·ty (swä′və-tē, swav′ə-) *n. pl.* **·ties** **1** The state of being suave; urbanity. **2** Something that is suave, bland, or agreeable. [<F *suavité* <L *suavitas, -tatis* < *suavis* sweet]

sub (sub) *n. Colloq.* Short for: **1** A substitute. **2** A subordinate or subaltern. **3** A subway. **4** A submarine.

sub- *prefix* **1** Under; beneath; below; as in:

subaquatic	subfloor
subastral	subfluvial
subcoastal	subsurface
subcurrent	subtext

2 *Anat.* Situated under or beneath, or on the ventral side of; as in:

subabdominal	submuscular
subalar	subnasal
subapical	subneural
subauricular	subnodal
subaxial	subocular
subcerebellar	suboptic
subclavicular	suboral
subcortical	suborbital
subcostal	subpelvic
subcranial	subphrenic
subcuticular	subpleural
subdental	subpubic
subdermal	subpulmonary
subdiaphragmatic	subrectal
subdorsal	subretinal
subepiglottic	subspinal
subgenital	subspinous
subgingival	substernal
subglottic	subungual
subintegumental	suburethral
subintestinal	subvaginal
submammary	subvertebral

3 Almost; nearly; slightly; imperfectly: chiefly in scientific terms; as in:

subacid	subfluid
subacidity	subhorizontal
subacidulous	subinflammation
subacrid	sublateral
subacuminate	sublinear
subalkaline	subluminous
subangular	submedial
subastringent	submetallic
subaudible	subnarcotic
subcalcareous	suboval
subcarbureted	subparallel
subcentral	subparalytic
subconcave	subpolar
subconchoidal	subsaline
subconical	subserrate
subcolumnar	subsibilant
subconvex	subtetanic
subcubical	subtypical
subdelirium	subvertical
subfebrile	subvirile

4 Lower in rank or grade; secondary; subordinate; as in:

subadministration	subholding
subadministrator	sub–idea
subagency	sublessee
subagent	sublessor
subassociation	sublieutenancy
subcantor	sublieutenant
subcause	submeaning
subchanter	submediator
subclerk	submortgage
subcommission	subofficer
subconstellation	subpart
sub–echo	subrector
sub–editor	subrent
sub–element	subsecretary
subflavor	subtone
subforeman	subvicar
subfunction	subworker

5 Forming a subdivision; as in:

sub–branch	submember
sub–bureau	suboffice
subcavity	subprovince
subclass	subscience
subclassification	subsegment
subcorporation	subseries
subcouncil	subshaft
subdepartment	subtype
subdialect	subunit
subdistrict	subzone

6 *Math.* Denoting a ratio, the inverse of a given ratio: The *subtriplicate* ratio is the inverse of the ratio of the cube. **7** *Chem.* **a** Present (in a compound) in less than normal amount: *subchloride, suboxide.* **b** Designating a basic salt compound: *subacetate, subcarbonate.* Also: *suc-* before *c,* as in *succumb; suf-* before *f,* as in *suffer; sug-* before *g,* as in *suggest; sum-* before *m,* as in *summon; sup-* before *p,* as in *support; sur-* before *r,* as in *surrogate; sus-* before *c, p, t,* as in *susceptible, suspect, sustain.* [<L *sub-* < *sub* under]

sub·a·cute (sub′ə-kyoot′) *adj.* **1** Somewhat acute. **2** Intermediate between acute and chronic: said of a disease. — **sub′a·cute′ly** *adv.*

sub·aer·i·al (sub′âr′ē-əl, -ā·ir′-) *adj.* Of, pertaining to, or formed at the earth's surface, in open air: contrasted with *aerial, submarine,* and *subterranean.*

su·bah (soo′bä) *n.* **1** A province or governmental district of India. **2** A subahdar. Also **su′ba.**

su·bah·dar (soo′bä-där′) *n.* The chief native officer of a company of sepoys in the former British East-Indian Army: also spelled *subah.* Also **su′ba·dar′.** [<Urdu *sūbahdār* <Persian <Arabic *sūbah* a province + Persian *dār* a possessor, master]

sub·al·pine (sub·al′pīn, -pin) *adj.* **1** Lower than alpine. **2** Of or pertaining to mountainous regions near but below the timber line.

sub·al·tern (sub-ôl′tərn) *adj.* **1** *Brit. Mil.* Ranking below a captain. **2** Of inferior rank or position; subordinate, as a species to a genus, or as a particular proposition under a universal. — *n.* **1** A person of subordinate rank or position. **2** *Brit. Mil.* An officer ranking below a captain. **3** (sub′əl·tûrn) *Logic* A specific class as included under a general one, or a particular statement as deducible from a

sub·al·ter·nant (sub·ôl·tûr′nənt) *adj.* Universal, as opposed to *particular.* — *n.* A universal proposition in its relation to the particular proposition containing the same terms. [<NL *subalternans, -antis,* ppr. of Med. L *subalternare* subordinate <LL *subalternus.* See SUBALTERN.]

sub·al·ter·nate (sub·ôl′tər·nit, -al′-) *adj.* 1 Subordinate; subaltern. 2 Successive, or succeeding by turns. 3 *Bot.* Alternate, with a tendency to become opposite. — *n.* A particular as opposed to a universal proposition. [< Med. L *subalternatus,* pp. of *subalternare.* See SUBALTERNANT.]

sub·al·ter·na·tion (sub·ôl′tər·nā′shən, -al′-) *n.* A succession; a subordination.

sub·ant·arc·tic (sub′ant·ärk′tik, -är′tik) *adj.* Denoting or pertaining to a region contiguous to that within the Antarctic Circle.

sub·a·qual (sub·ā′kwəl) *adj.* Situated below the level of the water table, as soils formed on lake or river beds. [<SUB- + L *aqua* water]

sub·a·que·ous (sub·ā′kwē·əs) *adj.* 1 Being, formed, or operating under water; submarine. 2 Occurring under or in water; adapted for use under water. 3 Having an appearance like that produced under water.

sub·arc·tic (sub·ärk′tik, -är′tik) *adj.* Denoting or pertaining to a region contiguous to that within the Arctic Circle.

sub·ar·cu·ate (sub·är′kyōō·it) *adj.* Moderately arched or bent. Also **sub·ar′cu·at·ed** (-ā′tid).

sub·ar·e·a (sub·âr′ē·ə) *n.* A small part of a given area, as in a field of gunfire.

sub·ar·id (sub·ar′id) *adj.* Partly arid; moderately dry.

sub·a·tom·ic (sub′ə·tom′ik) *adj.* Within the atom.

sub·au·di·tion (sub′ô·dish′ən) *n.* 1 The understanding or supplying of something not expressed. 2 A thought thus understood or supplied.

sub·ax·il·lar·y (sub·ak′sə·ler′ē) *adj.* 1 *Bot.* Lying under or beneath the axil. 2 *Anat.* Beneath the armpit.

sub–base (sub′bās′) *n.* 1 *Archit.* The lowest member of a base or pedestal. 2 A subdivision of a main base, as in a field of military operations. 3 The section of a base line between two fixed points, as the line connecting two microphones in a sound-ranging system.

sub–base·ment (sub′bās′mənt) *n.* An underground story, or any one of several below the first or true basement.

sub–bass (sub′bās′) *n.* In an organ, a 16-foot or 32-foot pedal stop. Also **sub′–base′**.

sub·cal·i·ber (sub·kal′ə·bər) *adj. Mil.* Of smaller caliber than the firearm from which it is to be fired: said of a projectile. A tube or disk is used to make up the deficit. Also **sub·cal′i·bre.**

sub·car·bide (sub·kär′bīd) *n. Chem.* A carbide containing less than the usual amount of carbon.

sub·car·ti·lag·i·nous (sub·kär′tə·laj′ə·nəs) *adj. Anat.* 1 Beneath cartilage or under tissue. 2 Partly cartilaginous.

sub·ce·les·tial (sub′si·les′chəl) *adj.* 1 Lower than celestial; beneath the heavens; mundane. 2 Directly beneath the zenith. — *n.* A subcelestial being.

sub·cel·lar (sub·sel′ər) *n.* A cellar under another cellar.

sub·chlo·ride (sub·klôr′īd, -klō′rid, -rid) *n.* A basic chloride: copper *subchloride,* Cu_2Cl_2.

sub·cla·vi·an (sub·klā′vē·ən) *Anat. adj.* 1 Situated beneath the clavicle. 2 Of or pertaining to the subclavian vessels. — *n.* A subclavian nerve, muscle, vein, etc. [<NL *subclavius* <L *sub-* under + *clavis* a key]

subclavian artery *Anat.* The large main artery that passes under the clavicle to convey blood to the arm.

subclavian groove *Anat.* A groove made by the subclavian artery or vein on the first rib.

subclavian vein *Anat.* That portion of the main venous trunk of the arm that lies under the clavicle.

sub·cli·max (sub·klī′maks) *n.* 1 A stage prior to or below the climax. 2 *Ecol.* a Any stage in the development of a plant or animal community determined by agencies other than climate which prevent attainment of the normal climax. b Any community so acted upon. — **sub·cli·mac·tic** (sub·klī·mak′tik) *adj.*

sub·com·mit·tee (sub′kə·mit′ē) *n.* An undercommittee; part of a committee appointed for special work.

sub·con·scious (sub·kon′shəs) *adj.* 1 Only dimly conscious; not clearly discerned by the conscious subject; lacking intellectual clearness. 2 *Psychol.* Denoting such phenomena of mental life as are not attended by full consciousness, as the many automatic processes involved in the performance of familiar actions. — *n.* 1 That portion of mental activity not directly in the focus of consciousness but sometimes susceptible to recall by the proper stimulus. 2 *Psychoanal.* The preconscious. — **sub·con′scious·ly** *adv.* — **sub·con′scious·ness** *n.*

sub·con·ti·nent (sub·kon′tə·nənt) *n.* A great land mass forming part of a continent but having considerable geographical independence, as India.

sub·con·tract (sub·kon′trakt) *n.* A contract subordinate to another contract and assigning part of the work to a third party. — *v.t.* & *v.i.* (sub′kən·trakt′) To make a subcontract (for); arrange for part or all of (work) to be performed by a third party. — **sub′con·tract′ed** *adj.*

sub·con·trac·tor (sub′kən·trak′tər, -kon′trak-) *n.* One who enters into a contract with a contractor to do work embraced in the latter's contract.

sub·cor·tex (sub·kôr′teks) *n.* *pl.* **·ti·ces** (-tə·sēz) *Anat.* That part of the brain which underlies the cortex. [<NL <L *sub-* under + *cortex* bark]

sub·cul·ture (sub·kul′chər) *n. Bacteriol.* A culture of bacteria or other material derived from a preexisting culture.

sub·cu·ta·ne·ous (sub′kyōō·tā′nē·əs) *adj.* 1 Situated, found, or applied beneath the skin. 2 Hypodermic. [<LL *subcutaneus* <L *sub-* under + *cutis* skin] — **sub′cu·ta′ne·ous·ly** *adv.*

sub·dea·con (sub·dē′kən) *n.* A member of the order of the ministry next below that of deacon, who assists at the Eucharist. [<AF *soudiake, subdiacne* <Med. L *subdiaconus* (< *sub-* under + *diaconus* deacon), trans. of LGk. *hypodiakonos*] — **sub·dea′con·ate** (-it) *n.*

sub·dean (sub·dēn′) *n.* An assistant or substitute dean. [<OF *soudeien* < *sou-* SUB- + *deien* a dean]

sub·dean·er·y (sub·dē′nər·ē) *n.* *pl.* **·er·ies** The office of a subdean.

sub·deb·u·tante (sub′deb·yōō·tänt′, -deb′yōō·tant) *n.* A young girl the year before she becomes a debutante. — *adj.* In a youthful style suitable for girls of this age. Also *U.S. Colloq.* **sub′deb′.**

sub·del·e·gate (sub·del′ə·gāt, -git) *n.* One who represents a delegate. — *v.t.* (-gāt) **·gat·ed, ·gat·ing** 1 To appoint as a subdelegate. 2 *Obs.* To delegate (authority, etc.) to another.

sub·de·pot (sub·dē′pō, -dep′ō) *n.* An auxiliary depot located near a base of operations, as for the supply of troops.

sub·di·ac·o·nate (sub′dī·ak′ə·nit, -nāt) *adj.* Of or pertaining to the office, rank, or order of subdeacon: also **sub′di·ac′o·nal.** — *n.* The office, rank, or order of subdeacon. [<Med. L *subdiaconatus* < *subdiaconus.* See SUBDEACON.]

sub·di·vide (sub′di·vīd′) *v.t.* & *v.i.* **·vid·ed, ·vid·ing** 1 To divide (a part) resulting from a previous division, divide again. 2 To divide (land) into lots for sale or improvement. [< LL *subdividere* <L *sub-* under + *dividere* DIVIDE]

sub·di·vi·sion (sub′di·vizh′ən) *n.* 1 Division following upon division. 2 A part, as of land, resulting from subdividing. See synonyms under PART.

sub·dom·i·nant (sub·dom′ə·nənt) *n. Music* The tone next below the dominant; fourth tone or degree of a major or minor scale. — *adj.* Less important than the dominant.

sub·duce (sub·dōōs′, -dyōōs′) *v.t.* **duced, ·duc·ing** *Obs.* To withdraw; take away. 2 To take as a part from a whole; subtract. Also **sub·duct** (-dukt′). [< L *subducere* <*sub-* from + *ducere* lead] — **sub·duc′tion** (-duk′shən) *n.*

sub·due (sub·dōō′, -dyōō′) *v.t.* **·dued, ·du·ing** 1 To gain dominion over, as by war or force; subjugate; vanquish. 2 To overcome by training, influence, or persuasion; tame. 3 To repress (emotions, impulses, etc.). 4 To reduce the intensity of; soften, as a color or sound. 5 To bring (land) under cultivation. [<OF *soduire* seduce <L *subducere* SUBDUCE; infl. in meaning by L *subdere* overcome] — **sub·du′a·ble** *adj.* — **sub·du′al** *n.* — **sub·du′er** *n.*
Synonyms: beat, break, bridle, conquer, control, crush, master, overbear, overcome, overpower, overwhelm, reduce, repress, subject, suppress, train, vanquish. See CHASTEN, CONQUER, REPRESS.

sub·el·a·phine (sub·el′ə·fin, -fin) *adj. Zool.* Designating a modified form of elaphine antlers. For illustration see ANTLER.

sub·e·qua·to·ri·al (sub·ē′kwə·tôr′ē·əl, -tō′rē-) *adj.* 1 Nearly equatorial. 2 Denoting or belonging to a region adjoining the equatorial region.

su·ber (sōō′bər) *n.* Cork. [<L] — **su·be·re·ous** (sōō·bir′ē·əs) *adj.*

su·ber·ic (sōō·ber′ik) *adj.* Of, pertaining to, or derived from cork.

suberic acid *Chem.* A white crystalline diacid, $C_8H_{14}O_4$, obtained by the action of nitric acid on cork and on various fatty oils.

su·ber·in (sōō′bər·in) *n.* A waxlike, fatty substance formed in cork cells.

su·ber·i·za·tion (sōō′bər·ə·zā′shən, -ī·zā′-) *n.* *Bot.* The transformation of plant cell walls into suberin or cork tissue.

su·ber·ize (sōō′bə·rīz) *v.t.* **·ized, ·iz·ing** To make corky, as cell walls.

su·ber·ose (sōō′bər·ōs) *adj.* 1 Corky. 2 Of or pertaining to suberin. Also **su′ber·ous** (-əs).

sub·fam·i·ly (sub·fam′ə·lē, -fam′lē) *n.* *pl.* **·lies** 1 A division of plants or animals next below a family but above the genus. 2 *Ling.* A division of languages below a family and above a branch.

sub·ge·nus (sub·jē′nəs) *n.* *pl.* **·gen·e·ra** (-jen′ər·ə) A primary subdivision of a genus including one or more species with common characters. — **sub′ge·ner′ic** (-ji·ner′ik) *adj.*

sub·gla·cial (sub·glā′shəl) *adj.* Deposited or formed at the bottom of or beneath a glacier.

sub·group (sub′grōōp′) *n.* 1 An inferior order, or one of the biological divisions of an order. 2 *Chem.* A group that is included within a superior group, as in the periodic table of the elements.

sub·head (sub′hed′) *n.* 1 A heading or title of a subdivision: also **sub·head′ing.** 2 An official next below the head in a college or school.

sub·hu·man (sub·hyōō′mən) *adj.* 1 Less than or imperfectly human. 2 *Anthropol.* Below the level of the primate type represented by *Homo sapiens.*

sub·hu·mid (sub·hyōō′mid) *adj.* Intermediate between semiarid and humid: said especially of a climate with sufficient precipitation to support a moderate to dense growth of tall and short grasses.

Su·bic (sōō′bik) A municipality of central Luzon, Philippines, at the head of Subic Bay, an inlet of the South China Sea near Bataan Peninsula; site of U. S. landing in World War II, January, 1945.

sub·in·ci·sion (sub′in·sizh′ən) *n.* 1 A cutting beneath or under. 2 Among certain primitive peoples, a slitting open of the urethra of the penis.

sub·in·dex (sub·in′deks) *n.* *pl.* **·in·dices** (-in′də·sēz) An indicative figure, letter, or sign following and slightly underneath a figure, letter, or sign: in M_n, X_2, Y_4, the subindices are *n*, 2, and 4.

sub·in·fec·tion (sub′in·fek′shən) *n. Pathol.* 1 Infection of cells weakened by prolonged resistance to toxin. 2 Infection by the toxic wastes of destroyed bacteria.

sub·in·feu·date (sub′in·fyōō′dāt) *v.t.* & *v.i.* **·dat·ed, ·dat·ing** To sublet by subinfeudation. Also **sub′in·feud′.**

sub·in·feu·da·tion (sub′in·fyōō·dā′shən) *n.* 1 The granting of lands by a feudal vassal to a tenant who thus becomes his vassal. 2 The feud or fief resulting from subinfeudation. — **sub′in·feu′da·to·ry** (sub′in·fyōō·dā′tôr·ē, -tō′rē) *adj.*

sub·in·trant (sub·in′trənt) *adj.* 1 Occurring

or entering secretly. 2 *Pathol.* Anticipating a recurrence of, as a paroxysm, a malarial fever, etc. [<L *subintrans, -antis,* ppr. of *subintrare* <*sub-* secretly + *intrare* enter] — **sub·in′·trance** (-trəns) *n.*

sub·ir·ri·gate (sub·ir′ə·gāt) *v.t.* **·gat·ed, ·gat·ing** To irrigate through underground pipes, etc. — **sub′ir·ri·ga′tion** *n.*

su·bi·to (sōō′bē·tō) *adv. Music* Quickly; suddenly. [<Ital. <L *subitus,* pp. of *subire* come or go stealthily <*sub-* secretly + *ire* go]

sub·ja·cent (sub·jā′sənt) *adj.* 1 Situated underneath. 2 Being at a lower elevation. [<L *subjacens, -entis,* ppr. of *subjacere* <*sub-* under + *jacere* lie] — **sub·ja′cen·cy** *n.*

sub·ject (sub′jikt) *adj.* 1 Being under the power of another; owing or yielding obedience to sovereign authority. 2 Exposed to some agency or tendency: *subject* to headache; a climate *subject* to storms. 3 Being under discretionary authority: a treaty *subject* to ratification. — *n.* 1 One who is under the governing power of another, as of a ruler or government, especially of a monarch. 2 One who or that which is employed or treated in a specified way, as a body for dissection, a person used in hypnotic experiments, one attacked by or liable to any disease. 3 Something upon which thought or the artistic constructive faculty is employed, as a theme of consideration or the general idea or plan of an artistic work. 4 *Gram.* The word, phrase, or clause of a sentence about which something is stated or asked in the predicate. 5 *Music* The melodic phrase on which a composition or a part of it is based. 6 A branch of learning. 7 The originating clause or motive. 8 The ego or self; that of which qualities or attributes are affirmed; substance; essential being; the thinking, feeling agent. 9 *Logic* In a proposition, that term about which something is affirmed or denied. See PROPOSITION. See synonyms under TOPIC. — *v.t.* (səb·jekt′) 1 To bring under dominion or control; subjugate. 2 To cause to undergo some experience or action. 3 To offer for consideration or approval; submit. 4 To make liable; expose: His inheritance was *subjected* to heavy taxation. 5 *Obs.* To place beneath. See synonyms under CONQUER, SUBDUE. [<OF *suget, sujet* <L *subjectus,* pp. of *subjicere* <*sub-* under + *jacere* throw; refashioned after L]

Synonyms (adj.): dependent, disposed, exposed, inferior, liable, obnoxious, prone, subordinate. *Antonyms*: clear, exempt, free, supreme, uncontrolled, unrestrained.

sub·jec·tion (səb·jek′shən) *n.* The act of making subject or bringing into a state of subjection.

sub·jec·tive (səb·jek′tiv) *adj.* 1 Relating to, or conditioned by, mental states or the ego; proceeding from or taking place within the thinking subject: opposed to *objective.* 2 Pertaining to the real nature or essence or substance of a person or thing; inherent; essential. 3 Peculiar to an individual; fanciful; illusory. 4 Inclined to be submissive; obedient. 5 *Gram.* Designating that case of the substantive used to denote its function as subject of a finite verb. 6 In literature and art, giving prominence to the subject or author as treating of his inner experience and emotion. 7 Introspective. — **sub·jec′tive·ly** *adv.* — **sub·jec′·tive·ness, sub·jec·tiv·i·ty** (sub′jek·tiv′ə·tē) *n.*

◆ *Subjective* and *objective,* paired words, are strictly speaking neither synonyms nor antonyms. In scholasticism and philosophies of idealism they are both concerned with the object perceived, but represent different approaches to it. *Objective* signifies the relating of mental states to an object, that is, to something outside the perceiving mind which is recognized as having an existence outside that mind. *Subjective* relates to a feeling, attitude, or cognition that is recognized as being a construct within the mind of the perceiver, even though it takes the external object as its point of departure. Different individuals may receive different *subjective* impressions from the same *objective* fact. See INHERENT, OBJECTIVE.

sub·jec·tiv·ism (səb·jek′tiv·iz′əm) *n.* 1 The doctrine that knowledge is merely subjective and relative and is derived from one's own consciousness. 2 The doctrine that we know directly no external object. 3 The doctrine that there is no objective standard, test, or measure of truth; relativism. 4 The doctrine that individual feeling is the standard by which to judge right and wrong. — **sub·jec′tiv·ist** *n.* — **sub·jec′tiv·is′tic** *adj.*

subject matter The object of consideration or study; the subject of thought.

sub·join (sub·join′) *v.t.* To add at the end; attach; affix. See synonyms under ADD. [<MF *subjoindre* <L *subjungere* <*sub-* in addition + *jungere* join]

sub·join·der (sub·join′dər) *n.* Something subjoined. [<SUBJOIN, on analogy with *rejoinder*]

sub ju·di·ce (sub jōō′di·sē) *Latin* Under judicial consideration.

sub·ju·gate (sub′jōō·gāt) *v.t.* **·gat·ed, ·gat·ing** 1 To bring under dominion; conquer; subdue. 2 To make subservient in any way; enslave. See synonyms under CONQUER. [<L *subjugatus,* pp. of *subjugare* <*sub-* under + *jugum* a yoke] — **sub′ju·ga′tion** *n.* — **sub′ju·ga′·tor** *n.*

sub·junc·tion (səb·jungk′shən) *n.* 1 The act of subjoining, or the state of being subjoined. 2 That which is subjoined. [<LL *subjunctio, -onis* <L *subjungere* SUBJOIN]

sub·junc·tive (səb·jungk′tiv) *Gram. adj.* Of or pertaining to that mood of the finite verb that is used to express a future contingency, a supposition implying the contrary, a mere supposition with indefinite time, or a wish or desire. In English the forms of the subjunctive mood are introduced by conjunctions of condition, doubt, contingency, possibility, etc., as *if, though, lest, unless, that, till,* or *whether,* but verbs in conditional clauses are not always in the subjunctive mood, for the use of these conjunctions with the indicative is very common. — *n.* 1 The subjunctive mood. 2 A verb form or construction in this mood. [<L *subjunctivus* <*subjunctus,* pp. of *subjungere* SUBJOIN]

sub·king·dom (sub·king′dəm) *n.* A phylum.

sub·lap·sar·i·an (sub′lap·sâr′ē·ən) *n.* A believer in the predestinarian view held by moderate Calvinists that God foresaw the fall of man and decreed to save some by election. — *adj.* Relating to the sublapsarians or to their tenets. [<NL *sublapsarius* <L *sub-* consequent upon, under + *lapsus* a fall] — **sub′·lap·sar′i·an·ism** *n.*

sub·la·tion (sub·lā′shən) *n. Med.* The detachment, displacement, or removal of a part. [<L *sublatio, -onis* <*sublatus,* pp. to *tollere* lift up, take away]

sub·lease (sub·lēs′) *v.t.* **·leased, ·leas·ing** To obtain or let (property) on a sublease. — *n.* (sub′lēs′) A lease of property from a tenant or lessee.

sub·let (sub·let′, sub′let′) *v.t.* **·let, ·let·ting** 1 To let to another (property held on a lease); underlet. 2 To let (work that one has contracted to do) to a subordinate contractor.

sub·le·thal (sub·lē′thəl) *adj.* Having an effect short of death: a *sublethal* dose of poison.

sub·li·mate (sub′lə·māt) *v.* **·mat·ed, ·mat·ing** *v.t.* 1 *Chem.* To convert from a solid to a vapor by heat, and then solidify again by cooling, with no apparent intermediate liquefaction. 2 To refine; purify. 3 *Psychol.* To convert the energy of (primitive impulses) into acceptable social and cultural manifestations. — *v.i.* 4 To undergo or engage in sublimation. — *adj.* Sublimated; refined. — *n. Chem.* The product of sublimation, especially when regarded as purified by the process. [<L *sublimatus,* pp. of *sublimare* <*sublimis* SUBLIME]

sub·li·ma·tion (sub′lə·mā′shən) *n.* 1 The act or process of sublimating. 2 That which has been sublimated; the pure essence of a thing. 3 *Psychol.* The transfer of psychic energy into socially acceptable channels of endeavor.

sub·lime (sə·blīm′) *adj.* 1 Characterized by elevation, nobility, or awe; grand; solemn. 2 Preeminent for nobility of character or attainment; majestic; noble: said of persons. 3 Being of the highest degree; supreme; utmost. 4 *Poetic* Of lofty bearing; haughty; proud; elated. — *n.* That which is sublime, in any sense: usually with the definite article. — *v.* **·limed, ·lim·ing** *v.t.* 1 To make sublime; ennoble. 2 To purify by sublimating. — *v.i.* To become sublimated. [<L *sublimis* lofty, prob. <*sub-* up to, under + *limen* a lintel] — **sub·lime′ly** *adv.* — **sub·lim′er** *n.* — **sub·lim·i·ty** (sə·blim′ə·tē), **sub·lime′ness** *n.*

Synonyms (adj.): beautiful, exalted, grand, lofty, magnificent, majestic, stately. *Sublime* represents the ultimate, the quintessence, and is seldom applied to persons. What is *beautiful* attracts, but what is *sublime* transcends the beautiful and inspires awe rather than simple delight. *Majestic* refers exclusively to superficial effect which makes an impression but has no connection with moral greatness. *Magnificent* denotes the possession at once of greatness, splendor, and richness; as, *magnificent* array. See GRAND. *Antonyms*: base, contemptible, insignificant, little, mean, petty, ridiculous.

Sublime Porte See PORTE.

sub·lim·i·nal (sub·lim′ə·nəl) *adj. Psychol.* 1 Below the threshold of consciousness: opposed to *supraliminal*: said of psychophysical changes of too small intensity to produce definite sensations or a clear awareness: a *subliminal* stimulus. 2 Belonging to the subconscious. [<SUB- + L *limen, liminis* a threshold, trans. of G *unter der Schwelle (des Bewusstseins)* under the threshold (of consciousness)]

sub·lin·gual (sub·ling′gwəl) *adj.* 1 Situated beneath the tongue. 2 Of or pertaining to the salivary gland situated beneath the tongue.

sub·lit·to·ral (sub·lit′ər·əl) *adj.* 1 Close to the seashore. 2 Pertaining to or designating the area between low-tide mark and a depth of 20 fathoms or of 40 meters.

sub·lu·nar·y (sub′lōō·ner′ē, sub·lōō′nər·ē) *adj.* 1 Situated beneath the moon: also **sub·lu·nar** (sub·lōō′nər). 2 Terrestrial; earthly. [<NL *sublunaris* <L *sub-* under + *luna* the moon]

sub·ma·chine gun (sub′mə·shēn′) A lightweight, gas-operated gun, automatic or semi-automatic in action, designed for firing from the shoulder or hip. — **Thompson submachine gun** An air-cooled, .45-caliber submachine gun with automatic firing action: also called *Tommy gun*: named for its inventor, John T. *Thompson,* 1860–1940, U.S. Army officer.

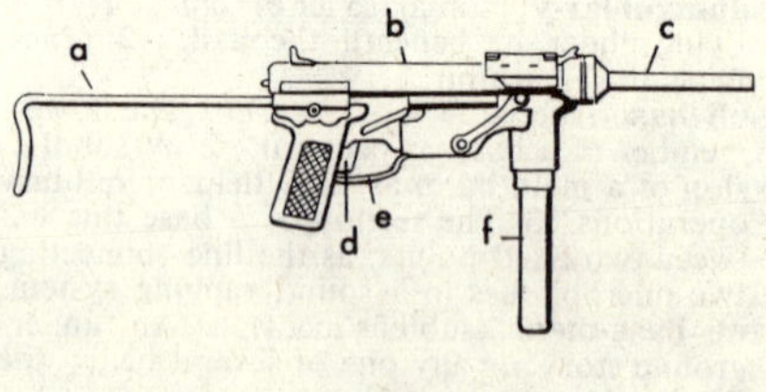

SUBMACHINE GUN
a. Stock. *b.* Housing. *c.* Barrel.
d. Trigger. *e.* Trigger guard. *f.* Clip.

sub·mar·gin·al (sub·mär′jən·əl) *adj.* 1 Below the margin. 2 Below economic sufficiency: *submarginal* land. 3 *Biol.* Situated close to the margin of an organ or structure.

submarginal land *Agric.* Land of such low degree of fertility or productivity as to be incapable of furnishing an economic return.

sub·ma·rine (sub′mə·rēn′) *adj.* Existing, done, or operating beneath the surface of the sea: a *submarine* mine: contrasted with *subaerial.* — *n.* (sub′mə·rēn) A boat designed to operate both on, and at various depths below, the surface of the sea.

submarine chaser A small patrol vessel designed and equipped for action against submarines.

sub·mar·i·ner (sub·mar′ə·nər) *n.* A trained and qualified member of a submarine crew.

sub·max·il·lar·y (sub·mak′sə·ler′ē) *Anat. adj.* 1 Of, pertaining to, or situated beneath the lower jaw. 2 Of or pertaining to one of the salivary glands situated near the angle of the lower jaw. — *n. pl.* **·lar·ies** The lower jaw bone: also **sub·max·il·la** (sub′mak·sil′ə).

sub·me·di·ant (sub·mē′dē·ənt) *n. Music* The sixth tone of a major or minor scale.

sub·men·tal (sub·men′təl) *adj.* 1 *Anat.* Situated beneath the chin: the *submental* artery. 2 Of or pertaining to the submentum.

sub·men·tum (sub·men′təm) *n. Entomol.* The basal sclerite of the labium of an insect, between the gula and the mentum. [<NL <L *sub-* under + *mentum* the chin]

sub·merge (səb·mûrj′) *v.* **·merged, ·merg·ing** *v.t.* 1 To place under or plunge into water. 2 To cover; hide. — *v.i.* 3 To sink or dive beneath the surface of water. Also **sub·merse′**

(-mûrs′). See synonyms under IMMERSE. [<L *submergere*, var. of *summergere* <*sub-* under + *mergere* plunge] —**sub·mer′gence, sub·mer′sion** (-mûr′shən, -zhən) *n.*

sub·mer·gi·ble (səb·mûr′jə·bəl) *adj.* Capable of being submerged. —**sub·mer′gi·bil′i·ty** *n.*

sub·mersed (səb·mûrst′) *adj.* 1 *Bot.* Growing under water. 2 Submerged. [<L *submersus*, pp. of *submergere* SUBMERGE]

sub·mers·i·ble (səb·mûr′sə·bəl) *adj.* That may be submerged. —*n.* A submarine.

sub·mi·cron (sub·mī′kron) *n.* A particle of from 50 to 1,000 angstroms in diameter.

sub·mi·cro·scop·ic (sub·mī′krə·skop′ik) *adj.* Below the limit of vision in a microscope.

sub·mine (sub′mīn′) *n.* A small, electrically actuated mine located near a submarine mine and used in the training of navy personnel.

sub·min·i·a·ture camera (sub′min′ē·ə·chŏŏr) A miniature camera using 16-mm. film for taking still photographs.

sub·min·i·a·tur·ize (sub·min′ē·ə·chə·rīz′) *v.t.* **·ized, ·iz·ing** To reduce, as certain delicate instruments, to the smallest size compatible with efficient use and service, as in the design and production of hearing aids. —**sub·min′i·a·tur·i·za′tion** *n.*

sub·miss (səb·mis′) *adj. Archaic* Submissive; soft; subdued. [<L *submissus*, pp. of *submittere* SUBMIT]

sub·mis·sion (səb·mish′ən) *n.* 1 The act of submitting; a yielding to the power or authority of another; obedience. 2 The state or quality of being submissive; the spirit of subjection or obedience; an acquiescent temper; humility; resignation; meekness. 3 The act of referring, or the agreement to refer, a matter of controversy to arbitration. 4 *Archaic* Acknowledgment of error.

Synonyms: obedience, patience, resignation, subjection, submissiveness. See PATIENCE.

sub·mis·sive (səb·mis′iv) *adj.* Willing or inclined to submit; yielding; obedient; docile. See synonyms under DOCILE, HUMBLE, OBSEQUIOUS, PASSIVE, SUPPLE. —**sub·mis′sive·ly** *adv.* —**sub·mis′sive·ness** *n.*

sub·mit (səb·mit′) *v.* **·mit·ted, ·mit·ting** *v.t.* 1 To place under or yield to the authority, will, or power of another; surrender. 2 To present for the consideration, decision, or approval of others; refer. 3 To present as one's opinion; suggest. —*v.i.* 4 To give up; surrender. 5 To be obedient or submissive; be acquiescent. See synonyms under BEND[1], DEFER, OBEY. [<L *submittere*, var. of *summittere* <*sub-* underneath + *mittere* send] —**sub·mit′tal** *n.* —**sub·mit′ter** *n.*

sub·mon·tane (sub·mon′tān) *adj.* 1 Situated at the foot of a mountain or mountain range. 2 Beneath a mountain. —**sub·mon′tane·ly** *adv.*

sub·mul·ti·ple (sub·mul′tə·pəl) *n. Math.* A number or quantity that is contained in another without remainder; an aliquot part. —*adj.* Contained in something an exact number of times. [<LL *submultiplus* <*sub-* opposite of, lesser + *multiplus* MULTIPLE]

sub·nor·mal (sub·nôr′məl) *adj.* 1 Below the normal. 2 *Psychol.* Of less than normal intelligence. —*n.* 1 *Math.* That portion of the axis of a curve included between the ordinate of one of its points and the normal to that point. 2 A subnormal individual. —**sub·nor·mal·i·ty** (sub′nôr·mal′ə·tē) *n.*

sub·o·ce·an·ic (sub′ō·shē·an′ik) *adj.* Occurring, formed, or happening beneath the ocean floor.

sub·or·der (sub′ôr′dər) *n.* 1 *Biol.* A category of animals or plants next below an order. 2 A subordinate architectural order modifying the principal order, generally for decoration. —**sub·or·di·nal** (sub·ôr′də·nəl) *adj.*

sub·or·di·nar·y (sub·ôr′də·ner′ē) *n. pl.* **·nar·ies** *Her.* One of a class of armorial charges usually considered less honorable than the ordinaries. Among them are the *bordure, flanch, orle, tressure*, etc.

sub·or·di·nate (sə·bôr′də·nit) *adj.* 1 Belonging to an inferior order in a classification; secondary; minor. 2 Subject or subservient to another; inferior in any way. 3 Dependent; joining dependent words to others. See synonyms under AUXILIARY, SUBJECT. —*n.* One who is subordinate; an inferior in rank or official position. —*v.t.* (-nāt) **·nat·ed, ·nat·ing** 1 To make subordinate; assign to a lower order or rank; hence, to hold as of less importance. 2 To make subject or subservient. [<L *subordinatus*, pp. of *subordinare* <*sub-* under + *ordinare* order] —**sub·or′di·nate·ly** *adv.* —**sub·or′di·nate·ness** *n.* —**sub·or′di·na′tion** *n.*

subordinate conjunction See under CONJUNCTION.

sub·or·di·na·tion·ism (sə·bôr′də·nā′shən·iz′əm) *n. Theol.* The doctrine that the second and third persons of the Trinity are inferior to the first person. —**sub·or′di·na′tion·ist** *n.*

sub·or·di·na·tive (sə·bôr′də·nā′tiv) *adj.* Having a tendency to or expressive of subordination.

sub·orn (sə·bôrn′) *v.t.* 1 To bribe or procure (someone) to commit perjury. 2 To incite or instigate to an evil act, especially a criminal act. 3 *Obs.* To decorate or adorn. [<L *subornare* <*sub-* secretly + *ornare* equip] —**sub·orn′er** *n.* —**sub·or·na·tion** (sub′ôr·nā′shən) *n.*

Su·bo·ti·ca (sōō′bô′ti·tsä) A city in northern Serbia, Yugoslavia: German *Maria Theresiopel*, Hungarian *Szabadka*. Also **Su′bo′ti·tsa.**

sub·ox·ide (sub·ok′sīd) *n. Chem.* An oxide having the minimum amount of oxygen.

sub·phy·lum (sub·fī′ləm) *n. Biol.* A primary division of a phylum, superior to the class.

sub·plinth (sub′plinth′) *n. Archit.* A block or base supporting a plinth; a second or lower plinth.

sub-plot (sub′plot′) *n.* A plot subordinate to the principal one in a novel, play, etc.

sub·poe·na (sə·pē′nə, səb-) *n.* A judicial writ requiring a person to appear at a specified time and place under penalty for default. —*v.t.* To notify or summon by writ or subpoena. Also **sub·pe′na.** [<Med. L <L *sub poena* <*sub* under + *poena* penalty]

sub-port (sub′pôrt′, -pōrt′) *n.* An auxiliary port, equipped to handle traffic diverted from the main port.

sub-post (sub′pōst′) *n.* An administrative subdivision of a military post.

sub·pre·fect (sub·prē′fekt) *n.* A subordinate prefect; in France, the administrative officer of an arrondissement. —**sub·pre′fec·ture** (-fek·chər) *n.*

sub·prin·ci·pal (sub·prin′sə·pəl) *n.* 1 A vice principal. 2 A rafter or brace next to or auxiliary to one of the main timbers of the frame. 3 *Music* An open diapason sub-bass in an organ.

sub·ra·mose (sub·rā′mōs) *adj. Bot.* 1 Branching moderately, as a plant. 2 Having few branches. [<NL *subramosus* <L *sub-* somewhat, under + *ramosus* RAMOSE]

sub·re·gion (sub′rē′jən) *n.* A subdivision of a region, especially with reference to the distribution of animals. —**sub·re′gion·al** *adj.*

sub·rep·tion (səb·rep′shən) *n.* 1 A procuring of some favor or reward by means of a fraudulent concealment or suppression of the truth. 2 Inference resulting from concealment, or misrepresentation of essential elements or facts. [<L *subreptio, -onis* <*subreptus*, pp. of *subripere* <*sub-* secretly + *rapere* snatch, seize]

sub·ro·gate (sub′rō·gāt) *v.t.* **·gat·ed, ·gat·ing** 1 To substitute (one thing) for another. 2 To substitute (one person) for another when attributing or assigning rights or appointing to an office. [<L *subrogatus*, pp. of *subrogare* substitute <*sub-* in place of + *rogare* ask]

sub·ro·ga·tion (sub′rō·gā′shən) *n.* 1 The succession or substitution of one person or thing by or for another. 2 *Law* The putting of a person who (as a surety) has paid the debt of another in the place of the creditor to whom he has paid it.

sub ro·sa (sub rō′zə) *Latin* Confidentially; in secret: literally, under the rose: because, in Egypt, the rose was the emblem of Horus, (Roman Harpocrates), mistakenly regarded by the Greeks and Romans as the god of silence, for he was often depicted as a child with finger on mouth.

sub·scap·u·lar (sub·skap′yə·lər) *adj. Anat.* Situated underneath the scapula. Also **sub·scap′u·lar′y** (-ler′ē). [<NL *subscapularis* <L *sub-* under + *scapula* a shoulder blade]

sub·scribe (səb·skrīb′) *v.* **·scribed, ·scrib·ing** *v.t.* 1 To write, as one's name, at the end of a document; sign. 2 To sign one's name to as an expression of assent, acceptance, etc.; attest to by signing. 3 To promise, especially in writing, to pay or contribute (a sum of money). —*v.i.* 4 To write one's name at the end of a document. 5 To give sanction, support, or approval; agree. 6 To promise to pay or contribute money. 7 To agree to receive and pay for an article, as a periodical, usually by written agreement: with *to*. [<L *subscribere* <*sub-* underneath + *scribere* write] —**sub·scrib′er** *n.*

sub·script (sub′skript) *adj.* 1 Written following and slightly beneath, as a small letter: iota *subscript*. 2 *Math.* Of a subindex. —*n.* A subscript sign, symbol, or letter. Compare SUPERSCRIPT. [<L *subscriptus*, pp. of *subscribere*. See SUBSCRIBE.]

sub·scrip·tion (səb·skrip′shən) *n.* 1 The act of subscribing; signature; hence, consent, confirmation, or agreement. 2 That which is subscribed; a signed paper or statement. 3 A signature written at the end of a document. 4 A signed acceptance of religious articles. 5 The individual or total sum or number subscribed for any purpose. 6 A formal agreement or undertaking evinced by signature, as payment of a certain price for the receipt of a magazine, book, ticket, etc. 7 *Archaic* Submission; obedience. 8 The part of a doctor's prescription which gives directions for compounding the ingredients. 9 The sale of books, magazines, tickets, etc., by mail or by personal canvass. —**to take up a subscription** To collect money (for some special purpose or cause) from a large number of people. —**sub·scrip′tive** *adj.* —**sub·scrip′tive·ly** *adv.*

subscription list A list of the names of people and the amounts they have subscribed, as for a periodical, a charity, or other cause.

sub·sec·tion (sub·sek′shən, sub′sek′shən) *n.* A subdivision of a section.

sub·sec·tor (sub·sek′tər, sub′sek′tər) *n.* A portion of a military sector or coastal frontier marked out for convenience in operations.

sub·se·quence (sub′sə·kwəns) *n.* 1 The condition of being subsequent. 2 The act of following. Also **sub′se·quen·cy.**

sub·se·quent (sub′sə·kwənt) *adj.* 1 Following in time, place, or order, or as a result. 2 Succeeding; consequent. [<L *subsequens, -entis*, ppr. of *subsequi* <*sub-* next below + *sequi* follow] —**sub′se·quent·ly** *adv.* —**sub′se·quent·ness** *n.*

sub·serve (səb·sûrv′) *v.t.* **·served, ·serv·ing** 1 To be of use or help in furthering (a process, cause, etc.); serve; promote. 2 To serve as a subordinate to (a person). See synonyms under SERVE. [<L *subservire* <*sub-* under + *servire* SERVE]

sub·ser·vi·ent (səb·sûr′vē·ənt) *adj.* 1 Adapted to promote some end or purpose; being of service; useful as a subordinate. 2 Hence, acting in the interests of another; servile; obsequious; truckling. —*n.* One who or that which subserves. See synonyms under BASE[2]. [<L *subserviens, -entis*, ppr. of *subservire* SUBSERVE] —**sub·ser′vi·ent·ly** *adv.* —**sub·ser′vi·ent·ness, sub·ser′vi·ence, sub·ser′vi·en·cy** *n.*

sub-shrub (sub′shrub′) *n.* An undershrub or very small shrub. —**sub′-shrub′by** *adj.*

sub·side (səb·sīd′) *v.i.* **·sid·ed, ·sid·ing** 1 To sink to a lower level. 2 To become less violent or agitated; become calm or quiet; abate. 3 To sink to the bottom, as sediment; settle. See synonyms under ABATE, FALL. [<L *subsidere* <*sub-* under + *sidere* settle <*sedere* sit]

sub·sid·ence (səb·sīd′ns, sub′sə·dəns) *n.* 1 The settling of heavy parts to the bottom; precipitation. 2 The sinking of water or other liquids to a lower or usual level: *subsidence* of a flood. 3 A gradual settling into a quiet or inactive state. 4 A gradual settling of the earth to a lower level, because of ground movements or underground workings. [<L *subsidentia* sediment <*subsidere* SUBSIDE]

sub·sid·i·ar·y (səb·sid′ē·er′ē) *adj.* 1 Assisting in an inferior capacity; supplementary; auxiliary; secondary. 2 Of, pertaining to, or in the nature of a subsidy; helping by a subsidy. —*n. pl.* **·ar·ies** 1 One who or that which furnishes supplemental aid or supplies; an auxiliary; assistant. 2 *Music* A theme subordinate to or dependent on the main theme or subject. [<L *subsidiarius* <*subsidium* <*subsidere* SUBSIDE] — **sub·sid'i·ar'i·ly** *adv.*

subsidiary coin Coin of small denomination, legal tender only to a limited amount; in the United States, any coin worth less than a dollar.

subsidiary company A company controlled by another company which owns the greater part of its shares.

sub·si·dize (sub'sə·dīz) *v.t.* **·dized, ·diz·ing** 1 To furnish with a subsidy; grant a regular allowance or pecuniary aid to. 2 To obtain the assistance of by a subsidy: now often implying bribery. Also *Brit.* **sub'si·dise.** — **sub'si·di·za'tion** *n.* — **sub'si·diz'er** *n.*

sub·si·dy (sub'sə·dē) *n. pl.* **·dies** 1 Pecuniary aid directly granted by government to an individual or private commercial enterprise deemed beneficial to the public. 2 Formerly, an aid or tax granted by the House of Commons to the king for urgent needs of the kingdom. 3 Any financial assistance afforded by one individual or government to another. [<AF *subsidie,* OF *subside* <L *subsidium* auxiliary forces, aid <*subsidere* SUBSIDE] *Synonyms*: aid, allowance, bonus, bounty, gift, grant, indemnity, pension, premium, reward, support, subvention, tribute. A nation grants a *subsidy* to an ally, pays a *tribute* to a conqueror. An *indemnity* is a single reparation demanded for a specific injury, while a *tribute* may be exacted indefinitely. A nation may also grant a *subsidy* to its own citizens as a means of promoting the public welfare; as, a *subsidy* to a steamship company. The somewhat rare term *subvention* is especially applied to a *grant* of governmental aid to a literary or artistic enterprise. The word *bounty* may be applied to almost any regular or stipulated *allowance* by a government to a citizen or citizens; as, a *bounty* for enlisting in the army, a *bounty* for killing wolves, a land *bounty* to encourage settlement of sparsely populated areas. A *bounty* is reward for a single act; a *pension* is earned by long service.

sub·sist (səb·sist') *v.i.* 1 To have existence or reality; continue to exist. 2 To remain alive; manage to live. 3 To continue unchanged; abide. 4 To have existence in or by something; inhere. — *v.t. Obs.* 5 To provide with food and clothing; support. See synonyms under LIVE. [<MF *subsister* <L *subsistere* <*sub-* under + *sistere* cause to stand <*stare* stand] — **sub·sist'er** *n.*

sub·sis·tence (səb·sis'təns) *n.* 1 The act of subsisting. 2 That on which one subsists; sustenance; means of support; livelihood. 3 The state of being subsistent; inherent quality. 4 That which subsists; real being. 5 A basis; a logical substance; hypostasis. Also **sub·sis'ten·cy.** [<LL *subsistentia* <*subsistere* SUBSIST]

subsistence department A former department of the army that provided and had charge of subsistence stores: these are now purchased and issued by the Quartermaster Corps: also called *commissary department.*

sub·sis·tent (səb·sis'tənt) *adj.* 1 That subsists or is inherent. 2 Existing; having real being or action. 3 Having subsistence.

sub·soil (sub'soil') *n.* The stratum of earth next beneath the surface soil. — *v.t.* To turn up the subsoil of; plow with a subsoil plow. — **sub'·soil'er** *n.*

subsoil plow *Agric.* A plow specially designed for loosening or turning up the subsoil.

sub·so·lar (sub·sō'lər) *adj.* 1 Situated directly beneath the sun, as at high noon at the equinoxes; also, between the tropics. 2 Mundane; earthly.

sub·son·ic (sub·son'ik) *adj.* 1 Designating those sound waves beyond the lower limits of human audibility, or with frequencies of less than about 25 cycles per second; infrasonic. 2 Of, pertaining to, characterized, or operated by such waves. Compare SUPERSONIC.

sub·spe·cies (sub·spē'shēz, -shiz) *n. Biol.* A subdivision of a species, variously ranked but usually distinguished by minor differences in characteristics and by having a particular geographic range within a larger area. [<NL <L *sub-* under + *species* an appearance, sort]

sub·stance (sub'stəns) *n.* 1 The material of which anything is made or constituted. 2 The essential part of anything said or written, put into a brief, condensed statement; the gist or purport. 3 The vital part of that which is spiritual or emotional. 4 Material possessions; wealth; property. 5 That which gives stability or solidity; confidence; ground. 6 *Philos.* The essential nature that underlies phenomena; the permanent cause underlying outward manifestations; that in which qualities or attributes inhere. 7 In Christian Science, Spirit. 8 Any particular kind of material. 9 Essential components or characteristic elements of ideas: The tenets are the same in *substance.* See synonyms under MASS[1]. [<OF <L *substantia* <*substare* be present <*sub-* under + *stare* stand]

sub·stan·dard (sub·stan'dərd) *adj.* 1 Below the standard. 2 Lower than the established rate or authorized requirements.

sub·stan·tial (səb·stan'shəl) *adj.* 1 Solid; strong; firm. 2 Of real worth and importance; of considerable value; valuable. 3 Considerable and sure. 4 Possessed of wealth or sufficient means; responsible. 5 Of or pertaining to substance; having real existence; not illusory; actual; permanent; lasting. 6 Containing or conforming to the essence of a thing; giving the correct idea; essential; material; fundamental. 7 Ample and nourishing. — *n.* 1 That which has substance; a reality. 2 The more important part. — **sub·stan'ti·al'i·ty** (-shē·al'ə·tē), **sub·stan'tial·ness** *n.* — **sub·stan'tial·ly** *adv.*

sub·stan·tial·ism (səb·stan'shəl·iz'əm) *n. Philos.* The doctrine that substantial realities are the sources or underlying ground of all phenomena, material and mental; the doctrine that matter is a real substance. — **sub·stan'tial·ist** *n.*

sub·stan·ti·ate (səb·stan'shē·āt) *v.t.* **·at·ed, ·at·ing** 1 To establish, as a position or a truth, by substantial evidence; verify. 2 To give form to; embody. 3 To make substantial, existent, or real; give substance to. See synonyms under CONFIRM, RATIFY. [<NL *substantiatus,* pp. of *substantiare* establish <L *substantia* SUBSTANCE] — **sub·stan'ti·a'tion** *n.* — **sub·stan'ti·a'tive** *adj.*

sub·stan·ti·val (sub'stən·tī'vəl) *adj.* 1 Of or pertaining to a substantive. 2 Self-existent. — **sub'stan·ti'val·ly** *adv.*

sub·stan·tive (sub'stən·tiv) *n.* 1 A noun. 2 Anything used in place of a noun, as a verbal form, phrase, or clause. 3 One who or that which is independent; a self-subsisting person or thing. — *adj.* 1 Capable of being used as a noun. 2 Expressive of or denoting existence: The verb "to be" is called the *substantive* verb. 3 Having substance or reality; hence, lasting. 4 Being an essential part or constituent. 5 Relating to what is essential. 6 Having distinct individuality. 7 Independent in resources; self-supporting, as a country. 8 Of considerable amount; substantial. 9 In dyeing, not needing a mordant. [<OF *substantif* <LL *substantivus* <L *substantia* SUBSTANCE] — **sub'stan·tive·ness** *n.*

substantive dye See under DYE.

sub·stan·tiv·ize (sub'stən·tiv·īz') *v.t.* **·ized, ·iz·ing** To treat or use as a substantive: The adjective "meek" is *substantivized* in "Blessed are the meek."

sub·sta·tion (sub'stā'shən) *n.* A subsidiary station, as an electric power station for switching, transforming, or converting purposes, a branch post office, etc.

sub·stit·u·ent (səb·stich'ōō·ənt) *n. Chem.* A radical, atom, or group, substituting or replacing another in a chemical reaction. — *adj.* Of a substituting atom or molecule. [<L *substituens, -entis,* ppr. of *substituere* SUBSTITUTE]

sub·sti·tute (sub'stə·tōōt, -tyōōt) *v.* **·tut·ed, ·tut·ing** *v.t.* 1 To put in the place of another person or thing. 2 To take the place of. — *v.i.* 3 To act as a substitute. 4 *Chem.* To exchange one constituent of a compound for, or replace it with, another. See synonyms under CHANGE. — *n.* 1 One who or that which takes the place or serves in lieu of another. 2 In the American Civil War, one hired to serve in the place of a man drafted into military service. 3 Any substance or material adapted to replace another in a given product or process, or for a specified purpose: Gelatin is a *substitute* for agar, synthetic rubber for cork, etc.: also called *alternative, replacement.* See synonyms under DELEGATE. [<L *substitutus,* pp. of *substituere* <*sub-* in place of + *statuere* set up]

sub·sti·tu·tion (sub'stə·tōō'shən, -tyōō'-) *n.* 1 The act of substituting, or the state of being substituted. 2 *Chem.* Any reaction which involves the replacement of certain elements or radicals by others: said especially of organic compounds. — **sub'sti·tu'tion·al** *adj.* — **sub'·sti·tu'tion·al·ly** *adv.*

sub·sti·tu·tive (sub'stə·tōō'tiv, -tyōō'-) *adj.* Acting or tending to act as a substitute; admitting of substitution.

sub·strate (sub'strāt) *n.* 1 *Biochem.* The material or substance acted upon by an enzyme or ferment. 2 A substratum. [<SUBSTRATUM]

sub·stra·tum (sub·strā'təm, -strat'əm) *n. pl.* **·stra·ta** (-strā'tə, -strat'ə) 1 An underlying stratum or layer, as of earth or rock; subsoil. 2 That which forms the foundation or groundwork. 3 Matter or mind considered as the ground of qualities and phenomena; the substance possessing attributes. 4 The substance in which something takes root, as vegetable or animal tissue. [<NL <L, pp. neut. of *substernere* spread underneath <*sub-* underneath + *sternere* strew] — **sub·stra'tive** *adj.*

sub·struc·tion (sub·struk'shən) *n.* A foundation. [<F <L *substructio, -onis* <*substruere* <*sub-* underneath + *struere* build] — **sub·struc'tion·al** *adj.*

sub·struc·ture (sub·struk'chər, sub'struk'-) *n.* 1 A structure serving as a foundation of a building, etc. 2 Groundwork. 3 The earthen roadway supporting railroad tracks. — **sub·struc'tur·al** *adj.*

sub·sume (səb·sōōm') *v.t.* **·sumed, ·sum·ing** 1 To place in some particular class; classify. 2 To include, as the specific or individual in the general. [<NL *subsumere* <L *sub-* underneath + *sumere* take] — **sub·sum'a·ble** *adj.*

sub·sump·tion (səb·sump'shən) *n.* 1 The act of subsuming. 2 That which is subsumed; an assumption; especially, the minor premise of a syllogism as stated after the major premise. 3 Formerly, a narrative of an alleged crime giving minute particulars. [<NL *subsumptio, -onis* <*subsumere* SUBSUME] — **sub·sump'·tive** *adj.*

sub·tan·gent (sub·tan'jənt) *n. Geom.* The portion of the axis of a curve cut off between the tangent to a given point and the ordinate of that point. [<NL *subtangens, -entis* <L *sub-* under + *tangens,* ppr. of *tangere* touch]

sub·tem·per·ate (sub·tem'pər·it) *adj.* 1 Pertaining to the colder parts of the temperate zone. 2 Slightly temperate.

sub·ten·ant (sub·ten'ənt) *n.* A person who rents or leases from a tenant; a sublessee. — **sub·ten'an·cy** *n.*

sub·tend (sub·tend') *v.t.* 1 *Geom.* To extend under or opposite to, as the chord of an arc or the side of a triangle opposite to an angle. 2 *Bot.* To enclose in its axil: A leaf *subtends* a bud. [<L *subtendere* <*sub-* underneath + *tendere* stretch]

sub·tense (sub·tens') *Geom. n.* 1 A line that subtends an arc or angle. 2 The chord of an arc. — *adj.* Pertaining to or used in estimating distance by measuring the subtended angle. [<NL *subtensa (linea)* (a) subtended (line), pp. fem. of L *subtendere* SUBTEND]

subter- *prefix* Under; less than: opposed to *super-*: *subteraqueous.* [<L *subter* below, beneath]

sub·ter·a·que·ous (sub'tə·rā'kwē·əs, -rak'wē-) *adj.* Situated beneath the surface of the water. Also **sub'ter·ra'que·ous.** [<L (assumed) *subteraqueus* <*subter-* beneath + *aqua* water]

sub·ter·fuge (sub'tər·fyōōj) *n.* That to which one resorts for escape or concealment; an evasion of an issue; a plan to avoid censure; a false excuse. See synonyms under ARTIFICE, SOPHISTRY. [<L *subterfugium* <*subterfugere* <*subter-* below, in secret + *fugere* flee, take flight]

sub·ter·nat·u·ral (sub'tər·nach'ər·əl) *adj.* Below the norms of nature.

sub·ter·rane (sub'tə·rān) *n.* 1 A basal or underlying terrane. 2 An underground room; a cave. [<L *subterraneus* <*sub-* under + *terra* the earth]

sub·ter·ra·ne·an (sub'tə·rā'nē·ən) *adj.* 1 Situated or occurring below the surface of the earth: contrasted with *subaerial* and *surficial*; underground. 2 Hidden. Also **sub'ter·ra'ne·al, sub'ter·ra'ne·ous, sub'ter·rene'** (-tə·rēn').

sub·ter·res·tri·al (sub'tə·res'trē·əl) *adj.* Subterranean; lower than the terrestrial. —*n.* A creature that lives underground.

sub·tile (sut'l, sub'til) *adj.* **1** Having fine structure; delicately formed; ethereal. **2** Characterized by material rarity; rarefied; refined; hence, penetrating; pervasive. **3** Subtle. [< OF *subtil*, alter of *soutil* SUBTLE; refashioned after L] —**sub'tile·ly** *adv.* —**sub'tile·ness** *n.*

Synonym: subtle. *Subtile* and *subtle* have been constantly used as interchangeable by good writers; but there is a present tendency to distinguish them by making *subtile* an attribute of things and *subtle* a characteristic of mind. *Subtle*, the later form of the word, is used preferably when the derogatory sense of crafty is to be expressed. See ACUTE, ASTUTE, FINE[1].

sub·til·i·ty (sub·til'ə·tē) *n.* The quality or state of being subtile; thinness; fineness.

sub·til·ize (sut'l·īz, sub'tə·līz) *v.* **·ized, ·iz·ing** *v.t.* **1** To make subtle or subtile; refine. **2** To make acute; sharpen, as the senses. **3** To discuss or argue subtly. —*v.i.* **4** To make subtle distinctions; use subtlety. [<Med. L *subtilizare* <L *subtilis* SUBTLE] —**sub'til·i·za'tion** *n.*

sub·til·ty (sut'l·tē, sub'təl·tē) *n. pl.* **·ties** **1** Refinement or niceness, or an instance of it; a nicety. **2** Subtlety.

sub·ti·tle (sub'tīt'l) *n.* A subordinate or explanatory title, as in a book, play, or document; a book title repeated, as on top of the first page of the text.

sub·tle (sut'l) *adj.* **1** Characterized by cunning, craft, or artifice; wily; crafty. **2** Keen; penetrative; discriminating: *subtle* humor; overrefined. **3** Apt; skilful. **4** Executed with nice art; ingenious; clever. **5** Insidious; secretly active. **6** Hard to understand; abstruse. **7** Of delicate texture. **8** Subtile. See synonyms under ACUTE, ASTUTE, FINE[1], INSIDIOUS, SUBTILE. [<OF *soutil* <L *subtilis* fine, orig. closely woven < *sub-* under + *tela* a web] —**sub'tle·ness** *n.* —**sub'tly** *adv.*

sub·tle·ty (sut'l·tē) *n. pl.* **·ties** **1** The state or quality of being subtle. **2** The ability to make fine distinctions; keenness of perception. **3** Something subtle, as a nice distinction.

sub·ton·ic (sub·ton'ik) *adj. Phonet.* Sonant or voiced, as certain consonants. —*n.* **1** *Phonet.* A subtonic sound. **2** *Music* The seventh of the scale; a semitone below the tonic.

sub·tor·rid (sub·tôr'id, -tor'-) *adj.* Subtropical.

sub·tract (səb·trakt') *v.t.* & *v.i.* To take away or deduct, as a portion from the whole, or one quantity from another. [<L *subtractus*, pp. of *subtrahere* < *sub-* away + *trahere* draw] —**sub·tract'er** *n.*

sub·trac·tion (səb·trak'shən) *n.* **1** The act or process of subtracting; a deducting; something deducted. **2** *Math.* The operation of finding the difference between two quantities (symbol −).

sub·trac·tive (səb·trak'tiv) *adj.* **1** Serving or tending to diminish. **2** *Math.* Having the minus sign; to be subtracted.

subtractive process *Phot.* A method of making two or more negatives through filters which exclude all but a desired color: used in color printing and engraving.

sub·tra·hend (sub'trə·hend) *n. Math.* That which is to be subtracted from a number or quantity (the minuend) to give the difference. [<L *subtrahendus (numerus)* (the number) to be subtracted, gerundive of *subtrahere* SUBTRACT]

sub·trans·lu·cent (sub'trans·lōō'sənt, -tranz-) *adj.* Not fully translucent, as certain gemstones and other minerals.

sub·treas·ur·y (sub·trezh'ər·ē) *n. pl.* **·ur·ies** **1** A branch of the U. S. Treasury Department maintained for receipt and safekeeping of government revenues: established in 1840 and abolished in 1920. **2** The building that housed such a branch. —**sub·treas'ur·er** *n.*

sub·trop·i·cal (sub·trop'i·kəl) *adj.* **1** Of, pertaining to, or designating regions adjacent to the tropical zone. **2** Designating either of two irregular belts of high atmospheric pressure roughly between 30° and 40° latitude, north and south. Also **sub·trop'ic.**

sub·trop·ics (sub·trop'iks) *n. pl.* Subtropical regions.

su·bu·late (sōō'byə·lāt, -lit) *adj. Biol.* Shaped like an awl; slender and tapering to a point. [<NL *subulatus* <L *subula* an awl]

sub·um·brel·la (sub'um·brel'ə) *n. Zool.* The under surface of the swimming bell of a jellyfish, or that surface situated in the region of the mouth. —**sub·um·bral** (sub·um'brəl), **sub'·um·brel'lar** *adj.*

sub·urb (sub'ûrb) *n.* A place adjacent to a city; in the plural, collectively, environs; outskirts; outlying residential districts; purlieus. [<OF *suburbe* <L *surburbium* < *sub-* near to + *urbs, urbis* a city]

sub·ur·ban (sə·bûr'bən) *adj.* Of or pertaining to a suburb; dwelling or located in a place which is a combination of the rural and urban. —*n.* A suburbanite.

sub·ur·ban·ite (sə·bûr'bən·īt) *n.* A resident of a suburb.

sub·ur·bi·a (sə·bûr'bē·ə) *n.* **1** The social and cultural world of suburbanites. **2** Suburbs or suburbanites collectively.

sub·ur·bi·car·i·an (sə·bûr'bə·kâr'ē·ən) *adj.* Being in the suburbs (of Rome): applied to the six sees that compose the province of the pope as metropolitan. [<LL *suburbicarius* <L *suburbium* SUBURB]

sub·vene (səb·vēn') *v.i.* **·vened, ·ven·ing** To come or happen so as to be of aid or support, especially by preventing something; intervene. [<L *subvenire* come to one's assistance < *sub-* up from under + *venire* come]

sub·ven·tion (səb·ven'shən) *n.* **1** The act of subvening; giving of succor; aid. **2** That which aids, especially a grant, as of money; subsidy. See synonyms under SUBSIDY. [<OF *subvencion* <LL *subventio, -onis* <L *subvenire.* See SUBVENE.] —**sub·ven'tion·ar'y** (-er'ē) *adj.*

sub·ver·sion (səb·vûr'shən, -zhən) *n.* **1** The act of subverting, or the state of being subverted; a demolition; overthrow. **2** A cause of ruin. Also **sub·ver'sal** (-səl). See synonyms under RUIN. [<OF <LL *subversio, -onis* <L *subvertere* SUBVERT]

sub·ver·sive (səb·vûr'siv) *adj.* Tending to subvert or overthrow. —*n.* A person who engages in subversion.

sub·vert (səb·vûrt') *v.t.* **1** To overthrow from the very foundation; destroy utterly. **2** To corrupt; undermine the morals or character of. [<OF *subvertir* <L *subvertere* overturn < *sub-* up from under + *vertere* turn] —**sub·vert'er** *n.* —**sub·vert'i·ble** *adj.*

Synonyms: destroy, extinguish, overthrow, overturn, supersede, supplant. To *supersede* implies the putting of something that is preferred in the place of that which is removed; to *subvert* does not imply substitution. To *supplant* is more often personal, signifying to take the place of another, usually by underhand means; one is *superseded* by authority, *supplanted* by a rival. See ABOLISH. *Antonyms*: conserve, perpetuate, preserve, sustain, uphold.

sub·vit·re·ous (sub·vit'rē·əs) *adj.* Having a luster resembling that of glass, but less brilliant.

sub·way (sub'wā) *n.* **1** An artificial passage below the surface of the ground; specifically, one for traffic, water and gas mains, electric cables, etc. **2** An underground railroad, usually electrically operated; also, a tunnel for such a railroad.

suc– Assimilated var. of SUB–.

suc·ce·da·ne·um (suk'si·dā'nē·əm) *n. pl.* **·ne·ums** or **·ne·a** (-nē·ə) One who or that which is a substitute. [<NL, neut. sing. of L *succedaneus* < *succedere* succeed, replace] —**suc'ce·da'ne·ous** *adj.*

suc·ceed (sək·sēd') *v.i.* **1** To come next in order or sequence; follow; ensue. **2** To come after another into office, ownership, etc.; be the successor: often with *to.* **3** To be successful; accomplish what is attempted or intended; also, formerly, to achieve an end in a specified manner: They *succeeded* badly. **4** *Law* To devolve: said of an estate. —*v.t.* **5** To be the successor or heir of. **6** To come after in time or sequence; follow. [<OF *succeder* <L *succedere* go under, follow after < *sub-* under + *cedere* go] —**suc·ceed'er** *n.*

Synonyms: achieve, attain, flourish, prevail, prosper, thrive, win. To *win* implies that someone loses, but one may *succeed* where no one fails. A solitary swimmer *succeeds* in reaching the shore; if we say he *wins* the shore we place him in competition with the water. Many students may *succeed* in study; a few *win* the special prizes for which all compete. See FOLLOW.

suc·cen·tor (sək·sen'tər) *n.* **1** A deputy precentor; subcantor; subchanter. **2** The leading bass or bass soloist in a church or cathedral choir. [<LL <L *succinere* sing to < *sub-* subordinately + *canere* sing]

suc·cès d'es·time (sük·se' des·tēm') *French* Success marked by the praise of critics but not by widespread popular approval: said of a play, book, etc.

suc·cess (sək·ses') *n.* **1** A favorable or prosperous course or termination of anything attempted; prosperous or advantageous issue. **2** A successful person or affair. **3** *Obs.* The outcome or result, favorable or unfavorable. **4** *Obs.* Succession. See synonyms under VICTORY. [<L *successus* < *succedere* SUCCEED]

suc·cess·ful (sək·ses'fəl) *adj.* **1** Of persons, obtaining what one desires or intends; especially, having reached a high degree of worldly prosperity. **2** Of things, terminating in or meeting with success; resulting favorably: said of a course of action, etc. See synonyms under AUSPICIOUS, FORTUNATE, HAPPY. —**suc·cess'ful·ly** *adv.* —**suc·cess'ful·ness** *n.*

suc·ces·sion (sək·sesh'ən) *n.* **1** The act of following in order, or the state of being successive; a following consecutively. **2** A group of things that succeed in order; a series, either in time or in place; sequence. **3** The act or right of legally or officially coming into a predecessor's office, possessions, etc.; also, the order of so succeeding, or that which is or is to be so taken. **4** The right or act of succeeding to a throne. **5** Descendants collectively; issue. See synonyms under TIME. —**suc·ces'sion·al** *adj.* —**suc·ces'sion·al·ly** *adv.*

suc·ces·sive (sək·ses'iv) *adj.* Following in succession; consecutive. —**suc·ces'sive·ly** *adv.* —**suc·ces'sive·ness** *n.*

suc·ces·sor (sək·ses'ər) *n.* One who or that which follows in succession; especially, a person who succeeds to a throne, property, or office.

suc·ci·nate (suk'si·nāt) *n. Chem.* A salt of succinic acid. [<SUCCIN(IC) + -ATE[3]]

suc·cinct (sək·singkt') *adj.* **1** Reduced or comprised within a narrow compass; terse; concise. **2** Supported by an encircling silken thread, as a butterfly chrysalis. **3** *Archaic* Encircled or held in position by or as by a girdle. See synonyms under TERSE. [<L *succinctus*, pp. of *succingere* < *sub-* underneath + *cingere* gird] —**suc·cinct'ly** *adv.* —**suc·cinct'ness** *n.*

suc·cinc·to·ri·um (suk'singk·tôr'ē·əm, -tō'rē-) *n. pl.* **·to·ri·a** (-tôr'ē·ə, -tō'rē·ə) A band or scarf embroidered with an Agnus Dei, worn pendent from the girdle: used by the pope on solemn occasions. [<LL <L *sub-* under + *cinctorium* a girdle < *cinctus*, pp. of *cingere* gird]

suc·cin·ic (sək·sin'ik) *adj.* Derived from or found in amber. [<F *succinique* <L *succinum* amber]

succinic acid *Chem.* Either of two white crystalline isomeric compounds, $C_4H_6O_2$, contained in amber and in certain plants, and also made synthetically.

suc·cor (suk'ər) *n.* **1** Help or relief rendered in danger, difficulty, or distress. **2** One who or that which affords relief. —*v.t.* To go to the aid of; help; rescue. See synonyms under AID, HELP, SERVE. Also *Brit.* **suc'cour.** [<OF *sucurs* <Med. L *succursus* <L *succurrere* < *sub-* up from under + *currere* run] —**suc'cor·a·ble** *adj.* —**suc'cor·er** *n.*

suc·co·ry (suk'ər·ē) *n.* Chicory. [Alter. of *cicoree, sichorie*, earlier vars. of CHICORY; infl. in form by MDu. *sukerie* chicory]

suc·co·tash (suk'ə·tash) *n.* A dish of Indian corn kernels and beans boiled together. [<Algonquian (Narraganset) *misickquatash* an ear of corn]

Suc·coth (sōōk'ōth, sōōk'ōs) See SUKKOTH.

suc·cu·bus (suk'yə·bəs) *n. pl.* **·bi** (-bī) One of a class of demons in female form fabled to have intercourse with men in their sleep. [<Med. L <LL *succuba* a strumpet <L *succubare* < *sub-* underneath + *cubare* lie]

suc·cu·lent (suk'yə·lənt) *adj.* **1** *Bot.* Juicy;

add,āce,câre,pälm; end,ēven; it,īce; odd,ōpen,ôrder; tŏŏk,pōōl; up,bûrn; ə = a in *above*, e in *sicken*, i in *clarity*, o in *melon*, u in *focus*; yōō = u in *fuse*; oi,oil; ou,pout; ch,check; g,go; ng,ring; th,thin; th,this; zh,vision. Foreign sounds à,œ,ü,kh,ṅ; and ◆: see page xx. < from; + plus; ? possibly.

succumb 1252 **Suez**

fleshy, as the tissues of certain plants. 2 Rich or vigorous: a *succulent* theme. [< *succulentus* < *succus* juice] — **suc'cu·lence**, **suc'cu·len·cy** n. — **suc'cu·lent·ly** adv.

suc·cumb (sə·kum') v.i. 1 To give way; yield, as to force or persuasion. 2 To die. [< OF *succomber* < L *succumbere* < *sub-* underneath + *cumbere* lie] — **suc·cum'bent** (-bənt) adj.

suc·cuss (sə·kus') v.t. To shake suddenly or forcibly. [< L *succussus*, pp. of *succutere* < *sub-* up from under + *quatere* shake] — **suc·cus'sive** adj.

suc·cus·sion (sə·kush'ən) n. 1 The act of shaking. 2 *Med.* A vigorous shaking of the patient to detect liquids in the thorax or other cavities of the body. Also **suc·cus·sa·tion** (suk'ə·sā'shən). — **suc·cus'sa·to·ry** (sə·kus'ə·tôr'ē, -tō'rē) adj.

such (such) adj. 1 Of that kind; of the same or like kind: often with *as* or *that* completing the comparison: *Such* wit *as* this is rare. 2 Specifically, being the same as what has been mentioned or indicated: *Such* was the king's command. 3 Being the same in quality: Let the truthful continue *such*. 4 Being the same as something understood by the speaker or the hearer, or purposely left indefinite: a concise and elliptical use by which specification is avoided: the chief of *such* a clan. 5 So extreme, unpleasant, or the like: an emphatic or expletive use: We have come to *such* a pass. — *pron.* 1 Such a person or thing, or (more commonly) such persons or things: by ellipsis of the noun: The friend of *such* as are in trouble. 2 The same; the aforesaid: I bring good tidings, for *such* the general sent. — adv. So: *such* destructive criticism. [OE *swelc*, *swilc*, *swylc*]

such-and-such (such'ən·such') adj. Being a particular person, thing, or time, not specifically named: He visited *such-and-such* a place. Also **such and such.**

such·like (such'līk') adj. Of a like or similar kind. — pron. Persons or things of that kind: mosses, ferns, and *suchlike*.

Sü·chow (shü'jō') 1 A city in SW Shantung province, China: formerly (1912-45) *Tungshan.* 2 A former name for IPIN.

suck (suk) v.t. 1 To draw into the mouth by means of a partial vacuum created by action of the lips and tongue. 2 To draw in or take up in a manner resembling this; inhale; absorb: The sponge *sucked* the water up. 3 To draw liquid or nourishment from with the mouth: to *suck* a lemon; also, to take into and hold in the mouth as if to do this: to *suck* one's thumb. 4 To consume by licking, or by holding in the mouth: to *suck* candy. 5 To bring to a specified state or condition by sucking: He *sucked* the lemon dry. — v.i. 6 To draw in liquid, air, etc., by suction. 7 To suckle. 8 To draw in air instead of water, as a defective pump does. 9 To make a sucking sound. — n. 1 The act of sucking; suction. 2 That which is sucked or comes by sucking. 3 A slight draft or drink. 4 A mother's milk. 5 A whirlpool or powerful eddy. [OE *sūcan*] — **suck'ing** adj. — **suck'ing·ly** adv.

suck·er (suk'ər) n. 1 One who or that which sucks; a suckler, as a sucking pig or a newly born whale. 2 A North American fresh-water fish (family *Catostomidae*), related to the cyprinoids, having the mouth usually protractile with thick and fleshy lips adapted for sucking in food. 3 *Zool.* An organ by which an animal adheres to other bodies; a suctorial organ. 4 *Slang* A toady; sponger; parasite; hanger-on. 5 *U.S. Slang* A foolish fellow; dolt; one easily deceived; a gull. 6 A piston, as of a syringe or a suction pump; a tube or pipe used for suction. 7 *Bot.* A shoot or branch originating on a subterranean portion of a stem. b A shoot or sprout arising from the root near or remote from the trunk of certain trees. 8 A haustorium. 9 A sweetmeat; also, sugar. — v.t. 10 To strip of suckers or shoots. — v.i. To form or send out suckers or shoots. [< SUCK]

Sucker State Nickname for ILLINOIS.

suck·fish (suk'fish') n. pl. **·fish** or **·fish·es** 1 A remora. 2 A fish (*Caularchus maeandricus*) of the Pacific coast, with a ventrally placed sucker by which it attaches itself to stones, shells, etc.

suck·le (suk'əl) v. **·led**, **·ling** v.t. 1 To give suck to, as at the breast. 2 To bring up; nourish. — v.i. 3 To take nourishment at the breast; suck. [ME *sucklen*, freq. of *suken* SUCK] — **suck'ler** n.

suck·ling (suk'ling) n. 1 An unweaned mammal. 2 A young, inexperienced person.

Suck·ling (suk'ling), **Sir John**, 1609-42, English poet and dramatist.

su·crate (soo'krāt) n. A compound in which sucrose or some analogous carbohydrate combines with a base to form a salt: calcium *sucrate*. [< F *sucre* sugar + *-ate* -ATE³]

su·cre (soo'krā) n. The monetary unit of Ecuador. [< Sp., after Antonio José de *Sucre*]

Su·cre (soo'krā) The nominal capital of Bolivia. La Paz is the seat of government.

Su·cre (soo'krā), **Antonio José de**, 1795?-1830, South American soldier; first president of Bolivia.

su·crose (soo'krōs) n. *Chem.* 1 Any one of the group of carbohydrates, including cane sugar, milk sugar, maltose, etc., having the common composition $C_{12}H_{22}O_{11}$, and deviating the plane of polarized light to the right. 2 Cane sugar as obtained from the sugarcane, maple, beet, etc. Also called saccharose. [< F *sucre* sugar + *-ose* -OSE²]

suc·tion (suk'shən) n. 1 The act or process of sucking. 2 The production of a partial vacuum in a space connected with a fluid or gas under pressure; also, any like act or effect. [< OF < L *suctio*, *-onis* < *sugere* suck]

suction pump A pump operating by suction, consisting of a piston working up and down in a cylinder, both equipped with valves: the most common form of house pump. Compare illustration under FORCE PUMP.

suction stop *Phonet.* A click, as in the Bushman and Hottentot languages. See CLICK (def. 3).

Suc·to·ri·a (suk·tôr'ē·ə, -tō'rē·ə) n. pl. A class or subclass of aquatic protozoans having in the adult stage long hollow tentacles for piercing and sucking. [< NL < L *suctus*, pp. of *sugere* suck]

suc·to·ri·al (suk·tôr'ē·əl, -tō'rē·əl) adj. 1 Adapted for sucking or for adhesion. 2 *Zool.* Living by sucking; having organs for sucking.

sud (sood) v. *Scot.* Should.

su·dan (soo·dan') adj. *Chem.* Designating any of a class of diazo compounds widely used as red and yellow dyes. [from *Sudan*]

Su·dan (soo·dan') A region extending across Africa from the Atlantic Ocean to the Red Sea, south of the Sahara: formerly *Nigritia*.

Sudan, Republic of the An independent country in NE Africa; 967,500 square miles; capital, Khartoum: formerly Anglo-Egyptian Sudan.

Su·da·nese (soo'də·nēz', -nēs') adj. Of or pertaining to the Sudan or its people. — n. pl. ·nese One living in the Sudan; the people of the Sudan collectively, including Negro and Negroid peoples, Hamites and certain Arab tribes.

Su·dan·ic (soo·dan'ik) n. A family of languages spoken in central Africa from the Atlantic to the Indian oceans, including Dinka, Ewe, Nubian, and Yoruba. — adj. Of or pertaining to this family.

su·dan·o·phil (soo·dan'ə·fil) adj. *Biol.* Staining readily with sudan dyes, as certain tissues or cells. Also **su·dan'o·phil'ic**, **su·dan·oph·i·lous** (soo'dən·of'ə·ləs). [< *sudano-* (< SUDAN) + -PHIL]

su·dar·i·um (soo·dâr'ē·əm) n. pl. ·dar·i·a (-dâr'ē·ə) 1 A handkerchief or cloth for drying or removing perspiration; specifically, the sweat cloth or handkerchief of St. Veronica, said to have been miraculously impressed with the features of Jesus when she wiped his face on his way to crucifixion. 2 The napkin about the head of Christ in the tomb. *John* xx 7. 3 Any miraculous picture of Christ; a veronica. 4 A sudatory (def. 2). Also **su·da·ry** (soo'dər·ē). [< L *sudor*, *-oris* sweat]

su·da·tion (soo·dā'shən) n. Morbid or excessive sweating. [< L *sudatio*, *-onis* < *sudatus*, pp. of *sudare* sweat]

su·da·to·ry (soo'də·tôr'ē, -tō'rē) adj. 1 Producing perspiration; sudorific. 2 Perspiring. — n. pl. ·ries 1 An agent that causes sweating; a sudorific. 2 A sweating bath; specifically, a hot-air room in a Roman bath: also **su·da·to·ri·um** (soo'də·tôr'ē·əm). [< L *sudatorius*]

sudd (sud) n. A floating mass of vegetation that frequently obstructs navigation on the White Nile. [< Arabic < *sudd* obstruct]

sud·den (sud'n) adj. 1 Happening quickly and without warning: *sudden* death. 2 Hurriedly or quickly contrived, used, or done; hasty. 3 Come upon unexpectedly; causing surprise. 4 Quick-tempered; precipitate; rash. See synonyms under IMPETUOUS, SWIFT¹. — n. The state of being sudden, or that which is sudden: obsolete except in a few phrases: **all of a sudden**, **all on a sudden**, **on a sudden** Without warning; on the spur of the moment. [< AF *sodein*, OF *soudain* < L *subitaneus* < *subitus*, pp. of *subire* come or go stealthily < *sub-* secretly + *ire* go] — **sud'den·ly** adv. — **sud'den·ness** n.

Su·der·mann (zoo'dər·män), **Hermann**, 1857-1928, German dramatist and novelist.

Su·de·ten·land (soo·dāt'n·land, *Ger.* zoo·dā'tən·länt) The border district of Bohemia and Moravia, Czechoslovakia; 8,976 square miles.

Su·de·tes (soo·dē'tēz) A mountainous system along the German-Czechoslovak and Polish-Czechoslovak border; highest point, 5,259 feet. Also **Su·det·ic Mountains** (soo·det'ik).

su·dor (soo'dôr) n. Visible perspiration; sweat. [< L] — **su·dor·al** (soo'dər·əl) adj.

su·dor·if·er·ous (soo'də·rif'ər·əs) adj. Secreting or producing sweat. [< NL *sudoriferus* < L *sudor*, *-oris* sweat + *ferre* carry] — **su'dor·if'er·ous·ness** n.

su·dor·if·ic (soo'də·rif'ik) adj. Causing perspiration. — n. A medicine that produces or promotes sweating. [< NL *sudorificus* < L *sudor*, *-oris* + *facere* make]

suds (sudz) n. pl. 1 Soapy water worked up into bubbles and froth; foam; lather. 2 *Slang* Beer: so called from its foamy properties. [Prob. < MDu. *sudde*, *sudse* a marsh, marsh water] — **suds'y** adj.

sue (soo) v. **sued**, **su·ing** v.t. 1 *Law* a To institute proceedings against for the recovery of some right or the redress of some wrong. b To prosecute (an action). c To seek a grant from (a court). 2 To endeavor to persuade by entreaty; beg; urge; petition. 3 To seek to win in marriage; woo. — v.i. 4 To institute legal proceedings. 5 To make entreaty. 6 *Archaic* To pay court; woo. [< AF *suer*, OF *sivre*, ult. < L *sequi* follow] — **su'er** n.

Sue (soo) Diminutive of SUSANNA.

Sue (soo, *Fr.* sü), **(Marie Joseph) Eugène**, 1804-57, French novelist.

suède (swād) n. Undressed kid: often attributively: *suède* gloves. [< F *Suède* Sweden, in phrase *gants de Suède* Swedish gloves]

suède fabric A woven or knitted fabric of cotton, rayon, or wool, finished to resemble suède leather: used for sports coats and jackets, linings, gloves, etc.

su·et (soo'it) n. The fatty tissues about the loins and kidneys of sheep, oxen, etc.: used in cookery and for making tallow. [Dim. of AF *sue*, OF *seu* < L *sebum* tallow, fat] — **su'et·y** adj.

Sue·to·ni·us (swi·tō'nē·əs), **Gaius Tranquillus**, A.D. 70?-140?, Roman historian.

Su·ez (soo·ez', soo'ez) A port in NE Egypt at the northern end of the **Gulf of Suez**, the NW arm of the Red Sea (about 180 miles long, 20 miles wide).

Suez, Isthmus of The neck of land joining Asia and Africa; between the Gulf of Suez and the Mediterranean; 72 miles wide at its narrowest point; traversed by the **Suez Canal**, a ship canal 107 miles long, 197 feet wide, constructed (1859–69) by Ferdinand de Lesseps.

suf- Assimilated var. of SUB-.

suf·fa·ri (sə·fä′rē) See SAFARI.

suf·fer (suf′ər) *v.i.* 1 To feel pain or distress. 2 To be affected injuriously; suffer loss or injury. 3 To undergo punishment; especially, to be put to death. 4 *Archaic* To tolerate or endure pain, injury, etc. — *v.t.* 5 To have inflicted on one; sustain, as an injury or loss. 6 To undergo; pass through, as change. 7 To bear; endure: He cannot *suffer* more pain. 8 To allow; permit: Will he *suffer* us to leave? See synonyms under ALLOW, ENDURE, PERMIT. [<AF *suffrir*, OF *sofrir*, ult. <L *sufferre* < *sub-* up from under + *ferre* bear] — **suf′·fer·er** *n.*

suf·fer·a·ble (suf′ər·ə·bəl, suf′rə-) *adj.* Such as can be suffered or endured; endurable. — **suf′·fer·a·ble·ness** *n.* — **suf′fer·a·bly** *adv.*

suf·fer·ance (suf′ər·əns, suf′rəns) *n.* 1 Permission given or implied by failure to prohibit; negative consent. 2 In customs, a permit for the shipment of certain kinds of goods to specified ports. 3 The act or state of suffering; wretchedness; experience of pain or evil; power to endure. 4 Patience or endurance under suffering; submission; submissiveness. 5 *Rare* Loss; injury; damage. See synonyms under PATIENCE. [<AF, OF *suffrance* <LL *sufferentia* < *sufferre* SUFFER]

suf·fer·ing (suf′ər·ing, suf′ring) *n.* 1 The state of anguish or pain of one who suffers; the bearing of pain, injury, or loss. 2 The pain so borne; distress; loss; injury. See synonyms under AGONY, PAIN. — *adj.* Inured to pain and loss; submissive. — **suf′fer·ing·ly** *adv.*

suf·fice (sə·fīs′) *v.* **·ficed, ·fic·ing** *v.i.* To be sufficient or adequate; meet the requirements or answer the purpose. — *v.t.* To be satisfactory or adequate for; satisfy. See synonyms under SATISFY, SERVE. [<OF *suffis-*, stem of *suffire* <L *sufficere* < *sub-* under + *facere* make] — **suf·fic′er** *n.*

suf·fi·cien·cy (sə·fish′ən·sē) *n. pl.* **·cies** 1 The state of being sufficient. 2 That which is sufficient; especially, adequate pecuniary means or income; a competency. 3 Full capability or qualification; efficiency. 4 Conceit; self-sufficiency. See synonyms under COMFORT.

suf·fi·cient (sə·fish′ənt) *adj.* 1 Being all that is needful; adequate; enough. 2 *Archaic* Capable; competent. 3 *Obs.* Financially competent; responsible. See synonyms under ADEQUATE, AMPLE, ENOUGH. [<OF <L *sufficiens, -entis*, ppr. of *sufficere* SUFFICE] — **suf·fi′cient·ly** *adv.*

suf·fix (suf′iks) *n.* 1 *Ling.* A letter or letters added to the end of a word or root, and functioning as a formative, derivative, or inflectional element, as *-er* in shorter, *-ful* in faithful, *-s* and *-es* in dogs, boxes, *-ed* in loved, *-ness* in kindness, etc. Compare COMBINING FORM, PREFIX. 2 Any added title or the like. 3 *Math.* A subindex. — *v.t.* To add as a suffix; append. [<NL *suffixum* <L *suffixus*, pp. of *suffigere* < *sub-* underneath + *figere* fix. Doublet of SOFFIT.] — **suf′fix·al** *adj.* — **suf·fix·ion** (sə·fik′shən) *n.*

suf·flate (sə·flāt′) *v.t.* **·flat·ed, ·flat·ing** *Obs.* To blow up or inflate. [<L *sufflatus*, pp. of *sufflare* < *sub-* up from under + *flare* blow] — **suf·fla′tion** *n.*

suf·fo·cant (suf′ə·kənt) *n.* Any substance or agent that produces suffocation.

suf·fo·cate (suf′ə·kāt) *v.* **·cat·ed, ·cat·ing** *v.t.* 1 To kill by obstructing respiration in any manner. 2 To obstruct or oppress, as by an inadequate supply of air. 3 To stifle; extinguish; smother, as a fire. — *v.i.* 4 To become choked or stifled; die from suffocation. [<L *suffocatus*, pp. of *suffocare* < *sub-* under + *fauces* throat] — **suf′fo·cat′ing·ly** *adv.* — **suf′fo·ca′tion** *n.* — **suf′fo·ca′tive** *adj.*

Suf·folk (suf′ək) *n.* 1 A breed of hardy, chestnut-colored English working horse, smaller and freer from feather than the Shire and Clydesdale breeds. It is heavy in body and has rather short legs. Also **Suffolk punch.** 2 A breed of hornless, short-wool Southdown sheep, with black face and legs: preeminent for the quality of its mutton. [from *Suffolk*, England]

Suf·folk (suf′ək) A county in eastern England; 1,507 square miles; administratively divided into **East Suffolk** (879 square miles; county town, Ipswich) and **West Suffolk** (628 square miles; county town, Bury St. Edmunds).

suf·fra·gan (suf′rə·gən) *Eccl. n.* An auxiliary or assistant bishop, who assists a bishop in the administration of the diocese, or is consecrated for service in a limited portion of the diocese: also **suffragan bishop.** — *adj.* Of or pertaining to a suffragan; assisting; auxiliary; subordinate to an archiepiscopal see. [<AF, OF <Med. L *suffraganeus* <L *suffragari* vote for, support] — **suf′fra·gan·ship′** *n.*

suf·frage (suf′rij) *n.* 1 A vote in support of some measure or candidate; hence, approbation; assent. 2 Voting; also, the right or privilege of voting; franchise: also **political suffrage.** 3 *Eccl.* Any short intercessory prayer or petition. — **woman suffrage** Political suffrage as belonging to or exercised by women. In the United States suffrage was granted to women in 1920 by the 19th amendment to the Constitution: also **female suffrage.** [<OF <L *suffragium* a voting tablet, vote]

suf·fra·gette (suf′rə·jet′) *n. Colloq.* A woman who advocated female suffrage; specifically, a member of a militant organization demanding it. [<SUFFRAGE + -ETTE] — **suf′fra·get′·tism** *n.*

suf·fra·gist (suf′rə·jist) *n.* 1 A voter. 2 An advocate of some particular form of suffrage, especially of woman suffrage.

suf·fru·tex (suf′rə·teks) *n. Bot.* 1 An undershrub; a small plant having a decidedly woody stem. 2 An herb with a permanent woody base. [<NL <L *sub-* under, less than + *frutex, -icis* a shrub] — **suf′fru·tes′cent** (-tes′ənt) *adj.*

suf·fru·ti·cose (sə·froo′tə·kōs) *adj. Bot.* Somewhat shrubby; woody; shrubby or woody at base and herbaceous above. [<NL *suffruticosus* < *suffrutex, -icis.* See SUFFRUTEX.]

suf·fu·mi·gate (sə·fyoo′mə·gāt) *v.t.* **·gat·ed, ·gat·ing** To fumigate from or as from underneath. [<L *suffumigatus*, pp. of *suffumigare* < *sub-* up from under + *fumigare* FUMIGATE]

suf·fu·mi·ga·tion (sə·fyoo′mə·gā′shən) *n.* 1 The act of suffumigating. 2 The act of burning perfumes. 3 A fume or vapor.

suf·fuse (sə·fyooz′) *v.t.* **·fused, ·fus·ing** To overspread, as with a vapor, fluid, or color. [<L *suffusus*, pp. of *suffundere* < *sub-* underneath, up from under + *fundere* pour] — **suf·fu·sive** (sə·fyoo′siv) *adj.*

suf·fu·sion (sə·fyoo′zhən) *n.* 1 The act of welling up or spreading over. 2 The state of being suffused; a blush. 3 That which suffuses: a *suffusion* of blood.

Su·fi (soo′fē) *n.* A follower of a system of Moslem philosophical and devotional mysticism, especially in Persia. [<Arabic *sufi*, lit., a man of wool < *sūf* wool] — **Su′fic, Su·fis·tic** (soo·fis′tik) *adj.*

Su·fism (soo′fiz·əm) *n.* The doctrine of the Sufis, which has inspired a mass of symbolical religious poetry.

sug- Assimilated var. of SUB-.

sug·ar (shoog′ər) *n.* 1 A sweet crystalline disaccharide having the formula $C_{12}H_{22}O_{11}$, obtained chiefly from the juice of the sugarcane or sugar beet; called, according to its source, **beet sugar, cane sugar, date sugar, grape sugar, maple sugar,** etc. ◆ Collateral adjective: *saccharine.* 2 Any of a large class of sweet, soluble, optically active carbohydrates which are ketone or aldehyde derivatives of the higher alcohols. They are widely distributed in plants and animals, play an important role in nutrition, and are generally classified on the basis of chemical structure as monosaccharides, disaccharides, trisaccharides. 3 Flattering or honeyed words, especially if used to disguise or soften an unpleasant or severe reality. 4 *Slang* Sweet one: a pet name. — *v.t.* 1 To sweeten, cover, or coat with sugar. 2 To make agreeable or less distasteful, as by flattery. — *v.i.* 3 *U.S. & Can.* To make maple sugar. 4 To form or produce sugar; granulate. [<OF *sucre* < Med. L *succarum*, ult. <Arabic *sukkar.* Prob. related to SACCHARIN.]

sugar apple The sweetsop.

sugar beet Any sugar-producing variety of the common garden beet.

sug·ar·ber·ry (shoog′ər·ber′ē) *n. pl.* **·ries** The hackberry.

sug·ar·bird (shoog′ər·bûrd′) *n.* 1 Any bird that sucks the nectar of flowers, as the honey creepers, honey-eaters, sunbirds, etc. 2 The evening grosbeak (*Hesperiphona vespertina*): so named by North American Indians from its fondness for maple sugar.

sugar bush A grove of sugar-maple trees: sometimes designating a grove of 200 or more trees.

sugar camp The collection of cabins and other buildings in a sugar bush where the maple sap is boiled.

sug·ar·cane (shoog′ər·kān′) *n.* A tall, stout, perennial grass (*Saccharum officinarum*) of tropical regions with a solid jointed stalk rich in sugar.

sug·ar-coat (shoog′ər·kōt′) *v.t.* 1 To cover with sugar. 2 To cause to appear attractive or less distasteful. — **sug′ar-coat′ed** *adj.* — **sug′ar-coat′ing** *n.*

sugar corn Sweet corn (def. 1).

sug·ar-cured (shoog′ər·kyoord′) *adj.* Cured by using sugar in the curing process: said of ham and pork.

sugar daddy *U.S. Slang* A wealthy old man who gives a young woman presents in return for her favors.

sug·ared (shoog′ərd) *adj.* Sugar-coated; honeyed; pleasant; sweetened.

sug·ar·house (shoog′ər·hous′) *n.* 1 A building in which the juices are extracted from sugarcane, sugar beets, etc., and made into raw sugar; a sugar refinery. 2 A building in which sugar is stored. 3 A building in a sugar camp in which maple sap is boiled.

sugaring off 1 The boiling of maple sap until it crystallizes into sugar. 2 The time of year at which this is done. 3 A community social gathering to take part in making maple sugar.

sugaring over Making palatable the unpalatable, especially facts.

sugar loaf 1 A conical mass of hard refined sugar. 2 A conical hat or hill. — **sug′ar-loaf′** *adj.*

Sugar Loaf Mountain A peak in Río de Janeiro, Brazil, at the entrance to Guanabara Bay; 1,296 feet: Portuguese *Pão de Açúcar.*

sugar maple The maple (*Acer saccharum*) of eastern North America from the sap of which maple sugar is made: also called *hard maple, rock maple.*

sugar of lead Lead acetate.

sugar of milk Lactose.

sugar orchard An orchard of sugar maples.

sugar pine A tall pine (*Pinus lambertiana*) of the Pacific coast, bearing very large cones and having wood much used in construction work.

sug·ar·plum (shoog′ər·plum′) *n.* 1 A small sweetmeat; a small ball or disk of candy; a bon-bon. 2 The shadbush.

sugar tree The sugar maple.

sug·ar·y (shoog′ər·ē) *adj.* 1 Composed of or as of sugar; sweet. 2 Fond of sugar. 3 Figuratively, honeyed; alluring. 4 Granular. — **sug′ar·i·ness** *n.*

sug·gan (sug′ən) *n.* 1 A type of rope made of twisted straw. 2 A saddle, collar, or bolster so made. 3 A heavy bed coverlet. [<Irish *sūgān*]

sug·gest (səg·jest′, sə·jest′) *v.t.* 1 To bring or put forward for consideration, action, or approval; propose. 2 To arouse in the mind by association or connection; connote: Hallowe'en *suggests* witches and black cats. 3 To give a hint or indirect suggestion of; intimate: This poem *suggests* a great deal of care and thought. 4 To act as or provide a motive for; prompt: The success of his novel *suggested* a sequel. See synonyms under ALLUDE, IMPORT. [<L *suggestus*, pp. of *suggerere* < *sub-* underneath + *gerere* carry] — **sug·gest′er** *n.*

sug·gest·i·bil·i·ty (səg·jes′tə·bil′ə·tē, sə-) *n.* 1 *Psychol.* Responsiveness to suggestion, normal in children and diminishing in adults, but heightened or abnormal in hypnosis, light sleep, and certain nervous conditions. 2 Readiness to believe and agree without reflection; compliancy of mind and will.

sug·gest·i·ble (səg·jes′tə·bəl, sə-) *adj.* 1 That can be suggested. 2 Easily led; yielding, especially to hypnosis: a *suggestible* patient.

sug·ges·tion (səg·jes′chən, sə·jes′-) *n.* **1** The act of suggesting. **2** A hint; insinuation. **3** The spontaneous calling up of an idea in the mind by a connected idea. **4** *Psychol.* **a** The inducing in a person of some idea, impulse, action, or mode of behavior through a stimulus, verbal or other, coming from another person but independent of critical argument or rational persuasion, as in hypnosis. **b** The idea, impulse, etc., so induced.

Synonyms: hint, innuendo, insinuation, intimation. A *suggestion* brings something before the mind less directly than by formal or explicit statement, as by a partial statement, an incidental allusion, an illustration, a question, or the like. *Suggestion* is often used of an unobtrusive statement of one's views or wishes to another, leaving consideration and any consequent action entirely to that person's judgment, and is hence, in many cases, the most respectful way in which to convey one's views to a superior or a stranger. An *intimation* is a *suggestion* in brief utterance, or sometimes by significant act, gesture, or token, of one's meaning or wishes; in the latter case it is often the act of a superior. A *hint* is still more limited in expression and more remote, and is always covert, but frequently with good intent; as, to give one a *hint* of danger or of opportunity. *Insinuation* and *innuendo* usually imply discredit; an *insinuation* is a covert or partly veiled injurious utterance; an *innuendo* is commonly secret as well as sly, as if pointing to something derogatory. See COUNSEL.

sug·ges·tive (səg·jes′tiv, sə-) *adj.* **1** Fitted or tending to suggest; stimulating to thought or reflection. **2** Hinting at indecent thoughts; suggesting the improper. — **sug·ges′tive·ly** *adv.* — **sug·ges′tive·ness** *n.*

sugh (so͞okh) *n. Scot.* A rushing sound; sough.

su·i·ci·dal (so͞o′ə·sīd′l) *adj.* Self-destructive; ruinous; pertaining to, or leading to, suicide; fatal to one's prospects or interests. — **su′i·ci′dal·ly** *adv.*

su·i·cide (so͞o′ə·sīd) *n.* **1** The intentional taking of one's own life. **2** Self-inflicted political, social, or commercial ruin. **3** One who commits self-murder. — *v.i.* **·cid·ed, ·cid·ing** *Colloq.* To commit suicide. [<NL *suicidium* <L *sui* of oneself + *caedere* kill]

su·i gen·e·ris (so͞o′ī jen′ər·is) *Latin* Literally, of his (her, its) particular kind; forming a kind by itself; unique.

su·i ju·ris (so͞o′ī jo͝or′is) *Latin* In one's own right; having legal capacity to act for oneself.

su·int (so͞o′int, swint) *n.* Natural wool grease from wool-washings: it consists of fatty substances combined with potash salts. [<F < *suer* sweat <L *sudare*]

Suisse (swēs) The French name for SWITZERLAND.

Sui·sun Bay (sə·so͞on′) The easternmost arm of San Francisco Bay.

suit (so͞ot) *n.* **1** A set of outer garments or armor to be worn together. **2** A set of garments consisting of a coat and trousers or skirt, made of the same fabric. **3** A group of things of like kind or pattern composing a series or set: now usually *suite.* **4** In card-playing, any one of the four sets of thirteen cards each that make up a pack, as spades, hearts, diamonds, or clubs. **5** *Law* A proceeding in a court of law or chancery in which a plaintiff demands the recovery of a right or the redress of a wrong: a term rarely applied to criminal prosecution. **6** *Archaic* Entreaty; petition; supplication. **7** The courting or courtship of a woman. **8** *U.S. Colloq.* A growth of hair or whiskers. See synonyms under PRAYER. — **to follow suit** **1** To play a card identical in suit to the card led. **2** To do as somebody or something else has done; follow an example. — *v.t.* **1** To meet the requirements of, or be appropriate to; be in accord with; befit. **2** To please; satisfy. **3** To render appropriate or accordant; accommodate; adapt. **4** *Archaic* To furnish with clothes. — *v.i.* **5** To be befitting; agree; correspond. **6** To be or prove satisfactory. **7** *Obs.* To clothe oneself. See synonyms under ACCOMMODATE, ADAPT. [< AF *siwte,* OF *sieute,* ult. <L *sequi* follow. Doublet of SUITE.]

suit·a·ble (so͞o′tə·bəl) *adj.* Capable of suiting; appropriate; applicable; proper. See synonyms under APPROPRIATE, BECOMING, CONVENIENT, EXPEDIENT, GOOD. — **suit′a·bil′i·ty, suit′a·ble·ness** *n.* — **suit′a·bly** *adv.*

suit·case (so͞ot′kās′) *n.* A flat, rectangular valise used for carrying clothing, etc.

suite (swēt) *n.* **1** A succession of things forming a series; a set of things having a certain dependence upon each other and intended to go or be used together. **2** A number of connected apartments or a set of furniture. **3** A collection of pictures illustrating consecutive events. **4** *Music* A form of instrumental composition formerly consisting of a series of dance movements intended for one instrument, but now often written for an orchestra and varying freely in its construction and movements. **5** A retinue; a company of attendants or followers. ◆ Homophone: *sweet.* [<F <OF *sieute.* Doublet of SUIT.]

suit·ing (so͞o′ting) *n.* Cloth from which to make entire suits of clothes.

suit·or (so͞o′tər) *n.* **1** One who institutes a suit in court. **2** A wooer. **3** A petitioner. [<AF *seutor* <LL *secutor, -oris* <L *secutus,* pp. of *sequi* follow]

Sui·yü·an (swā′yü·än′) A former province of northern China, incorporated in the Inner Mongolian Autonomous Region, June, 1954; 135,000 square miles; capital, Kweisui.

Su·khu·mi (so͞o′kho͞o·mē) A port on the Black Sea, capital of Abkhaz Autonomous S.S.R.

su·ki·ya·ki (so͞o′kē·yä′kē, -yak′ē) *n.* A Japanese dish of meat cooked with vegetables to which sake, sugar, and a little soya sauce are added. [<Japanese]

Suk·koth (so͝ok′ōth, so͝ok′ōs) *n. pl.* The feast of Tabernacles, a Jewish holiday beginning on the 15th of Tishri (late September–October): originally a harvest festival: also spelled *Succoth.* Also **Suk′kos, Suk′kot.** [<Hebrew *sūkōth* tabernacles, booths]

Suk·kur (so͝ok′o͝or) A city in SE central West Pakistan, on the Indus.

Su·ky (so͞o′kē) Diminutive of SUSANNA.

Su·la Islands (so͞o′lə) An Indonesian island group between the Banggai and the Obi island groups; total, 1,873 square miles. *Dutch* **Soe·la** (so͞o′lä).

sul·cate (sul′kāt, -kit) *adj. Biol.* Having long narrow furrows or channels; grooved; fluted. Also **sul′cat·ed.** [<L *sulcatus,* pp. of *sulcare* plow < *sulcus* a furrow] — **sul·ca′tion** *n.*

sul·cus (sul′kəs) *n. pl.* **·ci** (-sī) **1** A narrow channel or furrow. **2** *Anat.* One of a large number of shallow grooves on the surface of the mammalian brain. [<L]

sul·dan (sul′dən) See SOLDAN.

Su·lei·man (sü′lā·män′) Name of three Turkish rulers: also spelled *Solyman.* — **Suleiman the Magnificent,** 1496?–1566, Ottoman sultan 1520–66; added extensive territories to the Turkish Empire in Europe and encouraged arts and science.

sulfa- *combining form Chem.* Sulfur; related to or containing sulfur: also spelled *sulpha-.* Also, before vowels, **sulf-,** as in *sulfarsenide.* See also SULFO-. [<SULFUR]

sul·fa·di·a·zine (sul′fə·dī′ə·zēn) *n. Chem.* A white, crystalline, relatively non-toxic derivative of sulfanilamide, $C_{10}H_{10}N_4SO_2$, used in the treatment of infections due to streptococci, pneumococci, and staphylococci.

sul·fa drug (sul′fə) *Chem.* Any of a group of organic compounds consisting mainly of substituted sulfanilamide derivatives and having a wide range of therapeutic effects in the treatment of bacterial infections.

sul·fa·gua·ni·dine (sul′fə·gwä′nə·din, -dēn, -gwan′ə-) *n. Chem.* A white, crystalline, relatively non-toxic sulfonamide, $C_7H_{10}N_4O_2S$, used in the treatment of certain infections.

sul·fal·de·hyde (sul·fal′də·hīd) *n. Chem.* An oily liquid compound, C_3H_6S, sometimes used as a hypnotic.

sul·fa·nil·a·mide (sul′fə·nil′ə·mīd, -mid) *n. Chem.* A colorless, crystalline sulfonamide, $C_6H_8N_2O_2S$, originally widely developed and used as a chemotherapeutic agent in the treatment of various bacterial infections. [< SULF(A)- + ANIL(INE) + AMIDE]

sul·fa·pyr·i·dine (sul′fə·pir′ə·dēn, -din) *n. Chem.* A white, crystalline derivative of sulfanilamide, $C_{11}H_{11}N_3O_2S$, once widely used in the treatment of various infections.

sulf·ar·se·nide (sulf·är′sə·nīd, -nid) *n. Chem.* A compound of the arsenide and sulfide of a metal or metals.

sulf·ars·phen·a·mine (sulf′ärs·fen′ə·mēn, -fen·am′in) *n. Chem.* A yellow, almost odorless, water-soluble powder containing from 18 to 20 percent arsenic: used in the treatment of syphilis.

sul·fate (sul′fāt) *n. Chem.* A salt of sulfuric acid. Sulfates are widely distributed in nature and are important in the arts and in medicine. — *v.* **·fat·ed, ·fat·ing** *v.t.* **1** To form a sulfate of; treat with a sulfate or sulfuric acid. **2** *Electr.* To form a coating of lead sulfate on (the plate of a secondary battery). **3** To make (red lead) into lead sulfate by the action of sulfuric acid. — *v.i.* **4** To become sulfated. Also **sul′phate.** [<F <NL *sulfas, -atis* a sulfate <L *sulfur* sulfur]

sulfate process A method for manufacturing tough kraft paper by introducing sulfate of soda in the digesters containing the wood pulp.

sul·fa·thi·a·zole (sul′fə·thī′ə·zōl) *n. Chem.* A sulfanilamide derivative, $C_9H_9N_3O_2S_2$, considered particularly effective in treating certain pneumococcal and staphylococcal infections.

sul·fa·tize (sul′fə·tīz) *v.t.* **·tized, ·tiz·ing** To turn (ores, etc.) into sulfate, by roasting. Also **sul′pha·tize.**

sulf·hy·dryl (sulf·hī′dril) *n. Chem.* The univalent thiol radical SH: also called *mercapto.* [<SULF(A)- + HYDR- + -YL]

sul·fide (sul′fīd) *n. Chem.* A compound of sulfur with an element or radical. Also **sul′fid** (-fid), **sul′phide, sul′phid.** [<SULF(A)- + -IDE]

sul·fi·nyl (sul′fə·nil) *n. Chem.* Thionyl. [<*sulfine,* var. of SULFONIUM + -YL]

sul·fite (sul′fīt) *n. Chem.* A salt or ester of sulfurous acid. Also **sul′phite.** [<SULF(A)- + -ITE[2]] — **sul·fit′ic** (-fit′ik) *adj.*

sulfite process The production of chemical wood pulp by the use of calcium sulfite.

sulfo- *combining form Chem.* **1** Sulfur; containing sulfur. **2** Denoting the replacement of oxygen by sulfur in a compound. **3** Indicating the presence of the sulfonic or sulfonyl group. Also spelled *sulpho-.* Compare THIO-. [<SULFUR]

Sul·fo·nal (sul′fə·nal, sul′fə·nal′) *n.* Proprietary name for a brand of sulfonmethane.

sul·fon·a·mide (sul·fon′ə·mīd, sul′fən·am′īd, -id) *n. Chem.* Any group of organic compounds containing the univalent radical SO_2NH_2, especially those derived from para-aminobenzene-sulfonamide, $p\text{-}H_2N{\cdot}C_6H_4{\cdot}SO_2NH_2$, used in the treatment of certain bacterial infections. [<SULFON(E) + AMIDE]

sul·fo·nate (sul′fə·nāt) *v.t.* **·nat·ed, ·nat·ing** *Chem.* **1** To form into a sulfonic acid. **2** To subject to the treatment of sulfonic acid. — *n.* A salt or ester of sulfonic acid. [<SULFON(E) + -ATE[3]] — **sul′fo·na′tion** *n.*

sul·fone (sul′fōn) *n. Chem.* Any of several compounds consisting of two organic radicals in combination with the sulfonyl group and corresponding to the formula R_2SO_2. [<G *sulfon*] — **sul·fon·ic** (sul·fon′ik) *adj.*

sul·fon·eth·yl·meth·ane (sul′fōn·eth′il·meth′ān) *n. Chem.* A colorless crystalline compound, $C_8H_{18}O_4S_2$, used as a hypnotic and sedative: a form of sulfonmethane. [<SULFON(E) + ETHYL + METHANE]

sulfonic acid *Chem.* Any of several compounds consisting of an organic radical in combination with the sulfonic radical and corresponding to the formula $R{\cdot}SO_2OH$: used in organic synthesis.

sul·fo·ni·um (sul·fō′nē·əm) *n. Chem.* The ion, H_3S, resulting from the addition of a proton to hydrogen sulfide. [<SULF(A) + (AMM)ONIUM]

sul·fon·meth·ane (sul′fōn·meth′ān) *n. Chem.* A white, crystalline, organic compound, $C_7H_{16}O_4S_2$, used in medicine as a sedative and hypnotic. [<SULFON(E) + METHANE]

sul·fo·nyl (sul′fə·nil) *n. Chem.* The bivalent radical SO_2: also called *sulfuryl.* [<SULFON(E) + -YL]

sul·fur (sul′fər) *n.* **1** A pale-yellow, nonmetallic element (symbol S) in the oxygen group, found both free and combined in the native state. It exists in several allotropic forms, of which the more important are the two crystalline modifications, one orthorhombic (native) and the other monoclinic. Sulfur burns to form sulfur dioxide, which has a penetrating odor; it is used for making matches, gunpowder, vulcanized rubber, sulfuric acid, etc. See ELEMENT. **2** Any one of various yellowish pieridine butterflies, as the

common North American **clouded sulfur** (*Colias philodice*) or the **cloudless sulfur** (*Callidryas eubule*). — **flowers of sulfur** A fine yellow powder obtained by the distillation of sulfur. — *v.t.* To treat or fume, as a wine cask or a hive, with sulfur or with sulfurous acid. Also **sul′phur.** [< AF *sulfre*, OF *soufre* < L *sulfur*, *-uris*]
sul·fu·rate (sul′fyə-rāt, -fə-) *v.t.* **·rat·ed**, **·rat·ing** To sulfurize. [< SULFUR + -ATE[1]]
sul·fur–bot·tom (sul′fər-bot′əm) *n.* A very large baleen whale (*Sibbaldius musculus*) found in Atlantic and Pacific waters, having a yellowish belly and attaining an average length of 60–80 feet, with a maximum of about 100 feet.

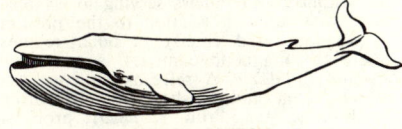

SULFUR–BOTTOM

sulfur dioxide *Chem.* A colorless, gaseous compound, SO_2, with a sharp odor and readily soluble in water: used in the manufacture of sulfuric acid, in bleaching, as a preservative, etc.
sul·fu·re·ous (sul-fyŏor′ē-əs) *adj.* Of or like sulfur. [< L *sulfureus* < *sulfur* sulfur]
sul·fu·ret (sul′fyə-ret) *v.t.* **·ret·ed** or **·ret·ted**, **·ret·ing** or **·ret·ting** To sulfurize. — *n.* (·rit) A sulfide. [< F *sulfuret* a sulfide < NL *sulfuretum* < L *sulfur*] — **sul′fu·ret′ed** or **sul′fu·ret′ted** *adj.*
sul·fu·ric (sul-fyŏor′ik) *adj. Chem.* Pertaining to or derived from sulfur, especially in its higher valence.
sulfuric acid *Chem.* A colorless, exceedingly corrosive, oily liquid, H_2SO_4, essentially a combination of sulfur trioxide and water, extensively employed in the manufacture of soda, batteries, guncotton, and in almost all chemical operations. Formerly called *vitriol.*
sul·fur·ize (sul′fyə-rīz, -fə-) *v.t.* **·ized**, **·iz·ing** 1 To impregnate, treat with, or subject to the action of sulfur. 2 To bleach or fumigate with sulfur. — **sul′fur·i·za′tion** *n.*
sul·fur·ous (sul′fər-əs, sul-fyŏor′əs) *adj.* 1 *Chem.* Of, pertaining to, or derived from sulfur: specifically applied to compounds that contain sulfur in its lower valence. 2 Fiery; hellish; blasphemous, as language.
sulfurous acid *Chem.* A compound corresponding to the formula H_2SO_3, and known only in solution and by its salts.
sulfur point *Physics* The boiling point of pure liquid sulfur at standard atmospheric pressure, 444.60° C.: one of the fixed points of the international temperature scale.
sulfur trioxide *Chem.* A compound, SO_3, formed by the union of sulfur dioxide and oxygen in the presence of a catalytic agent. With water, sulfur trioxide forms sulfuric acid; hence, it is often called **sulfuric anhydride.**
sul·fur·y (sul′fər-ē) *adj.* Resembling or suggesting sulfur; sulfureous.
sulfur yellow A light greenish-yellow color of very high brilliance, like the color of refined sulfur.
sul·fur·yl (sul′fər·il, -fyə-ril) *n.* Sulfonyl. [< SULFUR + -YL]
sulfuryl chloride *Chem.* A colorless, very pungent liquid compound, SO_2Cl_2, used in the manufacture of dyes, drugs, and poison gas.
sulk (sulk) *v.i.* To be sulky or morose. — *n.* A sulky mood or humor: often plural. [Back formation < SULKY]
sulk·y[1] (sul′kē) *adj.* **sulk·i·er**, **sulk·i·est** 1 Sullenly cross; doggedly or resentfully ill-humored. 2 Stunted; sluggish; dismal. See synonyms under MOROSE. [? OE (*ā*)*solcen* slothful, orig. pp. of (*ā*)*seolcan* be weak, slothful] — **sulk′i·ly** *adv.* — **sulk′i·ness** *n.*
sulk·y[2] (sul′kē) *n. pl.* **sulk·ies** A light, two-wheeled, one-horse vehicle for one person. — *adj.* Resembling this vehicle; a *sulky* plow. [< SULKY[1]; so called because one rides alone]
sull (sul) *n. Dial.* To sulk. [< SULLEN]

Sul·la (sul′ə), **Lucius Cornelius**, 138–78 B.C., Roman general and dictator.
sul·lage (sul′ij) *n.* 1 Mud or silt deposited by flowing water. 2 Refuse; sewage. [< AF *souiller*, *soillier* SOIL[2]; infl. in form by *sully*]
sul·len (sul′ən) *adj.* 1 Obstinately and gloomily ill-humored; morose; glum. 2 Depressing; somber: *sullen* clouds. 3 Slow; sluggish: a *sullen* tread. 4 Melancholy. 5 Ill-omened; threatening. See synonyms under GRIM. [Earlier *solein*, appar. < AF < *sol* SOLE[3]] — **sul′len·ly** *adv.* — **sul′len·ness** *n.*
Sul·li·van (sul′ə-vən), **Sir Arthur Seymour**, 1842–1900, English composer (often in collaboration with W. S. Gilbert). — **Harry Stack**, 1892–1949, U.S. psychiatrist, editor, and writer. — **John L(awrence)**, 1858–1918, U.S. pugilist. — **Louis Henri**, 1856–1924, U.S. architect.
Sul·li·vant (sul′ə-vənt), **William Starling**, 1803–73, U.S. botanist.
sul·ly (sul′ē) *v.* **·lied**, **·ly·ing** *v.t.* 1 To mar the brightness or purity of; soil; defile; tarnish. — *v.i.* To become soiled or tarnished: also figuratively. See synonyms under DEFILE, STAIN. — *n. pl.* **·lies** Anything that tarnishes; a stain; spot; blemish. [< MF *souiller* SOIL[2]]
Sul·ly (sul′ē, *Fr.* sü·lē′), **Duc de**, 1560–1641, Maximilien de Béthune, French statesman. — **Thomas**, 1783–1872, U.S. portrait painter born in England.
Sul·ly–Prud·homme (sü·lē′prü·dôm′), **René François Armand**, 1839–1907, French poet and critic.
sulph– For all words so spelled, see the forms beginning SULF–.
sulpha– Var. of SULFA–.
sulpho– Var. of SULFO–.
sul·tan (sul′tən) *n.* 1 The ruler of a Moslem country. 2 A gallinule with deep-blue or purple plumage and white lower tail coverts. *Ionornis martinica* is the purple gallinule or sultan of the warmer parts of America. 3 A small white-crested variety of the domestic fowl, originating in Turkey, having heavily feathered legs and feet. 4 Formerly, any ruler. — **the Sultan** The title of the sovereign of Turkey: office abolished, 1922. [< Med. L *sultanus* < Arabic *sulṭān* a sovereign, dominion]
sul·tan·a (sul·tan′ə, -tän′ə) *n.* 1 A sultan's wife, daughter, sister, or mother: also **sul·tan·ess** (sul′tən·is). 2 The mistress of a king or prince. 3 A variety of raisin from the district of Smyrna, Asia Minor. 4 Sultan (def. 3): also **sul·tan′a–bird′.** [< Ital., fem. of *sultano* a sultan < Arabic *sulṭān*]
sul·tan·ate (sul′tən·āt, -it) *n.* The authority or territorial jurisdiction of a sultan. Also **sul′tan·ship.**
sul·try (sul′trē) *adj.* **·tri·er**, **·tri·est** 1 Hot, moist, and still; close: said of weather. 2 Emitting an oppressive heat; burning; hot with anger. 3 Showing or suggesting passion; sensual. [< obs. *sulter*, var. of SWELTER] — **sul′tri·ly** *adv.* — **sul′tri·ness** *n.*
Su·lu (soo′loo) *n.* 1 A member of the chief Moro tribe occupying the Sulu Archipelago. 2 The Indonesian language of this tribe, closely related to Tagalog. — **Su·lu′an** *adj. & n.*
Sulu Archipelago An island group between Basilan and the NE coast of Borneo, comprising a province of the Philippines; 1,086 square miles; capital, Jolo.
Sulu Sea An arm of the Pacific Ocean between the SW Philippines and Borneo; over 400 miles long, east to west.
sum (sum) *n.* 1 The result obtained by addition. 2 The entire quantity, number, or substance; the whole; all: the *sum* total of my means; the *sum* and substance of the case. 3 Any indefinite amount: said chiefly of money. 4 A problem in arithmetic propounded for solution. 5 The summit; topmost or highest point; also, the maximum; the complement. 6 A summary; the pith or essence. See synonyms under AGGREGATE. — *v.* **summed**, **sum·ming** *v.t.* 1 To present in brief; recapitulate succinctly: usually with *up*: to *sum up* evidence. 2 To add into one total; ascertain the sum of: often with *up*. 3 To ascertain the sum of (the terms of a series). — *v.i.* 4 To make a summation or recapitula-

tion: generally with *up*. ◆ Homophone: *some.* [< AF, OF *summe*, *somme* < L *summa* (res) highest (thing), fem. of *summus* highest]
sum– Assimilated var. of SUB–.
su·mac (soō′mak, shoō′-) *n.* 1 Any of a genus (*Rhus*) of woody, erect, or root-climbing plants (family *Anacardiaceae*), with panicles of small flowers, small drupaceous fruits, and yielding a resinous or milky juice; especially, the **smooth sumac** (*R. glabra*) used in medicine. 2 The poison sumac. 3 The dried and powdered leaves of certain species of sumac, used for tanning and dyeing, especially of **tanner's sumac** (*Rhus coriaria*). Also **su′mach.** [< OF < Med. L *sumach* < Arabic *summāq*]
Su·ma·tra (soō·mä′trə) An Indonesian island south of the Malay Peninsula, comprising, with adjacent islands, six provinces of Indonesia; 163,557 square miles; chief city, Palembang. — **Su·ma′tran** *adj. & n.*
Sum·ba (soōm′bä) One of the Lesser Sunda Islands, SE of Sumbawa, in south central Nusa Tenggara province, Indonesia; 4,300 square miles; formerly *Sandalwood Island*: Dutch *Soemba.*
Sum·bar (soōm′bär) A river in SW Turkmen S.S.R. and Iran, flowing 150 miles west to its confluence with the Atrek river, on the U.S.S.R.–Iran border.
Sum·ba·wa (soōm·bä′wä) One of the Lesser Sunda Islands, east of Lombok, in NW central Nusa Tenggara province, Indonesia; 5,965 square miles: Dutch *Soembawa.*
sum·bul (soōm′bul) *n.* Muskroot. Also **sum′bal, sum′bul–root′** (-roōt′, -roōt′). [< F < Arabic *sunbul*] — **sum·bu·lic** (sum·boō′lik) *adj.*
sumbul tree See under AMMONIAC[2].
Su·mer (soō′mər) A region and ancient country of Mesopotamia, later the southern division of Babylonia; the sites of its once great cities were in south central Iraq.
Su·me·ri·an (soō·mir′ē-ən) *adj.* Of or pertaining to ancient Sumer, its people, or their language. — *n.* 1 One of an ancient non-Semitic people formerly occupying a part of lower Babylonia: culturally important in the Near East from about 3300–1800 B.C. 2 The agglutinative, unclassified language of these people, written in cuneiform characters and preserved on rocks and clay tablets the earliest of which date from about 4000 B.C. Also **Su·mir′i·an.**
sum·less (sum′lis) *adj.* Too great for computation; incalculable; without number.
sum·ma cum lau·de (sum′ə kum lô′dē, soōm′ə koōm lou′de) See under CUM LAUDE.
sum·mand (sum′and) *n.* That which is added; any of the numbers forming part of a sum. [< Med. L *summandus* (*numerus*) (the number) to be added < *summare* add < L *summa.* See SUM.]
sum·ma·rize (sum′ə-rīz) *v.t.* **·rized**, **·riz·ing** To make a summary of; sum up; epitomize. Also *Brit.* **sum′ma·rise.** — **sum′ma·ri·za′tion** *n.* — **sum′ma·riz′er** *n.*
sum·ma·ry (sum′ər-ē) *adj.* 1 Giving the substance or sum; greatly condensed; concise. 2 Performed without ceremony or delay; instant; offhand: used specifically in law. — *n. pl.* **·ries** An abridgment or epitome; abstract; compendium. See synonyms under ABRIDGMENT. [< Med. L *summarius* < *summarium* a summary < *summa.* See SUM.] — **sum·ma·ri·ly** (sum′ə-ral·ē, emphatic sə-mer′ə-lē) *adv.* — **sum′ma·ri·ness** *n.* — **sum′ma·rist** *n.*
sum·mate (sum·āt′) *v.t. & v.i.* **·mat·ed**, **·mat·ing** 1 To arrive at the sum of (a series). 2 To sum up. [Back formation < SUMMATION]
sum·ma·tion (sum·ā′shən) *n.* 1 The act or operation of obtaining a sum; the computation or statement of an aggregate sum or result; addition. 2 A speech or a portion of a speech summing up the principal points. [< NL *summatio*, -*onis* < Med. L *summare* add < *summa.* See SUM.]
sum·mer[1] (sum′ər) *n.* 1 The hottest or warmest season of the year: including June, July, and August, in the northern hemisphere. In the southern hemisphere the summer occurs during the months of the northern winter. ◆ Collateral adjective: *estival.* 2 Figuratively, a year of life, especially of early or happy life; a bright and prosperous period. — **Indian summer** A period of mild weather occurring in

the autumn, with hazy atmosphere usually along the horizon, and a clear sky. It corresponds to the English St. Luke's or St. Martin's summer. — **St. Luke's summer** or **little summer of St. Luke** A short period of warm weather in England expected for a few days beginning with St. Luke's day, the 18th of October. — **St. Martin's summer** A season of mild weather about St. Martin's day, the 11th of November, corresponding to the American Indian summer. — *v.t.* To keep or care for through the summer. — *v.i.* To pass the summer. — *adj.* Of, pertaining to, or occurring in summer. [OE *sumor, sumer*] — **sum'mer·ly** *adj. & adv.*
sum·mer[2] (sum'ər) *n. Archit.* 1 A heavy horizontal timber or girder serving as a support for some superstructure in a building, etc.; a lintel. 2 A large stone, as on a column or pilaster, for supporting one or more arches, or any similar structure. 3 A horizontal beam resting upon the walls or external frame of a building, and supporting the ends of joists: also **sum'mer·beam'**. [<OF *somier* a pack horse, beam <LL *saumarius* <L *sagmarius* <*sagma* a pack saddle <Gk.]

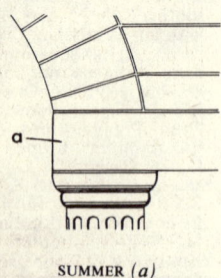
SUMMER (*a*)

sum·mer[3] (sum'ər) *n.* One who sums or adds skilfully.
summer flounder A flounder (*Paralichthys dentatus*) of the Atlantic coast of North America.
sum·mer·house (sum'ər·hous') *n.* A rustic structure, as in a garden, for rest or shade.
summer house A house or cottage in the country or at the seashore, used during the summer.
summer kitchen A kitchen in a separate building or in a separate room attached to a house, for use in hot weather.
sum·mer·sault (sum'ər·sôlt), **sum·mer·set** (sum'ər·set) See SOMERSAULT.
summer squash Any garden squash or pumpkin, varieties of *Cucurbita pepo*, grown principally for home consumption, as the crookneck.
sum·mer·time (sum'ər·tīm') *n.* Summer; the summer season. Also **sum'mer·tide'**.
sum·mer·y (sum'ər·ē) *adj.* Pertaining to or resembling summer.
sum·mit (sum'it) *n.* 1 The highest part; top; vertex. 2 The highest degree; maximum. [<OF *sommette*, dim. of *som* a summit, top <L *summum*, neut. of *summus* highest] — **sum'mit·al** *adj.* — **sum'mit·less** *adj.*
Synonyms: acme, apex, cap, climax, crown, height, peak, pinnacle, top, vertex. *Antonyms:* abyss, base, bottom, chasm, deep, depth, gorge, gulf, pit, vale, valley.
sum·mon (sum'ən) *v.t.* 1 To order to come; send for. 2 To call together; cause to convene, as a legislative assembly. 3 To order (a person) to appear in court by a summons. 4 To call forth or into action; arouse: usually with *up*; to summon up courage. 5 To bid or call on for a specific act: The garrison was *summoned* to surrender. See synonyms under ARRAIGN, CONVOKE. [<AF, OF *somondre* <L *summonere* suggest, hint <*sub-* secretly + *monere* warn]
sum·mon·er (sum'ən·ər) *n.* 1 A person who summons. 2 *Archaic* An officer who summons persons to appear in court.
sum·mons (sum'ənz) *n.* 1 A call to attend or act at a particular place or time. 2 *Law* A notice to a defendant summoning him to appear in court: either a judicial writ or process, or a notice signed by the plaintiff or his attorney; any citation issued to a party to an action to appear before a court or judge at chambers. See WRIT OF SUMMONS. 3 A notice to a person requiring him to appear in court as a witness or as a juror. 4 A military demand to surrender. 5 Any signal or sound that is a peremptory call. [<AF *somonse*, OF *sumunse* <*somonder* SUMMON]
sum·mum bo·num (sum'əm bō'nəm) *Latin* The chief, supreme, or highest good.

Sum·ner (sum'nər), **Charles**, 1811–74, U.S. statesman and abolitionist. — **James Batcheller**, born 1887, U.S. biochemist. — **William Graham**, 1840–1910, U.S. sociologist.
sump (sump) *n.* 1 *Mining* a A depression sunk below the lowest level in a mine shaft, to receive water and form a pool from which it may be pumped. b A sump winze. 2 *Mech.* The lowest part of the crankcase of an internal-combustion engine, acting as a reservoir for lubricating oil. 3 A cesspool or other reservoir for drainage. [<MDu. *somp, sump* a marsh. Akin to SWAMP.]
sumph (sumpf) *Scot. v.i.* To be sullen or stupid. — *n.* A simpleton; sullen or lubberly fellow; chump.
sump·ter (sump'tər) *n.* A pack animal; beast of burden. [<OF *sometier* a driver of a pack horse, ult. <L *sagma*. See SUMMER[2].]
sump·tu·ar·y (sump'chōō·er'ē) *adj.* Pertaining to expense; limiting or regulating expenditure, as some laws. [<L *sumptuarius* <*sumptus* expenditure <*sumere* take]
sumptuary law 1 A law limiting or regulating expenditure in order to prevent extravagance and inflation. 2 A law regulating private life on moral or religious grounds; a blue law.
sump·tu·ous (sump'chōō·əs) *adj.* Involving or showing lavish expenditure; hence, luxurious; magnificent. [<OF *sumptueux, somptueux* <L *sumptuosus* <*sumptus.* See SUMPTUARY.] — **sump'tu·ous·ly** *adv.* — **sump'tu·ous·ness** *n.*
sump·weed (sump'wēd') *n.* Marsh elder.
sump winze A winze sunk in the lowest level of a mine in order to explore the lode below.
sun (sun) *n.* 1 The heavenly body that is the center of attraction and the main source of light and heat in the solar system, with a mean distance from the earth of about 93,000,000 miles and a diameter of 864,000 miles. Its mass is 332,000 times that of the earth, but its density only about one-fourth. 2 Any star, especially one that is the center of a system revolving around it. 3 The light and heat radiated from the sun; sunshine. 4 Anything brilliant and magnificent, or that is a source of splendor. 5 The time of the earth's revolution round the sun; a year. 6 The daily appearance of the sun; a day; also, the time of its appearance or shining; sunrise. — **a place in the sun** A dominant position in international affairs; hence, a position in the spotlight; publicity. — *v.* **sunned, sun·ning** *v.t.* 1 To expose to the light or heat of the sun. 2 To warm or dry (something) in the sun. — *v.i.* 3 To bask in the sun; expose oneself to the light or heat of the sun. ◆ Homophone: *son.* [OE *sunne*]
Sun may appear as a combining form in hyphemes or solidemes, or as the first element in two-word phrases, with the following meanings:
1 Of the sun; of sunshine:

sun blaze	sun-loving
sun-eclipsing	sun-worship
sun glare	sun-worshiper
sunland	sun-worshiping

2 By or with the sun:

sun-bake	sun-filled
sun-baked	sun-flooded
sun-blind	sun-gilt
sun-blinded	sun-heated
sun-blistered	sun-kissed
sun-brown	sunlit
sun-browned	sun-scorched
sun-cracked	sun-scorching
sun-dappled	sun-streaked
sun-dried	sun-warmed
sun-dry	sun-withered

sun bath Exposure to the direct rays of the sun.
sun·bathe (sun'bāth') *v.i.* **-bathed, -bath·ing** To bask in the sun, especially as a method of tanning the skin. — **sun'-bath'er** *n.* — **sun'-bath'ing** *n.*
sun·beam (sun'bēm') *n.* 1 A ray or beam of the sun; light from the sun in a visible path. 2 *pl.* Sunlight.
sun·bird (sun'bûrd') *n.* 1 A brilliantly colored oriental singing bird (family *Nectariniidae*) resembling the hummingbird. 2 A sun bittern.
sun bittern Either of two birds of Central and South America (genus *Eurypyga*) related to the rails and herons, having a slender neck and bill, long wings and tail, and moderately long legs.
sun·bon·net (sun'bon'it) *n.* A bonnet of light material with projecting brim and sometimes a cape covering the neck.
sun·bow (sun'bō') *n.* A rainbow formed by the sun, as opposed to a lunar bow.
sun·burn (sun'bûrn') *n.* Discoloration or inflammation of the skin, produced by exposure to the sun. — *v.t. & v.i.* To affect or be affected with sunburn. — **sun'burnt', sun'burned'** *adj.*
sun·burst (sun'bûrst') *n.* 1 A strong burst of sunlight, as through rifted clouds. 2 A brooch or pin with jewels so set around a larger central gem as to suggest sun rays.
sun compass A compass serving to establish Greenwich time in relation to the position of the sun: used chiefly in polar regions, where the magnetic compass is unreliable.
sun·dae (sun'dē) *n.* A refreshment consisting of ice-cream and crushed fruit, flavoring, sirup, nuts, etc. [Prob. <*Sunday;* prob. so called because orig. sold only on that day]
Sun·da Islands (sun'də, *Du.* sōōn'dä) An Indonesian island group of the Malay Archipelago, between the Indian Ocean and the Java Sea; divided into *Greater Sunda Islands,* including Sumatra, Java, Borneo, and Celebes, and *Lesser Sunda Islands,* the smaller islands east of Java. *Dutch* Soen'da.
sun dance The greatest ceremonial dance of the Plains Indians, usually a summer solstice ceremony, comprising fast days, dance days, secret rites, and a public performance.
Sun·da Strait (sun'də, *Du.* sōōn'dä) The channel between Java and Sumatra, connecting the Java Sea with the Indian Ocean 16 to 70 miles wide.
Sun·day (sun'dē, -dā) *n.* The first day of the week; the Lord's day; the Christian Sabbath: sometimes used attributively. See synonyms under SABBATH. [OE *sunnan dæg* <*sunnan* of the sun + *dæg* a day; trans. of LL *dies solis* day of the sun]
Sun·day (sun'dē, -dā), **Billy,** 1862–1935, U.S. preacher and evangelist: full name William Ashley Sunday.
Sun·day-go-to-meet·ing (sun'dē·gō'tə·mē'ting) *adj. Colloq.* Best: *Sunday-go-to-meeting* clothes or manners.
Sunday school A school, generally attached to some church, in which religious instruction is given on Sunday, especially to the young; also, the pupils, or teachers and pupils, collectively: also called *Sabbath school.*
sun·der (sun'dər) *v.t.* To break apart; disunite; sever. — *v.i.* To be parted or severed. See synonyms under BREAK, CUT, REND, SEPARATE. — *n.* Division into parts; separation. — **in sun·der** Apart; separate from other parts. Compare ASUNDER. [OE *syndrian, sundrian*] — **sun'der·ance** *n.*
Sun·der·land (sun'dər·lənd) A port and county borough in NE Durham, England.
sun·dew (sun'dōō, -dyōō') *n.* Any of a genus (*Drosera*) of marsh plants that exude a viscid liquid from the tips of the hairs on the leaves. Insects are caught by the secretions and are utilized by the plant for its own nutrition.
sun·di·al (sun'dī'əl) *n.* A device that measures time and shows the time of day by means of the shadow of a style or gnomon thrown on a dial.
sun disk The winged disk. See under DISK.
sun·dog (sun'dôg', -dog') *n.* 1 A parhelion, appearing near the sun, sometimes with a luminous train, due to the presence of ice crystals in the air; a mock sun. 2 A small rainbow lying near the horizon.

SUNDIAL

sun·down (sun'doun') *n.* 1 Sunset: originally colloquial, like *sunup,* but now in good literary usage. 2 A broad-brimmed hat worn by women. [? Contraction of *sun-go-down*]
sun·down·er (sun'dou'nər) *n.* 1 *Colloq.* A tramp. 2 *Austral.* A vagrant who seeks food and lodging at back-country ranches, often about the time of sundown. 3 *Slang* A strict,

sundries — **rigidly** uncompromising ship's officer; originally, a ship's captain who granted liberty only until sundown.
sun·dries (sun′drēz) *n. pl.* Items or things too small or too numerous to be separately specified. [<SUNDRY]
sun·drops (sun′drops′) *n.* Any of several American species of evening primrose (genus *Oenothera*), having large yellow flowers, and blooming in the daytime.
sun·dry (sun′drē) *adj.* Of an indefinite small number; various; several; miscellaneous. See synonyms under MANY. [OE *syndrig* separate, private]
sune (soon) *adv. Scot.* Soon.
sun·fish (sun′fish′) *n. pl.* ·**fish** or ·**fish·es** 1 A large pelagic plectognath fish (genus *Mola*), having a deep compressed body truncate behind, as *Mola mola* of warm and tropical seas. It has tough and leathery flesh. 2 Any of several North American freshwater perchlike fishes (family *Centrarchidae*) of the genus *Lepomis*, as the pumpkinseed.

SUNFISH *(def. 1)*
(Up to 8 feet in length)

sun·flow·er (sun′flou′ər) *n.* Any of a genus (*Helianthus*) of tall, stout, rough herbs of the composite family, with large leaves and circular heads of flowers, those in the center tubular and usually purple, and those on the margin strap-shaped and bright-yellow; especially, the common sunflower (*H. annuus*), the source of an edible oil, and the State flower of Kansas.
Sunflower State Nickname of KANSAS.
sung (sung) Past participle and occasional past tense of SING.
Sung (sŏŏng) *n.* A dynasty in Chinese history, 960 to 1280, noted for its achievements in art and philosophy.
Sun·ga·ri (sŏŏng′gä·rē′) The largest river of Manchuria, flowing 1,150 miles NW to the Amur river. Chinese **Sung·hwa** (sŏŏng′hwä′).
Sung·kiang (sŏŏng′jyäng′) A former province of NE Manchuria region, NE China, incorporated in Heilungkiang province; 75,000 square miles; capital, Harbin.
sun·glass (sun′glas′, -gläs′) *n.* 1 A burning glass; a glass used for concentrating the rays of the sun. 2 *pl.* Spectacles that protect the eyes from the glare of the sun by their colored lenses.
sun·glow (sun′glō′) *n.* 1 The rose tint or faint yellow of the sky that precedes sunrise or follows sunset. 2 The warm glow of the sun.
sun-god (sun′god′) *n.* In the religions of some primitive agricultural peoples, a deity conceived of as life-giving and beneficent, and symbolized by the sun, as the ancient Egyptian Ra, ancient Irish Lug, Inti of the Incas, etc.: not to be confused with personifications of the sun in many cosmogonic myths (Greek Helios, for instance) which are mere explanatory etiological tales and do not posit a sun cult.
sunk (sungk) Past participle and alternative past tense of SINK.
sunk·en (sung′kən) Obsolete past participle of SINK. — *adj.* 1 Deeply depressed or fallen in; hollow: a *sunken* cheek. 2 Located beneath the surface of the ground or the water; also, marshy. 3 At a lower level: *sunken* gardens.
sun·ket (sung′kit, sŏŏng′-) *n. Scot.* A dainty; tidbit.
sunk fence A ditch having a retaining wall on one side to divide lands; a ha-ha.
sunk·ie (sungk′ē) *n. Scot.* A low stool or small seat.
sunk panel A panel so depressed as to form a recess below the surface of its frame.
sun lamp 1 A lamp giving illumination of high intensity, usually reflected by parabolic mirrors: used in motion-picture studios. 2 A lamp radiating ultraviolet rays: used for therapeutic treatments and as a protection against airborne bacteria in operating rooms, etc.
sun·less (sun′lis) *adj.* Dark; cheerless. — **sun′less·ness** *n.*
sun·light (sun′līt′) *n.* The light of the sun.
sunn (sun) *n.* An East Indian shrub (*Crotalaria juncea*) of the bean family, with bright-yellow flowers and tough, durable fiber: used for making cordage, bagging, and other coarse textiles: also called *Bombay* or *Madras hemp.* Also **sunn hemp.** [<Hind. *san*]
Sun·na (sŏŏn′ə) *n.* A path or manner of life; that part of the orthodox Moslem creed or law based on traditions of the Prophet's words and deeds: regarded by a numerous sect as of equal importance with the Koran, which it supplements; hence, the theory and practice of orthodox Islam. Also **Sun′nah**. [<Arabic *sunnah,* lit., a form, way]
Sun·nite (soon′īt) *n.* An orthodox Moslem of the sect accepting Sunna (tradition) and the Koran as of equal authority, and acknowledging the first four caliphs as rightful successors of the Prophet: opposed to *Shiah.* Also **Sun·ni** (sŏŏn′ē).
sun·ny (sun′ē) *adj.* ·**ni·er**, ·**ni·est** 1 Filled with the light and warmth of the sun; exposed to the sun. 2 Bright like the sun; of the sun or sunshine; hence, genial; cheery: a *sunny* smile. See synonyms under BRIGHT, CHEERFUL, HAPPY. — **sun′ni·ly** *adv.* — **sun′ni·ness** *n.*
sunny side 1 The side, as of a hill, facing the sun. 2 The cheerful view of any situation, question, etc.
sun parlor A room enclosed in glass and having a sunny exposure.
sun pillar A column of variously tinted light sometimes seen projecting vertically above or below the sun at sunrise or sunset. It is caused by the reflection of sunlight from small snow crystals.
sun·rise (sun′rīz′) *n.* 1 The daily first appearance of the sun above the horizon, with the atmospheric phenomena just preceding and following. 2 The time at which the sun rises; the region where it rises. 3 The east; Orient.
sun·room (sun′rŏŏm′, -rŏŏm′) *n.* A room built to admit a profusion of sunlight.
sun·scald (sun′skôld′) *n.* A diseased condition of plants induced by exposure to intense sunlight.
sun·scorch (sun′skôrch′) *n.* A scorched or burnt condition of plants.
sun·set (sun′set′) *n.* 1 The apparent daily descent of the sun below the horizon. 2 The time when the sun sets; the early evening. 3 The colors in the sky when the sun sets. 4 The west; Occident. 5 Figuratively, the ending or decline, as of life.
sun·shade (sun′shād′) *n.* Something used as a shade or protection from the rays of the sun, as a parasol, an awning, etc.
sun·shine (sun′shīn′) *n.* 1 The shining light of the sun; the direct rays of the sun. 2 The warmth of the sun's rays. 3 The place where the rays fall. 4 Figuratively, brightness; any cheering influence. — **sun′shin′y** *adj.*
Sunshine State Nickname of NEW MEXICO.
sun·spot (sun′spot′) *n.* 1 *Astron.* One of many dark irregular spots appearing periodically on the surface of the sun: believed to have connection with terrestrial magnetic storms. 2 An incandescent sun lamp used in color photography.
sun·stone (sun′stōn′) *n.* A variety of feldspar; aventurine.
sun·stroke (sun′strōk′) *n. Pathol.* A sudden onset of high fever induced by exposure to the sun and often marked by convulsions and coma; insolation. — **sun′struck′** (-struk′) *adj.*
sun tan A bronze-colored condition of the skin, produced by exposure to the sun. — **sun′-tanned′** (-tand′) *adj.*
sun·tans (sun′tanz′) *n. pl.* The lightweight summer uniform made of khaki worn by U. S. Army personnel: officially known as *cotton khakis,* and often called *khakis.* They are worn in the Navy by officers.
sun-up (sun′up′) *n.* Sunrise. [<SUN + UP; on analogy with *sundown*]
Sun Valley A resort village in south central Idaho; altitude, 6,000 feet.
sun·ward (sun′wərd) *adj.* Facing toward the sun. — *adv.* Toward the sun: also **sun′wards.**
sun·wise (sun′wīz′) *adv.* With the sun; in the direction of the sun; clockwise.
Sun Yat-sen (sŏŏn′ yät′sen′), 1865?-1925, Chinese statesman; president of China 1911-1912.
su·o ju·re (soo′ō joor′ē) *Latin* In one's own right.
su·o lo·co (soo′ō lō′kō) *Latin* In its own or proper place.
Su·o·mi (soo′ō·mē) *n. pl.* 1 The people of Finland; the Finns. 2 The language of the Finns; Finnish. — **Su·o′mic** *adj. & n.*
Su·o·mi (soo′ō·mē) The Finnish name for FINLAND.
sup[1] (sup) *v.t. & v.i.* **supped, sup·ping** To take (fluid food) in successive mouthfuls, a little at a time; sip. — *n.* A mouthful or taste of liquid or semiliquid food. [OE *sūpan* drink]
sup[2] (sup) *v.* **supped, sup·ping** *v.i.* To eat supper. — *v.t. Obs.* To furnish with or invite to supper. [<OF *soper, super;* ult. origin unknown]
sup- Assimilated var. of SUB-.
supe (soop) *n. Slang* A supernumerary actor. [Short for SUPERNUMERARY]
su·per[1] (soo′pər) *n. Colloq.* Shortened form of SUPERINTENDENT.
su·per[2] (soo′pər) *n. Slang* Shortened form of SUPERNUMERARY (def. 2).
su·per[3] (soo′pər) *n.* 1 An article of superior size or quality; also, such size or quality. 2 In bookbinding, a thin, starched cotton fabric used in reinforcement. — *adj.* 1 *Slang* First-rate; superfine. 2 Showing excessive loyalty: a *super* American. — *v.t.* To reinforce (a book) with super. [Short for SUPERIOR, SUPERFINE, etc.]
super- *prefix* 1 Above in position; over: *superstructure, superimpose.* 2 *Anat. & Zool.* Situated above, or on the dorsal side of: *superorbital.* 3 Above or beyond; more than: *supersonic, supersensible.* 4 Excessively: *supersaturate.* 5 *Med.* Exceeding the normal: *superacidity.* 6 *Chem.* Denoting a high proportion of the ingredient indicated (now superseded by PER-, BI-): *superphosphate.* 7 Surpassing in power or size all others of its class: *superhighway, supermarket.* In this sense the prefix is sometimes doubled to intensify the degree of superiority: a *super-supernavy* a navy far superior to any other. 8 Extra; additional: *supertax.* [<L *super-* < *super* above, beyond]
In the following list of words *super-* denotes excess or superiority, as *supercritical* excessively critical, *superexcellence* superior excellence.

superabhor	superbold
superabominable	superbrave
superabsurd	superbusy
superaccession	supercandid
superaccommodating	supercapable
superaccomplished	supercatastrophe
superaccumulate	supercatholic
superachievement	supercaution
superacquisition	superceremonious
superacute	superchivalrous
superadaptable	supercivil
superadequate	supercivilized
superadmiration	superclassified
superadorn	supercolossal
superaffluence	supercombination
superagency	supercommendation
superaggravation	supercommercial
superagitation	supercompetition
superambitious	supercomplex
superangelic	supercomprehension
superappreciation	supercompression
superarbitrary	superconfident
superarduous	superconformist
superarrogant	superconformity
superaspiration	superconfusion
superastonish	supercongestion
superattachment	superconservative
superattraction	supercontrol
superattractive	supercordial
superbelief	supercritic
superbeloved	supercritical
superbenefit	supercultivated
superbenevolent	supercurious
superbenign	supercynical
superbias	superdainty
superblessed	superdanger
superblunder	superdeclamatory

add, āce, cãre, päm; end, ēven; it, īce; odd, ōpen, ôrder; tŏŏk, pōōl; up, bûrn; ə = a in *above*, e in *sicken*, i in *clarity*, o in *melon*, u in *focus*; yōō = u in *fuse*; oi, oil; ou, pout; ch, check; g, go; ng, ring; th, thin; t̷h, this; zh, vision. Foreign sounds á, œ, ü, kh, ṅ; and ♦: see page xx. < from; + plus; ? possibly.

superdeficit
superdejection
superdelicate
superdemand
superdemonic
superdesirous
superdevelopment
superdevilish
superdevotion
superdiabolical
superdifficult
superdiplomacy
superdistribution
superdividend
superdonation
supereconomy
supereffective
supereffluence
superelastic
superelated
superelegance
supereligible
supereloquent
superemphasis
superendorsement
superendow
superenforcement
superenrolment
superestablishment
superesthetic
superethical
superevident
superexacting
superexalt
superexaltation
superexcellence
superexcellent
superexcitation
superexcited
superexcitement
superexiguity
superexpansion
superexpectation
superexpenditure
superexpressive
superexquisiteness
superextension
superfecundity
superfeminine
superfervent
superfoliation
superfolly
superformal
superformation
superformidable
superfriendly
superfructified
superfulfilment
supergaiety
supergallant
supergenerosity
superglorious
supergoodness
supergovernment
supergratification
supergravitation
superhandsome
superhearty
superhero
superheroic
superhistorical
superhypocrite
superideal
superignorant
superillustrate
superimpending
superimpersonal
superimportant
superimprobable
superimproved
superincentive
superinclination
superinclusive
superincomprehensible
superindependent
superindifference
superindignant
superindividualism
superindividualist
superindulgence
superindustrious
superinference
superinfinite
superinfirmity
superinfluence
superingenious
superinitiative

superinjustice
superinquisitive
superinsistent
superintellectual
superintolerable
superjurisdiction
superjustification
superknowledge
superlaborious
superlenient
superlie
superlogical
superloyal
superlucky
superluxurious
supermagnificently
supermanhood
supermarvelous
supermasculine
supermechanical
supermediocre
supermental
supermentality
supermetropolitan
supermishap
supermodest
supermoisten
supermorose
supermundane
supermystery
supernecessity
supernegligent
supernotable
supernumerous
superobedience
superobese
superobjectionable
superobligation
superobstinate
superoffensive
superofficial
superofficiousness
superopposition
superoratorical
superordinary
superorganize
superornamental
superoutput
superpatient
superpatriotic
superpatriotism
superperfection
superpious
superplease
superpolite
superpositive
superpraise
superprecise
superpreparation
superproduce
superprosperous
superpublicity
superpure
superpurgation
superradical
superrational
superrefined
superreform
superreliance
superremuneration
superrespectable
superresponsible
superrestriction
superreward
superrighteous
superromantic
supersacrifice
supersafe
supersagacious
supersanguine
supersarcastic
supersatisfaction
superscholarly
superscientific
supersensitive
supersensitiveness
supersensuousness
supersentimental
superserious
supersevere
supersignificant
supersimplify
supersmart
supersolemn
supersolemnly
supersolicitation
superspecialize

superspiritual
superspirituality
superstimulation
superstoical
superstrain
superstrenuous
superstrict
superstrong
superstylish
supersufficient
supersurprise
supersweet
supertension
superthankful
superthorough

supertoleration
supertragic
supertrivial
superugly
superunity
superurgent
supervexation
supervigilant
supervigorous
supervirulent
supervital
superwise
superworldly
superwrought
superzealous

su·per·a·ble (sōō′pər·ə·bəl) *adj.* That can be surmounted, overcome, or conquered. [<L *superabilis* < *superare* overcome < *super* over]

su·per·a·bound (sōō′pər·ə·bound′) *v.i.* To abound to excess or to an unusual extent. [<LL *superabundare* <L *super-* exceedingly + *abundare* overflow]

su·per·a·bun·dant (sōō′pər·ə·bun′dənt) *adj.* Excessive; more than sufficient. See synonyms under REDUNDANT. [<LL *superabundans, -antis,* ppr. of *superabundare.* See SUPERABOUND.] — **su′per·a·bun′dance** *n.* — **su′per·a·bun′dant·ly** *adv.*

su·per·a·cid·i·ty (sōō′pər·ə·sid′ə·tē) *n. Med.* An excess of acid, especially in the gastric juices; hyperacidity.

su·per·add (sōō′pər·ad′) *v.t.* To add in addition to something already added. [<LL *superaddere* < *super-* over and above + *addere* ADD] — **su′per·ad·di′tion** (-ə·dish′ən) *n.*

su·per·al·tar (sōō′pər·ôl′tər) *n. Eccl.* 1 A consecrated slab laid on an unconsecrated altar when mass is said in oratories or temporary chapels. 2 Sometimes, incorrectly, a retable. [<Med. L *superaltare* <L *super-* over + *altare* an altar]

su·per·an·nu·ate (sōō′pər·an′yōō·āt) *v.t.* **·at·ed, ·at·ing** 1 To retire or retire and pension on account of age: chiefly in past participle. 2 To set aside or discard as obsolete or too old. [<Med. L *superannuatus* more than a year old (said of cattle) <L *super annum* < *super* beyond + *annus* a year] — **su′per·an′nu·at′ed** *adj.* — **su′per·an·nu·a′tion** *n.*

su·per·aq·ual (sōō′pər·ak′wəl, -ā′kwəl) *adj.* Of, pertaining to, or denoting those soils lying just above the water table, from which they derive the greater part of their moisture. [<L *super* above + *aqua* water]

su·perb (sōō·pûrb′, sə-) *adj.* 1 Having grand, impressive beauty; majestic; imposing: a *superb* edifice. 2 Luxurious; rich and costly; elegant. 3 Very good; supremely fine. [<L *superbus* proud < *super-* over] — **su·perb′ly** *adv.* — **su·perb′ness** *n.*

su·per·bomb (sōō′pər·bom′) *n.* A hydrogen bomb.

su·per·cal·en·der (sōō′pər·kal′ən·dər) *n.* A calender having a number of polished rollers for giving a high finish to paper. See CALENDER[1]. — *v.t.* To give a high finish to (paper). — **su′per·cal′en·dered** *adj.*

su·per·car·go (sōō′pər·kär′gō) *n. pl.* **·goes** or **·gos** An agent on board ship in charge of the cargo and its sale and purchase. [Alter. of obs. *supracargo* <Sp. *sobrecargo* < *sobre-* over (<L *super-*) + *cargo* CARGO]

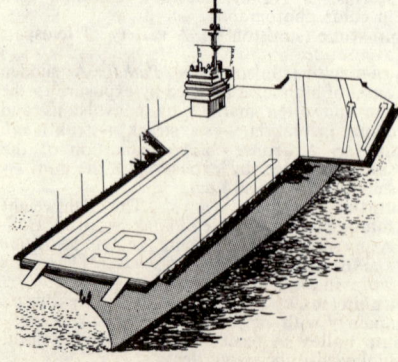

SUPERCARRIER OF THE FORRESTAL CLASS

su·per·car·ri·er (sōō′pər·kar′ē·ər) *n.* An aircraft-carrier of exceptional size.

su·per·charge (sōō′pər·chärj′) *v.t.* **·charged, ·charg·ing** 1 To adapt (an engine) to develop more power, as by fitting with a supercharger. 2 To charge to excess; overload. — *n.* (sōō′pər·chärj′) 1 An excess charge, in any sense. 2 *Her.* One charge or device borne on another.

su·per·charg·er (sōō′pər·chär′jər) *n.* A compressor for supplying air or combustible mixture to an internal-combustion engine at a pressure greater than that developed by the suction of the pistons alone.

su·per·cil·i·ar·y (sōō′pər·sil′ē·er′ē) *adj.* 1 Of or pertaining to the eyebrow. 2 Situated over the eyebrow; supraorbital: the *superciliary* arches. [<NL *superciliaris* <L *supercilium* an eyebrow < *super-* above + *cilium* an eyelid]

su·per·cil·i·ous (sōō′pər·sil′ē·əs) *adj.* Exhibiting haughty contempt or indifference; arrogant. See synonyms under HAUGHTY. [<L *superciliosus* < *supercilium.* See SUPERCILIARY.] — **su′per·cil′i·ous·ly** *adv.* — **su′per·cil′i·ous·ness** *n.*

su·per·class (sōō′pər·klas′, -kläs′) *n. Biol.* A division of plants or animals below a phylum but above a class.

su·per·col·um·nar (sōō′pər·kə·lum′nər) *adj. Archit.* 1 Erected above a colonnade or another column. 2 Having one order placed above another.

su·per·con·duc·tiv·i·ty (sōō′pər·kon′duk·tiv′ə·tē) *n. Electr.* The property, exhibited by certain metals and alloys, of becoming almost perfect conductors of electricity when their temperatures fall below transition points in the neighborhood of absolute zero. — **su′per·con·duc·tive** (-kən·duk′tiv) *adj.* — **su′per·con·duc′tor** *n.*

su·per·cool (sōō′pər·kōōl′) *v.t.* To cool, as a liquid, below the freezing point without solidification.

su·per·dom·i·nant (sōō′pər·dom′ə·nənt) *n. Music* The tone just above the dominant; the sixth or submediant.

su·per-du·per (sōō′pər·dōō′pər) *Slang adj.* Superlative: an intensive formation. — *n.* Anything especially fine. [Reduplication of SUPER[3]]

su·per·e·go (sōō′pər·ē′gō, -eg′ō) *n. Psychoanal.* A largely unconscious element of the personality, regarded as dominating the conscious ego, for which it acts principally in the role of conscience and critic.

su·per·em·i·nent (sōō′pər·em′ə·nənt) *adj.* Excelling or surpassing others; of a superior or remarkable quality; supremely exalted. [<L *supereminens, -entis,* ppr. of *supereminere* rise above < *super-* above + *eminere* rise. See EMINENT.] — **su′per·em′i·nence** *n.* — **su′per·em′i·nent·ly** *adv.*

su·per·er·o·gate (sōō′pər·er′ə·gāt) *v.i.* **·gat·ed, ·gat·ing** To do more than is required or ordered. [<L *supererogatus,* pp. of *supererogare* < *super-* over and above + *erogare* pay out < *ex-* out + *rogare* ask]

su·per·er·o·ga·tion (sōō′pər·er′ə·gā′shən) *n.* The performance of an act in excess of the demands or requirements of duty. — **works of supererogation** Good deeds done by saints of the Roman Catholic Church in excess of the requirements of divine law; also, voluntary good deeds performed by men over and above God's commandments.

su·per·e·rog·a·to·ry (sōō′pər·ə·rog′ə·tôr′ē, -tō′rē) *adj.* Of, pertaining to, or of the nature of supererogation; superfluous. Also **su′per·e·rog′a·tive.**

su·per·fam·i·ly (sōō′pər·fam′ə·lē, -fam′lē) *n. pl.* **·lies** *Biol.* A division of plants or animals ranking next above the family but below an order or superorder.

su·per·fe·cun·da·tion (sōō′pər·fē′kən·dā′shən, -fek′ən-) *n. Physiol.* The successive impregnation of two or more ova.

su·per·fe·male (sōō′pər·fē′māl) *n. Biol.* A supersexual organism, characterized in the fruit fly by a ratio of 3 X-chromosomes to 2 sets of autosomes.

su·per·fe·tate (sōō′pər·fē′tāt) *v.i.* **·tat·ed, ·tat·ing** *Physiol.* To conceive again prior to the birth of an embryo or fetus already conceived. [<L *superfetatus,* pp. of *superfetare* < *super-* over and above + *fetus* a foetus]

su·per·fe·ta·tion (sōō′pər·fi·tā′shən) *n.* 1 *Physiol.* **a** The second impregnation of a female already pregnant. **b** The progeny resulting from such second impregnation; hence, any unusual additional growth. 2 *Bot.* Fertilization of the same ovule by two or more kinds of pollen. Also **su'per·foe·ta'tion.**

su·per·fi·cial (sōō'pər·fish'əl) *adj.* 1 Of, pertaining to, lying near, or forming the surface; affecting only the surface. 2 Of or pertaining to only the ordinary and the obvious; not profound; shallow: a *superficial* writer. 3 Marked by partial knowledge; cursory; hasty; slight: *superficial* treatment of a subject. 4 Not real or genuine. 5 Square: said of measure. [<LL *superficialis* <L *superficies* SUPERFICIES] — **su'per·fi'ci·al'i·ty** (-fish'ē·al'ə·tē), **su'per·fi'cial·ness** *n.* — **su'per·fi'cial·ly** *adv.*

su·per·fi·ci·ar·y (sōō'pər·fish'ē·er'ē) *adj.* 1 Belonging or pertaining to the superficies; superficial. 2 *Law* Situated on another's land, or resulting from such situation.

su·per·fi·ci·es (sōō'pər·fish'i·ēz, -fish'ēz) *n. pl.* **·ci·es** 1 A surface or its area; superficial area. 2 External appearance; exterior part. [<L *super-* over + *facies* a face]

su·per·fine (sōō'pər·fīn') *adj.* 1 Of surpassing fineness and delicacy; of the best quality. 2 Overrefined; unduly elaborate; overnice. [<MF *superfin* < *super-* over (<L) + *fin* FINE¹] — **su'per·fine'ness** *n.*

su·per·flu·id (sōō'pər·flōō'id) *n. Physics* A peculiar state of matter noted in helium cooled to within a degree of absolute zero: it is characterized by an exceptional heat conductivity, a ready permeation of very dense substances, and the ability to flow upward against gravity: also called *quantum liquid.* — *adj.* (sōō'pər·flōō'id) Of or pertaining to such a state.

su·per·flu·i·ty (sōō'pər·flōō'ə·tē) *n. pl.* **·ties** 1 The state of being superfluous; superabundance. 2 That, or that part, which is superfluous. See synonyms under EXCESS. [<OF *superfluité* <Med. L *superfluitas, -tatis* <L *superfluus* excessive < *super-* over + *fluere* flow]

su·per·flu·ous (sōō·pûr'flōō·əs) *adj.* 1 Exceeding what is needed; excessively abundant; surplus. 2 *Music* Augmented: sometimes said of an interval. 3 *Archaic* Supererogatory; officious. 4 *Obs.* Overfed, overequipped, or oversupplied. See synonyms under REDUNDANT, WASTE. [<L *superfluus.* See SUPERFLUITY.] — **su·per'flu·ous·ly** *adv.* — **su·per'flu·ous·ness** *n.*

Su·per·for·tress (sōō'pər·fôr'tris) *n.* A heavy, four-engine bombing plane; the B-29: a trade name. Also **Su'per·fort'.**

su·per·fuse (sōō'pər·fyōōz') *v.* **·fused, ·fus·ing** *v.t.* To pour so as to cover something else, as cod-liver oil on wine. — *v.i.* To be poured over or on something. [<L *superfusus,* pp. of *superfundere* < *super-* over + *fundere* pour] — **su'per·fu'sion** (-fyōō'zhən) *n.*

su·per·gla·cial (sōō'pər·glā'shəl) *adj. Geol.* Resting upon or deposited from the surface of a glacier.

su·per·heat (sōō'pər·hēt') *v.t.* 1 To heat to excess; overheat. 2 To raise the temperature of (a vapor not in contact with its liquid) above the saturation point for a given pressure. 3 To heat (a liquid) above the boiling point for a given pressure, but without conversion into vapor. — *n.* (sōō'pər·hēt') The degree to which steam has been superheated, or the heat so imparted.

su·per·heat·er (sōō'pər·hē'tər) *n.* A mechanical contrivance for superheating steam, as by causing it to traverse small tubes in the lower part of a chimney.

su·per·het·er·o·dyne (sōō'pər·het'ər·ə·dīn') *adj. Electronics* Pertaining to or designating a type of radio reception in which the modulated incoming signals have the frequency of their carrier waves changed to an intermediate (inaudible) frequency, and are then rectified to reproduce the original sounds. — *n.* A radio receiving set for this method of reception. [<SUPER(SONIC) + HETERODYNE]

su·per·high·way (sōō'pər·hī'wā') *n.* A highway for high-speed traffic, generally with four or more traffic lanes divided by a safety strip.

su·per·hu·man (sōō'pər·hyōō'mən) *adj.* 1 Above the range of human power or skill; above and beyond what is human; miraculous; divine. 2 Beyond normal human ability or power. See synonyms under SUPERNATURAL. — **su'per·hu·man'i·ty** (-hyōō·man'ə·tē) *n.* — **su'per·hu'man·ly** *adv.*

su·per·im·pose (sōō'pər·im·pōz') *v.t.* **·posed, ·pos·ing** 1 To lay or impose upon something else. 2 To add to something else. — **su'per·im'po·si'tion** (-im'pə·zish'ən) *n.*

su·per·in·cum·bent (sōō'pər·in·kum'bənt) *adj.* Resting or lying upon something else. [<L *superincumbens, -entis,* ppr. of *superincumbere* < *super-* over + *incumbere* rest on. See INCUMBENT.] — **su'per·in·cum'bence** or **·ben·cy** *n.*

su·per·in·duce (sōō'pər·in·dōōs', -dyōōs') *v.t.* **·duced, ·duc·ing** To introduce additionally; bring in or cause as an addition. [<LL *superinducere* cover over, add <L *super-* over + *inducere* INDUCE] — **su'per·in·duc'tion** (-duk'shən) *n.*

su·per·in·tend (sōō'pər·in·tend') *v.t.* To have the charge and direction of; manage; supervise. [<LL *superintendere* < *super-* over + *intendere* aim at. See INTEND.]

su·per·in·ten·dence (sōō'pər·in·ten'dəns) *n.* Direction and management; guiding and controlling supervision. See synonyms under OVERSIGHT.

su·per·in·ten·den·cy (sōō'pər·in·ten'dən·sē) *n. pl.* **·cies** 1 The office or rank of a superintendent. 2 Superintendence.

su·per·in·ten·dent (sōō'pər·in·ten'dənt) *n.* 1 One whose function is to superintend some particular work, office, or undertaking: a school *superintendent,* road *superintendent.* 2 The person charged with supervising maintenance and repair in an office or apartment building. — *adj.* Of or pertaining to superintendence or a superintendent; superintending. [<LL *superintendens, -entis,* ppr. of *superintendere* superintend]

Synonyms (noun): conductor, curator, custodian, director, guardian, inspector, intendant, manager, master, overseer, superior, supervisor, warden.

su·pe·ri·or (sə·pir'ē·ər, sōō-) *adj.* 1 Surpassing in quantity, quality, or degree; more excellent; preferable; in an absolute sense, of great excellence: a *superior* man. 2 Of higher grade, rank, or dignity. 3 Too great or dignified to be under the influence of something specified; serenely unaffected or indifferent: with *to*: *superior* to envy. 4 Locally higher; more elevated; upper. 5 Situated relatively nearer the top of the head when the body is standing erect: opposed to *inferior.* 6 *Bot.* Situated above or over another organ or part, as an ovary when free from the calyx, or, in an axillary flower, a petal or lip which is the one next to the main axis of the plant. 7 *Printing* Set above the level of the line: said of type; thus, in C^4D^n, 4 and n are *superior.* 8 *Logic* Of wider application; generic: said of terms, conceptions, and propositions. 9 Supercilious; affecting superiority: a *superior* smile. See synonyms under EXCELLENT, PARAMOUNT, PREDOMINANT. — *n.* 1 One who surpasses another in rank or excellence. 2 The ruler of an ecclesiastical order or house, as an abbey, convent, or monastery. 3 *Printing* A superior letter or character. See synonyms under SUPERINTENDENT. [<OF <L, compar. of *superus* on high, above < *super* above] — **su·pe·ri·or·i·ty** (sə·pir'ē·ôr'ə·tē, -or'-, sōō-) *n.* — **su·pe'ri·or·ly** *adv.*

Su·pe·ri·or (sə·pir'ē·ər, sōō-) A port and industrial city in Wisconsin at the western end of Lake Superior.

Superior, Lake The northernmost, westernmost, and largest of the Great Lakes, in the United States and Canada; 31,820 square miles; length, 350 miles; width, 160 miles.

superior court See under COURT.

su·per·ja·cent (sōō'pər·jā'sənt) *adj.* Lying or resting immediately upon or above something else; overlying. [<LL *superjacens, -entis,* ppr. of *superjacere* < *super-* above + *jacere* lie]

su·per·la·tive (sə·pûr'lə·tiv, sōō-) *adj.* 1 Elevated to the highest degree; consummate; of supreme excellence or eminence. 2 *Gram.* Expressing or involving the extreme degree: said of a form of comparison of adjectives or adverbs: The *superlative* degree of "wise" is "wisest." See COMPARISON (def. 2). 3 Excessive. — *n.* 1 That which is of the highest possible excellence or superior to all others. 2 *Gram.* The highest degree of comparison of the adjective or adverb; any word or phrase in the superlative degree. [<OF *superlatif* <LL *superlativus* <L *superlatus* excessive < *super-* above + *latus,* pp. to *ferre* carry] — **su·per'la·tive·ly** *adv.* — **su·per'la·tive·ness** *n.*

su·per·lu·nar (sōō'pər·lōō'nər) *adj.* Being above or beyond the moon; celestial. Also **su'per·lu'na·ry.**

su·per·male (sōō'pər·māl') *n. Biol.* A supersexual individual having, in the fruit fly, a ratio of 1 X-chromosome to 3 sets of autosomes.

su·per·man (sōō'pər·man') *n. pl.* **·men** (-men') 1 A hypothetical superior being, characterized by perfection of physique, capacity for power, and a moral nature beyond good and evil, regarded as the product of evolutionary survival of the fittest; the *Ubermensch* of Nietzsche. 2 An intellectually and morally improved man; a superior man; one possessing superhuman powers. [Trans. of G *übermensch*]

su·per·mar·ket (sōō'pər·mär'kit) *n.* A large store or market selling food and household supplies and operating generally on a self-service, cash-and-carry basis. Also **super market.**

su·per·mo·ron (sōō'pər·môr'on, -mō'ron) *n. Psychiatry* A mentally deficient person ranking above the moron; one only slightly deficient mentally.

su·per·nal (sōō·pûr'nəl) *adj.* 1 Heavenly; celestial. 2 Placed or located above; lofty; overhead; towering. 3 Coming from above or from the sky. [<OF <L *supernus* < *super* over] — **su·per'nal·ly** *adv.*

su·per·na·tant (sōō'pər·nā'tənt) *adj.* 1 Floating uppermost, above something, or on the surface. 2 *Chem.* Denoting a liquid from which a precipitate has been thrown down. [<L *supernatans, -antis* < *super-* above + *natare* swim] — **su'per·na·ta'tion** (-nā·tā'shən) *n.*

su·per·na·tion·al (sōō'pər·nash'ən·əl) *adj.* Pertaining to all mankind, rather than to one nation only. — **su'per·na'tion·al·ism** *n.* — **su'per·na'tion·al·ist** *n.*

su·per·nat·u·ral (sōō'pər·nach'ər·əl) *adj.* 1 Existing or occurring through some agency beyond the known forces of nature. 2 Lying outside the sphere of natural law, whether psychic or physical. 3 Believed to be miraculous or caused by the immediate exercise of divine power. 4 Pertaining to the miraculous. — *n.* That which is outside the accepted and known order of nature; that which transcends nature. [<Med. L *supernaturalis* <L *super-* above + *natura* NATURE] — **su'per·nat'u·ral·ly** *adv.* — **su'per·nat'u·ral·ness** *n.*

Synonyms (adj.): miraculous, preternatural, superhuman. The *supernatural* is above or superior to the known powers of nature; the *preternatural* is aside from or beyond what we have been accustomed to regard as the result of natural law, often in the sense of inauspicious; as, a *preternatural* gloom. *Miraculous* is more emphatic and specific than *supernatural,* as referring to the direct personal intervention of divine power. *Miraculous* might be termed "extranatural," rather than *supernatural.* All that is beyond human power is *superhuman;* as, Prophecy gives evidence of *superhuman* knowledge; the word is sometimes applied to remarkable manifestations of human power, surpassing all that is ordinary. *Antonyms*: common, natural, ordinary, usual.

su·per·nat·u·ral·ism (sōō'pər·nach'ər·əl·iz'əm) *n.* 1 The quality of being supernatural. 2 Belief in the supernatural; especially, the doctrine that there is a power not to be identified with nature, but which is the ground of its existences and is manifested in its forces, laws, and events: opposed to *naturalism.* 3 The doctrine of spiritual revelation together with the belief in Providence, the efficacy of prayer, and related doctrines: opposed to *rationalism.* Also spelled *supranaturalism.* — **su'per·nat'u·ral·ist** *adj.* & *n.* — **su'per·nat'u·ral·is'tic** *adj.*

su·per·nor·mal (sōō'pər·nôr'məl) *adj.* 1 *Psychol.* Above the normal in characteristics, properties, or intelligence: a *supernormal* child. 2 Pertaining to or designating phenomena incapable of rigorous scientific explanation but conceivably in accord with still undiscovered natural laws.

su·per·nu·mer·ar·y (sōō′pər·nōō′mə·rer′ē, -nyōō′-) *adj.* 1 Being beyond a fixed or standard number. 2 Beyond a customary or necessary number; superfluous. — *n. pl.* **·ar·ies** 1 A person or thing in excess of the regular, necessary, or customary number. 2 A stage performer, as in mob scenes or processions, without any speaking part: often contracted to *supe* or *super.* [<LL *supernumerarius* a soldier added to a legion after it is complete <L *super numerum* < *super* over + *numerus* a number]

su·per·or·der (sōō′pər·ôr′dər) *n. Biol.* A plant or animal division intermediate between a class and an order.

su·per·or·gan·ic (sōō′pər·ôr·gan′ik) *adj.* 1 Not affected by the structure or activities of the organism; superadded to the organic; hence, psychic. 2 Belonging to society regarded as a higher organism than the individual: social term used by Herbert Spencer.

su·per·phos·phate (sōō′pər·fos′fāt) *n. Chem.* 1 An acid phosphate. 2 Any fertilizing material mostly consisting of soluble phosphates: *superphosphate* of lime.

su·per·phys·i·cal (sōō′pər·fiz′i·kəl) *adj.* Beyond or above the physical.

su·per·pose (sōō′pər·pōz′) *v.t.* **·posed, ·pos·ing** 1 To lay over or upon something else, as one layer upon another. 2 *Geom.* To suppose (one figure) to be placed upon another so that all like parts coincide. Compare SUPERIMPOSE. [<F *superposer* < *super-* over + *poser* POSE[1]] — **su′per·pos′a·ble** *adj.* — **su′per·posed′** *adj.* — **su′per·po·si′tion** (-pə·zish′ən) *n.*

su·per·pow·er (sōō′pər·pou′ər) *n.* 1 A theoretical entity conceived as having political power over other very powerful states or countries. 2 The combined sources of electric power in a given area, operating as a connected system to give greater service at a more economical cost to consumers.

su·per·pres·sure (sōō′pər·presh′ər) *n.* 1 Excessive pressure under given conditions. 2 *Aeron.* The amount by which the pressure within the gas cell of a dirigible exceeds atmospheric pressure.

su·per–roy·al (sōō′pər·roi′əl) *n.* A size of ledger paper, 20 by 28 inches.

su·per·sat·u·rate (sōō′pər·sach′ōō·rāt) *v.t.* **·rat·ed, ·rat·ing** 1 To saturate to excess or beyond the normal point. 2 To cause (a solution) to contain more of a dissolved substance than can be held under normal conditions of temperature. — **su′per·sat′u·ra′tion** *n.*

su·per·scribe (sōō′pər·skrīb′) *v.t.* **·scribed, ·scrib·ing** To write or engrave on the outside or on the upper part of; inscribe with a name or address; specifically, to address, as a letter. [<LL *superscribere* <L *super-* over + *scribere* write]

su·per·script (sōō′pər·skript′) *adj.* Written above or overhead: opposed to *subscript.* — *n.* 1 Superscription. 2 *Math.* An index or other mark following and above a letter or figure, as a^3, c'. [<LL *superscriptus,* pp. of *superscribere* SUPERSCRIBE]

su·per·scrip·tion (sōō′pər·skrip′shən) *n.* 1 The act of superscribing an address on a letter. 2 An upper or outer inscription, as a title or a direction; especially, an address on a letter. 3 That portion of a medical prescription that begins with the word *recipe* (generally abbreviated ℞, and meaning "take"). [<OF <LL *superscriptio, -onis* < *superscribere.* See SUPERSCRIBE.]

su·per·sede (sōō′pər·sēd′) *v.t.* **·sed·ed, ·sed·ing** 1 To take the place of, as by reason of superior worth, right, or appropriateness; replace; supplant. 2 To put something in the place of; set aside; suspend; annul. See synonyms under SUBVERT. [<OF *superceder* <L *supersedere* sit over, forbear < *super-* above + *sedere* sit] — **su′per·sed′er** *n.* — **su′per·se′dure** (-sē′jər), **su′per·ses′sion** (-sesh′ən) *n.*

su·per·se·de·as (sōō′pər·sē′dē·əs) *n. Law* A proceeding, as a writ, that operates to supersede or check proceedings. [<L, you shall desist]

su·per·sen·si·ble (sōō′pər·sen′sə·bəl) *adj.* Being above or beyond the range of the senses; supersensual; psychical. — **su′per·sen′si·bly** *adv.*

su·per·sen·su·al (sōō′pər·sen′shōō·əl) *adj.* 1 Being above the senses; supersensible. 2 Spiritual; ideal. Also **su′per·sen′so·ry** (-sen′sər·ē).

su·per·ser·vice·a·ble (sōō′pər·sûr′vis·ə·bəl) *adj.* Trying needlessly or disagreeably to be of service; officious. — **su′per·ser′vice·a·bly** *adv.*

su·per·sex (sōō′pər·seks′) *n. Biol.* A sterile organism having a mixture of male and female characteristics due to a disturbed ratio of autosomes to X–chromosomes, as in the fruit fly. — **su′per·sex′u·al** (-sek′shōō·əl) *adj.*

su·per·son·ic (sōō′pər·son′ik) *adj. Aeron.* Of, pertaining to, or characterized by a speed greater than that of sound: distinguished from *ultrasonic.*

su·per·son·ics (sōō′pər·son′iks) *n. pl. (construed as singular)* The science which treats of the phenomena of supersonic speed, with especial reference to their practical applications to aircraft, guided missiles, rockets, etc.: distinguished from *ultrasonics.*

su·per·state (sōō′pər·stāt′) *n.* A state established as the governing power of a union or federation of subordinate states.

su·per·sti·tion (sōō′pər·stish′ən) *n.* 1 A belief founded on irrational feelings, especially of fear, and marked by credulity; also, any rite or practice inspired by such belief. 2 Specifically, a belief in a religious system regarded (by others than the believer) as without reasonable support; also, any of its rites. 3 Credulity regarding or reverence for the occult or supernatural, as belief in omens, charms, and signs; loosely, any unreasoning or unreasonable belief or impression. 4 *Obs.* Undue scrupulousness. See synonyms under FANATICISM. [<OF <L *superstitio, -onis* excessive fear of the gods, amazement, dread < *superstare* < *super-* over + *stare* stand still]

su·per·sti·tious (sōō′pər·stish′əs) *adj.* 1 Disposed to believe in or be influenced by superstitions. 2 Of, pertaining to, or manifesting superstition. — **su′per·sti′tious·ly** *adv.* — **su′per·sti′tious·ness** *n.*

su·per·stra·tum (sōō′pər·strā′təm, -strat′əm) *n. pl.* **·stra·ta** (-strā′tə, -strat′ə) A layer superimposed upon another; a superficial stratum.

su·per·struct (sōō′pər·strukt′) *v.t.* To build or erect upon a foundation; build over or on something else. [<L *superstructus,* pp. of *superstruere* < *super-* over + *struere* build]

su·per·struc·tion (sōō′pər·struk′shən) *n.* 1 The act of building on a foundation or other structure. 2 A superstructure.

su·per·struc·ture (sōō′pər·struk′chər) *n.* 1 Any structure or any part of a structure above the basement or considered in relation to its foundation. 2 The sleepers, rails, etc., of a railway, as distinguished from the roadbed. 3 *Naut.* The parts of a ship's structure, especially of a warship, above the main deck. Compare SUBSTRUCTURE.

su·per·sub·tle (sōō′pər·sut′l) *adj.* Extremely subtle; oversubtle.

su·per·tax (sōō′pər·taks′) *n.* An extra tax in addition to the normal tax; especially, a graded additional tax on incomes above certain amounts; a surtax.

su·per·ton·ic (sōō′pər·ton′ik) *n. Music* The tone above the tonic or keynote; the second.

su·per·vene (sōō′pər·vēn′) *v.i.* **·vened, ·ven·ing** 1 To follow closely upon something; come as something extraneous or additional. 2 To take place; happen. See synonyms under HAPPEN. [<L *supervenire* < *super-* over and above + *venire* come] — **su′per·ven′ient** (-vēn′yənt) *adj.* — **su′per·ven′tion** (-ven′shən) *n.*

su·per·vise (sōō′pər·vīz) *v.t.* **·vised, ·vis·ing** To have a general oversight of; superintend; oversee. [<Med. L *supervisus,* pp. of *supervidere* <L *super-* over + *videre* see]

su·per·vi·sion (sōō′pər·vizh′ən) *n.* 1 The act of supervising; superintendence. 2 The authority to direct or supervise.

su·per·vi·sor (sōō′pər·vī′zər) *n.* 1 One who supervises or oversees; a superintendent; an inspector. 2 *U.S.* A township officer in administrative charge of its business; one of a board of such officers constituting a body having charge of the business of a county; a borough officer who has charge of road repairs, etc. 3 A person supervising teachers of special subjects in a school. 4 *Obs.* A beholder. — **su′per·vi′sor·ship** *n.* — **su′per·vi′so·ry** (-zər·ē) *adj.*

su·pi·nate (sōō′pə·nāt) *v.t.* & *v.i.* **·nat·ed, ·nat·ing** 1 To make or become supine. 2 To turn, as the hand or forelimb, so that the palm is upward or forward. [<L *supinatus,* pp. of *supinare* throw (someone) on the back < *supinus* SUPINE]

su·pi·na·tion (sōō′pə·nā′shən) *n. Physiol.* 1 The act of turning the palm of the hand, or the corresponding surface of the forelimb, upward. 2 The position of a limb so turned: opposed to *pronation.* 3 The act or state of lying supine.

su·pi·na·tor (sōō′pə·nā′tər) *n. Anat.* A muscle of the forearm by which supination is effected.

su·pine[1] (sōō·pīn′) *adj.* 1 Lying on the back, or with the face turned upward. 2 Having no interest or care; inactive; indolent; negligent; indifferent; listless. 3 Having an inclined position; sloping, as a hill. [<L *supinus* < *sup-,* root of *super* above] — **su·pine′ly** *adv.* — **su·pine′ness** *n.*

su·pine[2] (sōō′pīn) *n.* In Latin grammar, one of two parts of the verb, generally regarded as verbal nouns. The **first** or **former supine,** an accusative form in *-um,* is used after verbs of motion to express purpose, as in *Processit libatum* He went forth to sacrifice; the **second** or **latter supine,** an ablative form in *-u,* is used for specification, as in *Mirabile dictu!* Wonderful to relate! [<L *supinum (verbum)* (a) supine (word), neut. of *supinus* SUPINE[1]]

sup·per (sup′ər) *n.* The last meal of the day: frequently used of an evening banquet. [<OF *soper, super* sup, dine] — **sup′per·less** *adj.*

sup·plant (sə·plant′, -plänt′) *v.t.* 1 To take the place of; displace. 2 To take the place of (someone) by scheming, treachery, etc. 3 To replace (one thing) with another; remove; uproot. See synonyms under ABOLISH, SUBVERT. [<OF *supplanter* <L *supplantare* trip up < *sub-* up from below + *planta* the sole of the foot] — **sup·plan·ta·tion** (sup′lan·tā′shən) *n.* — **sup·plant′er** *n.*

sup·ple (sup′əl) *adj.* 1 Easily bent; flexible; pliant: a *supple* bow. 2 Yielding to the humor or wishes of others; especially, servilely compliant; obsequious. 3 Of the mind, showing adaptability; elastic; easily changing. — *v.t.* & *v.i.* **·pled, ·pling** To make or become supple. [<OF *supple, sople* <L *supplex, -icis* submissive, lit., bending under < *sub-* under + stem of *plicare* fold] — **sup′ple·ly** *adv.* — **sup′ple·ness** *n.*

Synonyms (adj.): compliant, elastic, fawning, flexible, limber, lissom, lithe, lithesome, obsequious, pliable, pliant, soft, submissive, willowy, yielding. See ACTIVE, OBSEQUIOUS. *Antonyms*: firm, fixed, inflexible, obstinate, pertinacious, rigid, stiff, stubborn, unbending, unyielding.

sup·ple·jack (sup′əl·jak) *n.* 1 Any of various woody climbers with tough and lithe stems; specifically, a high–climbing vine (genus *Berchemia*) of the southern United States. 2 A walking stick made from the wood of such a plant.

sup·ple·ment (sup′lə·ment) *v.t.* To make additions to; provide for what is lacking in. — *n.* (-mənt) 1 Something added that supplies a deficiency; especially, an addition to a publication. 2 A supplementary angle. See synonyms under APPENDAGE. [<L *supplementum* < *supplere* SUPPLY[1]]

sup·ple·men·tal (sup′lə·men′təl) *adj.* Like a supplement; supplementing; additional. Also **sup·ple·to·ry** (sup′lə·tôr′ē, -tō′rē).

sup·ple·men·ta·ry (sup′lə·men′tər·ē) *adj.* Supplemental.

supplementary angle See under ANGLE.

sup·pli·ance (sup′lē·əns) *n.* The act of supplicating; an urgent petition or prayer

sup·pli·ant (sup′lē·ənt) *adj.* 1 Entreating earnestly and humbly; beseeching. 2 Manifesting entreaty or submissive supplication. — *n.* One who supplicates. [<MF, ppr. of *supplier* <L *supplicare* SUPPLICATE] — **sup′pli·ant·ly** *adv.* — **sup′pli·ant·ness** *n.*

sup·pli·cant (sup′lə·kənt) *n.* One who supplicates; a suppliant. — *adj.* Asking or entreating humbly; beseeching. [<L *supplicans, -antis,* ppr. of *supplicare* SUPPLICATE]

sup·pli·cate (sup′lə·kāt) *v.* **·cat·ed, ·cat·ing** *v.t.* 1 To ask for humbly or by earnest prayer. 2 To beg something of; entreat. — *v.i.* 3 To beg or pray humbly; make an earnest request. See synonyms under ASK, PRAY. [<L *supplicatus,* pp. of *supplicare* supplicate < *sub-* under + *plicare* bend, fold] — **sup′pli·ca′tion** *n.* — **sup′pli·ca·to′ry** (-kə·tôr′ē, -tō′rē) *adj.*

sup·ply[1] (sə·plī′) *v.* **·plied, ·ply·ing** *v.t.* 1 To give or furnish (something needful or desirable):

to *supply* milk for a city. 2 To furnish with what is needed: to *supply* an army with ammunition. 3 To provide for adequately; satisfy: to *supply* a demand. 4 To make up for; make good or compensate for, as a loss or deficiency. 5 To fill (the place of another); also, to fill (an office, etc.) or occupy (a pulpit) as a substitute. — *v.i.* 6 To take the place of another temporarily. See synonyms under ACCOMMODATE, GIVE, PROVIDE. — *n. pl.* **·plies** 1 That which is or can be supplied; the available aggregate of things needed or demanded. 2 The amount of a commodity offered at a given price or available for meeting a demand. 3 Accumulated stores reserved for distribution, as for an army or a fleet: usually in the plural: He was cut off from his base of *supplies.* 4 A grant of money to the crown or for the public service; appropriation: usually in the plural. 5 An amount sufficient for a given use; store or quantity on hand. 6 A substitute or temporary incumbent. 7 *Obs.* Reinforcements for an army or navy. 8 The act of supplying. See synonyms under STOCK. [<OF *sopleer, soupleier* <L *supplere* <*sub-* up from under + *ple-*, root of *plenus* full]

sup·ply[2] (sup′lē) *adv.* In a supple manner; supplely. [<SUPPLE]

sup·port (sə·pôrt′, -pōrt′) *v.t.* 1 To bear the weight of, especially from underneath; hold in position; keep from falling, sinking, etc. 2 To bear or sustain (weight, etc.). 3 To keep (a person, the mind, etc.) from failing or declining; strengthen. 4 To serve to uphold or corroborate (a statement, theory, etc.); substantiate; verify. 5 To provide (a person, institution, etc.) with maintenance; provide for. 6 To give approval or assistance to; uphold; advocate; aid. 7 To endure or tolerate: I cannot *support* his insolence. 8 To carry on; keep up; maintain: to *support* a war. 9 In the theater: **a** To act (a role or part). **b** To act in a subordinate role to. — *n.* 1 The act of supporting. 2 One who or that which supports. 3 Subsistence. See synonyms under SUBSIDY. [<OF *supporter* <L *supportare* convey <*sub-* up from under + *portare* carry]

Synonyms (verb): bear, carry, maintain, prop, sustain, uphold. *Support* and *sustain* alike signify to hold up or keep up, to prevent from falling or sinking; but *sustain* has a special sense of continuous exertion or strength, as when we speak of *sustained* endeavor or a *sustained* note; a flower is *supported* by the stem or a temple roof by arches; the foundations of a great building *sustain* an enormous pressure; to *sustain* life implies a greater exigency and need than to *support* life; to say one is *sustained* under affliction emphasizes the severity of the trial and the completeness of the *upholding* more than if we say he is *supported.* To *bear* is the most general word, denoting all holding up or keeping up of any object, whether in rest or motion; it refers to something that is a tax upon strength or endurance; as, to *bear* a strain; to *bear* pain or grief. To *maintain* is to keep in a state or condition, especially in an excellent and desirable condition; as, to *maintain* health, reputation, position, etc. *Maintain* is a word of more dignity than *support*; a man *supports* his family; a state *maintains* an army or navy. To *prop* is always partial, signifying to add support to something that is insecure. See ABET, AID, ENDURE, KEEP, LEAN, PROP. *Antonyms*: abandon, betray, demolish, desert, destroy, drop, overthrow, wreck.

sup·port·a·ble (sə·pôr′tə·bəl, -pōr′-) *adj.* That may be supported or borne; bearable; endurable. — **sup·port′a·ble·ness, sup·port′a·bil′i·ty** *n.* — **sup·port′a·bly** *adv.*

sup·port·er (sə·pôr′tər, -pōr′-) *n.* 1 One who or that which supports, in any sense. 2 One who countenances or supports; an adherent. 3 *Her.* One of a pair representing living objects, standing on the dexter and sinister sides of a shield, as if supporting it. 4 An elastic or other support for some part of the body.

SUPPORTER (*def. 3*)

sup·pos·a·ble (sə·pō′zə·bəl) *adj.* That may be supposed. — **sup·pos′a·ble·ness** *n.* — **sup·pos′a·bly** *adv.*

sup·pos·al (sə·pō′zəl) *n.* The act or an instance of supposing; supposition.

sup·pose (sə·pōz′) *v.* **·posed, ·pos·ing** *v.t.* 1 To think or imagine to oneself as true; believe or believe probable; think; presume. 2 To assume as true for the sake of argument or illustration. 3 To require to exist as true; imply as cause or consequence; involve as an inference: Design in creation *supposes* the existence of a God. 4 To expect: I am *supposed* to follow. 5 To presuppose; assume. — *v.i.* 6 To make a supposition. [<OF *suposer* <*sup-* under (<L *sub-*) + *poser* POSE[1]] — **sup·pos′er** *n.*

Synonyms: conjecture, deem, guess, imagine, surmise, think. To *suppose* is temporarily to assume a thing as true, *either* with the expectation of finding it so or for the purpose of ascertaining what would follow if it were so. To *conjecture* is to put together the nearest available materials for a provisional opinion, always with some expectation of finding the facts to be as *conjectured.* To *imagine* is to form a mental image of something as existing, while its actual existence may be unknown, or even impossible. To *think,* in this application, is to hold as the result of thought what is admitted not to be matter of exact or certain knowledge; as, I do not know, but I *think* this to be the fact: a more conclusive statement than would be made by the use of *conjecture* or *suppose.* See GUESS. *Antonyms*: ascertain, conclude, discover, know, prove.

sup·posed (sə·pōzd′) *adj.* Accepted as genuine; believed; often, falsely imagined. — **sup·pos·ed·ly** (sə·pō′zid·lē) *adv.*

sup·po·si·tion (sup′ə·zish′ən) *n.* 1 The act of supposing, or that which is supposed; conjecture. 2 A hypothetical proposition made for the purpose of explaining certain facts, relating them, or of deducing consequences from them; hypothesis. See synonyms under FANCY, GUESS, HYPOTHESIS, IDEA, THOUGHT. [<Med. L *suppositio, -onis* <L, a substitute <*suppositus,* pp. of *supponere* suppose, substitute <*sub-* under + *ponere* place] — **sup′po·si′tion·al** *adj.* — **sup′po·si′tion·al·ly** *adv.*

sup·po·si·tious (sup′ə·zish′əs) *adj.* Supposed or assumed; hypothetical; also, imaginary.

sup·pos·i·ti·tious (sə·poz′ə·tish′əs) *adj.* Put in the place of or made to represent, in order to deceive or defraud; spurious. See synonyms under COUNTERFEIT. [<L *suppositus.* See SUPPOSITION.] — **sup·pos′i·ti′tious·ly** *adv.* — **sup·pos′i·ti′tious·ness** *n.*

sup·pos·i·tive (sə·poz′ə·tiv) *adj.* Including or implying supposition; supposed. — *n.* A conjunction introducing a supposition, as *if,* or *provided.* — **sup·pos′i·tive·ly** *adv.*

sup·pos·i·to·ry (sə·poz′ə·tôr′ē, -tō′rē) *n. pl.* **·ries** *Med.* A solid, readily fusible, medicated preparation for introduction into some canal, cavity, or internal organ. [<LL *suppositorium,* orig. neut. sing. of *suppositorius* placed underneath or up <L *suppositus.* See SUPPOSITION.]

sup·press (sə·pres′) *v.t.* 1 To put an end or stop to; quell; crush, as a rebellion. 2 To stop or prohibit the activities of, as a rival political group; abolish. 3 To withhold from knowledge or publication, as a book, news, etc. 4 To repress, as a groan or sigh. 5 To stop (a hemorrhage, etc.). See synonyms under ABOLISH, HIDE, REPRESS, RESTRAIN, SUBDUE. [<L *suppressus,* pp. of *supprimere* <*sub-* under + *premere* press] — **sup·press′er, sup·pres′sor** *n.* — **sup·press′i·ble** *adj.*

sup·pres·sion (sə·presh′ən) *n.* 1 The act of suppressing, or the state of being suppressed. 2 *Psychoanal.* The deliberate exclusion from consciousness and action of ideas, memories, or emotions, especially those regarded as unpleasant or as socially unacceptable.

sup·pres·sive (sə·pres′iv) *adj.* Tending to suppress.

sup·pu·rate (sup′yə·rāt) *v.i.* **·rat·ed, ·rat·ing** To form or generate pus; maturate. [<L *suppuratus,* pp. of *suppurare* <*sub-* under + *pus, puris* pus]

sup·pu·ra·tion (sup′yə·rā′shən) *n.* 1 The act or process of suppurating. 2 Pus.

sup·pu·ra·tive (sup′yə·rā′tiv) *adj.* Tending to or producing suppuration. — *n.* A remedy promoting suppuration.

supra– *prefix* Above; beyond: *supraliminal.* Used to form adjectives and often the equivalent of *super–* which is preferred in general words. [<L *supra-* <*supra* above, beyond] In anatomical and zoological terms *supra–* means above in position, on the dorsal side of; as in:

supra–abdominal	**suprahepatic**	**supranasal**
supra–auditory	**supralabial**	**supra–ocular**
supracaudal	**supramaxillary**	**supraspinal**

su·pra·lap·sar·i·an (sōō′prə·lap·sâr′ē·ən) *n.* A high Calvinist or holder of the doctrine that predestination preceded creation and the fall of man in the divine order of decrees. See INFRALAPSARIAN. [<NL *supralapsarius* <L *supra-* before + *lapsus* a fall] — **su′pra·lap·sar′i·an·ism** *n.*

su·pra·lim·i·nal (sōō′prə·lim′ə·nəl) *adj. Psychol.* Above the threshold of normal consciousness or sensation: opposed to *subliminal.*

su·pra·mo·lec·u·lar (sōō′prə·mə·lek′yə·lər) *adj.* 1 Containing more than one molecule. 2 Of greater complexity than a molecule.

su·pra·mun·dane (sōō′prə·mun′dān, -mun·dān′) *adj.* Being or placed beyond, or superior to, the world; supernatural; celestial. [Appar. <NL *supramundanus* <L *supra-* above + *mundus* the world]

su·pra·nat·u·ral·ism (sōō′prə·nach′ər·əl·iz′əm) See SUPERNATURALISM.

su·pra·or·bi·tal (sōō′prə·ôr′bi·təl) *adj. Anat.* Situated above the orbit of the eye. [<NL *supraorbitalis* <L *supra-* above + *orbita* ORBIT]

su·pra·pro·test (sōō′prə·prō′test) *n. Law* Acceptance or payment of a bill of exchange by one not a party to it after protest for non–acceptance or non–payment. [<Ital. *sopra protesta* upon protest <*sopra* (<L *supra* above) + *protesta* <L *protestari* PROTEST]

su·pra·re·nal (sōō′prə·rē′nəl) *Anat. adj.* Situated above the kidneys, or pertaining to the ductless glands above the kidneys. — *n.* A suprarenal gland. [<NL *suprarenalis* <L *supra-* above + *renalis* RENAL]

suprarenal gland *Anat.* A ductless gland lying outside the upper or anterior part of either kidney in most vertebrates: also called *adrenal gland.*

su·pra·ren·a·lin (sōō′prə·ren′ə·lin) *n.* Epinephrine. [<SUPRARENAL + -IN]

su·pra·ster·ol (sōō′prə·ster′ōl, -ol) *n. Biochem.* One of two inactive sterols produced as end products by the ultraviolet irradiation of ergosterol. [<SUPRA- + STEROL]

su·pra·tem·po·ral (sōō′prə·tem′pər·əl) *Anat. adj.* Situated in the upper part of the temporal bone or region. — *n.* A supratemporal bone.

su·prem·a·cy (sə·prem′ə·sē, sōō-) *n. pl.* **·cies** The state of being supreme; supreme power or authority. See synonyms under PRECEDENCE, VICTORY. — **royal supremacy** The judicial and executive supremacy of a sovereign as the supreme earthly head of the Christian church within his realm: used especially of English sovereigns.

su·preme (sə·prēm′, sōō-) *adj.* 1 Highest in power or authority; dominant. 2 Highest in degree, importance, or estimation; most extreme or momentous; utmost: *supreme* devotion. 3 Ultimate; last and greatest. See synonyms under ABSOLUTE, FIRST, IMPERIAL, PARAMOUNT, PREDOMINANT. — **the Supreme Being** God; the Deity. — *n.* 1 The supreme or highest point; acme. 2 One who is above the rest; a superior; chief. [<L *supremus* highest, superl. of *superus* that is above <*super* above] — **su·preme′ly** *adv.* — **su·preme′ness** *n.*

su·prême (sü·prem′) *n. French* 1 An especially choice portion of breast of fowl, fish, etc.: a culinary term. 2 A rich cream sauce.

Supreme Bench The United States Supreme Court.

Supreme Court See under COURT.

supreme sacrifice The sacrifice of one's life.

Supreme Soviet The Russian Congress consisting of two legislative chambers, the Soviet (Council) of the Union and the Soviet (Council) of Nationalities, which have equal rights and whose members are elected for a period of four years.

Supreme War Council An international body with headquarters at Versailles, composed of representatives of the Entente nations, of their allies, and of the United States, established at the end of World War I, to determine the terms of peace. This was accomplished by the signing of the Treaty of Versailles, June 28, 1919.

Sur (sŏŏr, sür) A port of SW Lebanon, on the Mediterranean; site of ancient Tyre: also *Tyre. Arabic* **El Sur.**

sur-[1] *prefix* A form of the Latin *super-* found in words which came into English through Old French. [<OF *sur-* <L *super-* SUPER-]

sur-[2] Assimilated var. of SUB-.

su·ra (sŏŏr′ə) *n.* A chapter or section of the Koran. [<Arabic *sūrah*, lit., a step, degree]

Su·ra·ba·ya (sōō′rä·bä′yä) A port and industrial city of NE Java: Dutch *Soerabaja*. Also **Su′ra·ba′ja.**

su·rah (sŏŏr′ə) *n.* A soft, usually twilled, silk fabric, used for women's wear, ties, etc.: now sometimes mixed with rayon. Also **surah silk.** [from *Surat*, India]

Su·ra·jah Dow·lah (sə·rä′jə dou′lə), 1728?–1757, nawab of Bengal; executed by the British. Also *Siraj-ud-daula*.

Su·ra·kar·ta (sŏŏr′ə·kär′tə) Former name for SOLO.

su·ral (sŏŏr′əl) *adj. Anat.* Of or pertaining to the calf of the leg. [<NL *suralis* <L *sura* the calf of the leg]

sur·ance (sŏŏr′əns, shŏŏr′-) *n. Obs.* Assurance. [<OF < *sur* SURE; infl. in meaning by ASSURANCE]

Su·rat (sōō·rat′, sōō′rət) A port on the Gulf of Cambay, northern Bombay State, India; the first British settlement in India, 1612.

sur·base (sûr′bās′) *n. Archit.* A molding or border above the dado and base of a pedestal or above the baseboard of a room. [<SUR-[1] + BASE[1]]

sur·based (sûr′bāst′) *adj. Archit.* 1 Having a surbase, as a pedestal. 2 Flattened; depressed. 3 Having the rise of the curve less than half the span: a *surbased* arch.

sur·cease (sûr·sēs′, sûr′sēs) *n.* Absolute cessation; end. —*v.t.* & *v.i.* **·ceased, ·ceas·ing** To cease entirely or finally; end. [<AF *sursise* omission, orig. pp. of *surseoir* refrain <L *supersedere* SUPERSEDE]

sur·charge (sûr′chärj′) *n.* 1 An excessive burden, load, or charge. 2 In chancery law, the showing of an omission of items in an account for which credit ought to be allowed: opposed to *falsification*. 3 An additional or excessive amount charged, especially an unlawful charge; an overcharge. 4 A new valuation or something additional printed on a postage or revenue stamp; also, a stamp so imprinted. —*v.t.* (sûr·chärj′) **·charged, ·charg·ing** 1 To charge (a person) too much; overcharge. 2 To show an omission of credits in (an account), or of something for which credit should have been allowed. 3 To overload. 4 To fill to excess. 5 To imprint a surcharge on (postage stamps). [<F < *surcharger* < *sur-* over + *charger* <OF *chargier* CHARGE] —**sur·charg′er** *n.*

sur·cin·gle (sûr′sing·gəl) *n.* 1 A girth or strap encircling the body of a beast of burden, for holding a saddle, etc. 2 A girdle, as of a cassock. —*v.t.* **·gled, ·gling** To gird or fasten with a surcingle. [<OF *surcengle* < *sur-* over + L *cingulum* a belt]

sur·coat (sûr′kōt′) *n.* An outer coat or garment; in the Middle Ages, a loose robe or cloaklike garment worn over armor. [<OF *surcot* < *sur-* over + *cot, cote* a coat]

sur·cu·lose (sûr′kyə·lōs) *adj. Bot.* Producing or having suckers: said of plants. [<L *surculosus* < *surculus* a twig, sucker, dim. of *surus* a twig]

surd (sûrd) *n.* 1 *Math.* An irrational number or quantity, especially an indicated root that can only be approximated, as $\sqrt{2}$. 2 *Phonet.* A speech sound made without vibration of the vocal cords. —*adj.* 1 *Math.* Incapable of being expressed in rational numbers; irrational. 2 *Phonet.* Voiceless: opposed to *sonant, voiced*. [<L *surdus* deaf, silent]

sure (shŏŏr) *adj.* 1 Not liable to change or failure; firm; unyielding; stable; infallible. 2 Fit, proper, or deserving to be depended on; reliable; trustworthy. 3 Free from doubt; certain; positive. 4 Certain of obtaining, attaining, or retaining something: with *of*. 5 Safe; secure from danger or harm. 6 Bound to happen. —*adv. Colloq.* Surely; certainly. —**to be sure** Indeed; certainly. —**to make sure** To make certain; secure. [<OF *sur* <L *securus*. Doublet of SECURE.] —**sure′ness** *n.*

Synonyms (adj.): actual, assured, aware, certain, clear, confident, indisputable, positive, real. See AUTHENTIC, FAITHFUL, SECURE.

sure-e·nough (shŏŏr′i·nuf′) *U.S. Colloq. adj.* Real; genuine. —*adv.* Really; surely.

sure-fire (shŏŏr′fīr′) *adj. Colloq.* Reliable; sure or certain to succeed, win, or come out as expected.

sure-foot·ed (shŏŏr′fŏŏt′id) *adj.* Not liable to fall or stumble; figuratively, not liable to err.

sure·ly (shŏŏr′lē) *adv.* 1 Without doubt; certainly. 2 Securely; safely.

sure thing A certainty; any project or undertaking bound to succeed.

sure·ty (shŏŏr′tē, shŏŏr′ə·tē) *n. pl.* **·ties** 1 A person who engages to be responsible for the debt, default, or miscarriage of another; bail. 2 An individual or corporation that, in consideration of the payment of a premium, acts as security for a principal (as a State, city, bank, etc.), against possible loss through the act of an associate or employee who is required to furnish such security. 3 A pledge of money deposited, or of credit given, to secure against loss or damage; security for payment or performance. 4 That which gives security or confidence; ground or basis of certainty or security. 5 The state of being sure; sureness; security; safety; certainty. 6 A sponsor. See synonyms under CERTAINTY, SECURITY. [<OF *surte* <L *securitas, -atis* < *securus* SECURE] —**sure′ty·ship** *n.*

surf (sûrf) *n.* 1 The swell of the sea that breaks upon a shore. 2 The foam caused by the billows. ◆ Homophone: *serf*. [Earlier *suff*, ? var. of SOUGH] —**surf′y** *adj.*

sur·face (sûr′fis) *n.* 1 The exterior part or face of anything that has length, breadth, and thickness. 2 That which has length and breadth, but not thickness; a superficies. 3 A superficial aspect; external view or appearance. 4 That portion of the side of a fortification which is bounded by the angle of the nearest bastion and the prolongation of the flank. —*v.* **·faced, ·fac·ing** *v.t.* 1 To put a surface on; especially, to make smooth, even, or plain. —*v.i.* 2 To mine at or near the surface. 3 To rise to the surface, as a submarine. [<F < *sur-* above + *face* FACE]

sur·face-ac·tive (sûr′fis·ak′tiv) *adj. Chem.* Pertaining to or denoting any of a class of substances which have the property of reducing the surface tension of a liquid in which they are dissolved: said especially of detergents.

surface boat A sink boat.

surface noise The mechanical noise produced by friction of the needle against the granular surface of a phonograph record.

surface plate *Mech.* A plate having a very accurate surface: used for testing other surfaces.

sur·fac·er (sûr′fis·ər) *n.* 1 A planing machine for giving a true surface to lumber; a surface planer: also **surfacing machine.** 2 A miner who works near the surface.

surface tension *Physics* That property of a liquid by virtue of which the surface molecules exhibit a strong inward attraction, thus forming an elastic skin which tends to contract to the minimum area.

sur·fac·tant (sûr·fak′tənt) *n. Chem.* A surface-active agent or a solute which tends to reduce the surface tension of the solvent, as a soap or detergent. [<SURF(ACE)-ACT(IVE) + -ANT]

surf·bird (sûrf′bûrd′) *n.* A ploverlike bird (*Aphriza virgata*) of the Pacific coast of America from Alaska to Chile.

surf·board (sûrf′bôrd′, -bōrd′) *n.* A long, narrow board used in surf-riding, as in Hawaii. Compare AQUAPLANE.

surf·boat (sûrf′bōt′) *n.* A boat of extra strength and buoyancy, for launching and landing through surf. —**surf′boat′man** (-mən) *n.*

surf duck One of various scoters or sea ducks, especially the surf scoter. Also **surf coot.**

sur·feit (sûr′fit) *v.t.* 1 To feed to fullness or satiety; overfeed. 2 To supply to satiety. —*v.i.* 3 To partake of food or drink to excess; overeat. 4 To overindulge. See synonyms under SATISFY. —*n.* 1 The act of surfeiting oneself; excess in eating or drinking; also, the excessive quantity partaken of. 2 The result of such excess; satiety; superfluity. 3 The state of being surfeited; oppressive fullness of the system caused by excess in eating or drinking. [<OF *sorfait* < *surfaire* overdo < *sur-* above + *faire* make <L *facere*] —**sur′·feit·er** *n.*

surf fish Any of a family (*Embiotocidae*) of viviparous sea fishes, perchlike in form, numerous in shallow places near shore all along the northern Pacific coast of North America.

sur·fi·cial (sûr·fish′əl) *adj. Geol.* Originally belonging to or being on the surface, as of the earth: contrasted with *subterranean*. [< SURFACE]

surf-rid·ing (sûrf′rī′ding) *n.* A water sport in which a person standing on a surfboard is borne by the breakers toward the shore.

surf scoter A North American scoter (*Melanitta perspicillata*). The adult male is black with a white spot on the forehead and the nape. Also **surf′er.**

surge (sûrj) *v.* **surged, surg·ing** *v.i.* 1 To rise high and roll onward, as waves; swell or heave. 2 To move or go in a manner suggestive of this: The mob *surged* through the square. 3 To increase or vary suddenly, as an electric current. 4 To slip, as a rope on a windlass. —*v.t.* 5 To cause to move in surges. 6 To let go suddenly, as a rope or cable. —*n.* 1 A large swelling wave; billow; also, such billows collectively. 2 The act of surging; a heaving and rolling motion, as of great waves. 3 *Naut.* The tapered drum of a capstan or windlass around which the rope surges. 4 *Electr.* A sudden fluctuation of voltage due to lightning, switching, etc. See synonyms under WAVE. ◆ Homophone: *serge*. [<OF *sourge-*, stem of *sourdre* rise <L *surgere*] —**surg′er** *n.* —**surg′y** *adj.*

sur·geon (sûr′jən) *n.* 1 One who practices surgery. 2 A medical officer in the military or naval service; a ship's doctor. 3 A surgeon fish. [<AF *surgien*, var. of OF *cirugien*. See CHIRURGEON.]

sur·geon·cy (sûr′jən·sē) *n. pl.* **·cies** The office, duties, or rank of a surgeon.

surgeon fish A West Indian fish (*Teuthis hepatus*) having erectile lancetlike spines at the sides of the tail.

Surgeon General 1 Chief officer of the Medical Department in the United States Army or Navy. 2 Chief medical officer of the United States Public Health Service. 3 *Brit.* One of the ten members of the Army Medical Staff.

surgeon's knot A knot used in tying ligatures, stitching up wounds, etc. See illustration under KNOT.

sur·ger·y (sûr′jər·ē) *n. pl.* **·ger·ies** 1 The branch of medical science that relates to body injuries, deformities, and morbid conditions that require being remedied by operations or instruments. 2 A place where surgical treatment or advice is regularly given; a surgeon's office; an operating room. 3 The work of a surgeon. 4 The treatment of diseases or injuries to nonhuman organisms by like methods: tree *surgery*. [<OF *surgerie*, contraction of *serurgerie*, ult. <LL *chirurgia* <Gk. *cheirourgia* a handicraft < *cheir, cheiros* the hand + *ergein* work]

sur·gi·cal (sûr′ji·kəl) *adj.* 1 Of or pertaining to surgery. 2 Designating a degree of anesthesia deep enough to permit major surgical operations. —**sur′gi·cal·ly** *adv.*

Su·ri·ba·chi (sŏŏr′ə·bä′chē), **Mount** An extinct volcano (546 feet) in southern Iwo Jima; scene of an American victory over the Japanese in World War II, February, 1945.

su·ri·cate (sŏŏr′ə·kāt) *n.* A small burrowing viverrine carnivore (*Suricata tetradactyla*) of South Africa, having only four toes: often domesticated. [<F *surikate* <Afrikaans, ? < a native South African name]

Su·ri·nam (sŏŏr′ə·näm′) Part of the Kingdom of the Netherlands, on the NE coast of South America; 55,129 square miles; capital, Paramaribo: also called *Dutch Guiana, Netherlands Guiana. Dutch* **Su·ri·na·me** (sü′rē·nä′mə).

Surinam River A river in central Surinam, flowing 300 miles north to the Atlantic near Paramaribo.

sur·loin (sûr′loin) *n.* Sirloin: the older spelling.

sur·ly (sûr′lē) *adj.* **·li·er, ·li·est** 1 Persistently rude and ill-humored; crabbed; cross; gruff. 2 Characterized by rudeness or gruffness, as a reply. 3 *Obs.* Haughty. See synonyms under HAUGHTY, MOROSE. [Earlier *sirly* like a lord

< *sir* a lord + *-ly* like] **—sur′li·ly** *adv.* **—sur′li·ness** *n.*

Sur·ma (sōōr′mä) A river in Manipur and SE Assam, NE India, and East Pakistan, flowing 320 miles north and west to the Meghna River, and forming numerous arms, especially in Assam.

sur·mise (sər·mīz′) *v.* **·mised, ·mis·ing** *v.t.* To infer on slight evidence; guess. **—** *v.i.* To make a conjecture thus. See synonyms under GUESS, SUPPOSE, SUSPECT. **—** *n.* (sər·mīz′, sûr′mīz) A conjecture made on slight evidence; supposition. See synonyms under GUESS, HYPOTHESIS. [<OF, an accusation, pp. fem. of *surmettre* accuse < *sur-* upon + *mettre* put <L *mittere* send]

sur·mount (sər·mount′) *v.t.* **1** To overcome; prevail over (a difficulty, etc.). **2** To mount to the top or cross to the other side of; get over, as an obstacle or mountain. **3** To be or lie over or above. **4** To place something above or on top of; cap. **5** *Obs.* To surpass; exceed. See synonyms under CONQUER. [<OF *surmunter* <Med. L *supermontare* <L *super-* over + *mons, montis* a hill, mountain] **—sur·mount′a·ble** *adj.* **—sur·mount′a·ble·ness** *n.* **—sur·mount′er** *n.*

sur·mul·let (sər·mul′it) See MULLET[1] (def. 2).

sur·name (sûr′nām′) *n.* A name subjoined to a given or Christian name; hence, a family name. **—** *v.t.* (sûr′nām′, sûr·nām′) **·named, ·nam·ing** To give a surname to; call by a surname. [Alter. of obs. *surnoun* <OF *surnom* < *sur-* above, beyond + *nom* a name <L *nomen, -inis;* infl. in form by NAME] **—sur′·nam′er** *n.*

sur·pass (sər·pas′, -päs′) *v.t.* **1** To go beyond or past in degree or amount; exceed; excel. **2** To transcend; be beyond the reach or powers of. [<MF *surpasser* < *sur-* above + *passer* PASS] **—sur·pass′a·ble** *adj.*

Synonyms: eclipse, outdo, outstrip, transcend. See BEAT, LEAD. *Antonyms*: fail, yield.

sur·pass·ing (sər·pas′ing, -päs′-) *adj.* Preeminently excellent. **—** *adv. Poetic* Surpassingly. **—sur·pass′ing·ly** *adv.* **—sur·pass′ing·ness** *n.*

sur·plice (sûr′plis) *n. Eccl.* A loose white vestment with full sleeves, worn over the cassock by the clergy of the Anglican, Moravian, and Roman Catholic churches, and also by choristers in a vested choir. [<AF *surpliz,* OF *sourpeliz* < Med. L *superpellicium (vestimentum)* an overgarment < *super-* over + *pellicia* a fur garment < *pellis* skin]

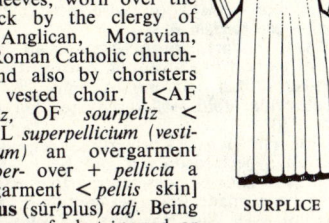

SURPLICE

sur·plus (sûr′plus) *adj.* Being in excess of what is used or needed. **—** *n.* **1** That which remains over and above what has been used or is required; overplus; residue. **2** Assets in excess of liabilities. **3** Excess of net assets above the face value of shares of a corporation. **4** A small unorganized tract of land in Maine set apart by State authority. See synonyms under EXCESS. [<OF <Med. L *superplus* < *super-* over and above + *plus* more]

sur·plus·age (sûr′plus·ij) *n.* **1** That which is over and above; surplus; overplus. **2** *Law* Matter in an instrument not necessary to the meaning; irrelevant matter.

sur·print (sûr′print′) *v.t.* To print again on or over (matter once printed). **—** *n.* That which is surprinted.

sur·pris·al (sər·prī′zəl) *n.* The act of surprising; surprise. Also *Rare* **sur·priz′al.**

sur·prise (sər·prīz′) *v.t.* **·prised, ·pris·ing 1** To cause to feel wonder or astonishment because unusual or unexpected. **2** To come upon suddenly or unexpectedly; take unawares. **3** To attack suddenly and without warning; capture by surprise. **4** To lead unawares, as into doing something not intended: with *into.* **5** To elicit in this manner: They *surprised* the truth from him. **—** *n.* **1** The act of surprising; a coming upon unawares. **2** A surprised state; astonishment. **3** Something that causes surprise, as a sudden and unexpected event, fact, or gift. Also *Rare* **sur·prize′.** [<OF *surpris,* pp. of *surprendre* < Med. L *superprendere* <L *super-* over + *prehendere* take] **—sur·pris′er** *n.*

surprise party A prearranged social gathering of persons, usually at a friend's home, but without previous notice to him.

sur·pris·ing (sər·prī′zing) *adj.* Causing wonder or astonishment; amazing. **—sur·pris′ing·ly** *adv.* **—sur·pris′ing·ness** *n.*

sur·re·al·ism (sə·rē′əl·iz′əm) *n.* A movement in 20th century literature and art which attempts to express and exhibit the workings of the subconscious mind, especially as manifested in dreams and uncontrolled by the reason or any conscious process: characterized by the incongruous and startling arrangement and presentation of subject matter. [<F *surréalisme* < *sur-* beyond, above + *réalisme* realism < *réal* REAL] **—sur·re′al·ist** *adj.* & *n.* **—sur·re′al·is′tic** *adj.* **—sur·re′al·is′ti·cal·ly** *adv.*

sur·re·but·tal (sûr′ri·but′l) *n. Law* A plaintiff's evidence or presentation of evidence, to support or maintain a surrebutter.

sur·re·but·ter (sûr′ri·but′ər) *n.* In common-law pleading, the plaintiff's reply to a defendant's rebutter. [<SUR-[1] + REBUTTER; on analogy with *surrejoinder*]

sur·reined (sə·rānd′) *adj. Obs.* Overridden; worn out. [? <SUR-[1] + *reined,* pp. of REIN]

sur·re·join·der (sûr′ri·join′dər) *n. Law* The plaintiff's answer to the defendant's rejoinder. [<SUR-[1] + REJOINDER]

sur·ren·der (sə·ren′dər) *v.t.* **1** To yield possession of or power over to another; give up because of demand or compulsion. **2** To give up; abandon, as hope. **3** To give up or relinquish, especially in favor of another; resign. **4** To give (oneself) over to a passion, influence, etc. **—** *v.i.* **5** To give oneself up, as to an enemy in warfare; yield. **—** *n.* The act of surrendering one's person to another, or the possession of something to another. [< AF *surrender,* OF *surrendre* < *sur-* over + *rendre* RENDER] **—sur·ren′der·er,** *Law* **sur·ren′der·or** *n.*

Synonyms (verb): abandon, alienate, capitulate, cede, give, relinquish, sacrifice, yield. A state *cedes* territory for a consideration, *surrenders* it to a conqueror; a military commander *abandons* an untenable position or unavailable stores. We *relinquish* a claim, *sacrifice* something precious through error, friendship, or duty, *yield* to convincing reasons, a stronger will, winsome persuasion, or superior force. To *yield* is to give place or give way under pressure, and hence under compulsion; it implies more softness or concession than *surrender.* See ABANDON.

surrender value The reserve value of an insurance policy payable to the insured or to the beneficiary when the policy is discontinued.

sur·rep·ti·tious (sûr′əp·tish′əs) *adj.* **1** Accomplished by secret or improper means; clandestine. **2** Acting secretly or by stealth. [<L *surreptitius, subrepticius* < *subreptus,* pp. of *subripere* steal < *sub-* secretly + *rapere* snatch] **—sur′rep·ti′tious·ly** *adv.* **—sur′rep·ti′tious·ness** *n.*

sur·rey (sûr′ē) *n.* A light pleasure vehicle, having two seats, both facing forward, four wheels, and sometimes a top. [Prob. from *Surrey,* England]

Sur·rey (sûr′ē) A county in SE England; 722 square miles; county town, Guildford.

Sur·rey (sûr′ē), **Earl of,** 1516?–47, Henry Howard, English courtier, soldier, and poet: executed for treason.

sur·ro·gate (sûr′ə·gāt) *n.* **1** One who or that which is substituted for another; a substitute. **2** *Brit.* A deputy appointed by an ecclesiastical judge to act in his place. **3** A probate judge. **—** *v.t.* **·gat·ed, ·gat·ing 1** To put in the place of another; substitute; subrogate. **2** To appoint (another) to succeed oneself. [<L *surrogatus* < *subrogatus,* pp. of *subrogare* < *sub-* in place of another + *rogare* ask]

sur·round (sə·round′) *v.t.* **1** To extend completely around; be on all sides of; encircle: Chairs *surrounded* the table. **2** To place something completely around; enclose. **3** To shut in or enclose, as enemy troops, on all sides so as to cut off communication or retreat; beset; invest. **—** *n.* That which surrounds; the surrounding area. [<AF *surunder,* OF *soronder* overflow <LL *superundare* < *super-* over + *undare* rise in waves < *unda* a wave]

Synonyms: compass, encompass, environ, invest. See EMBRACE.

sur·round·ing (sə·roun′ding) *n.* **1** That which surrounds, or any part of it; environment; conditions of life: usually in the plural. **2** The act of one who surrounds. **—** *adj.* Encompassing; enveloping.

sur·sum cor·da (sûr′səm kôr′də) *Latin* **1** *Eccl.* **a** Lift up your hearts: the opening words of the Preface in the mass. **b** A translation of this, used in other eucharistic liturgies. **2** A cry of encouragement, exhortation, etc.

sur·tax (sûr′taks′) *n.* An extra or additional tax; specifically, a graduated income tax over and above the usual or fixed income tax, levied on the amount by which net income exceeds a certain sum. **—** *v.t.* To assess with an extra or additional tax. [<F *surtaxe* < *sur-* above + *taxe* < *taxer* TAX]

Sur·tees (sûr′tēz), **Robert Smith,** 1803–1864, English novelist and editor.

sur·tout[1] (sər·tōōt′, -tōō′; *Fr.* sür·tōō′) *n.* A long, close-fitting overcoat. [<F < *sur-* above + *tout* all <L *totus*]

sur·tout[2] (sür·tōō′) *adv. French* Above all; chiefly; especially.

Su·ru·ga Bay (sōō·rōō·gä) An inlet of the Philippine Sea in central Honshu, Japan; 35 miles long, 15 to 35 miles wide.

sur·veil·lance (sər·vā′ləns, -vāl′yəns) *n.* The act of watching, or the state of being watched; a very close watch; a spying supervision. See synonyms under OVERSIGHT. [<F *surveiller* superintend < *sur-* over + *veiller* watch <L *vigilare*]

sur·veil·lant (sər·vā′lənt, -vāl′yənt) *adj.* Exercising surveillance; watching; watchful. **—** *n.* One who keeps watch so as to control; an overseer or a spy.

sur·vey (sər·vā′) *v.t.* **1** To look at in its entirety; view as from a height. **2** To look at carefully and minutely; scrutinize; inspect. **3** To determine accurately the area, contour, or boundaries of by measuring lines and angles according to the principles of geometry and trigonometry. **—** *v.i.* **4** To survey land. See synonyms under LOOK. **—** *n.* (sûr′vā, sər·vā′) **1** The operation, act, process, or results of finding the contour, area, boundaries, etc., of a surface. **2** A department or corps for carrying on such operations; also, an area that has been surveyed. **3** A general or comprehensive view; an overlooking. **4** A scrutinizing view; inspection. [<AF *survey-,* stem of *surveier,* OF *sorveir* <Med. L *supervidere* < *super-* over + *videre* look]

sur·vey·ing (sər·vā′ing) *n.* **1** The science and art of determining the area and configuration of portions of the surface of the earth and representing them on maps. **2** The work of one who makes surveys.

sur·vey·or (sər·vā′ər) *n.* **1** One who surveys lands, roads, mines, oil fields, etc.; especially, one engaged in the business of land surveying. **2** One who examines a thing for the purpose of ascertaining its condition, quality, or character; an inspector, as of customs. **3** A customs officer who examines merchandise brought into a port.

sur·vey·or·ship (sər·vā′ər·ship) *n.* The office of a surveyor.

surveyor's level A form of spirit level with telescope and tripod attachment, for use in surveying.

surveyor's measure A system of measurement used in surveying and based on the chain as a unit.

sur·viv·al (sər·vī′vəl) *n.* **1** The act of surviving; an outliving. **2** Something surviving. **3** *Sociol.* The persistence in a society of customs and beliefs originating under circumstances not fully understood or no longer valid. **4** One who or that which lives longer than others. Also *Archaic* **sur·viv′ance.**

survival of the fittest Natural selection.

sur·vive (sər·vīv′) *v.* **·vived, ·viv·ing** *v.i.* To live or continue beyond the death of another, the occurrence of an event, etc.; remain alive or in existence. **—** *v.t.* To live or exist beyond the death, occurrence, or end of; outlive; outlast. See synonyms under LIVE.

[<AF *survivre,* OF *sorvivre* <LL *supervivere* <*super-* above, beyond + *vivere* live] —**sur·viv′ing** *adj.* —**sur·vi′vor, sur·viv′er** *n.*

sur·vi·vor·ship (sər·vī′vər·ship) *n.* 1 The state of surviving. 2 *Law* The right of a surviving party, having a joint interest with others in property, to take the whole estate.

sus– Assimilated var. of SUB–.

Su·sa (sōō′sə) 1 An ancient city of Persia, capital of Elam, the site of which is in SW Iran: Old Testament *Shushan. Persian* **Shush** (shōōsh). 2 A former name for SOUSSE, Tunisia.

Su·san·na (sōō·zan′ə; *Dan., Du., Sw.* sōō·zä′nä, *Ital.* sōō·zän′nä, *Sp.* sōō·sä′nä) A feminine personal name. Also **Su·san** (sōō′zən), **Su·san·nah** (-zä′nə). [Hebrew, a lily]
—**Susanna** A Jewish captive in Babylon, falsely accused of adultery, whose life Daniel saved; also, the book of the Old Testament Apocrypha containing the story of Susanna.

sus·cep·ti·bil·i·ty (sə·sep′tə·bil′ə·tē) *n. pl.* **·ties** 1 The state or quality of being susceptible to influences or of easily receiving impressions. 2 The ability to receive or be impressed by deep emotions or strong feelings; sensibility. 3 *Physics* The ratio of the magnetization of a material to the magnetic force producing it.

sus·cep·ti·ble (sə·sep′tə·bəl) *adj.* 1 Yielding readily; capable of being influenced, acted on, or determined; unresistant; open; liable: usually with *of* or *to.* 2 Having delicate sensibility; sensitive; impressionable; easily affected. [<Med. L *susceptibilis* <L *suscipere* receive, undertake <*sub-* under + *capere* take] —**sus·cep′ti·ble·ness** *n.* —**sus·cep′ti·bly** *adv.*

sus·cep·tive (sə·sep′tiv) *adj.* Receptive; sensitive to; susceptible. —**sus·cep′tive·ness, sus·cep·tiv·i·ty** (sus′ep·tiv′ə·tē) *n.*

Su·si·an (sōō′zē·ən) *n.* The Elamite language.

Su·si·a·na (sōō′zē·ā′nə, -an′ə) See ELAM.

Su·sie, Su·sy (sōō′zē) Diminutives of SUSANNA.

sus·lik (sōōs′lik) *n.* A sciuroid rodent (*Citellus citellus*) of NE Europe and NW Asia, with a very short tail; a pouched marmot; spermophile. [<Russian]

sus·pect (sə·spekt′) *v.t.* 1 To think (a person) guilty as specified on little or no evidence. 2 To have distrust of; doubt: They *suspected* my motives. 3 To have an inkling or suspicion of; think possible: The police *suspect* arson. —*v.i.* 4 To have suspicions. —*adj.* (sus′pekt) Suspected; exciting suspicion. —*n.* (sus′pekt) A person suspected of a crime or other action. [<F *suspecter* <L *suspectus,* pp. of *suspicere* look under, mistrust <*sub-* from under + *specere* look] —**sus·pect′er** *n.*
Synonyms (verb): conjecture, distrust, doubt, mistrust, surmise. See DOUBT, GUESS.

sus·pend (sə·spend′) *v.t.* 1 To bar for a time from a privilege, office, or function as a punishment; debar. 2 To cause to cease for a time; interrupt; withhold temporarily: to *suspend* payments on a debt. 3 To hold in a state of indecision or abeyance; withhold or defer action on: to *suspend* a sentence. 4 To hang from a support so as to allow free movement. 5 To sustain in a body of nearly the same specific gravity; keep in suspension, as dust motes in the air. —*v.i.* 6 To stop for a time. 7 To fail to meet obligations; stop payment. [<OF *suspendre* <L *sub-* under + *pendere* hang]
Synonyms: debar, defer, delay, discontinue, fail, hang, hinder, intermit, interrupt, stay, stop, withhold. See ADJOURN. *Antonyms:* begin, continue, expedite, prolong, protract.

suspended animation Temporary loss of a vital force, simulating death.

sus·pend·er (sə·spen′dər) *n.* 1 One who or that which suspends. 2 One of a pair of straps for supporting the trousers: usually in the plural, **pair of suspenders.** 3 *Brit.* A garter.

sus·pense (sə·spens′) *n.* 1 The state of being uncertain, undecided, or insecure; anxiety. 2 The state of being suspended or stopped temporarily. 3 *Obs.* Cessation. See synonyms under DOUBT. [<OF *suspens, suspense* delay, abeyance <Med. L *suspensum,* orig. pp. neut. of L *suspendere* SUSPEND]

suspense account An account in which charges or credits are entered temporarily pending determination of their proper place.

sus·pen·sion (sə·spen′shən) *n.* 1 The act of suspending or hanging. 2 The state of deferment. 3 *Physics* A uniform dispersion of the fine particles of a solid in a liquid which does not dissolve them. Compare BROWNIAN MOVEMENT, COLLOID. 4 Cessation of payments in business; a going into liquidation: the *suspension* of a bank. 5 Any device used for the purpose of suspension, as in a compass. 6 *Mech.* A system of flexible or absorbent members, as springs in a vehicle, intended to insulate the chassis and body against road shocks transmitted by the wheels. 7 *Music* The prolongation of any note of a chord into the succeeding chord, causing at first dissonance which disappears by resolution; the note so prolonged. 8 The act of debarring from an office or its privileges.

suspension bridge See under BRIDGE.

suspension point One of a series of dots used to indicate the omission of words or sentences.

sus·pen·sive (sə·spen′siv) *adj.* 1 Tending to suspend or to keep in suspense. 2 Having the power of suspending operation: a *suspensive* veto. —**sus·pen′sive·ly** *adv.*

sus·pen·sor (sə·spen′sər) *n.* 1 A suspensory bandage. 2 *Bot.* The thread or chain of cells, in flowering plants and certain cryptogams, which produces at its extremity the developing embryo.

sus·pen·so·ry (sə·spen′sər·ē) *adj.* Suspending; sustaining; delaying. —*n. pl.* **·ries** A truss, bandage, or supporter.

suspensory ligament *Anat.* A fibrous membrane sustaining the lens of the eye.

sus·pi·cion (sə·spish′ən) *n.* 1 Conjecture; doubt; mistrust; imagining something wrong without proof or clear evidence. 2 *Colloq.* The least particle, as of a flavor. See synonyms under DOUBT. —*v.t. Dial.* To suspect. [<AF *suspecioun,* OF *sospeçon* <Med. L *suspectio, -onis* <L *suspicere.* See SUSPECT.] —**sus·pi′cion·al** *adj.*

sus·pi·cious (sə·spish′əs) *adj.* 1 Inclined to suspect. 2 Questionable. 3 Indicating suspicion. See synonyms under ENVIOUS, EQUIVOCAL. —**sus·pi′cious·ly** *adv.* —**sus·pi′cious·ness** *n.*

sus·pire (sə·spīr′) *v.i.* **·pired, ·pir·ing** 1 To sigh. 2 To breathe. [<L *suspirare* <*sub-* up from below + *spirare* breathe] —**sus·pi·ra·tion** (sus′pə·rā′shən) *n.*

Sus·que·han·na (sus′kwə·han′ə) A river in New York, Pennsylvania, and Maryland, flowing 444 miles south to Chesapeake Bay.

Sus·sex (sus′iks) 1 A county in SE England; administratively divided into **East Sussex** (829 square miles; county town, Lewes) and **West Sussex** (628 square miles; county town, Chichester). 2 A former Anglo-Saxon kingdom in southern England.

Sussex spaniel See under SPANIEL.

sus·tain (sə·stān′) *v.t.* 1 To keep from sinking or falling, especially by bearing up from below; uphold; support. 2 To endure without yielding; withstand. 3 To have inflicted on one; undergo; suffer, as loss or injury. 4 To keep up the courage, resolution, or spirits of; comfort. 5 To keep up or maintain; keep in effect or being: to *sustain* friendly relations. 6 To maintain by providing with food, drink, or other necessities; support. 7 To uphold or support as being true or just. 8 To prove the truth or correctness of; corroborate; confirm. See synonyms under AID, ASSENT, CARRY, CONFIRM, ENDURE, HELP, KEEP, PRESERVE, PROP, SUPPORT. [<OF *sustein-,* stem of *sustenir, sostenir* <L *sustinere* <*sub-* up from under + *tenere* hold] —**sus·tain′a·ble** *adj.* —**sus·tain′·er** *n.* —**sus·tain′ment** *n.*

sustaining program A radio or television program that has no commercial sponsor but is paid for by the network or station.

sus·te·nance (sus′tə·nəns) *n.* 1 The act or process of sustaining; especially, maintenance of life or health; subsistence. 2 That which sustains; especially, that which supports life; food. 3 Livelihood; means of support. See synonyms under FOOD, NUTRIMENT. [<AF *sustenaunce,* OF *sostenance* <*sostenir* SUSTAIN]

sus·ten·tac·u·lar (sus′ten·tak′yə·lər) *adj. Anat.* Supporting; sustaining. [<L *sustentaculum* a support <*sustentare* hold up, intens. of *sustinere* SUSTAIN]

sus·ten·ta·tion (sus′ten·tā′shən) *n.* 1 The act or process of sustaining; specifically, support of life; maintenance. 2 That which provides the means of support. 3 Upkeep or maintenance of an estate, building, etc. 4 Physical support. 5 Preservation on a certain level. [<OF *sustentacion* <L *sustentatio, -onis* <*sustentatus,* pp. of *sustentare* hold up <*sustinere* SUSTAIN] —**sus′ten·ta′tive** *adj.*

sus·ten·tion (sə·sten′shən) *n.* 1 Support. 2 The act of being sustained. [<SUSTAIN; on analogy with *retention, detention,* etc.]

su·sur·rant (sōō·sûr′ənt) *adj.* Softly murmuring; rustling; whispering. [<L *susurrans, -antis,* ppr. of *susurrare* whisper <*susurrus* a humming, whispering]

su·sur·rate (sōō·sûr′āt) *v.i.* **·rat·ed, ·rat·ing** To speak softly; whisper. [<L *susurratus,* pp. of *susurrare.* See SUSURRANT.] —**su·sur·ra·tion** (sōō′sə·rā′shən) *n.*

su·sur·rus (sōō·sûr′əs) *n.* A gentle sibilant murmur; whisper; rustling. [<L, a humming, whispering]

Suth·er·land (suth′ər·lənd) A county in northern Scotland; 2,028 square miles; county seat, Dornoch. Also **Suth′er·land·shire** (-shir).

Sutherland Falls Falls in SW South Island, New Zealand; 1,904 feet.

Suth·ron (suth′rən) See SOUTHRON.

Sut·lej (sut′lej) A river in SW Tibet, central Himachal Pradesh, and northern Punjab State, northern India and West Pakistan, flowing 850 miles SW to the Indus river on the former SW border of Bahawalpur.

sut·ler (sut′lər) *n.* A peddler who follows an army to sell goods and food to the soldiers. [<Du. *soeteler* a petty tradesman <*soetelen* perform mean duties] —**sut′ler·ship** *n.*

su·tra (sōō′trə) *n.* 1 A formulated doctrine, often so short as to be unintelligible without a key; literally, a rule or precept. 2 In Sanskrit literature, a short grammatical rule. 3 *pl.* A collection of writings or aphorisms, as the dialogs of the Buddha, the Laws of Manu. 4 In Buddhism, an extended writing, usually in verse, and often in dialog form, embodying important religious and philosophical propositions, sometimes directly, sometimes in highly allegorical or metaphorical language. Also **sut·ta** (sōōt′ə). [<Skt. *sūtra* a thread, rule <*siv* sew]

sut·tee (su·tē′, sut′ē) *n.* Formerly, the sacrifice of a Hindu widow on the funeral pyre of her husband: now forbidden; also, the widow so immolated. [<Hind. *satī* <Skt., a faithful wife, fem. of *sat* good, wise, orig. ppr. of *as* be] —**sut·tee′ism** *n.*

Sut·ter (sut′ər), **John Augustus,** 1803–80, U. S. pioneer in California, born in Germany.

Sut·ter's Mill (sut′ərz) A mill in eastern California: gold was discovered on its site in 1848.

sut·tle (sut′l) *adj.* Formerly, taken after the tare has been deducted and before the tret has been allowed; designating that allowance has been made for the container: said of weight. —*n.* Suttle weight. [Earlier var. of SUBTLE]

su·ture (sōō′chər) *n.* 1 The junction of two contiguous surfaces or edges along a line by or as by sewing. 2 *Anat.* The interlocking of two bones at their edges, as in the skull. 3 *Zool.* The line of junction between contiguous parts. 4 *Bot.* The line of dehiscence in plants. 5 *Surg.* **a** The act or operation of uniting parts by or as by stitching. **b** The sewing together of the cut or cleft edges of divided parts. **c** The thread, silver wire, or other material used in this operation. —*v.t.* **·tured, ·tur·ing** To unite by means of sutures; sew together. [MF <L *sutura* <*sutus,* pp. of *suere* sew] —**su′tur·al** *adj.* —**su′tur·al·ly** *adv.*

su·um cui·que (sōō′əm kī′kwē, kwī′-) *Latin* To each his own.

Su·va (sōō′vä) A port on the south coast of Viti Levu, capital of the Fiji Islands.

Su·vo·rov (sōō·vô′rôf), **Count Alexander Vasilievich,** 1729–1800, Russian field marshal.

Su·wal·ki (sōō·vä′ōō·kē) A town of NE Poland; formerly in Russia.

Su·wan·nee River (sōō·wô′nē, -won′ē) A river in Georgia and Florida, flowing 250 miles south, west, and SW to the Gulf of Mexico: also *Swanee.*

su·ze·rain (sōō′zə·rān, -rin) *n.* 1 One invested with superior or paramount authority; formerly, a feudal lord. 2 A nation having paramount control over a locally autonomous region. —*adj.* Sovereign; supreme. [<F *sus* above <L *susum, sursum* upwards; on analogy with *souverain* a sovereign] —**su′ze·rain·ty** *n.*

Su·zu·ki (soo-zoo'kē), **Daisetz Teitaro,** born 1870, Japanese scholar, author, and teacher, active in the U. S.; leading authority on Zen Buddhism, especially as its expositor to the West.

Su·zy (soo'zē) Diminutive of SUSANNA.

Sval·bard (sväl'bär) An archipelago in the Arctic Ocean, including Spitsbergen and other smaller islands and comprising a possession of Norway; 23,951 square miles; administrative capital, Longyear City on West Spitsbergen.

sva·raj (svä-räj') See SWARAJ.

Sved·berg (svā'berkh), **The (Theodor),** born 1884, Swedish chemist.

svelte (svelt) *adj.* Slender; slim; willowy. [<F *svelte* <Ital. *svelto* <L *ex-* out + *vellere* pluck]

Sverd·lovsk (sverd·lôfsk') A city in the central Ural Mountains in western Asiatic Russian S.F.S.R.: formerly *Ekaterinburg.*

Sver·drup (svar'droop), **Otto,** 1855–1930, Norwegian Arctic explorer.

Sver·drup Islands (svar'droop) An archipelago of northern Franklin District, Northwest Territories, in the Arctic Ocean.

Sve·rige (svā'rya) The Swedish name for SWEDEN.

Sviz·ze·ra (svēt·tsä'rä) The Italian name for SWITZERLAND.

swab (swob) *n.* **1** One of various utensils consisting essentially of a soft absorbent substance on the end of a handle: used for cleaning, etc. **2** A mop for cleaning decks, floors, etc. **3** A sailor who uses such a mop; a menial; a worthless person. **4** A cylindrical brush for cleaning firearms. **5** *Med.* **a** A bit of sponge or cloth for cleansing the mouth of, or used as a means of applying nourishment or medicine to, a sick person. **b** A specimen of mucus, etc., taken for examination; also, the cotton-wound wire used in obtaining it. — *v.t.* **swabbed, swab·bing** To clean or apply with a swab. Also spelled *swob.* [Back formation <SWABBER]

swab·ber (swob'ər) *n.* **1** One who uses a swab. **2** One fit only for swabbing. **3** A swab. [<MDu. *zwabber* <*zwabben* do dirty work, swab]

Swa·bi·a (swā'bē-ə) A region and former duchy of SW West Germany, which contains the Black Forest; the eastern section comprises an administrative province of SW Bavaria; 3,818 square miles; capital, Augsburg: German *Schwaben.* — **Swa'bi·an** *adj.* & *n.*

swad·dle (swod'l) *v.t.* **·dled, ·dling** To wrap with a bandage; especially, to wrap (an infant) with a long strip of linen or flannel; swathe. — *n.* A swaddling band. [OE *swæthel* swaddling clothes, a bandage <*swathian* swathe]

swaddling clothes **1** Bands or strips of linen or cloth wound around a newborn infant. **2** A time of immaturity, or the limitations that restrict the immature. Also **swaddling bands, swaddling clouts.**

Swa·de·shi (swə·dā'shē) *n.* A former political movement originating in Bengal, India, advocating the boycott of British goods as one means of obtaining swaraj or home rule. [<Skt. *svadeśin* native, national <*svadeśa* native country]

swag (swag) *n.* **1** *Slang* Property obtained by robbery or theft; plunder; booty. **2** *Austral.* A swagman's bundle or pack. **3** Baggage; luggage. **4** A swaying; a lurch. — *v.i.* **swagged, swag·ging** **1** *Brit. Dial.* To swing heavily. **2** *Austral.* To tramp, bearing a swag. **3** To sag; sway; lurch. [Prob. <Scand. Cf. dial. Norw. *svagga* sway.]

swag·bel·ly (swag'bel'ē) *n.* A person having a protuberant abdomen. [<SWAG, *v.* (def. 3) + BELLY] — **swag'bel'lied** *adj.*

swage (swāj) *n.* **1** A tool or form, often one of a pair, for shaping metal by hammering or pressure. **2** An ornamental border or molding. **3** A groove on an anvil for use in shaping metal. **4** A swage block. — *v.t.* **swaged, swag·ing** To shape (metal) with or as with a swage or swage block. [<OF *souage*; ult. origin uncertain]

swage block A heavy iron block or anvil having various grooves or holes for shaping metal, heading bolts, etc.: also called *swage.*

swag·ger (swag'ər) *v.i.* **1** To walk with a proud or insolent air; strut. **2** To boast; bluster. — *n.* Braggadocio; superiority in words or deeds. — *adj.* Showy or ostentatious in style, manner, or appearance. [Appar. freq. of SWAG] — **swag'ger·er** *n.* — **swag'ger·ing·ly** *adv.*

swagger coat A sports coat without a belt.

swagger stick A short canelike stick; specifically, one carried by a British soldier when off duty: also called *swanking stick.*

swagger suit A short flared coat and a skirt that matches.

swag·man (swag'mən) *n. pl.* **·men** (-men') *Austral.* One who seeks work, carrying his bundle or swag.

Swa·hi·li (swä·hē'lē) *n. pl.* **·li** **1** One of a Bantu people of Zanzibar and the adjacent coast, having an admixture of Arab blood. **2** These people collectively: with *the.* **3** The agglutinative language of the Swahili, belonging to the Bantu family of languages. [<Arabic, coastal <*sawāḥil,* pl. of *sāḥil* a coast] — **Swa·hi'li·an** *adj.*

swain (swān) *n.* **1** A youthful rustic; a lover. **2** *Obs.* A squire; a male servant. [<ON *sveinn* a boy, servant] — **swain'ish** *adj.* — **swain'ish·ness** *n.*

sward (swârd) *n. Scot.* Sward.

swale[1] (swāl) *n.* **1** Low, marshy ground. **2** *Dial.* Shade; a shady place. Also **swail.** [Prob. <Scand. Cf. ON *svalr* cool.]

swale[2] (swāl) See SWEAL.

swall (swäl) *v.* & *n. Scot.* Swell.

swal·low[1] (swol'ō) *v.t.* **1** To cause (food, etc.) to pass from the mouth into the stomach by means of muscular action of the gullet or esophagus. **2** To take in or engulf in a manner suggestive of this; absorb; envelop: often with *up.* **3** To put up with or endure; submit to, as insults. **4** *Colloq.* To believe credulously. **5** To refrain from expressing or giving vent to; suppress. **6** To take back; recant: to *swallow* one's words. — *v.i.* **7** To perform the act or the motions of swallowing. See synonyms under ABSORB. — *n.* **1** That which is swallowed at once; a small amount; a mouthful. **2** The gullet; throat; gorge. **3** The act of swallowing; appetite; inclination. **4** The channel in a hoisting block for the passage of the rope. **5** An abyss; whirlpool; also, a pit. [OE *swelgan* swallow] — **swal'low·er** *n.*

swal·low[2] (swol'ō) *n.* **1** Any of various small, widely distributed passerine birds (family *Hirundinidae*) with short, broad, depressed bill, long, pointed wings, and forked tail: noted for swiftness of flight and migratory habits, as the **bank swallow** (*Riparia riparia*), the American **tree swallow** (*Iridoprocne bicolor*), and the **barn swallow** (*Hirundo erythrogaster*). ◆ Collateral adjective: *hirundine.* **2** A similar bird, as the swift. [OE *swealwe*]

swallow dive A swan dive.

swal·low·tail (swol'ō·tāl') *n.* **1** *Colloq.* A man's dress coat with two long, tapering skirts or tails. **2** A butterfly (family *Papilionidae*) having a posterior, tail-like prolongation on each hind wing. — **swal'low-tailed'** *adj.*

swal·low·wort (swol'ō-wûrt') *n.* **1** A twining perennial herb (*Cynanchum vincetoxicum*) with greenish-white flowers and roots, the latter formerly used in medicine. **2** The common celandine: said to blossom with the arrival of the swallows and to wither when they depart. **3** One of several plants of the milkweed family.

swam (swam) Past tense of SWIM.

swa·mi (swä'mē) *n.* **1** Master; lord: used by Hindus as a title of respect. **2** A Hindu teacher, especially a religious teacher; a pundit. **3** Loosely, a yogi or fakir. Also **swa'my.** [<Hind. *svāmi* lord, master <Skt. *svāmin*]

swamp (swomp, swômp) *n.* A tract or region of low land saturated with water; a wet bog. Also **swamp'land'** (-land'). ◆ Collateral adjective: *paludal.* — *v.t.* **1** To drench or submerge with water or other liquid. **2** To overwhelm with difficulties; crush; ruin. **3** *Naut.* To sink in water, a swamp, etc. **4** To sink (a vessel) with water. — *v.i.* **5** To sink in water, in a swamp, etc. [Cf. LG *swampen* quake (said of a bog). Akin to SUMP.] — **swamp'y, swamp'ish** *adj.*

swamp angel **1** A person who lives in a swamp; a swamper. **2** The hermit thrush.

Swamp Angel A 200-pound Parrott gun used in the siege of Charleston, S.C., in 1863: so called because it was mounted in a swamp.

swamp blackbird The redwing (def. 1).

swamp boat A small, flat-bottomed, blunt-prowed boat powered by an engine with an airplane propeller mounted high in the stern: used in swampy or boggy areas.

swamp cabbage Skunk cabbage.

swamp·er (swom'pər, swôm'-) *n.* **1** One who lives in a swamp or in a swampy district. **2** One who clears a way in a swamp or forest for skidding logs; also, one who clears away underbrush, fallen trees, and other debris for logging operations.

swamp fever **1** Malaria. **2** An infectious anemia of equine animals, caused by a filtrable virus.

Swamp Fox Sobriquet of FRANCIS MARION.

swamp hare A rabbit (*Sylvilagus aquaticus*) frequenting the swamps of the southern United States.

swamp honeysuckle The swamp azalea (*Azalea viscosa*) of the SE United States.

swamp land **1** Land covered with swamps. **2** Fertile, arable land in a swamp.

swamp law Lynch law.

swamp locust The water locust.

swamp maple The red maple of North America (*Acer rubrum*).

swamp oak **1** An oak (*Quercus bicolor*) common in swamps of the eastern United States: also **swamp white oak.** **2** The pin oak.

swamp owl **1** The short-eared owl. **2** The barred owl. See under OWL.

swamp pine Any of certain pines common in swamps or swampy regions; especially, the loblolly pine and the slash pine.

swamp privet See under PRIVET.

swamp sparrow An American sparrow (*Melospiza georgiana*) resembling the song sparrow, inhabiting the swamps of the southern and eastern United States.

swamp willow The pussy willow.

swan[1] (swon, swôn) *n.* **1** A large, web-footed, long-necked bird (subfamily *Cygninae*), allied to but heavier than the goose, and noted for its grace on the water, as the whooper, the trumpeter swan, and the common North American whistling swan (*Cygnus columbiana*). The male is a *cob,* and the female is a *pen.* **2** Figuratively, a poet or singer. [OE]

TRUMPETER SWAN (Body length from 4 to 4 1/2 feet)

swan[2] (swon, swôn) *v.i. U.S. Dial.* Swear: chiefly in the phrase *I swan,* an exclamation of amazement. [Prob. <dial. E (Northern) *Is' wan,* lit., I shall warrant, used as euphemism for *swear*]

Swan (swon, swôn) The constellation Cygnus. See under CONSTELLATION.

swan dive A fancy dive performed with head tilted back and arms held like the wings of a swallow until near the water: also called *swallow dive.*

Swa·nee River (swô'nē, swon'ē) See SUWANNEE RIVER.

swang (swang) Dialectal past tense of SWING.

swan·herd (swon'hûrd', swôn'-) *n.* One who tends swans; especially, a royal officer of England having charge of marking the swans on the Thames which belong to the crown. Also **swan'mas'ter.** Compare SWAN-UPPING.

swank (swangk) *v.* & *n. Slang* Swagger; bluster. — *adj. Slang* **1** Ostentatiously fashionable; pretentious. **2** *Scot.* Slim; pliant; agile; jolly; lively. Also **swank'y.** [<dial. E. Appar. akin to MLG *swank* flexible, MHG *swanken* sway.] — **swank'i·ly** *adv.* — **swank'i·ness** *n.*

swank·ie (swangk'ē) *n. Scot.* An active, clever lad: sometimes said of a lass. Also **swank'y.**

swanking stick A swagger stick.

swan maiden In many ancient folk myths, a beautiful fairy maiden able to transform herself into a swan by means of a magic robe,

swan-neck

ring, or chain, and living under an enchantment or tabu affecting her life with a human lover.
swan-neck (swon'nek', swôn'-) *n.* Any of several mechanical contrivances resembling in outline the neck of a swan.
swan·ner·y (swon'ər·ē, swôn'-) *n. pl.* ·ner·ies A place where swans are bred or kept.
swan·pan (swän'pän') *n.* A Chinese abacus or frame of beads to aid reckoning: also spelled *schwanpan, shwanpan.* [< Chinese *suan p'an* a reckoning board]
Swan River (swon, swôn) A river in SW Western Australia, flowing 240 miles NW and SW to the Indian Ocean at Fremantle.
swan's-down (swonz'doun', swônz'-) *n.* 1 The down of a swan: used for trimming, powder puffs, etc. 2 Canton or cotton flannel. 3 A soft, thick, fine woolen cloth resembling down. Also **swans'down'**.
Swan·sea (swon'sē) A port and industrial county borough in SW Glamorganshire, southern Wales.
swan·skin (swon'skin', swôn'-) *n.* 1 The unplucked skin of a swan. 2 A soft, fine-twilled flannel or cotton fabric having a soft nap. — *adj.* Made of swanskin.
swan-song (swon'sông', -song', swôn'-) *n.* A last or dying work, as of a poet or composer: in allusion to the ancient fable that the swan sings a last song before dying.
swan-up·ping (swon'up'ing, swôn'-) *n. Brit.* The annual inspection and marking on the beak of the royal and other privileged young swans or cygnets on the Thames; also, the annual expedition for this purpose.
swap (swop) *v.t. & v.i.* **swapped, swap·ping** *Colloq.* To exchange (one thing for another); trade. — **to swap lies** To exchange tales; tell stories. — *n.* The act of swapping. Also spelled *swop.* [ME *swappen* strike (a bargain), slap; prob. ult. imit. of the sound of clapping the hands, as in bargaining]
swa·raj (swə·räj') *n.* 1 Formerly in British India, self-government; by extension, cultural and political development under native influence as distinguished from such development under British influence. 2 Home rule: the watchword of the Indian Nationalists; the party itself. Also spelled *swaraj.* [Skt. *svarāj* self-ruling < *sva-* own + *rāj* rule] — **swa·raj'ist** *n.* — **swa·raj'ism** *n.*
sward (swôrd) *n.* 1 Land thickly covered with grass; turf. 2 *Obs.* A skin; rind. Also **swarth** (swôrth). — *v.t. & v.i.* To cover or become covered with sward. [OE *sweard* a skin]
sware (swâr) Obsolete past tense of SWEAR.
swarm[1] (swôrm) *n.* 1 A large number or body of insects or small living things of any kind. 2 A hive of bees; also, a large number of bees leaving the parent stock at one time, to set up new lodgings, accompanied by a queen. 3 A crowd or throng of persons, animals, or things, especially when in motion or advancing under pressure. 4 *Biol.* A collection of free-swimming unicellular organisms, especially zoospores. See synonyms under FLOCK. — *v.i.* 1 To leave the hive in a swarm: said of bees. 2 To come together, move, or occur in great numbers. 3 To be crowded or overrun; teem: with *with.* 4 *Biol.* To come forth in a swarm. — *v.t.* 5 To fill with a swarm or crowd; throng. [OE *swearm*]
swarm[2] (swôrm) *v.t. & v.i.* To climb (a tree, etc.) by clasping it with the hands and limbs. [Orig. nautical cant. Prob. akin to SWARM[1].]
swarm·er (swôr'mər) *n.* 1 A swarm spore. 2 An insect that swarms, as a bee or gnat; one who or that which swarms.
swarm spore *Biol.* 1 A zoospore. 2 A flagellate spore. 3 A ciliated sponge embryo.
swart (swôrt) *adj.* 1 Swarthy; also, poetically, absolutely black. 2 Malignant; gloomy. Also **swarth** (swôrth). [OE *sweart*] — **swart'ness** *n.*
swart-back (swôrt'bak') *n. Scot.* The great black-backed gull (*Larus marinus*).
swarth[1] (swôrth) *n.* 1 *Obs.* Sward (*n.* def. 1). 2 *Dial.* An unripe crop of hay. [Var. of *swearth*]
swarth[2] (swôrth) *n. Dial.* The apparition of a person about to die; a wraith. [? Var. of SWART]
swarth·y (swôr'thē) *adj.* **swarth·i·er, swarth·i·est** Having a dark hue; of dark or sunburned complexion; tawny. Also **swart'y**. See synonyms under DARK. [Var. of obs. *swarty* <SWART] — **swarth'i·ly** *adv.* — **swarth'i·ness, swarth'ness** *n.*

swarve[1] (swôrv) *v.t. & v.i. Obs.* To swerve. [< dial. var. of SWERVE]
swarve[2] (swôrv) *v.t. & v.i.* To climb. [Origin uncertain. Prob. akin to SWARM[2].]
swarve[3] (swôrv) *v.i.* To swoon. [? <ON *svarfa* upset]
swash (swosh, swôsh) *v.i.* 1 To move or wash noisily, as waves. 2 To swagger. — *v.t.* 3 To splash (water, etc.). 4 To splash or dash water, etc., upon or against. — *n.* 1 The splash of a liquid. 2 A narrow channel through which tides flow. 3 A bar over which the waves pass freely. 4 Swill or wet refuse for pigs. 5 A swaggerer or his behavior. 6 *Slang* Worthless sentimental literature; trash. [Imit.]
swash-buck·ler (swosh'buk'lər, swôsh'-) *n.* A swaggering soldier; a bravo[2]. [<SWASH + BUCKLER; with ref. to striking one's own or one's opponent's shield with a sword] — **swash'buck'ler·ing** *n.* — **swash'buck'ling** *adj. & n.*
swash·er (swosh'ər, swôsh'-) *n.* A blusterer; braggart; bully.
swash·ing (swosh'ing, swôsh'-) *adj.* 1 Splashing. 2 Swaggering; blustering. 3 Crushing; violent.
swash letters Italic special letters having a top or bottom flourish on the side where there is most blank space.
swas·ti·ka (swos'ti·kə) *n.* 1 A primitive religious ornament or symbol, originally in the form of a gammation, but variously modified, the most typical being a Greek cross with the ends of the arms bent at right angles, and prolonged to the length of the upright arms, clockwise, or counterclockwise. See *b* in illustration. It dates back to the Bronze Age in Europe, and still exists as a religious symbol in India, Persia, China, Japan, and among North, Central, and South American Indians: believed to be a token of good luck or blessing. 2 The emblem of the Nazis: as *b* in illustration. See HAKENKREUZ. Compare FYLFOT. Also **swas'ti·ca**. [<Skt. *svastika* < *svasti* well-being, fortune < *sú* good + *astí* being < *as* be]

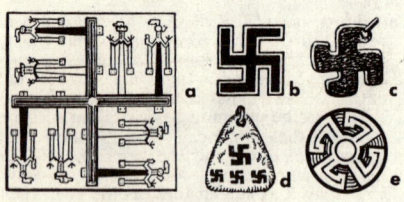

SWASH LETTERS

SWASTIKA
a. Navaho Indian.   c. Caucasian.
b. Indian.   d. Siberian.
         e. Pima Indian.

swat (swot) *v.t.* **swat·ted, swat·ting** To hit with a sharp blow. — *n.* A smart blow. Also spelled *swot.* [Var. of SQUAT, in dial. sense of "squash"]
Swat (swot) *n. pl.* **Swa·ti** (swä'tē) One of an East Indian Moslem people of Indo–European linguistic stock, dwelling in northern West Pakistan. Also **Swa'ti**.
Swat (swot) 1 A former princely state, in the former North–West Frontier Province, northern West Pakistan; 4,000 square miles; capital, Saidu. 2 A river in northern West Pakistan, flowing about 200 miles south, SW, and SE to the Kabul.
swatch (swoch) *n.* 1 A strip, as of cloth, especially one cut off for a sample. 2 [<dial. E (Northern), a cloth tally]
swath (swoth, swôth) *n.* 1 A row or line of cut grass. 2 The space cut by a machine or implement in a single course. 3 The width of grass cut by the sweep of a scythe. Also spelled *swathe.* — **to cut a wide swath** To accomplish much; hence, to make a fine impression. [OE *swæth* a track]
swathe[1] (swäth) *v.t.* **swathed, swath·ing** 1 To bind or wrap, as in bandages; swaddle. 2 To envelop; enwrap; surround. — *n.* A bandage for swathing. [OE *swathian*] — **swath'er** *n.*
swathe[2] (swäth) See SWATH.
Swa·tow (swä'tou') An industrial city and former treaty port on the South China Sea in eastern Kwangtung province, China.
swats (swots) *n. Scot.* New ale; small beer; drink.

swat·ter[1] (swot'ər) *n.* 1 One who or that which crushes with a blow. 2 A perforated rubber or meshed wire device for killing flies. 3 A hard-hitting baseball player.
swat·ter[2] (swot'ər) *v.i. Dial.* To splash water about, as geese and ducks in drinking. [Imit.]
sway (swā) *v.i.* 1 To swing from side to side or to and fro; oscillate. 2 To bend or incline to one side; lean; veer. 3 To tend in opinion, sympathy, etc. 4 To have influence or control; rule. — *v.t.* 5 To cause to swing from side to side. 6 To cause to bend or incline to one side. 7 *Naut.* To swing into place; hoist, as a yard or mast. 8 To cause (a person, opinion, etc.) to tend in a given way; influence. 9 To cause to swerve; deflect or divert, as from a course of action. 10 *Archaic* a To wield, as a weapon or, especially, a scepter. b To rule over; govern. See synonyms under GOVERN, INFLUENCE, SHAKE. — *n.* 1 Power exercised in governing; dominion; control. 2 The act of swaying, literal or figurative; a sweeping, swinging, or turning from side to side. 3 Momentum; inclination; bias. 4 Overpowering force or influence. [Prob. fusion of ON *sveigja* bend and LG *swajen* be moved to and fro by the wind]
sway-back (swā'bak') *n.* 1 A hollow or unnaturally sagging condition of the back, as in a horse. 2 An animal whose back sags abnormally.
sway-backed (swā'bakt') *adj.* 1 Having a sagged or hollow back. 2 Hence, strained or weakened, as by overwork.
Swa·zi (swä'zē) *n.* One of a tribe belonging to the Bantu peoples, and dwelling in Swaziland, Africa.
Swa·zi·land (swä'zē·land) A British protectorate in SE Africa between Mozambique and the Union of South Africa; 6,704 square miles; administrative capital, Mbabane.
sweal (swēl) *v.i. Brit. Dial.* 1 To melt and run down, as the tallow of a candle. 2 To burn away slowly; waste away. Also spelled *swale.* [OE *swelan, swælan* burn]
swear (swâr) *v.* **swore** (*Obs.* **sware**), **sworn, swear·ing** *v.i.* 1 To make a solemn affirmation with an appeal to God or to some deity, or with invocation of something held sacred, as in attestation of truth or proof of good intentions: He *swore* by all the gods. 2 To make a vow; utter a solemn promise. 3 To use profanity; invoke or mention sacred beings or things irreverently or blasphemously; curse. 4 *Law* To give testimony under oath. — *v.t.* 5 To affirm or assert solemnly by invoking sacred beings or things. 6 To promise with an oath or solemn affirmation; vow. 7 To declare or affirm upon oath: to *swear* treason against a man. 8 To take or utter (an oath). 9 To administer a legal oath to. — **to swear by** 1 To appeal to by oath. 2 To have complete confidence in. — **to swear in** To administer a legal oath to. — **to swear off** *Colloq.* To promise to renounce or give up: to *swear off* drink. — **to swear out** To obtain (a warrant for arrest) by making a statement or charge under oath. [OE *swerian*] — **swear'er** *n.*
sweat (swet) *v.* **sweat** or **sweat·ed, sweat·ing** *v.i.* 1 To exude or excrete sensible moisture from the pores of the skin; perspire. 2 To exude moisture in drops; ooze. 3 To gather and condense moisture in drops on its surface. 4 To pass through pores or interstices in drops. 5 To ferment, as tobacco leaves. 6 *Colloq.* To work hard; toil; drudge. 7 *Colloq.* To suffer: You will *sweat* for that! — *v.t.* 8 To exude (moisture) from the pores. 9 To gather or condense drops of (moisture). 10 To soak or stain with sweat. 11 To cause to sweat. 12 To cause to work hard. 13 *Colloq.* To force (employees) to work for low wages and under unfavorable conditions. 14 *Slang* To extort money from. 15 To heat (solder, etc.) until it melts. 16 To join, as metal objects, by applying heat after binding together with solder. 17 *Metall.* To heat so as to extract an element that is easily fusible; also, to extract thus. 18 To force moisture from, as wood in a charcoal kiln. 19 To subject to fermentation, as hides or tobacco. 20 To remove particles of (coins) illegally, as

by shaking them in a bag. **21** *Slang* To subject to torture or rigorous interrogation for the purpose of extracting information; put through the third degree. — **to sweat (something) out** *Slang* To wait through anxiously and helplessly: to *sweat out* a long delay. — *n.* **1** Sensible perspiration of animals, or any gathering of moisture in minute drops like those of perspiration on the skin. **2** The act or state of sweating; specifically, sweating induced by drugs or artificial means. **3** Figuratively, hard labor; drudgery. **4** *Colloq.* Fuming impatience; worry; hurry. **5** The act or process of causing to sweat, as a short rapid exercise given to a horse or the process of sweating hides or bricks. **6** *Obs.* The sweating sickness. [OE *swǣtan* < *swāt* sweat] — **sweat′i·ly** *adv.* — **sweat′i·ness** *n.* — **sweat′y** *adj.*

sweat·band (swet′band′) *n.* A band, usually of leather, inside the crown of a hat to protect it from sweat.

sweat·box (swet′boks′) *n.* **1** A device for sweating such products as hides and dried fruits. **2** Any very hot, close room. **3** *Colloq.* Formerly, a narrow cell where an unruly prisoner was confined; now, any place of confinement; specifically, a place where a prisoner is questioned or put through the third degree.

sweat·ed (swet′id) *adj.* **1** Saturated or covered with sweat; that has been made to perspire. **2** Employed in hard work for low pay; overworked and underpaid: a *sweated* industry.

sweat·er (swet′ər) *n.* **1** One who or that which sweats; specifically, an employer who underpays and overworks his employees. **2** A jerseylike knitted garment with or without sleeves. **3** A medicine that induces sweating; a sudorific.

sweat gland *Anat.* One of the convoluted tubules that secrete sweat, found in subcutaneous tissue and terminating externally in a small orifice or pore.

sweating sickness *Pathol.* A febrile infective disease epidemic in England in the 15th and 16th centuries, characterized by profuse sweating; miliaria. Also **sweating fever.**

sweat shirt A collarless pull-over sweater, sometimes lined with fleece: used by athletes.

sweat·shop (swet′shop′) *n.* A place where work is done under poor conditions, for insufficient wages, and for long hours.

Swede (swēd) *n.* **1** A native or naturalized inhabitant of Sweden. **2** A Swedish turnip; the rutabaga.

Swe·den (swēd′n) A kingdom in NE Europe, in the eastern part of the Scandinavian peninsula; 173,577 square miles; capital, Stockholm: Swedish *Sverige.*

Swe·den·borg (swēd′n·bôrg, *Sw.* svā′dən·bôr′y′), **Emanuel,** 1688–1772, Swedish mystic, philosopher, and scientist. — **Swe′den·bor′gi·an** (-bôr′jē·ən) *adj.* & *n.*

Swe·den·bor·gi·an·ism (swē′dən·bôr′jē·ən·iz′·əm) *n.* The system of philosophy or the theology developed by Emanuel Swedenborg, or from his writings, which teaches that Jesus Christ is the only God, and emphasizes a symbolic interpretation of the Bible. The Swedenborgian church, first organized in London in 1783, is called the *New Church,* or the *New Jerusalem Church.* Also **Swe′den·borg′ism** (-bôrg′iz·əm).

Swed·ish (swē′dish) *adj.* Pertaining to Sweden, the Swedes, or their language. — *n.* **1** The North Germanic language of Sweden, including Old Swedish (the pre-Reformation language), Modern Swedish, and several dialects. **2** The inhabitants of Sweden collectively: with *the.*

Swedish clover Alsike.

Swedish massage Massage given in combination with Swedish movements.

Swedish movements A system of muscular movements employed in treating certain diseases or developing the body.

Swedish turnip The rutabaga.

swee·ny (swē′nē) *n.* Atrophy of the shoulder muscles of a horse. [Perhaps < dial. G *schweine* atrophy < *schwinen* become emaciated]

sweep (swēp) *v.* **swept, sweep·ing** *v.t.* **1** To collect, remove, or clear away with a broom, brush, etc. **2** To clear or clean with or as with a broom or brush: to *sweep* a floor; to *sweep* the plains of buffalo. **3** To touch or brush with a motion as of sweeping: Her dress *swept* the ground; to *sweep* the strings of a harp. **4** To pass over or through swiftly, as in searching: His eyes *swept* the sky. **5** To cause to move with an even, continuous action: He *swept* the cape over her shoulders. **6** To move, carry, bring, etc., with strong or continuous force: The flood *swept* the bridge away. **7** To move over or through with strong or steady force: The gale *swept* the bay. **8** To drag the bottom of (a body of water, etc.). — *v.i.* **9** To clean or brush a floor or other surface with a broom, etc. **10** To move or go strongly and evenly, especially with speed: The train *swept* by. **11** To walk with or as with trailing garments: She *swept* into the room. **12** To trail, as a skirt. **13** To extend with a long reach or curve: The road *sweeps* along the lake shore on the north. See synonyms under CLEANSE. — *n.* **1** The act or result of sweeping. **2** The motion of a long stroke or movement: a *sweep* of the hand. **3** The act of clearing out or getting rid of; hence, removal from office or place: a clean *sweep* of the office-holders; also, a clearance. **4** A turning of the eye or of optical instruments over the field of vision. **5** The winning of a great success, as in an election. **6** The range, area, or compass reached by sweeping, as extent of stroke, range of vision, etc.; direction or extent of motion; hence, a curve or bend, as of a scythe blade, etc. **7** One who or that which sweeps. **8** A piece, as of a machine, along which something sweeps. **9** *Brit.* A chimneysweeper. **10** A long, heavy oar. **11** A well sweep. **12** A curved roadway or approach before a building. **13** *pl.* Sweepings, as of a place where precious metals are worked. **14** *Physics* An irreversible process in which a substance settles to thermal equilibrium or tends to do so. **15** In card games, a winning of all the points in a hand, as by taking of all the tricks in whist; in casino, the taking or capture of all the cards on the table. **16** *Colloq.* Sweepstakes. [ME *swepen,* alter. of *swopen* brush away <OE *swāpen*] — **sweep′er** *n.*

sweep·back (swēp′bak′) *n. Aeron.* **1** The backward inclination of the leading edge of an airplane wing. **2** The acute angle between the line of this inclination and the lateral axis of the airplane.

sweep·ing (swē′ping) *adj.* **1** Carrying off or clearing away with a driving movement. **2** Carrying all before it; covering a wide area; comprehensive. **3** General and thoroughgoing. — *n.* **1** The action of one who or that which sweeps. **2** *pl.* Things swept up; refuse. — **sweep′ing·ly** *adv.* — **sweep′ing·ness** *n.*

sweep·stakes (swēp′stāks′) *n. pl.* **·stakes** **1** A gambling arrangement by which all the sums staked may be won by one or by a few of the betters, as in a horse race. **2** A race for all the stakes. **3** A prize in a sporting contest comprising several stakes. **4** A lottery which offers sweepstakes as prizes. Also **sweep′stake′.**

sweep ticket A ticket which gives the holder a chance to win in a sweepstakes.

sweep·y (swē′pē) *adj.* **sweep·i·er, sweep·i·est** **1** Having a sweeping, swaying, or trailing motion. **2** Sweeping in curves, as a river.

sweer (swir) *adj. Dial.* **1** Heavy; lazy; indolent. **2** Reluctant; unwilling. [OE *swǣr*]

sweet (swēt) *adj.* **1** Agreeable to the sense of taste; having a flavor like that of sugar; especially, containing or due to sugar in some form. **2** Fresh, as opposed to *salt, sour,* or *rancid*; not fermented or decaying. **3** Gently pleasing to the senses; agreeable to the smell; pleasing in sound; melodious; fair; restful. **4** Agreeable or delightful to the mind; arousing gentle, pleasant emotions. **5** Having gentle, pleasing, and winning qualities; marked by kindness and amiability; dear; beloved. **6** Easy; smooth; noiseless: said of machines or contrivances. **7** Sound; rich; productive: said of soil. **8** Not dry: said of wines. **9** *Chem.* Free from acid, etc. — *n.* **1** The quality of being sweet; sweetness. **2** Something sweet: chiefly in the plural, as confections, preserves, candy. **3** A beloved person; darling. **4** Something agreeable or pleasing; pleasure. **5** A sweet smell; perfume. **6** *Brit.* A dessert. ◆ Homophone: *suite.* [OE *swēte*] — **sweet′ly** *adv.* — **sweet′ness** *n.*

Synonyms (adj.): honeyed, luscious, nectared, saccharine, sugared, sugary. See AMIABLE, LOVELY.

Sweet (swēt), **Henry,** 1845–1912, English philologist.

sweet alyssum A perennial Mediterranean herb *(Lobularia maritima)* of the mustard family, having very fragrant white blossoms.

sweet basil Basil (def. 1).

sweet bay **1** Laurel (def. 1). **2** A highly ornamental tree or shrub *(Magnolia virginiana),* with evergreen or deciduous leaves and large handsome flowers.

sweet·bread (swēt′bred′) *n.* The pancreas (**stomach sweetbread**) or the thymus gland (**neck sweetbread** or **throat sweetbread**) of a calf or other animal, when used as food. [<SWEET + BREAD, in obs. sense of "a morsel"]

sweet·bri·er (swēt′brī′ər) *n.* A stout prickly rose *(Rosa eglanteria)* with aromatic leaves. Also **sweet′bri′ar.**

sweet cicely **1** A small European perennial *(Myrrhis odorata)* having white fragrant flowers. **2** A related American herb (genus *Osmorhiza*) with white or purplish flowers and fleshy aromatic root.

sweet clover Melilot.

sweet corn **1** Any of several varieties of Indian corn rich in sugar, and shriveling when ripe. **2** Indian corn in the milky stage.

sweet·en (swēt′n) *v.t.* **1** To make sweet or sweeter. **2** To make more endurable; lighten. **3** To make pleasant or gratifying. **4** In poker, to increase the chips in (the pot). **5** To add gilt-edge securities to others so as to increase the value of (collateral for a loan). — *v.i.* **6** To become sweet or sweeter. — **sweet′en·er** *n.*

sweet·en·ing (swēt′n·ing) *n.* **1** The act of making sweet. **2** That which sweetens. — **long sweetening** Molasses; treacle. — **short sweetening** Sugar.

sweet fennel Finochio.

sweet fern **1** A shrub of the northern United States and Canada (genus *Comptonia*) with long, fernlike, fragrant leaves. **2** Any of several ferns (genus *Dryopteris*).

sweet·flag (swēt′flag′) *n.* A marsh-dwelling plant *(Acorus calamus),* with sword-shaped leaves and a thick creeping rootstock with an aromatic flavor; the calamus.

sweet·gale (swēt′gāl′) *n.* A branching shrub *(Myrica gale),* with both fertile and sterile flowers in short scaly catkins, and resinous, dotted, fragrant leaves. [<SWEET + GALE[2]]

sweet·gum (swēt′gum′) *n.* **1** A balsamiferous tree *(Liquidambar styraciflua)* of Atlantic North America, the wood of which is sometimes used to imitate mahogany. **2** The balsam or gum yielded by it.

sweet·heart (swēt′härt′) *n.* One who is particularly loved by or as a lover; a lover.

sweet·ing (swē′ting) *n.* **1** A sweet apple. **2** A sweetheart; dear one; darling.

sweet·ish (swē′tish) *adj.* Somewhat sweet; slightly sweet; also, nauseatingly sweet. — **sweet′ish·ly** *adv.* — **sweet′ish·ness** *n.*

sweet·leaf (swēt′lēf′) *n.* The horse sugar.

sweet marjoram Marjoram.

sweet·meat (swēt′mēt′) *n.* **1** A confection, preserve, or the like. **2** A candy or crystallized fruit. **3** *pl.* Very sweet candy, cakes, etc.

sweetness and light The essence of esthetic and moral culture, consisting of sympathy, appreciation, open-mindedness, and capacity to enlighten or be enlightened: phrase taken from Swift and popularized by Matthew Arnold.

sweet pea An ornamental annual climber *(Lathyrus odoratus)* of the bean family cultivated for its fragrant, varicolored flowers.

sweet pepper A mild variety of capsicum used for pickling and as a vegetable.

sweet potato **1** A perennial tropical vine *(Ipomoea batatas)* of the morning-glory family, with rose-violet or pink flowers and a fleshy tuberous root. **2** The root itself,

sweets (swēts) *n. pl.* **1** Sweet things to eat, as puddings, cakes, tarts, jellies, etc. **2** The pleasures and gratifying things in life: the *sweets* of success.

sweet-scent·ed (swēt′sen′tid) *adj.* Having a fragrant scent; sweet-smelling.

sweet·sop (swēt′sop′) *n.* **1** A tropical American tree (*Annona squamosa*) allied to the custard apple. **2** Its egg-shaped, scaly fruit; the sugar apple.

sweet tooth *Colloq.* A fondness or appetite for candy or sweets.

sweet william A perennial species of pink (*Dianthus barbatus*) with large lanceolate leaves and closely clustered, showy flowers.

swell (swel) *v.* **swelled**, **swelled** or **swol·len**, **swell·ing** *v.i.* **1** To increase in bulk or dimension, as by inflation with air or by absorption of moisture; dilate; expand. **2** To increase in size, amount, degree, etc. **3** To grow in volume or intensity, as a sound. **4** To rise in waves or swells, as the sea. **5** To bulge; protrude or belly, as a sail. **6** To become puffed up with pride. **7** To grow within one: My anger *swells* at the sight. — *v.t.* **8** To cause to increase in size or bulk. **9** To cause to increase in amount, extent, or degree. **10** To cause to bulge; belly. **11** To puff with pride. **12** *Music* To sing or play with combined crescendo and diminuendo. — *n.* **1** The act, process, or effect of swelling; expansion. **2** The long continuous body of a wave; a billow; hence, a rise of, or undulation in, the land. **3** A bulge or protuberance. **4** *Music* The union of crescendo and diminuendo; also, the signs (< >) indicating it. **5** A device by which the loudness of a musical instrument, as an organ, may be increased or diminished. **6** *Slang* A person of the ultra-fashionable set. See synonyms under WAVE. — *adj. Slang* **1** Of or pertaining to swells or ultrafashionable people; hence, in the height of fashion; smart. **2** First-rate; distinctive. [OE *swellan*]
  Synonyms (verb): bulge, dilate, distend, enlarge, expand, increase, inflate. See PUFF.
  Antonyms: contract, decrease, dwindle, shrink, shrivel, wither.

swell-box (swel′boks′) *n.* A chamber containing the pipes of the organ and having a front of movable slats which muffle the sound or allow it to be heard clearly.

swell-fish (swel′fish′) *n. pl.* **·fish** or **·fish·es** A fish of the eastern coast of the United States that inflates its body with air; a puffer or globefish.

swell-head (swel′hed′) *n. Slang* A conceited person.

swell·ing (swel′ing) *n.* **1** The act of expanding, inflating, or augmenting. **2** *Pathol.* Morbid enlargement of a part of the body. **3** A protuberance. — *adj.* Increasing; bulging.

swell mob *Slang* Well-dressed pickpockets collectively.

swel·ter (swel′tər) *v.i.* **1** To suffer from oppressive heat; perspire from heat. — *v.t.* **2** To cause to swelter. **3** *Obs.* To exude. — *n. Rare* A hot, sweltering condition; oppressive humid heat. [Freq. of obs. and dial. *swelt* be faint, die <OE *sweltan* die]

swel·ter·ing (swel′tər·ing) *adj.* **1** Oppressive; overpoweringly hot. **2** Overcome by heat. Also **swel′try** (-trē). — **swel′ter·ing·ly** *adv.*

swept (swept) Past tense and past participle of SWEEP.

swept·back (swept′bak′) *adj. Aeron.* Having the front edge (of a wing) tilted backward at an angle with the lateral axis of an airplane. Also called *backswept*.

swerve (swûrv) *v.t. & v.i.* **swerved**, **swerv·ing** To turn or cause to turn aside from a course or purpose; deflect. See synonyms under FLUCTUATE, WANDER. — *n.* The act of swerving; a sudden turning aside. [OE *sweorfan* file or grind away]

swev·en (swev′ən) *n. Obs.* A dream. [OE *swefn* sleep, a dream]

swift[1] (swift) *adj.* **1** Traversing space or performing movements in a brief time; rapid; quick. **2** Capable of quick motion; fleet; speedy. **3** Passing rapidly as time or events; also, coming without warning; unexpected. **4** Acting with readiness; prompt. — *adv.* Quickly; a poetic use. [OE] — **swift′ly** *adv.* — **swift′ness** *n.*
  Synonyms (adj.): expeditious, fast, fleet, flying, hasty, quick, rapid, speedy, sudden. See IMPETUOUS, NIMBLE. *Antonyms:* deliberate, dilatory, dull, lingering, slow, sluggish, tardy.

swift[2] (swift) *n.* **1** A bird of swallowlike form (family *Micropodidae*), possessing extraordinary powers of flight, including the builders of edible birds' nests (genus *Collocalia*) and the common American **chimney swift** (*Chaetura pelagica*). **2** One of various small lizards (genera *Sceloporus* and *Uta*) common in the western United States. **3** A reel having an adjustable diameter for winding yarn, etc. **4** The main cylinder of a carding machine; also, a similar part in other machines. [<SWIFT[1]]

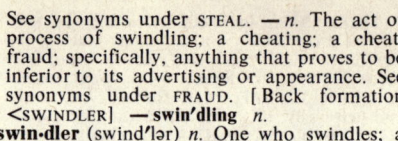
CHIMNEY SWIFT
(About 9 1/2 inches long)

Swift (swift), **Jonathan**, 1667–1745, English satirist born in Dublin: called "Dean Swift."

swift·er (swif′tər) *v.t.* *Naut.* To make taut, as shrouds of a ship, by means of a block and tackle. — *n. Naut.* **1** A rope around the extremities of the capstan bars to connect and steady them, and to give a hold for extra men. **2** One of the forward lower shrouds. **3** A rope for encircling a boat, to strengthen her or prevent chafing of her sides. [< obs. *swift* tie with ropes drawn tight, prob. <Scand. Cf. ON *svifta* reef (a sail).]

swift lizard The fence lizard.

swig[1] (swig) *n. Colloq.* A deep draft. — *v.t. & v.i.* **swigged**, **swig·ging** *Colloq.* To drink swigs (of). [Origin unknown]

swig[2] (swig) *Naut.* *v.t.* **swigged**, **swig·ging** To tighten (a rope that is fast at both ends) by hauling at right angles to its lead. — *n.* **1** A hauling on the bight of a rope fast at both ends. **2** A tackle having diverging ropes. [Akin to SWAG]

swill (swil) *v.t.* **1** To drink greedily and to excess. **2** *Brit.* To drench, as with water; rinse; wash. — *v.i.* **3** To drink to excess; tope. — *n.* **1** Liquid food for domestic animals; especially, the mixture of liquid and solid food given to swine; garbage. **2** Liquor drunk greedily or grossly; loosely, liquor in general. [OE *swillan, swillian* wash]

swim[1] (swim) *v.* **swam** (*Dial.* **swum**), **swum**, **swim·ming** *v.i.* **1** To move through water by working the legs, arms, fins, etc. **2** To be supported on water or other liquid; float. **3** To move with a smooth or flowing motion, as if swimming in water. **4** To be immersed in or covered with liquid; be flooded; overflow. — *v.t.* **5** To cross or traverse by swimming. **6** To cause to swim. See synonyms under FLOAT. — *n.* **1** The action or pastime of swimming. **2** A gliding, swaying motion or movement. **3** The air bladder of a fish; the sound: also **swim bladder**, **swimming bladder**. **4** *Colloq.* The current of affairs, especially of fashionable life: in the *swim*. [OE *swimman*] — **swim′mer** *n.*

swim[2] (swim) *v.i.* To be dizzy; reel; have a giddy sensation; seem to go round. — *n.* A sudden dizziness; temporary unconsciousness; swoon. [OE *swima* dizziness]

swim·mer·et (swim′ə·ret) *n. Zool.* One of a series of fringed, typically biramous abdominal appendages of a crustacean, adapted for swimming, for aid in respiration, and for carrying the eggs on females. [Dim. of *swimmer*]

swim·ming[1] (swim′ing) *n.* The act of one who swims. — *adj.* **1** Used for swimming; having the capacity of swimming; natatorial. **2** Watery; flooded with tears, as the eyes. [<SWIM[1]]

swim·ming[2] (swim′ing) *n.* A state of dizziness or vertigo. — *adj.* Affected by dizziness. [<SWIM[2]]

swimming hole A deep hole in a shallow running stream, used for swimming.

swim·ming·ly (swim′ing·lē) *adv.* In a swimming manner; easily, rapidly, and successfully.

Swin·burne (swin′bûrn) **Algernon Charles**, 1837–1909, English poet and critic.

swin·dle (swin′dəl) *v.* **·dled**, **·dling** *v.t.* **1** To cheat of money or property by deliberate fraud; defraud. **2** To obtain by such means. — *v.i.* **3** To practice fraud; be a swindler. See synonyms under STEAL. — *n.* The act or process of swindling; a cheating; a cheat; fraud; specifically, anything that proves to be inferior to its advertising or appearance. See synonyms under FRAUD. [Back formation <SWINDLER] — **swin′dling** *n.*

swin·dler (swind′lər) *n.* One who swindles; a rogue. [<G *schwindler* giddy-minded person, cheat < *schwindeln* act thoughtlessly, be giddy]

swindle sheet *U.S. Slang* An expense account.

swine (swīn) *n. pl.* **swine** **1** An omnivorous mammal (family *Suidae*) having a long mobile snout and cloven hoofs. **2** A domesticated hog. **3** A low, greedy, stupid, or vicious person: used contemptuously. [OE *swīn*]

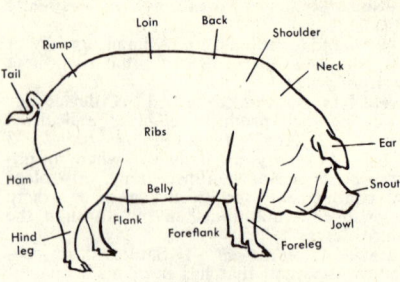
SWINE
Nomenclature of anatomical parts.

swine fever Hog cholera.

swine·herd (swīn′hûrd) *n.* A tender of swine.

Swi·ne·mün·de (svē′nə·mün′də) The German name for ŚWINOUJŚCIE.

swine-pox (swīn′poks′) *n.* A form of chicken pox affecting swine, in which the pustules are deep-seated.

swing (swing) *v.* **swung** (*Dial.* **swang**), **swung**, **swing·ing** *v.i.* **1** To move to and fro or backward and forward rhythmically, as something suspended; oscillate. **2** To move in a swing (def. 3). **3** To move with an even, swaying motion; walk with vigorous strides. **4** To turn; pivot: We *swung* around and went home. **5** To be suspended; hang. **6** *Colloq.* To be executed by hanging. — *v.t.* **7** To cause to move to and fro or backward and forward. **8** To cause to move with a sweeping or circular motion, as a sword, ax, etc.; brandish; flourish. **9** To cause to turn on or as on a pivot or central point. **10** To lift or hoist: They *swung* the mast into place. **11** *Colloq.* To bring to a successful conclusion; manage successfully. **12** *Colloq.* To arrange, sing, or play in the style of swing music. — **to swing round the circle** To make a political tour of any extended area or any local constituency: specifically said of a presidential candidate's tour of the whole United States. — *n.* **1** The action of swinging. **2** A free swaying motion. **3** A contrivance of hanging ropes with a seat on which a person may move to and fro through the air as a pastime. **4** Free course or scope; full liberty or license. **5** Compass; sweep. **6** The movement or rhythm characterizing certain styles of prose and poetry. **7** That which swings or is swung; a swinging blow or stroke. **8** The course of a career or period of activity; main current of business. **9** Swing music. [OE *swingan* scourge, beat up] — **swing′er** *n.*

swing back *Phot.* **1** A camera back provided with a hinge to allow free movement in any direction so as to minimize distortion of perspective or focus. **2** A camera so equipped.

swing bridge A bridge constructed to rotate in a horizontal plane through an angle of 90 degrees from its normal position, to permit the passage of large vessels, etc.

swinge[1] (swinj) *v.t.* **swinged**, **swinge·ing** *Archaic* To flog; chastise. [OE *swengan* shake, beat] — **swing′er** *n.*

swinge[2] (swinj) *v.t.* **swinged**, **swinge·ing** *Dial.* To singe. [? Alter. of SINGE; ? infl. in form by SWEAL]

swinge·ing (swin′jing) *adj. Colloq.* Very large; heavy; extravagant. [<SWINGE[1]]

swinging door A door so hung that it will open in either direction and swing shut when not held.

swin·gle (swing′gəl) *n.* **1** A large, knifelike

swingletree — wooden implement for beating flax: also **swing′-knife**. 2 The short wooden bar of a flail; a swiple. —v.t. **·gied, ·gling** To cleanse, as flax, by beating with a swingle; scutch. [<MDu. *swinghel*. Akin to SWING.]

swin·gle·tree (swing′gəl·trē′) n. A horizontal crossbar, to the ends of which the traces of a harness are attached; a whiffletree or singletree. See illustration under HARNESS. Also **swing·tree** (swing′trē′), **swin′gle·bar′** (-bär′).

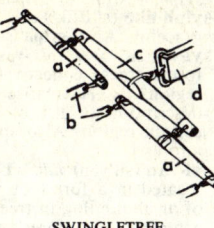

SWINGLETREE
a, a. Swingletrees.
b. Traces.
c. Double-tree.
d. Plow beam.

swing music 1 A development of jazz after about 1935 which achieved its effects by large bands of musicians, contrapuntal styles, and arranged ensemble playing rather than improvised solo performances. 2 The particular rhythmic quality in such music. Also called *swing*.

swing shift An evening work shift, usually lasting from about 4 p.m. to midnight.

swin·ish (swī′nish) adj. Of or like swine; degraded; beastly. See synonyms under BRUTISH. — **swin′ish·ly** adv. — **swin′ish·ness** n.

swink (swingk) v.i. Archaic & Brit. Dial. To toil hard; drudge. [OE *swincan*]

Swin·ner·ton (swin′ər·tən), **Frank (Arthur)**, born 1884, English novelist and critic.

Świ·no·ujś·cie (shvē′nô·ōō′əsh·che) A port of NW Poland: German *Swinemünde*.

swipe (swīp) v.t. **swiped, swip·ing** 1 Colloq. To give a strong blow; strike with a full swing of the arm. 2 Slang To steal; snatch. —n. Colloq. 1 A hard blow, especially in field games. 2 A well sweep, lever, pump handle, or the like. [Var. of SWEEP]

swipes (swīps) n. pl. Brit. Slang Poor, spoiled, or weak beer; small beer; beer in general. [<SWIPE, in obs. sense of "drink hastily"]

swip·le (swip′əl) n. That part of a threshing flail that strikes the grain; a swingle. Also **swip′ple**. [ME *swepelles* a broom. Akin to SWEEP.]

swirl (swûrl) v.t. & v.i. To move or cause to move along in irregular eddies; whirl. —n. 1 A whirling along, as in an eddy; whirl. 2 A curl or twist; spiral. [<dial. E (Scottish) *swyrle*. Prob. akin to dial. Norw. *svirla* whirl.]

swirl·y (swûr′lē) adj. 1 Full of swirls. 2 Scot. Tangled; knotty; gnarled: also **swirl′ie**.

swish (swish) v.i. 1 To move with a sweeping motion and a whistling sound, as a whip. —v.t. 2 To cause to swish. 3 To thrash; flog. —n. 1 A hissing, swishing sound, as of a lash through the air, or the swing of a silk skirt. 2 A movement producing such a sound. 3 An implement, as a broom, used with such a movement. [Imit.]

swiss (swis) n. Often cap. A sheer, crisp cotton fabric, similar to muslin, and often dotted or figured, when it is called *dotted swiss*.

Swiss (swis) adj. Pertaining to Switzerland; characteristic of Switzerland. —n. pl. **Swiss** A native or naturalized inhabitant of Switzerland.

Swiss chard Chard (def. 2).

Swiss cheese A pale-yellow cheese with many large holes, made in, or similar to that made in, Switzerland.

Swiss franc The monetary unit of Switzerland.

Swiss guards Mercenary soldiers from Switzerland formerly used as bodyguards by European monarchs; now guards at the Vatican.

Swiss steak A thick cut of steak floured and cooked, often with a sauce of tomatoes and onions.

switch (swich) n. 1 A small flexible rod; light whip. 2 A tress of human or false hair, fastened together at one end and used by women in building a coiffure. 3 A mechanism for shifting a railway train or other rail vehicles from one track to another. 4 The act or operation of switching, shifting, or changing. 5 The end of the tail in certain animals, as a cow. 6 Electr. A device to make or break a circuit, or transfer a current from one conductor to another. 7 A connecting trench between two lines of defensive trenches. 8 A blow with a switch. See synonyms under STICK. —v.t. 1 To whip or lash with or as with a switch. 2 To move, jerk, or whisk suddenly or sharply: The woman *switched* her skirts aside. 3 To turn aside or divert; shift. 4 To exchange: They *switched* plates. 5 To shift, as a railroad car, to another track. 6 Electr. To connect or disconnect with a switch. —v.i. 7 To turn aside; change; shift. 8 To be shifted or turned. 9 Dial. To walk with a jerky or uneven gait. [Earlier *swits*. Akin to LG *zwuske* a thin rod.]

switch·back (swich′bak′) n. 1 A railway ascending or descending a steep incline in a series of zigzag tracks. 2 A zigzag mountain road. 3 A railroad at amusement resorts in which the cars are hoisted to a starting point and descend along a circuitous route by gravity.

switch·board (swich′bôrd′, -bōrd′) n. A panel or arrangement of panels bearing switches for connecting and disconnecting electric circuits, as a telephone exchange.

switch·er (swich′ər) n. 1 A switch-tender. 2 One who or that which switches.

switch·man (swich′mən) n. pl. **·men** (-mən) One who handles railway switches.

switch plant A plant in which green shoots take the place of absent or reduced leaves.

switch·yard (swich′yärd′) n. A railroad yard for the assembling and breaking up of trains.

swith (swith) adv. Scot. or Obs. Strongly; very much; quickly. — interj. Begone! quick! Also **swithe** (swith).

swith·er (swith′ər) Dial. & Scot. v.i. To doubt; hesitate; fear. —n. 1 A state of doubt or hesitation. 2 A fright; perspiration; faint.

Swith·in (swith′in), **Saint**, died A.D. 862, bishop of Winchester. See under ST. SWITHIN'S DAY.

Swit·zer (swit′sər) n. 1 A Swiss. 2 Specifically, a Swiss mercenary soldier. Also **Swiss·er** (swis′ər).

Swit·zer·land (swit′sər·lənd) A republic in central Europe; 15,940 square miles; capital, Bern: French *Suisse*, German *Schweiz*, Italian *Svizzera*, Latin *Helvetia*. Also **Swiss Confederation**.

swiv·el (swiv′əl) n. 1 A coupling device, link, ring, or pivot that permits either half of a mechanism, as a chain, to rotate independently. 2 A rest on a boat's gunwale, on which a gun may be swept or swung in a horizontal plane. 3 Anything that turns on a pin or headed bolt. 4 A cannon that swings on a pivot: also **swivel gun**. 5 The shuttle of a loom. —v.t. & v.i. **·eled** or **·elled, ·el·ing** or **·el·ling** 1 To turn on or as on a swivel. 2 To provide with or secure by a swivel. —v.i. 3 To turn or swing on or as on a swivel. [ME *swyuel* <OE *swif-*, stem of *swīfan* move]

SWIVEL (a)

swiv·et (swiv′it) n. Colloq. Hurry; anxiety; eager, nervous haste or excitement: Don't be in such a *swivet*. Also **swiv′vet**. [Cf. obs. *swive* copulation <OE *swīfan* move] — **swiv′et·ty** adj.

swiz·zle (swiz′əl) n. One of various compounded intoxicating drinks; specifically, a drink made with rum or other spirit, sugar, bitters, and ice. —v.t. & v.i. **·zled, ·zling** Slang To guzzle. [Origin unknown] — **swiz′zler** n.

swizzle stick 1 A stick, usually with prongs set at right angles to one end, used to mix swizzle by whirling between the palms of the hands. 2 A slender rod of glass, plastic, etc., used to mix drinks.

swob (swob), **swob·ber** (swob′ər) See SWAB, etc.

swol·len (swō′lən) Alternative past participle of SWELL.

swoon (swoon) v.i. To fall in a faint; faint. —n. The act of swooning; a fainting fit. See synonyms under STUPOR. Also, Obs., *swoun*, *swound*. [ME *swounen*, back formation <*swowening*e SWOONING]

swoon·ing (swoo′ning) n. A fainting fit; swoon. —adj. Fainting. [ME *swoweninge* <OE *geswōgen* unconscious]

swoop (swoop) v.i. 1 To drop or descend suddenly, as a bird pouncing on its prey. —v.t. 2 To take or seize suddenly, as with a swoop. —n. A sweeping down or pouncing down, as by a bird of prey: often figuratively. [Var. of obs. *swope* <OE *swāpan* sweep; prob. infl. in form by dial. E *soop* sweep <ON *sōpa*]

swop (swop) See SWAP.

sword (sôrd, sōrd) n. 1 A weapon consisting of a long blade fixed in a hilt: used for cutting or thrusting, as a rapier, scimitar, or claymore. 2 The power of the sword; sovereignty; the power of life and death; especially, military as opposed to civil power. 3 War; also, the cause of death or ruin. 4 An end bar from which the lay of a hand loom hangs; also, the upright support of the lay of a power loom. — **at swords' points** Very unfriendly; hostile; ready for a fight. — **to put to the sword** To kill with a sword; slaughter in battle. [OE *sweord*]

sword bayonet A bayonet having the shape of a sword and used like one. See illustration under BAYONET.

sword·bill (sôrd′bil′, sōrd′-) n. A tropical American hummingbird (genus *Ensifera*) with a very long, slender bill.

sword cane A cane made to carry a sword or dagger.

sword·craft (sôrd′kraft′, -kräft′, sōrd′-) n. 1 Dexterity or skill in the use of the sword. 2 Exercise of authority by the sword, or by military power.

sword dance 1 A dance among or over naked swords laid on the ground. 2 A dance in which the female dancers pass under a double line of swords crossed over their heads by the men.

sword·er (sôr′dər, sōr′-) n. Obs. One skilled in the use of, or who fights with, a sword; hence, a cut-throat.

sword·fish (sôrd′fish′, sōrd′-) n. pl. **·fish** or **·fish·es** A large fish of the open sea (genus *Xiphias*) having the bones of the upper jaw consolidated to form an elongated swordlike process.

SWORDFISH
(Up to 20 feet in length)

sword·grass (sôrd′gras′, -gräs′, sōrd′-) n. 1 Any of several varieties of grasses or sedges (especially genus *Mariscus*, formerly *Cladium*) with sharp or serrated edges. 2 The sword lily.

sword·knot (sôrd′not′, sōrd′-) n. Formerly, a loop of leather used to fasten the hilt of a sword to the wrist; now, a tassel of cord or ribbon tied to a sword hilt.

sword lily A gladiolus.

sword play 1 Attack and defense with the sword. 2 Art or skill in fighting with the sword or in fencing; fencing. — **sword′-play′-er** n.

swords·man (sôrdz′mən, sōrdz′-) n. pl. **·men** (-mən) 1 One skilled in the use of or armed with a sword. 2 A soldier. Also **sword′man**. — **swords′man·ship, sword′man·ship** n.

swore (swôr, swōr) Past tense of SWEAR.

sworn (swôrn, swōrn) Past participle of SWEAR.

swot[1] (swot) Brit. Slang v.i. **swot·ted, swot·ting** To sweat or work hard over a task; grind. —n. Hard work; also, one who works hard, especially in studying. [Dial. var. of SWEAT]

swot[2] (swot) See SWAT[1].

swoun (swoon), **swound** (swoond) See SWOON.

swounds (zwoundz, zoundz), **swouns** (zwounz, zoonz) See ZOUNDS.

swum (swum) Past participle and dialectal past tense of SWIM.

swung (swung) Past tense and past participle of SWING.

sy- Var. of SYN-.

Syb·a·ris (sib′ə·ris) An ancient Greek city on the Gulf of Tarentum in southern Italy, famous as a center of luxurious living; founded in 720 B.C.; destroyed, 510 B.C.

syb·a·rite (sib′ə·rīt) *n.* A luxurious person; epicure; voluptuary. [<L *Sybarita* <Gk. *Sybaritēs* <*Sybaris* Sybaris]

Syb·a·rite (sib′ə·rīt) *n.* A native or citizen of Sybaris.

Syb·a·rit·ic (sib′ə·rit′ik) *adj.* 1 Of or pertaining to Sybaris or the Sybarites. 2 Hence, given to luxury; voluptuous. Also **Syb′a·rit′i·cal.** — **Syb′a·rit′i·cal·ly** *adv.* — **Syb·a·rit·ism** (sib′ə·rīt·iz′əm) *n.*

sy·bo (sī′bō) *n. pl.* **·boes** The cibol or Welsh onion. [<dial. E (Scottish), var. of CIBOL]

syc·a·mine (sik′ə·min) *n.* The mulberry tree (*Morus nigra*) of the New Testament. [<LL *sycaminus* <Gk. *sykaminos* a mulberry tree <Aramaic *shiqmīn*, pl. <Hebrew *shiqmah*]

syc·a·more (sik′ə·môr, -mōr) *n.* 1 A medium-sized bushy tree of Syria and Egypt (*Ficus sycomorus*) allied to the common fig. 2 Any of various plane trees widely distributed in the United States, especially the American sycamore (*Platanus occidentalis*) and the buttonwood of California. 3 An ornamental shade tree of Europe and Asia (*Acer pseudo-platanus*); the sycamore maple. Also *Obs.* **syc′o·more.** [<OF *sicamor* <LL *sycomorus* <Gk. *sykomoros* <*sykon* a fig + *moron* a mulberry]

syce (sīs) *n.* A groom; a man servant: also spelled *sice, saice.* [<Hind. *sā'is* <Arabic <*sūs* tend a horse]

sy·cee (sī·sē′) *n.* Pure uncoined silver ingots of various weight and size: used by the Chinese as a medium of exchange. Also **sycee silver.** — *adj.* Pure; unalloyed. [<dial. Chinese (Cantonese) *sai sze*, var. of Chinese *si szĕ* fine silk; so called because if pure it may be drawn out into fine threads]

sy·con (sī′kon) *adj. Zool.* Designating a type of sponge having an infolded body wall provided with radial canals for the reception of water, as in the typical genus *Sycon.* [<NL <Gk. *sykon* a fig]

sy·co·ni·um (sī·kō′nē·əm) *n. pl.* **·ni·a** (-nē·ə) *Bot.* An aggregate or multiple fruit in which many flowers have been developed on a fleshy receptacle, which is a flattened disk or forms a nearly closed cavity, as in the fig. [<NL <Gk. *sykon* a fig]

syc·o·phan·cy (sik′ə·fən·sē) *n. pl.* **·cies** The practices of a sycophant; base flattery; fawning.

syc·o·phant (sik′ə·fənt) *n.* 1 A servile flatterer; parasite. 2 *Obs.* An informer; accuser: the original meaning. 3 *Obs.* An impostor; deceiver. [<L *sycophanta* <Gk. *sykophantēs* an informer <*sykon* a fico + *phan-*, stem of *phainein* show] — **syc′o·phan′tic** (-fan′tik) or **·ti·cal** *adj.* — **syc′o·phan′ti·cal·ly** *adv.*

Syc·o·rax (sik′ō·raks) In Shakespeare's *Tempest*, Caliban's mother, a witch.

sy·co·sis (sī·kō′sis) *n. Pathol.* An inflamed staphylococcic infection of the skin involving the hair follicles, generally of the face and scalp. ◆ Homophone: *psychosis.* [<NL <Gk. *sykōsis* a fig-shaped ulcer <*sykon* a fig]

Syd·ney (sid′nē) A masculine personal name. Also *Sidney.* [from a surname, orig. <AF *St. Denis.* See DENIS]

Syd·ney (sid′nē) 1 The chief port and capital of New South Wales, Australia. 2 A port on Cape Breton Island, NE Nova Scotia, Canada.

Syd·ney (sid′nē), **Sir Philip** See SIDNEY.

Sy·e·ne (sī·ē′nē) The ancient name for ASWAN. — **Sy·e·nit·ic** (sī′ə·nit′ik) *adj.*

sy·e·nite (sī′ə·nīt) *n.* An igneous granular rock composed principally of feldspar and containing little or no quartz: also spelled *sienite.* [<F *syénite* <L *syenites* (*lapis*) (stone) of Syene <*Syene* Syene <Gk. *Syēnē*] — **sy′e·nit′ic** (-nit′ik) *adj.*

syke (sīk) *n. Scot.* A small stream from a bog: also spelled *sike.*

Syk·tyv·kar (sik′tif·kär′) The capital of Komi Autonomous S.S.R., in NE central European Russian S.F.S.R.

syl- Assimilated var. of SYN-.

syl·la·bar·y (sil′ə·ber′ē) *n. pl.* **·bar·ies** A list of characters representing syllables; the syllabic characters, collectively, of a language, as Chinese or Japanese, answering the function of an alphabet in writing. [<NL *syllabarium*, neut. of Med. L *syllabarius* <*syllaba* SYLLABLE]

syl·lab·ic (si·lab′ik) *adj.* 1 Of, pertaining to, or consisting of a syllable or syllables. 2 *Phonet.* Designating a consonant capable of forming a complete syllable without a vowel, as *l* in *middle* (mid′l) and *n* in *sudden* (sud′n). See SONORANT. 3 Having every syllable distinctly pronounced. 4 Designating a type of poetry based on a definite number of syllables per line rather than on stress or rhythm. Also **syl·lab′i·cal.** — *n. Phonet.* A syllabic consonant; a sonorant. — **syl·lab′i·cal·ly** *adv.*

syl·lab·i·cate (si·lab′ə·kāt) *v.t.* **·cat·ed, ·cat·ing** To form or divide into syllables. Also **syl·lab′i·fy.** — **syl·lab′i·ca′tion, syl·lab′i·fi·ca′tion** *n.*

syl·la·bism (sil′ə·biz′əm) *n.* 1 The use of characters representing syllables instead of letters in a written language. 2 The theory of syllables; division into syllables.

syl·la·bist (sil′ə·bist) *n.* One skilled in syllabicating.

syl·la·bize (sil′ə·bīz) *v.t.* **·bized, ·biz·ing** To divide (words) or form (letters) into syllables.

syl·la·ble (sil′ə·bəl) *n.* 1 *Phonet.* A word or part of a word uttered in a single vocal impulse, and consisting of a vowel (or diphthong) alone or with one or more consonants, or of a syllabic consonant. An **open syllable** is one ending in a vowel, as the first syllable of *si·lent* (sī′lənt); a **closed syllable** is one ending in a consonant, as the first and third syllables of *cat·a·pult* (kat′ə·pult). 2 A part of a written or printed word corresponding, more or less, to the spoken division. In this dictionary, syllable breaks are indicated by centered dots. 3 The smallest particle of expression; the least detail, mention, or trace: Please don't repeat a *syllable* of what you've heard here. — *v.* **·bled, ·bling** *v.t.* 1 To pronounce the syllables of; utter; speak. 2 *Obs.* To syllabicate. — *v.i.* 3 To pronounce syllables. [<AF *sillable*, OF *sillabe* <L *syllaba* <Gk. *syllabē* <*syllambanein* <*syn-* together + *lambanein* take]

syl·la·bub (sil′ə·bub) See SILLABUB.

syl·la·bus (sil′ə·bəs) *n. pl.* **·bus·es** or **·bi** (-bī) A concise statement of the main points of a subject; outline, as of a course of study; schedule; epitome; abstract; specifically, a short statement at the beginning of a brief of the legal points involved. [<NL <Med. L *syllabos*, a misprint for L *sittybas*, accusative pl. of *sittyba* label on a book <Gk.]

syl·lep·sis (si·lep′sis) *n. pl.* **·ses** (-sēz) A figure of speech, common in classical Greek and Roman literature, by which an adjective or a verb is made to modify or govern two nouns, but must be understood in a different sense for each noun. This figure conveys a double meaning, often with humorous effect, as in Pope's comment on Queen Anne: Dost sometimes *counsel take* —and sometimes *tea.* Compare ZEUGMA. [<LL *syllepsis* <Gk. *syllēpsis* <*syn-* together + *lēpsis* a taking <*lēb-, lab-*, stem of *lambanein* take] — **syl·lep′tic** *adj.*

syl·lo·gism (sil′ə·jiz′əm) *n.* 1 *Logic* **a** A formula of argument consisting of three propositions. The first two propositions, called *premises*, have one term in common furnishing a logical connection between the two other terms, which are then linked in the third proposition, called the *conclusion.* Example: All men are mortal (*major premise*); kings are men (*minor premise*); therefore, kings are mortal (*conclusion*). In this example, the *major term* is "mortal," the *minor term* is "kings," and the *middle term* is "men." **b** Deductive reasoning. 2 A subtle or crafty argument. [<OF *silogime* <L *syllogismus* <Gk. *syllogismos* <*syllogizesthai* SYLLOGIZE]

syl·lo·gis·tic (sil′ə·jis′tik) *adj.* Pertaining to, or having the nature or form of, a syllogism: also **syl′lo·gis′ti·cal.** — *n.* The art of reasoning by syllogism; the department of logic dealing with syllogisms: also **syl′lo·gis′tics.** — **syl′lo·gis′ti·cal·ly** *adv.*

syl·lo·gize (sil′ə·jīz) *v.t.* & *v.i.* **·gized, ·giz·ing** To reason or argue by syllogisms. [<OF *silogiser* <Med. L *syllogizare* <Gk. *syllogizesthai* <*syn-* together + *logizesthai* calculate, infer <*logos* discourse] — **syl′lo·gi·za′tion** (-jə·zā′shən) *n.*

sylph (silf) *n.* 1 Originally, in the system of Paracelsus, a being, male or female, mortal but without a soul, living in and on the air, and intermediate between material and immaterial beings. 2 A slender, graceful young woman or girl. 3 A South American hummingbird (*Cyanolesbia gorgo*), with a long, forked, brilliantly colored tail. [<NL *sylphes*, pl., ? coined by Paracelsus]

sylph·id (sil′fid) *n.* A young or diminutive sylph. — *adj.* Having qualities suggesting a sylph: also **sylph·i·dine** (sil′fə·din, -dīn). [<F *sylphide*, dim. of *sylphe* <NL *sylphes* SYLPH]

sylph·like (silf′līk′) *adj.* Like a sylph; slender; graceful. Also **sylph′ish, sylph′y.**

syl·va (sil′və) *n. pl.* **·vas** or **·vae** (-vē) 1 The forest trees, collectively, of a territory or region. 2 A treatise on forest trees, or a description or list of the forest trees of a certain region. Also spelled *silva.* [<L *silva* a forest]

syl·van (sil′vən) *adj.* 1 Of, pertaining to, or located in a forest or woods. 2 Composed of or abounding in trees or woods. 3 Characteristic of a forest or wood; rustic. — *n.* 1 In mythology, a spirit or deity of the forest. 2 *Archaic* or *Poetic* A person or animal dwelling in the woods. [<MF *sylvain* a sylvan <L *sylvanus, silvanus* <*silva* a wood]

syl·van·ite (sil′vən·īt) *n.* A metallic, steel-gray to silver-white telluride of gold or silver, crystallizing in the monoclinic system: when the crystals are arranged in patterns suggesting runic symbols, it is called *graphic gold, graphic tellurium.* [from (TRAN)SYLVAN(IA) + -ITE[1]]

Syl·va·nus (sil·vā′nəs) See SILVANUS.

Syl·ves·ter (sil·ves′tər) Silvester; a masculine personal name. Also *Sp.* **Syl·ves·tre** (sēl·ves′·trā). [<L, living in the wood]

syl·ves·tral (sil·ves′trəl) *adj.* Adapted to growing in woody and shady places, as certain plants; also, relating to the woods; wild. [<L *silvester, silvestris* <*silva* a forest]

Syl·vi·a (sil′vē·ə) A feminine personal name. [<L, of the forest]

Syl·vi·an fissure (sil′vē·ən) *Anat.* A deep fissure that separates the temporal lobe of the cerebrum from the parietal and frontal lobes. [<F *sylvien*, after François de la Boë *Sylvius*, 1614–72, Flemish anatomist]

syl·vite (sil′vīt) *n.* A vitreous, native potassium chloride, crystallizing in the isometric system. Also **syl′vin** (-vin), **syl′vine** (-vin, -vīn), **syl′·vin·ite.** [<NL (*sal digestivus*) *sylvii* (digestive salt) of Sylvius + -ITE[1]]

sym- Assimilated var of SYN-.

sym·bi·ont (sim′bī·ont, -bē-) *n. Biol.* An organism living in a state of symbiosis. Also **sym′·bi·on.** [<Gk. *symbioōn, -ontos*, ppr. of *bioein.* See SYMBIOSIS.] — **sym′bi·on′tic** *adj.*

sym·bi·o·sis (sim′bī·ō′sis, -bē-) *n. Biol.* The consorting together or partnership of dissimilar organisms, as of the algae and fungi in lichens. The term ordinarily connotes an association which is mutually advantageous. Compare CONSORTISM. [<NL <Gk. *symbiōsis* a living together, companionship <*symbioein* live together <*symbios* a companion, living together <*syn-* together + *bios* life] — **sym′bi·ot′ic** (-ot′ik) or **·i·cal** *adj.* — **sym′bi·ot′i·cal·ly** *adv.*

sym·bol (sim′bəl) *n.* 1 Something chosen to stand for or represent something else, usually because of a resemblance in qualities or characteristics; an object used to typify a quality, abstract idea, etc.: The oak is a *symbol* of strength. 2 A character, mark, abbreviation, conventional sign, or letter indicating something, as a quantity in mathematics, a substance in chemistry, a planet or celestial body, a quality, operation, relationship, etc. 3 A confession of faith; creed. 4 The disguised representation of an unconscious trend involving a person, object, act, etc. See synonyms under EMBLEM, LETTER, MARK, SIGN, SIMILE. ◆ Homophone: *cymbal.* [<LL *symbolum* <Gk. *symbolon* a mark, token <*symballein* put together <*syn-* together + *ballein* throw]

sym·bol·ae·og·ra·phy (sim′bəl·ē·og′rə·fē) *n.* The drawing up or framing of legal instruments. Also **sym′bol·e·og′ra·phy.** [<Gk. *symbolaiographia* <*symbolaiographos* a notary <*symbolaion* a mark, contract + *graphein* write]

sym·bol·ic (sim·bol′ik) *adj.* 1 Of or pertaining to a symbol or symbols; expressed by a symbol. 2 Serving as signs of relation or connection; relational; connective: distinguished from *presentive*: said of certain classes of words, as prepositions and conjunctions. 3 Characterized by or involving the use of symbols: *symbolic* poetry. Also **sym·bol′i·cal.** — **sym·bol′i·cal·ly** *adv.* — **sym·bol′i·cal·ness** *n.*

symbolical books Books containing the symbols or confessions of faith of a church, religious body, or inspired writer.

symbolic logic A development of formal logic in which the ambiguity of verbal propositions and of operations upon them is reduced to a minimum by the rigorous use of symbols each of which has only one referent within the given context. Also called *mathematical logic.*

sym·bol·ics (sim·bol′iks) n. pl. (construed as singular) The science or study of symbols or of ancient symbolic rites or creeds.

sym·bol·ism (sim′bəl·iz′əm) n. 1 Representation by symbols; treatment or interpretation of things as symbolic; also, the quality of being symbolic. 2 A system of symbols or symbolical representation. 3 The theories and practice of a group of symbolists. 4 Artistic imitation as a means of suggesting or expressing ideal or intangible states or ideas; also, the expression or representation of the invisible by conventional signs or figures.

sym·bol·ist (sim′bəl·ist) n. 1 One who uses symbols; one versed or ardent in the interpretation or use of symbols; especially, one who regards the elements in the Eucharist as mere symbols. 2 One of a class of French and Belgian writers and artists of the late 19th century, including Verlaine, Mallarmé and Maeterlinck, who sought to exalt the metaphysical by suggesting ideas and emotions by patterns of color and form and by symbolic meanings of objects, words, and sound.

sym·bol·is·tic (sim′bəl·is′tik) adj. 1 Expressed by symbols; characterized by the use of symbols. 2 Of or pertaining to symbolism; symbolic. Also **sym′bol·is′ti·cal.**

sym·bol·ize (sim′bəl·īz) v. ·ized, ·iz·ing v.t. 1 To be a symbol of; represent symbolically; typify. 2 To represent by a symbol or symbols. 3 To treat as symbolic or figurative. — v.i. 4 To use symbols. 5 Psychol. To transfer emotional values from one person, object, or act to another. Also Brit. **sym′bol·ise.** — **sym′bol·i·za′tion** n.

sym·bol·o·gy (sim·bol′ə·jē) n. The art of representing by, or of interpreting, symbols. [< SYMBOL(L) + -LOGY]

sym·met·al·ism (sim·met′l·iz′əm) n. A money system in which the unit of coinage is composed of two or more metals combined. [< SYM- + METAL + -ISM]

sym·met·ri·cal (si·met′ri·kəl) adj. 1 Exhibiting symmetry; having harmonious proportions or a correspondence in shape and size of parts; well-balanced; regular: a *symmetrical* structure. 2 Biol. Having parts or organs on one side corresponding to those on the other. 3 Bot. Regular as to number or shape of parts: said especially of a flower when the parts or divisions in each cycle (that is, the sepals, petals, stamens, and pistils) are of the same number or multiples of the same. 4 Chem. Denoting an arrangement of atoms of a molecule at equal relative intervals when graphically represented. 5 Med. Affecting corresponding organs or parts similarly. Also **sym·met′ric.** [<SYMMETRY] — **sym·met′ri·cal·ly** adv. — **sym·met′ri·cal·ness** n.

sym·me·trist (sim′ə·trist) n. A student or advocate of symmetry.

sym·me·trize (sim′ə·trīz) v.t. ·trized, ·triz·ing To make symmetrical or proportional. — **sym′me·tri·za′tion** n.

sym·me·try (sim′ə·trē) n. pl. ·tries 1 Corresponding arrangement or balancing of the parts or elements of a whole in respect to size, shape, and position on opposite sides of an axis or center; hence, loosely, congruity; harmony; also, an instance of such arrangement. 2 The element of beauty in nature or art that results from such arrangement and balancing. 3 Biol. Regular arrangement of parts or organs in an animal body so that a division will give halves corresponding in shape, size, function, relative position, etc.; similarity of structure. 4 Bot. Equality of number in the whorls of a flower, as of sepals, petals, etc. 5 Math. An arrangement of pairs of points in a general system such that of set of lines joining them together is divided into equal parts by a line, a plane, or a point. 6 Mineral. The symmetrical distribution of non-parallel but equivalent direc-

tions (faces, edges, etc.) in a crystal with reference to certain planes or lines called **planes** or **axes of symmetry.** [<MF *symmetrie* <LL *symmetria* <Gk. < *symmetros* measured together < *syn-* together + *metron* a measure]

Synonyms: agreement, conformity, harmony, order, parity, proportion, regularity, shapeliness. See HARMONY. *Antonyms:* deformity, discordance, disproportion, shapelessness.

Sym·onds (sim′əndz), **John Addington,** 1840–1893, English author.

Sy·mons (sī′mənz), **Arthur,** 1865–1945, English poet and critic born in Wales.

sym·pa·thec·to·my (sim′pə·thek′tə·mē) n. Surg. The operation of interrupting some portion of the sympathetic nervous system, as by transection or resection of a nerve pathway. [<SYMPATH(ETIC) + -ECTOMY]

sym·pa·thet·ic (sim′pə·thet′ik) adj. 1 Pertaining to, expressing, or proceeding from sympathy. 2 Having a fellow feeling for others; sympathizing; compassionate. 3 Being in accord or harmony; congenial. 4 Referring to sounds produced by responsive vibrations. 5 Anat. Designating the entire autonomic nervous system. Also **sym′pa·thet′i·cal.** See synonyms under HUMANE. [<NL *sympatheticus* <Gk. *sympathētikos* < *sympatheia.* See SYMPATHY.] — **sym′pa·thet′i·cal·ly** adv.

sympathetic ink An ink that is colorless and invisible until brought out by heat, light, or chemical action: also called *invisible ink.*

sympathetic nervous system Anat. That part of the autonomic nervous system which serves the viscera, glands, heart, blood vessels, and smooth muscles. It consists of a chain of ganglia on each side of the spinal column between the cervical and sacral regions, connected with nerve plexuses, and in general produces effects opposite to those coming from the parasympathetic system.

sym·path·i·co·to·ni·a (sim·path′i·kō·tō′nē·ə) n. Physiol. Increased dominance of the sympathetic nervous system over other body functions, marked by vascular spasm and high blood pressure. [<NL <E *sympathic,* var. of SYMPATHETIC + Gk. *tonos* tension] — **sym·path′i·co·ton′ic** (-ton′ik) adj.

sym·pa·thin (sim′pə·thin) n. Biochem. A substance liberated by the stimulation of certain fibers of the sympathetic nervous system and acting as a chemical mediator in associated nerve impulses. [<SYMPATH(ETIC) + -IN]

sym·pa·thism (sim′pə·thiz′əm) n. Suggestibility; the state of being susceptible to hypnotic or other influences.

sym·pa·thize (sim′pə·thīz) v.i. ·thized, ·thiz·ing 1 To share the sentiments or ideas of another; have the same feelings as another: with *with.* 2 To feel or express compassion, as for another's sorrow or affliction: with *with.* 3 To be in harmony or agreement. Also Brit. **sym′pa·thise.** See synonyms under CONSOLE. — **sym′pa·thiz′er** n. — **sym′pa·thiz′ing·ly** adv.

sym·pa·thy (sim′pə·thē) n. pl. ·thies 1 The quality of being affected by the state of another with feelings correspondent in kind; a fellow feeling; a mutual affinity or susceptibility; reaction to such relationship. 2 A feeling of compassion for another's sufferings; pity; commiseration. 3 An agreement of affections or inclinations, or a conformity of natural temperaments, which makes persons agreeable to one another; congeniality; accord. 4 That quality of inanimate things by virtue of which they attract or influence one another, or are supposed to do so; affinity: a sense once much used in alchemy and astrology: the *sympathy* of the lodestone for iron. See synonyms under BENEVOLENCE, PITY. [<L *sympathia* <Gk. *sympatheia* < *sympathēs* feeling compassion with another < *syn-* together + *pathos* a feeling, passion]

sympathy strike A strike in which the strikers support the demands of another group of workers but demand nothing for themselves.

sym·pa·try (sim′pə·trē) n. Ecol. The distribution of plant and animal species in coextensive areas. [<SYM- + L *patria* fatherland]

sym·pet·al·ous (sim·pet′l·əs) adj. Bot. Gamopetalous. [<NL *Sympetalae,* a division of dicotyledons <Gk. *syn-* together + *petalon* a leaf, petal]

sym·phon·ic (sim·fon′ik) adj. 1 Relating to or having the form of a symphony: also **sym·pho·net·ic** (sim′fə·net′ik). 2 Agreeing in sound; harmonious.

symphonic poem *Music* A composition in free form for symphony orchestra, composed either as a unit (as Liszt's *Les Préludes* or Strauss's *Death and Transfiguration*) or as a short series of pieces (as Debussy's *La Mer*), and following a descriptive, literary, or "program" outline; a tone poem: a form developed by Liszt in the 19th century.

sym·pho·ni·ous (sim·fō′nē·əs) adj. According in sound; harmonious; concordant; agreeing; sounding together or in harmony. — **sym·pho′ni·ous·ly** adv.

sym·pho·nize (sim′fə·nīz) v.t. & v.i. ·nized, ·niz·ing To harmonize.

sym·pho·ny (sim′fə·nē) n. pl. ·nies 1 A harmonious or agreeable mingling of sounds, whether vocal, instrumental, or both; figuratively, any concord or agreeable blending: *symphonies* in gray. 2 *Music* A composition for orchestra, consisting usually of four movements, of which one or more generally follow sonata form, and which are of diverse individuality united by homogeneous elements. 3 A symphony orchestra. [<OF *simphonie* <L *symphonia* <Gk. *symphōnia* < *syn-* together + *phōnē* a sound]

symphony orchestra A large orchestra composed usually of the string, brass, woodwind, and percussion sections needed to present symphonic works.

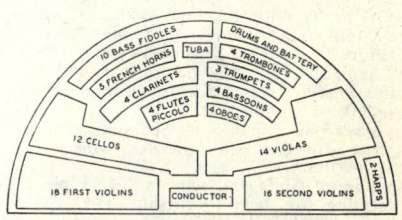

TRADITIONAL SEATING PLAN OF MODERN SYMPHONY ORCHESTRA

sym·phy·sis (sim′fə·sis) n. pl. ·ses (-sēz) 1 Anat. A junction of two parts of the skeleton, formed either by a growing together (*synostosis*) or by the intervention of cartilage (*synchondrosis*). 2 Bot. The union of similar parts, or of parts normally separate. [<NL <Gk., a growing together, esp. of the bones < *syn-* together + *phyein* grow]

sym·plec·tic (sim·plek′tik) adj. Geol. Denoting a rock texture formed by the intermingling of two different minerals. [<Gk. *symplektikos* plaiting together < *symplekein* < *syn-* together + *plekein* plait]

Sym·pleg·a·des (sim·pleg′ə·dēz) In Greek mythology, twin rocks forming a gateway to the Black Sea and supposed to swing together and crush whatever tried to pass between them. [<L <Gk. (*petrai*) *Symplēgades* the clashing (rocks) < *symplēgas, -ados* striking together < *syn-* together + *plēssein* strike]

sym·po·di·um (sim·pō′dē·əm) n. pl. ·di·a (-dē·ə) Bot. A false axis or stem of a plant, morphologically made up of a series of superposed branches imitating a simple stem; a pseudaxis. [<NL <Gk. *syn-* together + *podion,* dim. of *pous, podos* a foot] — **sym·po′di·al** adj. — **sym·po′di·al·ly** adv.

sym·po·si·ac (sim·pō′zē·ak) adj. Pertaining to, of the nature of, or occurring at a symposium; specifically, denoting convivial songs, glees, etc.: also **sym·po′si·al.** — n. A symposium. [<LL *symposiacus* <Gk. *symposiakos* < *symposion.* See SYMPOSIUM.]

sym·po·si·arch (sim·pō′zē·ärk) n. 1 The master or director of an ancient Greek symposium; hence, the master of a feast; a toastmaster. 2 Familiarly, a ruling spirit of a social or convivial company. [<Gk. *symposiarchos* a symposium (def. 3) + *archos* a ruler]

sym·po·si·um (sim·pō′zē·əm) n. 1 A meeting for discussion of a particular subject. 2 A collection of comments or opinions brought together; especially, a series of several brief essays or articles on the same subject by different writers, as in a magazine. 3 In ancient Greece, an after-dinner drinking party, characterized by conversation, music, dancing, and other amusements. 4 Any similar social gathering. Also called *symposiac.* Also **sym·po′si·on** (-zē·on). [<L <Gk. *symposion* <*syn-* together + *posis* a drinking <*po-*, stem of *pinein* drink]

symp·tom (simp′təm) *n.* 1 *Pathol.* An organic or functional condition indicating the presence of disease, especially when regarded as an aid in diagnosis. 2 That which serves to point out the existence of something else; any sign, token, or indication. See synonyms under SIGN. [<L *symptoma* <Gk. *symptōma* a chance, a disease <*sympiptein* happen to <*syn-* together + *piptein* fall]

symp·to·mat·ic (simp′tə·mat′ik) *adj.* 1 Pertaining to, of the nature of, or constituting a symptom or symptoms; indicative: Fever is *symptomatic* of inflammation. 2 According to symptoms: a *symptomatic* classification of diseases. Also **symp′to·mat′i·cal.** [<F *symptomatique* <LL *symptomaticus* <Gk. *symptōmatikos* <*symptōma, -atos* a symptom] — **symp′to·mat′i·cal·ly** *adv.*

symp·tom·a·tol·o·gy (simp′təm·ə·tol′ə·jē) *n.* 1 The branch of medicine that has for its object the observation and classification of symptoms. 2 The combined symptoms of a disease: also *semeiology, semeiotics.* Compare DIAGNOSIS. [<NL *symptomatologia* <Gk. *symptōma, -atos* a symptom + *logos* study]

syn- *prefix* With; together; associated with or accompanying: *syntax, syndrome.* Also: *sy-* before *sc, sp, st,* and *z,* as in *system; syl-* before *l,* as in *syllable; sym-* before *b, p,* and *m,* as in *sympathy; sys-* before *s,* as in *syssarcosis.* [<Gk. <*syn* together]

syn·aer·e·sis (si·ner′ə·sis) See SYNERESIS.

syn·aes·the·sia (sin′is·thē′zhə, -zhē·ə) See SYNESTHESIA.

syn·a·gog (sin′ə·gôg, -gog) *n.* 1 A place of meeting for Jewish worship and religious instruction. 2 A Jewish congregation or assemblage for religious instruction and observances. 3 The Jewish religion or communion. Also **syn′a·gogue.** [<OF *sinagoge* <LL *synagoga* <Gk. *synagōgē* an assembly, synagog <*synagein* bring together <*syn-* together + *agein* lead, bring] — **syn′a·gog′i·cal** (-goj′i·kəl), **syn′a·gog′al** (-gôg′əl, -gog′əl) *adj.*

syn·a·le·pha (sin′ə·lē′fə) *n.* The blending into a single syllable of two successive vowels of different syllables; especially, the suppression of a final vowel or diphthong before one that begins the next word: *th' Omnipotent* for *the Omnipotent.* Compare APOCOPE. Also **syn′a·le′phe** (-lē′fē), **syn′a·loe′pha, syn′a·loe′phe.** [<LL <Gk. *synaloiphē* <*synaleiphein* smear together <*syn-* together + *aleiphein* anoint]

syn·al·gi·a (si·nal′jē·ə) *n. Pathol.* Sympathetic pain transmitted to a remote organ through associated nerves. [<NL <Gk. *synalgeein* share in suffering <*syn-* together + *algeein* feel bodily pain <*algos, -eos* bodily pain] — **syn·al′gic** *adj.*

syn·an·ther·ous (si·nan′thər·əs) *adj. Bot.* Having the stamens cohering by their anthers, as in composite flowers. [<NL *Synanthereae,* former family name <Gk. *syn-* together + NL *anthera* an anther]

syn·apse (si·naps′) *n. Physiol.* The junction point of two neurons, across which a nerve impulse passes. [<NL *synapsis* <Gk., a junction <*syn-* together + *hapsis* a joining <*haptein* join]

syn·ap·sis (si·nap′sis) *n.* 1 *Biol.* The conjugation of maternal and paternal chromosomes preceding maturation, or the reduction division in the nucleus; syndesis. 2 A synapse. [<NL. See SYNAPSE.] — **syn·ap′tic** *adj.* — **syn·ap′ti·cal·ly** *adv.*

syn·ar·thro·sis (sin′är·thrō′sis) *n. pl.* **·ses** (-sēz) *Anat.* A joint that permits no motion between the parts articulated. Also **syn′ar·thro′di·a** (-dē·ə). [<NL <Gk. *synarthrōsis* <*syn-* together + *arthrōsis* a jointing <*arthron* a joint] — **syn′ar·thro′di·al** *adj.* — **syn′ar·thro′di·al·ly** *adv.*

syn·ax·is (si·nak′sis) *n.* A congregation assembled for public worship, especially for celebrating the Lord's Supper. [<LL <Gk. <*synagein.* See SYNAGOG.]

sync (singk) *n. Colloq.* Synchronization: a truncated form used in telecommunication.

syn·carp (sin′kärp) *n. Bot.* An aggregate fruit composed of several more or less coherent carpels, as in the blackberry, or a multiple fruit, as in the fig. Also **syn·car·pi·um** (sin·kär′pē·əm). [<NL *syncarpium* <Gk. *syn-* together + *karpos* a fruit]

syn·car·pous (sin·kär′pəs) *adj. Bot.* Characterized by or characteristic of a syncarp; consisting of united carpels: contrasted with *apocarpous.*

syn·cat·e·gor·e·mat·ic (sin·kat′ə·gôr′ə·mat′ik, -gor′-) *adj.* Pertaining to words that can only form parts of terms, as adverbs, prepositions, and conjunctions: opposed to *categorematic.* Also **syn·cat′e·gor′e·mat′i·cal.** [<Gk. *synkatēgorēmatikos* <*synkatēgorēma* <*synkatēgorein* predicate jointly <*syn-* together + *katēgoreein.* See CATEGORY.]

syn·chon·dro·sis (sing′kən·drō′sis) See under SYMPHYSIS. [<NL <Gk. *synchondrōsis* <*syn-* together + *chondros* cartilage]

syn·chor·ol·o·gy (sing′kô·rol′ə·jē) *n.* The study of the geographic occurrence and distribution of plant communities. Also spelled *synechorology.* [<SYN- + Gk. *chōros* a region + -LOGY] — **syn′chor·o·log′ic** (-kôr·ə·loj′ik) *adj.*

syn·chro·mesh (sing′krō·mesh′) *n. Mech.* 1 A gear system by which driving and driven members are brought to the same speed before engaging. 2 The mechanism by which this uniform speed of gears is obtained. [<SYNCHRO(NIZED) + MESH]

syn·chron·ic (sin·kron′ik) *adj.* 1 Synchronous. 2 *Ling.* Pertaining to the study of some aspect of a language at a given stage in its development: *synchronic* grammar. Also **syn·chron′i·cal.** See DIACHRONIC. [<LL *synchronus* SYNCHRONOUS] — **syn·chron′i·cal·ly** *adv.*

syn·chro·nic·i·ty (sing′krə·nis′ə·tē) *n.* The temporal coincidence of two or more events linked together by meaning, but without any causal connection; meaningful cross-connection between separate causal chains. [Trans. of G *synchronizität;* used by C. G. Jung]

syn·chro·nism (sing′krə·niz′əm) *n.* 1 The state of being synchronous. 2 Coincidence in time of different events or phenomena; simultaneousness. 3 A tabular grouping of historic personages or events according to their dates. 4 In art, representation in the same picture of events having differing dates. [<LL *synchronismus* <Gk. *synchronismos* <*synchronos* SYNCHRONOUS] — **syn′chro·nis′tic** or **·ti·cal** *adj.* — **syn′chro·nis′ti·cal·ly** *adv.*

syn·chro·nize (sing′krə·nīz) *v.* **·nized, ·niz·ing** *v.i.* 1 To occur at the same time; coincide. 2 To move or operate in unison. — *v.t.* 3 To cause (timepieces) to agree in keeping or indicating time. 4 To cause to operate in unison. 5 To assign the same date or period to; make contemporaneous. [<SYNCHRONISM] — **syn′chro·ni·za′tion** *n.* — **syn′chro·niz′er** *n.*

synchronized shifting A change in the speed of an automotive vehicle by means of synchromesh gearing.

syn·chron·o·scope (sin·kron′ə·skōp) *n.* A synchroscope. [<SYNCHRON(ISM) + -(O)SCOPE]

syn·chro·nous (sing′krə·nəs) *adj.* 1 Occurring at the same time; coincident. 2 Happening at the same rate. 3 *Physics* Having the same period or rate of vibration: *synchronous* currents. Also **syn′chro·nal.** [<LL *synchronus* <Gk. *synchronos* <*syn-* together + *chronos* time] — **syn′chro·nous·ly** *adv.* — **syn′chro·nous·ness** *n.*

synchronous converter *Electr.* A machine adapted for the conversion of direct into alternating current or vice versa.

synchronous machine *Electr.* A machine whose normal speed of operation is exactly proportional to the frequency of the current to which it is connected, as a motor or generator.

synchronous speed *Electr.* The speed of an alternating-current machine as determined by the frequency of the circuit.

syn·chro·scope (sing′krə·skōp) *n. Electr.* An apparatus for visually indicating the degree of synchronization in the working, speed, etc., of two or more engines, as in an airplane. Also *synchronoscope.* [<SYNCHRO(NISM) + -SCOPE]

syn·chro·tron (sing′krə·tron) *n. Physics* An atom-smashing machine combining features of the cyclotron and betatron to permit synchronizing the acceleration of atomic particles to the level of 300 million electron volts or greater. [<SYNCHRO(NIZE) + (ELEC)-TRON]

syn·clas·tic (sin·klas′tik) *adj.* Having the same kind of curvature in all directions; concave or convex in every direction: said of a surface: opposed to *anticlastic.* [<SYN- + Gk. *klastos* broken <*klaein* break]

syn·cli·nal (sin·klī′nəl) *adj.* 1 Sloping downward on each side toward a common line or point. 2 *Geol.* Dipping downward on each side toward the axis of the fold, as rock strata: opposed to *anticlinal.* Also **syn·clin·i·cal** (sin·klin′i·kəl). — *n.* A syncline. [<Gk. *synklinein* <*syn-* together + *klinein* incline]

syn·cline (sing′klīn) *n. Geol.* 1 A trough or structural basin toward which rocks dip. 2 A synclinal fold. [<Gk. *synklinein.* See SYNCLINAL.]

syn·clit·ism (sing′klə·tiz′əm) *n. Med.* The lateral turning of the fetal head in a natural presentation at childbirth, thus bringing the cranial planes into parallelism with the planes of the maternal pelvis. [<Gk. *syn-* together + *klitikos* <*klinein* incline, turn aside] — **syn·clit′ic** *adj.*

syn·co·pate (sing′kə·pāt) *v.t.* **·pat·ed, ·pat·ing** 1 To contract, as a word, by syncope. 2 *Music* To treat or modify, as a tone, by syncopation. [<LL *syncopatus,* pp. of *syncopare* affect with syncope <*syncope* SYNCOPE]

syn·co·pa·tion (sing′kə·pā′shən) *n.* 1 The act of syncopating or state of being syncopated; also, that which is syncopated; a dance or rhythm in syncopated time. 2 *Music* The beginning of a tone on an unaccented beat and its continuation through the following accented beat, or the beginning of a tone on the last half of a beat and continuing it through the first half of the next beat; also, the tone so treated, generally receiving an accent. 3 Any music featuring syncopation, as ragtime, jazz, etc. 4 Syncope of a word, or an example of it.

syn·co·pe (sing′kə·pē) *n.* 1 The elision of a sound or syllable in the middle part of a word, as *e'er* for *ever.* 2 *Music* Syncopation. 3 *Pathol.* Sudden faintness; swooning, with loss of sensation, motion, and consciousness. See synonyms under STUPOR. [Earlier *sincopis* <OF *sincopin,* ult. <LL *syncope* <Gk. *synkopē* <*syn-* together + *kop-,* stem of *koptein* strike, cut; refashioned after LL] — **syn′co·pal, syn·cop·ic** (sin·kop′ik) *adj.*

syn·cra·sy (sing′krə·sē) *n.* The blending, harmonizing, or massing of different or antagonistic elements. [<Gk. *synkrasis* a commixture <*syn-* together + *krasis* a mixing <*kerannynai* mix]

syn·cre·tism (sing′krə·tiz′əm) *n.* 1 A tendency or effort to reconcile and unite various systems of philosophy or religious opinion on the basis of tenets common to all and against a common opponent. 2 *Ling.* The fusion of two or more inflectional forms which were originally different, as of two cases. [<F *syncrétisme* <NL *syncretismus* <Gk. *synkrētismos* a union of two parties against a third <*synkrētizein* combine] — **syn′cre·tist** *n.* — **syn′cre·tis′tic** or **·ti·cal, syn·cret·ic** (sin·kret′ik) *adj.*

syn·cre·tize (sing′krə·tīz) *v.t.* **·tized, ·tiz·ing** To attempt to blend and reconcile, as various religions or philosophies. [<NL *syncretizare* <Gk. *synkrētizein* combine]

syn·cri·sis (sing′krə·sis) *n.* A figure of speech formed by comparison of opposite persons or things. [<LL <Gk. *synkrisis* <*synkrinein* compare <*syn-* together + *krinein* separate]

sync signal (singk) *Telecom.* The electromagnetic signal pulses by which the scanning process in television is synchronized for proper transmission and reception.

syn·cyt·i·um (sin·sit′ē·əm, -sish′əm) *n. pl.* **·cyt·i·a** (-sit′ē·ə, -sish′ə) *Biol.* 1 A multinucleate cell, or a mass of non-cellular, undifferentiated protoplasm. 2 Plasmodium. [<NL <Gk. *syn-* together + *kytos* a hollow] — **syn·cyt′i·al** *adj.*

syn·dac·tyl (sin·dak′til) *adj. Anat.* Having two or more digits either of the hand or of the foot wholly or partly united; web-footed: also **syn·dac′tyle, syn·dac′ty·lous.** — *n.* A mammal or bird which is syndactyl. [<F *syndactyle* <Gk. *syn-* together + *daktylos* a finger]

syn·dac·tyl·ism (sin·dak′til·iz′əm) *n.* **1** The condition of being syndactyl. **2** The union of two or more digits or toes.

syn·de·sis (sin′də·sis) *n. pl.* **·ses** (-sēz) Synapsis.

syndesmo– *combining form Anat.* A ligament; of or pertaining to a ligament or ligaments: *syndesmology.* Also, before vowels, **syndesm–.** [<Gk. *syndesmos* a ligament]

syn·des·mol·o·gy (sin′des·mol′ə·jē) *n.* The study of the anatomy and physiology of the ligaments.

syn·des·mo·sis (sin′des·mō′sis) *n. Anat.* The joining of two portions of the skeleton by means of ligamentous tissue. [<NL <Gk. *syndesmos* a ligament] —**syn′des·mot′ic** (-mot′ik) *adj.*

syn·det·ic (sin·det′ik) *adj.* Serving to unite or connect; connective, as a word. Also **syn·det′i·cal.** [<Gk. *syndetikos* <*syndeein* bind together <*syn-* together + *deein* bind] —**syn·det′i·cal·ly** *adv.*

syn·dic (sin′dik) *n.* A civil magistrate or officer representing a government or a community; also, one chosen to transact business for others: used also collectively for a body of officers or a council. [<F *syndic, syndique* a delegated representative <LL *syndicus* an advocate, delegate <Gk. *syndikos* a defendant's advocate <*syn-* together + *dikē* judgment] —**syn′di·cal** *adj.*

syn·di·cal·ism (sin′di·kəl·iz′əm) *n.* A social and political theory proposing the taking over of the means of production by syndicates of workers, preferably by means of the general strike, with consequent political control and the disappearance of the bourgeois state. [<F *syndicalisme* <*syndical* of a labor union <*(chambre) syndicale* a labor union <*syndic* a syndic] —**syn′di·cal·ist** *n.* —**syn′di·cal·is′tic** *adj.*

syn·di·cate (sin′də·kit) *n.* **1** An association of individuals united to negotiate some business or to prosecute some enterprise requiring large capital. **2** A combination of persons associated for purchasing manuscripts and selling them again to a number of periodicals, as newspapers, for simultaneous publication. **3** The office or jurisdiction of a syndic; syndics collectively: the original meaning. —*v.t.* (-kāt) **·cat·ed, ·cat·ing 1** To combine into or manage by a syndicate. **2** To sell for publication in many newspapers or magazines. [<F *syndicat* office of a syndic <*syndic.* See SYNDIC.]

syn·drome (sin′drōm) *n. Med.* An aggregate or set of concurrent symptoms together indicating the presence and nature of a disease. [<NL <Gk. *syndromē* <*syn-* together + *dramein* run] —**syn·drom·ic** (sin·drom′ik) *adj.*

syne (sīn) *adv. Scot.* **1** Since; ago: auld lang *syne.* **2** Afterward. **3** Then; moreover. Also **syn.**

sy·nec·do·che (si·nek′də·kē) *n.* A figure of speech in which a part is put for a whole or a whole for a part, an individual for a class, or a material for the thing, as a *roof* for a *house, marble* for a *statue.* [<LL <Gk. *synekdochē* <*synekdechesthai* take something with something else <*syn-* together + *ekdechesthai* take from <*ek-* from + *dechesthai* take] —**syn·ec·doch·ic** (sin′ek·dok′ik), **syn′·ec·doch′i·cal** *adj.*

syn·e·chor·ol·o·gy (sin′ə·kô·rol′ə·jē) See SYNCHOROLOGY.

sy·ne·cious (si·nē′shəs) See SYNOECIOUS.

syn·e·col·o·gy (sin′ə·kol′ə·jē) *n.* The study of plant and animal communities in relation to their environment; the ecology of organisms taken collectively. [<SYN- + ECOLOGY]

Syn·e·dri·on (sin·ē′drē·ən), **Syn·e·dri·um** (-drē·əm) See SANHEDRIN.

syn·er·e·sis (si·ner′ə·sis) *n.* **1** The coalescence of two vowels or syllables generally pronounced separately, as *seest* for *see-est*: opposed to *dieresis*; crasis. Compare SYNIZESIS. **2** *Chem.* The contraction of a gel, with the expulsion of water or other liquids, as in the clotting of blood. Also spelled *synaeresis.* [<LL *synaeresis* <Gk. *synairesis* a drawing together <*syn-* together + *haireein* take]

syn·er·get·ic (sin′ər·jet′ik) *adj.* Working together; cooperative, as the flexor muscles of the leg. [<Gk. *synergētikos* <*synergeein* cooperate <*syn-* together + *ergeein* work]

syn·er·gism (sin′ər·jiz′əm) *n.* **1** The doctrine that human effort cooperates with divine grace in the salvation of the soul. **2** *Med.* The mutually cooperating action of separate substances which together produce an effect greater than that of any component taken alone, as certain drug mixtures. [<NL *synergismus* <Gk. *synergos* working together <*synergeein.* See SYNERGETIC.]

syn·er·gist (sin′ər·jist) *n.* **1** One holding to synergism. **2** A cooperating organ, part, or medicine. —**syn′er·gis′tic** or **·ti·cal** *adj.*

syn·er·gy (sin′ər·jē) *n.* **1** Combined and correlated force; united action. **2** *Med.* Correlation or concurrence of action between different organs in health or disease, or between different drugs. Also **syn·er·gi·a** (si·nûr′jē·ə). [<NL *synergia* <Gk. *synergos.* See SYNERGISM.] —**syn·er′gic** *adj.*

syn·e·sis (sin′ə·sis) *n. Gram.* Construction in accordance with the sense rather than the syntax, as the use of a plural form of a verb with a collective noun to emphasize the individuals in the group. [<Gk., a joining together, understanding <*synienai* perceive <*syn-* together + *hienai* send]

syn·es·the·sia (sin′is·thē′zhə, -zhē·ə) *n. Physiol.* **1** Transferred sensation; sensation produced at a point different from the point of stimulation. **2** The producing of a subjective response normally associated with one sense by stimulation of another sense, as of a color from hearing a certain sound. Also spelled *synaesthesia.* [<NL *synaesthesia* <Gk. *synaisthēsis* joint perception <*synaisthanesthai* perceive simultaneously <*syn-* together + *aisthanesthai* perceive, feel] —**syn′es·thet′ic** (-thet′ik) *adj.*

syn·ga·my (sing′gə·mē) *n. Biol.* The union of male and female gametes in fertilization. [<SYN- + -GAMY] —**syn·gam·ic** (sin·gam′ik), **syn′ga·mous** *adj.*

Synge (sing), **John Millington,** 1871–1909, Irish dramatist and poet.

syn·gen·e·sis (sin·jen′ə·sis) *n. Biol.* **1** Sexual reproduction. **2** The theory that the sexually fertilized germ contains within itself the germs of all future generations: opposed to *epigenesis.* [<NL <Gk. *syn-* together + *genesis* GENESIS] —**syn·ge·net·ic** (sin′jə·net′ik) *adj.*

syn·i·ze·sis (sin′ə·zē′sis) *n.* **1** In Greek prosody, the union in pronunciation of two vowels that cannot form a diphthong, so as to pass for one syllable: differing from *contraction* in not being made in the written word, but only in pronunciation. Compare SYNERESIS. **2** *Biol.* The contractile massing of the chromatin during meiotic cell division: associated with *synapsis.* **3** *Med.* Contraction of the pupil of the eye. Also **syn′e·zi′sis** (-zī′sis). [<LL <Gk. *synizēsis* <*synizanein* sink down <*syn-* together + *izanein* settle down, sit <*izein* seat, sit]

syn·od (sin′əd) *n.* **1** An ecclesiastical council, stated or special, local or general; hence, any deliberative assembly. **2** *Astron.* A conjunction (def. 2). [OE *synoth* <LL *synodus* <Gk. *synodos,* lit., a coming together <*syn-* together + *hodos* a way; refashioned after MF *synode* <LL]

Syn·od (sin′əd) *n.* **1** One of certain ecclesiastical councils distinguished by their extent or locality. **2** In the Presbyterian churches, a council intermediate between presbyteries and General Assembly. **3** In the Dutch Reformed, German Reformed, and Lutheran churches in the United States, a supreme council, known as **General Synod,** and also a more limited one, known as **Particular** or **District Synod.**

sy·nod·i·cal (si·nod′i·kəl) *adj.* **1** Of, pertaining to, or of the nature of a synod; transacted in a synod. **2** *Astron.* Pertaining to the conjunction of two heavenly bodies one of which revolves round the other, or to the interval between two successive conjunctions: a *synodical* month. Also **syn·od·al** (sin′ə·dəl), **sy·nod′ic.** —**sy·nod′i·cal·ly** *adv.*

sy·noe·cious (si·nē′shəs) *adj. Bot.* Having male and female organs, either stamens and pistils or antheridia and archegonia, in the same inflorescence or receptacle, as in most composite plants and many mosses: also spelled *synecious.* [<Gk. *synoikia* living together <*syn-* together + *oikos* a house; formed on analogy with *dioecious, monoecious,* etc.]

syn·o·nym (sin′ə·nim) *n.* **1** A word having the same or almost the same meaning as some other; hence, one of a number of words that have one or more meanings in common: opposite of *antonym.* **2** The equivalent of a word in another language. **3** *Biol.* A scientific name, as of a genus or species, superseded or discarded, as by the law of priority or because of incorrect application. Also **syn′o·nyme.** [<LL *synonymum* <Gk. *synōnymon,* neut. of *synōnymos* having like meaning or name <*syn-* together + *onyma, onoma* a name] —**syn′o·nym′ic** or **·i·cal** *adj.* —**syn′·o·nym′i·ty** *n.*

sy·non·y·mize (si·non′ə·mīz) *v.t.* **·mized, ·miz·ing** To give the synonyms of; express by words of similar or equivalent meaning.

sy·non·y·mous (si·non′ə·məs) *adj.* Being a synonym or synonyms; equivalent or similar in meaning; closely related or nearly alike in significance. Also **syn·o·ny·mat·ic** (sin′·ə·ni·mat′ik), **syn′o·nym′ic** (-nim′ik) or **·i·cal.** —**sy·non′y·mous·ly** *adv.*

Synonyms: alike, correspondent, corresponding, equivalent, identical, interchangeable, like, same, similar, synonymic. In the strictest sense, *synonymous* words scarcely exist; rarely, if ever, are any two words in any language *equivalent* or *identical* in meaning; where a difference in meaning cannot be easily shown, a difference in usage, often involving connotation, usually exists, so that the words are not *interchangeable.* By *synonymous* words we usually understand words that coincide or nearly coincide in some part of their meaning, and may hence within certain limits be used interchangeably, while outside of these limits they may differ very greatly in meaning and use. To consider *synonymous* words *identical* is fatal to accuracy; to forget that they are *similar,* to some extent *equivalent,* and sometimes *interchangeable,* is destructive of freedom and variety.

sy·non·y·my (si·non′ə·mē) *n. pl.* **·mies 1** The quality of being synonymous; the expressing or extending of an idea by the use of synonyms. **2** The science or systematic collection and study of synonyms; the use and nice discrimination of synonyms: also **syn·o·nym·ics** (sin′ə·nim′iks). **3** A book treating of or discriminating the meaning of synonyms or of allied terms. **4** An index, list, or collection of synonyms, as in scientific nomenclature. [<LL *synonymia* <Gk. *synōnymia* <*synōnymos.* See SYNONYM.]

sy·nop·sis (si·nop′sis) *n. pl.* **·ses** (-sēz) A general view, as of a subject or its treatment; an abstract; syllabus; a summary. See synonyms under ABRIDGMENT. [<LL <Gk., a general view <*syn-* together + *opsis* a view]

sy·nop·tic (si·nop′tik) *adj.* **1** Giving a general view. **2** Presenting the same or a similar point of view; containing parts that, when compared, are virtually identical: said of the first three Gospels (**Synoptic Gospels**) as distinguished from the fourth. Also **sy·nop′·ti·cal.** [<NL *synopticus* <Gk. *synoptikos* <*synopsis* a synopsis] —**sy·nop′ti·cal·ly** *adv.*

syn·os·to·sis (sin′os·tō′sis) See under SYMPHYSIS. Also **syn·os·te·o·sis** (si·nos′tē·ō′sis). [Contraction of *synosteosis* <NL <Gk. *syn-* together + *osteon* a bone]

sy·nou·si·acs (si·noo′shē·aks, -nou′-) *n.* That branch of knowledge pertaining to societies: a term used in cataloging, as in libraries. [<Gk. *synousia* society <*synousa,* ppr. fem. of *syneinai* be with <*syn-* together + *einai* be]

sy·no·vi·a (si·nō′vē·ə) *n. Physiol.* The viscid, transparent, albuminous fluid secreted in the interior of joints and at other points where lubrication is necessary. [<NL *sinovia, synovia, synophia*; coined by Paracelsus, appar. <Gk. *syn-* together + L *ovum* an egg <Gk. *ōon*] —**sy·no′vi·al** *adj.*

syn·o·vi·tis (sin′ō·vī′tis) *n. Pathol.* Inflammation of a synovial membrane.

syn·sep·a·lous (sin·sep′ə·ləs) *adj. Bot.* Gamosepalous. [<SYN- + SEPAL + -OUS]

syn·tax (sin′taks) *n.* **1** The arrangement and interrelationship of words in grammatical constructions. 2 The branch of linguistics dealing with this. [<F *syntaxe* <LL *syntaxis* <Gk. <*syntassein* join together <*syn-* together + *tassein* arrange] —**syn·tac·tic** (sin·tak′tik) or **·ti·cal** *adj.* —**syn·tac′ti·cal·ly** *adv.*

syn·tech·nic (sin·tek′nik) *adj. Ecol.* Denoting resemblance among dissimilar animal forms due to influences of common environment. See CONVERGENCE (def. 7). [<Gk. *syntechnos* practicing the same art <*syn-* together + *technē* an art]

syn·the·sis (sin′thə·sis) *n. pl.* **·ses** (-sēz) 1 The assembling of separate or subordinate parts into a new form; also, the complex whole resulting from this. 2 *Ling.* The combination of radical and formative or inflectional elements in one word, as in *un-think-ing, home-wards.* 3 *Logic* **a** Combination of separate elements into a whole, as of species into genera: contrasted with *analysis.* **b** A process of reasoning from the whole to a part, from the general to the particular; deductive reasoning. 4 *Surg.* The operation of reuniting broken or divided parts, as of bones. 5 *Chem.* **a** The building up of compounds from a series of reactions involving elements, radicals, or simpler compounds. **b** The preparation by such means of organic compounds which have specific properties or are identical in certain respects with naturally occurring substances. Compare ANALYSIS. [<L <Gk. <*syntithenai* <*syn-* together + *tithenai* place] —**syn′the·sist** *n.*

syn·the·size (sin′thə·sīz) *v.t.* **·sized, ·siz·ing** 1 To unite or produce by synthesis. 2 To apply synthesis to. Also *Brit.* **syn′the·sise.**

syn·thet·ic (sin·thet′ik) *adj.* 1 Pertaining to or of the nature of synthesis; characterized by or consisting in synthesis; specifically, tending to reduce particulars to inclusive wholes: a *synthetic* mind. 2 *Chem.* Produced by the synthesis of simpler materials or substances: *synthetic* rubber. 3 Artificial; spurious. 4 *Ling.* Describing a language that utilizes inflectional affixes for the expression of relationships between words, as in Latin; inflectional: opposed to *analytic.* Also **syn·thet′i·cal.** —*n.* 1 Anything produced by synthesis. 2 *Chem.* A synthesized compound adapted for use as a substitute for some other material or substance. [<F *synthétique* <NL *syntheticus* <Gk. *synthetikos* <*synthetos* compounded <*syntithenai* <*syn-* together + *tithenai* place] —**syn·thet′i·cal·ly** *adv.*

synthetic philosophy Spencerism: so called by Spencer as being an attempt to combine all the sciences into a connected whole.

syn·to·nize (sin′tə·nīz) *v.t.* **·nized, ·niz·ing** *Electr.* 1 To place in resonance with each other, as radio frequencies. 2 To tune or tone together, as electrical instruments. [<SYNTON(Y) + -IZE] —**syn·ton·ic** (sin·ton′ik) or **·i·cal** *adj.* —**syn·ton′i·cal·ly** *adv.* —**syn′·to·ni·za′tion** *n.*

syn·to·ny (sin′tə·nē) *n. Electr.* 1 The harmonizing or tuning of particular transmitters and receivers each to the other. 2 Resonance. [<Gk. *syntonia* agreement <*syn-* together + *tonos* a tone]

syn·u·ra (sin·yo͝or′ə) *n. pl.* **·u·rae** (-yo͝or′ē) Any of a genus (*Synura*) of flagellate protozoans, uniting in subspherical clusters and discharging oil globules. They are common in swamp waters and render drinking water unpalatable by giving it a cucumberlike flavor. [<NL <Gk. *synouros, synoros* bordering on <*syn-* together + *oros* a boundary]

sy·pher (sī′fər) *v.t.* To make a lap joint with (two chamfered or beveled plank edges) so as to leave a flush surface. [Var. of CIPHER] —**sy′pher·ing** *n.*

syph·i·lis (sif′ə·lis) *n. Pathol.* An infectious, chronic, venereal disease caused by a spirochete (*Treponema pallidum*) transmissible by direct contact or congenitally. It usually progresses by three stages of increasing severity: primary, secondary, and tertiary. [after *Syphilis, sive Morbus Gallicus,* a Latin poem by Fracastoro, published in 1530, the hero of which, *Syphilus,* a shepherd, was the first sufferer from the disease] —**syph′i·loid, syph′·i·lous** *adj.*

syph·i·lit·ic (sif′ə·lit′ik) *adj.* Relating to or affected with syphilis. —*n.* A person suffering from syphilis. [<NL *syphiliticus* <*syphilis* SYPHILIS]

syph·i·lol·o·gy (sif′ə·lol′ə·jē) *n.* The science of syphilis, its cognate diseases, and their treatment. [<SYPHIL(IS) + -(O)LOGY] —**syph′i·lol′o·gist** *n.*

syph·i·lo·pho·bi·a (sif′ə·lə·fō′bē·ə) *n. Psychiatry* A morbid fear of syphilis. [<SYPHIL(IS) + -(O)PHOBIA] —**syph′i·lo·pho′bic** *adj.*

syphon (sī′fən) See SIPHON.

Syr·a·cuse (sir′ə·kyo͞os) 1 A port of SE Sicily: Italian *Siracusa.* Ancient **Syr·a·cu·sae** (sir′ə·kyo͞o′sē, -zē). 2 A city in central New York. —**Syr′a·cu′san** *adj.* & *n.*

Syr Dar·ya (sir där′yä) A river in SW Asiatic U.S.S.R. flowing about 1,327 miles NW to the Aral Sea: ancient *Jaxartes.*

sy·ren (sī′rən) See SIREN.

Syr·ette (si·ret′) *n.* A miniature syringe; especially, a small disposable tube for the emergency administration of morphine, for use on the battlefield, by paratroopers, etc.: a trade name.

Syr·i·a (sir′ē·ə) *n.* 1 A former republic, 1941–1958, south of Asia Minor on the NE coast of the Mediterranean, incorporated into the United Arab Republic in 1958; 72,234 square miles; capital, Damascus. *Arabic* **Esh Shan.** 2 A former French mandated territory, 1920–1941, roughly comprising Syria (def. 1) and Lebanon. 3 An ancient country including Syria (def. 1), Lebanon, Palestine (def. 2), and adjacent districts of western Asia.

Syr·i·ac (sir′ē·ak) *n.* The language of the Syrians, belonging to the eastern Aramaic subgroup of the Northwest Semitic languages. [<L *Syriacus* <Gk. *Syriakos* <*Syria* Syria]

Syr·i·an (sir′ē·ən) *adj.* Of or pertaining to Syria, ancient or modern. —*n.* 1 A native of Syria, especially one of the native Semitic people of Arabic, Phoenician, and Aramean descent. 2 One who is a member of the Christian church in Syria. [<OF *sirien* <L *Syrius* a Syrian <Gk. *Syrios* <*Syria* Syria]

Syrian Desert An arid wasteland of SW Asia between the lands along the eastern Mediterranean and the Euphrates valley.

sy·rin·ga (si·ring′gə) *n.* 1 Any of a genus (*Philadelphus*) of ornamental shrubs of the saxifrage family having cream-colored flowers resembling those of the orange in form and fragrance, especially the Lewis mock orange (*P. lewisii*), the State flower of Idaho. 2 Any of a genus (*Syringa*) of ornamental shrubs of the olive family having panicles of showy white or purple flowers; the lilacs. [<NL <Gk. *syrinx, -ingos* a pipe]

SYRINGA
(Plant to 10 feet)

syr·inge (sir′inj, si·rinj′) *n. Med.* An instrument used to withdraw a fluid from a reservoir and eject it in one or more jets or streams. The simplest forms are valveless single-acting devices; other forms consist of an elastic bag supplied with flexible inlet and outlet pipes each having a suitable check valve. —*v.t.* **·inged, ·ing·ing** To spray or inject by a syringe; cleanse or treat with injected fluid. [<Med. L *siringa* <Gk. *syrinx, -ingos* a tube, a pipe]

sy·rin·go·my·e·li·a (si·ring′gō·mī·ē′lē·ə) *n. Pathol.* A morbid condition of the spinal cord, due to the presence of liquid in abnormally formed cavities. [<NL <Gk. *syrinx, -ingos* a tube + *myelos* marrow]

syr·inx (sir′ingks) *n.* 1 *Ornithol.* A special modification of the windpipe serving as the song organ in birds. 2 A tube, pipe, or fistula. 3 Panpipes. [<Gk., a pipe] —**sy·rin·ge·al** (si·rin′jē·əl) *adj.*

Syr·inx (sir′ingks) In Greek mythology, a nymph pursued by Pan and changed into a reed, from which Pan made his pipes.

Sy·ros (sī′rəs) One of the Cyclades group, SW of Tenos; 33 square miles. Also *Siros.* Latin **Sy′rus.**

syr·phus fly (sûr′fəs) A fly of *Syrphus* or a related genus (family *Syrphidae*). The group is large and widely distributed, and contains many species which deceptively resemble bees and wasps. The larvae of many feed upon harmful plant lice. For illustration see INSECTS (beneficial). Also **syr·phid** (sûr′fid), **syr′phi·an.** [<NL <Gk. *syrphos* a gnat]

Syr·tis Ma·jor (sir′tis mā′jər) The ancient name for the GULF OF SIDRA.

Syrtis Minor An ancient name for the GULF OF GABÈS.

syr·up (sir′əp), **syr·up·y** See SIRUP, etc.

sys- Assimilated var. of SYN-.

sys·sar·co·sis (sis′är·kō′sis) *n. Anat.* The union of bones by means of muscles. [<NL <Gk. *syssarkōsis* <*syssarkoein* unite by or cover over with flesh] —**sys′sar·co′sic, sys′sar·cot·ic** (-kot′ik) *adj.*

sys·tal·tic (sis·tal′tik) *adj. Physiol.* Alternately contracting and dilating: the *systaltic* motion of the heart; pulsatory. Compare PERISTALSIS. [<LL *systalticus* <Gk. *systaltikos* depressing <*systellein* draw together <*syn-* together + *stellein* send]

sys·tem (sis′təm) *n.* 1 Orderly combination or arrangement, as of parts or elements, into a whole; specifically, such combination according to some rational principle; any methodical arrangement of parts. 2 In science and philosophy, an orderly collection of logically related principles, facts, or objects. 3 Any group of facts and phenomena regarded as constituting a natural whole and furnishing the basis and material of scientific investigation and construction: the solar *system.* 4 The connection or manner of connection of parts as related to a whole, or the parts collectively so related; a whole as made up of constitutive parts: a railroad *system.* 5 The state or quality of being in order or orderly; orderliness; method: He works with *system.* 6 *Physiol.* An assemblage of organic structures composed of similar elements and combined for the same general functions: the nervous *system*; also, the entire body, taken as a functional whole. 7 *Physics* An aggregation of matter in, or tending to approach, equilibrium. 8 *Mineral.* One of the six divisions into which all crystal forms may be grouped, depending upon the relative lengths and mutual inclinations of the assumed crystal axes. 9 *Geol.* A category of rock strata next below a group and above a series and corresponding with a period in the time scale. [<LL *systema* a musical interval <Gk. *systēma, -atos* an organized whole <*syn-* together + *histanai* stand, set up]

Synonyms: manner, method, mode, order, regularity, rule. *Order* in this connection denotes a fact or a result; as, These papers are in *order. Method* denotes a process; *rule* an authoritative requirement or an established course of things; *system,* not merely a law of action or procedure, but a comprehensive plan; *manner* refers to the external qualities of actions, and to those often as settled and characteristic; we speak of a *system* of taxation, a *method* of collecting taxes, the *rules* by which assessments are made; or we say, As a *rule* the payments are heaviest at a certain time of year; a just tax may be made odious by the *manner* of its collection. *Regularity* applies to even disposition of objects or uniform recurrence of acts in a series. There may be *regularity* without *order,* as in the recurrence of paroxysms of disease or insanity; there may be *order* without *regularity,* as in the arrangement of furniture in a room, where the objects are placed at varying distances. *Order* commonly implies the design of an intelligent agent or the appearance or suggestion of such design; *regularity* applies to an actual uniform disposition or recurrence with no suggestion of purpose, and as applied to human affairs is less intelligent and more mechanical than *order.* See BODY, FRAME, HABIT, HYPOTHESIS. *Antonyms*: chaos, confusion, derangement, disarrangement, disorder, irregularity.

sys·tem·at·ic (sis′tə·mat′ik) *adj.* 1 Of, pertaining to, of the nature of, or characterized by system. 2 Acting by system or method; methodical: *systematic* thieving. 3 Forming a system; systematized. 4 Carried out with organized regularity. 5 Taxonomic: *systematic* botany. Also **sys′tem·at′i·cal.** [<LL *systematicus* <LGk. *systēmatikos* <*systēma, -atos* a system] —**sys′tem·at′i·cal·ly** *adv.*

sys·tem·at·ics (sis′tə·mat′iks) *n. pl.* (*construed as singular*) 1 The art or principles of classification and nomenclature. 2 *Biol.* The science of the classification of organisms; taxonomy.

sys·tem·a·tism (sis′tə·mə·tiz′əm) *n.* 1 Systematic arrangement or classification. 2 Adherence to or reduction of principles, etc., to a system.

systematist 1275 **tablespoonful**

sys·tem·a·tist (sis′tə·mə·tist) *n.* **1** One who reduces things to systems, as a taxonomist. **2** One who forms or adheres to a system or to a systematic view of things.
sys·tem·a·tize (sis′tə·mə·tīz′) *v.t.* **·tized, ·tiz·ing** To reduce to a system; dispose methodically. See synonyms under REGULATE. Also **sys′tem·ize,** *Brit.* **sys′tem·a·tise′.** — **sys′tem·a·ti·za′tion, sys′tem·i·za′tion** *n.* — **sys′tem·a·tiz′er, sys′tem·iz′er** *n.*
sys·tem·ic (sis·tem′ik) *adj.* **1** Of or pertaining to system or a system; systematic. **2** *Physiol.* Pertaining to or affecting the body as a whole: a *systemic* poison. — **sys·tem′i·cal·ly** *adv.*
sys·to·le (sis′tə·lē) *n.* **1** *Physiol.* The regular contraction of the heart, especially of the ventricles, that impels the blood outward. Compare DIASTOLE. **2** The shortening of a syllable that is naturally or by position long. [< NL < Gk. *systolē* a contraction < *systellein*. See SYSTALTIC.] — **sys·tol·ic** (sis·tol′ik) *adj.*
syz·y·gy (siz′ə·jē) *n. pl.* **·gies 1** *Astron.* One of two opposite points in the orbit of a celestial body when it is in conjunction with or opposition to the sun; especially, the points on the moon's orbit when the moon is most nearly in line with the earth and the sun. **2** The union of parts or organisms. **3** A dipody or group of two feet in one verse. [< LL *systolē* a contraction < *systellein*. See SYSTALTIC.] — **sys·tol·ic** (sis·tol′ik) *adj.*

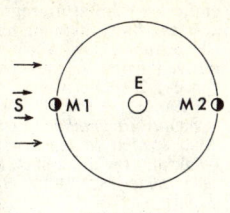

SYZYGY
S. Sun's rays. E. Earth. M1, M2. Syzygy of the moon.

syz·y·gi·a < Gk., a yoke, conjunction < *syzy·gos* yoked, paired < *syn-* together + *zeugnynai* yoke < *zygon* a yoke] — **sy·zyg·i·al** (si·zij′ē·əl) *adj.*
Sza·bad·ka (sô′bôd·kô) The Hungarian name for SUBOTICA.
Sze·chwan (se′chwän′, su′-) A province of SW and central western China; 338,136 square miles; capital, Chengtu.
Sze·ged (se′ged) A city of southern Hungary. Formerly **Sze·ge·din** (se′ge·din).
Sze·ming (su′ming′) A former name for AMOY.
Szent–Gyor·gyi von Nagy·ra·polt (sent′dyûr′dyē fon nod′y′ro′pōlt), **Albert,** born 1893, Hungarian biochemist.
Szi·lard (si·lärd′), **Leo,** born 1898, U. S. physicist born in Hungary.
Szom·bat·hely (som′bôt·hāy′) A city of western Hungary: German *Steinamanger.*

T

t, T (tē) *n. pl.* **t's, T's** or **ts, Ts, tees** (tēz) **1** The twentieth letter of the English alphabet, from Greek *tau* (a modification of Phoenician *tau*) and Latin *T.* **2** The sound of the letter *t,* the voiceless alveolar stop. See ALPHABET. — *symbol* Anything shaped like a T. — **to a T** Precisely; with exactness: probably in allusion to a T-square.
T (tē) *adj.* Shaped or having a cross-section like a T, as *T-beam, T-pipe,* etc. — *n.* Anything having the shape of a T.
't Contraction for IT: used initially, as in *'tis,* and finally, as in *on't.*
-t Inflectional ending used to indicate past participles and past tenses, as in *bereft, lost, spent:* equivalent to *-ed.*
Taal (täl) *n.* A form of Dutch spoken in South Africa; Afrikaans. [< Du., speech, language. Cf. TALE.]
Ta·al (tä·äl′), **Mount** An active volcano (984 feet) on an island in **Lake Taal** (94 square miles) in southern Luzon, Philippines.
tab (tab) *n.* **1** A flap, strip, tongue, or appendage of something, as a garment. **2** *Colloq.* Tally: to keep *tab.* **3** *Aeron.* An auxiliary airfoil attached to the control surface of an airplane. [Origin uncertain]
tab·a·cin (tab′ə·sin) *n. Chem.* A waxy, lemon-yellow, poisonous glycoside extracted from the leaves of Kentucky tobacco. [< F *tabac* tobacco + -IN]
tab·a·nid (tab′ə·nid) *n.* Any of a family (*Tabanidae*) of large, bloodsucking insects; a horsefly or deerfly. — *adj.* Of or pertaining to the *Tabanidae.* [< NL, family name < L *tabanus* a horsefly]
tab·ard (tab′ərd) *n.* **1** Formerly, a short, sleeveless or short-sleeved, outer garment. **2** A knight's cape or cloak, worn over his armor and emblazoned with his own arms; also, a similar garment worn by a herald and embroidered with his lord's arms. **3** A banner attached to a trumpet or bugle. [< OF *tabard,* ult. < L *tapete* tapestry]
tab·a·ret (tab′ə·rit) *n.* A strong, silk upholstery fabric with varicolored stripes of satin or moiré. [Prob. < TABBY]
Ta·bas·co (tə·bas′kō) *n.* A pungent sauce made from the red-pepper plant (genus *Capsicum*): a trade name.
Ta·bas·co (tə·bas′kō, *Sp.* tä·väs′kō) A state in SE Mexico; 9,782 square miles; capital, Villahermosa.
tab·by (tab′ē) *n. pl.* **·bies 1** Any of several plain-woven fabrics, as a striped or watered taffeta, or a moreen. **2** A garment made of a watered fabric. **3** A brindled or striped cat; popularly, any domestic cat, especially a female. **4** A gossiping old maid. **5** A building material of equal parts of lime and shells, and gravel, mixed with water. — *adj.* **1** Watered; mottled, as a fabric; also, brindled, as a cat. **2** Made of tabby. **3** Woven in the same way as fabric that is to be watered. — *v.t.* **·bied, ·by·ing** To give a wavy or watered appearance to (silk, etc.) by pressure between hot rollers; water; calender. [< F *tabis, atabis* < Arabic *uttābi* < *'Attābi,* name of a quarter of Baghdad where it was manufactured]
ta·ber (tā′bər) See TABOR.
tab·er·nac·le (tab′ər·nak′əl) *n.* **1** A tent or similar structure; slight shelter, fixed or portable. **2** Specifically, the portable sanctuary used by the Jews in the wilderness; later, the Jewish temple; hence, any house of worship, especially one of large size and not of especially ecclesiastical architecture: in England, the place of worship of some nonconformists. **3** The human body as the dwelling place of the soul. **4** The ornamental receptacle for the consecrated eucharistic elements, or for the pyx. **5** An ornamental recess or a structure sheltering something. **6** A socket or hinged post to unstep or lower a mast. — *v.i.* **·led, ·ling** To dwell in a tent; hence, to dwell transiently: The soul *tabernacles* in the body. [< OF < L *tabernaculum,* dim. of *taberna* shed] — **tab·er·nac·u·lar** (tab′ər·nak′yə·lər) *adj.*
ta·bes (tā′bēz) *n. Pathol.* **1** Emaciation with general languor, progressive atrophy, and hectic fever; a decline. **2** Locomotor ataxia: also **tabes dor·sa·lis** (dôr·sā′lis). [< L, a wasting away < *tabere* waste away] — **ta·bes·cence** (tə·bes′əns) *n.* — **ta·bes′cent** *adj.* — **ta·bet′ic** (-bet′ik) *n.* & *adj.* — **tab·id** (tab′id) *adj.*
ta·bet (tā′bit) *n. Scot.* Bodily sensation or strength. — **ta′bet·less** *adj.*
Tab·i·tha (tab′i·thə) A feminine personal name. [< Aramaic, gazelle]
— **Tabitha** A woman of Joppa noted for her good works. *Acts* ix 36.
Tab·las (täb′läs) The largest island in the Romblon group, central Philippines; 265 square miles.
tab·la·ture (tab′lə·chər) *n.* **1** *Anat.* One of the plates of bony tissue that form the walls of the cranium. **2** A tablelike painting or design. [< F < L *tabula* board]
ta·ble (tā′bəl) *n.* **1** An article of furniture with a flat horizontal top upheld by one or more supports. **2** Such a table around which persons sit for a meal; to set the *table.* **3** The food served or entertainment provided at a meal or dinner. **4** The company of persons at a table. **5** A gaming table, as for roulette, dice, etc. **6** A collection of related numbers, values, signs, or items of any kind, arranged for ease of reference or comparison, often in parallel columns: a *table* of logarithms; a *table* of statistics. **7** A synoptical statement; list: *table* of contents. **8** A tableland; plateau. **9** *Geol.* A horizontal stratum of rock. **10** The flat facet cut across the top of a precious stone. **11** *Archit.* **a** A raised horizontal surface or band of molding on a wall; a stringcourse. **b** A raised or sunken panel on a wall. **12** In palmistry, the quadrangle formed by four lines of the hand. **13** In backgammon: **a** Either of the two leaves of a backgammon board. **b** *pl. Obs.* Backgammon. **14** *Anat.* One of the flat bony plates forming the inner or outer part of the cranium. **15** A tablet or slab bearing an inscription: especially, one of those which bore the Ten Commandments or certain Roman laws. — **to turn the tables** To thwart an opponent's action and turn the situation to his disadvantage. — *v.t.* **·bled, ·bling 1** To place on a table, as a playing card. **2** To postpone discussion of (a resolution, bill, etc.) until a future time, or for an indefinite period. **3** *Rare* To make into or enter in a list or table; tabulate. [Fusion of OF *table* and OE *tabule,* both < L *tabula* board]
tab·leau (tab′lō, ta·blō′) *n. pl.* **·leaux** (-lōz) or **·leaus** (-lōz) **1** Any picture or picturesque representation; especially, an unexpected situation produced suddenly and dramatically. **2** A tableau vivant. [< F, dim. of *table.* See TABLE.]
ta·bleau vi·vant (ta·blō′ vē·vän′) *pl.* **ta·bleaux vi·vants** (ta·blō′ vē·vän′) *French* A picturelike scene represented by silent and motionless persons standing in appropriate attitudes: also called *living picture.*
Table Bay An inlet of the Atlantic in SW Cape of Good Hope Province, Union of South Africa: the harbor of Cape Town.
table book An ornamental book to be kept on a table.
ta·ble–chair (tā′bəl·châr′) *n.* An armchair or small bench having a hinged back which when tilted up forms a table-top: also called *chair-table.*
ta·ble·cloth (tā′bəl·klôth′, -kloth′) *n.* A cloth, often white, covering a table at meals; table cover.
ta·ble d'hôte (tab′əl dōt′, tä′bəl) *pl.* **ta·bles d'hôte** (tā′bəlz dōt′, tä′bəlz) **1** A common table for guests, as at a hotel. **2** A complete meal of several specified courses, served in a restaurant at a fixed price: used also attributively: a *table d'hôte* breakfast, dinner, etc. [< F, lit., table of the host]
ta·ble·land (tā′bəl·land′) *n.* **1** A broad, level, elevated region, usually treeless; a plateau: specifically, a precipitous mesa. **2** Level, elevated land.
table linen Tablecloths, napkins, doilies, etc., made of linen, cotton, etc.
Table Mountain A flat-topped mountain in SW Cape of Good Hope Province, near Cape Town, Union of South Africa; 3,549 feet.
ta·ble·spoon (tā′bəl·spōon′, -spōon′) *n.* A large spoon, larger than a dessertspoon, with a capacity of 15 cc. or three times the capacity of a teaspoon: used for serving food.
ta·ble·spoon·ful (tā′bəl·spōon·fool′, -spōon′) *n. pl.* **·fuls** As much as a tablespoon will hold:

add, āce, câre, pälm; end, ēven; it, īce; odd, ōpen, ôrder; tŏŏk, pōōl; up, bûrn; ə = *a* in *above, e* in *sicken, i* in *clarity, o* in *melon,* u in *focus;* yōō = u in *fuse;* oi, oil; ou, pout; ch, check; g, go; ng, ring; th, thin; ᵺ, this; zh, vision. Foreign sounds á, œ, ü, kh, ᴎ; and ◆: see page xx. < from; + plus; ? possibly.

usually reckoned as equivalent to half a fluid ounce, 15 cc., or three teaspoonfuls.

tab·let (tab′lit) *n.* **1** A thin leaf or sheet of solid material, as ivory or wood, for writing, painting, or drawing. **2** One of a set of leaves pivoted or joined together at one end and used for writing; also, a set of such leaves; hence, a pad, as of writing paper or note paper. **3** A small table or flat surface, especially one designed for or containing an inscription or design. **4** A small, flat or nearly flat piece of some prepared substance, as chocolate or soap. **5** A definite portion or weight of drug brought by pressure and the addition of a gum into a solid form; a troche or lozenge; also, an electuary. **6** A flat or tablelike surface. [<OF *tablete,* dim. of *table.* See TABLE.]

table tennis A game resembling tennis in miniature, played indoors with a small celluloid ball and wooden paddles on a large table; Ping-pong.

ta·ble·ware (tā′bəl·wâr′) *n.* Ware for table use; dishes, knives, forks, spoons, etc., collectively: called **table furniture** when napery is included.

tab·loid (tab′loid) *n.* A newspaper, one half the size of an ordinary newspaper, in which the news is presented by means of pictures and concise reporting. —*adj.* **1** Compact; concise; condensed. **2** Sensational: *tabloid* journalism. [<TABL(ET) + -OID]

Tab·loid (tab′loid) *n.* Proprietary name for any of various medical preparations and drugs in concentrated or condensed tablet form.

ta·boo (tə·bo͞o′, ta-) See TABU.

ta·bor (tā′bər) *n.* A small drum or tambourine on which a fife-player beats his own accompaniment; a timbrel. —*v.i.* To beat or play on a timbrel or small drum; beat lightly and repeatedly. Also spelled *taber.* Also **ta′bour.** [<OF *tabour,* prob. <Persian *tabīrah* drum] —**ta′bor·er** *n.*

Ta·bor (tā′bər), **Mount** A mountain near Nazareth in Galilee, northern Israel; 1,929 feet.

tab·o·ret (tab′ər·it, tab′ə·ret′) *n.* **1** A small tabor. **2** A stool or small seat, usually without arms or back. **3** An embroidery frame. **4** A needle case. Also **tab′ou·ret.** [<F *tabouret,* dim. of *tabour* TABOR]

tab·o·rine (tab′ə·rēn, tab·ə·rēn′) *n.* A small tabor, tambourine, or side drum. Also **tab′o·rin, tab′ou·rine.** [<OF *tabourin,* dim. of *tabour* TABOR]

Ta·briz (tä·brēz′) A city in NW Iran: ancient *Tauris:* also *Tebriz.*

ta·bu (tə·bo͞o′, ta-) *n.* **1** Among primitive peoples, especially the Polynesians, a religious and social interdict against the touching or mentioning of a certain person, thing, or place, the uttering of a certain name, or the performing of a certain action, because it is considered sacred, protective, dangerous, unclean, or possessed of mysterious powers. **2** The system or practice of such interdicts or prohibitions. **3** Any restriction or ban founded on custom or social convention. —*adj.* **1** Consecrated or prohibited by tabu. **2** Banned or forbidden by social authority or convention. —*v.t.* **1** To place under tabu. **2** To exclude; ostracize. Also spelled *taboo.* [<Tonga]

tab·u·lar (tab′yə·lər) *adj.* **1** Pertaining to or consisting of a table or list. **2** Computed from or with a mathematical table. **3** Having a flat surface; tablelike. [<L *tabularis* <*tabula* table] —**tab′u·lar·ly** *adv.*

tab·u·la ra·sa (tab′yo͞o·lə rä′sə) *Latin* **1** An empty or clean tablet; a clean slate. **2** The concept of the mind of a newborn child as a blank, to be written on by experience.

tab·u·lar·ize (tab′yə·lə·rīz′) *v.t.* **·ized, ·iz·ing** To arrange in tabular form, or in a table or tables; tabulate. —**tab′u·lar·i·za′tion** *n.*

tab·u·late (tab′yə·lāt) *v.t.* **·lat·ed, ·lat·ing** **1** To arrange in a table or list: to *tabulate* results. **2** To form with a tabular surface. —*adj.* **1** Having a flat surface or surfaces; broad and flat. **2** *Zool.* Having tabulated horizontal plates extending across the visceral cavity: said of certain corals. [<L *tabula* table + -ATE¹] —**tab′u·la′tion** *n.*

tab·u·la·tor (tab′yə·lā′tər) *n.* **1** One who or that which tabulates. **2** A device built into a typewriter with which statistical matter may be speedily written in tabulated form. **3** An automatic high-speed accounting machine for tabulating reports.

tac·a·ma·hac (tak′ə·mə·hak′) *n.* **1** A yellowish resinous substance with a strong odor, derived from various trees and used as incense. **2** Any of the trees producing this substance, especially the balsam poplar (*Populus tacamahaca*) of the United States. Also spelled *tacmahack.* Also **tac′a·ma·hac′a** (-hak′ə), **tac′a·ma·hack′.** [<Sp. *tacamaca, tacamahaca* <Nahuatl *tecomahca,* lit., fetid copal]

ta·can (ta·kan′) *n. Aeron.* A system for indicating the distance and bearing of an aircraft from a known fixed point by means of ultrahigh-frequency signals transmitted from the aircraft to a ground station. [<TAC(TICAL) A(IR) N(AVIGATION)]

tace (tās) *n.* Tasset.

ta·cet (tā′set) *Latin* Literally, it is silent: a musical direction for silence.

tache (tach) *n. Archaic* A hook or fastening; a clasp; buckle. Also **tach.** [<OF *tache* nail, fastening. Doublet of TACK.]

tach·e·om·e·ter (tak′ē·om′ə·tər) *n.* **1** A tachymeter. **2** A tachometer. [<Gk. *tachos, tacheos* speed + -METER] —**tach′e·om′e·try** *n.*

tach·i·na fly (tak′ə·nə) A fly (family *Tachinidae*) often resembling the house fly, whose larvae develop as parasites in the caterpillar or other insect. For illustration see under INSECTS (beneficial). [<NL *tachina* <Gk. *tachinos* swift]

Ta Ch′ing (dä′ jing′) The Manchu dynasty of China. See MANCHU.

tach·i·nid (tak′ə·nid) *n.* A tachina fly. —*adj.* Of or pertaining to the *Tachinidae.*

ta·chis·to·scope (tə·kis′tə·skōp) *n.* An apparatus for giving a brief but accurately measurable exposure to visual objects, for the purpose of determining the speed and conditions of their apperception. [<Gk. *tachistos* swiftest + -SCOPE]

tach·o·graph (tak′ə·graf, -gräf) *n.* **1** A registering tachometer. **2** The record it makes. [<Gk. *tachos* swiftness + -GRAPH]

ta·chom·e·ter (tə·kom′ə·tər) *n.* **1** An instrument for measuring linear and angular velocity, as of a machine, the flow of a current, blood, etc. **2** A device for indicating the speed of rotation of an engine, etc. [See TACHEOMETER]

ta·chom·e·try (tə·kom′ə·trē) *n.* The art or science of using a tachometer. —**tach·o·met·ric** (tak′ə·met′rik) *adj.*

tachy- *combining form* Speed; swiftness: *tachycardia.* [<Gk. *tachys* swift]

tach·y·car·di·a (tak′i·kär′dē·ə) *n. Pathol.* Abnormal rapidity of the heartbeat, usually indicating a pulse rate above 100 per minute. —**tach′y·car′di·ac** *adj.* & *n.*

tach·y·graph (tak′ə·graf, -gräf) *n.* **1** A tachygraphic manuscript or symbol. **2** A tachygrapher.

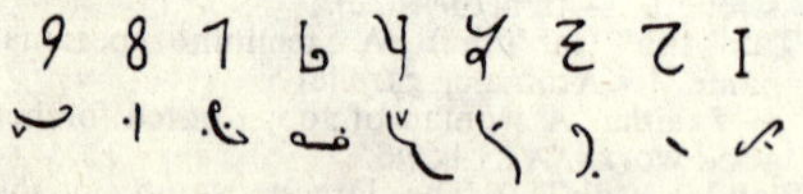

TACHYGRAPHS
Numerals: *Upper* Arabic, A.D. 976; *Lower* Modern Shorthand.

ta·chyg·ra·pher (tə·kig′rə·fər) *n.* **1** One who writes in shorthand; a stenographer. **2** One of the shorthand writers of the ancient Greeks and Romans. Also **ta·chyg′ra·phist.**

ta·chyg·ra·phy (tə·kig′rə·fē) *n. Archaic* Stenography; shorthand. —**tach·y·graph·ic** (tak′ə·graf′ik) or **·i·cal** *adj.* —**tach′y·graph′i·cal·ly** *adv.*

tach·y·lyte (tak′ə·līt) *n.* A pitch-black basaltic glass which is rapidly decomposed by acids. [<TACHY- + -LYTE¹; so called because easily decomposed] —**tach′y·lyt′ic** (-lit′ik) *adj.*

ta·chym·e·ter (tə·kim′ə·tər) *n.* **1** A surveying instrument for stadia surveying, having a level, telescope, vertical arc or circle, horizontal compass, and stadia wires. **2** A tachometer.

ta·chym·e·try (tə·kim′ə·trē) *n.* The art or science of using, or measuring with, the tachymeter. —**tach·y·met·ric** (tak′ə·met′rik) *adj.*

ta·chys·ter·ol (tə·kis′tə·rōl, -rol) *n. Biochem.* One of the substances formed by the ultraviolet irradiation of ergosterol, and the immediate predecessor of calciferol. [<TACHY- + STEROL]

tac·it (tas′it) *adj.* **1** Existing, inferred, or implied without being directly stated; implied by silence or silent acquiescence. **2** *Law* Not expressed but understood by provision or operation of the law. **3** Silent; emitting no sound; noiseless. ◆ Homophone: *tasset.* [<F *tacite* <L *tacitus,* pp. of *tacere* be silent] —**tac′it·ly** *adv.* —**tac′it·ness** *n.*

Synonyms: implicit, implied, understood, unexpressed, unspoken.

tacit mortgage A lien in the nature of a mortgage created by operation of law.

tac·i·turn (tas′ə·tûrn) *adj.* Habitually silent or reserved; disinclined to conversation. [<L *taciturnus* <L *tacere* be silent] —**tac′i·tur′ni·ty** *n.* —**tac′i·turn·ly** *adv.*

Synonyms: close, dumb, mute, reserved, reticent, silent, uncommunicative. *Dumb, mute,* and *silent* refer to fact or state; *taciturn* refers to habit and disposition. The talkative person may be stricken *dumb* with terror; the obstinate may remain *mute*; one may be *silent* through preoccupation or set purpose; but the *taciturn* person is averse to the utterance of thought or feeling and to communication with others. One who is *silent* does not speak at all; one who is *taciturn* speaks when compelled, but in a grudging way. *Reserved* suggests more of method and intention than *taciturn,* applying often to some special time or topic. *Reserved* is thus closely equivalent to *uncommunicative,* but is a somewhat stronger word, often suggesting pride or haughtiness, as when we say one is *reserved* toward strangers. *Antonyms:* communicative, free, garrulous, loquacious, talkative, unreserved.

Tac·i·tus (tas′ə·təs), **Gaius Cornelius,** A.D. 55?–after 117, Roman historian.

tack¹ (tak) *n.* **1** A small sharp-pointed nail, commonly with tapering sides and a flat head. **2** *Naut.* **a** A rope which holds down the weather clew of a course. **b** The weather clew of a square sail. **c** The lower forward corner of a fore-and-aft sail. **d** A rope by which the lower outer corner of a studdingsail is pulled to the end of the boom. **e** The direction in which a vessel sails when sailing close-hauled, considered in relation to the position of her sails: the starboard *tack* when the wind is coming from the right-hand side. **f** The distance or the course run at one time in such direction. **g** The act of tacking. **h** Any veering of a vessel to one side, as to take advantage of a side wind. **3** A change of policy; a new course of action. **4** A fastening; in needlework, a temporary stitch. **5** In Scots law, a contract; a lease; also, leased land. **6** The saddle, bridle, martingale, etc., used in riding horseback. —*v.t.* **1** To fasten or attach with tacks. **2** To secure temporarily, as with tacks or long stitches. **3** To attach as supplementary; append. **4** *Naut.* **a** To bring (a vessel) momentarily into the wind so as to go on the opposite tack. **b** To navigate (a vessel) to windward by making a series of tacks. —*v.i.* **5** *Naut.* **a** To tack a vessel. **b** To go on the opposite tack, or sail to windward by a series of tacks: said of vessels. **6** To change one's course of action; veer. [<AF *taque,* OF *tache* a nail <Gmc. Doublet of TACHE.] —**tack′er** *n.*

tack² (tak) *n.* Food in general: usually used contemptuously, and often in compounds: *hardtack.* [Origin uncertain]

tack·et (tak′it) *n. Scot.* A hobnail or clout.

tack hammer A small hammer for driving tacks.

tack·le (tak′əl, *in nautical usage* tā′kəl) *n.* **1** A rope, pulley, or combination of ropes and pulleys, used for hoisting or moving objects. **2** *Naut.* A mechanism for raising and lowering heavy weights, or managing sails and spars, as on shipboard. **3** A windlass or winch, together with ropes and hooks. **4** The instruments collectively used in any work or sport; gear; equipment: fishing *tackle.* **5** Formerly, the implements of war; weapons. **6** The act of tackling, or seizing and stopping, especially in

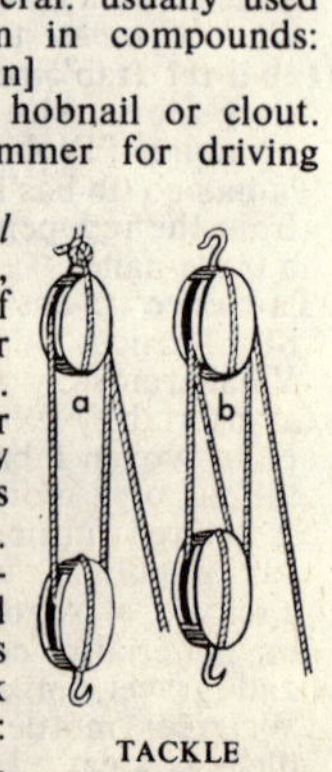

TACKLE
a. Gun.
b. Luff.

tackling

football. 7 In football, either of two linemen stationed between the guard and end: called **right** and **left tackle**. 8 A ship's rigging collectively. — v.t. **·led, ·ling** 1 To harness (a horse). 2 To deal with; undertake to master, accomplish, or solve: to *tackle* a task or a problem. 3 In football, to seize and stop (an opponent carrying the ball). [< MLG *takel* < *taken* seize] — **tack′ler** n.
tack·ling (tak′ling) n. Tackle collectively, or material for it.
tack·y[1] (tak′ē) *adj.* **tack·i·er, tack·i·est** Having adhesive properties; sticky: said especially of surfaces covered with partly dried varnish. Also **tack′ey**. [Prob. < TACK[1], v. (def. 2)]
tack·y[2] (tak′ē) *adj. U.S. Colloq.* 1 Unfashionable; plain; in bad taste; common. 2 Describing a party to which the guests come dressed in ridiculous costumes. [Cf. dial. G *taklig* untidy]
tac·ma·hack (tak′mə·hak) See TACAMAHACK.
Tac·na (täk′nä) A department of southern Peru on the Pacific; 4,920 square miles.
tac·node (tak′nōd) n. *Math.* A point of osculation. [< L *tactus*, pp. of *tangere* touch + NODE]
Ta·co·ma (tə·kō′mə) A port on Puget Sound in western Washington.
Ta·con·ic Mountains (tə·kon′ik) A range of the Appalachian system in New England and New York; highest point, 3,816 feet.
tac·o·nite (tak′ə·nīt) n. *Geol.* A variously tinted ferruginous chert enclosing the iron ores of the Mesabi district in Minnesota. [from *Tacon(ic Mountains)* + -ITE[1]]
tact (takt) n. 1 A quick or intuitive appreciation of what is fit, proper, or right; fine or ready mental discernment shown in saying or doing the proper thing, or especially in avoiding what would offend or disturb; skill or facility in dealing with men or emergencies; adroitness; cleverness; address. 2 The sense of touch; feeling; also, a touch or touching. 3 A perception or feeling, other than tactile, of the qualities of things. See synonyms under ADDRESS. [< L *tactus* a touching < *tangere* touch]
tact·ful (takt′fəl) *adj.* Possessing or manifesting tact; considerate. — **tact′ful·ly** *adv.* — **tact′ful·ness** n.
tac·tic (tak′tik) n. Tactics; also, a branch or detail of tactics. — *adj.* Tactical; of or pertaining to arrangement or tactics. [See TACTICS]
tac·ti·cal (tak′ti·kəl) *adj.* 1 Pertaining to or of the nature of tactics. 2 Exhibiting adroit maneuvering. — **tac′ti·cal·ly** *adv.*
tactical unit A military combat unit, running in size from the squad through the army group.
tac·ti·cian (tak·tish′ən) n. An expert in tactics; an adroit maneuverer.
tac·tics (tak′tiks) n. pl. 1 The science and art of military and naval evolutions; specifically, the art of handling troops in the presence of the enemy or for immediate objectives, as distinguished from *strategy*: construed as singular. 2 Any maneuvering or adroit management to effect an object. 3 Artful devices or their application. [< Gk. *taktika*, pl. of *taktikos* suitable for arranging or organizing < *tassein*, *tattein* arrange, order]
tac·tile (tak′til, -təl) *adj.* 1 Pertaining to the organs or sense of touch; caused by or consisting of contact; tactual. 2 That may be touched; tangible. [< F < L *tactilis* < *tactus* touch. See TACT.]
tac·til·i·ty (tak·til′ə·tē) n. Tangibility.
tac·tion (tak′shən) n. 1 The act of touching. 2 The state of being in contact. [< L *tactio*, *-onis* < *tactus*, pp. of *tangere* touch]
tact·less (takt′lis) *adj.* Without tact. — **tact′less·ly** *adv.* — **tact′less·ness** n.
tac·tom·e·ter (tak·tom′ə·tər) n. An esthesiometer. [< L *tactus* touch + -METER]
tac·tu·al (tak′chōō·əl) *adj.* 1 Pertaining to the sense or the organs of touch. 2 Derived from or caused by touch. [< L *tactus* touch. See TACT.] — **tac′tu·al·ly** *adv.*
Ta·cu·ba·ya (tä′kōō·bä′yä) A western section of Mexico City; site of the national astronomical observatory.
tad (tad) n. A little boy or girl; young child. [Prob. short for TADPOLE]
Tad·de·o (täd·dā′ō) Italian form of THADDEUS.
Ta·de·o (tä·thā′ō) Spanish form of THADDEUS.

tad·pole (tad′pōl) n. The aquatic larva of an amphibian, as a frog or toad, breathing by external gills and having a tail with extended membrane giving it a fishlike form. See FROG. [ME *taddepol* < *tadde* toad + *poll* head]
Ta·dzhik (tä·jēk′, -jik′) n. pl. **·dzhik** One of a people of Iranian descent inhabiting the Tadzhik S.S.R. and adjacent regions: also spelled *Tajik*.
Tadzhik Soviet Socialist Republic (tä·jēk′, -jik′) A constituent republic of the U.S.S.R. in central Asia; 55,043 square miles; capital, Stalinabad: also *Tajik S.S.R.* Also **Tad·zhik·i·stan** (tä·jē′kə·stän, -jik′ə-).
tae (tā) n. *Scot.* 1 A toe. 2 The prong of a fork.
tae·di·um vi·tae (tē′dē·əm vī′tē) *Latin* Weariness of life.
Tae·dong (tī′dông) A river of central and SW North Korea, flowing 245 miles SW to the Yellow Sea at Korea Bay near Chinnampo.
Tae·gu (tī′gōō) A city in SE South Korea. Japanese **Tai·kyu** (tī′kyōō).
Tae·jon (tī′jôn) A city of SW central South Korea. Japanese **Tai·den** (tī′den).
tael (tāl) n. 1 An Oriental weight varying from 1 to 2 1/2 ounces, commonly about 1 1/3 ounces. 2 A Chinese monetary unit of varying value. [< Pg. < Malay *tahil*]
ta′en (tān) *Scot.* Taken: a contraction.
tae·ni·a (tē′nē·ə) n. 1 In classical antiquity, a band, ribbon, or fillet for containing the hair. 2 *Archit.* A band or fillet between the Doric frieze and the architrave. 3 *Anat.* A band or strip of tissue, especially one of several ribbonlike arrangements of white substance in the brain, or one of the three longitudinal muscular bands of the colon. 4 *Zool.* A tapeworm (genus *Taenia*). Also spelled *tenia*. [< L < Gk. *tainia* fillet, tape]
tae·ni·a·cide (tē′nē·ə·sīd′), **tae·ni·a·fuge** (tē′nē·ə·fyōōj′), etc. See TENIACIDE, etc.
taf·fer·el (taf′ər·əl, -rel) n. *Naut.* 1 A taffrail. 2 Originally, the upper part of a vessel's stern. [< MDu. *tafereel* panel, picture, dim. of *tafel* table, panel < L *tabula* board]
taf·fe·ta (taf′ə·tə) n. A fine, glossy, uncorded, somewhat stiff silk fabric; a term variously applied at different times, as to certain silk-and-linen or silk-and-wool mixtures, now also to rayon. — *adj.* Made of or resembling taffeta; also, lacy; filmy; delicate. [< OF *taffetas* < Med. L *taffeta* < Persian *tāftah* < *tāftan* twist]
taff·rail (taf′rāl′) n. *Naut.* 1 The rail around a vessel's stern. 2 The upper part of a vessel's stern. [Alter. of TAFFEREL, after RAIL[1]]
taffrail log A nautical log.
taf·fy (taf′ē) n. 1 A confection made of brown sugar or molasses, mixed with butter, boiled down, and pulled into long ropes until it cools sufficiently to hold its shape: also spelled *toffee*, *toffy*. 2 *Colloq.* Flattery; blarney. [Origin unknown]
taf·i·a (taf′ē·ə) n. A spirituous liquor resembling rum, distilled in the West Indies from impure molasses or from refuse sugar. Also **taf′fi·a**. [< native name, prob. ult. < Malay *tāfīa* spirit distilled from molasses]
Ta·fi·lelt (tä·fē′lelt) The largest Saharan oasis in SE Morocco; 200 square miles. Also **Ta·fi′lalt**, **Ta·fi·let** (-let).
Taft (taft), **Lorado**, 1860–1936, U. S. sculptor. — **Robert A.**, 1899–1953, U. S. lawyer and politician; son of W. H. Taft. — **William Howard**, 1857–1930, U. S. statesman; president of the United States 1909–13.
tag[1] (tag) n. 1 Something tacked on or attached to something else; appendage. 2 A label tied or tacked on, as to a trunk; loosely, any label. 3 A loose, ragged edge of anything; tatter. 4 The tail or tip of the tail of any animal. 5 A matted and ragged lock of wool on a sheep; a loose lock of hair. 6 A worthless leaving; remnant; ort. 7 A flap or loop, as for drawing on a boot. 8 An aglet. 9 A decorative flourish, as on a signature. 10 In angling, a piece of bright material surrounding the shank of the hook in an artificial fly. 11 A lamb or yearling sheep. 12 A well-known quotation or saying, as in a song, poem, or book. 13 The refrain of a song or poem; also, the final lines of a speech in a

play; catchword; cue. 14 The crowd; rabble: often in the phrases **rag and tag** and **rag, tag, and bobtail**. — v. **tagged, tag·ging** v.t. 1 To supply, adorn, fit, mark, or label with a tag. 2 To shear away tags from (sheep). 3 To follow closely or persistently. — v.i. 4 To follow closely at one's heels: *The little boy tagged along.* [Prob. < Scand. Cf. Sw. *tagg* spike, tooth, Norw. *tagge* tooth.]
tag[2] (tag) n. A juvenile game in which the object of the players is to keep from being caught or touched by one, the tagger (usually called "it"), who chases them for that purpose. — v.t. **tagged, tag·ging** To overtake and touch, as in the game of tag. [< TAG[1]]
Ta·ga·log (tä·gä′log, tag′ə·log, -lōg) n. 1 A member of a Malay people native to the Philippines, especially Luzon. 2 One of the principal native languages and, since 1940, the official language of the Philippines, belonging to the Indonesian subfamily of the Austronesian family of languages. Also **Ta·gal** (tä·gäl′).
Ta·gan·rog (tä′gän·rôk′) A port in SW European Russian S.F.S.R., on the Gulf of Taganrog, a NE arm of the Sea of Azov.
tag day A day on which contributions are solicited for eleemosynary and other institutions: so called from the custom of giving a tag to each donor.
tagged atom *Physics* A radioisotope which betrays its presence in any part of a system into which it has been introduced, especially one used in the study, diagnosis, and treatment of disease.
tag·ger (tag′ər) n. 1 One who or that which tags or is tagged. 2 pl. Very thin tin plate.
tag·lock (tag′lok′) n. Daglock. [Var. of DAGLOCK]
Ta·gore (tə·gôr′, -gōr′, tä′gôr), **Sir Rabindranath**, 1861–1941, Hindu philosopher and poet.
ta·gua nut (tä′gwä) The ivory nut. [< native Colombian name]
Ta·gus (tā′gəs) A river in west central Spain and central Portugal, flowing 566 miles SW to the Atlantic at Lisbon: Spanish *Tajo*, Portuguese *Tejo*.
Ta·hi·ti (tä·hē′tē, tī′tē) The largest island of the Society group; 600 square miles; capital, Papeete; formerly *Otaheite*. French **Ta·i·ti** (tä·ē·tē′).
Ta·hi·ti·an (tä·hē′tē·ən, -shən) *adj.* Of or relating to Tahiti, its people, or their language. — n. 1 One of the native Polynesian people of Tahiti. 2 The Polynesian language of the Tahitians.
Ta·hoe (tä′hō, tä′-), **Lake** A lake on the boundary between California and Nevada; 195 square miles; elevation, 6,225 feet.
tah·sil·dar (tä′sēl·där′) n. An Indian officer of customs; a tax-collector. Also **tah′seel·dar**. [< Hind. *tahṣīldār* < Arabic *taḥṣīl* a collection + Persian *dār* holder]
Tai (tī) See THAI.
Tai·an (tī′än′) A city of western Shantung province, China.
Tai·chung (tī′chōong′) A city in west central Taiwan; an agricultural and industrial center.
tai·ga (tī′gə) n. The far northern coniferous forest of Siberia and by extension of Eurasia and America, extending to the northern limit of trees. [< Russian]
Tai·ho·ku (tī′hō·kōō) The Japanese name for TAIPEH.
tail[1] (tāl) n. 1 The hindmost part or rear end of an animal's body, especially when prolonged beyond the rest of the body. ◆ Collateral adjective: *caudal*. 2 Any slender, flexible, terminal prolongation of the body of a structure: the *tail* of a shirt or kite. 3 *Astron.* The luminous sheaf extending from the nucleus of a comet. 4 The hind, back, or inferior portion of anything. 5 pl. *Colloq.* The reverse side of a coin. 6 The lower end of a stream or pool. 7 Anything of tail-like appearance; a body of persons in single file; a queue; also, a retinue or suite. 8 A pigtail. 9 *Aeron.* One of several fixed horizontal or vertical surfaces of an airplane structure placed at some distance to the rear of the main bearing surfaces. 10 The rear portion of a bomb, projectile, rocket, or guided missile, usually equipped with vanes. 11

The bottom of a printed page. **12** *pl. Colloq.* A man's full-dress suit; also, a swallow-tailed coat. **13** The back end of a wagon. — *v.t.* **1** To furnish with a tail. **2** To cut off the tail of. **3** To be the tail or end of: to *tail* a procession. **4** To join (one thing) to the end of another. **5** To insert and fasten by one end, as a beam into a wall: with *in* or *on.* **6** *Colloq.* To follow secretly and stealthily; shadow. — *v.i.* **7** To extend or proceed in a line. **8** *Colloq.* To follow close behind. **9** To diminish gradually: His voice *tailed* off. **10** To be inserted and fastened at one end, as a beam. **11** *Naut.* To swing or go aground stern foremost: The ship *tailed* into the wind. — *adj.* **1** Rearmost; hindmost: the *tail* end. **2** Following; coming from behind: a *tail* wind. ◆ Homophone: *tale.* [OE *tægl*] — **tail′less** *adj.*

tail[2] (tāl) *Law adj.* Restricted; limited; abridged; restricted in succession to particular heirs: an estate *tail.* — *n.* A cutting off, abridgment, or limitation of ownership; an entail: an estate in *tail.* ◆ Homophone: *tale.* [<OF *taillié,* pp. of *taillier* cut]

Tai Lake (tī) One of China's largest lakes, on the Kiangsu-Chekiang border, between Shanghai and Nanking; 40 miles long, 30 miles wide. *Chinese* **T'ai Hu** (tī′hōō′).

tail beam Tailpiece (def. 4).

tail·board (tāl′bôrd′, -bōrd′) *n.* The hinged or vertically sliding board or gate closing the back end of a wagon body or truck body.

tail boom *Aeron.* An outrigger of an airplane to support the empennage.

tail coverts *Ornithol.* The feathers that lie at the base of the tail feathers above and below.

tail-first (tāl′fûrst′) *adv.* Backward; with the hind side foremost. Also **tail′fore′most** (-fôr′-mōst, -fōr′-).

tail·gate (tāl′gāt′) *n.* **1** One of the gates at the lower level of a canal lock. **2** A tailboard.

tail gun A gun mounted in the tail section of an airplane.

tail-heav·y (tāl′hev′ē) *adj.* Having too much weight at the rear: a *tail-heavy* airplane: opposed to *nose-heavy.*

tail·ing (tā′ling) *n.* **1** Refuse or residue from grain after milling, or from ground ore after washing: usually plural. **2** The inner, covered portion of a projecting brick or stone in a wall.

taille (tāl, *Fr.* tä′y′) *n.* In feudal France, a tax levied by a king or lord from which nobles and clergy were exempt. [<OF < *taillier* cut]

tail light A light attached to the rear of a vehicle. Also **tail lamp.**

tai·lor (tā′lər) *n.* One who makes to order or repairs men's or women's outer garments. — *v.i.* **1** To do a tailor's work. — *v.t.* **2** To fit with garments: He is well *tailored.* **3** To work at or make by tailoring: to *tailor* a coat. ◆ Collateral adjective: *sartorial.* [<OF *tailleor* < *taillier* cut <LL *taliare* split, cut, prob. <L *talea* rod]

tailor bee Any of certain leaf-cutting bees (family *Megachilidae*) that line their nests with pieces of leaves.

tai·lor·bird (tā′lər·bûrd′) *n.* A bird (genus *Sutoria*) of Asia and Africa, related to the warblers, that stitches leaves together to form a receptacle for its nest. See illustration under NEST.

tai·lor·ing (tā′lər·ing) *n.* **1** A tailor's trade or occupation. **2** The making or altering of a garment by a tailor. **3** The style and fit resulting from the work of a tailor.

tai·lor-made (tā′lər·mād′) *adj.* Made by a tailor: said especially of women's clothes of a plain, close-fitting, usually heavier type, as for walking, etc.: opposed to *ready-made.* — *n. Colloq.* A commercially prepared cigarette, as opposed to one which is rolled by hand.

tail·piece (tāl′pēs′) *n.* **1** Any endpiece or appendage. **2** In a violin or similar instrument, a piece of wood, as ebony, at the sounding-board end, having the strings fastened to it. **3** *Printing* An ornamental design on the lower blank portion of a short page. **4** A piece inserted by tailing, as a floor timber, one end of which rests on the wall, while the other is mortised into a trimmer or header.

tail·race (tāl′rās′) *n.* **1** That part of a millrace below the water wheel, bearing away the spent water. **2** *Mining* The channel for water to remove tailings.

tail·skid (tāl′skid′) *n. Aeron.* A runner fixed beneath the tail of an airplane.

tail·spin (tāl′spin′) *n.* **1** *Aeron.* The descent of an airplane along a helical path at a steep angle, either by accident or with power of recovery by manipulation of the controls. **2** *Colloq.* A sudden, sharp, emotional upheaval, often resulting in loss of control: He went into a *tailspin* over her.

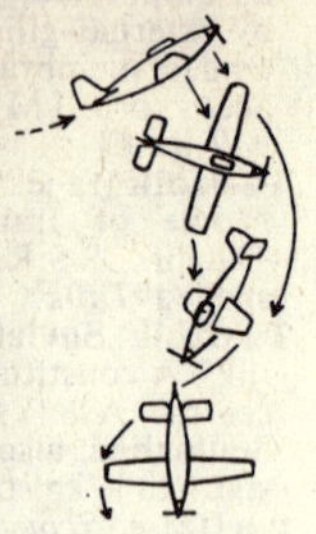

TAILSPIN

tail·stock (tāl′stok′) *n.* That standard or stock of a lathe through which passes the non-rotating spindle or dead center.

tail wind A wind blowing in the same general direction as the flight of an aircraft or course of a ship.

Tai·mir Peninsula (tī·mir′, tī′mir) A large peninsula of northern Asiatic Russian S.F.S.R., extending 700 miles NE to SW, between the Kara and Laptev seas. Also **Tai·myr′.**

tain[1] (tān) *n.* **1** Very thin plate. **2** Tinfoil suitable for backing mirrors. [Prob. aphetic var. of F *étain* tin]

tain[2] (tôn) *n.* Literally, a cattle raid; by extension, any of numerous Old Irish epics about a cattle raid. [<Irish *táin*]

Tai·nan (tī′nän′) A city of west central Taiwan.

Tai·na·ron (te′nä·rôn), **Cape** See MATAPAN, CAPE.

Tain Bo Cuail·gne (tôn bō kōōl′nye) The Cattle Raid of Cooley: title of the most famous epic in the Ulster cycle of Old Irish literature, and the oldest epic of all western Europe, embodying the life and exploits of Cuchulain and his single-handed defense of Ulster against the hosts of Connacht. [<Irish *táin* cattle raid + *bo* cow (<L *bos* ox) + *Cuailgne* Cooley, a hill district in County Louth]

Taine (tān, *Fr.* ten), **Hippolyte Adolphe,** 1828–93, French literary critic and historian.

Tai·no (tī′nō) *n. pl.* **·nos 1** A member of an extinct tribe of Indian aborigines of the West Indies, especially Haiti, probably the first encountered by Columbus. **2** The Arawakan language of this tribe.

taint (tānt) *v.t.* **1** To imbue with an offensive, noxious, or deteriorating quality or principle; infect with decay; render corrupt or poisonous. **2** To render morally corrupt or vitiated; contaminate; pollute. **3** *Obs.* To tincture; tinge. — *v.i.* **4** To be or become tainted. See synonyms under POLLUTE. — *n.* **1** A trace or germ of decay; a cause or result of corruption. **2** A moral stain or blemish; spot. [Fusion of aphetic form of ATTAINT and F *teint,* pp. of *teindre* tinge, color <L *tingere*]

Tai·peh (tī′pā′) The capital of Taiwan, in the northern part: Japanese *Taihoku.* Also **Tai′·pei′, T'ai′-pei′.**

Tai·ping (tī′ping′) *n.* An insurgent in the **Taiping Rebellion** in China (1850–64) led by one Hung-siu-tsuen, who sought to replace the Manchu dynasty with a native dynasty called the T'ai-p'ing Chao (Great Peace Dynasty): suppressed with the aid of a corps of Chinese led by Charles George Gordon. [<Chinese, great peace]

Tai·ping (tī′ping′) The capital of Perak state, Federation of Malaya, near the NW coast of the Federation; a tin-mining center.

Tai·sho (tī·shō) The title of the reign (1912–26) of Yoshihito, emperor of Japan. [<Japanese, great righteousness]

Tai·wan (tī′wän′) An island off the coast of SE China, comprising a province, and, together with the Pescadores, the National Republic of China; 13,890 square miles; capital, Taipeh; ceded to Japan, 1895–1945: formerly *Formosa.*

Tai·yü·an (tī′yü·än′) The capital of Shansi province, China.

Ta'iz (ta·iz′) **1** A province of SW Yemen. **2** The capital of Ta'iz province, residence of the Imam of Yemen and the second capital of the country.

taj (täj) *n. Persian* A diadem or crown; a headdress of distinction; specifically, a tall cap worn by Moslem dervishes.

Ta·jik (tä·jēk′, -jik′) See TADZHIK.

Ta·jik S.S.R. (tä·jēk′, -jik′) See TADZHIK S.S.R.

Taj Ma·hal (täj′ mə·häl′, täzh′) A mausoleum of white marble built (1631–45) by the emperor Shah Jehan at Agra, India, containing the tombs of his favorite wife and of himself.

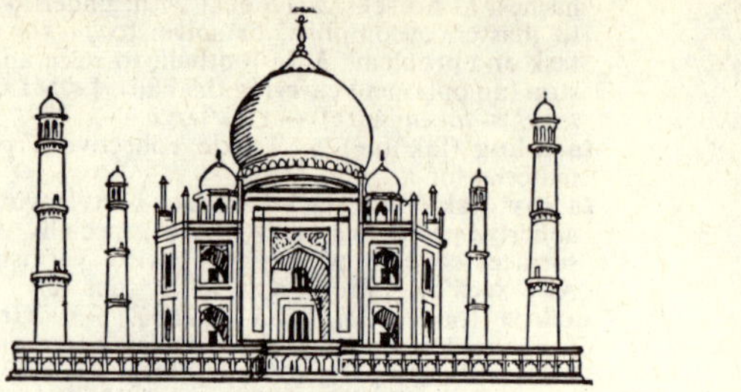

TAJ MAHAL, AGRA, INDIA.

Ta·jo (tä′hō) The Spanish name for the TAGUS.

Ta·ju·mul·co (tä′hōō·mōōl′kō) An extinct volcano in SW Guatemala; highest point in central America; 13,816 feet.

Ta·ka·mat·su (tä·kä·mät·sōō) A port in northern Shikoku, Japan.

Ta·ka·o·ka (tä·kä·ō·kä) A port of north central Honshu, Japan, a rice-trading and manufacturing center.

take (tāk) *v.* **took, tak·en, tak·ing** *v.t.* **1** To lay hold of; grasp. **2** To get possession of; seize. **3** To seize forcibly; capture; catch. **4** To catch in a trap or snare. **5** To gain in competition; win. **6** To choose; select. **7** To obtain by purchase; buy. **8** To rent or hire; lease: to *take* lodgings. **9** To receive regularly by payment; subscribe to, as a periodical. **10** To assume occupancy of: to *take* a chair. **11** To assume the responsibilities or duties of: to *take* office. **12** To bring or accept into some relation to oneself: He *took* a wife. **13** To assume as a symbol or badge: to *take* the veil. **14** To impose upon oneself; subject oneself to: to *take* a vow. **15** To remove or carry off: with *away.* **16** To remove from the proper place; misappropriate; steal. **17** To remove by death. **18** To subtract or deduct. **19** To be subjected to; undergo: to *take* a beating. **20** To submit to; accept passively: to *take* an insult. **21** To become affected with; contract: He *took* cold. **22** To affect: The fever *took* him at dawn. **23** To captivate; charm or delight: The dress *took* her fancy. **24** To conduct oneself in response to; react to: How did she *take* the news? **25** To undertake to deal with; contend with; handle: to *take* an examination. **26** To consider; deem: I *take* him for an honest man. **27** To understand; comprehend. **28** To strike in a specified place; hit: The blow *took* him on the forehead. **29** *Colloq.* To aim or direct: He *took* a shot at the target. **30** To carry with one; transport; convey: *Take* your umbrella! **31** To lead: This road *takes* you to town. **32** To escort; conduct: Who *took* her to the dance? **33** To receive into the body, as by eating, inhaling, etc.: *Take* a deep breath. **34** To accept, as something offered, due, or given; have conferred on one: to *take* a bribe; to *take* a degree. **35** To let in; admit: The ship is *taking* water; The car will *take* only six people. **36** To indulge oneself in; enjoy: to *take* a nap. **37** To perform, as an action: to *take* a stride. **38** To avail oneself of (an opportunity, etc.). **39** To put into effect; adopt: to *take* measures; to *take* advice. **40** To use up or consume; require as necessary; demand: The piano *takes* too much space; That *takes* a lot of nerve. **41** To make use of; apply: They *took* clubs to him; to *take* pains. **42** To travel by means of: to *take* a train to Boston. **43** To go to; seek: to *take* cover. **44** To ascertain or obtain by measuring, computing, etc.: to *take* a census. **45** To obtain or derive from some source; adopt or copy. **46** To obtain by writing; write down or copy: to *take* notes. **47** To obtain a likeness or representation of, as by drawing or photographing; also, to obtain (a likeness, picture, etc.) in such a manner. **48** To experience; feel: to *take* pride in an achievement. **49** To conceive or feel: She *took* a dislike to him. **50** To become impregnated with; absorb: The cloth will not *take* the pattern. **51** *Slang* To cheat; deceive. **52** *Gram.* To require by construction or usage: The verb

takes a direct object. —*v.i.* **53** To get possession. **54** To engage; catch, as mechanical parts. **55** To begin to grow; germinate. **56** To have the intended effect: The vaccination *took.* **57** To become popular; gain favor or currency, as a play. **58** To admit of being photographed: His face *takes* well. **59** To detract: with *from.* **60** To become (ill or sick). **61** To make one's way; go. See synonyms under ABSTRACT, ASSUME, CARRY, CATCH. —**to take after** **1** To resemble. **2** To follow as an example. —**to take amiss** To be offended by. —**to take at one's word** To believe. —**to take back** **1** To regain. **2** To retract. —**to take breath** To pause, as from working. —**to take down** **1** To pull down, as a building. **2** To dismantle; disassemble. **3** To humble. **4** To write down; make a record of. —**to take heart** To gain courage or confidence. —**to take in** **1** To admit; receive. **2** To lessen in size or scope. **3** To furl or brail (sail). **4** To include; embrace. **5** To understand. **6** To cheat or deceive. **7** To visit as part of a tour: Did you *take in* the Louvre? **8** To receive into one's home for pay, as lodgers or work. —**to take in vain** To use profanely or blasphemously, as the name of a deity. —**to take it** **1** To assume; understand. **2** To endure hardship, abuse, etc. —**to take off** **1** To remove, as a coat. **2** To carry away. **3** To kill. **4** To deduct. **5** To mimic; burlesque. **6** To rise from the ground or water in starting a flight, as an airplane. **7** To leave; depart. —**to take on** **1** To hire; employ. **2** To undertake to deal with; handle. **3** *Colloq.* To exhibit violent emotion. —**to take out** **1** To extract; remove. **2** To obtain from the proper authority, as a license or patent. **3** To lead or escort. —**to take over** **1** To assume control of. **2** To convey. —**to take place** To happen. —**to take stock** **1** To make an inventory. **2** To estimate probability, position, etc.; consider. —**to take the field** To begin a campaign or game. —**to take to** **1** To betake oneself to: to *take to* one's bed. **2** To develop the practice of, or an addiction to: He *took to* drink. **3** To become fond of; be attracted by. —**to take to heart** To be deeply affected by. —**to take up** **1** To raise or lift. **2** To make smaller or less; shorten or tighten. **3** To pay, as a note or mortgage. **4** To accept as stipulated: to *take up* an option. **5** To begin or begin again; resume. **6** To reprove or criticize. **7** To occupy, engage, or consume, as space or time. **8** To acquire an interest in or devotion to: to *take up* a cause. —**to take up with** *Colloq.* To become friendly with; associate with. —*n.* **1** The act of taking, or that which is taken. **2** An uninterrupted run of the camera and sound apparatus in recording any portion of a motion picture. **3** *Slang* The money collected; receipts. **4** The quantity collected at one time: the *take* of fish. [OE *tacan* <ON *taka*]

take-down (tāk′doun′) *adj.* Fitted for being taken apart or down easily: a *take-down* shack; a *take-down* rifle. —*n.* **1** Any article so constructed as to be taken apart easily. **2** The part of a take-down mechanism by means of which it is taken apart or down. **3** *Colloq.* The act of humiliating any one; humiliation.

take-home pay (tāk′hōm′) The remainder of one's wages or salary after tax and other payroll deductions.

take-in (tāk′in′) *n. Colloq.* An act of cheating or hoaxing.

take-off (tāk′ôf′, -of′) *n.* **1** *Colloq.* A satirical representation; caricature. **2** In horsemanship and athletics, the spot at which the feet leave the ground in leaping. **3** *Aeron.* The act of rising from and leaving the ground or water in an aircraft flight.

tak·er (tā′kər) *n.* One who takes; specifically, one who accepts a wager; also, a collector: a ticket *taker.*

take-up (tāk′up′) *n.* **1** *Mech.* A device for taking up lost motion or drawing in the slack of a thing, as in a loom. **2** The act of tightening or taking up.

tak·ing (tā′king) *adj.* **1** Fascinating; captivating. **2** *Colloq.* Contagious; infectious. —*n.* **1** The act of one who takes. **2** The thing or things taken; in fishing, a catch; haul; in the plural, receipts, as of money. **3** *Obs.* Agitation; perplexity; distress. —**tak′ing·ly** *adv.* —**tak′ing·ness** *n.*

Ta·kla·ma·kan (tä′klä′mä′kän′) A desert in the SW third of the Sinkiang-Uigur Autonomous Region, NW China. Also **Ta′kla′ Ma′kan′.**

Ta·ku (tä′kōō′) A port on the Gulf of Chihli, eastern Hopeh province, NE China.

tal·a·poin (tal′ə·poin) *n.* **1** A Buddhist priest or monk. **2** A West African monkey (*Cercopithecus talapoin*) of the guenon group, smallest of the Old World monkeys. [<Pg. *talapões*, pl. of *talapão* <Burmese *tala poi* our master]

ta·lar (tā′lər) *n. Archaic* A cloak or robe reaching to the ankles. [<L *talaris* of the ankles <*talus* ankle]

ta·lar·i·a (tə·lâr′ē·ə) *n. pl. Latin* Winged boots or sandals, or wings springing directly from the ankles: used in antique art as attributes of Mercury, Perseus, etc.

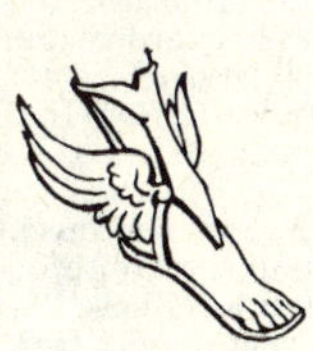

TALARIA

Ta·laud Islands (tä′lout) An Indonesian island group NE of Celebes; 495 square miles. Also **Ta·laur Islands** (tä′·lour).

Ta·la·ve·ra de la Rei·na (tä′lä·vā′rä thā lä rā′nä) A city of central Spain on the Tagus river; scene of a battle between Wellington and Joseph Bonaparte, 1809.

tal·bot (tôl′bət, tal′-) *n.* A sleuthhound, supposed to be related to the bloodhound. [after *Talbot,* English family name]

talc (talk) *n.* A soft, hydrous magnesium silicate, $H_2Mg_3(SiO_3)_4$, used in making paper, soap, toilet powder, lubricants, etc. Soapstone and French chalk are varieties of talc. —*v.t.* **talcked** or **talced, talck·ing** or **talc·ing** To treat with talc: to *talc* a photographic plate. [<F <Med. L *talcum* <Arabian *talq* <Persian *talk*]

Tal·ca (täl′kä) A city of central Chile; birthplace of Chile's independence, 1818; destroyed by earthquake, 1928.

talc·ose (tal′kōs) *adj.* Composed of or containing talc. Also **talc′ous** (tal′kəs).

tal·cum (tal′kəm) *n.* Talc. [<Med. L. See TALC.]

talcum powder Finely powdered and purified talc, used as a dusting agent, filter, and for the relief of chafed skin and prickly heat.

tale (tāl) *n.* **1** That which is told or related; a story; recital. **2** Hence, a connected narrative or account, whether oral or written, of an actual, legendary, or fictitious event or series of events. **3** An idle or malicious report; a piece of gossip. **4** A deliberately untrue story; a lie; falsehood. **5** *Archaic* A counting or enumeration; reckoning; numbering. **6** *Archaic* That which is counted; an amount; total; sum. **7** *Obs.* Speech; talk; also, the language of a country. ◆ Homophone: *tail.* [OE *talu* speech, narrative. Akin to TELL, TALK.]

tale·bear·er (tāl′bâr′ər) *n.* One who tells mischievous tales about other persons. —**tale′·bear′ing** *adj.* & *n.*

tal·ent (tal′ənt) *n.* **1** Mental endowments or capacities of a superior character; marked mental ability; also, mental ability in general. **2** A particular and uncommon aptitude for some special work or activity; a faculty or gift: a usage founded on a Scriptural parable (*Matt.* xxv 14–30), mental power being considered as a divine trust. **3** People of skill or ability, collectively: the *talent* of stage, screen, and radio. **4** *U.S. Slang* In horse-racing circles, those who make bets or take odds on their individual judgment and responsibility: distinguished from the bookmakers. **5** An ancient weight and denomination of money, varying in weight and value among different nations and in different periods. **6** *Obs.* Inclination; disposition. See synonyms under ABILITY, GENIUS. [OE *talente* appetite, will, inclination <L *talentum,* a sum of money <Gk. *talanton* weight, thing weighed]

tal·ent·ed (tal′ən·tid) *adj.* Having mental ability; gifted. See synonyms under CLEVER.

talent scout One whose business it is to discover talented or exceptionally gifted people, especially those suitable for dramatic or motion-picture careers.

ta·ler (tä′lər) *n.* A former German silver coin, the prototype of all dollars, issued in Bohemia and first dated 1518; a dollar: also spelled *thaler.* [<G. See DOLLAR.]

ta·les (tā′lēz) *n. pl.* **·les** (-lēz) *Law* **1** Persons to be summoned for jury duty to make up a deficiency when the regular panel is exhausted by challenges. **2** The writ for summoning such persons. [<L *tales (de circumstantibus)* such (of the bystanders), pl. of *talis* such a one; the phrase is from the writ summoning them]

tales·man (tālz′mən) *n. pl.* **·men** (-mən) One summoned to make up a jury when the regular panel is exhausted. [<TALES + MAN]

tale·tel·ler (tāl′tel′ər) *n.* **1** One who tells stories, etc.; a raconteur. **2** A talebearer. —**tale′tell′ing** *adj.* & *n.*

Ta·lien (dä′lyen′) The Chinese name for DAIREN.

tal·i·grade (tal′ə·grād) *adj. Zool.* Walking on the outer surface of the foot. [<L *talus* ankle + -GRADE]

tal·i·on (tal′ē·ən) *n.* Retaliation, as a form of justice. [<F <L *talio, -onis* <*talis* such]

tal·i·ped (tal′ə·ped) *adj.* Suffering from or afflicted with talipes; clubfooted. —*n.* A clubfooted person. [See TALIPES]

tal·i·pes (tal′ə·pēz) *n. Pathol.* **1** Malformation of the foot. **2** A clubfoot. [<NL <L *talus* ankle + *pes, pedis* foot]

tal·i·pom·a·nus (tal′ə·pom′ə·nəs) *n. Pathol.* Clubhand. [<*talipo-* (<TALIPES) + L *manus* a hand]

tal·i·pot (tal′ə·pot) *n.* A stately and valuable East Indian palm (*Corypha umbraculifera*) crowned by large leaves often used as fans, umbrellas, and as a house covering. Also **talipot palm.** [<Bengali *tālipāt* palm leaf <Skt. *tālī* fan palm + *pattra* leaf]

tal·is·man (tal′is·mən, -iz-) *n. pl.* **·mans** **1** Something supposed to produce or capable of producing extraordinary effects; a charm. **2** An astrological charm or symbol supposed to benefit or protect the possessor, especially by exerting magical or occult influence; in a wider sense, any amulet. [<F <Sp. <Arabic *ṭilsam, ṭilasm* magic figure <LGk. *telesma* a sacred rite <Gk. *teleein* initiate <*telos* end, completion]

Synonyms: amulet, charm. An *amulet* or *talisman* is strictly a material object; a *charm* may be a movement or a form of words. An *amulet* is ordinarily worn upon the person as a protection against disease, injury, or death. A *talisman* is any object supposed to work wonders, like Aladdin's lamp, whether kept in one's possession or not.

tal·is·man·ic (tal′is·man′ik) *adj.* Exerting magical or occult power. Also **tal′is·man′i·cal.**

talk (tôk) *v.i.* **1** To express or exchange thoughts in audible words; communicate by speech; speak or converse. **2** To communicate by means other than speech: to *talk* with one's fingers. **3** To speak irrelevantly; prate; chatter. **4** To confer; consult. **5** To gossip. **6** To make sounds suggestive of speech. **7** *Colloq.* To give information; inform. —*v.t.* **8** To express in words; utter. **9** To use in speaking; converse in: to *talk* Spanish. **10** To converse about; discuss: to *talk* business. **11** To bring to a specified condition or state by talking: to *talk* one into doing something. **12** To pass or spend, as time, in talking. —**to talk back** To answer impudently. —**to talk big** *Slang* To brag; boast. —**to talk down** To silence by talking; outtalk. —**to talk down to** To speak to (an audience of lower or supposedly lower intelligence than one's own) in simple, obvious words; speak to patronizingly. —**to talk shop** To talk about one's work. —**to talk up** **1** To discuss, especially so as to promote; praise; extol. **2** *Colloq.* To speak loudly or boldly. —*n.* **1** The act of talking; conversation; speech, especially when informal. **2** Report, rumor: We heard *talk* of war. **3** That which is talked about; a topic; theme; subject of conversation. **4** A conference for discussion or deliberation; a council. **5** Mere words; verbiage. **6** A language, dialect, or lingo; an argot:

talk·a·tive baseball *talk*. See synonyms under CONVERSATION. [ME *talken*, prob. freq. of *talen*, OE *talian* reckon, speak. Related to TELL, TALE.]
— *Synonyms (verb)*: chat, chatter, converse, discourse, speak. To *talk* is to utter a succession of connected words, ordinarily with the expectation of being listened to. To *speak* is to give articulate utterance even to a single word; the officer *speaks* the word of command, but does not *talk* it. To *chat* is ordinarily to utter in a familiar, conversational way; to *chatter* is to *talk* in an empty, ceaseless way like a magpie. See SPEAK.
talk·a·tive (tô′kə·tiv) *adj.* Given to much talking. See synonyms under GARRULOUS. — **talk′a·tive·ly** *adv.* — **talk′a·tive·ness** *n.*
talk·er (tô′kər) *n.* One who talks; also, a loquacious person.
talk·ie (tô′kē) *n. Colloq.* A motion picture with spoken words and sound effects. Also **talking picture**.
talk·ing (tô′king) *adj.* 1 Having power of speech or of imitating speech. 2 Talkative. — *n.* Conversation; talk.
talking machine A phonograph.
talk·ing-to (tô′king·tōō′) *n. pl.* **·tos** *Colloq.* A scolding; berating.
talk·y (tô′kē) *adj.* **talk·i·er, talk·i·est** Talkative.
tall (tôl) *adj.* 1 Having more than average height; high or lofty: a *tall* building. 2 Having specified height: He is five feet *tall*. 3 *Colloq.* Inordinate; extravagant; boastful: *tall* talk; also, unbelievable; remarkable: a *tall story*. 4 *Colloq.* Large; excellent; grand: a *tall dinner*. 5 *Obs.* Handsome; fine; proud. 6 *Obs.* Brave; sturdy; spirited. — *adv. Colloq.* Proudly; handsomely: He walks *tall*. [OE *getæl* swift, prompt] — **tall′ness** *n.*
tal·lage (tal′ij) *n.* In old English law, any form of assessment or taxation for raising revenue, including subsidies and customs. — *v.t.* **·laged, ·lag·ing** To tax. [< OF *taillage* < *taille* a tax, a cutting < *taillier* cut. See TAILOR.]
Tal·la·has·see (tal′ə·has′ē) The capital of Florida, in the northern part.
tall·boy (tôl′boi′) *n.* 1 *Brit.* A highboy. 2 A variety of chimney pot.
Tal·ley·rand-Pé·ri·gord (tal′ē·rand·pā′ri·gôr, *Fr.* tȧ·le·rän′pā·rē·gôr′), **Charles Maurice de**, 1754–1838, French statesman.
Tal·linn (tȧl′lin) The capital of Estonia, a port in the NW part, on the Gulf of Finland: German *Reval*, Russian *Revel*. Also **Tal′lin**.
tall·ish (tô′lish) *adj.* Rather tall.
tal·lith (tal′ith, täl′is) *n.* A fringed mantle of fine linen, originally covering the head and falling over the shoulders, now worn around the shoulders, sometimes as a scarf, by Jews engaged in prayer. [< Hebrew *tallīth* cover, sheet, robe]
tall oil A fatty resinous liquid obtained as a by-product from wood pulp: it is used as an emulsifying agent in various manufacturing processes. Also **tal·lol** (tal′ôl). [< Sw. *tallöl* pine oil]
tal·low (tal′ō) *n.* 1 A mixture of the harder animal fats, as of beef or mutton, refined for use in candles, soaps, oleomargarine, etc. 2 A vegetable fat obtained from the bayberry. — *v.t.* 1 To smear with tallow. 2 To fatten. [ME *talgh*, prob. < MLG *talg, talch*] — **tal′low·y** *adj.*
tal·low-chan·dler (tal′ō·chand′lər) *n.* A maker or vender of tallow candles.
tal·ly (tal′ē) *n. pl.* **·lies** 1 A piece of wood on which notches or scores are cut as marks of number. 2 A score or mark; hence, a reckoning; account. 3 A counterpart; duplicate. 4 A mark indicative of tale or number: used to denote one in a series. 5 A label; tag. — *v.* **·lied, ·ly·ing** 1 To score on a tally; mark; record. 2 To reckon; count; estimate: often with *up*. 3 To mark or cut corresponding notches in; cause to correspond. — *v.i.* 4 To correspond; agree precisely; fit: His story tallies with yours. 5 To keep score. [< AF *tallie* < L *talea* cutting] — **tal′li·er** *n.*
tal·ly-ho (tal′ē·hō′) *interj.* A huntsman's cry to hounds when the quarry is sighted. — *n.* 1 The cry of "tallyho." 2 A four-in-hand coach. — *v.t.* To urge on, as hounds, with the cry of "tallyho." — *v.i.* To cry "tallyho." [Alter. of F *taïaut*, a hunting cry]
tal·ly·man (tal′ē·mən) *n. pl.* **·men** (-mən) 1 One who keeps a count or a tally, especially of votes. 2 One who keeps a record of number, volume, and measurement, as of timber.
Tal·mi gold (tal′mē) Gold shell. [< G, orig. a trade name]
Tal·mud (tal′mud, täl′mŏŏd) *n.* The body of Jewish civil and religious law (and related commentaries and discussion) not comprised in the Pentateuch, commonly including the Mishna and the Gemara, but sometimes limited to the latter. [< Hebrew *talmūdh* instruction < *lāmadh* learn] — **Tal·mud′ic** or **·i·cal** *adj.* — **Tal′mud·ist** *n.*
tal·on (tal′ən) *n.* 1 The claw of a bird or other animal, especially of a bird of prey: often applied figuratively, as to a grasping human hand. 2 A projection on the bolt of a lock on which the key presses in shooting the bolt. 3 In card games, the part of a pack left on the table after the deal; the stock. 4 The heel of a sword blade. [< OF, spur < L *talus* heel] — **tal′oned** *adj.*
Ta·los (tā′los) In Greek mythology: 1 A giant man of brass presented by Zeus to Minos, king of Crete, who used him as a watchman. 2 A Greek inventor killed by his uncle, Daedalus, because of jealousy. Also **Ta′lus** (-ləs).
ta·luk (tȧ·lŏŏk′) *n.* In parts of India, a government district from which a revenue is derived; also, a tract of proprietary land; an estate. [< Arabic *ta'alluq* estate]
ta·lus (tā′ləs) *n. pl.* **·li** (-lī) 1 *Anat.* The astragalus. 2 A slope, as of a tapering wall. 3 *Geol.* The sloping mass of rock fragments below a cliff. Compare SCREE. 4 The slope given to the face of an earthwork or other fortification. [< L, ankle, heel]
tam (tam) *n.* A tam-o′-shanter.
tam·a·ble (tā′mə·bəl) *adj.* Capable of being tamed. Also **tame′a·ble**.
ta·ma·le (tə·mä′lē) *n.* A Mexican dish made of crushed Indian corn and meat, seasoned with red pepper, wrapped in corn husks, dipped in oil, and cooked by steam. Also **ta·mal** (tə·mäl′). [< Am. Sp. *tamales*, pl. of *tamal* < Nahuatl *tamalli*]
Tam·al·pais (tam′al·pī′əs), **Mount** A peak in western California, across the Golden Gate from San Francisco; 2,604 feet.
ta·man·dua (tə·man′dwə, tä′män·dwä′) *n.* A small arboreal ant-eater (*Tamandua tetradactyla*) of Central and South America. Also **ta·man·du** (tam′ən·dōō). [< Pg. < Tupian < *taixi* ant + *mondé* catch]
tam·a·rack (tam′ə·rak) *n.* 1 The American larch (*Larix laricina*) common all over northern North America. 2 Its wood: also called *hackmatack*. 3 The lodgepole pine of the Pacific coast. [< Algonquian]
ta·ma·rau (tä′mä·rou′) *n.* A small, dark-brown, short-horned buffalo (genus *Anoa*) of the island of Mindoro, standing about 40 inches high. Also spelled *timarau*. [< native name]
tam·a·rin (tam′ə·rin) *n.* One of various squirrel-like marmosets of Guiana and the Amazon valley; especially, the **silky tamarin** (*Leontocebus rosalia*). [< F < native Cariban name]
tam·a·rind (tam′ə·rind) *n.* 1 A tropical tree (*Tamarindus indica*) of the bean family, with hard yellow wood, pinnate leaves, and showy yellow flowers striped with red. 2 The fruit of this tree, a flat pod with soft acid pulp used in preserves and as a laxative drink. [< Sp. *tamarindo* < Arabic *tamr hindi* Indian date]
tam·a·risk (tam′ə·risk) *n.* An evergreen shrub (genus *Tamarix*) of the Mediterranean region, western Asia, and India, with slender branches bearing small, pinkish-white flowers in racemes. [< LL *tamariscus*, var. of L *tamarix* a tamarisk]
ta·ma·sha (tə·mä′shə) *n.* In India, any form of public procession, display, or entertainment; a show. [< Arabic *tamâsha* sightseeing, walking around]
Ta·ma·tave (tä′mä·täv′) A port of Madagascar, on the Indian Ocean, chief port of Malagasy Republic.
Ta·mau·li·pas (tä′mä·ōō·lē′päs) A state of NE Mexico, bordering on the United States and the Gulf of Mexico; 30,731 square miles; capital, Ciudad Victoria.
Ta·ma·yo (tä·mä′yō), **Rufino**, born 1899, Mexican painter.
tam·bac (tam′bak) See TOMBAC.
Tam·bo·ra (täm′bō·rä) A volcano on northern Sumbawa; 9,255 feet.
tam·bour (tam′boor) *n.* 1 A drum. 2 A light wooden frame, usually circular, on which material for embroidery may be stretched; also, a fabric embroidered on such a frame. 3 A palisade for defending an entrance to a fortified work. — *v.t. & v.i.* To embroider on a tambour. [< F < Arabic *ṭambūr* a stringed instrument; prob. infl. in meaning by OF *tabour* a tabor]

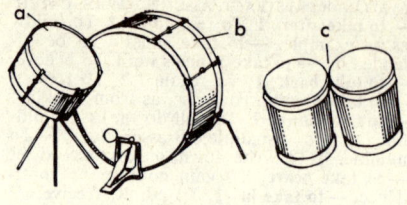

TAMBOURS
a. Snare drum. *b.* Bass drum. *c.* Bongo drums.

tam·bou·rin (tam′bə·rin) *n.* 1 A long, narrow, oblong drum, originating in Provence. 2 A gay, 18th century Provençal dance, or the music accompanying it. [< F, dim. of *tambour*]
tam·bou·rine (tam′bə·rēn′) *n.* A musical instrument like the head of a drum, with jingles in the rim, played by striking it with the hand; a timbrel. [< F]
Tam·bov (täm·bôf′) A city in south central European Russian S.F.S.R.
tame (tām) *adj.* **tam·er, tam·est** 1 Having lost its native wildness or shyness; domesticated. 2 In agriculture, brought under or produced by cultivation: *tame* hay or land. 3 Docile; tractable; hence, subdued or subjugated; spiritless; also, gentle; harmless. 4 Lacking in effectiveness; uninteresting; dull; insipid. See synonyms under DOCILE, FLAT, MEAGER. — *v.t.* **tamed, tam·ing** 1 To make tame; domesticate. 2 To bring into subjection or obedience; conquer or take the spirit or heart from; render spiritless. 3 To tone down; soften, as glaring colors. See synonyms under RECLAIM. [OE *tam*] — **tame′ly** *adv.* — **tame′ness** *n.* — **tam′er** *n.*
ta·mein (tä·mīn′) *n.* A draped garment, similar to an Indian sari, worn by Burmese women. [< Burmese *thamein*]
tame·less (tām′lis) *adj.* Untamable. — **tame′less·ness** *n.*
Tam·er·lane (tam′ər·lān), 1336?–1405, Tatar conqueror of Asia; also called *Timour, Timur*. Also **Tam·bur·laine** (tam′bər·lān).
Tam·il (tam′əl, tum′əl) *n.* 1 One of an ancient Dravidian people, and still the most numerous of the inhabitants of southern India and northern Ceylon. 2 Their language, the oldest and most widely used of the Dravidian languages.
tam·is (tam′is) *n.* 1 A strainer of cloth or gauze. 2 A fabric used for straining. Also **tam′my**. [< F, sieve]
Tam·ma·ny (tam′ə·nē) *n.* A fraternal society in New York City (founded 1789) serving as the central organization of the city's Democratic party: more commonly **Tammany Hall**, from its meeting place. The name has often been associated with political bossism. Also called **Tammany Society**. [Alter. of *Tamanend*, lit., the affable, name of a 17th c. Delaware Indian chief noted for his friendliness toward white men]
Tam·mer·fors (täm′mər·fôrs′) The Swedish name for TAMPERE.
Tam·mer·kos·ki (täm′mər·kôs′kē) The Finnish name for TAMPERE.
Tam·muz (täm′mŏŏz, *in Biblical usage* tam′uz) 1 In Babylonian mythology, the husband of Ishtar and god of agriculture, whose annual death and resurrection symbolize the cycle of months. 2 A Hebrew month. See CALENDAR (Hebrew). Also spelled *Thammuz*. Also **Tam·uz** (tam′uz). [< Hebrew]
tam-o′-shan·ter (tam′ə·shan′tər) *n.* A Scottish cap with a tight headband and a full, flat top, sometimes with a pompon or tassel. [after TAM O′ SHANTER]
Tam o′ Shan·ter (tam′ ə shan′tər) In Robert Burns's poem *Tam o′ Shanter*, the hero, a drunken farmer, who fancies himself pursued by witches.
tamp (tamp) *v.t.* 1 To force down or pack closer by firm, repeated blows. 2 To ram

Tampa — down, as a packing on a charge in a blasthole, in order to increase the explosive effect. — *n.* A tamper. [Back formation <TAMPION]

Tam·pa (tam′pə) A port of entry on **Tampa Bay,** an arm of the Gulf of Mexico in central western Florida.

tam·pa·la (tam′pə·lə) *n.* A horticultural variety of an Asian plant (*Amaranthus tricolor*), cultivated in the United States and esteemed for its edible, spinachlike leaves. [<Hind.]

tam·pan (tam′pan) *n.* A soft-bodied tick (genus *Argas*) of cosmopolitan distribution, a dangerous bloodsucking parasite of poultry whose bite is often injurious to men. Also called *miana bug*. [<native S. African name]

tam·per[1] (tam′pər) *v.i.* 1 To meddle; interfere: usually with *with*. 2 To make changes, especially so as to damage, corrupt, etc.: with *with*: to *tamper* with a manuscript. 3 To use corrupt measures, as bribery; scheme or plot. [Var. of TEMPER] — **tam′per·er.**

tamp·er[2] (tam′pər) *n.* 1 One who tamps. 2 An instrument for tamping. 3 *Physics* A reflector (def. 4).

Tam·pe·re (täm′pe·re) A city in SW Finland: Swedish *Tammerfors,* Finnish *Tammerkoski.*

Tam·pi·co (tam·pē′kō, *Sp.* täm·pē′kō) A port on the Gulf of Mexico in SE Tamaulipas state, NE Mexico.

Tampico fiber Istle.

tam·pi·on (tam′pē·ən) *n.* A tompion. [<F *tampon,* nasal var. of *tapon, tape* a bung < Gmc.]

tam·pon (tam′pon) *n. Med.* A plug of cotton or lint for insertion in a wound or body cavity. — *v.t.* To plug up, as a wound, with a tampon. [See TAMPION.]

tam-tam (tum′tum′) *n.* 1 A type of drum, used in the East Indies and western Africa. See TOM-TOM. 2 A Chinese gong. — *v.i.* To play on a tam-tam. [<Hind.; imit. in origin]

tan (tan) *v.* **tanned, tan·ning** *v.t.* 1 To convert into leather, as hides or skins, by treatment with an infusion of tannin obtained from the bark of the oak, hemlock, etc. 2 To make durable or hard, as fishnets or sails. 3 To bronze, as the skin, by exposure to sunlight. 4 *Colloq.* To thrash; flog. — *v.i.* 5 To become tanned, as hides or the skin. — *n.* 1 *Chem.* a Tanbark. b Tannin. 2 A yellowish-brown color tinged with red. 3 A dark or brown coloring of the skin, resulting from exposure to the sun: a coat of *tan.* — *adj.* 1 Of a yellowish- or reddish-brown; tan-colored. 2 Used in or pertaining to tanning. [OE *tannian* <Med. L *tannare* < *tanum* tanbark, prob. <Celtic. Cf. Breton *tann* oak.]

Ta·na (tä′nä) A river in SE Kenya, Africa, flowing about 500 miles east and south to the Indian Ocean.

Ta·na (tä′nä), **Lake** A lake in northern Ethiopia, source of the Blue Nile; about 1,400 square miles: also *Tsana.*

tan·a·ger (tan′ə·jər) *n.* Any of a family (*Thraupidae*) of arboreal oscine American birds related to the finches and noted for the brilliant plumage of the male. Most of the species are tropical, but a few migrate to the United States, especially the **scarlet tanager** (*Piranga erythromelas*) and the **western tanager** (*P. ludoviciana*). [<NL *tanagra* <Pg. *tangara* <Tupian] — **tan′a·grine** (-grēn) *adj.*

Tan·a·gra (tan′ə·grə, tə·nag′rə) A village in eastern Boeotia, east central Greece; known for the terra-cotta figurines excavated there.

Tan·a·is (tan′ə·is) An ancient name for the DON (def. 1).

Ta·na·na·rive (tä·nä′nä·rēv′) The capital of Malagasy Republic, in the central part: English *Antananarivo.* Also **Ta·na′na·ri′vo** (-rē′vō).

Ta·na·na River (tä′nä·nä′) A river of western Yukon and central and eastern Alaska, flowing 600 miles from near the Alaskan border to the Yukon River in central Alaska.

tan·bark (tan′bärk′) *n.* 1 The bark of certain trees, especially oak or hemlock, containing tannin in quantity, and used in tanning leather. 2 Spent bark from the tan vats, used on circus arenas, racetracks, etc.

Tan·cred (tang′krid), 1078?–1112, Norman hero of the first crusade.

tan·dem (tan′dəm) *adv.* One in front of the another: said of two or more persons or things so arranged, and of horses harnessed in single file instead of abreast. — *n.* 1 Two or more horses harnessed and driven in single file; also, such a turnout, including both horses and vehicle. 2 A bicycle with seats for two persons, one behind the other: also **tandem bicycle.** 3 Any arrangement of two or more persons or things placed one before another. — *adj.* Consisting of or being two arranged one before another. [<L, at length (of time); used in puns in sense of "lengthwise"]

Tan·djung·pi·nang (tän·jōōng′pē·näng′) A seaport SE of Singapore, capital of the Riouw Archipelago province of Indonesia.

Tan·djung·pri·ok (tän′jōōng·prē′ōk) See TANJUNGPRIOK.

Ta·ney (tä′nē), **Roger Brooke,** 1777–1864, U.S. jurist; chief justice of Supreme Court 1836–64.

tang[1] (tang) *n.* 1 A slender shank or tongue projecting from some metal part, as the end of a sword blade or of a chisel, for inserting into or fixing upon a handle, hilt, etc.; also, a tonguelike part, as of a belt buckle. 2 A penetrating taste, flavor, or odor, sometimes a disagreeable one; also, a trace; hint: a *tang* of pepper. 3 Any distinct quality, other than one that is sweet. — *v.t.* To provide with a tang. [<ON *tongi* a point, dagger]

tang[2] (tang) See TWANG.

Tang (täng) A Chinese dynasty, 618–907, under which China enjoyed its greatest period of literature and art.

Tan·gan·yi·ka (tan′gən·yē′kə, tang′-) A United Nations Trust Territory in eastern Africa, administered by Great Britain; 362,688 square miles; capital, Dar-es-Salaam; formerly part of German East Africa.

Tan·gan·yi·ka (tan′gən·yē′kə, tang′-), **Lake** A lake in the Great Rift Valley of east central Africa, SW of Victoria Nyanza; 12,700 square miles; 400 miles long; the longest and deepest (4,700 feet) lake in Africa, second deepest fresh-water body in the world.

tan·ge·lo (tan′jə·lō) *n. pl.* **·los** 1 A loose-skinned orangelike fruit, a hybrid of the tangerine and the pomelo. 2 The tree (genus *Citrus*) on which it grows. [<TANG(ERINE) + (POM)ELO]

tan·gen·cy (tan′jən·sē) *n. pl.* **·cies** The state of being tangent. Also **tan′gence.**

tan·gent (tan′jənt) *adj.* 1 *Geom.* Meeting at a point or along a line without further coincidence or intersection: said of either or both of two lines or surfaces so touching. 2 Touching; in contact. — *n.* 1 *Geom.* **a** A line tangent to a curve at any point. **b** The straight line through two coincident points of a curve. **c** The length of a tangent line from the point of contact to the axis of abscissas. 2 *Trig.* One of the functions of an angle; the quotient of the ordinate divided by the abscissa. 3 A sharp change in course or direction. [<L *tangens, -entis,* ppr. of *tangere* touch]

tan·gen·tial (tan·jen′shəl) *adj.* 1 Of, pertaining to, or moving in the direction of a tangent. 2 Touching slightly. 3 Divergent. Also **tan·gen′tal** (-jen′təl). — **tan·gen′ti·al′i·ty** (-shē·al′ə·tē) *n.* — **tan·gen′tial·ly** *adv.*

tan·ger·ine (tan′jə·rēn′) *n.* 1 A small, juicy orange with a loose, easily removed skin; a variety of mandarin (def. 2). 2 A slightly burnt-orange color, like the color of the tangerine. [from *Tangier*]

Tan·ger·ine (tan′jə·rēn′) *adj.* Of or pertaining to Tangier, Morocco. — *n.* A native or inhabitant of Tangier.

tan·gi·ble (tan′jə·bəl) *adj.* 1 Perceptible by touch; also, within reach by touch. 2 Figuratively, capable of being apprehended by the mind; of definite shape; not elusive or unreal: *tangible* evidence. 3 *Law* Perceptible to the senses; corporeal; material: *tangible* property. See synonyms under EVIDENT, PHYSICAL. — *n.* 1 That which is tangible. 2 *pl.* Material assets. [<F <L *tangibilis* < *tangere* touch] — **tan′gi·bil′i·ty, tan′gi·ble·ness** *n.* — **tan′gi·bly** *adv.*

Tan·gier (tan·jir′) A port on the northernmost coast of Morocco; formerly an international zone (**Tangier International Zone;** 225 square miles). French **Tan·ger** (tän·zhā′).

tan·gle[1] (tang′gəl) *v.* **·gled, ·gling** *v.t.* 1 To twist or involve in a confused and not readily separable mass. 2 To complicate; ensnare as in a tangle; trap; enmesh. — *v.i.* 3 To be or become entangled. — **to tangle with** *Colloq.* To embroil oneself with. — *n.* 1 A confused intertwining, as of threads or hairs; a snarl. 2 Hence, a state of confusion or complication; a jumbled mess. 3 A state of perplexity or bewilderment. [Nasalized var. of obs. *tagle* <Scand. Cf. dial. Sw. *taggla* disorder.] — **tan′gler** *n.*

tan·gle[2] (tang′gəl) *n.* 1 An edible seaweed (genus *Laminaria*). 2 *Scot.* A tall, lean person. [<ON *thöngull*]

tan·gle·ber·ry (tang′gəl·ber′ē) *n. pl.* **·ries** The blue huckleberry (*Gaylussacia frondosa*) of the eastern United States: also *dangleberry.*

tan·gly (tang′glē) *adj.* Consisting of or being in a tangle.

tan·go (tang′gō) *n. pl.* **·gos** 1 Any of several Latin-American dances, originally from Argentina, in 2/4 time and characterized by deliberate gliding steps and low dips. 2 Any syncopated tune or melody to which the tango may be danced. — *v.i.* To dance the tango. [<Am. Sp., fiesta <Sp., gipsy dance]

tan·gram (tang′grəm) *n.* A Chinese puzzle consisting of a square card or board cut by straight incisions into different-sized pieces (five triangles, a square, and a lozenge) to be combined into a variety of figures. [Arbitrary coinage, after ANAGRAM]

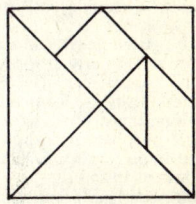
TANGRAM

tang·y (tang′ē) *adj.* **tang·i·er, tang·i·est** Having a tang in taste or odor; pungent.

Ta·nim·bar Islands (tə·nim′bär, tan′im·bär) An Indonesian island group in the Banda Sea; total, 2,172 square miles: also *Timorlaut.*

Ta·nis (tä′nis) An ancient city of Lower Egypt in the Nile delta: Old Testament *Zoan.*

tan·ist (tän′ist, thŏn′-) *n.* Among the ancient Celts, the heir apparent to a chieftainship, elected in the lifetime of a chief from among the chief's kinsmen. [<Irish *tānaiste* second, heir presumptive]

tan·ist·ry (tän′ist·rē, thŏn′-) *n.* The succession and life tenure relating to a tanist.

tan·jib (tun·jēb′) *n.* A kind of fine muslin fabric made in India. [<Bengali <Persian *tan-zib,* lit., ornament of the body]

Tan·jore (tan·jōr′, -jôr′) A city in SE Madras State, India.

Tan·jung·pri·ok (tän′jōōng·prē′ōk) A port of NW Java, the principal port of Indonesia and port for Jakarta: also *Tandjungpriok.*

U. S. ARMY TANK M48A2

tank (tangk) *n.* 1 A large vessel, basin, or receptacle for holding a fluid. 2 Any natural pool or pond. 3 *Mil.* A heavily armored combat vehicle of the Caterpillar tractor type, propelled by internal-combustion engines and mounting guns of various calibers.

— *v.t.* To place or store in a tank. [<Pg *tanque,* aphetic var of *estanque* <L *stagnum* pool] — **tank'less** *adj.* — **tank'like'** *adj.*

tan·ka[1] (tang'kə) *n.* **1** A Japanese verse form consisting of five lines, of which the first and third have five syllables, and the rest seven. **2** A poem imitating the Japanese tanka in verse form. [<Japanese]

tan·ka[2] (tang'kə) *n. pl.* **·ka** A descendant of an aboriginal race living on watercraft at Canton, China. [<Chinese *tankia* < *tan* egg + *chia* family, people]

tank·age (tangk'ij) *n.* **1** The act, process, or operation of putting in tanks. **2** The price for storage in tanks. **3** The capacity or contents of a tank. **4** Slaughterhouse waste, as bones and entrails, from which the fat has been rendered: used, when dried, as a fertilizer or coarse feed.

tank·ard (tangk'ərd) *n.* A large, one-handled drinking cup, usually made of pewter or silver, often with a cover. [<MDu. *tanckaert* <Med. L *tancardus,* prob. metathetic var. of L *cantharus* tankard, large goblet]

tank destroyer A motor vehicle equipped with an anti-tank gun.

tank·er (tangk'ər) *n.* A cargo vessel especially constructed for the transport of oil and gasoline.

tank farming Hydroponics. — **tank farmer**

tank·ful (tangk'fŏŏl') *n.* The quantity that fills a tank.

tank town *U. S. Colloq.* A small town where trains formerly stopped to refill from a water tank.

tank trap A camouflaged ditch excavated along the probable route of enemy tanks for the purpose of trapping them.

tan·nage (tan'ij) *n.* The act, process, or operation of tanning.

tan·nate (tan'āt) *n. Chem.* A salt or ester of tannic acid.

tanned (tand) Past tense and past participle of TAN.

Tan·nen·berg (tän'ən·berkh) A village in NE Poland; scene of major Russian defeat by German forces, 1914: before 1945 in East Prussia, Germany. *Polish* **Ste·bark** (stenm'·bärk).

tan·ner[1] (tan'ər) *n.* One who tans hides.

tan·ner[2] (tan'ər) *n. Brit. Colloq.* A sixpence. [Origin unknown]

tan·ner·y (tan'ər·ē) *n. pl.* **·ner·ies** A place where leather is tanned.

Tann·häu·ser (tän'hoi·zər) A German minnesinger and crusader of the 13th century, identified with a legendary knight who gives himself up to revelry with Venus and her court in the Venusberg, then makes a trip to Rome to seek absolution; hero of an opera by Wagner.

tan·nic (tan'ik) *adj.* Pertaining to or derived from tannin or tanbark.

tan·nif·er·ous (ta·nif'ər·əs) *adj.* Having or yielding tannin. [<TANNI(N) + -FEROUS]

tan·nin (tan'in) *n. Chem.* Any of a group of amorphous, brownish-white, astringent compounds that form shiny scales when extracted, as with water, from gallnuts, sumac, etc. Their principal applications in the arts are in the preparation of ink and the manufacture of leather. Also **tannic acid.** [<F *tanin* < *tan* tan]

tan·ning (tan'ing) *n.* **1** The art or process of converting hides into leather. **2** A bronzing, as of the skin, by exposure to the sun, wind, etc.

Tan·nu-tu·va People's Republic (tan'ōō·tōō'·və) A former name for TUVA AUTONOMOUS REGION.

tan·rec (tan'rek) See TENREC.

tan·sy (tan'zē) *n. pl.* **·sies** Any of a genus (*Tanacetum*) of coarse perennial herbs; especially, a species (*T. vulgare*) with yellow flowers and a strongly aromatic and bitter taste, used in medicine for its tonic properties. [<OF *tanesie,* aphetic var. of *athanasie* <LL *athanasia* <Gk., immortality]

Tan·ta (tän'tä) A city in Lower Egypt in the Nile delta.

tan·ta·late (tan'tə·lāt) *n. Chem.* A salt of tantalic acid.

tan·tal·ic (tan·tal'ik) *adj.* **1** Pertaining to tantalum; containing tantalum in its higher valence. **2** Designating a colorless crystalline acid, $HTaO_3$, derived from tantalum oxide, Ta_2O_5.

tan·ta·lite (tan'tə·līt) *n.* An iron-black ferrous tantalate, $FeTa_2O_6$, having a submetallic luster.

tan·ta·lize (tan'tə·līz) *v.t.* **·lized, ·liz·ing** To tease or torment by repeated frustration of hopes or desires. Also *Brit.* **tan'ta·lise.** [from *Tantalus*] — **tan'ta·li·za'tion** *n.* — **tan'·ta·liz'er** *n.* — **tan'ta·liz'ing·ly** *adv.*

tan·ta·lum (tan'tə·ləm) *n.* A silver-white, ductile, metallic element (symbol Ta) occurring in tantalite and other rare minerals. It becomes hard when hammered, and forms alloys with tungsten, molybdenum, and iron. See ELEMENT. [<TANTALUS; from its inability to absorb water]

Tan·ta·lus (tan'tə·ləs) In Greek mythology, a rich king, son of Zeus and father of Pelops and Niobe, who was punished in Hades for revealing the secrets of Zeus by being made to stand in water that receded when he tried to drink, and under fruit-laden branches he could not reach.

tan·ta·mount (tan'tə·mount) *adj.* Having equivalent value, effect, or import; equivalent: with *to.* [<AF *tant amunter* amount to as much <L *tantus* as much + OF *amonter* amount. See AMOUNT.]

tan·ta·ra (tan'tə·rä', tan·tar'ə, -tä'rə) *n.* A quick succession of notes from a horn; also, a hunting cry. [Imit.]

tan·tiv·y (tan·tiv'ē) *adj.* Swift; rapid. — *n. pl.* **·tiv·ies** **1** A hunting cry indicating that the chase is at full speed. **2** *Obs.* A rapid, rushing movement. — *adv.* Swiftly; with all speed. [Prob. imit. of the horse's gallop]

tant mieux (tän myœ') *French* So much the better.

tan·to (tän'tō) *adv. Italian* So much; too much: especially in the musical direction **non tanto,** not too much.

tant pis (tän pē') *French* So much the worse.

tan·trum (tan'trəm) *n.* A petulant fit of passion. [Origin unknown]

Tao·ism (dou'iz·əm, tou'-) *n.* One of the principal religions or philosophies of China, founded about 500 B.C. by Lâo-tse, who taught that happiness could be acquired through obedience to the requirements of man's nature and the simplification of social and political relations, in accordance with the Tao, or Way, the basic principle of the cosmos from which all of nature proceeds. [<Chinese *tao* way, road] — **Tao'ist** *adj.* & *n.* — **Tao·is'tic** *adj.*

Ta·os (tä'ōs) A resort town in northern New Mexico.

tap[1] (tap) *n.* **1** An arrangement for drawing out liquid, as beer from a cask. **2** A faucet or cock; spigot; also, a plug or stopper to close an opening in a cask or other vessel. **3** Liquor drawn from a tap; also, a particular liquor or quality of liquor contained in casks. **4** *Brit.* A place where liquor is served; a bar; taproom. **5** A tool for cutting internal screw threads. **6** A point of connection for an electrical circuit. — **on tap** **1** Contained in a cask; ready for tapping: beer *on tap.* **2** Provided with a tap. **3** Available; ready. — *v.t.* **tapped, tap·ping** **1** To provide with a tap or spigot. **2** To pierce or open so as to draw liquid from: to *tap* a sugar-maple tree. **3** To draw (liquid) from a container. **4** To make connection with: to *tap* a gas main. **5** To make connection with secretly: to *tap* a telephone wire. **6** To make an internal screw thread in with a tap: to *tap* a nut. [OE *tæppa*]

tap[2] (tap) *v.* **tapped, tap·ping** *v.t.* **1** To touch or strike gently. **2** To make or produce by tapping. **3** To apply leather to (the sole or heel of a shoe) in repair. — *v.i.* **4** To strike a light blow or blows, as with the finger tip. — *n.* **1** A gentle or playful blow. **2** Leather, etc., affixed to a shoe sole or heel; also, a metal plate on the toe or heel of a tap-dancer's shoe. **3** *pl.* A military signal by trumpet or beat of drum, sounded after tattoo, for the extinguishing of all lights in soldiers' quarters: often played after a military burial. [<OF *taper*]

ta·pa (tä'pä) *n.* The bark of the Asian paper-mulberry tree (*Broussonetia papyrifera*), used in making a kind of cloth, **tapa cloth.** [< native Polynesian name]

tap·a·der·a (tap'ə·dâr'ə) *n.* The leather hood of the stirrup of a Mexican saddle. Also **tap'a·der'o, tap'i·der'o.** [<Sp., cover < *tapar* stop up]

Ta·pa·jós (tä'pə·zhôs') A river of central and NE Brazil, flowing 500 miles NE from SW Pará state to the Amazon at Santerém.

tap·a·lo (tap'ə·lō) *n. pl.* **·los** A scarf or shawl of coarse cloth worn in Latin-American countries. [<Am. Sp., lit., cover it, imperative of *tapar* cover + *lo* it]

tap-dance (tap'dans', -däns') *v.i.* To dance or perform a tap dance. — **tap'-danc'er** *n.*

tap dance A dance, usually solo, in which the dancer emphasizes his steps by tapping the floor with the heels or toes of shoes or clogs designed to make audible the rhythm.

tape (tāp) *n.* **1** A narrow, stout strip of woven fabric. **2** Any long, narrow, flat strip of paper, metal, or the like, as the magnetic strip used in a tape recorder. **3** A tapeline. **4** Red tape. **5** A string or thread stretched breast-high across the finishing point of a racetrack and broken by the winner of the race. — *v.t.* **taped, tap·ing** **1** To wrap or secure with tape; also, to bandage: to *tape* a boxer's hands. **2** To measure with or as with a tapeline. **3** *Colloq.* To record on tape. **4** *Scot.* To use sparingly; stint. [OE *tæppe* strip of cloth] — **tape'less** *adj.* — **tape'like'** *adj.* — **tap'er** *n.*

tape·line (tāp'līn') *n.* A tape for measuring distances. Also **tape measure.**

ta·per (tā'pər) *n.* **1** A small candle; a burning wick or other light substance giving but feeble illumination. **2** A gradual diminution of size in an elongated object: the *taper* of a mast; also, any tapering object, as a cone. — *v.t.* & *v.i.* **1** To make or become smaller or thinner toward one end. **2** To lessen gradually; diminish: with *off.* — *adj.* Growing small by degrees in one direction; slender and conical or pyramidal. ◆ Homophone: *tapir.* [OE, dissimilated var. of Med. L *papyrus* taper, wick <L, papyrus; from the use of the pith of the papyrus as a wick] — **ta'per·ing·ly** *adv.*

tape recorder An electromagnetic apparatus which records by the effect of sound waves upon the particles adhering to a magnetic tape: in the playback the magnetic patterns are reconverted into the original electrical impulses and sound waves.

tap·es·try (tap'is·trē) *n. pl.* **·tries** **1** A loosely woven, ornamental fabric used for hangings, in which the woof is supplied by a spindle, the design being formed by stitches across the warp. **2** Loosely, a fabric imitating this process. — *v.t.* **·tried, ·try·ing** **1** To hang or adorn with tapestry. **2** To make or weave as tapestry. [<OF *tapisserie* < *tapis* carpet <L *tapete* <Gk. *tapētion,* dim. of *tapēs* rug]

tapestry carpet A carpet in which the fabric is woven after the designs are first printed.

ta·pe·tum (tə·pē'təm) *n. pl.* **·ta** (-tə) **1** *Bot.* A cell or layer of cells just outside the spore case of a plant, lining the cavity of an anther or sporangium. **2** *Zool.* A portion of the choroid coat of the eye in cats and certain other animals. **3** *Anat.* The fibers of the corpus callosum. [<LL <L *tapete* carpet. See TAPESTRY.]

tape·worm (tāp'wûrm') *n.* Any of various cestode worms (class *Cestoda*) with segmented, ribbonlike bodies, parasitic on the intestines of vertebrates; especially, the common pork tapeworm (*Taenia solium*) of man.

taph·e·pho·bi·a (taf'ə·fō'bē·ə) *n. Psychiatry* A morbid fear of being buried alive. [<Gk. *taphē* a grave + -PHOBIA] — **taph'e·pho'bic** *adj.*

tap house An inn; tavern; also, a barroom.

tap·i·o·ca (tap'ē·ō'kə) *n.* A nutritious starchy substance having irregular grains, obtained from cassava. [<Sp. <Tupi *tipioca* juice of the cassava < *ty* juice + *pýa* heart + *ocô* be removed]

ta·pir (tā'pər) *n.* A large, ungulate, herbivorous, typically nocturnal mammal (family *Tapiridae*), having short stout limbs and flexible proboscis, with the nostrils near the end. The tapir of South and Central America is brownish-black, that of the Malay Peninsula black and white. ◆ Homophone: *taper.* [<Sp. <Tupi *tapy'ra* tapir]

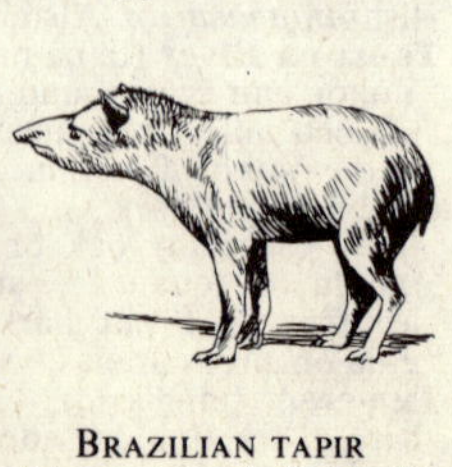

BRAZILIAN TAPIR
(From 3 to 3 1/2 feet high)

tap·is (tap'ē, tap'is; *Fr.* tȧ·pē') *n.* Tapestry, formerly used as a cover of a council table: now only in the phrase **on the tapis** (up for consideration). [<F. See TAPESTRY.]

ta·pis·sier (tȧ·pē·syā') *n. French* **1** A tapestry-maker. **2** An upholsterer. — **ta·pis·sière** (tȧ·pē·syâr') *n. fem.*

Tap·pan Zee (tap'ən zā') An expansion of the Hudson River in SE New York above New York City; 10 miles long and 3 miles wide.

tap·per (tap'ər) *n.* One who or that which taps, in any sense.

tap·pet (tap'it) *n. Mech.* A projecting arm of a mechanism, to operate an unattached part automatically, as to impart the motion of a cam to a valve. [<TAP²]

tappet rod *Mech.* A reciprocating rod bearing one or more tappets.

tap·ping (tap'ing) *n.* **1** The act of one who or that which taps in any sense. **2** Something taken by tapping, or running from a tap.

tap·pit (tap'it) *adj. Scot.* Having a tuft; crested.

tap·pit-hen (tap'it·hen') *n.* **1** A hen having a topknot. **2** An English pewter measure for liquors, holding three quarts: named for the knob on the lid resembling a hen's topknot. [<dial. E (Scottish) *tappit* topped + HEN]

tap·poon (ta·po͞on') *n.* A semicircular gate of heavy sheet iron, serving as a temporary dam for a small irrigating ditch. [<Sp. *tapón* plug < *tapar* stop up]

tap·room (tap'ro͞om', -ro͝om') *n.* A bar; bar-room.

tap·root (tap'ro͞ot', -ro͝ot') *n. Bot.* The principal descending root of a plant. — **tap'root'ed** *adj.*

taps (taps) See TAP² (*n.* def. 3).

tap·sal·tee·rie (tap'səl·tir'ē) *adv. Scot.* Upside down and in confusion; topsy-turvy.

tap·ster (tap'stər) *n.* One who draws and serves liquor; a bartender. [OE *tæppestre* barmaid]

Tap·ti (tap'tē) A river of west central India, flowing 450 miles west from west central Madhya Pradesh to the Gulf of Cambay just below Surat.

Ta·pu·ya (tȧ·po͞o'yä) *n.* A Tapuyan Indian.

Ta·pu·yan (tȧ·po͞o'yən) *n.* A large linguistic stock of South American Indians; Ge. — *adj.* Of or pertaining to this stock.

Ta·qua·rí (tä·kwȧ·rē') **1** A river in south central Mato Grosso, Brazil, flowing 350 miles SW from the Goiás border to the Paraguay river near the Bolivian border. **2** A river in NE Río Grande do Sul, Brazil, flowing 200 miles west and south to the Jacuí river.

tar¹ (tär) *n.* **1** A dark, oily, viscid mixture of hydrocarbons, especially phenols, obtained by the dry distillation of resinous woods, coal, etc. **2** Coal tar. Compare ASPHALT, PITCH¹. — *v.t.* **tarred, tar·ring** To cover with or as with tar. — **to tar and feather** To smear with tar and then cover with feathers: an old form of punishment. — *adj.* Made of, derived from, or resembling tar. [OE *teru*]

tar² (tär) *n. Colloq.* A sailor. [Short for TAR-PAULIN]

Tar·a (tär'ə) A village of central County Meath, Ireland; seat of the ancient Irish kings until the sixth century.

ta·ra·did·dle (tar'ə·did'l) See TARRADIDDLE.

Ta·ra·na·ki (tä'rä·nä'kē) The Maori name for EGMONT.

tar·an·tass (tar'ən·tas') *n.* A large four-wheeled vehicle on longitudinal bars in place of springs and mounted on a sledge in winter. Also **tar'an·tas'**. [<Russian *tarantas*]

tar·an·tel·la (tar'ən·tel'ə) *n.* A lively Neapolitan dance in 6/8 time: once thought to be a remedy for tarantism; also, the music written for it. [<Ital., dim. of *Taranto* Taranto; infl. by *tarantola* a tarantula]

tar·ant·ism (tar'ən·tiz'əm) *n.* A nervous and hysterical disorder characterized by stupor and hypochondria which, it was supposed, could be cured only by inordinate dancing and music; dancing disease. Formerly prevalent in southern Italy, it was believed to follow the bite of a tarantula. [<Ital. *tarantismo* <*Taranto* Taranto]

Ta·ran·to (tä'rän·tō) A port in SE Italy on the **Gulf of Taranto**, an arm of the Ionian Sea forming the instep of the Italian boot: ancient *Tarentum*.

ta·ran·tu·la (tə·ran'chə·lə) *n. pl.* **·las** or **·lae** (-lē) **1** A large, hairy, venomous spider (*Lycosa tarentula*) of southern Europe, still popularly but erroneously supposed to cause tarantism by its bite. **2** Any of various large, hairy American spiders (family *Theraphosidae*), especially of the genus *Eurypelma* of the SW United States, dreaded for their painful but not dangerous bite. [<Med. L <Ital. *tarantola* <*Taranto* Taranto]

TARANTULA
(Body from 2 to 3 1/2 inches)

tarantula hawk A large wasp (genus *Pepsis*) which paralyzes tarantulas with its sting and places them in its nest as food for its young. Also **tarantula killer.**

Ta·ra·pon (tä'rä·pōn) *n.* Micronesian.

Ta·ra·wa (tä·rä'wä, tä'rä·wä) The island headquarters of the Gilbert Islands and capital of the Gilbert and Ellice Islands colony; 8 square miles; scene of a United States victory over Japanese forces in World War II, November, 1943.

ta·rax·a·cum (tə·rak'sə·kəm) *n.* **1** Any of a genus (*Taraxacum*) of composite plants that includes the dandelion. **2** A medicinal preparation from the dried root of the common dandelion, used as a diuretic and laxative. [<NL <Arabic *tarakhshaqūq* bitter herb]

Tar·bell (tär'bel), **Ida Minerva,** 1857–1944, U. S. author.

tar·boosh (tär·bo͞osh') *n.* A brimless, usually red, felt cap with colored silk tassel, worn by Moslems. Also **tar·bush'**. [<Arabic *ṭarbūsh*]

tar camphor Naphthalene.

Tar·de·noi·si·an (tär'də·noi'zē·ən) *adj.* Of, pertaining to, or designating a subdivision of the Mesolithic culture epoch of the late Paleolithic period, related to the Azilian and characterized by small flint implements. [from *Fère-en-Tardenois*, town in NE France where remains were discovered]

Tar·dieu (tȧr·dyœ'), **André Pierre Gabriel Amédée,** 1876–1945, French statesman: pseudonym *George Villiers.*

tar·di·grade (tär'də·grād) *adj.* **1** Slow in motion or action; stepping or walking slowly. **2** Of or pertaining to a group (*Tardigrada*) of slow-moving microscopical arthropods, the water bears, found especially in water and damp moss. — *n.* One of the *Tardigrada*. [<F <L *tardigradus* < *tardus* slow + *gradi* walk]

tar·do (tär'dō) *adj. Music* Slow: a direction to performers. [<Ital.]

tar·dy (tär'dē) *adj.* **·di·er, ·di·est** **1** Not coming at the appointed time; dilatory; late. **2** Slow; reluctant. See synonyms under SLOW, TEDIOUS. [<F *tardif* <L *tardus* slow] — **tar'·di·ly** *adv.* — **tar'di·ness** *n.*

tare¹ (târ) *n.* **1** An unidentified weed that grows among wheat, supposed to be the darnel; hence, a seed of wickedness. *Matt.* xiii 25. **2** Any one of various species of vetch; especially, the common vetch (*Vicia sativa*). ◆ Homophone: *tear¹*. [? <F *tare* defect, rejectable thing. See TARE².]

tare² (târ) *n.* **1** An allowance made to a buyer of goods by deducting from the gross weight of his purchase the weight of the container. **2** *Chem.* An empty flask or vessel used as a counterweight. — *v.t.* **tared, tar·ing** To weigh, as a vessel or package, in order to determine the amount of tare. ◆ Homophone: *tear¹*. [<F <Arabic *ṭarḥah* < *ṭaraḥa* reject, throw away]

Ta·ren·tum (tə·ren'təm) Ancient name for TARANTO.

targ (tärg) *n.* A device for indicating on a plotting board the changing positions of a target. [Back formation <TARGET]

targe¹ (tärj) *n.* A shield; rarely, a target. [<OF <OE *targa* <ON]

targe² (tärj) *v.t. Scot.* **1** To censure severely; thrash. **2** To cross-question rigidly. **3** To subject to strict discipline.

tar·get (tär'git) *n.* **1** An object presenting a surface that may be used as a mark or butt, as in rifle or archery practice; anything that is shot at. **2** One who or that which is made an object of attack or a center of attention or observation; a butt: He was the *target* of the crowd's sneers. **3** A small, variously shaped and colored signal, usually placed near a railroad track, to indicate the position of the switches. **4** The vane or sliding sight on a surveyor's rod. **5** *Electronics* That electrode of a vacuum tube on which cathode rays are focused and from which X-rays are emitted. **6** A small round shield or buckler; a targe. [OE *targette, targuete*, dim. of *targe* shield. See TARGE¹.]

tar·get·eer (tär'gə·tir') *n.* A soldier armed with a shield.

Tar·gum (tär'gum, *Hebrew* tär·go͞om') *n. pl.* **Tar·gums** or *Hebrew* **Tar·gu·mim** (tär'go͞o·mēm') One of various ancient paraphrases of portions of the Hebrew scriptures in Aramaic or Chaldee. [<Aramaic *targūm* interpretation] — **Tar'gum·ic** or **·i·cal** *adj.* — **Tar'gum·ist** *n.*

Tar·heel (tär'hēl') *n. Colloq.* A native of the pine barrens of North Carolina; hence, any North Carolinian. Also **Tar Heel.**

Tarheel State Nickname of NORTH CAROLINA.

Ta·ri·fa (tä·rē'fä) A port on the Strait of Gibraltar in southern Spain.

tar·iff (tar'if) *n.* **1** A schedule of articles of merchandise with the rates of duty to be paid for their importation or exportation. **2** A duty, or duties collectively. **3** The law by which duties are imposed; also, the principles governing their imposition. **4** Any schedule of charges. — *v.t.* **1** To make a list or table of duties or customs on. **2** To fix a price or tariff on. [<Ital. *tariffa* <Arabic *ta'rif* information < '*arafa* know, inform]

Ta·rim (tä·rēm') The principal river of the Sinkiang-Uigur Autonomous Region, NW China, flowing 1,300 miles east from the west central part, forming the northern boundary of the Taklamakan desert.

Tark·ing·ton (tär'king·tən), **Booth,** 1869–1946, U. S. novelist.

tar·la·tan (tär'lə·tən) *n.* A thin, open-mesh transparent muslin, slightly stiffened and often rather coarse. [<F *tarlatane*; ult. origin unknown]

Tar·mac (tär'mak) *n.* A paving material made from coal tar: a trade name.

tarn (tärn) *n.* A small mountain lake. [ME *terne* <ON *tjörn*]

Tarn (tärn) A river in south central France, flowing 235 miles west from the Cévennes to the Garonne below Montauban.

tar·nal (tär'nəl) *U.S. Slang* Eternal; infernal; hence, damned. — *adv.* Very; damn. [Alter. of ETERNAL] — **tar'nal·ly** *adv.*

tar·na·tion (tär·nā'shən) *interj. & n. Dial.* Damnation: a euphemism. [Blend of TAR(NAL) + (DAM)NATION]

tar·nish (tär'nish) *v.t.* **1** To dim the luster of. **2** To dim the purity of; stain; disgrace. — *v.i.* **3** To lose luster, as by oxidation; become blemished. See synonyms under DEFILE¹, STAIN. — *n.* **1** Loss of luster; hence, a blemish. **2** The thin film of color on the exposed surface of a metal or mineral. [<OF *terniss-*, stem of *ternir* < *terne* dull, wan] — **tar'nish·a·ble** *adj.*

tarnished plant bug A common, brown-marked hemipterous insect (*Lygus pratensis*) of North America, which attacks many fruits and vegetables. For illustration see INSECTS (injurious).

Tar·no·pol (tär·nô'pôl) The Polish name for TERNOPOL.

Tar·nów (tär'no͞of) A city in southern Poland, 45 miles east of Cracow.

ta·ro (tä'rō) *n. pl.* **·ros** **1** Any one of several tropical plants (genus *Colocasia*) of the arum family, grown for their edible, cormlike rootstocks. **2** The rootstock of this plant. [<native Polynesian name]

tar·ot (tar'ō, -ət) *n.* One of a set of playing

TARBOOSH

cards with grilled or checkered backs used in Italy as early as the 14th century; also, a game played with such cards in which 22 are trumps and the other 56 are the usual Italian playing cards: used by fortune-tellers and gipsies in foretelling future events. [<F <Ital. *tarocco* < *taroccare* wrangle, play at cards; ult. origin obscure]

tar·pau·lin (tär·pô′lin, tär′pə-) *n.* 1 A waterproof canvas, impregnated with tar, for covering merchandise. 2 A sailor's wide-brimmed storm hat. 3 *Rare* A sailor. [<TAR[1] + PALL[1] + -ING[1]]

Tar·pe·ia (tär·pē′ə) The daughter of the governor of the citadel of Rome, who treacherously opened its gates to the Sabines on condition of receiving what they wore on their arms, meaning their golden bracelets. As they entered they crushed her with their shields instead. — **Tar·pe′ian** *adj.*

Tarpeian Rock A cliff upon the Capitoline Hill at Rome, from which state criminals were hurled to their death.

tar·pon (tär′pon, -pən) *n. pl.* **·pon** or **·pons** A large marine game fish with conspicuous silvery scales (*Tarpon atlanticus*) of the West Indies and the coast of Florida. [Origin unknown]

Tar·quin (tär′kwin) Anglicized name of two legendary kings of Rome, **Lucius Tarquinius Priscus** and **Lucius Tarquinius Superbus,** respectively fifth and seventh kings, of the sixth century B.C.

tar·ra·did·dle (tar′ə·did′l) *n. Colloq.* A prevarication; lie: also spelled *taradiddle.* [Origin uncertain]

tar·ra·gon (tar′ə·gon) *n.* 1 A European perennial plant (*Artemisia dracunculus*) allied to wormwood, and cultivated for its aromatic leaves which are used as seasoning. 2 The leaves of this plant. [<Sp. *taragona* <Arabic *tarkhun* <Gk. *drakōn* dragon]

Tar·ra·go·na (tar′ə·gō′nə) 1 A province in Catalonia, NE Spain; 2,425 square miles. 2 A manufacturing city, capital of Tarragona province, and formerly of a Roman province.

tar·ri·ance (tar′ē·əns) *n. Archaic* A tarrying; delay. [<TARRY + -ANCE]

tar·ri·er (tar′ē·ər) *n.* One who or that which tarries.

tar·row (tar′ō) *v.i. Scot.* 1 To show reluctance or hesitation; delay; tarry. 2 To feel loathing.

tar·ry[1] (tar′ē) *v.* **·ried, ·ry·ing** *v.i.* 1 To put off going or coming; linger. 2 To remain in the same place; abide; stay. 3 To wait. — *v.t.* 4 *Archaic* To wait for; await: to *tarry* his coming. See synonyms under ABIDE. — *n.* Sojourn; stay. [ME *tarien* vex, hinder, delay, fusion of OE *tergan* vex + OF *targer* delay <LL *tardicare* <L *tardare* delay < *tardus* slow]

tar·ry[2] (tär′ē) *adj.* Covered with tar; like tar.

tar·sal (tär′səl) *adj.* 1 Of, pertaining to, or situated near the tarsus or ankle. 2 Of or pertaining to the tarsi of the eye. See TARSUS.

Tar·shish (tär′shish) In the Bible, an ancient maritime country, often identified with Tartessus, in southern Spain. I *Kings* x 22.

tar·si·er (tär′sē·ər) *n.* A small, arboreal, insectivorous East Indian primate (*Tarsius spectrum*) of nocturnal habits, with large eyes and ears, long tail, and adhesive pads on elongated digits: it is the sole member of the suborder *Tarsioidea.* [<F < *tarse* tarsus; so called from its unusually long tarsal bones]

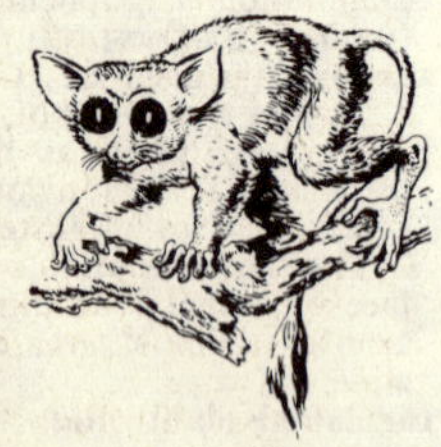

TARSIER
(Size of a small rat)

tarso- *combining form* 1 The tarsus; pertaining to the tarsus. 2 The tarsus of the eye; pertaining to the tarsal plate: *tarsoplasty,* plastic surgery of the eyelid. Also, before vowels, **tars-.** [<Gk. *tarsos* flat of the foot, edge of the eyelid]

tar·so·met·a·tar·sus (tär′sō·met′ə·tär′səs) *n. pl.* **·si** (-sī) *Ornithol.* The so-called tarsus of birds; the bone reaching from the tibia to the toes, consisting of the confluent proximal tarsal and metatarsal bones. [<NL]

tar·sus (tär′səs) *n. pl.* **·si** (-sī) 1 *Anat.* **a** The ankle, or, in man, the group of seven bones of which it is composed. **b** A plate of connective tissue in the eyelid. 2 *Zool.* **a** The shank of a bird's leg. **b** The distal part of the leg of certain arthropods. [<NL <Gk. *tarsos* flat of the foot, any flat surface]

Tar·sus (tär′səs) 1 A port near the NE Mediterranean in southern Turkey in Asia; anciently, the capital of Cilicia, and the birthplace of St. Paul. 2 A river in southern Turkey, flowing 95 miles south from the Taurus mountains to the Mediterranean below Tarsus: ancient *Cydnus.*

tart[1] (tärt) *adj.* 1 Having a sharp, sour taste. 2 Figuratively, severe; cutting; caustic: a *tart* remark. See synonyms under BITTER. [OE *teart*] — **tart′ly** *adv.* — **tart′ness** *n.*

tart[2] (tärt) *n.* 1 A small pastry shell with fruit or custard filling, and without a top crust, as distinguished from a pie. 2 In England, an uncovered fruit pie. 3 *Slang* A girl or woman of loose morality. [<OF *tarte*]

tar·tan[1] (tär′tən) *n.* 1 A woolen fabric having varicolored lines or stripes at right angles, forming a distinctive pattern; a woolen plaid; the characteristic dress of the Scottish Highlanders, each clan having its particular pattern or patterns; hence, any similar pattern; a plaid. 2 A garment made of tartan. — *adj.* Made of tartan; also, striped or checkered in a manner similar to the Scottish tartans. [? <OF *tiretaine* linsey-woolsey]

tar·tan[2] (tär′tən) *n.* 1 A Mediterranean vessel having one mast with a large lateen sail. 2 A variety of long, covered carriage. [<F *tartane* <Arabic *tarīdah,* kind of ship]

tar·tar (tär′tər) *n.* 1 An acid substance deposited from grape juice during fermentation as a pinkish sediment; crude potassium bitartrate. See ARGOL, CREAM OF TARTAR. 2 *Dent.* A yellowish incrustation on the teeth, chiefly calcium phosphate. [<F *tartre* <LL *tartarum* <Med. Gk. *tartaron,* ? <Arabic]

Tar·tar (tär′tər) *n.* 1 Tatar. 2 A person of intractable or savage temper; also, especially in the phrase **to catch a Tartar,** an opponent who turns out to be unexpectedly formidable: also **tar′tar.** — *adj.* Of or pertaining to the Tatars of Tartary. [<F *Tartare* <LL *Tartarus* <Persian *Tātar* Tatar; prob. infl. by L *Tartarus* Hell]

Tar·tar (tär′tər) *Obs.* Tartarus.

Tar·tar·e·an (tär·târ′ē·ən) *adj.* Of or pertaining to Tartarus.

tartar emetic *Chem.* A white, crystalline, poisonous tartrate of antimony and potassium, $K(SbO)C_4H_4O_6 \cdot 1/2H_2O$, with a sweet, afterward disagreeable, metallic taste: used in medicine, chiefly as an emetic, and in dyeing as a mordant.

tar·tar·e·ous (tär·târ′ē·əs) *adj.* Resembling tartar.

tar·tare sauce (tär′tər) A fish sauce consisting of mayonnaise, capers, chopped olives, and pickles. Also **tar′tar sauce.**

Tar·tar·i·an (tär·târ′ē·ən) *adj.* Of or pertaining to the Tatars or Tartary.

tar·tar·ic (tär·tar′ik, -tär′ik) *adj.* Pertaining to or derived from tartar or tartaric acid.

tartaric acid *Chem.* Any one of four isomeric organic compounds, $HOOC(CHOH)_2COOH$, differing from each other in their optical properties, especially the dextrorotatory form, occurring in the free state or as a potassium or calcium salt, as in grape juice, various unripe fruits, etc.

tar·tar·ize (tär′tə·rīz) *v.t.* **·ized, ·iz·ing** To impregnate or treat with tartar, cream of tartar, or tartar emetic. — **tar′tar·i·za′tion** *n.*

tar·tar·ous (tär′tər·əs) *adj.* Pertaining to or derived from tartar.

Tartar sable The kolinsky.

Tar·ta·rus (tär′tər·əs) 1 In Greek mythology, the abyss below Hades where Zeus confined the Titans. 2 Hades.

Tar·ta·ry (tär′tər·ē) A region of Asia and eastern Europe, mostly in central and western Asiatic Russian S.F.S.R., Soviet Central Asia, southern European Russian S.F.S.R., and Ukrainian S.S.R., ruled by the Tatars, under Mongol leadership, in the 13th and 14th centuries A.D. At its greatest extent, under Genghis Khan, it reached the Pacific; after his death, the Asian portion became known as **Great Tartary,** or **Asiatic Tartary,** while the European portion, ruled by the Golden Horde, became **Little Tartary,** or **European Tartary**: also *Tatary.*

Tar·tes·sus (tär·tes′əs) An ancient city and region in the SW part of the Iberian Peninsula near the Pillars of Hercules, often identified with Biblical Tarshish.

tart·let (tärt′lit) *n.* A small tart.

tar·trate (tär′trāt) *n. Chem.* A salt or ester of tartaric acid.

tar·trat·ed (tär′trā·tid) *adj. Chem.* Containing or combined with tartaric acid.

Tar·tu (tär′tōō) A city in SE central Estonia: German *Dorpat,* Russian *Yurev.*

tar·tufe (tär·tōōf′, *Fr.* tàr·tüf′) *n.* Any hypocrite or toady. Also **tar·tuffe′.** [after TARTUFE]

Tar·tufe (tär·tōōf′, *Fr.* tàr·tüf′) In Molière's comedy of the same name, the chief character, a person of pretended devoutness. Also **Tar·tuffe′.**

Tar·ve·si·um (tär·vē′sē·əm) The ancient name for TREVISO.

Tar·zan (tär′zan, tär·zan′) The hero of a series of novels by Edgar Rice Burroughs (1875–1950): an English child of noble birth abandoned in the African jungle, raised by apes, and possessing incredible strength, agility, and a knowledge of the speech of animals. Also **Tarzan of the Apes.**

Tash·kent (täsh·kent′) The largest city of Soviet Central Asia, capital of Uzbek S.S.R. Also **Tash·kend′** (-kend′).

ta·sim·e·ter (tə·sim′ə·tər) *n.* An electrical apparatus for detecting changes in pressure by the resulting variations in the conductivity of a solid, and so measuring changes, as in length, temperature, or moisture, that produce alteration of pressure. [<Gk. *tasis* extension (< *teinein* stretch) + -METER] — **tas·i·met·ric** (tas′ə·met′rik) *adj.* — **ta·sim′e·try** *n.*

task (task, täsk) *n.* 1 A specific amount of labor or study imposed by authority or required by duty or necessity. 2 Any work voluntarily undertaken and imposed on oneself. 3 An exhausting or vexatious employment; burden. 4 A specific military mission. 5 *Obs.* A tax; duty. — **to take to task** To reprove; lecture. — *v.t.* 1 To assign a task to. 2 To overtax with labor; burden. 3 To censure; reprimand. 4 *Obs.* To tax. [<AF *tasque* <LL *tasca, taxa* <L *taxare* appraise. Related to TAX.]

Synonyms (noun): business, drudgery, job, labor, lesson, stint, toil, work. See TOIL[1].

task·er (tas′kər, täs′-) *n.* 1 A reaper. 2 A thresher of grain. 3 A laborer who performs allotted work.

task force *Mil.* A tactical unit consisting of elements drawn from different branches of the armed services and devised and assigned to execute a specific mission.

task·mas·ter (task′mas′tər, täsk′mäs′tər) *n.* One who assigns tasks; figuratively, one who or that which loads with heavy burdens.

Tas·lan (taz′lan) *n.* A mechanical process for imparting bulk and texture to any standard textile yarn without altering basic chemical properties: a trade name.

Tas·man (täs′män), **Abel Janszoon,** 1603?–1659?, Dutch navigator who discovered Tasmania and New Zealand.

Tas·ma·nia (taz·mā′nē·ə) An island state in the Commonwealth of Australia, south of Victoria; 26,215 square miles; capital, Hobart: formerly *Van Diemen's Land.* — **Tas·ma′ni·an** *adj.* & *n.*

Tasmanian devil A ferocious burrowing carnivorous marsupial (*Sarcophilus ursinus*) of the dasyure family, with white markings on the black hide.

Tasmanian wolf The thylacine. Also **Tasmanian tiger.**

Tas·man Sea (taz′mən) The arm of the South Pacific Ocean between SE Australia and Tasmania on the west and New Zealand on the east.

tass (tas) *n. Scot.* A drinking cup, or its contents.

Tass (täs, tas) *n.* Russian news agency: from the initials of *Telegrafnoe Agentstvo Sovetskovo Soyuza* (Telegraph Agency Soviet Union).

tas·sel[1] (tas′əl) *n.* 1 A pendent ornament, for curtains, cushions, and the like, consisting of a tuft of loosely hanging threads or cords; formerly, a clasp for holding a cloak. 2 Something resembling a tassel, as the pendent head of some plants or flowers, or the pyramidal inflorescence on a stalk of Indian corn. —*v.* **·seled** or **·selled, ·sel·ing** or **·sel·ling** *v.t.* **1** To provide or adorn with tassels. **2** To form in a tassel or tassels. **3** To remove the tassels from (Indian corn). —*v.i.* **4** To put forth tassels, as Indian corn. [<OF, clasp]

tas·sel² (tas′əl) *n.* The tercel.

tas·set (tas′it) *n.* One of a series of overlapping metal plates pendent from the cuirass to protect the waist and thighs: often called *tace.* Also **tasse.** ◆ Homophone: *tacit.* [<F *tassette,* dim. of OF *tasse* a pouch]

tas·sie (tas′ē) *n. Scot.* A drinking cup.

Tas·so (täs′sō), **Torquato,** 1544–95, Italian epic poet.

taste (tāst) *v.* **tast·ed, tast·ing** *v.t.* **1** To perceive the flavor of (something) by taking into the mouth or touching with the tongue. **2** To take a little of (food or drink); eat or drink a little of. **3** To test the quality of (a product) thus: His business is *tasting* tea. **4** *Archaic* To have a relish for; like. **5** *Obs.* To prove or try by or as by touch. —*v.i.* **6** To take a small quantity into the mouth; take a taste: usually with *of.* **7** To have experience or enjoyment; be or become acquainted through experience: with *of*: to *taste* of great sorrow. **8** To have specified flavor when in the mouth: Sugar *tastes* sweet. —*n.* **1** The sensation excited when a soluble substance comes into contact with any of the taste buds; also, the quality thus perceived; flavor. **2** *Physiol.* Any of the four fundamental sensations, salt, sweet, bitter, or sour, excited alone or in any combination by the sole action of the gustatory nerves. **3** A small quantity tasted, eaten or sipped; a sample: often used figuratively. **4** Special fondness and aptitude for a pursuit; bent; inclination: a *taste* for music. **5** The power or faculty of apprehending and appreciating the beautiful in nature, art, and literature; critical perception or discernment. **6** Style or form with respect to the rules of propriety or etiquette: She behaves in very poor *taste.* **7** Individual preference or liking: That tie suits my *taste.* **8** The act of tasting. **9** *Obs.* The act of examining or testing. See synonyms under RELISH, SAVOR. [<OF *taster* taste, try, feel, prob. ult. <L *taxare* touch, handle, appraise] —**tast′a·ble** *adj.*

taste bud *Physiol.* One of the clusters of cells situated in the mucous membrane chiefly of the tongue and containing sensitive receptors for the discriminatory perception of taste.

taste·ful (tāst′fəl) *adj.* **1** Conforming to taste. **2** Possessing good taste. **3** Savory: a rare use. —**taste′ful·ly** *adv.* —**taste′ful·ness** *n.*
Synonyms: artistic, dainty, delicate, delicious, elegant, esthetic, esthetical, exquisite, fastidious, fine, nice. That which is *elegant* is made so not merely by nature, but by art and culture. *Nice* and *delicate* both refer to exact adaptation to some standard; as regards matters of taste, *delicate* is a higher and more discriminating word than *nice,* and is always used in a favorable sense; a *delicate* distinction is one worth observing; a *nice* distinction may be so, or may be overstrained and unduly subtle. *Esthetic* or *esthetical* refers to beauty or the appreciation of the beautiful, especially from the philosophic point of view. *Exquisite* denotes the utmost perfection of the *elegant* in minute details; we speak of an *elegant* garment, an *exquisite* lace. *Exquisite* is also applied to intense keenness of any feeling; as, *exquisite* pain. *Antonyms*: clumsy, coarse, deformed, disgusting, displeasing, distasteful, fulsome, gaudy, grotesque, harsh, hideous, horrid, inartistic, inharmonious, meretricious, offensive, rude, tawdry.

taste·less (tāst′lis) *adj.* **1** Having no flavor; insipid; dull. **2** Having lost the sense of taste. **3** Devoid of esthetic taste. **4** Lacking, or showing a lack of, good taste. —**taste′less·ly** *adv.* —**taste′less·ness** *n.*

tast·er (tās′tər) *n.* **1** One who tastes; specifically, one who tests the quality of for trade: a tea–*taster.* **2** A device to assist in testing or sampling. **3** A pipette, or a small, flat, circular metal vessel used in testing wines.

tast·y (tās′tē) *adj.* **tast·i·er, tast·i·est** *Colloq.* **1** Having a fine flavor; savory. **2** Tasteful. —**tast′i·ly** *adv.* —**tast′i·ness** *n.*

tat¹ (tat) *v.* **tat·ted, tat·ting** *v.t.* To make, as an edging, by tatting. —*v.i.* To make tatting. [Back formation <TATTING] —**tat′ter** *n.*

tat² (tat) *n.* A tap or blow: in the phrase **tit for tat.** [? Var. of TAP², *n.*]

Ta·tar (tä′tər) *n.* **1** One belonging to any of the Turkic peoples of eastern, western, and Ural Asiatic Russian S.F.S.R. and Soviet Central Asia; also, one of the Turkic Tatars of the Tatar Republic, the Crimea, the Kalmuck area, and the northern Caucasus. **2** Any of the Turkic languages of the Tatars, as Uzbek. **3** Originally, any of the Tungus of Manchuria and Mongolia. —*adj.* Of or pertaining to the Tatars. Also *Tartar.* [<Persian]

Ta·tar Autonomous Soviet Socialist Republic (tä′tər) An administrative division of east central European Russian S.F.S.R.; 26,100 square miles; capital, Kazan.

Ta·tar·i·an (tä·târ′ē·ən) *adj.* Of or pertaining to the Tatars: also **Ta·tar·ic** (tä·tar′ik). —*n.* A Tatar.

Ta·ta·ry (tä′tər·ē) See TARTARY.

tate (tāt) *n. Scot.* A wisp or tuft, as of hay or hair.

Tate (tāt), **Nahum,** 1652–1715, English dramatist; poet laureate 1692–1715.

ta·tie (tā′tē) *n. Brit. Dial.* A potato. Also **ta′ter, ta′ty.**

Tat·ler (tat′lər), **The** An English periodical, published thrice weekly by Sir Richard Steele from 1709 to 1711, chiefly written by Steele, occasionally by Addison: predecessor of *The Spectator.*

tat·ou·ay (tat′o͞o·ā, tä·to͞o′ī) *n.* A large South American armadillo (genus *Cabassous*). [<Sp. *tatuay* <Guarani *tatu–aí* < *tatu* armadillo + *aí* worthless; so called because it is inedible]

Ta·tra (tä′trä), **High** The highest group of the central Carpathian Mountains, in northern Czechoslovakia; highest peak, 8,737 ft. Also **Tatra Mountains.**

tat·ter (tat′ər) *n.* **1** A torn and hanging shred; rag. **2** *pl.* Ragged clothing. —*v.t.* To make ragged; tear into tatters. —*v.i.* To become ragged. [<Scand. Cf. ON *tötturr* rags.]

tat·ter·de·mal·ion (tat′ər·di·māl′yən, -mal′-) *n.* A person wearing ragged clothes; a ragamuffin. —*adj.* Ragged. [Origin unknown]

tat·tered (tat′ərd) *adj.* **1** Torn into tatters. **2** Clothed in rags; ragged.

Tat·ter·sall check (tat′ər·sôl) A check or plaid design of dark lines on a light ground: used especially in men's vests. Also **Tattersall plaid.** [From a pattern on blankets used in the London market of Richard *Tattersall,* 18th century horse merchant]

tat·ting (tat′ing) *n.* A lacelike threadwork, made by hand; also, the act or process of making it. [Origin unknown]

tat·tle (tat′l) *v.* **·tled, ·tling** *v.i.* **1** To talk idly; prate; chatter. **2** To tell tales about others; gossip. —*v.t.* **3** To reveal by gossiping. See synonyms under BABBLE. —*n.* **1** Idle talk or gossip. **2** Prattling speech, as of children. [Prob. <MDu. *tatelen*] —**tat′tling·ly** *adv.*

tat·tler (tat′lər) *n.* **1** One who tattles; a talebearer; tattletale. **2** Any long-billed bird of the genus *Totanus,* as the redshank and the yellowlegs. **3** The willet. **4** The wandering tattler (*Heteroscelus incanus*), a shore bird of the Pacific coast of the United States.

tat·tle·tale (tat′l·tāl′) *n.* A talebearer; tattler. —*adj.* Revealing; betraying.

tat·too¹ (ta·to͞o′) *v.t.* **1** To prick and mark (the skin) in patterns with indelible pigments. **2** To mark the skin with (designs, etc.) in this way. —*n. pl.* **·toos** A pattern or picture so made. [<Polynesian. Cf. Tahitian, Tongan *tatau,* Marquesan *tatu* < *ta* mark.] —**tat·too′er** *n.* —**tat·too′·ing** *n.*

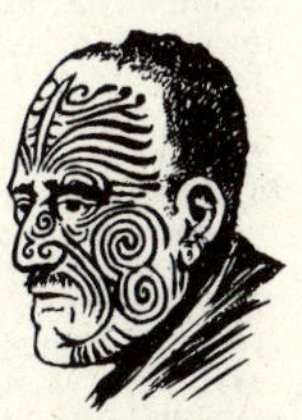

TATTOOING ON MAORI CHIEFTAIN

tat·too² (ta·to͞o′) *n.* **1** A continuous beating or drumming. **2** In military or naval usage, a signal by drum or bugle to repair to quarters, usually occurring about 9 p.m. [Var. of earlier *taptoo* <Du. *taptoe* < *tap* tap, faucet + *toe* shut]

tat·ty (tat′ē) *n. pl.* **·ties** *Anglo-Indian* An East Indian matting usually hung in doorways and window openings, and kept wet to cool the air. Also **tat′tie.** [<Hind. *ṭaṭṭī*] —**tat′tied** *adj.*

tau (tou) *n.* The nineteenth letter in the Greek alphabet: (Τ,τ) equivalent to the English *t.* As a numeral it denotes 300. [<Gk.]

Tauch·nitz (toukh′nits), **Baron Christian,** 1816–95, German publisher.

tau cross See under CROSS.

taught (tôt) Past tense and past participle of TEACH.

Taungs skull (toungz) A fossil skull assumed to represent Australopithecus. [from *Taungs,* South Africa, where remains were discovered]

taunt¹ (tônt) *n.* **1** A sarcastic, biting speech or remark; insulting reproach. **2** *Obs.* A butt of contemptuous reproach. See synonyms under SCORN, SNEER. —*v.t.* **1** To reproach with sarcastic or contemptuous words; mock; upbraid. **2** To tease in any way; provoke with taunts. See synonyms under MOCK, RIDICULE, SCOFF. [? <OF *tanter,* var. of *tenter* provoke, tempt. See TEMPT.] —**taunt′er** *n.* —**taunt′·ing·ly** *adv.*

taunt² (tônt) *adj. Naut.* Unusually tall: said of masts. [Aphetic var. of ATAUNT]

Taun·ton (tän′tən, tôn′-) The county town of Somerset, England; in the west central part.

tau particle *Physics* A rare, unstable atomic particle of the meson group, positively charged and with a mass about 1,000 times that of the electron.

taupe (tōp) *n.* **1** A mole. **2** The color of moleskin; dark gray, often tinged with brown, purple, or yellow. ◆ Homophone: *tope.* [<F <L *talpa* mole]

Tau·ric Cher·so·nese (tô′rik kûr′sō·nēz, -nēs) An ancient name for the CRIMEA.

tau·ri·form (tôr′ə·fôrm) *adj.* Shaped like a bull. [<L *tauriformis* < *taurus* bull + *forma* shape]

tau·rine¹ (tôr′ēn) *adj.* **1** Of or like a bull. **2** Related to or connected with the constellation or sign Taurus. [<L *taurinus* < *taurus* a bull]

tau·rine² (tôr′ēn, -in) *n. Chem.* A colorless crystalline compound, $C_2H_7NSO_3$, contained in the bile and muscles of oxen and other animals: also derived synthetically. [<L *taurus* bull + -INE²]

Tau·ris (tôr′is) **1** The Tauric Chersonese. **2** The ancient name for TABRIZ.

tauro– *combining form* Bull; ox; bovine. Also, before vowels, **taur–.** [<Gk. *tauros* a bull]

tau·ro·cho·lic acid (tôr′ə·kō′lik, -kol′ik) *Chem.* A bitter crystalline compound, $C_{26}H_{45}NSO_7$, contained in the bile of man and some animals, as the ox. [<TAURO- + Gk. *cholē* bile]

tau·rom·a·chy (tô·rom′ə·kē) *n.* The art of bullfighting. Also **tau·ro·ma·chi·a** (tôr′ə·mā′kē·ə). [<Gk. *tauromachia* < *tauros* a bull + *machesthai* fight]

Tau·rus (tôr′əs) **1** A zodiacal constellation, the Bull, containing the Hyades, the Pleiades, and Aldebaran. **2** The second sign of the zodiac, which the sun enters April 20. See CONSTELLATION, ZODIAC. [<L, bull]

Tau·rus (tôr′əs) A mountain range in southern Turkey; highest point, 12,251 feet: Turkish *Toros Dağlari.*

Taus·sig (tou′sig), **Frank William,** 1859–1940, U. S. political economist.

taut (tôt) *adj.* **1** Hard-drawn; stretched tight. **2** In proper shape; ready; tidy. **3** Tense; tight: *taut* muscles. **4** *Obs.* Filled to distention; firm. [ME *toyt, toht*; origin uncertain] —**taut′ly** *adv.* —**taut′ness** *n.*

taut·ed (tô′tid) *adj. Scot.* Tangled; tousled; matted: said of wool or hair.

taut·en (tôt′n) *v.t.* & *v.i.* To make or become taut; tighten.

tauto– *combining form* Same; identical: *tautomerism.* Also, before vowels, **taut–.** [<Gk. *tauto* the same]

tau·tog (tô·tôg′, -tog′) *n.* A blackish, edible, labroid fish (*Tautoga onitis*) of the North American Atlantic coast. Also **tau·taug′.** [<Algonquian *tautauog,* pl. of *tautau,* a kind of blackfish]

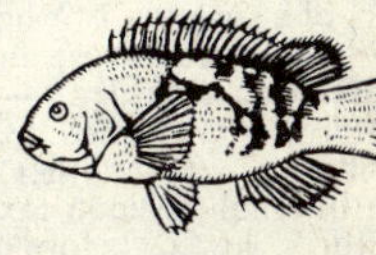

TAUTOG (About 16 inches long)

tau·tol·o·gism (tô·tol′ə·jiz′əm) *n.* Use of needlessly repetitious speech, or an instance of it; pleonasm. — **tau·tol′o·gist** *n.*

tau·tol·o·gize (tô·tol′ə·jīz) *v.i.* **·gized, ·giz·ing** To repeat needlessly the same idea in different words.

tau·tol·o·gy (tô·tol′ə·jē) *n. pl.* **·gies** Unnecessary repetition of the same idea in different words; pleonasm; also, an instance of such repetition; as, He is writing his own autobiography. See REDUNDANCE. [<LL *tautologia* <Gk. <*tauto* the same + *logos* discourse] — **tau·to·log·ic** (tô′tə·loj′ik) or **·i·cal** *adj.* — **tau′to·log′i·cal·ly** *adv.*

tau·to·mer·ic (tô′tə·mer′ik) *adj.* Having the property of tautomerism.

tau·tom·er·ism (tô·tom′ə·riz′əm) *n. Chem.* The property, exhibited by certain substances and compounds when subjected to appropriate chemical reaction, of assuming either of two interconvertible atomic structures, **tau·to·mers** (tô′tə·mərz), which are in equilibrium with each other. [<TAUTO- + Gk. *meros* part]

tau·tom·er·i·za·tion (tô·tom′ər·ə·zā′shən, -ī·zā′-) *n. Chem.* Conversion into a tautomeric structure.

tau·to·nym (tô′tə·nim) *n.* An instance of tautonymy.

tau·ton·y·my (tô·ton′ə·mē) *n. pl.* **·mies** *Biol.* **1** The possession by two or more distinct plants or animals of the same generic and specific names: prohibited by the rules of scientific nomenclature. **2** Identity of the generic, specific, and subspecific names of a given plant or animal, as *Bison bison bison*: a permitted practice. [<TAUTO- + Gk. *onyma* name] — **tau·to·nym·ic** (tô′tə·nim′ik) *adj.*

tav (täv) *n.* The twenty-second Hebrew letter. Also **taw.** See ALPHABET.

tav·ern (tav′ərn) *n.* **1** A public house where travelers and other guests are accommodated with lodging, food, and drink. **2** A house licensed to retail liquors to be drunk on the premises. [<OF *taverne* <L *taberna* hut, booth]

tav·ern·er (tav′ər·nər) *n. Archaic* A tavern-keeper; also, one who frequents taverns.

taw[1] (tô) *v.t.* **1** To convert into leather by some process other than soaking in tanning liquor, as by using alum and salt. **2** *Brit. Dial.* To beat; torture; vex; also, to harden or prepare. [OE *tawian* prepare, harass] — **taw′er** *n.*

taw[2] (tô) *n.* **1** A game of marbles. **2** The line from which marble-players shoot. **3** A marble used for shooting. — *v.i.* To shoot a marble or come to the mark before shooting. [<Scand. Cf. ON *taug* string.]

taw·dry (tô′drē) *adj.* **·dri·er, ·dri·est** Showy without elegance; excessively ornamental; gaudy. — *n.* Cheap, pretentious finery. [Short for *tawdry lace,* alter. of *St. Audrey's lace,* a type of silk neckpiece sold at St. Audrey's Fair at Ely, England] — **taw′dri·ly** *adv.* — **taw′dri·ness** *n.*

taw·ie (tô′ē) *adj. Scot.* Docile; tame: said of a horse, etc.

Taw·ney (tô′nē), **Richard Henry,** born 1880, English economist and historian.

taw·ny (tô′nē) *adj.* **·ni·er, ·ni·est** Tan-colored; brownish-yellow. Also **taw′ney.** [<AF *taune* <OF *tanné,* pp. of *tanner* tan] — **taw′ni·ness** *n.*

taw·pie (tô′pē) *n. Scot.* A foolish young woman.

taws (tôz) *Scot. n.* A whip made of a leather strap cut into thongs or of several thongs on a handle. — *v.t.* To flog; scourge. Also **tawse.**

tax (taks) *n.* **1** A compulsory contribution levied upon persons, property, or business for the support of government; by extension, any proportionate assessment, as on the members of a society. **2** A heavy demand on one's powers or resources; an onerous duty or requirement; a burden. — **direct tax** A tax, as on property or income, which the taxpayer cannot shift to another person. — **excise tax** An internal-revenue tax on domestic manufactures, levied before they are sold to the consumer. The term has been extended to include license duties. — **income tax** A tax levied on the income or profits of individuals and of corporations. See CAPITAL LEVY. — **indirect tax** A tax, such as a customs duty, paid by one person but ultimately shifted to the consumer. — **nuisance tax** A tax which yields little benefit in proportion to the amount of discontent it causes. — **single tax** A tax to be obtained from a single source, especially from a levy on land and natural resources, as a substitute for all other forms of taxation. The theory was first proposed by John Locke, and was elaborated and popularized in the 19th century by Henry George. [<*v.*] — *v.t.* **1** To impose a tax on; subject to taxation. **2** *Law* To settle or fix (amounts) as duly chargeable in any judicial matter: to *tax* costs. **3** To subject to a severe demand; impose a burden or load upon; task: He *taxes* my patience. **4** To make an accusation against; charge; also, to blame; censure: usually with *with.* [<OF *taxer* <L *taxare* estimate, appraise. Related to TASK.] — **tax′·a·bil′i·ty, tax′a·ble·ness** *n.* — **tax′a·ble** *adj.* — **tax′a·bly** *adv.* — **tax′er** *n.*

Synonyms (noun): assessment, custom, demand, duty, exaction, excise, impost, rate, rating, toll, tribute.

Tax may appear as a combining form in hyphemes or solidemes, or as the first element in two-word phrases, with the meaning of definition 1:

tax-assessor	**tax-evader**	**tax payment**
tax burden	**tax-evading**	**tax proposal**
tax-burdened	**tax-exempt**	**tax receipt**
tax claim	**tax-free**	**tax-repeal**
tax-collecting	**tax-laden**	**tax revenue**
tax-collector	**tax law**	**tax-ridden**
tax-cut	**tax levy**	**tax-supported**
tax-dodger	**taxman**	**tax system**
tax-dodging	**tax-paid**	**taxwise**

tax·a·ceous (tak·sā′shəs) *adj. Bot.* Designating or belonging to a widely distributed family (*Taxaceae*) of typically evergreen shrubs and trees, the yew family, having one or two integuments and drupelike or, rarely, cone fruits. [<NL <L *taxus* yew]

tax·a·tion (tak·sā′shən) *n.* The act of taxing; the amount assessed as a tax.

Tax·co (tas′kō, *Sp.* täs′kō) A resort city in Guerrero, SW Mexico. Officially **Tax·co de Al·ar·cón** (täs′kō thā äl′är·kōn′).

tax·gath·er·er (taks′gath′ər·ər) *n.* A collector of taxes. — **tax′gath′er·ing** *n.* & *adj.*

tax·i (tak′sē) *n.* A taxicab. — *v.* **tax·ied, tax·i·ing** or **tax·y·ing** *v.i.* **1** To ride in a taxicab. **2** To move along the ground or on the surface of the water under its own power, as an airplane before taking off or after landing. — *v.t.* **3** To cause (an airplane) to taxi. [<TAXI(CAB)]

tax·i·arch (tak′sē·ärk) *n.* The commander of a division of an ancient Greek army. [<Gk. *taxiarchēs* <*taxis* division of an army + *archos* leader <*archein* rule]

tax·i·cab (tak′sē·kab′) *n.* A passenger vehicle, usually an automobile fitted with a taximeter, available for hire. [Short for *taximeter cab*]

taxi dancer *U.S.* A girl employed by a dance hall or cabaret to dance with patrons for a certain fee. [<*taxi-* hired, as in *taxicab* + DANCER]

tax·i·der·mist (tak′sə·dûr′mist) *n.* One who practices taxidermy.

tax·i·der·my (tak′sə·dûr′mē) *n.* The art or process of stuffing and mounting the skins of dead animals for preservation or exhibition. [<Gk. *taxis* arrangement + *derma* skin] — **tax′i·der′mal, tax′i·der′mic** *adj.*

tax·i·me·ter (tak′si·mē′tər) *n.* **1** An instrument for measuring distances and recording fares. **2** A taxicab equipped with a taximeter. [<F *taximètre* <*taxe* tariff + *mètre* a meter]

tax·ine (tak′sēn, -sin) *n. Chem.* A yellow-white, poisonous alkaloid, $C_{37}H_{51}O_{10}N$, from the needles and seed of the English yew (*Taxus baccata*). It produces convulsion and paralyzes the heart. Also **tax·in** (tak′sin). [<L *taxus* yew + -INE[2]]

tax·i·plane (tak′sē·plān′) *n.* An airplane available for hire as a public vehicle.

tax·is (tak′sis) *n.* **1** *Surg.* A methodical application of manual pressure, as on a hernial tumor, for restoring the parts to their normal place. **2** *Zool.* The involuntary movement of an organism or cell, as a zoospore, in response to an external stimulus; specifically, a movement involving locomotion or change of place. Compare TROPISM. **3** In ancient Greece, a body of troops of varying size. **4** *Obs.* Order; arrangement, as of words in a sentence. [<Gk., arrangement <*tassein* arrange]

-taxis *combining form* Order; disposition; arrangement: *thermotaxis.* Also spelled **-taxy.** [<Gk. *taxis* arrangement]

tax·ite (tak′sīt) *n.* A volcanic rock which has crystallized in such manner as to have a clastic appearance. [<Gk. *taxis* arrangement + -ITE[1]] — **tax·it·ic** (tak·sit′ik) *adj.*

tax·on·o·mist (tak·son′ə·mist) *n.* One versed in taxonomy. Also **tax·on′o·mer.**

tax·on·o·my (tak·son′ə·mē) *n.* **1** The department of knowledge that embodies the laws and principles of classification. **2** *Biol.* The systematic arrangement of plant and animal organisms according to accepted diagnostic criteria which determine their assignment to each of the following major groups, beginning with the most inclusive: kingdom, phylum or division, class, order, family, genus, and species. [<F *taxonomie* <Gk. *taxis* arrangement + *nomos* law] — **tax·o·nom·ic** (tak′sə·nom′ik) or **·i·cal** *adj.* — **tax′o·nom′i·cal·ly** *adv.*

tax·pay·er (taks′pā′ər) *n.* **1** One who pays any tax. **2** A building, the rental from which is intended to cover merely the taxes on the land.

tax title The title conveyed to a purchaser of property sold for non-payment of taxes.

Ta·yg·e·ta (tā·ij′ə·tə) One of the Pleiades.

Tay·lor (tā′lər), **Bayard,** 1825–78, U. S. writer. — **Frederick Winslow,** 1856–1915, U. S. engineer; developed scientific shop management. — **Jeremy,** 1613–67, English bishop and author. — **(Joseph) Deems,** born 1885, U. S. composer. — **Laurette,** 1887–1946, *née* Cooney, U.S. actress. — **Myron,** 1874–1959, U.S. lawyer, businessman, and diplomat. — **Tom,** 1817–80, English dramatist. — **Zachary,** 1784–1850, U. S. general; president of the United States 1849–50.

Tay River (tā) The largest river in Scotland, flowing 118 miles SW and SE, from eastern Perthshire, near the Argyll border, to the **Firth of Tay,** an estuary of the North Sea extending 25 miles into central Scotland.

taz·za (tät′tsä) *n. Italian* A flat ornamental cup, especially one supported on a high foot.

T-base (tē′bās′) *n.* Two strips of wood nailed together in the form of a T and serving as a base for the tripod of a machine-gun.

Tbi·li·si (tpi′li·sē) The Georgian name for TIFLIS.

Tchad (chȧd) The French name for CHAD.

Tchai·kov·sky (chī·kôf′skē), **Peter (Pëtr) Ilich,** 1840–93, Russian composer.

tchick (chik) *n.* A sound made by pressing the tongue against the roof of the mouth and sucking it back, as in urging a horse. — *v.i.* To make a *tchick.* [Imit.]

tea (tē) *n.* **1** An evergreen Asian shrub or small tree (*Thea sinensis*), having a compact head of leathery, toothed leaves and white or pink flowers. **2** The prepared leaves of this plant, or an infusion of them used as a beverage. The difference between **black tea** and **green tea** is the result of manipulation, the latter being withered by steaming, thus retaining the green color, while leaves simply dried turn black. **3** Any infusion, decoction, solution, or extract to be used as a beverage or medicinally: beef *tea.* **4** The leaves of a particular variety of plant, prepared for making a beverage, or for medicinal purposes: senna *tea.* **5** A light evening or afternoon meal; also, a social gathering at which tea is served. [<Chinese *ch'a,* dial. Chinese *t'e*]

Tea may appear as a combining form in hyphemes or solidemes, or as the first element in two-word phrases; as in:

tea-blending	**tea-making**
tea bowl	**tea merchant**
teabox	**tea-packer**
teacart	**tea-packing**
tea china	**tea plant**
tea-colored	**tea-planter**
tea crop	**tea-planting**
tea dealer	**tea-producer**
tea-drinker	**tea-producing**
tea-drinking	**tea table**
tea-farming	**tea-taster**
tea-grower	**tea-tasting**
tea-growing	**teatime**
tea jar	**tea trade**
tea leaf	**tea tray**
tea-loving	**tea tree**
tea-maker	**teaware**

tea bag A small porous sack of cloth or paper

containing tea leaves, which is immersed in water to make tea.
tea ball 1 A perforated metal ball, filled with tea leaves, to be dropped or suspended in boiling water to make tea. 2 A tea bag.
tea·ber·ry (tē′ber′ē) n. pl. ·ries 1 The wintergreen, whose leaves are sometimes mixed with or used as tea. 2 The berry of this plant.
tea biscuit A biscuit or cracker, usually short and sweetened, served with tea.
tea caddy See CADDY[1] (def. 1).
teach (tēch) v. **taught, teach·ing** v.t. 1 To impart knowledge to by lessons; give instruction to; guide by precept or example; instruct: to *teach* a class. 2 To give instruction in; make known; communicate the knowledge of: to *teach* French. 3 To train by practice or exercise. — v.i. 4 To follow the profession of teaching. 5 To impart knowledge or skill. [OE *tǣcan, tǣcean*]
Synonyms: discipline, drill, educate, enlighten, indoctrinate, inform, initiate, instruct, nurture, school, train, tutor. To *teach* is to communicate knowledge; to *instruct* is to impart knowledge with special method and completeness; *instruct* has also an authoritative sense nearly equivalent to command. To *educate* is to draw out or develop the mental powers. To *train* is to direct to a certain result powers already existing. *Train* is used in preference to *educate* when the reference is to the inferior animals or to the physical powers of man; as, to *train* a horse; to *train* the hand or eye. To *discipline* is to bring into habitual and complete subjection to authority. To *nurture* is to furnish the care and sustenance necessary for physical, mental, and moral growth; *nurture* is a more tender word than *educate*. See INFORM[1], LEARN.
teach·a·ble (tē′chə·bəl) adj. 1 Capable of being taught; willing to learn; docile. 2 Capable of being imparted by teaching. See synonyms under DOCILE. — **teach′a·bil′i·ty, teach′a·ble·ness** n. — **teach′a·bly** adv.
teach·er (tē′chər) n. One who teaches; specifically, one whose occupation is to teach others; an instructor. See synonyms under MASTER.
teacher bird 1 The ovenbird. 2 The North American red-eyed vireo. [Imit.; because its cry sounds like "teacher"]
teachers' institute See under INSTITUTE.
teach·ing (tē′ching) n. 1 The act or business of a teacher. 2 That which is taught. See synonyms under DOCTRINE, EDUCATION, NURTURE.
tea cozy See COZY, n.
tea·cup (tē′kup′) n. 1 A small cup suitable for serving tea. 2 As much as a teacup will hold, usually four fluid ounces: also **tea′cup·ful′** (-fŏŏl′).
tea·house (tē′hous′) n. In the Orient, a public place serving tea and other light refreshments.
teak (tēk) n. 1 A large East Indian tree (*Tectona grandis*) of the vervain family, yielding a very hard, durable timber highly prized for shipbuilding. 2 The wood of this tree. [<Malayalam *tēkka*]
tea·ket·tle (tē′ket′l) n. A kettle with a spout, used for boiling water for culinary purposes.
teal (tēl) n. 1 Any of several small, short-necked river ducks (genera *Nettion* and *Querquedula*); especially, the common teal (*N. crecca*) of the Old World and the similar North American **green-winged teal** (*N. carolinense*) having grayish wing coverts and the head slightly crested. 2 A darkish, dull-blue color with a greenish cast. [ME *tele*]
team (tēm) n. 1 Two or more beasts of burden harnessed together: often including harness and vehicle; also, a single horse and vehicle. 2 A set of workers, or players competing in a game: a baseball *team*. 3 *Dial.* A flock; brood. 4 *Obs.* Race; lineage. — v.t. 1 To convey with a team. 2 To harness together in a team. — v.i. 3 To drive a team as a business. 4 To form a team; work as a team: to *team up*. — v.t. Of or pertaining to a team. ♦ Homophone: *teem*. [OE *tēam* offspring, succession, row. Related to TEEM[1].] — **team′ing** n.
team boat A paddle-wheel ferryboat propelled by horse power.
team·mate (tēm′māt′) n. A fellow player on a team.

team play Cooperation.
team·ster (tēm′stər) n. 1 One who drives or owns a team. 2 One who drives a truck or other commercial vehicle.
team·work (tēm′wûrk′) n. 1 Work done by or requiring to be done by or with a team of horses: distinguished from manual labor. 2 Unity of action by the players on an athletic team to further the success of the team. 3 Cooperation.
tea party A social gathering at which tea and light sandwiches or cakes are the principal refreshments.
tea·pot (tē′pot′) n. A vessel with a spout and handle in which tea is made and from which it is served.
tea·poy (tē′poi) n. A small three- or four-legged table for holding a tea service; originally, any small, ornamental, three-legged stand or table. [<Hind. *tipāī* < *tīn* three + Persian *pāē* foot]
tear[1] (târ) v. **tore, torn, tear·ing** v.t. 1 To pull apart, as cloth; part or separate by pulling; rip; rend. 2 To make by rending or tearing: to *tear* a hole in a dress. 3 To injure or lacerate, as skin. 4 To divide; disrupt: a party *torn* by dissension. 5 To distress or torment; anguish: The sight *tore* his heart. — v.i. 6 To become torn or rent. 7 To move with haste and energy. See synonyms under REND. — n. 1 A fissure made by tearing; a rent; an act of tearing. 2 *Slang* A carouse; a spree; frolic. 3 A rushing motion: to start off with a *tear*; also, any violent outburst, as of anger, enthusiasm, etc. ♦ Homophone: *tare*. [OE *teran*]
tear[2] (tir) n. 1 ♦ A drop of the saline liquid secreted by the lacrimal gland, for moistening the eye. 2 Something resembling or suggesting a drop of the lacrimal fluid. 3 A drop of any liquid. 4 A droplike portion, as of glass, amber, etc. 5 pl. Sorrow; lamentation. ♦ Homophone: *tier*. [OE *tēar*] — **tear′less** adj. — **tear′y** adj.
Tear may appear as a combining form in hyphemes or solidemes, with the meaning of definition 1:

tear-arresting	tear-lined
tear-baptized	tear-marked
tear-blinded	tear-mocking
tear-compelling	tear-moistened
tear-creating	tear-mourned
tear-dimmed	tearproof
tear-dropped	tear-provoking
tear-falling	tear-salt
tear-filled	tear-shedding
tear-freshened	tear-stained
tear-glistening	tear-swollen
tear-kissed	tear-wrung

tear·drop (tir′drop′) n. A tear.
tear·er (târ′ər) n. 1 One who or that which tears or rends. 2 *Slang* Anything remarkable for its size, intensity, violence, etc., as a storm.
tear·ful (tir′fəl) adj. 1 Weeping abundantly. 2 Causing tears. — **tear′ful·ly** adv. — **tear′ful·ness** n.
tear gas (tir) A lacrimator.
tear·ing (târ′ing) adj. *Colloq.* 1 Rushing along as in a hurry or rage. 2 Tremendous; mighty.
tear-jerk·er (tir′jûr′kər) n. *U. S. Slang* A story or play charged with sentimental sadness.
tea·room (tē′rōōm′, -rŏŏm′) n. A restaurant serving tea and other refreshments.
tea rose 1 Any of numerous garden roses thought to be tea-scented, primarily hybrids bred from the Chinese *Rosa odorata*. 2 A yellowish-pink color of many hues.
tear·proof (târ′prŏŏf′) v.t. To render less subject to shear: to *tearproof* paper.
tear-sheet (târ′shēt′) n. A page torn or cut from a magazine, book, or newspaper, containing matter of particular interest.
Teas·dale (tēz′dāl), **Sara**, 1884-1933, U. S. poet.
tease (tēz) v. **teased, teas·ing** v.t. 1 To annoy or harass with continual importunities, raillery, etc.; pester. 2 To scratch or dress in order to raise the nap, as cloth with teasels. 3 To tear or pull apart with instruments, as tissues in examination. 4 To comb or card, as wool or flax; also, to pick or shred, as hard-packed tobacco. — v.i. 5 To annoy a person in a facetious or petty manner. See

synonyms under AFFRONT. — n. 1 One who or that which teases. 2 The act of teasing or the state of being teased. [OE *tǣsan* tease, pluck, pull about] — **teas′ing** n. & adj. — **teas′ing·ly** adv.
tea·sel (tē′zəl) n. 1 A coarse, prickly Old World herb (genus *Dipsacus*) of which the flower head is covered with hooked bracts, especially the **fuller's teasel** (*D. fullonum*). 2 The rough bur of this plant, or a mechanical substitute used in dressing cloth. — v.t. **·seled** or **·selled**, **·sel·ing** or **·sel·ling** To use a teasel on; raise the nap of with a teasel. Also **tea′zel, tea′zle**. [OE *tǣsel*] — **tea′sel·er** or **tea′sel·ler** n.

TEASEL
(Plant to 5 feet or more)

teas·er (tē′zər) n. 1 One who or that which teases, as a machine used for teasing wool. 2 Anything tempting or whetting the appetites. 3 The barrier at the front of the stage. Compare BORDER.
tea service The articles used in serving tea: a silver *tea service*. Also **tea set**.
tea·shop (tē′shop′) n. 1 A tearoom. 2 *Brit.* A lunchroom.
tea·spoon (tē′spōōn′, -spŏŏn′) n. 1 A small spoon used for stirring tea, etc. 2 As much as a teaspoon will hold, 1/3 of a tablespoon, usually 1 1/3 fluid drams: also **tea′spoon·ful′** (-fŏŏl′).
teat (tēt) n. The protuberance on the breast or udder of most female mammals, through which the milk is drawn; a nipple; pap; dug. [<OF *tete* <Gmc.]
tea wagon A table on wheels for use in serving tea or refreshments.
Te·bet (tā-vāth′, tā′ves) A Hebrew month. Also **Te·beth′**. See CALENDAR (Hebrew).
Te·briz (tə-brēz′) See TABRIZ.
tech·ne·ti·um (tek-nē′shē·əm) n. The chemical element of atomic number 43 (symbol Tc), artificially produced by the bombardment of molybdenum with neutrons or deuterons: it displaces the hypothetical element *masurium*. [<NL <Gk. *technētos* artificial]
tech·nic (tek′nik) n. 1 Technique. 2 pl. The theory of an art or of the arts; specifically, the study of the techniques of an art. 3 pl. Technical rules, methods, etc. 4 pl. Technology. — adj. Technical.
tech·ni·cal (tek′ni·kəl) adj. 1 Pertaining to some particular art, science, or trade. 2 Peculiar to or from a specialized field of knowledge. 3 Of or pertaining to the mechanical arts. 4 Employing a specialized vocabulary, as in a treatise or textbook. 5 Considered in terms of an accepted body of rules and regulations: a *technical* defeat. 6 Designating a money market in which prices are for the most part determined by speculation or manipulation. [<Gk. *technikos* <*technē* art] — **tech′ni·cal·ly** adv. — **tech′ni·cal·ness** n.
tech·ni·cal·i·ty (tek′ni·kal′ə·tē) n. pl. **·ties** 1 The state of being technical. 2 The use of technical terms. 3 A technical point peculiar to some profession, art, trade, etc. 4 A petty distinction; quibble. Also **tech′nism**.
technical knockout In boxing, a victory awarded when one fighter has been beaten so severely that the referee discontinues the fight. Abbr. *t.k.o.*, T.K.O., or TKO.
tech·ni·cian (tek-nish′ən) n. 1 One skilled in the handling of instruments or in the performance of tasks requiring specialized training. 2 A rating in the armed services including those qualified for technical work; also, one having such a rating.
Tech·ni·col·or (tek′ni·kul′ər) n. A motion-picture process by which two or more sets of film of the same scene are photographed through different color filters and assembled in one positive film that reproduces the colors of the original scene: a trade name.
tech·nique (tek-nēk′) n. Working methods or manner of performance, as in art, science, etc. [<F <Gk. *technikos*. See TECHNICAL.]
tech·no- combining form 1 Art; skill; craft: *technology*. 2 Technical; technological. Also, before vowels, **techn-**. [<Gk. *technē* an art; skill]
tech·noc·ra·cy (tek·nok′rə·sē) n. pl. **·cies** 1 A

technography (tek·nog′rə·fē) n. 1 Description of the arts and crafts. 2 The scientific study of the development and geographic distribution of technical processes.

tech·nol·a·tor (tek·nol′ə·tər) n. One who has an excessive admiration for or belief in technology, especially in relation to social problems; immoderate worship of techniques, gadgets, machinery, and the like. [<TECHNO- + Gk. *latris* servant < *latron* pay, hire] — **tech·nol′a·try** n.

tech·no·lith·ic (tek′nə·lith′ik) adj. *Anthropol.* Pertaining to or designating those stone implements which were deliberately fashioned for some intended purpose.

tech·no·log·i·cal (tek′nə·loj′i·kəl) adj. Of, pertaining to, associated with, produced or affected by technology, especially in relation to improvements resulting from the application of technical advances in industry, manufacturing, commerce, and the arts. Also **tech′no·log′ic**. — **tech′no·log′i·cal·ly** adv.

tech·nol·o·gy (tek·nol′ə·jē) n. 1 Theoretical knowledge of industry and the industrial arts. 2 The application of science to the arts. 3 That branch of ethnology which treats of the development of the arts. — **tech·nol′o·gist** n.

tech·y (tech′ē) adj. **tech·i·er**, **tech·i·est** Peevishly sensitive; irritable; touchy. Also spelled *tetchy*. [< OF *teche* mark, quality] — **tech′i·ly** adv. — **tech′i·ness** n.

tec·tol·o·gy (tek·tol′ə·jē) n. The branch of morphology that treats of the manner in which organic forms are built up. [<Gk. *tektōn* carpenter, builder + -LOGY] — **tec·to·log′i·cal** (tek′tə·loj′i·kəl) adj.

tec·ton·ic (tek·ton′ik) adj. 1 Of or pertaining to building or construction. 2 *Geol.* a Characteristic of or relating to the structure of the earth's crust, especially as due to deformation. b Denoting the forces producing such structures. [<L *tectonicus* <Gk. *tektonikos* < *tektōn* carpenter]

tec·ton·ics (tek·ton′iks) n. pl. (construed as singular) 1 The science or art of constructing functionally beautiful buildings or things. 2 The geology of earth structure.

tec·tri·ces (tek·trī′sēz, tek′tri-) n. pl. of *tectrix* (tek′triks) *Ornithol.* The wing coverts of a bird. [<NL <L *tectus*, pp. of *tegere* cover] — **tec·tri′cial** (-trish′əl) adj.

Te·cum·seh (ti·kum′sə), 1768?–1813, Shawnee chief, an ally of Britain in the War of 1812, during which he was killed.

ted (ted) v.t. **ted·ded**, **ted·ding** To turn over and strew about, or spread loosely for drying, as newly mown grass. [Prob. <Scand. Cf. ON *tethja* spread manure.]

Ted (ted), **Ted·dy** (ted′ē) Diminutives of EDWARD, THEODORE.

ted·der (ted′ər) n. 1 One who or that which teds. 2 *Agric.* A machine for spreading hay to dry.

ted·dy (ted′ē) n. pl. **·dies** A short undergarment combining chemise and drawers in one. [Origin unknown]

teddy bear (ted′ē) A toy bear, usually covered with plush. Also **Teddy bear**. [after *Teddy*, a nickname of Theodore Roosevelt]

Te Deum (tē dē′əm) 1 An ancient Christian hymn beginning with these words. 2 The music to which this hymn is set. 3 Any thanksgiving service in which it is sung. [<L *Te Deum* (*laudamus*) (we praise) Thee, O God]

te·di·ous (tē′dē·əs) adj. 1 Causing weariness; wearisome; boring. 2 *Obs.* Moving slowly. [<LL *taediosus* <L *taedium* tedium, weariness] — **te′di·ous·ly** adv. — **te′di·ous·ness** n. — *Synonyms*: dilatory, dreary, dull, fatiguing, irksome, monotonous, slow, sluggish, tardy, tiresome, wearisome. See WEARISOME. *Antonyms*: active, alert, animated, brilliant, energetic, exciting, fresh, quick, stirring, vigorous, vivid.

te·di·um (tē′dē·əm) n. Tediousness; wearisomeness. [<L *taedium* <*taedere* vex, weary]

tee[1] (tē) n. 1 The letter T. 2 Something resembling the form of the letter T. 3 *Mining* The point of the meeting of two veins lying nearly at right angles to each other without intersecting. — adj. T-shaped. [OE *te* <L *te*, name of the letter T]

tee[2] (tē) n. 1 A little cone, as of damp sand or of wood, on which a golf ball is placed in making the first play to a hole. 2 The teeing ground in golf. — v.t. & v.i. **teed**, **tee·ing** To place (the ball) on a tee before striking it. — **to tee off** To strike (the ball) in starting play. [Prob. <TEE[3]]

tee[3] (tē) n. In certain games, a mark toward which the balls, quoits, etc., are directed, as in curling. — **to a tee** Exactly; as precisely as possible. [? <TEE[1]]

tee[4] (tē) n. A finial in the form of a conventionalized umbrella, used on pagodas, etc. [<Burmese *h'ti* umbrella]

teeing ground An area defined by marks within which the golf tee must be built or placed.

teem[1] (tēm) v.i. 1 To be full, as if at the point of producing; be full to overflowing; abound. 2 *Obs.* To bear young. — v.t. 3 To produce or bring forth, as offspring: often figuratively. ◆ Homophone: *team*. [OE *tīeman*, prob. <*tēam* progeny. Related to TEAM.] — **teem′er** n.

teem[2] (tēm) v.i. To pour; come down heavily: said of rain. — v.t. *Obs.* To pour out; empty. ◆ Homophone: *team*. [<ON *tœma* empty] — **teem′er** n.

teem·ing[1] (tē′ming) adj. 1 Prolific. 2 Full; overflowing. 3 Produced in great quantity. See synonyms under FERTILE.

teem·ing[2] (tē′ming) adj. Raining heavily.

teen (tēn) n. *Scot. & Brit. Dial.* Grief; trouble; also, provocation; vexation; anger. [OE *tēona* injury, vexation] — **teen′ful** adj. — **teen′ful·ly** adv.

-teen suffix Plus ten: used in cardinal numbers from 13 to 19 inclusive: *fifteen*. [OE -*tēne* < *tēn* ten]

teen age The age from 13 to 19 inclusive; hence, adolescence. — **teen′-age′** adj.

teen-ag·er (tēn′ā′jər) n. A person of teen age.

teens (tēnz) n. pl. The numbers that end in -*teen*; the years of one's age from 13 to 19 inclusive.

tee·ny (tē′nē) adj. **·ni·er**, **·ni·est** *Colloq.* Tiny. [Var. of TINY]

tee·pee (tē′pē) See TEPEE.

tee shirt (tē) See T-SHIRT.

Tees River (tēz) A river in northern England, flowing 70 miles east from eastern Cumberland, between Durham and York, to the North Sea.

tee·ter (tē′tər) v.i. 1 To see-saw. 2 To walk or move with a swaying or tottering motion. 3 To vacillate; waver. — v.t. 4 To cause to teeter. — n. 1 An oscillating motion. 2 A see-saw. 3 The spotted sandpiper: so called from its jerky motions. [<dial. E *titter*, prob. <ON *titra* tremble, shiver]

teeter board A see-saw.

tee·ter-tot·ter (tē′tər·tot′ər) n. A see-saw. [TEETER + TOTTER]

teeth (tēth) Plural of TOOTH.

teethe (tēth) v.i. **teethed**, **teeth·ing** To cut or develop teeth.

teeth·ing (tē′thing) n. The process of developing and cutting teeth, especially the milk teeth; dentition.

teething ring A ring of hard rubber, bone, or ivory for a teething baby to bite on.

tee·to·tal (tē·tōt′l) adj. 1 Pertaining to total abstinence from intoxicants. 2 Total; entire. [<TOTAL, with emphatic repetition of initial letter] — **tee·to′tal·ism** n. — **tee·to′tal·ly** adv.

tee·to·tal·er (tē·tōt′l·ər) n. One who abstains totally from intoxicants as beverages. Also **tee·to′tal·ist**, **tee·to′tal·ler**.

tee·to·tum (tē·tō′təm) n. 1 A kind of top having lettered and numbered sides: used in the game of put and take. 2 A child's toy, often four-sided, pierced by a peg and spun by the fingers. Also spelled *tetotum*: sometimes called *toddle-top*. [<T-*totum* <T + L *totus* all, from the fact that the side marked with a T wins the entire stake]

Teflon (tef′lon) n. A chemically resistant, heat-stable plastic polymer of fluorine and ethylene having wide application in industry and electronics: a trade name.

teg·men (teg′mən) n. pl. **·mi·na** (-mə·nə) 1 A covering or coat. 2 *Bot.* The soft inner covering of a seed. Also **teg′u·men** (-yə·min). [<L <*tegere* cover] — **teg′mi·nal** adj.

Teg·nér (teng·nâr′), **Esaias**, 1782–1846, Swedish poet. also **Teng′ner**.

Te·gu·ci·gal·pa (tā·gōō′sē·gäl′pä) The capital of Honduras, in the SW part of the country.

teg·u·la (teg′yə·lə) n. pl. **·lae** (-lē) A tile. [<L *tegula* roof tile < *tegere* cover]

teg·u·lar (teg′yə·lər) adj. 1 Pertaining to or resembling tiles. 2 Arranged like tiles. 3 Formed of overlapping plates or scales. Also **teg′u·lat′ed**. — **teg′u·lar·ly** adv.

teg·u·ment (teg′yə·mənt) n. A covering or envelope; an integument. [<L *tegumentum* < *tegere* cover] — **teg·u·men·ta·ry** (teg′yə·men′tər·ē), **teg′u·men′tal** adj.

te·hee (tē·hē′) v.i. **·heed**, **·hee·ing** To laugh frivolously or with derision; titter; giggle. — interj. An imitative exclamation. — n. A restrained laugh; titter. [Imit.]

Te·he·ran (tə·ə·rän′, -ran′; *Persian* te·hrän′) The capital of Iran, in the north central part; Roosevelt, Churchill, and Stalin conferred here in November, 1943. Also **Te·hran′**.

Teh·ri (tā′rē) A district of northern Uttar Pradesh State, India; 4,516 square miles; before 1949 a princely state; capital, Tehri. Also **Teh′ri–Garh·wal′** (-gûr′wäl′).

Te·huan·te·pec (te·wän′te·pek′), **Isthmus of** The narrowest part of southern Mexico (125 miles wide) between the Gulf of Tehuantepec, an arm of the Pacific Ocean, about 300 miles long, NW to SE, on the coast of southern Mexico, and the Gulf of Campeche.

Te·huan·te·pec Winds (te·wän′te·pek′) n. pl. Violent NE winds striking the Gulf of Tehuantepec in winter and early spring.

Te·huel·che (te·wel′che) n. One of a group of tribes of South American Indians inhabiting Patagonia, noted for their great height. — **Te·huel′che·an** (-chē·ən) adj.

te ig·i·tur (tē ij′ə·tər) The prayer or paragraph beginning the canon of the mass in Latin liturgies. [<L, thee therefore]

teil (tēl) n. 1 The linden. 2 The terebinth pistache (*Pistacia terebinthus*) of the Bible. Also called *teyl tree*. [<OF <L *tilia* lime tree]

teind (tēnd) n. *Scot.* A tithe or tithes.

Te·jo (tā′zhoō) The Portuguese name for the TAGUS.

Te·ju·co (tə·zhoō′koō) A former name for Diamantina, Brazil.

tek·non·y·my (tek·non′ə·mē) n. *Anthropol.* The custom of renaming a parent after his or her child. [<Gk. *teknon* child + *onyma*, *onoma* name]

tek·tite (tek′tīt) n. *Geol.* One of several kinds of rounded, glasslike objects, variously named and of unknown origin, found in widely scattered parts of the world and believed by some to be fragments of a shattered planet. [<Gk. *tēktos* molten + -ITE[1]]

tel- Var. of TELO-[1].

te·la (tē′lə) n. pl. **·lae** (-lē) 1 A tissue or weblike membrane. 2 *Anat.* One of the thin membranes (**tela choroidea**), prolongations of the pia mater, that cover the third and fourth ventricles of the brain. [<L, web]

tel·aes·the·sia (tel′əs·thē′zhə, -zhē·ə) See TELESTHESIA.

tel·a·mon (tel′ə·mon) n. pl. **tel·a·mo·nes** (tel′ə·mō′nēz) *Archit.* A male figure used as a pillar to support an entablature, etc. Compare ATLANTES, CARYATID. [<L <Gk. *telamōn* < *tlēnai* bear]

Tel·a·mon (tel′ə·mon) In Greek legend, the father of Ajax.

tel·an·gi·ec·ta·sia (tel·an′jē·ek·tā′zhə, -zhē·ə) n. *Pathol.* Permanent dilatation of the small arteries or capillaries, producing a vascular tumor: often seen in the form of maternal birthmarks; wine spots. Also **tel·an·gi·ec·ta·sis** (tel·an′jē·ek′tə·sis). [<NL <Gk. *telos* end + *angeion* vessel + *ekstasis* dilatation] — **tel·an′gi·ec·tat′ic** (-tat′ik) adj.

TELAMON

tel·au·to·gram (tel·ô′tə·gram) *n.* A record made by a telautograph.
tel·au·to·graph (tel·ô′tə·graf, -gräf) *n.* An electromagnetically operated device for reproducing writing or drawings at a distance.
Tel A·viv (tel′ ä·vēv′) The largest city of Israel, on the Mediterranean: since 1950 includes Jaffa.
tel·e (tel′ē) *n. Psychoanal.* The development between two or more persons of a relationship based on the gradual recognition of mutual attractions or repulsions either within or between social groups. [<Gk. *têle* far off]
tele- *combining form* 1 Far off; operating at a distance: *telegraph.* 2 Television; related to or transmitted by television: *telecast.* Also spelled *telo-.* Also, before vowels, **tel-.** [<Gk. *têle* far]
tel·e·car·di·o·gram (tel′ə·kär′dē·ə·gram′) *n.* A cardiogram electrically produced at a distance from the subject.
tel·e·cast (tel′ə·kast, -käst) *v.t. & v.i.* **·cast** or **·cast·ed**, **·cast·ing** To broadcast by television. — *n.* A program broadcast by television.
tel·e·com·mu·ni·ca·tion (tel′ə·kə·myoō′nə·kā′shən) *n.* 1 The art and science of communicating at a distance, especially by means of electromagnetic impulses, with or without wires, as in radio, radar, television, telegraphy, telephony, etc. Also **tel′e·com·mu·ni·ca′tions.** 2 Any message so transmitted.
tel·e·du (tel′ə·doō) *n.* A small, short-tailed East Indian mammal (genus *Mydaus*) which resembles the skunk in color and in its ability to emit a fetid odor when disturbed. [<Malay]
tel·e·fi·nal·ist (tel′ə·fī′nəl·ist) *n.* One who believes in final causes or the working out of a final purpose in life or the universe; a teleologist, especially one who seeks scientific proof of the existence of God.
tel·e·ga (te·le′gä) *n. Russian* A rude four-wheeled wagon with springs, used in Russia.
tel·e·gen·ic (tel′ə·jen′ik) *adj.* Videogenic.
tel·eg·no·sis (tel′əg·nō′sis) *n.* Knowledge of remote happenings by other than normal sensory means, as by clairvoyance. [<TELE- + Gk. *gnōsis* knowing]
Te·leg·o·nus (tə·leg′ə·nəs) In Greek legend, the son of Odysseus and Circe, who unknowingly killed his father in Ithaca and married Penelope, his father's wife.
te·leg·o·ny (tə·leg′ə·nē) *n. Biol.* The alleged influence of a previous sire on the progeny of the same mother from subsequent matings with other males. [<TELE- + -GONY] — **tel·e·gon·ic** (tel′ə·gon′ik), **te·leg′o·nous** *adj.*
tel·e·gram (tel′ə·gram) *n.* A message sent by telegraph. [<TELE- + -GRAM]
tel·e·graph (tel′ə·graf, -gräf) *n.* Any of various devices, systems, or processes for transmitting messages or signals to a distance, especially any form of such apparatus utilizing electromagnetic impulses transmitted by conducting wires between sending and receiving points. — *v.t.* 1 To send (a message) by telegraph. 2 To communicate with by telegraph. — *v.i.* 3 To transmit a message by telegraph. [<TELE- + -GRAPH]
te·leg·ra·pher (tə·leg′rə·fər) *n.* One who is employed in sending telegrams or is skilled in telegraphy. Also **te·leg′ra·phist.**
tel·e·graph·ic (tel′ə·graf′ik) *adj.* Of or pertaining to the telegraph; transmitted by means of telegraphy. Also **tel′e·graph′i·cal.** — **tel′e·graph′i·cal·ly** *adv.*
tel·e·graph·one (tel′ə·leg′rə·fōn) *n.* An instrument for recording and reproducing sound, similar in principle to the tape recorder but adapted for connection with a transmitter or microphone. [<TELE- + -GRA(PH) + -PHONE]
tel·e·graph·o·scope (tel′ə·graf′ə·skōp) *n.* An instrument for transmitting and reproducing a picture telegraphically. [<TELE- + GRAPHO- + -SCOPE]
te·leg·ra·phy (tə·leg′rə·fē) *n.* 1 The process of conveying messages by telegraph. 2 The art or science of the construction and operation of telegraphs.
Tel·e·gu (tel′ə·goō) See TELUGU.
tel·e·ki·ne·sis (tel′ə·ki·nē′sis) *n.* 1 Movement of an object or inanimate body without apparent external cause. 2 The alleged power of a spiritualist medium to bring about such movements without direct or observable contact. — **tel′e·ki·net′ic** (-net′ik) *adj.*
tel·e·lec·tric (tel′i·lek′trik) *adj.* Denoting the transmission, as of music, to a distance by electricity. [<TEL(E)- + ELECTRIC]
Te·lem·a·chus (tə·lem′ə·kəs) In Greek legend, son of Odysseus and Penelope, who helped his father kill his mother's suitors.
tel·e·mark (tel′ə·märk) *n.* In skiing, a turn effected by shifting the weight to one advanced ski and turning its tip inward: used to change direction or stop quickly. [from *Telemark,* Norway]
tel·e·me·chan·ics (tel′ə·mə·kan′iks) *n.* 1 The theory and practice of operating mechanisms from a distance. 2 Remote control operation, as by electromagnetic and radio impulses.
tel·e·me·ter (tə·lem′ə·tər) *n.* 1 An apparatus for determining distances by the measurement of angles. 2 An electrical apparatus for indicating or measuring various quantities and for transmitting the data to a distant point. — **te·lem′e·try** *n.* — **tel·e·met·ric** (tel′ə·met′rik) *adj.*
tel·e·mo·tor (tel′ə·mō′tər) *n.* A hydraulic or electrical device by which power is applied at a distance, especially in operating the steering gear of a vessel by turning the wheel on the bridge.
tel·en·ceph·a·lon (tel′en·sef′ə·lon) *n. Anat.* The terminal division of the neural tube of the embryo from which are developed the cerebral hemispheres and olfactory lobes; the endbrain. [<TELE- + ENCEPHALON] — **tel·en·ce·phal·ic** (tel′en·si·fal′ik) *adj.*
teleo- Var. of TELO-[1].
tel·e·ol·o·gy (tel′ē·ol′ə·jē, tē′lē-) *n.* 1 The branch of cosmology that treats of final causes. See FINAL CAUSE. 2 The philosophical and biological doctrine of design which holds that the phenomena of organic life and development can be explained by conscious or purposive causes directed to definite ends and not by mechanical causes; vitalism as opposed to *mechanism.* 3 The explanation of nature in terms of utility or purpose, especially divine purpose; the study of a creative design in the processes of nature. [<NL *teleologia* <Gk. *telos, teleos* end + *logos* discourse] — **tel·e·o·log·i·cal** (-ə·loj′i·kəl) or **tel′e·o·log′ic** *adj.* — **tel′e·o·log′i·cal·ly** *adv.* — **tel′e·ol′o·gist** *n.*
tel·e·ost (tel′ē·ost, tē′lē-) *n.* Any of a large and widely distributed group or order (*Teleostei*) of fishes having true bones: distinguished from cyclostomes and elasmobranchs. — *adj.* Of, pertaining to, or having the characteristics of the teleosts. Also **tel′e·os′te·an.** [<Gk. *telos* end + *osteon* bone]
te·lep·a·thy (tə·lep′ə·thē) *n.* The supposed communication of one mind with another at a distance by other than normal sensory means; thought-transference. [<TELE- + -PATHY] — **tel·e·path·ic** (tel′ə·path′ik) *adj.* — **tel′e·path′i·cal·ly** *adv.* — **te·lep′a·thist** *n.*
tel·e·phone (tel′ə·fōn) *n.* An instrument for reproducing sound or speech at a distant point, by the electromagnetic transmission of variable audio frequencies over a conducting wire or other communication channel. — **wireless telephone** A radiotelephone. — *v.* **·phoned**, **·phon·ing** — *v.t.* 1 To send by telephone, as a message. 2 To communicate with by telephone. — *v.i.* 3 To communicate by telephone. [<TELE- + -PHONE] — **tel′e·phon′er** *n.*
telephone receiver That part of a telephone in which a diaphragm is caused to vibrate by electric impulses, converting the varying current into sound.
tel·e·phon·ic (tel′ə·fon′ik) *adj.* 1 Of or pertaining to the telephone. 2 Conveying sound to a great distance. Also **tel′e·phon′i·cal.** — **tel′e·phon′i·cal·ly** *adv.*
tel·e·pho·no·graph (tel′ə·fō′nə·graf, -gräf) *n.* A combination of a phonograph and a telephone receiver by which telephone messages can be recorded and then reproduced. — **tel′e·pho′no·graph′ic** *adj.*
te·leph·o·ny (tə·lef′ə·nē) *n.* The art or process of communicating by telephone, with or without wires directly connecting the terminal points.
tel·e·pho·to (tel′ə·fō′tō) *adj.* 1 Denoting a combination of lenses which produces a large image of a distant object in a camera; telephotographic. 2 Pertaining to telephotography.
tel·e·pho·to·graph (tel′ə·fō′tə·graf, -gräf) *n.* 1 A picture transmitted by wire or radio. 2 A picture made with a telephoto lens. — **tel′e·pho′to·graph′ic** *adj.*
tel·e·pho·tog·ra·phy (tel′ə·fə·tog′rə·fē) *n.* 1 The art of producing photographic images of distant objects on a larger scale than is possible with an ordinary camera. 2 The reproduction of photographs or other picture material by radio or wire communication.
tel·e·plasm (tel′ə·plaz′əm) *n.* Ectoplasm (def. 2). [<TELE- + -PLASM]
tel·e·print·er (tel′ə·prin′tər) *n.* A teletypewriter.
Tel·e·promp·ter (tel′ə·promp′tər) *n.* A prompting device for television whereby a prepared script, unseen by the audience, is shown to a speaker or performer, enlarged line by line: a trade name.
tel·e·ra·di·o (tel′ə·rā′dē·ō) *n. pl.* **·di·os** Television and radio taken collectively, especially with reference to their use as advertising media. — *adj.* Pertaining to or by means of teleradio. [<TELE- (def. 2) + RADIO]
tel·e·ran (tel′ə·ran) *n. Telecom.* A system of air navigation which combines the principles of television and radar, the information being gathered by ground stations and transmitted to all aircraft within range. [<TELE- (def. 2) + R(ADAR) A(IR) N(AVIGATION)]
tel·e·scope (tel′ə·skōp) *n.* 1 An optical instrument for enlarging the image of a distant object, consisting of an object glass for collecting light beams from the object and an eyepiece for viewing the image. The **refracting telescope** transmits the rays to a focus through a combination of lenses called the object glass; the **reflecting telescope** brings them to a focus by reflection from a concave mirror. 2 A valise or traveling bag that shuts with one section inside the other, and thus can be extended, like a telescope. — *v.* **·scoped**, **·scop·ing** *v.t.* 1 To drive or slide together so that one part fits into another in the manner of the sections of a small telescope. 2 To crush by driving something into or upon. — *v.i.* 3 To crash or be forced into one another, as railroad cars in a collision. [<TELE- + -SCOPE]

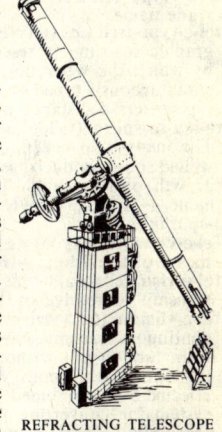

REFRACTING TELESCOPE
Yerkes Observatory
30-inch.

telescope word A blend (def. 2).
tel·e·scop·ic (tel′ə·skop′ik) *adj.* 1 Pertaining to the telescope. 2 Visible only through a telescope. 3 Far-seeing. 4 Having sections that slide within or over one another. Also **tel′e·scop′i·cal.** — **tel′e·scop′i·cal·ly** *adv.*
tel·e·scop·tics (tel′ə·skop′tiks) *n. pl.* (construed as singular) The art of designing, constructing, and using telescopes, regarded as the hobby of skilled amateurs.
te·les·co·py (tə·les′kə·pē) *n.* The art of using or making telescopes. — **te·les′co·pist** *n.*
tel·e·script (tel′ə·skript) *n.* A script written or adapted for a television program. [<TELE- (def. 2) + SCRIPT]
tel·e·set (tel′ə·set) *n.* A television receiving set.
tel·e·sis (tel′ə·sis) *n. Sociol.* Satisfactory progress toward an intended purpose, especially as the result of skilled direction of forces and intelligent planning. [<NL <Gk. *telein* fulfil < *telos, teleos* end]
tel·e·spec·tro·scope (tel′ə·spek′trə·skōp) *n.* 1 A combined telescope and spectroscope. 2 A spectroscope for attachment to a telescope.
tel·e·ster·e·o·scope (tel′ə·ster′ē·ə·skōp, -stir′-) *n.* An optical instrument that presents images of objects at a distance from the observer in enhanced relief.

tel•es•the•sia (tel′is·thē′zhə, -zhē·ə) *n.* Susceptibility to stimuli coming from a distance and beyond the normal range of the senses: also spelled *telaesthesia.* [<NL <Gk. *tēle* far + *aisthēsis* feeling] —**tel′es•thet′ic** (-thet′ik) *adj.*

tel•e•stich (tel′ə·stik, tə·les′tik) *n.* An acrostic in which the significant letters are at the ends of the lines. [<Gk. *telos* end + *stichos* line]

tel•e•ther•a•py (tel′ə·ther′ə·pē) *n. Med.* **1** Treatment by radiation administered in massive doses at a distance from the body. **2** The prescribing of medical treatment by telephone, letter, etc.: also called **absent treatment.**

tel•e•ther•mo•graph (tel′ə·thûr′mə·graf, -gräf) *n.* **1** A self-registering telethermometer. **2** A record made by this instrument. [<TELE- + THERMOGRAPH]

tel•e•ther•mom•e•ter (tel′ə·thûr·mom′ə·tər) *n.* Any apparatus used to indicate the temperature of a distant point, as a thermocouple. —**tel′e•ther•mom′e•try** *n.*

tel•e•tran•scrip•tion (tel′ə·tran·skrip′shən) *n.* A method for transcribing television programs on films for subsequent presentation; also, the transcription itself.

tel•e•type (tel′ə·tīp) *v.t.* & *v.i.* **•typed, •typ•ing** To communicate (with) by teletypewriter or Teletype. —*n.* A teletypewriter. —**tel′e•typ′er** *n.*

Tel•e•type (tel′ə·tīp) *n.* A teletypewriter: a trade name.

tel•e•type•writ•er (tel′ə·tīp′rī′tər) *n.* A telegraphic instrument resembling a typewriter, by which the work done on one machine is simultaneously typed on electrically connected typewriters a distance away.

te•leu•to•spore (tə·lōō′tə·spôr, -spōr) *n. Bot.* The one- or two-celled, usually stalked, thick-walled spore produced as the final stage in the growth of rust fungi. [<Gk. *teleutē* fulfilment + SPORE] —**te•leu′to•spor′ic** (-spôr′ik, -spor′ik) *adj.*

tel•e•view (tel′ə·vyōō) *v.t.* & *v.i.* To observe by means of television. —**tel′e•view′er** *n.*

tel•e•vise (tel′ə·vīz) *v.t.* & *v.i.* **•vised, •vis•ing** To transmit or receive by television.

tel•e•vi•sion (tel′ə·vizh′ən) *n.* The exact and continuous transmission of visual images, still or in motion but without permanent recording, for instantaneous viewing at a distance: effected by a combined optical and electrical system for converting light waves into corresponding electrical impulses which are reconverted into their visual form in a receiving set. —**tel′e•vi′sion•al, tel′e•vi′sion•ar′y** (-vizh′·ən·er′ē) *adj.*

tel•fer (tel′fər) See TELPHER.

tel•ford (tel′fərd) *adj.* Designating a road made of large broken stone packed with smaller pieces, covered with a layer of finely broken stone or gravel, and rolled hard and smooth. —*n.* A road having such a surface. [after Thomas *Telford,* 1757–1834, Scottish engineer]

tel•ford•ize (tel′fər·dīz) *v.t.* **•ized, •iz•ing** To make or cover (a road) with a telford surface.

tel•har•mo•ni•um (tel′här·mō′nē·əm) *n.* An instrument by which an operator at a central station playing on a keyboard controlling alternating electric currents is able to produce music at a distance. [<TEL(E)- + HARMONIUM] —**tel′har•mon′ic** (-mon′ik) *adj.* —**tel•har′mo•ny** (-här′mə·nē) *n.*

tel•ic (tel′ik, tē′lik) *adj.* Connected with, tending toward, or denoting a purpose; teleological. [<Gk. *telikos* < *telos* end] —**tel′i•cal•ly** *adv.*

te•li•o•stage (tē′lē·ə·stāj′, tel′ē-) *n. Bot.* The last stage in the life cycle of rust fungi. [<TELIUM + STAGE]

te•li•um (tē′lē·əm, tel′ē-) *n. Bot.* The sorus of the teliostage of the rust fungus. [<NL <Gk. *telos, teleos* end] —**te′li•al** *adj.*

tell (tel) *v.* **told, tell•ing** *v.t.* **1** To relate in detail; narrate, as a story. **2** To make known by speech or writing; communicate. **3** To make known; reveal; disclose: to *tell* secrets. **4** To decide; ascertain: I cannot *tell* who is to blame. **5** To utter; express in words: to *tell* a lie. **6** To give a command to; bid; order: I *told* him to go home. **7** To let know; inform. **8** *Colloq.* To inform or assure emphatically: It's cold out, I *tell* you! **9** To count; enumerate: to *tell* one's beads. —*v.i.* **10** To give an account or description: usually with *of.* **11** To disclose something; inform: with *on.* **12** To serve as indication or evidence: with *of*: Their rags *told* of their poverty. **13** To produce a marked effect: Every blow *told.* See synonyms under AFFIRM, ASSERT, INFORM[1], PUBLISH, RELATE, SPEAK. —**all told** Everyone or everything being counted; in all. —**to tell off 1** To count and set apart. **2** *Colloq.* To reprimand severely. —*n. Dial.* **1** Something told; story; say. **2** Account; story; explanation: according to his *tell.* [OE *tellan.* Akin to TALE, TALK.] —**tell′•a•ble** *adj.*

Tell (tel), **William** A legendary Swiss hero in the struggle for independence from Austria. He refused to salute the governor's cap, which had been set up as a symbol of Austrian authority, and was forced to shoot an apple off his son's head with bow and arrow.

Tell el A•mar•na (tel el ə·mär′nə) Site of the ruins of an ancient city on the east bank of the Nile, Upper Egypt; Ikhnaton's capital, built about 1360 B.C.

Tell el Ke•bir (tel el ke·bir′) A village of NE Egypt; scene of a British victory over Egyptian forces, 1882.

tell•er (tel′ər) *n.* **1** One who relates or informs. **2** A person who receives or pays out money, as in a bank. **3** A person appointed to collect and count ballots in a legislative body or other assembly.

Tel•ler (tel′ər), **Edward,** born 1908, U.S. atomic physicist born in Hungary.

tell•ing (tel′ing) *adj.* Producing a great effect; impressive; effective; striking. See synonyms under VIVID. —**tell′ing•ly** *adv.*

tell•tale (tel′tāl′) *adj.* **1** Tattling; talebearing. **2** Betraying. —*n.* **1** One who improperly gives information concerning the private affairs of others; a tattler. **2** That which conveys information, especially in an involuntary way; a token. **3** An instrument or device, usually automatic, for giving information as to number, position, condition, etc. **4** A pointer or piece moving in a slot and indicating the degree of inflation of an organ bellows. **5** A row of dangling straps or ropes suspended above a railway track to warn anyone standing on a car roof of the approach of a low overhead structure. **6** A clock to record the times of coming and going, as of workmen, or as a watchman's clock. **7** An index showing the position of a vessel's helm. **8** A yellowlegs or tattler.

telltale sandpiper The yellowlegs.

tel•lu•rate (tel′yə·rāt) *n. Chem.* A salt of telluric acid.

tel•lu•ri•an (te·lōōr′ē·ən, tel·yōōr′-) *adj.* Of or pertaining to the earth or its inhabitants. —*n.* An inhabitant of the earth. [<L *tellus, -uris* the earth]

tel•lu•ric (te·lōōr′ik, tel·yōōr′-) *adj.* **1** Of or pertaining to the earth; terrestrial; earthly. **2** *Chem.* Derived from or containing tellurium, especially in its higher valence.

telluric acid *Chem.* A weak acid, H_2TeO_4, obtained by oxidizing tellurium. It is analogous to sulfuric acid.

tel•lu•ride (tel′yə·rīd, -rid) *n. Chem.* A compound of tellurium with an element or an organic radical: *telluride* of lead.

tel•lu•rite (tel′yə·rīt) *n.* **1** A white or yellow native tellurium dioxide, TeO_2. **2** *Chem.* A salt of tellurous acid.

tel•lu•ri•um (te·lōōr′ē·əm, tel·yōōr′-) *n.* A rare non-metallic element (symbol Te) resembling sulfur and selenium, occasionally found native as tin-white, rhombohedral crystals, but usually combined with metals, as telluride of gold. See ELEMENT. [<NL <L *tellus, -uris* the earth]

tel•lu•rize (tel′yə·rīz) *v.t.* **•rized, •riz•ing** To cause to combine with tellurium.

tel•lur•nick•el (tel′ər·nik′əl) *n.* Melonite. [< TELLUR(IUM) + NICKEL]

tel•lu•rous (tel′yər·əs, te·lōōr′əs, tel·yōōr′-) *adj. Chem.* Of, pertaining to, or derived from tellurium, especially in its lower valence: *tellurous* acid, H_2TeO_3.

Tel•lus (tel′əs) In Roman mythology, the goddess of the earth: identified with the Greek *Gaea.* Also **Tellus Mater.**

telo-[1] *combining form* Final; complete; perfect: *telophase:* also, before vowels, *tel-.* Also *teleo-.* [<Gk. *telos* end]

telo-[2] Var. of TELE-.

tel•o•blast (tel′ə·blast) *n. Zool.* A large cell at the growing end of the embryo, in annelids, etc., which produces rows of smaller cells. [<TELO-[1] + -BLAST]

tel•o•dy•nam•ic (tel′ə·dī·nam′ik, -di-) *adj.* Of, related to, or employed in the transmission of power to a distance, specifically by cables and pulleys. [<TELO-[2] + DYNAMIC]

tel•o•lec•i•thal (tel′ə·les′ə·thəl) *adj. Biol.* Having the nutritive part of the yolk at one pole: said of ova, as of birds, with unequal or partial segmentation. [<TELO-[1] + LECITHAL]

tel•o•phase (tel′ə·fāz) *n. Biol.* The closing phase of mitosis, when the cell divides and the daughter nuclei are formed. [<TELO-[1] + PHASE]

tel•pher (tel′fər) *n.* A light car suspended from cables and usually propelled by electricity: used for aerial transportation. —*v.t.* To transport by telpher. Also spelled *telfer.* [<TEL(E)- + Gk. *pherein* bear] —**tel′pher•ic** *adj.* —**tel′pher•age** (-ij) *n.*

tel•son (tel′sən) *n. Zool.* The last abdominal segment of the body of an arthropod, as of a lobster, shrimp, or scorpion. [<Gk. *telson* boundary]

Tel•u•gu (tel′ōō·gōō) *n. pl.* **•gu 1** A Dravidian language, spoken by more than 30 million people, in the vernacular more widespread than Tamil, and with Tamil, most important in literary culture. **2** One of a Dravidian people of Telugu speech, inhabiting NW Andhra Pradesh, India. —*adj.* Of or pertaining to the Telugu or to Telugu. Also spelled *Telegu.*

tem•blor (tem·blôr′) *n. pl.* **•blors** or **•blo•res** (-blô′rās) An earthquake. [<Sp.]

Tem•bu•land (tem′bōō·land) A district of eastern Cape of Good Hope Province, Union of South Africa; 3,448 square miles; capital, Umtata.

tem•e•rar•i•ous (tem′ə·râr′ē·əs) *adj.* Unreasonably adventurous; rash; reckless. [<L *temerarius* < *temere* rashly] —**tem′e•rar′i•ous•ly** *adv.* —**tem′e•rar′i•ous•ness** *n.*

te•mer•i•ty (tə·mer′ə·tē) *n.* Venturesome or foolish boldness; rashness; disregard of personal danger or consequences. [<L *temeritas, -tatis* < *temere* rashly]

Synonyms: audacity, foolhardiness, hardihood, hastiness, heedlessness, precipitancy, precipitation, presumption, rashness, recklessness, venturesomeness. *Rashness* applies to the actual impulsive rushing into danger without counting the cost; *temerity* denotes the needless exposure of oneself to peril because of lack of foresight. *Rashness* is used chiefly of bodily acts, *temerity* often of mental or social matters. We say it is amazing that one should have had the *temerity* to make a statement which could be readily proved a falsehood; in such use *temerity* is often closely allied to *hardihood, audacity,* or *presumption. Venturesomeness* dallies on the edge of danger and experiments with it; *foolhardiness* rushes in for want of sense, *heedlessness* for want of attention, *rashness* for want of reflection, *recklessness* from disregard of consequences. *Antonyms*: care, caution, circumspection, cowardice, hesitation, timidity, wariness.

Tem•es•vár (te′mesh·vär) The Hungarian name for TIMIȘOARA.

tem•i•ak (tem′ē·ak) *n.* An Eskimo jacket or coat. [<Eskimo *tingmiag, tingmeak,* lit., bird]

Tem•pe (tem′pē), **Vale of** A valley, about 5 miles long, between Mount Olympus and Mount Ossa in Thessaly, Greece: famous for its beauty and in ancient times regarded as sacred to Apollo. *Greek* **Tem•be** (tem′bē).

Tem•pel•hof (tem′pəl·hof, *Ger.* tem′pəl·hōf) A southern district of West Berlin, Germany; site of the city's chief airport.

tem•per (tem′pər) *n.* **1** Heat of mind or passion; disposition to become angry; passion; irritation. **2** Quality of mind with reference to the passions, emotions, or affections; disposition. **3** Composure of mind; equanimity; self-command; calmness: used only in the phrases **to keep, or to lose, one's temper. 4** *Metall.* The condition of a metal as regards hardness and brittleness, especially when due to heating and sudden cooling. **5** Consistency due to mixture, as of mortar, etc. **6** Lime or an equivalent used in clarifying sugar. **7** An alloy, as that added to tin to make pewter. **8** *Obs.* Constitutional condition, resulting, according to the

ancients, from the proportion in which the four humors were mixed. 9 *Archaic* A mean; medium. [<v.] — *v.t.* 1 To bring to a state of moderation or suitability, as by addition of another quality; free from excess; moderate; mitigate: to *temper* justice with mercy. 2 To bring to the proper consistency, texture, etc., by moistening and working: to *temper* clay. 3 To bring (metal) to a required hardness and elasticity by heating and suddenly cooling. 4 *Music* To adjust the tones of (an instrument) by temperament; tune. 5 *Obs.* To adjust. — *v.i.* 6 To be or become tempered. [Fusion of OE *temprian* mingle, regulate and of OF *temprer, tremper* soak, temper (steel), both <L *temperare* combine in due proportion. For sense development of noun defs. 1, 2, and 3, see def. 8.] — **tem′per·a·bil′i·ty** *n.* — **tem′per·a·ble** *adj.* — **tem′per·er** *n.*
Synonyms (noun): constitution, disposition, frame, grain, humor, mood, nature, organization, temperament. See ANGER, CHARACTER.
Synonyms (verb): accommodate, adapt, adjust, appease, assuage, attemper, calm, fit, moderate, modify, mollify, pacify, qualify, restrain, soften, soothe.
tem·per·a (tem′pər·ə, *Ital.* tem′pä·rä) *n.* 1 A painting medium which is essentially an emulsion prepared by any of numerous recipes, and composed characteristically of oil usually thickened, with or without a resin such as dammar varnish, and egg and water. 2 The method of painting by this medium, which falls into three principal divisions: *unvarnished tempera, varnished tempera,* and *tempera,* as underpainting for oil glazes: widely used in the Renaissance and revived in modern times, sometimes in combination with oil techniques. [<Ital. < *temperare* temper <L]
tem·per·a·ment (tem′pər·ə·mənt, -prə-) *n.* 1 The characteristic physical and mental peculiarities of an individual as manifested in his reactions. 2 *Music* The tuning of an instrument so that the intervals of the scale shall follow a suitable law of succession. 3 Mental constitution; make-up; disposition. 4 Adjustment or compromise. 5 *Obs.* Temperature. See synonyms under CHARACTER, TEMPER. [<L *temperamentum* proper mixture < *temperare* mix in due proportions]
tem·per·a·men·tal (tem′pər·ə·men′təl, -prə-) *adj.* 1 Of or pertaining to temperament. 2 Having a strongly marked temperament. 3 Sensitive; easily excited; changeable. — **tem′per·a·men′tal·ly** *adv.*
tem·per·ance (tem′pər·əns) *n.* 1 The state or quality of being temperate; habitual moderation, especially in the indulgence of any appetite. 2 Specifically, the principle and practice of total abstinence from intoxicants. 3 *Obs.* Calmness; self-control. See synonyms under ABSTINENCE. — *adj.* 1 Of or pertaining to public places where alcoholic beverages are not sold. 2 Of, relating to, practicing, or promoting total abstinence from intoxicants. [<OF <L *temperantia,* orig. neut. pl. of *temperans, -antis,* ppr. of *temperare* mix in due proportions]
temperance pledge A pledge not to indulge in alcoholic drinks.
tem·per·ate (tem′pər·it) *adj.* 1 Observing moderation or self-control; specifically, by extension, not indulging in intoxicating liquors. 2 Moderate as regards temperature; free from extremes of heat or cold; mild. 3 Characterized by moderation or the absence of extremes; not excessive. 4 Calm; restrained; self-controlled. 5 *Music* Tempered: said of an interval or scale. See synonyms under SOBER. [<L *temperatus,* pp. of *temperare* mix in due proportions] — **tem′per·ate·ly** *adv.* — **tem′per·ate·ness** *n.*
temperate zone See ZONE (def. 1).
tem·per·a·ture (tem′pər·ə·chər, -prə-) *n.* 1 Condition as regards heat or cold. 2 The degree of heat in a body or substance, as measured on the graduated scale of a thermometer. See table below. 3 Sensible heat of the human body; also, excess of this above the normal. 4 *Obs.* Constitution; temperament; mixture; temperateness; temperance. [<L *temperatura* due measure < *temperatus.* See TEMPERATE.]

To convert from Fahrenheit to Celsius (Centigrade): Subtract 32 from the Fahrenheit reading, multiply by 5, and divide the product by 9. *Example:* 65° F. − 32 = 33; 33 × 5 = 165; 165 ÷ 9 = 18.3° C. To convert from Celsius to Fahrenheit: Multiply the Celsius reading by 9, divide the product by 5, add 32. *Example:* 30° C. × 9 = 270; 270 ÷ 5 = 54; 54 + 32 = 86° F.

CONVERSION TABLE

Fahrenheit	Celsius	Fahrenheit	Celsius
500	260.0	−10	−23.3
400	204.4	−20	−28.9
300	149.0	−30	−34.4
212	100.0	−40	−40.0
200	93.3	−50	−45.6
100	37.8	−60	−51.1
90	32.2	−70	−56.7
80	26.7	−80	−62.2
70	21.1	−90	−67.8
60	15.6	−100	−73.3
50	10.0	−200	−129.0
40	4.4	−300	−184.0
32	0.0	−400	−240.0
30	−1.1	*−459.4	−273.0
20	−6.6		
10	−12.2		
0	−17.8	*Absolute zero	

temperature coefficient *Physics* The amount of change in some specified physical quantity per unit change in temperature: it may be positive or negative and is usually expressed as the quotient of the change observed after a rise of 1° C. divided by the constant value of the quantity at 0° C.
temperature gradient The rate of change in temperature with change in altitude or other variable factors.
tem·pered (tem′pərd) *adj.* 1 Having temper or a temper, in any sense; mostly in compounds: quick-*tempered,* ill-*tempered.* 2 *Music* Adjusted in pitch to some mean temperament. 3 Moderated by admixture. 4 Having the right degree of hardness and elasticity: well-*tempered* steel.
tem·per·pin (tem′pər·pin′) *n.* 1 A wooden screw used to regulate the motion of a spinning wheel. 2 A tuning peg of a violin.
tem·pest (tem′pist) *n.* 1 An extensive and violent wind, usually attended with rain, snow, or hail. 2 A violent commotion or agitation; a fierce tumult. See synonyms under STORM. — *v.t.* To agitate violently; affect as a tempest does. [<OF *tempeste* <L *tempestas* space of time, weather < *tempus* time]
tem·pes·tu·ous (tem·pes′chōō·əs) *adj.* Stormy; turbulent; violent. [<OF *tempestueux* <LL *tempestuosus* <L *tempestas* weather. See TEMPEST.] — **tem·pes′tu·ous·ly** *adv.* — **tem·pes′tu·ous·ness** *n.*
tem·plar (tem′plər) *n.* A law student or a barrister who has apartments in the buildings known as the Inner and the Middle Temple in London. [<OF *templier* <Med. L *templarius* <L *templum.* See TEMPLE.]
Tem·plar (tem′plər) *n.* A Knight Templar.
tem·plate (tem′plit) *n.* 1 A pattern or gage, as of wood or metal, used as a guide in shaping something or in checking the accuracy of work. 2 In building, a stout stone or timber for distributing weight or thrust. 3 A wedge for a building block under a ship's keel. Also spelled *templet.* [<F *templette* stretcher, dim. of *temple* small timber <L *templum*]
tem·ple¹ (tem′pəl) *n.* 1 A stately edifice consecrated to one or more deities and forming a seat of their worship. 2 An edifice dedicated to public worship; especially, in the United States, a Reform synagog. 3 In France, a Protestant church. 4 Figuratively, any place considered as occupied by God; specifically, a sanctified human body. 5 A building erected and dedicated for the administration of Mormon ordinances; a Mormon church. — **the Temple** 1 Either of two medieval establishments in London and Paris, once occupied by the Knights Templar. In London, since 1185, the district lying between Fleet Street and the Thames river, the site of the **Inner** and **Middle Temple.** See INNS OF COURT. 2 Any of three successive sacred edifices built in Jerusalem for the worship of Jehovah. [OE *tempel* <L *templum* temple]

tem·ple² (tem′pəl) *n.* The region on each side of the head above the cheek bone. [<OF <L *tempora,* pl. of *tempus* temple]
tem·ple³ (tem′pəl) *n.* An attachment to a loom that serves to keep the last woven part of the fabric stretched and to prevent chafing of the warp. [<F <L *templum,* a small timber]
Tem·ple (tem′pəl), **Sir William,** 1628–99, English statesman, diplomat, and writer. — **William,** 1881–1944, English prelate; archbishop of Canterbury 1942–44.
Temple Bar A historic three-arched gateway in London marking the western boundary of the city proper and on which the heads of traitors and other malefactors were exposed. It was dismantled in 1878 but re-erected at Waltham Cross, in Essex, in 1888.
tem·pled (tem′pəld) *adj.* Honored with or enshrined in a temple: a *templed* god.
tem·plet (tem′plit) See TEMPLATE.
tem·po (tem′pō) *n.* *pl.* ·**pos** or ·**pi** (-pē) 1 *Music* Relative speed at which a composition is rendered; time; rhythm of a tune. 2 Characteristic manner or style; rate of speed or activity in general. [<Ital. <L *tempus* time]
tem·po·la·bile (tem′pō·lā′bil) *adj. Biol.* Subject to decay or destruction within a certain period of time, as a serum. [<L *tempus* time + *labilis* perishable]
tem·po·ral¹ (tem′pər·əl) *adj.* 1 Pertaining to affairs of the present life, as contrasted with those of a future life; earthly, as opposed to heavenly. 2 Pertaining to or limited by time; transitory, as opposed to eternal. 3 Related to or concerned with worldly affairs; worldly; material, as opposed to spiritual. 4 Pertaining to civil law or authority; lay; secular: contrasted with *clerical.* 5 *Gram.* Of, pertaining to, or denoting time: *temporal* conjunctions. See synonyms under PROFANE. — **lords temporal** English, Scottish, and Irish lay peers with seats in the House of Lords. [<OF *temporel* <L *temporalis* <*tempus, temporis* time] — **tem′po·ral·ly** *adv.* — **tem′po·ral·ness** *n.*
tem·po·ral² (tem′pər·əl) *adj. Anat.* Of, pertaining to, or situated at the temple or temples: the *temporal* bone. [<L *temporalis* <*tempora.* See TEMPLE².]
tem·po·ral³ (tem′pər·äl) *n. SW U.S.* A field or portion of land; a farm, especially one not requiring irrigation. [<Sp. *temporal* storm, tempest; ? < *terreno de temporal* land where heavy rains fall]
temporal bone *Anat.* A compound bone situated at the side of the head in man and other mammals, and containing the organ of hearing.
tem·po·ral·i·ty (tem′pə·ral′ə·tē) *n.* *pl.* ·**ties** 1 *Usually pl.* A temporal or material matter, interest, revenue, etc.; specifically, an ecclesiastical possession or revenue. 2 The state of being temporal or temporary: opposed to *perpetuity.*
tem·po·ra mu·tan·tur (tem′pər·ə myōō·tan′tər) *Latin* The times are changed.
tem·po·rar·y (tem′pə·rer′ē) *adj.* 1 Lasting or intended to be used for a short time only; transitory; of passing interest: opposed to *permanent.* 2 *Obs.* Contemporary. See synonyms under TRANSIENT. [<L *temporarius* <*tempus, temporis* time] — **tem′po·rar′i·ly** *adv.* — **tem′po·rar′i·ness** *n.*

TEMPLE OF HORUS, EDFU, BEGUN 237 B.C.
Greco-Egyptian style.

tem·po·rize (tem′pə·rīz) *v.i.* ·**rized,** ·**riz·ing** 1 To act evasively so as to gain time or put off decision or commitment. 2 To give real or apparent compliance to the circumstances;

comply. 3 To parley so as to gain time: with *with.* 4 To effect a compromise; negotiate: with *with* or *between.* Also *Brit.* **tem'po·rise.** [<F *temporiser* <L *tempus, temporis* time] —**tem'po·ri·za'tion** *n.* —**tem'po·riz'er** *n.* —**tem'po·riz'ing·ly** *adv.*

tempt (tempt) *v.t.* 1 To attempt to persuade (a person) to do wrong, as by promising pleasure or gain. 2 To be attractive to; invite: Your offers do not *tempt* me. 3 To provoke or risk provoking: to *tempt* fate. 4 *Obs.* To test; prove. See synonyms under ALLURE. [<OF *tempter, tenter* <L *temptare, tentare* test, try, prob. intens. of *tendere* stretch] —**tempt'a·ble** *adj.* —**tempt'er** *n.* —**tempt'ress** *n. fem.*

temp·ta·tion (temp·tā'shən) *n.* 1 That which tempts, especially to evil. 2 The state of being tempted, or enticed to evil; the act of tempting or testing. 3 A state of mental conflict between heavenly and infernal influences.

tempt·ing (temp'ting) *adj.* Alluring; attractive; seductive. —**tempt'ing·ly** *adv.* —**tempt'·ing·ness** *n.*

tem·pus fu·git (tem'pəs fyōō'jit) *Latin* Time flies.

tem·u·line (tem'yə·lēn, -lin) *n. Chem.* A colorless, sirupy, poisonous alkaloid, $C_7H_{12}ON_2$, from the seeds of the darnel. [<L *temulentus* drunken + -INE[2]]

ten (ten) *n.* 1 The cardinal number following nine and preceding eleven, or any of the symbols or combinations of symbols (10, x, X) used to represent it. 2 Anything containing or representing ten units or members; a playing card marked with ten pips; also, a ten-dollar bill. —*adj.* Being or consisting of one more than nine; decennary. [OE]

ten- Var. of TENO-.

ten·a·ble (ten'ə·bəl) *adj.* Capable of being held, maintained, or defended. [<F <*tenir* hold <L *tenere*] —**ten'a·bil'i·ty, ten'a·ble·ness** *n.* —**ten'a·bly** *adv.*

ten·ace (ten'ās) *n.* The combination in the same hand of the best and third best cards (**major tenace**) or of the second and fourth best cards (**minor tenace**) of any suit. [<Sp. *tenaza* pincers, tongs <*tenaz* tenacious <L *tenax, tenacis.* See TENACIOUS.]

te·na·cious (ti·nā'shəs) *adj.* 1 Having great cohesiveness of parts; tough. 2 Adhesive; sticky. 3 Holding or tending to hold strongly, as opinions, rights, etc.: followed by *of*; hence, stubborn; obstinate; unyielding; persistent. 4 Apt to retain; strongly retentive, as memory. See synonyms under STRONG. [<L *tenax, tenacis* holding fast <*tenere* hold, grasp, embrace] —**te·na'cious·ly** *adv.* —**te·na'·cious·ness** *n.*

te·nac·i·ty (ti·nas'ə·tē) *n.* 1 The state or quality of being tenacious. 2 That quality of a body in consequence of which it resists being pulled or forced apart.

te·nac·u·lum (ti·nak'yə·ləm) *n. pl.* **·la** (-lə) *Surg.* A hooked instrument for seizing and holding parts of the body, as arteries, during surgical operations. [<LL, holder <L *tenax, tenacis.* See TENACIOUS.]

te·naille (te·nāl') *n.* A low outwork, usually with one or two reentering angles, in the main ditch between two bastions. —*v.t.* To equip with tenailles. Also **te·nail'.** [<F <LL *tenacula,* pl. of *tenaculum.* See TENACULUM.]

ten·an·cy (ten'ən·sē) *n. pl.* **·cies** 1 The holding of lands or tenements by any form of title; occupancy. 2 The period of holding or occupying lands, tenements, or office; temporary possession. 3 A habitation or dwelling place held of another.

ten·ant (ten'ənt) *n.* 1 One who holds or possesses lands or property by any kind of title; especially, one who holds under another; a lessee. 2 A defendant in an action concerning real property. 3 A dweller in any place; an occupant. —*v.t.* To hold as tenant; occupy. —*v.i.* To be a tenant. [<F, orig. ppr. of *tenir* hold <L *tenere*] —**ten'ant·a·ble** *adj.* —**ten'ant·less** *adj.*

tenant farmer One who farms land owned by another and pays rent usually in a share of the crops.

ten·ant-right (ten'ənt·rīt') *n.* A customary right belonging to a tenant, even if not specifically stipulated, as a right to continuous occupancy without increase of rent, or a right to compensation for improvements.

ten·ant·ry (ten'ən·trē) *n. pl.* **·ries** 1 Tenants collectively. 2 Tenantship; tenancy.

Te·nas·se·rim (tə·nas'ər·im) A former administrative division of SE Burma extending in a narrow strip of coast 400 miles down the Malay Peninsula to the Isthmus of Kra; 31,588 square miles; capital, Moulmein.

ten-cent store (ten'sent') See FIVE- AND TEN-CENT STORE.

tench (tench) *n.* A European fresh-water cyprinoid fish (*Tinca tinca*), very tenacious of life, and having small, deeply embedded scales. [<F *tenche* <LL *tinca* tench]

Ten Commandments See under COMMANDMENT.

tend[1] (tend) *v.i.* 1 To have an aptitude, tendency, or disposition; incline: He *tends* to talk too much. 2 To have influence toward a specified result; lead or conduce: Education *tends* to refinement. 3 To go in a certain direction. [<OF *tendre* <L *tendere* extend, tend]

tend[2] (tend) *v.t.* 1 To attend to the needs or requirements of; take care of; minister to: to *tend* a fire. 2 To watch over; look after: to *tend* children. 3 To watch (a vessel at anchor) with the intention of so managing her when the tide changes as to prevent fouling the anchor and chain. —*v.i.* 4 To be in attendance; serve or wait: with *on* or *upon.* 5 *Colloq.* To give attention or care: with *to.* [Aphetic var. of ATTEND]

ten·dance (ten'dəns) *n.* 1 The act of tending; attendance; service. 2 *Archaic* Attendants collectively. Also **ten'dence.**

Ten Degree Channel A passage from the Bay of Bengal to the Andaman Sea between the Andaman and Nicobar Islands, along 10° N; about 90 miles wide.

ten·den·cy (ten'dən·sē) *n. pl.* **·cies** 1 The state of being directed toward some purpose, end, or result; inclination; bent; aptitude. 2 That which tends to produce some specified effect. 3 Bias; propensity. 4 Trend of a speech; purpose of a story. See synonyms under AIM, DIRECTION, INCLINATION. [<Med. L *tendentia,* orig. neut. pl. of *tendens, -entis,* ppr. of *tendere* extend, tend]

ten·den·tious (ten·den'shəs) *adj.* Having a purposed aim or intentional tendency. [<G *tendenziös* <*tendenz* tendency <Med. L *tendentia.* See TENDENCY.] —**ten·den'tious·ly** *adv.* —**ten·den'tious·ness** *n.*

Ten·denz (ten·dens') *n. German* Tendency or drift; partisan or biased attitude, as in a work of literature or art; angle; slant.

ten·der[1] (ten'dər) *adj.* 1 Yielding easily to force that tends to crush, bruise, break, or injure; soft or delicate. 2 Easily chewed or cut: said of food, especially meat. 3 Delicate or weak; not strong or hardy. 4 Youthful and delicate; not strengthened by maturity: a *tender* age. 5 Characterized by or expressive of a delicate sensibility; kind; affectionate; gentle: *tender* mercy; a *tender* father. 6 Capable of arousing sensitive feelings; touching: *tender* memories; a *tender* sight. 7 Susceptible to spiritual or moral feelings: a *tender* conscience. 8 Painful if touched; easily pained: a *tender* sore. 9 Of delicate effect or quality; soft: a *tender* light. 10 Requiring deft or delicate treatment; ticklish; touchy: a *tender* subject. 11 *Naut.* Careening too easily under sail: said of a ship. See synonyms under BLAND, FRAGILE, FRIENDLY, HUMANE, MERCIFUL. —*v.t.* To make tender; soften. [<OF *tendre* <L *tener, teneris*] —**ten'der·ly** *adv.* —**ten'·der·ness** *n.*

ten·der[2] (ten'dər) *v.t.* 1 To present for acceptance, as a resignation; offer. 2 *Law* To proffer, as money, in payment, in discharge of a debt, or to fulfil a contract. —*n.* 1 The act of tendering; an offer; specifically, in law, a formal offer of satisfaction. 2 That which is offered as payment, especially money: legal *tender.* [<F *tendre* <L *tendere* extend, tend] —**ten'der·er** *n.*

tend·er[3] (ten'dər) *n.* 1 A vessel used to bring supplies, passengers, and crew back and forth between a larger vessel and a nearby shore; also, a vessel which services another at sea. 2 A boat used to carry provisions, etc., to whalers and lighthouses. 3 A vehicle attached to the rear of a steam locomotive to carry fuel and water for it. 4 One who tends or ministers to. [<TEND[2]]

ten·der·foot (ten'dər·fōōt') *n. pl.* **·foots** or **·feet** (-fēt') *U.S.* 1 A newcomer in the West; one not yet inured to the hardships of or not yet experienced in the life of the plains, the mining camp, etc.; a greenhorn: opposed to *longhorn.* 2 Any inexperienced person. 3 A boy scout in the beginning class or group. —*adj.* Inexperienced; also, made up of inexperienced people: a *tenderfoot* gang.

ten·der-heart·ed (ten'dər·här'tid) *adj.* Having deep or quick sensibility, as to love, pity, etc.; compassionate; sympathetic; easily impressed by sorrow or pain. —**ten'der-heart'·ed·ly** *adv.* —**ten'der-heart'ed·ness** *n.*

ten·der·ize (ten'də·rīz) *v.t.* **·ized, ·iz·ing** To make tender, as meat.

ten·der·iz·er (ten'də·rī'zər) A substance, as papain, for softening the tough fibers and connective tissues of meat in order to make it more palatable.

ten·der·loin (ten'dər·loin') *n.* The tender part of the loin of beef, pork, etc., lying close to the ventral side of the lumbar vertebrae. —**the tenderloin district** 1 A former district of New York City, coinciding with a certain police precinct from 23rd to 42nd streets, west of Broadway, where vice flourished and police corruption was common. 2 Hence, any district in any city which is noted for its night life, a high incidence of crime, and police leniency.

ten·di·nous (ten'də·nəs) *adj.* 1 Of, pertaining to, resembling, or formed by a tendon. 2 Having or full of tendons; sinewy. [<F *tendineux* <Med. L *tendo, -inis.* See TENDON.]

ten·don (ten'dən) *n. Anat.* One of the bands of tough, fibrous connective tissue forming the termination of a muscle and serving to transmit its force to some other part; a sinew. [<F <Med. L *tendo, -inis* <Gk. *tenōn* a sinew <*tenein* stretch]

tendon of Achilles *Anat.* Achilles' tendon.

ten·dril (ten'dril) *n. Bot.* One of the slender, leafless, coiling organs which serve a climbing plant as a means of attachment to a wall, tree trunk, or other supporting surface. [<F *tendrillon,* dim. of *tendron* sprout <*tendre* tender; infl. in meaning by F. *tendre* stretch] —**ten'dril·lar, ten'dril·ous** *adj.*

ten·e·brae (ten'ə·brē) *n. pl.* The matins and lauds of Thursday, Friday, and Saturday of Holy Week, sung on the afternoon or evening of the preceding days. [<L, shadows]

ten·e·brif·ic (ten'ə·brif'ik) *adj.* Making dark or gloomy. [<L *tenebrae* darkness + -FIC]

ten·e·brous (ten'ə·brəs) *adj.* Gloomy; dark; obscure. [<L *tenebrosus* <*tenebrae* darkness] —**ten'e·bros'i·ty** (-bros'ə·tē) *n.*

Ten·e·dos (ten'ə·dos, *Gk.* ten'ə·dôs) A Turkish island in the Aegean near the western entrance to the Dardanelles; 15 square miles: Turkish *Bozcaada.*

ten·e·ment (ten'ə·mənt) *n.* 1 A room, or set of rooms, designed for one family. See TENEMENT HOUSE. 2 *Law* Anything of a permanent nature that may be held by one person of another as property, as land, houses, offices, rents, franchises, etc. 3 A house or building; especially, a dwelling house rented or intended for rent; a tenement house. 4 Figuratively, an abode. [<OF <LL *tenementum* tenure <L *tenere* hold] —**ten'e·men'ta·ry** (-men'tər·ē), **ten'e·men'tal** (-men'təl) *adj.*

tenement house A building or house, usually of inferior type and situated in the poorer sections of a city, rented, leased, or let, to be occupied as the home of three or more families living independently of one another, or by more than two families on a floor, all having a common right in stairways, yards, etc.

te·nen·dum (ti·nen'dəm) *n. Law* The clause in a deed in which, before the abolition of feudal tenures, the tenure was defined: now part of the habendum clause. See HABENDUM. [<L, that which must be held, gerundive of *tenere* hold]

Ten·er·ife (ten'ə·rif', -rēf'; *Sp.* tā'nā·rē'fā) The largest of the Canary Islands; 794 1/2 square miles; capital, Santa Cruz de Tenerife; contains the **Peak of Tenerife** (also *Teyde*), a dormant volcano and the highest peak on Spanish soil; 12,200 feet. Also **Ten'er·iffe'.**

te·nes·mus (ti·nes'məs, -nez'-) *n. Pathol.* A painful straining and ineffectual effort to evacuate the bladder or the bowels. [<NL <L *tenesmus* a straining <Gk. *teneismos* <*teinein* stretch] —**te·nes'mic** *adj.*

ten·et (ten'it, tē'nit) *n.* An opinion, principle,

tenfold

dogma, or doctrine that a person or organization believes or maintains as true. See synonyms under DOCTRINE. [< L, he holds < *tenere* hold]

ten·fold (ten'fōld') *adj.* Made up of ten; ten times as many or as much; ten times repeated; decuplicate. — *adv.* In a tenfold manner or degree.

Ten·gri Khan (teng'grē khän') The second highest peak of the Tien Shan, in NE Kirghiz S.S.R.; 22,949 feet. Also **Khan Tengri**.

Ten·gri Nor (teng'grē nôr', nōr') The Mongolian name for NAM TSO.

te·ni·a (tē'nē·ə) See TAENIA.

te·ni·a·cide (tē'nē·ə·sīd') *n.* A substance which destroys tapeworms, as the oleoresin of certain ferns, carbon tetrachloride, etc.: also spelled *taeniacide.* Also **te'ni·a·fuge'** (-fyōōj') [< L *taenia* < Gk. *tainia* tapeworm + -CIDE] — **te·ni·a·ci'dal** *adj.*

te·ni·a·sis (ti·nī'ə·sis) *n. Pathol.* Any morbid or toxemic condition due to the presence of tapeworms in the body: also spelled *taeniasis.* [< Gk. *tainia* tapeworm + -IASIS]

Ten·iers (ten'yərz, *Flemish* te·nīrs'), **David**, 1582–1649, Flemish painter: called "the Elder." — **David**, 1610–90, Flemish painter: called "the Younger"; son of preceding.

Ten·nes·se·an (ten'ə·sē'ən) *n.* A native or inhabitant of Tennessee. — *adj.* Of or pertaining to Tennessee.

Ten·nes·see (ten'ə·sē') A State in the SE United States; 42,246 square miles; capital, Nashville; entered the Union June 1, 1796; nicknamed *Volunteer State:* abbr. *Tenn.*

Tennessee River A river of the east central United States, rising in eastern Tennessee and flowing 652 miles SW, NW, and north through Alabama, Tennessee, and Kentucky to the Ohio River at Paducah, Kentucky.

Tennessee Valley Authority A Federal corporation established in 1933 by the U.S. government to take custody of the Wilson Dam and associated plants at Muscle Shoals in Tennessee, developing and operating them in the national interest, with special reference to electric power, irrigation, fertilizers, and flood control. Abbr. *TVA, T.V.A.*

Ten·niel (ten'yəl), **Sir John**, 1820–1914, English illustrator and cartoonist.

ten·nis (ten'is) *n.* A game played by striking a ball to and fro with rackets over a net stretched perpendicularly across a space called a court. It has two forms, **court tennis**, played indoors in a specially prepared building, and **lawn tennis**, played out-of-doors on a court of grass, clay, concrete, etc. [< AF *tenetz* take, receive, imperative of *tenir* hold; from the call of the server]

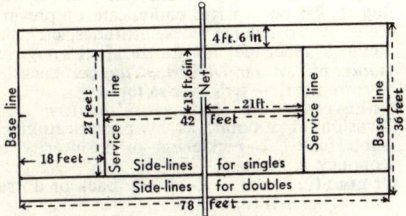

TENNIS COURT — PLAN AND DIMENSIONS

Ten·ny·son (ten'ə·sən), **Alfred**, 1809–92, Lord Tennyson, English poet laureate 1850–92.

Ten·ny·so·ni·an (ten'ə·sō'nē·ən) *adj.* Relating to or characteristic of Alfred Tennyson, or his verse or style.

teno- combining form *Med.* Tendon; related to a tendon, or to tendons: *tenotomy:* also, before vowels, *ten-*. Also **tenonto-**. [< Gk. *tenōn* a tendon]

Te·noch·ti·tlán (tā·nōkh'tē·tlän') The capital of the ancient Aztec Empire, on the site of Mexico City.

ten·on (ten'ən) *n.* A projection on the end of a timber, etc., for inserting in a socket to form a joint. — *v.t.* **1** To form a tenon on. **2** To join by a mortise and tenon. [< F < *tenir* hold]

ten·o·ni·tis (ten'ə·nī'tis) *n. Pathol.* Inflammation of a tendon. [< NL < Gk. *tenōn* a tendon]

ten·or (ten'ər) *n.* **1** A settled course or manner of progress. **2** Course of thought; general purport. **3** *Law* The purport or substance and effect of a document; an exact transcript, as of a record. **4** General character and tendency; nature. **5** The highest adult male voice (except the falsetto); a singer having such a voice, or a part to be sung by it. **6** An instrument playing the part intermediate between the bass and the alto; especially, the viola. **7** In bellringing, the lowest bell, irrespective of peal. — *adj.* **1** Of or pertaining to a tenor. **2** Having a relation to other instruments that the tenor bears to other musical parts: a *tenor* violin. [< OF *tenour* < L *tenor* a course < *tenere* hold; in def. 5, so called because this voice originally sang or "held" the melody]

ten·or·ite (ten'ə·rīt) *n.* Native oxide of copper, occurring in minute black scales; black copper. [after Prof. G. *Tenore,* president (1841) of Naples Academy]

te·nor·rha·phy (ti·nôr'ə·fē, -nor'-) *n. Surg.* Suture of the ends of a divided tendon. [< TENO- + -RRHAPHY]

te·not·o·my (ti·not'ə·mē) *n. Surg.* The operation of cutting a tendon. [< TENO- + -TOMY]

ten·pen·ny (ten'pen'ē, -pə·nē) *adj.* **1** Valued at tenpence. **2** Designating the size of nails three inches long. See -PENNY.

ten·pin (ten'pin') *n.* One of the pins used in the game of tenpins.

ten·pins (ten'pinz') *n.* A game, played in a bowling alley, in which the players attempt to bowl down ten pins set up at the far end of the alley.

ten·rec (ten'rek) *n.* One of several insectivorous mammals of Madagascar; especially, the spiny-coated, tailless *Tenrec ecaudatus,* from 12 to 16 inches long: also spelled *tanrec.* [< F < Malagasy *trandraka*]

Ten·sas River (ten'sô) A river in eastern Louisiana, flowing 175 miles south and SW to the Ouachita River in east central Louisiana.

tense¹ (tens) *adj.* **1** Stretched tight; taut. **2** Under mental or nervous strain; strained. **3** *Phonet.* Pronounced with the tongue and its muscles taut, as (ē) and (ōō); narrow: opposed to *lax.* — *v.t.* & *v.i.* **tensed, tens·ing** To make or become strained or drawn tight. [< L *tensus,* pp. of *tendere* stretch] — **tense'ly** *adv.* — **tense'ness** *n.*

tense² (tens) *n.* A form of a verb that relates it to time viewed either as finite past, present, or future, or as non-finite. — **sequence of tenses** In inflected languages, the customary choice of tense for a verb that follows another in a sentence, particularly in reported or indirect discourse. ◆ The general principle of sequence of tenses in English is that present follows present and past follows past. Thus, the tense of the subordinate clause tends to shift back to agree with the tense of the main verb. "He *wants* to go," becomes, in indirect discourse, "They said that he *wanted* to go." However, if continued, habitual, future, or universal action is expressed, the present tense may be retained in the subordinate clause: They told me that he *is* still in town; Columbus proved that the world *is* round. The present tense is also retained in the subordinate clause for emphasis: They just learned he *is* going after all. In subordinate clauses of purpose the general rule of tense sequence holds true: We *are working* so that we *can* go to Europe; We *worked* so that we *could* go to Europe. In conditional sentences expressing a simple fact or open question, the main and subordinate verbs remain independent: If he *said* that, I *can't* prove it. However, sequence of tenses is strictly observed in a highly improbable or contrary-to-fact statement. Time present is then expressed by the use of the past tense: If he *had* any sense, he *wouldn't drive* that car. Time past is expressed by the past perfect tense: If I *had had* my wits about me, I *would have* telephoned immediately. [< OF *tens* < L *tempus* time, tense]

ten·si·ble (ten'sə·bəl) *adj.* **1** Extensible. **2** Capable of being made tense; tensile.

ten·sile (ten'sil, *Brit.* ten'sīl) *adj.* **1** Of or pertaining to tension. **2** Capable of extension. **3** Producing tones from stretched strings of instruments. [< NL *tensilis* < L *ten-*

tentation

sus. See TENSE¹.] — **ten·sil·i·ty** (ten·sil'ə·tē) *n.*

tensile strength *Physics* The resistance of a material to forces of rupture and longitudinal stress: usually expressed in pounds or tons per square inch.

ten·sim·e·ter (ten·sim'ə·tər) *n.* An instrument for measuring the tension of gases; a manometer. [< *tensi-* (< TENSION) + -METER]

ten·si·om·e·ter (ten'sē·om'ə·tər) *n.* A device for determining tensile strength. [See TENSIMETER.]

ten·sion (ten'shən) *n.* **1** The act of stretching; the condition of being stretched tight. **2** Mental strain; intense nervous anxiety. **3** Any strained relation, as between governments. **4** *Physics* **a** Stress on a material caused by pulling: opposed to *compression,* and distinguished from *torsion.* **b** The condition of a body when acted on by such stress. **5** The expansive force of a gas. **6** A regulating device, as that on a sewing machine to regulate the tightness of the thread. **7** *Electr.* Electromotive force; also, electric potential. [< L *tensio, -onis* < *tensus.* See TENSE¹.] — **ten'sion·al** *adj.*

ten·si·ty (ten'sə·tē) *n.* The state of being tense; tension.

ten·sive (ten'siv) *adj.* **1** Caused by or causing tension. **2** Causing a sensation of stiffness or contraction.

ten·sor (ten'sər, -sôr) *n.* **1** *Anat.* A muscle that stretches a part. **2** *Math.* A vector quantity which may be fully described only with reference to more than three components. [< NL < L *tensus.* See TENSE¹.]

ten-strike (ten'strīk') *n.* **1** In bowling, the knocking down by a player of all the pins at one bowl: also called *strike.* **2** *U.S. Colloq.* Hence, a stroke of unexampled success; a very profitable bargain.

tent¹ (tent) *n.* A shelter of canvas or the like, supported by poles and fastened by cords to pegs (called **tent pegs**) driven into the ground. — *v.t.* To cover with or as with a tent. — *v.i.* To pitch a tent; camp out. [< F *tente* < LL *tenta,* orig. neut. pl. of *tentus,* pp. of *tendere* stretch. Cf. L *tentorium* awning.]

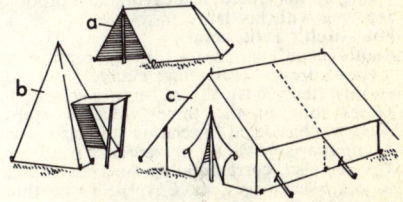

TENTS
a. Pup tent. b. Pyramid tent. c. Wall tent.

tent² (tent) *Surg. n.* A small roll, as of lint, placed in a wound or orifice to prevent its closing. — *v.t.* To keep open with a tent; also, to probe. [< F *tente* < *tenter* test, probe < L *tentare*]

tent³ (tent) *Scot. v.t.* **1** To pay attention to; observe. **2** To hinder; prevent. **3** To attend upon; look after. — *n.* **1** Attention; note; heed. **2** An open-air wooden pulpit. — **tent'·less** *adj.*

tent⁴ (tent) *n.* A deep-red wine obtained chiefly from Spain. [< Sp. *tinto* deep-colored < L *tinctus* dyed. See TINT.]

ten·ta·cle (ten'tə·kəl) *n.* **1** *Zool.* A protruding flexible process or appendage (usually of the head) of invertebrate animals, functioning as an organ of touch, prehension, or motion. Some examples are the hollow fleshy processes about the mouth of a polyp communicating with the body cavity, the eyestalks of a gastropod, and the arms of a cuttlefish, especially one of the two longer arms of a decapod. **2** *Bot.* A sensitive glandular hair, as on the leaves of the sundew. **3** Something resembling a tentacle; a tendril. Also **ten·tac·u·lum** (ten·tak'yə·ləm). [< L *tentaculum* < *tentare* touch, try] — **ten·tac'u·lar** *adj.*

tent·age (ten'tij) *n.* **1** The supply of tents available for any purpose. **2** Tents collectively.

ten·ta·tion (ten·tā'shən) *n.* The act or process of adjusting by experimentation until a desired effect is secured. [<F <L *tentatio, -onis* < *tentare* try. See TEMPT.]

ten·ta·tive (ten′tə·tiv) *adj.* **1** Used in making a trial; provisional or conjectural; experimental and subject to change. **2** *Med.* Based on subjective and objective symptoms: said of a diagnosis subject to change. —*n.* An experiment; conjecture. [<Med. L *tentativus* <L *tentatus,* pp. of *tentare* try, probe] —**ten′ta·tive·ly** *adv.* —**ten′ta·tive·ness** *n.*

tent caterpillar The gregarious larva of several North American moths (family *Lasiocampidae)* that spins a large silken web which shelters the colony, especially the **orchard caterpillar** (genus *Malacosoma).*

tent·ed (ten′tid) *adj.* **1** Overspread or covered with or sheltered by tents: the *tented* field. **2** Resembling a tent.

tented arch A fingerprint pattern in which the skin ridges have an upward thrust in the shape of a tent, arranging themselves on both sides of a spine or axis.

ten·ter[1] (ten′tər) *n.* **1** A frame or machine for stretching cloth to prevent shrinkage while drying. **2** *Obs.* A tenterhook. —*v.t.* To stretch on or as on a tenter. —*v.i.* To be or admit of being stretched thus. [<L *tentus* extended. See TENT[1].]

tent·er[2] (ten′tər) *n. Brit.* One who especially attends to anything; particularly, one who attends to machinery in a factory. [<TENT[3]]

ten·ter·hook (ten′tər·ho͝ok′) *n.* A sharp hook for holding cloth while being stretched on a tenter. —**to be on tenterhooks** To be in a state of anxiety or suspense.

tenth (tenth) *adj.* **1** Next in order after the ninth. **2** Designating one of ten equal parts. —*n.* **1** One of ten equal parts. **2** *Music* An interval compounded of an octave and a third; a note separated from another by this interval. **3** An organ stop tuned a tenth above the diapasons. **4** A tax of one tenth of one's income; a tithe. [ME *tenthe*] —**tenth′ly** *adv.*

ten·tie (ten′tē) *adj. Scot.* Attentive; cautious. Also **ten′ty.**

tent pegging A cavalry exercise in British military tournaments in which the horseman, riding at full speed, endeavors to uproot a tent peg with his lance.

tent stitch Petit point.

te·nue (tə·nü′) *n. French* **1** Appearance or style of dress. **2** Bearing; manner.

ten·u·is (ten′yo͞o·is) *n. pl.* **·u·es** (-yo͞o·ēz) In Greek, one of the three voiceless stops, κ, π, τ, considered in relation to their voiced counterparts, γ, β, δ, or voiceless fricatives, χ, φ, θ; also, corresponding voiceless sounds in other languages, as *k, t, p.* [<L, thin, trans. of Gk. *psilos* bare, unaspirated]

ten·u·ous (ten′yo͞o·əs) *adj.* **1** Thin; slim; delicate; also, weak; flimsy. **2** Having slight density; rare: opposed to *dense.* See synonyms under FINE. [<L *tenuis* thin] —**ten′u·ous·ly** *adv.* —**ten′u·ous·ness, ten·u·i·ty** (ten·yo͞o′ə·tē, ti·no͞o′-) *n.*

ten·ure (ten′yər) *n.* **1** A holding, as of land. **2** The act of holding in general, or the state of being held. **3** The term during which a thing is held, as an office. **4** The conditions or manner of holding. See synonyms under OCCUPATION. [<F < *tenir* hold <L *tenere*] —**ten·u·ri·al** (ten·yo͝or′ē·əl) *adj.* —**ten·u′ri·al·ly** *adv.*

te·nu·to (te·no͞o′tō) *adj. Music* Sustained; held for the full time. [<Ital.]

te·nu·to–mark (te·no͞o′tō·märk′) *n. Music* A horizontal stroke over a note or chord that is to be held for its full value.

Te·o·bal·do (tā′ō·bäl′dō) Italian form of THEOBALD.

te·o·cal·li (tē′ə·kal′ē, *Sp.* tā′ō·kä′yē) *n.* **1** A temple peculiar to the ancient Mexicans and Central Americans, usually erected on a truncated pyramid. **2** A mound of similar form. Also **te·o·pan** (tā′ō·pän′). [<Sp. < Nahuatl, house of the god < *teotl* a god + *calli* house]

Te·o·do·ri·co (*Ital.* tā′ō·dō·rē′kō, *Sp.* -thō-) Italian and Spanish form of THEODORIC.

Te·o·do·ro (*Ital.* tā′ō·dō′rō, *Sp.* -thō′rō) Italian and Spanish form of THEODORE. —**Te′o·do′ra** (-rä) *fem.*

Te·o·fi·lo (tā·ō′fē·lō) Italian and Spanish form of THEOPHILUS.

te·o·sin·te (tē′ə·sin′tē) *n.* A stout, hardy perennial grass *(Euchlaena mexicana),* closely allied to Indian corn, and used for fodder. [<Sp. <Nahuatl *teocentli,* lit., divine maize < *teotl* a god + *centli* corn]

te·pee (tē′pē) *n.* A conical tent of the North American Plains Indians, usually covered with skins or other material: also spelled *teepee, tipi.* [<Dakota *tipi* < *ti* dwell + *pi* used for]

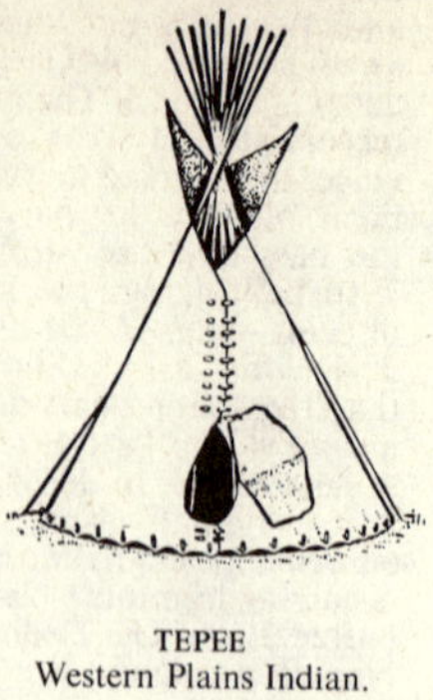

TEPEE
Western Plains Indian.

tep·e·fy (tep′ə·fī) *v.t.* & *v.i.* **·fied, ·fy·ing** To make or become tepid. [<L *tepefacere* make tepid < *tepere* be lukewarm + *facere* make] —**tep′e·fac′tion** (-fak′·shən) *n.*

teph·rite (tef′rīt) *n.* An ash-gray to black volcanic rock, essentially an alkaline andesite, with either nepheline or leucite. [<L *tephritis* <Gk. *tephra* ashes] —**te·phrit·ic** (tə·frit′ik) *adj.*

te·phro·sin (tə·frō′sin) *n. Chem.* A white crystalline compound, $C_{23}H_{22}O_7$, extracted from the leaves of a leguminous plant *(Tephrosia vogeli),* from derris, and cube: used as a fish poison. [<NL <Gk. *tephros* ash-colored < *tephra* ashes]

te·phro·sis (tə·frō′sis) *n.* Cremation; incineration. [<NL <Gk. *tephrōsis* < *tephra* ashes]

tep·id (tep′id) *adj.* Moderately warm; lukewarm, as a liquid. [<L *tepidus* < *tepere* be lukewarm] —**te·pid·i·ty** (tə·pid′ə·tē), **tep′id·ness** *n.* —**tep′id·ly** *adv.*

tep·i·dar·i·um (tep′ə·dâr′ē·əm) *n. pl.* **·dar·i·a** (-dâr′ē·ə) In the Roman baths, the intermediate apartment between the cold- and the hot-bath rooms. [<L < *tepidus.* See TEPID.]

Te·quen·da·ma Falls (tā′ken·dä′mä) A falls in the Bogotá river, west of Bogotá, central Colombia; 482 feet high.

te·qui·la (tə·kē′lə) *n.* A Mexican alcoholic liquor made from the maguey. [from *Tequila,* Jalisco, Mexico]

ter- *combining form* Three; third; threefold; three times: *tercentenary.* [<L *ter* thrice]

ter·a·phim (ter′ə·fim) *n. pl. sing.* **ter·aph** (ter′·əf) or **ter·a·phim** Images, small idols, or household gods consulted as oracles by some of the ancient Hebrews: used as a plural or collective singular in the Bible. [<Hebrew *terāphīm*]

ter·a·tism (ter′ə·tiz′əm) *n. Biol.* A monstrosity; especially, a malformed human or animal fetus. [<Gk. *teras* monster]

terato- *combining form* A wonder; monster: *teratogeny.* Also, before vowels, **terat-.** [< Gk. *teras, teratos* a wonder]

ter·a·tog·e·ny (ter′ə·toj′ə·nē) *n. Biol.* The production of monsters or other abnormal organisms. Also **ter′a·to·gen′e·sis** (-tō·jen′ə·sis). [<TERATO- + -GENY] —**ter′a·to·gen′ic** (-tō·jen′ik) *adj.*

ter·a·toid (ter′ə·toid) *adj.* Like a monstrosity; abnormal. [<TERAT(O)- + -OID]

ter·a·tol·o·gy (ter′ə·tol′ə·jē) *n.* The branch of biology and medicine treating of abnormal growths or monstrosities. [<TERATO- + -LOGY] —**ter′a·to·log′ic** (-tō·loj′ik) or **·i·cal** *adj.* —**ter′a·tol′o·gist** *n.*

Te·ra·u·chi (te·rä·o͞o·chē), **Count Juichi,** 1879–1946, Japanese general.

ter·bi·a (tûr′bē·ə) *n. Chem.* Oxide of terbium, Tb_2O_3.

ter·bi·um (tûr′bē·əm) *n.* A metallic element (symbol Tb) belonging to the lanthanide series, found in gadolinite and other rare-earth minerals. See ELEMENT. [<NL < *Ytterby,* a town in Sweden] —**ter′bic** *adj.*

terbium metal One of a group in the lanthanide series of elements, including gadolinium, europium, and terbium.

Ter Borch (tûr bôrkh), **Gerard,** 1617?–81, Dutch painter.

Ter·cei·ra (tûr·sâr′ə) The easternmost island of the central Azores; 153 square miles.

ter·cel (tûr′səl) *n.* A male falcon, especially the peregrine falcon: also spelled *tassel.* Also **terce·let** (tûrs′lit). [<OF <L *tertius* third; said to be so called because every third egg in a falcon's nest was thought to produce a male]

ter·cen·te·nar·y (tûr·sen′tə·ner′ē, tûr′sen·ten′·ər·ē) *adj.* Of or pertaining to a period of 300 years or to a 300th anniversary. —*n. pl.* **·nar·ies** The 300th anniversary. Also *tricentennial.* Also **ter·cen·ten·ni·al** (tûr′sen·ten′ē·əl).

ter·cet (tûr′sit, tûr·set′) *n.* **1** *Music* A triplet. **2** A group of three lines riming together or connected with adjacent triplets by double or triple rime. [<F <Ital. *terzetto,* dim. of *terzo* <L *tertius* a third]

ter·e·bene (ter′ə·bēn) *n. Chem.* A colorless, aromatic liquid hydrocarbon mixture of terpenes from oil of turpentine: used as an antiseptic and expectorant. [<TEREB(INTH) + (TERP)ENE]

te·reb·ic (te·reb′ik, -rē′bik) *adj. Chem.* Of, pertaining to, or derived from a white crystalline acid, $C_7H_{10}O_4$, derived from oil of turpentine. [<TEREBINTH]

ter·e·binth (ter′ə·binth) *n.* A small tree *(Pistacia terebinthus)* with winged pinnate leaves resembling those of the common ash but smaller: the original source of turpentine. [<L *terebinthus* <Gk. *terebinthos*]

ter·e·bin·thine (ter′ə·bin′thin) *adj.* Of or pertaining to the terebinth or turpentine. Also **ter′e·bin′thic.**

te·re·do (tə·rē′dō) *n.* One of a genus *(Teredo)* of marine mollusks (family *Teredinidae*); a shipworm. [<L, borer <Gk. *terēdōn* < *terein* rub hard, bore]

Te·rek (te′rek, *Russian* tye′rik) A river in northern Caucasus, Russian S.F.S.R., flowing 367 miles north and NE from north central Georgian S.S.R. to the NW Caspian Sea.

Ter·ence (ter′əns), 190?–159 B.C., Roman playwright: full name *Publius Terentius Afer.*

Te·re·sa (tə·rē′sə; *Ital., Sp.* tā·rā′sä) Italian and Spanish form of THERESA.

Te·re·sian (ti·rē′shən) *n.* A Carmelite friar or nun of the order founded by St. Teresa of Ávila in 1562.

Te·re·si·na (tā′rə·zē′nə) The capital of Piauí state, Brazil.

te·rete (tə·rēt′, ter′ēt) *adj.* Cylindrical and slightly tapering; round in cross-section. [<L *teres, teretis* round, rounded off < *terere* rub]

Te·reus (tir′yo͞os, tir′ē·əs) In Greek mythology, a Thracian king who was transformed into a hoopoe by the gods after he had raped his sister-in-law. See PHILOMELA.

ter·fa (tûr′fə) *n.* An edible fungus (genera *Terfezia* and *Tirmania*) of the deserts of North Africa, having a subterranean fruit body resembling truffles and eaten by the Arabs. [<Arabic *tirfāsh* truffle]

ter·gal (tûr′gəl) *adj.* Of or pertaining to the tergum; dorsal.

ter·gem·i·nate (tər·jem′ə·nit) *adj. Bot.* Having three pairs of forked leaflets. [<TER- + GEMINATE]

ter·gi·ver·sate (tûr′ji·vər·sāt′) *v.i.* **·sat·ed, ·sat·ing** **1** To be evasive; equivocate or prevaricate. **2** To change sides, attitudes, etc.; become a renegade; apostatize. [<L *tergiversatus,* pp. of *tergiversari* < *tergum* back + *versare* turn] —**ter′gi·ver·sa′tor** *n.*

ter·gi·ver·sa·tion (tûr′ji·vər·sā′shən) *n.* **1** Evasion of a point, as by prevarication or subterfuge. **2** Fickleness or insincerity of conduct; shiftiness.

ter·gum (tûr′gəm) *n. Zool.* The back or dorsal part of an arthropod. [<L]

Ter·hune (tər·hyo͞on′), **Albert Payson,** 1872–1942, U. S. author.

term (tûrm) *n.* **1** A word or expression used to designate some definite thing; a technical expression: a scientific *term.* **2** Any word or expression conveying some conception or thought: a *term* of reproach; to speak in general *terms.* **3** *pl.* The conditions or stipulations according to which something is to be done or acceded to: the *terms* of sale; peace *terms.* **4** *pl.* Mutual relations; footing: usually preceded by *on* or *upon*: England was on friendly *terms* with France. **5** *Math.* **a** The antecedent or consequent of a ratio. **b** The numerator or denominator of a fraction. **c** One of the quantities of an algebraic expression that are connected by the plus and minus signs. **d** One of the quantities which compose a series or progression. **6** *Logic* **a** In a proposition, either of the two parts, the subject and predicate, which are joined by a copula. **b** Any of the three elements of a syllogism, each of which appears twice. In a syllogism, the **major term** is the predicate of both the major premise and the conclusion. The **minor term** is the subject of both the minor premise and the conclusion. See SYLLOGISM. 7 A fixed period or definite length of time: a *term* of office. 8 One of the periods of the year appointed for holding instruction in colleges and schools. 9 *Law* **a** One of the prescribed periods of the year during which a court may hold a session. **b** A specific extent of time during which a termor may hold an estate. **c** A space of time allowed a debtor to meet his obligation. **10** *Med.* The time for childbirth. **11** *Archaic* An utmost limit; boundary. **12** *Archit.* A pillar of tapering form, ending in a sculptured head or bust. — *v.t.* To designate by means of a term; name or call. [<OF *terme* <L *terminus* a limit]

Synonyms (noun): article, condition, expression, member, name, phrase, word. *Term* in its figurative use always retains something of its literal sense of a boundary or limit. The *articles* of a contract or other instrument are simply the portions into which it is divided for convenience; the *terms* are the essential statements on which its validity depends—as it were, the landmarks of its meaning or power; a *condition* is a contingent *term*, which may become fixed upon the happening of some contemplated event. In logic a *term* is one of the essential members of a proposition, the boundary of statement in some one direction. Thus in general use *term* is more restricted than *word, expression,* or *phrase*; a *term* is a *word* that limits meaning to a fixed point of statement or to a special class of subjects; as, when we speak of the definition of *terms*, that is of the key *words* in any discussion; or we say "that is a legal or scientific *term*." See BOUNDARY, DICTION.

ter·ma·gant (tûr′mə·gənt) *n.* A scolding or abusive woman; shrew. — *adj.* Violently abusive and quarrelsome; vixenish. [<TERMAGANT] — **ter′ma·gan′cy** *n.*

Ter·ma·gant (tûr′mə·gənt) An idol or imaginary deity of very turbulent, overbearing character that the medieval romances represented Moslems as worshiping. Also **Ter′·ma·gaunt, Ter′ma·gund** (-gənd).

Ter·man-Mer·rill test (tûr′mən·mer′il) *Psychol.* An extension of the Stanford revision intelligence test which includes ages down to two years and adds many more items at upper levels, with a choice of two alternative scales.

term day 1 A designated day; specifically, quarter-day. 2 At hiring fairs, the day from which the contract of service dates.

term·er (tûr′mər) *n.* 1 *Law* A termor. 2 *Colloq.* A prisoner serving a certain term: usually with an ordinal: a first-*termer.*

ter·mes (tûr′mēz) *n.* A termite.

ter·mi·na·ble (tûr′mə·nə·bəl) *adj.* That may be terminated; limitable; not perpetual. — **ter′·mi·na·bil′i·ty, ter′mi·na·ble·ness** *n.* — **ter′mi·na·bly** *adv.*

ter·mi·nal (tûr′mə·nəl) *adj.* 1 Pertaining to or creative of a boundary, limit, or terminus: a *terminal* railroad station. 2 Pertaining to the delivery or storage of freight or baggage: *terminal* charges. 3 Pertaining to a term or name. 4 Situated at or forming the end of a series or part. 5 *Bot.* Borne at the end of a stem or branch. 6 Of, pertaining to, or occurring in or at the end of a period of time; of a fixed period. — *n.* 1 That which terminates; a terminating point or part; termination; end. 2 *Electr.* One of the two free ends of a conductor, particularly if proceeding from an electric source, as a battery or dynamo. 3 *Archit.* A terminal figure or pedestal; terminus. 4 The edges or planes that form the end of a crystal. 5 A railroad terminus. 6 *pl.* Charges for the use of terminal facilities, or for the handling of freight at railroad terminuses. 7 *Physiol.* The end structure or end of a neuron or nerve fiber. [<F <LL *terminalis* <L *terminus* boundary] — **ter′mi·nal·ly** *adv.*

Ter·mi·na·li·a (tûr′mə·nā′lē·ə) *n. pl. Latin* The ancient Roman festival of Terminus, celebrated on Feb. 23 by the decoration of the boundary markers between private properties, and the offering of sacrifices.

terminal rime The riming of a word or group of syllables at the end of a verse with that at the end of another verse in the same stanza or poem.

terminal velocity *Physics* The velocity acquired by a freely falling body when the resistance of the medium equals the weight of the body.

ter·mi·nate (tûr′mə·nāt) *v.* **·nat·ed, ·nat·ing** *v.t.* 1 To put an end or stop to. 2 To form the conclusion of; finish. 3 To bound or limit. — *v.i.* 4 To have an end; come to an end. See synonyms under ABOLISH, CEASE, END. [<L *terminatus,* pp. of *terminare* end, limit <*terminus* a limit]

ter·mi·na·tion (tûr′mə·nā′shən) *n.* 1 The act of setting bounds or limits. 2 The act of ending or concluding. 3 That which bounds or limits; close; end; limit in time or space. 4 Outcome; result; conclusion. 5 The final letters or syllable of a word; a suffix. See synonyms under BOUNDARY, END.

ter·mi·na·tion·al (tûr′mə·nā′shən·əl) *adj.* Of, pertaining to, or formative of a syllable or other termination; formed by suffixes.

ter·mi·na·tive (tûr′mə·nā′tiv) *adj.* Designed or tending to terminate; determining; definitive; bounding; conclusive. — **ter′mi·na′tive·ly** *adv.*

ter·mi·na·tor (tûr′mə·nā′tər) *n.* 1 One who or that which terminates. 2 *Astron.* The boundary between the illuminated and dark portions of the moon or of a planet.

ter·mi·ner (tûr′mə·nər) *n. Law* The act or function of determining. See OYER AND TERMINER. [<AF *terminour* <F *terminer* end <L *terminare. See* TERM.]

ter·mi·nism (tûr′mə·niz′əm) *n.* 1 *Theol.* The doctrine that God has ordained a limit in the life of each man and of mankind beyond which the opportunity for salvation is lost. 2 A form of nominalism; specifically, the doctrine of William of Ockham, who stated that universals are abstract terms or predicables, rather than either real existents or mere vocal sounds. [<L *terminus* term]

ter·mi·nol·o·gy (tûr′mə·nol′ə·jē) *n.* 1 The study or the use of terms. 2 The technical terms used in a science, art, trade, etc. 3 Nomenclature. [<L *terminus* + -LOGY] — **ter′mi·no·log′i·cal** (-nə·loj′i·kəl) *adj.* — **ter′·mi·no·log′i·cal·ly** *adv.* — **ter′mi·nol′o·gist** *n.*

ter·mi·nus (tûr′mə·nəs) *n. pl.* **·nus·es** or **·ni** (-nī) 1 The final point or goal; end; terminal. 2 The farthermost station on a railway; also, the town in which such station is situated. 3 A boundary or border; also, a boundary mark. See synonyms under END. [<L]

Ter·mi·nus (tûr′mə·nəs) In Roman mythology, the god of boundaries and landmarks.

ter·min·us ad quem (tûr′mə·nəs ad kwem) *Latin* The end or limit to which; the goal; the terminating point of an argument, period, etc.

ter·mi·nus a quo (tûr′mə·nəs ā kwō) *Latin* The starting point.

ter·mite (tûr′mīt) *n.* A white ant. Also *termes.* For illustration see INSECTS (injurious). [<L *termes, termitis*]

term·less (tûrm′lis) *adj.* 1 Of boundless extent or duration. 2 Independent of conditions; unconditional. 3 *Archaic* Incapable of being expressed by terms; indescribable.

term·ly (tûrm′lē) *adj.* Happening or done every term. — *adv.* Periodically.

term·or (tûr′mər) *n. Law* A person who holds lands or tenements for a definite number of years or for life.

tern[1] (tûrn) *n.* Any of several gull-like birds (subfamily *Sterninae*), having the bill pointed and the mandibles co-terminal, smaller than most gulls, with wings more pointed, and the tail usually deeply forked; especially, the common tern (*Sterna hirundo*) of the Atlantic coasts, white with a black cap, and the **least** or **minute tern** (*S. antillarum*). ◆ Homophones: *terne, turn.* [< Scand. Cf. Dan. *terne* tern.]

BLACK TERN (Body length about 10 inches; wingspread, 25 inches)

tern[2] (tûrn) *n.* 1 That which is composed of three; specifically, three numbers in a lottery that, when drawn together, secure a large prize. 2 In New England, a three-masted schooner. ◆ Homophones: *terne, turn.* [<L *terni* by threes <*ter* thrice]

ter·na·ry (tûr′nər·ē) *adj.* 1 Formed or consisting of three; grouped in threes. 2 *Math.* Containing three variables; also, pertaining to systems of notation, having three as a base, or radix. 3 *Chem.* Having three separate parts, as atoms, elements, etc. 4 *Metall.* Made of an alloy which contains three metals. — *n. pl.* **·ries** A group of three; a triad. [<L *ternarius* <*terni* by threes]

ter·nate (tûr′nāt) *adj.* 1 Classified or arranged in threes. 2 *Bot.* Trifoliolate; consisting of threes. [<NL *ternatus* <L *terni* by threes] — **ter′nate·ly** *adv.*

Ter·na·te (ter·nä′tā) An Indonesian island of the northern Moluccas; 41 square miles.

terne (tûrn) *v.t.* **terned, tern·ing** To cover with a thin layer of lead and tin. — *n.* Terne plate. ◆ Homophones: *tern, turn.* [<F *terne* dull; from the resulting finish]

terne plate Steel plate with a coating of lead and tin, having a dull finish and inferior in quality to standard tin plate.

Ter·ni (ter′nē) A city in Umbria, central Italy.

ter·ni·on (tûr′nē·ən) *n.* 1 A set of three. 2 A section of a book composed of three sheets in double folds, or 12 pages. [<L *terni* by threes]

Ter·no·pol (tûr·nō′pəl, *Russian* tyir·nō′pəl) A city in western Ukrainian S.S.R.: *Polish* **Tar·no·pol** (tär·nō′pəl).

Ter·pan·der (tər·pan′dər) Greek poet and musician of the seventh century B.C.

ter·pene (tûr′pēn) *n. Chem.* Any of a class of isomeric hydrocarbons, $C_{10}H_{16}$, contained chiefly in the essential oils of coniferous plants. [<*terp(entin),* earlier form of TURPENTINE + -ENE]

ter·pin·e·ol (tər·pin′ē·ōl, -ol) *n. Chem.* A colorless, unsaturated, tertiary alcohol, $C_{10}H_{17}OH$, derived from the essential oils of various plants and also made synthetically: it has an odor of lilacs and is used in perfumery. [< *terpin,* earlier form of TERPENE + -OL[1]]

ter·pi·nol (tûr′pə·nōl, -nol) *n.* An oily, colorless, liquid mixture of various terpenes, having an odor of hyacinth. [See TERPINEOL]

Terp·sich·o·re (tûrp·sik′ə·rē) The Muse of dancing. [<Gk. *Terpsichorē* <*terpsichoros* delighting in the dance <*terpsis* enjoyment + *choros* dance] — **Terp·si·cho·re·an** (tûrp′·si·kə·rē′ən) *adj.*

terp·si·cho·re·an (tûrp′si·kə·rē′ən) *adj.* Of or relating to dancing: also **terp′si·cho·re′al.** — *n. Colloq.* A dancer.

ter·ra (ter′ə) *n. Latin* The earth; earth.

ter·ra al·ba (ter′ə al′bə) 1 Pipe clay. 2 The pigment made from ground gypsum. 3 Magnesia. 4 A grade of kaolin used as an adulterant of paints. [<L, white earth]

ter·race (ter′is) *n.* 1 An artificial raised level space, as of lawn, having one or more vertical or sloping sides; also, such levels collectively. 2 A raised level supporting a row of houses, or the houses occupying such a position. 3 The flat roof of an Oriental or Spanish house. 4 A relatively narrow step in the face of a steep natural slope. 5 An open gallery; balcony. — *v.t.* **·raced, ·rac·ing** To form into or provide with a terrace or terraces. [<OF <Ital. *terraccia* <L *terra* earth]

ter·ra cot·ta (ter′ə kot′ə) 1 A hard, durable, kiln-burnt clay, reddish-brown in color and usually unglazed: widely used as a structural material and also, in glazed and colored forms, for tiles, building façades, etc. 2 A statue or figure made of this clay. 3 A brownish-orange color resembling that of terra cotta. [<Ital., cooked earth]

ter·ra fir·ma (ter′ə fûr′mə) Solid ground, as distinguished from the sea or the air. [<L]

ter·rain (te·rān′, ter′ān) *n.* 1 Battleground, or a region suited for defense, fortifications, etc. 2 A piece or plot of ground; a region or territory viewed with regard to its suitability for some particular purpose. 3 A terrane. [<F <L *terrenum* <*terrenus* earthen <*terra* earth]

ter·ra in·cog·ni·ta (ter′ə in·kog′nə·tə) 1 An unknown land or region. 2 An unexplored field of study or knowledge. [<L]

Ter·ra·my·cin (ter′ə·mī′sin) *n.* Proprietary name for an antibiotic isolated from a soil mold (*Streptomyces rimosus*), of value in the

treatment of a wide variety of pathogenic infections.
ter·rane (te·rān′, ter′ān) n. 1 Geol. A continuous formation or continuous series of related formations; an area of particular rocks. 2 A tract or region considered with reference to some special purpose. [<F *terrain*. See TERRAIN.]
ter·ra·pin (ter′ə·pin) n. One of the several North American edible tortoises (family *Testudinidae*); especially, the diamond-back terrapin. [<Algonquian]

DIAMOND-BACK TERRAPIN
(Shell from 4 to 7 inches)

ter·ra·que·ous (te·rā′kwē·əs) adj. Composed of, living in, or consisting of, both land and water. [<L *terra* earth, land + AQUEOUS]
ter·rar·i·um (te·râr′ē·əm) n. pl. **·rar·i·ums** or **·rar·i·a** (-râr′ē·ə) 1 A small enclosure or box with glass sides for live lizards, growing plants, etc. 2 A vivarium for land animals. [<L *terra* earth + -ARIUM, on analogy with *aquarium*]
ter·raz·zo (ter·rät′sō) n. Flooring made of small pieces of marble or colored stone set in concrete. Also **ter·raz′zo Ve·ne·zia·no** (vā′nā·tsyä′nō). [<Ital. <L *terra* earth]
terre (târ) n. French Earth. See TERRA.
Terre·bonne (ter′bon′, -bôn′), **Bayou** A lagoon in SE Louisiana, flowing 55 miles south to **Terrebonne Bay**, a shallow inlet of the Gulf of Mexico SW of New Orleans.
ter·reen (te·rēn′) See TERRINE.
Terre Haute (ter′ə hōt′) A city in western Indiana, on the Wabash River.
ter·rene[1] (te·rēn′) adj. 1 Pertaining to earth; earthy. 2 Earthly; worldly; mundane. —n. 1 The surface of the earth. 2 The earth; a land or terrain. [<L *terrenus* <*terra* earth]
ter·rene[2] (te·rēn′) See TERRINE.
terre·plein (ter′plān) n. 1 The upper surface of a rampart behind the parapet, on which the guns are mounted. 2 An embankment with a level top. [<F *terre* earth + *plein* level]
ter·res·tri·al (tə·res′trē·əl) adj. 1 Belonging to the earth: opposed to *celestial* or *cosmic*. 2 Pertaining to land or earth: *terrestrial* magnetism. 3 Biol. Living on or growing in the earth or land: opposed to *aquatic*, *aerial*, etc. 4 Belonging to or consisting of land, as distinct from water, trees, etc. 5 Worldly; mundane. —n. An inhabitant of the earth. [<L *terrestris* <L *terra* land] —**ter·res′tri·al·ly** adv. —**ter·res′tri·al·ness** n.
ter·ret (ter′it) n. 1 One of two metal rings projecting from the saddle of a harness, through which the reins are passed. 2 A ring for attaching a leash to a dog's collar, etc. Also **ter′rit**. [ME *toret* <F *touret* small wheel, dim. of *tour* a turn]
terre–ten·ant (ter′ten′ənt) n. Law 1 The person who is in actual possession of lands. 2 The owner or holder of the legal estate in lands. Also spelled *ter–tenant*. [<AF *terre tenaunt* holding land <F *terre* (<L *terra* land) + *tenaunt* holding, ppr. of *tenir* hold <L *tenere*]
terre–verte (ter′vert′) n. 1 An earthy silicate resembling glauconite and used as a green pigment by artists. 2 Glauconite. [<F *terre verte* green earth]
ter·ri·ble (ter′ə·bəl) adj. 1 Of a nature to excite terror; appalling. 2 *Colloq.* Characterized by excess; severe; extreme. 3 Inspiring awe. See synonyms under AWFUL, FORMIDABLE, FRIGHTFUL, GRIM. [<F <L *terribilis* <*terrere* terrify] —**ter′ri·ble·ness** n. —**ter′ri·bly** adv.
ter·ric·o·lous (te·rik′ə·ləs) adj. *Biol.* Living on or in the ground. Also **ter·ric′o·line** (-lēn, -lin). [<L *terricola* earth dweller <*terra* earth + *colere* dwell]
ter·ri·er[1] (ter′ē·ər) n. A small, active, wiry dog of several breeds, formerly used to hunt burrowing animals and noted for the courage and eagerness with which it "goes to earth" in pursuit of its quarry. See AIREDALE, DANDIE DINMONT, SCHNAUZER. [<OF <L *terrarius* pertaining to earth. See TERRIER[2].]
— **Bedlington terrier** A liver-colored or blue terrier with muscular body, long neck, narrow skull, and thick coat. It is very game in attacking badgers, foxes, or vermin. [<*Bedlington*, shire in Northumberland County, England]
— **Boston terrier** A small, non-sporting terrier of dark brindle color marked with white, crossbred from the English bulldog and the white English terrier, and having a square skull, short tail, and short, smooth coat.
— **bull terrier** A white terrier first crossbred from the bulldog and the white English terrier, then crossed with the Spanish pointer. It has a muscular, well-balanced body, long head, flat skull, and short, stiff coat.
— **Cairn terrier** A small, stocky, alert terrier of Scotland, having a broad head, a rough outer coat of any color except white, pointed

Welsh Terrier   Airedale   Lakeland Terrier   Irish Terrier

West Highland White Terrier   Cairn Terrier   Yorkshire Terrier   Skye Terrier

Norwich Terrier   Dandie Dinmont   Boston Terrier   Bull Terrier

Scottish Terrier   Schnauzer   Wire-haired Fox Terrier   Smooth-haired Fox Terrier

Bedlington Terrier   Manchester Terrier   Sealyham Terrier   Kerry Blue Terrier

ears, and a black nose: used as a retriever and to exterminate vermin.
— **Clydesdale terrier** A straight-eared, silky-haired dog, with tiny, erect ears and short legs, bred from but smaller than the Skye terrier.
— **fox terrier** A small white terrier, either smooth or wire-haired: formerly bred for bringing the fox out of his burrow, now usually a pet.
— **Irish terrier** A small, red or golden-red, rough-haired terrier having a rather long body built on racing lines: used for hunting small or big game and vermin and also for retrieving in water.
— **Kerry blue terrier** A breed of terrier originating in County Kerry, Ireland, having a long straight head, straight legs, long head, and soft, wavy, bluish-black coat: used to hunt and retrieve small game and as a herd dog, watchdog, and companion.
— **Lakeland terrier** A courageous breed of terrier having a dense, harsh coat of black, blue, or grizzle and tan: originally from Cumberland County, England, and used in hunting the otter or the fox.
— **Lhasa terrier** A breed of terrier native to Tibet, with a heavy yellow, black, white, or brown coat, straight forelegs, and tail curled over its back.
— **Manchester terrier** A small, speedy, short-haired, black-and-tan terrier, originally bred in Manchester, England, and known at one time as **Black-and-Tan terrier**.
— **Norwich terrier** A breed of small, wire-haired terrier, native in England, and usually red, black-and-tan, or grizzled: used in hunting.

— Scottish terrier A Scotch breed of small, wire-haired, alert, and intelligent terrier, having a compact body, short legs, small eyes and skull, and gray, brindled, grizzled, black, sandy, or wheaten coat. Also **Scotch terrier, Scottie.**

— Sealyham terrier A terrier of mixed ancestry, native of Sealyham, Wales, with short legs, a wide skull, square jaws, and wiry coat, usually solid white, but sometimes marked with lemon or brown on ears and head: used in hunting badger, fox, and otter.

— Skye terrier The smallest, lowest-set, and longest-bodied of all useful terriers, unrivaled for acute scent, hearing, sight, and alacrity. Its coat is long and straight, usually blue, gray, or fawn with black points.

— Welsh terrier An old breed of rough-haired, black-and-tan terrier, a native of Wales, having a broader head than a fox terrier and a flat skull: used for hunting otter, fox, and badger.

— West Highland white terrier A breed of small terrier, with long, low, compact body, short legs, and a pure white, coarse, wiry outer coat: said to have existed in Scotland prior to 1600 A.D.

— Yorkshire terrier A toy breed, among the smallest of all varieties of terriers. It has semi-erect ears and a coat of long, silky, dark steel-blue hair, with golden tan on chest and head. At first a pet of the working classes, especially of weavers, it later became a fashionable pet.

ter·ri·er[2] (ter′ē·ər) *n. Law* **1** A land survey setting forth in detail the number of acres, names of tenants, etc., in a given district: the *terrier* of glebe lands. **2** A book containing the lists of the lands either of a private person or a corporation; a rent roll. [<OF, list of tenants <LL *terrarius* a roll describing landed property <L, pertaining to land < *terra* land]

ter·rif·ic (tə·rif′ik) *adj.* **1** Arousing or calculated to arouse great terror or fear. **2** *Colloq.* Excessive; extreme; tremendous. See synonyms under AWFUL, FRIGHTFUL. **— ter·rif′i·cal·ly** *adv.*

ter·ri·fy (ter′ə·fī) *v.t.* **·fied, ·fy·ing** To fill with extreme terror. See synonyms under FRIGHTEN. [<L *terrificare* < *terrificus* causing fear < *terrere* frighten + *facere* make]

ter·rig·e·nous (te·rij′ə·nəs) *adj.* **1** Produced from or of the earth. **2** *Geol.* Derived from the land: said of marine deposits formed of material washed from the land, as contrasted with those of organic, chemical, or other origin, formed in the sea. **3** Earthborn. Also **ter·ri·gene** (ter′ə·jēn). [<L *terrigenus* < *terra* earth + *gignere* be born]

ter·rine (te·rēn′) *n.* **1** An earthenware jar containing some delicacy for the table and sold with its contents: a *terrine* of preserved ginger. **2** A kind of ragout or stew. Also spelled *terreen, terrene.* [<F <LL *terrineus* made of earth <L *terra* earth. Doublet of TUREEN.]

ter·ri·to·ri·al (ter′ə·tôr′ē·əl, -tō′rē-) *adj.* **1** Pertaining to a territory or territories; limited to a particular territory. **2** Designating military forces intended for territorial defense. **3** Belonging to a particular locality. **4** Organized or intended primarily for national defense: a *territorial* reserve. **— ter′ri·to′ri·al·ly** *adv.*

Ter·ri·to·ri·al (ter′ə·tôr′ē·əl, -tō′rē-) *adj.* Of or pertaining to any or all of the Territories of the United States: the *Territorial* system. **— *n.*** A member of the Territorial Army in Great Britain who enlisted for home defense but volunteered for overseas service in World War I.

ter·ri·to·ri·al·ism (ter′ə·tôr′ē·əl·iz′əm, -tō′rē-) *n.* The organizations, theories, or doctrines of the territorial systems. **— ter′ri·to′ri·al·ist** *n.*

ter·ri·to·ri·al·i·ty (ter′ə·tôr′ē·al′ə·tē, -tō′rē-) *n.* Territorial condition, status, or position.

ter·ri·to·ri·al·ize (ter′ə·tôr′ē·əl·īz′, -tō′rē-) *v.t.* **·ized, ·iz·ing** **1** To enlarge by annexation of territory. **2** To reduce to the political status of a territory. **3** To distribute among certain territories. **— ter′ri·to′ri·al·i·za′tion** *n.*

territorial jurisdiction *Law* The sovereign jurisdiction exercised by a state over all lands, waters, persons, and properties within its boundaries.

territorial system **1** A system of church government in which all inhabitants of a territory are required to belong to the same religion as the civil ruler. **2** Local organization for militia service. **3** Landlordism; a system giving predominance to landowners.

territorial waters The belt of sea under a state's territorial jurisdiction: formerly, the range of a cannon shot, or three miles: now often controversial. Also **territorial sea.**

ter·ri·to·ry (ter′ə·tôr′ē, -tō′rē) *n. pl.* **·ries** **1** The domain over which a sovereign state exercises jurisdiction. **2** Any considerable tract of land; a region; district; figuratively, sphere; province. **3** An area assigned for a special purpose: the *territory* of a commercial traveler. [<L *territorium* < *terra* earth]

Ter·ri·to·ry (ter′ə·tôr′ē, -tō′rē) *n. U.S.* A region having a certain degree of self-government, but not having the status of a State. Alaska and Hawaii were formerly Territories.

ter·ror (ter′ər) *n.* **1** An overwhelming impulse of fear; extreme fright or dread. **2** That which or one who causes extreme fear. **3** *Colloq.* An intolerable nuisance: That child is a holy *terror.* See synonyms under ALARM, FEAR, FRIGHT. [<F *terreur* <L *terror* fright < *terrere* frighten]

ter·ror·ism (ter′ə·riz′əm) *n.* **1** The act of terrorizing. **2** A system of government that seeks to rule by intimidation. **3** Unlawful acts of violence committed in an organized attempt to overthrow a government.

ter·ror·ist (ter′ər·ist) *n.* **1** One who adopts or supports a policy of terrorism. **2** A Jacobin or Republican of the French Revolution of 1789, especially during the Reign of Terror. **3** A member of political extremist groups in czarist Russia. **4** An alarmist; a scaremonger. **— ter′ror·is′tic** *adj.*

ter·ror·ize (ter′ə·rīz) *v.t.* **·ized, ·iz·ing** **1** To reduce to a state of terror; terrify. **2** To coerce through intimidation. Also *Brit.* **ter′·ror·ise. — ter′ror·i·za′tion** *n.* **— ter′ror·iz′er** *n.*

ter·ry (ter′ē) *n. pl.* **·ries** **1** The loop raised for the nap in weaving pile fabrics. **2** A pile dressmaking fabric in which the loops are uncut: also **terry cloth.** **3** A looped cotton fabric, very water-absorbent, used chiefly for towels and beach robes. [Prob. <F *tiré*, pp. of *tirer* draw <L *trahere*]

Ter·ry (ter′ē), **Dame Ellen,** 1848–1928, English actress.

terse (tûrs) *adj.* **1** Elegantly concise; short and to the point. **2** Rubbed to a polish; clean; polished; refined. [<L *tersus*, pp. of *tergere* rub off, rub down] **— terse′ly** *adv.* **— terse′ness** *n.*

Synonyms: brief, compact, compendious, concise, condensed, laconic, pithy, sententious, short, succinct. Anything *short* or *brief* is of relatively small extent. That which is *concise* is trimmed down, and that which is *condensed* is, as it were, pressed together, so as to include as much as possible within a small space. That which is *compendious* gathers the substance of a matter into a few weighty and effective words. *Succinct* writing is taut and lean without extraneous detail. *Summary* implies compression to the utmost, often to the point of abruptness; as, a *summary* statement or a *summary* dismissal. That which is *terse* has an elegant and finished completeness within the smallest possible compass. A *sententious* style is one abounding in maxims or short, pithy phrases. A *pithy* utterance gives the gist of a matter effectively, whether in rude or elegant style. *Antonyms*: diffuse, lengthy, long, prolix, tedious, verbose, wordy.

ter-ten·ant (ter′ten′ənt) See TERRE-TENANT.

ter·tial (tûr′shəl) *Ornithol. adj.* Of or pertaining to the third row of flight feathers in a bird's wing. **— *n.*** A tertiary feather. [<L *tertius* third < *ter* thrice]

ter·tian (tûr′shən) *adj.* Recurring every third day, reckoned inclusively, hence every alternate day. **— *n. Pathol.*** A disease, the paroxysms of which return every other day; a tertian fever. [<L *(febris) tertiana* tertian (fever) < *tertius* third]

ter·ti·ar·y (tûr′shē·er′ē, -shə·rē) *adj.* **1** Third in point of time, number, degree, or standing. **2** Tertial. **3** *Eccl.* Pertaining to the third order of a religious body. **4** *Chem.* **a** Having three substituted atoms or radicals: a *tertiary* amine. **b** Denoting a radical in which three bonds of the combining carbon atoms are directly connected with three other carbon atoms: *tertiary* butyl. **— *n. pl.*** **·ar·ies** **1** *Ornithol.* One of the feathers attached to the humerus joint of the wing of a bird. **2** Any member of the third order of a monastic body. [<L *tertiarius* < *tertius* third]

Ter·ti·ar·y (tûr′shē·er′ē, -shə·rē) *Geol. adj.* Of or pertaining to the earlier of the two geological periods or systems comprising the Cenozoic era, following the Cretaceous and succeeded by the Quaternary. **— *n.*** The Tertiary period or system, characterized by the rise of mammals.

ter·ti·um quid (tûr′shē·əm kwid) *Latin* **1** A third something; an indefinite or undefined thing related in some way to two definite or known things. **2** A mediating factor between essentially opposite things.

Ter·tul·li·an (tər·tul′ē·ən) Anglicized name of Quintus Septimius Florens Tertullianus, A.D. 160?–230?, Latin church father.

Te·ru·el (tä′rōō·el′) **1** A province in NE Spain; 6,710 square miles. **2** The capital city of Teruel province; scene of fierce fighting in the Spanish Civil War, 1937.

ter·va·lent (tûr′və·lənt, tər·vā′lənt) *adj. Chem.* Trivalent.

ter·za-ri·ma (ter′tsä·rē′mä) *n. pl.* **ter·ze-ri·me** (ter′tsā·rē′mā) A form of Italian triplet, in iambic decasyllables or hendecasyllables, in which the middle line of the first triplet rimes with the first and third lines of the following triplet: used by Dante in the *Divine Comedy.* [<Ital., third or triple line]

ter·zet·to (ter·tset′tō) *n. pl.* **·ti** (-tē) *Music* A short composition for three performers or singers; a trio. [See TERCET]

Tesch·en (tesh′ən) A territory and former principality in southern Poland and eastern Silesia, Czechoslovakia; 850 square miles; incorporated in Germany, 1939–45. *Czech* **Tě·šín** (tye′shēn), *Polish* **Cie·szyn** (che′shin).

Te·sho La·ma (te′shō lä′mə) See under DALAI LAMA.

Tes·la (tes′lə), **Nikola,** 1857–1943, U. S. electrical inventor born in Yugoslavia.

Tes·lin Lake (tez′lin, tes′-) A lake between NW British Columbia and southern Yukon, Canada; about 200 square miles; 80 miles long, 1 to 3 miles wide.

tes·sel·late (tes′ə·lāt) *v.t.* **·lat·ed, ·lat·ing** To construct in the style of checkered mosaic; lay or adorn with squares or tiles, as pavement. [<L *tessellatus* checkered < *tessella*, dim. of *tessera* cube. See TESSERA.] **— tes′sel·lat′ed** *adj.*

tes·sel·la·tion (tes′ə·lā′shən) *n.* **1** Tessellated work. **2** The art or act of doing such work.

tes·ser·a (tes′ər·ə) *n. pl.* **·ser·ae** (-ər·ē) **1** A small square, as of stone, glass, etc., used in mosaic work. **2** A small object, often a square or cube, as of bone or wood, used as a die in gambling or as a token, voucher, or the like. [<L <dial. Gk. (Ionic) *tesseres* four]

tes·ser·act (tes′ər·akt) *n. Math.* **1** A construct intended to illustrate graphically or in the form of a model the general appearance of a four-dimensional figure. **2** A hypercube bounded by 8 cubes or cells, with 16 vertices, 24 faces, and 32 edges. [<dial. Gk. (Ionic) *tesseres* four + *aktis* ray]

Tes·sin (te·sēn′) See TICINO.

test[1] (test) *v.t.* **1** To subject to a test or trial; try. **2** *Chem.* **a** To refine, as gold or silver, by means of lead, as in the process of cupellation. **b** To examine by means of some reagent, as in testing for sulfuric acid. **— *v.i.* 3** *Chem.* To undergo testing; also, to show specified qualities or properties under testing: The alcohol *tested* 75 percent. See synonyms under EXAMINE. [<*n.*] **— *n.* 1** Subjection to conditions that disclose the true character of a person or thing in relation to some particular quality. **2** An examination made for the purpose of proving or disproving some matter in doubt, as mental condition. **3** A criterion or standard of judgment. **4** An oath or other confirmatory evidence of principles or belief. **5** *Chem.* **a** A reaction by means of which the identity of a compound or one of its constituents may be determined. **b** Its agent or the result. **6** An earthen vessel similar to a cupel, formerly used in testing metals. **7** A series of questions, problems, etc., intended to measure the extent of knowledge, aptitudes, intelligence, and other mental traits: an intelligence *test.* See synonyms under PROOF. [<OF, a cupel, pot <L *testum* an earthen vessel < *testa* potsherd, shell] — **test'a·ble** *adj.*

test[2] (test) *n.* **1** *Zool.* A rigid external case or covering of many invertebrates, as a sea urchin or mollusk; a shell. **2** *Bot.* A testa. [<L *testa* shell]

test[3] (test) *v.t.* To attest. [<OF *tester* bequeath <L *testari* be a witness. See TESTAMENT.]

tes·ta (tes'tə) *n. pl.* **·tae** (-tē) **1** *Bot.* The outer, usually hard and brittle coat or integument of a seed. **2** *Zool.* A test. [See TEST[2]]

tes·ta·ce·an (tes·tā'shē·ən, -shən) *adj.* Pertaining or belonging to an order *(Testacea)* of rhizopods enclosed in a single-chambered cell. [<NL <L *testaceum* shellfish < *testaceus.* See TESTACEOUS.]

tes·ta·ceous (tes·tā'shəs) *adj.* **1** Of or derived from shells or shellfish. **2** Having a hard shell. **3** Dull brick-red or brownish-yellow. [<L *testaceus* of shell, brick < *testa* a shell]

tes·ta·cy (tes'tə·sē) *n. Law* The state of being testate or of having left a will at death: opposed to *intestacy.*

tes·ta·ment (tes'tə·mənt) *n.* **1** The written declaration of one's last will: usually **last will and testament.** In strictness, a testament differs from a will in that it bequeaths personal property only, but the words are commonly used interchangeably. **2** In Biblical use, a covenant; dispensation. [<F <L *testamentum* < *testari* testify < *testis* a witness] — **tes'ta·men'tal** *adj.*

Tes·ta·ment (tes'tə·mənt) *n.* **1** One of the two volumes of the Bible, distinguished as the **Old** and the **New Testament.** **2** Specifically, a volume containing the New Testament.

tes·ta·men·ta·ry (tes'tə·men'tər·ē) *adj.* **1** Derived from, bequeathed by, or set forth in a will. **2** Appointed or provided by, or done in accordance with, a will. **3** Pertaining to a will, or to the administration or settlement of a will; testamental. **4** *Often cap.* Pertaining to a Testament.

tes·tate (tes'tāt) *adj.* Having made a will before decease. [<L *testatus,* pp. of *testari* be a witness. See TESTAMENT.]

tes·ta·tor (tes·tā'tər, tes'tā·tər) *n.* **1** The maker of a will. **2** One who has died leaving a will. [<L] — **tes·ta'trix** (-triks) *n. fem.*

test·er[1] (tes'tər) *n.* One who tests; a device for testing.

tes·ter[2] (tes'tər) *n.* A flat canopy over a tomb, pulpit, or bed. [<OF *testiere* < *teste* head <L *testa* shell, skull]

TESTER

tes·ter[3] (tes'tər) *n. Obs.* A silver coin of the Tudor period, originally equal to twelve pence, later worth sixpence. [<OF *teston* coin < *teste* head. See TESTER[2].]

tes·ti·cle (tes'ti·kəl) *n. Biol.* One of the two genital glands of the male in which the spermatozoa and certain internal secretions are formed; a testis. [<L *testiculus,* dim. of *testis* testicle]

tes·tic·u·late (tes·tik'yə·lit, -lāt) *adj.* **1** Shaped or formed like a testicle. **2** Solid and ovate, like the roots of certain orchids. **3** Having organs like testicles. [<L *testiculus* + -ATE[1]]

tes·tif·i·cate (tes·tif'ə·kāt) *n.* In Scots law, a solemn written assertion. [<L *testificatus,* pp. of *testificari* bear witness. See TESTIFY.]

tes·ti·fi·ca·tion (tes'tə·fə·kā'shən) *n.* **1** The act of testifying or the giving of testimony. **2** The testimony given. — **tes'ti·fi·ca'tor** *n.*

tes·ti·fy (tes'tə·fī) *v.* **·fied, ·fy·ing** *v.i.* **1** To make solemn declaration of truth or fact. **2** *Law* To give testimony; bear witness. **3** To serve as evidence or indication: Her rags *testified* to her poverty. — *v.t.* **4** To bear witness to; affirm positively. **5** *Law* To state or declare on oath or affirmation. **6** To be evidence or indication of. **7** To make known publicly; declare. See synonyms under AFFIRM, AVOW. [<L *testificari* < *testis* witness + *facere* make] — **tes'ti·fi'er** *n.*

tes·ti·mo·ni·al (tes'tə·mō'nē·əl) *n.* **1** A formal token of regard. **2** A written certificate; an acknowledgment of services or worth; a letter of recommendation. — *adj.* Pertaining to or constituting testimony or a testimonial. [<L *testimonialis* < *testimonium.* See TESTIMONY.]

tes·ti·mo·ny (tes'tə·mō'nē) *n. pl.* **·nies** **1** A statement or affirmation of a fact, as before a court; evidence; proof. **2** The aggregate of proof offered in a case. **3** The act of testifying; attestation. **4** Public declaration regarding some experience. **5** The Decalog; the Old Testament Scriptures. [<L *testimonium* < *testis* a witness] — **tes'ti·mo'nied** *Obs. adj.*

Synonyms: affidavit, affirmation, attestation, deposition, proof, witness. *Testimony,* in legal as well as in common use, denotes the statements of witnesses. *Deposition* and *affidavit* denote *testimony* reduced to writing. The *deposition* differs from the *affidavit* in that the latter is voluntary and without cross-examination, while the former is made under interrogatories and subject to cross-examination. *Evidence* is a broader term, including the *testimony* of witnesses and all facts of every kind that tend to prove a thing true; we have the *testimony* of a traveler that a fugitive passed this way; his footprints in the sand are additional *evidence* of the fact. Compare PROOF.

tes·tis (tes'tis) *n. pl.* **·tes** (-tēz) A testicle. [<L]

test meal A meal of prescribed materials and quantity taken as a preliminary to a subsequent examination of the contents of the stomach.

tes·ton (tes'tən, tes·tōōn') *n. Obs.* **1** A European silver coin: so called from the head on the obverse side. **2** A French coin of the 16th century. **3** An English silver coin; a tester. Also **tes·toon** (tes·tōōn'). [<F <Ital. *testone,* aug. of *testa* head <L, skull]

tes·tos·ter·one (tes·tos'tə·rōn) *n. Biochem.* A male sex hormone, $C_{19}H_{28}O_2$, isolated as a white crystalline substance from the testes, and also made synthetically. [<TESTIS + STER(OL) + -ONE]

test paper **1** *Chem.* A paper saturated with some reagent that readily changes color when exposed to certain others, as litmus paper. **2** A list of questions, problems, etc., for the testing of students.

test pilot An aviator who flies airplanes of new design to test their performance under various conditions.

test tube A glass tube, open at one end, and usually with a rounded bottom, used in making chemical or biological tests.

tes·tu·di·nal (tes·tōō'də·nəl, -työō'-) *adj.* Pertaining to or like a turtle or tortoise shell: also **tes·tu'di·nate.** [<L *testudo, -inis* tortoise]

Tes·tu·din·i·dae (tes'tōō·din'i·dē, -työō-) *n. pl.* A family of reptiles having a dorsal shell or carapace constituted chiefly by the vertebrae and ribs and a ventral shell or plastron; tortoises and turtles. [<NL <L *testudo, -inis* tortoise]

tes·tu·do (tes·tōō'dō, -työō'-) *n. pl.* **·di·nes** (-də·nēz) **1** A shed or screen used by the Romans for the protection of soldiers in siege operations. **2** A protecting cover formed by soldiers in ranks by overlapping their shields above their heads. [<L < *testa* shell]

Tes·tu·do (tes·tōō'dō, -työō'-) *n.* A genus typical of land tortoises. [<L]

tes·ty (tes'tē) *adj.* **·ti·er, ·ti·est** Having an irritable disposition; touchy. See synonyms under FRETFUL. [<AF *testif* heady <OF *teste* head <L *testa* skull] — **tes'ti·ly** *adv.* — **tes'ti·ness** *n.*

te·tan·ic (ti·tan'ik) *adj.* Relating to or productive of tetanus: also **te·tan'i·cal.** — *n.* A drug capable of causing convulsions, as strychnine or nux vomica.

tet·a·nize (tet'ə·nīz) *v.t.* **·nized, ·niz·ing** To affect with tetanic spasms. — **tet'a·ni·za'tion** *n.*

tet·a·nus (tet'ə·nəs) *n.* **1** *Pathol.* An acute infectious disease caused by a bacillus *(Clostridium tetani)* and characterized by rigid spasmodic contraction of various voluntary muscles, especially that form affecting the muscles of the jaw, called *lockjaw.* **2** *Physiol.* A state of contraction in a muscle excited by a rapid series of shocks. [<L <Gk. *tetanos* spasm < *teinein* stretch]

tet·a·ny (tet'ə·nē) *n. Pathol.* **1** Intermittent tetanic spasms, usually due to defective metabolism. **2** Tetanus.

tetarto- *combining form* Four; fourth. Also, before vowels, **tetart-.** [<Gk. *tetartos* fourth < *tettares* four]

te·tar·to·he·dral (ti·tär'tō·hē'drəl) *adj.* Possessing one fourth of the planes necessary for true symmetry: said of crystals.

tetch·y (tech'ē) See TECHY.

tête-à-tête (tāt'ə·tāt', *Fr.* tet·à·tet') *adj.* Being face to face; literally, head to head; hence, confidential, as between two persons. — *n.* **1** A private interview; a confidential chat between two persons. **2** An S-shaped sofa on which two persons may face each other. — *adv.* In private or personal talk. [<F]

tête-bêche (tet·besh') *adj. French* Literally, head to foot: said of a pair of stamps so printed that one is reversed in relation to the other.

tête-de-pont (tet·də·pôn') *n. pl.* **têtes-de-pont** (tet·də·pôn') *French* A bridgehead.

teth (teth) *n.* The ninth Hebrew letter. See ALPHABET.

teth·er (teth'ər) *n.* **1** Something used to check or confine, as a rope for fastening an animal. **2** The range, scope, or limit of one's powers or field of action. — **at the end of one's tether** At the extreme end or limit of one's resources. — *v.t.* To fasten or confine by a tether. [ME *tethir* <Scand. Cf. ON *tiodhr* a tether.]

Te·thys (tē'this) In Greek mythology, a Titaness, sister and wife of Oceanus and mother of the Oceanids.

Te·ton Range (tē'ton, tēt'n) A range of the Rocky Mountains, chiefly in NW Wyoming; highest peak, 13,776 feet.

te·to·tum (te·tō'təm) See TEETOTUM.

tetra- *combining form* Four; fourfold: *tetrachord.* Also, before vowels, **tetr-.**

tet·ra·ba·sic (tet'rə·bā'sik) *adj. Chem.* **1** Containing four atoms of hydrogen replaceable by a base or basic radicals: said of certain acids. **2** Denoting a compound with four atoms of a univalent metal or the equivalent.

tet·ra·brach (tet'rə·brak) *n.* A Greek or Latin word or foot made up of four short syllables. [<Gk. *tetrabrachys* < *tessares, tettares* four + *brachys* short]

tet·ra·cene (tet'rə·sēn) *n. Chem.* A yellow, solid, nitrogen compound, $C_2H_8N_{10}$, used as a sensitizer or combustion initiator in priming compositions. Also *tetrazine.*

tet·ra·chlo·ride (tet'rə·klôr'īd, -id, -klō'rīd, -rid) *n. Chem.* A compound containing four atoms of chlorine. Also **tet'ra·chlo'rid** (-klôr'id, -klō'rid).

tet·ra·chord (tet'rə·kôrd) *n. Music* **1** A scale series of half an octave. **2** The interval of a perfect fourth. [<Gk. *tetrachordon* a musical instrument < *tetras* group of four + *chordē* string] — **tet'ra·chor'dal** *adj.*

te·trac·id (te·tras'id) *Chem. adj.* Denoting a base which is capable of combination with four molecules of a monobasic acid to form a salt or ester. — *n.* A base having four replaceable hydroxyl radicals. [<TETR(A)- + ACID]

tet·ra·cy·cline (tet'rə·sī'klin) *n. Chem.* A nitrogenous compound, $C_{22}H_{24}N_2O_8$, isolated as a yellow, odorless, crystalline powder from certain species of a soil bacillus (genus *Streptomyces*). It forms the base of several antibiotics, as Aureomycin and Terramycin. [< *tetracyclic,* containing four atomic rings + -INE[2]]

tet·rad (tet'rad) *n.* **1** A collection of four, or the number four. **2** An atom, radical, or element that is quadrivalent. **3** *Biol.* The group of four chromatids into which two bivalent chromosomes divide in the last stages of meiosis. **4** A crystal having an axis showing fourfold symmetry. [<Gk. *tetras, -ados* group of four]

te·trad·y·mite (te·trad'ə·mīt) *n.* A soft, metallic, pale steel-gray, bismuth telluride, Bi_2Te_3, crystallizing in the rhombohedral system. [<G *tetradymit* <Gk. *tetradymos* fourfold; from its occurring in compound twin crystals]

tet·ra·dyn·a·mous (tet'rə·din'ə·məs, -dī'nə-) *adj. Bot.* Having six stamens, of which four, arranged in opposite pairs, are longer than the other two and inserted above them, as in flowers of the mustard family. [<TETRA- + Gk. *dynamis* power]

tet·ra·eth·yl lead (tet′rə·eth′il led) Lead tetraethyl.

tet·ra·gon (tet′rə·gon) *n. Geom.* A plane figure having four angles; a quadrangle. [<Gk. *tetragōnon* a quadrangle < *tetra-* four + *gōnia* angle]

tet·rag·o·nal (tet·rag′ə·nəl) *adj.* 1 Being or pertaining to a tetragon; having four angles; quadrangular. 2 Belonging to or designating a crystal system characterized by four alternately dissimilar planes of symmetry intersecting at angles of 45 degrees and a fifth symmetrical plane at right angles to the others.

tet·ra·gram (tet′rə·gram) *n.* A word of four letters.

Tet·ra·gram·ma·ton (tet′rə·gram′ə·ton) *n.* In Hebrew texts, the group of four letters (JHVH, JHWH, YHVH, or YHWH) representing the holy and ineffable name of God. The common transliteration Jehovah is the result of a combination of the Tetragrammaton with the vowel points of *Adonai* "my Lord," which is substituted in reading the name. [<Gk. *tetragrammaton* < *tetra-* four + *gramma* a letter < *graphein* write]

tet·ra·he·dral (tet′rə·hē′drəl) *adj.* 1 Of or pertaining to a tetrahedron. 2 Made up of or having four sides. [<Gk. *tetraedros*. See TETRAHEDRON.]

tet·ra·he·drite (tet′rə·hē′drīt) *n.* A steel-gray, fine-grained mineral, usually a sulfide of copper and antimony but having other elements, found in tetrahedral crystals. [<TETRAHEDRON]

tet·ra·he·dron (tet′rə·hē′drən) *n. pl.* **·dra** (-drə) 1 *Geom.* A solid bounded by four plane triangular faces. 2 An anti-tank obstacle shaped like a pyramid. [<Gk. *tetraedron*, neut. of *tetraedros* < *tetra-* four + *hedra* base]

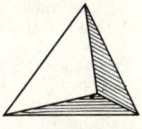
TETRAHEDRON

tet·ral·o·gy (te·tral′ə·jē) *n. pl.* **·gies** 1 A group of four dramas, three tragic and one satyric, presented together at the festivals of Dionysus at Athens. 2 Hence, any series of four related dramatic or operatic works. [<Gk. *tetralogia* < *tetra-* four + *logos* word, speech]

tet·ram·er·ous (te·tram′ər·əs) *adj.* 1 Having four parts. 2 *Bot.* Having the parts or organs in four; arranged in fours or multiples of four: often written *4-merous*. 3 *Zool.* Having four joints; having four-jointed tarsi. Also **te·tram′-er·al.**

tet·ram·e·ter (te·tram′ə·tər) *adj.* Having four measures. In classical trochaic, iambic, and anapestic verse a measure consists of two feet (a dipody); hence, a trochaic tetrameter contains eight feet to the line. In English, a tetrameter has four feet or measures. — *n.* A verse (line) thus composed. [<LL *tetrametrus* <Gk. *tetrametros* < *tetra-* four + *metron* measure]

tet·ra·morph (tet′rə·môrf) *n.* The union of the four attributes of the four Evangelists in one composite figure, winged, and standing on winged wheels of fire, the wings being full of eyes. [<Gk. *tetramorphon* four-shaped < *tetra-* four + *morphē* form]

tet·ra·pet·al·ous (tet′rə·pet′l·əs) *adj. Bot.* Having four petals.

tet·ra·pod (tet′rə·pod) *n.* Four-footed. [<NL *tetrapodus* <Gk. *tetrapous, tetrapodos* four-footed < *tetra-* four + *pous* foot]

tet·rap·o·dy (te·trap′ə·dē) *n. pl.* **·dies** A group of four feet, as a colon, meter, or verse containing that number. [<Gk. *tetrapodia* < *tetrapous*. See TETRAPOD.] — **tet·ra·pod·ic** (tet′rə·pod′ik) *adj.*

tet·rap·ter·ous (te·trap′tər·əs) *adj. Biol.* Having four wings, as certain fruits and insects. [<NL *tetrapterus* <Gk. *tetrapteros* four-winged < *tetra-* four + *pteron* wing]

tet·ra·py·lon (tet′rə·pī′lon) *n. Archit.* A structure having four gateways or penetrated by two intersecting passages, as some arches. [<Gk. *tetrapylos* with four gates < *tetra-* four + *pylē* a gate]

tet·rarch (tet′rärk, tē′trärk) *n.* 1 The governor of one of four divisions of a country or province. 2 A tributary prince under the Romans; a subordinate ruler. 3 Anciently, in the Greek army, the commander of a subdivision of a phalanx. [<LL *tetrarcha* <L *tetrarches* <Gk. *tetrarchēs* < *tetra-* four + *archos* ruler] — **tet·rar·chy** (tet′rär·kē, tē′trär-), **tet·rarch·ate** (tet′rär·kāt, -kit, tē′trär-) *n.*

tet·ra·seme (tet′rə·sēm) *n.* A long syllable or a foot equal to four short syllables. [<TETRA- + Gk. *sēma* sign] — **tet′ra·se′mic** *adj.*

tet·ra·spore (tet′rə·spôr, -spōr) *n. Bot.* An asexual spore produced by certain algae: named from the fact that often four are produced together from a mother cell.

tet·ra·stich (tet′rə·stik) *n.* A poem or stanza of four lines; a quatrain. [<TETRA- + Gk. *stichos* row, line] — **tet′ra·stich′ic** *adj.*

tet·ras·ti·chous (te·tras′tə·kəs) *adj. Bot.* Four-ranked; having organs, as leaves on a stem, arranged in four vertical rows or ranks.

tet·ra·style (tet′rə·stīl) *adj.* Having four pillars. — *n. Archit.* 1 A temple having four columns in the front or end row. 2 Any building or structure having four pillars in a row or rows. [<L *tetrastylos* <Gk. < *tetra-* four + *stylos* column]

tet·ra·syl·la·ble (tet′rə·sil′ə·bəl) *n.* A word of four syllables. — **tet′ra·syl·lab′ic** or **·i·cal** *adj.*

tet·ra·tom·ic (tet′rə·tom′ik) *adj. Chem.* 1 Containing four atoms. 2 Containing four replaceable univalent atoms or molecules. 3 Quadrivalent.

tet·ra·va·lent (tet′rə·vā′lənt) *adj. Chem.* Quadrivalent.

tet·ra·zine (tet′rə·zēn, -zin) *n. Chem.* Tetracene.

Te·traz·zi·ni (tā′trät·tsē′nē), **Luisa,** 1874?-1940, Italian coloratura soprano active in the United States.

tet·rode (tet′rōd) *n. Electronics* A vacuum tube containing four elements, the fourth usually being an additional grid interposed between the first grid and the plate. [<TETR(A)- + -ODE[1]]

te·trox·ide (te·trok′sīd, -sid) *n. Chem.* An oxide containing four atoms of oxygen to the molecule. Also **te·trox′id** (-sid). [<TETR(A)- + OXIDE]

tet·ryl (tet′ril) *n. Chem.* A yellowish, crystalline nitrogen compound, $C_7H_5N_5O_8$, used as an explosive in boosters and detonators. [<TETR(A)- + -YL]

tet·ter (tet′ər) *n. Pathol.* A vesicular skin disease, as eczema. [OE *teter*]

Te·tuán (tā·twän′) A port on the Mediterranean Sea in NE Morocco; former capital of Spanish Morocco.

Tet·zel (tet′səl), **Johann,** 1465-1519, German Dominican monk, opponent of Luther. Also spelled *Tezel*.

Teu·cer (tōo′sər, tyōo′-) In Greek legend: 1 The half-brother of Ajax, who founded Salamis in Cyprus; noted as an archer. 2 The first king of Troy.

Teu·cri·a (tōo′krē·ə, tyōo′-) See TROAS.

Teu·cri·an (tōo′krē·ən, tyōo′-) *adj.* 1 Trojan. 2 Of or pertaining to Teucer.

teugh (tōokh, tyōokh) *adj. Scot.* Tough. Also **teuch.** — **teugh′ly** *adv.* — **teugh′ness** *n.*

Teu·to·bur·ger Wald (toi′tō·bŏŏr′gər vält) A range of hills in western Germany; highest point, 1,465 feet; scene of a German victory by Arminius over the Roman army commanded by Varus, 9 A.D.

Teu·ton (tōot′n, tyōot′n) *n.* 1 One of an ancient German tribe that dwelt in Jutland north of the Elbe, appearing in history as **Teu·to·nes** (tōo′tə·nēz, tyōo′-), together with the **Cim·bri** (sim′brī), a possibly related tribe, in 113 B.C. 2 One belonging to any of the Teutonic peoples; especially, a German.

Teu·ton·ic (tōo·ton′ik, tyōo-) *adj.* 1 Of or pertaining to the Teutons; especially, designating the blond peoples of northern Europe, formerly including the Angles, Saxons, Danes, Normans, Norwegians, the Goths, Franks, Lombards, Vandals, etc.; now embracing also the English, Germans, Dutch, etc. 2 Of or pertaining to that subfamily of Indo-European languages now called *Germanic,* including Gothic, the Scandinavian languages, and all the High and Low German languages and dialects, among which are German, Dutch, Flemish, and English. — *n.* The Germanic subfamily of languages.

Teutonic Knights The Knights of St. Mary's Hospital at Jerusalem, an order of military monks deriving their name and office from a German hospital founded at Jerusalem in 1128. One of the three great military orders founded during the Crusades to convert the heathen, help pilgrims, and nurse the sick, the knights later moved to eastern Europe and during the Middle Ages became the spearhead of German expansion toward Slavic and Baltic territories. Their costume was a white mantle with a black cross. See HOSPITALER, KNIGHT TEMPLAR.

Teu·ton·ism (tōōt′n·iz′əm, tyōōt′n-) *n.* 1 A custom or mode of expression peculiar to Germans or Teutons; Germanism: also **Teu·ton·i·cism** (tōō·ton′ə·siz′əm, tyōō-). 2 A belief in the superiority of the Teutonic race. 3 Teutonic character and civilization. — **Teu′ton·ist** *n.*

Teu·ton·ize (tōōt′n·īz, tyōōt′n-) *v.t. & v.i.* **·ized, ·iz·ing** To make or become Teutonic or German. — **Teu·ton·i·za′tion** *n.*

Te·ve·re (tā′vä·rā) The Italian name for the river TIBER.

tew (tōō, tyōō) *Brit. Dial. v.i.* To work hard; fuss or bustle. — *n.* A state of excitement, worry, or bustling.

Tewkes·bur·y (tōōks′ber·ē, -bər·ē, tyōōks′-) A municipal borough in NW Gloucestershire, England; scene of the final defeat of the Lancastrian forces in the Wars of the Roses, 1471.

Tex·ar·kan·a (teks′är·kan′ə) A dual city in NE Texas and SW Arkansas, with two municipal governments.

tex·as (tek′səs) *n. U.S.* The uppermost structure on a river steamboat, containing the pilot house, officers' cabins, etc.; often, a row of staterooms behind the pilot house, or having the pilot house set on top of it. [from *Texas*; so called from the former custom of naming staterooms after the States, those of the officers being the largest]

Tex·as (tek′səs) *n.* A Caddo Indian.

Tex·as (tek′səs) *n.* A State in the SW United States, bordering on Mexico and the Gulf of Mexico; 267,339 square miles; capital, Austin; entered the Union Dec. 29, 1845; nicknamed *Lone Star State:* abbr. *Tex.* — **Tex′an** *n. & adj.*

Texas cattle Cattle bred from the old longhorn stock of early Texas.

Texas fever A destructive cattle disease caused by a blood parasite transmitted by the cattle tick, *Margaropus annulatus.* Also **Texan fever.**

Texas leaguer *U.S. Colloq.* In baseball, a looping fly ball that falls safe between an infielder and an outfielder.

Texas Ranger 1 A member of the mounted State police force of Texas. 2 Originally, one of a band of armed and mounted men organized in Texas to fight Indians and keep order on the frontiers.

Texas sparrow A plain, olive-backed fringilline bird (*Arremonops rufivirgatus*) found in Mexico and southern Texas.

Texas tower A radar station having several tall towers erected on a platform which may be moored permanently in the sea as part of a radar-warning network.

Texas trail The old Chisholm cattle trail from Red River, Texas, to Abilene, Kansas.

Tex·o·ma (tek·sō′mə), **Lake** A reservoir in northern Texas and southern Oklahoma created by damming the Red River; one of the largest in the U.S.; 227 square miles.

text (tekst) *n.* 1 The actual or original words of an author; the body of matter on a written or printed page, as distinguished from notes, commentary, illustrations, etc. 2 A written or printed version of the matter of an author's works: the folio *text* of Shakespeare. 3 Any one of various recensions that are taken to represent the authentic words, or portion of the words, of the original Scriptures. 4 A verse of Scripture, particularly when cited as the basis of a discourse or sermon. 5 Any subject of discourse; a topic; theme. 6 One of several styles of letters or types. 7 A textbook. [<OF *texte* <L *textus* fabric, structure < *texere* weave]

text·book (tekst′bo͝ok′) *n.* A book used as a standard work or basis of instruction in any branch of knowledge; schoolbook; manual.
tex·tile (teks′til, -tīl) *adj.* 1 Pertaining to weaving or woven fabrics. 2 Such as may be woven; manufactured by weaving. —*n.* 1 A woven fabric; textile material. 2 Material capable of being woven. [<L *textilis* < *textus* fabric. See TEXT.]
tex·tu·al (teks′cho͞o·əl) *adj.* 1 Pertaining to, contained in, or based on the text of a book, especially of the Scriptures; literal; word for word. 2 Versed in texts. [<OF *textuel* < *texte.* See TEXT.] —**tex′tu·al·ly** *adv.*
tex·tu·al·ism (teks′cho͞o·əl·iz′əm) *n.* 1 Rigid adherence to the letter of a text. 2 The method or principles of textual criticism.
tex·tu·al·ist (teks′cho͞o·əl·ist′) *n.* 1 A close adherent to the letter of a text. 2 One who is versed in or cites texts readily.
tex·tu·ar·y (teks′cho͞o·er′ē) *adj.* 1 Contained in a text. 2 Of, belonging to, or adhering to a text. —*n. pl.* **·ar·ies** A textualist.
tex·ture (teks′chər) *n.* 1 The arrangement or character of the threads, etc., of a woven fabric. 2 The mode of union or disposition of elementary constituent parts, as in a photograph, or surface of paper, etc.; minute structure or make; structural order. 3 The structure, especially as regards detail, of a work of art. 4 Any woven fabric; a web. [<L *textura* < *textus* fabric. See TEXT.] —**tex′tur·al** *adj.* —**tex′tur·al·ly** *adv.*
Tey·de (tā′thā) See PEAK OF TENERIFE under TENERIFE.
teyl tree (tēl) See TEIL.
Tez·cat·li·po·ca (tes·kät′lē·pō′kä) In Aztec mythology, the creator, and god of night and fruitfulness.
Tez·el (tet′səl) See TETZEL.
-th[1] *suffix of nouns* 1 The act or result of the action expressed in the root word: *growth.* 2 The state or quality of being what is indicated in the root word: *health.*
-th[2] *suffix* Used in ordinal numbers: *tenth.* Also, after vowels, *-eth*, as in *fortieth.* [OE *-tha, -the*]
-th[3] See -ETH[1].
thack (thak) *n.* & *v.t. Scot.* Thatch; roof.
Thack·er·ay (thak′ər·ē), **William Makepeace,** 1811–63, English novelist.
Thad·de·us (thad′ē·əs) A masculine personal name. Also **Thad′dae·us,** *Ger.* **Thad·dä·us** (tä·dā′o͞os), *Polish* **Ta·de·usz** (tä·dā′o͞osh), *Pg.* **Thad·de·o** (täd·dā′ō). [<Aramaic, praise]
thae (thā) *adj.* & *pron. Brit. Dial.* Those; these. [OE *tha,* pl. of *sē, thē* the, this, that]
Thai (tī) *n.* 1 The people collectively of Thailand, Laos, and parts of Burma, including the Laos, Shan, and Siamese. 2 A family of languages spoken by these people: considered by some to be a branch of the Sino-Tibetan family. —*adj.* Of or pertaining to the Thai, their culture, or their languages. Also spelled *Tai.*
Thai·land (tī′land) A constitutional monarchy in SE Asia; 198,404 square miles; capital, Bangkok: formerly *Siam. Thai* **Mu·ang Thai** (mo͞o′äng tī′).
thairm (thârm) See THARM.
Tha·is (thā′is) 1 An Athenian courtesan, mistress of Alexander the Great, whom she accompanied to Asia. 2 (*Fr.* tà·ēs′) An opera by Massenet based on a novel by Anatole France, which deals with the conversion of an Alexandrian courtesan by a monk.
thal·a·men·ceph·a·lon (thal′ə·men·sef′ə·lon) *n. Anat.* Diencephalon. [<THALAM(US) + ENCEPHALON]
tha·lam·ic (thə·lam′ik) *adj.* Of or pertaining to a thalamus, especially to the optic thalamus.
thal·a·mus (thal′ə·məs) *n. pl.* **·mi** (-mī) 1 *Anat.* The optic thalamus. 2 *Bot.* The receptacle of a flower. [<L <Gk. *thalamos* chamber]
tha·las·sic (thə·las′ik) *adj.* 1 Of or pertaining to the seas, sometimes as opposed to the oceans. 2 Pelagic; oceanic. [<Gk. *thalassa* sea]
thalasso- *combining form* The sea; of or pertaining to the sea: *thalassophobia.* Also, before vowels, **thalass-.** Also **thalassi-.** [<Gk. *thalassa* the sea]
thal·as·sog·ra·phy (thal′ə·sog′rə·fē) *n.* Oceanography. [<THALASSO- + -GRAPHY]
tha·las·so·pho·bi·a (thə·las′ə·fō′bē·ə) *n.* Morbid fear of the sea. [<THALASSO- + -PHOBIA]
tha·ler (tä′lər) See TALER.
Tha·les of Miletus (thā′lēz), 640?–546? B.C., Greek philosopher and scientist.
Tha·li·a (thə·lī′ə) 1 The Muse of comedy and pastoral poetry. 2 One of the three Graces. [<L <Gk. *Thaleia* < *thallein* bloom]
Tha·lic·trum (thə·lik′trəm) *n.* A genus of perennial herbs of the crowfoot family, the meadow rues. [<L <Gk. *thaliktron* < *thallein* bloom]
thal·lic (thal′ik) *adj. Chem.* Of, pertaining to, or derived from thallium, especially in its higher valence.
thal·line (thal′ēn, -in) *n. Chem.* A white, crystalline, synthetic alkaloid, $C_{10}H_{13}NO$: its salts are used as antipyretics and antiseptics.
thal·li·um (thal′ē·əm) *n.* A soft, white, crystalline metallic element (symbol Tl) closely resembling lead: its salts are used in rat poison, insecticides, and in making optical glass. See ELEMENT. [<NL <Gk. *thallos* a green shoot; from the bright green line in its spectrum, which led to its discovery]
thal·loid (thal′oid) *adj.* Resembling a thallus. Also **thal·loi·dal** (thə·loid′l).
thal·lo·phyte (thal′ə·fīt) *n.* Any plant belonging to a major division or phylum of plants (*Thallophyta*), comprising the bacteria, fungi, algae, and lichens. Many of the forms are unicellular and those more highly developed are without true roots, stems, or leaves. [< *thallo-* (<THALLUS) + -PHYTE] —**thal′lo·phyt′ic** (-fit′ik) *adj.*
thal·lous (thal′əs) *adj. Chem.* Derived from thallium, especially in its lower valence. Also **thal·li·ous** (thal′ē·əs).
thal·lus (thal′əs) *n. pl.* **·lus·es** or **·li** (-ī) *Bot.* A plant body without true root, stem, or leaf, as in thallophytes. [<L, a shoot <Gk. *thallos* < *thallein* bloom]
Thames (temz) A river of southern England, rising in the Cotswold Hills, south central Gloucestershire, and flowing 209 miles east through London to the North Sea between northern Kent and southern Essex; the principal river of England.
Thames River 1 (temz) A river in SE Ontario, Canada, flowing SW 160 miles to Lake St. Clair. 2 (thāmz, tāmz) A river and estuary in SE Connecticut, flowing about 15 miles south to Long Island Sound.
Tham·muz (täm′mo͞oz, tam′uz) See TAMMUZ.
than (than, *unstressed* thən) *conj.* 1 When, as, or if compared with: after an adjective or adverb to express comparison between what precedes and what follows: I am stronger *than* he (is); I know her better *than* (I know) him. 2 Except; but: used after *other, else,* etc: no other *than* you. *Than* is sometimes considered a preposition in the one phrase, *than whom*: an eminent judge *than whom* no other is more just. [OE *thanne* then]
than·age (thā′nij) *n.* 1 In early English law, the state, jurisdiction, or office of a thane. 2 The land held by a thane or the tenure by which he held it. Also spelled *thenage.* [<AF *thaynage* <OE *thegn* a thane]
thanato- *combining form* Death; of or pertaining to death: *thanatophobia.* Also, before vowels, **thanat-.** [<Gk. *thanatos* death]
than·a·toid (than′ə·toid) *adj.* Resembling death; deadly.
than·a·to·pho·bi·a (than′ə·tə·fō′bē·ə) *n.* Morbid fear of death. [<THANATO- + PHOBIA] —**than′a·to·pho′bic** *adj.*
than·a·top·sis (than′ə·top′sis) *n.* A musing or meditation upon death; a view of death. [<THANAT(O)- + -OPSIS]
Than·a·tos (than′ə·tos) 1 In Greek mythology, the god of death: identified with the Roman *Mors.* 2 *Psychoanal.* The death instinct: opposed to *Eros,* the life instinct.
thane (thān) *n.* 1 Originally, a warrior companion of an English king before the Conquest. 2 Later, a man who ranked above an ordinary freeman or ceorl (churl) but below an earl or nobleman. 3 *Scot.* The chief of a clan; a baron; one of the old nobility in the service of the king. Also spelled *thegn.* [OE *thegn*]
Than·et (than′it), **Isle of** An island comprising the NE corner of Kent county, England; 10 miles long, 5 miles wide.
thank (thangk) *v.t.* 1 To express gratitude to; give thanks to. 2 To hold responsible; blame: often used ironically. [OE *thancian* < *thanc* thanks, thought]
thank·ful (thangk′fəl) *adj.* 1 Deeply sensible of favors received; grateful. 2 Done or made to express thanks; manifesting thanks. See synonyms under GRATEFUL. —**thank′ful·ly** *adv.* —**thank′ful·ness** *n.*
thank·it (thangk′it) *Scot.* Past participle of THANK.
thank·less (thangk′lis) *adj.* 1 Not feeling or expressing gratitude; ungrateful; unresponsive. 2 Not gaining or likely to gain thanks; unthanked; unappreciated. —**thank′less·ly** *adv.* —**thank′less·ness** *n.*
thanks (thangks) *n. pl.* Expressions of gratitude; grateful acknowledgment. —*interj.* My thanks to you; I thank you. —**thanks to** 1 Thanks be given to. 2 Because of.
thanks·giv·er (thangks′giv′ər) *n.* One who gives thanks.
thanks·giv·ing (thangks′giv′ing) *n.* 1 The act of giving thanks, as to God; the expression of gratitude. 2 A form of words or worship in recognition of divine mercies. 3 A public celebration in recognition of divine favor. 4 A day set apart for such celebration.
Thanksgiving Day *U.S.* Usually, the last Thursday in November, set apart as an annual festival of thanksgiving to God for the year's blessings. Also **Thanksgiving.**
thank·wor·thy (thangk′wûr′thē) *adj.* Deserving or worthy of thanks; meritorious.
Thap·sus (thap′səs) An ancient ruined town on the coast of eastern Tunisia; scene of Julius Caesar's defeat of Cato the Younger, 46 B.C.
Thar Desert (tär) A sandy waste in NW India (much of western Rajasthan and southern Punjab) and West Pakistan; over 15,000 square miles: also *Indian Desert.*
tharm (thärm) *n. Obs.* 1 The belly; intestines. 2 Twisted gut. Also spelled *thairm.* [OE]
Tha·sos (thā′sos) A Greek island in the north Aegean Sea; 170 square miles.
that (that, *unstressed* thət) *adj. pl.* **those** 1 Pertaining to some person or thing previously mentioned, understood, or specifically designated: *that* man. 2 Denoting something more remote in place, time, or thought: correlative to *this.* —*pron.* 1 As a demonstrative, the person or thing implied, mentioned, or understood; or the person or thing there or in the second place: *That* is the dress I like. 2 As a relative, who or which: used as a correlative to *such* or *so.* ◆ In earlier English, *that* was the relative pronoun, *who, what,* and *which* being only interrogatives until they gradually assumed the force of relatives, and in some uses superseded *that.* When the relative clause qualifies or makes an addition to the main clause, *who* or *which* is generally preferred, whereas *that* usually introduces a restrictive clause. Thus we say: Washington, *who* was the first president, is often called Father of his Country. But: *The* Washington *that* emigrated to this country was his ancestor. —*adv.* 1 *Colloq.* In such a manner or degree; so. 2 To that extent: I can't see *that* far. —*conj. That* is used primarily to connect a subordinate clause with its principal clause, with the following meanings: 1 As a fact that: introducing a fact: I tell you *that* it is so. 2 So that; in order that: I tell you *that* you may know. 3 For the reason that; seeing that; because: She wept *that* she was growing old. 4 As a result: introducing a result, consequence, or effect: He bled so profusely *that* he died. 5 At which time; when: It was only yesterday *that* I saw him. 6 Introducing an exclamation: O *that* he would come! See synonyms under BUT. —**so that** 1 To the end that. 2 With the result that. 3 Provided. See THOSE. [OE *thæt,* neut. of *sē* the]
thatch (thach) *n.* 1 A covering of reeds, straw, etc., arranged on a roof so as to shed water. 2 Any of various palms whose leaves are used for thatching, especially those of the genera *Thrinax* and *Sabal.* 3 Any of certain tall, coarse American grasses (genus *Spartina*) of the northern Atlantic coasts. —*v.t.* To cover with a thatch. [OE *thæc* cover] —**thatch′er** *n.* —**thatch′y** *adj.*
thatch·ing (thach′ing) *n.* 1 The act or process of covering a roof with a thatch. 2 Material used for a thatch.
thaumato- *combining form* A wonder; a miracle: *thaumatology.* Also, before vowels, **thaumat-.** [<Gk. *thauma, -atos* a wonder]

thau·ma·tol·o·gy (thô′mə·tol′ə·jē) *n.* The scientific study of miracles. [<THAUMATO- + -LOGY]

thau·ma·trope (thô′mə·trōp) *n.* An optical toy or instrument in which pictures on opposite sides of a card appear to blend together when the card is rapidly twirled. [<Gk. *thauma* wonder + -TROPE]

thau·ma·turge (thô′mə·tûrj) *n.* One who performs wonders or miracles; a wonder-worker; magician. Also **thau′ma·tur′gist.** [<Gk. *thaumatourgos* < *thauma* wonder + *ergon* work]

thau·ma·tur·gy (thô′mə·tûr′jē) *n.* Magic; the performance or working of wonders or miracles. **— thau′ma·tur′gic** or **·gi·cal** *adj.*

thaw (thô) *v.i.* **1** To melt or dissolve; become liquid or semi-liquid, as snow or ice. **2** To rise in temperature so as to melt ice and snow: said of weather and used impersonally. **3** To become less cold and unsociable. *— v.t.* **4** To cause to thaw. See synonyms under MELT. *— n.* **1** The act of thawing, or the state of being thawed. **2** Warmth of weather such as melts things frozen; also, figuratively, state of warmer feeling or expression. [OE *thawian*] **— thaw′er** *n.*

Thax·ter (thaks′tər), **Celia,** 1835–94, *née* Laighton, U. S. poet.

Thay·er (thā′ər), **Sylvanus,** 1785–1872, U. S. Army officer: called "Father of West Point." **— William Roscoe,** 1859–1923, U. S. historian and biographer.

the[1] (*stressed* thē; *unstressed before a consonant* thə; *unstressed before a vowel* thi) *definite article* or *adj. The* is opposed to the indefinite article *a* or *an,* and is used, especially before nouns, to render the modified word more particular or individual. It is used specifically: **1** When reference is made to a particular person, thing, or group: *The* natives are getting restless; He left *the* room. **2** To give an adjective substantive force, or render a notion abstract: *the* quick and *the* dead; *the* doing of the deed. **3** Before a noun to make it generic: *The* dog is a friend of man. **4** With the force of a possessive pronoun: He kicked me in *the* (my) leg. **5** To give distributive force: equivalent to *a, per, each,* etc.: a dollar *the* volume. **6** *Scot.* & *Irish* To designate the head of a clan or group: *the* MacIntosh. **7** To designate a particular one as emphatically outstanding: usually stressed in speech and italicized in writing: He is *the* officer for the command. **8** As part of a title: *The* Duke of York. [OE *the,* later form of *sē*]

the[2] (thə) *adv.* By that much; by so much; to this extent: *the* more, *the* merrier: used to modify words in the comparative degree. [OE *thȳ,* oblique case of *sē* the]

the- Var. of THEO-.

the·a·ceous (thē·ā′shəs) *adj. Bot.* Designating a family *(Theaceae)* of shrubs and trees having alternate, simple leaves, large flowers, and a typically capsular fruit; the tea family. [<NL < *Thea,* genus name < dial. Chinese *t'e* tea; incorrectly taken by Linnaeus as "divine herb" <Gk. *thea* a goddess]

the·an·throp·ic (thē′ən·throp′ik) *adj.* **1** Being both divine and human. **2** Having or pertaining to a nature both divine and human. Also **the′an·throp′i·cal.** [<Gk. *theanthrōpos* < *theos* god + *anthrōpos, -ōpou* man]

the·an·thro·pism (thē·an′thrə·piz′əm) *n.* **1** The doctrine of the manifestation of God in man, or of the union of the divine and human in Christ. **2** The ascription of human characteristics to a deity; anthropomorphism. **3** Belief in the possibility of the combination in one being of a nature both human and divine. **— the·an′thro·pist** *n.*

the·ar·chy (thē′är·kē) *n. pl.* **·chies 1** Government by God or by a god. **2** A theocracy. **3** A body or class of deities. [<Gk. *thearchia* < *theos* god + *archein* rule]

the·a·ter (thē′ə·tər) *n.* **1** A building especially adapted to dramatic, operatic, or spectacular representations; playhouse. **2** The theatrical world and everything relating to it. **3** A room or hall arranged with seats that rise as they recede from a platform, especially adapted to lectures, surgical demonstrations, etc. **4** Any place of semicircular form with seats rising by easy gradations. **5** Any place or region that is the scene of events: a *theater* of operations in war. Also **the′a·tre.** [<OF *theatre* <Gk. *theatron* < *theasthai* behold]

the·a·ter-go·er (thē′ə·tər·gō′ər) *n.* One who frequents theaters. Also **the′a·tre-go′er.** **— the′a·ter-go′ing, the′a·tre-go′ing** *n.*

the·a·ter-in-the-round (thē′ə·tər·in·thə·round′) *n.* An arena theater.

The·a·tin (thē′ə·tin) *n.* **1** A member of a congregation founded in 1524 by Bishop Carafa and St. Cajetan. **2** A member of an order of nuns founded by Ursula Benincasa, who died in 1618. Also **The′a·tine** (-tēn, -tin). [<NL *theatinus,* from *Teate,* ancient name of Chieti, Italy]

the·at·ri·cal (thē·at′ri·kəl) *adj.* **1** Pertaining to the theater or to dramatic performances. **2** Designed for show, display, or effect; showy; artificial. **3** Suited to dramatic presentation. **4** Like the manner of actors; histrionic. Also **the·at′ric.** *— n. pl.* Dramatic performances: especially when given by amateur performers. [<LL *theatricus* <Gk. *theatrikos* < *theatron.* See THEATER.] **— the·at′ri·cal·ly** *adv.* **— the·at′ri·cal·ness** *n.*

the·at·ri·cal·ism (thē·at′ri·kəl·iz′əm) *n.* Theatrical or melodramatic manner or style.

the·at·rics (thē·at′riks) *n. pl. (construed as singular)* The art of bringing about effects appropriate for dramatic performances.

The·ba·id (thē′bā·id, thi·bā′-) *n.* A Latin epic by Statius, narrating the story of the siege of Boeotian Thebes.

The·ba·id (thē′bā·id, thi·bā′-) The territory about Thebes in either Egypt or Greece.

the·ba·ine (thē′bə·ēn, thi·bā′ēn, -in) *n. Chem.* A silvery-white, poisonous, crystalline alkaloid, $C_{19}H_{21}O_3N$, found in opium and resembling strychnine in action: also called *paramorphine.* Also **the′ba·in** (-in). [from Egyptian *Thebes,* where a kind of opium was produced + -INE[2]]

Thebes (thēbz) **1** The ancient capital of Upper Egypt; Luxor and Karnak occupy part of its site on the Nile: Greek *Diospolis.* **2** The chief city of ancient Boeotia, Greece; destroyed in 336 B.C. by Alexander the Great: also **The·bae** (thē′bē). **3** A commercial city in east central Greece on the site of ancient Thebes; important in the Middle Ages: *Greek* **The·vai** or **Thi·vai** (thē′vā). **— The·ban** (thē′bən) *adj.* & *n.*

the·ca (thē′kə) *n. pl.* **·cae** (-sē) **1** A sheath or case. **2** *Anat.* The investment of the spinal cord formed by the dura mater, sometimes called **theca vertebralis. 3** *Bot.* A spore case, sac, or capsule. [<L <Gk. *thēkē* case] **— the′cal** *adj.*

the·cate (thē′kit, -kāt) *adj.* Having a sheath; sheathed.

thé dan·sant (tā′ dän·sän′) *pl.* **thés dan·sants** (tā′ dän·sän′) *French* Literally, a dancing tea; an afternoon tea at which there is dancing.

thee (thē) *pron.* **1** The objective case of *thou.* **2** Thou: used generally by Quakers with a verb in the third person singular: *Thee* knows my mind. [OE *thē,* accusative case of *thū* thou]

theek (thēk) *v.t. Brit. Dial.* To thatch. [Scottish var. of THATCH] **— theek′ing** *n.*

Thee·lin (thē′lin) *n.* Proprietary name for a brand of estrone.

Thee·lol (thē′lōl, -lol) *n.* Proprietary name for a brand of estriol.

theft (theft) *n.* **1** The act of thieving; larceny. **2** *Rare* That which is stolen. [OE *thēoft, thīefth*]

the·gith·er (thi·gith′ər) *adv. Scot.* Together.

thegn (thān) See THANE.

the·ine (thē′ēn, -in) *n. Chem.* The alkaloid found in the tea plant: chemically identical with caffeine. Also **the′in** (-in). [<F *théine* <NL *thea* tea <dial. Chinese *t'e*]

their (thâr) *pronominal adj.* The possessive case of the pronoun *they* employed attributively; belonging or pertaining to them: *their* homes. [ME <ON *theirra* of them]

theirs (thârz) *pron.* **1** The possessive case of *they,* used predicatively; belonging or pertaining to them: That house is *theirs.* **2** The things or persons belonging or relating to them: our country and *theirs.* **— of theirs** Belonging or pertaining to them; their: the double possessive. [<THEIR + *-s,* on analogy with *his*]

the·ism[1] (thē′iz·əm) *n.* **1** Belief in, or in the existence of, God, a god, or gods: opposed to *atheism.* **2** Belief in a personal God as creator and supreme ruler of the universe, who transcends his creation but works in and through it in revealing himself to men. Compare DEISM, PANTHEISM. **3** Belief in one god; monotheism: opposed to *polytheism.* **4** Formerly, deism. **5** *Philos.* The doctrine that one supreme reality, intrinsically complete and perfect, is the final ground and source of everything other than itself: a doctrine resembling monotheism and some types of monism, but opposed to atheism, agnosticism, deism, materialism, pantheism, and polytheism. [<Gk. *theos* god] **— the′ist** *n.* **— the·is′tic** or **·ti·cal** *adj.* **— the·is′ti·cal·ly** *adv.*

the·ism[2] (thē′iz·əm) *n. Pathol.* The toxic effects of excessive tea-drinking. [<NL *thea* tea. See THEINE.]

Theiss (tīs) The German name for TISZA.

Thé·lème (tā·lem′), **Abbey of** An abbey described by Rabelais in *Gargantua,* and having only one rule, "Do what you like."

the·li·tis (thi·lī′tis) *n. Pathol.* Inflammation of the nipple. [<NL <Gk. *thēlē* teat + -ITIS]

Thel·ma (thel′mə) A feminine personal name. [<Gk., nursling]

The·lon River (thi·lon′) A river in Northwest Territories, Canada, rising in eastern Mackenzie district and flowing about 550 miles north, NE, and east to NW Hudson Bay on the central eastern coast of Keewatin district.

them (them, *unstressed* thəm) *pron.* The objective case of *they.* [ME *theim* <ON, to them]

the·mat·ic (thē·mat′ik) *adj.* **1** Of, constituting, or pertaining to a theme or themes. **2** *Ling.* Constituting a stem. Also **the·mat′i·cal. — the·mat′i·cal·ly** *adv.*

theme (thēm) *n.* **1** A subject of discourse; a topic to be discussed or developed in speech or writing; hence, any topic. **2** An essay or dissertation; loosely, a brief composition in any form, written as an exercise. **3** *Ling.* The stem of a word, to which are attached the inflectional endings, consisting of the root unmodified or with some internal change and, often, a thematic vowel common to the particular stem class. **4** A melodic subject usually developed with variations in a musical composition. **5** One of the administrative divisions of the Byzantine Empire. See synonyms under TOPIC. [<OF *teme* <L *thema* <Gk. *the-,* stem of *tithenai* place]

theme song 1 A melody used throughout a dramatic presentation to furnish the key to the mood. **2** A strain of music which, from repetition, identifies a daily or periodical radio or television presentation, a dance band, etc.

The·mis (thē′mis) In Greek mythology, a goddess of law and justice, daughter of Uranus and Gaea. [<Gk., law]

The·mis·to·cles (thi·mis′tə·klēz), 527?–460? B.C., Athenian statesman and soldier.

them·selves (them′selvz′, *unstressed* thəm-) *pron.* Emphatic or reflexive form of THEY, THEM: the plural of HIMSELF, HERSELF, ITSELF.

then (then) *adv.* **1** At that time. **2** Soon or immediately afterward; next in space or time. **3** At another time: often introducing a sequential statement following *now, at first,* etc. *— conj.* **1** For that reason; as a consequence; accordingly. **2** In that case: I will *then,* since you won't. *— adj.* Being or acting in, or belonging to, that time: the *then* secretary of state. *— n.* A specific time already mentioned or understood; that time. [OE *thanne*]

then·age (then′ij) See THANAGE

the·nar (thē′när) *n. Anat.* **1** The palm of the hand. **2** The prominence on the palm at the base of the thumb. *— adj.* Of or pertaining to the palm of a hand or the sole of a foot: also **the′nal.** [<Gk. *thenar* palm of the hand]

thence (thens) *adv.* **1** From that place. **2** From the circumstance, fact, or cause; therefore. **3** From that time; after that time. **4** *Archaic* Away from there; elsewhere; absent. [ME *thannes* <OE *thanon* from there + *-s*[3]]

thence·forth (thens′fôrth′, -fōrth′, thens′fôrth′, -fōrth′) *adv.* From that time on; thereafter.

thence·for·ward (thens′fôr′wərd) *adv.* 1 Thenceforth. 2 From that place or time forward. Also **thence′for′wards.**

theo- *combining form* God; of or pertaining to God, a god, or gods: *theophany, theodicy.* Also, before vowels, *the-.* [<Gk. *theos* a god]

The·o·bald (thē′ə-bôld, tib′əld; *Dan.* tā′ō-bäl, *Ger.* tā′ō-bält, *Sw.* tā′ō-bäld) A masculine personal name. Also *Lat.* **The·o·bal·dus** (thē′ə-bôl′dəs), *Pg.* **The·o·bal·do** (tā′ōō-bäl′thōō), *Sp.* **The·u·de·bal·do** (tā′ōō-thä·väl′thō). [<Gmc., bold patriot]

The·o·bro·ma (thē′ə-brō′mə) *n.* A genus of small trees indigenous in tropical America, especially *Theobroma cacao,* source of the cocoa and chocolate of commerce, now cultivated also in the Old World tropics. [<NL <Gk. *theos* god + *brōma* food]

the·o·bro·mine (thē′ə-brō′mēn, -min) *n. Chem.* A bitter, colorless, crystalline alkaloid, $C_7H_8N_4O_2$, resembling caffeine, contained in cacao beans: used in medicine as a diuretic and myocardial stimulant. [<THEOBROM(A) + -INE²]

the·o·cen·tric (thē′ə-sen′trik) *adj.* Having God for its center; proceeding from and returning to God.

the·oc·ra·cy (thē·ok′rə-sē) *n. pl.* **·cies** 1 A state, polity, or group of people that claims a deity as its ruler, as ancient Israel after the Exodus. 2 Government of a state by a god, or by a priestly class claiming to have divine authority, as in the Papacy. [<Gk. *theokratia* < *theos* god + *kratein* rule] — **the·o·crat·ic** (thē′ə-krat′ik) or **·i·cal** *adj.*

the·o·cra·sy (thē·ok′rə-sē) *n.* 1 The mingling of several deities or divine attributes in one personality. 2 The mystical intimacy or union of the soul with God. [<LGk. *theokrasia* <Gk. *theos* god + *krasis* mingling]

the·o·crat (thē′ə-krat) *n.* 1 A theocratic or divine ruler. 2 An advocate of theocracy.

The·oc·ri·tus (thē·ok′rə-təs) Greek pastoral poet of the third century B.C.

the·od·i·cy (thē·od′ə-sē) *n. pl.* **·cies** 1 Justification of the divine providence by the attempt to reconcile the existence of evil with the goodness and sovereignty of God: a term established by Leibnitz in 1710. 2 The branch of philosophy that treats of the being, perfections, and government of God and the immortality of the soul. [<F *théodicée* <Gk. *theos* god + *dikē* justice]

the·od·o·lite (thē·od′ə-līt) *n.* One of several surveying and astronomical instruments for measuring horizontal and vertical angles by means of a small telescope turning on both a horizontal and a vertical axis. [An arbitrary formation] — **the·od·o·lit·ic** (-lit′ik) *adj.*

The·o·dore (thē′ə-dôr, -dōr) A masculine personal name. Also *Dan., Ger., Sw.* **The·o·dor** (tā′ō-dôr), *Fr.* **Thé·o·dore** (tā-ô-dôr′), *Gk.* **The·o·do·ros** (thē·o′dō·ros), *Du.* **The·o·do·rus** (tā′ō-dô′rəs). [<Gk., gift of God] — **The·o·do·ra** (thē′ə-dôr′ə) *fem.*

The·od·o·ric (thē·od′ər·ik) A masculine personal name. Also **The·od′er·ick, The·od′o·rick,** *Fr.* **Thé·o·do·ric** (tā-ō-dō-rēk′), *Ger.* **The·o·do·rich** (tā-ō′dō-rikh), *Lat.* **The·o·do·ri·cus** (tā′ō-də-rī′kəs). [<Gmc., ruler of the people] — **Theodoric,** 454?–526, king of the Ostrogoths; invaded and conquered Italy.

The·o·do·si·us (thē′ə-dō′shē-əs), 346?–395, Roman emperor 379–395.

the·og·o·ny (thē·og′ə-nē) *n.* The generation or genealogy of the gods, especially as recited in ancient poetry. [<Gk. *theogonia* < *theos* god + *gonos* generation < *gignesthai* be born] — **the·o·gon·ic** (thē′ə-gon′ik) *adj.* — **the·og′o·nist** *n.*

the·o·log (thē′ə-log, -log) *n.* A theological student. Also **the′o·logue.** [<L *theologus* <Gk. *theologos* one who treats of the gods < *theos* god + *logos* discourse < *legein* speak]

the·o·lo·gian (thē′ə-lō′jē·ən, -jən) *n.* One versed in theology, especially a professor of the Christian church; a professor of divinity; a divine.

the·o·log·i·cal (thē′ə-loj′i-kəl) *adj.* 1 Pertaining or relating to theology. 2 Linked to, based on, or referring to divine revelation. 3 Pertaining to the exposition or expounders of theology. Also **the·o·log′ic.** — **the·o·log′i·cal·ly** *adv.*

theological virtues See under VIRTUE.

the·ol·o·gize (thē·ol′ə-jīz) *v.* **·gized, ·giz·ing** *v.t.* To devise or fit (something) into a system of theology. — *v.i.* To reason theologically. Also *Brit.* **the·ol′o·gise.**

the·ol·o·gy (thē·ol′ə-jē) *n. pl.* **·gies** 1 The study of religion, culminating in a synthesis or philosophy of religion; also, a critical survey of religion, especially of the Christian religion. 2 A body of doctrines concerning God, including his attributes and relations with man; especially, such a body of doctrines as set forth by a particular church or religious group: Catholic *theology.* [<OF *theologie* <LL *theologia* <Gk. < *theos* god + *logos* discourse]

the·om·a·chy (thē·om′ə-kē) *n.* 1 A combat with the gods, as that waged by the Titans. 2 A battle among the gods. [<Gk. *theomachia* < *theos* god + *machē* combat]

the·o·mor·phic (thē′ə-môr′fik) *adj.* Having the form or likeness of God. [<Gk. *theomorphos* < *theos* god + *morphē* form]

the·o·mor·phism (thē′ə-môr′fiz-əm) *n.* The doctrine that man has the likeness or form of God.

the·o·pa·thet·ic (thē′ō-pə-thet′ik) *adj.* Pertaining to or of the nature of theopathy: *theopathetic* mysticism. Also **the·o·path′ic** (thē′ə-path′ik).

the·op·a·thy (thē·op′ə-thē) *n.* Religious emotion aroused by meditation on God; mystical ecstasy. [<Gk. *theopathia* the suffering of God < *theos* a god + *path-,* stem of *paschein* suffer]

the·oph·a·ny (thē·of′ə-nē) *n. pl.* **·nies** A manifestation or appearance of a deity or of the gods to man. [<L *theophania* <Gk. < *theos* god + *phainein* show]

The·oph·i·lus (thē·of′ə-ləs; *Du., Ger.* tā-ō′fē-lōōs) A masculine personal name. Also *Fr.* **Thé·o·phile** (tā-ō-fēl′), *Pg.* **The·o·phi·lo** (tā-ō′fē-lōō). [<Gk., lover of God]

The·o·phras·tus (thē′ə-fras′təs), 372?–287? B.C., Greek philosopher.

the·o·phyl·line (thē′ə-fil′ēn, -in) *n. Chem.* A white, bitter, crystalline alkaloid, $C_7H_8O_2N_4$, obtained from tea leaves and also made synthetically: it is an isomer of theobromine. [<NL *thea* tea + Gk. *phyllon* leaf + -INE²]

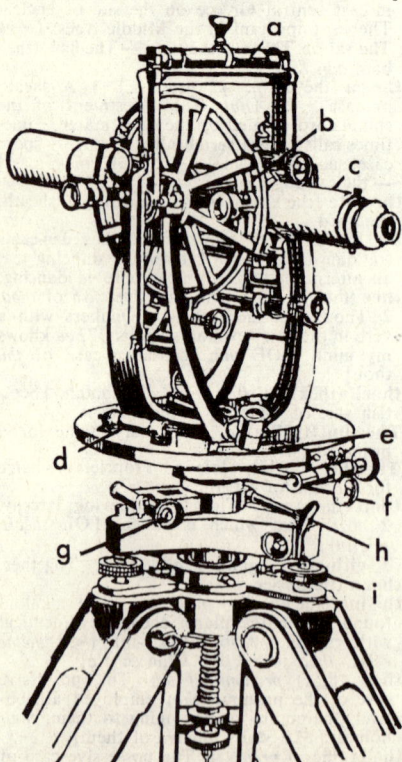

THEODOLITE
a. Striding level. *b.* Vertical limb and vernier. *c.* Telescope. *d.* Plate bubble. *e.* Horizontal limb and vernier. *f.* Clamp and tangent screw. *g.* Lower clamp screw. *h.* Tangent screw. *i.* Leveling screw.

the·or·bo (thē·ôr′bō) *n.* A 17th century lute having two necks. [<F *théorbe* <Ital. *tiorba,* prob. after the name of the inventor]

the·o·rem (thē′ə-rəm, thir′əm) *n.* 1 A proposition demonstrably true or acknowledged as such. 2 *Math.* **a** A proposition setting forth something to be proved. **b** A proposition that has been proved or assumed to be true. **c** A rule or statement of relations formulated in symbols. [<F *théorème* <Gk. *theōrēma* sight, theory < *theōreein* look at] — **the·o·re·mat·ic** (thē′ər·ə·mat′ik), **the′o·rem′ic** (-rem′ik) *adj.*

the·o·ret·ic (thē′ə·ret′ik) *adj.* 1 Theory, as distinct from practice. 2 *pl.* Theoretical matters; specifically, the theoretical aspect of a science. — *adj.* Theoretical.

the·o·ret·i·cal (thē′ə·ret′i·kəl) *adj.* 1 Of, relating to, or consisting of theory. 2 Relating to knowledge or pure science without reference to its application: compare EXPERIMENT (def. 3). 3 Existing only in theory; hypothetical. 4 Addicted to theorizing; unaffected by practical considerations; hence, impractical; visionary. Also **theoretic.** — **the·o·ret′i·cal·ly** *adv.*

the·o·re·ti·cian (thē′ər·ə·tish′ən) *n.* One who deals with the speculative, hypothetical, or ideal rather than with the practical and executive aspects of a subject.

the·o·rist (thē′ər·ist) *n.* One who theorizes.

the·o·rize (thē′ə·rīz) *v.i.* **·rized, ·riz·ing** To form or express theories; speculate. Also *Brit.* **the′o·rise.** — **the′o·ri·za′tion** *n.* — **the′o·riz′er** *n.*

the·o·ry (thē′ər·ē, thir′ē) *n. pl.* **·ries** 1 A plan or scheme existing in the mind only, but based on principles verifiable by experiment or observation. 2 A body of the fundamental principles underlying a science or the application of a science: the *theory* of relativity. 3 Abstract knowledge of any art, as opposed to the practice of it. 4 A proposed explanation or hypothesis designed to account for any phenomenon. 5 Loosely, mere speculation or hypothesis; an individual idea or guess. 6 *Math.* An arrangement of results, or a body of theorems, presenting a systematic view of some subject: the *theory* of functions. 7 The science of musical composition, as distinguished from the art of execution. See synonyms under HYPOTHESIS, IDEA. [<F *théorie* <Gk. *theōria* view, speculation < *theōreein* look at]

the·os·o·phy (thē·os′ə·fē) *n.* Mystical speculation applied to deduce a philosophy of the universe. In its modern phase, a system that claims to embrace the essential truth underlying all systems of religion, science, and philosophy. Its doctrines resemble closely those of Buddhism and Brahmanism, teaching the existence of an omnipotent, infinite, eternal, and immutable principle transcending the power of human conception, and the identity of all souls, through the cycle of incarnation with a universal spirit. [<Med. L *theosophia* <Gk. < *theosophos* wise in divine matters < *theos* god + *sophos* wise] — **the·o·soph·ic** (thē′ə·sof′ik) or **·i·cal** *adj.* — **the′o·soph′i·cal·ly** *adv.* — **the·os′o·phist** *n.*

The·o·to·co·pu·li (tā′ə·tō·kō′pōō·lē), **Domenico** See GRECO, EL.

The·ra (thir′ə, thē′rə) The southernmost island of the Cyclades in the Aegean; 31 square miles; formerly *Santorin.*

ther·a·peu·tic (ther′ə·pyōō′tik) *adj.* 1 Having healing qualities; curative. 2 Pertaining to therapeutics. Also **ther′a·peu′ti·cal.** [<NL *therapeuticus* <Gk. *therapeutikos* < *therapeutēs* an attendant < *therapeuein* serve, take care of < *therapōn* an attendant] — **ther′a·peu′ti·cal·ly** *adv.*

therapeutic dose That quantity of a drug which will produce the greatest beneficial effect in the given instance; the optimal dose.

ther·a·peu·tics (ther′ə·pyōō′tiks) *n. pl.* (construed as singular) 1 The department of medical science that treats of remedies for disease and their application. 2 The art and science of healing. — **ther′a·peu′tist** *n.*

ther·a·py (ther′ə·pē) *n. pl.* **·pies** 1 The treatment of disease by drugs or other curative processes: chiefly used in compounds: *hydrotherapy.* 2 Healing or curative quality. [<NL *therapia* <Gk. *therapeuein* < *therapōn* an attendant, a servant] < care of. See THERAPEUTIC. — **ther′a·pist** *n.*

there (thâr) *adv.* 1 At or in that place; in a place other than that of the speaker; opposed to *here.* 2 To, toward, or into that place; thither. 3 At that stage or point of action

there·a·bout (thâr′ə·bout′) *adv.* Near that number, quantity, degree, place, or time; approximately. Also **there′a·bouts′**.
there·af·ter (thâr·af′tər, -äf′-) *adv.* 1 Afterward; from that time on. 2 Accordingly.
there·a·gainst (thâr′ə·genst′) *adv.* Against or in opposition to that thing; on the other hand.
there·at (thâr·at′) *adv.* At that event, place, or time; at that incentive; upon that.
there·by (thâr′bī′) *adv.* 1 Through the agency of that. 2 Connected with that. 3 Conformably to that. 4 Nearby; thereabout. 5 By it or that; into possession of it or that: How did you come *thereby?*
there·for (thâr·fôr′) *adv.* For this, that, or it; in return or requital for this or that: We return thanks *therefor.*
there·fore (thâr′fôr′, -fōr′) *adv. & conj.* For that or this reason; on that ground or account; hence; consequently: He did not run fast enough; *therefore* he lost the race.
— *Synonyms (conj.):* accordingly, because, consequently, hence, since, then, thence, whence, wherefore. *Therefore* is the most precise and formal word for expressing the direct conclusion of a chain of reasoning; *then* carries a similar but slighter sense of inference, which it gives incidentally rather than formally; as, If this is true, *then* we can go. *Consequently* denotes a direct result, but more frequently of a practical than a theoretical kind; as, Important matters demand my attention; *consequently* I shall not sail today. *Accordingly* denotes correspondence, which may or may not be consequence; it is often used in narration; as, The soldiers were eager and confident; *accordingly* they sprang forward at the word of command. *Thence* is a word of more sweeping inference than *therefore*, applying not merely to a single set of premises but often to all that has gone before, including the reasonable inferences that have not been formally stated. *Wherefore* is the correlative of *therefore*, and *whence* of *hence* or *thence*, appending the inference or conclusion to the previous statement without a break. Compare synonyms for BECAUSE.
there·from (thâr·frum′, -from′) *adv.* From this, that, or it; from this or that time, place, state, event, or thing.
there·in (thâr·in′) *adv.* 1 In that place. 2 In that time, matter, or respect.
there·in·af·ter (thâr′in·af′tər, -äf′-) *adv.* In a subsequent part of that (book, document, speech, etc.).
there·in·to (thâr′in·tōō′) *adv.* Into this, that, or it.
Ther·e·min (ther′ə·min) *n.* A musical instrument played by manual interference with two sets of radio-frequency waves issuing from a pair of oscillators adapted for tone variation and volume control: a trade name. [after Léon Thérémin, born 1896, Russian-French inventor]
there·of (thâr·uv′, -ov′) *adv.* 1 Of or relating to this, that, or it. 2 From or because of this or that cause or particular; therefrom.
there·on (thâr·on′, -ôn′) *adv.* 1 On this, that, or it. 2 Thereupon; thereat.
there's (thârz) There is: a contraction.
The·re·sa (tə·rē′sə, -res′ə) A feminine personal name. Also *Fr.* **Thérèse** (tā·rez′). [<Gk., harvester]
— **Saint Theresa of Ávila**, 1515–82, Spanish Carmelite nun and mystic: also *Teresa*.
there·to (thâr·tōō′) *adv.* 1 To this, that, or it. 2 In addition; furthermore. Also **there′·un·to′** (-un·tōō′).
there·to·fore (thâr′tə·fôr′, -fōr′) *adv.* Before this or that; previously to that.
there·un·der (thâr·un′dər) *adv.* Under this or that. 2 Less, as in number. 3 In a lower or lesser status or rank.
there·up·on (thâr′ə·pon′, -ə·pôn′) *adv.* 1 Upon that; upon it. 2 Following upon or in consequence of that. 3 Immediately following; at once.
there·with (thâr·with′, -with′) *adv.* 1 With this, that, or it. 2 Thereupon; thereafter; immediately afterward.
there·with·al (thâr′with·ôl′) *adv.* 1 With all this or that; besides. 2 *Obs.* Therewith; with this, that, or it.
the·ri·a·ca (thi·rī′ə·kə) *n.* 1 An ancient antidote for the bite of venomous creatures, containing numerous drugs mixed with honey. 2 Molasses; treacle. Also **the·ri·ac** (thir′ē·ak). [< LL *theriaca*, an antidote for poison < Gk. *thēriakos* pertaining to wild beasts < *thērion*, dim. of *thēr* wild beast] — **the·ri′a·cal** *adj.*
the·ri·an·thro·pism (thir′ē·an·thrə·piz′əm) *n.* Representation of preternatural beings in combined forms of man and beast, especially in primitive polytheistic worship: the religions of *therianthropism.* [< Gk. *thērion* wild beast + *anthrōpos*, -*opou* man] — **the′ri·an·throp′ic** (-an·throp′ik) *adj.*
the·ri·o·mor·phic (thir′ē·ə·môr′fik) *adj.* Beastlike in form: *theriomorphic* gods. Also **the′ri·o·mor′phous**. [< Gk. *thērion* wild beast + *morphē* form]
therm (thûrm) *n.* 1 A unit of heat used as a basis for the sale of illuminating gas in England, equal to 100,000 British thermal units. 2 One thousand great calories. 3 The great calorie. 4 The lesser calorie. Also **therme**. [< Gk. *thermē* heat]
therm- Var. of THERMO-.
Ther·ma (thûr′mə) Ancient name for THESSALONIKE.
ther·mae (thûr′mē) *n. pl.* 1 Hot springs or baths. 2 Specifically, the public baths of the ancient Romans; also, the bathhouses. [< L < Gk. *thermai*, pl. of *thermē* heat]
ther·mal (thûr′məl) *adj.* 1 Pertaining to, determined by, or measured by heat. 2 Hot or warm. Also **ther′mic**. — **ther′mal·ly** *adv.*
thermal barrier *Aeron.* The limit imposed upon the operating speed of jet engines, rockets, motors, and the like by temperatures above the melting point of their materials.
thermal death Heat death.
thermal diffusion 1 The diffusion of heat. 2 *Physics* A method for the separation of isotopes by passing a gas through a vertical tube containing an electrically heated wire which produces a concentration of the heavier components at the bottom and of the lighter components at the top.
therm·an·es·the·sia (thûr′mən·is·thē′zhə, -zhē·ə) *n. Pathol.* Loss of ability to recognize sensations of heat or cold; absence of temperature sense. Also **therm′an·aes·the′sia**. [< THERM(O)- + ANESTHESIA]
therm·el (thûr′mel) *n.* A thermocouple or group of thermocouples when used to determine temperatures. [< Gk. *thermē* heat]
therm·es·the·sia (thûr′mis·thē′zhə, -zhē·ə) *n. Physiol.* The ability to recognize changes of temperature; temperature sensitivity. Also **therm′aes·the′sia**. [< THERM(O)- + ESTHESIA]
Ther·mi·dor (thûr′mə·dôr′, *Fr.* ter·mē·dôr′) See under CALENDAR (Republican). [< F < Gk. *thermē* heat + *dōron* gift]
therm·i·on (thûrm′ī′ən, thûr′mē·ən) *n. Physics* An electrically charged particle emitted by a heated body: it may be either positive or negative. [< THERM(O)- + ION] — **therm·i·on·ic** (thûr′mē·on′ik) *adj.*
therm·i·on·ics (thûr′mē·on′iks) *n. pl.* (construed as singular) The science and practical application of thermionic phenomena.
thermionic tube A vacuum tube emitting thermions from a heated electrode. Also *Brit.* **thermionic valve**.
therm·is·tor (thûr·mis′tər) *n. Electr.* A small, compact thermometric device consisting of a semiconducting material having a large temperature coefficient of resistance: widely used in the measurement of microwave power, of temperatures, and as a protective device in circuits. [< THERM(O)- + (RES)ISTOR]
ther·mit (thûr′mit) *n.* A mixture composed of finely divided aluminum and oxide of iron, chromium, or manganese. When such a mixture is brought to a sufficient temperature, the oxygen of the oxide unites with the aluminum, producing an intense heat. Also **ther′mite** (-mīt). [< Gk. *thermē* heat]
thermo- *combining form* Heat; of, related to, or caused by heat: *thermolysis, thermostat*. Also, before vowels, **therm-**. [< Gk. *thermos* heat, warmth]
ther·mo·bar·o·graph (thûr′mō·bar′ə·graf, -gräf) *n.* An apparatus for measuring the pressure and temperature of a gas simultaneously.
ther·mo·ba·rom·e·ter (thûr′mō·bə·rom′ə·tər) *n.* 1 An apparatus for measuring atmospheric pressure by the boiling point of water: used in determining altitudes. 2 A form of barometer that can be inverted and made to serve as a thermometer.
ther·mo·cau·ter·y (thûr′mō·kô′tər·ē) *n.* Cautery by means of heated wires or points.
ther·mo·chem·is·try (thûr′mō·kem′is·trē) *n.* The branch of chemistry that treats of the relations between chemical reactions and the evolution and absorption of heat observed to accompany them. — **ther′mo·chem′i·cal** (-kem′i·kəl) *adj.* — **ther′mo·chem′ist** *n.*
ther·mo·cline (thûr′mō·klīn) *n.* A gradient indicating marked changes in temperature with depth, especially between discontinuous layers of ocean waters.
ther·mo·cou·ple (thûr′mō·kup′əl) *n.* A pair of dissimilar metals so joined as to produce a thermoelectric effect when the contact surfaces are at different temperatures. Also **ther′mo·e·lec′tric couple**.
ther·mo·dy·nam·ics (thûr′mō·dī·nam′iks, -di-) *n. pl. (construed as singular)* That branch of physical science which treats of the relations between heat and energy, especially the convertibility of one into the other and the mechanical work involved. — **ther′mo·dy·nam′ic** or **·i·cal** *adj.* — **ther′mo·dy·nam′i·cist** (-nam′ə·sist) *n.*
ther·mo·e·lec·tric (thûr′mō·i·lek′trik) *adj.* Of or pertaining to thermoelectricity. Also **ther′mo·e·lec′tri·cal**. — **ther′mo·e·lec′tri·cal·ly** *adv.*
ther·mo·e·lec·tric·i·ty (thûr′mō·i·lek′tris′ə·tē) *n.* Electricity generated by differences of temperature, especially between two different metals in contact when one of the junctions is heated.
ther·mo·e·lec·tro·mo·tive (thûr′mō·i·lek′trə·mō′tiv) *adj.* Of, pertaining to, or designating electromotive force caused by difference of temperature.
ther·mo·gal·va·nom·e·ter (thûr′mō·gal′və·nom′ə·tər) *n.* A combination of a galvanometer and a thermocouple used to measure minute variations of temperature.
ther·mo·gen·e·sis (thûr′mō·jen′ə·sis) *n.* The production of heat, especially of animal heat by organic action. — **ther′mo·gen′ic, ther′mog·e·nous** (thər·moj′ə·nəs), **ther′mo·ge·net′ic** (-jə·net′ik) *adj.*
ther·mo·gram (thûr′mə·gram) *n.* The record made by a thermograph.
ther·mo·graph (thûr′mə·graf, -gräf) *n.* An instrument for recording temperature variations; a self-registering thermometer.
ther·mog·ra·phy (thər·mog′rə·fē) *n.* 1 Photography by means of heat waves emitted by an object which has been coated with luminescent paint and exposed to ultraviolet light. 2 *Printing* Any process of reproducing written or printed characters that employs heat. — **ther·mo·graph·ic** (thûr′mə·graf′ik) *adj.*
ther·mo·hal·ine (thûr′mō·hal′ēn, -īn, -in) *adj.* Pertaining to or characterized by variations in the temperature and salinity of sea water. [< THERMO- + Gk. *hals* salt + -INE¹]
ther·mo·junc·tion (thûr′mō·jungk′shən) *n.* The point of contact between the pair of conductors forming a thermocouple.
ther·mo·kin·e·mat·ics (thûr′mō·kin′ə·mat′iks) *n. pl. (construed as singular)* The study of heat in motion or of the motive power of heat.
ther·mo·la·bile (thûr′mō·lā′bil) *adj. Biochem.* Decomposed, destroyed, affected, or liable to be adversely affected by heat, as some enzymes and toxins: opposed to *thermostable*. [< THERMO- + LABILE]
ther·mo·lu·mi·nes·cence (thûr′mō·lōō′mə·nes′əns) *n.* 1 The emission of light from a substance or material under the action of

heat. 2 A luminous effect in rock crystals from which electrons displaced by radioactivity have been released at definite temperatures, with or without pressure: sometimes indicative of the age of sedimentary rocks. **— ther'mo·lu'mi·nes'cent** *adj.*

ther·mol·y·sis (thər·mol'ə·sis) *n.* 1 *Chem.* The resolution of a compound substance into its component elements by the application of heat. 2 *Physiol.* The dissipation of heat from the animal body by physical processes. [< THERMO- + -LYSIS] **— ther·mo·lyt·ic** (thûr'mə·lit'ik) *adj.*

ther·mo·mag·net·ic (thûr'mō·mag·net'ik) *adj.* Of or pertaining to the relations between heat and magnetism.

ther·mom·e·ter (thər·mom'ə·tər) *n.* An instrument for measuring the temperature of a substance, body, or space. The ordinary thermometer consists of a graduated glass capillary tube or stem with a bulb containing mercury which expands or contracts as the temperature rises or falls. The **differential thermometer** has two air bulbs connected by a U-tube, containing colored liquid, so that when the bulbs are exposed to different temperatures a shifting of the liquid in the tube will be caused by the difference of expansion of air in the bulbs. A **resistance thermometer** indicates, by means of the change in electrical conductivity of wires with temperature, the temperature of any given wire or its environment. **— clinical thermometer** A thermometer accurately calibrated for determining body temperature, especially of a person. [<THERMO- + METER]

ther·mom·e·try (thər·mom'ə·trē) *n.* The measurement of temperature, or the art thereof, by means of the thermometer; specifically, the use of the thermometer in medical diagnosis. **— ther·mo·met·ric** (thûr'mō·met'rik) or **·ri·cal** *adj.* **— ther'mo·met'ri·cal·ly** *adv.*

ther·mo·mo·tor (thûr'mō·mō'tər) *n.* A heat engine; especially, a hot-air engine. Compare MOTOR.

ther·mo·nu·cle·ar (thûr'mō·nōō'klē·ər, -nyōō'-) *adj. Physics* Pertaining to or characterized by the mass-energy reactions involving the fusion of light atomic nuclei subjected to very high temperatures, especially with reference to stellar energy and the hydrogen bomb.

ther·mo·pen·e·tra·tion (thûr'mō·pen'ə·trā'·shən) *n.* Diathermy.

ther·mo·phil·ic (thûr'mō·fil'ik) *adj.* Fond of heat: used mainly of certain bacteria. Also **ther'mo·phile** (-fil, -fīl). [<THERMO- + Gk. *philos* loving]

ther·mo·pile (thûr'mō·pīl) *n.* A group of thermocouples acting jointly to produce electric energy, especially when used with a galvanometer to measure heat.

ther·mo·plas·tic (thûr'mō·plas'tik) *adj.* Plastic in the presence of or under the application of heat: said especially of certain synthetic molding materials. **—** *n.* A thermoplastic substance or material.

Ther·mop·y·lae (thər·mop'ə·lē) A narrow mountain pass in Greece; scene of a battle, 480 B.C., in which the Spartans under the command of Leonidas held off the Persians under Xerxes and finally died to the last man rather than yield.

Ther·mos bottle (thûr'məs) A container shaped like a bottle or flask, having two walls separated by a vacuum which serves to insulate the contents so that they retain their temperature: a trade name.

ther·mo·scope (thûr'mə·skōp) *n.* An instrument for detecting changes or differences of temperature without accurately measuring them. [<THERMO- + -SCOPE] **— ther'mo·scop'ic** (-skop'ik) or **·i·cal** *adj.*

ther·mo·set·ting (thûr'mō·set'ing) *adj.* Having the property of assuming a fixed shape after being molded under heat, as certain phenol and other synthetic resins.

ther·mo·si·phon (thûr'mō·sī'fən) *n.* A device consisting of siphon tubes to increase or induce circulation by making use of temperature differential in a water-cooling system, as in that of an internal-combustion engine.

ther·mo·sta·ble (thûr'mō·stā'bəl) *adj.* 1 Resistant to heat, as certain plastics and chemicals. 2 *Biochem.* Unaffected by moderate heats; denoting immune substances, as certain toxins or ferments, which may be heated to 55° C. without loss of special properties: opposed to *thermolabile.* Also **ther'mo·sta'bile.** **— ther'·mo·sta·bil'i·ty** (-stə·bil'ə·tē) *n.*

ther·mo·stat (thûr'mə·stat) *n.* A device for the automatic regulation of temperature by means of a relay utilizing the expansion and contraction caused by temperature changes in certain metals: used for actuating fire alarms, opening or closing dampers, regulating steam pressures, etc. [<THERMO- + Gk. *statos* standing] **— ther'mo·stat'ic** *adj.* **— ther'mo·stat'i·cal·ly** *adv.*

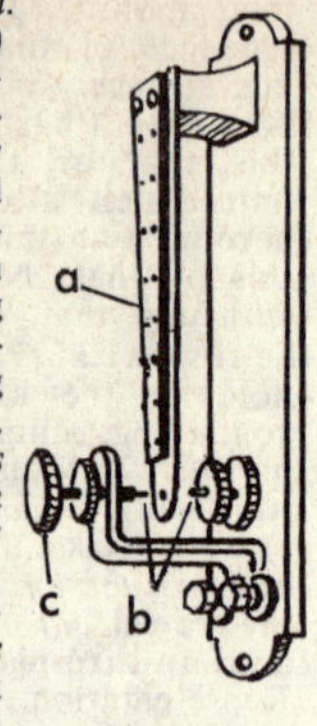

THERMOSTAT
a. Bimetal bar.
b. Contact points.
c. Control knob.

ther·mo·stat·ics (thûr'mō·stat'iks) *n. pl. (construed as singular)* The science that deals with the equilibrium of heat.

ther·mo·tank (thûr'mō·tangk') *n.* A tank or box in which steam, water, air, or the like circulates through pipes and thus heats or cools the air passing through the tank.

ther·mo·tax·is (thûr'mō·tak'sis) *n. Biol.* 1 The regulation or normal adjustment of the animal heat in an organism. 2 The determination of movement by heat. **— ther'mo·tax'ic, ther'mo·tac'tic** (-tak'tik) *adj.*

ther·mo·ten·sile (thûr'mō·ten'sil) *adj.* Relating to variation of tensile strength caused by temperature.

therm·o·ther·a·py (thûr'mō·ther'ə·pē) *n. Med.* The treatment of disease by the application of heat.

ther·mot·ics (thər·mot'iks) *n. pl. (construed as singular)* The science of heat. [<Gk. *thermotēs* heat]

ther·mot·ro·pism (thər·mot'rə·piz'əm) *n. Biol.* 1 The property or phenomenon of movement in growing plants or other organisms brought about by the influence of heat or cold. 2 The attraction or repulsion from a source of heat evinced by some bacteria. **— ther·mo·trop·ic** (thûr'mō·trop'ik) *adj.*

the·roid (thir'oid) *adj.* Resembling or like a beast. [<Gk. *thēroeidēs* < *thēr, thēros* a wild beast + *eidos* form]

the·ro·phyte (thir'ə·fīt) *n. Bot.* An annual plant which completes its life cycle in one vegetative season. [<Gk. *theros* summer + -PHYTE]

the·ro·pod (thir'ə·pod) *n.* Any of a suborder *(Theropoda)* of saurischian dinosaurs of the Triassic and Cretaceous periods, including the true carnivorous types, as *Allosaurus* and *Tyrannosaurus.* **—** *adj.* Of or pertaining to the *Theropoda.* [<NL *Theropoda* <Gk. *thēr, thēros* a wild beast + *pous, podos* foot] **— the·rop·o·dan** (thi·rop'ə·dən) *adj.* & *n.*

Ther·si·tes (thər·sī'tēz) In the *Iliad,* an ugly and scurrilous Greek soldier in the Trojan War, killed by Achilles for troublemaking.

ther·sit·i·cal (thər·sit'i·kəl) *adj.* Characteristic of Thersites; hence, loud and scurrilous; abusive.

the·sau·ric (thi·sôr'ik) *adj.* Encyclopedic; having or containing large stores of miscellaneous information.

the·sau·ro·sis (thē'sô·rō'sis) *n. Pathol.* A condition marked by the storage in the body of excessive amounts of normal or foreign substances. [<Gk. *thēsauros* treasure + -OSIS]

the·sau·rus (thi·sôr'əs) *n. pl.* **·sau·ri** (-sôr'ī) 1 A place where treasure is laid up; a storehouse. 2 A repository of words or knowledge; hence, a lexicon or cyclopedia. [<L <Gk. *thēsauros* treasure house. Doublet of TREASURE.]

these (thēz) Plural of THIS.

The·seus (thē'sōōs, -sē·əs) In Greek mythology, the chief hero of Attica, son of Aegeus and king of Athens, celebrated for many adventures, chiefly the killing of the Minotaur, and for unifying Attica with Athens as its capital. See ARIADNE, HIPPOLYTUS, PHAEDRA, PIRITHOUS. **— The'se·an** (-sē·ən) *adj.*

the·sis (thē'sis) *n. pl.* **·ses** (-sēz) 1 A proposition. 2 Specifically, a formal proposition, advanced and defended by argumentation. 3 A formal treatise on a particular subject, especially, a dissertation presented by a candidate for an academic degree. 4 In early prosody, that part of a foot which had the ictus or stress. 5 In later Roman usage and in modern prosody, the unaccented part of a foot; also, the depression of the voice in pronouncing it. See ARSIS. 6 *Logic* An affirmative proposition; a premise or postulate, as opposed to a hypothesis. 7 *Music* The down beat; the accented part of a measure. [<L <Gk., a placing, proposition < *tithenai* put, place]

Thes·pi·an (thes'pē·ən) *adj.* 1 Of or relating to Thespis. 2 Of or relating to drama; dramatic; tragic. **—** *n.* An actor or actress.

Thes·pis (thes'pis) Greek poet of the sixth century B.C.; reputed father of Greek tragedy.

Thes·sa·lo·ni·an (thes'ə·lō'nē·ən) *n.* 1 A native or inhabitant of modern Thessalonike or of ancient Thessalonica. 2 *pl.* Either of two epistles in the New Testament **(First** and **Second Thessalonians)** written by St. Paul to the Christians of Thessalonica. **—** *adj.* Of or pertaining to Thessalonike.

Thes·sa·lo·ni·ke (thes'ä·lô·nē'kē) The Greek name for SALONIKA. Ancient **Thes·sa·lo·ni·ca** (thes'ə·lō·nī'kə, -lon'i·kə).

Thes·sa·ly (thes'ə·lē) A division of north central Greece; 5,399 square miles; chief town, Larissa. **— Thes·sa·li·an** (the·sā'lē·ən) *adj.* & *n.*

the·ta (thā'tə, thē'tə) *n.* The eighth letter in the Greek alphabet (Θ, ϑ, θ): equivalent in classical Greek to *t* + *h,* as in *right-hand,* but in modern Greek to spirant *th,* as in *thin.* [<Gk. *thēta*]

thet·ic (thet'ik) *adj.* 1 In ancient prosody, beginning with, bearing, relating to, or of the nature of a thesis. 2 Characterized by positive statement; arbitrary; dogmatic. Also **thet'i·cal.** [<Gk. *thetikos* fit for placing < *thetos* placed < *the-,* stem of *tithenai* place] **— thet'i·cal·ly** *adv.*

The·tis (thē'tis) In Greek mythology, a Nereid, wife of Peleus and mother of Achilles: by dipping Achilles into the Styx, she made him invulnerable, except in the right heel, by which she had held him.

the·ur·gy (thē'ûr·jē) *n. pl.* **·gies** 1 Divine or supernatural intervention in human affairs. 2 The working of miracles through divine or supernatural aid. 3 Magic, as practiced by the Neo-Platonists, by means of which miraculous effects were supposedly produced through the intervention of beneficent spirits; white magic. [<Gk. *theourgia* < *theourgos* divine worker < *theos* god + *ergon* work] **— the·ur·gic** (thē·ûr'jik), **the·ur'gi·cal** *adj.* **— the·ur'gi·cal·ly** *adv.* **— the'ur·gist** *n.*

thew (thyōō) *n.* 1 A sinew or muscle, especially when strong or well-developed. 2 *pl.* Bodily strength or vigor. [ME *theawes* good qualities, strength <OE *thēau* habit, characteristic quality] **— thew'y** *adj.*

thew·less (thyōō'lis) *adj. Scot.* 1 Having no thews; inactive; weak. 2 Spiritless; inert.

they (thā) *pron.* 1 The persons, beings, or things previously mentioned or understood: the nominative plural of *he, she, it.* 2 People in general; men: *They* say rain is expected. [<ON *their,* pl. of *sā* this, that]

they'd (thād) Contraction of: 1 They had. 2 They would.

they'll (thāl) They will: a contraction.

they're (thâr) They are: a contraction.

they've (thāv) They have: a contraction.

thi- Var. of THIO-.

thi·a·mine (thī'ə·mēn, -min) *n. Biochem.* A white crystalline compound, $C_{12}H_{18}ON_4SCl_2$; vitamin B_1, found in various natural sources, as cereal grains, green peas, liver, egg yolk, etc., and also made synthetically. Thiamine is the anti-beriberi vitamin. Also **thi'a·min** (-min). [<THI- + -AMINE]

thi·a·zine (thī'ə·zēn, -zin) *n. Chem.* One of a class of organic ring compounds of one atom of nitrogen, one of sulfur, and four of carbon. Also **thi'a·zin** (-zin). [<THI- + -AZINE]

thi·a·zole (thī'ə·zōl) *n. Chem.* A colorless, stable, liquid compound, C_3H_3NS, whose derivatives yield dyestuffs and certain sulfa drugs. Also **thi'a·zol** (-zōl, -zol). [<THI- + AZOLE]

Thi·bault (tē·bō'), **Jacques Anatole** See FRANCE, ANATOLE.

Thi·baut (tē·bō') French form of THEOBALD.

Thi·bet (ti·bet') See TIBET.

thick (thik) *adj.* 1 Having relatively large depth or extent from one surface to its opposite; having the dimension that is commonly least, comparatively great; not thin: distinguished from *long* and *broad.* 2 Having a specified dimension of this kind, whether great or small: an inch *thick.* **3** Arranged compactly; close: a *thick* forest; also, following at brief intervals; frequent, as blows, raindrops, etc. **4** Set or furnished closely or abundantly with objects; abounding. **5** Having considerable density or consistency; dense; hence, turbid; impure; heavy. **6** Overcharged with vapor; foggy; misty. **7** Lacking quickness of apprehension; dull; stupid. **8** Indistinct; muffled: a *thick* sound; also, guttural; husky; throaty. **9** *Colloq.* Very friendly; intimate. **10** *Colloq.* Excessive; going too far; being beyond the bounds of what is tolerable. — *adv.* In a thick manner; placed or following closely. — **to lay it on thick** *Colloq.* **1** To overstate; exaggerate. **2** To praise fulsomely. — *n.* **1** The dimension of thickness; the thickest part. **2** The thickest or most intense time or place of anything: the *thick* of the fight. — **through thick and thin** Through good times and bad; loyally; through good fortune and adversity. [OE *thicce*] — **thick'ly** *adv.*

Synonyms (adj.): close, cloudy, compact, condensed, dense, dull, foggy, gross, hazy, inspissate, misty, muddy, turbid. See BLUNT.

thick·en (thik'ən) *v.t.* & *v.i.* **1** To make or become thick or thicker. **2** To make or become more intricate or intense: The plot *thickens.* — **thick'en·er** *n.*

thick·en·ing (thik'ən·ing) *n.* **1** The act of making or becoming thick. **2** Something added to a liquid to increase its consistency. **3** That or that part which is or has been thickened.

thick·et (thik'it) *n.* A thick growth, as of underbrush, through which a passage is not easily effected; a coppice; jungle. [OE *thiccet* < *thicce* thick]

thick·head (thik'hed') *n.* A stupid person; numskull. Also **thick'skull'** (-skul').

thick·ish (thik'ish) *adj.* Somewhat thick.

thick·leaf (thik'lēf') *n.* Any plant of a genus *(Crassula)* of herbs and shrubs having large fleshy leaves and white, rose, or yellow flowers.

thick·ness (thik'nis) *n.* **1** The state or quality of being thick. **2** The dimension or measure of a solid other than its length or width. **3** A sheet, layer, etc., as of paper.

thick–set (thik'set') *adj.* **1** Having a short, thick body; stout. **2** Set like a thicket; closely planted. — *n.* **1** A thicket; also, a thick hedge. **2** A fustianlike fabric having a nap like velveteen.

thick·skin (thik'skin') *n.* A thick–skinned person.

thick–skinned (thik'skind') *adj.* **1** Having a thick skin; pachydermatous. **2** Insensitive; callous to hints or insults.

thick–wit·ted (thik'wit'id) *adj.* Stupid; obtuse; dense.

thief (thēf) *n. pl.* **thieves** (thēvz) **1** One who takes something belonging to another; one who steals. **2** *Law* One guilty of simple or compound larceny, embezzlement, or swindling. **3** That which causes loss: Procrastination is the *thief* of time. See synonyms under ROBBER. [OE *thēof*]

Thiers (tyâr), **Louis Adolphe,** 1797–1877, French statesman and historian.

thieve (thēv) *v.* **thieved, thiev·ing** *v.t.* To take by theft; purloin; steal. — *v.i.* To be a thief; commit theft. [OE *thēofian*]

thieve·less (thēv'lis) *adj. Scot.* **1** Ungracious; hard. **2** Listless.

thiev·er·y (thē'vər·ē) *n. pl.* **·er·ies** The practice or act of thieving; theft; also, an instance of thieving.

thiev·ish (thē'vish) *adj.* **1** Addicted to thieving. **2** Acting by stealth; furtive. **3** Relating to or like a thief. **4** Partaking of the nature of theft. — **thiev'ish·ly** *adv.* — **thiev'ish·ness** *n.*

thig (thig) *v.t.* & *v.i. Scot.* To beg or solicit (gifts, etc.). — **thig'ger** *n.*

thigh (thī) *n.* **1** The leg between the hip and the knee of man or the corresponding portion in other animals. ◆ Collateral adjective: *femoral.* **2** The femur of an insect. [OE *thēoh*]

thigh bone The femur.

thig·mo·tax·is (thig'mə·tak'sis) *n. Biol.* Stereotropism. [<Gk. *thigma* touch + -TAXIS] — **thig'mo·tac'tic** (-tak'tik) *adj.* — **thig'mo·tac'·ti·cal·ly** *adv.*

thig·mot·ro·pism (thig·mot'rə·piz'əm) *n. Biol.* Involuntary response to mechanical stimulation of any kind, as displayed by many insects and by the tendrils, leaves, etc., of certain plants. [<Gk. *thigma* touch + TROPISM] — **thig·mo·trop·ic** (thig'mə·trop'ik) *adj.*

thill (thil) *n.* One of the shafts of a vehicle, between which a horse is harnessed. [OE *thille* board]

thim·ble (thim'bəl) *n.* **1** A caplike cover with a pitted surface, worn in sewing to protect the end of the finger that pushes the needle. **2** *Mech.* A sleeve through which a bolt passes, or which unites two rods, tubes, or the like. **3** *Naut.* **a** A metal anti–chafing ring forming a guard over a loop or eye in a sail. **b** The metal piece about which a rope is bent and spliced to the main body of the rope to form an eye. [OE *thȳmel* < *thūma* thumb]

thim·ble·ber·ry (thim'bəl·ber'ē) *n. pl.* **·ries** Any of certain American raspberries or blackberries having a thimble–shaped fruit; especially, the blackcap raspberry, the **fragrant thimbleberry** *(Rubus odoratus),* and the **western thimbleberry** *(R. parviflorus).*

thim·ble·rig (thim'bəl·rig') *n.* **1** A swindling trick in which a pea or ball is shifted by sleight of hand from one to another of three inverted thimble–shaped cups. **2** A gambler who operates a thimblerig. — *v.t.* **·rigged, ·rig·ging** To cheat by or as by thimblerig. — **thim'ble·rig'ger** *n.*

thim·ble·weed (thim'bəl·wēd') *n.* Any of various plants (genus *Rudbeckia*) with thimble–shaped receptacles, as the rudbeckia and the American wood anemone.

thin (thin) *adj.* **thin·ner, thin·nest** **1** Having opposite surfaces relatively close to each other; being of little depth or width; not thick. **2** Lacking roundness or plumpness of figure; lean; slender. **3** Having the component parts or particles scattered or diffused; not dense or abundant; sparse; rare: *thin* ranks, *thin* gas. **4** Having little body or substance; of a loose texture; hence, insufficient to conceal or cover: *thin* clothing; flimsy: a *thin* excuse. **5** Having little or no consistency, as a liquid: *thin* molasses. **6** Lacking in essential ingredients or qualities: *thin* blood. **7** Having little volume or richness; shrill or metallic, as a voice. **8** Not abundantly supplied or furnished; bare; scant: a *thin* table. **9** Not having sufficient contrasts of shade to print well: said of a photographic negative. **10** Lacking vigor or force; feeble; superficial: *thin* wit. See synonyms under FINE[1], GAUNT, MEAGER. — *v.t.* & *v.i.* **thinned, thin·ning** To make or become thin or thinner. [OE *thynne*] — **thin'ly** *adv.* — **thin'ness** *n.*

thine (thīn) *pron.* **1** The possessive case of thou, used predicatively; belonging or pertaining to thee: *Thine* is the kingdom. **2** The things or persons belonging or pertaining to thee. — **of thine** Belonging or relating to thee; thy: the double possessive. — *pronominal adj. Archaic* Thy: *thine* eyes. [OE *thīn,* genitive of *thū* thou]

thing[1] (thing) *n.* **1** That which exists or is conceived to exist as a separate entity; an entity; being. **2** That which is designated, as contrasted with the word or symbol used to denote it. **3** A matter or circumstance; an affair; concern: *Things* have changed. **4** An act or deed; transaction: That was a shameless *thing* to do. **5** A statement or expression; utterance: to say the right *thing.* **6** An idea; opinion; notion: Stop putting *things* in her head. **7** A quality; attribute; characteristic: Kindness is a precious *thing.* **8** An inanimate object, as distinguished from a living organism. **9** An organic being: usually with a qualifying word: Every living *thing* dies. **10** An object that is not or cannot be described or particularized: The *thing* disappeared in the shadows. **11** A person, regarded in terms of pity, affection, or contempt: that poor *thing;* You stupid *thing!* **12** *pl.* Possessions; belongings: to pack one's *things.* **13** *pl.* Clothes; especially, outer garments: Take off your *things* and stay awhile. **14** A piece of literature, art, music, etc.: He read a few *things* by Byron. **15** The proper or befitting act or result: with *the*: That was not the *thing* to do. **16** The important or remarkable point: with *the*: The *thing* we learned from the war was this. **17** *Law* A subject or property or dominion, as distinguished from a person. — **to see things** To have hallucinations. [OE, thing, cause, assembly. Akin to THING[2].]

thing[2] (ting) *n.* A Scandinavian legislative or judicial body: the *Storthing,* the Norwegian parliament: also spelled *ting.* [<ON, assembly. Akin to THING[1].]

thing·a·ma·bob (thing'ə·mə·bob') *n. Colloq.* A thing the specific name of which is unknown or forgotten; a dingus. Also **thing'um·a·bob', thing'um·bob.**

thing·a·ma·jig (thing'ə·mə·jig') *n. Colloq.* A thingamabob. Also **thing'um·a·jig'.**

T–hinge (tē'hinj') *n.* A hinge the two sections of which have the form of the letter T. See illustration under HINGE.

thing in itself *Philos.* A noumenon; the ultimate, metaphysical reality behind the physical phenomena perceived by the senses, which, according to Kant, can never be known: the English rendering of the German *Ding an sich.*

think[1] (thingk) *v.* **thought** (thôt), **think·ing** *v.t.* **1** To produce or form in the mind; conceive mentally: to *think* evil thoughts. **2** To examine in the mind; meditate upon, or determine by reasoning: He was *thinking* what to do next; to *think* a plan through. **3** To believe; consider: I *think* him guilty. **4** To expect; anticipate: They did not *think* to meet us. **5** To bring to mind; remember; recollect: I cannot *think* what he said. **6** To have the mind preoccupied by: to *think* business morning, noon, and night. **7** To intend; purpose: Do they *think* to rob me? — *v.i.* **8** To use the mind or intellect in exercising judgment, forming ideas, etc.; engage in rational thought; reason. **9** To have a particular opinion, sentiment, or feeling: I don't *think* so. — **to think better of** **1** To abandon a course of action; alter one's intentions: I was going to call but I *thought better of* it. **2** To form a better opinion of. — **to think fit, proper, right,** etc. To regard as worth doing. — **to think nothing of** **1** To have a low opinion of; ignore. **2** To consider easy to do. — **to think of** **1** To bring to mind; remember; recollect. **2** To conceive in the mind; invent; imagine. **3** To have a specified opinion or attitude toward; regard. **4** To be considerate of; have regard for. — **to think over** To reflect upon; ponder. — **to think up** To devise, arrive at, or invent by thinking. — *n.* An act of thinking; a thought. [OE *thencean*; influenced in form by THINK[2]].

think[2] (thingk) *v.i.* To seem; appear: now obsolete except with the pronoun as indirect object in the combinations *methinks, methought.* [OE *thyncan* seem]

think·a·ble (thingk'ə·bəl) *adj.* Susceptible of being thought; conceivable; hence, possible to be believed.

think·er (thingk'ər) *n.* **1** One who thinks. **2** A person of powerful mind who devotes himself to abstract thought.

think·ing (thingk'ing) *adj.* **1** Exercising the mental capacities. **2** Capable of such exercise; rational. — *n.* **1** Mental action; thought. **2** The product of such action, as an idea. See synonyms under REFLECTION, THOUGHT. — **think'ing·ly** *adv.*

thin·ner (thin'ər) *n.* **1** One who or that which thins. **2** A liquid, as turpentine or petroleum spirits, mixed with paint in order to give it a proper consistency for working.

thin·nish (thin'ish) *adj.* Somewhat thin.

thin–skinned (thin'skind') *adj.* **1** Having a thin skin. **2** Hence, easily hurt or offended; sensitive.

thio– *combining form Chem.* Containing sulfur; denoting a compound of sulfur, especially one in which sulfur has displaced oxygen: *thiocyanic.* Also, before vowels, sometimes *thi–.* Compare SULFURO–. [<Gk. *theion* sulfur]

thi·o·a·ce·tic (thī'ō·ə·sē'tik, -ə·set'ik) *adj. Chem.* Designating a yellow, fuming, pungent acid, C_2H_4OS, used in ammonia solutions as a precipitant of metals. [<THIO– + ACETIC]

thi·o·al·co·hol (thī'ō·al'kə·hôl) *n. Chem.* Thiol.

thi·o·al·de·hyde (thī'ō·al'də·hīd) *n. Chem.* An aldehyde containing sulfur as a substitute for oxygen.

thi·o·bac·te·ri·um (thī'ō·bak·tir'ē·əm) *n.* Any

add,āce,câre,pälm; end,ēven; it,īce; odd,ōpen,ôrder; tŏŏk,pōōl; up,bûrn; ə = a in *above,* e in *sicken,* i in *clarity,* o in *melon,* u in *focus;* yōō = u in *fuse;* oi,oil; ou,pout; ch,check; g,go; ng,ring; th,thin; th,this; zh,vision. Foreign sounds à,œ,ü,kh,ṅ; and ◆: see page xx. < from; + plus; ? possibly.

of an order *(Thiobacteriales)* of bacteria which utilize the sulfur of decaying organic matter.

thi·o·car·bam·ide (thī′ō·kär·bam′īd, -id, -kär′·bə·mīd) *n.* Thiourea. [<THIO- + CARBAMIDE]

thi·o·cy·a·nate (thī′ō·sī′ə·nāt) *n. Chem.* A salt or ester of thiocyanic acid.

thi·o·cy·an·ic (thī′ō·sī·an′ik) *adj. Chem.* Designating or pertaining to a colorless liquid acid, HSCN, soluble in water and having a pungent odor. [<THIO- + CYANIC]

thi·o·gen (thī′ə·jen) *n.* A bacterial organism producing sulfur. [<THIO- + -GEN]

Thi·o·kol (thī′ə·kōl, -kol) *n.* A synthetic material consisting of organic polysulfides and resembling natural rubber in its physical properties: a trade name.

thi·ol (thī′ōl, -ol) *n. Chem.* Any of a class of sulfur compounds which are analogs of the alcohols and have the general formula RSH, in which R is a hydrocarbon radical: used largely in compounding, as *ethanethiol,* C_2H_5SH. Formerly called *mercaptan.* [<THI- + -OL[1]]

thi·on·ic (thī·on′ik) *adj. Chem.* **1** Of, pertaining to, containing, or derived from sulfur. **2** Denoting any of a group of unstable acids having the general formula $H_2S_nO_6$. [<Gk. *theion* sulfur]

thi·o·nine (thī′ə·nēn, -nin) *n. Chem.* A dark-green thiazine derivative, $C_{12}H_9N_3S$, made by synthesis, with a glistening metallic luster that yields purplish colors to silk and wool. Also **thi′o·nin** (-nin). [<Gk. *theion* sulfur + -INE[2]]

thi·o·nyl (thī′ə·nil) *n. Chem.* The bivalent sulfur radical SO: also called *sulfinyl.* [<Gk. *theion* sulfur + -YL]

thi·o·phene (thī′ə·fēn) *n. Chem.* A colorless liquid hydrocarbon, C_4H_4S, with an odor resembling that of benzene, found in coal tar and also made by synthesis. Also **thi′o·phen** (-fen). [<THIO- + PH(ENYL) + -ENE]

thi·o·sin·am·ine (thī′ō·sin·am′in, -sin′ə·mēn) *n. Chem.* A crystalline compound, $C_4H_8N_2S$, formed by the union of allyl mustard oil and alcohol with ammonia: used in photography. Also **thi′o·sin·am′in** (-am′in). [<THIO- + Gk. *sin(api)* mustard + AMINE]

Thi·o·spi·ril·lum (thī′ō·spī·ril′əm) *n.* A genus of motile, sulfur-containing bacteria found in fresh or salt water. [<THIO- + SPIRILLUM]

thi·o·sul·fate (thī′ō·sul′fāt) *n. Chem.* A salt of thiosulfuric acid.

thi·o·sul·fu·ric (thī′ō·sul·fyŏŏr′ik) *adj. Chem.* Designating or pertaining to an unstable acid, $H_2S_2O_3$, known chiefly by its salts, which have extensive applications in bleaching and photography.

thi·o·u·re·a (thī′ō·yŏŏ·rē′ə) *n. Chem.* A white, solid compound, NH_2CSNH_2, prepared from urea by replacement of oxygen by sulfur: used in organic synthesis, in photography, and as an insecticide: also called *thiocarbamide.* [<THIO- + UREA]

thir (thûr, thir) *pron. Scot.* These.

third (thûrd) *adj.* **1** Next in order after second: the ordinal of *three.* **2** Being one of three equal parts. —*n.* **1** One of three equal parts of anything. **2** The person or thing coming after the second, as in a series. **3** *pl. Law* The third part of a husband's personal estate, allotted to the widow in case of his dying intestate and leaving an heir; also, loosely, a dower. **4** A unit of time or of an arc, equal to one sixtieth of a second. **5** *Music* **a** The interval between any note and the next note but one above it on a diatonic scale, known as a **major third** when such interval is two whole steps or degrees of the staff, and as a **minor third** when it is a step and a half. **b** A note separated by this interval from any other, considered in relation to that other; specifically, the third above the keynote. **c** Two notes at this interval written or sounded together, or the consonance so produced. **6** In baseball, the third base. —*adv.* In the third order, rank, or place: also, in formal discourse, **third′ly.** [OE *thridda* < *thrī* three]

third base In baseball, the third base reached by the runner, at the left-hand angle of the infield. Compare illustration under BASEBALL.

third class **1** In the U. S. postal system, a classification of mail that includes all miscellaneous printed matter but not newspapers and periodicals legally entered as second class. **2** A classification of accommodations on some ships and trains, usually the cheapest and least luxurious available; formerly, on a ship, steerage; also, the passengers traveling in this classification.

third degree **1** *Colloq.* Severe or brutal examination of a prisoner by the police for the purpose of securing information or a confession; hence, any brutal treatment. **2** In Freemasonry, the degree of Master Mason.

third estate The commons or common people; the third political class of a kingdom, following the nobility and the clergy. See under ESTATE.

third eyelid The nictitating membrane.

Third Order *Eccl.* A confraternity, generally for laymen, associated with a religious order and following a modified rule. [after the *Third Order* of St. Francis, founded 1221]

third person The person or thing spoken of, or the grammatical form indicating such person or thing.

third rail An insulated rail placed as a conductor on the track of an electric railway, from which the current is taken by means of a contact device, the running rails acting as return conductors. —**third′-rail′** *adj.*

Third Reich See under REICH.

thirl[1] (thûrl) *v.t. Scot.* & *Brit. Dial.* **1** To thrill. **2** To drill or bore. [OE *thyrlian* < *thȳrel* hole < *thurh* through]

thirl[2] (thûrl) *v.t. Scot.* To bind, as by a lease. —*n.* Thirlage.

thirl·age (thûr′lij) *n.* A feudal obligation upon certain tenants or the inhabitants of certain districts to bring their grain to a certain mill for grinding; also, the fee for such grinding. Also *thirl.* [Metathetic var. of obs. *thrillage* < obs. *thrill* enthrall < OE *thræl* thrall]

thirst (thûrst) *n.* **1** A distressful feeling of dryness in the throat and mouth, accompanied by an increasingly urgent desire for liquids. **2** The physiological condition which produces this feeling. **3** Any eager desire; a longing or craving: a *thirst* for glory. See synonyms under APPETITE. —*v.i.* **1** To feel thirst; be thirsty. **2** To have an eager desire or craving; long; yearn. [OE *thurst*] —**thirst′er** *n.*

thirst·y (thûrs′tē) *adj.* **thirst·i·er, thirst·i·est** **1** Affected with thirst. **2** Lacking moisture; arid; parched. **3** Eagerly desirous. **4** *Colloq.* Causing thirst. [OE *thurstig*] —**thirst′i·ly** *adv.* —**thirst′i·ness** *n.*

thir·teen (thûr′tēn′) *n.* The cardinal number preceding fourteen and following twelve, or any of the symbols (13, xiii, XIII) which represent it. —*adj.* Consisting of or being one more than twelve. [OE *thrēotēne*]

thirteen fires The original thirteen States of the United States: a North American Indian term.

thir·teenth (thûr′tēnth′) *adj.* **1** Third in order after the tenth: the ordinal of *thirteen.* **2** Being one of thirteen equal parts. —*n.* **1** One of thirteen equal parts. **2** The next one after the twelfth.

thir·ti·eth (thûr′tē·ith) *adj.* **1** Tenth in order after the twentieth: the ordinal of *thirty.* **2** Being one of thirty equal parts. —*n.* **1** One of thirty equal parts of anything. **2** The tenth in order after the twentieth.

thir·ty (thûr′tē) *n.* The cardinal number preceding thirty-one and following twenty-nine; thrice ten; also, any of the symbols (30, xxx, XXX) used to represent it. —*adj.* Consisting of or being ten more than twenty, or thrice ten; tricennial. [OE *thrītig*]

thir·ty-sec·ond (thûr′tē·sek′ənd) *adj.* **1** Being the second after the thirtieth. **2** Being one of thirty-two equal parts. —*n.* A thirty-second note.

thirty-second note *Music* A note having one thirty-second of the time of a whole note; a demisemiquaver.

thir·ty-two·mo (thûr′tē·tōō′mō) *n. pl.* **·mos** A sheet of paper folded so as to make 32 leaves about 3 1/8 by 4 3/4 inches; hence a book or pamphlet having 32 leaves to the sheet. —*adj.* Having 32 leaves to a sheet. Commonly written *32mo.*

Thirty Years' War See table under WAR.

this (this) *adj. pl.* **these** **1** That is near or present, either actually or in thought: *This* house is for sale; I shall be there *this* evening. **2** That is understood or has just been mentioned: *This* offense justified my revenge. **3** That is nearer than or contrasted with something else: opposed to *that*: *This* tree is still alive, but that one is dead; He ran *this* way and that. **4** These: used of a number or collection considered as a whole: He has been dead *this* fourteen nights. —*pron.* **1** The person or thing near or present, being understood or just mentioned: *This* is where I live; *This* is the guilty man. **2** The person or thing nearer than or contrasted with something else: opposed to *that*: *This* is a better painting than that. **3** The idea, statement, etc., about to be made clear: I will say *this*: he is a hard worker. —*adv.* To this degree; thus or so: I was not expecting you *this* soon. [OE]

This·be (thiz′bē) See PYRAMUS AND THISBE.

this·tle (this′əl) *n.* **1** One of various vigorous prickly plants (genera *Carduus, Cirsium, Cnicus,* and *Onopordum*) with cylindrical or globular heads of tubular purple flowers; especially, the **bull thistle** *(Cirsium lanceolatum)* of Scotland, and the **Canada thistle** *(Cirsium canadense).* **2** Any of several prickly plants of other genera. [OE *thistel*] —**this′·tly** *adj.*

thistle butterfly A butterfly *(Vanessa cardui)* resembling the painted beauty but having usually four eyespots on the under side of each wing: also called *painted lady.*

this·tle·down (this′əl·doun′) *n.* The pappus of a thistle; the ripe silky fibers from the dry flower of a thistle.

thith·er (thith′ər, thith′-) *adv.* **1** To that place; in that direction: opposed to *hither.* **2** *Archaic* To that end, point, or result. —*adj.* Situated or being on the other side; farther; more distant: the *thither* bank of the river. [OE *thider*]

thith·er·to (thith′ər·tōō′, thith′-) *adv.* Up to that time.

thith·er·ward (thith′ər·wərd, thith′-) *adv.* In that direction; toward that place. Also **thith′·er·wards.**

thix·ot·ro·py (thik·sot′rə·pē) *n. Chem.* The property possessed by certain gels of liquefying under the action of vibrating forces. [<Gk. *thixis* touch + *tropē* turning] —**thix·o·trop·ic** (thik′sə·trop′ik) *adj.*

tho (thō) See THOUGH.

thole[1] (thōl) *n.* A pin or pair of pins serving as a fulcrum for an oar in rowing. Also **thole pin.** [OE *thol* pin]

thole[2] (thōl) *v.t.* & *v.i. Archaic* To endure; suffer; tolerate. [OE *tholian* suffer]

Thom·as (tom′əs; *Dan., Du., Ger., Sw.* tō′mäs; *Fr.* tō·mä′) A masculine personal name. [< Hebrew, twin]

—**Thomas** One of the Twelve Apostles, known for his doubting disposition. *John* xx 25.

—**Thomas à Becket** See BECKET.

—**Thomas à Kempis** See KEMPIS.

—**Thomas of Er·cel·doune** (ûr′səl·dōōn), 1220?–97, Scottish seer and poet: best known as *Thomas the Rhymer.*

Tho·mas (tō·mä′), **Ambroise,** 1811–96, French composer.

Tho·mas (tom′əs), **Dylan,** 1914–53, Welsh poet and author. —**George Henry,** 1816–70, U. S. general. —**Norman,** born 1884, U. S. socialist leader and writer. —**Seth,** 1785–1859, U. S. clock manufacturer. —**Theodore,** 1831?–1905, U. S. orchestra conductor born in Germany.

Tho·mism (tō′miz·əm, thō′-) *n.* The doctrine of St. Thomas Aquinas, who attempted to combine Aristotelian metaphysics, ontology, logic, and method with Christian theology into one comprehensive system, including theology, natural philosophy, esthetics, ethics, psychology, and politics. He held that human reason was the faculty by which men apprehended many truths, but that the divinely revealed truths necessary for salvation could be known only through faith; that reason was distinct from faith, though not opposed to it when rightly used; and that reason served faith by preparing men's minds to receive revealed truth, by expounding and systematizing that truth, and by defending it against attack. The system of dogmatic theology constructed by St. Thomas remains the standard within the Roman Catholic Church, and has had a wide influence in many other communions. —**Tho′·mist** *adj.* & *n.* —**Tho·mis′·tic** or **·ti·cal** *adj.*

Thomp·son (tomp′sən), **Benjamin** See RUMFORD, COUNT. —**Francis,** 1859–1907, English poet.

Thomp·son River (tomp′sən) A river in southern British Columbia, Canada, flowing 304

Thompson submachine gun   1307   **thought**

miles west and south to the Fraser River.
Thompson submachine gun See under SUB-MACHINE GUN.
Thom·son (tom'sən), **George Paget,** born 1892, English physicist; son of Sir Joseph John. — **James,** 1700-48, Scottish poet. — **Sir John Arthur,** 1861-1933, Scottish biologist. — **Sir Joseph John,** 1856-1940, English physicist.
thong (thông, thong) n. 1 A narrow strip, properly of leather, as for tying or fastening. 2 A whiplash. [OE *thwang* thong]
Thor (thôr, tōr) In Norse mythology, the god of war, thunder, and strength, and son of Odin: he destroyed the enemies of the gods with his magic hammer.
tho·rac·ic (thō·ras'ik, thô-) adj. Of, relating to, or situated in or near the thorax. [<NL *thoracicus* <Gk. *thōrax* the chest]
thoracic duct Anat. The canal emptying into the left subclavian vein which collects the lymph from parts of the body below the diaphragm.
thoraco- *combining form* Med. & Surg. The thorax or the chest; of or related to the thorax: *thoracotomy.* Also, before vowels, **thorac-.** [<Gk. *thōrax* the chest]
tho·ra·co·plas·ty (thôr'ə·kō·plas'tē, thō'rə-) n. Surg. An operation for the removal and replacement of several ribs in order to provide a thoracic cavity within which the underlying lung is kept permanently collapsed: used in the treatment of tuberculosis. [<THORACO- + -PLASTY]
tho·ra·cot·o·my (thôr'ə·kot'ə·mē, thō'rə-) n. Surg. Incision of the wall of the chest. [<THORACO- + -TOMY]
tho·rax (thôr'aks, thō'raks) n. pl. **tho·rax·es** or **tho·ra·ces** (thôr'ə·sēz, thō'rə-) 1 Anat. The part of the body between the neck and the abdomen, enclosed by the ribs. 2 Entomol. The middle region of the body of an insect, between the head and the abdomen. 3 Zool. The corresponding region of the body in other arthropods. [<L <Gk. *thōrax*]
Tho·reau (thôr'ō, thō'rō, thə·rō'), **Henry David,** 1817-1862, U.S. author.
Tho·rez (tō·rez'), **Maurice,** born 1900, French Communist.
tho·ri·a (thôr'ē·ə, thō'rē·ə) n. A white, very heavy oxide of thorium, ThO₂, used with zirconia and other earths in the mantle of Welsbach's incandescent lamp. [<NL <*thorium* THORIUM]
tho·ri·a·nite (thôr'ē·ə·nīt, thō'rē-) n. A black radioactive mineral composed chiefly of thorium, cerium, and uranium oxides.
tho·rite (thôr'īt, thō'rīt) n. A vitreous, yellow to black, thorium silicate, ThSiO₄. [<THOR(IUM) + -ITE]
tho·ri·um (thôr'ē·əm, thō'rē-) n. A gray, radioactive, metallic element (symbol Th) of the actinide series, found only in small quantities in certain rare minerals. Its oxide, ThO₂, is used in the manufacture of gas mantles, and its isotope of mass 232 has been used in the generation of atomic energy. See ELEMENT. [after *Thor*] — **tho'ric** adj.
thorium series Physics The group of radioactive elements beginning with thorium of mass 232 and a half-life of 1.39×10^{10} years, with successive disintegrations terminating in the stable isotope of lead of mass 208.
thorn (thôrn) n. 1 An indurated, leafless spine or sharp-pointed process from a branch. 2 One of various other sharp processes, as the spine of a porcupine. 3 Any of various thorn-bearing shrubs or trees; especially, a genus (*Crataegus*) of rosaceous plants, the hawthorn. 4 Anything or anyone that occasions discomfort, pain, or annoyance; a vexation. 5 The name of the Old English rune Þ; also, the corresponding Icelandic character: equivalent originally to *th*, both voiced and unvoiced, but finally only to the unvoiced sound, as in *thorn*, from which it derives its name. *Y* or *y* is sometimes used as a makeshift for it in early English, as in the contraction yᵉ. Compare EDH. — *v.t.* To pierce or prick with a thorn. [OE] — **thorn'less** adj.
Thorn (thôrn) The German name for TORUŃ.
thorn apple 1 Jimsonweed: so called from its spiny seedcase. 2 Any plant of the same genus. 3 The fruit of the hawthorn; a haw.
thorn·back (thôrn'bak') n. 1 A European ray (*Raia clavata*) whose back is studded with short stout spines. 2 The common European spider crab (*Maia squinado*). 3 Any of certain American skates or sticklebacks.
thorn·bill (thôrn'bil') n. Any of certain brightly-colored hummingbirds of South America (genera *Rhamphomicron* and *Chalcostigma*) characterized by a long, sharp bill.
thorn broom The furze.
thorn tree 1 The hawthorn. 2 The honey locust.
thorn·y (thôr'nē) adj. **thorn·i·er, thorn·i·est** 1 Full of thorns; spiny. 2 Sharp like a thorn, literally or figuratively; painful; vexatious; presenting difficulties or trials. — **thorn'i·ness** n.
Thorn·dike (thôrn'dīk), **Ashley,** 1871-1933, U.S. educator. — **Edward Lee,** 1874-1949, U.S. psychologist; brother of the preceding. — **Lynn,** born 1882, U.S. historian; brother of the preceding. — **Dame Sybil,** born 1882, English actress.
tho·ron (thôr'on, thō'ron) n. A gaseous radioactive emanation produced during the atomic disintegration of thorium: it has a half-life of 54.5 seconds. [<NL <THOR(IUM) + -on, as in *neon*]
thor·ough (thûr'ō) adj. 1 Carried to completion; thoroughgoing: a *thorough* search; also, carrying (a task) to completion; persevering: a very *thorough* worker. 2 Marked by careful attention throughout; not superficial; hence, complete; perfect. 3 Completely (such and such); through and through: a *thorough* nincompoop. 4 Painstakingly conforming to a standard. 5 Obs. Going or passing through. See synonyms under RADICAL. — *adv. & prep. Obs.* Through. Also *Obs.* **thor'o.** [Emphatic var. of THROUGH] — **thor'ough·ly** adv. — **thor'ough·ness** n.
Thor·ough (thûr'ō) n. The administrative policy of Charles I's minister, the Earl of Strafford: so called by himself as being a method of carrying through his ideas in spite of all opposition.

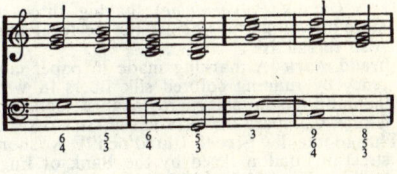

THOROUGH–BASS
The numbers under the bass indicate the notes of the chords in the treble.

thor·ough-bass (thûr'ō·bās') n. Music 1 A bass part accompanied by shorthand marks, as numerals, below the staff, to indicate the general harmony: now disused. 2 Loosely, the science of harmony or the art of harmonic composition.
thorough brace A strong leather strap extending under each side of the body of a carriage and serving as a support and a spring. — **thor'ough-braced'** adj.
thor·ough·bred (thûr'ō·bred', thûr'ə-) n. 1 Pure stock. 2 *Colloq.* A person of culture and good breeding. — *adj.* 1 Belonging to the strain of horses known as Thoroughbred. 2 Bred from pure stock. 3 Possessing the traits of a thoroughbred.
Thor·ough·bred (thûr'ō·bred') n. A horse whose ancestry is recorded in the English Stud Book, and which is therefore descended from one of three Eastern sires: the Byerly Turk, the Darley Arabian, or the Godolphin.
thor·ough·fare (thûr'ō·fâr', thûr'ə-) n. 1 A frequented way or course; especially, a road or street through which the public have unobstructed passage; highway. 2 A traveling or passing through, or the right or possibility of doing so; a passage: now chiefly in the phrase *no thoroughfare.* 3 An outlet to an enclosed place, as to a court. 4 Any place through which much traffic passes, as a strait, river, or other waterway. See synonyms under ROAD, WAY. [ME *thurghfare* <OE *thurh* through + *faru* going]
thor·ough·go·ing (thûr'ō·gō'ing, thûr'ə-) adj. 1 Characterized by extreme thoroughness or efficiency. 2 Unmitigated: a *thoroughgoing* scoundrel.
thor·ough-paced (thûr'ō·pāst', thûr'ə-) adj. 1 Perfectly trained, as a horse. 2 Hence, thoroughgoing; accomplished: a *thoroughpaced* villain.
thor·ough-pin (thûr'ō·pin') n. Dropsical swelling of the sheath of the tendon of a flexor muscle connected with the hock of a horse: it appears on both sides of the leg, as if the latter had been pierced by a pin. Also **thor'ough-shot'** (-shot').
thor·ough·wort (thûr'ō·wûrt', thûr'ə-) n. 1 A stout, hairy herb, the boneset, 2 to 5 feet high, with white flowers, common in the United States and Canada. 2 Any other eupatorium.
thorp (thôrp) n. A hamlet; small cluster of houses in the country: now chiefly in names of places. Also **thorpe.** [OE. Akin to DORP.]
Thors·havn (tôrs'houn') Capital of the Faeroe Islands, in the central part of the group.
Thor·vald·sen (tôr'väl·sən), **Albert Bertel,** 1770-1844, Danish sculptor. Also **Thor'wald·sen.**
those (thōz) adj. & pron. Plural of THAT. [OE *thās*]
Thoth (thōth, tōt) In Egyptian mythology, the god of wisdom, inventor of art, science, and letters: identified with the Greek Hermes Trismegistus: represented with the head of an ibis or a dog.
Thoth·mes (thōth'mēz, tōt'mes) Any of several Egyptian kings, between 1587-1328 B.C.: also **Thuthmose.**
thou (thou) *pron.* The person spoken to, as denoted in the nominative case: archaic except in Biblical, homiletic, elevated, or poetic language, in prayers to a deity, or in certain dialects. [OE *thū*]

THOTH

though (thō) *conj.* 1 Notwithstanding the fact that: introducing a clause expressing an actual fact. 2 Conceding or granting that; even if: introducing a clause assumed or admitted as supposedly true. 3 And yet; still; however: introducing a modifying clause or statement added as an afterthought: I am well, *though* I do not feel very strong. 4 Notwithstanding what has been done or said; nevertheless: But they have, *though.* As used in this sense, *though* is sometimes regarded as a conjunctive adverb. Also spelled *tho.* Compare HOWEVER. See synonyms under BUT¹. [Prob. fusion of OE *thēah* and ON *thō*]
thought¹ (thôt) n. 1 The act or process of using the mind actively and deliberately; meditation; cogitation. 2 The product of thinking; an idea, concept, judgment, opinion, or the like. 3 Intellectual activity of a specific kind: Greek *thought.* 4 Consideration; attention; heed: to take *thought* on how to do something. 5 Intention or idea of doing something; plan; design: All *thought* of returning was abandoned. 6 Expectation; an-

---

add, āce, câre, päm; end, ēven; it, īce; odd, ōpen, ôrder; tōōk, pōōl; up, bûrn; ə = a in *above*, e in *sicken*, i in *clarity*, o in *melon*, u in *focus*; yōō = u in *fuse*; oi, oil; ou, pout; ch, check; g, go; ng, ring; th, thin; ᴛʜ, this; zh, vision. Foreign sounds á, œ, ü, ń, ḣ; and ♦: see page xx. < from; + plus; ? possibly.

ticipation: He had no *thought* of finding her there. 7 A trifle; a small amount: Be a *thought* more cautious. [<THOUGHT[2]]
Synonyms: cogitation, conception, conclusion, consideration, contemplation, deliberation, fancy, idea, imagination, judgment, meditation, musing, notion, opinion, reflection, reverie, speculation, study, supposition, thinking, view. See IDEA, MIND, REFLECTION.

thought[2] (thôt) Past tense and past participle of THINK. [OE *thōht*]

thought·ful (thôt′fəl) *adj.* 1 Full of thought; meditative: a *thoughtful* face. 2 Showing, characterized by, or employed in thought; promotive of thought. 3 Attentive; careful; especially, manifesting regard for others; considerate: often with *of* or an infinitive: *thoughtful* of one's reputation; *thoughtful* to lay up a store for winter. — **thought′ful·ly** *adv.* — **thought′ful·ness** *n.*
Synonyms: attentive, careful, circumspect, considerate, heedful, mindful, provident. An *attentive* person waits upon another to supply what is needed or desired. A *thoughtful* person provides in advance for needs and wishes not yet manifested. A *considerate* person carefully spares another all that would harm, grieve, or annoy; one who is *circumspect* carefully avoids all that might compromise himself. See SEDATE. *Antonyms*: careless, gay, giddy, heedless, inadvertent, inattentive, inconsiderate, neglectful, negligent, reckless, remiss.

thought·less (thôt′lis) *adj.* 1 Manifesting lack of thought or care; heedless; also, giddy. 2 Stupid. See synonyms under IMPROVIDENT, IMPRUDENT. — **thought′less·ly** *adv.* — **thought′·less·ness** *n.*

thought–trans·fer·ence (thôt′trans·fûr′əns) *n.* Telepathy.

thou·sand (thou′zənd) *n.* The cardinal number following 999; one hundred times ten, or any of the symbols (1,000, m, M) used to represent it; also, loosely, an indefinitely large number. — *adj.* Consisting of a hundred times ten; millenary. [OE *thusend*] — **thou′sand·fold′** (-fōld′) *adj.* & *adv.*

Thousand Islands A group of 1,500 islets in an expansion of the upper St. Lawrence River, near Lake Ontario.

thou·sandth (thou′zəndth) *adj.* 1 Last in a series of a thousand: an ordinal numeral. 2 Being one of a thousand equal parts. — *n.* 1 One of a thousand equal parts. 2 The next in order after the 999th.

thowe (thō) *n. Scot.* Thaw. Also **thow.**

thow·less (thou′lis) *adj. Scot.* Inactive; lazy; without ambition or energy. See THEWLESS.

Thrace (thrās) An ancient region, later a Roman province, NE of Macedonia in the eastern part of the Balkan Peninsula: modern Thrace is divided into a Greek division (3,315 square miles); and a Turkish division corresponding to Turkey in Europe. Ancient **Thra·cia** (thrā′shə).

Thra·cian (thrā′shən) *adj.* Pertaining to Thrace or its people. — *n.* 1 One of the people of Thrace. 2 The Indo-European language of the ancient Thracians, related to Phrygian. 3 A gladiator who fought in the native dress of the Thracians. See illustration under GLADIATOR.

Thracian Chersonese An ancient name for GALLIPOLI PENINSULA.

thral·dom (thrôl′dəm) *n.* 1 The state of being a thrall. 2 Figuratively, any sort of bondage or servitude. See synonyms under BONDAGE. Also **thrall′dom.**

Thrale (thrāl), **Mrs.** See PIOZZI, HESTER LYNCH.

thrall (thrôl) *n.* 1 A person in bondage; a slave; serf; hence, figuratively, one controlled by a passion or vice. 2 The condition of bondage; thraldom. — *v.t. Archaic* To reduce to thraldom; enslave. — *adj.* Held in subjection; enslaved. [OE *thræl* <ON]

thrang (thräng) *Scot. adj.* Occupied fully; busy. — *n.* A throng; crowd.

thrash (thrash) *v.t.* 1 To thresh, as grain. 2 To beat as if with a flail; flog; whip. 3 To defeat utterly. — *v.i.* 4 To move or swing about with flailing, violent motions. 5 *Naut.* To work to windward, against the tide, etc. See synonyms under BEAT. — **to thrash out** To discuss fully and to a conclusion. — *n.* 1 The act of thrashing. 2 In swimming, a kick used with the crawl and back strokes. [Dial. var. of THRESH]

thrash·er[1] (thrash′ər) *n.* 1 One who or that which thrashes. 2 *Agric.* A threshing machine. 3 The thresher shark.

thrash·er[2] (thrash′ər) *n.* Any of several long-tailed American songbirds (genus *Toxostoma*) resembling the thrushes and related to the mockingbirds, especially the common eastern **brown thrasher** (*T. rufum*), colored foxy-red with black spots. [<dial. E *thresher* <THRUSH[1]]

thrash·ing (thrash′ing) *n.* A sound beating or whipping.

thra·son·i·cal (thrā·son′i·kəl) *adj.* Characterized by boasting or ostentation; bragging; boastful. [<L *Thraso*, a braggart soldier in Terence's *Eunuch* <Gk. *Thrason* < *thrasus* rash] — **thra·son′i·cal·ly** *adv.*

Thras·y·bu·lus (thras′ə·byōō′ləs) Greek patriot and naval commander, died 389 B.C.

thrave (thrāv) *n. Scot.* & *Brit. Dial.* 1 Twenty-four sheaves of grain. 2 An indefinite number; a company; throng; also, a bundle.

thraw[1] (thrô) *Scot. n.* 1 A wrench or twist. 2 A throe. — *v.t.* 1 To twist or wrench. 2 To thwart; frustrate. — *adj.* Awry.

thraw[2] (thrô) *v.* & *n. Scot.* & *Brit. Dial.* Throw.

thrawn (thrôn) *adj. Scot.* 1 Wrenched; awry; twisted; crooked. 2 Obstinate; contrary.

thread (thred) *n.* 1 A very slender cord or line composed of two or more yarns or filaments, as of flax, cotton, silk, or other fibrous substance, twisted together. 2 A filament of any substance, as of metal, glass, or tissue; a hair. 3 A fine stream or beam: a *thread* of light. 4 A fine line of color. 5 Anything suggestive of a thread; something that runs a continuous course through a series, serving to give sequence to the whole: the *thread* of his discourse. 6 *Mining* A very thin seam or vein of ore. 7 *Mech.* The spiral ridge of a screw. 8 Thread of life. — *v.t.* 1 To pass a thread through the eye of: to *thread* a needle. 2 To arrange or string on a thread, as beads. 3 To cut a thread on or in, as a screw. 4 To make one's way through or over: to *thread* a maze. 5 To make (one's way) carefully. 6 To be present throughout; pervade. — *v.i.* 7 To make one's way carefully; step. 8 To drop from a fork or spoon in a fine thread: said of boiling sirup when it has reached a certain consistency. — *adj.* Pertaining to, resembling, or made of thread; filar. [OE *thrǣd*] — **thread′er** *n.* — **thread′like′** *adj.*

thread·bare (thred′bâr′) *adj.* 1 Worn so that the threads show, as a rug or garment. 2 Clad in worn garments. 3 Commonplace; hackneyed. See synonyms under COMMON, TRITE. — **thread′bare′ness** *n.*

thread feather *Ornithol.* An extremely slender feather, having the vane rudimentary or absent; filoplume.

thread·fin (thred′fin′) *n.* A fish of tropical seas (family *Polynemidae*), having three or more threadlike rays below the pectoral fins. Also **thread′fish′.**

thread mark A marking made in paper currency by running colored silk fibers in with the pulp, as a safeguard against counterfeiting. Compare GRANITE PAPER, SILK PAPER.

Thread·nee·dle Street (thred′nēd′l) A short street in London faced by the Bank of England. — **The Old Lady of Threadneedle Street** The Bank of England.

thread of life The course of existence, represented by the ancient Greeks and Romans as a thread being spun and cut off by the three Fates, Atropos being the one who cut it.

thread·worm (thred′wûrm′) *n.* A threadlike nematode worm; a pinworm or filaria.

thread·y (thred′ē) *adj.* 1 Resembling a thread; filamentous; tenuous. 2 Consisting of, containing, or covered with thread.

threap (thrēp) *v.t.* & *v.i. Scot.* & *Brit. Dial.* To contradict; dispute; also, to rebuke; insist. Also **threep.**

threat (thret) *n.* 1 A declaration of an intention to inflict injury or pain; a menace. 2 An announcement or omen of impending danger or evil. 3 A menace or danger of any sort. — *v.t. Archaic* To threaten. [OE *thrēat* crowd, oppression]

threat·en (thret′n) *v.t.* 1 To utter menaces or threats against. 2 To be menacing or dangerous to. 3 To be ominous or portentous of. 4 To utter threats of (injury, vengeance, etc.). — *v.i.* 5 To utter threats. 6 To have a menacing aspect; lower: The rising waters seemed to *threaten.* [OE *thrēatnian* urge, compel] — **threat′en·er** *n.* — **threat′en·ing·ly** *adv.*
Synonym: menace. *Threaten* is applied alike to vast and trivial matters; *menace* only to those of moment. Either persons or things may *threaten*; *menace* is chiefly used of persons or of things personified. One may *threaten* by word or act; *menace* is for the most part limited to actions or concrete things; one *threatens* another with death; he *menaces* him with a revolver.

three (thrē) *n.* 1 The cardinal number following two and preceding four, or any of the symbols (3, iii, III) used to represent it. 2 Any group of three persons or things; a playing card with three pips. — *adj.* Being one more than two; ternary. [OE *thrī*]

three-base hit (thrē′bās′) A fair hit in baseball that enables the batter to reach third base without the help of an error. Also **three′-bag′ger** (-bag′ər).

three-cent piece (thrē′sent′) A copper and nickel coin of the United States from 1865–1890.

three-col·or (thrē′kul′ər) *adj.* Pertaining to or denoting a process of color printing based on three primary colors, each of which is transferred to the printing surface from a separate, accurately registered plate.

three-deck·er (thrē′dek′ər) *n.* 1 A vessel having three decks or gun decks. 2 Any structure having three levels. 3 A sandwich made with three slices of bread.

three·fold (thrē′fōld′) *adj.* Made up of three; three times as many or as great; triplicate. — *adv.* Triply; in a threefold manner or degree.

three-mile limit (thrē′mīl′) See under LIMIT.

three·pence (thrip′əns, threp′-, thrup′-) *n. Brit.* 1 The sum of three pennies. 2 A small coin of Great Britain, made of alloy, formerly of silver, worth three pennies: also **threepenny bit.**

three·pen·ny (thrip′ə·ni, threp′-, thrup′-, thrē′pen′ē) *adj. Brit.* 1 Worth or costing threepence. 2 Hence, of little value.

three-phase (thrē′fāz′) *adj. Electr.* Designating a combination of alternating currents or circuits each of which differs in phase by one third of a cycle or 120 degrees.

three-piled (thrē′pīld′) *adj.* 1 Having a triple pile or nap: said of velvet; also, figuratively, costly or extravagant. 2 Clad in or wearing such velvet; hence, wealthy. 3 Piled in a set or sets of three.

three-ply (thrē′plī′) *adj.* Consisting of three thicknesses, strands, layers, etc.

three-point landing (thrē′point′) 1 *Aeron.* A perfect airplane landing, with the front wheels and tail skid or wheel touching the ground simultaneously. 2 Any successful outcome.

three-quar·ter binding (thrē′kwôr′tər) A style of bookbinding having the strip of leather over the back and corners projecting to a greater width than in half-binding.

Three Rivers The English name for TROIS RIVIÈRES, Canada.

three R's See under R.

three·score (thrē′skôr′, -skōr′) *adj.* & *n.* Sixty.

three·some (thrē′səm) *adj.* Performed by three; triple: a *threesome* reel. — *n.* A golf match in which one plays against two, the latter playing one ball between them alternately.

three-square (thrē′skwâr′) *adj.* Having three plane faces of equal width: said especially of certain files of triangular cross-section.

threm·ma·tol·o·gy (threm′ə·tol′ə·jē) *n.* The science of breeding animals and plants. [< Gk. *thremma, -atos* a nursling + -LOGY]

thren·o·dy (thren′ə·dē) *n. pl.* **·dies** An ode or song of lamentation; a dirge. Also **thren′ode** (-ōd). [<Gk. *thrēnōidia* < *thrēnos* lament + *ōidē* song] — **thre·no·di·al** (thri·nō′dē·əl), **thre·nod·ic** (thri·nod′ik) *adj.* — **thren′o·dist** *n.*

thre·o·nine (thrē′ə·nēn, -nin) *n. Biochem.* A crystalline amino acid, $C_4H_9NO_3$, isolated as a product of the hydrolysis of certain proteins and regarded as an essential to proper nutrition.

thresh (thresh) *v.t.* 1 To beat stalks of (ripened grain) with a flail or machine so as to separate the grain from the straw or husks. 2 To beat; flog. — *v.i.* 3 To thresh grain. 4 To move or thrash about. — **to thresh out** (or **over**) To discuss fully and to a conclusion. — *n.* The act of threshing; a threshing. [OE *therscan*] — **thresh′ing** *n.*

thresh·er (thresh′ər) n. 1 One who or that which threshes; specifically, a machine for threshing. 2 A large shark (*Alopias vulpes*) of warm seas, having the dorsal lobe of the tail extremely long, supposedly for splashing the water to round up its prey: also **thresher shark.**

thresh·old (thresh′ōld, -hōld) n. 1 The plank, timber, or stone lying under the door of a building; doorsill. 2 The entrance, entering point, or beginning of anything: the *threshold* of the 20th century. 3 *Physiol.* The point at which a stimulus, as of a nerve or muscle, just produces a response; especially, the minimum degree of stimulation necessary for conscious perception: also called *limen.* ◆ Collateral adjective: *liminal.* [OE *therscold*]

threw (thrōō) Past tense of THROW.

thrice (thrīs) adv. 1 Three times. 2 In a three-fold manner; hence, fully; repeatedly. [ME *thries* <OE *thriwa* thrice + *-s*³]

thrift (thrift) n. 1 Care and wisdom in the management of one's resources; frugality. 2 A flourishing condition; vigorous growth, as of a plant. 3 Any of a genus (*Armeria*, formerly *Statice*) of tufted herbs of the north temperate zone growing on mountains and the seashore; especially, the common thrift (*A. maritima*), having white or pink flower heads. 4 *Obs.* The state of one who thrives; prosperity. 5 *Scot. & Brit. Dial.* Effort; occupation; work. [<ON. Akin to THRIVE.] — **thrift′less** adj. — **thrift′less·ly** adv. — **thrift′less·ness** n.
— Synonyms: gain, profit, prosperity. See FRUGALITY.

thrift·y (thrif′tē) adj. **thrift·i·er, thrift·i·est** 1 Displaying thrift or good management; economical; frugal. 2 Prosperous; thriving. 3 Growing vigorously. See synonyms under PRUDENT. — **thrift′i·ly** adv. — **thrift′i·ness** n.

thrill¹ (thril) v.t. 1 To cause to feel a sudden wave of emotion; move to great or tingling excitement. 2 To cause to vibrate or tremble. — v.i. 3 To feel a sudden wave of emotion or excitement. 4 To vibrate or tremble; quiver. See synonyms under SHAKE. — n. 1 A tremor of feeling or excitement. 2 A pulsation. 3 *Med.* An abnormal vibratory or tremulous resonance perceived in auscultation; fremitus. [Metathetic var. of THIRL¹] — **thrill′ing** adj. — **thrill′ing·ly** adv.

thrill² (thril) See TRILL¹.

thrill·er (thril′ər) n. 1 One who or which thrills. 2 *Colloq.* An exciting book, play, or motion picture.

thrip (thrip) n. *Brit. Slang* A threepenny piece.

thrips (thrips) n. A small insect (order *Thysanoptera*), many species of which are injurious to grain and plants. [<L <Gk., wood-worm]

thrive (thrīv) v.i. **throve** (thrōv) or **thrived, thrived** or **thriv·en** (thriv′ən), **thriv·ing** 1 To prosper; be successful, especially by being thrifty. 2 To grow with vigor; flourish. See synonyms under FLOURISH, SUCCEED. [<ON *thrífast,* orig. reflexive of *thrífa* grasp. Akin to THRIFT.] — **thriv′er** n. — **thriv′ing·ly** adv.

throat (thrōt) n. 1 The anterior part of the neck, extending from the back of the mouth and containing the epiglottis, larynx, trachea, and pharynx. 2 Anything resembling a throat; an entrance, inlet, or orifice: the *throat* of a bottle. 3 *Naut.* The end of a gaff nearest the mast. — v.t. 1 *Rare* To utter in a guttural tone. 2 To provide with a throat; channel; groove. [OE *throte*]

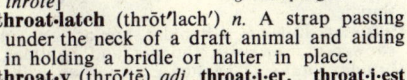

HUMAN THROAT
a. Soft palate.
b. Tonsils.
c. Pharynx.
d. Epiglottis.
e. Vocal cords.
f. Larynx.
g. Esophagus.

throat·latch (thrōt′lach′) n. A strap passing under the neck of a draft animal and aiding in holding a bridle or halter in place.

throat·y (thrōt′ē) adj. **throat·i·er, throat·i·est** Uttered in the throat; guttural. — **throat′i·ly** adv. — **throat′i·ness** n.

throb (throb) v.i. **throbbed, throb·bing** 1 To beat or pulsate rhythmically, as the heart; especially, to beat rapidly or violently; palpitate. 2 To feel or show emotion. — n. 1 The act or state of throbbing. 2 A pulsation or beat, especially one caused by excitement or emotion. [? Imit.] — **throb′ber** n.

throe (thrō) n. 1 A violent pang or pain; agony: said especially of the pains of death and childbirth. 2 Any agonized or agonizing activity. See synonyms under AGONY, PAIN. — v.t. & v.i. **throed, throe·ing** *Rare* To put in, suffer, or undergo agony. ◆ Homophone: *throw.* [ME *throwe,* prob. fusion of OE *throwian* suffer and *thrāwan* twist, throw]

throm·bin (throm′bin) n. *Biochem.* The enzyme present in blood serum that reacts with fibrinogen to form fibrin in the process of clotting. [<THROMBUS]

throm·bo·cyte (throm′bə·sīt) n. A blood platelet. [<Gk. *thrombos* clot + -CYTE]

throm·bo·gen (throm′bə·jen) n. *Biochem.* Prothrombin. [<Gk. *thrombos* clot + -GEN]

throm·bo·plas·tin (throm′bō·plas′tin) n. *Biochem.* A complex substance present in the blood and other animal tissues, which reacts with calcium ions to produce prothrombin. Also **throm·bo·kin·ase** (-kin′ās, -ki′nās). [<Gk. *thrombos* clot + -PLAST + -IN] — **throm′bo·plas′tic** adj.

throm·bo·sis (throm·bō′sis) n. *Pathol.* Local coagulation of blood in the heart, arteries, veins, or capillaries, forming by its clot an obstruction to circulation. [<NL <Gk. *thrombōsis* < *thrombos* clot] — **throm·bot′ic** (-bot′ik) adj.

throm·bus (throm′bəs) n. pl. **·bi** (-bī) *Pathol.* The blood clot formed in thrombosis. [<Gk. *thrombos* clot, lump]

throne (thrōn) n. 1 The royal chair occupied by a sovereign on state occasions. 2 The chair of state of a pope or of some other dignitary, as a cardinal, archbishop, or bishop. 3 Royal estate or dignity; sovereign power. 4 One invested with sovereign power; sometimes, the rank or authority of any high dignitary. 5 *pl.* The third of the nine orders of angels in the celestial hierarchy. — v.t. & v.i. **throned, thron·ing** To place or sit on a throne; enthrone; exalt. [<OF *trone* <L *thronus* <Gk. *thronos* seat]

throng (throng, thrông) n. 1 A multitude of people crowded closely together. 2 Any numerous collection. — v.t. 1 To crowd into and occupy fully; jam. 2 To press or crowd upon. — v.i. 3 To collect or move in a throng. See synonyms under JAM¹. [OE *gethrang*]
— Synonyms (noun): concourse, crowd, host, jam, mass, multitude, press. A *crowd* is a company of persons filling to excess the space they occupy and pressing inconveniently upon one another; the total number in a *crowd* may be great or small. *Throng* implies that the persons are numerous as well as pressed or pressing closely together; there may be a dense *crowd* in a small room, but there cannot be a *throng.* *Host* and *multitude* both imply vast numbers, but a *multitude* may be diffused over a great space so as to be nowhere a *crowd*; *host* is a military term, and properly denotes an assembly too orderly for crowding. *Concourse* signifies a spontaneous gathering of many persons moved by a common impulse, and suggests less massing and pressure than is indicated by the word *throng.* Compare ASSEMBLY, COMPANY.

throp·ple (throp′əl) n. *Scot. & Brit. Dial.* The windpipe or throttle.

thros·tle (thros′əl) n. 1 *Scot.* A thrush, especially the song thrush. 2 A machine for twisting and winding fibers from roves. [OE. Related to THRUSH¹.]

throt·tle (throt′l) n. 1 The throat or windpipe. 2 *Mech.* A valve controlling the supply of steam to a steam engine, or of vaporized fuel to the cylinders of an internal-combustion engine: also **throttle valve.** 3 The lever which operates the throttle: also **throttle lever.** — v.t. **·tled, ·tling** 1 To press or constrict the windpipe or throat; strangle; choke or suffocate. 2 To silence, stop, or suppress by or as by choking. 3 To reduce or shut off the flow of (steam, or fuel in an internal-combustion engine). 4 To reduce the speed of by means of a throttle; slow down. — v.i. 5 To suffocate; choke. [Dim. of ME *throte* throat] — **throt′tler** n.

through (thrōō) prep. 1 From end to end, side to side, or limit to limit of; into at one side, end, or point, and out of at another. 2 Covering, entering, or penetrating all parts of; throughout; also, over the surface of. 3 From the first to the last of; during the time or period of. 4 In the midst of; here and there upon or in. 5 By way of: He departed *through* the door. 6 By means of; by the instrumentality or aid of. 7 Having reached the end of, especially with success: He got *through* his examinations easily. 8 On account of; because or as a result of. See synonyms under BY. — adv. 1 From one end, side, surface, etc., to or beyond another. 2 From beginning to end. 3 To a termination or conclusion, especially a successful one: to pull *through.* 4 Completely; entirely: He is wet *through.* — **through and through** Thoroughly; completely. — adj. 1 Going from beginning to end without stops or with very few stops, and without reshipment or change: a *through* train; also, pertaining to or serving an entire distance or route: a *through* ticket. 2 Extending from one side or surface to another. 3 Unobstructed; open; clear: a *through* road. 4 Arrived at an end; finished: Are you *through* with my pen? 5 At the end of all relations or dealings: He is *through* with his old friends. Also spelled *thru.* [OE *thurh*]

through·ith·er (thrōō′ith′ər) *Scot.* adj. Disorderly; harum-scarum. — adv. Pell-mell. Also **through′-oth′er** (-uth′ər), **throu·ther** (thrōō′thər).

through·ly (thrōō′lē) adv. *Archaic* Thoroughly.

through·out (thrōō·out′) adv. Through or in every part: The house was searched *throughout.* — prep. All through; everywhere in: *throughout* the nation.

through·put (thrōō′poot′) n. The quantity of raw materials which may be processed for intended final use in a given time, as in an oil refinery or a chemical plant.

throve (thrōv) Past tense of THRIVE.

throw (thrō) v. **threw** (thrōō), **thrown, throw·ing** v.t. 1 To propel through the air by means of a sudden straightening or whirling of the arm. 2 To propel or hurl: The mortar *threw* shells into the town. 3 To put hastily or carelessly: He *threw* a coat over his shoulders. 4 To direct or project (light, shadow, a glance, etc.). 5 To bring to a specified condition or state by or as by throwing: to *throw* the enemy into a panic. 6 To cause to fall; overthrow: The horse *threw* its rider. 7 In wrestling, to force the shoulders of (an opponent) to the ground. 8 To cast (dice). 9 To make (a specified cast) with dice. 10 To cast off or shed; lose: The horse *threw* a shoe. 11 *Colloq.* To lose purposely, as a race. 12 To give birth to (young): said of domestic animals. 13 To move, as a lever or switch, in connecting or disconnecting a circuit, mechanism, etc.; also, to connect or disconnect in this manner. 14 *Slang* To give (a party, etc.). 15 In card games, to play or discard. 16 In ceramics, to shape on a potter's wheel. 17 To spin (filaments, as of silk) into thread. — v.i. 18 To cast or fling something. — **to throw away.** 1 To cast off; discard. 2 To waste; squander. — **to throw back** To revert to ancestral characteristics. — **to throw cold water on** To discourage. — **to throw in** 1 To cause (gears or a clutch) to mesh or engage. 2 To contribute; add. 3 To join with others. — **throw off** 1 To cast aside; reject; spurn. 2 To rid oneself of. 3 To do or utter in an offhand manner. 4 To disconnect, as a machine; release. — **to throw oneself at** To strive to gain the affections or love of. — **to throw oneself into** To engage or take part in vigorously. — **to throw oneself on (upon)** To entrust oneself to; rely on. — **to throw open** 1 To open suddenly or completely, as a door. 2 To free from restrictions or obstacles. — **to throw out** 1 To put forth; emit. 2 To cast out or aside; discard; reject. 3 To utter as if accidentally: to *throw out* hints. 4 In baseball, to retire (a runner) by throwing the ball to the base toward which he is advancing. — **to throw over** 1 To overturn. 2 To discard. — **to throw together** To put together hastily or roughly. — **to throw up** 1 To erect hastily. 2 To give up; relinquish. 3 To vomit. 4 *Colloq.* To mention or repeat, as a fault or taunt. — *n.* 1 An act of throwing or hurling; a cast; fling. 2 The distance over which a missile may be thrown: a stone's *throw*. 3 A cast of dice, or the resulting number; hence, a hazard; venture. 4 *Mech.* **a** The radius of the circle described by a crank, cam, or the like. **b** The travel or extent of reciprocating motion obtainable, as from a crank, piston, slide valve, etc. 5 A scarf used for draping an easel or picture frame; also, a woman's scarf or boa. 6 *Geol.* **a** A faulting, or dislocation of rock strata. **b** The amount of vertical displacement produced by dislocation of strata. 7 The sudden fluctuation of a magnetic needle when the force is suddenly changed. 8 The distance from a motion-picture projector to the screen. 9 In wrestling, a flooring of one's opponent so that both his shoulders touch the mat simultaneously for ten seconds. ♦ Homophone: throe. [OE *thrāwan* turn, twist, curl] — **throw'er** *n.*

throw·back (thrō'bak') *n.* 1 *Biol.* **a** Reversion to an earlier ancestral or primitive type, phase, or condition of physical being or development. **b** An example of such reversion. 2 Anything returned for revision, correction, redirection, etc.

throw·ster (thrō'stər) *n.* 1 A thrower of dice; gamester. 2 One who throws silk.

thru (thrōō) See THROUGH.

thrum[1] (thrum) *v.* **thrummed, thrum·ming** *v.t.* 1 To play on or finger (a stringed instrument) idly and without expression. 2 To drum or tap monotonously or listlessly. 3 To recite or repeat in a droning, monotonous way. — *v.i.* 4 To thrum a stringed instrument. 5 To sound when played thus, as a guitar. 6 *Scot.* To purr. — *n.* Any monotonous drumming. [Prob. imit.]

thrum[2] (thrum) *n.* 1 The fringe of warp threads remaining on a loom beam after the web has been cut off; also, one of such threads. 2 Any loose thread or fringe, or a tuft of filaments or fibers; a tassel. 3 *pl.* Coarse or waste yarn. 4 *pl. Naut.* Bits of rope yarn for sewing on canvas to make chafing gear or collision mats. 5 *Bot.* A threadlike organ or part of a flower; stamen. 6 *Scot.* A bit; particle; I don't care a *thrum*. 7 *Scot.* A tangle. — *v.t.* **thrummed, thrum·ming** 1 To cover or trim with thrums or similar appendages. 2 *Naut.* To insert bits of rope yarn (in canvas) to produce a rough surface or mat to be used to prevent chafing. [OE *-thrum* ligament, as in *tungethrum* the ligament of the tongue]

thrum·my (thrum'ē) *adj.* **·mi·er, ·mi·est** Made of or with thrums or resembling a thrum; shaggy; rough.

thrush[1] (thrush) *n.* Any one of many migratory, passerine birds of the family Turdidae, having typically a long and slightly graduated tail, long wings, and spotted under parts. The robin, **hermit thrush** (*Hylocichla guttata*), **wood thrush** (*H. mustelina*), and the European **song thrush** (*Turdus philomelus*) are examples. ♦ Collateral adjective: *turdine*. [OE *thrysce*]

WOOD THRUSH
(To 8 1/2 inches in length)

thrush[2] (thrush) *n.* 1 *Pathol.* A vesicular disease of the mouth, lips, and throat caused by a fungus (*Monilia albicans*): generally confined to infants. 2 A disease of a horse's foot characterized by suppuration. Also called *sprue*. [Cf. Dan. *tröske*, Sw. *trosk* a mouth disease]

thrust (thrust) *v.* **thrust, thrust·ing** *v.t.* 1 To push or shove with force or sudden impulse. 2 To pierce with a sudden forward motion; stab, as with a sword or dagger. 3 To interpose; put in. — *v.i.* 4 To make a sudden push or thrust. 5 To force oneself on or ahead; push one's way; crowd: with *through, into, on,* etc. See synonyms under DRIVE, PUSH. — *n.* 1 A sudden and forcible push, especially with a long, pointed weapon: distinguished from *cut.* 2 A vigorous attack; sharp onset. 3 *Engin.* A stress or strain tending to push a member of a structure outward or sidewise: the *thrust* of an arch. 4 In steam vessels, the pushing strain exerted by a propeller shaft. 5 The pulling effect of a tractor-airplane propeller. 6 A crushing of coal-mine pillars by the weight of the roof. 7 *Geol.* A rock fault due to horizontal compression; also, the plane of such a fault. [<ON *thrȳsta*] — **thrust'er** *n.*

thrust fault *Geol.* A fault resulting from horizontal compression in which the hanging wall appears to have moved upward, with a corresponding shortening of the entire rock mass: opposed to *gravity fault.* Also called *reverse fault.*

thru·way (thrōō'wā') *n.* A long-distance express highway.

Thu·cyd·i·des (thōō·sid'ə·dēz, thyōō-), 471?-399? B.C., Athenian statesman and historian. — **Thu·cyd'i·de'an** (-dē'ən) *adj.*

thud (thud) *n.* A dull, heavy sound, as of a hard body striking upon a comparatively soft one; also, the blow causing such a sound; a thump. — *v.i.* **thud·ded, thud·ding** To make a thud. [OE *thyddan* strike, thrust, press]

thug (thug) *n.* 1 Formerly, one of an organization of religious, professional assassins in northern India. 2 Any cutthroat or ruffian. [<Hind. *ṭhag* <Skt. *sthaga* swindler] — **thug'ger·y** *n.* — **thug'gish** *adj.*

thug·gee (thug'ē) *n.* The system of secret assassination formerly practiced by thugs in India. [<Hind. *ṭhagī*]

thu·ja (thōō'jə) *n.* Any of a genus (*Thuja*) of evergreen trees and shrubs of the pine family, including the arborvitae, source of the medicinal **oil of thuja**. Also spelled *thuya.* [<NL <Gk. *thyia,* an African tree]

Thu·le (thōō'lē, tōō'-) 1 In ancient geography, the northernmost limit of the habitable world: identified with Iceland or Mainland in the Shetland Islands. See ULTIMA THULE. 2 A settlement in NW Greenland; site of a major United States military installation.

thu·li·a (thōō'lē·ə) *n. Chem.* Oxide of thulium, Tm_2O_3, found in samarskite. [<THULIUM]

thu·li·um (thōō'lē·əm) *n.* A metallic element (symbol Tm) of the erbium family in the lanthanide series. See ELEMENT. [from THULE]

thumb (thum) *n.* 1 The inner digit of a limb when set apart from and apposable to the other fingers; especially, the short, thick digit on the radial side of the human hand; the pollex. 2 *Ornithol.* The first radial digit of the wing of certain birds. 3 The division in a glove or mitten that covers the thumb. 4 *Archit.* An ovolo. — **all thumbs** *Colloq.* Clumsy with the hands; not deft. — **thumbs down** A sign of negation or disapproval. — **under one's thumb** Under one's influence or power. — *v.t.* 1 To press, rub, soil, or wear with the thumb in handling, as the pages of a book. 2 To perform with or as with the thumbs; hence, to do or handle clumsily. 3 To run through the pages of (a book, manuscript, etc.) rapidly and perfunctorily. 4 To solicit or obtain (a ride) in an automobile by standing by the road and indicating with the thumb the direction one wishes to go; also, to make (one's way) thus: He *thumbed* his way from Boston to New York. [OE *thūma*]

thumb-in·dex (thum'in'deks) *v.t.* To provide with a thumb index.

thumb index A series of scalloped indentations cut along the right-hand edge of a book and labeled to indicate its various sections.

thumb·kin (thum'kin) *n.* A thumbscrew or pair of thumbscrews.

thumb·ling (thum'ling) *n.* A diminutive being; dwarf. Compare FINGERLING.

thumb·nail (thum'nāl') *n.* 1 The nail of the thumb. 2 Anything as small and essentially complete as a thumbnail. — *adj.* Small and essentially complete: a *thumbnail* sketch.

thumb·nut (thum'nut') *n.* A threaded nut having one or more wings or projections for screwing by the thumb and fingers; wing-nut. See illustration under NUT.

thumb·print (thum'print') *n.* An impression or print made by the thumb.

thumb·screw (thum'skrōō) *n.* 1 A screw to be turned by thumb and fingers. See illustration under SCREW. 2 An instrument of torture for compressing the thumb or thumbs.

thumb·stall (thum'stôl') *n.* A covering or sheath, as of leather, for the thumb.

thumb·tack (thum'tak') *n.* A broad-headed tack that may be pushed in with the thumb.

Thum·mim (thum'im) See under URIM.

thump (thump) *n.* A blow with a blunt or heavy object; also, the sound made by such a blow; a dull thud. See synonyms under BLOW. — *v.t.* 1 To beat or strike so as to make a heavy thud or thuds. 2 *Colloq.* To beat or defeat severely. — *v.i.* 3 To strike with a thump. 4 To make a thump or thumps; pound or throb. [Imit.] — **thump'er** *n.*

thump·ing (thum'ping) *adj.* 1 That thumps. 2 *Colloq.* Huge; whopping.

thumps (thumps) *n. pl.* 1 Hiccups in a horse. 2 A lung disease in swine, caused by infestation with the larvae of a roundworm (genus *Ascaris*). [<THUMP; from the sound of the contractions of the diaphragm]

Thun (tōōn) A town on the Aar river in central Switzerland. *French* **Thoune** (tōōn).

Thun (tōōn), **Lake of** An expansion of the Aar river in central Switzerland; 10 miles long; 18 square miles. *German* **Thu·ner·see** (tōō'nər·zā).

thun·der (thun'dər) *n.* 1 The sound that accompanies lightning, caused by the sudden heating and expansion of the air along the path of the lightning flash. 2 Any loud, rumbling or booming noise, suggestive of thunder. 3 An awful denunciation or threat; a vehement or powerful utterance, oratorical or other. 4 *Rare* A lightning stroke; thunderbolt. — **to steal one's thunder** To take for one's own use anything especially popular or effective originated by another: said especially of an argument. — *v.i.* 1 To give forth a peal or peals of thunder: used impersonally: It *thunders*. 2 To make a noise like thunder. 3 To utter vehement denunciations or threats. — *v.t.* 4 To utter or express with a noise like or suggestive of thunder: The cannon *thundered* defiance. [OE *thunor*] — **thun'der·er** *n.*

thun·der·a·tion (thun'də·rā'shən) *interj. & n. Slang* Damnation: a euphemism. [<THUNDER + (DAMN)ATION]

thun·der·bird (thun'dər·bûrd') *n.* An enormous bird believed to produce thunder by flapping its wings, lightning by opening and closing its eyes, and rain by allowing a huge lake to run off its back: common to the folklore of the North American Indians of the Plains and the Canadian forests.

thun·der·bolt (thun'dər·bōlt') *n.* 1 One electric discharge accompanied by a clap of thunder: formerly conceived of as a molten ball or bolt hurled by the lightning flash. 2 Any person or thing acting with or as with the force and speed or destructiveness of lightning.

thun·der·clap (thun'dər·klap') *n.* 1 A sharp, violent detonation of thunder. 2 Anything having the violence or suddenness of a clap of thunder.

thun·der·cloud (thun'dər·kloud') *n.* A dark, heavy mass of cloud highly charged with electricity.

thun·der·head (thun'dər·hed') *n.* A rounded mass of cumulus cloud, either silvery-white or dark with silvery edges, often developing into a thundercloud.

thun·der·ing (thun'dər·ing) *adj.* 1 Giving forth, or accompanied by, thunder. 2 Unusually great or extreme; superlative.

thun·der·ous (thun'dər·əs) *adj.* Producing or emitting thunder or a sound like thunder. Also **thun'drous** (-drəs). — **thun'der·ous·ly** *adv.*

thun·der·peal (thun'dər·pēl') *n.* A clap of thunder.

thun·der·show·er (thun'dər·shou'ər) *n.* A shower of rain with thunder and lightning.

thunder snake The house snake: so called because forced out of its hole by heavy rain.

thun·der·squall (thun'dər·skwôl') *n.* A squall accompanied by thunder.

thun·der·stone (thun′dər·stōn′) *n.* **1** *Archaic* A thunderbolt. **2** A stone or rock supposed to have accompanied a thunderbolt. **3** A belemnite.
thun·der·storm (thun′dər·stôrm′) *n.* A local storm accompanied by lightning and thunder.
thun·der·stroke (thun′dər·strōk′) *n.* A stroke of lightning.
thun·der·struck (thun′dər·struk′) *adj.* **1** Struck by lightning. **2** Amazed, astonished, or confounded, as with fear, surprise, or the like. Also **thun·der·strick·en** (-strik′ən).
thun·der·y (thun′dər·ē) *adj. Colloq.* Thunderous; indicative of, or accompanied by thunder.
Thur·ber (thûr′bər), **James Grover**, born 1894, U.S. humorous artist and writer.
thu·ri·ble (thōō′rə·bəl) *n.* A censer. [< L *thuribulum* < *thus, thuris* frankincense]
thu·ri·fer (thōō′rə·fər) *n.* A censer-bearer; an acolyte or altar boy who carries a thurible. [< L < *thus, thuris* frankincense + *ferre* bear, carry]
thu·rif·er·ous (thōō·rif′ər·əs) *adj.* Yielding or bearing incense.
Thu·rin·gi·a (thōō·rin′jē·ə, -jə) A former state of central Germany in southwestern East Germany; 6,022 square miles; capital, Weimar; formerly a region of central Germany including several duchies and principalities. German **Thü·rin·gen** (tü′ring·ən).
Thu·rin·gi·an (thōō·rin′jē·ən) *adj.* **1** Of or relating to Thuringia or its inhabitants. **2** *Geol.* Denoting the upper division of the Permian in Europe. — *n.* **1** One of an ancient Teutonic tribe which occupied a kingdom in central Germany until the sixth century, when they were conquered by the Franks. **2** A citizen or inhabitant of modern Thuringia.
Thuringian Forest A wooded mountain range of central Germany; highest point, 3,222 feet. German **Thü·rin·ger Wald** (tü′ring·ər vält).
Thurs·day (thûrz′dē, -dā) *n.* The fifth day of the week. [Fusion of OE *Thunres dæg* day of Thunor and ON *Thōrsdagr* day of Thor; trans. of LL *dies Jovis* day of Jove]
Thursday Island An island of NE Australia in Torres Strait, comprising a municipality of Queensland; 1 1/2 miles long, 1 mile wide.
thus (thus) *adv.* **1** In this or that or the following way or manner. **2** To such degree or extent; so: *thus* far. **3** In these circumstances or conditions; in this case; therefore. [OE]
thu·ya (thōō′yə) See THUJA.
thwack (thwak) *v.t.* To strike with something flat; whack. — *n.* A blow with some flat or blunt instrument. [Prob. OE *thaccian* smack; infl. in form by *whack*] — **thwack′er** *n.*
thwart (thwôrt) *v.t.* **1** To prevent the accomplishment of, as by interposing an obstacle; also, to prevent (one) from accomplishing something; foil; frustrate; balk. **2** *Obs.* To move or place over or across. See synonyms under BAFFLE, HINDER[1]. — *n.* An oarsman's seat extending athwart a boat. — *adj.* **1** Lying, moving, or extending across something; transverse. **2** *Obs.* Perverse or cross-grained; ill-natured. — *adv. & prep.* Athwart. [< ON *thvert*, neut. of *thverr* transverse] — **thwart′er** *n.*
Thwing (twing), **Charles Franklin**, 1853–1937, U.S. educator.
thy (thī) *pronominal adj.* The possessive case of the pronoun *thou* used attributively; belonging or pertaining to thee: *Thy* kingdom come. [Apocopated var. of THINE]
Thy·es·te·an banquet (thī·es′tē·ən) A cannibal feast: so called from the feast at which Thyestes was served his own sons. See ATREUS.
Thy·es·tes (thī·es′tēz) In Greek legend, a son of Pelops and brother of Atreus. — **Thy·es′te·an, Thy·es′ti·an** *adj.*
thy·la·cine (thī′lə·sīn, -sin) *n.* A nearly extinct, carnivorous, doglike marsupial (*Thylacinus cynocephalus*) of Tasmania, grayish-brown with dark transverse bands on the hinder part of the back: also called **Tasmanian wolf**. [< NL < Gk. *thylax, thylakos* pouch]

THYLACINE
(About 18 inches high at the shoulder)

thyme (tīm) *n.* Any of a genus (*Thymus*) of small shrubby plants of the mint family, having aromatic leaves and cultivated for seasoning in cookery; especially, the **wild thyme** (*T. serpyllum*). [< F *thym* < L *thymum* < Gk. *thymon*] — **thym′y** *adj.*
thym·e·la·e·a·ceous (thim′ə·lē·ā′shəs) *adj. Bot.* Designating a family (*Thymelaeaceae*) of apetalous trees or shrubs having very tough bark; the mezereon family. [< NL, family name < L *thymelaea* < Gk. *thymelaia* < *thymon* thyme + *elaia* olive tree]
thym·ic[1] (thī′mik) *adj.* Pertaining to or derived from thyme.
thy·mic[2] (thī′mik) *adj.* Of, pertaining to, or derived from the thymus.
thy·mol (thī′mōl, -mol, thī′-) *n. Chem.* A crystalline compound, $C_{10}H_{13}OH$, contained in certain volatile oils, as those of thyme and horsemint, and made synthetically: used as an antiseptic. [< THYME + -OL[2]]
thymol iodide *Chem.* A reddish-brown mixture of iodine derivatives and thymol, used as a deodorant and antiseptic.
thy·mus (thī′məs) *n. Anat.* A lymphoid organ of glandular character and unknown function, developed in the region of the neck in many vertebrates. In man and other mammals it lies at the root of the neck, just above the heart, and is most prominent in the young. It is the neck sweetbread of calves and lambs. [< NL < Gk. *thymos*]
thy·re·oid (thī′rē·oid) *adj.* Thyroid.
thyro- combining form *Med. & Surg.* The thyroid; of or related to the thyroid: *thyrotropin*. Also, before vowels, **thyr-**. Also **thyreo-**. [< Gk. *thyreoeidēs* thyroid]
thy·ro·hy·oid (thī′rō·hī′oid) *adj. Anat.* Having a relationship to the thyroid gland and the hyoid bone: the *thyrohyoid* ligament. See illustration under LARYNX. [< THYRO- + HYOID]
thy·roid (thī′roid) *adj.* **1** Relating or pertaining to the thyroid cartilage or the thyroid gland. **2** Shaped like a shield; also, having a shield-shaped marking. — *n.* **1** The thyroid cartilage or gland. **2** The dried and powdered thyroid gland of certain domesticated food animals, used in the treatment of myxedema, goiter, obesity, and other disorders. [< Gk. *thyreoeidēs* shield-shaped < *thyreos* large shield + *eidos* form]
thyroid cartilage *Anat.* The largest cartilage of the larynx, composed of two blades whose juncture in front forms the Adam's apple.
thy·roid·ec·to·my (thī′roid·ek′tə·mē) *n. Surg.* Excision of the thyroid gland. [< THYROID + -ECTOMY]
thyroid gland *Anat.* A bilobate endocrine gland situated in front of and on each side of the trachea, close to the larynx. It secretes thyroxin, vitally important in growth and in the prevention of such disorders as goiter, cretinism, etc.
thy·roid·i·tis (thī′roid·ī′tis) *n. Pathol.* Inflammation of the thyroid gland.
thy·ro·tox·i·co·sis (thī′rō·tok′sə·kō′sis) *n. Pathol.* A morbid or diseased condition resulting from excessive activity of the thyroid gland, as in exophthalmic goiter. [< THYRO- + TOXICOSIS] — **thy′ro·tox′ic** (-tok′sik) *adj.*
thy·ro·tro·pin (thī·rot′rə·pin) *n. Biochem.* A hormone from the anterior lobe of the pituitary gland, regarded as having an affinity for the thyroid gland. [< THYRO- + -TROP(E) + -IN]
thy·rox·in (thī·rok′sin) *n. Biochem.* A white, odorless, crystalline compound, $C_{15}H_{11}O_4NI_4$, obtained as the hormone of the thyroid gland and also made synthetically: used in the treatment of thyroid disorders. Also **thy·rox′ine** (-sēn, -sin). [< THYR(O)- + OXY- + -IN] — **thy·rox·in·ic** (thī′rok·sin′ik) *adj.*
thyrse (thûrs) *n.* A thyrsus.
thyr·soid (thûr′soid) *adj. Bot.* Resembling or shaped like a thyrsus. Also **thyr·soi·dal** (thûr·soid′l).
thyr·sus (thûr′səs) *n. pl.* **·si** (-sī) **1** A staff wreathed in ivy and crowned with a pine cone or a bunch of ivy leaves with grapes or berries: an attribute of Dionysus and the satyrs. **2** *Bot.* A branched panicle in which the middle branches are longer than those above or below them, as in the lilac and grape. [< Gk. *thyrsos*]

thy·sa·nu·ran (thī′sə·noor′ən, -nyoor′-, this′ə-) *adj.* Designating or belonging to an order (*Thysanura*) of primitive wingless insects, including the silverfish and the firebrat. — *n.* One of the *Thysanura*. [< NL, name of the order < Gk. *thysanos* fringe + *oura* tail] — **thy·sa·nu′rous** *adj.*
thy·self (thī·self′) *pron.* Emphatic or reflexive form of the second person singular pronouns *thee* and *thou*: I love thee for *thyself*.
Thys·sen (tis′ən), **Fritz**, 1873–1951, German industrialist.
ti[1] (tē) *n. Music* In solmization, a syllable representing the seventh note of the diatonic scale: formerly called *si*. [See GAMUT.]
ti[2] (tē) *n.* One of several Asian trees (genus *Cordyline*) of the lily family, especially the **ti palm** (*C. terminalis*) of eastern Asia, having many foliage forms. [< Polynesian]
Ti·a Jua·na (tē′ə wä′nə) See TIJUANA.
ti·a·ra (tī·âr′ə, tē·är′ə, -ar′ə) *n.* **1** The pope's triple crown, emblematic of his claim to spiritual and temporal authority; hence, the papal dignity. Compare MITER. **2** The upright headdress worn by the ancient Persian kings. **3** A coronet or form of headdress denoting princely rank; also, anything in imitation of it worn for personal adornment. **4** A Phrygian cap for men and women, long, conical, and falling over the brow: found in Greco-Roman art as the attribute of Paris, Mithras, and others. [< L < Gk. *tiara* Persian headdress]

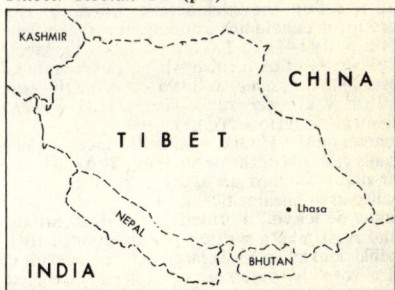
PAPAL TIARA

Tib·bett (tib′it), **Lawrence**, 1896–1960, U.S. baritone.
Ti·ber (tī′bər) A river of central Italy, flowing 251 miles south from the Apennines through Rome to the Tyrrhenian Sea, SE of Rome: Italian *Tevere*.
Ti·be·ri·as (tī·bir′ē·əs), **Lake** See GALILEE, SEA OF.
Ti·be·ri·us (tī·bir′ē·əs), 42 B.C.–A.D. 37, second emperor of Rome, A.D. 14–37: full name Tiberius Claudius Nero Caesar.
Ti·bes·ti Massif (ti·bes′tē) The highest mountain group of the Sahara and of French Equatorial Africa, lying mostly in NW Chad, partly in Libya and French West Africa; highest point, 11,204 feet.
Ti·bet (ti·bet′) A former independent theocracy of central Asia, south of the Sinkiang-Uigur Autonomous Region, China, and north of India, Nepal, Sikkim, and Bhutan; incorporated, 1950–57, in China, as the **Tibetan Autonomous Region**; about 470,000 square miles; capital, Lhasa: Chinese *Sitsang*: also *Thibet*. Tibetan **Pö** (pœ).

Ti·bet·an (ti·bet′n) *adj.* Of or pertaining to Tibet, the Tibetans, or to their language, religion, or customs. — *n.* **1** One of the native Mongoloid people of Tibet, now intermixed with Chinese and various peoples of India. **2** The Sino-Tibetan language of Tibet. Also spelled *Thibetan*.
Tibetan lion dog The Lhasa apso.
tib·i·a (tib′ē·ə) *n. pl.* **tib·i·ae** (tib′ē·ē) or **tib·i·as** **1** *Anat.* The inner and larger of the two bones of the leg below the knee; the shin bone. See illustration under FOOT. **2** *Entomol.* The fourth or penultimate joint of the leg of an insect, between the femur and the tarsus. **3** An ancient flute or pipe provided with holes

for the fingers, originally made of an animal's leg bone. [<L] —**tib'i·al** *adj.*

Ti·bul·lus (ti·bul'əs), **Albius,** 54?–18 B.C., Roman elegiac poet.

Ti·bur (tī'bər) Ancient name of TIVOLI.

tic (tik) *n.* **1** An involuntary spasm or twitching of muscles, usually of the face and sometimes of neurotic origin. **2** Tic douloureux. [<F]

ti·cal (ti·käl', -kôl', tē'kəl) *n.* **1** The former name for the baht, a Thai unit of currency. **2** A Thai unit of weight, equivalent to about half an ounce: also called *baht.* [<Malay *tikal*]

tic dou·lou·reux (tik dōō'lōō·rōō', *Fr.* tēk' dōō·lōō·rœ') *Pathol.* An acutely painful neuralgia of the face with paroxysmal muscular twitchings. [<F, painful tic]

Ti·ci·no (tē·chē'nō) A river of Switzerland and Italy, flowing 150 miles south through Lake Maggiore to the Po below Pavia. Ancient **Ti·ci·nus** (ti·sī'nəs).

tick[1] (tik) *n.* **1** A light recurring sound made by a watch, clock, or similar mechanism. **2** *Brit. Colloq.* The length of time occupied by one tick of a watch or clock: I'll be through in five *ticks.* **3** A mark, as a dot or dash, used in checking off something. —*v.i.* To make or sound a tick or ticks; make a recurrent clicking sound, as a running watch or clock. —*v.t. Brit.* To mark or check with ticks. —**to tick off** *Brit. Colloq.* To tell off. [Prob. imit.]

tick[2] (tik) *n.* **1** One of numerous flat, leathery, bloodsucking arachnids (order *Acarida*) that attack the skin of man and other animals; especially, the **cattle tick** *(Margaropus annulatus),* causative agent of Texas fever. **2** Any of certain two-winged or wingless parasitic insects (family *Hippoboscidae*), as the **sheep ticks** and **bat ticks.** [Cf. LG *tieke,* G *zecke* a tick]

tick[3] (tik) *n.* **1** The stout outer covering of a mattress; also, the material for such covering. **2** *Colloq.* Ticking. [Earlier *teke, tyke,* ult. <L *teca, theca* <Gk. *thēke* a case]

tick[4] (tik) *n. Brit. Colloq.* Credit; trust: to buy something on *tick.* [Short for TICKET]

tick·er (tik'ər) *n.* **1** One who or that which ticks. **2** A telegraphic receiving instrument which records stock quotations on a paper ribbon (**ticker tape**). **3** *Slang* A watch. **4** *Slang* The heart.

tick·et (tik'it) *n.* **1** A note or notice; a memorandum; also, a slip of paper containing a notice or memorandum. **2** A card with words or characters on it showing that the holder is entitled to something, as transportation in a public vehicle, admission to a theater, or the like. **3** A certificate or license, as of an airplane pilot or the captain of a ship. **4** A label or tag for attachment or identification. **5 a** A list of candidates of a single party on a ballot: the Democratic *ticket.* **b** The group of candidates running for the offices of a party. —*v.t.* **1** To fix a ticket to; label. **2** To present or furnish with a ticket or tickets. [<MF *etiquet* a little note <OF *estiquette* <*estiquer* stick, fix <OLG *stekan.* Doublet of ETIQUETTE.]

ticket agent 1 One who sells tickets, especially railroad or theater tickets. **2** An agency, or one who runs an agency, for the sale of railroad or theater tickets.

ticket of leave Formerly, in Great Britain and Australia, a written permit granted to a penal convict to be at large before the expiration of his sentence on certain specified conditions.

tick fever Any of several fevers caused by ticks, especially the Texas fever of cattle, and Rocky Mountain spotted fever, transmitted to man by the bite of a wood tick.

tick·ing (tik'ing) *n.* A strong, closely woven cotton or linen fabric: used for mattress covering, awnings, etc. [<TICK[3] + -ING[1]]

tick·le (tik'əl) *v.* **·led, ·ling** *v.t.* **1** To excite the nerves of by touching or scratching on some sensitive spot, producing a thrilling sensation resulting in spasmodic laughter or twitching; titillate. **2** To arouse or excite agreeably; please: Compliments *tickle* our vanity. **3** To amuse or entertain; delight. **4** To move, stir, or get by or as by tickling. —*v.i.* **5** To have or experience a thrilling or tingling sensation: My foot *tickles.* —*n.* The sensation produced by tickling; titillation; also, the touch or action producing such sensation. [ME *tikelen,* ? metathetic var. of ON *kitla* tickle]

tick·le-grass (tik'əl·gras', -gräs') *n.* Rough bent grass *(Agrostis hiemalis).*

tick·ler (tik'lər) *n.* **1** One who or that which tickles. **2** A memorandum book or file, as of bills receivable, notes due, etc.

tickler coil In radio, a coil of the regenerative type coupled in series with the plate circuit and employed to intensify sound on a receiving circuit by means of a feedback action.

tick·lish (tik'lish) *adj.* **1** Sensitive to tickling. **2** Liable to be upset; unstable; also, easily offended; sensitive. **3** Attended with risk; difficult; delicate. —**tick'lish·ly** *adv.* —**tick'·lish·ness** *n.*

Tick·nor (tik'nər), **George,** 1791–1871, U.S. historian and educator.

tick·seed (tik'sēd') *n.* **1** The coreopsis. **2** The tick trefoil. [<TICK[2] + SEED]

tickseed sunflower A square-stemmed species of bur marigold (genus *Bidens*), with a panicle of large-rayed yellow flowers.

tick-tack (tik'tak') *n.* **1** A recurrent sound like that of the ticking of a clock. **2** Anything that makes a tapping or rattling noise; specifically, a device for making a rattling noise against a window or door, consisting of a small weight hung by a string, and worked from a distance by pulling a long cord attached: used in playing pranks. [Imit. reduplication of TICK[1]]

tick-tack-toe (tik'tak·tō') *n.* **1** A game for two players, who alternately put circles or crosses in the spaces of a figure formed by two sets of parallel lines crossing at right angles. Each player tries to get a row of three circles or three crosses before his opponent does. **2** The prank or trick of using a tick-tack. Also **tick'-tack-too'** (-tōō').

tick-tock (tik'tok') *n.* The oscillating sound of a clock. —*v.i.* To make this sound. [Imit.]

tick trefoil Any of several leguminous plants (genus *Desmodium*) whose leaves and pods cling to the coats of animals and to clothing. [<TICK[2] + TREFOIL]

Ti·con·der·o·ga (tī'kon·də·rō'gə) A town on Lake George in New York; site of **Fort Ticonderoga,** captured from the French in 1759 and from the British by the American revolutionists in 1775.

tid·al (tīd'l) *adj.* **1** Of, pertaining to, or influenced by the tides; periodically flowing and ebbing: a *tidal* river. **2** Dependent on the rise of the tide as to time of starting or leaving: a *tidal* steamship.

tidal wave 1 Any great incoming rise of waters along a shore, caused by windstorms at sea or by excessively high tides. **2** A tsunami. **3** A great movement in popular feeling or in the affairs of men.

tid·bit (tid'bit') *n.* A choice bit, as of food. Also, *Brit., titbit.* [<dial. E *tid* a small object + BIT[1]]

tid·dle·dy·winks (tid'l·dē·wingks') *n.* A game in which the players attempt to snap little disks of bone, ivory, or the like, from a plane surface into a cup. Also **tid'dly·winks'** (tid'·lē-). [Prob. <*tiddly,* a child's word for *little*]

tide[1] (tīd) *n.* **1** The periodic rise and fall of the surface of the ocean, and of the waters connected with the ocean, caused by the attraction of moon and sun. In each lunar day of 24 hours and 51 minutes there are two high tides and two low tides, alternating at equal intervals of flood and ebb. **Spring tides** are high tides above the average, occurring when the moon is new or full; **neap tides** are high tides below the average, occurring when the moon is in the first or third quarter. See FLOOD, EBB. **2** Anything that comes like the tide at flood; the time at which something is most flourishing. **3** The natural drift or tendency of events; also, a current; stream. **4** Season; time; especially, a season of the ecclesiastical year: used chiefly in composition and in the phrase *time and tide*: *Christmastide.* **5** *Archaic* A suitable or favorable occasion; opportunity. See synonyms under STREAM. —*v.* **tid·ed, tid·ing** *v.i.* **1** To ebb and flow like the tide. **2** To float with the tide. —*v.t.* **3** To carry or help like a boat buoyed up by the tide: Charity *tided* us over the depression. **4** To surmount; survive; endure, as a difficulty: with *over*: to *tide* over hard times. [OE *tīd* a period, season] —**tide'less** *adj.*

tide[2] (tīd) *v.i. Archaic* To betide; happen. [OE *tīdan*]

tide·land (tīd'land') *n.* Land alternately covered and uncovered by the tide.

tide·rip (tīd'rip') *n.* Riptide. [<TIDE[1] + RIP[2]]

tide·wait·er (tīd'wā'tər) *n.* **1** A customs officer who boards vessels entering port, to enforce customs regulations. **2** A politician who waits for a strong indication of public opinion before determining his action.

tide·wa·ter (tīd'wô'tər, -wot'ər) *n.* Water which inundates land at high tide; also, water affected by the tide on the seacoast or in a river; hence, loosely, the seacoast. —*adj.* Pertaining to the tidewater; also, situated on the seacoast: the *tidewater* country.

tide·way (tīd'wā') *n.* A channel where the tide runs.

ti·dings (tī'dingz) *n. pl. (sometimes construed as singular)* A report or information; news. [OE *tīdung*; infl. in meaning by ON *tīthindi* news, a message]

Synonyms: advice, information, intelligence, news. *News* is the most general of these words, signifying something that has either just happened or just become known. *Advices* are communications of fact by a trusted informant with the design of guiding or influencing the action of the recipient; the word signifies *news* with a practical purpose and value. *Intelligence* is *news* or *information,* often secret *information,* specifically communicated, usually in certain form. See NEWS.

ti·dy (tī'dē) *adj.* **·di·er, ·di·est 1** Marked by neatness and order; trim. **2** Of an orderly disposition. **3** *Colloq.* Moderately large; considerable: a *tidy* sum. **4** *Colloq.* Tolerable; fairly good. See synonyms under NEAT[1]. —*v.t.* & *v.i.* **ti·died, ti·dy·ing** To make (things) tidy; put (things) in order. —*n. pl.* **·dies** A light, detachable covering, to protect the back or arms of a chair or sofa from dirt and wear. [ME *tidi* <OE *tīd* time] —**ti'di·ly** *adv.* —**ti'di·ness** *n.*

ti·dy·tips (tī'dē·tips') *n. pl* **·tips** Any of a genus *(Layia)* of ornamental annual plants of California, having yellow flower heads tipped with white; especially, *L. elegans.*

tie (tī) *v.* **tied, ty·ing** *v.t.* **1** To fasten with cord, rope, etc., the ends of which are then drawn into a knot. **2** To draw the parts of together or into place by a cord or band fastened with a knot: to *tie* one's shoes. **3** To form (a knot). **4** To form a knot in, as string. **5** To fasten, attach, or join in any way. **6** To restrain or confine; restrict; bind. **7 a** To equal (a competitor) in score or achievement. **b** To equal (a competitor's score). **8** *Colloq.* To unite in marriage. **9** *Music* To unite by a tie. —*v.i.* **10** To make a tie or connection. **11** To make the same score; be equal. See synonyms under BIND. —**to tie down** To hinder; restrict. —**to tie up 1** To fasten with rope, string, etc. **2** To wrap, as with paper, and then fasten with string, cord, etc. **3** To moor (a vessel). **4** To block; hinder. **5** To have or be already committed, in use, etc., so as to be unavailable. —*n.* **1** A flexible bond or fastening secured by drawing the ends into a knot or loop. **2** Any bond or obligation, mental, moral, or legal: *ties* of affection. **3** An exact equality in number, as of a score, votes, etc.; hence, a contest which neither side wins; a draw. **4** Something that is tied or intended for tying, as a shoelace, necktie, or the like. **5** *Engin.* A structural member fastening parts together and receiving tensile stress: distinguished from a *strut.* **6** *Music* A curved line placed over or under two musical notes of the same pitch on the staff to make them represent one tone length. **7** *pl.* Low shoes fastened with lacings: Oxford *ties.* **8** One of a set of timbers laid crosswise on the ground as supports for railroad tracks. [OE *tīgan* bind <*tēah, tēag* a rope]

tie beam A timber that serves as a tie in a roof, etc.

Tie·bout (tē'bout) Dutch form of THEOBALD.

Tieck (tēk), **Ludwig,** 1773–1853, German poet and critic.

tie-in (tī'in') *n.* Connection; association; relation.

tie-in sale A sale in which the buyer, in order to get the article he wants, is required to buy a second article.

tie·man·nite (tē'mə·nīt) *n.* A metallic, steel-to lead-gray, opaque mercuric selenide, HgSe.

[<G *tiemannit,* after W. *Tiemann,* 19th c. German mineralogist, its discoverer]

Tien Shan (tyen' shän') A mountain chain of central Asia, chiefly in the Tadzhik S.S.R., Kirghiz S.S.R., and Sinkiang–Uigur Autonomous Region, China; highest point, 24,406 feet.

Tien·tsin (tin'tsin', *Chinese* tyen'jin') A port near the Gulf of Chihli, NE China; formerly included in Hopeh Province; since 1935, an independent municipality under direct control of the central government; the leading transportation and industrial center of northern China.

Tie·po·lo (tye'pō-lō), **Giovanni Battista,** 1696?–1770?, Venetian painter.

tier[1] (tir) *n.* A rank or row in a series of things placed one above another. — *v.t.* & *v.i.* To place or rise in tiers. ◆ Homophone: *tear.* [Earlier *tire* <OF, a sequence < *tirer* draw, elongate]

ti·er[2] (tī'ər) *n.* 1 One who or that which ties; also, that used for or in tying. 2 A child's apron.

tierce (tirs) *n.* 1 A former liquid measure equivalent in the United States to 42 wine gallons; a third of a pipe or butt. 2 A cask holding this amount, intermediate between a hogshead and a barrel. 3 In card games, a sequence of three cards of the same suit. 4 In fencing, the third standard position from which a guard, parry, or thrust can be made. 5 *Eccl.* The third canonical hour, nine a.m., or the office or service of that hour: often called *undersong.* 6 *Music* An interval of a third. 7 A set of three. [<OF *tierce, terce* a third <L *tertia,* fem. of *tertius*]

Tier·ra del Fue·go (tyer'ä del fwā'gō) 1 An archipelago at the southern tip of South America, belonging to Chile and to Argentina; separated from the mainland by the Strait of Magellan; total, 27,476 square miles: 7,996 square miles in Argentina, the rest in Chile. 2 The largest island of the group: 18,000 square miles: 7,750 square miles in Argentina, the rest in Chile.

tiers é·tat (tyâr zä·tä') *French* The third estate, especially in prerevolutionary France.

tie-up (tī'up') *n.* 1 A situation, resulting from a strike, the breakdown of machinery, etc., in which further progress or operation is impossible: a *tie-up* in traffic. 2 *Dial.* The part of a barn where cows and oxen are kept.

tiff[1] (tif) *n.* 1 A peevish display of irritation; a pet; huff. 2 A light quarrel; a spat. — *v.i.* To be in or have a tiff. [Prob. imit.]

tiff[2] (tif) *Obs. n.* A small draft of liquor; a sip; drink. — *v.t.* To sip; taste. [Cf. ON *thefr* a smell, taste]

tiff[3] (tif) *v.i. Anglo-Indian* To take tiffin or lunch. [Back formation < *tiffing* TIFFIN]

tif·fa·ny (tif'ə·nē) *n. pl.* ·nies 1 A very thin transparent cotton gauze. 2 Formerly, a very thin silk. [<OF *tifinie, tiphanie* Epiphany <LL *theophania* THEOPHANY; ? so called because its transparency manifests the wearer]

Tif·fa·ny (tif'ə·nē), **Charles Lewis,** 1812–1902, U.S. jeweler.

tif·fin (tif'ən) *Anglo-Indian n.* Midday luncheon. — *v.i.* To lunch; tiff. [Appar. < *tiffing,* ppr. of TIFF[2]]

Tif·lis (tif'lis, *Russian* tif-lēs') The capital of Georgian S.S.R., on the Kura River: Georgian *Tbilisi.*

ti·ger (tī'gər) *n.* 1 A large carnivorous feline mammal (*Felis tigris*) of Asia, with vertical black wavy stripes on a tawny body and black bars or rings on the limbs and tail. 2 One of several other large ferocious animals, as the South American jaguar or the African leopard; also, the thylacine of Tasmania. 3 A fierce, cruel person. 4 *U.S.* An additional cheer or yell (often the word "tiger") given at the conclusion of a round of cheering. [<OF *tigre* <L *tigris*

BENGAL TIGER
(About 6 1/2 feet long; tail, 3 feet)

<Gk., ?< Avestan *tīghri* an arrow, a dart]

tiger beetle Any of certain very active, predacious beetles (genus *Cicindela*) having spotted or striped wings, which dart upon their prey from a concealment. For illustration see INSECTS (beneficial).

tiger cat 1 A wildcat, resembling, but smaller than, the tiger, as the Asian **marbled tiger cat** (*Felis marmorata*), the African serval, the American ocelot, and the margay. 2 A domestic cat having striped markings.

ti·ger-eye (tī'gər·ī') *n.* 1 A gemstone, usually the mineral crocidolite altered by oxidation, showing a beautiful chatoyant luster. One variety is called *hawk's-eye.* Also **ti'ger's-eye'.** 2 A tiger cat.

ti·ger·ish (tī'gər·ish) *adj.* Of, pertaining to, or resembling the tiger or its habits; predacious; bloodthirsty: also spelled *tigrish.*

tiger lily 1 A tall cultivated lily (*Lilium tigrinum*) from China, with nodding orange flowers spotted with black. 2 Any of various lilies with similar flowers, especially the leopard lily (*L. pardalinum*).

tiger moth A stout-bodied moth (family *Arctiidae*) with striped or spotted wings.

tight (tīt) *adj.* 1 So closely held together or constructed as to be impervious to fluids; not leaky: a *tight* roof; a *tight* vessel. 2 Firmly fixed or fastened in place; secure. 3 Fully stretched, so as not to be slack; taut; tense: *tight* as a drum. 4 Strict; stringent: a *tight* schedule. 5 Fitting closely; especially, fitting too closely: said of a garment, shoe, cork, etc. 6 *Colloq.* Difficult to cope with; troublesome: a *tight* spot; a *tight* squeeze. 7 *Colloq.* Parsimonious; tight-fisted; close. 8 Characterized by a feeling of constriction: a *tight* cough. 9 *Slang* Drunk; intoxicated. 10 Evenly matched: said of a race or contest. 11 Difficult to obtain because of scarcity or financial restrictions: said of money or commodities. 12 Straitened from lack of money or commodities: a *tight* market. 13 Yielding very little or no profit: said of a bargain. — *adv.* 1 Firmly; securely; Hold me *tight.* 2 Closely; with much constriction: The dress fits too *tight.* — **to sit tight** To remain firm in one's position; refrain from budging. [ME *thight,* appar. <Scand. Cf. ON *théttr* dense.] — **tight'·ly** *adv.* — **tight'ness** *n.*

-tight *combining form* Impervious to: *watertight.*

tight·en (tīt'n) *v.t.* & *v.i.* To make or become tight or tighter. — **tight'en·er** *n.*

tight-fist·ed (tīt'fis'tid) *adj.* Stingy; parsimonious.

tight-lipped (tīt'lipt') *adj.* Having the lips held tightly together; hence, unwilling to talk; reticent or secretive.

tight·rope (tīt'rōp') *n.* A tightly stretched rope on which acrobats perform. — *adj.* Pertaining to or performing on a tightrope: a *tightrope* walker.

tights (tīts) *n. pl.* Skin-fitting garments, commonly for the legs and lower torso.

tight·wad (tīt'wod') *n. U.S. Slang* A skinflint; miser. [<TIGHT + WAD[1]]

Tig·lath-pi·le·ser (tig'lath·pi·lē'zər, -pī-) Any of several Assyrian kings and conquerors; especially, Tiglath-pileser III, reigned 745–727 B.C.

tig·lic (tig'lik) *adj. Chem.* 1 Derived from croton oil. 2 Designating a white, crystalline, poisonous acid, $C_5H_8O_2$, contained as an ester in croton oil. Also **tig·lin·ic** (tig·lin'-ik). [<NL (*Croton*) *tiglium* the croton oil plant, prob. ult. <Gk. *tilos* thin feces; so called because of its purgative properties]

Ti·gré (tē·grā') *n.* A modern Semitic language of Ethiopia, descended from the ancient Ethiopian.

Ti·gré (tē·grā') A province of northern Ethiopia, bordering on Eritrea; formerly an independent kingdom; about 26,000 square miles; capital, Makale. Also **Ti·gre** (tēg'r').

ti·gress (tī'gris) *n.* A female tiger.

Ti·gri·ña (tē·grē'nyä) *n.* A Southwest Semitic language spoken in Ethiopia.

Ti·gris (tī'gris) A river of SW Asia, flowing about 1,150 miles SE from east central Turkey in Asia through Iraq to the Euphrates NW of Basra.

ti·grish (tī'grish) See TIGERISH.

Ti·hwa (dē'hwä') See URUMCHI.

Ti·jua·na (tē·hwä'nä) A border town in NW Lower California, Mexico: also *Tia Juana.*

tike (tīk) *n.* 1 A low-bred dog; a cur. 2 *Scot.* An uncouth fellow; a boor. 3 *Colloq.* A small child. Also spelled *tyke.* [<ON *tík* a bitch]

Ti·ki (tē'kē) In Maori mythology, the creator of the first man.

til (til, tēl) *n.* Sesame. [<Hind. <Skt. *tilá*]

Til·burg (til'bûrg, *Du.* til'bûrkh) A city in North Brabant province, south Netherlands.

til·bur·y (til'ber·ē) *n. pl.* ·bur·ies A form of gig seating two persons. [after *Tilbury,* an early 19th c. London coachmaker who invented it]

Til·da (til'də) Diminutive of MATILDA.

til·de (til'də, -dē) *n.* 1 A sign (~) used in Spanish over *n* to indicate nasal palatalization or the sound of *ny,* as in *cañon,* canyon. 2 The same sign (usually called **til**) used in Portuguese over a vowel or the first vowel of a diphtong to indicate nasalization, as in *lã, Camões.* [<Sp. <L *titulus* superscription, title]

Til·den (til'dən), **Samuel Jones,** 1814–86, U.S. statesman.

Til·dy (til'dē), **Zoltán,** born 1889, Hungarian politician; president of Hungary 1946–48.

tile (tīl) *n.* 1 A thin piece or plate of baked clay, sometimes decorated, used for covering roofs, floors, etc., and as an ornament. 2 A short earthenware pipe, used in forming sewers. 3 Tiles collectively; tiling. 4 *Colloq.* A high silk hat. — *v.t.* **tiled, til·ing** 1 To cover with tiles. 2 To secure against intrusion; specifically, in Freemasonry, to place the doorkeeper or tiler at the door of (a lodge) to keep out unauthorized persons. [OE *tigule, tigele,* ult. <L *tegula* <*tegere* cover]

tile·fish (tīl'fish') *n. pl.* **·fish** or **·fish·es** A large marine fish (*Lopholatilus chamaeleonticeps*) of the western Atlantic, marked with large yellow spots, and esteemed as food. [<NL (*Lophola*)*til*(*us*), genus name; infl. by *tile,* because its markings resemble ornamental tiles]

til·er (tī'lər) *n.* 1 A maker or layer of tiles. 2 The doorkeeper of a Masonic lodge.

til·i·a·ceous (til'ē·ā'shəs) *adj. Bot.* Designating or belonging to a widely distributed family (*Tiliaceae*) of trees, shrubs, and herbs, the linden family, having clusters of often fragrant flowers. [<NL <L *tiliaceus* <*tilia* the linden tree]

til·ing (tī'ling) *n.* 1 The act, operation, or system of using tiles for roofing or drainage. 2 Tiles collectively. 3 Something made of or faced with tiles.

till[1] (til) *v.t.* & *v.i.* To put and keep (soil) in order for the production of crops, as by plowing, harrowing, hoeing, sowing, etc.; cultivate. [OE *tilian* strive, acquire] — **till'a·ble** *adj.*

till[2] (til) *prep.* 1 To the time of; up to; until: He slept *till* noon. 2 Before: with the negative: He couldn't leave *till* today. 3 *Scot. & Brit. Dial.* To; unto; as far as. — *conj.* 1 Up to such time as; until: *till* death do us part. 2 Before: with the negative: They couldn't go *till* the carriage came for them. [OE *til* <ON, to]

till[3] (til) *n.* A drawer, compartment, or tray; a money drawer. [Earlier *tille,* prob. <ME *tillen, tyllen* draw]

till[4] (til) *n. Geol.* An unassorted, commingled, and chiefly unstratified mass of clay, sand, pebbles, and boulders, deposited by masses of ice. [Var. of ME *thill,* ? <OE *thille* a board, flooring]

till·age (til'ij) *n.* The cultivation of land. See synonyms under AGRICULTURE. [<TILL[1] + -AGE]

til·land·si·a (ti·land'zē·ə) *n.* Any of a genus (*Tillandsia*) of mainly epiphytic bromeliaceous plants of tropical America and the southern United States, having narrow, entire, often scurfy leaves, and flowers in a terminal spike. [<NL, after Elias *Tillands,* 18th c. Swedish botanist]

till·er[1] (til'ər) *n.* One who or that which tills; a plowman; a farmer. [<TILL[1]]

till·er[2] (til'ər) *n.* 1 A lever to turn a rudder. 2 A means of guidance. [<OF *telier* stock

of a crossbow <Med. L *telarium* a weaver's beam <L *tela* a web; prob infl. in meaning by ME *tillen* draw]

till·er[3] (til′ər) *n.* **1** A shoot from the base of a stem; sucker. **2** A sapling. —*v.i.* To put forth stems from the root; send forth new shoots. [Prob. OE *telgor* a twig <*telga* a branch]

til·lot (til′ət) *n. Brit.* A type of cloth used for wrapping fabric. Also **til′let.** [Earlier *tillet,* appar. <OF *tellette,* var. of *teilete, toilete* a wrapper of cloth. See TOILET.]

Til·lot·son (til′ət·sən), **John,** 1630–94, English theologian.

Til·ly (til′ē), **Count von,** 1559–1632, Johann Tserklaes, Flemish general of the imperial league in the Thirty Years' War.

til·ly-val·ly (til′ē·val′ē) *interj. Brit.* Nonsense; bosh. Also **til′ly-fal′ly** (-fal′ē). [Origin unknown]

Til·sit (til′zit) A former name for SOVETSK.

tilt[1] (tilt) *v.t.* **1** To cause to rise at one end or side; incline at an angle; slant; lean; tip. **2** To aim or thrust, as a lance. **3** To charge or overthrow in a tilt or joust. **4** To hammer or forge with a tilt hammer. —*v.i.* **5** To incline at an angle; lean. **6** To contend with the lance; engage in a joust. See synonyms under TIP[1]. —*n.* **1** An inclination from the vertical or horizontal position; slant; slope; also, the act of inclining, or the state of being inclined. **2** A medieval sport in which mounted knights, charging with lances, endeavored to unseat each other. **3** Any encounter resembling or suggestive of that between two tilting knights; hence, a quarrel; dispute; altercation; also, a thrust or blow, as with a lance. **4** A tilt hammer. **5** A seesaw. **6** The American black-necked stilt. —**at full tilt** At full speed; at full charge. [ME *tylten* be overthrown, totter <OE *tealt* unsteady] —**tilt′er** *n.*

tilt[2] (tilt) *n.* A canvas canopy or awning on a boat, wagon, booth, or the like. —*v.t.* To furnish or cover with an awning or tilt. [Var. of ME *tild, teld,* OE *teld* a tent]

tilth (tilth) *n.* **1** The act of tilling; cultivation of soil; tillage. **2** That part of the surface soil affected by tillage; cultivated land. [OE < *tilian* till]

tilt hammer A trip hammer.

tilt roof A round-topped roof: so called from its resemblance to the canopy or tilt of a covered wagon.

tilt-up (tilt′up′) *n.* The spotted sandpiper: so called from its teetering habits.

tilt·yard (tilt′yärd′) *n.* A courtyard or other place for tilting.

Tim (tim) Diminutive of TIMOTHY.

Ti·ma·ga·mi (ti·mä′gə·mē), **Lake** A lake in east central Ontario, Canada; 90 square miles.

ti·ma·rau (tē′mə·rou′) See TAMARAU.

Tim·a·ru (tim′ə·rōō) A port on eastern South Island, New Zealand.

tim·bal (tim′bəl) *n.* **1** A kettledrum. **2** *Entomol.* The drumlike, sound-producing, folding membrane of the shrilling organ of a male cicada or harvest fly. Also spelled *tymbal.* [<F *timbale,* appar. alter. of *attabale* <Sp. *atabal* ATABAL]

tim·bale (tim′bəl, *Fr.* tan̈·bȧl′) *n.* **1** A dish made of chicken, fish, cheese, or vegetables, pounded fine and mixed with the white of eggs, sweet cream, etc., cooked in a drum-shaped mold, then turned out and served with sauce. **2** A small cup made of fried pastry, in which food may be served. [<F. See TIMBAL.]

tim·ber (tim′bər) *n.* **1** Wood suitable for building purposes, prepared for use. **2** Growing or standing trees; also, woodland. **3** A single piece of squared wood prepared for use or already in use. **4** Any principal beam in a vessel's framing. **5** The wooden part or handle of any implement. **6** Loosely, the materials for any structure; hence, also, human material: That boy has good *timber* in him. See synonyms under STICK. —*v.t.* To provide or shore with timber. [OE] —**tim′ber·er** *n.*

tim·bered (tim′bərd) *adj.* **1** Covered with growing trees; wooded. **2** Constructed of timber.

tim·ber·head (tim′bər·hed′) *n. Naut.* **1** An end of a timber projecting above the deck, and used for attaching lines, etc. **2** An upright post fastened to the deck at the point where a timber's end would come.

timber hitch *Naut.* A knot by which a rope is fastened around a spar.

tim·ber·ing (tim′bər·ing) *n.* **1** Timberwork; timbers collectively. **2** The act or process of furnishing with timber.

tim·ber·land (tim′bər·land′) *n.* Land covered with forests.

timber line **1** The upper limit of tree growth on mountains and in arctic regions; the line above which no trees grow. **2** The boundary line of a tract of timber. —**tim′ber-line′** (-līn′) *adj.*

timber wolf The large gray or brindled wolf (*Canis occidentalis*) of the forests of the northern United States and Canada: distinguished from the *coyote* or *prairie wolf.*

TIMBER WOLF
(About 4 feet long; 26 inches high)

tim·ber·work (tim′bər·wûrk′) *n.* Work constructed of wood, especially the framing of a structure.

tim·bre (tim′bər, tam′-; *Fr.* tan̈′br′) *n.* **1** The inherent quality of tone which serves to distinguish one musical instrument or voice from another and renders it unique: sometimes called *tone color.* **2** In acoustics, the character or quality of a sound that is produced by the relative number and strength of its harmonics: distinguished from *intensity* (amplitude of vibrations) and *pitch* (frequency of vibrations). **3** *Phonet.* The degree of resonance of a voiced sound, especially a vowel. [<F <OF, a small bell, sound of a bell, orig. a timbrel <L *tympanum* a kettledrum <Gk. *tympanon*]

tim·brel (tim′brəl) *n.* An ancient Hebrew instrument resembling a tambourine. [Dim. of earlier *timbre* a timbrel <OF. See TIMBRE.]

tim·breled (tim′brəld) *adj.* Chanted to the accompaniment of a timbrel. Also **tim′brelled.**

Tim·buk·tu (tim·buk′tōō, tim′buk·tōō′) A town of central French Sudan, French West Africa, near the Niger; formerly a major center of the slave trade: French *Tombouctou.*

time (tīm) *n.* **1** The general idea, relation, or fact of continuous or successive existence; infinite duration or its measure. **2** A definite portion of duration; a moment; period; season. **3** A considerable period marked off by some special characteristics; era. **4** The portion of duration allotted to some specific purpose, as that allotted to human life or to any particular life, military service, a prison sentence, etc. **5** The length of an apprenticeship. **6** Period of gestation. **7** A portion of duration available or sufficient for, or allotted to, some special purpose or event; also, leisure: I have no *time* to read. **8** Indefinite duration viewed in the concrete as measurable and terminable, but not precisely limited: You build for *time,* we for eternity. **9** A general term indicating a subdivision of one of the grander divisions of geological history. **10** A point in duration; date; occasion; especially, the hour of death or of travail: Your *time* has come! **11** A portion of duration considered as having some quality or experience of its own, personal or general: in the latter sense usually in the plural: *Times* are hard. **12** A system of reckoning or measuring duration, especially with reference to the rotation and revolution of the earth, or to the movements of the celestial bodies. See also DAYLIGHT-SAVING TIME, STANDARD TIME, and lists given below. **13** A case of recurrence or repetition: many a *time,* three *times* a day. **14** The temporal relation of a verb. **15** *Music* **a** The characteristic tempo suited to a particular style of composition. **b** The division of musical composition into measures of equal length; rhythm: common *time,* triple *time.* Rhythms which are divisible by two are called **duple** or **common time,** as 2/2, 2/4, 2/8, 4/2, 4/4, 4/8, etc. Rhythms which are divisible by three are called **triple time,** as 3/2, 3/4, 3/8. **Compound triple times** are 9/4, 9/8, 9/16, 5/4, and 5/8. **16** A measured interval in verse; a unit of duration in rhythmical utterance; a mora. **17** One of the Aristotelian unities of the drama. See under UNITY. **18** Period during which work has been, or remains to be done; also, the amount of pay due one, especially on an hourly rate: *time* and a half for overtime. **19** Rate of movement, as in dancing, marching, etc.; tempo. **20** *pl.* In arithmetic, the fact or process of being multiplied or added to or by: Five *times* four is twenty; also, the multiplication sign ×. **21** Fit or proper occasion: This is no *time* to quibble. —**at the same time** **1** At the same moment. **2** Despite that; however; nevertheless. —**at times** Now and then. —**to bring to time** To call to account; discipline; force to conform. —**to have a time** To experience unusual pleasure, difficulty, etc. —**high time** The expiration of, or a time past the expiration of, a period of which something should have been accomplished. —**in time** **1** While time permits or lasts; before it is too late. **2** In the progress of time; ultimately. —**to keep time** **1** To indicate time correctly, as a clock; run in time, as a train. **2** To make regular or rhythmic movements in unison with another or others. **3** To render a musical composition in proper time or rhythm. **4** To make a record of the number of hours worked by an employee or employees. —**to make a time** To make a fuss or to-do. —**to make time** **1** To gain time; especially, to make up for lost time by extra speed, as a train. **2** To perform, achieve, or arrive in a certain time: to *make* good *time.* **3** *Slang* To impress or influence favorably: with *with.* —**on time** **1** Promptly; according to schedule: The train left *on time.* **2** Paid for, or to be paid for, later or in instalments. —*adj.* **1** Of or pertaining to time. **2** Devised so as to operate, explode, etc., at a specified time: a *time* bomb, *time* lock. **3** Payable at, or to be paid for at, a future date. —*v.t.* **timed, tim·ing** **1** To regulate as to time. **2** To cause to correspond in time: They *timed* their steps to the music. **3** To choose or arrange the time or occasion for: He *timed* his arrival for five o'clock. **4** To mark the rhythm or measure of. **5** To assign metrical or rhythmic qualities to (a syllable or note). **6** To ascertain or record the speed or duration of: to *time* a horse or a race. [OE *tīma*]

—**astronomical time** Prior to Jan. 1, 1925, the 24-hour period reckoned from noon to noon: since that date reckoned from midnight to midnight in order to bring civil and navigational practice into conformity with each other.

—**civil time** (or **civil day**) The 24-hour period extending from midnight to midnight: generally divided into two sections of 12 hours each, but in navigation, aeronautics, and other technical uses reckoned from 0 (midnight) to 24 hours. The same reckoning now applies to *astronomical time.*

—**Greenwich mean time** See CIVIL TIME.

—**Greenwich time** Time as reckoned from the zero meridian of Greenwich, England. To each hour in advance of, or behind, Greenwich time there corresponds a difference of 15 degrees longitude east or west of the Greenwich meridian.

—**local time** Time, whether sidereal or solar, as reckoned from a local meridian other than the standard meridian.

—**mean time** Time reckoned from the hour angle of the mean sun; the *mean solar day* is the 24-hour interval between two successive lower transits of the mean sun across the meridian of a place and corresponds exactly with civil time.

—**sidereal time** Time computed from the hour angle of a fixed point on the celestial sphere known as the first point in Aries, coincident with the vernal equinox; the *sidereal day* is the interval between two successive upper transits of the vernal equinox across the meridian.

—**solar time** Time reckoned from the hour angle of the central point of the sun's disk; the *apparent solar day* is the slightly variable interval between two successive lower transits of the sun across the meridian of a place, noon being the moment of upper transit or the hour angle plus 12 hours.

—**zone time** Time corresponding to that within a zone of 7 1/2 degrees on either side of a meridian; used in the determination of a ship's longitude.

Synonyms (noun): age, date, duration, epoch, era, period, season, sequence, succession. *Sequence* and *succession* apply to events viewed as following one another; *time* and *duration* denote something conceived of as enduring while events take place and acts are

time and again

done. According to the necessary conditions of human thought, events are contained in *time* as objects are in space, *time* existing before the event, measuring it as it passes, and still existing when the event is past. *Duration* and *succession* are more general words than *time*; we can speak of infinite or eternal *duration* or *succession*, but *time* is commonly contrasted with eternity. *Time* is measured or measurable *duration*.

time and again Frequently. Also **time after time**.

time belt A time zone.

time-card (tīm′kärd′) n. A card for recording the time of arrival and departure of an employee.

time clock A clock equipped for automatically recording times of arrival and departure, or for actuating release mechanisms, as on vault doors, etc.

time exposure *Phot.* A film exposure made at spaced intervals by two separate manual operations of the shutter instead of automatically.

time-hon·ored (tīm′on′ərd) *adj.* Observed or honored from former times; claiming veneration as of long existence. See synonyms under ANCIENT[1]. Also *Brit.* **time′-hon′oured**.

time immemorial A considerable and indefinite length of time; specifically, in law, time beyond legal memory, now reckoned at twenty years: the period of the statute of limitations relating to realty; formerly, "a time whereof the memory of man runneth not to the contrary," fixed, in England, as the commencement of the reign of Richard I.

time·keep·er (tīm′kē′pər) n. 1 One who or that which keeps time. 2 One who declares the time in a race, game, athletic match, etc., or records the hours worked by employees. 3 A railroad train starter. 4 A timepiece.

time·less (tīm′lis) *adj.* 1 Independent of, or unaffected by, time; unending. 2 *Archaic* Untimely. 3 Not assigned or limited to any special time, era, or epoch; without a date. See synonyms under ETERNAL. — **time′less·ly** *adv.* — **time′less·ness** *n.*

time lock A lock, having a clock mechanism attached, so devised as to prevent its being unlocked before a specified time.

time·ly (tīm′lē) *adj.* **·li·er, ·li·est** 1 Being or occurring in good or proper time; opportune; seasonable; also, well-timed. 2 *Archaic* Early. — *adv.* Opportunely; seasonably; early. — **time′li·ness** *n.*

ti·me·ma (ti·mē′mə) n. A stick insect. [<NL]

time·o Dan·a·os et do·na fe·ren·tes (tī·mē′ō dan′ā·ōs et dō′nə fə·ren′tēz) *Latin* I fear the Greeks, even when they bring gifts; hence, the motives of a foe offering a gift are suspect.

time·ous (tī′məs) *adj. Scot.* Seasonable; timely.

time-out (tīm′out′) n. 1 A short recess requested by a team during play. 2 Any interval of rest taken during the course of a regular period of work. Also **time out**.

time out of mind Longer than is known or can be remembered; time immemorial.

time·piece (tīm′pēs′) n. A chronometer; a clock, or watch.

tim·er (tī′mər) n. 1 A timekeeper, or one who gives or officially records time. 2 A stopwatch, as for timing a race. 3 A device attached in an adjustable form to an internal-combustion engine so as to time the spark automatically.

time-sav·ing (tīm′sā′ving) *adj.* Calculated or devised to save time by facilitating work: Vacuum cleaners are *time-saving* devices.

time-serv·er (tīm′sûr′vər) n. One who yields to the apparent demands of the time, without reference to principle; a temporizer. Also **time′-pleas′er** (-plē′zər). — **time′-serv′ing** *adj. & n.*

time signature *Music* A sign placed at the beginning of a composition, immediately after the key signature, to indicate the rhythm or time.

Times Square A square in New York City formed by the intersection of Broadway and Seventh Avenue, extending from 42nd to 45th street; by extension, the area around it, the city's entertainment district.

time·ta·ble (tīm′tā′bəl) n. A tabular statement of the times at which certain things, as arrivals and departures of trains, boats, high and low tides, etc., are to take place.

time·work (tīm′wûrk′) n. Work paid for on the basis of a set wage per hour, day, week, etc. — **time′work′er** *n.*

time-worn (tīm′wôrn′, -wōrn′) *adj.* Showing the ravages of time; affected by time.

time zone One of the 24 established divisions or sectors into which the globe is divided for convenience in reckoning standard time from the meridian of Greenwich: each sector represents 15 degrees of longitude, or a time interval of 1 hour. See table below. See also STANDARD TIME.

WORLD TIME ZONES

Each zone comprises (with certain geographic adjustments) an area 7 1/2 degrees on each side of the reference longitude from Greenwich, and the zone number is equivalent to the number of hours later (−) or earlier (+) than Greenwich time. The places given in parentheses are for convenience of reference.

Zone No. East of Greenwich	Longitude from Greenwich	Zone No. West of Greenwich
0 (Greenwich)	0°	0 (Greenwich)
− 1 (Berlin)	15°	+ 1 (Iceland)
− 2 (Leningrad)	30°	+ 2 (Azores)
− 3 (Baghdad)	45°	+ 3 (Rio de Janeiro)
− 4 (Bokhara)	60°	+ 4 (Halifax)
− 5 (Bombay)	75°	+ 5 (Washington)
− 6 (Lhasa)	90°	+ 6 (Chicago)
− 7 (Singapore)	105°	+ 7 (Denver)
− 8 (Manila)	120°	+ 8 (Vancouver)
− 9 (Kyoto)	135°	+ 9 (Dawson)
−10 (Melbourne)	150°	+10 (Tahiti)
−11 (Kamchatka)	165°	+11 (Nome)
−12 (Fiji Is.)	180°	+12 (Samoa)

(*International Date Line*)

Tim·gad (tim·gäd′) An ancient ruined city in NE Algeria.

tim·id (tim′id) *adj.* Shrinking from danger or publicity; easily frightened; shy; lacking self-confidence. See synonyms under FAINT, PUSILLANIMOUS. [<L *timidus* <*timere* fear] — **ti·mid·i·ty** (ti·mid′ə·tē), **tim′id·ness** *n.* — **tim′id·ly** *adv.*

tim·ing (tī′ming) n. 1 In music, oratory, acting, etc., the act or art of regulating the speed of performance, utterance, etc., so as to accentuate the impressiveness of certain parts; also, the effect produced by such regulation. 2 In certain sports, as swimming, boxing, etc., the regulation of the speed of a blow or stroke so that it reaches its highest effectiveness at just the right moment.

WORLD TIME ZONES
The system of keeping standard time at sea has been adopted by most of the world's navies.

Ti·mi·șoa·ra (tē′mē·shwä′rä) A city in western Rumania: Hungarian *Temesvár*.

ti·moc·ra·cy (tī·mok′rə·sē) n. pl. **·cies** 1 A state in which the honor attaching to the position of ruler becomes an object of contention, and is sought by the ambitious with intrigue, rather than accepted as a trust. 2 A state in which honors are bestowed according to property owned. [<OF *tymocracie* <Med. L *timocratia* <Gk. *timokratia* <*timē* honor + *krateein* rule] — **ti·mo·crat·ic** (tī′mə·krat′ik) *adj.*

ti·mol·o·gy (tī·mol′ə·jē) n. *Philos.* A study or theory of value or excellence, especially of inherent rather than relative value. [<Gk. *timē* honor, valuation + -LOGY] — **ti·mo·log·ic** (tī′mə·loj′ik) or **·i·cal** *adj.* — **ti·mol′o·gist** *n.*

Ti·mon (tī′mən) An Athenian of the fifth century B.C.; called "the Misanthrope"; hero of Shakespeare's *Timon of Athens*.

Ti·mor (tē′môr, ti·môr′) The largest and easternmost of the Lesser Sunda Islands; divided into *Indonesian Timor*, included in Nusa Tenggara province, in the western portion; 5,765 square miles; capital, Kupang: formerly *Netherlands Timor*; and into *Portuguese Timor*, a Portuguese province, in the eastern portion; 5,761 square miles, including the exclave of Ambeno and the islands of Atauro and Jaco; capital, Dili.

Timor Archipelago See LESSER SUNDA ISLANDS.

Ti·mor-laut (tē′môr-lout) See TANIMBAR ISLANDS. Also **Ti·mor-laoet** (tē′môr-lout).

tim·or·ous (tim′ər·əs) *adj.* 1 Fearful of danger; timid. 2 Indicating or produced by fear. See synonyms under PUSILLANIMOUS. [<OF *timoureus, temeros* <Med. L *timorosus*, ult. <L *timor, -oris* fear] — **tim′or·ous·ly** *adv.* — **tim′or·ous·ness** *n.*

Timor Sea An arm of the Indian Ocean between northern Australia and Timor.

Ti·mo·shen·ko (tē′mō·sheng′kō), **Semion Konstantinovich**, born 1895, Russian marshal in World War II.

tim·o·thy (tim′ə·thē) n. A perennial fodder grass (*Phleum pratense*) having its flowers in a long, dense, cylindrical, spikelike panicle. Also **timothy grass**. [after *Timothy* Hanson, who took the seed from New York to the Carolinas about 1720]

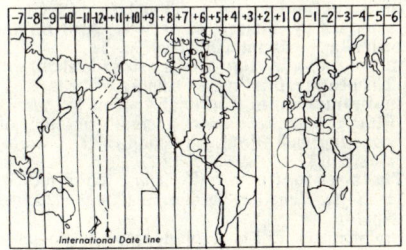

TIMOTHY
(Plant to 6 feet high)

Tim·o·thy (tim′ə·thē) n. A masculine personal name. Also *Dan., Du., Ger., Sw.* **Ti·mo·the·us** (tē·mō′tē·ōōs), *Fr.* **Ti·mo·thée** (tē·mō·tā′), *Ital., Sp.* **Ti·mo·te·o** (tē·mō·tā′ō), *Pg.* **Ti·mo·the·o** (tē′mō·tā′ōō). [<Gk. *honoring a god*]
— **Timothy** A convert and companion of the apostle Paul, either of two pastoral epistles in the New Testament, addressed to Timothy and attributed to Paul.

Ti·mour, Ti·mur (tē·mōōr′, tē-) See TAMERLANE.

tim·pa·ni (tim′pə·nē) n. pl. sing. **·pa·no** (-pə·nō) Kettledrums; a set of kettledrums in an orchestra: also spelled *tympani*. [<Ital., pl. of *timpano* <L *tympanum* a drum <Gk. *tympanon*] — **tim′pa·nist** *n.*

tim·pa·num (tim′pə·nəm) See TYMPANUM.

tin (tin) n. 1 A white, malleable, metallic element (symbol Sn) of low tensile strength and crystalline structure, found chiefly in combination and extensively used in making alloys. See ELEMENT. 2 Tin plate. 3 An article of tinware; a container or box made of tin. 4 *Brit.* A tin container for preserved foods; a can. 5 *Slang* Money. — *v.t.* **tinned, tin·ning** 1 To coat or cover with tin or tin plate. 2 To pack or put up in tins. — *adj.* Made of tin. [OE]

Ti·na (tē′nə) Diminutive of CHRISTINA.

tin·a·mou (tin′ə·mōō) n. Any of certain South American birds (family *Tinamidae*), resembling quails, and hunted as game birds. [<F <Cariban *tinamu*]

tin·cal (ting′kal, -käl, -kôl) n. Native borax. [<Malay *tiṅkal* <Persian *tiṅkāl, tiṅkar* <Skt. *ṭaṅkaṇa* borax]

tinct (tingkt) v.t. To tinge; tint. — *adj. Poetic* Slightly tinged. — n. 1 *Poetic* A tint. 2 *Obs.* A tincture; specifically, the elixir vitae. [<L *tinctus*, pp. of *tingere* dye, color]

tinc·to·ri·al (tingk·tôr′ē·əl, -tō′rē-) *adj.* 1 Of or pertaining to color or hue. 2 Affording or imbuing with tint or color. [<L *tinctorius* <*tinctus*. See TINCT.]

tinc·ture (tingk′chər) n. 1 A solution, usually in alcohol, of some principle used in medicine. 2 A tinge of color; tint. 3 A slight flavor superadded; modicum; spice. 4 That part of a substance which is extracted by a solvent. 5 One of the metals, colors, or furs used in heraldic description. — *v.t.* **·tured, ·tur·ing** 1 To impart a slight hue or tinge to. 2 To imbue with flavor, odor, etc. 3 To imbue with a specified moral or mental quality. [<L *tinctura* a dyeing <*tinctus*. See TINCT.]

tin·der (tin′dər) n. Any readily combustible substance, as charred linen or touchwood, that will ignite (without explosion) on contact with a spark. [OE *tynder*] —**tin'der·y** *adj.*

tin·der·box (tin'dər·boks') *n.* **1** A portable metallic box containing tinder, and usually flint and steel to ignite it. **2** A highly inflammable mass of material. **3** A person with an easily excitable temper.

tine (tīn) *n.* A spike or prong, as of a fork or of an antler. [OE *tind*] —**tined** *adj.*

tin·e·a (tin'ē·ə) *n.* **1** Any of a genus (*Tinea*) of small, narrow-winged moths, including the case-making clothes moth (*T. pellionella*). **2** *Pathol.* Ringworm; any fungous skin disease. [<NL <L, a moth, gnawing worm]

tin·e·id (tin'ē·id) *adj.* Of or pertaining to a family (*Tineidae*) of moths. —*n.* One of the *Tineidae.* [<NL <*Tinea* TINEA]

tin·foil (tin'foil') *n.* Tin or an alloy of tin made into thin sheets for use as wrapping material and in decoration.

ting[1] (ting) *n.* A single high metallic sound, as of a small bell. —*v.t.* & *v.i.* To give forth or cause to give forth a ting. [Imit.]

ting[2] (ting) See THING[2].

ting-a-ling (ting'ə·ling') *n.* The sound of a little bell.

tinge (tinj) *v.t.* **tinged, tinge·ing** or **ting·ing** **1** To imbue with a faint trace of color; impart a tint to. **2** To impart a slight characteristic quality of some other element to. See synonyms under STAIN. —*n.* **1** A faint trace of added color. **2** A quality or peculiar characteristic imparted to something by the slight admixture of some foreign element. [<L *tingere* dye]

tin·gle (ting'gəl) *v.* **·gled, ·gling** *v.i.* **1** To experience a prickly, stinging sensation, as the skin from exposure to cold, or the ears from a sharp blow. **2** To cause such a sensation. —*v.t.* **3** To cause to tingle. —*n.* **1** A prickly, stinging sensation; a tingling. **2** A jingle or tinkling. [Appar. var. of TINKLE] —**tin'gler** *n.* —**tin'gly** *adj.*

Ting·ley (ting'lē), **Katherine**, 1847–1929, *née* Wescott, U.S. theosophist.

tin·horn (tin'hôrn') *Slang n.* **1** A pretentious person without any real ability, power, influence, etc. **2** A gambler who bets with low stakes. —*adj.* **1** Resembling or characteristic of a cheap gambler. **2** Pretentious. [With ref. to the fine appearance, but poor quality, of a tin horn]

Tin·i·an (tin'ē·ən, tē'nē·än') One of the southern Marianas Islands; 39 square miles.

tink (tingk) *v.i.* To make a single or separate tinkling sound; chink. —*n.* A tinkle or tinkling. [Imit.]

tink·er (tingk'ər) *n.* **1** An itinerant mender of domestic tin utensils, as pots and pans. **2** Loosely, one who does repairing work of any kind; a jack-of-all-trades. **3** A clumsy workman; a botcher. **4** The act of roughly repairing; hasty workmanship. **5** A young mackerel about two years old. **6** The chub mackerel. **7** The razor-billed auk. —*v.i.* **1** To work as a tinker. **2** To work in a clumsy, makeshift fashion on anything. **3** To potter; fuss. —*v.t.* **4** To mend as a tinker. **5** To repair clumsily or inexpertly. [Var. of earlier *tinekere* a worker in tin]

tinker's damn *Slang* Any useless or worthless article: commonly in the phrase *not worth a tinker's damn.* Also **tinker's dam.** [<TINKER + DAMN; with ref. to the reputed profanity of tinkers]

tin·kle (ting'kəl) *v.* **·kled, ·kling** *v.i.* **1** To produce slight, sharp, metallic sounds, as a small bell. —*v.t.* **2** To cause to tinkle. **3** To summon or signal by a tinkling. —*n.* A sharp, clear, tinkling sound. [Freq. of TINK] —**tin'kling** *adj.* & *n.* —**tin'kly** *adj.*

tin lizzie *U.S. Slang* The Model T automobile.

tin·ner (tin'ər) *n.* **1** A miner employed in tin mines. **2** A maker of or dealer in tinware; a tinsmith.

tin·ni·tus (ti·nī'təs) *n. Pathol.* A subjective ringing, rushing, or buzzing sound in the ears, not caused by any external stimulus. [<NL <L <*tinnire* ring]

tin·ny (tin'ē) *adj.* **·ni·er, ·ni·est** **1** Pertaining to, composed of, or abounding in tin. **2** Sounding as if a tin pan were being struck: a *tinny* sound. **3** Tasting of tin, as food from a can. —**tin'ni·ly** *adv.* —**tin'ni·ness** *n.*

tin-pan (tin'pan') *adj.* Noisy; clanging; inharmonious; tinny. Also **tin'-pan'ny.**

tin-pan alley (tin'pan') **1** A street or section of a city frequented by musicians and song writers and occupied by publishers of popular music: originally used to designate a section of New York where cheap, tinny-sounding pianos were supposedly heard in publishers' offices. **2** The composers and publishers of popular music, collectively.

tin-plate (tin'plāt') *v.t.* **-plat·ed, -plat·ing** To plate with tin. —**tin'-plat'er** *n.*

tin plate Sheet iron or steel plated with tin.

tin·sel (tin'səl) *n.* **1** Very thin glittering bits of brass, copper, and other cheap metals, used for display and to ornament articles of dress; also, the thin metal from which they are cut. **2** A fabric in which such spangles or bits of metal are woven, or to which they are attached; also, a fabric or yarn containing gold or silver thread. **3** Anything sparkling and showy, with little real worth; superficial adornment and brilliancy. —*adj.* **1** Made or covered with tinsel. **2** Of tinsel-like qualities; superficially brilliant; tawdry. —*v.t.* **·seled** or **·selled, ·sel·ing** or **·sel·ling** **1** To adorn or decorate with or as with tinsel. **2** To give a metallic appearance to (ceramic ware) by washing with a metallic substance. [<MF *étincelle* <OF *estincelle* <L *scintilla* a spark]

tin·smith (tin'smith') *n.* One who works with tin or tin plate.

tin spirits *Chem.* A solution of a tin salt in acid, used in dyeing.

tin·stone (tin'stōn') *n.* Cassiterite.

tint[1] (tint) *n.* **1** A variety of color; tincture; specifically, a tendency toward or slight admixture of a different color; tinge: red with a blue *tint.* **2** A gradation or shading of a color made by dilution with white to lessen its chroma and saturation. **3** Any color having a brilliance higher than that of median gray. **4** In engraving, an effect of light, shade, texture, etc., produced by the spacing of lines or by hatching. **5** An impression from a block bearing a design to be printed in a faint color as a background: used on checks as a safeguard against erasure. —*v.t.* **1** To give a tint to; tinge. **2** In engraving, to form a tint upon. See synonyms under STAIN. [Alter. of TINCT; ? infl. in form by Ital. *tinta* color] —**tint'er** *n.*

tint[2] (tint) *Scot.* Past tense and past participle of TINE[2].

Tin·tag·el Head (tin·taj'əl) A promontory with castle ruins in western Cornwall, England; traditionally, the birthplace of King Arthur.

Tin·tern Abbey (tin'tərn) The ruins of a Cistercian abbey founded in the 12th century in Monmouth, England, on the Wye.

tin·tin·nab·u·lar (tin'ti·nab'yə·lər) *adj.* Characterized by tinkling, as of bells. Also **tin'tin·nab'u·lar'y, tin'tin·nab'u·lous.**

tin·tin·nab·u·la·tion (tin'ti·nab'yə·lā'shən) *n.* The pealing, tinkling, or ringing of bells.

tin·tin·nab·u·lum (tin'ti·nab'yə·ləm) *n. pl.* **·la** (-lə) A bell; especially, a small tinkling or signaling bell. [<L, a small bell <*tintinnare* ring]

Tin·to·ret·to (tin'tə·ret'ō, *Ital.* tēn'tō·ret'tō), 1518–94, Venetian painter: real name *Jacopo Robusti.*

tin·type (tin'tīp') *n.* A photograph taken on a sensitized film supported on a thin sheet of enameled tin or iron; a ferrotype.

tin·ware (tin'wâr') *n.* Household articles, collectively, made of tin plate.

tin·work (tin'wûrk') *n.* **1** Articles made of tin; work with tin. **2** *pl.* A place or establishment where tin is manufactured or mined.

ti·ny (tī'nē) *adj.* **·ni·er, ·ni·est** Very small; minute; wee. See synonyms under LITTLE, MINUTE[2], SMALL. [< obs. *tine* a small amount, bit + -Y[3]; ult. origin unknown]

-tion *suffix of nouns* **1** Action or process of: *rejection.* **2** Condition or state of being: *completion.* **3** Result of: *connection.* Also *-ation, -cion, -ion, -sion, -xion.* [<F *-tion* <OF *-cion* <L *-tio, -tionis*]

tip[1] (tip) *n.* A slanting or inclined position; a tilt. —*v.* **tipped, tip·ping** *v.t.* **1** To cause to lean by lowering or raising one end or side; cant; tilt. **2** To overturn or upset: often with *over.* —*v.i.* **3** To become tilted; slant. **4** To overturn; topple: with *over.* [ME *tipen* overturn; origin uncertain] —**tip'per** *n.*

Synonyms (verb): cant, careen, heel, incline, lean, list, slant, slope, tilt. To *tilt* or *tip* is to throw out of a horizontal position by raising one side or end or lowering the other. *Slant* and *slope* are said of things somewhat fixed or permanent in a position out of the horizontal or perpendicular: the roof *slants,* the hill *slopes. Incline* is a more formal word for *tip,* and also for *slant* or *slope.* To *cant* is to set slantingly; in many cases *tip* and *cant* might be interchanged, but *tip* is more temporary, often momentary; one *tips* a pail so that the water flows over the edge; a mechanic *cants* a table by making or setting one side higher than the other. *Careen, heel,* and *list* are used of vessels which from any cause, as leakage, shifting of cargo, etc., are off an even keel.

tip[2] (tip) *v.t.* **tipped, tip·ping** **1** To strike lightly, or with something light; tap. **2** In baseball, to strike (the ball) a light, glancing blow. —*n.* A tap; light blow. [Earlier *tippe,* prob. <LG. Cf. Du. *tippen* tap.]

tip[3] (tip) *n.* **1** A small gift of money for services rendered, given to a servant, waiter, porter, or the like. **2** A friendly, helpful hint; specifically, secret information presumed to increase a better's or speculator's chance of winning. [< *v.*] —*v.* **tipped, tip·ping** *v.t.* **1** To give a small gratuity to. **2** *Colloq.* To give secret information to, as in betting and speculation: often with *off.* —*v.i.* **3** To give tips. [Orig. <thieves' cant, ? <TIP[2]] —**tip'·per** *n.*

tip[4] (tip) *n.* **1** The point or extremity of anything tapering; end: the *tip* of the tongue. **2** A piece or part made to form the end of anything, as a nozzle, ferrule, etc. **3** The upper part of a hat crown; also, the lining in the upper part of the crown. —*v.t.* **tipped, tip·ping** **1** To furnish with a tip. **2** To form the tip of. **3** To cover or adorn the tip of. [Prob. <MDu., a point]

ti palm See TI[2].

tip·cart (tip'kärt') *n.* A cart having a body that can be tipped for unloading.

tip·cat (tip'kat') *n.* A game played with a stick or bat and a small piece of wood pointed at the ends and called a *cat,* which the batter hits lightly into the air and then hits again, trying to drive it as far as possible; also, the cat. [<TIP[1] + CAT]

ti·pi (tē'pē) See TEPEE.

tip-off (tip'ôf', -of') *n. Colloq.* A hint or warning.

Tip·pe·ca·noe (tip'ē·kə·nōō') The nickname of William Henry Harrison: from his victory over Tecumseh's Indians at Tippecanoe River in 1811. The name provided the presidential campaign slogan, **Tippecanoe and Tyler too,** for Harrison and his vice-presidential running mate, John Tyler, in 1840.

Tip·pe·ca·noe River (tip'ē·kə·nōō') A river in north central Indiana, flowing 166 miles NW, west, and SW to the Wabash River near Lafayette; scene of General W. H. Harrison's victory over Indians, 1811.

Tip·per·ar·y (tip'ə·râr'ē) A county of NE Munster province, Ireland; 1,643 square miles; county town, Clonmel.

tip·pet (tip'it) *n.* **1** An outdoor covering for the neck, or neck and shoulders, hanging well down in front. **2** *Eccl.* A long scarf worn by clergymen in the Anglican Church. **3** A ruff of feathers on birds, etc. [Prob. dim. of TIP[4]]

tip·ple[1] (tip'əl) *v.t.* & *v.i.* **·pled, ·pling** To drink (alcoholic beverages) frequently and habitually. —*n.* Liquor consumed in tippling. [Cf. Norw. *tipla* drip, tipple] —**tip'pler** *n.*

tip·ple[2] (tip'əl) *n.* **1** An apparatus for tipping loaded cars. **2** The place where such tipping is done. [<dial. E *tipple* topple, freq. of TIP[1]]

tip·py (tip'ē) *adj.* **·pi·er, ·pi·est** *Colloq.* Shaky; unsteady; apt to tip over. [<TIP[1] + -Y[3]]

tip·staff (tip'staf', -stäf') *n.* **1** *pl.* **·staffs** In England, a sheriff's subordinate; bailiff; constable; also, a court crier. **2** *pl.* **·staves** (-stāvz') A staff having a metal tip: a badge of office. [<TIP(PED) STAFF]

tip·ster (tip'stər) *n. Colloq.* One who sells tips for betting, as on a race. [<TIP[3]]

tip·sy (tip'sē) *adj.* **·si·er, ·si·est** **1** Befuddled with drink, but not really drunk; partially intoxicated; high. **2** Tippy; shaky; also, crooked; askew. [<TIP[1]] —**tip'si·ly** *adv.* —**tip'si·ness** *n.*

tip·toe (tip'tō') *n.* **1** The tip of a toe, or the tips of all the toes collectively. **2** Topmost height; also, alertness of expectation: usually in the phrase **to be on tiptoe** or **a-tiptoe,** to be eagerly expectant. —*v.i.* **·toed, ·toe·ing** To walk on tiptoe; go stealthily. —*adj.* **1** Standing on tiptoe. **2** Quiet; gentle; stealthy. —*adv.* On tiptoe, in any sense.

tip–top (tip′top′) *Colloq. adj.* Best of its kind; first-rate. —*n.* The highest point, quality, or degree; the very top; the best. —*adv.* In a tip-top manner. [<TIP[4] + TOP[1]] —**tip′-top′per** *n.*

Ti·pu Sa·hib (ti′pōō sä′hib), 1753?–99, sultan of Mysore; fought against the British 1775–79. Also **Tip′poo Sa′hib.**

Ti·rach Mir (tē′rəch mēr′) See TIRICH MIR.

ti·rade (tī′rād, tə·rād′) *n.* **1** A prolonged declamatory outpouring, as of censure. **2** *Music* A diatonic run, filling the interval between two musical notes. [<F <Ital. *tirata* a volley, pp. of *tirare* fire, pull]

ti·rail·leur (tir′ə·lûr′, *Fr.* tē·rà·yœr′) *n.* A sharpshooter; skirmisher. [<F]

Ti·ra·na (tē·rä′nə) The capital of Albania, in the central part. Also **Ti·ra′në.**

tire[1] (tīr) *v.* **tired, tir·ing** *v.t.* **1** To reduce the strength of, as by toil; weary; fatigue. **2** To reduce the interest or patience of, as with tediousness. —*v.i.* **3** To become weary or exhausted. **4** To lose patience, interest, etc. —**to tire of** To become weary of or impatient with. —**to tire out** To weary completely. —*n. Dial.* The sensation of fatigue; weariness. [OE *tiorian, teorian*]

Synonyms (verb): exhaust, fag, fatigue, harass, jade, weary. To *tire* is to reduce one's strength in any degree by exertion; one may be *tired* just enough to make rest pleasant, or even unconsciously *tired,* becoming aware of the fact only when he ceases the exertion. One who is *fatigued* suffers from painful lack of strength as the result of overtaxing; an invalid may be *fatigued* with very slight exertion; when one is *wearied,* the painful lack of strength is the result of long-continued demand or strain; one is *exhausted* when the strain has been so severe and continuous as utterly to consume the strength, so that further exertion is for the time impossible. One is *fagged* by drudgery; he is *jaded* by incessant repetition of the same act until it becomes increasingly difficult or well-nigh impossible; as, a horse is *jaded* by a long and unbroken journey. See WEAR[1].

tire[2] (tīr) *n.* **1** A band or hoop surrounding the rim of a wheel. **2** A flexible tube, usually of inflated rubber, set in a rim and protected by an outer covering: used on automobiles, bicycles, etc., to reduce vibration. —*v.t.* **tired, tir·ing** To furnish with a tire; put a tire on. Also, *Brit., tyre.* [Special use of TIRE[4]]

tire[3] (tīr) *Archaic v.t.* **1** In falconry, to rend and devour; draw; pull. —*v.i.* **2** To prey. **3** To be preoccupied; dote; gloat. [<OF *tirer;* ult. origin uncertain]

tire[4] (tīr) *Obs. v.t.* To attire; dress; adorn. —*n.* **1** A tiara; headdress. **2** Attire. [Aphetic var. of ATTIRE]

tire[5] (tīr) *n.* A volley of cannon; a broadside. [<OF *tir* < *tirer* draw, shoot; ult. origin uncertain]

tired (tīrd) *adj.* Weary; exhausted; jaded; fatigued. [Orig. pp. of TIRE[1]] —**tired′ly** *adv.* —**tired′ness** *n.*

tire·less (tīr′lis) *adj.* Proof against fatigue; untiring. See synonyms under INDEFATIGABLE. [<TIRE[1] + -LESS] —**tire′less·ly** *adv.* —**tire′less·ness** *n.*

Ti·re·si·as (tī·rē′sē·əs) In Greek mythology, a Theban soothsayer, blinded by Athena whom he saw bathing: in recompense she gave him power to foretell the future. —**Ti·re′si·an** *adj.*

tire·some (tīr′səm) *adj.* Tending to tire, or causing one to tire; tedious. See synonyms under TEDIOUS, TROUBLESOME, WEARISOME. —**tire′some·ly** *adv.* —**tire′some·ness** *n.*

tire·wom·an (tīr′wŏŏm′ən) *n. pl.* **·wom·en** (-wim′in) *Obs.* A lady's maid; an abigail. Also **tir′ing-wom′an.** [<TIRE[4] + WOMAN]

Ti·rich Mir (tē′rich mēr′) The highest mountain in the Hindu Kush, in extreme NW West Pakistan near the border of Afghanistan; 25,263 feet: also *Tirach Mir.*

tiring room *Archaic* A dressing-room, especially in a theater. [<*tiring,* ppr. of TIRE[4] + ROOM]

tirl (tûrl) *Scot. v.t.* To cause to produce a vibrating or thrilling noise, as by plucking a string. —*n.* A vibrating or thrilling noise.

tirl·ing pin (tûr′ling) *Scot.* A vertical twisted iron bar passed through a loose ring and fastened to a door: formerly used as a knocker.

Tir·no·vo (tir′nô·vô) A city in northern Bulgaria; scene of the declaration of the country's independence, 1908: also *Trnovo.*

ti·ro (tī′rō) See TYRO.

Ti·rol (ti·rōl′, tir′ōl, tī′rōl) An autonomous province of western Austria in the eastern Alps north of Italy; 4,883 square miles; capital, Innsbruck: also *Tyrol.* —**Ti·ro′le·an** *adj. & n.*

Tir·o·lese (tir′ō·lēz′, -lēs′) *adj.* Of or pertaining to Tirol or its inhabitants. —*n. pl.* **·lese** A native of Tirol. Also *Tyrolese.*

Tir·pitz (tir′pits), **Alfred von,** 1849–1930, German admiral.

tir·ri·vee (tir′ə·vē) *n. Scot.* A burst of ill-humor; fit of passion; tantrum.

Tir·so de Mo·li·na (tir′sō thā mō·lē′nä), 1571?–1648, Spanish dramatist: real name Gabriel Téllez.

Ti·ruch·i·rap·pal·li (ti·rŏŏch′ē·räp′ə·lē) A city in east central Madras State, India: also *Trichinopoly.*

'tis (tiz) It is: a contraction.

Ti·sa (tē′sä) The Czech and Rumanian name for the Tisza.

ti·sane (ti·zan′, *Fr.* tē·zàn′) *n.* A slightly medicated decoction, usually of herbs, prepared for the sick; a ptisan. [<F <L *ptisana* a ptisan]

Tish·ri (tish′rē) The first month of the Hebrew calendar. See CALENDAR (Hebrew). Also **Tis·ri** (tiz′rē). [<Hebrew <Aramic *tishrī* < *sherā* begin; infl. by Babylonian *tashrītu* the seventh month, first month of the second half of the year]

Ti·siph·o·ne (ti·sif′ə·nē) In Greek mythology, one of the three Furies.

tis·sue (tish′ōō) *n.* **1** Any light or gauzy textile fabric, usually of silk; originally, cloth interwoven with gold or silver thread. **2** *Biol.* One of the elementary aggregates of cells and their products, developed by plants and animals for the performance of a particular function: connective *tissue.* **3** A connected or interwoven series; chain; fabrication: a *tissue* of lies. **4** Tissue paper. —*v.t.* **·sued, ·su·ing** *Rare* **1** To make into tissue. **2** To adorn with tissue; weave. [<OF *tissu* a rich stuff, orig. pp. of *tistre* weave <L *texere*]

tissue culture The science and art of growing body tissues in a culture medium.

tis·sued (tish′ōōd) *adj.* **1** Clad in tissue. **2** Variegated.

tissue paper Very thin, unsized, almost transparent paper for wrapping delicate articles, protecting engravings, etc.

Ti·sza (ti′so) A river flowing 800 miles from the Carpathian Mountains, in SW Ukrainian S.S.R., south through Hungary and Yugoslavia to the Danube north of Belgrade: German *Theiss,* Czech and Rumanian *Tisa.*

tit[1] (tit) *n.* **1** A titmouse. **2** One of various other small birds, as a titlark, etc. [Short for TITMOUSE, TITLARK, etc]

tit[2] (tit) *n.* A light blow; tap: chiefly in the phrase *tit for tat.* [Var. of TIP[2]]

tit[3] (tit) *n.* Teat; breast; nipple. [OE *titt*]

tit[4] (tit) *n.* **1** A small or worn-out horse; a nag. **2** *Slang* A young woman or girl: a disrespectful term. [ME, a little thing, ? <Scand. Cf. dial. Norw. *titta* little girl.]

ti·tan (tīt′n) *n.* Any person having gigantic strength or size; a giant. —*adj.* Titanic. [after *Titan*] —**tī′tan·ess** *n. fem.*

Ti·tan (tīt′n) **1** In Greek mythology, one of a race of giant gods, children of Uranus and Gaea, who were vanquished and succeeded by the Olympian gods, who imprisoned them in Tartarus. **2** Helios: so called by some Latin poets.

ti·tan·ate (tī′tə·nāt) *n. Chem.* A salt or ester of titanic acid. [<TITAN(IC)[2] + -ATE[3]]

Ti·tan·esque (tīt′n·esk′) *adj.* Of or befitting the Titans; gigantic.

Ti·tan·ess (tīt′n·is) A female Titan.

Ti·ta·ni·a (ti·tā′nē·ə, tī-) Queen of fairyland and wife of Oberon in Shakespeare's *A Midsummer Night's Dream.*

ti·tan·ic[1] (tī·tan′ik) *adj.* Gigantic; huge; tremendous. [<Gk. *titanikos* <*Titanes* the Titans]

ti·tan·ic[2] (tī·tan′ik, ti-) *adj. Chem.* Of or pertaining to titanium, especially in its higher valence. [<TITAN(IUM) + -IC]

Ti·tan·ic (tī·tan′ik) *adj.* Pertaining to the Titans, characteristic of, or resembling the Titans.

titanic acid *Chem.* **1** A white pulverulent titanium dioxide, TiO_2, found native as rutile, etc.: a common constituent of iron ores: also **titanic oxide. 2** One of various weak acids derived from titanium dioxide.

ti·tan·if·er·ous (tī′tən·if′ər·əs) *adj.* Containing or yielding titanium. [<TITAN(IUM) + -(I)FEROUS]

Ti·tan·ism (tīt′n·iz′əm) *n.* Defiance of, or rebellion against, constituted authority or social conventions: a characteristic attributed to the Titans in Greek mythology.

ti·tan·ite (tī′tən·īt) *n.* Sphene. [<G *titanit* < *titanium* titanium]

ti·ta·ni·um (ti·tā′nē·əm, ti-) *n.* A widely distributed dark-gray metallic element (symbol Ti) resembling tin and silicon, found in small quantities in many minerals and used to toughen steel alloys. See ELEMENT. [<NL <L *Titani* the Titans <Gk. *Titanes*; named on analogy with *uranium*]

titanium tetrachloride *Chem.* A colorless liquid compound, $TiCl_4$, used as a smoke-producing agent in warfare.

Ti·tan·om·a·chy (tī′tən·om′ə·kē) *n.* In Greek mythology, the war of the Titans against the Olympian gods. [<Gk. *Titanomachia* <*Titan* a Titan + *machē* a battle]

ti·tan·o·there (tī′tən·ə·thir′, tī·tā′nə-, ti-) *n. Paleontol.* Any of an extinct family (*Titanotheriidae*) of large, odd-toed ungulates resembling the rhinoceros and common in the Lower Eocene of the Tertiary period. [<NL <Gk. *Titan* a Titan + *thērion,* dim. of *thēr* a wild beast]

ti·tan·ous (tī′tən·əs, tī·tan′əs, ti-) *adj. Chem.* Of or pertaining to titanium, especially in its lower valence. [<TITAN(IUM) + -OUS[2]]

tit·bit (tit′bit′) See TIDBIT.

Tite (tēt) French form of TITUS.

ti·ter (tī′tər, tē′-) *n. Chem.* **1** The strength or concentration of a solution as determined by titration. **2** The temperature at which a molten fatty acid or wax solidifies. Also spelled *titre.* [<F *titre* the fineness of gold or silver alloy]

tit for tat Retaliation in kind; blow for blow. [? Alter. of *tip for tap*; ? infl. in form by MF *tant pour tant* tit for tat]

tith·a·ble (tī′thə·bəl) *adj.* Liable to be tithed, as property.

tithe (tīth) *n.* **1** A tax or assessment of one tenth, especially when payable in kind; loosely, any ratable tax. **2** Specifically, in England, a tenth part of the yearly proceeds arising from lands and from the personal industry of the inhabitants, for the support of the clergy and the church. **3** The tenth part of anything; hence, a small part. —*v.t.* **tithed, tith·ing 1** To give or pay a tithe, or tenth part of. **2** To tax with tithes. [ME *tithe, tethe,* OE *tēotha, tēogotha* a tenth] —**tith′er** *n.*

tith·ing (tī′thing) *n.* **1** The act of levying tithes. **2** A tenth part. **3** In old English law, a civil division composed of ten freeholders and their families.

tith·ing·man (tī′thing·mən) *n. pl.* **·men** (-mən) **1** Anciently, in England, the chief of a tithing; more recently, a constable. **2** In the New England colonies, an officer for enforcing Sunday observance and order.

Ti·tho·nus (ti·thō′nəs) In Greek mythology, a son of Laomedon who was loved by Eos. She persuaded Zeus to grant him immortality but neglected to request for him eternal youth, so that Tithonus shriveled as he grew older and older, and was finally changed into a grasshopper. [<L <Gk. *Tithōnos*]

ti·ti[1] (tē′tē) *n.* **1** An evergreen shrub or small tree (*Cliftonia monophylla*) with fragrant white flowers, native in swamps of the southern United States. **2** Any of a genus (*Cyrilla*) of related trees of tropical America; especially, the **white titi** (*C. racemiflora*). [<Sp. <Aymaran]

ti·ti[2] (tē·tē′) *n.* One of several small South American monkeys (genus *Callicebus*). [<Sp. *titi* <Guarani *titi*]

ti·tian (tish′ən) *n.* A reddish-yellow color

add,āce,câre,pälm; end, ēven; it,īce; odd,ōpen,ôrder; tŏŏk,pōōl; up,bûrn; ə = a in *above,* e in *sicken,* i in *clarity,* o in *melon,* u in *focus;* yōō = u in *fuse;* oi,oil; ou,pout; ch,check; g,go; ng,ring; th,thin; th,this; zh,vision. Foreign sounds à,œ,ü,kh,ṅ; and ◆: see page xx. < from; + plus; ? possibly.

much used by Titian, especially in painting women's hair. —*adj.* Having or pertaining to the color titian. [after *Titian*]

Ti·tian (tish′ən), 1477–1576, Venetian painter: real name *Tiziano Vecellio.*

Ti·ti·ca·ca (tē′tē·kä′kä), **Lake** The largest lake in South America, in the Andes between SE Peru and west central Bolivia; 3,200 square miles; elevation, 12,500 feet; the highest large lake in the world.

tit·il·lant (tit′ə·lənt) *n.* An excitant. [<L *titillans, -antis,* ppr. of *titillare* tickle]

tit·il·late (tit′ə·lāt) *v.t.* **·lat·ed, ·lat·ing** 1 To cause a tickling sensation in. 2 To excite pleasurably in any way. [<L *titillatus,* pp. of *titillare* tickle]

tit·il·la·tion (tit′ə·lā′shən) *n.* 1 The act of titillating, or the state of being titillated. 2 Any momentary exciting or gratifying sensation. —**tit′il·la′tive** *adj.*

tit·i·vate (tit′ə·vāt) *v.t.* & *v.i.* **·vat·ed, ·vat·ing** *Colloq.* To put on decorative touches; smarten; dress up: also spelled *tittivate.* [Earlier *tidivate, tiddivate,* ? <TIDY, on analogy with *cultivate*] —**tit′i·va′tion** *n.*

tit·lark (tit′lärk′) *n.* A pipit. [ME *tit* a little thing + LARK]

ti·tle (tīt′l) *n.* 1 *Law* **a** The means whereby the owner of lands has the just possession of his property; the union of possession, the right of possession, and the right of property in lands and tenements; also, the legal evidence of one's right of property, or the means by or source from which one's right to property has accrued: *title* by purchase. **b** The distinguishing form of words that heads or opens a legal document or statute; also, the opening clause containing the name of the court in which any action is pending, together with the names of the parties, etc. 2 A claim based on an acknowledged or alleged right: What is his *title* to credence? 3 A section or division of a statute, legal document, treatise, or the like. 4 An inscription that serves as a name for designating something, as a book or legal document. 5 A name; descriptive designation. 6 An appellation significant of office, rank, etc.; especially, a designation of nobility. 7 In or near Rome, a church or parish headed by a cardinal: so called because dedicated to or named after the title of some martyr or saint. 8 A source of maintenance, as a patrimony, or a place of duty, especially with income attached, a right or nomination to which is a canonical prerequisite to ordination. 9 In some sports, supremacy; championship: to play for the *title.* See synonyms under NAME. —*v.t.* **·tled, ·tling** 1 To give a name to; entitle; call. 2 To confer an honorary title upon; ennoble. [<OF <L *titulus* a label, an inscription. Doublet of TITTLE.] —**ti′tle·less** *adj.*

ti·tled (tīt′ld) *adj.* Having a title, especially of nobility.

ti·tle·hold·er (tīt′l·hōl′dər) *n.* One who possesses a title, especially a championship title. Also **ti′tlist.** —**ti′tle·hold′ing** *adj.*

title page A page containing the title of a work and the names of its author and its publisher.

title role The character in a play, opera, or motion picture for whom it is named.

tit·man (tit′mən) *n. pl.* **·men** (-mən) *U.S. Colloq.* 1 The smallest pig in a litter; the runt of a litter of pigs. 2 A man small or stunted either physically or mentally. [ME *tit* a little thing + MAN]

tit·mouse (tit′mous′) *n. pl.* **·mice** (-mīs′) Any of several small oscine birds (family *Paridae*) related to the nuthatches; especially, the **tufted titmouse** (*Baeolophus bicolor*) of the United States, having a conspicuous crest. [Alter. of ME *titmose* < *tit-* little + *mose,* alter. of OE *mase* a titmouse; infl. in form by MOUSE]

TITMOUSE (About 5 1/2 inches long)

Ti·to (tē′tō) Italian, Spanish, and Portuguese form of TITUS.

Ti·to (tē′tō), **Marshal,** born 1891?, Yugoslav guerrilla leader in World War II; premier 1945–53; president 1953–: real name *Josip Broz.*

Ti·to·grad (tē′tô·gräd) The capital of Montenegro in southern Yugoslavia: formerly **Pod·go·ri·ca** (pod′gô·rē′tsä).

Ti·to·ism (tē′tō·iz′əm) *n.* The assertion by a Communist state of its national interests in opposition to Soviet domination, such as occurred under Marshal Tito in Yugoslavia.

ti·trate (tī′trāt, tit′rāt) *v.t.* & *v.i.* **·trat·ed, ·trat·ing** *Chem.* To determine the strength of (a solution) by means of standard solutions or by titration. [<F *titrer* < *titre.* See TITER.]

ti·tra·tion (tī·trā′shən, ti-) *n. Chem.* The process of determining the strength or concentration of a given solution by adding to it measured amounts of a standard solution until the desired chemical reaction has been effected.

ti·tre (tī′tər, tē′-) See TITER.

tit·ter (tit′ər) *v.i.* To laugh in a suppressed way, as from nervousness or in ridicule; snicker; giggle. —*n.* The act of tittering; a giggling. [Imit.] —**tit′ter·er** *n.* —**tit′ter·ing·ly** *adv.*

tit·tie (tit′ē) *n. Scot.* A sister. Also **tit′ty.**

tit·tle (tit′l) *n.* 1 The minutest quantity; iota. 2 Originally, a very small mark in writing, as the dot over an *i,* etc.; any diacritical mark. [<L *titulus.* Doublet of TITLE.]

tit·tle-tat·tle (tit′l·tat′l) *n.* 1 Foolish or trivial talk; gossip. 2 An idle, trifling or tattling talker. —*v.i.* **·tled, ·tling** To talk foolishly or idly; gossip; chatter. [Reduplication of TATTLE]

tit·tup (tit′əp) *v.i.* **·tuped** or **·tupped, ·tup·ing** or **·tup·ping** To act in a restless or lively manner; dance along; prance. —*n.* A prancing or curveting action, indicating gaiety or frolicsomeness; a caper. [Appar. imit. of hoof beats]

tit·u·ba·tion (tich′ōō·bā′shən, tit′yə-) *n. Pathol.* A stumbling; tottering; a disturbance of equilibrium resulting in the stumbling gait characteristic of spinal disease. [<L *titubatio, -onis* < *titubatus,* pp. of *titubare* stagger]

tit·u·lar (tich′ōō·lər, tit′yə-) *adj.* 1 Existing in name or title only; nominal. 2 Pertaining to a title. 3 Bestowing or taking title. See TITLE (def. 8). —*n.* One having a title in virtue of which he holds an office or benefice, whether he performs its duties or not; in ecclesiastical law, one holding a sinecure title. Also **tit′u·lar′y** (-ler′ē). [<L *titulus* a title] —**tit′u·lar·ly** *adv.*

Ti·tus (tī′təs) A masculine personal name. [<L, safe]

—**Titus** A disciple of the apostle Paul; also, the epistle in the New Testament addressed to Titus and attributed to Paul.

—**Titus,** A.D. 40?–81, emperor of Rome A.D. 79–81: full name *Titus Flavius Sabinus Vespasianus.*

Ti·u (tē′ōō) In Teutonic mythology, god of war and sky: identified with the Norse *Tyr.*

Tiv·o·li (tiv′ə·lē, *Ital.* tē′vô·lē) A town in central Italy NE of Rome: ancient *Tibur.*

tiv·y (tiv′ē) *adv.* With great speed: a hunting cry. [Appar. short for TANTIVY]

tiz·zy[1] (tiz′ē) *n. pl.* **·zies** *Slang* A bewildered or excited state of mind; a dither. [Origin unknown]

tiz·zy[2] (tiz′ē) *n. pl.* **·zies** *Brit. Slang* A sixpence. [Prob. alter. of TESTER[3]; infl. in form by slang *tilbury* a sixpence]

Tji·la·tjap (chē·lä′chäp) A port of southern Java, Indonesia. Also **Chi·la′chap.**

Tji·re·bon (chē′re·bôn′) A port of NW central Java, Indonesia, SE of Jakarta: also *Cheribon.*

Tlax·ca·la (tläs·kä′lä) A state of central Mexico; 1,555 square miles; capital, Tlaxcala.

Tlem·cen (tlem·sen′) A city of NW Algeria. Also **Tlem·sen′.**

Tlin·git (tling′git) *n. pl.* North American Indians belonging to any of eighteen tribes comprising the Koluschan linguistic stock, and inhabiting the Alexander Archipelago of SE Alaska. They are a seafaring people of fairly advanced culture. Also **Tlin·kit** (tling′kit).

tme·sis (tmē′sis, mē′sis) *n.* The separation of the elements of a compound word by an intervening word, as in the phrase *to us ward,* meaning "toward us." [<L <Gk. *tmēsis* a cutting < *temnein* cut]

TNT (tē′en′tē′) *n.* 1 Trinitrotoluene. 2 *Colloq.* Any explosive and dangerous circumstance, force, or person. Also **T.N.T.** [<T(RI)N(ITRO)-T(OLUENE)]

to (tōō, *unstressed* tə) *prep.* 1 In a direction toward or terminating in: going *to* town. 2 Opposite, in contact with, or near: face *to* face; Hold me *to* your breast. 3 Intending or aiming at; having as an object or purpose: Come *to* my rescue. 4 Resulting in; having as a condition or effect: frozen *to* death; flattered *to* his ruin. 5 Belonging in connection or accompaniment with; denoting the relation of things made to go together or between which there is correspondence: the key *to* the barn; March *to* the music. 6 In honor of: Drink *to* me only with thine eyes. 7 In comparison, correspondence, or agreement with: often denoting ratio: 9 is *to* 3 as 21 *to* 7; four quarts *to* the gallon. 8 Until; approaching as a limit; denoting the end of a period of time, or a time not reached: *to* my dying day; five minutes *to* one. 9 For the utmost duration of; as far as: a miser *to* the end of his days. 10 In respect of; concerning: blind *to* her charms; a speech *to* the point. 11 In close application toward: Buckle down *to* work; Fall *to* dinner. 12 For; with regard for: The contest is open *to* everyone. 13 Noting an indirect or limiting object after verbs, adjectives, or nouns, and designating the recipient of the action: taking the place of the dative case in other languages: Give the ring *to* me; That fact is not apparent *to* me. 14 By: known *to* the world. 15 From the point of view of: It seems *to* me. 16 *Dial.* At or in (a place): He is not *to* home now. 17 *Colloq.* With: The land was planted *to* potatoes. 18 About; involved in: That's all there is *to* it. ◆ *To* also serves to indicate the infinitive, and is often used elliptically for it: You may come if you care *to.* See synonyms under AT, INTO. —*adv.* 1 To or toward something. 2 In a direction, position, or state understood or implied; especially, shut or closed: Pull the door *to.* 3 Into a normal condition; into consciousness: She soon came *to.* 4 *Naut.* With head to the wind: said of a sailing vessel: to lie *to.* 5 Upon the matter at hand; into action or operation: They fell *to* with good will. 6 Nearby; at hand. —**to and fro** In opposite or different directions; back and forth. [OE *tō*]

toad (tōd) *n.* 1 A tailless, jumping, insectivorous amphibian (family *Bufonidae*), resembling the frog but without teeth in the upper jaw, and resorting to water only to breed. 2 Some similar amphibian; especially, the **Surinam toad** (*Pipa pipa*) or the European **midwife toad** (*Alytes obstetricans*). 3 Any person regarded scornfully or contemptuously. [OE *tādige*]

TOAD (Species vary from 2 to 6 inches)

toad-eat·er (tōd′ē′tər) *n.* A fawning parasite; a sycophant. [Orig. an assistant to a charlatan, who ate, or pretended to eat, toads (held to be poisonous) to show the efficacy of a patent medicine]

toad·fish (tōd′fish′) *n. pl.* **·fish** or **·fish·es** Any of a family (*Batrachoididae*) of fishes with scaleless skin and mouth and head resembling those of a toad.

toad·flax (tōd′flaks′) *n.* 1 A common, showy perennial weed (*Linaria vulgaris*) of the figwort family, having terminal spikes of spurred yellow flowers marked with an orange spot: also called *butter-and-eggs.* 2 Any other plant of the genus *Linaria.* [So called because spotted like toads and having a flaxlike foliage]

toad spit Cuckoo spit. Also **toad spittle.**

toad·stone[1] (tōd′stōn′) *n. Dial.* A volcanic rock, generally decomposed, occurring in limestone in Derbyshire, England. [? So called from a resemblance of its markings to those of a toad]

toad·stone[2] (tōd′stōn′) *n.* A natural or artificial stone resembling a toad in color and form, and long believed to be formed in a toad: worn as a talisman. [<TOAD + STONE; trans. of L *batrachites* <Gk.]

toad·stool (tōd′stōōl′) *n.* 1 Any one of many umbrella-shaped fungi, growing on decaying vegetable matter, common in woods and damp places; a mushroom. 2 *Colloq.* A poisonous mushroom.

toad·y (tōd′ē) *n.* *pl.* **toad·ies** An obsequious flatterer; a fawning, servile person; a toad-eater. — *v.t.* & *v.i.* **toad·ied, toad·y·ing** To act the toady (to). ◆ Homophone: *tody*. [Short for TOAD-EATER] — **toad′y·ish** *adj.* — **toad′y·ism** *n.*

to-and-fro (tōō′ən-frō′) *adj.* Moving back and forth; undulating; alternating. — *n.* Motion back and forth.

toast[1] (tōst) *v.t.* 1 To brown before or over a fire; especially, to brown (bread or cheese) before a fire or in a toaster. 2 To warm thoroughly before a fire. — *v.i.* 3 To become warm or toasted. — *n.* Sliced bread browned in a toaster or at a fire; toasted bread. [< OF *toster* roast, grill < L *tostus* < *torrere* parch, roast]

toast[2] (tōst) *n.* 1 The act of drinking to someone's health or to some sentiment. 2 The person or sentiment named in thus drinking: She was the *toast* of the town. — *v.t.* To drink to the health of or in honor of. — *v.i.* To drink a toast or toasts. [< TOAST[1] in obs. sense of "a spiced piece of toast put in a drink to flavor it"]

toast·er[1] (tōs′tər) *n.* A device for making toast.

toast·er[2] (tōs′tər) *n.* One who proposes a toast.

toast·mas·ter (tōst′mas′tər, -mäs′tər) *n.* A person who, at public dinners, announces the toasts, calls upon the various speakers, etc. — **toast′mis′tress** (-mis′tris) *n. fem.*

to·bac·co (tə-bak′ō) *n.* *pl.* **·cos** or **·coes** 1 An annual plant of the nightshade family (genus *Nicotiana*), especially *N. tabacum*, the chief source of the tobacco of commerce, originally of tropical America, but now cultivated in various parts of the world. 2 Its leaves prepared in various ways, as for smoking, chewing, snuffing, etc. 3 The use of tobacco for smoking. 4 The various products prepared from tobacco leaves, as cigarettes, cigars, etc. [< Sp. *tabaco* < Cariban, a tube or pipe in which the natives smoked tobacco]

TOBACCO (Plant to 8 feet or more)

tobacco heart *Pathol.* A cardiac disorder brought about by excessive smoking and characterized by a rapid or uneven pulse; nicotinism.

to·bac·co·nist (tə-bak′ə-nist) *n. Brit.* One who deals in tobacco.

tobacco worm Either of two large green worms (*Protoparce sexta* and *P. quinquemaculata*) with white stripes and a slender horn at the rear end of the body, destructive to tobacco plants.

To·ba·go (tō-bā′gō) See TRINIDAD AND TOBAGO.

To·bi·as (tə-bī′əs, tō-; *Dan., Du., Ger., Sp.* tō-bē′äs) A masculine personal name. Also **To·bi′ah** (-bī′ə), *Fr.* **To·bie** (tō-bē′), *Ital.* **To·bi·a** (tō·bē′ä). [<Hebrew, the Lord is (my) good]

To·bit (tō′bit) A pious Hebrew captive in Nineveh, hero of the Apocryphal book of the Old Testament bearing his name.

to·bog·gan (tə-bog′ən) *n.* A light sledlike vehicle, consisting of a long thin board or boards curved upward at the forward end; used for transporting goods or coasting, especially on prepared slides. — *v.i.* 1 To coast on a toboggan. 2 To move downward swiftly: Wheat prices *tobogganed*. [< dial. F (Canadian) *tabagan* a sleigh < Algonquian. Cf. Micmac *tobākūn*.] — **to·bog′gan·er, to·bog′gan·ist** *n.*

toboggan slide A slope prepared for coasting with toboggans: often a winding track with banked curves.

To·bol (tō-bôl′y′) A river in northern Kazakh S.S.R. and SW Asiatic Russian S.F.S.R., flowing 1,042 miles NE from the Ural Mountains of NW central Kazakh S.S.R., near the Russian border, to the Irtish River at Tobolsk.

To·bolsk (tō-bôlsk′) A city in SW Asiatic Russian S.F.S.R., at the junction of the Tobol and Irtish rivers.

To·bruk (tō-brōōk′, tō′brōōk) A port of eastern Cyrenaica, Libya; scene of several battles of World War II, 1941–42. *Italian* **To·bruch** (tō′-brōōk).

to·by (tō′bē) *n.* *pl.* **·bies** 1 A mug or jug for ale or beer, often made in the form of an old man wearing a three-cornered hat. 2 *Colloq.* A form of stogie cigar. [< TOBY]

TOBY JUG

To·by (tō′bē) Diminutive of TOBIAS.

To·can·tins (tō′kän·tēns′) A river in north central and north Brazil, flowing 1,640 miles north to the Pará River.

toc·ca·ta (tə-kä′tə, *Ital.* tôk·kä′tä) *n. Music* A rapid free composition for piano, organ, or other keyboard instrument, often preceding a fugue. [< Ital., lit., a touching, orig. pp. fem. of *toccare* touch]

To·char·i·an (tō-kâr′ē-ən, -kär′-) *n.* 1 One of an ancient cultured people known to the Greeks and Chinese as having inhabited central Asia in the first Christian millennium; conquered by the Uigurs. 2 The language of the Tocharians, belonging to the centum division of the Indo-European language family: unknown before 1904, when it was brought to light through manuscripts of the seventh century found in ruined temples in Chinese Turkestan. Two dialects have been distinguished, usually referred to as *Tocharian A* and *Tocharian B*. Also spelled *Tokharian*.

toch·er (tokh′ər) *Scot.* & *Brit. Dial.* *n.* The dowry of a bride. — *v.t.* To give a dowry to; dower. [<Irish *tochar* an assigned portion < *tochuirim* I put to, assign < *chuirim* I put]

toco- combining form Child; pertaining to children or to childbirth: *tocology*. Also, before vowels, **toc-**. [< Gk. *tokos* child, childbirth]

to·col·o·gy (tō-kol′ə-jē) *n.* The science and art of midwifery; obstetrics: also spelled *tokology*. [< TOCO- + -LOGY]

to·coph·er·ol (tō-kof′ə-rōl, -rol) *n. Biochem.* Any of three closely related alcohols, widely distributed in nature and also made synthetically; especially, alpha-tocopherol, $C_{29}H_{50}O_2$, an active principle of vitamin E. [< TOCO- + Gk. *pherein* bear + -OL[1]; so called because thought to be effective against sterility]

Tocque·ville (tōk·vēl′), **Alexis Charles Henri Maurice Clérel de,** 1805–59, French statesman and political writer.

toc·sin (tok′sin) *n.* 1 A signal sounded on a bell; alarm. 2 An alarm bell. [< MF < OF *toquassen* < Provençal *tocasenh* < *tocar* strike, touch + *senh* a bell < LL *signum* a signal bell < L, a sign]

tod[1] (tod) *n.* 1 A bushy clump. 2 A former weight for wool, about 28 pounds. [ME *todde*, prob. <LG. Cf. East Frisian *todde* small load, bundle.]

tod[2] (tod) *n. Scot.* & *Brit. Dial.* A fox.

to·day (tə-dā′) *adv.* 1 On or during this present day. 2 At the present time; nowadays. — *n.* The present day, time, or age. Also **to-day**′. ◆ Collateral adjective: *hodiernal*. [OE *tō dæg* < *tō* to + *dæg* a day]

Todd (tod), **Sir Alexander R.,** born 1907, English chemist.

tod·dle (tod′l) *v.i.* **·dled, ·dling** To walk unsteadily and with short steps, as a little child. — *n.* The act of toddling; a child's walk; also, a stroll. [? Freq. of TOTTER] — **tod′dler** *n.*

tod·dle-top (tod′l-top′) *n.* Teetotum.

tod·dy (tod′ē) *n.* *pl.* **·dies** 1 A drink made with spirits, hot water, sugar, and a slice of lemon. 2 The sap or juice that flows from the incised spathes of certain East Indian palms; also, a spirituous liquor distilled from it. The principal palms yielding toddy are called **toddy palms,** as the wild date of India (*Phoenix sylvestris*). [<Hind. *tārī* toddy (def. 2) < *tār* palm tree <Skt. *tāla* a palmyra]

Tod·le·ben (tōt′lä·bən) See TOTLEBEN.

to-do (tə-dōō′) *n. Colloq.* Confusion or bustle, as on account of something disturbing; a demonstration; a fuss. [OE *tō-dōn* < *tō* asunder + *dōn* do, put]

Todt (tōt), **Fritz,** 1891–1942, German military engineer.

to·dy (tō′dē) *n.* *pl.* **·dies** Any of numerous very small insectivorous West Indian birds (genus *Todus*) related to the kingfishers; especially, the **green tody** (*Todus godus*) of Jamaica,

bright green with a scarlet throat. ◆ Homophone: *toady*. [< F *todier* < L *todus*, a kind of small bird]

toe (tō) *n.* 1 One of the digits of the foot; also, the forward part of the foot, as distinguished from the *heel*. 2 That portion of a shoe, boot, sock, stocking, skate, or the like that covers, or corresponds in position with, the toes. 3 The lower end or projection of something, resembling or suggestive of a toe. 4 *Mech.* **a** A pivot or journal in a bearing. **b** A horizontally projecting arm on a stem, as for operating a valve, raised by a cam or lifted. 5 The end of the head of a golf club. 6 In a railroad switch, the space between the rails at the unchanneled end of a frog. — **on one's toes** Alert; wide-awake. — **to tread on (someone's) toes** To offend (a person); trespass on (someone's) feelings, opinions, prejudices, etc. — *v.* **toed, toe·ing** *v.t.* 1 To touch with the toes: to *toe* the line. 2 To kick with the toe. 3 To furnish with a toe. 4 To drive (a nail or spike) obliquely; also, to attach (beams, etc.) end to end, by nails driven thus. 5 To strike (a golf ball) with the toe of the club. — *v.i.* 6 To stand or walk with the toes pointing in a specified direction: to *toe* out. — **to toe the mark** To touch a certain line or mark with the toes preparatory to starting a race; hence, to abide by the rules; conform to discipline or a standard. [OE *tā*] — **toe′less** *adj.*

toe cap A cap covering for the tip or toe of a boot or shoe. See illustration under SHOE.

toe crack A sandcrack.

toed (tōd) *adj.* 1 Having toes: chiefly in composition: pigeon-*toed*. 2 Fastened or fastening by obliquely driven nails; also, driven obliquely, as a nail.

toe-dance (tō′dans′, -däns′) *v.i.* **-danced, -danc·ing** To dance on tiptoe; perform a toe dance. — **toe′-danc′er** *n.*

toe dance A dance performed on tiptoe.

toe·hold (tō′hōld′) *n.* 1 In climbing, a small space which supports the toes. 2 Any means of entrance, support, or the like; a footing: The Marines gained a *toehold* on the island. 3 A hold in which a wrestler bends back the foot of his opponent.

toe·nail (tō′nāl′) *n.* 1 A nail growing on the toe. 2 A nail driven obliquely to hold the foot of a stud or brace. — *v.t.* To fasten with obliquely driven nails.

toff (tof, tôf) *n. Brit. Slang* A dandy; also, a gentleman. [Earlier *toft* <TUFT (def. 3)]

tof·fee, tof·fy (tof′ē, tôf′ē) See TAFFY.

toft (tôft, toft) *n. Brit.* 1 Land once occupied as a messuage, on which the buildings have decayed or been burned; a homestead. 2 A hillock or knoll. [OE, a homestead <ON *topt*, *tupt*]

tog (tog) *Colloq.* *n.* 1 A coat. 2 *pl.* Clothes; outfit: football *togs*. — *v.t.* **togged, tog·ging** To dress; clothe: often with *up* or *out.* [Short for vagabond's cant *togemans, togman* coat, cloak <F *toge* a toga <L *toga*]

to·ga (tō′gə) *n.* *pl.* **·gas** or **·gae** (-jē) 1 The distinctive outer garment worn in public by a citizen of ancient Rome. 2 Any gown or cloak characteristic of a calling or profession: the lawyer's *toga*. [< L < *tegere* cover]

to·gaed (tō′gəd) *adj.* Robed in the toga; hence, classical and stately. Also **to·gat·ed** (tō′gā·tid).

to·ga vi·ri·lis (tō′gə vi·rī′lis) *Latin* The toga assumed by a male citizen of ancient Rome at the age of 14 as a token of manhood.

ROMAN TOGA

to·geth·er (tōō·geth′ər, tə-) *adv.* 1 Into union or contact with each other; conjointly. 2 In the same place or at the same spot; with each other; in company. 3 At the same moment of time; simultaneously. 4 Without cessation or intermission. 5 With one another; mutually. [OE *tōgædere, tōgadore* < *tō* to + *gædre* together. Akin to GATHER.]

to·geth·er·ness (tōō·geth′ər·nis, tə-) *n.* The state of being associated or united.

tog·ger·y (tog′ər·ē) *n. pl.* **·ger·ies** *Colloq.* Togs collectively; clothes.
tog·gle (tog′əl) *n.* **1** A pin, or short rod, properly attached in the middle, as to a rope, and designed to be passed through a hole or eye and turned. **2** A toggle iron. **3** A toggle joint. —*v.t.* **·gled, ·gling** To fix, fasten, or furnish with a toggle or toggles. [Prob. nautical var. of dial. *tuggle,* appar. freq. of TUG]
toggle iron A harpoon, as for killing whales, so arranged as to turn crosswise when it enters the animal's body. Also **toggle harpoon.**
toggle joint *Mech.* A joint having a central hinge like an elbow, and operable by applying the power at the junction, thus changing the direction of motion and giving indefinite mechanical pressure.

TOGGLE JOINT
Level Type

toggle switch *Electr.* A switch in the form of a projecting lever whose movement through a small arc opens or closes an electric circuit.
To·gliat·ti (tō·lyät′tē), **Palmiro,** born 1893, leader of the Italian Communist party.
To·go (tō′gō) An independent republic in western Africa; 22,008 square miles; capital, Lomé: formerly French Togoland, a United Nations Trust Territory. —**To′go·lese′** (-lēs, -lēz) *adj.* & *n.*
To·go (tō·gō), **Count Heihachiro,** 1847–1934, Japanese admiral; defeated the Russian fleet at the battle of Tsushima, 1905.
To·go·land (tō′gō·land), **British** See GHANA.
Togoland, French See TOGO.
toil[1] (toil) *n.* **1** Fatiguing work; labor; hence, any oppressive task. **2** Any notable work accomplished by labor. **3** *Obs.* Strife; struggle. —*v.i.* **1** To work arduously; labor painfully and tiringly. **2** To progress or make one's way with slow and labored steps. —*v.t.* **3** To accomplish or obtain by toil. See synonyms under STRUGGLE. [<AF *toil* a dispute, OF *tooil* trouble <AF *toiler* strive, OF *tooillier* soil, agitate <L *tudiculare* stir about < *tudicula* a machine for bruising olives, dim. of *tudes* a mallet] —**toil′er** *n.*
Synonyms (noun): drudgery, labor, stent, stint, task, travail, work. *Work* is exertion of body or mind that taxes the powers for the accomplishment of some end. The term is a broad one; *work* may be light and pleasant, or severe and exhausting. *Labor* is always strenuous; it is hard *work. Toil* is still more severe. One may enjoy *work* and be cheerful in *labor,* but *toil* oppresses. *Drudgery* is often applied to menial service, but also to any *work* that is not only hard, but dull and mechanical. A *task* is a definite amount of *work* appointed and required by another; yet we sometimes speak of a *task* which one imposes upon himself; this in popular language is called a *stint* or *stent.* See TASK, WORK. *Antonyms*: amusement, ease, idleness, leisure, play, recreation, relaxation, repose, rest.
toil[2] (toil) *n.* A net, snare, or other trap: now generally used figuratively and commonly in the plural. [<MF *toiles* nets < *toile* cloth <OF *teile* <L *tela* a web]
toile (twäl) *n.* A sheer linen fabric; also, a fine cretonne with scenic designs printed in one color. [<F. See TOIL[2].]
toi·let (toi′lit) *n.* **1** The act or process of dressing oneself; formerly, especially of dressing the hair. **2** Attire; toilette; also, a toilette or costume. **3** A dressing-table. **4** A lavatory or watercloset; also, a bathroom. [<F *toilette* orig. a cloth dressing gown, dim. of *toile* cloth. See TOIL[2].]
toi·let·ry (toi′lit·rē) *n. pl.* **·ries** Any of the several articles used in making one's toilet, as soap, comb, brush, etc.
toi·lette (toi·let′, *Fr.* twà·let′) *n.* **1** The act or process of grooming oneself, usually including bathing, hair-dressing, application of cosmetics and perfume, and costuming. **2** A person's actual dress or style of dress; also, any specific costume or gown: an elaborate *toilette.* [<F. See TOILET.]
toilet water A scented liquid containing a small amount of alcohol, used in or after the bath, applied to the face after shaving, etc.
toil·ful (toil′fəl) *adj.* Replete with toil; laborious. —**toil′ful·ly** *adv.*
toil·some (toil′səm) *adj.* Accomplished with fatigue; involving toil. See synonyms under ARDUOUS, DIFFICULT. —**toil′some·ly** *adv.* —**toil′some·ness** *n.*
toil·worn (toil′wôrn′, -wōrn′) *adj.* Exhausted by toil; showing the effects of toil.
toit (toit) *v.i. Brit. Dial.* **1** To dawdle; saunter. **2** To totter. Also spelled *toyte.* [Origin uncertain]
To·jo (tō·jō), **Hideki,** 1885–1948, Japanese general and statesman in World War II.
To·kay (tō·kā′) *n.* **1** A white or reddish-blue grape from Tokay, Hungary. **2** A wine made from it. [from *Tokay,* a town in northern Hungary]
To·ke·lau (tō′kə·lou′) A New Zealand island group north of Samoa; 4 square miles: also *Union Islands.*
to·ken (tō′kən) *n.* **1** Anything indicative of some other thing; a visible sign; indication; evidence: in *token* of respect. **2** A symbol: This gift is a *token* of my affection. **3** *Obs.* A signal. **4** Some tangible proof or evidence of a statement or of one's identity, authority, etc. **5** A memento; keepsake; souvenir. **6** A characteristic mark or feature. **7** A piece of metal issued as currency and having a face value greater than its actual value. **8** A piece of metal issued by a transportation company and good for one fare. See synonyms under EMBLEM, MARK[1], SIGN, TRACE[1]. —*v.t.* To evidence by a token; betoken. —*adj.* Done or given as a token, especially in partial fulfillment of an obligation or engagement: a *token* payment. [OE *tācen, tācn*]
to·kened (tō′kənd) *adj. Obs.* Marked by spots: the *tokened* pestilence. [<TOKEN, in obs. sense "a spot on the body indicating disease"]
To·khar·i·an (tō·kâr′ē·ən, -kär′-) See TOCHARIAN.
to·kol·o·gy (tō·kol′ə·jē) See TOCOLOGY.
To·ku·shi·ma (tō·ko͞o·shē·mä) A port of eastern Shikoku, Japan.
To·ky·o (tō′kē·ō, *Japanese* tō·kyō) The capital of Japan, a port on **Tokyo Bay,** an inlet of the Philippine Sea in central Honshu, Japan: formerly *Edo* or *Yedo.* Also **To′ki·o.**
to·la (tō′lä) *n. Anglo-Indian* A weight, about 180 grains, for gold and silver; the weight of one rupee. [<Hind. <Skt. *tulā* a balance, weight < *tul-* weigh]
to·lan (tō′lan) *n.* A white crystalline unsaturated hydrocarbon, $C_{14}H_{10}$, prepared by synthesis. Also **to·lane** (tō′lān). [<TOL(UENE) + -ANE[2]]
tol·booth (tōl′bo͞oth′, -bo͞oth′) See TOLLBOOTH.
told (tōld) Past tense and past participle of TELL.
tole[1] (tōl) *v.t.* **toled, tol·ing 1** *Dial.* To draw as with a lure; entice; decoy. **2** *Obs.* To pull; drag; draw. Also spelled *toll.* [Var. of TOLL[2]]
tole[2] (tōl) *n.* A metalware, enameled or lacquered in various colors and frequently gilded: used for tea trays and utensils, and also esteemed as an ornamental material for lamps and decorative objects. Also **tôle.** ◆ Homophone: *toll.* [<F *tôle* sheet iron, dial. var. of *table* a table]
To·le·do (tə·lē′dō) *n. pl.* **·dos** A sword or sword blade from Toledo, Spain. Also **to·le′do.**
To·le·do (tə·lē′dō) **1** A city in NW Ohio near Lake Erie. **2** (*Sp.* tō·lā′thō) An ancient city of central Spain on the Tagus.
tol·er·a·ble (tol′ər·ə·bəl) *adj.* **1** Passably good; commonplace. **2** Endurable; capable of being borne. **3** Allowable. **4** *Colloq.* In passably good health. [<OF <L *tolerabilis* able to endure < *tolerare* endure] —**tol′er·a·ble·ness** *n.* —**tol′er·a·bly** *adv.*
tol·er·ance (tol′ər·əns) *n.* **1** The character, state, or quality of being tolerant. **2** Indulgence or forbearance in judging the opinions, customs, or acts of others; freedom from bigotry or from racial or religious prejudice. **3** The act of enduring, or the capacity for endurance. **4** *Mech.* A fractional allowance for variations from the specified standard weight, dimensions, etc., of mechanical constructions. **5** A legally permissible variation from the standard of weight, fineness, etc., of coins: also called *remedy.* **6** *Med.* Natural or acquired ability to endure without ill effects large or increasing amounts of specified substances, particularly drugs.
tol·er·ant (tol′ər·ənt) *adj.* **1** Of a long-suffering disposition. **2** Indulgent; liberal. **3** *Med.* Capable of taking with impunity unusual or excessive doses of dangerous drugs. [<F <L *tolerans, -antis,* ppr. of *tolerare* endure] —**tol′er·ant·ly** *adv.*
tol·er·ate (tol′ə·rāt) *v.t.* **·at·ed, ·at·ing 1** To allow to be or be done without active opposition. **2** To concede, as the right to opinions or participation. **3** To bear, sustain, or be capable of enduring or sustaining. **4** *Med.* To endure, as a poisonous amount of dose, with impunity. See synonyms under ABIDE, ALLOW, ENDURE, PERMIT. [<L *toleratus,* pp. of *tolerare* endure] —**tol′er·a′tive** *adj.* —**tol′er·a′tor** *n.*
tol·er·a·tion (tol′ə·rā′shən) *n.* **1** The act or practice of tolerance. **2** The recognition of the rights of the individual to his own opinions and customs, as in matters pertaining to religious worship, when they do not interfere with the rights of others or with decency and order. **3** The spirit and desire to be tolerant in matters of opinion; forbearance; freedom from bigotry or race prejudice.
tol·i·dine (tol′ə·dēn, -din) *n. Chem.* One of several isomeric bases, $(CH_3 \cdot C_6H_3 \cdot NH_2)_2$, derived from dimethyl benzidine: one form is used in making dyes. Also **tol′i·din** (-din). [<TOL(UOL) + (BENZ)IDINE]
To·li·ma (tō·lē′mä), **Ne·va·da del** (nā·vä′thä thel) A volcano in the Andes Mountains of west central Colombia; 18,438 feet; last eruption, 1829.
toll[1] (tōl) *n.* **1** A fixed compensation for some privilege granted or service rendered, especially for one granted in a general or public way, as passage on a bridge or turnpike, or that taken by a miller for grinding grain (commonly a portion of the grain). **2** The right to levy such charge. **3** Something taken or elicited like a toll; price: The train wreck took a heavy *toll* of lives. **4** A due charged for the privilege of shipping or landing goods. **5** A charge for transportation of goods, especially by rail or canal. **6** A charge for a long-distance telephone call. See synonyms under TAX. —*v.t.* To take as a toll. —*v.i.* To take or exact a toll. ◆ Homophone: *tole.* [OE, ? <LL *toloneum* <L *telonium* <Gk. *telōnion* a customhouse < *telōnes* a tax collector < *telos* a tax]
toll[2] (tōl) *v.t.* **1** To cause (a bell) to sound slowly and at regular intervals. **2** To announce thus; especially, to announce (a death, funeral, etc.) by tolling. **3** To call or summon by tolling. **4** To decoy (game, especially ducks). **5** *Rare* To entice. —*v.i.* **6** To sound slowly and at regular intervals. —*n.* The sound of a bell rung slowly and with single, regularly repeated strokes. ◆ Homophone: *tole.* [Prob. <TOLL[1], in obs. sense of "pull, draw"]
toll·age (tō′lij) *n.* **1** A charge in the nature of a toll. **2** The toll itself.
toll·bar (tōl′bär′) *n.* A tollgate, properly one with a single bar.
toll·booth (tōl′bo͞oth′, -bo͞oth′) *n.* **1** *Scot.* A jail; prison: also spelled *tolbooth.* **2** A tollhouse.
toll bridge A bridge at which toll for passage is paid.
toll call A long-distance telephone call, charged for at more than local rates.
toll collector A collector of tolls.
toll·er (tō′lər) *n.* **1** One who tolls a bell. **2** A bell used for tolling. **3** A small dog trained to toll or decoy ducks.
Tol·ler (tôl′ər), **Ernst,** 1893–1939, German dramatist and politician.
toll·gate (tōl′gāt′) *n.* A gate at the entrance to a bridge, or on a road, at which toll is paid.
toll·house (tōl′hous′) *n.* A toll collector's lodge adjoining a tollgate.
toll·keep·er (tōl′kē′pər) *n.* One who keeps a tollgate.
toll line A telephone line or channel, as between two central offices in different exchanges, for the use of which a toll is charged; a long-distance circuit.
Tol·stoy (tol′stoi, tōl′-; *Russian* tol·stoi′), **Count Leo Nikolaevich,** 1828–1910, Russian novelist and social reformer. Also **Tol′stoi.**
Tol·tec (tol′tek, tōl′-) *n.* One of certain ancient Nahuatlan tribes that dominated central and southern Mexico about A.D. 900–1100 and through contact with Mayan culture founded the highly civilized Nahua culture of the Aztecs: referred to in Aztec and Mayan legend. See NAHUA. —*adj.* Of or pertaining to the Toltecs. [<Nahuatl *Tolteca*] —**Tol′tec·an** *adj.*
to·lu (tə·lo͞o′) *n.* Balsam of Tolu. [<Sp. *tolú,*

toluate 1321 **tone**

from Santiago de *Tolu*, a seaport in Colombia.]
tol·u·ate (tol′yōō-āt) *n. Chem.* A salt or ester of a toluic acid. [<TOLU(IC) + -ATE³]
To·lu·ca (tō-lōō′kä) The capital of Mexico state, central Mexico. Also **Toluca de Ler·do** (thā ler′thō).
tol·u·ene (tol′yōō-ēn) *n. Chem.* A limpid hydrocarbon, $C_6H_5CH_3$, of the aromatic series, homologous with benzene, and obtained from coal tar by distillation: it is used in making dyestuffs, explosives, and other substances. [<TOLU + -ENE; so called because orig. obtained from tolu]
to·lu·ic (tə-lōō′ik, tol′yōō-ik) *adj. Chem.* Designating or pertaining to any one of four isomeric acid derivatives of toluene, $C_8H_8O_2$, occurring as white crystalline compounds. [<TOLU(ENE) + -IC]
tol·u·ide (tol′yōō-īd, -id) *n. Chem.* One of a series of compounds obtained from toluene by substituting a tolyl radical for hydrogen in the amino group. Also **tol′u·id** (-id). [<TOLU(ENE) + -IDE]
to·lu·i·dine (tə-lōō′ə-dēn, -din) *n. Chem.* One of three isomeric compounds, C_7H_9N, homologous with aniline and derived from the nitro-compounds of toluene. Also **to·lu′i·din** (-din). [<TOLUID(E) + -INE²]
tol·u·ol (tol′yōō-ōl, -ol) *n. Chem.* Crude commercial toluene. Also **tol′u·ole** (-ōl). [<TOLU + (BENZ)OL]
tol·u·yl (tol′yōō-il) *n. Chem.* The univalent acid radical C_8H_7O. [<TOLU(IC) + -YL]
tol·yl (tol′il) *n. Chem.* The univalent radical $C_6H_4CH_3$, derived from toluene; cresyl. [<TOL(UIC) + -YL]
tom (tom) *n.* The male of various animals, especially the cat. — *adj.* Male: a *tom* pheasant. [from the personal name *Tom*]
Tom (tom) Diminutive of THOMAS.
Tom (tom) A river in SW central Asiatic Russian S.F.S.R., flowing 440 miles west and NW, from east Stalinsk to the Ob river NW of Tomsk.
tom·a·hawk (tom′ə-hôk) *n.* A war weapon used by the Algonquian Indians of North America, originally a carved club about three feet long, having a knob of solid wood on the end in which a piece of bone or metal was inserted; later, the light ax or hatchet-shaped weapon with an iron blade obtained in trade with Europeans. Tomahawks were either thrown or wielded in the hand. — *v.t.* To strike or kill with a tomahawk. [<Algonquian *tamahak*, short for *tamahaken* a cutting utensil < *tamahaken* he uses for cutting < *tamaham* he cuts]

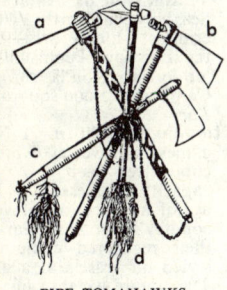
PIPE TOMAHAWKS
a. Cree. b. Iroquois.
c. Omaha. d. Osage.

tom·al·ley (tom′al-ē) *n.* The liver of the lobster, turning green when cooked: considered a great delicacy. Also **to·mal·ly** (tə-mal′ē). [Prob. <Cariban]
to·man (tō·män′) *n.* A Persian gold coin of varying value; formerly a money of account. [<Persian *tūmān, tumān, tuman* <Turki, lit., ten thousand]
Tom-and-Jer·ry (tom′ən-jer′ē) *n.* A drink made with brandy, rum, beaten egg, hot milk or water, sugar, and nutmeg. [after Corinthian *Tom* and *Jerry* Hawthorn, two main characters in *Life in London*, 1821, by Pierce Egan, 1772–1849, English writer on sports and sports jargon]
To·más (tō-mäs′) Spanish form of THOMAS.
to·ma·tin (tō-mā′tin, -mä′-) *n.* An antibiotic extracted from the leaves and plants of the tomato plant and also from the leaf juices of potatoes and green peppers. [<TOMAT(O) + -IN]
to·ma·to (tə-mā′tō, -mä′-) *n. pl.* **·toes** 1 The pulpy edible berry, yellow or red when ripe, of a tropical American perennial plant (*Lycopersicon esculentum*) of the nightshade family, highly esteemed as a vegetable. 2 The plant itself. 3 *U.S. Slang* A girl or woman. [<Sp. *tomate* <Nahuatl *tomatl*]
tomato fruitworm The bollworm.
tomb (tōōm) *n.* 1 A place for the burial of the dead; a vault; grave. 2 A place where the dead lie. 3 Death itself. 4 A tombstone. — *v.t.* To entomb; bury; inter. [<AF *tumbe*, OF *tombe* <LL *tumba* <Gk. *tymbos* a mound]
tom·bac (tom′bak) *n.* Any of several copper-and-zinc alloys used to make gongs and bells in the East, and cheap jewelry in Europe: often spelled *tambac*. Also **tom′bak, tom′-bak.** [<F <Pg. <Malayan *tambāga* copper <Skt. *tāmraka*]
Tom·big·bee River (tom·big′bē) A river in NE Mississippi and SW Alabama, flowing 384 miles SE and south to a junction with the Alabama River, forming the Mobile River, 30 miles north of Mobile Bay.
Tom·bouc·tou (tôń·bōōk·tōō′) The French name for TIMBUKTU.
tom·boy (tom′boi) *n.* A girl of romping and boisterous conduct; hoyden. [<TOM + BOY] — **tom′boy·ish** *adj.* — **tom′boy·ish·ness** *n.*
Tombs (tōōmz), **the** Formerly, the New York City police prison: so called from the funereal appearance of the building in which it was housed until 1948; also, loosely, the prison which replaced the original Tombs.
tomb·stone (tōōm′stōn) *n.* A stone, usually inscribed, marking a place of burial.
Tomb·stone (tōōm′stōn) A city in SE Arizona; formerly the site of the richest gold mines in Arizona.
tom·cat (tom′kat) *n.* A male cat. [after *Tom*, a male cat, hero of *The Life and Adventures of a Cat*, 1760, a very popular anonymous work]
tom·cod (tom′kod) *n.* Any of several small edible fishes (genus *Microgadus*) common on the Atlantic coast of North America. [<TOM + COD]
Tom Collins A drink consisting of gin, lemon or lime juice, sugar, and carbonated water.
Tom, Dick, and Harry Any persons taken at random from the crowd: used disparagingly, and often preceded by *every*.
tome (tōm) *n.* A volume, particularly if large; originally, one of a series of volumes. [<MF <L *tomus* <Gk. *tomos* a fragment, volume <*temnein* cut]
-tome *combining form* A cutting instrument (of a specified kind): *microtome*. [<Gk. *tomos* a cutting <*temnein* cut]
to·men·tose (tō·men′tōs, tō′men·tōs) *adj. Biol.* Covered with matted woolly hairs; flocculent. Also **to·men′tous** (-təs). [<L *tomentosus* <*tomentum* a stuffing for cushions]
to·men·tum (tə-men′təm) *n. pl.* **·ta** (-tə) 1 *Anat.* A network of small blood vessels of the pia mater where applied to the brain or spinal cord. 2 *Bot.* A form of pubescence composed of matted woolly hairs. [<L. See TOMENTOSE.]
tom·fool (tom′fōōl′) *n.* 1 An idiotic or silly person. 2 An amusing trifler. — *adj.* Ridiculous; very stupid. [after *Tom Fool*, a name formerly applied to mental defectives]
tom·fool·er·y (tom′fōō′lər-ē) *n. pl.* **·er·ies** 1 Nonsensical behavior. 2 Kickshaws. Also **tom′fool′ish·ness.**
tom-ful·ler (tom′fōōl′ər) *n.* Sour or fermented hominy prepared as food: originally a Choctaw Indian dish. Also **tom′ful′la** (-fōō′lə). **tom fuller.** [<Choctaw *tahfula* hominy]
Tom·ma·si·ni (tôm′mä·zē′nē), **Vicenzo,** 1880–1950, Italian composer.
tom·my¹ (tom′ē) *n. pl.* **·mies** *Slang* A roll; a loaf or piece of bread. [<brown *Tommy* <*Tommy Brown*, appar. a personification of brown bread]
tom·my² (tom′ē) *n. pl.* **·mies** Provisions or goods given instead of money in payment of wages; also, the system of paying workmen partly or entirely in kind. [Short for *tommy-shop*, a store run on the truck system]
tom·my³ (tom′ē) *n. pl.* **·mies** A Tommy Atkins; a British soldier: also **Tom′my.** [Short for TOMMY ATKINS]
tom·my (tom′ē) Diminutive of THOMAS.
Tommy At·kins (at′kinz) A British private of the regular 'army. [after *Thomas Atkins*, a name used on specimen forms in the official regulations of the British Army after 1815]
Tommy gun A Thompson submachine gun. [<*Tommy*, dim. of *Thompson* + GUN]
to·mo·dro·mic (tō′mə-drō′mik, -drom′ik) *adj.* Having a flight path which cuts athwart a moving target; heading to cut or intercept: said of guided missiles. [<Gk. *tomos* cutting (<*temnein* cut) + *dromos* a running <*dramein* run]
to·mog·ra·phy (tō·mog′rə-fē) *n. Med.* X-ray photography of a predetermined plane of the body, with a blurring or elimination of details in other planes. [<Gk. *tomos* a slice (<*temnein* cut) + (PHOTO)GRAPHY]
to·mor·row (tə-môr′ō, -mor′ō) *adv.* On or for the next day after today. — *n.* The next day after today; the morrow. Also **to·mor′row.** [OE *tō morgen* <*tō* to + *morgen* morning, morrow]
tom·pi·on (tom′pē-ən) *n. Mil.* A stopper, as the plug put into the mouth of a cannon, to exclude moisture, etc.: also called *tampion*. [Var. of TAMPION]
Tomsk (tomsk, *Russian* tômsk) A city in west central Siberia, Russian S.F.S.R.
Tom Thumb In English folklore, the son of a plowman, as big as his father's thumb, who undergoes many adventures, including being swallowed by a cow and a giant.
— **General Tom Thumb** The stage name of Charles Sherwood Stratton, 1838–83, a dwarf exhibited by P. T. Barnum.
tom·tit (tom′tit′) *n.* 1 A tit; titmouse. 2 Any of various small birds, as a chickadee or a wren. [<TOM + TIT¹]
tom-tom (tom′tom′) *n.* 1 The native drum of India, Africa, etc., variously shaped, and usually beaten with the hands. 2 A percussion instrument of monotonous tone, used in some modern orchestras for special effects. 3 A copper or copper-alloy disk-shaped instrument sounded with a felt-covered hammer or stick; a Chinese gong. Also spelled *tam-tam*. [<Hind. *tamtam*, imit. of the instrument's sound]
-tomy *combining form* 1 *Surg.* A cutting of a (specified) part or tissue: *osteotomy*. 2 A (specified) kind of cutting or division: *dichotomy*. [<Gk. *tomē* a cutting <*temnein* cut]
ton¹ (tun) *n.* 1 Any of several large measures of weight; particularly, the **short ton** of 2000 pounds avoirdupois, commonly used in the United States and Canada; the **long ton** of 2240 pounds of Great Britain; or the **metric ton** of 1000 kilograms. 2 A unit for reckoning the displacement or weight of vessels, 35 cubic feet of sea water weighing about one long ton: called **displacement ton.** 3 A unit for reckoning the freight-carrying capacity of a ship, usually equivalent to 40 cubic feet of space but varying with the cargo: called **freight ton, measurement ton.** 4 A unit for reckoning the internal capacity of merchant vessels for purposes of registration, equivalent to 100 cubic feet or 2.832 cubic meters: called **register ton.** [Var. of TUN; infl. in form by OF *tonne* a cask]
ton² (tôn) *n. French* Tone; style; the prevailing fashion; vogue.
ton- Var. of TONO-.
-ton *suffix* Town: used in place names: *Charleston, Brockton*. [OE *-tun* <*tūn* a town]
to·nal (tō′nəl) *adj.* Of or pertaining to tone or tonality. — **to′nal·ly** *adv.*
to·nal·ite (tō′nəl-īt) *n.* A quartz-mica diorite. Also **to′nal·yte.** [from *Tonale*, in the Tirol, where it was first described]
to·nal·i·ty (tō-nal′ə-tē) *n. pl.* **·ties** 1 *Music* The quality and peculiarity of a tonal system; the melodic and harmonic relations between the tones of a scale or system of tones; a key or mode. 2 The general color scheme or collective tones of a painting. 3 Tonicity.
to-name (tōō′nām′) *n. Scot.* 1 Some special distinguishing name; nickname. 2 A surname. [OE *tō-nama*]
to·na·pha·si·a (tō′nə-fā′zhē-ə, -zhə) *n. Psychiatry* Inability to recall a familiar tune; musical aphasia. [<NL <L *tonus* TONE + Gk. *aphasia* inability to speak]
tone (tōn) *n.* 1 Sound in relation to quality, volume, duration, and pitch. 2 *Physics* A sound having a definite pitch, and due to vibration of a sounding body. The pitch of a

tone depends on rate of vibration and its force on amplitude of vibration; its timbre is a complex resultant of concomitant vibration. If the vibration is simple harmonic motion the tone is pure; if there are complex components, the one of lowest pitch is the **fundamental tone** and the other components, in a simple ratio to the lowest, are **partial tones** or *overtones.* The combined result of all the partial tones gives the quality or *timbre* of the tone. **3** *Music* **a** The timbre, or peculiar characteristic sound, as of a voice or instrument. **b** The interval corresponding to one degree of the scale or staff; two semitones: sometimes called a **major tone** or **whole tone**, in distinction from a *semitone.* **4** A predominating disposition; especially, a frame or condition of mind; mood. **5 a** Characteristic style or tendency; tenor; quality: a want of moral *tone.* **b** Style or distinction; elegance: The party had *tone.* **6** Vocal inflection as expressive of feeling: a *tone* of pity. **7** *Ling.* A musical intonation or modulation of the voice by which a word or phrase may be changed in meaning or function: Peking Chinese distinguishes four *tones.* **8** *Phonet.* **a** The acoustical pitch, or change in pitch, of a phrase or sentence: In English, a questioning is indicated by a rising *tone.* **b** Special stress or accent given to one syllable of a word, or to one of the words in a sentence or phrase. **9** The prevailing effect of a picture, due to the management of chiaroscuro and to the effect of light upon the quality of color. **10** A shade, hue, tint, or degree of a particular color, or some slight modification of it: a deep *tone* of yellow; red with a purplish *tone.* **11** *Phot.* The shade or color of a photographic positive picture; also, the color of a negative film. **12** *Physiol.* The general condition of the body with reference to the vigorous and healthy discharge of its functions. See synonyms under SOUND[1]. —*v.* **toned, ton·ing** *v.t.* **1** To give tone to; modify in tone. **2** To tune or modify with reference to musical quality, as an instrument. **3** To intone in monotonous recitative; intone. **4** To alter the color or increase the brilliancy of (a photographic print) by a chemical bath. —*v.i.* **5** To assume a certain tone or hue. **6** To blend or harmonize, as in tone or shade. **—to tone down 1** To subdue the tone of (a painting). **2** To moderate in quality or tone. **—to tone up 1** To raise in quality or strength. **2** To elevate in pitch. **3** To gain in vitality. [<OF *ton* <L *tonus* <Gk. *tonos* a pitch of voice, a stretching < *teinein* stretch] **—ton′er** *n.*

Tone (tōn), **Wolfe,** 1763–98, Irish revolutionist and author.

tone color Timbre.

tone·less (tōn′lis) *adj.* Having no tone; without tone. **—tone′less·ly** *adv.* **—tone′less·ness** *n.*

tone poem A symphonic poem.

to·net·ic (tō·net′ik) *adj. Ling.* Tonic. [<TONE + (PHON)ETIC]

tong[1] (tông, tong) *v.t.* To gather, collect, or seize with tongs. —*v.i.* To use or fish with tongs. [<TONGS]

tong[2] (tông, tong) *n.* A Chinese closed society; in the United States, a secret society composed of Chinese. [<Chinese *t'ang* a hall, meeting place]

ton·ga (tong′gə) *n. Anglo-Indian* A light two-wheeled cart for four persons, in use in the country districts of India. [<Hind. *tāṅgā*]

Ton·ga (tong′gə) *n.* A Polynesian language spoken in the Tonga Islands.

Ton·ga Islands (tong′gə) An island group SE of the Fiji Islands in the South Pacific, comprising an independent Polynesian kingdom under British protection; total, 270 square miles; capital, Nukualofa: also *Friendly Islands.*

Ton·ga·land (tong′gə·land) A region of Zululand on the Mozambique border.

Ton·ga·re·va (tông′ä·rä′vä) See PENRHYN.

tongs (tôngz, tongz) *n. pl.* (*sometimes construed as singular*) **1** An implement for grasping, holding, or lifting objects, consisting usually of a pair of pivoted levers: also called **pair of tongs. 2** One of various grasping mechanisms. [OE *tang, tange*]

tongue (tung) *n.* **1** A protrusile, freely moving organ situated in the mouth of most vertebrates and supported by the hyoid bone: most completely developed in mammals, where it is important in taking in and masticating food, as one of the organs of taste, and in man as an organ of speech. ◆ Collateral adjective: *lingual.* **2** An organ or part of the mouth of various insects and fishes, having a similar shape or function. **3** An animal's tongue, as of beef, prepared as food. **4** The power of speech or articulation: to lose one's *tongue.* **5** Manner or style of speaking: a smooth *tongue.* **6** Mere speech, as contrasted with fact or deed. **7** Utterance; talk; discourse. **8** A language, vernacular, or dialect. **9** *Archaic* A people or race, regarded as having its own language: a Biblical use. **10** Anything resembling an animal tongue in appearance, shape, or function. **11** A slender projection of land, as a cape or small promontory. **12** A long narrow bay or inlet of water. **13** A jet of flame. **14** A strip of leather for closing the gap in the front of a laced shoe. **15** The fastening pin of a brooch or buckle. **16** *Music* The free or vibrating end of a reed in a wind instrument. **17** The clapper of a bell. **18** The harnessing pole of a horse-drawn vehicle. **19** The pointed, movable rail in a street railway switch. **20** *Mech.* Any flange or projecting part of a machine or mechanical device. **21** A projecting edge or tenon of a board for insertion into a corresponding groove of another board, thus forming a **tongue-and-groove joint. 22** A spike on a sword blade on which the hilt is secured. **23** The movable arm of a bevel. **24** A small, young sole. See synonyms under LANGUAGE. **—gift of tongues** See under GIFT. **—to hold one's tongue** To keep silent. **—with tongue in cheek** With mental reservations; facetiously; insincerely. —*v.* **tongued, tongu·ing** *v.t.* **1** To use the tongue in playing (a wind instrument) so as to produce marcato or staccato effects; also, to modify the sound of (a flute, cornet, etc.) by the use of the tongue. **2** To touch or lap with the tongue. **3 a** To cut a tongue on (a board). **b** To join or fit by a tongue-and-groove joint. **4** *Poetic* To utter; articulate. **5** *Archaic* To reproach; chide. —*v.i.* **6** To use the tongue in playing a wind instrument. **7** To talk or prattle. **8** To extend as a tongue. [OE *tunge.* Akin to LANGUAGE.]

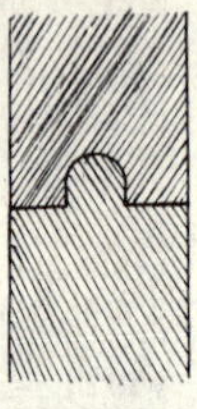

TONGUE AND GROOVE

tongued (tungd) *adj.* Having a tongue: chiefly in compounds: four-*tongued.*

tongue·grass (tung′gras′, -gräs′) *n.* Peppergrass.

tongue·less (tung′lis) *adj.* Having no tongue; hence, speechless.

tongue-tie (tung′tī′) *n.* Abnormal shortness of the frenum of the tongue, whereby its motion is impeded or confined. —*v.t.* **1** To deprive of speech or the power of speech, or of distinct articulation. **2** To bewilder or amaze so as to render speechless. **—tongue′-tied′** *adj.*

tongue-twist·er (tung′twis′tər) *n.* A word or phrase difficult to articulate quickly: "Miss Smith's fish-sauce shop" is a *tongue-twister.*

tongue worm A hemichordate animal.

ton·ic (ton′ik) *adj.* **1** Having power to invigorate or build up; bracing. **2** Pertaining to tone or tones; specifically, in music, pertaining to the keynote. **3** In art, denoting the general color effect and the light and shade in a picture or scene. **4** *Physiol.* **a** Of or pertaining to tension, especially muscular tension. **b** Rigid; unrelaxing: *tonic* spasm. **5** *Ling.* **a** Of or pertaining to musical intonations or modulations of words, sentences, etc. **b** Designating languages which distinguish words of identical or very similar form by variations in tone or pitch, as Chinese. **6** *Phonet.* **a** Stressed, as a syllable. **b** *Obs.* Voiced. —*n.* **1** *Med.* A drug that gradually restores the normal tone of organs from a condition of debility. **2** Whatever imparts vigor or tone. **3** The basic note of a key; keynote. [<Gk. *tonikos* < *tonos* sound, tone]

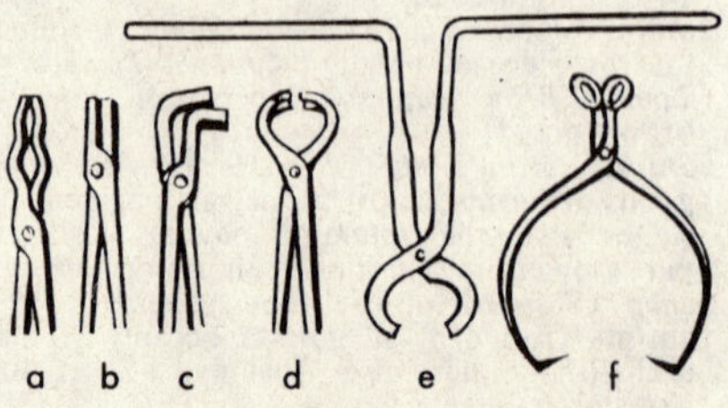

TONGS
a, b, c, d. Blacksmith's tongs.
e. Rail tongs. *f.* Ice tongs.

tonic accent 1 An accent that is spoken or pronounced rather than written. **2** *Phonet.* Emphasis placed on a syllable or sound by raising or changing the pitch of the voice.

to·nic·i·ty (tō·nis′ə·tē) *n.* **1** The state of being tonic; tone. **2** *Physiol.* The peculiar elastic condition of healthy tissue; tonus. **3** Health and vigor generally.

tonic sol-fa A system of teaching, writing, and reading music, especially vocal music, that lays particular stress on the tonal relations of the various elements of the key. The initials of the syllables used in solmization are employed to write its scale. [<TONIC + SOL[1] + FA]

to·night (tə·nīt′) *adv.* **1** In or during the present or coming night. **2** *Obs.* Last night. —*n.* The night that follows this day; also, the present night. Also **to-night′.** [OE *tō niht* < *tō* to + *niht* night]

Ton·ite (tōn′īt) *n.* A blasting explosive of the guncotton class, with addition of barium nitrate and dinitrobenzene: a trade name.

Tonk (tongk) **1** A former princely state of the Rajputana States, India; merged with Rajasthan, 1948; 2,543 square miles. **2** A city of SE Rajasthan, India, formerly the capital of Tonk state.

Ton·ka bean (tong′kə) **1** An odoriferous seed obtained from a tropical American tree (*Dipteryx odorata*), and used for the adulteration of vanilla, flavoring of tobacco, snuff, etc. **2** The tree from which it is obtained. [Prob. <Negro name for the bean in Guiana]

Ton·kin (ton′kin, tong′-) A former name for northern North Vietnam, once a powerful independent kingdom, later a French protectorate; 44,670 square miles; capital, Hanoi. Also **Tong′king, Ton′king. —Ton′kin·ese′** *adj.* & *n.*

Tonkin, Gulf of An arm of the South China Sea between North Vietnam and the southernmost Chinese areas of Hainan island and the Luichow Peninsula.

Ton·le Sap (ton′lā sap) A lake in central Cambodia; 1,000 square miles: French *Grand Lac.*

ton·nage (tun′ij) *n.* **1** The cubic capacity of a merchant vessel expressed in tons of 100 cubic feet each. **2** The freight-carrying capacity of a vessel. **3** The aggregate freightage of a collection of vessels, especially of a country's merchant marine, as represented by their registered cubic capacity. **4** A tax levied on vessels at a given rate per ton. **5** The total weight of materials produced, mined, or transported. [<OF < *tonne* a ton, tun]

ton·neau (tu·nō′) *n. pl.* **·neaus** (-nōz′) or **·neaux** (-nōz′) The rear part of an early type of automobile or vehicle, with low sides enclosing the seats, and a door at the rear or the side; also, the whole body of an automobile having such a rear part. [<F, lit., a barrel]

tono- *combining form* **1** Tension; pressure: *tonoplast.* **2** *Music* Tone; pitch: *tonometer* (def. 2). Also, before vowels, *ton-.* [<Gk. *tonos* tension < *teinein* stretch]

ton·o·graph (ton′ə·graf, -gräf, tō′nə-) *n.* A recording tonometer. [<TONO- + GRAPH]

to·nom·e·ter (tō·nom′ə·tər) *n.* **1** An instrument to measure strains within a liquid that tend to pull the particles asunder. **2** An accurately pitched tuning fork or set of forks; any instrument for determining the pitch of a tone. **3** An instrument for measuring tension in the eyeball or varying pressure of the blood.

to·nom·e·try (tō·nom′ə·trē) *n.* The art of using a tonometer. **—ton·o·met·ric** (ton′ə·met′rik, tō′nə-) *adj.*

ton·o·plast (ton′ə·plast) *n. Biol.* An inner plasmic membrane lining the vacuole of a cell and controlling the osmotic pressure. [<TONO- + -PLAST]

ton·o·scope (ton′ə·skōp, tō′nə-) *n.* An instrument by which a player or singer can observe departures from pitch or tone.

ton·sil (ton′səl) *n. Anat.* One of two oval lymphoid organs situated on either side of the passage from the mouth to the pharynx. [<L *tonsillae* the tonsils] **—ton′sil·lar, ton′sil·ar** *adj.*

ton·sil·lec·to·my (ton'sə-lek'tə-mē) n. Surg. Removal of a tonsil. [<TONSIL + -ECTOMY]

ton·sil·li·tis (ton'sə-lī'tis) n. Pathol. Inflammation of the tonsils. —**ton'sil·lit'ic** (-lit'ik) adj.

ton·sil·lo·tome (ton-sil'ə-tōm) n. An instrument used for cutting away a portion of the tonsils. [< tonsillo- (<TONSIL) + -TOME]

ton·sil·lot·o·my (ton'sə-lot'ə-mē) n. Surg. The operation of cutting away the tonsils or a part of them. [< tonsillo- (<TONSIL) + -TOMY]

ton·so·ri·al (ton-sôr'ē-əl, -sō'rē-) adj. Pertaining to a barber or to barbering: chiefly used in the humorous phrase, tonsorial artist. [<L tonsorius < tonsor, -oris a barber < tonsus, pp. of tondere shear, clip]

ton·sure (ton'shər) n. 1 The shaving of the head, or of the crown of the head, as of a priest or monk, or the state of being thus shaven; hence, the priestly office. 2 That part of a priest's or monk's head left bare by shaving. —v.t. ·sured, ·sur·ing To shave the head of. [<OF <L tonsura a shearing < tonsus. See TONSORIAL.] —**ton'sured** adj.

ton·tine (ton'tēn, ton-tēn') n. 1 A form of collective life annuity, the individual profits of which increase as the number of survivors diminishes, the final survivor taking the whole. 2 The subscribers to such an annuity, collectively. 3 The share of a single subscriber. [<F, after Lorenzo Tonti, a Neapolitan banker who introduced it into France in about 1653]

to·nus (tō'nəs) n. 1 Tonicity. 2 Physiol. **a** The ability of a muscle to contract in response to a stimulus. **b** A condition of prolonged muscular spasm. [<L, TONE]

ton·y (tō'nē) adj. ton·i·er, ton·i·est Colloq. Aristocratic; high-toned; fashionable; stylish; swell. [<TONE (def. 5b)]

too (tōō) adv. 1 In addition; likewise; also: beautiful and good too. 2 In excessive quantity or degree; more than sufficiently: too long and too technical. 3 In a degree beyond expression or endurance; extremely: I am too happy for you. 4 Colloq. Indeed: an intensive, often used to reiterate a contradicted statement: You are too going! [Stressed var. of OE tō to]

took (tōōk) Past tense of TAKE.

Tooke (tōōk), (**John**) **Horne,** 1736–1812, English politician and philologist.

tool (tōōl) n. 1 A simple mechanism or implement, as a hammer, saw, spade, or chisel, used chiefly in the direct manual working, moving, shaping, or transforming of material. 2 A power-driven apparatus, as a lathe, used for cutting and shaping the parts of a machine. 3 The cutting or shaping part of such an apparatus. 4 A bookbinder's hand stamp used in lettering or ornamenting book covers. 5 A person used to carry out the designs of others or another; a dupe. 6 Law Any instrument or apparatus necessary to the efficient prosecution of one's profession or trade. —v.t. 1 To shape, mark, or ornament with a tool. 2 To provide with tools. 3 Colloq. To drive, as an automobile, or convey (a person) by driving. 4 In bookbinding, to ornament or impress designs upon with a roller bearing a pattern. —v.i. 5 To work with a tool or tools. 6 Colloq. To drive or travel in a vehicle. [OE tōl]

— Synonyms (noun): apparatus, appliance, implement, instrument, machine, mechanism, utensil. A tool is both contrived and used for extending the force of an intelligent agent to something that is to be operated upon. An instrument is anything through which power is applied and a result produced; in general usage, the word is of considerably wider meaning than tool; as, a piano is a musical instrument. Instruments is the word usually applied to tools used in scientific pursuits; as, we speak of a surgeon's or an optician's instruments. An implement is a mechanical agency considered with reference to some specific purpose to which it is adapted; as, an agricultural implement, implements of war. Implement is a less technical term than tool. A utensil is that which may be used for some special purpose; the word is especially applied to articles used for domestic or agricultural purposes; as, kitchen utensils, farming utensils.

Mechanism is a word of wide meaning, denoting any combination of mechanical devices for united action. A *machine* in the most general sense is any mechanical *instrument* for the conversion of motion; in this sense a lever is a *machine*; but in more commonly accepted usage a *machine* is distinguished from a *tool* by its complexity, and by the combination and coordination of powers and movements to produce results.

tool·ing (tōō'ling) n. 1 The ornamentation or work done with tools. 2 The application of a tool or tools to any work.

tool·mak·er (tōōl'mā'kər) n. A maker of tools.

toom (tōōm) adj. Scot. & Brit. Dial. Empty; void; futile. [OE tōm]

Toombs (tōōmz), **Robert Augustus,** 1810–85, American Confederate general.

toon[1] (tōōn) n. 1 The fine, close-grained red wood of an East Indian tree (*Toona ciliata*) of the mahogany family, used for furniture, boxes, and construction. 2 The tree itself. [<Hind. tun <Skt. tunna]

toon[2] (tōōn) n. Scot. Hamlet; town.

toot (tōōt) v.i. 1 To blow a horn, whistle, etc., especially with short blasts. 2 To give forth a blast or toot, as a horn. 3 To make a similar sound. —v.t. 4 To sound (a horn, etc.) with short blasts. 5 To sound (a blast, etc.). —n. 1 A short note or blast on a horn. 2 Slang A spree; especially, a drinking spree. [? <MLG tuten; prob. orig. imit.] —**toot'er** n.

tooth (tōōth) n. pl. **teeth** (tēth) 1 One of the hard, dense structures in the mouth of a vertebrate, used for seizing and chewing food, as offensive and defensive weapons, etc. It consists chiefly of dentine or ivory, invested on the outer surface and crown with enamel, and a root embedded in the gum, with a small opening leading into a pulp cavity richly supplied with blood vessels and nerves. ◆ Collateral adjective: *dental*. 2 One of various hard calcareous or chitinous bodies of the oral or gastric regions of invertebrates. 3 Any one of various small toothlike projections

TEETH OF HUMAN ADULT
A. Cross-section of a molar.
B. Left upper jaw. C. Left lower jaw.
A. a. Crown. b. Enamel. c. Pulp cavity. d. Dentine. e. Cement. f. Roots.
B. & C. g. Incisors. h. Canines. i. Bicuspids. j. Molars. k. Wisdom teeth.

4 Zool. A process near the hinge of a bivalve shell. 5 Bot. One of the processes in the peristome of a moss. 6 Something resembling a tooth in form or use; specifically, a projecting point, pin, tine, or cog, as on a saw, comb, fork, rake, or gearwheel. 7 Appetite, liking, or taste (for something): She has a sweet *tooth*. 8 pl. That part which opposes, as in the gnawing, biting, or piercing manner of a tooth; the face of opposition, especially when involving resistance or risk: the *teeth* of the wind; He disobeyed them in their *teeth*. 9 pl. Means of enforcement: to put *teeth* into a law. 10 In paper or painting grounds, coarseness; irregularity of surface. —**armed to the teeth** Completely or heavily armed. —**by the skin of one's teeth** Barely; by the narrowest possible margin. —**in the teeth of** Directly against, counter to, or in defiance of. —**to put teeth into** To provide (something) with strength or power. —**to show one's teeth** To display a disposition to fight; threaten. —**to throw (or cast) in one's teeth** To fling at one, as a challenge or taunt. —v.t. 1 To supply with teeth, as a rake or saw. 2 To give a serrated edge to; indent. —v.i. 3 To become interlocked, as gearwheels; to gear. [<ON tōth, tōdh]

tooth·ache (tōōth'āk') n. 1 Pain in a tooth or the teeth, generally due to caries exposing the nerve. 2 Neuralgia of the teeth or of the jaw bone.

tooth and nail By biting and scratching; hence, fiercely; with all possible strength and effort: to fight *tooth and nail*.

tooth·brush (tōōth'brush') n. A small brush used for cleaning the teeth.

toothed (tōōtht, tōōthd) adj. 1 Having teeth, notches, cogs, or jags. 2 Bot. Dentate. 3 Coarse; irregular of surface: said of paper or painting grounds.

tooth·less (tōōth'lis) adj. 1 Being without teeth. 2 Incapable of biting; harmless.

tooth·paste (tōōth'pāst') n. A paste used in cleaning the teeth.

tooth·pick (tōōth'pik') n. 1 A small sliver of wood or metal, used for removing particles of food from between the teeth. 2 U.S. Slang A bowie knife: sometimes **Arkansas toothpick**.

tooth·pow·der (tōōth'pou'dər) n. A powder used in cleaning the teeth.

tooth rash Strophulus.

tooth·shell (tōōth'shel') n. A burrowing mollusk (genus *Dentalium*), having a long, very slender tubular shell.

tooth·some (tōōth'səm) adj. Having a pleasant taste. —**tooth'some·ly** adv. —**tooth'some·ness** n.

tooth·wort (tōōth'wûrt') n. 1 Any of a genus (*Dentaria*) of spring-blooming herbs of the mustard family, with compound toothed leaves and terminal clusters of white or purplish flowers. 2 Any of a genus (*Lathraea*) of small, parasitic plants having rootstocks covered with white scales instead of leaves.

tooth·y (tōōth'ē) adj. tooth·i·er, tooth·i·est Having large or prominent teeth.

too·tle (tōōt'l) v.i. ·tled, ·tling To toot lightly or continuously, especially on the flute, as in double-tonguing. —n. The act of, or sound produced by, tootling. [Freq. of TOOT]

toot·sy (tōōt'sē) n. pl. ·sies Slang The foot of a child or woman: an endearing or humorous term. Also **toot'sy–woot'sy** (-wōōt'sē). [Child's term for a foot]

too·zie (tōō'zē) See TOWZIE.

top[1] (top) n. 1 The uppermost or highest part, end, side, or surface of anything. 2 That end or part of anything, regarded as the higher or upper extremity: the *top* of the street. 3 A lid or cover: a bottle *top*. 4 The roof of a vehicle, as an automobile. 5 The crown of the head: from *top* to toe. 6 pl. The aboveground part of a plant producing root vegetables. 7 The highest degree or reach: at the *top* of one's voice; the *top* of one's ambition. 8 The highest or most prominent place or rank: at the *top* of one's profession. 9 One who is highest in rank or position: the *top* of one's class. 10 The choicest or best part: the *top* of the crop. 11 In bridge, the highest card in a suit. 12 In billiards, tennis, golf, etc.: **a** A stroke in which the player hits the ball above the center or on the upper half. **b** The forward spinning motion imparted to the ball by such a stroke. 13 Naut. A platform at the head of the lower section of a ship's mast, used as a place to stand and for extending the topmast rigging. 14 Chem. The most volatile part of a substance in distillation. 15 Scot. **a** The hair on one's head. **b** A bird's crest. **c** A horse's forelock. **d** A bunch of hair, wool, flax, etc. See synonyms under SUMMIT. —**to blow one's top** Slang 1 To break out in a rage; flare up. 2 To go insane. —adj. 1 Of or pertaining to the top. 2 Forming or comprising the top or upper part. 3 Highest in rank or quality; chief: *top* authors. 4 Greatest in amount or degree: *top* prices. —v. **topped, top·ping** v.t. 1 To remove the top or upper end of; prune. 2 To provide with a top, cap, etc. 3 To form the top of. 4 To reach or pass over the top of; surmount. 5 To surpass or exceed. 6 Chem. To take away the most volatile part of by distillation. 7 In golf, tennis, etc.: **a** To hit the upper part of (the ball) in making a stroke. **b** To make (a stroke) thus. —v.i. 8 To top someone or something. —**to top off** 1 To put something on the top of. 2 To complete; finish. [OE]

top[2] (top) n. A toy of wood or metal, with a point on which it is made to spin, as by the unwinding of a string, a spring, etc. [OE]

top– Var. of TOPO–.

to·paz (tō′paz) *n.* 1 A native fluosilicate of aluminum, often found in yellow prismatic crystals valued as gemstones. 2 The yellow sapphire, a highly prized corundum of Ceylon: also called **Oriental topaz.** 3 Citrine (def. 2). 4 Either of two large tropical American hummingbirds (*Topaza pyra* and *T. pella*) with brilliant green–and–gold plumage. 5 A brownish–gold color, the color of the mineral. [<OF *topaze, topace* <L *topazus* <Gk. *topazos*]

to·paz·o·lite (tō·paz′ə·līt) *n.* A variety of andradite, yellow or sometimes green. [<Gk. *topazos* topaz + -LITE]

top·boot (top′bo͞ot′) *n.* A boot with a high top, sometimes ornamented with materials different from the rest of the boot. —**top′–boot′ed** *adj.*

top buggy A buggy with a top that may be raised or folded back.

top·coat (top′kōt′) *n.* A lightweight overcoat.

top–drawer (top′drôr′) *adj. Colloq.* Of the highest standard or merit.

top–dress (top′dres′) *v.t. Agric.* To apply manure on the top of, instead of plowing it into, a field.

top dressing *Agric.* A dressing of manure not to be plowed under the surface of a field.

tope[1] (tōp) *v.t.* **toped, top·ing** To drink (alcoholic beverages) excessively and frequently. ◆ Homophone: *taupe.* [? Related to earlier *top* tilt, turn over]

tope[2] (tōp) *n. Dial.* A small European shark or dogfish (genus *Galeorhinus*). ◆ Homophone: *taupe.* [? <dial. E (Cornish); ult. origin unknown]

tope[3] (tōp) *n. Anglo–Indian* A grove, especially a mango grove. ◆ Homophone: *taupe.* [< Tamil *tōppu*]

tope[4] (tōp) *n. Anglo–Indian* A round Buddhist shrine, dome, or tower, constructed to contain relics of the Buddhas, to indicate some sacred site, or for the burial of priests: also called *stupa.* ◆ Homophone: *taupe.* [<Hind. *top,* prob. <Pali *thūpo* <Skt. *stūpa*]

to·pec·to·my (tō·pek′tə·mē, tə-) *n. Surg.* An operation in which certain prefrontal cortical areas of the brain are removed. [<TOP(O)- + -ECTOMY]

to·pee (tō·pē′, tō′pē) See TOPI.

to·pek (tō′pek) *n.* A North American Indian or Eskimo hut of weeds, twigs, and animal skins. [<Eskimo *toopik, tupek* a tent]

To·pe·ka (tə·pē′kə) The capital of Kansas, on the Kansas River in the NE part.

to·pep·o (tə·pep′ō) *n. pl.* **·pep·oes** 1 A hybrid plant obtained by crossing the Chinese pepper with a variety of tomato, cultivated for its edible fruit. 2 The fruit itself. [<TO(MAT)O + PEP(PER)]

top·er (tō′pər) *n.* A habitual drunkard; sot. [<TOPE[1]]

top·flight (top′flīt′) *adj.* Of the highest quality; outstanding; superior.

top·full (top′fo͝ol′) *adj. Rare* Brimful.

top·gal·lant (tə·gal′ənt, top′gal′ənt) *n. Naut.* 1 The mast, sail, yard, or rigging immediately above the topmast and topsail. 2 The parts of a deck that are higher than the rest. —*adj.* Pertaining to the topgallants. [<TOP[1] + GALLANT; with ref. to "making a gallant show" compared with the lower tops]

toph (tōf) *n.* Tufa. Also **tophe.** [<L *tophus, tofus*]

top·ham·per (top′ham′pər) *n. Naut.* 1 Spars and rigging usually kept aloft. 2 The light upper sails and rigging. 3 Casks, cables, rigging, etc., encumbering the deck. [<TOP[1] + HAMPER[1], *n.*] —**top′–ham′pered** *adj.*

top hat A high silk hat for men.

top·heav·y (top′hev′ē) *adj.* Having the top or upper part too heavy for the lower part; ill–proportioned; impracticable. —**top′heav′i·ness** *n.*

To·phet (tō′fet) 1 In the Old Testament, a place in the valley of Hinnom, near Jerusalem, where the Jews were said to sacrifice their children to Moloch: later used as a place for burning the city's refuse. 2 A place of endless perdition; hell. Also **To′pheth** (-fet). [<Hebrew *tōpheth,* ? an altar]

top–hole (top′hōl′) *adj. Brit. Slang* First–rate; excellent.

to·phus (tō′fəs) *n. pl.* **·phi** (-fī) 1 *Dent.* Tartar of the teeth. 2 *Pathol.* A deposit of urates around and at the surface of joints in persons affected with gout. 3 *Mineral.* Any natural calcareous tufa. [<L, tufa]

to·pi (tō·pē′, tō′pē) *n.* A hat or helmet, especially a light helmet made of pith: also spelled *topee.* [<Hind. *topī*]

to·pi·ar·y (tō′pē·er′ē) *adj.* Arranged or trimmed in, or making use of, fantastic shapes of shrubs and evergreen trees, as in gardening, etc. —*n. pl.* **·ar·ies** A topiary garden. [<L *topiarius* concerning ornamental gardening < *topia opera* ornamental gardening <Gk. *topion,* dim. of *topos* a place]

top·ic (top′ik) *n.* 1 A subject of discourse or of a treatise; any matter treated of in speech or writing; a theme for discussion. 2 *pl.* In rhetorical invention, the part that treats of the selection and arrangement of the proofs; also, the places or classes in which the various kinds of proofs are to be found. 3 A subdivision of an outline or a treatise. —*adj. Obs.* Topical. [<L *topica* <Gk. (*ta*) *topica,* lit., (matters) concerning commonplaces, title of a work by Aristotle, neut. pl. of *topikos* of a place < *topos* a place, commonplace]

Synonyms (*noun*): division, head, issue, matter, motion, point, proposition, question, subject, theme. Since a *topic* for discussion is often stated in the form of a *question, question* has come to be extensively used to denote a debatable *topic,* especially of a practical nature; as, the labor *question.* In deliberative assemblies the *motion* or other matter for consideration is known as the *question*; a member is required to speak to the *question.* In speaking or writing the general *subject* or *theme* may be termed the *topic,* but it is more usual to apply the latter term to the subordinate *divisions, points,* or *heads* of discourse; as, To enlarge on this *topic* would carry me far from my *subject.*

top·i·cal (top′i·kəl) *adj.* 1 Pertaining to a topic. 2 Of the nature of merely probable argument. 3 Belonging to a place or spot; local. 4 Pertaining to matters of present interest: a *topical* song. 5 *Med.* Local. —**top′i·cal·ly** *adv.*

top kick *Slang* A top sergeant.

top·knot (top′not′) *n.* 1 A crest, tuft, or knot on the top of the head, as of feathers on the head of a bird. 2 The hair of the human head when worn as a high knot. 3 A knot or bow worn by women, as a headdress, etc.

top–loft·y (top′lôf′tē, -lof′tē) *adj.* 1 Towering very high. 2 Very proud or haughty; inflated; pompous. —**top′–loft′i·ness** *n.*

top·mast (top′mast, top′mast′, -mäst′) *n. Naut.* The mast next above the lower mast.

top minnow Any of a family (*Poeciliidae*) of small, typically viviparous fishes which feed near the surface of the water, especially *Gambusia affinis,* widely used to combat mosquitoes.

top·most (top′mōst′) *adj.* Being at the very top.

top–notch (top′noch′) *adj. Colloq.* Excellent; best. —**top′–notch′er** *n.*

topo– *combining form* A place or region; regional: *topography.* Also, before vowels, *top–.* [<Gk. *topos* a place]

to·pog·ra·pher (tə·pog′rə·fər) *n.* An expert in topography.

topographic anatomy The anatomy of parts of the body, especially in relation to the surrounding parts.

to·pog·ra·phy (tə·pog′rə·fē) *n.* 1 The detailed description of particular places. 2 The art of representing on a map the physical features of a place. 3 The physical features, collectively, of a region. 4 Topographic surveying. [<TOPO- + -GRAPHY] —**top·o·graph·ic** (top′ə·graf′ik) or **·i·cal** *adj.* —**top′o·graph′i·cal·ly** *adv.*

to·pol·o·gy (tə·pol′ə·jē) *n.* 1 The branch of geometry which studies those properties of figures or solid bodies which remain invariant under all continuous deformation: also called *analysis situs.* 2 Topographic anatomy. 3 *Med.* The relation between the forward part of the fetus and the birth canal. [<TOPO- + -LOGY] —**top·o·log·ic** (top′ə·loj′ik) or **·i·cal** *adj.*

top·o·nym (top′ə·nim) *n.* 1 *Anat.* The name of a region of the body, as distinguished from an organ. 2 *Biol.* A name based on the place of origin of a zoological or botanical type. 3 Any name derived from the name of a place. [<TOPO- + Gk. *onoma, onyma* a name] —**top′o·nym′ic** or **·i·cal** *adj.*

to·pon·y·my (tə·pon′ə·mē) *n. pl.* **·mies** 1 The nomenclature of anatomical regions. 2 The science or study of place names, or a register of place names.

top·o·type (top′ə·tīp) *n. Biol.* A plant or animal specimen selected from the locality typical of the species. [<TOPO- + TYPE]

top·per (top′ər) *n.* 1 One who or that which cuts off the top of something. 2 *Slang* One who or that which is of supreme quality. 3 *Slang* A high silk hat.

top·ping (top′ing) *adj.* 1 Towering high above; eminent; distinguished. 2 Making great pretensions; arrogant; domineering. 3 *Brit. Colloq.* Excellent; first–rate. —*n.* 1 The act of one who tops, in any sense. 2 That which forms the top of anything.

topping lift *Naut.* A rope extending from the lower masthead to the outer end of a boom, for hoisting or supporting the boom.

top·ple (top′əl) *v.* **·pled, ·pling** *v.t.* 1 To push over and cause to totter or fall by its own weight; overturn. —*v.i.* 2 To totter and fall, as by its own weight. 3 To lean or jut out, as if about to fall. [Freq. of TOP[1], *v.*]

tops (tops) *adj. Slang* Excellent; first–rate.

top·sail (top′səl, top′sāl′) *n. Naut.* 1 In a square–rigged vessel, a square sail set next above the lowest sail of a mast. 2 In a fore–and–aft–rigged vessel, a square or triangular sail carried above the gaff of a lower sail.

top–se·cret (top′sē′krit) *adj. U.S.* Designating defense information requiring the strictest measures of secrecy and safeguard. Compare SECRET (*adj.* def. 5), CONFIDENTIAL (def. 4).

top sergeant *Colloq.* The first sergeant of a company, battery, or troop.

top·side (top′sīd′) *n. Naut.* The portion of a ship above the main deck. —*adv.* To or on the upper parts of a ship: He is going *topside.*

top·soil (top′soil′) *n.* The surface soil of land: distinguished from *subsoil.* —*v.t.* To remove the surface soil of (an area or region).

top·stone (top′stōn′) *n.* A capstone.

Top·sy (top′sē) In *Uncle Tom's Cabin,* a young Negro slave girl who, when questioned about her origins, replied that she had "just growed."

top·sy–tur·vy (top′sē·tûr′vē) *adv.* Upside–down; hind side before; in utter confusion. —*adj.* Being in an upset or disordered condition; upside–down. —*n.* A state of confusion; disorder; chaos. [Earlier *topsy–tervy, topsy–tirvy,* prob. <TOP[1] + obs. *terve, tirve* turn, overturn] —**top′sy–tur′vi·ly** *adv.* —**top′·sy–tur′vi·ness** *n.* —**top′sy–tur′vy·dom** *n.*

toque (tōk) *n.* 1 A small, close–fitting, brimless hat worn by women. 2 The tall conical headdress formerly worn by the doges of Venice. 3 A black velvet cap, ornamented with eagle's plumes and furnished with a band and brim: worn by both sexes in France before the Restoration. Also **to·quet** (tō·kā′). [<F, a cap <Sp. *toca* <Basque *tauka,* a kind of cap]

tor (tôr) *n.* A high, rocky hill; a jutting rock. [OE *torr* <Celtic]

to·rah (tôr′ə, tō′rə) *n.* In Hebrew literature, a law; also, counsel or instruction proceeding from a specially sacred source. Also **to′ra.** [<Hebrew *tōrāh* an instruction, law < *yārāh* throw, show, instruct]

To·rah (tôr′ə, tō′rə) *n.* The Mosaic law; the Pentateuch.

tor·bern·ite (tôr′bərn·īt) See under URANITE. [<G *torbernit, torberit* <NL *torbernus,* after *Torber* Bergmann, 18th c. Swedish chemist]

torc (tôrk) See TORQUE[2].

torch (tôrch) *n.* 1 A source of light, as from flaming pine knots, or from some material dipped in tallow or oil, and fixed at the end of a handle or pole. 2 Anything that illuminates or brightens: the *torch* of science. 3 A portable device giving off an intensely hot flame and used for burning off paint, melting solder, etc. 4 A flashlight. —**to carry a (or the) torch for** *Slang* To continue to love (someone), though the love is unrequited. [<OF *torche,* ult. <L *torquere* twist; so called because early torches were made of twisted tow dipped in pitch]

torch·bear·er (tôrch′bâr′ər) *n.* 1 One who carries a torch. 2 One who imparts knowledge, truth, etc. 3 *Colloq.* One loud in his praise of a friend.

torch·light (tôrch′līt′) *n.* The light of a torch or torches. —*adj.* Lighted by torches: a *torchlight* rally.

torchlight procession A parade of persons

carrying torches, usually a political demonstration.

tor·chon lace (tôr′shon, *Fr.* tôr·shôṅ′) **1** A coarse, durable bobbin lace in simple geometrical designs made of linen thread. **2** An imitation of this made by machine. [<F *torchon* a dishcloth < *torcher* wipe]

torch singer One who sings torch songs.

torch song A popular love song, slow and melancholy, expressing sadness and hopeless yearning. [< phrase "carry a torch for." See under TORCH.]

torch·wood (tôrch′wo͝od′) *n.* **1** Any of a genus *(Amyris)* of tropical American shrubs and small trees, especially *A. balsamifera.* **2** Its bright-burning, fragrant wood.

Tor·de·sil·las (tôr′thā·sē′lyäs) A village in NW Spain; scene of the signing of a treaty between Spain and Portugal setting the line of demarcation for colonial expansion, 1494.

tore[1] (tôr, tōr) Past tense of TEAR[1].

tore[2] (tôr, tōr) *n.* Torus (defs. 1 and 4). [<F *tore* <L *torus* a torus]

tor·e·a·dor (tôr′ē·ə·dôr′, *Sp.* tō′rā·ä·thôr′) *n.* One who engages in a bullfight, especially on horseback; a bullfighter: also spelled *toreador.* [<Sp. < *torear* fights bulls < *toro* a bull <L *taurus*]

to·re·ro (tō·rā′rō) *n. pl.* **·ros** (-rōs) *Spanish* A bullfighter, usually on foot.

to·reu·tics (tə·ro͞o′tiks) *n. pl.* *(construed as singular)* The art of working in ornamental relief or intaglio, especially in metal. [<Gk. *toreutikos* < *toreuein* work in relief, bore] **—to·reu′tic** *adj.*

tor·ic (tôr′ik, tor′-) *adj.* Of, pertaining to, or resembling a torus; segmental.

toric lens *Optics* A lens in which one of the surfaces is a segment of a torus: used for eyeglasses because of its special refracting powers.

to·ri·i (tôr′i·ē, tō′ri·ē) *n.* The gateway of a Shinto temple or of a shrine: properly comprising two uprights with one straight crosspiece, and another above with a concave lintel. [<Japanese]

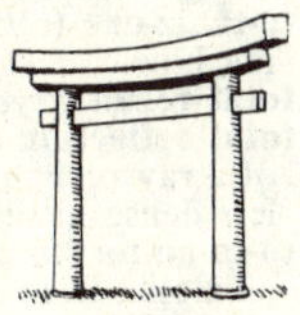

JAPANESE TORII

To·ri·no (tō·rē′nō) The Italian name for TURIN.

tor·ment (tôr′ment) *n.* **1** Intense bodily pain or mental anguish; agony; torture. **2** One who or that which torments. **3** The inflicting of torture. **4** *Archaic* Any device for inflicting torture, as the rack; also, the torture inflicted. **5** Hell. See synonyms under AGONY, PAIN. **—*v.t.*** (tôr·ment′) **1** To subject to excruciating physical or mental suffering; torture. **2** To make miserable; afflict or vex grievously. **3** To harass or tease. **4** To distort; also, to throw into violent agitation. See synonyms under PERSECUTE. [<AF *turment*, OF *torment, tourment* <L *tormentum* a rack, orig. a machine for hurling missiles by means of torsion < *torquere* twist] **—tor·ment′ing·ly** *adv.* **—tor·ment′ing·ness** *n*

tor·men·til (tôr′men·til) *n.* A slender, trailing, Old World herb *(Potentilla erecta)*, with yellow flowers. Its root, a powerful astringent, has been used in treating diarrhea and dysentery, and also in tanning. [<OF *tormentille* < Med. L *tormentilla*, dim. of L *tormentum* TORMENT; so called because used as a pain killer]

tor·men·tor (tôr·men′tər) *n.* **1** One who or that which torments. **2** A movable panel of sound-insulating material for controlling the acoustics on a sound stage outside of the field of the camera. **3** A movable piece of theater scenery at either side and back of the proscenium arch to mask sidelights and downstage entrances and exits. Also **tor·ment′er.**

torn (tôrn, tōrn) Past participle of TEAR[1].

tor·na·do (tôr·nā′dō) *n. pl.* **·does** or **·dos** **1** A whirling wind of exceptional violence, usually associated with thunderstorms and accompanied by a pendulous, funnel-shaped cloud marking the narrow path of greatest destruction. **2** A violent thunderstorm or squall of the west coast of Africa. **3** A hurricane or violent windstorm of the tropical Atlantic. See synonyms under CYCLONE. [Alter. of *ternado*, prob. alter. of Sp. *tronada* a thunderstorm < *tronar* thunder <L *tonare*, infl. in form by Sp. *tornar* turn, because characterized by shifting or whirling winds] **—tor·nad′ic** (-nad′ik) *adj.*

Tor·ne (tôr′nə, tōr′-) A river in northern Sweden and northern Finland, flowing 250 miles SE and south from Lake Torne to its confluence with the Muonio, and thence along the Swedish-Finnish border to the Gulf of Bothnia.

Torne, Lake A lake near the Norwegian border in extreme NW Sweden; 124 square miles; 40 miles long, 1 to 6 miles wide.

to·roid (tôr′oid, tō′roid) *n.* **1** *Geom.* **a** A surface generated by the rotation of any closed plane curve, as a circle or ellipse, about an axis lying in its plane. **b** The solid produced by such a surface. **2** *Electr.* An electromagnetic coil wound upon a ring of circular cross-section. [<TOR(US) + -OID] **—to·roi′dal** *adj.*

To·ron·to (tə·ron′tō) The capital of Ontario province, Canada, on Lake Ontario; a leading industrial center.

To·ros Dağ·la·ri (tō·rôs′ dä′lä·rē′) The Turkish name for the Taurus range.

to·rose (tôr′ōs, tō′rōs, tō·rōs′, tō-) *adj.* **1** Having protuberances; bulging. **2** *Bot.* Knobby; cylindrical and swollen at intervals. Also **to·rous** (tôr′əs, tō′rəs). [<L *torosus* < *torus* a swelling] **—to·ros·i·ty** (tō·ros′ə·tē) *n.*

tor·pe·do (tôr·pē′dō) *n. pl.* **·dos** or **·does** **1** A device or apparatus containing an explosive to be fired by concussion or otherwise. **2** A self-propelling, cigar-shaped projectile for carrying a powerful detonating charge under water to a hostile vessel. **3** A submarine mine. **4** A cartridge placed on a railway track and exploded by the weight of a train passing over it, the report serving as a warning signal to the train crew. **5** A cartridge exploded in an oil or gas well to start or increase the flow. **6** A toy of gravel and a fulminating powder wrapped in paper, and exploded by being dashed against some hard surface. **7** A ray fish *(Torpedo ocellata)* having an electric apparatus with which it stuns or kills its prey; a crampfish; numbfish. **8** *Colloq.* A gangster, especially an armed bodyguard prepared to attack or kill without warning. **—aerial torpedo** A torpedo projectile, moving under its own power and usually released from low-flying aircraft at fixed or floating targets. **—*v.t.*** **·doed, ·do·ing** To damage or sink (a vessel) with a torpedo or torpedos. [<L, stiffness, numbness < *torpere* be numb]

torpedo boat A small, swift, lightly armed and armored surface vessel equipped with one or more tubes for the discharge of torpedos.

tor·pe·do-boat destroyer (tôr·pē′dō·bōt′) A small, swift, lightly armed war vessel; a destroyer.

torpedo tube A tube in a torpedo boat or other war vessel, through which torpedos are launched.

tor·pid[1] (tôr′pid) *adj.* **1** Having lost sensibility or power of motion, partially or wholly, as a hibernating animal. **2** Dormant; numb. **3** Sluggish; apathetic; dull. See synonyms under LIFELESS, NUMB. [<L *torpidus* < *torpere* be numb] **—tor·pid·i·ty** (tôr·pid′ə·tē), **tor′pid·ness** *n.* **—tor′pid·ly** *adv.*

tor·pid[2] (tôr′pid) *n.* **1** An eight-oared, clinker-built racing boat for the second crew at Oxford University; also, one of its crew. **2** *pl.* The Lenten races in which such boats take part. [<TORPID[1]; so called because the second crew consisted of awkward or very young oarsmen]

tor·por (tôr′pər) *n.* **1** Complete or partial insensibility; stupor. **2** Apathy; torpidity. [<L < *torpere* be numb] **—tor·po·rif·ic** (tôr′pə·rif′ik) *adj.*

tor·quate (tôr′kwit, -kwāt) *adj. Zool.* Having a torque or ring, as of color, about the neck; collared. [<L *torquatus* having a collar < *torques.* See TORQUES.]

Tor·quay (tôr·kē′) A port and municipal borough in southern Devon, England.

torque[1] (tôrk) *n.* **1** *Mech.* **a** Anything that causes or tends to cause torsion in a body; the moment of forces that causes rotation or twisting. **b** The rotary force in a mechanism. **c** The degree of smoothness in the conversion of reciprocating into rotary motion. **2** *Optics* The rotatory effect upon the plane of polarization produced by the passage of light through certain liquids and crystals. [<L *torquere* twist]

torque[2] (tôrk) *n.* A necklace, armlet, or collar of wire, usually twisted: worn especially by ancient Gauls and Britons: also spelled *torc.* [<L *torques.* See TORQUES.]

Tor·que·ma·da (tôr′kwə·mä′də, *Sp.* tôr′kā·mä′thä), **Tomás de,** 1420–98, Dominican monk; first inquisitor general of Spain.

tor·ques (tôr′kwēz) *n. Zool.* A natural ring or collar, of feathers or hair, on the neck of a bird or other animal. [<NL <L, a twisted collar < *torquere* twist]

tor·re·a·dor (tôr′ē·ə·dôr′) See TOREADOR.

Tor·re An·nun·zia·ta (tôr′rā än′no͞on·tsyä′tä) A port on the Bay of Naples, southern Italy; destroyed by an eruption of Vesuvius, 1631, and rebuilt.

tor·re·fy (tôr′ə·fī, tor′-) *v.t.* **·fied, ·fy·ing** To dry or roast by exposure to heat, as ores or drugs. Also **tor′ri·fy.** [<MF *torréfier* <L *torrefacere* < *torrere* dry, parch + *facere* make] **—tor′re·fac′tion** (-fak′shən) *n.*

Tor·rens (tôr′ənz, tor′-), **Lake** A salt lake in SE central South Australia; 120 miles long; 2,230 square miles.

tor·rent (tôr′ənt, tor′-) *n.* **1** A stream of water flowing with great velocity or turbulence. **2** Any similar stream, as of lava. **3** Any abundant or tumultuous flow: a *torrent* of rain; a *torrent* of abuse. **—*adj.*** Like a torrent; pouring forth with violence. [<OF <L *torrens, -entis,* lit., boiling, burning, ppr. of *torrere* parch]

tor·ren·tial (tô·ren′shəl, to-) *adj.* **1** Of, pertaining to, or resulting from the action of a torrent or torrents. **2** Figuratively, suggestive of a torrent in rapidity and volume; outpouring; overpowering: *torrential* passion. **—tor·ren′tial·ly** *adv.*

Tor·re·ón (tôr′rā·ôn′) A city in Coahuila, northern Mexico; an industrial center.

Tor·res Strait (tôr′əs, -iz, tor′-) A strait between Australia and New Guinea; 95 miles wide; connects the Arafura and Coral seas.

Tor·res Ve·dras (tôr′rizh vā′thrəsh) A town north of Lisbon in western Portugal; Wellington's headquarters in the Peninsula campaign, 1810.

Tor·rey (tôr′ē, tor′ē), **John,** 1796–1873, U.S. botanist.

Tor·ri·cel·li (tôr′rē·chel′lē), **Evangelista,** 1608–1647, Italian physicist; discovered the principle of the barometer. **—Tor·ri·cel·li·an** (tôr′·i·sel′ē·ən, -chel′ē·ən) *adj.*

Torricellian tube A vertical glass tube containing mercury or other fluid, sealed at the top and having the lower end in a container of the same fluid. [after E. TORRICELLI]

Torricellian vacuum The vacuum above the fluid in a Torricellian tube or in the top of a barometer tube.

tor·rid (tôr′id, tor′-) *adj.* **1** Exposed to the full force of the sun's heat; sultry. **2** Having power to parch or burn; scorching; burning; hot and dry. [<L *torridus* < *torrere* parch] **—tor·rid·i·ty** (tô·rid′ə·tē, to-), **tor′rid·ness** *n.* **—tor′rid·ly** *adv.*

torrid zone See under ZONE.

tor·sade (tôr·säd′) *n.* **1** A molded ornament resembling a twisted cable. **2** A twisted cord for draperies. [<F <Med. L *torsus,* var. of L *tortus,* pp. of *torquere* twist]

tor·si·bil·i·ty (tôr′sə·bil′ə·tē) *n.* Capacity for undergoing torsion, measured by the amount of torsion produced.

tor·sion (tôr′shən) *n.* **1** The act of twisting, or the state of being twisted. **2** *Mech.* Deformation of a body, as a thread or rod, by twisting, one end being held fast while the other is subjected to a torque around its length as an axis. **3** The force with which a twisted cord or cable tends to return to its former position: distinguished from *tension.* [<OF <LL *torsio, -onis,* var. of L *tortio, -onis* < *tortus,* pp. of *torquere* twist] **—tor′sion·al** *adj.* **—tor′sion·al·ly** *adv.*

torsion balance An instrument for determining very minute forces by measuring the angle through which an arm turns before the resisting force of torsion acts upon the supporting wire or filament.

torsion bar A solid or laminated bar or rod, anchored on one end, which acts as a spring when subjected to torsion (def. 2). — **tor′sion-bar′** adj.
torsk (tôrsk) n. 1 A gadoid fish; the cusk. 2 The codfish. [<Norw. <ON *thorskr*, prob. < base of *thurr* dry. Akin to THIRST.]
tor·so (tôr′sō) n. pl. **·sos** or **·si** (-sē) 1 The trunk of a human body. 2 In sculpture, a statue deprived of head and limbs. 3 Any fragmentary or defective thing. [<Ital., a stalk, core, trunk of a body <L *thyrsus* a stalk <Gk. *thyrsos* a thyrsus]
tort (tôrt) n. *Law* Any private or civil wrong by act or omission for which a civil suit can be brought, but not including breach of contract. [<OF <L *tortus* twisted. See TORSION.]
torte (tôrt, *Ger.* tôr′tə) n. A rich cake variously made of butter, eggs, fruits, and nuts. [<G]
tort-fea·sor (tôrt′fē′zər) n. *Law* One who has committed a tort; a wrongdoer. [<OF *tortfesor*, *tortfaiseur* < *tort* a wrong, tort + *fesor*, *faiseur* a doer < *faire* do <L *facere*]
tor·ti·col·lis (tôr′tə-kol′is) n. *Pathol.* A spasmodic affection of the muscles of the neck which draws the head to one side; wryneck. [<NL <L *tortus* twisted + *collum* neck] — **tor′ti·col′lar** adj.
tor·tile (tôr′til) adj. Twisted up into a coil. [<L *tortilis* <*tortus* twisted. See TORSION.]
tor·til·la (tôr-tē′yä) n. A flat cake made of coarse cornmeal and baked on a hot sheet of iron or a slab of stone: the customary substitute for bread in Mexico. [<Sp., dim. of *torta* a cake <LL, a twisted loaf <L, pp. fem. of *torquere* to twist]
tor·tious (tôr′shəs) adj. *Law* Of the nature of or implying a tort; wrongful. [<AF *torcious* < *torcion*, *tortion*, var. of OF *torsion* torsion; infl. in meaning by TORT] — **tor′tious·ly** adv.
tor·tive (tôr′tiv) adj. *Obs.* Twisted. [<L *tortivus* < *tortus*. See TORSION.]
tor·toise (tôr′təs) n. 1 A turtle; chelonian; specifically, one of a terrestrial or fresh-water species, or a terrestrial as distinguished from an aquatic species. 2 A testudo. — **giant tortoise** Any of several species of very large herbivorous land tortoises (family *Testudinidae*), especially those found on the Galápagos Islands, which may reach a length of four feet and weigh 600 pounds. [Earlier *tortuce* < Med. L *tortuca*, ult. <L *tortus* twisted; so called from its crooked feet]

GIANT TORTOISE
(Largest specimens: up to 5 1/2 feet long by 4 1/2 feet wide)

tortoise beetle A small, iridescent beetle (family *Chrysomelidae*) having a tortoiselike form.
tortoise plant Elephant foot.
tortoise shell 1 The shell of a marine turtle, especially of the hawkbill, valuable in the arts. 2 A cat having fur mottled with black and yellow like the shell of a tortoise. — **tor′toise-shell′** adj.
Tor·to·la (tôr-tō′lə) The chief island of the British Virgin Islands; 21 square miles; capital, Road Town.
tor·tri·cid (tôr′trə-sid) n. Any of a large family (*Tortricidae*) of small, usually bright-colored moths with rectangular fore wings, including many important pests of fruit and forest trees. — adj. Of or pertaining to the *Tortricidae*. [<NL <*Tortrix*, type genus <L *tortus*. See TORSION.]
Tor·tu·ga (tôr-tōō′gə) An island off the northern coast of Haiti, to which it belongs; 70 square miles; a 17th century pirate stronghold. French **Île de la Tor·tue** (ēl′ də lä tôr-tü′).
tor·tu·os·i·ty (tôr′chōō-os′ə-tē) n. 1 The quality or state of being tortuous; a twist. 2 A bend or twist; winding.

tor·tu·ous (tôr′chōō-əs) adj. 1 Consisting of or abounding in irregular bends or turns; twisting. 2 Figuratively, morally irregular or crooked; not straightforward; devious. [<AF <L *tortuosus* < *tortus*. See TORSION.] — **tor′tu·ous·ly** adv. — **tor′tu·ous·ness** n.
tor·ture (tôr′chər) n. 1 Infliction of or subjection to extreme physical pain. 2 A former judicial mode of getting evidence by inflicting pain. 3 Great mental suffering; agony. 4 Something that causes severe pain. 5 A violent perversion or straining. See synonyms under AGONY, PAIN. — v.t. **·tured, ·tur·ing** 1 To inflict extreme pain upon; cause to suffer keenly in body or mind; specifically, to put to judicial torture. 2 To twist or turn into an abnormal form; distort; wrench. [<OF <L *tortura*, lit., a twisting <*tortus*. See TORSION.] — **tor′tur·er** n.
tor·u·lose (tôr′yə-lōs) adj. *Bot.* Having alternate swellings and constrictions like the vegetative growth of *Torula*, a genus of fungus. Also **tor′u·lous** (-ləs). [<NL *Torula*, dim. of *torus* a torus]
To·ruń (tô′rōōn′y) A port on the Vistula in north central Poland: German *Thorn*.
to·rus (tôr′əs, tō′rəs) n. pl. **to·ri** (tôr′ī, tō′rī) 1 *Archit.* A large convex molding, nearly semicircular in cross-section: used in bases as the lowest molding, or in columns above the plinth. 2 *Anat.* A rounded ridge, as on the occipital bone of the skull. 3 *Bot.* The swollen end of a flowerstalk which bears the floral leaves; the receptacle. 4 *Geom.* The surface or solid generated by the rotation of a conic section about an axis in its own plane. [<L, lit., a swelling]
to·ry (tôr′ē, tō′rē) n. pl. **·ries** *Obs.* 1 A freebooter among the outlawed Irish in the 17th century. 2 Any outlaw or bandit. [<Irish *tōruidhe* a robber, a pursuer < *tōir* pursue]
To·ry (tôr′ē, tō′rē) n. pl. **To·ries** 1 A historical English political party, successor to the Cavaliers and opponent of the Whigs; since about 1832 called the Conservative party. 2 One who at the period of the American Revolution adhered to the cause of British sovereignty over the colonies. 3 A very conservative person: also **tory**. — **To′ry·ism** n.
Tos·ca·na (tôs-kä′nē) The Italian name for TUSCANY.
Tos·ca·ni·ni (tos′kə-nē′nē, tôs′-; *Ital.* tôs′kä-nē′nē), **Arturo**, 1867–1957, Italian orchestra conductor active in the United States.
tosh (tosh) n. *Brit. Colloq.* Nonsense; rubbish; bosh. [? Alter of BOSH]
toss (tôs, tos) v.t. 1 To throw, pitch, or fling about. 2 To make restless; agitate; disturb. 3 To throw with the hand, especially with the palm of the hand upward; pitch. 4 To lift with a quick motion, as the head. 5 To bandy about, as something discussed. 6 To toss up with. See TO TOSS UP, below. — v.i. 7 To be moved or thrown about; be flung to and fro, as a ship in a storm. 8 To throw oneself from side to side; roll about restlessly, as in sleep. 9 To go quickly or angrily, as with a toss of the head. 10 To toss up a coin. — **to toss** (or **peak**) **oars** To raise the oars out of the rowlocks to a vertical position. — **to toss off** 1 To drink at one draft. 2 To utter, write, or do in an offhand manner. — **to toss up** To throw a coin into the air to decide a matter or choice by the way in which it falls. — n. 1 The act of tossing; specifically, a gentle throwing from the hand; a pitch; also, the distance over which a thing is tossed. 2 A quick upward or backward movement of the head; any quick jerk. 3 The state of being tossed about; excitement; agitation. 4 A toss-up or wager. 5 *Scot.* A belle; a toast. [Prob. <Scand. Cf. dial. Norw. *tossa* spread, strew.] — **toss′er** n.
toss·pot (tôs′pot′, tos′-) n. A toper; drunkard.
toss-up (tôs′up′, tos′-) n. *Colloq.* 1 The throwing up, as of a coin, to decide a bet, etc. 2 An even or fair chance.
tot[1] (tot) n. A little child; toddler; also, anything small or trifling. [Prob. <Scand. Cf. ON *tuttr* a dwarf.]
tot[2] (tot) v.t. *Colloq.* To add; total: with *up* or *together*. [Short for TOTAL]
to·tal (tōt′l) n. 1 The whole sum or amount; the whole, especially the whole considered as an aggregate of parts or elements. See synonyms under AGGREGATE, MASS[1]. — adj. 1 Constituting or comprising a whole, without diminution or division; being a total: the sum *total*. 2 Extending throughout the whole; comprising everything; complete; perfect: a *total* loss. — v. **·taled** or **·talled, ·tal·ing** or **·tal·ling** v.t. 1 To ascertain the total of. 2 To come to or reach as a total; amount to. — v.i. 3 To amount: often with *to*. [<OF <Med. L *totalis* <L *totus* all] — **to′tal·ly** adv.
total abstinence See under ABSTINENCE.
total depravity The condition defined by the doctrine that human nature has no tendency to piety or spirituality, but has the opposite tendency, every faculty having an innate taint: one of the five points of Calvinism. Compare ORIGINAL SIN.
total emission *Physics* The maximum emission of electrons from the cathode of a thermionic or vacuum tube.
to·tal·i·sa·tor (tōt′l-ə-zā′tər, -ī-zā′-) n. *Brit.* A machine used at racetracks for totaling the bets, reckoning the resulting pay-off odds, and recording these on a large scoreboard visible to the grandstand; a pari-mutuel. Also **to′tal·i·za′tor**.
to·tal·i·tar·i·an (tō-tal′ə-târ′ē-ən) adj. Designating or characteristic of a government controlled exclusively by one party or faction, which suppresses all opposition and criticism and controls and regiments all social, cultural, and economic activity in the country to advance its political aims. — n. An adherent of totalitarian government. [<TOTALIT(Y) + -ARIAN] — **to·tal′i·tar′i·an·ism** n.
to·tal·i·ty (tō-tal′ə-tē) n. 1 An aggregate of parts or individuals. 2 The state of being total or entire. 3 *Astron.* The state or period of an eclipse while it is total. Also *totality of an eclipse*. See synonyms under AGGREGATE, MASS[1].
to·tal·ize (tōt′l-īz) v.t. **·ized, ·iz·ing** To collect into or ascertain as an aggregate; make total. — **to′tal·i·za′tion** n.
to·tal·iz·er (tōt′l-ī′zər) n. A pari-mutuel machine.
total recall *Psychol.* Hypermnesia.
total reflection *Optics* The complete reflection of a ray of light passing from a denser to a less dense medium.
to·ta·quine (tō′tə-kwin) n. A mixture of the alkaloids from cinchona bark, including an effective percentage of quinine: used in the treatment of malaria. [<NL *totaquina* <L *tota*, fem. of *totus* all + Quechua *(quin)quina* cinchona bark]
tote (tōt) *Colloq.* v.t. **tot·ed, tot·ing** 1 To carry or bear on the person, as a burden. 2 To carry, transport, or haul, as supplies. 3 In arithmetic, to carry. 4 To wear habitually: He *totes* a gun. — n. 1 The act of toting. 2 A load or haul. [Prob. <West African] — **tot′er** n.
to·tem (tō′təm) n. 1 Among many primitive peoples, especially the North American Indians, an animal, plant, or other natural object believed to be ancestrally related to a tribe, clan, or family group or to be its tutelar spirit. 2 The representation of such an animal, plant, or object taken as an emblem or symbol. 3 The name or symbol of a person, clan, or tribe. [<Algonquian. Cf. Ojibwa *ototeman* his relations.] — **to·tem·ic** (tō-tem′ik) adj.
to·tem·ism (tō′təm-iz′əm) n. 1 Belief in totems and the practices associated therewith. 2 The system of dividing a tribe into sibs or clans according to their totems. — **to′tem·ist** n. — **to′tem·is′tic** adj.
totem pole A post or pole, usually of cedar and sometimes as much as 50 feet high, carved or painted with totemic symbols, erected outside an Indian house or as a memorial to a deceased, especially among the Indians of the NW American coast. Also **totem post**.

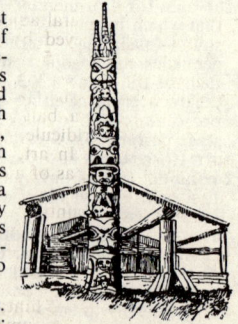

TOTEM POLE
Haida Indians, Queen Charlotte Islands, B.C.

toth·er (tuth′ər) pron. *Colloq.* The other; other. Also **t'oth′er**. [ME *the tother* < *thet other* the other]
toti- *combining form* Whole; wholly: *totipalmate*. [<L *totus* whole]

to·ti·pal·mate (tō′ti·pal′māt) *adj. Ornithol.* Wholly webbed; having all four toes joined by a web, as pelicans. [<TOTI- + PALMATE] —**to′ti·pal·ma′tion** (-pal·mā′shən) *n.*

to·tip·o·tence (tō·tip′ə·təns) *n. Biol.* Power to regenerate the whole of an organism, or some one part, from a fragment. [<TOTI- + *potence,* var. of POTENCY] —**to·tip′o·tent** *adj.*

Tot·le·ben (tôt′lā·bən, tot·lā′-), **Count Franz Eduard Ivanovich,** 1818–84, Russian general and engineer. Also *Todleben.*

Tot·ten·ham (tot′n·əm) A municipal borough of SE Middlesex, and northern suburb of London, England.

tot·ter (tot′ər) *v.i.* **1** To walk feebly and unsteadily. **2** To shake or waver, as if about to fall; be unsteady. —*n.* The act of tottering. See synonyms under SHAKE. [Prob. <Scand. Cf. Norw. *totra, tutra* quiver.] —**tot′ter·er** *n.* —**tot′ter·y** *adj.*

tot·ter·ing (tot′ər·ing) *adj.* Unsteady; that totters; variable. —**tot′ter·ing·ly** *adv.*

to·tum (tō′təm) *Latin* The whole; all.

tou·can (to͞o′kan, to͞o·kän′) *n.* A large, fruit-eating bird of tropical America (family *Rhamphastidae*) with brilliant plumage and an immense thin-walled beak. [<F <Pg. *tucano* <Tupian *tucana*]

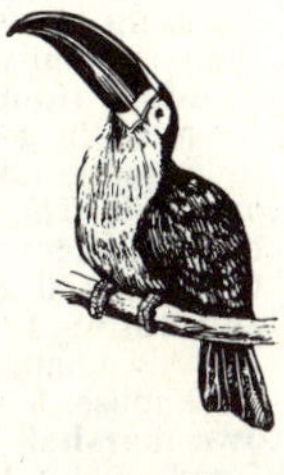

TOUCAN (About 12 inches over-all)

touch (tuch) *v.t.* **1** To place the hand, finger, etc., in contact with. **2** To be in or come into contact with. **3** To bring into contact with something else. **4** To hit or strike lightly; tap. **5** To lay the hand or hands on. **6** To border on; adjoin. **7** To come to; reach. **8** To attain to; equal. **9** To mark or delineate lightly, as with a brush or pen. **10** To modify by adding fine strokes or lines; retouch. **11** To color slightly: The sun *touched* the clouds with gold. **12** To affect injuriously; taint: Vegetables *touched* by frost. **13** To affect by contact; act upon: The drill could not *touch* the steel. **14** To affect the emotions of; soften; move. **15** To move to anger; irritate. **16** To strike the strings or keys of (a musical instrument); play on. **17** To play (a tune). **18** To relate to; concern: This quarrel *touches* you. **19** To treat or discuss in passing; deal with. **20** To have to do with, use, or partake of: I will not *touch* this food. **21** *Slang* To borrow money from. **22** *Slang* To steal. **23** *Geom.* To be tangent to. **24** *Obs.* To test, as gold with a touchstone. —*v.i.* **25** To touch someone or something. **26** To come into or be in contact. See synonyms under REACH. —**to touch at** To stop briefly at (a port or place) in the course of a journey or voyage. —**to touch off 1** To cause to explode; detonate; fire. **2** To cause to happen or occur. —**to touch on** (or **upon**) **1** To relate to; concern. **2** To treat briefly or in passing. —**to touch up 1** To strike or prod gently; rouse. **2** To add finishing touches or corrections to. —*n.* **1** The act or process of touching or coming in contact with (something). **2** The act or state of being touched. **3** That one of the special senses that gives the impression of contact with external material objects or their impact upon the body. ◆ Collateral adjective: *tactile.* **4** The sensation conveyed by touching something: a smooth *touch.* **5** *Med.* **a** Examination by feeling; palpation. **b** Digital examination of the vagina in obstetrics. **6** A stroke; hit; blow: to give a ball a slight *touch.* **7** A stroke of wit, ridicule, etc.: He felt the *touch* of her wit. **8** In art, any slight or delicate effort or effect, as of a brush, pen, or chisel; a light stroke or mark: to apply the finishing *touches* to a painting. **9** Any slight detail or effort given to anything, as to a literary work. **10** The manner or style in which an artist, workman, or author executes his work: a master's *touch*; a freedom of *touch.* **11** A trace; tinge; hint; infusion: a *touch* of irony; a *touch* of autumn. **12** A slight attack or twinge: a *touch* of rheumatism; a *touch* of remorse. **13** A small quantity or dash: to apply a *touch* of perfume. **14** Close communication, contact, or sympathy: to keep in *touch* with; to lose *touch* with. **15** A test; trial: to put something to the *touch* of proof. **16** *Music* **a** In the pianoforte, the resistance made to the fingers by the keys. **b** The manner in which a player presses the keyboard. **17** In Rugby football and soccer, the ground just outside the touch lines. **18** An official stamp impressed upon ware made of gold, silver, or pewter, to testify to its fineness. **19** *Obs.* A touchstone, or the method of assaying by the use of a touchstone. **20** *Slang* A sum of money obtained, usually from a friend or acquaintance, by borrowing or mooching. **21** *Slang* A request for such a sum of money: to make a *touch.* **22** *Slang* A person who is an easy mark for a loan or gift of money: usually with an attributive word: a soft *touch*; an easy *touch.* [<OF *tochier, tuchier*; prob. ult. imit.] —**touch′a·ble** *adj.* —**touch′a·ble·ness** *n.* —**touch′er** *n.*

touch-and-go (tuch′ən·gō′) *adj.* **1** Risky; precarious. **2** Hasty and casual; perfunctory.

touch and go 1 An uncertain, risky, or precarious state of things; a narrow escape. **2** An instantaneous or rapid action.

touch·back (tuch′bak′) *n.* In football, the act of touching the ball to the ground behind the player's own goal line when the impetus that sent the ball over the goal line was given to it by an opponent.

touch·down (tuch′doun′) *n.* A scoring play in football in which the ball is held on or over the opponent's goal line and is there declared dead.

tou·ché (to͞o·shā′) *French adj.* In fencing, touched by the point of an opponent's foil. —*interj.* You've scored a point! That argument struck home!: an exclamation used to indicate an opponent's success.

touched (tucht) *adj.* **1** That has been subjected to contact. **2** Slightly unbalanced in mind; crack-brained.

touch·hole (tuch′hōl′) *n.* The orifice in old-fashioned cannon or firearms through which the powder was ignited.

touch·ing (tuch′ing) *adj.* Appealing to the susceptibilities; affecting; pathetic. See synonyms under PITIFUL. —*n.* **1** The act of one who touches. **2** The sense of touch. —*prep.* With regard to; concerning; with respect to. —**touch′ing·ly** *adv.* —**touch′ing·ness** *n.*

touch lines The side boundary lines of a Rugby football or soccer field.

touch-me-not (tuch′mē·not′) *n.* **1** Any plant of the genus *Impatiens,* as the garden balsam (*I. balsamina*): so called from the explosive discharge of the seeds by the ripe capsules when touched. **2** The squirting cucumber. **3** *Pathol.* Lupus.

touch paper Paper made slow-burning by saturation with saltpeter: used for firing explosives, as in pyrotechny.

touch·stone (tuch′stōn′) *n.* **1** A fine-grained dark stone, as jasper, formerly used to test the fineness of gold by the color of the streak made on the stone. **2** A criterion or standard by which the qualities of something are tested.

Touch·stone (tuch′stōn) A witty clown in Shakespeare's *As You Like It.*

touch-up (tuch′up′) *n.* A finishing touch or retouch.

touch·wood (tuch′wo͝od′) *n.* **1** Wood, decayed or thoroughly dried, for use as tinder; punk. **2** Dried fungi or fungous growth; amadou.

touch·y (tuch′ē) *adj.* **touch·i·er, touch·i·est 1** Apt to take offense on very little provocation; irascible; also, apt or liable to take fire, as tinder. **2** In art, done with short, light touches of the brush or pencil instead of with firm, unbroken lines. See synonyms under FRETFUL. —**touch′i·ly** *adv.* —**touch′i·ness** *n.*

tough (tuf) *adj.* **1** Susceptible of great tension or strain without breaking; also, of a close texture. **2** Not easily separated; tenacious; viscid; ropy. **3** Possessing great physical endurance: a *tough* constitution. **4** Possessing moral or intellectual endurance; steadfast; persistent; also, stubborn. **5** Irreclaimably vicious; disreputable; vulgar. **6** Difficult to accomplish; laborious; also, severe. **7** Hard to believe; incredible. —*n.* A lawless person; a rowdy; ruffian. [OE *tōh*] —**tough′ly** *adv.* —**tough′ness** *n.*

tough·en (tuf′ən) *v.t.* & *v.i.* To make or become tough or tougher. —**tough′en·er** *n.*

Toul (to͞ol) A town in NE France; besieged and captured by the Germans in the Franco-Prussian War, 1870.

Tou·lon (to͞o·lôṅ′) A French port and naval base on the Mediterranean 29 miles SE of Marseille.

Tou·louse (to͞o·lo͞oz′) A city of southern France, on the Garonne, capital of the Haute-Garonne department.

Tou·louse-Lau·trec (to͞o·lo͞oz′lō·trek′), **Henri Marie Raymond de,** 1864–1901, French painter and lithographer.

toun (to͞on) *n. Scot.* A town; also, a farmhouse.

tou·pee (to͞o·pā′, -pē′) *n.* **1** A little tuft or lock of hair. **2** A curl or lock of hair worn as a false front or at the top of a wig. **3** A wig worn to cover baldness or a bald spot. [<F *toupet* <OF *toup, top* a tuft of hair, prob. <Gmc.]

tour (to͝or) *n.* **1** A round trip or journey or a rambling excursion. **2** A passing around; circuit for inspection or sightseeing. **3** A turn or shift, as of service. See synonyms under JOURNEY. —**grand tour** A tour of the principal cities of Europe, customary in the 17th and 18th centuries for young English gentlemen as a supplement to their education: chiefly used in the expression *to make the grand tour.* —*v.t.* **1** To make a tour of; travel. **2** To present on a tour: to *tour* a play. —*v.i.* **3** To go on a tour. [<MF <OF *tor, tors* <L *tornus* a lathe <Gk. *tornos;* infl. in meaning by OF *tourner* TURN]

tou·ra·co (to͝or′ə·kō) See TURACOU.

Tou·raine (to͞o·ren′) A region and former province of west central France; capital, Tours.

tour·bil·lion (to͝or·bil′yən) *n.* **1** A whirling wind or a vortex, or something resembling them. **2** A kind of rocket with a spiral flight. [<MF *tourbillon* a whirlwind <OF *torbeillon* <L *turbo, -inis*]

Tour·coing (to͞or·kwaṅ′) A city in northern France near the Belgian border in the Nord department.

tour de force (to͞or də fôrs′) *French* A feat of remarkable strength or skill.

tour·ing (to͝or′ing) *adj.* Used for touring; that tours.

touring car A large, open automobile with a capacity for five or more passengers and baggage, built especially for touring. Also *Brit.* **tour′er.** —**convertible touring car** A touring car with folding top and disappearing or removable windows.

tour·ism (to͝or′iz·əm) *n.* **1** Traveling as a recreation. **2** Touring groups; tourists. **3** The organization and guidance of tourists. —**tour·is′tic** *adj.*

tour·ist (to͝or′ist) *n.* One who makes a tour or a pleasure trip. —*adj.* Of or suitable for tourists.

tourist camp A roadside group of cabins for the accommodation of transients, usually automobilists.

tourist class A class of accommodations for steamship passengers, lower than cabin class.

tour·ma·line (to͝or′mə·lēn, -lin) *n.* A complex borosilicate of aluminum, with a vitreous to resinous luster and found commonly black or brownish or bluish-black, but sometimes blue, green, red, or colorless. The transparent variety, when cut, is esteemed as a gemstone. Also spelled *turmaline.* Also **tour′ma·lin** (-lin). [<F, ult. <Singhalese *tōramalli* carnelian]

Tour·nai (to͞or·nā′) A town in SW Belgium on the Sabeldt; a manufacturing and quarrying center. Also **Tour·nay′.**

tour·na·ment (tûr′nə·mənt, to͝or′-) *n.* **1** In medieval times, a pageant in which two opposing parties of men in armor contended on horseback, with blunted weapons, in mock combat. **2** The jousts, sports, or contests in which such combatants engaged. **3** A comparatively recent sport of skilled horsemen, who tilt at rings suspended in the air, seeking to bear them off on their lances. **4** Any contest of skill involving a number of competitors and a series of games: a chess *tournament.* **5** An encounter, as of arms: Don Quixote's

add,āce,câre,pälm; end,ēven; it,īce; odd,ōpen,ôrder; to͝ok,po͞ol; up,bûrn; ə = a in *above,* e in *sicken,* i in *clarity,* o in *melon,* u in *focus*; yo͞o = u in *fuse*; oi,oil; ou,pout; ch,check; g,go; ng,ring; th,thin; th,this; zh,vision. Foreign sounds à,œ,ü,kh,ṅ; and ◆: see page xx. < from; + plus; ? possibly.

tournament with the barber. Also *tourney.* [<OF *torneiement, tornoiement* < *torneier, tornoier* tourney, ult. <L *tornare.* See TURN.]

Tour·neur (tûr′nər), **Cyril,** 1575?–1626, English dramatist.

tour·ney (tûr′nē, to͝or′-) *v.i.* To take part in a tournament; tilt. —*n.* A tournament. [<OF *torneier.* See TOURNAMENT.]

tour·ni·quet (to͝or′nə·ket, -kā, tûr′-) *n. Surg.* A bandage, etc., for stopping the flow of blood through an artery by compression. [<F < *tourner* TURN]

tour·nure (to͝or·nür′) *n. French* **1** The curving shape of a figure; outline; contour. **2** A light pad or cushion formerly worn by women to give the effect of well-rounded hips; a bustle; also, the drapery at the back of a gown.

tour of duty *Mil.* The hours or period of time during which a member of the armed services is on official duty, or assigned to a particular duty: the 24-hour *tour of duty* as officer of the day.

Tours (to͝or) A city of west central France between the Loire and the Cher above their confluence; scene of Charles Martel's defeat of the Saracens, 732.

touse (touz) *Dial.* & *Scot. v.t.* **1** To stir up, as a row. **2** To tousle; dishevel; rumple. —*n.* Disturbance. [ME *tusen, tousen,* prob. <Gmc.] —**tous′er** *n.*

tou·sle (tou′zəl) *v.t.* **·sled, ·sling** To disarrange or disorder, as the hair or dress. —*n.* **1** *Scot.* A tussle; also, a rude dalliance. **2** A tousled mass or mop of hair. Also **tou′zle.** [Freq. of TOUSE]

tous-les-mois (to͞o·lā·mwä′) *n.* The edible, starchlike tubers of a perennial herb of the West Indies and South America *(Canna edulis),* used in making baby food and as a substitute for arrowroot. [<F, all the months, every month; so called because edible the year round]

Tous·saint l'Ou·ver·ture (to͞o·saṅ′ lo͞o·ver·tür′), **Dominique François,** 1743–1803, Negro general; liberator of Haiti.

tou·sy (tou′zē) See TOWSY.

tout (tout) *Colloq. v.i.* **1** To solicit patronage, customers, votes, etc. **2** To spy on a race horse so as to gain information for betting; act as a tout. —*v.t.* **3** To solicit; importune. **4** **a** To spy on (a race horse) to gain information for betting. **b** To sell information concerning (a race horse). —*n.* **1** One who touts. **2** In horse-racing, a spy who sells information regarding horses entered for a race. **3** One who solicits business. **4** A spy for a robber. [OE *tōtian, tȳtan* peep, look out]

tout à fait (to͞o tà fe′) *French* Entirely; quite.

tout à l'heure (to͞o tà lœr′) *French* Instantly; just now; presently.

tout au con·traire (to͞o tō kôn·trâr′) *French* Quite to the contrary; quite the reverse.

tout à vous (to͞o tà vo͞o′) *French* Wholly yours; sincerely yours; at your service.

tout de suite (to͞ot swēt′) *French* Immediately; at once.

tout en·sem·ble (to͞o tän säṅ′bl′) *French* **1** All in all; everything considered. **2** The general effect.

tout·er (tou′tər) *n.* **1** One who plies or solicits customers or supporters obtrusively: a *touter* for a candidate for election. **2** *Colloq.* A runner.

tout le monde (to͞o lə mônd′) *French* All the world; everybody.

to·va·risch (to·vä′rish) *n. Russian* Comrade.

tow[1] (tō) *n.* A short, coarse hemp or flax fiber prepared for spinning. [Prob. OE *tōw-* for spinning, as in *tōwlic* pertaining to spinning]

tow[2] (tō) *v.t.* To pull or drag by a rope or chain; drag or pull along. See synonyms under DRAW. —*n.* **1** The act of towing, or the state of being towed. **2** That which is towed, as barges by a tugboat. **3** That which tows. **4** A rope or cable used in towing; towline. —**to take in tow** To take in charge for or as for towing; take under protection; take charge of. [OE *togian*]

tow·age (tō′ij) *n.* **1** The service of, or charge for, towing. **2** The act of towing. [<TOW[1]]

to·ward (tôrd, tōrd, tə·wôrd′) *prep.* **1** In the direction of; facing. **2** With respect to; regarding: his attitude *toward* women. **3** In anticipation of or as a contribution to; for: He is saving *toward* his education. **4** Near in point of time; approaching; about: arriving *toward* evening. **5** Tending to result in; designed or likely to achieve: a struggle *toward* mutual understanding. Also **to·wards′.** See synonyms under AT. —*adj.* (tôrd, tōrd) *Archaic* or *Rare* **1** Ready to do or learn; apt. **2** Docile. **3** In progress: used predicatively. **4** Impending or imminent. [OE *tōweard* < *tō* to + *-weard* -ward] —**to·ward′ness** *n.*

to·ward·ly (tôrd′lē, tōrd′-) *Archaic adj.* **1** Ready to do or learn; compliant; docile. **2** Unusually advanced; forward. **3** Favorable; promising; propitious. —**to·ward′li·ness** *n.*

tow·boat (tō′bōt′) *n.* A tugboat.

tow·el (toul, tou′əl) *n.* **1** A cloth or paper for drying anything by wiping. **2** An altar cloth. —*v.t.* **·eled** or **·elled, ·el·ing** or **·el·ling** To wipe or dry with a towel. [<OF *toaille,* prob. < OHG *dwahila* a washcloth < *dwahan* wash]

tow·el·ing (tou′ling, tou′əl·ing) *n.* Material, as crash, for towels. Also *Brit.* **tow′el·ling.**

tow·er (tou′ər) *n.* **1** A structure very tall in proportion to its other dimensions, and frequently forming part of a large building; properly, a structure larger than a pinnacle, and less tapering than a steeple. **2** A tall, wooden, movable structure from which besiegers formerly stormed a fortress. **3** A place of security or defense; fortified place; citadel. —*v.i.* **1** To rise or stand like a tower; extend to a great height. **2** To fly directly upward, as some birds. [Fusion of OE *torr* (<L *turris*) and OE *tūr* <OF *tor, tur* <L *turris*]

tow·ered (tou′ərd) *adj.* **1** Furnished with towers for ornament or defense. **2** Rising like a tower.

tow·er·ing (tou′ər·ing) *adj.* **1** Like a tower; lofty; hence, very high or great: also **tow′er·y.** **2** Rising or increasing to a high pitch of violence or intensity; furious. See synonyms under HIGH.

Tower of London A group of buildings comprising a fortress and palace on the north bank of the Thames, built in 1078 around the original tower (the White Tower) and used as a royal residence, a political prison, and a museum.

tower of silence A circular tower with central well, having a high outer wall, and inner platform on which the Parsees expose the bodies of their dead to be eaten by vultures, so that the bodies may be dissipated without polluting the earth: also called *dakhma, dokhma.*

tow·head (tō′hed′) *n.* **1** A head of very light-colored or flaxen hair, or a person having such hair. **2** *Rare* Tousled hair, or a person with tousled hair. **3** *U.S.* A wooded sandbar or newly formed island in a river: a navigational hazard. [<TOW[1] + HEAD] —**tow′-head′ed** *adj.*

tow·hee (tou′hē, tō′-) *n.* An American bird related to the buntings and the sparrows, especially the **Alabama towhee** *(Pipilo erythrophthalmus)* and the **green-tailed towhee** *(Oberholseria chlorura)* of the western United States. Also **towhee bunting.** [Imit. of one of its notes]

tow·line (tō′lin′) *n.* A line, rope, or chain used in towing.

tow·mont (tō′mənt) *n. Scot.* A twelvemonth. Also **tow′mond** (-mənd).

town (toun) *n.* **1** Any considerable collection of dwellings and other buildings larger than a village and comprising a geographical and political community unit, but not incorporated as a city. **2** The local government of such a community; also, the voters, the representatives, or the inhabitants collectively. **3** A subdivision of a county, usually rural, that may include a number of villages and towns; a township. **4** In New England, a local unit governing itself through a town meeting. **5** *Brit.* Originally, a collection of dwellings enclosed for security within some form of fortification; subsequently, any collection of dwelling houses larger than a village. ◆ Collateral adjective: *oppidan.* **6** A closely settled urban district as contrasted with the open country: *town* and country. **7** The city or town nearest to where one lives: a trip to *town;* also, the downtown or business section of a city or town. **8** A group of prairie-dog burrows. —**on the town** **1** Dependent on municipal charity. **2** *Slang* On a round of pleasure in the city. —**to go to town** *Slang* To succeed in the highest degree. —**to paint the town red** *Slang* To carouse. —*adj.* **1** Of or pertaining to, like, situated in, or for use in town: *town* clothes. **2** Supported by town funds: a *town* library. [OE *tūn, tuun* an enclosure, group of houses]

Town may appear as a combining form in hyphemes or solidemes, or as the first element in two-word phrases:

town-absorbing	**town-hating**
town-born	**town-imprisoned**
town-bound	**town jail**
town-bred	**town-keeping**
town bridge	**town life**
town car	**town lot**
town church	**town-loving**
town-dotted	**town-made**
town dweller	**town park**
town-dwelling	**town sick**
town-flanked	**townsickness**
town-frequenting	**town-tied**
town-goer	**town-trained**
town-going	**town-weary**

town clerk An official who keeps the records of a town.

town crier A person appointed to make proclamations through the streets of a town.

town farm A farm maintained by town or township funds for the poor or indigent of the community.

town·folk (toun′fōk′) *n.* People who live in towns or in a particular town or city. Also **towns′folk′, towns′peo′ple** (-pē′pəl).

town hall The building containing the public offices of a town and used for meetings of the town council and other official business.

town house **1** A residence in a town or city. **2** A town hall. **3** *U. S. Obs.* **a** An almshouse; workhouse. **b** A town prison.

town marshal **1** An officer of a town police force. **2** In the American colonies, a town officer who levied and collected taxes, fines, etc.

town meeting **1** A general assemblage of the people of a town. **2** An assembly of qualified voters for the purpose of transacting town business; also, the voters assembled.

town·ship (toun′ship) *n.* **1** *U.S.* **a** A territorial subdivision of a county with certain corporate powers of municipal government for local purposes; also, the corporation or government thereof. **b** In New England, a local political unit governed by a town meeting. **2** A unit of area in surveys of U. S. public lands, normally six miles square, subdivided into 36 sections of one square mile each. **3** *Brit.* Anciently, an organized group of families forming the political unit of early society which existed prior to the parish. [OE *tūnscipe* < *tūn* a village, group of houses]

towns·man (tounz′mən) *n. pl.* **·men** (-mən) **1** A resident of a town; also, a fellow citizen. **2** In New England, a town officer; a selectman. **3** In a school or college town, one who lives in the town as contrasted with a student or teacher in the school or college.

towns·wom·an (tounz′wo͝om′ən) *n. pl.* **·wom·en** (-wim′in) A woman living in a town.

tow·path (tō′path′, -päth′) *n.* A path along a river or canal used by men, horses, or mules towing boats; a towing path.

tow·rope (tō′rōp′) *n.* A heavy rope or cable used in towing. Also called *towline.*

tow·y (tō′ē) *adj.* Composed of, like, or containing tow. [<TOW[1]]

tow·zie (tou′zē) *adj. Scot.* Disheveled; rumpled; shaggy: also spelled *toozie, tousy.* Also **tow′sie.**

tox·al·bu·min (tok′sal·byo͞o′min) *n. Biochem.* Any protein substance having toxic properties, as snake venom, ricin, certain bacterial cultures, etc. [<TOX(IC) + ALBUMIN]

tox·e·mi·a (tok·sē′mē·ə) *n. Pathol.* A poisoned condition of the body caused by the absorption of bacterial toxins from a local source of infection and their distribution by the blood. Also **tox·ae′mi·a.** [<NL <Gk. *toxicon* a poison + *haima* blood] —**tox·e′mic, tox·ae′·mic** *adj.*

tox·ic (tok′sik) *adj.* **1** Pertaining to poison; poisonous. **2** Due to or caused by poison or a toxin. Also **tox′i·cal.** [<Med. L *toxicus* poisoned, poisonous <L *toxicum* a poison, orig. a poison for arrows <Gk. *toxicon (pharmakon)* (a poison) for arrows < *toxa* arrows < *toxon* a bow] —**tox′i·cal·ly** *adv.*

tox·i·cant (tok′sə·kənt) *adj.* **1** Possessing poisonous qualities. **2** Producing a poisonous effect. —*n.* A toxic substance; poison; also, an intoxicant. [<LL *toxicans, -antis,* ppr. of

toxicare smear with poison <L *toxicum.* See TOXIC.]

tox·i·ca·tion (tok′sə·kā′shən) *n.* 1 The act of poisoning. 2 The state of being poisoned. 3 Poisoning.

tox·ic·i·ty (tok·sis′ə·tē) *n.* 1 The quality of being toxic. 2 The degree or intensity of virulence of a poison.

toxico- *combining form* Poison; of or pertaining to poison, or to poisons: *toxicology.* Also, before vowels, **toxic-.** [<Gk. *toxicon* poison]

tox·i·co·gen·ic (tok′sə·kō·jen′ik) *adj.* 1 Producing poisons or toxins. 2 Generated or formed by toxic matter.

tox·i·col·o·gy (tok′sə·kol′ə·jē) *n.* The science that treats of the origin, nature, properties, and effects of poisons, of their detection in the organs or tissues, of their antidotes, and of the treatment of diseases due to poisoning. [<F *toxicologie*] —**tox′i·co·log′i·cal** (-kō·loj′i·kəl) *adj.* —**tox′i·co·log′i·cal·ly** *adv.* —**tox′i·col′o·gist** *n.*

tox·i·co·ma·ni·a (tok′sə·kō·mā′nē·ə, -mān′yə) *n.* A morbid desire to take poison. [<TOXICO- + -MANIA]

tox·i·co·pho·bi·a (tok′sə·kō·fō′bē·ə) *n.* A morbid fear of poison or of being poisoned: also called *iophobia, toxiphobia.* [<TOXICO- + -PHOBIA] —**tox′i·pho′bic** *adj.*

tox·i·co·sis (tok′sə·kō′sis) *n. pl.* **·ses** (-sēz) *Pathol.* A morbid condition due to the effect of toxins generated within the system or administered from without. [<NL <L *toxicum* poison]

tox·in (tok′sin) *n.* 1 Any of a class of more or less unstable poisonous compounds elaborated by animal, vegetable, or bacterial organisms and acting as causative agents in many diseases, usually after an incubation period. 2 Any toxic matter generated in living or dead organisms. Also **tox·ine** (tok′sēn). [<TOX(IC) + -IN]

tox·is·ter·ol (tok·sis′tər·ōl, -ol) *n. Biochem.* A toxic compound produced by the excessive irradiation of ergosterol and intermediate between calciferol and suprasterol. [<TOXI(C) + STEROL]

tox·o·phil (tok′sə·fil) *adj. Biol.* Having an affinity for or being in harmony with a toxin. Also **tox′o·phile** (-fīl, -fil). [<*toxo-* (<TOXIN) + -PHIL]

tox·o·plas·mo·sis (tok′sō·plaz·mō′sis) *n.* A diseased condition resulting from the presence of or infection by sporozoan parasites (genus *Toxoplasma*) which act principally upon the nervous system of certain animals and sometimes of man. [<NL <*Toxoplasma,* genus name]

toy (toi) *n.* 1 An article constructed for the amusement of children; a plaything; hence, any trifling or diverting object; an ornament; trinket. 2 Any diminutive object imitating a larger one and fitted for entertainment and instruction. 3 *Obs.* Wanton play; dalliance. 4 A small dog bred to extreme smallness and kept as a pet: also **toy dog.** 5 *Scot.* A head covering for women that hangs loosely over the shoulders; a toy-mutch. 6 *Obs.* A dance tune. 7 *Archaic* A quaint utterance, idle tale, or anecdote; fancy; jest. See synonyms under GAUD. —*v.i.* To trifle; play. —*adj.* Resembling a toy; of miniature size. [Prob. fusion of ME *toye* flirtation, sport + Du. *tuig* tools, stuff] —**toy′er** *n.* —**toy′ish** *adj.*

To·ya·ma (tō·yä·mä) A port of north central Honshu, Japan.

toy-mutch (toi′much) *n. Scot.* Toy (def. 5).

Toyn·bee (toin′bē), **Arnold Joseph,** born 1889, English historian.

to·yo (tō′yō) *n.* A shiny, rice-paper straw. [<Japanese]

To·yo·ha·shi (tō·yō·hä·shē) A city of southern Honshu, Japan; a manufacturing center.

to·yon (tō′yən) *n.* An evergreen shrub (*Photinia arbutifolia*) indigenous to the Pacific coast of North America, having white flowers, followed by persistent berries of a bright red color; California holly. [<Sp. *tollón* < N. Am. Ind. (Mexican)]

toy shop A shop where toys are displayed for sale.

tra·be·at·ed (trā′bē·ā′tid) *adj. Archit.* 1 Having an entablature. 2 Having beams or long stones as lintels instead of an arch. Also **tra′·be·ate** (-it, -āt). [Irregularly formed <L *trabs, trabis* a beam]

tra·be·a·tion (trā′bē·ā′shən) *n. Archit.* 1 The state of being trabeated. 2 An entablature.

tra·bec·u·la (trə·bek′yə·lə) *n. pl.* **·lae** (-lē) 1 A small supporting band or bar. 2 *Anat.* The interwoven bands of connective tissue that form the supporting framework of an organ, as the spleen. 3 *Bot.* A row or plate of sterile cells extending across the cavity in the sporangium of a moss. [<L, dim. of *trabs, trabis* a beam] —**tra·bec′u·lar** *adj.*

Tra·ben-Trar·bach (trä′bən·trär′bäkh) A town in Rhineland-Palatinate, West Germany, on the Mosel; a wine center.

Trab·zon (träb·zôn′) The Turkish name for TREBIZOND.

trace[1] (trās) *n.* 1 A vestige or mark left by some past event or agent, especially when regarded as a sign or clue. 2 A barely detectable quantity, quality, token, or characteristic; touch. 3 *Chem.* A proportion or ingredient too small to be weighed (often abbreviated *tr.*): a *trace* of soda. 3 An imprint or mark indicating the passage of a person or thing, as a footprint, etc. 4 A path or trail through woods or forest beaten down by men or animals. 5 A lightly drawn line; something traced. 6 The point or line on a map or on the ground indicating the position of a trench, an aircraft flight path, etc. 7 The path of a tracer bullet. 8 *Psychol.* An engram. —*v.* **traced, trac·ing** *v.t.* 1 To follow the tracks, course, or development of. 2 To follow (tracks, a course of development, etc.). 3 To discover or ascertain by examination or investigation; find out or determine. 4 To draw; sketch. 5 To copy (a drawing, etc.) on a superimposed transparent sheet. 6 To form (letters, etc.) with careful strokes. 7 To mark with an impressed design; chase. 8 To imprint (a pattern or design). 9 To mark or record by a curved or broken line. 10 To go or move over, along, or through. —*v.i.* 11 To make one's way; proceed. 12 To have its origin; go back in time. [<OF *tracier,* ult. <L *tractus* a dragging, a track <*trahere* draw] —**trace′a·ble** *adj.* —**trace′a·bil′i·ty, trace′a·ble·ness** *n.* —**trace′a·bly** *adv.* —**trace′·less** *adj.*

Synonyms (noun): footmark, footprint, footstep, mark, memorial, remains, remnant, sign, token, track, vestige. A *vestige* is always slight compared with that whose existence it recalls; as, Scattered mounds containing human implements are *vestiges* of a former civilization. A *vestige* is always a part of that which has passed away; a *trace* may be merely the *mark* it has made, or some slight evidence of its presence or of the effect it has produced; as, *Traces* of game were observed by the hunter. See CHARACTERISTIC, MARK[1].

trace[2] (trās) *n.* 1 One of two side straps or chains for connecting the collar of a harness with the swingletree. 2 *Mech.* A link or connecting bar hinged at each end to other pieces of a mechanism, to transmit motion from one part to another. —**to kick over the traces** To throw off control; become unmanageable. —*v.t.* **traced, trac·ing** To fasten, as with traces. [<OF *traiz, trais,* pl. of *trait* a dragging, a leather harness <L *tractus.* See TRACE[1].]

trac·er (trā′sər) *n.* 1 One who or that which traces. 2 One of various instruments used in tracing drawings, etc. 3 An inquiry forwarded from one point to another, to trace missing mail matter, etc. 4 *Surg.* An instrument for laying bare and tracing the course of nerves, muscles, etc. 5 One who searches for lost property, as on railroads. 6 *Mil.* **a** A chemical incorporated in certain types of ammunition used for ranging, signaling, or incendiary purposes. **b** A tracer bullet. 7 *Med.* A radioisotope introduced into the body for the purpose of following the processes of metabolism, the course or location of a disease, etc. 8 A message that describes a person or thing wanted, as by the police. [<TRACE[1]]

tracer bullet A bullet which leaves a line of smoke or fire in its wake, thus indicating its course for correction of aim.

trac·er·y (trā′sər·ē) *n. pl.* **·er·ies** 1 Ornamental stonework formed of ramifying lines. 2 Any work resembling this.

tra·che·a (trā′kē·ə) *n. pl.* **·che·ae** (-ki·ē) 1 *Anat.* The duct, composed of membrane and incomplete cartilaginous rings, by which air passes from the larynx to the bronchi and the lungs; the windpipe. 2 *Zool.* One of the passages by which air is conveyed from the exterior in air-breathing arthropods, as insects and arachnids. 3 *Bot.* A duct or vessel in plants, particularly one having spiral markings. [<Med. L <LL *trachia* <Gk. *(artēria) tracheia* a rough (artery), fem. of *trachys* rough] —**tra′che·al** *adj.*

tracheal tissue *Bot.* Plant tissue consisting of tracheae or tracheids: one of the chief constituents of xylem.

tra·che·id (trā′kē·id) *n. Bot.* An elongated, taper-pointed, woody plant cell, especially when marked with bordered pits and serving for support, as in the pine family. [<G *tracheïde* <Med. L *trachea* the trachea] —**tra·che·i·dal** (trə·kē′ə·dəl) *adj.*

tra·che·i·tis (trā′kē·ī′tis) *n. Pathol.* Inflammation of the trachea or windpipe. [<NL <Med. L *trachea* the trachea]

tracheo- *combining form* The trachea; of or pertaining to the trachea: *tracheotomy.* Also, before vowels, **trache-.** [<TRACHEA]

tra·che·os·co·py (trā′kē·os′kə·pē) *n. Med.* Instrumental inspection of the windpipe. —**tra′che·o·scop′ic** (-ō·skop′ik) *adj.* —**tra′che·os′co·pist** *n.*

tra·che·ot·o·my (trā′kē·ot′ə·mē) *n. Surg.* The operation of making an incision into the windpipe. —**tra′che·ot′o·mist** *n.*

trach·le (träkh′əl) *Scot. v.t.* To trail or draggle; fatigue; exhaust. —*n.* Any exercise involving or resulting in unusual fatigue; a burdensome work. Also spelled *trauchle.* [Corruption of DRAGGLE]

tra·cho·ma (trə·kō′mə) *n. Pathol.* A contagious virus disease of the eye characterized by hard papillary elevations or granular excrescences on the inner surface of the eyelids, with inflammation of the lining; granular conjunctivitis. [<NL <Gk. *trachōma, -atos* roughness <*trachys* rough] —**tra·chom·a·tous** (trə·kom′·ə·təs) *adj.*

trachy- *combining form* Rough; uneven: *trachycarpous,* bearing rough fruit. Also, before vowels, **trach-.** [<Gk. *trachys* rough]

tra·chyte (trā′kīt, trak′īt) *n.* A light-colored, rough volcanic rock having a porphyritic texture, composed essentially of alkaline feldspar and one or more secondary minerals. [<F <Gk. *trachytēs* ruggedness <*trachys* rough]

trac·ing (trā′sing) *n.* 1 The act of one who traces. 2 An ornamentation produced by etching, drawing, or tracing. 3 A copy made by tracing on transparent paper. 4 A record made by a self-registering instrument.

track (trak) *n.* 1 A mark or trail left by the passage of anything: the *track* of a storm. 2 A footprint or series of footprints. 3 Any regular path; course: the *track* of a comet round the sun. 4 Any kind of racecourse; also, sports performed on such a course; track athletics. 5 A set of rails or a rail on which something may travel; specifically, the pair of metal rails on which a railway train or tramway runs; also, the rail or pair of rails with its ties, bolts, etc.; by extension, the whole trackway. 6 A trace or vestige. 7 A sequence of events; a succession of ideas. 8 Awareness of the progress or sequence; count; record: to keep *track* of. 9 Tread (def. 2). See synonyms under MARK[1], ROAD, TRACE[1], WAY. —**to make tracks** To hurry; run away in haste. —**in one's tracks** Right where one is; on the spot. —**to jump the track** 1 To leave the rails, as a railroad engine or car. 2 To depart from any usual course or procedure. —*v.t.* 1 To follow the tracks of; trail. 2 To discover and follow up or out, by means of marks or indications. 3 To make tracks upon or with: to *track* snow through a house. 4 To traverse, as on foot: to *track* the wild forests. 5 To furnish with rails or tracks. —*v.i.* 6 To measure a certain distance between wheels. 7 To have the wheels equal in span or gage to the wheels of another vehicle. 8 To run in the same track; be in alinement. —*adj.* Pertaining to or performed on a track. [<OF *trac*, prob. <Gmc. Cf. Du. *trek* pull.] —**track′er** *n.* —**track′a·ble** *adj.*

track·age (trak′ij) *n.* 1 Railroad tracks collectively. 2 The right of one company to use the track system of another company. 3 The charge for this right. 4 A towing, especially of a vessel in a canal, with a rope from the towpath.

track boat A boat towed from a path along the shore.

track events The races at an athletic meet: distinguished from *field events.* Also **track athletics.**

track·less (trak′lis) *adj.* 1 Unmarked by footsteps; pathless: the *trackless* desert. 2 Leaving no traces: a *trackless* fugitive. 3 Not running on tracks or rails: a *trackless* trolley.

trackless trolley A trolley bus.

track·man (trak′mən) *n. pl.* **·men** (-mən) *U.S.* A person employed to inspect regularly the condition of a section of railroad track; trackwalker.

track meet An athletic contest made up of track events.

track road A tow path.

track·walk·er (trak′wô′kər) *n. U.S.* A trackman.

track·way (trak′wā′) *n.* The permanent way of a railroad.

tract[1] (trakt) *n.* 1 An extended area, as of land or water. 2 Continued duration, as of time. 3 *Anat.* An extensive region of the body, especially one comprising a system of parts or organs: the alimentary *tract.* [<L *tractus* a drawing out, duration < *trahere* draw. Doublet of TRAIT.]

tract[2] (trakt) *n.* 1 A short treatise, as on some question of religion or morals; a propaganda leaflet. 2 An anthem sometimes substituted for the Alleluia: so styled because, instead of being treated antiphonally, it is sung *tractim* (continuously) and as a solo: also **trac′tus.** [Short for L *tractatus* a tractate]

tract·a·ble (trak′tə·bəl) *adj.* 1 Easily led or controlled; manageable; docile. 2 Readily worked or handled; malleable. See synonyms under DOCILE. [<L *tractabilis* < *tractare* handle, freq. of *trahere* draw] —**tract′a·bly** *adv.* —**tract′a·ble·ness, tract′a·bil′i·ty** *n.*

Trac·tar·i·an (trak·târ′ē·ən) *n.* One of the authors of the series of 90 pamphlets called *Tracts for the Times,* or one with their views; loosely, any one of the High Church group in the Church of England. —*adj.* Pertaining to the Tractarians or to their teachings.

Trac·tar·i·an·ism (trak·târ′ē·ən·iz′əm) *n.* The tenets or principles expressed in *Tracts for the Times* (1833–41) by the leaders of the religious movement known as the Oxford Movement which made an effort to bring to the Church of England a new awareness of its historic claim to be a branch of the Catholic Church by reaffirming the validity of its orders and the efficacy of its sacraments, by opposing rationalism in theology, and by seeking to restore its ancient liturgical practices.

trac·tate (trak′tāt) *n.* A short treatise; a tract. [<L *tractatus* a handling, treatise, pp. of *tractare.* See TRACTABLE.]

trac·tile (trak′til) *adj.* That can be drawn out; ductile. [<L *tractilis* < *tractus.* See TRACE[1].] —**trac·til′i·ty** *n.*

trac·tion (trak′shən) *n.* 1 The act of drawing, as by motive power over a surface. 2 The state of being drawn, or the power employed. 3 *Physiol.* Contraction, as of a muscle. 4 Adhesive or rolling friction, as of wheels on a track. [<Med. L *tractio, -onis* <L *tractus.* See TRACE[1].] —**trac′tion·al** *adj.*

traction engine A locomotive for hauling on roads or ground, as distinguished from one used on a railway.

trac·tive (trak′tiv) *adj.* Having or exerting traction.

trac·tor (trak′tər) *n.* 1 A machine or instrument for pulling or drawing. 2 A powerful, motor-driven vehicle, usually having heavy treads, used, as on farms, to draw a plow, reaper, etc. 3 An automotive vehicle with a driver's cab, used to haul trailers, etc. 4 A traction engine. 5 *Aeron.* **a** An airplane with the propeller or propellers situated in front of the supporting surface to pull it through the air: also **tractor airplane. b** The propeller of a tractor airplane. [<NL <L *tractus.* See TRACE[1].]

trade (trād) *n.* 1 A business, particularly a skilled or specialized handicraft; a craft. 2 Mercantile traffic; commerce. 3 A bargain; deal; also, an exchange; specifically, a corrupt bargain in patronage between political-party leaders. 4 The people following a particular calling. 5 The amount of business or exchange done in a particular place; a firm's customers. 6 Customary pursuit; occupation. 7 *Brit.* **a** The submarine service of the Royal Navy. **b** The liquor traffic. 8 *Obs.* A trail or track. 9 *Obs.* A course, path, passage, or way. 10 *Obs.* Custom, habit, or practice. 11 A trade wind: usually in the plural. See synonyms under BUSINESS, SALE, TRAFFIC. —*v.* **trad·ed, trad·ing** *v.t.* To dispose of by bargain and sale; now, especially, to barter; exchange. —*v.i.* To engage in commerce or in business transactions of bargain and sale. —**to trade in** To give in exchange as payment or part payment. —**to trade off** To get rid of by exchange or trading. —**to trade on** To take advantage of. [<MLG, a track. Akin to TREAD.]

trade acceptance A bill of exchange drawn by the seller of goods on the purchaser who accepts the draft by writing across the face of it when and where it is payable.

trade book An edition of a book designed for ordinary sale to the general public, as distinguished from a textbook, limited or de luxe edition, etc.

trade dollar See under DOLLAR.

trade-in (trād′in′) *n.* Something given or accepted in payment or part payment for something else; an exchange.

trade journal A periodical publishing news and discussions of a particular trade or business.

trade-last (trād′last′, -läst′) *n. Colloq.* A favorable remark that one has heard and offers to repeat to the person complimented in return for a similar remark.

trade·mark (trād′märk′) *n.* A name, symbol, design, device, or word, or any combination thereof, used by a merchant or manufacturer to identify his goods and distinguish them from those made or sold by others. A trademark may or may not be legally registered as such.—*v.t.* 1 To label with a trademark. 2 To register as a trademark. —**trade′marked′** *adj.*

trade name 1 The name by which an article, process, service, or the like is designated in trade. 2 A name given by a manufacturer to designate a proprietary article, sometimes having the status of a trademark or of a copyrighted and patented proprietary name. 3 A style or a name of a business house acquired by purchase from a retiring firm or trader.

trad·er (trā′dər) *n.* 1 One who trades. 2 Any vessel employed in a particular trade. 3 A member of a stock exchange who trades for himself, and not for customers.

trade rat A pack rat.

trad·es·can·ti·a (trad′əs·kan′shē·ə, -shə) *n.* Any of a genus (*Tradescantia*) of perennial American herbs, often having grasslike leaves and showy flowers with ephemeral petals. [<NL, after John *Tradescant,* died in 1638, English traveler and naturalist]

trade school See under SCHOOL.

trades·folk (trādz′fōk′) *n. pl.* People engaged in trade; shopkeepers. Also **trades′peo′ple** (-pē′pəl).

trades·man (trādz′mən) *n. pl.* **·men** (-mən) 1 A retail dealer; shopkeeper. 2 *Brit.* A mechanic.

trades·wom·an (trādz′wŏŏm′ən) *n. pl.* **·wom·en** (-wim′in) A woman engaged in trade or the sale of goods.

trade union An organized association of workmen formed for the protection and promotion of their common interests, especially with regard to wages, hours, and working conditions. Also **trades union.** —**trade′-un′ion·ism** *n.* —**trade′-un′ion·ist** *n.*

trade wind Either of two steady winds blowing in the same course toward the equator from about 30° N and S latitude, one from the northeast on the north, the other from the southeast on the south side of the equatorial line.

trad·ing (trā′ding) *adj.* 1 Carrying on trade. 2 Corrupt; venal: said of officials. 3 *Obs.* Pursuing a steady course.

trading post A building or small settlement in unsettled territory where a trader or trading company has set up a station for barter (usually in furs) with North American Indians or other natives.

trading stamp See under STAMP.

tra·di·tion (trə·dish′ən) *n.* 1 The transmission of knowledge, opinions, doctrines, customs, practices, etc., from generation to generation, originally by word of mouth and by example. 2 That which is so transmitted; a body of beliefs and usages handed down from generation to generation; also, any particular story, belief, or usage so handed down; hence, remembrance, or recollection existing as by transmission. 3 That body of Christian doctrine, handed down through successive generations and held by some churches to belong to the deposit of faith, even if it may not be found in the Holy Scripture. 4 Among the Jews, an unwritten code said to have been revealed to Moses on Mount Sinai at the time of the delivery of the Decalog and handed down through the oral teaching of prophets and doctors of the law. 5 The record of the acts and utterances of Mohammed, known as the *Sunna,* which, forming no part of the Koran, was formerly communicated only by verbal utterance from father to son. 6 In literature, the drama, and the fine arts, the accumulated knowledge, taste, and experience handed down from one generation of writers, actors, or artists to another; the historic conceptions and usages of a school, collectively, or any one such conception or usage: the *traditions* of the stage. 7 A custom so long continued that it has almost the force of a law. 8 *Law* Delivery of possession. [<OF *tradicion* <L *traditio, -onis* a delivery, surrender < *traditus,* pp. of *tradere* deliver < *trans-* across + *dare* give. Doublet of TREASON.] —**tra·di′tion·er, tra·di′tion·ist** *n.*

tra·di·tion·al (trə·dish′ən·əl) *adj.* 1 Relating to or depending on tradition. 2 Characterizing a school of English Biblical critics who hold the Greek texts of the New Testament to be the foundation of the true text. Also **tra·di′tion·ar′y** (-er′ē). —**tra·di′tion·al·ist** *adj.* & *n.* —**tra·di′tion·al·is′tic** *adj.* —**tra·di′tion·al·ly** *adv.*

tra·di·tion·al·ism (trə·dish′ən·əl·iz′əm) *n.* 1 A system of faith founded on tradition. 2 Adherence to tradition; especially, undue reverence for tradition in religious matters.

trad·i·tive (trad′ə·tiv) *adj. Obs.* Traditional. [Appar. <MF *traditif* <L *traditus.* See TRADITION.]

trad·i·tor (trad′ə·tər) *n. pl.* **trad·i·to·res** (trad′ə·tôr′ēz, -tō′rēz) A traitor among the early Christians at the time of the Roman persecutions. [<L, a deliverer, betrayer < *tradere.* See TRADITION.]

tra·duce (trə·dōōs′, -dyōōs′) *v.t.* **·duced, ·duc·ing** To misrepresent wilfully the conduct or character of; defame; slander. See synonyms under ASPERSE, REVILE. [<L *traducere* transport, bring into disgrace < *trans-* across + *ducere* lead] —**tra·duc′er** *n.* —**tra·duc′i·ble** *adj.* —**tra·duc′ing·ly** *adv.* —**tra·duc·tion** (trə·duk′shən) *n.*

tra·du·cian·ism (trə·dōō′shən·iz′əm, -dyōō′-) *n.* The doctrine that the soul, equally with the body, is produced and begotten by the parent or parents: distinguished from *creationism* and *preexistence.* [<LL *traducianus* <L *tradux, -icis* a shoot for propagation < *traducere.* See TRADUCE.] —**tra·du′cian·ist** *n.* —**tra·du′cian·is′tic** *adj.*

Tra·fal·gar (trə·fal′gər, *Sp.* trä′fäl·gär′), **Cape** A headland on the Atlantic coast of SW Spain; scene of a naval battle in which Nelson, though fatally wounded, defeated the French and Spanish fleets, 1805.

traf·fic (traf′ik) *n.* 1 The exchange of goods, wares, etc.; the business of buying and selling, between individuals or communities; trade. 2 The business of transportation, as by railroad. 3 The subjects of transportation collectively; the things carried. 4 A business procedure; transaction; hence, intercourse. 5 The passing of pedestrians and vehicles along a road; the flow of telephone messages, etc. 6 Unlawful or improper trade: *traffic* in stolen goods. —*v.i.* **·ficked, ·fick·ing** 1 To engage in buying and selling; do business, especially

illegally: with *in.* 2 To have dealings: with *with.* [<MF *trafic, trafique* <Ital. *traffico* < *trafficare* <L *trans-* across + Ital. *ficcare* thrust in <L *figere* fasten] —**traf′fick·er** *n.*
Synonyms (noun): business, commerce, trade. *Commerce* is the broadest and noblest term of this group. *Trade* may be local; *commerce* is always extended and is between members of distinct communities, states, or nations; as, foreign, interstate, or intrastate *commerce*; foreign, domestic, or free-port *trade*. *Traffic* is local, as between different parts of one city or between two or more cities. *Trade* may be largely by letter or telegram, etc.; *traffic* involves the actual passing to and fro of persons or commodities and may be applied directly to persons when considered as in some way a source of gain: the passenger *traffic* of a railroad. *Traffic* always suggests stir and bustle: the din of *traffic*; one may say dull *trade*, but scarcely dull *traffic*. Compare synonyms for BUSINESS.
Traffic may appear as a combining form in hyphemes or as the first element in two-word phrases, with the following meanings:
1 Of or pertaining to the flow of roadway traffic:

traffic accident	**traffic congestion**
traffic artery	**traffic-laden**
traffic-congested	**traffic lane**

2 Of or pertaining to the laws or regulation of roadway traffic:

traffic cop	**traffic signal**
traffic court	**traffic violation**
traffic policeman	**traffic violator**

traf·fic·a·tor (traf′ə·kā′tər) *n.* A traffic signal. [<TRAFFIC + -ATOR]
traffic circle A circular intersection, where traffic is maintained in one direction, so constructed as to allow vehicles to enter or leave it at any of the converging roads, or to change course, without interruption of the flow of traffic.
traffic light A signal light which, by changing color, directs the flow of traffic along a road or highway.
trag·a·canth (trag′ə·kanth) *n.* 1 A white or reddish gum obtained from various species of Old World leguminous herbs (genus *Astragalus*), especially *A. gummifer* of SW Asia: used in pharmacy and the arts. 2 Any of the shrubs yielding this gum. [<MF *tragacante* <L *tragacantha* <Gk. *tragakantha* a tragacanth shrub < *tragos* a male goat + *akantha* a thorn]
tra·ge·di·an (trə·jē′dē·ən) *n.* 1 An actor in tragedy. 2 A writer of tragedies.
tra·ge·di·enne (trə·jē′dē·en′) *n.* An actress of tragedy. [<F]
trag·e·dy (traj′ə·dē) *n. pl.* **·dies** 1 A form of drama in which the protagonist, having some quality of greatness (and, in Greek, Roman, and Renaissance tragedy, in high place) comes to disaster through some flaw (which may be a noble fault) in his nature that interacts with the fabric of events (the plot) to bring about his inevitable downfall or death, the action being managed in a way to produce pity and fear in the spectator and to effect a catharsis of these feelings. The failure to achieve this leads to **tragedy manquée,** which falls short of true tragedy. To the outcome of death or madness usual in ancient and Renaissance tragedy, modern tragedy adds the possibility of frustration and unfulfilment from which there seems no escape. Opposed to *comedy.* 2 A fatal event or course of events; murder, especially one involving dramatic incidents. 3 A very terrible or sorrowful fate or end. 4 The art or theory of acting or composing tragedy. [<OF *tregedie, tragedie* <L *tragoedia* <Gk. *tragōidia* appar. < *tragos* a goat + *ōidē* a song; semantic development uncertain]
Trag·e·dy (traj′ə·dē) Tragedy personified, especially as Melpomene.
trag·ic (traj′ik) *adj.* 1 Involving death or calamity; causing suffering; fatal; terrible. 2 Pertaining to or having the nature of tragedy. 3 Appropriate to or like tragedy, especially in drama. Also **trag′i·cal.** [<L *tragicus* <Gk. *tragikos* pertaining to tragedy < *tragos* a goat] —**trag′i·cal·ly, trag′ic·ly** *adv.* —**trag′·i·cal·ness** *n.*
trag·i·com·e·dy (traj′i·kom′ə·dē) *n. pl.* **·dies** A drama in which tragic and comic scenes are intermingled. [<MF *tragi-comédie* <LL *tragicomoedia* <L *tragico-comoedia* < *tragicus* TRAGIC + *comoedia* COMEDY] —**trag′i·com′ic** or **·i·cal** *adj.* —**trag′i·com′i·cal·ly** *adv.*
trag·o·pan (trag′ə·pan) *n.* An Asian pheasant (genus *Tragopan*) of which the horned pheasant, having gorgeous ocellated plumage, is a variety. [<NL <L, a fabulous bird <Gk. < *tragos* a goat + *Pan* Pan]
tra·gus (trā′gəs) *n. pl.* **·gi** (-jī) *Anat.* A flattened, somewhat conical eminence of the auricle in front of the opening of the external ear. [<LL <Gk. *tragos* the hairy part of the ear, a he-goat; so called because of the hairs on it]
tra·heen (trə·hēn′) *n. Irish* A stocking without a sole.
traik (trāk) *Scot. v.i.* 1 To wander idly or with fatigue; tramp; trudge. 2 To go astray. —*n.* 1 The flesh of sheep that have died from disease or accident. 2 A stroll or saunter; also, a wearisome tramp or journey: also spelled *trake.*
traik·et (trā′kit) *adj. Scot.* Overfatigued; tired out. Also **traik′it.**
trail (trāl) *v.t.* 1 To draw along lightly over a surface; also, to drag or draw after: to *trail* a robe. 2 To follow the trail of; trace; track. 3 *Mil.* To carry, as a rifle, by grasping it in the right hand just above the balance, with the muzzle to the front and the butt nearly touching the ground. 4 To tread or force down, as grass into a pathway. 5 *Naut.* To allow (the oars) to drift alongside the boat. —*v.i.* 6 To hang or float loosely. 7 To grow along the ground or over rocks, bushes, etc., in a loose, creeping way. 8 To follow behind loosely; stream. 9 To saunter leisurely along; move heavily. 10 To lag behind; straggle; remain in the rear. —*n.* 1 The track left by anything that has moved or been drawn or dragged over any surface. 2 The track or indications followed by a huntsman or by a dog in hunting; the scent. 3 The path worn by persons or by animals; particularly, a route made by repeated passage through a wilderness. 4 Anything drawn behind or in the wake of something; a train; specifically, the train of a dress or gown. 5 *Mil.* The inclined stock of a gun carriage, or extension of the stock that rests on the ground when the piece is not limbered up: when divided longitudinally into two parts, it is called a **split trail.** —**to hit** (or **take**) **the trail** To set out on a journey. [<AF *trailler* haul, tow a boat <L *tragula* a dragnet < *trahere* draw]
trail·er (trā′lər) *n.* 1 One who or that which trails. 2 A vehicle drawn by another having motive power. 3 A trailing plant. 4 A caravanlike vehicle used as a dwelling place and drawn on the highway by a motorcar or truck.
trailing arbutus An evergreen perennial (*Epigaea repens*) of the heath family, bearing clusters of fragrant pink flowers; the mayflower: the State flower of Massachusetts.
trailing edge *Aeron.* The rear edge of an airfoil or propeller blade.
trail rope 1 A guiderope. 2 A rope used for dragging or towing. 3 A rope attached to a horse's halter or tied around its neck, but allowed to drag while the horse grazes. 4 *Mil.* A prolonge.
train (trān) *n.* 1 Anything drawn out to a length, or any series of things drawn along. 2 A continuous line of coupled railway cars. 3 A series, succession, or set of connected things; a sequence; especially, an assemblage of people or objects drawn up processionally or in orderly disposition. 4 A retinue or body of retainers; suite. 5 Something pulled along with and in the track of another. 6 An extension of a dress skirt, trailing behind the wearer. 7 Proper order; due course. 8 *Mech.* A series of parts acting upon each other, as for transmitting motion: also called *drive train, power train.* 9 *Mil.* **a** The variation of the axis of a gun in a horizontal plane. **b** Collectively, the men, animals, and vehicles attached to a military body for the transportation of its ammunition, supplies, etc. 10 A succession or line of wagons and pack animals en route. 11 A line of gunpowder or other combustible laid to conduct fire to a charge, mine, or the like. See synonyms under PROCESSION. —*v.t.* 1 To bring to a requisite standard, as of conduct or skill, by protracted and careful instruction; specifically, to mold the character of; educate; instruct: sometimes with *up.* 2 To render skilful or proficient, as a mechanic or soldier. 3 To make obedient to orders or capable of performing tricks, as an animal. 4 To bring into a required physical condition by means of a course of diet and exercise: to *train* a man for a boat race. 5 To lead into taking a particular course; develop into a fixed shape: to *train* a plant on a trellis. 6 To put or point in an exact direction; bring to bear; aim, as a cannon. 7 *Obs.* To mislead; entice. 8 *Obs.* To draw along; trail. —*v.i.* 9 To undergo a course of training. 10 To give a course of training; drill. See synonyms under LEARN, SUBDUE, TEACH. [Fusion of OF *traîne* a dragging and *train* a series, procession, both < *traïner, trahiner* draw <L *trahere*] —**train′·a·ble** *adj.* —**train′less** *adj.*
Train may appear as a combining form in hyphemes or solidemes, or as the first element in two-word phrases, with the meaning of definition 2:

train caller	**train schedule**
train conductor	**train service**
train crew	**train signal**
train flagman	**train staff**
train foreman	**train stop**
train inspector	**train ticket**
train-lighting	**traintime**
train line	**train trip**
train recorder	**trainway**
train reporter	**train whistle**
train robber	**train wreck**

train·a·si·um (trā·nā′zē·əm) *n.* A structure of bars crossing and intersecting one another to form ladders, tunnels, etc.: used in developing the muscles, as in military training. [<TRAIN + (GYMN)ASIUM]
train·band (trān′band′) *n.* A militia organization, especially one in London, England, during the Stuart period (17th century). [Short for *trained band*]
train·bear·er (trān′bâr′ər) *n.* An attendant who holds the long train of a dress or robe.
trained nurse One who has been trained in and graduated from a nurses' training school.
train·ee (trā·nē′) *n.* One who undergoes training.
train·er (trā′nər) *n.* 1 One who trains. 2 One who directs and superintends a course of physical training, or who supervises the physical condition of members of an athletic team. 3 An apparatus or device used in training: a Link *trainer.* 4 One who trains a cannon; specifically, in the U. S. Navy, the member of the gun's crew who gives direction to the gun. 5 One who trains animals for shows, contests, animal acts, etc.
train·ing (trā′ning) *n.* 1 Systematic instruction and drill. 2 The condition of being physically fit for the performance of an athletic exercise or contest; also, the act or science of bringing one to such a condition. See synonyms under EDUCATION, LEARNING, NURTURE.
training school A school for practical instruction and drill; specifically, a school in which students receive special vocational or technical instruction and practice.
training ship A vessel on which apprentice seamen and cadets are educated in seamanship, navigation, etc.
train·man (trān′mən) *n. pl.* **·men** (-mən) A railway employee serving on a train; especially, a brakeman.
train·mas·ter (trān′mas′tər, -mäs′-) *n.* A railroad official supervising some division or subdivision of a rail line.
train oil Oil obtained from the fat of whales, especially from the right whale, and from cod livers, etc. [Earlier *trane* <MDu. *traen* extracted oil]
traipse (trāps) *v.i.* **traipsed, traips·ing** *Colloq.* To walk about in an idle or aimless manner; go on foot: also spelled *trapes.* [Earlier *trapass,*

prob. <OF *trapasser*, var. of *trespasser* TRESPASS]

trait (trāt) *n.* **1** A distinguishing feature or quality of mind or character. **2** A line, stroke, or touch. See synonyms under CHARACTERISTIC. [<F <MF *traict* <L *tractus*. Doublet of TRACT[1].]

trai·tor (trā′tər) *n.* **1** One who betrays a trust; especially, one who commits treason. **2** Hence, one who acts deceitfully and falsely. [<OF *traitre, traitor* <L *traditor*. See TRADITOR.] —**trai′tor·ism** *n. Obs.* —**trai′tress** (-tris) *n. fem.*

trai·tor·ous (trā′tər·əs) *adj.* **1** Inclined to treason. **2** Involving treason. See synonyms under PERFIDIOUS. —**trai′tor·ous·ly** *adv.* —**trai′tor·ous·ness** *n.*

Tra·jan (trā′jən), A.D. 56–117, Roman emperor 98–117: full name *Marcus Ulpius Trajanus*. —**Tra·jan·ic** (trā·jan′ik) *adj.*

tra·ject (trə·jekt′) *v.t.* To throw or cast over, through, or across, as a beam of light; transmit. [<L *trajectus*, pp. of *trajicere* < *trans-* over + *jacere* throw] —**tra·jec′tion** *n.*

tra·jec·to·ry (trə·jek′tər·ē) *n. pl.* **·ries** **1** The path described by an object or body moving in space. **2** The path of a projectile after leaving the muzzle of a gun. **3** *Geom.* **a** A curve which cuts a set of curves at the same angle. **b** A surface which passes through a given set of points. [<Med. L *trajectorius* <L *trajectus*. See TRAJECT.]

trake (trāk) See TRAIK (*n.* def. 2).

tral·a·ti·tion (tral′ə·tish′ən) *n. Obs.* The use of a word or expression in a figurative sense; metaphor. [<L *tralatio, -onis* < *tralatus, translatus*, pp. to *transferre* TRANSFER]

tral·a·ti·tious (tral′ə·tish′əs) *adj.* **1** Traditional; legendary. **2** Not literal; figurative; metaphorical.

Tra·lee (trə·lē′) The county town of County Kerry, Ireland, in the western part.

tram[1] (tram) *n.* **1** *Brit.* A tramway. **2** A street railway car for passengers; a tramcar. **3** A four-wheeled vehicle for conveying coals to or from a pit's mouth. —*v.t.* **trammed, tram·ming** To convey in a tramcar. [Short for TRAMROAD]

tram[2] (tram) *n.* **1** A trammel. **2** *Mech.* Accuracy or trueness of adjustment. Compare TRAMMEL. —*v.t.* **trammed, tram·ming** To use a trammel in adjusting (any part). [Short for TRAMMEL]

tram[3] (tram) *n.* A thick silk thread used for the cross threads of the best silks and velvets. Also **trame.** [<F *trame* <OF *traime* a woof, machination <L *trama* a woof]

tram·car (tram′kär′) *n. Brit.* A car or carriage that runs on a tramway; particularly, a street car; a tram. [<TRAM[1] + CAR]

tram·line (tram′līn′) *n. Brit.* A street-car line.

tram·mel (tram′əl) *n.* **1** That which limits freedom or activity; an impediment; hindrance. **2** A fetter, shackle, or bond, particularly one of such kind as is used in teaching a horse to amble. **3** An instrument whose parts slide on a rod, especially one bearing pointers, for use as a compass, or for describing ellipses. **4** A gage for adjusting machine parts. **5** A two-piece hook, adjustable for length, used to suspend cooking pots from a fireplace crane. **6** A net formed of three layers, the central one being of finer mesh in order to catch the fish which pass through either of the others: also **trammel net.** —*v.t.* **·meled** or **·melled, ·mel·ing** or **·mel·ling** **1** To hinder or obstruct; restrict. **2** To entangle in or as in a snare; imprison. Also **tram′el** or **tram′ell.** [<OF *tramail* a net <LL *tramaculum, tremaculum* <L *tri-* three + *macula* a mesh] —**tram′mel·er** or **tram′mel·ler** *n.*

tra·mon·tane (trə·mon′tān, tram′ən·tān) *adj.* **1** Situated beyond the mountains; ultramontane; hence, barbarous; foreign. **2** Coming from the other side of the mountains. —*n.* A foreigner or barbarian; originally, a resident beyond the mountains. [<Ital. *tramontana* north wind, polestar <L *transmontanus* beyond the mountains < *trans-* over + *mons, montis* a mountain]

tramp (tramp) *v.i.* **1** To walk or wander, especially as a tramp or vagabond. **2** To walk heavily or firmly. —*v.t.* **3** To walk or wander through. **4** To walk on heavily; trample. —*n.* **1** A heavy continued tread. **2** The sound produced by continuous and heavy marching or walking. **3** A long stroll on foot. **4** One who walks from place to place; a vagrant; vagabond. **5** A steam vessel that goes from port to port picking up freight wherever it can be obtained: also **tramp steamer.** **6** A metal plate on a shoe to protect it from wear or from a spade in digging. [ME *trampen* <Gmc. Cf. LG *trampen*.]

tramp·er (tram′pər) *n.* One who or that which tramps; specifically, a vagabond.

tram·ple (tram′pəl) *v.* **·pled, ·pling** *v.t.* To tread on heavily; injure, violate, or encroach upon by or as by tramping. —*v.i.* To tread heavily or ruthlessly; tramp. —*n.* The act or sound of treading under foot. [ME *trampelen*, freq. of *trampen* TRAMP] —**tram′pler** *n.*

tram·po·line (tram′pə·lin) *n.* **1** An acrobatic performance on stilts. **2** A heavy mat or net used in acrobatic exhibitions. Also **tram′po·lin.** [<Ital. *trampoli* stilts]

trampoline trainer A section of strong canvas stretched on a frame, on which a person may bound or spring: used in training for body control and acrobatics.

tram·road (tram′rōd′) *n.* A road with wheel tracks of stone, wood, or metal; especially, a railroad in a mine. [<dial. E *tram* a rail, wagon shaft (prob. <LG *traam* a beam, shaft) + ROAD]

tram·way (tram′wā′) *n. Brit.* **1** A street railroad. **2** A roadway having plates or rails on one part of it on which wheeled vehicles may run. **3** A system of cars suspended from cables, often operating in counterbalancing pairs: also called **aerial tramway.**

trance[1] (trans, träns) *n.* **1** A state in which the soul seems to have passed out of the body; an ecstasy; rapture. **2** *Psychol.* A condition between sleep and waking characterized by dissociation, involuntary movements, and automatisms of behavior, as in hypnosis and mediumistic seances. **3** A dreamlike state marked by bewilderment and an insensibility to ordinary surroundings. **4** A state of deep abstraction. See synonyms under DREAM. —*v.t.* **tranced, tranc·ing** To entrance, usually in a figurative sense; enchant. [<OF *transe* passage, dread of coming evil < *transir* pass, die, benumb <L *transire*. See TRANSIENT.]

trance[2] (trans, träns) *n. Scot.* A passage or hallway; an alley, courtyard, or close.

tran·gam (trang′gəm) *n. Obs.* A worthless person or thing; a knick-knack or trinket. Also **tran·kum** (trang′kəm). [Origin uncertain]

tran·quil (trang′kwil) *adj.* **·quil·er** or **·quil·ler, ·quil·est** or **·quil·lest** **1** Free from agitation or disturbance; calm: said of persons. **2** Quiet and motionless: said of things. See synonyms under CALM, PACIFIC, SEDATE. [<L *tranquillus* quiet] —**tran′quil·ly** *adv.* —**tran′quil·ness** *n.*

tran·quil·ize (trang′kwəl·īz) *v.t.* & *v.i.* **·ized, ·iz·ing** To make or become tranquil. Also **tran′quil·lize**, *Brit.* **tran′quil·lise.** —**tran′quil·i·za′tion** *n.*

Synonyms: allay, appease, assuage, calm, compose, hush, lull, moderate, pacify, quell, quiet, soothe, still. See ALLAY. *Antonyms*: agitate, alarm, arouse, disturb, excite, inflame, rouse, stimulate, stir.

tran·quil·iz·er (trang′kwəl·ī′zər) *n.* **1** One who or that which tranquilizes. **2** *Med.* An ataractic drug. Also **tran′quil·liz′er.**

tran·quil·li·ty (trang·kwil′ə·tē) *n.* The state of being tranquil; rest; quiet. Also **tran·quil′i·ty.** See synonyms under APATHY, REST.

trans- *prefix* **1** Across; beyond; through; on the other side of; as in:

transarctic	**transequatorial**
transborder	**transfrontier**
transchannel	**transisthmian**
transdesert	**transpolar**

In adjectives and nouns of place, the prefix may signify "on the other side of" (opposed to *cis-*) or "across; crossing." Through long usage, certain of these are written as solid words, as *transalpine, transatlantic*; otherwise, words in this class, unless by contrary official usage, are properly written with a hyphen, as in:

trans-American	**trans-Germanic**
trans-Andean	**trans-Himalayan**
trans-Arabian	**trans-Iberian**
trans-Baltic	**trans-Mediterranean**
trans-Canadian	**trans-Siberian**

2 Through and through; changing completely; as in:

transcolor	**transfashion**

3 Surpassing; transcending; beyond; as in:

transconscious	**transmundane**
transempirical	**transnational**
transhuman	**transphysical**
transmaterial	**transrational**
transmental	

4 *Anat.* Across; transversely; as in:

transcortical	**transocular**
transduodenal	**transthoracic**
transfrontal	**transuterine**

[<L <*trans* across, beyond, over]

trans·act (trans·akt′, tranz-) *v.t.* To carry through; accomplish; do. —*v.i. Rare* To do business. [<L *transactus*, pp. of *transigere* drive through, accomplish < *trans-* through + *agere* drive, do] —**trans·ac′tor** *n.*

Synonyms: accomplish, act, conduct, do, negotiate, perform, treat. There are many acts that one may *do, accomplish*, or *perform* unaided; what he *transacts* is by means of or in association with others; one may *do* a duty, *perform* a vow, *accomplish* a task, but he *transacts* business, since that always involves the agency of others. To *negotiate* and to *treat* are likewise collective acts, but *negotiate* implies deliberation with adjustment of mutual claims and interests, while *transact* implies execution. Nations may *treat* of peace without result, but when a treaty is *negotiated* peace is secured; the citizens of the two nations are then free to *transact* business with one another.

trans·ac·tion (trans·ak′shən, tranz-) *n.* **1** The management of any affair. **2** Something transacted; an affair; a business deal. **3** *pl.* Published reports, as of a society. —**trans·ac′tion·al** *adj.*

Synonyms: act, action, affair, business, deed, doing, proceeding. A man's *acts* or *deeds* may be exclusively his own; his *transactions* involve the agency or participation of others. A *transaction* is something completed; a *proceeding* is or is viewed as something in progress; but since *transaction* is often used to include the steps leading to the conclusion, while *proceedings* may result in *action*, the dividing line between the two words becomes sometimes quite faint. Both *transactions* and *proceedings* are used of the records of a deliberative body, especially when published. See ACT.

Trans-A·lai Range (trans′ä·lī′) A branch of the Pamir-Alai mountain system on the Kirgiz-Tadzhik S.S.R. border; highest point, 23,382 feet.

trans·al·pine (trans·al′pin, -pīn, tranz-) *adj.* **1** On the other side of the Alps, especially from Rome. **2** Crossing or extending across the Alps. **3** Of or pertaining to the country or the people beyond the Alps. —*n.* A native of or a resident beyond the Alps. [<L *transalpinus* < *trans-* across + *alpinus* alpine < *Alpes* the Alps]

Transalpine Gaul The section of Gaul on the northern side of the Alps.

trans·at·lan·tic (trans′ət·lan′tik, tranz′-) *adj.* **1** On the other side of the Atlantic. **2** Across or crossing the Atlantic.

trans·berke·li·an (trans·bûrk′lē·ən) *adj. Physics* Of or pertaining to unstable radioactive elements beyond berkelium, atomic No. 97, as californium, einsteinium, fermium, mendelevium, and nobelium. [<TRANS- + BERKEL(IUM) + -IAN]

trans·ca·lent (trans·kā′lənt) *adj.* Permitting or facilitating the passage of heat. [<TRANS- + L *calens, -entis*, ppr. of *calere* be hot] —**trans·ca′len·cy** *n.*

Trans·cas·pi·a (trans·kas′pē·ə) A former administrative division of Russian Turkestan, roughly coextensive with the Turkmen S.S.R., in which it was incorporated in 1924; capital, Ashkhabad. Also **Trans·cas′pi·an Region.**

Trans·cau·ca·sia (trans′kô·kā′zhə, -shə) A region of southeastern U.S.S.R., between the Caucasus mountains on the north and Iran and Turkey in Asia on the south, comprising the republics of Armenia, Azerbaijan, and Georgia, and from 1922 to 1936 constituting the **Transcaucasian Socialist Federated Soviet Republic.** —**Trans′cau·ca′sian** *adj.* & *n.*

trans·ceiv·er (tran·sē′vər) *n. Electronics* A radio unit, usually for portable or mobile service, containing equipment for both transmission and reception. [<TRANS(MITTER) + (RE)CEIVER]

tran·scend (tran·send′) *v.t.* 1 To rise above in excellence or degree. 2 To overstep or exceed, as a limit. — *v.i.* 3 To be surpassing; excel. See synonyms under SURPASS. [<L *transcendere* surmount < *trans-* beyond, over + *scandere* climb] — **tran·scend′i·ble** *adj.*

tran·scen·dent (tran·sen′dənt) *adj.* 1 Of very high and remarkable degree; surpassing; superexcellent. 2 *Philos.* In Kantianism, lying beyond the bounds of all possible human experience; hence, beyond knowledge. 3 *Theol.* Pertaining to God as exalted above the universe; beyond limitation; hence, perfect. See synonyms under EXCELLENT, TRANSCENDENTAL. — *n.* That which is transcendent or surpassingly great or remarkable. [<L *transcendens, -entis,* ppr. of *transcendere* TRANSCEND] — **tran·scen′dence** *n.* — **tran·scen′dent·ly** *adv.* — **tran·scen′dent·ness** *n.*

tran·scen·den·tal (tran′sen·den′təl) *adj.* 1 Of very high degree; transcendent. 2 Pertaining to or being a transcendent; not included in any of the categories. See CATEGORY. 3 *Philos.* **a** In Kant's system, of an a priori character; transcending experience but not knowledge. **b** Rising above the common notions of men; with the Cartesians, pertaining to body and spirit alike. 4 Wildly speculative; above, beyond, or contrary to common sense. 5 *Math.* That cannot be formed by the five fundamental operations of algebra, each performed a finite number of times. — **tran′scen·den′tal·ly** *adv.*

Synonyms: instinctive, intuitive, original, primordial, transcendent. *Intuitive* truths are those which are in the mind independently of all experience, not being derived from experience nor limited by it. All *intuitive* truths or beliefs are *transcendental.* But *transcendental* is a wider term than *intuitive,* including all within the limits of thought that is not derived from experience, as the ideas of space and time. *Transcendent, transcendental,* and *intuitive* are opposed to *empirical*; or, according to the philosophy of Kant, *transcendent* is opposed to *immanent,* and *transcendental* to *empirical.* See MYSTERIOUS.

tran·scen·den·tal·ism (tran′sen·den′tal·iz′əm) *n.* 1 The state or quality of being transcendental. 2 In common usage, that which, in philosophy or religion, is vague, visionary, or sublimated. 3 *Philos.* The doctrine that man can attain knowledge which goes beyond or transcends appearances or phenomena. In the Kantian sense, transcendentalism affirmed the existence of a priori principles of cognition. The New England movement, as represented by Emerson and others, has been characterized by the absence of a formal system of thought, the exaltation of the spiritual in a general sense over the material, and the immanence of the divine in all creation, especially as set forth in Emerson's "The Over-Soul." — **tran′scen·den′tal·ist** *n.* & *adj.*

transcendental number See under NUMBER.

trans·con·ti·nen·tal (trans′kon·tə·nen′təl) *adj.* Extending or passing across a continent.

tran·scribe (tran·skrīb′) *v.t.* **·scribed, ·scrib·ing** 1 To write over again; copy or recopy in handwriting or typewriting from an original or from shorthand notes. 2 *Telecom.* To make an electrical recording of for use on a later radio program. 3 To adapt (a musical composition) for a change of instrument or voice. [<L *transcribere* < *trans-* over + *scribere* write] — **tran·scrib′a·ble** *adj.* — **tran·scrib′er** *n.*

tran·script (tran′skript) *n.* 1 A copy made directly from an original. 2 Any copy. See synonyms under DUPLICATE. [Fusion of OF *transcrit* (pp. of *transcrire* transcribe <L *transcribere*) and L *transcriptus,* pp. of *transcribere* TRANSCRIBE]

tran·scrip·tion (tran·skrip′shən) *n.* 1 The act of transcribing; a copying. 2 A copy; transcript. 3 *Telecom.* An electrical recording made for the purpose of a later radio broadcast. 4 *Music* The adaptation of a composition for some instrument or voice other than that for which it was written. — **tran·scrip′tion·al, tran·scrip′tive** *adj.*

trans·cul·tu·ra·tion (trans·kul′chə·rā′shən) *n.* *Anthropol.* 1 The processes, resulting in the development of new cultural phenomena and the disappearance of old, involved in the transition of a group or a people from one culture context to another. 2 The transition itself. [<TRANS- + *culturation* development of a culture <CULTUR(E) + -ATION] — **trans·cul′tu·ra′tive** *adj.*

trans·cur·rent (trans·kûr′ənt) *adj.* Passing or extending transversely.

trans·duc·er (trans·dōō′sər, -dyōō′-, tranz-) *n.* *Physics* Any device whereby the energy of one power system may be transmitted to another system, whether of the same or a different type. [<L *transducere,* var. of *traducere.* See TRADUCE.]

tran·sect (tran·sekt′) *v.t.* To dissect transversely. [<TRANS- + L *sectus,* pp. of *secare* cut] — **tran·sec′tion** (-sek′shən) *n.*

tran·sept (tran′sept) *n.* *Archit.* One of the lateral members or projections between the nave and choir of a cruciform church: commonly distinguished as the *north* and *south transepts.* [<Med. L *transeptum,* short for L *transversum septum* < *transversus* TRANSVERSE + *septum* an enclosure] — **tran·sep′tal** *adj.* — **tran·sep′tal·ly** *adv.*

trans·e·unt (tran′sē·ənt) *adj.* Proceeding from and operating beyond itself on another, as a physical cause: opposed to *immanent.* [<L *transiens, transeuntis.* See TRANSIENT.]

trans·fer (trans·fûr′, trans′fər) *v.* **·ferred, ·fer·ring** *v.t.* 1 To carry, or cause to pass, from one person, place, etc., to another. 2 To make over possession of to another. 3 To convey (a drawing) from one surface to another, as by specially prepared paper. — *v.i.* 4 To transfer oneself. 5 To be transferred. 6 To change from one car or line to another on a transfer. See synonyms under CONVEY. — *n.* (trans′fər) 1 The act of transferring, or the state of being transferred. 2 That which is transferred; specifically, in art, lithography, etc., a design conveyed or to be conveyed, as by copying ink or pressure, in reverse, from one surface to another. 3 A place, method, or means of transfer. 4 A ticket, entitling a passenger on one car or boat to ride on another, as on a connected line, with or without paying an additional fare; also, the place where such transfer is made. 5 A delivery of title or property from one person to another. 6 The exchange of a person from one organization to another, from one military division to another, from one school to another, etc. 7 An order transferring money or securities. [<OF *transferer* <L *transferre* < *trans-* across + *ferre* carry] — **trans·fer′a·bil′i·ty** *n.* — **trans·fer′a·ble** *adj.*

trans·fer·ee (trans′fə·rē′) *n.* 1 *Law* The person to whom a transfer is made. 2 One who is transferred.

trans·fer·ence (trans·fûr′əns) *n.* 1 Transfer. 2 *Psychoanal.* **a** The reproduction of the repressed or forgotten experiences of early childhood, accompanied by a transfer of emotions from the original object or person to another. **b** Displacement (def. 7). [<NL *transferentia* <L *transferens, -entis,* ppr. of *transferre* TRANSFER] — **trans·fer·en·tial** (trans′fə·ren′shəl) *adj.*

trans·fer·or (trans·fûr′ər) *n.* The vender or conveying party in a transfer.

trans·fer·rer (trans·fûr′ər) *n.* One who or that which transfers.

Trans·fig·ur·a·tion (trans′fig·yə·rā′shən) *n.* 1 The supernatural transformation of Christ on the mount as recorded in the Gospels. *Matt.* xvii 1–9. 2 A festival commemorating this: August 6.

trans·fig·ure (trans·fig′yər) *v.t.* **·ured, ·ur·ing** 1 To change the outward form or appearance of. 2 To make glorious; idealize. See synonyms under CHANGE. [<L *transfigurare* change the shape of < *trans-* across + *figura* shape] — **trans′fig·ur·a′tion, trans·fig′ure·ment** *n.*

trans·fi·nite (trans·fī′nīt) *adj.* 1 Beyond the finite. 2 *Math.* Of, pertaining to, or characterizing the properties of a set of numbers whose cardinality is not expressible by any finite number.

trans·fix (trans·fiks′) *v.t.* 1 To pierce through; impale. 2 To fix in place by impaling. 3 To make motionless, as with horror, amazement, etc. See synonyms under PIERCE. [<L *transfixus,* pp. of *transfigere* < *trans-* through, across + *figere* fasten] — **trans·fix′ion** (-fik′shən) *n.*

trans·flu·ent (trans′flōō·ənt) *adj.* 1 Flowing across or through. 2 *Her.* Flowing through the arches of a bridge. [<L *transfluens, -entis,* ppr. of *transfluere* < *trans-* across + *fluere* flow]

trans·flux (trans′fluks) *n.* A flowing or running through, across, or beyond.

trans·form (trans·fôrm′) *v.t.* 1 To give a different form to; change the character of. 2 To alter the nature of; convert. 3 *Math.* To change (one expression or operation) into another equivalent to it or having similar properties. 4 *Electr.* **a** To change the potential or the type of, as a current from higher to lower voltage, or from alternating to direct. **b** To alter the energy form of, as electrical into mechanical. 5 In alchemy, to transmute. — *v.i.* 6 To be or become changed in form or character. See synonyms under CHANGE. [<L *transformare* < *trans-* over + *formare* form < *forma* a form] — **trans·form′a·ble** *adj.*

trans·for·ma·tion (trans′fər·mā′shən) *n.* 1 A change. 2 The act of transforming or the state of being transformed. 3 A wig or partial wig worn by a woman.

trans·form·a·tive (trans·fôr′mə·tiv) *adj.* Having power or a tendency to transform.

trans·form·er (trans·fôr′mər) *n.* 1 One who or that which transforms. 2 *Electr.* A device for altering the strength and potential of a current; especially, a form of induction coil used in alternating-current systems of electrical distribution, by which a current of high voltage is transformed to one of lower voltage, or vice versa: classed accordingly either as **step-down** or **step-up transformers.**

trans·form·ism (trans·fôr′miz·əm) *n.* *Biol.* 1 The theory of the development of one species from another through gradual modifications and without the intervention of special acts of creation. 2 Any doctrine or example of evolution.

trans·fuse (trans·fyōōz′) *v.t.* **·fused, ·fus·ing** 1 To pour, as a fluid, from one vessel to another. 2 To cause to be imparted or instilled. 3 *Med.* To transfer (blood) from one person or animal to another. [<L *transfusus,* pp. of *transfundere* < *trans-* across + *fundere* pour] — **trans·fus′er** *n.* — **trans·fus′i·ble, trans·fu·sive** (trans·fyōō′siv) *adj.*

trans·fu·sion (trans·fyōō′zhən) *n.* 1 The act of pouring from one vessel into another; hence, transference; transmission. 2 *Med.* **a** The transfer of blood from one person or animal to the veins or arteries of another. **b** A similar transfer of any other fluid, as a saline solution.

trans·gress (trans·gres′, tranz-) *v.t.* 1 To break over the bounds of, as a law; violate. 2 To pass beyond or over (limits); exceed; trespass. — *v.i.* 3 To break a law; sin. See synonyms under BREAK. [Appar. <OF *transgresser* <L *transgressus,* pp. of *transgredi* < *trans-* across + *gradi* step] — **trans·gress′i·ble** *adj.* — **trans·gress′ing·ly** *adv.* — **trans·gres′sor** *n.*

trans·gres·sion (trans·gresh′ən, tranz-) *n.* 1 The act of transgressing; sin. 2 An overpassing. 3 *Geol.* An overlap. See synonyms under OFFENSE, SIN[1].

trans·gres·sive (trans·gres′iv, tranz-) *adj.* Apt to transgress; faulty; culpable. — **trans·gres′sive·ly** *adv.*

tran·shape (tran·shāp′) See TRANSSHAPE.

tran·ship (tran·ship′), **tran·ship·ment** (tran·ship′mənt) See TRANSSHIP, etc.

trans·hu·mance (trans·hyōō′məns) *n.* The moving of cattle or other animals to more suitable places as the seasons change, especially of herds to and from mountain pastures. [<F < *transhumer* <Sp. *trashumar* <L *trans-* across + *humare* cover with earth < *humus* earth] — **trans·hu′mant** *adj.*

tran·sience (tran′shəns) *n.* The quality of existing for a short time only; also, something that is transient: the *transience* of life. Also **tran′sien·cy.**

tran·sient (tran′shənt) *adj.* 1 Passing before the vision in a brief time; of short duration; brief; hasty. 2 Not permanent; temporary; casual. 3 *Obs.* Proceeding from one place or object to another; imparted. — *n.* One who

add,āce,câre,pälm; end,ēven; it,īce; odd,ōpen,ôrder; tŏŏk,pōōl; up,bûrn; ə = a in *above,* e in *sicken,* i in *clarity,* o in *melon,* u in *focus*; yōō = u in *fuse*; oi,oil; ou,pout; ch,check; g,go; ng,ring; th,thin; th,this; zh,vision. Foreign sounds à,œ,ü,kh,ṅ; and ◆: see page xx. < from; + plus; ? possibly.

or that which is transient; specifically, a lodger or boarder who remains for a short time. [<L *transiens, -euntis,* ppr. of *transire* < *trans-* across + *ire* go] —**tran′sient·ly** *adv.* —**tran′sient·ness** *n.*

Synonyms (adj.): brief, ephemeral, evanescent, fleeting, flitting, flying, fugitive, momentary, passing, short, temporary, transitory. A thing is *transient* which in fact is not lasting; a thing is *transitory* which by its very nature must soon pass away; a thing is *temporary* which is intended to last or be made use of but a little while; as, a *transient* joy; this *transitory* life; a *temporary* chairman. That which is *ephemeral,* literally lasting but for a day, is looked upon as at once slight and perishable, and the word carries often a suggestion of contempt; with no solid qualities or worthy achievement a pretender may sometimes gain an *ephemeral* popularity. That which is *fleeting* is viewed as in the act of passing swiftly by, and that which is *fugitive* as eluding attempts to detain it; that which is *evanescent* is in the act of vanishing even while we gaze, as the hues of the sunset. *Antonyms*: abiding, enduring, eternal, everlasting, immortal, imperishable, lasting, permanent, perpetual, persistent, undying, unfading.

tran·si·gent (tran′sə·jənt) *n.* A person who is willing to compromise or to be brought to terms. [<L *transigens, -entis,* ppr. of *transigere* settle. See TRANSACT.]

tran·sil·i·ent (tran·sil′ē·ənt) *adj.* Leaping or passing abruptly from one thing or condition to another; saltatory; spanning; extending over. [<L *transiliens, -entis,* ppr. of *transilire* < *trans-* across + *salire* leap] —**tran·sil′i·ence** *n.*

trans·il·lu·mi·nate (trans′i·lo͞o′mə·nāt, tranz′-) *v.t.* **·nat·ed, ·nat·ing** *Med.* To cause light to pass through (an organ or part of the body) to reveal its condition.

tran·sis·tor (tran·zis′tər, -sis′-) *n. Electronics* A miniature, compact device for the control and amplification of an electron current by passing it from a fine tungsten wire through a semiconductor to a second wire, without the use of a vacuum and at low power consumption. [<TRANS(FER) (RES)ISTOR]

tran·sis·tor·ize (tran·zis′tər·īz, -sis′-) *v.t.* **·ized, ·iz·ing** To equip with transistors instead of vacuum tubes, as a radio, hearing aid, etc.

tran·sit (tran′sit, -zit) *n.* **1** The act of passing over or through; passage. **2** The act of carrying across or through; conveyance. **3** A specific passage or route; also, a traveler through a country. **4** *Astron.* **a** The passage of one heavenly body over the disk of another. **b** The moment of passage of a celestial body across the meridian: when in that half of the meridian containing the zenith it is *superior* or *upper* transit; when in that half containing the nadir it is *inferior* or *lower* transit. **5** A transit compass. See synonyms under JOURNEY, MOTION. [<L *transitus* < *transire* cross. See TRANSIENT.]

transit compass A surveying instrument resembling a theodolite, for measuring horizontal angles. Also **transit theodolite.**

transit instrument **1** An astronomical telescope mounted in the plane of the meridian and turning on a fixed east–and–west axis: used to determine the time of passage of an object over the meridian. **2** A transit compass.

tran·si·tion (tran·zish′ən) *n.* **1** Passage from one place, condition, or action to another; change. **2** *Music* A passing modulation, an abrupt change of key, or a passage leading from one theme to another. **3** The time, period, or place of such passage; also, its product or result. See synonyms under CHANGE, MOTION. —**tran·si′tion·al, tran·si′tion·ar′y** (-er′ē) *adj.* —**tran·si′tion·al·ly** *adv.*

transition point *Physics* A single point or temperature at which different phases of a substance can exist together.

tran·si·tive (tran′sə·tiv) *adj.* **1** *Gram.* Having, requiring, or terminating upon a direct object; also, expressing an action performed by a subject or agent, that passes over to or takes effect on some person or thing as its object. **2** Having the power of passing; effecting transition. —*n. Gram.* A transitive verb. [<LL *transitivus* <L *transitus* transit. See TRANSIT.] —**tran′si·tive·ly** *adv.* —**tran′si·tive·ness, tran′si·tiv′i·ty** *n.*

transitive verb A verb whose action, performed by a subject or agent, requires or terminates upon a direct object. The verbs in the following are transitive: *catch* the ball; he *shot* the gun; they *shot* the traitor; cats *climb* trees; we *speak* French.

tran·si·to·ry (tran′sə·tôr′ē, -tō′rē) *adj.* Existing for a short time only; transient. See synonyms under TRANSIENT. [<OF *transitoire* <L *transitorius* having, allowing passage through < *transitus.* See TRANSIT.] —**tran′si·to′ri·ly** *adv.* —**tran′si·to′ri·ness** *n.*

Trans–Jor·dan (trans·jôr′dən, tranz-) An Arab territory included in the Hashemite Kingdom of the Jordan; 34,758 square miles; capital, Amman; a former British mandate. Formerly called **Trans·jor·da·ni·a** (trans′jôr·dā′·nē·ə, tranz′-).

Trans·kei (trans·kā′) A native reserve district of Transkeian Territories; 2,504 square miles.

Trans·kei·an Territories (trans·kā′ən) A division of eastern Cape of Good Hope Province, Union of South Africa; 16,554 square miles; capital, Umtata.

trans·late (trans·lāt′, tranz-, trans′lāt, tranz′-) *v.* **·lat·ed, ·lat·ing** *v.t.* **1** To give the sense or equivalent of, as a word or an entire work, in another language; change into another language. **2** To interpret; explain in other words. **3** To remove, as an ecclesiastic, from one office to another. **4** To change into another form; transform. **5** To convey or remove from one place to another, as a human being from earth to heaven without natural death. **6** *Archaic* To transport; enrapture. **7** *Mech.* To impart to (any body) motion in which all the parts follow the same direction. **8** To retransmit, as a message, by means of a telegraphic relay. —*v.i.* **9** To act as translator; also, to admit of translation: This book *translates* easily. **10** To give form to ideas. See synonyms under INTERPRET. [? <OF *translater* <L *translatus,* pp. to *transferre* TRANSFER] —**trans·lat′a·ble** *adj.* —**trans·lat′a·ble·ness** *n.*

trans·la·tion (trans·lā′shən, tranz-) *n.* **1** The act of translating, or the state of being translated. **2** A transfer from one language to another; a turning of a foreign literary composition into the vernacular; a reproduction of a work in a language different from the original. **3** *Mech.* Motion in which all the parts of a body follow the same direction: distinguished from *rotation.* **4** Automatic resending of a telegraphic message to a more distant point. See synonyms under DEFINITION. —**trans·la′tion·al** *adj.*

trans·la·tor (trans·lā′tər, tranz-, trans′lā·tər, tranz′-) *n.* **1** One who translates; also, an interpreter. **2** A telegraph repeater. —**trans·la·to·ri·al** (trans′lə·tôr′ē·əl, -tō′rē-, tranz′-) *adj.*

Trans·lei·tha·ni·a (trans′lī·thā′nē·ə, -lī·tä′-) A region in Hungary east of the Leitha river.

trans·lit·er·ate (trans·lit′ə·rāt, tranz-) *v.t.* **·at·ed, ·at·ing** To represent, as a word, by the alphabetic characters of another language having the same sound: distinguished from *translate.* [<TRANS- + L *litera* a letter] —**trans·lit′er·a′tion** *n.*

trans·lo·cate (trans·lō′kāt, tranz-) *v.t.* **·cat·ed, ·cat·ing** To cause to shift from one place or position to another.

trans·lo·ca·tion (trans′lō·kā′shən, tranz′-) *n.* **1** A shift in position. **2** *Genetics* The attachment of a part of a chromosome to another chromosome, with resulting changes in the arrangement of the genes.

trans·lu·cent (trans·lo͞o′sənt, tranz-) *adj.* Allowing the passage of light, but not permitting a clear view of any object; semitransparent. See synonyms under CLEAR, TRANSPARENT. [<L *translucens, -entis,* ppr. of *translucere* < *trans-* through, across + *lucere* shine] —**trans·lu′cence, trans·lu′cen·cy** *n.* —**trans·lu′cent·ly** *adv.*

trans·lu·nar (trans·lo͞o′nər, tranz-) *adj.* **1** Situated beyond the moon. **2** Ethereal; visionary. Also **trans·lu′na·ry** (-nər·ē). [<TRANS- + L *luna* the moon]

trans·lu·vi·al (trans·lo͞o′vē·əl, tranz-) *adj.* Pertaining to or characterized by progressive leaching, with some erosion: said of soils. [<TRANS- + (AL)LUV(IUM) + -IAL]

trans·ma·rine (trans′mə·rēn′, tranz′-) *adj.* **1** Beyond the sea. **2** Born or found overseas. **3** Crossing the sea. [<L *transmarinus* < *trans-* across + *mare* the sea]

trans·mi·grant (trans·mī′grənt, tranz-, trans′·mə-, tranz′-) *adj.* Passing from one place or condition to another. —*n.* An emigrant or an immigrant. [<L *transmigrans, -antis,* ppr. of *transmigrare* TRANSMIGRATE]

trans·mi·grate (trans·mī′grāt, tranz-, trans′·mə-, tranz′-) *v.i.* **·grat·ed, ·grat·ing** **1** To migrate, as from one place or condition to another; pass from one country or jurisdiction to another. **2** To pass into another body, as the soul at death. [<L *transmigratus,* pp. of *transmigrare* < *trans-* across + *migrare* migrate] —**trans·mi′gra·tor** *n.* —**trans·mi·gra·to·ry** (trans·mī′grə·tôr′ē, -tō′rē, tranz-) *adj.*

trans·mi·gra·tion (trans′mī·grā′shən, -mə-, tranz′-) *n.* The act of transmigrating; especially, the assumed passing of the soul from one body, after death, to another; metempsychosis. —**trans′mi·gra′tion·ism** *n.*

transmigration of souls The doctrine that souls pass into other bodies after death.

trans·mis·si·ble (trans·mis′ə·bəl, tranz-) *adj.* That may be transmitted. Also **trans·mit′ti·ble** (-mit′ə·bəl). —**trans·mis′si·bil′i·ty** *n.*

trans·mis·sion (trans·mish′ən, tranz-) *n.* **1** The act of transmitting. **2** The state of being transmitted. **3** That which is transmitted. **4** *Mech.* **a** A device that transmits power from the engine of an automobile to the driving wheels and varies the speed ratios between them. The principal types are **automatic transmission,** in which the speed ratios are automatically selected and engaged (see also FLUID DRIVE), and **manual transmission,** in which the speed ratios are selected and engaged by hand. **b** The gears for changing speed. [<L *transmissio, -onis* < *transmissus,* pp. of *transmittere* TRANSMIT]

Trans–Mis·sis·sip·pi (trans′mis′ə·sip′ē, tranz′-) *adj.* Of or pertaining to the region west of the Mississippi River.

trans·mis·sive (trans·mis′iv, tranz-) *adj.* **1** Derivable. **2** Tending to transmit; capable of sending or being sent through. **3** Derived; transmitted.

trans·mit (trans·mit′, tranz-) *v.t.* **·mit·ted, ·mit·ting** **1** To send from one place or person to another; forward or convey; dispatch. **2** To pass on by heredity; transfer. **3** To serve as a medium of passage for; conduct. **4** To send out by means of radio waves. **5** To cause (light, sound, etc.) to pass through a medium. **6** *Mech.* To convey (force, motion, etc.) from one part or mechanism to another. See synonyms under CARRY, CONVEY, SEND[1]. [<L *transmittere* < *trans-* across + *mittere* send] —**trans·mit′tal** *n.*

trans·mit·tance (trans·mit′ns, tranz-) *n.* **1** The act or process of transmitting. **2** *Physics* That proportion of radiant energy transmitted by a body upon which it is impinging. Compare OPACITY.

trans·mit·ter (trans·mit′ər, tranz-) *n.* **1** One who or that which transmits. **2** A telegraphic sending instrument. **3** That part of a telephone into which a person talks. **4** That part of a radio or television system which produces, modulates, and transmits radio-frequency waves.

trans·mog·ri·fy (trans·mog′rə·fī, tranz-) *v.t.* **·fied, ·fy·ing** To convert into a different shape; transform. [A humorous coinage; ? alter. of TRANSMIGRATE] —**trans·mog′ri·fi·ca′tion** *n.*

trans·mon·tane (trans·mon′tān, tranz-, trans′·mon·tān′, tranz′-) *adj.* Situated beyond a mountain. [Fusion of OF *transmontane,* alter. of *tramontane* polestar, north pole and L *transmontanus* TRAMONTANE]

trans·mu·ta·tion (trans′myo͞o·tā′shən, tranz′-) *n.* **1** The act of transmuting. **2** In alchemy, the supposed change of a baser metal into one of greater value, as of lead into gold. **3** *Physics* The change of one element into another through alteration of its nuclear structure, as in radioactivity or by bombardment with high-energy particles, etc. **4** *Biol.* Successive change of form; transformism. See synonyms under CHANGE. —**trans′mu·ta′tion·al, trans·mut·a·tive** (trans·myo͞o′tə·tiv, tranz-) *adj.*

trans·mute (trans·myo͞ot′, tranz-) *v.t.* **·mut·ed, ·mut·ing** To change in nature or form; alter in essence. Also **trans·mu′tate.** See synonyms under CHANGE. [<L *transmutare* < *trans-* across + *mutare* change] —**trans·mut′a·ble** *adj.* —**trans·mut′a·bil′i·ty, trans·mut′a·ble·ness** *n.* —**trans·mut′a·bly** *adv.* —**trans·mut′er** *n.*

trans·nep·tu·ni·an (trans′nep·to͞o′nē·ən, -tyo͞o′-, tranz′-) *adj. Astron.* Beyond the planet Neptune. [<TRANS- + NEPTUN(E) + -IAN]

trans·nor·mal (trans·nôr′məl, tranz-) *adj.* Supernormal.

trans·o·ce·an·ic (trans′ō·shē·an′ik, tranz′-) *adj.* 1 Lying beyond or over the ocean. 2 Crossing the ocean.

tran·som (tran′səm) *n.* 1 A horizontal piece framed across an opening; a lintel. 2 A window above such a bar, especially a small window above a door. 3 A horizontal construction dividing a window into stages. 4 A tie beam. 5 *Naut.* A beam running across and forming part of the stern frame of a ship. 6 The horizontal crossbar of a gallows or cross. [<L *transtrum* a crossbeam < *trans* across] —**tran′somed** *adj.*

transom window 1 A window divided into stages by transoms. 2 A window over a door transom and often hinged to it.

tran·son·ic (tran·son′ik) *adj. Aeron.* Of, pertaining to, or characterized by speeds between the subsonic and supersonic.

transonic barrier *Aeron.* A barrier to flight encountered by aircraft not designed to exceed subsonic speed: caused by turbulence of the airflow around different parts of the plane. Also called *sonic barrier, sound barrier.*

Trans·ox·i·an·a (trans·ok′sē·an′ə) See SOGDIANA.

trans·pa·cif·ic (trans′pə·sif′ik) *adj.* 1 Crossing the Pacific Ocean. 2 Situated across or beyond the Pacific.

trans·pa·dane (trans′pə·dān) *adj.* Being beyond the river Po, from Rome as a standpoint. [<L *transpadanus* < *trans-* across + *padanus* of the Po <*Padus* the river Po]

Transpadane Gaul The section of Gaul in Italy north of the river Po.

trans·par·en·cy (trans·pâr′ən·sē, -par′-) *n. pl.* **·cies** 1 The quality of being transparent. 2 Something, as a picture on glass, intended to be viewed by shining a light through it. 3 *Phot.* The light-transmitting power of a sensitized negative. 4 Simplicity. Also **trans·par′ence.**

trans·par·ent (trans·pâr′ənt, -par′-) *adj.* 1 Admitting the passage of light, and of clear views of objects beyond; pervious to light: *transparent* glass: distinguished from *translucent.* 2 Figuratively, easy to see through or understand; hence, without guile; frank. 3 Diaphanous. 4 Luminous; bright. [<Med. L *transparens, -entis* <L *trans-* across + *parere* appear, be visible] —**trans·par′ent·ly** *adv.* —**trans·par′ent·ness** *n.*

Synonyms: clear, diaphanous, limpid, lucid, pellucid, translucent. Whatever offers no obstruction to the vision is *clear*; *limpid, lucid,* and *pellucid* refer to a shining, sparkling clearness. A *transparent* body allows the forms and colors of objects beyond to be seen through it; a *translucent* body allows light to pass through, but may not permit forms and colors to be distinguished; plate glass is *transparent,* ground glass is *translucent. Limpid* refers to a liquid clearness, or that which suggests it; as, *limpid* streams. See CANDID, CLEAR, EVIDENT, MANIFEST, PLAIN[1]. *Antonyms*: cloudy, dark, dim, obscure, opaque, turbid.

transparent velvet A soft, lightweight velvet suitable for draping, having a silk or rayon back and a rayon pile.

tran·spic·u·ous (tran·spik′yo͞o·əs) *adj.* Transparent. [<Med. L *transpicuus* <L *transpicere* look, see through < *trans-* through + *specere* look]

trans·pierce (trans·pirs′) *v.t.* **·pierced, ·pierc·ing** To pierce through; penetrate completely. [<MF *transpercer* < *trans-* (<L, across, through) + *percer* pierce]

tran·spi·ra·tion (tran′spə·rā′shən) *n.* A transpiring or exhalation, as through a porous substance or through the tissues of a plant.

tran·spire (tran·spīr′) *v.* **·spired, ·spir·ing** *v.t.* 1 *Physiol.* To send off through the excretory organs, as of the skin and lungs; exhale. —*v.i.* 2 *Physiol.* To be emitted, as through the skin; be exhaled, as moisture or odors. 3 To become known. 4 *Colloq.* To happen; occur. [<F *transpirer* <L *trans-* across, through + *spirare* breathe]

trans·plant (trans·plant′, -plänt′) *v.t.* 1 To remove and plant in another place. 2 To remove and settle or establish for residence in another place. 3 *Surg.* To transfer (a portion of tissue) from its original site to another part of the same individual, or to another individual. —*n.* (trans′plant′, -plänt′) 1 That which is transplanted, as a seedling. 2 A transplanting. [<LL *transplantare* <L *trans-* across + *plantare* plant] —**trans′plan·ta′tion** *n.* —**trans·plant′er** *n.* —**trans·plant′ing** *n.*

trans·pond·er (trans·pon′dər) *n. Electronics* A device that receives a signal from one telecommunication circuit and transmits the corresponding signal to another circuit: used in conjunction with an interrogator. Also called *pulse repeater.* [<TRANS(MITTER) + (RES)PONDER]

trans·po·ni·ble (trans·pō′nə·bəl) *adj.* Transposable. [<L *transponere* transpose (< *trans-* across + *ponere* put) + -IBLE] —**trans·po′ni·bil′i·ty** *n.*

trans·pon·tine (trans·pon′tin, -tīn) *adj.* Situated on the other side of a bridge: said of London south of the Thames. [<TRANS- + L *pons, pontis* a bridge]

trans·port (trans·pôrt′, -pōrt′) *v.t.* 1 To carry or convey from one place to another. 2 To carry into banishment, especially beyond the sea. 3 To carry away with emotion. 4 *Obs.* To take out of the world; kill. See synonyms under CARRY, CONVEY, RAVISH. —*n.* (trans′pôrt, -pōrt) 1 The state of being transported, as with rapture. 2 *pl.* The varied and recurrent emotions that characterize such a state. 3 Transportation. 4 A vessel, rolling stock, or other means of conveyance used by a government to transport troops, military supplies, etc. 5 The act of transporting convicts. 6 A deported convict. 7 *Aeron.* An airplane used to transport passengers, mail, etc. See synonyms under ENTHUSIASM, RAPTURE. [<MF *transporter* <L *transportare* < *trans-* across + *portare* carry] —**trans·port′er** *n.*

trans·port·a·ble (trans·pôr′tə·bəl, -pōr′-) *adj.* 1 That may be transported. 2 Rendering liable to transportation (def. 2). —**trans·port′a·bil′i·ty** *n.*

trans·por·ta·tion (trans′pər·tā′shən) *n.* 1 The act of transporting; conveyance. 2 The sending away of a convict to a remote place. 3 Vehicles used in transporting; also, charge for conveyance. 4 A ticket, pass, or other printed matter entitling a passenger to travel on a railroad train, street car, etc.

trans·port·ing (trans·pôr′ting, -pōr′-) *adj.* Enrapturing; ravishing; ecstatic. —**trans·port′ing·ly** *adv.*

trans·pose (trans·pōz′) *v.t.* **·posed, ·pos·ing** 1 To reverse the order or change the place of; interchange. 2 *Math.* To transfer (a term) with a changed sign from one side of an algebraic equation to the other, so as not to destroy the equality of the members. 3 To change in place or order, as a word in a sentence. 4 *Music* To write or play in a different key. 5 To transport. 6 *Obs.* To transform. [<OF *transposer* <L *trans-* over + OF *poser.* See POSE[1].] —**trans·pos′a·ble** *adj.* —**trans·pos′er** *n.*

trans·po·si·tion (trans′pə·zish′ən) *n.* 1 The act of transposing, or the state of being transposed. 2 That which has been transposed. Also **trans·po·sal** (trans·pō′zəl). —**trans′po·si′tion·al** *adj.*

trans·shape (trans·shāp′) *v.t.* **·shaped, ·shap·ing** To change the shape of: also spelled *transhape.*

trans·ship (trans·ship′) *v.t.* & *v.i.* **·shipped, ·ship·ping** To transfer from one conveyance or line to another: also spelled *tranship.* —**trans·ship′ment** *n.*

tran·sub·stan·ti·ate (tran′səb·stan′shē·āt) *v.t.* **·at·ed, ·at·ing** 1 To change from one substance into another; transmute; transform. 2 *Theol.* To change the substance of (the bread and wine of the Eucharist) into the body and blood of Christ. [<Med. L *transubstantiatus,* pp. of *transubstantiare* <L *trans-* over + *substantia* substance]

tran·sub·stan·ti·a·tion (tran′səb·stan′shē·ā′shən) *n.* 1 *Theol.* The conversion of the substance of the eucharistic elements into that of Christ's body and blood: a doctrine of the Greek and Roman Catholic churches. Compare CONSUBSTANTIATION, IMPANATION. 2 A change of anything into something essentially different. —**tran′sub·stan′ti·a′tion·al·ist** *n.*

tran·su·date (tran′so͞o·dāt) *n.* 1 The fluid that transudes. 2 The act or process of transuding. Also **tran′su·da′tion** (-dā′shən). [<NL *transudatus,* pp. of *transudare* TRANSUDE]

tran·sude (tran·so͞od′) *v.i.* **·sud·ed, ·sud·ing** To pass through the pores or tissues, as of a membrane. [<NL *transudare* <L *trans-* across, through + *sudare* sweat] —**tran·su′da·to′ry** (-də·tôr′ē, -tō′rē) *adj.*

trans·u·ra·ni·an (trans′yo͞o·rā′nē·ən, tranz′-) *adj. Physics* Of or pertaining to those radioactive elements having an atomic number greater than that of uranium. Also **trans′u·ran′ic** (-ran′ik), **trans′u·ra′ni·um.** [<TRANS- + URAN(IUM) + -IAN]

Trans·vaal (trans·väl′, tranz-) A province of NE Union of South Africa; 110,450 square miles; seat of government, Pretoria; formerly an independent republic.

trans·val·ue (trans·val′yo͞o, tranz-) *v.t.* **·ued, ·u·ing** 1 To appraise the value of, as conduct, morals, beliefs, and the like, in accordance with principles at variance with accepted or conventional standards. 2 *Psychoanal.* To attach to an idea or complex of ideas a disproportionate emotional value, as in dreams, schizophrenia, etc. —**trans′val·u·a′tion** *n.*

trans·ver·sal (trans·vûr′səl, tranz-) *adj.* Transverse. —*n. Geom.* A straight line intersecting a system of lines.

trans·verse (trans·vûrs′, tranz-) *adj.* 1 Lying or being across; athwart. 2 *Anat.* Placed across the long axis of a part: a *transverse* muscle. —*n.* (*also* trans′vûrs, tranz′-) 1 That which is transverse. 2 *Geom.* That axis of a conic section which passes through its foci. [<L *transversus* lying across, pp. of *transvertere* < *trans-* across + *vertere* turn] —**trans·verse′ly** *adv.*

transverse process *Anat.* A long process extending laterally from a vertebra.

transverse wave *Physics* A wave whose component particles oscillate in a direction perpendicular to the line of propagation.

trans·ves·ti·tism (trans·ves′tə·tiz′əm, tranz-) *n. Psychiatry* A morbid craving to dress in the garments appropriate to members of the opposite sex. Compare EONISM. [<TRANS- + L *vestitus,* pp. of *vestire* clothe] —**trans·ves′tite** (-tīt) *n.*

Trans–vi·sion (trans·vizh′ən, tranz-) *n.* A method of preparing illustrations in which successive parts of an object, as a machine or anatomical structure, are printed on a series of closely alined layers of transparent acetate, in order to show the relationship of details to each other and to the whole: a trade name.

Tran·syl·va·ni·a (tran′sil·vā′nē·ə) A region and former province in central Rumania; 24,000 square miles; formerly the eastern part of Hungary. —**Tran′syl·va′ni·an** *adj.* & *n.*

trap[1] (trap) *n.* 1 A device for catching game or other animals, as a pitfall or a baited device so arranged that a slight disturbance causes it to close or fall and thus kill or capture the victim. 2 A contrivance for hurling clay pigeons or glass balls into the air for sportsmen to shoot at. 3 Any artifice by which a person may be betrayed or taken unawares. 4 *Mech.* A U- or S-bend in a pipe, etc., for stopping return flow, as of noxious gas. 5 A trap door. 6 *Colloq.* A light, two-wheeled carriage suspended by springs. 7 A rattletrap. 8 *pl.* Traps. 9 a The game of trap ball. b A pivoted piece of wood, resembling a low shoe, used in the game of trap ball to throw a ball into the air. 10 In some games, especially golf, an obstacle or hazard: a water *trap,* sand *trap.* 11 *U.S. Slang* The mouth: Shut your *trap.* —*v.* **trapped, trap·ping** *v.t.* 1 To catch in a trap; ensnare. 2 To provide with a trap. 3 To stop or hold by some obstruction: said of a liquid. —*v.i.* 4 To set traps for game; be a trapper. [OE *treppe, træppe*]

trap[2] (trap) *n.* 1 *pl. Colloq.* Personal effects, as luggage; also, household goods. 2 A trapping. —*v.t.* **trapped, trap·ping** To adorn with trappings; bedeck. [Orig. a cloth covering for a horse, alter. of OF *drap* a cloth, covering <Med. L *drappus*; ult. origin uncertain]

trap[3] (trap) *n. Geol.* A dark, fine-grained igneous rock, often of columnar structure, as basalt, dolerite, etc.: also called *traprock*. [<Sw. *trapp* <*trappa* a stair; so called from the steplike arrangement of this rock in other rock]

tra·pan (trə·pan′) See TREPAN[2].

Tra·pa·ni (trä′pä·nē) An Italian port on the NW tip of Sicily: ancient *Drepanum*.

trap ball 1 A game in which a player strikes one end of a trap with a bat and thus flips a ball into the air for other players to try to catch. 2 The ball used in this game. See TRAP[1] (*n.* def. 9).

trap door A door, hinged or sliding, to cover an opening, as in a floor or roof.

trap-door spider (trap′dôr′, -dōr′) A large spider (family *Ctenizidae*) that inhabits a vertical, tubular pit in the ground, covered by a circular trap door hinged at one side to the silken lining of the tube, especially *Bothriocyrtum californica* of the SW United States.

trapes (trāps) See TRAIPSE.

tra·peze (trə·pēz′, tra-) *n.* 1 A short swinging bar, suspended by two ropes, for various gymnastic exercises. 2 *Geom.* A trapezium. [<F *trapèze* <NL *trapezium* a trapezium]

tra·pe·zi·form (trə·pē′zə·fôrm) *adj.* Having the form of a trapezium. [<TRAPEZI(UM) + -FORM]

tra·pe·zi·um (trə·pē′zē·əm) *n. pl.* **·zi·a** (-zē·ə) 1 *Geom.* **a** A four-sided plane figure of which no two sides are parallel. **b** Formerly, in England, a trapezoid (def. 1b). 2 *Anat.* **a** The bone of the distal row of the carpus situated on the radial side at the base of the thumb. **b** A band of transverse fibers found in the pons Varolii of the brain. 3 *Astron.* The four brightest stars in the nebula of Orion, at the angles of a trapezium. [<NL <Gk. *trapezion*, dim. of *trapeza* a table, lit., a four-footed (bench) < *tetra-* four + *peza* foot]

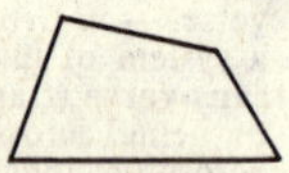

TRAPEZIUM

trap·e·zo·he·dron (trap′ə·zō·hē′drən, trə·pē′-) *n. pl.* **·dra** (-drə) A crystal figure bounded by six, eight, or twelve faces, each having unequal intercepts on all axes. [<NL < *trapezium* a trapezium + Gk. *hedra* a base]

trap·e·zoid (trap′ə·zoid) *n.* 1 *Geom.* **a** A quadrilateral of which two sides are parallel. **b** Formerly, in England, a plane quadrilateral having no parallel sides; a trapezium. 2 *Anat.* An irregular bone in the second row of the carpus, at the end of the forefinger. [<NL *trapezoïdes* <Gk. *trapezoeidēs* tablelike < *trapeza* a table + *eidos* a form] —**trap′e·zoi′dal** *adj.*

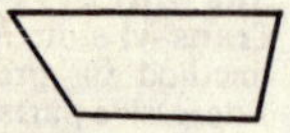

TRAPEZOID

trap·fall (trap′fôl′) *n.* A trap door yielding under pressure of feet.

trap line 1 The ensnaring filament in a spider's web. 2 A series of traps set out at approximately equal distances.

trap net A fishing net having a funnel-shaped entrance into an oblong net pen from which egress is almost impossible.

trap·pe·an (trap′ē·ən, trə·pē′ən) *adj.* Of or pertaining to traprock. Also **trap′pous, trap·pose** (trap′ōs). [<TRAP[3]]

trap·per (trap′ər) *n.* One whose occupation is the trapping of fur-bearing animals.

trap·ping (trap′ing) *n.* 1 An ornamental housing or harness for a horse. 2 *pl.* Adornments of any kind; embellishments; superficial dress. See synonyms under CAPARISON. [<TRAP[2]]

trap·pist (trap′ist) *n.* A nunbird. [<TRAPPIST]

Trap·pist (trap′ist) *n.* A member of an ascetic order of monks, a branch of the Cistercians, founded at Soligny-la-Trappe, France, and noted for silence and abstinence. [<F *Trappiste*, from *La Trappe*, name of their first abbey, established 1664]

trap·rock (trap′rok′) *n.* Trap[3].

traps (traps) *n. pl.* Percussion instruments, such as drums, cymbals, etc. [<TRAP[1] (def. 8)]

trap·shoot·ing (trap′shōō′ting) *n.* The sport of shooting pigeons, or artificial substitutes sent up from spring traps. See TRAP[1] (*n.* def. 2). —**trap′shoot′er** *n.*

trash[1] (trash) *n.* 1 Worthless or waste matter of any kind; rubbish. 2 That which is broken or lopped off, as loppings of trees. 3 The lowest grade of tobacco. 4 The dry refuse of sugarcane after the juice has been expressed. 5 A worthless person, or one of ill repute. —*v.t.* 1 To free from trash. 2 To strip of leaves; prune; lop. 3 To regard as trash; discard. [Cf. dial. Norw. *trask* lumber, trash, baggage]

trash[2] (trash) *n.* 1 Something fastened to an animal's neck to serve as a check. 2 A clog; collar; leash; any hindrance. —*v.t.* To keep in check with a leash, trash, or halter. [<OF *trachier*, var. of *tracier*. See TRACE[1].]

trash·trie (trash′trē) *n. Scot.* Trash.

trash·y (trash′ē) *adj.* **trash·i·er, trash·i·est** 1 Consisting of or like trash; worthless. 2 Cheap; inferior: said of literature. 3 Covered with underbrush or waste: said of land. —**trash′i·ly** *adv.* —**trash′i·ness** *n.*

Tra·si·me·no (trä′sē·mē′nō), **Lake** A lake in central Italy, 10 miles in diameter; 50 square miles; here Hannibal defeated the Romans, 217 B.C.: also *Lake of Perugia*. Ancient **Tras·i·me·nus** (tras′i·mē′nəs).

tras·ko (träs′kō) *n.* A Swedish dance in 4/4 time but having a polka step: also called *wooden-shoe dance*. [<Sw. < *traska* patter, trot]

trass (tras) *n.* A gray, yellow, or whitish earth, related to pozzuolana, common in volcanic districts: used in preparation of a hydraulic cement. [<G <Du. *tras* <earlier *taras*. Akin to TERRACE.]

trauch·le (troukh′lə) See TRACHLE.

trau·ma (trô′mə, trou′-) *n. pl.* **·mas** or **·ma·ta** (-mə·tə) 1 *Pathol.* Any injury to the body caused by shock, violence, etc.; a wound. 2 *Psychiatry* A severe emotional shock having a deep, often lasting effect upon the personality. 3 A traumatism. [<NL <Gk. *trauma, -atos* a wound]

trau·mat·ic (trô·mat′ik) *adj.* 1 Of or pertaining to trauma. 2 Connected with or resulting from shock, a wound, or wounds. [<LL *traumaticus* <Gk. *traumatikos* < *trauma, -atos* a wound] —**trau·mat′i·cal·ly** *adv.*

trau·ma·tism (trô′mə·tiz′əm) *n. Pathol.* 1 The general condition of the system resulting from a severe wound or external injury. 2 The injury or wound itself; a trauma. Also **trau′ma·to′sis** (-tō′sis).

trau·ma·to·pho·bi·a (trô′mə·tə·fō′bē·ə) *n.* A morbid fear of injury. [< *traumato-* (<Gk. *trauma* a wound) + -PHOBIA]

trau·mat·ro·pism (trô·mat′rə·piz′əm) *n. Biol.* The growth or involuntary movement of an organism as determined by an injury. [<TRAUMA + TROPISM]

trav·ail[1] (trav′āl, trə·vāl′) *v.i.* 1 To weary. —*v.i.* 2 To suffer the pangs of childbirth. 3 To toil; labor. —*n.* 1 Labor in childbirth. 2 Anguish or distress encountered in achievement. 3 Hard or agonizing labor. 4 Physical agony. See synonyms under TOIL[1]. [<OF < *travaillier* labor, toil, ult. <LL *trepalium* a three-pronged instrument of torture < *tres, tria* three + *palus* a stake]

tra·vail[2] (trà·và′y′) *n. pl.* **·vails** (-và′y′) *French* A travois.

Trav·an·core (trav′ən·kôr′) A former administrative division and princely state of Travancore-Cochin State, India; 7,662 square miles.

Trav·an·core-Co·chin (trav′ən·kôr′kō′chin, koch′in) A former constituent state of SW India; mostly incorporated in Kerala State, 1956; 9,144 square miles; capital, Trivandrum.

trave (trāv) *n. Obs.* 1 A frame to confine a beast of burden while being shod. 2 A crossbeam; transom. [<OF <L *trabs, trabis* a beam]

trav·el (trav′əl) *v.* **trav·eled** or **·elled, trav·el·ing** or **·el·ling** *v.i.* 1 To go from one place to another or from place to place; make a journey or tour. 2 To proceed; advance. 3 To go about from place to place as a traveling salesman. 4 *U. S. Colloq.* To move with speed. 5 To pass or be transmitted, as light, sound, etc. 6 *Mech.* To move in a fixed path, as part of a mechanism. —*v.t.* 7 To move or journey across or through; traverse. —*n.* 1 The act of traveling; a journeying: chiefly in the plural. 2 *pl.* A narration of things experienced or observed in traveling. 3 A moving or progress of any kind. 4 *Mech.* Movement or length of stroke. 5 The passage of people and vehicles to, over, or past a certain place. 6 Tourists, collectively. 7 Distance traveled; mileage. See synonyms under JOURNEY. [Var. of TRAVAIL[1]] —**trav′el·ing** or **·el·ling** *adj.* & *n.*

trav·eled (trav′əld) *adj.* 1 Having made many journeys, especially to distant lands. 2 Experienced as the result of travel. 3 Frequented or used by travelers: a *traveled* district. Also **trav′elled.**

trav·el·er (trav′əl·ər, trav′lər) *n.* 1 One who travels or journeys from place to place. 2 An animal or thing considered with reference to its mode or speed of movement. 3 A traveling salesman; specifically, a drummer: also **commercial traveler.** 4 *Naut.* **a** A metal ring or thimble running freely on a rope, rod, or spar. **b** A bar affixed to the deck, along which a ring or thimble slides. 5 A traveling crane or other moving device for transporting heavy objects. 6 In the theater, an overhead rod or pipe in the flys of the stage from which small spotlights are suspended and made available for unusual lighting effects. 7 The rings and track for drawn curtains. Also **trav′el·ler.**

traveler's tree Ravenala.

trav·el·ing (trav′əl·ing, trav′ling) *adj.* 1 Designed or used for travel: a *traveling* bag. 2 Itinerant: a *traveling* tinker. 3 Portable; movable. 4 *Mech.* **a** Running or sliding along a fixed course, as a ring or thimble. **b** Constructed with a part that travels.

traveling crane A hoisting and transporting apparatus which moves along a supporting frame or bridge, the frame itself moving on tracks. Compare illustration under GANTRY.

traveling man A commercial traveler; a traveling salesman.

trav·e·log (trav′ə·lôg, -log) *n.* A lecture or discourse on or an account of travel, usually illustrated pictorially. Also **trav′e·logue.** [< TRAVEL, on analogy with *monolog, dialog,* etc.]

trav·erse (trav′ərs, trə·vûrs′) *v.* **·ersed, ·ers·ing** *v.t.* 1 To pass over, across, or through. 2 To move back and forth over or along. 3 To examine carefully; survey or scrutinize. 4 To oppose; thwart. 5 To turn (a gun, lathe, etc.) to right or left; swivel. 6 *Law* To make denial of; in legal pleading, to deny and tender issue upon, as a matter of fact alleged by the opposite party; impeach the validity of an inquest of office. 7 *Naut.* To brace (a yard) fore and aft. —*v.i.* 8 To move back and forth. 9 To move across; cross. 10 To turn; swivel. 11 In fencing, to slide one's blade toward the hilt of an opponent's sword while maintaining pressure on it. —*n.* (trav′ərs) 1 A part, as of a machine or structure, that traverses, as a crosspiece, crossbeam, transom, or the like. 2 *Archit.* A gallery or loft communicating with opposite sides of a building. 3 Something serving as a screen or barrier. 4 *Geom.* A transversal. 5 The act of traversing or traveling; a journey; passage. 6 *Mech.* Sidewise travel, as of the tool in a slide rest. 7 The act of traversing or denying; a denial; in legal pleading, a formal denial. 8 *Naut.* A zigzag track of a vessel while beating to windward. 9 A short line surveyed from a main line, to establish the position of a side point. 10 *Mil.* A bank of earth thrown up, as from a trench, to afford protection from gunfire. 11 Something that obstructs, vexes, or thwarts. 12 A path cut transversely in the side of a cliff or mountain; also, the cliff across which a path is cut. 13 A sled having a long board connecting two or more sleds or two or more sets of runners. —*adj.* (trav′ərs) Transverse; lying or being across. —*adv.* (trav′ərs, trə·vûrs′) Transversely; crosswise. [<OF *traverser* <LL *traversare, transversare* <L *transversus* TRANSVERSE] —**trav′ers·a·ble** *adj.* —**trav·er·sal** (trav′ər·səl, trə·vûr′səl) *n.* —**trav′ers·er** *n.*

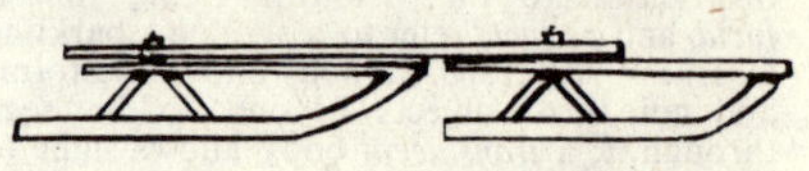

TRAVERSE *(def. 13)*

Trav·erse (trav′ərs), **Lake** A lake on the

boundary of South Dakota and Minnesota; 26 miles long, 3 miles wide.

traverse jury A trial jury. [<TRAVERSE (*v.* def. 3) + JURY]

trav·er·tine (trav′ər·tin, -tēn, -tin) *n.* A porous, light-yellow, crystalline calcium carbonate deposited in solution from ground or surface waters: a form of limestone used for building purposes. Also **trav′er·tin** (-tin). [<Ital. *travertino, tivertino* <L *Tiburtinus* Tiburtine <*Tibur*s, *-urtis* of Tibur <*Tibur* Tibur]

trav·es·ty (trav′is·tē) *n. pl.* **·ties** 1 A grotesque imitation; burlesque. 2 In literature, a burlesque treatment of a lofty subject. See synonyms under CARICATURE. —*v.t.* **·tied, ·ty·ing** To make a travesty on; burlesque; parody. [<MF *travesti*, pp. of *(se) travestir* disguise (oneself) <Ital. *travestire* disguise]

tra·vois (tra·voi′) *n. pl.* **·vois** (-voiz′) or **·vois·es** (-voi′ziz) A primitive sled constructed of two poles which serve as shafts for a dog or other draft animal and which drag on the ground, bearing a frame for the load: used by North American Indians and lumbermen in logging: also spelled *travail.* Also **tra·voise′** (-voiz′). [<dial. F (Canadian), alter. of F *travail,* a frame in which horses are held while being shod <OF]

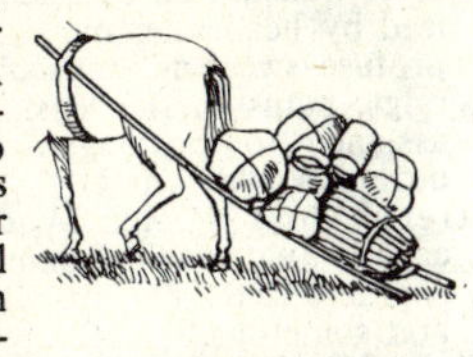

TRAVOIS

trawl (trôl) *n.* 1 A stout line, sometimes over a mile long, anchored and buoyed, and having hanging from it many lines frequently spaced and bearing baited hooks: also called *trotline.* 2 A great net shaped like a flattened bag, for towing on the bottom of the ocean by a boat. —*v.t.* To drag, as a net to catch fish. —*v.i.* To fish with a trawl line, trawl net, or the like. [Cf. MDu. *traghel* a dragnet; prob. infl. by *trail*] —**trawl′ing** *n.*

trawl·er (trô′lər) *n.* 1 A person engaged in trawling. 2 A vessel used for trawling.

trawl·ey (trô′lē) See TROLLEY (def. 4).

tray (trā) *n.* 1 A flat shallow utensil or bowl with raised edges, for various uses. 2 A shallow box without a cover, used in trunks and otherwise. 3 A kind of flat board with a low rim, made of wood, metal, or other material, and used for carrying or holding articles; also, its contents. ◆ Homophone: *trey.* [OE *treg, trig* a wooden board]

treach·er·ous (trech′ər·əs) *adj.* 1 Traitorous; perfidious. 2 Having a good appearance, but bad in character or nature; untrustworthy; affording unsafe footing: a *treacherous* path. See synonyms under INSIDIOUS, PERFIDIOUS, ROTTEN. —**treach′er·ous·ly** *adv.* —**treach′er·ous·ness** *n.*

treach·er·y (trech′ər·ē) *n. pl.* **·er·ies** Violation of allegiance, confidence, or plighted faith; perfidy; treason. See synonyms under FRAUD. [<OF *trecherie, tricherie* <*tricher, trechier* cheat]

trea·cle (trē′kəl) *n.* 1 The sirup obtained in refining sugar. 2 Molasses. 3 A saccharine fluid of certain plants. 4 Originally, a compound used as an antidote. 5 *Obs.* A panacea. [<OF *triacle* <L *theriaca* <Gk. *thēriakē* a remedy for poisonous bites <*thērion*, dim. of *thēr* a wild beast] —**trea′cly** *adj.*

tread (tred) *v.* **trod** (*Archaic* **trode**), **trod·den** or **trod, tread·ing** *v.t.* 1 To step or walk on, over, along, etc.: to *tread* the floor. 2 To press with the feet; trample: to *tread* grass. 3 To accomplish in walking or in dancing: to *tread* a measure. 4 To copulate with: said of male birds. —*v.i.* 5 To place the foot down; walk. 6 To press the ground or anything beneath the feet: usually with *on.* —**to tread water** In swimming, to keep the body erect and the head above water by moving the feet up and down as if walking. —*n.* 1 The act or manner of treading; a walking or stepping. 2 That on which something treads or rests in moving, or which affords space for or as for treading. 3 The part of a wheel that bears upon the ground or rails. 4 The outer surface of an automobile tire, or the distance between opposite wheels. 5 The part of a rail on which the wheels bear. 6 The cicatricle or chalaza of an egg. 7 The impression made by a foot, a tire, etc. 8 The flat part of a step in stairs. [OE *tredan*] —**tread′er** *n.* —**tread′ing** *n.*

trea·dle (tred′l) *n.* A lever operated by the foot, usually to cause rotary motion. —*v.i.* **·led, ·ling** To work a treadle. Also spelled *treddle.* [OE *tredel* <*tredan* tread] —**tread′ler** *n.*

tread·mill (tred′mil′) *n.* 1 A mechanism rotated by the walking motion of one or more persons: formerly used as a prison punishment. 2 A somewhat similar mechanism operated by a quadruped. 3 Toilsome effort; monotonous routine.

tread·way (tred′wā′) *n.* The roadway in certain types of bridges.

trea·son (trē′zən) *n.* 1 Betrayal, treachery, or breach of allegiance or of obedience toward the sovereign or government. Treason against the United States is declared by the Constitution (Article 3, section 3) to "consist only in levying war against them, or in adhering to their enemies, giving them aid and comfort." 2 A breach of faith; treachery. See synonyms under FRAUD. [<AF *treyson*, OF *traïson* <L *traditio, -onis* a betrayal, delivery. Doublet of TRADITION.]

trea·son·a·ble (trē′zən·ə·bəl) *adj.* Of, involving, or characteristic of treason. —**trea′son·a·ble·ness** *n.* —**trea′son·a·bly** *adv.*

trea·son·ous (trē′zən·əs) *adj.* Full of treason; treasonable. —**trea′son·ous·ly** *adv.*

treas·ure (trezh′ər) *n.* 1 The precious metals; money; jewels. 2 Riches accumulated or possessed; a stock or store of anything; wealth. 3 Something very precious. See synonyms under WEALTH. —*v.t.* **·ured, ·ur·ing** 1 To lay up in store; accumulate. 2 To retain carefully, as in the mind: generally with *up.* 3 To set a high value upon; prize. See synonyms under CHERISH. [<OF *tresor* <L *thesaurus.* Doublet of THESAURUS.]

Treasure Island An artificial island in San Francisco Bay, used as a naval base; 400 acres.

treas·ur·er (trezh′ər·ər) *n.* 1 One who has the care of treasure or of a treasury. 2 An officer legally authorized to receive, care for, and disburse public revenues upon lawful orders. 3 A similar custodian of the funds of a society or a corporation.

Treasure State Nickname of MONTANA.

treas·ure-trove (trezh′ər·trōv′) *n.* 1 *Law* Money, plate, or the like, found hidden in the earth, etc., the owner being unknown. 2 Any wealth-yielding discovery; loosely, treasure; riches. [<AF *tresor trové* <*tresor* TREASURE + *trové*, pp. of *trover* find]

treas·ur·y (trezh′ər·ē) *n. pl.* **·ur·ies** 1 The place of receipt and disbursement of public revenue, or of funds belonging to a corporation. 2 A repository, especially of words, as a dictionary or thesaurus. —**Department of the Treasury** An executive department of the U. S. government (established in 1789), headed by the Secretary of the Treasury, which superintends and manages the national finances, controls the coinage and printing of money, and supervises the Coast Guard (except when it is a part of the Navy, in wartime or when the president directs), the Bureau of Narcotics, and the Secret Service. Also **Treasury Department.** [<OF *tresorie* <*tresor* TREASURE]

Treasury note A demand note on which the face value is printed, issued by the Treasury: a legal tender for all debts, public and private, unless otherwise expressly stipulated.

treat (trēt) *v.t.* 1 To conduct oneself toward in a specified manner: He *treated* her shamefully. 2 To look upon or regard in a specified manner: They *treat* the matter as a joke. 3 To subject to chemical or physical action, as for altering or improving. 4 To give medical or surgical attention to. 5 To deal with in writing or speaking; handle. 6 To deal with or develop (a subject in art) in a specified manner or style. 7 To pay for the entertainment, food, or drink of. —*v.i.* 8 To handle a subject in writing or speaking: usually with *of.* 9 To carry on negotiations; negotiate. 10 To pay for another's entertainment. See synonyms under TRANSACT. —*n.* 1 Something that gives unusual pleasure. 2 Entertainment of any kind furnished gratuitously to another. 3 *Colloq.* One's turn to pay for refreshment or entertainment, especially for drinks. [<OF *tretier, traitier* <L *tractare.* See TRACTABLE.] —**treat′a·ble** *adj.* —**treat′er** *n.* —**treat′ing** *n.*

trea·tise (trē′tis) *n.* 1 An elaborate, formal, and systematic literary composition presenting a serious subject in all its parts: distinguished from an *essay* in being longer, more exhaustive, and less popular, and from a *monograph* in being less full and complete. 2 *Obs.* A story; tale. [<AF *tretiz,* OF *traitier* TREAT]

treat·ment (trēt′mənt) *n.* 1 The act, mode, or process of treating anything, as a raw material, substance, or product. 2 *Med.* The management of illness, by the use of drugs, dieting, or other means designed to bring relief or effect a cure. 3 In motion pictures and television, an expanded synopsis of a story, used in planning, writing, or marketing a play or scenario.

trea·ty (trē′tē) *n. pl.* **·ties** 1 A formal agreement or compact, duly concluded and ratified, between two or more states. 2 *Obs.* The act of negotiating for an agreement; also, the agreement so made. 3 *Obs.* An entreaty. [<AF *treté,* OF *traitié*, pp. of *traitier* TREAT]

treaty port Any of several sea and river ports, especially in China, in which foreigners were permitted to reside, purchase property, and erect business establishments: abolished by various treaties.

Treb·bia (treb′byä) A river in NW Italy, flowing 70 miles NE from the Apennines NE of Genoa to the Po river near Piacenza; on its banks Hannibal defeated the Romans, 218 B.C.

Treb·i·zond (treb′i·zond) 1 A province of NW Turkey; 1,753 square miles; an empire of Asia Minor from 1204 to 1461. 2 Its capital, a port on the Black Sea. Turkish *Trabzon.*

tre·ble (treb′əl) *v.t. & v.i.* **·led, ·ling** To multiply by three; triple. —*adj.* 1 Threefold; triple. 2 Soprano. —*n.* 1 *Music* The soprano; the highest register of the compass of an instrument; a soprano singer. 2 High, piping sound. 3 A musical instrument of treble pitch; a violin. [<OF <L *triplus.* Doublet of TRIPLE.] —**treb′le·ness** *n.* —**treb′ly** *adv.*

treb·u·chet (treb′yōō·shet) *n.* A medieval catapultlike device for throwing heavy missiles. The missile, on the long arm of a lever, was hurled with great force by the sudden descent of a heavy weight on the short arm. Also **treb′uck·et** (-uk·it). [<OF <*trebucher* trip, fall]

tre·cen·to (trā·chen′tō) *n. Italian* The 14th century, as producing a particular style of Italian literature and art (in literature, Petrarchism); the early Italian style.

tred·dle (tred′l) See TREADLE.

tree (trē) *n.* 1 A perennial woody plant having usually a single self-supporting trunk, with branches and foliage growing at some distance above the ground, the whole ranging from about ten feet to as high as 300 feet. ◆ Collateral adjective: *arboreal.* 2 Any shrub or plant that assumes treelike shape or dimensions. 3 Something whose outline resembles that of a tree: a genealogical *tree*; a branching diagram; a treelike group of crystals. 4 A timber or heavy piece of wood, as in a framing: usually in composition: *axletree,* boot-*tree,* etc. 5 A gibbet; also, a cross. —**up a tree** *Colloq.* In a position from which there is no retreat; cornered; caught; also, in an embarrassing position. —*v.t.* **treed, tree·ing** 1 To force to climb or take refuge in a tree: to *tree* an opossum. 2 *Colloq.* To get the advantage of; corner. 3 To stretch, as a boot, on a boot-tree. [OE *trēow, triow, trēo*]

Tree may appear as a combining form in hyphemes or solidemes, or as the first element in two-word phrases; as in:

tree-bordered	**tree-clad**	**tree-covered**
tree-boring	**tree-climbing**	**tree-crowned**

add,āce,câre,pälm; end, ēven; it,īce; odd,ōpen,ôrder; to͝ok,po͞ol; up,bûrn; ə = a in *above,* e in *sicken,* i in *clarity,* o in *melon,* u in *focus*; yo͞o = u in *fuse*; oi,oil; ou,pout; ch,check; g,go; ng,ring; th,thin; th,this; zh,vision. Foreign sounds à,œ,ü,kh,ṅ; and ◆: see page xx. < from; + plus; ? possibly.

tree–dotted	**tree–hopping**	**tree protector**
tree–dwelling	**tree–inhabiting**	**tree–pruning**
tree–feeding	**tree insulator**	**tree–ripened**
tree–fringed	**tree–lined**	**tree–sawing**
tree–garnished	**tree–locked**	**tree–shaded**
tree–girt	**tree–loving**	**tree–skirted**
tree guard	**tree–marked**	**tree–spraying**
tree–haunting	**tree–planted**	**tree tag**
tree–hewing	**tree planter**	**treetop**
tree holder	**tree–planting**	**tree–trimmer**

Tree (trē), **Sir Herbert Beerbohm,** 1853–1917, English actor and impresario.

tree fern Any of various ferns (families *Cyatheaceae* and *Dicksoniaceae*) with large fronds and woody trunks that often attain a treelike size.

tree frog An arboreal amphibian (family *Hylidae*), having the toes dilated with viscous, adhesive disks. Also **tree toad.**

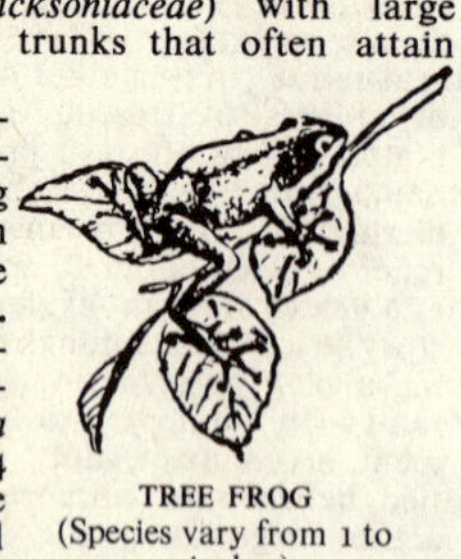

TREE FROG
(Species vary from 1 to 5 inches)

tree heath An evergreen shrub of southern Europe (*Erica arborea*) about 4 feet high, with white flowers: also called *brier.*

tree·nail (trē′nāl′, tren′əl, trun′əl) *n.* A wooden peg or nail of dry, hard wood which swells when wet, used for fastening timbers, especially in shipbuilding: also *trenail, trunnel.*

tree of heaven A large ornamental tree (*Ailanthus altissima*) of eastern Asia. It has large green flowers, those on the male or staminate trees being very ill–scented; ailanthus.

tree of knowledge of good and evil In the Bible, a tree in Eden whose fruit Adam and Eve were forbidden to eat. *Gen.* iii 3, 6. Also **tree of knowledge.**

tree of life **1** Arborvitae. **2** In the Bible: **a** A tree in the garden of Eden whose fruit conferred immortality. *Gen.* iii 22. **b** A similar tree in heaven. *Rev.* xxii 2. **3** Any source of vitality or life.

Tree–Plant·er State (trē′plan′tər, -plän′-) Nickname of NEBRASKA.

tree ring A growth ring.

tree sparrow A North American sparrow (*Spizella arborea*) which nests in Canada and migrates southward in winter: also called *Canada sparrow.*

tree surgeon One skilled in the science of tree surgery.

tree surgery The treatment of disease conditions and decay in trees by operative methods.

tref (trāf) *adj.* Unclean. See KOSHER. Also **tre·fa** (trā′fə). [<Yiddish *trēf* impure, forbidden <Hebrew *ṭerēphāh* an animal torn by wild beasts, lit., that which is torn < *ṭāraf* tear]

tre·foil (trē′foil) *n.* **1** Any one of the clovers (genus *Trifolium*), so called from the trifoliolate leaves. **2** Certain other plants with trifoliolate leaves, as the black medic. **3** A three–lobed architectural ornamentation. [< AF *trifoil,* OF *trefeuil* <L *trifolium*]

tre·ha·la (tri·hä′lə) *n. Biochem.* A carbohydrate substance forming the pupal case of certain weevils (genus *Larixus*) and deposited upon Asian plants of the genus *Echinops.* [<NL <Turkish *tīqālah*]

tre·ha·lose (trē′hə·lōs) *n. Biochem.* A crystalline disaccharide, $C_{12}H_{22}O_{11}$, elaborated by many fungi and stored as a food reserve instead of starch. [TREHAL(A) + -OSE[2]]

treil·lage (trā′lij) *n.* A trellis. [<MF < *treille* a bower, trellis, arbor <L *trichila, tricla*]

Treitsch·ke (trīch′kə), **Heinrich von,** 1834–96, German historian and political writer.

trek (trek) *v.* **trekked, trek·king** *v.i.* **1** In South Africa, to travel by ox wagon. **2** To travel; migrate. — *v.t.* **3** In South Africa, to draw (a vehicle or load): said of an ox. — *n.* **1** An organized migration, as for the founding of a colony. **2** A journey; also, a stage in a journey. **3** The act of pulling. — **Great Trek** The migration of the Boers of Cape Colony across the Vaal, Orange, and Drakenburg rivers (1835–38) which led to the formation of the Orange Free State and the South African Republic. Also spelled *treck.* [<Du. *trekken* draw, travel <MDu. *trecken,* intensive of *trēken* <OHG *trechan* draw] — **trek′ker** *n.*

trel·lis (trel′is) *n.* **1** A crossbarred grating or lattice, used as a screen or a support for vines, etc. **2** A summerhouse or other structure of trelliswork. — *v.t.* **1** To interlace so as to form a trellis. **2** To furnish with or fasten on a trellis. [<OF *treliz, trelis* <L *trilix, trilicis* of three threads < *tri-* three + *licium* a thread]

trel·lis·work (trel′is·wûrk′) *n.* Latticework.

trem·a·tode (trem′ə·tōd) *n.* One of a class (*Trematoda*) of typically parasitic flatworms, including the liver flukes. [<NL <Gk. *trēmatōdēs* perforated < *trēma, -atos* a hole + *eidos* form] — **trem′a·toid** (-toid) *adj.*

trem·ble (trem′bəl) *v.i.* **·bled, ·bling** **1** To shake involuntarily, as with fear or weakness; be agitated. **2** To have slight, irregular vibratory motion, as from some jarring force; quiver; shake. **3** To feel anxiety or fear. **4** To quaver, as the voice. See synonyms under QUAKE, SHAKE. — *n.* **1** The act or state of trembling. **2** *pl.* A debilitating disease of cattle and sheep, possibly caused by eating certain plants, and communicated to man as the milk sickness. [<OF *trembler* <LL *tremulare* < *tremulus* tremulous < *tremere* tremble, shake] — **trem′bler** *n.* — **trem′bling** *adj.* & *n.* — **trem′bling·ly** *adv.* — **trem′bly** *adj.*

tre·men·dous (tri·men′dəs) *adj.* **1** Causing or fitted to cause astonishment by its magnitude, force, etc.: a *tremendous* blow; awe–inspiring; terrible. **2** *Colloq.* Extraordinarily big; remarkable. See synonyms under FORMIDABLE. [<L *tremendus* to be trembled at < *tremere* tremble] — **tre·men′dous·ly** *adv.* — **tre·men′dous·ness** *n.*

trem·e·tol (trem′ə·tōl, -tol) *n.* An oily, poisonous alcohol isolated from certain plants, as the white snakeroot, and believed to be the cause of trembles in sheep. [<L *tremere* tremble + -OL[1]]

trem·o·lite (trem′ə·līt) *n.* A light–colored calcium–magnesium amphibole, $CaMg_3Si_4O_{12}$. [from *Tremola,* Switzerland, where it was first found + -ITE[1]]

trem·o·lo (trem′ə·lō) *n. pl.* **·los** *Music* **1** A vibrating, beating, or throbbing sound produced vocally or instrumentally. **2** The mechanism for causing this effect in organ tones. [<Ital., trembling <L *tremulus.* See TREMBLE.]

trem·or (trem′ər, trē′mər) *n.* **1** A quick, vibratory movement caused by an external impulse; a shaking; also, a succession of such movements. **2** Any involuntary quivering or trembling of the body or limbs; a shiver. **3** *Pathol.* An involuntary and continued quivering or shaking of the whole or some part of the body: a form of paralysis. **4** Any trembling, quivering effect. See synonyms under FEAR. [<OF, fear, a trembling <L < *tremere* tremble]

trem·u·lant (trem′yə·lənt) *adj.* Trembling; tremulous: also **trem′u·lent.** — *n.* A tremolo. [<LL *tremulans, -antis,* ppr. of *tremulare* TREMBLE]

trem·u·lous (trem′yə·ləs) *adj.* **1** Characterized or affected by trembling: *tremulous* speech. **2** Showing timidity and irresolution. **3** Characterized by mental excitement. [<L *tremulus.* See TREMBLE.] — **trem′u·lous·ly** *adv.* — **trem′u·lous·ness** *n.*

tre·nail (trē′nāl, tren′əl, trun′əl) See TREENAIL.

trench (trench) *n.* **1** A long narrow excavation in the ground; ditch. **2** A long irregular ditch, lined with a parapet of the excavated earth, to protect troops: often with a descriptive word: *communication, reserve, shelter,* or *supply trench.* — *v.t.* **1** To dig a trench or trenches in. **2** *Mil.* To fortify with trenches; construct trenches against. **3** To cut deep furrows in; ditch. **4** To confine in a trench, as water; entrench. — *v.i.* **5** To cut or dig trenches. **6** To cut; carve. **7** To encroach. [<OF *trenche* a cutting, gash < *trenchier* cut, ult. <L *truncare* lop off < *truncus* a tree trunk]

Trench (trench), **Richard Chenevix,** 1807–86, English prelate, poet, and philologist.

trench·ant (tren′chənt) *adj.* **1** Cutting deeply and quickly; sharp: a *trenchant* sword. **2** Figuratively, clear, vigorous, and effective; cutting, as sarcasm. [<OF, ppr. of *trenchier.* See TRENCH.] — **trench′an·cy** *n.* — **trench′ant·ly** *adv.*

trench coat A loose–fitting overcoat of rainproof fabric with removable lining, several pockets, and a belt.

trench·er[1] (tren′chər) *n.* **1** A wooden plate formerly used at table; originally, a square piece of board used to cut food on. **2** *Archaic* The food served on trenchers; hence, the table or its pleasures. **3** *Obs.* A thick slice of bread used as a platter. [<AF *trenchour,* OF *tranchouoir* < *trenchier.* See TRENCH.]

trench·er[2] (tren′chər) *n.* **1** One who digs trenches. **2** One who carves. [<TRENCH, *v.* + -ER[1]]

trench·er·man (tren′chər·mən) *n. pl.* **·men** (-mən) **1** A feeder; eater; especially, one who enjoys food. **2** A table companion: also **trench′er·mate′** (-māt′). **3** A hanger–on; parasite.

trench fever *Pathol.* A remittent rickettsial fever transmitted by body lice and characterized by headache, nausea, high temperature, profuse sweating, muscular pains, and neuralgic pains in the legs. It attacked soldiers assigned to prolonged service in trenches during World War I.

trench foot *Pathol.* A disease of the feet caused by continued dampness and cold, and characterized by discoloration, weakness, and sometimes gangrene.

trench knife A double–edged steel knife with a long blade, used in hand–to–hand combat.

trench mortar Any of various portable, muzzleloading mortars designed for firing a projectile at a high trajectory. Also **trench gun.**

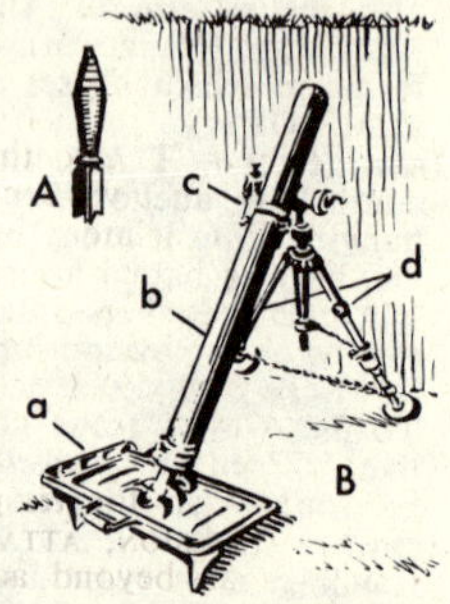

TRENCH MORTAR
A. Shell. *B.* 8 mm. mortar. *a.* Base plate. *b.* Tube. *c.* Sight. *d.* Bipod.

trench mouth *Pathol.* A mildly contagious disease of the mouth, gums, and sometimes the larynx and tonsils, caused by a soil bacillus; Vincent's angina.

trend (trend) *v.i.* To have or take a general course or direction; incline. — *n.* General course or direction; bent. [OE *trendan* roll]

Treng·ga·nu (treng·gä′nōō) A state on the South China Sea in the Federation of Malaya; 5,050 square miles; capital, Kuala Trengganu.

Trent (trent) A city in northern Italy on the Adige: ancient *Tridentum. Italian* **Tren·to** (tren′tō).

Trent (trent) The third longest river of England, flowing 170 miles SE and NE from NW Staffordshire, to a confluence with the Ouse, forming the Humber.

Trent, Council of A council of the Roman Catholic Church, held at intervals in Trent, Italy, from 1545 to 1563: it condemned the leading doctrines of the Reformation.

trente–et–qua·rante (träṅ·tā·kä·räṅt′) *n.* A gambling game played with cards laid out in two rows on a table, the top row representing "black" and the lower, "red." All court cards count as 10, and the dealing of each row is discontinued when the pips total between 31 and 40, inclusive. The players, who play against the bank, win if the row of cards appropriated to the color they have chosen totals, in pips, nearer 31 than the other. Compare ROUGE ET NOIR. [<F, thirty and forty]

Tren·ti·no (tren·tē′nō) A district around Trent, Italy: that part of the southern Tirol under Italian control.

Tren·ton (tren′tən) The capital of New Jersey, on the Delaware River at the west central border of the State.

Trent River (trent) A river in SE Ontario, Canada, flowing 150 miles south and east to Lake Ontario.

tre·pan[1] (tri·pan′) *n.* **1** An early form of the trephine. **2** A large rock–boring tool. — *v.t.* **·panned, ·pan·ning** **1** *Mech.* To use a trepan upon. **2** *Surg.* To subject to the operation of trephining. **3** To cut a hole partly through, as the back of a brush, for the insertion of bristles. Also **trep·a·nize** (trep′ə·nīz). [<OF, a borer <Med. L *trepanum* a crown saw <Gk. *trypanon* a borer < *trypaein*

bore] —**trep·a·na·tion** (trep′ə·nā′shən) *n.* —**tre·pan′ner** *n.*
tre·pan[2] (tri·pan′) *Obs. & Archaic v.t.* To ensnare. —*n.* A snare; trick; also, a trickster. Also spelled *trapan.* [< thieves' cant *trapan* <TRAP; prob. infl. in form by *trepan*[1]]
tre·pang (tri·pang′) *n.* An East Indian holothurian or sea cucumber, especially *Holothuria marmorata* or a related species: a Chinese delicacy esteemed for soups: also called *bêche-de-mer.* [<Malay *trīpang*]
tre·phine (tri·fīn′, -fēn′) *n. Surg.* A cylindrical saw for removing a piece of bone from the skull, to relieve pressure, etc. —*v.t.* **·phined, ·phin·ing** To operate on with a trephine. [Earlier *trafine* <L *tres fines* three ends; infl. in form by *trepan*[1]]
trep·i·da·tion (trep′ə·dā′shən) *n.* 1 A state of agitation from fear. 2 An involuntary trembling. 3 *Obs.* Confused haste. 4 *Obs.* A vibrating or vibration, as of leaves. Also **tre·pid·i·ty** (tri·pid′ə·tē). See synonyms under FEAR. [<L *trepidatio, -onis* < *trepidatus,* pp. of *trepidare* hurry, be alarmed < *trepidus* alarmed]
trep·o·neme (trep′ə·nēm) *n.* Any of a genus (*Treponema*) of corkscrew-shaped bacteria parasitic in the blood and tissues of animals. *T. pallidum* is the morbific agent of syphilis. [<NL <Gk. *trepein* turn + *nēma* thread] —**trep′o·nem′a·tous** (-nem′ə·təs) *adj.*
tres·pass (tres′pəs, -pas′) *v.i.* 1 *Law* To violate wilfully and forcibly the personal or property rights of another; commit a trespass: with *on* or *upon.* 2 To pass the bounds of propriety or rectitude, to the injury of another; intrude offensively; encroach: with *on* or *upon.* 3 To violate a positive law, rule, or custom: with *against.* —*n.* 1 Any voluntary transgression of law or rule of duty; any offense done to another. 2 *Law* Any wrongful act accompanied with force, either actual or implied, as wrongful entry on another's land, whereby another is injuriously treated; also, an action for trespass. See synonyms under AGGRESSION, ATTACK, OFFENSE. [<OF *trespasser* pass beyond, across <Med. L *transpassare* <L *trans-* across, beyond + *passare* PASS] —**tres′pass·er** *n.*
tress (tres) *n.* 1 A lock, curl, or ringlet of human hair, especially when abundant: applied also, figuratively, to adornment suggesting tresses. 2 *pl.* The hair of a woman or girl, especially when worn loose. [<OF *tresce* <LL *tricia;* ult. origin uncertain] —**tress′y** *adj.*
-tress *suffix* Used in feminine nouns corresponding to masculine nouns in *-ter, -tor:* a contracted form of *-teress, -toress.* Compare -ER[1], -ESS.
tressed (trest) *adj.* Wearing or arranged in tresses; braided; also, curled.
tres·sure (tresh′ər) *n. Her.* A bearing around the edge of a shield; modified or double orle, generally ornamented with fleurs-de-lis. Also **tres′sour.** [<OF *tresseor, tressure* < *tresce* a tress] —**tres′sured** *adj.*
tres·tle (tres′əl) *n.* 1 A beam or bar supported by four divergent legs, for bearing platforms, etc. 2 An open braced framework for supporting the horizontal stringers of a railway bridge, etc. 3 In carpentry, an intervening stud. 4 A trestletree. 5 *pl.* The props of a vessel on the ways. [<OF *trestel* <L *transtrum.* See TRANSOM.]
tres·tle·tree (tres′əl·trē′) *n. Naut.* One of a pair of pieces at right angles to a lower mast, to support the crosstrees, etc.
tres·tle·work (tres′əl·wûrk′) *n.* 1 Trestles collectively. 2 A bridge made of trestles or braced framework, especially of wood. Also **tres′tling.**
tret (tret) *n.* A former allowance to purchasers for waste due to transportation. [<AF, OF *tret,* var. of *traict.* See TRAIT.]
Tre·vel·yan (tri·vel′yən), **George Macaulay,** born 1876, English historiân; son of the following. —**Sir George Otto,** 1838–1928, English historian and statesman.
Treves (trēvz) The English name for TRIER. *French* **Trèves** (trev).
trev·et (trev′it) See TRIVET.
trev·is (trev′is) *n.* 1 A bar or beam. 2 A crosspiece; partition. [Var. of TRAVERSE]
Tre·vi·so (trā·vē′zō) A city 16 miles NW of Venice in NE Italy: ancient *Tarvisium.*
trews (trōōz) *n. Scot.* Trousers, especially those worn under the kilt. Also spelled *trooz.*
trey (trā) *n.* A card, domino, or die having three spots or pips. ◆ Homophone: *tray.* [<OF *trei, treis* <L *tres* three]
trez tine (trez) The royal tine. Also **tres tine, trey tine.** [Prob. <L *tres* three + TINE[1]]
tri- *prefix* 1 Three; threefold; thrice: *tricycle, trisect.* 2 *Chem.* Containing three (specified) atoms, radicals, groups, etc.: *trioxide, trisulfide.* 3 Occurring every three (specified) intervals, or three times within a (specified) interval: *triweekly.* [<L *tri-* threefold < *tres* three]
tri·a·ble (trī′ə·bəl) *adj.* 1 That may be tried or tested. 2 *Law* That may undergo a judicial examination or determination. —**tri′a·ble·ness** *n.*
tri·ac·id (trī·as′id) *n. Chem.* An acid containing three hydroxyl radicals which are replaceable by acid radicals.
tri·ad (trī′ad) *n.* 1 A group of three persons or things. 2 *Music* A chord of three tones or notes; often the common chord, consisting of a fundamental tone with its third and fifth higher. A **major triad** has a major third and a perfect fifth; a **minor triad** has a minor third and a perfect fifth. 3 *Chem.* **a** A trivalent atom or radical. **b** One of a group of three elements having similar chemical properties, as chlorine, bromine, and iodine. [<L *trias, -adis* <Gk. *trias, -ados* < *treis* three] —**tri·ad′ic** *adj. & n.*
tri·age (trī′ij, trē·äzh′) *n. Med.* The sorting out of a group of sick and wounded persons and classifying them according to a system of priorities for the treatment of mass casualties under conditions of limited medical resources and personnel. [<OF < *trier* pick out, sort]
tri·ag·o·nal (trī·ag′ə·nəl) *adj.* Having three angles; triangular. [Var. of TRIGONAL, on analogy with *tetragonal, pentagonal,* etc.]
tri·al (trī′əl, trīl) *n.* 1 The act of testing or proving by experience or use. 2 The state of being tried or tested by suffering: the hour of *trial.* 3 Experimental treatment or action performed to determine a result: to learn by *trial* and error. 4 An experience, person, or thing that puts strength, patience, or faith to the test. 5 An attempt or effort to do something; a try: to make a *trial.* 6 The examination, before a tribunal having assigned jurisdiction, of the facts or law involved in an issue in order to determine that issue. 7 A former method of determining guilt or innocence by subjecting the accused to physical tests of endurance, as by ordeal or by combat with his accuser. 8 *Brit.* An academic or licensing examination. See synonyms under ENDEAVOR, MISFORTUNE, PROOF. —**on trial** In the process of being tried or tested. —*adj.* 1 Of or pertaining to a trial or trials. 2 Made or performed in the course of trying or testing: a *trial* trip. 3 Used in testing: a *trial* specimen. [<AF < *trier* TRY]
trial balance In double-entry bookkeeping, a draft or statement of the debit and credit footings or balances of each account in the ledger.
trial balloon 1 A balloon released in order to test atmospheric and meteorological conditions, as wind velocities, air currents, etc. 2 Any tentative plan or scheme advanced to test public reaction.
trial jury A jury impaneled to try a civil or criminal case: also called *petit* or *petty jury.*
tri·a·morph (trī′ə·môrf) *n.* Any mineral or other substance that crystallizes in three different forms. [<L *tres, tria* three + Gk. *morphē* form] —**tri′a·mor′phous** *adj.*
tri·an·gle (trī′ang′gəl) *n.* 1 *Geom.* A figure,

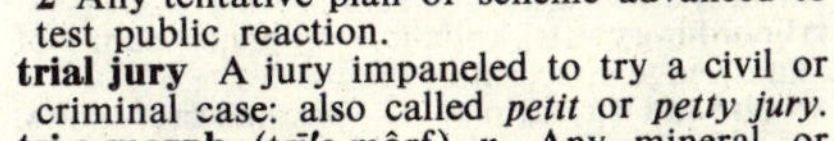

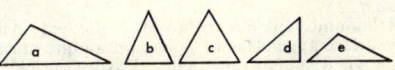

TRIANGLE
a. Scalene. *b.* Isosceles. *c.* Equilateral. *d.* Right-angled. *e.* Obtuse.

especially a plane figure, bounded by three sides, and having three angles. 2 Something resembling such a figure in shape or arrangement. 3 A flat drawing implement for making parallel or diagonal lines, etc. 4 A group or set of three; a triad. 5 A situation involving three persons: the eternal *triangle.* 6 *Music* A musical instrument of percussion, consisting of a resonant bar bent into a triangle and open at one corner, sounded by being struck with a small metal rod. [<OF <L *triangulum* < *triangulus* three-cornered < *tri-* three + *angulus* an angle]
tri·an·gu·lar (trī·ang′gyə·lər) *adj.* 1 Pertaining to, like, or bounded by a triangle: also **tri′an·gled.** 2 Concerned with or pertaining to three things, parties, or persons. [<LL *triangularis* <L *triangulum* a triangle] —**tri·an′gu·lar′i·ty** (-lar′ə·tē) *n.* —**tri·an′gu·lar·ly** *adv.*
triangular number See under NUMBER.
tri·an·gu·late (trī·ang′gyə·lāt) *v.t.* **·lat·ed, ·lat·ing** 1 To divide into triangles. 2 To survey by triangulation. 3 To give triangular shape to. —*adj.* Marked with triangles. [<L *triangulum* a triangle + -ATE[1]]
tri·an·gu·la·tion (trī·ang′gyə·lā′shən) *n.* The laying out and accurate measurement of a network of triangles, especially on the surface of the earth, as in surveying.
Tri·an·gu·lum (trī·ang′gyə·ləm) *Astron.* A zodiacal constellation. See CONSTELLATION. [<L, a triangle]
tri·ap·si·dal (trī·ap′sə·dəl) *adj. Archit.* Distinguished by or constructed with three apses. [<TRI- + L *apsis, -idis* an apse + -AL]
tri·ar·chy (trī′är·kē) *n. pl.* **·chies** Government by three persons, or a country so governed; a triumvirate. [<Gk. *triarchia* < *tri-* three + *archein* rule]
Tri·as·sic (trī·as′ik) *adj. Geol.* Of or pertaining to the lowest of the three geological periods comprised in the Mesozoic era. —*n.* The Triassic period or rock system, following the Permian and succeeded by the Jurassic. Also **Tri·as** (trī′əs). [<LL *trias.* See TRIAD.]
tri·at·ic stay (trī·at′ik) *Naut.* A device consisting of two pendants connected by a span, and attached respectively to the foremast head and mainmast head of a ship: used principally for hoisting boats in and out of a vessel.
tri·a·tom·ic (trī′ə·tom′ik) *adj. Chem.* 1 Containing only three atoms in the molecule. 2 Containing three replaceable univalent atoms. 3 Trivalent.
tri·ax·i·al (trī·ak′sē·əl) *adj.* Having three axes.
tri·a·zine (trī′ə·zēn, -zin, trī·az′ēn, -in) *n. Chem.* 1 One of three heterocyclic compounds, $C_3H_3N_3$, each having three carbon and three nitrogen atoms in the ring. 2 Any of their derived compounds. Also **tri·a·zin** (trī′ə·zin, trī·az′in). [<TRI- + AZ(O)- + -INE[2]]
tri·a·zo·ic (trī′ə·zō′ik) *adj. Chem.* Hydrazoic. [<TRI- + -AZ(O)- + -IC]
tri·a·zole (trī′ə·zōl, trī·az′ōl) *n.* One of four five-membered ring compounds, $C_2H_3N_3$, in which nitrogen atoms have replaced two CH groups. [<TRI- + AZ(O)- + OLE[1]]
trib·ade (trib′əd) *n.* A female homosexual, especially one who assumes the role of the male; a Lesbian. [<MF <L *tribas, -adis* <Gk. *tribas, -ados* < *tribein* rub]
trib·a·dism (trib′ə·diz′əm) *n.* Homosexual practices between females; Lesbianism.
tri·bal·ism (trī′bəl·iz′əm) *n.* Tribal organization, culture, or relations.
tri·ba·sic (trī·bā′sik) *adj. Chem.* 1 Containing three atoms of hydrogen replaceable by a base or basic radical: said of certain acids. 2 Having three hydroxyl groups in the molecule.
tribe (trīb) *n.* 1 A division, class, or group of people, varying ethnologically according to the circumstances from which their separation or distinction is supposed to originate. 2 Among primitive peoples, a group or aggregation of persons, usually consanguineous and endogamous, under one chief, characterized by its own culture, and having a name, a dialect, a government, and usually a territory of its own: Kaffir *tribes.* 3 In ancient states, an ethnic, hereditary, or political division of a united people: the *tribes* of Athens or of Israel. 4 A division of freeholders with a right to vote in certain of the ancient Roman councils. The Latins, Sabines, and Etruscans

probably represented primitive clan divisions, to which Servius Tullius added a fourth when making his territorial division of Rome. Outside the city the spread of tribal organizations was coincident with the founding of new colonies. 5 A number of persons of any class or profession taken together: often derogatory or contemptuous: the theatrical *tribe.* 6 *Biol.* A group of plants or animals of indefinite rank. 7 Among stockbreeders, the descendants of a particular female bearer through females. See synonyms under PEOPLE. [Fusion of OF *tribu* (<L *tribus* a tribe) and L *tribus*] — **tri'bal** *adj.* — **tri'bal·ly** *adv.*

tribes·man (trībz'mən) *n. pl.* **·men** (-mən) A member of a tribe.

trib·o·e·lec·tric (trib'ō·i·lek'trik) *adj.* Of, pertaining to, or characterized by frictional electricity, as when a glass rod is rubbed with flannel. [< *tribo-* (<Gk. *tribein* rub) + ELECTRIC] — **trib'o·e·lec·tric'i·ty** (-i·lek'tris'ə·tē) *n.*

triboelectric series A grouping of substances in such order that each one may be positively electrified by rubbing with those below it in the series.

trib·o·lu·mi·nes·cence (trib'ō·lōō'mə·nes'əns) *n.* Luminescence produced by crushing or grinding certain substances, as glass. [< *tribo-* (<Gk. *tribein* rub) + LUMINESCENCE] — **trib'o·lu'mi·nes'cent** *adj.*

tri·brach (trī'brak, trib'rak) *n.* In ancient prosody, a foot composed of three short syllables, two of which belong to the thesis and one to the arsis. [<L *tribrachys* <Gk. < *tri-* three + *brachys* short]

tri·brom·eth·a·nol (trī'brom·eth'ə·nōl, -nol) *n. Chem.* A white crystalline compound, $C_2H_3Br_3O$, with an ethereal odor, used as a general anesthetic. Also **tri·bro'mo·eth'a·nol** (trī·brō'mō-).

tri·brom·phe·nol (trī'brom·fē'nōl, -nol) *n. Chem.* A colorless crystalline compound, $C_6H_3Br_3O$: used in medicine as an antiseptic.

trib·u·la·tion (trib'yə·lā'shən) *n.* A condition of affliction and distress; suffering; also, that which causes it. See synonyms under GRIEF, MISFORTUNE. [<OF *tribulacion* <LL *tribulatio, -onis* <L *tribulatus,* pp. of *tribulare* thrash < *tribulum* a threshing floor < *tri-,* root of *terere* rub, grind]

tri·bu·nal (tri·byōō'nəl, trī-) *n.* 1 A court of justice; any judicial body, as a board of arbitrators. 2 The seat set apart for judges, magistrates, etc. [<L < *tribunus* TRIBUNE]

trib·u·nate (trib'yə·nit, -nāt) *n.* The office or dignity of a tribune. Also **trib·une·ship** (trib'yōōn·ship).

trib·une[1] (trib'yōōn, *Brit.* trī'byōōn) *n.* 1 In Roman history, a magistrate chosen by the plebeians to protect them against patrician oppression. 2 One of various civil or military officers of later times; any champion of the people: as the title of a newspaper, often pronounced tri·byōōn'. [<L *tribunus,* lit., head of a tribe < *tribus* a tribe] — **trib'u·nar'y** (-yə·ner'ē), **trib'u·ni'cial** (-yə·nish'əl), **trib'u·ni'cian** *adj.*

trib·une[2] (trib'yōōn) *n.* 1 A raised floor for a Roman magistrate's chair. 2 A bishop's throne. 3 A rostrum or platform. [<MF <Ital. *tribuna* <L *tribunal* a tribunal]

trib·u·tar·y (trib'yə·ter'ē) *adj.* 1 Bringing supply; contributory; subsidiary: a *tributary* stream. 2 Offered or due as tribute; having the character of tribute: a *tributary* payment. 3 Paying tribute; hence, subordinate, as a state. — *n. pl.* **·tar·ies** 1 A person or state paying tribute; a dependent. 2 A stream flowing into another; an affluent. [<L *tributarius* < *tributum.* See TRIBUTE.] — **trib'u·tar'i·ly** *adv.* — **trib'u·tar'i·ness** *n.*

trib·ute (trib'yōōt) *n.* 1 Money or other valuables paid by one state or ruler to another as an acknowledgment of submission or as the price of peace and protection, or by virtue of some treaty; also, the taxes imposed to raise money to make such payment. 2 The obligation or necessity of making such gift or payment; the state of being tributary. 3 Anything given, paid, or rendered as by a subordinate to a superior; figuratively, that which is due to worth, affection, or duty; contribution; tax; gift; offering; meed: I must render my *tribute* of praise. See synonyms under SUBSIDY, TAX. [<L *tributum,* neut. of *tributus,* pp. of *tribuere* pay, allot]

trice (trīs) *v.t.* **triced, tric·ing** To raise with a rope; also, to tie or lash: usually with *up.* — *n.* An instant: only in the phrase *in a trice.* [< MDu. *trisen* hoist]

tri·cen·ni·al (trī·sen'ē·əl) *adj.* Of or pertaining to the number thirty; taking place every thirtieth year. [<L *tricennium* a period of thirty years < *tricies* thirty times + *annus* a year]

tri·cen·ten·ni·al (trī'sen·ten'ē·əl) *adj.* & *n.* Tercentenary.

tri·ceps (trī'seps) *n. Anat.* A muscle having three heads; specifically, the large muscle at the back of the upper arm, of which the function is to extend the forearm. [<L *triceps, -cipitis* three-headed < *tri-* three + *caput, capitis* a head]

trich- Var. of TRICHO-.

tri·chi·a·sis (tri·kī'ə·sis) *n. Pathol.* 1 A condition of ingrowing hairs about an orifice, especially ingrowing eyelashes. 2 The presence of hairlike filaments in the urine. [<LL *trichiasis* <Gk. < *trichiaein* be hairy < *thrix, trichos* hair]

tri·chi·na (tri·kī'nə) *n. pl.* **·nae** (-nē) A small nematode parasitic worm (*Trichinella spiralis*) that in its larval stage sometimes infests the muscles of man, swine, and other mammals. [<NL <Gk. *trichinos* of hair < *thrix, trichos* hair]

trich·i·nize (trik'ə·nīz) *v.t.* **·nized, ·niz·ing** To infect with trichinae. — **trich'i·ni·za'tion** *n.*

Trich·i·nop·o·ly (trich'ə·nop'ə·lē) See TIRUCHIRAPPALI.

trich·i·no·sis (trik'ə·nō'sis) *n. Pathol.* The disease produced by trichinae in the intestines and muscles of the body. Also **trich'i·ni'a·sis** (-nī'ə·sis). [<TRICHINA + -OSIS] — **trich'i·nosed, trich'i·not'ic** (-not'ik), **trich'i·nous** *adj.*

trich·ite (trik'īt) *n.* 1 *Mineral.* A microscopic crystallite, curved, bent, or zigzag in form, found in volcanic rocks. 2 *Bot.* One of the needle-shaped, radial crystals occurring in starch grains. 3 *Zool.* A rodlike organ surrounding the mouth and gullet of certain ciliate protozoa. [<G *trichit* <Gk. *thrix, trichos* hair] — **tri·chit·ic** (tri·kit'ik) *adj.*

tri·chlo·ride (trī·klôr'īd, -id, -klō'rīd, -rid) *n. Chem.* Any compound having three chlorine atoms in its molecule. Also **tri·chlo·rid** (trī·klôr'id, -klō'rid).

tri·chlo·ro·eth·yl·ene (trī·klôr'ō·eth'əl·ēn, -klō·rō-) *n. Chem.* A colorless, odorless, volatile liquid, C_2HCl_3, used in organic synthesis, in chemical manufactures, and as a general anesthetic. Also **tri'chlor·eth'yl·ene** (trī'klôr-).

tricho- *combining form* Hair; of or resembling a hair or hairs: *trichocyst:* also, before vowels, *trich-.* Also **trichi-.** [<Gk. *thrix, trichos* a hair]

trich·o·bac·te·ri·a (trik'ō·bak·tir'ē·ə) *n.* A group of bacteria which includes forms possessing flagella.

trich·o·cyst (trik'ə·sist) *n. Biol.* 1 A stinging capsule containing a protrusible hairlike body: found in various protozoans. 2 A thread cell. — **trich'o·cys'tic** *adj.*

trich·o·gyne (trik'ə·jin, -jīn) *n. Bot.* The slender threadlike portion of the procarp in red algae which receives the male fertilizing bodies. [<TRICHO- + Gk. *gynē* a woman, female]

trich·oid (trik'oid) *adj.* Having the form or appearance of hair. [<Gk. *trichoeidēs* < *thrix, trichos* a hair + *eidos* form]

tri·chol·o·gy (tri·kol'ə·jē) *n.* The sum of knowledge concerning the hair.

tri·cho·ma (tri·kō'mə) *n. pl.* **·ma·ta** (-mə·tə) 1 *Pathol.* a Entropion. b Matted and crusted hair; plica polonica. 2 *Bot.* One of the threads or filaments of filamentous algae: also spelled *trichome.* [<NL <Gk. *trichōma* growth of hair < *trichoein* cover with hair < *thrix, trichos* hair] — **tri·chom'ic** (-kom'ik) *adj.*

trich·ome (trik'ōm, trī'kōm) *n. Bot.* 1 Any surface appendage or epidermal outgrowth in a plant, comprising hairs, bristles, prickles, scales, root hairs, etc. 2 A trichoma (def. 2). [<Gk. *trichōma.* See TRICHOMA.]

Tri·chop·ter·a (tri·kop'tər·ə) *n. pl.* An order of insects including the caddis flies. The aquatic larvae of most species construct cases of sand or other material; the adults are mothlike, with two pairs of hairy wings and well-developed compound eyes. [<NL <Gk. *thrix, trichos* hair + *pteron* a wing]

tri·cho·sis (tri·kō'sis) *n. Pathol.* Any morbid condition of the hair. [<NL <Gk. *trichōsis* growth of hair < *trichoein.* See TRICHOMA.]

tri·chot·o·my (trī·kot'ə·mē) *n.* 1 Division into three parts. 2 *Logic* The threefold division of a genus or class. 3 *Theol.* The division of human nature into body, soul, and spirit. [<Gk. *tricha* threefold + -TOMY] — **trich·o·tom·ic** (trik'ə·tom'ik), **tri·chot'o·mous** *adj.* — **tri·chot'o·mous·ly** *adv.*

tri·chro·ism (trī'krō·iz'əm) *n.* The property of a crystal of transmitting light of different colors in three different directions. [<Gk. *trichroos* of three colors < *tri-* three + *chroia* color, skin] — **tri·chro·ic** (trī·krō'ik) *adj.*

tri·chro·mat·ic (trī'krō·mat'ik) *adj.* Of, pertaining to, having, or using three colors, as the normal eye, the three-color process in photography and printing, etc. Also **tri'chrome, tri·chro·mic** (trī·krō'mik). [<TRI- + CHROMATIC] — **tri·chro'ma·tism** (-mə·tiz'əm) *n.*

trick (trik) *n.* 1 A device for getting an advantage by deception; a petty artifice. 2 A malicious, injurious, or annoying act: a dirty *trick.* 3 A practical joke; prank: the *tricks* of schoolboys. 4 A particular habit or manner; characteristic; trait; also, a vicious habit. 5 A peculiar skill or knack. 6 An act of legerdemain; a feat of jugglery: conjurer's *tricks.* 7 In card games, the whole number of cards played in one round. 8 The turn of one sailor at the helm; a turn or spell of duty; a railroad or factory shift. 9 *Colloq.* A toy; trifle; plaything; a child. See synonyms under ARTIFICE, FRAUD. — **to do** (or **turn**) **the trick** *Slang* To produce the desired result. — *v.t.* 1 To deceive or cheat; delude. 2 To dress or array; adorn: with *up* or *out.* — *v.i.* 3 To practice trickery or deception. See synonyms under DECEIVE. [<AF *trique,* OF *triche* deceit < *trichier* cheat, prob. ult. <L *tricare, tricari* trifle, play tricks < *tricae* trifles, tricks] — **trick'er** *n.* — **trick'less** *adj.*

trick·er·y (trik'ər·ē) *n. pl.* **·er·ies** 1 The practice of tricks; artifice; stratagem; wiles. 2 Dressing up; decorations. See synonyms under DECEPTION.

trick·ing (trik'ing) *n.* The act of dressing up; also, ornaments. — *adj. Obs.* Given to tricks; tricky.

trick·ish (trik'ish) *adj.* Apt to be tricky; partaking of trickery. — **trick'ish·ly** *adv.* — **trick'ish·ness** *n.*

trick·le (trik'əl) *v.* **·led, ·ling** *v.i.* 1 To flow or run drop by drop or in a very thin stream. 2 To move, come, go, etc., slowly or bit by bit. — *v.t.* 3 To cause to trickle. — *n.* The act or state of trickling, or that which trickles. [ME *triklen,* prob. alter. of *striklen,* freq. of *striken* strike] — **trick'ly** *adj.*

trick·let (trik'lit) *n.* A tiny rill. [Dim. of TRICKLE]

trick·ster (trik'stər) *n.* One who plays tricks; a cheat.

trick·sy (trik'sē) *adj.* 1 Fond of tricks or pranks; mischievous; playful. 2 Given to artifice or stratagem; cunning; crafty. 3 Tending to elude or deceive; illusory. 4 Neat; trim; spruce; smartly attired. — **trick'si·ness** *n.*

trick-track (trik'trak') *n.* A form of backgammon; specifically, an old form in which pegs as well as pieces were used. Also **tric'-trac'.** [<F *trictrac* <MF, a clicking noise; imit. of the sound of the pieces during a game]

trick·y (trik'ē) *adj.* **trick·i·er, trick·i·est** 1 Disposed to or characterized by trickery; deceitful. 2 Vicious, as an animal. 3 Intricate; requiring or showing adroitness or skill in making: *tricky* clothes. See synonyms under INSIDIOUS. Also *Scot.* **trick'ie.** — **trick'i·ly** *adv.* — **trick'i·ness** *n.*

tri·clin·ic (trī·klin'ik) *adj.* Describing a crystal form having three unequal and dissimilar axes with oblique intersections. [<TRI- + Gk. *klinein* incline + -IC]

tri·clin·i·um (trī·klin'ē·əm) *n. pl.* **·i·a** (-ē·ə) 1 In Roman antiquity, a dining table of four sides, three sides of which were provided with low couches upon which guests could recline. 2 The Roman dining-room. [<L <Gk. *triklinion,* dim. of *triklinos* a dining-room with three couches < *tri-* three + *klinē* a couch]

tri·col·or (trī'kul'ər) *adj.* Having or characterized by three colors: also **tri'col'ored.** — *n.* 1 A flag of three colors; the French national flag of blue, white, and red vertical bands.

2 The tricolor cockade of the French Revolutionists. Also *Brit.* **tri'col'our.** [<F *tricolore* <LL *tricolor* <L *tri-* three + *color* color]

tri·corn (trī'kôrn) *n.* A hat with the brim turned up on three sides, worn during the 17th and 18th centuries by both men and women: used improperly in the form *tricorne* for the two-cornered hat of the French gendarmes. See BICORN. — *adj.* Three-horned; three-pronged; having three hornlike processes. [<F *tricorne* <L *tricornis* three-horned < *tri-* three + *cornu* a horn]

tri·cor·nered (trī'kôr'nərd) *adj.* Three-cornered.

tri·cos·tate (trī·kos'tāt) *adj. Biol.* Having three ribs or costae. [<TRI- + L *costa* a rib + -ATE[1]]

tri·cot (trē'kō, *Fr.* trē·kō') *n.* 1 A hand-knitted or woven fabric, or a machine-made imitation thereof. 2 A soft ribbed cloth. 3 A tight-fitting garment worn by ballet dancers. [<F, knitting < *tricoter* knit, ? ult. <LG *striken* make movements]

tri·crot·ic (trī·krot'ik) *adj. Med.* Having three distinct rhythmic waves in succession, as the pulse: also **tri·cro·tous** (trī'krə·təs). [<TRI- + Gk. *kroteein* knock, beat] — **tri·crot·ism** (trī'krə·tiz'əm) *n.*

tri·cus·pid (trī·kus'pid) *adj.* 1 Having three cusps or points, as a molar tooth or a valve of the heart. 2 Of or pertaining to the tricuspid valve. Also **tri·cus'pi·dal.** [<L *tricuspis, -idis* three-pointed < *tri-* three + *cuspis, -idis* a point]

tri·cus·pi·date (trī·kus'pə·dāt) *adj.* Three-pointed: a *tricuspidate* leaf.

tricuspid valve *Anat.* A three-segmented valve which controls the flow of blood from the right atrium to the right ventricle of the heart.

tri·cy·cle (trī'sik·əl) *n.* 1 A three-wheeled vehicle of the velocipede class. 2 A motorcycle with three wheels. [<F <*tri-* three + Gk. *kyklos* a circle]

tri·cy·clic (trī·sī'klik, -sik'lik) *adj.* Having or characterized by three cycles or identical units of structure: a *tricyclic* chemical compound.

tri·dac·tyl (trī·dak'til) *adj. Anat.* Possessing three fingers or toes. [<Gk. *tridaktylos* < *tri-* three + *daktylos* a digit]

tri·dec·ane (trī·dek'ān) *n. Chem.* A light, colorless, liquid hydrocarbon, $C_{13}H_{28}$, of the methane series, having an odor like turpentine. [<TRI- + DECANE]

tri·dent (trīd'nt) *n.* 1 A three-pronged implement or weapon, the emblem of Neptune (Poseidon); hence, dominion over the sea. 2 The three-pronged spear with which the Roman retiarius was armed. 3 A fishspear with three prongs. 4 *Geom.* A plane cubic curve somewhat resembling a three-pronged spear. — *adj.* Having three teeth or prongs: also **tri·den·tate** (trī·den'tāt), **tri·den'tat·ed.** [<L *tridens, -dentis* < *tri-* three + *dens, dentis* a tooth]

Tri·den·tine (trī·den'tin, -tīn, tri-) *adj.* 1 Pertaining to Trent or to the Council of Trent. 2 Adhering to the decrees of the Council of Trent. — *n.* A Roman Catholic: from the fact that the creed (**Tridentine Creed**) of the Roman Catholic Church as now held was formulated by the Council of Trent. [<Med. L *Tridentinus* <*Tridentum* Trent]

Tri·den·tum (trī·den'təm) The ancient name for TRENT, Italy.

tri·di·men·sion·al (trī'di·men'shən·əl) *adj.* Of three dimensions; having length, breadth, and thickness. — **tri'di·men'sion·al'i·ty** *n.*

tri·di·ur·nal (trī'dī·ûr'nəl) *adj.* Occurring every three days or lasting three days.

tri·e·cious (trī·ē'shəs) See TRIOECIOUS.

tried (trīd) *adj.* 1 Tested; trustworthy, as a friend or a formula. 2 Freed of impurities, as metal or oil. 3 Rendered, as fat.

tri·en·ni·al (trī·en'ē·əl) *adj.* 1 Taking place every third year. 2 Lasting three years. — *n.* 1 A ceremony or event observed or celebrated every three years; a third anniversary. 2 A plant lasting three years. — **tri·en'ni·al·ly** *adv.*

tri·en·ni·um (trī·en'ē·əm) *n. pl.* **·en·ni·ums** or **·en·ni·a** (-en'ē·ə) A period of three years. [<L < *tri-* three + *annus* a year]

Trier (trir) A city of Rhineland-Palatinate, West Germany, on the Moselle river in the central western part near Luxembourg: English *Treves,* French *Trèves.*

tri·er·arch (trī'ər·ärk) *n.* In Greek antiquity, the captain of a trireme; also, at Athens, one who alone or with others fitted out and maintained a trireme. [<L *trierarchus* <Gk. *triērarchos* < *triērēs* a trireme + *archein* rule]

tri·er·ar·chy (trī'ər·är'kē) *n. pl.* **·chies** 1 The command of a trireme. 2 The fitting out and maintaining of a trireme. 3 The body of trierarchs collectively. [<Gk. *triērarchia* < *triērarchos* TRIERARCH]

Tri·este (trē·est', *Ital.* trē·es'tā) An Italian port on the **Gulf of Trieste,** a NE inlet of the Gulf of Venice; formerly an Austrian port and later part of the Free Territory of Trieste.

Trieste, Free Territory of A former free territory, including the city of Trieste and adjoining portions of Istria, constituted by the Italian-Allied peace treaty of 1947; 298 square miles; in 1954 the smaller part (81 square miles, including the city of Trieste) reverted to Italy, and the larger (217 square miles) passed to Yugoslavia.

tri·e·ter·ic (trī'ə·ter'ik) *adj.* Happening every other year. [<L *trietericus* <Gk. *trietērikos* < *trietēris* a festival celebrated every other year < *tri-* three + *etos* a year]

tri·fa·cial (trī·fā'shəl) *adj.* Trigeminal (def. 2).

tri·fid (trī'fid) *adj.* Divided into three parts or sections; three-cleft. [<L *trifidus* < *tri-* three + *fid-,* stem of *findere* split]

tri·fle (trī'fəl) *v.* **·fled, ·fling** *v.i.* 1 To treat something as of no value or importance; dally: with *with.* 2 To act or speak frivolously or idly; jest. 3 To play; toy. 4 To pass time idly; idle. — *v.t.* 5 To pass (time) in an idle and purposeless way. — *n.* 1 Anything of very little value or importance. 2 A light confection, usually made of alternate layers of macaroons or ladyfingers with sugared fruit, covered with a custard and topped with meringue or whipped cream. 3 A variety of pewter. — **a trifle** Slightly; to a small extent: a *trifle* short. [<OF *truffler,* var. of *truffer* deceive, jeer at < *trufle,* dim. of *trufe* a cheating, mockery; ult. origin unknown] — **tri'·fler** *n.*

tri·fling (trī'fling) *adj.* 1 Frivolous. 2 Insignificant. See synonyms under CHILDISH, IDLE, INSIGNIFICANT, LITTLE, RIDICULOUS, VAIN. — **tri'fling·ly** *adv.*

tri·fo·cal (trī·fō'kəl) *adj.* 1 Having three foci. 2 *Optics* Pertaining to or describing a lens ground in three segments, for near, intermediate, and far vision respectively.

tri·fold (trī'fōld) *adj.* Triple.

tri·fo·li·ate (trī·fō'lē·it, -āt) *adj. Bot.* Having three leaves or leaflike processes. Also **tri·fo'·li·at'ed.** [<TRI- + FOLIATE]

tri·fo·li·o·late (trī·fō'lē·ə·lāt') *adj. Bot.* Having three leaflets.

tri·fo·li·um (trī·fō'lē·əm) *n.* Any of a genus (*Trifolium*) of small plants of the bean family, the clovers, with trifoliolate leaves, and purple, red, white, or yellow flowers. [<NL <L < *tri-* three + *folium* a leaf]

tri·fo·ri·um (trī·fôr'·ē·əm, -fō'rē-) *n. pl.* **·fo·ri·a** (-fôr'ē·ə, -fō'·rē·ə) *Archit.* A gallery above the arches of the nave in a church. [<Med. L <L *tri-* three + *foris* a door] — **tri·fo'ri·al** *adj.*

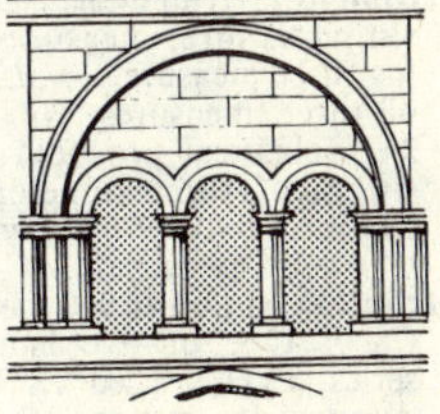
TRIFORIUM

tri·formed (trī'·fôrmd') *adj.* 1 Having three forms or shapes. 2 Consisting of three parts or divisions. Also **tri'·form.** — **tri·form'i·ty** *n.*

tri·fur·cate (trī·fûr'kāt) *adj.* Three-forked; trichotomous. Also **tri·fur'cat·ed.** — **tri·fur·ca·tion** (trī'fər·kā'shən) *n.*

trig[1] (trig) *adj.* 1 Characterized by tidiness; trim; neat. 2 Strong; sound; firm. 3 In a depreciative sense, correct; precise; prim. 4 Faithful; trustworthy; dependable. 5 Active; alert. 6 Full; inflated. — *v.t.* **trigged, trig·ging** To make trig or neat; dress finely or smartly: often with *out* or *up.* [<ON *tryggr* true, trusty] — **trig'ly** *adv.* — **trig'ness** *n.*

trig[2] (trig) *v.t.* **trigged, trig·ging** 1 To check, as with a skid; obstruct; stop. 2 To shore; prop. — *n.* A check or brake, as a skid or drag for a wheel. [? <ON *tryggja* make firm]

tri·gem·i·nal (trī·jem'ə·nəl) *adj.* 1 Being in three parts; threefold; triple. 2 Of or pertaining to the trigeminus: *trigeminal* neuralgia. — *n.* The trigeminus. [<L *trigeminus* born three at a time < *tri-* three + *geminus* a twin]

tri·gem·i·nus (trī·jem'ə·nəs) *n. pl.* **·ni** (-nī) *Anat.* The fifth cranial or trifacial nerve, the great nerve of sensation for the face and head. [<NL <L. See TRIGEMINAL.]

trig·ger[1] (trig'ər) *n.* 1 The fingerpiece of a gunlock or pistol-lock, for releasing the hammer. 2 A catch or small lever doing similar service in a trap or other mechanism. — **quick on the trigger** 1 Quick to shoot. 2 Quick to act in response to a suggestion; quick-witted; alert. — *v.t.* To cause or precipitate. [Earlier *tricker* <Du. *trekker* < *trekken* pull, tug at]

trig·ger[2] (trig'ər) *n. Dial.* A skid or trig. [< TRIG[2] + -ER]

trig·ger·fish (trig'ər·fish') *n. pl.* **·fish** or **·fish·es** A plectognath fish (genus *Balistes*) found mainly in the tropical Pacific region, with an ovate body covered with large, rough scales: named from the triggerlike second spine of the dorsal fin.

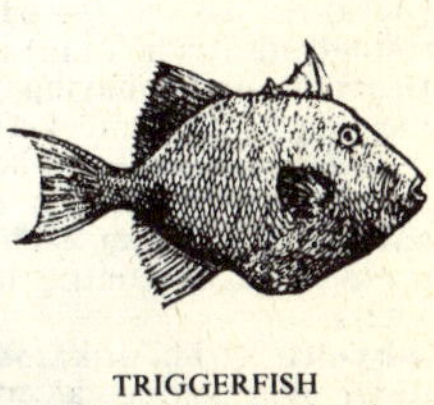
TRIGGERFISH
(About 12 inches long)

tri·glyph (trī'glif) *n. Archit.* An ornament in a Doric frieze consisting of a tablet with three parallel vertical channels or glyphs, and standing on each side of the metopes. [<L *triglyphus* <Gk. *triglyphos* thrice grooved < *tri-* three + *glyphē* a carving < *glyphein* carve, engrave]

tri·glyph·ic (trī·glif'ik) *adj.* 1 Pertaining to or consisting of triglyphs. 2 Having three groups of characters or carvings. Also **tri'·glyph·al, tri·glyph'i·cal.**

tri·go (trē'gō) *n. Spanish* Wheat.

tri·gon (trī'gon) *n.* 1 A triangle; especially, the triangle of reference used in trilinear coordinates. 2 One of four parts of the zodiac, each consisting of three signs. 3 In Greek and Roman antiquity, a lyre or harp of triangular form. [<L *trigonum* <Gk. *trigōnon,* orig. neut. of *trigōnos* three-angled < *tri-* three + *gōnia* an angle]

trig·o·nal (trig'ə·nəl) *adj.* 1 Pertaining to or in the form of a trigon; triangular; three-cornered. 2 Characterized, in the hexagonal crystal system, by having a principal (vertical) axis of threefold symmetry. Also **trig'o·nous.** — **trig'o·nal·ly** *adv.*

trig·o·nom·e·ter (trig'ə·nom'ə·tər) *n.* 1 An instrument for solving triangles mechanically. 2 *Obs.* An expert in trigonometry.

trigonometric functions Certain functions of an angle or arc used in trigonometry. The most commonly used are: sine, cosine, tangent, cotangent, secant, cosecant. In illustration *A* the functions of the angle θ are defined as ratios or fractions. They are:

$$\text{sine } \theta = \frac{AB}{AC} \qquad \text{cosecant } \theta = \frac{AC}{AB}$$
$$\text{cosine } \theta = \frac{CB}{AC} \qquad \text{secant } \theta = \frac{AC}{CB}$$
$$\text{tangent } \theta = \frac{AB}{CB} \qquad \text{cotangent } \theta = \frac{CB}{AB}$$

The functions may be represented also as lines by constructing the reference triangle in a circle whose radius is taken as unity, and drawing additional lines as in *B.* The *sine* is then AB, the *cosine* CB, the *tangent* ED, the *cotangent* GF, the *secant* CE, and the *cosecant* CF. These are all spoken of as the sine, cosine, etc., of the arc AD as well as of the angle θ. Other, less common, trigonometric functions are: versed sine, coversed sine, exsecant, and

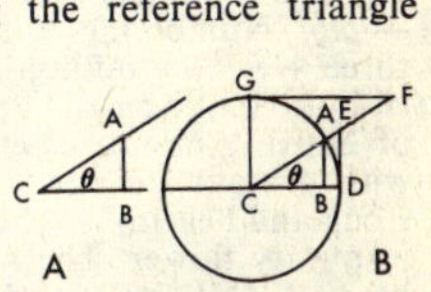

TRIGONOMETRIC FUNCTIONS

add,āce,câre,pälm; end,ēven; it,īce; odd,ōpen,ôrder; to͝ok,po͞ol; up,bûrn; ə = a in *above,* e in *sicken,* i in *clarity,* o in *melon,* u in *focus;* yo͞o = u in *fuse;* oi,oil; ou,pout; ch,check; g,go; ng,ring; th,thin; th,this; zh,vision. Foreign sounds à,œ,ü,kh,ṅ; and ♦: see page xx. < from; + plus; ? possibly.

haversine. These functions are expressed as follows:

versed sine θ (or versine θ) = 1 − cosine θ
coversed sine θ (or versed cosine θ) = 1 − sine θ
exsecant θ = secant θ − 1
haversine θ = 1/2 versine θ

trig·o·nom·e·try (trig′ə·nom′ə·trē) *n.* The branch of mathematics that treats of the relations of the sides and angles of triangles and of the methods of applying these relations in the solution of problems involving triangles: widely used in navigation, surveying, etc. [<NL *trigonometria* <Gk. *trigōnon* a triangle + *metron* measure] —**trig·o·no·met·ric** (trig′ə·nō·met′rik) or **·ri·cal** *adj.* —**trig′o·no·met′ri·cal·ly** *adv.*

tri·graph (trī′graf, -gräf) *n.* A group of three letters representing one articulate sound: *eau* in *beau*; also, the sound thus represented. [<TRI- + -GRAPH] —**tri·graph′ic** *adj.*

tri·he·dron (trī·hē′drən) *n.* *pl.* **·dra** (-drə) *Geom.* A figure having three plane surfaces meeting at a point. [<NL <Gk. *tri-* three + *hedra* a base] —**tri·he′dral** *adj.*

tri·hy·brid (trī·hī′brid) *n. Biol.* A hybrid whose parents differ from each other in respect to three pairs of contrasting Mendelian characters.

tri·hy·dric (trī·hī′drik) *adj. Chem.* Pertaining to or designating a compound containing three hydroxyl groups. Also **tri·hy·drox·y** (trī′hī·drok′sē). [<TRI- + HYDR(OXYL) + -IC]

tri·ju·gate (trī′jōō·gāt, trī·jōō′gāt, -git) *adj. Bot.* Having three pairs of leaflets. Also **tri·ju·gous** (trī′jōō·gəs, trī·jōō′gəs). [<L *trijugus* threefold <*tri-* three + *jugum* a yoke]

tri·lat·er·al (trī·lat′ər·əl) *adj.* Having three sides. [<L *trilaterus* <*tri-* three + *latus, lateris* a side] —**tri·lat′er·al·ly** *adv.*

tril·by (tril′bē) *n.* A soft felt hat with indented crown: informal wear for men in Great Britain. [after *Trilby*, a novel (1894) by George Du Maurier, in which such hats are described]

tri·lin·e·ar (trī·lin′ē·ər) *adj.* Pertaining to, referring to, or bounded by three lines.

tri·lin·gual (trī·ling′gwəl) *adj.* Derived from, composed of, or using three languages: a *trilingual* discourse. Also **tri·lin′guar.** [<L *trilinguis* <*tri-* three + *lingua* a tongue]

tri·lit·er·al (trī·lit′ər·əl) *adj.* Consisting of three letters. —**tri·lit′er·al·ism** *n.*

trill[1] (tril) *v.t.* **1** To sing or play in a quavering or tremulous tone. **2** *Phonet.* To articulate with a trill. —*v.i.* **3** To utter, make, or give forth a quavering or tremulous sound. **4** *Music* To execute a trill or shake. —*n.* **1** A tremulous utterance of successive tones, as of certain insects or birds; a warble. **2** *Music* A quick alternation of two notes either a tone or a semitone apart; shake. **3** *Phonet.* A rapid vibration of a speech organ, as of the tip of the tongue against the alveolar ridge or the uvula against the back of the tongue, as in the articulation of *rr* in Spanish. **4** A consonant or word so uttered. Also spelled *thrill.* [<Ital. *trillare*, prob. <Gmc.]

trill[2] (tril) *v.t.* & *v.i. Archaic* **1** To flow or cause to flow in a trickle, as tears. **2** To turn or roll; also, to quiver. [ME *trillen* <Scand. Cf. Sw. & Norw. *trilla* roll.]

tril·ling (tril′ing) *n.* A compound crystal made up of three individuals. [Cf. Dan. *trilling* a triplet]

tril·lion (tril′yən) *n.* A cardinal number; in the French and United States system of numeration, 1 followed by 12 zeros; in the English and German system, 1 followed by 18 zeros. —*adj.* Numbering a trillion. [<MF <*tri-* three + *million* million] —**tril′lionth** *adj.* & *n.*

tril·li·um (tril′ē·əm) *n.* Any of a genus (*Trillium*) of North American herbs of the lily family, with a stout stem, rising from a short rootstock and bearing a whorl of three leaves and a solitary flower. The fruit is a red or purple berry. [<NL <L *tri-* three; so called because of its three leaves]

tri·lo·bate (trī·lō′bāt, trī′lə·bāt) *adj.* **1** Three-lobed. **2** *Bot.* Having three lobes, as some leaves. Also **tri·lo′bal, tri·lo′bat·ed, tri′lobed.**

tri·lo·bite (trī′lə·bīt) *n. Paleontol.* Any of a subclass or group (*Trilobita*) of early Paleozoic marine arthropods related to the crustaceans, having a flattened body divided into a variable number of segments covered by a hard dorsal shield marked in three lobes. [<NL *Trilobites* <Gk. *tri-* three + *lobos* a lobe] —**tri′lo·bit′ic** (-bit′ik) *adj.*

tri·loc·u·lar (trī·lok′yə·lər) *adj.* Having three cells or chambers. [<TRI- + L *loculus* a small receptacle, dim. of *locus* a place]

tril·o·gy (tril′ə·jē) *n.* *pl.* **·gies** A group of three literary or dramatic compositions, each complete in itself, but continuing the same general subject. [<Gk. *trilogia* <*tri-* three + *logos* a discourse]

trim (trim) *v.* **trimmed, trim·ming** *v.t.* **1** To put in or restore to order; make neat by clipping, pruning, etc. **2** To remove by cutting: usually with *off* or *away.* **3** To put ornaments on; decorate. **4** In carpentry, to smooth; dress. **5** *Colloq.* **a** To chide; rebuke. **b** To punish or thrash; beat. **c** To defeat. **d** To cheat; victimize. **6** *Naut.* **a** To adjust (sails or yards) for sailing. **b** To cause (a ship) to sit well in the water by adjusting cargo, ballast, etc. **7** *Aeron.* To bring (an airplane) to level or balanced flight by adjusting control surfaces. **8** *Obs.* To furnish; equip. —*v.i.* **9** *Naut.* **a** To be or remain in equilibrium: said of a ship. **b** To adjust sails or yards for sailing. **10** To act so as to appear to favor opposing sides in a controversy. —*n.* **1** State of adjustment or preparation; fitting condition; orderly disposition: All was in good *trim.* **2** Condition as to general appearance; dress; style. **3** *Naut.* Fitness for sailing: said of a vessel in reference to disposition of ballast, masts, cargo, etc. **4** *Naut.* Actual or comparative degree of immersion: said of a vessel. **5** Particular character or nature; kind; stripe. **6** The moldings, etc., as about the doors of a building; also, the hardware trimmings of a house, such as hinges, window fastenings, etc. **7** Ornament; trapping; dress. **8** Material rejected or cut out, as sections from a motion-picture film. **9** In advertising, window dressing or display. **10** The interior furnishings of an automobile body. **11** *Aeron.* The position of an aircraft relative to balanced flight. —*adj.* **trim·mer, trim·mest** **1** Adjusted to a nicety; being in perfect order; handsomely equipped or of stylish and smart appearance; spruce; precise; jaunty. **2** Excellently fit; nice; pretty; fine. See synonyms under NEAT[1], STAUNCH. —*adv.* In a trim manner: also **trim′ly.** [OE *trymman* arrange, strengthen <*trum* steadfast, strong] —**trim′ness** *n.*

tri·mer (trī′mər) *n. Chem.* A compound formed by the union of three molecules of another compound or substance, as benzene from acetylene. [<TRI- + Gk. *meros* a part]

trim·er·ous (trim′ər·əs) *adj.* **1** Composed of three similar parts. **2** *Bot.* Three-parted. **3** *Entomol.* Having three joints, as the tarsus of an insect: often written *3-merous.* [<TRI- + Gk. *meros* a part]

tri·mes·ter (trī·mes′tər) *n.* A three-month period; quarter. [<F *trimestre* <L *trimestris* <*tri-* three + *mensis* a month] —**tri·mes′·tral, tri·mes′tri·al** *adj.*

trim·e·ter (trim′ə·tər) *adj.* In prosody, consisting of three measures or of lines containing three measures. —*n.* **1** A verse consisting of three measures, as the iambic trimeter. **2** In classical prosody, a line or verse consisting of three dimeters, or six feet. [<L *trimetrus* <Gk. *trimetros* <*tri-* three + *metron* a measure]

tri·meth·yl·pen·tane (trī′meth·il·pen′tān) *n. Chem.* One of five isomers of the pentane series, C_8H_{18}, used as a solvent and high-compression motor fuel. Also called *isooctane.* Compare OCTANE NUMBER. [<TRI- + METHYL + PENTANE]

tri·met·ric (trī·met′rik) *adj.* **1** Trimeter. **2** Orthorhombic. Also **tri·met′ri·cal.**

trimetric projection *Geom.* A three-dimensional geometric projection in which each dimension is measured on a separate scale and according to arbitrarily assigned angles.

tri·met·ro·gon (trī·met′rə·gon) *n.* A high-speed system of aerial topographic photography, in which a unit of three cameras takes simultaneous pictures of one area, from three positions, one vertical and two at matching oblique angles. —*adj.* Of or pertaining to this system or camera unit. [<TRI- + METRO- + -GON]

trim·mer (trim′ər) *n.* **1** One who or that which trims. **2** A time-server. **3** *Brit.* One who keeps the balance between opposing political parties by throwing his support from one to the other. **4** A small horizontal beam, as in a floor, into which the ends of one or more joists are framed. **5** A tool or machine with which to trim; specifically, a large table with power saw used to trim lumber for buildings.

trim·ming (trim′ing) *n.* **1** Something added for ornament or to give a finished appearance or effect. **2** Material attached to a garment, etc., for ornamentation or effect. **3** *pl.* Articles or equipment; fittings, as the hardware of a house. **4** *pl.* The usual or proper accompaniments or condiments of an article or food. **5** *pl.* That which is removed by trimming, cutting, or clipping; in shearing, wool from the shanks. **6** A severe reproof or a chastisement; flogging; beating. **7** *Colloq.* A defeat. **8** The act of one who trims.

tri·mo·lec·u·lar (trī′mə·lek′yə·lər) *adj. Chem.* Having, consisting of, or pertaining to three molecules.

tri·month·ly (trī·munth′lē) *adj.* & *adv.* Done or occurring every third month.

tri·morph (trī′môrf) *n.* **1** A substance existing or occurring in three forms. **2** One of the forms in which such a substance exists. [<Gk. *trimorphos* having three forms <*tri-* three + *morphē* a form]

tri·mor·phism (trī·môr′fiz·əm) *n.* **1** *Bot.* The existence on the same plant of three distinct forms of flowers as regards the relative lengths of stamens and pistils. **2** *Mineral.* The property of crystallizing in three series of fundamentally different forms with the same ultimate chemical composition. **3** *Zool.* Difference of species in form, color, etc., characterizing three distinct types. —**tri·mor′phic, tri·mor′phous** *adj.*

Tri·mur·ti (tri·mōōr′tē) *n.* In Hindu mythology, the triad of the Vedas, consisting of Brahma (the Creator), Vishnu (the Preserver), and Siva (the Destroyer). [<Skt. *trimūrti* <*tri-* three + *mūrti* shape]

Tri·na·cri·a (tri·nā′krē·ə, trī-) Ancient name for SICILY. —**Tri·na′cri·an** *adj.*

tri·nal (trī′nəl) *adj.* **1** Of or pertaining to three. **2** Having three parts; threefold. [<LL *trinalis* <L *trinus* three each <*tres, tria* three]

tri·na·ry (trī′nər·ē) *adj.* Made up of three parts or proceeding by threes; ternary. [<LL *trinarius* of three kinds <L *trinus.* See TRINAL.]

Trin·co·ma·lee (tring′kō·mə·lē′) **1** A port of NE Ceylon, capital of its eastern province and of Trincomalee district. **2** A district of NE Ceylon; 1,165 square miles.

trin·dle (trin′dəl) *n.* **1** One of several forked pieces of wood or metal between the cords and boards of a book to flatten its front and back edges before cutting. **2** *Brit. Dial.* A wheel, especially of a barrow. **3** A large wooden tub. —*v.t.* **1** *Brit. Dial.* To trundle; roll. —*v.i. Brit. Dial.* **2** To travel easily and rapidly. **3** To roll; advance by rolling. Also **trin′tle** (-təl). [ME *trindel*, var. of OE *trendel* a circle]

trine[1] (trīn) *adj.* **1** Threefold; triple: also *trinal.* **2** In astrology, relating to or situated in trine; auspicious. —*n.* **1** A compound in three parts or elements; a trio; triad. **2** *Her.* A charge composed of three objects. **3** In astrology, the aspect of two planets when 120° apart. —*v.t. Obs.* In astrology, to place or join in trine. [<OF *trin, trine* <L *trinus.* See TRINAL.]

trine[2] (trīn) *v.i. Obs.* To proceed; go. [<Scand. Cf. OSw. *trina* tramp.]

Trine (trīn) *n.* The Trinity.

Trin·i·dad and To·ba·go (trin′ə·dad, tō·bā′gō, *Span.* trē′nē·thäth′) A British colony in the West Indies NE of Venezuela, comprising the islands of **Trinidad;** 1,864 square miles; and **Tobago;** 116 square miles; capital, Port-of-Spain, on Trinidad; a member of The West Indies since January, 1958.

Tri·nil man (trī′nil) *Paleontol.* Pithecanthropus. [from *Trinil*, Java, where remains were found]

Trin·i·tar·i·an (trin′ə·târ′ē·ən) *adj.* **1** Of or pertaining to the Trinity. **2** Holding or professing belief in the Trinity: distinguished from *Unitarian.* —*n.* A believer in the doctrine of the Trinity. [<NL *trinitarius* <LL *trinitas* TRINITY] —**Trin′i·tar′i·an·ism** *n.*

tri·ni·trate (trī·nī′trāt) *n. Chem.* A nitrate containing three nitric-acid radicals in combination: bismuth *trinitrate.*

tri·ni·tro·ben·zene (trī·nī′trō·ben′zēn) *n. Chem.* A yellow crystalline compound, $C_6H_3(NO_2)_3$, occurring in three forms, one of which is highly explosive.

tri·ni·tro·cre·sol (trī·nī'trō·krē'sōl, -sol) *n. Chem.* A yellow crystalline organic compound, $C_8H_8N_3O_7$, used as an explosive. [<TRI- + NITRO- + CRESOL]

tri·ni·tro·phe·nol (trī·nī'trō·fē'nōl, -nol) *n.* Picric acid. [<TRI- + NITROPHENOL]

tri·ni·tro·tol·u·ene (trī·nī'trō·tol'yōō·ēn) *n. Chem.* A high explosive, $C_7H_5N_3O_6$, made by treating toluene with nitric acid: used for filling high explosive shells, for it melts readily and can be poured safely and rapidly: also called *TNT, trotyl.* Also **tri·ni'tro·tol'u·ol** (-yōō·ōl, -ol). [<TRI- + NITRO- + TOLUENE]

trin·i·ty (trin'ə·tē) *n. pl.* **·ties** **1** In art, a symbolic representation of the Trinity. **2** The state or character of being three; also, any union of three parts or elements in one; a trio; triad. [<OF *trinite* <LL *trinitas* <L, a triad < *trinus.* See TRINAL.]

Trin·i·ty (trin'ə·tē) *n.* **1** *Theol.* A threefold personality existing in the one divine being or substance; the union in the one God of Father, Son, and Holy Spirit as three infinite persons. **2** Trinity Sunday.

Trinity, Cape A cliff on the lower Saguenay River, SE central Quebec, Canada; 1,500 feet high.

Trinity River A river in Texas, flowing 510 miles SE to **Trinity Bay,** the NE arm of Galveston Bay.

Trinity Sunday *Eccl.* The eighth Sunday after Easter, observed as a festival in honor of the Trinity. Also *Trinity.*

trin·ket (tring'kit) *n.* **1** Any small ornament, as of jewelry. **2** Any small article forming part of an outfit. **3** A trifle; a trivial object; a toy. **4** *Obs.* A knife. See synonyms under GAUD. [<AF *trenquet,* OF *trenchet* a toy knife, ornament, < *trenchier.* See TRENCH.]

trin·kums (tring'kəmz) *n. pl. Scot.* & *Brit. Dial.* Small ornaments; trinkets. Also **trin'·kum-tran'kums** (-trang'kəmz). [Appar. alter. of TRINKET]

tri·no·dal (trī·nōd'l) *adj. Bot.* Having three nodes or nodal points.

tri·no·mi·al (trī·nō'mē·əl) *adj.* **1** *Biol.* Of, having, or employing three terms or names— the generic, the specific, and the subspecific or varietal, as *Lynx rufus texensis,* the Texas bobcat. **2** *Math.* Consisting of three terms connected by plus or minus signs or both. —*n.* **1** An algebraic expression consisting of three terms connected by plus or minus signs or both, as $3x + y - 27z$. **2** A trinomial name. Also **tri·nom'i·nal** (-nom'ə·nəl), **tri·on'·y·mal** (-on'ə·məl). [<TRI- + (BI)NOMIAL]

tri·o (trē'ō, *for def. 1 also* trī'ō) *n. pl.* **tri·os** **1** Any three things grouped or associated together. **2** *Music* **a** A composition for three performers. **b** The second part of a minuet or scherzo, of a march, and of dance forms generally. **c** A group of three musicians who render trios. [<F <Ital. < *tre* three <L *tres, tria*]

tri·ode (trī'ōd) *n. Electronics* A three-element vacuum tube, containing an anode, cathode, and a control grid or electrode. [<TRI- + (ELECTR)ODE]

tri·oe·cious (trī·ē'shəs) *adj. Bot.* Having in different plants of the same species male, female, and hermaphrodite flowers: also spelled *triecious.* Also **tri·oi'cous** (-oi'kəs). [<NL *Trioecia,* order name <Gk. *tri-* three + *oikos* a house] —**tri·oe'cious·ly** *adv.*

tri·o·let (trī'ə·lit) *n.* A stanza of eight lines on two rimes, the first line repeated as the fourth and seventh and the second as the eighth. Its rime scheme is *abaaabab.* [<F, dim. of *trio* TRIO]

Tri·o·nal (trī'ə·nal) *n.* Proprietary name for a brand of sulfonethylmethane.

tri·ose (trī'ōs) *n. Biochem.* A monosaccharide whose molecule contains three atoms of carbon and three of oxygen. [<TRI- + -OSE[2]]

tri·ox·ide (trī·ok'sīd, -sid) *n. Chem.* An oxide containing three atoms of oxygen in combination: iron *trioxide,* Fe_2O_3. Also **tri·ox'·id** (-sid).

trip (trip) *n.* **1** A short journey; excursion; jaunt. **2** A misstep or stumble occasioned by losing the balance or striking the foot against an object. **3** An active, nimble step or movement. **4** The number of fish caught in an excursion. **5** A single tack to windward. **6** *Mech.* A pawl or similar device that trips, or the action of such a device. **7** A sudden catch, especially of the legs and feet, as of a wrestler. **8** A blunder; mistake. See synonyms under JOURNEY. —*v.* **tripped, trip·ping** *v.i.* **1** To stumble. **2** To move quickly with light or small steps; saunter. **3** To commit an error; make a false step; go astray. **4** *Mech.* To run past the nicks or dents in the ratchet escape wheel of a timepiece. **5** *Rare* To make a journey. —*v.t.* **6** To cause to stumble: often with *up.* **7** To detect and expose in an error; defeat the purpose of. **8** To perform (a dance) lightly. **9** *Mech.* To set free or in operation by releasing a stay, catch, trigger, etc. **10** *Naut.* **a** To loosen, as an anchor, from the bottom by a long rope or cable. **b** To hoist (the topmast) so as to prepare it for being lowered. **c** To tilt (a yard) similarly. [<OF *treper, triper* leap, trample, ? <MDu. *trippen* trip, hop]

tri·pal·mi·tin (trī·pal'mə·tin) *n.* Palmitin.

tri·par·tite (trī·pär'tīt) *adj.* **1** Divided into three parts or divisions; threefold: a *tripartite* leaf: also **tri·part·ed** (trī'pär·tid). **2** *Law* Pertaining to or executed between three parties. **3** *Math.* Homogeneous in three sets of variables. [<L *tripartitus* < *tri-* three + *partitus,* pp. of *partiri* divide] —**tri·par'tite·ly** *adv.*

tri·par·ti·tion (trī'pär·tish'ən) *n.* Division into three parts, into thirds, or among three.

tripe (trīp) *n.* **1** A part of the stomach of a ruminant, as the ox, used for food. **2** *Colloq.* Contemptible or worthless stuff; nonsense; an inferior, mean, or offensive thing, person, or matter. [<OF *tripe, trippe* <Arabic *tharb* entrails, a net]

tri·pe·dal (trī'pə·dəl, trī·pēd'l, trip'ə·dəl) *adj.* Having three feet; three-footed. [<L *tripedalis* < *tri-* three + *pes, pedis* foot]

tri·per·son·al (trī·pûr'sən·əl) *adj.* Consisting of or relating to three persons.

tri·per·son·al·i·ty (trī·pûr'sən·al'ə·tē) *n. Theol.* The state or quality of existing in three persons in one Godhead; trinity.

tri·pet·al·ous (trī·pet'l·əs) *adj. Bot.* Having three petals.

trip hammer A heavy power hammer that is raised or tilted by a cam and then allowed to drop: also called *tilt hammer.*

tri·phase (trī'fāz) *adj. Electr.* Having or employing three phases, as in an alternating current.

tri·phen·yl·meth·ane (trī·fen'əl·meth'ān) *n. Chem.* A hydrocarbon, $(C_6H_5)_3CH$, occurring in colorless leaflets: used in organic synthesis and in the manufacture of dyes. [<TRI- + PHENYL + METHANE]

tri·phib·i·an (trī·fib'ē·ən) *adj.* Describing a joint military and naval operation which utilizes terrestrial, marine, and aerial weapons simultaneously; literally, capable of life in three elements—air, earth, and water. Also **tri·phib'i·ous.** [<TRI- + (AM)PHIBIAN]

triph·thong (trif'thông, -thong, trip'-) *n.* **1** A combination of three vowel sounds in one syllable, as in one pronunciation of *fire.* **2** A trigraph composed of vowels, as in *beau.* [<TRI- + (DI)PHTHONG] —**triph·thon'gal** *adj.*

triph·y·lite (trif'ə·līt) *n.* A greenish-gray, bluish, transparent to translucent phosphate of iron and lithium, crystallizing in the orthorhombic system. Also **triph'y·line** (-lin, -lēn). [<TRI- + Gk. *phylē* a tribe + -ITE[1]; so called because it contains three bases]

tri·pin·nate (trī·pin'āt) *adj. Bot.* Thrice pinnate, as when the pinnae of a bipinnate leaf become again pinnate in certain ferns. Also **tri·pin'nat·ed.** —**tri·pin'nate·ly** *adv.*

tri·pin·nat·i·fid (trī'pə·nat'ə·fid) *adj. Bot.* Tripinnately cleft. [<TRIPINNATE + -FID]

tri·plane (trī'plān') *n.* An airplane having three supporting surfaces arranged one above the other.

trip·le (trip'əl) *v.* **·led, ·ling** *v.t.* **1** To make threefold in number or quantity. —*v.i.* **2** To be or become three times as many or as large. **3** In baseball, to make a triple. —*adj.* **1** Consisting of three things united or of three parts; threefold. **2** Multiplied by three; thrice said or done. **3** *Archaic* Third. —*n.* **1** A set or group of three. **2** In baseball, a three-base hit. [<MF <L *triplus* <Gk. *triploos* threefold. Doublet of TREBLE.] —**trip'ly** *adv.*

Triple Alliance **1** An alliance between England, Holland, and Sweden against Louis XIV of France, formed in 1668. **2** A league between England, France, and Holland, formed in 1717, but called **Quadruple** when joined by Austria in 1718, designed to secure the succession to the crown of England for the house of Hanover, that of France for the house of Bourbon, and to prevent the union of France and Spain under one crown. **3** An alliance formed in 1795 between Austria, Great Britain, and Russia against France. **4** A Dreibund.

Triple Entente A friendly understanding formed between Great Britain, France, and Russia prior to World War I to counteract the Dreibund.

trip·le-ex·pan·sion (trip'əl·ik·span'shən) *adj.* Designating a compound steam engine constructed with three cylinders of graduated sizes in which the steam is successively expanded.

triple measure *Music* A measure of three beats, the first accented, the second and third unaccented.

trip·le-nerved (trip'əl·nûrvd') *adj. Bot.* Three-nerved; having three principal nerves arising from or near the base, as certain leaves.

triple play In baseball, a play during which three men are put out.

trip·let (trip'lit) *n.* **1** A group of three of a kind. **2** One of three children born at one birth. **3** A group of three rimed lines. **4** *Music* A group of three notes performed in the time of two. **5** A bicycle for three. [<TRIPLE, on analogy with *doublet*]

trip·le·tail (trip'əl·tāl') *n.* A large edible marine fish (*Lobotes surinamensis*) of warm seas, with soft dorsal and anal fins extended backward, suggesting additional tails.

triple time See under TIME.

triple voile Ninon.

tri·plex (trī'pleks, trip'leks) *adj.* Having three parts; threefold. —*n. Music* Triple measure. [<L < *tri-* three + *plicare* fold]

trip·li·cate (trip'lə·kit) *adj.* Threefold; made in three copies. —*n.* A third thing corresponding to two others of the same kind or three similar things collectively: a document signed in *triplicate.* —*v.t.* (-kāt) **·cat·ed, ·cat·ing** To make three times as much or as many; treble. [<L *triplicatus,* pp. of *triplicare* triple < *triplex* TRIPLEX] —**trip'li·cate·ly** *adv.*

trip·li·ca·tion (trip'lə·kā'shən) *n.* **1** The act of triplicating. **2** That which is triplicated or made threefold, or is in three layers.

tri·plic·i·ty (tri·plis'ə·tē) *n. pl.* **·ties** **1** Threefold character. **2** A group or combination of three; a triad; a triplet. **3** In astrology a combination of three of the twelve signs of the zodiac. [<LL *triplicitas, -tatis* <L *triplex, -icis* TRIPLEX]

trip·lite (trip'līt) *n.* A brown or black, translucent to opaque, fluophosphate of iron and manganese. [<G *triplit* <Gk. *triploos* triple; with ref. to its three cleavages]

trip·lo·blas·tic (trip'lə·blas'tik) *adj. Biol.* Having or characterized by three germ layers, as the embryos of the higher animals. [<Gk. *triploos* triple + BLASTIC]

trip·loid (trip'loid) *adj.* **1** Trebled. **2** *Genetics* Noting the occurrence in certain cells of three times the basic number of chromosomes. —*n.* A triploid cell or organism. [<NL *triploides* <Gk. *triploos* threefold + *eidos* form]

trip·loi·dy (trip·loi'dē) *n.* The condition of being triploid.

trip·lo·pi·a (trip·lō'pē·ə) *n. Pathol.* A defect of vision in which objects are seen tripled. [<NL <Gk. *triploos* threefold + *ōps, ōpos* eye]

tri·pod (trī'pod) *n.* **1** A utensil or article having three feet or legs. **2** A three-legged stand, as for supporting a camera, compass, or other instrument. [<L *tripus, -podis* <Gk. *tripous* < *tri-* three + *pous* foot]

trip·o·dal (trip'ə·dəl) *adj.* **1** Of the nature or form of a tripod. **2** Having three feet or legs. Also **tri·po·di·al** (tri·pō'dē·əl, trī-), **tri·pod'ic** (-pod'ik).

trip·o·dy (trip′ə·dē) *n. pl.* **·dies** A verse or meter having three feet. [<TRI- + (DI)PODY]

trip·o·li (trip′ə·lē) *n.* Rottenstone. [from *Tripoli,* Libya, where it is found]

Trip·o·li (trip′ə·lē) **1** One of the two capitals (with Bengasi) and the largest city of Libya, a port on the central Mediterranean and the capital of Tripolitania province: Phoenician *Oea.* **2** A port of NW Lebanon on the Mediterranean: ancient **Trip·o·lis** (trip′ə·lis). — **Tri·pol·i·tan** (tri·pol′ə·tən) *adj.* & *n.* — **Trip′o·line** (-lin) *adj.*

Trip·o·li·ta·ni·a (trip′ə·li·tā′nē·ə) The western province of Libya, on the Mediterranean; 82,990 square miles; capital, Tripoli; a former Barbary State. Ancient **Trip′o·lis.** — **Trip′o·li·ta′ni·an** *adj.* & *n.*

tri·pos (trī′pos) *n.* **1** An honors examination held at Cambridge University, England, especially in mathematics. **2** *Obs.* A tripod. [Appar. alter. of L *tripus* TRIPOD]

trip·per (trip′ər) *n.* **1** One who trips in any sense. **2** *Brit. Colloq.* One who makes trips; a tourist or traveler. **3** *Mech.* A trip or tripping mechanism, as a device on a railroad track which operates a catch on a passing train to give a signal or alarm.

trip·pet (trip′it) *n. Mech.* A cam, toe, or projecting piece, designed to strike some other piece at fixed intervals. [<TRIP, *v.*]

trip·ping (trip′ing) *n.* **1** The act of one who or that which trips. **2** A light dance. — *adj.* Light; nimble; easy; stepping. — **trip′ping·ly** *adv.*

trip·tane (trip′tān) *n. Chem.* A hydrocarbon compound, C_7H_{16}, derived from butane and having a very high octane number. [Contraction of *tripentane* <TRI- + PENTANE]

triptane number An improved measure of the efficiency of a motor fuel, expressed in terms of a blend of normal heptane and triptane, each containing a specified amount of tetraethyl lead.

trip·ter·ous (trip′tər·əs) *adj. Bot.* Having three wings or winglike processes, as certain seeds. [<TRI- + Gk. *pteron* a wing, on analogy with *dipterous*]

Trip·tol·e·mus (trip·tol′ə·məs) In Greek mythology, a hero said to have given mankind the secret of the cultivation of grain. Also **Trip·tol′e·mos** (-mos).

trip·tote (trip′tōt) *n.* A substantive having but three cases. [<L *triptota,* pl. <Gk. *triptōta,* neut. pl of *triptōtos* having three cases]

trip·tych (trip′tik) *n.* **1** A picture, carving, or work of art on three panels side by side. **2** Three pictures associated in their subjects and placed side by side in compartments. **3** A writing tablet in three sections, made of various laminate materials. Also **trip′ty·ca** (-ti·kə), **trip′ty·chon** (-ti·kon). [<Gk. *triptychos* threefold <*tri-* three + *ptyx, ptychos* a fold <*ptyssein* fold]

tri·pu·di·ate (trī·pyōō′dē·āt) *v.i.* **·at·ed, ·at·ing** To dance, especially in a measured way. [<L *tripudiatus,* pp. of *tripudiare* <*tripudium* a religious dance, prob. <*tri-* three + *.pes, pedis* foot] — **tri·pu′di·a′tion** *n.*

Tri·pu·ra (tri′pōō·rə) A union territory of NE India; 4,032 square miles; capital, Agartala.

tri·quet·rous (trī·kwet′rəs, -kwē′trəs) *adj.* **1** Three-sided. **2** Having three acute or salient angles. **3** Three-cornered, as certain stems and bones. [<L *triquetrus*]

tri·ra·di·ate (trī·rā′dē·āt) *adj.* Having three rays or radiate branches: the *triradiate* sulcus of the brain. Also **tri·ra′di·al, tri·ra′di·at′ed.** — **tri·ra′di·al·ly, tri·ra′di·ate′ly** *adv.*

tri·reme (trī′rēm) *n.* An ancient Greek or Roman warship with three banks of oars. [<L *triremis* <*tri-* three + *remus* an oar]

tri·sac·cha·ride (trī·sak′ə·rīd, -rid) *n. Biochem.* Any of a class of saccharides which yield three monosaccharide molecules when subjected to hydrolysis, as raffinose. Also **tri·sac′cha·rid** (-rid).

Tris·ag·i·on (tris·ag′ē·on, -ā′gē-) *n.* A hymn, probably of Hebrew origin, in the liturgy of the Greek and Oriental churches, beginning with a threefold invocation of the Deity as Holy. Also **Tris·ag′i·um, Tris·hag′i·on.** [<Gk. *trisagion,* orig. neut. of *trisagios* thrice holy <*tris* thrice (<*treis* three) + *hagios* holy]

tri·sect (trī·sekt′) *v.t.* To divide into three parts, especially, as in geometry, into three equal parts. [<TRI- + L *sectus,* pp. of *secare* cut] — **tri·sect′ed** *adj.* — **tri·sec′tion** (-sek′shən) *n.* — **tri·sec′tor** *n.*

tri·sec·trix of MacLaurin (trī·sek′triks) *Math.* The plane curve of the equation $x^3 + xy^2 + ay^2 - 3ax^2 = 0$. Symmetric about the x-axis, passing through the origin and asymptotic to the line $x = -a$, the trisectrix is so named because it can be employed to trisect an angle. [after Colin *MacLaurin,* 1698–1746, Scottish mathematician]

tri·seme (trī′sēm) *n.* A syllable or foot consisting of or equivalent to three morae or short syllables, as the tribrach, iambus, and trochee. — *adj.* Consisting of or equal to three morae or short syllables: also **tri·se·mic** (trī·sē′mik). [<Gk. *trisēmos* <*tri-* three + *sēma* a sign]

tri·sep·al·ous (trī·sep′əl·əs) *adj. Bot.* Having three sepals.

tri·sep·tate (trī·sep′tāt) *adj. Biol.* Having three septa.

tri·se·ri·al (trī·sir′ē·əl) *adj.* **1** Arranged in three series or rows. **2** *Bot.* Tristichous. Also **tri·se′ri·ate** (-it, -āt). — **tri·se′ri·al·ly, tri·se·ri·a·tim** (trī·sir′ē·ā′tim) *adv.*

tris·kel·i·on (tris·kel′ē·ən) *n. pl.* **·kel·i·a** (-kel′ē·ə) A symbolic figure charaterized by three lines or three human legs radiating from a common center. It is used as the arms of the Isle of Man. Also **tris·cele** (tris′sēl), **tris·kele** (tris′kēl). [<Gk. *triskelēs* of three legs <*tri-* three + *skelos* a leg]

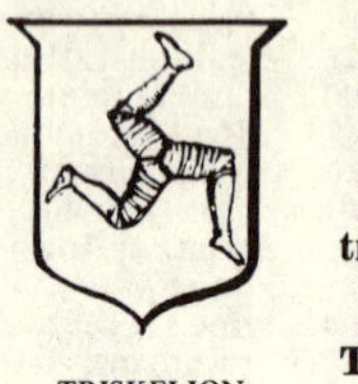
TRISKELION

Tris·me·gis·tus (tris′mə·jis′təs, triz′-) See HERMES TRISMEGISTUS.

tris·mus (triz′məs, tris′-) *n. Pathol.* Tetanic spasm causing rigid closure of the jaws; lockjaw. [<NL <Gk. *trismos* a gnashing of teeth, a grinding] — **tris′mic** *adj.*

tris·oc·ta·he·dron (tris·ok′tə·hē′drən) *n. pl.* **·dra** (-drə) **1** A solid having 24 equal faces corresponding by threes to the faces of an octahedron. **2** A holohedral isometric crystal included under 24 equal isosceles triangular faces with eight planes meeting at the extremities of the rectangular axes: also **trigonal trisoctahedron. 3** An isometric holohedron included under 24 similar and equal trapeziform faces; a trapezohedron: also **tetragonal trisoctahedron.** [<Gk. *tris* thrice (<*treis* three) + OCTAHEDRON.] — **tris·oc′ta·he′dral** *adj.*

tri·sper·mous (trī·spûr′məs) *adj. Bot.* Having three seeds.

tri·spo·rous (trī·spôr′əs, -spō′rəs) *adj. Bot.* Having three spores. Also **tri·spor′ic** (-spôr′ik, -spor′ik). [<TRI- + -SPOROUS]

Tris·tan (tris′tän, -tən) In medieval legend, a knight sent to Ireland to bring back the princess Iseult the Beautiful as a bride for his uncle, King Mark of Cornwall. Iseult and Tristan mistakenly drink a magic love potion, and ultimately die together. In some versions, Tristan is later married to Iseult of the White Hand, daughter of the Duke of Brittany. Also **Tris·tram** (tris′trəm).

Tris·tan da Cun·ha (tris·tän′ dä kōōn′yä) A British island group in the South Atlantic, midway between South America and the Cape of Good Hope; administered with St. Helena; 40 square miles.

triste (trēst) *adj. French* Sorrowful; sad.

tris·tesse (trēs·tes′) *n. French* Sadness; melancholy.

trist·ful (trist′fəl) *adj. Archaic* Sad; gloomy; sorrowful. [<obs. *trist* sad <OF *triste* <L *tristis*] — **trist′ful·ly** *adv.*

tris·tich (tris′tik) *n.* A strophe or system of three lines; triplet. Compare COUPLET, DISTICH. [<TRI- + (DI)STICH]

tris·ti·chous (tris′tə·kəs) *adj.* **1** Three-ranked. **2** *Bot.* Having parts, as leaves, arranged in three vertical rows. [<Gk. *tristichos* three-rowed <*tri-* three + *stichos* a row]

tri·stim·u·lus (trī·stim′yə·ləs) *adj.* **1** Having, pertaining to, or caused or characterized by three distinct stimuli. **2** In color analysis, designating an instrument or method for measuring a color stimulus in terms of three selected primary stimuli.

tri·sty·lous (trī·stī′ləs) *adj. Bot.* Having three styles.

tri·sul·fide (trī·sul′fīd, -fid) *n. Chem.* A sulfide containing three atoms of sulfur in combination. Also **tri·sul′fid** (-fid), **tri·sul′phide, tri·sul′phid.**

tri·syl·la·ble (trī·sil′ə·bəl) *n.* A word of three syllables. — **tri·syl·lab·ic** (trī′si·lab′ik) or **·i·cal** *adj.* — **tri′syl·lab′i·cal·ly** *adv.*

tri·tag·o·nist (trī·tag′ə·nist) *n.* In Greek drama, the actor who played the third part; hence, also, a third-rate actor. [<Gk. *tritagōnistēs* <*tritos* third + *agōnistēs* a contender, actor <*agōnizesthai* contend <*agōn* a contest]

trit·an·o·pi·a (trit′ən·ō′pē·ə) *n. Pathol.* Impairment of vision for blue and yellow; blue blindness. Also **trit′an·op′si·a** (-op′sē·ə). [<NL <Gk. *tritos* third + ANOPIA] — **trit′an·op′tic** (-op′tik) *adj.*

trite (trīt) *adj.* **1** Used so often as to be hackneyed; made commonplace by repetition. **2** *Archaic* Worn-out; frayed. [<L *tritus,* pp. of *terere* rub] — **trite′ly** *adv.* — **trite′ness** *n.*

Synonyms: common, commonplace, hackneyed, musty, rusty, stale, stereotyped, threadbare, worn. See COMMON. *Antonyms*: bright, brilliant, fresh, new, original, racy, striking, telling, vivid.

tri·the·ism (trī′thē·iz′əm) *n. Theol.* The doctrine of the separate existence of three Gods: sometimes opprobriously applied to belief in the distinct personality of the Father, the Son, and the Holy Spirit. [<TRI- + Gk. *theos* a god] — **tri′the·ist** *n.* — **tri′the·is′tic** or **·ti·cal** *adj.*

tri·thing (trī′thing) *n.* In English law, a riding. See RIDING[2]. [OE *thrithing* <ON *thridhungr* a third part]

Trit·i·cum (trit′ə·kəm) *n.* A widely distributed and important genus of cereal grasses, the wheats, especially *T. aestivum* and its numerous cultivated varieties. [<NL <L, wheat]

trit·i·um (trit′ē·əm, trish′ē·əm) *n.* The rare hydrogen isotope of atomic mass 3, whose nucleus contains one proton and two neutrons. [<NL <Gk. *tritos* third]

tri·ton[1] (trīt′n) *n.* A marine gastropod (genus *Triton*) with many gills and a trumpet-shaped shell. [<NL <L, Triton]

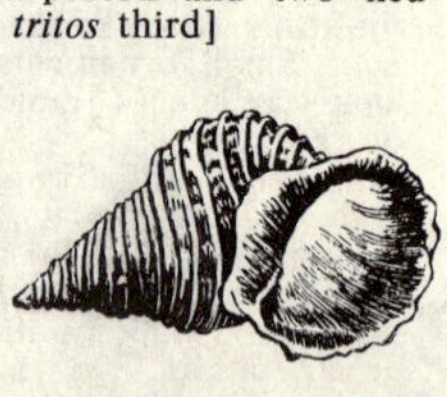
TRITON

tri·ton[2] (trī′ton) *n.* The nucleus of an atom of tritium. [<TRIT(IUM) + (ELECTR)ON]

Tri·ton (trīt′n) **1** In Greek mythology: **a** A son of Poseidon (Neptune) and Amphitrite, represented with a man's head and upper body and a dolphin's tail. **b** One of a race of attendants of the sea gods. **2** *Her.* A merman; also, a Neptune holding a trident. — **Tri′ton·ess** *n. fem.*

tri·tone (trī′tōn) *n. Music* An augmented fourth, as containing three whole tones. [<Med. L *tritonus* <Gk. *tritonos* <*tri-* three + *tonos.* See TONE.]

trit·u·rate (trich′ə·rāt) *v.t.* **·rat·ed, ·rat·ing** To reduce to a fine powder or pulp by grinding or rubbing; pulverize. — *n.* **1** That which has been triturated. **2** A trituration (def. 3). [<LL *trituratus,* pp. of *triturare* thresh <L *tritura* a rubbing, threshing <*tritus.* See TRITE.] — **trit·u·ra·ble** (trich′ər·ə·bəl) *adj.* — **trit′u·ra′tor** *n.*

trit·u·ra·tion (trich′ə·rā′shən) *n.* **1** The act of triturating; reduction to a very fine powder by grinding or rubbing, as in a mortar. **2** The process of reducing to a pulp. **3** A triturated preparation, especially one in which 10 parts of a medicinal substance are triturated with 90 parts of milk sugar: also *triturate.*

tri·umph (trī′əmf) *v.i.* **1** To win a victory; be victorious. **2** To be successful. **3** To rejoice over a victory; exult. **4** To celebrate a triumph, as a victorious Roman general. — *v.t.* **5** *Obs.* To conquer. See synonyms under REJOICE. — *n.* **1** In Roman antiquity, the religious pageant of the entry of a victorious consul, dictator, or pretor into Rome: given only for a decisive victory over a foreign enemy. **2** Exultation over victory. **3** The condition of being victorious; victory. **4** *Obs.* A trump card. **5** *Obs.* Any public spectacular display, procession, or pageant. See synonyms under HAPPINESS, VICTORY. [<OF *triumpher* <L *triumphare* <*triumphus* a triumph <Gk. *thriambos* a processional hymn to Dionysus] — **tri′umph·er** *n.*

tri·um·phal (trī·um′fəl) *adj.* 1 Of, pertaining to, or of the nature of a triumph. 2 Celebrating a victory.

triumphal arch A large monumental arch erected in ancient or modern times to commemorate any great victory or achievement.

tri·um·phant (trī·um′fənt) *adj.* 1 Exultant for or as for victory. 2 Crowned with victory; victorious. 3 *Obs.* Of supreme magnificence or beauty; glorious. 4 *Obs.* Triumphal. [<L *triumphans, -antis,* ppr. of *triumphare* TRIUMPH] — **tri·um′phant·ly** *adv.*

tri·um·vir (trī·um′vər) *n.* *pl.* **·virs** or **·vi·ri** (-və·rī) One of three men united in public office or authority, as in ancient Rome. [<L < *trium virorum* of three men < *tres, trium* three + *vir* a man] — **tri·um′vi·ral** *adj.*

tri·um·vi·rate (trī·um′vər·it, -və·rāt) *n.* 1 A group or coalition of three men who unitedly exercise authority or control; government by triumvirs. 2 The office of a triumvir; also, the triumvirs collectively. 3 A group of three men; a trio. [<L *triumviratus* < *triumvir* TRIUMVIR]

tri·une (trī′yōōn) *adj.* Three in one: said of the Godhead. — *n.* A group of three things united; a triad; a trinity in unity. [<TRI- + L *unus* one]

Tri·u·ni·tar·i·an (trī·yōō′nə·târ′ē·ən) *n.* A Trinitarian. [<TRIUNIT(Y) + -ARIAN]

tri·u·ni·ty (trī·yōō′nə·tē) *n.* Trinity.

tri·va·lent (trī·vā′lənt, triv′ə·lənt) *adj.* *Chem.* Having a valence or combining value of three. [<TRI- + L *valens, -entis,* ppr. of *valere* be strong] — **tri·va′lence, tri·va′len·cy** *n.*

tri·valve (trī′valv′) *adj.* Having three valves, as a shell. — *n.* A trivalve shell.

Tri·van·drum (tri·van′drəm) A port on the Malabar Coast, capital of Kerala State, SW India, in the SW part of the State.

triv·et (triv′it) *n.* A three-legged stand for holding cooking vessels in a fireplace, a heated iron, or a hot dish on a table: also *trevet.* [OE *trefet* <L *tripes, -pedis* three-footed < *tri-* three + *pes, pedis* a foot]

triv·i·a (triv′ē·ə) *n. pl.* Insignificant or unimportant matters; trifles. [<NL <L *trivialis* TRIVIAL]

triv·i·al (triv′ē·əl) *adj.* 1 Of little value or importance; trifling; insignificant. 2 Such as is found everywhere or every day; ordinary; commonplace. 3 Occupied with trifles; of low ability or wit; unscholarly. See synonyms under CHILDISH, INSIGNIFICANT, LITTLE, RIDICULOUS, VAIN, VENIAL. [<L *trivialis* of the crossroads, commonplace < *trivium* a crossing of three roads < *tri-* three + *via* a road] — **triv′i·al·ism** *n.* — **triv′i·al·ly** *adv.*

triv·i·al·i·ty (triv′ē·al′ə·tē) *n.* *pl.* **·ties** 1 The state or quality of being trivial: an age of *triviality*: also **triv′i·al·ness.** 2 A trivial matter; a trivialism.

triv·i·um (triv′ē·əm) *n.* In medieval schools, the course in the liberal arts embracing grammar, logic, and rhetoric. Compare QUADRIVIUM. [<Med. L <L. See TRIVIAL.]

tri·week·ly (trī·wēk′lē) *adj.* & *adv.* 1 Occurring three times a week. 2 Sometimes, done or occurring every third week.

-trix *suffix* A feminine termination of agent nouns the masculine form of which is *-tor*: *testatrix.* See -OR[1]. [<L *-trix*]

Tr·no·vo (tûr′nô·vô) See TIRNOVO.

troak (trōk) See TROKE.

Tro·as (trō′as) The region of western Asia Minor on the Aegean surrounding the ancient city of Troy: also *Teucria.* Also **the Tro′ad** (-ad).

Tro·bri·and Islands (trō′brē·änd) A volcanic island group off the eastern tip of New Guinea; a dependency of the Australian Trust Territory of Papua and New Guinea; total, 175 square miles.

tro·car (trō′kär) *n.* *Surg.* A sharp-pointed instrument used with a cannula to drain off internal fluids. Also **tro′char.** [<F *troquart, trois-quarts* < *trois* three + *carre* face; so called because of its triangular shape]

tro·cha (trō′chä) *n.* 1 A path; road. 2 An obstruction on a road, to hinder an enemy; a military cordon. [<Sp.]

tro·cha·ic (trō·kā′ik) *adj.* Pertaining to, containing, or composed of trochees: a *trochaic* foot or verse. — *n.* A trochaic verse or line. [<MF *trochaïque* <L *trochaicus* <Gk. *trochaikos* < *trochaios* TROCHEE]

tro·chal (trō′kəl) *adj.* 1 Shaped like a wheel; rotiform. 2 Trochilic. [<Gk. *trochos* a wheel]

tro·chan·ter (trō·kan′tər) *n.* 1 *Anat.* One of several bony processes on the upper thigh bone. 2 *Entomol.* The small second segment of an insect's leg. [<MF <Gk. *trochantēr* < *trechein* run]

tro·che (trō′kē) *n.* A medicated lozenge, usually circular. [Alter. of obs. *trochisk* <MF *trochisque* a lozenge <L *trochiscus* <Gk. *trochiskos* a small wheel, a lozenge < *trochos* a wheel < *trechein* run]

tro·chee (trō′kē) *n.* In prosody, a foot comprising a long and short syllable (—⏑), or, in modern verse, an accented syllable followed by an unaccented one. [<L *trochaeus* <Gk. *trochaios (pous)* a running (foot) < *trechein* run]

Troch·el·min·thes (trok′əl·min′thēz) *n. pl.* A phylum of minute, transparent, aquatic protozoans which move by means of cilia, including the wheel animalcules. [<NL <Gk. *trochos* a wheel + *helmins, helminthos* a worm]

tro·chil·ic (trō·kil′ik) *adj.* 1 Of the nature of or pertaining to rotary motion. 2 Capable of such motion. [<Gk. *trochilos* a pulley, taken as var. of *trochos* a wheel < *trechein* run]

troch·i·lus (trok′ə·ləs) *n.* *pl.* **·li** (-lī) 1 The crocodile bird: also **tro·chil** (trō′kil, trok′il), **troch′i·los** (-los). 2 A hummingbird (family *Trochilidae*). 3 One of various small warblers or warblerlike birds. [<L *trochilus* a crocodile bird <Gk. *trochilos* < *trechein* run]

troch·le·a (trok′lē·ə) *n.* *pl.* **·le·ae** (-li·ē) *Anat.* A grooved pulleylike surface, permitting smooth motion, as between the humerus and ulna. [<L, a pulley <Gk. *trochilia, trochileia* < *trechein* run]

troch·le·ar (trok′lē·ər) *adj.* 1 *Anat.* Of, pertaining to, or situated near a trochlea. 2 Of the nature of a pulley. 3 Short, cylindrical, compressed, and contracted in the middle of its circumference like a pulley block. [<NL *trochlearis* <L *trochlea* TROCHLEA]

troch·le·ar·i·form (trok′lē·ar′ə·fôrm) *adj.* Having the form of a pulley; trochlear.

tro·choid (trō′koid) *adj.* Rotating upon its own axis; pivotal: also **tro·choi′dal.** — *n.* *Math.* A plane curve traced by a point on a circle or on its extended radius as the circle rolls, without slipping, on a straight line: when the point is on the circumference of the circle, the curve traced is a cycloid. [<Gk. *trochoeidēs* round, wheel-like < *trochos* a wheel + *eidos* form, shape] — **tro·choi′dal·ly** *adv.*

troch·o·phore (trok′ə·fôr, -fōr) *n.* *Zool.* A pear-shaped larval form of certain aquatic invertebrates, as annelids, brachiopods, and mollusks. Also **troch′o·sphere** (-sfir). [<Gk. *trochos* a wheel + -PHORE]

trock (trok) See TROKE.

trod (trod) Past tense and alternative past participle of TREAD.

trod·den (trod′n) Past participle of TREAD.

trode (trōd) Archaic past tense of TREAD.

trog·lo·dyte (trog′lə·dīt) *n.* 1 A prehistoric cave man. 2 Figuratively, a hermit; anyone of primitive or degenerate habits. 3 An anthropoid ape, as the gorilla. 4 The wren. [<L *troglodyta* <Gk. *trōglodytēs* < *trōglē* a hole + *dyein* go into] — **trog′lo·dyt′ic** (-dit′ik), **trog′lo·dyt′i·cal** *adj.*

tro·gon (trō′gon) *n.* A tropical American bird (family *Trogonidae*) noted for its resplendent plumage. [<NL <Gk. *trōgōn,* ppr. of *trōgein* gnaw]

troi·ka (troi′kə) *n.* A Russian vehicle drawn by a team of three horses driven abreast; also, the team, or both team and vehicle together. [<Russian]

Troi·lus (troi′ləs, trō′i·ləs) In Greek legend, a son of Priam killed by Achilles; in medieval legend, Chaucer's *Troilus and Criseyde,* and Shakespeare's *Troilus and Cressida,* Cressida's lover.

troilus butterfly The green-clouded or spicebush swallowtail butterfly (*Papilio troilus*) of eastern North America. [after *Troilus*]

Trois Ri·vières (trwä rē·vyâr′) A city on the St. Lawrence River at the mouth of the St. Maurice River in southern Quebec, Canada: English *Three Rivers.*

Tro·jan (trō′jən) *n.* 1 A native of Troy. 2 A brave, persevering person; one who works earnestly or suffers courageously. 3 *Colloq.* A jolly fellow; boon companion. — *adj.* Of or pertaining to ancient Troy. Also called *Dardan, Dardanian.* [Earlier *Troyan, Troian* <L *Troianus* < *Troja* Troy]

Trojan horse 1 In classical legend, a large, hollow wooden horse, described in Vergil's *Aeneid,* filled with Greek soldiers and left at the Trojan gates: when it was brought within the walls the soldiers emerged at night and admitted the Greek army, who burned the city: also called *wooden horse.* 2 *Mil.* The infiltration of military men into a potentially hostile region for the purpose of nullifying resistance against attack: compare FIFTH COLUMN.

Trojan War In Greek legend, the ten years' war waged by the confederated Greeks under their king, Agamemnon, against the Trojans to recover Helen, the wife of Menelaus, who had been abducted by Paris: celebrated especially in the *Iliad* and the *Odyssey.* See APPLE OF DISCORD.

troke (trōk) *Scot.* *n.* 1 Exchange; also, articles of trade; small wares; truck. 2 Familiar intercourse or acquaintance. — *v.t.* & *v.i.* To exchange; barter. Also spelled *troak, trock.*

troll[1] (trōl) *v.t.* 1 To cause to roll; revolve. 2 To sing in succession, as in a round or catch. 3 To sing in a full, hearty manner. 4 To fish for with a moving lure, as from a moving boat. 5 To move (the line or lure) in fishing. 6 *Obs.* To pass around, as a bottle or decanter. — *v.i.* 7 To roll; turn. 8 To sing a tune, etc., in a full, hearty manner. 9 To be uttered in such a way. 10 To fish with a moving lure. 11 *Obs.* To move about; ramble. — *n.* 1 A catch or round. 2 A rolling movement or motion; hence, repetition or routine. 3 In fishing, a spoon or other lure. [? <OF *troller* quest, wander <Gmc. Cf. MHG *trollen* walk with short steps.] — **troll′er** *n.*

troll[2] (trōl) *n.* In Scandinavian folklore, a giant; later, a friendly but often mischievous dwarf. Also **trold** (trōld). [<ON]

trol·ley (trol′ē) *n.* *pl.* **·leys** 1 A grooved metal wheel for rolling in contact with an electric conductor (the **trolley wire**), to convey the current to a motor car; a trolley pole. 2 In a subway system, a bow or shoe adapted to the same purpose attached to a current-taker operating through a slot in the track: also **trolley wheel.** 3 A car or system so operated. 4 A small truck or car for conveying material, as in a factory or around a furnace: sometimes applied to trucks in mines, etc.: also spelled *trawley.* 5 *Brit. Dial.* A small hand or donkey cart. 6 A parcels carrier. 7 The mechanism of a traveling crane. 8 A small car running on tracks and worked by a lever operated by hand: used by workmen on a railway. — *v.t.* & *v.i.* To convey or travel by trolley. Also spelled *trolly.* [<TROLL[1]]

trolley bus A passenger conveyance operating without rails, propelled electrically by current taken from an overhead wire by means of a trolley: also called *trackless trolley.* Also **trolley coach.**

trolley car A car arranged with a trolley and motor for use on an electric railway operated by the trolley system.

trolley line A system of street cars propelled on the trolley system; also, the road itself.

trol·ley·man (trol′ē·man′) *n.* *pl.* **·men** (-men′) A man who operates a trolley; especially, a conductor or motorman.

trolley pole A pole on a trolley car carrying the trolley wheel.

troll·ing (trō′ling) *n.* The method or act of fishing by dragging a hook and line, as behind a boat and near the surface: usually with a spoon bait or the like. [<TROLL[1]]

trolling bait Spoon bait. Also **trolling hook, trolling spoon.**

trolling rod A strong fishing rod for trolling.

trol·lop (trol′əp) *n.* 1 A slatternly woman. 2 A prostitute. [<dial. E (Scottish) <ME

trollen roll about; prob. infl. in meaning by *trull*] — **trol′lop·ish, trol′lop·y, trol′lop·ing** *adj.*

Trol·lope (trol′əp), **Anthony**, 1815–82, English novelist.

trol·ly (trol′ē) *n. pl.* **·lies** *n.* & *v.* Trolley.

Trom·be·tas (trōn̄m·bā′təs) A river of NW Pará state, Brazil, flowing 470 miles south and east to the Amazon.

trom·bic·u·li·a·sis (trom·bik′yə·lī′ə·sis) *n. Pathol.* Infestation with mites of the genus *Trombicula*, the chiggers. Also **trom·bic′u·lo′·sis** (-lō′sis). [<NL *Trombicula* + -IASIS]

trom·bone (trom′bōn, trom·bōn′) *n.* A powerful brass wind instrument of the trumpet family possessing a complete chromatic scale. It consists of a cupped mouthpiece and a long tube bent twice upon itself, the outer bend being a U-shaped slide, by the motion of which the length of the vibrating air column may be so adjusted as to produce any note within its compass. [<Ital., aug. of *tromba* a trumpet] — **trom′bon·ist** *n.*

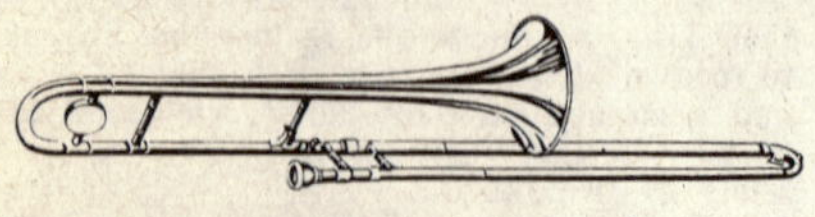

TROMBONE

trom·mel (trom′əl) *n. Metall.* A perforated steel plate, usually cylindrical in form, used for sifting or screening rock, ore, etc. [<G, a drum]

trom·o·ma·ni·a (trom′ə·mā′nē·ə, -mān′yə) *n.* Delirium tremens. [<NL <Gk. *tromos* a trembling + *mania* madness]

Tromp (trômp), **Cornelius van**, 1629–91, Dutch admiral; son of the following. — **Marten Harpertzoon**, 1597–1653, Dutch admiral.

trompe (tromp) *n.* **1** *Metall.* An apparatus that supplies a blast of air, as to a forge, by the action of a thin column of water falling through a large, long tube and thus carrying air by entanglement. **2** An arched and vaulted structure that supports a portion of a building. Also **tromp.** [<F, lit., a trumpet]

Trom·sö (trom′zō, *Norw.* trôms′œ, trōōms′œ) A port on eastern **Tromsö Island**, an islet of NW Norway (8 square miles).

tro·na (trō′nə) *n.* A vitreous, gray or white, monoclinic hydrous sodium carbonate, $Na_2CO_3HNaCO_3 \cdot 2H_2O$. [<Sw., appar. <Arabic *trōn*, short for *natrūn* NATRON]

Trond·heim (trôn′hām) A port on **Trondheim Fiord**, in Norway: formerly *Nidaros*. Formerly spelled **Trond·hjem** (trôn′yem).

troop (trōōp) *n.* **1** An assembled company; gathering; a herd or flock. **2** *Usually pl.* A body of soldiers; soldiers collectively. **3** The cavalry unit of formation, corresponding to a company of infantry. **4** A body of Boy Scouts consisting of four patrols of eight scouts each. **5** Formerly, a troupe; a company of actors. See synonyms under ARMY. — *v.i.* **1** To move along or gather as a troop or as a crowd. **2** *Archaic* To associate; consort. — *v.t.* **3** To form into troops. **4** *Brit. Mil.* To carry ceremoniously before troops: to *troop* the colors. ◆ Homophone: *troupe*. [<OF *trope* <LL *troppus* a flock <Gmc.]

troop·er (trōō′pər) *n.* **1** A cavalryman. **2** A mounted policeman. **3** A troop horse; charger. **4** A troopship.

troop·i·al (trōō′pē·əl) *n.* Any American bird of the family *Icteridae*, including the blackbirds, orioles, bobolinks, and meadowlarks, especially *Icterus icterus* of South America and the West Indies, mostly black varied with yellow and white. Also spelled *troupial*. [<F *troupiale* < *troupe* <OF *trope* TROOP; so called because it goes in flocks]

troop·lift (trōōp′lift′) *n.* **1** The troop-carrying capacity of a nation's passenger ships, merchant marine, or aviation. **2** The actual transport of troops.

troop·ship (trōōp′ship′) *n.* A ship for carrying troops; a transport.

troost·ite (trōōs′tīt) *n.* A variety of willemite in large reddish crystals. [after Gerhard *Troost*, 1776–1850, U.S. mineralogist]

trooz (trōōz) See TREWS.

trop (trō) *adv. French* Too much; too many; too.

tro·pa·co·caine (trō′pə·kō·kān′, -kō′kān, -kō′·kə·ēn) *n. Chem.* A white crystalline compound, $C_{15}H_{19}O_2N$, obtained from Java coca leaves and also made synthetically from atropine and hyoscine: used as an anesthetic. [<*(benzoyl-pseudo)trop(eine)*, its chemical name + COCAINE]

Tro·pae·o·lum (trō·pē′ə·ləm) *n.* A genus of tropical American plants with alternate leaves and bright-colored flowers supposed to resemble ancient trophies. Many species, known as *nasturtiums*, are cultivated. [<NL <Gk. *tropaion* a trophy; so called from the resemblance of the leaf and flower to a shield and helmet]

tro·pae·um (trō·pē′əm) *n.* In Greek antiquity, a monument of victory, composed of captured arms, set up by the Greeks at a place where they had defeated an enemy. Also **tro·pai·on** (trō·pā′on). [<L, TROPHY]

-tropal *combining form* -tropic.

tro·par·i·on (trō·pâr′ē·on) *n. pl.* **·par·i·a** (-pâr′·ē·ə) In the Greek Church, a stanza of, or the several stanzas constituting, a hymn. [<Gk., dim. of *tropos*. See TROPE.]

trope (trōp) *n.* **1** The figurative use of a word. **2** Loosely and less properly, a figure of speech; figurative language in general. **3** A short distinguishing cadence interpolated in Gregorian melodies. **4** An interpolated phrase that was occasionally inserted in various parts of the mass prior to the 16th century. [<F <L *tropus* a figure of speech <Gk. *tropos* a turn < *trepein* turn]

-trope *combining form* **1** One who or that which turns or changes: *allotrope*. **2** Turning; turned in a (specified) way: *hemitrope*. [<Gk. *tropos* a turning < *trepein* turn]

tro·pe·ine (trō′pē·in, -ēn) *n. Chem.* An ester of tropine, from which it is formed by the action of certain organic acids. Also **tro′pe·in** (-in). [Alter. of TROPINE]

tro·pe·o·lin (trō·pē′ə·lin) *n. Chem.* Any of several orange azo dyes formed by the action of diazosulfuric acids on phenols. Also **tro·pae′·o·lin.** [<TROPAEOL(UM) + -IN; so called because their hues resemble those of the flower]

troph·al·lax·is (trof′ə·lak′sis) *n. Biol.* The free exchange of food substances among individuals, considered as an essential factor in the life cycle of certain insects, especially army ants. [<NL <Gk. *trophē* food + *allaxis* an exchange] — **troph′al·lac′tic** (-lak′tik) *adj.*

troph·ic (trof′ik) *adj.* Pertaining to nutrition and its processes. Also **troph′i·cal.** [<Gk. *trophikos* < *trophē* nourishment < *trephein* nourish] — **troph′i·cal·ly** *adv.*

tro·phied (trō′fēd) *adj.* Adorned with trophies.

tropho- *combining form* Nutrition; nourishment; of or pertaining to food or nutrition: *trophoplasm*. Also, before vowels, **troph-**. [<Gk. *trophē* food, nourishment < *trephein* feed, nourish]

troph·o·blast (trof′ə·blast) *n. Biol.* The ectodermal layer of cells in the embryo that establishes relation with the uterus and is concerned in the nutrition of the embryo and fetus. Also **troph′o·derm** (-dûrm). [<TROPHO- + -BLAST] — **troph′o·blas′tic** *adj.*

troph·o·gen·e·sis (trof′ə·jen′ə·sis) *n. Biol.* The production of variations among plants and animals by differences in food and nutrition, as distinguished from genetic factors. Also **tro·phog·e·ny** (trō·foj′ə·nē). — **troph′o·gen′ic** *adj.*

troph·o·plasm (trof′ə·plaz′əm) *n. Biol.* **1** The nutritive or vegetative substance of the cell, as distinguished from the idioplasm. **2** Formerly, a cytoplasmic substance distinguished from the archiplasm. — **troph′o·plas′mic** *adj.*

troph·o·ther·a·py (trof′ə·ther′ə·pē) *n. Med.* The treatment of disease by diet therapy.

tro·phot·ro·pism (trō·fot′rə·piz′əm) *n. Bot.* The movement or curvature, as toward or away from nutrient substances, induced in a growing plant by the influence of the chemical nature of its surroundings. — **troph·o·trop·ic** (trof′ə·trop′ik) *adj.*

troph·o·zo·ite (trof′ə·zō′īt) *n. Zool.* A parasitic sporozoan at the stage of entering the blood cell of its host, feeding on the nutritive material in the blood. [<TROPHO- + Gk. *zōion* an animal + -ITE[1]]

tro·phy (trō′fē) *n. pl.* **·phies** **1** Anything taken from an enemy and displayed or treasured in proof of victory; hence, a momento of victory or success: *trophies* of the chase. **2** An ancient Roman memorial of victory in imitation of the Greek tropaeum, but a permanent structure, decorated with arms or beaks of ships suspended over the undecorated parts. **3** An ornamental group of objects hung together on a wall, or any collection of objects typical of some event, art, industry, or branch of knowledge. **4** A memento or memorial. **5** *Archit.* A group of arms and armor carved in marble or cast in bronze rising from a circular or quadrangular stepped base. [<MF *trophée* <L *trophaeum, tropaeum* <Gk. *tropaion* < *tropē* a defeat, turning < *trepein* turn, rout]

-trophy *combining form* A (specified) kind of nutrition or nurture: *hypertrophy*. Corresponding adjectives end in *-trophic*. [<Gk. *trophē*. See TROPHO-.]

trop·ic (trop′ik) *n.* **1** *Geog.* Either of two parallels of latitude at a distance from the equator, north and south, equal to the obliquity of the ecliptic, or 23° 27′, on which the sun is seen in the zenith on the days of its greatest declination: called respectively **tropic of Cancer** and **tropic of Capricorn.** **2** *Astron.* **a** Either of two corresponding parallels of declination in the celestial sphere similarly named, and respectively 23° 27′ north or south from the celestial equator. **b** Either of the two points in the celestial sphere where the sun reaches its maximum distance north or south of the celestial equator; a solstice. **3** *pl.* The regions of the earth's surface between the tropics of Cancer and Capricorn, where the sun crosses the zenith twice in the course of the year: with the definite article; the torrid zone. — *adj.* Of or pertaining to the tropics; tropical. [<L *tropicus* <Gk. *tropikos (kyklos)* the tropical (circle), pertaining to the turning of the sun at the solstice < *tropē*. See TROPHY.]

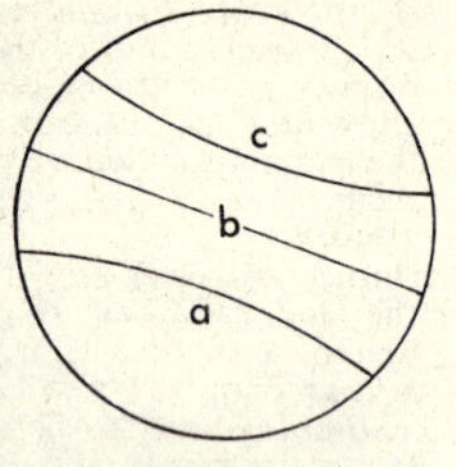

TROPICS
a. Tropic of Capricorn.
b. Equator.
c. Tropic of Cancer.

-tropic *combining form* Having a (specified) tropism; turning or changing in a (particular) way, or in response to a (given) stimulus: *chemotropic, phototropic.*

trop·i·cal (trop′i·kəl) *adj.* **1** Of, pertaining to, or characteristic of the tropics. **2** Of the nature of a trope or metaphor; changed from the original to a figurative meaning. — **trop′i·cal·ly** *adv.*

trop·i·cal·ize (trop′i·kəl·īz′) *v.t.* **·ized, ·iz·ing** To adapt, as clothing, war equipment, ships, etc., for service in tropical areas. — **trop′i·cal·i·za′tion** *n.*

tropic bird A long-winged, oceanic, ternlike bird (genus *Phaëthon*), found mostly in the tropics, having the two middle tail feathers elongated.

tro·pine (trō′pēn, -pin) *n. Chem.* A colorless crystalline alkaloid, $C_8H_{15}NO$, with a tobacco odor, formed when atropine is hydrolyzed. Also **tro′pin** (-pin). [<ATROPINE]

tro·pism (trō′piz·əm) *n. Biol.* **1** The involuntary response of an organism, or of any of its parts, to an external stimulus. **2** Any automatic reaction to a stimulus. [<Gk. *tropē* a turning] — **tro·pis·tic** (trō·pis′tik) *adj.*

-tropism *combining form* A (specified) tropism; a tendency to turn or change in response to a (given) stimulus: *chemotropism, phototropism*. Corresponding adjectives end in *-tropic*. [<TROPISM]

tro·pist (trō′pist) *n.* **1** One given to the use of tropes. **2** One who interprets and explains a text, especially Scripture, tropically or figuratively.

tro·pol·o·gy (trō·pol′ə·jē) *n.* **1** The use of tropical or figurative language. **2** Consideration or treatment of the Scriptures both literally and figuratively, or as having a double sense. **3** A treatise on figures of speech. [<LL *tropologia* <Gk. < *tropos* TROPE + *logos* discourse] — **trop·o·log·ic** (trop′ə·loj′ik) or **·i·cal** *adj.* — **trop′o·log′i·cal·ly** *adv.*

trop·o·pause (trop′ə·pôz) *n. Meteorol.* A transition zone in the atmosphere between the troposphere and the stratosphere at which the fall of temperature with increasing height abruptly ceases. [<TROPO(SPHERE) + Gk. *pausis* a ceasing]

tro·poph·i·lous (trō·pof′ə·ləs) *adj. Ecol.* Adapted to extreme conditions of moisture or of heat: said of plants. [<Gk. *tropos* a turning, change + *philos* loving; with ref. to adaptation to seasonal changes]

trop·o·phyte (trop′ə·fīt) *n. Ecol.* Any of the plants that adapt themselves to seasonal changes of dryness or cold and also of moisture: they form the highest type of temperate-zone plants, as the deciduous trees. [<Gk. *tropos* a turning, change + -PHYTE] — **trop′o·phyt′ic** (-fit′ik) *adj.*

trop·o·sphere (trop′ə·sfir) *n. Meteorol.* The region of the atmosphere from the earth's surface to the tropopause, having a height of from six to twelve miles and characterized by decreasing temperature with increasing altitude. [<F *troposphère* <Gk. *tropos* a turning + F *sphère* <L *sphaera* SPHERE]

trop·po (trop′ō, *Ital.* trôp′pō) *adv. Music* Too much: andante ma non *troppo* (andante but not too much). [<Ital.]

-tropous *combining form* Turned in a specified way: *anatropous.* Corresponding nouns end in *-tropy.*

-tropy *combining form* 1 -tropism. 2 A state of being turned. See -TROPOUS. [<Gk. *tropē* a turning < *trepein* turn]

Tros·sachs (tros′aks, -əks) A valley in Perthshire, central Scotland: scene of Scott's *The Lady of the Lake.*

trot[1] (trot) *n.* 1 A progressive motion of a quadruped, in which each diagonal pair of legs is alternately lifted, thrust forward, and placed upon the ground almost simultaneously, the body of the animal being entirely unsupported twice during each stride; the sound of this gait. 2 A race for trotters. 3 A little child; toddler: a term of endearment. 4 Steady going or movement, implying persistence and diligence: I have been on the *trot* all day. 5 *Colloq.* A literal translation of a foreign-language text, used as an aid in study or in examination; a crib; pony. — *v.* **trot·ted, trot·ting** *v.i.* 1 To go at a trot. 2 To go quickly; hurry. — *v.t.* 3 To cause to trot. 4 To ride at a trotting gait. — **to trot out** To bring forth for inspection, approval, etc. [<OF < *troter* <OHG *trottōn* tread]

trot[2] (trot) *n. Archaic* An old woman: a derogatory term. [<AF *trote*; ult. origin uncertain]

troth (trôth, trōth) *n.* 1 Good faith; fidelity; also, the act of pledging fidelity; especially, betrothal. 2 Truth; verity. — *v.t. Archaic* To betroth; pledge. [ME *trowthe, trouthe,* var. of OE *trēowth* truth]

troth·plight (trôth′plīt′, trōth′-) *Archaic v.t.* To betroth; affiance. — *n.* Betrothal. — *adj.* Betrothed: also **troth′-plight′ed.** [<TROTH + PLIGHT[2]]

trot·line (trot′līn′) *n.* A trawl line.

Trot·sky (trot′skē), **Leon,** 1879–1940, Russian Bolshevist leader; exiled 1929; murdered: real name *Lev Davidovitch Bronstein.*

Trot·sky·ism (trot′skē·iz′əm) *n.* The doctrines of Trotsky and his followers; especially, his belief in "permanent revolution" or the theory that Communism to succeed must be international. — **Trot′sky·ist** *n.*

Trot·sky·ite (trot′skē·īt) *n.* An adherent of any of the various factions of the Communist party originally led by Leon Trotsky, who opposed Stalinism and supported international Communism.

trot·ter (trot′ər) *n.* 1 One who or that which trots; a trotting horse; specifically, a horse trained to trot for speed. 2 *Colloq.* An animal's foot: a pig's *trotters.*

tro·tyl (trō′til) *n.* Trinitrotoluene. [<(TRINI)-TROT(OLUENE) + -YL]

trou·ba·dour (trōō′bə·dôr, -dōr, -dōōr) *n.* One of a class of lyric poets, sometimes including wandering minstrels and jongleurs, originating in Provence in the 11th century and flourishing in southern France, northern Italy, and eastern Spain during the 12th and 13th centuries. Compare TROUVÈRE. See synonyms under POET. [<MF <Provençal *trobador* < *trobar* compose, invent, find; ult. origin uncertain]

Trou·betz·koy (trōō·bets′koi), **Princess** See RIVES, AMÉLIE.

troub·le (trub′əl) *n.* 1 The state of being distressed, annoyed, or confused; also, grief; affliction; disturbance. 2 A person, circumstance, or event that occasions difficulty or perplexity; the vexation thus occasioned; annoyance; worry; civil unrest or agitation. 3 Toilsome exertion; pains. 4 Any serious or permanent diseased condition: lung *trouble.* See synonyms under ANXIETY, CARE, GRIEF, MISFORTUNE, PAIN. — *v.* **·led, ·ling** *v.t.* 1 To cause mental agitation to; distress; worry. 2 To agitate or disturb; stir up or roil, as water. 3 To inconvenience or incommode. 4 To annoy or pester; bother. 5 To cause physical pain or discomfort to; afflict. — *v.i.* 6 To take pains; bother. 7 To worry. See synonyms under PERPLEX. [<OF *truble, turble* < *turbler* <L *turbula* a mob, dim. of *turba* a crowd] — **troub′ler** *n.* — **troub′ling·ly** *adv.*

troub·le·mak·er (trub′əl·mā′kər) *n.* One who habitually stirs up trouble.

troub·le-shoot·er (trub′əl·shōō′tər) *n.* 1 A mechanic; a repairman. 2 One who locates difficulties and seeks to remove them. 3 A person trained to find and eliminate trouble in the operation of a machine, process, or the like; a maintenance man. — **troub′le-shoot′·ing** *n.*

troub·le·some (trub′əl·səm) *adj.* 1 Causing trouble; vexatious; burdensome; trying; afflictive: a *troublesome* business. 2 Marked by violence; tumultuous. 3 Greatly agitated or disturbed; troublous. — **troub′le·some·ly** *adv.* — **troub′le·some·ness** *n.*

Synonyms: afflictive, annoying, arduous, burdensome, difficult, galling, harassing, hard, importunate, intrusive, irksome, laborious, painful, perplexing, teasing, tiresome, trying, vexatious, wearisome. *Antonyms*: amusing, cheering, easy, entertaining, grateful, gratifying, helpful, light, pleasant.

troub·lous (trub′ləs) *adj.* 1 Marked by commotion or tumult; full of trouble: *troublous* times. 2 Uneasy; restless. 3 *Obs.* Troublesome.

trou-de-loup (trōō′də·lōō′) *n. pl.* **trous-de-loup** (trōō′də·lōō′) *Usually pl.* A conical pit having a vertical central stake with a pointed top, used as a defense against cavalry. [<F *trou* a hole + *de* of + *loup* a wolf]

trough (trôf, trof; *Dial.* trôth, troth) *n.* 1 A long, narrow, open receptacle for conveying a fluid or for holding food or water for animals. 2 A long, narrow channel or depression, as between ridges on land or waves at sea. 3 A gutter for rain water fixed under the eaves of a building. [OE *trog*]

trounce (trouns) *v.t.* **trounced, trounc·ing** 1 To beat or thrash severely; punish. 2 *Colloq.* To defeat. [<OF *tronce* a thick piece of wood <L *truncus* stem, trunk] — **trounc′ing** *n.*

troupe (trōōp) *n.* A company of actors or other performers. — *v.i.* **trouped, troup·ing** To travel as one of a company of actors or entertainers. ◆ Homophone: *troop.* [<MF <OF *trope* TROOP] — **troup′er** *n.*

troup·i·al (trōō′pē·əl) See TROOPIAL.

trou·sers (trou′zərz) *n. pl.* A man's garment, covering the body from the waist to the ankles or knees and divided so as to make a separate covering for each leg. Also **trow′sers.** [Blend of obs. *trouse* breeches (<Irish *triubhas*) and DRAWERS]

trousse (trōōs) *n.* 1 A collection of small implements in a sheath or case. 2 A case containing knives, tweezers, etc., fastened to the belt: a surgeon's *trousse.* [<F. See TRUSS.]

trous·seau (trōō·sō′, trōō′sō) *n. pl.* **·seaux** (-sōz′, -sōz) 1 A bride's outfit, especially of clothing. 2 *Obs.* A bundle; truss. [<F < *trousse* a packed collection of things. See TRUSS.]

trout (trout) *n.* 1 A salmonoid fish mostly found in fresh waters and highly esteemed as a game and food fish. The **brown trout** or **river trout** *(Salmo trutta),* attaining a length of 30 inches, is common in Europe; the **cut-throat trout** *(S. clarkii),* and the **rainbow trout** or steelhead *(S. gairdnerii)* are species of western North America. The **speckled trout** or **brook trout** *(Salvelinus fontinalis)* is common in eastern North America. 2 A fish resembling, or supposed to resemble, the above, as the greenling. [OE *truht* <LL *tructus, tructa* <Gk. *trōktēs* a nibbler < *trōgein* gnaw]

trou·vère (trōō·vâr′) *n.* One of a class of poets flourishing in northern France from the 11th to the 14th centuries, distinguished from the troubadours of southern France by the prevailingly narrative and epic character of their works, which include chansons de geste, fabliaux, romances, and chronicles. Also **trou·veur** (trōō·vûr′). [<F <OF *trovere* < *trover* find, compose; ult. origin uncertain]

Trou·ville (trōō·vēl′) A port and resort in northern France, 9 miles south of Le Havre. Also **Trou·ville′-sur-Mer′** (-sür·mâr′).

tro·ver (trō′vər) *n. Law* An action to recover the value of personal property of the plaintiff wrongfully withheld or converted by another to his own use: originally an action of trespass against one who found the goods of another, and refused to give them up; the finding, however, became a fiction. [<OF, find; ult. origin uncertain]

trow (trō) *v.t.* & *v.i.* 1 *Archaic* To suppose; think; believe. 2 *Obs.* To wonder. [Fusion of OE *truwian* < *truwa* faith and *trēowan* believe < *trēowe* true]

trow·el (trou′əl, troul) *n.* 1 A flat-bladed, sometimes pointed implement having an offset handle: used by masons, plasterers, and molders. 2 A small concave scoop with a handle: used in digging about small plants, potting them, etc. 3 A molder's smoothing tool. — *v.t.* **·eled** or **·elled, ·el·ing** or **·el·ling** To apply, dress, or form with a trowel. [<OF *truele* <LL *truella* <L *trulla,* dim. of *trua* a stirring spoon, ladle] — **trow′el·er** or **trow′el·ler** *n.*

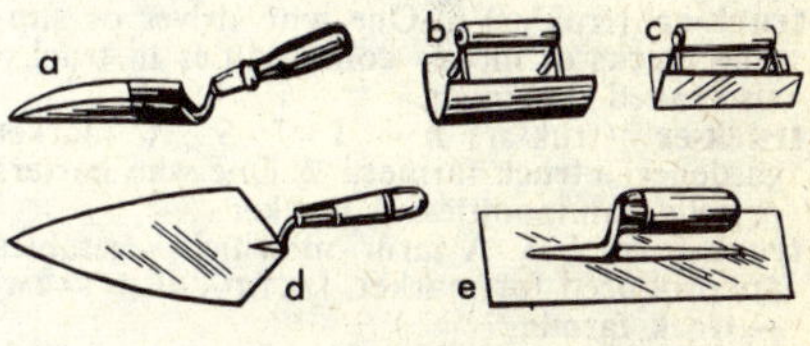

TYPES OF TROWELS
a. Garden. *b.* Circle. *c.* Corner.
d. Brick. *e.* Plastering.

trowel bayonet A spade-shaped bayonet.

trowth (trōth) *n. Scot.* 1 Truth. 2 Troth.

troy (troi) *n.* A system of weights in which 12 troy ounces make a pound, used by jewelers in England and the United States. See under WEIGHT. Also **troy weight.** [from *Troyes;* with ref. to a weight used at a fair held there]

Troy (troi) 1 The site of nine superimposed ruined cities in NW Asia Minor: the seventh stratum, a Phrygian city of perhaps about 1200 B.C., the scene of the *Iliad,* was also called *Ilium, Ilion.* 2 A city on the Hudson River in eastern New York.

Troy coach A type of passenger coach commonly used in travel in the United States before the building of railroads, seating nine inside, having room for driver, six passengers, and baggage outside, and drawn by four to six horses. [from *Troy,* N. Y.]

Troyes (trwä) A city in NE central France, on the Seine river; a major textile center.

tru·an·cy (trōō′ən·sē) *n. pl.* **·cies** The state or habit of being truant; an act of playing truant. Also **tru′ant·ry.**

tru·ant (trōō′ənt) *n.* One who absents himself, especially from school, without leave. — *v.i.* To play the truant. — *adj.* 1 Playing the truant; idle. 2 Relating to or characterizing a truant. [<OF, a vagabond, prob. <Celtic]

truant officer *U. S.* An official who investigates truancy from school.

truce (trōōs) *n.* 1 An agreement between belligerents for a temporary suspension of hostilities; an armistice. 2 Temporary cessation or intermission. [Plural of ME *trew,* OE *trēowa* faith, a promise. Akin to TRUE, TRUST.]

Tru·cial O·man (trōō′shəl ō·män′) A region on the eastern coast of the Arabian peninsula extending along the **Trucial Coast,** a nearly 400-mile section between Oman and Qatar, and consisting largely of seven **Trucial Sheikdoms** bound by treaties with Great Britain; about 32,300 square miles.

truck[1] (truk) *n.* **1** One of several forms of strong vehicles, variously constructed, for moving bulky articles, freight, etc.; a dray; a stout automotive vehicle on rubber tires able to carry heavy loads. **2** A two-wheeled barrowlike vehicle with a forward lip and no sides, for use in moving barrels, boxes, etc., by hand. **3** A two-, three-, four-, or sometimes six-wheeled vehicle used about railway stations, for moving trunks, etc.: distinguished as **baggage truck, freight truck,** or **wagon truck. 4** Any of numerous small, flat-topped cars moved by pushing or pulling and used in stores. **5** *Brit.* An open or platform freight car. **6** *Naut.* A disk at the upper extremity of a mast or flagpole through which the halyards of signals are run. **7** A wheel: the original sense, now rare, and usually implying a small tireless wheel. — *v.t.* **1** To carry on a truck. — *v.i.* **2** To carry goods on a truck. **3** To drive a truck. [Appar. <L *trochus* a hoop <Gk. *trochos* a wheel < *trechein* run]

truck[2] (truk) *v.t. & v.i.* To exchange or barter; also, to peddle. — *n.* **1** Commodities for sale. **2** *U.S.* Garden produce for market: often in compounds: *truck* farming, etc. **3** *Colloq.* Rubbish; worthless articles collectively. **4** Barter. **5** *Colloq.* Intercourse; dealings: I will have no *truck* with him. [<OF *troquer* barter; origin unknown]

truck·age[1] (truk′ij) *n.* **1** Money paid for conveyance of goods on trucks. **2** Such conveyance. [<TRUCK[1] + -AGE]

truck·age[2] (truk′ij) *n.* Exchange; barter. [<TRUCK[2] + -AGE]

truck·er[1] (truk′ər) *n.* One who drives or supplies trucks or moves commodities in trucks: also called *truckman.*

truck·er[2] (truk′ər) *n.* **1** *U.S.* A market gardener; a truck farmer. **2** One who barters or sells commodities; a hawker.

truck farm *U.S.* A farm on which vegetables are produced for market. [<TRUCK[2] + FARM] — **truck farming**

truck·head (truk′hed′) *n.* The terminal to which supplies are brought by truck and from which they are distributed to the required points. [< TRUCK[1] + HEAD; on analogy with *railhead*]

truck house Formerly, a building used to store articles used in trading with the Indians. Also **trucking house.**

truck·ing[1] (truk′ing) *n.* The act or business of transportation by trucks.

truck·ing[2] (truk′ing) *n.* **1** Exchanging or bartering; dealings; intercourse. **2** *U.S.* Cultivation of vegetables for market; truck farming.

truck·le (truk′əl) *v.* **·led, ·ling** *v.i.* **1** To yield meanly or weakly: with *to.* **2** To roll on truckles or casters. — *v.t.* **3** To cause to roll on truckles or casters. [<*n.*] — *n.* **1** A small wheel. **2** *Dial.* A trundle bed. [<AF *trocle, trokle* <L *trochlea.* See TROCHLEA.] — **truck′ler** *n.* — **truck′ling·ly** *adv.*

truckle bed A trundle bed.

truck·man[1] (truk′mən) *n.* **1** A truck driver. **2** One engaged in the business of trucking.

truck·man[2] (truk′mən) *n. pl.* **·men** (-mən) A dealer in truck; one who trucks or trades.

truck system The practice of paying wages to workmen in goods instead of money.

truc·u·lence (truk′yə·ləns) *n.* Savageness of character, behavior, or aspect. Also **truc′u·len·cy.**

truc·u·lent (truk′yə·lənt) *adj.* **1** Of savage character; awakening terror; cruel; ferocious. **2** Scathing; harsh; violent: said of writing or speech. [<L *truculentus* < *trux, trucis* fierce] — **truc′u·lent·ly** *adv.*

Tru·deau (tro͞o′dō), **Edward Livingston,** 1848–1915, U.S. physician, pioneer in tuberculosis treatment.

trudge (truj) *v.i.* **trudged, trudg·ing** To walk wearily or laboriously; plod. — *n.* A tiresome walk or tramp. [Earlier *tredge, tridge*; origin uncertain] — **trudg′er** *n.*

trudg·en stroke (truj′ən) In swimming, a former racing stroke similar to the crawl stroke but performed with a frog kick or a scissors kick. Also **trudgen, trudgeon stroke.** [after John *Trudgen,* 19th c. British swimmer, who introduced the stroke into England, 1873]

Tru·dy (tro͞o′dē) Diminutive of GERTRUDE.

true (tro͞o) *adj.* **tru·er, tru·est 1** Faithful to fact or reality; not false or erroneous: a *true* judgment or proposition. In this sense *true* is often used elliptically for *it is true* or *that is true*: *True,* I hated him. **2** Being real or natural; genuine, not counterfeit: a *true* specimen, *true* gold. **3** Faithful to friends, promises, or principles; loyal; steadfast: *true* love, a *true* friend. **4** Conformable to an existing standard type or pattern; exact: a *true* copy. **5** Faithful to the requirements of law or justice; legitimate: the *true* king. **6** Faithful to truth; trustful; honest: a *true* man. **7** Faithful to the promise or predicted event; correctly indicative: a *true* sign. **8** *Biol.* **a** Possessing all the attributes of a developed organ or structure of its class; complete. **b** Fulfilling a given function, homologous with organs performing a like function, or being the essential part of an organ. **c** Of pure strain or pedigree: a *true* collie dog. **d** Conformed to the structure of the type; properly so called: said of a plant or animal, as distinguished from others improperly so called: a *true* locust. **9** Exactly correspondent in pitch or key; in perfect tune: His voice is *true.* See synonyms under AUTHENTIC, CORRECT, FAITHFUL, GOOD, HONEST, JUST[1], MORAL, RIGHT, PURE. — *n.* **1** Truth; covenant; pledge. **2** *pl.* **trues** or **truce** *Obs.* An armistice or truce. — **in** (or **out of**) **true** In (or not in) line of adjustment: said of a mark or part, as in a drawing or a machine. — *adv.* **1** In truth; truly. **2** In a true and accurate manner: The wheel runs *true.* **3** Conformably to the ancestral type: in the phrase *to breed true.* — *v.t.* **trued, tru·ing** To bring to conformity with a standard or requirement; form or adjust, as with geometrical precision: to *true* a frame or a tool. [OE *trēowe.* Akin to TRUCE, TRUST.] — **true′ness** *n.*

true bill *Law* **1** The endorsement by a grand jury on a bill of indictment which they find to be sustained by the evidence. **2** A bill so endorsed.

true·blue (tro͞o′blo͞o′) *n.* **1** Originally, a fast blue color or dye; hence, constancy or unchangingness. **2** In the 17th century, a Scotch Presbyterian or Covenanter, so called from the blue adopted as the distinctive color of his political party. **3** A person of uncompromising faithfulness or loyalty as to party, sect, friendship, or principle. — *adj.* Staunch; faithful; dependable; genuine.

true copy An exact, verbatim transcript of any document, report, etc.; especially, one certified as correct by a qualified authority.

true level The geoid.

true·love (tro͞o′luv′) *n.* **1** One truly beloved; a sweetheart: used also adjectively. **2** *Obs.* Truelovers' knot. **3** The herb-Paris, so called because its four leaves are set together in the form of a truelovers' knot.

true·lov·ers' knot (tro͞o′luv′·ərz) A complicated double knot, a symbol of fidelity in love.

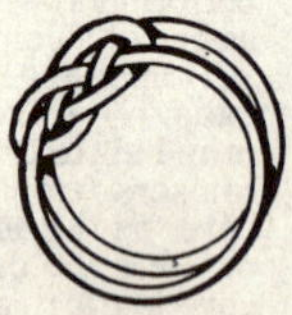

TRUELOVERS' KNOT
One of several so called.

true·pen·ny (tro͞o′pen′ē) *n.* *Archaic* **1** Originally, a coin of genuine metal. **2** A trusty or genuine person; an honest fellow.

true rib See under RIB.

true time Mean time, or mean solar time.

true toxin An endotoxin.

truf·fle (truf′əl, tro͞o′fəl) *n.* Any of various fleshy underground fungi (genus *Tuber*), regarded as a choice table delicacy. [<OF *trufe, truffe,* prob. <Ital. *truffa,* ult. <L *tuber* a tuber]

tru·ism (tro͞o′iz·əm) *n.* An obvious or self-evident truth; a platitude. See synonyms under AXIOM.

Tru·jil·lo (tro͞o·hē′yō), **Ciudad** See CIUDAD TRUJILLO.

Tru·jil·lo Mo·li·na (tro͞o·hē′yō mō·lē′nä), **Rafael,** born 1891, Dominican general; president of the Dominican Republic 1930–38, 1942–57.

Truk (truk, tro͝ok) An island group in the eastern Caroline Islands; total, 40 square miles.

trull (trul) *n.* A prostitute; drab. [<G *trulle, trolle.* ? Akin to TROLL[1].]

tru·ly (tro͞o′lē) *adv.* **1** In conformity with fact. **2** With accuracy. **3** With loyalty or fidelity. **4** *Archaic* Surely; verily. **5** Lawfully; legally.

Tru·man (tro͞o′mən), **Harry S,** born 1884, president of the United States 1945–1953.

Trumbull (trum′bəl), **John,** 1750–1831, American poet and satirist. — **John,** 1756–1843, American painter. — **Jonathan,** 1710–85, American statesman.

trump[1] (trump) *n.* **1** In various card games, a card of the suit selected to rank above all others temporarily. **2** The suit thus determined: usually in the plural. **3** *Colloq.* A very acceptable and agreeable person; good fellow. — *v.t.* **1** To take (another card) with a trump. **2** To surpass; excel; beat. — *v.i.* **3** To play a trump. — **to trump up** To make up or invent for a fraudulent purpose. [Alter. of TRIUMPH]

trump[2] (trump) *n.* **1** *Poetic* A trumpet. **2** *Scot.* A jew's-harp. [<OF *trompe* <Gmc.]

trump·er·y (trum′pər·ē) *n. pl.* **·er·ies 1** Worthless finery. **2** Rubbish; nonsense. **3** Deceit; trickery. See synonyms under GAUD. — *adj.* Having a showy appearance, but valueless. [<OF *tromperie* < *tromper* TRUMP[1], *v.*]

trum·pet (trum′pit) *n.* **1** A soprano wind instrument with a flaring bell and a long metal tube. The tube was formerly always straight, but now may recurve singly or doubly. **2** A powerful reed stop in an organ. **3** Something resembling a trumpet in form. **4** A tube for collecting and conducting sounds to the ear; an ear trumpet. **5** A loud penetrating sound like that of a trumpet; trumpeting. **6** *pl.* A pitcherplant (*Sarracenia flava*) of the southern United States having trumpet-shaped leaves. **7** *Obs.* A trumpeter. — *v.t.* **1** To sound or proclaim by or as by trumpet; publish abroad. — *v.i.* **2** To blow a trumpet. **3** To give forth a sound as if from a trumpet. [<OF *trompette,* dim. of *trompe* TRUMP[2]]

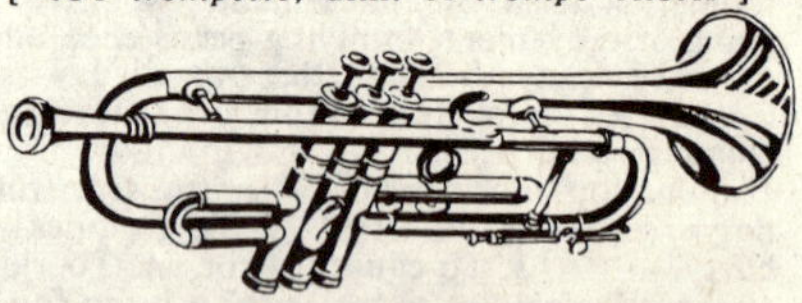

TRUMPET

trumpet creeper A woody vine (*Campsis radicans*) of the southern United States, with scarlet trumpet-shaped flowers. Also **trumpet vine.**

trum·pet·er (trum′pit·ər) *n.* **1** One who sounds a trumpet. **2** One who publishes something loudly abroad. **3** A large South American bird, related to the cranes; especially, the golden-breasted trumpeter (*Psophia crepitans*), often domesticated. **4** A large North American wild swan (*Cygnus buccinator*), having a clarionlike cry: now very scarce: also **trumpeter swan. 5** One of a breed of domestic pigeons.

trumpet flower Any of various plants having trumpet-shaped flowers, as the trumpet creeper, the trumpet honeysuckle.

trumpet honeysuckle A twining honeysuckle (*Lonicera sempervirens*) with oblong leaves and trumpet-shaped flowers, scarlet without and yellow within.

trumpet tree A West Indian and South American tree (*Cecropia peltata*) whose hollow branches are used for musical instruments. Also **trum′pet·wood′** (-wo͝od′).

trum·pet·weed (trum′pit·wēd′) *n.* **1** The joe-pye weed. **2** The boneset.

trun·cate (trung′kāt) *v.t.* **·cat·ed, ·cat·ing** To cut the top or end from. — *adj.* **1** Truncated. **2** *Biol.* Appearing as though cut or broken squarely off, as the end of certain leaves and shells, the tail of certain birds, the caudal fin of some fishes, etc. [<L *truncatus,* pp. of *truncare* < *truncus* TRUNK] — **trun·ca·tion** (trung·kā′shən) *n.*

trun·cat·ed (trung′kā·tid) *adj.* **1** Cut off; shortened. **2** Describing a cone or pyramid whose vertex is cut off by a plane usually parallel to the base. **3** *Mineral.* Having the edges or angles cut off, as certain crystals. **4** *Biol.* Truncate.

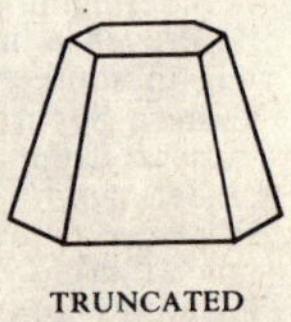

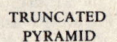

TRUNCATED PYRAMID

trun·cheon (trun′chən) *n.* **1** A short, heavy stick; a club; staff. **2** The baton of a military officer or marshal. **3** A tree whose branches have been lopped off to hasten growth; tree trunk: the original meaning. **4** *Obs.* A short club or cudgel; a spear shaft. **5** *Brit.* A policeman's club. — *v.t.* To beat as with a truncheon; cudgel. [<OF *truncun, tronchon* a stump, ult. <L *truncus* TRUNK]

trun·dle (trun′dəl) *n.* **1** A small broad wheel, as of a caster. **2** The act, motion, or sound of trundling. **3** A trundle bed. **4** A lantern wheel. **5** *Obs.* A small low-wheeled vehicle; truck. — *v.t.* & *v.i.* **·dled, ·dling 1** To roll along, as a hoop. **2** To rotate. [Var. of TRINDLE] — **trun′dler** *n.*

trundle bed A bed with very low frame resting upon casters, so that it may be rolled under another bed: also called *truckle bed.*

trun·dle·tail (trun′dəl-tāl′) *n. Obs.* A curly-tailed dog; also, a curly tail.

trunk (trungk) *n.* **1** The main stem or stock of a tree, as distinguished from its branches or roots. **2** The human body, apart from the head, neck, and limbs; the torso. **3** *Entomol.* The thorax. **4** *Anat.* The main stem of a nerve, blood vessel, or lymphatic. **5** The main line of a communication or transportation system. **6** The circuit connecting two telephone exchanges. **7** The main body, line, or stem of anything, as distinct from its appendages. **8** A proboscis, as of an elephant. **9** A large box or case used for packing and carrying clothes or other articles, as for a journey. **10** A large compartment at the rear of an automobile, used for storage. **11** *pl.* A close-fitting garment covering the loins and often part of the thighs, worn by male swimmers, athletes, etc. **12** *pl. Obs.* Trunk hose. **13** *Mech.* **a** A trough, chute, or conduit. **b** A large hollow piston in which a connecting rod moves. **14** *Naut.* **a** The well for the centerboard of a vessel. **b** A casing connecting the hatchways of two or more decks and forming a shaft. **c** Any structure placed on the upper deck of a ship, as for shelter. **15** *Archit.* The shaft of a column. See synonyms under BODY. — *adj.* Being or belonging to a trunk or main body: a *trunk* railroad. [<OF *tronc* <L *truncus* stem, trunk, orig. adj., mutilated; def. 8 infl. in meaning by F *trompe* a trumpet]

trunk engine A steam engine having a trunk or open cylinder attached to the piston in place of the usual piston rod, permitting direct attachment of the connecting rod to the piston head.

trunk·fish (trungk′fish′) *n. pl.* **·fish** or **·fish·es** A plectognath fish (family *Ostraciidae*) of warm seas, characterized by a body covering of hard, bony plates.

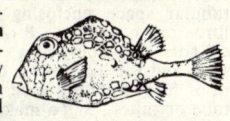

TRUNKFISH
(Rarely to 10 inches)

trunk hose Full breeches worn by gentlemen in the 16th and early 17th centuries, extending from the waist to the middle of the thigh: originally one piece with the hose. Also **trunk breeches.**

trunk line The main line of a transportation or communication system, as distinguished from a branch line.

trunk sleeve A sleeve made very full at the top after the manner of trunk hose.

trun·nel (trun′əl) *n.* A treenail. [Var. of TREENAIL]

trun·nion (trun′yən) *n.* **1** One of two opposite cylindrical projections from the sides of a cannon, forming an axis on which it is elevated or depressed. **2** A similar support on which the cylinders of some engines oscillate. [<F *trognon* the core of a fruit, a stump, trunk; ult. origin unknown]

Tru·ro (trŏŏr′ō) **1** A municipal borough in SW Cornwall, England. **2** A port of central Nova Scotia, Canada.

truss (trus) *n.* **1** *Med.* A bandage or support for a rupture. **2** A braced framework of ties, beams, or bars, usually arranged in a series of triangles, as for the support of a roof, airplane, or bridge. **3** A bundle, especially of hay or straw. In England, 56 pounds of old or 60 pounds of new hay make a *truss*; 36 pounds make a *truss* of straw. **4** *Naut.* A heavy iron piece by which a lower yard is attached to a mast. **5** *Bot.* A compact terminal cluster of flowers. **6** *Archit.* A projection from the face of a wall, used to support a cornice; a large corbel; a bracket or modillion. **7** A pack; package. — *v.t.* **1** To tie or bind; fasten with *up.* **2** To support by a truss; brace, as a roof. **3** To fasten the wings of (a fowl) with skewers or twine before cooking. **4** To fasten, tighten, or tie around one, as a garment or laces. **5** To hang, as a criminal: with *up.* [< OF *trusse, trousse* < *trousser, trusser* pack up, bundle, prob. <L *torca* a bundle < *torques.* See TORQUES.] — **truss′er** *n.*

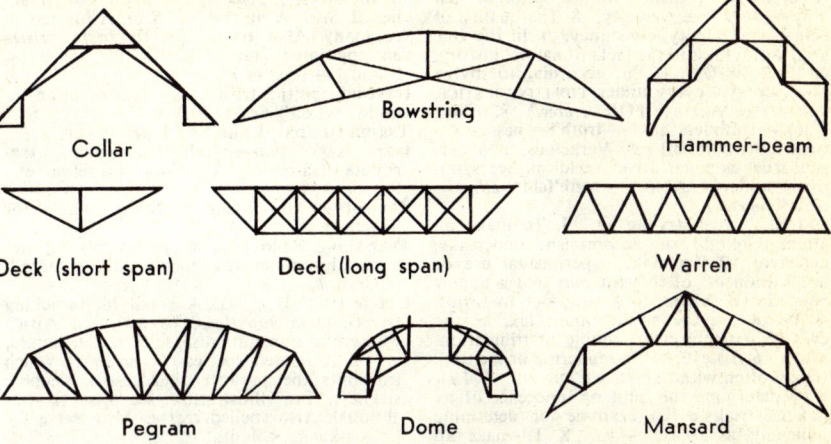

Collar   Bowstring   Hammer-beam

Deck (short span)   Deck (long span)   Warren

Pegram   Dome   Mansard

TYPES OF TRUSSES

truss bridge A bridge stiffened, supported, or formed by a truss or construction of trusses.

truss·ing (trus′ing) *n.* **1** A system of diagonal tension rods and struts for strengthening or stiffening a structure, as a railway car or a vessel's hull. **2** Trusses collectively. **3** The act of one who trusses. **4** A bracing with ties, struts, or the like.

trust (trust) *n.* **1** A confident reliance on the integrity, veracity, or justice of another; confidence; faith; also, the person or thing so trusted. **2** Something committed to one's care for use or safekeeping; a charge; responsibility. **3** The state or position of one who has received an important charge. **4** A confidence in the reliability of persons or things without careful investigation. **5** Credit, in the commercial sense. **6** *Law* The confidence, or the obligation arising from the confidence, reposed in a person (called the *trustee*) to whom the legal title to property is conveyed for the benefit of another (the *cestui que trust*), that he will faithfully apply the property according to such confidence; also, the beneficial title or ownership of property of which the legal title is in another. ◆ Collateral adjective: *fiducial.* **7** The property or thing held in trust; also, the relation subsisting between the holder and the property so held. **8** A permanent combination, now illegal, for the purpose of controlling the production, price, etc., of some commodity or the management, profits, etc., of some business; also, a trust company. Compare CARTEL, CORNER, MONOPOLY, POOL², SYNDICATE. **9** Confident expectation; belief; hope. **10** Custody; care; keeping. **11** *Obs.* Trustworthiness. — *v.t.* **1** To have trust in; rely upon. **2** To commit to the care of another; entrust. **3** To commit something to the care of: with *with.* **4** To allow to do something without fear of the consequences. **5** To expect with confidence or with hope. **6** To believe. **7** To allow business credit to. — *v.i.* **8** To place trust or confidence; rely: with *in.* **9** To hope: with *for.* **10** To allow business credit. — **to trust to** To depend upon; confide in. — *adj.* Held in trust: *trust* property, *trust* money. [<ON *traust,* lit., firmness. Akin to TRUCE, TRUE.] — **trust′er** *n.* — **trust′less** *adj.*
Synonyms (noun): assurance, belief, confidence, credence, expectation, faith, hope. See ASSURANCE, BELIEF, FAITH.
Synonyms (verb): believe, commit, confide, hope. See COMMIT, LEAN¹. Antonyms: despair, disbelieve, discredit, distrust, mistrust, suspect.

trust company An incorporated institution empowered by its charter to accept and execute trusts, as provided by law, to receive deposits of money and other personal property and issue obligations therefor, and to lend money on real and personal securities.

trus·tee (trus-tē′) *n.* **1** One who holds property in trust; especially, in popular usage, one of a body of men, often elective, who hold the property and manage the affairs of a church or public institution. **2** One in whose hands property is attached by a trustee process. — *v.t.* **·teed, ·tee·ing 1** *Law* To attach by trustee process (the property of a debtor in the hands of a third person). **2** To place (property) in the care of a trustee.

trustee process A statutory remedy whereby a creditor may reach property or assets of his debtor in the hands of a third person.

trus·tee·ship (trus-tē′ship) *n.* **1** The post or function of a trustee. **2** Supervision and control of a trust territory by a country or countries commissioned by the United Nations; also, the territory so controlled.

trust·ful (trust′fəl) *adj.* Disposed to trust. — **trust′ful·ly** *adv.* — **trust′ful·ness** *n.*

trust fund Money, securities, or similar property held in trust.

trust·ing (trus′ting) *adj.* Having trust; trustful. — **trust′ing·ly** *adv.* — **trust′ing·ness** *n.*

trust officer An administrator of a trust company or the trust department of a bank.

Trust Territory An area, usually a former colonial possession, governed by a member state of the United Nations as an Administering Authority reporting to the United Nations Trusteeship Council functioning under the authority of the General Assembly with the exception that the trusteeships of the areas designated as strategic are supervised by the Security Council after first having approved the trust agreements. Trust Territories include former League of Nations mandates. Also **trust territory, UN Trust Territory.** The independence of the Cameroons, Somaliland, and Togoland in 1960–61 reduced the Trust Territories to:

Trust Territory	Administered by
Nauru	Australia on behalf of Australia, New Zealand, and the United Kingdom
(Papua and) New Guinea	Australia
Ruanda-Urundi	Belgium
West Samoa	New Zealand
Tanganyika	The United Kingdom
Trust Territory of the Pacific Islands	United States (a strategic territory)

trust·wor·thy (trust′wûr′thē) *adj.* Worthy of confidence; reliable. See synonyms under AUTHENTIC, FAITHFUL, HONEST, RELIABLE, STAUNCH. — **trust′wor′thi·ly** (-wûr′thə-lē) *adv.* — **trust′wor′thi·ness** *n.*

trust·y (trus′tē) *adj.* **trust·i·er, trust·i·est 1** Faithful to duty or trust. **2** Staunch; firm. **3** *Obs.* Trustful. See synonyms under FAITHFUL, HONEST, JUST¹, RELIABLE, STAUNCH. — *n. pl.* **trust·ies** A trustworthy person; especially, a convict who has been found serviceable and reliable and to whom special liberties are granted. —**trust'i·ly** *adv.* —**trust'i·ness** *n.*

truth (trōōth) *n. pl.* **truths** (trōōthz, trōōths) **1** The state or character of being true in relation to being, knowledge, or speech. **2** Conformity to fact or reality. **3** Conformity to rule, standard, model, pattern, or ideal. **4** Conformity to the requirements of one's being or nature; steadfastness; sincerity. **5** That which is true; a statement or belief which corresponds to the reality. **6** A fact as the object of correct belief; reality. **7** A tendency or disposition to speak or tell only what is true; veracity. **8** The quality of being true; fidelity; constancy. **9** In the fine arts, faithfulness to the facts of nature, history, or life. **10** *Obs.* Right, according to divine law. See synonyms under FIDELITY, JUSTICE, VERACITY, VIRTUE. [OE *trēowth* < *trēowe* true] —**truth'less** *adj.* —**truth'less·ness** *n.*

truth·ful (trōōth'fəl) *adj.* Veracious, as a person; true, as a narrative; veridical. See synonyms under CANDID. —**truth'ful·ly** *adv.* —**truth'ful·ness** *n.*

try (trī) *v.* **tried, try·ing** *v.t.* **1** To make an attempt to do or accomplish; undertake; endeavor. **2** To make experimental use or application of: often with *out:* to *try* a new pen. **3** To subject to a test; put to proof. **4** To put severe strain upon; tax, as the eyes. **5** To subject to trouble or tribulation; afflict. **6** To extract by rendering or melting; refine: often with *out:* to *try* out oil. **7** *Law* **a** To determine the guilt or innocence of by judicial trial. **b** To examine or determine judicially, as a case. —*v.i.* **8** To make an attempt; put forth effort. **9** To make an examination or test. See synonyms under CHASTEN, ENDEAVOR, EXAMINE, STRUGGLE. —**to try on** To put on (a garment) to test it for fit or appearance. —**to try out** To attempt to qualify: He *tried out* for the football team. —*n. pl.* **tries** **1** The act of trying; trial; experiment. **2** In Rugby football, the act of touching the ball down behind an opponent's goal, which scores three points. [<OF *trier* sift, pick out, prob. <LL *tritare* thresh <L *tritus.* See TRITE.] —**tri'er** *n.*

try·ing (trī'ing) *adj.* Testing severely; hard to endure. See synonyms under ARDUOUS, DIFFICULT, TROUBLESOME.

trying plane A long plane used to true up the edges of boards to be joined; a jointer. Also **try plane.**

try·lon (trī'lon) *n.* A three-sided pylon: used as part of the main gateway to the New York World's Fair, 1939. [<TR(I)- + (P)YLON]

try·ma (trī'mə) *n. pl.* **·ma·ta** (-mə·tə) *Bot.* A drupelike, commonly two-celled fruit with a bony nucleus and a fleshy, leathery, or fibrous dehiscent or separating exocarp, as the hickory nut and walnut. [<NL <Gk. *tryma, trymē* a hole < *tryein* wear away]

try-out (trī'out') *n. U.S. Colloq.* A test of ability, as of an actor or athlete, often in competition with others.

tryp·a·no·some (trip'ə·nə·sōm') *n.* Any of a genus *(Trypanosoma)* of flagellate infusorians infesting the blood of man and some lower animals. They destroy the red corpuscles, and cause serious and even fatal diseases, as the sleeping sickness. Also **tryp'a·no·so'ma** (-sō'mə). [<Gk. *trypanon* a borer + -SOME[2]]

tryp·a·no·so·mi·a·sis (trip'ə·nō·sō·mī'ə·sis) *n. Pathol.* Any disease caused by the presence in the body of trypanosomes. Also **tryp'a·no·so'ma·to'sis** (-sō'mə·tō'sis). [<TRYPANOSOME + -IASIS]

tryp·ars·am·ide (trip'är·sam'id, -īd, trip·är'sə·mid, -mīd) *n. Chem.* A colorless crystalline compound, $C_8H_{10}O_4N_2AsNa$, used in the treatment of trypanosomiasis and certain forms of syphilis. [<TRYP(ANOSOME) + ARS(ENIC) + AMIDE]

tryp·sin (trip'sin) *n. Biochem.* A proteolytic enzyme contained in the pancreatic juice. [<G <Gk. *tripsis* a rubbing (< *tribein* rub) + (PEP)SIN] —**tryp'tic** (-tik) *adj.*

tryp·sin·o·gen (trip·sin'ə·jen) *n. Biochem.* The substance secreted by the pancreas and converted into trypsin by the action of intestinal enzymes. [<*trypsino-* <TRYPSIN + -GEN]

tryp·to·phan (trip'tə·fan) *n. Biochem.* A crystalline amino acid, $C_{11}H_{12}O_2N_2$, contained in variable amounts in most proteins and associated with the digestive functions. Also **tryp'to·phane** (-fān). [< *tryptic* (<TRYPSIN) + *-phan,* var. of -PHANE]

try·sail (trī'səl, -sāl') *n. Naut.* A small sail bent to a gaff abaft the foremast and mainmast of a ship: also called *spencer.* [< nautical phrase *(at) try* lying to in a storm + SAIL]

try square A carpenter's square having usually a wooden stock and a steel blade.

tryst (trist, trīst) *v.t.* **1** To agree to meet. **2** To appoint (a time), as for meeting. **3** To arrange for in advance; engage. **4** *Obs.* To trust. —*v.i.* **5** To agree upon some place or time of meeting. —*n.* **1** An appointment to meet, or the meeting place agreed upon: also **tryst'·ing.** **2** *Scot.* A market. **3** *Scot.* A journey in company. Also **tryste.** [<OF *triste, tristre* an appointed station in hunting, prob. < Scand.] —**tryst'er** *n.*

tryst·ed (tris'tid, trī'stid) *adj.* Agreed upon.

tsa·de (tsä·dā') See SADE.

Tsa·na (tsä'nä), **Lake** See TANA, LAKE.

tsar (tsär), **tsar·e·vitch** (tsär'ə·vich), **tsa·rev·na** (tsä·rev'nä), **tsa·ri·na** (tsä·rē'nä) etc. See CZAR, etc.

Tsa·ri·tsyn (tsä·rē'tsin) A former name for STALINGRAD.

Tsar·sko·e Se·lo (tsär'skə·yə sye·lô') Former imperial summer residence near Leningrad: modern *Pushkin.*

tset·se (tset'sē) *n.* **1** A small bloodsucking fly *(Glossina morsitans)* of southern Africa whose bite transmits disease in cattle, horses, etc. **2** A related species *(G. palpalis),* which transmits the parasite that causes sleeping sickness. For illustration see INSECTS (injurious). Also spelled *tzetze.* Also **tsetse fly.** [<Afrikaans <Bantu]

Tshi (chwē, chē) See TWI.

T-shirt (tē'shûrt') *n.* **1** A cotton undershirt with short sleeves: so called because T-shaped. **2** A sleeveless jersey or sweater of similar cut for outer wear. Also spelled *tee shirt.* Also **T shirt.**

Tsi·nan (jē'nän') A port in NE China, capital of Shantung province; a former treaty port: also *Chinan.*

Tsing·hai (ching'hī') A province of NW China; 318,450 square miles; capital, Sining: also *Chinghai.*

Tsing·tao (ching'dou') A port in eastern Shantung province, China.

Tsi·tsi·har (tsē'tsē'här') A city in former north central Manchuria, NE China, former capital of Heilungkiang province.

Tso·ne·can (tsō·nā'kən) *n.* A linguistic stock of South American Indians, including all the Tehuelchan tribes, and, possibly, the Onas of Tierra del Fuego. By some linguists called **Cho·ne·an** (chō'nē·ən).

T-square (tē'skwâr') *n.* An instrument by which to measure or lay out right angles or parallel lines, consisting usually of a flat strip with a shorter head at right angles to it and slightly offset so that it may be slid along the edge of a drawing board.

Tsu·ga·ru Strait (tsōō·gä·rōō) The passage from the Sea of Japan to the Pacific between Honshu and southern Hokkaido, Japan; 15 to 25 miles wide.

tsu·na·mi (tsōō·nä'mē) *n.* An extensive and often very destructive ocean wave caused by a violent submarine earthquake: erroneously called a tidal wave. [<Japanese, a storm wave]

Tsu·shi·ma (tsōō·shē·mä) A Japanese island in Korea Strait; 271 square miles, including 42 offshore islets; scene of a naval battle in the Russo-Japanese War in which the Russian fleet was destroyed, 1905.

tsu·tsu·ga·mu·shi disease (tsōō·tsōō'gä·mōō'·shē) *Pathol.* A rickettsial fever endemic in Japan and the Orient, caused by a micro-organism *(Rickettsia orientalis)* transmitted to man by the infected larvae of a mite (genus *Trombicula*): also called *Japanese river fever, river fever, scrub typhus.* [<Japanese *tsutsugamush,* a small Japanese mite < *mushi* a bug]

Tu·a·mo·tu Archipelago (tōō'ä·mō'tōō) An island chain extending 1,300 miles south of the Marquesas Islands in eastern French Oceania; 330 square miles: also *Low Archipelago:* formerly *Paumotu Archipelago.*

Tuan (twän) *n.* Sir; mister: courteous Malayan form of address for a European. [< Malay]

Tua·reg (twä'reg) *n.* **1** A member of the nomadic Berber tribes of the central and western Sahara. **2** The Berber dialect spoken by these people.

tu·a·ta·ra (tōō'ä·tä'rä) *n.* A sphenodon. Also **tu'a·te'ra** (-tā'rä). [<Maori < *tua* on the farther side, the back + *tara* the spine]

Tu·a·tha De Da·naan (thōō'ə·hə dā dä'nôn) In ancient Irish mythology, a race of gods who ruled in Ireland until their defeat by the Milesians: now conceived of as fairies. See DANU. [<OIrish, people of Danu]

tub (tub) *n.* **1** A broad, open-topped vessel, usually of wood, and formed with staves, bottom, hoops, and handles on the side. **2** A bathtub. **3** *Brit. Colloq.* A bath taken in a tub. **4** The amount that a tub contains. **5** Anything resembling a tub, as a broad, clumsy boat: contemptuous or humorous. **6** A small cask. **7** A bucket for bringing ore or coal up a shaft; also, an underground tram. **8** A keeve. **9** A sweating in a tub. —*v.t.* & *v.i.* **tubbed, tub·bing** To wash, bathe, or place in a tub. [<MDu. *tubbe*] —**tub'ba·ble** *adj.* —**tub'ber** *n.*

tu·ba (tōō'bə, tyōō'-) *n. pl.* **·bas** or **·bae** (-bē) **1** A large bass instrument of the saxhorn family. **2** An ancient Roman war trumpet. **3** A powerful reed stop in an organ. [<L, a war trumpet]

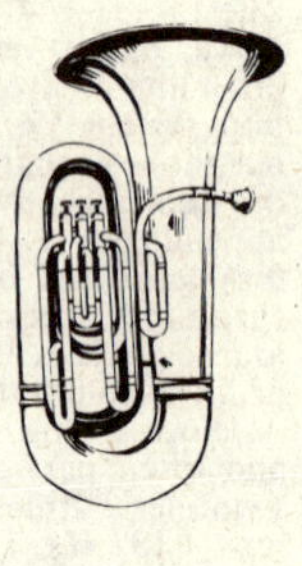

TUBA

tu·bal (tōō'bəl, tyōō'-) *adj.* **1** Relating to a tube. **2** *Anat.* Pertaining to the Fallopian tube.

Tu·bal-cain (tōō'bəl·kān, tyōō'-) The first artificer in brass and iron. *Gen.* iv 22.

tu·bate (tōō'bāt, tyōō'-) *adj.* Of the form of or provided with a tube; tubular. [<NL *tubatus* <L *tubus* a pipe]

tub·by (tub'ē) *adj.* **·bi·er, ·bi·est** **1** Resembling a tub in form; round and fat; corpulent. **2** Lacking resonance when struck; sounding dull or wooden, as a musical instrument.

tube (tōōb, tyōōb) *n.* **1** A long hollow cylindrical body of metal, glass, rubber, etc., generally used for the conveyance of something through it; a pipe. **2** The principal part of a gun. **3** Any similar device having a tube or tubelike part, as a telescope. **4** *Biol.* Any elongated hollow part or organ, as the united part of a gamopetalous corolla or a gamosepalous calyx. **5** A subway or a tunnel. **6** An electron, thermionic, or vacuum tube. **7** The tubular space enclosing lines of magnetic force or induction. **8** A collapsible metal cylinder for containing paints, toothpaste, glue, and the like. —*v.t.* **tubed, tub·ing** **1** To fit or furnish with a tube. **2** To enclose in a tube or tubes. **3** To make tubular. [<F <L *tubus* a tube] —**tube'less** *adj.* —**tub'er** *n.*

Tube may appear as a combining form in hyphemes or solidemes, with the meaning of noun definition 1:

tube-drawing	**tube-rolling**
tube-drilling	**tube-scraping**
tube-fed	**tube-shaped**
tube-filling	**tubesmith**
tubemaker	**tube-straightening**
tubemaking	**tubework**

tube foot *Zool.* An ambulacral sucker; one of the small vascular locomotor processes exserted through the ambulacral pores of echinoderms.

tu·ber (tōō'bər, tyōō'-) *n.* **1** *Bot.* A short, thickened portion of an underground stem, as in the potato or artichoke. **2** *Anat.* A swelling or prominence; tubercle. [<L, a swelling]

tu·ber·cle (tōō'bər·kəl, tyōō'-) *n.* **1** A small rounded eminence or nodule. **2** *Bot.* A minute swelling on the roots of leguminous plants, which contains a micro-organism believed to absorb nitrogen from the air for the use of the plant. **3** *Pathol.* A small granular tumor formed within an organ from morbid or infected matter: in the lungs, the seat of pulmonary consumption. **4** *Anat.* A small knoblike excrescence, especially on the skin or on a bone. [<L *tuberculum,* dim. of *tuber* a swelling] —**tu·ber·cu·loid** (tōō·bûr'kyə·loid, tyōō-) *adj.*

tubercle bacillus The rod-shaped, Gram-positive bacterium *(Mycobacterium tuberculosis)* which is the cause of tuberculosis in man.

tu·ber·cu·lar (tōō·bûr′kyə·lər, tyōō-) *adj.* 1 Affected with tubercles; nodular. 2 Tuberculous. — *n.* One affected with tuberculosis.
tu·ber·cu·late (tōō·bûr′kyə·lit, -lāt) *adj.* 1 Nodular. 2 Affected with tubercles; tuberculous. Also **tu·ber′cu·lat′ed**. [<NL *tuberculatus* <L *tuberculum* TUBERCLE] — **tu·ber′cu·la′tion** *n.*
tu·ber·cu·lin (tōō·bûr′kyə·lin, tyōō-) *n. Bacteriol.* A sterile liquid prepared from attenuated cultures of the tubercle bacillus, used especially as a test for tuberculosis in children and animals. Also **tu·ber′cu·line** (-lin, -lēn). [<L *tuberculum* TUBERCLE + -IN]
tuberculo- *combining form* 1 Tuberculosis; of or pertaining to tuberculosis; tuberculous. 2 The tubercle bacillus. Also, before vowels, **tubercul-**. [<L *tuberculum*, dim. of *tuber* a swelling]
tu·ber·cu·lo·sis (tōō·bûr′kyə·lō′sis, tyōō-) *n. Pathol.* A communicable disease caused by infection with the tubercle bacillus, characterized by the formation of tubercles within some organ or tissue: when affecting the lungs, known as **pulmonary tuberculosis**. [<NL <L *tuberculum* TUBERCLE + -OSIS]
tu·ber·cu·lous (tōō·bûr′kyə·ləs, tyōō-) *adj.* Of, pertaining to, or affected with tuberculosis.
tu·ber·if·er·ous (tōō′bə·rif′ər·əs, tyōō-) *adj.* Bearing or producing tubers. [<TUBER + -(I)FEROUS]
tu·ber·oid (tōō′bər·oid, tyōō′-) *adj.* Resembling a tuber.
tube·rose¹ (tōōb′rōz′, tyōōb′-, tōō′bə·rōs′, tyōō′-) *n.* A bulbous plant (*Polianthes tuberosa*) of the amaryllis family, bearing a long raceme of fragrant white flowers. [<NL *Tuberosa*, species name <L *tuberosus* knobby <*tuber* a swelling]
tu·ber·ose² (tōō′bər·ōs, tyōō′-) *adj.* Tuberous. [<TUBER + -OSE]
tu·ber·os·i·ty (tōō′bə·ros′ə·tē, tyōō′-) *n. pl.* **·ties** 1 The state of being tuberous. 2 A swelling or protuberance. 3 *Anat.* A large, rough eminence on a bone, as for the attachment of a muscle.
tu·ber·ous (tōō′bər·əs, tyōō′-) *adj.* 1 Bearing projections or prominences. 2 Resembling tubers. 3 *Bot.* Bearing tubers.
tuberous root *Bot.* One of the tuberlike parts of a multiple or fascicled fleshy root, as in the dahlia.
tu·bi·form (tōō′bə·fôrm, tyōō′-) *adj.* Having the form of a tube; tubular. [<*tubi-* (<TUBE) + -FORM]
tub·ing (tōō′bing, tyōō′-) *n.* 1 Tubes collectively. 2 A piece of tube or material for tubes. 3 Material for pillowcases. 4 The act of making tubes.
Tü·bing·en (tü′bing·ən) A university town in the state of Baden-Württemberg, SW West Germany.
Tub·man (tub′mən), **William Vacanarat**, born 1895, Liberian statesman; president of Liberia 1944-.
tub·thump·er (tub′thum′pər) *n. U.S. Colloq.* A noisy speaker; a soapbox orator.
Tu·bu·ai Islands (tōō′bōō·ī′) An island group south of the Society Islands, comprising a part of French Oceania; 115 square miles: also *Austral Islands*.
tu·bu·lar (tōō′byə·lər, tyōō′-) *adj.* 1 Having the form of a tube; tube-shaped. 2 Made up of or provided with tubes. 3 Pertaining to or sounding as if produced in a tube. [<L *tubulus* TUBULE]
tu·bu·late (tōō′byə·lāt, tyōō′-) *v.t.* **·lat·ed, ·lat·ing** 1 To shape or fashion into a tube. 2 To furnish with a tube. — *adj.* Shaped like or into a tube; also, provided with a tube: also **tu′bu·lat′ed**. [<L *tubulatus* tubular <*tubulus* TUBULE] — **tu′bu·la′tor** *n.*
tu·bu·la·tion (tōō′byə·lā′shən, tyōō′-) *n.* 1 The formation of a tube. 2 The arrangement of a set of tubes.
tu·bule (tōō′byōōl, tyōō′-) *n.* A minute tube. [<L *tubulus*, dim. of *tubus* a tube] — **tu′bu·li·form′** *adj.*
tu·bu·li·flo·rous (tōō′byə·lə·flôr′əs, -flō′rəs, tyōō′-) *adj. Bot.* Having tubular florets: said of composite plants with all the florets tubular. [<*tubuli* (<TUBULE) + -FLOROUS]
tu·bu·lous (tōō′byə·ləs, tyōō′-) *adj.* 1 Tubeshaped; tubular. 2 *Bot.* Having tubular florets. 3 Consisting of or containing small tubes. Also **tu′bu·lose** (-lōs).
tu·bu·lure (tōō′byə·lər, tyōō′-) *n.* The short open tube of a retort, receiver, or bell jar. [<F <L *tubulus* TUBULE]
Tu·ca·na (tōō·kā′nə, tyōō-) *Astron.* A southern constellation. See CONSTELLATION.
tu·chun (dōō′jün′) *n. Chinese* Formerly, the military governor of a Chinese province. — **tu′chun·ate** *n.* — **tu′chun·ism** *n.*
tuck¹ (tuk) *n.* 1 A fold made in a garment, usually horizontal. 2 A flap forming a continuation of one side of a book cover, and inserted in a loop or pocket in the other side. 3 *Naut.* That part of a vessel's hull where the after planks meet. 4 *Brit. Slang* Food. 5 *U.S. Slang* Stamina; determination. [<*v.*] — *v.t.* 1 To fold under; thrust or press in the ends or edges of. 2 To wrap or cover snugly. 3 To thrust or press into a close place; cram; hide. 4 To make tucks in, by folding and stitching. — *v.i.* 5 To contract; draw together. 6 To make tucks. [Fusion of OE *tūcian* ill-treat, lit., tug and MDu. *tucken* pluck]
tuck² (tuk) *Scot. & Obs. n.* 1 A stroke; tap; beat, as of a drum. 2 A flourish, as of a trumpet. — *v.t. & v.i.* To beat; tap, as a drum. [<AF *toker*, OF *toucher* touch]
tuck³ (tuk) *n. Archaic* A long narrow sword; rapier. [<AF *etoc*, OF *estoc* <*estoquier* <Du. *stocken* pierce]
Tuck (tuk), **Friar** A jovial priest, associate and confessor of Robin Hood.
tuck·a·hoe (tuk′ə·hō) *n.* An underground fungus (*Poria cocos*) with a rough brown edible sclerotium: found in the southern United States: also called *Indian bread* or *Virginia truffle*. [<Algonquian (Virginian) *tockawhoughe*]
tuck·er¹ (tuk′ər) *n.* 1 One who or that which tucks: the *tucker* of a sewing machine. 2 A covering, formerly worn over the neck and shoulders by women.
tuck·er² (tuk′ər) *v.t. Colloq.* To weary completely; exhaust: usually with *out*. [Freq. of TUCK¹, *v.*]
tuck·et (tuk′it) *n. Archaic* A flourish on a trumpet. [Dim. of TUCK²]
Tuc·son (tōō·son′, tōō′son) A city in SE Arizona; a manufacturing, rail, and mining center.
Tu·cu·mán (tōō′kōō·män′) A city in NW Argentina; an agricultural and industrial center.
-tude *suffix of nouns* Condition or state of being: *gratitude*. [<F *-tude* <L *-tudo*]
Tu·dor (tōō′dər, tyōō′-) *adj.* 1 Of or pertaining to the **Tudors**, an English royal family descended from Sir Owen Tudor, a Welshman who married Catherine of Valois, widow of Henry V. See the table of sovereigns under ENGLAND. 2 Designating or pertaining to the architecture, poetry, etc., developed during the reigns of the Tudors.
Tudor architecture The latest phase of the Perpendicular style, developed under the Tudors to make houses more livable. It employed large windows, many fireplaces, large bays, steep roofs, flattened arches, much carving, and paneling. The house plan was generally a quadrangle, an H or an E.
Tues·day (tōōz′dē, -dā, tyōōz′-) *n.* The third day of the week; the day after Monday. [OE *tīwesdæg* day of Tiw <*Tīw*, an ancient Teutonic deity + *dæg* a day; trans. of LL *dies Martis* Mars's day]
tu·fa (tōō′fə, tyōō′-) *n.* 1 A variety of calcium carbonate with cellular structure, as deposited from springs and streams. 2 Tuff. [<Ital. *tufa, tufo* <L *tofus, tophus*] — **tu·fa·ceous** (tōō·fā′shəs, tyōō-) *adj.*
tuff (tuf) *n.* A fragmentary volcanic rock composed of material varying in size from fine sand to coarse gravel: often used for structural purposes. [<MF *tufe, tuffe* <Ital. *tufo* TUFA] — **tuff·a′ceous** *adj.*
tuft (tuft) *n.* 1 A collection or bunch of small, flexible parts, as hair, grass, or feathers, held together at the base. 2 A clump or knot; frequently, a cluster of threads drawn tightly through a quilt, mattress, or upholstery to secure the stuffing. 3 A gold tassel formerly worn by titled undergraduates at Oxford and Cambridge universities; also, a student who wears such a tuft. — *v.t.* 1 To separate or form into tufts. 2 To cover or adorn with tufts. — *v.i.* 3 To form tufts. [<OF *tuffe*, prob. <Gmc.] — **tuft′er** *n.* — **tuft′y** *adj.*
tuft·ed (tuf′tid) *adj.* 1 Having, or adorned with, a tuft; crested: the *tufted* duck. 2 Forming a tuft or dense cluster; cespitose.
tuft-hunt·er (tuft′hun′tər) *n. Archaic* 1 Originally, a student at Oxford or Cambridge who sought association with titled students distinguished by gold tufts on their hats. 2 One who seeks the acquaintance of persons of rank; a snob; sycophant; parasite. — **tuft′·hunt′ing** *n. & adj.*
tug (tug) *v.* **tugged, tug·ging** *v.t.* 1 To pull at with effort; strain at. 2 To pull, draw, or drag with effort. 3 To tow with a tugboat. — *v.i.* 4 To pull strenuously: to *tug* at an oar. 5 To strive; toil; struggle. See synonyms under DRAW. — *n.* 1 An act of tugging; a violent pull. 2 A strenuous contest; a struggle; wrestle. 3 A tugboat. 4 A trace of a harness; also, *Scot.*, rawhide: formerly used in making traces. 5 *Brit.* A colleger or member of the king's foundation at Eton College. 6 *Brit. Dial.* A high-wheeled cart for carrying logs, etc., slung beneath its axles. [ME *toggen*, intens. of OE *tēon* tow; infl. by ON *toga* draw] — **tug′ger** *n.*
tug·boat (tug′bōt′) *n.* A small, compact, ruggedly built vessel operated by steam or other power and designed for towing: also called *towboat*.
Tu·ge·la (tōō·gā′lə) A river in Natal, Union of South Africa, flowing 300 miles east from near the Basutoland and Orange Free State borders, through the **Tugela Falls** (2,810 feet) to the Indian Ocean.
Tug·gurt (tōō·gōōrt′) A territory of NE Southern Territories, Algeria; 52,094 square miles; capital, Tuggurt: French *Touggourt*.
tug of war 1 A contest in which a number of persons at one end of a rope pull against a like number at the other end, each side endeavoring to drag the other across a line marked between. 2 A laborious effort; supreme contest.
Tui·le·ries (twē′lər·ēz, *Fr.* twēl·rē′) A palace of the French kings in Paris; begun in 1564 and burned in 1871; site occupied by the **Tuileries Gardens**, a public park near the Louvre.
tuille (twēl) *n.* In armor, a steel protection for the thighs, attached by straps to the tassets. [<MF <OF *tieule* <L *tegula* a tile]
tu·i·tion (tōō·ish′ən, tyōō-) *n.* 1 The act or business of teaching any branch of learning; instruction. 2 The charge or payment for instruction. 3 *Archaic* Guardianship; care. See synonyms under EDUCATION, LEARNING, NURTURE. [<AF *tuycioun*, OF *tuicion* <L *tuitio, -onis* a guard, guardianship <*tuitus*, pp. of *tueri* look at, watch] — **tu·i′tion·al, tu·i′tion·ar′y** (-er′ē) *adj.*
Tu·la (tōō′lä) 1 A city of west central European Russian S.F.S.R. 2 A city in Hidalgo state, central Mexico; site of the ancient capital of the Toltecs: officially **Tula de Al·len·de** (thä ä·yen′dä).
tu·la·re·mi·a (tōō′lə·rē′mē·ə) *n.* A disease of rodents, especially rabbits, caused by a micro-organism (*Pasteurella tularensis*) which may be transmitted to man by flies and certain insects, producing an undulant fever; rabbit fever. Also **tu′la·rae′mi·a**. [<NL, from *Tulare* County, California + Gk. *haima*
tu·le (tōō′lē) *n.* A large bulrush (*Scirpus acutus*) of the sedge family growing on damp or flooded land in the southwestern United States. [<Sp. <Nahuatl *tullin*]
tu·lip (tōō′lip, tyōō′-) *n.* 1 Any of numerous hardy bulbous herbs (genus *Tulipa*) of the lily family, bearing variously colored bellshaped flowers. 2 A bulb or flower of this plant. [<F *tulipe* <OF *tulipan* <Turkish *tuliband* <Persian *dulband* a turban]
tu·lip·o·ma·ni·a (tōō′lip·ō·mā′nē·ə, -mān′yə, tyōō′-) *n.* A craze for the acquisition or cultivation of tulips; specifically, that which arose in Holland early in the 17th century and which spread into wild speculation like

tuliptree / **Tungusian**

an epidemic. [< *tulipo*- (<TULIP) + -MANIA]
— tu'li·po·ma'ni·ac n.
tu·lip·tree (tōō'lip·trē', tyōō'-) n. 1 A large magnoliaceous tree (*Liriodendron tulipifera*) of the eastern United States, with greenish cup-shaped flowers. 2 Any of various other trees having tuliplike flowers.
tu·lip·wood (tōō'lip·wŏŏd', tyōō'-) n. 1 The wood of the tuliptree. 2 Any of several ornamental cabinet woods yielded by various trees: so called from their color or markings. 3 Any of the trees themselves.
Tul·la·more (tul'ə·môr', -mōr') The county town of County Offaly, Ireland.
tulle (tōōl, *Fr.* tül) n. A fine, silk, open-meshed material, used for veils, etc. [<F, from *Tulle*, a city in SW France, where first made]
Tul·ly (tul'ē) See CICERO, MARCUS TULLIUS.
Tul·sa (tul'sə) A city in NE Oklahoma, on the Arkansas River.
tum·ble (tum'bəl) v. ·bled, ·bling v.i. 1 To roll or toss about. 2 To perform acrobatic feats, as somersaults, etc. 3 To fall violently or awkwardly. 4 To move in a careless or headlong manner; stumble. 5 *Colloq.* To understand; comprehend; with *to.* 6 *Metall.* To smooth, clean, or polish, as castings, by friction with each other or with a polishing material, in a rotating box or barrel. — v.t. 7 To toss carelessly; cause to fall. 8 To throw into disorder or confusion; disturb; rumple. — n. 1 The act of tumbling; a fall. 2 A state of disorder or confusion. [ME *tumbel*, freq. of *tumben*, OE *tumbian* fall, leap] — **tum'bling** n.
tum·ble·bug (tum'bəl·bug') n. A scarabaeid beetle that rolls up a ball of dung to enclose its eggs.
tum·ble-down (tum'bəl·doun') adj. Rickety, as if about to fall in pieces; dilapidated.
tumble gear *Mech.* A type of reversing gear comprising a rocking frame adapted to bring either of two idlers into mesh with the driving gear.
tum·ble-home (tum'bəl·hōm') n. *Naut.* The inward inclination of a vessel's hull above the line of extreme breadth.

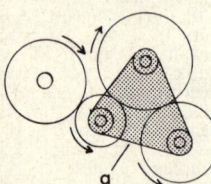
TUMBLE GEAR
a. Rocking frame.

tum·bler (tum'blər) n. 1 A drinking glass without a foot; also, its contents. The base was formerly rounded, so that the glass would not stand upright. 2 One who or that which tumbles; especially, an acrobat or contortionist. 3 One of a breed of domestic pigeons noted for the habit of turning forward somersaults during flight. 4 A greyhound used formerly in coursing. 5 In a lock, a latch that engages a bolt and prevents its being shot in either direction unless the tumbler is raised by the key bit. 6 In a firearm lock, a piece attached to the hammer and receiving the thrust of the mainspring. 7 A tumbling box. 8 *Mech.* a A piece of metal that projects from a revolving or rocking shaft and communicates motion to another piece. b The rocking frame in a tumble gear. 9 *Scot.* A light cart. 10 A child's toy, so formed and weighted as to rock at the slightest touch.
tum·ble·weed (tum'bəl·wēd') n. Any of various plants which, when withered, break from the root and are driven about by the wind, widely scattering their seed.
tumbling box *Metall.* A box, usually cylindrical and mounted on a horizontal shaft, in which articles, as castings, are cleaned by friction against each other and the walls of the box. Also **tumbling barrel.**
tum·brel (tum'bril) n. 1 *Obs.* A two-wheeled military covered cart for carrying tools, ammunition, etc. 2 A farmer's cart; especially, a boxlike cart for carrying and dumping dung. 3 A rude cart in which prisoners were taken to the guillotine during the French Revolution. 4 Formerly, a ducking stool set

TUMBREL

on wheels. Also **tum'bril.** [<OF *tomberel* < *tomber* fall, ult. <Gmc.]
tu·me·fa·cient (tōō'mə·fā'shənt, tyōō'-) adj. Producing or tending to produce tumefaction. [<L *tumefaciens*, ppr. of *tumefacere* TUMEFY]
tu·me·fac·tion (tōō'mə·fak'shən, tyōō'-) n. 1 Any puffing up of a part, especially as in a tumor. 2 A swelling; puffiness. 3 The act of tumefying; state of being tumefied. [<MF <L *tumefactus*, pp. of *tumefacere* TUMEFY]
tu·me·fy (tōō'mə·fī, tyōō'-) v.t. & v.i. ·fied, ·fy·ing To swell or puff up. [<MF *tuméfier* <L *tumefacere* < *tumere* swell + *facere* make]
Tu·men (tōō'mun') A river on the Korea-Manchuria border, flowing 324 miles east to the Sea of Japan.
tu·mer·os·i·ty (tōō'mə·ros'ə·tē) n. The state or quality of being swollen. [<L *tumere* swell + -OSITY]
tu·mes·cence (tōō·mes'əns, tyōō'-) n. 1 The state or quality of being swollen. 2 The act or process of becoming tumid, as an organ or part of the body. 3 That which is swollen.
tu·mes·cent (tōō·mes'ənt, tyōō'-) adj. 1 Swelling; somewhat tumid. 2 Beginning to swell. [<L *tumescens*, *-entis*, ppr. of *tumescere*, inceptive of *tumere* swell]
tu·mid (tōō'mid, tyōō'-) adj. 1 Swollen; enlarged; protuberant. 2 Inflated or pompous in style; bombastic. 3 Bursting; teeming. [<L *tumidus* < *tumere* swell] — **tu'mid·ly** adv.
tu·mid·i·ty (tōō·mid'ə·tē, tyōō-) n. The state or character of being tumid. Also **tu·mid·ness** (tōō'mid·nis, tyōō'-).
tu·mor (tōō'mər, tyōō'-) n. 1 *Pathol.* A local swelling on or in any part of the body, especially from some autonomous morbid growth of tissue which may or may not become malignant; a neoplasm. 2 *Obs.* High-sounding words or style; bombast. 3 *Obs.* A swelling of any kind, as of water. Also *Brit.* **tu'mour.** — **fatty tumor** Lipoma. [<OF *tumour* <L *tumor* a swelling < *tumere* swell] — **tu'mor·ous** adj.
tump (tump) n. *Dial.* A little mound or hill, as about a plant; a barrow. [Cf. Welsh *twmp*] — **tump'y** adj.
Tu·muc-Hu·mac Mountains (tōō·mōōk'ōō·mäk') A range on the border between Brazil and Dutch and French Guiana; highest point, 2,800 feet.
tu·mu·lar (tōō'myə·lər, tyōō'-) adj. Having the form of a mound.
tu·mu·lose (tōō'myə·lōs, tyōō'-) adj. Full of mounds or hills. Also **tu'mu·lous.** [<L *tumulosus* < *tumulus* a mound] — **tu·mu·los'i·ty** (-los'ə·tē) n.
tu·mult (tōō'mult, tyōō'-) n. 1 The commotion, disturbance, or agitation of a multitude; an uproar; turbulence; hubbub. 2 Any violent commotion or agitation, as of the mind. [<OF *tumulte* <L *tumultus* < *tumere* swell] — Synonyms: agitation, bluster, bustle, commotion, confusion, disorder, disturbance, ferment, flurry, hubbub, hurly-burly, noise, outbreak, racket, riot, turbulence, turmoil, uproar. See NOISE, QUARREL[1], REVOLUTION. Antonyms: calmness, peace, quiet, repose, tranquillity.
tu·mul·tu·ous (tōō·mul'chōō·əs, tyōō'-) adj. 1 Characterized by tumult; disorderly. 2 Causing or affected by tumult or agitation; agitated or disturbed. Also **tu·mul'tu·ar'y** (-er'ē). See synonyms under NOISY, TURBULENT, VIOLENT. — **tu·mul'tu·ous·ly** adv. — **tu·mul'tu·ous·ness, tu·mul'tu·ar'i·ness** n.
tu·mu·lus (tōō'myə·ləs, tyōō'-) n. pl. ·li (-lī) A sepulchral mound, often of great size. Compare BARROW[2], CAIRN. [<L, a mound < *tumere* swell]
tun (tun) n. 1 A large cask. 2 A brewers' fermenting vat. 3 The amount of malt liquor fermented at one operation; a brew. 4 A varying measure of capacity, usually equal to 252 gallons. — v.t. **tunned, tun·ning** 1 To put into a cask or tun. 2 To add to a liquor, as for flavoring. [OE *tunne*]
tu·na[1] (tōō'nə) n. A tunny. Also **tuna fish.** [<Am. Sp., ult. <L *thunnus* TUNNY]
tu·na[2] (tōō'nə) n. 1 A tropical American prickly pear (*Opuntia tuna*), or its edible fruit. 2 One of a number of other prickly pears. [<Am. Sp., prob. <Taino]
tun·a·ble (tōō'nə·bəl, tyōō'-) adj. 1 That may be put in tune. 2 Being in tune. 3 *Obs.*

Tuneful; musical. Also **tune'a·ble.** — **tun'a·ble·ness** n. — **tun'a·bly** adv.
Tun·bridge Wells (tun'brij) A municipal borough in SW Kent, England.
tun·dra (tun'drə, tōōn'-) n. A rolling, treeless, often marshy plain of Siberia, arctic North America, etc. [<Russian <Lapp]
tune (tōōn, tyōōn) n. 1 A melodious succession of musical tones adjusted to some measure and constituting one whole; a melody or air. 2 A setting for a hymn or psalm used in worship. 3 The state or quality of being in the proper pitch or key. 4 Concord or unison. 5 Suitable temper or humor; state of mind. 6 *Obs.* A musical tone or sound. — **to change one's tune** To assume a different manner or style. — **to the tune of** To the serious or exorbitant amount of: *to the tune of* a thousand dollars. — v. **tuned, tun·ing** v.t. 1 To adjust to a musical standard; put in tune; attune. 2 To adapt to a particular tone, expression, or mood. 3 To bring into harmony or accord. 4 To utter or express musically; sing. — v.i. 5 To be in harmony. — **to tune in** To adjust a radio receiver to the frequency of (a station, broadcast, etc.). — **to tune out** To adjust a radio receiver to exclude (interference, a station, etc.). — **to tune up** 1 To bring (musical instruments) to a common pitch. 2 To adjust (a machine, engine, etc.) to proper working order. [Var. of TONE] — **tun'ing** adj. & n.
Synonyms (noun): air, cadence, concord, harmony, measure, melody, strain.
tune·ful (tōōn'fəl, tyōōn'-) adj. Musically disposed; melodious; musical. — **tune'ful·ly** adv. — **tune'ful·ness** n.
tune·less (tōōn'lis, tyōōn'-) adj. 1 Not being in tune. 2 Not employed in making music; silent. 3 Lacking in rhythm, melody, etc. — **tune'less·ly** adv. — **tune'less·ness** n.
tun·er (tōō'nər, tyōō'-) n. 1 One who or that which tunes. 2 One who puts musical instruments, as pianos, in tune.
tune-up (tōōn'up', tyōōn'-) n. *Colloq.* An adjustment to bring a motor or other device into proper operating condition.
Tung (dŏŏng) A river of SE China, flowing 250 miles SW and west from southern Kiangsi province to a delta in east central Kwangtung province, emptying into the Canton River.
Tung·chow (tōōng'jō') A former name for NANTUNG.
tung oil (tung) A yellow, ill-smelling oil extracted from the seeds of the Chinese **tung tree** (*Aleurites fordii*), now cultivated in the U.S.: used in paints, varnishes, etc., as a highly effective drying agent, and also as a waterproofing agent. [<Chinese *t'ung* the tung tree]
tung·state (tung'stāt) n. *Chem.* A salt of tungstic acid: sodium *tungstate.*
tung·sten (tung'stən) n. A steel-gray, brittle, heavy metallic element of the chromium group (symbol W), occurring in sheelite and in wolframite, having a high melting point and much used in the manufacture of filaments for electric lamps and tungsten-steel tools. Also called *wolfram.* See ELEMENT. [<Sw. <*tung* weighty + *sten* stone] — **tung·sten·ic** (tung·sten'ik) adj.
tungsten lamp An incandescent electric lamp having a filament of metallic tungsten.
tungsten steel A hard, tenacious steel that contains tungsten.
tung·stic (tung'stik) adj. *Chem.* Of, pertaining to, derived from, or containing tungsten, especially in its highest valence. [<TUNGST(EN) + -IC]
tungstic acid *Chem.* Either of two acids consisting of tungsten oxide combined with water, and uniting with bases to form salts; especially, the yellow crystalline monohydrate, H_2WO_4.
tung·stite (tung'stīt) n. A yellow or yellowish-green native tungsten trioxide, WO_3. Also **tungstic ocher.**
Tung·ting (dŏŏng'ting'), **Lake** A lake in NE Hunan province, SE central China; one of the largest lakes in China; 1,450 square miles.
Tun·gus (tŏŏn·gŏŏz') n. 1 One of a Mongoloid people of the Tungusic group inhabiting eastern Siberia. 2 The language of the Tungus, belonging to the Manchu-Tungusic subfamily of Altaic languages: also *Tungusic.* Also **Tun·guz'.**
Tun·gus·i·an (tŏŏn·gŏŏz'ē·ən) adj. Of or per-

Tungusic 1353 **tureen**

taining to the Tungus or their language. — n. One of the Tungus.
Tun·gus·ic (toong-gōōz′ik) adj. 1 Of, pertaining to, or denoting a group of tribes including the Tungus and Manchus. 2 Tungusian. — n. The Tungus language.
Tun·gus·ka (toong-gōōs′kä) Any of three rivers in north central Asiatic Russian S.F.S.R., known as the **Upper Tunguska** (see ANGARA), the **Stony** (or **Middle**) **Tunguska**, flowing 975 miles NW and west to the Yenisei, and the **Lower Tunguska**, flowing 1,587 miles north and NW to the Yenisei.
tu·nic (tōō′nik, tyōō′-) n. 1 Among the ancient Greeks and Romans, a body garment, with or without sleeves, reaching to the knees: worn usually without a girdle. 2 A modern outer garment gathered at the waist, as a short overskirt or a blouse. 3 A surcoat worn over armor. 4 A tunica. 5 Bot. Any loose membranous skin enveloping an organ, as a seed coat. 6 Brit. The undercoat worn by soldiers, policemen, etc. 7 A bishop's tunicle; a dalmatic. [< F tunique < L tunica < Semitic]
tu·ni·ca (tōō′nə·kə, tyōō′-) n. pl. **·cae** (-sē) Biol. A covering or investing part; a mantle of tissue, as of the kidney, ovaries, etc.; tunic. [< NL < L, a tunic]
tu·ni·cate (tōō′nə·kit, -kāt, tyōō′-) adj. 1 Bot. Covered with a tunic, as the bulb of an onion. 2 Zool. Having a tunic; of, or pertaining to the tunicates. Also **tu′ni·cat′ed**. — n. Any of a subphylum (Tunicata or Urochordata) of small marine chordates covered with a transparent tunic, as the ascidians: called sea squirts. [< NL tunicata < L tunicata (animalia) coated (animals), neut. pl. of tunicatus, pp. of tunicare clothe with a tunic < tunica a tunic]
tu·ni·cle (tōō′ni·kəl, tyōō′-) n. 1 A light or fine tunic. 2 A slight natural covering. 3 A short ecclesiastical vestment. [< L tunicula, dim. of tunica a tunic]
tuning fork A fork-shaped piece of steel which vibrates with a definite frequency when struck: used to measure the pitch of musical tones.
Tu·nis (tōō′nis, tyōō′-) 1 A former Barbary state of northern Africa. 2 The capital and chief port of Tunisia, on the Mediterranean.
Tu·ni·sia (tōō·nish′ə, -nish′ē·ə, -nē′zhə, tyōō′-) A republic in northern Africa, proclaimed July, 1957; 48,195 square miles; capital, Tunis; formerly, Tunis.
Tu·ni·sian (tōō·nish′ən, -nē′zhən, tyōō′-) adj. Of or relating to Tunisia, or Tunis, or their inhabitants. — n. 1 An inhabitant or native of Tunisia or Tunis. 2 The speech of Tunisia, a North Arabic dialect.
tunk·et (tung′kit) n. U.S. Dial. Hell: a euphemistic expletive. [Origin unknown]
tun·nage (tun′ij) n. Brit. Tonnage.
tun·nel (tun′əl) n. 1 An artificial subterranean passageway or gallery, especially one under a hill, etc., as for a railway. 2 Any similar passageway under or through something. 3 A funnel. 4 The main flue or shaft of a chimney or the like. 5 An adit or level in a mine. — v. **·neled** or **·nelled**, **·nel·ing** or **·nel·ling** v.t. 1 To make a tunnel through. 2 To shape or make in the form of a tunnel: to tunnel a passage. — v.i. 3 To make a tunnel. [Fusion of OF tonnelle a partridge net and tonel, dim. of tonne a cask] — **tun′nel·er** or **tun′nel·ler** n.
tunnel disease Caisson disease.
tun·ny (tun′ē) n. pl. **·nies** 1 A large, oily, marine fish (family Thunnidae) related to

GREAT TUNNY

the mackerel, especially the **great tunny** (Thunnus thynnus) of warm seas, sometimes weighing 1,500 pounds. 2 One of various related

fishes, as the albacore and the California horse mackerel (Trachurus symmetricus). [< OF thon < L thunnus < Gk. thynnos]
Tu·ol·um·ne River (tōō·ol′ə·mē) A river in central California, flowing 110 miles west to the San Joaquin River.
tup (tup) n. 1 A ram, or male sheep. 2 The striking part of a power hammer. — v.t. & v.i. **tupped, tup·ping** To copulate with (a female): said of the ram. [ME tupe, tope, prob. < Scand. Cf. Norw. & Sw. tupp a cock.]
tu·pe·lo (tōō′pə·lō) n. pl. **·los** 1 One of several trees of Asia and the southeastern United States (genus Nyssa), especially the sourgum or blackgum. 2 The wood of any of these trees. [< Muskhogean]
Tu·pi (tōō·pē′) n. pl. **Tu·pis** or **Tu·pi** 1 A member of any of a group of South American Indian tribes, comprising the northern branch of the Tupian stock, and occupying the Amazon, Tapajós, and Xingú valleys. 2 The language spoken by the Tupis, used as a lingua franca along the Amazon: also called Neengatu. [< Tupian, a comrade]
Tu·pi·an (tōō·pē′ən) adj. Of or pertaining to the Tupis or their language. — n. A large linguistic stock of South American Indians of some one hundred tribes of the Tupis and Guaranis, scattered throughout the continent (except in Venezuela): also **Tu·pi′-Gua·ra·ni′** (-gwä′rä·nē′).
Tu·pun·ga·to (tōō′pōōng·gä′tō) A peak in the Andes on the Argentine-Chile border; 21,940 feet.
tuque (tōōk, tyōōk) n. A Canadian cap consisting of a knitted cylindrical bag with tapered ends, worn by thrusting one end inside the other, for tobogganing, etc. [< dial. F (Canadian) < F toque TOQUE]
tu quo·que (tōō kwō′kwē, tyōō′) Latin A retort in kind from a person assailed: also used attributively; literally, thou also.
tu·ra·cou (tōō′rä·kōō′) n. An African bird (Turacus fischeri) related to the cuckoo, remarkable for its red-and-green plumage: also called **touraco**. [< F touraco < native West African name]
Tu·ra·ni·an (tōō·rā′nē·ən, tyōō·) adj. Of or pertaining to a large family of agglutinative languages of Europe and northern Asia, neither Indo-European nor Semitic, specifically known as the Ural-Altaic languages, or any of the people who speak them. — n. 1 One whose mother tongue is a Ural-Altaic language; a person of Ural-Altaic stock. 2 The Ural-Altaic languages collectively. 3 Theoretically, one of an unknown nomadic people who antedated the Aryans in Europe and Asia. [< Persian Tūrān, a country north of the Oxus River]
tur·ban (tûr′bən) n. 1 An Oriental head covering consisting of a sash or shawl, twisted about the head or about a cap. 2 Any similar headdress. 3 A round-crowned brimless hat for women or children. [< F turban, turbant < Pg. turbante < Turkish tülbend, dial. alter. of dulbend < Persian dulband < dul a turn + band a band] — **tur′baned** (-bənd) adj.
tur·ba·ry (tûr′bər·ē) n. pl. **·ries** 1 In English law, the liberty of digging turf or peat upon another's ground. 2 A place where turf or peat is dug. [< OF turberie, OF tourberie < tourbe peat < LG turf, turv turf]
tur·bel·lar·i·an (tûr′bə·lâr′ē·ən) n. Any of a class (Turbellaria) of motile aquatic flatworms having a ciliated epidermis and sometimes brilliantly colored: includes the planarians. [< NL < L turbellae a tumult, pl. dim. of turba a crowd]
tur·beth (tûr′bəth), **tur·bith** See TURPETH.
tur·bid (tûr′bid) adj. 1 Having the sediment or lees stirred up; cloudy; muddy. 2 Being in a state of confusion; disturbed. See synonyms under THICK. [< L turbidus < turbare trouble < turba a crowd] — **tur′bid·ly** adv. — **tur′bid·ness**, **tur·bid·i·ty** (tûr·bid′ə·tē) n.
tur·bi·dim·e·ter (tûr′bə·dim′ə·tər) n. An instrument for measuring the turbidity of a liquid. [< TURBIDI(TY) + -METER]
tur·bi·nal (tûr′bə·nəl) adj. Spirally coiled; turbinate; top-shaped. — n. A turbinate bone or cartilage. [< L turbo, -inis a whirlwind, top]

tur·bi·nate (tûr′bə·nit, -nāt) adj. 1 Top-shaped; also, spinning like a top. 2 Zool. Tapering from a broad base to the apex, as certain spiral shells. 3 Anat. Pertaining to one of the thin, curved bones on the walls of the nasal passages. Also **tur′bi·nat′ed**. [< L turbinatus < turbo, -inis a whirlwind]
tur·bi·na·tion (tûr′bə·nā′shən) n. 1 A conelike formation. 2 The act, state, or condition of spinning like a top.
tur·bine (tûr′bin, -bīn) n. An engine consisting of one or more rotary units, mounted on a shaft and usually provided with a series of curved vanes, actuated by the reaction, impulse, or suction of steam, water, gas, or other fluid under pressure. [< F < L turbo, -inis a whirlwind, top]

TURBINE
Type used in an electric-light plant.

tur·bit (tûr′bit) n. One of a breed of domestic pigeons having a small head with the feathers at the back curled upward. [Appar. < L turbo, -inis a top; so called with ref. to its shape]
turbo- combining form A turbine; related to or operated by a turbine or turbines: turbojet. [< L turbo a top]
tur·bo·fan (tûr′bō·fan′) n. Aeron. 1 A compressor having ducted fans which supply air to a jet engine. 2 The engine using such a fan.
tur·bo·gen·er·a·tor (tûr′bō·jen′ə·rā′tər) n. An electric power-generating machine adapted for direct coupling to a steam turbine.
tur·bo·jet (tûr′bō·jet′) n. Aeron. 1 A gas turbine which drives the air compressor and auxiliaries of certain types of jet engines. 2 The engine itself.
tur·bo·prop (tûr′bō·prop′) n. Aeron. 1 A gas turbine connecting directly with the propeller. 2 An engine having such a turbine. [< TURBO- + PROP(ELLER)]
tur·bo·su·per·charg·er (tûr′bō·sōō′pər·chär′jər) n. Aeron. A compact, highly efficient supercharging device utilizing exhaust gases, for use on aircraft engines operating at very high altitudes.
tur·bot (tûr′bət) n. pl. **·bot** or **·bots** 1 A large European flatfish (Psetta maxima) esteemed as food. 2 One of various related flatfishes. [< AF turbut, OF tourbout, ? < OSw. törnbut < törn a thorn + but the butt]
tur·bu·lence (tûr′byə·ləns) n. 1 The state or condition of being violently disturbed, restless, or confused. 2 Physics The irregular flow of a gas or fluid caused by an obstacle or by friction, as of a ship or airplane in rapid motion. 3 Meteorol. A disturbed condition of the atmosphere due to irregular wind currents. Also **tur′bu·len·cy**.
tur·bu·lent (tûr′byə·lənt) adj. 1 Being in violent agitation or commotion. 2 Inclined to rebel; insubordinate. 3 Having a tendency to disturb or throw into confusion. [< MF < L turbulentus full of disturbance < turbare. See TURBID.] — **tur′bu·lent·ly** adv. Synonyms: agitated, blustering, boisterous, disorderly, disturbed, insurgent, mutinous, obstreperous, rebellious, refractory, riotous, seditious, tumultuous, wild. See NOISY, VIOLENT.
Tur·co (tûr′kō) n. pl. **·cos** An Algerian light-infantryman serving in the French army; an Algerian tirailleur. [< F < Sp., a Turk]
Tur·co·man (tûr′kə·mən) See TURKOMAN.
tur·di·form (tûr′də·fôrm) adj. Thrushlike in form or structure. [< L turdus a thrush + -FORM]
tur·dine (tûr′din, -dīn) adj. 1 Belonging or pertaining to a large and widely distributed family (Turdidae) of singing birds, including thrushes and bluebirds. 2 Pertaining to the subfamily (Turdinae) which includes the true thrushes. [< NL, subfamily name < L turdus a thrush]
tu·reen (tōō·rēn′, tyōō-) n. A deep, covered dish, as for soup. [Earlier terrene < F terrine. Doublet of TERRINE.]

---

add, āce, câre, päm; end, ēven; it, īce, odd, ōpen, ôrder; tōōk, pōōl; up, bûrn; ə = a in above, e in sicken, i in clarity, o in melon, u in focus; yōō = u in fuse; oi, oil; ou, pout; ch, check; g, go; ng, ring; th, thin; th, this; zh, vision. Foreign sounds å, œ, ü, kh, ṅ; and ♦: see page xx. < from; + plus; ? possibly.

Tu·renne (tü·ren′), **Viscount de,** 1611–75, Henri de la Tour d'Auvergne, French general and marshal.

turf (tûrf) *n. pl.* **turfs** (*Archaic* **turves**) **1** A mass of matted roots of grass and other fine plants filling the upper stratum of certain soils; a sod. **2** Peat. **3** Loosely, a grass plot. **4** A racecourse; horse-racing: in the phrase **the turf.** — *v.t.* To cover with turf; sod. [OE]

Tur·fan (to͝or′fän′) A depression in eastern Sinkiang–Uigur Autonomous Region, NW China, lowest point of the Chinese mainland (940 feet below sea level); center of an ancient Indo-Iranian civilization (A.D. 200–400) and site of the capital of a Uigur empire (about 800–1200).

turf·man (tûrf′mən) *n. pl.* **·men** (-mən) A man who is devoted to or connected with horse-racing.

turf·y (tûr′fē) *adj.* **turf·i·er, turf·i·est 1** Covered with turf. **2** Resembling turf in character or appearance. **3** Pertaining to the turf, to horse-racing, or to a racecourse; horsy. — **turf′i·ness** *n.*

Tur·ge·nev (to͝or·gā′nyef), **Ivan Sergeyevich,** 1818–83, Russian novelist. Also **Tur·ge′niev.**

tur·gent (tûr′jənt) *adj. Obs.* Turgid. [<L *turgens, -entis,* ppr. of *turgere* swell] — **tur′·gent·ly** *adv.*

tur·ges·cence (tûr·jes′əns) *n.* **1** The process of swelling up; the state of being swollen. **2** Hence, empty pompousness; inflation. Also **tur·ges′cen·cy.** [<Med. L *turgescentia* <L *turgescens, -entis,* ppr. of *turgescere,* inceptive of *turgere* swell] — **tur·ges′cent** *adj.* — **tur·ges′cent·ly** *adv.*

tur·gid (tûr′jid) *adj.* **1** Unnaturally distended, as by contained air or liquid; swollen. **2** Figuratively, inflated; bombastic; tumid: a *turgid* tale of woman wronged. [<L *turgidus* < *turgere* swell] — **tur′gid·ly** *adv.*

tur·gid·i·ty (tûr·jid′ə·tē) *n.* **1** The state or quality of being turgid. **2** *Biol.* The internal pressure of a cell against its enclosing membrane. Also **tur·gid·ness** (tûr′jid·nis).

tur·gite (tûr′jīt) *n.* A fibrous, earthy iron ore, found as a reddish-black or dark-red ferric hydroxide. [from *Turginsk,* a copper mine in the Ural Mountains]

tur·gor (tûr′gər) *n.* **1** The state of being turgid; turgidity. **2** *Physiol.* The normal condition of the blood vessels and of cells distended by their protoplasmic contents: also **vital turgor.** [<LL <L < *turgere* swell]

Tur·got (tür·gō′), **Anne Robert Jacques,** 1727–81, French statesman, financier, and economist.

Tu·rin (to͝or′in, tyo͝or′-, to͞o·rin′, tyo͞o-) A city on the Po river in NW Italy; capital of the kingdom of Sardinia until 1860 and of Italy until 1864; a major industrial and transportation center: Italian *Torino.*

Turk (tûrk) *n.* **1** A native or inhabitant of Turkey; an Ottoman. **2** One of any of the peoples speaking any of the Turkic languages, and ranging from the Adriatic to the Sea of Okhotsk: believed to be of the same ultimate extraction as the Mongols. **3** Loosely, a Moslem. **4** A Turkish horse.

Turk·cap lily (tûrk′kap′) See under LILY.

Tur·ke·stan (tûr′kə·stan′, -stän′) A region of central Asia extending from the Caspian Sea to the Gobi Desert and divided by the Pamir and Tien Shan mountain systems into **Russian** (or **Western**) **Turkestan,** comprising the Kazakh, Kirghiz, Tadzhik, Turkmen, and Uzbek S.S.R. (also called *Soviet Central Asia*); and **Chinese** (or **Eastern**) **Turkestan,** comprising the Sinkiang–Uigur Autonomous Region.

tur·key (tûr′kē) *n. pl.* **·keys 1** A large American bird (family *Meleagridae*) related to the pheasant, having the head naked and the tail extensible upward and sideward; especially, the American domesticated turkey (*Meleagris gallopavo*): much esteemed as food. **2** A guinea fowl. **3** *U. S. Slang* A play (occasionally, a motion picture) that is a failure. [Short for *turkey cock* the guinea fowl, from *Turkey;* later applied erroneously to the American bird]

Tur·key (tûr′kē) A republic of Asia Minor and SE Europe; 296,108 square miles (Turkey in Asia, known as *Anatolia,* 287,043 square miles, Turkey in Europe, 9,065 square miles); capital, Ankara.

turkey buzzard A sooty-black vulture of tropical America (*Cathartes aura*), with a naked red head and neck.

TURKEY BUZZARD (Length, 30 inches; wingspread, 6 feet)

Turkey carpet A hand-made Turkish carpet or one having a Turkish design and texture. Also **Turkish carpet.**

turkey cock 1 A male turkey. **2** One who struts and behaves in a pompous, conceited manner.

turkey gobbler A turkey cock.

turkey red 1 A brilliant red pigment, or its color. **2** Cotton cloth dyed with this permanent bright red.

tur·key-trot (tûr′kē·trot′) *n.* A dance consisting of a stop-step, glide, and turn to syncopated music, the feet being kept well apart and a swinging motion being given to the shoulders: popular in the early 20th century.

turkey vulture The turkey buzzard.

Tur·ki (to͝or′kē) *adj.* **1** Of or pertaining to any of the languages included in the Turkic subfamily of Altaic languages. **2** Of or pertaining to any of the peoples speaking any of these languages, as the Osmanlis and Chuvashes of Turkey, NW Persia, Transcaucasia, etc., and the Asian Tatar tribes, as the Uigurs, Uzbeks, Kipchaks, Turkomans, etc., of Mongolia and Turkestan. — *n.* **1** The Turkic languages. **2** A member of any of the Turki peoples.

Turk·ic (tûr′kik) *n.* A subfamily of the Altaic family of languages, including Osmanli or Turkish, Azerbaijani, Uzbek, Chuvash, Yakut, etc. — *adj.* Pertaining to this linguistic subfamily, or to any of the peoples speaking these languages.

Turk·ish (tûr′kish) *adj.* **1** Of or pertaining to Turkey or the Turks. **2** Of or relating to the Turkic subfamily of Altaic languages, especially to Osmanli. — *n.* The Altaic language of Turkey; Osmanli.

Turkish bath A bath originating in the East in which sweating is induced by exposure to high temperature, usually in a room heated by steam, followed by washing, rubbing, kneading, or the like.

Turkish delight A sweetmeat of Turkish origin, usually consisting of cubes having a gelatinous consistency coated with powdered sugar and having any of various fruit flavors. Also **Turkish paste.**

Turkish Empire The Ottoman Empire.

Turkish pound A Turkish lira. Symbol £T.

Turkish towel A heavy, rough towel with loose, uncut pile. Also **turkish towel.**

Turk·ism (tûr′kiz·əm) *n.* **1** The religion, or the social or political system, characteristic of the Turks. **2** Any distinctive peculiarity of Turkish speech or custom.

Turk·men (tûrk′men) *n.* The Turkic language of the Turkomans.

Turk·men Soviet Socialist Republic (tûrk′·men) A constituent republic of the U. S. S. R. in Central Asia; 189,370 square miles; capital, Ashkhabad. Also **Turk·me·ni·stan** (tûrk′me·ni·stan′, -stän′; *Russian* to͝ork′mye·nyi·stän′).

tur·kois (tûr′koiz) See TURQUOISE.

Tur·ko·man (tûr′kə·mən) *n. pl.* **·mans 1** A member of any of the Turki peoples dwelling in those parts of Turkestan comprising the Russian Turkmen, Uzbek, and Kazakh Soviet Republics. **2** Turkmen. Also spelled *Turcoman.* Also **Turk′man.** [<Persian *Turkumān* one like a Turk] — **Turk·me·ni·an** (tûrk·mē′·nē·ən) *adj.* & *n.* — **Turk·o·man·ic** (tûr′kə·man′ik) *adj.*

Turks and Cai·cos Islands (tûrks, kā′kəs) An archipelago SE of the Bahamas, comprising a dependency of Jamaica in The West Indies; 202 square miles.

Turk's-cap lily (tûrks′kap′) See under LILY.

Turk's-head (tûrks′hed′) *n.* An ornamented knot of turbanlike form.

Tur·ku (to͝or′ko͞o) A city of SW Finland, on the Gulf of Bothnia: Swedish *Åbo.*

tur·ma·line (tûr′mə·lēn) See TOURMALINE.

tur·mer·ic (tûr′mər·ik) *n.* **1** The root of an East Indian plant (*Curcuma longa*) of the ginger family, used as a condiment, aromatic stimulant, dyestuff, etc. **2** The plant. **3** Any of several plants resembling turmeric. — *adj.* Of, pertaining to, or saturated with turmeric. [Earlier *tarmaret,* ? <F *terre mérite* deserving earth <Med. L *terra merita;* ult. origin uncertain]

turmeric paper A paper, yellow from saturation with the extract of turmeric, used as a test for alkalis, turning it brown, and for boric acid, turning it red-brown: also called *curcuma paper.*

tur·moil (tûr′moil) *n.* Confused motion; disturbance; tumult. See synonyms under TUMULT. — *v.t.* & *v.i. Archaic* To be or cause to be in a state of turmoil. [? <OF *tremouille* hopper of a mill <L *tremere* tremble; prob. infl. in form by *turn* and *moil*]

turn (tûrn) *v.t.* **1** To give a rotary motion to; cause to rotate, as about an axis. **2** To change the position of, as by rotating: to *turn* a trunk on its side. **3** To move so that the upper side becomes the under: to *turn* a page. **4** To bring the subsoil of to the surface, as by plowing or spading. **5** To alter (a garment) by reversing the material: to *turn* a cuff. **6** To reverse the arrangement or order of; cause to be upside down. **7** To upset mentally; dement or distract; infatuate. **8** To revolve mentally; ponder: often with *over.* **9** To sprain or strain: to *turn* one's ankle in running. **10** To nauseate (the stomach). **11** To shape (an object revolving in a lathe, etc.) in rounded form by application of a cutting tool. **12** To give rounded or curved form to. **13** To give graceful or finished form to: to *turn* a phrase. **14** To perform by revolving: to *turn* cartwheels. **15** To bend, curve, fold, or twist. **16** To bend or blunt (the edge of a knife, etc.). **17** To change or transform; convert: to *turn* water into wine. **18** To translate: to *turn* French into English. **19** To exchange for an equivalent: to *turn* stocks into cash. **20** To adapt to some use or purpose; apply: to *turn* information to good account. **21** To cause to become as specified: The sight *turned* him sick. **22** To change the color of. **23** To make sour or rancid; ferment or curdle. **24** To change the direction of. **25** To direct or aim; point. **26** To change the direction or focus of (thought, attention, etc.). **27** To deflect or divert: to *turn* a blow. **28** To repel: to *turn* a charge. **29** To go around or to the other side of: to *turn* a corner. **30** To pass or go beyond: to *turn* twenty-one. **31** To cause or compel to go; send; drive: to *turn* a beggar from one's door. **32** To keep circulating in trade: to *turn* goods or money. **33** *Obs.* To pervert. — *v.i.* **34** To move around an axis or center; rotate; revolve. **35** To move completely or partially on or as if on an axis: He *turned* and ran. **36** To change position; also, to roll from side to side, as in bed. **37** To take a new direction: We *turned* north. **38** To reverse position; become inverted. **39** To reverse direction or flow: The tide has *turned.* **40** To change the direction or focus of one's thought, attention, etc.: Let us *turn* to the next problem. **41** To depend; hinge: with *on* or *upon.* **42** To be affected with giddiness; whirl, as the head. **43** To become upset or nauseated, as the stomach. **44** To change attitude, sympathy, or allegiance: to *turn* on one's neighbors. **45** To rebel; act in retaliation: The worm *turns;* to *turn* on one's persecutors. **46** To become transformed; change: The water *turned* into ice. **47** To become as specified: His hair *turned* gray. **48** To change color: said especially of leaves. **49** To become sour, rancid, or fermented, as milk or wine. **50** *Naut.* To tack or put about. **51** *Obs.* To vacillate. See synonyms under BEND, CHANGE, REVOLVE. — **to turn against** To become or cause to become opposed or hostile to. — **to turn an honest dollar** (or **penny**) To earn money honestly. — **to turn down 1** To diminish the flow, volume, etc., of: *Turn down* the gas. **2** *Colloq.* **a** To reject or refuse, as a proposal, or request. **b** To refuse the request, proposal, etc., of. — **to turn in 1** To fold or double. **2** To bend or incline inward. **3** To deliver; hand over. **4** *Colloq.* To go to bed. — **to turn off 1** To stop the operation, flow, etc., of. **2** To leave the direct road; make a turn. **3** To deflect or divert. **4** *Brit.* To dismiss; discharge. — **to turn on** To set in operation, flow, etc.: to *turn on* an engine. — **to turn**

out 1 To turn inside out. 2 To eject or expel; put out. 3 To dismiss or discharge. 4 To turn off (def. 1). 5 To bend or incline outward. 6 To produce by work or toil; make. 7 To come or go out, as for duty or service. 8 To prove (to be); be found. 9 To become or result. 10 To equip or fit; dress. 11 *Colloq.* To get out of bed. —**to turn over** 1 To change the position of; invert. 2 To upset; overturn. 3 To hand over; transfer or relinquish. 4 To do business to the amount of. 5 To invest and get back (capital). 6 To use in trade or exchange; buy and then sell: to *turn over* merchandise. —**to turn tail** To run away; flee. —**to turn to** 1 To set to work. 2 To seek aid from. 3 To refer or apply to. —**to turn up** 1 To bring or fold the under side upward. 2 To bend or incline upward. 3 To bring or be brought to view by plowing, digging, etc.; find or be found. 4 To increase the flow, volume, etc., of. 5 To put in an appearance; arrive. —*n.* 1 The act of turning, or the state of being turned. 2 A change to another direction, motion, or position: a *turn* of the tide. 3 A deflection or deviation from a course; a bend; a change in policy or trend: a *turn* of fortune. 4 The point at which a change takes place: a *turn* for the better in a crisis or an illness. 5 Motion about or as about a center; a rotation or revolution: the *turn* of a crank. 6 Favorable, fitting, or regular time or chance in some succession or rotation, or the work it offers; a job; also, a round; spell: one's *turn* to read, a *turn* of work. 7 Characteristic form or style; distinguishing shape; mold; cast: the *turn* of an ankle or sentence. 8 Disposition; tendency; manner: a humorous *turn*; a knack or special ability: a *turn* for study. 9 A deed performed, regarded as aiding or injuring another: an ill *turn*; also, an advantage proposed or gained: It served his *turn.* 10 A walk, drive, or trip to and fro; promenade: a *turn* in the park. 11 A trip back and forth in taking a load of anything: *turns* made to a mill; also, the load so taken. 12 A round in a skein or coil. 13 *Music* An instrumental or vocal embellishment formed by a group of four notes rapidly performed, the first a tone above and the third a tone below the principal tone, which occupies the second and fourth positions. In an **inverted turn** the tones are reversed in order. 14 *Colloq.* A spell of dizziness or faintness; a shock to the nerves, as from alarm: It gave her quite a *turn.* 15 A variation or difference in type or kind. 16 A short theatrical act of any description; also, in sport, a contest; a bout. 17 A twist, as of a rope, around a tree or post. 18 A transaction on the stock exchange, involving purchase and sale, or the reverse; also, any business transaction. 19 In infantry drill, a maneuver in which a line of troops changes the direction of its front, usually in preparation for marching. —**at every turn** On every occasion; constantly. —**by turns** 1 In alternation or sequence. 2 At intervals. —**in turn** One after another; in proper order or sequence. —**out of turn** Not in proper order or prescribed order or sequence. —**to a turn** Just right; perfectly or exactly: said especially of cooked food, in allusion to the turning of the spit in roasting. —**to take turns** To act, play, etc. in proper order. ◆ Homophones: *tern, terne.* [Fusion of OE *tyrnan* and *turnian* and OF *turner,* all <L *tornare* turn in a lathe < *turnus* a lathe <Gk. *tornos*]

turn·a·bout (tûrn′ə·bout′) *n.* 1 One who overturns things; a radical. 2 A merry-go-round. 3 A turn-about-face.

turn about 1 Alternately; in proper succession or order: also **turn and turn about.** 2 A turn-about-face.

turn-a·bout-face (tûrn′ə·bout′fās′) *n.* A change from one loyalty or viewpoint to another; adoption of new opinions or policy: also *turnabout.*

turn·a·round (tûrn′ə·round′) *n.* 1 The act of returning a ship, aircraft, or other vehicle to its original point of departure. 2 A round trip. 3 The temporary shutdown of an industrial plant to permit the inspection, repair, and replacement of essential operating machinery.

turn·buck·le (tûrn′·buk′əl) *n. Mech.* A form of coupling so threaded that when connected lengthwise between two metal rods it may be turned so as to regulate the distance between them.

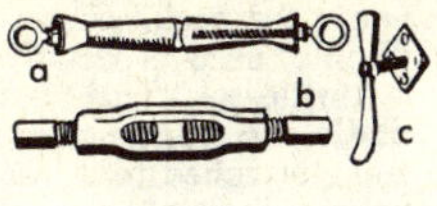

TURNBUCKLES
a. Insulated for electric wires.
b. Type for metal tie rods.
c. For window shutters.

turn·coat (tûrn′kōt′) *n.* One who goes over to the opposite side or party; a renegade.

turn-down (tûrn′doun′) *adj.* Folded down, as a collar; also, capable of being turned down.

turned comma *Printing* An inverted comma, as a single initial quotation mark.

turn·er[1] (tûr′nər) *n.* One who turns; specifically, one who fashions objects with a lathe.

turn·er[2] (tûr′nər) *n.* A gymnast; a member of a turnverein. [<G < *turnen* engage in gymnastics <F *turner* turn]

Tur·ner (tûr′nər), **Frederick Jackson,** 1861–1932, U. S. historian. —**Joseph Mallord William,** 1775–1851, English painter.

turn·er·y (tûr′nər·ē) *n. pl.* **·er·ies** 1 A place where lathework, especially ornamental work, is carried on. 2 The act or process of turning, or articles and ornamentation made with a lathe.

turn·hall (tûrn′hôl′) *n.* A building in which gymnasts, especially members of a turnverein, practice; a gymnasium. Also *German* **Turn·hal·le** (to͝orn′häl′ə). [<G *turnhalle* < *turnen* exercise + *halle* a hall]

turn indicator 1 *Aeron.* An instrument of the gyroscope type used to indicate any turning movement of an airplane in flight: also **turn meter.** 2 In motor vehicles, any device, as a flashing light, which enables the driver to signal his intention to turn: also **turning signals.**

turn·ing (tûr′ning) *n.* 1 The act of one who turns. 2 The art of shaping wood, metal, etc., in a lathe. 3 Any deviation from a straight or customary course; a winding; bend. 4 The point where a road forks. 5 A movement placing a military force on the flank or rear of an enemy's position.

turning point 1 The point of a decisive change in direction of action; a crisis. 2 The point at which the direction of a motion is reversed. 3 A marked object toward which a surveying instrument is sighted from each of two positions in a direct line with each other in the process of leveling.

tur·nip (tûr′nip) *n.* 1 The fleshy globular edible root of either of two brassicaceous biennial herbs, *Brassica rapa* and the rutabaga. 2 Either of the plants. [Earlier *turnepe,* ?<F *tour* a turn, rotation (<L *tornus* a lathe) + ME *nepe* <L *napus* a turnip; with ref. to its round shape]

tur·nix (tûr′niks) *n.* One of a genus *(Turnix)* of small three-toed birds with short tails found in warm regions of the Old World. [<NL, short for L *coturnix* a quail]

turn·key (tûrn′kē′) *n.* 1 One who has charge of the keys of a prison; a jailer. 2 An instrument formerly used for extracting teeth.

turn-off (tûrn′ôf′, -of′) *n. Colloq.* A road, path, or way branching off from a main thoroughfare.

turn·out (tûrn′out′) *n.* 1 An act of turning out or coming forth. 2 An assemblage of persons; attendance. 3 A quantity produced; output. 4 Array; equipment; outfit. 5 A railroad siding. 6 The movement of a vehicle from a line of traffic to pass other vehicles. 7 A section of narrow road widened to permit vehicles to pass one another. 8 A carriage or wagon with its horses and equipage 9 *Brit.* A labor strike; also, a striker.

turn·o·ver (tûrn′ō′vər) *n.* 1 The act or process of turning over; an upset or overthrow, as of a vehicle. 2 A change or revolution: a *turnover* in affairs. 3 A small pie or tart made by covering half of a circular crust with fruit, jelly, or the like, and turning the other half over on top. 4 The amount of business accomplished, or of work achieved; turnout. 5 A completed commercial transaction or course of business; also, the money receipts of a business for a given period. 6 The rate at which persons hired by a given establishment within a given period are replaced by others; also, the number of persons hired. —*adj.* 1 Designed for turning over or reversing. 2 Capable of being turned over or folded down. 3 Made with a part folded down: a *turnover* collar.

turn·pike (tûrn′pīk′) *n.* 1 A road on which there are tollgates. 2 Loosely, any highway: also **turnpike road.** 3 A tollbar or tollgate. 4 *Obs.* A turnstile. See synonyms under ROAD. [ME *turnpyke* a spiked road barrier <TURN, *v.* + *pyke* PIKE[1]]

turn·sole (tûrn′sōl′) *n.* 1 Any of several plants supposed to turn their flowers toward the sun; especially, the heliotrope and the sunflower. 2 Litmus. 3 One of various other blue coloring matters obtained from certain lichens and herbs. [<OF *tournesole* <Ital. *tornasole* < *tornare* turn (<L) + *sole* the sun <L *sol*]

turn·spit (tûrn′spit′) *n.* 1 One who turns a spit; a menial. 2 A dog formerly used in a treadmill to turn a roasting spit.

turn·stile (tûrn′stīl′) *n.* 1 A kind of gate or closure consisting of a vertical post and horizontal arms which by revolving permit persons, but not cattle, to pass; also, one that permits persons to pass in one direction only. 2 A similar device for registering the number of persons entering a building or for automatically admitting passengers to subways, buses, etc., on the deposit of fares.

turn·stone (tûrn′stōn′) *n.* A ploverlike migratory bird (genus *Arenaria*) of northern regions: so called from its habit of turning over stones to obtain its food; especially, the **ruddy turnstone** *(A. interpres)* and the **black turnstone** *(A. melanocephala)* of North America.

turn·ta·ble (tûrn′tā′bəl) *n.* 1 A rotating platform arranged to turn a section of a bridge in order to open a passage for ships. 2 Such a platform to turn a locomotive, car, etc.: also *Brit.* **turn′plate′** (-plāt′). 3 A small rotating disk in a microscope. 4 A rotating table in a show window. 5 The disk which carries a phonograph record.

turn·up (tûrn′up′) *n.* 1 That which is turned up, as part of a garment. 2 A particular card or die turned up in games of chance. 3 Pure chance; a toss-up; an unexpected phenomenon. 4 *Colloq.* A boxing contest; hence, a fight or row; commotion. —*adj.* Turned up.

turn·ver·ein (to͝orn′fe·rīn) *n.* An association of turners or gymnasts; an athletic club. [<G < *turnen* exercise + *verein* a club]

tur·pen·tine (tûr′pən·tīn) *n.* 1 A resinous, oily mixture of various pinenes exuding from any one of several coniferous trees, especially the longleaf pine *(Pinus palustris).* 2 The semifluid resin of the terebinth: also **Chian turpentine.** —**oil of turpentine** The colorless essential oil formed when turpentine is distilled with steam and consisting of a mixture of terpenes: widely used in industry, medicine and the arts. —*v.t.* **·tined, ·tin·ing** 1 To put turpentine with or upon; saturate with turpentine. 2 To obtain crude turpentine from (a tree). [<OF *turbentine* <L *terebinthinus* of the terebinth tree < *terebinthus* the terebinth <Gk. *terebinthos*]

turpentine tree The peebeen.

tur·peth (tûr′pith) *n.* 1 The root of an East Indian plant (genus *Ipomoea*) allied to the one yielding the common jalap, similar to it in properties: also **vegetable turpeth.** 2 A lemon-yellow basic mercuric sulfate, $HgSO_4 \cdot 2HgO$, used in medicine as an emetic: also **turpeth mineral.** Also called *turbeth, turbith.* [<OF *turbit* <Med. L *turbithum* <Arabic *turbid* <Persian]

Tur·pin (tûr′pin), **Richard,** 1706–39, English highwayman: known as *Dick Turpin.*

tur·pi·tude (tûr′pə·to͞od, -tyo͞od) *n.* Inherent baseness; vileness; depravity, or any action showing depravity. [<MF <L *turpitudo, -inis* < *turpis* vile]

tur·quoise (tûr′koiz, -kwoiz) *n.* 1 A blue or green hydrous aluminum phosphate, $H_5Al_2PO_8$, colored by copper: found massive, and in its highly polished blue varieties esteemed as a gemstone. 2 A light, greenish blue, the color of the turquoise: also **turquoise blue**. Also spelled *turkois*. Also *Obs*. **tur·quois** (tûr′koiz′). [< MF *(pierre) turquoise* Turkish (stone) < OF *turqueise*, fem. of *turqueis* Turkish; so called because first imported through Turkey]

tur·rel (tûr′əl) *n. Obs.* An auger used by coopers. [? Dim. of OF *tour* a turn < L *tornus* a lathe]

tur·ret (tûr′it) *n.* 1 A small tower, often merely ornamental, rising above a larger structure, as on a castle. 2 *Mil.* A rotating armed tower, large enough to contain a powerful gun or guns and gunners, forming part of a man-of-war or of a fort; a similar enclosed structure in a tank or a bombing or combat airplane. 3 The clerestory of a railway car. 4 In ancient warfare, a high wooden structure, supported on slides or wheels, intended to enable besiegers to surmount the walls against which it was pushed. 5 *Mech.* In a lathe, a cylinder fitted with sockets or chucks for the reception of various tools, any one of which may be presented in succession in the axial line of the work: also **turret head**. [< OF *torete*, dim. of *tor* TOWER]

tur·ret·ed (tûr′it·id) *adj.* 1 Provided with turrets. 2 Having the form of a turret. 3 *Zool.* Having a long spire, as certain shells.

turret lathe A power-driven metalworking machine having a rotating turret head holding various tools, each of which in turn processes the material.

tur·ri·cal (tûr′i·kəl) *adj.* Of, pertaining to, or like a turret. [< L *turris* a tower]

tur·ric·u·late (tə·rik′yə·lit, -lāt) *adj.* 1 Having or resembling a turret or turrets. 2 Turreted or having a spire: said of shells. [< L *turricula*, dim. of *turris* a tower + -ATE²]

tur·tle¹ (tûr′təl) *n.* 1 Any of numerous reptiles (order *Chelonia*) having a horny, toothless beak, and characterized by a short, stout body covered above and below with a bony carapace and plastron respectively, into which all the members may be drawn for protection; a tortoise; specifically, a marine species as distinguished from a terrestrial or fresh-water species. 2 The flesh of certain varieties of turtle, served as food. 3 A stout frame in the form of a segment of a cylinder, used to hold the type in a type-revolving web press. — **green turtle** An important food turtle (*Chelonia mydas*) of wide distribution in tropical and semitropical seas: so called from the greenish color of its flesh. — **to turn turtle** To overturn; capsize. — *v.i.* **·tled, ·tling** To hunt for or catch turtles. [Appar. alter. of Sp. *tortuga* < Med. L *tortuca* TORTOISE; infl. in form by TURTLE²]

tur·tle² (tûr′təl) *n. Archaic* A turtle dove. [OE < L *turtur*]

tur·tle·back (tûr′təl·bak′) *n.* 1 *Naut.* An arched covering, resembling the shell of a turtle, built over the main deck of a ship as protection against heavy seas: usually at the bow or stern. Also **turtle deck**. 2 *Archeol.* A rude, chipped stone implement whose facets resemble the sculptured carapace of a turtle. [< TURTLE¹ + BACK]

turtle dove 1 An Old World dove (genus *Streptopelia*), noted for its affection for its mate and young. 2 One of other pigeons, as the mourning dove. [< TURTLE² + DOVE]

tur·tle·head (tûr′təl·hed′) *n.* Any species of a genus (*Chelone*) of hardy American herbs of the figwort family, with large white or purple flowers.

turtle neck A high collar that fits snugly about the neck, usually rolled or turned over double: used especially on athletic sweaters. — **tur′tle·neck′** *adj.*

tur·tle·peg (tûr′təl·peg′) *n.* A small, sharp, steel spike attached to a line and loosely mounted upon a shaft which is thrown like a harpoon to capture sea turtles.

tur·tle·stone (tûr′təl·stōn′) *n.* Septarium.

turves (tûrvz) Archaic plural of TURF.

Tus·ca·loo·sa (tus′kə·loo′sə) A city of west central Alabama on the Black Warrior River.

Tus·can (tus′kən) *adj.* 1 Pertaining to Tuscany. 2 Designating the Tuscan order of architecture. — *n.* 1 A native or naturalized inhabitant of Tuscany. 2 Any Italian dialect used in Tuscany; especially, the one spoken in Florence, used by Dante, who thus set the pattern for what has become the standard literary Italian language.

Tuscan Archipelago An Italian island group in the Tyrrhenian Sea between the coast of Tuscany and Corsica; total, 115 square miles.

Tuscan order A Roman order of architecture resembling Roman Doric but having bolder moldings, no decorated details, and no triglyphs.

Tus·ca·ny (tus′kə·nē) A region and former duchy of west central Italy; 8,876 square miles; chief city, Florence: Italian *Toscana*.

Tus·ca·ro·ra (tus′kə·rôr′ə, -rō′rə) *n. pl.* **·ra** or **·ras** One of a tribe of North American Indians of Iroquoian stock formerly living in North Carolina, now surviving in New York and Ontario. They joined the Five Nations in 1722.

Tus·cu·lum (tus′kyə·ləm) An ancient ruined city of Latium, SE of Rome. — **Tus′cu·lan** *adj.*

tush¹ (tush) *interj.* An exclamation expressing disapproval, impatience, etc.

tush² (tush) See TUSK.

tushed (tusht) *adj.* Having tushes or tusks.

TUSCAN ORDER
a. Cornice.
b. Frieze.
c. Architrave.
d. Capital.
e. Shaft.
f. Base.

tusk (tusk) *n.* 1 A long, pointed tooth, as in the boar, walrus, or elephant. 2 A sharp, projecting, toothlike point. 3 A shoulder on a tenon, to strengthen it at its base; also, a tenon having such a shoulder. — *v.t.* 1 To gore with the tusks. 2 To root up with the tusks. Also *tush*. [Metathetic var. of OE *tux*] — **tusked** (tuskt) *adj.* — **tusk′less** *adj.* — **tusk′like′** *adj.*

tusk·er (tus′kər) *n.* An elephant or wild boar with developed tusks.

tusk tenon A tenon strengthened by a step or steps, or by a shoulder. [< TUSK (def. 3) + TENON]

tus·sah (tus′ə) *n.* 1 A wild and semi-domesticated Asian silkworm (*Antheraea paphia*) which spins large cocoons of brownish or yellowish silk. 2 The silk, or the tough, durable fabric woven from it. Also **tus′sa, tus·sar** (tus′ər), **tus′seh, tus′ser, tus·sore** (tus′ôr, -ōr), **tus′sur**. [< Hind. *tasar* < Skt. *tasara*, *trasara*, lit., a shuttle]

tus·sis (tus′is) *n. Pathol.* A cough: bronchial *tussis*. [< NL < L] — **tus′sal, tus′sive** *adj.*

tus·sle (tus′əl) *v.t.* & *v.i.* **·sled, ·sling** To fight or struggle in a vigorous, determined way; engage in a tussle. — *n.* A disorderly struggle, as in sport; scuffle. [Var. of TOUSLE]

tus·sock (tus′ək) *n.* 1 A tuft or clump of grass or sedge. 2 A tuft, as of hair or feathers. Also **tus′suck**. [Prob. dim. of obs. *tusk* a tuft of hair, ? < TUSK] — **tus′sock·y** *adj.*

tussock moth Any of various robust, medium-sized moths (family *Lymantriidae*) whose larvae bear tufts of hairs and are very destructive of broad-leaved deciduous trees, as the gipsy moth.

tut (tut) *interj.* An exclamation to check rashness or express impatience: often repeated.

Tut-ankh-a-men (toot′ängk·ä′min) Egyptian pharaoh who reigned 1358-1350 B.C.

tu·te·lage (too′tə·lij, tyoo′-) *n.* 1 The state of being under a tutor or guardian. 2 The act or office of a guardian; guardianship. 3 The act of tutoring; instruction. [< L *tutela* a watching, guardianship < *tutus* safe < *tueri* watch, guard]

tu·te·lar (too′tə·lər, tyoo′-) *adj.* 1 Invested with guardianship. 2 Pertaining to a guardian. Also **tu′te·lar·y** (-ler·ē).

tu·te·nag (too′tə·nag, tyoo′-) *n.* 1 A white alloy, with varying proportions of copper, zinc, and nickel. 2 Zinc or spelter. Also **tu′te·nague**: sometimes spelled *teutenag*. [< Pg. *tutanaga, tutenaga* < Marathi *tuttināg* < Skt. *tuttha* copper sulphate + *nāga* tin, lead]

Tu·to·caine (too′tə·kān) *n.* Proprietary name for an ivory-colored, odorless, crystalline compound, $C_{14}H_{22}O_2N_2$, used as a local anesthetic.

tu·tor (too′tər, tyoo′-) *n.* 1 One who instructs another in one or more branches of knowledge; a private teacher. 2 A college teacher who gives individual instruction. 3 *Brit.* A college official entrusted with the tutelage and care of undergraduates assigned to him. 4 *Law* A guardian of a minor or of a woman. — *v.t.* 1 To act as tutor to; instruct; teach; train. 2 To have the guardianship of. 3 To treat severely or sternly, as a tutor might; discipline. — *v.i.* 4 To do the work of a tutor. 5 To be tutored or instructed. See synonyms under TEACH. [< AF, OF *tutour* < L *tutor* a watcher, guardian < *tutus*. See TUTELAGE.] — **tu·to·ri·al** (too·tôr′ē·əl, -tō′rē-, tyoo-) *adj.*

tu·tor·age (too′tər·ij, tyoo′-) *n.* The office of a tutor; tutorship.

tutorial system A system of education, generally collegiate, in which each student is assigned to a tutor, who directs his studies and has general supervision over his instruction.

tu·tor·ship (too′tər·ship, tyoo′-) *n.* 1 The office of a tutor or of a guardian. 2 Tutelage.

tu·toy·er (tü·twä·yā′) *v.t. French* To speak to with the French singular pronoun *tu, te, toi* instead of the more formal plural pronoun *vous*; address intimately.

tut·ti (toot′ē) *Music adj.* All: a term used to indicate that all performers are to take part: contrasted with *solo*. — *n.* A composition, piece, movement, or passage to be performed by all the voices and instruments together: contrasted with *solo*. [< Ital., pl. of *tutto* all]

tut·ti-frut·ti (too′tē·froo′tē) *n.* A confection, chewing gum, ice-cream, etc., made with different fruits. — *adj.* Having fruit flavors. [< Ital., all fruits]

tut·ty (tut′ē) *n.* An impure zinc oxide obtained as a sublimate in the flues of zinc-smelting furnaces: used as a polishing powder. [< OF *tutie* < Arabic *tūtiya* oxide of zinc, ? < Persian]

tu·tu (tü·tü′) *n. French* A short, full, projecting skirt consisting of many layers of sheer fabric, worn by ballet dancers.

Tu·tu·i·la (too′too·ē′lä) The chief island of the American Samoan group; 40 square miles; capital, Pago Pago.

tu·um (too′əm) *pron. Latin* Yours; thine: used in the phrase *meum and tuum*, mine and thine.

Tu·va Autonomous Region (too′və) An administrative division of southern Asiatic Russian S.F.S.R.; 66,100 square miles: formerly *Tannu-Tuva People's Republic*. Also **Tu-in-i an Autonomous Region** (too·vin′ē·ən).

tu·whit tu·whoo (too·hwit′ too·hwoo′) The cry of an owl. [Imit.]

tux·e·do (tuk·sē′dō) *n. pl.* **·dos** 1 A man's semi-formal dinner coat without tails. 2 The suit of which the coat is a part. Also **Tux·e′do**. [from *Tuxedo* Park, N.Y.; so called because first worn at the country club there]

Tux·tla (toost′lä) The capital of Chiapas state, southern Mexico, in the west central part of the state. Also **Tuxtla Gu·tiér·rez** (goo·tyär′räs).

tu·yère (twē·yâr′, twir; *Fr.* tü·yâr′) *n. Metall.* The pipe through which air is forced into a furnace or forge: also spelled *twyere*. [< F, a nozzle < *tuyau* a pipe]

Tver (tver) A former name for KALININ.

twa (twä, twô) *adj. Scot.* Two.

twad·dle (twod′l) *v.t.* & *v.i.* **·dled, ·dling** To talk foolishly and pretentiously. See synonyms under BABBLE. — *n.* Pretentious, silly talk; also, a twaddler. [Prob. alter. of TWATTLE] — **twad′dler** *n.*

twain (twān) *adj. Archaic* Two: rare except in poetic usage. See synonyms under BOTH. — *n.* 1 A couple; two. 2 In river navigation, two

TURTLES
a. Wood turtle. b. Emyd. c. Trionychid.

Twain

fathoms or twelve feet. [OE *twēgen*, masculine of *twa* two]
Twain (twān), **Mark** See MARK TWAIN.
twal (twäl, twôl) *adj. Scot.* Twelve. Also **twall**.
twal·pen·nies (twäl'pen'ēz, twôl'-) *n. pl. Scot.* Twelvepence: in old Scots currency, equal to one penny sterling. — **twal'pen'nie, twal'pen'ny** *adj.*
twang (twang) *v.t. & v.i.* **twanged, twang·ing** 1 To make or cause to make a sharp, vibrant sound, as a bowstring. 2 To utter or speak with a harsh, nasal sound. — *n.* 1 A sharp, vibrating sound, as of a tense string plucked. 2 A sharp, nasal sound of the voice. 3 A sound resembling either of the foregoing. Also *tang*. [Imit.] — **twang'y** *adj.*
twan·gle (twang'gəl) *v.t. & v.i.* **·gled, ·gling** To twang. — *n.* A twang. [Freq. of TWANG] — **twan'gler** *n.*
Twan·kay (twang'kā) *n.* A variety of green tea. Also **twan·ky** (twang'kē). [from Chinese *T'un ch'i*, a town in Anwhei province, where originally grown]
twa·some (twä'səm, twô'-) *Scot. adj.* Twosome. — *n.* Two persons in company; a pair.
twat·tle (twot'l) *v.t. & v.i.* **·tled, ·tling,** *n.* Twaddle. [Short for *twittle-twattle*, var. of TITTLE-TATTLE]
tway·blade (twā'blād) *n.* Any one of various hardy terrestrial orchids (genera *Listera* or *Liparis*) with two radical leaves: also spelled *twyblade*. [< archaic *tway* two, var. of TWAIN + BLADE]
tweak (twēk) *v.t.* To pinch and twist sharply; twitch. — *n.* A twisting pinch; twitch. [Var. of dial. *twick*, OE *twiccan* twitch] — **tweak'y** *adj.*
tweed (twēd) *n.* 1 A soft woolen fabric with a homespun surface: often woven in two or more colors to effect a check or plaid pattern. 2 A tweed suit or coat. — **Harris tweed** A homespun woolen cloth, usually of mixed colors, made at Harris in the Hebrides. [Alter. of dial. E (Scottish) *tweel*, var. of TWILL; prob. infl. in form by *Tweed* river, which flows through the district where it is woven]
Tweed (twēd) A river in Peeblesshire, Scotland, forming part of the boundary of England and Scotland and flowing 97 miles NE to the North Sea.
Tweed (twēd), **William Marcy,** 1823–78, U.S. politician: called "Boss Tweed."
Tweed·dale (twēd'dāl) See PEEBLES.
twee·dle¹ (twēd'l) *v.* **·dled, ·dling** *v.t.* 1 To play (a musical instrument) casually or carelessly. 2 To wheedle; cajole. — *v.i.* 3 To produce a series of shrill tones. 4 To play a musical instrument casually or carelessly. — *n.* A sound resembling the tones of a violin. [Imit. of the sound of a reed pipe]
twee·dle² (twēd'l) *v.* **·dled, ·dling** *v.t.* To handle carelessly. — *v.i.* To wriggle. [Var. of TWIDDLE]
twee·dle·dum and twee·dle·dee (twēd'l-dum', twēd'l-dē') Two things between which there is the slightest possible distinction: from John Byrom, *On the Feuds between Handel and Bononcini* (1723). [Orig. imit. of low- and high-pitched musical instruments, respectively]
Tweedledum and Tweedledee Twin brothers of almost identical appearance in Lewis Carroll's *Through the Looking-Glass*.
Tweed Ring The political group, headed by William M. ("Boss") Tweed and other Tammany Hall politicians, which controlled New York city government (1865–71) and plundered millions of dollars. — **Tweed'ism** *n.*
Tweeds·muir (twēdz'myŏŏr), **Baron** See BUCHAN, JOHN.
'tween (twēn) Contraction of BETWEEN.
tweet (twēt) *v.i.* To utter a thin, chirping note. — *n.* A twittering or chirping. Also **tweet'tweet',** *n.* [Imit.]
tweet·er (twē'tər) *n. Electronics* A loudspeaker used to reproduce the treble register in high-fidelity sound equipment: distinguished from *woofer*. [<TWEET]
tweeze (twēz) *v.t.* **tweezed, tweez·ing** *Colloq.* To handle, pinch, pluck, etc., with tweezers. [Back formation <TWEEZERS]
tweez·ers (twē'zərz) *n.pl.* Small pincers for tiny objects: often called **a pair of tweezers.**

2 *Obs.* A set of surgeon's instruments; also, a surgeon's instrument case. [Alter. of *tweezes*, pl. of *tweeze*, earlier *etweese* a case of small instruments <F *étuis*, pl. of *étui* ÉTUI]
twelfth (twelfth) *adj.* 1 Second in order after the tenth: the ordinal of *twelve*. 2 Being one of twelve equal parts. — *n.* 1 One of twelve equal parts; the quotient obtained by dividing by twelve. 2 *Music* An interval compounded of an octave and a fifth. [OE *twelfta*]
Twelfth-cake (twelfth'kāk') *n.* A cake prepared for a Twelfth-night festival. [Short for *Twelfth-night cake*]
Twelfth-day (twelfth'dā') *n.* The festival of the Epiphany, the twelfth day after Christmas.
Twelfth-night (twelfth'nīt') *n.* The eve (Jan. 5th) of Twelfth-day, or the evening before Epiphany; sometimes, the evening (Jan. 6th) of Epiphany. — *adj.* Of or pertaining to Twelfth-night.
Twelfth·tide (twelfth'tīd') *n.* Twelfth-day; the twelfth day after Christmas; Epiphany.
twelve (twelv) *adj.* Consisting of twice six: a cardinal numeral. — *n.* The sum of ten and two, or the symbols (12, xii, XII) representing it. — **the Twelve** The twelve apostles. See APOSTLE (def. 1). [OE *twelf*]
Twelve Apostles 1 A governing body of the Mormon Church, composed of twelve high officials. 2 The twelve disciples of Jesus: more commonly *the Twelve*.
twelve·mo (twelv'mō) *adj. & n.* Duodecimo.
twelve·month (twelv'munth') *n.* A year.
twelve-tone (twelv'tōn') *adj. Music* Of, pertaining to, or composed in a system or technique developed by Arnold Schönberg, in which any particular series of twelve tones containing all twelve of the tones of the chromatic scale is used in various permutations as the basis of composition, usually without reference to a fixed tonal center; dodecaphonic.
twen·ti·eth (twen'tē·ith) *adj.* 1 Tenth in order after the tenth: the ordinal of *twenty*. 2 Being one of twenty equal parts. — *n.* One of twenty equal parts; the quotient of a unit divided by twenty. [OE *twentigotha* < *twentig* twenty]
twen·ty (twen'tē) *adj.* 1 Consisting of twice ten; vicenary. 2 *Archaic* A considerable but indefinite number. — *n. pl.* **·ties** The sum of ten and ten, or the symbols (20, xx, XX) representing it. [OE *twentig*] — **twen'ty·fold'** *adj.*
twen·ty-one (twen'tē·wun') *n.* Vingt-et-un: a card game.
twerp (twûrp) *n. Slang* A small, contemptible person. [Cf. obs. *twirk* twitch, var. of TWIRL]
Twi (twē) *n.* A Sudanic language spoken by African Negroes in Ghana: also called *Ashanti*: also spelled *Tshi*.
twi- *prefix* Two; double; twice: *twibil*. Also spelled *twy-*. [OE, double < *twa* two]
twi·bil (twī'bil) *n.* 1 An ax with two cutting edges. 2 A double-bladed battle-ax. 3 A garden tool like an ax; a mattock. Also **twi'bill.** [OE < *twi-* two + *bill* an ax]
twice (twīs) *adv.* 1 Two times. 2 In double measure; doubly. [OE *twiges*, gen. of *twiga* twice]
twice-laid (twīs'lād') *adj.* 1 Made from the yarns of old or used rope. 2 Made from remnants or refuse.
twic·er (twī'sər) *n. Brit.* A printer who is both compositor and pressman.
Twick·en·ham (twik'ən·əm) A municipal borough on the Thames, 11 miles SW of London, England; home of Alexander Pope.
twid·dle (twid'l) *v.* **·dled, ·dling** *v.t.* 1 To twirl idly; toy or play with. — *v.i.* 2 To revolve or twirl. 3 To toy with something idly. 4 To be busy about trifles. — *n.* A gentle twirling, as of the fingers. [Prob. <ON *tridla* stir; ? infl. in meaning by TWIRL and FIDDLE] — **twid'dler** *n.*
twi·er (twī'ər) Altered form of TUYÈRE.
twig¹ (twig) *n.* A small shoot or branchlet of a tree. ♦ Collateral adjective: *viminal*. [OE *twigge*] — **twig'less** *adj.*
twig² (twig) *v.* **twigged, twig·ging** *Slang v.t.*

TWIBIL (*def. 2*)

1 To observe closely; notice or watch. 2 To comprehend; understand. — *v.i.* 3 To understand. [Cf. Irish *tuigim* I understand]
twig³ (twig) *n. Archaic* The fashion: an old fop in good twig. [? <obs. *twig* act vigorously; ult. origin uncertain]
twig blight 1 Any of various bacterial or fungous infections of plants which attack the twigs, resulting in extreme decay. 2 Dieback.
twig borer The larva of a lepidopterous insect (*Anarsia lineatella*) which bores into the twigs of certain fruit trees, as the peach, plum, and apricot, with destructive effect.
twigged (twigd) *adj.* Having shoots or twigs.
twig·gen (twig'ən) *adj.* Made of twigs; wicker.
twig girdler See GIRDLER (def. 3).
twig·gy (twig'ē) *adj.* Like, or abounding in, twigs.
twi·light (twī'līt') *n.* 1 The light diffused over the sky after sunset and before sunrise (especially, in popular use, the former) which is caused by the reflection of sunlight from the higher portions of the atmosphere. 2 Any faint light; shade; obscurity: the *twilight* of the groves. 3 Indistinct apprehension or perception: the *twilight* of doubt or barbarism. 4 A hazy or obscure condition following the waning of past glory, achievements, etc.: the *twilight* of the gods. — *adj.* 1 Pertaining or peculiar to twilight; crepuscular. 2 Imperfectly or faintly lighted; shaded; dim. [ME *twylight* <OE *twi-* (<*twa* two) + LIGHT; used in sense of "the light between the two," i.e., between day and night]
twilight arch The arch that bounds the brightest region of twilight.
twilight of the gods See RAGNARÖK.
twilight sleep *Med.* A light or partial anesthesia, induced artificially, as by injection of morphine and scopolamine, in which the patient loses the power to remember present events and sensations: sometimes used to relieve childbirth pains. [Trans. of G *dämmerschlaf*]
twill (twil) *n.* 1 One of the three foundation systems of weaves, in which the shuttle carries the woof thread over one and under two or more warp threads, producing the characteristic diagonal ribs or lines in fabrics. 2 A fabric woven with a twill; twilled cloth.

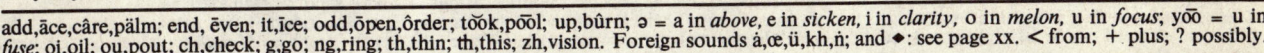
TWILL
Enlarged to show weave.

— *v.t.* To weave (cloth) so as to produce diagonal lines or ribs on the surface. [Var. of ME *twile*, OE *twili* a twilled fabric < *twi-* <*twa* two, partial trans. of L *bilix* having a double thread]
twilled (twild) *adj.* Woven so as to produce a diagonal rib or line; ribbed or ridged.
twin (twin) *n.* 1 One of two young produced at the same birth. 2 The counterpart or exact mate of another. 3 An intergrowth of two or more crystals of the same substance according to some definite law, a single plane or axis usually being common to the different individuals. — **the Twins** Castor and Pollux, the two brightest stars in the constellation Gemini; also, the constellation. — *adj.* 1 Being, or standing in the relation of, a twin or twins. 2 Consisting of, forming, or being one of a pair of similar and closely related objects; double; twofold. — **the Twin Cities** St. Paul and Minneapolis. — *v.* **twinned, twin·ning** *v.i.* 1 To bring forth twins. 2 To be matched or equal; agree. 3 *Archaic* To be born as a twin. — *v.t.* 4 To bring forth as twins. 5 To couple; match. 6 *Scot.* To separate: also **twine** (twīn). [Fusion of OE *twinn, getwinn* (< *twi-* <*twa* two) and ON *tvinnr, tvennr* double]
twin bed One of a pair of single beds.
twin·ber·ry (twin'ber'ē, -bər-ē) *n. pl.* **·ries** 1 The partridgeberry. 2 A North American shrub (*Lonicera involucrata*) with elliptic leaves, yellowish-red flowers, and shining black berries.
twin-born (twin'bôrn') *adj.* Brought forth at the same birth; born as a twin or twins.

twine[1] (twīn) *v.* **twined, twin·ing** *v.t.* 1 To twist together, as threads. 2 To form by such twisting. 3 To coil or wrap about something. 4 To encircle by winding or wreathing. 5 To enfold; embrace. —*v.i.* 6 To interlace; become twined. 7 To proceed in a winding course; meander. See synonyms under BEND, TWIST. —*adj.* Of or like twine. —*n.* 1 A string composed of two or more strands twisted together; loosely, any small cord. 2 The act of twining or entwining. 3 A form or conformation produced by twining. 4 An interweaving or interlacing. 5 *Obs.* A twisting about rapidly; spin. [OE *twīn* a twisted double thread < *twi-* double < *twa* two] —**twin'er** *n.*

twin·flow·er (twin'flou'ər) *n.* A trailing evergreen plant (genus *Linnaea*) of the honeysuckle family, as the Old World *L. borealis*, with fragrant rose or white bell-shaped flowers growing in pairs, and its American variety, *L. borealis americana*.

twinge (twinj) *v.t.* & *v.i.* **twinged, twing·ing** To affect with or suffer a sudden pain or twinge. —*n.* A sharp, darting, local pain; twitch; also, a mental pang. See synonyms under PAIN. [OE *twengan* pinch]

twi-night (twī'nīt') *adj.* Beginning in the late afternoon and continuing, under artificial light, into the night, as baseball games or other outdoor contests. [Blend of TWILIGHT and NIGHT]

twin·kle (twing'kəl) *v.* **·kled, ·kling** *v.i.* 1 To shine with fitful, intermittent gleams, as a star. 2 To be bright, as with amusement: Her eyes *twinkled.* 3 To wink or blink; open and shut with a quick, involuntary motion. 4 To move rapidly to and fro; flicker: *twinkling* feet. —*v.t.* 5 To emit or cause to flash out, as gleams of light. 6 To move (the eyelids) quickly and repeatedly. —*n.* 1 A tremulous gleam of light; sparkle; glimmer. 2 A quick or repeated movement of the eyelids; also, a wink or sparkle of the eye. 3 An instant; a twinkling. See synonyms under LIGHT[1]. [OE *twinclian*] —**twin'kler** *n.*

twin·kling (twing'kling) *n.* 1 The act of scintillating. 2 A wink or twinkle. 3 The act of winking, or the time required for it. 4 A moment. See synonyms under LIGHT.

twin-leaf (twin'lēf') *n.* A small perennial herb (*Jeffersonia diphylla*) of the barberry family native in eastern North America, having solitary white flowers and leaves divided into kidney-shaped leaflets.

twinned (twind) *adj.* 1 Produced at one birth; twin. 2 Formed by twinning, as a crystal.

twin·ning (twin'ing) *n.* 1 The production of two young at one birth; the bearing of twins. 2 Close union or combination; coupling of two related objects. 3 The formation of twin crystals, each the counterpart of the other.

twin-screw (twin'skrōō') *adj.* Of a vessel, having two propeller shafts, one on each side of the keel, and two propellers, normally turning in opposite directions. —*n.* (twin'skrōō') Such a vessel.

twirl (twûrl) *v.t.* & *v.i.* 1 To whirl or rotate. 2 In baseball, to pitch. —*n.* 1 A whirling motion, or a quick twisting action, as of the fingers. 2 A curl; twist; coil. [Alter. of ME *tirlen*, var. of *trillen* TRILL[2]; appar. infl. by *whirl*] —**twirl'er** *n.*

twist (twist) *v.t.* 1 To wind (strands, etc.) around each other. 2 To form by such winding: to *twist* thread. 3 To give spiral, circular, or semicircular form to, as by turning at either end. 4 To force out of natural shape; distort or contort. 5 To distort the meaning of. 6 To confuse; perplex. 7 To wreathe, twine, or wrap. 8 To cause to revolve or rotate. 9 To impart spin to (a ball) so that it moves in a curve. —*v.i.* 10 To become twisted. 11 To move in a winding course; meander or bend. 12 To squirm; writhe. —*n.* 1 The act, manner, or result of twisting or turning on an axis. 2 The state of being twisted. 3 *Physics* **a** A torsional strain. **b** The angle of torsion, as of a rod or bar. 4 A curve; turn; bend; winding: This path is full of *twists* and turns. 5 A contortion or twisting of a facial or bodily feature: a smile with a certain *twist.* 6 A wrench; strain, as of a joint or limb: He fell and gave his ankle a *twist.* 7 A peculiar or perverted inclination, bent, or attitude: the *twist* of a criminal's mind. 8 A distortion; deviation; wresting: a *twist* of meaning. 9 Thread or cord made of tightly twisted or braided strands. 10 *Naut.* One of the strands of a rope. 11 A twisted roll of bread. 12 Tobacco twisted in the form of a large cord. 13 In baseball, billiards, tennis, etc.: **a** A spin or whirling motion given to a ball by a certain stroke or throw. **b** The stroke or throw producing such a spin. **c** The act or knack of imparting such a spin. [ME *twisten* divide in two, combine two, prob. <OE *-twist* a rope, as in *mæst-twist* a rope to stay a mast < *twi-* double < *twa* two]

Synonyms (verb): bend, contort, crook, encircle, entwine, twine, wreathe. To *twist* is to bend a thing somewhat spirally upon itself. To *twine* is to *bend* it around some other object. Wrestlers *twine* their arms about each other, but if a combatant's arm is *twisted* it is likely to disable him. An iron shaft may be *twisted* out of shape, but not *twined*; the groove of a rifle barrel is *twisted*, not *twined*; a wreath is *twined* around one's temples, but not *twisted.* Compare BEND, PERVERT.

twist drill *Mech.* A drill or bit whose body is cut with deep spiral grooves to carry out the chips.

twisted pine The lodge-pole pine.

twist·er (twis'tər) *n.* 1 One who or that which twists. 2 A ball, as in cricket, bowled with a twist. 3 In baseball, a curve; also, one who pitches a curve. 4 *U. S.* A tornado.

Twist·or (twis'tər) *n. Electronics* A device for increasing the storage capacity of digital computers, consisting of a grid of copper wires and segments of magnetic wire twisted to change the magnetization from longitudinal to helical: a trade name.

twit (twit) *v.t.* **twit·ted, twit·ting** To taunt, reproach, or annoy by reminding of a mistake, fault, etc. —*n.* A taunting allusion; reproach. [Aphetic var. of ME *atwite*, OE *ætwitan* taunt < *æt-* at + *witan* accuse] —**twit'ter** *n.*

twitch (twich) *v.t.* 1 To pull sharply; pluck with a jerky movement. 2 In lumbering, to drag or skid (logs) along the ground with a chain. —*v.i.* 3 To tug or move with a quick, spasmodic jerk, as a muscle. —*n.* 1 A sudden involuntary contraction of a muscle. 2 A sudden jerk or pull. [ME *twicchen.* Akin to OE *twiccian* pluck.] —**twitch'ing·ly** *adv.*

twitch·grass (twich'gras', -gräs') *n.* Couch-grass.

twit·ter[1] (twit'ər) *v.i.* 1 To utter a series of light chirping or tremulous notes, as a bird. 2 To titter. 3 *Brit. Dial.* To be excited; tremble. —*v.t.* 4 To utter or express with a twitter. —*n.* 1 A succession of light, tremulous sounds. 2 *Bot.* A disease of plants caused by insects. [Imit.]

twit·ter[2] (twit'ər) *v.t.* To taunt; upbraid. [Freq. of TWIT]

twit·ter[3] (twit'ər) *v.t.* Spin or twist unevenly. [<earlier *twit*, a fault or entanglement in thread; ult. origin uncertain]

twixt (twikst) *prep. Poetic* Betwixt: an abbreviated form.

two (tōō) *adj.* Being one more than one, or a unit taken once again; binary: a cardinal numeral. See synonyms under BOTH. —*n.* 1 The sum of one and one: a cardinal number. 2 Any symbol or set of symbols (2, ii, II) for this number. —**in two** Bisected; bipartite; asunder; apart. [OE *twā, tū*]

two-base hit (tōō'bās') In baseball, a hit in which the batter reaches second base without benefit of an error. Also **two'-bag'ger.**

two-bit (tōō'bit') *adj. U.S. Slang* Cheap; small-time: a *two-bit* gambler.

two bits *U.S. Colloq.* Twenty-five cents.

two-by-four (tōō'bi·fôr', -fōr') *adj.* 1 Measuring two inches by four inches. 2 *U.S. Slang* Of trifling size or significance; narrow or limited. —*n.* (tōō'bi·fôr', -fōr') A piece of lumber measuring two inches by four inches before finishing: much used in building.

two-cy·cle (tōō'sī'kəl) *adj.* Designating a type of internal-combustion engine in which the piston completes its work in two strokes.

two-edged (tōō'ejd') *adj.* Having an edge on each side; cutting both ways.

two-faced (tōō'fāst') *adj.* 1 Having two faces. 2 Double-dealing; insincere; of dissimulating tendency. —**two'-fac'ed·ly** (-fā'sid·lē, -fāst'lē) *adv.*

two·fer (tōō'fər) *U.S. Slang n.* 1 An article advertised or sold at two for the price of one. 2 A free coupon which entitles the holder to two theater tickets for the price of one if presented at the box office of the designated attraction. —*adj.* Offering two of anything for the price of one: a *twofer* sale. [Alter. of *two for (one)*]

two·fold (tōō'fōld') *adj.* Double. —*adv.* In a twofold manner or degree; doubly.

two-hand·ed (tōō'han'did) *adj.* 1 Requiring both hands at once. 2 Constructed for use by two persons. 3 Ambidextrous. 4 Having two hands.

two-mast·er (tōō'mas'tər, -mäs'-) *n.* A ship with two masts.

two-name (tōō'nām') *adj.* Bearing two names or signatures.

two-name paper A negotiable paper, bearing either two signatures or one signature and one endorsement, two persons thus being responsible for the payment of the instrument.

two·pence (tup'əns) *n. Brit.* 1 Money of account of the value of two pennies. 2 A silver coin of the same value, now issued only for alms money, distributed by order of the British sovereign on Maundy Thursday. 3 A trifle; small amount: She doesn't care *twopence* about him.

two·pen·ny (tup'ən·ē) *adj. Brit.* 1 Of the price or value of twopence. 2 Cheap; worthless.

two-phase (tōō'fāz') *adj. Electr.* Diphase.

two-ply (tōō'plī') *adj.* 1 Made of two united webs; woven double: a *two-ply* carpet. 2 Made of two strands or two thicknesses of material.

Two Sic·i·lies (sis'ə·lēz), **The** A kingdom formed by the union of Sicily with Naples in 1130; incorporated with Italy in 1861.

two·some (tōō'səm) *n.* 1 Two persons together. 2 A match with one player on each side. —*adj.* 1 Performed or participated in by two, as a dance. 2 Comprising two or a pair.

two-spot (tōō'spot') *n.* 1 A playing card having two pips; a deuce. 2 *U.S. Slang* A small or unimportant person. 3 *U. S. Slang* A two-dollar bill. 4 *U. S. Slang* A prison sentence of two years.

two-step (tōō'step') *n.* A round dance consisting of a sliding step in 2/4 time; also, the music for it.

two-time (tōō'tīm') *v.t.* **-timed, -tim·ing** *Slang* To be unfaithful to (someone), especially in love; delude; deceive. —**two'-tim'er** *n.*

two-way (tōō'wā') *adj.* 1 Having an arrangement that will permit a fluid to be directed in either of two channels: specifically said of cocks and valves. 2 *Math.* Having a double mode of variation. 3 Permitting traffic in either direction: a *two-way* street.

twy- See TWI-.

twy·blade (twī'blād) See TWAYBLADE.

twy·ere (twī·ir') See TUYÈRE.

-ty[1] *suffix of nouns* The state or condition of being: *sanity.* [<F *-té* <L *-tas*]

-ty[2] *suffix* Ten; ten times: used in numerals, as *thirty, forty,* etc. [OE *-tig* ten]

Tyb·alt (tib'əlt) A masculine personal name. See THEOBALD.

—**Tybalt** In Shakespeare's *Romeo and Juliet*, nephew to Lady Capulet; kills Mercutio and is killed by Romeo.

Ty·burn (tī'bərn) A former place of execution in London, England.

Ty·che (tī'kē) In Greek mythology, the goddess of chance: identified with the Roman *Fortuna.*

ty·coon (tī·kōōn') *n.* 1 *U. S. Colloq.* A wealthy and powerful industrial or business leader. 2 A shogun. [<Japanese *taikun* a mighty lord <Chinese *ta* great + *kiun* a prince]

Ty·deus (tī'dyōōs, -dē·əs) The father of Diomedes and one of the Seven against Thebes.

Ty·di·des (ti·dī'dēz) In Greek mythology, Diomedes, son of Tydeus.

ty·ing (tī'ing) *n.* The act of fastening, or a fastening, as a ribbon or cord.

tyke (tīk) *n.* 1 A tike. 2 *Brit. Dial.* A man from the county of Yorkshire. [Var. of TIKE]

Ty·ler (tī'lər), **John,** 1790–1862, president of the United States 1841–45. —**Wat,** died 1381, English rebel; opposed taxation.

ty·lo·sis (tī·lō'sis) *n. pl.* **·ses** (-sēz) 1 *Bot.* A bladderlike enlargement of a plant cell, intruding within the cavity of a vessel from the wall of a contiguous growing cell. 2 The formation of calluses, especially on the skin. 3 The callus so formed. [<Gk. *tylōsis* < *tylos* a lump, callus]

tym·bal (tim'bəl) See TIMBAL.

tymp (timp) *n. Metall.* A water-cooled block

of refractory material or of cast iron, as the top of the opening between the crucible and the forehearth of a blast furnace. [Short for TYMPAN]

tym·pan (tim′pən) *n.* **1** *Printing* A thickness (or, more usually, several thicknesses), as of paper, on the impression surface of a printing press: used to improve the quality of the presswork. **2** *Archit.* A tympanum. **3** A membrane or other thin sheet tightly stretched. **4** A drum. [<OF <L *tympanum.* See TYMPANUM.]

tym·pa·ni (tim′pə·nē) See TIMPANI.

tym·pan·ic (tim·pan′ik) *adj.* **1** Like or of the nature of a drum. **2** Of or pertaining to the middle ear.

tympanic bone *Anat.* An incomplete bony ring that surrounds the external auditory canal.

tympanic membrane *Anat.* The drumhead membrane separating the middle ear from the external ear; the eardrum. See illustration under EAR.

tym·pa·nist (tim′pə·nist) *n.* One who beats or plays upon a tympan; a drummer.

tym·pa·ni·tes (tim′pə·nī′tēz) *n. Pathol.* Swelling of the abdomen due to accumulation of gas. [<LL <Gk. *tympanitēs* < *tympanon.* See TYMPANUM.] **—tym′pa·nit′ic** (-nit′ik) *adj.*

tym·pa·ni·tis (tim′pə·nī′tis) *n. Pathol.* Inflammation of the mucous membrane lining the tympanum. [<NL <L *tympanum* a drum]

tym·pa·num (tim′pə·nəm) *n. pl.* **·na** (-nə) **1** *Anat.* The middle ear; also, the tympanic membrane. **2** *Archit.* An ornamental space, as over a doorway, bounded by an arch or within the coping of a pediment. **3** A large drum wheel fitted with buckets for raising water from a flowing stream. **4** An ancient form of drum. **5** *Electr.* The diaphragm in a telephone. Also spelled *timpanum.* [<NL <L, a drum <Gk. *tympanon* < *typtein* beat]

tym·pa·ny (tim′pə·nē) *n. pl.* **·nies** Tympanites. [<Med. L *tympanias* <Gk. < *tympanon* a drum]

Tyn·dale (tin′dəl), **William,** 1484–1536, English priest and religious reformer; translated New Testament; executed for heresy.

Tyn·dall (tin′dəl), **John,** 1820–93, English physicist born in Ireland.

Tyndall effect *Physics* The scattering of light due to its passage through a medium containing minute suspended particles in continuous rapid motion. Also **Tyndall cone.** [after John *Tyndall*]

Tyn·dal·li·za·tion (tin′dəl·ə·zā′shən, -ī·zā′-) *n. Med.* A form of sterilization in which heat is applied intermittently in order to destroy spores in their less resistant adult form. [after John *Tyndall*]

Tyn·dar·e·us (tin·dâr′ē·əs) In Greek mythology, a king of Sparta and husband of Leda.

Tyne (tīn) A river in Northumberland and Durham, England, flowing 30 miles south to the North Sea at **Tyne·mouth** (tīn′məth, tin′-), a port with a 12th century priory.

typ– Var. of TYPO–.

ty·pal (tī′pəl) *adj.* Typical.

type (tīp) *n.* **1** Something that represents or symbolizes something else; an image; emblem; symbol. **2** *Theol.* That by which something is prefigured. **3** An object representative of, or embodying the characteristics of, a class or group. **4** *Biol.* **a** The general plan of an organism, with special reference to those structural and physiological characteristics which make it representative of a group, species, class, etc. **b** An individual considered as representative of members of the next higher category in a biological system of classification: the *type* of a genus, family, order, etc. **5** A variety of some physiological substance as determined by specific differences in properties and in mode of action when compared with another variety of the same substance: a blood *type.* **6** *Printing* A piece or block of metal or of wood, bearing on its upper surface, usually in relief, a letter or character for use in printing; also, such pieces collectively. See AGATE, PICA, POINT SYSTEM. **7** A distinctive sign; stamp; mark. **8** A plan to which proposed work or action should conform, as in fine arts; a standard or model. **9** In coinage, the characteristic device on either side of a medal or coin. See synonyms under EMBLEM, EXAMPLE, LETTER, MODEL, SIGN. **—*v.*** **typed, typ·ing** *v.t.* **1** To assign to a particular type or role, as an actor. **2** To determine the type of; identify: to *type* a blood sample. **3** To typewrite. **4** To represent; typify. **5** To prefigure. **—*v.i.*** **6** To typewrite. [<MF <L *typus* <Gk. *typos* an impression, figure, type < *typtein* strike]

–type *combining form* **1** Representative form; stamp; type: *prototype.* **2** Used in or produced by printing, photography, or other duplicating processes, or by type: *Linotype, collotype.* [<Gk. *typos* stamp]

type foundry An establishment in which metal type is made. **—type founder —type founding**

type genus *Biol.* A genus that combines the essential characteristics of the higher group (as a family) to which it belongs; the representative genus after which a family is named.

type–high (tīp′hī′) *adj. Printing* Designating the standard height of type (*height–to–paper*) from base to the level of the printing surface; in the United States, 0.918 of an inch: also *letter–high.*

type line One of the innermost ridges that circumscribe the pattern area of a fingerprint and assist in the identification of the fingerprint type.

type metal The alloy of which type is made, usually of lead, tin, and antimony, in various proportions.

type·script (tīp′skript′) *n.* Matter which has been typewritten: also called *typoscript.* [< TYPE(WRITTEN) + SCRIPT]

type·set·ter (tīp′set′ər) *n.* **1** A compositor. **2** A machine for composing type. **—type′·set′ting** *n.* & *adj.*

type species *Biol.* The plant or animal species regarded as most typical of the genus to which its name is given; a genotype.

type specimen *Biol.* The individual plant or animal on whose description the distinguishing characters of a species are based.

type·write (tīp′rīt′) *v.t.* & *v.i.* **·wrote, ·writ·ten, ·writ·ing** To write with a typewriter. Also *type.*

type·writ·er (tīp′rī′tər) *n.* **1** A machine for producing printed characters as a substitute for writing: it usually has a keyboard, depression of the keys serving to impress a type upon the paper through the medium of an inked ribbon. **2** A typist.

type·writ·ing (tīp′rī′ting) *n.* **1** The act or operation of using a typewriter. **2** Work done by such process.

ty·pha (tī′fə) *n.* **1** The cat–tail. **2** A fiber resembling kapok prepared from the spikes of the cat–tail, for use in life preservers, pillows, etc. [<NL < *typhē* a cat–tail]

typh·li·tis (tif·lī′tis) *n. Pathol.* Inflammation of the cecum. [<NL <Gk. *typhlos* blind] **—typh·lit′ic** (-lit′ik) *adj.*

typhlo– *combining form* **1** Blindness; of or pertaining to blindness, or to the blind: *typhlology.* **2** *Anat.* & *Med.* The cecum; related to the cecum: *typhlotomy,* a cutting into the cecum. Also, before vowels, **typhl–.** [<Gk. *typhlos* blind]

typh·lol·o·gy (tif·lol′ə·jē) *n.* The branch of medicine and pathology that deals with blindness. [<TYPHLO– + -LOGY]

typh·lo·sis (tif·lō′sis) *n.* Blindness. [<NL <Gk. *typhlōsis* < *typhlos* blind]

typho– *combining form* Typhus; typhoid: *typhogenic.* Also, before vowels, **typh–.** [<Gk. *typhos* smoke, stupor]

Ty·pho·eus (tī·fō′yōōs) In Greek mythology, a giant with a hundred snake heads, killed by Zeus's thunderbolt. **—Ty·pho·e·an** (tī·fō′ē·ən) *adj.*

ty·pho·gen·ic (tī′fə·jen′ik) *adj.* Producing typhus.

ty·phoid (tī′foid) *adj.* **1** Pertaining to or resembling typhoid fever: also **ty·phoi′dal, ty′·phose** (-fōs). **2** Resembling typhus. **—*n.*** Typhoid fever. [<TYPH(US) + -OID]

typhoid bacillus A motile, flagellated, Gram–negative bacterium (*Eberthella* or *Salmonella typhosa*), usually introduced into the body by food or drink: the bacillus that causes typhoid fever.

typhoid carrier A person who, with few or none of the clinical symptoms of infection, carries the typhoid bacillus and can communicate it to others in its active form. Also **ty·pho·phore** (tī′fə·fôr, -fōr).

typhoid fever *Pathol.* An acute, infectious fever caused by the typhoid bacillus and characterized by severe intestinal disturbances, a typical eruption of bright rose–red spots on the chest and abdomen, and great physical prostration.

ty·phoi·din (tī·foi′din) *n. Bacteriol.* A culture of the typhoid bacillus, used as a test for passive or active infection. [<TYPHOID + -IN]

ty·pho·ma·lar·i·al (tī′fō·mə·lâr′ē·əl) *adj. Pathol.* Describing a fever resembling that of typhoid but believed to be malarial in origin. [<TYPHO– + MALARIAL]

ty·pho·ma·ni·a (tī′fə·mā′nē·ə, -mān′yə) *n. Pathol.* The delirious state associated with typhoid fever or typhus. Also **ty·pho·ni·a** (tī·fō′nē·ə).

Ty·phon (tī′fon) In Greek mythology, a monster overcome and buried by Zeus under Mount Etna.

ty·phoon (tī·fōōn′) *n.* A tropical storm of cyclonic force and peculiar violence, occurring in the western Pacific and the China Sea. See synonyms under CYCLONE. [< dial. Chinese *tai feng,* lit., big wind; infl. by obs. *typhon* a whirlwind (<Gk. *typhōn* a hurricane) and by obs. *tuphan, tufan* a typhoon <Arabic *ṭūfān,* ? ult. from the same Gk. source]

ty·phus (tī′fəs) *n. Pathol.* An acute, contagious, rickettsial disease caused by a micro–organism (*Rickettsia prowazeki*) and marked by high fever with eruption of red spots, cerebral disorders, and extreme prostration; typhus fever: also called *Brill's disease.* **Epidemic typhus** is transmitted by the bite of the body louse, and **endemic** or **murine typhus** by the bite of the rat flea. [<NL <Gk. *typhos* smoke, a stupor < *typhein* smoke] **—ty′·phous** *adj.*

typ·i·cal (tip′i·kəl) *adj.* **1** Having the nature or character of a type; constituting a type or pattern; symbolic. **2** Conforming to the essential features of a species, group, class, etc.; characteristic. Also **typ′ic;** also *typal.* See synonyms under NORMAL. [<Med. L *typicalis* <L *typicus* <Gk. *typikos* < *typos* TYPE] **—typ′i·cal·ly** *adv.* **—typ′i·cal·ness** *n.*

typ·i·fy (tip′ə·fī) *v.t.* **·fied, ·fy·ing** **1** To represent by a type; signify, as by an image or token. **2** To constitute a type or serve as a characteristic example of. **—typ′i·fi·ca′tion** (-fə·kā′shən) *n.* **—typ′i·fi′er** *n.*

typ·ist (tī′pist) *n.* One who uses a typewriting machine.

typo– *combining form* Type; of or related to type: *typography.* Also, before vowels, *typ–* [<Gk. *typos* stamp, type]

ty·po·graphed (tī′pə·graft, -gräft) *adj.* Printed from type, or from plates in which the design is raised above the level of the body of the plate.

ty·pog·ra·pher (tī·pog′rə·fər) *n.* A printer.

ty·po·graph·i·cal (tī′pə·graf′i·kəl) *adj.* Pertaining to, concerned with, or effected by typography or printing. Also **ty′po·graph′ic.** **—ty′·po·graph′i·cal·ly** *adv.*

ty·pog·ra·phy (tī·pog′rə·fē) *n.* **1** The arrangement of composed type. **2** The style and appearance of printed matter. **3** The act or art of composing and printing from types. [<TYPO– + -GRAPHY]

ty·pol·o·gy (tī·pol′ə·jē) *n.* The study of types, as in systems of classification. [<TYPO– + -LOGY]

ty·po·script (tī′pō·skript) See TYPESCRIPT.

Ty·poth·e·tae (tī·poth′ə·tē, tī′pə·thē′tē) *n. pl.* An association of master printers; hence, by extension, **ty·poth′e·tae,** printers collectively: used in the names of organized groups of printers. [<NL <Gk. *typos* TYPE + *tithenai* set, put]

Tyr (tür, tir) In Norse mythology, the god of war and son of Odin: identified with the Teutonic *Tiu:* also spelled *Tyrr.*

ty·ra·mine (tī′rə·mēn′, tir′ə-) *n. Chem.* A white, crystalline, nitrogenous compound, $C_8H_{11}ON$, found in ergot, ripe cheese, and putrefying animal tissue: the hydrochloride is used in medicine. [<TYR(OSINE) + AMINE]

ty·ran·ni·cal (ti·ran′i·kəl, tī-) *adj.* Of or like a

tyrant; harsh; despotic; arbitrary. Also **ty·ran′·nic.** See synonyms under ABSOLUTE, ARBITRARY. — **ty·ran′ni·cal·ly** *adv.* — **ty·ran′ni·cal·ness** *n.*
ty·ran·ni·cide (ti·ran′ə·sīd, tī-) *n.* 1 The slayer of a tyrant. 2 The slaying of a tyrant. [<F <L *tyrannicida* < *tyrannus* a tyrant + *caedere* kill; def. 2 <L *tyrannicidium*]
tyr·an·nize (tir′ə·nīz) *v.* **·nized, ·niz·ing** *v.i.* 1 To exercise power cruelly or unjustly. 2 To rule as a tyrant; have absolute power. — *v.t.* 3 To treat tyrannically; domineer. Also *Brit.* **tyr·an·nise.** [<MF *tyranniser* <LL *tyrannizare* <Gk. *tyrannizein* < *tyrannos* a tyrant] — **tyr′·an·niz′er** *n.*
ty·ran·no·saur·us (ti·ran′ə·sôr′əs, tī-) *n. Paleontol.* A carnivorous dinosaur (*Tyrannosaurus rex*) inhabiting North America in the Cretaceous period: it was characterized by its huge bulk, massive jaws, and ability to walk erect on its hind legs. [<NL <Gk. *tyrannos* a tyrant + *sauros* a lizard]
tyr·an·nous (tir′ə·nəs) *adj.* Despotic, tyrannical. See synonyms under ARBITRARY. — **tyr′an·nous·ly** *adv.* — **tyr′an·nous·ness** *n.*
tyr·an·ny (tir′ə·nē) *n. pl.* **·nies** 1 Absolute power arbitrarily or unjustly administered; despotism. 2 An arbitrarily cruel exercise of power; a tyrannical act. 3 In Greek history, the office or the administration of a tyrant. 4 Severity; roughness. [<OF *tirannie* <L *tyrannia* < *tyrannus* a tyrant]

ty·rant (tī′rənt) *n.* 1 One who rules oppressively or cruelly; a despot. 2 One who exercises absolute power without legal warrant, whether ruling well or ill: the original meaning in ancient Greece. [<OF *tiran, tyran* <L *tyrannus* <Gk. *tyrannos* a master, a usurper]
tyrant flycatcher Any American flycatcher (family *Tyrannidae*), as the kingbird, pewee, etc.
Ty·ras (tī′rəs) The ancient name for BELGOROD-DNESTROVSKI.
tyre (tīr) See TIRE².
Tyre (tīr) A port and capital of ancient Phoenicia, on the site of modern Sur in SW Lebanon.
Tyr·i·an (tir′ē·ən) *adj.* 1 Of or pertaining to Tyre. 2 Having the color of Tyrian dye: purple. — *n.* A native of Tyre.
Tyrian dye 1 A purple or crimson dyestuff obtained by the ancient Greeks and Romans from certain mollusks of the genus *Murex*. 2 A violet-purple color of high saturation and low brilliance. Also **Tyrian purple.**
ty·ro (tī′rō) *n. pl.* **·ros** One who is in the rudiments of any study or the preliminary stage of any occupation; a beginner; novice: also spelled *tiro*. [<Med. L <L *tiro* a recruit]
Ty·rol (ti·rōl′, tir′ōl, tī′rōl) See TIROL.
Tyr·o·lese (tir′ə·lēz′, -lēs′) See TIROLESE.
Ty·ro·lienne (tē·rô·lyen′) *n.* A ländler. [<F, fem. of *tyrolien* Tyrolean]

Ty·rone (ti·rōn′) A county of Ulster, western Northern Ireland; 1,218 miles; county town, Omagh.
ty·ro·sin·ase (tī′rō·si·nās′, tir′ō-) *n. Biochem.* A plant and animal enzyme which converts tyrosine into dark pigments, as melanin. [<TYROSIN(E) + -ASE]
ty·ro·sine (tī′rō·sēn, -sin, tir′ō-) *n. Biochem.* A white crystalline amino acid, $C_9H_{11}O_3N$, formed by the hydrolysis of many plant and animal proteins. [<Gk. *tyros* cheese + -INE²]
ty·ro·sin·o·sis (tī′rō·sin·ō′sis, tir′ō-) *n. Pathol.* A disorder caused by defective metabolism of tyrosine in the body. [<TYROSIN(E) + -OSIS]
ty·ro·thri·cin (tī′rō·thrī′sin, -thris′in) *n.* An antibiotic isolated from a soil bacterium (*Bacillus brevis*): similar to gramicidin and used therapeutically in localized infections. [<TYRO(SINE) + Gk. *thrix, trichos* a hair + -IN]
Tyrr (tür, tir) See TYR.
Tyr·rhe·ni·an Sea (ti·rē′nē·ən) The part of the Mediterranean between Italy, Sardinia, Corsica, and Sicily. [<Gk. *Tyrrhēnia* Tuscany]
Tyr·tae·us (tûr·tē′əs) Greek poet of the seventh century B.C.
Tyu·men (tyōō·men′, *Russian* tyōō·myän′y′) A city in western Asiatic Russian S.F.S.R.
Tyu·zen·zi (chōō·zen·jē) See CHUZENJI.
tzar (tsär), **tza·ri·na** (tsä·rē′nä), etc. See CZAR, etc.
tzet·ze (tset′sē) See TSETSE.

U

u, U (yōō) *n. pl.* **u's, U's** or **Us** (yōōz) 1 The twenty-first letter of the English alphabet: from Greek *upsilon*. In Roman it was written V and had both consonant and vowel value. In English U was formerly the uncial or cursive form of V; gradually V came to be preferred in initial position in writing, and, as the sound at the beginning of a word is ordinarily consonantal, U was finally restricted to vowel use. 2 Any sound of the letter *u*. See ALPHABET. — *symbol* 1 *Chem.* Uranium (symbol U). 2 Anything shaped like a U.
Uau·pés (wou·pās′) A river in SE Colombia and NW Brazil, flowing 500 miles SE to the Río Negro: also *Vaupés.*
U·ban·gi (ōō·bäng′gē) A river of central Africa, flowing 1,400 miles from NE Belgian Congo to the Congo river and forming part of the boundary between the Belgian Congo and French Equatorial Africa.
U·ban·gi-Sha·ri (ōō·bäng′gē-shä′rē) See CENTRAL AFRICAN REPUBLIC.
U·be (ōō′bē, *Japanese* ōō·be) A city of SW Honshu island, Japan.
Ü·ber·mensch (ü′bər·mensh) *n. German* The superman, in Nietzsche's terminology.
u·bi·e·ty (yōō·bī′ə·tē) *n.* The state of being in a place; local relation. [<NL *ubietas, -tatis* <L *ubi* where]
u·biq·ui·tar·i·an (yōō·bik′wə·târ′ē·ən) *n.* One who has ubiquitous existence.
U·biq·ui·tar·i·an (yōō·bik′wə·târ′ē·ən) *n.* A believer in the omnipresence of the human nature of Christ and, as a consequence, in his necessary actual bodily presence in the Eucharist. Also **U·bi·quar·i·an** (yōō′bə·kwâr′ē·ən), **U′bi·quist, U·biq′ui·tist.**
U·biq·ui·tar·i·an·ism (yōō·bik′wə·târ′ē·ən·iz′əm) *n.* The tenets of the Ubiquitarians. Also **U·biq·ui·tar·i·an·iz′əm).**
u·biq·ui·tous (yōō·bik′wə·təs) *adj.* Existing, or seeming to exist, everywhere at once; omnipresent. Also **u·biq′ui·tar′y** (-ter′ē). — **u·biq′ui·tous·ly** *adv.* — **u·biq′ui·tous·ness** *n.*
u·biq·ui·ty (yōō·bik′wə·tē) *n.* 1 The state of being in an indefinite number of places at once; omnipresence real or seeming. 2 The state of existing always without beginning or end. [<F *ubiquité* <L *ubique* everywhere]
u·bi su·pra (yōō′bī sōō′prə) *Latin* Where (mentioned) above.

U-boat (yōō′bōt′) *n.* A German submarine. [<G *U-boot*, contraction of *Unterseeboot*, lit., undersea boat]
U-bolt (yōō′bōlt′) *n.* A bolt bent like the letter U, and fitted with a screw and nut at each end.
U·che·an (yōō·chē′ən) *n.* A North American Indian linguistic stock, consisting only of the Yuchi tribe.
U·dai·pur (ōō·dī′pŏŏr, ōō′dī-) 1 A former princely state of the Rajputana States, India; since 1948 merged with the State of Rajasthan; 13,170 square miles: also *Mewar.* 2 A city of southern Rajasthan, India; formerly capital of Udaipur state.
U·dall (yōōd′l), **Nicholas**, 1506?-56, English scholar and dramatist: also *Uvedale.*
U·day Shan·kar (ōō′dī shän·kär′), born 1900, Indian dancer.
ud·der (ud′ər) *n.* A large, pendulous, milk-secreting gland provided with nipples or teats for the suckling of offspring, as in cows. [OE *ūder*]
U·di·ne (ōō′dē·nä) A city in NE Italy.
Ud·murt Autonomous Soviet Socialist Republic (ōōd′mŏŏrt, ōōd·mōōrt′) An administrative division of east central European Russian S.F.S.R.; 16,300 square miles; capital, Izhevsk.
u·do (ōō′dō) *n.* A bushy plant (*Aralia cordata*) of Japan and China which, when young, yields edible shoots. [<Japanese]
u·dom·e·ter (yōō·dom′ə·tər) *n.* A pluviometer. [<L *udus* moist + -METER] — **u·do·met·ric** (yōō′də·met′rik) *adj.* — **u·dom′e·try** *n.*
Ue·le (wel′ä) A river of NE Belgian Congo, flowing 700 miles north, NW, and west to a confluence with the Bomu at the border of French Equatorial Africa, forming the Ubangi; also *Welle.*
U·fa (ōō·fä′) 1 A city on the Byelaya river in eastern European Russian S.F.S.R.; capital of Bashkir Autonomous S.S.R. 2 A river in eastern European Russian S.F.S.R., flowing 599 miles NW and SW from the southern Urals to the Byelaya river at Ufa.
UFO Unidentified flying object: an official U.S. Air Force designation. Compare FLYING SAUCER.
U·gan·da (yōō·gan′də, ōō·gän′dä) A British protectorate in east central Africa; 93,981 square miles; capital, Entebbe.

ugh (ukh, u, ŏŏkh, ŏŏ) *interj.* An exclamation of repugnance or disgust. [Imit.]
ug·li·fy (ug′lə·fī) *v.t.* **·fied, ·fy·ing** To make ugly. — **ug′li·fi·ca′tion** (-fə·kā′shən) *n.*
ug·ly (ug′lē) *adj.* **·li·er, ·li·est** 1 Displeasing to the esthetic feelings, as from lack of grace or proportion; distasteful in appearance; ill-looking; unsightly. 2 Repulsive to the moral sentiments; revolting. 3 Bad in character or consequences, as a rumor or a wound. 4 *Colloq.* Ill-tempered; quarrelsome. 5 Portending storms; threatening: said of the weather. [<ON *uggligr* dreadful < *uggr* fear] — **ug′li·ly** *adv.* — **ug′li·ness** *n.*
ugly duckling 1 In Hans Christian Andersen's story, *The Ugly Duckling*, a young swan hatched by a duck, belittled and persecuted by all the ducks for his strange appearance, until he grew into the most beautiful bird on the pond. 2 Any ill-favored or unpromising child who unexpectedly grows into a beauty or a wonder.
U·go (ōō′gō) Italian form of HUGH. Also **U′go·li′no** (-lē′nō).
U·gri·an (ōō′grē·ən, yōō′-) *n.* 1 A member of any of the Finno-Ugric peoples of Hungary and western Siberia, including the Ostyaks, Voguls, and Magyars. 2 Ugric. — *adj.* Of or pertaining to the Ugrians, their culture, or their languages.
U·gric (ōō′grik, yōō′-) *n.* A branch of the Finno-Ugric subfamily of Uralic languages, comprising Magyar (Hungarian), Ostyak, and Vogul. — *adj.* Of or pertaining to any of these languages.
U·gro-Al·ta·ic (ōō′grō·al·tā′ik, yōō′grō-) *n. & adj.* Ural-Altaic.
ug·some (ug′səm) *adj. Scot.* Disgusting.
uh·lan (ōō′län, ōō·län′, yōō′lən) *n.* 1 A cavalryman and lancer of a type originating in eastern Europe, formerly prominent in European armies, notably the German. 2 One of a body of Tatar militia. Also spelled *ulan.* [<G <Polish <Turkish *ōghlān* lad, servant]
Uh·land (ōō′länt), **Johann Ludwig**, 1787-1862, German poet.
Ui·gur (wē′gŏŏr) *n.* 1 One of a Turkic people who ruled in Mongolia and East Turkestan from the eighth to the twelfth century, now the majority of the population of the Sinkiang-Uigur Autonomous Region, NW China.

2 The Turkic language of these people. — **Ui·gu·ri·an** (wē-gōor′ē-ən), **Ui·gu·ric** (wē-gōor′ik) *adj.*

u·in·tah·ite (yōō-in′tə-īt) *n.* A variety of asphalt common in Utah: often called *gilsonite.* Also **u·in′ta·ite.** [from *Uinta* Mountains]

U·in·ta Mountains (yōō-in′tə) A range in NE Utah and SW Wyoming; highest point, 13,498 feet (the highest point in Utah).

uit (oit, œit) *prep. Afrikaans* Out; out of.

uit·land·er (it′lan-dər, oit′-; *Afrikaans* œit′-län-dər) *n. Afrikaans* A foreigner; formerly, in the South African Republic, a foreign white resident.

uit·span (œit′spän) *v.* & *n. Afrikaans* Outspan.

U·ji·ji (ōō-jē′jē) A port on Lake Tanganyika in western Tanganyika, Africa.

Ú·j·pest (ōō′ē-pesht) A city on the Danube in central Hungary: German *Neupest.*

u·kase (yōō′kās, yōō-kāz′) *n.* 1 Formerly, an edict or decree of the imperial Russian government. 2 Any official decree. [<Russian *ukaz*]

U·kraine (yōō-krān′, yōō′krān, yōō-krīn′) A rich agricultural region in SW European U.S.S.R., comprised in the Ukrainian S.S.R. Also **U·krain·i·a** (yōō-krā′nē-ə, -krī′-). *Ukrainian* **U·krai·na** (ōō-krä-yē′nä).

U·krain·i·an (yōō-krā′nē-ən, -krī′-) *adj.* Of or pertaining to the Ukraine, its people, or their language. — *n.* 1 A native or inhabitant of the Ukraine. 2 An East Slavic language spoken in the Ukraine. See under RUSSIAN.

Ukrainian S.S.R. A constituent republic of SW European U.S.S.R.; 231,986 square miles; capital, Kiev: also *Ukraine, Ukrainia.*

u·ku·le·le (yōō′kə-lā′lē, *Hawaiian* ōō′kōō-lā′lā) *n.* A guitarlike musical instrument having four strings. [<Hawaiian, flea <*uku* insect + *lele* jump; from the movements of the fingers in playing]

U·lan (ōō′län, ōō-län′, yōō′lən) See UHLAN.

UKULELE

U·lan Ba·tor (ōō′län bä′tôr) The capital of the Mongolian People's Republic: formerly *Urga:* Chinese *Kulun.*

U·lan Ho·to (ōō′län khō′tō) The capital of Inner Mongolian Autonomous Region, northern China, in the NE part of the Region: Chinese *Wulanhaote:* formerly *Wangyehmiao.* Also *Khoto.*

U·lan-U·de (ōō′län-ōō′de) The capital of Buryat-Mongol Autonomous S.S.R., in SE central Asiatic Russian S.F.S.R., SE of Baikal lake.

ul·cer (ul′sər) *n.* 1 *Pathol.* An open sore on an external or internal surface of the body, usually accompanied by disintegration of tissue with the formation of pus. 2 Figuratively, a corroding fault or vice; corruption; evil. [<L *ulcus, ulceris*]

ul·cer·ate (ul′sə-rāt) *v.t.* & *v.i.* **·at·ed, ·at·ing** To make or become ulcerous. [<L *ulceratus,* pp. of *ulcerare* <*ulcus, ulceris* ulcer] — **ul′cer·a′tive** *adj.*

ul·cer·a·tion (ul′sə-rā′shən) *n.* 1 The forming of an ulcer, or the condition of being affected with ulcers. 2 An ulcer, or ulcers collectively.

ul·cer·ous (ul′sər-əs) *adj.* 1 Resembling an ulcer. 2 Affected with ulcers. — **ul′cer·ous·ly** *adv.* — **ul′cer·ous·ness** *n.*

-ule *suffix of nouns* Small; little: used to form diminutives: *granule.* [<F *-ule* <L *-ulus, -ula, -ulum,* diminutive suffix]

u·le·ma (ōō′lə-mä′) *n.* 1 In Moslem countries, a council or college of learned officials (priests, judges, or scholars) who are trained in Moslem religion and law, and interpret the Koran. 2 Hence, any Moslem scholar. [<Turkish *'ulema* <Arabic *'ulamā,* pl. of *'alim* wise <*'alama* know]

-ulent *suffix of adjectives* Abounding in; full of (what is indicated in the main element): *opulent, truculent.* Corresponding nouns are formed in **-ulence,** as in *opulence, truculence.* [<L *-ulentus*]

Ul·fi·las (ul′fi-ləs), A.D. 311?-383, bishop of the Goths; translated the Bible into Gothic: also spelled *Wulfila.* Also **Ul′fi·la** (-lə).

U·li·thi (ōō-lē′thē) An atoll of the western Caroline Islands; 19 miles long, 10 miles wide.

ul·lage (ul′ij) *n.* The quantity that a vessel, as a wine cask, lacks of being full; wantage. [<AF *ulliage,* OF *ouillage* <*ouiller* fill up (to the bunghole) <*ueil* eye, bunghole <L *oculus* eye]

Ulls·wa·ter (ulz′wô-tər, -wot·ər) The second largest English lake, in Cumberland and Westmoreland; 3 square miles.

Ulm (ŏŏlm) A city on the Danube River in central eastern Baden-Württemberg, SW West Germany.

ul·ma·ceous (ul-mā′shəs) *adj. Bot.* Designating or belonging to a family (*Ulmaceae*) of shrubs and trees of the order *Urticales,* the elm family, widely distributed in temperate and tropical regions, and characterized by alternate simple leaves, apetalous bisexual or unisexual flowers, and a compressed fruit. [<NL, family name <L *ulmus* elm]

ul·na (ul′nə) *n. pl.* **·nae** (-nē) or **·nas** *Anat.* In vertebrates above fishes, that one of the two long bones of the forearm or foreleg which articulates with the radius and is on the same side as the little finger or fifth digit. [<L, elbow] — **ul′nar** *adj.*

-ulose *suffix of adjectives* Marked by or abounding in: widely used in scientific and technical terms: *ramulose.* Compare -ULOUS (def. 2). [<L *-ulosus,* adjective suffix]

U·lot·ri·chi (yōō-lot′rə-kī) *n. pl.* In the classification of Huxley, a subdivision of the human species, characterized by woolly or crispy hair. Also **U·lot′ri·ches** (-kēz). [<NL *Ulotriches* <Gk. *oulothrix, oulotrichos* woolly-haired <*oulos* woolly + *thrix* hair] — **u·lot′ri·chous** (-kəs) *adj.*

-ulous *suffix of adjectives* 1 Tending to do or characterized by (what is indicated by the main element): *tremulous, ridiculous.* 2 Full of: *meticulous, populous.* Compare -ULOSE. [<L *-ulus* and *-ulosus,* adjective suffixes]

Ul·pi·an (ul′pē-ən), A.D. 170?-228, Roman jurist; full name *Domitius Ulpianus.*

ul·ster (ul′stər) *n.* A very long, loose overcoat, sometimes belted at the waist: made originally of frieze from Ulster, Ireland.

Ul·ster (ul′stər) A former province of northern Ireland comprising the nine counties listed below; 8,331 square miles. In 1925 six of the counties (Antrim, Armagh, Downe, Fermanagh, Londonderry, Tyrone: 5,238 square miles) became Northern Ireland and three (Cavan, Donegal, Monaghan: 3,093 square miles) the Province of Ulster of the Republic of Ireland. — **Ul′ster·man** (-mən) *n.*

Ulster cycle The older and more famous of the two cycles of Old Irish epic and romance. The manuscripts date from the seventh and eighth centuries, but celebrate the Ireland and Irish heroes of the first century, and depict a civilization of barbaric splendor that dates back centuries earlier. See FENIAN CYCLE, TAIN BO CUAILGNE.

ul·te·ri·or (ul-tir′ē-ər) *adj.* 1 More remote; not so pertinent as something else to the matter spoken of: applied to immaterial things: *ulterior* considerations; also, intentionally unrevealed; hidden: *ulterior* motives. 2 Following; succeeding; later in time, or secondary in importance. 3 Lying beyond or on the farther side of a certain bounding line. [<L, compar. of *ulter* beyond] — **ul·te′ri·or·ly** *adv.*

ul·ti·ma (ul′ti-mə) *n.* The last syllable of a word. [<L, fem. of *ultimus* last]

ul·ti·mate (ul′tə-mit) *adj.* 1 Beyond which there is no other; last of a series; final. 2 Fundamental or essential; hence, not susceptible of further analysis; elementary; primary. 3 Most distant; farthest; extreme. 4 *Mech.* Designating the maximum strength of a body, or a strain of the least intensity sufficient to cause rupture. — *n.* 1 The final result; last step; conclusion. 2 A fundamental or final fact. [<LL *ultimatus,* orig. pp. of *ultimare* come to an end <*ultimus* farthest, last, superl. of *ulter* beyond] — **ul′ti·mate·ness** *n.*

ul·ti·mate·ly (ul′tə-mit·lē) *adv.* In the end; at last; finally.

ul·ti·ma Thu·le (ul′tə-mə thōō′lē, tōō′lē) 1 Farthest Thule: in ancient geography, the northernmost habitable regions of the earth. 2 Any distant, unknown region. 3 The farthest possible point, degree, or limit.

ul·ti·ma·tum (ul′tə-mā′təm, -mä′-) *n. pl.* **·tums** or **·ta** (-tə) 1 A final statement, as concerning terms or conditions; in diplomacy, the final terms offered by one party, as during negotiations concerning a treaty, the rejection of which by the other party will result in breaking off all negotiation; loosely, a last proposal, offer, concession, or demand. 2 Anything ultimate. [<NL <LL, neut. of *ultimatus.* See ULTIMATE.]

ul·ti·mo (ul′tə-mō) *adv. Latin* In the last month: shortened to *ult.,* following a date: the 15th *ult.:* distinguished from *proximo* (*prox.*) or *instant* (*inst.*).

ul·ti·mo·gen·i·ture (ul′tə-mō-jen′ə-chər) *n.* The rule whereby the youngest son takes the inheritance: the opposite of *primogeniture.* [<L *ultimus* last + GENITURE]

ul·tra (ul′trə) *adj.* Going beyond the bounds of moderation; extreme; extravagant. — *n.* One who holds extreme opinions; a radical. [<L, beyond, on the other side]

ultra- *prefix* 1 On the other side of; beyond in space (compare TRANS-); as in:

ultra-Arctic	ultra-Neptunian
ultra-equinoctial	ultra-stellar
ultra-galactic	ultra-terrene
ultra-lunar	ultra-terrestrial
ultra-Martian	ultra-zodiacal

2 Going beyond the limits of; surpassing; as in:

ultra-atomic	ultra-molecular
ultra-centenarian	ultra-natural
ultra-human	ultra-total

3 Beyond what is usual or natural; excessively; as in:

ultra-affected	ultra-moderate
ultra-agnostic	ultra-modest
ultra-ambitious	ultra-mulish
ultra-Anglican	ultra-nominalistic
ultra-believing	ultra-ornate
ultra-benevolent	ultra-orthodox
ultra-Christian	ultra-orthodoxy
ultra-classical	ultra-partisan
ultra-confident	ultra-physical
ultra-conservatism	ultra-positivistic
ultra-conservative	ultra-precision
ultra-cooperative	ultra-Protestant
ultra-cosmopolitan	ultra-Protestantism
ultra-credulous	ultra-prudent
ultra-democratic	ultra-purist
ultra-despotic	ultra-Puritan
ultra-discipline	ultra-radical
ultra-educationist	ultra-refined
ultra-episcopal	ultra-refinement
ultra-evangelical	ultrareligious
ultra-exclusive	ultra-revolutionary
ultrafashionable	ultra-revolutionist
ultra-fastidious	ultra-ritualism
ultra-federalist	ultra-romanticist
ultra-feudal	ultra-royalism
ultra-filtration	ultra-royalist
ultra-Gallican	ultra-scientific
ultra-German	ultra-sensual
ultra-honorable	ultra-sentimental
ultra-intellectual	ultra-servile
ultra-legality	ultra-Spartan
ultra-liberal	ultra-spiritual
ultra-liberalism	ultra-splendid
ultra-logical	ultra-sterile
ultra-loyal	ultra-strict
ultra-manners	ultra-theological
ultra-maternal	ultra-virtuous

ULNA
A. Front view.
B. Back view.
a. Elbow joint.
b. Ulna.
c. Radius.
d. Wrist.

ul·tra·cen·tri·fuge (ul′trə·sen′trə·fyōōj) *n.* A centrifuge whose rotor, sometimes driven by blasts of hydrogen, will exert a force of about one million times gravity: used for high precision scientific and laboratory work. — **ul′tra·cen′tri·fu·ga′tion** (-fyōō·gā′shən) *n.*

ul·tra·crit·i·cal (ul′trə·krit′i·kəl) *adj.* Unduly critical.

ul·tra·fil·ter (ul′trə·fil′tər) *n. Chem.* A filter having extremely minute pores, as a living membrane or a film of gelatin on filter paper: used to sift out colloidal particles which pass through ordinary filters. — **ul′tra·fil·tra′tion** (-fil·trā′shən) *n.*

ultrahigh frequency *Electronics* A band of wave frequencies in radio between 300 and 3,000 megacycles per second. Abbr. *uhf.*

ul·tra·ist (ul′trə·ist) *n.* One who in opinions or conduct goes beyond moderation; a radical; an extremist. — *adj.* Radical; extreme: also **ul′tra·is′tic.** — **ul′tra·ism** *n.*

ul·tra·ma·rine (ul′trə·mə·rēn′) *n.* 1 A deep, usually purplish-blue, permanent pigment made by treating the powdered mineral lapis lazuli. 2 A similar pigment made largely by synthesis from kaolin, silica, soda, sulfur, and charcoal: also called *new blue, French blue.* 3 The color of ultramarine. — *adj.* Being beyond or across the sea. [<Med. L *ultramarinus* <L *ultra* beyond + *marinus* marine]

ul·tra·mi·crobe (ul′trə·mī′krōb) *n.* A microorganism that is invisible in the optical microscope.

ul·tra·mi·crom·e·ter (ul′trə·mī·krom′ə·tər) *n.* A micrometer designed for measurements requiring a high order of precision and accuracy.

ul·tra·mi·cro·scope (ul′trə·mī′krə·skōp) *n.* An optical instrument for detecting objects too small to be seen with an ordinary microscope, by means of an intense beam of light thrown from the side upon the spot to be examined.

ul·tra·mi·cro·scop·ic (ul′trə·mī′krə·skop′ik) *adj.* 1 Too minute to be seen by an ordinary microscope. 2 Relating to the ultramicroscope. Also **ul′tra·mi′cro·scop′i·cal.** — **ul′tra·mi·cros′co·py** (-mi·kros′kə·pē) *n.*

ul·tra·mod·ern (ul′trə·mod′ərn) *adj.* Excessively or inordinately new or modern; extreme in modern tendencies or ideas. — **ul′tra·mod′ern·ism** *n.* — **ul′tra·mod′ern·ist** *n.* — **ul′tra·mod′ern·is′tic** *adj.*

ul·tra·mon·tane (ul′trə·mon′tān) *adj.* 1 Situated beyond the mountains: opposed to *cismontane*; beyond or south of the Alps, i.e., Italian or papal. 2 In politics or ecclesiastical matters, supporting the policy of the papal court. — *n.* 1 One who resides beyond the Alps. 2 One who supports the papal policy in political or ecclesiastical matters. [<Med. L *ultramontanus* <L *ultra* beyond + *montanus* pertaining to a mountain <*mons, montis* mountain]

Ul·tra·mon·ta·nism (ul′trə·mon′tə·niz′əm) *n.* The policy of Roman Catholics who wish to see all power in the church in the hands of the pope, in opposition to those desiring a more independent development of the national churches; curialism: opposed to *Gallicanism.*

ul·tra·mun·dane (ul′trə·mun′dān) *adj.* Extending beyond the world, the solar system, or the present life. [<L *ultramundanus*]

ul·tra·na·tion·al·ism (ul′trə·nash′ən·əl·iz′əm) *n.* Extreme devotion to or support of national, as opposed to international, interests or considerations. — **ul′tra·na′tion·al** *adj.* — **ul′tra·na′tion·al·ist** *n.* & *adj.*

ul·tra·pho·tic (ul′trə·fō′tik) *adj. Physics* Denoting wavelengths of radiant energy beyond the visible region of the spectrum, as ultraviolet and infrared.

ul·tra·red (ul′trə·red′) *adj.* Infrared.

ul·tra·son·ic (ul′trə·son′ik) *adj. Physics* Pertaining to or designating sound waves having a frequency above the limits of audibility, or in excess of about 15 kilocycles per second: distinguished from *supersonic.*

ul·tra·son·ics (ul′trə·son′iks) *n. pl. (construed as singular)* The study of acoustic phenomena in the frequency range above that of audibility.

ul·tra·trop·i·cal (ul′trə·trop′i·kəl) *adj.* 1 Situated beyond the tropics. 2 Hotter than the tropics.

ul·tra·vi·o·let (ul′trə·vī′ə·lit) *adj. Physics* Lying beyond the violet end of the visible spectrum: said of high-frequency light waves more refrangible than the violet and having wavelengths ranging from about 3,900 angstroms to the upper limits of X-rays. Compare INFRARED.

ul·tra vi·res (ul′trə vī′rēz) *Latin* 1 *Law* Beyond the lawful capacity or powers: said especially of corporations as to acts or contracts not within the scope of the powers conferred upon them and which are *ipso facto* void: applied also to acts which although within their powers have been done without their required consent, as in the case of powers delegated to directors. 2 Figuratively, not permissible; forbidden: a colloquial use.

ul·tra·vi·rus (ul′trə·vī′rəs) *n.* A filtrable virus. [<NL]

U·lugh Muz·tagh (ōō′lōō mōōz·tä′) A peak on the east central border of the Sinkiang-Uigur and the Tibetan autonomous regions, western China; 25,340 feet.

ul·u·lant (yōōl′yə·lənt, ul′-) *adj.* Howling; hooting. [<L *ululans, -antis,* ppr. of *ululare* howl]

ul·u·late (yōōl′yə·lāt, ul′-) *v.i.* **·lat·ed, ·lat·ing** To howl, hoot, or wail. [<L *ululatus,* pp. of *ululare* howl] — **ul′u·la′tion** *n.*

Ul·ya·novsk (ōōl·yä′nôfsk) A port on the Volga in east central European Russian S.F.S.R.

U·lys·ses (yōō·lis′ēz, *Ger.* ōō·lü′ses) A masculine personal name. Also *Fr.* **U·lysse** (ü·lēs′), *Ital.* **U·lis·se** (ōō·lēs′sā). [<Gk., the hater] — **Ulysses** Odysseus.

um·bel (um′bəl) *n. Bot.* An indeterminate inflorescence in which a number of nearly equal pedicels radiate from a small area at the top of a very short axis, giving an umbrellalike appearance. [<L *umbella* a parasol, dim. of *umbra* shadow. Related to UMBRELLA.]

um·bel·late (um′bə·lit, -lāt) *adj.* Disposed in or resembling umbels. Also **um′bel·lar, um′bel·lat′ed.** [<NL *umbellatus*]

um·bel·let (um′bə·lit) *n.* An umbellule.

um·bel·lif·er·ous (um′bə·lif′ər·əs) *adj.* 1 Bearing umbels. 2 Designating or pertaining to an important and widely distributed family *(Umbelliferae)* of herbs and some shrubs, the parsley or carrot family, comprising many plants used as food, for flavoring, and in medicine. [<NL, family name <L *umbella* parasol + *ferre* bear]

um·bel·lu·late (um·bel′yə·lit, -lāt) *adj.* Having or disposed in umbellules.

um·bel·lule (um′bəl·yōōl, um·bel′-) *n. Bot.* A small or secondary umbel. [<NL *umbellula,* dim. of L *umbella* parasol]

um·ber[1] (um′bər) *n.* A chestnut- to liver-brown hydrated ferric oxide, containing some manganese oxide and clay: used as a pigment; also, the color. When in its natural state it is known as **raw umber,** and when heated, so as to produce a reddish-brown, as **burnt umber.** — *adj.* Of or pertaining to umber; of a dusky hue; brownish. — *v.t.* To color with umber; darken, as by staining. [<F *(terre d')ombre* <Ital. *ombra,* prob. <L *Umbra,* fem. of *Umber* of Umbria, where originally found; ? infl. in Ital. by *ombra* shadow, shade <L *umbra*]

um·ber[2] (um′bər) *n.* 1 Shade; hence, some indefinite dark color. 2 The umbrette: also **umber bird.** 3 The grayling. [<F *ombre* <L *umbra* shade]

Um·ber·to (ōōm·ber′tō) Italian form of HUMBERT.

um·ber·y (um′bər·ē) *adj.* Pertaining to or like umber; dusky.

um·bil·i·cal (um·bil′i·kəl) *adj.* 1 Pertaining to or situated near the umbilicus. 2 Placed near the navel; central. [<LL *umbilicalis* <L *umbilicus* navel]

umbilical cord *Anat.* The ropelike tissue connecting the navel of the fetus with the placenta.

um·bil·i·cate (um·bil′ə·kit, -kāt) *adj.* 1 Resembling a navel, as by having a central depression or mark. 2 Having an umbilicus or navel-shaped depression, as a shell. Also **um·bil′i·cat′ed.**

um·bil·i·ca·tion (um·bil′ə·kā′shən) *n.* 1 The state of being umbilicate. 2 An umbiliform depression.

um·bil·i·cus (um·bil′ə·kəs, um′bə·lī′kəs) *n. pl.* **·ci** (-sī) 1 *Anat.* The depression at the middle of the abdomen where the umbilical cord of the fetus was attached; the navel. 2 *Zool.* An indention or depression at the axial base of a spiral shell, as in many gastropods. 3 *Ornithol.* Either of the apertures (inferior and superior) of the calamus of a feather. 4 *Bot.* A navel-shaped depression; a hilum. [<L]

um·bil·i·form (um·bil′ə·fôrm) *adj.* Navel-shaped. [<*umbili-* (<UMBILICUS) + -FORM]

um·ble pie (um′bəl) See HUMBLE PIE.

um·bles (um′bəlz) *n. pl.* The entrails of a deer; humbles. [Var. of NUMBLES]

um·bo (um′bō) *n. pl.* **um·bo·nes** (um·bō′nēz) or **·bos** 1 The boss or projecting spike in the center of a shield. 2 *Zool.* An elevation, boss, or knob, as the prominence of a bivalve shell near the hinge, or the plate of an echinoderm. 3 *Bot.* The top of the cap of certain fungi. 4 *Anat.* The surface of the tympanic membrane at the point of attachment to the malleus. [<L] — **um′bo·nal, um·bon·ic** (um·bon′ik) *adj.*

um·bo·nate (um′bə·nit, -nāt) *adj.* Having an umbo or bosslike protuberance. Also **um′bo·nat′ed.**

um·bra (um′brə) *n. pl.* **·brae** (-brē) 1 That region of a shadow from which the direct light is entirely cut off. 2 *Astron.* **a** In an eclipse, that part of the shadow of the earth or moon within which the moon or the sun is entirely hidden. See PENUMBRA. **b** The inner dark portion of a sunspot. [<L, shadow]

um·brage (um′brij) *n.* 1 Resentment, as at being obscured by another. 2 A sense of injury; offense: now usually in **to give** (or **take) umbrage.** 3 That which gives shade, as a leafy tree. 4 *Poetic* Shade or shadow cast. 5 *Rare* Mere shadowy appearance; semblance. 6 *Obs.* Disgrace; odium. See synonyms under OFFENSE, PIQUE. [<F *ombrage* <L *umbraticus* shady <*umbra* shade]

um·bra·geous (um·brā′jəs) *adj.* 1 Shady or shaded; forming or providing shade. 2 Quick to take offense; peevish; suspicious. 3 *Obs.* Obscure. [<F *ombrageux* <*ombrage.* See UMBRAGE.] — **um·bra′geous·ly** *adv.* — **um·bra′geous·ness** *n.*

um·brel·la (um·brel′ə) *n.* 1 A light portable canopy on a folding frame, carried as a protection against sun or rain. 2 *Zool.* The contractile, jellylike portion of the body of a medusa expanded like a bell or umbrella. [<Ital. *ombrella,* alter. (after *ombra* shade) of L *umbella* parasol. Related to UMBEL.]

umbrella bird Any of several South American birds (genus *Cephalopterus*), the male of which has a broad crest likened to an umbrella. *C. ornatus* has lustrous black plumage with an umbrellalike crest of blue, hairlike feathers.

umbrella leaf A smooth perennial herb *(Diphylleia cymosa)* of the barberry family, with a single large peltate leaf, one to two feet across, and a terminal cyme of white flowers. It is found in the southern United States.

umbrella palm A palm *(Hedyscepe canterburyana)* having pinnate leaves, native to Lord Howe Island in the British Solomons.

umbrella tree 1 A small magnolia *(Magnolia tripetala)* of the southern United States, with fragrant white flowers and oval leaves 16 to 30 inches long, crowded in an umbrellalike whorl at the ends of the branches. 2 Any one of several other trees with large, round cordate leaves.

um·brel·la·wort (um·brel′ə·wûrt′) *n.* A typically North American herb (genus *Allionia*) with the flowers enclosed in a three- or four-parted involucre.

um·brette (um·bret′) *n.* A dusky brown African wading bird *(Scopus umbretta),* related to the herons and storks. Also called *umber.* [<F *ombrette,* orig. dim. of *ombre* a shadow]

Um·bri·a (um′brē·ə) An ancient and modern region of central Italy between the Tiber and the Adriatic; 3,270 square miles; capital, Perugia.

Um·bri·an (um′brē·ən) *adj.* Of Umbria, or its people. — *n.* 1 A native or inhabitant of Umbria. 2 The extinct language of ancient Umbria, belonging to the Osco-Umbrian branch of the Italic languages.

um·brif·er·ous (um·brif′ər·əs) *adj.* Affording or making a shade; umbrageous. [<L *umbrifer* <*umbra* shade + *ferre* bear] — **um·brif′er·ous·ly** *adv.*

u·mi·ak (ōō'mē·ak) *n.* *Eskimo* A large, open boat, about 30 feet long and 8 feet wide, made by drawing skins over a wooden frame, and frequently used by Eskimo women: also spelled *oomiak*. Also **u'mi·ack.**

um·laut (ōōm'lout) *n.* **1** *Ling.* **a** The change in quality of a vowel sound caused by its partial assimilation to a vowel or semi-vowel (often later lost) in the following syllable; vowel mutation: primarily a phenomenon of the Germanic languages. English plurals showing internal vowel modification, such as *feet* and *geese*, are a result of this process. **b** A vowel which has been so altered, as *ä*, *ö*, and *ü* in German. **2** In German, the two dots (¨) put over a vowel modified by umlaut: short for **umlaut-mark.** — *v.t.* To modify by umlaut or mutation. [< G, change of sound < *um* about + *laut* sound]

Um·nak Island (ōōm'nak) One of the Fox Islands in the Aleutian Islands; 83 miles long, 2 to 18 miles wide.

um·pir·age (um'pīr·ij, -pə·rij) *n.* The office, function, or decision of an umpire. Also **um'pire·ship.**

um·pire (um'pīr) *n.* **1** *Law* A person called upon to settle a disagreement in opinion between arbitrators. **2** In general, anything by which a question in controversy is settled. **3** In various games, as baseball, a person chosen to enforce the rules of the game, and in case of controversy to settle disputed points. See synonyms under JUDGE. — *v.t.* & *v.i.* **·pired, ·pir·ing** To decide as umpire; act as umpire (of or in). [Aphetic alter. of ME *noumpere* < OF *nonper* odd, uneven (i.e., third) < *non* not + *per* even, equal]

Um·ta·ta (ōōm·tä'tə) A city in NE Cape of Good Hope Province, South Africa, capital of Transkeian Territories and of Tembuland district.

UN See UNITED NATIONS.

un-[1] *prefix* Not; opposed to. [OE] ♦ **Un-**[1] is used to express negation, lack, incompleteness or opposition. It is freely attached to adjectives and adverbs, less often to nouns. See UN-[2].

un-[2] *prefix* Back. [OE *un-, on-, and-*] ♦ **Un-**[2] is used to express reversal of the action of verbs, or to form verbs from nouns indicating removal from the state or quality expressed by the noun, or sometimes to intensify the force of negative verbs. Beginning at the foot of this page will be found a partial list of words which are formed with un-[1] and un-[2]. Other compounds of these prefixes, with strongly positive, specific, or special meanings, will be found in vocabulary place. In the verbs in the list *un-* gives the sense of reversal: *unchain* "to loose the chains of." In the nouns and the adjectives usually it has negative or privative force. Thus, *unburdened* may be regarded as an adjective meaning "not burdened," or as a participle of the verb *unburden*, meaning "relieved of a burden." *In-* as a prefix of adjectives expresses in usage more of negation, *un-* more of mere lack or privation: a child's *unartistic* speech, a writer's *inartistic* diction. In general, *in-* is more confined to words of Latin origin.

Pronunciations may be ascertained by consulting the second element in its vocabulary place.

un·a·ble (un·ā'bəl) *adj.* **1** Lacking the necessary power or resources; not able: usually used with an infinitive: *unable to walk.* **2** Lacking mental capacity; incompetent.

3 *Obs.* Feeble; helpless. — **un·a·bil·i·ty** (un'ə·bil'ə·tē) *Obs.* *n.* — **un·a'bly** *Obs.* *adv.*

un·a·bridged (un'ə·brijd') *adj.* Not abridged; not being a shorter or condensed version of another work; original and complete in itself: an *unabridged* dictionary.

un·ac·com·mo·dat·ed (un'ə·kom'ə·dā'tid) *adj.* **1** Not made suitable; ill-adapted or -adjusted. **2** Being without accommodations or conveniences.

un·ac·com·mo·dat·ing (un'ə·kom'ə·dā'ting) *adj.* Not disposed to accommodate; unobliging.

un·ac·com·plished (un'ə·kom'plisht) *adj.* **1** Having fallen short of accomplishment; not done or finished. **2** Lacking accomplishments.

un·ac·count·a·ble (un'ə·koun'tə·bəl) *adj.* **1** Impossible to be accounted for; inexplicable; hence, remarkable; extraordinary. **2** Exempt from supervision or control; irresponsible. — **un'ac·count'a·ble·ness** *n.* — **un'ac·count'a·bly** *adv.*

un·ac·count·ed-for (un'ə·koun'tid·fôr') *adj.* Unexplained; not accounted for.

un·ac·cus·tomed (un'ə·kus'təmd) *adj.* **1** Not made familiar by use or by practice: *unaccustomed* to hardship. **2** Not familiar or well known; strange: an *unaccustomed* sight. — **un'·ac·cus'tomed·ness** *n.*

un·ad·vised (un'əd·vīzd') *adj.* **1** Not advised; not having received advice. **2** Rash or imprudent; lacking consideration. — **un'ad·vis'ed·ly** (-vī'zid·lē) *adv.* — **un'ad·vis'ed·ness** *n.*

un·af·fect·ed (un'ə·fek'tid) *adj.* **1** Not showing affectation; natural; sincere; real. **2** Not influenced or changed. See synonyms under SIMPLE. — **un'af·fect'ed·ly** *adv.* — **un'af·fect'ed·ness** *n.*

Un·a·las·ka Island (un'ə·las'kə, ōō'nə-) One of the SW Fox Islands of the Aleutian Islands; 30 miles long, 6 to 30 miles wide.

un·al·ien·a·ble (un·āl'yən·ə·bəl) *adj.* *Obs.* Inalienable.

un·al·loyed (un'ə·loid') *adj.* Free from alloy or admixture; pure; also, figuratively, perfectly complete; absolute: *unalloyed* content.

un-A·mer·i·can (un'ə·mer'ə·kən) *adj.* Not having characteristics of persons or things native to the United States; lacking in patriotism and national feeling toward the United States; not consistent with American ideals, objectives, spirit, etc.

U·na·mu·no y Ju·go (ōō'nä·mōō'nō ē hōō'gō), **Miguel de,** 1864-1936, Spanish philosopher and novelist.

un·a·neled (un'ə·nēld') *adj.* *Obs.* Not having received extreme unction. [< UN-[1] + ANELE]

u·na·nim·i·ty (yōō'nə·nim'ə·tē) *n.* The state of being unanimous; complete agreement in opinion or purpose. See synonyms under HARMONY. [< OF *unanimité* < L *unanimitas, -tatis* < *unanimus*. See UNANIMOUS.]

u·nan·i·mous (yōō·nan'ə·məs) *adj.* **1** Sharing the same views or sentiments; consentient; harmonious. **2** Establishing or expressive of unanimity; showing or resulting from the assent of all concerned: the *unanimous* voice of the jury. [< L *unanimus, unanimis* < *unus* one + *animus* mind] — **u·nan'i·mous·ly** *adv.* — **u·nan'i·mous·ness** *n.*

un·ap·peal·a·ble (un'ə·pē'lə·bəl) *adj.* **1** Admitting no appeal to a higher court: an *unappealable* case. **2** That cannot be appealed from; conclusive; final.

un·ap·pro·pri·at·ed (un'ə·prō'prē·ā'tid) *adj.* Not set apart for special use; not taken possession of by or formally granted to a particular person or company.

un·ap·proved (un'ə·prōōvd') *adj.* **1** Not re-

garded with approval; not approved. **2** *Obs.* Not verified by proof; not proved.

un·apt (un·apt') *adj.* **1** Not likely or inclined. **2** Not suitable or qualified. **3** Not ready-witted. — **un·apt'ly** *adv.* — **un·apt'ness** *n.*

un·ar·gued (un·är'gyōōd) *adj.* **1** Not argued; undebated. **2** Undisputed. **3** *Obs.* Not censured: a Latinism.

un·arm (un·ärm') *v.t.* To disarm; deprive of weapons.

un·armed (un·ärmd') *adj.* **1** Not armed; without weapons. **2** Having no sharp, hard projections, as spines, prickles, plates, etc.: said of plants and animals.

un·as·sum·ing (un'ə·sōō'ming) *adj.* Unpretentious; modest. — **un'as·sum'ing·ly** *adv.*

un·at·tached (un'ə·tacht') *adj.* **1** Not attached. **2** *Law* Not held or seized, as in satisfaction of a judgment. **3** In the armed forces, not assigned to a regiment or company.

u·nau (yōō·nō', -nô', ōō·nou') *n.* The common two-toed sloth of Brazil (genus *Choloepus*). [< F < Tupian]

u·na vo·ce (yōō'nə vō'sē) *Latin* Unanimously; with one voice.

un·a·void·a·ble (un'ə·voi'də·bəl) *adj.* **1** That cannot be avoided; inevitable. **2** That cannot be made null and void; not voidable. See synonyms under NECESSARY.

un·a·ware (un'ə·wâr') *adj.* **1** Giving no heed; not cognizant, as of something specified. **2** *Poetic* Carelessly unmindful; inattentive; heedless. — *adv.* *Obs.* Unawares.

un·a·wares (un'ə·wârz') *adv.* **1** Unexpectedly. **2** Without premeditation; unwittingly.

un·backed (un·bakt') *adj.* **1** Never having borne a rider, as a horse; unbroken. **2** Left without backers or support; not supported financially; also, in sports, not wagered on. **3** Without a back, as a stool.

un·baked (un·bākt') *adj.* **1** Not baked; insufficiently baked. **2** Immature; crude.

un·bal·ance (un·bal'əns) *v.t.* **·anced, ·anc·ing** **1** To deprive of balance. **2** To disturb or derange.

un·bal·anced (un·bal'ənst) *adj.* **1** Not in a state of equilibrium. **2** In bookkeeping, not adjusted so as to balance. **3** Lacking mental balance; unsound; erratic.

un·bal·last·ed (un·bal'əs·tid) *adj.* **1** Not steadied by ballast. **2** Not firm; wavering.

un·bar (un·bär') *v.* **·barred, ·bar·ring** *v.t.* To remove the bar from. — *v.i.* To become unlocked or unbarred; open.

un·barbed (un·bärbd') *adj.* **1** Not fitted or made with barbs. **2** *Obs.* Untrimmed; unbarbered.

un·bat·ed (un·bā'tid) *adj.* *Archaic* **1** Not bated or blunted by having a button on the point, as a lance or other thrusting weapon. **2** Unabated; undiminished.

un·bear (un·bâr') *v.t.* **·beared, ·bear·ing** To free from the pressure of the checkrein, as a horse.

un·be·com·ing (un'bi·kum'ing) *adj.* **1** Not becoming; unsuited to the wearer, place, or surroundings: an *unbecoming* robe. **2** Not befitting; not worthy of. **3** Not decorous; improper. — **un'be·com'ing·ly** *adv.* — **un'be·com'ing·ness** *n.*

un·be·known (un'bi·nōn') *adj.* Unknown: used with *to*. Also **un'be·knownst'** (-nōnst').

un·be·lief (un'bi·lēf') *n.* **1** Absence of positive belief; incredulity. **2** A refusal to believe; belief in a contrary proposition; disbelief, as in religion. **3** In Scriptural use, lack of faith in God's promises. See synonyms under DOUBT.

unabashed	unacquitted	unallowable	unanswered	unasked	unattested	unawaked
unabated	unadaptable	unalterable	unappalled	unaspirated	unattracted	unawakened
unabetted	unadjustable	unaltered	unapparent	unaspiring	unattractive	unawed
unabolished	unadjusted	unaltering	unappeasable	unassailable	unauspicious	unbaptized
unabsolved	unadorned	unambiguous	unappeased	unassailably	unauthentic	unbearable
unacademic	unadulterated	unambitious	unappetizing	unassailed	unauthentical	unbeaten
unaccented	unadvisable	unamiable	unappreciative	unassignable	unauthenticated	unbefitting
unacceptable	unadvisably	unamusing	unapproachable	unassigned	unauthorized	unbelievable
unaccepted	unaesthetic	unanalytic	unapproachably	unassisted	unavailable	unbeloved
unacclimated	unafraid	unanalyzable	unapproached	unsoiled	unavailably	unbeneficed
unacclimatized	unaggressive	unanimated	unarmored	unassumed	unavailing	unbenighted
unaccompanied	unagitated	unannealed	unarrested	unattainable	unavenged	unbenign
unaccredited	unaided	unannounced	unartful	unattained	unavouched	unbeseeming
unacknowledged	unalleviated	unanswerable	unartistic	unattempted	unavowed	unbesought
unacquainted		unanswerably	unashamed	unattended	unavowedly	unbespoken

add, āce, câre, pälm; end, ēven; it, īce; odd, ōpen, ôrder; tōōk, pōōl; up, bûrn; ə = a in *above*, e in *sicken*, i in *clarity*, o in *melon*, u in *focus*; yōō = u in *fuse*; oi, oil; ou, pout; ch, check; g, go; ng, ring; th, thin; t͟h, this; zh, vision. Foreign sounds á, œ, ü, kh, ṅ; and ♦: see page xx. < from; + plus; ? possibly.

un·be·liev·er (un′bi·lē′vər) *n.* 1 One who withholds belief. 2 One who has no religious faith. 3 One having a religion different from that of the speaker or writer; specifically, a non–Christian. See synonyms under SKEPTIC.
un·be·liev·ing (un′bi·lē′ving) *adj.* 1 Doubting; skeptical; incredulous. 2 Disbelieving, particularly in regard to religious matters. — **un′·be·liev′ing·ly** *adv.* — **un′be·liev′ing·ness** *n.*
un·belt (un·belt′) *v.t.* 1 To remove the belt of. 2 To remove from the belt; ungird.
un·bend (un·bend′) *v.* **·bent**, **·bend·ing** *v.t.* 1 To relax, as from exertion or formality: to *unbend* the mind. 2 To straighten (something bent or curved). 3 To relax, as a bow, from tension. 4 *Naut.* **a** To loose; untie, as a rope. **b** To detach or remove (a sail from a spar or stay. — *v.i.* 5 To become free of restraint or formality; relax. 6 To become straight or nearly straight again.
un·bend·ing (un·ben′ding) *adj.* Not bending easily; stiff; hence, unyielding; resolute; firm, as character. — *n.* Relaxation. — **un·bend′·ing·ly** *adv.* — **un·bend′ing·ness** *n.*
un·bi·ased (un·bī′əst) *adj.* Having no bias; especially, having no mental bias; not prejudiced or warped; impartial. Also **un·bi′assed**. — **un·bi′ased·ly** *adv.* — **un·bi′ased·ness** *n.*
un·bid·den (un·bid′n) *adj.* 1 Not commanded; not invited: an *unbidden* guest. 2 Not called forth; spontaneous: *unbidden* thoughts. Also **un·bid′**.
un·bind (un·bīnd′) *v.t.* **·bound**, **·bind·ing** 1 To free from bindings; undo; hence, to release. 2 To remove, as something that binds; unfasten. See synonyms under RELEASE. [OE *unbindan*]
un·bit·ted (un·bit′id) *adj.* Not furnished with or restrained by a bit or bridle; uncontrolled.
un·blenched (un·blencht′) *adj. Obs.* Not dismayed or confounded.
un·blessed (un·blest′) *adj.* 1 Not having been blessed or admitted to blessedness or divine favor. 2 Unhappy. 3 Unhallowed or unholy; evil.
un·blood·y (un·blud′ē) *adj.* 1 Not stained by blood; hence, not attended with slaughter, as a conflict. 2 Not of a bloodthirsty disposition.
un·blush·ing (un·blush′ing) *adj.* Not blushing; immodest; shameless. — **un·blush′ing·ly** *adv.*
un·bod·ied (un·bod′ēd) *adj.* 1 Having no body; immaterial. 2 Disembodied.
un·bolt (un·bōlt′) *v.t.* To release, as a door, by withdrawing a bolt; unlock; open. — *v.i. Obs.* To remove a bolt or bar; hence, to expose something to view; make explanation.
un·bolt·ed[1] (un·bōl′tid) *adj.* Not fastened by bolts; not bolted.
un·bolt·ed[2] (un·bōl′tid) *adj.* 1 Not separated by bolting; not sifted: *unbolted* flour. 2 *Obs.* Gross; coarse.
un·boned (un·bōnd′) *adj.* 1 Without bones. 2 Not having had the bones removed.
un·bon·net (un·bon′it) *v.t. & v.i.* To remove the bonnet or other covering from (the head); uncover. — **un·bon′net·ed** *adj.*
un·born (un·bôrn′) *adj.* 1 Not yet born; of a future time or generation; future. 2 Not in existence.
un·bos·om (un·bŏŏz′əm, -bōō′zəm) *v.t.* To reveal, as one's thoughts or secrets; disclose or give vent to: often used reflexively. — *v.i.* To say what is troubling one; tell one's thoughts, feelings, etc. — **un·bos′om·er** *n.*
un·bound·ed (un·boun′did) *adj.* 1 Having no bounds; of unlimited extent; very great; boundless. 2 Having no boundary, as a line that returns into itself or a closed surface. 3 Going beyond bounds; unrestrained. — **un·bound′ed·ly** *adv.* — **un·bound′ed·ness** *n.*

un·bowed (un·boud′) *adj.* Not bent; not bowed or subdued; proud in defeat or adversity.
un·brace (un·brās′) *v.t.* **·braced**, **·brac·ing** 1 To free from bands or braces. 2 To free from tension; loosen. 3 To weaken; make feeble.
un·breathed (un·brēthd′) *adj.* 1 Not breathed; hence, not whispered, or spoken; not communicated to another. 2 *Obs.* Unexercised; not practiced.
un·bred (un·bred′) *adj.* 1 Devoid of good breeding; ill–bred. 2 Not taught; untrained: sometimes followed by *to*: *unbred* to spinning. 3 *Obs.* Unbegotten; not born.
un·bri·dled (un·brīd′ld) *adj.* 1 Having no bridle on: an *unbridled* horse. 2 Without restraint; unrestrained; unruly: an *unbridled* tongue; *unbridled* license. — **un·bri′dled·ly** *adv.* — **un·bri′dled·ness** *n.*
un·bro·ken (un·brō′kən) *adj.* 1 Not broken; whole; entire: an *unbroken* seal. 2 Unviolated: *unbroken* faith; an *unbroken* promise. 3 Uninterrupted; regular; smooth: *unbroken* sleep; an *unbroken* prairie. 4 Not weakened; strong; firm. 5 Not broken to harness or service, as a draft animal. 6 Not disarranged or thrown out of order. Also *Obs.* **un·broke′**. — **un·bro′ken·ly** *adv.* — **un·bro′ken·ness** *n.*
un·buck·le (un·buk′əl) *v.t. & v.i.* **·led**, **·ling** To unfasten the buckle or buckles (of).
un·build (un·bild′) *v.t.* **·built**, **·build·ing** To demolish; destroy.
un·bur·den (un·bûr′dən) *v.t.* To free from a burden; relieve. Also *Archaic* **un·bur′then** (-thən).
un·but·ton (un·but′n) *v.t.* To unfasten the button or buttons of.
un·caged (un·kājd′) *adj.* 1 Not locked up in a cage; free. 2 Released from a cage; freed.
un·called (un·kôld′) *adj.* Not in response to a summons; without being asked or demanded.
un·called–for (un·kôld′fôr′) *adj.* Unnecessary; gratuitous; not justified by circumstances; discourteous.
un·can·ny (un·kan′ē) *adj.* 1 Exciting superstitious fear; weird; unnatural; eerie. 2 So good as to seem almost supernatural in origin: *uncanny* accuracy. 3 *Scot.* Dangerous; severe, as a wound. — **un·can′ni·ly** *adv.* — **un·can′ni·ness** *n.*
un·cap (un·kap′) *v.* **·capped**, **·cap·ping** *v.t.* To take off the cap or covering of. — *v.i.* To remove the hat or cap, as in respect.
un·ca·pa·ble (un·kā′pə·bəl) *adj. Obs.* Incapable.
un·caused (un·kôzd′) *adj.* Existing without a cause; not caused; not created: an *uncaused* deity.
un·cer·e·mo·ni·ous (un′ser·ə·mō′nē·əs) *adj.* Informal; abrupt; discourteous. — **un′cer·e·mo′ni·ous·ly** *adv.*
un·cer·tain (un·sûr′tən) *adj.* 1 Not certain; that cannot be relied upon; variable; changeful; fitful; erring: an *uncertain* friend; *uncertain* weather; an *uncertain* shot. 2 That cannot be certainly predicted; being of doubtful issue. 3 Not having certain knowledge or assured conviction. 4 Not surely or exactly known: a lady of *uncertain* age. 5 Having no exact or precise significance: *uncertain* phraseology. See synonyms under EQUIVOCAL, PRECARIOUS, VAGUE. — **un·cer′tain·ly** *adv.*
un·cer·tain·ty (un·sûr′tən·tē) *n. pl.* **·ties** 1 The state of being uncertain; doubt: also **un·cer′tain·ness**. 2 A doubtful matter; a contingency. See synonyms under DOUBT.
uncertainty principle *Physics* A statement of the impossibility of exactly determining at any given instant or by a single operation more than one magnitude or quantity, as the

velocity, position, etc., of an electron: also called *indeterminacy principle*.
un·chain (un·chān′) *v.t.* To release from a chain; set free.
un·chanc·y (un·chan′sē) *adj. Scot.* 1 Unpropitious; unlucky. 2 Ill–timed; inopportune. 3 Unsafe; dangerous.
un·charged (un·chärjd′) *adj.* 1 Not loaded. 2 Not attacked or accused. 3 Not required or asked to pay a price or meet an expense. 4 Having no electrical charge.
un·char·i·ta·ble (un·char′i·tə·bəl) *adj.* Not charitable; harsh in judgment; censorious. — **un·char′i·ta·ble·ness** *n.* — **un·char′i·ta·bly** *adv.*
un·chris·tian (un·kris′chən) *adj.* 1 Unbecoming to a Christian. 2 Foreign to Christianity; hence, uncharitable, ungracious, rude, etc. 3 Non–Christian; pagan.
un·church (un·chûrch′) *v.t.* 1 To deprive of membership in a church; expel from a church. 2 To excommunicate. 3 To deny the validity of the sacraments and order of, as a sect.
un·cial (un′shəl, -shē·əl) *adj.* Pertaining to or consisting of a form of letters found in manuscripts from the fourth to the eighth century, and resembling modern capitals but more rounded. — *n.* 1 An uncial letter. 2 An uncial manuscript. [<L *uncialis* inch–high <*uncia* inch, ounce]

a ⲉ̄ⲛⲥⲟⲩⲕⲁⲓⲁⲩⲧⲱⲙⲟⲛⲱⲁⲛ
b **ETCONLOQUEBANTUR**

UNCIALS

a. Greek uncials — fifth century.
b. Latin uncials — circa A.D. 700.

un·ci·form (un′sə·fôrm) *adj.* Shaped like a hook; hooklike. — *n.* The unciform bone. [<L *uncus* hook + -FORM]
unciform bone *Anat.* A bone of the distal row of the wrist on the ulnar side, articulating with the fourth and fifth metacarpals.
unciform process *Anat.* 1 A projection upon the anterior surface of the unciform bone. 2 The uncinate process.
un·ci·na·ri·a·sis (un′si·nə·rī′ə·sis) *n. Pathol.* Ancylostomiasis. [<NL <*Uncinaria*, genus name <L *uncinus* a hook, barb, dim. of *uncus* hook]
un·ci·nate (un′sə·nit, -nāt) *adj. Biol.* Hooked or bent at the end; having a hooked appendage. Also **un′ci·nal**, **un′ci·nat·ed**. [<L *uncinatus* <*uncinus*, dim. of *uncus* hook]
uncinate process *Anat.* A hooklike process on the ethmoid bone.

UNCINATE APPENDAGES

un·cir·cum·cised (un·sûr′kəm·sīzd) *adj.* Not circumcised; Gentile; heathen.
un·cir·cum·ci·sion (un′sûr·kəm·sizh′ən) *n.* 1 The state of being uncircumcised. 2 Those not circumcised; in Scripture, the Gentiles.
un·civ·il (un·siv′əl) *adj.* 1 Wanting in civility; discourteous; ill–bred. 2 *Obs.* Uncivilized. See synonyms under BLUFF, HAUGHTY. — **un·civ′il·ly** *adv.*
un·civ·i·lized (un·siv′ə·līzd) *adj.* Destitute of civilization; barbarous. See synonyms under BARBAROUS.
un·clad (un·klad′) *adj.* Without clothes; naked.
un·clasp (un·klasp′, -kläsp′) *v.t.* 1 To release

unbetrayed	unbought	unbrotherly	uncanonical	unchanged	uncholeric	unclothe
unbetrothed	unbound	unbruised	uncarbureted	unchanging	unchosen	unclothed
unbewailed	unboundable	unbrushed	uncared–for	unchangingly	unchristened	uncloud
unbias	unbraid	unburied	uncarpeted	unchaperoned	unclaimed	unclouded
unblamable	unbranched	unburnt	uncastrated	uncharted	unclassed	uncloyed
unblamably	unbranching	unbusinesslike	uncaught	unchartered	unclassic	uncoated
unblamed	unbranded	unbuttoned	unceasing	unchary	unclassifiable	uncocked
unbleached	unbreakable	uncage	uncensored	unchaste	unclassified	uncoerced
unblemished	unbreathable	uncalculate	uncensured	unchastened	uncleaned	uncoffined
unblest	unbreech	uncalculating	uncertified	unchastised	uncleansed	uncollectable,
unblissful	unbreeched	uncalendered	unchainable	unchastity	uncleared	uncollectible
unboastful	unbribable	uncanceled	unchained	unchecked	uncleavable	uncollected
unbookish	unbridgeable	uncandid	unchallenged	unchewed	unclipped	uncolonized
unborrowed	unbridged	uncandidly	unchambered	unchilled	unclog	uncolored
unbottomed	unbridle	uncanonic	unchangeable	unchivalrous	unclogged	uncombed

from a clasp. 2 To release the clasp of. —v.i. 3 To become released from a clasp.

un·cle (ung'kəl) n. 1 The brother of one's father or mother; also, the husband of one's aunt. ◆ Collateral adjective: *avuncular*. 2 An elderly man: used in direct address. 3 *Colloq.* A pawnbroker. [<F *oncle* <L *avunculus* a mother's brother, orig. dim. of *avus* grandfather]

un·clean (un·klēn') adj. 1 Not clean; foul. 2 Characterized by impure thoughts; unchaste; depraved. 3 Ceremonially impure. See synonyms under FOUL. — **un·clean'ness** n.

un·clean·ly[1] (un·klen'lē) adj. 1 Lacking cleanliness. 2 Impure; indecent; not chaste. [< UN-[1] + CLEANLY, adj.] — **un·clean'li·ness** n.

un·clean·ly[2] (un·klēn'lē) adv. In an unclean manner. [<UNCLEAN + -LY[2]]

un·clench (un·klench') v.t. & v.i. To relax or open from a clenched condition. Also **un·clinch'** (-klinch').

Uncle Re·mus (rē'məs) In Joel Chandler Harris's series of Afro-American Negro folk tales and folk songs, the old southern Negro who tells the stories of Br'er Rabbit, Br'er Fox, and others to a small white boy.

Uncle Sam The personification of the government of the United States or of the people of the United States: represented as a tall, lean man with chin whiskers, wearing a plug hat, blue swallow-tailed coat, and red-and-white striped pants. See BROTHER JONATHAN.

Uncle Tom The chief character in Harriet Beecher Stowe's *Uncle Tom's Cabin*, a faithful, elderly Negro slave, who, on the death of a kind master, was sold to the brutal Simon Legree and subjected to cruelties from which he died.

un·clew (un·kloō') v.t. 1 *Naut.* To unfurl. 2 *Archaic* To unroll; undo; also, to ruin.

un·cloak (un·klōk') v.t. 1 To remove the cloak or covering from. 2 To unmask; expose. —v.i. 3 To remove one's cloak or outer garments.

un·close (un·klōz') v.t. & v.i. ·closed, ·clos·ing 1 To open or set open. 2 To reveal; disclose. — **un·closed'** adj.

un·co (ung'kō) *Scot. & Brit. Dial.* adj. Being out of the ordinary; strange; weird; reserved. —n. 1 Anything out of the common or surprising; hence, a strange person or thing. 2 pl. News. —adv. Remarkably or excessively; very: *unco gude*.

un·cock (un·kok') v.t. 1 To release and let down the hammer of (a firearm) without exploding the charge. 2 To restore to usual position, as a hat.

un·coft (un·koft') adj. *Scot.* Unbought.

un·coil (un·koil') v.t. & v.i. To unwind or become unwound.

un·coined (un·koind') adj. 1 Not fabricated; natural. 2 Not minted.

un·com·fort·a·ble (un·kum'fər·tə·bəl, -kumpf'tə·bəl) adj. 1 Not at ease; feeling discomfort. 2 Causing uneasiness or disquietude, physical or mental; disquieting. — **un·com'fort·a·bly** adv.

un·com·mer·cial (un'kə·mûr'shəl) adj. 1 Not engaged or versed in commerce. 2 Conflicting with the spirit of commerce.

un·com·mit·ted (un'kə·mit'id) adj. 1 Not committed; specifically, not performed or done. 2 Not entrusted. 3 Not bound by a pledge.

un·com·mon (un·kom'ən) adj. Unusual; remarkable. See synonyms under EXTRAORDINARY, ODD, RARE. — **un·com'mon·ly** adv.

un·com·mu·ni·ca·tive (un'kə·myoō'nə·kə·tiv, -nə·kā'tiv) adj. Not communicative; not disposed to talk, either to express oneself or to give information; reserved; taciturn.

Un·com·pah·gre Peak (un'kəm·pä'grē) A mountain in SW central Colorado, the highest of the San Juan Mountains; 14,306 feet.

un·com·pro·mis·ing (un·kom'prə·mī'zing) adj. Making or admitting of no compromise; inflexible; strict. — **un·com'pro·mis'ing·ly** adv. — **un·com'pro·mis'ing·ness** n.

un·con·cern (un'kən·sûrn') n. Absence of or freedom from concern or anxiety; indifference. See synonyms under APATHY.

un·con·cerned (un'kən·sûrnd') adj. Undisturbed; not anxious; indifferent. — **un·con·cern'ed·ly** (-sûr'nid·lē) adv. — **un'con·cern'ed·ness** n.

un·con·di·tion·al (un'kən·dish'ən·əl) adj. Limited by no conditions; absolute. See synonyms under ABSOLUTE. — **un'con·di'tion·al·ly** adv.

unconditional surrender The unconditional acceptance of military defeat by a warring enemy power, subject only to terms to be subsequently imposed by the victors.

un·con·di·tioned (un'kən·dish'ənd) adj. 1 Not restricted; unconditional. 2 In metaphysics, not limited by conditions of space or time; free from relation; unrelated; absolute. 3 *Psychol.* Not having a reaction or reflex developed by a specified condition or conditions; not acquired; natural. 4 Admitted to a school, college, or higher class without condition.

un·con·form·a·ble (un'kən·fôr'mə·bəl) adj. 1 Not conforming or conformable; inconsistent. 2 *Geol.* Showing unconformity. — **un'con·form'a·bil'i·ty, un'con·form'a·ble·ness** n. — **un'con·form'a·bly** adv.

un·con·for·mi·ty (un'kən·fôr'mə·tē) n. pl. ·ties 1 Want of conformity; nonconformity. 2 *Geol.* a A lack of continuity between groups of stratified rocks in contact, indicative of a gap in the stratigraphic record. b The contact layer between such groups.

un·con·scion·a·ble (un·kon'shən·ə·bəl) adj. 1 Going beyond customary or reasonable bounds. 2 Not governed by sense or prudence; unconscientious; devoid of conscience. 3 *Law* Inequitable. — **un·con'scion·a·ble·ness** n. — **un·con'scion·a·bly** adv.

un·con·scious (un·kon'shəs) adj. 1 Temporarily deprived of consciousness. 2 Not cognizant; unaware: with *of*: *unconscious* of his charm. 3 Not known or felt to exist; not produced or accompanied by conscious effort: *unconscious* thought. 4 Not endowed with consciousness or a mind. —n. *Psychoanal.* That extensive area of the psyche which is not in the immediate field of awareness and whose content, when consisting of repressed material, may affect the personality through dreams, morbid fears and compulsions, forms of behavior, etc.: with *the*. — **un·con'scious·ly** adv. — **un·con'scious·ness** n.

un·con·sol·i·dat·ed (un'kən·sol'ə·dā'tid) adj. *Geol.* Not compact or solid, as rock or soil material in a form of loose aggregation.

un·con·sti·tu·tion·al (un'kon·sti·too'shən·əl, -tyoō'-) adj. Contrary to or violative of the constitution or fundamental law of a state. — **un'con·sti·tu'tion·al'i·ty** n. — **un'con·sti·tu'tion·al·ly** adv.

un·con·trol·la·ble (un'kən·trō'lə·bəl) adj. Beyond control; ungovernable. See synonyms under REBELLIOUS, VIOLENT. — **un'con·trol'la·ble·ness, un'con·trol·la·bil'i·ty** n. — **un'con·trol'la·bly** adv.

un·con·ven·tion·al (un'kən·ven'shən·əl) adj. Not adhering to conventional rules; informal; free. — **un'con·ven'tion·al'i·ty** n. — **un'con·ven'tion·al·ly** adv.

un·con·vert·ed (un'kən·vûr'tid) adj. 1 Not converted. 2 *Theol.* Impenitent; without saving faith.

un·cork (un·kôrk') v.t. To draw the cork from.

un·count·ed (un·koun'tid) adj. 1 Not counted. 2 Beyond counting; innumerable.

un·cou·ple (un·kup'əl) v. ·led, ·ling v.t. 1 To disconnect or unfasten. 2 To set loose; unleash (dogs). —v.i. 3 To break loose. — **un·coup'led** (-kup'əld) adj.

un·couth (un·koōth') adj. 1 Marked by awkwardness or oddity; outlandish; ungainly; unrefined; rough. 2 Not common; not well-known. 3 Mysterious; alarming. See synonyms under AWKWARD, BARBAROUS, RUSTIC. [OE *uncūth* unknown <*un-* not + *cūth*, pp. of *cunnan* know] — **un·couth'ly** adv. — **un·couth'ness** n.

un·cov·e·nant·ed (un·kuv'ə·nən·tid) adj. 1 Not bound by a covenant or promise; not having entered into a covenant or league. 2 Not guaranteed by a covenant: used specifically to describe divine grace or mercy not promised by a covenant.

un·cov·er (un·kuv'ər) v.t. 1 To remove the covering from. 2 To make known; reveal; disclose. 3 In military tactics, to expose successively, as lines of formation. —v.i. 4 To remove a covering; raise or remove the hat, as in token of respect.

un·cov·ered (un·kuv'ərd) adj. 1 Not covered; devoid of covering. 2 Not covered by collateral security.

un·cre·ate (un'krē·āt') v.t. ·at·ed, ·at·ing To deprive of existence.

un·cre·at·ed (un'krē·ā'tid) adj. 1 Not yet created or brought into being. 2 *Philos.* Not created; self-existent.

unc·tion (ungk'shən) n. 1 The act of anointing, as with oil. 2 *Eccl.* a A ceremonial anointing with oil, as in consecration or dedication. b The sacramental rite of anointing the sick, reserved in the Roman Catholic Church for those in danger of death: also called **extreme unction**. 3 The act of treating medicinally by anointing. 4 A substance used in anointing, as an unguent or a salve; something that soothes or palliates. 5 The quality or characteristic of speech, especially in religious discourse, that awakens or is intended to awaken deep sympathetic feeling; sometimes, effusive or affected emotion. [<F *onction* <L *unctio*, *-onis* <*ungere* anoint] — **unc'tion·less** adj.

unc·tu·ous (ungk'choō·əs) adj. 1 Having the characteristics of an unguent; greasy. 2 Characterized by deep sympathetic feeling. 3 Characterized by affected emotion; hence, oily-tongued; unduly suave. 4 Being greasy or soapy to the touch, as certain minerals. 5 Soft; rich in organic matter, as certain soils. 6 Having plasticity, as clay. [<Med. L *unctuosus* <L *unctum* ointment, orig. neut. pp. of *ungere* anoint] — **unc'tu·ous·ly** adv. — **unc'tu·ous·ness** n. — **unc'tu·os'i·ty** (-choō·os'ə·tē) n.

un·cut (un·kut') adj. 1 Not cut. 2 In bookbinding, having untrimmed margins. 3 Unground, as a gem.

un·damped (un·dampt') adj. *Physics* Pertaining to or designating those electromagnetic oscillations which continue without change in amplitude: *undamped* radio waves.

un·daunt·ed (un·dôn'tid, -dän'-) adj. Not daunted; fearless; intrepid. See synonyms

uncombinable	uncomplimentary	unconfinedly	unconquered	uncontradictable	uncorrected	uncrossed
uncombinably	uncompounded	unconfirmed	unconscientious	uncontradicted	uncorroborated	uncrowded
uncombined	uncomprehended	unconfused	unconsecrated	uncontrite	uncorrupt	uncrown
uncomely	uncomprehending	unconfusedly	unconsenting	uncontrolled	uncorrupted	uncrowned
uncomforted	uncomprehensible	unconfuted	unconsidered	uncontrolledly	uncorruptly	uncrystalline
uncomforting	uncompressed	uncongeal	unconsoled	uncontroverted	uncorruptness	uncrystallizable
uncommanded	uncompromised	uncongealable	unconsonant	uncontrovertible	uncountable	uncrystallized
uncommissioned	uncomputed	uncongealed	unconstant	uncontrovertibly	uncourteous	uncultivable
uncompanionable	unconcealable	uncongenial	unconstituted	unconversant	uncourtliness	uncultivated
uncomplaining	unconcealed	uncongeniality	unconstrained	unconvinced	uncourtly	uncultured
uncomplaisant	unconceded	uncongenially	unconstricted	unconvincing	uncredited	uncumbered
uncomplaisantly	unconcerted	unconnected	unconsumed	uncooked	uncrippled	uncurb
uncompleted	unconciliated	unconnectedly	uncontaminated	uncooperative	uncritical	uncurbable
uncompliable	uncondemned	unconquerable	uncontending	uncoordinated	uncriticizable	uncurbed
uncomplicated	unconquerably	uncontested	uncorked	uncross	uncurdled

add, āce, câre, päim; end, ēven; it, īce; odd, ōpen, ôrder; tōōk, pōōl; up, bûrn; ə = a in *above*, e in *sicken*, i in *clarity*, o in *melon*, u in *focus*; yōō = u in *fuse*; oi, oil; ou, pout; ch, check; g, go; ng, ring; th, thin; th, this; zh, vision. Foreign sounds à, œ, ü, kh, ñ; and ◆: see page xx. < from; + plus; ? possibly.

under BRAVE. — **un·daunt′ed·ly** adv. — **un·daunt′ed·ness** n.

un·dé (un′dā) adj. Her. Wavy; undulating: said of an ordinary or of the lines dividing the shield. Also **un′dée, un′dy** (-dē). [< OF < L unda wave]

un·dec·a·gon (un·dek′ə·gon) n. A figure that has eleven angles and eleven sides. [< L undecim eleven + -GON]

un·de·ceiv·a·ble (un′di·sē′və·bəl) adj. 1 That cannot be deceived. 2 Obs. Not deceitful.

un·de·ceive (un′di·sēv′) v.t. ·ceived, ·ceiv·ing To free from deception, error, or illusion.

un·de·ceived (un′di·sēvd′) adj. 1 Not deceived. 2 Freed from error or deception.

un·de·cen·ni·al (un′di·sen′ē·əl) adj. 1 Pertaining to a period of eleven years or to the eleventh year. 2 Lasting eleven years, or occurring or celebrated on the eleventh year or every eleven years. Also **un′de·cen′na·ry** (-sen′ər·ē). [< L undecim eleven + annus year]

un·de·cid·ed (un′di·sī′did) adj. 1 Not having the mind made up. 2 Not decided upon; not determined. See synonyms under IRRESOLUTE. — **un′de·cid′ed·ly** adv.

un·decked (un·dekt′) adj. 1 Having no ornaments; not decked out. 2 Having no deck, as a vessel.

un·dec·u·ple (un·dek′yə·pəl) adj. 1 Consisting of eleven. 2 Having eleven parts or members; elevenfold. 3 Taken by elevens. — n. A number or sum eleven times as great as another. — v.t. & v.i. ·pled, ·pling To multiply by eleven; make or become eleven times as large. [< L undecim eleven, on analogy with decuple]

un·de·cu·pli·cate (un′də·kyōō′plə·kit, -kāt) adj. 1 Elevenfold. 2 Raised to the eleventh power. — v.t. & v.i. (-kāt) ·cat·ed, ·cat·ing To multiply by eleven; undecuple. — n. One of eleven like things. — **un′de·cu′pli·cate·ly** adv. — **un′de·cu′pli·ca′tion** n.

un·de·mon·stra·tive (un′di·mon′strə·tiv) adj. Not demonstrative; not characterized by show of feeling.

un·de·ni·a·ble (un′di·nī′ə·bəl) adj. 1 That cannot be denied; indisputably true; obviously correct: an undeniable fact. 2 Unquestionably good; excellent: His reputation was undeniable. — **un′de·ni′a·bly** adv.

un·der (un′dər) prep. 1 Beneath, so as to have something directly above; covered by: layer under layer. 2 In a place lower than; at the foot or bottom of: under the hill. 3 Beneath the shelter of: under the paternal roof. 4 Beneath the concealment, guise, or assumption of: under a false name. 5 Less than in number, degree, age, value, or amount: under 10 tons. 6 Inferior to in quality, character, or rank. 7 Beneath the domination of; owing allegiance to; subordinate or subservient to: under the Nazi flag. 8 Subject to the guidance, tutorship, or direction of: He studied under Mendelssohn. 9 Subject to the moral obligation of: a statement under oath; subject to the sanction of; with the liability or certainty of incurring: under penalty of the law. 10 Subject to the influence or pressure of: under the circumstances; swayed or impelled by: under fear of death. 11 Driven or propelled by: under sail, under steam. 12 Included in the group or class of; found in the matter titled or headed: See under History. 13 Being the subject of: under medical treatment. 14 During the period of; in the reign of; pending the administration of. 15 By virtue of; authorized, substantiated, attested, or warranted by: under his own signature. 16 In conformity to or in accordance with; having regard to. 17 Planted or sowed with: an acre under wheat. See synonyms under BENEATH. — adv. 1 In or into a position below something; underneath. 2 In or into an inferior or subordinate degree or rank. 3 So as to be covered or hidden; in or into concealment. 4 Less than the required or appointed amount. — **to go under** To fail or collapse, as a business venture. — adj. 1 Situated or moving under something else; lower or lowermost: an under layer. 2 Zool. Ventral: the under side of a rattlesnake. 3 Subordinate; lower in rank or authority. 4 Insufficient; less than usual, standard, or prescribed. 5 Held in subjection or restraint: used predicatively: Hold your emotions under. [OE]

under- combining form 1 Below in position; situated or directed beneath; on the underside; as in:

underarch	underjaw
underbody	underlip
underbridge	undermark
underbud	undernamed
undercasing	underpart
undercellar	underpier
undercurved	underprop
underdraw	undershore
undereaten	undersole
underfeathering	underspread
underfill	understroke
underfire	undersurface
undergnaw	undersweep
undergore	underthrust

2 Below a surface or covering; lower; as in:

underbodice	undergarb
undercloth	underglow
undercrust	undergown
underdish	undergrove
underdrawers	underjacket
underdress	underlife
underearth	underpetticoat
underflooring	underregion

3 Inferior in rank or importance; subordinate; subsidiary; as in:

under-actor	under-officer
under-agent	under-official
under-captain	under-secretary
under-chief	under-secretaryship
under-clerk	under-servant
under-god	under-steward
under-king	under-teacher
under-kingdom	undertitle
under-man	under-treasurer

4 Insufficient; less than is usual or proper; as in:

underact	underpowered
underbill	underpraise
undercapitalize	underprize
underclothed	underproportioned
underconsumption	underripe
underdeveloped	undersailed
underexercise	undersaturated
undergrow	underspecified
underload	understaffed
undermanned	understimulus
underniceness	understocked
underofficered	undertaxed
underpeopled	undertrained
underpopulated	

5 At a lower rate; less in degree or amount; as in:

underprice	underspend

6 Subdued; hidden; as in:

underbreath	undernote
underfeeling	underthought
undermelody	undervoice

under a cloud Overshadowed by reproach or distrust.

un·der-age (un′dər-āj′) adj. Not of a requisite age; immature.

un·der·arm (un′dər-ärm′) adj. Situated or placed under the arm: the underarm section of a blouse. — n. The armpit.

un·der·arm (un′dər-ärm′) adj. In various sports, as tennis, baseball, etc., delivered with the hand lower than the elbow.

un·der·bel·ly (un′dər-bel′ē) n. pl. ·lies 1 The lower region of the belly. 2 Any similar unprotected part: the soft underbelly of Europe.

un·der·bid (un′dər-bid′) v.t. ·bid, ·bid·ding 1 To bid lower than, as in a competition. 2 In auction bridge, to fail to bid the full value of (a hand). — **un′der·bid′der** n.

un·der·bred (un′dər-bred′) adj. 1 Of impure breed; not thoroughbred. 2 Lacking in good breeding; ill-bred. See synonyms under VULGAR.

un·der·brush (un′dər-brush′) n. Small trees and shrubs growing beneath forest trees; undergrowth. Also **un′der·bush′** (-bŏŏsh′).

un·der·buy (un′dər-bī′) v.t. ·bought, ·buy·ing 1 To buy at a price lower than that paid by (another). 2 To pay less than the value for.

under canvas Naut. Under sail; propelled by the wind.

un·der·car·riage (un′dər-kar′ij) n. 1 The framework supporting the body of a structure, as an automobile. 2 The principal landing gear of an aircraft.

un·der·charge (un′dər-chärj′) v.t. ·charged, ·charg·ing 1 To make an inadequate charge for. 2 To load with an insufficient charge, as a gun. — n. (un′dər-chärj′) An inadequate or insufficient charge.

un·der·class·man (un′dər-klas′mən, -kläs′-) n. pl. ·men (-mən) A member of the freshman or of the sophomore class in a school or college.

un·der·clay (un′dər-klā′) n. A layer of clay underlying a coal seam, often containing the roots of ancient coal-forming plants: also called seatstone.

un·der·clothes (un′dər-klōz′, -klōthz′) n. pl. Clothes designed for underwear, or to be worn next the skin. Also **un′der·cloth′ing**.

un·der·coat (un′dər-kōt′) n. 1 A coat worn under another. 2 Underfur. 3 A layer of black, soft material sprayed on the underparts of an automobile to prevent rust, absorb sound, etc.: also **un′der·coat′ing**. — v.t. To provide with an undercoat (def. 3).

un·der·cool (un′dər-kōōl′) v.t. To supercool.

un·der·cov·er (un′dər-kuv′ər) adj. Secret; surreptitious; specifically, engaged in spying or secret investigation: an undercover man.

under cover Secretively; surreptitiously.

un·der·cov·ert (un′dər-kuv′ərt) n. Ornithol. A wing covert.

un·der·croft (un′dər-krôft′, -kroft′) n. A subterranean chamber, vault, or passage; the crypt of a church. [< UNDER- + obs. croft vault, ult. < L crypta crypt]

un·der·cur·rent (un′dər-kûr′ənt) n. 1 A current, as of water or air, below another or below the surface. 2 A hidden drift or tendency, as of popular sentiments.

un·der·cut (un′dər·kut′) n. 1 The act or result of cutting under. 2 The tenderloin. 3 A slanting cut in a sawed log. 4 A notch cut in the side of a tree so that it will fall toward that side when sawn through. 5 Any part that is cut away below: the undercut of a carriage. 6 In sports, a cut or backspin imparted to the ball by an underhand or downward stroke. — v.t. (un′dər·kut′) ·cut, ·cut·ting 1 To cut under. 2 To cut away a lower portion of so as to leave a part overhanging: The river undercut its banks. 3 To work or sell for lower

uncured	undecaying	undefended	undenominational	undeserved	undestroyed	undiminishable
uncurious	undecipherable	undefensible	undenounced	undeservedly	undetachable	undiminished
uncurl	undecipherably	undefiled	undependable	undeservedness	undetached	undimmed
uncurled	undeciphered	undefinable	undeplored	undeserving	undetected	undiplomatic
uncurrent	undeclared	undefined	undeposed	undesignated	undeterminable	undisbanded
uncursed	undeclinable	undeformed	undepraved	undesigned	undeterred	undiscerned
uncurtained	undeclined	undelayable	undepreciated	undesignedly	undeveloped	undiscernedly
uncushioned	undecomposable	undelayed	undepressed	undesignedness	undeviating	undiscernible
undamaged	undecomposed	undelineated	undeputed	undesirability	undevoured	undiscernibly
undangered	undecorated	undeliverable	underived	undesirable	undifferentiated	undiscerning
undated	undefaceable	undelivered	underivedness	undesirably	undiffused	undischarged
undaughterly	undefaced	undemocratic	underogating	undesired	undigested	undisciplined
undazzled	undefacedness	undemonstrable	underogatory	undesirous	undignified	undisclosed
undebatable	undefeatable	undemonstrably	undescribed	undesisting	undilated	undisconcerted
undecayed	undefeated	undenied	undescried	undespairing	undiluted	undiscordant

payment than (a rival). **4** In golf, to impart backspin to (the ball) by striking it obliquely downward. **5** In tennis, to use an underhand stroke in cutting (the ball). —*adj.* **1** Having the parts in relief cut under. **2** Done by undercutting.

un·der·do (un′dər·do͞o′) *v.t.* & *v.i.* **·did, ·done, ·do·ing** To do less than is expected or needed.

un·der·dog (un′dər·dôg′, -dog′) *n.* **1** The dog that is losing, has lost, or is at a disadvantage in a dogfight. **2** The weaker or worsted person. **3** Anyone in a position of inferiority.

un·der·done (un′dər·dun′) *adj.* **1** Insufficiently done. **2** Not cooked to the full.

un·der·dose (un′dər·dōs′) *n.* A dose less than that prescribed, customary, or requisite. —*v.t.* (un′dər·dōs′) **·dosed, ·dos·ing** To administer too small a dose to.

un·der·drain (un′dər·drān′) *n.* A subsurface drain built to carry off water percolating through the soil above. —**un′der·drain′age** (-drā′nij) *n.*

un·der·drive (un′dər·drīv′) *n. Mech.* A gearing device which turns a drive shaft at a speed less than that of the engine: opposed to *overdrive.*

un·der·es·ti·mate (un′dər·es′tə·māt) *v.t.* **·mat·ed, ·mat·ing** To put too low an estimate or valuation upon (things or people). See synonyms under DISPARAGE. —*n.* (-mit) **1** An insufficiently high opinion. **2** An estimate below the just value or expense. —**un′der·es′ti·ma′tion** *n.*

un·der·ex·pose (un′dər·ik·spōz′) *v.t.* **·posed, ·pos·ing** *Phot.* To expose (a film) less than is required for proper development. —**un′der·ex·posed′** *adj.* —**un′der·ex·po′sure** (-spō′zhər) *n.*

un·der·feed (un′dər·fēd′) *v.t.* **·fed, ·feed·ing** **1** To feed insufficiently. **2** To supply fuel for (an engine) from beneath.

under fire Engaged in a battle; exposed to fire; being attacked: said of troops.

un·der·foot (un′dər·fo͝ot′) *adv.* **1** Beneath the feet; down on the ground; immediately below. **2** In the way.

un·der·fur (un′dər·fûr′) *n.* The coat of dense, fine hair forming the main part of a pelt, as in seals.

un·der·gar·ment (un′dər·gär′mənt) *n.* A garment to be worn under the ordinary outer garments.

un·der·gird (un′dər·gûrd′) *v.t.* **·girt** or **·gird·ed, ·gird·ing** To fasten or gird, as by something that passes underneath.

un·der·glaze (un′dər·glāz′) *adj.* Used in or suitable for porcelain decoration: said of painting in vitrifiable pigment before the glaze is applied.

un·der·go (un′dər·gō′) *v.t.* **·went, ·gone, ·go·ing** **1** To be subjected to; have experience of; suffer. **2** To bear up under; endure. **3** *Obs.* To exist under. See synonyms under ENDURE.

un·der·grad·u·ate (un′dər·graj′o͞o·it) *n.* A student of a university or college who has not taken the bachelor's degree.

un·der·ground (un′dər·ground′) *adj.* **1** Situated, done, or operating beneath the surface of the ground. **2** Hence, done in secret; clandestine. —*n.* **1** That which is beneath the surface of the ground, as a passage or space. **2** A railway operated in a system of tunnels beneath the ground. **3** A group secretly organized to resist or oppose those in control of a government or country. —*adv.* (un′dər·ground′) Beneath the surface of the ground; hence, secretly.

Underground Railroad A system of cooperation among anti–slavery people, before 1861, for assisting fugitive slaves to escape to Canada and the free States.

un·der·grown (un′dər·grōn′) *adj.* Not fully grown; undersized.

un·der·growth (un′dər·grōth′) *n.* **1** A growth of smaller plants among larger ones; specifically, a thicket or coppice in or as in a forest. **2** Condition of being undergrown. **3** A close growth of hair beneath and finer than the outer growth of a pelt.

un·der·hand (un′dər·hand′) *adj.* **1** Done or acting in a treacherously secret manner; unfair; sly. **2** In baseball, cricket, etc., underarm. —*adv.* Underhandedly; slily; clandestinely.

un·der·hand·ed (un′dər·han′did) *adj.* Clandestinely carried on; underhand. —**un′der·hand′ed·ly** *adv.* —**un′der·hand′ed·ness** *n.*

un·der·hung (un′dər·hung′) *adj.* **1** *Anat.* Protruding from beneath, as a lower jaw: said of persons, dogs, etc., with a jaw protruding either beyond the upper jaw or unusually far. **2** Running on rollers on a rail below it, as a sliding door. **3** Underslung.

un·der·laid (un′dər·lād′) *adj.* **1** Laid underneath; supporting. **2** Supported by or having something lying or placed underneath.

un·der·lay (un′dər·lā′) *v.t.* **·laid, ·lay·ing** **1** To place (one thing) under another. **2** To furnish with a base or lining. **3** *Printing* To support or raise by underlays. —*n.* (un′dər·lā′) **1** *Printing* A piece of paper, etc., placed under certain parts of a printing form, to bring them to the proper level. **2** *Mining* An inclination, as of a lode. **3** A wager made at odds unfavorable to the better: opposed to *overlay.*

un·der·lease (un′dər·lēs′) *n.* A lease of premises by a lessee; sublease.

un·der·let (un′dər·let′) *v.t.* **·let, ·let·ting** **1** To lease (premises already held on lease); sublet. **2** To lease at less than the usual rate.

un·der·lie (un′dər·lī′) *v.t.* **·lay, ·lain, ·ly·ing** **1** To lie below or under. **2** To be the ground or support of: the principle that *underlies* a scheme. **3** To constitute a first or prior claim or lien over: A first mortgage *underlies* a second. **4** To be subject, answerable, or liable to. [OE *underlicgan*]

un·der·line (un′dər·līn′) *v.t.* **·lined, ·lin·ing** **1** To mark with a line underneath; underscore. **2** To emphasize. **3** To advertise the performance of (a play or other show) in a line subjoined to a current announcement: *Faust* is *underlined* for Thursday. —*n.* **1** A line underneath, as beneath a printed or written word or syllable to indicate emphasis or stress. **2** An announcement of a performance subjoined to the advertisement of a current one.

un·der·lin·en (un′dər·lin′ən) *n.* Linen underwear; any underwear.

un·der·ling (un′dər·ling) *n.* A subordinate; an inferior; a servile person.

un·der·ly·ing (un′dər·lī′ing) *adj.* **1** Lying under: *underlying* strata. **2** Hence, figuratively, fundamental: *underlying* principles. **3** Prior in claim or lien. See UNDERLIE (def. 3).

un·der·men·tioned (un′dər·men′shənd) *adj.* Mentioned below in a writing.

un·der·mine (un′dər·mīn′, un′dər·mīn) *v.t.* **·mined, ·min·ing** **1** To excavate beneath; dig a mine or passage under: to *undermine* a fortress. **2** To weaken by wearing away at the base. **3** To weaken or impair secretly or by degrees: to *undermine* the influence or the health of someone. See synonyms under WEAKEN. —**un′der·min′er** *n.*

un·der·most (un′dər·mōst′) *adj.* Having the lowest place or position.

un·der·neath (un′dər·nēth′, -nēth′) *adv.* **1** In a place below. **2** On the under or lower side. See synonyms under BENEATH. —*prep.* **1** Beneath; under; below. **2** Under the form or appearance of. **3** Under the authority of; in the control of. —*adj.* Lower. —*n.* The lower or under part or side. [OE *underneothan*]

un·der·nour·ish (un′dər·nûr′ish) *v.t.* To provide with nourishment insufficient in amount or quality for proper health and growth. —**un′der·nour′ish·ment** *n.*

un·dern·song (un′dərn·sông′, -song′) *n.* Tierce (def. 4). [OE *undern* midday, midday meal + SONG]

un·der·pants (un′dər·pants′) *n. pl.* An undergarment worn over the loins and sometimes extending over the thighs or lower legs.

un·der·pass (un′dər·pas′, -päs′) *n.* A passage beneath; the section of a way or road that passes under railway tracks or under another road.

un·der·pay (un′dər·pā′) *v.t.* **·paid, ·pay·ing** To pay insufficiently.

un·der·pin (un′dər·pin′) *v.t.* **·pinned, ·pin·ning** **1** To support, as a wall or structure, from below, especially when a previous support is removed, by inserting a prop or pier. **2** To corroborate; support.

un·der·pin·ning (un′dər·pin′ing) *n.* **1** Material or framework used to support a wall or building from below. **2** *pl. Colloq.* The legs.

un·der·pitch (un′dər·pich′) *adj. Archit.* Designating a main vault intersected by another at a lower level. [<UNDER- + PITCH², *n.* (def. 3)]

un·der·plant (un′dər·plant′, -plänt′) *v.t. Rare* To plant young trees under (existing trees).

un·der·plot (un′dər·plot′) *n.* **1** A subsidiary literary or dramatic plot; an episode. **2** A piece of roguery or trickery; an underhand action. —**un′der·plot′ter** *n.*

un·der·priv·i·leged (un′dər·priv′ə·lijd) *adj.* At a social or economic disadvantage; specifically, through economic cause, not privileged to enjoy certain rights theoretically possessed by all members of a community or state.

un·der·pro·duc·tion (un′dər·prə·duk′shən) *n.* Production below capacity or below requirements; abnormally low production. Compare OVERPRODUCTION.

un·der·proof (un′dər·pro͞of′) *adj.* Having less strength than proof spirit.

un·der·prop (un′dər·prop′) *v.t.* **·propped, ·prop·ping** To prop from below; support.

un·der·quote (un′dər·kwōt′) *v.t.* **·quot·ed, ·quot·ing** **1** To undersell or offer to undersell, as goods or stocks. **2** To underbid.

un·der·rate (un′dər·rāt′) *v.t.* **·rat·ed, ·rat·ing** To rate too low; underestimate. See synonyms under DISPARAGE.

un·der·run (un′dər·run′) *v.t.* **·ran, ·run, ·run·ning** **1** To run or pass beneath. **2** *Naut.* To examine (a line, hawser, etc.) from below by drawing a boat along beneath it.

un·der·score (un′dər·skôr′, -skōr′) *v.t.* **·scored, ·scor·ing** To draw a line below, as for indicating emphasis; underline. —*n.* (un′dər·skôr′, -skōr′) A line drawn beneath a word, etc., as for emphasis.

un·der·sea (un′dər·sē′) *adj.* Existing, carried on, or adapted for use beneath the surface of the sea: *undersea* exploration; an *undersea* oil well. —*adv.* Beneath the surface of the sea: also **un′der·seas′.**

un·der·sell (un′dər·sel′) *v.t.* **·sold, ·sell·ing** **1** To sell at a lower price than. **2** To sell for less than the real value. —**un′der·sell′er** *n.*

un·der·set¹ (un′dər·set′) *v.t.* **·set, ·set·ting** **1** To prop up; support. **2** *Brit.* To underlet; sublet.

undiscouraged	**undismissed**	**undistracted**	**undouble**	**undutiful**	**uneliminated**	**unendorsed**
undiscoverable	**undispatched**	**undistraught**	**undoubting**	**undutifully**	**unelucidated**	**unendowed**
undiscoverably	**undispelled**	**undistressed**	**undrained**	**undutifulness**	**unemancipated**	**unendurable**
undiscovered	**undispensed**	**undistributed**	**undramatic**	**undyed**	**unembarrassed**	**unenduring**
undiscredited	**undisputable**	**undisturbed**	**undramatical**	**uneatable**	**unembellished**	**unenforceable**
undiscriminating	**undisputed**	**undisturbedly**	**undramatized**	**uneaten**	**unemotional**	**unenforced**
undiscussed	**undissected**	**undisturbedness**	**undrape**	**unecclesiastic**	**unemphatic**	**unenfranchised**
undisguised	**undissembling**	**undiversified**	**undraped**	**uneclipsed**	**unemphatical**	**unengaged**
undisguisedly	**undisseminated**	**undiverted**	**undreaded**	**uneconomic**	**unemptied**	**unengaging**
undisheartened	**undissolved**	**undivested**	**undreamed,**	**uneconomical**	**unenclosed**	**un–English**
undishonored	**undissolving**	**undivided**	**undreamt**	**unedible**	**unencumbered**	**unenjoyable**
undisillusioned	**undistilled**	**undivorced**	**undreamed–of**	**unedifying**	**unendangered**	**unenjoyed**
undismantled	**undistinguishable**	**undivulged**	**undried**	**uneducable**	**unendeared**	**unenlightened**
undismayed	**undistinguished**	**undomestic**	**undrilled**	**uneducated**	**unended**	**unenlivened**
undismembered	**undistinguishing**	**undomesticated**	**undrinkable**	**uneffaced**	**unending**	**unenriched**

add,āce,câre,pälm; end,ēven; it,īce; odd,ōpen,ôrder; to͝ok,po͞ol; up,bûrn; ə = a in *above,* e in *sicken,* i in *clarity,* o in *melon,* u in *focus;* yo͞o = u in *fuse;* oi,oil; ou,pout; ch,check; g,go; ng,ring; th,thin; th,this; zh,vision. Foreign sounds à,œ,ü,kh,ṅ; and ◆: see page xx. < from; + plus; ? possibly.

un·der·set² (un′dər·set′) *n.* An undercurrent.
un·der·set·ter (un′dər·set′ər) *n.* 1 An underpinning prop or support. 2 *Brit.* One who sublets.
un·der·set·ting (un′dər·set′ing) *n.* Any underpinning; also, a base or pedestal.
un·der·shap·en (un′dər·shā′pən) *adj.* Below normal size; imperfectly formed.
un·der–sher·iff (un′dər·sher′if) *n.* A deputy sheriff, especially one upon whom the sheriff's duties devolve in his absence.
un·der·shirt (un′dər·shûrt′) *n.* A garment worn beneath the shirt, generally of cotton, wool and cotton, or silk.
un·der·shot (un′dər·shot′) *adj.* 1 Propelled by water that flows underneath: said of a water wheel. 2 Projecting; having a projecting lower jaw or teeth: said especially of a bulldog.
un·der·shrub (un′·dər·shrub′) *n.* A small shrub or plant, with shrubby base.

UNDERSHOT WATER WHEEL

un·der·side (un′dər·sīd′) *n.* The lower or under side or surface.
un·der·sign (un′dər·sīn′) *v.t.* To sign at the foot of; subscribe: used chiefly in the past participle. — **the undersigned** The subscriber or subscribers to a document.
un·der·sized¹ (un′dər·sīzd′) *adj.* Of less than the normal or average size.
un·der·sized² (un′dər·sīzd′) *adj.* Insufficiently sized, as paper.
un·der·skirt (un′dər·skûrt′) *n.* 1 A skirt worn beneath another; a petticoat. 2 The foundation skirt of a draped gown.
un·der·sleeve (un′dər·slēv′) *n.* A sleeve worn beneath another, especially when of contrasting color and showing through slashes or openings.
un·der·slung (un′dər·slung′) *adj.* Having the springs fixed to the axles from below, instead of resting upon them: said of certain automobiles.
un·der·soil (un′dər·soil′) *n.* Subsoil.
un·der·song (un′dər·sông′, -song′) *n.* 1 A subordinate strain or subdued melody. 2 An underlying meaning.
un·der·sparred (un′dər·spärd′) *adj. Naut.* Having too few, too short, or too slight spars or masts.
un·der·spin (un′dər·spin′) *n.* In golf, a backward spin imparted to the ball.
un·der·stand (un′dər·stand′) *v.* **·stood**, **·stand·ing** *v.t.* 1 To come to know the meaning or import of; apprehend. 2 To perceive the nature or character of: I do not *understand* her. 3 To have comprehension or mastery of: Do you *understand* German? 4 To be aware of; realize: She *understands* her position. 5 To have been told; believe: I *understand* that she went home. 6 To take or suppose to mean; infer: How am I to *understand* that remark? 7 To accept as a condition or stipulation: It is *understood* that the tenant will provide his own heat. 8 To supply in thought when unexpressed, as the subject of a sentence. — *v.i.* 9 To have understanding; comprehend. 10 To be informed; believe. See synonyms under APPREHEND, KNOW, PERCEIVE, SOLVE. — **to understand each other** To be in agreement; be privately in sympathy with each other. [OE *understandan* < *under-* under + *standan* stand] — **un′der·stand′a·ble** *adj.* — **un′·der·stand′a·bly** *adv.*

un·der·stand·ing (un′dər·stan′ding) *n.* 1 The act of one who understands, or the resulting state; intellectual apprehension; mental discernment; comprehension. 2 The power by which one understands. 3 The sum of the mental powers by which knowledge is acquired, retained, and extended; the power of apprehending relations and making inferences from them. 4 The facts or elements of a case as apprehended by any one intelligence; an individual view of a case; opinion. 5 An agreement between two or more persons; an informal or confidential compact; also, the subject of such compact; the thing agreed on; sometimes, an arrangement or settlement of differences, or of disputed points: That was not our *understanding*; They have come to an *understanding*. — *adj.* Possessing comprehension and good sense. — **un′der·stand′ing·ly** *adv.* — **un′der·stand′ing·ness** *n.*
Synonyms (noun): apprehension, comprehension, discernment, intellect, intelligence, judgment, mind, perception, reason. See INTELLECT, MIND, WISDOM.
un·der·state (un′dər·stāt′) *v.* **·stat·ed**, **·stat·ing** *v.t.* 1 To state with less force than the truth warrants or allows. 2 To state, as a number or dimension, as less than the true one. — *v.i.* 3 To make an understatement.
un·der·state·ment (un′dər·stāt′mənt) *n.* A statement covering less than the truth or fact.
un·der·stood (un′dər·stood′) Past tense and past participle of UNDERSTAND. — *adj.* Taken for granted; agreed upon by all.
un·der·strap·per (un′dər·strap′ər) *n.* An underling; a subordinate agent.
un·der·strap·ping (un′dər·strap′ing) *adj.* Subordinate; inferior.
un·der·stra·tum (un′dər·strā′təm, -strat′əm) *n. pl.* **·stra·ta** (-strā′tə, -strat′ə) An underlying stratum; substratum, literal or figurative.
un·der·stud·y (un′dər·stud′ē) *v.t. & v.i.* **·stud·ied**, **·stud·y·ing** 1 To study (a part) in order to be able, if necessary, to take the place of the actor playing it. 2 To act as an understudy to (another actor). — *n. pl.* **·stud·ies** 1 An actor or actress who can take the place of another actor in a given role when necessary. 2 A person prepared to perform the work or fill the position of another.
un·der·take (un′dər·tāk′) *v.* **·took**, **·tak·en**, **·tak·ing** *v.t.* 1 To take upon oneself; agree or attempt to do; begin. 2 To contract to do; pledge oneself to. 3 To guarantee or promise. 4 To take under charge or guidance. 5 *Obs.* To enter into combat with. — *v.i.* 6 To make oneself responsible or liable; be surety: with *for.* See synonyms under ENDEAVOR.
un·der·tak·er (un′dər·tā′kər *for def. 1*; un′dər·tā′kər *for def. 2*) *n.* 1 One who undertakes any work or enterprise; especially, a contractor. 2 One whose business it is to arrange for burying the dead and to conduct funerals.
un·der·tak·ing (un′dər·tā′king; *for def. 3* un′dər·tā′king) *n.* 1 The act of one who undertakes any task or enterprise. 2 The thing undertaken; an enterprise; task. 3 The management of funerals; the business of an undertaker. 4 An engagement, promise, or guaranty.
under tenant A tenant of a tenant; one who holds premises by a lease from one who is himself a lessee.
under the rose Sub rosa.
under the weather *Colloq.* 1 Depressed by unpleasant weather; hence, somewhat ill; indisposed. 2 Inebriated; drunk. 3 In financial straits.
under the yoke In subjection.
un·der·thrust (un′dər·thrust′) *n. Geol.* 1 A deformation of the earth's crust in which a mass of rock is pushed beneath an overlying mass. 2 The intruded rock mass itself.
un·der·tint (un′dər·tint′) *n.* A subdued tint.
un·der·tone (un′dər·tōn′) *n.* 1 A tone of lower pitch or loudness than is usual; the tone of a subdued voice; sometimes, a whisper. 2 A subdued shade of a color, as when spread thinly on a white surface; also, a color upon which other colors have been imposed and which is seen through them, modifying their effect. 3 A meaning or suggestion implied but not expressed. 4 An underlying stability in the price level of some stocks.
un·der·took (un′dər·took′) Past tense of UNDERTAKE.
un·der·tow (un′dər·tō′) *n.* 1 The flow of water beneath and in a direction opposite to the surface current. 2 The backward undercurrent below the surf.
un·der·trick (un′dər·trik′) *n.* In certain card games, a trick required to make the number declared, but not taken.
un·der·trump (un′dər·trump′) *v.t.* To play to (a previous card in the same trick) a trump lower than one already played by another player; also, to trump with too low a trump, and so be overtrumped.
un·der·val·ue (un′dər·val′yōō) *v.t.* **·ued**, **·u·ing** 1 To value too lightly; underrate; underestimate. 2 *Obs.* To hold inferior: with *to* before the object compared. See synonyms under DISPARAGE. — **un′der·val·u·a′tion** *n.*
un·der·vest (un′dər·vest′) *n. Brit.* An undershirt.
un·der·waist (un′dər·wāst′) *n.* 1 A waist to be worn under another waist. 2 An undergarment worn by children; a skeleton undergarment to which other garments may be buttoned.
un·der·wa·ter (un′dər·wô′tər, -wot′ər) *adj.* Lying or situated below water level; specifically, below the water line of a ship.
un·der·wear (un′dər·wâr′) *n.* Garments worn underneath the ordinary outer garments; underclothes.
un·der·weight (un′dər·wāt′) *adj.* Having less than the normal weight. — *n.* Insufficiency of weight; also, weight below normal.
un·der·went (un′dər·went′) Past tense of UNDERGO.
un·der·wing (un′dər·wing′) *n. Entomol.* One of the posterior pair of wings in an insect.
underwing moth A large noctuid moth (genus *Catocala*), whose front wings are an inconspicuous brown or gray, while in most species the hind wings, usually concealed, are banded with black and crimson.
un·der·wood (un′dər·wōōd′) *n.* Low trees and brush growing among large forest trees.
un·der·work (un′dər·wûrk′) *v.t.* 1 To work for lower wages than. 2 To exact too little work from. 3 *Obs.* To weaken or injure by underhand contrivances; undermine. (un′·dər·wûrk′) 4 To do too little work. — *n.* (un′dər·wûrk′) Subordinate, inferior, unimportant, or routine work.
un·der·world (un′dər·wûrld′) *n.* 1 In Greek and Roman mythology, the abode of the dead; Hades; Orcus. 2 In later folklore, sometimes a beautiful country under the earth or sea; also, fairyland, sometimes entered through a well. 3 The antipodes; also, all beneath the horizon. 4 The sublunary world; the earth. 5 The debased, criminal, or degenerate components of the social order; the world of crime and vice; gangsterdom.
un·der·write (un′dər·rīt′) *v.* **·wrote**, **·writ·ten**, **·writ·ing** *v.t.* 1 To write beneath; subscribe. 2 In finance, to execute and deliver (a policy of insurance on specified property, especially

unenrolled	unescapable	unexcused	unexplored	unfashionable	unfeigningly	unfittingly
unenslaved	unessayed	unexecuted	unexported	unfashioned	unfelt	unfixed
unentangled	unestablished	unexercised	unexposed	unfastened	unfeminine	unfixedness
unentered	unesthetic	unexhausted	unexpressed	unfatherly	unfenced	unflagging
unenterprising	unestimated	unexpanded	unexpunged	unfathomable	unfermented	unflattered
unentertaining	unethical	unexpectant	unexpurgated	unfathomed	unfertile	unflattering
unenthralled	unexaggerated	unexpended	unextended	unfatigued	unfertilized	unflavored
unenthusiastic	unexalted	unexpert	unexterminated	unfavored	unfetter	unflickering
unentitled	unexamined	unexpiated	unextinguishable	unfeared	unfettered	unfoiled
unenviable	unexcavated	unexpired	unextinguished	unfearing	unfilial	unforbearing
unenvied	unexcelled	unexplainable	unfadable	unfeasible	unfilled	unforbid
unenvious	unexchangeable	unexplained	unfaded	unfed	unfilmed	unforbidden
unenvying	unexcited	unexplicit	unfading	unfederated	unfiltered	unforced
unequipped	unexciting	unexploded	unfallen	unfeignedly	unfinished	unforcedly
unerased	unexcluded	unexploited	unfaltering	unfeignedness	unfired	unfordable

underwriter   1369   **unfailing**

marine property); insure; assume (a risk) by way of insurance. **3** To engage to buy, at a determined price and time, all or part of the stock in (a new enterprise or company) that is not subscribed for by the public; loosely, to guarantee or assume responsibility for, as an enterprise. **4** To undertake to pay, as a subscription or written pledge of money. — *v.i.* **5** To act as an underwriter; especially, to issue a policy of insurance. [OE *underwrītan*, trans. of L *subscribere*]

un·der·writ·er (un'dər·rī'tər) *n.* **1** A body corporate or a person in the insurance business; one who sets up the premium for a risk. **2** One who underwrites (def. 3) an issue of stocks, bonds, or the like.

un·de·sign·ing (un'di·zī'ning) *adj.* Without ulterior purpose or selfish plan; artless; sincere.

un·de·ter·mined (un'di·tûr'mind) *adj.* **1** Not decided or fixed. **2** Not determined.

un·did (un·did') Past tense of UNDO.

un·dine (un·dēn', un'dēn, -dīn) *n. Med.* A small glass cup or flask for irrigating the eye. [< L *unda* wave + -INE¹; from its wavy profile]

Un·dine (un·dēn', un'dēn, -dīn) A water nymph without a soul, which she later received by marrying a mortal and bearing a child: heroine of a book (1811) by Baron de la Motte Fouqué, German author. [< NL *Undina* < L *unda* wave]

un·di·rect·ed (un'di·rek'tid, -dī-) *adj.* **1** Unguided, or uninformed as to direction. **2** Not addressed: said of a letter.

un·dis·posed (un'dis·pōzd') *adj.* **1** Not sold, settled, placed, or otherwise decided: frequently with *of*. **2** *Obs.* Disinclined. — **un'·dis·pos'ed·ness** (-pō'zid·nis) *n.*

un·do (un·doō') *v.t.* **·did**, **·done**, **·do·ing** **1** To cause to be as if never done; reverse, annul, or cancel. **2** To loosen or untie. **3** To unfasten and open. **4** To bring to ruin; destroy. **5** *Obs.* To solve, as a riddle. [OE *undōn*] — **un·do'er** *n.*

un·do·ing (un·doō'ing) *n.* **1** Reversal of what has been done. **2** Destruction; ruin; cause of ruin. **3** The action of unfastening, loosening, opening, etc. **4** *Psychoanal.* The abolition of painful experiences by the unconscious, resulting in obliviousness to the unacceptable fact.

un·done¹ (un·dun') *adj.* **1** Untied; unfastened. **2** Ruined. [Orig. pp. of UNDO]

un·done² (un·dun') *adj.* Not done. [< UN-¹ + DONE]

un·doubt·ed (un·dou'tid) *adj.* **1** Assured beyond question; being beyond a doubt. **2** Not viewed with distrust; unsuspected. See synonyms under INCONTESTABLE. — **un·doubt'ed·ly** *adv.*

un·draw (un·drô') *v.t. & v.i.* **drew**, **drawn**, **·draw·ing** To draw open, away, or aside.

un·dress (un·dres') *v.t.* **1** To divest of clothes; strip. **2** To remove the dressing or bandages from, as a wound. **3** To divest of special attire; disrobe. — *v.i.* **4** To remove one's clothing. — *n.* Ordinary attire; negligée, as opposed to full or evening dress; specifically, the military or naval uniform worn by officers when not on parade or at functions necessitating full dress. The term has fallen into disuse in the United States Army. — *adj.* (un'dres') Pertaining to everyday attire; hence, informal.

un·dressed (un·drest') *adj.* **1** Not dressed. **2** Not treated or dressed: said of kid leather.

Und·set (ŏn'set), **Sigrid**, 1882–1949, Norwegian novelist.

und so wei·ter (ŏont zō vī'tər) *German* And so forth; et cetera; abbreviated *u.s.w.*

un·due (un·doō', -dyoō') *adj.* **1** Excessive; disproportionate. **2** Not justified by law; illegal. **3** Not due; in process of becoming due, but not yet demandable. **4** Not appropriate; improper.

un·du·lant (un'dyə·lənt, -də-) *adj.* Undulating; fluctuating. [< L *undul(atus)* + -ANT]

undulant fever *Pathol.* A persistent and wasting infectious disease of wide distribution, caused by a bacterium (genus *Brucella*) which is usually transmitted to man in the milk of cows and goats. The disease is marked by fluctuating or recurrent fever, with swelling of the joints, neuralgic pains, profuse perspiration, and enlargement of the spleen: also called *brucellosis, Malta fever, Mediterranean fever*.

un·du·late (un'dyə·lāt, -də-) *v.* **·lat·ed**, **·lat·ing** *v.t.* **1** To cause to move like a wave or in waves. **2** To give a wavy appearance to. — *v.i.* **3** To move like a wave or waves. **4** To have a wavy form or appearance. — *adj.* (-lit, -lāt) **1** Wavy. **2** Having wavelike markings, as of color. See synonyms under FLUCTUATE. [< L *undulatus* undulated, ult. < *unda* wave]

un·du·lat·ing (un'dyə·lā'ting, -də-) *adj.* Having the appearance of waves; vibrating; waving; wavy.

un·du·la·tion (un'dyə·lā'shən, -də-) *n.* **1** The act of undulating; a waving or sinuous motion; a wave. **2** An appearance as of waves; a gentle rise and fall. **3** *Physics* The continuous propagation of waves through a medium; wave motion. — **un'du·la·to·ry** (-lə·tôr'ē, -tō'rē) *adj.*

un·du·la·tus (un'dyə·lā'təs, -də-) *n. Meteorol.* A variety of stratocumulus cloud characterized by elongated wavelike undulations, sometimes in different directions. [< NL < L. See UNDULATE.]

un·du·lous (un'dyə·ləs, -də-) *adj.* Undulatory; undulating.

un·du·ly (un·doō'lē, -dyoō'-) *adv.* **1** Excessively. **2** In violation of a moral or of a legal standard; unjustly.

un·dy·ing (un·dī'ing) *adj.* Immortal.

un·earned (un·ûrnd') *adj.* Not earned by labor; also, undeserved.

unearned increment See under INCREMENT.

un·earth (un·ûrth') *v.t.* **1** To dig or root up from the earth. **2** To reveal; discover.

un·earth·ly (un·ûrth'lē) *adj.* **1** Not earthly; sublime. **2** Supernatural; terrifying; weird; terrible. **3** Ridiculously unconventional; inconvenient, or unpleasant; preposterous: at this *unearthly* hour. — **un·earth'li·ness** *n.*

un·eas·y (un·ē'zē) *adj.* **·eas·i·er**, **·eas·i·est** **1** Deprived of ease; disturbed; unquiet. **2** Not affording ease or rest; uncomfortable; causing discomfort. **3** Showing embarassment or constraint; strained. **4** *Obs.* Difficult. — **un·eas'i·ly** *adv.* — **un·eas'i·ness** *n.*

un·eath (un·ēth') *adv. Obs.* Scarcely; hardly; not easily. [OE *unēathe* not easily]

un·em·ploy·a·ble (un'əm·ploi'ə·bəl) *adj.* Not employable. — *n.* A person who, because of illness, age, mental or physical incapacity, or other reason, cannot be employed.

un·em·ployed (un'əm·ploid') *adj.* **1** Having no occupation: out of work. **2** Not put to use or turned to account; uninvested: *unemployed* resources. See synonyms under IDLE, VACANT. — *n.* A jobless person: with *the, unemployed persons* collectively. — **un'em·ploy'ment** *n.*

un·en·cum·bered funds (un'en·kum'bərd) **1** Funds not designated for any specific use; general funds. **2** Funds not pledged in connection with present or future obligations.

un·e·qual (un·ē'kwəl) *adj.* **1** Not having equivalent or equal extension, duration, or properties; not equal in strength, ability, wealth, status, or other respects. **2** Inadequate for the purpose; insufficient: with *to*. **3** Not balanced; disproportioned; inequitable; unfair. **4** Wanting in uniformity; varying; irregular. **5** *Bot.* Unsymmetrical; *unequal* distribution. **6** Involving poorly matched competitors or contestants: an *unequal* contest. — **un·e'qual·ly** *adv.*

un·e·qualed (un·ē'kwəld) *adj.* Not equaled or matched; unrivaled; supreme. Also **un·e'qualled**.

un·e·quiv·o·cal (un'i·kwiv'ə·kəl) *adj.* Understandable in only one way; distinct; plain. See synonyms under ABSOLUTE, CLEAR, PLAIN. — **un'e·quiv'o·cal·ly** *adv.*

un·err·ing (un·ûr'ing, -er'-) *adj.* Making no mistakes; not erring: also, sure; accurate; infallible. — **un·err'ing·ly** *adv.*

UNESCO (yōō·nes'kō) The United Nations Educational, Scientific and Cultural Organization, established November, 1946, to "advance mutual knowledge and understanding of peoples," promote popular education, and assist in the diffusion of knowledge. Also **Unesco**.

un·es·sen·tial (un'ə·sen'shəl) *adj.* **1** Not absolutely required; not of prime importance. **2** Unimportant. **3** Void of essence, real or apparent. — **un'es·sen'tial·ly** *adv.*

un·e·vac·u·a·ble (un'i·vak'yōō·ə·bəl) *adj.* Not capable of being removed or evacuated, as in a military action or air raid.

un·e·ven (un·ē'vən) *adj.* **1** Not even, smooth, or level; rough. **2** Not level, parallel, or perfectly horizontal. **3** Not divisible by two without remainder; odd: said of numbers. **4** Not uniform; variable; spasmodic. **5** *Obs.* Not having correspondence; not balanced; not fair or just; also, ill-suited; not matched. See synonyms under IRREGULAR, ROUGH. — **un·e'ven·ly** *adv.* — **un·e'ven·ness** *n.*

un·e·vent·ful (un'i·vent'fəl) *adj.* Devoid of noteworthy events; quiet.

un·ex·am·pled (un'ig·zam'pəld) *adj.* So great, remarkable, or striking as to have no precedent or analogy; without a parallel example.

un·ex·cep·tion·a·ble (un'ik·sep'shən·ə·bəl) *adj.* That cannot be objected to; irreproachable. — **un'ex·cep'tion·a·ble·ness** *n.* — **un'ex·cep'tion·a·bly** *adv.*

un·ex·cep·tion·al (un'ik·sep'shən·əl) *adj.* **1** Being no exception; ordinary. **2** Subject to no exception: *unexceptional* orders.

un·ex·pect·ed (un'ik·spek'tid) *adj.* **1** Coming without warning; not expected: said especially of things of such a kind that one would not naturally expect them; sudden; strange and unforeseen. — **the unexpected** Unexpected things or events collectively; that which is unforeseen. — **un'ex·pect'ed·ly** *adv.* — **un'ex·pect'ed·ness** *n.*

un·ex·pe·ri·enced (un'ik·spir'ē·ənst) *adj.* **1** Not experienced; not had, undergone, possessed, or known: *unexperienced* pain. **2** Lacking experience; inexperienced.

un·ex·pres·sive (un'ik·spres'iv) *adj.* **1** Not having expression; inexpressive. **2** *Obs.* Inexpressible. — **un'ex·pres'sive·ly** *adv.*

un·fail·ing (un·fā'ling) *adj.* **1** Giving or constituting a supply that never fails; inexhaustible: an *unfailing* spring. **2** Always fulfilling

unforeboding	unformulated	unfrozen	ungird	ungraded	unhandled	unhealthful
unforeknown	unforsaken	unfulfilled	ungirded	ungrafted	unhandsome	unheated
unforeseeable	unfortified	unfurnished	ungirt	ungrained	unhang	unheeded
unforeseeing	unfought	unfurrowed	ungladdened	ungrammatical	unhanged	unheedful
unforeseen	unfound	ungallant	unglazed	ungrammatically	unharassed	unheeding
unforetold	unframed	ungalled	unglossed	ungratified	unhardened	unheedingly
unforfeited	unfranchised	ungarnished	unglove	ungrounded	unharmed	unhelped
unforged	unfraternal	ungartered	ungloved	unharming	unhelpful
unforgetful	unfraught	ungathered	unglue	unguided	unharmonious	unheralded
unforgetting	unfree	ungenial	ungot	unhackneyed	unharnessed	unheroic
unforgivable	unfreedom	ungenteel	ungoverned	unhailed	unharrowed	unhesitating
unforgiven	unfreezable	ungentle	ungown	unhalved	unharvested	unhewn
unforgiving	un-French	ungentlemanly	ungowned	unhammered	unhasty	unhindered
unforgot	unfrequent	ungently	ungraced	unhampered	unhatched	unhired
unforgotten	unfrequently	un-get-at-able	ungraceful	unhandicapped	unhealed	unhistoric

add, āce, cȃre, pälm; end, ēven; it, īce; odd, ōpen, ôrder; tŏŏk, pōōl; up, bûrn; ə = a in *above*, e in *sicken*, i in *clarity*, o in *melon*, u in *focus*; yōō = u in *fuse*; oi, oil; ou, pout; ch, check; g, go; ng, ring; th, thin; th, this; zh, vision. Foreign sounds à, œ, ü, kh, ṅ; and ◆: see page xx. < from; + plus; ? possibly.

un·fair (un·fâr′) *adj.* 1 Marked by dishonesty or fraud; showing partiality or prejudice; not fair: *unfair* dealing. 2 Not compatible with law and justice; illegal: *unfair* competition. 3 *Obs.* Not pleasing or comely. See synonyms under BAD. [OE *unfæger* ugly] —**un·fair′ly** *adv.* —**un·fair′ness** *n.*

un·faith·ful (un·fāth′fəl) *adj.* 1 Manifesting lack or absence of faith; unworthy of trust; perfidious; faithless; not true to marriage vows: an *unfaithful* husband. 2 Not true to a standard or to an original; not accurate or exact: an *unfaithful* description. 3 *Obs.* Not having religious faith; unbelieving; infidel. See synonyms under PERFIDIOUS. —**un·faith′ful·ly** *adv.* —**un·faith′ful·ness** *n.*

un·fa·mil·iar (un′fə·mil′yər) *adj.* 1 Not familiarly knowing: I am *unfamiliar* with it. 2 Not familiarly known: an *unfamiliar* face. —**un′·fa·mil′i·ar′i·ty** (-mil′ē·ar′ə·tē) *n.* —**un′fa·mil′iar·ly** *adv.*

un·fas·ten (un·fas′ən, -fäs′-) *v.t.* To untie; loosen; open. —*v.i.* To become untied or loosened.

un·fa·thered (un·fä′thərd) *adj.* 1 Having no acknowledged father; hence, illegitimate. 2 Unauthenticated.

un·fa·vor·a·ble (un·fā′vər·ə·bəl) *adj.* Not favorable; unpropitious; adverse. Also *Brit.* **un·fa′vour·a·ble.** —**un·fa′vor·a·ble·ness** *n.* —**un·fa′vor·a·bly** *adv.*

Unfederated Malay States Former collective name for the states of Perlis, Kedah, Kelanton, Trengganu, and Johore, now members of the Federation of Malaya.

un·feel·ing (un·fē′ling) *adj.* 1 Not sympathetic; hard; cruel. 2 *Obs.* Destitute of feeling or sensation. See synonyms under HARD. —**un·feel′ing·ly** *adv.* —**un·feel′ing·ness** *n.*

un·feigned (un·fānd′) *adj.* Not feigned; not pretended; sincere; genuine.

un·fel·lowed (un·fel′ōd) *adj.* 1 Unequaled; unmatchable. 2 Alone; without a companion.

un·fit (un·fit′) *v.t* **·fit·ted** or **·fit, ·fit·ting** To deprive of requisite fitness, skill, etc.; disqualify. —*adj.* 1 Having no fitness; unsuitable. 2 Not appropriate; improper. 3 Not completely trained; not in best condition: said of race horses. —**un·fit′ly** *adv.* —**un·fit′ness** *n.*

un·fix (un·fiks′) *v.t.* 1 To unfasten; loosen; detach. 2 To unsettle.

un·fledged (un·flejd′) *adj.* 1 Not yet fledged; immature, as a young bird. 2 Inexperienced: an *unfledged* orator.

un·flesh·ly (un·flesh′lē) *adj.* Not corporeal, worldly, or sensual; ethereal; spiritual.

un·flinch·ing (un·flin′ching) *adj.* Performing or suffering without shrinking; steadfast; brave. —**un·flinch′ing·ly** *adv.* —**un·flinch′ing·ness** *n.*

un·fold¹ (un·fōld′) *v.t.* 1 To open or spread out (something folded). 2 To lay open to view. 3 To make clear by detailed explanation; explain: to *unfold* a plan. 4 To develop. —*v.i.* 5 To become opened; expand. 6 To become manifest. See synonyms under AMPLIFY, INTERPRET, SOLVE. [OE *unfealdan*] —**un·fold′er** *n.*

un·fold² (un·fōld′) *v.t.* To free or let loose from a fold or pen.

un·for·get·ta·ble (un′fər·get′ə·bəl) *adj.* Not forgettable; memorable. —**un′for·get′ta·bly** *adv.*

un·formed (un·fôrmd′) *adj.* 1 Devoid of shape or form; not fully developed in character; crude. 2 Unorganized.

un·for·tu·nate (un·fôr′chə·nit) *adj.* 1 Having ill fortune; not prosperous; unsuccessful. 2 Causing or attended by ill fortune; disastrous. See synonyms under BAD. —*n.* 1 One who is unfortunate. 2 Specifically, one who has lapsed from virtue; a prostitute. —**un·for′tu·nate·ly** *adv.* —**un·for′tu·nate·ness** *n.*

un·found·ed (un·foun′did) *adj.* 1 Resting on no solid foundation; groundless; baseless. 2 Not founded or established. —**un·found′ed·ly** *adv.*

un·fre·quent·ed (un′fri·kwen′tid) *adj.* Rarely or never visited or frequented.

un·friend·ed (un·fren′did) *adj.* Without friends. —**un·friend′ed·ness** *n.*

un·friend·ly (un·frend′lē) *adj.* 1 Unkindly disposed; inimical; hostile. 2 Not favorable or propitious. See synonyms under INIMICAL. —*adv.* In an unfriendly manner. —**un·friend′li·ness** *n.*

un·frock (un·frok′) *v.t.* 1 To divest of a frock or gown. 2 To depose, as a monk or priest, from ecclesiastical rank.

un·fruit·ful (un·frōōt′fəl) *adj.* 1 Bearing no fruit; having no offspring; barren. 2 Having no useful results; fruitless: an *unfruitful* line of thought. —**un·fruit′ful·ly** *adv.* —**un·fruit′ful·ness** *n.*

un·fumed (un·fyōomd′) *adj.* 1 Not fumigated. 2 *Obs.* Undistilled.

un·fund·ed (un·fun′did) *adj.* Not funded: said of a debt.

un·furl (un·fûrl′) *v.t. & v.i.* 1 To unroll, as a flag; spread out; expand. 2 To unfold. —**un·furled′** *adj.*

un·gain·ly (un·gān′lē) *adj.* Lacking grace or ease; clumsy. See synonyms under AWKWARD. —*adv.* In an awkward manner. —**un·gain′li·ness** *n.*

Un·ga·va (ung·gä′və, -gā′və) A district of northern Quebec province, extending south of Ungava Bay, and including part of Labrador; 239,780 square miles: also *New Quebec*.

Ungava Bay An inlet of Hudson Strait in northern Quebec province; 200 miles long, 160 miles wide at the mouth.

Ungava Peninsula A peninsula of northern Quebec province between Ungava Bay and Hudson Bay; 400 miles long, 350 miles wide.

un·gen·er·ous (un·jen′ər·əs) *adj.* 1 Not generous; illiberal; niggardly; unkind or harsh in judging others. —**un·gen′er·ous·ly** *adv.*

un·gift·ed (un·gif′tid) *adj.* 1 Not gifted or endowed with talent. 2 Not having received gifts.

un·god·ly (un·god′lē) *adj.* 1 Having no reverence for God; impious; wicked. 2 Unholy; sinful. 3 *Colloq.* Outrageous. —*adv.* In an ungodly manner. —**un·god′li·ness** *n.*

un·got·ten (un·got′n) *adj.* 1 Not begotten. 2 Not obtained; not acquired.

un·gov·ern·a·ble (un·guv′ər·nə·bəl) *adj.* That cannot be governed; refractory; unruly. See synonyms under PERVERSE, REBELLIOUS, VIOLENT. —**un·gov′ern·a·ble·ness** *n.* —**un·gov′ern·a·bly** *adv.*

un·gra·cious (un·grā′shəs) *adj.* 1 Lacking in graciousness of manner; unmannerly. 2 Not pleasing; offensive; unacceptable. 3 *Obs.* Odious. —**un·gra′cious·ly** *adv.* —**un·gra′cious·ness** *n.*

un·grate·ful (un·grāt′fəl) *adj.* 1 Feeling or showing a lack of gratitude; not thankful. 2 Not pleasant; disagreeable. 3 Unrewarding; yielding no return. —**un·grate′ful·ly** *adv.* —**un·grate′ful·ness** *n.*

un·gual (ung′gwəl) *adj.* Having, resembling, or pertaining to a hoof, claw, or nail. [<L *unguis* hoof, claw, nail]

un·guard (un·gärd′) *v.t.* To deprive of a guard; expose.

un·guard·ed (un·gär′did) *adj.* 1 Having no guard; being without protection. 2 Done or spoken without proper caution; careless: *unguarded* speech. —**un·guard′ed·ly** *adv.* —**un·guard′ed·ness** *n.*

un·guent (ung′gwənt) *n.* Any ointment for local application; a salve or cerate. [<L *unguentum* < *unguere* anoint]

un·guen·tar·y (ung′gwən·ter′ē) *adj.* Of, for, like, or pertaining to unguents.

un·guic·u·late (ung-gwik′yə·lit, -lāt) *adj.* 1 *Zool.* Having claws, as a carnivorous mammal. 2 *Bot.* Having a stalklike or clawlike base, as the petals of pinks. —*n.* A mammal having claws, as distinguished from an ungulate or cetacean. [<NL *unguiculatus* <L *unguiculus* fingernail, dim. of *unguis* nail]

un·gui·form (ung′gwi·fôrm) *adj.* Claw-shaped; hooked; unciform. [<L *unguis* nail + -FORM]

un·gui·nous (ung′gwi·nəs) *adj.* Resembling, containing, or consisting of oil or fat; unctuous. [<L *unguinosus* <*unguen, -inis* ointment]

un·guis (ung′gwis) *n.* *pl.* **·gues** (-gwēz) 1 A nail, claw, hoof, or talon. 2 A structure resembling a nail. 3 *Bot.* A claw or lower contracted part of a petal. [<L, nail]

un·gu·la (ung′gyə·lə) *n.* *pl.* **·lae** (-lē) 1 *Zool.* A hoof, claw, nail, or talon. 2 *Surg.* An instrument for removing a dead fetus from the womb. 3 *Geom.* That which is left of a cone or cylinder when the top is cut off by a plane oblique to the base: so called from its resemblance to a horse's hoof. 4 *Bot.* An unguis. [<L, hoof, <*unguis* nail]

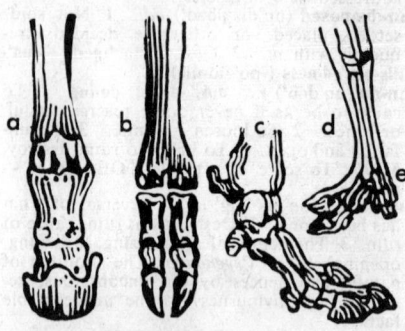

UNGULATE FEET
a. Hind foot of horse.
b. Foot of a stag.
c. Forefoot of Indian rhinoceros.
d. Side view of stag foot, showing false hoof at e.

un·gu·lar (ung′gyə·lər) *adj.* Of, pertaining to, or bearing a nail, hoof, or claw; ungual.

un·gu·late (ung′gyə·lit, -lāt) *adj.* 1 Having hoofs; hoof-shaped. 2 Designating, pertaining to, or belonging to a large division (*Ungulata*) of hoofed, herbivorous mammals, including the elephant, rhinoceros, horse, cony, hog, and all the ruminants. —*n.* A hoofed mammal. [<LL *ungulatus* <L *ungula* hoof]

un·gu·li·grade (ung′gyə·lə·grād′) *adj.* Walking on hoofs, as a horse or cow. [<L *ungula* hoof + -GRADE]

un·hair (un·hâr′) *v.t. & v.i.* To free or become free of hair, as hides by soaking and scraping.

un·hal·low (un·hal′ō) *v.t.* To profane; desecrate.

un·hal·lowed (un·hal′ōd) *adj.* 1 Left secular. 2 Not sacred. 3 Unholy; wicked.

un·halsed (un·hôlst′) *adj.* *Scot.* Not saluted or greeted.

un·hand (un·hand′) *v.t.* To remove one's hand from; release from the hand or hands; let go.

un·hand·y (un·han′dē) *adj.* 1 Inconvenient;

unhomogeneous	unidentified	unimportance	uninfested	uninstructive	uninterpolated	unjaded
unhonored	unidiomatic	unimportant	uninflammable	uninsurable	uninterpreted	unjoined
unhood	unilluminated	unimposing	uninflected	uninsured	uninterrupted	unjointed
un-Horatian	unillumined	unimpressed	uninfluenced	unintellectual	unintimidated	unjoyful
unhostile	unillustrated	unimpressible	uninfluential	unintelligibility	unintoxicated	unjudged
unhoused	unimaginable	unimpressionable	uninformed	unintelligible	uninvaded	unjustifiable
unhuman	unimaginably	unimpressive	uninfringed	unintelligibly	uninvented	unjustifiably
unhumanize	unimaginative	uninaugurated	uninhabitable	unintended	uninventive	unkempt
unhung	unimagined	unincurred	uninhabited	unintentional	uninverted	unkindled
unhurt	unimbued	uncorporated	uninhibited	unintentionally	uninvested	unkindliness
unhurtful	unimitated	uncumbered	uniniitated	uninteresting	uninvited	unkingly
unhygienic	unimpaired	unindemnified	uninjured	uninterestingly	uninviting	unkissed
unhyphenated	unimpassioned	unindicated	uninspired	unintermitted	uninvoked	unkneeled
unhyphened	unimpeded	unindorsed	uninspiring	unintermittent	uninvolved	unknightly
unideal	unimplored	uninfected	uninstructed	unintermitting	unissued	unknit

unhappy hard to handle. 2 Clumsy; lacking in manual skill. — **un·hand′i·ly** *adv.*
un·hap·py (un-hap′ē) *adj.* **·pi·er, ·pi·est** 1 Subject to conditions that prevent or destroy happiness; sad; depressed. 2 Causing or constituting misery, unrest, or dissatisfaction: *unhappy* circumstances. 3 Characterized by or exhibiting ill fortune; unfortunate; unpropitious. 4 Exhibiting lack of tact or judgment; inappropriate; inopportune. 5 *Obs.* Evil. See synonyms under BAD, SAD. — **un·hap′pi·ly** *adv.* — **un·hap′pi·ness** *n.*
un·har·bored (un-här′bərd) *adj.* 1 Having no harbor, shelter, or cover. 2 *Obs.* Not affording shelter. Also *Brit.* **un·har′boured.**
un·har·ness (un-här′nis) *v.t.* 1 To remove the harness from; unyoke; release. 2 To remove the armor from.
un·hat (un-hat′) *v.* **·hat·ted, ·hat·ting** *v.i.* To take off one's hat, especially to show respect or in worship. — *v.t.* To remove the hat from.
un·health·y (un-hel′thē) *adj.* **·health·i·er, ·health·i·est** 1 Lacking health, vigor, or wholesomeness; sickly; unsound: *unhealthy* animals or plants; also, indicating such a condition: *unhealthy* signs. 2 Loosely, insalubrious; injurious to health. 3 Morally or spiritually unsound, defective, or pernicious: *unhealthy* fiction. — **un·health′i·ly** *adv.* — **un·health′i·ness** *n.*
un·heard (un-hûrd′) *adj.* 1 Not perceived by the ear. 2 Not granted a hearing. 3 Obscure; unknown.
un·heard-of (un-hûrd′uv′, -ov′) *adj.* Not known of before; unknown or unprecedented.
un·helm (un-helm′) *v.t.* To remove the helmet or helm of. — *v.i.* To remove one's helmet.
un·hinge (un-hinj′) *v.t.* **·hinged, ·hing·ing** 1 To take from the hinges. 2 To remove the hinges of. 3 To detach; dislodge. 4 To throw into confusion; disorder. 5 To make unstable; unsettle, as the mind.
un·hitch (un-hich′) *v.t.* To unfasten.
un·ho·ly (un-hō′lē) *adj.* **·ho·li·er, ·ho·li·est** 1 Not hallowed. 2 Lacking purity; wicked; sinful. See synonyms under PROFANE, SINFUL. [OE *unhālig*] — **un·ho′li·ly** *adv.* — **un·ho′li·ness** *n.*
un·hook (un-hŏŏk′) *v.t.* 1 To remove from a hook. 2 To unfasten the hook or hooks of. — *v.i.* 3 To become unhooked.
un·hoped (un-hōpt′) *adj.* Not hoped (for); unexpected; exceeding hope: chiefly in the compound **un·hoped′-for′.**
un·horse (un-hôrs′) *v.t.* **·horsed, ·hors·ing** 1 To throw from a horse. 2 To dislodge; overthrow. 3 To remove a horse or horses from: to *unhorse* a vehicle.
un·hou·seled (un-hou′zəld) *adj.* *Obs.* Not having received the last sacraments. [<UN-¹ + HOUSEL + -ED²]
un·hur·ried (un-hûr′ēd) *adj.* Leisurely; not hurried.
un·husk (un-husk′) *v.t.* 1 To strip the husk from. 2 To expose; lay open.
uni- *combining form* One; single; one only: *unifoliate.* [<L *unus* one]
U·ni·at (yoo′nē-at) *n.* A member of any community of Eastern Christians that acknowledges the supremacy of the pope at Rome, but retains its own liturgy, ceremonies, and rites: also called **United Armenian**, **United Greek**. Compare LATIN CHURCH. — *adj.* Of the Uniats or their faith. Also **U′ni·ate** (-it, -āt). [<Russian *uniyat* <*uniya* union <L *unus* one; from being in union with the Roman Catholic Church]

u·ni·ax·i·al (yoo′nē-ak′sē-əl) *adj.* 1 Having one axis. 2 Doubly refracting and having only a single optical axis, as crystals of the tetragonal and hexagonal systems. 3 *Bot.* Unbranched, as a primary stem terminating in a flower.
u·ni·cam·er·al (yoo′nə-kam′ər-əl) *adj.* Consisting of but one chamber, as a legislature.
u·ni·cel·lu·lar (yoo′nə-sel′yə-lər) *adj. Biol.* Consisting of a single cell, as a protozoan; one-celled.
u·ni·col·or (yoo′nə-kul′ər) *adj.* Of one color.
u·ni·corn (yoo′nə-kôrn) *n.* 1 A fabulous horselike animal with one horn. 2 A two-horned animal, identified with the urus, so called in the early English versions of the Bible to render the Latin and Greek mistranslations of the Hebrew *re′ ēm*: translated as *wild ox* in the Revised Version. *Deut.* xxxiii 17. [<OF *unicorne* <L *unicornis* one-horned <*unus* one + *cornu* a horn]

UNICORN

u·ni·cos·tate (yoo′nə-kos′tāt) *adj.* 1 Having a single principal costa, rib, or nervure. 2 *Bot.* Having a midrib, as a leaf.
u·ni·cy·cle (yoo′nə-sī′kəl) *n.* A cycle or velocipede having a single wheel propelled by pedals.
un·i·de·aed (un′ī-dē′əd) *adj.* Not having ideas; frivolous.
u·ni·di·rec·tion·al (yoo′nə-di-rek′shən-əl, -dī-) *adj.* Moving in the same direction. 2 Designed or equipped to operate best in only one direction, as a radio antenna. 3 *Electr.* Of or pertaining to a direct current.
u·ni·fi·a·ble (yoo′nə-fī′ə-bəl) *adj.* That can be unified.
u·nif·ic (yoo-nif′ik) *adj.* Unifying.
unified field theory *Physics* 1 Any mathematically rigorous generalization which will combine two or more physical theories in a form permitting accurate inclusive predictions not deducible from one theory alone, as the electromagnetic theory of Maxwell. 2 Such a generalization, as tentatively formulated by Einstein, to unify the theories of electromagnetism, gravitation, and relativity.
u·ni·fi·lar (yoo′nə-fī′lər) *adj.* 1 Possessing but a single thread. 2 Utilizing only one suspending thread.
u·ni·flo·rous (yoo′nə-flôr′əs, -flō′rəs) *adj. Bot.* One-flowered.
u·ni·fo·li·ate (yoo′nə-fō′lē·it, -āt) *adj. Bot.* Having one leaf.
u·ni·fo·li·o·late (yoo′nə-fō′lē·ə-lit, -lāt) *adj. Bot.* Having a single leaflet, as the compound leaves of the orange.
u·ni·form (yoo′nə-fôrm) *adj.* 1 Being the same or alike, as in form, appearance, quantity, quality, degree, or character; not varying: *uniform* temperature. 2 Agreeing with each other; harmonious; accordant; consonant: *uniform* tastes. See synonyms under ALIKE. — *n.* A dress or suit of uniform style and appearance worn by members of the same organization, service, etc., as soldiers, sailors, postmen, etc. See synonyms under DRESS. — **dress uniform** A military or naval uniform worn at social or ceremonial events. — *v.t.*

To put into or clothe with a uniform. [<F *uniforme* <L *uniformis* <*unus* one + *forma* form] — **u′ni·form·ness** *n.* — **u′ni·form·ly** *adv.*
u·ni·for·mal·ize (yoo′nə-fôr′məl-īz) *v.t.* **·ized, ·iz·ing** To bring into a uniform system; render uniform.
Uniform Code of Military Justice The code of laws and related procedures enacted in 1951 by the U. S. Congress for the government of the personnel of the armed services: supersedes the former Army *Articles of War* and the *Articles for the Government of the Navy.*
u·ni·formed (yoo′nə-fôrmd) *adj.* Dressed in uniform.
u·ni·form·i·tar·i·an·ism (yoo′nə-fôr′mə-târ′ē-ən-iz′əm) *n. Geol.* The doctrine that essential uniformity in causes and effects, forces and phenomena, has prevailed in all ages of the world's physical history, and that the activities of the past were similar in mode and intensity to those of the present: opposed to *catastrophism.* — **u′ni·form′i·tar′i·an** *adj. & n.*
u·ni·form·i·ty (yoo′nə-fôr′mə-tē) *n. pl.* **·ties** 1 The state or quality of being uniform, or an instance of it; consistency throughout; lack of diversity. 2 Conformity or compliance, as in opinions or religion. 3 Monotony; sameness.
u·ni·fy (yoo′nə-fī) *v.t.* **·fied, ·fy·ing** To cause to be a unit; make uniform; unite; cause to be one. [<F *unifier* <LL *unificare* <L *unus* one + *facere* make] — **u′ni·fi·ca′tion** (-fə-kā′shən) *n.* — **u′ni·fi′er** *n.*
u·ni·gen·i·ture (yoo′nə-jen′ə-chər) *n.* The state of being an only child, or, in theology, of being the only begotten Son.
u·ni·ju·gate (yoo′nə-joo′gāt, -jōō′git, -gāt) *adj. Bot.* Having one pair, as of leaflets: said especially of a pinnate leaf. [<UNI- + JUGATE]
u·ni·lat·er·al (yoo′nə-lat′ər·əl) *adj.* 1 One-sided; relating to one side only; made, undertaken, done, or signed by only one of two or more people or parties. 2 *Law* Binding or obligatory on one party only. 3 Arranged or growing on one side only, as a plant or animal organ. 4 *Med.* Affecting but one side of the body. 5 Relating to or tracing ancestry on one side only.
u·ni·lit·er·al (yoo′nə-lit′ər·əl) *adj.* Comprising but one letter.
u·ni·loc·u·lar (yoo′nə-lok′yə-lər) *adj. Biol.* Having or consisting of one cell or chamber, as an anther, ovary, etc.
U·ni·mak Island (yoo′nə-mak) The most northeasterly of the Fox Islands in the NE Aleutian Islands; 70 miles long.
un·im·peach·a·ble (un′im-pē′chə-bəl) *adj.* Not to be called in question as regards truth, honesty, etc.; faultless; blameless. — **un′im·peach′a·bly** *adv.*
un·im·proved (un′im-proovd′) *adj.* 1 Not improved; not bettered or advanced: *unimproved* health. 2 Having no improvements; not cleared, cultivated, or built upon: *unimproved* land. 3 Not made anything of; unused: *unimproved* opportunities. 4 *Obs.* Not proved or tried.
un·in·gen·ious (un′in-jēn′yəs) *adj.* 1 Lacking ingenuity; not possessed of inventiveness. 2 *Obs.* Uningenuous.
un·in·gen·u·ous (un′in-jen′yoo·əs) *adj.* Not ingenuous; sly or designing.
un·in·tel·li·gent (un′in-tel′ə·jənt) *adj.* 1 Not intelligent; characterized by lack of intelligence. 2 Unwise; ignorant. — **un′in·tel′li·gence** *n.*
un·in·ter·est·ed (un·in′tər·is·tid, -tris-) *adj.* 1 Having no interest in, as in property.

unknowable	unlevel	unlocated	unmangled	unmaternal	unmentionability	unmodified
unknowing	unlevied	unlocked	unmanifested	unmatted	unmentioned	unmodish
unlabeled,	unlibidinous	unlovable	unmanipulated	unmatured	unmercenary	unmoistened
unlabelled	unlicensed	unloved	unmannerly	unmeant	unmerchantable	unmold
unladylike	unlifelike	unloveliness	unmannish	unmeasurable	unmerited	unmolested
unlamented	unlighted	unloverlike	unmannishly	unmeasured	unmethodical	unmollified
unlash	unlikable,	unloving	unmanufacturable	unmechanical	unmilitary	unmolten
unlashed	unlikeable	unlubricated	unmanufactured	unmedicated	unmilled	unmortgaged
unlaundered	unlined	unlying	unmarketable	unmeditated	unmingle	unmotivated
unleased	unlink	unmagnified	unmarred	unmelodious	unmingled	unmounted
unled	unliquefiable	unmaidenly	unmarriageable	unmelted	unmirthful	unmourned
unlessened	unliquefied	unmailable	unmarried	unmenaced	unmistaken	unmovable
unlessoned	unlit	unmalleable	unmastered	unmendable	unmitigable	unmoved
unlet	unliveliness	unmanageable	unmatched	unmended	unmixed,	unmown
unletted	unlively	unmanful	unmated	unmensurable	unmixt	unmurmuring

add, āce, câre, päm; end, ēven; it, īce; odd, ōpen, ôrder; tŏŏk, pōōl; up, bûrn; ə = a in *above*, e in *sicken*, i in *clarity*, o in *melon*, u in *focus*; yōō = u in *fuse*; oi, oil; ou, pout; ch, check; g, go; ng, ring; th, thin; th, this; zh, vision. Foreign sounds å, œ, ü, kh, ñ; and ◆: see page xx. <from; + plus; ? possibly.

2 Taking no interest in; indifferent; unconcerned.

un·ion (yōōn′yən) *n.* **1** The act of uniting, or the state of being united; a joining; coalescence; junction. **2** That which is constituted as one by the combination of elements previously separate; a coalition; confederation; league. **3** A combination of co-laborers for the joint and mutual protection of their common interests. See TRADE UNION. **4** *Brit.* An amalgamation of parishes for administration of poor relief; also, a workhouse administered by such a union. **5** Agreement in sentiment or action; harmony; concord; unanimity. **6** The joining of two persons in marriage, or the resulting state of wedlock. **7** A device emblematic of union borne in the canton of a flag, as the three crosses in a British ensign; the canton itself containing such device, sometimes used separately as a flag, as the blue canton with white stars in the flag of the United States, and the Union Jack of Great Britain. **8** A coupling or connection for pipes or rods. **9** A fabric made of two or more materials, as cotton and wool. **10** *Obs.* A pearl of extraordinary worth. — *adj.* Of, pertaining to, or adhering to a union, particularly a political or trade union. [< F < LL *unio, -onis* < L *unus* one. Doublet of ONION.]

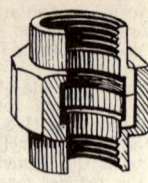

PIPE UNION

Synonyms (noun): coalition, combination, conjunction, junction, juncture, oneness, unification, unity. *Unity* is oneness, the state of existing as essentially one, especially of that which never has been divided or of that which cannot be conceived of as resolved into parts; as, the *unity* of the human soul. *Union* is a bringing together of things that have been distinct, so that they combine or coalesce to form a new whole, or the state or condition of things thus brought together; in a *union* the separate individuality of the things united is never lost sight of; we speak of the *union* of the parts of a fractured bone. See ALLIANCE, ASSOCIATION, ATTACHMENT, HARMONY, MARRIAGE. *Antonyms*: analysis, contrariety, decomposition, disconnection, disjunction, dissociation, disunion, division, divorce, separation, severance.

Un·ion (yōōn′yən) *n.* **1** The United States regarded as a national unit: with *the.* **2** The Union of South Africa. — *adj.* Of, pertaining to, or loyal to the United States, especially the Federal government during the Civil War: a *Union* soldier; He was *Union* to the core.

union card 1 A card certifying that the person named belongs to a certain labor union. **2** A card certifying that the shop named hires only union labor.

union down Reversed, as a flag, so as to have the union or canton at the lower edge: a signal of distress.

Union Islands See TOKELAU ISLANDS.

un·ion·ism (yōōn′yən-iz′əm) *n.* **1** The principle of combination for unity of purpose and action. **2** Trade-unionism. **3** Adherence to or advocacy of political union between states, as opposed to secession. **4** Advocacy of or adherence to the principles of the Unionists of Great Britain who opposed Home Rule for Ireland. — **un′ion·is′tic** *adj.*

un·ion·ist (yōōn′yən-ist) *n.* **1** An advocate of union or unionism. **2** A member of a trade union.

Un·ion·ist (yōōn′yən-ist) *n.* **1** One who before and during the Civil War in the United States supported the Union cause and opposed secession; a Union man. **2** One of those opposed to loosening the formal ties between Great Britain and Ireland, whether belonging to the Conservatives or the branch of the Liberals (**Liberal Unionists**) that separated from their party in 1886 in opposition to the advocates of Home Rule for Ireland. From this time the term *Unionists* began to come into use, at first to signify both the Conservative and the Liberal Unionist parties, and later, as the distinction between the two wings gradually grew smaller, the Conservative Party.

un·ion·ize (yōōn′yən-īz) *v.* **·ized, ·iz·ing** *v.t.* To cause to join, or to organize into a union, especially a trade union. — *v.i.* To become a member of or organize a trade union. — **un′ion·i·za′tion** *n.*

union jack A flag consisting of the union or canton only.

Union Jack The British national flag. It is a combination of the flags of England, Scotland, and Ireland.

Union of South Africa See SOUTH AFRICA, UNION OF.

Union of Soviet Socialist Republics A federal union of 16 constituent republics of eastern Europe and northern Asia, extending from the Arctic Ocean to the Black Sea and east to the Pacific; 8,646,400 square miles; capital, Moscow: formerly *Russia*: also *Soviet Russia, Soviet Union.* Abbr. *U.S.S.R., USSR.*

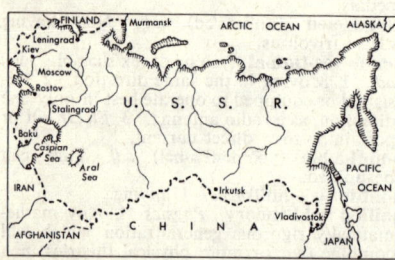

union shop An industrial establishment in which only members of a trade union are employed.

union station A railroad station or depot used by two or more railroad lines.

union suit A one-piece undergarment consisting of shirt and drawers.

Union Territory In India, under the provisions of the States Reorganization Act, which came into effect November 1, 1956, a division of India administered by the central government, rather than governing itself as a State. The six Union Territories of India are the Andaman and Nicobar Islands, Delhi, Himachal Pradesh, the Laccadive Islands, Manipura, and Tripura; Pondicherry is also temporarily administered as a Union Territory.

u·ni·pa·ren·tal (yōō′nə·pə·ren′təl) *adj.* Having or produced by one parent only; asexual.

u·nip·a·rous (yōō·nip′ər·əs) *adj.* **1** *Bot.* Having but one axis or stem. **2** Bringing forth but one offspring at a time, or not having borne more than one. [< UNI- + -PAROUS]

u·ni·per·son·al (yōō′nə·pûr′sən·əl) *adj.* **1** Manifested or existing in but one person. **2** *Gram.* Used in only one person, especially the third person singular; impersonal.

u·ni·pet·al·ous (yōō′nə·pet′l·əs) *adj. Bot.* Having only one petal.

u·ni·pla·nar (yōō′nə·plā′nər) *adj.* Lying or taking place in one plane.

u·ni·po·lar (yōō′nə·pō′lər) *adj.* **1** *Physics* Showing only one kind of polarity. **2** *Anat.* Having, or operating by means of, one pole: said especially of nerve cells having only one process.

u·nique (yōō·nēk′) *adj.* **1** Being the only one of its kind; being without equal; singular; uncommon; rare. **2** Not complicated with other things. **3** Sole; single. See synonyms under ODD, QUEER, RARE. [< F < L *unicus* < *unus* one] — **u·nique′ly** *adv.* — **u·nique′ness** *n.*

u·ni·sep·tate (yōō′nə·sep′tāt) *adj.* Having a single septum or partition.

u·ni·sex·u·al (yōō′nə·sek′shōō·əl) *adj.* Of one sex: specifically said of flowers and animals having one kind of sexual organs only.

u·ni·son (yōō′nə·sən, -zən) *n.* **1** A condition of perfect agreement and accord; harmony. **2** *Music* A state in which the instruments or voices perform the same part; unity of pitch; also, the interval of one or more octaves. See synonyms under HARMONY, MELODY. [< L *unisonus* having a single sound < *uni-* one + *sonus* a sound]

u·nis·o·nal (yōō·nis′ə·nəl) *adj.* Being in unison. Also **u·nis′o·nant.** — **u·nis′o·nal·ly** *adv.*

u·nis·o·nance (yōō·nis′ə·nəns) *n.* Accordance in sound or pitch.

u·nis·o·nous (yōō·nis′ə·nəs) *adj.* **1** Unisonal. **2** Sounding alone. [< Med. L *unisonus*]

u·nit (yōō′nit) *n.* **1** A single person or thing regarded as an individual or belonging to an entire group. **2** A body or group considered as a single whole among a plurality of similars. **3** A standard quantity with which others of the same kind are compared for purposes of measurement and in terms of which their magnitude is stated. **4** *Math.* A quantity whose measure is represented by the number 1; a least whole number; specifically, in arithmetic, the number 1 itself; unity. **5** *Med.* The quantity of a drug, vaccine, serum, or antigen required to produce a given effect or yield a particular result. **6** A fundamental quantity used in calculating how much scholastic work a school or college student has finished. [Short for UNITY]

unit angle A radian.

u·ni·tar·i·an (yōō′nə·târ′ē·ən) *n.* One who rejects the doctrine of the Trinity; a non-Trinitarian monotheist. — *adj.* Pertaining to a unit. [< NL *unitarius* unitary]

U·ni·tar·i·an (yōō′nə·târ′ē·ən) *n.* A member of a Protestant denomination which rejects the doctrine of the Trinity, but accepts the ethical teachings of Jesus and emphasizes complete freedom of religious opinion, the importance of personal character, and the independence of each local congregation. — *adj.* Pertaining to the Unitarians, or to their teachings.

u·ni·tar·i·an·ism (yōō′nə·târ′ē·ən·iz′əm) *n.* Any unitary system.

U·ni·tar·i·an·ism (yōō′nə·târ′ē·ən·iz′əm) *n.* *Theol.* The doctrine of the Unitarians.

u·ni·tar·y (yōō′nə·ter′ē) *adj.* **1** Pertaining to a unit; characterized by, based on, or pertaining to unity. **2** Having the nature of a unit; whole.

unit character *Genetics* One of two or more contrasting characters which is transmitted as a unit and without modification.

u·nite (yōō·nīt′) *v.* **·nit·ed, ·nit·ing** *v.t.* **1** To join together so as to form a whole; combine; compound. **2** To bring into close connection, as by legal, physical, marital, social, or other tie; join in action, interest, etc. **3** To attach permanently or solidly; cause to

unmusical	unnoted	unobtruding	unornamental	unparental	unperceivable	unphonetic
unmuzzle	unnoticeable	unobtrusive	unornate	unperceived	unpicked
unmuzzled	unnoticed	unobtrusiveness	unorthodox	unparted	unperceiving	unpicturesque
unmystified	unnurtured	unoccasioned	unorthodoxy	unpartisan,	unperfected	unpierced
unnail	unobjectionable	unoffended	unpasteurized	unperformed	unpile
unnamable,	unobliged	unoffending	unowned	unpatched	unperplexed	unpitying
unnameable	unobliging	unoffensive	unoxidized	unpatented	unpersuadable	unplaced
unnamed	unobnoxious	unoffered	unpacified	unpatriotic	unpersuaded	unplagued
unnaturalized	unobscured	unofficious	unpacker	unpaved	unpersuasive	unplait
unnavigable	unobservant	unoiled	unpainful	unpeaceable	unperturbed	unplanned
unnavigated	unobserved	unopen	unpalatable	unpeaceful	unperused	unplanted
unneeded	unobserving	unopened	unpalatably	unpedigreed	unphilanthropic	unplayed
unneedful	unobstructed	unopposed	unpardonable	unpen	unphilological	unpleasing
unnegotiable	unobtainable	unoppressed	unpardonably	unpenetrated	unphilosophic	unpledged
unneighborly	unobtained	unordained	unpardoned	unpensioned	unphilosophical	unpliable

adhere; combine. — v.i. **4** To become or be merged into one; be consolidated; combine. **5** To join together for action; act in conjunction; concur. [< LL *unitus*, pp. of *unire* make one < L *unus* one]
Synonyms: amalgamate, associate, attach, blend, cement, coalesce, cohere, combine, compound, conjoin, connect, consolidate, fuse, incorporate, join, link, merge. See MIX. Compare ADD, COMPLEX. *Antonyms:* analyze, decompose, disconnect, disjoin, disrupt, dissever, dissociate, dissolve, disunite, divide, resolve, separate, sever, sunder.
u·nit·ed (yōō·nī′tid) *adj.* Incorporated into one; allied; combined; harmonious. — **u·nit′ed·ly** *adv.* — **u·nit′ed·ness** *n.*
United Arab Republic An integral republic formed in 1958 by the merger of the non-contiguous former republics of Egypt and Syria which became its constituent provinces; 458,432 square miles; capital, Cairo.
United Arab States A federation formed in 1958 by the United Arab Republic and the Kingdom of Yemen, each of which retains its sovereignty; 533,432 square miles; capitals, Cairo, Sanaa, and Tai′z.
United Armenian A Uniat.
United Brethren 1 Originally, the Moravians: also **Old United Brethren. 2** A Christian denomination founded in 1800 by P. W. Otterbein and Martin Boehm: officially, **United Brethren in Christ.**
United Church of Christ A Protestant denomination formed in 1957 by a union of the Congregational Christian Churches and the Evangelical and Reformed Church.
United Greek A Uniat.
United Kingdom 1 The kingdom of the British Isles, comprising Great Britain, Northern Ireland, the Isle of Man, and the Channel Islands; 94,284 square miles; capital, London: officially **United Kingdom of Great Britain and Northern Ireland. 2** Formerly, Great Britain and Ireland (1801–1922).
United Nations 1 A coalition of 26 national states formed on January 2, 1942 to resist the military aggression of Germany, Italy, Japan, and other Axis powers in World War II. **2** An international organization of sovereign states (originally called the **United Nations Organization**) created by the United Nations Charter drafted in September-October, 1944, at Dumbarton Oaks and adopted at San Francisco in May and June, 1945: the 26 states of the United Nations coalition and 25 other states constitute the original membership.
Original Members: Argentina, Australia, Belgium, Belorussian S.S.R., Bolivia, Brazil, Canada; Chile, China, Colombia, Costa Rica, Cuba, Czechoslovakia, Denmark, Dominican Republic, Ecuador, Egypt, El Salvador, Ethiopia, France, Greece, Guatemala, Haiti, Honduras, India, Iraq, Lebanon, Liberia, Luxembourg, Mexico, Netherlands, New Zealand, Nicaragua, Norway, Panama, Paraguay, Persia, Peru, Philippines, Poland, Saudi Arabia, Syria, Turkey, Ukrainian S.S.R., Union of South Africa, United Kingdom, United States, Uruguay, U.S.S.R., Venezuela, Yugoslavia. Other nations joined in subsequent years.
1946 Afghanistan, Iceland, Sweden, Thailand.
1947 Pakistan, Yemen.
1948 Burma.
1949 Israel.
1950 Indonesia.
1955 Albania, Austria, Bulgaria, Cambodia,
Ceylon, Finland, Hungary, Irish Republic, Italy, Jordan, Laos, Libya, Nepal, Portugal, Rumania, Spain.
1956 Japan, Morocco, Sudan, Tunisia.
1957 Ghana, Malaya.
1958 United Arab Republic, Guinea.
1960 Cameroun, Central African Republic, Chad, Congo Republic, Cyprus, Dahomey, Gabon Republic, Ivory Coast, Malagasy Republic, Mali, Niger, Nigeria, Republic of the Congo, Senegal, Somalia, Togo, and Republic of the Upper Volta.
Also called **UN.**
United Nations Trust Territory See TRUST TERRITORY.
United Presbyterian Church in the United States of America The largest U. S. Presbyterian body, formed in 1958 by the merger of two separate churches.
United Press A news-collecting and -distributing organization, merged in 1958 with the International News Service to form the **United Press International.**
United Provinces of Agra and Oudh A former British province in north central India; became the State of Uttar Pradesh, 1950.
United Society of Believers in Christ's Second Appearing See SHAKER.
United States The United States of America. — **to talk United States** To speak out forcefully; speak in a manner that is supposedly characteristically American.
United States Army, Navy, Air Force, etc. See under ARMY, NAVY, AIR FORCE, etc.
United States of America A federal republic of North America, including 50 states, and the District of Columbia, the Canal Zone, Puerto Rico, the Virgin Islands of the United States, American Samoa, Guam, Wake, and several other scattered islands of the Pacific; total area, about 3,720,407 square miles; capital, Washington, in the District of Columbia: also *America, the States, United States:* abbr. *U.S.A., U.S., US.*

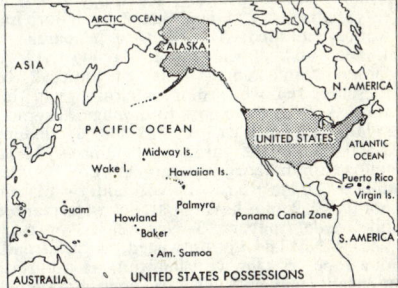

United States Office of Education A branch of the Department of Health, Education, and Welfare which directs activities relating to the promotion of education in all fields.
unit factor *Biol.* A gene controlling the inheritance of a unit character.
unit fraction See under FRACTION.
u·ni·tive (yōō′nə·tiv) *adj.* Productive of or promoting union; having power to unite; characterized by union.
u·ni·tize (yōō′nə·tīz) *v.t.* ·tized, ·tiz·ing To form into a whole; unify.
unit modifier A conventional or improvised compound used adjectively before a substantive. Examples: *blue-green* algae, *bitter-sweet* chocolate, *suit-coat* pattern, *situation-comedy* plot, *storm-window* installation, *contour-plowing* method, *most-favored-nation* clause.
◆ The use of the hyphen in the unit modifier is to avoid ambiguity in a word sequence where the relationship is not immediately apparent from context: The house had faded red-brick walls (faded walls of red brick, *not* faded red walls of brick). The hyphen here is to be considered a nonce use and not as a spelling form or variant.
unit rule *U.S.* A rule in a national convention of the Democratic party, requiring that, if so instructed by a State party convention, the vote of an entire delegation shall be determined by a majority of its members.
u·ni·ty (yōō′nə·tē) *n. pl.* ·ties **1** The state, property, or product of being united, physically, socially, or morally; oneness: opposed to *division, plurality.* **2** Union, as of constituent parts or elements: national *unity.* **3** Agreement of parts; harmonious adjustment of constituent elements; sameness of character: the *unity* of two writings. **4** The fact of something's being a whole that is more than or different from its parts or their sum. **5** Singleness of purpose or action. **6** A state of general good feeling; mutual understanding; concord: brethren dwelling together in *unity.* **7** *Math.* The number one; the ratio of two equal quantities; by extension, an operator that leaves unchanged the quantity or thing on which it operates, as does the number one when used as a factor. **8** In literature and the arts, combination into a homogeneous artistic whole, exhibiting oneness of purpose, thought, spirit, and style, with subordination of all parts to the general effect. **9** In the drama, observance, complete or partial, of the law of dramatic unities. **10** Identity. See synonyms under HARMONY, UNION. — **the law of dramatic unities** The law of Aristotle that in a drama there must be unity of action, unity of time, and unity of place. These unities were strictly observed by the French classical dramatists of the 17th century, but were violated by Shakespeare and certain of the German playwrights. [< F *unité* < L *unitas* < *unus* one]
U·ni·ty (yōō′nə·tē) *n.* A religious philosophy, based on the teachings of Jesus Christ, stating that man is inseparable from the spirit of God within him, and that through prayerful realization of this spirit he can obtain healing of all life's inharmonies in mind, body, etc.: founded in 1889.
Unity of the Brethren See MORAVIAN.
u·ni·va·lent (yōō′nə·vā′lənt) *adj.* **1** *Chem.* Having a valence or combining value of one; monovalent. **2** *Biol.* Pertaining to or designating a single unpaired chromosome. — **u′ni·va′lence, u′ni·va′len·cy** *n.*
u·ni·valve (yōō′nə·valv′) *adj.* Having only one valve, as a mollusk. Also **u′ni·val′vate, u′ni·valved′.** — *n.* **1** A mollusk having a univalve shell; a gastropod. **2** A shell of a single piece. — **u′ni·val′vu·lar** (-val′vyə·lər) *adj.*
u·ni·ver·sal (yōō′nə·vûr′səl) *adj.* **1** Prevalent or common everywhere or among all things or persons specified or implied: a *universal* belief; *universal* suffrage; a *universal* language. **2** Of or including everyone: a *universal* church. **3** Applicable to all cases: a *universal* law; a *universal* cure. **4** Accomplished or interested in a vast variety of subjects or activities: Leonardo da Vinci was a *universal* genius. **5** Of, pertaining to, or occurring throughout the universe: a *universal* being. **6** *Mech.* Adapted or adaptable to a great variety of uses, shapes, etc., as certain machines or machine parts. **7** *Logic* **a** Including all the

unpliant	unposted	unprimed	unpropitiable	unpunctual	unquenched	unreasoningly
unplighted	unpractical	unprincely	unpropitiated	unpunishable	unquestioning	unrebukable
unplowed,	unpredictable	unprinted	unpropitious	unpunished	unquotable	unrebuked
unploughed	unpreoccupied	unprivileged	unproportionate	unpurchasable	unraised	unreceipted
unplucked	unprepossessing	unprized	unproportioned	unpure	unransomed	unreceivable
unplugged	unprescribed	unprobed	unproposed	unpurged	unrated	unreceived
unpoetic	unpresentable	unprocessed	unprosperous	unpurified	unratified	unreceptive
unpoetical	unpreserved	unprocurable	unprotected	unpurposed	unravaged	unreciprocated
unpointed	unpressed	unprofaned	unproved	unpursuing	unrazored	unreclaimable
unpolarized	unpresumptuous	unprofited	unproven	unpuzzle	unreachable	unreclaimed
unpolished	unpretending	unprogressive	unprovoked	unquaffed	unreached	unrecognizable
unpolitical	unpretentious	unprohibited	unprovoking	unquailing	unreadable	unrecognized
unpolluted	unprevailing	unpromising	unpruned	unqualifying	unrealizable	unrecommended
unpondered	unpreventable	unprompted	unpublished	unquelled	unrealized	unrecompensed
unpopulated	unprevented	unpronounced	unpucker	unquenchable	unreasoned	unreconcilable

add, āce, câre, pälm; end, ēven; it, īce; odd, ōpen, ôrder; tŏŏk, pōōl; up, bûrn; ə = a in *above*, e in *sicken*, i in *clarity*, o in *melon*, u in *focus*; yōō = u in *fuse*; oi, oil; ou, pout; ch, check; g, go; ng, ring; th, thin; ṯẖ, this; zh, vision. Foreign sounds à, œ, ü, kh, ṅ; and ◆: see page xx. < from; + plus; ? possibly.

individuals of a class or genus; generic. **b** In a proposition, predicable of all the individuals denoted by the subject: opposed to particular: "All men are mortal" is a *universal* proposition. See synonyms under COMMON, GENERAL. —*n.* **1** *Logic* **a** A universal proposition. **b** One of the five predicables, that is, genus, species, difference, property, and accident, known collectively as the *universals.* **c** A general or abstract concept considered as having absolute reality or mental or nominal existence. **2** Any general or universal notion or idea. **3** A metaphysical being which preserves its identity in spite of the changes through which it passes, as the ego. [<OF <L *universalis* <*universus.* See UNIVERSE.] —**u'ni·ver'sal·ly** *adv.* —**u'ni·ver'sal·ness** *n.*

universal developer *Phot.* A developer adapted for use with various types of films, plates, or papers.

universal donor One whose blood belongs to group O and may be transfused with little or no danger of agglutination into a person belonging to any of the four blood groups.

U·ni·ver·sal·ism (yōō'nə·vûr'səl·iz'əm) *n. Theol.* The doctrine that all souls will finally be saved and that good will triumph universally.

U·ni·ver·sal·ist (yōō'nə·vûr'səl·ist) *adj.* Pertaining to Universalism or Universalists. —*n.* A believer in the doctrines of Universalism, or a member of the Universalist denomination.

u·ni·ver·sal·i·ty (yōō'nə·vər·sal'ə·tē) *n.* **1** The state of being all-embracing. **2** Unrestricted fitness or adaptability.

u·ni·ver·sal·ize (yōō'nə·vûr'səl·īz) *v.t.* **·ized, ·iz·ing** To make universal.

universal joint *Mech.* A joint that permits both connected parts of a machine to be turned in any direction within definite limits; specifically, a coupling for connecting two shafts, etc., so as to permit angular motion in all directions. Also **universal coupling.**

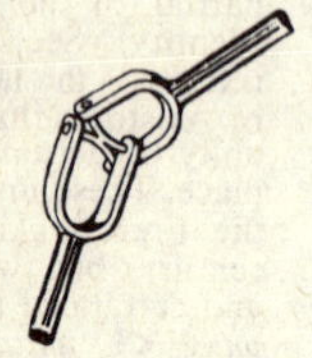

UNIVERSAL JOINT

u·ni·verse (yōō'nə·vûrs) *n.* **1** The aggregate of all existing things; the whole creation embracing all celestial bodies and all of space; the cosmos. **2** In restricted sense, the earth. **3** Human beings collectively; mankind. **4** *Logic* All objects, collectively, that are the subjects of consideration at once: also **universe of discourse.** **5** *Stat.* All the instances in a given class: contrasted with *sample.* [<F *univers* <L *universum,* neut. of *universus* turned, combined into one, all collectively <*unus* one + *versus,* pp. of *vertere* turn]

u·ni·ver·si·ty (yōō'nə·vûr'sə·tē) *n. pl.* **·ties** **1** An educational institution for higher instruction or for the examination of students already instructed. Universities arose in Europe in the Middle Ages and were first essentially ecclesiastic. Their functions gradually became specialized, some dividing into several *faculties,* each of which took charge of some one great branch of instruction, or into *colleges,* as now in the older English universities, where the relation of the university to the college is similar to that of a federal government to its component states. In the United States the word has been used loosely, chiefly to mean a collection of educational associations including a college (which offers degrees in general subjects) and several more advanced and specialized faculties, either professional, as law, medicine, etc., or academic, as history or mathematics. **2** All the students of such an institution. **3** *Brit. Colloq.* A university team or crew.

u·ni·ver·son (yōō'nə·vûr'son) *n. Physics* A hypothetical primordial entity originally containing the entire mass of the universe and whose splitting simultaneously gave rise to the cosmos and to its balanced opposite, the anticosmos. [<UNIVERSE]

u·niv·o·cal (yōō·niv'ə·kəl) *adj.* Having but one proper sense or meaning. —*n.* A word that has but one meaning. [<LL *univocus* <L *unus* one + *vox* voice]

un·just (un·just') *adj.* **1** Not legitimate, fair, or just; wrongful. **2** Unrighteous; acting contrary to right and justice. **3** *Archaic* Faithless. **4** *Archaic* Dishonest. —**un·just'ly** *adv.* —**un·just'ness** *n.*

un·kempt (un·kempt') *adj.* **1** Not combed; neglected; untidy. **2** Without polish; rough. [<UN-[1] + *kempt* combed, pp. of dial. *kemb,* var. of COMB] —**un·kempt'ness** *n.*

un·kenned (un·kend') *adj. Scot. & Brit. Dial.* Unknown. Also **un·kend', un·kent'** (-kent').

un·ken·nel (un·ken'əl) *v.t.* **1** To drive or release from a kennel or lair. **2** To bring to light; disclose.

un·kind (un·kīnd') *adj.* Showing lack of kindness; unsympathetic; harsh; cruel. [OE *uncynde* strange, unnatural] —**un·kind'ly** *adv.* —**un·kind'ness** *n.*

un·known (un·nōn') *adj.* **1** Not known; not apprehended mentally; not recognized, as a fact or person. **2** Not ascertained; incalculable. See synonyms under MYSTERIOUS, SECRET. —*n.* An unknown person or quantity. —**the Great Unknown** Life after death; future life.

Unknown Soldier One of the unidentified dead of World War I who is honored as a symbol of all his compatriots who died in action; extended to include unknown dead of World War II and the Korean conflict.

un·la·bored (un·lā'bərd) *adj.* **1** Produced without strain or effort; seemingly free and easy; natural. **2** Uncultivated by labor; unworked; untilled. Also *Brit.* **un·la'boured.**

un·lace (un·lās') *v.t.* **·laced, ·lac·ing** **1** To loosen or unfasten the lacing of; untie. **2** To loosen or remove (armor or clothing) in this way. **3** *Obs.* To expose to damage; disgrace.

un·lade (un·lād') *v.t. & v.i.* **·lad·ed, ·lad·ing** **1** To unload the cargo of (a ship). **2** To unload or discharge (cargo, etc.).

un·laid (un·lād') *adj.* **1** Not laid or placed; not fixed. **2** Not having parallel watermarked lines: *unlaid* paper. **3** Not allayed or pacified. **4** Not laid up; untwisted, as the strands of a rope. **5** *Obs.* Not laid out, as a corpse.

un·latch (un·lach') *v.t.* To open or unlock by releasing the latch. —*v.i.* To come open or unlocked.

un·law·ful (un·lô'fəl) *adj.* Contrary to or in violation of law; illegal; illicit; also, illegitimate. See synonyms under CRIMINAL. —**un·law'ful·ly** *adv.* —**un·law'ful·ness** *n.*

un·lay (un·lā') *v.t. & v.i.* **·laid, ·lay·ing** To untwist: said specifically of the strands of a rope. [<UN-[2] + LAY[1] (def. 21)]

un·lead (un·led') *v.t.* **1** To strip of lead. **2** *Printing* To remove the leads from between (lines of type matter).

un·lead·ed (un·led'id) *adj.* **1** Not supplied or weighted with lead. **2** *Printing* Having no leads between the lines; not spaced with leads.

un·learn (un·lûrn') *v.t.* **·learned** or **·learnt, ·learn·ing** To dismiss from the mind (something learned); forget.

un·learn·ed (un·lûr'nid) *adj.* **1** Not learned; not possessed of or characterized by learning; illiterate; ignorant; untaught. **2** Unworthy or unsuggestive of a scholar; not like the production of a learned man. **3** (un·lûrnd') Not acquired by learning or study. See synonyms under IGNORANT.

un·leash (un·lēsh') *v.t.* To set free from or as from a leash.

un·leav·ened (un·lev'ənd) *adj.* Not leavened: said specifically of the bread used at the feast of the Passover.

un·less (un·les') *conj.* **1** If it be not a fact that; supposing that . . . not; except that: *Unless* we persevere, we shall lose. **2** *Obs.* For fear that; lest. —*prep.* Save; except; excepting: with an implied verb: *Unless* a miracle, he'll not be back in time. See synonyms under BUT. [Earlier *onlesse (that)* in a less case <ON + LESS]

un·let·tered (un·let'ərd) *adj.* Not educated; not lettered; illiterate.

un·like (un·līk') *adj.* Having little or no resemblance; different. See synonyms under ALIEN, CONTRARY, HETEROGENEOUS. —*adv.* In another manner: with *to* expressed or implied. By the ellipsis of *to* it approaches prepositional use. [ME *unliche*] —**un·like'·ness** *n.*

un·like·ly (un·līk'lē) *adj.* **1** Improbable. **2** Not inviting or promising success. —*adv.* Improbably. —**un·like'li·ness, un·like'li·hood** *n.*

un·lim·ber (un·lim'bər) *v.t. & v.i.* To disconnect (a gun or caisson) from its limber; prepare for action.

un·lim·it·ed (un·lim'it·id) *adj.* **1** Having no limits in space, number, or time; unbounded; endless; unnumbered. **2** Not limited by restrictions; unconfined: *unlimited* authority. **3** Not limited by exceptions or qualifications; undefined. —**un·lim'it·ed·ly** *adv.* —**un·lim'it·ed·ness** *n.*

un·liq·ui·dat·ed (un·lik'wə·dā'tid) *adj.* **1** *Law* Unascertained as to amount; undetermined or not settled: *unliquidated* damages. **2** Not yet eliminated from the living: said of the survivors of attempted genocide.

un·list·ed (un·lis'tid) *adj.* Not listed; specifically, noting stocks quoted in the unlisted department of a stock exchange, but not admitted to dealings on the floor of the New York Stock Exchange.

un·live (un·liv') *v.t.* **·lived, ·liv·ing** To live so as to wipe out the effects of (a former period of life); undo by living; live down.

un·load (un·lōd') *v.t.* **1** To remove the load or cargo from. **2** To take off or discharge (cargo, etc.). **3** To relieve of something burdensome or oppressive. **4** To withdraw the charge of ammunition from. **5** *Colloq.* To dispose of, especially by selling in large quantities. —*v.i.* **6** To discharge freight, cargo, or other burden.

un·load·er (un·lō'dər) *n.* One who or that which unloads; specifically, a contrivance for unloading something, as hay or coal.

un·lock (un·lok') *v.t.* **1** To unfasten (something locked). **2** To open or undo; release. **3** To lay open; reveal or disclose. —*v.i.* **4** To become unlocked.

un·looked-for (un·lo͝okt'fôr') *adj.* Not anticipated; unexpected.

un·loose (un·lo͞os') *v.t.* **·loosed, ·loos·ing** To release from fastenings; set loose or free. See synonyms under RELEASE.

un·loos·en (un·lo͞o'sən) *v.t.* To loose; unloose.

un·love·ly (un·luv'lē) *adj.* Unattractive; disagreeable; ugly.

un·luck·y (un·luk'ē) *adj.* **·luck·i·er, ·luck·i·est** **1** Not favored by luck; unfortunate. **2** Resulting in or attended by ill luck; causing misfortune; disastrous. **3** Ill-omened; inauspicious: an *unlucky* day. See synonyms under

unreconciled
unrecorded
unrecounted
unrecoverable
unrecruited
unrectified
unredeemed
unredressed
unreelable
unrefined
unreflected
unreflecting
unreformable
unreformed
unreformedness
unrefreshed
unrefreshing
unregarded
unregistered
unregretted
unregulated
unrehearsed
unrelated
unrelatedness
unrelaxed
unrelaxing
unrelievable
unrelieved
unrelished
unremarked
unremedied
unremembered
unremittable
unremitted
unremorseful
unremovable
unremoved
unremunerated
unremunerative
unrendered
unrenewed
unrenounced
unrenowned
unrent
unrented
unrepaid
unrepairable
unrepaired
unrepealed
unrepentant
unrepented
unrepenting
unrepining
unreplaced
unreplenished
unreported
unrepresentative
unrepresented
unrepressed
unreprievable
unreprieved
unreprovable
unrequested
unrequited
unresented
unresigned
unresistant
unresisted
unresisting
unresistingly
unresolved
unrespectable
unrespectful
unrested
unresting
unrestrainable
unrestraint
unrestricted
unretarded
unretentive
unretracted
unretrieved
unreturned
unrevealed
unrevenged
unreversed
unrevised
unrevoked
unrewarded
unrhetorical
unrhymed
unrhythmic
unrhythmical
unrighted
unrightful
unrimed
unripened
unrisen
unroasted
unrobe
unromantic
unromantically
unroof
unrough
unruled

un·make (un-māk') v.t. ·made, ·mak·ing 1 To reverse the making of; reduce to the original condition or form. 2 To ruin; destroy. 3 To depose, as from a position of authority.

un·man (un-man') v.t. ·manned, ·man·ning 1 To cause to lose courage or fortitude; dishearten. 2 To render unmanly or effeminate. 3 To deprive of virility; emasculate; castrate. 4 To remove the men from, as a ship or fortress.

un·man·ly (un-man'lē) adj. 1 Not masculine; effeminate; not virile; not courageous. 2 Not gentlemanly; not honorable. —**un·man'li·ness** n.

un·manned (un-mand') adj. 1 Not manned. 2 Deprived of virility or manhood. 3 Obs. Unaccustomed to men; untamed: said of hawks. 4 Uninhabited.

un·man·ner·ly (un-man'ər-lē) adj. Lacking manners; rude. —adv. Impolitely. —**un·man'ner·li·ness** n.

un·marked (un-märkt') adj. 1 Bearing no mark; having no distinctive mark. 2 Electr. Denoting that pole of a magnet which points south. 3 Not noticed. 4 Not examined; hence, uncorrected; ungraded: unmarked test papers.

un·mask (un-mask', -mäsk') v.t. 1 To remove a mask from. 2 To reveal; disclose. —v.i. 3 To remove one's mask or disguise.

un·mean·ing (un-mē'ning) adj. 1 Having no meaning: an unmeaning speech or look. 2 Having no expression; not displaying intelligence. —**un·mean'ing·ly** adv. —**un·mean'ing·ness** n.

un·meet (un-mēt') adj. Not meet, adapted, or suitable; not proper or fit; unbecoming. [OE unmǣte] —**un·meet'ly** adv. —**un·meet'ness** n.

un·men·tion·a·ble (un-men'shən-ə-bəl) adj. Not proper to be mentioned or discussed; embarrassing; shameful; disgraceful. —**un·men'tion·a·ble·ness** n. —**un·men'tion·a·bly** adv.

un·men·tion·a·bles (un-men'shən-ə-bəlz) n. pl. Things or articles not ordinarily discussed or mentioned; usually, undergarments; formerly, trousers; pants.

un·mer·ci·ful (un-mûr'sə-fəl) adj. 1 Showing no mercy; cruel; pitiless; unconscionable. 2 Extreme; exorbitant. —**un·mer'ci·ful·ly** adv. —**un·mer'ci·ful·ness** n.

un·mew (un-myōō') v.t. To release from confinement; set free.

un·mind·ful (un-mīnd'fəl) adj. Not keeping in mind; neglectful; inattentive. —**un·mind'ful·ly** adv. —**un·mind'ful·ness** n.

un·mis·tak·a·ble (un'mis-tā'kə-bəl) adj. That cannot be mistaken for something else; evident; clear; obvious. See synonyms under CLEAR, EVIDENT, MANIFEST. —**un'mis·tak'a·bly** adv.

un·mi·ter (un-mī'tər) v.t. To divest of a miter; deprive of the office of bishop. Also **un·mi'tre**.

un·mit·i·gat·ed (un-mit'ə-gā'tid) adj. 1 Not mitigated or lightened in effect; unabated; unassuaged: unmitigated sorrow. 2 As bad as can be; unconscionable: an unmitigated rogue. —**un·mit'i·gat'ed·ly** adv.

un·mod·u·lat·ed (un-moj'ōō-lā'tid) adj. 1 Without modulation. 2 Telecom. Denoting a carrier wave of constant amplitude, as during a pause in broadcasting.

un·moor (un-mōōr') Naut. v.t. 1 To loose the moorings for; release from moorings: to unmoor a ship. 2 To release all but one anchor of (a vessel formerly moored by two or more). —v.i. 3 To cast off moorings.

un·mor·al (un-môr'əl, -mor'-) adj. Having no moral sense or relation; not pertaining to morality: distinguished from amoral and immoral. —**un·mo·ral·i·ty** (un'mə-ral'ə-tē) n.

un·mor·tise (un-môr'tis) v.t. ·tised, ·tis·ing 1 To loosen; loosen the mortised joints of. 2 To separate.

un·muf·fle (un-muf'əl) v. ·fled, ·fling v.t. 1 To take the covering from. 2 To remove the muffling of (a drum, oar, etc.). —v.i. 3 To remove that which muffles.

un·nat·u·ral (un-nach'ər-əl) adj. 1 Contrary to the laws of nature; opposed to what is natural. 2 Contrary to the common laws of morality or decency; monstrous; inhuman: unnatural crimes. 3 Destitute of natural feeling. 4 Not consistent with nature; artificial: unnatural acting. See synonyms under FACTITIOUS, IRREGULAR. —**un·nat'u·ral·ly** adv. —**un·nat'u·ral·ness** n.

un·nec·es·sar·y (un-nes'ə-ser'ē) adj. Not required; not necessary. —**un·nec'es·sar'i·ly** adv. —**un·nec'es·sar'i·ness** n.

un·nerve (un-nûrv') v.t. ·nerved, ·nerv·ing To deprive of strength, firmness, self-control, or courage; unman. See synonyms under WEAKEN.

un·num·bered (un-num'bərd) adj. 1 Not counted. 2 Innumerable. 3 Not assigned a number; not marked with a number.

u·no an·i·mo (yōō'nō an'ə·mō) Latin With one mind; unanimously; in agreement.

un·oc·cu·pied (un-ok'yə-pīd) adj. 1 Empty; not dwelt in; uninhabited: an unoccupied house. 2 Idle; unemployed; not put to use: an unoccupied day.

un·of·fi·cial (un'ə-fish'əl) adj. 1 Not of an official character. 2 Not in an official capacity. 3 Not in the regular list of the pharmacopoeia.

un·or·gan·ized (un-ôr'gən·īzd) adj. 1 Not organized. 2 Not living; inorganic; structureless. 3 Not unionized. Also Brit. **un·or'gan·ised**.

un·o·rig·i·nal (un'ə-rij'ə-nəl) adj. Not original.

un·pack (un-pak') v.t. 1 To open and take out the contents of. 2 To take out of the container, as something packed. 3 To remove a load or pack from; unload. —v.i. 4 To unpack a trunk, goods, etc.

un·paid (un-pād') adj. 1 Not met or discharged, as a debt. 2 Not receiving pay; serving without pay. 3 Having wages remaining due.

un·paired (un-pârd') adj. 1 Not paired; not forming one of a pair; not matched. 2 Anat. a Having no corresponding part in the opposite half of the body. b Situated in the median plane of the body.

un·par·al·leled (un-par'ə-leld) adj. Without parallel; unmatched; unprecedented.

un·par·lia·men·ta·ry (un'pär-lə-men'tər-ē) adj. Not parliamentary; contrary to the rules that govern deliberative or legislative bodies. —**un'par·lia·men·ta'ri·ly** adv. —**un'par·lia·men'·ta·ri·ness** n.

un·peg (un-peg') v.t. ·pegged, ·peg·ging To remove the peg or pegs from; unfasten.

un·peo·ple (un-pē'pəl) v.t. ·pled, ·pling To depopulate.

un·peo·pled (un-pē'pəld) adj. 1 Uninhabited. 2 Depopulated.

un·per·fo·rat·ed (un-pûr'fə-rā'tid) adj. 1 Not perforated. 2 In philately, imperforate.

un peu (œn pœ) French A little; somewhat.

un·pick (un-pik') v.t. 1 To undo by removing the stitches; also, to remove (stitches). 2 To open with a pick or picklock.

un·pin (un-pin') v.t. ·pinned, ·pin·ning 1 To remove the pins from. 2 To unfasten by removing pins.

un·pleas·ant (un-plez'ənt) adj. Disagreeable; objectionable; not pleasing. —**un·pleas'ant·ly** adv.

un·pleas·ant·ness (un-plez'ənt-nis) n. 1 The quality, character, or condition of being unpleasant or disagreeable. 2 Any disagreeable experience or event; a disagreement or quarrel. —**the late unpleasantness** The American Civil War: now chiefly humorous; by extension, any recent war.

un·plumbed (un-plumd') adj. 1 Not sounded; not explored fully; unfathomed. 2 Not furnished with plumbing.

un·poised (un-poizd') adj. Not poised or balanced.

un·pol·i·cied (un-pol'ə·sēd) adj. 1 Having no established system of civil polity. 2 Obs. Unguided by reason or prudence; impolitic.

un·po·lit·ic (un-pol'ə-tik) adj. Impolitic.

un·polled (un-pōld') adj. Not registered: an unpolled vote or voter; not having voted at an election.

un·pop·u·lar (un-pop'yə-lər) adj. Having no popularity; generally disliked or condemned. —**un·pop'u·lar·ly** adv. —**un·pop'u·lar·i·ty** (-lar'ə-tē) n.

un·prac·ticed (un-prak'tist) adj. 1 Being without practice; inexperienced. 2 Not carried out in practice; not used. 3 Not yet tried.

un·prec·e·dent·ed (un-pres'ə-den'tid) adj. Being without precedent; preceded by no similar case; unexampled. See synonyms under EXTRAORDINARY. —**un·prec'e·dent'ed·ly** adv.

un·prej·u·diced (un-prej'ōō-dist) adj. 1 Free from prejudice or bias; impartial. 2 Not impaired, as a right. See synonyms under CANDID.

un·pre·med·i·tat·ed (un'pri·med'ə-tā'tid) adj. 1 Not planned beforehand; undesigned: unpremeditated assault. 2 Not previously considered or thought of. —**un'pre·med'i·tat'ed·ly** adv. —**un'pre·med'i·tat·ed·ness** n.

un·pre·pared (un'pri·pârd') adj. 1 Having made no preparations: an unprepared student. 2 Not brought into a state of preparation; not yet ready: Dinner is still unprepared. 3 Done or carried out without preparation; impromptu: an unprepared speech. —**un·pre·par·ed·ly** (un'pri-pâr'id-lē) adv. —**un'pre·par'ed·ness** n.

un·priced (un-prīst') adj. 1 Having no fixed price. 2 Priceless.

un·prin·ci·pled (un-prin'sə-pəld) adj. Destitute of conscientious scruples; unscrupulous; wicked. See synonyms under BAD, IMMORAL. —**un·prin'ci·pled·ness** n.

un·print·a·ble (un-prin'tə-bəl) adj. Not fit to be printed.

un·priz·a·ble (un-prī'zə-bəl) adj. Obs. 1 Of worth beyond estimation; invaluable. 2 Not prized; valueless.

un·pro·duc·tive (un'prə·duk'tiv) adj. 1 Producing little or nothing; barren, literally or figuratively. 2 Econ. Not adding to exchangeable value: unproductive labor. —**un'pro·duc'·tive·ly** adv. —**un'pro·duc'tive·ness** n.

un·pro·fes·sion·al (un'prə-fesh'ən·əl) adj. 1 Having no profession; also, lay; amateur. 2 Violating the rules or ethical code of a profession; not up to the standard of a profession: unprofessional work. —**un'pro·fes'sion·al·ly** adv.

un·prof·it·a·ble (un-prof'it-ə·bəl) adj. Productive of no profit; serving no desirable purpose; fruitless; futile: unprofitable conversation; an

unsafe	unsatiating	unscholarlike	unseconded	unsew	unshorn	unsized
unsafely	unsatisfactory	unscholarly	unsectarian	unsewn	unshrinkable	unskeptical
unsaintly	unsatisfied	unschooled	unsecured	unsexual	unshrinking	unslacked
unsalability,	unsatisfying	unscientific	unseeing	unshaded	unshriven	unslaked
unsaleability	unsaved	unscorched	unsegmented	unshakable	unshrouded	unsleeping
unsalable,	unsawed	unscorned	unselected	unshaken	unshrunk	unslumbering
unsaleable	unsawn	unscoured	unselective	unshamed	unshunned	unsmiling
unsalaried	unsayable	unscourged	unselfish	unshapely	unshut	unsmirched
unsalted	unscabbarded	unscratched	unselfishly	unshared	unsifted	unsmoked
unsanctified	unscaled	unscreened	unsent	unshaved	unsigned	unsoaked
unsanctioned	unscanned	unscriptural	unsentimental	unshaven	unsilenced	unsober
unsanitary	unscarred	unsculptured	unserved	unshed	unsimilar	unsocial
unsated	unscented	unsealed	unserviceable	unshelled	unsingable	unsoiled
unsatiable	unscepical	unseaworthiness	unserviceably	unsheltered	unsinkable	unsold
unsatiated	unscheduled	unseaworthy	unset	unshod	unsisterly	unsoldierlike

add, āce, câre, pälm; end, ēven; it, īce; odd, ōpen, ôrder; tōōk, pōōl; up, bûrn; ə = a in above, e in sicken, i in clarity, o in melon, u in focus; yōō = u in fuse; oi, oil; ou, pout; ch, check; g, go; ng, ring; th, thin; th, this; zh, vision. Foreign sounds à, œ, ü, kh, н; ◆: see page xx. < from; + plus; ? possibly.

unprofitable transaction. —**un·prof′it·a·ble·ness** n. —**un·prof′it·a·bly** adv.
un·pro·nounce·a·ble (un′prə-noun′sə-bəl) adj. 1 Not easy to pronounce, especially properly. 2 Not fit to be mentioned.
un·pro·vid·ed (un′prə-vī′did) adj. 1 Not furnished or provided: with *with*, formerly with *of*: to be *unprovided* with suitable raiment. 2 Not fittingly prepared; not ready: *unprovided* for a sudden change. —**un′pro·vid′ed·ly** adv.
un·qual·i·fied (un-kwol′ə-fīd) adj. 1 Being without the proper qualifications; unfit. 2 Having failed to qualify; lacking legal power or authority. 3 Without limitation or restrictions; absolute; entire: *unqualified* approval. —**un·qual′i·fied′ly** adv. —**un·qual′i·fied′ness** n.
un·ques·tion·a·ble (un-kwes′chən-ə-bəl) adj. Too certain or sure to admit of question; being beyond a doubt; indisputable. See synonyms under INCONTESTABLE, NOTORIOUS. —**un·ques′tion·a·bly** adv.
un·ques·tioned (un-kwes′chənd) adj. 1 Not called in question; undoubted. 2 Not to be frustrated or opposed; indisputable. 3 Not interrogated.
un·qui·et (un-kwī′ət) adj. 1 Not at rest; disturbed; restless, in mind or physically. 2 Causing unrest or discomfort. —**un·qui′et·ly** adv. —**un·qui′et·ness** n.
un·quote (un-kwōt′) v.t. & v.i. **·quot·ed, ·quot·ing** To close (a quotation).
un·rav·el (un-rav′əl) v. **·eled** or **·elled, ·el·ing** or **·el·ling** v.t. 1 To separate the threads of, as a tangled skein or knitted article. 2 To free from entanglement; unfold; explain, as a mystery or a plot. — v.i. 3 To become unraveled. See synonyms under INTERPRET.
un·read (un-red′) adj. 1 Not informed by reading; ignorant. 2 Not yet perused.
un·read·y (un-red′ē) adj. 1 Being without readiness or alertness; not apt or quick to see or appreciate. 2 Not in a condition to act effectively; unprepared. —**un·read′i·ly** adv. —**un·read′i·ness** n.
un·re·al (un-rē′əl, -rēl′) adj. Having no reality, actual existence, or substance; having no genuineness; insincere; artificial; also, fanciful; visionary. —**un·re·al·i·ty** (un′rē·al′ə·tē) n. —**un·re′al·ly** adv.
un·rea·son (un-rē′zən) n. Lack or absence of reason; irrationality; also, absurdity; nonsense.
un·rea·son·a·ble (un-rē′zən-ə-bəl) adj. 1 Acting without or contrary to reason. 2 Not according to reason; irrational. 3 Exceeding what is reasonable; immoderate; exorbitant. See synonyms under ABSURD, IMMODERATE. —**un·rea′son·a·ble·ness** n. —**un·rea′son·a·bly** adv.
un·rea·son·ing (un-rē′zən-ing) adj. So intolerant, or so unaccompanied by reason or control, as to be obstinate, blind, or wild.
un·reck·on·a·ble (un-rek′ən-ə-bəl) adj. That cannot be reckoned or computed; unlimited.
un·reel (un-rēl′) v.t. & v.i. To unwind, as from a reel.
un·reeve (un-rēv′) v. **·reeved** or **·rove** (*for pp. also* **·rov·en), ·reev·ing** Naut. v.t. 1 To take out or withdraw (a rope) from a block, thimble, deadeye, etc. —v.i. 2 To become unreeved. 3 To unreeve a rope.
un·re·flec·tive (un′ri-flek′tiv) adj. Not given to reflection; not thoughtful.
un·re·gen·er·ate (un′ri-jen′ər-it) adj. Not having been changed spiritually by regeneration; remaining unreconciled to God; loosely, sinful. Also **un′re·gen′er·at·ed** (-ā′tid). —**un′re·gen′er·a·cy** (-ə-sē) n. —**un′re·gen′er·ate·ly** adv.
un·re·lent·ing (un′ri-len′ting) adj. 1 Continuing to be severe; pitiless; inexorable. 2 Not diminishing, or not changing, in pace, effort, speed, etc. —**un′re·lent′ing·ly** adv.

un·re·li·a·ble (un′ri-lī′ə-bəl) adj. That cannot be relied upon; not dependable. —**un′re·li·a·bil′i·ty, un′re·li′a·ble·ness** n. —**un′re·li′a·bly** adv.
un·re·li·gious (un′ri-lij′əs) adj. 1 Irreligious; hostile to religion. 2 Having no religion; not connected with religion.
un·re·mit·ting (un′ri-mit′ing) adj. Incessant; not relaxing. —**un′re·mit′ting·ly** adv. —**un′re·mit′ting·ness** n.
un·re·proved (un′ri-prōōvd′) adj. 1 Not censured or blamed; not reproved. 2 Not liable to reproof; above reproach.
un·re·serve (un′ri-zûrv′) n. Absence of reserve; freedom of style or manner.
un·re·served (un′ri-zûrvd′) adj. 1 Given or done without reserve. 2 Having no reserve of manner; informal; open; frank. See synonyms under CANDID, IMPLICIT. —**un′re·serv′ed·ly** (un′ri-zûr′vid-lē) adv. —**un′re·serv′ed·ness** n.
un·re·spec·tive (un′ri-spek′tiv) adj. Obs. 1 Undiscriminating. 2 Inattentive; heedless. 3 Common; not restricted.
un·res·pit·ed (un-res′pit-id) adj. 1 Not postponed; not respited, as from a sentence of the law. 2 Obs. Having no intermission.
un·re·spon·sive (un′ri-spon′siv) adj. Showing no reaction or response; unsympathetic. —**un′re·spon′sive·ly** adv. —**un′re·spon′sive·ness** n.
un·rest (un-rest′) n. 1 Restlessness, especially of the mind. 2 Trouble; turmoil: used with regard to public or political conditions and suggesting premonitions of revolt.
un·re·strained (un′ri-strānd′) adj. Not restrained; free; not controlled. —**un·re·strain′ed·ly** (un′ri·strā′nid-lē) adv.
un·rid·dle (un-rid′l) v.t. **·dled, ·dling** To solve, as a mystery.
un·ri·fled[1] (un-rī′fəld) adj. Smooth-bored, as a gun.
un·ri·fled[2] (un-rī′fəld) adj. Not rifled, seized, or plundered.
un·rig (un-rig′) v.t. **rigged, rig·ging** Naut. To strip of rigging.
un·right·eous (un-rī′chəs) adj. 1 Not righteous; wicked; sinful. 2 Contrary to the law of right; unjust. See synonyms under SINFUL. —**un·right′eous·ly** adv. —**un·right′eous·ness** n.
un·rip (un-rip′) v.t. **·ripped, ·rip·ping** To separate by ripping; rip or cut open.
un·ripe (un-rīp′) adj. 1 Not arrived at maturity; not ripe; immature. 2 Premature. 3 Not ready; not prepared. [OE, *untimely*] —**un·ripe′ness** n.
un·ri·valed (un-rī′vəld) adj. Having no rival or competitor; unequaled; matchless. Also **un·ri′valled.**
un·roll (un-rōl′) v.t. 1 To spread or open (something rolled up). 2 To exhibit to view. 3 *Rare* To remove from a roll or register. —v.i. 4 To become unrolled.
un·root (un-rōōt′, -rōōt′) v.t. To uproot.
un·ruf·fled (un-ruf′əld) adj. 1 Not disturbed or agitated emotionally; calm. 2 Not ruffled or made rough physically.
un·ru·ly (un-rōō′lē) adj. Disposed to resist rule or discipline; intractable; ungovernable. See synonyms under RESTIVE. —**un·ru′li·ness** n.
un·sad·dle (un-sad′l) v.t. **·dled, ·dling** 1 To remove a saddle from. 2 To remove from the saddle; unhorse.
un·said (un-sed′) adj. Not said; not spoken.
un·sat·u·rat·ed (un-sach′ə-rā′tid) adj. 1 Falling short of saturation, as a solution. 2 *Chem.* Not combined to the greatest possible extent; capable of uniting with certain other elements or compounds to form additional compounds.
un·sa·vor·y (un-sā′vər-ē) adj. 1 Having a disagreeable taste or odor. 2 Suggesting something disagreeable, offensive, or unclean;

also, morally bad: an *unsavory* reputation. 3 *Obs.* Having no savor; tasteless; odorless. Also *Brit.* **un·sa′vour·y.** —**un·sa′vor·i·ly** adv. —**un·sa′vor·i·ness** n.
un·say (un-sā′) v.t. **·said, ·say·ing** To retract (something said).
un·scathed (un-skāthd′) adj. Uninjured.
un·scram·ble (un-skram′bəl) v.t. **·bled, ·bling** *Colloq.* To resolve the confused, scrambled, or disordered condition of.
un·screw (un-skrōō′) v.t. 1 To remove the screw or screws from. 2 To remove or detach by withdrawing screws, or by turning. —v.i. 3 To permit of being unscrewed.
un·scru·pu·lous (un-skrōō′pyə-ləs) adj. Not scrupulous; having no scruples; unprincipled. —**un·scru′pu·lous·ly** adv. —**un·scru′pu·lous·ness** n.
un·seal (un-sēl′) v.t. 1 To break or remove the seal of. 2 To open (that which has been sealed or closed).
un·seam (un-sēm′) v.t. To open the seam or seams of.
un·search·a·ble (un-sûr′chə-bəl) adj. That cannot be searched or explored; hidden; mysterious. —**un·search′a·bly** adv.
un·sea·son·a·ble (un-sē′zən-ə-bəl) adj. Not being in the proper season or not being in time; inappropriate; ill-timed. —**un·sea′son·a·ble·ness** n. —**un·sea′son·a·bly** adv.
un·sea·soned (un-sē′zənd) adj. 1 Not seasoned; not flavored. 2 Immature; unripe; not properly aged. 3 Not habituated. —**un·sea′soned·ness** n.
un·seat (un-sēt′) v.t. 1 To remove from a seat or fixed position. 2 To unhorse. 3 To deprive of office or rank; depose.
un·seem·ly (un-sēm′lē) adj. **·li·er, ·li·est** Unbecoming; indecent; not handsome. —adv. In an unseemly fashion. —**un·seem′li·ness** n.
un·seen (un-sēn′) adj. Not seen; not evident; invisible; not previously seen or prepared, as a passage for translation.
un·set·tle (un-set′l) v. **·tled, ·tling** v.t. 1 To move from a fixed or settled condition. 2 To confuse; disturb. —v.i. 3 To become unsteady or unfixed. See synonyms under DISPLACE.
un·sex (un-seks′) v.t. 1 To deprive of the distinctive qualities of a sex; especially, to render unfeminine or unwomanly. 2 To castrate.
un·shack·le (un-shak′əl) v.t. **·led, ·ling** To unfetter; free from shackles. —**un·shack′led** adj.
un·shap·en (un-shā′pən) adj. Not shaped; imperfectly formed; badly shaped. Also **un·shaped′.**
un·sheathe (un-shēth′) v.t. **·sheathed, ·sheath·ing** To take from or as from a scabbard or sheath; bare.
un·ship (un-ship′) v.t. **·shipped, ·ship·ping** 1 To unload from a ship or other vessel; also, to dismiss from a ship. 2 To remove from the place where it is fixed or fitted, as a rudder or oar.
un·sick·er (un-sik′ər) adj. *Scot.* Insecure; unreliable; undependable. —**un·sick′er·ly** adv. —**un·sick′er·ness** n.
un·sight·ed (un-sī′tid) adj. 1 Not sighted; not in view. 2 Having no sight, as a cannon. 3 Not aimed with the assistance of a sight, as a shot.
un·sight·ly (un-sīt′lē) adj. **·li·er, ·li·est** Offensive to the sight; ugly. —**un·sight′li·ness** n.
unsight unseen Sight unseen: former usage.
un·skaithed (un-skāthd′) adj. *Scot.* Unscathed.
un·skil·ful (un-skil′fəl) adj. 1 Lacking or not evincing skilfulness; awkward. 2 *Obs.* Lacking in discernment; ignorant. Also **un·skill′ful.** —**un·skil′ful·ly** adv. —**un·skil′ful·ness** n.

unsoldierly	unspeculative	unsquandered	unsterilized	unsuggestive	unsustained	untalented
unsolicited	unspelled	unsquared	unstick	unsuited	unswayed	untalked-of
unsolicitous	unspent	unstack	unstigmatized	unsullied	unsweetened	untamable,
unsolid	unspilled	unstainable	unstinted	unsunk	unswept	untameable
unsoluble	unspilt	unstained	unstitched	unsunned	unswerving	untame
unsolvable	unspiritual	unstamped	unstrained	unsupportable	unsworn	untamed
unsolved	unspirituality	unstandardized	unstressed	unsupported	unsymmetrical	untangled
unsophistication	unspiritually	unstarched	unstripped	unsupportedly	unsympathetic	untanned
unsorted	unspiritualness	unstarred	unstuffed	unsuppressed	unsympathizing	untapped
unsought	unspoiled	unstatesmanlike	unstung	unsure	unsystematic	untarnished
unsounded	unspoilt	unsteadfast	unsubdued	unsurmountable	unsystematized	untasted
unsoured	unspoken	unsteadily	unsubmissive	unsurpassable	untack	untaxable
unsowed	unsportsmanlike	unsteadiness	unsubscribed	unsurpassed	untactful	untaxed
unsown	unsprinkled	unsteady	unsubstantiated	unsusceptible	untainted	unteachable
unspecified	unsprung	unstemmed	unsuccess	unsuspicious	untaken	untechnical

un·skilled (un·skild′) *adj.* 1 Destitute of skill or dexterity in artisan's work; good only for common labor: an *unskilled* workman. 2 Produced without or not requiring special skill or training; untrained. 3 Destitute of practical knowledge; unskilful.
un·sling (un·sling′) *v.t.* **·slung**, **·sling·ing** 1 To remove, as a rifle, from a slung position. 2 *Naut.* To take the slings from.
un·snap (un·snap′) *v.t.* **·snapped**, **·snap·ping** To undo the snap or snaps of; unfasten.
un·snarl (un·snärl′) *v.t.* To disentangle.
un·so·cia·ble (un·sō′shə·bəl) *adj.* 1 Not sociable; not inclined to seek the society of others. 2 Not congenial or in accord: an *unsociable* group. 3 Not encouraging social intercourse. — **un·so′cia·bil′i·ty**, **un·so′cia·ble·ness** *n.* — **un·so′cia·bly** *adv.*
un·sol·der (un·sod′ər) *v.t.* 1 To disunite or take apart (something soldered). 2 To separate; sunder.
un·son·sie (un·son′sē) *adj. Scot.* Unlucky; disagreeable. Also **un·son′cy**, **un·son′sy**.
un·so·phis·ti·cat·ed (un′sə·fis′tə·kā′tid) *adj.* 1 Not sophisticated; showing inexperience or naiveté; artless; simple. 2 Free from adulteration; genuine; pure. See synonyms under CANDID, RUSTIC. — **un′so·phis′ti·cat′ed·ly** *adv.* — **un′so·phis′ti·cat′ed·ness** *n.*
un·sound (un·sound′) *adj.* 1 Lacking in soundness; not having material strength and solidity; weak; rotten. 2 Not sound in health; diseased. 3 Not logically valid; erroneous; in religion, heterodox. 4 Disturbed; not profound: said of sleep. — **un·sound′ly** *adv.* — **un·sound′ness** *n.*
un·spar·ing (un·spâr′ing) *adj.* 1 Not sparing or saving; lavish; liberal. 2 Showing no mercy. — **un·spar′ing·ly** *adv.* — **un·spar′ing·ness** *n.*
un·speak (un·spēk′) *v.t.* **·spoke**, **·spo·ken**, **·speak·ing** To retract (something said); take back.
un·speak·a·ble (un·spē′kə·bəl) *adj.* 1 That cannot be expressed; unutterable: *unspeakable* joy. 2 Extremely bad or objectionable: an *unspeakable* crime. 3 Mute. — **un·speak′a·ble·ness** *n.* — **un·speak′a·bly** *adv.*
un·spe·cial·ized (un·spesh′əl·īzd) *adj.* 1 Not specialized. 2 *Biol.* Not set apart for a special function or purpose; generalized. Also *Brit.* **un·spe′cial·ised**.
un·sphere (un·sfir′) *v.t.* **·sphered**, **·spher·ing** To take out of its sphere or place.
un·spot·ted (un·spot′id) *adj.* 1 Not marked or marred with spots. 2 Not morally tainted; immaculate; free from blemishes; perfect. 3 Ceremonially clean. See synonyms under PURE. — **un·spot′ted·ness** *n.*
un·sprung weight (un·sprung′) In automobiles, the weight of components not supported by the suspension, as the wheel assemblies: opposed to *sprung weight*.
un·sta·ble (un·stā′bəl) *adj.* 1 Lacking in stability or firmness; not stable. 2 Having no fixed purposes; easily influenced; inconstant: an *unstable* character. 3 *Chem.* Readily decomposable, as certain compounds. 4 Subject to a radical change by the application of a slight force: *unstable* equilibrium. See FICKLE, PRECARIOUS. — **un·sta′ble·ness** *n.* — **un·sta′bly** *adv.*
un·state (un·stāt′) *v.t.* **·stat·ed**, **·stat·ing** 1 To divest of statehood. 2 To deprive of dignity, rank, or office.
un·steel (un·stēl′) *v.t.* To deprive of steel-like quality; disarm; soften.
un·step (un·step′) *v.t.* **·stepped**, **·step·ping** To take out of a step or socket: to *unstep* a mast.

un·stop (un·stop′) *v.t.* **·stopped**, **·stop·ping** 1 To remove a stop or stopper from. 2 To open by removing obstructions; clear. 3 To open the stops of (an organ).
un·stopped (un·stopt′) *adj.* 1 Not stopped; unobstructed. 2 *Phonet.* Of consonants, not stopped; capable of being prolonged: said of the continuants, as (z) and (l).
un·stowed (un·stōd′) *adj.* 1 Not stowed or filled, as a ship, with cargo. 2 Lying loose in the hold or on deck, as cargo.
un·strap (un·strap′) *v.t.* **·strapped**, **·strap·ping** To unfasten or loosen the strap or straps of.
un·strat·i·fied (un·strat′ə·fīd) *adj. Geol.* Not deposited in beds or strata, as igneous rocks.
un·stri·at·ed (un·strī′ā·tid) *adj.* Without striations; smooth-textured, as certain muscles.
un·string (un·string′) *v.t.* **·strung**, **·string·ing** 1 To remove from a string, as pearls. 2 To take the string or strings from. 3 To loosen the string or strings of, as a bow or guitar. 4 To relax, as if by loosening; weaken: usually in the passive: Her nerves were *unstrung*.
un·striped (un·strīpt′) *adj.* 1 Not striped; unstriated. 2 *Anat.* Denoting certain muscles that act independently of the will, as the heart muscles. See under MUSCLE.
un·strung (un·strung′) *adj.* 1 Having the strings removed or relaxed. 2 Unnerved; emotionally upset; weakened; relaxed.
un·stud·ied (un·stud′ēd) *adj.* 1 Not planned; unpremeditated. 2 Not stiff or artificial; natural. 3 Not acquainted through study; unversed: with *in*. See synonyms under SIMPLE.
un·sub·stan·tial (un′səb·stan′shəl) *adj.* 1 Lacking solidity, strength, or weight. 2 Having no valid basis. 3 Having no bodily existence; fanciful. — **un′sub·stan′tial·ly** *adv.* — **un′sub·stan′ti·al′i·ty** (-shē·al′ə·tē) *n.*
un·suc·cess·ful (un′sək·ses′fəl) *adj.* Having or meeting with no success: said of persons or their acts: *unsuccessful* in business, an *unsuccessful* attempt. — **un′suc·cess′ful·ly** *adv.* — **un′suc·cess′ful·ness** *n.*
un·suit·a·ble (un·soō′tə·bəl) *adj.* Not suitable; unfitting. — **un·suit·a·bil·i·ty** (un′soō·tə·bil′ə·tē), **un·suit′a·ble·ness** *n.* — **un·suit′a·bly** *adv.*
un·sung (un·sung′) *adj.* 1 Not celebrated in song or poetry; obscure. 2 Not yet sung, as a song.
un·sus·pect·ed (un′sə·spek′tid) *adj.* 1 Not suspected, as of evil; not under suspicion. 2 Not imagined or known to exist.
un·sus·pect·ing (un′sə·spek′ting) *adj.* Having no suspicion; trusting.
un·swathe (un·swäth′) *v.t.* **·swathed**, **·swath·ing** To remove swathings from; free from swathings.
un·swear (un·swâr′) *v.t.* **swore**, **sworn**, **·swear·ing** To revoke (an oath); retract; abjure.
un·tan·gle (un·tang′gəl) *v.t.* **·gled**, **·gling** To free from entanglement or embarrassment; resolve.
un·taught (un·tôt′) *adj.* Not having been instructed; ignorant.
un·teach (un·tēch′) *v.t.* **·taught**, **·teach·ing** 1 To cause to forget or to disbelieve what has been taught. 2 To cause to be forgotten or disbelieved.
un·ten·a·ble (un·ten′ə·bəl) *adj.* 1 That cannot be maintained: *untenable* theories. 2 Incapable of being defended or held, as a fortress. — **un·ten′a·ble·ness** *n.*
un·tent·ed (un·ten′tid) *adj.* 1 Having no tents. 2 *Obs.* Not kept open or dressed, as a wound.
Un·ter den Lin·den (ŏon′tər den lin′dən) A famous avenue in East Berlin; literally, under the lindens.

Un·ter·wal·den (ŏon′tər·väl′dən) A canton of central Switzerland; 296 square miles.
un·thanked (un·thangkt′) *adj.* 1 Not thanked. 2 *Obs.* Not received with thankfulness.
un·thank·ful (un·thangk′fəl) *adj.* 1 Not grateful. 2 Not received with thanks; unwelcome. — **un·thank′ful·ly** *adv.* — **un·thank′ful·ness** *n.*
un·think (un·thingk′) *v.t.* **·thought**, **·think·ing** To retract in thought; change the mind concerning.
un·think·ing (un·thingk′ing) *adj.* 1 Not having the power of thought. 2 Lacking thoughtfulness, care, or attention; heedless; inconsiderate. See synonyms under IMPRUDENT. — **un·think′ing·ly** *adv.* — **un·think′ing·ness** *n.*
un·thread (un·thred′) *v.t.* 1 To remove the thread from, as a needle. 2 To find one's way out of, as a maze.
un·ti·dy (un·tī′dē) *adj.* **·di·er**, **·di·est** Showing or characterized by lack of tidiness. [ME *untīdī*] — **un·ti′di·ly** *adv.* — **un·ti′di·ness** *n.*
un·tie (un·tī′) *v.* **·tied**, **·ty·ing** *v.t.* 1 To loosen or undo, as a knot or knotted rope. 2 To free from that which binds or restrains. — *v.i.* 3 To become untied. See synonyms under RELEASE. [OE *untīgan*] — **un·tied′** *adj.*
un·til (un·til′) *prep.* 1 Up to the time of; till: We will wait *until* midnight. 2 Before: used with a negative: The music doesn't begin *until* nine. 3 *Scot. & Brit. Dial.* Unto. — *conj.* 1 To the time when: *until* I die. 2 To the place or degree that: Walk east *until* you reach the river. 3 Before: with a negative: He couldn't leave *until* the car came for him. [ME *untill* < *und-* up to, as far as + TILL]
un·time·ly (un·tīm′lē) *adj.* Coming before time or not in proper time; unseasonable; ill-timed: also *Scot.* **un·time′ous**. — *adv.* Before the proper time; inopportunely.
un·ti·tled (un·tīt′ld) *adj.* 1 Having no right (to a throne). 2 Having no title, as a book. 3 Having no title of distinction: *untitled* nobility.
un·to (un′toō) *prep.* 1 *Poetic & Archaic* To: used in all senses except to indicate the infinitive. 2 *Archaic* Until. — *conj. Obs.* Up to the extent or time that; until. [ME *un-*, *und-* up to, as far as + TO, on analogy with *until*]
un·told (un·tōld′) *adj.* 1 That cannot be told, revealed, or described; inexpressible: *untold* misery. 2 That cannot be numbered or estimated; hence, of great number or extent: *untold* numbers; *untold* treasure. 3 Not told.
un·touch·a·bil·i·ty (un′tuch·ə·bil′ə·tē) *n.* The character or state of being untouchable.
un·touch·a·ble (un·tuch′ə·bəl) *adj.* 1 Inaccessible to the touch; out of reach; intangible; unrivaled; unapproachable. 2 Forbidden to the touch. 3 Unpleasant, disgusting, vile, or dangerous to touch; that should not be touched. — *n.* In India, a member of the lowest caste; one whose touch was formerly counted as pollution by Hindus of higher station.
un·to·ward (un·tôrd′, -tōrd′) *adj.* 1 Causing annoyance or hindrance; vexatious. 2 Not yielding readily; refractory; perverse. 3 *Obs.* Uncouth; ungraceful. See synonyms under PERVERSE. — **un·to′ward·ly** *adv.* — **un·to′ward·ness** *n.*
un·trav·eled (un·trav′əld) *adj.* 1 Not passed over, as a road. 2 Not having traveled; hence, narrow in ideas; provincial. Also **un·trav′elled**.
un·tread (un·tred′) *v.t.* **trod**, **·trod·den** or **·trod**, **·tread·ing** To retrace.
un·tried (un·trīd′) *adj.* 1 Not tried or tested. 2 Not tried in court.
un·trimmed (un·trimd′) *adj.* 1 Not adorned

untempered	untillable	untranslatable	untuneful	unvail	unvitrified	unweakened
untenanted	untilled	untranslated	unturned	unvalidated	unvocal	unweaned
untended	untinged	untransmitted	untwilled	unvanquished	unvolatilized	unwearable
unterrified	untired	untrapped	untwisted	unvaried	unvulcanized	unweary
untested	untiring	untraversable	untypical	unvarying	unwakened	unwearying
untether	untouched	untraversed	un-uniformed	unveiled	unwalled	unweathered
unthatched	untraceable	untreasured	un-uniformly	unventilated	unwanted	unweave
untheatrical	untraced	untrim	un-united	unveracious	unwarlike	unwed
unthinkable	untracked	untroubled	unurged	unverifiable	unwarmed	unwedded
unthoughtful	untractable	untrustiness	unusable	unverified	unwarned	unweeded
unthought-of	untrained	untrusty	unutilizable	unversed	unwashed	unwelded
unthrift	untrammeled,	untuck	unutilized	unvexed	unwasted	unwetted
unthriftiness	untrammelled	untufted	unuttered	unvexext	unwasting	unwhetted
unthrifty	untransferable	untunable	unvaccinated	unvisited	unwatched	unwhipped
unthrone	untransferred	untuned	unvacillating	unvitiated	unwavering	unwhipt

with trimmings: an *untrimmed* hat. 2 Not made trim or orderly; not clipped or pruned: an *untrimmed* tree. 3 *Obs.* Virgin.

un·trod·den (un·trod′n) *adj.* Not having been trodden upon; hence, unfrequented. Also **un·trod′.**

un·true (un·trōō′) *adj.* 1 Lacking truth; not true; not corresponding with fact. 2 Not conforming to rule or standard. 3 Not adhering to faith, pledge, or duty; disloyal. See synonyms under BAD, PERFIDIOUS. —**un·tru′ly** *adv.*

un·truss (un·trus′) *v.t.* 1 To loosen or free from or as from a truss; unfasten; undo. 2 *Obs.* To take off (breeches); undress.

un·trust·ful (un·trust′fəl) *adj.* 1 Not trusting or trustful. 2 Not to be trusted; untrustworthy.

un·trust·wor·thy (un·trust′wûr′thē) *adj.* Worthy of no trust; unreliable. See synonyms under BAD, PERFIDIOUS. —**un·trust′wor′thi·ness** *n.* —**un·trust′wor′thi·ly** *adv.*

un·truth (un·trōōth′) *n. pl.* **·truths** (-trōōths′, -trōōthz′) 1 The quality or character of being untrue; want of veracity. 2 *Obs.* Lack of fidelity; disloyalty. 3 Something that is not true; a falsehood; lie. See synonyms under DECEPTION, LIE. [OE *untrēowth*]

un·truth·ful (un·trōōth′fəl) *adj.* Not truthful; untrue; not veracious. —**un·truth′ful·ly** *adv.* —**un·truth′ful·ness** *n.*

un·tu·tored (un·tōō′tərd, -tyōō′-) *adj.* Having had no tutor or teacher; hence, uninstructed; raw. See synonyms under IGNORANT.

un·twine (un·twīn′) *v.* **·twined, ·twin·ing** *v.t.* To undo (something twined); unwind by disentangling. —*v.i.* To become untwined.

un·twist (un·twist′) *v.t.* & *v.i.* To separate or open by a movement the reverse of twisting; unwind or untwine.

U·nun·gun (ōō·nōōng′gōōn) *n. pl.* Literally, people: the collective name for two Eskimo tribes, the Unalaskans and the Atkans, inhabiting the Aleutian Islands; the Aleuts.

un·used (un·yōōzd′) *adj.* 1 Not made use of; disused; also, never having been used. 2 Not accustomed or wont: with *to.*

un·u·su·al (un·yōō′zhōō·əl) *adj.* Of a character, kind, number, or size not usually met with. See synonyms under EXTRAORDINARY, ODD, RARE. —**un·u′su·al·ly** *adv.* —**un·u′su·al·ness** *n.*

un·ut·ter·a·ble (un·ut′ər·ə·bəl) *adj.* 1 That cannot be uttered; too great or deep for verbal expression; ineffable: *unutterable* bliss. 2 Unpronounceable. —**un·ut′ter·a·ble·ness** *n.* —**un·ut′ter·a·bly** *adv.*

un·val·ued (un·val′yōōd) *adj.* 1 Not valued; neglected; unappreciated. 2 Not having a fixed value; not appraised. 3 *Obs.* Inestimable.

un·var·nished (un·vär′nisht) *adj.* 1 Having no covering of varnish. 2 Having no embellishment; plain: the *unvarnished* truth.

un·veil (un·vāl′) *v.t.* To remove the veil or covering from; disclose to view; reveal. —*v.i.* To remove one's veil; reveal oneself.

un·voice (un·vois′) *v.t.* **·voiced, ·voic·ing** *Phonet.* To pronounce (a voiced sound) without vibration of the vocal cords; devocalize.

un·voiced (un·voist′) *adj.* 1 Not expressed. 2 *Phonet.* Not voiced; rendered voiceless: The final (v) in "have" is often heard *unvoiced* in "have to": also *devoiced.*

un·voic·ing (un·voi′sing) *n. Phonet.* The change of a voiced consonant to its unvoiced counterpart, as of (b) to (p).

un·war·rant·a·ble (un·wôr′ən·tə·bəl, -wor′-) *adj.* That cannot be warranted; unjustifiable; indefensible. —**un·war′rant·a·bly** *adv.*

un·war·rant·ed (un·wôr′ən·tid, -wor′-) *adj.* 1 Having no warrant; unwarrantable; unjustifiable. 2 Being without warranty or guarantee.

un·war·y (un·wâr′ē) *adj.* Taking no precautions against accident or danger; especially, not realizing the necessity of such precautions; incautious. —**un·war′i·ly** *adv.* —**un·war′i·ness** *n.*

un·wea·ried (un·wir′ēd) *adj.* 1 Not tired. 2 Indefatigable.

un·wel·come (un·wel′kəm) *adj.* 1 Not welcome; not desired: an *unwelcome* guest. 2 Causing no satisfaction: *unwelcome* news. —**un·wel′come·ly** *adv.* —**un·wel′come·ness** *n.*

un·well (un·wel′) *adj.* 1 Somewhat ill; ailing. 2 Menstruating; indisposed by reason of menstruation: a euphemism. See synonyms under SICKLY. —**un·well′ness** *n.*

un·wept (un·wept′) *adj.* 1 Not lamented or wept for, as a deceased person. 2 Not shed, as tears.

un·whole·some (un·hōl′səm) *adj.* 1 Deleterious to physical or mental health. 2 Unsound in quality or condition; diseased or decayed: *unwholesome* provisions. 3 Impaired in health; sickly in appearance: an *unwholesome* look. 4 Not contributing to moral health; pernicious: *unwholesome* literature. See synonyms under BAD, NOISOME. —**un·whole′some·ly** *adv.* —**un·whole′some·ness** *n.*

un·wield·y (un·wēl′dē) *adj.* Moved or managed with difficulty, as from great size or awkward shape; bulky; clumsy. —**un·wield′i·ly** *adv.* —**un·wield′i·ness** *n.*

un·willed (un·wild′) *adj.* 1 Not willed or intended; spontaneous. 2 Being without, or deprived of, purpose or will.

un·will·ing (un·wil′ing) *adj.* 1 Not willing; reluctant; loath. 2 Done with reluctance. 3 *Obs.* Not intended; involuntary. See synonyms under RELUCTANT. —**un·will′ing·ly** *adv.* —**un·will′ing·ness** *n.*

un·wind (un·wīnd′) *v.* **·wound, ·wind·ing** *v.t.* 1 To reverse the winding of; untwist or wind off; uncoil. 2 To disentangle. —*v.i.* 3 To become unwound.

un·wise (un·wīz′) *adj.* Acting with, or showing, lack of wisdom; injudicious; foolish. [OE *unwīs*] —**un·wise′ly** *adv.* —**un·wis′dom** (-wiz′dəm) *n.*

un·wish (un·wish′) *v.t.* 1 To retract (something wished); stop wishing. 2 To wish (something) not to be. 3 *Obs.* To destroy or do away with by wishing.

un·wished (un·wisht′) *adj.* 1 Not desired or wished. 2 Unwelcome.

un·wit·ting (un·wit′ing) *adj.* 1 Having no knowledge or consciousness of the thing in question; unknowing or unconscious. 2 Unintentional. [OE *unwitende*] —**un·wit′ting·ly** *adv.*

un·wont·ed (un·wun′tid, -won′-) *adj.* 1 Not according to wont or custom; unusual; uncommon. 2 *Obs.* Not accustomed; unfamiliar. See synonyms under EXTRAORDINARY. —**un·wont′ed·ly** *adv.* —**un·wont′ed·ness** *n.*

un·world·ly (un·wûrld′lē) *adj.* 1 Not motivated by worldly values or interests; spiritually minded. 2 Unearthly; spiritual; not belonging to this world. —**un·world′li·ness** *n.*

un·wor·thy (un·wûr′thē) *adj.* 1 Not worthy or deserving of something specified: usually with *of.* 2 Not befitting or becoming: often with *of*; wrong; improper: conduct *unworthy* of a gentleman. 3 Lacking worth or merit; unfit; wrong; contemptible. See synonyms under BAD, SINFUL. —**un·wor′thi·ly** *adv.* —**un·wor′thi·ness** *n.*

un·wrap (un·rap′) *v.* **·wrapped, ·wrap·ping** *v.t.* To take the wrapping from; open; undo. —*v.i.* To become unwrapped.

un·wrin·kle (un·ring′kəl) *v.t.* **·kled, ·kling** To free from wrinkles; smooth.

un·writ·ten (un·rit′n) *adj.* 1 Not reduced to writing; not written down; oral; traditional. 2 Having no writing upon it; blank.

unwritten law 1 A rule or custom established by general usage: an *unwritten law* of gentlemanly decorum. 2 Law which rests on custom and judicial decision, and not on a written command, decree, or statute. See COMMON LAW under LAW. 3 A custom in some communities granting a measure of immunity to those who commit criminal acts of revenge in support of personal or family honor, especially in cases of seduction, adultery, etc.

un·yoke (un·yōk′) *v.* **·yoked, ·yok·ing** *v.t.* 1 To release from a yoke. 2 To separate; part. —*v.i.* 3 To become unyoked. 4 To stop work; cease. [OE *ungeocian*]

un·yoked (un·yōkt′) *adj.* 1 Not subjected to or not wearing a yoke. 2 Freed from a yoke. 3 *Obs.* Unrestrained; licentious.

up (up) *adv.* 1 Toward a higher place or level: opposed to *down.* 2 In or on a higher place; above the horizon. 3 Toward that which is figuratively or conventionally higher: **a** To or at a higher price: Barley is *up.* **b** To or at a higher rank: people who have come *up* in the world. **c** To or at a greater size or larger amount: to swell *up.* **d** To or at a higher musical pitch. **e** To or at a place that is locally or arbitrarily regarded as higher: *up* north. 4 To a vertical position; standing; on one's feet. 5 Risen from bed. 6 So as to be level (to) or even (with) in space, time, degree, or amount: *up* to date; *up* to the brim. 7 In or into commotion or activity; in progress: They were stirred *up* to mutiny; to be *up* in arms. 8 Into existence: to draw *up* a document; to turn *up.* 9 In or into prominence; under consideration: The question was *up* for debate. 10 Into or in a place of safekeeping; aside: Fruits are put *up* in glass jars. 11 At an end or close: Your time is *up.* 12 Completely; wholly: Houses were burned *up*; The brooks dried *up.* 13 In baseball and cricket, at bat: He made but one hit in three times *up.* 14 In tennis and other sports: **a** In the lead; ahead: said of a player or team. **b** Apiece; alike: said of a score. 15 Bound for: said of a ship: *up* for Panama. 16 Running for as a candidate: Jones is *up* for mayor. 17 On trial before a magistrate: *up* for manslaughter. 18 *Naut.* Shifted to windward, as a tiller. —**all up with** At an end for; no further hope for. —**to be up against** *Colloq.* To meet with; be face to face with. —**to be up against it** *Colloq.* To be in difficulty; have financial trouble. —**to be up to** 1 *Colloq.* To be doing or plotting; be about to do: What is he *up to*? 2 To be equal to; to be capable of: I'm not *up to* moving all this furniture today. 3 To be incumbent upon; be dependent upon: It's *up to* him to save us. —*adj.* 1 Moving, sloping, or directed upward or in a direction arbitrarily regarded as upward. 2 At stake, as in gambling: to have money *up* on a horse race. 3 *Colloq.* Going on; taking place: What's *up*? 4 *Colloq.* In a state acquainted (with), equal (to), or a match (for); of a kind or character capable (of): He is *up* in that subject. 5 In golf: **a** In advance of the opponent or opponents: with a number indicating the extent (in holes) of such advance: three *up* and four to play: opposed to *down.* **b** Struck so as to travel as far as or beyond the hole: said of the ball. 6 Rising, risen, overflowing, or at flood: The moon is *up*; The river is *up.* —**up and around** *Colloq.* Sufficiently recovered to walk, as following an illness or injury; on one's feet again; convalescent and ambulatory. —**up to no good** Engaged in or contemplating some mischief or improper act. —*prep.* 1 From a lower to a higher point or place of, on, or along; toward a higher condition or rank on or in: *up* the social ladder. 2 To or at a point farther above or along: The farm is *up* the road. 3 From the coast toward the interior of (a country, as being higher); from the mouth toward the source of (a river): to sail *up* a river. 4 At, on, or near the height or top of: said of position or situation. —*n.* One who or that which is up, as elevated ground, an ascent or upward movement, state of prosperity, etc.: usually plural. —**ups and downs** Changes of fortune or circumstance. —*v.* **upped, up·ping** *Colloq. v.t.* 1 To increase; make larger; cause to rise. 2 To put or take up. —*v.i.* 3 To rise. [OE]

up– *combining form* As an element in solidemes *up* has adverbial force with various meanings, as in the following examples:

1 To a higher place or level:

upbear
upbearer
upborne
upbuilder
upbuilding
upclimb
upcoil
upcurl
upcurve
updart
updive
upfling
upflow
upgaze
upgoing
upgrow
upheap
upleap
uppile
upraiser
upreach
upreaching
uprise
uprush
upsend
upshoot
upsoar
upstare
upstep
upsurge
upswell
uptilt
uptoss
upvomit
upwaft
upwreathe

---

unwifelike
unwifely
unwincing
unwinking
unwinning
unwithered
unwithering
unwitnessed
unwomanlike
unwomanly
unwon
unwooded
unwooed
unworkable
unworked
unworkmanlike
unworn
unworshiped,
unworshipped
unwound
unwounded
unwoven
unwreathe
unwrought
unwrung
unyielding
unyouthful
unzealous

2 To a greater size or larger amount:
  upbulging   upflashing   uplight
  upflaring   upflooding   upswell

3 To a vertical position:
  upprop   upstand   upsticking

4 In or into commotion or activity:
  upboil   upbubbling   upstir

5 Completely; wholly:
  upbind   upfold   upgird
  updry   upgather   uphoard

up-and-com·ing (up′ənd-kum′ing) *adj.* Enterprising; energetic; promising.

U·pan·i·shad (ōō·pan′ə-shad, -pä′nə-shäd) *n. Sanskrit* Literally, a philosophical treatise; one of the treatises forming the third division of the Vedas, dealing with the nature of man and the universe.

u·par·na (ōō-pär′nə) *n.* A silk or muslin scarf, interwoven with gold or silver threads, worn as a shawl by men and as a shawl or veil by women in India. [< Hind.]

u·pas (yōō′pəs) *n.* **1** A tall evergreen moraceous tree (*Antiaris toxicaria*) of the island of Java, with an acrid, milky, poisonous juice. **2** The poisonous sap of this tree, used by natives in the manufacture of arrow poison; also, a similar poison from the **upas–tieute**, a climbing shrub (genus *Strychnos*) of the family *Loganiaceae*. **3** Hence, something morally deadly. [< Malay (*pohon*) *upas* poison (tree)]

up·beat (up′bēt′) *n. Music* An unaccented beat; the beat at which the hand is raised.

up·bow (up′bō′) *n.* An upward stroke of the violin bow, indicated in score by the symbol **V**: opposed to *down–bow*.

up·braid (up-brād′) *v.t.* To reproach for some wrongdoing; scold or reprove. — *v.i.* To utter reproaches. See synonyms under REPROVE, REVILE. [OE *upbregdan* < *up–* up + *bregdan* weave, twist] — **up·braid′er** *adj.* — **up·braid′ing** *n.* — **up·braid′ing·ly** *adv.*

up·bring·ing (up′bring′ing) *n.* Rearing and training received by a person during childhood.

up·bye (up′bī) *adv. Scot.* A little farther on; up the way. Also **up′by**.

up·cast (up′kast′, -käst′) *adj.* Cast, turned, or directed upward. — *n.* **1** A casting or throwing upward; that which is so cast. **2** An airshaft in a mine. **3** An upward current of air, as in a mine shaft. **4** *Scot.* An upset, or a reproach.

up–coun·try (up′kun′trē) *Colloq. n.* Country somewhat distant from the seashore or from lowlands; inland country. — *adj.* Living in, from, or characteristic of inland places. — *adv.* (up′kun′trē) In, into, or toward the interior: to move *up–country*.

up·date (up-dāt′) *v.t.* ·dat·ed, ·dat·ing To bring up to date; to revise, with corrections, additions, etc., as a textbook or manual.

up–end (up′end′) *v.t. & v.i.* To set or stand on end.

U·per·ni·vik (ōō-pûr′ni-vēk) A Danish settlement on an islet in Baffin Bay, western Greenland. Also **U·per′na·vik**.

up·grade (up′grād′) *n.* An upward incline or slope. — *v.t.* (up-grād′) ·grad·ed, ·grad·ing **1** To improve the breed of (animals) by the introduction of a higher strain. **2** To raise to a higher grade, rank, or responsibility, as an employee.

up·growth (up′grōth′) *n.* **1** The process of growing up. **2** That which grows or has grown up.

up·heav·al (up-hē′vəl) *n.* **1** The act of upheaving, or the state of being upheaved. **2** *Geol.* An elevation of the earth's surface due to a warping of large rock masses. **3** Overthrow or violent disturbance of the established social order.

up·heave (up-hēv′) *v.* ·heaved or ·hove, ·heav·ing *v.t.* To heave or raise up. — *v.i.* To be raised or lifted.

up·held (up-held′) Past tense and past participle of UPHOLD.

up·hill (up′hil′) *adv.* Up or as up a hill or ascent; against difficulties. — *adj.* **1** Going up a hill or an ascent; sloping upward. **2** Attended with difficulty or exertion. — *n.* (up′hil′) An upward slope; rising ground; a steep rise.

up·hold (up-hōld′) *v.t.* ·held, ·hold·ing **1** To hold up; raise. **2** To keep from falling or sinking. **3** To give aid or support to; encourage. **4** To regard with approval. See synonyms under ABET, AID, ASSENT, CONFIRM, HELP, JUSTIFY, PRESERVE, SUPPORT. — **up·hold′er** *n.*

up·hol·ster (up-hōl′stər) *v.t.* **1** To fit, as furniture, with coverings, cushioning, etc. **2** To provide or adorn with hangings, curtains, etc., as an apartment. **3** To furnish with a covering of any kind. [Back formation < UPHOLSTERER]

up·hol·ster·er (up-hōl′stər-ər) *n.* One who furnishes upholstery; one who upholsters. [< obs. *upholster*, alter. of ME *upholder* a tradesman + -ER¹]

up·hol·ster·y (up-hōl′stər-ē, -strē) *n. pl.* ·ster·ies **1** Goods used in upholstering. **2** Textile decoration of an apartment. **3** The act, art, or business of upholstering.

u·phroe (yōō′frō, yōōv′rō) See EUPHROE.

up·keep (up′kēp′) *n.* The act or state of maintenance; also, means or cost of maintenance.

up·land (up′lənd, -land′) *n.* **1** The higher portions of a region, district, farm, etc. **2** The country in the interior. — *adj.* **1** Pertaining to an upland; higher in situation. **2** Pertaining to or situated in inland districts.

upland plover See under PLOVER.

up·lift (up-lift′) *v.t.* **1** To lift up, or raise aloft; elevate. **2** To raise the tone of; put on a higher plane, mentally or morally. See synonyms under HEIGHTEN, RAISE. — *adj.* (up′lift′) Uplifted: a rare form. — *n.* (up′lift′) **1** The act of raising; the fact of being raised. **2** A movement upward. **3** *Geol.* An upheaval. **4** Mental or spiritual stimulation or elevation. **5** A social movement aiming to improve, morally or esthetically, the condition of the underprivileged. **6** A brassière designed to lift and support the breasts. — **up·lift′er** *n.*

up·most (up′mōst′) *adj.* Uppermost.

U·po·lu (ōō-pō′lōō) One of the chief islands of Western Samoa; 430 square miles; capital, Apia.

up·on (ə-pon′, ə-pôn′) *prep.* **1** On, in all its meanings. **2** On, in an elevated position: *upon* the throne. **3** On, by motion upward: to get *upon* a roof. — *adv.* On: completing a verbal idea: The paper has been written *upon*. Also *Scot.* **up·o′**. See synonyms under ABOVE. [ME]
◆ *Upon* now differs little in use from *on*, the former being sometimes used for reasons of euphony and also preferably when motion into position is involved, the latter when merely rest or support is to be indicated. When *upon* has its original meaning of *up* and *on*, it is written as two words, *up* having its adverbial force: Let us go *up on* the roof.

up·per (up′ər) *adj.* **1** Higher than something else; being above. **2** Higher in place: opposed to *lower*. **3** Higher in station or dignity; superior: opposed to *inferior*: the *upper* house. — **to get the upper hand** To get the advantage. — *n.* **1** That part of a boot or shoe above the sole: the vamp. **2** *pl.* Cloth gaiters. — **on one's uppers** *Colloq.* Having worn out the soles of one's shoes; hence, at the end of one's resources; destitute. [ME, orig. compar. of UP]

Up·per (up′ər) *adj. Geol.* Designating a later period or a later formation of a specified period: the *Upper* Cambrian.

Upper Austria A province of northern Austria; 4,624 square miles; capital, Linz. German **O·ber·ös·ter·reich** (ō′bər·œs′tə·rīkh).

Upper Bavaria An administrative division of southern Bavaria, West Germany; 6,308 square miles; capital, Munich. German **O·ber·bay·ern** (ō′bər·bī′ərn).

upper berth The top berth in a ship, railroad sleeping-car, cabin, etc., where two bunks or beds are built one above the other.

Upper Burma See BURMA, UNION OF.

Upper Canada A former British province (1791 to 1840) in the southern part of Ontario province, Canada.

upper case 1 Case² (def. 3). **2** Capital letters.

upper class The socially or economically superior group in society. — **up′per–class′** (-klas′, -kläs′) *adj.*

upper classman A junior or senior in a school or college.

upper crust *Colloq.* That portion of society assuming or thought of as having more social standing by reason of wealth or ancestry.

up·per·cut (up′ər-kut′) *n.* In boxing, a blow upward from the waist or hip, delivered under or inside the opponent's guard. — *v.t. & v.i.* ·cut, ·cut·ting To strike with an uppercut.

Upper Egypt See under EGYPT.

Upper Franconia An administrative division of NE Bavaria, Germany; 2,897 square miles; capital, Bayreuth. German **O·ber·fran·ken** (ō′bər·fräng′kən).

upper hand The advantage.

Upper House The branch, in a bicameral legislature, where membership is more restricted, as the U. S. Senate and the English House of Lords. Also **upper house**.

up·per·most (up′ər-mōst′) *adj.* **1** Highest in place, rank, authority, or vantage ground. **2** First to come into the mind: one's *uppermost* thoughts. Also *upmost*. — *adv.* In the highest place; also, first, as in time.

Upper New York Bay An arm of the Atlantic at the junction of the Hudson and the East River, joined to Newark Bay and Long Island Sound.

Upper Peninsula The northern part of Michigan, between Lake Superior and Lake Michigan; 16,538 square miles; 320 miles long, 125 miles wide.

Upper Silesia A former province of eastern Germany, now in western Poland.

Upper Vol·ta (vol′tə), **Republic of the** An independent republic of the French Community in western Africa; 105,900 square miles; capital, Ouagadougou; formerly the French overseas territory of Upper Volta.

up·pish (up′ish) *adj. Colloq.* Inclined to be self–assertive; assuming; pretentious; snobbish. Also **up′pi·ty**. — **up′pish·ly** *adv.* — **up′pish·ness** *n.*

up·raise (up-rāz′) *v.t.* ·raised, ·rais·ing To lift up; elevate. Also **up·rear′** (-rir′).

up·right (up′rīt′) *adj.* **1** Being in a vertical position; erect. **2** Morally correct; especially, just and honest. See synonyms under GOOD, HONEST, INNOCENT, JUST, MORAL, PURE, VIRTUOUS. — *n.* **1** Something having a vertical position, as an upright timber or piano. **2** The state of being upright: a post out of *upright*. **3** In football, one of the goal posts. — *adv.* Vertically; honestly; sincerely; justly. [OE *upriht* < *up–* up + *riht* right] — **up′right′ly** *adv.* — **up′right′ness** *n.*

upright piano See under PIANO.

up·rise (up-rīz′) *v.i.* ·rose, ·ris·en, ·ris·ing **1** To get up; rise, as from a seat or from sleep. **2** To be or become erect. **3** To go upward; ascend. **4** To increase; swell. **5** To rise into view. **6** To rise in revolt. — *n.* (up′rīz′) **1** The act of rising; ascent. **2** An upward slope; upgrade.

up·ris·ing (up-rī′zing, up′rī′zing) *n.* **1** The act of rising. **2** Revolt; insurrection. **3** An ascent; a slope; acclivity.

up–riv·er (up′riv′ər) *adj. & adv.* On or toward the upper part of a river. — *n.* A region located up–river.

up·roar (up′rôr′, -rōr′) *n.* Violent disturbance and noise; tumult. See synonyms under NOISE, TUMULT. — *v.* (up-rôr′, -rōr′) *v.t. & v.i.* To make an uproar. — *v.t.* To throw into uproar or confusion. [< Du. *oproer* < *op–* up + *roeren* stir]

up·roar·i·ous (up-rôr′ē-əs, -rō′rē-) *adj.* Accompanied by or making uproar. See synonyms under NOISY. — **up·roar′i·ous·ly** *adv.* — **up·roar′i·ous·ness** *n.*

up·root (up-rōōt′, -rōōt′) *v.t.* To tear up by the roots; eradicate; destroy utterly. See synonyms under EXTERMINATE. — **up·root′al** *n.* — **up·root′er** *n.*

up·rouse (up-rouz′) *v.t.* ·roused, ·rous·ing To rouse up, as from sleep.

Up·sa·la (up-sä′lə, *Sw.* ōōp′sä·lä) A city in eastern Sweden; site of Upsala University, founded 1477, one of world's oldest universities. Also **Upp·sa′la**.

up·set (up-set′) *v.* ·set, ·set·ting *v.t.* **1** To overturn. **2** To throw into confusion or disorder. **3** To disconcert; derange or disquiet. **4** To defeat, especially unexpectedly: Navy *upset* Army. 5 To shorten and thicken (metal) by hammering or by pressure: to *upset* a bolt to form a head or to *upset* the metal tire of a wheel. 6 To swage (the ends of the teeth of a saw). —*v.i.* 7 To become overturned. —*adj.* (*also* up′set′) Set up; required: in the phrase **upset price,** a price at which property is offered for sale, as by an auctioneer, as the lowest selling price. —*n.* (up′set′) The act of upsetting, or the state of being upset. —**up·set′ter** *n.*

up·set·ting (up·set′ing) *adj.* 1 Overturning; disturbing; disconcerting. 2 *Scot.* Conceited; ambitious; forward: an *upsetting* boy. —*n.* The act of upsetting or the condition of being upset; disturbance; disquiet.

up·shot (up′shot′) *n.* 1 The final outcome. 2 *Obs.* The final shot in an archery match; the final or parting shot. See synonyms under CONSEQUENCE.

up·side (up′sīd′) *n.* The upper side or part. —**to be upsides with** *Brit. Dial.* To be even with.

up·side–down (up′sīd′doun′) *adj.* Having the upper side down; in disorder. —*adv.* With the upper side down; in disorder. [Alter. of ME *up so down* up as if down]

up·si·lon (yo͞op′sə·lon, *Brit.* yo͞op·sī′lən) *n.* The twentieth letter and sixth vowel in the Greek alphabet (Υ, υ): having the sound of French *u,* Latin and Old English *y.* It is transliterated in English as *u* or *y.* See Y. [<Gk. *ypsilon* smooth y]

up·spring (up′spring′) *n.* 1 A leap up into the air. 2 *Obs.* An upstart. —*v.i.* **·sprang** or **·sprung, ·spring·ing** To spring up.

up·stage (up′stāj′) *n.* The half of a stage, from left to right, extending from the center to the backdrop. —*adj.* 1 Pertaining to the back half of a stage. 2 *Colloq.* Conceited; haughty; stuck–up; supercilious. —*adv.* Toward or on the back half of a stage. —*v.t.* **·staged, ·stag·ing** To steal a scene from (another actor).

up·stairs (up′stârz′) *adj.* Pertaining to an upper story. —*n.* The upper story; the part of a building above the ground floor. —*adv.* In, to, or toward an upper story. —**to kick upstairs** To promote so as to get out of the way.

up·stand·ing (up·stan′ding) *adj.* Standing up; erect; hence, honest; upright; straightforward.

up·start (up·stärt′) *v.i.* To start or spring up suddenly. —*adj.* (up′stärt′) 1 Suddenly raised to prominence, wealth, or power. 2 Characteristic of a parvenu; vulgar; pretentious. —*n.* (up′stärt′) One who or that which springs up suddenly; especially, one who has suddenly risen from a humble position to consequence; a parvenu, especially one who assumes an arrogant tone or bearing.

up·state (up′stāt′) *U.S. adj.* Of, from, or designating that part of a State lying outside, usually north, of the principal city. —*n.* The outlying, usually northern, sections of a State; especially, the northern part of New York State. —*adv.* In or toward the outlying or northern sections of a State. —**up′stat′er** *n.*

up·stream (up′strēm′) *adv.* Toward the upper part of a stream; against the current; toward or at a place nearer the source.

up·stroke (up′strōk′) *n.* An upward stroke, as of a pen or pencil.

up·sweep (up′swēp′) *n.* 1 A sweeping up or upward. 2 The upturning of the lower jaw, as in the bulldog. —*v.t.* & *v.i.* (up·swēp′) **·swept, ·sweep·ing** To brush or sweep upward or up.

up–swept (up′swept′) *adj.* Of or pertaining to a style of hairdressing in which the hair is swept upward smoothly in the back and piled high on the top of the head.

up·swing (up′swing′) *n.* 1 A swinging upward. 2 An improvement. —*v.i.* (up·swing′) **·swung, ·swing·ing** To swing upward; improve.

up·take (up′tāk′) *n.* 1 The act of lifting or taking up. 2 A boiler flue that unites the combustion gases and carries them toward the smokestack. 3 An upward ventilating shaft in a mine. 4 Mental comprehension; understanding.

up·throw (up′thrō′) *n.* 1 A throwing upward; an upheaval. 2 *Geol.* An upward displacement of the rock on one side of a fault.

up·thrust (up′thrust′) *n.* 1 An upward thrust. 2 *Geol.* An upheaval (usually violent) of rocks in the earth's crust.

up to See under UP.

up–to–date (up′tə·dāt′) *adj.* Having the latest information, fashion, manner, or improvement: an *up-to-date* dictionary.

up to date To the present time.

up·town (up′toun′) *adv.* In or toward the upper part of a town. —*adj.* Pertaining to or resident in the upper part of a town or city, or that part which is conventionally regarded as the upper part, usually the residence section.

up·turn (up·tûrn′) *v.t.* To turn up or over, as sod with the plow; hence, to overturn; upset. —*n.* (up′tûrn′) A turning upward; an increase; an improvement.

up·ward (up′wərd) *adv.* 1 In or toward a higher place; in an ascending course or direction; toward the source: to look *upward;* to trace a stream *upward.* 2 With increase or advancement; toward a higher price: Prices tended *upward.* 3 In excess; more: children five years old and *upward.* 4 Toward that which is better, nobler, or holier. 5 In the upper parts. Also **up′wards.** —**upward of** or **upwards of** Higher than or in excess of. —*adj.* Turned or directed toward a higher place. [OE *upweard* < *up-* up + *-weard* -WARD] —**up′ward·ly** *adv.*

Ur (ûr) An ancient city of Sumer, southern Mesopotamia, the site of which is on the Euphrates in SE Iraq. Old Testament **Ur of the Chal·dees** (kal·dēz′, kal′dēz).

ur–[1] Var. of URO–[1].

ur–[2] Var. of URO–[2].

u·ra·chus (yo͝or′ə·kəs) *n. Anat.* A canal connecting the bladder of the fetus with the allantois. [<NL <Gk. *ourachos* urinary canal of a fetus < *ouron* urine + *echein* hold]

u·rae·mi·a (yo͞o·rē′mē·ə), **u·rae·mic** (yo͞o·rē′·mik) See UREMIA, etc.

u·rae·us (yo͞o·rē′əs) *n.* The emblem of the sacred serpent (haje) in the headdress of Egyptian divinities and kings: a symbol of sovereignty. [<NL <Gk. *ouraios* of a tail < *oura* tail]

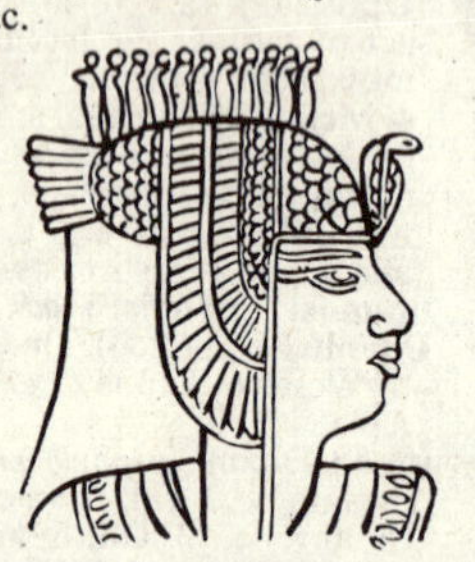

URAEUS

U·ral (yo͝or′əl, *Russian* o͞o·räl′) A river in SE European Russian S.F.S.R., and SW Kazakh S.S.R., formerly considered as forming part of the boundary between Asia and Europe and flowing 1,574 miles south and west from the Ural Mountains to the Caspian Sea.

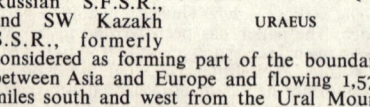

U·ral–Al·ta·ic (yo͝or′əl·al·tā′ik) *n.* A hypothesized family of languages embracing almost all the agglutinative languages of Europe and northern Asia, comprising the Uralic (Finno–Ugric, Samoyedic) and Altaic (Turkic, Mongolian, Manchu–Tungusic) subfamilies. —*adj.* 1 Of or pertaining to the Ural and Altai mountain ranges. 2 Of, pertaining to, or designating the Ural–Altaic languages, or any of the peoples natively speaking any of these languages. Also called *Turanian, Ugro–Altaic.*

U·ral·ic (yo͞o·ral′ik) *n.* A family of agglutinative languages comprising the Finno–Ugric and Samoyedic subfamilies: by some linguists classified with Altaic in one great Ural–Altaic family. —*adj.* Of or pertaining to this linguistic family. Also **U·ra·li·an** (yo͞o·rā′lē·ən).

u·ral·ite (yo͝or′əl·īt) *n.* A pyroxene altered to amphibole. [from *Ural* Mountains] —**u′ral·it′ic** (-it′ik) *adj.*

U·ral Mountains (yo͝or′əl, *Russian* o͞o·räl′) A mountain system in the Russian S.F.S.R., the traditional boundary between Asia and Europe; 1,300 miles long, from the Arctic Ocean to the Kazakh S.S.R. border; highest peak, 6,184 feet.

U·ralsk (o͞o·rälsk′) A city on the Ural river in NW Kazakh S.S.R.

u·ra·nal·y·sis (yo͝or′ə·nal′ə·sis) See URINALYSIS.

U·ra·ni·a (yo͞o·rā′nē·ə) 1 The Muse of astronomy. 2 The heavenly one: an epithet of Aphrodite. [<L <Gk. *Ourania* < *ouranios* heavenly < *ouranos* heaven]

U·ra·ni·an (yo͞o·rā′nē·ən) *adj.* 1 Of or pertaining to the planet Uranus. 2 Celestial.

u·ran·ic (yo͞o·ran′ik) *adj.* 1 Relating to the heavens; astronomical. 2 *Chem.* Pertaining to or derived from uranium, especially in its higher valence. [<Gk. *ouranos* heaven]

u·ran·i·nite (yo͞o·ran′ə·nīt) *n.* A greenish–black, opaque uranium mineral containing also lead, nitrogen, helium, thorium, radium, and certain rare earths, occurring in octohedral crystals. In the massive form it is called *pitchblende.* [<URANIUM]

u·ran·ism (yo͝or′ən·iz′əm) *n.* Homosexuality: opposed to *dionism.* [<Gk. *Ourania* URANIA (def. 2)]

u·ra·nite (yo͝or′ə·nīt) *n.* Any of several uranium minerals, especially uranium phosphates, torbernite or copper uranite, autunite or lime uranite. [<URANIUM] —**u′ra·nit′ic** (-nit′ik) *adj.*

u·ra·ni·um (yo͞o·rā′nē·əm) *n.* A heavy, nickel–white, radioactive, metallic element (symbol U) of the actinide series, found only in combination: it is a principal source of radium, and one of its isotopes is important in the generation of atomic energy. See ELEMENT. [<URANUS]

uranium series *Physics* A series of radioactive elements beginning with uranium of mass 238 and a half–life of 4.5×10^{10} years and continuing through successive disintegrations to the stable isotope of lead of mass 206.

urano– *combining form Astron.* The heavens; of or pertaining to the heavens, or to celestial bodies: *uranography.* Also, before vowels, **uran–.** [<Gk. *ouranos* heaven]

u·ra·nog·ra·phy (yo͝or′ə·nog′rə·fē) *n.* Scientific description of the celestial bodies; the making of celestial globes and maps: also spelled *ouranography.* —**u′ra·nog′ra·pher** or **·phist** *n.* —**u′ra·no·graph′ic** (-nō·graf′ik) or **·i·cal** *adj.*

u·ra·nous (yo͝or′ə·nəs) *adj. Chem.* Of or pertaining to uranium, especially in its lower valence.

U·ra·nus (yo͝or′ə·nəs) 1 In Greek mythology, the son and husband of Gaea (Earth) and father of the Titans, Furies, and Cyclops: overthrown by his son Kronos: also spelled *Ouranos.* 2 *Astron.* A planet of the solar system; seventh in distance from the sun. Its mean distance from the sun is 1,781 millions of miles, its sidereal period about 84 terrestrial years, and its diameter about 31,000 miles. It has four satellites. See PLANET. [<L <Gk. *Ouranos* < *ouranos* heaven]

u·ra·nyl (yo͝or′ə·nil) *n. Chem.* The bivalent radical UO_2, found in many uranium compounds. [<URANIUM + -YL]

u·ra·re (yo͞o·rä′rē) *n.* Curare: also called *oorali.* Also **u·ra′ri.**

U·ra·ri·coe·ra (o͞o·rä′rē·kwä′rə) A river in northern Brazil, flowing 300 miles east from the Venezuela border to the Río Branco.

u·rase (yo͝or′ās) See UREASE.

u·rate (yo͝or′āt) *n. Chem.* A salt of uric acid.

U·ra·wa (o͞o·rä·wä) A city of central Honshu island, Japan.

ur·bac·i·ty (ûr·bas′ə·tē) *n.* The character or quality of being so narrow that one's interests do not extend beyond the limits of one's own city. [<L *urbs, urbis* a city]

ur·ban (ûr′bən) *adj.* Pertaining to, characteristic of, including, or constituting a city; situated or dwelling in a city. [<L *urbanus.* See URBANE.]

Ur·ban (ûr′bən) Name of eight popes. —**Urban II,** 1042?–99, pope 1088–99; preached the First Crusade.

urban district A subdivision, for administrative purposes, of a shire of England, Wales, or Northern Ireland, usually comprising several thickly populated communities.

ur·bane (ûr·bān′) *adj.* 1 Characterized by or having refinement, especially in manner; polite; courteous; suave: opposed to *rustic.* 2 *Obs.* Urban. See synonyms under POLITE. [<L *urbanus* of a city < *urbs, urbis* a city] —**ur·bane′ly** *adv.* —**ur·bane′ness** *n.*

ur·ban·i·ty (ûr·ban′ə·tē) *n. pl.* **·ties** 1 The character or quality of being urbane; refined or elegant courtesy; strictly, the city quality, from the assumption that life in the city results in superior refinement. 2 *Obs.* Polished humor or wit. [<L *urbanitas, -tatis* < *urbs, urbis* a city]

ur·ban·ize (ûr′bən·īz) *v.t.* **·ized, ·iz·ing** 1 To render urban. 2 *Rare* To render urbane. —**ur′ban·i·za′tion** *n.*

ur·bi·cul·ture (ûr′bə·kul′chər) *n.* The study of the proper development, planning, and use of cities, especially in relation to the needs of their inhabitants. [<L *urbs, urbis* a city + CULTURE]

ur·bi et or·bi (ûr′bī et ôr′bī) *Latin* To the city (Rome) and to the world: used in official announcements, as papal bulls.

Urbs Ve·tus (ûrbz vē′təs) The ancient Latin name for ORVIETO.

ur·ce·o·late (ûr′sē·ə·lit, -lāt′) *adj. Bot.* Pitcher- or urn-shaped, as a corolla. [<L *urceolus*, dim. of *urceus* pitcher]

ur·chin (ûr′chin) *n.* 1 A roguish, mischievous boy. 2 A cylinder in a carding machine. 3 A hedgehog. 4 A sea urchin. 5 *Obs.* An elf, as often assuming the form of a hedgehog. — *adj. Obs.* Elfish; mischievous. [ME *irchoun* <OF *irechon, ireçon* <L *ericius* hedgehog < *er* hedgehog]

Ur·du (ŏŏr′dŏŏ, ôr·dŏŏ′, ûr′dŏŏ) *n.* A variety of Hindustani used by the Moslems, containing many Arabic and Persian elements and written in the Arabic alphabet: the official language of Pakistan. Also spelled *Oordoo*. [<Hind. *urdū*, short for (*zaban-i-*) *urdū* (language of the) camp <Turkish *ordū* camp <Persian *urdū*. Related to HORDE.]

-ure *suffix of nouns* 1 The act, process, or result of: *pressure*. 2 The function, rank, or office of: *prefecture*. 3 The means or instrument of: *ligature*. [<F <L *-ura*]

u·re·a (yŏŏ·rē′ə) *n. Biochem.* A very soluble colorless crystalline compound, $CO(NH_2)_2$, formed by the oxidation of nitrogenous compounds in the body, and also made synthetically: used in medicine and in the making of plastics. Also called *carbamide*. [<NL <F *urée* <Gk. *ouron*] — **u·re′al** *adj.*

urea resin Any of a class of thermosetting resins obtained by the reaction of urea and formaldehyde in the presence of certain modifying agents.

u·re·ase (yŏŏr′ē·ās, -āz) *n. Biochem.* An enzyme which promotes the hydrolysis of urea, with the formation of ammonium carbonate. [<UREA + -ASE]

U·red·i·na·les (yŏŏ·red′ə·nā′lēz) *n. pl.* An order of fungi characterized by a branched, septate mycelium and the formation of reddish or yellow spores; the rust fungi. [<NL <L *uredo, -inis* blast, blight <*urere* burn]

u·re·din·i·um (yŏŏr′ə·din′ē·əm) *n. Bot.* The spore fruit of a rust fungus which produces the uredospores. Also **u·re·di·um** (yŏŏ·rē′dē·əm), **u′re·do·so′rus** (-dō·sôr′əs, -sō′rəs). [<NL <L *uredo, -inis* blight. See UREDO.]

u·re·do (yŏŏ·rē′dō) *n.* 1 The uredo stage. 2 Urticaria. [<L, blight <*urere* burn]

u·re·do·spore (yŏŏ·rē′dō·spôr, -spōr) *n. Bot.* A unicellular thin-walled spore produced as a repeating generation in summer as part of the life cycle of a rust fungus. Also **u·re·din·i·o·spore** (yŏŏr′ə·din′ē·ə·spôr′, -spōr′).

uredo stage *Bot.* The stage in the life history of certain rust fungi during which uredospores are produced.

u·re·ide (yŏŏr′ē·īd, -id) *n. Chem.* Any of several nitrogenous compounds derived from urea and an acid or aldehyde by the removal of water.

u·re·mi·a (yŏŏ·rē′mē·ə) *n. Pathol.* An abnormal condition of the blood due to the presence of urea with other urinary constituents ordinarily excreted by the kidneys. Also *uraemia, urinemia*. [<NL <UR-¹ + -EMIA] — **u·re′mic** *adj.*

-uret *suffix Chem.* Used to denote a compound: now replaced by *-ide*. [<F <-ure -URE]

u·re·ter (yŏŏ·rē′tər) *n. Anat.* The duct by which urine passes from the kidney to the bladder or the cloaca. [<NL <Gk. *ourētēr* < *ourein* urinate] — **u·re′ter·al, u·re·ter·ic** (yŏŏr′ə·ter′ik) *adj.*

u·re·ter·ec·to·my (yŏŏ·rē′tə·rek′tə·mē) *n. Surg.* Excision of all or part of the ureter. [<URETER(O)- + -ECTOMY]

uretero- *combining form Med.* A ureter; of or related to a ureter. Also, before vowels, **ureter-**. [<Gk. *ourētēr* <*ourein* urinate]

u·re·than (yŏŏr′ə·than′, yŏŏ·reth′ən) *n. Chem.* 1 A white crystalline compound, $C_3H_7NO_2$, derived from carbamic acid by substituting ethyl for the hydrogen of the hydroxyl group: some of its derivatives are used as hypnotics and sedatives: also called *ethylurethane*.

2 Any ester of carbamic acid. Also **u·re·thane** (yŏŏr′ə·thān′, yŏŏ·reth′ān). [<UR(EA) + ETHAN(E)]

u·re·thra (yŏŏ·rē′thrə) *n. Anat.* The duct by which urine is discharged from the bladder of most mammals, and which, in males, carries the seminal discharge. [<LL <Gk. *ourēthra* < *ouron* urine] — **u·re′thral** *adj.*

u·re·thri·tis (yŏŏr′ə·thrī′tis) *n. Pathol.* Inflammation of the urethra: *urethroscope*. Also, before vowels, **urethr-**. [<Gk. *ourēthra* the urethra]

urethro- *combining form Med.* The urethra; of or pertaining to the urethra: *urethroscope*. Also, before vowels, **urethr-**. [<Gk. *ourēthra* the urethra]

u·re·thro·scope (yŏŏ·rē′thrə·skōp) *n. Med.* An instrument for examining the urethra. — **u·re′thro·scop′ic** (-skop′ik) *adj.* — **u·re·thros·co·py** (yŏŏr′ə·thros′kə·pē) *n.*

u·ret·ic (yŏŏ·ret′ik) *adj. Med.* 1 Diuretic. 2 Of or pertaining to the urine; urinary. [<LL *ureticus* <Gk. *ourētikos* < *ouron* urine]

U·rey (yŏŏr′ē), **Harold Clayton**, born 1893, U.S. chemist.

Ur·fa (ŏŏr·fä′) A city in southern Turkey in Asia, near the Syrian border: ancient *Edessa*.

Ur·fé (dür·fā′), **Honoré d'**, 1568–1625, French novelist.

Ur·ga (ŏŏr′gä) The former name for ULAN BATOR.

urge (ûrj) *v.* **urged, urg·ing** *v.t.* 1 To drive or force forward; impel; push. 2 To plead with or entreat earnestly, as with arguments or explanations: He *urged* them to accept the plan. 3 To press or argue the doing, consideration, or acceptance of; advocate earnestly. 4 To move or force to some course or action; constrain. 5 To stimulate or excite; incite; intensify. 6 To ply or use vigorously, as oars. — *v.i.* 7 To present or press arguments, claims, etc. 8 To exert an impelling or prompting force. See synonyms under ACTUATE, PERSUADE, PIQUE, PLEAD, PUSH, QUICKEN. — *n.* 1 A strong impulse to perform a certain act. 2 The act of urging; the state of being urged. [<L *urgere* drive, urge]

Ur·gel (ŏŏr·hel′) A city in Lérida, NE Spain, in the Pyrenees SW of Andorra; seat of a bishop who is joint suzerain of Andorra.

ur·gen·cy (ûr′jən·sē) *n. pl.* **·cies** 1 The quality of being urgent. 2 Pressure by entreaty; pressure of necessity. 3 The act of urging. 4 Something urgent. See synonyms under NECESSITY.

ur·gent (ûr′jənt) *adj.* 1 Characterized by urging or importunity; requiring prompt attention; pressing; imperative. 2 Eagerly importunate or insistent. [<F <L *urgens, -entis*, ppr. of *urgere* drive] — **ur′gent·ly** *adv.*

Synonyms: importunate, pertinacious, pressing, solicitous.

-urgy *combining form* Development of or work with a (specified) material or product: *metallurgy, chemurgy, zymurgy*. [<Gk. *-ourgia* < *ergon* work]

U·ri (ŏŏr′ē) A canton in central Switzerland; 415 square miles; capital, Altdorf.

-uria *combining form Pathol.* A (specified) condition of the urine: usually used to indicate disease or abnormality: *hematuria, dysuria*. [<NL <Gk. *-ouria* < *ouron* urine]

U·ri·ah (yŏŏ·rī′ə) A masculine personal name. Also *Ital.* **U·ri·a** (ŏŏ·rē′ä), **U·ri·as** (*Ger.* ŏŏ·rē′äs, *Lat.* yə·rī′əs), *Fr.* **U·rie** (ü·rē′). [<Hebrew, God is light]

— **Uriah** A Hittite captain in the Israelite army, husband of Bathsheba, treacherously sent to his death by David. II *Sam.* xi 15–17.

Uriah Heep An unctuous, fawning, scheming character in Dickens's *David Copperfield*; hence, an odious hypocrite.

u·ric (yŏŏr′ik) *adj.* Of, pertaining to, or derived from urine. [<F *urique*]

uric acid *Biochem.* A white, almost insoluble dibasic acid, $C_5H_4N_4O_3$, of varying crystalline forms, found in small quantity in human urine. It is a product of the incomplete oxidation of animal tissue and animal diet, and forms the nucleus of most urinary and renal calculi.

urico- *combining form* Uric acid; of or related to uric acid: *uricolysis*, the splitting up of uric acid. Also, before vowels, **uric-**. [<URIC]

U·ri·el (yŏŏr′ē·əl) A masculine personal name. [<Hebrew, light of God]

— **Uriel** One of the seven archangels of Christian legend: in Milton's *Paradise Lost*, represented as "regent of the sun."

U·rim (yŏŏr′im) *n. pl.* 1 Objects mentioned in the Old Testament (*Ex.* xxviii 30, etc.) in connection with the breastplate of the high priest: generally in the phrase **Urim and Thummim**, supposed to have been precious stones used in casting lots, one signifying an affirmative and the other a negative answer. 2 In Mormon theology, with the Thummim, the sacred objects used by seers under divine direction, especially those used by Joseph Smith in translating the *Book of Mormon*. [<Hebrew *ūrīm* fires < *ūr* shine]

u·ri·nal (yŏŏr′ə·nəl) *n.* 1 A toilet or closet convenience or fixture for men's use in urination; also, a private place containing such conveniences for public use, as in a park. 2 A receptacle for urine; a glass receptacle, as a bottle, used in the inspection of urine. [<OF <Med.L *urinale*, orig. neut. of L *urinalis* pertaining to urine < *urina* urine]

u·ri·nal·y·sis (yŏŏr′ə·nal′ə·sis) *n. pl.* **·ses** (-sēz) Chemical analysis of the urine: also spelled *uranalysis*. [<NL <URIN(O)- + (AN)ALYSIS]

u·ri·nar·y (yŏŏr′ə·ner′ē) *adj.* Of, pertaining to, or concerned in the production and excretion of urine: the *urinary* organs. — *n. pl.* **·nar·ies** 1 A reservoir for storing urine, etc., for use as manure. 2 A urinal.

urinary calculus *Pathol.* A concretion formed in the urinary passages; the stone.

u·ri·nate (yŏŏr′ə·nāt) *v.i.* **·nat·ed, ·nat·ing** To void or pass urine. [<Med. L *urinatus*, pp. of *urinare* pass urine < *urina* urine] — **u′ri·na′tion** *n.*

u·rine (yŏŏr′in) *n.* A pale-yellow fluid secreted from the blood of mammals by the kidneys, stored in the bladder, and voided through the urethra: the principal vehicle by which nitrogenous and saline matters are removed from the system. [<F <L *urina*]

u·ri·ne·mi·a (yŏŏr′ə·nē′mē·ə) *n.* Uremia. Also **u′ri·ne′mi·a**. [<URIN(O)- + -EMIA] — **u′ri·ne′mic** or **·nae′mic** *adj.*

u·ri·nif·er·ous (yŏŏr′ə·nif′ər·əs) *adj.* Concerned in the conveyance of urine.

urino- *combining form* Urine. Also, before vowels, **urin-**, as in *urinalysis*. [<L *urina* urine]

u·ri·no·gen·i·tal (yŏŏr′ə·nō·jen′ə·təl) *adj.* Urogenital.

u·ri·nos·co·py (yŏŏr′ə·nos′kə·pē) *n. pl.* **·pies** Uroscopy.

u·ri·nous (yŏŏr′ə·nəs) *adj.* Of, pertaining to, containing, or resembling urine. Also **u′ri·nose** (-nōs).

Ur·mi·a (ŏŏr′mē·ə), **Lake** The largest lake of Iran, between Tabriz and the Turkish border; in summer, 1,500 square miles; in winter, 2,300 square miles; 90 miles long, 30 miles wide: also *Rizaiyeh*. Persian **U·ru·mi·yeh** (ŏŏ·rŏŏ·mē·ye′).

urn (ûrn) *n.* 1 A rounded or angular vase having a foot, variously used in antiquity as a receptacle for the ashes of the dead, a water vessel, measure, etc. 2 A vessel for preserving the ashes of the dead; a grave. 3 In ancient Rome, a receptacle used to hold lots drawn in voting. 4 A vase-shaped receptacle having a faucet, and designed for keeping tea, coffee, etc., hot, as by means of a spirit lamp. ◆ *Homophones: earn, erne*. [<F *urne* <L *urna*]

URN
In park at Versailles.

uro-¹ *combining form* Urine; pertaining to urine or to the urinary tract: *urology*. Also, before vowels, **ur-**. [<Gk. *ouron* urine]

uro-² *combining form* A tail; of or related to the tail; caudal: *uropod*. Also, before vowels, **ur-**. [<Gk. *oura* a tail]

u·ro·bi·lin (yŏŏr′ə·bī′lin, -bī′lin) *n. Biochem.* A brownish, resinous bile pigment, found in urine and sometimes in the blood. [<URO-¹ + BILE + -IN]

---

add, āce, câre, pälm; end, ēven; it, īce; odd, ōpen, ôrder; tŏŏk, pōōl; up, bûrn; ə = a in *above*, e in *sicken*, i in *clarity*, o in *melon*, u in *focus*; yŏŏ = u in *fuse*; oi, oil; ou, pout; ch, check; g, go; ng, ring; th, thin; t͡h, this; zh, vision. Foreign sounds à, œ, ü, kh, ñ; and ◆: see page xx. < from; + plus; ? possibly.

u·ro·chord (yŏŏr'ə·kôrd) n. Zool. The notochord or central axis of larval ascidians and certain adult tunicates. [<URO-² + CHORD²] —**u'ro·chor'dal** adj.

U·ro·chor·da·ta (yŏŏr'ō·kôr·dā'tə) n. pl. The tunicates. [<NL <URO-² + CHORDATA]

u·ro·chrome (yŏŏr'ə·krōm) n. Biochem. The yellow pigment which gives to urine its characteristic color.

u·rochs (yŏŏr'oks) n. The urus. [<G]

U·ro·de·la (yŏŏr'ō·dē'lə) n. pl. Caudata. [<NL <URO-² + Gk. dēlos visible]

u·ro·gen·i·tal (yŏŏr'ō·jen'ə·təl) adj. Of or pertaining to the urinary and genital organs and their functions.

u·ro·gen·i·tals (yŏŏr'ō·jen'ə·təlz) n. pl. The urogenital organs.

u·rog·e·nous (yŏŏ·roj'ə·nəs) adj. Producing or promotive of the urinary secretion. [<URO-¹ + -GENOUS]

u·ro·lith (yŏŏr'ə·lith) n. Pathol. A urinary calculus. [<URO-¹ + -LITH¹] —**u'ro·lith'ic** adj.

u·ro·li·thi·a·sis (yŏŏr'ō·li·thī'ə·sis) n. Pathol. Any diseased condition due to the formation of urinary calculi. [<NL <URO-¹ + LITHIASIS]

u·rol·o·gy (yŏŏ·rol'ə·jē) n. The branch of medical science that relates to the urine and to the genitourinary tract in health and in disease. —**u·ro·log·ic** (yŏŏr'ə·loj'ik) or **·i·cal** adj. —**u·rol'o·gist** n.

u·ro·pod (yŏŏr'ə·pod) n. Zool. An abdominal or caudal limb or appendage of an arthropod, especially one of the posterior pairs of pleopods in a crustacean. [<URO-² + -POD] —**u·rop·o·dal** (yŏŏ·rop'ə·dəl), **u·rop'o·dous** adj.

u·ro·pyg·i·al (yŏŏr'ə·pij'ē·əl) adj. Of or pertaining to the uropygium.

uropygial gland Ornithol. The gland at the base of a bird's tail, secreting an oily substance used to preen the feathers.

u·ro·pyg·i·um (yŏŏr'ə·pij'ē·əm) n. Ornithol. The terminal part of the body supporting the tail feathers of a bird; rump. [<NL <Gk. ouropygion, alter. (after oura tail) of orrhopygion < orrhos end of the os sacrum + pygē rump]

u·ros·co·py (yŏŏ·ros'kə·pē) n. Med. Diagnosis by examination of the urine. [<URO-¹ + -SCOPY] —**u·ro·scop·ic** (yŏŏr'ə·skop'ik) adj. —**u·ros'co·pist** n.

U·ro·tro·pin (yŏŏ·rō'trō·pin) n. Proprietary name for a brand of methenamine.

u·ro·xan·thin (yŏŏr'ə·zan'thin) n. Indican (def. 2). [<URO-¹ + XANTHIN]

Ur·quhart (ûr'kərt), **Sir Thomas**, 1611–60, Scottish author; translator of Rabelais.

ur·sa (ûr'sə) n. Latin A she-bear: used in the phrases Ursa Major and Ursa Minor.

Ursa Major Astron. The Great Bear, a large northern constellation containing the seven conspicuous stars called the Septentriones, including the two Pointers, Dubhe and Merak, which point to the polestar: also called Big Dipper, the Dipper, Charles's Wain. See CONSTELLATION.

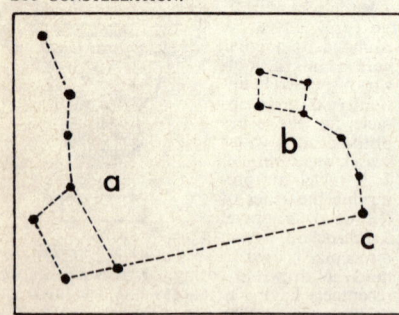

URSA MAJOR AND URSA MINOR
a. Ursa Major. b. Ursa Minor. c. Polestar.

Ursa Minor Astron. The Little Bear, a northern constellation including the polestar: also called Little Dipper, Dog's Tail. See CONSTELLATION.

ur·si·form (ûr'sə·fôrm) adj. Having the form of a bear. [<L ursus bear + -FORM]

ur·sine (ûr'sīn, -sin) adj. 1 Pertaining to or like a bear. 2 Clothed with dense bristles, as certain caterpillars. [<L ursinus <ursus bear]

ursine howler See HOWLER (def. 3).

Ur·spra·che (ŏŏr'shprä'khə) n. German A primitive, original, or parent language; particularly, a hypothetical primitive Indo-European language.

Ur·su·la (ûr'syə·lə, -sə-, Du. ŏŏr'sŏŏ·lä) A feminine personal name. Also Fr. **Ur·sule** (ür·sül'), Ger., Sw. **Ur·sel** (ŏŏr'səl), Sp. **Ur·so·la** (ŏŏr'sō·lä). [<L, little she-bear]
— **Ursula, Saint** A Cornish princess of the fourth or fifth century, martyred, according to legend, with eleven thousand virgins at Cologne by the Huns.

Ur·su·line (ûr'syə·lin, -sə-, -līn) adj. Pertaining to St. Ursula or to an order of nuns founded in 1537 by St. Angela Merici: they are engaged chiefly in the education of girls. — n. An Ursuline nun.

Ur·text (ŏŏr'tekst) n. German Earliest or primary form of a written text.

ur·ti·ca·ceous (ûr'ti·kā'shəs) adj. Bot. Belonging to a widely distributed family (Urticaceae) of trees, shrubs, or herbs, the nettle family, some of which are provided with sharp, stinging hairs. [<NL <L urtica nettle]

ur·ti·car·i·a (ûr'tə·kâr'ē·ə) n. Pathol. A disease of the skin, variously caused, characterized by evanescent, rounded elevations resembling wheals raised by a whip, and attended with intense itching; nettle rash; hives. [<NL <L urtica nettle] —**ur'ti·car'i·al** or **·i·ous** adj.

ur·ti·cate (ûr'tə·kāt) v.t. & v.i. ·cat·ed, ·cat·ing To sting, as with nettles. [<Med. L urticatus, pp. of urticare sting <urtica nettle]

ur·ti·ca·tion (ûr'tə·kā'shən) n. Med. 1 Formerly, the act, process, or effect of whipping with nettles as a stimulant, as in paralysis. 2 A tingling or burning sensation. 3 The development of urticaria.

U·ru·bam·ba (ŏŏ·rŏŏ·väm'bä) A river in southern Peru, flowing about 450 miles NW and north from the Andes of SE Peru to the Ucayali in central Peru.

U·ru·guay (yŏŏr'ə·gwā, Sp. ŏŏ'rŏŏ·gwī') 1 A republic of SE South America, on the Atlantic; 72,172 square miles; capital, Montevideo. 2 A river in SE South America, flowing 1,000 miles SW to the Río de la Plata. **U'ru·guay'an** adj. & n.

U·rum·chi (ŏŏ·rŏŏm'chē') The capital of the Sinkiang-Uigur Autonomous Region, NW China. Also **U'rum'tsi'**.

U·run·di (ŏŏ·rŏŏn'dē) A former district of German East Africa; since 1923, the southern county of Ruanda-Urundi, 10,658 square miles.

u·rus (yŏŏr'əs) n. An extinct, long-horned, wild ox of Germany (Bos primigenius), so named by Julius Caesar: also called aurochs, urochs. [<L <Gmc. Cf. OHG ur.]

u·ru·shi·ol (ŏŏ·rŏŏ·shē·ōl', -ol') n. A poisonous, irritant liquid, the active principle of poison ivy and the Japanese lac tree. [<Japanese urushi lacquer + -OL²]

us (us) pron. The objective case of WE. [OE ūs]

us·a·ble (yŏŏ'zə·bəl) adj. 1 Capable of being used. 2 That can be used conveniently. Also **use'a·ble**. —**us'a·ble·ness** n. —**us'a·bly** adv.

us·age (yŏŏ'sij, -zij) n. 1 The manner of using or treating a person or thing; treatment; also, the act of using. 2 Customary or habitual practice, or something permitted by it or done in accordance with it; custom or a custom: an act permitted by usage; ancient usages. 3 Law Uniform practice. 4 The way of using words, speech patterns, etc., that is general and established among the majority of the native speakers and writers of a language. 5 Obs. Conduct; behavior. See synonyms under HABIT. — **nonjurors' usages** In English and Scottish history, certain ceremonies, including mixing wine with water, prayer for the dead, trine immersion at baptism, the chrism at confirmation, anointing of the sick, etc., adopted by the nonjurors. [<OF <Med. L usaticum <L usus. See USE.]

us·ance (yŏŏ'zəns) n. 1 A period of time, variable as between various countries, which, by commercial usage, is allowed, exclusive of days of grace, for payment of bills of exchange, especially foreign. 2 Econ. An income derived from the possession of wealth in any way it may be invested. 3 Obs. Employment; use. 4 Obs. Interest on money. 5 Obs. Custom. Also Obs. **us'aunce**. [<OF <us <L usus. See USE.]

use (yŏŏz) v. used, us·ing v.t. 1 To employ for the accomplishment of a purpose; make use of. 2 To put into practice or employ habitually; make a practice of: to use diligence in business. 3 To expend the whole of; consume: often with up. 4 To conduct oneself toward; treat: to use one badly. 5 To make familiar by habit or practice; accustom; inure: usually in the past participle: He is used to exposure. 6 To partake of; smoke or chew: He does not use tobacco. — v.i. 7 To do something customarily or habitually; be accustomed or wont: now only in the past tense as an auxiliary to form a phrase equivalent to a frequentative past tense: I used to go there. See synonyms under EMPLOY, OCCUPY. — n. (yŏŏs) 1 The act of using; application or employment to an end, particularly a good or useful end; the fact or condition of being employed. 2 Suitableness or adaptability to an end; serviceableness: the uses of adversity. 3 Way or manner of using. 4 Occasion or need to employ; necessity: I have no use for it; purpose; function. 5 Habitual practice or employment; the fact of being habitually used; custom; usage. 6 Any special form, ceremony, or ritual of public worship, or any individual service that arose in or was perpetuated by a church, diocese, or branch of a church: Sarum use, Roman use, York use. Compare LITURGY. 7 Law The permanent equitable right that a beneficiary has to the enjoyment of the rents and profits of lands and tenements of which the legal title and possession are vested in another in trust for the beneficiary. 8 Obs. Ordinary experience or occurrence. 9 Usury. See synonyms under CUSTOM, HABIT, OCCUPATION, SERVICE, UTILITY. [<OF user <L usus, pp. of uti use]

use·ful (yŏŏs'fəl) adj. Serviceable; serving a use or purpose, especially a valuable one; productive of good; beneficial. —**use'ful·ly** adv. —**use'ful·ness** n.
Synonyms: adapted, advantageous, available, beneficial, conducive, convenient, favorable, good, helpful, profitable, salutary, serviceable, suitable, suited. See CONVENIENT, EXPEDIENT, GOOD. Compare UTILITY. Antonyms: see synonyms for USELESS.

use·less (yŏŏs'lis) adj. Unserviceable; being of no use; not serving, or not capable of serving, any beneficial purpose. —**use'less·ly** adv. —**use'less·ness** n.
Synonyms: abortive, bootless, fruitless, futile, ineffectual, nugatory, null, profitless, unavailing, unprofitable, unserviceable, vain, valueless, worthless. That which is bootless, fruitless, or profitless fails to accomplish any valuable result; that which is abortive, ineffectual, or unavailing fails to accomplish a result that it was, or was supposed to be, adapted to accomplish. That which is useless, futile, or vain is inherently incapable of accomplishing a specified result. Useless in the widest sense signifies not of use for any valuable purpose, and is thus closely similar to valueless and worthless. Fruitless is more final than ineffectual, as applying to the sum or harvest of endeavor. That which is useless lacks fitness for a purpose; that which is vain lacks imaginable fitness. See VAIN, WASTE. Antonyms: see synonyms for USEFUL.

us·er (yŏŏ'zər) n. 1 One who or that which uses. 2 Law The exercise or enjoyment of a right.

Ush·ant (ush'ənt) An island off NW France, comprising the westernmost point of France; 5 miles long, 2 miles wide: French Île d'Ouessant.

U·shas (ŏŏ'shəs, ŏŏ·shäs') In Hindu mythology, the goddess of the dawn.

ush·er (ush'ər) n. 1 One who acts as doorkeeper, as of a court or other assemblyroom. 2 An officer whose duty it is to introduce strangers or walk before a person of rank. 3 One who conducts persons to seats, as in a church or theater. 4 Brit. An underteacher. — v.t. 1 To act as an usher to; escort; conduct. 2 To precede as a harbinger; be a forerunner of. [<OF uissier <L ostiarius doorkeeper < ostium door]

ush·er·ette (ush'ə·ret') n. A female usher, as in a theater.

Usk (usk) A river in SW England and SE

Wales, flowing 60 miles east, SE, and south from the eastern border of Carmarthenshire to the Bristol Channel at Newport.

Üs·küb (üs·küp′) The Turkish name for SKOPLJE.

Us·ku·da·ma (ōōs′kōō·dä′mə) An ancient name for ADRIANOPLE.

Üs·kü·dar (üs′kü·där′) A Turkish city on the Asian side of the Bosporus opposite Istanbul: also *Scutari*.

Us·nach (ōōsh′nə) In old Irish legend, a famous warrior, father of three even more famous sons.

Us·pal·la·ta (ōōs′pä·yä′tä) A pass over the Andes between Santiago, Chile, and Mendoza, Argentina; elevation, 12,650 feet: also *La Cumbre*.

us·que·baugh (us′kwə·bô) *n.* A distilled spirit, as whisky: so called in Ireland and Scotland. Also **us′qua·bae, us′que, us′que·bae**. [< Irish and Scottish Gaelic *uisge-beatha* < *uisge* water + *beatha* life]

Ussh·er (ush′ər), **James,** 1581–1656, Irish bishop and theologian.

Us·su·ri (ōō·sōō′rē) A river forming part of the boundary between northeasternmost China and southeasternmost U.S.S.R., and flowing 365 miles north from the SW Sikhotè-Alin Range to the Amur at Khabarovsk.

U·sta·ši (ōō·stä′shē) A Croat fascist party in World War II supported by the German and Italian governments. Also **U·sta′chi**.

Us·ti·la·go (us′tə·lā′gō) *n.* A genus of smut fungi (order *Ustilaginales*) which attack the tissues of many plants, especially cereals, as *U. zeae*, destructive of corn, *U. tritici*, parasitic on wheat, etc. [< NL < LL < L *ustulatus* scorched. See USTULATE.]

Us·ti nad La·bem (ōō′styĕ näd lä′bem) A city on the Elbe in NW Bohemia, Czechoslovakia: German *Aussig*.

us·tion (us′chən) *n.* 1 The act of burning, or the state of being burnt. 2 *Med.* Cauterization by burning. [< L *ustio, -onis* < *ustus*, pp. of *urere* burn]

us·tu·late (us′chōō·lit, -lāt) *adj.* Scorched, burned, or colored as if by burning or scorching. [< L *ustulatus*, pp. of *ustulare* scorch, freq. of *urere* burn]

us·tu·la·tion (us′chōō·lā′shən) *n.* 1 The act of burning or searing. 2 In pharmacy, the drying of substances by heat preparatory to pulverization. 3 The burning of wine.

u·su·al (yōō′zhōō·əl) *adj.* Such as occurs in the ordinary course of events; frequent; common. [< OF < LL *usualis* < L *usus* use. See USE.] — **u′su·al·ly** *adv.* — **u′su·al·ness** *n.*
Synonyms: accustomed, common, customary, everyday, familiar, frequent, general, habitual, normal, ordinary, prevailing, prevalent, regular, wonted. In strictness, *common* and *general* apply to the greater number of individuals in a class; but both words are in good use as applying to the greater number of instances in a series, so that it is possible to speak of one person's *common* practice or *general* custom, but *ordinary* or *usual* would in such case be preferable. See COMMON, FREQUENT, GENERAL, HABITUAL, NORMAL. *Antonyms:* exceptional, extraordinary, infrequent, out-of-the-way, rare, singular, strange, uncommon, unusual.

u·su·fruct (yōō′zyōō·frukt, yōō′syōō-) *n. Law* The right of using the property of another and of drawing the profits it produces without wasting its substance. [< LL *usufructus* < L *ususfructus* < *usus et fructus* use and fruit]

u·su·fruc·tu·ar·y (yōō′zyōō·fruk′chōō·er′ē, yōō′·syōō-) *n. pl.* **·ar·ies** One who holds property for use by usufruct, as a tenant. — *adj.* Of, pertaining to, or having the nature of a usufruct. [< LL *usufructuarius* < *usufructus*. See USUFRUCT.]

u·su·rer (yōō′zhər·ər) *n.* 1 One who practices usury; one who lends money, especially at an exorbitant or illegal rate. 2 *Obs.* One who lends money on interest; any moneylender. [< OF *usurier* < Med. L *usurarius* < L *usura* use, usury. See USURY.]

u·su·ri·ous (yōō·zhōōr′ē·əs) *adj.* Practicing usury; having the nature of usury. — **u·su′ri·ous·ly** *adv.* — **u·su′ri·ous·ness** *n.*

u·surp (yōō·zûrp′, -sûrp′) *v.t.* 1 To seize and hold (the office, rights, or powers of another) without right or legal authority; take possession of by force. 2 To take arrogantly, as if by right. — *v.i.* 3 To practice usurpation; encroach with *on* or *upon*. See synonyms under ASSUME. [< OF *usurper* < L *usurpare* make use of, usurp, ? < *usus* use + *rapere* seize] — **u·surp′er** *n.* — **u·surp′ing·ly** *adv.*

u·sur·pa·tion (yōō′zər·pā′shən, -sər-) *n.* 1 The act of usurping: said especially of unlawful or forcible seizure of kingly power. 2 *Law* The wrongful intrusion into or unjust exercise of the privileges of any office, franchise, or right of another.

u·su·ry (yōō′zhər·ē) *n. pl.* **·ries** 1 The act or practice of exacting a rate of interest beyond what is allowed by law. 2 *Obs.* The lending of money at interest; interest in general. 3 *Law* A premium paid for the use of money beyond the rate of interest established by law. [< OF *usure* < L *usura* < *usus*, pp. of *uti* use]

u·sus lo·quen·di (yōō′səs lō·kwen′dī) *Latin* Usage in speaking.

ut (ōōt) *n.* The first note in the Guido scale: now commonly *do.* [See GAMUT]

U·tah (yōō′tô, -tä) A State of the western United States; 84,916 square miles; capital, Salt Lake City; entered the Union Jan. 4, 1896; nickname, *Beehive State*: abbr. *Ut.* — **U′tah·an** *adj. & n.*

U·ta·ma·ro (ōō·tä·mä·rō), **Kitagama,** 1754?–1806, Japanese engraver and designer of color prints.

ut dic·tum (ut dik′təm) *Latin* As said or directed.

Ute (yōōt, yōō′tē) *n.* One of a group of tribes of North American Indians of Shoshonean stock, including the Uncompahgre, Kaviawach, and Uinta, formerly living in Utah, Colorado, and New Mexico: now on reservations in Colorado and Utah.

u·ten·sil (yōō·ten′səl) *n.* A vessel, tool, implement, etc., serving a useful purpose, especially for domestic or farming use. See synonyms under TOOL. [< OF *utensile* < L *utensilis* fit for use < *utens*, ppr. of *uti* use]

u·ter·ine (yōō′tər·in, -īn) *adj.* 1 Pertaining to the uterus. 2 Born of the same mother, but having a different father. [< LL *uterinus* born of the same mother]

u·ter·i·tis (yōō′tə·rī′tis) *n. Pathol.* Metritis. [< NL]

utero- combining form The uterus; of or pertaining to the uterus. Also, before vowels, **uter-**, as in *uteritis*. [< L *uterus* the uterus]

u·ter·us (yōō′tər·əs) *n. pl.* **u·ter·i** (yōō′tər·ī) 1 *Anat.* The organ of a female mammal in which the young are protected and developed before birth; the womb. In the higher mammals the uterus is single, but in the lower, as marsupials and monotremes, it is double. 2 *Zool.* Any differentiated portion of an oviduct found in various animals, other than mammals, serving as a repository for the development and nourishment of the eggs or the young during the embryonic stage. [< L]

Ut·gard (ōōt′gärd) In Norse mythology, the abode of Utgard-Loki.

Ut·gard–Lo·ki (ōōt′gärd·lō′kē) In Norse mythology, an invulnerable giant.

U·ther (yōō′thər) A legendary king of Britain; father of Arthur. See ARTHUR, KING; IGRAINE; PENDRAGON.

U·ti·ca (yōō′tə·kə) 1 An ancient city, 20 miles NW of Carthage in northern Africa; site 18 miles north of modern Tunis. 2 A city of central New York, on the Mohawk River.

u·tile (yōō′til) *adj.* Useful: now rare. [< L *utilis* < *uti* use]

u·til·i·tar·i·an (yōō·til′ə·târ′ē·ən) *adj.* 1 Relating to utility; especially, placing utility above beauty or the amenities of life. 2 Pertaining to or advocating utilitarianism. — *n.* 1 An advocate of utilitarianism. 2 One devoted to mere material utility.

u·til·i·tar·i·an·ism (yōō·til′ə·târ′ē·ən·iz′əm) *n.* 1 *Philos.* A system that holds usefulness to be the end and criterion of action; specifically, the ethical doctrine that actions derive their moral quality from their usefulness as means to some end, especially as means productive of happiness or unhappiness. Jeremy Bentham, James Mill, and John Stuart Mill (who coined the word *utilitarianism*), understood by it the ethical theory which makes the pleasure or happiness of the individual or of mankind the end and criterion of the morally good and right. 2 The doctrine, in civics and politics, that the greatest happiness of the greatest number should be the sole end and criterion of all public action. 3 Devotion to mere material interests and aims.

u·til·i·ty (yōō·til′ə·tē) *n. pl.* **·ties** 1 Fitness for some desirable, practical purpose; serviceableness; also, that which is necessary. 2 Fitness to supply the natural needs of man. 3 In philosophy, the happiness of mankind; the greatest happiness of the greatest number; the utilitarianism expounded by J. S. Mill. 4 *Obs.* Use; profit. 5 A public service, as gas, water, or other service. 6 *pl.* Shares of utility company stocks. [< F *utilité* < L *utilitas* < *utilis* useful < *uti* use]
Synonyms: advantage, advantageousness, avail, benefit, expediency, policy, profit, serviceableness, use, usefulness. *Utility* is somewhat more abstract and philosophical than *usefulness* or *use*, and is often employed to denote adaptation to produce a valuable result, while *usefulness* denotes the actual production of such result. We contrast beauty and *utility*. We say of an invention its *utility* is questionable, or, on the other hand, its *usefulness* has been proved by ample trial, or, I have found it of *use. Expediency* (literally, the getting the foot out) refers primarily to escape from or avoidance of some difficulty or trouble. *Policy* is often used in a kindred sense, more positive than *expediency*, but narrower than *utility*, as in the proverb "Honesty is the best policy." See PROFIT, SERVICE. *Antonyms:* disadvantage, folly, futility, impolicy, inadequacy, inexpediency, inutility, unprofitableness, worthlessness.

utility man 1 A regular member of a theatrical company who must be prepared, on short notice, to go on in any of the less important parts. 2 In baseball, a member of a team who acts as a substitute.

u·til·ize (yōō′təl·īz) *v.t.* **·ized, ·iz·ing** To make useful or serviceable; turn to practical account; make use of. Also *Brit.* **u′til·ise.** — **u′til·iz′a·ble** *adj.* — **u′til·i·za′tion** *n.* — **u′til·iz′er** *n.*

ut in·fra (ut in′frə) *Latin* As below.

u·ti pos·si·de·tis (yōō′tī pos′ə·dē′təs) *Latin* In international law, the principle that the parties to a war retain what they possessed at its close, unless otherwise provided by treaty; literally, as you possess.

ut·most (ut′mōst) *adj.* 1 Of the highest degree or the largest amount or number; greatest; uttermost. 2 Being at the farthest limit or most distant point; most remote; last. — *n.* The greatest possible extent; the most possible. See synonyms under END. [OE *utmest*]

U·to-Az·tec·an (yōō′tō·az′tek·ən) *n.* One of the chief linguistic stocks of North American Indians, formerly occupying two large regions of the NW and SW United States, comprising three branches (Shoshonean, Piman, and Nahuatlan) and embracing about fifty tribes: still surviving in the United States and Mexico. — *adj.* Of or pertaining to this linguistic stock.

u·to·pi·a (yōō·tō′pē·ə) *n.* 1 Any state, condition, or place of ideal perfection. 2 A visionary, impractical scheme for social improvement. [from *Utopia*]

U·to·pi·a (yōō·tō′pē·ə) An imaginary island described as the seat of a perfect social and political life in a romance by Sir Thomas More, published in 1516. [< NL < Gk. *ou* not + *topos* place]

u·to·pi·an (yōō·tō′pē·ən) *adj.* Excellent, but existing only in fancy or theory; ideal. See synonyms under IMAGINARY. — *n.* One who advocates impractical reforms; a visionary.

U·to·pi·an (yōō·tō′pē·ən) *adj.* Pertaining to or like Utopia. — *n.* A dweller in Utopia.

u·to·pi·an·ism (yōō·tō′pē·ən·iz′əm) *n.* Highly idealistic and impractical views, especially about social problems.

U·trecht (yōō′trekt, *Du.* ü′trekht) 1 A province of central Netherlands; 511 square miles. 2 Its capital, scene of the signing of a treaty (1713) ending the War of the Spanish Succession.

u·tri·cle (yōō′tri·kəl) *n.* 1 *Anat.* A small saclike cavity, especially the larger of two found in the bony vestibule of the inner ear. 2 *Bot.* **a** A small fruit having an inflated pericarp, as in the pigweed. **b** An air cell, as in certain aquatic plants. [< L *utriculus*, dim. of *uter* skin bag]

u·tric·u·lar (yōō·trik′yə·lər) *adj.* 1 Resembling a utricle or small sac. 2 Bladderlike; bearing or provided with utricles. Also **u·tric′u·late** (-lit, -lāt).

u·tric·u·li·tis (yōō·trik′yə·lī′tis) *n. Pathol.* Inflammation of a utricle, as of the inner ear. [< NL]

u·tric·u·lus (yōō·trik′yə·ləs) *n. pl.* **·li** (-lī) Utricle. [< L]

U·tril·lo (ōō·trē′lyō, ōō·tril′ō; *Fr.* ü·trē·lō′) **Maurice**, 1883–1955, French painter.

ut su·pra (ut sōō′prə) *Latin* As above: abbreviated *ut sup.*

Ut·tar Pra·desh (ōōt′ər prə·dāsh′) A constituent State of northern India, formed in 1950; 113,409 square miles; capital, Lucknow: formerly *United Provinces of Agra and Oudh.*

ut·ter[1] (ut′ər) *v.t.* 1 To give out or send forth with audible sound; express; say. 2 *Law* To put in circulation; now, especially, to deliver or offer (something forged or counterfeit) to another. 3 *Obs.* To give vent to in any way; give forth; emit. 4 *Obs.* To issue or deliver, as merchandise, in the course of trade. See synonyms under SPEAK. [ME *outre*, freq. of obs. *out* say, speak out] — **ut′ter·a·ble** *adj.* — **ut′ter·er** *n.*

ut·ter[2] (ut′ər) *adj.* 1 Realized or developed to the last degree; absolute; total: *utter misery.* 2 Being or done without conditions or qualifications; unqualified; final; peremptory; absolute: *utter denial.* 3 *Obs.* Outer; remote. [OE *ūttra*, orig. compar. of *ūt* out]

ut·ter·ance[1] (ut′ər·əns) *n.* 1 The act of uttering; vocal expression; manner of speaking; also, the power of speech. 2 A thing uttered or expressed. See synonyms under REMARK.

ut·ter·ance[2] (ut′ər·əns) *n. Obs.* The bitter end; the uttermost; the last extremity; death: in the phrase **to the utterance**. [Var. of OUTRANCE]

ut·ter·ly (ut′ər·lē) *adv.* In a complete manner; entirely; thoroughly.

ut·ter·most (ut′ər·mōst′) *adj. & n.* Utmost.

U–tube (yōō′tōōb′, -tyōōb′) *n.* A tube bent into U form, especially such a tube made of glass for laboratory use.

U–turn (yōō′tûrn′) *n. Colloq.* A continuous turn which reverses the direction of a vehicle on a road.

u·va (yōō′və) *n. Bot.* A succulent fruit having a central placenta, as a grape. [< L]

u·var·o·vite (ōō·vär′ōf·īt) *n.* An emerald-green calcium-chromium garnet. [after Count S. *Uvarov*, 1785–1855, Russian nobleman]

u·va·ur·si (yōō′və·ûr′sī) *n.* A trailing plant, the bearberry (def. 1). [< L, bear's grape]

u·ve·a (yōō′vē·ə) *n. Anat.* 1 The inner, colored layer of the iris. 2 The iris, ciliary muscle, and choroid coat. [< Med. L < L *uva* grape] — **u′ve·al** *adj.*

U·vé·a (ōō·vā′ä) The largest island of the Wallis archipelago, capital of the protectorate; 7 miles long, 4 miles wide.

u·ve·i·tis (yōō′vē·ī′tis) *n. Pathol.* Inflammation of the uvea or iris. [< NL <UVEA] — **u′ve·it′ic** (-it′ik) *adj.*

u·ve·ous (yōō′vē·əs) *adj.* 1 Resembling a grape or a cluster of grapes. 2 Uveal. [< L *uva* grape]

u·vu·la (yōō′vyə·lə) *n. pl.* **·las** or **·lae** (-lē) *Anat.* 1 The pendent fleshy portion of the soft palate. 2 Either of two other similar processes, one at the neck of the bladder and the other on the under side of the cerebellum. [< LL, dim. of *uva* grape]

u·vu·lar (yōō′vyə·lər) *adj.* 1 Pertaining to or of the uvula. 2 *Phonet.* Produced by vibration of, or with the back of the tongue near or against, the uvula. — *n. Phonet.* A uvular sound.

u·vu·li·tis (yōō′vyə·lī′tis) *n. Pathol.* Inflammation of the uvula. [< NL]

Ux·bridge (uks′brij) An urban district of Middlesex, England, NW of London.

Ux·mal (ōōz·mäl′, ōōsh-, ōōs-) An ancient Mayan city of Yucatán, SE Mexico; site, 40 miles south of Mérida.

ux·or (uk′sôr) *n. Latin* Wife: abbreviated *ux.*

ux·o·ri·al (uk·sôr′ē·əl, -sō′rē-, ug·zôr′ē·əl, -zō′rē-) *adj.* 1 Of, pertaining to, characteristic of, or becoming to a wife. 2 Uxorious. [< L *uxorius* < *uxor* wife]

ux·o·ri·cide (uk·sôr′ə·sīd, -sō′rə-, ug·zôr′ə-, -zō′rə-) *n.* 1 The act of murdering or killing one's wife; wife-murder. 2 One who murders his wife. [< L *uxor* wife + -CIDE] — **ux·or′i·ci′dal** (-sīd′l) *adj.*

ux·o·ri·ous (uk·sôr′ē·əs, -sō′rē-, ug·zôr′ē-, -zō′rē-) *adj.* Fatuously or foolishly devoted to one's wife; showing extreme or foolish fondness for one's wife. [< L *uxorius* < *uxor* wife] — **ux·o′ri·ous·ly** *adv.* — **ux·o′ri·ous·ness** *n.*

Uz·bek (ōōz′bek, uz′-) *n.* 1 A member of a Turkic people dominant in Turkestan; a native or inhabitant of the Uzbek S.S.R. 2 The Turkic language of the Uzbeks. Also **Uz′beg.**

Uz·bek Soviet Socialist Republic (ōōz′bek, uz′-) A constituent republic of the U.S.S.R. in Central Asia; 154,014 square miles; capital, Tashkent. Also **Uz′bek·i·stan′.**

U·zhok (ōō′zhôk) A pass of the Carpathians in SW Ukrainian S.S.R. *Polish* **U·żok** (ōō′zhôk). *Hungarian* **U·zsok** (ōō′zhôk).

V

v, V (vē) *n. pl.* **v's** or **V's, vs** or **Vs, vees** (vēz) 1 The twenty-second letter of the English alphabet; ultimately from Phoenician *vau*, vocalized by the Greeks into *upsilon*, and used by the Romans in the form V with the value of a semivowel (w) and, later, a consonant (v). In English it was used interchangeably with the character *u* until fairly modern times. Compare U, W. 2 The sound of the letter *v*, the voiced, labiodental fricative. See ALPHABET. — *n.* 1 A V-shaped piece, or two pieces at an acute angle, as part of a construction: also **vee**. 2 *Colloq.* A five-dollar bill. — *symbol* 1 The Roman numeral five. See under NUMERAL. 2 *Chem.* Vanadium (symbol V). 3 *Electr.* Volt. 4 Anything shaped like a V.

V–1 (vē′wun′) *n.* The robot bomb used against England by the Germans in World War II. See ROBOT BOMB under BOMB. [< G *vergeltungswaffe eins* retaliation weapon 1]

V–2 (vē′tōō′) *n.* A jet-propelled rocket bomb carrying a bomb load of one ton or more, and able to travel about 200 miles from its launching site: used against England by the Germans in World War II. [< G *vergeltungswaffe zwei* retaliation weapon 2]

Vaal (väl) A river of the Union of South Africa, forming part of the boundary between the Orange Free State and the Transvaal and flowing 750 miles SW and west, from near the Swaziland border in SE Transvaal, to the Orange River near Kimberley.

Vaa·sa (vä′sä) A port of western Finland on the Gulf of Bothnia. *Swedish* **Va′sa.**

va·can·cy (vā′kən·sē) *n. pl.* **·cies** 1 The state of being vacant; vacuity; emptiness; specifically, emptiness of mind. 2 That which is vacant, empty, or unoccupied; empty space. 3 An interruption of continuity of thought or space; a gap; chasm. 4 An unoccupied post, place, or office; a place destitute of an incumbent. 5 *Rare* Unoccupied time; leisure.

va·cant (vā′kənt) *adj.* 1 Containing or holding nothing; being without contents or occupants; especially, devoid of occupants; empty. 2 Occupied with nothing; unemployed; unencumbered; free. 3 Being or appearing without intelligence; inane. 4 Having no incumbent; unfilled: a *vacant* office. 5 *Law* Unoccupied or unused, as land; also, abandoned; having neither claimant nor heir, as an estate. 6 Free from cares. 7 Devoid of thought; unreflecting. [< F < L *vacans, -antis*, ppr. of *vacare* be empty] — **va′cant·ly** *adv.*

Synonyms: blank, empty, unemployed, unfilled, unoccupied, vacuous, void, waste. That is *empty* which contains nothing; that is *vacant* which is without that which has filled or might be expected to fill it; *vacant* has extensive reference to rights or possibilities of occupancy. A *vacant* room may not be *empty*, and an *empty* house may not be *vacant*. *Void* and *devoid* are rarely used in the literal sense, but are for the most part confined to abstract relations, *devoid* being followed by *of*, and having with that addition the effect of a prepositional phrase: The article is *devoid* of sense; The contract is *void* for want of consideration. *Waste*, in this connection, applies to that which is made so by devastation or ruin, or gives an impression of desolation, especially as combined with vastness, probably from association of the words *waste* and *vast*; *waste* is applied also to uncultivated or unproductive land, if of considerable extent; we speak of a *waste* tract or region. *Vacuous* refers to the condition of being *empty* or *vacant*, regarded as continuous or characteristic. See BLANK, IDLE. *Antonyms*: brimful, brimmed, brimming, busy, crammed, crowded, full, gorged, inhabited, jammed, occupied, overflowing, packed, replete.

va·can·ti·a bo·na (vā·kan′shē·ə bō′nə) *Latin* Goods without an owner; escheated goods.

va·cate (vā′kāt) *v.* **·cat·ed, ·cat·ing** *v.t.* 1 To make vacant; surrender possession of by removal. 2 To set aside; annul. 3 To give up (a position or office); quit. — *v.i.* 1 To leave an office, position, place, etc. 5 *Colloq.* To go away; leave. See synonyms under CANCEL. [< L *vacatus*, pp. of *vacare* be empty]

va·ca·tion (və·kā′shən) *n.* 1 An intermission of activity, employment, or stated exercises, as for recreation or rest; a holiday. 2 *Law* The period of time intervening between stated terms of court. 3 The intermission of the course of studies and exercises in an educational institution. 4 The act of vacating. 5 *Obs.* The time during which an office is vacant. — *v.i.* To take a vacation. [< F < L *vacatio, -onis* freedom from duty < *vacatus*. See VACATE.] — **va·ca′tion·er** *n.*

va·ca·tion·ist (və·kā′shən·ist) *n.* One who is taking a vacation or staying at a resort; a tourist.

vac·ci·nal (vak′sə·nəl) *adj.* Of the nature of or relating to vaccine or vaccination.

vac·ci·nate (vak′sə·nāt) *v.* **·nat·ed, ·nat·ing** *Med. v.t.* To inoculate with a vaccine as a preventive or therapeutic measure; especially, to inoculate against smallpox. — *v.i.* To perform the act of vaccination. [< VACCIN(E) + -ATE[2]]

vac·ci·na·tion (vak′sə·nā′shən) *n. Med.* The act or process of vaccinating, especially against smallpox.

vac·ci·na·tion·ist (vak′sə·nā′shən·ist) *n. Med.* An advocate of vaccination.

vac·ci·na·tor (vak′sə·nā′tər) *n. Med.* 1 One who vaccinates. 2 An instrument used for vaccination.

vac·cine (vak′sēn, -sin) *n.* 1 The virus of cowpox, as prepared for or introduced by vaccination: usually lymph, dried or fluid, or part of the crust from a pustule. 2 Any in-

vaccine point *Med.* A sharp-pointed piece of bone, ivory, or the like, coated with vaccine for inoculation purposes.

vac·cin·i·a (vak·sin'ē·ə) *n. Pathol.* Cowpox. Also **vac·ci·na** (vak·sī'nə). [< NL < L *vaccinus.* See VACCINE.]

vac·cin·i·a·ceous (vak·sin'ē·ā'shəs) *adj. Bot.* Pertaining or belonging to a genus (*Vaccinium*; family *Ericaceae* or *Vacciniaceae*) of shrubs with cylindrical or globular flowers and small blue, black, or red berries, including the blueberry, huckleberry, and cranberry. [< NL < L *vaccinium* blueberry]

vac·ci·ni·za·tion (vak'sə·nə·zā'shən, -nī·zā'-) *n. Med.* Repeated inoculation with a vaccine.

vac·cin·o·ther·a·py (vak'sən·ō·ther'ə·pē) *n. Med.* Treatment by bacterial vaccines.

vac·il·late (vas'ə·lāt) *v.i.* **·lat·ed, ·lat·ing** 1 To sway one way and the other; totter; waver. 2 To fluctuate. 3 To waver in mind; be irresolute. See synonyms under FLUCTUATE. [< L *vacillatus,* pp. of *vacillare* waver] — **vac'il·la'tion** *n.*

vac·il·lat·ing (vas'ə·lā'ting) *adj.* Inclined to waver; uncertain; wavering. Also **vac'il·lant**, **vac'il·la·to·ry** (-lə·tôr'ē, -tō'rē). — **vac'il·lat'ing·ly** *adv.*

vac·u·a (vak'yōō·ə) Plural of VACUUM.

va·cu·i·ty (va·kyōō'ə·tē) *n. pl.* **·ties** 1 The state of being a vacuum; emptiness. 2 Vacant space; a void. 3 Freedom from mental exertion; idleness. 4 Lack of intelligence; stupidity. 5 Nothingness. 6 An inane or idle thing or statement: His speech was weakened by *vacuities.* [< F *vacuité* < L *vacuitas, -tatis* < *vacuus* empty]

vac·u·o·lat·ed (vak'yōō·ə·lā'tid) *adj. Biol.* Having one or more vacuoles. — **vac'u·o·la'tion** *n.*

vac·u·ole (vak'yōō·ōl) *n. Biol.* A minute cavity containing air, a watery fluid, or a chemical secretion of the protoplasm, found in an organ, tissue, or cell. [< F < L *vacuum,* neut. of *vacuus* empty]

vac·u·ous (vak'yōō·əs) *adj.* 1 Having no contents; containing no matter; empty. 2 Lacking intelligence; blank. 3 Idle; unoccupied. See synonyms under VACANT. [< I *vacuus*] — **vac'u·ous·ly** *adv.* — **vac'u·ous·ness** *n.*

vac·u·um (vak'yōō·əm, -yōōm) *n. pl.* **·ums** or **·u·a** (-yōō·ə) 1 *Physics* **a** A space absolutely devoid of matter. **b** A space from which air or other gas has been exhausted to a very high degree. 2 A partial diminution of the normal atmospheric pressure. 3 A void; an empty feeling. — *adj.* 1 Of, or used in the production of, a vacuum. 2 Exhausted or partly exhausted of gas, air, or vapor. 3 Operated by suction to produce a vacuum. — *v.t. Colloq.* To use a vacuum cleaner on. [< L, neut. of *vacuus* empty]

vacuum bottle A bottle having a double wall separated by a vacuum which permits the contents to be kept cold or hot for an appreciable period. Also **vacuum flask.**

vacuum cleaner A machine for cleaning floors, carpets, furnishings, etc., by the suction of an air current.

vacuum fan A fan producing suction or an incomplete vacuum.

vacuum gage A gage containing mercury for testing the pressure consequent on producing a vacuum, as in a condenser. Also **vacuum gauge.**

vacuum pump A pulsometer.

vacuum tube *Electronics* 1 A sealed glass tube exhausted of air to a high degree and containing electrodes between which electric discharges may be passed. 2 An electron tube.

vacuum valve *Brit.* A vacuum tube.

va·de in pa·ce (vā'dē in pā'sē) *Latin* Go in peace.

va·de me·cum (vā'dē mē'kəm) *Latin* Go with me; hence, anything carried for constant use, as a guidebook, manual, or bag. Also **va'de·me'cum, va·de-me'cum.**

Va·duz (fä·dōōts') The capital of the principality of Liechtenstein, near the Rhine, SE of St. Gall, Switzerland.

vae vic·tis (vē vik'təs) *Latin* Woe to the vanquished.

vag·a·bond (vag'ə·bond) *n.* 1 One who wanders from place to place without visible means of support; a tramp. 2 One without a settled home; a wanderer; nomad. 3 A worthless fellow; rascal. — *adj.* 1 Pertaining to a vagabond; nomadic. 2 Having no definite residence; wandering; irresponsible. 3 Driven to and fro; aimless. [< F < L *vagabundus* < *vagus* wandering] — **vag'a·bond'age** *n.* — **vag'a·bond'ish** *adj.* — **vag'a·bond'ism** *n.*

vagabond neurosis Dromomania.

va·gar·y (və·gâr'ē) *n. pl.* **·gar·ies** A wild fancy; extravagant notion. See synonyms under FANCY, WHIM. [< obs. *vagary, v.,* wander < L *vagari*]

va·gi·na (və·jī'nə) *n. pl.* **·nas** or **·nae** (-nē) 1 *Anat.* **a** A sheath or sheathlike covering. **b** The canal leading from the external genital orifice in female mammals to the uterus. 2 *Zool.* The terminal portion of the oviduct of various invertebrates. 3 *Bot.* A tubular part surrounding another, as the basal portion of a leaf around a stem. [< L, a sheath]

vag·i·nal (vaj'ə·nəl, və·jī'-) *adj.* 1 Pertaining to or like a sheath; thecal. 2 Pertaining to. the vagina.

vag·i·nate (vaj'ə·nit, -nāt) *adj.* 1 Having a sheath. 2 Formed into a sheath; tubular. Also **vag'i·nat'ed.** [< NL *vaginatus* < L *vagina* sheath]

vag·i·nec·to·my (vaj'ə·nek'tə·mē) *n. Surg.* 1 Removal or obliteration of the vaginal canal. 2 Resection of the serous membrane of the testis: also **vag'i·na·lec'to·my** (-nə·lek'tə·mē). [< VAGIN(O)- + -ECTOMY]

vag·i·nis·mus (vaj'ə·niz'məs, -nis'-) *n. Pathol.* Spasm of the sphincter muscle of the vagina with extreme sensitivity of the adjacent parts. [< NL]

vag·i·ni·tis (vaj'ə·nī'tis) *n. Pathol.* Inflammation of the vagina. [< NL]

vagino- combining form *Med.* The vagina; of or pertaining to the vagina. Also, before vowels, **vagin-,** as in *vaginectomy.* [< L *vagina* a sheath, the vagina]

va·gi·tus (və·jī'təs) *n.* The first cry of the newborn infant. [< L, pp. of *vagire* cry, squall]

va·go·to·ni·a (vā'gə·tō'nē·ə) *n. Pathol.* Excessive or morbid excitability of the vagus nerve, characterized by vasomotor instability, involuntary spasms, sweating, and constipation. [< NL < *vagus* the vagus nerve + Gk. *tonos* tone, tension] — **va'go·ton'ic** (-ton'ik) *adj.*

va·gran·cy (vā'grən·sē) *n. pl.* **·cies** The state, condition, or action of a vagrant. Also **va'grant·ness.**

va·grant (vā'grənt) *n.* 1 A person without a settled home; an idle wanderer; vagabond; tramp. 2 A roving person; wanderer. — *adj.* 1 Wandering about as a vagrant. 2 Pertaining to one who or that which wanders; nomadic. 3 Having a wandering course; capricious; wayward. [ME *vagaraunt,* alter. of AF *wakerant* < OF *wacrant,* ppr. of *wacrer* walk, wander < Gmc.; infl. in form by L *vagari* wander] — **va'grant·ly** *adv.*

va·grom (vā'grəm) *adj. Obs.* Vagrant. [Alter. of VAGRANT; used by Dogberry in Shakespeare's *Much Ado About Nothing*]

vague (vāg) *adj.* 1 Lacking definiteness or precision. 2 Of uncertain source or authority: a *vague* rumor. 3 Not clearly recognized, understood, stated, or felt. 4 *Obs.* Roving, vagrant. 5 Shadowy; hazy. [< F < L *vagus* wandering] — **vague'ly** *adv.* — **vague'ness** *n.* Synonyms: ambiguous, doubtful, dreamy, indefinite, indeterminate, indistinct, lax, loose, obscure, uncertain, undetermined, unsettled.

va·gus (vā'gəs) *n. pl.* **·gi** (-jī) *Anat.* Either of the tenth pair of cranial nerves originating in the medulla oblongata and sending branches to the lungs, heart, stomach, and most of the abdominal viscera; the pneumogastric nerve. Also **vagus nerve.** [< L, wandering]

Váh (väkh) A river in western Slovakia, Czechoslovakia, flowing 245 miles SW to the Danube: German **Waag.**

vail[1] (vāl) *n. & v.t. Obs.* Veil.

vail[2] (vāl) *Obs. v.i.* To be of use; avail. — *n.* 1 Usually pl. A gratuity or tip; a perquisite, often corrupt. 2 A windfall; find. 3 Advantage; proceeds; profit. ◆ Homophones: *vale, veil.* [Aphetic var. of AVAIL] — **vail'a·ble** *adj.*

vail[3] (vāl) *v.t. Archaic* 1 To let fall; lower, as the topsail, in salute or submission. 2 To take off (the hat, etc.) in respect or submission. ◆ Homophones: *vale, veil.* [Aphetic form of obs. *avale* < F *avaler* lower < *à val* down < L *ad vallem,* lit., to the valley]

vain (vān) *adj.* 1 Elated with self-admiration; greedy of applause. 2 Characterized by frivolity. 3 Ostentatious; showy: said of things. 4 Unproductive; worthless; fruitless; useless. 5 Without any substantial foundation; empty; unreal. — **in vain** To no purpose; without effect. ◆ Homophones: *vane, vein.* [< F < L *vanus* empty] — **vain'ly** *adv.* — **vain'ness** *n.* Synonyms: abortive, baseless, bootless, deceitful, delusive, dreamy, empty, fruitless, futile, idle, inconstant, ineffectual, light, profitless, shadowy, trifling, trivial, unavailing, unimportant, unreal, unsatisfying, unsubstantial, useless, vapid, worthless. *Vain* keeps the etymological idea through all changes of meaning; a *vain* endeavor is *empty* of result, or of adequate power to produce a result, a *vain* pretension is *empty* or destitute of support, a *vain* person has a conceit that is *empty* or destitute of adequate cause or reason. See USELESS. Antonyms: effective, efficient, firm, potent, powerful, real, solid, sound, substantial, valid, valuable, worthy.

vain·glo·ry (vān·glôr'ē, -glō'rē) *n.* Excessive or groundless vanity; also, vain pomp; boastfulness. See synonyms under PRIDE. [< OF *vaine gloire* < Med. L *vana gloria* empty pomp, show] — **vain·glo'ri·ous** *adj.* — **vain·glo'ri·ous·ly** *adv.* — **vain·glo'ri·ous·ness** *n.*

vair (vâr) *n.* 1 *Her.* One of the furs represented by rows of small shield-shaped figures. 2 *Obs.* A fur used for the garments of the nobility (14th century). [< F < LL *varius* ermine < L *varius* parti-colored, various]

Va·lais (vȧ·le') A canton in southern Switzerland, in the upper Rhône valley; 2,021 square miles; capital, Sion: German **Wallis.**

val·ance (val'əns) *n.* 1 A drapery hanging from the tester of a bedstead. 2 A short, full drapery across the top of a window. 3 A damask used for upholstering. — *v.t.* **·anced, ·anc·ing** To furnish with or as with drapery, or a valance. [Prob. < OF *avalant,* ppr. of *avaler* descend] — **val'anced** *adj.*

Val·dai Hills (väl·dī') A low plateau and group of hills in western European Russian S.F.S.R.; maximum height, 1,053 feet.

Val·de·mar (väl'də·mär) See WALDEMAR.

Val d'A·os·ta (väl dä·ōs'tä) An autonomous region of NW Italy bordering on France and Switzerland; 1,260 square miles; capital, Aosta.

Val·di·vi·a (val·dīv'ē·ə, *Sp.* bäl·dē'vyä), **Pedro de,** 1500?-54, Spanish conqueror of Chile.

vale[1] (vāl) *n.* 1 A valley; a low-lying tract of land: now chiefly poetic. 2 A trough or channel. See synonyms under VALLEY. ◆ Homophones: *vail, veil.* [< OF *val* < L *vallis*]

va·le[2] (vā'lē) *interj. Latin* Farewell; literally, be in good health.

val·e·dic·tion (val'ə·dik'shən) *n.* A bidding farewell. See synonyms under FAREWELL. [< L *valedictus,* pp. of *valedicere* say farewell < *vale* farewell, orig. imperative of *valere* be well + *dicere* say]

val·e·dic·to·ri·an (val'ə·dik·tôr'ē·ən, -tō'rē-) *n.* One who delivers a valedictory; specifically, a student who delivers a valedictory at the graduating exercises of an educational institution: usually the member of the graduating class whose rank in scholarship is highest.

val·e·dic·to·ry (val'ə·dik'tər·ē) *adj.* Pertaining to a leave-taking. — *n. pl.* **·ries** A parting address, as by a member (ordinarily the first in rank) of a graduating class. See synonyms under FAREWELL.

va·lence (vā'ləns) *n. Chem.* 1 The property possessed by an element or radical of combining with or replacing other elements or radicals in definite and constant proportion. 2 The number of atoms of hydrogen (or its equivalent), taken as unity, with which an atom or radical can combine, or which it can replace. It varies with different elements, and with certain elements in different compounds. 3 *Med.* The combining power of certain substances or bodies, as serums, chromosomes, and the like. Also **va′len·cy.** [<LL *valentia* strength, orig. neut. pl. of L *valens, -entis,* ppr. of *valere* be well, be strong]

valence electron *Chem.* One of the electrons in the outermost shell of an atom, regarded as being responsible for the chemical reaction of an element.

va·len·ci·a (və·len′shē·ə, -shə) *n.* A woven fabric with wool weft and silk or cotton warp. [from VALENCIA]

Va·len·ci·a (və·len′shē·ə, -shə; *Sp.* bä·len′·thyä) 1 A region and former Moorish kingdom of eastern Spain on the Mediterranean; 8,996 square miles. 2 A province of eastern Spain, center of the Valencia region; 4,155 square miles. 3 A port of eastern Spain, the chief city of Valencia region and capital of Valencia province. 4 A city of north central Venezuela.

Va·len·ci·ennes (və·len′sē·enz′, *Fr.* và·läṅ·syen′) *n.* A kind of bobbin lace with a floral pattern, originally made at Valenciennes. Also **Valenciennes lace, Val lace.**

Va·len·ciennes (và·läṅ·syėn′) A city of northern France on the Escaut in Nord department.

Va·lens (vā′lenz), 328?–378, Roman emperor of the East 364–378.

val·en·tine (val′ən·tīn) *n.* 1 A letter or token of affection sent, often anonymously, to a person of the opposite sex on St. Valentine's Day (Feb. 14), the anniversary of the beheading of this martyr by the Romans. 2 A sweetheart. [<OF]

Val·en·tine (val′ən·tīn) A masculine personal name. Also **Va·len·tin** (*Russian* vä·lyen·tyin′; *Ger.* vä′len·tēn; *Sw.* vä′len·tēn′; *Fr.* và·läṅ·täṅ′), *Sp.* **Va·len·tín** (bä′len·tēn′), *Pg.* **Va·len·tim** (vä′leṅ·tēṅ′), *Ital.* **Va·len·ti·no** (vä′len·tē′nō), *Lat.* **Va·len·ti·nus** (val′ən·tī′nəs), *Du.* **Va·len·tijn** (vä′len·tīn). [<L, well, healthy] — **Valentine, Saint,** Christian martyr of the third century A.D.

Val·en·tin·i·an (val′ən·tin′ē·ən) Name of three Roman emperors.
— **Valentinian I,** 321–375, reigned 364–375.
— **Valentinian II,** 372?–392, reigned 375–392.
— **Valentinian III,** 419–455, reigned 425–455: full name *Flavius Placidus Valentianus.*

Va·le·ra (və·ler′ə), **Eamon de** See under DE VALERA.

val·er·ate (val′ə·rāt) *n. Chem.* A salt of valeric acid. Also **va·le·ri·an·ate** (və·lir′ē·ən·āt′).

Va·le·ri·a (və·lir′ē·ə, *Ital.* vä·ler′yä) A feminine personal name. Also *Fr.* **Va·lé·rie** (và·lā·rē′), *Ger.* **Va·le·ri·e** (vä·lā′rē·ə). [<L, strong]

va·le·ri·an (və·lir′ē·ən) *n.* 1 Any of a genus (*Valeriana*) of Old World perennial herbs; especially, one species (*V. officinalis*), with small pink or white flowers and a strong odor. 2 Its root, used in medicine as a carminative and sedative. [<OF *valeriane* <Med. L *valeriana,* appar. ult. <*Valerius,* a personal name]

Va·le·ri·an (və·lir′ē·ən) Anglicized name of *Publius Licinius Valerianus,* 193?–260?, Roman emperor 254?–260?.

va·le·ri·a·na·ceous (və·lir′ē·ə·nā′shəs) *adj. Bot.* Pertaining to a family (*Valerianaceae*) of herbs, including valerian. [<NL <Med. L *valeriana* VALERIAN]

va·ler·ic (və·ler′ik, -lir′-) *adj.* 1 Of, pertaining to, or derived from valerian. 2 *Chem.* Pertaining to or designating one of four isomeric acids, $C_5H_{10}O_2$, of which two are found in valerian: all are made synthetically. Also **va·le·ri·an·ic** (və·lir′ē·an′ik).

Va·lé·ry (và·lā·rē′), **Paul Ambroise,** 1871–1945, French poet and philosopher.

val·et (val′it, val′ā; *Fr.* và·le′) *n.* 1 A gentleman's personal servant. 2 A man servant in a hotel who performs personal services for patrons. —*v.t.* & *v.i.* To serve or act as a valet. [<F, a groom <OF *vaslet, varlet,* dim. of *vasal* vassal. Doublet of VARLET.]

va·let de cham·bre (và·le′ də shäṅ′br′) *pl.* **va·lets de cham·bre** (và·le′) *French* A valet.

Va·let·ta (və·let′ə) See VALLETTA.

val·e·tu·di·nar·i·an (val′ə·to͞o′də·nâr′ē·ən, -tyo͞o′-) *n.* A chronic invalid; one unduly solicitous about his health. —*adj.* Seeking to recover health; infirm. Also **val′e·tu′di·nar′y.** [<L *valetudinarius* infirm < *valetudo, -inis* health, ill health < *valere* be well] —**val′e·tu′di·nar′·i·an·ism** *n.*

val·gus (val′gəs) *adj.* Knock-kneed or bowlegged. —*n. Pathol.* An abnormal eversion of the foot, as by a depression of the arch. [<L, bowlegged]

Val·hal·la (val·hal′ə) 1 In Norse mythology, the great hall into which the souls of heroes fallen bravely in battle were borne by the valkyries and received and feasted by Odin. 2 An edifice wherein the remains or memorials of deceased heroes of a nation are placed. Also **Val·hall** (val·hal′): also spelled *Walhalla.* [<NL <ON *valhöll,* genitive of *valhallar* hall of the slain < *valr* the slain + *höll* hall]

val·iant (val′yənt) *adj.* 1 Strong and intrepid; powerful and courageous. 2 Performed with valor; bravely conducted; heroic. See synonyms under BRAVE. [<OF *vailant,* ppr. of *valoir* be strong <L *valere*] —**val′iant·ly** *adv.* —**val′iant·ness, val′iance, val′ian·cy** *n.*

val·id (val′id) *adj.* 1 Based on evidence that can be supported; sound; just; sufficient and effective in law. 2 *Obs.* Strong. 3 *Stat.* Having a high degree of correlation with its criterion: distinguished from *reliable.* [<F *valide* <L *validus* powerful < *valere* be strong] —**val′id·ly** *adv.* —**val′id·ness** *n.*

Synonyms: cogent, conclusive, convincing, efficacious, efficient, good, incontestable, irrefragable, irrefutable, just, logical, solid, sound, substantial, sufficient, undeniable, weighty. See POWERFUL. *Antonyms*: see synonyms for VAIN.

val·i·date (val′ə·dāt) *v.t.* **·dat·ed, ·dat·ing** 1 To make valid; ratify and confirm. 2 To declare legally valid; legalize. See synonyms under RATIFY. —**val′i·da′tion** *n.*

va·lid·i·ty (və·lid′ə·tē) *n.* 1 The state or quality of being valid; soundness, as in law or reasoning; efficacy. 2 *Archaic* Health; strength. 3 *Obs.* Worth.

val·ine (val′ēn) *n. Biochem.* Either of two isomeric amino acids, $C_5H_{11}O_2N$, found in small amounts in casein and edestin. [<VAL(ERIC) + -INE[2]]

va·lise (və·lēs′) *n.* A portable receptacle for clothes and toilet articles; traveling bag. [<F <Ital. *valigia;* ult. origin uncertain]

val·kyr·ie (val·kir′ē, val′kir·ē) *n.* In Norse mythology, one of the maidens who ride through the air and choose heroes from among those slain in battle, and carry them to Valhalla. Also **val′kyr, Val·kyr′ie.** [<ON *valkyrja,* lit., chooser of the slain < *valr* the slain + stem of *kjósa* choose, select] —**val·kyr′i·an** *adj.*

Val·la·do·lid (val′ə·dō′lid, *Sp.* bä′lyä·thō′lēth) 1 A province of north central Spain; 3,221 square miles. 2 A city of central Spain, former capital of Castile and capital of Valladolid province.

Val·lar·ta (bä·lyär′tä), **Manuel Sandoval,** born 1899, Mexican physicist.

val·la·tion (və·lā′shən) *n.* 1 The art of planning or erecting fortifications. 2 A rampart. [<LL *vallatio, -onis* <L *vallare* protect with a wall < *vallum* a wall] —**val·la·to·ry** (val′ə·tôr′ē, -tō′rē) *adj.*

val·lec·u·la (və·lek′yə·lə) *n. pl.* **·lae** (-lē) 1 A furrow or depression. 2 *Anat.* A deep sulcus (**vallecula cerebelli**) enclosing the median lobe on the inferior surface of the cerebellum; also, a depression on the back of the tongue on either side of the epiglottis. 3 *Bot.* A groove or furrow, as those between the ridges on the fruit of plants of the parsley family. [<NL, var. of L *vallicula,* dim. of *vallis* a valley] —**val·lec′u·lar, val·lec′u·late** (-lit, -lāt) *adj.*

Val·let·ta (və·let′ə) The capital of Malta, a port on the SE coast; site of a major British naval base: also *Valetta.*

val·ley (val′ē) *n.* 1 A depression of the earth's surface, as one through which a stream flows; level or low land between mountains, hills, or high lands; also, the people who inhabit a valley. 2 *Archit.* **a** The gutter or angle formed by the meeting of the two roof slopes. **b** An interval in a vault, or the space between vault ridges as seen from above. 3 A vallecula. [<OF *valee* < *val* <L *vallis* a valley]

Synonyms: canyon, dale, dell, dingle, glen, gorge, gulch, gully, ravine, vale.

Valley Forge A village in SE Pennsylvania, scene of Washington's winter encampment, 1777–78, in the American Revolution.

Valley of Ten Thousand Smokes A region in Katmai National Monument, southern Alaska, punctuated by thousands of small volcanoes; 72 square miles.

Val·lom·bro·sa (väl′lôm·brō′zä) A resort town near Florence, Italy, in the Apennines.

Val·my (vȧl·mē′) A village in NE France, near Reims; scene of a French victory over the Prussians, 1792.

Va·lois (và·lwà′) A medieval county and former duchy of northern France.

Va·lois (và·lwà′) A French dynasty; began 1328 with Philip VI of Valois, ended with Henry III, 1589.

Va·lo·na (vä·lô′nä) An ancient port on the **Bay of Valona,** an inlet of the Strait of Otranto in SW Albania (15 miles long, 3 miles wide): formerly *Avlona*: Albanian *Vlona.*

va·lo·ni·a (və·lō′nē·ə) *n.* The dried acorn cups of the Old World **valonia oak** (*Quercus macrolepis*), used as a tanning material. [<Ital. *vallonea* <Modern Gk. *balania* an evergreen oak, pl. of *balani* an acorn <Gk. *balanos*]

val·or (val′ər) *n.* Intrepid courage, especially in warfare; personal bravery. Also *Brit.* **val′·our.** See synonyms under COURAGE, PROWESS. [<OF *valour* <LL *valor* worth < *valere* be strong]

val·or·i·za·tion (val′ər·ə·zā′shən, -ī·zā′-) *n.* The maintenance by governmental action of an artificial price for any product. [<Pg. *valorização* < *valor* value <LL. See VALOR.]

val·or·ize (val′ə·rīz) *v.t.* **·ized, ·iz·ing** To subject to valorization. Also *Brit.* **val′or·ise.**

val·or·ous (val′ər·əs) *adj.* Courageous; valiant. —**val′or·ous·ly** *adv.* —**val′or·ous·ness** *n.*

Val·pa·rai·so (val′pə·rā′zō, -sō, -rī′-) A port of central Chile; the most important port on the west coast of South America. *Spanish* **Val·pa·ra·í·so** (bäl′pä·rä·ē′sō).

val·u·a·ble (val′yo͞o·ə·bəl, val′yə·bəl) *adj.* 1 Having financial worth, price, or value; costly. 2 Of a nature or character capable of being valued or estimated: These goods are *valuable* by money. 3 Having moral worth, value, or importance; very serviceable; worthy; estimable: a *valuable* friend. See synonyms under EXCELLENT, GOOD, IMPORTANT. —*n. Usually pl.* An article of worth or value, as a piece of jewelry. —**val′u·a·ble·ness** *n.* —**val′u·a·bly** *adv.*

val·u·a·tion (val′yo͞o·ā′shən) *n.* 1 The act of valuing. 2 Estimated worth or value; appraisement; price. 3 Personal estimation; judgment of merit or character: to set a high *valuation* on one's skill or power. —**val′u·a′·tion·al** *adj.*

val·u·a·tor (val′yo͞o·ā′tər) *n.* One who makes appraisals; an appraiser.

val·ue (val′yo͞o) *n.* 1 The desirability or worth of a thing; intrinsic worth; utility. 2 The rate at which a commodity is potentially exchangeable for others; a fair return in service, goods, etc.; worth in money; market price; also, the ratio of utility to price; a bargain. 3 Attributed or assumed valuation; esteem or regard. 4 Exact meaning; signification; import: the *value* of the words "will" and "shall." 5 *Music* The relative length of a tone as signified by a note. 6 *Math.* The quantity, magnitude, or number an algebraic symbol or expression is supposed to denote. 7 Rank in a system of classification. 8 In the graphic arts, the relation of the elements of a picture, as light and shade, to one another, especially with reference to their distribution and interdependence, apart from the idea of hue. 9 *Phonet.* The special quality of the sound represented by a written character: the *values* of the letter *e.* See synonyms under PRICE, PROFIT. —**book value** The value of property or stock as shown by the books of the company that owns it; the value of stock of a corporation based on the profit or loss shown by its books, and distinguished from the face, market, or artificially created value. —**face value** 1 The value stated on the face of a bond, coin, note, etc. 2 Seeming or apparent value: a promise taken at its *face value.* —**par value** The nominal value, or value printed on a security or stock: not necessarily the market value of the shares. —*v.t.* **·ued, ·u·ing** 1 To estimate the value or worth of; assess; appraise. 2 To regard highly; esteem; prize. 3 To place a relative estimate of value or desirability upon: to *value* honor more than life. See synonyms under APPRECIATE, CHERISH. [<OF *valu,* pp. of *valoir* be worth <L *valere*] —**val′ue·less** *adj.* —**val′u·er** *n.*

val·ued (val′yōōd) *adj.* **1** Regarded or estimated; hence, much or highly esteemed: a *valued* friend. **2** Having a value: a many-*valued* function.
valued policy **1** A policy in which the value of a ship or cargo is agreed on and inserted as the amount of damages in case of total loss. **2** A policy requiring an insurance company to pay the insured the full amount of his policy, regardless of the actual value of the property, if it is totally destroyed.
val·val (val′val) *adj.* Of or pertaining to a valve. Also **val′var** (-vər).
val·vate (val′vāt) *adj.* **1** Serving as or like a valve; having a valve; valvular. **2** *Bot.* Touching by contiguous edges but not overlapping: applied to most dehiscent capsules in which the component parts separate like valves, to certain anthers, and to the petals or sepals of many flowers in estivation. [< L *valvatus* with folding doors < *valva*. See VALVE.]
valve (valv) *n.* **1** *Mech.* Any contrivance or arrangement that permits the flow of a liquid, gas, vapor, or loose material in either of two directions, and closes against its return. **2** *Obs.* One of a pair of folding doors. **3** *Anat.* A structure formed by one or more loose folds of the lining membrane of a vessel or other organ, preventing or retarding the flow of a fluid in one direction and allowing it in another. **4** *Zool.* **a** One of the parts of a shell, as of a mollusk. **b** A covering plate or one of two or more external pieces forming a sheath, as for an ovipositor. **5** *Bot.* **a** One of the parts into which a capsule splits in dehiscence. **b** One of the halves of an anther after its opening. **6** *Electr.* A device for controlling the direction of flow of a current, as an electrolytic cell, or a vacuum tube. **7** *Brit.* A radio tube. **8** A device in certain brass-wind instruments for lengthening the air column and lowering the pitch of the instrument's scale, by turning the air current from the main tube into an additional side tube. — *v.t.* **valved, valv·ing** To furnish with valves; control the flow of by means of a valve. [< L *valva* leaf of a door] — **valve′less** *adj.*

GATE VALVE
Cross-section.
a. Screw.
b. Gate closed.

valve-in-head engine (valv′in-hed′) An internal-combustion engine having overhead valves; an overhead-valve engine. See OVERHEAD VALVE.
valve·let (valv′lit) *n.* A little valve; a valvule, as of a pericarp.
val·vu·lar (val′vyə-lər) *adj.* **1** Pertaining to or of the nature of a valve, as of the heart. **2** Having valves; acting as a valve.
val·vule (val′vyōōl) *n.* A small valve; a structure like a small valve. Also **val′vu·la** (-vyə-lə). [< F < Med. L *valvula,* dim. of L *valva* a door]
val·vu·li·tis (val′vyə-lī′tis) *n. Pathol.* Inflammation of any membrane that serves as a valve in the organs or channels of circulation. [< NL < Med. L *valvula* VALVULE + -ITIS]
vam·brace (vam′brās) *n.* Armor for the forearm from the elbow to the wrist: also, *Obs.,* **vantbrace.** [Var. of *vantbrace* < AF *vantbras,* OF *avant-bras* < *avant* in front of + *bras* arm] — **vam′braced** *adj.*
va·moose (va-mōōs′) *v.t. & v.i.* **·moosed, ·moos·ing** *U.S. Slang* To leave hastily or hurriedly; quit. Also **va·mose′** (-mōs′). [< Sp. *vamos* let us go]
vamp[1] (vamp) *n.* **1** The piece of leather forming the upper front part of a boot or shoe. **2** Something added to give an old thing a new appearance. **3** *Music* A simple improvised accompaniment. — *v.t.* **1** To provide with a vamp. **2** To repair or patch. **3** *Music* To improvise an accompaniment to. — *v.i.* **4** *Music* To improvise accompaniments. [< OF *avampie* forepart of the foot < *avant* before + *pied* foot] — **vamp′er** *n.*
vamp[2] (vamp) *Slang v.t.* **1** To seduce or prey upon (a man) by utilizing one's feminine charms. — *v.i.* **2** To play the vamp. — *n.* An unscrupulous flirt or coquette. See VAMPIRE (def. 2). [Short for VAMPIRE]
vam·pire (vam′pīr) *n.* **1** A living corpse that rises from its grave at night to feed upon the living, usually by sucking the blood: a widespread folk belief originating in primitive cannibalism but developed primarily in Slavic folklore. It is not a demon or a ghost, but the physical body of one who has died; it cannot be exorcised, but must be disinterred and either burned or fastened in the grave with a stake through its heart. Belief in vampires still exists in Slavic Europe, Hungary, Greece, and Iceland. Bram Stoker's *Dracula* is a famous treatment of the subject. **2** A man or woman who preys upon persons of the opposite sex; especially, a woman who brings her lover to a state of poverty or degradation. **3** A large bat (genera *Desmodus* and *Diphylla*) of South or Central America, which drinks the blood of horses, cattle, and, sometimes, men: more fully **true vampire.** **4** An insectivorous or frugivorous bat (genera *Phyllostomus* and *Vampyrum*) formerly supposed to suck blood: a **false vampire.** [< F < G *vampir* <Slavic] — **vam·pir′ic** (-pir′ik), **vam·pir′ish** (-pir′ish) *adj.*

ALBINO VAMPIRE
(Bats vary from 2 to 28 inches in body length)

vam·pir·ism (vam′pī·riz′əm, -pə-) *n.* **1** Belief in vampires. See VAMPIRE (def. 1). **2** The act or practice of a vampire; bloodsucking. **3** The practice of extortion or of preying upon others.
van[1] (van) *n.* **1** A large covered wagon or vehicle, for removing furniture, household goods, etc.; a caravan. **2** *Brit.* A closed railway car for luggage, etc.; also, a vehicle, open or covered, used for carrying light goods. [Short for CARAVAN]
van[2] (van) *n.* **1** An advance guard, as of an army, or foremost division of a fleet. **2** The leaders of a movement; those at the front of any line or unit. [Short for VANGUARD]
van[3] (van) *n.* A fan or winnowing machine; hence, a wing. [Dial. var. of FAN[1]]
van[4] (vän) *prep. Dutch* Of; from: used with Dutch family names, originally designating where the family came from or received its name.
Van (vän) Singular of VANIR.
Van (vän) A town on the eastern shore of **Lake Van** (1,453 square miles) in western Turkey in Asia.
van·a·date (van′ə-dāt) *n. Chem.* A salt or ester of vanadic acid. Also **va·na·di·ate** (və-nā′-dē-āt).
va·nad·ic (və-nad′ik) *adj. Chem.* Of, pertaining to, or derived from vanadium, especially in its higher valence.
vanadic acid *Chem.* Any of several acids known only in their salts, as *meta*-vanadic acid, a yellow compound, HVO₃, used as a pigment and a substitute for gold bronze.
va·nad·i·nite (və-nad′ə-nīt) *n.* A native vanadate and chloride of lead, found in opaque prismatic crystals of red and yellow color. [<VANAD(IUM) + -IN + -ITE[1]]
va·na·di·um (və-nā′dē-əm) *n.* A rare, silver-white metallic element (symbol V) of the phosphorus group, difficult to extract from the vanadates and minerals in which it is found. It is useful in an alloy steel to increase tensile strength. See ELEMENT. [< NL <ON *Vanadis,* a name of the Norse goddess Freya]
vanadium steel Steel containing from .1 to .25 percent of vanadium to increase its tensile strength.
vanadium toner *Phot.* A substance containing vanadium dichloride, VCl₂, and used to impart a green tone to bromide prints.
van·a·dous (van′ə-dəs) *adj. Chem.* Of, pertaining to, or derived from vanadium, especially in its lower valence. Also **va·na·di·ous** (və-nā′dē-əs).
Van·brugh (van·brōō′, van′brə) **Sir John,** 1664–1726, English playwright and architect.
Van Bu·ren (van byōōr′ən), **Martin,** 1782–1862, president of the United States 1837–41.
Van·cou·ver (van-kōō′vər) **1** A port of SW British Columbia opposite **Vancouver Island,** the largest island off the western coast of North America, comprising part of British Colombia; 12,408 square miles. **2** A city of SW Washington on the Columbia River.
Van·cou·ver (van-kōō′vər), **George,** 1758?–1798, English navigator.
Vancouver, Mount A peak on the Yukon-Alaska border in the St. Elias Mountains; 15,700 feet.
van·dal (van′dəl) *n.* A ruthless plunderer; wilful destroyer of what is beautiful or artistic. — *adj.* Being a vandal; barbarous. [< VANDAL] — **van·dal·ic** (van·dal′ik) *adj.*
Van·dal (van′dəl) *n.* One of a Germanic people from a region between the Vistula and Oder rivers, south of the Baltic, who invaded the western Roman Empire in the fourth century. At the beginning of the fifth century, they ravaged Gaul and overran Spain and North Africa. In 455 they pillaged the city of Rome, destroying many artistic and literary treasures. Their kingdom, established in North Africa with Carthage as its capital, was overthrown in 534 by the Romans under Belisarius. — **Van·dal·ic** (van·dal′ik) *adj.* — **Van′dal·ism** *n.*
van·dal·ism (van′dəl·iz′əm) *n.* Hostility to or wilful destruction of artistic works, or of property in general.
Van·den·berg (van′dən·bûrg), **Arthur Hendrick,** 1884–1951, U. S. statesman. — **Hoyt,** 1899–1954, U. S. Air Force general.
Van·der·bilt (van′dər·bilt), **Cornelius,** 1794–1877, U. S. capitalist: called "Commodore Vanderbilt."
Van Die·men Gulf (van dē′mən) An arm of the Timor Sea between Northern Territory and Melville Island, Australia; 90 miles long, 50 miles wide. [after Anthony *Van Diemen,* 1593–1645, Dutch admiral]
Van Die·men's Land (van dē′mənz) The former name for TASMANIA.
Van Do·ren (van dôr′ən, dō′rən), **Carl,** 1885–1950, U. S. writer and editor. — **Mark,** born 1894, U. S. poet, writer, and critic; brother of preceding.
Van Dyck (van dīk′), **Anthony,** 1599–1641, Flemish painter. Also **Van-dyke.**
Van·dyke (van·dīk′) *adj.* Of or pertaining to Anthony Van Dyck (or Vandyke), or to his style; also, of or pertaining to the dress or fashions represented in the paintings of Van Dyck. — *n.* **1** A painting by Van Dyck. **2** A Vandyke cape, collar, or beard.
Van Dyke (van dīk′), **Henry,** 1852–1933, U. S. clergyman, educator, and author.
Vandyke beard A peaked or pointed beard resembling those depicted in Van Dyck's paintings.
Vandyke brown A deep-brown pigment, a kind of bog-earth or peat color used by the painter Van Dyck; any of the various brown pigments, as those resembling burnt umber.
Vandyke collar A broad, deep collar or cape of fine linen and lace resembling those represented in portraits by Van Dyck. Also **Vandyke cape.**
vane (vān) *n.* **1** A thin plate, pivoted out of center, on a vertical rod, to indicate the direction of the wind; a weathercock: also **wind vane.** **2** A slender flag or streamer used for the same purpose. **3** An arm or blade, as of a windmill, propeller, projectile, etc. **4** *Ornithol.* The web of a feather. **5** The target on a leveling rod. **6** The sight on a quadrant, compass, or similar instrument, by which the direction of the object viewed is determined. **7** One of the plates or strips of metal fixed in the tail of a bomb, guided missile, or the like, to provide stability or guidance. **8** *Obs.* A flag; pennon. ◆ Homophones: *vain, vein.* [Dial. var. of *fane* a small flag, OE *fana* a flag] — **vaned** *adj.*

WINDMILL VANES

Vane (vān), **Sir Henry,** 1613–62, English

Puritan statesman; executed on charge of treason.

Vä·ner (vā′nər), **Lake** See VENER, LAKE.

van Eyck (van īk′) See EYCK, VAN.

vang (vang) *n. Naut.* One of two guy ropes from the end of a gaff to the deck: used to steady the gaff. [<Du., a catch <*vangen* catch]

van Gogh (van gō′, gôkh′; *Du.* vän khokh′), **Vincent,** 1853–90, Dutch painter.

van·guard (van′gärd) *n.* **1** The advance guard of an army; the van. **2** Hence, one who or that which is foremost. [<OF *avangarde,* var. of *avantgarde* <*avant* before + *garde* guard]

Vanguard *n.* The second U. S. artificial satellite, launched from Cape Canaveral, Fla., on March 17, 1958, to a maximum altitude of 2,513 miles: diameter, 6.4 inches; weight, 3.25 pounds; equipment, two radio transmitters.

va·nil·la (və·nil′ə) *n.* **1** Any of a genus (*Vanilla*) of tall climbing orchids of tropical America. **2** The long dehiscent capsule of one species (*V. planifolia*) of this genus. **3** A flavoring extract made from these capsules. [<NL <Sp. *vainilla,* dim. of *vaina* sheath, pod <L *vagina* sheath; so called from the little pods that contain its seeds]

vanilla plant An erect perennial herb (*Trilisa odoratissima*) of the composite family growing in SE United States: the leaves have a vanilla odor when bruised: also called *hound's-tongue.*

va·nil·lic (və·nil′ik) *adj.* Of, pertaining to, or derived from vanilla or vanillin.

va·nil·lin (və·nil′in) *n. Chem.* A colorless crystalline compound, $C_8H_8O_3$, contained in vanilla, of which it is the odoriferous principle: also made synthetically. Also **va·nil′·line** (-in, -ēn).

Va·nir (vä′nir) *sing.* **Van** (vän) In Norse mythology, an early race of fertility deities, of whom the names Njord, Frey, and Freya survive: later combined with the Aesir.

van·ish (van′ish) *v.i.* **1** To disappear from sight; fade away; depart. **2** To pass out of existence; be annihilated. **3** *Math.* To become equal to zero. —*n. Phonet.* The slight terminal sound of certain vowels, as the faint (o͞o) heard after the (ō) in *go.* [Aphetic var. of OF *esvanniss-,* stem of *esvanir* <L *evanescere* fade away. See EVANESCE.] —**van′ish·er** *n.*

van·i·tas van·i·ta·tum (van′ə·tas van′ə·tā′təm) *Latin* Vanity of vanities.

van·i·ty (van′ə·tē) *n. pl.* **·ties** **1** The condition or character of being vain; a feeling of shallow pride; conceit; ambitious display; ostentation; show. **2** The quality or state of being vain or empty, or destitute of reality, etc. **3** That which is vain or unsubstantial. **4** A vanity bag or box. **5** A dressing-table. See synonyms under ARROGANCE, EGOTISM, LEVITY, PRIDE. [<OF *vanité* <L *vanitas, -tatis* <*vanus* empty, vain]

vanity bag or **box** A bag or box containing face powder, rouge, puff, mirror, etc. Also *vanity.*

Vanity Fair **1** In Bunyan's *Pilgrim's Progress,* a fair depicting the world as a scene of vanity and folly. **2** A novel by W. M. Thackeray, satirizing the weaknesses and follies of human nature. **3** The world of fashion and frivolity.

Van Loon (van lōn), **Hendrik Willem,** 1882–1944, U. S. author and lecturer born in Holland.

van·ner (van′ər) *n. Brit.* A truck driver; one who drives a van.

van·quish (vang′kwish, van′-) *v.t.* **1** To defeat in battle; overcome; conquer. **2** To suppress or overcome (a feeling or emotion): to *vanquish* lust. **3** To defeat, as in argument; confute. See synonyms under BEAT, CONQUER, SUBDUE. [<OF *veinquiss-,* stem of *veinquir* conquer <L *vincere*] —**van′quish·a·ble** *adj.* —**van′quish·er** *n.*

Van Rens·se·laer (van ren′sə·lər, -lir), **Stephen,** 1764–1839, U. S. general and politician.

Van·sit·tart (van·sit′ərt), **Robert Gilbert,** 1881–1957, first Baron Vansittart, British diplomat.

van·tage (van′tij) *n.* **1** Superiority over a competitor, as in means of attack; advantage. **2** In lawn tennis, the state of the game when either player has scored a point after deuce. **3** An opportunity; chance. [Aphetic var. of OF *avantage* ADVANTAGE] —**van′tage·less** *adj.*

vantage ground A position or condition which gives one an advantage.

vant·brace (vant′brās) See VAMBRACE.

van't Hoff (vänt hôf), **Jacobus Henricus,** 1852–1911, Dutch physical chemist.

van't Hoff's law *Chem.* A statement that the osmotic pressure of a substance in solution approximately equals the pressure it would have in a gaseous state at the same temperature and volume as the solution. [after J. H. *van't Hoff*]

Va·nu·a Le·vu (vä′no͞o·ä lā′vo͞o) Second largest of the Fiji Islands; 2,137 square miles.

Van Vech·ten (van vek′tən), **Carl,** born 1880, U. S. novelist, photographer, and music critic.

van·ward (van′wərd) *adj.* Pertaining to or situated in the van or front: *vanward* regiments.

Van·zet·ti (van·zet′ē), **Bartolomeo,** 1888–1927, Italian anarchist active in the United States. See SACCO, NIKOLA.

vap·id (vap′id) *adj.* **1** Having lost sparkling quality and flavor. **2** Flat; dull; insipid. See synonyms under FLAT, VAIN. [<L *vapidus* insipid] —**va·pid·i·ty** (və·pid′ə·tē), **vap′id·ness** *n.* —**vap′id·ly** *adv.*

va·por (vā′pər) *n.* **1** Moisture in the air; especially, visible floating moisture, as light mist. **2** Any light, cloudy substance in the air, as smoke or fumes. **3** Any substance in the gaseous state, which, under ordinary conditions, is usually a liquid or solid. **4** A gas below its critical temperature. **5** That which is fleeting and unsubstantial. **6** A remedial agent applied by inhalation; also, a substance vaporized for use in industries. **7** Boastful swagger; vaporing. **8** *pl.* Depression of spirits; hypochondria. —*v.t.* **1** To vaporize. —*v.i.* **2** To emit vapor. **3** To evaporate. **4** To make idle boasts; brag. Also *Brit.* **va′pour.** [<AF *vapour,* OF *vapeur* <L *vapor* steam] —**va′por·a·bil′i·ty** *n.* —**va′por·a·ble** *adj.* —**va′por·er** *n.*

vapor density *Physics* **1** The density of a substance in the state of vapor, reaching its maximum before the substance passes into the liquid state. **2** The density of a gas or vapor as compared with that of hydrogen at the same temperature and pressure.

va·por·es·cence (vā′pə·res′əns) *n.* The process of forming mist or vapor. —**va′por·es′cent** *adj.*

vapori- *combining form* Vapor; of or related to vapor, steam, etc.: *vaporimeter.* Also, before vowels, **vapor-.** [<L *vapor* steam]

va·por·if·ic (vā′pə·rif′ik) *adj.* Producing vapors. [<NL *vaporificus* <L *vapor, -oris* steam + *facere* make]

va·por·im·e·ter (vā′pə·rim′ə·tər) *n.* An instrument for determining vapor pressure.

va·por·ing (vā′pər·ing) *adj.* Boasting; swaggering. —*n.* The act of boasting or talking pretentiously. —**va′por·ing·ly** *adv.*

va·por·ish (vā′pər·ish) *adj.* **1** Somewhat like vapor. **2** Somewhat hypochondriac.

va·por·i·za·tion (vā′pər·ə·zā′shən, -ī·zā′-) *n.* **1** The act or process of vaporizing, or the state of being vaporized. **2** *Med.* Treatment with vapors.

va·por·ize (vā′pə·rīz) *v.t.* & *v.i.* **·ized, ·iz·ing** To convert or be converted into vapor. —**va′·por·iz′a·ble** *adj.*

va·por·iz·er (vā′pə·rī′zər) *n.* **1** One who or that which vaporizes. **2** An atomizer.

va·por·ous (vā′pər·əs) *adj.* **1** Of or like vapor; foggy; misty; ethereal. **2** Full of vapors; hypochondriac; also, producing vapors; flatulent. **3** Vainly imaginative; whimsical. Also **va′por·y.** —**va·por·os·i·ty** (vā′pə·ros′ə·tē) *n.* —**va′por·ous·ly** *adv.* —**va′por·ous·ness** *n.*

vapor pressure *Physics* The pressure of a confined vapor when it is in equilibrium with its liquid at any specific temperature. Also **vapor tension.**

va·que·ro (vä·kā′rō) *n. pl.* **·ros** (-rōz, *Sp.* -rōs) A herdsman; cowboy. [<Sp. <*vaca* cow <L *vacca*]

var (vär) *n. Electr.* The reactive volt-ampere. [<V(OLT) + A(MPERE) + R(EACTIVE)]

va·ra (vä′rä) *n.* A Spanish and Portuguese measure of length, varying from 2.7 to 3.6 feet. —**square vara** A varying measure of surface. [<Sp. and Pg., lit., a rod <L, a forked pole <*varus* bent]

Va·rang·er Fjord (vä·rang′ər) An inlet of the Barents Sea in NE Norway; 60 miles long, 3 to 35 miles wide.

Va·ran·gi·an (və·ran′jē·ən) *n.* A Norse rover; one of a group of predatory Scandinavian seamen who, in the ninth century, sailed down the Volga into the Caspian Sea and established a dynasty in Caucasia. [<Med. L *Varangus* <Med. Gk. *Barangos* <Slavic, ult. <ON *Væringi* an ally <*vārar* pledges]

va·ra·ni·an (və·rā′nē·ən) *n.* One of a family of lizards (*Varanidae*) with tongue sheathed at the root and forked at the tip; a monitor: also **var·a·nid** (var′ə·nid). —*adj.* Of or pertaining to the *Varanidae.* [<NL *Varanus* <Arabic *waran* a monitor lizard]

Var·dar (vär′där) A river of southern Yugoslavia and NE Greece, flowing 230 miles NE and SE to the Gulf of Salonika near Salonika: Greek *Axios.*

va·reuse (va·rœz′) *n. French* A loose, woolen jacket, similar to a peajacket.

Var·gas (vär′gəs), **Getulio,** 1883–1954, provisional president of Brazil 1930–34; president 1934–45 and 1951–54; forced out of office.

vari- *combining form* Various; different: *variform, varicolored.* Also *vario-.* [<L *varius* varied]

var·i·a·ble (vâr′ē·ə·bəl) *adj.* **1** Having the capacity of varying; alterable; mutable. **2** Having a tendency to change; not constant; fickle. **3** Having no definite value as regards quantity. **4** *Biol.* Prone to variation from a normal or established type: said of plants and animals. See synonyms under FICKLE, IRREGULAR, MOBILE[1]. —*n.* **1** That which is liable to change. **2** *Math.* **a** A quantity susceptible of fluctuating in value or magnitude under different conditions. **b** A symbol representing one of a group of objects. **3** *Meteorol.* A shifting wind or winds; also, in the plural, places where such winds are common. [<OF <L *variabilis* <*variare* VARY] —**var′i·a·bil′·i·ty, var′i·a·ble·ness** *n.* —**var′i·a·bly** *adv.*

variable star See under STAR.

variable zone A temperate zone.

var·i·ance (vâr′ē·əns) *n.* **1** The act of varying, or the state of being variant; difference; discrepancy; hence, dissension; discord. **2** *Law* **a** A disagreement between the allegations in the pleadings and the proof in an essential matter. **b** A material disagreement between the writ beginning an action and the declaration or complaint. **3** *Stat.* The square of the standard deviation. **4** *Chem.* Degree of freedom. See synonyms under QUARREL[1].

var·i·ant (vâr′ē·ənt) *adj.* **1** Having or showing variation; varying; differing. **2** Tending to vary; variable; changing. **3** Restless; fickle; inconstant. **4** Differing from a standard or type; discrepant. See synonyms under HETEROGENEOUS. —*n.* **1** A thing that differs from another in form only; especially, a different spelling, pronunciation, or form of the same word. **2** A variate. [<OF <L *varians, -antis,* ppr. of *variare* VARY]

var·i·ate (vâr′ē·āt) *v.t.* & *v.i. Obs.* To vary. —*n.* **1** That which varies; a variable. **2** *Stat.* The magnitude or value of a variable. [<L *variatus,* pp. of *variare* VARY]

var·i·a·tion (vâr′ē·ā′shən) *n.* **1** The act, process, state, or result of varying; modification; diversity. **2** The extent to which a thing varies. **3** Inflection, as of declensions or conjugations; also, change in certain vowel sounds. **4** A repetition of the essential features of a musical theme or melody with fanciful embellishments. **5** *Astron.* **a** An inequality in the moon's motion. **b** Any change in the elements of an orbit. **6** *Biol.* Deviation in structure or function from the type or parent form of an organism, as by heredity or in response to conditions of environment. **7** *Stat.* Dispersion. See synonyms under CHANGE, DIFFERENCE. [<F] —**var′i·a′tion·al** *adj.*

var·i·cel·la (var′ə·sel′ə) *n. Pathol.* Chicken pox. [<NL, dim. of *variola.* See VARIOLA.]

var·i·cel·late (var′ə·sel′it, -āt) *adj. Zool.* Marked with small varices, as certain shells. [<NL *varicella,* dim. of L *varix* a varicose vein]

var·i·cel·loid (var′ə·sel′oid) *adj.* Resembling varicella: *varicelloid* smallpox.

varico- *combining form Med.* A vein; of or related to veins, especially to varicose veins: *varicotomy.* Also, before vowels, **varic-.** [<L *varix, varicis* a varicose vein]

var·i·ces (var′ə·sēz) Plural of VARIX.

var·i·co·cele (var′ə·kō·sēl′) *n. Pathol.* A tumor formed by varicose veins of the spermatic cord. [<NL <L *varix, -icis* a varicose vein + Gk. *kēlē* a tumor]

var·i·col·ored (vâr′i·kul′ərd) *adj.* Variegated in color; parti-colored; diversified; of various colors. Also *Brit.* **var′i·col′oured.**

var·i·cose (var′ə·kōs) *adj. Pathol.* Abnormally dilated or contorted, as veins. [<L *varicosus* < *varix, -icis* a varicose vein]

var·i·co·sis (var′ə·kō′sis) *n. Pathol.* A condition in which there are varicose veins; varicosity. [<NL]

var·i·cos·i·ty (var′ə·kos′ə·tē) *n. pl.* **·ties** *Pathol.* 1 The condition of being varicose. 2 A varix.

var·i·cot·o·my (var′ə·kot′ə·mē) *n. Surg.* Excision of a varix or of a varicose vein. [< VARICO- + -TOMY]

var·ied (vâr′ēd) *adj.* 1 Partially or repeatedly altered. 2 Consisting of diverse sorts. 3 Differing from one another. 4 Varicolored. See VARY. — **var′ied·ly** *adv.*

varied thrush A robinlike bird of the western United States (*Ixoreus naevius*) with a plump, rust-colored body, black or gray wings, and a dark band across the breast.

var·i·e·gate (vâr′ē·ə·gāt′) *v.t.* **·gat·ed, ·gat·ing** 1 To mark with different colors or tints; dapple; spot; streak. 2 To make varied; diversify. — *adj.* Variegated. [<LL *variegatus,* pp. of *variegare* variegate < *varius* various + *agere* drive, do]

var·i·e·gat·ed (vâr′ē·ə·gā′tid) *adj.* 1 Having diverse colors; varied in color, as with streaks or blotches. 2 Having or exhibiting different forms, styles, or varieties. 3 *Bot.* Designating a type of inflorescence: see illustration under INFLORESCENCE.

var·i·e·ga·tion (vâr′ē·ə·gā′shən) *n.* 1 The act of variegating, or the state of being variegated. 2 Diversity of colors.

va·ri·e·tal (və·rī′ə·təl) *adj. Biol.* Of, pertaining to, or of the nature of a distinct variety: opposed to *specific* or *generic.* — **va·ri′e·tal·ly** *adv.*

va·ri·e·ty (və·rī′ə·tē) *n. pl.* **·ties** 1 The state or character of being various or varied; diversity. 2 A collection of diverse things; an assortment of unlike objects. 3 The possession of different characteristics by one individual. 4 A limited class of things that differ in certain common peculiarities from a larger class to which they belong; sometimes, an example of such a sort or kind. 5 *Biol.* An individual, or a group of individuals of a species, that differs from the type in certain characters capable of perpetuation, and that is usually fertile with any other member of the species; a subdivision of a species; subspecies. See synonyms under CHANGE. [<MF *variété* <L *varietas* < *varius* various]

variety shop 1 A store having a great variety of merchandise for sale, as hardware, dry-goods, notions, toys, and other small wares. 2 A general store. Also **variety store.**

variety show A theatrical show consisting of a series of short acts or numbers, including songs, dances, dramatic sketches, acrobatic feats, animal acts, etc.; a vaudeville show.

var·i·form (vâr′ə·fôrm) *adj.* Of diverse form; having different shapes.

vario– Var. of VARI–.

var·i·o–coup·ler (vâr′ē·ō·kup′lər) *n. Electr.* A tuning coil in which the secondary winding is mounted to rotate within the primary winding. It resembles the variometer.

va·ri·o·la (və·rī′ə·lə) *n. Pathol.* Smallpox. [< Med. L, a pustule <L *varius* speckled] — **va·ri′o·lar, va·ri′o·lous** *adj.*

var·i·o·late (vâr′ē·ə·lāt′) *v.t.* **·lat·ed, ·lat·ing** To vaccinate with smallpox virus. — **var′i·o·la′tion, var′i·o·li·za′tion** *n.*

var·i·ole (vâr′ē·ōl) *n.* 1 A foveola. 2 A spherulite or variolite. [<F <Med. L *variola.* See VARIOLA.]

var·i·o·lite (vâr′ē·ə·līt′) *n.* A dense, finely crystalline variety of basalt, characterized by whitish spheroid granules. [<G *variolit* < Med. L *variola.* See VARIOLA.]

var·i·o·lit·ic (vâr′ē·ə·lit′ik) *adj.* 1 Of, pertaining to, or containing variolite. 2 Spotted.

var·i·o·loid (vâr′ē·ə·loid′) *Pathol. n.* A mild form of smallpox, occurring after vaccination or in persons who have had smallpox. — *adj.* 1 Resembling smallpox. 2 Pertaining to varioloid.

var·i·om·e·ter (vâr′ē·om′ə·tər) *n. Electr.* 1 An instrument used to determine the variation of magnetic force at different times or at different places, usually by means of a needle suspended within the magnetic field. 2 A variable inductance device composed of two coils connected in series, one of which revolves within the other, and capable of controlling the strength of a current. [<VARIO- + -METER]

var·i·o·rum (vâr′ē·ôr′əm, -ō′rəm) *adj.* Having notes or comments by different critics or editors. — *n.* 1 An edition containing various versions of a text, usually with notes and commentary. 2 A text or edition, especially the complete works of a classical author, containing various notes and comments: also **variorum edition.** [<L *(cum notis) variorum* (with the notes) of various persons]

var·i·ous (vâr′ē·əs) *adj.* 1 Characteristically different from one another; diverse. 2 Being more than one and easily distinguishable; several. 3 Many-sided; variform. 4 Having a changeable or inconstant nature; unfixed. 5 Having a diversity of appearance; variegated. See synonyms under HETEROGENEOUS, MANY. [<L *varius*] — **var′i·ous·ly** *adv.* — **var′i·ous·ness** *n.*

Var·i–Typ·er (vâr′i·tī′pər) *n.* A compact, electrically operated, self-justifying composing machine for the rapid preparation of copy for all kinds of printing reproduction: a trade name.

var·ix (vâr′iks) *n. pl.* **var·i·ces** (vâr′ə·sēz) 1 *Pathol.* **a** A Permanent dilatation of a vein or other vessel of circulation. **b** A vessel thus distorted, as a varicose vein. 2 *Zool.* A ridge marking the former position of the outer lip of certain univalve shells. [<L, a varicose vein]

var·let (vär′lit) *n.* 1 *Archaic* A low menial or subordinate; formerly, a page. 2 A knave or scoundrel. [<OF, a groom. Doublet of VALET.]

var·let·ry (vär′lit·rē) *n.* The rabble; the mob.

var·mint (vär′mənt) *n. Dial.* Any person or animal considered as troublesome; vermin. [Alter. of VERMIN] — **var′mint·ry** *n.*

Var·na (vär′nä) A major seaport on the Black Sea in eastern Bulgaria: formerly (1949–58) called *Stalin.*

var·nish (vär′nish) *n.* 1 A solution of certain gums or resins in alcohol, linseed oil, etc., used to produce a shining, transparent coat on a surface. 2 Any natural or artificial product resembling varnish. 3 Outward show, or any superficial polish, as of politeness. — *v.t.* 1 To cover with varnish. 2 To give a smooth or glossy appearance to. 3 To improve the appearance of; polish. 4 To hide by a deceptive covering or appearance; gloss over. [<OF *vernis* <Med. L *vernicium* sandarac <Med. Gk. *bernikē,* prob. from Gk. *Berenikē,* a city in Cyrenaica] — **var′nish·a·ble** *adj.* — **var′nish·er** *n.* — **var′nish·ing** *n.*

varnish tree 1 A tree of China and Japan (*Toxicodendron verniciffuum*) yielding a milky juice suitable for making varnish. 2 The candlenut tree.

Var·ro (var′ō), **Marcus Terentius,** 116–27? B.C., Roman scholar and author.

var·si·ty (vär′sə·tē) *n. pl.* **·ties** *Colloq.* 1 *Brit.* University. 2 The team that represents a university, college, or school in any activity, as in football, debating, etc. [Aphetic alter. of UNIVERSITY]

Var·u·na (var′ōō·nə, vûr′-) In the earliest (Vedic) Hindu mythology, the god of the sky, creator and supreme god of the universe; later, in the Puranas, the god of the waters.

var·us (vâr′əs) *n. Pathol.* A malformation in which a bone or joint is turned away from its normal position. [<NL <L, growing inward, bandy-legged]

Var·us (vâr′əs), **Publius Quintilius,** died A.D. 9, Roman general, commander in Germany; defeated by Arminius.

varve (värv) *n.* One of a series of finely stratified seasonal deposits, as of clay or shale: often useful in determining the age of geological formations. [<Sw. *varv* a layer]

var·y (vâr′ē) *v.* **var·ied, var·y·ing** *v.t.* 1 To change the form, nature, substance, etc., of; modify. 2 To cause to be different from one another. 3 To impart variety to; diversify. 4 *Music* To embellish (a melody) by changes of rhythm, harmony, etc. — *v.i.* 5 To become changed in form, nature, substance, etc. 6 To be diverse or different; differ. 7 To deviate; depart: with *from.* 8 To change in succession; alternate. 9 *Math.* To be subject to continual change. 10 *Biol.* To undergo variation. See synonyms under CHANGE, FLUCTUATE. [<OF *varier* <L *variare* < *varius* various, diverse] — **var′i·er** *n.*

varying hare Any of certain hares whose coats turn white in the winter, specifically the American *Lepus americanus.*

vas (vas) *n. pl.* **va·sa** (vā′sə) *Biol.* A vessel or duct. [<L, vessel, dish]

vas– Var. of VASO–.

Va·sa·ri (vä·zä′rē), **Giorgio,** 1511–74, Italian painter, architect, and biographer of artists.

vas·cu·lar (vas′kyə·lər) *adj. Biol.* 1 Of, pertaining to, consisting of, or containing vessels or ducts, as blood vessels, etc. 2 Having vessels. 3 Richly supplied with blood vessels. Also **vas′cu·lose** (-lōs), **vas′cu·lous** (-ləs). [<L *vasculum,* dim. of *vas* vessel] — **vas′cu·lar′i·ty** (-lar′ə·tē) *n.* — **vas′cu·lar·ly** *adv.*

vascular bundle Bundle (def. 4).

vascular tissue *Bot.* Plant tissue made up of vessels or ducts through which the sap is conveyed.

vas·cu·lum (vas′kyə·ləm) *n. pl.* **·la** (-lə) 1 A small box used in plant collecting. 2 *Bot.* An ascidium. [<L, little vessel]

vas def·er·ens (vas def′ər·ənz) *Anat.* The duct by which semen is conveyed from the epididymis to the seminal vesicles. [<NL <L *vas* vessel + *deferens* leading down]

vase (vās, vāz, väz) *n.* An urnlike vessel, usually rounded and generally of greater height than width, ordinarily used as an ornament or for holding flowers. [<F <L *vas* vessel]

va·sec·to·my (və·sek′tə·mē) *n. Surg.* Removal of a portion of the vas deferens. [<VAS- + -ECTOMY]

Vas·e·line (vas′ə·lēn, -lin) *n.* Proprietary name for various semisolid hydrocarbons derived from petroleum: a brand of petrolatum.

Vash·ti (vash′tī) In the Bible, the queen of Ahasuerus, of Persia, whom he divorced because she refused to come to a royal banquet as commanded. *Esther* i 10–21.

Va·si·lev·ski (vä′sē·lyef′skē), **Alexander Mikhailovich,** born 1901, Russian marshal and chief of general staff in World War II.

vaso– *combining form* 1 *Physiol.* A vessel, especially a blood vessel: *vasomotor.* 2 *Med.* The vas deferens: *vasosection,* the severing of the vas deferens. Also, before vowels, *vas–.* [<L *vas* a vessel]

vas·o·con·stric·tor (vas′ō·kən·strik′tər) *adj. Physiol.* Causing constriction of a blood vessel when stimulated. — *n.* A nerve or drug causing such constriction.

vas·o·den·tine (vas′ō·den′tēn, -tin) *n.* Dentine in which the capillaries have remained wide enough to give passage to the blood.

vas·o·di·la·tor (vas′ō·dī·lā′tər) *adj. Physiol.* Causing dilatation of a blood vessel, as certain nerves or drugs.

vas·o·mo·tor (vas′ō·mō′tər) *adj. Physiol.* Producing movement, either of contraction or dilatation, in the walls of vessels.

vas·sal (vas′əl) *n.* 1 One who held land of a superior lord by a feudal tenure; a liegeman or feudal tenant. 2 A dependent, retainer, or servant of any kind; a slave or bondman. — *adj.* Having the character of or pertaining to a vassal; tributary; hence, servile. [<OF <Med. L *vassallus* <LL *vassus* a servant <Celtic]

vas·sal·age (vas′əl·ij) *n.* 1 The condition, duties, and obligations of a vassal; also, the feudal system. 2 Servitude in general. 3 Lands held by feudal tenure; a fief. 4 Vassals collectively.

vas·sal·ize (vas′əl·īz) *v.t.* **·ized, ·iz·ing** To reduce to vassalage; use as a vassal.

vast (vast, väst) *adj.* 1 Of great extent; immense; enormous; huge; also, very spacious. 2 Very great in number, quantity, or amount. 3 Very great in degree, intensity, or importance. 4 *Obs.* Wide and waste; desolate; desert. See synonyms under IMMENSE, LARGE. — *n.* 1 A boundless space; immensity. 2 *Brit. Dial.* A great quantity. [<L, waste, empty, vast. Related to WASTE.] — **vast′ly** *adv.*

add, āce, câre, pälm; end, ēven; it, īce; odd, ōpen, ôrder; tŏŏk, pōōl; up, bûrn; ə = a in *above,* e in *sicken,* i in *clarity,* o in *melon,* u in *focus;* yōō = u in *fuse;* oi, oil; ou, pout; ch, check; g, go; ng, ring; th, thin; th, this; zh, vision. Foreign sounds à, œ, ü, kh, ṅ; and ♦: see page xx. < from; + plus; ? possibly.

—**vast'ness, vas·ti·tude** (vas'tə·tōōd, -tyōōd, väs'-) n.
vas·ta·tion (vas·tā'shən) n. 1 Purification by the spiritual burning away of evil. 2 Obs. Devastation. [< L vastatio, -onis < vastare lay waste < vastus waste, empty]
Väs·ter·ås (ves·tər·ōs') A city of east central Sweden; an industrial center: formerly Vesterås.
vas·ti·ty (vas'tə·tē, väs'-) n. pl. **·ties** Vastness; immensity. [< F vastité < vaste large]
vast·y (vas'tē, väs'-) adj. Poetic Vast. [< VAST]
vat (vat) n. A large vessel, tub, or cistern, especially for holding liquids and dyeing materials. —v.t. **vat·ted, vat·ting** To put into a vat; treat in a vat. [OE fæt]
vat dye Chem. A dye produced by oxidation, and resistant to sunlight and washing.
Va·té (vä'tē) See EFATE.
vat·ic (vat'ik) adj. Pertaining to or proceeding from a prophet or seer; oracular; prophetic; inspired: vatic dicta, vatic lips. Also **vat'i·cal**. [< L vates prophet]
Vat·i·can (vat'ə·kən) n. 1 The papal palace in Vatican City, Rome. 2 The papal government, as distinguished from the Quirinal, or Italian civil government. — **Council of the Vatican** An ecumenical council, 1869–70, at the Vatican which declared the pope's infallibility, when speaking ex cathedra, to be a dogma of the church. [from L Vaticanus Vatican Hill in Rome]
Vatican City A sovereign papal state within the city of Rome, established June 10, 1929; 108.7 acres, including the Vatican and St. Peter's Church, along with the square in front of it; twelve buildings outside this area, both in and outside Rome, enjoy extraterritorial rights: Italian Città del Vaticano.
Vat·i·can·ism (vat'ə·kən·iz'əm) n. The ecclesiastical system and the tenet based on the supremacy and infallibility of the pope.
vat·i·cide (vat'ə·sīd) n. The killing of a prophet; also, a prophet-slayer. [< L vates prophet + -CIDE]
va·tic·i·nal (və·tis'ə·nəl) adj. Prophetic.
va·tic·i·nate (və·tis'ə·nāt) v.t. & v.i. **·nat·ed, ·nat·ing** To prophesy; foretell. [< L vaticinatus, pp. of vaticinari prophesy < vates prophet] — **va·tic'i·na'tion** n. — **va·tic'i·na'tor** n. — **va·tic'i·na·to·ry** (-nə·tôr'ē, -tō'rē) adj.
Vät·ter (vet'ər), **Lake** The second largest lake in Sweden, in the south central part; 733 square miles: formerly Vetter. Swedish **Vät·tern** (vet'ərn).
Va·tu·tin (vä·tōō'tin), **Nikolai**, 1901?–44, Russian general in World War II.
vau (väv) See VAV.
Vau·ban (vō·bän'), **Marquis de**, 1633–1707, Sébastien le Prestre, French military engineer.
vau·che·ri·a·ceous (vō·kir'ē·ā'shəs) adj. Bot. Belonging or pertaining to a genus (Vaucheria) of green algae consisting of long and usually branched filaments which grow in feltlike masses in shallow water and on muddy banks: often called **green felt**. [< NL, after Jean Pierre Vaucher, 1763–1841, Swiss botanist]
Vau·cluse (vō·klüz') A department in Provence, SE France; 1,381 square miles; capital, Avignon.
Vaud (vō) A canton in west central Switzerland; 1,239 square miles; capital, Lausanne: German Waadt.
vaude·ville (vōd'vil, vô'də·vil) n. 1 A miscellaneous theatrical entertainment consisting of a slight dramatic sketch or pantomime interspersed with songs and dances; a series of short sketches, songs, dances, acrobatic feats, etc., having no dramatic connection; a variety show; also, a theater presenting such shows. 2 A street ballad; originally, a satirical or topical popular song. [< F, alter. of (chanson de) Vau de Vire (song of) the valley of the Vire river (in Normandy), where Basselin, the best-known composer of such songs, lived]
Vau·dois (vō·dwä') n. pl. **·dois** (-dwä') 1 An inhabitant, or the inhabitants collectively, of the Swiss canton of Vaud. 2 The dialect of this canton.
Vau·dois (vō·dwä') n. pl. The Waldenses.
Vaughan (vôn), **Henry**, 1622–95, English poet.
Vaughan Williams, Ralph, born 1872, English composer.
vault¹ (vôlt) n. 1 An arched apartment or chamber; also, any subterranean compartment; cellar. 2 An arched structure; arched

ceiling or roof. 3 Any vaultlike covering; the sky. 4 An arched roof of a cavity. 5 An underground room or compartment for storing wine, valuables, etc. 6 A burial chamber. —v.t. 1 To form with a vaulted roof; cover with or as with a vault. 2 To construct in the form of a vault. [< OF volte, vaute, ult. < L volutus, pp. of volvere turn about, roll] — **vault'ed** adj.

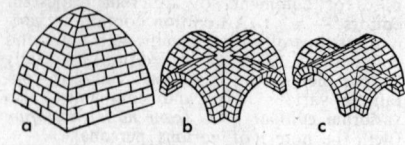

TYPES OF VAULTS
a. Cloister or cove. b. Groin.
c. Welsh or underpitch.

vault² (vôlt) v.t. 1 To leap over, especially with the aid of a pole or with the hands resting on something. 2 To mount with a leap, as a horse. —v.i. 3 To leap; spring. 4 To do a curvet. See synonyms under LEAP. —n. 1 A leap or bound; a springing leap, as one made with the aid of a pole. 2 The curvet of a horse. [< OF volter leap, gambol, ? ult. < L volutus, pp. of volvere turn about, roll] — **vault'er** n.
vault·ing¹ (vôl'ting) n. 1 Vaulted work, or vaults collectively. 2 The work or art of building a vault.
vault·ing² (vôl'ting) adj. 1 That overleaps; hence, unduly confident or presumptuous: vaulting ambition. 2 That can be used in vaulting, as in gymnastics.
vaunt (vônt, vänt) v.i. To speak boastfully. —v.t. To boast of. See synonyms under FLAUNT. —n. Boastful assertion or ostentatious display. See synonyms under OSTENTATION. [< OF vanter << LL vanitare brag < L vanus empty, vain] — **vaunt'er** n. — **vaunt'ing** n. — **vaunt'ing·ly** adv.
vaunt-cou·ri·er (vänt'kōōr'ē·ə, vônt'-) n. 1 Archaic A horseman or soldier sent in advance of an army. 2 A forerunner; precursor; herald. [< F avant-courier]
vaunt·ie (vôn'tē) adj. Scot. Boastful. Also **vaunt'y, vawnt'ie**.
Vau·pés (vou'pās, Sp. bou·pās') See UAUPÉS.
Vaux (vō) A village west of Château-Thierry in north central France; point of furthest German advance on the road to Paris in World War I, 1918.
vav (väv) n. The sixth Hebrew letter: also spelled vau. Also **vaw**. See ALPHABET.
vav·a·sor (vav'ə·sôr, -sōr) n. 1 The rank of a principal vassal next below a baron. 2 A vassal holding lands from a great vassal, and having other vassals under him. Also **vav'a·sour** (-sōōr). [< OF vavassour << LL vassus vassorum vassal of vassals]
va·ward (vä'wərd) adj. Obs. Vanward. [Alter. of obs. avantward < AF avantwarde, OF avant-garde vanguard]
V-day (vē'dā') n. A day of victory; in World War II, **V-E Day** (vē'ē'dā') (officially May 8, 1945), the date of victory of the United Nations in Europe, and **V-J Day** (vē'jā'dā') (officially Sept. 2, 1945, Tokyo time), the date of their victory in the Pacific.
Ve·a·dar (vē'ä·där, vā'-) A Hebrew month. See CALENDAR (Hebrew).
veal (vēl) n. 1 The flesh of a calf considered as food. 2 Obs. **bob veal** The flesh of a calf so young as to be unfit for food. [< OF viel calf < L vitellus, dim. of vitulus calf]
Veb·len (veb'lən), **Thorstein Bunde**, 1857–1929, U.S. economist and sociologist.
vec·tion (vek'shən) n. Med. The carrying of a disease organism from an infected to a well person. [< L vectio, -onis a conveyance, carrying < vehere carry]
vec·tor (vek'tər) n. 1 Math. **a** A line representing a physical quantity that has magnitude and direction in space, as velocity, acceleration or force: distinguished from scalar. **b** A radius vector. 2 Med. A carrier of pathogenic micro-organisms from one host to another: The anopheles mosquito is a vector of the malaria parasite. [< L, carrier < vehere carry] — **vec·to'ri·al** (vek·tôr'ē·əl, -tō'-) adj.
vectorial angle Math. A polar angle.
vec·tur·ism (vek'chə·riz'əm) n. U.S. The

hobby of collecting old transportation tokens from bus and trolley lines. [< L vecturus fare, passage money < vehere carry] — **vec'tur·ist** n.
Ve·da (vā'də, vē'-) n. The body of ancient Indian sacred writings, dating from the second millennium B.C., which form the Hindu scriptures; specifically, the four major collections included in this literature: **Rigveda**, containing sacrificial hymns addressed to the gods; **Yajurveda**, containing liturgical formulas; **Samaveda**, a group of hymns chiefly in honor of Indra; and **Aharvaveda**, a large collection of charms and incantations. [< Skt., knowledge] — **Ve·da·ic** (vi·dā'ik) adj. — **Ve·da·ism** (vā'də·iz'əm, vē'-) n.
Ve·dan·ta (vi·dän'tə, -dan'-) n. Any of several schools of Hindu religious philosophy based on the Upanishads; especially, the monistic system of Shankara which teaches the worship of Brahma as the creator and soul of the universe. [< Skt. < Veda Veda + anta end] — **Ve·dan'tic** adj. — **Ve·dan'tism** n. — **Ve·dan'tist** n.
Ved·da (ved'ə) n. One of a primitive people of Ceylon of doubtful classification, slender, dark, small, with heavy, wavy hair, having both Caucasoid and Australoid traits, but not typically either: by some anthropologists thought to be remnants of an original Indo-Australoid race. Also **Ved'dah**. [< Singhalese, hunter]
Ved·der (ved'ər), **Elihu**, 1836–1923, U.S. painter and illustrator.
ve·dette (vi·det') n. 1 A mounted sentinel placed in advance of an outpost. 2 A small vessel used to watch the movements of the enemy: also **vedette boat**. 3 Colloq. In France, a female movie star. Also spelled vidette. [< F < Ital. vedetta, alter. (after vedere see) of veletta, dim. of Sp. vela vigil < L vigilare watch]
Ve·dic (vā'dik, vē'-) adj. Of or pertaining to the Vedas or the language in which they were written. —n. Vedic Sanskrit.
Vedic Sanskrit See under SANSKRIT.
vee (vē) n. 1 The sound or the shape of the letter V. 2 Anything shaped like the letter V. 3 U.S. Slang A five-dollar bill. —adj. V-shaped.
Veep (vēp) n. U.S. Slang The vice president (V. P.): applied especially to Alben W. Barkley, vice president of the United States 1948–1952.
veer¹ (vir) v.i. 1 Naut. To turn to another course; wear ship. 2 To change direction by a clockwise motion, as the wind. 3 To shift from one position to another; be variable or fickle. —v.t. 4 To change the direction of. See synonyms under CHANGE, FLUCTUATE, WANDER. —n. A change in direction; a swerve. [< F virer turn]
veer² (vir) v.t. & v.i. Naut. To let out or allow (a rope, anchor chain, etc.) to run out to a certain length. [< MDu. vieren slacken]
veer·y (vir'ē) n. pl. **veer·ies** A melodious tawny thrush (Hylocichla fuscescens) of eastern North America: also called Wilson's thrush. [Prob. imit.]
Ve·ga (vē'gə, vā'-) A star, Alpha in the constellation Lyra; 0.14 magnitude. [< Med. L < Arabic (al-Nasr) al-Waqi the falling (vulture)]
Ve·ga (vā'gə, Sp. bā'gä), **Lope de**, 1562–1635, Spanish dramatist and poet: full name Lope Felix de Vega Carpio.
veg·e·ta·ble (vej'ə·tə·bəl, vej'tə-) n. 1 The edible part of any herbaceous plant, raw or cooked, chiefly when served with an entree, or before the dessert. 2 Any member of the vegetable kingdom; a plant. See synonyms under FRUIT. —adj. 1 Pertaining to plants, especially garden or farm vegetables. 2 Derived from, of the nature of, or resembling plants. 3 Made from or consisting of vegetables. [< OF << LL vegetabilis full of life < L vegetare animate < vegetus vigorous, lively < vegere be lively] — **veg'e·ta·bil'i·ty** n. — **veg'e·ta·bly** adv.
vegetable butter See BUTTER¹ (def. 2).
vegetable fibers Textile fibers such as cotton, flax, kapok, jute, ramie, etc.
vegetable ivory Ivory nut.
vegetable kingdom The domain of nature that includes all organisms classified as plants. Compare ANIMAL KINGDOM, MINERAL KINGDOM.
vegetable marrow 1 A plant of the gourd family (Cucurbita pepo), having a tender,

edible fruit. **2** The fruit, esteemed as a vegetable: also called *marrow squash.*
vegetable oyster The salsify.
vegetable silk A cottonlike material obtained from the seed pods of a Brazilian tree *(Chorisia speciosa)* and used for stuffing cushions, etc.
vegetable sponge A luffa.
vegetable tallow Any of several fatty vegetable substances, variously derived, resembling tallow, and used locally for making candles, soap, etc.
vegetable wax Any wax derived from a plant.
veg·e·tal (vej'ə·təl) *adj.* **1** Of or pertaining to plants or vegetables; vegetative. **2** Characterizing those vital processes which are common to plants and animals, especially as distinguished from sensation and volition. [<L *vegetus* lively, vigorous. See VEGETABLE.]
veg·e·tant (vej'ə·tənt) *adj.* **1** Invigorating; vivifying; stimulating growth. **2** Vegetating; plantlike. [<L *vegetans, -antis,* ppr. of *vegere* be active, lively]
veg·e·tar·i·an (vej'ə·târ'ē·ən) *adj.* **1** Pertaining to or advocating the eating of only vegetable foods. **2** Exclusively vegetable, as a diet. —*n.* One who holds or practices vegetarianism: also **veg·e·tist** (vej'ə·tist).
veg·e·tar·i·an·ism (vej'ə·târ'ē·ən·iz'əm) *n.* The theory or practice of eating only vegetables and fruits. Also **veg'e·tism.**
veg·e·tate (vej'ə·tāt) *v.i.* **·tat·ed, ·tat·ing 1** To grow, as a plant. **2** To live in a monotonous, passive way. **3** *Pathol.* To increase in size. [<L *vegetatus,* pp. of *vegetare* animate. See VEGETABLE.]
veg·e·ta·tion (vej'ə·tā'shən) *n.* **1** The process of vegetating. **2** Plant life in the aggregate. **3** *Pathol.* An excrescence on the body; an abnormal or fibrous growth. **4** A plantlike growth. —**veg'e·ta'tion·al** *adj.*
veg·e·ta·tive (vej'ə·tā'tiv) *adj.* **1** Of, pertaining to, or exhibiting the processes of plant life. **2** Growing or capable of growing as plants; productive. **3** Having a mere physical existence; showing but little mental activity. **4** Asexual. **5** Concerned with growth and nutrition. **6** Functioning involuntarily or unconsciously: a *vegetative* process. Also **veg·e·tive** (vej'ə·tiv). —**veg'e·ta'tive·ly** *adv.* —**veg'e·ta'tive·ness** *n.*
Ve·glia (ve'lyä) The Italian name for KRK.
ve·he·ment (vē'ə·mənt) *adj.* **1** Arising from or marked by impetuosity of feeling or passion; ardent. **2** Acting with great force or energy; energetic; violent; furious. See synonyms under ARDENT, EAGER[1], HOT, VIOLENT. [<OF <L *vehemens, -entis* impetuous, rash; ult. origin uncertain] —**ve'he·mence, ve'he·men·cy** *n.* —**ve'he·ment·ly** *adv.*
ve·hi·cle (vē'ə·kəl) *n.* **1** That in or on which anything is carried; especially, a contrivance fitted with wheels or runners for carrying something; a conveyance, as a car or sled. **2** *Med.* A medium, as a liquid, with which is mixed some other substance that it may be applied or administered more easily; an excipient. **3** The medium with which pigments are mixed in painting. **4** Anything by means of which something else, as power, thought, etc., is transmitted or communicated; a device used to transmit an effect. [<F *véhicule* <L *vehiculum* <*vehere* carry, ride] —**ve·hic·u·lar** (vi·hik'yə·lər) *adj.*
Vehm·ge·richt (fām'gə·rikht) *n. pl.* **·rich·te** (-rikh'tə) An institution peculiar to Germany, especially Westphalia, from about 1150 to 1568, consisting of irregular tribunals. Civil cases were tried openly, but serious crimes, such as heresy, witchcraft, murder, etc., were tried by night in secret session. [<G <*fehm* judgment, punishment + *gericht* law, court]
Ve·ii (vē'yī) An ancient Etruscan city in central Italy, destroyed by Romans, 396 B.C.; site, 10 miles NW of Rome.
veil (vāl) *n.* **1** A piece of thin and light fabric, worn over the face or head for concealment, protection, or ornament. **2** Any piece of fabric used to conceal an object; a screen; curtain; mask. **3** Figuratively, that which conceals from inspection; a disguise; pretext. **4** A velum. **5** A caul. **6** The life of a nun; vows made by a nun. —**to take the veil** To become a nun. —*v.t.* **1** To cover with a veil. **2** To hide; disguise. See synonyms under HIDE[1], MASK[1], PALLIATE. Also, *Obs., vail.* ◆ Homophones: *vail, vale.* [<OF *veile* <L *velum* piece of cloth, sail] —**veiled** *adj.* —**veil'er** *n.*
veil·ing (vā'ling) *n.* **1** The act of covering with a veil. **2** Material for veils. **3** A veil.
vein (vān) *n.* **1** *Anat.* One of the muscular tubular vessels that convey blood to the heart; loosely, any blood vessel. ◆ Collateral adjective: *venal.* **2** *Entomol.* One of the radiating supports of an insect's wing; a rib or nerve. **3** *Bot.* One of the slender vascular bundles that form the framework of a leaf. **4** *Geol.* The filling of a fissure or fault in a rock, particularly if deposited by aqueous solutions. **5** A lode. **6** A bed or shoot of ore parallel with the fault. **7** A long, irregular, colored streak, as in wood, marble, etc. **8** A distinctive trait; a specific tendency or disposition. **9** A temporary state of mind; humor; mood. **10** A cavity; cleft; fissure. **11** A crevice or natural channel through which water trickles. —*v.t.* **1** To furnish or fill with veins. **2** To streak or ornament with veins. **3** To extend over or throughout as veins. ◆ Homophones: *vain, vane.* [<OF *veine* <L *vena* blood vessel] —**vein'less** *adj.* —**vein'y** *adj.*

VEINING OF SILVER MAPLE LEAF

veined (vānd) *adj.* **1** Having veins. **2** Marked with or abounding in veins. **3** Marked with streaks of another color.
vein·ing (vā'ning) *n.* **1** A vein or network of veins. **2** A streaked or veined surface.
vein·let (vān'lit) *n.* A small vein.
vein·stone (vān'stōn') *n.* Gangue.
Ve·la (vē'lə) A southern constellation, formerly part of the larger one, Argo Navis. [<L, veil]
ve·la·men (və·lā'mən) *n. pl.* **·lam·i·na** (-lam'ə·nə) **1** *Anat.* Any membrane, covering, or integument. **2** *Bot.* An envelope consisting of several layers of empty cells, forming the outer covering of the aerial roots of certain orchids and arums. Also **vel·a·men·tum** (vel'ə·men'təm). [<L, covering <*velare* veil]
ve·lar (vē'lər) *adj.* **1** Of or pertaining to a velum, especially the soft palate. **2** *Phonet.* Formed with the back of the tongue touching or near the soft palate, as (k) in *cool,* (g) in *go.* —*n. Phonet.* A velar sound. [<L *velaris* <*velum* sail, curtain]
ve·lar·ize (vē'lə·rīz) *v.* **·ized, ·iz·ing** *Phonet. v.t.* To modify (a sound) by raising the back of the tongue toward the soft palate. —*v.i.* To be modified to a velar sound.
ve·lar·i·um (və·lâr'ē·əm) *n. pl.* **·lar·i·a** (-lâr'ē·ə) *Latin* The awning spread over the seats in ancient Roman theaters or amphitheaters.
Ve·lás·quez (və·las'kwiz, *Sp.* bä·läs'kāth), **Diego Rodriguez de Silva y,** 1599–1660, Spanish painter. Also **Ve·láz·quez** (bä·läth'·kāth).
ve·late (vē'lāt, -lit) *adj. Biol.* Having a velum or veil. [<L *velatus,* pp. of *velare* veil]
ve·la·tion (vē·lā'shən) *n.* **1** The forming of a velum. **2** The act of veiling, or the state of being veiled; hence, concealment; mystery.
veldt (velt, felt) *n.* In South Africa, open country or pasture land; grassland having few shrubs or trees. Low-lying wooded land is known as **bush veldt,** and the high treeless plains as **high veldt.** Also **veld.** [<Afrikaans *veld* <Du., field]
vel·i·ger (vel'ə·jər) *n. Zool.* The larva of a mollusk at the stage succeeding the trochophore and when it has a ciliated swimming membrane or membranes. [<LL *veliger* sail-bearing <L *velum* sail + *gerere* bear]
ve·lig·er·ous (və·lij'ər·əs) *adj.* Bearing a velum or membranous partition.
Ve·li·ki Kvar·ner (ve'li·kē kvär'nər) An arm of the Adriatic Sea, SE of Istria, in NW Croatia, Yugoslavia: also *Gulf of Quarnero. Italian* **Gol·fo di Quar·ne·ro** (gôl'fō dē kwär·nā'rō).
vel·i·ta·tion (vel'ə·tā'shən) *n.* A petty skirmish; a wordy controversy. [<L *velitatio, -onis* <*velitatus,* pp. of *velitari* skirmish <*veles, velitis* a foot soldier]
ve·li·tes (vē'lə·tēz) *n. pl.* Light-armed Roman soldiers used as skirmishers in ancient legions. [<L, pl. of *veles* foot soldier]
vel·le·i·ty (ve·lē'ə·tē) *n. pl.* **·ties** A very low degree of desire or volition, not leading to action; a mere wish. [<Med. L *velleitas, -tatis* <L *velle* wish]
vel·li·cate (vel'ə·kāt) *v.t.* & *v.i.* **·cat·ed, ·cat·ing** To twitch or pluck. [<L *vellicatus,* pp. of *vellicare* twitch, freq. of *vellere* pluck] —**vel'li·ca'tion** *n.* —**vel'li·ca'tive** *adj.*
vel·lum (vel'əm) *n.* **1** Fine parchment made from the skins of calves: used for expensive binding, printing, etc. **2** A manuscript written on such parchment. **3** Paper made to resemble parchment. [<OF *velin, vellin* <*veel, viel* calf. See VEAL.]
ve·lo·ce (vā·lō'chā) *adv. Music* Rapidly; in quick tempo; swiftly. [<Ital., swift]
ve·loc·i·pede (və·los'ə·pēd) *n.* **1** An early form of bicycle or tricycle; also, a child's tricycle. **2** A light handcar or vehicle propelled by hands or feet and used along railroad tracks. [<L *velox, velocis* swift + -PEDE]
ve·loc·i·ty (və·los'ə·tē) *n. pl.* **·ties 1** The state of moving swiftly; rapid motion; celerity; speed. **2** *Physics* **a** The rate of change of position in a moving object. **b** The rate of motion in a stated direction: a vector quantity: distinguished from *speed.* [<L *velocitas, -tatis* <*velox* swift]
velocity of escape *Physics* The minimum velocity at which any particle or object would permanently escape the gravitational field of a body of stated mass: on the earth this velocity is approximately 7 miles per second. See table under PLANET.
ve·lo·drome (vē'lə·drōm) *n.* A racecourse, as for bicycles. [<L *velox* speedy + -DROME]
ve·lours (və·lo͝or') *n. pl.* **·lours** A soft, velvetlike, closely woven cotton or wool fabric having a short, thick pile. Also **ve·lour'.** [<F. See VELURE.]
ve·lou·té (və·lo͞o·tā') *n. French* A rich white sauce made by thickening chicken or veal stock with flour and butter. Also **sauce velouté.**
ve·lum (vē'ləm) *n. pl.* **·la** (-lə) **1** *Biol.* A thin membranous covering or partition, as in certain jellyfishes, and in mushrooms. **2** *Anat.* The soft palate. See PALATE. [<L]
ve·lure (və·lo͝or') *n.* **1** Velvet, or a fabric resembling velvet; specifically, a heavy fabric of linen, silk, or jute, used for hangings, table covers, and the like. **2** A velvet or silk pad for smoothing a silk hat. —*v.t.* **·lured, ·lur·ing** To smooth with a soft pad, as a hat. [<F *velours* <L *villosus* shaggy <*villus* shaggy hair]
ve·lu·ti·nous (və·lo͞o'tə·nəs) *adj. Bot.* Covered with close, soft hairs, like the pile of velvet; velvety. [<NL *velutinus* <Med. L *velutum,* var. of *velvetum.* See VELVET.]
vel·ver·et (vel'və·ret') *n.* A velvet fabric with cotton backing.
vel·vet (vel'vit) *n.* **1** A fabric, properly of silk, now sometimes made of cotton or one of the synthetics, closely woven and having on one side a thick, short, smooth pile: called **pile velvet** when the pile is formed of loops, and **cut velvet** when the pile is of single threads. **2** The furry skin covering a growing antler. —**chiffon velvet** A very soft, lightweight velvet having the pile pressed flat. —*adj.* **1** Made of velvet. **2** Smooth and soft to the touch; velvety. [<Med. L *velvetum,* ult. <L *villus* shaggy hair]
velvet carpet A carpet having a pile longer than a Brussels carpet but cut in the manner of a Wilton carpet: also **tapestry velvet carpet.**
vel·vet·een (vel'və·tēn') *n.* **1** A cotton fabric, with a short, close pile like velvet. **2** *pl.* Clothes, especially trousers, made of this material. [<VELVET] —**vel'vet·eened'** *adj.*
vel·vet·leaf (vel'vit·lēf') *n.* **1** Any one of several plants, especially the Indian mallow. **2** A tropical climbing shrub *(Cissampelos pareira)* of the moonseed family, the bark of which yields a variety of pareira brava.
vel·vet·y (vel'vit·ē) *adj.* **1** Like velvet; smooth and soft to appearance or touch. **2**

Mild and smooth to the taste: *velvety* liqueur.

ve·na (vē′nə) *n. pl.* **·nae** (-nē) *Anat.* A vein. [<L]

ve·na ca·va (vē′nə kā′və) *n. pl.* **ve·nae ca·vae** (vē′nē kā′vē) *Anat.* Either of the two great venous trunks (called *superior* and *inferior*) emptying into the right auricle of the heart. See illustration under HEART. [<L, hollow vein]

ve·nal[1] (vē′nəl) *adj.* **1** Ready to sell honor or principle, or to accept a bribe; mercenary; purchasable: said of persons. **2** Subject to sordid bargaining or to corrupt influences; salable. **3** Characterized by corruption and venality. [<L *venalis* < *venum* sale] — **ve′nal·ly** *adv.*

Synonyms: hireling, mercenary, purchasable, salable. *Mercenary* has especial application to character or disposition; as, a *mercenary* spirit; *mercenary* motives—that is, a spirit or motives to which money is the chief consideration or the moving principle. Thus, etymologically, the *mercenary* can be hired, while the *venal* are openly or actually for sale; *hireling* signifies serving for hire or pay, or having the spirit or character of one who works or of that which is done directly for hire or pay. The *hireling*, the *mercenary*, and the *venal* are alike in making principle, conscience, and honor of less account than gold or sordid considerations; but the *mercenary* and *venal* may be simply open to the bargain and sale which the *hireling* has already consummated. A public officer who makes his office tributary to private speculation is *mercenary*; if he receives a stipulated recompense for administering his office at the behest of some leader, faction, corporation, or the like, he is both *hireling* and *venal*; if he sells essential advantages, without subjecting himself to any direct domination, his course is *venal*, but not *hireling*. *Antonyms*: disinterested, honest, honorable, incorruptible, patriotic, unpurchasable.

ve·nal[2] (vē′nəl) *adj.* Of or pertaining to the veins; venous. [<L *vena* vein]

ve·nal·i·ty (vē·nal′ə·tē) *n. pl.* **·ties** The state or character of being basely or improperly influenced by sordid considerations; prostitution, as of talents, office, etc., for gain or reward; willingness to accept bribes. [<L *venalitas, -tatis*]

ve·nat·ic (vē·nat′ik) *adj.* **1** Of, used in, or pertaining to hunting. **2** Living by or fond of hunting. Also **ven·a·to·ri·al** (ven′ə·tôr′ē·əl, -tō′rē-). [<L *venaticus* < *venatus*, pp. of *venari* hunt] — **ve·nat′i·cal** *adj.* — **ve·nat′i·cal·ly** *adv.*

ve·na·tion (vē·nā′shən) *n. Biol.* **1** The particular arrangement of veins, as in a leaf. **2** The distribution of veins in an organism or part, as an insect wing, etc. [<L *vena* a vein]

vend (vend) *v.t.* **1** To sell. **2** To utter (an opinion); publish. — *v.i.* **3** To be a vender. **4** To be sold. [<F *vendre* <L *vendere* < *venum* sale + *dare* give] — **ven·di·tion** (ven·dish′ən) *n.*

ven·dace (ven′dis) *n.* A small whitefish *(Coregonus vandesius)* of some British lakes. Also **ven′dis.** [<F *vandoise* dace]

ven·dee (ven·dē′) *n.* The person or party to whom something, especially land, is sold.

Ven·dée (väṅ·dā′) A region and department in western France on the Bay of Biscay; scene of a royalist revolt, 1793–95; 2,708 square miles. — **Ven·de·an** (ven·dē′ən) *adj.* & *n.*

Ven·dé·miaire (väṅ·dā·myâr′) See under CALENDAR (Republican).

vend·er (ven′dər) *n.* One who sells; a peddler or hawker; vendor.

ven·det·ta (ven·det′ə) *n.* Private warfare or feud, as in revenge for a murder, injury, etc.; a blood feud in which the relatives of the killed or injured person take vengeance on the offender or his relatives. It is still prevalent in Sicily, Corsica, and Montenegro, and, to some extent, in certain other districts. [<Ital. <L *vindicta* vengeance]

vend·i·ble (ven′də·bəl) *adj.* Capable of being vended or sold; marketable. — *n.* A thing exposed for sale. [<L *vendibilis* < *vendere* sell] — **vend′i·bil′i·ty, vend′i·ble·ness** *n.* — **vend′i·bly** *adv.*

Ven·dôme (väṅ·dôm′) A town and former duchy in north central France.

ven·dor (ven′dər) The common legal spelling of VENDER.

ven·due (ven·do͞o′, -dyo͞o′) *n.* A public sale or auction. [<F, orig. fem. pp. of *vendre* sell]

ve·neer (və·nir′) *n.* **1** A thin layer, as of choice wood, upon a commoner surface; a layer of superior material for overlaying a cheaper one. **2** Any of the thin layers glued together to strengthen plywood. **3** Figuratively, mere outside show or elegance. — *v.t.* **1** To cover (a surface) with veneers; overlay for decoration or finer finish. **2** To glue together to form plywood. **3** To conceal, as something disagreeable or coarse, with an attractive or deceptive surface. [Earlier *fineer* <G *furnieren* inlay <F *fournir* furnish] — **ve·neer′er** *n.*

ve·neer·ing (və·nir′ing) *n.* **1** The art of applying veneer. **2** Material used for veneers. **3** A facing or surface of veneer.

ven·e·punc·ture (ven′ə·pungk′chər) See VENIPUNCTURE.

Ve·ner (ve′nər), **Lake** The largest lake of Sweden, in the SW part; 2,141 square miles; 90 miles long, 5 to 46 miles wide: also *Vaner. Swedish* **Vä·nern** (ve′nərn).

ven·er·a·ble (ven′ər·ə·bəl) *adj.* **1** Meriting or commanding veneration; worthy of reverence: now usually implying age. **2** Exciting reverential feelings because of sacred or elevated associations. **3** Revered: used as a title for an archdeacon in Anglican churches, and for those beatified in the Roman Catholic Church. See synonyms under ANCIENT[1]. [<OF <L *venerabilis* < *venerari* revere] — **ven′er·a·ble·ness, ven′er·a·bil′i·ty** *n.* — **ven′er·a·bly** *adv.*

ven·er·ate (ven′ə·rāt) *v.t.* **·at·ed, ·at·ing** To look upon or regard with respect and deference; revere. [<L *veneratus*, pp. of *venerari* revere]

Synonyms: adore, honor, respect, revere, reverence. In the highest sense, to *revere* or *reverence* is to hold in mingled love and honor with something of sacred fear; to *revere* is a wholly spiritual act; to *reverence* is often, but not necessarily, to give outward expression to the reverential feeling; we *revere* or *reverence* the divine majesty. *Revere* is a stronger word than *reverence* or *venerate*. To *venerate* is to hold in exalted honor without fear, and is applied to objects less removed from ourselves than those we *revere*, being said especially of aged persons, of places or objects having sacred associations, and of abstractions; we *venerate* an aged friend or some great cause, as that of civil or religious liberty; we do not *venerate* God, but *revere* or *reverence* him. We *adore* with a humble yet free outflowing of soul. See ADMIRE, DEFER. *Antonyms*: contemn, despise, disdain, dishonor, disregard, scorn, slight, spurn.

ven·er·a·tion (ven′ə·rā′shən) *n.* **1** The act of venerating; reverence; profound respect combined with awe, evoked by the high character or wisdom of a person. **2** The act of worshiping; worship.

Synonyms: adoration, awe, dread, reverence. *Awe* is inspired by that in which there is sublimity or majesty so overwhelming as to awaken a feeling akin to fear; in *awe*, considered by itself, there is no element of esteem or affection, but a sense of the vastness, power, or grandeur of the object. *Dread* is a shrinking apprehension or expectation of possible harm awakened by any one of many objects or causes; in its higher uses *dread* approaches the meaning of *awe*, but with more of chilliness and cowering, and without that subjection of soul to the grandeur and worthiness of the object that is involved in *awe*. *Reverence* and *veneration* are less overwhelming than *awe* or *dread*, and suggest something of esteem, affection, and personal nearness. We may feel *awe* of that which we cannot reverence, as a grandly terrible ocean storm; *awe* of the divine presence is more distant and less trustful than *reverence*. *Veneration* is commonly applied to things which are not subjects of *awe*. *Adoration*, in its full sense, is loftier than *veneration*, less restrained and awed than *reverence*, and with more of the spirit of direct, active, and joyful worship. See REVERENCE. Compare VENERATE. *Antonyms*: contempt, disdain, dishonor, disregard, scorn.

ve·ne·re·al (və·nir′ē·əl) *adj.* **1** Pertaining to or proceeding from sexual intercourse. **2** Communicated by sexual relations with an infected person: a *venereal* disease. **3** Pertaining to or curative of diseases so communicated. **4** Infected with venereal disease. [<L *venereus* <*Venus, -eris*, the goddess of love]

venereal disease *Pathol.* One of several diseases propagated directly or indirectly by sexual intercourse, as syphilis, gonorrhea, and chancroid.

ve·ne·re·ol·o·gy (və·nir′ē·ol′ə·jē) *n.* The study and treatment of venereal diseases. — **ve·ne′re·ol′o·gist** *n.*

ven·er·y[1] (ven′ər·ē) *n.* Sexual indulgence, especially when excessive. [<L *Venus, -eris*, the goddess of love]

ven·er·y[2] (ven′ər·ē) *n. pl.* **·er·ies** The hunting of game; the sport of hunting; the chase. [<F *venerie* < *vener* hunt <L *venari* hunt]

ven·e·sec·tion (ven′ə·sek′shən) *n. Surg.* Phlebotomy. [<Med. L *venae sectio* cutting of a vein]

Ve·ne·ti·a (və·nē′shē·ə, -shə) **1** An ancient division of the Roman Empire comprising that part of Italy between the Po river and the Alps. **2** Venezia. **3** Veneto.

Ve·ne·tian (və·nē′shən) *adj.* Pertaining to Venice, or to the medieval school of architecture developed there. — *n.* **1** A native of Venice. **2** *Colloq.* A Venetian blind. **3** A heavy braid or tape used on Venetian blinds.

Venetian blind A flexible window screen that may be raised or lowered, having overlapping horizontal slats so fastened on webbing or tape as to regulate, exclude, or admit light.

Venetian carpet A worsted carpet for stairs and hallways, commonly of a simple striped pattern.

Venetian glass A delicate and fine glassware originally made at or near Venice.

Venetian red Red ocher. See under OCHER.

Venetian school **1** A school of painting originating in and near Venice in the 15th century and distinguished by richness of coloring, as in the work of Titian, Tintoretto, Giorgione, etc. **2** A school of Italian architecture.

Ve·net·ic (və·net′ik) *n.* **1** A member of an ancient people of NE Italy. **2** Their Indo-European language, possibly related to Illyrian and Messapian. — *adj.* Of or pertaining to these people or their language.

Ve·ne·to (vā′nā·tō) A region of northern and NE Italy; 7,093 square miles; capital, Venice.

Ve·ne·zia (vā·nā′tsyä) **1** The Italian name for VENICE. **2** A province in Veneto, northern Italy; 949 square miles; capital, Venice.

Venezia Giu·lia (jo͞o′lyä) A former region of NE Italy including Trieste, Rijeka, and Istria; 3,370 square miles; divided 1947 between Italy (180 square miles), Yugoslavia (2,891 square miles), and the Free Territory of Trieste (298 square miles, later also divided).

Ven·e·zue·la (ven′ə·zwē′lə, -zo͞o·ā′lə; *Sp.* bā′·nā·swā′lä) A republic in northern South America; 352,143 square miles; capital, Caracas. — **Ven′e·zue′lan** *adj.* & *n.*

Venezuela, Gulf of See under MARACAIBO.

ven·geance (ven′jəns) *n.* **1** The infliction of a deserved penalty; retributive punishment. **2** In a bad sense, wrathful avenging of a wrong; revenge. **3** *Obs.* Mischief; evil. See synonyms under REVENGE. — **with a vengeance** With great force or violence; extremely; to an unusual extent. [<AF <OF *venger* avenge <L *vindicare* defend, avenge < *vindex, vindicis* claimant, protector]

venge·ful (venj′fəl) *adj.* Prone to inflict vengeance; vindictive. — **venge′ful·ly** *adv.* — **venge′ful·ness** *n.*

ve·ni·al (vē′nē·əl, vēn′yəl) *adj.* That may be pardoned or overlooked; excusable. [<OF <L *venialis* < *venia* forgiveness, mercy] — **ve′·ni·al′i·ty** (-al′ə·tē), **ve′ni·al·ness** *n.* — **ve′ni·al·ly** *adv.*

Synonyms: excusable, pardonable, slight, trivial. Aside from its technical ecclesiastical use, *venial* is always understood as marking some fault comparatively *slight* or *trivial*. A *venial* offense is one readily overlooked; a *pardonable* offense requires more serious consideration, but on deliberation is found to be susceptible of pardon. *Excusable* is scarcely applied to offenses, but to matters open to doubt or criticism rather than direct censure; so used, it often falls little short of justifiable; as, I suppose, under those circumstances, his action was *excusable.* *Antonyms*: inexcusable, mortal, unpardonable.

venial sin *Theol.* A pardonable offense, or an unpremeditated one: opposed to *mortal* or *deadly* sin.

Ven·ice (ven′is) A port in NE Italy, built on 118 islands in the **Lagoon of Venice,** the NW part of the **Gulf of Venice,** the northern sector of the Adriatic between Istria and the Po river delta: Italian *Venezia.*

Ve·ni Cre·a·tor (vē′nī krē·ā′tər) *Latin* A hymn to the Holy Ghost: so called from its beginning, *Veni Creator Spiritus* (Come, Creator Spirit).

ven·i·punc·ture (ven′ə·pungk′chər) *n. Surg.* The operation of puncturing a vein: also spelled *venepuncture.* [<L *vena* vein + PUNCTURE]

ve·ni·re (vi·nī′rē) *n. Law* A writ issued to the sheriff for summoning persons to serve as a jury in court: from its phrase *venire facias* (that you cause to come). [<L]

ve·ni·re·man (vi·nī′rē·mən) *n. pl.* **·men** (-mən) A juryman; a person summoned to serve on a jury.

ven·i·son (ven′ə·zən, -sən; *Brit.* ven′zən) *n.* **1** Deer flesh used for food. **2** *Obs.* The flesh of any edible game. [<F *venaison* <L *venatio, -onis* hunting <*venatus,* pp. of *venari* hunt]

Ve·ni·te (vi·nī′tē) *n.* The 95th psalm, used as a canticle in various liturgies: from its first word. [<L, come, imperative of *venire* come]

ve·ni, vi·di, vi·ci (vē′nī, vī′dī, vī′sī; wā′nē, wē′dē, wē′kē) *Latin* I came, I saw, I conquered: words used by Julius Caesar to report his victory over Pharnaces, king of Pontus, to the Roman Senate.

Ve·ni·ze·los (ve′nē·ze′lôs), **Eleftherios,** 1864–1936, Greek statesman; premier 1917–20 and 1928–32.

ven·om (ven′əm) *n.* **1** The poisonous fluid that certain animals, as serpents and scorpions, secrete, and which produces toxic effects when introduced into the system by a bite or sting. **2** Something harmful; hence, malignity; spite. **3** Any poison. —*v.t.* To imbue with poison; envenom. [<OF *venim* <L *venenum* poison] —**ven′om·er** *n.*

ven·om·ous (ven′əm·əs) *adj.* **1** Having glands secreting venom. **2** Able to give a poisonous sting; virulent; noxious. **3** Working harm; baneful. **4** Malignant; spiteful. See synonyms under MALICIOUS. —**ven′om·ous·ly** *adv.* —**ven′om·ous·ness** *n.*

ve·nose (vē′nōs) *adj.* **1** Having numerous or prominent veins, as a leaf; veiny. **2** Venous. [<L *venosus*]

ve·nos·i·ty (vi·nos′ə·tē) *n.* **1** An excess of venous blood in a part. **2** A plentiful supply of blood vessels.

ve·nous (vē′nəs) *adj. Physiol.* **1** Of, pertaining to, contained, or carried in a vein or veins. **2** Designating the blood carried by the veins and distinguished from arterial blood by its darker color, absence of oxygen, and presence of carbon dioxide. **3** Marked with or having veins. [<L *venosus* <*vena* vein] —**ve′nous·ly** *adv.* —**ve′nous·ness** *n.*

vent (vent) *n.* **1** An opening, commonly small, for the passage of fluids, gases, etc.; hence, an outlet of any kind: also **vent hole.** **2** The act of giving utterance, as to passion; expression; escape; passage: now usually in the phrase **to give vent to.** **3** *Zool.* The external opening of the alimentary canal, especially of animals below mammals; the anus. **4** A touchhole of a gun. —*v.t.* **1** To give expression to: to *vent* one's rage. **2** To relieve, as by giving vent to emotion. **3** To permit to escape at a vent, as a gas. **4** To make a vent in, as a mold. [ME *fent* <OF *fente* cleft <*fendre* cleave <L *findere* split]

vent·age (ven′tij) *n.* **1** A small opening. **2** A finger hole in a musical instrument. [<VENT]

ven·tail (ven′tāl) *n.* The adjustable front of a helmet, permitting complete defense of the face in combat. [<OF *ventaile* <*vent* wind]

vent·er[1] (ven′tər) *n.* One who vents.

ven·ter[2] (ven′tər) *n.* **1** The belly or stomach. **2** Any protuberant part. **3** The womb. **4** A hollowed part, as of a bone. [<L, stomach]

ven·ti·duct (ven′tə·dukt) *n.* An air passage, especially a subterranean ventilating passage. [<L *ventus* wind + DUCT]

ven·ti·late (ven′tə·lāt) *v.t.* **·lat·ed, ·lat·ing** **1** To produce a free circulation of air in, as by means of open shafts, windows, doors, etc.; admit fresh air into. **2** To provide with a vent. **3** To make widely known; expose to examination and discussion. **4** To oxygenate, as blood. **5** *Obs.* To winnow; fan, as wheat. [<L *ventilatus,* pp. of *ventilare* fan <*ventus* wind] —**ven′ti·lat′ing, ven′ti·la′tion** *n.* —**ven′ti·la′tive** *adj.*

ven·ti·la·tor (ven′tə·lā′tər) *n.* **1** One who or that which ventilates. **2** A device or arrangement for supplying fresh air. —**ven′ti·la·to′ry** (-lə·tôr′ē, -tō′rē) *adj.*

Ven·ti·mi·glia (ven′tē·mē′lyä) An Italian port on the Gulf of Genoa NE of Nice; an important international railroad station.

Ven·tôse (väṅ·tōz′) See under CALENDAR (Republican).

ven·trad (ven′trad) *adv. Biol.* Toward the belly or undersurface. [<L *venter* belly]

ven·tral (ven′trəl) *adj.* **1** *Biol.* **a** Of, pertaining to, or situated on or near the abdomen or abdominal surface of an animal. **b** On or toward the lower surface of the body: the *ventral* plates of a serpent: opposed to *dorsal.* **2** *Bot.* Pertaining to the surface of a petal, carpel, etc., that faces the center of a flower. —*n.* One of the paired fins on the underside of fishes, homologous with the hind limb of higher vertebrates: in full, **ventral fin.** [<L *ventralis* <*venter, ventris* belly] —**ven′tral·ly** *adv.*

ven·tri·cle (ven′trə·kəl) *n. Anat.* **1** Any of various cavities in the body, as of the brain, the spinal cord, or between the true and false vocal cords in the larynx. **2** One of the two lower chambers of the heart, from which blood received from the atria is forced into the arteries. [<L *ventriculus,* dim. of *venter, ventris* belly]

ven·tri·cose (ven′trə·kōs) *adj.* **1** Having a protruding belly. **2** Swelling out or inflated on one side or in the middle; bellied; distended. Also **ven′tri·cous** (-kəs). [<NL *ventricosus* <L *venter, ventris* belly] —**ven′tri·cos′i·ty** (-kos′ə·tē) *n.*

ven·tric·u·lar (ven·trik′yə·lər) *adj.* **1** Of, pertaining to, or of the nature of a ventricle. **2** Swollen and distended; ventricose.

ven·tric·u·lose (ven·trik′yə·lōs) *adj.* Slightly ventricose. [<L *ventriculosus* <*ventriculus.* See VENTRICLE.]

ven·tri·lo·qui·al (ven′trə·lō′kwē·əl) *adj.* Pertaining to, resembling, or practicing ventriloquism. Also **ven·tril·o·qual** (ven·tril′ə·kwəl), **ven·tril′o·quous.** —**ven′tri·lo′qui·al·ly** *adv.*

ven·tril·o·quism (ven·tril′ə·kwiz′əm) *n.* The art or practice of speaking in such a manner that the sounds seem to come from some source other than the person speaking. Also **ven·tril′o·quy** (-kwē). [<L *venter* belly + *loqui* speak] —**ven·tril′o·quist** *n.* —**ven·tril′o·quis′tic** *adj.*

ven·tril·o·quize (ven·tril′ə·kwīz) *v.t.* & *v.i.* **·quized, ·quiz·ing** To speak as a ventriloquist. Also *Brit.* **ven·tril′o·quise.**

ventro- *combining form Anat.* The abdomen; related to or near the abdomen; ventral. [<L *venter, ventris* the belly, abdomen]

Vents·pils (vents′pils) A port on the Baltic Sea in NW Latvia: German *Windau.*

ven·ture (ven′chər) *v.* **·tured, ·tur·ing** *v.t.* **1** To expose to chance or risk; hazard; stake. **2** To run the risk of; brave. **3** To express at the risk of denial or refutation: to *venture* a suggestion. **4** To place or send on a chance, as in a speculative business enterprise. **5** *Obs.* To trust as an agent or doer; rely on. —*v.i.* **6** To take a risk; dare. —*n.* **1** The staking of a thing upon a contingency; a risk; hazard. **2** An undertaking attended with risk; a business speculation. **3** That which is ventured; especially, property risk. **4** That which is unforeseen and hazardous; chance; fortune: a rare usage. See synonyms under HAZARD. —**at a venture** At hazard; at random; without aim or thought. [Aphetic form of ADVENTURE] —**ven′tur·er** *n.*

ven·ture·some (ven′chər·səm) *adj.* **1** Bold; daring. **2** Involving hazard; risky. See synonyms under BRAVE, IMPRUDENT. —**ven′ture·some·ly** *adv.* —**ven′ture·some·ness** *n.*

ven·tur·ous (ven′chər·əs) *adj.* **1** Adventurous; willing to take risks and brave dangers; bold. **2** Hazardous; risky; dangerous. See synonyms under IMPRUDENT. —**ven′tur·ous·ly** *adv.* —**ven′tur·ous·ness** *n.*

ven·ue (ven′yōō) *n. Law* **1** The place or neighborhood where a crime is committed or a cause of action arises; the county or political division from which the jury must be summoned and in which the trial must be held. **2** The clause, usually at the beginning of a declaration or indictment, indicating the county in which the proceeding is pending. **3** A clause in an affidavit, stating where it was made and sworn to. —**change of venue** The change of the place of trial, for good cause shown, from one county to another. [<OF, orig. fem. pp. of *venir* come <L *venire*]

ven·ule (ven′yōōl) *n.* A small vein; veinlet, as of an insect. [<L *venula,* dim. of *vena* vein] —**ven′u·lar** (-yə·lər) *adj.*

ven·u·lose (ven′yə·lōs) *adj.* Having numerous veinlets, as a leaf. Also **ven′u·lous** (-ləs). [<VENULE]

ve·nus (vē′nəs) *n.* A bivalve having three hinge teeth in each valve, as the quahaug. [<VENUS; from the resemblance of the lunula of the closed valve to the vulva]

Ve·nus (vē′nəs) **1** In Roman mythology, the goddess of spring, bloom, and beauty: identified with the Greek *Aphrodite.* **2** A statue or painting of Venus. **3** A lovely woman. **4** *Astron.* The second planet from the sun, the most brilliant object in the heavens except the sun and the moon. It moves in an orbit between those of Mercury and Earth at a mean distance from the sun of about 67,000,000 miles, completing a revolution in 224.7 days. Its diameter is about 7,700 miles and it has no satellites. **5** *Obs.* In alchemy, the metal copper. [<L]

Ve·nus·berg (vē′nəs·bûrg, *Ger.* vā′nōōs·berkh) In medieval German legend, a mountain in the dark recesses of which Venus lured men to sensuous pleasures. See TANNHÄUSER.

Venus flytrap A plant (*Dionaea muscipula*), with clustered leaves the blades of which instantly close upon insects or other objects lighting upon them: found native chiefly in the sandy bogs of eastern North and South Carolina. Also **Ve·nus's-fly·trap** (vē′nəs·iz·flī′trap′).

VENUS FLYTRAP
(From 4 to 14 inches tall)

Venus of Mi·lo (mē′lō) A marble statue of Venus, nude above the thighs and with the arms missing, discovered in 1820 on the island of Milo: now in the Louvre. Also **Venus de Milo.**

Ve·nus's-comb (vē′nəs·iz·kōm′) *n.* A European plant (*Scandix pecten-veneris*) with white flowers in numerous umbels, and lobed leaves suggestive of a comb: often called *shepherd's-needle, devil's-darning-needle.*

Ve·nus's-gir·dle (vē′nəs·iz·gûr′dəl) *n.* A ctenophore of warm seas, having a transparent body that shimmers with blue, green, or violet colors.

Ve·nus's-hair (vē′nəs·iz·hâr′) *n.* A maidenhair fern (*Adiantum capillus-veneris*) having a black stipe and branches.

ver·a (ver′ə, var′ə) *adj.* & *adv. Scot.* Very.

Ve·ra (vir′ə) A feminine personal name. [<Slavic, faith]

ve·ra·cious (və·rā′shəs) *adj.* **1** Habitually disposed to speak the truth; truthful. **2** Conforming to or expressing truth; true; accurate. [<L *verax, veracis* <*verus* true] —**ve·ra′cious·ly** *adv.* —**ve·ra′cious·ness** *n.*

ve·rac·i·ty (və·ras′ə·tē) *n. pl.* **·ties** **1** The habitual regard for truth; truthfulness; honesty. **2** Agreement with truth; accuracy, or fact; trueness. **3** That which is true; truth. [<F *véracité* <L *verax.* See VERACIOUS.]

Synonyms: candor, fact, frankness, honesty, ingenuousness, reality, truth, truthfulness, verity. *Truth* is primarily and *verity* is always a quality of thought or speech, especially of speech, as in exact conformity to *fact.* *Veracity* is properly a quality of a person, the habit of speaking and the disposition to

add, āce, câre, pälm; end, ēven; it, īce; odd, ōpen, ôrder; tŏŏk, pōōl; up, bûrn; ə = a in *above,* e in *sicken,* i in *clarity,* o in *melon,* u in *focus;* yōō = u in *fuse;* oi, oil; ou, pout; ch, check; g, go; ng, ring; th, thin; th, this; zh, vision. Foreign sounds à, œ, ü, kh, ṅ; and ◆: see page xx. < from; + plus; ? possibly.

speak the *truth*. *Truthfulness* is a quality that may inhere either in a person or in his statements or beliefs. *Candor, frankness, honesty*, and *ingenuousness* are closely allied with *veracity*, and *fact, reality*, and *verity* with *truth*, while *truthfulness* may accord with either. *Truth* in a secondary sense may be applied to intellectual action or moral character, in the former case becoming a close synonym of *veracity*: She knows him to be a man of *truth*. *Antonyms*: deceit, deception, delusion, duplicity, error, fabrication, fallacy, falsehood, falsity, fiction, guile, imposture, lie, untruth.

Ve·ra·cruz (ver′ə·krōōz′, *Sp.* bä′rä·krōōs′, -krōōth′) 1 A coast state in eastern Mexico; 27,752 square miles; capital, Jalapa. 2 Its chief city, a port on the Gulf of Mexico: officially **Veracruz Lla·ve** (yä′vā).

ve·ran·da (və·ran′də) *n.* An open portico, gallery, or balcony, usually roofed, along the outside of a building; a porch or stoop. Compare LOGGIA. Also **ve·ran′dah.** [<Hind. *varandā* <Pg. *varanda* railing, balustrade, prob. < *vara* rod, pole <L *vara* forked pole]

ve·ra·no (və·rä′nō) *n.* The dry midwinter season in tropical America. [<Sp., lit., summer]

ve·rat·ric acid (və·rat′rik) *Chem.* A colorless crystalline acid, $C_9H_{10}O_4$, contained in sabadilla seeds and also made synthetically. [<L *veratrum* hellebore]

ve·rat·ri·dine (və·rat′rə·dēn, -din) *n. Chem.* A yellowish, amorphous alkaloid, $C_{36}H_{51}O_{11}N$, contained in sabadilla seeds. Also **ve·rat′ri·din** (-din). [<L *veratrum* hellebore + -ID(E) + -INE[2]]

ver·a·trine (ver′ə·trēn, -trin) *n. Chem.* A white or grayish-white, amorphous (rarely crystalline), extremely poisonous mixture of alkaloids, contained in sabadilla seeds: used in medicine as an analgesic in neuralgia and rheumatism. Also **ve·ra·tri·a** (və·rā′trē·ə), **ver′a·trin** (-trin), **ver′a·tri′na** (-trī′nə). [<L *veratrum* hellebore + -INE[2]]

ver·a·trize (ver′ə·trīz) *v.t.* **·trized, ·triz·ing** To treat with veratrine so as to produce its toxic effects.

ve·ra·trum (və·rā′trəm) *n.* Hellebore (def. 2). [<L]

verb (vûrb) *n. Gram.* 1 One of a class of words which assert, declare, or predicate something; that part of speech which expresses existence, action, or occurrence, as the English words *be, collide, think*. 2 Any word or construction functioning similarly. [<F *verbe* <L *verbum* word. Akin to WORD.]

ver·bal (vûr′bəl) *adj.* 1 Of, pertaining to, or connected with words; concerned with words rather than the ideas they convey: *verbal* distinctions. 2 Uttered by the mouth; expressed in words orally; not written: a *verbal* communication; a *verbal* contract or agreement. 3 Having word corresponding with word; literal: a *verbal* translation. 4 *Gram.* **a** Partaking of the nature of or derived from a verb: a *verbal* noun. **b** Used to form verbs: a *verbal* prefix. — *n. Gram.* A noun directly derived from a verb, in English often having the form of the present participle, and signifying the act or process of what is expressed in the verb root; as, there shall be *weeping* and *wailing* and *gnashing* of teeth; also, an infinitive used as a noun, as, *to err* is human: also **verbal noun.** Compare GERUND. [<F <LL *verbalis* <L *verbum* word] — **ver′bal·ly** *adv.*

Synonyms (adj.): literal, oral, vocal. These words, whose etymology would make them similar in meaning, are differentiated in usage by their applications. *Oral* (L *os* the mouth) signifies uttered through the mouth or (in common phrase) by word of mouth; *vocal* (L *vox* the voice) signifies of or pertaining to the voice, uttered or modulated by the voice, and especially uttered with or sounding with full, resonant voice; *literal* (L *litera* a letter) signifies consisting of or expressed by letters, or according to the letter in the broader sense of the exact meaning or requirement of the words used; what is called "the letter of the law" is its *literal* meaning without going behind what is expressed by the letters on the page. Thus *oral* applies to that which is given by spoken words in distinction from that which is written or printed; as, *oral* tradition; an *oral* examination. By this rule we should in strictness speak of an *oral* contract or an *oral* message, but *verbal* contract and *verbal* message, as indicating that which is spoken rather than by written word, have become fixed in the language. A *verbal* translation may be *oral* or written, so that it is word for word; a *literal* translation follows the construction and idiom of the original as well as the words; thus a *literal* translation is more than one that is merely *verbal*; both *verbal* and *literal* are opposed to *free*. In the same sense, of attending to words only, we speak of *verbal* criticism, a *verbal* change. *Vocal* has primary reference to the human voice; as, *vocal* sounds, *vocal* music; *vocal* may be applied within certain limits to inarticulate sounds given forth by other animals than man; as, The woods were *vocal* with the songs of birds; *oral* is never so applied.

ver·bal·ism (vûr′bəl·iz′əm) *n.* A verbal remark or expression; sometimes, a meaningless form of words; wordiness.

ver·bal·ist (vûr′bəl·ist) *n.* One who deals with words or is skilled in the use and meanings of words; a critic of words.

ver·bal·ize (vûr′bəl·īz) *v.* **·ized ·iz·ing** *v.t.* 1 *Gram.* To make a verb of; change into a verb. 2 To express in words. — *v.i.* 3 To speak or write verbosely. — **ver′bal·i·za′tion** *n.* — **ver′bal·iz′er** *n.*

ver·ba·tim (vər·bā′tim) *adv.* In the exact words; word for word. [<LL <L *verbum* word]

ver·ba·tim et lit·e·ra·tim (vər·bā′tim et lit′ə·rā′tim) *Latin* Word for word and letter for letter.

ver·be·na (vər·bē′nə) *n.* Any of a genus (*Verbena*) of American garden plants having dense terminal spikes of showy, often fragrant flowers. [<L, foliage, vervain. Doublet of VERVAIN.]

ver·be·na·ceous (vûr′bə·nā′shəs) *adj.* Belonging to a family (*Verbenaceae*) of herbs, shrubs, and trees, the verbena family, having opposite or whorled leaves and more or less two-lipped or irregular corollas. [<NL, family name <L *verbena* vervain]

ver·bi·age (vûr′bē·ij) *n.* Use of many words without necessity; wordiness; verbosity. See synonyms under CIRCUMLOCUTION, DICTION. [<F < *verbier* gabble < *verbe*. See VERB.]

verb·i·fy (vûr′bə·fī) *v.t.* **·fied, ·fy·ing** To form into or use as a verb.

ver·big·er·ate (vər·bij′ə·rāt) *v.i.* **·at·ed, ·at·ing** *Psychiatry* To repeat meaningless words, phrases, or sentences over and over, as in certain forms of schizophrenia. [<L *verbigerare* chatter, babble < *verbum* word + *gerere* carry on, conduct] — **ver·big′er·a′tion** *n.*

ver·bose (vər·bōs′) *adj.* Using or containing a wearisome and unnecessary number of words; wordy. See synonyms under GARRULOUS. [<L *verbosus* < *verbum* word] — **ver·bose′ly** *adv.* — **ver·bose′ness** *n.*

ver·bos·i·ty (vər·bos′ə·tē) *n.* *pl.* **·ties** The state or quality of being verbose; wordiness.

ver·bo·ten (fər·bōt′n) *adj. German* Forbidden; authoritatively prohibited.

ver·bum sat sa·pi·en·ti (vûr′bəm sat sā′pē·en′·tī) *Latin* A word to the wise is sufficient: abbr. *verbum sap.*

Ver·cin·get·o·rix (vûr′sin·jet′ər·iks) Gallic chieftain, leader of rebellion against Julius Caesar, who put him to death in 45 B.C.

Ver·dan·di (vər·dän′dē) One of the Norns.

ver·dant (vûr′dənt) *adj.* 1 Green with vegetation; covered with grass or green leaves; fresh. 2 Immature in experience; unsophisticated. See synonyms under FRESH, RUSTIC. [<F *verdoyant*, ppr. of *verdoyer* grow green, ult. <L *viridis* green] — **ver′dan·cy** *n.* — **ver′·dant·ly** *adv.*

verd antique (vûrd) 1 A variety of serpentine. 2 Dark-green andesite porphyry containing crystals of feldspar: also **Oriental verd antique.** 3 A green coating that forms on ancient bronzes. [<OF *verd antique* ancient green]

Verde (vûrd), **Cape** The westernmost point of Africa, in Senegal, French West Africa, a peninsula about 20 miles long, up to 7 miles wide: also *Cape Vert.*

ver·der·er (vûr′dər·ər) *n.* An officer in charge of the royal forests in early England. Also **ver′der·or.** [<AF *verder*, OF *verdier* <Med. L *viridarius* <L *viridis* green]

Ver·di (ver′dē), **Giuseppe,** 1813–1901, Italian composer.

ver·dict (vûr′dikt) *n.* 1 The decision of a jury in an action. 2 A conclusion expressed; an opinion. [<AF *verdit*, OF *voirdit* <L *vere dictum* truly said < *verus* true + *dictum*, pp. of *dicere* say; later refashioned after L]

ver·di·gris (vûr′də·grēs, -gris) *n.* 1 *Chem.* A green basic acetate of copper obtained by treating copper with acetic acid: used as a pigment, for dyeing and calico printing, in medicine, and in the preparation of other copper pigments. 2 The green or bluish patina formed on copper, bronze, or brass surfaces after long exposure to the air. [<OF *verd de Grice, vert de Grece*, lit., green of Greece]

ver·din (vûr′din) *n.* A small, brightly-colored titmouse (*Auriparus flaviceps*) with a yellow head, of the southwestern United States and northern Mexico. [<F, yellowhammer]

ver·di·ter (vûr′də·tər) *n.* 1 A pigment made by grinding a basic copper carbonate; bice: azurite yields **blue verditer,** malachite yields **green verditer.** 2 Verdigris. [<OF *verd de terre*, lit., green of earth]

Ver·dun (vâr·dun′, *Fr.* ver·dœṅ′) A town on the Meuse in NE France; scene of heroic French resistance to German attack during several battles of World War I, 1916. Also Verdun-sur-Meuse (-sür·mœz′).

ver·dure (vûr′jər) *n.* 1 The fresh greenness of growing vegetation, or such vegetation itself. 2 A tapestry representing trees and other vegetation. [<F < *verd* green <L *viridis*] — **ver′dure·less** *adj.*

ver·dur·ous (vûr′jər·əs) *adj.* Covered with verdure; verdant. — **ver′dur·ous·ness** *n.*

ver·e·cund (ver′ə·kund) *adj. Rare* Modest; bashful; coy; shy. [<L *verecundus*]

Ve·ree·ni·ging (fə·rē′nə·khing) A city in the southern Transvaal, Union of South Africa, on the Vaal: site of the signing of the treaty concluding the Boer War, 1902.

Ver·ein (fer·īn′) *n. German* A society; association: often compounded, as in *Turnverein.*

Ve·re·shcha·gin (vyi·ryish·chä′gin), **Vasili Vasilevich,** 1842–1904, Russian genre painter.

verge[1] (vûrj) *n.* 1 The extreme edge of something having defined limits; brink; margin. 2 A bounding or enclosing line; a circlet; ring; also, the space enclosed. 3 A stick or rod, or something having this shape; a wand or staff as a symbol of authority or emblem of office. 4 *Obs.* In England, a stick or wand which tenants held in the hand while swearing fealty to their lord. 5 The spindle of a balance wheel, especially in an old-fashioned vertical escapement. 6 *Archit.* **a** A column shaft. **b** The projecting edge of the tiling on a gable. 7 In old English law, the area over which the authority of an official extended. See synonyms under BOUNDARY, MARGIN. — *v.i.* **verged, verg·ing** 1 To be contiguous or adjacent. 2 To form the limit or verge. [<F, rod, stick <L *virga* twig]

verge[2] (vûrj) *v.i.* **verged, verg·ing** 1 To come near; approach; border: often with *on*: His speech *verges* on the chaotic. 2 To tend; slope; incline. [<L *vergere* bend, turn]

verg·er (vûr′jər) *n.* 1 An official who carries a verge before a scholastic, legal, or ecclesiastical dignitary; specifically, in English cathedrals and collegiate churches, one who carries the mace before the dean or canons. 2 *Brit.* One in charge of the interior of a cathedral or church; usher. 3 *Obs.* A master of ceremonies. [<F < *verge* rod]

Ver·gil (vûr′jil) Anglicized name of *Publius Vergilius Maro*, 70–19 B.C., Roman epic poet. Also spelled *Virgil.*

Ver·gil·i·an (vər·jil′ē·ən) *adj.* Pertaining to or in the style of Vergil: also spelled *Virgilian.*

ver·glas (vər·gläs′) *n. French* A thin, slippery coating of ice on rock: a mountaineering term.

ve·rid·i·cal (və·rid′i·kəl) *adj.* Telling or expressing the truth; truthful; accurate. Also **ve·rid′ic.** [<L *veridicus* speaking the truth < *verus* true + *dicere* say] — **ve·rid′i·cal′i·ty** (-kal′ə·tē) *n.* — **ve·rid′i·cal·ly** *adv.*

ver·i·fi·ca·tion (ver′ə·fə·kā′shən) *n.* 1 The act of verifying, or the state of being verified. 2 *Law* An oath appended to an account, petition, or plea, as to the truth of the facts stated in it; also, at common law, the formal statement at the end of a plea, "and this he is ready to verify."

ver·i·fi·ca·tive (ver′ə·fə·kā′tiv) *adj.* Aiding or resulting in verification.

ver·i·fy (ver′ə·fī) v.t. ·fied, ·fy·ing 1 To prove to be true or accurate; substantiate; confirm. 2 To test or ascertain the accuracy or truth of. 3 Law a To affirm under oath. b To add a confirmation to. [<OF verifier <Med. L verificare make true <verus true + facere make] — ver′i·fi′a·ble adj. — ver′i·fi′er n.

ver·i·ly (ver′ə·lē) adv. 1 In truth; assuredly; certainly. 2 Sincerely and truly; really; confidently. [<VERY]

ver·i·sim·i·lar (ver′ə·sim′ə·lər) adj. Appearing or seeming to be true; likely; probable. [<L verisimilis <verus true + similis like] — ver′i·sim′i·lar·ly adv.

ver·i·si·mil·i·tude (ver′ə·si·mil′ə·tōōd, -tyōōd) n. 1 Appearance of truth; likelihood. 2 That which resembles truth. See synonyms under PROBABILITY. [<L verisimilitudo <verisimilis. See VERISIMILAR.]

ver·ism (ver′iz·əm) n. A style in art and literature that follows the theory that reality should be rigidly represented, even when it is ugly or vulgar. [<L verus true] — ver′ist n. & adj. — ve·ris·tic (və·ris′tik) adj.

ver·i·ta·ble (ver′ə·tə·bəl) adj. Conforming to truth or fact; genuine; true; real. See synonyms under AUTHENTIC. [<F vérité. See VERITY.] — ver′i·ta·ble·ness n. — ver′i·ta·bly adv.

ver·i·tas (ver′ə·tas) n. Latin Truth.

ver·i·ty (ver′ə·tē) n. pl. ·ties 1 The quality of being correct or true as a statement or representation of reality. 2 A true statement; a fact; truth. See synonyms under VERACITY. [<F vérité <L veritas truth <verus true]

ver·juice (vûr′jōōs) n. 1 The sour juice of green fruit, as of unripe grapes. 2 Sharpness or sourness of disposition or manner; acidity. [<OF verjus <vert green + jus juice]

Ver·kho·yansk Range (vir·khō·yänsk′) A mountain system in northern Yakut Autonomous S.S.R. north of the Arctic Circle; highest point, 8,000 feet.

Ver·laine (ver·len′), **Paul**, 1844-96, French poet.

Ver·meer (vər·mâr′), **Jan**, 1632-75, Dutch painter.

ver·meil (vûr′mil) n. 1 Silver or bronze gilt. 2 A transparent water varnish. 3 An orange-red garnet. 4 Poetic Vermilion, or the color of vermilion. — adj. Of a bright-red color. [<OF <L vermiculus, dim. of vermis worm, the cochineal insect]

vermi- combining form A worm; of or related to a worm, or to worms: vermiform. [<L vermis a worm]

ver·mi·cel·li (vûr′mə·sel′ē, Ital. ver′mē·chel′lē) n. A food paste made into slender wormlike cords thinner than spaghetti or macaroni. [<Ital., lit., little worms, pl. of vermicello <L vermiculus. See VERMEIL.]

ver·mi·ci·dal (vûr′mə·sīd′l) adj. Destructive of intestinal worms; anthelmintic.

ver·mi·cide (vûr′mə·sīd) n. Any substance that kills worms; specifically, any medicine or drug destructive of intestinal worms. [<VERMI- + -CIDE]

ver·mic·u·lar (vər·mik′yə·lər) adj. 1 Pertaining to a worm; having the form or motion of a worm. 2 Like the tracks of a worm. [<L vermicularis <vermiculus, dim. of vermis worm] — ver·mic′u·lar·ly adv.

vermicular work 1 A form of rusticated masonry simulating worm tracks. 2 Ornamental work consisting of winding tracks in mosaic work. Also **vermiculated work**.

ver·mic·u·late (vər·mik′yə·lāt) v.t. ·lat·ed, ·lat·ing 1 To adorn with tracery simulating the tracks of worms. 2 To make worm-eaten; infest with worms. — adj. 1 Wormlike or covered with wormlike markings. 2 Having the motions of a worm; insinuating; wavy. 3 Worm-eaten. [<L vermiculatus, pp. of vermiculari be worm-eaten <vermiculus, dim. of vermis a worm]

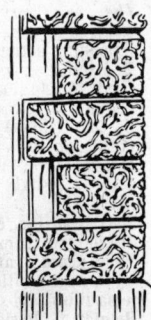
VERMICULAR WORK

ver·mic·u·la·tion (vər·mik′yə·lā′shən) n. 1 Wormlike motion, as of the intestines. 2 Vermicular ornamentation. 3 The state of being wormy. 4 A track left by worms. 5 A fine wavy color marking, as on a bird.

ver·mic·u·lite (vər·mik′yə·līt) n. A laminated hydrous silicate, derived chiefly as an alteration product of biotite, phlogopite, and other micaceous minerals. [<L vermiculus, dim. of vermis worm + -ITE¹]

ver·mic·u·lose (vər·mik′yə·lōs) adj. 1 Worm-eaten; wormy. 2 Worm-shaped; wormlike. Also **ver·mic′u·lous** (-ləs). [<LL vermiculosus <L vermiculus, dim. of vermis worm]

ver·mi·form (vûr′mə·fôrm) adj. Like a worm in shape. [<Med. L vermiformis <L vermis a worm + forma a form]

vermiform appendix Anat. A slender, wormlike diverticulum, 3 to 6 inches long, protruding from the end of the cecum in man and certain other mammals.

vermiform process Anat. 1 Either surface of the median lobe of the cerebellum. 2 The vermiform appendix.

ver·mi·fuge (vûr′mə·fyōōj) n. Any remedy that destroys intestinal worms. — adj. Anthelmintic. [<L vermis a worm + fugare expel]

ver·mil·ion (vər·mil′yən) n. 1 A brilliant, durable red pigment consisting of mercuric sulfide, obtained native by grinding the mineral cinnabar to a fine powder, or artificially, as by treating a mixture of mercury and sulfur with potassium hydroxide. 2 The color of the pigment, an intense orange red. — adj. Of a bright-red color. — v.t. To color with vermilion; dye bright red. [<OF vermeilon, vermillon <vermeil VERMEIL]

ver·min (vûr′min) n. pl. ·min 1 Noxious small animals or parasitic insects, as lice, fleas, worms, rats, mice, etc. 2 Brit. Certain animals injurious to game, as weasels, owls, etc. 3 A repulsive or noxious human being, or such persons collectively. [<OF <L vermis a worm]

ver·mi·nate (vûr′mə·nāt) v.i. ·nat·ed, ·nat·ing To produce or breed vermin, especially parasitic vermin. — ver′mi·na′tion n.

ver·mi·nous (vûr′mə·nəs) adj. 1 Infested with vermin. 2 Affected with intestinal worms, or caused, as a disease, by vermin. 3 Of the nature of vermin.

Ver·mont (vər·mont′) A State in NE United States; 9,609 square miles; capital, Montpelier; entered the Union March 4, 1791; nickname, Green Mountain State: abbr. Vt. — **Ver·mont′er** n.

ver·mou·lu (ver·mōō·lü′) adj. French Worm-eaten.

ver·mouth (vûr′mōōth, vər·mōōth′) n. A liqueur made from white wine flavored with aromatic herbs. Also **ver′muth**. [<F vermout <G wermuth wormwood]

ver·nac·u·lar (vər·nak′yə·lər) n. 1 The native language of a locality. 2 The common daily speech of the people, as opposed to the literary language. 3 The vocabulary or jargon of a particular profession or trade: to speak the medical vernacular. 4 An idiomatic word or phrase. 5 The common name of a plant or animal as distinguished from its scientific designation. See synonyms under LANGUAGE. — adj. 1 Originating in or belonging to one's native land; indigenous: said of a language, idiom, etc. 2 Using the colloquial native tongue, rather than the literary language: vernacular poets. 3 Written in the language indigenous to a country or people: a vernacular translation of the Bible. 4 Characteristic of a specific locality or country; local: vernacular arts. 5 Rare Peculiar to a particular region; endemic: a vernacular disease. 6 Designating the common name of a plant or animal. [<L vernaculus domestic, native <verna a home-born slave, a native] — ver·nac′u·lar·ly adv.

ver·nac·u·lar·ism (vər·nak′yə·lə·riz′əm) n. 1 A vernacular term or idiom. 2 The use of the vernacular as opposed to classic or literary language.

ver·nal (vûr′nəl) adj. 1 Belonging to, appearing in, or appropriate to spring. 2 Pertaining to youth; having a springlike freshness. [<L vernalis <vernus belonging to spring <ver spring] — ver′nal·ly adv.

vernal equinox See under EQUINOX.

ver·nal·ize (vûr′nəl·īz) v.t. ·ized, ·iz·ing To accelerate the growth of (a plant) by subjecting the seeds to artificial treatment, as by moistening at a low temperature. — ver′nal·i·za′tion n.

vernal point The vernal equinox. See under EQUINOX.

ver·na·tion (vər·nā′shən) n. Bot. The disposition of leaves within the leaf bud, as regards their folding, coiling, etc. [<NL vernatio, -onis <vernare flourish <ver spring]

Verne (vûrn, Fr. vern), **Jules**, 1828-1905, French writer of scientific novels.

Ver·ner (vûr′nər), **Karl Adolph**, 1846-96, Danish philologist.

Verner's Law A law regarding certain consonant changes in Germanic languages, set forth by Karl Verner in 1876, stating that certain exceptions to Grimm's Law are due to a still wider law, namely, the position of the primary accent in the parent language. It shows that original Indo-European voiceless plosives p, t, k, which, according to Grimm's Law became in Germanic the voiceless fricatives f, th (as in thin), h, became, instead, except in specific combinations, the corresponding voiced fricatives and, ultimately, the voiced plosives b, d, g, if the original Indo-European tonic stress was not on the immediately preceding syllable. This statement can be illustrated in English by the distinction in final consonant between death and dead. The same process operated for the Indo-European s, which became z, and finally r, as can be seen in English was and were, raise and rear.

ver·ni·cose (vûr′nə·kōs) adj. Bot. Appearing as if varnished, as some leaves. [<NL vernicosus <Med. L vernicium VARNISH]

ver·ni·er (vûr′nē·ər) n. 1 The small, movable, auxiliary scale for obtaining fractional parts of the subdivisions of a fixed scale on a theodolite, barometer, sextant, gage, or other measuring instrument. 2 Mech. An auxiliary device to insure fine adjustments in precision instruments. [after Pierre Vernier]

Ver·nier (ver·nyā′), **Pierre**, 1580?-1637, French mathematician.

ver·nis·sage (ver′nē·säzh′) n. The opening day of an exhibition of oil paintings to which critics are often invited: also called varnishing day. [<F <vernir varnish; because the painters varnish their works on this day]

Ver·no·le·ninsk (vyer′nə·lyä′nyinsk) A former name for NIKOLAEV.

Ver·non (vûr′nən), **Edward**, 1684-1757, English admiral.

Ver·nyi (vyer′nē) A former name for ALMA-ATA.

Ve·ro·na (və·rō′nə, Ital. vā·rō′nä) A city in Veneto, NE Italy. — **Ve·ro·nese** (ver′ə·nēz′, -nēs′) adj. & n.

Ve·ro·nal (ver′ə·nəl) n. Proprietary name for a brand of barbital.

Ve·ro·ne·se (vā′rō·nā′zä), **Paolo**, 1528-88, Venetian painter; real name Cagliari.

ve·ron·i·ca¹ (və·ron′i·kə) n. The speedwell. [<Med. L, appar. after St. Veronica]

ve·ron·i·ca² (və·ron′i·kə) n. A cloth said to have been miraculously impressed with the face of Christ on his way to Calvary, handed to him by a woman named Veronica to wipe the perspiration from his face; also, the representation of the face on this handkerchief; hence, a cloth or handkerchief having on it a representation of Christ's face. See SUDARIUM. [<Med. L <LL veraiconica, prob. <L verus true + Gk. eikōn an image, likeness]

ve·ron·i·ca³ (və·ron′i·kə, Sp. bā·rō′nē·kä) n. In bullfighting, a maneuver in which the torero faces the bull and holds the cape directly in front of himself. [<Sp.]

Ve·ron·i·ca (və·ron′i·kə, Ital. vā′rō·nē′kä) A feminine personal name. Also Fr. **Véronique** (vā·rō·nēk′). [See VERONICA²]

— **Veronica, Saint** A legendary follower of Christ, upon whose handkerchief a picture of Christ's features is said to have appeared.

Ver·ra·za·no (ver′rä·tsä′nō), **Giovanni da**, 1480?-1528?, Italian navigator.

Ver·roc·chio (ver·rôk′kyō), **Andrea del**, 1435-88, Florentine sculptor and painter.

ver·ru·ca (ve·rōō′kə) n. pl. ·cae (-sē) 1 Med.

A wart. 2 *Biol.* A wart or wartlike elevation on animals or plants. [<L, a wart, orig. a steep place]

ver·ru·ca·no (ver′ə·kä′nō) *n.* A hard conglomerate of quartz cemented by various siliceous materials, usually colored. [<Ital., from Mount *Verruca* near Pisa]

ver·ru·cose (ver′ə·kōs) *adj.* Abounding in wartlike elevations; warty. Also **ver′ru·cous** (-kəs). [<L *verrucosus* < *verruca* a wart] — **ver′ru·cos′i·ty** (-kos′ə·tē) *n.*

Ver·sailles (vər·sī′, -sālz′; *Fr.* ver·sä′y′) A city of north central France, 11 miles SW of Paris; site of the palace of Louis XIV; scene of the signing of a treaty (1919) between the Allies and Germany after World War I.

ver·sant (vûr′sənt) *n. Geog.* 1 An entire area having a general slope in one direction. 2 The general aspect or slope of any portion of country; inclination. [<F, ppr. of *verser* overturn, pour <L *versare*, freq. of *vertere* turn]

ver·sa·tile (vûr′sə·til) *adj.* 1 Having an aptitude for new tasks or occupations; many-sided. 2 Subject to change; inconstant; variable. 3 Freely swinging or turning: said of an anther part so slightly attached to its support that it readily swings to and fro. 4 Capable of being turned forward or backward, as the toe of a bird; movable in every direction, as insect antennae. [<F <L *versatilis* < *versare*, freq. of *vertere* turn] — **ver′sa·tile·ly** *adv.* — **ver′sa·til′i·ty, ver′sa·tile·ness** *n.*

vers de so·cié·té (ver′ də sō·syā·tā′) *French* A form of light verse characterized by grace, elegance, and wit.

verse (vûrs) *n.* 1 A single metrical or rhythmical line made up of a number of feet, arranged according to a specific rule. 2 A group of metrical lines; a stanza. 3 Metrical composition as distinguished from prose; poetry. 4 a Composition in meter; versification. b A specified type of metrical composition; type of meter or metrical structure: iambic *verse.* 5 One of the short divisions of a chapter of the Bible; also, a short division of any composition. 6 The solo part of a song, anthem, or other piece. See synonyms under METER[2], POETRY. — *v.t.* & *v.i.* **versed, vers·ing** *Rare* To versify. [Fusion of OE *fers* and OF *vers*, both <L *versus* a turning, a verse < *vertere* turn]

versed (vûrst) *adj.* 1 Thoroughly acquainted; having ready skill; proficient. 2 Turned about; reversed. [<L *versatus*, pp. of *versari* occupy oneself]

versed sine, ver·sine (vûr′sīn) See under SINE[1]. [<NL *sinus versus* < *sinus* a sine + L *versus*, pp. of *vertere* turn]

verse·mon·ger (vûrs′mung′gər, -mong′-) *n.* A writer of inferior verses; poetaster.

ver·si·cle (vûr′si·kəl) *n.* 1 A little verse. 2 One of a series of lines said or sung alternately by minister and people. [<L *versiculus*, dim. of *versus* VERSE]

ver·si·col·or (vûr′si·kul′ər) *adj.* 1 Showing a variety of colors; variegated. 2 Changing from one color to another in different lights; iridescent. Also *Brit.* **ver′si·col′our.** [<L < *versus*, pp. of *vertere* turn + *color* color]

ver·sic·u·lar (vər·sik′yə·lər) *adj.* Relating to verses, especially Biblical verses; marking the division into verses. [<L *versiculus* VERSICLE]

ver·sie·ra (ver·syä′rä) *n. Math.* The witch of Agnesi. See WITCH OF AGNESI. [<Ital., a ghost, hobgoblin]

ver·si·fy (vûr′sə·fī) *v.* **·fied, ·fy·ing** *v.t.* 1 To change from prose into verse. 2 To narrate or treat in verse. — *v.i.* 3 To write poetry; make verses. [<OF *vercifier, versifier* <L *versificare* < *versus* VERSE + *facere* make] — **ver′si·fi·ca′tion** (-fə·kā′shən) *n.* — **ver′si·fi′er** *n.*

ver·sion (vûr′zhən, -shən) *n.* 1 That which is translated or rendered from one language into another; a translation; a translation of the original Hebrew and Greek of the Old and New Testaments, or any part of them, into some other tongue. 2 A description of something as modified by the relator. 3 *Med.* a The manual turning of a fetus in the womb so as to secure proper delivery. b Displacement of the uterus in which the organ is deflected without bending upon itself. 4 *Obs.* A transformation; conversion. [<MF <Med. L *versio, -onis* a turning <L *vertere* turn] — **ver′·sion·al** *adj.*

vers li·bre (ver lē′br′) *French* Free verse.

ver·so (vûr′sō) *n. pl.* **·sos** 1 A left-hand page of a book, piece of music, or sheet of folded paper: also called *reverso.* Compare RECTO. 2 The reverse of a coin or medal. Compare OBVERSE. [<L *verso (folio)* a turned (leaf), ablative neut. sing. pp. of *vertere* turn]

verst (vûrst) *n.* A Russian measure of distance: about two thirds of a mile, or 1.067 kilometers. [<F *verste* and G *werst* <Russian *versta*, orig. a line]

ver·sus (vûr′səs) *prep.* 1 Against: used in naming or entitling actions in courts: plaintiff *versus* defendant; in contests: Dempsey *versus* Tunney: usually contracted to *v.* or *vs.* 2 Considered as the alternative of: free trade *versus* high tariffs. [<L, toward, turned toward, orig. pp. of *vertere* turn]

vert (vûrt) *n.* 1 In English forest law, anything that grows and bears green leaves within a forest, especially thick coverts; also, the right to cut green or growing wood in a forest. 2 *Her.* The color or tincture green. [<MF *vert, verd* <L *viridis* green]

Vert (ver), **Cape** See VERDE, CAPE.

ver·te·bra (vûr′tə·brə) *n. pl.* **·brae** (-brē) or **·bras** *Anat.* One of the segmented bones of the spinal column. In man, and the higher vertebrates, each vertebra, with its semicylindrical central body and attached processes, articulates with those on either side by means of elastic fibrous pads. [<L, a joint, a vertebra < *vertere* turn]

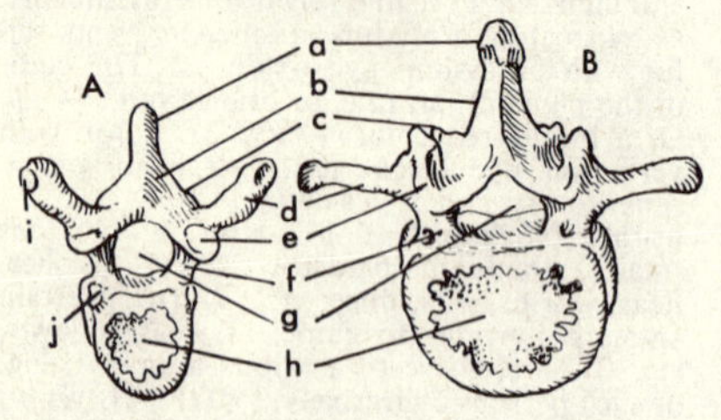

HUMAN VERTEBRAE
A. Sixth thoracic vertebra.
B. Third lumbar vertebra.

a.	Spinous process.	*f.*	Pedicle.
b.	Lamina.	*g.*	Vertebral foramen.
c.	Inferior articular process.	*h.*	Body.
d.	Transverse process.	*i.*	Facet for rib.
e.	Superior articular process.	*j.*	Demi-facet for rib.

ver·te·bral (vûr′tə·brəl) *adj.* 1 Pertaining to or of the nature of a vertebra. 2 Having, or composed of, vertebrae.

vertebral column The spinal column; the backbone.

ver·te·brate (vûr′tə·brāt, -brit) *adj.* 1 Having a backbone or spinal column. 2 Pertaining to or characteristic of vertebrates. 3 Vertebral. — *n.* Any of a primary division or subphylum *(Vertebrata)* of animals, of the phylum *Chordata*, characterized by a spinal column, as fishes, birds, mammals, and a few primitive forms in which a notochord represents the backbone. [<L *vertebratus* jointed < *vertebra* VERTEBRA]

ver·te·bra·tion (vûr′tə·brā′shən) *n.* 1 The formation of vertebrae. 2 Segmentation like that of the spinal column.

ver·tex (vûr′teks) *n. pl.* **·tex·es** or **·ti·ces** (-tə·sēz) 1 The highest point or summit of anything; apex; top. 2 *Astron.* a The zenith. b The point in the sky toward or from which a group of stars appears to be moving. 3 The top of the head; also, in craniometry, the top of the arch of the skull. 4 *Math.* a The point of intersection of the sides of an angle. b The point opposite to, and farthest from, the base. c The intersection of three or more edges of a polyhedron. [<L, the top < *vertere* turn]

ver·ti·cal (vûr′ti·kəl) *adj.* 1 Of or pertaining to the vertex. 2 Occupying a position directly above or overhead; being at the highest point. 3 Directed perpendicularly to the plane of the horizon; upright; plumb. 4 Of or pertaining to the crown of the head. 5 *Bot.* a Perpendicular to the surface or to the axis of support. b In the direction of the axis of growth; lengthwise. 6 *Econ.* Of or pertaining to a business concern that undertakes a process from raw material to consumer: a *vertical* trust. — *n.* 1 A vertical line, plane, or circle. 2 An upright beam or rod in a truss. [<MF <L *verticalis* < *vertex, -icis* VERTEX] — **ver′ti·cal′i·ty** (-kal′ə·tē), **ver′ti·cal·ness** *n.* — **ver′ti·cal·ly** *adv.*

vertical circle *Astron.* A great circle perpendicular to the plane of the horizon.

ver·ti·ces (vûr′tə·sēz) Plural of VERTEX.

ver·ti·cil (vûr′tə·sil) *n. Biol.* A set of organs, as leaves or tentacles, disposed in a circle around an axis; whorl; a volution of a spiral shell. [<L *verticillus* a whorl, dim. of *vertex, -icis* VERTEX]

ver·ti·cil·las·ter (vûr′tə·si·las′tər) *n. Bot.* An inflorescence or flower cluster with the flowers seemingly in a whorl, but composed of a pair of dense sessile cymes in the axils of opposite leaves, as in most mints. [<NL <L *verticillus* VERTICIL + -ASTER]

ver·tic·il·late (vər·tis′ə·lit, -lāt, vûr′tə·sil′it, -āt) *adj.* 1 Arranged in a verticil or whorl. 2 Having parts so arranged; whorled. Also **ver·tic′il·lat′ed.** [<NL *verticillatus* <L *verticillus* VERTICIL] — **ver·tic′il·late·ly** *adv.* — **ver·tic′il·la′tion** *n.*

ver·tic·i·ty (vər·tis′ə·tē) *n.* Tendency to move toward the north, as manifested by a magnetic needle. [<NL *verticitas, -tatis* <L *vertex, -icis* VERTEX]

ver·tig·i·nous (vər·tij′ə·nəs) *adj.* 1 Affected by vertigo; dizzy. 2 Turning round; whirling; revolving. 3 Liable to cause giddiness; making dizzy. [<L *vertiginosus* < *vertigo, -inis* VERTIGO] — **ver·tig′i·nous·ly** *adv.* — **ver·tig′i·nous·ness** *n.*

ver·ti·go (vûr′tə·gō) *n. pl.* **·goes** or **ver·tig·i·nes** (vər·tij′ə·nēz) *Pathol.* Any of a group of disorders, variously caused, in which a person feels as if he or his surroundings are whirling around; dizziness. [<L, lit., a turning around < *vertere* turn]

Ver·tum·nus (vər·tum′nəs) In Roman mythology, the god of the changing seasons and growing plants; husband of Pomona: also *Vortumnus.*

Ver·u·la·mi·um (ver′yo͝o·lā′mē·əm) The ancient Roman name for ST. ALBANS.

Ve·rus (vir′əs), **Lucius Aurelius,** 130–169, Roman emperor 161–169.

ver·vain (vûr′vān) *n.* Any of various plants (genus *Verbena*), congeners of the common cultivated ornamental verbenas, as the American blue vervain *(V. hastata)*, or the common European vervain *(V. officinalis)*. [<OF *verveine* <L *verbena.* Doublet of VERBENA.]

verve (vûrv) *n.* 1 Enthusiasm or energy, especially as manifested in artistic production; hence, spirit; vigor. 2 *Rare* Special bent or talent. [<F, prob. <L *verba*, pl. of *verbum* a word]

ver·vet (vûr′vit) *n.* A South African monkey (genus *Cercopithecus*), grayish-green speckled with black, and with reddish-white cheeks and belly. [<F < *ver(t)* green (<L *viridis*) + *(gri)vet* a grivet; so called because of its color]

ver·y (ver′ē) *adv.* In a high degree; in large measure; extremely; exceedingly: *very* generous. ◆ Some grammarians feel that *very* may properly be used before a participle only when the latter precedes the noun it modifies, as in *a very agitated speech*; when the participle is used after some form of the verb *to be*, another adverb is interposed, as in *He was very much* (or *greatly*) *agitated.* However, the construction without the adverb is now widely accepted with participles of emotion or feeling, and is often found, as well, with some participles describing physical condition, as in *They were very disturbed*; *His face is very changed in aspect.* — *adj.* **ver·i·er, ver·i·est** 1 Absolute; actual; simple; utter: said of truth. 2 Exclusive; peculiar: the *very* wisdom of God. 3 Unqualified; utter; complete: a *very* rogue. 4 Selfsame; identical: my *very* words. 5 The (thing) itself: used as an intensive equivalent to *even*: The *very* stones cry out. 6 *Obs.* True: *very* God; also, truthful; veracious. [<AF *verrai*, OF *verai* <L *verus* true]

Ver·y light (ver′ē, vir′ē) A brilliant signal flare discharged from a special type of pistol, the **Very pistol.** [after E. W. *Very*, 1847–1910, U. S. naval officer and inventor]

Ve·sa·li·us (vi·sā′lē·əs), **Andreas,** 1514–64, Belgian physician; founder of modern anatomy.

ve·si·ca (vi·sī′kə) *n. pl.* **·cae** (-sē) A bladder. [<L]

ves·i·cal (ves′i·kəl) *adj.* Of, pertaining to, supplying, or affecting the bladder.

ves·i·cant (ves′i·kənt) *adj.* Blister-producing. — *n.* 1 That which produces blisters; a

blister. 2 A chemical warfare agent which attacks the skin, as mustard gas or lewisite: also called *blister gas*. [<NL *vesicans, -antis,* ppr. of *vesicare* raise blisters <L *vesica* a blister, bladder]

Ve·si·ca pis·cis (vi·sī′kə pis′is, pī′sis) The pointed oval aureole used by medieval sculptors and painters to enclose the figure of Christ, the Virgin Mary, or an apostle.

ves·i·cate (ves′i·kāt) *v.* **·cat·ed, ·cat·ing** *v.t.* To raise blisters on. — *v.i.* To become blistered; blister, as the skin. [<NL *vesicatus,* pp. of *vesicare.* See VESICANT.] — **ves′i·ca′tion** *n.*

ves·i·ca·to·ry (ves′i·kə·tôr′ē, və·sik′ə·tôr′ē, -tō′rē) *adj.* Capable of producing blisters; vesicant. — *n. pl.* **·ries** Any substance, as an ointment or plaster, that causes a blister.

ves·i·cle (ves′i·kəl) *n.* 1 Any small bladderlike cavity, cell, or cyst. 2 A small sac, containing gas or fluid. 3 *Pathol.* Any small rounded elevation of the cuticle containing a clear liquid; a blister. 4 *Bot.* A small bladderlike cavity filled with air. 5 *Geol.* A small spherical cavity found commonly in volcanic rocks. [<L *vesicula,* dim. of *vesica* a bladder]

vesico- combining form *Med.* The urinary bladder; of or pertaining to the urinary bladder: *vesicotomy,* a cutting into the bladder. Also, before vowels, **vesic-.** [<L *vesica* a bladder]

ve·sic·u·la (və·sik′yə·lə) *n. pl.* **·lae** (-lē) A little bladder; vesicle. [<L, VESICLE]

ve·sic·u·lar (və·sik′yə·lər) *adj.* 1 Of, pertaining to, composed of, or resembling vesicles. 2 Bearing or containing vesicles or air bladders. [<L *vesicula* VESICLE] — **ve·sic′u·lar·ly** *adv.*

ve·sic·u·late (və·sik′yə·lāt) *v.t. & v.i.* **·lat·ed, ·lat·ing** To make or become vesicular or vesiculate. — *adj.* (-lit, -lāt) Full of or having vesicles; also, vesicular. [Back formation < *vesiculated* <NL *vesiculatus* <L *vesicula* VESICLE] — **ve·sic′u·la′tion** *n.*

Vesle (vel) A river in NE France, flowing 90 miles NW from the badlands of Champagne, NE of Châlons-sur-Marne, through Reims to the Aisne east of Soissons.

Ve·son·ti·o (və·son′shē·ō) The ancient name for BESANÇON.

Ve·soul (və·zo͞ol′) A city in eastern France, capital of Haute-Saône department.

Ves·pa·sian (ves·pā′zhən) Anglicized name of *Titus Flavius Vespasianus,* A.D. 9–79, Roman emperor 69–79.

ves·per (ves′pər) *n.* 1 A bell that calls to vespers: also **vesper bell.** 2 An evening service, prayer, or song. 3 *Obs.* Evening. — *adj.* Pertaining to or suitable for evening or vespers. [<L, the evening star]

Ves·per (ves′pər) The evening star; Hesperus; the planet Venus when an evening star. [<L]

ves·per·al (ves′pər·əl) *adj.* Pertaining to evening or the service of vespers. — *n.* 1 A book of the music and office of vespers. 2 A cover for an altar cloth.

ves·pers (ves′pərz) *n. pl.* Often cap. 1 *Eccl.* The sixth in order of the canonical hours. 2 *Eccl.* A service of worship in the late afternoon or evening; specifically, in the Anglican church, Evening Prayer; in the Roman Catholic Church, a public service on Sundays or holy days at which the office of vespers is said or sung. 3 The hour of vespers, usually about sunset. [<OF *vespres* <Med. L *vesperae* <L *vespera* evening]

vesper sparrow An American sparrow (*Pooecetes gramineus*) distinguished by the partial whiteness of its outer tail feathers: so called from its evening song.

ves·per·til·i·o·nine (ves′pər·til′ē·ə·nīn′, -nin) *adj.* Belonging to a family (*Vespertilionidae*) of insectivorous bats. [<L *vespertilio, -onis* a bat + -INE[1]] — **ves′per·til′i·o·nid** *adj. & n.*

ves·per·tine (ves′pər·tin, -tīn) *adj.* 1 Pertaining to or occurring in the evening. 2 Flying, opening, etc., in the evening, as a bat, flower, etc. 3 Descending toward the horizon at the sunset hour. Also **ves′per·ti′nal** (-tī′nəl). [<L *vespertinus* < *vesper* VESPER]

ves·pi·ar·y (ves′pē·er′ē) *n. pl.* **·ar·ies** A nest of social wasps or its colony. [<L *vespa* a wasp + (AP)IARY]

ves·pid (ves′pid) *n.* Any of a large family (*Vespidae*) of hymenopterous insects, including social wasps and hornets. — *adj.* Of or pertaining to the *Vespidae.* [<NL <L *vespa* a wasp] — **ves′pi·form** (-pə·fôrm) *adj.*

ves·pine (ves′pīn, -pin) *adj.* Of or pertaining to wasps. [<L *vespa* a wasp + -INE[1]]

Ves·puc·ci (ves·po͞ot′chē), **Amerigo,** 1451–1512, Italian navigator for whom America was named.

ves·sel (ves′əl) *n.* 1 A hollow receptacle of any form or material, especially one capable of holding a liquid. 2 A ship or craft designed to float on the water: usually one larger than a rowboat; also, an airship. 3 *Biol.* A duct or canal for containing or transporting a body fluid, as an artery, vein, etc. 4 *Bot.* A water-conducting tube in plants. 5 Figuratively, a person viewed as having capacity or fitness to receive or contain something; one who receives: chiefly in religious use: a *vessel* of mercy or of wrath. [<OF <L *vascellum,* dim. of *vas* a vessel]

vest (vest) *n.* 1 A short sleeveless jacket worn by men and sometimes by women under the coat; waistcoat; originally, a kind of cassock: in England chiefly a trade term. 2 A close jacket formerly worn by women; now, an extra piece of trimming on the front of the body or waist of a woman's dress, usually V-shaped. 3 An undervest or undershirt. 4 Clothing of any kind; vesture; array; dress. 5 *Obs.* An ecclesiastical vestment. — *v.t.* 1 To confer (ownership, authority, etc.) upon some person or persons: usually with *in.* 2 To place ownership, control, or authority with (a person or persons). 3 To clothe or robe, as with vestments. — *v.i.* 4 To clothe oneself, as in vestments. 5 To be or become vested; devolve. [<F *veste* <Ital. <L *vestis* clothing, a garment]

ves·ta (ves′tə) *n.* A friction match of wax; a short wax taper; a wooden match. [after *Vesta*]

Ves·ta (ves′tə) 1 In Roman mythology, the goddess of the hearth and the hearth fire, protectress of the state and custodian of the sacred fire tended by the vestals: identified with the Greek *Hestia.* 2 A minor planet.

ves·tal (ves′təl) *n.* 1 One of the virgin priestesses of Vesta: also **vestal virgin.** 2 A woman of pure character; a virgin; nun. — *adj.* 1 Pertaining to Vesta. 2 Suitable for a vestal or a nun; hence; chaste. [<L *vestalis* <*Vesta* Vesta]

ves·ted (ves′tid) *adj.* 1 Having vestments; robed. 2 *Law* Held by a tenure subject to no contingency; complete; established by law as a permanent right: *vested* interests.

vest·ee (ves·tē′) *n.* 1 An imitation blouse-front worn in the front of a suit or dress. 2 A broadcloth garment without sleeves worn with a formal riding habit. [Dim. of VEST]

Ves·ter·a·len (ves′tər·ô′lən) A Norwegian archipelago in the Norwegian Sea north of the Lofoten Islands; total, about 1,200 square miles. Also **Ves′ter·aa′len.**

ves·ti·ar·y (ves′tē·er′ē) *adj.* Pertaining to clothes. — *n. pl.* **·ar·ies** *Obs.* A vestry; robing-room. [<OF *vestiairie* <Med. L *vestiarium.* See VESTRY.]

ves·tib·u·lar (ves·tib′yə·lər) *adj.* Pertaining to or like a vestibule. Also **ves·tib′u·late** (-lit, -lāt).

ves·ti·bule (ves′tə·byo͞ol) *n.* 1 A small antechamber between the outer door of a building and an interior one; an entrance hall; lobby. 2 An enclosed passage from one railway passenger car to another. 3 Formerly, a walled place before the entrance to a Roman or Greek house; later, a porch. 4 *Anat.* Any one of several chambers or channels adjoining or communicating with others: the *vestibule* of the ear. — *v.t.* **·buled, ·bul·ing** 1 To provide with a vestibule or vestibules. 2 To couple (railroad cars) and connect by vestibules. [<L *vestibulum* an entrance hall] — **ves′ti·buled** *adj.*

vestibule train A passenger train with enclosed platforms connected by flexible walls and roof, forming a weatherproof passageway between connected cars, called **vestibule cars.**

ves·tige (ves′tij) *n.* 1 A visible trace, impression, or a sensible evidence or sign, of something absent, lost, or gone; trace; originally, a footprint; track. 2 *Biol.* A part or organ, small or degenerate, but well developed and functional in ancestral forms of organisms. See synonyms under MARK[1], TRACE[1]. [<F <L *vestigium* a footprint]

ves·tig·i·al (ves·tij′ē·əl) *adj.* Of, or of the nature of a vestige; surviving in small or degenerate form. — **ves·tig′i·al·ly** *adv.*

ves·tig·i·um (ves·tij′ē·əm) *n. pl.* **·tig·i·a** (-tij′ē·ə) A vestigial part; vestige. [<L, a footprint]

vest·ment (vest′mənt) *n.* 1 An article of dress; clothing or covering; particularly, a garment or robe of state or office. 2 *Eccl.* **a** One of the ritual garments of the clergy, especially one worn at the Eucharist. **b** A chasuble. See synonyms under DRESS. [<OF *vestement* <L *vestimentum* clothes < *vestire* clothe] — **vest′ment·al** *adj.*

vest-pock·et (vest′pok′it) *adj.* 1 Small enough to fit in a vest pocket; very small; diminutive: a *vest-pocket* edition. 2 Much smaller than standard or usual size: a *vest-pocket* battleship.

ves·try (ves′trē) *n. pl.* **·tries** 1 A room where vestments are put on or kept. 2 A room for altar linens, sacred vessels, etc., attached to a church and often called a *sacristy.* 3 A room in a church used for Sunday school, meetings, as a chapel, etc. 4 In the Anglican church: **a** A body administering the affairs of a parish or congregation; also, a meeting of such a body. **b** In English parishes, a business meeting of all the parishioners or their representatives. 5 A place of meeting for the parish vestry; a vestry hall. [<AF *vestrie,* OF *vestiarie* <Med. L *vestiarium* a wardrobe <L *vestis* a garment]

ves·try·man (ves′trē·mən) *n. pl.* **·men** (-mən) A member of a vestry.

ves·ture (ves′chər) *n.* 1 Something that covers; garments; clothing; a robe. 2 *Law* All that covers land, except trees. 3 A covering or envelope. See synonyms under DRESS. — *v.t.* **·tured, ·tur·ing** To cover or clothe with vesture; vest; robe; envelop: usually in the past participle. [<OF <*vestir* cloth <L *vestire* clothe]

ve·su·vi·an (və·so͞o′vē·ən) *n.* 1 Vesuvianite. 2 A kind of match or fusee which burns with a spluttering flame; used for lighting cigars, etc. [from *Vesuvius*]

ve·su·vi·an·ite (və·so͞o′vē·ən·īt′) *n.* A vitreous, brown to green, translucent hydrous silicate of calcium and aluminum, with traces of iron and magnesium: also called *idocrase.* [<VESUVIAN + -ITE[1]]

Ve·su·vi·us (və·so͞o′vē·əs) The only active volcano on the European mainland, on the Bay of Naples, Italy; 3,891 feet. *Italian* **Ve·su·vio** (vā·zo͞o′vyō). — **Ve·su′vi·an** *adj.*

vet (vet) *Colloq. n.* A veterinary surgeon. — *v. vet·ted, vet·ting* *v.t.* 1 To treat as a veterinarian does. 2 *U.S. Slang* To criticize or emend: to *vet* a manuscript. — *v.i.* 3 To treat animals medically. [Short for VETERINARIAN]

vetch (vech) *n.* 1 Any of a genus (*Vicia*) of climbing herbaceous vines of the bean family, especially the common broadbean, grown for fodder. 2 A leguminous European plant (*Lathyrus satirus*) yielding edible seeds. — **bitter vetch** A species of vetch (*V. ervilia*) of which the seeds contain a bitter, poisonous alkaloid: also called *ers.* [<AF *veche, vecce* <L *vicia*]

vetch·ling (vech′ling) *n.* A plant (genus *Lathyrus*), nearly allied to the vetches; especially, a European species (*L. pratensis*), naturalized in the United States and Canada. [Dim. of VETCH]

vet·er·an (vet′ər·ən, vet′rən) *n.* 1 One who is much experienced in any service, especially a soldier or an ex-soldier. 2 A member of the armed forces who has been in active service. — *adj.* 1 Having had long experience or practice; old in service. 2 Belonging to or suggestive of a veteran. 3 Long continued; extending over a long period. [<MF <L *veteranus* < *vetus, veteris* old]

Veterans Administration An agency of the U.S. government which administers all federal laws relating to the relief of former members of the military and naval services.

Veterans Day A U.S. national holiday honoring veterans of the armed forces, held on November 11, the anniversary of the date in 1918 when the Allies granted an armistice to the Central Powers in World War I. Formerly called *Armistice Day*.

Veterans of Foreign Wars A society of ex-servicemen who have served in the United States Army, Navy, or Marine Corps in a war with and in a foreign country: founded 1899.

vet·er·i·nar·i·an (vet′ər·ə·nâr′ē·ən, vet′rə-) *n.* A practitioner of veterinary medicine or surgery. [<L *veterinarius* VETERINARY]

vet·er·i·nar·y (vet′ər·ə·ner′ē, vet′rə-) *adj.* Pertaining to the diseases or injuries of animals, and to their treatment by medical or surgical means. —*n. pl.* **·nar·ies** A veterinarian; a veterinary surgeon. [<L *veterinarius* pertaining to beasts of burden < *veterinus* < *veterina* beasts of burden, ult. < *vehere* carry]

vet·i·ver (vet′ə·vər) *n.* **1** An Asian grass (*Vetiveria zizanioides*) grown in Florida and the SE United States. **2** Its aromatic roots, used for weaving mats, fans, etc., and as a source of **vetiver oil**, an ingredient of perfumes. [<F *vétyver* <Tamil *vettivēru*, lit., a root that is dug up < *vēr* a root]

Vet·lu·ga (vet·lōō′gə) A river in central European Russian S.F.S.R., flowing 500 miles west, north, and south to the Volga.

ve·to (vē′tō) *v.t.* **·toed, ·to·ing 1** To refuse executive approval of (a bill passed by a legislative body). **2** To forbid or prohibit authoritatively; refuse consent to. —*n. pl.* **·toes 1** The prerogative in a chief executive of refusing to approve a legislative enactment by withholding his signature. **2** The act of vetoing; also, the official communication containing a refusal to approve a bill. **3** Any authoritative prohibition. [<L, I forbid] —**ve′to·er** *n.*

veto message A message giving the reasons of the chief executive for refusing his approval of a proposed law.

veto power 1 The right or power possessed by a branch of the government to forbid or refuse approval of projects proposed by another department. **2** A power vested in the chief executive to prevent the enactment of bills passed by the legislature.

Vet·ter (vet′ər), **Lake** A former spelling for VÄTTER, LAKE.

vex (veks) *v.t.* **1** To provoke to anger or displeasure by small irritations; irritate; annoy. **2** To trouble or afflict. **3** To throw into commotion; agitate. **4** To make a subject of dispute: a *vexed* question. See synonyms under AFFRONT, PIQUE[1]. [<OF *vexer* <L *vexare* shake]

vex·a·tion (vek·sā′shən) *n.* **1** The act of vexing, or the state of being vexed; irritation. **2** That which vexes; annoyance; affliction; cause of trouble or distress. See synonyms under CHAGRIN, IMPATIENCE.

vex·a·tious (vek·sā′shəs) *adj.* **1** Being a source of vexation. **2** Full of vexation; harassing; annoying. See synonyms under TROUBLESOME, WEARISOME. —**vex·a′tious·ly** *adv.* —**vex·a′tious·ness** *n.*

vexed (vekst) *adj.* **1** Harassed; troubled; irritated; agitated; disturbed. **2** Much debated; contested: a *vexed* question. —**vex·ed·ly** (vek′sid·lē) *adv.* —**vex′ed·ness** *n.*

vex·il (vek′sil) *n.* A vexillum (def. 2). [Short for VEXILLUM]

vex·il·lar·y (vek′sə·ler′ē) *n. pl.* **·lar·ies** A standard-bearer. —*adj.* **1** Of or pertaining to a vexillum: also **vex′il·lar** (-lər). **2** Of or pertaining to a standard or ensign. [<L *vexillarius* a standard-bearer < *vexillum* VEXILLUM]

vex·il·late (vek′sə·lit, -lāt) *adj.* Having a vexillum or vexilla.

vex·il·lum (vek·sil′əm) *n. pl.* **vex·il·la** (vek·sil′ə) **1** In Roman antiquity, a square flag, or standard; hence, a company or troop of soldiers serving under a separate standard. **2** *Bot.* The large upper petal of a papilionaceous flower. **3** *Ornithol.* The web of a feather. [<L < *vehere* carry]

V-for·ma·tion (vē′fôr·mā′shən) *n. Aeron.* A V-shaped flight formation of three or more aircraft.

vi·a (vī′ə, vē′ə) *prep.* By way of; by a road passing through: He went to Boston *via* New Haven. [<L, ablative sing. of *via* a way]

Vi·a Ap·pi·a (vī′ə ap′ē·ə) The ancient Latin name for the APPIAN WAY.

vi·a·ble (vī′ə·bəl) *adj.* Capable of living and developing normally, as a newborn infant, a seed, etc. [<F < *vie* life <L *vita*] —**vi′a·bil′i·ty** *n.*

Vi·a Dol·o·ro·sa (vī′ə dol′ō·rō′sə) *Latin* The road traveled by Jesus to Golgotha; literally, the sorrowful way.

vi·a·duct (vī′ə·dukt) *n.* A bridgelike structure, especially a large one of arched masonry, to carry a roadway or the like over a valley or ravine. Compare AQUEDUCT. [<L *via* a way + (AQUE)DUCT]

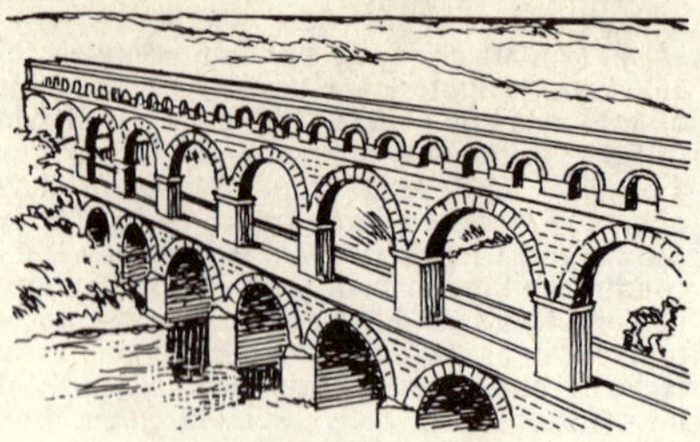

VIADUCT
Roman aqueduct, Nîmes, France.

Vi·a Fla·min·i·a (vī′ə flə·min′ē·ə) The Latin name for the FLAMINIAN WAY.

vi·al (vī′əl) *n.* A small bottle for liquids, commonly cylindrical; also, more widely, any bottle: also spelled *phial*. —**to pour out the vials of wrath upon** To inflict retribution or vengeance on. See *Rev.* xvi. —*v.t.* **·aled** or **·alled, ·al·ing** or **·al·ling** To put or keep in or as in a vial. [<OF *viole* <L *phiala* a saucer <Gk. *phialē* a shallow cup]

vi·a me·di·a (vī′ə mē′dē·ə) *Latin* A middle way.

vi·and (vī′ənd) *n.* **1** An article of food, especially meat. **2** *pl.* Victuals; provisions; food. See synonyms under FOOD. [<AF *viaunde*, OF *viande*, ult. <L *vivenda*, neut. pl. gerundive of *vivere* live]

vi·at·ic (vī·at′ik) *adj.* Of or pertaining to a journey or to traveling. Also **vi·at′i·cal**. [<L *viaticus* < *via* a way]

vi·at·i·cum (vī·at′ə·kəm) *n. pl.* **·ca** (-kə) or **·cums 1** *Eccl.* The Eucharist, as given on the verge of death. **2** In ancient Rome, the provision of necessaries for an official journey of a magistrate; later, provisions for any journey. [<L, traveling money, neut. sing. of *viaticus* < *via* a way. Doublet of VOYAGE.]

vi·a·tor (vī·ā′tər) *n. pl.* **vi·a·to·res** (vī′ə·tôr′ēz, -tō′rēz) A traveler; wayfarer. [<L < *via* a way]

Vi·borg (vē′bôr·y′) The Swedish name for VYBORG.

vi·brac·u·lum (vī·brak′yə·ləm) *n. pl.* **·la** (-lə) *Zool.* One of the slender, whiplike defensive organs of the cells of many polyzoans. [<NL <L *vibrare* shake] —**vi·brac′u·lar** *adj.* —**vi·brac′u·loid** *adj.*

vi·bran·cy (vī′brən·sē) *n. pl.* **·cies** The state or character of being vibrant; resonance.

vi·brant (vī′brənt) *adj.* **1** Having, showing, or resulting from vibration; vibrating; resonant. **2** Throbbing; pulsing: a nation *vibrant* with enthusiasm. **3** Energetic; vigorous. **4** *Phonet.* Produced with vibration of the vocal cords; voiced. —*n. Phonet.* A speech sound made with vibration of the vocal cords; a voiced sound. [<L *vibrans, -antis*, ppr. of *vibrare* shake] —**vi′brant·ly** *adv.*

vi·bra·phone (vī′brə·fōn) *n.* A type of marimba in which a pulsating sound is produced by motor-driven valves in the resonators. Also **vi′bra-harp′** (-härp′). [<VIBRA(TO) + -PHONE]

vi·brate (vī′brāt) *v.* **·brat·ed, ·brat·ing** *v.i.* **1** To move or swing back and forth, as a pendulum. **2** To move back and forth rapidly; quiver. **3** To sound: The note *vibrates* on the ear. **4** To be emotionally moved; thrill. **5** To vacillate; waver, as between choices. —*v.t.* **6** To cause to move or swing back and forth. **7** To cause to quiver or tremble. **8** To send forth (sound, etc.) by vibration. **9** To measure by each vibration: a pendulum *vibrating* seconds. See synonyms under QUAKE, SHAKE. [<L *vibratus* < *vibrare* shake]

vi·bra·tile (vī′brə·til, -tīl) *adj.* **1** Adapted to, having, or used in vibratory motion. **2** Pertaining to or resembling vibration.

vi·bra·til·i·ty (vī′brə·til′ə·tē) *n.* Capability, or quality, of being vibratile.

vi·bra·tion (vī·brā′shən) *n.* **1** The act of vibrating; oscillation. **2** *Physics* **a** A periodic, usually rapid back-and-forth motion of a particle, as of electrons in an atom, or the parts of an elastic or rigid body suddenly released from tension. **b** Any physical process characterized by cyclic variations in amplitude, intensity, or the like, as wave motion or an electric field. **c** A single complete oscillation. See synonyms under WAVE. —**vi·bra′tion·al** *adj.*

vi·bra·to (vē·brä′tō) *n. Music* A trembling or pulsating effect, not confined to vocal music, caused by rapid variation of emphasis on the same tone: properly distinguished from *tremolo*, where there is an alternation of tones. [<Ital., pp. of *vibrare* vibrate <L, shake]

vi·bra·tor (vī′brā·tər) *n.* **1** One who or that which vibrates. **2** An electrically operated massaging apparatus. **3** *Electronics* **a** An electromagnetic switch mechanism for converting direct into alternating current by continuously vibrating impulses. **b** An oscillator (def. 3).

vi·bra·to·ry (vī′brə·tôr′ē, -tō′rē) *adj.* Pertaining to, causing, or characterized by vibration. Also **vi′bra·tive** (-tiv).

vib·ri·o (vib′rē·ō) *n.* Any of a genus (*Vibrio*) of motile, rodlike bacteria in which the cells are but slightly sinuous and have one or more flagellae at each end; especially, the Gram-negative **comma vibrio** (*V. comma*), found in the intestines of cholera victims. [<NL <L *vibrare* shake]

vib·ri·oid (vib′rē·oid) *adj.* Resembling a vibrio. —*n.* A vibrioid body. [<VIBRI(O) + -OID]

vibrioid body *Bot.* A minute organ of the cell of some fungi and algae usually seen embedded in the peripheral layers of the protoplasm, rodlike in shape and showing independent movements.

vi·bris·sa (vī·bris′ə) *n. pl.* **·bris·sae** (-bris′ē) *Biol.* **1** One of the stiff, coarse hairs found in the nostrils of man and about the mouth of many other mammals, as the cat: they often function as tactile organs. **2** One of the vaneless, hairlike rictal feathers of many insectivorous birds, especially flycatchers. [<L *vibrissae* hairs in a man's nostrils < *vibrare* shake]

vi·bro·scope (vī′brə·skōp) *n.* A device for observing and recording vibrations, especially those of harmonic character. [< *vibro-* (<VIBRATE) + -SCOPE]

vi·bro·tro·pism (vī′brə·trō′piz·əm) *n. Biol.* The involuntary response of an organism to a vibratory stimulus. [< *vibro-* (<VIBRATE) + -TROPISM] —**vi′bro·trop′ic** (-trop′ik) *adj.*

vi·bur·num (vī·bûr′nəm) *n.* Any of a large and widely distributed genus (*Viburnum*) of shrubs or small trees of the honeysuckle family, bearing small flowers and berrylike fruit; especially, the dockmackie, the sheepberry, and the hobble-bush. [<L, the wayfaring tree]

vic·ar (vik′ər) *n.* **1** In general, one who is authorized to perform functions, especially religious ones, in the stead of another; a substitute in office. **2** Hence, an agent; deputy. **3** *Brit.* The priest of a parish of which the main revenues are appropriated or impropriated by a layman, the priest himself receiving but a stipend; any incumbent of a parish who is not a rector. **4** In the Roman Catholic Church, a substitute or representative of an ecclesiastical person; in a strict sense, one whose jurisdiction is confined to the external forum. **5** In some parishes of the Protestant Episcopal Church, the clergyman who is the head of a chapel; also, a clergyman having charge of a church or mission as the bishop's deputy. [<AF *vikere, vicare*, OF *vicaire* <L *vicarius* a substitute < *vicis* a change]

vic·ar·age (vik′ər·ij) *n.* **1** The benefice, office, or duties of a vicar. **2** A vicar's residence or household.

vicar apostolic In the Roman Catholic Church, formerly, a bishop or archbishop appointed by the pope to act in his stead in a given district; more recently, a titular bishop exercising episcopal jurisdiction where there is no see canonically.

vicar fo·rane (fō·rān′, fō-) In the Roman Catholic Church, a clergyman appointed by a bishop, having a limited jurisdiction over the inferior clergy in the parishes constituting the deanery; a rural dean. [<VICAR + Med. L *foraneus* outside the episcopal city, rural <L *foras* out of doors]

vicar general 1 In the Roman Catholic Church, a functionary appointed by the bishop as assistant or representative in certain

vicarial   1399   **Victoria**

matters of jurisdiction, but without power to perform the specific function of the episcopal order. 2 In the Church of England, an official assisting the bishop or archbishop in ecclesiastical causes. 3 In English history, the ecclesiastical vicegerent of the king: a title bestowed on Thomas Cromwell by Henry VIII.

vi·car·i·al (vī-kâr′ē·əl, vi-) *adj.* 1 Vicarious; delegated. 2 Belonging to, relating to, or acting as a vicar.

vi·car·i·ate (vī-kâr′ē·it, -āt, vi-) *n.* A delegated office or power; specifically, the office or authority of a vicar. Also **vic·ar·ate** (vik′ər·it).

vi·car·i·ous (vī-kâr′ē·əs, vi-) *adj.* 1 Made or performed by substitution; suffered or done in place of another; substitutionary: a *vicarious sacrifice*; also, enjoyed or felt by a person as a result of his imagined participation in an experience that is not his own: *vicarious gratification*. 2 Filling the office of or acting for another. 3 Of, pertaining to, or belonging to a vicar or deputy; deputed; delegated. 4 *Med.* Performing, as an organ, the functions of another; substitutive; also, occurring in an abnormal situation: *vicarious menstruation*. [<L *vicarius*. See VICAR.] — **vi·car′i·ous·ly** *adv.* — **vi·car′i·ous·ness** *n.*

vic·ar·ly (vik′ər·lē) *adj.* Resembling or pertaining to a vicar.

vicar of Christ The pope, regarded as Christ's representative on earth.

vic·ar·ship (vik′ər·ship) *n.* The office or position of a vicar.

vice[1] (vīs) *n.* 1 A moral blemish or taint; an immoral habit or trait: the *vice* of intemperance. 2 Habitual indulgence in degrading or harmful appetites; deviation from moral rectitude; depravity. 3 Something that mars; a defect; blemish. 4 A physical deformity, taint, or imperfection. 5 A bad trick, as of a horse. See synonyms under SIN[1]. — **inherent vice** In insurance, a hazard arising from a preexistent condition not manifest when the commodity was insured and therefore not covered by the insurance policy: Eggs of worms were the *inherent vice* that ruined the cargo of hides. ◆ Homophone: *vise*. [<OF <L *vitium* a fault]

vice[2] (vīs) See VISE.

vice[3] (vīs) *adj.* Acting in the place of; substitute; deputy: *vice president*. — *n.* One who acts in the place of another; a substitute; deputy.

— **vi·ce** (vī′sē) *prep.* Instead of; in the place of. [<L, ablative of *vicis* change]

Vice may appear as a combining form in hyphemes or as the first element in two-word phrases:

vice-chair	vice-ministry
vice chairman	vice principal
vice-chairmanship	vice-principalship
vice dean	vice rector
vice-government	vice-rectorship
vice governor	vice-reign
vice-governorship	vice-wardenship

vice admiral A commissioned officer in the Navy or Coast Guard who ranks next above a rear admiral and next below an admiral. [<AF *visadmirail*, OF *visamiral* < *vis-* in place (<L *vice*) + *admirail*, *amiral* an admiral]

vice-ad·mir·al·ty (vīs′ad′mər·əl·tē) *n.* The area under or office of a vice admiral.

vice chancellor 1 *Law* A judge in equity courts subordinate to the chancellor. 2 A deputy chancellor in a university. [<OF *vichancelier* <Med. L *vicecancellarius* <L *vice* in place + LL *cancellarius* a chancellor] — **vice′-chan′cel·lor·ship**′ *n.*

vice consul One who exercises consular authority, either as the substitute or as the subordinate of a consul. — **vice-con·su·lar** (vīs′kon′sə·lər) *adj.* — **vice–con·su·late** (vīs′·kon′sə·lit) *n.* — **vice′-con′sul·ship** *n.*

vice·ge·ren·cy (vīs·jir′ən·sē) *n. pl.* **·cies** 1 The office or authority of a vicegerent; the fact of ruling as a vicegerent or by a vicegerent. 2 A district ruled by a vicegerent.

vice·ge·rent (vīs·jir′ənt) *n.* One duly authorized to exercise the powers of another; a deputy; vicar. — *adj.* Acting in the place of another, usually that of a superior. [<Med. L *vicegerens, -entis* <L *vice* in place + *gerens,*

-entis, ppr. of *gerere* carry, manage] — **vice·ge′ral** *adj.*

vi·ce·nar·y (vis′ə·ner′ē) *adj.* 1 Consisting of or pertaining to twenty. 2 Relating to a system of notation based upon twenty. [<L *vicenarius* < *viceni* twenty each < *viginti* twenty]

vi·cen·ni·al (vī-sen′ē·əl) *adj.* Occurring once in twenty years; also, lasting or existing twenty years. [<L *vicennium* a twenty-year period < *vicies* twenty times + *annus* a year]

Vi·cen·te (*Pg.* vē-seń′tə, *Sp.* bē-then′tä) Portuguese and Spanish form of VINCENT.

Vi·cen·za (vē-chen′tsä) 1 A province in Veneto, northern Italy; 1,051 square miles. 2 A city in NE Italy, capital of Vicenza province; ancient **Vi·cen·ti·a** (vī-sen′shē·ə).

vice president An officer ranking next below a president, and acting, on occasion, in his place. The vice president of the United States is elected at the same time and in the same manner as the president, and is designated by the Constitution to be president of the Senate and to succeed the president in case of that officer's death, resignation, removal, or inability. — **vice–pres·i·den·cy** (vīs′prez′ə·dən·sē) *n.* — **vice′–pres′i·den′tial** (-prez′ə·den′shəl) *adj.*

vice·re·gal (vīs·rē′gəl) *adj.* Of or relating to a viceroy, his office, or his jurisdiction. Also **vice–roy′al** (-roi′əl). — **vice·re′gal·ly** *adv.*

vice regent A deputy regent. — **vice′–re′gent** (-rē′jənt) *adj.*

vice·roy (vīs′roi) *n.* 1 One who rules a country, colony, or province by the authority of his sovereign or king. 2 A North American nymphalid butterfly (*Basilarchia archippus*), orange-red with black markings and a row of white marginal spots. The larva feeds on the willow, poplar, and certain other trees. [<MF *viceroy, visroy* < *vice-*, *vis-* in place (<L *vice*) + *roy* a king, ult. <L *rex, regis*]

vice·roy·al·ty (vīs·roi′əl·tē) *n.* 1 The office or authority of a viceroy. 2 The term of office of a viceroy. 3 A district ruled or governed by a viceroy. Also **vice′roy·ship**.

vice squad A police division charged with combating illegal prostitution, perversion, gambling, and other vices.

vi·ce ver·sa (vī′sē vûr′sə, vīs′) The order being changed; the relation of terms being reversed; conversely. [<L]

Vi·cha·da (bē-chä′thä) A river in eastern Colombia, flowing 400 miles east to the Orinoco at the Venezuela border.

Vi·chy (vē·shē′) A resort city in central France; provisional capital of France during German occupation, World War II.

vi·chy·ssoise (vē′shē·swäz′) *n.* A potato cream soup flavored with leeks, celery, etc., usually served cold with a sprinkling of chives. [<F, of Vichy <*Vichy* Vichy]

Vi·chy water (vish′ē) The effervescent mineral water from the springs at Vichy, France; any mineral water resembling it. Also **Vi′chy, vi′chy**.

vic·i·nage (vis′ə·nij) *n.* 1 Neighboring places collectively; vicinity. 2 The state of being a neighbor or neighbors. See synonyms under NEIGHBORHOOD. [<OF *visenage, vicenage* <L *vicinus* nearby]

vic·i·nal (vis′ə·nəl) *adj.* 1 Neighboring; adjoining; near. 2 *Mineral.* Designating a crystal form closely approximating one of the fundamental forms. 3 *Chem.* Designating a benzene derivative in which the substituted elements or radicals are in consecutive order on the benzene ring. [<L *vicinalis* < *vicinus* a neighbor, orig. nearby]

vicinal planes In crystallography, crystal planes which may approximate or take the place of the fundamental planes.

vicinal road A local road, as distinguished from one between towns.

vic·i·nism (vis′ə·niz′əm) *n. Ecol.* Plant variation resulting from the proximity of other plants.

vi·cin·i·ty (vi-sin′ə·tē) *n. pl.* **·ties** 1 Nearness in space or relationship; proximity. 2 A region adjacent to or near; neighborhood; vicinage. See synonyms under NEIGHBORHOOD. [<L *vicinitas, -tatis* < *vicinus* nearby]

vi·cious (vish′əs) *adj.* 1 Addicted to vice; corrupt in conduct or habits; wicked; depraved. 2 Partaking of what is base, low, and vile; morally injurious; evil. 3 Unruly

or dangerous; refractory, as an animal. 4 Defective or faulty: *vicious* arguments. 5 Impure or incorrect; corrupted, as a text, manuscript, etc. 6 Noxious; poisonous; foul, as water, air, etc. 7 *Colloq.* Marked by malice or spite; malignant: a *vicious* lie. See synonyms under CRIMINAL, IMMORAL, IRREGULAR, RESTIVE. [<OF <L *vitiosus* < *vitium* a fault] — **vi′cious·ly** *adv.* — **vi′cious·ness** *n.*

vicious circle 1 The process or predicament that arises when the solution of a problem creates a new problem and each successive solution adds another problem. 2 *Logic* Argument in a circle. See under CIRCLE. 3 *Med.* The accelerating effect of one disease upon another when the two are coexistent.

vi·cis·si·tude (vi-sis′ə·tōōd, -tyōōd) *n.* 1 *pl.* Irregular changes or variations, as of conditions or fortune: the *vicissitudes* of life. 2 A change; especially, a complete change; mutation or mutability. 3 Alternating change or succession, as of the seasons. See synonyms under CHANGE. [<MF <L *vicissitudo* < *vicis* a turn, change]

vi·cis·si·tu·di·nar·y (vi-sis′ə·tōō′də·ner′ē, -tyōō′-) *adj.* Marked by or subject to change or alternation. Also **vi·cis′si·tu′di·nous**.

Vicks·burg (viks′bûrg) A city in western Mississippi on the Mississippi River; besieged and taken by the Union army in the Civil War, 1863.

vi·con·ti·el (vī·kon′tē·əl) *adj. Obs.* Of or pertaining to a viscount or sheriff. [<AF, OF *vicontal* < *viconte* a viscount]

vic·tim (vik′tim) *n.* 1 A living creature sacrificed to some deity or as a religious rite. 2 A person sacrificed in the pursuit of some object; one who is injured or killed, as by misfortune or calamity. 3 A sufferer from any diseased condition or morbid feeling. 4 One who is swindled; a dupe. [<L *victima* a beast for sacrifice]

vic·tim·ize (vik′tim·īz) *v.t.* **·ized, ·iz·ing** To make a victim of, especially by defrauding or swindling; dupe; cheat. See synonyms under ABUSE. — **vic′tim·i·za′tion** *n.* — **vic′tim·iz′er** *n.*

vic·tor (vik′tər) *n.* One who vanquishes an enemy; one who is successful in any struggle or contest; winner; conqueror. — *adj.* Pertaining to a victor; victorious; triumphant: the *victor* nation. [<AF *victor, victour*, OF *victeur* <L *victus*, pp. of *vincere* conquer]

Synonyms (noun): conqueror, master, vanquisher, winner. A *victor* wins in a single battle or contest; a *conqueror* wins by subjugating his opponents in many battles or campaigns.

Vic·tor (vik′tər, *Fr.* vēk·tôr′) A masculine personal name. [<L, a conqueror]

Vic·tor Em·man·u·el (vik′tər i·man′yōō·əl) Name of three Italian kings.

— **Victor Emmanuel I**, 1759–1824, king of Sardinia 1820–21.

— **Victor Emmanuel II**, 1820–78, king of Sardinia 1849–61, and first king of Italy 1861–78.

— **Victor Emmanuel III**, 1869–1947, king of Italy 1900–46.

vic·to·ri·a (vik·tôr′ē·ə, -tō′rē·ə) *n.* 1 A low, light, four-wheeled carriage, with a calash top, a seat for two persons over the rear axle, and a raised driver's seat. 2 A passenger automobile with a calash top which usually covers the rear seat only. [after Queen *Victoria*]

VICTORIA

Vic·to·ri·a (vik·tôr′ē·ə, -tō′rē·ə) A feminine personal name.

— **Victoria**, 1819–1901, queen of Great Britain 1837–1901.

Victoria 1 A state of SE Australia; 87,884 square miles; capital, Melbourne. 2 The capital of British Columbia, a port at the southern extremity of Vancouver Island. 3 A port on Hong Kong Island, capital of Hong Kong colony, China. 4 A province of SE Southern Rhodesia; 21,028 square miles.

Victoria, Lake The largest lake in Africa

and the second largest fresh-water body in the world, in British East Africa, between Uganda and Tanganyika, with a NE portion in Kenya; 26,828 square miles. Also *Victoria Nyanza.*

Victoria, Mount 1 The highest peak of the Owen Stanley Range, SE New Guinea; 13,240 feet. 2 The highest peak of the Chin Hills, Upper Burma; 10,018 feet.

Victoria Cross See under CROSS.

Victoria Desert The southern belt of the Western Australian desert, south of the Gibson Desert: also *Great Victoria Desert.*

Victoria Falls A cataract on the Zambesi River between Northern and Southern Rhodesia; 343 ft. high; over a mile wide; discovered by Livingstone in 1855.

Victoria Island An island in SW Franklin district, Northwest Territories, Canada; 80,340 square miles.

Victoria Land A part of Antarctica south of New Zealand, east of the Ross Sea, and west of Wilkes Land, consisting of a series of snow-covered mountains; highest point, 13,350 feet.

Vic·to·ri·an (vik·tôr′ē·ən, -tō′rē-) *adj.* 1 Of or relating to Queen Victoria, or to her reign. 2 Pertaining to or characteristic of the ideals and standards of morality and taste prevalent during the reign of Queen Victoria; prudish; conventional; narrow. 3 Of or pertaining to Victoria, Australia. — *n.* 1 Anyone, especially an author, contemporary with Queen Victoria. 2 An article of furniture, dress, or the like, identified with or dating from the Victorian age.

Victoria Nile See under NILE.

Vic·to·ri·an·ism (vik·tôr′ē·ən·iz′əm, -tō′rē-) *n.* The state or quality of being Victorian, as in style or moral outlook.

Victoria Ny·an·za (nī·an′zə, nyän′zä) See VICTORIA, LAKE.

Victoria River A river in western Northern Territory, Australia, flowing 350 miles NE, north, and west to the Timor Sea.

Victoria waterlily Any of a genus (*Victoria*) of very large tropical American waterlilies, having leaves often five feet in diameter, and huge, showy, crimson-and-white flowers.

vic·to·ri·ous (vik·tôr′ē·əs, -tō′rē-) *adj.* 1 Having won victory; conquering; triumphant. 2 Bringing victory; distinguished by victory; instrumental in bringing victory. 3 Relating to victory. — **vic·to′ri·ous·ly** *adv.* — **vic·to′ri·ous·ness** *n.*

vic·to·ry (vik′tər·ē) *n. pl.* **·ries** 1 The state of being a victor. 2 The overcoming of an enemy or of any difficulty. [<OF *victorie, victoire* <L *victoria* < *victor.* VICTOR]

Synonyms: achievement, advantage, conquest, mastery, success, supremacy, triumph. *Victory* is the state resulting from the overcoming of an opponent or opponents in any contest, or from the overcoming of difficulties, obstacles, evils, etc., considered as opponents or enemies. In the latter sense any hard-won *achievement, advantage,* or *success* may be termed a *victory.* In *conquest* and *mastery* there is implied a permanence of state that is not implied in *victory. Triumph,* originally denoting the public rejoicing in honor of a *victory,* has come to signify also an exultant, complete, and glorious *victory.* Compare CONQUER. *Antonyms*: defeat, disappointment, disaster, failure, frustration, miscarriage, overthrow, retreat, rout.

Victory Medal Either of two bronze medals awarded to all who served in the U. S. armed forces in World War I or World War II, worn with the **Victory Ribbon,** combining six colors of the rainbow.

Vic·tro·la (vik·trō′lə) *n.* A make of phonograph: a trade name.

vict·ual (vit′l) *n.* 1 Food for human beings, as prepared for eating: usually in the plural, and, except in dialect, seldom used in any but a humorous or depreciatory sense. 2 *Obs.* Provisions of any kind. See synonyms under FOOD. — *v.* **·ualed** or **·ualled, ·ual·ing** or **·ual·ling** *v.t.* 1 To furnish with victuals. — *v.i.* 2 To lay in supplies of food. 3 *Rare* To eat; feed. [<OF *vitaile* <LL *victualia* provisions, neut. pl. of L *victualis* of food < *victus* food]

vict·ual·age (vit′l·ij) *n. Rare* Victuals; victualing.

vict·ual·er (vit′l·ər) *n.* 1 One who supplies or sells victuals; specifically, one engaged in supplying an army, navy, or ship with provisions; a commissary; sutler. 2 An innkeeper. 3 A victualing ship. Also **vict′ual·ler.**

vi·cu·ña (vi·kōōn′yə, -kyōō′nə) *n.* 1 A small ruminant (*Lama vicugna*) of the high Andes related to the llama and alpaca, having fine and valuable wool. 2 A fiber and textile made from this wool, or some substitute. Also **vi·cu′·gna.** [<Sp. <Quechua]

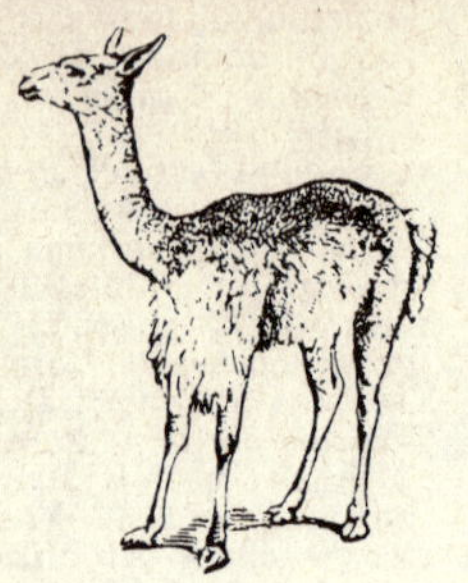

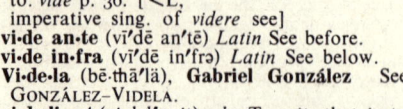

VICUÑA
(Up to 3 feet high at the shoulder)

vicuña cloth Soft cloth made of vicuña wool.

vi·de (vī′dē) See: used to make a reference or direct attention to: *vide* p. 36. [<L, imperative sing. of *videre* see]

vi·de an·te (vī′dē an′tē) *Latin* See before.

vi·de in·fra (vī′dē in′frə) *Latin* See below.

Vi·de·la (bē·thā′lä), **Gabriel González** See GONZÁLEZ-VIDELA.

vi·de·li·cet (vi·del′ə·sit) *adv.* To wit; that is to say; namely: abbr. *viz.* [<L < *videre licet* it is permitted to see]

vid·e·o (vid′ē·ō) *adj.* Of or pertaining to television, especially to the picture portion of a program. Compare AUDIO. [<L, I see]

vid·e·o·gen·ic (vid′ē·ō·jen′ik) *adj.* Having such characteristics, as coloration, form, etc., as appear effectively in television: also *telegenic.* [<VIDEO + -GENIC]

vid·e·og·no·sis (vid′ē·og·nō′sis) *n. Med.* The diagnosis of disease by means of X-ray pictures transmitted by television to a radiologic center. [<VIDEO + (DIA)GNOSIS]

vi·de post (vī′dē pōst′) *Latin* See after; see what follows.

vide·ruff (vīd′ruf) *n.* An old card game. Compare RUFF[2]. [< *vide,* var. of *vied,* pp. of VIE, in obs. sense of "wager money at cards" + RUFF[2]]

vi·de su·pra (vī′dē sōō′prə) *Latin* See above.

vi·dette (vi·det′) See VEDETTE.

vi·de ut su·pra (vī′dē ut sōō′prə) *Latin* See what is written above.

vid·u·age (vid′yōō·ij) *n.* Widowhood; also, widows collectively. [<L *vidua* a widow, orig. fem. of *viduus* bereft]

vie (vī) *v.* **vied, vy·ing** *v.i.* 1 To strive for superiority; put forth effort to excel or outdo others, as in a race: with *with* or *for.* — *v.t.* 2 *Rare* To put forth in competition; match. 3 *Obs.* To wager; bet. See synonyms under CONTEND, STRUGGLE. [<MF *envier* invite, challenge <L *invitare* invite]

Vied·ma (vyed′mä, *Sp.* byeth′mä) The capital of Río Negro province, in SE central Argentina near the Atlantic.

Vi·en·na (vē·en′ə) A city on the Danube, capital of Austria, in the NE part: German *Wien.*

Vienne (vyen) 1 A city in SE France, on the Rhône. 2 A river of west central France, flowing 230 miles west and north to the Loire. 3 A department of west central France; 2,719 square miles; capital, Poitiers.

Vi·en·nese (vē′ə·nēz′, -nēs′) *adj.* Belonging or relating to Vienna, or to its inhabitants. — *n. pl.* **·nese** A native or citizen of Vienna.

Vien·nois (vye·nwä′) An ancient county of SE France.

Vien·tiane (vyaṅ·tyän′) A city on the Mekong river, the administrative capital of Laos, in the NW central part on the Thailand border.

Vie·ques Island (byā′käs) An island belonging to and east of Puerto Rico; 52 square miles.

Vier·wald·stät·ter See (fir′vält·shtet′ər zā) The German name for the LAKE OF LUCERNE.

vi et ar·mis (vī et är′mis) *Latin* With force and arms.

Vi·et Minh (vē·et′ min′) A popular name for the DEMOCRATIC REPUBLIC OF VIETNAM. See under VIETNAM.

Vi·et·nam (vē·et·näm′) A country in SW Indochina, comprising South Vietnam (formerly Cochin China), Central Vietnam (formerly Annam), and North Vietnam (formerly Tonkin); total, 129,093 square miles; divided into the **Democratic Republic of Vietnam (Viet Minh** or **North Vietnam)** north of the 17th parallel; 63,344 square miles; capital, Hanoi; and the **Republic of Vietnam (South Vietnam),** formerly (1949–54), as the **National State of Vietnam,** an associated state within the French Union; 65,749 square miles; capital Saigon. Also **Viet Nam.** — **Vi·et′nam·ese′** (-ēz′, -ēs′) *n.* & *adj.*

Vi·e·tor (fē′ā·tôr), **Wilhelm,** 1850–1918, German philologist and writer.

view (vyōō) *n.* 1 The act of seeing; survey; inspection. 2 Mental examination or inspection. 3 Power of seeing, or range of vision; reach of perception or insight; range or scope of thought. 4 That which is seen; a spectacle; prospect. 5 A representation of a scene; especially, a landscape; also, a sketch; design; plan. 6 Reference to something regarded as the object of action; intention; purpose. 7 Manner of looking at things; opinion; judgment; belief: What are your *views* on this subject? 8 *Obs.* Appearance; aspect; show. See synonyms under PURPOSE, SCENE, THOUGHT[1]. — **in view of** In consideration of. — **on view** Open to the public; set up for public inspection. — **with a view to** With the aim or purpose of. — *v.t.* 1 To look at; see; behold. 2 To look at carefully; scrutinize; examine. 3 To survey mentally; consider. See synonyms under EXAMINE, LOOK. [<OF *veue,* orig. pp. of *veoir* see <L *videre*] — **view′er** *n.*

view-hal·loo (vyōō′hə·lōō′) *n.* A shout uttered by a huntsman when a fox breaks cover. Also **view′-hal·lo′** (-lō′), **view′-hal·loa′.**

view·less (vyōō′lis) *adj.* 1 Devoid of a view; that cannot be viewed. 2 Having no views or opinions. 3 Invisible; unseen. — **view′less·ly** *adv.* — **view′less·ness** *n.*

view·point (vyōō′point′) *n.* Point of view.

view·y (vyōō′ē) *adj.* 1 *Colloq.* Having visionary ideas or peculiar views; visionary. 2 Appearing good at first sight; showy.

Vi·gée-Le·brun (vē·zhā′lə·brœṅ′), **Marie Anne,** 1755–1842, French painter.

vi·ges·i·mal (vī·jes′ə·məl) *adj.* 1 Twentieth. 2 Of or pertaining to twenty; proceeding by twenties. [<L *vigesimus,* var. of *vicesimus* < *viceni.* See VICENARY.]

vig·il (vij′əl) *n.* 1 The act or state of keeping awake; a nightlong watch; watchfulness. 2 *Eccl.* **a** The eve of a holy day, especially the eve of a fast day. **b** *pl.* Religious devotions on such an eve. 3 *Usually pl.* Any nocturnal devotions. [<AF *vigile* <L *vigilia* < *vigil* awake]

vig·il·am·bu·lism (vij′əl·am′byə·liz′əm) *n. Psychol.* A state in which a person, while awake, is unconscious of his surroundings: a condition resembling somnambulism. [<L *vigil* awake + *ambulare* walk]

vig·i·lance (vij′ə·ləns) *n.* 1 The quality of being vigilant; alertness; watchfulness in guarding against danger or providing for safety. 2 A morbid watchfulness; insomnia. See synonyms under CARE.

vigilance committee 1 A body of men self-organized for the maintenance of order and the administration of summary justice in communities where regular authority is lacking or inefficient, especially in lawless sections of the western United States. 2 Formerly, in the southern United States, a group of white citizens organized to terrify and control Negroes and abolitionists.

vig·i·lant (vij′ə·lənt) *adj.* Characterized by vigilance; being on the alert; watchful; heedful; wary. [<MF <L *vigilans, -antis,* ppr. of *vigilare* keep awake <*vigil* awake] — **vig′i·lant·ly** *adv.* — **vig′i·lant·ness** *n.*

Synonyms: active, alert, awake, careful, cautious, circumspect, heedful, mindful, sleepless, wakeful, wary, watchful, wide-awake. *Vigilant* implies more sustained activity and more intelligent volition than *alert.* One is *vigilant* against danger; he may be *alert* or *watchful* for good as well as against evil; he is *wary* in view of suspected stratagem, trickery, or treachery. A person may be *wakeful* because of some merely physical excitement or excitability, as through insomnia; yet he may be utterly careless and negligent in his wakefulness, the reverse of *watchful*; a person who is truly *watchful* must keep himself *wakeful* while on watch, in which case *wakeful* has something of mental quality. *Watchful,* from the English, and *vigilant,* from the Latin, are almost exact equivalents; but *vigilant* has a

somewhat sharper definiteness and somewhat more of a suggestion of volition; one may be habitually *watchful*; one is *vigilant* of set purpose and for direct cause. See ALERT. *Antonyms*: careless, drowsy, dull, heedless, inattentive, incautious, inconsiderate, neglectful, negligent, oblivious, thoughtless, unwary.

vig·i·lan·te (vij′ə·lan′tē) *n.* One who belongs to a vigilance committee. Also **vigilance man.** [<Sp., vigilant <L *vigilans* VIGILANT]

vi·gnette (vin·yet′) *n.* **1** Originally, a running ornament of leaves and tendrils, as in Gothic architecture. **2** A decorative or illustrative design placed on or before the title page of a book, at the end or beginning of a chapter, etc.; also, in medieval manuscript, an ornamented capital letter. **3** An engraving, photograph, or the like, having a background that shades off gradually. **4** A word-picture which delineates something subtly and delicately. — *v.t.* **·gnet·ted, ·gnet·ting** **1** To make with a gradually shaded background or border, as a photograph. **2** To ornament with vignettes. [<F, dim. of *vigne* a vine]

vi·gnet·ter (vin·yet′ər) *n.* **1** A device, as a shaded paper with an oval hole in the center, used by photographers in printing vignettes. **2** One who makes vignettes: also **vi·gnet′tist.**

Vi·gno·la (vē·nyō′lä), **Giacomo da** See BAROZZI.

Vi·gny (vē·nyē′), **Comte Alfred Victor de,** 1799–1863, French poet, dramatist, and novelist.

Vi·go (vē′gō, *Sp.* bē′gō) A port of Pontevedra province, on **Vigo Bay**, an inlet of the Atlantic in NW Spain (18 miles long, 1/2 to 10 miles wide).

vig·or (vig′ər) *n.* **1** Active strength or force, physical or mental. **2** Vital or natural power, as in a healthy animal or plant. **3** Forcible exertion of strength; energy; intensity. **4** Legal force; validity. Also *Brit.* **vig′our.** — **in vigor** *Law* In operation; effective. [<AF *vigur, vigour,* OF *vigor* <L < *vigere* be lively, thrive]

vi·go·ro·so (vē′gō·rō′sō) *adj. Music* Vigorous; energetic: a direction. [<Ital.]

vig·or·ous (vig′ər·əs) *adj.* **1** Full of physical or mental vigor; robust. **2** Marked by or accompanied by vigor; performed or done with vigor; showing vigor; energetic. See synonyms under ACTIVE, FRESH, HEALTHY, POWERFUL, STRONG, VIVID. — **vig′or·ous·ly** *adv.* — **vig′or·ous·ness** *n.*

Vii·pu·ri (vē′po͝o·rē) The Finnish name for VYBORG.

Vi·ja·ya·na·gar (vij′ə·yə·nug′ər) **1** A former princely state in the Rajputana States, India; since 1949 a part of Bombay State; 135 square miles. **2** A village of northern Bombay State, India, formerly capital of Vijayanagar state.

vi·king (vī′king) *n.* One of the Scandinavian warriors who harried the coasts of Europe from the eighth to the tenth centuries; a pirate; sea rover. Also **Vi′king.** [<ON *víkingr* a pirate, ? <OE and Frisian *wicing* < *wīc* a camp <L *vicus* a village]

Vi·la (vē′lə) The capital of the New Hebrides condominium, on Efate.

vi·la·yet (vē′lä·yet′) *n.* An administrative division of Turkey. [<Turkish *vilāyet* <Arabic *wilāyat* < *wāli* a governor]

vile (vīl) *adj.* **vil·er, vil·est** **1** Morally base, despicable, or loathsome; shamefully wicked; sinful; corrupt; filthy; disgusting. **2** Of little worth or account; mean. **3** Objectionable in any way; disagreeable: a general term of derogation. See synonyms under BAD[1], BASE[2], BRUTISH, COMMON, CRIMINAL, IMMORAL, INFAMOUS, SINFUL, VULGAR. [<AF, OF, fem. of *vil* <L *vilis* cheap] — **vile′ly** *adv.* — **vile′ness** *n.*

vil·i·a·co (vil′ē·ä′kō) *n. Obs.* A villain; scoundrel. [<Ital., ult. <L *vilis* cheap]

vil·i·fy (vil′ə·fī) *v.t.* **·fied, ·fy·ing** **1** To speak of as vile; defame; slander; traduce. **2** To make base or vile; degrade. [<LL *vilificare* <L *vilis* cheap + *facere* make] — **vil′i·fi·ca′tion** (-fə·kā′shən) *n.* — **vil′i·fi′er** *n.*

vil·i·pend (vil′ə·pend) *v.t.* **1** To think or speak of disparagingly; depreciate. **2** To vilify; defame. [<OF *vilipender* <L *vilipendere* < *vilis* cheap + *pendere* weight]

vill (vil) *n.* In old English law, a village; hamlet; township; also, a manor. [<AF *vill,* OF *vile, ville* a country house, village <L *villa.* See VILLA.]

vil·la (vil′ə) *n.* Originally, a country house with some suggestion of opulence; now, a suburban or rural residence. See synonyms under HOUSE. [<Ital. <L, a country house, farm, dim. of *vicus* a village]

Vil·la (vē′yä, *Sp.* bē′yä), **Francisco,** 1877–1923, Mexican revolutionary leader called "Pancho": real name *Doroteo Arango.*

Vil·la Bens (bē′lyä bāns) Capital of the Southern Protectorate of Morocco, Spanish West Africa, on the SW coast; until 1940, capital of Spanish Sahara: formerly *Cabo Jubi.*

Vil·la Cis·ne·ros (bē′lyä thēs·nā′rōs) The capital of Río de Oro, Spanish West Africa, on the central western coast near the tropic of Cancer.

vil·la·dom (vil′ə·dəm) *n. Brit.* Villas collectively; also, their occupants; the world of suburban villas.

vil·lage (vil′ij) *n.* **1** A collection of houses in a rural district, smaller than a town but larger than a hamlet, and usually arranged according to a regular plan. Villages may or may not be incorporated. **2** In some States, a municipality smaller than a city. Compare TOWNSHIP. **3** A collection of habitations of animals: a gopher *village.* **4** The inhabitants of a village, collectively; the villagers. **5** An encampment or community of North American Indians or Eskimos: permanent, or sometimes temporary, during a migration or for a season. — *adj.* Of, pertaining to, or characteristic of a village. [<OF <L *villaticum,* neut. sing. of *villaticus* pertaining to a villa < *villa* a villa]

village community An agricultural community with a simple organization, such as was found in early England, Germany, Russia, India, etc.; specifically, a free, self-dependent, communal group, regarded by many writers as the political unit out of which the modern state developed.

vil·lag·er (vil′ij·ər) *n.* One who lives in a village.

vil·lage·ry (vil′ij·rē) *n. Obs.* A collection of villages.

Vil·la·her·mo·sa (bē′yä·er·mō′sä) The capital of Tabasco state, SE Mexico.

vil·lain (vil′ən) *n.* **1** One who has committed or is disposed to commit any flagitious or disgraceful crime or series of crimes; a scoundrel; rogue: often used jocosely: He's a little *villain.* **2** A character in a novel, play, etc., who represents such a person and is the opponent of the hero or protagonist; also, an actor who regularly portrays such a character. **3** A villein. **4** *Obs.* A countryman; boor; clown; rustic. — *adj.* **1** Base; vile. **2** Of low birth; occupying a low station in life. ◆ Homophone: *villein.* [<AF, OF *vilein, vilain* a farm servant <LL *villanus* <L *villa* a villa]

vil·lain·age (vil′ən·ij) See VILLEINAGE.

vil·lain·ess (vil′ən·is) *n.* A female villain.

vil·lain·ous (vil′ən·əs) *adj.* **1** Having the nature of a villain. **2** Marked by extreme depravity. **3** *Colloq.* Very bad; disgusting; abominable: said of things: *villainous* words. See synonyms under BAD[1], INFAMOUS. — **vil′lain·ous·ly** *adv.* — **vil′lain·ous·ness** *n.*

vil·lain·y (vil′ən·ē) *n. pl.* **·lain·ies** **1** The quality or condition of being villainous; moral depravity. **2** Conduct befitting a villain; a villainous act; a crime. **3** *Obs.* Villeinage; servitude. **4** *Obs.* A low or miserable condition or state. See synonyms under ABOMINATION.

Vil·la-Lo·bos (vē′lə·lō′bo͝osh, -bo͞os), **Heitor,** 1887–1959, Brazilian composer and conductor.

vil·lan·age (vil′ən·ij) *n.* **1** Villeinage. **2** *Obs.* Villainy. [Var. of VILLEINAGE]

vil·la·nel·la (vil′ə·nel′ə, *Ital.* vēl′lä·nel′lä) *n. pl.* **·nel·le** (-nel′ē, *Ital.* -nel′lā) **1** A light, rustic part song, or dance accompanying it. **2** An early form of madrigal, popular in Naples during the sixteenth century. [<Ital., fem. dim. of *villano* <LL *villanus.* See VILLAIN.]

vil·la·nelle (vil′ə·nel′) *n.* A verse form, originally French, in 19 lines and 2 rimes, arranged in five tercets and a concluding quatrain. [<F <Ital. *villanella* a villanella]

Vil·lard (vi·lärd′), **Oswald Garrison,** 1872–1949, U.S. journalist.

Vil·lars (vē·lår′), **Duc Claude Louis Hector de,** 1653–1734, French marshal.

vil·lat·ic (vi·lat′ik) *adj.* Of or pertaining to a villa, farm, or village; rural. [<L *villaticus.* See VILLAGE.]

vil·lein (vil′ən) *n.* In the manorial system of feudal times, a member of any of the classes of freemen ranking below the thanes; more specifically, a free peasant ranking below a socman but above a cotter. By the 13th century the term *villein* was applied to a class of serfs who were regarded as freemen in respect to their legal relations with all persons except their lord, whose slaves they were. — *adj. Obs.* Relating to villeins; low-born. ◆ Homophone: *villain.* [<AF, OF *vilein, vilain.* See VILLAIN.]

vil·lein·age (vil′ən·ij) *n.* In feudal law, the tenure by which villeins held land; also, the status or condition of a villein: also spelled *villanage.* Also **vil′len·age.**

Ville·neuve (vēl·nœv′), **Pierre Charles Jean Baptiste Silvestre de,** 1763–1806, French admiral.

Vil·liers (vil′ərz), **George** See BUCKINGHAM.

vil·li·form (vil′ə·fôrm) *adj.* **1** Having the form of a villus. **2** Resembling nap, as of plush, as the teeth of fishes when numerous, small, and close together in velvety bands. [<NL *villiformis* <L *villus* tuft of hair + *forma* form]

Vil·lon (vē·yôṅ′), **François,** 1431–85?, French poet: real name *François de Montcorbier.*

vil·los·i·ty (vi·los′ə·tē) *n. pl.* **·ties** **1** The state or condition of being villous. **2** A villous surface or coating. **3** A villus.

vil·lous (vil′əs) *adj.* **1** Covered with short, soft hairs; nappy. **2** Covered with or having villi. Also **vil′lose** (-ōs). [<L *villosus* < *villus* tuft of hair] — **vil′lous·ly** *adv.*

vil·lus (vil′əs) *n. pl.* **vil·li** (vil′ī) **1** *Anat.* One of the short, hairlike processes found on certain membranes, as of the small intestine, where they aid in the digestive process. **2** *Bot.* One of the long, close, rather soft hairs on the surface of certain plants. [<L, a tuft of hair, shaggy hair, var. of *vellus* a fleece, wool]

Vil·na (vil′nə, *Russian* vel′nä) The capital of Lithuania, in the SE part: Polish *Wilno.* Also **Vil·ni·us** (vil′nē·əs), **Vil·nyus** (vil′nyəs).

Vi·lyui (vyē·lyo͞o′ē) A river in western Yakut Autonomous S.S.R., flowing 1,512 miles east to the Lena.

vim (vim) *n.* Force or vigor; energy; spirit. [<L, accusative of *vis* power]

vi·men (vī′mən) *n. pl.* **vim·i·na** (vim′ə·nə) *Bot.* A long, flexible shoot or branch. [<L *vimen, -inis* a twig < *viere* bend together, plait]

vim·i·nal (vim′ə·nəl) *adj. Rare* Pertaining to twigs; made of or producing twigs. [<L *vimen, -inis.* See VIMEN.]

Vim·i·nal (vim′ə·nəl) One of the seven hills on which ancient Rome was built.

vi·min·e·ous (vī·min′ē·əs) *adj.* **1** Having or resembling long, flexible shoots or branches. **2** Composed of twigs. [<L *vimineus* < *vimen, -inis.* See VIMEN.]

Vi·my (vē·mē′) A town in Pas-de-Calais department, northern France, near **Vimy Ridge,** scene of fierce fighting in World War I, 1915–1917.

vin (vaṅ) *n. French* Wine.

vin- Var. of VINI-.

vi·na (vē′nä) *n.* An East Indian musical instrument with seven steel strings stretched on a long, fretted fingerboard over two gourds. [<Hind. *vīṇā* <Skt.]

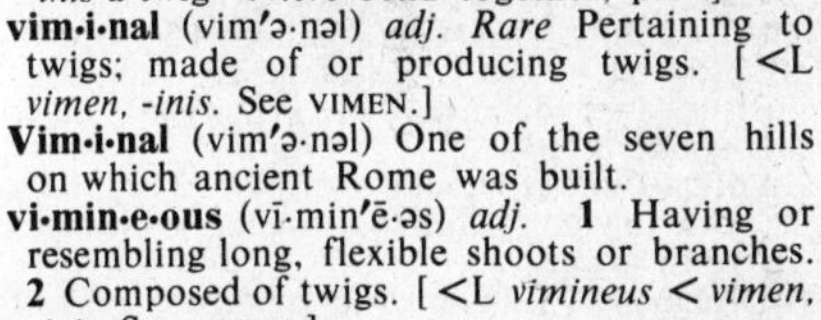

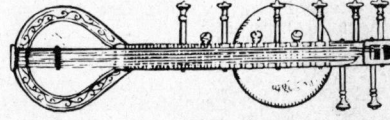

VINA OF BENARES

vi·na·ceous (vī·nā′shəs) *adj.* **1** Of or pertaining to wine or grapes. **2** Of the color of red wine. [<L *vinaceus* < *vinum* wine]

Vi·ña del Mar (bē′nyä thel mär′) A city on the Pacific in central Chile; a beach resort and industrial and agricultural center.

vin·ai·grette (vin′ə·gret′) *n.* **1** An ornamental

box or bottle, with a perforated top, for holding vinegar, smelling salts, or a pungent drug: also *vinegarette.* **2** Vinaigrette sauce. [<F, dim. of *vinaigre* vinegar]

vinaigrette sauce A vinegar and savory herb sauce served with fish and cold meats.

vi·nasse (vi·nas′) *n.* A residual product containing potassium salts, obtained from the winepress or from beets after the sugar has been extracted. [<F]

Vin·cennes (vin·senz′, *Fr.* vaṅ·sen′) **1** A city on the Wabash River in SW Indiana; site of French mission established in 1702. **2** A city just east of Paris, France.

Vin·cent (vin′sənt, *Fr.* vaṅ·säṅ′) A masculine personal name. Also *Ger.* **Vin·cenz** (vin′sents), *Ital.* **Vin·cen·zo** (vin·chen′tsō). [<L, conquering]

— **Vincent de Paul, Saint,** 1574?–1660, French Roman Catholic priest, founder of several charitable organizations.

Vin·cen·tian (vin·sen′shən) *n.* A member of a Roman Catholic order founded in 1625 by St. Vincent de Paul. See LAZARIST.

Vincent's infection *Pathol.* Trench mouth. Also **Vincent's angina, Vincent's disease.** [after J. H. *Vincent,* 1862–1950, French physician]

Vin·ci (vēn′chē), **Leonardo da** See LEONARDO DA VINCI.

vin·ci·ble (vin′sə·bəl) *adj.* That may be conquered or overcome; conquerable. [<L *vincibilis* < *vincere* conquer] — **vin′ci·bil′i·ty, vin′·ci·ble·ness** *n.*

vin·cu·lum (vingk′yə·ləm) *n. pl.* **·la** (-lə) **1** A bond of union. **2** *Anat.* A confining band of fascia. **3** *Math.* A straight line drawn over several algebraic terms, or a brace uniting them to show that all are to be operated on together. [<L < *vincire* bind]

Vin·dhya Pra·desh (vind′hyə prə·dāsh′) A former State of central India, incorporated in Madhya Pradesh State, 1956; 24,600 square miles; capital, Rewa.

Vin·dhya Range (vind′hyə) A chain of hills in central India; highest point, 3,400 feet.

vin·di·ca·ble (vin′də·kə·bəl) *adj.* That may be vindicated; justifiable.

vin·di·cate (vin′də·kāt) *v.t.* **·cat·ed, ·cat·ing** **1** To clear of accusation, censure, suspicion, etc. **2** To support or maintain, as a right or claim, against denial, opposition, etc. **3** To serve to justify. **4** *Rare* To lay claim to. **5** *Obs.* To avenge; punish. **6** *Obs.* To set free; rescue. See synonyms under AVENGE, JUSTIFY. [<L *vindicatus,* pp. of *vindicare* avenge, claim] — **vin′di·ca′tor** *n.*

vin·di·ca·tion (vin′də·kā′shən) *n.* The act of vindicating, or the state of being vindicated; justification; defense. See synonyms under APOLOGY, DEFENSE.

vin·di·ca·tive (vin′də·kā′tiv) *adj.* That contributes to vindication; that vindicates or serves to vindicate.

vin·di·ca·to·ry (vin′də·kə·tôr′ē, -tō′rē) *adj.* **1** Bringing to vindication; justificatory. **2** Punitive; avenging.

vin·dic·tive (vin·dik′tiv) *adj.* **1** Having a revengeful spirit; of a revengeful character. **2** *Obs.* Punitive. [<L *vindicta* a revenge] — **vin·dic′tive·ly** *adv.* — **vin·dic′tive·ness** *n.*

vine (vīn) *n.* **1** Any of a large and widely distributed group of plants having a slender, weak stem that may clasp or twine about a support by means of tendrils, leaf petioles, etc. **2** The stem itself. **3** A grapevine. [<OF *vigne, vine* <L *vinea* vineyard < *vinum* wine]

vine·dress·er (vīn′dres′ər) *n.* One who trims or prunes grapevines.

vin·e·gar (vin′ə·gər) *n.* **1** An acid liquid obtained by the acetous fermentation of alcoholic liquids, as cider, beer, wine, etc., and used as a condiment and preservative. **2** *Med.* A preparation of dilute acetic acid. **3** Anything metaphorically sour or soured, as a face; acerbity, as of speech. [<OF *vyn egre, vinaigre* < *vin* wine (<L *vinum*) + *aigre, egre* sour <L *acer* sharp] — **vin′e·gar·ish** *adj.*

vinegar eel A small nematode worm (*Anguillula aceti*) common in vinegar, sour paste, and similar fermenting liquids. Also **vinegar worm.**

vin·e·gar·ette (vin′ə·gə·ret′) See VINAIGRETTE (def. 1).

vinegar fly A fruit fly (def. 2).

vin·e·gar·roon (vin′ə·gə·rōōn′) *n.* The whip-tailed scorpion (*Mastigoproctus giganteus*) of the SW United States and Mexico, so called from its odor when alarmed: erroneously supposed to be venomous. Also **vin′e·ge·rone′** (-rōn′). [<Sp. *vinagre* vinegar < *vino* wine (<L *vinum*) + *agrio* sour <L *acer* sharp]

vin·e·gar·y (vin′ə·gər·ē) *adj.* **1** Being like or suggestive of vinegar; sour; acid. **2** Crabbed; of a sour disposition.

Vine·land (vīn′lənd) See VINLAND.

vin·er·y (vī′nər·ē) *n. pl.* **·er·ies** **1** A greenhouse for grapes; grapery. **2** Vines in general.

vine·yard (vin′yərd) *n.* **1** A large collection of cultivated grapevines. **2** Figuratively, a field for labor, especially spiritual culture or labor. [Earlier *wineyard,* OE *wīngeard;* infl. in form by VINE]

vine·yard·ist (vin′yər·dist) *n.* One who grows or cultivates grapevines.

vingt-et-un (vaṅ·tā·œṅ′) *n.* A game of cards played with a full pack, the object being to draw cards on which the aggregate number of spots shall reach as near as possible to but not exceed 21. Also called *twenty-one, blackjack.* [<F, twenty-one]

vini- *combining form* Wine; of or pertaining to wine or to wine grapes: *viniculture, viniferous:* also, before vowels, *vin-.* Also **vino-.** [<L *vinum* wine]

vi·nic (vī′nik, vin′ik) *adj.* Of, pertaining to, or derived from wine: *vinic* alcohol. [<L *vinum* wine]

vin·i·cul·ture (vin′ə·kul′chər) *n.* The cultivation of grapes for winemaking. — **vin′i·cul′tur·al** *adj.*

vin·i·cul·tur·ist (vin′ə·kul′chər·ist) *n.* One engaged in viniculture.

vi·nif·er·ous (vī·nif′ər·əs) *adj.* Producing wine. [<VINI- + -FEROUS]

vin·i·fi·ca·tor (vin′ə·fə·kā′tər) *n.* An apparatus for receiving and condensing the vapor of alcohol that rises from the fermenting must during the making of wine. [<VINI- + L *-ficator* a maker < *facere* make]

Vin·land (vin′lənd) A name given to part of the coast of North America by Norse voyagers: also *Vineland.*

Vin·ni·tsa (vin′it·sə, *Russian* vyēn′nyē·tsə) A city on the Bug river in SW central Ukrainian S.S.R.

Vi·no·gra·doff (vē′nə·grä′dôf), **Sir Paul Gavrilovich,** 1854–1925, Russian jurist and medieval historian, active in England.

vi·nom·e·ter (vi·nom′ə·tər, vī-) *n.* A hydrometer for measuring the percentage of alcohol in wine. [<VINO- + -METER]

vin or·di·naire (vaṅ ôr·dē·nâr′) *French* A cheap wine; literally, ordinary wine.

vi·nos·i·ty (vī·nos′ə·tē) *n.* **1** The state or quality of being vinous. **2** The general character of a wine, including the bouquet, flavor, body, etc. **3** Addiction to or fondness for wine. [<LL *vinositas, -tatis* <L *vinosus* VINOUS]

vi·nous (vī′nəs) *adj.* **1** Pertaining to, characteristic of, or having the qualities of wine. **2** Caused by, affected by, or addicted to wine. **3** Wine-colored. [<L *vinosus* < *vinum* wine]

Vin·son (vin′sən), **Fred M.,** 1890–1953, U.S. administrator and jurist; chief justice of the United States 1946–53.

vin·tage (vin′tij) *n.* **1** The yield of a vineyard or wine-growing district for one season. **2** The visible fruit of vineyards. **3** The harvesting of a vineyard and the first steps in the making of wine. **4** Wine, especially wine of high quality. **5** The year or the region in which a particular wine is produced. **6** *Colloq.* The type or kind current or appropriate at a particular time or in a particular season of the past: a joke of ancient *vintage.* [<AF *vintage,* alter. of *vindage, vendage,* OF *vendange* <L *vindemia* < *vinum* wine + *demere* remove < *de-* off + *emere* take; infl. in form by *vintner*]

vin·tag·er (vin′tij·ər) *n.* A harvester of grapes.

vintage wine Wine of an exceptionally good year, especially a dated champagne or port.

vint·ner (vint′nər) *n.* A wine merchant, especially at wholesale. [<OF *vinetier, vinotier* < *vinot,* dim. of *vin* wine <L *vinum*]

vin·y (vī′nē) *adj.* Pertaining to, like, of, full of, or yielding vines.

vi·nyl (vī′nəl) *n. Chem.* The univalent radical, CH_2:CH, derived from ethylene, especially when used in organic synthesis. [<L *vinum* wine + -YL]

vinyl acetate *Chem.* A colorless liquid, $C_4H_6O_2$, used as a starting point in the synthesis of various resins and plastics.

vinyl alcohol *Chem.* A hypothetical unstable alcohol, C_2H_4O, derived from acetylene.

vinyl chloride *Chem.* A compound of vinyl and chlorine, C_2H_3Cl, used in the production of synthetic fibers.

vinyl polymer *Chem.* Any of a class of organic compounds obtained by the polymerization of vinyl compounds.

vi·ol (vī′əl) *n.* **1** Any member of a family of stringed musical instruments, predecessors of the violin family, originating in the later Middle Ages and passing out of use in the 18th century, having usually six strings, and played with a bow. **2** A stringed instrument of the violin class. See BASS VIOL. [Earlier *vielle* <AF, OF <Med. L *vidula, vitula* <Gmc.; infl. in form by OF *viole*]

vi·o·la (vē·ō′lə, vī-; *Ital.* vyō′lä) *n.* **1** A four-stringed musical instrument of the violin family, somewhat larger than the violin, and tuned a fifth lower, with a graver and less brilliant tone. Its four strings are tuned in fifths. **2** A medieval viol. **3** An organ stop of eight-foot length and tone, producing stringlike tones. [<Ital., orig. a viol <Med. L *vidula* <Gmc.]

Vi·o·la (vī′ō·lə, vē′-, vī·ō′lə) A feminine personal name. Also **Vi·o·lan·te** (*Pg.* vē′ō·läṅ′tə, *Sp.* bē′ō·län′tā), *Ger.* **Vi·o·le** (vē·ō′lə). [<L, a violet]

— **Viola** The heroine in Shakespeare's *Twelfth Night.*

vi·o·la·ble (vī′ə·lə·bəl) *adj.* That may be violated. [<L *violabilis* < *violare* VIOLATE] — **vi′o·la·ble·ness, vi′o·la·bil′i·ty** *n.* — **vi′o·la·bly** *adv.*

vi·o·la·ceous (vī′ə·lā′shəs) *adj.* **1** Having a violet hue. **2** *Bot.* Of or pertaining to the violet or the violet family (*Violaceae*) of herbs, shrubs, and trees. [<L *violaceus* < *viola* a violet]

vio·la da gam·ba (vyō′lä dä gäm′bä) **1** The bass of the viol family, held between the legs, and having a range and tone similar to those of the violoncello: also *bass viol.* **2** An organ stop producing tones akin to those of the viola da gamba and usually having an eight-foot length and tone. [<Ital., viola of the leg]

vi·o·late (vī′ə·lāt) *v.t.* **·lat·ed, ·lat·ing** **1** To break or infringe, as a law, oath, agreement, etc. **2** To treat irreverently; profane, as a holy place. **3** To break in upon; disturb. **4** To ravish; rape. **5** To do violence to; offend grossly; outrage. **6** *Obs.* To treat roughly; abuse. [<L *violatus,* pp. of *violare* use violence < *vis* force] — **vi′o·la′tor** *n.*

Synonyms: abuse, debauch, defile, deflower, desecrate, hurt, injure, outrage, pollute, profane, rape, ravish. See ABUSE, POLLUTE.

vi·o·la·tion (vī′ə·lā′shən) *n.* The act of violating, or the state of being violated.

vi·o·la·tive (vī′ə·lā′tiv) *adj.* Having a tendency to violate; violating; involving violation.

vi·o·lence (vī′ə·ləns) *n.* **1** The quality or state of being violent; intensity; fury; also, an instance of violent action. **2** Violent or unjust exercise of power; injury; outrage; desecration; profanation. **3** *Law* Physical force unlawfully exercised; an act tending to intimidate or overawe by causing apprehension of bodily injury. **4** The perversion or distortion of the meaning of a text, word, or the like; unjustified alteration of wording. [<AF, OF <L *violentia* < *violentus* violent]

Synonyms: acuteness, boisterousness, eagerness, fierceness, force, fury, impetuosity, injury, intensity, outrage, passion, poignancy, rage, severity, sharpness, vehemence, violation, wildness, wrath. See OUTRAGE. *Antonyms*: calmness, feebleness, forbearance, gentleness, meekness, mildness, patience, self-command, self-control, self-restraint.

vi·o·lent (vī′ə·lənt) *adj.* **1** Proceeding from or marked by great physical force or roughness; sudden; forcible. **2** Caused by or exhibiting intense emotional or mental excitement; passionate; impetuous; fierce. **3** Characterized by intensity of any kind; extreme: *violent* heat. **4** Marked by unjust exercise of force; harsh; severe: to take *violent* measures. **5** Resulting from external force or injury; not in the ordinary course of nature: a *violent* death. **6** Tending to

violent presumption   1403   **Virgin Islands**

pervert the meaning or sense: a *violent construction*. [<OF <L *violentus* < *vis* force] — **vi′o·lent·ly** *adv.*
Synonyms: acute, boisterous, fierce, forceful, frantic, frenzied, fuming, furious, immoderate, impetuous, intense, irate, mad, maniacal, outrageous, passionate, poignant, raging, raving, severe, sharp, tumultuous, turbulent, uncontrollable, ungovernable, vehement, wild. See FIERCE, HOT, IMMODERATE, TURBULENT.

violent presumption *Law* An inference based on evidence that is so strong as to be almost conclusive.

vi·o·les·cent (vī′ə·les′ənt) *adj.* Having a tinge of violet color. [<L *viola* a violet + -ESCENT]

vi·o·let (vī′ə·lit) *n.* **1** Any of a widely distributed genus (*Viola*) of herbaceous perennial herbs, bearing flowers typically of a purplish-blue color; especially, the common **garden violet** (*V. odorata*). The violet is the State flower of Illinois, New Jersey, Rhode Island, and Wisconsin. **2** Any of several similar plants: the dog's-tooth *violet*. **3** A color seen at the end of the spectrum, opposite the red and beyond the blue; also, a pigment of this color. — *adj.* Of the color of violet. [<OF *violette*, dim. of *viole* <L *viola* a violet]

violet rays High-frequency radiation from the violet end of the visible spectrum: distinguished from *ultraviolet*.

vi·o·lin (vī′ə·lin′) *n.* **1** A musical instrument having four strings and a sounding box of seasoned wood, and played by means of a bow; a fiddle. It is the treble member of the **violin family**, which includes the viola, violoncello, and double-bass, and is distinguished in its modern form by its fully molded belly and back. **2** A violinist, especially in an orchestra: He is second *violin*. [<Ital. *violino*, dim. of *viola* a viola]

VIOLIN
a. Scroll.
b. Peg box.
c. Peg.
d. Nut.
e. Fingerboard.
f. Neck plate.
g. Sound holes.
h. Bridge.
i. Tailpiece.
j. Chin rest.
k. Button.

vi·o·lin·ist (vī′ə·lin′ist) *n.* One who plays the violin.

vi·ol·ist (vī′ə·list) *n.* One who plays the viol or viola.

Viol·let-le-Duc (vyô·le′lə·dük′), **Eugène Emmanuel**, 1814-79, French architect and archeologist.

vi·o·lon·cel·list (vē′ə·lən·chel′ist) *n.* One who plays the violoncello: usually abbreviated to *cellist* or *'cellist*.

vi·o·lon·cel·lo (vē′ə·lən·chel′ō) *n. pl.* **·los** A bass instrument of the violin family, having four strings tuned an octave lower than the viola, and held between the performer's knees when played: commonly called *cello* or *'cello*. [<Ital., dim. of *violone* a bass viol, aug. of *viola*. See VIOLA.]

vi·o·lo·ne (vyō·lō′nā) *n.* **1** The double-bass of the viol family, playing an octave lower than the viola da gamba: the immediate ancestor of the modern double-bass, which replaces the true double-bass of the violin family. **2** An organ stop with stringlike tone quality, having a 16-foot length and tone. **3** A small-scaled organ stop of eight-foot length and tone. [<Ital., aug. of *viola* a viol. See VIOLA.]

vi·os·ter·ol (vī·os′tər·ol, -ōl) *n.* Irradiated ergosterol, a vitamin D preparation variously used in medicine. [<(ULTRA)VIO(LET) + (ERGO)STEROL]

vi·per (vī′pər) *n.* **1** Any of a family (*Viperidae*) of venomous Old World snakes, especially the common European viper or adder (*Vipera berus*), about two feet long and variously colored; also, the African puff viper, and the horned viper. **2** One of a family (*Crotalidae*) of typically American poisonous snakes, the **pit vipers**, including the rattlesnake, copperhead, and fer-de-lance, which are characterized by a small depression between the nostril and the eye. **3** Any poisonous or allegedly poisonous snake. **4** A venomous, malicious, treacherous, or spiteful person. **5** *U.S. Slang* A marihuana smoker. [<OF *vipere*, *vipre* <L *vipera*, contraction of *vivipara* < *vivus* living + *parere* bring forth] — **vi′per·ine** (vī′pər·in, -pə·rīn) *adj.* — **vi′per·ish** *adj.*

vi·per·ous (vī′pər·əs) *adj.* **1** Snakelike; viperine. **2** Venomous. — **vi′per·ous·ly** *adv.*

vi·per's-bu·gloss (vī′pərz·byoo′glôs, -glos) *n.* Blueweed.

vi·ra·gin·i·ty (vir′ə·jin′ə·tē) *n. Psychiatry* The assumption by a woman of male characteristics and reactions. [<L *virago*, -*inis* VIRAGO]

vi·ra·go (vi·rā′gō, vī-) *n. pl.* **·goes** or **·gos** **1** A turbulent woman; vixen. **2** *Obs.* A woman of extraordinary size and courage; a female warrior; Amazon. [<L, manlike woman < *vir* a man]

vi·ral (vī′rəl) *adj. Med.* Of, pertaining to, caused by, or resembling a virus.

Vir·chow (vir′kō), **Rudolf**, 1821-1902, German pathologist.

vir·e·lay (vir′ə·lā) *n.* A form of old French verse, arranged in any of various arbitrary orders; especially, a verse form having only two rimes throughout; also, a form in which each stanza has two rimes, one repeated from the preceding stanza and a new one that will be repeated in the next. Also *French* **vire·lai** (vēr′lē′). [<OF *virelai*, prob. alter. of *vireli*, *virli* a refrain of old dance songs]

vir·e·o (vir′ē·ō) *n. pl.* **·os** Any of various small, insectivorous birds (family *Vireonidae*), predominantly dull-green and grayish, which make slight, cup-shaped, pensile nests; a greenlet. The **red-eyed vireo** (*Vireo olivaceus*), the **yellow-throated vireo** (*V. flavifrons*), the **white-eyed vireo** (*V. griseus*), the **blue-headed** or **solitary vireo** (*V. solitarius*), and the **warbling vireo** (*V. gilvus*) are common in the United States. Many of the species are noted for their song. [<L, a kind of small bird, ? the greenfinch]

vir·e·o·nine (vir′ē·ə·nīn′, -nin) *adj.* Characteristic of or pertaining to a vireo and related birds. — *n.* A vireo or related bird. [<L *vireo*, -*onis*. See VIREO.]

vi·res·cence (vī·res′əns) *n.* **1** The state or condition of becoming green. **2** *Bot.* Abnormal assumption of green by the usually bright-colored organs of plants, as when petals become green like normal leaves.

vi·res·cent (vī·res′ənt) *adj.* Greenish or becoming green. [<L *virescens*, -*entis*, ppr. of *virescere* grow green < *virere* be green]

vir et ux·or (vir et uk′sôr) *Latin* Husband and wife.

vir·ga (vûr′gə) *n. Meteorol.* Drooping streamers or wisps of precipitation from clouds, usually of the altocumulus and altostratus types. [<L, twig, streak in the sky]

vir·gate[1] (vûr′git, -gāt) *adj.* **1** Long, straight, and slender like a wand. **2** *Bot.* Bearing or producing many small twigs. [<L *virga* a twig, rod]

vir·gate[2] (vûr′git, -gāt) *n.* An early English measure of land, varying greatly (15, 20, 24, 30, and sometimes 40 acres) in different parts of England. [<Med. L *virgata* (*terrae*) a virgate (of land) <L *virga* a rod]

Vir·gil (vûr′jəl), **Vir·gil·i·an** (vər·jil′ē·ən) See VERGIL, etc.

vir·gin (vûr′jin) *n.* **1** A person, especially a young woman, who has never had sexual intercourse; a maiden. **2** A chaste young girl or unmarried woman; a spinster. **3** *Eccl.* A member of a religious community who has taken a vow of chastity; a nun. **b** A chaste, unmarried woman honored for her piety or virtue: used as an epithet of saints: St. Cecilia, *virgin* and martyr. **4** Any female animal before its first copulation. **5** *Entomol.* A female insect producing fertile eggs by parthenogenesis. — *adj.* **1** Being a virgin. **2** Consisting of virgins: a *virgin* band. **3** Pertaining or suited to a virgin; chaste; maidenly. **4** Uncorrupted; pure; undefiled: *virgin* whiteness. **5** Not hitherto used, touched, tilled, or worked upon by man: *virgin* soil; *virgin* forest. **6** Not previously processed, manufactured, or acted upon; new: *virgin* rubber; *virgin* wool. **7** Obtained from the first pressing (of olives, nuts, etc.) without the use of heat: said of an oil. **8** *Metall.* Produced directly from ore, or at the primary smelting: *virgin* silver. **9** *Mining* Occurring in native form; unalloyed; unmixed: *virgin* gold. **10** First: a ship's *virgin* voyage. **11** Untrained; lacking experience or contact with: waters *virgin* of ships. **12** *Zool.* Parthenogenetic. [<OF *virgine* <L *virgo*, -*inis* a maiden]

Vir·gin (vûr′jin) **1** Mary, the mother of Jesus: usually with *the*: also, **the Virgin Mary**, **the Blessed Virgin**. **2** The constellation Virgo. See CONSTELLATION.

vir·gin·al[1] (vûr′jin·əl) *adj.* Related to, like, or suited to a virgin; pure; modest; maidenly. [<OF <L *virginalis* < *virgo*, -*inis* a virgin]

VIRGINAL—LATE 16TH CENTURY

vir·gin·al[2] (vûr′jin·əl) *n.* A legless keyboard musical instrument of the 16th and 17th centuries, predecessor of the harpsichord: often in the plural, sometimes called a **pair of virginals**. [<OF, <VIRGINAL[1]; ? so called from its use by young men and girls]

virgin birth 1 *Zool.* Parthenogenesis. **2** *Usually cap. Theol.* The doctrine that Jesus Christ was conceived by divine agency and born without impairment of the virginity of his mother Mary.

vir·gin·hood (vûr′jin·hŏŏd) *n.* Virginity.

Vir·gin·ia (vər·jin′yə) A middle Atlantic State of the United States; 40,815 square miles; capital, Richmond; entered the Union June 25, 1788, one of the original thirteen States; nickname, *Old Dominion*: abbr. *Va.*, *Virg.* Original name: *Commonwealth of Virginia*.

Virginia cowslip A smooth perennial herb (*Mertensia virginica*) of the eastern United States, with clusters of blue or purple tubular flowers. Also **Virginia bluebell**.

Virginia creeper A common American climbing vine (*Parthenocissus* or *Ampelopsis quinquefolia*) of the grape family, with compound toothed leaves, small green flowers, and inedible blue berries: also called *American ivy*, *woodbine*.

Virginia deer A large, graceful, white-tailed deer (*Odocoileus virginianus*), native in the eastern United States and as far west as the Great Plains.

Virginia dogwood The flowering dogwood: State flower of Virginia. See under DOG-WOOD.

Virginia Key Northernmost of the Florida Keys, one mile south of Miami Beach; about 2 miles long.

Vir·gin·ian (vər·jin′yən) *adj.* **1** Of, pertaining to, or from Virginia. **2** Of, pertaining to, or designating the language of certain Algonquian North American Indians of eastern Virginia, North Carolina, and Maryland, especially of the Powhatan confederacy, formerly dwelling on the James River, Virginia. — *n.* A native or citizen of Virginia.

Virginia nightingale The cardinal bird.

Virginia rail fence A worm fence; a stake-and-rider.

Virginia reel A country dance in which the performers stand in two parallel lines facing one another and perform various figures, usually at the direction of a caller.

Virginia truffle Tuckahoe.

Virginia trumpet flower The trumpet creeper.

Virgin Islands A group of islands in the West Indies, east of Puerto Rico; divided into the **Virgin Islands of the United States**, an unincorporated territory comprising the

add, āce, cāre, pälm; end, ēven; it, īce; odd, ōpen, ôrder; tŏŏk, pōōl; up, bûrn; ə = a in *above*, e in *sicken*, i in *clarity*, o in *melon*, u in *focus*; yōō = u in *fuse*; oi, oil; ou, pout; ch, check; g, go; ng, ring; th, thin; ᵺ, this; zh, vision. Foreign sounds à, œ, ü, kh, ṅ; and ◆: see page xx. < *from*; + *plus*; ? *possibly*.

islands of St. Thomas, St. John, and St. Croix, and adjacent islets, purchased from Denmark in 1917; 133 square miles; capital, Charlotte Amalie, on St. Thomas: formerly *Danish West Indies*; and the BRITISH VIRGIN ISLANDS; total area, 200 square miles.

vir·gin·i·ty (vər·jin′ə·tē) *n. pl.* **·ties** 1 The state of being a virgin; maidenhood; virginal chastity. 2 The state of being unsullied or unused.

vir·gin·i·um (vər·jin′ē·əm) *n.* Former name of an element now identified as francium. [from the State of Virginia]

Virgin Mary Mary, the mother of Jesus.

Virgin River A river in Utah, Arizona, and Nevada, flowing 200 miles SW to Lake Mead.

vir·gin's-bower (vûr′jinz·bou′ər) *n.* A species of clematis (*Clematis virginiana*) bearing white flowers in leafy panicles.

Vir·go (vûr′gō) 1 A zodiacal constellation south of Ursa Major and Boötes; the Virgin. See CONSTELLATION. 2 The sixth sign of the zodiac. See illustration under ZODIAC. [<L, a virgin]

vir·gu·late (vûr′gyə·lit, -lāt) *adj.* Diminutively virgate; like a small rod. [<L *virgula.* See VIRGULE.]

vir·gule (vûr′gyōōl) *n.* A slanting line (/) used to indicate a choice between two alternatives, as in the phrase *and/or.* See SOLIDUS. [<L *virgula*, dim. of *virga* a rod]

vir·i·des·cent (vir′ə·des′ənt) *adj.* Greenish, or becoming slightly green. [<LL *viridescens, -entis*, ppr. of *viridescere* become green <*viridis* green] — **vir′i·des′cence** *n.*

vi·rid·i·an (və·rid′ē·ən) *n.* A durable bluish-green pigment consisting of hydrated chromic oxide. [<L *viridis* green]

vi·rid·i·ty (və·rid′ə·tē) *n.* Fresh greenness, as of vegetation; verdure. [<L *viriditas, -tatis* greenness, verdure <*viridis* green]

vir·ile (vir′əl) *adj.* 1 Having the characteristics of manhood. 2 Having the vigor or strength of manhood; sturdy, intrepid, and forceful; masculine. 3 Capable of procreation. See synonyms under MASCULINE. [<OF <L *virilis* <*vir* a man]

vir·il·ism (vir′əl·iz′əm) *n.* 1 The appearance in a woman of secondary male sexual and physical characteristics. 2 Female hermaphroditism.

vi·ril·i·ty (və·ril′ə·tē) *n. pl.* **·ties** The state, character, or quality of being virile.

vi·rip·o·tent (və·rip′ə·tənt) *adj.* 1 Sexually mature. 2 Nubile. [<LL *viripotens, -entis* <L *vir, viri* a man + *potens, -entis* able, powerful]

virl (vûrl) *n. Scot.* A ring around a column; a band; a ferrule.

vi·rol·o·gy (və·rol′ə·jē, vī-) *n.* The study of viruses, especially in their relation to disease. [<*viro-* (<VIRUS) + -LOGY] — **vi·rol′o·gist** *n.*

Vir·ta·nen (vir′tä·nen), **Artturi Ilmari,** born 1895, Finnish biochemist.

vir·tu (vər·tōō′, vûr′tōō) *n.* 1 Rare, curious, or beautiful quality: generally in the phrase **objects** or **articles of virtu.** 2 A taste for such objects. 3 Such objects collectively. [<Ital. *virtù* <L *virtus.* See VIRTUE.]

vir·tu·al (vûr′chōō·əl) *adj.* 1 Being in effect, but not in form or appearance; having potency, validity, or essential qualities: opposed to *apparent* or *nominal.* 2 *Obs.* Potent; effective; energizing. [<Med. L *virtualis* <L *virtus.* See VIRTUE.] — **vir′tu·al′i·ty** (-al′ə·tē) *n.* — **vir′tu·al·ly** *adv.*

virtual focus See under FOCUS.

virtual image See under IMAGE.

vir·tue (vûr′chōō) *n.* 1 The disposition to conform to the law of right; moral excellence; rectitude. 2 The practice of moral duties and the abstinence from immorality and vice: a life devoted to *virtue.* 3 Sexual purity; chastity, especially in women. 4 A particular moral excellence, especially one of those considered to be of special importance and classified by Plato as the four **cardinal virtues** (justice, temperance, prudence, and fortitude), to which the Christian scholastic moralists added the three **theological virtues** (faith, hope, and charity or love). The latter are sometimes called the **supernatural** or **Christian virtues** and the former the **natural virtues,** and all seven are opposed to the Seven Deadly Sins. 5 Any admirable quality, merit, or accomplishment: Patience is a *virtue.* 6 Active quality; power; efficacy; especially, medical efficacy; potency. 7 The quality of manliness; strength; valor. 8 *pl.* The fifth of the nine orders of angels in the celestial hierarchy. — **by** (or **in**) **virtue of** By or through the fact, quality, force, or authority of. [<OF *vertu* <L *virtus* strength, bravery <*vir* man]

Synonyms: chastity, duty, excellence, faithfulness, goodness, honesty, honor, integrity, justice, morality, probity, purity, rectitude, righteousness, rightness, truth, uprightness, virtuousness, worth, worthiness. *Virtue* is *goodness* that is victorious through trial, perhaps through temptation and conflict. *Goodness* may be much less than *virtue*, as lacking the strength that comes from trial and conflict, or it may be more than *virtue*, as rising above the possibility of temptation and conflict. *Virtue* is human; we do not predicate it of God. *Morality* is conformity to the moral law in action, whether in matters concerning ourselves or others, whether with or without right principle. *Honesty* and *probity* are used especially of one's relations to his fellow men, *probity* being to *honesty* much what *virtue* is to *goodness*; *probity* is *honesty* tried and proved, especially in those things that are beyond the reach of legal requirement; above the commercial sense, *honesty* may be applied to the highest truthfulness of the soul to and with itself. *Integrity*, in the full sense, is moral wholeness; when used of contracts and dealings, it has reference to inherent character and principle, and denotes more than conventional *honesty*. *Purity* is freedom from all admixture, especially of that which debases. *Duty*, the rendering of what is due to any person or in any relation, is the fulfilment of moral obligation. *Rectitude* and *righteousness* denote conformity to the standard of right; *righteousness* is used especially in the religious sense. *Uprightness* refers especially to conduct. Compare INNOCENCE, JUSTICE, RELIGION. *Antonyms*: evil, vice, viciousness, wrong. See synonyms for SIN.

vir·tu·os·i·ty (vûr′chōō·os′ə·tē) *n. pl.* **·ties** 1 The state of being a virtuoso; the technical mastery of an art, as music. 2 A taste for the fine arts, especially the taste of a dilettante. 3 Virtuosi collectively.

vir·tu·o·so (vûr′chōō·ō′sō) *n. pl.* **·si** (-sē) or **·sos** 1 A master of technique, as a skilled musician; one who displays virtuosity. 2 A connoisseur; a collector or lover of curios or works of art. 3 *Obs.* A savant; a scientist; learned person. [<Ital., skilled, learned <LL *virtuosus* full of excellence <L *virtus.* See VIRTUE.]

vir·tu·ous (vûr′chōō·əs) *adj.* 1 Characterized by, exhibiting, or having the nature of virtue; morally pure and good; chaste: now said especially of women. 2 Potent; efficacious. — **vir′tu·ous·ly** *adv.* — **vir′tu·ous·ness** *n.*

Synonyms: blameless, chaste, correct, dutiful, equitable, estimable, excellent, exemplary, good, honest, just, pure, right, righteous, upright, worthy. See GOOD, INNOCENT, JUST, MODEST, MORAL, PURE. *Antonyms*: see synonyms for CRIMINAL, SINFUL.

vir·tu·te of·fi·ci·i (vər·tōō′tē ə·fish′ē·ī, vər·tyōō′tē) *Latin* By virtue of office.

vir·u·lence (vir′yə·ləns, vir′ə-) *n.* 1 The quality of being virulent. 2 Extreme bitterness or malignity. 3 The power of bacteria and other micro-organisms to overcome the resistance of the host.

vir·u·lent (vir′yə·lənt, vir′ə-) *adj.* 1 Manifesting or partaking of the nature of virus; exceedingly noxious. 2 Very bitter in enmity. 3 *Med.* Actively poisonous or infective; malignant. 4 Having or exhibiting virulence. See synonyms under BITTER[1], MALICIOUS. [<L *virulentus* full of poison <*virus* a poison] — **vir′u·lent·ly** *adv.*

vi·rus (vī′rəs) *n.* 1 Venom; snake poison. 2 Any virulent substance developed by morbid processes within an animal body, and capable of transmitting a specific disease, as smallpox: when inoculated in an attenuated form it is called a *vaccine.* 3 Any of a class of filter-passing, pathogenic agents, chiefly protein in composition but often reducible to crystalline form, and typically inert except when in contact with certain living cells: also **filtrable virus.** 4 Figuratively, a moral taint; a corrupting influence. 5 Bitterness of mind; acrimony; malice. [<L, poison, slime]

vis (vis) *n. pl.* **vi·res** (vī′rēz) *Latin* Force; potency.

vi·sa (vē′zə) *n.* 1 An official endorsement, as on a passport, certifying that it has been found correct and that the bearer may proceed. 2 A signature of approval, as by an authorized inspecting officer. — *v.t.* **·saed, ·sa·ing** 1 To put a visa on. 2 To give a visa to. Also spelled *visé.* [<F <L, fem. sing. pp. of *videre* see]

vis·age (viz′ij) *n.* The face or look of a person, or of an animal; distinctive aspect. [<OF <*vis* a face <L *visus* a look <*videre* see]

vis·aged (viz′ijd) *adj.* Having a visage of some character indicated.

vis·ard (viz′ərd) See VIZARD.

vis-à-vis (vē′zə·vē′, *Fr.* vē·zà·vē′) *n.* 1 One of two persons or things that face each other, as in dancing. 2 A seat having an S-shaped back so arranged that two persons can sit side by side, but facing in opposite directions. — *adj.* & *adv.* Face to face. [<F, face to face]

Vi·sa·yan (vē·sä′yən) *n.* 1 One of the native people of the Philippines, occupying the Visayan Islands and northern Mindanao. 2 The language of these people, belonging to the Indonesian subfamily of Austronesian languages. — *adj.* Of or pertaining to the Visayans or their language. Also spelled *Bisayan.*

Vi·sa·yan Islands (vē·sä′yən) A group of the central Philippines, comprising Bohol, Cebu, Leyte, Masbate, Negros, Panay, Samar, Romblon, and the islets adjacent to them; total 23,621 square miles: also *Bisayan Islands.* Also **Vi·sa′yas** (-yəs).

Visayan Sea A part of the Pacific in the central Philippines, bounded by the Visayan Islands.

Vis·by (vēs′bü) A port on western Gotland island, SE Sweden: German *Wisby.*

vis·ca·cha (vis·kä′chə) *n.* 1 A large burrowing rodent (*Lagostomus maximus*) of the South American pampas, related to the chinchilla, with three-toed hind feet. 2 An allied genus (*Lagidium*) of the Andes, resembling the gray squirrel but with large rabbitlike ears. [<Sp. <Quechuan *uiscacha*]

vis·cer·a (vis′ər·ə) *n. pl. sing.* **vis·cus** (vis′kəs) 1 *Anat.* The internal organs, especially those of the great cavities of the body, as the stomach, lungs, heart, etc. 2 Commonly, the intestines. [<L, pl. of *viscus, visceris* an internal organ]

vis·cer·al (vis′ər·əl) *adj.* 1 Pertaining to the viscera. 2 Abdominal.

Visch·er (fish′ər), **Peter,** 1455?–1529, German sculptor.

vis·cid (vis′id) *adj.* Sticky or adhesive; mucilaginous; viscous. See synonyms under ADHESIVE. [<LL *viscidus* <L *viscum* birdlime, mistletoe] — **vis′cid·ly** *adv.* — **vis′cid·ness** *n.*

vis·cid·i·ty (vi·sid′ə·tē) *n.* The quality or state of being viscid.

vis·coi·dal (vis·koid′l) *adj.* Somewhat viscid.

Vis·con·ti (vēs·kôn′tē) A Lombard family which ruled Milan from 1277 to 1447.

vis·cose (vis′kōs) *n.* A thick, honeylike substance produced by the action of caustic soda and carbon disulfide upon cellulose: an important source of rayon yarns and fabrics. — *adj.* 1 Viscous. 2 Of, pertaining to, containing, or made from viscose. [<LL *viscosus* VISCOUS]

viscose rayon Rayon formed from fibers composed of regenerated cotton or wood-pulp cellulose which has been coagulated or solidified from a solution of cellulose xanthate.

vis·co·sim·e·ter (vis′kə·sim′ə·tər) *n.* An apparatus for determining the viscosity of liquids. Also **vis·com·e·ter** (vis·kom′ə·tər). [<VISCOSI(TY) + -METER]

vis·cos·i·ty (vis·kos′ə·tē) *n. pl.* **·ties** 1 The state, quality, property, or degree of being viscous. 2 *Physics* That property of fluids by virtue of which they offer resistance to flow or to any change in the arrangement of their molecules. Compare POISE[2].

vis·count (vī′kount) *n.* 1 In England, a title of nobility between earl and baron. 2 In continental Europe, a title next below that of count; also, the son or younger brother of a count. 3 Formerly, a representative

viscountcy

or deputy of a count or earl in the government of a district; specifically, in English use, a sheriff. [<AF *viscounte*, OF *visconte* < *vis-* in place (<L *vice*) + *counte*, *conte* COUNT²]

vis·count·cy (vī'kount-sē) *n. pl.* **·cies** The rank, title, or dignity of a viscount. Also **vis'count·ship**, **vis'count·y**.

vis·count·ess (vī'koun-tis) *n.* The wife of a viscount, or a peeress holding the title in her own right.

vis·cous (vis'kəs) *adj.* 1 Glutinous; semifluid; sticky. 2 Imperfectly fluid, as warm tar. See synonyms under ADHESIVE. [<LL *viscosus* <L *viscum* birdlime, mistletoe] — **vis'cous·ly** *adv.* — **vis'cous·ness** *n.*

vis·cus (vis'kəs) Singular of VISCERA.

vise (vīs) *n.* A clamping device, usually of two jaws made to be closed together with a screw, lever, or the like, for grasping and holding a piece of work. — *v.t.* **vised**, **vis·ing** To hold, force, or squeeze in or as in a vise. ◆ Homophone: *vice*. Also spelled *vice*.
[<OF *vis* a screw <L *vitis* vine; with ref. to the spiral growth of vine tendrils]

vi·sé (vē'zā) See VISA.

Vish·nu (vish'nōō) In Hindu theology, the second god of the trinity (Brahma, Vishnu, and Siva), known as "the Preserver"; of his many incarnations the most famous is as Krishna.

vis·i·bil·i·ty (viz'ə-bil'ə-tē) *n. pl.* **·ties** 1 Condition, capability or degree of being seen. 2 *Meteorol.* The condition of the atmosphere as affecting the distance at which objects can be seen and identified. 3 *Physics* The ratio of the luminous flux of a given wavelength to the radiant energy producing it.

vis·i·ble (viz'ə-bəl) *adj.* 1 Perceivable by the eye; capable of being seen. 2 Apparent at sight; evident. 3 At hand; available; manifest. See VISIBLE SUPPLY. 4 Accessible to visitors; prepared or disposed to be seen or visited. 5 Constructed so that certain parts can be seen by the operator: a *visible* typewriter. See synonyms under EVIDENT, MANIFEST. [<OF <L *visibilis* <*visus*, pp. of *videre* see] — **vis'i·ble·ness** *n.* — **vis'i·bly** *adv.*

visible speech Phonetic symbols devised by Alexander Melville Bell to represent every possible utterance of the organs of speech.

visible supply The total of the known available supply of any commodity, as wheat in elevators and in shipment.

Vis·i·goth (viz'ə-goth) *n.* One of the western Goths, a Teutonic people that invaded the Roman Empire in the third and fourth centuries and settled in France and Spain. See OSTROGOTH. [<LL *Visigothus* <Gmc.; ? lit., the western Goths] — **Vis'i·goth'ic** *adj.*

vi·sion (vizh'ən) *n.* 1 The faculty or sense of sight, localized in the eye, which, with its receptors and associated organs, is normally adapted to receive the stimulus of radiant energy within a certain range of wavelengths. 2 That which is or has been seen; also, something or someone beautiful or delightful: She is a *vision* of loveliness. 3 A mental representation of or as of external objects or scenes, as in sleep; an apparition; dream; fantasy; specifically, an inspired revelation. 4 Some product of the fancy or imagination; an imaginary or unreal thing: *visions* of sugarplums. 5 The ability to anticipate and make provision for future events; foresight. 6 Insight; imagination. See synonyms under DREAM. — *v.t.* To see in or as in a vision. [<OF <L *visio*, *-onis* <*visus*, pp. of *videre* see]

vi·sion·al (vizh'ən-əl) *adj.* Of, pertaining to, or consisting of vision or a vision. — **vi'sion·al·ly** *adv.*

vi·sion·ar·y (vizh'ən-er'ē) *adj.* 1 Not founded on fact; imaginary; impracticable. 2 Affected by fantasies; dreamy; impractical. 3 Associated with apparitions, dreams, etc. See synonyms under FANCIFUL, IDEAL, IMAGINARY, ROMANTIC. — *n. pl.* **·ar·ies** 1 One who has visions. 2 A dreamer; an impractical schemer. — **vi'sion·ar'i·ness** *n.*

vis·it (viz'it) *v.t.* 1 To go or come to see (a person) from friendship, courtesy, on business, etc.; make a call on. 2 To go or come to (a place, etc.), as for transacting business or for touring: to *visit* the Louvre. 3 To be a guest of; stay with temporarily: I *visited* them for several days. 4 To go or come to so as to make official inspection or inquiry: to *visit* a military school. 5 To come upon or afflict. 6 To inflict punishment upon or for. 7 To comfort or bless: The Lord hath *visited* His people. — *v.i.* 8 To make a visit; pay a call or calls. 9 To inflict punishment or vengeance. See synonyms under AVENGE. — *n.* 1 The act of going or coming to see a person or thing, especially with some formality and with the intention of staying some time; a sojourn in a place or with a person; a call or stay. 2 *Colloq.* A talk or friendly chat. 3 An authoritative personal call for inspection and examination or discharge of an official or professional duty. — **right of visit** See RIGHT OF SEARCH. [<OF *visiter* <L *visitare* go to see, freq. of *visare* <*visus*, pp. of *videre* see]

vis·it·a·ble (viz'it-ə-bəl) *adj.* 1 Subject to visitation or punishment. 2 Agreeable to visitors, as a country or region. 3 Having a social position.

vis·it·ant (viz'ə-tənt) *n.* 1 A visitor; that which comes and goes or makes a transient appearance. 2 A migratory animal or bird at a particular region. 3 A visitor as if from another sphere; a supernatural being. — *adj.* Acting as a visitor; paying visits. [<MF <L *visitans*, *-antis*, ppr. of *visitare* VISIT]

vis·i·ta·tion (viz'ə-tā'shən) *n.* 1 The act or fact of visiting; a visit; also, the state or circumstance of being visited. 2 The visit of a bishop to his diocese; an official or authoritative inspection and examination of a foundation, institution, or establishment to set affairs to rights, correct abuses, enforce laws or rules, etc. 3 In Biblical and religious use, a visit of blessing or affliction: a blessed *visitation* from on high; a dreadful *visitation* of famine. 4 *Obs.* The purpose or object of a visit. 5 The resorting of birds or animals to unusual places. See synonyms under MISFORTUNE. — **vis'i·ta'tion·al** *adj.*

Vis·i·ta·tion (viz'ə-tā'shən) *n.* A religious festival held on July 2 in honor of the visit of the Virgin Mary to Elizabeth. Luke i 40.

vis·i·ta·to·ri·al (viz'ə-tə-tôr'ē-əl, -tō'rē-) *adj.* Of or pertaining to visitation; done under an official right of visitation. Also **vis'i·to'ri·al**.

visiting card A calling card.

vis·i·tor (viz'ə-tər) *n.* One who visits. Also **vis'it·er**.

Vis·la (vēs'lə) The Russian name for the VISTULA.

vis major (vis mā'jər) *Latin* 1 Irresistible or uncontrollable force; inevitable accident. 2 *Law* An unavoidable accident: in civil law, nearly the same as, but broader than, an act of God.

vis med·i·ca·trix na·tu·rae (vis med'ə-kā'triks nə-choor'ē) *Latin* The curative power of nature.

vi·sor (vī'zər, viz'ər) *n.* 1 A projecting piece on a cap shielding the eyes. 2 In ancient armor, the front piece of a helmet which protected the upper part of the face and could be raised. — *v.t.* To mask; cover with a visor. Also spelled *vizor.* [<AF *viser*, OF *visiere* <*vis* face. See VISAGE.]

vis·ta (vis'tə) *n.* 1 A view or prospect, as along an avenue; an outlook. 2 A mental view embracing a series of events. [<Ital. <L *visus*, pp. of *videre* see]

Vis·tu·la (vis'chōō-lə) The longest river in Poland, flowing 678 miles north from the Carpathian Mountains to the **Vistula Lagoon** (German *Frisches Haff*), a coastal inlet (332 square miles; about 60 miles long, 7 to 11 miles wide) of the Gulf of Danzig: German *Weichsel*, Russian *Visla*.

vis·u·al (vizh'ōō-əl) *adj.* 1 Pertaining to, resulting from, or serving the sense of sight; ocular. 2 Perceptible by sight; visible. 3 Optical: the *visual* focus of a lens. 4 Produced or induced by mental images: a *visual* conception. [<MF <LL *visualis* <L *visus* a sight <*videre* see]

visual field The total area visible to the unmoving eye or eyes at any given moment.

vis·u·al·i·ty (vizh'ōō-al'ə-tē) *n. pl.* **·ties** 1 The quality or condition of being visual; mental visibility. 2 That which is or may be perceived by or as by vision.

vis·u·al·ize (vizh'ōō-əl-īz') *v.* **·ized**, **·iz·ing** *v.t.* To form a mental image of; picture in the mind. — *v.i.* To form mental images. Also *Brit.* **vis'u·al·ise'**. — **vis'u·al·ism** *n.* — **vis'u·al·i·za'tion** *n.*

vis·u·al·iz·er (vizh'ōō-əl-ī'zər) *n.* 1 One who visualizes. 2 One whose mental images are formed chiefly by visualization: also **vis'u·al·ist** (-ist).

visual purple *Biochem.* A complex reddish-purple protein present in the rods of the vertebrate retina; it is an important factor in the process of vision, especially at night: also called *rhodopsin*.

visual yellow *Biochem.* The pigmented protein into which visual purple is changed by the action of light: heat acts upon it to produce vitamin A: also called *retinene*.

vis vi·tae (vis vī'tē) *Latin* The force of life; vitality. Also **vis vi·ta·lis** (vī-tā'lis).

vi·ta·ceous (vī-tā'shəs) *adj.* *Bot.* Designating or belonging to a family (*Vitaceae*) of mostly woody and climbing vines, the grape family, having alternate leaves, inconspicuous greenish flowers in clusters, and berry-like fruit. [<NL, family name <L *vitis* vine]

vi·tal (vīt'l) *adj.* 1 Pertaining to life. 2 Essential to or supporting life. 3 Affecting life; fatal to life: a *vital* error or wound. 4 Necessary to existence or continuance; necessary; essential; life-sustaining. 5 Relating to the facts of life, as births, deaths, etc.: *vital* statistics. [<OF <L *vitalis* <*vita* life] — **vi'tal·ly** *adv.*

vital force A form of energy regarded as acting independently of all physical and chemical forces in the causation of life and in the development of living phenomena. Also **vital principle**.

vi·tal·ism (vīt'l·iz'əm) *n.* 1 The doctrine that life had its origin and support in some principle that is neither material nor organic. 2 *Philos.* A movement represented by Henri Bergson, which upholds the principles of freedom and self-determination and the creative power of the human consciousness. It places intuition above intellect, and considers the universe as living and self-evolving without predestined development or end. Compare BERGSONISM, ÉLAN VITAL. 3 *Biol.* The theory that organic growth is due to forces that operate only in living organisms and differ in kind from the chemical and physical forces at work in the inorganic world: opposed to *mechanism.* — **vi'tal·ist** *n.* — **vi'tal·is'tic** *adj.*

vi·tal·i·ty (vī-tal'ə-tē) *n.* 1 The state of being vital; vital force; the principle of life; animation; life. 2 Power of continuing in force or effect. See synonyms under LIFE.

vi·tal·ize (vīt'l-īz) *v.t.* **·ized**, **·iz·ing** To make vital; endow with life or animation; animate. — **vi'tal·i·za'tion** *n.* — **vi'tal·iz'er** *n.*

vi·tals (vīt'lz) *n. pl.* 1 The parts necessary to life, as the heart and brain: used also figuratively. 2 The parts essential to the health, maintenance, etc., of anything.

vital statistics Quantitative data relating to certain aspects and conditions of human life, especially in relation to large population groups.

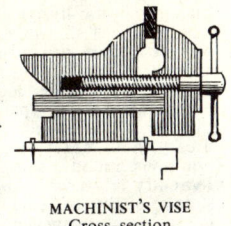

VISHNU

MACHINIST'S VISE Cross-section.

---

add, āce, câre, pälm; end, ēven; it, īce; odd, ōpen, ôrder; tōōk, pōōl; up, bûrn; ə = a in *above*, e in *sicken*, i in *clarity*, o in *melon*, u in *focus*; yōō = u in *fuse*; oi, oil; ou, pout; ch, check; g, go; ng, ring; th, thin; ŧħ, this; zh, vision. Foreign sounds á, œ, ü, kh, ń; and ◆: see page xx. < from; + plus; ? possibly.

vi·ta·mer (vī′tə·mər) *n. Biochem.* Any dietary factor or other substance that possesses the activity of a given vitamin or acts to counteract a vitamin deficiency, as carotenoid in human subjects. [<VITA(MIN) + Gk. *meros* a part]

vi·ta·min (vī′tə·min) *n. Biochem.* Any of a group of complex organic substances found in minute quantities in most natural foodstuffs, and closely associated with the maintenance of normal physiological functions in man and animals. Numerous forms have been isolated and described under special names. Also **vi′ta·mine** (-mēn, -min). [<L *vita* life + AMINE] — **vi′ta·min′ic** *adj.*

vitamin A The fat-soluble vitamin occurring in green and yellow vegetables, dairy products, liver oil, and fish oil: it prevents atrophy of epithelial tissue and protects against night blindness.

vitamin B complex A group of water-soluble vitamins widely distributed in plants and animals, most members of which have special names.

vitamin B_1 Thiamine.

vitamin B_2 Riboflavin.

vitamin B_6 Pyridoxine.

vitamin B_{12} A vitamin extracted from liver and believed to be protective against pernicious anemia.

vitamin C Ascorbic acid.

vitamin D The anti-rachitic vitamin occurring chiefly in fish-liver oils. Many closely related forms are known.

vitamin D_1 An impure mixture of calciferol and lumisterol.

vitamin D_2 Calciferol.

vitamin D_3 A form of vitamin D_2 found principally in fish-liver oils.

vitamin E The anti-sterility vitamin, found in whole grain cereals, seeds of legumes, corn and cottonseed oils, egg yolks, meat, and milk: known to be a mixture of alpha-, beta-, and gamma-tocopherols.

vitamin G Riboflavin.

vitamin H Biotin.

vitamin K_1 A vitamin, found in green leafy vegetables, which promotes the clotting of blood: also called *phylloquinone.*

vitamin K_2 A form of vitamin K_1 prepared from fishmeal.

vi·ta·min·ol·o·gy (vī′tə·min·ol′ə·jē) *n.* The scientific study of vitamins. [<VITAMIN + -(O)LOGY]

vitamin P The factor present in citrus juices along with vitamin C; citrin. It promotes the normal permeability of capillary walls.

vi·ta·scope (vī′tə·skōp) *n.* A device by which pictures taken by the kinetoscope are enlarged and exhibited on a screen. [<L *vita* life + -SCOPE]

Vi·tebsk (vē′tepsk) A city on the Western Dvina river in NE Belorussian S.S.R.

vi·tel·lin (vi·tel′in, vī-) *n. Biochem.* A phosphoprotein occurring in the yolk of eggs. [<VITELL(US) + -IN]

vi·tel·line (vi·tel′in, vī-) *adj.* **1** Of or pertaining to the food yolk of an egg. **2** Of a dull yellow, approaching red; of the color of the yolk of eggs. — *n.* The yolk of an egg. [<Med. L *vitellinus* <L *vitellus* VITELLUS]

vi·tel·lus (vi·tel′əs, vī-) *n.* The egg yolk. [<L, orig. dim. of *vitulus* a calf]

vi·tesse (vē·tes′) *n. French* Speed: used especially in the phrases **grande vitesse** (gränd), fast express, and **pe·tite vitesse** (pə·tēt′), ordinary express, or freight, etc.

vi·ti·ate (vish′ē·āt) *v.t.* **·at·ed, ·at·ing** **1** To impair the use or value of; spoil. **2** To debase or corrupt. **3** To render legally ineffective: Fraud *vitiates* a contract. See synonyms under CORRUPT, DEFILE[1], POLLUTE. [<L *vitiatus,* pp. of *vitiare* <*vitium* a fault] — **vi·ti·a·ble** (vish′ē·ə·bəl) *adj.* — **vi′ti·a′tion** *n.* — **vi′ti·a′·tor** *n.*

vi·ti·at·ed (vish′ē·ā′tid) *adj.* Contaminated; rendered defective; invalidated.

vit·i·cul·ture (vit′ə·kul′chər, vī′tə-) *n.* **1** The science and art of grape-growing. **2** The culture of the vine. [<L *vitis* a vine + CULTURE] — **vit′i·cul′tur·al** *adj.* — **vit′i·cul′tur·er, vit′i·cul′tur·ist** *n.*

Vi·ti Le·vu (vē′tē lā′vōō) The largest of the Fiji Islands; 4,010 square miles; capital, Suva.

vit·i·li·go (vit′ə·lī′gō) *n. Pathol.* A skin disease characterized by a partial privation of color in spots, with a tendency to increase in size; piebald skin; leukoderma. [<L *vitiligo* tetter <*vitium* a fault]

Vi·tim (vi·tēm′, *Russian* vē·tyēm′) A river in NE Buryat-Mongol Autonomous S.S.R., flowing 1,132 miles to the Lena.

Vi·to·ri·a (vē·tôr′ē·ə, *Sp.* bē·tō′ryä) A city in north central Spain, capital of a Basque province.

Vi·tó·ria (vē·tô′ryə) The capital of Espírito Santo state, SE central Brazil; a port on the Altantic coast.

vit·rain (vit′rān) *n.* A variety of bituminous coal having a vitreous appearance and a structure characterized by narrow, compact, crystalline bands. [<L *vitrum* glass, on analogy with *fusain* (def. 2)]

vit·re·ous (vit′rē·əs) *adj.* **1** Pertaining to glass; glassy. **2** Obtained from glass. **3** Resembling glass in some property or properties; vitriform. **4** Pertaining to the vitreous humor. [<L *vitreus* <*vitrum* glass] — **vit′re·os′i·ty** (-os′ə·tē), **vit′re·ous·ness** *n.*

vitreous electricity Electricity generated by rubbing glass with silk: regarded as positive.

vitreous humor *Anat.* The transparent jellylike tissue that fills the ball of the eye and is enclosed by the hyaloid membrane. Also **vitreous body.**

vi·tres·cence (vi·tres′əns) *n.* The state of becoming vitreous.

vi·tres·cent (vi·tres′ənt) *adj.* **1** Capable of being turned into glass. **2** Tending to become glass. [<L *vitrum* glass + -ESCENT]

vi·tres·ci·ble (vi·tres′ə·bəl) *adj.* Capable of forming a viscous, glasslike layer under the action of great heat, as certain crushed minerals. [<VITRESC(ENT) + -IBLE]

vitri- *combining form* Glass; of or pertaining to glass; crystalline: *vitriform.* Also, before vowels, **vitr-.** [<L *vitrum* glass]

vit·ric (vit′rik) *adj.* Pertaining to or like glass. Compare CERAMIC.

vit·ri·fac·ture (vit′rə·fak′chər) *n.* The manufacture of vitreous or vitrified wares, as glass. [<VITRI- + (MANU)FACTURE]

vit·ri·fi·ca·tion (vit′rə·fə·kā′shən) *n.* **1** The process of vitrifying. **2** The state of being vitrified. **3** A vitrified object. Also **vit′ri·fac′tion** (-fak′shən).

vit·ri·form (vit′rə·fôrm) *adj.* Having a glassy appearance; glasslike.

vit·ri·fy (vit′rə·fī) *v.t.* & *v.i.* **·fied, ·fy·ing** To change into glass or a vitreous substance; make or become vitreous. [<MF *vitrifier* <L *vitrum* glass + *facere* make] — **vit′ri·fi′a·ble** *adj.*

vit·rine (vit′rin) *n.* A glass showcase for art objects. [<F <*vitre* glass <L *vitrum*]

vit·ri·ol (vit′rē·ōl, -əl) *n.* **1** *Chem.* **a** Sulfuric acid, originally made from green vitriol: more commonly called **oil of vitriol. b** Any sulfate of a heavy metal, as **green vitriol,** from iron; **blue vitriol,** from copper; **white vitriol,** from zinc. **2** Anything sharp or caustic, as sarcasm. — *v.t.* **·oled** or **·olled, ·ol·ing** or **·ol·ling** **1** To injure (a person) with vitriol. **2** To subject (anything) to the agency of vitriol. [<OF <Med. L *vitriolum* <L *vitrum* glass; so called because of its glassy appearance]

vit·ri·ol·ic (vit′rē·ol′ik) *adj.* **1** Derived from a vitriol. **2** Corrosive, burning, or caustic.

vit·ri·ol·ize (vit′rē·əl·īz′) *v.t.* **·ized, ·iz·ing** **1** To corrode, injure, or burn with sulfuric acid. **2** To convert into or impregnate with vitriol. — **vit′ri·ol·i·za′tion** *n.*

Vi·tru·vi·us (vi·trōō′vē·əs) Roman architect, military engineer, and writer of the first century B.C.; full name, *Marcus Vitruvius Pollio.* — **Vi·tru′vi·an** *adj.*

vit·ta (vit′ə) *n. pl.* **vit·tae** (vit′ē) **1** A fillet or band for the head; specifically, a sacred or sacrificial headband or chaplet worn by brides, vestals, priests, poets, and sacrificial victims. **2** *Bot.* An oil tube; a tube or canal in the fruit of plants of the parsley family, containing an aromatic oil. **3** *Zool.* A band or stripe, as of color. [<L <*viere* plait]

vit·tate (vit′āt) *adj.* **1** Having or bearing vittae or a vitta. **2** Striped.

Vit·to·rio (vit·tô′ryō) Italian form of VICTOR.

Vit·to·rio E·ma·nue·le (vit·tô′ryō ā·mä·nwā′lā) See VICTOR EMMANUEL.

Vit·to·rio Ve·ne·to (vit·tô′ryō ve′nā·tō) A city in NE Italy: scene of an Italian victory and armistice in World War I, November 3, 1918.

vit·u·line (vich′ōō·līn, -lin) *adj.* Pertaining to, of, or like a calf or veal. [<L *vitulinus* <*vitulus* a calf]

vi·tu·per·ate (vī·tōō′pə·rāt, -tyōō′-, vi-) *v.t.* **·at·ed, ·at·ing** To find fault with abusively; rail at; berate; scold. See synonyms under ABUSE. [<L *vituperatus,* pp. of *vituperare* blame, scold <*vitium* a fault + *parare* prepare] — **vi·tu′per·a′tion** *n.* — **vi·tu′per·a′tive** *adj.* — **vi·tu′per·a′tive·ly** *adv.* — **vi·tu′per·a′·tor** *n.*

vi·va (vē′vä) *interj.* Live! Long live!: a shout of applause; an acclamation or salute. [<Ital., 3rd person sing. present subjunctive of *vivere* live <L]

vi·va·ce (vē·vä′chā) *adv. Music* Lively; quickly; briskly. Also **vi·va′ce·men′te** (-mān′tā). [<Ital. <L *vivax.* See VIVACIOUS.]

vi·va·cious (vi·vā′shəs, vī-) *adj.* **1** Full of life and spirits; lively; active. **2** *Obs.* Tenacious of life. [<L *vivax, vivacis* <*vivere* live] — **vi·va′cious·ly** *adv.* — **vi·va′cious·ness** *n.*

Synonyms: animated, brisk, cheerful, frolicsome, gay, jocose, jocund, lively, merry, mirthful, pleasant, sparkling, spirited, sportive. See ALIVE, SPRIGHTLY. *Antonyms*: dead, dreary, dull, heavy, inanimate, lifeless, monotonous, moody, spiritless, stolid, stupid.

vi·vac·i·ty (vi·vas′ə·tē, vī-) *n. pl.* **·ties** **1** The state or quality of being vivacious. **2** Sprightliness, as of temper or behavior; liveliness. **3** A vivacious act, expression, etc.

Vi·val·di (vē·väl′dē), **Antonio,** 1675?–1743, Italian violinist and composer.

vi·van·dière (vē·vän·dyâr′) *n.* Formerly, a woman who supplied provisions and liquors to troops in the field, as in the French army. [<F, fem. of *vivandier* a sutler, ult. <L *vivenda.* See VIAND.]

vi·var·i·um (vī·vâr′ē·əm) *n. pl.* **·var·i·a** (-vâr′ē·ə) or **·var·i·ums** A place for keeping or raising live animals, fish, or plants, as a park, pond, aquarium, cage, etc. Also **viv·a·ry** (viv′ər·ē). [<L, orig. neut. of *vivarius* concerning live things <*vivus* alive <*vivere* live]

vi·va vo·ce (vī′və vō′sē) *Latin* By spoken word; orally: used both as an adverb and adjective.

vive (vēv) *French interj.* Live! Long live!: used in acclamation: opposed to *à bas.*

vive la ré·pu·blique (vēv là rā·pü·blēk′) *French* Long live the republic!

vive le roi (vēv lə rwà′) *French* Long live the king!

vi·ver·rine (vī·ver′īn, -in, vi-) *adj.* Belonging or pertaining to a family *(Viverridae)* of small carnivores including civets and mongooses. — *n.* A civet. [<NL *viverrinus* <L *viverra* a ferret]

vi·vers (vī′vərz) *n. pl. Scot.* Food; provisions.

vives (vīvz) *n. pl.* A morbid enlargement of the submaxiliary glands of the horse: also called *fives.* [Earlier *avives* <OF <Sp. *avivas* <Arabic *addhība* <*al* the + *dhība* a she-wolf]

Viv·i·an (viv′ē·ən, *Ger.* vē′vē·än) A personal name. Also **Viv·i·en** (viv′ē·ən, *Fr.* vē·vyan′), *Fr. fem.* **Vi·vienne** (vē·vyen′), *Ital. fem.* **Vi·via·na** (vē·vyä′nä). [<L, lively]

— **Vivian** In Arthurian romance, the wily mistress of Merlin, who imprisons him by his own magic: also known as *the Lady of the Lake, Nimue.* Also **Vivien, Viviane.**

viv·id (viv′id) *adj.* **1** Having an appearance of vigorous life; intense: said of colors having intense luminosity. **2** Producing or fitted to evoke lifelike imagery or suggestion. **3** Acting or exercised with lively interest; keen; clearly felt; strongly expressed. [<L *vividus* lively <*vivere* live] — **viv′id·ly** *adv.* — **viv′id·ness** *n.*

Synonyms: animated, bright, brilliant, clear, graphic, intense, keen, lively, luminous, quick, sprightly, stirring, telling, vigorous. See GRAPHIC. *Antonyms*: dim, dreary, dull, gloomy, heavy, lifeless, prosy, spiritless, stupid.

viv·i·fy (viv′ə·fī) *v.t.* **·fied, ·fy·ing** **1** To give life to; animate; vitalize. **2** To make more vivid or striking. [<OF *vivifier* <LL *vivificare* <L *vivus* alive + *facere* make] — **viv′i·fi·ca′tion** (-fə·kā′shən) *n.* — **viv′i·fi′er** *n.*

vi·vip·a·rous (vī·vip′ər·əs) *adj.* **1** *Zool.* Bringing forth living young, as most mammals: contrasted with *oviparous.* **2** *Bot.* Producing bulbs or seeds that germinate while still attached to the parent plant; proliferous. [<L *viviparus* <*vivus* alive + *parere* bring forth] — **vi·vip′a·rous·ly** *adv.* — **vi·vip′a·rous·ness, viv·i·par·i·ty** (viv′ə·par′ə·tē) *n.*

viv·i·sect (viv′ə-sekt) v.t. To dissect or operate upon (a living animal), with a view to exposing its physiological processes. — v.i. To practice vivisection. [Back formation < VIVISECTION] — viv′i·sec′tor n.

viv·i·sec·tion (viv′ə-sek′shən) n. 1 The dissection of a living animal. 2 Experimentation on living animals by means of operations designed to promote knowledge of physiological and pathological processes. [<L vivus living, alive + sectio, -onis a cutting. See SECTION.] — viv′i·sec′tion·al adj. — viv′i·sec′tion·ist n. & adj.

vix·en (vik′sən) n. 1 A turbulent, quarrelsome woman; shrew. 2 A female fox. [ME fixen a she-fox, fem. of OE fox] — vix′en·ish adj. — vix′en·ly adj. & adv.

viz·ard (viz′ərd) n. A mask; visor: also spelled visard. [Alter. of VISOR]

viz·ard·ed (viz′ərd·id) adj. Masked; disguised or protected by a vizard.

Viz·ca·ya (vēs·kä′yä, Sp. bēth·kä′yä) The Spanish name for BISCAY.

Viz·e·tel·ly (viz′ə·tel′ē), **Frank Horace**, 1864-1938, U.S. lexicographer and encyclopedist born in England.

vi·zier (vi·zir′, viz′yər) n. A high official of a Moslem country, especially of the old Turkish Empire; a minister of state. Also **vi·zir′**. — **grand vizier** The highest dignitary in Moslem countries; the prime minister. [<Turkish vezir <Arabic wazīr a counselor, orig. a porter < wazara carry]

vi·zier·ate (vi·zir′it, -āt, viz′yər·it, -yə·rāt) n. The office or dignity of a vizier. Also **vi·zier′·al·ty, vi·zier′ship, vi·zir′ate, vi·zir′ship.**

vi·zor (vī′zər, viz′ər) n. The movable upper front piece of a helmet protecting the eyes. See VISOR.

Vlad·i·mir (vlad′ə·mir, Russian vlä·dyē′mir), 956?-1015, first Christian Russian ruler.

Vlad·i·mir (vlad′ə·mir, Russian vlä·dyē′mir) A city in central European Russian S.F.S.R., NE of Moscow.

Vla·di·vos·tok (vlad′ə·vos·tok′, -vos′tok; Russian vlä·dye·vos·tôk′) A port on the Sea of Japan in extreme SE Asiatic Russian S.F.S.R.

Vla·minck (vlà·maṅk′), **Maurice de**, 1876-1958, French painter.

Vlis·sing·en (vlis′ing·ən) The Dutch name for FLUSHING.

Vlo·na (vlō′nä) The Albanian name for VALONA.

Vl·ta·va (vul′tà·va) A river in central Bohemia, Czechoslovakia, flowing 267 miles north from the Bohemian Forest, through Prague, to the Elbe: German Moldau.

V-mail (vē′māl′) n. Mail written on special forms, transmitted overseas in World War II on microfilm, and enlarged at point of reception for final delivery. [<V(ICTORY) + MAIL¹]

vo·ca·ble (vō′kə·bəl) n. 1 A word, chiefly as regarded in relation to its sound or combination of sounds instead of its meaning. 2 A vocal sound. — adj. Utterable. [<F <L vocabulum a name, appellation < vocare call < vox voice]

vo·cab·u·lar·y (vō·kab′yə·ler′ē) n. pl. ·lar·ies 1 A list of words or of words and phrases, especially one arranged in alphabetical order and defined or translated; a lexicon; glossary. 2 All the words of a language. 3 A sum or aggregate of the words used or understood by a particular person, class, etc., or employed in some specialized field of knowledge: the vocabulary of Shakespeare. [<LL vocabularius <L vocabulum. See VOCABLE.]

vocabulary entry 1 A word or term given in a vocabulary. 2 A word, term, or phrase entered in a dictionary, in some readily distinguishable type, for purposes of definition or identification. Vocabulary entries may be listed in alphabetical place (main entries), run in within a main entry (additional parts of speech, inflected forms, idioms, etc.), run on at the end of an entry (derivatives and related words), listed under a word, prefix, or combining form (self-explanatory compounds and two-word phrases), or entered in a special section of the book. In this dictionary, all vocabulary entries are shown in boldface type or preceded by a boldface em-dash.

vo·cal (vō′kəl) adj. 1 Of or pertaining to the voice; uttered by the voice; oral: vocal protests. 2 Having voice; endowed with the power of utterance: vocal creatures. 3 Composed for or performed by the voice: a vocal score. 4 Concerned in the production of voice: the vocal organs. 5 Full of voices or sounds; resounding: The air was vocal with their cries. 6 Eloquent without need of speech: the vocal beauty of the Parthenon. 7 Freely expressing oneself in speech; readily given to voicing opinions: the vocal segment of the populace. 8 Phonet. a Voiced; sonant, as b, d, g, distinguished from p, t, k. b Pertaining to or like a vowel; vocalic. See synonyms under VERBAL. — n. Phonet. 1 A vowel. 2 A voiced consonant. [<L vocalis speaking, sounding < vox, vocis a voice. Doublet of VOWEL.] — vo′cal·ly adv.

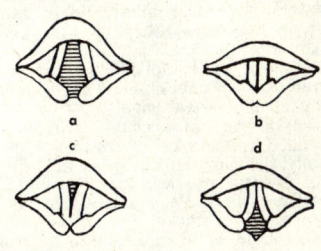

VOCAL CORDS
a. Open. b. Closed. c. Voice. d. Whisper.

vocal cords Two membranous bands extending from the thyroid cartilage of the larynx. The edges of these bands, when drawn tense, are caused to vibrate by the passage of air from the lungs, thereby producing voice; the degree of tension of the cords controls the pitch of the voice.

vo·cal·ic (vō·kal′ik) adj. Consisting of, like, or relating to vowel sounds.

vo·ca·lise (vō·kä·lēz′) n. Music A practice exercise for singers designed to develop flexibility and control of pitch and tonal beauty, usually employing vowels or Italian syllables. [<F]

vo·cal·ism (vō′kəl·iz′əm) n. 1 Vocalization. 2 A vocalic sound; also, a vowel system. 3 Singing; also, the technique of singing.

vo·cal·ist (vō′kəl·ist) n. A singer, especially one who has a cultivated voice.

vo·cal·ize (vō′kəl·īz) v. ·ized, ·iz·ing v.t. 1 To make vocal; utter, say, or sing; make sonant. 2 To provide a voice for; render articulate. 3 To mark with vowel points, as a Hebrew text. 4 Phonet. a To change to or use as a vowel: to vocalize y. b To voice. — v.i. 5 To produce sounds with the voice, as in speaking or singing. 6 Phonet. To be changed to a vowel. — vo′cal·i·za′tion n. — vo′cal·iz′er n.

vo·ca·tion (vō·kā′shən) n. 1 A stated or regular occupation; a calling. 2 A call to, or fitness for, a certain career, especially a religious position. 3 The work or profession for which one has a sense of special fitness. See synonyms under BUSINESS. [<L vocatio, -onis <vocatus, pp. of vocare call] — vo·ca′tion·al adj. — vo·ca′tion·al·ly adv.

vocational adviser One who diagnoses the personal characteristics of people with the view of suggesting suitable vocations for them; a specialist in vocational guidance. Also **vocational expert.**

vocational school See under SCHOOL.

voc·a·tive (vok′ə·tiv) adj. 1 Pertaining to or used in the act of calling. 2 Gram. In some inflected languages, denoting the case of a noun, pronoun, or adjective used in direct address: The name "Brutus" is in the vocative case in "Et tu, Brute." — n. Gram. 1 The vocative case. 2 A word in this case. [<F, fem. of vocatif <L vocativus <vocare call]

vo·ces (vō′sēz) Plural of vox.

vo·cif·er·ant (vō·sif′ər·ənt) adj. Vociferous; clamorous; uttering loud cries. — n. A vociferous person. [<L vociferans, -antis, ppr. of vociferari. See VOCIFERATE.] — vo·cif′er·ance n.

vo·cif·er·ate (vō·sif′ə·rāt) v.t. & v.i. ·at·ed, ·at·ing To cry out with a loud voice; exclaim noisily; shout; bawl. See synonyms under CALL. [<L vociferatus, pp. of vociferari cry out < vox, vocis a voice + ferre carry] — vo·cif′er·a′tion n. — vo·cif′er·a′tor n.

vo·cif·er·ous (vō·sif′ər·əs) adj. Making a loud outcry; clamorous; noisy. See synonyms under NOISY. — vo·cif′er·ous·ly adv. — vo·cif′er·ous·ness n.

vod·ka (vod′kə, Russian vôd′kä) n. An alcoholic liquor, originally Russian, usually made from a fermented mash of wheat and neither flavored nor aged. [<Russian, dim. of voda water]

voe (vō) n. Scot. A small bay, creek, or inlet.

Vo·gel·kop (vō′gəl·kôp′) A peninsula of NW Netherlands New Guinea, connected to the mainland by an isthmus 20 miles wide; about 225 miles east to west, about 135 miles north to south.

vo·gie (vō′gē) adj. Scot. 1 Merry; cheerful. 2 Vain; conceited.

vogue (vōg) n. 1 The prevalent way or fashion; popular temporary usage: often preceded by in. 2 Popular favor; general acceptance; popularity. [<F, fashion, orig. rowing <voguer row <Ital. vogare <MHG wogen sail <woge a wave]

Vo·gul (vō′gool) n. 1 One of a Finno-Ugric people of the Ural Mountains. 2 The Ugric language of these people.

voice (vois) n. 1 The sound produced by the vocal organs of a person or animal; also, the quality or character of such sound: a melodious voice. 2 The power or faculty of vocal utterance; speech. 3 A sound suggesting vocal utterance or speech: the voice of the wind. 4 Opinion or choice expressed; also, the right of expressing a preference or judgment: to have a voice in the affair. 5 Instruction; admonition; teaching: the voice of nature. 6 A speaker; also, a person or agency by which the thought, wish, or purpose of another is expressed: This journal is the voice of the teaching profession. 7 Expression of thought, opinion, feeling, etc.: to give voice to one's ideals. 8 Phonet. The sound produced by vibration of the vocal cords, as heard in the utterance of vowels and certain consonants, as (g), (m),(v): distinguished from whisper, and also from breath, as heard in (k), (sh), (f). 9 Musical tone produced by vibration of the vocal cords and resonating in the cavities of the throat and head; also, the ability to sing, or the state of the vocal organs with regard to this ability: to be in poor voice. 10 Gram. The relation of the action expressed by the verb to the subject, or the form of the verb indicating this relationship. In English, as in most Indo-European languages, a distinction between an active and a passive voice is made, indicating, respectively, that the subject of the sentence is either performing the action or is being acted upon. (Active: He wrote the letter. Passive: The letter was written by him.) In Greek and Sanskrit verbs, there is, in addition, a middle voice, representing the subject as acting upon himself directly, or in his own interest. 11 Obs. Report; rumor; fame. — **with one voice** With one accord; unitedly; unanimously. — v.t. **voiced, voic·ing** 1 To put into speech; give expression to; utter. 2 Music To regulate the tones of; tune, as the pipes of an organ. 3 Phonet. To utter with voice or sonance. [<OF vois <L vox, vocis]

voiced (voist) adj. 1 Having a voice; expressed by voice. 2 Phonet. Uttered with vibration of the vocal cords, as (b), (d), (z); sonant: opposed to surd, voiceless.

voice·ful (vois′fəl) adj. Having vocal quality; vocal; sounding. — **voice′ful·ness** n.

voice·less (vois′lis) adj. 1 Having no voice, speech, or suffrage. 2 Phonet. Produced without voiced breath, as (p), (t), (s); surd: opposed to sonant, voiced. — **voice′less·ly** adv. — **voice′less·ness** n.

voice part A single part, as a melody written for the voice and either sung, or played by a solo instrument, in a concerted composition.

void (void) adj. 1 Not occupied by matter or by visible matter; empty. 2 Destitute; clear or free: with of: void of reason, void of offense.

VIZOR
15th century.

voidable 1408 **voltaic**

3 Unoccupied, as a house or room; having no incumbent. 4 Having no legal force or validity; incapable of confirmation or ratification; invalid; null. 5 Producing no effect; useless. See synonyms under VACANT. — n. 1 An empty space; a vacuum. 2 A breach of surface or matter; a disconnecting space. 3 Empty condition or feeling; a blank. — v.t. 1 To make void or of no effect; annul. 2 To empty or remove (contents); evacuate, as urine. 3 Archaic To leave empty or vacant. [<OF voide, fem. of voit, ult. <LL vocuus empty <L vacuus] — void′er n.

void·a·ble (voi′də·bəl) adj. 1 Capable of being made void: A voidable contract is valid unless annulled. 2 That may be evacuated. — void′a·ble·ness n.

void·ance (void′ns) n. 1 The act of voiding, evacuating, ejecting, or emptying. 2 The state or condition of being void; vacancy: voidance of a benefice. [<AF voidaunce, OF vuidance <voider empty <voit VOID]

void·ed (voi′did) adj. 1 Made empty or void; cleared of contents; having a vacant space. 2 Her. Having the central area removed, so as to leave only an outline through which the field is visible: said of a charge, as a cross.

voi·là (vwä·lä′) interj. French There! behold! literally, see there.

voi·là tout (vwä·lä tōō′) French That is all; there is the whole matter.

voile (voil, Fr. vwäl) n. A fine, sheer cotton, silk, wool, or rayon fabric like heavy veiling: used for summer dresses and curtains. [<F, a veil <OF veile VEIL]

voir dire (vwär dēr′) Law A legal oath administered to a witness to be examined, to make true answers to the questions to be asked him regarding his competency. [<OF voir truth + dire say]

voix (vwä) n. French The voice.

voix cé·leste (vwä sā·lest′) French An organ stop consisting of two ranks of soft flue stops which produce a waving effect; literally, heavenly voice.

Voj·vo·di·na (voi′vô·di·nä) An autonomous province of NE Yugoslavia, included in Serbia as its northern part and bordering on Hungary and Rumania; 8,683 square miles; capital, Novi Sad. Also **Voy′vo·di·na, Voi′vo·di·na.**

vo·lant (vō′lənt) adj. 1 Passing through the air; flying, or able to fly. 2 Characterized by lightness and quickness; nimble. 3 Her. Flying, as a bird or bee. [<OF, ppr. of voler fly <L volare]

vo·lan·te (vō·län′tā) adj. Music Swift and light. [<Ital., ppr. of volare fly <L]

Vo·la·pük (vō′lə·pük′) n. A proposed universal language, invented in 1879 by Johann M. Schleyer, a German priest. [<Volapük vol world + pük speech] — **Vo′la·pük′ist** n.

vo·lar[1] (vō′lər) adj. Used in flying; pertaining to flight. [<L volare fly]

vo·lar[2] (vō′lər) adj. Pertaining to the sole of the foot or palm of the hand. [<L vola sole, palm]

vol·a·tile (vol′ə·til) adj. 1 Evaporating rapidly at ordinary temperatures on exposure to the air; capable of being vaporized. 2 Easily influenced; fickle; changeable. 3 Transient; fleeting; ephemeral. 4 Obs. Flying, or able to fly. See synonyms under MOBILE[1]. [<OF volatil <L volatilis <volare fly]

volatile oil Any oil that may be readily vaporized, especially one distilled from plants: distinguished from fixed oil.

volatile salts Salts that volatilize without residue; sal volatile.

vol·a·til·i·ty (vol′ə·til′ə·tē) n. 1 The state or quality of being volatile. 2 The property of being freely or rapidly diffused in the atmosphere. Also **vol′a·tile·ness.**

vol·a·til·ize (vol′ə·til·īz′) v.t. & v.i. ·ized, ·iz·ing 1 To make or become volatile. 2 To pass off or cause to pass off in vapor; evaporate. — **vol′a·til·iz′a·ble** adj. — **vol′a·til·i·za′tion** n. — **vol′a·til·iz′er** n.

vol-au-vent (vô·lō·vän′) n. French A patty shell of light puff paste filled with a ragout of meat, fowl, or fish.

vol·can·ic (vol·kan′ik) adj. 1 Of, pertaining to, or characteristic of a volcano or volcanoes. 2 Produced by a volcano or by igneous action: distinguished from plutonic. — **vol·can·ic·i·ty** (vol′kə·nis′ə·tē) n. — **vol·can′i·cal·ly** adv.

volcanic glass An igneous rock of volcanic origin and glassy texture having cooled too quickly to crystallize, as obsidian.

volcanic rocks Rocks formed by the consolidation of lava from volcanoes.

vol·can·ism (vol′kən·iz′əm) n. The conditions, phenomena, or science of volcanoes or volcanic action.

vol·can·ist (vol′kən·ist) n. One who studies, or is expert on volcanoes; a volcanologist.

vol·can·ize (vol′kən·īz) v.t. ·ized, ·iz·ing To subject to the action and effects of volcanic heat. — **vol′can·i·za′tion** n.

vol·ca·no (vol·kā′nō) n. pl. ·noes or ·nos Geol. An opening in the earth's surface surrounded by an accumulation of ejected material, forming a hill or mountain, from which heated matter is or has been ejected: known in the former case as active, and in the latter as dormant or extinct. [<Ital. <L Volcanus, Vulcanus Vulcan]

Volcano Islands Three small islands, including Iwo Jima, in the western Pacific south of the Bonin Islands, administered by the United States; total, 11 square miles; held by Japan, 1887–1945. Japanese **Ka·zan Ret·to** (kä·zän ret·tō).

vol·can·ol·o·gy (vol′kən·ol′ə·jē) n. The scientific study of volcanoes. — **vol′can·o·log′i·cal** (-ə·loj′i·kəl) adj. — **vol′can·ol′o·gist** n.

vole[1] (vōl) n. Any of a genus (Microtus) of short-tailed, mouselike or ratlike rodents; especially, the common European field mouse or the North American meadow mouse. [Short for earlier vole mouse <vole a field (<Norw. voll) + MOUSE]

vole[2] (vōl) n. In some card games, as écarté, a winning of all the tricks in a deal; hence, the entire range; a slam. — **to go the vole** To risk all for great gains. [<F, appar. <voler fly <L volare]

vol·er·y (vōl′ər·ē) n. pl. ·er·ies A large bird cage; aviary; also, the birds in it. [<F volerie a flying <voler. See VOLE[2].]

Vol·ga (vol′gə, Russian vôl′gä) The longest river in Europe, in central European Russian S.F.S.R., flowing 2,290 miles east and south from the Valdai Hills to its delta, the **Volga Basin** (approximately 2,500 square miles), on the Caspian Sea.

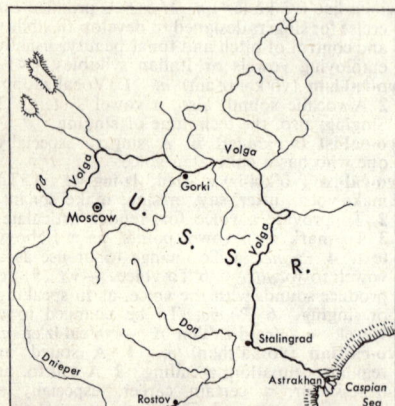

Vol·hyn·i·a (vol·hin′ē·ə, vo·lin′ē·ə) A historical region, formerly in Poland, in NW Ukrainian S.S.R.; about 27,230 square miles. Russian **Vo·lyn** (vo·lin′y), Polish **Wo·łyń** (vô′win·y). — **Vol·hyn′ian** (-ē·ən) adj.

vol·i·tant (vol′ə·tənt) adj. Flying, or having power to fly; volant. [<L volitans, -antis, ppr. of volitare, freq. of volare fly]

vol·i·ta·tion (vol′ə·tā′shən) n. The act or power of flying; flight. [<L volitatus, pp. of volitare. See VOLITANT.] — **vol′i·ta′tion·al** adj.

vo·li·tient (və·lish′ənt) adj. Exercising the will, or having freedom of will; voluntary. [<VOLITI(ON) + -ENT] — **vo·li′tien·cy** n.

vo·li·tion (və·lish′ən) n. 1 The act or faculty of willing; exercise of the will; especially, the termination of a process of deliberation or vacillation of purpose by a decision or choice. 2 The faculty of will by which the powers are directed toward the attainment of a chosen end; willpower. 3 That which is specifically willed or determined upon. [<F <Med. L volitio, -onis <L vol-, stem of velle will] — **vo·li′tion·al** adj. — **vo·li′tion·al·ly** adv.

vo·li·tive (vol′ə·tiv) adj. 1 Of, pertaining to, or originating in the will. 2 Expressing a wish or permission.

Vol·khov (vôl′khof) A river of NW European Russian S.F.S.R., flowing 140 miles NE from Lake Ilmen, through Novgorod, to Lake Ladoga. Also **Vol′khof.**

Volks·lied (fôlks′lēt′) n. pl. ·lied·er (-lē′dər) German A folk song; popular song.

vol·ley (vol′ē) n. 1 A simultaneous discharge of many missiles; also, the missiles so discharged. 2 Any discharge of many things at once: a volley of oaths. 3 In tennis, a return of the ball before it touches the ground. 4 In soccer, a kick given the ball before its rebound. 5 In cricket, a ball bowled so that it strikes the head of the wicket before it touches the ground. — v.t. & v.i. ·leyed, ·ley·ing 1 To discharge or be discharged in a volley; let fly together. 2 In tennis, to return (the ball) without allowing it to touch the ground. 3 In soccer, to kick (the ball) before its rebound; in cricket, to bowl (a ball) full pitch. [<MF volée, pp. fem. of voler fly <L volare]

volley ball A game in which a number of players on both sides of a high net endeavor to keep a large ball in motion with the hands from side to side without letting it drop; also, the ball used. Also **vol′ley·ball** (vol′ē·bôl′).

Vo·log·da (vô′lə́g·də) A city in north central European Russian S.F.S.R.; an industrial and dairying center; capital of a 15th century principality.

Vo·los (vō′los) A port city on the Gulf of Volos, an inlet of the Aegean in SE Thessaly, Greece (about 20 miles long and wide); the principal port of Thessaly and capital of Magnesia nome.

vo·lost (vō′lost) n. A district having one joint administrative assembly; a rural soviet; formerly, a canton. [<Russian volost′]

vol·plane (vol′plān) v.i. ·planed, ·plan·ing To glide in an airplane. — n. An airplane glide. [<F vol plané gliding flight <vol flight + plané, pp. of planer glide, soar] — **vol′plan·ist** n.

Vol·sci·an (vol′shən) adj. Of or pertaining to the **Vol·sci** (vol′sī), a warlike people of ancient Italy, subdued by the Romans about 350 B.C. — n. 1 One of the Volsci. 2 Their language, belonging to the Sabellian branch of the Italic languages.

Vol·stead Act (vol′sted) An act to enforce the Eighteenth (Prohibition) Amendment to the Constitution of the United States, and defining intoxicating liquors as those containing more than one half of one percent of alcohol by volume: effective 1920–33. [after Representative Andrew J. Volstead, 1860–1947, of Minnesota] — **Vol′stead·ism** n.

Vol·sun·ga Sa·ga (vol′sŏong·gə sä′gə) A prose version of the Icelandic legends of the dwarf race, the Nibelungs, and Sigurd, the grandson of Volsung. See NIBELUNGENLIED. [<ON Völsunga saga, lit., saga of the Volsungs]

volt[1] (vōlt) n. The unit of electromotive force, or that difference of potential which, when steadily applied to a conductor whose resistance is one ohm, will produce a current of one ampere. Abbr. v., V. [after Alessandro Volta]

volt[2] (vōlt) n. 1 In horse-training, a gait in which the horse moves partially sidewise round a center with the head turned out; a circular tread. 2 In fencing, a sudden leap to avoid a thrust. [<F volte a turn <Ital. volta, orig. pp. fem. of volvere turn <L]

vol·ta (vōl′tə, Ital. vôl′tä) n. pl. ·te (-tā) Music A turning; a time: used mainly in phrases. — **prima volta** First time. — **seconda volta** Second time. — **una volta** Once. [<Ital. See VOLT[2].]

Vol·ta (vōl′tä) The principal river of Ghana, formed by the confluence, in north central Ghana, of the **Black Volta** and **White Volta**, and flowing 300 miles (with Black Volta about 800 miles) SE to the Bight of Benin.

Vol·ta (vōl′tä), **Count Alessandro**, 1745–1827, Italian physicist and pioneer in electricity.

volt·age (vōl′tij) n. Electromotive force expressed in volts: the voltage of a current.

vol·ta·ic (vol·tā′ik) adj. 1 Pertaining to electricity developed through chemical action or contact; galvanic: a voltaic battery; voltaic cell. 2 Of or pertaining to Alessandro Volta. [after Alessandro Volta]

voltaic battery An assembly of voltaic cells which operate as a unit in generating an electric current.

voltaic cell Cell (def. 5).

voltaic couple A pair of dissimilar, usually metallic, substances which will produce an electric current when immersed in an electrolyte.

voltaic pile An arrangement of dissimilar metal disks, placed alternately and having between them paper moistened with acids for the generation of an electric current. Also *galvanic pile.*

Vol·taire (vol·târ′, *Fr.* vôl·târ′), **François Marie Arouet de,** 1694–1778, French author and philosopher.

vol·ta·ism (vol′tə·iz′əm) *n.* The act of producing an electric current by the chemical action of a liquid on dissimilar metals; galvanism. [after Alessandro *Volta*]

volt·am·e·ter (vol·tam′ə·tər) *n.* A coulometer. [<VOLTA(IC) + -METER]

volt·am·me·ter (vōlt′am′mē′tər) *n.* A wattmeter. [<VOLT(AGE) + AM(PERAGE) + -METER]

volt–am·pere (vōlt′am′pir) *n.* A watt: so called because it is the rate of working in an electric circuit when the current is one ampere and the potential one volt.

volte–face (volt·fäs′, *Fr.* vôlt·fàs′) *n.* **1** A turning about so as to face in the opposite direction. **2** A complete change of attitude or reversal of opinion. [<F <Ital. *volta faccia* < *volta* a turning + *faccia* a face <L *facies*]

vol·ti (vôl′tē) *interj. Music* Turn; specifically, a direction to turn the leaf. [<Ital., imperative sing. of *voltare* turn < *volta.* See VOLT[2].]

vol·ti·geur (vôl·tē·zhœr′) *n.* One who vaults; a tumbler; formerly, in the French army, a skirmisher in a light infantry regiment. [<F < *voltiger* hover, vault <Ital. *volteggiare* < *volta.* See VOLT[2].]

volt·me·ter (vōlt′mē′tər) *n.* An instrument for determining the voltage or potential difference existing between any two points, generally consisting of a calibrated galvanometer wound with a coil of high resistance.

Vol·tur·no (vôl·to͞or′nō) The chief river of southern Italy, arising in the Apennines and flowing 109 miles to the Tyrrhenian Sea NW of Naples.

vol·u·ble (vol′yə·bəl) *adj.* **1** Having a flow of words or fluency in speaking; talkative; garrulous. **2** Turning readily or easily; revolving; apt or formed to roll. **3** Twining, as a plant. [<MF <L *volubilis* easily turned < *volutus,* pp. of *volvere* turn] **— vol′u·bil′i·ty, vol′u·ble·ness** *n.* **— vol′u·bly** *adv.*

vol·ume (vol′yo͞om, -yəm) *n.* **1** A collection of sheets of paper bound together; a book; a separately bound part of a work; anciently, a written roll, a scroll, as of papyrus or vellum. **2** Sufficient matter to fill a volume. **3** Something of a swelling form; coil; fold or turn. **4** A large quantity; a considerable amount. **5** Space occupied, as measured by cubic units, that is, cubic centimeters, cubic feet, etc. **6** The amount of space included by the bounding surfaces of a solid. **7** *Music* Fullness or quantity of sound or tone. **— to speak volumes** To be full of meaning; express a great deal. [<OF *volum* <L *volumen* a roll, scroll < *volutus.* See VOLUBLE.]

vol·umed (vol′yo͞omd, -yəmd) *adj.* **1** Rounded or swelling in form: *volumed* mists. **2** Having bulk or quantity. **3** Being in one or more volumes: a two–*volumed* history.

vo·lu·me·ter (və·lo͞o′mə·tər) *n.* An instrument for measuring the volume of a gas by the amount of liquid displaced by it in a graduated vessel, under known conditions of pressure and temperature. [<VOLU(ME) + -METER]

vol·u·met·ric (vol′yə·met′rik) *adj. Chem.* Of or pertaining to measurement of substances by comparison of volumes or by volumetric analysis. Also **vol′u·met′ri·cal. — vol′u·met′ri·cal·ly** *adv.* **— vo·lu·me·try** (və·lo͞o′mə·trē) *n.*

volumetric analysis *Chem.* The quantitative analysis of a substance by determining the amount of a standard solution required to effect a reaction in a known quantity of the substance.

vo·lu·mi·nos·i·ty (və·lo͞o′mə·nos′ə·tē) *n.* The state or quality of being voluminous; especially, copiousness or prolixity.

vo·lu·mi·nous (və·lo͞o′mə·nəs) *adj.* **1** Consisting of many volumes; capable of filling several volumes; also, of great bulk. **2** Writing or having written much; productive.. **3** Having coils, folds, convolutions, or windings. [<LL *voluminosus* <L *volumen, -inis* a roll] **— vo·lu′mi·nous·ly** *adv.* **— vo·lu′mi·nous·ness** *n.*

vol·un·ta·rism (vol′ən·tə·riz′əm) *n.* **1** The theory that will is the ultimate principle or constituent of reality, both in experience and development of the individual, and in the constitution and evolution of the universe. **2** The theory that will is the fundamental psychic factor. Compare VITALISM. **— vol′un·ta·rist** *n.* **— vol′un·ta·ris′tic** *adj.*

vol·un·tar·y (vol′ən·ter′ē) *adj.* **1** Proceeding from the will or from one's own free choice: *voluntary* murder; specifically, unconstrained; intentional; volitional. **2** Endowed with, possessing, or exercising will or free choice: a *voluntary* donor. **3** Effected by choice or volition; acting without constraint. **4** Subject to or directed by the will, as a muscle or movement. **5** Of or relating to voluntaryism. **6** *Law* Unconstrained of will; done without compulsion; performed without legal obligation; also, done without valuable consideration; gratuitous. See synonyms under SPONTANEOUS. **—** *n. pl.* **·tar·ies 1** Any work or performance not compelled or imposed by another. **2** *Music* **a** An organ solo, often improvised, played before, during, or after a service. **b** *Rare* A piece of music, usually spontaneous, played or sung as a prelude. **3** *Obs.* A volunteer. [<OF *voluntaire* <L *voluntarius* < *voluntas* will] **— vol′un·tar′i·ly** *adv.* **— vol′un·tar′i·ness** *n.*

vol·un·tar·y·ism (vol′ən·ter′ē·iz′əm) *n.* The principle that religious and educational institutions should be supported by voluntary contributions. **— vol′un·tar′y·ist** *n.*

voluntary system A system of freely given support in distinction from state support of religious or educational institutions.

vol·un·teer (vol′ən·tir′) *n.* **1** One who enters into any service of his own free will. **2** One who voluntarily enters military service, but is then subject to the same regulations and discipline as other soldiers: opposed to *conscript.* **3** *Law* One who takes title under a deed made without valuable consideration; also, a voluntary agent or actor in a transaction. **—** *adj.* **1** Pertaining to or composed of volunteers; voluntary. **2** Springing up naturally or spontaneously, as from fallen or self–sown seed: a *volunteer* growth. **—** *v.t.* To offer to give or do. **—** *v.i.* To enter or offer to enter into some service or undertaking of one's free will; enlist. [< obs. F *voluntaire* <OF, VOLUNTARY]

Volunteers of America A religious and philanthropical organization founded in the United States in 1896 by Commander and Mrs. Ballington Booth, who resigned from the Salvation Army for that purpose.

Volunteer State Nickname of TENNESSEE.

vo·lup·tu·ar·y (və·lup′cho͞o·er′ē) *adj.* Pertaining to or promoting sensual indulgence and luxurious pleasures. **—** *n. pl.* **·ar·ies** One addicted to sensual pleasures; a sensualist. [<L *voluptuarius* < *voluptas* pleasure]

vo·lup·tu·ous (və·lup′cho͞o·əs) *adj.* **1** Belonging to, producing, exciting, or yielding sensuous gratification. **2** Pertaining to or devoted to the enjoyment of pleasures or luxuries; luxurious; sensual. **3** Having a full and beautiful form, as a woman. [<OF *voluptueux* <L *voluptuosus* full of pleasure < *voluptas* pleasure] **— vo·lup′tu·ous·ly** *adv.* **— vo·lup′tu·ous·ness** *n.*

vo·lute (və·lo͞ot′) *n.* **1** *Archit.* A spiral scroll–like ornament, as in Corinthian capitals; a scroll. **2** *Zool.* One of the whorls or turns of a spiral shell. **—***adj.* **1** Rolled up; forming spiral curves. **2** Having a spiral form, as a machine part. [<F <L *voluta* a scroll, orig. fem. pp. of *volvere* turn]

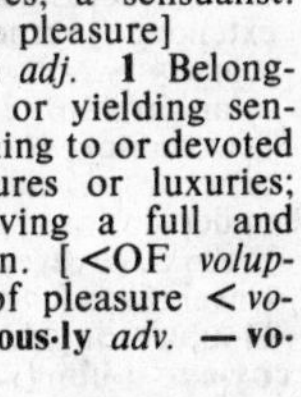

VOLUTE

vo·lut·ed (və·lo͞o′tid) *adj.* Having a volute or flat spiral scroll.

volute spring *Mech.* A flat metallic spring coiled in a spiral conical form.

vo·lu·tion (və·lo͞o′shən) *n.* **1** A spiral turn or twist; convolution. **2** A whorl of a spiral shell. **3** A revolving movement.

vol·va (vol′və) *n. Bot.* That part of the sheath enclosing certain young mushrooms which, on being ruptured in the course of growth, forms a cuplike appendage at the base of the stem. [<L < *volvere* wrap, turn]

vol·vu·lus (vol′vyə·ləs) *n. pl.* **·li** (-lī) *Pathol.* Obstruction of the intestines caused by twisting. [<NL <L *volvere* turn]

vo·mer (vō′mər) *n. Anat.* **1** A bone of the face situated between the nasal passages on the median line in vertebrates above fishes. **2** A bone of the roof of the mouth in fishes, behind the premaxillaries. [<NL <L *vomer* a plow] **— vo·mer·ine** (vō′mər·in, vom′ər-) *adj.*

vom·i·ca (vom′i·kə) *n. pl.* **·cae** (-sē) *Pathol.* **1** A collection of purulent matter within an organ. **2** An ulcerous cavity, especially in the lungs. **3** Expectoration of putrid matter. [<L, a boil, ulcer < *vomere* vomit]

vom·it (vom′it) *v.i.* **1** To throw up or eject the contents of the stomach through the mouth. **2** To issue with violence from any hollow place; be ejected. **—** *v.t.* **3** To throw up or eject from the stomach, as food. **4** To discharge or send forth copiously or forcibly: The volcano *vomited* smoke. **—** *n.* **1** Matter that is ejected, as from the stomach in vomiting. **2** A sickness which is characterized by vomiting. **3** An emetic. **4** The act of vomiting. [<L *vomitare,* freq. of *vomere* vomit] **— vom′it·er** *n.*

vomiting gas Chlorpicrin.

vom·i·tive (vom′ə·tiv) *adj.* Causing vomiting. **—** *n.* An emetic.

vom·i·to (vom′ə·tō, *Sp.* vō′mē·tō) *n. Pathol.* Yellow fever; black vomit. Also **vomito ne·gro** (nē′grō, *Sp.* nā′grō). [<Sp. *vómito* <L *vomitus,* pp. of *vomere* vomit]

vom·i·to·ry (vom′ə·tôr′ē, -tō′rē) *adj.* Efficacious in producing vomiting. **—** *n. pl.* **·ries 1** An emetic. **2** An opening or vent through which matter is discharged. **3** In a Roman amphitheater, one of the entrances from the encircling arcades to the passages leading to the seats.

vom·i·tu·ri·tion (vom′ə·cho͞o·rish′ən) *n. Pathol.* **1** Violent vomiting with the ejection of but little matter; retching. **2** Vomiting with but small effort; repeated vomiting. [<F <L *vomitus* a vomiting < *vomere* vomit]

von (von, *Ger.* fôn, *unstressed* fən) *prep. German* Of; from: used in German and Austrian family names as an attribute of nobility, corresponding to the French *de.*

voo·doo (vo͞o′do͞o) *n.* **1** A primitive religion of West African origin, found among Haitian and West Indian Negroes and the Negroes of the southern United States, characterized by belief in sorcery and the use of charms, fetishes, witchcraft, etc. **2** A witch doctor; a Negro conjurer who practices voodoo. **3** A voodoo charm or fetish. **—** *adj.* Of or pertaining to the beliefs, ceremonies, or practices of voodoo. **—** *v.t.* To put a spell upon after the manner of a voodoo; bewitch. [<Creole *voudou* <Ewe *vodu*]

voo·doo·ism (vo͞o′do͞o·iz′əm) *n.* **1** The religion of voodoo. **2** Belief in or practice of this religion. **— voo′doo·ist** *n.* **— voo′doo·is′tic** *adj.*

-vora *combining form Zool.* Used to denote orders or genera when classified according to their food: *Carnivora.* An individual member of such an order or genus is denoted by –**vore.** [<NL <L *-vorus.* See –VOROUS.]

vo·ra·cious (vô·rā′shəs, vō-, və-) *adj.* **1** Eating with greediness; ravenous. **2** Greedy; rapacious. **3** Ready to swallow up or engulf. **4** Insatiable; immoderate. See synonyms under GREEDY. [<L *vorax, -acis* < *vorare* devour] **— vo·ra′cious·ly** *adv.* **— vo·rac·i·ty** (vô·ras′ə·tē, vō-, və-), **vo·ra′cious·ness** *n.*

Vor·arl·berg (fōr′ärl·berkh) An autonomous province of western Austria, bordering on Germany, Switzerland, and Liechtenstein; 1,004 square miles; capital, Bregenz.

Vor·i·ai Spor·a·des (vô′rē·e spô·rä′thes) The Greek name for the NORTHERN SPORADES. See under SPORADES.

vor·la·ge (fōr′lä·gə) *n.* In skiing, a posture in which the body leans forward, beyond the perpendicular to the incline. [<G, lit., a lying

forward < *vorlagern* extend forward < *vor-* forward + *lagern* lie, lay]

Vo·ro·nezh (vo·rô'nesh) A city in SW European Russian S.F.S.R.; a major industrial center.

Vo·ro·shi·lov (vo'ro·shē'lôf), **Klement Efremovich,** born 1881, Russian politician; marshal in World War II.

Vo·ro·shi·lov·grad (vo'ro·shē'lôf·grät) See LUGANSK.

-vorous *combining form* Consuming; eating or feeding upon: *omnivorous, carnivorous.* [<L *-vorus* < *vorare* devour]

vor·tex (vôr'teks) *n. pl.* **·tex·es** or **·ti·ces** (-tə·sēz) **1** A mass of rotating or whirling fluid, especially when sucked spirally toward the center; a whirlpool; an eddy. **2** A portion of fluid whose particles have rotary motion. [<L, var. of *vertex* top, point]

vor·ti·cal (vôr'ti·kəl) *adj.* Of, like, or causing a vortex. [<L *vortex, -icis* a vortex] **—vor'·ti·cal·ly** *adv.*

vor·ti·cose (vôr'tə·kōs) *adj.* Rotating rapidly; whirling; vortical. [<L *vorticosus* < *vortex, -icis* a vortex]

vor·tig·i·nous (vôr·tij'ə·nəs) *adj.* Moving as in a vortex. [<L *vortigo, -inis,* var. of *vertigo* a spinning]

Vor·tum·nus (vôr·tum'nəs) See VERTUMNUS.

Vosges Mountains (vōzh) A mountain chain in eastern France; highest peak, 4,672 feet.

Vos·toch·no-Si·bir·sko·ye Mo·re (vos·toch'nô·sē·bir'skô·yə mô'rə) See SIBERIAN SEA, EAST.

vo·ta·ry (vō'tər·ē) *n. pl.* **·ries 1** One devoted to some particular worship, pursuit, study, etc. **2** A worshiper, as of an idol. Also **vo'ta·rist. —** *adj.* Consecrated by a vow or promise; votive. [<L *votus,* pp. of *vovere* vow] **—vo'ta·ress, vo'tress** *n. fem.*

vote (vōt) *n.* **1** A formal expression of will or opinion in regard to some question submitted for decision, as in electing officers, passing resolutions, etc. **2** That by which such choice is expressed, as a show of hands, or ballot. **3** The result of an election; also, votes in the aggregate: the foreign *vote.* **4** The right to vote. **5** A voter. **6** *Obs.* A wish, vow, or prayer. **— casting vote** A deciding vote given by the chairman of an assembly in cases where the votes of the members tie. **—** *v.* **vot·ed, vot·ing** *v.t.* **1** To enact or determine by vote. **2** To cast one's vote for: to *vote* a straight ticket. **3** *Colloq.* To declare by general agreement: to *vote* a concert a success. **—** *v.i.* **4** To cast one's vote; express opinion or preference by or as by a vote. **— to vote down** To defeat or suppress by voting against. **— to vote in** To elect. [<L *votum* a vow, wish, orig. pp. neut. of *vovere* vow. Doublet of VOW.] **—vot'er** *n.*

vote-get·ter (vōt'get'ər) *n.* **1** A person with ability to win votes. **2** A campaign slogan, platform, etc., that draws votes. **—vote'-get'ting** *n.* & *adj.*

voting precinct An election district.

vo·tive (vō'tiv) *adj.* Dedicated by a vow; performed in fulfilment of a vow. [<L *votivus* < *votum.* See VOTE.] **—vo'tive·ly** *adv.* **—vo'·tive·ness** *n.*

votive mass A mass not rubrically assigned to a particular day, but said at the choice of the priest.

vouch (vouch) *v.i.* **1** To give one's own assurance or guarantee; bear witness: with *for*: I will *vouch* for their honesty. **2** To serve as assurance or proof: with *for*: The evidence *vouches* for his innocence. **—** *v.t.* **3** To bear witness to; attest or affirm. **4** To cite as support or justification, as a precedent, authority, etc. **5** To uphold by satisfactory proof or evidence; substantiate. **6** *Law* To call upon or summon (a person) to defend a title. **7** *Obs.* To call to witness. **—** *n.* A declaration that attests; an assertion. [<OF *vocher, voucher* <L *vocare* call < *vox, vocis* a voice]

vouch·ee (vou·chē') *n. Law* A person who is called into an action to warrant or defend a title.

vouch·er (vou'chər) *n.* **1** Any material thing (as a writing) that serves to vouch for the truth of something, or attest an alleged act, especially the receipt of money. **2** One who vouches for another; a witness. **3** In early English law, the calling in of a person, or the person called in, as warrantor, to defend a title.

vouch·safe (vouch'sāf') *v.* **·safed, ·saf·ing** *v.t.* **1** To grant, as with condescension; permit; deign. **2** *Obs.* To assure or guarantee. **—** *v.i.* **3** To condescend; deign. [<VOUCH + SAFE] **— vouch'safe'ment** *n.*

vous·soir (vōō·swär') *n. Archit.* A stone in an arch shaped to fit its curve. [<OF *vausoir, volsoir* curvature of a vault, ult. <L *volutus.* See VOLUBLE.]

vow (vou) *n.* **1** A solemn promise to God or to a deity or saint to perform some act or make some gift or sacrifice: generally made in a time of peril or need, and on the condition of the fulfilment of some petition or in return for special divine favor: the *vow* of Jephthah. **2** A solemn engagement to adopt a certain course of life, pursue some end, observe some moral precept, or surrender oneself to a higher life of holiness; also, a pledge of faithfulness: marriage *vows.* **3** A solemn and emphatic affirmation. See synonyms under OATH. **— to take vows** To enter a religious order. **—** *v.t.* **1** To promise solemnly; especially, to promise to God or to some deity. **2** To declare with assurance or solemnity. **3** To make a solemn promise or threat to do, inflict, etc. **—** *v.i.* **4** To make a vow. [<AF *vu,* OF *vo, vou* <L *votum.* Doublet of VOTE.] **—vow'er** *n.*

vow·el (vou'əl) *n.* **1** *Phonet.* A voiced speech sound produced by the relatively unimpeded passage of air through the mouth, altering in quality according to the shape of the resonance cavity: distinguished from *consonant.* Vowels may be characterized by the height of the tongue (high, mid, low), the place of articulation (front, central, back), the tension of the tongue muscles (tense, lax), and the presence of lip rounding; as, (ōō) is a high, back, tense, rounded vowel. **2** A letter indicating such a sound, as *a, e, i, o,* or *u.* **—** *adj.* Of or pertaining to a vowel; vocal. [<OF *vouele* <L *vocalis (littera)* vocal (letter) < *vox, vocis* a voice, sound. Doublet of VOCAL.]

vow·el·ize (vou'əl·iz) *v.t.* **·ized, ·iz·ing** To supply with vowel points or signs: to *vowelize* a Hebrew text. **—vow'el·i·za'tion** *n.*

vowel point One of a system of diacritical marks written above or below the consonants in Hebrew and certain other Semitic languages to indicate the vowel sound following the consonant.

vox (voks) *n. pl.* **vo·ces** (vō'sēz) Voice; especially, in music, a voice; part. [<L]

vox an·gel·i·ca (voks an·jel'i·kə) **1** An organ stop of two ranks of pipes, one of which is tuned slightly sharper than the other, so that beats are produced giving a tremulous effect; voix céleste: also **vox cae·les·tis** (si·les'tis). **2** A single-rank stop of soft, sweet quality. [<L, an angelic voice]

vox clan·des·ti·na (voks klan'des·tī'nə) *Latin* A secret voice; a whisper.

vox hu·ma·na (voks hyōō·mä'nə) A reed stop with very short pipes used for clarinet tones in an organ. [<L, a human voice]

vox po·pu·li (voks pop'yə·lī) *Latin* The voice of the people; public sentiment.

voy·age (voi'ij) *n.* **1** A journey by water, especially by sea: commonly used of a somewhat extended journey by water; formerly, any journey: a *voyage* across the sea. **2** A journey in an airship. **3** A book describing a voyage or voyaging: Hakluyt's *Voyages.* **4** Any enterprise or project; also, course. See synonyms under JOURNEY. **—** *v.* **·aged, ·ag·ing** *v.i.* To make a voyage; journey by water. **—** *v.t.* To travel over. [<OF *veiage, voiage* <L *viaticum.* Doublet of VIATICUM.] **—voy'ag·er** *n.*

voy·age·a·ble (voi'ij·ə·bəl) *adj.* Navigable.

vo·ya·geur (vwä·yà·zhœr') *n. pl.* **·geurs** (-zhœr') *French* An employee of Hudson's Bay Company, engaged in carrying men, supplies, etc., between remote trading posts; also, a Canadian boatman or fur trader.

vo·yeur (vwä·yûr') *n.* One who is sexually gratified by looking at sexual objects or acts; one given to scopophilia. [<F < *voir* see] **—vo·yeur'ism** *n.*

V-par·ti·cle (vē'pär'ti·kəl) *n. Physics* A hyperon.

vrai·sem·blance (vre·sän·bläns') *n. French* A show or appearance of truth; verisimilitude. [<F < *vrai* true + *semblance* appearance]

Vry·burg (frī'bûrg) The capital of Bechuanaland, Union of South Africa, an agricultural and livestock center near the Transvaal border.

VT fuze A proximity fuze. [<V(ARIABLE) T(IME)]

Vuel·ta A·ba·jo (vwel'tä ä·bä'hō) A region including all Cuba west of Havana.

vug (vug, vŏŏg) *n. Mining* An opening in a mineral vein into which crystals often project. Also **vugg, vugh.** [<Cornish *vooga* a cave] **—vug'gy** *adj.*

Vuil·lard (vwē·yàr'), **Jean Édouard,** 1868–1940, French painter.

Vul·can (vul'kən) In Roman mythology, the god of fire and of metallurgy: identified with the Greek *Hephaestus.*

vul·ca·ni·an (vul·kā'nē·ən) *adj.* **1** Volcanic: also **vul·can·ic** (vul·kan'ik). **2** Of or pertaining to Plutonism; plutonic. [<L *Vulcanius* pertaining to Vulcan < *Vulcanus* Vulcan]

Vul·ca·ni·an (vul·kā'nē·ən) *adj.* **1** Relating to Vulcan or to the art of working in metals. **2** Wrought by Vulcan or by Vulcan's art. Also **Vul·can·ic** (vul·kan'ik).

vul·can·ite (vul'kən·īt) *n.* A dark-colored hard variety of India rubber that has been subjected to vulcanization: also called *hard rubber.* **—** *adj.* Made of vulcanite. [after *Vulcan*]

vul·can·i·za·tion (vul'kən·ə·zā'shən, -ī·zā'-) *n.* **1** The process of treating crude India rubber with sulfur or sulfur compounds in varying proportions and at different temperatures, thereby increasing its strength and elasticity, yielding either soft rubber or vulcanite. **2** A similar process applied to other substances.

vul·can·ize (vul'kən·īz) *v.t.* & *v.i.* **·ized, ·iz·ing** To subject to or undergo the process of vulcanization. [after *Vulcan*] **—vul'can·iz'·a·ble** *adj.* **—vul'can·iz'er** *n.*

vulcanized fiber A cellulose material made from cotton and linen rags passed through a solution of zinc chloride, or sometimes of sulfuric acid.

vul·can·ol·o·gy (vul'kən·ol'ə·jē) *n.* Volcanology. [<L *Vulcanus* of Vulcan + -(O)LOGY] **—vul'·can·o·log'i·cal** (-ə·loj'i·kəl) *adj.* **—vul'can·ol'·o·gist** *n.*

vul·gar (vul'gər) *adj.* **1** Pertaining to the common people; plebeian; general; popular. **2** Pertaining to or characteristic of the people at large, as distinguished from the privileged or educated classes; coarse; boorish; offensive to good taste or sensitive feelings; low. **3** Written in or translated into the common language or dialect; vernacular. **—** *n. Obs.* **1** The common people. **2** The vernacular tongue. [<L *vulgaris* < *vulgus* the common people] **—vul'gar·ly** *adv.*

Synonyms (adj.): base, broad, coarse, gross, ignoble, inelegant, inferior, loose, low, mean, obscene, obscure, offensive, rude, unauthorized, underbred, vile. See COMMON. *Antonyms*: aristocratic, chaste, choice, cultivated, cultured, dainty, elegant, high-bred, learned, literary, lofty, polite, refined, select, stylish.

vulgar fraction A common fraction. See under FRACTION.

vul·gar·i·an (vul·gâr'ē·ən) *n.* A person of vulgar tastes or manners; especially, a wealthy person with coarse ideas or low standards.

vul·gar·ism (vul'gə·riz'əm) *n.* **1** Vulgarity. **2** A word, phrase, or expression that is in common colloquial or unrefined usage, though not necessarily coarse or gross: distinguished from those in literary or standard usage.

vul·gar·i·ty (vul·gar'ə·tē) *n. pl.* **·ties 1** The quality or character of being vulgar; low condition in life; commonness. **2** Lack of refinement in conduct or speech, or an instance of it; coarseness. Also **vul'gar·ness.**

vul·gar·ize (vul'gə·rīz) *v.t.* **·ized, ·iz·ing** To make vulgar. Also *Brit.* **vul'gar·ise. —vul'gar·i·za'tion** *n.* **—vul'gar·iz'er** *n.*

Vulgar Latin See under LATIN.

vul·gate (vul'gāt) *adj.* Common; popular; usual; generally accepted; in common use. **—** *n.* **1** The vulgar tongue; colloquial everyday speech. **2** Any commonly accepted text. [<L *vulgatus* common, orig. pp. of *vulgare* make common < *vulgus* the common people]

Vul·gate (vul'gāt) *n.* **1** St. Jerome's Latin version of the Bible, now revised and used as the authorized version by the Roman Catholics. Jerome translated the Gospels into Latin, then the vernacular or vulgar tongue, about A.D. 383, the remaining New Testament somewhat later, and the Old Testament from the Hebrew between 390 and 405. The Sistine edition of the Vulgate, published under Pope Clement VIII in 1592–93, is the source of the modern revision of the Douai version ordered by Pius X in 1908. **—** *adj.* Belonging

vulgo or relating to the Vulgate. [< Med. L *vulgata (editio)* the popular (edition), fem. of L *vulgatus* common]

vul·go (vul′gō) *adv. Latin* Commonly; popularly.

vul·ner·a·ble (vul′nər·ə·bəl) *adj.* 1 That may be wounded; capable of receiving injuries. 2 Liable to attack; assailable. 3 In contract bridge, having won one game of a rubber, and hence subject to doubled penalties if contract is not fulfilled. [< LL *vulnerabilis* wounding < L *vulnerare* wound < *vulnus, -eris* a wound] — **vul·ner·a·bil′i·ty, vul′ner·a·ble·ness** *n.* — **vul′ner·a·bly** *adv.*

vul·ner·ar·y (vul′nə·rer′ē) *adj.* Tending to cure wounds. — *n. pl.* **·ries** A healing application for wounds, as a preparation of medicinal plants. [< L *vulnerarius* < *vulnus, -eris* a wound]

Vul·pec·u·la (vul·pek′yə·lə) A small northern constellation lying between Cygnus and Aquila; the Fox: sometimes called **Vulpecula cum An·se·re** (kum an′sə·rē) (*The Little Fox with the Goose*). See CONSTELLATION. [< L, dim. of *vulpes* a fox]

vul·pec·u·lar (vul·pek′yə·lər) *adj.* Of or pertaining to a fox, especially a young one; vulpine.

vul·pi·cide (vul′pə·sīd) *n.* 1 One who kills a fox otherwise than by hunting. 2 The act of killing a fox when not hunting it with hounds. [< L *vulpes, -is* a fox + -CIDE] — **vul′pi·ci′dal** (-sīd′l) *adj.*

vul·pine (vul′pin, -pīn) *adj.* 1 Pertaining to a fox; resembling foxes. 2 Like a fox; sly; crafty; cunning. [< L *vulpinus* < *vulpes* a fox]

vul·pi·nite (vul′pə·nīt) *n.* A scaly variety of anhydrite from Vulpino, Italy.

vul·ture (vul′chər) *n.* 1 Any of various large birds of prey (family *Cathartidae* or *Vulturidae*) related to the eagles, hawks, and falcons, having the head and neck naked or partly naked, and feeding mostly on carrion; especially, the common **turkey vulture**, or buzzard, and the tropical American **king vulture** (*Gypagus papa*), strikingly colored, with black wings and tail. 2 Something or someone that preys upon a person in the manner of a vulture. [< AF *vultur*, OF *voltour* < L *vultur, vulturius*] — **vul·tur·ine** (vul′chə·rīn, -chər·in), **vul′tur·ous** *adj.*

VULTURE
(From 30 to 55 inches; wingspread from 7 to 11 feet)

vul·va (vul′və) *n. pl.* **·vae** (-vē) *Anat.* The external genital parts of the female, including the labia majora and minora, the clitoris, and the area between the clitoris and the labia minora. [< L, a covering, womb] — **vul′val, vul′var** *adj.* — **vul′vi·form** (-və·fôrm) *adj.*

Vyat·ka (vyät′kə) 1 A former name for KIROV. 2 A river in east central European Russian S.F.S.R., flowing 849 miles north, SW, and SE from the central Ural foothills to the Kama.

Vy·borg (vē′bôrg, *Russian* vi′berk) A port on the Gulf of Finland in NW Russian S.F.S.R. near the Finnish border. Swedish *Viborg*, Finnish *Viipuri*.

Vy·cheg·da (vi′chəg·də) A river in northern European Russian S.F.S.R., flowing 700 miles south and west to the Northern Dvina.

vy·ing (vī′ing) *adj.* That vies or contends. — **vy′ing·ly** *adv.*

Vy·shin·ski (vi·shin′skē), **Andrei**, 1883–1954, U.S.S.R. lawyer and politician: also *Vishinski*.

W

w, W (dub′əl·yōō, -yōō) *n. pl.* **w's, W's** or **ws, Ws** or **dou·ble·yous** 1 The 23rd letter of the English alphabet; double u: a ligature of vv or uu. It first came into English writing as a substitution by Norman scribes of the 11th century for the Old English rune *wen*, which later dropped completely out of use. 2 The sound of the letter *w*, a voiced bilabial velar semivowel before vowels (*we, wage, worry*), and a *u*-glide in diphthongs (*how, allow, dew, review*). It is silent before *r* (*wrist, write, wrong*), and is often lost internally (*two, sword, answer*). ◆ The combination *wh-* (in Old English spelled *hw-*) is pronounced in this dictionary as (hw); many educated speakers of English, however, use simple (w) instead, and this pronunciation should be inferred as an acceptable variant in every case. Some speakers normally use still a third sound here, a voiceless allophone of (w) heard also after voiceless consonants, as in *sweet, twin*, etc. See ALPHABET. — *symbol* 1 *Chem.* Tungsten (symbol W, for *wolfram*). 2 *Electr.* Watt.

wa′ (wä) *n. Scot.* Wall.

Waadt (vät) The German name for VAUD.

Waag (väkh) The German name for the VAH.

Waal (väl) The southern branch of the Rhine in the Netherlands, flowing 52 miles west from the Rhine proper to the Maas.

Waals (väls), **Johannes Diderik van der**, 1837–1923, Dutch physicist.

wab (wäb) *n. Scot.* A web.

Wa·bash (wô′bash) A river in western Ohio and north central and western Indiana, forming part of the boundary between Indiana and Illinois and flowing 475 miles NW, west, SW, and south from western Ohio to the Ohio River at the SW corner of Indiana.

wab·ble[1] (wob′əl) *n., v.t.* & *v.i.* Wobble. [Var. of WOBBLE] — **wab′bler** *n.* — **wab′bly** *adj.*

WAC (wak) *n.* A member of the Women's Army Corps. [< W(OMEN'S) A(RMY) C(ORPS)]

Wace (wās, wäs), 1100?–75, Anglo-Norman poet.

wack·e (wak′ə) *n.* A brown earthy or clayey variety of basaltic rock. [< G < MHG, a large stone < OHG *waggo* a pebble]

wack·y (wak′ē) *adj.* **wack·i·er, wack·i·est** *Slang* Extremely irrational or impractical; erratic; screwy. [Prob. < WHACK; with ref. to the mental impairment caused by repeated blows on the head]

Wa·co (wā′kō) A manufacturing city in central Texas.

wad[1] (wod) *n.* 1 A small compact mass of any soft or flexible substance, especially as used for stuffing, packing, or lining; also, a lump; mass: a *wad* of hair; also, a chew of tobacco, or a portion the right size for chewing. 2 A piece of paper, cloth, or leather used to hold in a charge of powder in a muzzleloading gun; also, a pasteboard or paper disk to hold powder and shot in place in a shotgun shell. 3 Fibrous material for stopping up breaks, leakages, etc.; wadding. 4 *Colloq.* A large amount. 5 *Colloq.* A roll of banknotes; hence, money; wealth. 6 A hydrated oxide of manganese and other metals. — *v.* **wad·ded, wad·ding** *v.t.* 1 To press (fibrous substances, as cotton) into a mass or wad. 2 To roll or fold into a tight wad, as paper. 3 To pack with wadding for protection, as valuables, or to stuff or line with wadding. 4 To place a wad in, as a gun; hold in place with a wad. — *v.i.* 5 To form into a wad. [Cf. Sw. *vadd* a wad] — **wad′dy** *adj.*

wad[2] (wod) *n. Scot.* A pledge; wager.

wad[3] (wod) *v.t.* & *v.i. Scot.* To wed.

wad[4] (wäd, wod) *v. Scot.* Would.

Wa·dai (wä·dī′) A former independent sultanate of north central Africa; now part of central and eastern Chad territory, French Equatorial Africa; 94,225 square miles: French *Ouadi*.

wad·ding (wod′ing) *n.* 1 Wads collectively. 2 Any substance, as carded cotton, used as material for wads. 3 The act of applying a wad or wads.

Wad·ding·ton (wod′ing·tən), **Mount** The highest mountain in British Columbia, in the SW part; 13,260 feet.

wad·dle (wod′l) *v.i.* **·dled, ·dling** 1 To walk with short steps, swaying from side to side. 2 To move clumsily; totter. — *n.* The act of waddling; a clumsy rocking walk, like that of a duck. [Freq. of WADE] — **wad′dler** *n.* — **wad′dly** *adj.*

wad·dy (wod′ē) *n. pl.* **·dies** *Austral.* 1 A thick war club used by the aborigines. 2 A walking stick; piece of wood. — *v.t.* **·died, ·dy·ing** To strike with a waddy.

wade (wād) *v.* **wad·ed, wad·ing** *v.i.* 1 To walk through water or, by extension, any substance more resistant than air, as mud, sand, etc. 2 To proceed slowly or laboriously: to *wade* through a lengthy book. 3 *Obs.* To go; proceed. — *v.t.* 4 To pass or cross, as a river, by walking on the bottom; walk through; ford. — **to wade in** (or **into**) *Colloq.* To attack or begin energetically or vigorously. — *n.* 1 The act of wading. 2 A ford. [OE *wadan* go]

wad·er (wā′dər) *n.* 1 One who wades. 2 A long-legged wading bird, as a snipe, plover, or stork. 3 *pl.* High waterproof boots, worn especially by anglers.

wa·di (wä′dē) *n. pl.* **·dies** 1 In Arabia and northern Africa, a river or valley; a ravine containing the bed of a watercourse, usually dry except in the rainy season. 2 An oasis. Also **wa′dy**. [< Arabic *wādī*]

Wa·di Hal·fa (wä′dē häl′fə) A city on the Nile in the northern Sudan near the Egyptian border; the northern gateway to the Sudan.

Wad·jak (wä′jək) A village in central Java.

Wadjak man A hominid (*Homo wadjakensis*), probably of the third interglacial period, and represented solely by two skulls found near Wadjak, Java, in 1891: of larger cranial capacity than earlier types but otherwise controversial because of the crushed condition of the remains and inadequate investigation at the site.

wad·mal (wod′məl) *n. Obs.* A thick, coarse, hairy, durable woolen cloth, used by the poor of northern Europe for garments. Also **wad′maal, wad′mol**. [< ON *vathmál* woolen fabric]

wad·na (wod′nə) *Scot.* Would not.

wad·set (wod′set′) *n.* In Scots law, a pledge, as of land, as security for a debt. — *v.t.* **·set·ted, ·set·ting** In Scots law, to mortgage.

wad·set·ter (wod′set′ər) *n. Scot.* One receiving a wadset.

wae (wā) *n. Scot.* Woe. — **wae′ness** *n.*

wae·ful (wā′fōōl) *adj. Scot.* Woeful; sad. Also **wae′fu** (-fōō).

wae·sucks (wā′suks) *interj. Scot.* Alas! Also **wae′suck.** [< WAE + alter. of SAKE[1]]

WAF (waf, wäf) *n.* A member of the Women in the Air Force. [< W(OMEN IN THE) A(IR) F(ORCE)]

Wafd (woft) *n.* A nationalist party in Egypt founded about 1919. [< Arabic, a deputation] — **Wafd′ist** *n.* & *adj.*

wa·fer (wā′fər) *n.* 1 A very thin crisp biscuit, cooky, or cracker; also, a small disk of candy. 2 *Eccl.* A small flat disk of unleavened bread stamped with a cross or the letters IHS, and used in the Eucharist in some churches; the sacred host. 3 A thin hardened disk of gelatin, flour, isinglass, or other suitable substance, used for sealing letters, attaching papers, or receiving the impression of a seal. 4 *Med.* **a** A thin double layer of dried paste enclosing a pill or capsule. **b** A suppository. 5 A disk of priming material used in early artillery. — *v.t.* To attach, seal, or fasten with a wafer. [<AF *wafre* <MLG *wafel.* Akin to WAFFLE.]

waff[1] (waf, wäf) *Scot. & Brit. Dial. v.t. & v.i.* To wave. — *n.* 1 The act of waving. 2 A light ailment. 3 A gust; puff. 4 A glimpse; sight. 5 A spirit or ghost. [Var. of WAVE]

waff[2] (waf, wäf) *Scot. adj.* 1 Low-born; worthless; inferior. 2 Strayed; solitary. — *n.* A tramp; vagrant.

Waf·fen Schutz·staf·feln (vä′fən shōōts′shtä·feln) *German* The divisions of the *Schutzstaffel* used by the Nazis to curb disturbances inside Germany; literally, protection echelon in arms: usually written **Waffen SS.**

waff·ie (wä′fē) *n. Scot.* A tramp.

waf·fle (wof′əl, wô′fəl) *n.* A batter cake, crisper than a pancake, baked in a waffle iron marked with regular indentations. [< Du. *wafel* a wafer. Akin to WAFER.]

waffle iron A type of utensil for cooking waffles, consisting of two metal griddles, hinged together, and usually marked with indentations so as to give a large heating surface when closed on each other: now usually made of aluminum and heated electrically.

waft[1] (waft, wäft) *v.t.* 1 To carry or bear gently or lightly over air or water; float. 2 To convey as if on air or water. — *v.i.* 3 To float, as on the wind. — *n.* 1 The act of one who or that which wafts. 2 A breath or current of air; also, a passing odor. [Back formation < *wafter,* in obs. sense, "an escort ship" <Du. *wachter* a guard < *wachten* guard]

waft[2] (waft, wäft) *n.* 1 A signal flag or pennant, sometimes used to indicate the direction of the wind to a ship's helmsman. 2 A signal made with a flag or pennant. — *v.t. Obs.* 1 To signal or beckon to with the hand. 2 To turn; direct, as a glance. [Alter. of dial. E *waff,* var. of WAVE]

waft[3] (waft, wäft) *n. Scot.* Woof; weft.

waft·age (waf′tij, wäf′-) *n.* Conveyance by wafting.

waft·er (waf′tər, wäf′-) *n.* 1 One who or that which wafts. 2 A form of fan or revolving disk used in a blower.

waf·ture (waf′chər, wäf′-) *n.* 1 A wafting or waving motion. 2 Conveyance by wafting. 3 That which is wafted, as an odor.

wag[1] (wag) *v.* **wagged, wag·ging** *v.t.* 1 To cause to move lightly and quickly from side to side or up and down; oscillate; swing: The dog *wags* its tail. 2 To move (the tongue) in talking. — *v.i.* 3 To move lightly and quickly from side to side or up and down. 4 To move busily in animated talk or gossip: said of the tongue. 5 To proceed at a regular pace: Life *wags* on. 6 To waddle. 7 *Brit. Slang* To play truant. — *n.* The act or motion of wagging: a *wag* of the head. [ME *waggen,* prob. <Scand. Cf. Sw. *vagga* rock a cradle. Akin to OE *wagian* oscillate.]

wag[2] (wag) *n.* A droll or humorous fellow; a wit; joker. [Short for obs. *waghalter* <WAG[1] + HALTER[1]]

wage (wāj) *v.t.* **waged, wag·ing** 1 To engage in and maintain vigorously; carry on: to *wage* war. 2 *Obs.* To pledge; put down as security; hence, to wager; bet. 3 *Obs.* To attempt; risk. 4 *Brit. Dial.* To pay a salary to; hire; employ. — *n.* 1 Payment for service rendered, especially the pay of artisans or laborers receiving a fixed sum by the day, week, or month, or for a certain amount of work; hire. 2 *pl.* The remuneration received by labor as distinguished from that received by capital, including the expenses incurred for superintendence and management, called respectively **wages of superintendence** and **wages of management.** 3 Figuratively, produce; yield. 4 *Obs.* A pledge; gage; also, the state of being pledged: to lay one's fortune in *wage.* See synonyms under SALARY. See LIVING WAGE, MINIMUM WAGE. [<AF *wagier,* OF *guagier* pledge <*gage* a pledge. Doublet of GAGE[2].] ◆ The plural of **wage** is sometimes construed as a singular: The *wages* of sin *is* death.

Wage may appear as a combining form in hyphemes or solidemes, or as the first element in two-word phrases:

wage board	**wage floor**
wage ceiling	**wage–freeze**
wage–control	**wage–labor**
wage differential	**wage law**
wage–driver	**wage level**
wage–incentive	**wage–slave**
wage–increase	**wage–slavery**
wage–paying	**wage structure**
wage rate	**wagework**

wage–earn·er (wāj′ûr′nər) *n.* One who works for wages.

wa·ger (wā′jər) *v.t. & v.i.* To stake (something) on an uncertain event; bet. — *n.* 1 An agreement between persons that something, as money, shall be delivered over to one of them on the happening or not happening of an uncertain event; a bet. 2 The thing so pledged. 3 The act of giving a pledge. [<AF *wageure* < *wagier.* See WAGE.] — **wa′ger·er** *n.*

wager of law Anciently, a mode of trial whereby a defendant acquitted himself of a debt by taking his oath that he owed the plaintiff nothing, and having eleven compurgators present to swear that they believed his oath to be true.

wage scale 1 A scale or series of amounts of wages paid for similar duties. 2 The scale of wages paid by a single employer.

wage·worker (wāj′wûr′kər) *n.* An employee receiving wages.

wag·ger·y (wag′ər·ē) *n. pl.* **·ger·ies** 1 Mischievous jocularity; drollery. 2 A jest; joke. See synonyms under WIT[1]. [<WAG[2] + -ERY]

wag·gish (wag′ish) *adj.* 1 Being or acting like a wag. 2 Said or done in waggery. See synonyms under JOCOSE. — **wag′gish·ly** *adv.* — **wag′gish·ness** *n.*

wag·gle (wag′əl) *v.* **·gled, ·gling** *v.t.* To cause to move with rapid to–and–fro motions; wag; swing: The duck *waggles* its tail. — *v.i.* To totter; wobble. — *n.* The act of waggling or wagging. [Freq. of WAG[1]] — **wag′gling·ly** *adv.* — **wag′gly** *adj.*

Wag·ner (väg′nər), **(Wilhelm) Richard,** 1813–83, German composer, poet, and critical writer.

Wag·ner·esque (väg′nə·resk′) *adj.* Similar to or suggestive of the works or style of Richard Wagner.

Wag·ne·ri·an (väg·nir′ē·ən) *adj.* Relating to Richard Wagner or to his style, theory, or works. — *n.* One who advocates or accepts the theories of Richard Wagner; also, one who admires his works.

Wag·ner·ism (väg′nə·riz′əm) *n.* The theory of Richard Wagner regarding music drama, as exemplified in the construction and rendition of his own works. Its chief point, especially that in which it differs from the method of the old Italian composers of opera, is its abundant use of the leitmotif for cumulative dramatic effect and its insistence on the equal participation of music, both vocal and orchestral, poetry, scenic effect, and dramatic action, no one of these being subordinate.

Wag·ner–Jau·regg (väg′nər·you′rek), **Julius,** 1857–1940, Austrian neurologist and psychiatrist. Also **Wag′ner von Jau′regg.**

wag·on (wag′ən) *n.* 1 A strong four–wheeled vehicle used to carry heavy loads of freight. Compare DRAY, WAIN. 2 An open four–wheeled vehicle for carrying hay, corn, etc.: a farm *wagon.* 3 A light four–wheeled vehicle used for various business purposes, as a grocer's *wagon.* 4 *Brit.* A railway freight car. 5 A covered four–wheeled vehicle used as living quarters by gipsies, traveling showmen, etc. 6 *Obs.* A chariot. 7 *Colloq.* A patrol wagon. 8 A station wagon. 9 *Slang* An automobile. 10 A child's four–wheeled toy cart. 11 *Astron.* Charles's Wain. 12 A stand on wheels or casters for serving food or drink: a tea *wagon.* — **on the (water) wagon** *Colloq.* Abstaining from alcoholic beverages. — **to fix (someone's) wagon** *Slang* To even scores with; obtain revenge on: I'll *fix your wagon!* — *v.t.* To carry or transport in a wagon. Also *Brit.* **wag′gon.** [<Du. *wagen.* Akin to WAIN.]

wag·on·age (wag′ən·ij) *n.* 1 The amount paid for conveyance in a wagon. 2 Wagons collectively. Also *Brit.* **wag′gon·age.**

wagon bed The body of a wagon.

wag·on·er (wag′ən·ər) *n.* 1 One whose business is driving wagons. 2 *Obs.* A charioteer. Also *Brit.* **wag′gon·er.**

Wag·on·er (wag′ən·ər) 1 Charles's Wain. See under WAIN. 2 The constellation Auriga.

wag·on·ette (wag′ən·et′) *n.* A light wagon, with or without a cover, with lengthwise seats facing inward and a crosswise seat in front for the driver. Also **wag′on·et.** [Dim. of WAGON]

wag·on–head·ed (wag′ən·hed′id) *adj. Archit.* Having a semicylindrical head or top, resembling the top of a covered wagon; having a round–arched roof.

wa·gon–lit (và·gôn·lē′) *n. pl.* **–lits** (-lē′) *French* A sleeping–car on a French railway.

wag·on·load (wag′ən·lōd′) *n.* The amount that a wagon can carry.

wagon train 1 A train or line of wagons. 2 A group of wagons and families typical of those which formerly traveled together to settle new regions, especially in the western United States. 3 The equipment of a military force for the carriage of ammunition, provisions, etc.

Wa·gram (vä′gräm) A village NE of Vienna, Austria; scene of Napoleon's victory over the Austrians, 1809.

wag·some (wag′səm) *adj.* Mischievous; waggish: a rare use. [<WAG[2] + -SOME[1]]

wag·tail (wag′tāl′) *n.* 1 Any of several small singing birds (genus *Motacilla*), named from their habit of jerking the tail; especially, the **yellow wagtail** *(M. flava)* of Asia and eastern Alaska. 2 Any of certain American birds that wag the tail when walking on the ground, especially the ovenbird and the water thrush.

Wa·ha·bi (wä·hä′bē) *n.* A believer in Wahabiism. Also **Wa·ha′bee, Wah·ha′bi.**

Wa·ha·bi·ism (wä·hä′bē·iz′əm) *n.* A puristically orthodox Moslem sect of Arabia, related to the Sunnites, founded by Abdul–Wahhab; the religion of the ruling family of Saudi Arabia.

wah·con·da (wä·kon′dä) See WAKANDA.

wa·hoo[1] (wä·hōō′, wä′hōō) *n.* A deciduous North American shrub or small tree *(Euonymus atropurpureus)* with finely toothed leaves, purple flowers, and scarlet fruit: also called *burningbush.* [<Siouan (Dakota) *wanhu,* lit., arrowwood]

wa·hoo[2] (wä·hōō′, wä′hōō) *n.* 1 The American winged elm *(Ulmus alata).* 2 The white basswood *(Tilia heterophylla).* 3 The cascara buckthorn. [<Muskhogean (Creek) *uhawhu* the winged elm]

Wai·chow (wī′jō′) A former name for WAI-YEUNG.

waif (wāf) *n.* 1 A homeless, neglected wanderer; a stray. 2 *Law* Something stolen and then abandoned by the thief in his flight to avoid arrest. 3 Anything found and unclaimed, the owner being unknown. 4 A nautical signal; a waft. — *v.t.* To throw away; cast off, as a waif. — *adj.* Stray; wandering; homeless. [<AF *waif,* OF *gaif,* prob. <Scand. Cf. ON *veif* something flapping < *veifa* wave.]

Wai·ki·ki (wī′kē·kē, wī′kē·kē′) A section of Honolulu; site of a resort beach on Honolulu harbor, SE Oahu, Hawaii.

wail (wāl) *v.t. & v.i.* To grieve with mournful cries; lament; cry out in sorrow. — *n.* A prolonged, high–pitched sound of lamentation; a shrill moan of grief; also, any mournful sound, as of the wind. ◆ Homophone: *wale.* [<ON *vǣla* wail < *væ, vei* woe] — **wail′er** *n.*

wail·ful (wāl′fəl) *adj.* 1 Deeply sorrowing; mournful. 2 Making a mournful sound.

Wailing Place of the Jews A courtyard in Jerusalem, one wall of which reputedly contains fragments of Solomon's temple. Here the Jews assemble on Fridays to mourn and pray. Also **Wailing Wall.**

wain (wān) *n.* An open, four–wheeled wagon for hauling heavy loads. — **Charles's Wain** Seven bright stars in Ursa Major; the Dipper: also **the Wain:** sometimes called the *Wagoner.* ◆ Homophone: *wane.* [OE *wægn, wǣn.* Akin to WAGON.]

wain·scot (wān′skət, -skot) *n.* 1 A facing for inner walls, usually of wood, but sometimes of marble or other material: usually paneled and of elaborate workmanship. 2 *Brit.* A superior quality of imported oak used for paneling; also, a piece of such wood. 3 The lower part of an inner wall, when finished with material different from the rest of the wall. — *v.t.* **·scot·ed** or **·scot·ted, ·scot·ing** or **·scot·ting** To face or panel with wainscot. [<MLG *wagenschot* < *wagen* a wagon + *schot* a wooden partition]

wain·scot·ing (wān′skət·ing, -skot-) *n.* Material

for a wainscot; a wainscot; wainscots collectively. Also **wain′scot·ting.**

wain·wright (wān′rīt) *n.* A maker of wagons.

Wain·wright (wān′rīt), **Jonathan,** 1883–1953, U. S. general.

Wai·pa·hu (wī·pä′ho͞o) A city of southern Oahu, Hawaii, on Pearl Harbor, NW of Honolulu.

wair (wâr) See WARE[3].

waist (wāst) *n.* **1** The narrow part of the body between the chest and the hips. **2** The middle part or section of any object, especially if of less diameter than the ends: the *waist* of a violin. **3** *Naut.* That section of a ship between the quarter-deck and the forecastle. **4** The central section of an airplane. **5** That part of a woman's dress or other garment covering the body from the waistline to the neck or shoulders; a bodice; also, an undergarment for children, to which other garments may be buttoned. **6** A waistband. ◆ Homophone: *waste.* [ME *wast.* Akin to OE *wæstm* growth.]

waist·band (wāst′band′, -bənd) *n.* A band encircling the waist, especially a band inside the top of a skirt or the upper part of trousers.

waist·cloth (wāst′klôth′, -kloth′) See LOINCLOTH.

waist·coat (wāst′kōt′, wes′kit) *n.* **1** A man's garment, now commonly sleeveless, buttoning in front and extending just below the waistline; a vest. **2** A similar garment worn by women. **3** A long vest formerly worn with trunk and hose under a slip doublet.

waist·coat·ing (wāst′kō′ting, wes′kit·ing) *n.* A textile fabric specially designed for men's waistcoats.

waist·er (wās′tər) *n. Rare* An apprentice or new hand on a whaling vessel, placed at work in the ship's waist to learn his duties.

waist·ing (wās′ting) *n.* Any material suitable for making waists.

waist·line (wāst′līn′) *n.* The line of the waist, between the ribs and the hips; in dressmaking, the line at which the skirt of a dress meets the waist.

wait (wāt) *v.i.* **1** To stay or remain in expectation, as of an anticipated action or event: with *for, until,* etc. **2** To be or remain in readiness. **3** To remain temporarily neglected or undone. **4** To perform duties of personal service or attendance; especially, to act as a waiter or waitress: She *waits* at table. — *v.t.* **5** To stay or remain in expectation of: to *wait* one's turn. **6** *Colloq.* To put off or postpone; defer; delay: Don't *wait* breakfast for me. **7** *Obs.* To attend; escort. **8** *Obs.* To attend as a result or consequence. See synonyms under ABIDE, LINGER. — **to wait on** (or **upon**) **1** To act as a servant or attendant to. **2** To go to see; call upon; visit. **3** To attend as a result or consequence. — **to wait up** To delay going to bed in anticipation of the arrival of someone. — *n.* **1** The act of waiting, or the time spent in waiting; delay. **2** An ambush or trap; snare: to lie in *wait* for a victim. **3** A member of a musical band organized to play and sing in the streets, at night or dawn: now applied only to those who sing carols in the streets at Christmastime. **4** *Obs.* A watchman or guard. [<AF *waitier,* OF *guaitier* <OHG *wahten* watch < *wahta* a guard]

wait–a–bit (wāt′ə·bit′) *n.* Any one of various plants with sharp or hooked thorns that catch and tear the clothing, and thus detain those who would pass through them, as the greenbrier or the prickly ash. [Trans. of Afrikaans *wacht–een–beetje*]

Waite (wāt), **Morrison Remick,** 1816–88, U. S. jurist; chief justice of the Supreme Court 1874–88.

wait·er (wā′tər) *n.* **1** One who waits upon others, as in a restaurant. **2** One who awaits something. **3** A tray for dishes, etc. **4** *Obs.* A watchman or keeper.

wait·ing (wā′ting) *n.* The act or business of a waiter; attendance. — **in waiting** In attendance, especially at court. — *adj.* That waits; expecting.

waiting list A list of people waiting to be admitted to some institution, as a school or club, or to some privilege or opportunity.

waiting room A room for the use of persons waiting, as for a railroad train, a doctor, dentist, or the like.

wait·ress (wā′tris) *n.* A woman or girl employed to wait on guests at table, as in a restaurant.

waive (wāv) *v.t.* **waived, waiv·ing 1** To give up or relinquish a claim to. **2** To refrain from insisting upon or taking advantage of; forgo. **3** To put off; postpone; delay. **4** *Law* To surrender, abandon, or relinquish voluntarily, either expressly or by implication, as a claim, privilege, or right. **5** *Obs.* To reject; cast off; abandon; desert. ◆ Homophone: *wave.* [<AF *weyver,* OF *gaiver* abandon <AF *weyf, waif* WAIF]

waiv·er (wā′vər) *n. Law* The voluntary relinquishment of a right, privilege, or advantage; also, the instrument which evidences such relinquishment. [<AF, var. of *weyver* abandon. See WAIF.]

Wai·yeung (wī′yüng′) A city in eastern Kwangtung province, China; a river port on the Tung, east of Canton: formerly *Waichow.*

wa·kan·da (wä·kän′dä) *n.* Among the Sioux, the great power or supreme being behind the world: identical with Algonquian *manito*: also spelled *wahconda.* [<Siouan]

Wa·ka·ya·ma (wä·kä·yä·mä) A port of southern Honshu island, Japan.

wake[1] (wāk) *v.* **woke** or **waked, waked** (*Dial.* **wok·en**), **wak·ing** *v.i.* **1** To be roused from sleep or slumber. **2** To be or remain awake. **3** To become active or alert after being inactive or dormant. **4** *Dial.* To keep watch or guard at night; especially, to hold a wake (def. 1). **5** *Obs.* To feast or revel late into the night. — *v.t.* **6** To rouse from sleep or slumber; awake. **7** To rouse or stir up; excite: to *wake* evil passions. **8** *Dial.* To keep a vigil over; especially, to hold a wake over. See synonyms under STIR[1]. — *n.* **1** A watch over the body of a dead person through the night, just before the burial, by the relatives and friends: common among the Irish, and often accompanied by conviviality. **2** Formerly, in the Anglican Church, a dedication festival or anniversary celebration of a parish church, preceded by a night vigil in the church. **3** The act of refraining from sleep, especially on a festive or solemn occasion. **4** *Obs.* The act of waking, or the state of being awake; vigil. [Fusion of OE *wacan* awake and *wacian* be awake. Akin to WATCH.]

◆ **awake, awaken, wake, waken** These four verbs, so similar in basic meaning, offer a confusing variety of choices in actual use. In the imperative, *Wake up!* is the familiar and homely form; the other three would be felt as poetic. *Awake* and *wake* have checkered form–histories. In the King James Bible, Shakespeare, and Milton, only the inflected forms in *-ed* are found. But *awake* and *wake* each had a strong verb as well as a weak one in its ancestry, and in the late seventeenth century the alternative inflected forms *awoke, awoken,* and *woke, woken* emerged, reinforced by analogy with *break, broke, broken,* etc. These alternative forms have led to uncertainty and confusion in usage. For the past tense of *awake, awoke* is usual; *awaked* tends to be felt as Biblical. *Awoke* as the past participle is rare, and *awaked* seems awkward to some; what happens in practice is a borrowing of the past participle from *awaken.* For *wake,* the more usual past is *woke* (or *woke up*); *waked* in the intransitive is also standard, but in the transitive sense it is dialectal, referring to holding a vigil or wake: They *waked* old Tim on Thursday night. Real uncertainty arises over the form to choose for the past participle of *wake.* In British (or dialectal American) usage, *woken,* or, for the phrasal verb, *woken up* is used: He had *woken* (or *woken up*) early. American usage here employs *waked (up),* or if this is felt as awkward, particularly in the passive, the past participle is borrowed from *waken* or *awaken*: What *woke* her? She was *awakened* (or *wakened*) by the noise of the crash.

wake[2] (wāk) *n.* **1** The track left by a vessel passing through the water. **2** Any course passed over. — **in the wake of 1** Following close behind. **2** In consequence of. [<ON *vök* an opening in ice]

Wake·field (wāk′fēld) **1** A county borough in southern Yorkshire, England; scene of Lancastrian victory in the Wars of the Roses, 1460. **2** The birthplace of George Washington, an estate on the Potomac River in SE Virginia.

wake·ful (wāk′fəl) *adj.* **1** Remaining awake, especially at the ordinary time of sleep; not sleeping or sleepy. **2** Watchful; alert. **3** Unable to sleep; restless; suffering from insomnia. **4** Arousing from or as from sleep. See synonyms under VIGILANT. — **wake′ful·ly** *adv.* — **wake′ful·ness** *n.*

Wake Island (wāk) A U. S. naval and air base, comprising a coral atoll and three islands in the North Pacific, acquired by the United States in 1898; 4 square miles; about 45 miles long, 2 1/4 miles wide; occupied by Japanese forces, 1941–45.

wake·less (wāk′lis) *adj.* Uninterrupted; unbroken: a *wakeless* sleep.

wak·en (wā′kən) *v.t.* **1** To rouse from sleep; awake. **2** To rouse to alertness or activity. — *v.i.* **3** To cease sleeping; wake up. **4** *Obs.* To keep awake; also, to keep watch. [OE *waecnan, wacnian*]

wak·en·er (wā′kən·ər) *n.* One who or that which awakens.

wak·er (wā′kər) *n.* **1** A wakener. **2** One who stays awake; a watcher.

wake·rife (wāk′rīf) *adj. Scot.* or *Obs.* Wakeful; alert. — **wake′rife·ness** *n.*

wake·rob·in (wāk′rob′in) *n.* **1** The cuckoo pint. **2** Any species of trillium; the mooseflower. **3** The jack–in–the–pulpit.

wakf (wukf) *n.* In Moslem law, the inalienable dedication of property in trust for the service of God or charitable uses; also, the property so dedicated. [<Arabic *waqf*]

wa·kif (wu′kif) *n.* One who makes a wakf.

wa·ki·ki (wä′ki·kē) *n.* Shell money of the South Sea Islands. [<Melanesian]

Waks·man (waks′mən), **Selman Abraham,** born 1888, U. S. microbiologist; discovered streptomycin.

Wa·la·chi·a (wo·lā′kē·ə) See WALLACHIA.

Wał·brzych (vä′o͞o·bzhikh) A city in SW Poland, in former Lower Silesia; a coal-mining and manufacturing center. *German* **Wal·den·burg** (väl′dən·bo͝orkh).

Wal·che·ren (väl′khə·rən) An island at the mouth of the Scheldt river in SW Netherlands; the westernmost island of Zeeland province; 80 square miles.

Wal·deck (väl′dek) An administrative district of Hesse; 420 square miles; formerly a principality of western Germany, a state of the German Republic (1918–29), and a Prussian province (1929–45).

Wal·de·mar (väl′də·mär) Name of four kings of Denmark: also *Valdemar.*

— **Waldemar I,** 1131–82, king 1157–82: called "Waldemar the Great."

— **Waldemar II,** 1170–1241, king 1202–41; greatly extended Danish territory: called "Waldemar the Victorious."

Wal·den·ses (wol·den′sēz) *n. pl.* A sect of religious dissenters founded about 1170 by Peter Waldo or Valdo, a rich merchant of Lyons, France. Waldo and his disciples sought to restore the church to its early purity and poverty, but were excommunicated by Pope Alexander III, and severely persecuted. [<Med. L *Waldenses* of (Peter) *Waldo*] — **Wal·den′si·an** *adj.* & *n.*

wald·grave (wôld′grāv) *n.* **1** An old German title of nobility. **2** Originally, the lord or intendant of a forest. Compare LANDGRAVE, MARGRAVE. [<G *waldgraf* < *wald* a wood + *graf* a count]

Wal·do (wôl′dō), **Peter** Late 12th century French religious reformer: also *Valdo, Valdez.* See WALDENSES.

Wal·dorf salad (wôl′dôrf) A salad made of chopped celery, apples, and walnuts, and garnished with lettuce and mayonnaise. [from the first *Waldorf*–Astoria Hotel, New York City]

Wald·stät·ter, Die Vier (dē fēr vält′shtet′ər) Lucerne, Schwyz, Unterwalden and Uri, the forest cantons of Switzerland: the original cantons of the Swiss federation.

Wald·teu·fel (väl′toi·fəl), **Émile,** 1837–1915, French composer born in Alsace.

wale[1] (wāl) *n.* 1 A stripe or ridge made on living flesh by a rod, whip, or stick; a wheal. 2 *Naut.* One of certain strakes of outer planking running fore and aft on a vessel: the channel *wales*. 3 A ridge or rib on the surface of cloth; hence, texture; grain. — *v.t.* **waled, wal·ing** 1 To raise wales or stripes on by striking, as with a lash; flog; beat. 2 To manufacture, as cloth, with a ridge or rib. 3 To weave, as wickerwork, with several rods together. 4 To protect, fasten, or hold with wales. ◆ Homophone: *wail.* [OE *walu*]

wale[2] (wāl) *Dial. & Scot. n.* A choice or preference of one thing from among others; also, the best; the cream. — *adj.* Well-selected; choice. — *v.t.* **waled, wal·ing** To choose; select; hence, to woo. ◆ Homophone: *wail.* [<ON *val* choice]

Wal·er (wā′lər) *n. Anglo-Indian* A horse of New South Wales exported to India for cavalry service; also, any horse from Australia.

Wales (wālz) A peninsula of SW Britain, comprising a principality of England, with which it has been politically united since 1536; 8,016 square miles.

Wal·fish Bay (wŏl′fish) See WALVIS BAY.

Wal·hal·la (wal·hal′ə, -häl′ä, val-) See VALHALLA.

walk (wôk) *v.i.* 1 To advance on foot in such a manner that one part of a foot is always on the ground; of quadrupeds, to advance in such a manner that two or more feet are always on the ground. 2 To move or go on foot for exercise or amusement. 3 To proceed or advance slowly. 4 To move in a manner suggestive of walking, as a piece of masonry subjected to wind pressure. 5 To act or live in some manner: to *walk* in peace. 6 To return to earth and appear, as a ghost. 7 In baseball, to advance to first base on balls. 8 In basketball, to take more than two steps while holding the ball. 9 *Obs.* To be in continual motion. — *v.t.* 10 To pass through, over, or across at a walk: to *walk* the floor. 11 To cause to go at a walk; lead, ride, or drive at a walk: to *walk* a horse. 12 To force or help to walk. 13 To accompany on a walk. 14 To bring to a specified condition by walking. 15 To measure or survey by traversing on foot: to *walk* a boundary. 16 To cause to move with a motion resembling a walk: to *walk* a trunk on its corners. 17 In baseball, to allow to advance to first base on balls. 18 In basketball, to take more than two steps while holding (the ball). — **to walk off** 1 To depart, especially abruptly or without warning. 2 To get rid of (fat, drunkenness, etc.) by walking. — **to walk off with** 1 To win. 2 To steal. — **to walk out** *Colloq.* 1 To go out on strike. 2 To keep company: with *with* or *together.* — **to walk out on** *Colloq.* To forsake; desert. — **to walk over** 1 In certain sports, to walk over the course without a competitor so as to perform the technicality of winning; hence, to gain an easy victory. 2 To defeat easily; overwhelm. — **to walk Spanish** To be compelled to walk on tiptoe by being seized by the scruff of the neck and the seat of the trousers, as in some boys' games; hence, to walk gingerly; also, to be discharged. — *n.* 1 The act of walking, as for enjoyment or recreation; a stroll. 2 Manner of walking; gait; specifically, the gait of a horse in which two or more feet are always on the ground. 3 Method or way of living; behavior. 4 Chosen profession or habitual sphere of action: the different *walks* of life. 5 Distance as measured by the time taken by one who walks: It's an hour's *walk* to my house. 6 A place laid out or set apart for walking or resorted to by those who walk; a path, avenue, promenade, or sidewalk for pedestrians. 7 A ropewalk. 8 The formation of, or space between, two lines or rows of plants or trees, as in a coffee plantation. 9 A piece of ground set apart for the feeding and exercise of domestic animals; range; pasture: a *sheepwalk*. 10 A hawker's or vender's district or beat; a beat. 11 A contest of speed in walking. 12 In baseball slang, a base on balls. [OE *wealcan* roll, toss] — **walk′er** *n.* — **walk′ing** *adj. & n.*

walk-a·round (wôk′ə·round′) *n.* 1 A rhythmic Negro dance performed by a group walking around in a large circle; also, the music composed for this dance. 2 A dance of this kind performed on the stage; also, the music for it.

walk-a·way (wôk′ə·wā′) *n.* A contest won without serious opposition.

Walk·er (wô′kər), **John**, 1732–1807, English lexicographer and actor. — **William,** 1824–60, U.S. filibuster in Lower California and Nicaragua.

walk·ie-talk·ie (wô′kē·tô′kē) *n. Telecom.* A portable radio set, equipped for both sending and receiving, and light enough to be carried by one man. Also spelled *walky-talky*.

walking bass An insistently reiterated bass figure, usually in eighth notes, used in boogie-woogie music.

walking beam *Mech.* In a vertical engine, a horizontal beam that transmits power to the crankshaft through the connecting rod.

WALKING BEAM

walking delegate See under DELEGATE.

walking fern A tufted evergreen fern (*Camptosorus rhizophyllus*) with fronds ending in long tapering tips which take root and thus give rise to new plants. Also **walking leaf**.

walking papers Notice of dismissal from employment, office, position, etc.

walking stick 1 A staff or cane carried in the hand. 2 Any of a family (*Phasmidae*) of insects having legs, body, and wings resembling one of the twigs among which it lives.

walk-on (wôk′on′, -ôn′) *n.* An actor who plays a bit part or merely walks on the stage; also, the part.

walk-out (wôk′out′) *n. Colloq.* 1 The act of walking out. 2 A workmen's strike.

walk-o·ver (wôk′ō′vər) *n.* 1 A horse race in which there is only one horse entered, and which can thus be won by going over the course at a mere walk. 2 An easy or unopposed success.

walk-up (wôk′up′) *Colloq. n.* An apartment house having no elevator. — *adj.* Having no elevator.

Wal·kü·re (väl·kü′rə), **Die** *German* The second music drama in Richard Wagner's tetralogy, *Der Ring des Nibelung*. See RING OF THE NIBELUNG, VALKYRIE.

walk·way (wôk′wā′) *n.* A sidewalk; a passage; a garden path.

wal·kyr·ie (wal·kir′ē, val-) *n.* A valkyrie. [OE *wælcyrie*, lit., a chooser of the slain]

walk·y-talk·y (wôk′ē·tô′kē) See WALKIE-TALKIE.

wall (wôl) *n.* 1 A continuous structure, as of stone or brick, designed to enclose an area, to provide defense or security, or to be the surrounding exterior of a house or a partition between rooms or halls; also, a fence of stone or brickwork, surrounding or separating yards, fields, etc. ◆ Collateral adjective: *mural*. 2 A barrier or rampart constructed for defense: in the plural, fortifications. 3 A sea wall; levee. 4 The side of any cavity, vessel, or receptacle; a parietal surface: the *walls* of the abdomen. 5 Something suggestive of a wall or barrier: a *wall* of bayonets. See synonyms under RAMPART. — **to drive, push,** or **thrust to the wall** To force (one) to an extremity; crush. — **to go to the wall** To be pressed or driven to an extremity; be forced to yield. — **to take the wall** To take the inner side of the walk; hence, to take a rude advantage. — *v.t.* 1 To provide, surround, protect, etc., with or as with a wall or walls. 2 To fill or block with a wall: often with *up.* — *adj.* Of or pertaining to a wall; hanging or growing on a wall. [OE *weall, wall* <L *vallum* a rampart < *vallus* a stake, palisade]

Wall may appear as the first element in two-word phrases:

wall arcade	wall bracket	wall crane
wall arch	wall case	wall engine
wall berry	wall casing	wall face
wall border	wall clock	wall garden
wall box	wall coping	wall map
wall mosaic	wall plant	wall tower
wall moss	wall plug	wall tree
wall nook	wall top	wall vase

wal·la·by (wol′ə·bē) *n. pl.* **·bies** One of the smaller kangaroos frequenting forests and regions of brush growth in Australia, as the **rock wallaby** (genus *Petrogale*). [<Australian *wolabā*]

Wal·lace (wol′is), **Alfred Russel,** 1823–1913, English naturalist. — **(Richard Horatio) Edgar,** 1875–1932, English novelist. — **Henry Agard,** born 1888, U.S. vice president 1941–1944; agriculturist, editor, and politician. — **Lewis,** 1827–1905, U.S. general, administrator, and author: known as *Lew Wallace*. — **Sir William,** 1272?–1305, Scottish national hero; executed by the English.

Wal·lach (väl′äkh), **Otto,** 1847–1931, German chemist.

Wal·la·chi·a (wo·lā′kē·ə) A historic region and former principality in southern and SE Rumania; 29,575 square miles; chief city, Bucharest: also *Walachia*. — **Wal·la′chi·an** *adj. & n.*

wal·lah (wä′lä) *n. Anglo-Indian* A person engaged in a specified occupation or activity, as a merchant, vender, agent, worker, or servant; popularly and somewhat contemptuously, a man or fellow. Also **wal′la. — punka wallah** The servant whose job it is to keep the punka in motion. [<Hind. *-vālā*, suffix indicating a personal agent]

wal·la·roo (wol′ə·rōō′) *n.* A species of kangaroo (*Macropus robustus*) of great size. Also called **wallaroo kangaroo**. [<Australian *wolarū*]

Wal·la·sey (wol′ə·sē) A county borough in NW Cheshire, England, on the Mersey river opposite, and forming part of the port of, Liverpool.

Wal·la-wal·la (wol′ə·wol′ə) *n.* One of a small tribe of North American Indians of Shahaptian linguistic stock of the NW Pacific coast: now on a reservation in Oregon.

Wal·la Wal·la (wol′ə wol′ə) A city in SE Washington near the Oregon border on the Walla Walla River.

Walla Walla River A river in NE Oregon and SW Washington, flowing 60 miles NW to the Columbia River.

wall·board (wôl′bôrd′, -bōrd′) *n.* A material composed of several layers of compressed wood chips and pulp, molded and sized for use as a substitute for wooden boards and plaster.

wall·creep·er (wôl′krē′pər) *n.* A small, brilliantly colored Old World bird (*Tichodroma muraria*) that obtains its insect prey by creeping on cliffs and walls.

Wal·len·stein (wol′ən·stīn, *Ger.* vol′ən·shtīn), **Albrecht Wenzel Eusebius von,** 1583–1634, Duke of Friedland; Austrian general in the Thirty Years' War.

Wal·ler (wol′ər), **Edmund,** 1606–87, English poet.

wal·let (wol′it) *n.* 1 A pocketbook, usually of leather, for holding unfolded banknotes, personal papers, etc.; a billfold. 2 A leather or canvas bag for tools, etc. 3 A knapsack. 4 *Obs.* Any baggy protuberance hanging loosely. [ME *walet,* ? metathetic var. of *watel* a bag, basket <OE *watul* a wattle]

wall·eye (wôl′ī′) *n.* 1 An eye in which the iris is light-colored or white. 2 An eye in which the cornea is opaque and whitish; also, leukoma of the cornea. 3 A large staring eye, usually one showing much white, because of divergent strabismus. 4 Any of several walleyed fishes, as the walleyed pike or perch, the alewife, or the walleyed pollack. [Back formation <WALLEYED]

wall·eyed (wôl′īd′) *adj.* 1 Affected with divergent strabismus. 2 Having a whitish or grayish eye; also, affected with leukoma of the cornea. 3 Squinting. 4 Having large, staring eyes, as a fish. 5 *Slang* Drunk. [<ON *valdeygthr,* alter. of *vagl eygr* < *vagl* a film on the eye + *eygr* having eyes < *auga* eye]

walleyed pike An American fresh-water percoid fish (genus *Stizostedion*) of the Great Lakes, having large eyes, esteemed as a game fish. Also **walleyed perch**.

walleyed pollack A coal-black North American pollack (*Pollachius fucensis*) of Pacific waters.

walleyed surf fish A sooty fish (*Hyperprosopon argenteus*) common in California waters.
wall fern The common polypody.
wall·flow·er (wôl′flou′ər) *n.* 1 Any of a genus (*Cheiranthus*) of European herbs of the mustard family, in particular the popular garden perennial *C. cheiri*, having fragrant yellow, orange, or red flowers. 2 An Australian desert shrub (genus *Gastrolobium*). 3 *Colloq.* A man or woman at a party who remains sitting or standing by the wall, presumably for want of a dancing partner.
wall fruit Fruit grown and ripened close to a wall or fence.
wal·lie (wol′ē) *n. Scot.* A valet.
Wal·lis and Fu·tu·na Islands (wol′is, foo̅·too̅′nə) Two closely connected protectorates NE of Fiji Islands, both dependencies of New Caledonia, including the chief islands of Uvéa, Futuna, and Alofi; total, 75 square miles.
wall lizard A gecko.
Wal·lo·ni·an (wo·lō′nē·ən) *adj.* Of or pertaining to the Walloons or the dialect spoken by them. — *n.* 1 A Walloon. 2 The French dialect of the Walloons.
Wal·loon (wo·loon′) *n.* 1 One of a people inhabiting southern and southeastern Belgium and the adjoining regions of France, originally descended from the ancient Belgae. 2 Their language, a dialect of French. 3 One of the Huguenot colonists who came to the United States from Artois, France. — *adj.* Of or pertaining to the Walloons or their dialect.
wal·lop (wol′əp) *v.t. Colloq.* 1 To beat soundly; thrash. 2 To hit with a hard blow. 3 To defeat soundly. — *v.i. Dial.* or *Colloq.* 4 To move quickly and strenuously; gallop. 5 To move in an awkward, floundering manner; waddle. — *n.* 1 *Brit. Dial. & Scot.* A lively rolling motion; a gallop. 2 *Colloq.* A severe blow. [<AF *waloper*, OF *galoper*. Doublet of GALLOP.]
wal·lop·er (wol′əp·ər) *n. Colloq.* 1 One who wallops. 2 Something astounding or amazing; an extraordinary statement or act; a whopper.
wal·lop·ing (wol′əp·ing) *Colloq. adj.* Extraordinarily large; whopping: a *walloping* lie. — *n.* A beating; whipping.
wal·low (wol′ō) *v.i.* 1 To roll about, as in mud, snow, etc.; flounder: The hippopotamus *wallows* in the mud. 2 To move with a heavy, rolling motion, as a ship in a storm. 3 To live or indulge complacently or wantonly: to *wallow* in sensuality or wealth. — *n.* 1 The act of wallowing. 2 A pool, mudhole, or slough in which animals wallow; also, any depression or hollow made by or suggesting such use. [OE *wealwian*] — **wal′low·er** *n.*
wall·pa·per (wôl′pā′pər) *n.* Paper specially prepared and printed in colors and designs, for covering walls and ceilings of rooms. — *v.t.* To cover or provide with wallpaper.
wall pellitory See PELLITORY.
wall plate 1 A horizontal timber on a wall, for bearing the ends of joists, girders, etc. 2 *Mech.* A plate for attaching a bearing or the like to a wall.
wall rock *Mining* The non-metalliferous rock between two lodes.
wall rocket A British perennial (*Diplotaxis tenuifolia*) of the mustard family, with large yellow flowers.
wall rue A small delicate spleenwort (*Asplenium ruta-muraria*) growing on walls and cliffs.
Walls·end (wôlz′end) *n.* A size or grade of coal for household purposes. [from *Wallsend*, England]
Walls·end (wôlz′end) A municipal borough in SE Northumberland, England, on the Tyne just NE of Newcastle-on-Tyne.
Wall Street 1 A street in lower Manhattan, New York City; the financial center of the United States. 2 American financiers collectively, their interests, power, etc., or the American financial world.

wall tent A tent having vertical sides and peaked top.
wall·ly (wä′lē, wol′ē) See WALY¹.
wal·ly-drai·gle (wä′lē-drā′gəl, wol′ē-) *n. Scot.* 1 The youngest in a family; also, a young bird in the nest. 2 Any feeble or ill-grown creature. Also **wal′ly·drag′** (-drag′, -dräg′).
wal·nut (wôl′nut′, -nət) *n.* 1 Any of various deciduous, typically European and Asian trees (genus *Juglans*), cultivated as ornamental shade trees and valued for their timber and their edible nuts; especially, the **black walnut** (*J. nigra*) of the eastern United States, and the **English, Persian, Circassian,** or **Caucasian walnut** (*J. regia*). 2 The wood or nut of any of these trees. 3 The shagbark hickory, or its nut. 4 The color of the wood of any of these trees, especially of the black walnut, a very dark brown; also, the color of the shell of the English walnut, a dull, medium yellowish brown: also called **walnut brown**. [OE *walhhnutu*, *wealh hnutu* < *wealh* foreign + *hnutu* a nut]

BLACK WALNUT
a. Catkin.
b. Shuck, nut inside.
c. Nut, shuck removed.

Wal·pole (wôl′pōl, wol′-), **Horace**, 1717–97, fourth earl of Orford; English author and wit; son of Sir Robert Walpole. — **Sir Hugh Seymour**, 1884–1941, English novelist. — **Sir Robert**, 1676–1745, first earl of Orford; English statesman.
Wal·pur·gis Night (väl·pŏor′gis) The night before May 1, originally dedicated to St. Walpurga, an English nun of the eighth century who founded religious houses in Germany: associated in German folklore with a witches' Sabbath on the Brocken. Also German **Wal·pur′gis·nacht′** (-näkht′). [<G *Walpurgisnacht*]
wal·rus (wôl′rəs, wol′-) *n.* A large, marine, seal-like mammal (family *Odobenidae*) of arctic seas, with flexible hind limbs, tusklike canines in the upper jaw, and a thick, heavy neck; especially, the common Atlantic walrus (*Odobenus rosmarus*). — *adj.* 1 Belonging or pertaining to a walrus. 2 Designating a type of mustache suggestive of the coarse bristles on the muzzle of a walrus. [<Du. *walrus* <Scand. Cf. Dan. *hyalros*, ? <ON *hrossvalr*, lit., a horse whale.]
Wal·sall (wôl′sôl) A county borough in southern Staffordshire, England.
Wal·sing·ham (wôl′sing·əm), **Sir Francis**, 1530?–90, English statesman.
Wal·ter (wôl′tər, *Ger. Sw.* väl′tər) A masculine personal name. Also *Ger., Sw.* **Wal·ther** (väl′tər). [<Gmc., ruler of the army]
Wal·ter (väl′tər), **Bruno**, born 1876, German orchestra conductor active in the United States: real name Bruno Schlesinger.
Wal·ter (wôl′tər), **John**, 1739–1812, English journalist; founder of the London *Times*.
Wal·tham (wôl′thəm) An industrial city in eastern Massachusetts on the Charles River west of Boston.
Wal·tham·stow (wôl′thəm·stō, -təm-) A municipal borough of SW Essex, England, NE of London.
Wal·ther von der Vo·gel·wei·de (väl′tər fôn der fō′gəl·vī′də), 1170?–1230?, German minnesinger.
Wal·ton (wôl′tən), **Izaak**, 1593–1683, English

WALRUS
(Body to 10 feet; weight to 3,000 pounds)

author. — **William Turner**, born 1902, English composer.
waltz (wôlts) *n.* 1 A round dance executed to music in triple time. 2 The music for such a dance, or any composition written in the triple time characteristic of the waltz. — *v.i.* 1 To dance a waltz. 2 To move quickly: He *waltzed* out of the room. — *v.t.* 3 To cause to waltz. — *adj.* Pertaining to, or typical of, the waltz: *waltz* time. [<G *walzer* < *walzen* waltz, roll] — **waltz′er** *n.*
Wal·vis Bay (wôl′vis) 1 An inlet of the Atlantic in South-West Africa. 2 An enclave in South-West Africa, administered by that territory, but an integral part of the Cape of Good Hope Province, Union of South Africa; on Walvis Bay; 374 square miles. 3 A port in this enclave: also *Walfish Bay. Afrikaans* **Wal·vis·bi′** (-vis·bī′).
wa·ly¹ (wä′lē, wol′ē) *Scot. adj.* 1 Beautiful; pleasing; excellent. 2 Strong; robust; vigorous. — *n. pl.* **·lies** 1 Something pleasing to the eye; a toy; ornament. 2 Good luck. 3 *pl.* Finery. Also spelled *wally*.
wa·ly² (wä′lē) *interj. Dial. & Scot.* Alas!: an expression of sorrow or lament.
wam·ble (wom′əl, wam′-) *Dial. v.i.* **·bled**, **·bling** 1 To move unsteadily; roll. 2 To twist or turn; writhe. 3 *Obs.* To feel nausea; be giddy or faint. — *n.* 1 A rolling gait. 2 A rolling or upheaving of the stomach; nausea. [ME *wamlen*. Cf. Dan. *vamle* feel nausea, Norw. *vamla* stagger.] — **wam′bling·ly** *adv.* — **wam′bly** *adj.*
wame (wām) *n. Scot.* The abdomen; belly; womb.
wame·fou (wām′foo) *n. Scot.* A bellyful. Also **wame′fu′**, **wame′ful** (-fool).
wamp·ish (wom′pish) *v.t. Scot.* To toss or throw about; wave; brandish.
wam·pum (wom′pəm, wôm′-) *n.* 1 Beads made of the interior parts of shells, formerly used as currency among North American Indians and between the Indians and white settlers: used loose, strung on strings, and also made into belts, scarfs, etc. The strings were often worn as ornaments, necklaces, bracelets, etc. The belts, woven with symbolic designs, were used in rituals, official communications, proposals, ratification of treaties, alliances, etc. The beads were either black, dark-purple, or white, the last being specifically **wam′pum·peag** (-pēg). The dark beads were double the value of the white. See SEAWAN. 2 *Colloq.* Money. [<Algonquian *wampum(peage)*, lit., a white (string of beads)]

WAMPUM
The historic Pennwampum — Iroquois Indian.

wampum snake The hoop snake: so called from its coloring.
wa·mus (wô′məs, wom′əs) *n.* A cardigan; a heavy outer jacket of strong, coarse cloth, worn in the United States. Also **wam′mus**, **wam′pus** (-pəs). [<Du. *wammes*, short for *wambuis* <OF *wambois* a leather doublet <OHG *wamba* the belly]
wan¹ (won) *adj.* 1 Pale, as from sickness or anxiety; pallid; livid; careworn; of a sickly hue. 2 Having a gloomy aspect; dismal; dark: said of scenes or landscapes. 3 *Obs.* Sad; mournful. 4 Faint; feeble: a *wan* smile. See synonyms under GHASTLY, PALE². — *v.t.* & *v.i.* **wanned**, **wan·ning** To make or become wan. — *n. Rare* The quality of being wan; paleness. [OE *wann* dark, gloomy] — **wan′ly** *adv.* — **wan′ness** *n.*
wan² (won) Obsolete past tense of WIN.
Wan·a·mak·er (won′ə·mā′kər), **John**, 1838–1922, U.S. merchant.
wand (wond) *n.* 1 A slender, flexible rod waved by a magician, conjurer, or legerdemain artist; also, any rod indicating an office or function of the bearer, as a scepter. 2 A musician's baton. 3 A thin, flexible stick or twig; also, a willow shoot; osier. 4 In archery, a slat used as a mark and placed at varying distances for men and women.

See synonyms under STICK. [<ON *vŏndr,* Akin to WIND².]

wan·der (won'dər) *v.i.* **1** To move or travel about without destination or purpose; roam; rove. **2** To go casually or by an indirect route; idle; stroll. **3** To extend in an irregular course; twist or meander. **4** To turn from a true or direct course; stray. **5** To deviate in conduct or opinion; go astray. **6** To think or speak deliriously or irrationally. — *v.t.* **7** *Poetic* To wander through or across. — *n.* The act of wandering; a ramble. [OE *wandrian*] — **wan·der·er** *n.* — **wan·der·ing** *adj.* — **wan'der·ing·ly** *adv.*
— **Synonyms** *(verb)*: deviate, digress, diverge, err, ramble, range, roam, rove, stray, swerve, veer. To *wander* is to move in an indefinite or indeterminate way which may or may not be a departure from a prescribed way; to *deviate* is to turn from a prescribed or right way, physically, mentally, or morally, usually in an unfavorable sense; to *diverge* is to turn from a course previously followed or that something else follows, and has no unfavorable implication; to *digress* is used only with reference to speaking or writing; to *err* is used of intellectual or moral action. To *swerve* or *veer* is to turn suddenly from a prescribed or previous course, and often but momentarily; *veer* is more capricious and repetitious; the horse *swerves* at the flash of a sword; the wind *veers*; the ship *veers* with the wind. To *stray* is to go in a somewhat purposeless way aside from the regular path or usual haunts or abode, usually with unfavorable implication; cattle *stray* from their pastures; an author *strays* from his subject. *Stray* is in most uses a lighter word than *wander*. *Ramble* in its literal use is always a word of pleasant suggestion, but in its figurative use somewhat contemptuous; as, *rambling* talk. See RAMBLE.

wandering albatross A large, whitish, black-winged, web-footed sea bird (*Diomedea exulans*), having extraordinary powers of flight.
wandering jew 1 A perennial trailing herb (*Tradescantia fluminensis*) of the spiderwort family, with hairy white flowers and vivid green leaves sometimes striped with yellow. **2** A related plant (*Zebrina pendula*) with red or white flowers and striped leaves.
Wandering Jew See under JEW.
wandering kidney A floating kidney.
wan·der·lust (won'dər·lust', *Ger.* vän'dər·lōōst) *n.* An impulse to travel; restlessness combined with a sense of adventure. [<G <*wandern* travel + *lust* joy]
wan·der·oo (won'də·rōō') *n.* **1** A large black monkey (*Macaca silenus*) of western India, having a heavy whitish mane. **2** A Ceylonese langur (*Presbytis cephalopterus*). [<Singhalese *vanduru*, pl. of *vandurā* the Ceylonese langur <Skt. *vānara* a monkey]
wan·dle (won'dəl, -əl) *adj. Dial.* Supple; nimble. [Back formation <OE *wandlung* changeableness]
Wands·worth (wondz'wûrth) A metropolitan borough in SW London, England.
wane (wān) *v.i.* **waned, wan·ing 1** To diminish in size and brilliance: opposed to *wax.* **2** To decline or decrease gradually; draw to an end. — *n.* **1** Decrease, as of power, prosperity, or reputation. **2** The decrease of the moon's visible illuminated surface; also, the period of such decrease. **3** The beveled edge of a board sawn from a log; also, the bark or defective portion on the edge or corner of a board. ◆ Homophone: *wain.* [OE *wanian* lessen]
wan·ey (wā'nē) *adj.* Having a beveled edge, as the wane of a plank: also spelled *wany.* [<WANE, *n.* (def. 3)]
Wang·a·nu·i (wŏng'ə·nōō'ē) A port of southern North Island, New Zealand.
wan·gle (wang'gəl) *v.* **·gled, ·gling** *Colloq. v.t.* **1** To obtain or make by indirect or irregular methods; contrive: to *wangle* an introduction to a celebrity. **2** To manipulate or adjust, especially dishonestly. **3** To wriggle or wag. — *v.i.* **4** To resort to indirect, irregular, or dishonest methods. **5** To wriggle. [? Alter. of WAGGLE] — **wan'gler** *n.*
Wan·hsien (wän'shyen') A city on the Yangtze River, eastern Szechwan province, central China; a major commercial port NE of Chungking.
wan·i·gan (won'ə·gən) *n.* **1** In American log-

ging camps, a place for the storage of small supplies or reserve stock; also, a large chest for the lumbermen's clothing, shoes, tobacco, etc. **2** A raft of square timber, on which is a shanty fitted with sleeping and cooking accommodations: used by Maine lumbermen. **3** The accountant's shack in a lumber camp. Also **wan·gan** (won'gən), **wan'gun, wan'ni·gan.** [Earlier *wangan* <Algonquian *atawangan* <*atawan* buy, sell]
wan·ion (won'yən) *n. Archaic* Disaster, or bad luck; a curse: used only in the phrases **in a wanion, with a wanion,** etc. [Alter of dial. ME (Northern) *waniand,* ppr. of *wanien* wane]
Wan·ne-Eick·el (vän'ə·ī'kəl) A city in west central North Rhine-Westphalia NW of Bochum, West Germany; a Ruhr chemical and coal-mining center.
Wan·stead and Wood·ford (won'sted, -stid, wŏŏd'fərd) A municipal borough in Essex, England, NE of London.
want (wont, wônt) *v.t.* **1** To feel a desire or wish for. **2** To wish; desire: used with the infinitive: Your friends *want* to help you. **3** To be deficient in; lack; be without. **4** To be lacking to the extent of: He *wants* three inches of six feet. **5** *Brit.* To need; require. — *v.i.* **6** To have need: usually with *for.* **7** To be needy or destitute. **8** *Rare* To be lacking or absent. — *n.* **1** Lack or absence of something; scarcity; shortage. **2** Privation; indigence; destitution; need. **3** Something that is lacking or needed; a need. **4** A conscious or felt need of something; a craving. [Prob. <ON *vanta* be lacking] — **want'er** *n.*
— **Synonyms** *(noun)*: absence, dearth, default, defect, deficiency, lack, necessity, need, privation, scantiness, scarceness, scarcity. See NECESSITY, POVERTY. Antonyms: abundance, affluence, fullness, luxury, plenty, profusion, riches, wealth.
wa'n't (wont, wônt) Was not: a dialectal contraction.
want ad *Colloq.* An advertisement in a newspaper for something wanted, as hired help, a job, a lodging, etc.
want·age (won'tij, wôn'-) *n.* Whatever is lacking; deficiency.
want column A column of want ads in a newspaper or other periodical.
want·ing (won'ting, wôn'-) *adj.* **1** Not at hand; missing; lacking: One juror is still *wanting.* **2** Marked by lack or deficiency; not coming up to need or expectation: He was found *wanting.* **3** *Colloq.* Deficient in intellect; feeble-minded. — **wanting in** Deficient in. — *prep.* With the exception of; less; save; minus.
wan·ton (won'tən) *adj.* **1** Dissolute; unchaste; licentious; lewd; lustful. **2** Recklessly inconsiderate, heartless, or unjust; evincing a malicious nature: *wanton* savagery; also, unprovoked: a *wanton* murder. **3** Of vigorous and abundant growth; rank. **4** Extravagant; running to excess; unrestrained: *wanton* speech. **5** Not bound or tied; loose: *wanton* curls; also, frolicsome; prankish. **6** *Obs.* Refractory; rebellious. — *v.i.* **1** To act wantonly or playfully; revel or sport. **2** To grow luxuriantly. — *v.t.* **3** To waste wantonly. — *n.* **1** A lewd or licentious person, especially a woman. **2** A playful or frolicsome person or animal. **3** A trifler; dallier. **4** *Obs.* A person who has been much indulged; a pet. [ME *wantoun* <OE *wan* deficient + ME *towen,* OE *togen,* pp. of *tēon* bring up, educate] — **wan'ton·ly** *adv.* — **wan'ton·ness** *n.*
— **Synonyms** *(adj.)*: airy, free, frisky, frolicsome, gay, loose, merry, playful, reckless, sportive, unbridled, uncurbed, unrestrained, wandering, wild. See IMMODEST. Antonyms: austere, demure, discreet, sedate, serious, thoughtful.
wan·y (wā'nē) See WANEY.
wap¹ (wop, wap) *Dial.* or *Archaic v.t.* & *v.i.* **wapped, wap·ping 1** To whip; beat; strike. **2** To flutter or flap, as wings. — *n.* **1** A stroke; blow. **2** A quarrel; fight. **3** A storm. [Prob. var. of WHOP]
wap² (wop, wap) *Dial. v.t.* **wapped, wap·ping** To wrap; tie; bind. — *n.* A wrapping. [? Alter. of WARP]
wap·en·shaw (wop'ən·shô, wap'-) *Scot.* A show of weapons; review of weapons. Also **wap'in·schaw, wap'pen·schaw'ing.**
wap·en·take (wop'ən·tāk, wap'-) *n.* An old administrative and judicial subdivision of some English counties, equivalent to the

hundred of most counties. [OE *wǣpengetæc* <ON *vāpnatak* a (symbolical) flourish of weapons denoting confirmation of the decisions of an assembly < *vāpna,* genitive pl. of *vāpn* a weapon + *tak* a taking]
wap·i·ti (wop'ə·tē) *n.* A large North American deer (*Cervus canadensis*): usually *elk.* Cf. Algonquian. Shawnee *wapiti* pale, white.]
wap·per-jawed (wop'ər·jôd') *adj. U. S. Dial.* **1** Having a wry or undershot jaw. **2** Out of true; askew.

WAPITI
(About 5 feet high at the shoulders; antler spread to 3 feet)

war¹ (wôr) *n.* **1** A contest between or among nations or states, or between different parties in the same state, carried on by force and with arms. **2** Any act or state of hostility; enmity; strife; also, a contest or conflict. **3** *Poetic* **a** A battle. **b** The supplies and paraphernalia of war. **c** Armed troops; an army. See table MAJOR WARS OF HISTORY on page 1417. **4** The science or art of military operations; strategy. — *v.i.* **warred, war·ring 1** To wage war; fight or take part in a war. **2** To be in any state of active opposition; contend; strive. — *adj.* Of or pertaining to, used in, or resulting from war. [OE *wyrre, werre* <AF *werre* <OHG *werra* strife, confusion]
War may appear as a combining form in hyphemes or solidemes, or as the first element in two-word phrases:

war-blasted	war-making
war-born	war march
war-breeder	war-marked
war-breeding	war neurosis
war bride	war office
war-broken	war party
war budget	war prisoner
war chant	war-production
war chief	war-proof
war cloud	war-ridden
war code	war-risk
warcraft	war service
war-debt	war-shaken
war-disabled	war song
war dog	war-stirring
war drum	war-swept
war-famed	war talk
war-footing	war tax
war gains	wartime
war-god	war-torn
war-goddess	war-tossed
war-hardened	war traitor
war-impoverished	war vessel
war insurance	war-wasted
war law	war-wearied
war leader	war-weary
war loan	war-work
war-loving	war-worker
war-machine	war-worn
war-made	warworthy
war-maimed	war-wounded
war-maker	war zone

war² (wär) *Dial. v.t.* To guard against; ware. — *adj.* Cautious; wary. [Var. of WARE²]
war³ (wär) *adj. & adv. Scot. & Brit. Dial.* Worse.
War·beck (wôr'bek), **Perkin,** 1474–99, Walloon impostor and pretender to the English throne; hanged.
war belt Among certain North American Indians, a belt of wampum bearing symbolic figures or designs, sent by one tribe to another or passed from tribe to tribe, as a message declaring war, summoning a group of tribes to war, invoking aid in war, etc.
War between the States The United States Civil War: used especially in the former Confederate States.
war bird Among certain North American Indians, the golden eagle: so called because its feathers were worn in the war bonnet.
war·ble (wôr'bəl) *v.* **·bled, ·bling** *v.t.* **1** To sing with trills and runs, or with tremulous vibrations. **2** To celebrate in song. — *v.i.* **3** To sing with trills, etc. **4** To make a liquid,

MAJOR WARS OF HISTORY

Name	Contestants (victor shown first)	Notable Battles	Treaties
Greco-Persian Wars 499-478 B.C.	Greek states — Persia	Marathon, 490; Thermopylae, Salamis, 480; Plataea, 479	
Peloponnesian War 431-404 B.C.	Sparta — Athens	Syracuse, 415; Cyzicus, 410; Aegospotami, 405	Peace of Nicias, 421
First Punic War 264-241 B.C.; **Second Punic War** 218-201 B.C.; **Third Punic War** 149-146 B.C.	Rome — Carthage	Drepanum, 249; Aegates, 241; Lake Trasimene, 217; Cannae, 216; Zama, 202	
Islamic Invasion of Europe 630-19th century	Christianity — Islam	Constantinople, 717-718; Tours, 732; Manzikert, 1071; Hattin, 1187; Lepanto, 1571; Vienna, 1524, 1683; Zenta, 1697	Pruth, 1711; Kutchuk-Kanardji, 1774; Sistova, 1791
Norman Conquest 1066	Normandy — England	Hastings, 1066	
Crusades 1096-1291	Christianity — Islam (indecisive)	Jerusalem, 1099; Acre, 1191	
Hundred Years' War 1338-1453	England — France	Crécy, 1346; Poitiers, 1356; Agincourt, 1415; Siege of Orléans, 1428-39	
Wars of the Roses 1455-85	Lancaster — York (indecisive)	St. Albans, 1455	
Thirty Years' War 1618-48	Catholics — Protestants	Leipzig, Breitenfeld, 1631; Lützen, 1632	Westphalia, 1648
Civil War (English) 1642-46	Roundheads — Cavaliers	Marston Moor, 1643; Naseby, 1645	
Second Great Northern War 1700-1721	Russia — Sweden and Baltic allies	Poltava, 1709	Nysted, 1721
War of the Spanish Succession 1701-14	England, Austria, Prussia, Netherlands — France, Spain	Blenheim, 1704	Utrecht, 1713
War of the Austrian Succession 1740-48	France, Prussia, Sardinia, Spain — Austria, England	Dettingen, 1743; Fontenoy, 1745	Aix-la-Chapelle, 1748
French & Indian War 1755-63	England — France	Plains of Abraham, 1759; Montreal, 1760	
Seven Years' War 1756-63	Prussia — Austria, France, Russia	Rossbach, Leuthen, 1757	Hubertusberg, 1763
Revolutionary War 1775-83	American Colonies — England	Lexington, Concord, Bunker Hill, 1775; Saratoga, 1777; Yorktown, 1781	Paris, 1783
Napoleonic Wars 1796-1815	England, Austria, Russia, Prussia, etc. — France	Nile, 1798; Trafalgar, 1805; Jena, Auerstädt, 1806; Leipzig, 1813; Waterloo, 1815	Campoformio, 1797; Tilsit, 1807; Schönbrunn, 1809; Paris, 1814-15; Vienna, 1815
War of 1812 1812-15	United States — England	Lake Erie, 1813; New Orleans, 1815	Ghent, 1814
War of Independence (Greek) 1821-29	Greece, England, Sweden, Russia — Turkey	Navarino, 1827	London, 1827
Mexican War 1846-48	United States — Mexico	Resaca de la Palma, 1846; Chapultepec, 1847	Guadalupe Hidalgo, 1848
Crimean War 1854-56	Turkey, England, France, Sardinia — Russia	Sevastopol, 1854	Paris, 1856
Civil War (United States) 1861-65	Union (North) — Confederate States (South)	Bull Run, 1861; Antietam, 1862; Chancellorsville, Gettysburg, Vicksburg, Chattanooga, 1863; Wilderness, 1864	
Franco-Prussian War 1870-71	Prussia — France	Sedan, 1870	Versailles, 1871
Spanish-American War 1898	United States — Spain	Manila Bay, Santiago, 1898	Paris, 1898
Boer War 1899-1902	England — Transvaal Republic & Orange Free State	Ladysmith, 1899	Vereeniging, 1902
Russo-Japanese War 1904-1905	Japan — Russia	Port Arthur, Mukden, Tsushima, 1905	Portsmouth, 1905
First Balkan War 1912-13; **Second Balkan War** 1913	Bulgaria, Serbia, Greece, Montenegro — Turkey	Scutari, 1912; Salonika, 1912; Adrianople, 1912	London, 1913; Bucharest, 1913
World War I 1914-18	Allies — Central Powers	Dardanelles, 1915; Verdun, Somme, Jutland, 1916; Caporetto, 1917; Vittorio Veneto, Amiens, Marne, Ypres, 1918	Versailles, Saint-Germain, Neuilly, 1919; Trianon, Sèvres, 1920; Lausanne, 1923
Civil War (Spanish) 1936-39	Insurgents — Loyalists	Teruel, 1937; Ebro River, 1938	
World War II 1939-45	United Nations — Axis 1939-45	Dunkirk, 1940; Crete, 1941; El Alamein, 1942; Tunis, 1943; Stalingrad, 1942-43; Kharkov, 1943; Cassino, 1943-44; Saint-Lô, 1944; Rhine, Ruhr, Berlin, 1945	Potsdam, 1945
	United Nations — Japan 1941-45	Pearl Harbor, 1941; Bataan, 1941-42; Singapore, Coral Sea, Midway Island, Guadalcanal, 1942; Bismarck Sea, Tarawa, 1943; Leyte Gulf, 1944; Philippines, 1944-45; Okinawa, 1945	San Francisco, 1951
Korean War 1950-52	United Nations — North Korea	Inchon, Pyongyang, 1950; Seoul, 1951	Panmunjom, 1953

murmuring sound, as a stream. **5** *U.S.* To yodel. See synonyms under SING. —*n.* The act of warbling; a carol; song. [<AF *werbler,* OF *guerbler* < *werble* a warble <OHG *werbel* something that revolves. Akin to WHIRL.]

war·ble[2] (wôr′bəl) *n.* **1** A hard swelling on the back of a horse, caused by the chafing of the saddle. **2** A boil or swelling under the hide of a horse, cow, deer, or the like, caused by the maggot of a botfly or warblefly. **3** A warblefly. [Cf. obs. Sw. *varbulde* < *var* pus + *bulde* a tumor] —**war′bled** *adj.*

war·ble·fly (wôr′bəl·flī′) *n. pl.* **·flies** Any of a family *(Hypodermatidae)* of dipterous insects resembling the botflies, whose larvae produce swellings under the hides of cattle, horses, etc. [<WARBLE[2] + FLY[2]]

war·bler (wôr′blər) *n.* **1** One who or that which warbles; a songster. **2** Any of a family *(Sylviidae)* of plain-colored, mostly Old World birds allied to the kinglets and noted for their song, as the whitethroat. **3** Any of a large and varied family *(Compsothlypidae)* of small American insectivorous birds, usually brilliantly colored and with little powers of song, as the **summer** or **yellow warbler** *(Dendroica aestiva),* the redstart, ovenbird, and water thrush. Also **wood warbler.**

war bonnet The ceremonial head dress of the North American Plains Indians, consisting of a rawhide cap fitting the head and extending down the back to the heels, the crown and the extension being decorated with feathers of the golden eagle.

War·burg (vär′bŏŏrkh), **Otto Heinrich,** born 1883, German physiologist and chemist.

War College One of four colleges in the United States giving advanced instruction to experienced military, naval, and air officers; specifically, the **Army War College,** Carlisle Barracks, Pennsylvania, under the Department of the Army; the **Naval War College,** Newport, Rhode Island, under the Navy Department; and the **Air War College,** near Montgomery, Alabama, under the Department of the Air Force. The **National War College,** Washington, D.C., operating under the Joint Chiefs of Staff, prepares officers of the armed services, the State Department, and other executive departments for duties on the highest level concerned with the national security.

war correspondent A newspaper reporter or representative of some other periodical engaged to write up the scenes of combat from direct observation.

war cry A rallying cry used by combatants in a war, or by participants in any contest.

ward (wôrd) *n.* **1** The act of guarding; protection. **2** The state of being under a guard or guardian; custody; confinement; also, guardianship; control. **3** A guarded or protected place; a prison; jail; also, a division or subdivision of a jail or hospital: the maternity *ward.* **4** A territorial division of a city, made for convenience of government; also, in certain northern counties of England, a division equivalent to a hundred or wapentake. **5** A person who is in the charge or under the protection of a guardian. **6** An instrument or means of defense; a protection. **7** A defensive attitude or movement, as in fencing; guard. **8** A projection inside a lock, designed to obstruct the turning of any key other than the proper one; also, a corresponding notch in the bit of a key. **9** In feudal law, a minor under the care or protection of a guardian. **10** A warden; overseer. **11** *Obs.* A company of men detailed to defend or guard; a garrison; watch. See synonyms under SHELTER. —*v.t.* **1** To repel or turn aside, as a thrust or blow: usually with *off.* **2** To put in a ward; keep in safety. **3** *Archaic* To guard; protect. [OE *weard* a watching < *weardian* watch, guard; infl. in some senses by AF *warde,* OF *garde* <Gmc.]

-ward *suffix* Toward; in the direction of: *upward, homeward.* Also **-wards.** [OE *-weard, -weardes* at, toward]

Ward (wôrd), **Artemas,** 1727–1800, American Revolutionary general. —**Artemus** Pseudonym of Charles Farrar Browne, 1834–67, U.S. humorist. —**Mary Augusta,** 1851–1920, *née* Arnold, English novelist: known as *Mrs. Humphrey Ward.*

war dance A dance of savage tribes before going to war or in celebration of a victory.

war·den[1] (wôr′dən) *n.* **1** One who keeps ward; a warder or gatekeeper. **2** A chief officer, as in a prison. **3** In England, the head of certain colleges. **4** In Connecticut, the chief executive of a borough. **5** A churchwarden. See synonyms under SUPERINTENDENT. [<AF *wardein,* OF *gardein, guarden* <Gmc. Doublet of GUARDIAN.]

war·den[2] (wôr′dən) *n.* A variety of pear used chiefly for cooking. Also **War′den.** [ME *wardon,* prob. <AF *warder,* OF *garder* keep <Gmc.]

war·den·ry (wôr′dən·rē) *n. pl.* **·ries** The office, functions, or jurisdiction of a warden. Also **war′den·ship** (-ship).

War Department A former executive department of the U.S. government (1789–1947) in charge of matters relating to the Army and (later) the Army Air Force: now absorbed into the Department of Defense.

ward·er (wôr′dər) *n.* **1** A keeper; guard; sentinel; watchman. **2** An official staff or baton; a truncheon. **3** A prison official; warden. [<AF *wardere* < *warder,* var. of OF *guarder* guard, keep]

ward-heel·er (wôrd′hē′lər) *n.* *U.S. Slang* A hanger-on of a political boss, who does minor tasks, canvasses votes, etc. [<WARD (def. 4) + HEELER (def. 1)]

ward-hold·ing (wôrd′hōl′ding) *n.* The holding of lands by military tenure: distinguished from *feu.*

ward·ress (wôrd′ris) *n.* A female warden.

ward·robe (wôrd′rōb′) *n.* **1** A large upright cabinet for wearing apparel; formerly, a large clothes closet or room, where clothes were also made and repaired. **2** All the garments of any one person. **3** In a noble or royal household, the department responsible for clothing, jewelry, etc. **4** The costumes of a theater or theatrical troupe. **5** The styles of a particular season taken collectively: the spring *wardrobe.* [<AF *warderobe,* OF *garderobe* < *warder* keep + *robe* a robe, dress]

ward·room (wôrd′rōōm′, -rŏŏm′) *n.* On a warship, the quarters allotted to the commissioned officers above the rank of ensign, excepting the commander, who has his own quarters; especially, the dining-room of these officers; also, these officers regarded as a group.

ward·ship (wôrd′ship) *n.* **1** The state of a ward; pupilage. **2** In feudal law, the right by which the lord had the custody of the bodies, and the custody and profits of the lands, of minor heirs of a deceased tenant.

ware[1] (wâr) *n.* **1** Articles of the same class; especially, manufactured articles: used collectively, often in composition: *tableware, glassware.* **2** *pl.* Articles of commerce; goods; merchandise; products. **3** Pottery; ceramic articles; earthenware. ◆ Homophone: *wear.* [OE *waru*]

ware[2] (wâr) *v.t.* **wared, war·ing** To beware of: used mainly in the imperative: *Ware* the dog. —*adj. Obs.* Conscious; aware; hence, on one's guard; cautious. ◆ Homophone: *wear.* [Fusion of OE *warian* beware and AF *warer,* OF *garer* <Gmc. Akin to WARN.]

ware[3] (wâr) *v.t. Scot.* To expend; lay out; also, to lavish; squander: also spelled *wair.* ◆ Homophone: *wear.*

ware·house (wâr′hous′) *n.* **1** A storehouse for goods or merchandise. **2** *Brit.* A large wholesale shop. —*v.t.* **·housed** (-houzd′), **·hous·ing** (-hou′zing) To place or store in a warehouse, especially in a bonded warehouse.

ware·house·man (wâr′hous′mən) *n. pl.* **·men** (-mən) One who makes a business of storing goods.

ware·room (wâr′rōōm′, -rŏŏm′) *n.* A room for the storage, exhibition, or sale of goods or wares.

war·fare (wôr′fâr′) *n.* **1** The waging or carrying on of war; conflict with arms; war. **2** Struggle; strife.

War·field (wôr′fēld), **David,** 1866–1951, U.S. actor.

War for Southern Independence See CIVIL WAR (AMERICAN) in table under WAR.

war game **1** Kriegspiel. **2** *pl.* Practice maneuvers imitating the conditions of actual warfare.

war hawk One who advocates war; a jingo.

war·head (wôr′hed′) *n. Mil.* **1** An ogive-shaped chamber in the nose of a torpedo, containing the charge of high explosive. **2** A similar chamber in a bomb, guided missile, or the like.

war horse **1** A heavy horse used in warfare; a charger. **2** *Colloq.* A veteran; especially, an aggressive or veteran politician.

war·i·son (war′ə·sən) *n.* **1** A signal for assault: an erroneous use. **2** Reward; healing. [<AF *warison,* OF *garison* wealth, possession]

wark[1] (wärk) *n. Scot.* Work.

wark[2] (wärk) *Scot. & Brit. Dial. n.* Ache; pain. —*v.i.* To suffer pain; ache; throb.

war·like (wôr′līk′) *adj.* **1** Disposed to engage in war; belligerent. **2** Relating to, used in, or suggesting war. **3** Threatening war; belligerent; hostile.

Synonyms: martial, military, soldierlike, soldierly. *Antonyms*: civil, effeminate, meek, pacific, peaceful, unmilitary, unsoldierlike, unsoldierly, unwarlike.

war·lock[1] (wôr′lok′) *n.* A wizard; sorcerer; also, a demon. [OE *wǣrloga* a traitor, foe, devil < *wǣr* a covenant + *lēogen* lie, deny]

war·lock[2] (wôr′lok′) *n.* A scalp lock worn by the warriors of certain North American Indian tribes. [<WAR + LOCK[2]]

war·lord (wôr′lôrd′) *n.* **1** A leader or high-ranking officer in a militaristic nation. **2** The warlike ruler or leader of a local region or group of bandits, especially in the Orient.

warm (wôrm) *adj.* **1** Moderately hot; having, or characterized by, heat somewhat greater than temperate: *warm* water; a *warm* climate. **2** Imparting heat: a *warm* fire. **3** Imparting, promoting, or preserving warmth; preventing loss of bodily heat: a *warm* coat. **4** Having a feeling of heat somewhat greater than ordinary: *warm* from exertion. **5** Possessing or marked by ardor, zeal, liveliness, enthusiasm, or cordiality: a *warm* argument; *warm* wishes. **6** Excited; agitated; also, vehement; passionate: a *warm* temper. **7** United by ardent affection: *warm* friends; also, amorous; loving. **8** Having predominating tones of red or yellow: opposed to *cool.* **9** Recently made; fresh: a *warm* trail; hence, near a hidden object, as in certain games of children. **10** *Colloq.* Uncomfortable by reason of annoyances or danger: They made the town *warm* for him. **11** Characterized by brisk activity: a *warm* skirmish. **12** *Colloq.* Rich; wealthy. —*v.t.* **1** To make warm; heat slightly: often with *up.* **2** To make ardent or enthusiastic; interest. **3** To fill with kindly feeling: The sight *warms* my heart. —*v.i.* **4** To become warm. **5** To become ardent or enthusiastic: often with *up* or *to.* **6** To become kindly disposed or friendly: with *to* or *toward.* —*n.* *Colloq.* The state or sensation of being or becoming warm; warmth; a heating. [OE *wearm*] —**warm′ly** *adv.* —**warm′ness** *n.*

warm-blood·ed (wôrm′blud′id) *adj.* **1** Having warm blood: said of animals, as mammals and birds, that preserve a nearly uniform and high body temperature, whatever the surrounding medium; homoiothermal. **2** Enthusiastic; ardent; passionate.

warm·er (wôr′mər) *n.* One who or that which warms.

warm front *Meteorol.* The irregular boundary line between an advancing mass of warm air and the underlying colder air mass.

warm-heart·ed (wôrm′här′tid) *adj.* Kind; affectionate.

warming pan A closed metal pan with a long handle, containing live coals or hot water, for warming a bed.

warm·ish (wôr′mish) *adj.* Rather warm.

war·mon·ger (wôr′mung′gər, -mong′-) *n.* One who propagates warlike ideas; a jingo. —**war′·mon′ger·ing** *adj. & n.*

Warm Springs A resort town in western Georgia; site of an institution for the study and treatment of poliomyelitis; here Franklin D. Roosevelt died, 1945.

warmth (wôrmth) *n.* **1** The state, quality, or sensation of being warm. **2** Ardor or fervidness of disposition or feeling; excitement of temper or mind. **3** The effect produced by warm colors. [ME *wermthe,* ult. <OE *wærm* warm]

Synonyms: animation, ardor, cordiality, eagerness, earnestness, emotion, energy, enthusiasm, excitement, fervidness, fervor, geniality, glow, heat, intensity, irascibility, life, passion, vehemence, zeal. Compare ENTHUSIASM. *Antonyms*: coldness, coolness, frigidity, iciness, indifference, insensibility, torpor.

warm-up (wôrm′up′) *n. Colloq.* The act of exercising or limbering up just before a game, contest, etc.

warn (wôrn) *v.t.* **1** To make aware of impending or possible harm; put on guard; caution. **2** To advise; admonish; counsel. **3** To inform; give notice in advance. **4** To notify (a person) to stay, go, or keep: with *off, away,* etc. See synonyms under ADMONISH. [OE *warenian, wearnian.* Akin to WARE[2].] **— warn′er** *n.*

warn·ing (wôr′ning) *n.* **1** The act of one who warns, or that which he communicates; notice of danger. **2** That which warns or admonishes. See synonyms under COUNSEL, EXAMPLE. **—** *adj.* Serving as a warning. **— warn′·ing·ly** *adv.*

war nose The end of a projectile or shell which carries the detonating device.

War of American Independence *Brit.* The American Revolution.

War of Independence The American Revolution.

War of Secession The Civil War in the United States.

War of the Rebellion The Civil War in the United States: used especially in the States that adhered to the Union.

War of the Spanish Succession See table under WAR.

War of 1812 See table under WAR.

warp (wôrp) *v.t.* **1** To turn or twist out of shape, as by shrinkage or heat. **2** To turn from a correct or proper course; give a twist or bias·to; corrupt; pervert. **3** To stretch or arrange (yarn) so as to form a warp. **4** *Naut.* To move (a vessel) by hauling on a rope or cable, which is usually fastened to something stationary, as a pier or anchor. **5** *Aeron.* To change the curvature of (an airfoil or wing) by twisting, so as to bring the airplane into balance. **—** *v.i.* **6** To become turned or twisted out of shape, as wood in drying. **7** To turn or deviate from a correct or proper course; go astray. **8** *Naut.* To move by means of ropes fastened to a pier, anchor, etc. See synonyms under BEND[1]. **—** *n.* **1** The state of being warped or twisted out of shape; a twist or distortion, especially in a piece of wood. **2** A mental or moral deviation or aberration; bias. **3** The threads that run the long way of a fabric, crossing the woof. **4** The heavy cords forming the carcass of a pneumatic tire. **5** *Naut.* A light cable used for warping a ship or boat; a towline or towrope. **6** A length of rope yarn or rope. [OE *weorpan* throw] **— warp′er** *n.*

war paint 1 Paint applied to faces and bodies by North American Indians and other primitive peoples in token of going to war. **2** Hence, any preparation for battle. **3** *Colloq.* Any front assumed to intimidate an adversary or increase self-confidence. **4** *Colloq.* Rouge and other cosmetics applied to the person; hence, full dress and personal adornment; finery; also, official garb or regalia.

war·path (wôr′path′, -päth′) *n.* The route taken by an attacking party of American Indians; the state of war; also, a war expedition. **— on the warpath 1** On a warlike expedition; at war. **2** Ready for a fight; thoroughly angry; ready to begin hostilities.

warp beam The roller or beam in a loom on which the warp is wound.

war·plane (wôr′plān′) *n.* An airplane equipped for fighting.

war·pow·er (wôr′pou′ər) *n.* The armed potential of a country; capacity of a nation's manpower and resources for waging war.

war powers Certain powers granted under the Constitution of the United States to the national government or to the chief executive in time of war, to prosecute war and act in all contingent emergencies.

war·ra·gal (wär′ə·gəl) *n.* The dingo. Also **war′ri·gal.** [<Australian *warregal* a dog, savage]

war·rant (wôr′ənt, wor′-) *n.* **1** *Law* A judicial writ or order authorizing arrest, search, seizure, or any other designated act in aid of the administration of justice. **2** Something which assures or attests; a voucher; evidence; guarantee. **3** That which gives authority for some course or act; sanction; justification: What *warrant* have you for that statement? **4** A certificate of appointment given to army and navy officers of rank lower than commissioned officers. See under OFFICER. **5** A document giving a certain authority; specifically, a document authorizing receipt or payment of money: a dividend *warrant.* See synonyms under PRECEDENT. **—** *v.t.* **1** To assure or guarantee the quality, accuracy, certainty, or sufficiency of: to *warrant* a title to property. **2** To assure or guarantee the character or fidelity of; pledge oneself for. **3** To guarantee against injury, loss, etc. **4** To be sufficient grounds for; justify: The facts did not *warrant* your action. **5** To give legal authority or power to, so as to secure against harm; empower; authorize. **6** *Colloq.* To say confidently; feel sure. See synonyms under JUSTIFY. [<AF *warant,* OF *guarant* <Gmc.] **— war′rant·a·ble** *adj.* **— war′rant·a·bly** *adv.* **— war′rant·er** *n.*

war·ran·tee (wôr′ən·tē′, wor′-) *n. Law* The person to whom a warranty is given.

warrant officer See under OFFICER.

war·rant·or (wôr′ən·tôr, wor′-) *n. Law* One who makes or gives a warranty to another.

war·ran·ty (wôr′ən·tē, wor′-) *n. pl.* **·ties 1** *Law* An assurance or undertaking by the seller of property, express or implied, that the property is or shall be as it is represented or promised to be. **2** In conveyancing, a covenant in a deed whereby the grantor binds himself and his heirs to secure to the grantee the estate conveyed to him. **3** In insurance law, a stipulation or engagement on the part of the insured that the facts in relation to the risk are as stated by him. **4** Authorization; warrant. **5** *Dial.* Security; guaranty. [<AF *warantie,* OF *guarantie* <OF *guarant* a warrant. Doublet of GUARANTY.]

War·re·go River (wor′i·gō) A river in east central Australia, flowing 495 miles SW to the Darling River.

war·ren (wôr′ən, wor′-) *n.* **1** A place where rabbits live and breed in communities. **2** An enclosure for keeping small game; also, a place for keeping fish in a river. **3** An obscure crowded place of habitation. **4** In English law, a franchise, either by prescription or royal grant, to keep in an enclosure "beasts and fowls of warren," that is, animals that are by nature wild. See also FREEWARREN. [<AF *warenne* a game park, a rabbit warren < *warir* preserve <Gmc.]

War·ren (wôr′ən, wor′-), **Earl,** born 1891, U. S. administrator; chief justice of the U. S. Supreme Court 1953–. **— Joseph,** 1741–75, American physician and general. **— Robert Penn,** born 1905, U. S. poet, novelist, and educator.

war·ren·er (wôr′ən·ər, wor′-) *n.* The keeper of a warren.

Warren hoe A garden hoe with the sides tapering to a point: used to make furrows for seeds: a trade name. See illustration under HOE.

War·ring·ton (wôr′ing·tən, wor′-) A county borough in southern Lancashire, England, on the Mersey east of Liverpool.

war·ri·or (wôr′ē·ər, -yər, wor′-) *n.* A man engaged in or experienced in warfare; one devoted to a military life. **—** *adj.* Military; martial. [<AF *werreieor* < *werreier* make war < *werre* WAR]

war-risk insurance (wôr′risk′) Insurance written by the government of the United States for military and naval personnel.

war·saw (wôr′sô) *n.* **1** A fish, the black grouper *(Garrupa nigrita)* of the South Atlantic and Gulf of Mexico. **2** A jewfish *(Promicrops guttatus)* of tropical American waters. [Alter. of Sp. *guasa*; prob. infl. in form by *Warsaw*]

War·saw (wôr′sô) The capital of Poland, on the Vistula, in the east central part of the country. *Polish* **War·sza·wa** (vär·shä′vä).

war·ship (wôr′ship′) *n.* Any vessel used in naval combat; especially, an armored vessel.

war·sle (wär′səl) *n., v.t. & v.i. Scot.* Wrestle. Also **war′stle. — war′sler** *n.*

Wars of the Roses See table under WAR.

wart (wôrt) *n.* **1** A small, usually hard and non-malignant excrescence formed on and rooted in the skin. **2** A spongy excrescence found on the pasterns of a horse. **3** A hard glandular protuberance on a plant. [OE *wearte*]

War·ta (vär′tä) A river in NW Poland, flowing 492 miles north and west to the Oder. *German* **War·the** (vär′tə).

Wart·burg (värt′bŏŏrkh) A castle in the former state of Thuringia, SW of Eisenach, SW East Germany, where Luther translated the New Testament (1521–22).

wart·hog (wôrt′hôg′, -hog′) *n.* An African veldt wild hog *(Phacochoerus aethiopicus)* having warty excrescences on the face and large tusks in both jaws.

WARTHOG
(From 2 to 2 1/2 feet at the shoulder)

War·ton (wôr′tən), **Thomas,** 1728–90, English literary historian, critic, and poet laureate.

wart·y (wôr′tē) *adj.* **wart·i·er, wart·i·est 1** Characterized by having warts: *warty*-flowered panic grass. **2** Of the nature of warts.

war whoop A yell made by American Indians, as a signal for attack or to terrify their opponents in battle.

War·wick (wôr′ik, wor′-) **1** A county of central England; 983 square miles. Also **War′·wick·shire** (-shir). **2** A municipal borough of central Warwick on the Avon; county town of Warwick.

War·wick (wôr′ik, wor′-), **Earl of,** 1428–71, Richard Neville, Earl of Salisbury, English statesman and soldier: called the "King-maker."

war·y (wâr′ē) *adj.* **war·i·er, war·i·est 1** Carefully watching and guarding. **2** Shrewd; wily. See synonyms under POLITIC, VIGILANT. [<WARE[2], *adj.*] **— war′i·ly** *adv.* **— war′i·ness** *n.*

was (woz, wuz, *unstressed* wəz) First and third person singular, past indicative of BE. [OE *wæs,* first and third person sing. of *wesan* be]

Wa·satch Plateau (wô′sach) A high tableland of central Utah at the southern end of the Wasatch Range; highest point 12,300 feet.

Wasatch Range A section of the Rocky Mountains in SE Idaho and northern Utah; highest point, 12,008 feet.

wase (wāz) *n. Obs.* or *Dial.* A wisp or bundle of hay, straw, or the like; especially, a cushion of such material for use between the head and a load borne thereon.

wash (wosh, wôsh) *v.t.* **1** To cleanse by immersing in or applying water or other liquid, often with rubbing or scrubbing. **2** To purify from pollution, defilement, or guilt. **3** To wet or cover with water or other liquid. **4** To flow against or over; lave: a beach *washed* by the ocean. **5** To carry away or remove by the action of water: with *away, off, out,* etc. **6** To form or wear by erosion: The storm *washed* gulleys in the hillside. **7** To purify, as gas, by passing through a liquid. **8** To coat with a thin or watery layer of color. **9** To cover with a thin coat of metal. **10** *Mining* **a** To subject (gravel, earth, etc.) to the action of water so as to separate the ore, etc. **b** To separate (ore, etc.) thus. **11** *Aeron.* To warp. **—** *v.i.* **12** To wash oneself. **13** To wash clothes, etc., in water or other liquid. **14** To withstand the effects of washing: That calico will *wash.* **15** *Brit. Colloq.* To undergo testing successfully: That story won't *wash.* **16** To flow with a lapping sound, as waves. **17** To be carried away or removed by the action of water: with *away, off, out,* etc. **18** To be eroded by the action of water. See synonyms under CLEANSE, PURIFY. **— to wash out** *Slang* **1** To fail and be dropped from a course, especially in military flight training. **2** To damage (an aircraft) irreparably, especially in landing. **—** *n.* **1** The act or process of washing; cleansing; ablution. **2** A number of articles, as of clothing, set apart for washing or being washed at one time; a washing; laundry. **3** Liquid or semi-liquid refuse; especially, waste food from the kitchen; swill. **4** A preparation used in washing or coating; specifically, a liquid cosmetic or a mouthwash; also, a water-color or India-ink pigment for spreading lightly and evenly on a drawing or picture. **5** The breaking of a body of water upon the shore, or the sound made by waves breaking or surging against a surface; swash. 6 Erosion of soil or earth by the action of rain or running water. 7 Backwash. 8 *Aeron.* Local air currents set up by the passing of an airplane. 9 An area washed by a sea or river; also, the shallow part of a river or an arm of the sea; a marsh; bog. 10 Material collected and deposited by water, as in the bed of a river or along its banks. 11 *U.S.* The dry bed of a stream; an arroyo. 12 Fermented liquor ready for the distillery. — *adj.* Washable; that may be washed without injury: *wash* fabrics. [OE *wæscan, wascan*]

Wash (wosh, wôsh), **The** An inlet of the North Sea on the eastern coast of England between Norfolk and Lincolnshire; 20 miles long, 15 miles wide.

wash·a·ble (wosh′ə·bəl, wôsh′-) *adj.* That may be washed without fading or injury.

wash·board (wosh′bôrd′, -bōrd′, wôsh′-) *n.* 1 A board or frame having a corrugated surface on which to rub clothes while washing them. 2 *Naut.* A thin plank adjusted to turn the wash of the sea from a deck or port of a ship.

wash bowl A basin or bowl, either portable or stationary, used for washing the hands and face. Also **wash basin.**

wash·cloth (wosh′klôth′, -kloth′, wôsh′-) *n.* A small cloth used for washing the body.

wash·day (wosh′dā′, wôsh′-) *n.* A day of the week set aside for doing household washing.

washed-out (wosht′out′, wôsht′-) *adj.* 1 Faded; colorless; pale. 2 *Colloq.* Exhausted; worn-out; tired.

washed-up (wosht′up′, wôsht′-) *adj. Slang* Finished; done with; through; also, washed-out; tired.

wash·er (wosh′ər, wô′shər) *n.* 1 One who washes. 2 *Mech.* A small, flat, perforated disk of metal, leather, or wood, used for placing beneath a nut or at an axle bearing or joint, to serve as a cushion, to relieve friction, etc. 3 A machine for washing (ore or clothes). 4 A device for purifying gases; a scrubber.

wash·er·man (wosh′ər·mən, wô′shər-) *n. pl.* **·men** (-mən) A laundryman.

wash·er·wom·an (wosh′ər·wŏŏm′ən, wô′shər-) *n. pl.* **·wom·en** (-wim′in) A laundress.

wash·in (wosh′in′, wôsh′-) *n. Aeron.* An increase in the angle of incidence of an airplane wing relative to the tip.

wash·ing (wosh′ing, wô′shing) *n.* 1 The act of one who washes. 2 Things (as clothing) washed on one occasion, or collected during a certain time. 3 That which is retained after being washed: a *washing* of ore. 4 A thin coating of metal: The forks had received only one *washing* of silver. 5 The sale of stock or other securities at a stock exchange between parties of one interest, in order to create a fictitious activity. — *adj.* Used in or intended for washing.

washing soda Sodium carbonate.

Wash·ing·ton (wosh′ing·tən, wô′shing-) 1 A State in NW United States, adjoining Canada; 68,192 square miles; capital, Olympia; entered the Union Nov. 11, 1889; nickname, *Evergreen State*: abbr. *Wash.* 2 A city coextensive with the District of Columbia and capital of the United States. — **Wash′ing·to′ni·an** (-tō′nē·ən) *adj.* & *n.*

Wash·ing·ton (wosh′ing·tən, wô′shing-), **Booker Taliaferro,** 1856–1915, U.S. Negro educator. — **George,** 1732–99, American patriot, soldier, and statesman; first president of the United States 1789–97. — **Martha,** 1731–1802, *née* Dandridge (Mrs. Daniel Parke Custis 1749–57), wife of George Washington.

Washington, Lake A lake in west central Washington, near Seattle; 20 miles long.

Washington, Mount The highest peak of the White Mountains of New Hampshire; 6,288 feet.

Washington palm The fan palm (*Washingtonia filifera*) of California and the Colorado desert.

Washington pie A layer cake with a filling of cream or jam.

Washington's Birthday The anniversary of George Washington's birth February 22: a legal holiday in most States of the United States.

Wash·i·ta River (wosh′ə·tô, wô′shə-) 1 See OUACHITA RIVER. 2 A river in Texas and Oklahoma, flowing 450 miles SE and east from the Texas Panhandle near the Oklahoma border to Lake Texoma; formerly flowed 40 miles further to the Red River.

wash leather Any soft, washable leather; especially, chamois leather, or an imitation.

wash·out (wosh′out′, wôsh′-) *n.* 1 A considerable erosion of earth by the action of water; also, the excavation thus made; a gully or gulch. 2 *Aeron.* A decrease in the angle of incidence of an airplane wing toward the tip. 3 *Slang* A hopeless or total failure; wreck.

wash·rag (wosh′rag′, wôsh′-) *n.* A washcloth.

wash·room (wosh′rŏŏm′, -rŏŏm′, wôsh′-) *n.* A lavatory.

wash sale On a stock exchange, the buying of stock by the seller's agents, to mislead as to the real demand.

wash·stand (wosh′stand′, wôsh′-) *n.* A piece of furniture used for holding the utensils for ablutions; a stand for wash bowl, pitcher, etc.

wash·tub (wosh′tub′, wôsh′-) *n.* A tub used for washing.

wash·wom·an (wosh′wŏŏm′ən, wôsh′-) *n. pl.* **·wom·en** (-wim′in) A washerwoman.

wash·y (wosh′ē, wô′shē) *adj.* **wash·i·er, wash·i·est** 1 Overly wet; sodden; water-logged. 2 Bringing rain: said of weather or wind. 3 Wanting in substance, solidity, stamina, or force; wishy-washy; feeble. 4 Sweating: said of horses. — **wash′i·ness** *n.*

was·n't (woz′ənt, wuz′-) Was not: a contraction.

wasp (wosp, wôsp) *n.* Any of numerous hymenopterous insects, chiefly of the superfamilies *Sphecoidea* and *Vespoidea*, of which the workers and females are provided with effective stings. The typical social wasps construct papery nests of masticated vegetable material; they feed on fruits, the nectar of flowers, and on insects. The solitary wasps construct nests of mud or sand. ◆ Collateral adjective: *vespine.* [OE *wæsp*]

wasp·ish (wos′pish, wôs′-) *adj.* 1 Having a nature like a wasp; irritable; irascible. 2 Having a wasplike form or slender waist. See synonyms under FRETFUL. — **wasp′ish·ly** *adv.* — **wasp′ish·ness** *n.*

wasp waist A person's waist, so slender as to suggest that of a wasp. — **wasp-waist·ed** (wosp′wās′tid, wôsp′-) *adj.*

wasp·y (wos′pē, wôs′-) *adj.* **wasp·i·er, wasp·i·est** Like a wasp; waspish.

was·sail (wos′əl, was′-, wo·sāl′) *n.* 1 An ancient salutation or toast; an expression of good will in festivities, especially when pledging someone's health. See DRINK-HAIL. 2 The liquor prepared for a wassail; especially, a mixture of ale and wine with sugar, roasted apples, spices, etc. 3 A festivity at which healths are drunk; a carousal. 4 *Brit.* A convivial song. — *v.i.* To take part in a wassail; carouse. — *v.t.* To drink the health of; toast. [ME *wæs hæil* <ON *ves heill* be whole (*i.e.*, in good health)] — **was′sail·er** *n.*

Was·ser·mann (väs′ər·män), **August von,** 1866–1925, German physician and bacteriologist. — **Jakob,** 1873–1934, German novelist.

Wasserman reaction A diagnostic test for syphilis, based on testing the serum of the blood for syphilitic antibodies. Also **Wassermann test.** [after August von *Wasserman*]

wast[1] (wost, *unstressed* wəst) Archaic second person singular, past indicative of BE: used with *thou.*

wast[2] (wast) *adj. Scot.* West.

wast·age (wās′tij) *n.* That which is lost by leakage, wear, waste, etc.

waste (wāst) *adj.* 1 Cast aside as worthless or of no practical value; used; worn out; discarded. 2 Excreted; cast out of an animal body, as food, etc. 3 Not under cultivation; untilled; hence, unproductive; unoccupied. 4 Made desolate; ruined; dismal; gloomy. 5 Containing or conveying waste products. 6 Produced in excess of consumption; superfluous: *waste* energy. 7 *Obs.* Wasteful; lavish. — *v.* **wast·ed, wast·ing** *v.t.* 1 To use or expend thoughtlessly, uselessly, or without return; be prodigal or extravagant of; squander. 2 To cause to lose strength, vigor, or bulk; make weak or feeble. 3 To use up; exhaust; consume. 4 To fail to use or take advantage of, as an opportunity. 5 To lay waste; desolate; devastate. — *v.i.* 6 To lose strength, vigor, or bulk; become weak or feeble: often with *away.* 7 To diminish or dwindle gradually. 8 To pass gradually: said of time. See synonyms under SQUANDER, WEAR[1]. — *n.* 1 The act of wasting or squandering, or the state of being wasted; useless or unnecessary expenditure. 2 A place or a region that is devastated or made desolate; wilderness; desert. 3 A continuous, gradual diminishing of strength, vigor, or substance by use or wear. 4 The act of laying waste or devastating; ravage: the *waste* of war. 5 Something rejected as worthless or unneeded; specifically, tangled spun cotton thread, the refuse of a textile factory; also, steam or other fluid that escapes without being used. 6 Garbage; rubbish; trash. 7 The waste products of the soil due to erosion by chemical or human action and carried out to sea by running water. 8 A wasting disease; specifically, consumption. ◆ Homophone: *waist.* [<AF *waster,* ult. <L *vastare* lay waste <*vastus* desert, desolate. Related to VAST.]

Synonyms (adj.): excess, extra, redundant, refuse, superfluous, useless, valueless, worthless. See BLEAK, VACANT. *Antonyms*: choice, good, precious, useful, valuable.

Synonyms (noun): chaff, debris, dregs, dross, leavings, offal, offscouring, refuse, remains, scum, sediment. See EXCESS, LOSS.

Waste may appear as a combining form in hyphemes or solidemes, or as the first element in two-word phrases, with the meaning: containing or conveying refuse or waste; as in:

waste bin	**waste sluice**
waste-collector	**waste trap**
waste gate	**waste-water**
waste heap	**wasteway**
waste pit	**wasteyard**

waste·bas·ket (wāt′bas′kit, -bäs′-) *n.* A basket for paper scraps and other waste.

waste·ful (wāst′fəl) *adj.* 1 Prone to waste; extravagant. 2 Causing waste; ruinous. — **waste′ful·ly** *adv.* — **waste′ful·ness** *n.*

waste·land (wāst′land′) *n.* A barren or desolate land.

waste paper Paper thrown away as worthless. Also **waste·pa·per** (wāst′pā′pər). — **waste′-pa′per** *adj.*

waste-paper basket A wastebasket.

waste pipe A pipe for carrying off waste-water, etc.

wast·er (wās′tər) *n.* One who wastes; a wastrel.

wast·ing (wās′ting) *adj.* 1 Producing emaciation; sapping the strength; enfeebling: a *wasting* fever. 2 Laying waste; devastating.

wast·rel (wās′trəl) *n.* 1 An abandoned child; a waif. 2 A waster; a profligate; spendthrift. [Dim. of WASTER]

wast·ry (wās′trē) *Scot.* or *Obs. adj.* Wasteful. — *n.* Wastefulness: also **waste′rie, wast′·rie, wast′rife.**

wat[1] (wat) *adj. Scot.* 1 Intemperate. 2 Wet.

wat[2] (wot) *n.* A hare. [Prob. from *Wat,* nickname for WALTER]

wa·tap (wä·täp′) *n.* Roots of the spruce, cedar, pine, etc., used by North American Indians to sew bark for canoes and other objects. Also **wa·ta·pe** (wä·tä′pe). [<Algonquian (Narraganset) *wattap* a root of a tree]

watch (woch) *v.i.* 1 To be constantly on the alert; give earnest heed; be observant, vigilant, or attentive. 2 To look attentively; observe. 3 To wait expectantly for something; be in a state of expectation: with *for.* 4 To do duty as a guard or sentinel; serve as a watchman. 5 To have in one's care or keeping; guard; tend. 6 To be awake; go without sleep; keep vigil. — *v.t.* 7 To keep under observation; look at steadily and attentively; observe. 8 To follow the course of mentally; keep informed concerning. 9 To be alert for; wait for expectantly: to *watch* one's opportunity. 10 To keep watch over; guard. See synonyms under ABIDE, LOOK. — *n.* 1 The act of watching; wakefulness with close and continuous attention; careful observation; vigil. 2 One of the divisions of the night made in ancient times: with the Hebrews, one third; with the Romans, one fourth; hence, any indefinite waking period which marks the passage of the night. 3 Position or service as a guard or sentry. 4 *Obs.* Vigilance; a vigil; wake. 5 One or more persons set to watch; a watchman or set of watchmen; sentinel; guard. 6 The place occupied by or assigned to a guard. 7 The period of time during which a guard is on duty. 8 *Naut.* **a** One of the two divisions of a ship's officers and crew, performing duty in alternation. **b** The period of time during which each division is on duty: four hours, except the dog-watches, from 4 to 6 and from 6 to 8 p.m., which are interposed daily to shift night duty from one watch to the other alternately. 9 A small, portable timepiece, actuated by a coiled spring, for keeping and indicating time. 10 *Obs.* A candle marked into equal sections, each of which burns a known length of time. 11 *Obs.* The cry of a watchman. 12 *Obs.* Wakefulness; the state of staying or being awake. See synonyms under OVERSIGHT. [OE *waeccan*. Akin to WAKE¹.]

watch cap In the U.S. Navy, a small, knitted woolen cap of navy blue worn by enlisted men during cold weather.

watch·case (woch'kās') *n.* 1 The protecting case of a watch: usually of gold or silver. 2 *Obs.* A sentry box.

watch·cry (woch'krī') *n. pl.* **·cries** A slogan; a watchword.

watch·dog (woch'dôg', -dog') *n.* A dog kept to guard a building or other property.

watch·er (woch'ər) *n.* 1 One who watches; especially, one who watches by a sickbed, deathbed, or corpse. 2 One who watches the voting at the polls on election day to detect dishonest practices.

watch·ful (woch'fəl) *adj.* 1 Vigilant. 2 *Obs.* Wakeful. See synonyms under ALERT, VIGILANT. — **watch'ful·ly** *adv.* — **watch'ful·ness** *n.*

watch·guard (woch'gärd') *n.* A chain, cord, or ribbon attached to a watch and fastened to the clothing.

watch·mak·er (woch'mā'kər) *n.* One who makes or repairs watches.

watch·man (woch'mən) *n. pl.* **·men** (-mən) 1 Formerly, one of a group of men appointed to keep watch or patrol the streets of a town or village at night. 2 Anyone who keeps watch or guard; especially, a man employed to guard a building, etc., at night.

watch night New Year's Eve.

watch·tow·er (woch'tou'ər) *n.* A tower upon which a sentinel is stationed.

watch·word (woch'wûrd') *n.* 1 A secret password. 2 A rallying cry or maxim.

Wa·ten·stedt-Salz·git·ter (vä'tən-shtet-zälts'-git-ər) A city in SE Lower Saxony, north central West Germany.

wa·ter (wô'tər, wot'ər) *n.* 1 A colorless limpid liquid compound of hydrogen and oxygen, H₂O, in the proportion of two volumes of hydrogen to one of oxygen, or by weight of approximately 2 parts of hydrogen to 16 of oxygen. Water has its maximum density at 4° C. or 39° F., one cubic centimeter weighing a gram. It freezes at 0° C. or 32° F., and boils at 100° C. or 212° F. 2 Any body of water, as a lake, river, or a sea; in Scotland, a small river. 3 Any one of the aqueous or liquid secretions of animals; also, perspiration, tears, urine, etc. 4 Any preparation of water holding a gaseous or volatile substance in solution. 5 The transparency or luster of a precious stone or a pearl; hence, excellence; purity. 6 An undulating sheen given to certain fabrics, as silk, etc. 7 In commerce and finance, stock issued without increase of paid-in capital to represent it. — **above water** Out of danger; secure. — **hard water** Water containing in solution salts of calcium and magnesium, especially the sulfates or bicarbonates of these elements: so called because of the difficulty of obtaining a soap lather with such water. — **soft water** Water free from the salts of calcium and magnesium, as rain water and water found in sandstone districts. — *v.t.* 1 To pour water upon; irrigate. 2 To provide with water for drinking; give water to. 3 To dilute or weaken with water: often with *down*. 4 To give an undulating sheen to the surface of (silk, linen, etc.) by uneven pressure after damping and heating. 5 To enlarge the number of shares of (a stock company) without increasing the paid-in capital in proportion. 6 To provide with streams: used in the passive participle. — *v.i.* 7 To secrete or discharge water, tears, etc. 8 To fill with saliva, as the mouth, from desire for food. 9 To drink water. 10 To take in water, as a locomotive. [OE *wæter*. Akin to OTTER.]

Water may appear as a combining form in hyphemes or solidemes, or as the first element in two-word phrases:

water-analysis	water-laden
water barge	water-locked
water-bearing	water pail
water bottle	waterplane
water-bound	water plant
water bucket	water police
water-carrier	water problem
water-carrying	water project
water cask	water pump
water channel	water-quenched
water content	water-resistant
water-deposited	water resources
water diver	water-rolling
water-drain	water-rot
water-drawer	water-rotted
water-drinker	water-route
water-drinking	water-scarcity
water flow	water-sealed
water-flushed	water service
water fountain	water-soaked
waterfree	water-sodden
water-girt	water source
water-gray	water tap
water-green	water trough
water heater	water turbine
waterhole	water-walled
water insect	water-washed
water jar	water-wasting

water adder 1 The water moccasin. 2 The water snake.

wa·ter·age (wô'tər-ij, wot'ər-) *n. Brit.* Conveyance of merchandise by water; also, the fee paid for such transportation.

wa·ter·back (wô'tər-bak') *n.* A coil or chamber for heating water in the back of a range or other stove.

water balance *Biol.* The preservation of a nearly uniform water content in an organism, especially a plant.

water bear See under TARDIGRADE.

Wa·ter-bear·er (wô'tər-bâr'ər, wot'ər-) The constellation Aquarius.

water bearing *Mech.* A journal bearing in which water under pressure does the work of a lubricant.

water beetle Any of several aquatic beetles (especially the families *Dystiscidae*, *Hydrophilidae* or *Gyrinidae*), having legs flattened and fringed with hairs for swimming.

water bird Any bird living on or near the water.

water biscuit A plain cracker or biscuit of flour, shortening, and water.

water blink In arctic regions, a cloud or spot on the horizon arising from and indicating the presence of open water: a sign of the breaking up of winter.

water blister A blister containing limpid watery matter.

wa·ter·bloom (wô'tər-blōōm', wot'ər-) *n.* The sudden appearance of large masses of blue-green algae in bodies of fresh water.

wa·ter-borne (wô'tər-bôrn', -bōrn', wot'ər-) *adj.* 1 Floating on water. 2 Transported or carried by water: *water-borne* commerce.

wa·ter·brain (wô'tər-brān') *n.* A disease of sheep characterized by staggering as from giddiness; gid.

water brake *Mech.* A brake, formerly used on steam locomotives, formed by injecting a jet of water into a cylinder when the engine is running, the impedance of the water providing a braking effect.

wa·ter·brash (wô'tər-brash') *n. Pathol.* Pyrosis; heartburn.

wa·ter·buck (wô'tər-buk') *n.* 1 Either of two large African antelopes (genus *Kobus*), frequenting the neighborhood of rivers and swimming with ease; especially, *K. ellipsiprymnus* of south central Africa. 2 Any of several similar antelopes. [<Du. *waterbok*]

water buffalo 1 A buffalo (*Bubalus bubalus*) of India, the largest of wild cattle, attaining a height of 6 feet at the withers and a very wide spread of horns. When domesticated it becomes a useful draft animal. 2 The carabao. Also called *Indian buffalo*.

WATER BUFFALO
(Spread of horns up to 9 feet)

water bug 1 The Croton bug. 2 Any of various hemipterous bugs (family *Belostomatidae*) which live in the water, especially the large species (*Lethocerus americanus*) common in North America. 3 The water scorpion.

Wa·ter·bu·ry (wô'tər-ber'ē, wot'ər-) A city in SW Connecticut; an industrial center, especially of the brass industry.

water chestnut 1 The hard horned edible fruit of an aquatic plant (*Trapa natans*). 2 The plant itself: also **water caltrop, wa'ter·nut'**.

water chinkapin 1 The American or yellow lotus (*Nelumbium pentapetalum*). 2 One of its edible nutlike seeds. Also **water chinquapin**.

wa·ter·clock (wô'tər-klok', wot'ər-) *n.* Any device, as a clepsydra, for measuring time by the fall or flow of water.

WATER CHINKAPIN
a. Flower. *b.* Leaf. *c.* Fruit.

wa·ter·clos·et (wô'tər-kloz'it, wot'ər-) *n.* A room or closet having a hopper flushed and discharged by means of water, used as a privy; also, the hopper and its trap.

wa·ter-col·or (wô'tər-kul'ər, wot'ər-) *adj.* Of, pertaining to, used with, or executed in water colors.

water color 1 A color prepared for painting with water as the medium, as distinguished from one to be used with oil, tempera, etc., as the medium, and characterized by the fact that the result may be either transparent or opaque. 2 That branch of painting in which water colors are used, or the method of using them. 3 A picture or painting done in water colors.

wa·ter-cool (wô'tər-kōōl', wot'ər-) *v.t.* To cool by means of water; as by using a water jacket on an internal-combustion engine. — **wa'ter-cooled'** *adj.* — **wa'ter-cool'ing** *adj. & n.*

water cooler A vessel or apparatus for cooling and dispensing drinking water: often operated electrically.

wa·ter·course (wô'tər-kôrs', -kōrs', wot'ər-) *n.* 1 A stream of water; river; brook; a stream having a bed and banks. 2 The course or channel of a stream of water; a canal. See synonyms under STREAM.

wa·ter·craft (wô'tər-kraft', -kräft', wot'ər-) *n.* 1 Skill in sailing boats or in aquatic sports. 2 Any boat or ship; also, sailing vessels collectively.

water crake 1 The spotted crake. 2 The water ouzel. See under OUZEL.

wa·ter·cress (wô'tər-kres', wot'ər-) *n.* A creeping perennial herb (*Rorippa nasturtium-aquaticum*) of the mustard family, having pinnate leaves and white flowers. It grows in springs and clear cool streams and is cultivated for use as salad.

water culture Hydroponics.

water cure 1 *Med.* Hydropathy. 2 *Colloq.* A kind of torture in which large quantities of water are put forcibly down the victim's throat.

water cushion A pool of water maintained to absorb the impact of water, as from the spillway of a dam.

water dog 1 A dog that takes readily to the water, as the water spaniel. 2 A dog trained to retrieve water fowl. 3 *Colloq.* An old sailor.

Wa·ter·ee (wô'tə-rē') The lower course of the

water elm The planer tree.
wa·ter·er (wô′tər·ər, wot′ər-) n. 1 One who waters, in any sense. 2 Any contrivance used for watering.
wa·ter·fall (wô′tər·fôl′, wot′ər-) n. 1 A cataract; cascade. 2 Colloq. A chignon suggesting a cascade. [OE *wætergefeall*]
water fence A fence built into or across a stream, or one extending into the water on the shore of a lake or the sea, to prevent cattle, horses, etc., from passing around it.
wa·ter-find·er (wô′tər·fīn′dər, wot′ər-) n. A dowser who tries to locate underground water with a divining rod. See RHABDOMANCY.
wa·ter·flea (wô′tər·flē′, wot′ər-) n. Any of numerous minute, fresh-water crustaceans (family *Daphniidae*), about the size of a flea, which swim with a jumping motion.
Wa·ter·ford (wô′tər·fərd, wot′ər-) 1 A maritime county in eastern Munster province, Ireland; 710 square miles. 2 Its county town, a port on **Waterford Harbor**, an inlet of the Atlantic in southern Ireland; 15 miles long.
water fowl 1 A bird that lives on or about the water, especially a swimming game bird. 2 Such birds collectively.
wa·ter·front (wô′tər·frunt′, wot′ər-) n. 1 Real property abutting on or overlooking a natural body of water. 2 That part of a town which fronts on a body of water. 3 A coil or chamber for heating water in the front of a range or other stove.
water gage A gage indicating the level of water in a boiler, etc. Also **water gauge.**
water gall 1 A hollow in the earth made by a flood, etc.; a washout. 2 A partial rainbow: also, *Scot.*, *weather gall*. [< WATER + GALL²]
water gap A deep ravine in a mountain ridge giving passage to a stream.
water gas A highly poisonous mixture of hydrogen and carbon monoxide produced by forcing steam over white-hot carbon (as coal or coke): used for cooking and heating, and when carbureted, as an illuminant. — **wa·ter-gas** (wô′tər·gas′, wot′ər-) adj.
wa·ter·gate (wô′tər·gāt′, wot′ər-) n. Floodgate (def. 1).
wa·ter·glass (wô′tər·glas′, -gläs′, wot′ər-) n. 1 A waterclock; clepsydra. 2 A glass-bottomed tube or box for examining objects lying or moving under water. 3 A substance composed of sodium silicate, potassium silicate, or both, soluble in hot water: used in preserving eggs, as a facing for walls, etc. 4 A water gage on a steam boiler, etc. 5 A vessel for holding water; a drinking glass.
wa·ter·gum (wô′tər·gum′, wot′ər-) n. 1 The American sourgum or tupelo tree. 2 A tall, slender, ornamental shrub (*Tristania laurina*) of the myrtle family, native in Australia, with opposite leaves and bright-yellow flowers.
water hammer 1 The concussion of confined water when its flow is suddenly arrested, as when a faucet is suddenly closed. 2 The hammering sound caused in pipes containing water when live steam is admitted. 3 A sealed tube void of air but containing water which strikes against the ends of the tube with a sharp knocking sound when shaken: used to demonstrate the equal rate of fall of solids and liquids in a vacuum.
water haul 1 In fishing, an empty haul of the net. 2 Any fruitless attempt or effort.
water hemlock Any of a genus (*Cicuta*) of poisonous, typically North American flowering herbs of the carrot family; especially, the **spotted water hemlock** (*C. maculata*) of the United States, highly injurious to livestock, and the Old World species (*C. virosa*).
water hen 1 Any of several coots or gallinules that frequent ponds and streams; especially, the moorhen. 2 The American coot (*Fulica americana*).
water hyacinth An aquatic herb of tropical America (*Eichornia crassipes*) with pendulous branched roots and a whorl of floating glossy leaves from whose center blooms a cluster of bluish-purple to lilac and white flowers.
water ice 1 An ice made with water, sugar, and fruit juice. 2 Ice formed by the freezing of water as distinguished from that formed by the packing together of snow.
wa·ter-inch (wô′tər·inch′, wot′ər-) n. An old unit of hydraulic measure based on the discharge of water from a round hole with a diameter of one inch: reckoned at fourteen pints a minute.
wa·ter·ing (wô′tər·ing, wot′ər-) n. 1 The act of one who waters. 2 The process of producing a wavy ornamental effect. — adj. 1 Sprinkling; irrigating; that waters. 2 Situated near the shore or near mineral springs: a *watering place*.
watering cart A cart carrying a barrel or large tank of water: used for sprinkling streets.
watering place 1 A place where water can be obtained, as a spring; also, a place by a road where horses can be watered. 2 A health resort having mineral springs; also, a pleasure resort near the water.
watering pot A tin can having a spout fitted with a perforated nozzle: used for watering flowers, etc.
wa·ter·ish (wô′tər·ish, wot′ər-) adj. Resembling water; watery; hence, weak.
wa·ter-jack·et (wô′tər·jak′it, wot′ər-) v.t. To encase in or fit with a water jacket.
water jacket A casing containing water and surrounding a cylinder or mechanism, especially the cylinder block of an internal-combustion engine, for keeping it cool.
water jump A water barrier, as a pool, stream, or ditch, to be jumped over by the horses in a steeplechase.
water leaf Any of a genus (*Hydrophyllum*) of delicate biennial or perennial herbs with white or blue flowers, growing in the woods of North America.
wa·ter·less (wô′tər·lis, wot′ər-) adj. Without water; arid; dry.
wa·ter-lev·el (wô′tər·lev′əl, wot′ər-) adj. Following the course of a river: a *water-level route*.
water level 1 The level of still water in the sea or in any other body of water. 2 A water table. 3 *Naut.* A ship's water line. 4 A leveling instrument in which water serves to determine the horizontal line.
wa·ter-lift (wô′tər·lift′, wot′ər-) n. The transportation of personnel, equipment, and supplies by water, with special reference to the 1950 military campaign in Korea.
wa·ter-lil·y (wô′tər·lil′ē, wot′ər-) n. pl. ·lil·ies 1 Any plant of a genus (*Nymphaea*) of showy aquatic herbs of temperate and tropical regions, with large floating leaves and flowers; especially, the fragrant **white water-lily** (*N. odorata*) of the eastern United States. 2 The yellow pond lily (*Nuphar luteum*) of the same family. 3 The Victoria waterlily.
water line 1 *Naut.* That part of the hull of a ship which corresponds with the water level at various loads. 2 Water level. 3 A river or system of waterways affording transportation.
water locust A small species of the American honey locust (*Gleditsia aquatica*) growing in southern swamps and boglands: also called *swamp locust*.
wa·ter·logged (wô′tər·lôgd′, -logd′, wot′ər-) adj. 1 Heavy and unmanageable on account of the leakage of water into the hold, as a ship. 2 Water-soaked; saturated with water. [< WATER + LOG v., in obs. sense of "to reduce to the condition of a log"]
Wa·ter·loo (wô′tər·lōō′, wô′tər·lōō′) A village in central Belgium; scene of Napoleon's final defeat by Wellington and Blücher, June 18, 1815; hence, final and decisive defeat; a complete reverse.
water lot 1 A building lot fronting on a body of water, as a river, harbor, etc. 2 A lot or piece of ground wholly or partially covered by water, or a piece of marsh or swamp land designated to be filled in for use.
water main A large conduit for carrying water, especially one laid underground.
wa·ter·man (wô′tər·mən, wot′ər-) n. pl. ·men (-mən) A man who plies for hire with a boat or small vessel on the water; a boatman. — **wa′ter·man·ship′** n.
water marigold An aquatic plant (*Bidens becki*) with terminal heads of yellow flowers.
wa·ter·mark (wô′tər·märk′, wot′ər-) n. 1 A mark showing the extent to which water rises; especially, the line marking the limit of the ebb and flow of the tide. 2 A series of translucent lines, letters, or designs made in paper by shaping the wires of the dandy rolls over which the paper passes while still in a pulpous state; also, the metal pattern which produces these markings. — v.t. 1 To impress (paper) with a watermark. 2 To impress as a watermark.
wa·ter·mel·on (wô′tər·mel′ən, wot′ər-) n. 1 The large edible fruit of a trailing plant (*Citrullus vulgaris*) of the gourd family, containing a many-seeded red or pink pulp and a refreshing sweet, watery juice. 2 The plant on which this fruit grows.
water meter An instrument for registering the amount of water flowing through a pipe, etc.
water milfoil Any of a genus (*Myriophyllum*) of aquatic herbs with graceful, feathery leaves.
water mill A mill operated by waterpower.
water moccasin The cottonmouth.
water motor 1 A turbine operated by waterpower. 2 A water wheel.
water nymph In classical mythology, any nymph or goddess living in or guarding a body of water; a naiad, Nereid, Oceanid, etc.
water oak A species of oak (*Quercus nigra*) growing near swamps and streams in the eastern United States.
water of Ayr See AYR STONE.
water of crystallization *Chem.* Water forming part of crystallized salts, from which it may be eliminated by heat, often with loss of crystalline structure. Also **water of hydration.**
water of life A rare and mysterious water that restores the dead to life. The human hope and belief that death can be overcome is expressed in the water-of-life motif in the peasant folklore of every European country, in the myths of the ancient Persians, Greeks, Romans, Hebrews, Hindus, in Japanese mythology, and in the folk tales of all primitive peoples, as the Polynesians, North and South American Indians, etc.
water ouzel See under OUZEL.
water ox A water buffalo.
wa·ter-part·ing (wô′tər·pär′ting, wot′ər-) n. A watershed.
wa·ter-pep·per (wô′tər·pep′ər, wot′ər-) n. Any of several species of knotweed, especially the common smartweed.
water pimpernel 1 The brookweed. 2 The common pimpernel.
water plantain Any of a genus (*Alsima*) of common, smooth, aquatic herbs with leaves like those of the plantain, especially the North American species (*A. plantago-aquatica*).
water polo A game played in a swimming pool by two teams of seven swimmers each, who push or throw a round, buoyant ball toward opposite goals.
wa·ter-pow·er (wô′tər·pou′ər, wot′ər-) n. 1 The power of water derived from its gravity or its momentum as applied to the driving of machinery. 2 A descent or fall in a stream from which motive power may be obtained.
water pox *Pathol.* Varicella.
wa·ter·proof (wô′tər·prōōf′, wot′ər-) adj. 1 Proof against water. 2 Impervious to water. 3 Coated with some substance, as rubber, which resists the passage of water. — n. 1 Material or fabric rendered impervious to water. 2 *Brit.* A raincoat or other garment made of such fabric. — v.t. To render waterproof.
water purslane 1 An herb (*Isnardia* or *Ludwigia palustris*) of the evening-primrose family, procumbent and creeping in muddy places and floating in water. 2 An aquatic plant (*Didiplis* or *Peplis diandra*) growing in swampy ground in the U.S.
water ram A hydraulic ram.
water rat 1 The American muskrat. 2 The European water vole (*Microtus amphibius*). 3 Any of a subfamily (*Hydromyinae*) of aquatic rodents of New Guinea, Australia, and the Philippines. 4 *Slang* A thief or tough who frequents the waterfront.
water repellent *Chem.* Any of various chemicals, as an emulsion of aluminum acetate, used to make textiles, leather, and other porous materials resistant to wetting by water but which does not waterproof them or impair their desirable properties. — **wa·ter-re·pel·lent** (wô′tər·ri·pel′ənt, wot′ər-) adj.
wa·ter-right (wô′tər·rīt′, wot′ər-) n. 1 The right to draw upon a water supply. 2 The

water sapphire A rich blue variety of iolite often worn as an ornament. [Trans. of F *saphir d'eau*]

wa·ter·scape (wô'tər·skāp, wot'ər-) *n.* A sea or other water view, as distinguished from a landscape. [<WATER + (LAND)SCAPE]

water scorpion Any of numerous hemipterous insects of aquatic habits (family *Nepidae*), having raptorial front legs and a long breathing tube at the end of the abdomen.

wa·ter·shed (wô'tər·shed', wot'ər-) *n.* 1 The line of separation between two contiguous drainage valleys. 2 The whole region from which a river receives its supply of water.

wa·ter·shield (wô'tər·shēld', wot'ər-) *n.* 1 An aquatic American herb (*Brasenia schreberi*) of the waterlily family, with the stems and the under sides of the leaves covered with a viscid jelly. 2 Any plant of a kindred genus (*Cabomba*), especially the fanwort.

wa·ter·sick (wô'tər·sik', wot'ər-) *adj.* Unproductive because of excessive irrigation: said of land.

wa·ter·side (wô'tər·sīd', wot'ər-) *n.* The shore of a body of water; the water's edge. — *adj.* 1 Of, pertaining to, or living or growing by the water's edge. 2 Working by the waterside, as a stevedore.

wa·ter·ski (wô'tər·skē') *v.i.* -skied, -ski·ing To glide over water on water-skis, while being towed by a motorboat. — *n.* A broad, skilike runner with a fitting to hold the foot: worn in the sport of water-skiing. — **wa'ter-ski'er** *n.* — **wa'ter-ski'ing** *n.*

water snake 1 A serpent of aquatic habits. 2 Any of a genus (*Natrix*) of harmless North American snakes that live chiefly in water.

wa·ter·soak (wô'tər·sōk', wot'ər-) *v.t.* To fill the pores or crevices of with water; soak in water.

wa·ter·sol·u·ble (wô'tər·sol'yə·bəl, wot'ər-) *adj.* Biochem. Soluble in water: said especially of certain organic compounds: a *water-soluble* vitamin: distinguished from *fat-soluble*.

water spaniel The Irish water spaniel. See under SPANIEL.

water speedwell A common plant (*Veronica anagallis-aquatica*) of the composite family, growing in damp places.

wa·ter·spout (wô'tər·spout', wot'ər-) *n.* 1 A moving, whirling column of spray and mist, with masses of water in the lower parts, accumulated because of a tornado at sea or on other large bodies of water. 2 A pipe for the free discharge of water, especially one connecting with the gutters of a roof.

water sprite A water nymph.

water starwort Any of a widely distributed genus (*Callitriche*) of herbaceous aquatic plants, especially *C. autumnalis*, common in the United States.

water station A place beside a railroad where there is a water tank for supplying locomotives with water.

water strider Any of a family (*Gerridae*) of hemipterous insects with elongate middle and hind legs adapted for darting over the surface of water.

water supply 1 The water available for the use of a community or region. 2 The means for supplying it, as reservoirs, lakes, etc. — **wa'ter-sup'ply** (wô'tər·sə·plī') *adj.*

water system 1 A river with all its tributaries, considered as a hydrologic unit. 2 Water supply.

water table 1 *Archit.* A projecting ledge, molding, or string-course, running along the sides of a building to shed the rain. 2 The surface marking the upper level of a water-saturated zone extending beneath the ground to depths determined by the thickness of the permeable strata.

water tank A large cistern of wood or metal, as upon an engine or building, for storing or supplying water; specifically, one built alongside a railroad track for supplying water to locomotives.

water thrush 1 Any of certain American warblers (genus *Seiurus*), frequenting swamps and streams; especially, the common or **northern water thrush** (*S. noveboracensis*), olive-brown above, yellowish beneath, with dusky streaks and a buffy superciliary line, or the **Louisiana water thrush** (*S. motacilla*), with a pure-white superciliary line. 2 The water ouzel.

water tiger The larva of the diving beetle.

wa·ter·tight (wô'tər·tīt', wot'ər-) *adj.* 1 So closely made that water cannot enter or leak through. 2 Constructed so as to be impermeable; without loopholes: a *watertight* legal document.

Wa·ter·ton Lakes National Park (wô'tər·tən) A national park in SW Alberta, Canada; 204 square miles; adjoining Glacier National Park, Montana, forming with it the **Waterton-Glacier International Peace Park**; total area, 1,800 square miles; highest U. S. point, 10,448 feet; highest Canadian point, 9,600 feet; established 1932.

water tower 1 A standpipe or tower, often of considerable height, used as a reservoir for a system of water distribution. 2 A vehicular towerlike structure having an extensible vertical pipe from which water can be played on a burning building from a great height.

wa·ter-tube boiler (wô'tər·tōōb', -tyōōb', wot'ər-) A type of boiler in which continuously heated water circulates through a series of tubes communicating with a steam chamber.

water turkey The snakebird.

water vapor The vapor of water, especially when found below the boiling point, as in the atmosphere. Compare STEAM.

water wave 1 An undulating effect of the hair, artificially produced when the hair is wet, and usually set by drying with heat. 2 A wave of water; a billow.

wa·ter·way (wô'tər·wā', wot'ər-) *n.* A river, channel, or other stream of water as a means of communication; water route.

wa·ter·weed (wô'tər·wēd', wot'ər-) *n.* 1 A submerged aquatic perennial (*Anacharis canadensis*), having whitish flowers. 2 Any of various other aquatic plants, as the pondweed.

water wheel 1 A wheel so equipped with floats, buckets, etc., that it may be turned by flowing water. See OVERSHOT WHEEL. 2 A noria.

water wings A waterproof, wing-shaped fabric device that may be inflated with air and used as a support for the body while swimming or learning to swim.

water witch 1 One who claims to discover underground springs with the use of a divining rod or hazel wand. 2 Any of various quick-diving water birds, as certain grebes.

water witching The use of a divining rod to discover water; rhabdomancy.

wa·ter·works (wô'tər·wûrks', wot'ər-) *n. pl.* 1 A display or pageant presented on floats; a display of fountains in operation. 2 A system of machines, buildings, and appliances for furnishing a water supply, especially for a city; also, any mill or factory run by waterpower. 3 *Slang* Tears: usually in the phrase *to turn on the waterworks.* b Rain.

wa·ter·worn (wô'tər·wôrn', -wōrn', wot'ər-) *adj.* Worn smooth by running or falling water.

wa·ter·y (wô'tər·ē, wot'ər·ē) *adj.* 1 Containing or discharging water; brimming; tearful; soft and flabby. 2 Resembling water; thin or liquid. 3 Consisting of or pertaining to water. — **wa'ter-i-ness** *n.*

Wat·ford (wot'fərd) A municipal borough in Hertford, England, NW of London.

Wat·ling Island (wot'ling) See SAN SALVADOR. Also **Wat'lings Island.**

Wat·son (wot'sən), **John,** 1850–1907, Scottish minister and author: pseudonym, **Ian Maclaren.** — **John Broadus,** born 1878, U. S. psychologist. — **Sir William,** 1858–1935, English poet.

Wat·son-Watt (wot'sən·wot'), **Sir Robert Alexander,** born 1892, Scottish physicist.

watt (wot) *n.* The practical unit of electric power, activity, or rate of work: equivalent to 10^7 ergs or one joule per second, or approximately 1/746 of a horsepower; a volt-ampere. [after James *Watt*]

Watt (wot), **James,** 1736–1819, Scottish inventor and engineer.

wat·tage (wot'ij) *n.* 1 Amount of electric power in terms of watts. 2 The total number of watts needed to operate an appliance.

Wat·teau (wä·tō', wot'ō) *adj.* Of or pertaining to Antoine Watteau, or the costumes shown in his pictures.

Wat·teau (wä·tō', *Fr.* vȧ·tō'), **Jean Antoine,** 1684–1721, French painter.

Watteau back A style of women's dress in which the fullness of the back is confined at the neck in plaits or gathers, and falls from there to the waistline.

Wat·ten·scheid (vät'ən·shit) A city of east central North Rhine-Westphalia, central West Germany; a coal-mining and manufacturing center in the Ruhr.

Wat·ter·son (wot'ər·sən), **Henry,** 1840–1921, U. S. editor and journalist.

watt-hour (wot'our') *n.* Electrical energy equivalent to that represented by one watt acting for one hour.

wat·tle (wot'l) *n.* 1 A frame of rods or twigs woven together; a hurdle or other wickerwork. 2 A twig or withe, especially as used for interweaving with others; also, collectively, material for fences, hurdles, roofs, etc. 3 A naked, fleshy process, often wrinkled and brightly colored, hanging from the throat of a bird or snake. 4 A pendent fold of skin on the throat or neck of some domestic swine. 5 A barbel of a fish. 6 Any one of various acacias of Australia, Tasmania, and South Africa: so called by the early colonists, who used the branches to make hurdles. 7 *pl.* Rods for supporting thatch on a roof. — *v.t.* **·tled, ·tling** 1 To weave or twist, as twigs, into a network. 2 To form, as baskets, by intertwining flexible twigs. 3 To bind together with wattles. — *adj.* Made of or covered with wattles; formed by wattling. [OE *watul*]

wat·tle·bird (wot'l·bûrd') *n.* Any of several large Australian honeyeaters (genus *Anthochaera*), having conspicuous wattles about the head and face.

wat·tled (wot'ld) *adj.* 1 Made with wattles. 2 Having a wattle, as a bird. 3 *Her.* Having wattles, comb, or gills of a tincture different from that of the body.

watt·less (wot'lis) *adj. Electr.* Denoting an alternating current, or the component of such a current, which is neutralized by the originating electromotive force and which for that reason does not produce any power.

watt·me·ter (wot'mē'tər) *n.* An instrument for measuring in watts the rate of doing electrical work: also called *voltammeter*.

Watts (wots), **George Frederick,** 1817–1904, English painter and sculptor. — **Isaac,** 1674–1748, English theologian and hymn writer.

Watts–Dun·ton (wots'dun'tən), **Theodore,** 1832–1914, English critic and poet.

Wa·tu·si (wä·tōō'sē) *n. pl.* **·si** One of a pastoral, Bantu-speaking people of east central Africa.

Wau (wou) A town in the Territory of New Guinea, NE New Guinea; a gold-mining center reached only by air.

wauch (wôkh, wäkh) *adj. Scot.* 1 Clammy; damp; nauseous. 2 Faint; languid; weak. Also **waugh** (wôf).

waucht (wôkht, wäkht, wäft) *Scot. n.* A large draft, as of liquor. — *v.t. & v.i.* To drink; quaff. Also **waught** (wôft, wäft).

Waugh (wô), **Alec,** born 1898, English novelist and travel writer. — **Evelyn,** born 1903, English novelist and critic; brother of preceding.

wauk[1] (wôk) *v.i. Scot.* To full cloth.

wauk[2] (wôk) *v.i. Scot.* To wake; watch over.

waul (wôl) *v.i.* To give a prolonged, plaintive cry like that of a cat: also spelled *wawl*. [Imit.]

waur (wôr) *adj. Scot.* Worse.

wave (wāv) *v.* **waved, wav·ing** *v.i.* 1 To move freely back and forth or up and down, as a flag in the wind; undulate or fluctuate. 2 To be moved back and forth or up and down as a signal; also, to make a signal by moving something thus. 3 To have an undulating shape or form; be sinuous: Her hair *waves*. — *v.t.* 4 To cause to move back and forth or up and down: to *wave* a banner. 5 To form with an undulating surface, edge, or outline. 6 To give a wavy appearance to; water, as silk. 7 To form into waves or undulations: to *wave* one's hair. 8 To signal by waving something: He *waved* me aside. 9 To express by waving something: to *wave* farewell. See synonyms under FLAUNT, SHAKE. —*n.* **1** A ridge or undulation moving on the surface of a liquid, the particles composing it having an oscillatory motion usually in the form of closed or nearly closed curves in a plane at right angles to the direction of movement of the ridge itself. **2** *Physics* One of the complete vibratory impulses set up by a disturbance propagated through the particles of a body or elastic medium, as a rope, air, or water. Each impulse, as in the transmission of light or sound, has characteristics of length, frequency, duration, etc., determined by the nature of the disturbance in the medium involved. **3** One of the rising curves on an undulatory edge or surface; one of a series of curves: amber *waves* of grain. **4** Something that comes, like a wave, with great volume or power; a flood; a period of marked activity or excitement: a *wave* of enthusiasm. **5** A wavelike stripe or undulation impressed on a surface, as on watered silk; also, a wavelike tress or curl of hair. **6** *Poetic* Any body of water; the sea. **7** The act of waving; a sweeping or undulating motion, as with the hand or a flag. **8** One of a series, as of groups or events, occurring or moving with wavelike fluctuations: He went ashore with the first *wave* of Marines. **9** A progressive change in temperature or in barometrical condition passing over a large area: a heat *wave.* ◆ Homophone: *waive.* [OE *wafian*] —**wav′er** *n.*

Synonyms (noun): billow, breaker, ripple, surge, swell, undulation, vibration.

Wave (wāv) *n.* A member of the WAVES. [Back formation from WAVES (taken as a pl.)]

wave cloud *Meteorol.* A cloud consisting of parallel bands or ridges, separated by strips of clear sky, due to air currents occurring at the bounding plane between two strata of the atmosphere.

wave front The leading surface of a wave as it advances through a medium.

wave–guide (wāv′gīd′) *n.* **1** Any system of material boundaries by which the direction of waves may be controlled. **2** *Electronics* A device, typically an arrangement of hollow metal pipes of varying size and cross–section, through which high–frequency electromagnetic waves may be guided as required.

wave·length (wāv′length′) *n. Physics* The distance, measured along the line of propagation, between two points representing similar phases of two consecutive waves. It is a fundamental unit in the study of radiant energy.

wave·less (wāv′lis) *adj.* Having no waves; tranquil. See synonyms under PACIFIC.

wave·let (wāv′lit) *n.* A little wave.

Wa·vell (wā′vəl), **Sir Archibald Percival,** 1883–1950, first Earl Wavell, Viscount Wavell of Cyrenaica and Tripolitania, British field marshal and administrator in World War II.

wa·vel·lite (wā′və·līt) *n.* A vitreous, translucent, hydrous aluminum phosphate, crystallizing in the orthorhombic system. [after Dr. William *Wavell,* died 1829, English physician, its discoverer]

wave mechanics The branch of physics which investigates the wave characteristics ascribed to the atom and its associated particles, and seeks to explain physical processes in terms of these characteristics as revealed by the quantum theory of atomic structure.

wave·me·ter (wāv′mē′tər) *n.* An apparatus for determining wavelengths and wave frequencies, as in a radio circuit.

wave number *Physics* The number of electromagnetic waves in a space of 1 centimeter, equal to the frequency of the wave divided by the velocity of light: it is the reciprocal of the wavelength.

wave·off (wāv′ôf′, -of′) *n. Aeron.* The act of denying landing privileges to an approaching aircraft, usually an aircraft making a faulty approach for landing on an aircraft carrier.

wa·ver (wā′vər) *v.i.* **1** To move one way and the other; sway; flutter. **2** To be uncertain or undecided; show irresolution; vacillate. **3** To show signs of falling back or giving way; reel; falter. **4** To flicker; gleam. **5** To quaver; tremble. See synonyms under FLUCTUATE, QUAKE, SHAKE. —*n.* A wavering. ◆ Homophone: *waiver.* [<ME *waveren,* freq. of OE *wafian* wave] —**wa′ver·er** *n.* —**wa′ver·ing** *adj.* —**wa′ver·ing·ly** *adv.*

Wa·ver·ley Novels (wā′vər·lē) A series of historical novels by Sir Walter Scott, published 1814–1831.

WAVES (wāvz) *n.* A corps of women in the U. S. Navy, which includes all women except nurses; officially, *Women in the United States Navy* (1946). [<*W(omen) A(ppointed for) V(oluntary) E(mergency) S(ervice),* an earlier name]

wave train A series of waves sent out at regular intervals from a vibrating body.

wave trap **1** *Telecom.* A device, usually connected with the antenna, for improving the selectivity of a radio receiver by cutting out undesired wave frequencies. **2** A widening inward of the distance between the sides of adjoining piers to allow for the spreading of storm waves.

wa·vey (wā′vē) *n.* The snow goose. [Var. of WAVY; so called because faintly streaked on head, neck, and back with darker plumage]

wav·y (wā′vē) *adj.* **wav·i·er, wav·i·est** **1** Full of waves; ruffled by or raised into waves. **2** Undulatory; waving. **3** Unstable; wavering. —**wav′i·ly** *adv.* —**wav′i·ness** *n.*

wawl (wôl) See WAUL.

wax[1] (waks) *n.* **1** A yellow fatty solid excreted from the abdominal rings of bees and used by them to build honeycombs; beeswax. It has a honeylike odor and a balsamic taste; becomes plastic with the heat of the hand; is insoluble in water, but is almost completely dissolved by boiling alcohol. **2** Any of a class of plant and animal substances consisting of the esters of fatty acids and alcohols other than glycerol, and including spermaceti and carnauba wax; specifically, a substance derived from the fruit of the wax myrtle or bayberry. **3** A solid mineral substance resembling wax, as ozocerite or paraffin. **4** A substance used for joining surfaces, sealing documents, etc.; sealing wax. **5** A mixture of pitch and tallow or some resinous composition used by shoemakers to wax their thread. **6** Earwax; cerumen. **7** *U. S.* The sap of the sugar maple after being boiled down and cooled. **8** A substance resembling beeswax secreted by certain scale insects. —*v.t.* To coat or treat with wax. —*adj.* Made of or pertaining to wax. [OE *weax*]

wax[2] (waks) *v.i.* **waxed, waxed** (*Poetic* **wax·en**), **wax·ing** **1** To become larger gradually; increase in size or numbers; grow: said specifically of the moon as it approaches fullness: opposed to *wane.* **2** To become as specified: to *wax* angry. [OE *weaxan* grow]

wax[3] (waks) *v.t. Colloq.* To record phonographically: to *wax* a folk song. [<WAX[1]; so called because wax was formerly used in making phonograph records]

wax[4] (waks) *n. Colloq.* A tantrum; fit of bad temper. [? < phrase *wax angry*]

wax bean A variety of string bean (*Phaseolus vulgaris*) cultivated in the United States: also called *butter bean.*

wax·ber·ry (waks′ber′ē) *n. pl.* **·ries** **1** The wax myrtle, or bayberry. **2** Its wax–covered fruit. **3** The snowberry.

wax·bill (waks′bil′) *n.* **1** Any of various small Old World seed–eating birds of the weaverbird family (genus *Estrilda*), having beaks resembling sealing wax. **2** The Java sparrow.

wax·en (wak′sən) *adj.* **1** Resembling wax. **2** Consisting wholly or in part of wax; covered with wax. **3** Pale; pallid: a *waxen* complexion; also, pliable or impressible as wax.

wax end A stout thread, or the end of a thread, made stiff and pointed with shoemakers' wax, or waxed and twisted with a bristle, as for the purpose of sewing shoes. Also **waxed end.**

wax myrtle Any of a genus (*Myrica*) of North American shrubs or small trees, especially *M. cerifera,* having fragrant leaves and small berries covered with white wax, often used in making candles: also called *bayberry, candleberry.*

wax palm **1** A South American palm (*Ceroxylon andicola*) with pinnate leaves, having a lofty straight trunk covered with a waxy, whitish, resinous substance. **2** A Brazilian palm (*Copernicia cerifera*) whose young leaves yield the carnauba wax of commerce.

wax paper Paper coated or treated with wax and used to protect against moisture. Also **waxed paper.**

wax plant The Indian pipe.

wax·weed (waks′wēd′) *n.* An annual, clammy, hairy herb (*Cuphea petiolata*) of the loosestrife family with irregular purplish axillary flowers.

wax·wing (waks′wing′) *n.* Any of various crested passerine birds (family *Bombycillidae*) of America and Asia, having soft, mainly brown plumage, and the tips of the secondary wing feathers tipped with horny appendages resembling red or yellow sealing wax; especially, the two best–known North American species, the **cedar waxwing** (*Bombycilla cedrorum*), and the larger **Bohemian waxwing** (*B. garrula pallidiceps*).

CEDAR WAXWING
(About 7 inches long)

wax·work (waks′wûrk′) *n.* **1** Work produced in wax; particularly, ornaments or life–size figures of wax. **2** *pl.* A collection of such figures.

wax·work·er (waks′wûr′kər) *n.* One who works in wax; one who makes waxwork.

wax·worm (waks′wûrm′) A honeycomb moth.

wax·y (wak′sē) *adj.* **wax·i·er, wax·i·est** **1** Resembling wax in appearance, consistency, or adhesive qualities; waxen; pliable; impressionable. **2** Having the dull whitish or yellowish color of wax; pale; pallid; bloodless. **3** Made of or abounding in wax; rubbed with wax; waxed. **4** *Pathol.* Characterized by the formation of an insoluble, waxlike protein in certain organs of the body, as the kidney; amyloid. —**wax′i·ness** *n.*

way (wā) *n.* **1** Direction; turn; route; line of motion or progress: Which *way* is the city? **2** A path, course, or track leading from one place to another or along which one goes; a road, street, highway, lane, path, or the like. **3** Space or room to advance or work: Make *way* for the king. **4** Length of space passed over; hence, distance in general: a little *way* off: often, popularly or dialectally, *ways.* **5** Passage from one place to another; hence, onward movement; headway; progress. **6** A customary or habitual manner or style; a manner peculiar to an individual, class, or people: the British *way* of doing things. **7** A chosen line or plan of action; a procedure; method: In what *way* will you accomplish this? **8** A point of relation; particular: He erred in two *ways.* **9** A course of life or experience: the *way* of sin. **10** *Colloq.* Vocation; line of business; profession. **11** *Colloq.* State of health: to be in a bad *way.* **12** A course wished for or resolved upon; something which one resolves to do: Have it your *way.* **13** The range of one's notice or observation: An accident threw it in his *way.* **14** *Naut.* **a** The movement of a vessel through the water; forward motion; headway. **b** *pl.* A tilted framework of timbers upon which a ship slides when launched. **15** The direction of the weave in textile goods. **16** *Law* A right of way. **17** *Mech.* A longitudinal guide for material being worked upon, or for a moving table bearing the work. **18** *Colloq.* Neighborhood, or route taken to go home: He lives out of my *way.* —**by the way** In passing; incidentally. —**by way of** **1** With the object or purpose of; to serve as: *by way of* introduction. **2** Through; via: We went home *by way of* Main Street. —**out of the way** **1** Removed, as an obstruction; unable to hinder or impede. **2** Out of the proper course; hence, remarkable; unusual; also, improper; wrong: Has he done anything *out of the way*? **3** Out of place; lost; mislaid; remote. —**under way** In motion; well along; making progress. —*adv. Colloq.* Away; very far; all the great distance: He went *way* to Denver. ◆ Homophone: *weigh.* [OE *weg*]

Synonyms (noun): alley, avenue, bridlepath, channel, course, driveway, highroad, highway, lane, pass, passage, path, pathway, road, route, street, thoroughfare, track. Wherever there is room for an object to proceed, there is a *way.* A *road* (originally a ride*way*) is a prepared way for traveling with horses or vehicles, a *way* suitable to be traversed only by foot–passengers or by animals being called a *path, bridlepath,* or *track*; as, The *roads* in that country are mere *bridlepaths.* A *road* may be private: a *highway* or *highroad* is public, *highway* being a specific name for a *road* legally set apart for the use of the public forever; a *highway* may be over water as well as over land. A *route* is a line of travel, and may be over many *roads*. A *street* is in some center of habitation, as a city, town, or village; when it passes between rows of dwellings, the country *road* becomes the village *street*. An *avenue* is a long, broad, and imposing or principal *street*. *Track* is a word of wide signification; we speak of a goat-*track* on a mountainside, a railroad *track*, a racetrack, the *track* of a comet; on a traveled *road* the line worn by regular passing of hoofs and wheels is called the *track*. A *passage* is between any two objects or lines of enclosure, a *pass* commonly between mountains. A *driveway* is within enclosed grounds, as of a private residence. A *channel* is a *waterway*. A *thoroughfare* is a *way* through. See AIR¹, DIRECTION, ROAD.

way back *Colloq.* Long ago. [Short for AWAY BACK]

way·bill (wā′bil′) *n.* A list describing or identifying goods or naming passengers carried by a common carrier, as a railroad, train, steamer, or other public vehicle.

way·far·er (wā′fâr′ər) *n.* One who journeys along a way on foot.

way·far·ing (wā′fâr′ing) *adj. & n.* Journeying; being on the road.

way freight Freight taken on or put off at way stations; also, a freight train stopping at way stations and handling such goods: distinguished from *through freight*.

way-go·ing (wā′gō′ing) *adj.* Pertaining to one's going away; going away; departing.

way-going crop *Law* A crop sown by a tenant during his term, but ripening after its expiration. [Short for *away-going crop*]

Way·land (wā′lənd) In Teutonic and English mythology, an invisible blacksmith with magical powers: in German folklore spelled *Wieland*. Also **Wayland (the) Smith**.

way·lay (wā′lā′) *v.t.* **·laid**, **·lay·ing** 1 To lie in ambush for and attack, as in order to rob. 2 To accost on the way. [< WAY + LAY¹, on analogy with MHG *wegelagen* < *wegelage* an ambush] — **way′lay′er** *n.*

Wayne (wān), **Anthony**, 1745–96, American Revolutionary general: called "Mad Anthony" Wayne.

way passenger A passenger getting on or off a public conveyance, as a train, steamship, bus, etc., at a way station; a local passenger.

-ways *suffix·of adverbs* In a (specified) manner, direction, or position: *noways, sideways*: often equivalent to -WISE. Also **-way**. [< WAY + -s³]

ways and means Means or methods of accomplishing an end or defraying expenses; specifically, in legislation, methods of raising funds for the use of the government.

way·side (wā′sīd′) *adj.* Pertaining to the side of a road; growing or being near the wayside. — *n.* The side or edge of the road or highway.

way station Any station between principal stations, especially on a railroad; a local station.

way train A train stopping at way stations; a local train.

way·ward (wā′wərd) *adj.* 1 Wandering away; wilful; froward. 2 Without definite way or course; unsteady; vacillating; capricious. 3 Unexpected or unwished for: a *wayward* fortune. See synonyms under PERVERSE. [ME *weiward*, short for *aweiward* < *awei* away + -WARD] — **way′ward·ly** *adv.* — **way′ward·ness** *n.*

way·worn (wā′wôrn′, -wōrn′) *adj.* Fatigued by travel.

Wa·zir·i·stan (wä-zir′i-stän′) A tribal region in NW central Pakistan on the Afghanistan border; 5,214 square miles.

we (wē) *pron.* 1 The persons speaking or writing as they denote themselves, or a single person writing or speaking when referring to himself and one or more others: the nominative case. 2 A single person denoting himself, as a sovereign, editor, writer, or speaker, when wishing to give his words an impersonal character. [OE]

weak (wēk) *adj.* 1 Lacking in physical strength; wanting in energy, activity, or vigor; feeble; debilitated. 2 Insufficiently resisting stress; incapable of supporting weight: a *weak* link or bridge. 3 Lacking in strength of will or stability of character; yielding easily to temptation; pliable. 4 Ineffectual, as from deficient supply: *weak* artillery support. 5 Lacking in power or sonorousness: a *weak* voice. 6 Lacking a specified component or components in the usual or proper amount; of less than customary strength or potency: *weak* tea, a *weak* tincture. 7 Lacking the power or ability to perform properly its function: a *weak* heart. 8 Lacking in mental or moral strength; liable to err or fail through feebleness of conception or vacillation of judgment. 9 Showing or resulting from poor judgment or a want of discretion or firmness: a *weak* plan; unable to persuade or convince: a *weak* argument. 10 Lacking in influence or authority: a *weak* state. 11 Deficient in strength, durability, skill, experience, or the like. 12 *Gram.* In Germanic languages: **a** Of verbs, forming the past tense and past participle by the addition of a dental suffix to the present stem; as, English *ask, asked; sight, sighted*; German *leben, lebte, gelebt*. Some weak verbs in English show vowel change in the stem (as in *leave, left*), but in such cases the change is due to factors other than ablaut. Also called *regular*. **b** Of nouns and adjectives (in German and Old English), inflected in the less full manner originally restricted to stems ending in *-n*. Weak nouns and adjectives in Old English characteristically terminate in *-a* in the masculine singular (*nama* name) and *-e* in the feminine singular (*tunge* tongue). In German, a descriptive adjective appears in the weak form when preceded by a limiting word, such as the definite article, having strong inflection (*der gute Mann*). Compare STRONG (def. 28). 13 *Phonet.* Unstressed; unaccented, as a syllable or sound. 14 *Phot.* Thin; wanting in contrast: a *weak* negative. 15 In prosody, indicating a verse ending in which the accent falls on a word or syllable otherwise without stress. 16 Declining in price; without an active market: The wheat market is *weak*. 17 Wanting in impressiveness or interest: a *weak* play or book. See synonyms under FAINT, FRAGILE, PUSILLANIMOUS, SICKLY. ◆ Homophone: *week*. [< ON *veikr*. Akin to OE *wac*.] — **weak′ly** *adv.* — **weak′ness** *n.*

Weak may appear as a combining form in hyphemes or as the first element in two-word phrases; as in:

weak-backed	weak-nerved
weak-bodied	weak point
weak-built	weak side
weak-eyed	weak-sighted
weak-growing	weak-spirited
weak-handed	weak-stemmed
weak-headed	weak-throated
weak-headedness	weak-toned
weak-hearted	weak-voiced
weak-limbed	weak-walled
weak-looking	weak-willed
weak-made	weak-winged
weak-mindedness	weak-witted

weak·en (wē′kən) *v.t. & v.i.* To make or become weak or weaker. — **weak′en·er** *n.*

Synonyms: debilitate, depress, enervate, enfeeble, impair, invalidate, lower, paralyze, reduce, relax, sap, undermine, unnerve. See IMPAIR.

weak·fish (wēk′fish′) *n. pl.* **·fish** or **·fish·es** Any of various American marine food fishes (genus *Cynoscion*), especially the common variety (*C. regalis*), frequenting coastal waters of the eastern United States.

weak-kneed (wēk′nēd′) *adj.* Weak in the knees; hence, without resolution, strong purpose, or energy; spineless.

weak·ling (wēk′ling) *n.* A feeble person or animal. — *adj.* Having no natural strength or vigor.

weak·ly (wēk′lē) *adj.* **·li·er**, **·li·est** Sickly; feeble; weak. See synonyms under SICKLY.

weak-mind·ed (wēk′mīn′did) *adj.* 1 Indecisive; unable to say no. 2 Feeble-minded.

weak·ness (wēk′nis) *n.* 1 The state, condition, or quality of being weak. 2 A characteristic indicating feebleness. 3 A slight failing; a fault.

weak sister *Colloq.* 1 The weakling in any group; specifically, one who cannot be depended on to stand firm against opposition. 2 Any ineffectual person.

weak spot 1 Any spot having less strength than the contiguous area, as in a fabric, fence, etc. 2 The most vulnerable part of an argument, proposition, etc. 3 The weakest or least dependable person on a team, in a group, etc.

weal¹ (wēl) *n.* 1 A sound or healthy state, either of persons or things; prosperity; welfare. 2 *Obs.* The body politic, state, or nation: now only in the phrase, *public weal*. 3 *Obs.* Wealth; worldly store. [OE *wela*. Akin to WELL.]

weal² (wēl) *n.* A discolored ridge or stripe on the skin, as from the blow of a whip; a wheal. [Var. of WALE¹; infl. in form by obs. *wheal* a pustule]

weald (wēld) *n.* An exposed forest area; waste woodland; also, an open region; down. [OE, a forest]

Weald (wēld), **The** A district of SE England between the North and South Downs in Kent, Surrey and Sussex counties; formerly forested, now primarily agricultural.

wealth (welth) *n.* 1 A large aggregate of real and personal property; an abundance of those material or worldly things that men desire to possess; riches; also, the state of being rich. 2 *Econ.* All material objects which have economic utility; also, in the private sense, all property possessing a monetary value. 3 Great abundance of anything: a *wealth* of learning. 4 *Obs.* Weal; well-being. — **personal wealth** Those faculties, energies, and habits which contribute to personal industrial efficiency. [ME *welthe* < *wele* weal, on analogy with *health*]

Synonyms: abundance, affluence, comfort, competence, competency, fortune, funds, goods, independence, lucre, mammon, money, opulence, pelf, plenty, possession, produce, property, riches, substance, treasure. See PROPERTY. *Antonyms:* see synonyms for POVERTY.

wealth·y (wel′thē) *adj.* **wealth·i·er**, **wealth·i·est** 1 Possessing wealth; affluent. 2 More than sufficient; abounding. — **wealth′i·ly** *adv.* — **wealth′i·ness** *n.*

wean¹ (wēn) *v.t.* 1 To transfer (the young of any animal) from dependence on its mother's milk to another form of nourishment. 2 To estrange from former habits or associations; alienate the affections of: usually with *from*. [OE *wenian* accustom]

wean² (wēn) *n. Scot.* A baby; infant.

wean·er (wē′nər) *n.* 1 One who weans. 2 A muzzle used in weaning a calf.

wean·ling (wēn′ling) *adj.* Freshly weaned. — *n.* A child or animal newly weaned.

weap·on (wep′ən) *n.* 1 Any implement of war or combat, as a sword, gun, etc. 2 Figuratively, any means that may be used against an adversary. 3 *pl.* The thorns or prickles of plants, or the stings, claws, etc., of animals. See synonyms under ARMS. [OE *wæpen*] — **weap′on·less** *adj.*

weap·oned (wep′ənd) *adj.* Furnished with weapons; bearing arms.

weap·on·eer (wep′ən·ir′) *n.* A person concerned with the design, improvement, production, and use of weapons, especially of the atomic and thermonuclear type.

wear¹ (wâr) *v.* **wore**, **worn**, **wear·ing** *v.t.* 1 To carry or have on the person as a garment, ornament, etc. 2 To have or bear on the person habitually or as a practice: He *wears* a derby. 3 To have in one's appearance or aspect; exhibit: He *wears* a scowl. 4 To bear habitually in a specified manner; carry: He *wears* his age well; She *wears* her hair in a chignon. 5 To display or fly: A ship *wears* its colors. 6 To impair, waste, or consume by use or constant action. 7 To cause or produce by scraping, rubbing, etc.: to *wear* a hole in a coat. 8 To bring to a specified condition by wear: to *wear* a sleeve to tatters. 9 To exhaust the strength or patience of; weary. — *v.i.* 10 To be impaired or diminished gradually by use, rubbing, etc. 11 To withstand the effects of use, wear, etc., as specified: The vest *wears* well. 12 To become as specified from use or attrition: His patience is *wearing* thin. 13 To pass gradually or

tediously: with *on* or *away*: The day *wears on.* — **to wear out 1** To make or become worthless by use: The cloak is *worn out.* **2** To waste gradually; use up: He *wears out* patience. **3** To tire or exhaust. — *n.* **1** The act of wearing, or the state of being worn: the worse for *wear.* **2** The material or articles of dress worn or made to be worn; a fashion: silk for summer *wear*; also in compounds: *footwear, underwear.* **3** The destructive effect of use or work; impairment from use or time. **4** Capacity for resistance to use or impairment; endurance; lasting quality; durability. ◆ Homophone: *ware.* [OE *werian*] — **wear′a·ble** *adj.* — **wear′er** *n.*
— *Synonyms* (verb): abrade, chafe, consume, deteriorate, diminish, fret, fritter, impair, rub, tire, waste.
wear² (wâr) *v.* **wore, worn, wear·ing** *Naut. v.t.* To change the course of (a vessel), so as to bring the wind to the other side, by turning it through an arc in which its head points momentarily directly to leeward. — *v.i.* To go about with the wind astern. Compare TACK¹. ◆ Homophone: *ware.* [Prob. alter. of VEER¹; infl. in form by *wear*¹]
wear·a·ble (wâr′ə·bəl) *adj.* That can be worn. — *n. pl.* Garments.
wear and tear Loss by the service, exposure, decay, or injury incident to ordinary use.
wea·ri·ful (wir′i·fəl) *adj.* Tiresome; wearisome. — **wea′ri·ful·ly** *adv.* — **wea′ri·ful·ness** *n.*
wea·ri·less (wir′i·lis) *adj.* Unwearying; untiring.
wear·ing (wâr′ing) *adj.* **1** Fatiguing; exhausting; wasting: *wearing* trials. **2** Capable of being, or designed to be, worn. — **wear′ing·ly** *adv.*
wearing apparel Clothing; garments.
wear·ish (wâr′ish) *adj.* *Obs.* or *Dial.* **1** Insipid; watery. **2** Wizened; shrunk; withered. [ME *werische*; origin uncertain] — **wear′ish·ly** *adv.* — **wear′ish·ness** *n.*
wea·ri·some (wir′i·səm) *adj.* Causing fatigue; tiresome. — **wea′ri·some·ly** *adv.* — **wea′ri·some·ness** *n.*
— *Synonyms*: annoying, fatiguing, irksome, laborious, tedious, tiresome, vexatious, wearing, weary. See TEDIOUS, TROUBLESOME. *Antonyms*: cheering, enlivening, inspiring, inspiriting, restful, reviving, rousing, soothing, stirring, thrilling.
wea·ry (wir′ē) *adj.* **·ri·er, ·ri·est 1** Worn with exertion, vexation, or suffering; tired; fatigued. **2** Discontented or vexed by continued endurance, or by something disagreeable: usually with *of*: *weary* of life. **3** Indicating or characteristic of fatigue: a *weary* sigh. **4** Causing weariness; wearisome. — *v.t.* & *v.i.* **·ried, ·ry·ing** To make or become weary; tire. See synonyms under TIRE¹. [OE *wērig*] — **wea′ri·ly** *adv.* — **wea′ri·ness** *n.*
wea·sand (wē′zənd) *n. Archaic* The windpipe; in general, the throat: often spelled *wizen.* Also *Scot.* **wea′son.** [OE *wǣsend*]
wea·sel (wē′zəl) *n.* Any of certain small, slender, reddish-brown, carnivorous mammals (genus *Mustela*) that prey on smaller mammals and birds. In northern regions their fur turns white in winter. [OE *wesle*]
weasel word A word that weakens a statement by rendering it ambiguous or equivocal: term popularized by Theodore Roosevelt.

WEASEL
(Head and body from 6 to 7 inches)

weath·er (weth′ər) *n.* **1** The general atmospheric condition, as regards temperature, moisture, winds, or other meteorological phenomena. **2** The common phenomena of wind, rain, cold, heat, cloudiness, or storm. **3** Bad weather; storm. — **to keep one's weather eye open** *Colloq.* To be alert. — **under the weather** *Colloq.* **1** Ailing; ill. **2** Somewhat intoxicated. — *v.t.* **1** To expose to the action of the weather. **2** To discolor, crumble, or otherwise affect by action of the weather. **3** To pass through and survive, as a crisis. **4** To slope, as a roof, so as to shed water. **5** *Naut.* To pass to windward of: to *weather* Cape Fear. — *v.i.* **6** To undergo changes resulting from exposure to the weather. **7** To resist the action of the weather. — *adj.* Facing the wind; windward: opposed to *lee.* ◆ Homophone: *wether.* [OE *weder*]
Weather may appear as a combining form in hyphemes or solidemes, or as the first element in two-word phrases:

weather-bitten	weather report
weather-bleached	weather-reporter
weather-blown	weather-reporting
weather-burnt	weather-rotted
weather-driven	weather-scarred
weather-eaten	weathersick
weather forecast	weather-tanned
weather-hardened	weathertight
weather-observer	weather-tough
weather-observing	weather-withstanding

weath·er·beat·en (weth′ər·bēt′n) *adj.* Bearing the effects of exposure to weather.
weath·er·board (weth′ər·bôrd′, -bōrd′) *n.* **1** A board for the outside covering of wooden buildings, usually feather-edged and nailed so as to form lap joints with the boards above and below and thus shed rain; a clapboard. **2** *Naut.* The windward side of a vessel. — *v.t.* To fasten weatherboards on.
weath·er·board·ing (weth′ər·bôr′ding, -bōr′-) *n.* **1** Weatherboards collectively, or material for making them. **2** The outer wooden covering of the walls and roof of a building.
weath·er·bound (weth′ər·bound′) *adj.* Detained by unfavorable weather, as a vessel in port.
Weather Bureau A bureau of the Department of Commerce in Washington, D.C., for meteorological observation, the diffusion of information concerning the weather, etc.
weath·er·cast (weth′ər·kast′, -käst′) *n.* A radio or television broadcast reporting on weather conditions. [<WEATHER + (BROAD)CAST] — **weath′er·cast′er** *n.*
weath·er·cock (weth′ər·kok′) *n.* **1** A vane, properly one in the semblance of a cock, which turns to indicate the direction of the wind; a weathervane. **2** A fickle person or variable thing.
weath·er·drome (weth′ər·drōm′) *n. Meteorol.* A large, floating structure resembling an airdrome, permanently anchored in offshore waters to serve as a weather station.

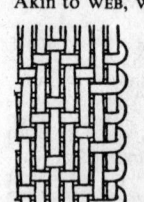
WEATHERCOCK

weath·ered (weth′ərd) *adj.* **1** Affected by exposure to the atmosphere; seasoned. **2** *Archit.* Sloped to prevent water lodging on the surface, as woodwork or stonework. **3** Worn, shaped, or stained by exposure in the atmosphere: said of rocks. **4** Denoting wood that has been artificially colored to represent weathering.
weather gage 1 *Naut.* The advantage to a ship or yacht of receiving the wind first; a position to the windward. **2** Any advantage gained.
weath·er·glass (weth′ər·glas′, -gläs′) *n.* A meteorological instrument for indicating the state of the weather; especially, a common barometer.
weath·er·ize (weth′ə·rīz) *v.t.* **·ized, ·iz·ing** To process (fabrics, leather, etc.) chemically or otherwise, so as to make impervious or highly resistant to moisture or other effects of severe weather.
weath·er·ly (weth′ər·lē) *adj. Naut.* Capable of keeping close into the wind without drifting to leeward.
weath·er·man (weth′ər·man′) *n. pl.* **·men** (-men′) *n. Colloq.* A meteorologist, especially one concerned with daily weather conditions and reports.
weather map *Meteorol.* A map or chart compiled periodically from official sources and indicating, for a given region and specified time, various components of the weather, as temperature, atmospheric pressure, wind velocity, rain, snow, cloud formations, etc.
weath·er·proof (weth′ər·prōōf′) *adj.* Capable of withstanding rough weather without appreciable deterioration. — *v.t.* To make weatherproof.

weather station A station or office where meteorological observations are taken and recorded.
weath·er·strip (weth′ər·strip′) *n.* A narrow strip of material to be placed over or in crevices, as at doors and windows, to exclude drafts, rain, etc.: also **weath′er·strip′ping.** — *v.t.* **-stripped, -strip·ping** To equip or fit with weather-strips.
weath·er·vane (weth′ər·vān′) *n.* A vane; weathercock.
weath·er·vi·sion (weth′ər·vizh′ən) *n.* A system which combines television and radar in the rapid dissemination of weather data to aircraft. [<WEATHER + (TELE)VISION]
weath·er·wise (weth′ər·wīz′) *adj.* Experienced in observing or predicting the weather.
weath·er·worn (weth′ər·wôrn′, -wōrn′) *adj.* Worn by exposure to the weather.
weave (wēv) *v.* **wove** or (*esp.* for defs. 7 and 10) **weaved, wo·ven** or *less frequently* **wove, weav·ing** *v.t.* **1** To form, produce, or manufacture as a textile, by interlacing threads or yarns; especially, to make by interlacing woof threads among warp threads in a loom. **2** To form by interlacing strands, strips, twigs, etc.: to *weave* a basket. **3** To produce by combining details or elements: to *weave* a story. **4** To bring together so as to form a whole: to *weave* fancies into theories. **5** To twist or introduce into, about, or through something else: to *weave* ribbons through one's hair. **6** To spin (a web). **7** To make or effect by side-to-side movements: to *weave* one's way. — *v.i.* **8** To make cloth, etc., by weaving. **9** To become woven or interlaced. **10** To move with a side-to-side motion. — *n.* A particular method or style of weaving. — **plain** or **taffeta weave** A weave in which each filling yarn passes successively over and under each warp yarn, forming an even surface. — **satin weave** An irregular weave in which the warp or filling yarns pass over a number of yarns of the other set before interweaving, thus forming a smooth, unbroken, lustrous surface. — **twill weave** A strong weave having a distinct diagonal line or rib, caused by the passage of the filling yarns over one warp yarn and under two or more. [OE *wefan.* Akin to WEB, WEFT.]

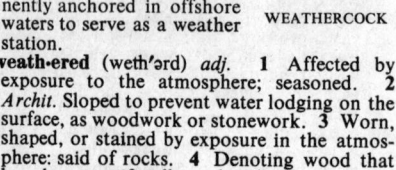
WEAVES
Simple figured.   Leno.   Five-shaft satin.

weav·er (wē′vər) *n.* **1** One who weaves. **2** A weaverbird.
weav·er·bird (wē′vər·bûrd′) *n.* Any of various finchlike birds (family *Ploceidae*) of the warmer parts of Asia, Africa, and Australia, that construct intricately woven nests.
weaver's bottom *Pathol.* An inflamed condition of the tissue over the ischium, or seat bone, arising from long sitting.
weaver's hitch A sheet bend. Also WEAVER'S KNOT.
web (web) *n.* **1** Textile fabric, especially as in the piece or as being woven in a loom. **2** A long sheet or roll of material formed like a web of cloth; especially, a roll of paper as it comes from the mill. **3** The network of delicate threads spun by a spider to entrap its prey; a cobweb. **4** Any complex network: a *web* of highways; anything artfully contrived or elaborated into a trap or snare: a *web* of espionage. **5** *Zool.* A membrane or fold of skin connecting the digits of an animal, as in aquatic birds, otters, bats, frogs, etc. **6** *Ornithol.* The series of barbs on either side of the shaft of a feather; the vane. **7** *Mech.* A plate or sheet, as of metal, connecting the heavier sections, ribs, frames, etc., of any tool or mechanical element. **8** The plate between the flange and head of a railroad rail. **9** *Archit.* The part of a ribbed vault between the ribs. **10** *Anat.* A membrane; tissue; tela. **11** A thin metal plate, as the blade of

a saw or sword, or the bit of a key. — **pin and web** A darkening speck on the cornea, with a film spreading fanwise from the cornea. — *v.t.* **webbed, web·bing** 1 To provide with a web. 2 To cover or surround with a web; entangle. [OE. Akin to WEAVE, WEFT.]
Webb (web), **Beatrice,** 1858-1943, *née* Potter, English economist and sociologist; wife of Sidney James Webb. — **Mary,** 1881-1927, *née* Meredith, English novelist. — **Sidney James,** 1859-1947, Baron Passfield, English economist and sociologist.
webbed (webd) *adj.* 1 Having a web. 2 Having the digits united by a membrane.
web·bing (web'ing) *n.* 1 A woven strip of strong fiber, as for girths, seat bottoms, etc. 2 Any woven texture; the structure of a web.
web·by (web'ē) *adj.* **·bi·er, ·bi·est** 1 Relating to or consisting of a web or membrane. 2 Palmate.
we·ber (vā'bər, wē'bər) *n. Physics* 1 The mks unit of magnetic flux, equal to 100,000,000 maxwells. 2 *Obs.* Coulomb. 3 *Obs.* Ampere. [after Wilhelm Eduard *Weber*]
We·ber (vā'bər), **Ernst Heinrich,** 1795-1878, German physiologist. — **Baron Karl Friedrich Ernst von,** 1786-1826, German composer. — **Wilhelm Eduard,** 1804-91, German physicist; brother of Ernst Heinrich Weber.
web·foot (web'foot') *n.* 1 A foot with webbed toes. 2 A web-footed bird or animal. 3 The condition of being web-footed.
web-foot·ed (web'foot'id) *adj.* Having the toes connected by a membrane, as many aquatic animals and birds.
web press A printing press which is fed from a continuous roll of paper instead of sheets.
web·ster (web'stər) *n. Obs.* A weaver. [OE *webbestre*, fem. of *webba* a weaver]
Web·ster (web'stər), **Daniel,** 1782-1852, U.S. statesman. — **John,** 1580?-1625, English dramatist. — **Noah,** 1758-1843, U.S. lexicographer.
Web·ste·ri·an (web·stir'ē·ən) *adj.* Of or pertaining to Daniel or Noah Webster.
web·worm (web'wûrm') *n.* Any of various caterpillars, usually gregarious and very destructive of foliage, which build large silken webs or tents for shelter; especially, the **common garden webworm** (*Loxostege similalis*).
wecht (wekht) *n. Scot.* Weight.
wed (wed) *v.* **wed·ded, wed·ded** or **wed, wed·ding** *v.t.* 1 To take as one's husband or wife; marry. 2 To unite or give in matrimony; join in wedlock. 3 To attach as if in marriage; join securely: chiefly in the past participle, with *to*: *wedded* to his job. — *v.i.* 4 To take a husband or wife; marry. [OE *weddian* pledge]
we'd (wēd) 1 We had. 2 We would.
Wed·dell Sea (wed'l) An embayment of the South Atlantic in Antarctica, SE of the Palmer Peninsula and South America.
wed·ding (wed'ing) *n.* 1 The ceremony of a marriage with the attendant nuptial festivities; also, the ceremony alone; originally, a betrothal. 2 The anniversary or celebration of a marriage. Such weddings are named from the character of the presents regarded as appropriate: golden *wedding* (50th); for list, see under ANNIVERSARY. See synonyms under MARRIAGE. [OE *weddung* < *weddian* pledge]
wedding cake A very rich fruit or pound cake served at a wedding reception, and also often divided among absent friends.
We·de·kind (vā'də·kint), **(Benjamin) Frank·(lin),** 1864-1918, German poet and playwright.
wedge (wej) *n.* 1 One of the so-called mechanical powers, consisting of a double inclined plane; specifically, a V-shaped piece of metal, wood, etc., used for splitting substances, raising weights, and the like. 2 Anything in the form of a wedge, as a piece of pie; specifically, a formation, as of soldiers or football players, arranged like a wedge. 3 A right triangular prism, having one very acute angle. 4 Any one of the triangular characters used in cuneiform writing. 5 *Meteorol.* **a** A wedge-shaped area of high barometric pressure as shown on a weather map. **b** An air mass advancing in the form of a wedge. 6 Any action or procedure which facilitates a change in policy, entrance, intrusion, etc.: also **entering wedge.** — *v.* **wedged, wedg·ing** *v.t.* 1 To force apart or split with or as with a wedge; rend; rive. 2 To compress or fix in place with a wedge. 3 To crowd or squeeze (something). — *v.i.* 4 To force oneself or itself in like a wedge. [OE *wecg*]
wedg·ie (wej'ē) *n. Colloq.* A kind of shoe worn by women, having a wedge-shaped piece making a solid sole, flat on the ground from heel to toe.
Wedg·wood (wej'wŏod') *n.* Any of various fine, hard earthenwares invented by Josiah Wedgwood, characterized by an unglazed, tinted clay background bearing small, finely detailed, classical figures in cameo relief applied in white paste. It is typically tinted either light or dark blue, but **Wedgwood bamboo ware** (yellow), **Wedgwood basalt** (black), and **Wedgwood queen's** (cream-colored) are also famous. Also **Wedgwood ware.**
Wedg·wood (wej'wŏod'), **Josiah,** 1730-95, English potter; inventor of the ware bearing his name. — **Josiah Clement,** 1872-1943, first Baron Wedgwood of Barlaston, English naval architect and statesman; great-great-grandson of the preceding: called the "Father of the Labour Party."
Wedgwood blue Either of two shades of blue, a light grayish blue and a dark reddish blue: the typical blues of the Wedgwood wares.
wedg·y (wej'ē) *adj.* Having the form or uses of a wedge; cuneal.
wed·lock (wed'lok) *n.* The ceremony of marriage, or the state of being married; matrimony. See synonyms under MARRIAGE. [OE *wedlāc* < *wed* a pledge + *-lāc*, suffix of nouns of action]
Wednes·day (wenz'dē, -dā) *n.* The fourth day of the week. [OE *Wōdnes dæg* day of Woden, trans. of LL *Mercurii dies* day of Mercury]
wee (wē) *adj.* **we·er, we·est** Very small; tiny. — *n. Scot.* A short time or space; a bit: bide a *wee*. [ME *wei*, OE *wēg, wēge* a quantity]
weed[1] (wēd) *n.* 1 Any unsightly or troublesome plant that grows in abundance; especially, any coarse, herbaceous plant growing to injurious excess on cultivated or fallow ground where it is not wanted, as dock, ragweed, etc. 2 *Colloq.* Tobacco: usually with *the*; also, a cigarette or cigar. 3 Any worthless animal or thing; specifically, a horse that is unfit for racing or breeding. 4 The stem and leaves of any useful plant as distinguished from its flower and fruit: The plant runs to *weed*. 5 *Obs.* Thick, luxuriant growth, as of underbrush or shrubs. — *v.t.* 1 To pull up and remove weeds from: to *weed* a garden. 2 To remove (a weed): often with *out*. 3 To remove (anything regarded as harmful or undesirable): with *out*. 4 To rid of anything harmful or undesirable. — *v.i.* 5 To remove weeds, etc. [OE *wēod*] — **weed'less** *adj.*
weed[2] (wēd) *n.* 1 A token of mourning, as a band of crape, worn as part of the dress: He wore a *weed* on his hat; especially in the plural, a widow's mourning garb. 2 *Obs.* Any article of clothing. [OE *wǣd, wǣde* a garment]
Weed (wēd), **Thurlow,** 1797-1882, U.S. journalist and politician.
weed·er (wē'dər) *n.* 1 One who weeds. 2 An implement for removing weeds.
weeding hoe A narrow-bladed hoe for weeding: usually with prongs on the end opposite the blade. See illustration under HOE.
weed·y (wē'dē) *adj.* **weed·i·er, weed·i·est** 1 Having or containing a growth of weeds; abounding in weeds. 2 Of or pertaining to a weed or weeds. 3 Resembling a weed; weedlike, as in rapid, ready growth. 4 *Colloq.* Gawky; awkward; ungainly: *weedy* youths. — **weed'i·ly** *adv.* — **weed'i·ness** *n.*
wee folk The fairies, elves, etc.
Wee·haw·ken (wē'hô·kən) A township in NE New Jersey opposite the Hudson River from New York City; scene of the fatal wounding of Alexander Hamilton in a duel with Aaron Burr, 1804.
week (wēk) *n.* 1 A period of seven successive days; especially, such a period beginning with Sunday. 2 The period of time within a week devoted to work: The office has a 35-hour *week*. 3 A period of seven days preceding or following any given day or date: a *week* from Tuesday. ◆ Homophone: *weak*. [OE *wicu, wice*]
week·day' (wēk'dā') *n.* Any day of the week except Sunday.
week-end (wēk'end') *n.* The end of the week; specifically, the time from Friday evening or Saturday noon to the following Monday morning. — *v.i.* To pass the week-end: We *week-ended* in the country.
week-end·er (wēk'en'dər) *n.* One who goes on week-end vacation trips.
Week·ley (wēk'lē), **Ernest,** 1865-1954, English lexicographer and etymologist.
week·ly (wēk'lē) *adv.* Once a week; especially, at regular seven-day intervals. — *adj.* 1 Of or pertaining to a week or to weekdays. 2 Done or occurring once a week; also, reckoned by the week; hebdomadal. — *n. pl.* **·lies** A publication issued once a week.
weel (wēl) *adj., adv., & interj. Scot.* Well.
Weems (wēmz), **Mason Locke,** 1759-1825, American clergyman and biographer: known as *Parson Weems.*
ween (wēn) *v.t.* & *v.i. Archaic* To suppose; guess; fancy. [OE *wēnan* think]
ween·di·jo (wēn'də·jō) See WINDIGO.
ween·ie (wē'nē) *n. Colloq.* A wiener.
weep[1] (wēp) *v.* **wept, weep·ing** *v.i.* 1 To manifest grief or other strong emotion by shedding tears: to *weep* for joy. 2 To mourn; lament: with *for*. 3 To give out or shed water or other liquid in drops, as the stems of some plants under pressure; bleed. — *v.t.* 4 To weep for; mourn or bewail. 5 To shed (tears, or drops of other liquid). 6 To bring to a specified condition by weeping: to *weep* oneself to sleep. — *n.* The act of weeping, or a fit of tears. [OE *wēpan*]
weep[2] (wēp) *n.* A lapwing; pewit. [Imit.]
weep·er (wē'pər) *n.* 1 One who weeps, as a hired mourner. 2 A long piece of black crape worn as a sign of mourning, customarily hanging down from the hat. 3 A pendant of moss, as from a branch. 4 A hole through which water may drip.
weep·ing (wē'ping) *adj.* 1 That weeps; crying; tearful. 2 Having slim, pendulous branches: the *weeping* ash.
weeping ash A variety of the common European ash (*Fraxinus excelsior pendula*) with drooping branches.

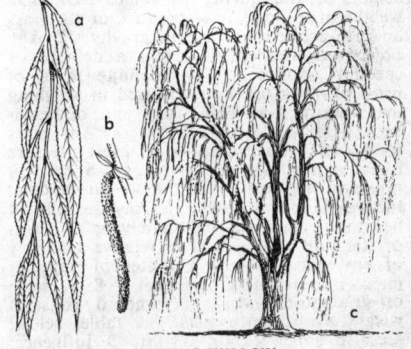

WEEPING WILLOW
a. Leaves. *b.* Catkin. *c.* Tree.

weeping willow An Old World willow (*Salix babylonica*), remarkable for its long, slender, pendulous branches.
weet (wēt) *n.* 1 The imitation of the call of various birds. 2 The peetweet, or common European sandpiper. [Imit.]
wee·ver (wē'vər) *n.* Any of various edible marine fishes (genus *Trachinus*), having upward-looking eyes and sharp dorsal and opercular spines, with which they can inflict serious wounds. [<AF *wivre*, OF *guivre*, orig. a serpent, dragon <L *vipera* a viper]
wee·vil (wē'vəl) *n.* 1 Any of numerous small beetles (family *Curculionidae*) with elongated snoutlike heads which bear the mouth parts at the end and the antennae along the sides: also called *curculio*. Weevil larvae feed on plants and plant products, especially flowers, fruits, and trees; many are serious pests. 2 Any insect injurious to stored grain. [OE *wifel* a beetle] — **wee'vil·y, wee'vil·ly** *adj.*
weft (weft) *n.* 1 The cross-threads in a web

of cloth; woof. 2 A woven fabric; web. [OE. Akin to WEAVE, WEB.]

Wehr·macht (vâr′mäkht) *n. German* The armed forces, collectively, of Germany: literally, defense force.

Wei (wā) A river in NW central China, flowing 540 miles east from SE Kansu province to the Yellow River at its bend on the Shansi border.

Weich·sel (vīkh′səl) The German name for the Vistula.

wei·ge·la (wī·gē′lə, -jē′-, wī′jə·lə) *n.* Any of a large genus *(Weigela)* of deciduous Asian shrubs of the honeysuckle family; especially, *W. florida,* cultivated extensively in the United States for its profusion of dark rose-purple flowers. [<NL, after Dr. C. E. *Weigel,* 1748–1831, German physician]

weigh[1] (wā) *v.t.* **1** To determine the weight of, as by measuring on a scale or balance. **2** To balance or hold in the hand so as to estimate weight or heaviness. **3** To measure (a quantity or quantities of something) according to weight: with *out.* **4** To consider carefully; estimate the worth or advantages of: to *weigh* a proposal. **5** To press or force down by weight or heaviness; burden or oppress: with *down.* **6** To raise or hoist: now only in the phrase *to weigh anchor.* **7** *Obs.* To think well of; esteem; regard. — *v.i.* **8** To have weight; be heavy to a specified degree: She *weighs* ninety pounds. **9** To have influence or importance: The girl's testimony *weighed* heavily with the jury. **10** To be burdensome or oppressive: with *on* or *upon*: What *weighs* on your mind? **11** *Naut.* **a** To raise anchor. **b** To begin to sail. See synonyms under CONSIDER, DELIBERATE, EXAMINE. — **to weigh in** Of a prize fighter or other contestant, to be weighed before a contest. — **to weigh one's words** To consider one's words carefully before speaking them. ◆ Homophone: *way.* [OE *wegan* weigh, carry, lift] — **weigh′er** *n.*

weigh[2] (wā) *n.* Way: used in the phrase *under weigh* by mistaken analogy with *aweigh.* See AWEIGH. ◆ Homophone: *way.* [Var. of WAY; infl. in form by *weigh*[1], in phrase "weigh anchor"]

weight (wāt) *n.* **1** The measure of the force with which bodies tend toward the earth's center, or the quality thus measured. The weight of a body is a product of its mass and the acceleration due to gravity. **2** Any object or mass which weighs a definite or specific amount. **3** A definite mass of brass, iron, or other metal, used in weighing machines as a standard; any unit of heaviness, as a pound, ounce, etc. **4** Any mass used as a counterpoise or to exert pressure by force of gravity: a *paperweight.* **5** Burden; pressure; oppressiveness: the *weight* of care; the *weight* of an attack. **6** Any quantity of heaviness, expressed indefinitely or in terms of standard units. **7** The relative tendency of any mass toward a center of superior mass: the *weight* of a planet. **8** A scale or graduated system of standard units of weight: avoirdupois *weight.* See tables below; see also under METRIC SYSTEM. **9** Influence; importance; consequence: a man of *weight.* **10** The comparative heaviness of clothes, as appropriate to the season: summer *weight.* **11** *Stat.* **a** The relative value of an item in a statistical compilation. **b** The frequency of its occurrence among related items, or the number used to express such frequency. — *v.t.* **1** To add weight to; make heavy. **2** To oppress or burden. **3** To adulterate or treat (fabrics or other merchandise) with cheap foreign substances. **4** *Stat.* To give weight to. [OE *wiht, gewiht*] — **weight′less** *adj.*

Synonyms (noun): burden, efficacy, gravity, heaviness, import, load, moment, ponderosity, ponderousness, power, pressure. See LOAD.

AVOIRDUPOIS WEIGHT

27.34+ grains (gr.)	=	1 dram (dr. av.)
16 drams av.	=	1 ounce (oz. av.)
16 ounces av.	=	1 pound (lb., lbs. av.)
2000 pounds av.	=	1 short ton (sh. tn.)
2240 pounds av.	=	1 long ton (l. tn.)

TROY WEIGHT

24 grains	=	1 pennyweight (dwt.)
20 pennyweight	=	1 ounce (oz. t.)
12 ounces	=	1 pound (lb., lbs. t.)

weight·y (wā′tē) *adj.* **weight·i·er, weight·i·est** **1** Having great weight; ponderous. **2** Having power to move the mind; cogent. **3** Of great importance. **4** Influential, as in public affairs. **5** Burdensome. See synonyms under HEAVY, IMPORTANT. — **weight′i·ly** *adv.* — **weight′i·ness** *n.*

Wei·hai (wā′hī′) A port and naval base in NE Shantung province, NE China; leased with the surrounding area (285 square miles) to Great Britain, 1898–1930. Formerly **Wei·hai·wei** (wā′hī′wā′).

Wei·mar (vī′mär) A city in SW East Germany, formerly capital of Thuringia.

Wei·mar·an·er (vī′mər·ä′nər) *n.* A breed of dog of the hound type, blue- or amber-eyed, gray in color, used for hunting and as a watchdog. [from *Weimar,* Germany, where the breed originated]

weir (wir) *n.* **1** An obstruction placed in a stream to raise the water, divert it into a millrace or irrigation ditches, or form a fish pond; a dam; also, that part of a dam which contains gates for discharging surplus water. **2** A series of wattled enclosures in a stream, to catch fish. **3** An aperture of a determined shape and size, made in a vertical plate: used to determine the quantity of water flowing through it. [OE *wer* < *werian* dam up]

Weir (wir), **Robert Walter,** 1803–89, and his sons, **John Ferguson,** 1841–1926, and **Julian Alden,** 1852–1919, U. S. painters.

weird (wird) *adj.* **1** Concerned with the unnatural or with witchcraft; unearthly; uncanny. **2** Pertaining to or having to do with fate or the Fates. — **the Weird Sisters** **1** The Fates. **2** The three witches in Shakespeare's *Macbeth.* — *n. Scot.* **1** One's allotted fate; fortune. **2** Destiny; fate. **3** One of the Fates. **4** A prophecy; prediction. **5** A spell; enchantment. [OE *wyrd* fate]

Weis·mann (vīs′män), **August,** 1834–1914, German biologist. — **Weis·man·ni·an** (vīs·män′ē·ən) *adj.* & *n.*

Weis·mann·ism (vīs′män·iz′əm) *n.* The theory of evolution, as propounded by August Weismann, which asserts the continuity of the germ plasm within but in isolation from the soma, and denies the heritability by offspring of characters acquired by the parents during their lifetime.

weiss beer (vīs) A light, whitish beer, brewed usually from wheat. [<G *weissbier* pale Berlin beer, lit., white beer]

Weis·sen·fels (vī′sən·fels) A city in south central East Germany, in the former state of Saxony-Anhalt.

Weiss·horn (vīs′hôrn) A peak in southern Switzerland; 14,804 feet.

Weiz·mann (wīts′mən, vīts′män), **Chaim,** 1874–1952, Israeli chemist and Zionist leader; first president of Israel 1948–52; born in Russian Poland.

we·jack (wē′jak) *n.* The fisher or pekan. [<Algonquian. Cf. Cree *otchek.*]

we·ka (wē′kə, wā′-) *n.* A wingless rail (genus *Ocydroma*) of New Zealand, now nearly extinct: also called *woodhen.* [<Maori]

welch (welch, welsh) See WELSH.

Welch (welch, welsh) See WELSH.

Welch (welch), **William Henry,** 1850–1934, U. S. pathologist and sanitarian; a founder of Johns Hopkins Medical School.

wel·come (wel′kəm) *adj.* **1** Admitted gladly to a place or festivity; received cordially: a *welcome* guest. **2** Producing satisfaction or pleasure; pleasing: *welcome* tidings. **3** Made free to use or enjoy: She is *welcome* to my purse. See synonyms under AGREEABLE, DELIGHTFUL. — *n.* The act of bidding or making welcome; a hearty greeting given or cordial reception accorded to a guest or visitor. — **to wear out one's welcome** To come so often or to linger so long as no longer to be welcome. — *v.t.* **·comed, ·com·ing** **1** To give a welcome to; greet gladly or hospitably. **2** To receive with pleasure: to *welcome* constructive advice. [OE *wilcuma* < *will-* will, pleasure + *cuma* a guest; infl. in form by WELL[2] and COME, on analogy with OF *bien venu*] — **wel′come·ly** *adv.* — **wel′come·ness** *n.* — **wel′com·er** *n.*

weld[1] (weld) *v.t.* **1** To unite, as two pieces of metal, with or without pressure, by the application of heat along the area of contact. **2** To bring into close association or connection. — *v.i.* **3** To admit of being welded. — *n.* The consolidation of pieces of metal by welding; also, the closed joint so formed. [Alter of WELL[1], *v.*] — **weld′a·bil′i·ty** *n.* — **weld′a·ble** *adj.* — **weld′er** *n.*

weld[2] (weld) *n.* **1** An erect Old World annual *(Reseda luteola),* formerly cultivated for dyers' use: also called *yellowweed.* **2** The yellow pigment obtained from it: also spelled *woald.* [ME *welde.* Cf. MLG *walde,* MDu. *woude.*]

wel·fare (wel′fâr) *n.* **1** The condition of faring well; exemption from pain or discomfort; prosperity; also, condition, as regards well-being: Inquire concerning thy brethren's *welfare.* **2** Organized efforts by a community or organization to improve the social and economic condition of a group or class: Negro *welfare*; more fully, **welfare work.** [ME *wel fare* < *wel* well + *fare* a going <OE *faran* go]

Welfare Island An island in the East River, New York City; 139 acres; site of two municipal hospitals: formerly *Blackwell's Island.*

welfare state A state or polity in which the government assumes a large measure of responsibility for the social welfare of its members, as through unemployment and health insurance, fair employment legislation, etc.

wel·kin (wel′kin) *n. Archaic* or *Poetic* **1** The vault of the sky; the heavens. **2** The air. [OE *wolcn, wolcen* a cloud]

well[1] (wel) *n.* **1** A hole or shaft sunk into the earth to obtain a fluid, as water, oil, brine, or natural gas. **2** A spring of water; a place where water issues from the ground; a fountain. **3** A source of continued supply, or that which issues forth continuously; a wellspring: a *well* of learning. **4** A depression, cavity, or vessel resembling a well: an *inkwell.* **5** A cavity in the lower part of some sorts of furnaces to receive falling metal. **6** In an English law court, the railed-in space between the bench and the bar, reserved for solicitors. **7** *Archit.* **a** The vertical opening contained within a winding staircase. **b** A similar opening descending through floors, or a deep enclosed space in a building for light or ventilation: an air *well*; an elevator *well.* **8** *Naut.* The boxed-in space in a vessel's hold, enclosing the pumps. **9** A compartment admitting water, in which fish are preserved alive. **10** A dangerous eddy or whirlpool in the sea. — *v.i.* To pour forth or flow up, as water in a spring. — *v.t.* To gush: Her eyes *welled* tears. [OE *wielle* < *weallan* boil, bubble up]

well[2] (wel) *adv.* **bet·ter, best** **1** Satisfactorily; favorably; according to one's wishes: Everything goes *well.* **2** In a good or correct manner; properly; excellently; expertly: to dance or speak *well.* **3** Suitably; befittingly; with reason or propriety: I cannot *well* remain here. **4** In a successful manner; prosperously; also, agreeably or luxuriously: He lives *well.* **5** Intimately: How *well* do you know him? **6** To a large or proper extent or degree; plentifully: a *well*-stocked larder. **7** Completely; wholly. **8** Far; at some distance: He lagged *well* behind us. — **as well** **1** Also; in addition. **2** With equal effect or consequence: He might just *as well* have sold it. — **as well as** **1** As satisfactorily as. **2** To the same degree as. **3** In addition to. — *adj.* **1** Satisfactory; rightly done or arranged; fortunate; suitable; gratifying: always in the predicate position: It is *well.* **2** Having physical health; free from ailment of mind or body. **3** Prosperous; comfortable. — *interj.* An exclamation used to express surprise, expectation, resignation, doubt, indignation, acquiescence, etc., or merely to preface a remark. [OE *wel.* Akin to WEAL.]

Synonyms (adj.): advantageous, beneficial, convenient, desirable, excellent, expedient, favorable, fortunate, good, happy, lucky, prosperous. See HEALTHY. *Antonyms*: see synonyms for BAD.

Well may appear as a combining form in hyphemes when joined to participles to form unit modifiers: thus, predicatively, His words were *well* chosen; but attributively, his *well-chosen* words. The following examples are self-explanatory:

well-accepted **well-acquainted**
well-accustomed **well-acted**
well-acknowledged **well-adjusted**

well-administered	well-knit
well-aimed	well-liked
well-aired	well-looking
well-armed	well-loved
well-armored	well-made
well-arranged	well-managed
well-assorted	well-mannered
well-assured	well-measured
well-attested	well-ordered
well-attired	well-paid
well-authenticated	well-phrased
well-behaved	well-placed
well-beloved	well-planned
well-built	well-pleased
well-chaperoned	well-pleasing
well-chosen	well-poised
well-considered	well-prepared
well-contented	well-preserved
well-covered	well-proportioned
well-cultivated	well-recognized
well-defended	well-regulated
well-defined	well-remembered
well-deserving	well-rooted
well-digested	well-seasoned
well-disciplined	well-selected
well-done	well-skilled
well-dressed	well-spent
well-earned	well-stocked
well-educated	well-swept
well-established	well-timed
well-financed	well-trained
well-fitted	well-trimmed
well-formed	well-understood
well-fortified	well-used
well-fought	well-versed
well-furnished	well-wooded
well-governed	well-worded
well-handled	well-worn
well-informed	well-woven
well-judged	well-written
well-kept	well-wrought

we'll (wēl) We shall; we will: a contraction.

Wel·land (wel′ənd) A city in southern Ontario, on the Welland Ship Canal, a waterway (28 miles long) connecting Lake Ontario with Lake Erie.

well-ap·point·ed (wel′ə·poin′tid) adj. Properly equipped; excellently furnished.

well-a·way (wel′ə·wā′) interj. Obs. Woe is me! alas! Also **well′a·day′**. [OE wei lā wei, alter. of wā lā wā woe! lo! woe!; infl. in form by ON vei woe]

well-bal·anced (wel′bal′ənst) adj. Evenly balanced; adjusted with reference to welfare.

well-be·ing (wel′bē′ing) n. A condition of happiness or prosperity; welfare.

well-born (wel′bôrn′) adj. Of good lineage.

well-bred (wel′bred′) adj. 1 Well brought up; polite. 2 Of good ancestry; of good or pure stock.

well-curb (wel′kûrb′) n. The frame or stone ring around the mouth of a well.

well-dis·posed (wel′dis·pōzd′) adj. Favorably inclined.

well-do·er (wel′dōō′ər) n. 1 A performer of moral and social duties. 2 Scot. & Brit. Dial. One who is prosperous or well-to-do. — **well′-do′ing** adj. & n.

Wel·le (we′lā) See UELE.

well enough Tolerably good or satisfactory. — **to let well enough alone** To leave things as they are lest the result of interference be worse.

Welles (welz), **Gideon,** 1802-78, U.S. politician and writer. — **(George) Orson,** born 1915, U.S. actor and producer. — **Sumner,** born 1892, U.S. diplomat.

Welles·ley (welz′lē), **Richard Colley,** 1760-1842, first Marquis of Wellesley, British statesman; brother of the Duke of Wellington.

well-fa·vored (wel′fā′vərd) adj. Of attractive appearance; comely; handsome. Also Brit. **well′-fa′voured.**

well-fed (wel′fed′) adj. Plump; fat; sleek.

well-fixed (wel′fikst′) adj. Colloq. Affluent; well-to-do.

well-found (wel′found′) adj. 1 Found to meet expectations. 2 Well equipped.

well-found·ed (wel′foun′did) adj. Based on fact: well-founded suspicions.

well-groomed (wel′grōōmd′) adj. 1 Carefully curried, as a horse. 2 Carefully dressed and scrupulously neat; having a fashionable, sleek appearance.

well-ground·ed (wel′groun′did) adj. 1 Adequately schooled in the elements of a subject. 2 Well-founded.

well-head (wel′hed′) n. A natural source supplying water to a spring or well.

well-heeled (wel′hēld′) adj. Slang Plentifully supplied with money. [<WELL² + HEEL¹, v. (def. 6)]

Wel·ling·ton (wel′ing·tən) The capital of New Zealand, a port on southern North Island.

Wel·ling·ton (wel′ing·tən), **Duke of,** 1769-1852, Arthur Wellesley, British general; defeated Napoleon at Waterloo; prime minister; born in Ireland.

Wellington boot A high boot covering the leg as far as the knee in front but cut away behind.

well-in·ten·tioned (wel′in·ten′shənd) adj. Having good intentions; well-meant: often with connotation of failure or of clumsy or harmful execution.

well-known (wel′nōn′) adj. Widely known; famous.

well-mean·ing (wel′mē′ning) adj. Having good intentions. — **well′-meant′** (-ment′) adj.

well met Welcome.

well-nigh (wel′nī′) adv. Very nearly; almost.

well-off (wel′ôf′, -of′) adj. In comfortable circumstances; wealthy; fortunate.

well-read (wel′red′) adj. Having a wide knowledge of literature or books; having read much: usually with in.

Wells (welz) A municipal borough in east central Somerset, England; noted for its 12th century cathedral.

Wells (welz), **Henry,** 1805-78, U.S. express operator; organized Wells, Fargo & Co. in 1852. — **H(erbert) G(eorge),** 1866-1946, English author.

wells·ite (welz′īt) n. A vitreous hydrated silicate of barium, calcium, potassium, and aluminum, crystallizing in the monoclinic system. [after H. L. Wells, 1855-1924, U.S. chemist]

well-spo·ken (wel′spō′kən) adj. 1 Fitly or excellently said. 2 Of gentle speech and manners.

well-spring (wel′spring) n. 1 An inexhaustible fountain. 2 A source of continual supply.

well sweep A long tapering pole swung on a pivot attached to a high post, and having the bucket suspended from one end, for use in drawing water.

well-thought-of (wel′thôt′uv′, -ov′) adj. In good repute; esteemed; respected.

WELL SWEEP

well-to-do (wel′tə·dōō′) adj. In prosperous circumstances; evincing a state of comfort or wealth.

well-wish·er (wel′wish′ər) n. One who wishes well, as to another. — **well′-wish′ing** adj. & n.

Wels·bach (welz′bak, Ger. vels′bäkh), **Baron Carl Auer von,** 1858-1929, Austrian chemist and inventor.

Welsbach burner A burner of the Bunsen type, having a cotton-gauze mantle impregnated with thoria and ceria, so arranged that upon ignition of a mixture of gases the mantle becomes incandescent. [after Baron Carl A. von Welsbach]

welsh (welsh, welch) v.t. & v.i. Slang 1 To cheat by failing to pay a bet or debt. 2 To avoid fulfilling (an obligation). Also spelled welch. [? Back formation < welsher, prob. <Welsher a Welshman, with ref. to supposed national traits]

Welsh (welsh, welch) adj. Pertaining to Wales, its people, or their language. — n. 1 The natives of Wales, a people of Celtic stock: with the: also called Cymry. 2 The language of Wales, belonging to the Brythonic or Cymric group of the Celtic subfamily of Indo-European languages: also called Cymric. Also spelled Welch. [OE Welisc < wealh a foreigner (one not of Saxon origin)] — **Welsh′man** n.

Welsh cor·gi (kôr′gē) n. Either of two ancient breeds of a Welsh working dog, characterized by a long body, short legs, and erect ears: the **Cardigan Welsh corgi** has a long tail, the **Pembroke Welsh corgi** a short tail. [< Welsh < corr dwarf + ci dog]

Welsh rabbit A concoction of melted cheese cooked in cream or milk, often with ale or beer added, and served hot on toast or crackers. ◆ The form rarebit was a later development and is the result of mistaken etymology.

welt (welt) n. 1 A strip of material, covered cord, etc., applied to a seam to cover or strengthen it. 2 In shoemaking, a strip of leather set into the seam between the edges of the upper and the outer sole. 3 In carpentry, a batten or strip made fast over a flush seam. 4 A wale or stripe raised on the skin by a blow. — v.t. 1 To sew a welt on or in; decorate with a welt. 2 Colloq. To flog severely, so as to raise welts or wales. [ME welte, walt. Cf. OE weltan roll.]

Welt·an·schau·ung (velt′än·shou′ŏŏng) n. German Literally, world viewing; philosophy of life: a comprehensive philosophy regarding the cosmos; ideology.

Welt·an·sicht (velt′än·zikht) n. German Literally, world view; a special view or interpretation of reality, seen as a whole.

wel·ter¹ (wel′tər) v.i. 1 To roll about; wallow. 2 To lie or be soaked in some fluid, as blood. 3 To surge or move tumultuously, as the sea. — n. 1 A rolling movement, as of waves; hence, commotion. 2 That in which weltering is done; a wallow. [<MDu. welteren]

wel·ter² (wel′tər) adj. Designating or pertaining to a horse race in which welterweights are carried. [< welter a heavyweight (horseman), ? <WELT, v. (def. 2)]

welter race A horse race in which heavy weights are imposed on the horses, in order to permit amateur jockeys to ride.

wel·ter·weight (wel′tər·wāt′) n. 1 The weight (regularly 28, sometimes 40 pounds, in addition to weight for age) borne by a horse running in a welter race; hence, loosely, a heavyweight. 2 A boxer whose fighting weight is between 135 and 147 pounds. [< welter a heavyweight + WEIGHT]

Welt·lit·e·ra·tur (velt′lit′ə·rä·tŏŏr′) n. German World literature.

Welt·po·li·tik (velt′pō·li·tēk′) n. German International politics; world policy.

Welt·schmerz (velt′shmerts) n. German World-weariness; melancholy pessimism over the state of the world; romantic discontent.

Wem·bley (wem′blē) A municipal borough of Middlesex, England, 8 miles NW of London.

Wemyss (wēmz) A parish on the Firth of Forth, central Fifeshire, Scotland.

wen¹ (wen) n. 1 Pathol. Any encysted tumor containing a suetlike substance, occurring commonly on the scalp. 2 Any protuberance. [OE wenn, wænn] — **wen′nish, wen′ny** adj.

wen² (wen) n. The old English rune ρ, replaced by modern English w. [OE, var. of wynn joy]

Wen·ces·laus (wen′səs·lôs), 1361-1419, Holy Roman Emperor 1378-1400; king of Bohemia 1378-1419. Also **Wen′ces·las** (-läs), Ger. **Wen·zel** (ven′tsəl), **Wen·zes·laus** (ven′tsəs·lous).

wench (wench) n. 1 A young peasant woman; also, a female servant; maid. 2 Any young woman; girl; maiden. 3 Archaic A prostitute; strumpet. — v.i. To keep company with strumpets. [ME wenche, short for wenchel <OE wencel a child, servant]

Wen·chow (wen′chou′, Chinese wun′jō′) A port on the Wu (def. 2) and chief city of SE Chekiang province, central eastern China.

wend (wend) v.t. & v.i. To direct (one's course); go. [OE wendan]

Wend (wend) n. One of a Slavic people now occupying the region between the Elbe and Oder rivers in Saxony and Prussia; a Sorb. [<G Wende, Winde]

Wen·dat (wen′dat) See WYANDOT.

Wen·dell (wen′dəl), **Barrett,** 1855-1921, U.S. scholar.

Wend·ish (wen′dish) adj. Of or pertaining to the Wends or their language; Sorbian. — n. The West Slavic language of the Wends; Sorbian. Also **Wend′ic.**

went (went) An obsolete past tense and past participle of wend, now used as past tense of GO.

wen·tle·trap (wen′təl·trap′) *n.* Any of a genus *(Epitonium)* or family *(Epitoniidae)* of mollusks, having a white, turreted, many-whorled shell. [<Du. *wenteltrap* a spiral staircase or shell]

wept (wept) Past tense and past participle of WEEP.

were (wûr, *unstressed* wər) Plural and second person singular past indicative, and past subjunctive singular and plural of BE. [OE *wǣre, wǣron,* pt. forms of *wesan* be]

we're (wir) We are: a contraction.

wer·en't (wûr′ənt) Were not: a contraction.

were·wolf (wir′wo͝olf′, wûr′-) *n. pl.* **·wolves** (-wo͝olvz′) In European folklore, a human being transformed into a wolf by bewitchment, or one having power to assume wolf form at will. Also **wer′wolf′**. [OE *werwulf* man-wolf < *wer* a man + *wulf* a wolf]

Wer·fel (ver′fəl), **Franz,** 1890–1945, German novelist, poet, and dramatist.

wer·geld (wûr′geld) *n.* In Anglo-Saxon and Teutonic law, a fine or pecuniary compensation for crime against the person, especially for homicide, paid by the kindred of the slayer to those of the slain. Also **were′gild** (-gild), **wer′gelt** (-gelt). [OE, lit., man-yield, *i.e.,* man-price < *wer* a man + *geld, gield* yield]

Wer·ner (ver′nər), **Alfred,** 1866–1919, Swiss chemist.

wer·ner·ite (wûr′nər·īt) *n.* Scapolite. [after A. G. *Werner,* 1750–1817, German mineralogist]

wert (wûrt, *unstressed* wərt) Archaic second person singular, past tense of both indicative and subjunctive of BE: used with *thou.*

Wes·cott (wes′kot), **Glenway,** born 1901, U. S. novelist and poet.

we'se (wēz) **1** *Dial.* We is: a mistake for *we are*: *We'se* going. **2** *Scot.* We shall.

We·ser (vā′zər) A river in east and north central West Germany, flowing 300 miles north to the North Sea.

Wes·ley (wes′lē, *Brit.* wez′lē), **Charles,** 1707?–1788, English clergyman and hymn writer; brother of John Wesley. **—John,** 1703–91, English clergyman; founder of Methodism.

Wes·ley·an (wes′lē·ən, *Brit.* wez′lē·ən) *adj.* Of or pertaining to the Wesleys, especially John Wesley, as the founder of Methodism. **—** *n.* A disciple of John Wesley; a Methodist. **— Wes′ley·an·ism** *n.*

Wes·sex (wes′iks) The ancient kingdom of the West Saxons, including modern Berkshire, Dorset, Hampshire, Somerset, and Wiltshire in southern England.

west (west) *n.* **1** The point of the compass at which the sun sets at the equinox, directly opposite *east.* See COMPASS CARD. **2** Any direction, region, or part of the horizon near that point. **— the West 1** The countries lying west of Asia and Turkey; the Occident. **2** The western hemisphere, discovered by explorers sailing westward from Europe. **3** The Western Roman Empire. **4** In the United States: **a** Formerly, the region west of the Allegheny Mountains. **b** The region west of the Mississippi, especially the northwestern part of this region. **—** *adj.* **1** To, toward, facing, or placed in the west; western. **2** Coming from the west: the *west* wind. **3** Designating or located in that part of a church directly opposite the altar. **—** *adv.* In or toward the west; in a westerly direction. **— go West, young man** Go settle in the unsettled western regions of the United States, a land of little competition and unusual opportunity: advice usually attributed to Horace Greeley. [OE]

West (west), **Benjamin,** 1738–1820, American painter. **— Rebecca** Pseudonym of Cicily Isabel Fairfield, born 1892, English novelist and critic.

West Bengal See under BENGAL.

west-bound (west′bound′) *adj.* Going westward. Also **west′bound′**.

West Brom·wich (brum′ich, -ij) A county borough in southern Stafford, England, just NW of Birmingham.

west by north One point north of west on the mariner's compass. See COMPASS CARD.

west by south One point south of west on the mariner's compass. See COMPASS CARD.

West·cott (west′kot), **Edward Noyes,** 1846–1898, U. S. banker and novelist.

West End The western part of London, England; includes parks and a fashionable shopping district and notable residential section.

west·er (wes′tər) *v.i.* To turn, trend, or shift to the west. **—** *n.* A wind, especially a storm, blowing from the west. [<WEST + -ER[5]]

west·er·ing (wes′tər·ing) *adj.* Moving or turning westward: the *westering* sun. **—** *n.* Movement or declension toward the west.

west·er·ly (wes′tər·lē) *adj.* **1** In, toward, or of the west. **2** From the west: a *westerly* wind. **—** *n. pl.* **·lies** A wind blowing from the west. **—** *adv.* **1** From the west. **2** Toward the west. **— west′er·li·ness** *n.*

Wes·ter·marck (ves′tər·märk), **Edward Alexander,** 1862–1939, Finnish anthropologist.

west·ern (wes′tərn) *adj.* **1** Being in the west; of, pertaining to, or directed toward the west. **2** Coming from the west: the *western* winds. **—** *n.* **1** A westerner. **2** A type of fiction or motion picture using cowboy and pioneer life in the western United States as its material. [OE *westerne* < *west*]

West·ern (wes′tərn) *adj.* **1** Proceeding from or characteristic of the West; Occidental. **2** Belonging or pertaining to the Western Church: *Western* ritual. **3** Of or pertaining to the western part of the United States of America. **—** *n.* A person identified with or belonging to the Western Church.

Western Australia The largest state of the Commonwealth of Australia, including all of the Australian continent west of 129° E.; about 975,920 square miles; capital, Perth.

Western Church The medieval church of the Western Roman Empire, now the Roman Catholic Church: distinguished from the church of the Eastern Empire, now the Greek or Eastern Church.

Western Dvina See DVINA (def. 2).

west·ern·er (wes′tər·nər) *n.* One who dwells in a western region, especially in the western United States.

western frontier Formerly, that part of the United States bordering on the west in still unsettled regions.

western hemisphere See under HEMISPHERE.

Western India States A former political agency in western India; by 1950 merged with Saurashtra, which merged with Bombay in 1956.

Western Islands The Hebrides.

west·ern·ism (wes′tər·niz′əm) *n.* An expression or practice peculiar to the West, especially the western United States.

west·ern·ize (wes′tər·nīz) *v.t.* **·ized, ·iz·ing** To make western in characteristics, habits, etc. **— west′ern·i·za′tion** *n.*

Western Ocean In ancient geography, the ocean lying westward of the known world; hence, the Atlantic Ocean.

Western Reserve A region now comprising ten counties in the NE portion of Ohio, on Lake Erie from the Pennsylvania border to near Sandusky, Ohio, reserved by Connecticut for her settlers when she ceded her western lands to the Federal Government in 1786, but relinquished in 1800.

Western (Roman) Empire The part of the Roman Empire west of the Adriatic, which existed as a separate empire from 395 A.D. until the fall of Rome in 476 A.D.

Western Samoa, Territory of See SAMOA.

Western Turkestan See under TURKESTAN.

West Flanders A province of western Belgium; 1,249 square miles; capital, Bruges.

West Germanic See under GERMANIC.

West Germany See under GERMANY.

West Ham A county borough in Essex, England; a NE suburb of London.

West Har·tle·pool (här′təl·po͞ol) A county borough on the North Sea in SE Durham, England.

West Indies A series of island groups separating the North Atlantic from the Caribbean, between North and South America, divided into the *Bahamas,* the *Greater Antilles,* and the *Lesser Antilles.* **— West Indian**

West Indies, The A federation comprising ten British colonies in the Caribbean, including Antigua, Barbados, Dominica, Grenada, Jamaica, Montserrat, St. Lucia, St. Vincent, Trinidad and Tobago, and St. Christopher, Nevis and Anguilla; formed January, 1958; 7,766 square miles; capital on Trinidad.

west·ing (wes′ting) *n.* **1** Distance accomplished toward the west. **2** *Naut.* The amount by which a ship has increased her west longitude from a specified meridian.

West·ing·house (wes′ting·hous), **George,** 1846–1914, U. S. inventor.

West Lo·thi·an (lō′thē·ən) A county in SE Scotland on the Firth of Forth; 120 square miles; county town, Linlithgow: formerly *Linlithgowshire.*

West·meath (west′mēth) An inland county of Leinster province, Ireland; 681 square miles; county town, Mullingar.

West·min·ster (west′min·stər) A city and metropolitan borough in the county of London, England on the north bank of the Thames; London's largest borough; site of the Houses of Parliament and Buckingham Palace.

Westminster Abbey A Gothic church in Westminster, London, begun in A.D. 1050; burial place of English kings and notables.

West·mor·land (west′môr·lənd, -mōr-; *Brit.* west′mər·lənd) A county in the Lake District, NW England; 789 square miles; county town, Appleby.

west-north·west (west′nôrth′west′, *in nautical usage* west′nôr·west′) *adj., adv., & n.* Midway between west and northwest. See COMPASS CARD.

Wes·ton su·per Ma·re (wes′tən so͞o′pər mā′rē, mâr′) A resort and municipal borough on Bristol Channel, in NE Somerset, England.

West Pakistan 1 A province of Pakistan comprising all of Pakistan west of the Republic of India with the exception of the **Federal District of Pakistan** around and including Karachi (812 square miles); formed in 1955 by the merger of the former provinces of Baluchistan, North-West Frontier Province, Punjab and Sind, along with Bahawalpur, Khairpur, and the other princely states in the area; 309,424 square miles; capital, Lahore. **2** Formerly, the region comprising all of the western portion of Pakistan, including the four former provinces, the princely states, and the present federal district; 310,236 square miles.

West·pha·li·a (west·fā′lē·ə) A former province of Prussia, since 1945 a part of North Rhine-Westphalia, West Germany; scene of the signing of a treaty by France, Sweden, and the Holy Roman Empire at the end of the Thirty Years' War, 1648; 7,806 square miles; capital, Münster. *German* **West·fa·len** (vest·fä′lən). **— West·pha′li·an** *adj. & n.*

West Point A U. S. military reservation on the Hudson River in SE New York; seat of the United States Military Academy.

West Prussia A former province of Prussia; since 1945 under Polish administration; capital, Danzig. *German* **West·preus·sen** (vest′·proi′sən).

West Quod·dy Head (kwod′ē) The easternmost point of continental United States, a promontory on the Atlantic coast of Maine near the Canadian border.

Wes·tra·li·a (wes·trā′lē·ə, -trāl′yə) A contraction of WESTERN AUSTRALIA. **— Wes·tra′li·an** *adj. & n.*

West Riding An administrative division of SW York, England; 2,936 square miles; capital, Wakefield.

West River The chief river of southern China, flowing 1,250 miles east from eastern Yünnan province (900 miles as the Hungshui to its confluence with the Yü) to the South China Sea: Chinese *Si Kiang.*

West Saxon The dialect of Old English spoken in Wessex: preserved in most of the literature of the period.

west-south·west (west′south′west′, *in nautical usage* west′sou·west′) *adj., adv., & n.* Midway between west and southwest. See COMPASS CARD.

West Spitsbergen The largest island of the Spitsbergen group, NW Svalbard; about 15,000 square miles.

West Virginia A State of the east central United States; 24,181 square miles; capital, Charleston; entered the Union June 20, 1863; nickname, *Panhandle State*: abbr. *W. Va.* **— West Virginian**

West·wall (west′wôl) See LIMES (def. 2).

west·ward (west′wərd) *adj.* Tending, moving, lying, or facing toward the west. **—** *adv.* Toward the west: also **west′wards.** [OE *westweard* < *west* the west] **— west′ward·ly** *adv.*

wet (wet) *adj.* **wet·ter, wet·test 1** Moistened

For pronunciation of WH- *see discussion under* W.

or saturated with water or other liquid; consisting of or covered with moisture. 2 Marked by showers or by heavy rainfall; rainy: the *wet* season. 3 Not dry: *wet* varnish. 4 *Colloq.* Favoring or not prohibiting the manufacture and sale of alcoholic beverages: a *wet* State; also, opposed to prohibition. 5 Preserved in liquid; also, bottled in alcohol, as laboratory specimens. 6 *Chem.* Treated or separated by means of liquid reagents. — **all wet** *Slang* Quite wrong; crazy; mistaken: He's *all wet*. — *n.* 1 Water; moisture; wetness. 2 Showery or rainy weather; rain. 3 *Colloq.* One opposed to prohibition. — *v.t.* & *v.i.* **wet** or **wet·ted, wet·ting** To make or become wet. — **to wet one's whistle** *Colloq.* To take a drink. [OE *wǣt*] — **wet′ly** *adv.* — **wet′ness.** *n.* — **wet′ta·ble** *adj.* — **wet′ter** *n.*
We·tar (wē′tär) One of the southern Molucca Islands, Indonesia, north of Portuguese Timor; 1,400 square miles.
wet·back (wet′bak′) *n.* *U.S. Colloq.* A Mexican laborer who enters the United States illegally. [So called because many cross the border by swimming or wading across the Rio Grande]
wet–blank·et (wet′blang′kit) *v.t.* To discourage; depress.
wet blanket A discouragement, or one who discourages any proceedings.
weth·er (weth′ər) *n.* A castrated ram. ♦ Homophone: *weather.* [OE]
wet–nurse (wet′nûrs′) *v.t.* A woman who is hired to suckle the child of another woman.
wet pack *Med.* A method of reducing fever or of relieving a disturbed psychoneurotic condition by wrapping the patient in wet sheets.
Wet·ter·horn (vet′ər-hôrn) A mountain of three peaks in the Bernese Alps, Switzerland; 12,153 feet.
wet·ting (wet′ing) *n.* 1 The act of one who wets; the state of being wetted. 2 A liquid, as water, used in moistening something, as flour in breadmaking.
wetting agent *Chem.* Any of a class of substances that, by reducing surface tension, enable a liquid to spread more readily over a solid surface to which it is applied: a form of detergent.
we've (wēv) We have: a contraction.
Wex·ford (weks′fərd) 1 A maritime county of SE Leinster province, Ireland; 908 square miles. 2 Its county town, a port on **Wexford Harbor**, an inlet of St. George's Channel in SE Ireland.
Wey·den (wī′dən), **Roger van der,** 1399?-1464, Flemish painter.
Wey·gand (ve·gän′), **Maxime,** born 1867, French general in World Wars I and II.
Wey·man (wā′mən), **Stanley,** 1855-1928, English novelist.
Wey·mouth (wā′məth) A port in southern Dorset, England; the old part of the present municipal borough of **Weymouth and Melcombe Regis.**
wha (hwä) *pron.* *Scot.* Who.
whack (hwak) *v.t.* & *v.i.* 1 *Colloq.* To strike sharply; beat; hit. 2 *Slang* To share: often with *up.* 3 *Slang* To drive (mules or oxen). — *n.* 1 *Colloq.* A sharp, resounding stroke or blow. 2 *Slang* A share; portion. 3 *Slang* A turn; a chance; a try. — **to have a whack at** *Slang* 1 To give a blow to. 2 To have a chance or turn at; to have a chance to try. — **out of whack** *Slang* Out of order. [? Var. of THWACK]
whack·er (hwak′ər) *n.* 1 One who whacks. 2 The driver of a mule team. 3 *Colloq.* A whopper.
whack·ing (hwak′ing) *Colloq.* *adj.* Strikingly large; whopping. — *adv.* Very; extremely.
whai·sle (hwā′zl) *v.i.* *Scot.* To breathe hard or roughly; wheeze. Also **whai′zle.**
whale[1] (hwāl) *n.* 1 A cetaceous mammal of fishlike form, especially one of the larger pelagic species, as distinguished from dolphins and porpoises. Whales have the fore limbs developed as broad flattened paddles, the hind limbs absent, and a thick layer of fat or blubber immediately beneath the skin. The principal types are the toothless or whalebone whales (suborder *Mysticeti*), and the toothed whales (suborder *Odontoceti*). 2 *Colloq.* Something extremely good or large: a *whale* of a party. — *v.i.* **whaled, whal·ing** To engage in the hunting of whales. [OE *hwæl*]
whale[2] (hwāl) *v.t.* **whaled, whal·ing** *Colloq.* To strike as if to produce wales or stripes; flog; wale. [Var. of WALE[1], *v.*]
whale·back (hwāl′bak′) *n.* A steamship having a convex main deck, used on the Great Lakes in passenger and freight traffic.
whale·boat (hwāl′bōt′) *n.* A long, deep rowboat, sharp at both ends, often steered with an oar: so called because first used in whaling, now carried on steamers as lifeboats.

WHALEBOAT

whale·bone (hwāl′bōn′) *n.* 1 The horny substance developed in plates from the palate of the whalebone whales; baleen. 2 A strip of whalebone, used in stiffening dress bodies, corsets, etc.
whale iron A harpoon.
whale·man (hwāl′mən) *n.* *pl.* **·men** (-mən) One who hunts whales; a whaler.
whal·er (hwā′lər) *n.* 1 A person or a vessel engaged in whaling. 2 A whaleboat.
Whales (hwālz), **Bay of** An inlet of the Antarctic Ocean in Ross Shelf Ice just north of Little America.
whale shark A very large pelagic shark (*Rhineodon typus*) somewhat resembling the basking shark in its habits but often reaching a length of 50 feet: it has a spotted body and very small teeth adapted for feeding on plankton.
whal·ing (hwā′ling) *n.* The industry of capturing whales. — *adj.* *Slang* Huge; whopping.
whaling station A place on shore to which whales are taken to be flensed and the oil tried out.
wham·my (hwam′ē) *n.* *pl.* **·mies** *U. S. Slang* 1 A gesture made by extending in parallel the index and little finger from the closed fist and pointing toward the person or object intended: an ancient form of hexing (*mano cornuta*, sign of the horns). If the fingers of both hands are used, the fists being in contact, it is a *double whammy.* 2 A jinx; hex: to put the *whammy* on someone. [<*wham*, colloq. interjection imit. of the sound of a hard blow]
Wham·po·a (hwäm′pō′ä′) The deep-water port for Canton in southern Kwangtung province, China, on an island in the Canton River.
whang[1] (hwang) *v.t.* & *v.i.* *Colloq.* To beat or sound with a resounding noise. — *n.* *Colloq.* A beating or banging; heavy blow; whack. [Imit.]
whang[2] (hwang) *n.* 1 A buckskin thong or one made of a deer sinew. 2 Whang leather. 3 *Scot.* A big slice, as of bread or cheese; a chunk. — *v.t.* 1 To beat as with a thong; lash. 2 To beat or strike violently. 3 *Scot.* & *Dial.* To fling; throw violently; hurl. 4 *Scot.* To slice, usually in large pieces. [Var. of OE *thwang* a thong]
whang·ee (hwang·ē′) *n.* 1 Any of a genus (*Phyllostachys*) of tall woody Asian grasses related to the bamboo. 2 A cane or stick made of the stalk of one of these plants. [<Chinese *huang* bamboo sprout]
whang leather A leather, usually of deerskin, made for lacings, thongs, etc. [<WHANG[2] + LEATHER]
Whang·poo (hwang′pōō′) See HWANGPOO.
whap (hwap), **whap·per** (hwap′ər), etc. See WHOP, etc.
wharf (hwôrf) *n.* *pl.* **wharves** (hwôrvz) or **wharfs** 1 A structure of masonry or timber erected on the shore of a harbor, river, or the like, alongside which vessels may lie to load or unload cargo, passengers, etc.; also, any landing place for vessels, as a pier or quay. 2 *Obs.* A river bank; also, the seashore. — *v.t.* 1 To moor to a wharf. 2 To provide or protect with a wharf or wharves. 3 To deposit or store on a wharf. [OE *hwearf, hwerf* a dam]
wharf·age (hwôr′fij) *n.* 1 Charge for the use of a wharf. 2 Wharf accommodations for shipping.
wharf boat A barge or float with a platform used as a landing stage for men and freight on rivers where the water level is changeful: usually connected with the shore or levee by a bridge.
wharf·in·ger (hwôr′fin·jər) *n.* One who keeps a wharf for landing goods and collects wharfage fees. [Earlier *wharfager* + intrusive *n*]
wharf rat 1 A rat that inhabits wharves; especially, the brown or Norway rat. 2 *U.S. Slang* A man or boy who loiters habitually about wharves, especially with thievish or other criminal intent.
Whar·ton (hwôr′tən), **Edith Newbold,** 1862-1937, *née* Jones, U. S. novelist.
wharve (hwôrv) See WHERVE.
wha's (hwäz), **whase** (hwāz) *pron.* *Scot.* Whose.
what (hwot, hwut) *adj.* 1 In interrogative construction, asking for information that will specify the person or thing qualified by it: Of *what* person do you speak? 2 How surprising, ridiculous, great, or the like: used in exclamation to express excess or something exceptional in the person or thing qualified: commendatory or the reverse according to circumstances: *What* genius! *What* a noise that boy is making! 3 How much: an ambiguous use: *What* cash has he? — *pron.* 1 Which circumstance, event, relation, or the like: asking for some specification concerning persons or things referred to: an interrogative pronoun used in absolute interrogation: Who and *what* is he? When used of persons, it ordinarily implies some shade of contempt. In this sense *what* is used elliptically for "What did you say?" or in surprise or indignation: *What!* did he really say that? Formerly it was used as a common introductory expletive like *well,* especially in a summons, as in the phrase *what ho!* 2 That which: a double relative, equivalent to a demonstrative followed by a simple relative: Tell me *what* it is; *What* followed occupied little time. 3 *Dial.* or *Illit.* That or which: a simple relative: a donkey *what* wouldn't go. — *adv.* 1 In what respect; to what extent: *What* are you profited? 2 In some measure; partly: usually followed by *with*: *What* with the heat, and *what* with the noise, it is distracting. 3 For what reason; why. 4 How extraordinarily! how!: an exclamatory or intensive use. — *conj.* 1 So far as; as well as: He did *what* he could at the time. 2 That: especially in the phrase *but what.* [OE *hwæt,* neut. of *hwa* who]
what–all (hwot′ôl′, hwut′-) *pron.* *Colloq.* Whatever; everything.
Whate·ly (hwāt′lē), **Richard,** 1787-1863, English prelate and logician.
what·ev·er (hwot′ev′ər, hwut′-) *pron.* 1 As a compound relative, the whole that; anything that; no matter what: often added for emphasis to a negative assertion: *whatever* makes life dear; I do not want anything *whatever.* 2 *Colloq.* What: usually interrogative: *Whatever* were you saying? Also *Poetic* **what′e′er′** (-âr′).
what–not (hwot′not′, hwut′-) *n.* 1 An ornamental set of shelves for holding bric-à-brac, etc. 2 Anything you please; something or other.
what·so·ev·er (hwot′sō·ev′ər, hwut′-) *adj.* & *pron.* Whatever: a slightly more formal usage. Also *Poetic* **what′so·e′er′** (-âr′).
whaup[1] (hwäp, hwôp) *n.* *Scot.* & *Brit. Dial.* A curlew. [Imit.]
whaup[2] (hwäp, hwôp) *Scot.* *v.i.* 1 To fuss about noisily. 2 To whine; whistle. — *n.* 1 A whistle or cry. 2 A pod; capsule. 3 A clumsy lout; also, a scamp. 4 An outcry; a fuss.
wheal (hwēl) *n.* A discolored ridge on the skin, as from hives or the stroke of a whip;

also, a whelk. ◆ Homophone: wheel. [Alter. of WALE¹]

wheat (hwēt) n. 1 A grain yielding an edible flour, the annual product of a cereal grass (genus *Triticum*): the most important of the cereals, it is excelled only by rice in the number of people by whom it is used as a staple food. 2 The plant that produces this grain, especially *Triticum aestivum* and varieties, a tall, slender annual or biennial of cosmopolitan distribution, bearing at its summit an imbricated spike of usually four-flowered spikelets culminating in the ear or head. 3 A wheatfield; a crop of wheat. [OE *hwǣte*]

WHEAT
a. Ear of bearded wheat.
b. Ear of beardless wheat.
c, d. Grain: front and back of b.

wheat-ear (hwēt'ir) n. A thrushlike bird (*Oenanthe oenanthe*) of the northern parts of the northern hemisphere, related to the whinchat, ash-gray above and white below, with the wings, sides of head, and tip of tail black. [Earlier *wheatears* <WHITE + *ers, eeres* rump]

wheat-en (hwēt'n) adj. Belonging to or made of wheat.

Wheat-ley (hwēt'lē), **Phillis**, 1753?–84, American Negro poet born in Africa.

Wheat-stone (hwēt'stōn), **Sir Charles**, 1802–1875, English physicist.

Wheat-stone bridge (hwēt'stōn) *Electr.* An instrument for the measurement of differential resistance in an electric current. Also **Wheatstone's bridge.** [after Sir Charles Wheatstone]

wheat-worm (hwēt'-wûrm) n. A threadworm (*Tylenchus tritici*) destructive of wheat. Also **wheat eelworm.**

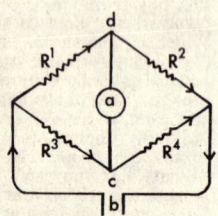

WHEATSTONE BRIDGE
a. Galvanometer.
b. Battery.
c, d. Bridge.
R¹, R² Resistances to be compared.
R³, R⁴ Known resistances which can be varied. When galvanometer shows no current, R¹R² = R³R⁴.

whee-dle (hwēd'l) v. **·dled, ·dling** v.t. 1 To persuade or try to persuade by flattery, cajolery, etc.; coax. 2 To obtain by cajoling or coaxing. — v.i. 3 To use flattery or cajolery. [? OE *wǣdlian* beg, be poor <*wǣdl* poverty] — **whee'dler** n. — **whee'dling·ly** adv.

wheel (hwēl) n. 1 A circular rim and hub connected by spokes or rays in one structure, or a disk, capable of rotating on a central axis and used to reduce friction and facilitate movement or transportation, as in vehicles, or to act with a rotary motion, as in machines. 2 Anything resembling or suggestive of a wheel; a disk or a circle, or any circular object or formation. 3 An instrument or device having a wheel or wheels as its distinctive characteristic, as a bicycle, a steering wheel or steering gear, or the like. 4 An old instrument of torture or execution, consisting of a wheel to which the limbs of the victim were tied and then broken with an iron bar; also, the death so inflicted. 5 The wheel with which the goddess of fortune is represented, symbolizing the vicissitudes and uncertainty of human fate. 6 A turning; revolution; rotation. 7 Figuratively, that which imparts or directs motion or controls activity; the moving force: the *wheels* of democracy. 8 A turning of a body of troops or a swinging of a line of ships in which a change of direction is accomplished while the different units keep in alinement. 9 A rotating firework; a pinwheel or catherine wheel. 10 A refrain of a song. 11 The rotating disk used in various gambling games, especially roulette; hence, roulette. — **Pelton wheel** A device consisting of a wheel which carries on it a succession of cupshaped buckets, and is made to rotate by the impingement of high-pressure jets of water on the buckets, the form of which is such as to prevent the accumulation of dead water. — **wheels within wheels** An intricate series of motives or influences, acting and reacting on one another. — v.t. 1 To move or convey on wheels. 2 To cause to turn on or as on an axis; pivot or revolve. 3 To perform with a circular movement. 4 To provide with a wheel or wheels. — v.i. 5 To turn on or as on an axis; pivot; rotate or revolve. 6 To take a new direction or course of action; change attitudes, opinions, etc.: often with *about.* 7 To move in a circular or spiral course. 8 To roll or move on wheels. — adj. 1 Pertaining to or shaped like a wheel. 2 Harnessed to a vehicle directly in front of the wheels: said of a draft animal when there is a leader or leaders in front. ◆ Homophone: *wheal.* [OE *hwēol*]

wheel and axle *Mech.* A wheel or drum mounted on an axle with a rope wound about the drum so that a slight pull on one end of the rope will raise a disproportionately heavy weight attached to the other: one of the so-called simple machines.

wheel animalcule n. A rotifer.

wheel·bar·row (hwēl'bar'ō) n. A boxlike vehicle ordinarily with one wheel and two handles, for moving small loads. — v.t. To convey in a wheelbarrow.

wheel·base (hwēl'bās') n. The distance from the center of the back axle to the center of the front axle, as in an automobile.

wheel·bug (hwēl'bug') n. A large hemipterous insect (*Arilus cristatus*) of the southern United States, which preys upon caterpillars and other soft-bodied insects: so called from a semicircular crest on the thorax resembling a cogwheel.

wheel·chair (hwēl'châr') n. A mobile chair mounted between large wheels, for the use of invalids. Also **wheel chair.**

wheeled (hwēld) adj. 1 Having wheels; furnished with a wheel or wheels: often in compounds: a two-*wheeled* cart. 2 Effected or borne by wheels: *wheeled* transportation.

wheel·er (hwē'lər) n. 1 One who wheels. 2 A wheelhorse or other draft animal working next the wheel. 3 Something furnished with a wheel or wheels: a side-*wheeler.*

Wheel·er (hwē'lər), **Joseph,** 1836–1906, American Confederate general; appointed U.S. general in 1898.

wheel·horse (hwēl'hôrs') n. 1 A horse harnessed to the pole or shafts when there is a leader or leaders in front; hence, one who does the heaviest work. 2 In politics, a person bearing great responsibility, or one to be greatly depended upon.

wheel·house (hwēl'hous') n. 1 A small house on the deck of a vessel in which the steering wheel is located: a pilothouse. 2 A paddle box.

wheel·ing (hwē'ling) n. 1 The act of one who wheels, especially of one riding a bicycle. 2 The condition of the roads, as regards traveling on wheels. 3 A rotating movement; a turning.

Wheel·ing (hwē'ling) A port on the Ohio River in NW West Virginia; an industrial center.

wheel lock 1 An old form of lock for small arms, in which a small steel wheel, actuated by a spring and released by a trigger, produced sparks by rotating against a flint. 2 A lock or catch for stopping a vehicle wheel.

wheel·man (hwēl'mən) n. pl. **·men** (-mən) 1 The man who steers a vessel. 2 A bicyclist. Also **wheels'man.**

Whee·lock (hwē'lok), **Eleazar,** 1711–79, U.S. clergyman and educator.

wheel window See ROSE WINDOW.

wheel·work (hwēl'wûrk') n. Mech. The gearing and arrangement of wheels in a machine or mechanical device.

wheel·wright (hwēl'rīt') n. A man whose business is making or repairing wheels and wheeled vehicles.

wheen (hwēn) n. Scot. & Dial. A few; also, an indefinite quantity.

whee·ple (hwē'pəl) Scot. v.i. To whistle, as a curlew or plover. — n. The whistle of a curlew or plover. Also **wheep.**

wheeze (hwēz) v.t. & v.i. **wheezed, wheez·ing** To breathe or utter with a husky, whistling sound. — n. 1 A wheezing sound. 2 A whispering sound so exaggerated as to give rise to the sound popularly called a "stage whisper." 3 *Colloq.* A popular tale, saying, or trick, especially an ancient one. [Prob. <ON *hwæsa* hiss] — **wheez'er** n. — **wheez'ing·ly** adv.

wheez·y (hwē'zē) adj. **wheez·i·er, wheez·i·est** Subject to wheezing, or making a wheezing sound. — **wheez'i·ly** adv. — **wheez'i·ness** n.

whelk¹ (hwelk) n. Any of various large marine mollusks (family *Buccinidae*), having whorled shells, that burrow in sand and prey on clams, etc. The common whelk (*Buccinum undatum*) is much eaten in Europe. [OE *weoloc*]

whelk² (hwelk) n. A swelling, protuberance, or pustule; wheal; especially, a pimple or eruption of pimples on the face. [OE *hwylca* a pustule <*hwelian* suppurate]

whelk·y¹ (hwel'kē) adj. **whelk·i·er, whelk·i·est** 1 Protuberant; rounded. 2 Shelly. Also spelled *welky.* [<WHELK¹]

whelk·y² (hwel'kē) adj. **whelk·i·er, whelk·i·est** Marked with pustules or whelks. [<WHELK²]

COMMON WHELK
(About 4 inches)

whelm (hwelm) v.t. 1 To cover with water or other fluid; submerge; engulf. 2 To overpower; overwhelm. — v.i. 3 To roll with engulfing force. [Prob. blend of OE *helmian* cover and *gehwelfan* bend over]

whelp (hwelp) n. 1 One of the young of a dog, wolf, lion, or other beast of prey; sometimes, a dog of any age. 2 A worthless young fellow; a cub; puppy: used contemptuously. 3 *Mech.* **a** One of a series of longitudinal ridges on a windlass or capstan. **b** One of the teeth of a sprocket wheel. — v.t. & v.i. To give birth (to): said of dogs, lions, etc. [OE *hwelp*]

when (hwen) adv. 1 Interrogatively, at what or which time: *When* did you arrive? 2 Conjunctively: **a** At which: the time *when* we went on the picnic. **b** At which or what time: They watched till midnight *when* they fell asleep. **c** As soon as: He laughed *when* he heard it; You may play *when* you finish work. **d** Although: He walks *when* he might ride. **e** At the time that; while: *when* you were in church; *when* we were young. **f** If; considering that: *When* in doubt, ask; How can I buy it *when* I have no money? **g** After which; then: We had just awakened *when* you called. — pron. What or which time: since *when*; till *when*. — n. The time; date: I don't know the *when* or the circumstances. [OE *hwanne, hwænne*]

when·as (hwen'az') conj. Obs. 1 Whereas; while. 2 When. Also **when that.**

whence (hwens) adv. 1 Interrogatively, from what place or source; of what origin: *Whence* and what art thou? *Whence* is the correlative of *thence.* 2 Conjunctively: **a** From what or which place, source, or cause; from which: the place *whence* these sounds arise. **b** To which place; where: Return *whence* you came. **c** For which reason; wherefore. [ME *whannes, whennes,* adverbial genitive of *whanne,* OE *hwanne* when]

whence·so·ev·er (hwens'sō·ev'ər) adv. & conj. From whatever place, cause, or source.

when·e'er (hwen·âr') adv. & conj. Poetic Whenever.

when·ev·er (hwen·ev'ər) adv. & conj. At whatever time.

when·so·ev·er (hwen'sō·ev'ər) adv. & conj. At what time soever; whenever.

where (hwâr) adv. 1 Interrogatively: **a** At or in what place, relation, or situation: *Where* is my book? **b** To what place or end; whither: *Where* are you going? **c** From what place; whence: *Where* did you get that hat? 2 Conjunctively: **a** At or in which or what place; at the place in which: *where* men gather. **b** To a place or situation in or to which; whither: Let us go *where* the mountains and the trees are. ◆ *Where* is the correlative of *there.* In composition with a preposition, *where*

For pronunciation of WH– see discussion under w.

whereabouts — whimper

where·a·bouts (hwâr′ə·bouts′) *adv.* 1 Near or at what place; about where. 2 *Obs.* About which; concerning which. Also *Rare* **where′·a·bout**. — *n.* The place in or near which a person or thing is.

where·as (hwâr·az′) *conj.* 1 Since the facts are such as they are; seeing that: often used in the preamble of a resolution, etc. 2 The fact of the matter being that; when in truth: implying opposition to a previous statement. — *n. pl.* **·as·es** A clause or item beginning with the word "whereas."

where·at (hwâr·at′) *adv.* 1 Interrogatively, at what: *Whereat* are you angry? 2 Conjunctively, at which; for which reason; whereupon: He won the race, *whereat* we were delighted.

where·by (hwâr·bī′) *adv.* 1 Interrogatively, by what; how. 2 Conjunctively, by, near, or through which.

wher·e'er (hwâr·âr′) *adv. Poetic* Wherever.

where·fore (hwâr′fôr′, -fōr′) *adv.* 1 Interrogatively, for what reason; what for; to what end; why: *Wherefore* didst thou doubt? 2 Conjunctively, for which reason. See synonyms under THEREFORE. — *n.* The cause; reason: the whys and wherefores. [< WHERE + FOR]

where·from (hwâr′frum′, -from′) *conj.* From which; whence.

where·in (hwâr·in′) *adv.* 1 Interrogatively, in what; in what particular or regard: *Wherein* is the error? 2 Conjunctively, in which thing, place, circumstance, etc.: a state *wherein* there is discord.

where·in·to (hwâr′in·tōō′) *adv.* 1 Interrogatively, into what. 2 Conjunctively, into which: the gulf *whereinto* he sailed.

where·of (hwâr·uv′, -ov′) *adv.* 1 Interrogatively, of or from what: *Whereof* did you partake? 2 Conjunctively, of which or whom: the house *whereof* he is the head.

where·on (hwâr·on′, -ôn′) *adv.* 1 Interrogatively, on what or whom. 2 Conjunctively, on which: a rock *whereon* to build.

where·so·ev·er (hwâr′sō·ev′ər) *adv. & conj.* 1 In or to whatever place; wherever. 2 Whithersoever. 3 Whencesoever.

where·som·ev·er (hwâr′səm·ev′ər) *adv. & conj. Dial.* Wherever; wheresoever. [< WHERE + SOMEVER < *som ever*, just (< Scand.) + EVER]

where·through (hwâr·thrōō′) *adv. & conj.* Through which.

where·to (hwâr·tōō′) *adv.* 1 Interrogatively, to what place or end: *Whereto* serves avarice? 2 Conjunctively, to which or to whom; whither: the grave *whereto* we haste. Also *Archaic* **where·un′to**.

where·up·on (hwâr′ə·pon′, -ə·pôn′) *adv.* 1 Interrogatively, upon what; whereon. 2 Conjunctively, upon which or whom; in consequence of which; after which: *whereupon* they took in sail.

wher·ev·er (hwâr·ev′ər) *adv. & conj.* In, at, or to whatever place; wheresoever.

where·with (hwâr·with′, -with′) *adv.* 1 Interrogatively, with what: *Wherewith* shall I do it? 2 Conjunctively, with which; by means of which: *wherewith* we abated hunger. — *pron.* That with or by which: with the infinitive: I have not *wherewith* to do it. — *n.* The requisites; wherewithal.

where·with·al (hwâr′with·ôl′) *n.* The necessary means or resources; especially, the necessary money: with the definite article. — *adv. & pron.* (hwâr′with·ôl′) Wherewithal.

wher·ry (hwer′ē) *n. pl.* **·ries** 1 A light, fast rowboat used on inland waters. 2 *Brit.* A decked fishing vessel with two sails. 3 An open rowboat for racing or exercise, built for one person. 4 *Brit.* A very broad, light barge. — *v.t. & v.i.* **·ried, ·ry·ing** To transport in or use a wherry. [? < WHIR; with ref. to rapid movement]

wherve (hwûrv) *n.* In spinning, a pulley on the spindle: also spelled *wharve*. [OE *hweorfa*]

whet (hwet) *v.t.* **whet·ted, whet·ting** 1 To sharpen, as a knife, by friction. 2 To make more keen or eager; excite; stimulate, as the appetite. — *n.* 1 The act of whetting. 2 Something that whets. [OE *hwettan*] — **whet′er** *n.*

wheth·er (hweth′ər) *conj.* As the first alternative; in case; if: introducing an alternative clause, followed by a correlative *or*, or *or whether*; sometimes also introducing a single alternative, the other, usually a negative, being implied: Tell us *whether* you are going (or not). — *pron.* Which: properly of two, less exactly of more than two: an archaism used interrogatively and relatively. — **whether or no** Regardless; in any case. [OE *hwæther, hwether*]

whet·slate (hwet′slāt′) *n.* A hard, fine-grained siliceous rock used for whetstones.

whet·stone (hwet′stōn′) *n.* A fine-grained stone for whetting cutting knives, axes, etc. [OE *hwetstān* < *hwettan* whet + *stān* a stone]

whew (hwyōō) *interj.* An exclamatory sound, expressive usually of amazement, dismay, relief, admiration (real or feigned), or discomfort from the heat. [Imit. of whistling]

Whew·ell (hyōō′əl), **William**, 1794-1866, English scientist and philosopher.

whey (hwā) *n.* A clear, straw-colored liquid that separates from the curd when milk is curdled, as in making cheese. [OE *hwæg, hweg*] — **whey′ey, whey′ish** *adj.*

whey-face (hwā′fās′) *n.* Formerly, a face or person pale as if from fear; now, one of pale, sallow complexion. — **whey′-faced′** *adj.*

which (hwich) *pron. & adj.* 1 Interrogatively, what individual person or thing, or group of persons or things collectively, of a certain number or class: asking for the indication or definite description. In this sense *which* is used both substantively and adjectively, singular and plural: Which shall I take? Which apple do you want? Which mammals are carnivorous? 2 As a relative pronoun, that particular one or ones of a certain number or class of impersonal beings or things: pointing out or definitely fixing upon that which is designated in the antecedent word, phrase, or clause to which it is related: now generally as a substantive, but sometimes as an adjective: He raised his hand, *which* gesture attracted my attention. Which as a relative now refers only to animals, without distinction of masculine or feminine, or to things without life; it was formerly used for persons, even in the most exalted sense, as "Our Father, *which* art in heaven." 3 Also relatively, the one that: often equivalent to the use of the interrogative in a dependent question: used substantively or adjectivally: Tell me which (or *which* apple) you prefer. [OE *hwelc, hwilc*]

which·ev·er (hwich·ev′ər) *pron. & adj.* Whether one or another (of two or of several); no matter which. Also **which′so·ev′er**.

whick·er (hwik′ər) *v.i. & n.* Whinny. [Imit.]

whid[1] (hwid) *Scot. n.* A brisk, nimble, scurrying movement. — *v.i.* **whid·ded, whid·ding** To move nimbly: said of small animals.

whid[2] (hwid) *Scot. n.* 1 A fib; lie. 2 A quarrel. 3 A word. — *v.i.* **whid·ded, whid·ding** To tell a lie; fib.

whid·ah bird (hwid′ə) An African weaverbird (subfamily *Viduinae*), the male of which has the tail greatly lengthened in the breeding season: formerly called *widow bird*. Also **whid′ah, whidah finch**: also spelled *whydah*. [Alter. of *widow bird*; infl. in form by *Whidah* former name of Ouidah, a seaport in French West Africa, near which this bird is commonly found]

whiff (hwif) *n.* 1 Any sudden or slight gust or puff of air. 2 A gust or puff of odor: a *whiff* of onions. 3 A sudden expulsion of breath or smoke from the mouth; a puff. 4 An inhalation of smoke. — *v.t.* 1 To drive or blow with a whiff or puff. 2 To exhale or inhale in whiffs. 3 To smoke, as a pipe. — *v.i.* 4 To blow or move in whiffs or puffs. 5 To exhale or inhale whiffs. [Alter. of ME *weffe* an offensive odor; imit.] — **whiff′er** *n.*

whif·fet (hwif′it) *n. Colloq.* 1 A trifling, useless person; whippersnapper: in slight contempt. 2 A small, snappish dog. 3 A little whiff. [? Dim. of WHIFF]

whif·fle (hwif′əl) *v.* **·fled, ·fling** *v.i.* 1 To blow with puffs or gusts; shift about, as the wind. 2 To vacillate; veer. — *v.t.* 3 To blow or dissipate with or as with a puff. [Freq. of WHIFF]

whif·fler (hwif′lər) *n.* 1 One who fluctuates or shuffles in argument; a trifler. 2 One who whiffs tobacco. 3 A piper; fifer. — **whiff′fler·y** *n.*

whif·fle·tree (hwif′əl·trē′) *n.* A swingletree: also called *whippletree*. [Var. of WHIPPLETREE]

whig[1] (hwig) *v.i. Scot.* To drive onward; move along easily; jog.

whig[2] (hwig) *n. Dial.* 1 Sour whey. 2 Buttermilk. [Var. of ME *hweg* whey]

Whig (hwig) *n.* 1 An American colonist who supported the Revolutionary War in the 18th century: opposed to *Tory*; later, a member of a party opposed to the Democratic and succeeded by the Republican party in 1856. 2 A member of the Liberal party in England in the 18th and 19th centuries, as opposed to a *Tory* or *Conservative*. 3 In earlier usage, a Presbyterian rebel of the west of Scotland in the 17th century: thus named in derision; also, after the Restoration (1660), a Roundhead, as opposed to a Cavalier. — *adj.* Consisting of or supported by Whigs. [Prob. short for WHIGGAMORE] — **Whig′gish** *adj.* — **Whig′gish·ly** *adv.* — **Whig′gish·ness** *n.*

Whig·ga·more (hwig′ə·môr, -mōr) *n.* 1 A member of a body of insurgents who in 1648 marched on Edinburgh and opposed the compromise with Charles I. 2 In the later 17th century, a Scotch Presbyterian; a Whig (def. 3). Also **Whig′a·more**. [Prob. < dial. E (Scottish) *whiggamaire* < *whig* a cry to urge on a horse + *mere* a horse]

Whig·ger·y (hwig′ər·ē) *n. pl.* **·ger·ies** The doctrines of Whigs. Also **Whig′gism**.

whig·ma·lee·rie (hwig′mə·lir′ē) *n. Scot.* A small or useless ornament; gewgaw; also, a whim. Also **whig′ma·lee′ry, whig′me·lee′rie**.

while (hwīl) *n.* 1 A short time; also, a period of time, or time in general: Stay and rest a *while*. 2 Time or pains expended on a thing; trouble; labor: only in the phrase *worth while* or *worth one's while*. — **between whiles** From time to time. — **the while** At the same time: He went about his work and sang *the while*. — *conj.* 1 During the time that; as long as. 2 At the same time that; although: *While* he found fault, he also praised. 3 *Colloq.* Whereas: This man is short, *while* that one is tall. 4 *Brit. Dial.* Until; till. — *v.t.* **whiled, whil·ing** To cause to pass lightly and pleasantly; spend; pass: usually with *away*: to *while* away the time. [OE *hwīl*]

whiles (hwīlz) *Archaic or Dial. adv.* Occasionally; at intervals. — *conj.* While; during the time that.

whi·lom (hwī′ləm) *Archaic adj.* Being once upon a time; former. — *adv.* 1 Formerly; at one time. 2 At times. [OE *hwīlum* at times, dative pl. of *hwīl* a while]

whilst (hwīlst) *conj.* While: an old form still widely used, especially in England. [ME *whilest* < *whiles*, genitive of WHILE + -*t*]

whim (hwim) *n.* 1 A sudden, unexpected, and unreasonable deviation of the mind from its usual or natural course; caprice; freak. 2 An old form of mine hoist, run by horsepower. [Short for earlier *whim-wham* a trifle, ? < Scand. Cf. ON *hvima* wander with the eyes.] — **Synonyms**: caprice, crotchet, fancy, freak, humor, kink, quirk, vagary, whimsy, wrinkle. See FANCY.

whim·brel (hwim′brəl) *n.* A small northern curlew with a white rump, especially a species (*Numenius phaeopus*) of northern portions of the eastern hemisphere. [? < obs. *whimp* whimper, prob. imit. of its cry]

whim·per (hwim′pər) *v.i.* To cry or whine with plaintive broken sounds. — *v.t.* 1 To utter with a whimper. — *n.* A low, broken, whining cry; whine. [Imit.] — **whim′per·er** *n.* — **whim′per·ing** *n.* — **whim′per·ing·ly** *adv.*

WHIDAH BIRD
(From 12 to 14 inches over all)

whim·si·cal (hwim′zi·kəl) *adj.* 1 Having eccentric ideas; capricious. 2 Oddly constituted; fantastic; quaint. See synonyms under FICKLE, ODD, QUEER. — **whim′si·cal·ly** *adv.* — **whim′·si·cal·ness** *n.*

whim·si·cal·i·ty (hwim′zi·kal′ə·tē) *n. pl.* **·ties** 1 Whimsicalness. 2 A singularity. 3 A quaint, fanciful, or odd idea or its expression.

whim·sy (hwim′zē) *n. pl.* **·sies** 1 A whim; caprice; freak. 2 Tenuously fanciful humor: His poems display more *whimsy* than imagination. Also **whim′sey.** See synonyms under WHIM. [Prob. related to WHIM]

whin[1] (hwin) *n.* Furze; gorse. [Prob. <Scand. Cf. Dan. & Norw. *hvine* a kind of grass.]

whin[2] (hwin) *n.* Whinstone. [<dial. E (Scottish) *quin*; ult. origin uncertain]

whin·chat (hwin′chat) *n.* A small, Old World, thrushlike singing bird *(Saxicola rubetra)*, streaked with brown above and rufous below. [<WHIN[1] + CHAT[1]]

whine (hwīn) *v.* **whined, whin·ing** *v.i.* 1 To utter a low, plaintive, nasal sound expressive of grief or distress. 2 To complain in a mean or childish way. — *v.t.* 3 To utter with a whine. — *n.* The act or sound of whining; any peevish complaint. [OE *hwīnan* whiz] — **whin′er** *n.* — **whin′ing·ly** *adv.* — **whin′y** *adj.*

whing·er (hwing′ər) *n. Brit. Dial.* A dirk, used at meals or as a weapon; a hanger. See HANGER. Also **whing′ar.** [Prob. var. of WHINYARD]

whin·ny[1] (hwin′ē) *v.* **·nied, ·ny·ing** *v.i.* To neigh, especially in a low or gentle way. — *v.t.* To express with a whinny. — *n. pl.* **·nies** The cry or call of a horse; a neigh, especially if low and gentle. [<WHINE]

whin·ny[2] (hwin′ē) *adj.* **·ni·er, ·ni·est** Abounding in whin or furze. [<WHIN[1]]

whin·stone (hwin′stōn′) *n.* Any very hard, dark-colored rock, as basalt or chert. [<WHIN[2] + STONE]

whin·yard (hwin′yərd) *n. Dial.* 1 One of certain ducks, especially the pochard. 2 A hanger or sword. [Earlier *whyneherd,* ? <OE *hwīnan* whiz; the duck is so called because of the swordlike shape of its bill]

whip (hwip) *v.* **whipped** or **whipt, whip·ping** *v.t.* 1 To strike with a lash, rod, strap, etc. 2 To punish by striking thus; flog. 3 To drive or urge with lashes or blows: with *on, up, off,* etc. 4 To strike in the manner of a whip: The wind *whipped* the trees. 5 To attack with scathing criticism; berate; flay. 6 To beat, as eggs or cream, to a froth. 7 To seize, move, jerk, throw, etc., with a sudden motion: with *away, in, off, out,* etc. 8 In fishing, to make repeated casts upon the surface of (a stream, etc.). 9 To wrap (rope, cable, etc.) with light line so as to prevent chafing or wear; serve. 10 To wrap or bind about something. 11 To form, as a flat seam, by laying two selvages of a fabric together and sewing with a loose overcast or overhand stitch. 12 *U.S. Colloq.* To defeat; overcome, as in a contest. 13 *Naut.* To hoist by means of a whip (def. 5). — *v.i.* 14 To go, come, move, or turn suddenly and quickly: with *away, in, off, out,* etc. 15 To thrash about in a manner suggestive of a whip: pennants *whipping* in the wind. 16 In fishing, to make repeated casts with rod and line. — **to whip in** 1 To keep from scattering, as hounds in a hunt. 2 To keep together or united, as a political party. — **to whip up** 1 To excite; arouse. 2 *Colloq.* To prepare quickly, as a meal. — *n.* 1 An instrument consisting of a lash attached to a handle, used for driving draft animals or for administering punishment. 2 One who handles a whip expertly; a driver. 3 A stroke, blow, or cut with a whip. 4 A member of a legislative body appointed unofficially to enforce the discipline and look after the interests of his party: often called **party whip;** also, a call made upon members of a legislature by such a person to bring or keep them in their places at a given time, as when a vote or division may be expected. 5 *Mech.* A simple form of hoisting apparatus, consisting of a rope passing over an elevated single pulley, and used for lifting light objects. 6 One who operates such an apparatus. 7 A huntsman who whips in the hounds to control them; a whipper-in. 8 *Electr.* A vibrating spring that whips back and forth, closing different circuits in electrical apparatus. 9 A dish or dessert containing cream or eggs whipped to a froth: prune *whip.* 10 A thrashing motion, as of a rope or wire suddenly broken. 11 Flexibility in the shaft of a golf club. 12 An arm of a windmill. 13 *Obs.* or *Scot.* An attack of illness; also, a sudden movement; a single swift attack or blow. [ME *wippen, hwippen.* Cf. MDu. *wippen* swing, leap, dance.]

whip·cord (hwip′kôrd′) *n.* 1 A strong, hard-twisted, sometimes braided hempen cord, used in making whiplashes. 2 A cord of catgut. 3 A twill-weave fabric, similar to gabardine, but with a more pronounced diagonal rib on the right side: used for riding habits and other outdoor garments.

whip·graft (hwip′graft′, -gräft′) *v.t. Bot.* To graft by fitting a tongue cut on the cion to a slit cut slopingly in the stock. — **whip′graft′·age, whip′graft′ing** *n.*

whip·hand (hwip′hand′) *n.* 1 The hand that wields the whip; in riding or driving, the right hand. 2 An instrument or means of mastery; advantage: She, not he, has the *whiphand.*

whip·lash (hwip′lash′) *n.* The flexible striking part of a whip.

whip·per (hwip′ər) *n.* 1 One who whips; especially, one appointed to inflict legal punishment by flogging. 2 A flagellant. 3 One who hoists coal, wood, merchandise, etc., with a whip, as from a ship's hold.

whip·per-in (hwip′ər·in′) *n.* 1 In hunting, one employed to assist the huntsman and to enforce obedience among the hounds. 2 A political or parliamentary whip.

whip·per·snap·per (hwip′ər·snap′ər) *n.* A pretentious but insignificant person. [? Extension of *whipsnapper* a cracker of whips]

whip·pet (hwip′it) *n.* 1 A swift dog resembling an English greyhound in miniature, characterized by a long, narrow head, long arched back, smooth, close coat, and a long, tapering tail. 2 A small, light, speedy tank used in World War I: also **whippet tank.** 3 Anything suggestive of a whippet, as in size, speed, etc. [Dim. of WHIP; so called with ref. to its rapid movement]

WHIPPET
(From 23 to 28 inches high at the shoulder)

whip·ping (hwip′ing) *n.* 1 The act of one who whips; castigation; state or fact of being flogged or defeated. 2 Material used to bind the head of a rope, or to bind the head to the shaft of a golf club.

whipping boy Formerly, a boy brought up as companion to a prince or other noble youth, and punished in his stead for all misdeeds; now, anyone who receives punishment deserved by another.

whipping post The fixture to which those sentenced to flogging are secured; hence, legal punishment by flogging.

Whip·ple (hwip′əl), **George Hoyt,** born 1878, U.S. pathologist. — **William,** 1730–85, American Revolutionary general; signed Declaration of Independence.

whip·ple·tree (hwip′əl·trē′) *n.* A swingletree. [Prob. <WHIP]

whip·poor·will (hwip′ər·wil) *n.* A small nocturnal bird *(Antrostomus vociferus),* allied to the goatsuckers, common in the eastern United States. [Imit. of its reiterated cry]

whip·saw (hwip′sô′) *n.* A thin, narrow, tapering ripsaw about six feet long. — *v.t.* **·sawed, ·sawed** or **·sawn, ·saw·ing** 1 To saw with a whipsaw. 2 In faro, to beat (an opponent) in two bets, one to win and one to lose, at the same time. 3 To get the best of (an opponent) in spite of every effort he makes.

whip scorpion Any of various scorpionlike arachnids (family *Thelyphonidae*) having an abdomen terminating in a slender appendage like a whiplash, and lacking a sting; especially, the vinegarroon.

whip·stall (hwip′stôl′) *Aeron. n.* The stalled condition of a sharply climbing airplane in which the nose whips violently downward. — *v.i.* To bring about or go into a whipstall.

whip·stitch (hwip′stich′) *v.t.* To sew or gather with overcast stitches, as the turned edge of a ruffle; overcast. — *n.* 1 An overcast stitch in whipping an edge or seam. 2 A tailor.

whip·stock (hwip′stok′) *n.* That part of a whip to which the lash is attached; a whip handle.

whipt (hwipt) Alternative past tense and past participle of WHIP.

whip·worm (hwip′wûrm′) *n.* A nematode *(Trichuris trichiura),* with the posterior part of the body thickened: found in the human cecum.

whir (hwûr) *v.t.* & *v.i.* **whirred, whir·ring** To fly, move, or whirl with a buzzing sound. — *n.* 1 A whizzing, swishing sound, as that caused by the sudden rising of birds. 2 Confusion; bustle. Also **whirr.** [Prob. <Scand. Cf. Dan. *hvivre.* Akin to WHIRL.]

whirl (hwûrl) *v.i.* 1 To turn or revolve rapidly, as about a center. 2 To turn away or aside quickly. 3 To move or go swiftly. 4 To have a sensation of spinning: My head *whirls.* — *v.t.* 5 To cause to turn or revolve rapidly. 6 To carry or bear along with a revolving motion: The wind *whirled* the dust into the air. 7 *Obs.* To hurl. — *n.* 1 A swift rotating or revolving motion. 2 Something whirling, as a cloud of dust. 3 Confusion; turmoil. [Prob. <ON *hvirfla* revolve. Akin to WARBLE[1].]

whirl-a-bout (hwûrl′ə·bout′) *n.* Anything that turns swiftly around or about; a whirligig.

whirl·er (hwûr′lər) *n.* 1 One who or that which whirls. 2 A rotating hook or reel used in ropemaking.

whirl·i·gig (hwûr′lə·gig′) *n.* 1 Any toy or small device that revolves rapidly on an axis. 2 A merry-go-round. 3 Anything that performs quick revolutions or moves in a cycle: the *whirligig* of time. 4 Any of a family *(Gyrinidae)* of water beetles that frequent the surface of smooth water and move in swift circles: also **whirligig beetle.** 5 A trifling ornament, as one used by printers; also, a fanciful notion. [< *whirly* (<WHIRL) + GIG[1] (def. 4)]

whirl·pool (hwûrl′po͞ol′) *n.* 1 An eddy or vortex where water moves with a gyrating sweep, as from the meeting of two currents. 2 Any disturbance from such causes, whether accompanied by vortical motion or not.

whirl·wind (hwûrl′wind′) *n.* 1 A moving atmospheric vortex; a funnel-shaped column of air, with a rapid circular and upward spiral motion around a vertical or inclined axis, causing waterspouts, sand pillars, and dust whirls. 2 Any violent rushing or rotatory movement. See synonyms under CYCLONE. — *adj.* Extremely swift or impetuous: a *whirlwind* courtship.

whirl·y·bird (hwûr′lē·bûrd′) *n. Colloq.* A helicopter. [< *whirly* (<WHIRL) + BIRD]

whir·ry (hwûr′ē) *v.t.* & *v.i.* **·ried, ·ry·ing** *Scot.* To hurry.

whish[1] (hwish) *v.i.* To move with a sibilant, whistling sound. — *n.* A swishing sound like that made by cutting the air with a pliant rod. [Imit.]

whish[2] (hwish) *interj.* Hush! silence! Also **whisht** (hwisht). [Alter. of HUSH; infl. in form by WHISHT]

whisht (hwisht, hwist, wisht; *Scot.* hwusht) *Scot. v.t.* To hush. — *v.i.* To be silent. — *n.* The slightest sound; a whisper. — **to hold one's whisht** To be or remain silent.

whisk (hwisk) *v.t.* 1 To bear along or sweep with light movements, as of a small broom or a fan: often with *away* or *off*: to *whisk* flies away. 2 To cause to move with a quick sweeping motion. 3 To beat or mix with a quick movement, as eggs, cream, etc. — *v.i.* 4 To move quickly and lightly. — *n.* 1 A light stroke; a sudden, sweeping movement. 2 A little broom or brush. 3 A little bunch, as of straw, feathers, etc.; wisp. 4 A small culinary instrument for rapidly whipping (cream, etc.) to a froth. 5 A neckerchief of lawn or lace formerly worn by women. [Prob. <Scand. Cf. Dan. *viske* wipe, rub.]

whisk·broom (hwisk′bro͞om′, -bro͝om′) *n.* A small, short-handled broom for brushing clothing, etc.

whisk·er (hwis′kər) *n.* 1 *pl.* The hair that grows on the sides of that of a man's face, as distinguished from that on his lips, chin, and throat; loosely, the beard or any part of the beard; also, formerly, a mustache. 2 A hair from the whiskers or beard. 3 One of the long, bristly hairs on the sides of the mouth of some animals, as the cat, or a similar formation of bristles, as about the mouth of a bird; a vibrissa. 4 One who or that which whisks; formerly, a switch. 5 One of two small projecting spars or booms on the side of a bowsprit, to extend the jib or flying-jib guys: also **whisker boom.** —**whisk'ered, whisk'er·y** *adj.* —**whisk'er·less** *adj.*

whisk-grass (hwisk'gras', -gräs') *n.* Zacatón.

whis·ky (hwis'kē) *n. pl.* **·kies** 1 An alcoholic liquor obtained by the distillation of a fermented starchy compound, usually a grain. Whisky is often named (sometimes improperly) from the substance from which it is made, as **corn whisky, rye whisky,** etc.; or from the place or country of production, as **Bourbon whisky, Irish whisky,** etc. 2 A drink or portion of whisky. Compare USQUEBAUGH. —*adj.* Pertaining to or made of whisky. Also **whis'·key.** [Short for *usquebaugh* <Irish *uisgebeatha,* lit., water of life <*uisge* water + *beatha* life]

whis·ky-jack (hwis'kē·jak') *n.* The gray or Canada jay *(Perisoreus canadensis),* common in the northern forests of North America, about lumber camps, etc. [Alter. of earlier *whisky-john,* alter. of Algonquian (Cree) *wiskatjan*]

whisky sour An alcoholic drink made with whisky, lemon juice, and sugar.

whis·per (hwis'pər) *n.* 1 A low, soft, sibilant voice; articulated but not sonant breath; also, a low, rustling sound, as of waves or leaves. See VOICE. 2 A whispered utterance; secret communication; hint; insinuation. —*v.i.* 1 To speak in a whisper. 2 To talk cautiously or furtively; plot or gossip. 3 To make a low, rustling sound, as leaves. —*v.t.* 4 To utter in a whisper. 5 To speak to in a whisper. [OE *hwisprian*] —**whis'per·er** *n.* —**whis'per·ing** *n.* & *adj.* —**whis'per·ing·ly** *adv.* —**whis'·per·y** *adj.*

whist[1] (hwist) *n.* A game of cards played by four persons with a full pack of 52 cards, opposite players being partners: all the cards are played in each hand, the highest card of the suit led played in each of the 13 tricks, or a card of the trump suit, or the highest trump played, winning such trick. Every trick above the sixth counts one point. See CONTRACT BRIDGE under BRIDGE[2]. [Alter. of earlier *whisk*; ult. origin unknown]

whist[2] (hwist) *interj.* Hush! be still! —*adj.* Silent or quiet; mute. See also WHISHT. [Prob. imit.]

whis·tle (hwis'əl) *n.* 1 A device for producing a shrill, musical sound, operated on the principle of forcing a current of air, steam, or the like, through a pipe or tube of narrowed aperture or against a thin edge. 2 A musical sound, more or less shrill, made without the use of the vocal cords, by sending the breath through a small orifice formed by contracting the lips; also, the act of making this sound. 3 The sound produced by a whistle, or any sound suggestive of it, as the sound of wind rushing by an object, or of a flying missile, or the shrill cry of some birds. 4 A summons or call made by a whistle: The dog comes at his master's *whistle.* 5 *Slang* The mouth and throat: to wet one's *whistle.* 6 The short, loud cry of a male moose or elk. —*v.* **·tled, ·tling** *v.i.* 1 To make a sound or series of sounds like a whistle. 2 To cause a sharp, shrill sound by swift passage through the air, or by passage past an edge or through an orifice: The bullets *whistled* over our heads. 3 To blow or sound a whistle. —*v.t.* 4 To produce, as a tune or melody, by whistling. 5 To call, manage, or direct by whistling. 6 To send with a whistling sound. —**to whistle for** To go without; fail to get. [OE *hwistle* a shrill pipe]

whis·tler (hwis'lər) *n.* 1 One who or that which whistles. 2 A large gray marmot *(Marmota caligata)* of NW North America. 3 One of various birds, as the American goldeneye or the English widgeon: so called from the noise of their wings in flight. 4 A broken-winded horse; a roarer.

Whis·tler (hwis'lər), **James Abbott McNeill,** 1834–1903, U.S. artist and etcher. —**Whis·tle·ri·an** (hwis·lir'ē·ən) *adj.*

whistle stop *U.S. Colloq.* A small town, where a train stops only on signal. —**whis·tle-stop** (hwis'əl·stop') *adj.*

whistling swan See under SWAN.

whit (hwit) *n.* The smallest particle; speck: usually with a negative: not a *whit* abashed. See synonyms under PARTICLE. [Var. of WIGHT[1], as used in phrases *any wight, no wight,* OE *ænig wiht, nān wiht* a little amount]

Whit·by (hwit'bē) A port on the North Sea in the North Riding, NE York, England.

white (hwīt) *adj.* **whit·er, whit·est** 1 Having the color produced by reflection of all the rays of the solar spectrum, as from a finely powdered surface; having the color of new snow: opposed to *black.* 2 Light or comparatively light in color; specifically, light-colored as opposed to *red*: *white* wine. 3 Bloodless; ashen: *white* with rage. 4 Very fair; blond. 5 Silvery, hoary, or gray, as with age. 6 Covered with snow; snowy. 7 Made of silver; also unburnished, as silverwork. 8 Habited in white clothing: *white* nuns. 9 Not intentionally wicked or evil; not malicious or harmful: a *white* lie. 10 Figuratively, free from spot or stain; innocent: a *white* soul. 11 Incandescent; being at white heat. 12 Blank; unmarked by ink: said of a space in an advertisement or the like. 13 Having a light-colored skin; Caucasian: opposed especially to *Negro,* but often to the yellow, brown, or red races of men. 14 Of, pertaining to, or governed by the white race: *white* supremacy. 15 *Colloq.* Fair and honorable; straightforward; honest. 16 Propitious; auspicious: a rare meaning. 17 In certain European countries, constitutional; conservative, as a party; opposed to the radicals or revolutionaries. See synonyms under PALE[2]. —*n.* 1 That color seen when sunlight is reflected without sensible absorption of any of the visible rays of the spectrum; the color in the scale of grays which is entirely without hue and is the opposite of *black.* 2 The state or condition of being white; whiteness; figuratively, innocence; truth. 3 The white or light-colored part of something; specifically, the albumen of an egg, or the white part of the eyeball. 4 Anything that is white or nearly white, as cloth or garments; in the plural, a white uniform or outfit: The sailor wore his summer *whites.* 5 White wine. 6 A white paint or pigment; hence, by comparison, a color approaching pure white in its effect. 7 In chess or checkers, the white or light men, or the player who has them. 8 *pl.* Flour made from the finest and whitest part of the wheat. 9 *pl. Printing* Blank spaces in a picture, plate, mold, etc. 10 In archery, the outermost ring of a target; also, a hit on that ring, scoring one point. 11 A member of a fair-skinned race; especially, one of the Caucasian race as distinguished from a Negro, an Indian, a Chinese, etc. 12 In some European countries, a member of a party opposed to the radicals or revolutionaries; a conservative. 13 *pl. Pathol.* Leukorrhea. —*v.t.* **whit·ed, whit·ing** 1 To make white; whiten; bleach. 2 *Printing* To make or leave blank spaces in, as between lines or about an illustration: often with *out*: to *white* out a column. [OE *hwīt*]

White (hwīt), **Andrew,** 1832–1918, U.S. educator, historian, and diplomat. —**Gilbert,** 1720–93, English naturalist and antiquary. —**Peregrine,** 1620–1704, first child of English parentage born in New England. —**Stanford,** 1853–1906, U.S. architect. —**William Allen,** 1868–1944, U.S. editor and author.

white alkali 1 The product obtained from soda ash during the manufacture of carbonate of soda, dissolved in water, clarified, and freed from moisture by evaporation. 2 Pure soda ash.

white ant A small, whitish, isopterous insect, the termite, closely resembling the true ant in general appearance and social habits: it exists in tropical and warmer temperate regions, and does much damage to wooden structures, furniture, etc., by boring. For illustration see INSECTS (injurious).

white·bait (hwīt'bāt') *n.* 1 The young of various clupeoid fishes, especially of sprat and herring, netted in great quantities, especially at the mouth of the Thames, England, and served as a delicacy. 2 One of various species of silversides of fresh and salt waters of the United States.

white bear The polar bear.

white·beard (hwīt'bird') *n.* An old man with a white or gray beard.

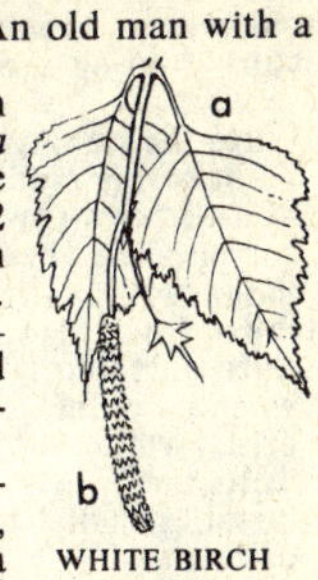

WHITE BIRCH
a. Leaf. *b.* Fruit.
(20 to 30 feet tall, rarely 40)

white birch 1 The North American birch *(Betula papyrifera)* with thin, white bark resembling paper. 2 The common European birch *(Betula pendula, B. pubescens),* having an ash-colored bark; also, a related Asian species *(B. platyphylla).*

white book In some European countries and in Japan, a formal report issued by a government on some special subject; in England an alternate of the bluebook: so called from the colors of the bookbinding.

white brant The snow goose.

white bryony A species of bryony *(Bryonia alba)* common in Europe.

white·cap (hwīt'kap') *n.* 1 A foam-crested wave. 2 One of several birds having white about the head.

White·cap (hwīt'kap') *n.* Formerly, in the Middle West and southward, one of a lawless, secret organization of men, who, under the pretense of regulating public morals, imposed lynch-law rule upon individuals who incurred their ill will: so named from their white caps or hoods.

white cedar 1 An evergreen tree *(Chamaecyparis thyoides)* of the cypress family, growing in moist places along the Atlantic coast. 2 Its soft, easily worked wood. 3 The arborvitae.

White·chap·el (hwīt'chap·əl) A district in Stepney borough, eastern London, England; the older Jewish quarter.

white clover A common variety *(Trifolium repens)* of clover, with white flowers.

white coal Water considered as a source of power.

white-col·lar (hwīt'kol'ər) *adj.* Pertaining to or designating salaried workers in occupations which demand a well-dressed appearance.

white comb A contagious disease of poultry, caused by a fungus *(Lophophyton gallinae),* and marked by the formation of grayish patches on the comb and a breaking off of the feathers.

white crane The whooping crane *(Grus americana)* of North America, which is pure white when adult.

white curlew The white ibis *(Guara alba)* of the southern United States.

whit·ed sepulcher (hwī'tid) A hypocrite; a person with a pleasing outward aspect, but corrupted thoughts. *Matt* xxiii 27.

white elephant 1 A rare pale-gray variety of Asian elephant held sacred by the Burmese and Siamese. 2 Anything rare, expensive, and difficult to keep; a burdensome possession.

white-eye (hwīt'ī') *n.* 1 The white-eyed vireo *(Vireo griseus)* of North America. 2 Any of numerous small singing birds *(Zosterops* and related genera), mostly of the Old World tropics: named from the circle of white feathers around the eye.

white-eyed (hwīt'īd') *adj.* Having the iris of the eye white or colorless, as an albino.

white-faced (hwīt'fāst') *adj.* 1 Pallid in countenance; pale. 2 Having a white mark or spot on the face or front of the head: the *white-faced* hornet. 3 Having a white facing or exposed surface, as a skirt.

white feather A mark of cowardice, full-blooded gamecocks being said to have no white feathers.

White·field (hwīt'fēld), **George,** 1714–70, English preacher; one of the founders of Methodism.

white·fish (hwīt'fish') *n. pl.* **·fish** or **·fish·es** 1 A salmonoid food fish (genus *Coregonus*) of North America, living mostly in lakes and having teeth minute or absent. 2 One of various other species of fish, as the menhaden, the European whiting, or the silver salmon *(Oncorhynchus kisutch).* 3 A tropical marine food fish of California *(Caulolatilus princeps).* 4 The young of the bluefish. 5 The beluga.

white flag 1 A flag of truce. 2 A signal of surrender when hoisted over a fortified position or a body of men.

add,āce,câre,pälm; end,ēven; it,īce; odd,ōpen,ôrder; to͝ok,po͞ol; up,bûrn; ə = a in *above,* e in *sicken,* i in *clarity,* o in *melon,* u in *focus*; yo͞o = u in *fuse*; oi,oil; ou,pout; ch,check; g,go; ng,ring; th,thin; th,this; zh,vision. Foreign sounds à,œ,ü,kh,ṅ; and ◆: see page xx. < from; + plus; ? possibly.

white flax Gold-of-pleasure.
white-foot·ed mouse (hwīt'foot'id) The deer mouse.
White Friar A Carmelite: so called from the color of his cloak.
White·fri·ars (hwīt'frī'ərz) The neighborhood surrounding the site of a former Carmelite monastery in Fleet Street, London.
white frost Hoar frost.
white gerfalcon The gerfalcon in the phase when its plumage is of a conspicuous, highly prized white color.
white gold An alloy of gold with a white metal, usually nickel and zinc, sometimes palladium.
white·gum (hwīt'gum') n. 1 An Australian eucalyptus with a white bark. 2 The American sweetgum.
White·hall (hwīt'hôl) 1 A former royal palace near Westminster Abbey. 2 A street in Westminster, London, where a number of government offices are located. 3 The British government.
White·head (hwīt'hed), **Alfred North**, 1861–1947, English mathematician and philosopher, active in the United States.
white-head·ed (hwīt'hed'id) adj. 1 Having white hair, feathers, etc., on the head; also, very blond; flaxen-haired. 2 Best-loved; favorite: an Irishism.
white heat 1 The temperature at which a body becomes incandescent. 2 A condition of extreme anger or emotional strain.
White·horse (hwīt'hôrs) The capital of Yukon Territory on the upper Yukon River. Also **White Horse**.
white-hors·es (hwīt'hôr'siz) n. pl. Foam-crested waves; white caps (def. 1).
white-hot (hwīt'hot') adj. 1 Exhibiting the condition of white heat. 2 Colloq. Extremely angry.
White House, The 1 The official residence of the president of the United States, at Washington, D.C.: a white building in American colonial style, officially called the *Executive Mansion*. 2 The executive branch of the United States government.
white lead 1 A poisonous white pigment composed of lead carbonate and hydrated lead oxide and prepared by several processes: also called *ceruse*. 2 Native carbonate of lead; cerusite. See LEAD.
white leather Whitleather.
white lie See under LIE.
white-liv·ered (hwīt'liv'ərd) adj. 1 Having a pale and feeble look. 2 Base; cowardly; envious.
white lupine A white-flowered variety (*Lupinus albus*) of lupine, grown in Europe for forage.
white·ly (hwīt'lē) adv. With a pale appearance; so as to look white.
white mahogany Primavera.
white man's burden The alleged duty of the white peoples to spread culture among the so-called backward peoples of the world: phrase originated by Rudyard Kipling.
white maple Any of certain maples having a whitish bark, as the silver maple (*Acer saccharinum*) and red maple (*A. rubrum*), both of North America.
white matter *Anat*. That portion of the brain and spinal cord that is composed mainly of medullated nerve fibers, giving it a white appearance: contrasted with *gray matter*.
white meat 1 The light-colored meat or flesh of animals, as veal or the breast of turkey. 2 *Obs*. Food made from milk, butter, cheese, eggs, and other animal products.
white metal See under METAL.
White Mountains 1 A range of the Appalachians in north central New Hampshire; highest peak, 6,288 feet. 2 A range of mountains in eastern Arizona; highest point, 11,590 feet. 3 A range of mountains in eastern California and SW Nevada; highest point, 14,242 feet.
whit·en (hwīt'n) v.t. & v.i. To make or become white; blanch; bleach. See synonyms under BLEACH. — **whit'en·er** n.
white·ness (hwīt'nis) n. 1 The state of being white; freedom from stains or darkness of surface. 2 Pallor from emotion or from illness. 3 Cleanness or pureness of heart; innocence.
White Nile See NILE.

white oak 1 A North American oak (*Quercus alba*) of the eastern United States, with long leaves having from five to nine entire, rounded lobes. 2 Either of two related species, the **swamp white oak** (*Q. bicolor*) and the **Oregon white oak** (*Q. garryana*). 3 The British oak (*Q. petraea*). 4 The wood of any species of white oak.

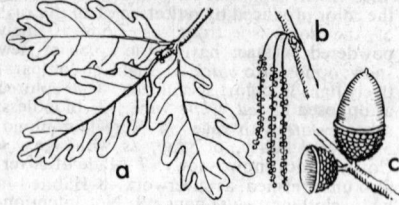

WHITE OAK
a. Leaf. *b*. Blossom. *c*. Acorn.

white of egg Egg white.
white·out (hwīt'out') n. *Meteorol*. An atmospheric condition in arctic regions in which a blending of clouds and snow cover produces a uniform milky whiteness characterized by the absence of shadow and the invisibility of all but very dark objects.
white paper A government publication on some subject of less importance than that treated in a white book or a bluebook. See WHITE BOOK, BLUEBOOK.
White Pass A pass in the Coast Mountains, on the border between SE Alaska and NW British Columbia; elevation 2,888 feet.
white pepper See under PEPPER.
white perch A small food fish (*Morone americana*) related to the sea basses, found in Atlantic coastal waters and sometimes landlocked in streams of the United States.
white pine 1 A pine (*Pinus strobus*) widely distributed in eastern North America, with soft, bluish-green leaves in clusters of five. The cone and tassel of this tree are the State emblem of Maine. 2 The light, soft wood of this tree. 3 Any of several varieties of this pine.
white-pine weevil (hwīt'pīn') A weevil (*Pissodes strobi*) of NE North America which feeds on the leading shoots of white pine and other conifers. For illustration see INSECTS (injurious).
white plague *Pathol*. Tuberculosis, especially of the lungs.
white poplar 1 A large, rapidly growing Old World tree (*Populus alba*), often planted in the United States for shade or for its ornamental leaves, which are green above and clothed with a silvery-white down beneath; the silver poplar. 2 The aspen.
white potato The common potato.
white rabbit The varying hare.
white race The Caucasoid ethnic division of mankind.
white rat 1 Any albino rat. 2 One of a special breed of albino Norway rats much used in biological and medical experimentation.
White River 1 A river in northern and eastern Arkansas and SW Missouri, flowing 690 miles to the Mississippi. 2 A river in Nebraska and southern South Dakota, flowing 507 miles NE to the Missouri.
White Russian See under RUSSIAN.
White Russian S.S.R. See BELORUSSIAN SOVIET SOCIALIST REPUBLIC. Also **White Russia**.
White Sands National Monument A government reservation in southern New Mexico; 219 square miles; established 1933.
white sapphire A variety of translucent, colorless corundum.
White Sea An inlet of the Barents Sea in NW European U.S.S.R.; 36,680 square miles. *Russian* **Be·lo·e Mo·re** (bye'lə·yə môr'yə).
white slave A girl forced into or held in prostitution. — **white-slave** (hwīt'slāv') adj.
White-slave Act The Mann Act.
white-slav·er (hwīt'slā'vər) n. One who procures for or engages in white-slavery.
white-slav·er·y (hwīt'slā'vər·ē) n. The business or practice of forced prostitution.
white·smith (hwīt'smith') n. 1 A worker in white metals, as a tinsmith or silversmith.

2 A finisher, polisher, or galvanizer of iron. Compare BLACKSMITH.
white spruce A spruce (*Picus glauca*) of Canada and the northern United States.
white squall *Meteorol*. A small whirlwind occurring in the tropics, having no accompanying cloud but a white patch above the storm center and often making ocean waters foam-white.
white·tail (hwīt'tāl) n. 1 The white-tailed deer. 2 The wheatear.
white-tailed deer (hwīt'tāld') The common North American deer (*Odocoileus virginianus*), having a moderately long tail white on the underside: also called *Virginia deer*.
white·throat (hwīt'thrōt') n. One of various Old World warblers, especially the common or **greater whitethroat** (*Sylvia cinerea*), with gray head, white throat, and rufous wings.
white-throat·ed sparrow (hwīt'thrō'tid) A common North American sparrow (*Zonotrichia albicollis*), with a prominent white patch on the throat.
white tie 1 A white bow tie, worn with a swallowtail coat. 2 A swallowtail coat and its correct accessories: the phrase is used on invitations, etc., to indicate formal attire will be worn.
white trash Poor whites.
white turnip The common turnip (*Brassica rapa*).
white vitriol Hydrated zinc sulfate, $ZnSO_4 \cdot 7H_2O$, widely used in medicine as an emetic, astringent, and antiseptic.
white·wash (hwīt'wosh', -wôsh') n. 1 A mixture of slaked lime and water, sometimes with salt, whiting, and glue added, used for whitening walls, etc. 2 A toilet preparation for whitening the skin. 3 Figuratively, a report falsely ascribing virtues, suppressing adverse evidence, etc. 4 A failure to score in a game. — v.t. 1 To coat with whitewash. 2 To gloss over; hide. 3 *Colloq*. In sports, to defeat without allowing the losing side to score. See synonyms under BLEACH. — **white'wash·er** n.
white wax Paraffin.
white·weed (hwīt'wēd') n. The oxeye daisy.
white whale The beluga.
white-wing (hwīt'wing') n. 1 One of the members of the Department of Sanitation of New York City: so called because they formerly wore white uniforms. 2 Any person who wears a white uniform. 3 The surf duck.
white-winged dove (hwīt'wingd') A dove (*Melopelia asiatica*) of the SW United States with a rounded, white-banded tail and a conspicuous white patch on the wings.
white·wood (hwīt'wood') n. 1 Any of various trees yielding a whitish timber, as the basswood, the tuliptree, the cottonwood, the wild cinnamon, etc. 2 The wood of these trees.
whith·er (hwith'ər) adv. 1 As a relative, to which or what: approaching a conjunctive use: the village *whither* we went. 2 As an interrogative, to which or to what place. 3 Wheresoever; whithersoever. 4 To what degree or extent. [OE *hwider*]
whith·er·so·ev·er (hwith'ər·sō·ev'ər) adv. To whatever place.
whit·ing[1] (hwī'ting) n. A pure white chalk, levigated and washed for use in making putty and whitewash, as a pigment, and for polishing.
whit·ing[2] (hwī'ting) n. 1 A small European gadoid food fish (*Merlangus merlangus*) without a barbel. 2 The hake (def. 1). 3 Any of several silvery sciaenoid fishes (genus *Menticirrhus*), especially the **Carolina whiting** (*M. americanus*), common on the coast of the southern United States. 4 The menhaden. [<MDu. *wijting* < *wit* white]
whit·ish (hwī'tish) adj. Somewhat white or, especially, very light gray. — **whit'ish·ness** n.
whit·leath·er (hwīt'leth'ər) n. Leather tawed with alum to render it pliable; white leather. [<WHITE + LEATHER]
whit·low (hwīt'lō) n. 1 *Pathol*. **a** An inflammatory tumor, especially on the terminal phalanx of a finger, seated between the epidermis and true skin; a run-round. **b** A small tumor within the sheath of a tendon, or between the bone and its enveloping membrane; a felon. 2 A disease of horses' feet. [ME *whitflaw*, appar. <WHITE + FLAW[1]]
Whit·man (hwīt'mən), **Marcus**, 1802–47, U.S. missionary massacred by Indians in Oregon. — **Walt**, 1819–92, U.S. poet.

For pronunciation of WH- *see discussion under* W.

Whit-Mon·day (hwit′mun′dē, -dā) *n.* The Monday next following Whitsunday: observed in England as a holiday. Also **Whit′mon′day, Whit′sun-Mon′day.** [On analogy with WHITSUNDAY]

Whit·ney (hwit′nē), **Eli,** 1765–1825, American inventor. — **Gertrude,** 1877?–1942, *née* Vanderbilt, U.S. sculptress. — **Josiah Dwight,** 1819–96, U.S. geologist. — **William Dwight,** 1827–94, U.S. philologist; brother of Josiah Dwight.

Whit·ney (hwit′nē), **Mount** A peak of the southern Sierra Nevada Range in eastern California; 14,496 feet; highest point in the United States.

whit·rack (hwit′rak) *n. Dial.* & *Scot.* A weasel. [ME *whitratt* <WHITE + RAT]

Whit·sun (hwit′sən) *n.* Whitsunday: frequently used in composition: *Whitsun*-ale, *Whitsun*-week. [ME *witsonen, whitsone* < *whitsondei* WHITSUNDAY]

Whit·sun·day (hwit′sun′dē, -dā) *n.* The seventh Sunday after Easter: a church festival commemorating Pentecost. [OE *Hwīta Sunnandæg,* lit., white Sunday; so called from the white robes worn by recently baptized persons on that day]

Whit·sun-week (hwit′sən·wēk′) *n.* The week that begins with Whitsunday. Also **Whit′·sun·tide′** (-tīd′).

Whit·ta·ker (hwit′ə·kər), **Samuel Estes,** born 1886, U.S. jurist; associate justice of the U.S. Supreme Court 1957–.

whit·ter (hwit′ər) *n. Scot.* **1** A copious draft of liquor, etc. **2** Anything weak. **3** Chatter; loquacity. **4** A token; sign.

Whit·ti·er (hwit′ē·ər) A port of southern Alaska on Prince William Sound.

Whit·ti·er (hwit′ē·ər), **John Greenleaf,** 1807–1892, U.S. poet.

Whit·ting·ton (hwit′ing·tən), **Richard,** 1358?–1423, English tradesman; lord mayor of London, 1397, 1406, and 1419.

whit·tle[1] (hwit′l) *v.* **·tled, ·tling** *v.t.* **1** To cut or shave bits from (wood, a stick, etc.). **2** To make or shape by carving or whittling. **3** To reduce or wear away by paring a little at a time: with *down, off, away,* etc.: to *whittle* down costs. — *v.i.* **4** To whittle wood, usually as an aimless diversion. See synonyms under CUT. [<*n.*] — *n. Dial.* & *Scot.* A knife; especially, a sheath knife worn at the belt, or any large knife. [Alter. of ME *thwitel* <OE *thwitan* cut] — **whit′tler** *n.*

whit·tle[2] (hwit′l) *n. Dial.* **1** A blanket. **2** A shaggy mantle formerly worn by country-women. Also **whittle shawl.** [OE *hwītel* < *hwīt* white]

whit·tlings (hwit′lingz) *n. pl.* The fine chips and shavings made with a whittle or by a whittler.

Whit-Tues·day (hwit′to͞oz′dē, -dā, -tyo͞oz′-) *n.* The day after Whit-Monday. Also **Whit′sun-Tues′day.** [On analogy with WHITSUNDAY]

whiz (hwiz) *v.* **whizzed, whiz·zing** *v.i.* **1** To make a hissing and humming sound while passing rapidly through the air. **2** To move or pass with such a sound. — *v.t.* **3** To cause to whiz. — *n.* **1** A sibilant sound with some sonant character, such as is produced by a missile passing through the air. **2** *Slang* Any person or thing of extraordinary excellence or ability. **3** *Slang* A bargain. **4** *Slang* A celebration; a spree. Also **whizz.** [Imit.]

whiz-bang (hwiz′bang′) *n. Slang* A high-explosive shell; also, a firecracker that explodes with a loud noise. Also **whizz′-bang′.**

who (ho͞o) *pron. possessive case* **whose;** *objective case* **whom** **1** As an interrogative, which or what person or persons. **2** As a relative, that: pointing out or fixing upon a particular person or persons, and identifying the subject or object in a relative clause with that of the principal clause. **3** As a compound relative, he, she, or they that: *Who* steals my purse steals trash. — **as who should say** As one who should say; as if one should say. [OE *hwa, hwā*]

◆ In modern usage, *who* as a relative is applied only to persons, *which* only to animals or to inanimate objects, *that* to persons or things indifferently. *Whose* is correctly used as the possessive of *which,* as well as of *who,* especially where the phrase *of which* would seem awkward: the man *whose* house was sold; a peak *whose* (*of which* the) summit seeks the sky. The use of *whom* as an interrogative pronoun in initial position, as in *Whom* did you see?, is supported by some grammarians, but the more natural *Who* did you see? *Who* did you give the book to? are in wider use and are now considered acceptable. However, when used after a verb or preposition, *whom* is still required, as in To *whom* did you give it? You saw *whom*? See also usage note under THAT (pronoun).

whoa (hwō) *interj.* Stop! stand still! [Var. of HO]

who·dun·it (ho͞o·dun′it) *n. Colloq.* A type of mystery fiction or dramatic production which challenges the reader or auditor to detect the perpetrator of a crime. [<WHO + DONE + IT; coined by Donald Gordon in 1930 in *American News of Books*]

who·ev·er (ho͞o·ev′ər) *pron.* Any one without exception; any person who.

whole (hōl) *adj.* **1** Containing all the parts necessary to make up a total; undivided and undiminished; entire; complete. **2** Having all the essential or original parts in their proper constitution; unbroken and uninjured; sound; intact. **3** Specifically, in or having regained sound health; hale. **4** Having the same parents; full, as opposed to *half-*: a *whole* brother. **5** *Colloq.* Each one of (something); all: He ate the *whole* batch of cookies. **6** *Math.* Integral. — **on the whole** Taking one thing with another. — **out of whole cloth** Fabricated; made up, without foundation in truth or fact, as a story or lie. — *n.* **1** All the parts or elements entering into and making up a thing. **2** An organization of parts making a unity or system; an organism. See synonyms under AGGREGATE, MASS[1]. ◆ Homophone: *hole.* [OE (Northumbrian) *hol,* var. of *hāl.* Related to HALE[2].]

whole blood **1** Full blood. **2** Blood as taken direct from the body, especially that used in transfusions.

whole brother See under BROTHER.

whole gale *Meteorol.* A gale of force 10 on the Beaufort scale.

whole·heart·ed (hōl′här′tid) *adj.* Done or experienced with all earnestness; characteristically sincere, sound, generous, or kind. — **whole′heart′ed·ly** *adv.* — **whole′heart′ed·ness** *n.*

whole-hog (hōl′hôg′, -hog′) *adj. Colloq.* Thoroughgoing.

whole hog *Colloq.* **1** The whole of anything: to believe in the *whole hog,* accept the *whole hog.* **2** Reliance or approval; trust: We don't put the *whole hog* on them. — **to go the whole hog** *Colloq.* To do something thoroughly; become involved without reservation.

whole milk Milk containing all its constituents: distinguished from *skim milk.*

whole·ness (hōl′nis) *n.* Entireness; completeness.

whole note *Music* A semibreve. See NOTE *n.* (def. 12).

whole number *Math.* A unit or a number composed of units; an integral number or integer: distinguished from *fraction* and *mixed number.*

whole·sale (hōl′sāl′) *n.* The sale of goods by the piece or in large bulk or quantity: opposed to *retail.* — *adj.* **1** Selling in quantity, not at retail: a *wholesale* druggist. **2** Done in buying and selling in quantity: the *wholesale* trade. **3** Pertaining to wholesale trade: the *wholesale* price. **4** Hence, made or done on a large scale; made or done indiscriminately: *wholesale* murder. — *adv.* In bulk or quantity; hence, indiscriminately: to berate the medical profession *wholesale.* — *v.t.* & *v.i.* **·saled, ·sal·ing** To sell at wholesale. [ME *holesale* < *by hole sale* in large quantities] — **whole′sal′er** *n.*

whole sister See under SISTER.

whole snipe See under SNIPE.

whole·some (hōl′səm) *adj.* **1** Tending to promote health; salubrious; healthful: *wholesome* air or food. **2** Favorable to virtue and well-being; salutary; sound; beneficial. **3** Healthy; physically, mentally, and morally sound: a *wholesome* girl. **4** Indicative or characteristic of health: *wholesome* red cheeks. **5** Safe; free from danger or risk: This is not a *wholesome* situation. **6** *Obs.* Auspicious; favorable. See synonyms under HEALTHY. [ME *holsum* < *hol* WHOLE + OE *-sum* -SOME[1]] — **whole′some·ly** *adv.* — **whole′some·ness** *n.*

whole-souled (hōl′sōld′) *adj.* Feeling or acting with one's whole heart; devoted; generous.

whole-wheat (hōl′hwēt′) *adj.* Made from wheat grain and bran.

who'll (ho͞ol) Who will; who shall: a contraction.

whol·ly (hō′lē, hōl′lē) *adv.* **1** Completely; totally. **2** Exclusively; only.

whom (ho͞om) *pron.* The objective case of WHO. [OE *hwam,* dative of *hwā* who]

whom·ev·er (ho͞om′ev′ər), **whom·so** (ho͞om′·sō′), **whom·so·ev·er** (ho͞om′sō·ev′ər) Objective cases of WHOEVER, WHOSO, etc.

whoop (ho͞op, hwo͞op, hwo͝op) *v.i.* **1** To utter loud cries, as of excitement, rage, or exultation. **2** To hoot, as an owl. **3** To make a loud, gasping inspiration, as after a paroxysm of coughing. — *v.t.* **4** To utter with a whoop or whoops. **5** To call, urge, chase, etc., with whoops; hoot. — **to whoop up** **1** To arouse enthusiasm in or for; ballyhoo. **2** To raise, as a price, or sum of money. — **to whoop it** (or **things**) **up** **1** *Slang* To make noisy revelry. **2** To arouse enthusiasm. — *n.* **1** A shout of excitement, encouragement, or exultation; also, a hoot of derision. **2** A signal halloo or a guiding call, as to incite dogs or men in the chase. **3** A loud, convulsive inspiration after a paroxysm of coughing in whooping cough; a sonorous indrawing of breath. **4** An owl's hoot. — *interj.* Hurrah! halloo! [Imit.]

whoop·ee (hwo͞o′pē, hwo͝op′ē) *Slang interj.* An exclamation of joy, excitement, etc. — *n.* A hilarious, festive time. — **to make whoopee** To have a noisy, festive time. [<WHOOP]

whoop·er (ho͞o′pər, hwo͞o′pər, hwo͝op′ər) *n.* **1** One who or that which whoops. **2** A large Old World swan *(Cygnus cygnus).* **3** The white crane: so called from its loud cry.

whoop·ing cough (ho͞o′ping, ho͝op′ing) *Pathol.* A contagious respiratory disease of bacterial origin chiefly affecting children, marked in its final stage by recurrent paroxysms of violent coughing, ending with a whoop; pertussis.

whooping crane See under CRANE.

whop (hwop) *Colloq. n.* A blow or fall, or the resulting noise. — *v.* **whopped, whop·ping** *v.t.* **1** To strike or beat. **2** To defeat convincingly. — *v.i.* **3** To drop or fall suddenly; flop. Also spelled *whap.* [Var. of WAP[1]]

whop·per (hwop′ər) *n. Colloq.* **1** One who whops. **2** Something large or remarkable, especially a surprising falsehood. Also spelled *whapper.*

whop·ping (hwop′ing) *adj.* Unusually large; excessively exaggerated.

whore (hôr, hōr) *n.* A prostitute. — *v.* **whored, whor·ing** *v.i.* **1** To have illicit sexual intercourse, especially with a prostitute. **2** To be a whore. — *v.t.* **3** To make a whore of; corrupt; debauch. ◆ Homophone: *hoar.* [OE *hōre,* prob. <ON *hōra*]

whore·dom (hôr′dəm, hōr′-) *n.* **1** The practice of illicit sexual intercourse. **2** Whores collectively. **3** In the Bible, idolatry. [Prob. <ON *hōrdōmr*]

whore·house (hôr′hous′, hōr′-) *n.* A house of prostitution.

whore·mas·ter (hôr′mas′tər, -mäs′-, hōr′-) *n.* **1** A procurer; pander. **2** A whoremonger.

whore·mon·ger (hôr′mung′gər, -mong′-, hōr′-) *n.* **1** A man who has intercourse with whores. **2** A pander.

whore·son (hôr′sən, hōr′-) *n. Obs.* The son of a whore: commonly, a term of contempt. [ME *hores son,* trans. of AF *fiz a putain*]

whor·ish (hôr′ish, hōr′ish) *adj.* Addicted to unlawful sexual indulgences; unchaste; lewd. — **whor′ish·ly** *adv.* — **whor′ish·ness** *n.*

whorl (hwûrl, hwôrl) *n.* **1** The flywheel of a spindle; wherve. **2** *Bot.* A set of leaves, etc., on the same plane with one another, distributed in a circle; a verticil. **3** *Zool.* A turn or volution, as of a spiral shell. **4** Any of the convoluted ridges of a fingerprint. [ME *wharwyl,*

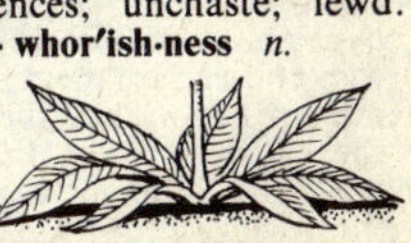

WHORL *(def. 2)*

add,āce,câre,pälm; end, ēven; it,īce; odd,ōpen,ôrder; to͝ok,po͞ol; up,bûrn; ə = a in *above,* e in *sicken,* i in *clarity,* o in *melon,* u in *focus;* yo͞o = u in *fuse;* oi,oil; ou,pout; ch,check; g,go; ng,ring; th,thin; th,this; zh,vision. Foreign sounds à,œ,ü,kh,ṅ; and ◆: see page xx. < from; + plus; ? possibly.

whorwhil, appar. vars. of WHIRL; infl. in form by *wharve*]
whorled (hwûrld, hwôrld) *adj.* Furnished with or arranged in whorls.
whort (hwûrt) *n.* The whortleberry, or its fruit. Also **whor·tle** (hwûr'təl). [OE *horta* a whortleberry]
whor·tle·ber·ry (hwûr'təl·ber'ē) *n. pl.* **·ries** 1 A European variety of blueberry (*Vaccinium myrtillus*); the bilberry. 2 Its blue-black fruit. 3 The huckleberry. [Dial. var. of HURTLEBERRY]
whose (hōōz) The possessive case of WHO and often of WHICH. See under WHO. [OE *hwæs*, genitive of *hwā* who]
whose·so·ev·er (hōōz'sō·ev'ər) Possessive case of WHOSOEVER.
who·so (hōō'sō) *pron.* Whoever; any person who. [Reduced form of OE *swā hwā swā*, generalized form of *hwā* who]
who·so·ev·er (hōō'sō·ev'ər) *pron.* Any person whatever; who; whoever.
why (hwī) *adv.* 1 For what cause, purpose, or reason; wherefore: used interrogatively: *Why* did you go? 2 Because of which; for which; the reason or cause for which: used relatively: I don't know *why* he went; I know no reason *why* he went. — *n. pl.* **whys** 1 An explanatory cause; reason; cause. 2 A puzzling problem; riddle; enigma. — *interj.* An introductory expletive, sometimes denoting surprise. [OE *hwī, hwȳ*, instrumental case of *hwæt* what]
whyd·ah (hwid'ə), **whydah bird** See WHIDAH BIRD.
wi (wi) *prep. Scot.* With.
wich (wich) See WITCH².
Wich·i·ta (wich'ə·tô) *n.* A member of a North American Indian confederacy of Caddoan linguistic stock, formerly inhabiting Oklahoma and Texas.
Wich·i·ta (wich'ə·tô) A city on the Arkansas River in south central Kansas; a center of the food and oil industries.
Wichita River A river in Texas, flowing 250 miles NE to the Red River.
wick¹ (wik) *n.* A band of loosely twisted or woven fibers, as in a candle or lamp, acting by capillary attraction to convey oil or other illuminant to a flame. [OE *wēoca*] — **wick'·ing** *n.*
wick² (wik) *Scot. v.t.* In curling, to strike (a stone) obliquely. — *n.* 1 In curling, an opening surrounded by stones already played. 2 A creek; inlet.
wick³ (wik) *n.* A village or town: now mostly in composition, often as **-wich**: *Woolwich*. [OE *wīc*, appar. <L *vicus*.]
wick·ed (wik'id) *adj.* 1 Evil in principle and practice; vicious; sinful; depraved. 2 Mischievous; roguish. 3 Noxious; pernicious. 4 Troublesome; painful. See synonyms under BAD, CRIMINAL, IMMORAL, INFAMOUS, PROFANE, SINFUL. [ME <*wikke, wicke*, appar. <OE *wicca* a wizard] — **wick'ed·ly** *adv.*
wick·ed·ness (wik'id·nis) *n.* 1 The quality of being wicked; moral depravity; sin; vice; crime: opposed to *goodness*. 2 A wicked thing or act; wicked conduct: to work *wickedness*.
wick·er (wik'ər) *adj.* Made of twigs, osiers, etc. — *n.* 1 A pliant young shoot or rod; twig; osier. 2 Ware made of such shoots. [Prob. <Scand. Cf. dial. Sw. *viker* <*vika* bend.]
wick·er·work (wik'ər·wûrk') *n.* A fabric or texture, as a basket, made of woven twigs, osiers, etc.; basketwork.
wick·et (wik'it) *n.* 1 A small door or gate subsidiary to or made within a larger entrance. 2 A small opening in a door. 3 A small sluicegate in a canal lock or at the end of a millrace. 4 In cricket, an arrangement of three upright rods called *stumps* set near together, with two crosspieces called *bails* laid over the top; also, the place at which the wicket is set up; the right or turn of each batsman at the wicket; the playing pitch between the wickets: a fast *wicket*; an inning that is not finished or not begun: The eleven won by three *wickets*, that is, the two men at bat and one yet to go in. 5 In croquet, an arch, usually of wire. [<AF *wiket*, OF *guichet*, prob. <Gmc.]
wick·et-keep·er (wik'it·kē'pər) *n.* In cricket, the fielder stationed immediately behind the wicket which is being bowled at.

wick·i·up (wik'ē·up) *n.* A loosely constructed hut of certain North American Indian tribes: distinguished from *tepee* or *wigwam*: also spelled *wikiup*. [<Algonquian. Cf. Sac and Fox *wikiyap* a lodge.]
Wick·liffe (wik'lif), **Wic·lif**, **Wic·liff·ite**, etc. See WYCLIF, etc.
Wick·low (wik'lō) A maritime county of eastern Leinster province, Ireland; 782 square miles; county town, Wicklow.
wic·o·py (wik'ə·pē) *n. pl.* **·pies** 1 The leatherwood. 2 The basswood. 3 Any of several species of willow herb. [<Algonquian. Cf. Cree *wikupiy*.]
wid·der·shins (wid'ər·shinz) See WITHERSHINS.
wid·dle (wid'l) *v.t. & n. Dial. & Scot.* Wriggle; struggle; waddle.
wid·dy¹ (wid'ē) *n. pl.* **·dies** *Scot.* A halter of withes; withy; hangman's noose; hence, the gallows. Also **wid'die**.
wid·dy² (wid'ē) *n. pl.* **·dies** *Dial.* Widow. [Var. of WITHY]
wide (wīd) *adj.* **wid·er, wid·est** 1 Having relatively great extent between sides; broad, as opposed to *narrow*. 2 Extended far in every direction; ample; spacious: a *wide* expanse. 3 Having a specified degree of width or breadth: an inch *wide*. 4 Distant from the desired or proper point by a great extent of space; remote; wild: *wide* of the mark. 5 Figuratively, having intellectual breadth; considering questions from all points of view; liberal: a man of *wide* views. 6 Fully open; expanded or extended: *wide* eyes. 7 *Phonet.* Lax. 8 Comprehensive; inclusive: *wide* learning. 9 Loose; ample; roomy: *wide* breeches. 10 In the stock exchange, exhibiting a considerable range between high and low, or bid and offered prices: a *wide* opening. See synonyms under LARGE. — *n.* 1 In cricket, a ball bowled too far over or on either side of the wicket to be within the batsman's reach. 2 Breadth of extent; also, a broad, open space. — *adv.* 1 To a great distance; extensively. 2 Far from the mark. 3 To the greatest extent; fully open. [OE *wīd*] — **wide'ly** *adv.* — **wide'ness** *n.*
♦ Various self-explaining compounds have *wide* as their first element: **wide'-arched'**, **wide'-branched'**, **wide'-brimmed'**, etc.
wide-angle lens (wīd'ang'gəl) *Phot.* A type of camera lens designed and ground to permit an angle of view wider than that of the ordinary lens, or more than 50 degrees.
wide-a·wake (wīd'ə·wāk') *adj.* Marked by vigilance and alertness; keen. See synonyms under ALERT, VIGILANT. — *n.* A soft, broad-brimmed felt hat: also **wide'-a·wake'** hat.
wide-eyed (wīd'īd') *adj.* With the eyes wide open, as if gazing intently in wonder or surprise.
wid·en (wīd'n) *v.t. & v.i.* To make or become wide or wider. See synonyms under AMPLIFY. — **wid'en·er** *n.*
Wide·ner (wīd'nər), **Peter Arrell Brown**, 1834-1915, U.S. businessman and philanthropist.
wide-o·pen (wīd'ō'pən) *adj.* 1 Opened wide: The gates are *wide-open*. 2 *Colloq.* Remiss in the enforcement of laws which regulate various forms of vice, as gambling, prostitution, etc.: a *wide-open* city.
wide·spread (wīd'spred') *adj.* Extending over a large space or territory; general: a *widespread* belief. Also **wide'spread'**.
widge·on (wij'ən) *n.* Any of a genus (*Mareca*) of river ducks with short bill and wedge-shaped tail; especially, the **American widgeon**, or **baldpate** (*M. americana*), esteemed as a game bird: also spelled *wigeon*. [Cf. MF *vigeon* a wild duck]
Wi·dor (vē·dôr'), **Charles Marie**, 1845-1937, French organist and composer.
wid·ow (wid'ō) *n.* 1 A woman who has lost her husband by death and has not remarried. 2 In some card games, an additional hand dealt to the table; also, a kitty. 3 *Printing* An incomplete line of type ending a paragraph; especially, a single line or less at the top of a page or column. — *v.t.* 1 To make a widow of; deprive of a husband: usually in the past participle: a woman *widowed* by war. 2 To deprive of something desirable; bereave. 3 *Rare* To survive as the widow of. 4 *Rare* To recognize as a widow; give the rights of

a widow to. — *adj.* Widowed. [OE *widewe, wuduwe*]
widow bird A whidah bird. [<NL *Vidua*, genus name, trans. of Pg. *viuva*, lit., a widow]
wid·ow·er (wid'ō·ər) *n.* A man whose wife is dead, and who has not married again. [ME *widwer* <*widwe*, OE *widewe* a widow]
wid·ow·hood (wid'ō·hood) *n.* The state or condition of being a widow, or, rarely, of being a widower; also, the period during which one is a widow.
widow's cruse An endless or inexhaustible supply: in allusion to the stories in I *Kings* xvii 10-16, and II *Kings* iv 1-7.
widow's mite See MITE².
widow's peak See PEAK¹.
width (width) *n.* 1 Dimension or measurement of an object taken from side to side, or at right angles to the length. 2 Wideness; the state or fact of being wide. 3 Something that has width; specifically, in dressmaking, one of the several pieces of material used in making a garment. [<WIDE, on analogy with *breadth*]
width·wise (width'wīz') *adv.* In the direction of the width; from side to side. Also **width'way'** (-wā'), **width'ways'**.
Wi·du·kind (vē'dōō·kint) See WITTEKIND.
wiel (wēl) *n. Scot.* An eddy; pool.
Wie·land (vē'länt), **Christoph Martin**, 1733-1813, German poet, novelist, and translator. — **Heinrich**, 1877-1957, German chemist.
wield (wēld) *v.t.* 1 To use or handle, as a weapon or instrument, especially with full command and effect. 2 To exercise (authority, power, influence, etc.). 3 *Obs.* To exercise authority over; command. [Fusion of OE *wealdan* cause, and OE *wildan* rule] — **wield'·a·ble** *adj.* — **wield'er** *n.*
wield·y (wēl'dē) *adj.* **wield·i·er, wield·i·est** Easily handled; wieldable: opposed to *unwieldy*.
Wie·licz·ka (vye·lech'kä) A town 7 miles SE of Cracow, Poland; a salt-mining center since the 11th century.
Wien (vēn) The German name for VIENNA.
Wien (vēn), **Wilhelm**, 1864-1928, German physicist.
wie·ner (wē'nər) *n. U.S.* A kind of sausage, often shorter than a frankfurter, made of beef and pork: often called *weenie*. Also **wie·ner·wurst** (wē'nər·wûrst', *Ger.* vē''nər·vōōrst'). [Short for G *wiener-(wurst)* Vienna (sausage)]
Wie·ner schnit·zel (vē'nər shnit'səl) A breaded veal cutlet, seasoned or garnished in any of several ways, as with capers, anchovies, a fried egg, or the like. [<G <*Wiener* Viennese + *schnitzel* a cutlet, dim. of *schnitz* a slice <*schneiden* cut]
Wieprz (vyepsh) A river in central Poland, flowing 194 miles NW to the Vistula.
Wies·ba·den (vēs'bä·dən) The capital of Hesse, West Germany, on the Rhine west of Frankfurt: site of a famous spa.
wife (wīf) *n. pl.* **wives** (wīvz) 1 A woman joined to a man in lawful wedlock; a spouse: the correlative of *husband*. ♦ Collateral adjective: *uxorial*. 2 A grown woman; adult female: usually in composition or in certain phrases: *housewife*, old *wives'* tales. — **to take (a woman) to wife** To marry (a woman). [OE *wīf*] — **wife'dom, wife'hood** *n.* — **wife'less** *adj.* — **wife'ly** *adj.*
wife-carl (wīf'kärl) *n. Scot.* A man who meddles with household affairs, especially such as belong naturally to women.
wig (wig) *n.* An artificial covering of hair for the head, so constructed as to form an imitation of the natural growth or to act as a coiffure. — *v.t.* **wigged, wig·ging** 1 To furnish with a wig or wigs. 2 *Brit. Colloq.* To censure severely; berate or scold, especially in public. [Short for PERIWIG]
wig·an (wig'ən) *n.* A stiff, canvaslike fabric used for stiffening the borders of garments. [from *Wigan*, where originally made]
Wig·an (wig'ən) A county borough in south central Lancashire, England.
wig·eon (wij'ən) See WIDGEON.
wigged (wigd) *adj.* Furnished with or wearing a wig.
wig·ger·y (wig'ər·ē) *n. pl.* **·ger·ies** 1 A peruke;

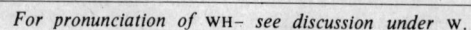
BARRISTER'S WIG

wig; also, wigs collectively. 2 Excessive formality; red-tapism. 3 The material of a wig; false hair.
Wig·gin (wig'in), **Kate Douglas**, *née* Smith, 1856-1923, U.S. educator and novelist.
wig·ging (wig'ing) *n. Brit. Colloq.* A rebuke; a scolding.
wig·gle (wig'əl) *v.t. & v.i.* ·gled, ·gling To move or cause to move quickly and irregularly from side to side; squirm; wriggle. — *n.* The act of wiggling. [? <MLG *wiggelen*] — **wig'gly** *adj.*
wig·gler (wig'lər) *n.* 1 One who or that which wiggles. 2 The larva of a mosquito; a wiggletail.
Wig·gles·worth (wig'əlz·wûrth), **Michael**, 1631-1705, American divine and poet.
wig·gle·tail (wig'əl·tāl) *n.* 1 The larva of a mosquito. 2 A tadpole.
wight[1] (wīt) *n.* A person; creature: usually an archaic or humorous term. [OE *wiht* a creature]
wight[2] (wīt) *adj. Obs.* Full of prowess; strong and valiant; active; swift. [<ON *vigt*, neut. of *vigr* able to fight]
Wight (wīt), **Isle of** An island in the English Channel just off the southern coast of England, comprising an administrative county of Hampshire; 147 square miles.
Wig·man (vikh'män), **Mary**, born 1886, German dancer; leading pioneer of modern dance.
Wig·ner (wig'nər), **Eugene Paul**, born 1902, U.S. physicist born in Hungary.
Wig·town (wig'tən) A county in SW Scotland; 487 square miles; county town, Wigtown. Also **Wig'town·shire** (-shir).
wig·wag (wig'wag') *v.t. & v.i.* ·wagged, ·wagging 1 To move briskly to and fro; wag. 2 To send (a message) by hand flags, torches, etc. — *n.* The act or art of signaling with such flags, etc., or the message so sent. [< dial. E *wig* wiggle + WAG[1]] — **wig'wag'ger** *n.*
wig·wam (wig'wom, -wôm) *n.* 1 A dwelling or lodge of the North American Indians of Algonquian stock, used in the area from Canada to North Carolina and in the Great Lakes regions: commonly an arbor-shaped or conical framework of poles covered with bark, rush matting, or hides. 2 By extension, a family of Indians. 3 A dwelling or lodge of North American Indians of other than Algonquian stock: a misuse by early travelers. 4 *Slang* A public building used for political gatherings, mass meetings, etc. — **the Wigwam** Tammany Hall. [<Algonquian (Ojibwa) *wigwaum*, lit., their dwelling]

WIGWAM
Eastern North
American Indian.

wik·i·up (wik'ē·up) See WICKIUP.
Wil·ber·force (wil'bər·fôrs, -fōrs), **William**, 1759-1833, English abolitionist and philanthropist.
Wil·bur (wil'bər) A masculine personal name. [<Gmc.,? resolute protection]
Wil·cox (wil'koks), **Ella**, *née* Wheeler, 1855?-1919, U.S. poet and author.
wild (wīld) *adj.* 1 Inhabiting the forest or open field; not domesticated or tamed; living in a state of nature: a *wild* horse; shy and easily startled: The deer are *wild*. 2 Growing or produced without care or culture; not cultivated: *wild* flowers. 3 Being in the natural state; being without civilized inhabitants or cultivation; desert; waste: *wild* prairies. 4 Living without any civilization and in a rude, savage way; uncivilized: the *wild* men of Borneo. 5 Boisterous; in a bad sense, dissolute; prodigal; in a milder sense, frolicsome and gay. 6 Affected with or originating violent disturbances, as of the elements or of human passions; stormy; turbulent: a *wild* night, a *wild* crowd. 7 Showing reckless want of judgment; rashly imprudent; extravagant: a *wild* speculation. 8 Fantastically irregular or disordered; odd in arrangement or effect; strange or weird: a *wild* imagination, *wild* dress. 9 Eager and excited, as by reason of joy, fear, desire, etc.: She was *wild* with delight. 10 Excited to frenzy or distraction; roused to fury or desperation; crazed or crazy: The mosquitoes are driving me *wild*. 11 Being or going far from the proper course or from the mark aimed at; erratic; wide of the mark: a *wild* ball, a *wild* guess. 12 In some card games, having its value arbitrarily determined by the dealer or holder: to play poker with fours *wild*. See synonyms under ABSURD, BLEAK[1], FIERCE, INSANE, ROMANTIC, TURBULENT, VIOLENT, WANTON. — *n.* An uninhabited or uncultivated place; a waste; wilderness: often in the plural: the *wilds* of Africa. — **the wild** The wilderness; also, the free, natural, wild life: the call of *the wild*. — *adv.* 1 Wildly. 2 Without control; unrestrainedly: The locomotive is running *wild*. [OE *wilde*] — **wild'ly** *adv.* — **wild'ness** *n.*
wild allspice The spicebush.
wild boar The native hog (*Sus scrofa*) of continental Europe, southern Asia, and North Africa, and formerly of Great Britain.
wild brier 1 Any species of rose in the wild state. 2 The dog rose. 3 The sweetbrier.
wild carrot An umbelliferous herb (*Daucus carota*) from which the cultivated carrot is derived; Queen Anne's lace.
wild·cat (wīld'kat') *n.* 1 A small, undomesticated feline carnivore (*Felis sylvestris*) of Europe, resembling the domestic cat, but larger and stronger. 2 The North American bobcat (genus *Lynx*). 3 One of several other small felines, as the ocelot and serval. 4 Figuratively, an aggressive, quick-tempered person. 5 An unattached locomotive and its tender, used on special work, as when sent out to haul a train, etc. 6 A successful oil well drilled in an area previously unproductive. 7 A tricky or unsound business venture; specifically, a worthless mine. Also **wild. cat.** — *adj.* 1 Unsound; risky; especially, financially unsound or risky: a *wildcat* venture. 2 Illegal; made, produced, or carried on without official sanction or authorization. 3 Not running on schedule time; also, running wild or without control, as a railroad train or engine. — *v.t. & v.i.* ·cat·ted, ·cat·ting To drill for oil in (an area not known to be productive). — **wild'cat'ting** *n. & adj.*
wildcat bank Prior to the passage of the National Bank Act of 1863-64, a bank operating with insufficient capital to redeem its circulating notes.
wildcat bill A note of a wildcat bank.
wildcat mine A worthless mine; especially, one represented to possible investors as being profitably productive.
wildcat strike A strike unauthorized by regular union procedure.
wild·cat·ter (wīld'kat'ər) *n.* 1 A promoter of mines of doubtful value. 2 One who develops oil wells in unproved territory. 3 One who manufactures illicit whisky.
wild cherry Any of certain species of cherry found growing wild; especially, the **wild black cherry** (*Prunus serotina*) and the chokecherry.
Wilde (wīld), **Oscar Fingall O'Flahertie Wills**, 1856-1900, Irish poet and playwright.
wilde·beest (wīld'bēst, wil'də-; *Du.* vil'də-bāst) A gnu. [<Afrikaans <Du. *wild* wild + *beeste* a beast]
wil·der (wil'dər) *Poetic v.t.* 1 To bewilder. 2 To lead astray; mislead. — *v.i.* 3 To be bewildered. 4 To wander; stray. [Prob. back formation <WILDERNESS] — **wil'der·ment** *n.*
Wil·der (wīl'dər), **Thornton Niven**, born 1897, U.S. novelist and playwright.
wil·der·ness (wil'dər·nis) *n.* 1 An uncultivated, uninhabited, or barren region. 2 A waste, as of an ocean. 3 A multitudinous and confusing collection: a *wilderness* of curiosities. 4 *Obs.* Wildness. [OE *wilder* a wild beast (< *wilde* wild + *deor* an animal, deer) + -NESS]
Wil·der·ness (wil'dər·nis), **The** A region in NE Virginia; scene of a Civil War battle, 1864.
wild·fire (wīld'fīr') *n.* 1 A raging, destructive fire: now generally in phrases like *to spread like wildfire*. 2 A composition of inflammable materials, or the flame produced by it, very hard to put out; Greek fire. 3 A phosphorescent luminousness; ignis fatuus. 4 *Obs.* A spreading inflammation of the skin; erysipelas. 5 A skin disease of sheep with inflammation.

wild flax 1 Toadflax. 2 Gold-of-pleasure.
wild·flow·er (wīld'flou'ər) *n.* 1 Any uncultivated flower. 2 The plant growing it. Also **wild flower.**
wild·fowl (wīld'foul') *n.* Wild game-birds, especially wild ducks and geese. Also **wild fowl.**
wild gean (gēn) 1 A European wild cherry (*Prunus avium*), yielding a fine cabinet wood. 2 Its small dark fruit, the mazzard cherry. [<WILD + dial. E *gean* the wild cherry <OF *guine*, prob. <Gmc.]
wild goose An undomesticated goose, as the English graylag, or the Canada goose.
wild-goose chase (wīld'gōōs') Pursuit of the unknown or unattainable; a bootless enterprise.
wild honeysuckle The pinkster flower.
wild hyacinth 1 The eastern camas (*Camassia scilloides*) of the United States. 2 The wood hyacinth.
wild indigo Any of a genus (*Baptisia*) of perennial North American herbs, especially one (*B. tinctoria*) having yellow flowers and a root which yields a purgative glycoside.
wild·ing (wīl'ding) *adj.* Growing wild; uncultivated; undomesticated. — *n.* 1 An uncultivated plant; a fruit tree on its own roots growing among grafted trees. 2 A cultivated plant that has sprung up spontaneously; an escape (def. 4). 3 A creature not conforming to type.
wild lettuce 1 A tall, yellow-flowered herb (*Lactuca virosa*) found in the northern United States. 2 The round-leaved wintergreen (*Pyrola rotundifolia*). 3 The prickly lettuce (*Lactuca serriola*).
wild·life (wīld'līf') *n.* Wild animals, trees, and plants collectively, especially as objects of government conservation. — *adj.* Pertaining to wild animals, trees, and plants collectively.
wild·ling (wīld'ling) *n.* An uncultivated plant or flower; a wild animal. [<WILD + -LING[1]]
wild madder 1 Madder (def. 1). 2 Either of two herbs (genus *Galium*) of the madder family, the white bedstraw (*G. mollugo*) and the dye bedstraw (*G. tinctorium*).
wild mandrake The May apple.
wild mare 1 A nightmare. 2 A see-saw.
wild mustard An annual herb (*Brassica kaber*) of the mustard family, frequently growing as a weed, whose seeds are sometimes used as a substitute for mustard and its leaves cooked as greens: also called *charlock*.
wild oat 1 An uncultivated grass (genus *Avena*); especially, the common European meadow weed (*A. fatua*). 2 *pl.* Indiscretions of youth.
wild olive Any of various trees resembling the olive or bearing an olivelike fruit.
wild pansy The European heartsease (*Viola tricolor*), from which the common garden pansy is derived.
wild parsley 1 Any of a genus (*Lomatium*) of perennial herbs of the carrot family, especially the nine-leaf species (*L. simplex*), valued as a forage plant in the western United States: also called *biscuitroot*. 2 Lovage.
wild parsnip 1 The parsnip in its uncultivated, weedlike form. 2 A perennial herb (*Angelica lyalli*) of the carrot family, resembling the water hemlock but non-poisonous and useful as a forage plant.
wild pink An American catchfly (*Silene caroliniana*) with white or rose-colored flowers and long spatulate or lanceolate flowers.
wild rice 1 A tall aquatic grass of North America (*Zizania aquatica*). 2 The grain of this plant; formerly used as food by North American Indians, now esteemed as a table delicacy: also called *Indian rice*.
wild rose Any of various uncultivated roses of the north temperate zone, as the sweetbrier.
wild rubber Rubber as extracted from the rubber tree (genus *Hevea*) in its wild state.
wild rye A tall perennial grass (genus *Elymus*), widely distributed in temperate regions.
wild sage 1 The sagebrush of the western United States. 2 The Old World vervain sage (*Salvia verbeneca*) known as wild clary.
wild spinach 1 A goosefoot sometimes used as a substitute for spinach. 2 One of several other spinaceous plants.
wild turkey A large North American turkey

(Meleagris gallopavo silvestris) formerly ranging east of the Rocky Mountains from southern Canada to Florida and Mexico, and first domesticated in Mexico: now rare in the wild state.

wild vanilla A smooth, erect, perennial herb *(Trilisa odoratissima)* of the composite family, found in the SE United States. Its leaves give off an odor of vanilla.

wild wall A soundproof movable wall used on motion-picture sets: also called *jockey wall.*

Wild West The western United States, especially in its early period of Indian fighting, pioneer conditions, and lawlessness.

Wild West show A circus or a feature of a circus presenting feats of Indian and cowboy horsemanship; also, a rodeo.

wild·wood (wīld′wo͝od′) *n.* Natural forest land.

wild yam An uncultivated species of yam *(Dioscorea villosa)* of the eastern United States; the colic root.

wile (wīl) *n.* 1 An act or a means of cunning deception; also, any beguiling trick or artifice. 2 Craftiness; cunning. See synonyms under ARTIFICE. —*v.t.* **wiled, wil·ing** 1 To lure, beguile, or mislead. 2 To pass divertingly, as time: usually with *away*: by confusion with *while.* [OE *wīl,* prob. <Scand. Cf. ON *vēl* an artifice.]

Wi·ley (wī′lē), **Harvey Washington,** 1844–1930, U. S. chemist.

Wil·fred (wil′frid) A masculine personal name. Also **Wil′frid.** [<Gmc., willing peace]

wil·ful (wil′fəl) *adj.* 1 Bent on having one's own way; headstrong; self-willed. 2 Resulting from the exercise of one's own will; voluntary; intentional. Also spelled *willful.* See synonyms under PERVERSE. —**wil′ful·ly** *adv.* —**wil′ful·ness** *n.*

Wil·helm (vil′helm) German form of WILLIAM. —**Wilhelm I,** 1797–1888, king of Prussia 1861–88 and emperor of Germany 1871–88. —**Wilhelm II,** 1859–1941, emperor of Germany 1888–1918.

Wil·hel·mi·na (wil′hel·mē′nə) A feminine personal name. Also **Wil·hel·mine** (wil′hel·mēn, *Fr.* vē·lel·mēn′, *Ger.* vil′hel·mē′nə). [Fem. of WILHELM]

—**Wilhelmina,** born 1880, queen of the Netherlands 1890–1948; abdicated in favor of her daughter Juliana: full name *Wilhelmina Helena Pauline Maria of Orange-Nassau.*

Wil·helms·ha·ven (vil′helms·hä′fən) A port on the North Sea in former Oldenburg state, Lower Saxony, NW West Germany.

Wil·helm·stras·se (vil′helm·shträ′sə) 1 A street in Berlin on which the German foreign office and government offices were formerly located. 2 Formerly, the German government, especially its foreign policies.

Wilkes (wilks), **Charles,** 1798–1877, U.S. admiral and Antarctic explorer. —**John,** 1727–97, English politician.

Wilkes-Bar·re (wilks′bar·ē) A city on the Susquehanna River in NE Pennsylvania, an industrial and coal-mining center.

Wilkes Land (wilks) Part of Antarctica on the Indian Ocean south of Australia, between Queen Mary Coast and George V Coast; site of the south magnetic pole in its eastern part.

Wil·kins (wil′kinz), **Sir George Hubert,** 1888–1958, Australian aviator and explorer. —**Mary Eleanor** See FREEMAN, MARY E. WILKINS.

Wil·kin·son (wil′kən·sən), **James,** 1757–1825, American Revolutionary general and politician.

will[1] (wil) *n.* 1 The power of conscious, deliberate action; the faculty by which the rational mind makes choice of its ends of action, and directs the energies in carrying out its determinations; in popular usage, choice, purpose, or directive effort. 2 The act or experience of exercising this faculty; a volition or a choice. 3 Strong determination; practical enthusiasm; energy of character: He works with a *will*; also, self-control. 4 That which has been resolved or determined upon; a purpose. 5 Power to dispose of a matter arbitrarily; discretion. 6 *Law* The legal declaration of a man's intentions as to his estate after his death; the written instrument by which someone declares his desires for the distribution of his property. 7 A conscious inclination toward any end or course; a wish. 8 A request or command. —**at will** As one pleases. —*v.* **willed, will·ing;** third person singular, present indicative **wills** *v.t.* 1 To decide upon; choose. 2 To resolve upon as an action or course; determine to do. 3 To give, devise, or bequeath by a will. 4 To control, as a hypnotized person, by the exercise of will. 5 *Archaic* To have a wish for; desire. —*v.i.* 6 To exercise the will. [OE *willa*] —**will′a·ble** *adj.*

Synonyms (noun): decision, desire, disposition, inclination, resolution, volition, wish. *Will* is a word of wide range of meaning, and both as faculty and act has been the subject of many and various theories; in popular language *will* is often equivalent to *desire* or *inclination,* as when we speak of doing something against our *will. Volition* is a word of scientific precision, denoting the determinative element of *will.*

will[2] (wil) *v.* Present *sing. & pl.*: **will** (*Archaic* **thou wilt**); past: **would** (*Archaic* **thou would·est** or **wouldst**) As an auxiliary verb *will* is used with the infinitive without *to,* or elliptically without the infinitive, to express: 1 Futurity: They *will* arrive by dark. ◆ See usage note under SHALL. 2 Willingness or disposition: Why *will* you not tell the truth? 3 Capability or capacity: The ship *will* survive any storm. 4 Custom or habit: He *will* sit for hours and brood. 5 *Colloq.* Probability or inference: I expect this *will* be the main street. —*v.t.* & *v.i.* To wish or have a wish; desire: What *wilt* thou? As you *will.* [OE *willan*]

Wil·lam·ette River (wi·lam′it) A river in NW Oregon, flowing 190 miles north to the Columbia River.

Wil·lard (wil′ərd), **Emma,** 1787–1870, *née* Hart, U. S. pioneer in education for women. —**Frances Elizabeth Caroline,** 1839–98, U. S. temperance advocate.

Will·cocks (wil′koks), **Sir William,** 1852–1932, English engineer.

willed (wild) *adj.* Having a will, especially one of a given character: mostly in composition: self-*willed.*

wil·lem·ite (wil′əm·īt) *n.* A vitreous or resinous orthosilicate of zinc, crystallizing in the hexagonal system and occurring in many colors. [<Du. *willemit,* after *Willem I* William of Orange]

Wil·lem·stad (wil′əm·stät, vil′-) A town on Curaçao, capital of Netherlands Antilles.

will·er (wil′ər) *n.* One who wills.

Willes·den (wilz′dən) A municipal borough in Middlesex, England, NW of London.

wil·let (wil′it) *n.* A large, light-colored shore bird *(Catoptrophorus semipalmatus)* of North America, related to the snipes. [Short for *pill-will-willet,* imit. of the cry of the bird]

will·ful (wil′fəl), **will·ful·ly, will·ful·ness** See WILFUL, etc.

will I, nill I or **will he, nill he** or **will ye, nill ye** Willingly or unwillingly; without choice. See WILLY-NILLY.

Wil·liam (wil′yəm) A masculine personal name. Also *Du.* **Wil·lem** (vil′əm). See also WILHELM. [<Gmc., resolute protection]

—**William I,** 1027–87, invaded England, 1066; king of England 1066–87: known as *William the Conqueror.*

—**William II,** 1056–1100, king of England 1087–1100: known as *William Rufus.*

—**William III,** 1650–1702, stadholder of Holland 1672–1702; invited to England; ruled 1689–1702 jointly with his wife Mary.

—**William IV,** 1765–1837, king of England 1830–37.

—**William of Malmesbury,** 1095?–1143? English historian.

—**William of Orange,** 1533–84, founded the Dutch republic; stadholder 1579–84: called "William the Silent."

Wil·liams (wil′yəms), **Roger,** 1603?–85, English clergyman; founded Rhode Island. —**Roger John,** born 1893, U. S. biochemist. —**Tennessee,** born 1914, U. S. playwright: original name *Thomas Lanier Williams.* —**William Carlos,** born 1883, U. S. poet, novelist, playwright, and physician.

Wil·liams·burg (wil′yəmz·bûrg) A town in eastern Virginia; founded in 1693; capital of Virginia (1699–1779); restored to condition of the colonial period.

wil·lies (wil′ēz) *n. pl. Slang* Nervousness; jitters; the creeps: with *the.* [? <WILLY-NILLY; with ref. to a state of indecision]

will·ie·waught (wil′ē·wäkht) *n. Scot.* A draft of liquor. Also **will′ie·waucht.**

will·ing (wil′ing) *adj.* 1 Having the mind favorably inclined or disposed. 2 Answering to demand or requirement; compliant. 3 Gladly proffered or done; hearty. 4 Of or pertaining to the faculty or power of choice; volitional. See synonyms under SPONTANEOUS. —**will′ing·ly** *adv.* —**will′ing·ness** *n.*

Wil·lis (wil′is), **Nathaniel Parker,** 1806–67, U. S. writer and editor.

wil·li·waw (wil′ē·wô) *n.* A sudden, violent blast of wind moving seaward down the slope of a mountainous coast, especially in the Strait of Magellan. [Origin unknown]

Will·kie (wil′kē), **Wendell Lewis,** 1892–1944, U. S. lawyer and political leader.

will-o'-the-wisp (wil′ə·thə·wisp′) *n.* 1 Ignis fatuus. 2 Any elusive or deceptive object. —*adj.* Deceptive; fleeting; misleading. [Earlier *Will with the wisp*]

wil·low (wil′ō) *n.* 1 Any of a large genus *(Salix)* of shrubs and trees related to the poplars, having generally smooth branches and often long, slender, pliant, and sometimes pendent branchlets. 2 The soft white wood of the willow. 3 *Colloq.* Something made of willow wood, especially a baseball or cricket bat. 4 A machine for giving a preliminary cleaning to cotton, flax, hemp, wool, etc., by means of long spikes projecting from a revolving cone or cylinder. —*v.t.* To clean, as cotton, wool, etc., with a willow. —*adj.* Of or pertaining to the willow; made of willow wood. [OE *wilige, welig*] —**wil′low·ish** *adj.*

wil·low·er (wil′ō·ər) *n.* One who or that which willows.

willow herb 1 Any of a genus *(Epilobium)* of perennial herbs of the evening-primrose family, especially the fireweed *(E. angustifolium),* having scattered, willowlike leaves and large, pink flowers. 2 The purple loosestrife *(Lythrum salicaria).*

willow oak 1 An oak *(Quercus phellos)* of the eastern United States, having long, slender, entire leaves resembling willow leaves. 2 The laurel oak *(Q. laurifolia).*

willow pattern A decorative design introduced on household china in England in 1780 and since extremely popular: so called from the willow tree, usually blue on a white background, which appears in the design.

wil·low·ware (wil′ō·wâr′) *n.* China decorated with the willow pattern.

wil·low·y (wil′ō·ē) *adj.* 1 Abounding in willows. 2 Having supple grace of form or carriage. See synonyms under SUPPLE.

will·pow·er (wil′pou′ər) *n.* Ability to control oneself; determination; strength or firmness of mind.

Will·stät·ter (vil′shtet·ər), **Richard,** 1872–1942, German organic chemist.

will·y[1] (wil′ē) *adj. Obs.* Willing; also, propitious. [Cf. ON *viljugr*]

wil·ly[2] (wil′ē) *v.t.* **·lied, ·ly·ing** To willow, as cotton, flax, hemp, etc.

will·yard (wil′yərd) *adj. Scot.* Wilful; also, abashed; bewildered. Also **will′yart** (-yərt).

wil·ly-nil·ly (wil′ē·nil′ē) *adj.* Having no decisiveness; uncertain; irresolute. —*adv.* Willingly or unwillingly. [Earlier *will I, nill I* whether I will or not]

wil·ly-wil·ly (wil′ē·wil′ē) *n. pl.* **·lies** *Austral.* 1 A violent storm of wind and rain on the NW coast of Australia. 2 A local duststorm.

Wil·ming·ton (wil′ming·tən) 1 A port of entry on the Delaware River in northern Delaware. 2 A port of entry in SE North Carolina.

Wil·no (vil′nô) The Polish name for VILNA.

Wil·son (wil′sən), **Alexander,** 1766–1813, American ornithologist born in Scotland. —**Charles Thomson Rees,** 1869–1959, Scottish physicist. —**Edmund,** born 1895, U. S. critic, author, and dramatist. —**Henry,** 1812–75 U. S. statesman. —**James,** 1742–98, American patriot, signed Declaration of Independence. —**John,** 1785–1854, Scottish poet: pseudonym *Christopher North.* —**(Thomas) Woodrow,** 1856–1924, U. S. educator and statesman; president of the United States 1913–21.

Wil·son (wil′sən), **Mount** A peak in SW California, near Pasadena; 5,710 feet; site of a famous observatory.

Wilson Dam A power dam in the Tennessee River at Muscle Shoals, NW Alabama; 137 feet high, 4,862 feet long; forms **Lake Wilson** (25 square miles; 15 1/2 miles long, 1 1/2 miles wide) over Muscle Shoals.

Wilson's petrel The storm petrel. [after Alexander *Wilson*]
Wilson's phalarope A shore bird (*Steganopus tricolor*) which breeds in northern North America and winters as far south as the Falkland Islands. [after Alexander *Wilson*]
Wilson's plover The ring plover (*Pagolla wilsonia*) of the southern United States and South America. [after Alexander *Wilson*]
Wilson's snipe Snipe (def. 1). [after Alexander *Wilson*]
Wilson's thrush The veery. [after Alexander *Wilson*]
Wilson's warbler A small, very active flycatcher (*Wilsonia pusilla*) of eastern North America, black-crowned with a yellow and olive-green body. [after Alexander *Wilson*]
wilt[1] (wilt) *v.i.* 1 To lose freshness; droop or become limp, as a flower that has been cut or that has not been watered. 2 To lose energy and vitality; become faint or languid: We *wilted* under the hot sun. 3 To lose courage or spirit; subside suddenly. — *v.t.* 4 To cause to droop or wither. 5 To cause to lose vitality and energy. — *n.* 1 The act of wilting; also, languor; faintness. 2 An infectious and virulent disease sometimes epidemic among certain caterpillars and insect larvae, which are reduced to a liquefied mass by its ravages: also **wilt disease**. [Prob. dial. var. of obs. *welk* wither. Cf. MDu. *welken* wither.]
wilt[2] (wilt) Archaic second person singular, present tense of WILL[2], used with *thou*.
Wil·ton (wil'tən) *n.* A kind of carpet resembling the Brussels carpet, but having the loops of the pile cut, thus giving it a velvety texture: originally made at Wilton, England. Also **Wilton carpet, Wilton rug**.
Wilt·shire (wilt'shir) *n.* One of a breed of longhorned sheep raised in Wiltshire, England.
Wilt·shire (wilt'shir) A county in southern England; 1,345 square miles; county town, Salisbury. Shortened form **Wilts**.
Wiltshire cheese A variety of Cheddar cheese.
wi·ly (wī'lē) *adj.* **·li·er, ·li·est** Full of or characterized by wiles; sly; cunning. See synonyms under INSIDIOUS, POLITIC. — **wi'li·ly** *adv.* — **wi'li·ness** *n.*
wim·ble (wim'bəl) *n.* Anything that bores a hole, especially if turned by hand, as a gimlet, auger, brace and bit, or the like. — *v.t.* **·bled, ·bling** To bore or pierce, as with a wimble. [< AF, OF *guimbel* < MLG *wiemel*. Akin to GIMLET.]
Wim·ble·don (wim'bəl·dən) A town and municipal borough SW of London in NE Surrey, England; scene of international tennis matches.
wim·ple (wim'pəl) *n.* 1 A cloth, as of linen or silk, wrapped in folds around the neck close under the chin and over the head, exposing only the face: formerly worn as a protection by women outdoors, and still by nuns. 2 *Scot.* A fold; plait; also, a curve; a winding turn, as in a river or road. — *v.* **·pled, ·pling** *v.t.* 1 To cover or clothe with a wimple; veil. 2 To make or fold into plaits, as a veil. 3 To cause to move with slight undulations; ripple. 4 *Obs.* To deceive; hoodwink. — *v.i.* 5 To lie in plaits or folds. 6 To ripple. [OE *wimpel*]

WIMPLE
14th century.

Wims·hurst machine (wimz'hûrst) A machine for the generation of static electricity by means of two insulated rotating disks carrying a number of equally spaced strips of conducting material which, by friction, build up an electrostatic charge. [after James *Wimshurst*, 1832-1903, English engineer, its inventor]
win[1] (win) *v.* **won** (*Obs.* **wan**), **won, win·ning** *v.i.* 1 To gain a victory; be victorious; prevail, as in a contest: May the best man *win*. 2 To succeed in an effort or endeavor. 3 To succeed in reaching or attaining a specified end or condition; get: often with *across, over, through,* etc.: The fleet *won* through the storm. 4 *Obs.* To fight; struggle. — *v.t.* 5 To be successful in; gain victory in: to *win* a game; to *win* an argument. 6 To gain in competition or contest: to *win* the blue ribbon. 7 To gain by effort, persistence, etc.: to *win* fame or fortune. 8 To influence so as to obtain the good will or favor of: often with *over*: His eloquence *won* the audience; We *won* him over to our side. 9 To secure the love of; gain in marriage: He wooed and *won* her. 10 To succeed in reaching; attain: to *win* the harbor. 11 To make (one's way), especially with effort. 12 To capture; take possession of. 13 To earn or procure, as a living: to *win* support from poor soil. 14 *Mining* **a** To extract, as ore or coal, or metal from ore. **b** To reach and open (a deposit, vein, etc.); prepare for mining. See synonyms under ALLURE, CONQUER, GAIN[1], GET, OBTAIN, PERSUADE, SUCCEED. — **to win out** *Colloq.* To succeed to the fullest extent or expectation. — *n.* 1 A victory; success. 2 Profit; winnings. [OE *winnan* contend, labor]
win[2] (win) *v.t. Scot. & Irish* 1 To winnow. 2 To cure, as hay.
win[3] (win) *n. Scot.* Wind.
win·cey (win'sē) *n.* A fabric woven with cotton or linen warp and woolen filling. [Short for *wincey-woolsey*, alter. of LINSEY-WOOLSEY]
wince[1] (wins) *v.i.* **winced, winc·ing** To shrink back or start aside, as from a blow or pain; flinch. — *n.* The act of wincing. [< AF *wenchier* (assumed), var. of OF *quenchier* avoid < Gmc.] — **winc'er** *n.*
wince[2] (wins) *n.* A dyer's winch or windlass. [Var. of WINCH[1]]
winch[1] (winch) *n.* 1 A windlass, particularly one used for hoisting, as on a truck or the mast of a crane, derrick, etc., having usually one or more hand cranks geared to a drum around which the rope or chain winds. 2 A crank with a handle, used to impart motion to a grindstone or the like. — *v.t.* To move, hoist, or haul with or as with a winch. [OE *wince*] — **winch'er** *n.*

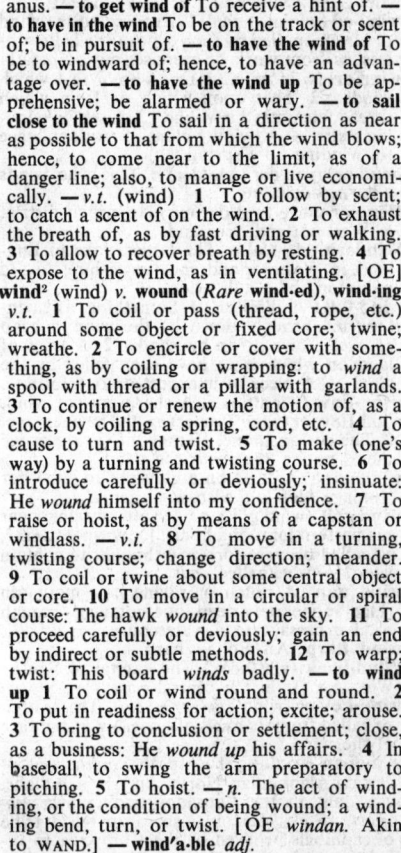
WINCH

winch[2] (winch) *v.i. Obs.* To wince; flinch. [See WINCE[1].]
Win·ches·ter (win'ches·tər) 1 The county town of Hampshire, England; known for its 11th century cathedral. 2 A city of northern Virginia near eastern West Virginia; scene of several Civil War battles, 1862 and 1864.
Winchester rifle A breechloading, lever-action, repeating rifle with a tubular magazine under the barrel, first produced in 1866: a trade name. Also **Winchester**. [after Oliver F. *Winchester*, 1810-80, U.S. industrialist]
Winck·el·mann (vingk'əl·män), **Johann Joachim,** 1717-68, German archeologist and art critic.
wind[1] (wind, *Poetic* wīnd) *n.* 1 Any movement of air, especially a natural horizontal movement; air in motion naturally. See BEAUFORT SCALE. 2 Any powerful or destructive wind; a tornado; hurricane. 3 The direction from which a wind blows; one of the cardinal points of the compass: They gathered from the four *winds*. 4 Air in motion by artificial means: the *wind* of a bullet, *wind* from a bellows. 5 Air pervaded by a scent: The deer got *wind* of the hunter; hence, figuratively, a suggestion or intimation: to get *wind* of a plot. 6 The power of breathing or respiring; breath: He lost his *wind* in the race. 7 Breath as expended in words, especially as having more sound than sense; idle chatter; also, vanity; conceit. 8 *pl.* The wind instruments of an orchestra; also, the players of these instruments. See WIND INSTRUMENT. 9 The gaseous product of indigestion; flatulence. 10 In pugilism, the pit of the stomach where a blow may cause temporary stoppage of breath: He was hit in the *wind*. — **in the wind** 1 Impending; astir; afoot. 2 Inebriated; drunk. — **in the wind's eye** Directly opposed to the point from which the wind blows. — **to break wind** To expel gas through the anus. — **to get wind of** To receive a hint of. — **to have in the wind** To be on the track or scent of; be in pursuit of. — **to have the wind of** To be to windward of; hence, to have an advantage over. — **to have the wind up** To be apprehensive; be alarmed or wary. — **to sail close to the wind** To sail in a direction as near as possible to that from which the wind blows; hence, to come near to the limit, as of a danger line; also, to manage or live economically. — *v.t.* (wind) 1 To follow by scent; to catch a scent of on the wind. 2 To exhaust the breath of, as by fast driving or walking. 3 To allow to recover breath by resting. 4 To expose to the wind, as in ventilating. [OE]
wind[2] (wīnd) *v.* **wound** (*Rare* **wind·ed**), **wind·ing** *v.t.* 1 To coil or pass (thread, rope, etc.) around some object or fixed core; twine; wreathe. 2 To encircle or cover with something, as by coiling or wrapping: to *wind* a spool with thread or a pillar with garlands. 3 To continue or renew the motion of, as a clock, by coiling a spring, cord, etc. 4 To cause to turn and twist. 5 To make (one's) way by a turning and twisting course. 6 To introduce carefully or deviously; insinuate: He *wound* himself into my confidence. 7 To raise or hoist, as by means of a capstan or windlass. — *v.i.* 8 To move in a turning, twisting course; change direction; meander. 9 To coil or twine about some central object or core. 10 To move in a circular or spiral course: The hawk *wound* into the sky. 11 To proceed carefully or deviously; gain an end by indirect or subtle methods. 12 To warp; twist: This board *winds* badly. — **to wind up** 1 To coil or wind round and round. 2 To put in readiness for action; excite; arouse. 3 To bring to conclusion or settlement; close, as a business: He *wound up* his affairs. 4 In baseball, to swing the arm preparatory to pitching. 5 To hoist. — *n.* The act of winding, or the condition of being wound; a winding bend, turn, or twist. [OE *windan*. Akin to WAND.] — **wind'a·ble** *adj.*
wind[3] (wīnd, wind) *v.t.* **wind·ed** (*erroneously* **wound**), **wind·ing** 1 To blow, as a horn; sound. 2 To give (a call or signal), as with a horn. [< WIND[1]; infl. by *wind*[2]]
wind·age (win'dij) *n.* 1 The rush of air caused by the rapid passage of an object, as a projectile or a railway train. 2 Deflection of an object, as a bullet, from its natural course due to wind pressure. 3 In a muzzleloading rifled gun, the difference between the diameter of a projectile and the bore through which it is discharged; also, in a smoothbore gun, the space between the surface of the bore and the projectile. 4 *Mech.* The free air space between any moving piece and the socket or bore in which it travels. 5 A contusion caused by sudden compression of air due to the passing of a ball or gunshot near the body. 6 *Naut.* The surface offered to the wind by a vessel.
Win·dau (vin'dou) The German name for VENTSPILS.
Win·daus (vin'dous), **Adolf,** born 1876, German chemist.
wind·bag (wind'bag') *n.* 1 A wordy talker. 2 A bellows. 3 *Slang* The chest.
wind-blown (wind'blōn') *adj.* 1 Tossed or blown by the wind. 2 Having a permanent direction of growth as determined by prevailing winds: said of plants and trees. 3 Pertaining to an irregular hair arrangement causing the hair in front to appear as if blown forward by the wind.
wind-borne (wind'bôrn', -bōrn') *adj.* Carried or transported by the wind.
wind-bound (wind'bound') *adj.* Delayed by contrary winds.
wind-break (wind'brāk') *n.* Anything, as a hedge, fence, etc., that protects from or breaks the force of the wind.
Wind-break·er (wind'brā·kər) *n.* A sturdy, warm sports jacket with fitted waistband: a trade name.
wind-bro·ken (wind'brō'kən) *adj.* Asthmatic; broken-winded: said of a horse.
Wind Cave National Park (wind) A region containing a large limestone cavern in the Black Hills in SW South Dakota; 41 square miles; established 1903.

wind-cone (wind′kōn′) n. A windsock.
wind·ed (win′did) adj. 1 Exposed to the wind or air, or roused by such exposure. 2 Breathless, as from work or exercise; out of breath.
wind·er[1] (wīn′dər) n. 1 One who or that which winds. 2 That upon which or from which thread, etc., may be wound. 3 A step in winding stairs. 4 A twining plant. 5 An appliance for winding up a spring.
wind·er[2] (wīn′dər, win′dər) n. One who winds a horn, bugle, etc.
Win·der·mere (win′dər·mir) An urban district in Westmorland, England.
Windermere, Lake The largest lake in England, in Westmorland and Lancashire; 10 1/2 miles long by 1 mile wide.
wind·fall (wind′fôl) n. 1 Something, as ripening fruit, brought down by the wind; a heap of trees blown down by wind. 2 A tract of land on which trees have been felled by the wind. 3 A piece of unexpected good fortune.
wind·flaw (wind′flô) n. A sharp gust of wind.
wind·flow·er (wind′flou′ər) n. 1 The anemone. 2 The rue anemone. [Trans. of Gk. *anemōnē* the anemone < *anemos* the wind]
wind gage A scale on a gunsight to allow for windage (def. 2). Also **wind gauge.**
wind·gall (wind′gôl′) n. A soft swelling near the pastern joint of a horse. [< WIND + GALL[2]; so called because formerly thought to contain wind] — **wind′galled′** adj.
wind gap A notch or ravine in a mountain ridge, moderately deep, but not deep enough to give passage to a watercourse.
wind harp An Eolian harp.
Wind·hoek (vint′hook) The capital of South-West Africa, in the central part.
wind·hov·er (wind′huv′ər) n. Brit. The kestrel: so called from its habit of hovering in the face of the wind.
win·di·go (win′di·gō) n. In the mythology of certain Algonquian North American Indians, especially in the Labrador and Ojibwa districts, an evil demon; also, a mythical tribe of cannibals believed by the Chippewa to inhabit an island in Hudson Bay: also spelled *weendijo.* [<Algonquian (Ojibwa) *weendigo* a cannibal]
wind·ing[1] (wīn′ding) n. 1 The act or condition of one who or that which winds; a spiral turning or coiling. 2 A bend or turn, or a series of them. 3 A warp or twist from a plane surface. 4 *Electr.* The manner in which the wire is wound in a coil, as on the armature of a dynamo. 5 A defective gait of horses in which one leg seems to wind around the other. — adj. 1 Turning spirally about an axis or core. 2 Having bends or lateral turns. 3 Twisting from a plane.
wind·ing[2] (wīn′ding) n. A boatswain's signal.
winding frame A device or machine for winding, as a reel.
wind·ing·ly (wīn′ding·lē) adv. In a winding manner.
winding sheet (wīn′ding) The sheet that wraps a corpse.
wind instrument (wind) A musical instrument whose sounds are produced by vibrations of air injected by the lungs or by mechanical bellows. Those blown by air from the lungs are known as **wood-wind instruments** or **woodwinds**, consisting of the flutes, oboes, clarinets, etc., and the **brass-wind instruments** or **brasses**, consisting of the horns, trumpets, trombones, tubas, etc. Those in which the vibration of the air column is induced by bellows are the various types of organ, accordion, etc.
wind·jam·mer (wind′jam′ər) n. 1 *Naut.* A merchant sailing vessel, as distinguished from a steamship. 2 A member of its crew. 3 *Slang* A chatterbox; a loquacious person.
wind·lass (wind′ləs) n. Any of several devices for hauling or lifting, especially that form familiar in well curbs, consisting of a drum or barrel on which the hoisting rope winds, and turned by means of cranking. — **Chinese** or **differential windlass** A horizontal wheel and axle having two drums of different di-

DIFFERENTIAL WINDLASS

ameters on the same axis, one of which pays out as the other winds up, the power being increased in inverse proportion to the difference between the diameters. — v.t. & v.i. To raise or haul with a windlass. [Alter. of ME *windas* <ON *vindass* < *vinda* wind + *ass* a beam; infl. in form by WINDLE[2]]
win·dle[1] (win′dəl) n. A basket. [OE *windel* a basket < *windan* plait, twist]
win·dle[2] (win′dəl) *Scot. & Brit. Dial.* v.t. & v.i. To wind. — n. Something used for winding or turning. [Freq. of WIND[2]]
wind·less (wind′lis) adj. 1 Without wind; breezeless; calm. 2 Being out of breath.
win·dle·straw (win′dəl·strô′) n. *Scot. & Brit. Dial.* A withered stalk of any one of several grasses, used in plaiting or ropemaking. 2 A feeble, unhealthy person. 3 The whitethroat warbler. Also **win′dle·strae′** (-strā′). [OE *windelstrēaw*, ? < *windel* basket + *strēaw* straw]
wind·ling (wind′ling) n. 1 *Dial.* That which is torn off by the wind, as a branch of a tree. 2 *Scot.* A bottle of straw. [<WIND[1] + -LING[1]]
wind·mill (wind′mil′) n. 1 A mill consisting of a tower within which is a shaft having at the top a horizontal axis which bears a rudder at one end and at the other a system of adjustable slats, wings, or sails which, in revolving, transmit motion to a pump, millstone, or the like. 2 Anything resembling a windmill. 3 An imaginary wrong, evil, or foe: usually in the phrase, **to fight** (or **tilt at**) **windmills**, in allusion to Don Quixote's combats with windmills, which he mistook for giants.
win·dow (win′dō) n. 1 An opening in the wall of a building, to admit light or air, capable of being opened and closed, and including, architecturally, the casement, sash, panes, etc.; in common usage, sometimes, the sash alone: Raise the *window.* 2 A windowpane. 3 Anything resembling or suggesting a window; a windowlike aperture: The eyes are the *windows* of the soul. 4 A transparent patch through which the address of an envelope can be read. — v.t. 1 To provide with a window or windows. 2 To fill with holes resembling windows. [<ON *vindauga* < *vindr* wind + *auga* an eye]
window box 1 One of the grooves along the sides of a window frame for the weights that counterbalance a lifting sash. 2 A box, generally long and narrow, along a window ledge or sill, for growing plants.
win·dow–dress·ing (win′dō·dres′ing) n. 1 The act or the art of arranging merchandise attractively in shop and store windows; also, the goods so displayed; hence, anything superficially attractive. 2 A business report that unduly stresses favorable conditions. 3 Anything added or done to make something else more attractive: The prosecution of the thieves was mere *window-dressing* for his campaign for governor. — **win′dow-dress′er** n.
win·dow·pane (win′dō·pān′) n. A single sheet of glass for a window. Also **window pane.**
window seat A seat in the recess of a window.
window shade A flexible fabric shade, usually mounted on a spring roller, used to regulate light at a window.
win·dow–shop (win′dō·shop′) v.i. -**shopped**, -**shop·ping** To look at goods shown in store windows without buying them. — **win′dow-shop′per** n. — **win′dow–shop′ping** n. & adj.
wind·pipe (wind′pīp′) n. The duct by which the breath is carried to and from the lungs; the trachea.
Wind River Range (wind) A range of the Rocky Mountains in west central Wyoming; highest point, 13,787 feet, highest point in Wyoming.
wind rose *Meteorol.* A diagram indicating the direction and relative velocities of the wind in a given locality by means of lines of varying length radiating from a common center.
wind·row (wind′rō′) n. 1 A long ridge or pile of hay or grain raked together preparatory to building into cocks. 2 A row of Indian corn set by setting two rows together. 3 A wind-swept line of dust, surf,

leaves, etc. 4 A deep furrow made for planting. 5 Land on which the trees have been felled by the wind; sometimes, a tornado track: also **wind slash.** — v.t. To rake or shape into a windrow. — **wind′row′er** n.
wind sail 1 *Naut.* A canvas tube or funnel with a spreading opening at one side of the top that may be stayed to face the wind: used to conduct fresh air below decks. 2 A sail on the arm of a windmill.
wind scale See BEAUFORT SCALE.
wind·shake (wind′shāk′) n. A defect in wood; anemosis.
wind·shield (wind′shēld′) n. 1 Any arrangement for breaking the force of the wind against an object. 2 A transparent screen of glass or similar material, attached in front of the occupants of an automobile, airplane, etc., as protection against wind and weather. 3 A covering for a chimney.
wind·sock (wind′sok′) n. *Meteorol.* A large, conical bag, open at both ends, mounted on a pivot, and used to indicate the direction of the wind by the current of air which blows through it; a drogue: also called *windcone.*
Wind·sor (win′zər) Name of the royal family of Great Britain since July 27, 1917, when it was officially changed from *Saxe-Coburg-Gotha.*
Wind·sor (win′zər) 1 A municipal borough in eastern Berkshire, England; site of **Windsor Castle,** a residence of the English sovereigns since the time of William the Conqueror. Officially **New Windsor.** 2 A city on the Detroit River in SE Ontario, Canada, opposite Detroit, Michigan.
Wind·sor (win′zər), **Duke of** See EDWARD VIII.
Windsor chair A wooden chair, with or without arms, common in England and America in the 18th century, typically with a spindle back, turned, slanting legs, and a flat or saddle seat.
Windsor tie A wide, soft necktie knotted loosely in a double bow, usually of black silk cut on the bias.

COMB-BACKED WINDSOR CHAIR

wind·storm (wind′stôrm′) n. A violent wind, usually with little or no precipitation.
wind·suck·er (wind′suk′ər) n. A horse that cribs. — **wind′suck′ing** n. & adj.
wind tee (wind) A T-shaped weathervane, especially one located on or near an aircraft landing field.
wind tunnel *Aeron.* A tunnel-like structure in which the effects of artificially produced winds may be investigated, as on airplane wings and other surfaces.
wind–up (wīnd′up′) n. 1 The act of concluding or closing. 2 A conclusion; a final act or part. 3 In baseball, the swing of the arm preparatory to pitching the ball.
wind·ward (wind′wərd) adj. Being on the side exposed to the wind. — n. The direction from which the wind blows. — **to windward of** Advantageously placed with respect to. — adv. In the direction from which the wind blows: opposed to *leeward.*
Wind·ward Islands (wind′wərd) A West Indies island group north of Trinidad, comprising four British colonies, all federating units of The West Indies, on the islands of Dominica, Grenada, St. Lucia, and St. Vincent, together with the Grenadines; 820 square miles; capital, St. George's, on Grenada. The French island of Martinique; Barbados, a British colony; and three islands of the Netherlands Antilles, Aruba, Bonaire, and Curaçao, are also sometimes included in this group.
Windward Passage The strait between Cuba and Hispaniola in the West Indies; 50 miles wide.
wind·y (win′dē) adj. **wind·i·er, wind·i·est** 1 Pertaining to, consisting of, or abounding in wind; stormy; tempestuous: *windy* weather. 2 Exposed to the wind; wind-swept: high on a *windy* hill. 3 Suggestive of wind; boisterous;

swift: *windy* emotions. **4** Producing, due to, or troubled with gas in the stomach or intestines; producing or affected with flatulence; flatulent: *windy* food. **5** Given to or expressed in bombast; pompous, loquacious, or bragging: *windy* talk, a *windy* orator. See synonyms under BLEAK[1]. — **wind'i·ly** *adv.* — **wind'i·ness** *n.*
Windy City A nickname for CHICAGO.
wine (wīn) *n.* **1** The fermented juice of the grape, containing alcohol to as high as 14 percent by volume, and various other volatile and non-volatile organic substances. The principal types are dry, sweet, red, white, still, and sparkling. Fortified wines have brandy added, and contain alcohol of from 16 to 23 percent. **2** By extension, the fermented juice of some fruit other than the grape: elderberry *wine*; sometimes, a fermented vegetable juice: dandelion *wine*. **3** The effects of drinking too much wine; intoxication. **4** A convivial gathering at which wine and other liquors are served; a wine party. **5** A medicinal preparation in which wine is used as the menstruum: *wine* of opium. **6** Any color resembling the color of wine, especially of a red wine, usually a dark, purplish red. — **Adam's wine** Water. — **new wine in old bottles** Any dynamic new thing, as a doctrine, theory, etc., which cannot be restricted by older forms or customs: with reference to *Matt.* ix 17. — *v.* **wined, win·ing** *v.t.* To entertain or treat with wine. — *v.i.* To drink wine. [OE *wīn* < L *vinum*]
wine·bib·bing (wīn'bib'ing) *adj.* Addicted to excessive drinking of wine. — *n.* The habitual, excessive drinking of wine. — **wine'bib'ber** *n.*
wine card The list of alcoholic drinks for sale at a hotel or restaurant.
wine cellar A storage space for wines; also, the wines stored.
wine-col·ored (wīn'kul'ərd) *adj.* Having the color of red wine.
wine fly Any fly (as of the genus *Piophila*) whose larva lives in wine or other fermented liquor.
wine gallon See under GALLON.
wine·glass (wīn'glas', -gläs') *n.* A small goblet from which to drink wine.
wine·glass·ful (wīn'glas·fo͝ol', -gläs-) *n.* *pl.* **·fuls** The amount a wineglass will hold, approximately equivalent to two fluid ounces or four tablespoonfuls.
wine·grow·er (wīn'grō'ər) *n.* One who cultivates a vineyard and makes wine; a viticulturist. — **wine'grow'ing** *adj.* & *n.*
wine measure A system of liquid measures formerly used for wines and spirits in which the gallon was equal to the present U.S. gallon.
wine palm Any palm from which palm wine is obtained.
wine·press (wīn'pres') *n.* An apparatus or a place where the juice of grapes is expressed. — **wine'press'er** *n.*
wine purple A hue of purple consisting of 50 percent red, 33 percent black, and 17 percent blue.
win·er·y (wī'nər·ē) *n.* *pl.* **·er·ies** **1** An establishment for making wine. **2** A room for fining and storing wines.
Wine·sap (wīn'sap) *n.* An American variety of red winter apple.
wine·skin (wīn'skin') *n.* The skin of some domestic quadruped kept as entire as possible and made into a tight bag for containing wine: much used in the Orient.
wine-sop (wīn'sop') *n.* Any farinaceous foodstuff steeped or sopped in wine, as bread or cake.
wine steward An attendant in a restaurant or hotel who takes orders for wines, and who is in charge of the wine cellar.
wine·tast·er (wīn'tās'tər) *n.* A person who tastes wine to judge its quality.
wine vinegar A vinegar made from wine.
wine whey *Brit.* A beverage made of wine and curdled milk.
Win·fre·da (win·frā'də) Latin form of WINIFRED. Also *Du.* **Win·fried** (vin'frēt), *Sw.* **Win·frid** (vin'frid).
wing (wing) *n.* **1** An organ of flight; specifically, one of the anterior movable pair of appendages of a bird or bat, homologous with the forelimbs of vertebrates but adapted for flight. **2** An analogous organ in insects and some other animals. **3** One of the pectoral fins of a flying fish. **4** *Slang* An arm; specifically, in baseball, the arm used for throwing or pitching. **5** Something regarded as conferring the power of swift motion or performing some function of a wing: on *wings* of song. **6** Flight or passage by or as by wings; also, the means or act of flying: to take *wing*. **7** Anything resembling or suggestive of a wing in form, function, or appearance; specifically, one of a pair of pneumatic devices for aid in swimming; a shoulder ornament. **8** The flare of a moldboard plowshare; also, the curved mudguard or fender of an automobile. **9** Something moved by or moving in the wind, as the vane of a windmill or a winnowing fan. **10** *Mil.* Either division of a military force on either side of the center. **11** An analogous formation in certain outdoor games, as hockey or football. **12** Either of two extremist groups or factions in a political or other organization: the left *wing*. **13** *Archit.* A part attached to a side; especially, a projection or extension of a building on the side of the main portion. **14** A sidepiece at the top of an armchair. **15** A side section of something that shuts or folds, as a double door, a screen, etc. **16** In fortifications, one of the sides connecting an outwork with the main fort. **17** *Anat.* An ala: a *wing* of the nose. **18** *Bot.* Any thin membranous or foliaceous expansion of an organ, as of certain stems, seeds, samaras, etc. **19** *Zool.* One of the lateral finlike expansions of the foot of a pteropod. **20** One of the sides of a stage; a small platform at either side of the stage; also, a piece of scenery for the side. **21** *Aeron.* One of the sustaining surfaces of an airplane. **22** A tactical and administrative unit of the U.S. Air Force, under the direction of a wing commander, larger than a group and smaller than a command. **23** A shore dam or jetty for narrowing a channel; also, an extension of a dam at either end, usually built at an angle. — **on** (or **upon**) **the wing** In flight; as, a bird *on the wing*; hence, just about to go; departing; also, journeying. — **to take wing** To fly away. — **under one's wing** Under one's protection. — *v.t.* **1** To pass over or through in flight. **2** To accomplish by flying: the eagle *winged* its way. **3** To enable to fly. **4** To cause to go swiftly; speed: Hope *winged* his steps. **5** To transport by flight. **6** To provide with wings for flight; also, to feather (an arrow). **7** To supply with a side body or part: The house was *winged* on both sides. **8** To wound (a bird) in a wing; hence, to disable by a minor wound: I *winged* him in the arm. — *v.i.* **9** To fly; soar. [< ON *vængr*]
wing and wing *Naut.* With sails spread or boomed out on each side like wings: said of a fore-and-aft vessel running downwind.
wing-back (wing'bak') *n.* In football, the position taken by one (**single wingback**) or two (**double wingback**) of the backs behind or beyond the ends; also, the back so posted.
wing-bow (wing'bō') *n.* A distinctive mark of color on the bend of the wing in a domestic fowl.
wing chair A large armchair, upholstered throughout, with high back and side pieces designed as protection from drafts.
wing cover The elytron of an insect. Also **wing case**.
wing covert *Ornithol.* One of the small close feathers clothing the bend of a bird's wing and covering the insertion of the flight feathers. Those of the lining of the wing are called *undercoverts*.

WING CHAIR

winged (wingd, *Poetic* wing'id) *adj.* **1** Having wings. **2** Passing swiftly; soaring; lofty; rapt. **3** Alive with creatures having wings. **4** (wingd) *Colloq.* Wounded or disabled in or as in the wing.
winged wolf A harpy (def. 2).
wing flap *Aeron.* A control surface hinged to an airplane wing, used primarily to increase lift and to retard the speed.
wing-foot·ed (wing'fo͝ot'id) *adj.* Rapid; swift.
wing·less (wing'lis) *adj.* Having no wings, or having aborted wings.
wing·let (wing'lit) *n.* An alula.
wing loading *Aeron.* The over-all weight of a fully loaded airplane divided by the area of the supporting surface, exclusive of the stabilizer and elevators. Also **wing load**.
wing-nut (wing'nut') *n.* A thumbnut.
wing·o·ver (wing'ō'vər) *n.* *Aeron.* A flight maneuver in which an airplane at the top of a climbing turn and just before stalling is put into a dive before resuming normal flight in the direction from which it started.
wing rail A guardrail, as at a railway switch.
wing skid *Aeron.* A device set beneath the wing tip of an airplane to guard the tip against contact with the ground.
wing·spread (wing'spred') *n.* The distance between the tips of the fully extended wings of a bird, insect, or airplane.
wing walk *Aeron.* A reinforced section of an airplane wing, used as a walking strip.
wing-wea·ry (wing'wir'ē) *adj.* Fatigued from flight or travel.
wing·y (wing'ē) *adj.* Winged; swift.
Win·i·fred (win'ə·frid) A feminine personal name. Also **Win'e·fred**, **Win'i·frid**. [< Welsh *Gwenfrewi* a white wave]
wink (wingk) *v.i.* **1** To close and open the eye or eyelids quickly. **2** To draw the eyelids of one eye together, as in conveying a hint or making a sign. **3** To shut one's eyes, especially in ignoring; pretend not to see: usually with *at*. **4** To emit fitful gleams; twinkle. — *v.t.* **5** To close and open (the eye or eyelids) quickly. **6** To move, force, etc., by winking: with *away*, *off*, etc. **7** To signify or express by winking. — *n.* **1** The act of winking. **2** The time necessary for a wink. **3** A twinkle. **4** A hint conveyed by winking. **5** A short nap: especially in the phrase **forty winks**. [OE *wincian* close the eyes]
Win·kel·ried (ving'kəl·rēt), **Arnold von** A 14th century Swiss patriot.
wink·er (wing'kər) *n.* **1** One who winks. **2** A blinder for a horse. **3** *Slang* An eyelash. **4** A small secondary bellows for use with an organ. **5** The nictitating membrane, as of a bird. **6** The muscle by which winking is done. **7** *pl.* *Slang* Spectacles.
win·kle (wing'kəl) *n.* A periwinkle[1]. [Short for PERIWINKLE[1]]
win·na (win'ə) *Scot.* Will not.
Win·ne·ba·go (win'ə·bā'gō) *n.* *pl.* **·gos** or **·goes** One of a tribe of North American Indians of Siouan linguistic stock, formerly occupying what is now eastern Wisconsin, south of Green Bay, where many still survive.
Win·ne·ba·go (win'ə·bā'gō), **Lake** The largest lake in Wisconsin, in the eastern part; 215 square miles; 30 miles long.
Win·ne·pe·sau·kee (win'ə·pə·sô'kē), **Lake** The largest lake in New Hampshire, in the east central part; 25 miles long, 12 miles wide. Also **Win'ni·pe·sau'kee**.
win·ner (win'ər) *n.* One who or that which wins.
Win·nie (win'ē) Diminutive of WINIFRED, WINSTON.
win·ning (win'ing) *adj.* **1** Successful in achievement, especially in competition. **2** Capable of winning or charming; attractive; winsome. — *n.* **1** The act of one who wins. **2** That which is won: usually in the plural. **3** A new opening in a mine; also, a section of a mine prepared for working. — **win'ning·ly** *adv.* — **win'ning·ness** *n.*
winning gallery In court tennis, the grille or square opening in the penthouse in the rear of the hazard court: so named because a ball played into it counts as a win.
winning hazard See HAZARD *n.* (def. 6).
winning post The post or goal at the end of a racecourse.
Win·ni·peg (win'ə·peg) The capital of Manitoba province, Canada, on the Red River in the SE part.
Winnipeg, Lake A lake in south central Manitoba, Canada; 240 miles long, 55 miles wide; 9,398 square miles.
Win·ni·pe·go·sis (win'ə·pə·gō'sis), **Lake** A lake in western Manitoba, Canada, west of Lake Winnipeg; 125 miles long, 25 miles wide; 2,086 square miles.
Winnipeg River A river in NW Ontario and

SE Manitoba, flowing 200 miles NW from Lake of the Woods to Lake Winnipeg.

win·nock (win′ək) *n. Scot.* A window.

win·now (win′ō) *v.t.* **1** To separate (grain, etc.) from the chaff by means of wind or a current of air. **2** To blow away (the chaff) thus. **3** To examine so as to separate good from bad; analyze minutely; sift. **4** To separate (what is valuable) from what is valueless, or to eliminate (what is valueless) from what is valuable; distinguish; sort: often with *out.* **5** To blow upon; cause to flutter. **6** To beat or fan (the air) with the wings. **7** To scatter by blowing; disperse. **8** *Rare* To proceed along (a course) by flapping the wings. —*v.i.* **9** To separate grain from chaff. **10** To fly; flap. —*n.* **1** Any device used in winnowing grain. **2** The act of winnowing; also, a vibrating motion caused by a current of air. [OE *windwian* < *wind* the wind] —**win′now·er** *n.*

win·o (wī′nō) *n. pl.* **·noes** or **·nos** *U.S. Slang* A drunkard who habitually drinks sweet, fortified wines. [<WINE]

Wins·low (winz′lō), **Edward,** 1595–1655, English Puritan, governor of Plymouth Colony.

win·some (win′səm) *adj.* Having a winning appearance or manner; pleasing; attractive; rarely, joyous. See synonyms under AMIABLE, LOVELY. [OE *wynsum* < *wyn* joy] —**win′·some·ly** *adv.* —**win′some·ness** *n.*

Win·sor (win′zər), **Justin,** 1831–97, U.S. historian and librarian.

Win·ston (win′stən) A masculine personal name. [Orig. from *Winston,* a hamlet near Circencester, England]

Win·ston-Sa·lem (win′stən·sā′ləm) A city in NW central North Carolina; one of the world's chief tobacco centers.

win·ter (win′tər) *n.* **1** The coldest season of the year, extending from the end of autumn to the beginning of spring: in the northern hemisphere, astronomically from the winter solstice, December 21, to the vernal equinox, March 21, but popularly regarded as including December, January, and February. ◆ Collateral adjectives: *hibernal, hiemal.* **2** Any time compared to winter, as being marked by lack of life, warmth, and cheer. **3** A year as including the winter season: used in reckoning the age of elderly persons: a man of ninety *winters.* —*v.i.* To pass the winter: We *wintered* in Bermuda. —*v.t.* To care for, feed, or protect during the winter: to *winter* animals or plants. —*adj.* **1** Pertaining to or taking place in winter; hibernal. **2** Suitable to or characteristic of winter. [OE] —**win′ter·er** *n.* —**win′ter·ish** *adj.* —**win′·ter·less** *adj.*

winter aconite A European tuberous-rooted, hardy, flowering garden herb *(Eranthis hyemalis)* of the crowfoot family, 5 to 8 inches high, with bright-yellow sessile flowers and oblong anthers.

win·ter·ber·ry (win′tər·ber′ē) *n. pl.* **·ries** Any of several North American shrubs (genus *Ilex*) of the holly family, bearing bright-red berrylike drupes about the size of a pea; especially, the smooth winterberry *(I. laevigata)* of the eastern United States.

win·ter·bourne (win′tər·bôrn′, -bōrn′, -bo͝orn′) *n.* A stream flowing only during excessive rainfall, as in winter, when water at the source rises above a certain level. [OE *winter burna* < *winter* + *burna* a stream]

win·ter-feed (win′tər·fēd′) *v.t.* **-fed, -feed·ing** To feed (stock) during the time when grazing is impossible.

win·ter·green (win′tər·grēn) *n.* **1** A small evergreen plant *(Gaultheria procumbens)* of eastern North America, bearing a cluster of aromatic oval leaves and white, bell-shaped flowers surrounded by red berries (often called *teaberries* or *checkerberries*). **2** Oil of wintergreen: a colorless, volatile oil extracted from the leaves of the true wintergreen, used as a flavor and in medicine: often called *Gaultheria oil.* **3** Any of various English low evergreen herbs (genus *Pyrola*). In the United States they are sometimes called *shinleaf* or **English** or **false wintergreen.** [On analogy with Du. *wintergroen*; so called because it is an evergreen]

winter itch Frost itch.

win·ter·ize (win′tə·rīz) *v.t.* **·ized, ·iz·ing** To prepare or equip for winter.

win·ter·kill (win′tər·kil′) *v.t.* & *v.i.* To die or kill by exposure to extreme cold: said of plants and grains. —**win′ter·kill′ing** *adj.* & *n.*

win·ter·ly (win′tər·lē) *adj.* Wintry; cheerless.

winter melon A hardy, cold-resistant muskmelon *(Cucumis melo,* variety *inodorus).*

Win·ter·thur (vin′tər·to͝or) A city of northern Switzerland NE of Zurich; a rail and industrial center.

win·ter·tide (win′tər·tīd′) *n. Poetic* Winter. Also **win′ter·time′** (-tīm′).

winter wheat Wheat planted before snowfall and harvested the following summer.

Win·throp (win′thrəp), **John,** 1588–1649, English Puritan; governor of Massachusetts Colony. —**John,** 1606–76, governor of Connecticut Colony; son of the preceding.

win·try (win′trē) *adj.* **·tri·er, ·tri·est** Belonging to winter; cold; frosty; brumal. Also **win′·ter·y** (-tər·ē). —**win′tri·ly** *adv.* —**win′tri·ness** *n.*

win·y (wī′nē) *adj.* **win·i·er, win·i·est** Having the taste or qualities of wine.

Win·yah Bay (win′yô) An estuary of the Pee Dee and other rivers in South Carolina NE of Charleston; 14 miles long.

winze[1] (winz) *n. Mining* A small inclined shaft from one level of a mine to another. [Earlier *winds,* ? <obs. *wind* a windlass, fusion of MDu. *winde* a windlass and WIND[2]]

winze[2] (winz) *n. Scot.* An oath.

wipe (wīp) *v.t.* **wiped, wip·ing** **1** To subject to slight friction or rubbing, usually with some soft, absorbent material. **2** To remove by rubbing lightly; brush: usually with *away* or *off.* **3** To move, apply, or draw for the purpose of wiping: He *wiped* his hand across his brow. **4** To apply solder to with a piece of greased cloth or leather; solder with a wiper or pad: to *wipe* a joint. See synonyms under CLEANSE. —**to wipe out** To remove or destroy utterly; annihilate. —*n.* **1** The act of wiping or rubbing. **2** *Slang* A sweeping blow or stroke; a swipe. **3** *Mech.* A wiper or cam. **4** *Slang* A handkerchief. **5** *Slang* A jeer; jibe. [OE *wīpian.* Akin to WISP.]

wip·er (wī′pər) *n.* **1** One who wipes. **2** An article designed or used for wiping. **3** *Mech.* A cam having one or more slightly curved projections serving, when mounted on a rock shaft or rotating shaft, to give a reciprocating (usually vertical) motion to another part. **4** *Electr.* A moving member of an electrical device which makes contact with the terminals. **5** One who cleans locomotives in a roundhouse.

wire (wīr) *n.* **1** A slender rod, strand, or thread of ductile metal, usually formed by drawing through dies or holes. **2** Something made of wire, as a fence, a bar of a cage, or a snare made for catching small animals. **3** A telegraph cable. **4** The telegraph system as a means of communication. **5** A telegram. **6** The screen of a papermaking machine. **7** A fine metallic thread, a cobweb, or one of a set of ruled lines, in the focus of a telescope. **8** *Ornithol.* A long slender filament of the plumage of various birds. **9** *pl.* A secret means of exerting influence: to pull the *wires*: from the analogy with the system of hidden wires by which puppets are operated. **10** An imaginary line marking the finish of a racecourse. —**to lay wires for** To prepare for. —**under wire** Fenced. —*v.* **wired, wir·ing** *v.t.* **1** To fasten with wire. **2** To furnish or equip with wiring: The studio was *wired* for sound. **3** In croquet, to place (a ball) so that the wire of an arch will be between it and another ball. **4** To catch, as a rabbit, with a snare of wire. **5** *Colloq.* To transmit or send by electric telegraph: to *wire* an order. **6** *Colloq.* To send a telegram to: Will you *wire* John? **7** To place on wire, as beads. —*v.i.* **8** *Colloq.* To telegraph. [OE *wīr*]

wire cloth A fabric of woven wire, as for strainers, window screens, etc.

wire coat An outer coat, as of some dogs, of dense stiff hair.

wire-danc·er (wīr′dan′sər, -dän′-) *n.* One who performs feats of balancing, etc., upon a wire stretched in mid-air: also called *wirewalker.* —**wire′-danc′ing** *n.*

wire-draw (wīr′drô′) *v.t.* **-drew, -drawn, -draw·ing** **1** To draw, as a metal rod, through a series of holes of diminishing diameter to produce a wire. **2** To treat (a subject) with excessive subtlety or overrefinement. —**wire′-draw′er** *n.* —**wire′-draw′ing** *n.*

wire entanglement *Mil.* An obstruction of barbed wire, set up to hinder enemy action in warfare.

wire gage **1** A gage for measuring the diameter of round wire, usually a round plate with slots on its periphery numbered according to an arbitrary standard, or a long graduated plate with a slot of diminishing width. **2** A standard system of sizes for wire. Also **wire gauge.**

wire gauze A material of a gauzelike structure made of interwoven strands of wire.

wire glass See under GLASS.

wire·grass (wīr′gras′, -gräs′) *n.* **1** A European grass *(Poa compressa)* having slender, compressed stems, cultivated in the United States and Canada: also called *Canada bluegrass.* **2** Any one of several similar grasses.

wire-haired griffon (wīr′hârd′) A griffon (def. 2).

wire·less (wīr′lis) *adj.* **1** Without wire or wires; having no wires. **2** *Brit.* Radio. —*n.* **1** The wireless telegraph or telephone system, or a message transmitted by either. **2** *Brit.* Radio. —*v.t.* & *v.i. Brit.* To communicate (with) by wireless telegraphy; radio.

wireless telegraphy or **telephony** Telegraphy or telephony without wires connecting the points of transmission and reception, the message being transmitted through space by electromagnetic waves; radio communication.

wire·man (wīr′mən) *n. pl.* **·men** (-mən) **1** A man who has to do with wire. **2** One who handles wire for telegraph lines, etc.; a wirer.

wire mark The faint impression left on paper by the wires of the mold during manufacture.

wire netting Netting made of wire, as window screens, fences, etc.

Wire·pho·to (wīr′fō′tō) *n. pl.* **·tos** An apparatus and method for transmitting and receiving photographs by wire: a trade name.

wire·pull·er (wīr′po͝ol′ər) *n.* One who pulls wires, as of a puppet; hence, one who uses secret means to control others or gain his own ends; an intriguer.

wire·pull·ing (wīr′po͝ol′ing) *n.* **1** The pulling of wires, as in a puppet show. **2** The use of secret influence to obtain an end.

wir·er (wīr′ər) *n.* **1** A trapper who snares with wire contrivances. **2** A wireman.

wire recorder A device for recording sounds on an uncoiling fine wire by magnetic registration of variations in the flow of electrical current from a microphone: these sounds are reproduced as the magnetized wire is passed back between the poles of the electromagnet.

wire rope A rope of wires firmly wound together.

wire·sonde (wīr′sond′) *n. Meteorol.* A type of radiosonde for use at low altitudes, the required data being transmitted by wire to ground stations. [<WIRE + (RADIO)SONDE]

wire-spun (wīr′spun′) *adj.* Wire-drawn; spun or drawn out too fine; overrefined.

wire·tap·ping (wīr′tap′ing) *n.* The act or process of tapping a wire for the purpose of diverting the current or of securing or sending information: often illicitly. —**wire′tap′per** *n.*

wire·walk·er (wīr′wô′kər) *n.* A wire-dancer.

wire wheel In automobiles, a wheel in which slender metal spokes, usually in a criss-cross pattern, connect the hub and rim.

wire·work (wīr′wûrk′) *n.* **1** Small articles made of wire cloth. **2** Wire fabrics in general. —**wire′work′er** *n.*

wire·works (wīr′wûrks′) *n. pl.* **·works** **1** A factory where wire or articles of wire are made. **2** A shop where wire is woven and manufactured into protective screens, filters, or the like.

wire·worm (wīr′wûrm′) *n.* **1** The cylindrical brown to whitish larva of a click beetle, with a stiff, wiry texture: some species are common in fields, where they damage the roots of plants. For illustration see INSECTS (injurious). **2** A millipede.

wire-wove (wīr′wōv′) *adj.* **1** Denoting a high grade of paper with a smooth writing surface. **2** Woven of wire.

wir·ing (wīr′ing) *n.* An entire system of wire installed for the distribution of electric power, as for lighting, heating, radio, engine ignition, or the like.

wit·ted (wit'id) *adj.* Having wit: used principally in compounds with the meaning having (a specified kind of) wit: quick-*witted*, half-*witted*.

Wit·te·kind (vit'ə-kint), died in battle 807?, leader of the Saxons against Charlemagne. Also spelled *Widukind*.

Wit·ten (vit'n) A city on the Ruhr in North Rhine-Westphalia, Germany.

Wit·ten·berg (wit'n-bûrg, *Ger.* vit'n-berkh) A city on the Elbe river in central East Germany, in the former state of Saxony-Anhalt; the Protestant Reformation originated here, 1517.

Witt·gen·stein (vit'gən-shtīn), **Ludwig,** 1889-1951, Austrian philosopher active in England.

Witt·gen·stein Island (vit'gən-shtīn) A former name for FAKARAVA.

wit·ti·cism (wit'ə-siz'əm) *n.* A witty saying. [<WITTY, on analogy with *criticism*; coined by Dryden]

wit·ting[1] (wit'ing) *adj.* Aware; done consciously, with knowledge and responsibility. [<WIT[2]]

wit·ting[2] (wit'ing) *n. Obs.* Knowledge; information. [<ON *vitand* consciousness <*vita* know]

wit·ting·ly (wit'ing-lē) *adv.* With knowledge and by design; knowingly and designedly.

wit·tol (wit'l) *n. Obs.* A contented cuckold; a husband who is aware of, but indifferent to, his wife's infidelity. [ME *wetewold* <*wete* know + (*coke*)*wold* cuckold]

wit·ty (wit'ē) *adj.* **·ti·er, ·ti·est** 1 Given to making original or clever speeches; quick at repartee; humorous. 2 Displaying or full of wit. See synonyms under HUMOROUS. [OE *wittig* wise] — **wit'ti·ly** *adv.* — **wit'ti·ness** *n.*

Wit·wa·ters·rand (wit-wä'tərs-ränt, -rand) A region of southern Transvaal on a rocky ridge near Johannesburg; 1,000 square miles; site of the world's richest gold fields: also *The Rand*.

witz·chour·a (wits-choor'ə) *n.* A mantle with large sleeves and a wide collar, worn in the early 19th century. [<F *vitchoura* <Polish *wilczura* a wolf-skin coat <*wilk* a wolf]

wive (wīv) *v.* **wived, wiv·ing** *v.t.* 1 To furnish with a wife. — *v.i.* 2 To marry a woman. [OE *wīfian* <*wīf* a wife, woman]

wi·vern (wī'vərn) *n. Her.* A two-legged, winged dragon, with barbed and knotted tail: often spelled *wyvern*. Also **wi'ver**. [<AF *wivre*, OF *guivre* a dragon, serpent, var. of *vivre* <L *vipera*]

wives (wīvz) Plural of WIFE.

wiz (wiz) *n. Slang* A wizard (def. 2). [Short for WIZARD]

wiz·ard (wiz'ərd) *n.* 1 One supposed to be in league with the devil; a male witch; sorcerer. 2 *Colloq.* A very skilful or clever person: He's a *wizard* with machinery. 3 *Obs.* A wise man; sage. — *adj.* 1 Having magical powers. 2 Fascinating; enchanting. [ME *wysard* <*wys*, OE *wīs* wise]

WIVERN

wiz·ard·ry (wiz'ərd-rē) *n.* The practice or methods of a wizard.

wiz·en[1] (wiz'ən) *v.t. & v.i.* To become or cause to become withered; shrivel. — *adj.* Wizened; shrunken; shriveled. [OE *wisnian* dry up, wither]

wiz·en[2] (wiz'ən) *n. Dial.* or *Obs.* The weasand. Also **wiz'zen**.

wiz·ened (wiz'ənd) *adj.* Shrunken; withered; dried up.

Wlo·cla·wek (vwô-tswä'vek) A city on the Vistula in central Poland; a manufacturing center. Russian **Vlo·tslavsk** (vlo·tsläfsk').

woad (wōd) *n.* 1 An Old World herb (*Isatis tinctoria*) of the mustard family; dyer's-weed. 2 The blue dyestuff obtained from its leaves. [OE *wād*] — **woad'ed** *adj.*

woad·wax·en (wōd'wak'sən) *n.* Dyer's-broom: also spelled *woodwaxen*.

woald (wōld) See WELD[2].

wob·ble (wob'əl) *v.* **·bled, ·bling** *v.i.* 1 To move or sway unsteadily, as a top while rotating at a low speed. 2 To show indecision or unsteadiness; waver; vacillate. — *v.t.* 3 To cause to wobble. — *n.* An unsteady motion, as that of unevenly balanced rotating bodies. Also spelled *wabble*. [? <LG *wabbeln*] —

wob'bler *n.* — **wob'bling** *adj.* — **wob'bling·ly** *adv.* — **wob'bly** *adj.*

wobble pump A hand pump.

wob·bly (wob'lē) *n. pl.* **·blies** *U.S. Slang* A member of the Industrial Workers of the World (I.W.W.). [Appar. mispronunciation of *w* in *I.W.W.*]

Wode·house (wood'hous, wōd'-), **P(elham) G(renville),** born 1881, English humorous novelist.

Wo·den (wōd'n) The Old English name for Odin, the chief Norse god. Wednesday is named for Woden. Also **Wo'dan**.

woe (wō) *n.* 1 Overwhelming sorrow; grief. 2 Heavy affliction or calamity; disaster: His *woes* are many. — *interj.* Alas! used to proclaim disaster or to express sorrow, to denounce, or invoke censure. Also **wo.** [OE *wa* misery]

woe·be·gone (wō'bi-gôn', -gon') *adj.* Overcome with woe; mournful; sorrowful. Also **wo'be-gone'**. See synonyms under SAD.

woe·ful (wō'fəl) *adj.* 1 Accompanied by or causing woe; direful. 2 Expressive of sorrow; doleful. 3 Deserving condemnation; paltry; miserable; mean. Also **woe'some, wo'ful**. See synonyms under PITIFUL, SAD. — **woe'ful·ly** *adv.* — **woe'ful·ness** *n.*

Woerth (vært) A town in NE France; scene of a French defeat in the Franco-Prussian War, 1870.

Woë·vre (vō·e'vr') A tableland in NE France, near Verdun; scene of severe fighting in World War I, 1915 and 1918.

Wof·fing·ton (wof'ing-tən), **Margaret,** 1714?-1760; English actress born in Dublin: commonly called "Peg Woffington."

Wöh·ler (vœ'lər), **Friedrich,** 1800-82, German chemist.

wo·kas (wō'kəs) *n.* A yellow waterlily (*Nuphar polysepalum*) of western North America, having small seeds formerly roasted and eaten by the Klamath Indians. Also **wo'cus.** [<Klamath]

woke (wōk) Past tense of WAKE[1].

wok·en (wō'kən) Dialectal past participle of WAKE[1].

Wol·cott (wool'kət), **Oliver,** 1726-97, American statesman; signed Declaration of Independence.

wold (wōld) *n.* 1 An undulating tract of open upland; down or moor. 2 *Obs.* A forest. [OE *wald* a forest]

Wolds (wōldz), **the** A range of hills in Lincolnshire and Yorkshire, England, parallel to the coast, north and south of the Humber; highest point, 800 feet.

wolf (wōolf) *n. pl.* **wolves** (woolvz) 1 Any of a genus (*Canis*) of large carnivorous mammals related to the dog, especially the common European species (*C. lupus*) or the timber wolf of North America. ◆ Collateral adjective: *lupine*. 2 Any ravenous, cruel, or rapacious person or thing; hence, popularly, a philanderer. 3 *Entomol.* The destructive larva of various beetles and moths. 4 The harsh, dissonant sound of certain chords on a keyed instrument, as an organ or piano, when tuned by a system of unequal temperament; in bowed instruments, a harsh, discordant sound caused by defective vibration of one or more notes of a scale. — **to cry wolf** To give a false alarm. — **to keep the wolf from the door** To avert want or starvation. — *v.t.* To devour ravenously; gulp down: He *wolfed* his food. [OE *wulf*]

Wolf (wōolf) The constellation Lupus.

Wolf (vôlf), **Friedrich August,** 1759-1824, German classical scholar. — **Hugo,** 1860-1903, Austrian composer.

wolf·ber·ry (wōolf'ber'ē) *n. pl.* **·ries** A shrub (*Symphoricarpos occidentalis*) of the honeysuckle family, with pinkish, bell-shaped flowers and white berries in spikes, growing in the western United States.

Wolf Cub A member of the division of Boy Scouts for boys between 8 and 11 years of age.

wolf dog 1 A large dog for hunting wolves. 2 A cross between a wolf and a dog.

Wolfe (woolf), **Charles,** 1791-1823, Irish poet. — **James,** 1727-59, English general; defeated the French under Montcalm at Quebec; both he and Montcalm were killed. — **Thomas Clayton,** 1900-38, U.S. novelist.

Wolff (vôlf), **Christian von,** 1679-1754, German philosopher. — **Kaspar Friedrich,** 1733-1794, German anatomist.

Wolf-Fer·ra·ri (vôlf'fer-rä'rē), **Ermanno,** 1876-1948, Italian composer.

Wolff·i·an (wool'fē·ən) *adj.* Pertaining to or named after the German anatomist Kaspar F. Wolff.

Wolffian body *Anat.* The mesonephros.

wolf fish A large fish (*Anarhichas lupus*) of the North Atlantic, with powerful teeth adapted for crushing shellfish.

Wolf·gang (vôlf'gäng) *German* A masculine personal name. [<G, a wolf's progress]

wolf·hound (wōolf'hound') *n.* Either of two breeds of large dogs, the **Russian wolfhound** or *borzoi* and the **Irish wolfhound**, a dog resembling the Great Dane, trained, or originally intended, to catch and kill wolves.

wolf·ish (wool'fish) *adj.* 1 Having the qualities of a wolf; rapacious; savage. 2 *Colloq.* Ravenously hungry. — **wolf'ish·ly** *adv.* — **wolf'ish·ness** *n.*

wolf pack A number of submarines which cooperate in making concerted attacks on enemy ships or convoys.

wolf·ram (wool'frəm) *n.* 1 Wolframite. 2 Tungsten. [<G, prob. <*wolf* a wolf + *rahm* cream, soot]

wolf·ram·ite (wool'frəm·īt) *n.* A submetallic, grayish-black or brown tungstate of iron and magnesium, crystallizing in the monoclinic system. It is important commercially as a source of tungsten and its compounds. [<G *wolframit* <*wolfram* tungsten]

Wol·fram von Esch·en·bach (vôl'främ fôn esh'ən-bäkh), 1165?-1220?, German poet.

wolf's-bane (wōolfs'bān') *n.* 1 A species of aconite; monkshood. 2 A species of European arnica (*Arnica montana*), used as a lotion for bruises. 3 The silk vine. [Trans. of NL *lycoctonum* <Gk. *lykoktonon*, lit., a wolf-slayer <*lykos* a wolf + *kteinein* kill]

Wol·las·ton (wool'əs·tən), **William Hyde,** 1766-1828, English chemist and physicist.

wol·las·ton·ite (wool'əs·tən·īt') *n.* A vitreous, white, translucent calcium silicate, crystallizing in the monoclinic system. [after Dr. W. H. *Wollaston*]

Wolse·ley (woolz'lē), **Garnet Joseph,** 1833-1913, first Viscount Wolseley, British general.

Wol·sey (wool'zē), **Thomas,** 1475?-1530, English cardinal and statesman.

wolv·er (wool'vər) *n.* One who hunts wolves.

Wol·ver·hamp·ton (wool'vər·hamp'tən) A county borough and industrial center in southern Stafford, England.

wol·ver·ine (wool'və·rēn') *n.* A rapacious and cunning carnivore (genus *Gulo*) of northern forests, with stout body and limbs and bushy tail. Also **wol'ver·ene'**. [Dim. of WOLF]

WOLVERINE (Body to 3 feet long; tail, 1 1/2 feet)

Wolverine State Nickname of MICHIGAN.

wolves (woolvz) Plural of WOLF.

wom·an (woom'ən) *n. pl.* **wom·en** (wim'in) 1 An adult human female. 2 The female part of the human race; women collectively. 3 Womanly character; femininity: usually with *the*. 4 As applied to a man, one who is effeminate, timid, or weak. 5 A female attendant or servant. 6 A paramour or kept mistress. 7 *Colloq.* A wife. — *adj.* 1 Feminine; characteristic of women. 2 Female: when used with a plural noun, usually *women*: *women* students. 3 Affecting or pertaining to women. — *v.t. Obs.* To play the part of a woman in or in reference to. [OE *wīfmann* <*wīf* a wife + *mann* a human being]

wom·an·hood (woom'ən·hood) *n.* 1 The state of a woman or of womankind. 2 Women collectively.

wom·an·ish (woom'ən·ish) *adj.* Characteristic of a woman; effeminate. See synonyms under FEMININE. — **wom'an·ish·ly** *adv.* — **wom'an·ish·ness** *n.*

wom·an·ize (woom'ən·īz) *v.* **·ized, ·iz·ing** *v.t.*

womankind   1448   **woodpecker**

To make effeminate or womanish. — *v.i. Colloq.* To consort with women illicitly.
wom·an·kind (woom′ən·kīnd′) *n.* Women collectively.
wom·an·ly (woom′ən·lē) *adj.* Having the qualities natural, suited, or becoming to a woman; feminine. — *adv.* In a feminine manner; like a woman. — **wom′an·li·ness** *n.*
woman suffrage See under SUFFRAGE. — **wom′an-suf′fra·gist** *n.*
womb (woom) *n.* 1 The organ in which the young of higher mammals are developed; the uterus. 2 The place where anything is engendered or brought into life. 3 A cavity viewed as enclosing something. 4 *Obs.* The belly or stomach. [OE *wamb, womb* the belly]
wom·bat (wom′bat) *n.* An Australian nocturnal marsupial (family *Phascolomidae*) resembling a small bear. [< Australian]
wombed (woomd) *adj.* Having a womb; hence, hollow; capacious; cavernous. Also **womb′y.**
wom·en (wim′in) Plural of WOMAN.
wom·en·folk (wim′in·fōk′) *n. pl.* Women collectively. Also **wom′en·folks′.**

WOMBAT
(From 3 to 4 feet in length)

Women in the Air Force A corps of women in the U.S. Air Force, including all women except nurses and medical specialists. Abbr. *WAF* or *W.A.F.*
Woman's Army Corps A corps of women in the U.S. Army, composed of all women except nurses and medical specialists. Abbr. *WAC* or *W.A.C.*
women's rights The rights of women to enjoy equal legal rights and privileges with men, as of suffrage, property, and education.
wom·er·a (wom′ər·ə) *n.* A device used by Australian aborigines for throwing javelins, spears, etc. [< Australian]
won[1] (wun) Past tense and past participle of WIN.
won[2] (wun) *v.i. Scot. & Brit. Dial.* To abide; dwell; live. [OE *wunian*]
won·der (wun′dər) *n.* 1 A feeling of mingled surprise and curiosity; astonishment. 2 That which causes wonder, a prodigy; a strange thing; a miracle. See synonyms under PRODIGY. — **nine days' wonder** Something that excites public wonder for a short time. — *v.t.* 1 To have a feeling of doubt and strong curiosity in regard to. — *v.i.* 2 To be affected or filled with wonder; marvel. 3 To be doubtful; query mentally; want to know. [OE *wundor*] — **won′der·er** *n.* — **won′der·ing** *adj.* — **won′der·ing·ly** *adv.*
won·der·ful (wun′dər·fəl) *adj.* Of a nature to excite wonder; marvelous. See synonyms under EXTRAORDINARY. — **won′der·ful·ly** *adv.* — **won′der·ful·ness** *n.*
won·der·land (wun′dər·land′) *n.* A realm of fairy romance or wonders.
won·der·ment (wun′dər·mənt) *n.* 1 The emotion of wonder; surprise. 2 Something wonderful; a marvel.
Wonder State Nickname for ARKANSAS.
won·der·strick·en (wun′dər·strik′ən) *adj.* Suddenly smitten with wonder or admiration. Also **won′der-struck′** (-struk′).
won·der·work (wun′dər·wûrk′) *n.* A work inspiring wonder; miracle. — **won′der-work′er** *n.* — **won′der-work′ing** *adj.*
won·drous (wun′drəs) *adj.* Commanding wonder; wonderful; marvelous. — *adv.* Surprisingly. [Alter. of ME *wonders*, genitive of WONDER.] — **won′drous·ly** *adv.* — **won′drous·ness** *n.*
won·ky (wong′kē) *adj. Brit. Slang* Unsteady; liable to break down; shaky; feeble. [Prob. OE *wancol* shaky]
won·ner (wun′ər) *n. Scot.* A prodigy; wonder.
Won·san (wœn·sän) A port in eastern Korea: Japanese *Gensan*.
wont (wunt, wōnt) *v.* **wont, wont** or **wont·ed, wont·ing** *v.t.* 1 To accustom or habituate: used reflexively. — *v.i.* 2 To be accustomed; be used. 3 *Obs.* To dwell. [< *adj.*] — *n.* Ordinary manner of doing or acting; habit.

See synonyms under HABIT. [OE *gewunod*, pp. of *gewunian* be accustomed]
won't (wōnt) Will not: a contraction of Middle English *woll not*. Also *Scot.* **won·na** (wun′nə).
wont·ed (wun′tid, wōn′-) *adj.* 1 Commonly used or done; habitual. 2 Habituated; accustomed; at ease; at home. See synonyms under HABITUAL, USUAL. — **wont′ed·ness** *n.*
woo[1] (woo) *v.t.* 1 To make love to, especially so as to marry; court. 2 To entreat earnestly; beg. 3 To invite; solicit; seek. — *v.i.* 4 To pay court; make love. See synonyms under ADDRESS. [OE *wōgian*]
woo[2] (woo) *n. Scot.* Wool.
wood[1] (wood) *n.* 1 A large and compact collection of trees; a forest; grove: also **woods.** 2 The hard, fibrous material between the pith and bark of a tree or shrub, and also occurring in some herbaceous plants; the xylem. 3 The hard substance of a tree or shrub, whether as growing or as cut for use, for building, fuel, etc.; lumber; timber. 4 Something made of wood. 5 *pl.* A rural district; backwoods. — **to knock wood** To tap on a piece of wood or a wooden object as a charm against bad luck, especially while making an optimistic statement. — *adj.* 1 Made of wood; wooden. 2 Made for using or holding wood: a *wood* stove. 3 Living or growing in woods: the *wood* anemone. — *v.t.* 1 To furnish with wood for fuel. 2 To convert into a forest; plant with trees. — *v.i.* 3 To take on a supply of wood. [OE *widu, wiodu*]
wood[2] (wood) *Obs. v.i.* To act like a maniac; rave. — *adj.* Furious; frantic; raging; mad. [OE *wōd* insane]
Wood (wood), **Grant,** 1892–1942, U.S. painter. — **Leonard,** 1860–1927, U.S. physician, army officer, and colonial administrator.
wood alcohol Methanol.
wood anemone Any of several small plants (genus *Anemone*), growing in woodlands and blooming in the early spring, especially the common American species (*A. quinquefolia*), and the common European species (*A. nemorosa*).
wood betony 1 The common lousewort (*Pedicularis canadensis*) of the eastern United States, with yellow or reddish flowers. 2 The common betony (*Stachys officinalis*).
wood·bin (wood′bin′) *n.* A box or crib for holding firewood.
wood·bine (wood′bīn′) *n.* 1 The common European honeysuckle. 2 The Virginia creeper. [OE *wudubind* < *wudu* wood + *bindan* bind]
wood·block (wood′blok′) *n.* 1 A block of wood prepared for engraving. 2 A woodcut.
wood block 1 A block of wood for paving, etc. 2 *Music* A percussion instrument consisting of a hollow block of wood struck with a drumstick.
wood·bor·er (wood′bôr′ər, -bō′rər) *n.* Any of a large family (*Buprestidae*) of brilliantly colored beetles whose larvae are very destructive of trees. For illustration see INSECTS (injurious).
wood·carv·ing (wood′kär′ving) *n.* 1 The art of carving wood, especially for decoration. 2 A carving in wood. — **wood′carv′er** *n.*
wood·chat (wood′chat) *n.* 1 A European butcherbird (*Lanius collorio*) with reddish plumage and a notched beak. 2 Any of several Asian birds (genera *Ianthia* and *Larvivora*) of the thrush family. [Prob. partial trans. of G *waldkatze* the butcherbird, lit., wood cat]
wood·chuck (wood′chuk) *n.* A marmot (*Marmota monax*) of eastern North America; a ground hog. [Prob. alter. of WEJACK; infl. in form by WOOD[1] and CHUCK[1]]
wood coal 1 Charcoal made from wood: also **wood charcoal.** 2 Lignite.
wood·cock (wood′kok′) *n.* 1 A small European gamebird (*Scolopax rusticola*), having the thighs entirely feathered. 2 A related North American bird (*Philohela minor*). 3 *Obs.* A dolt; fool.
wood·craft (wood′kraft′, -kräft′) *n.* 1 Skill in such things as belong to woodland life, such as hunting or trapping; the faculty of finding

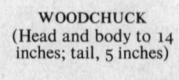
WOODCHUCK
(Head and body to 14 inches; tail, 5 inches)

one's way, and living comfortably in the wilderness. 2 Skill in woodwork or in constructing articles of wood. — **wood′crafts′man** (-krafts′mən, -kräfts′-) *n.*
wood·cut (wood′kut′) *n.* 1 An engraving on wood. 2 A print from such a block.
wood·cut·ter (wood′kut′ər) *n.* One who cuts or chops wood. — **wood′cut′ting** *n.*
wood·ed (wood′id) *adj.* Having a supply of wood; abounding with trees.
wood·en (wood′n) *adj.* 1 Made of wood: *wooden* tools. 2 Like a block of wood; stupid; mechanical; stiff; awkward. 3 Dull; spiritless. — **wood′en·ly** *adv.* — **wood′en·ness** *n.*
wood engraving 1 The art of cutting designs on wood for printing; the making of woodcuts. 2 A block thus engraved or a print therefrom. — **wood engraver**
wood·en·head (wood′n·hed′) *n. Colloq.* A stupid person; blockhead. — **wood′en·head′ed** *adj.*
wooden horse See TROJAN HORSE (def. 1).
wooden Indian 1 A carved and painted wooden figure of a North American Indian, usually in a standing position, formerly placed in front of cigar stores as an advertisement. 2 An inarticulate, sluggish, or dull person.
wooden nutmeg 1 An imitation nutmeg: proverbially used by New England (especially Connecticut) traders. 2 Any deceptive device.
wood·en-shoe dance (wood′n·shoo′) The trasko.
wood·en·ware (wood′n·wâr′) *n.* Dishes, vessels, bowls, etc., made of wood: said especially of household utensils.
Wood Green A municipal borough in Middlesex, England, 7 miles north of London.
wood grouse The capercaillie.
wood·hen (wood′hen′) *n.* A weka.
wood·house (wood′hous′) *n.* A house or shed for storing firewood. Also called *woodshed.*
wood hyacinth A small European squill (*Scilla nonscripta*), with clusters of bell-shaped blue, white, or pink flowers.
wood ibis A very large storklike bird (*Mycteria americana*) with a white body, glossy black tail, and naked head, common in wooded swamps of South America and the southern United States.
wood·ie (wood′ē) *n. Scot.* The gallows; a hangman's rope: used humorously.
wood·land (wood′lənd, -land) *n.* Land occupied by or covered with wood or trees; timberland. — *adj.* (-lənd) Belonging to or dwelling in the woods. — **wood′land′er** *n.*
woodland caribou See under CARIBOU.
wood·lark (wood′lärk′) *n.* A European passerine bird (*Lullula arborea*) resembling the skylark but with a sweeter note.
Woodlark Island A Papuan island in the Territory of Papua and New Guinea SE of New Guinea; 400 square miles.
wood lot A plot of land devoted to the growing of forest trees or consisting of woodland.
wood louse Any of numerous small terrestrial flat-bodied crustaceans (genera *Oniscus, Porcellion,* and others) commonly found under old logs; a sow bug or pill bug.
wood·man (wood′mən) *n. pl.* **·men** (-mən) 1 A woodcutter; lumberman. 2 A forester; also, a dweller in forests. 3 A hunter of forest game. Also *woodsman.*
wood·note (wood′nōt′) *n.* A simple, artless, or natural song, as of a wild bird.
wood nymph 1 A goddess or nymph of the forest; a dryad. 2 Any of several South American hummingbirds (genus *Thalurania*). 3 A butterfly of the family *Satyridae*, including species usually brown in color and with eyelike spots on the wings. They occur in woods and are not attracted to flowers.
wood·peck·er (wood′pek′ər) *n.* Any of a large family (*Picidae*) of birds related to the flickers, having stiff tail feathers to aid in climbing, strong claws, and a sharp, chisel-like bill for drilling holes in wood in search of insects; especially, the **red-headed woodpecker** (*Melanerpes erythrocephalus*) of North America, which has the head

RED-HEADED WOODPECKER
(9 to 9 1/2 inches long)

and upper breast deep red and the tail black, tipped with white; and the **pileated woodpecker** of North America (*Ceophloeus pileatus*) with a scarlet crest, white throat, white wing markings, and a large yellowish bill.

wood pewee Pewee (def. 1).

wood pigeon 1 The cushat. 2 The wild band-tailed pigeon (*Columba fasciata*) of the western United States.

wood·pile (wood'pīl') n. A pile of wood, especially of wood cut or split in sizes for burning in a fireplace or stove.

wood pitch The final residuum of wood tar.

wood·print (wood'print') n. A woodcut.

wood pulp Wood reduced to a pulp, as by grinding to a powder and digesting with chemicals: used for making paper.

wood pussy *Colloq.* A skunk.

wood rat A pack rat.

wood·ruff (wood'ruf') n. Any of several common European woodland herbs (genus *Asperula*) of the madder family, especially the **sweet-scented woodruff** (*A. odorata*), used to flavor wine and in perfumery. [OE *wudurofe* < *wudu* wood]

woods (woodz) n. pl. A forest or wooded area.

Woods, Lake of the See LAKE OF THE WOODS.

wood-screw (wood'skroo') n. A screw with a thread of coarse pitch, used for fastening pieces against wood. See illustration under SCREW.

wood·shed (wood'shed') n. A woodhouse.

woods·i·a (wood'zē·ə) n. Any of a genus (*Woodsia*) of small tufted ferns, found in rocky places. [after Joseph *Woods*, 1776–1864, English botanist]

woods·man (woodz·mən) n. pl. ·men (-mən) 1 A woodman. 2 A man skilled in woodcraft.

wood sorrel Oxalis.

wood spirit Wood alcohol or methanol.

Wood·stock (wood'stok) A municipal borough of central Oxfordshire, England.

wood sugar Xylose.

woods·y (wood'zē) adj. *Colloq.* Of, pertaining to, or dwelling in the woods; suggesting the woods: a *woodsy* fragrance.

wood tar A tar produced by the dry distillation of wood: it contains turpentine, resins, oils, creosote, and other hydrocarbons, and yields pyroligneous acid.

wood thrush 1 A large, common woodland thrush of North America (*Hylocichla mustelina*), noted for the vigor and sweetness of its song. 2 The missel thrush.

wood tick Any of certain ticks found in the woods, especially *Dermacentor variabilis* which transfers itself from underbrush to passing animals or human beings.

wood·turn·ing (wood'tûr'ning) n. The process or art of shaping blocks of wood into various forms by means of a lathe. — **wood'turn'er** n.

wood vinegar 1 Impure acetic acid from the distillation of wood. 2 Pyroligneous acid.

wood violet The bird's-foot violet.

wood·wax·en (wood'wak'sən) See WOADWAXEN.

wood·wind (wood'wind') n. A musical wind instrument made of wood. See WIND INSTRUMENT. — adj. Pertaining to or characteristic of a wooden wind instrument.

wood·work (wood'wûrk') n. 1 The wooden parts of any structure, especially interior wooden parts, as moldings or doors. 2 Work made of wood. — **wood'work'er** n. — **wood'work'ing** n.

wood·worm (wood'wûrm') n. A worm or larva dwelling in or that bores in wood.

wood·y (wood'ē) adj. **wood·i·er, wood·i·est** 1 Of the nature of wood; containing wood; ligneous. 2 Pertaining to wood; resembling wood. 3 Wooded; abounding with woods; sylvan. — **wood'i·ness** n.

woo·er (woo'ər) n. One who woos; a lover.

woof¹ (woof) n. 1 The weft of a woven fabric; the threads that are carried back and forth across the fixed threads of the warp in a loom. 2 The texture of a fabric. [OE *ōwef*]

woof² (woof) n. A sound made in imitation of the growl or low suppressed bark of a dog or bear. [Imit.]

woof·er (woof'ər) n. *Electronics* A loudspeaker used to reproduce the bass register in high-fidelity sound equipment, generally in connection with a tweeter. [<WOOF²]

wool (wool) n. 1 The soft, curly or crisped hair obtained from the fleece of sheep and some allied animals, especially that from domesticated sheep, noted for its felting properties and which provides the widest range of fibers for yarns and textiles. 2 The underfur of a fur-bearing animal. 3 Short kinky or crisp human hair, as of a Negro. 4 Material or garments made of wool. 5 Something resembling or likened to wool. — **all wool and a yard wide** Perfect in quality and quantity; hence, one hundred percent genuine. — **to pull the wool over one's eyes** To delude or deceive one. — adj. Made of or pertaining to wool or woolen material. [OE *wull*]

wool-clip (wool'klip') n. The amount of wool clipped from the sheep in one year.

wool-dyed (wool'dīd') adj. Dyed before the wool has been spun into yarn: said of fabrics.

wool·en (wool'ən) adj. 1 Consisting wholly or in part of wool; like wool. 2 Pertaining to wool or its manufacture. — n. Cloth or clothing made of wool: especially in the plural. Also **wool'len**.

Woolf (woolf), **(Adeline) Virginia**, 1882–1941, *née* Stephen, English novelist and essayist.

wool fat Lanolin. Also **wool grease**.

wool·fell (wool'fel') n. The pelt of a sheep or other wool-bearing animal with the wool still on it. [<WOOL + FELL⁴]

wool·gath·er·ing (wool'gath'ər·ing) n. Any trivial or purposeless employment; especially, idle reverie: from gathering wool caught on bushes, which required much wandering to collect even a little. — adj. Idly indulging in fancies. — **wool'gath'er·er** n.

wool·grow·er (wool'grō'ər) n. A person who raises sheep for the production of wool. — **wool'grow'ing** adj.

Wooll·cott (wool'kət), **Alexander**, 1887–1943, U.S. journalist and critic.

Wool·ley (wool'ē), **Sir (Charles) Leonard**, 1880–1960, English archeologist.

wool·ly (wool'ē) adj. ·li·er, ·li·est 1 Consisting of, covered with, or resembling wool; wool-bearing. 2 Soft and vaporous; lacking clearness; not sharply detailed; fuzzy; blurry. 3 Having a rounded and somewhat fleecy appearance, as clouds. 4 Having a growth of wool-like hairs. 5 Resembling the roughness and excitement of the West: usually in the phrase *wild and woolly*. — n. pl. ·lies A garment made of wool; especially, woolen underwear. Also **wool'y**. — **wool'li·ness, wool'i·ness** n.

woolly bear The larva of any of several tiger moths: so called because covered with long dense hairs.

wool·pack (wool'pak') n. 1 A bag or wrapper of canvas, cotton, etc., for packing a bale of wool. 2 A bale or bundle of wool. 3 *Meteorol.* A cumulus cloud.

wool·sack (wool'sak') n. 1 A sack of wool; specifically, the seat of the lord chancellor in the English House of Lords, a cushion stuffed with wool. 2 The office of lord high chancellor.

wool·sta·pler (wool'stā'plər) n. A dealer in or sorter of wool. — **wool'sta'pling** adj. & n.

Wool·wich (wool'ich, -ij) A metropolitan borough of London on the south bank of the Thames.

Wool·worth (wool'wûrth), **Frank Winfield**, 1852–1919, U.S. merchant; developed the five-and-ten-cent store.

Woon·sock·et (woon·sok'it) A city in NE Rhode Island on the Blackstone River.

woo·ra·li (woo·rä'lē) Curare. Also **woo·ra'ri** (-rē). [Var. of CURARE]

Woo·sung (woo'soong') The outer port of Shanghai, China, at the mouth of the Hwang-poo, north of Shanghai.

wooz·y (woo'zē) adj. *Slang* 1 Befuddled, especially with drink. 2 Fuzzy. [Prob. < *wooze*, var. of OOZE] — **wooz'i·ly** adv. — **wooz'i·ness** n.

wop (wop) n. *Slang* An Italian: a derogatory term. [? < dial. Ital. (Sicilian) *guapo* a dandy <Sp.]

Worces·ter (woos'tər) 1 A midland county in England; 699 square miles. Also **Worces'ter·shire** (-shir). 2 Its county town, famous for its 14th century cathedral. 3 A city in central Massachusetts, second largest in the State; an industrial, rail, and university center.

Worces·ter (woos'tər), **Joseph Emerson**, 1784–1865, U.S. lexicographer.

Worcester china A very fine china or porcelain made in Worcester, England, from 1751: also **Worcester porcelain**, and called **Royal Worcester** by royal warrant.

Worcestershire sauce A piquant sauce made originally in Worcester, England, from vinegar and many other ingredients. Also **Worcester sauce**.

word (wûrd) n. 1 A speech sound or combination of sounds which has come to signify and communicate a particular idea or thought, and which functions as the smallest meaningful unit of a language when used in isolation. There are **basic** or **radical words** as *master*, *man*, **derivative words** as *masterful*, *manly*, **inflectional words** as *masters*, *men*, and **compound words** as *masterpiece*, *manpower*, etc. In terms of modern linguistics, a word may be a single morpheme (a free form, as *master*) or a union of morphemes (free and bound forms, as *masters*, *masterful*, *masterpiece*). 2 The letters or characters that stand for a significant vocal sound. 3 A vocable considered only as a sound: ideas rather than *words*. 4 *Usually pl.* Conversation; talk: a man of few *words*. 5 A brief remark; hence, a short and pithy saying. 6 A communication or message: Send him *word*. 7 A command, signal, or direction: Give the *word* to start. 8 A promise; hence, good faith: a man of his *word*. 9 A party cry; watchword. 10 *pl.* Language used in anger, rebuke, or otherwise emotionally: They had *words*. See synonyms under TERM. — v.t. To express in a word or words, especially in selected words; phrase. [OE. Akin to VERB.]

Word (wûrd) n. 1 The Logos; the Son of God. 2 Divine Wisdom, as in *John* i. 3 The Scriptures as an embodiment of divine revelation.

word·age (wûr'dij) n. Words collectively.

word-blind·ness (wûrd'blīnd'nis) n. Alexia. — **word'-blind'** adj.

word·book (wûrd'book') n. 1 A collection of words; vocabulary; lexicon; dictionary. 2 An opera libretto.

word-coin·er (wûrd'koi'nər) n. One who makes up new words; a practitioner of logotechnics.

word deafness Inability to understand speech, resulting from disease of the cortical center: a form of aphasia.

word for word In the exact words; literally; verbatim.

word·i·ly (wûr'də·lē) adv. In a wordy manner; verbosely. — **word'i·ness, word'ish·ness** n.

word·ing (wûr'ding) n. The act or style of expressing in words; phraseology; also, words used; expression. See synonyms under DICTION.

word·less (wûrd'lis) adj. Having no words; dumb; silent.

word play 1 Repartee; fencing with words. 2 Subtle discussion on words and their meaning. 3 Play on words.

word square An arrangement of a set of words in rectangular form, so that they can be read in either horizontal or vertical lines, as in the accompanying example.

```
FRET
REAR
EASE
TREE
```

Words·worth (wûrdz'wûrth), **William**, 1770–1850, English poet; laureate 1843–50.

word-watch·er (wûrd'woch'ər) n. A close observer of words and their ways.

word·y (wûr'dē) adj. **word·i·er, word·i·est** 1 Of the nature of words; verbal. 2 Expressed in many words. 3 Given to the use of words; verbose; prolix.

wore (wôr, wōr) Past tense of WEAR¹ & WEAR².

work (wûrk) n. 1 Continued exertion or activity directed to some purpose or end; especially, manual labor; hence, opportunity for labor; occupation. 2 That upon which labor is expended; an undertaking; task. 3 That which is produced by or as by labor; specifically, an engineering structure; fortification; a design produced with a needle; also, a product of mental labor, as a book or opera. 4 A manufacturing or other industrial establishment: usually in the plural. 5 *pl.* Running gear or machinery, as of a watch. 6

Manner of working, or style of treatment; management; workmanship. 7 *pl.* Moral duties considered as external acts, especially as meritorious. 8 A froth or foam produced by fermentation in making vinegar, etc. 9 A feat or deed. 10 *Physics* A transference of energy from one body to another, resulting in the motion or displacement of the body acted upon, in the direction of the acting force: it is expressed as the product of the force and the amount of displacement in the line of its action. — *v.* **worked** (*Archaic* **wrought**), **work·ing** *v.i.* 1 To perform work; labor; toil. 2 To be employed in some trade or business. 3 To perform a function; operate: The machine *works* well. 4 To prove effective or influential; succeed: His stratagem *worked.* 5 To move or progress gradually or with difficulty: He *worked* up in his profession. 6 To become as specified, as by gradual motion: The bolts *worked* loose. 7 To have some slight improper motion in functioning: The wheel *works* on the shaft. 8 To move from nervousness or agitation: His features *worked* with passion. 9 To undergo kneading, hammering, etc.; be shaped: Copper *works* easily. 10 To ferment. 11 *Naut.* To labor in a heavy sea so as to loosen seams and fastenings: said of a ship. — *v.t.* 12 To cause or bring about; effect; accomplish: to *work* a miracle. 13 To cause to function; direct the operation of: to *work* a machine. 14 To make or shape by toil or skill. 15 To prepare, as by manipulating, hammering, etc.: to *work* dough. 16 To decorate, as with embroidery or inlaid work. 17 To cause to be productive, as by toil: to *work* a mine. 18 To cause to do work: He *works* his employees too hard. 19 To cause to be as specified, usually with effort: We *worked* the timber into position. 20 To make or achieve by effort: He *worked* his way to the top of his profession; to *work* one's passage on a ship. 21 To carry on some activity in (an area, etc.); cover: to *work* a stream for trout. 22 To solve, as a problem in arithmetic. 23 To cause to move from nervousness or excitement: to *work* one's jaws. 24 To excite; provoke: He *worked* himself into a passion. 25 To influence or manage, as by insidious means; lead. 26 To cause to ferment. 27 *Colloq.* To practice trickery upon; cheat; swindle. 28 *Colloq.* To make use of for one's own purposes; use. — **to work in** To put in; insert or be inserted. — **to work off** To get rid of, as extra flesh by exercise. — **to work on** (or **upon**) 1 To try to influence or persuade. 2 To influence or affect. — **to work out** 1 To make its way out or through. 2 To effect by work or effort; accomplish. 3 To exhaust, as a mineral vein or a subject of inquiry. 4 To discharge, as a debt, by labor rather than by payment of money. 5 To develop; form, as a plan. 6 To solve. 7 **a** To prove effective or successful. **b** To result as specified. — **to work up** 1 To excite; rouse, as rage or a person to rage. 2 To form or shape by working; develop. 3 To make one's or its way. [OE *weorc*]

Synonyms (noun): achievement, action, business, deed, doing, drudgery, employment, exertion, labor, occupation, performance, product, production, toil. *Work* is the generic term for any continuous application of energy toward an end; *work* may be hard or easy. *Labor* is hard and wearying *work*; *toil* is straining and exhausting *work*. *Work* is also used for any result of working, physical or mental; as, a *work* of art; a *work* of genius. In this connection, *work* has special uses, which *labor* and *toil* do not share. *Drudgery* is plodding, irksome, and often menial *work*. See ACT, BUSINESS, PRODUCTION, TASK, TOIL[1]. *Antonyms*: ease, idleness, leisure, recreation, relaxation, repose, rest, vacation.

-work *combining form* 1 A product made from a (specified) material: *paperwork, brickwork.* 2 Work of a (given) kind: *piecework.* 3 Work performed in a (specified) place: *housework.* [<WORK]

work·a·ble (wûr′kə·bəl) *adj.* 1 Of a nature to be operated, as a machine. 2 Practicable, as a plan. 3 That can be developed, as a mine. 4 Able to work. 5 That can be worked upon or influenced. — **work′a·bil′i·ty, work′a·ble·ness** *n.*

work·a·day (wûrk′ə·dā′) *adj.* 1 Of, pertaining to, or suitable for working days; everyday. 2 Commonplace; prosaic. [Alter. of ME *werkeday* < *werke,* OE *weorca* work + DAY; infl. in form by NOWADAYS]

work·bag (wûrk′bag′) *n.* A bag for holding tools or materials; a bag or reticule for needlework.

work·bench (wûrk′bench′) *n.* A bench for work, especially that of a carpenter, machinist, etc.

work·book (wûrk′bŏŏk′) *n.* 1 A booklet based on a course of study and containing problems and exercises which a student works out directly on the pages. 2 A manual containing operating instructions. 3 A book for recording work performed or planned.

work·box (wûrk′boks′) *n.* A small bag or box for needlework, etc.

work·day (wûrk′dā′) *n.* 1 Any day not a Sunday or holiday; a working day. 2 The part of the day or number of hours of one day spent in work. — *adj.* Workaday.

work·er (wûr′kər) *n.* 1 One who or that which performs work; specifically, a laborer as distinguished from a *capitalist.* 2 An individual female of an insect colony, as a true ant, a bee, or a white ant, with undeveloped sexual organs.

work·fel·low (wûrk′fel′ō) *n.* A companion in work.

work·folk (wûrk′fōk′) *n. pl.* Manual laborers. Also **work′folks′** (-fōks′).

work·house (wûrk′hous′) *n.* 1 *Brit.* A house for paupers able to work; an almshouse. 2 *Obs.* A workshop. 3 An industrial prison for petty offenders.

work·ing (wûr′king) *adj.* 1 Engaged actively in some employment. 2 That works, or performs its function: This is a *working* model. 3 Sufficient for use or action: They formed a *working* agreement. 4 Relating to or occupied by work: a *working* day. 5 Throbbing with pain; also, twitching: said especially of the face muscles. 6 Fermenting, as wine. — *n.* 1 The act or operation of any person or thing that works, in any sense. 2 That part of a mine or quarry where excavation is going on or has gone on.

working capital 1 That part of the finances of a business available for its operation. 2 The amount of quick assets which exceed current liabilities.

working day 1 A day not a legal holiday; a workday. 2 The number of hours constituting a day's work: a four-hour *working day.*

working drawing In engineering, etc., a drawing made to scale, as of a part of a machine or building, for the direction of workmen, contractors, etc.

work·ing·man (wûr′king·man′) *n.* *pl.* **·men** (-men′) A male worker; laborer.

working papers An age certificate and other official papers certifying that a minor may be legally employed.

working substance *Mech.* The fluid, as steam, or gasoline vapor, under pressure, that serves to operate a prime mover. Also **working fluid.**

work·ing·wom·an (wûr′king·wŏŏm′ən) *n.* *pl.* **·wom·en** (-wim′in) A female worker; laborer.

work·less (wûrk′lis) *adj.* Jobless; unemployed.

work·load (wûrk′lōd′) *n.* The amount of work apportioned to a person, machine, or department over a given period.

work·man (wûrk′mən) *n.* *pl.* **·men** (-mən) One who earns his bread by manual labor; an artisan; mechanic; workingman. — **work′·man·ly** *adj.*

work·man·like (wûrk′mən·līk) *adj.* Like or befitting a skilled workman; skilfully done. — **work′man·ly** *adv.*

work·man·ship (wûrk′mən·ship) *n.* 1 The art or skill of a workman, or the quality of work. 2 The work or result produced by a worker.

work of art A product of the fine arts, especially painting and sculpture, but including artistic, literary, and musical productions.

work·out (wûrk′out′) *n.* A test, trial, practice performance, etc., to discover, maintain, or increase ability for some work or competition, as a practice boxing bout, a fast turn around a track by a horse, runner, etc.

work·peo·ple (wûrk′pē′pəl) *n. pl.* People employed in work, especially in manual labor; working people.

work·room (wûrk′rōōm′, -rŏŏm′) *n.* A room where work is performed.

works (wûrks) *n.* 1 A manufacturing establishment including buildings and equipment: a gas *works.* 2 *Slang* The whole of anything; the kit and caboodle; everything: the whole *works.* — **to give (someone) the works** *Slang* To maul; kill by shooting. — **to shoot the works** *Slang* To make a supreme effort; risk one's all in one single attempt.

works council A committee of employed workers organized by an employer to discuss company and industrial problems and relations; a company union or similar group.

work·sheet (wûrk′shēt′) *n.* 1 A sheet of paper on which practice work or rough drafts of problems are written. 2 A sheet of paper used to record work schedules and operations.

work·shop (wûrk′shop′) *n.* A building or room where any work is carried on; workroom.

work·ta·ble (wûrk′tā′bəl) *n.* A table with drawers and other conveniences for use while working, especially while sewing.

work·week (wûrk′wēk′) *n.* The total number of hours worked in a week; also, the number of working hours in a week.

world (wûrld) *n.* 1 The earth; the terraqueous globe; the universe (of which the earth was once supposed to be the center); any similar orb; a part of the earth: the Old *World.* 2 A division of existing or created things belonging to the earth; natural grand division: the mineral, vegetable, or animal *world.* 3 The human inhabitants of the earth; mankind. 4 A definite class of people having certain interests or activities in common: the scientific *world*; a sphere or domain: the *world* of letters. 5 Man regarded socially; the public; hence, public or social life and intercourse. 6 The practices, usages, and ways of men: He knows the *world.* 7 A total of things as pertaining to or affecting an individual man; a career among men; one's experience in life: to begin the *world* anew. 8 The course of events as affecting one personally; individual condition or circumstances: How goes the *world* with you? Your *world* is changed then. 9 A scene of existence or of affairs regarded from a moral or religious point of view; secular affairs; worldly aims, pleasures, or people collectively; earthly existence; mortal life. 10 Figuratively, great quantity, number, or size: a *world* of trouble. — **for all the world** In every respect. — **on top of the world** *Colloq.* Elated. — **to bring into the world** To give birth to. [OE *weorold*]

World may appear as a combining form in hyphemes or as the first element in two-word phrases; as in:

world affairs	**world love**
world-alarming	**world-minded**
world battle	**world-old**
world builder	**world order**
world-changing	**world peace**
world citizen	**world politics**
world commerce	**world price**
world conflict	**world problem**
world-conquering	**world-rejoicing**
world-conscious	**world-renounced**
world-covering	**world-renouncing**
world destroyer	**world-renowned**
world-domination	**world report**
world dominion	**world revolution**
world-embracing	**world-roving**
world empire	**world sadness**
world-encircling	**world-shaking**
world esteem	**world sorrow**
world-famed	**world state**
world-famous	**world struggle**
world hero	**world trade**
world history	**world-wandering**
world leader	**world-winning**
world-long	**world-worn**

World Court See PERMANENT COURT OF INTERNATIONAL JUSTICE under COURT.

world·ling (wûrld′ling) *n.* One who lives merely for this world; a worldly-minded person.

world·ly (wûrld′lē) *adj.* **·li·er, ·li·est** 1 Pertaining to the world; mundane; earthly; not spiritual. 2 Devoted to temporal things; secular. 3 Sophisticated; worldly-wise. 4 *Obs.* Lay, as opposed to clerical. — *adv.* In a worldly manner. See synonyms under PROFANE. — **world′li·ness** *n.*

world·ly-mind·ed (wûrld′lē-mīn′did) *adj.* Absorbed in the things of this world. — **world′ly-mind′ed·ly** *adv.* — **world′ly-mind′ed·ness** *n.*

world·ly-wise (wûrld′lē·wīz′) *adj.* Wise in the

world power ways and affairs of the world; sophisticated.

world power A state or organization whose policy and action are of world-wide influence.

world series In baseball, the games played at the finish of the regular schedule between the champion teams of the American and National Leagues, the first team to win four games being adjudged world's champions. Also **world's series**.

world's fair An international exhibit of the folk crafts and arts, agricultural and industrial products, and scientific progress of various countries.

world soul 1 The hypothetical soul of the world; the All-Soul, conceived of after the analogy of the indwelling soul of man. 2 The principle that animates and informs the physical world. Also **world spirit**.

world's people Worldly people; those not belonging to some specific religious sect or group: a term used especially by the Friends.

World War See table under WAR.

world-wea·ry (wûrld'wir'ē) adj. ·ri·er, ·ri·est Dissatisfied with life and its conditions; weary and tired of this life.

world-wide (wûrld'wīd') adj. Extended throughout the world.

world without end Forever.

worm (wûrm) n. 1 A small, legless, invertebrate crawling animal, with an elongated, soft, and usually naked body, as a flatworm, roundworm, or annelid. ◆ Collateral adjective: vermicular. 2 A small creeping animal with short or undeveloped feet, as an insect larva, a grub, angleworm, etc. 3 Figuratively, that which suggests the action or habit of a worm as eating away or as an agent of decay or destruction, as remorse, death, etc. 4 A despicable or despised person; also, a feeble mortal. 5 Something conceived to be like a worm. 6 A screw thread. 7 A worm screw. 8 The spiral part of a corkscrew. 9 A spiral part in a still. 10 An organ or part that resembles a worm in shape, as the lytta of the dog or the vermiform process. 11 pl. An intestinal disorder due to the presence of parasitic worms. 12 The windings of a log road made to lessen the steepness of a grade. 13 The zigzag course of a log fence or a rail fence. — v.t. 1 To insinuate (oneself or itself) in a wormlike manner; effect as by crawling: with *in* or *into*; to *worm* one's way. 2 To draw forth by artful means, as a secret: with *out*. 3 To remove worms from. 4 To wind yarn, etc., along (a rope) so as to fill up the grooves between the strands. 5 To remove the lytta or worm from, as a dog. — v.i. 6 To move or progress slowly and stealthily. [OE *wyrm*] — **worm'er** n.

Worm (vôrm), **Olaus**, 1588-1654, Danish anatomist and physician.

worm-eat·en (wûrm'ēt'n) adj. Eaten or bored through by worms.

worm fence See under FENCE.

worm gear 1 Mech. A worm wheel having teeth shaped so as to mesh with a worm screw. 2 A worm wheel.

worm·hole (wûrm'hōl') n. The hole made by a worm or a wormlike animal, as in plants, earth, or stone. — **worm'holed'** adj.

WORM GEAR

Wor·mi·an (wôr'mē·ən) adj. Relating to or discovered by Olaus Worm.

Wormian bones Anat. Small bones occasionally lying along the lines of the cranial sutures.

wor·mil (wûr'məl) n. A warblefly or botfly larva. [Var. of dial. *warnel*, OE *wernægel*]

worm-root (wûrm'rōōt', -rŏŏt') n. Pinkroot.

Worms (wûrmz, *Ger*. vôrms) A city on the Rhine in Rhenish Hesse, SW West Germany; scene of the Diet of Worms (1521) by which Martin Luther was pronounced a heretic.

worm screw Mech. A short threaded portion of a shaft constituting an endless screw working with a worm wheel.

worm-seed (wûrm'sēd') n. 1 The seeds of any of various plants used as a vermifuge. 2 The plants themselves; especially, santonica, and a species of goosefoot (*Chenopodium ambrosioides*).

worm wheel Mech. A toothed wheel gearing with a worm screw.

worm·wood (wûrm'wŏŏd') n. 1 Any of a genus (*Artemisia*) of European herbs or small shrubs related to the sagebrush, especially a common species (*A. absinthium*), aromatic, tonic, bitter, and used in making absinthe. 2 That which embitters or makes bitter; bitterness. [Alter. of obs. *wermod* <OE; infl. in form by *worm* and *wood*[1]]

worm·y (wûr'mē) adj. **worm·i·er, worm·i·est** 1 Infested with or injured by worms. 2 Of or pertaining to worms; resembling a worm. 3 Earthy; groveling. — **worm'i·ness** n.

worn (wôrn, wōrn) Past participle of WEAR. — adj. 1 Affected by attrition or any similar continuous action. 2 Used, as a garment; showing the effects of anxiety, etc., as the mind; hackneyed, as phrases. 3 Exhausted or spent; used up.

worn-out (wôrn'out', wōrn'-) adj. 1 Used until without value for its purpose. 2 Thoroughly tired; exhausted.

wor·ri·cow (wûr'ē·kou') n. Scot. A hobgoblin; the devil; any hideous object or person; bugbear; a scarecrow.

wor·ri·some (wûr'ē·səm) adj. Causing worry or anxiety.

wor·rit (wûr'it) n. Colloq. Worry; vexation. [Appar. alter. of WORRY]

wor·ry (wûr'ē) v. ·ried, ·ry·ing v.i. 1 To be uneasy in the mind; feel anxiety about something; fret. 2 To pull or tear at something with the teeth: with *at*. 3 Colloq. To advance or manage despite trials or difficulties: with *along* or *through*. — v.t. 4 To cause to feel uneasy in the mind; trouble. 5 To bother; pester. 6 To mangle or kill by biting, shaking, or tearing with the teeth. 7 Scot. or Obs. To strangle; choke. See synonyms under PERSECUTE. — n. pl. ·ries A state of anxiety; vexation. See synonyms under ANXIETY, CARE. [OE *wyrgan* strangle] — **wor'ri·er** n. — **wor'ri·ment** n.

worse (wûrs) Used as comparative of *bad, ill, evil,* and the like. — adj. 1 Bad or ill in a greater degree; more evil, unworthy, etc. 2 Physically ill in a greater degree. 3 Less favorably situated as to means and circumstances. — n. Something worse; disadvantage; loss. — adv. 1 In a manner more evil or ill. 2 With greater intensity, severity, etc. 3 Consequently; less. [OE *wyrsa*]

wors·en (wûr'sən) v.t. & v.i. To make or become worse.

wors·er (wûr'sər) adj. & adv. Worse: a former redundant form of the comparative, on the analogy of *lesser*: now regarded as a vulgarism.

wor·set (wûr'sit) adj. & n. Scot. Worsted.

wor·ship (wûr'ship) n. 1 The act or feeling of adoration or homage; the paying of religious reverence, as in prayer, praise, etc. 2 The act or feeling of deference, respect, or honor toward virtue, power, or the like. 3 Excessive or ardent admiration; also, the object of such love or admiration. 4 A title of honor in addressing persons of station. See synonyms under RELIGION, REVERENCE. — v. ·shiped or ·shipped, ·ship·ing or ·ship·ping v.t. 1 To pay an act of worship to; venerate; adore. 2 To treat with intense or exaggerated admiration or affection. 3 Obs. To honor. — v.i. 4 To perform acts or have sentiments of worship. [OE *weorthscipe* < *weorth* worthy] — **wor'ship·er, wor'ship·per** n.

Synonyms (verb): adore, deify, exalt, honor, idolize, revere, reverence. See PRAISE. Antonyms: abhor, abjure, abominate, blaspheme, curse, denounce, detest, renounce, revile, scoff, scorn.

wor·ship·ful (wûr'ship·fəl) adj. 1 Worthy of honor; entitled to respect by reason of character or position: applied to dignitaries, as magistrates, etc. In Freemasonry, it is part of a specific official title, as of masters. 2 Esteemed; distinguished; honorable. 3 Giving reverence; adoring. — **wor'ship·ful·ly** adv. — **wor'ship·ful·ness** n.

worst (wûrst) Used as the superlative of *bad, ill,* or *evil*. — adj. Bad, ill, or evil in the highest degree. — **in the worst way** Slang Very much. — n. The most evil state or result. — **at worst** On the most pessimistic estimate. — **to get the worst of it** To be defeated or put at a disadvantage. — adv. In the worst or most extreme manner or degree. — v.t. To get the advantage over; defeat; vanquish. See synonyms under BEAT, CONQUER. [OE *wyrsta*]

wors·ted (wŏŏs'tid, wûr'stid) n. 1 Woolen yarn spun from long staple, with fibers combed parallel and twisted hard. 2 A lightly twisted woolen yarn. 3 A tightly woven or smooth fabric made from worsted yarns, as gabardine or serge. — adj. Consisting of or made from this yarn. [from *Worsted*, former name of a parish in Norfolk, north of Norwich, England]

wort (wûrt) n. 1 A plant or herb: usually in combination: *liverwort, navelwort*. 2 The sweet, unfermented infusion of malt that becomes beer when fermented. [OE *wyrt* a root, a plant]

worth[1] (wûrth) n. 1 That quality which renders a thing useful or desirable; value or excellence of any kind; hence, market value. 2 That quality or combination of qualities that makes one deserving of esteem; mental and moral excellence. 3 Wealth. — adj. 1 Having value; equal in value (to); exchangeable (for). 2 Deserving (of): in either a good or bad sense. 3 Having possessions to the value of: He is *worth* a million. — **for all it is worth** To the utmost. — **for all one is worth** With every effort possible; to the utmost of one's capacity. [OE *weorth*]

Synonyms (noun): character, desert, excellence, integrity, merit, preciousness, value. See PRICE, VIRTUE.

worth[2] (wûrth) v.i. To betide or befall: now only in phrases, as **woe worth the day**, etc. [OE *weorthan* come to be]

-worth combining form Of the value of: *pennyworth*. [OE *weorth* worth]

Wor·thing (wûr'thing) A municipal borough on the coast of southern Sussex, southern England, west of Brighton.

worth·less (wûrth'lis) adj. Having no worth; having no utility or value; destitute of dignity, virtue, or standing. See synonyms under BAD[1], BASE[2], USELESS, VAIN, WASTE. — **worth'less·ly** adv. — **worth'less·ness** n.

worth·while (wûrth'hwīl') adj. Sufficiently important to occupy the time; of enough value to repay the effort. ◆ This compound originated from the phrase *worth the while* and it is firmly established as a solideme in American usage. In British usage it is usually a hypheme: **worth-while**. — **worth'while'ness** n.

wor·thy (wûr'thē) adj. ·thi·er, ·thi·est 1 Possessing worth; deserving of respect or honor; having valuable or useful qualities. 2 Having such qualities as to be deserving of or adapted to some specified thing; fit; suitable: followed by *of* (rarely *for*), sometimes by an infinitive and rarely by the object directly. 3 Obs. Well deserved; fitting. See synonyms under BECOMING, EXCELLENT, GOOD, MORAL, VIRTUOUS. — n. pl. ·thies 1 A person of eminent worth. 2 Humorously, a person of local note; a character. [ME *wurthi, worthi*] — **wor'thi·ly** adv. — **wor'thi·ness** n.

-worthy combining form 1 Meriting or deserving: *trustworthy*. 2 Valuable as; having worth as: *newsworthy*. 3 Fit for: *seaworthy*. [OE *wyrthe* worthy]

wot (wot) Present tense, first and third person singular, of WIT[2].

Wot·ton (wot'n), **Sir Henry**, 1568-1639, English diplomat and poet.

would (wŏŏd) Past tense of WILL, expressing desire, condition, or what might be expected: used also to express determination: He would go, I couldn't stop him. [OE *wolde*, pt. of *willan* will]

◆ **would & should** Anybody dealing in ifs, as-ifs, promises, threats, hopes, wishes, or similar feelings and attitudes about the real or imagined future, will find himself making plentiful use of *would* and/or *should*, which do duty to express the vestigial and fast vanishing subjunctive mood. As to which one to use, the distinctions are subtle and fine-drawn and at several points British usage calls for *should* where American practice is

add, āce, cāre, päim; end, ēven; it, īce; ōpen, ôrder; tōōk, pōōl; up, bûrn; ə = a in *above*, e in *sicken*, i in *clarity*, o in *melon*, u in *focus*; yōō = u in *fuse*; oi, oil; ou, pout; ch, check; g, go; ng, ring; th, thin; th, this; zh, vision. Foreign sounds á, œ, ü, kh, ñ; and ◆ see page xx. < *from*; + *plus*; ? *possibly*.

would-be (wood'bē') *adj.* Desiring or professing to be: a *would-be* poet.
would·n't (wood'nt) Would not: a contraction.
wound[1] (woond, *Poetic* wound) *n.* 1 A hurt or injury caused by violence; especially, a cut, bruise, stab, etc.; a trauma. 2 A breach or cut of the bark or substance of a tree or plant. 3 Hence, any injury or cause of pain or grief, as to the feelings, honor, etc. — *v.t.* & *v.i.* To inflict a wound or wounds (upon); cause injury or grief (to); hurt. See synonyms under AFFRONT, HURT, PIQUE[1]. [OE *wund*] —**wound'ed** *adj.* —**wound'less** *adj.*
wound[2] (wound) Past tense and past participle of WIND[2].
Wou·ter (wou'tər) Dutch form of WALTER.
wove (wōv) Past tense and alternative past participle of WEAVE.
wo·ven (wō'vən) Past participle of WEAVE.
wove paper Paper carrying the marks of the wire gauze on which it was laid during finishing.
wow (wou) *interj.* An exclamation of wonder, surprise, pleasure, or pain. — *n. Slang* An extraordinary success. — *v.t. Slang* To be extraordinarily successful with.
wow·ser (wou'zər) *n. Australian Slang* One who is opposed to Sunday amusements, sports, etc.; a hypocritical censor of the lesser vices; a meddlesome puritan or sanctimonious reformer. [? < *wow*!, uttered in disapproval]
wrack[1] (rak) *n.* 1 Marine vegetation and floating material cast ashore by the sea, as seaweed or eelgrass; kelp. 2 The state of being wrecked; ruin; destruction: chiefly in the phrase *wrack and ruin.* 3 Shipwreck; a wrecked vessel; wreckage. 4 *Scot.* & *Brit. Dial.* Weeds. — *v.t.* & *v.i.* To wreck or be wrecked. ♦ Homophone: *rack.* [Fusion of OE *wræc* punishment, revenge and MDu. *wrak* a wreck]
wrack[2] (rak) *n.* A rack of clouds; any floating vapor. Compare RACK[3]. [Var. of RACK[3]]
wraith (rāth) *n.* 1 An apparition of a person thought to be alive, seen shortly before or shortly after his death. 2 Any specter, ghost, or apparition. [< dial. E (Scottish), alter. of *warth* < ON *vörthr* a guardian < *vartha* guard]
wrang (rang) *adj.* & *n. Scot.* Wrong.
Wran·gel (vrän'gil), **Ferdinand Petrovich von,** 1794?–1870, Russian explorer. Also **Wran'gell.**
Wran·gell (rang'gəl), **Mount** An active volcano (14,005 feet) in the western **Wrangell Mountains,** a range in SE Alaska; highest peak, 16,208 feet.
Wran·gell Island (rang'gəl) 1 An island of the Alexander Archipelago, SE Alaska; 30 miles long, 5 to 14 miles wide. 2 An island in the western Chukchi Sea off NE Siberia; part of Khabarovsk territory, Asiatic Russian S.F.S.R.; 75 miles long, 45 miles wide; 1,740 square miles. Also **Wran'gel Island.** Russian **O·strov Vran·ge·ly** (ô'strôf vrän'gi·lyə).
wran·gle (rang'gəl) *v.* **·gled, ·gling** *v.i.* 1 To argue or dispute noisily and angrily; brawl. — *v.t.* 2 To argue; debate. 3 To herd or round up, as livestock on a range. See synonyms under CONTEND. — *n.* An angry or noisy dispute; a quarrel. See synonyms under ALTERCATION, DISPUTE, QUARREL[1]. [Cf. LG *wrangeln* quarrel, freq. of *wrangen* struggle]
wran·gler (rang'glər) *n.* 1 One who wrangles. 2 At Cambridge University, England, one who has taken the highest mathematical honors. 3 A herdsman on a range.
wrap (rap) *v.* **wrapped** or **wrapt, wrap·ping** *v.t.* 1 To surround and cover by something folded or wound about; swathe; enwrap. 2 To cover with paper, etc., folded about and secured. 3 To wind or fold (a covering) about something. 4 To surround so as to obscure; blot out or conceal; envelop. 5 To fold, wind, or draw together. — *v.i.* 6 To be or become twined or coiled: with *about, around,* etc. — *n.* 1 An article of dress drawn or folded about a person; a wrapper. 2 *pl.* Outer garments collectively, as cloaks, scarfs, etc. 3 A blanket. [ME *wrappen*; origin uncertain]
wrap·a·round windshield (rap'ə·round) In automobiles, a windshield curving back into the sides of the body, thus providing a greater field of vision.
wrap·per (rap'ər) *n.* 1 A paper enclosing a newspaper, magazine, or similar packet for mailing or otherwise. 2 A detachable paper cover to protect the binding of a book. 3 A loose flowing outer garment; a dressing gown. 4 A tobacco leaf of high quality enclosing a cigar or plug of tobacco. 5 One who wraps articles.
wrap·ping (rap'ing) *n.* A covering; something in which an object is wrapped.
wrap·ras·cal (rap'ras'kəl) *n.* A long loose overcoat fashionable during the 18th century.
wrapt (rapt) Erroneous spelling of RAPT.
wrasse (ras) *n.* Any of a group of spiny-finned food fishes (family *Labridae*) of warm tropical seas, often highly colored; especially, the tautog. [<Cornish *wrach* < *gwrach,* orig., an old woman]
wrath (rath, räth; *Brit.* rôth) *n.* 1 Determined and lasting anger; extreme or violent rage; fury; vehement indignation. 2 An act done in violent rage. See synonyms under ANGER, VIOLENCE. — *v.t.* & *v.i. Obs.* To make or become angry. — *adj. Obs.* Wroth; angry. [OE *wræththu* < *wrath* wroth]
Wrath (rath, räth), **Cape** A promontory at the NW extremity of Scotland in NW Sutherland.
wrath·ful (rath'fəl, räth'-) *adj.* 1 Full of wrath; extremely angry. 2 Springing from or expressing wrath. —**wrath'ful·ly** *adv.* —**wrath'ful·ness** *n.*
wrath·y (rath'ē, räth'ē) *adj.* **wrath·i·er, wrath·i·est** 1 Disposed to wrath. 2 *Colloq.* Wroth. —**wrath'i·ly** *adv.* —**wrath'i·ness** *n.*
wreak (rēk) *v.t.* 1 To inflict or exact, as vengeance. 2 To satiate; give free expression to, as a feeling or passion. ♦ Homophone: *reek.* [OE *wrecan* drive, avenge]
wreath (rēth) *n.* 1 A twisted band, as of flowers, commonly circular, as for a crown or chaplet. 2 Any curled band of circular or spiral shape, as of smoke or snow. [OE *writha* < *writhan* writhe] —**wreath'y** *n.*
Wreath (rēth) See CORONA AUSTRALIS.
wreathe (rēth) *v.* **wreathed, wreath·ing** *v.t.* 1 To form into a wreath, as by twisting or twining. 2 To adorn or encircle with or as with wreaths. 3 To envelop; cover: His face was *wreathed* in smiles. — *v.i.* 4 To take the form of a wreath. 5 To twist, turn, or coil, as masses of cloud. See synonyms under TWIST. [Earlier *wrethe,* back formation <ME *wrethen,* var. of *writhen,* pp. of *writhen* writhe; infl. by *wreath*]
wreck (rek) *v.t.* 1 To cause the destruction or wreck of, as a vessel; shipwreck. 2 To bring ruin, damage, or destruction upon. 3 To tear down, as a building; dismantle. — *v.i.* 4 To suffer wreck; be ruined. 5 To engage in wrecking, as for plunder or salvage. See synonyms under RUIN. [< *n.*] — *n.* 1 The act of wrecking, or the state of being wrecked; the ruin of anything, especially if effected violently. 2 That which has been wrecked or ruined, as a vessel or an army; hence, an emaciated person. 3 Wreckage; shipwreck. 4 *Law* Property cast upon land by the sea, either broken portions of a wrecked vessel or cargo from it. ♦ Homophone: *reck.* [<AF *wrec, wrech,* OF *warec* <ON (assumed) *wrek* < *wrekan* drive]
wreck·age (rek'ij) *n.* 1 The act of wrecking, or the state of being wrecked; wrecked material. 2 Broken or disordered remnants or fragments from a wreck.
wreck·er (rek'ər) *n.* 1 One who causes wreck, destruction, or frustration of any sort. 2 One employed in tearing down and removing old buildings. 3 A person, train, car, or machine that clears away wrecks. 4 One employed to recover disabled vessels or wrecked cargoes for the owners; also, a vessel employed in this service; a salvager. 5 One who lures ships to destruction by false lights on the shore in order to plunder the wreck. 6 One who ruins something valuable, as a bank or a railroad, especially for his own profit.
wreck·ful (rek'fəl) *adj. Poetic* Causing wreck; involving ruin.
wrecking company 1 A business organization that salvages wrecked ships. 2 A business organization that tears down and removes old buildings.
wren (ren) *n.* 1 Any of numerous small passerine birds (family *Troglodytidae*) having short rounded wings and a short tail, including the common **house wren** (*Troglodytes aëdon*), the **Carolina wren** (*Thryothorus ludovicianus*), **Bewick's wren** (*Thryomanes bewicki*) of North America, and the European **wren** (*Nannus troglodytes*). 2 Any one of numerous similar birds. [OE *wrenna*]
Wren (ren), **Sir Christopher,** 1632–1723, English architect. —**Percival Christopher,** 1885–1941, English novelist.
wrench (rench) *n.* 1 A violent twist; hence, a twist causing pain or injury; sprain. 2 Any strain or sudden or violent tension; sudden and violent emotion. 3 Any perversion or distortion of an original meaning. 4 A tool for twisting or turning bolts, nuts, pipe, etc.

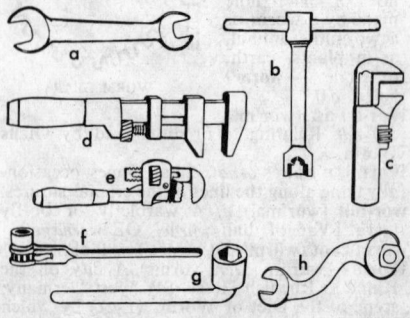

TYPES OF WRENCHES
a. Engineer's wrench. e. Pipe wrench.
b. Socket wrench. f. Ratchet wrench.
c. Bicycle wrench. g. Offset wrench.
d. Monkey wrench. h. S-wrench.

— *v.t.* 1 To twist violently; turn suddenly by force; wrest. 2 To twist forcibly so as to cause strain or injury; sprain. 3 To twist from the proper meaning, intent, or use. — *v.i.* 4 To give a twist or wrench. [OE *wrenc* a trick. Akin to WRINKLE[1].]

Wrens (renz) *n. pl. Brit. Colloq.* Women's Royal Naval Service, an organization to relieve men of certain shore duties connected with the Royal Navy: so called from the initial letters *W,R,N,S,* plus *E.*

wrest (rest) *v.t.* 1 To pull or force away by violent twisting or wringing; wrench. 2 To turn from the true meaning, character, intent, or application; distort; pervert. 3 To seize forcibly by violence, extortion, or usurpation. — *n.* 1 An act of wresting; a violent twist. 2 A misapplication or perversion. 3 A crooked act; wile. 4 A key for tuning a stringed instrument, as a harp. ◆ Homophone: *rest.* [OE *wrǣstan*] — **wrest′er** *n.*

wres·tle (res′əl) *v.* **·tled, ·tling** *v.i.* 1 To engage in wrestling. 2 To struggle, as for mastery; contend. — *v.t.* 3 To engage in (a wrestling match), or wrestle with. 4 To throw (a calf) and hold it down for branding. — *n.* A wrestling match; a hard struggle. [OE *wrǣstlian,* freq. of *wrǣsten* wrest]

wres·tler (res′lər) *n.* One who wrestles; especially, a person who competes in wrestling matches.

wres·tling (res′ling) *n.* A sport or exercise in which each of two unarmed contestants endeavors to throw the other to the ground or force him into a certain fallen position.

wretch (rech) *n.* 1 A base, vile, or contemptible person; despicable character. 2 A miserable or unhappy person; also, sometimes, any person or creature viewed with pity. ◆ Homophone: *retch.* [OE *wrecca* an outcast < *wrecan* drive]

wretch·ed (rech′id) *adj.* 1 Sunk in dejection; profoundly unhappy. 2 Causing misery or grief. 3 Mean; paltry; worthless; unsatisfactory in ability or quality. 4 Despicable; contemptible. See synonyms under BAD[1], BASE[2], PITIFUL. — **wretch′ed·ly** *adv.* — **wretch′·ed·ness** *n.*

wrig·gle (rig′əl) *v.* **·gled, ·gling** *v.i.* 1 To twist in a sinuous manner; squirm; writhe. 2 To proceed as by twisting or crawling. 3 To make one's way by evasive or indirect means. — *v.t.* 4 To cause to wriggle. — *n.* The motion of one who or that which wriggles; a squirm. [<MLG *wriggeln,* freq. of *wriggen* twist] — **wrig′gly** *adj.*

wrig·gler (rig′lər) *n.* 1 Someone or something that wriggles. 2 A mosquito larva.

wright (rīt) *n.* 1 One who does mechanical or constructive work. 2 An artificer or workman: used chiefly in compounds: *shipwright.* ◆ Homophones: *right, rite, write.* [OE *wyrhta*]

Wright (rīt), **Frank Lloyd,** 1869–1959, U.S. architect. — **Harold Bell,** 1872–1944, U.S. novelist. — **Joseph,** 1855–1930, English philologist and lexicographer. — **Orville,** 1871–1948, U.S. pioneer in aviation, with his brother **Wilbur,** 1867–1912. — **Richard,** 1908–1960, U.S. novelist. — **Willard Huntington,** 1888–1939, U.S. writer and art critic: pseudonym *S. S. Van Dine.*

wring (ring) *v.* **wrung** (*Rare* **wringed**), **wring·ing** *v.t.* 1 To squeeze or compress by twisting; turn and strain with force; pass (clothes) through a wringer. 2 To squeeze or press out, as water, by twisting. 3 To extort; acquire by extortion. 4 To distress; torment. 5 To twist or wrest violently out of shape or place: to *wring* his neck. 6 *Obs.* To pervert; distort. — *v.i.* 7 To writhe or squirm, as with anguish. 8 To perform the action of wringing. ◆ Homophone: *ring.* [OE *wringan*]

wring·bolt (ring′bōlt′) A ring bolt. [Earlier *wrainbolt,* var. of *ring bolt*]

wring·er (ring′ər) *n.* 1 One who or that which wrings. 2 A contrivance used to press water out of fabrics after washing; also, the operator of such a machine.

wrin·kle[1] (ring′kəl) *n.* 1 A small ridge or prominence, as on a smooth surface; a crease; fold. 2 Specifically, a small fold or crease in the skin, usually produced by age or by excessive exposure to the elements. 3 A ripple; little wave. — *v.t.* & *v.i.* **·kled, ·kling** To contract or be contracted into wrinkles or ridges; pucker. [OE *wrincle.* Akin to WRENCH.] — **wrin′kly** *adj.*

wrin·kle[2] (ring′kəl) *n. Colloq.* A curious or ingenious notion or device; happy thought; a novelty, as in dress. See synonyms under WHIM. [Prob. dim. of OE *wrenc* a trick]

wrist (rist) *n.* 1 The part of the arm immediately adjoining the hand; the carpus. ◆ Collateral adjective: *carpal.* 2 The part of a glove or garment that covers the wrist. 3 A wrist pin. [OE, prob. < *wrīthan* writhe]

wrist·band (rist′band′, -bənd, riz′-) *n.* The band of a sleeve that covers the wrist or ends a shirt sleeve.

wrist–drop (rist′drop′) *n. Pathol.* Paralysis of the forearm, usually due to lead poisoning.

wrist·let (rist′lit) *n.* 1 A flexible band worn on the wrist for ornament or warmth. 2 A bracelet. 3 *Slang* A handcuff.

wrist·lock (rist′lok′) *n.* In wrestling, a hold whereby an opponent is made helpless by twisting his arm with a grip at the wrist.

wrist pin *Mech.* 1 A pin holding together the piston and connecting rod of a steam engine. 2 A similar pin in the cross-head of an internal-combustion engine.

wrist watch A watch set in a leather or metal wristlet and worn at the wrist.

writ[1] (rit) *n.* 1 *Law* A mandatory precept, under seal, issued by a court, and commanding the person to whom it is addressed to do or not to do some act. 2 That which is written: now chiefly in the phrase **Holy Writ,** meaning the Bible. [OE, a writing < *wrītan* write]

writ[2] (rit) Archaic or dialectal past tense and past participle of WRITE.

write (rīt) *v.* **wrote** (*Archaic* or *Dial.* **writ**), **writ·ten** (*Archaic* or *Dial.* **writ**), **writ·ing** *v.t.* 1 To trace or inscribe (letters, words, numbers, symbols, etc.) on a surface with pen or pencil, or by other means. 2 To describe in writing: to *write* one's impressions of a journey. 3 To communicate by letter: Be sure to *write* all the news; He *writes* that he will be home soon. 4 To communicate with by letter: He *writes* her every day. 5 To produce by writing; be the author or composer of. 6 To draw up; draft: to *write* one's will; to *write* a check. 7 To cover or fill with writing: to *write* two full pages. 8 To leave marks or evidence of: Anxiety is *written* on his face. 9 To spell or inscribe as specified: He *writes* his name with two *n's.* 10 To entitle or designate in writing: He *writes* himself "General." 11 To underwrite: to *write* an insurance policy. — *v.i.* 12 To trace or inscribe letters, etc., on a surface, as of paper. 13 To write a letter or letters; communicate in writing. 14 To be engaged in the occupation of a writer or author. 15 To produce a specified quality of writing. See synonyms under INSCRIBE. — **to write down** 1 To put into writing. 2 To injure or depreciate in writing. — **to write in** 1 To insert in writing, as in a document. 2 To cast (a vote) for one not listed on a ballot by inserting his name in writing. — **to write off** 1 To cancel or remove (claims, debts, etc.) from an open account. 2 To acknowledge the loss or failure of. — **to write out** 1 To put into writing. 2 To write in full or complete form. — **to write up** 1 To describe fully in writing. 2 To praise fully or too fully in writing. 3 In accounting, to put an unusually high value upon. ◆ Homophones: *right, rite, wright.* [OE *wrītan*]

write–off (rīt′ôf′, -of′) *n.* 1 A cancellation. 2 An amount canceled or noted as a loss.

writ·er (rī′tər) *n.* 1 One who writes. 2 One who engages in literary composition. 3 Formerly, in England, a copying-clerk in a government office; also, sometimes, a clerk; a penman. 4 *Scot.* Formerly, an attorney. 5 That which writes or assists in writing: used in composition: *typewriter.*

writer's cramp *Pathol.* Spasmodic contraction of the muscles of the fingers and hand, caused by excessive writing. Also **writer's palsy** or **spasm.**

write–up (rīt′up′) *n. Colloq.* A written description, record, or account, usually laudatory, as of a town, manufacturing enterprise, or public institution: We shall give the place a *write-up* soon.

writhe (rīth) *v.t.* & *v.i.* **writhed, writh·ing** To twist with violence; wrench; distort, as the body, face, or limbs in pain. — *n.* An act of writhing; a contortion. See synonyms under STRUGGLE. [OE *wrīthan*] — **writh′er** *n.*

writh·en (rith′ən) Obsolete past participle of WRITHE. — *adj. Poetic* Twisted; distorted.

writ·ing (rī′ting) *n.* 1 The act of one who writes. 2 The characters so made; chirography; handwriting. 3 Anything written or expressed in letters, especially a literary production. 4 *Law* A written instrument: words, or characters that stand for words or ideas, traced on some substance, as paper, wood, or stone, with an implement, as a pen, pencil, or brush, or by some other device, as stamping, printing, or engraving. 5 The profession or occupation of a writer. 6 The practice, art, form, or style of literary composition.

writing machine A typewriter.

writ·ing–mas·ter (rī′ting·mas′tər, -mäs′-) *n.* A teacher of penmanship.

writing paper Paper prepared to receive ink in writing.

writ of error *Law* A commission by which the judges of one court are authorized to examine a record upon which a judgment was given in another court, and to affirm or reverse the judgment according to law.

writ of prohibition *Law* A writ issued by a superior court to an inferior court, commanding it to desist from proceeding in a matter not within its jurisdiction.

writ of right *Law* 1 Formerly, in England, a writ in an action for the purpose of establishing a title to real estate. 2 A similar common-law writ.

writ of summons *Law* The writ by which, in modern practice, a civil action is commenced; a written order to an authorized officer to notify a person to appear in court to answer a complaint.

writ·ten (rit′n) Past participle of WRITE.

Wro·cław (vrô′tswäf) A city in SW Poland, on the Oder; a German city from the 13th century; part of Prussia 1741–1945; a major industrial center: German *Breslau.*

wrong (rông, rong) *adj.* 1 Deviating from moral rectitude as prescribed by civil or divine law or by conscience; immoral. 2 Not just, proper, or equitable according to a standard, code, or convention; incongruous; improper. 3 Deviating from fact and truth; not according to reality; erroneous; mistaken: a *wrong* estimate. 4 Not in accordance with rule or appropriateness; improper; incorrect: to enter the *wrong* store. 5 Deviating from the proper design, intention, or requirement; unsuitable: the *wrong* side of cloth; the *wrong* letter in a word. 6 Unsatisfactory: a *wrong* reply. — **to go wrong** 1 To lapse from the strict path of rectitude. 2 To turn out badly; go astray. — **on the wrong side of** (**30, 40,** etc.) Older than (30, etc.). See synonyms under IMMORAL, SINFUL. — *adv.* In a wrong direction, place, or manner; awry or amiss; erroneously. — *n.* 1 That which is contrary to justice or rectitude; an injury; mischief. 2 Hence, some particular form of disobedience or non-conformity to lawful authority, human or divine. 3 *Law* An invasion or violation of one's legal rights; specifically, a crime; a tort. See synonyms under INJURY, INJUSTICE, SIN[1]. — *v.t.* 1 To violate the rights of; inflict injury or injustice upon. 2 To impute evil to unjustly; misrepresent: If you think so, you *wrong* him. 3 To seduce (a woman). 4 To treat dishonorably; malign. 5 *Scot.* To injure. See synonyms under ABUSE. [OE *wrang* twisted <ON *rangr* awry, unjust] — **wrong′·ness** *n.*

wrong·do·er (rông′dōō′ər, rong′-) *n.* One who commits a fault or crime. — **wrong′do′ing** *n.*

wrong·er (rông′ər, rong′-) *n.* One who commits an offense, injury, or trespass.

wrong font *Printing* The wrong font or type face: indicated by the abbreviation *w.f.* in marking printers' proofs.

wrong·ful (rông′fəl, rong′-) *adj.* 1 Characterized by wrong or injustice; injurious; unjust. 2 Unlawful; illegal. — **wrong′ful·ly** *adv.* — **wrong′ful·ness** *n.*

wrong–head·ed (rông′hed′id, rong′-) *adj.* Having perverted judgment; perverse; obstinate.

—wrong′-head′ed·ly adv. —wrong′-head′ed·ness n.
wrong·ly (rông′lē, rong′-) adv. In a wrong manner; erroneously; falsely.
wrote (rōt) Past tense of WRITE.
wroth (rôth) adj. Filled with anger; angry. —n. Obs. Anger. [OE wrāth]
wrought (rôt) Archaic past tense and past participle of WORK. —adj. 1 Beaten or hammered into shape by tools: wrought gold. 2 Worked; molded. 3 Made with delicacy; elaborated carefully. 4 Made; fashioned; formed: The cathedral was wrought by skilled hands. [ME wrogt, var. of worht, pp. of wirchen work]
wrought iron See under IRON.
wrought–i·ron casting (rôt′ī′ərn) Mitis casting.
wrought up Excited.
wrung (rung) Past tense and past participle of WRING.
wry (rī) adj. wri·er, wri·est 1 Bent to one side or out of position; contorted; askew; also, made by twisting or distorting the features: a wry smile. 2 Hence, deviating from that which is right or proper; perverted, as a course or an interpretation; warped. —v.t. wried, wry·ing To twist; contort. ♦ Homophone: rye. [ME wrye < OE wrigian move, tend] —wry′ly adv. —wry′ness n.
Wry may appear as a combining form in hyphemes, with the meaning of adjective definition 1:

wry-eyed   wry-looking   wry-set
wry-faced   wry-mouthed   wry-toothed

wry·neck (rī′nek′) n. 1 A bird (genus Jynx) resembling and allied to the woodpeckers, with the habit of twisting its head and neck. 2 A rheumatic affection in the muscles of the neck; torticollis. 3 One having a twisted neck; a person afflicted with torticollis.
Wu (wōō) The chief river of Kweichow province, south central China, comprising the upper course of the Kien and flowing over 5000 miles NE, north, and NW into the Kien proper in SE Szechwan province.
Wu (wōō), C. C., 1886-1934, Chinese statesman and diplomat: full name Wu Ch'ao-ch'u.
Wu·chang (wōō′chäng′) A formerly independent city and former capital of Hupeh province, China, on the Yangtze river, now part of Wuhan.
Wuch·er·er·i·a (vukh′ə·rer′ē·ə) n. A genus of parasitic nematode worms, especially W. bancrofti, the causative agent of elephantiasis. [after Otto Wucherer, 1820-73, German physician]
wud (wud) adj. Scot. Mad; insane.

Wu·han (wōō′hän′) A city comprising the three formerly independent cities of Hankow, Hanyang, and Wuchang on the Yangtze river, capital of Hupeh province, east central China: also Han Cities.
Wu·hu (wōō′hōō′) A port on the Yangtze river in central Anhwei province, central eastern China.
wul·fen·ite (wōōl′fən·īt) n. A resinous or hard, yellow, brown, or red molybdate of lead, PbMoO$_4$, usually occurring in tabular crystals. [after F. X. von Wulfen, 1728-1805, Austrian mineralogist]
Wul·fi·la (wōōl′fə·lə) See ULFILAS.
wun (wun) Scot. Won. —v.t. & v.i. To win. —n. Wind.
Wundt (vōōnt), Wilhelm Max, 1832-1920, German psychologist and physiologist. —Wundt′i·an adj.
wun·na (wun′nə) Scot. Will not.
Wup·per·tal (vōōp′ər·täl) A city in SW central North Rhine–Westphalia, west central West Germany, formed by the union of the cities of Barmen and Elberfeld.
Würm (vürm) See GLACIAL EPOCH.
Würm-see (vürm′zā′) See STARNBERGERSEE.
Würt·tem·berg (wûr′təm·bûrg, Ger. vür′təm·berkh) A former state of SW Germany; 7,532 square miles; capital, Stuttgart; divided after 1945 into Württemberg–Baden and Württemberg–Hohenzollern, which merged in 1951, along with Baden, to form the state of Baden–Württemberg.
Würt·tem·berg–Ba·den (wûr′təm·bûrg·bäd′n, Ger. vür′təm·berkh·bä′dən) A former state of SW Germany in the Federal Republic (1949) formed in 1945 by the union of northern Württemberg and northern Baden; 6,062 square miles; capital, Stuttgart; in 1951 the states Baden, Württemberg–Baden and Württemberg–Hohenzollern merged to form the state of Baden–Württemberg; 13,800 square miles; capital, Stuttgart.
Würt·tem·berg–Ho·hen·zol·lern (wûr′təm·bûrg·hō′ən·zol′ərn, Ger. vür′təm·berkh·hō′ən·tsōl′ərn) A state of SW Germany in the Federal Republic (1949); formed in 1945 by the union of southern Württemberg and the former Prussian province of Hohenzollern; 4,018 square miles; capital, Tübingen; merged into the state of Baden–Württemberg in 1951.
Würz·burg (würts′bûrg, Ger. vürts′bōōrkh) A city in Lower Franconia, NW Bavaria, SW central West Germany.
Wu·sih (wōō′shē′) A city of southern Kiangsu province, central eastern China.
Wu T'ing–fang (wōō′ ting′fäng′), 1841-1922, Chinese reformer and diplomat; father of C. C. Wu.

Wu·wei (wōō′wā′) A city in central Kansu province, north central China.
Wy·an·dot (wī′ən·dot) n. One of a tribe of North American Indians of Iroquoian stock, formerly very powerful in the Ohio valley and lake regions: descendants of a group of fugitive Hurons who called themselves Wendat: presently settled in Oklahoma. Also Wy′an·dotte.
Wy·an·dotte (wī′ən·dot) n. One of an American breed of domestic fowls. [after the Wyandot Indians]
Wy·att (wī′ət), Sir Thomas, 1503-42, English poet and diplomat.
wych (wich) See WITCH².
wych–elm (wich′elm′) n. 1 A wide-spreading elm (Ulmus glabra), with large, dull-green leaves, common in England, Ireland, and Scotland: also called Scotch elm. 2 Witch hazel. Also witch. [< wych, var. of WITCH² + ELM]
Wych·er·ley (wich′ər·lē), William, 1640?-1716, English dramatist and poet.
wych–ha·zel (wich′hā′zəl) n. 1 Witch hazel. 2 Wych-elm. [Var. of WITCH HAZEL]
Wyc·lif (wik′lif), John, 1324?-84, English reformer; first translator, with assistants, of the entire Bible into English. Also spelled Wiclif, Wickliffe, Wycliffe. —Wyc′lif·ite adj. & n.
wye (wī) n. The letter Y, or something Y-shaped.
Wye (wī) A river of SE Wales and SW England, flowing 130 miles SE from SW Montgomery to the Severn estuary.
Wyld (wīld), Henry Cecil Kennedy, 1870-1945, English philologist and lexicographer.
wyle (wīl) v.t. Scot. or Obs. To beguile; wile.
Wy·lie (wī′lē), Elinor Morton, 1885-1928, née Hoyt, U.S. poet and novelist: married name Mrs. William Rose Benét. —Philip Gordon, born 1902, U.S. author.
wy·lie–coat (wī′lē·kōt′, wil′ē-, wul′ē-) n. Scot. A boy's flannel underdress; also, a flannel petticoat.
Wy·o·ming (wī·ō′ming) A State in the NW United States; 97,914 square miles; capital, Cheyenne; entered the Union July 10, 1890; nickname, Equality State: abbr. Wyo., Wy. —Wy·o′ming·ite n.
Wyoming Valley A valley along the north branch of the Susquehanna River in NE Pennsylvania; scene of a massacre of settlers by Indians and Tories, 1778; chief city, Wilkes-Barre.
Wythe (with), George, 1726-1806, American jurist; signer of the Declaration of Independence.
wy·vern (wī′vərn) See WIVERN.

X

x, X (eks) n. pl. x's or X's, xs or Xs, ex·es (ek′siz) 1 The 24th letter of the English alphabet: from the ancient western Greek alphabets of Chalcis, Boeotia, and Elis, and Roman X. 2 The sound of the letter x: in English variously sounded as (ks), as in axle, box, next; (gz), as in executive, exert; (ksh), as in noxious; (gzh); as in luxurious; and initially, always (z) as in xenophobe, xylophone, Xanthippe. See ALPHABET. — symbol 1 The Roman numeral ten. See under NUMERAL. 2 Math. The principal unknown quantity; hence, anything unknown. 3 A mark shaped like an X, representing the signature of one who cannot write. 4 A mark used in diagrams, maps, etc., to place some event or substance, or to point out something to be emphasized. 5 A symbol used to indicate a kiss. 6 Anything shaped like an X. — X marks the spot Colloq. "Here": used in diagrams, maps, or the like, to indicate a specific locality.
xan·thate (zan′thāt) n. Chem. A salt or ester of xanthic acid. [<XANTH(IC) + -ATE³]
xan·the·in (zan′thē·in) n. Biochem. A water-soluble yellow coloring matter found in the cell sap of some plants. [<F xanthéine <Gk. xanthos yellow]
xan·the·las·ma (zan′thə·laz′mə) n. Pathol. A form of xanthoma marked by the appearance of small yellowish disks on the eyelids. [<NL <Gk. xanthos yellow + elasmos a metal plate]
Xan·thi·an (zan′thē·ən) adj. Relating to Xanthus.
xan·thic (zan′thik) adj. 1 Having a yellow or yellowish color. 2 Chem. Of or pertaining to xanthin or xanthine. [<F xanthique <Gk. xanthos yellow]
xanthic acid Chem. Any of a group of unstable, colorless, liquid thio compounds made by decomposing a xanthate with a dilute acid.
xan·thin (zan′thin) n. Biochem. An insoluble yellow pigment found in yellow flowers. [<G <Gk. xanthos yellow]
xan·thine (zan′thēn, -thin) n. Biochem. A white, crystalline, nitrogenous compound, C$_5$H$_4$N$_4$O$_2$, contained in blood, urine, and other animal secretions, and in some plants. It leaves a yellow residue when evaporated with nitric acid. [<F <Gk. xanthos yellow]
Xan·thip·pe (zan·tip′ē) The wife of Socrates; renowned as a shrew. Also Xan·tip′pe.

xantho- combining form Yellow: xanthophyll. Also, before vowels, xanth-. [<Gk. xanthos yellow]
xan·tho·car·pous (zan·thō·kär′pəs) adj. Bot. Yellow-fruited.
xan·tho·chroid (zan′thə·kroid) Anthropol. adj. Characterized by a light-colored or fair complexion. — n. One who exhibits xanthochroid characteristics. [<XANTHO- + Gk. chroa color + -OID]
xan·tho·ma (zan·thō′mə) n. Pathol. A skin disease marked by the presence of small yellowish disks formed by the deposit of lipoids. [<XANTH- + -OMA]
xan·tho·phyll (zan′thə·fil) n. Biochem. A yellow pigment, C$_{40}$H$_{56}$O$_2$, contained in plants and related to carotene. Also xan′tho·phyl. [<F xanthophylle <Gk. xanthos yellow + phyllon a leaf]
xan·thop·si·a (zan·thop′sē·ə) n. Pathol. A disorder of vision in which all objects appear yellow. [<NL <Gk. xanthos yellow + opsis a sight]
xan·thous (zan′thəs) adj. 1 Yellow. 2 Anthropol. a Of or pertaining to the yellow-skinned or Mongoloid type of mankind. b Of or re-

Xanthus 1455 **xylylene**

lating to that variety of mankind that has yellowish, brown, or auburn hair, including the Teutons and Scandinavians; blond. Opposed to *melanous*. [<XANTH(O)- + -OUS]

Xan·thus (zan′thəs) An ancient, ruined city of Lycia, SW Turkey in Asia, near the Mediterranean.

Xa·vi·er (zā′vē·ər, zav′ē-; *Sp.* hä·vyer′), **Saint Francis,** 1506-52, Spanish Jesuit missionary in the Orient; founder, with Ignatius Loyola, of the Society of Jesus: called "the Apostle of the Indies." — **Xa·ve·ri·an** (zā·vir′ē·ən) *adj. & n.*

X-chro·mo·some (eks′krō′mə·sōm) *n.* A sex chromosome.

xe·bec (zē′bek) *n.* A small, three-masted Mediterranean vessel, with both square and lateen sails: formerly used by Algerine pirates: also spelled *zebec*. [Earlier *chebec* <F <Sp. *jabeque, xabeque* <Arabic *shabbāk*]

xe·ni·a (zē′nē·ə) *n. Bot.* The influence of the pollen of one species upon the maternal tissues of another species after hybrid fertilization: a phenomenon observed particularly in maize, which often shows blue kernels in a yellow-seeded variety pollinated by a blue-seeded one. [<NL <Gk. *xenia* hospitality < *xenos* a guest]

xe·ni·al (zē′nē·əl) *adj.* Of or pertaining to hospitality. [<Gk. *xenia*. See XENIA.]

xeno- *combining form* Strange; foreign; different: xenophobia. Also, before vowels, **xen-**. [<Gk. *xenos* a stranger]

Xe·noc·ra·tes (zi·nok′rə·tēz), 396?-314 B.C., Greek philosopher.

xe·nog·a·my (zi·nog′ə·mē) *n. Biol.* Cross-fertilization. — **xe·nog′a·mous** *adj.*

xen·o·gen·e·sis (zen′ə·jen′ə·sis) *n. Biol.* **1** Abiogenesis. **2** Metagenesis. **3** The fancied production of an organism unlike either of its parents. Also **xe·nog·e·ny** (zi·noj′ə·nē). — **xen′o·ge·net′ic** (-jə·net′ik), **zen′o·gen′ic** *adj.*

xe·no·gloss·i·a (zē′nə·glôs′sē·ə, -glos′ē·ə) *n.* In psychic research, the alleged power of a person to communicate with others in a language which he has never learned. [<NL <Gk. *xenos* strange + *glōssa* a tongue]

xen·o·lith (zen′ə·lith) *n. Geol.* A rock fragment enclosed within a larger mass of igneous rock.

xen·o·mor·phic (zen′ə·môr′fik) *adj. Mineral.* Not having its own characteristic form, but having an irregular shape that is imposed by the interference of surrounding minerals: said of the constituents of a crystalline rock.

xe·non (zē′non) *n.* A heavy, inert, gaseous element (symbol Xe) occurring in extremely small quantities in the atmosphere. It solidifies at a very low temperature. See ELEMENT. [<Gk., neut. of *xenos* strange]

Xe·noph·a·nes (zi·nof′ə·nēz) Sixth century B.C. Greek philosopher and poet.

xen·o·phobe (zen′ə·fōb) *n.* A person who hates or distrusts strangers or foreigners.

xen·o·pho·bi·a (zen′ə·fō′bē·ə) *n.* Dislike of strangers or foreigners.

Xen·o·phon (zen′ə·fən), 435?-355? B.C., Greek historian and soldier.

Xe·res (hā′rās, *older* shā′rās, sher′es) The former name for JEREZ.

Xé·rez (hā′rās, -räth), **Francisco de,** 1504-1547?, Spanish historian of the conquest of Peru.

xer·ic (zer′ik, zir′ik) *adj.* Of, pertaining to, or characterized by extreme dryness; arid.

xero- *combining form* Dry; dryness: *xerophyte*. Also, before vowels, **xer-**. [<Gk. *xēros* dry]

xe·ro·chore (zir′ō·kôr, -kōr) *n. Ecol.* A region of extreme dryness; the desert areas of the earth, collectively. — **xe′ro·chor′ic** (-kôr′ik, -kor′ik) *adj.*

xe·ro·der·ma (zir′ō·dûr′mə) *n. Pathol.* Roughness and dryness of the skin, with scaly desquamation. [<NL <Gk. *xēros* dry + *derma* skin] — **xe′ro·der·mat′ic** (-dər·mat′ik), **xe′ro·der·ma·tous** (-dûr′mə·təs) *adj.*

Xe·ro·form (zir′ə·fôrm) *n.* Proprietary name for a yellow powder containing bismuth and tribromphenol in equal quantities: used as an intestinal and surgical antiseptic.

xe·rog·ra·phy (zi·rog′rə·fē) *n.* A method of printing by electrostatic attraction in which a negatively charged ink powder is sprayed upon the positively charged copy area of a metal plate, whence it is transferred to the positively charged printing surface. — **xe·ro·graph·ic** (zir′ō·graf′ik) *adj.* — **xe·rog′raph·er** *n.*

xe·ro·mor·phy (zir′ō·môr′fē) *n. Bot.* The form or structure of the plant by which it is protected from desiccation. [<XERO- + Gk. *morphē* form] — **xe′ro·mor′phic** *adj.*

xe·roph·i·lous (zir·rof′ə·ləs) *adj. Bot.* Growing in or adapted to drought: said of plants living in dry, hot climates, as the cactus.

xe·roph·thal·mi·a (zir′əf·thal′mē·ə) *n. Pathol.* Inflammation with thickening of the lining membrane of the eye, but without liquid discharge: associated with conjunctivitis and a lack of vitamin A. [<NL <Gk. *xēros* dry + *ophthalmos* an eye]

xe·ro·phyte (zir′ə·fit) *n. Bot.* A plant adapted to dry conditions of air and soil. — **xe′ro·phyt′ic** (-fit′ik) *adj.*

xe·ro·print·ing (zir′ō·prin′ting) *n.* A simplified variation of xerography which uses a suitably prepared plate on a rotating cylinder.

xe·ro·sere (zir′ə·sir) *n. Ecol.* The series of changes in the succession of the plant formation found upon dry soil. [<XERO- + SERE²]

xe·ro·sis (zi·rō′sis) *n. Pathol.* A condition of abnormal dryness of a part; specifically, a dry, harsh, thickened, and scaly condition of the skin or mucous membrane of a part. [<NL <Gk. *xēros* dry] — **xe·rot′ic** (-rot′ik) *adj.*

xe·ro·tro·pism (zir′ō·trō′piz·əm) *n. Bot.* The tendency of plants, or plant parts, to alter their position so as to protect themselves from desiccation. — **xe′ro·trop′ic** *adj.*

xe·rus (zir′əs) *n.* Any of a genus (*Xerus*) of African ground squirrels with long tails and coarse hair. [<NL <Gk. *xēros* dry]

Xerx·es (zûrk′sēz), 519?-465? B.C., Persian king 486?-465; invaded Greece, but was defeated at Salamis 480 B.C.

Xho·sa (kō′sä) *n.* The Bantu language of the Kaffirs, closely related to Zulu: also called *Kaffir*: also spelled *Xosa*.

xi (zī, sī; *Gk.* ksē) *n.* The fourteenth letter in the Greek alphabet (Ξ, ξ): equivalent to English *x* or *z*. [<Gk.]

Xin·gu (shing·gōō′) A river in northern and central Brazil, flowing 1,230 miles north from central Mato Grosso to the Amazon at the head of its delta.

-xion Var. of -TION.

xiphi- *combining form* A sword; of or pertaining to a sword: xiphisternum. Also, before vowels, **xiph-**. [<Gk. *xiphos* a sword]

xiph·i·ster·num (zif′ə·stûr′nəm) *n.*, *pl.* **-na** (-nə) *Anat.* The lower segment or ensiform process of the sternum. Also *xiphoid*. [<NL <Gk. *xiphos* a sword + *sternon* the breastbone]

xiph·oid (zif′oid) *adj.* Shaped like a sword: the *xiphoid* cartilage at the lower end of the breastbone. — *n.* The xiphisternum.

xiph·o·su·ran (zif′ə·sōōr′ən) *n.* Any of an order (*Xiphosura*) of primitive arachnids having a horseshoe-shaped carapace and a long swordlike tail; a king crab. — *adj.* Of or pertaining to the *Xiphosura*. [<NL <Gk. *xiphos* a sword + *oura* a tail]

Xmas Christmas: popular abbreviation. [<*X*, abbr. for *Christ* <Gk. *X*, chi, the first letter of *Christos* Christ + -MAS]

Xo·chi·mil·co (sō′chi·mēl′kō) A resort city south of Mexico City in Federal District, central Mexico; famous for its "floating gardens."

Xo·sa (kō′sä) See XHOSA.

XP Chi and rho: The first two letters of ΧΡΙΣΤΟΣ, the Greek word for Christ: introduced by Constantine the Great as an emblem of Christ.

X-ray (eks′rā′) *v.t.* To examine, diagnose, or treat with X-rays. — *n.* A picture made with X-rays; roentgenogram: also **X-ray photograph**.

X-rays (eks′rāz′) *n. pl.* Electromagnetic radiations of extremely short wavelength, emitted from a substance when it is bombarded by a stream of electrons moving at a sufficiently high velocity, as in a Coolidge tube. Their great penetrating power, ionizing effect, and property of acting on photographic plates have many useful applications, especially in the detection, diagnosis, and treatment of certain organic disorders, chiefly internal. Also called *Roentgen rays*. [Trans. of G *X-strahlen*, name coined by Roentgen, their discoverer, because their nature was unknown]

X-ray therapy Medical treatment by the use of X-rays.

Xu·thus (zōō′thəs) In Greek legend, son of Hellen and ancestor of the Ionians.

xy·lan (zī′lan) *n. Biochem.* A yellow, gummy hemicellulose found in straw, oat hulls, peanut shells, and other plant wastes: it yields xylose on hydrolysis. [<Gk. *xylon* wood]

xy·lem (zī′ləm) *n. Bot.* The portion of a vascular bundle in higher plants that is made up of woody tissue, parenchyma, and associated cells, etc. Compare PHLOEM. [<<Gk. *xylon* wood]

xy·lene (zī′lēn) *n. Chem.* Any one of three isomeric colorless hydrocarbons, $C_6H_4(CH_3)_2$, contained in coal tar and wood tar. A mixture of the three yields a colorless, inflammable liquid used as a solvent and in medicine as an antiseptic. Also **xy′lol** (-lōl, -lol). [<Gk. *xylon* wood + -ENE]

xy·lic (zī′lik) *adj.* Of, pertaining to, or derived from xylene. [<XYL(ENE) + -IC]

xylic acid *Chem.* One of six isomeric crystalline carboxyl derivatives of xylene, C_8H_9-COOH.

xy·li·dine (zī′lə·dēn, -din, zil′ə-) *n. Chem.* Any of six isomeric amino derivatives of xylene, $C_8H_{11}N$: they are homologs of aniline, and are used in the synthesis of certain dyes. Also **xy′li·din** (-din). [<XYL(ENE) + -ID(E) + -INE²]

xylo- *combining form* Wood; of or pertaining to wood; woody: *xylocarpous*. Also, before vowels, **xyl-**. [<Gk. *xylon* wood]

xy·lo·car·pous (zī′lō·kär′pəs) *adj. Bot.* Having a hard, woody fruit.

xy·lo·graph (zī′lə·graf, -gräf) *n.* **1** An engraving on wood, or a print from such engraving. **2** An impression obtained from the grain of wood, as used for surface decoration. — **xy′lo·graph′ic** or **·i·cal** *adj.* — **xy·log·ra·pher** (zī·log′rə·fər) *n.*

xy·log·ra·phy (zī·log′rə·fē) *n.* **1** Wood engraving, especially of the 15th century. **2** Printing with wood engravings. **3** Painting or printing on wood for decorative purposes. **4** The making of prints or impressions showing the grain of wood.

xy·loid (zī′loid) *adj.* Of, pertaining to, or resembling wood.

xy·loph·a·gous (zī·lof′ə·gəs) *adj.* Feeding on or boring in wood, as insect larvae. [<XYLO- + -PHAGOUS] — **xy·lo·phage** (zī′lə·fāj) *n.*

xy·lo·phone (zī′lə·fōn) *n.* A musical instrument consisting of a row of parallel wooden bars graduated in length to form a musical scale and struck by small mallets or sounded by rubbing. [<XYLO- + -PHONE] — **xy·loph·o·nist** (zī·lof′ə·nist) *n.*

XYLOPHONE

xy·lose (zī′lōs) *n. Chem.* A pentose, $C_5H_{10}O_5$, obtained by treating xylan with sulfuric acid; wood sugar: the levorotatory form is used in the synthesis of vitamin C. [<XYL(AN) + -OSE]

xy·lot·o·mous (zī·lot′ə·məs) *adj.* Adapted to cutting or boring wood, as an insect. [<XYLO- + Gk. *tomē* a cutting < *temnein* cut]

xy·lot·o·my (zī·lot′ə·mē) *n.* The preparation of wood for examination by microscope, as for scientific purposes. [<XYLO- + -TOMY] — **xy·lot′o·mist** *n.*

xy·lyl (zī′lil) *n. Chem.* The univalent radical, $(CH_3)_2C_6H_3$, derived from xylene. [<XYL(ENE) + -YL]

xy·ly·lene (zī′lə·lēn) *n. Chem.* The bivalent radical, C_8H_8, contained in xylene. [<XYLYL + -ENE]

xyst (zist) *n.* **1** In classical antiquity, a hall or covered portico used by athletes for their exercises: chiefly for use in stormy weather. **2** A garden walk or terrace. [< L *xystus* < Gk. *xystos*, orig. scraped, polished < *xyein* scrape, polish]

xys·ter (zis′tər) *n.* A surgical instrument for scraping bones. [< NL < Gk. *xystēr* a scraper, < *xyein* scrape]

Y

y, Y (wī) *n. pl.* **y's, Y's** or **ys, Ys** or **wyes** (wīz) **1** The 25th letter of the English alphabet: ultimately from Phoenician *vau*, Greek *upsilon*. The Romans took it from the Greek alphabet sometime in the first century B.C. and used it as a vowel. **2** The sound of the letter *y*. Initial *y* (introducing either a vowel or a syllable) is a voiced palatal semivowel, as in *yet, you, yonder, beyond*. Final *y* is either a vowel, pronounced (ē), as in *honey, pretty, steady*; a diphthong, pronounced (ī), as in *fly, my*; or the final glide of a diphthong, as in *gray, obey, annoy*. Internal *y* is pronounced as a vowel (i), as in *lyric, myth, syllable*; a diphthong (ī), as in *lyre, type, psychic*; an *r*-colored central vowel (ûr) or (ər), as in *myrtle, martyr*. See ALPHABET.
Y (wī) *n.* **1** Something similar to a Y in shape. **2** A branch pipe, forked pipe, or coupling in the shape of the letter Y. **3** A forked piece, often with the branches curved, usually one of a pair, serving as a rest or support, as for some part of a sighting instrument.
y- *prefix* Used in Middle English as a sign of the past participle, as an intensive, or without perceptible force: *yclad, yclept*. It survives (as *a–*) in such words as *alike, aware*, etc. Also spelled *i–*, as in *iwis*. [OE *ge-*]
-y[1] *suffix of adjectives* Being, possessing, or resembling what is expressed in the main element: *stony, rainy*. Also *-ey*, when added to words ending in *y*, as in *clayey, skyey*. [OE *-ig*]
-y[2] *suffix* The quality or state of being: *victory*: often used in abstract nouns formed from adjectives in *-ous* and *-ic*. [< F *-ie* < L *-ia*; also < Gk. *-ia, -eia*]
-y[3] *suffix* Little; small: *kitty*: often used in nicknames or to express endearment, as in *Tommy*. [Prob. < dial. E (Scottish) < OF *-i, -e*, dim. suffixes]
Ya·an (yä′än′) A city in west central Szechwan province, western China, formerly the capital of Sikang province, which was incorporated in Szechwan province in 1955.
yab·ber (yab′ər) *n. Austral. Colloq.* Speech; talk; jabber. [< Australian *yabba* < *ya* speak]
Ya·blo·noi Range (yi·blə·noi′) Part of the watershed between the Arctic and Pacific drainage areas, in SE Asiatic Russian S.F.S.R.; highest peak, 5,280 feet. Also **Ya·blo·no·vy** (yi′blə·nô′vē).
yacht (yot) *n.* A relatively small vessel specially built or fitted for private pleasure excursions, as distinguished from war or commerce. — *v.i.* To cruise, race, or sail in a yacht. [< Du. *jaghte*, short for *jaghtschip* a pursuit ship < *jaght* hunting (< *jagen* hunt) + *schip* a ship]
yacht club A club of yachtsmen.
yacht·ing (yot′ing) *n.* The act, practice, or pastime of sailing in or managing a yacht.
yachts·man (yots′mən) *n. pl.* **·men** (-mən) One who owns or sails a yacht; a devotee of yachting. Also **yacht′er, yacht′man.** — **yachts′wom′an** (-wŏŏm′ən) *n. fem.*
yachts·man·ship (yots′mən·ship) *n.* The art of managing a yacht; skill in yachting. Also **yacht′man·ship.**
Yad·kin River (yad′kin) The upper course of the Pee Dee River, flowing 204 miles NE and SE from NW to south central North Carolina.
yaff (yaf) *v.i. Brit. Dial.* To bark like a dog when excited; hence, to speak sharply; scold. [Imit.]
yaf·fle (yaf′əl) *n.* The green woodpecker. [Imit. of its cry]
yag·ger (yag′ər) *n. Scot.* An itinerant peddler; wanderer; ranger.
ya·gua·run·di (yä·gwə·run′dē) See JAQUARONDI.
yah[1] (yä, ya) *interj.* An exclamation of disgust; bah.

yah[2] (yä, yâ) *interj. Colloq.* Yes. [Alter. of YES; infl. in form by G *ja* yes]
Ya·ha·ta (yä·hä·tä) See YAWATA.
ya·hoo (yä′hōō, yä′-, yä·hōō′) *n.* **1** Any person of low or vicious instincts. **2** An awkward fellow; a bumpkin. [< YAHOO]
Ya·hoo (yä′hōō, yä′-, yä·hōō′) *n.* One of an imaginary race of brutes possessing human form and vices, described by Swift in *Gulliver's Travels*. See HOUYHNHNM.
yaird (yârd) *n. Scot.* **1** A yard (36 inches). **2** A garden; courtyard; churchyard.
Yah·weh (yä′we) In the Old Testament, the national god of Israel; God: a modern transliteration of the Tetragrammaton. See JEHOVAH. Also spelled *Jahveh, Jahwe*. Also **Yah·veh** (yä′ve). [< Hebrew YHWH]
Yah·wism (yä′wiz·əm) *n.* **1** The ancient Hebrew religion centered on the monotheistic worship of Yahweh. **2** The use of the name Yahweh for God. Also spelled *Jahvism, Jahwism*. Also **Yah′vism** (-viz·əm).
Yah·wist (yä′wist) *n.* In Biblical criticism, the writer supposed to have written those parts of the Hexateuch in which God is mentioned as Yahweh (erroneously Jehovah). Compare ELOHIST. Also spelled *Jahvist, Jahwist*. Also **Yah′vist** (-vist).
Yah·wis·tic (yä·wis′tik) *adj.* **1** Of or relating to Yahwist or Yahwism. **2** Characterized by the use of the name Yahweh (or Jehovah) for God. Compare ELOHISTIC. Also spelled *Jahvistic, Jahwistic*. Also **Yah·vis′tic** (-vis′-).
yak (yak) *n.* A large bovine ruminant (*Bos grunniens*) of the higher regions of central Asia: it has long hair fringing the shoulders, sides, and tail, and is often domesticated. [< Tibetan *gyag*]
Yak·i·ma (yak′ə·mə) A city on the Yakima River in southern Washington.
Yakima River A river in central and southern Washington, flowing 203 miles SE from the Cascade Range to the Columbia River.
Ya·ko (yä′kō) See JACO.
Ya·kof (yä′kôf) Russian form of JAMES.
Ya·kut (yä·kōōt′) *n.* **1** One of a people living in the Yakut Autonomous S.S.R. **2** The Turkic language of these people.
Ya·kut Autonomous Soviet Socialist Republic (yä·kōōt′) An administrative division of NE Asiatic Russian S.F.S.R.; 1,181,971 square miles; capital, Yakutsk.
Ya·kutsk (yä·kōōtsk′) A city on the Lena river; capital of Yakut Autonomous S.S.R.
yald[1] (yäd, yôd) See YELD.
yald[2] (yäd, yôd) *adj. Scot.* Athletic; supple; active: also spelled *yauld*.
Yale (yāl), **Elihu**, 1649–1721, English merchant; benefactor of Yale College (now Yale University, in New Haven, Connecticut); born in America. — **Linus**, 1821–68, U. S. locksmith.
Yal·ta (yäl′tə, yôl′-) A port on the Black Sea in the southern Crimea, U.S.S.R.; scene of a conference of Roosevelt, Churchill, and Stalin in February, 1945.
Ya·lu (yä′lōō′) A river forming part of the boundary between Manchuria, NE China, and Korea, and flowing 500 miles SW to the Yellow Sea: Japanese *Oryokko*.
Ya·lung (yä′lōōng′) A river in Szechwan province, China, flowing 800 miles south from SE Tsinghai province to the Yangtze, on the border of Yünnan province.

yam (yam) *n.* **1** The fleshy, edible, tuberous root of any of a genus (*Dioscorea*) of climbing tropical plants. **2** Any of the plants growing this root. **3** A large variety of the sweet potato. **4** *Scot.* A potato. [< Pg. *inhame* < Senegal *nyami* eat]
Ya·ma·ga·ta (yä·mä·gä·tä), **Prince Aritomo**, 1838–1922, Japanese admiral.
Ya·mal Peninsula (ye·mäl′) A peninsula of Asiatic Russian S.F.S.R. between the Kara Sea and Ob Gulf, about 400 miles long, up to 140 miles wide.
Ya·ma·mo·to (yä·mä·mō·tō), **Isoroku**, 1884–1943, Japanese admiral.
Ya·ma·shi·ta (yä·mä·shē·tä), **Tomoyuki**, 1885–1946, Japanese general; captured Singapore (February) and Philippines (May) 1942; executed as war criminal; called "the Tiger of Malaya."
Yam·bol (yäm′bôl) A city of east central Bulgaria: also *Jambol*. Turkish **Yam·bo·li** (yäm′·bô·lē).
yam·mer (yam′ər) *v.i. Colloq.* **1** To complain peevishly; whine; whimper. **2** To howl; roar; shout. — *v.t.* **3** To utter peevishly; complain. [OE *geōmrian* lament < *geōmor* sorrowful; infl. in form by MDu. *jammeren* complain]
ya·men (yä′mən) *n. Chinese* The office or official residence of a public functionary, as a mandarin; also, any department of the public service: the *yamen* of public justice. Also **ya′mun.**
Ya·nam (yə·num′) A city and former French settlement in NE Andhra Pradesh State, SE India. French **Ya·na·on** (yà·nà·ôn′).
yang (yang) *n.* In Chinese philosophy and art, the male element, source of life and heat, represented symbolically by a circular diagram bisected by an S-curve, one half red (*yang*), the other half black (*yin*): originated during the Han Dynasty: opposed to *yin*. Also **Yang**. [< Chinese]
Yang (yang), **C(hen) N(ing)**, born 1922, U. S. physicist born in China.
Yang·chow (yäng′jō′) A city in central Kiangsu province, eastern China.
Yang·tze (yang′tsē′, *Chinese* yäng′tse′) The longest river of Asia and China, flowing 3,430 miles from the Tibetan highlands to the East China Sea near Shanghai; forms border between Tibetan Autonomous Region and Szechwan province. Also **Yang′tze-Ki·ang′** (-kē·ang′, *Chinese* jē·äng′), **Yangtse-Kiang.**

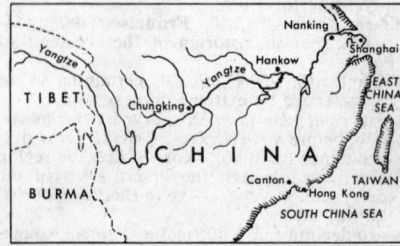
YAK
(From 5 to 5 1/2 feet high at the shoulder)

Ya·ni·na (yä′nē·nä) See IOANNINA.
yank (yangk) *v.t.* **1** To jerk or pull suddenly. — *v.i.* **2** To give a pull or jerk. **3** *Brit.* To be vigorously active. **4** *Brit.* To jabber; scold. — *n.* **1** *Colloq.* A sudden sharp pull; jerk. **2** *Scot.* A sharp blow or slap; buffet. [? < dial. E (Scottish) *yank* a sharp sudden blow]
Yank (yangk) *n. & adj. Colloq.* Yankee. [Short for YANKEE]
Yan·kee (yang′kē) *n.* **1** Originally, a native or inhabitant of New England. **2** A Northerner; especially, a Federal soldier during the Civil War: so called in the South. **3** Any

Yankeedom 1457 **year**

citizen of the United States; an American: a foreign, chiefly British, usage. —*adj.* **1** Of or pertaining to the Yankees. **2** *Brit.* American. [Prob. back formation <*Jan Kees* (taken as a plural), John Cheese, orig. a nickname for a Hollander; later applied by Dutch colonists in New York to English settlers in Connecticut]

Yan·kee·dom (yang′kē·dəm) *n.* **1** New England or the northern States as opposed to southern. **2** The United States as a whole. **3** Yankees collectively or as a class.

Yankee Doodle A song, of many humorous verses, popular in pre-Revolutionary times and one of the national airs of the United States.

Yan·kee·ism (yang′kē·iz′əm) *n.* **1** Yankee characteristics collectively. **2** A Yankee word, trait, or idiom, especially as restricted to New England.

Yan·kee·land (yang′kē·land′) *n. Colloq.* **1** The United States. **2** New England or the northern States as opposed to the southern.

yank·ing (yang′king) *adj.* **1** Inclined to jerk or pull sharply, as a horse. **2** *Scot.* Active; enterprising.

Yan·tra (yän′trä) A river in northern Bulgaria, flowing 168 miles NE to the Danube.

Ya·oun·dé (yä·ōōn·dā′) The capital of French Cameroons, central western Africa; a trading, manufacturing, and educational center.

yap (yap) *n.* **1** *Slang* Talk; jabber. **2** *Slang* A rowdy or bumpkin. **3** A bark or yelp. **4** A worthless dog. —*v.i.* **yapped, yap·ping 1** *Slang* To prate; jabber. **2** *Colloq.* To bark or yelp, as a cur. [Imit. of a dog's bark]

Yap (yäp, yap) An island group in the western Carolines; 80 square miles: formerly *Guap.*

ya·pon (yä′pon) See YAUPON.

Ya·qui (yä′kē) *n.* One of a tribe of North American Indians belonging to the Piman branch of the Uto-Aztecan linguistic stock, now living in southern Sonora, Mexico.

Ya·qui (yä′kē) A river in NW Mexico, flowing 420 miles SW and south from the Sierra Madre Occidental to the Gulf of California; the largest river of Sonora state.

Yar·bor·ough (yär′bûr·ō, *Brit.* yär′bər·ə) *n.* A whist or bridge hand with no card above a nine. [after an earl of *Yarborough,* who bet against the occurrence of such a hand]

yard[1] (yärd) *n.* **1** The standard English and American measure of length: 3 feet, or 36 inches, or 0.914 meter. **2** A yardstick. **3** *Naut.* A long, slender, tapering spar set crosswise on a mast and used to support sails. [OE *gyrd* a rod, a measure of length]

yard[2] (yärd) *n.* **1** A tract of ground enclosed or set apart. **2** An enclosure, usually small and near a residence or other building; by extension, the grounds near a house, college, or university, whether enclosed or not. **3** An enclosure used for some specific work: often in composition: a *brickyard, shipyard.* **4** An enclosure or piece of ground adjacent to a railroad station, used for making up trains and for storing the rolling stock. **5** The winter pasturing ground of deer and moose: a moose *yard.* **6** An enclosure for animals, poultry, etc. —*v.t.* To put or collect into or as into a yard. —*v.i.* To gather into an enclosure or yard. [OE *geard* an enclosure]

yard·age[1] (yär′dij) *n.* The amount or length of something in yards, as of silk. [<YARD[1]]

yard·age[2] (yär′dij) *n.* The use of or charge for a yard in handling cattle as they are moved to and from railway cars. [<YARD[2]]

yard·arm (yärd′ärm′) *n. Naut.* Either end of a yard of a square sail.

yard·grass (yärd′gras′, -gräs′) *n.* A coarse, widely distributed, annual grass (*Eleusine indica*); goosegrass.

yard·man[1] (yärd′mən) *n. pl.* **·men** (-mən) *Naut.* A sailor who works on the yards.

yard·man[2] (yärd′mən) *n. pl.* **·men** (-mən) A man employed in a yard, especially on a railroad.

yard·mas·ter (yärd′mas′tər, -mäs′-) *n.* A railroad official having charge of a yard.

yard·stick (yärd′stik′) *n.* **1** A graduated measuring stick a yard in length. **2** A measure or standard of comparision. Also **yard′·wand′** (-wond′).

yare (yâr) *adj. Archaic & Dial.* **1** Responding quickly to the helm; manageable: said of a ship. **2** Brisk; prompt. **3** Prepared; ready. —*adv. Obs.* With dispatch; quickly; soon. [OE *gearu* ready] —**yare′ly** *adv.*

Yar·kand (yär·kand′) **1** A town and oasis of SW Sinkiang-Uigur Autonomous Region, NW China: Chinese *Soche.* **2** A river of SW Sinkiang-Uigur Autonomous Region, NW China, flowing 500 miles NE from the Karakoram range to the Tarim.

Yar·mouth (yär′məth) **1** A port at the entrance to the Bay of Fundy in SW Nova Scotia, Canada. **2** See GREAT YARMOUTH, England.

yarn (yärn) *n.* **1** Any spun material, natural or synthetic, prepared for use in weaving, knitting, or crotcheting. **2** Continuous strands of spun fiber, as wool, cotton, linen, silk, jute, or rayon. **3** A quantity of such spun material. **4** *Colloq.* A long, exciting story of adventure, often of doubtful truth: a sailor spinning a *yarn.* —*v.i. Colloq.* To tell a yarn or yarns. [OE *gearn*]

Ya·ro·slavl (yä′rō·slav′əl) A city on the Volga in north central European Russian S.F.S.R.

yar·row (yar′ō) *n.* A genus (*Achillea*) of perennial carduaceous herbs of Europe and North America; especially, the common yarrow or milfoil, with small white flowers and a pungent odor and taste. [OE *gearwe*]

yar·rup (yar′əp) *n.* Flicker[2]. [Imit. of its song]

yash·mak (yäsh·mäk′, yash′mak) *n.* The double veil or covering for the face worn by Moslem women when in public. Also **yash·mac′, yas·mak′.** [<Arabic *yashmaq*]

Yass-Can·ber·ra (yäs′kan′bər·ə) The former name for the AUSTRALIAN CAPITAL TERRITORY before it was called the FEDERAL CAPITAL TERRITORY.

yat·a·ghan (yat′ə·gan, -gən; *Turkish* yä′tä·gän′) *n.* A Turkish sword or scimitar with a double-curved blade and a handle without a guard: often called *ataghan.* Also **yat′a·gan.** [<Turkish *yātāghan*]

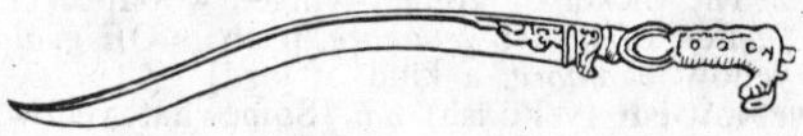

TURKISH YATAGHAN

yaud (yäd, yôd) *n. Scot.* An old mare. See JADE.

yauld[1] (yôd, yäd, yäld) See YALD[2].

yauld[2] (yäld) See YELD.

yaup (yôp) See YAWP.

yau·pon (yô′pən) *n.* A bushy evergreen shrub (*Ilex vomitoria*) of the holly family, found in the southern United States, where its leaves were used for tea and by the North Carolina Indians for their celebrated *black drink*: also spelled *yapon, youpon, yupon.* [<Siouan (Catawba) *yopún,* dim. of *yop* a bush]

Ya·va·ri (yä′vä·rē′) See JAVARI. Also **Ya′va·ry′.**

yaw (yô) *v.i.* **1** *Naut.* To steer wildly or out of its course, as a ship when struck by a heavy sea. **2** To move unsteadily or irregularly. **3** *Aeron.* To deviate from the flight path by angular displacement about the vertical axis; fishtail. —*v.t.* **4** To cause to yaw. —*n.* **1** A movement of a ship or aircraft by which it temporarily alters its course. **2** Irregular, unsteady, or deviating motion. [Cf. ON *jaga* move to and fro]

Ya·wa·ta (yä·wä·tä) A city of northern Kyushu island, Japan: also *Yahata.*

yawl[1] (yôl) See YOWL.

yawl[2] (yôl) *n.* **1** A small sailing vessel rigged like a cutter, with the addition of a jiggermast astern of the rudder post: thus distinguished from a ketch. **2** A ship's small boat; jollyboat. **3** A small fishing boat. [Appar. <Du. *jol,* orig. a boat used in Jutland]

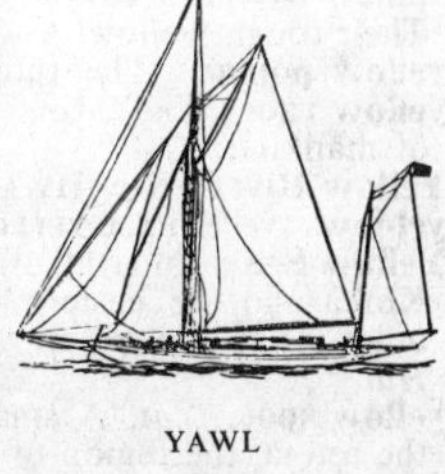

YAWL

yawl-rigged (yôl′rigd′) *adj. Naut.* Having two masts, the after one very small and stepped astern of the rudder post, and both rigged with fore-and-aft sails.

yaw·me·ter (yô′mē′tər) *n. Aeron.* An instrument for measuring the angle of yaw in an aircraft.

yawn (yôn) *v.i.* **1** To open the mouth wide, either voluntarily, as an animal seeking its prey, or involuntarily, with a long, full inspiration of the breath, usually as the result of drowsiness, fatigue, or boredom. **2** To be or stand wide open, especially as ready to engulf or receive something: A chasm *yawned* below. —*v.t.* **3** To express or utter with a yawn. —*n.* **1** A wide opening of the mouth, especially as from weariness. **2** The act of opening wide. [Prob. fusion of OE *geonian* yawn and *gānian* gape] —**yawn′er** *n.*

yawp (yôp) *v.i.* **1** To bark or yelp. **2** *Colloq.* To gape; yawn audibly. **3** *Brit. Colloq.* To shout; bawl; talk loudly. —*n.* **1** A bark or yelp. **2** A shout; noise; noisy talking; also, a loud, uncouth outcry. **3** *Scot.* The scream of a bird, especially when in distress. **4** *Scot.* A cough. Also spelled *yaup.* [Imit.] —**yawp′er** *n.*

yaws (yôz) *n. pl. Pathol.* Frambesia. [<Carib-an *yáya*]

yay (yā) *U. S. Dial. adj.* **1** This many; this much. **2** Ever so many: for *yay* years. —*adv.* **1** To this extent. **2** Ever so: *yay* big. [Cf. G *je* ever]

Yazd (yezd) See YEZD.

Yaz·oo River (yaz′ōō) A river in west central Mississippi, flowing 189 miles SW to the Mississippi at Vicksburg.

Y-car·ti·lage (wī′kär′tə·lij) *n. Anat.* A piece of cartilage shaped like the letter Y, situated at the bottom of the socket of the hip joint.

Y-chro·mo·some (wī′krō′mə·sōm) *n.* A sex chromosome.

y·clept (i·klept′) *adj. Archaic* Called; named: now a humorous term. Also **y·cleped′.** [OE *geclypod,* pp. of *clypian* call]

ye[1] (t͟hē) The: an archaic contraction in which the *y* represents the thorn (þ) of the Old and Middle English alphabet. Often printed y^e.

ye[2] (yē) *pron. Archaic* The persons addressed: now confined almost exclusively to poetic or formal pulpit style. Historically *ye* is only a nominative form: "Blessed are *ye* when men shall revile *you.*" *Matt.* v 11. [OE *ge,* nominative pl.]

yea (yā) *adv.* **1** Yes: used to express affirmation or assent: in this sense now superseded by *yes.* **2** Not only so, but more so: to intensify or amplify a meaning: fifty, *yea,* a hundred: an archaic term. **3** In reality; verily: a form of introduction in a sentence. **4** So as to be realized: All the promises of God in him are *yea* and Amen; truly; really: a use of the Authorized Version of the Bible. —*n.* An expression of affirmation; an affirmative vote; by extension, one who casts such a vote. [OE *gēa*]

ye·ah (ye′ə) *adv. Slang* Yes. [<YES]

yeal·ing (yē′ling) *n. Scot.* A contemporary; an equal in age: also spelled *yeelin.*

yean (yēn) *v.t. & v.i.* To bring forth (young), as a goat or sheep. [OE (assumed) *geēanian.* Akin to OE *geēan* pregnant.]

yean·ling (yēn′ling) *n.* The young of a goat or sheep. —*adj.* Young; newly born.

year (yir) *n.* **1** The period of time in which the earth completes a revolution around the sun: about 365 days, used as a unit of time, and divided into 12 months. It is now reckoned as beginning January 1 and ending December 31. **2** Any period of 12 months. **3** The period of time during which a planet revolves around the sun. **4** *pl.* Length or time of life; age; sometimes, old age: active for his *years.* —**astronomical year** The period between two passages of the sun through the same equinox, which determines the changing seasons. Its length is 365 days, 5 hours, 48 minutes, 46 seconds. Also **equinoctial, natural, solar,** or **tropical year.** —**calendar, civil,** or **legal year** The period of time from midnight of December 31 to the same hour twelve months thereafter. Formerly, in England, the legal year began with March 25, but historic years were counted from January 1. In 1751 the English Parliament prescribed that the legal year should begin with the

add,āce,câre,pälm; end,ēven; it,īce; odd,ōpen,ôrder; to͝ok,pōōl; up,bûrn; ə = a in *above,* e in *sicken,* i in *clarity,* o in *melon,* u in *focus*; yōō = u in *fuse*; oi,oil; ou,pout; ch,check; g,go; ng,ring; th,thin; t͟h,this; zh,vision. Foreign sounds à,œ,ü,kh,ṅ; and ◆: see page xx. < from; + plus; ? possibly.

first of January, 1752. —**common year** That of 365 days, approaching most nearly in the number of days to the astronomical year. The leap year has 366 days. —**fiscal year** A financial year of a national treasury or of a business at the end of which accounts are balanced; any twelve-month period used as a basis of business reckoning. —**lunar year** That of thirteen months, one month being added at intervals to make the mean length of the astronomical year, as in the Hebrew calendar. —**sidereal year** The period of 365 days, 6 hours, 9 minutes, 9 seconds, in which the sun apparently returns to the same position among the stars. It is longer than the astronomical year, owing to the precession of the equinoxes. —**Sothic year** The fixed solar year of the Egyptians, consisting of 365 days and 6 hours: so called because determined by the heliacal rising of the Dog Star (Sothis). [OE *gēar*]

year·book (yir′bŏŏk′) *n.* A book published annually, presenting information about the previous year.

year·ling (yir′ling) *n.* A young animal past its first year and not yet two years old; specifically, a colt or filly a year old dating from January 1 of the year of foaling. —*adj.* Being a year old.

year·long (yir′lông′, -long′) *adj.* Continuing through a year.

year·ly (yir′lē) *adj.* **1** Included within a year's time. **2** Occurring once a year; annual. **3** Continuing or lasting for a year: a *yearly* subscription. —*adv.* Once a year; annually.

yearn (yûrn) *v.i.* **1** To desire something earnestly; long: with *for.* **2** To be deeply moved; feel sympathy. [OE *giernan, geornan.* Akin to OE *georn* eager.]

yearn·ing (yûr′ning) *n.* A strong emotion of longing or desire, especially with tenderness. —**yearn′ing·ly** *adv.*

yeast (yēst) *n.* **1** A substance consisting of minute cells of ascomycetous fungi (genus *Saccharomyces*) that clump together, forming a yellow, frothy, viscous growth which, in contact with saccharine liquids, develops or increases by germination, producing fermentation by means of enzymes, in which process alcohol and carbon dioxide are produced, as in the brewing of beer and the raising of bread. **2** Such a substance mixed with flour or meal, and sold commercially. **3** Any of a family *(Saccharomycetaceae)* of yeast-forming fungi. **4** Froth or spume. **5** Figuratively, mental or moral ferment: the *yeast* of youth. —*v.i.* To foam; froth. [OE *gist*]

yeast cake A mixture of living yeast cells and starch in compressed form suitable for use in baking or brewing.

yeast powder Dried and powdered yeast used as a leavening agent.

yeast·y (yēs′tē) *adj.* **1** Of, pertaining to, or resembling yeast. **2** Causing or characterized by fermentation. **3** Restless; unsettled; frivolous. **4** Covered with or consisting mainly of froth or foam. **5** Light or unsubstantial. —**yeast′i·ness** *n.*

Yeats (yāts), **William Butler,** 1865–1939, Irish poet, dramatist, and essayist.

Ye·do (ye·dō) A former name for TOKYO.

yeel·in (yē′lin) See YEALING.

yegg (yeg) *n. Slang* An itinerant burglar; a criminal tramp; a safe-cracker; loosely, any burglar. Also **yegg′man.** [Prob. <earlier *yekkman* a beggar in San Francisco's Chinatown <dial. Chinese *yekk* a beggar]

Ye·gor·yevsk (yə·gôr′yəfsk) A city in west central European Russian S.F.S.R.; a cotton milling center: also *Egorevsk.*

Ye·hsien (ye′shyen′) A city of NE Shantung province, NE China, near the Gulf of Chihli.

yeld (yeld) *adj. Scot.* Not giving milk; barren: also spelled *yald, yauld.* Also **yell.**

Yel·ga·va (yel′gə·və) See JELGAVA.

yelk (yelk) *n. Dial.* Yolk.

yell (yel) *v.t.* & *v.i.* To shout; scream; roar; also, to cheer. See synonyms under CALL, ROAR. —*n.* **1** A sharp, loud, inarticulate cry, as of pain, terror, anger, etc. **2** A rhythmic cheer composed of a prearranged set of words and shouted by a group in unison. [OE *gellan, giellan*] —**yell′er** *n.*

yell·och (yel′əkh) *v.i.* & *n. Scot.* Yell; scream.

yel·low (yel′ō) *adj.* **1** Having the color of ripe lemons, or sunflowers. **2** Changed to a sallow color by age, sickness, or the like: a paper *yellow* with age. **3** Having a sallow complexion, as a member of the Mongoloid ethnic group. **4** Jaundiced; hence, melancholy; jealous. **5** Sensational, especially offensively so: said of newspapers: *yellow* journalism. **6** *Colloq.* Cowardly; mean; dishonorable. —*n.* **1** The color of the spectrum between green and orange, including wavelengths centering at about 5,890 angstroms; the color of ripe lemons. **2** Any pigment or dyestuff having such a color. **3** The yolk of an egg. **4** *pl.* Any of various unrelated plant diseases in which there is stunting of growth and yellowing of the foliage; especially, an infectious virus disease of peach, nectarine, apricot, and almond trees. **5** *pl.* Jaundice, especially a variety that affects domestic animals. **6** *pl. Obs.* Jealousy; hence, a jealous frame of mind. —*v.t.* & *v.i.* To make or become yellow. [OE *geolu*] —**yel′low·ly** *adv.* —**yel′low·ness** *n.*

yel·low·bark (yel′ō·bärk′) *n.* Calisaya.

yel·low–bel·lied (yel′ō·bel′ēd) *adj.* **1** *Slang* Cowardly; yellow. **2** Having a yellow underside: *yellow-bellied* sapsucker.

yel·low·bird (yel′ō·bûrd′) *n.* **1** The goldfinch (def. 2). **2** The yellow warbler.

yellow daisy The black-eyed Susan.

yel·low–dog contract (yel′ō·dôg′, -dog′) A contract with an employer in which an employee agrees not to join a labor union during his term of employment.

yellow fever *Pathol.* An acute, infectious intestinal disease of tropical and semitropical regions, caused by a filtrable virus transmitted by the bite of a mosquito (genus *Aëdes*). It is characterized by hemorrhages, jaundice, vomiting, and fatty degeneration of the liver. Also *yellow jack.*

yel·low·ham·mer (yel′ō·ham′ər) *n.* **1** An Old World bunting *(Emberiza citrinella)* with the sides of the head, neck, and breast bright yellow, the back yellow and black, and the top of the head and tail feathers blackish. **2** The flicker or golden-winged woodpecker. [Alter. of earlier *yelambre,* prob. <OE *geolo* yellow + *amore,* a kind of bird]

yel·low·ish (yel′ō·ish) *adj.* Somewhat yellow. —**yel′low·ish·ness** *n.*

yellow jack **1** A carangoid fish *(Caranx bartholomaei)* of the West Indies and Florida. **2** The flag of the quarantine service. **3** Yellow fever.

yellow jacket Any of various social wasps (genus *Vespa*) with bright-yellow markings.

yellow jasmine or **jessamine** A smooth twining shrub *(Gelsemium sempervirens)* with bright-yellow flowers.

yellow journal A cheaply sensational newspaper or other publication. [So called from the use of yellow ink in printing a cartoon strip, "The Yellow Kid," in the *New York Journal,* commencing Oct. 18, 1896]

yellow lead ore Wulfenite.

yel·low·legs (yel′ō·legz′) *n.* **1** Either of two North American sandpipers (genus *Totanus*) with long yellow legs: the **greater,** or **winter, yellowlegs** *(T. melanoleucus),* or the **lesser yellowlegs** *(T. flavipes).* **2** *U.S. Colloq.* Formerly, in the U.S. Army, a cavalry soldier.

yellow metal **1** A brass consisting of 60 parts copper and 40 parts zinc. **2** Gold.

yellow perch Perch² (def. 1).

yellow peril The political power of the peoples of eastern Asia, conceived of as threatening white supremacy.

yellow pine **1** Any of various American pines, as the Georgia or loblolly pine. **2** Their tough, yellowish wood.

yellow poplar The tuliptree.

yellow race The Mongoloid ethnic division of mankind.

Yellow River See HWANG HO.

yel·lows (yel′ōz) See YELLOW (*n.* defs. 4 and 5).

Yellow Sea An arm of the Pacific between Korea and the eastern coast of China; 400 miles long, 400 miles wide: Chinese *Hwang Hai.*

yellow spot *Anat.* A small yellowish spot in the retina, the region of most acute vision.

Yellowstone Falls Two waterfalls of the Yellowstone River in Yellowstone National Park: **Upper Yellowstone Falls,** 109 feet; **Lower Yellowstone Falls,** 308 feet.

Yellowstone National Park The largest and oldest of the United States national parks, at the junction of Wyoming, Montana, and Idaho, largely in NW Wyoming; 3,458 square miles; established, 1872.

Yel·low·stone River (yel′ō·stōn) A river in NW Wyoming, SE Montana, and NW North Dakota, flowing 671 miles NW to the Missouri River and passing through Yellowstone National Park where it forms **Yellowstone Lake,** 20 miles long, 14 miles wide; 140 square miles.

yellow streak A personality trait combining cowardice, treachery, and meanness.

yel·low·tail (yel′ō·tāl′) *n.* **1** Any of various fishes having a yellowish tail. **2** A carangoid fish (genus *Seriola*), especially the **California yellowtail** *(S. dorsalis).* **3** A California rockfish *(Sebastodes flavidus).* **4** The menhaden.

yel·low·throat (yel′ō·thrōt′) *n.* Any of various American warblers (genus *Geothlypis*), especially the **Maryland yellowthroat** *(G. trichas),* olive-green, with yellow throat and breast.

yel·low–throat·ed warbler (yel′ō·thrō′tid) A warbler *(Dendroica dominica)* of wooded regions of the southern United States.

yellow waterlily A yellow variety of pondlily (genus *Nuphar*).

yel·low·weed (yel′ō·wēd′) *n.* **1** Any of various goldenrods; especially, the Canada goldenrod *(Solidago canadensis).* **2** The bulbous crowfoot *(Ranunculus bulbosus).* **3** The European ragwort *(Senecio jacobaea).* **4** Weld² (def. 1).

yel·low·wood (yel′ō·wŏŏd′) *n.* **1** The yellow or yellowish wood of a medium-sized tree *(Cladrastis lutea)* of the southern United States, with smooth bark and showy white flowers; gopherwood. The wood yields a yellow dye. **2** The tree. **3** Any one of several other trees with yellowish wood, as the Osage orange, buckthorn, smoketree, or the like.

yel·low·y (yel′ō·ē) *adj.* Yellowish.

yellow yel·dring (yel′drin) *Dial.* The yellowhammer. Also **yellow yoldring.** [<YELLOW + var. of dial. E *yowlring* <*yowlo* yellow + RING]

yelp (yelp) *v.i.* To utter a sharp or shrill cry; give a yelp. —*v.t.* To express by a yelp or yelps. —*n.* **1** A sharp, shrill cry; a sharp, crying bark, as of a dog in distress. **2** The sharp, staccato cry of the turkey hen. [OE *gielpan* boast] —**yelp′er** *n.*

yelp·ing (yel′ping) *n.* The act of one who yelps; utterance of quick, sharp cries or barks, as of a dog; also, the sounds so uttered.

Yem·en (yem′ən) A kingdom of the SW Arabian peninsula; 75,000 square miles; capitals, Sana and Ta'iz; joined United Arab States, 1958. —**Yem·e·ni** (yem′ə·nē), **Yem·e·nite** (yem′ə·nīt) *adj.* & *n.*

yen¹ (yen) *Slang n.* An ardent longing or desire; an intense want; an infatuation. —*v.i.* **yenned, yen·ning** To yearn; long. [<Chinese, opium, smoke]

yen² (yen) *n.* The monetary unit of the Japanese, containing 100 sen. [<Japanese <Chinese *yüan* round, a dollar]

Ye·nan (ye′nän′) A city in northern Shensi province, north central China; a commercial center; headquarters of the Chinese Communist party, 1937–47: formerly (1913–1948) *Fushih.*

Yen·geese (yeng′gēz) *n. pl.* White people; specifically, English settlers in New England. [Appar. N. Am. Ind. alter. of ENGLISH]

Yen·i·sei (yen′ə·sā′) A river in central Asiatic Russian S.F.S.R., flowing 2,364 miles NW through **Yenisei Bay** to **Yenisei Gulf** (90 miles wide), its estuary in the Arctic Ocean: also *Enisei.*

Yen·i·seisk (yen′ə·sāsk′) A city on the Yenisei river in central Krasnoyarsk territory, central Asiatic Russian S.F.S.R.

Yen·tai (yen′tī′) A port on the Yellow Sea in NE Shantung province, NE China, on the northern coast of the Shantung peninsula: formerly *Chefoo.*

yeo·man (yō′mən) *n. pl.* **·men** (-mən) **1** *Brit.* A freeholder next under the rank of gentleman; in early times, one who owned a small landed estate; in modern usage, a farmer, especially one who cultivates his own farm; loosely, a man of the common people. **2** A petty officer in the U. S. Navy, Coast Guard, or Army Transport Service, who performs clerical duties. **3** *Brit.* One of the higher-class attendants in the service of a nobleman

yeomanly or of royalty: a *yeoman* of the crown; sometimes, a servitor of lower rank: a *yeoman* of the chamber, the buttery, etc. 4 *Brit.* A member of the yeomanry cavalry; also, a Yeoman of the (Royal) Guard. 5 *Obs.* One who acts as an assistant in a subordinate capacity; a helper; journeyman. [ME *yeman, yoman*, prob. contraction of *yengman* a young man <OE *geong* young + *mann* a man]

yeo·man·ly (yō'mən·lē) *adj.* Pertaining to or resembling a yeoman; of yeoman's rank; brave; rugged; staunch. — *adv.* Like a yeoman; bravely; staunchly.

Yeoman of the (Royal) Guard A member of the special bodyguard of the English royal household, consisting of one hundred yeomen chosen from the best rank below the gentry, and first appointed by Henry VII. See BEEFEATER.

yeo·man·ry (yō'mən·rē) *n.* 1 The collective body of yeomen; freemen; farmers. 2 *Brit.* A home guard of volunteer cavalry, created in 1761, consisting of gentlemen and gentlemen farmers, known since 1901 as the **Imperial yeomanry.** In 1907 it became a part of the Territorial Army.

yeoman's service Faithful and useful support or service; loyal assistance in need. Also **yeoman service.**

yep (yep) *adv. Colloq.* Yes. [Alter. of YES]

-yer Var. of -IER.

yer·ba (yâr'bə, yûr'-) *n.* Maté (def. 1). [<Sp. *yerba (maté)* the herb (maté)]

Yer·ba Bue·na Island (yâr'bə bwā'nə, yûr'-) An island of 300 acres in San Francisco Bay, California; mid–point of the San Francisco–Oakland Bay Bridge.

yerb tea (yûrb, yärb) *Dial.* Herb tea. [<*yerb*, dial. var. of HERB + TEA]

Ye·re·men·ko (yi'ryi·myen'kə), **Andrei Ivanovich**, born 1892, Russian general; broke the siege of Stalingrad in World War II.

Ye·re·van (ye're·vän') The Armenian name for ERIVAN.

yerk (yûrk) *Obs.* or *Dial. v.t. & v.i.* 1 To tie with a jerk; bind tightly. 2 To crack, as a whip. 3 To beat; lash; excite. 4 To jerk; to kick, as a horse. — *n.* A jerk; a smart blow.

Yer·kes (yûr'kēz), **Charles Tyson**, 1837–1905, U.S. financier. — **Robert Mearns**, 1876–1956, U. S. psychobiologist.

yes (yes) *adv.* As you say; truly; just so: a reply of affirmation or consent: opposed to *no*, and equivalent to a repetition of the words of a question or command in the form of an assertion. The word is sometimes used to enforce by repetition or addition something that precedes. — *n. pl.* **yes·es** or **yes·ses** A reply in the affirmative. — *v.t. & v.i.* **yessed, yes·sing** To say "yes" (to). [OE *gēse*, prob. < *gēa* yea + *sī*, third person sing. present subj. of *bēon* be]

ye'se (yēs) *Scot.* You shall; ye shall.

Ye·sil Ir·mak (ye·shēl' ir·mäk') A river in northern Turkey in Asia, flowing 260 miles NW to the Black Sea: ancient *Iris.*

Ye·sil·köy (ye'shēl·kœ·e') The Turkish name for SAN STEFANO.

yes man *Colloq.* One who agrees without criticism; a servile, acquiescent assistant or subordinate; a toady.

yester- *prefix* Pertaining to the day before the present; by extension of the preceding, used of longer periods than a day: *yesteryear*. [<YESTER(DAY)]

yes·ter·day (yes'tər·dē, -dā') *n.* 1 The day preceding today. 2 Loosely, the near past. — *adv.* 1 On the day last past. 2 At a recent time. [OE *geostran dæg* < *geostran* yesterday + *dæg* day]

yes·ter·eve·ning (yes'tər·ēv'ning) *n.* The evening of yesterday. Also **yes'ter·eve'**, **yes'ter·e'ven** (-ē'vən), **yes'treen** (yes·trēn').

yes·ter·morn·ing (yes·tər·môr'ning) *n.* The morning of yesterday. Also **yes'ter·morn'.**

yes·tern (yes'tərn) *adj. Archaic* Of or pertaining to yesterday. [<YESTER(DAY), on analogy with *eastern, western,* etc.]

yes·ter·night (yes'tər·nīt') *n. Archaic & Poetic* The night last past. — *adv.* In or during the night last past. [OE *geostran* yesterday + *niht* night]

yes·ter·noon (yes'tər·nōōn') *n.* The noon of yesterday.

yes·ter·week (yes'tər·wēk') *n.* Last week.

yes·ter·year (yes'tər·yir') *n.* Last year. [Trans. of F *antan*, coined by D. G. Rossetti]

yet (yet) *adv.* 1 In addition; besides; further: often with a comparative. 2 Before or at some future time; eventually: He will *yet* succeed. 3 In continuance of a previous state or condition; still: I can hear him *yet*. 4 At the present time; now: Don't go *yet*. 5 After all the time that has or had elapsed: Are you not ready *yet*? 6 Up to the present time; heretofore: commonly with a negative: He has never *yet* lied to me. 7 Than that which has been previously affirmed: with a comparative: It was hot yesterday; today it is hotter *yet*. 8 As much as; even: He did not believe the reports, nor *yet* the evidence. — **as yet** Up to now. — *conj.* 1 Nevertheless; notwithstanding: I speak to you peaceably, *yet* you will not listen. 2 But: He is willing, *yet* unable. 3 Although: active, *yet* ill. See synonyms under BUT[1], NOTWITHSTANDING. [OE *gīet, gīeta*]

Synonyms (adverb): besides, further, hitherto, now, still. *Yet* and *still* have many closely related senses, and, with verbs of past time, are often interchangeable; we may say "while he was *still* a child." *Yet*, like *still*, often applies to past action or state extending to and including the present time, especially when joined with *as*; we can say "He is feeble as *yet*," or "He is *still* feeble," with scarcely appreciable difference of meaning, except that the former statement implies somewhat more of expectation than the latter. *Yet* with a negative applies to completed action, often replacing a positive statement with *still*: "He has not gone *yet*" is nearly the same as "He is here *still*." *Yet* has a reference to the future which *still* does not share; "We may be successful *still*" implies that we may continue to enjoy in the future such success as we are winning now.

yett (yet) *n. Scot.* A gate.

yeuk (yōōk) *n. & v.i. Scot.* Itch. Also **yeuck, yewk.** — **yeuk'y** *adj.*

yew (yōō) *n.* 1 Any one of several evergreen trees or shrubs (genus *Taxus*), with flat, lanceolate, dark-green leaves and a red berrylike fruit; especially, the **European** or **English yew** (*T. baccata*), a medium-sized coniferous tree of slow growth and long life, with spreading horizontal branches and dense dark-green foliage. 2 The hard, fine-grained, durable wood of the common yew, of a purplish or deep-brown color. 3 A bow made from the wood of the yew tree. ◆ Homophone: *ewe*. [OE *ēow, īw*]

YEW

Yezd (yezd) A city in central Iran: also *Yazd.*

Ye·zo (ye·zō) The former name for HOKKAIDO.

Yg·dra·sil (ig'drə·sil) In Norse mythology, the huge ash tree whose roots and branches bind together heaven, earth, and hell: also spelled *Igdrasil.* Also **Yg'dra·sill, Ygg'dra·sill.**

Y·gerne (i·gûrn') See IGRAINE.

Y-gun (wī'gun') *n. Mil.* A gun having two barrels set at an angle, used for discharging depth bombs against enemy submarines, and mounted aft, usually on a destroyer. [So called because shaped like a Y]

YHWH Yahweh. See JEHOVAH.

Yid·dish (yid'ish) *n.* A Germanic language derived from the Middle High German spoken in the Rhineland in the thirteenth and fourteenth centuries, now spoken primarily by Jews in Poland, Lithuania, the Ukraine, and Rumania, and by Jewish immigrants from these regions in other parts of the world. It contains elements of Hebrew and the Slavic languages, and is written in slightly modified Hebrew characters. — *adj.* 1 Of or pertaining to Yiddish; written or spoken in Yiddish. 2 *Slang* Jewish. [<G *jüdisch* Jewish]

yield (yēld) *v.t.* 1 To give forth by a natural process, or as a result of labor or cultivation: The field will *yield* a good crop. 2 To give in return, as for investment; furnish: The bonds *yield* five percent interest. 3 To give up, as to superior power; surrender; relinquish: often with *up*: to *yield* a fortress; to *yield* oneself up to one's enemies. 4 To concede or grant: to *yield* precedence; to *yield* consent. 5 *Obs.* To pay, repay, or reward. — *v.i.* 6 To provide a return; produce; bear. 7 To give up; submit; surrender. 8 To give way, as to pressure or force; bend, collapse, etc. 9 To assent or comply, as under compulsion; consent: We *yielded* to their persuasion. 10 To give place, as through inferiority or weakness: with *to*: We will *yield* to them in nothing. See synonyms under ALLOW, BEND[1], DEFER[2], OBEY, PRODUCE, SURRENDER. — *n.* 1 The amount yielded; product; result, as of cultivation or mining. 2 The profit derived from invested capital. 3 The proceeds of a tax after the expenses of collection and administration have been deducted. 4 *Mil.* The explosive force of an atomic or thermonuclear bomb as expressed in kilotons or megatons. See synonyms under HARVEST, PRODUCT. [OE *gieldan, geldan* pay] — **yield'er** *n.*

yield·ing (yēl'ding) *adj.* Disposed to yield; flexible; obedient. See synonyms under DOCILE, SUPPLE. — **yield'ing·ly** *adv.* — **yield'ing·ness** *n.*

yield point *Physics* The amount of stress, measured in unit area, under which a given material, as a rod of metal, will exhibit permanent deformation; the point at which a stress or strain just exceeds the elastic strength of the material. Also **yield strength.**

yill (yil) *n. Scot.* Ale.

yin[1] (yin) *n. Scot.* One.

yin[2] (yin) *n.* In Chinese philosophy and art, the female element, which stands for darkness, cold, and death. Compare YANG. Also **Yin.** [<Chinese]

yince (yins) *adv. Scot.* Once.

Yin-chwan (yin'chwän') A city of NE Kansu province, NW central China; capital (1928–1954) of former Ningsia province: formerly (until 1945) *Ningsia.*

Ying·kow (ying'kō') A port on the Gulf of Liaotung in SW Liaoning province, NE China.

yip (yip) *n.* A yelp, as of a dog. — *v.i.* **yipped, yip·ping** To yelp. [Imit.]

yird (yûrd) *n. Scot.* Earth. Also **yirth** (yûrth).

yirr (yûr) *v.i. Scot.* To snarl; yell; growl, as a dog.

yit (yit) *adv. & conj. Dial. & Obs.* Yet. [Var. of YET]

-yl *suffix Chem.* Used to denote a radical, especially a univalent one: *ethyl, butyl.* [<Gk. *hylē* wood, matter]

y·lang-y·lang (ē'läng-ē'läng) *n.* 1 A tree (*Cananga odorata*) of Malaysia; the Malayan custard apple. 2 A perfume derived from the greenish-yellow flowers of this tree. Also spelled *ilang-ilang.* [<Tagalog *álang-ílang* flower of flowers]

Y-lev·el (wī'lev'əl) *n.* A combined telescope and spirit level on a Y-shaped mounting which may be rotated: used in surveying, etc.

Y-mir (ē'mir, ü'mir) In Norse mythology, the progenitor of the giants, formed of frost and fire, out of whose body the gods created the world. Also **Y'mer.**

y·nough (i·nuf') *adj. & adv. Obs.* Enough. Also **y·nough', y·now** (i·nou', i·nō'). [ME, enough, OE *genōg*]

yod (yōd, *Hebrew* yōōd) *n.* The tenth Hebrew letter. Also **yodh.** See ALPHABET. [<Hebrew *yōdh*, lit., a hand]

yo·del (yōd'l) *n.* A melody or refrain sung to meaningless syllables, with abrupt changes from chest to head tones and the reverse: common among Swiss and Tyrolese mountaineers. — *v.t. & v.i.* **·deled** or **·delled, ·del·ing** or **·del·ling** To sing with a yodel, changing the voice quickly from its natural tone to a falsetto and back. Also **yo'dle.** [<G *jodeln*, lit., utter the syllable *jo*] — **yo'del·er, yo'del·ler, yo'dler** *n.*

yo·ga (yō'gə) *n.* A Hindu system of mystical and ascetic philosophy which involves withdrawal from the world and abstract meditation upon any object, as the Supreme Spirit,

yogh

with the purpose of identifying one's consciousness with the object. [<Hind. <Skt., lit., union] — **yo′gic** adj.
yogh (yōkh) n. A Middle English letter which represented a voiced or voiceless palatal fricative, or a voiced velar fricative. It is variously spelled in Modern English as y, as in lay, w, as in law, and gh, as in daughter and enough.
yo·gi (yō′gē) n. A follower of the yoga philosophy; an ascetic or adept, supposed to possess magical powers. Also **yo′gee**, **yo′gin**. [<Hind. yogī <Skt. <yoga yoga]
yo·gurt (yō′gŏort) n. A thick, curdled milk treated with cultures of bacteria regarded as beneficial to the intestines; also called matzoon. Also **yo′ghurt**, **yo′ghourt**. [<Turkish yōghurt]
yoicks (yoiks) interj. A cry formerly used in foxhunting to urge on the hounds: also hoicks. [Earlier hoik, var. of hike; prob. imit.]
yoke (yōk) n. 1 A curved timber with attachments used for coupling draft animals, as oxen, usually having a bow at each end to receive the neck of the animal. 2 Any of many similar contrivances, as a frame fitted for a person's shoulders from the ends of which are suspended burdens intended to balance, as pails of milk. 3 Naut. A crosspiece on a rudder head, carrying yoke lines for steering. 4 Mech. A strap, clamp, clip, slotted piece, or the like, serving to confine, guide, or guard the movement of a part of a machine or mechanism. 5 A crossbar suspended from the collars in double harness for supporting the tongue or pole. 6 A part of a garment designed to support a plaited or gathered part, as at the hips or shoulders, giving shape to the garment. 7 That which binds or connects; a bond: the yoke of love. 8 In ancient Rome, a device consisting of two upright spears with a third laid transversely across them, under which a conquered army was made to march. 9 Servitude, or some visible mark of it; bondage. 10 sing. & pl. A couple; pair; team: a yoke of oxen. 11 Obs. The amount of land a yoke of oxen can plow in a day. 12 Scot. The time required for a yoke of oxen to accomplish a specified amount of work; hence, a part of the day. — v. **yoked**, **yok·ing** v.t. 1 To attach by means of a yoke, as draft animals; put a yoke upon. 2 To join with or as with a yoke; couple or link. 3 To join in marriage. 4 Rare To bring into bondage; enslave. — v.i. 5 To be joined or linked; unite. [OE geoc]

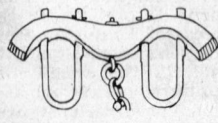

YOKE (def. 1)

yoke·fel·low (yōk′fel′ō) n. A mate or companion in labor. Also **yoke′mate**′ (-māt′).
yo·kel (yō′kəl) n. A countryman; country bumpkin: a contemptuous term. [? <dial. E, a green woodpecker, a yellowhammer] — **yo′kel·ish** adj.
yok·ing (yō′king) n. 1 The act of one who yokes. 2 Scot. As much work as is done by a yoke of draft animals at a time.
Yok·kai·chi (yōk·kī·chē) A city of central southern Honshu island, Japan, on Ise Bay.
Yo·ko·ha·ma (yō′kə·hä′mə) A port on Tokyo Bay in central Honshu island, Japan.
Yo·ko·su·ka (yō′kə·sōō′kə) A port at the entrance to Tokyo Bay, central Honshu island, Japan.
yol·dring (yōl′drin) n. Scot. & Brit. Dial. A species of bunting; the yellowhammer (def. 1). Also **yol′ding** (-ding), **yol′drin**. [Var. of earlier yowling <ME yowlow yellow + RING]
yolk (yōk, yōlk) n. 1 The yellow portion of an egg. 2 Biol. That portion of the contents or substance of the ovum which is used for the nourishment and formation of the embryo, consisting of fat or oil drops, etc., as distinguished from the albumen or white of an egg. ♦ Collateral adjective: vitelline. 3 A fine yellow soapy exudation in sheep's wool. [OE geol(o)ca, lit., (the) yellow part <geolu yellow]
yolk·y (yō′kē) **yolk·i·er**, **yolk·i·est** adj. 1 Of or pertaining to a yolk. 2 Affected with or containing yolk: yolky wool.
yom (yom, yōm) n. Hebrew Day: used in designating days of feast or fasting: Yom Kippur.
Yom Kip·pur (yom kip′ər, Hebrew yōm kip′- ŏor) The Jewish Day of Atonement: the 10th of Tishri (September–October). It is marked by continuous prayer and fasting for 24 hours from sundown on the evening previous. [<Hebrew yōm kipūr day of atonement]
yon (yon) adj. & adv. Archaic, Dial. & Poetic Yonder; that or those over there: yon fine house. [OE geon]
yond (yond) adj. & adv. Archaic & Dial. Yonder. [OE geond across; infl. in meaning by yon]
yon·der (yon′dər) adj. Being at a distance indicated. — adv. In that place; there. [ME, prob. extension of yone, OE geon yon]
yo·ni (yō′nē) n. The female organ of generation: the symbol under which Shakti is worshiped in India. [<Skt.]
yon·ker (yong′kər) See YOUNKER.
yont (yont) prep. Scot. Beyond.
yore (yôr, yōr) n. Old time; time long past: in days of yore. — adv. Obs. Long ago; in olden times. [OE geara formerly, prob. orig. genitive pl. of gear year]
Yor·ick (yôr′ik, yor′-) A court jester to the king of Denmark, mentioned in Shakespeare's Hamlet.
York (yôrk) A royal house of England; a branch of the Plantagenet line; reigned 1461–1485.
York (yôrk) 1 A maritime county in NE England; the largest county in England; 6,080 square miles; divided into East, West, and North Riding: also Yorkshire. 2 Its county town, a city on the Ouse, famous for its Norman cathedral: capital of Roman Britain as Eboracum.
York (yôrk), **Alvin Cullum**, born 1887, U.S. soldier and hero in World War I.
York, **Cape** 1 The northernmost point of Australia, in Queensland on Torres Strait. 2 A promontory of NW Greenland on Baffin Bay at the western end of Melville Bay; site of major meteorites, discovered by Peary.
Yorke Peninsula (yôrk) A promontory of southern South Australia; 160 miles long, 35 miles wide. Also **Yorke's Peninsula**.
York·ist (yôr′kist) n. An adherent of the house of York.
York River An estuary in SE Virginia, flowing into Chesapeake Bay; 40 miles long, 1 to 2 1/2 miles wide.
York·shire (yôrk′shir, -shər) See YORK (def. 1).
York·shire pudding (yôrk′shir, -shər) A batter pudding baked in the drippings of roasting meat, either in the same pan or under the spit of a roast.
York·town (yôrk′toun) A town in SE Virginia on the York River; scene of Cornwallis's surrender to Washington in the Revolutionary War, 1781.
Yo·ru·ba (yō′rōō·bä) 1 A Negro belonging to an extensive linguistic family of the African Slave Coast between the lower Niger and Dahomey rivers. Many North American Negroes are of Yoruba descent. 2 The language of the Yoruba, one of the dominant tongues of the Sudanic family. — **Yo′ru·ban** adj.
Yo·ru·ba (yō′rōō·bä) A former native state in SW Nigeria.
Yo·sem·i·te National Park (yō·sem′ə·tē) A government reservation in east central California noted for its scenic grandeur; 1,183 square miles; highest point, 13,095 feet; established in 1890.
Yosemite Valley A gorge in the western Sierra Nevada mountains in Yosemite National Park in east central California; 7 miles long, 1 mile wide; traversed by the Merced River which forms **Yosemite Falls**, a triple cataract (Upper Fall, 1,430 feet; Lower Fall, 320 feet; total drop, with intermediate cascades, 2,425 feet).
Yo·shi·da (yō·shē·dä), **Zengo**, born 1885, Japanese admiral.
Yo·shi·hi·to (yō·shē·hē·tō), 1879–1926, emperor of Japan 1912–26.
Yo·su (yō·sōō) A port of southern South Korea. Japanese **Rei·sui** (rā·syē).
you (yōō) pron. 1 The person or persons, animal or animals, personified thing or things addressed, in either the nominative or objective case: as a subject, always linked with a plural verb. 2 Colloq. One; anyone: You learn by trying. [OE ēow, dative and accusative of ge ye]
you'd (yōōd) You had; you would: a contraction.
you'll (yōōl) You will: a contraction.
young (yung) adj. **young·er** (yung′gər), **young·est** (yung′gist) 1 Being in the early period of life or growth; having existed a short or comparatively short time; not old. 2 Not having progressed far; newly formed: The day was young. 3 Pertaining to youth or early life. 4 Full of vigor or freshness. 5 Being without experience; immature. 6 Denoting the younger of two persons having the same name or title; junior. 7 Geol. Having the characteristics of an early stage in the geological cycle: said of a river or of certain land forms. 8 Radical or progressive in social or political aims: used with proper names: the Young Turks, Young Italy. See synonyms under FRESH, NEW, YOUTHFUL. — n. 1 Young persons as a group; youth collectively. 2 Offspring, especially of animals. — **with young** With child; pregnant. [OE geong]
Young (yung), **Arthur Henry**, 1866–1943, U.S. cartoonist. — **Brigham**, 1801–77, U.S. Mormon leader. — **Edward**, 1683–1765, English poet. — **Francis Brett**, 1884–1954, English novelist. — **Mahonri Mackintosh**, born 1877, U.S. sculptor. — **Owen D.**, born 1874, U.S. lawyer and industrialist. — **Thomas**, 1773–1829, English physicist.
young·ber·ry (yung′ber′ē) n. pl. **·ries** A type of large dark-red berry, hybridized from a trailing blackberry and a dewberry, found in the western United States. [after B. M. Young, U.S. horticulturist]
young blood Youth; young people.
younger hand In card games, the hand next to the leader: also called pone: opposed to eldest hand.
young–eyed (yung′īd′) adj. Having youthful eyes or fresh vision; bright-eyed.
young·ish (yung′ish) adj. Rather young.
young·ling (yung′ling) n. 1 A young person, animal, or plant. 2 An inexperienced person. — adj. Young. [OE geongling]
Young Plan The plan, adopted in 1929, whereby the amount of German reparations for World War I was finally determined. [after Owen D. Young]
Young Pretender See STUART, CHARLES EDWARD.
young·ster (yung′stər) n. 1 A young person; a child; youth; sometimes, also, a colt or other young animal. 2 Colloq. A junior military officer. [<YOUNG + -STER, infl. by younker]
Youngs·town (yungz′toun) A city in NE Ohio; a major steelmaking center.
youn·ker (yung′kər) n. 1 A German squire. 2 Colloq. A youngster. 3 A young gentleman; knight. Also spelled yonker. [<MDu. jonckher a young gentleman <jonc young + here a lord, master]
you·pon (yōō′pon) See YAUPON.
your (yŏor, yôr) pronominal adj. The possessive case of the pronoun you employed attributively; belonging or pertaining to you: your fate. [OE ēower, genitive of ge ye]
you're (yŏor, yôr) You are: a contraction.
yours (yŏorz, yôrz) pron. 1 The possessive case of you used predicatively; belonging or pertaining to you: This room is yours. 2 The things or persons belonging or pertaining to you: a home as quiet as yours; God bless you and yours. — **of yours** Belonging or relating to you; your: the double possessive. [ME youres]
your·self (yŏor·self′, yôr-) pron. pl. **·selves** (-selvz′) A reflexive and often emphatic form of the pronoun of the second person. Yourself is employed as a simple objective: This rests with yourself, or in apposition with you: You did it yourself. Its use as a subject nominative is obsolete. Yourself is also used reflexively: You've cut yourself, and, rarely, as a substantive: You're not yourself today. Also Scot. **your·sel′**.
youth (yōōth) n. pl. **youths** (yōōths, yōōthz) 1 The state or condition of being young. 2 The period when one is young; that part of life between childhood and manhood; adolescence. 3 The early period of being or development, as of a movement. 4 A young man: in this sense with plural: several youths: used, also, as a collective noun: the youth of the land. [OE geoguth]

youth·ful (yōōth′fəl) *adj.* 1 Pertaining to youth; characteristic of youth; hence, buoyant; fresh; vigorous. 2 Having youth; being still young; immature. 3 Not far advanced; early; new. 4 *Geol.* Young. — **youth′ful·ly** *adv.* — **youth′ful·ness** *n.*
— **Synonyms**: boyish, childish, childlike, girlish, juvenile, puerile, young. *Boyish, childish,* and *girlish* are used in a good sense of those to whom they properly belong, but in a bad sense of those from whom more maturity is to be expected; *childish* eagerness or glee is pleasing in a child, but unbecoming in a man; *puerile* in modern use is distinctly contemptuous. *Juvenile* and *youthful* are commonly used in a favorable and kindly sense in their application to those still *young; youthful* may have a favorable import as applied to any age, as when we say the old man still retains his *youthful* ardor, vigor, or hopefulness: *juvenile* in such use would belittle the statement. See FRESH, NEW.

you've (yōōv) You have: a contraction.
yow (you) See YOWL.
yowe (yō) *n. Obs. & Dial.* A ewe. Also **yow** (yō). [Dial. var. of EWE]
yow·ie (yō′ē) *n. Scot.* A small ewe. [Dim. of YOWE]
yowl (youl) *v.i.* To utter a yowl; howl; yell. — *n.* A loud, prolonged, wailing cry; a howl. Also spelled *yawl, yow.* [Cf. ON *gaula* howl, yell]
yo-yo (yō′yō′) *n. pl.* **-yos** A wheel-like toy with a deep central groove around which is looped a string connecting the toy with the operator's finger. As the toy spins up and down the string it may be put through a variety of movements by manipulation of the string. [Origin unknown]
Y·pres (ē′pr′) A town in NW Belgium; site of three major battles of World War I, 1914, 1915, 1917. Flemish **Ie·per** (yā′pər).
Yp·si·lan·ti (ip′sə·lan′tē, *Gk.* ēp′sē·län′tē), **Alexander**, 1792–1828, Greek patriot. — **Demetrios**, 1793–1832, Greek patriot; brother of the preceding.
Y·quem (ē·kem′) *n.* A highly esteemed Sauterne wine. [from Château *Yquem,* an estate in SW France]
Y·sa·bel (ē·sä·bel′) See SANTA ISABEL.
Y·ser (ē·zer′) A river in northern France and western Belgium, flowing 48 miles NE from near St. Omer through Nord department and West Flanders to the North Sea at Nieuport.
Y·seult (i·sōōlt′) See ISEULT.
Ys·sel (ī′səl) See IJSSEL.
Ys·trad·y·fod·wg (üs′träd·i·vod′ŏŏg) See RHONDDA.
Y–track (wī′trak′) A track at approximately right angles to a line of railroad, and connected with it by two switches: used in place of a turntable.
yt·ter·bi·a (i·tûr′bē·ə) *n. Chem.* White ytterbium oxide, Yb_2O_3.
yt·ter·bi·um (i·tûr′bē·əm) *n.* A rare metallic element (symbol Yb) occurring in minute amounts in gadolinite and certain other minerals which yield rare-earth elements. See ELEMENT. [<NL, from *Ytterby,* a town in Sweden where gadolinite was first found] — **yt·ter′bic** *adj.*
ytterbium metal Yttrium metal.

yt·tri·a (it′rē·ə) *n. Chem.* A white insoluble earth, yttrium sesquioxide, Y_2O_3. [<NL, from *Ytterby.* See YTTERBIUM.]
yt·tric (it′rik) *adj. Chem.* Of, pertaining to, or derived from yttrium, especially in its higher valence. [<YTTRI(UM) + -IC]
yt·trif·er·ous (i·trif′ər·əs) *adj.* Yielding or containing yttrium. [<YTTRI(UM) + -FEROUS]
yt·tri·um (it′rē·əm) *n.* A rare element (symbol Y) belonging to the lanthanide series. It is formed as a dark-gray powder upon the electrolysis of its double chloride with sodium, and occurs in gadolinite, samarskite, and other rare-earth minerals. See ELEMENT. [<NL <YTTRIA]
yttrium metal *Chem.* Any of the lanthanide series of elements related to yttrium, as dysprosium, erbium, holmium, thulium, ytterbium, and lutetium.
Yü (yü) A river of southern China, flowing 500 miles east from eastern Yünnan through southern Kwangsi to a confluence with the Hungshui, forming the West River proper: also *Siang.*
yu·an (yōō·än′, *Chinese* yü′än′) *n.* The former monetary unit of the Chinese Republic. Also **yuan dollar**. [<Chinese *yüan,* lit., a circle]
Yü·an (yü′än′) A river in NW Hunan province, SE central China, flowing 540 miles NE and east from western Kweichow province to Tungting Lake. Also **Yü·en** (yü′en′).
Yü·an Shih–k'ai (yü′än′ shē′kī′), 1859–1916, Chinese general; president of the Chinese Republic 1912–1916.
Yu·bi (yōō′bē), **Cape** See JUBY, CAPE.
Yu·ca·tán (yōō′kə·tan′, *Sp.* yōō′kä·tän′) 1 A peninsula of SE Mexico and NE Central America (including British Honduras and part of Guatemala); 70,000 square miles; separated from Cuba by **Yucatán Channel**, a strait between Yucatán and Cuba, connecting the Gulf of Mexico with the Caribbean; 135 miles wide. 2 A state in SE Mexico at the NW end of the peninsula; 13,706 square miles; capital, Mérida.
yuc·ca (yuk′ə) *n.* 1 Any of a large genus (*Yucca*) of liliaceous plants of the southern United States, Mexico, and Central America, generally found in dry, sandy places, having a woody stem, usually very short, but sometimes arborescent, which bears a large panicle of white, bell-shaped, drooping flowers emerging from a crown of sword-shaped leaves. 2 The flower of this plant, the State flower of New Mexico. [<NL <Sp. *yuca* < Taino]
yucca moth A moth (*Tegeticula* or *Pronuba yuccasella*) whose larvae feed on yucca seed pods.
Yu·chi (yōō′chē) *n.* One of a tribe of North American Indians, the one tribe comprising the Uchean linguistic stock, formerly dwelling along the Savannah River in eastern Georgia. In 1836 they migrated with the Creeks to what is now Oklahoma.

YUCCA (Plant from 2 to 10 feet tall)

Yu·ga (yōō′gə) *n.* An age; cycle; a period of long duration according to Hindu thought. Each **Maha–yuga** or great age of the world, consisting of 4,320,000 years, is subdivided into four *Yugas* or ages: **Krita–yuga** (1,728,000 years), **Treta–yuga** (1,296,000 years), **Dvâpara–yuga** (864,000 years), and **Kali–yuga** (432,000 years), which began in 3094 B.C. These ages decrease successively in excellence; the life of man is supposed to last for 400 years in the first, 300 years in the second, 200 years in the third, and 100 years in the fourth, or Kali age. Also **Yug** (yōōg). [<Skt., an age, yoke]
Yu·go·sla·vi·a (yōō′gō·slä′vē·ə), **Federal People's Republic of** A state of SE Europe on the Adriatic; the largest country of the Balkans; 98,538 square miles; capital, Belgrade; 1918–1929, the *Kingdom of the Serbs, Croats, and Slovenes,* formed by the union of Serbia and Montenegro with former Austro-Hungarian provinces; 1929–41 the *Kingdom of Yugoslavia;* 1941–45 occupied by Axis Powers; the present federal republic comprises six "people's republics": Bosnia and Herzegovina, Croatia, Macedonia, Montenegro, Serbia, and Slovenia: also *Jugoslavia.* — **Yu′go·slav** (-släv, -slav), **Yu′go·sla′vi·an** *adj. & n.* — **Yu′go·slav′ic** (-slä′vik, -slav′ik) *adj.*
Yu·it (yōō′it) *n.* One of the Eskimos inhabiting northeastern Siberia. Compare INNUIT. [<Eskimo, men]
Yu·ka·wa (yōō·kä·wä), **Hideki**, born 1907, Japanese physicist.
Yu·kon (yōō′kon) A territory in NW Canada between Alaska and the Northwest Territories; 207,076 square miles; capital, Whitehorse.
Yukon River (yōō′kon) A river in NW Canada and central Alaska, flowing 1,979 miles to the Bering Sea.
Yule (yōōl) *n.* Christmas time, or the feast celebrating it. [OE *geól(a)* Christmas day, Christmastide]
yule candle A large candle formerly used to light Christmas festivities.
Yule Day *Dial.* or *Scot.* Christmas Day.
yule log A large log or block of wood, brought in with much ceremony, and made the foundation of the Christmas Eve fire. Also **yule block, yule clog**.
Yule·tide (yōōl′tīd′) *n.* Christmas time.
Yu·ma (yōō′mə) *n.* One of a tribe of North American Indians, the dominant tribe of the Yuman linguistic stock, formerly living along the Gila and Colorado rivers in northern Mexico and Arizona and in SE California: now on a reservation in California.
Yu·man (yōō′mən) *n.* A North American Indian linguistic stock of the SW United States and NW Mexico, including the Mohave and Yuma tribes.
Yün·nan (yōō′nän′, *Chinese* yün′nän′) A province of SW China; 154,014 square miles; capital, Kunming.
yu·pon (yōō′pən) See YAUPON.
Yur·ev (yōōr′yəf) The Russian name for TARTU.
Yu·zov·ka (yōō′zəf·kə) The former name for STALINO.
Yve·tot (ēv·tō′) A town in northern France NW of Rouen; a monarchy in the 15th and 16th centuries; at present a textile center.
y·wis (i·wis′) See IWIS.

Z

z, Z (zē, *Brit.* zed) *n. pl.* **z's, Z's** or **zs, Zs** or **zees** (zēz) 1 The 26th letter of the English alphabet: from Phoenician *zayin,* Greek *zeta,* Roman Z. It was not used by the Romans until about the first century B.C. 2 The sound of the letter *z,* a voiced alveolar fricative corresponding to the voiceless *s.* See ALPHABET.
Z (zē) *n.* Something resembling a letter Z in shape: sometimes written *zee*.
Za·brze (zäb′zhe) A city in southern Poland,

formerly (1742–1945) in Upper Silesia: German *Hindenburg.*
Za·ca·te·cas (sä′kä·tā′käs) A state in central Mexico; 28,117 square miles; capital, Zacatecas.
za·ca·tón (sä·kä·tōn′, *Sp.* thä′kä·tōn′) *n.* A species of muhly grass (*Muhlenbergia macroura*) found in Mexico, the roots of which are often used in making brushes: also called *Mexican broomroot, whisk grass.* [<Sp.

< *zacate* forage, grass, hay <Nahuatl *zacatl*]
Zac·chae·us (za·kē′əs) A masculine personal name. Also **Zac·che·us** (za·kē′əs), *Fr.* **Zachée** (zá·shā′), *Ital.* **Za·che·o** (dzä·kā′ō). [< Hebrew, remembrance of the Lord]
— **Zacchaeus** A wealthy publican at whose house Jesus dined in Jericho. *Luke* xix 2.
Zach (zak) Diminutive of ZACHARIAH, ZACHARIAS.
Zach·a·ri·ah (zak′ə·rī′ə) A masculine personal

name. Also **Zach·a·ry** (zak'ər·ē), *Dan.*, *Du.*, *Sw.* **Za·cha·ri·as** (zä'kä·rē'äs), *Fr.* **Za·cha·rie** (zȧ·shȧ·rē'), *Ital.* **Zac·ca·ri·a** (dzäk'kä·rē'ä), *Lat.* **Zach·a·ri·as** (zak'ə·rī'əs), *Sp.* **Za·ca·rí·as** (thä'kä·rē'äs). [< Hebrew, remembrance of the Lord]
— **Zachariah** The last king of Israel of Jehu's race. *II Kings* xiv 29.
— **Zacharias** The father of John the Baptist. *Luke* i 5.
Za·cyn·thus (zə·sin'thəs) The ancient name for ZANTE.
Za·dar (zä'där) A port of western Croatia, Yugoslavia, on the Adriatic; formerly (1918–1947) in Venezia Giulia, Italy: Italian *Zara*.
Za·dok (zā'dok) A masculine personal name. Also *Fr.* **Za·doc** (zȧ·dôk'), *Lat.* **Za·do·cus** (zə·dō'kəs). [< Hebrew, the just]
— **Zadok** One of the two chief priests of the Davidic sanctuary in Jerusalem. *II Sam.* viii 17.
zaf·fer (zaf'ər) *n.* A blue pigment made by roasting cobalt ores to yield an impure cobalt oxide: used for enamel and for painting on glass and porcelain. Also **zaf'far**, **zaf'fir**, **zaf'·fre**. [< Ital. *zaffera*, prob. < Arabic *ṣufr* copper]
Zag·a·zig (zag'ə·zig, zä·gä·zēg') A city of SE Lower Egypt.
Zagh·lul Pa·sha (zag·lōōl' pä'shä), **Saad**, '860?–1927, Egyptian lawyer and statesman.
Za·greb (zä'greb) The capital of Croatia, in NW Yugoslavia; a major industrial center and the second largest city of Yugoslavia: German *Agram*.
Za·gre·us (zā'grē·əs, -grōōs) In Greek mythology, a son of Zeus and Persephone, slain by the Titans and revived as Dionysus. See ORPHIC MYSTERIES.
Zag·ros Mountains (zag'rəs) The chief mountain system of Iran, extending from Azerbaijan to Iranian Baluchistan; highest point, over 14,900 feet.
Za·ha·roff (zä·hä'rəf), **Sir Basil**, 1850?–1936, international financier and armament manufacturer born in Turkey of Greek and Russian parents.
Zah·ran (zä'rän) See DHAHRAN.
zai·bat·su (zī·bät·sōō) *n. Japanese* The wealthy clique of Japan, representing four or five dominant families.
Za·les·ki (zȧ·les'kē), **August**, born 1883, Polish statesman.
Za·ma (zä'mə) An ancient town in Numidia, northern Africa, SW of Carthage: scene of Hannibal's defeat by Scipio Africanus, 202 B.C., ending the strength of Carthage: site of a modern village of north central Tunisia.
za·mar·ra (zä·mär'ə, -mär'ə) *n.* A sheepskin coat worn by Spanish shepherds. Also **za·mar'ro** (-mär'ō, -mar'ō). [< Sp.]
Zam·be·zi (zam·bē'zē) A river in southern Africa, flowing 1,650 miles SE from northwesternmost Northern Rhodesia through Rhodesia (forming the border between Northern and Southern Rhodesia) to the Indian Ocean in Mozambique. Also **Zam·be'si**. *Portuguese* **Zam·be·ze** (zäm·bā'zə).
Zam·bo·an·ga (säm'bō·äng'gä) A port of SW Mindanao, Philippines.
za·mi·a (zā'mē·ə) *n.* Any of a genus (*Zamia*) of palmlike trees and low shrubs of the cycad family, having unbranched stems terminating in a tuft of thick, pinnate, often spiny-edged leaves. [< NL << Lat. *zamiae*, misreading of L (*nuces*) *azaniae* pine (nuts)]
za·min·dar (zə·mēn'där) See ZEMINDAR.
Za·mo·ra (thä·mô'rä) 1 An ancient city of NW Spain; capital of Zamora province. 2 A province of NW Spain, bordering on Portugal; 4,081 square miles.
Za·mo·ra y Tor·res (thä·mô'rä ē tôr'räs), **Niceto Alcalá**, 1877–1949, Spanish politician; president of Spain 1931–36.
za·na·na (zə·nä'nə) See ZENANA.
Zang·will (zang'gwil), **Israel**, 1864–1926, English novelist and dramatist.
Zan·te (zan'tē, *Ital.* dzän'tā) 1 The southernmost main island of the Ionian Islands, Greece; 157 square miles; ancient *Zacynthus*. *Greek* **Za·kyn·thos** (zȧ·kin'thos). 2 Its capital, a port on the SE coast.
Zan·thox·y·lum (zan·thok'sə·ləm) *n.* A genus of trees of the rue family with prickly leaves, of which some species have medicinal properties. [< NL < Gk. *xanthos* yellow + *xylon* wood]

za·ny (zā'nē) *n. pl.* **·nies** 1 In old comic plays, one who took the part of an awkward simpleton by imitating the other performers, especially the clown, with ludicrous failure. 2 A simpleton; buffoon; fool. 3 *Obs.* An attendant. [< F *zani* < Ital. *zanni* servants who act as clowns in early Italian comedy < dial. Ital. *Zanni*, var. of *Giovanni* John]
Zan·zi·bar (zan'zə·bär, zan'zə·bär') 1 An island in the Indian Ocean, 22 miles from the Tanganyika coast of eastern Africa; 640 square miles. 2 A sultanate including the islands of Zanzibar and Pemba, comprising a British protectorate since 1890; 1,020 square miles. 3 Its capital, a port on the west coast of Zanzibar island.
Za·pa·ta (sä·pä'tä), **Emiliano**, 1877?–1919, Mexican revolutionary leader 1911–1916.
za·pa·te·o (thä'pä·tā'ō) *n.* A Spanish folk dance. [< Sp. < *zapato* a shoe, clog]
Za·po·rozh·e (zȧ·pə·rôzh'yə) A city on the Dnieper in southern Ukrainian S.S.R.: formerly *Aleksandrovsk*, *Alexandrovsk*.
zap·ti·ah (zup·tē'ä) *n.* A Turkish policeman. Also **zap·ti'e**, **zap·ti'eh**. [< Turkish *ḍabtiyeh* < Arabic *ḍabṭ* administration, regulation]
Za·ra (zä'rä, *Ital.* dzä'rä) The Italian name for ZADAR.
Za·ra·go·za (thä'rä·gō'thä) The Spanish name for SARAGOSSA.
Za·ra·pe (sä·rä'pä) See SERAPE.
Za·ra·thus·tra (zä'rä·thōōs'trä, zar'ə·thōōs'trə) See ZOROASTER.
za·ra·tite (zä'rə·tīt) *n.* A massive, vitreous nickel carbonate found usually as an emerald-green incrustation: also called *emerald nickel*. [< Sp. *zaratita*, after a Señor *Zarate* of Spain]
za·re·ba (zȧ·rē'bȧ) *n.* 1 In the Sudan, a stockade, thorn hedge, or other palisaded enclosure for protecting a village or camp: used also as a means of military defense. 2 A village or camp so protected; by extension, any village. Also **za·ree'ba**. [< Arabic *zarībah*, a pen for cattle < *zarb* a sheepfold]
zarf (zärf) *n.* A metal cup-shaped holder, of open or ornamental filigree, for a hot coffee cup, used in the Levant. [< Arabic *ẓarf* a vessel, sheath]
zar·zue·la (thär·thwä'lä) *n. Spanish* A form of lyrical theater in which song is intermingled with spoken dialog; operetta.
za·stru·ga (zä·strōō'gə) *n. pl.* **·gi** (-jē) *Meteorol.* One of a series of long parallel snow ridges formed by the wind on the open plains of Russia: also spelled *sastruga*. [< Russian]
zax (zaks) *n. Dial.* A sax or slate ax. [Var. of SAX¹]
za·yin (zä'yin) *n.* The seventh Hebrew letter. See ALPHABET. [< Hebrew *zāyin*]
Z-bar (zē'bär), **Z-beam** (zē'bēm') *n.* A Z-iron.
Ze·a (zē'ə) *n.* A genus of tall annual cereal grasses which includes corn or maize. *Zea mays*, Indian corn, is the only species. [< NL, spelt < Gk. *zeia* one-seeded wheat]
Ze·a (zē'ə) The medieval name for KEOS.
zeal (zēl) *n.* Ardor for a cause, or, less often, for a person; enthusiastic devotion; fervor. See synonyms under ENTHUSIASM, WARMTH. [< OF *zele* < L *zelus* < Gk. *zēlos* < *zeein* be hot]
Zea·land (zē'lənd) A Danish island between the Kattegat and the Baltic Sea, on which Copenhagen is located; the largest island of Denmark, separated from Sweden by the Oresund; 2,709 square miles: German *Seeland*, Danish *Sjaelland*.
zeal·ot (zel'ət) *n.* One who is overzealous; a fanatic; immoderate partisan. [< LL *zelotes* < Gk. *zēlōtēs* < *zēloein* be zealous < *zēlos* zeal]
Zeal·ot (zel'ət) *n.* A member of a fanatical Jewish party (A.D. 6–70) in almost continual revolt against the Romans.
zeal·ot·ry (zel'ət·rē) *n.* The conduct or disposition of a zealot.
zeal·ous (zel'əs) *adj.* Filled with or incited by zeal; enthusiastic. See synonyms under EAGER.
— **zeal'ous·ly** *adv.* — **zeal'ous·ness** *n.*
ze·a·xan·thin (zē'ə·zan'thin) *n. Biochem.* A yellow pigment, $C_{40}H_{56}O_{2}$, related to carotene and obtained in the form of golden-orange leaflets from yellow corn, egg yolk, and green leaves. [< ZEA + XANTH- + -IN]
Zeb·a·di·ah (zeb'ə·dī'ə) A masculine personal name. [< Hebrew, God has bestowed]
ze·bec (zē'bek), **ze·beck** See XEBEC.

Zeb·e·dee (zeb'ə·dē) A masculine personal name. [Contraction of ZEBADIAH]
— **Zebedee** The father of James and John, disciples of Christ. *Matt.* iv 21.
ze·bra (zē'brə) *n.* Any of various African equine mammals resembling the ass, having a white or yellowish-brown body fully marked with variously patterned, dark-brown or blackish bands; especially, the true or **mountain zebra** (*Equus zebra*) of the Cape of Good Hope Province and Grevy's zebra (*E. grevyi*) of Abyssinia and northeast Africa. [< Pg. < Bantu (Congo)] — **ze'·brine** (-brēn, -brin), **ze'broid** (-broid) *adj.*
zebra wolf The thylacine.

ZEBRA
(From 10 1/2 to 13 hands high at the withers)

ze·bra·wood (zē'brə·wŏŏd') *n.* 1 The wood of a large tree (*Connarus guianensis*) of Guiana, light brown in color with dark stripes, used in making furniture. 2 The tree. 3 The striped or banded wood of various other trees.
ze·bu (zē'byōō) *n.* The domesticated ox (*Bos indicus*) of India, China, and East Africa, having a hump on the withers, a large dewlap, and short curved horns: there are many breeds, varying in color, some being reared for milk and flesh, and others for riding and draft. [< F *zébu* <Tibetan]

ZEBU
(From 3 to 4 1/2 feet high at the shoulder)

Zeb·u·lon (zeb'yə·lən) A son of Jacob and ancestor of the tribe of Israel bearing that name. *Gen.* xxx 20. Also **Zeb'u·lun**.
zec·chi·no (tsek·kē'nō) *n. pl.* **·ni** (-nē) A gold coin of the republic of Venice; the sequin. Also **zec·chin** (zek'in), **zech'in**. [< Ital. See SEQUIN.]
Zech·a·ri·ah (zek'ə·rī'ə) A masculine personal name. [Var. of ZACHARIAH]
— **Zechariah** Hebrew prophet of the sixth century B.C., who promoted the rebuilding of the Temple; also, the Old Testament book bearing his name. Also *Zacharias*.
zed (zed) *n. Brit.* The letter z: generally called *zee* in the United States. [< F *zède* < L *zeta* < Gk. *zēta*]
Zed·e·ki·ah (zed'ə·kī'ə) A masculine personal name. [< Hebrew, justice of the Lord]
— **Zedekiah** The last king of Judah, 597–586 B.C.; son of Josiah. *II Kings* xxiv 17.
zed·o·ar·y (zed'ō·er'ē) *n.* The root of a species of turmeric (*Curcuma zedoaria*), used in medicine as a stomachic and as a carminative. [< Med. L *zedoarium* <Arabic *zedwār*]
zee¹ (zē, *Du.* zā) *n. Dutch* Sea: used in geographic names: *Zuyder Zee*, *Tappan Zee*.
zee² (zē) *n.* The letter Z, z.
Zee·brug·ge (zē'brŏŏg·ə, *Flemish* zā'brŏŏkh·ə) A port of NW Belgium on the North Sea in West Flanders province.
Zee·land (zē'lənd, *Du.* zā'länt) A province of SE Netherlands bordering on Belgium and including Walcheren and other islands; 650 square miles.
Zee·man (zā'män), **Pieter**, 1865–1943, Dutch physicist.
Zeeman effect *Physics* The splitting of spectral lines when the source emitting them is placed in a strong magnetic field. [after Pieter *Zeeman*]
ze·in (zē'in) *n. Biochem.* A simple protein derived from corn: it is insoluble in water but soluble in 70 to 80 percent alcohol. [< ZEA + -IN]
Zeit·geist (tsīt'gīst) *n. German* The spirit of the time; the intellectual and moral tendencies that characterize any age or epoch. [< G *zeit* time + *geist* spirit]
Zeke (zēk) Diminutive of EZEKIEL.
Ze·lin·ski (zyi·lyen'skē), **Nikolai**, 1861–1953, Russian chemist.

ze·lo·typ·i·a (zel′ə·tip′ē·ə) *n. Psychiatry* A morbid or excessive zeal in the prosecution of any cause. [<NL <LL, jealousy <Gk. *zelotypia* <*zēlos* zeal + *typ-*, root of *typtein* strike]

ze·min·dar (zə·mēn·där′) *n.* In India, a tax farmer, required, under the Mogul rule, to pay a fixed sum for the tract of land assigned him; hence, later, especially in Bengal, a native landlord required to pay a certain land tax to the English government; an owner of the soil: also spelled *zamindar*. [<Hind. <Persian *zamīndār* <*zamīn* earth + *dār* a holder]

Zem·po·al·te·pec (sǎm′pō·äl′tā·pek′) A peak in Oaxaca, Mexico; 11,142 feet. Also **Zem·po·al·té·petl** (sǎm′pō·äl·tā′pet′l).

zemst·vo (zem′stvô, *Russian* zyem′stfô) *n.* A Russian elective district and provincial representative assembly; replaced in 1917 by the soviet system. [<Russian *zemlya* land]

Ze·mun (ze′mo͞on) The port section of Belgrade, Yugoslavia, on the Danube; formerly a separate city: German *Semlin*.

ze·na·na (zə·nä′nə) *n.* In India, the women's apartments; the East Indian harem: also spelled *zanana*. [<Hind. *zenāna* belonging to women <Persian *zanāna* <*zan* woman]

Zen Buddhism (zen) A form of meditative Buddhism whose adherents believe in and work toward abrupt enlightenment; much emphasis is placed on the identity of nirvana and samsara, and on direct transmission of the enlightened state from master to pupil, with a minimum of words; scriptures and ritual forms are minimized, while continual meditation and practical physical labor are stressed. It originated in China when late northern Indian Buddhism came into contact with Taoism around A.D. 500, whence it acquired many Taoist features; it then spread to Japan, where it greatly influenced Japanese culture in all areas, especially in the age of the samurai, whose feudal code, bushido, derived from Zen, as did judo and jiujitsu. [<Japanese *zen* meditation <Chinese *chan* <Skt. *dhyana*]

Zend (zend) *n.* 1 The ancient translation and commentary, in a literary form of Middle Persian (Pahlavi), of the Avesta, the sacred writings of the Zoroastrian religion. 2 Erroneously, the language of the Avesta; Avestan. [<F <Persian, interpretation] — **Zend′ic** *adj.*

Zend-A·ves·ta (zend′ə·ves′tə) *n.* The Avesta, including the later translation and commentary called the Zend. [Alter. of Persian *Avestā-va-Zend* the Avesta with its interpretation <Avestan *Avestā* a sacred text + Persian *zend* interpretation] — **Zend′-A·ves·ta′ic** (-ə·ves·tā′ik) *adj.*

zen·dik (zen·dēk′) *n.* In Eastern countries, an atheist or heretic; one who practices black magic. [<Arabic *zindīq* an atheist <Persian *zandīq* a fire worshiper]

Zeng·er (zeng′ər), **John Peter**, 1697–1746, American printer and publisher.

ze·nith (zē′nith) *n.* 1 The point in the celestial sphere that is exactly overhead: opposed to *nadir*. 2 The culminating point of prosperity, greatness, etc.; summit. [<OF *cenit* <Arabic *samt* (*ar-rās*) the path (over the head)]

Ze·no (zē′nō) Either of two ancient Greek philosophers:
— **Zeno of Elea**, 490?–430? B.C., early Greek philosopher of the Eleatic school, noted for his arguments (paradoxes) against motion and multiplicity.
— **Zeno the Stoic**, 342?–270? B.C., Greek philosopher; founder of the Stoic school.

Ze·no·bi·a (zi·nō′bē·ə) Queen of Palmyra in the third century; conquered and captured by the Roman emperor Aurelian.

ze·o·lite (zē′ə·līt) *n.* A secondary mineral occurring in cavities and veins in eruptive rocks, usually a hydrous silicate of aluminum and sodium: various forms are used as water softeners. [<Sw. *zeolit* <Gk. *zēein* boil + *lithos* stone] — **ze′o·lit′ic** (-lit′ik) *adj.*

Zeph·a·ni·ah (zef′ə·nī′ə) A masculine personal name. [<Hebrew, the Lord has hidden]
— **Zephaniah** Hebrew prophet of the seventh century B.C.; also, the book of the Old Testament bearing his name. Also *Sophonias*.

zeph·yr (zef′ər) *n.* 1 The west wind; poetically, any soft, gentle wind. 2 Worsted or woolen yarn of very light weight used for embroidery, shawls, etc.: also **zephyr worsted**. 3 Figuratively, anything very light and airy. [<L *zephyrus* <Gk. *zephyros*]

zephyr cloth Thin, fine cashmere used for women's clothing.

Zeph·y·rus (zef′ər·əs) In Greek mythology, the west wind: regarded as the mildest and gentlest of all sylvan deities.

Zep·pe·lin (zep′ə·lin, *Ger.* tsep′ə·lēn′) *n.* A large dirigible having a rigid, cigar-shaped body, as originally designed and constructed by Count Ferdinand von Zeppelin.

Zep·pe·lin (zep′ə·lin, *Ger.* tsep′ə·lēn′), **Count Ferdinand von**, 1838–1917, German general; aeronaut and airship builder.

Zer·matt (tser·mät′) A resort village of SE Valais canton, SW Switzerland; elevation, 5,315 feet.

ze·ro (zir′ō) *n. pl.* **ze·ros** or **ze·roes** 1 The numeral or symbol 0; a cipher. 2 *Math.* **a** A cardinal number indicating the absence of quantity. **b** The point where a continuous function changes its sign from plus to minus, or vice versa. 3 The point on a scale, as of a thermometer, from which measures are counted. 4 *Mil.* A setting for a gunsight which adjusts both for elevation and wind. 5 The lowest point in any standard of comparison; nullity. — *v.t.* **ze·roed**, **ze·ro·ing** To adjust (instruments) to an arbitrary zero point for synchronized readings. — **to zero in** 1 To bring an aircraft into a desired position, as for bombing or landing. 2 To adjust the sight of (a gun) by calibrated results of firings. — *adj.* Without value or appreciable change. [<F *zéro* <Ital. *zero* <Arabic *sifr*. Doublet of CIPHER.]

ze·ro-beat (zir′ō·bēt′) *adj. Electronics* Homodyne.

zero count The final time unit as called off in a count-down.

zero hour 1 The time set for attack or other military operations: also called *H-hour*. 2 The moment of undertaking something; any critical moment.

zest (zest) *n.* 1 Agreeable excitement and keen enjoyment of the mind accompanying exercise, mental or physical. 2 That which imparts such excitement and relish. 3 Specifically, an agreeable and piquant flavor in anything tasted, especially if added to the usual flavor, as that imparted to soups or wines by the essential oil of lemon peel, or by spice; figuratively, increase of enjoyment produced by the addition of any agreeable stimulant. 4 A piece of orange or lemon peel used to flavor anything, or the aromatic oil squeezed from it: a rare usage. See synonyms under APPETITE, RELISH. — *v.t.* To give zest or relish to; make piquant. [<F *zeste* lemon peel (for flavoring)] — **zest′ful** *adj.*

ze·ta (zā′tə, zē′-) *n.* The sixth letter (Z,ζ) in the Greek alphabet, corresponding to English *z*, in ancient Greek sounded *zd* or *dz*, in modern Greek *z*.

Ze·thus (zē′thəs) In Greek mythology, Amphion's twin brother. Also **Ze′thos**. See AMPHION.

Zet·land (zet′lənd) See SHETLAND.

zeug·ma (zo͞og′mə) *n.* A rhetorical figure in which an adjective is made to modify, or a verb to govern, two nouns, while applying properly only to one: *She was remembered but they forgotten*. Compare SYLLEPSIS. [<NL <Gk., a yoking <*zeugnymi* yoke]

Zeus (zo͞os) In Greek mythology, the supreme deity, ruler of the celestial realm, son of Kronos and Rhea and husband of Hera: identified with the Roman *Jupiter*.

Zeus-Am·mon (zo͞os′am′ən) See AMMON (def. 1).

Zeux·is (zo͞ok′sis) Greek painter of the late fifth century B.C.

Zhda·nov (zhdä′nôf) A port on the Sea of Azov, SE Ukrainian S.S.R.: formerly *Mariupol*.

Zhda·nov (zhdä′nôf), **Andrei**, 1896–1948, U.S.S.R. politician and general.

Zhi·to·mir (zhi·tô′mir) A city in west central Ukrainian S.S.R.

Zhu·kov (zho͞o′kôf), **Georgi**, born 1895, U.S.S.R. marshal and statesman.

zib·e·line (zib′ə·lin, -lin) *adj.* Pertaining to the sable; made of sable fur. — *n.* The fur of the sable. Also **zib′el·line**. [<F <OF *sebelin*, ult. <Slavic. Akin to SABLE.]

zib·et (zib′it) *n.* A carnivore, the Asian or Indian civet (*Viverra zibetha*), with the black markings less distinct and the tail more ringed than the common civet. It is often domesticated. Also **zib′eth**. [<Med. L *zibethum* <Arabic *zabād* a civet]

Zieg·feld (zēg′feld, zig′-), **Florenz**, 1869–1932, U.S. theatrical producer.

zig·gu·rat (zig′o͝o·rat) *n.* Among the Assyrians and Babylonians, a terraced temple tower pyramidal in form, each successive story being smaller than the one below, leaving a terrace around each of the floors. Also **zik′ku·rat** (zik′-). [<Assyrian *ziqquratu*, orig. a mountain top]

ZIGGURAT

zig·zag (zig′zag) *n.* A series of short, sharp turns or angles from one side to the other in succession, or something, as a path or pattern, characterized by such angles. — *adj.* Having a series of short alternating turns or angles from side to side. — *adv.* In a zigzag manner. — *v.t.* & *v.i.* **·zagged**, **·zag·ging** To form or move in zigzags. [<F <G *zickzack*, prob. reduplication of *zacke* a sharp point]

zig·zag·ger (zig′zag·ər) *n.* 1 One who or that which zigzags. 2 A sewing-machine attachment for stitching appliqué, joining lace and insertion to fabric, etc.

zil·lah (zil′ə) *n. Anglo-Indian* A provincial governmental district in India.

Zil·pah (zil′pə) The mother of Gad. *Gen.* xxx 10.

Zim·ba·bwe (zim·bä′bwā) The site of a ruined city (probably of a Bantu people, dating from the 15th century) of SE Southern Rhodesia; discovered about 1870.

Zim·ba·list (zim′bə·list, *Russian* zim′bə·lyest′), **Efrem**, born 1889, U.S. violinist born in Russia.

Zim·mern (zim′ərn), **Sir Alfred**, 1879–1957, English political scientist.

zinc (zingk) *n.* A bluish-white metallic element (symbol Zn) occurring mostly in combination: extensively used in the arts, as in the manufacture of brass, and for roofing, etc., also as the negative electrode in electric batteries. Its salts have various applications, as the oxide in painting and the chloride and sulfate in medicine, and, like other salts of heavy metals, are poisonous. See ELEMENT. — *v.t.* **zinced** or **zincked**, **zinc·ing** or **zinck·ing** To coat or cover with zinc; galvanize. [<G *zink*; ult. origin unknown] — **zinc′ic** *adj.* — **zinck′y**, **zinc′y**, **zink′y** *adj.*

zinc·al·ism (zingk′əl·iz′əm) *n. Pathol.* Chronic zinc poisoning.

zinc·ate (zir.gk′āt) *n. Chem.* A salt derived from zinc hydroxide by substitution of a metal for the hydrogen. [<ZINC + -ATE³]

zinc blende Sphalerite.

zinc chloride *Chem.* A white deliquescent compound, ZnCl₂, extensively used in medicine, industry, and the arts.

zinc·if·er·ous (zingk·if′ər·əs, zin·sif′ər·əs) *adj.* Yielding zinc, as ore. Also **zink·if′er·ous**. [<ZINC + -(I)FEROUS]

zinc·i·fy (zingk′ə·fī) *v.t.* **·fied**, **·fy·ing** To apply zinc to, as by coating or impregnating. [<ZINC- + -(I)FY] — **zinc′i·fi·ca′tion** (-fə·kā′shən) *n.*

zinc·ite (zingk′īt) *n.* A deep-red, translucent to subtranslucent zinc oxide, ZnO, crystallizing in the hexagonal system; zinc ore. [<ZINC + -ITE²]

zin·co·graph (zingk′ə·graf, -gräf) *n.* An etching on zinc; a picture obtained by zincography. Also **zin′co·type** (-tīp). [<ZINC + -(O)GRAPH] — **zin·cog′ra·pher** (zing·kog′rə·fər) *n.*

zin·cog·ra·phy (zing·kog′rə·fē) *n.* The art of etching on zinc to produce plates for printing. [<ZINC + -(O)GRAPHY] — **zinc·o·graph·ic** (zingk′ə·graf′ik) or **·i·cal** *adj.*

zinc ointment A medicated ointment containing zinc oxide.

zinc·ous (zingk'əs) *adj. Chem.* Pertaining to or derived from zinc; zincic.
zinc oxide *Chem.* White pulverulent oxide, ZnO, made by burning zinc in air. It is used as a pigment, chiefly as a substitute for white lead, and in medicine as a mild antiseptic.
zinc sulfate *Chem.* A crystalline compound, $ZnSO_4 \cdot 7H_2O$, obtained by the action of sulfuric acid on zinc; white vitriol.
zinc white Zinc oxide used as a pigment.
zin·fan·del (sin'fən·del) *n.* A dry, red or white, claret-type wine made in California. [? from a European place name]
zing (zing) *Colloq. n.* 1 A high-pitched buzzing or humming sound. 2 Energy; vitality; vigor. — *v.i.* To make a shrill, humming sound. [Imit.]
zin·ga·ro (tsēng'gä·rō) *n. pl.* **·ri** (-rē) Italian A gipsy. Also **zin'ga·no** (-nō). — **zin'ga·ra** (-rä) *n. fem.*
zin·gi·ber·a·ceous (zin'jə·bə·rā'shəs) *adj. Bot.* Of or pertaining to a family (*Zingiberaceae*) of monocotyledonous tropical plants, the ginger family, having aromatic rootstocks and large leaves, and including the ginger, cardamon, and turmeric plants. Also **zin'zi·ber·a'ceous** (zin'zə-). [<NL, family name <LL *zingiber* GINGER]
zink·en·ite (zingk'ən·īt) *n.* A metallic steel-gray mineral, $PbSb_2S_4$, crystallizing in the orthorhombic system. Also **zinck'en·ite.** [<G *zinkenit,* after J. K. L. Zinken, 1798–1862, German mine director]
zin·ni·a (zin'ē·ə) *n.* Any of a genus (*Zinnia*) of American, chiefly Mexican, herbs of the composite family, having opposite entire leaves and showy flowers; especially, the common zinnia (*Z. elegans*), the State flower of Indiana. [<NL, after J. G. Zinn, 1727–59, German professor of medicine]
Zi·nov·iev (zē·nôv'yif), **Grigori,** 1883–1936, U.S.S.R. political leader.
Zins·ser (zin'sər), **Hans,** 1878–1940, U. S. bacteriologist.
Zin·zen·dorf (tsin'tsən·dôrf), **Count Nicholas Ludwig von,** 1700–60, German theologian; founded sect of Moravians.
Zi·on (zī'ən) 1 A hill in Jerusalem, the site of the Temple and the royal residence of David and his successors: regarded by the Jews as a symbol for the center of Jewish national culture, government, and religion. 2 The Jewish people. 3 Any place or community considered to be especially under God's rule, as ancient Israel or the Christian church. 4 The heavenly Jerusalem; heaven. Also spelled *Sion.* [OE *Sion* <LL <Gk. *Seōn, Seiōn* <Hebrew *tsīyōn* a hill]
Zi·on·ism (zī'ən·iz'əm) *n.* A movement for a resettlement of the Jews in Palestine. The form which lays stress upon the political questions involved is sometimes called **political Zionism,** and the term **religious Zionism** is used by those Zionists who lay a special stress upon the regeneration of the Holy Land as a center of social and religious influence for Judaism. Also **Zion movement.** — **Zi'on·ist** *adj. & n.* — **Zi'on·is'tic** *adj.* — **Zi'on·ite** *n.*
Zion National Park A government reservation in SW Utah; 147 square miles; established in 1919; contains **Zion Canyon,** a gorge 1/2 mile deep, about 15 miles long.
Zi·on·ward (zī'ən·wərd) *adv.* Toward Zion; Godward; heavenward.
zip (zip) *n.* 1 A sharp, hissing sound, as of a bullet passing through the air. 2 *Colloq.* Energy; vitality; vim. — *v.* **zipped, zip·ping** *v.t.* 1 To fasten with a sliding fastener. — *v.i.* 2 *Colloq.* To be very energetic. 3 To move or fly with a zip. [Imit.]
Zi·pan·gu (zi·pang'gōō) Japan: name used by Marco Polo.
zip gun A home-made pistol consisting of a small pipe or other metal tube, usually accommodating a .22 caliber bullet, fastened to a block of wood and equipped with a firing pin actuated by a spring or rubber band.
Zi·phi·i·dae (zi·fī'i·dē) *n. pl.* A family of beaked whales with a dolphinlike form and generally only a pair of teeth in the lower jaw, as the bottlenose whale (*Hyperoodon ampullatus*). [<NL <*Ziphius* <L *xiphias* a swordfish <Gk. *xiphios* <*xiphos* a sword]
zip·per (zip'ər) *n.* A slide fastener.
Zip·per (zip'ər) *n.* An overshoe or boot in which buttons or laces are replaced by a slide fastener: a trade name.

zip·py (zip'ē) *adj.* **·pi·er, ·pi·est** *Colloq.* Brisk; energetic; lively; snappy.
zir·con (zûr'kon) *n.* 1 An adamantine, variously colored zirconium silicate, $ZrSiO_4$. The transparent reddish variety, called *hyacinth,* is used as a gem, as are also the leaf-green, yellowish, colorless, or smoky varieties called *jargon.* 2 A variety of this mineral having an artificially produced steely-blue color of high brilliance and luster: esteemed as a gem. [<F *zircone* <Arabic *zarqūn* cinnabar <Persian *zargūn* golden < *zar* gold + *gūn* color]
zir·con·ate (zûr'kən·āt) *n. Chem.* A salt formed by replacing hydrogen in zirconium hydroxide with a metal. [<ZIRCON(IUM) + -ATE³]
zir·co·ni·a (zûr·kō'nē·ə) *n. Chem.* A white pulverulent zirconium dioxide, ZrO_2, obtained by heating zirconium to redness in contact with air: when strongly heated it becomes luminous, and it is hence used in certain forms of incandescent burners. [<NL <ZIRCON]
zir·co·ni·um (zûr·kō'nē·əm) *n.* A metallic element (symbol Zr) chemically resembling titanium, prepared as a black amorphous powder, as steel-gray shining scales, or as crystalline laminae: used in alloys, as an opacifier of lacquers, and as an abrasive. See ELEMENT. [<NL <ZIRCON] — **zir·con'ic** *adj.*
Z-i·ron (zē'ī'ərn) *n.* An angle iron of Z form: also called **Z-bar, Z-beam.**
Žiš·ka (tsis'kä), **John,** 1360?–1424, Bohemian general; leader of the Hussites. Also **Žiž·ka** (zhish'kä).
zith·er (zith'ər) *n.* A simple form of stringed instrument, having a flat sounding board and from thirty to forty strings that are played by plucking with a plectrum. Also **zith'ern** (-ərn). [<G <L *cithara* <Gk. *kithara.* Doublet of CITHARA and GUITAR.]
zit·tern (zit'ərn) See CITHERN.
zi·zith (tsē·tsēt', tsī'sis) *n.* The fringe or tassel formerly worn by Jews on the outer garment (*Num.* xv 38), but now worn on the tallith during prayer. [<Hebrew *tsītsīth*]
ziz·zle (ziz'əl) *v.i.* **·zled, ·zling** *Brit. Dial.* To make a sputtering or hissing sound, as meat when cooking; sizzle. [Imit.]
Zla·to·ust (zlä'tə·ōōst') *n.* A city of SW Asiatic Russian S.F.S.R. in the southern Urals.
zlo·ty (zlô'tē) *n. pl.* **·tys** or **·ty** The monetary unit of Poland. [<Polish, lit., golden]
zo– Var. of ZOO–.
–zoa combining form *Zool.* Used to denote the names of groups: *Protozoa, Hydrozoa.* An individual in such a group is denoted by –zoan. [<NL <Gk. *zōion* an animal]
Zo·an (zō'an) The Old Testament name for TANIS.
zo·an·thro·py (zō·an'thrə·pē) *n.* The obsessive delusion that one has become a beast; lycanthropy. [<NL *zoanthropia* <Gk. *zōion* an animal + *anthrōpos* a man] — **zo·an·throp'ic** (zō'ən·throp'ik) *adj.*

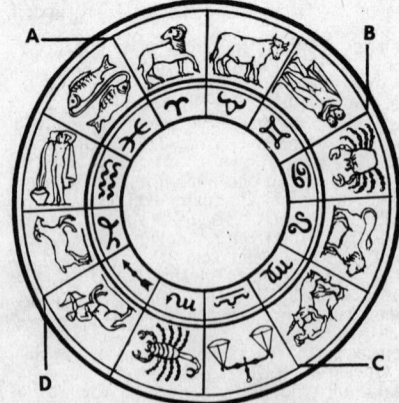

SIGNS OF THE ZODIAC
Reading clockwise:
A. Vernal equinox: Aries, Taurus, Gemini.
B. Summer solstice: Cancer, Leo, Virgo.
C. Autumnal equinox: Libra, Scorpio, Sagittarius.
D. Winter solstice: Capricorn, Aquarius, Pisces.

zo·di·ac (zō'dē·ak) *n.* 1 An imaginary belt encircling the heavens and extending about 8° on each side of the ecliptic, within which are the orbits of the moon, sun, and larger planets. It is divided into twelve parts, called **signs of the zodiac,** which formerly corresponded to twelve constellations bearing the same names. Now, owing to the precession of the equinoxes, each constellation is in the sign that has the name next following its own. 2 *Figuratively,* a complete circuit; round. 3 *Rare* A circle or halo; also, a girdle. [<OF *zodiaque* <L *zodiacus* <Gk. (*kyklos*) *zōdiakos* (circle) of animals <*zōdion* a sculptured animal, dim. of *zōion* an animal] — **zo·di·a·cal** (zō·dī'ə·kəl) *adj.*
zodiacal light *Astron.* A cone-shaped tract of faint light lying near the plane of the ecliptic: it may be seen in the west after twilight in winter and spring, or in the east before daybreak from September till January. It is attributed to the reflection of sunlight from a cloud of fine meteoric dust.
Zo·e (zō'ē) A feminine personal name. [<Gk., life]
zo·e·trope (zō'ə·trōp) *n.* A toy having a revolving cylinder with slits through which a series of pictures inside are seen in apparent motion. Compare PHENAKISTOSCOPE. [<Gk. *zōē* life + -TROPE] — **zo'e·trop'ic** (-trop'ik) *adj.*
zo·ic (zō'ik) *adj.* Pertaining to or characterized by animals or animal life. [<Gk. *zōikos* <*zōion* an animal]
zois·ite (zoi'sīt) *n.* A vitreous, transparent to subtranslucent silicate of aluminum and calcium, in which iron sometimes replaces the aluminum. [<G *zoisit,* after Baron Zois von Edelstein, 1747–1819, its discoverer]
Zo·la (zō'lə, zō·lä'; *Fr.* zô·lá'), **Émile,** 1840–1902, French writer. — **Zo'la·esque'** (-esk') *adj.*
zoll·ver·ein (tsôl'fer·īn) *n.* 1 A former trade league constituted by twenty-six German states. 2 Hence, a union of states for tariff purposes. [<G <*zoll* a tax, custom + *verein* a union]
Zom·ba (zom'bə) The capital of Nyasaland, Federation of Rhodesia and Nyasaland, in the SE part.
zom·bi (zom'bē) *n.* 1 In West African voodoo cults, the python deity; also, the snake deity of the voodoo cults of Haiti and of the Negroes of the southern United States. 2 The supernatural power by which a dead body is believed to be reanimated; specifically, a corpse reactivated by sorcery, but still dead. 3 Loosely, a ghost. Also **zom'bie.** [<West African. Cf. Bantu (Congo) *zumbi* fetish.] — **zom'bi·ism** *n.*
Zom·bie (zom'bē) *n.* A large, strong cocktail made from several kinds of rum, fruit juices, and liqueur. [<ZOMBI]
zo·nal (zō'nəl) *adj.* Of, pertaining to, exhibiting, or marked by a zone or zones; having the form of a zone. Also **zo'na·ry** (-nər·ē).
zo·nate (zō'nāt) *adj.* 1 Marked with zones or concentric colored bands. 2 *Bot.* Disposed in a single row, as certain tetraspores. Also **zo'nat·ed.** [<ZON(E) + -ATE¹] — **zo·na·tion** (zō·nā'shən) *n.*
zone (zōn) *n.* 1 One of five divisions of the earth's surface, enclosed between two parallels of latitude and named for the prevailing climate. These are the **torrid zone,** extending on each side of the equator 23° 27'; the **temperate** or **variable zones,** included between the parallels 23° 27' and 66° 33' on both sides of the equator; and the **frigid zones,** within the parallels 66° 33' and the poles. 2 In war, a region proscribed for neutrals as being within the range of military or naval operations: a defense *zone,* combat *zone*; also, any region neutralized by agreement of combatants: a demilitarized *zone.* 3 *Ecol.* A belt or area delimited from others by the character of its plant or animal life, its climate, geological formations, etc. 4 A region of land distinguished or set off by some special characteristic: a canal *zone.* 5 *Anat.* A beltlike area distinguished from its surroundings either by structure or appearance. 6 *Mineral.* Any series of faces upon a crystal whose planes form a prismatic surface. 7 A belt, stripe, etc., distinguished by color or

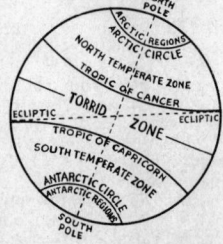

TERRESTRIAL ZONES

the like, encircling an object. **8** *Geom.* A portion of the surface of a sphere enclosed between two parallel planes. **9** Originally (now chiefly in poetry), a belt or girdle. **10** The total number of railroad stations situated in a certain area measured from a place whence traffic is shipped; also, a circular area within which a uniform fare is charged by the transportation companies. **11** In the United States parcel post system, any one of the concentric areas within each of which a uniform rate is charged. **12** A postal district in a city. — *v.t.* **zoned, zon·ing 1** To divide into zones; especially, to divide (a city, etc.) into zones which are restricted as to types of construction and activity, as residential, industrial, etc. **2** To encircle with a zone or belt. **3** To mark with or as with zones or stripes. [<MF <L *zona* <Gk. *zōnē* a girdle] — **zoned** *adj.*

zone axis *Mineral.* The imaginary line through a crystal, to which all the faces in a given zone, and the mutual intersections of those faces, are parallel.

zone·less (zōn′lis) *adj.* Having no zone or belt.

zone time See under TIME.

Zon·gul·dak (zông′gōōl·däk′) **1** A port on the Black Sea in NW Turkey in Asia, capital of Zonguldak province; a coal-shipping center. **2** A province of NW Turkey in Asia, bordering on the Black Sea; 2,876 square miles.

zo·nule (zōn′yōōl) *n.* A small zone, belt, or ring. Also **zo′nu·la.** [<NL *zonula,* dim. of L *zona* ZONE]

zoo (zōō) *n.* A menagerie. [Short for ZOOLOGICAL GARDEN]

zoo– *combining form* Animal; of or related to animals, or to animal forms: *zoology, zoophyte.* Also, before vowels, *zo–.* [<Gk. *zōion* an animal]

zo·o·chem·is·try (zō′ə·kem′is·trē) *n.* Animal chemistry; specifically, the chemistry of the solids and fluids contained in the animal organism. — **zo′o·chem′i·cal** (-kem′i·kəl) *adj.*

zo·o·chore (zō′ə·kôr, -kōr) *n.* A plant distributed by animals.

zo·o·ge·o·graph·ic (zō′ə·jē′ə·graf′ik) *adj.* Of, pertaining to, or engaged in zoogeography. Also **zo′o·ge′o·graph′i·cal.** — **zo′o·ge′o·graph′i·cal·ly** *adv.*

zoogeographic realm One of a series of major geographic areas characterized by the dominance of certain animal groups. The principal realms in the classification of A. R. Wallace are: the Palearctic, Nearctic, Neotropical, Ethiopian, Oriental, and Australian. The first two are often considered together as the Holarctic realm.

zo·o·ge·og·ra·phy (zō′ə·jē·og′rə·fē) *n.* **1** The systematic study of the distribution of animals and of the factors controlling it. **2** The study of the relations between special animal groups and the land or aquatic areas in which they predominate. — **zo′o·ge·og′ra·pher** *n.*

zo·o·gloe·a (zō′ə·glē′ə) *n. pl.* **·gloe·ae** (-glē′ē) A colony of bacteria forming a jellylike mass held together by a viscid sheath secreted by themselves. [<NL <Gk. *zōion* an animal + *gloios* sticky stuff]

zo·og·ra·phy (zō·og′rə·fē) *n.* The branch of zoology that describes animals; descriptive zoology. — **zo·og′ra·pher** or **·phist** *n.* — **zo·o·graph·ic** (zō′ə·graf′ik) or **·i·cal** *adj.*

zo·oid (zō′oid) *n.* **1** *Biol.* Any organism, usually very small, capable of spontaneous movement and independent existence, as a spermatozoon. **2** *Zool.* **a** One of the distinct members of a compound or colonial organism, as in a bryozoan. **b** A free-swimming organism produced as a stage in the life cycle of a jellyfish. — *adj.* Having essentially the nature of an animal: also **zo·oi·dal** (zō·oid′l). [<ZOO- + -OID]

zo·ol·a·try (zō·ol′ə·trē) *n.* Animal-worship. [<ZOO- + -LATRY] — **zo·ol′at·er** *n.* — **zo·ol′a·trous** *adj.*

zo·o·log·i·cal (zō′ə·loj′i·kəl) *adj.* **1** Of, pertaining to, or occupied with zoology. **2** Relating to or characteristic of animals. Also **zo′o·log′ic.** — **zo′o·log′i·cal·ly** *adv.*

zoological garden A park or garden in which wild animals are kept for exhibition.

zo·ol·o·gy (zō·ol′ə·jē) *n.* **1** The science that treats of animals with reference to their structure, functions, development, nomenclature, and classification. **2** The animal kingdom, or local examples of it, regarded biologically. **3** A scientific treatise on animals. [<NL *zoologia* <Gk. *zōion* an animal + *logos* a word, discourse] — **zo·ol′o·gist** *n.*

zoom (zōōm) *v.i.* **1** To make a low-pitched but loud humming or buzzing sound. **2** To climb sharply in an airplane, using the energy of momentum. — *v.t.* **3** To cause to zoom. — *n.* The act of zooming. [Imit.]

zo·om·e·try (zō·om′ə·trē) *n.* Measurement of the parts of animals and determination of their relative magnitude. [<ZOO- + -METRY] — **zo·o·met·ric** (zō′ə·met′rik) *adj.*

zo·o·mor·phism (zō′ə·môr′fiz·əm) *n.* **1** The conception, symbolization, or representation of a man or a god in the form of an animal; also, the attribution of divine or human qualities to animals. **2** The representation of animals or animal forms in art or symbolism. **3** Transformation into animals. In this sense compare BEAST MARRIAGE, SWAN MAIDEN. See also CIRCE. Also **zo′o·mor′phy.** — **zo′o·mor′phic** *adj.*

zo·on (zō′on) *n. pl.* **zo·ons** or **zo·a** (zō′ə) *Biol.* A developed individual of a compound animal or of a simple egg. [<NL <Gk. *zōion* an animal] — **zo·on·al** (zō·on′əl) *adj.*

zo·oph·a·gous (zō·of′ə·gəs) *adj.* Feeding on animals. [<ZOO- + -PHAGOUS]

zo·o·phile (zō′ə·fil, -fīl) *n.* **1** A zoophilous plant. **2** A lover of animals; specifically, one who objects to vivisection; also, one addicted to zoophilism: also **zo·oph·i·list** (zō·of′ə·list). [<ZOO- + -PHILE] — **zo′o·phil′ic** (-fil′ik) *adj.*

zo·oph·i·lism (zō·of′ə·liz′əm) *n.* **1** Fondness for animals. **2** The obtaining of sexual gratification by the fondling of animals. Also **zo·o·phil·i·a** (zō′ə·fil′ē·ə).

zo·oph·i·lous (zō·of′ə·ləs) *adj.* **1** Animal-loving. **2** Adapted for pollination by animals, as certain plants. [<ZOO- + -PHILOUS]

zo·o·pho·bi·a (zō′ə·fō′bē·ə) *n.* A morbid fear of animals. [<ZOO- + -PHOBIA] — **zo′o·pho′bic** *adj.*

zo·o·phyte (zō′ə·fīt) *n.* An invertebrate animal resembling a plant, as a coral or sea anemone. [<ZOO- + -PHYTE] — **zo′o·phyt′ic** (-fit′ik) or **·i·cal** *adj.*

zo·o·plas·ty (zō′ə·plas′tē) *n. Surg.* That operation by which a part of an animal body is grafted on some part of the human body. — **zo′o·plas′tic** *adj.*

zo·o·sperm (zō′ə·spûrm) *n.* **1** A zoospore. **2** A spermatozoon. — **zo′o·sper·mat′ic** (-spər·mat′ik) *adj.*

zo·o·spo·ran·gi·um (zō′ə·spə·ran′jē·əm) *n. pl.* **·gi·a** (-jē·ə) *Bot.* A sporangium producing zoospores. [<NL <Gk. *zōion* an animal + *spora* a seed + *angeion* a vessel] — **zo′o·spo·ran′gi·al** *adj.*

zo·o·spore (zō′ə·spôr, -spōr) *n.* **1** *Bot.* A motile spore destitute of any cell wall, produced particularly among some algae and fungi: they move sometimes in an ameboid manner, but more frequently they are provided with cilia, by the lashing of which the spore is propelled through the water. **2** *Zool.* A flagellate or ameboid motile cell in certain protozoa. [<ZOO- + SPORE] — **zo′o·spor′ic** (-spôr′ik, -spor′ik), **zo·os·po·rous** (zō·os′pər·əs) *adj.*

zo·os·ter·ol (zō·os′tər·ōl, -ol) *n. Biochem.* Any of a class of sterols found in animal tissues, as cholesterol. [<ZOO- + STEROL]

zo·ot·o·my (zō·ot′ə·mē) *n.* The anatomy or dissection of animals; comparative anatomy. [<NL *zootomia* <Gk. *zōion* an animal + *tomē* a cutting <*temnein* cut] — **zo·o·tom·ic** (zō′ə·tom′ik) or **·i·cal** *adj.* — **zo′o·tom′i·cal·ly** *adv.* — **zo·ot′o·mist** *n.*

zo·o·tox·in (zō′ə·tok′sin) *n.* A toxin derived from animals, as snake venom, the poison of bee stings, etc.

zoot suit (zōōt) *Slang* A suit having an extra-long coat and baggy trousers narrowing at the ankle. [Origin uncertain]

zoot-suit·er (zōōt′sōō′tər) *n.* A member of any of various groups wearing zoot suits. They were originally worn by metropolitan Negroes and jazz musicians, but later by the rowdier elements of large cities, such as hoodlums and members of boys' gangs.

zor·il (zôr′il, zor′-) *n.* An African musteline carnivore *(Ictonyx striata)* which can emit a noxious odor; the Cape polecat. Also **zo·ril·la.** (zə·ril′ə). [<F *zorille* <Sp. *zorrilla* a polecat, dim. of *zorra* a fox]

Zorn (sôrn), **Anders Leonhard,** 1860–1920, Swedish painter, etcher, and sculptor.

Zo·ro·as·ter (zō′rō·as′tər) The traditional founder of the ancient Persian religion, believed to have lived about 600 B.C.: also called *Zarathustra.* — **Zo′ro·as′tri·an** *adj. & n.*

Zo·ro·as·tri·an·ism (zō′rō·as′trē·ən·iz′əm) *n.* The religious system founded by Zoroaster on the old Aryan folk religion and taught in the Zend-Avesta. It recognizes two creative powers, one good and the other evil, includes the belief in life after death, and teaches the final triumph of good over evil. See AHRIMAN, ORMUZD. Also **Zo′ro·as′trism.**

Zor·ril·la y Mo·ral (thôr·re′lyä ē mō·räl′), **José,** 1817–93, Spanish poet and dramatist.

Zos·i·mus (zos′ə·məs, zō′sə-) Greek historian of the fifth century A.D.

zos·ter (zos′tər) *n.* **1** An ancient Greek belt or girdle worn especially by men. **2** *Pathol.* Shingles; herpes zoster. [<L <Gk. *zōstēr* a girdle <*zōnynai* gird]

Zou·ave (zōō·äv′, zwäv) *n.* **1** A light-armed French infantryman wearing a brilliant Oriental uniform, originally an Algerian recruit. **2** In the Civil War, a member of a volunteer regiment assuming the name and part of the dress of the French Zouaves. **3** A woman's short, gaily embroidered jacket: also **Zouave jacket.** [<F <Arabic *Zouaoua,* a Kabyle tribe; so called because orig. recruited from this tribe]

zounds (zoundz) *interj.* An exclamation denoting astonishment: also spelled *swounds.* [Short for *God's wounds*]

Zsig·mon·dy (zhig′môn·dē), **Richard,** 1865–1929, German chemist born in Austria.

zuc·chet·to (tsōōk·ket′tō) *n.* A skullcap worn by ecclesiastics in the Roman Catholic Church: black for a priest, purple for a bishop, red for a cardinal, and white for the pope. Also **zuc·chet′ta.** [Var. of Ital. *zucchetta,* orig. a small gourd, dim. of *zucca* a gourd]

zuc·chi·ni (zōō·kē′nē, *Ital.* dzōōk·kē′nē) *n.* A type of green summer squash (evolved from *Cucurbita pepo*) of a small cylindrical shape. Also called *Italian squash.* [<Ital., pl. of *zucchino,* dim. of *zucca* a gourd, squash]

Zug (tsōōkh) **1** The smallest canton of Switzerland; 92 square miles. **2** Its capital, a town on the **Lake of Zug** (15 square miles; about 9 miles long), in the north central part of the country. *French* **Zoug** (zōōg).

Zug·spit·ze (tsōōk′shpit·sə) The highest mountain in Germany, in the Bavarian Alps on the Austrian border; 9,722 feet.

Zu·lo·a·ga (thōō′lō·ä′gä), **Ignacio,** 1870–1945, Spanish painter.

Zu·lu (zōō′lōō) *n. pl.* **Zu·lus** or **Zu·lu 1** One of a Bantu nation of Natal, South Africa, sometimes included with the Kaffirs. **2** The language of the Zulus, belonging to the Bantu family of agglutinative languages. — *adj.* Of, pertaining to, or characteristic of the Zulus or their language.

Zu·lu·land (zōō′lōō·land′) A district of NE Natal, Union of South Africa, formerly a native kingdom; 10,362 square miles.

zum Bei·spiel (tsōōm bī′shpēl) *German* For example: abbreviated *z.B.*

Zu·ñi (zōō′nyē) *n.* **1** One of a tribe of North American Indians of pueblo culture but comprising a distinct linguistic stock: still occupying the big Zuñi pueblo in New Mexico, which is now a reservation. **2** The language of this tribe. — **Zu′ñi·an** *adj. & n.*

Zur·ba·rán (thōōr′bä·rän′), **Francisco de,** 1598–1662, Spanish painter.

Zu·rich (zōōr′ik) **1** A canton of NE Switzerland on the border of Baden-Württemberg, West Germany; 667 square miles. **2** Its capital, a city on the northern shore of the **Lake of Zurich** (35 square miles; 25 miles long) in NE Switzerland. *German* **Zü·rich** (tsü′rikh).

Zuy·der Zee (zī′dər zē, *Du.* zœi′dər zā) A former shallow inlet of the North Sea in NW Netherlands; 80 by 34 miles; enclosed by a dike; drainage projects have reclaimed much

of the land and formed Lake Ijssel. Also **Zui′- der Zee.**
Zweig (tsvīg, tsvīkh), **Arnold,** born 1887, German novelist. — **Stefan,** 1881-1942, Austrian dramatist and novelist.
Zwick·au (tsvik′ou) A city in the former state of Saxony, central southern East Germany.
zwie·back (tswē′bäk, swē′-, swī′-; *Ger.* tsvī′- bäk) *n.* A biscuit of wheaten bread or rusk baked yellow in the loaf and later sliced and toasted. [<G, twice baked <*zwie*- twice (<*zwei* two) + *backen* bake]
Zwing·li (tsving′lē), **Ulrich,** 1484-1531, Swiss Protestant reformer. Also *Huldreich Zwingli.*
Zwing·li·an (zwing′lē-ən, tsving′-) *adj.* Of or pertaining to the doctrines taught by Zwingli, especially to the doctrine that the Eucharist is simply a memorial or a symbolic commemoration of the death of Christ. — *n.* A follower of Zwingli. — **Zwing′li·an·ism** *n.*
zwit·ter·i·on (tsvit′ər·ī′ən) *n. Physics* An ion which carries both a negative and a positive charge, as in certain amino acids. [<G *zwitter* hybrid, hermaphrodite, mongrel + ION] — **zwit′ter·i·on′ic** (-ī·on′ik) *adj.*
Zwol·le (zvôl′ə) The capital of Overijssel province, north central Netherlands; a manufacturing and dairy center; site of many 15th century buildings.
Zwor·y·kin (zwôr′i·kin), **Vladimir Kosma,** born 1889, U.S. research engineer in electronics; born in Russia.
zyg·a·poph·y·sis (zig′ə·pof′ə·sis) *n. pl.* **·ses** (-sēz) *Anat.* One of the processes, usually disposed in pairs, by which a vertebra articulates with another; an articular process. [<NL <Gk. *zygon* a yoke + *apophysis* a branch. See APOPHYSIS.] — **zyg′a·po·phys′e·al** (-pō·fiz′ē-əl) or **·i·al** *adj.*
zygo- *combining form* Yoke; pair; resembling a yoke, especially in shape: *zygospore.* Also, before vowels, **zyg-.** [<Gk. *zygon* a yoke]
zy·go·dac·tyl (zī′gō·dak′til) *Zool. adj.* Having paired toes, one pair directed forward and the other pair backward, as in parrots and woodpeckers. — *n.* A zygodactyl bird. [<ZYGO- + Gk. *daktylos* a finger]
zy·go·ma (zī·gō′mə) *n. pl.* **·ma·ta** (-mə·tə) *Anat.* 1 The long arch that joins the temporal and malar bones on the side of the skull. 2 The zygomatic bone. 3 The zygomatic process. [<NL <Gk. *zygōma* <*zygon* a yoke] — **zy·go·mat·ic** (zī′gō·mat′ik) *adj.*
zygomatic arch *Anat.* The zygoma.
zygomatic bone *Anat.* The malar bone or cheek bone.
zygomatic process *Anat.* That process of the temporal bone which helps to form the zygomatic arch.
zy·go·mor·phic (zī′gō·môr′fik) *adj. Biol.* Bilaterally symmetrical: said of organisms or parts of organisms divisible into similar halves in only one plane. Also **zy′go·mor′- phous.** [<ZYGO- + -MORPHIC] — **zy′go·mor′- phism** *n.*
zy·go·phyl·la·ceous (zī′gō·fi·lā′shəs) *adj. Bot.* Designating or pertaining to a family (*Zygophyllaceae*) of herbs and shrubs, the caltrop family, having jointed branches, two-foliolate or pinnate stipulate leaves, and axillary white, red, or yellow flowers: mainly tropical in distribution. [<NL <Gk. *zygon* a yoke + *phyllon* a leaf]
zy·go·phyte (zī′gō·fīt) *n. Bot.* A plant in which reproduction is by means of zygospores. [<ZYGO- + -PHYTE]
zy·go·sis (zī·gō′sis) *n. Biol.* The union of gametes or cells; conjugation. [<NL <Gk. *zygōsis* a joining <*zygon* a yoke]
zy·go·spore (zī′gō·spôr, -spōr) *n. Bot.* A spore formed by the conjugation of two similar gametes, as in algae and fungi. Also **zy′go·sperm** (-spûrm). [<ZYGO- + SPORE]
zy·gote (zī′gōt, zig′ōt) *n. Biol.* 1 The product of the union of two gametes. 2 An individual developed from such a union. [<Gk. *zygōtos* yoked <*zygoein* yoke <*zygon* a yoke] — **zy·got·ic** (zī·got′ik) *adj.*
zy·mase (zī′mās) *n. Biochem.* An enzyme, obtained principally from yeast, which induces fermentation by breaking down glucose and related carbohydrates into alcohol and carbon dioxide. [<F <Gk. *zymē* leaven]
zyme (zīm) *n.* 1 A ferment. 2 A disease germ or virus supposed to be the specific cause of a zymotic disease. [<Gk. *zymē* leaven]
zy·mic (zī′mik) *adj.* Relating to or produced by fermentation.
zymo- *combining form* Fermentation; of or related to fermentation: *zymology.* Also, before vowels, **zym-.** [<Gk. *zymē* leaven]
zy·mo·gen (zī′mō·jən) *n.* 1 *Biochem.* A substance that develops into an enzyme when suitably activated, as in the stomach or pancreas. 2 *Biol.* A bacterial organism which produces enzymes or fermentation. Compare PATHOGEN. Also **zy′mo·gene** (-jēn). [<ZYMO- + -GEN]
zy·mo·gen·e·sis (zī′mō·jen′ə·sis) *n. Biochem.* The transformation of a zymogen into an enzyme.
zy·mo·gen·ic (zī′mō·jen′ik) *adj.* 1 Of, pertaining to, or relating to zymogen. 2 Capable of producing a ferment, as yeast. Also **zy·mog·e·nous** (zī·moj′ə·nəs).
zymogenic organism Any micro-organism which causes fermentation, as yeast.
zy·mol·o·gy (zī·mol′ə·jē) *n.* The study of the principles of fermentation and the action of enzymes. [<ZYMO- + -LOGY] — **zy·mo·log·ic** (zī′mə·loj′ik) or **·i·cal** *adj.* — **zy·mol′o·gist** *n.*
zy·mol·y·sis (zī·mol′ə·sis) *n.* Fermentation or the action of enzymes. [<ZYMO- + -LYSIS] — **zy·mo·lyt·ic** (zī′mə·lit′ik) *adj.*
zy·mom·e·ter (zī·mom′ə·tər) *n.* An instrument for measuring the degree of fermentation. [<ZYMO- + -METER]
zy·mo·sis (zī·mō′sis) *n.* 1 Any form of fermentation. 2 *Med.* **a** A fermentation giving rise to a morbid or diseased condition, as by the action of bacteria. **b** Any contagious or infectious disease produced by morbific fermentation; a zymotic disease. [<NL <Gk. *zymōsis* <*zymoein* leaven, ferment <*zymē* leaven]
zy·mot·ic (zī·mot′ik) *adj.* Relating to or produced by or from fermentation, as certain epidemic or contagious diseases. [<Gk. *zymōtikos* <*zymoein.* See ZYMOSIS.]
zy·mur·gy (zī′mûr·jē) *n.* A branch of chemistry treating of processes in which fermentation is the principal feature, as brewing, making of yeast, and winemaking. [<ZYM(O)- + -URGY]
Zy·ri·an Autonomous Region (zir′ē·ən) A former name for the KOMI AUTONOMOUS S.S.R.
Zyr·ya·novsk (zir·yä′nôfsk) A city of NE Kazakh S.S.R. near the Russian S.F.S.R. border.

THE ALFRED BERNHARD NOBEL PRIZES

PHYSICS

No awards were made in the years 1916, 1931, 1934, 1940-1942.

1901 WILHELM CONRAD ROENTGEN, 1845-1923, German: Roentgen rays.
1902 HENDRICK ANTOON LORENTZ, 1853-1928, Dutch; and PIETER ZEEMAN, 1865-1943, Dutch: influence of magnetism on electricity.
1903 HENRI ANTOINE BECQUEREL, 1852-1908, French: spontaneous radioactivity; and PIERRE CURIE, 1859-1906, French, with MARIE SKLODOWSKA CURIE, 1867-1934, Polish: phenomena of radiation.
1904 JOHN WILLIAM STRUTT, 1842-1919, English: density of gases.
1905 PHILIPP EDUARD ANTON VON LENARD, 1862-1947, German: cathode rays.
1906 JOSEPH JOHN THOMSON, 1856-1940, English: passage of electricity through gases.
1907 ALBERT ABRAHAM MICHELSON, 1852-1931, American: spectroscopic and metrological investigations.
1908 GABRIEL LIPPMANN, 1845-1921, French: color photography.
1909 GUGLIELMO MARCONI, 1874-1937, Italian; and FERDINAND BRAUN, 1850-1918, German: the development of wireless telegraphy.
1910 JOHANNES DIEDERIK VAN DER WAALS, 1837-1923, Dutch: equations of state for gases and liquids.
1911 WILHELM WIEN, 1864-1928, German: heat radiation.
1912 GUSTAF DALÉN, 1869-1937, Swedish: automatic light regulators.
1913 HEIKE KAMERLINGH ONNES, 1853-1926, Dutch: low temperatures.
1914 MAX VON LAUE, 1879-   , German: crystal diffraction of Roentgen rays.
1915 WILLIAM HENRY BRAGG, 1862-1942, English; and WILLIAM LAWRENCE BRAGG, 1890-   , English: crystal analysis.
1917 CHARLES GLOVER BARKLA, 1877-1944, English: Roentgen radiation of the elements.
1918 MAX PLANCK, 1858-1947, German: quantum theory.
1919 JOHANNES STARK, 1874-1957, German: Doppler effect in canal rays and spectrum investigations.
1920 CHARLES EDOUARD GUILLAUME, 1861-1938, French: anomalies in nickel steel alloys.
1921 ALBERT EINSTEIN, 1879-1955, German: photoelectric effect.
1922 NIELS BOHR, 1885-   , Danish: atomic structure.
1923 ROBERT ANDREWS MILLIKAN, 1868-1953, American: elementary electric charge and photoelectric phenomena.
1924 KARL MANNE GEORG SIEGBAHN, 1886-   , Swedish: X-ray spectroscopy.
1925 JAMES FRANCK, 1882-   , German; and GUSTAV HERTZ, 1887-   , German: electron impact on atoms.
1926 JEAN BAPTISTE PERRIN, 1870-1942, French: discontinuous structure of matter.
1927 ARTHUR HOLLY COMPTON, 1892-   , American: Compton effect; and CHARLES THOMSON REES WILSON, 1869-   , English: paths of electrically charged particles.
1928 OWEN WILLANS RICHARDSON, 1879-   , English: thermionic ions.
1929 LOUIS-VICTOR DE BROGLIE, 1892-   , French: wave character of electrons.
1930 CHANDRASEKHARA VENKATA RAMAN, 1888-   , Indian: diffusion of light and Raman effect.
1932 WERNER HEISENBERG, 1901-   , German: quantum mechanics.
1933 ERWIN SCHROEDINGER, 1887-   , German; and PAUL ADRIEN MAURICE DIRAC, 1902-   , English: atomic theory.
1935 JAMES CHADWICK, 1891-   , English: discovery of the neutron.
1936 VICTOR FRANZ HESS, 1883-   , Austrian: cosmic radiation; and CARL DAVID ANDERSON, 1905-   , American: discovery of the positron.
1937 GEORGE PAGET THOMSON, 1892-   , English; and CLINTON JOSEPH DAVISSON, 1881-   , American: crystal diffraction of electrons.
1938 ENRICO FERMI, 1901-1954, Italian: neutron investigations.
1939 ERNEST ORLANDO LAWRENCE, 1901-   , American: cyclotron.
1943 OTTO STERN, 1888-   , American: atomic rays and magnetic moment of the proton.

1944 ISIDOR ISAAC RABI, 1898- , American: magnetic properties of atomic nuclei.
1945 WOLFGANG PAULI, 1900- , Austrian: exclusion principle.
1946 PERCY WILLIAMS BRIDGMAN, 1882- , American: high-pressure physics.
1947 EDWARD VICTOR APPLETON, 1892- , English: ionosphere investigations.
1948 PATRICK MAYNARD STUART BLACKETT, 1897- , English: cosmic radiation.
1949 HIDEKI YUKAWA, 1907- , Japanese: the meson particle.
1950 CECIL FRANK POWELL, 1903- , English: the pi-meson.
1951 JOHN DOUGLAS COCKCROFT, 1897- , English: and ERNEST THOMAS SINTON WALTON, 1903- , English: transmutation of atomic nuclei.
1952 FELIX BLOCH, 1905- , American; and EDWARD MILLS PURCELL, 1912- , American: nuclear magnetic measurements.
1953 FRITS ZERNIKE, 1888- , Dutch: phase contrast method.
1954 MAX BORN, 1882- , English: quantum mechanics; and WALTHER BOTHE, 1891-1957, German: coincidence method.
1955 WILLIS EUGENE LAMB, 1913- , American: hydrogen spectrum; and POLYCARP KUSCH, 1911- , American: magnetic moment of the electron.
1956 WILLIAM SHOCKLEY, 1910- , American; JOHN BARDEEN, 1908- , American; and WALTER HOUSER BRATTAIN, 1902- , American: transistor effect.
1957 TSUNG DAO LEE, 1926- , American; and CHEN NING YANG, 1922- , American: parity laws.
1958 ILYA M. FRANK, 1908- , Russian; IGOR Y. TAMM, 1895- , Russian; and PAVEL A. CHERENKOV, 1904- , Russian: Cherenkov radiation.
1959 OWEN CHAMBERLAIN, 1920- , American; and EMILIO GINO SEGRÈ, 1905- , American: antiproton.
1960 DONALD ARTHUR GLASER, 1926- , American: invention of the bubble chamber.

CHEMISTRY

No awards were made in the years 1916, 1917, 1919, 1924, 1933, 1940-1942.

1901 JACOBUS HENRICUS VAN'T HOFF, 1852-1911, Dutch: osmotic pressure in solutions.
1902 EMIL HERMANN FISCHER, 1852-1919, German: sugar synthesis.
1903 SVANTE AUGUST ARRHENIUS, 1859-1927, Swedish: electrolysis.
1904 WILLIAM RAMSAY, 1852-1916, English: discovery of inert gases in the air.
1905 ADOLPH VON BAEYER, 1835-1917, German: organic chemistry and dyestuffs.
1906 HENRI MOISSAN, 1852-1907, French: isolation of fluorine and invention of electric furnace.
1907 EDUARD BUCHNER, 1860-1917, German: biological chemistry.
1908 ERNEST RUTHERFORD, 1871-1937, English: radioactive disintegration of elements.
1909 WILHELM OSTWALD, 1853-1932, German: catalysis and chemical equilibrium.
1910 OTTO WALLACH, 1847-1931, German: alicyclic organic compounds.
1911 MARIE SKLODOWSKA CURIE, 1867-1934, Polish: discovery of radium and polonium.
1912 VICTOR GRIGNARD, 1871-1935, French: Grignard reaction; and PAUL SABATIER, 1854-1941, French: hydrogenation.
1913 ALFRED WERNER, 1866-1919, Swiss: molecular linkage.
1914 THEODORE WILLIAM RICHARDS, 1868-1928, American: determination of atomic weights.
1915 RICHARD WILLSTAETTER, 1872-1942, German: coloring matter of plants, especially chlorophyll.
1918 FRITZ HABER, 1868-1934, German: synthetic production of ammonia.
1920 WALTHER HERMANN NERNST, 1864-1941, German: thermochemistry.
1921 FREDERICK SODDY, 1877-1956, English: origin and nature of isotopes.
1922 FRANCIS WILLIAM ASTON, 1877-1945, English: discovery of isotopes and of whole-number rule.
1923 FRITZ PREGL, 1869-1930, Austrian: microanalysis of organic substances.
1925 RICHARD ZSIGMONDY, 1865-1929, German: colloid chemistry.
1926 THE SVEDBERG, 1884- , Swedish: disperse systems.
1927 HEINRICH OTTO WIELAND, 1877-1957, German: constitution of bile acids.
1928 ADOLPH WINDAUS, 1876- , German: constitution of sterols and connection with vitamins.
1929 ARTHUR HARDEN, 1865-1940, English: enzyme action; and HANS KARL AUGUST SIMON VON EULER-CHELPIN, 1873- , German: sugar fermentation and enzyme action.
1930 HANS FISCHER, 1881-1945, German: coloring matter of blood and leaves, synthesis of hemin.
1931 CARL BOSCH, 1874-1940, German; and FRIEDRICH BERGIUS, 1884-1949, German: chemical high-pressure methods.
1932 IRVING LANGMUIR, 1881-1957, American: surface chemistry.
1934 HAROLD CLAYTON UREY, 1893- , American: heavy hydrogen.
1935 FRÉDÉRIC JOLIOT, 1900- , French; and IRÈNE JOLIOT-CURIE, 1897-1956, French: synthesis of new radioactive elements.
1936 PETER JOSEPH WILHELM DEBYE, 1884- , Dutch: molecular structure and X-ray diffraction in gases.
1937 WALTER NORMAN HAWORTH, 1887-1950, English: carbohydrates and vitamin C; and PAUL KARRER, 1889- , Swiss: carotenoids and vitamins A and B.
1938 RICHARD KUHN, 1900- , German: carotenoids and vitamins.
1939 ADOLPH FRIEDRICH JOHANN BUTENANDT, 1903- , German: sexual hormones; and LEOPOLD RUZICKA, 1887- , Yugoslav: polymethylenes and terpenes.
1943 GEORGE DE HEVESY, 1885- , Hungarian: use of isotopes.
1944 OTTO HAHN, 1879- , German: atomic disintegration.
1945 ARTTURI ILMARI VIRTANEN, 1895- , Finnish: nutritive and agricultural chemistry.
1946 JAMES BATCHELLER SUMNER, 1887-1955, American: crystallization of enzymes; and JOHN HOWARD NORTHROP, 1891- , American; and WENDELL MEREDITH STANLEY, 1904- , American: preparation of enzymes and viruses in pure form.
1947 ROBERT ROBINSON, 1886- , English: alkaloid investigations.
1948 ARNE WILHELM KAURIN TISELIUS, 1902- , Swedish: electrophoresis of proteins.
1949 WILLIAM FRANCIS GIAUQUE, 1895- , American: chemical thermodynamics and low temperatures.
1950 OTTO PAUL HERMANN DIELS, 1876-1954, German; and KURT ALDER, 1902-1958, German: -diene synthesis and production of odors.
1951 GLENN THEODORE SEABORG, 1912- , American; and EDWIN MATTISON MCMILLAN, 1907- , American: chemistry of transuranian elements.
1952 ARCHER JOHN PORTER MARTIN, 1910- , English; and RICHARD LAURENCE MILLINGTON SYNGE, 1914- , English: partition chromatography.
1953 HERMANN STAUDINGER, 1881- , German: macro-molecular chemistry.
1954 LINUS CARL PAULING, 1901- , American: chemical bond and complex structures.
1955 VINCENT DU VIGNEAUD, 1901- , American: sulfur compounds.
1956 CYRIL NORMAN HINSHELWOOD, 1897- , English; and NIKOLAI NIKOLAIEVICH SEMYONOV, 1896- , Russian: chemical reactions.
1957 ALEXANDER R. TODD, 1907- , English: nucleotides and nucleotide enzymes.
1958 FREDERICK SANGER, 1918- , English: structure of proteins, especially of insulin.
1959 JAROSLAV HEYROVSKY, 1890- , Czechoslovakian: polarographic method of analysis.
1960 WILLARD FRANK LIBBY, 1901- , American: development of a method for age determination in archeology, geology, geophysics, and other sciences.

PHYSIOLOGY AND MEDICINE

No awards were made in the years 1915-1918, 1921, 1925, 1940-1942.

1901 EMIL ADOLPH VON BEHRING, 1854-1917, German: serum therapy against diphtheria.
1902 RONALD ROSS, 1857-1932, English: malaria investigations.
1903 NIELS RYBERG FINSEN, 1860-1904, Danish: treatment of disease by light rays.
1904 IVAN PETROVICH PAVLOV, 1849-1936, Russian: physiology of digestion.
1905 ROBERT KOCH, 1843-1910, German: work on tuberculosis.
1906 CAMILLO GOLGI, 1843-1926, Italian; and SANTIAGO RAMÓN Y CAJAL, 1852-1934, Spanish: structure of nervous system.
1907 CHARLES LOUIS ALPHONSE LAVERAN, 1845-1922, French: protozoan cause of disease.
1908 PAUL EHRLICH, 1854-1915, German; and ILYA ILYICH METCHNIKOFF, 1845-1916, Russian: work on immunity.
1909 EMIL THEODOR KOCHER, 1841-1917, Swiss: researches on thyroid gland.
1910 ALBRECHT KOSSEL, 1853-1927, German: chemistry of the cell.
1911 ALLVAR GULLSTRAND, 1862-1930, Swedish: dioptries of the eye.
1912 ALEXIS CARREL, 1873-1944, American: vascular ligature and grafting of blood vessels and organs.
1913 CHARLES ROBERT RICHET, 1850-1935, French: anaphylaxis.
1914 ROBERT BÁRÁNY, 1876-1936, Austrian: physiology and pathology of vestibular system.
1919 JULES BORDET, 1870- , Belgian: researches on immunity.
1920 SCHACK AUGUST STEENBERG KROGH, 1874-1949, Danish: mechanism of the capillaries.
1922 ARCHIBALD VIVIAN HILL, 1886- , English: heat production of muscles; and OTTO MEYERHOF, 1884-1951, German: metabolism of muscles.
1923 FREDERICK GRANT BANTING, 1891-1941, Canadian; and JOHN JAMES RICHARD MACLEOD, 1876-1935, Canadian: discovery of insulin.
1924 WILLEM EINTHOVEN, 1860-1927, Dutch: discovery of the mechanism of the electrocardiogram.
1926 JOHANNES FIBIGER, 1867-1928, Danish: discovery of a carcinoma parasite.
1927 JULIUS WAGNER-JAUREGG, 1857-1940, Austrian: discovery of therapeutic value of malaria inoculation in paralytic dementia.
1928 CHARLES NICOLLE, 1866-1936, French: work on typhus.
1929 CHRISTIAAN EIJKMAN, 1858-1930, Dutch: discovery of antineuritic vitamins; and FREDERICK GOWLAND HOPKINS, 1861-1947, English: discovery of growth-promoting vitamins.
1930 KARL LANDSTEINER, 1866-1943, American: discovery of human blood groups.
1931 OTTO H. WARBURG, 1883- , German: discovery of character and mode of action of the respiratory ferment.
1932 CHARLES SCOTT SHERRINGTON, 1857-1952, English; and EDGAR DOUGLAS ADRIAN, 1889- , English: function of the neuron.
1933 THOMAS HUNT MORGAN, 1866-1945, American: hereditary function of the chromosomes.
1934 GEORGE HOYT WHIPPLE, 1878- , American; GEORGE RICHARDS MINOT, 1885-1950, American; and WILLIAM PARRY MURPHY, 1892- , American: liver therapy against anemias.
1935 HANS SPEMANN, 1869-1941, German: discovery of organizer effect in embryonic development.
1936 HENRY HALLETT DALE, 1875- , English; and OTTO LOEWI, 1873- , Austrian: chemical transmission of nerve impulses.
1937 ALBERT SZENT-GYÖRGI VON NAGYRAPOLT, 1893- , Hungarian: researches on biological combustion.
1938 CORNEILLE JEAN FRANÇOIS HEYMANS, 1892- , Belgian: sinus and aortic mechanisms in regulation of respiration.
1939 GERHARD DOMAGK, 1895- , German: discovery of properties of early sulfonamides.
1943 HENRIK CARL PETER DAM, 1895- , Danish: discovery of vitamin K; and EDWARD ADELBERT DOISY, 1893- , American: discovery of chemical nature of vitamin K.
1944 JOSEPH ERLANGER, 1874- , American; and HERBERT SPENCER GASSER, 1888- , American: functions of the single nerve fibers.
1945 ALEXANDER FLEMING, 1881-1955, English; ERNST BORIS CHAIN, 1906- , German; and HOWARD FLOREY, 1898- , English: discovery of penicillin.
1946 HERMANN JOSEPH MULLER, 1890- , American: discovery of X-ray production of genetic mutations.
1947 CARL FERDINAND CORI, 1896- , Czechoslovak; GERTI THERESA CORI, 1896-1957, Czechoslovak: catalytic metabolism of animal starch; and BERNARDO ALBERTO HOUSSAY, 1887- , Portuguese: significance of hormone of the hypophysis.
1948 PAUL HERMANN MÜLLER, 1899- , Swiss: discovery of effect of DDT on arthropods.
1949 WALTER RUDOLF HESS, 1881- , Swiss: organization of the diencephalon; and ANTONIO CAETANO DE ABREU FREIRE EGAS MONIZ, 1874-1955, Portuguese: discovery of therapeutic value of prefrontal lobotomy.
1950 EDWIN CALVIN KENDALL, 1886- , American; and PHILIP SHOWALTER HENCH, 1896- , American; and TADEUS REICHSTEIN, 1897- , Swiss: hormones of the adrenal complex.
1951 MAX THEILER, 1899- , American: researches on yellow fever.
1952 SELMAN ABRAHAM WAKSMAN, 1888- , American: discovery of streptomycin.
1953 HANS ADOLF KREBS, 1900- , English: citric acid cycle; and FRITZ ALBERT LIPMANN, 1899- , American: discovery of coenzyme A.
1954 JOHN FRANKLIN ENDERS, 1897- , American; THOMAS HUCKLE WELLER, 1915- , American; and FREDERICK CHAPMAN ROBBINS, 1916- , American: poliomyelitis viruses.

1955 AXEL HUGO THEODOR THEORELL, 1903- , Swedish: oxidation enzymes.
1956 ANDRÉ COURNAND, 1895- , American; WERNER FORSSMANN, 1904- , German; and DICKINSON W. RICHARDS JR., 1895- , American: heart and circulatory system.
1957 DANIEL BOVET, 1907- , Swiss: synthetic compounds acting on the vascular system and skeletal muscles.
1958 GEORGE WELLS BEADLE, 1903- , American; and EDWARD LAWRIE TATUM, 1909- , American: discovery that genes act by regulating definite chemical events; and JOSHUA LEDERBERG, 1925- , American: discoveries concerning genetic recombination and the organization of the genetic material of bacteria.
1959 SEVERO OCHOA, 1905- , American; and ARTHUR KORNBERG, 1918- , American: biological synthesis of ribonucleic acid and deoxyribonucleic acid.
1960 MACFARLANE BURNET, 1899- , Australian; and PETER BRIAN MEDAWAR, 1915- , British: immunological tolerance.

LITERATURE

No awards were made in the years 1914, 1918, 1935, 1940-1943.

1901 RENÉ FRANCOIS ARMAND SULLY PRUDHOMME, 1839-1907, French: exceptional merit as a writer, especially in his poetry.
1902 THEODOR MOMMSEN, 1817-1903, German: in special recognition of his *Römische Geschichte*.
1903 BJØRNSTJERNE BJØRNSON, 1832-1910, Norwegian: for devotion to poetic creation.
1904 FRÉDÉRIC MISTRAL, 1830-1914, French: Provençal poetry and scholarship; and JOSÉ ECHEGARAY Y EIZAGUIRRE, 1833-1916, Spanish: original dramatic work.
1905 HENRYK SIENKIEWICZ, 1846-1916, Polish: achievements as an epic writer.
1906 GIOSUÈ CARDUCCI, 1835-1907, Italian: poetry, scholarship, and criticism.
1907 RUDYARD KIPLING, 1865-1936, English: the body of his creative work.
1908 RUDOLF CHRISTOPH EUCKEN, 1846-1926, German: for the vigor and idealism of his philosophic works.
1909 SELMA LAGERLÖF, 1858-1940, Swedish: imaginative and spiritual qualities of her poetry.
1910 PAUL VON HEYSE, 1830-1914, German: for his work as a lyricist, dramatist, novelist, and story writer.
1911 MAURICE MAETERLINCK, 1862-1949, Belgian: for his versatile literary work, especially in drama.
1912 GERHART HAUPTMANN, 1862-1946, German: for the wide range of his creative work in realm of dramatic poetry.
1913 RABINDRANATH TAGORE, 1861-1941, Indian: for the beauty of his English rendering of Indian poetic thought.
1915 ROMAIN ROLLAND, 1866-1944, French: for the lofty idealism and deep human sympathy of his writings.
1916 VERNER VON HEIDENSTAM, 1859-1940, Swedish: for work representing a new epoch in Swedish literature.
1917 KARL GJELLERUP, 1857-1919, Danish: for his varied and rich poetry; and HENRIK PONTOPPIDAN, 1857-1943, Danish: for his vital description of contemporary life in Denmark.
1919 CARL SPITTELER, 1845-1924, Swiss: especially in recognition of his epic *Olympischer Frühling*.
1920 KNUT HAMSUN, 1859-1952, Norwegian: for his monumental work *Markens grøde*.
1921 ANATOLE FRANCE (Jacques Anatole Thibault), 1844-1924, French: for his brilliant work as an author, typical of the French genius.
1922 JACINTO BENAVENTE, 1866-1954, Spanish: for his success in perpetuating the honorable traditions of Spanish drama.
1923 WILLIAM BUTLER YEATS, 1865-1939, Irish: for his poetry.
1924 WŁADYSŁAW REYMONT, 1868-1925, Polish: for his great national epic *The Peasants*.
1925 GEORGE BERNARD SHAW, 1856-1950, Irish: for his total work, characterized by idealism and humanity.
1926 GRAZIA DELEDDA, 1875-1936, Italian: for the idealism, depth, and sympathy of her writings.
1927 HENRI BERGSON, 1859-1941, French: for his rich and life-giving ideas and the artistry of their presentation.
1928 SIGRID UNDSET, 1882-1949, Norwegian: for her powerful pictures of northern life in medieval times.
1929 THOMAS MANN, 1875-1955, German: principally for his great novel *Buddenbrooks*.
1930 SINCLAIR LEWIS, 1885-1951, American: for his great and living art of painting life and talent for creating types.
1931 ERIK AXEL KARLFELDT, 1864-1931, Swedish: for his poetry.
1932 JOHN GALSWORTHY, 1867-1933, English: for the power of his descriptions, especially in the *Forsyte Saga*.
1933 IVAN BUNIN, 1870-1953, Russian: for his skill in representing the classical tradition of Russian prose.
1934 LUIGI PIRANDELLO, 1867-1936, Italian: for his bold and ingenious recreation of dramatic and scenic art.
1936 EUGENE GLADSTONE O'NEILL, 1888-1953, American: for dramatic works of vital energy, sincerity, and intensity of feeling.
1937 ROGER MARTIN DU GARD, 1881- , French: for the artistic vigor and truth of his novel-cycle *Les Thibault*.
1938 PEARL SYDENSTRICKER BUCK, 1892- , American: for rich and genuinely epic pictures of Chinese peasant life.
1939 FRANS EEMIL SILLANPÄÄ, 1888- , Finnish: for his art in depicting Finnish peasant life and customs.
1944 JOHANNES V. JENSEN, 1873-1950, Danish: for the strength and fertility of his poetic imagination.
1945 GABRIELLA MISTRAL (Lucila Godoy y Alcayaga), 1889-1957, Chilean: for the splendor and idealism of her lyric poetry.
1946 HERMANN HESSE, 1877- , Swiss: for the humanistic vigor of his writings and the high quality of his style.
1947 ANDRÉ GIDE, 1869-1951, French: for the extensive and artistic worth of his literary work.
1948 THOMAS STEARNS ELIOT, 1888- , English: for his remarkable achievement as a pioneer in modern poetry.
1949 WILLIAM FAULKNER, 1897- , American: for his unique contribution to the modern American novel.
1950 BERTRAND ARTHUR WILLIAM RUSSELL, 1872- , English: for his varied and significant writings championing humanitarian ideals and freedom of thought.
1951 PÄR LAGERKVIST, 1891- , Swedish: for vigor and independence of mind, especially as shown in his novel *Barabbas*.
1952 FRANÇOIS MAURIAC, 1885- , French: for the spiritual insight and artistic integrity of his novels.
1953 WINSTON LEONARD SPENCER CHURCHILL, 1874- , English: for his historical and biographical description and defense of human values.
1954 ERNEST HEMINGWAY, 1898- , American: for his mastery of story-telling and influence on style.
1955 HALLDOR KILJAN LAXNESS, 1902- , Icelandic: for his epic power and renewal of Icelandic narrative art.
1956 JUAN RAMÓN JIMÉNEZ, 1881-1958, Spanish: for his lyrical poetry, an example of artistical purity.
1957 ALBERT CAMUS, 1913- , French: for his important literary production, illuminating the problems of the human conscience.
1958 BORIS LEONIDOVICH PASTERNAK, 1890-1960, Russian: for his important achievement in contemporary lyrical poetry and in the field of great Russian epic tradition.
1959 SALVATORE QUASIMODO, 1901- , Italian: for his lyrical poetry, which expresses the tragic experience of life in our times.
1960 SAINT-JOHN PERSE (pseudonym of *Alexis Léger*), 1887- , French: for the soaring flight and the evocative imagery of his poetry.

PEACE

No awards were made in the years 1914-1916, 1918, 1923, 1924, 1928, 1932, 1939-1943, 1948, 1955, 1956, 1960.

1901 HENRI DUNANT, 1828-1910, Swiss, founder of Red Cross; and FRÉDÉRIC PASSY, 1822-1912, French, peace advocate.
1902 ÉLIE DUCOMMUN, 1833-1906, Swiss, of the Bureau International Permanent de la Paix; and CHARLES ALBERT GOBAT, 1843-1914, Swiss, also of the Bureau International.
1903 WILLIAM RANDAL CREMER, 1838-1908, English, founder of the International Arbitration League.
1904 INSTITUT DE DROIT INTERNATIONAL, founded 1873 in Ghent.
1905 BERTHA VON SUTTNER, 1843-1914, Austrian writer.
1906 THEODORE ROOSEVELT, 1858-1919, President of the United States.
1907 ERNESTO TEODORO MONETA, 1833-1918, Italian, President of Lombard League of Peace; and LOUIS RENAULT, 1843-1918, French, professor of international law.
1908 KLAS PONTUS ARNOLDSON, 1844-1916, Swedish writer; and FREDERIK BAJER, 1837-1922, Danish member of Parliament.
1909 AUGUSTE MARIE FRANÇOIS BEERNAERT, 1829-1912, Belgian, member of Permanent Court of Arbitration; and PAUL HENRI BENJAMIN BALLUET D'ESTOURNELLES DE CONSTANT DE REBECQUE, 1852-1924, French, founder of French Parliamentary Arbitration Committee.
1910 BUREAU INTERNATIONAL PERMANENT DE LA PAIX, founded 1891 at Berne.
1911 TOBIAS MICHAEL CAREL ASSER, 1838-1913, Dutch Prime Minister and ALFRED HERMANN FRIED, 1864-1921, Austrian journalist and editor.
1912 ELIHU ROOT, 1845-1937, American, Senator and Secretary of State.
1913 HENRI LA FONTAINE, 1854-1943, Belgian senator.
1917 COMITÉ INTERNATIONAL DE LA CROIX-ROUGE, founded 1863.
1919 THOMAS WOODROW WILSON, 1856-1924, President of the United States.
1920 LÉON BOURGEOIS, 1851-1925, President of French Senate and of Council of League of Nations.
1921 KARL HJALMAR BRANTING, 1860-1925, Swedish Prime Minister; and CHRISTIAN LOUIS LANGE, 1869-1938, Swiss, Secretary General to Inter-Parliamentary Union.
1922 FRIDTJOF NANSEN, 1861-1930, Norwegian explorer and humanitarian.
1925 AUSTEN CHAMBERLAIN, 1863-1937, British Secretary of State for Foreign Affairs; and CHARLES GATES DAWES, 1865-1951, Vice President of the United States.
1926 ARISTIDE BRIAND, 1862-1932, French Secretary of State for Foreign Affairs; and GUSTAV STRESEMANN, 1878-1929, German Chancellor and Foreign Minister.
1927 FERDINAND BUISSON, 1841-1932, French professor; and LUDWIG QUIDDE, 1858-1941, German professor.
1929 FRANK BILLINGS KELLOGG, 1856-1937, American Secretary of State.
1930 LARS OLOF (JONATHAN) NATHAN SÖDERBLOM, 1866-1931, Archbishop of Upsala.
1931 JANE ADDAMS, 1860-1935, American social worker; and NICHOLAS MURRAY BUTLER, 1862-1947, President of Columbia University.
1933 NORMAN ANGELL, 1874- , English author and journalist.
1934 ARTHUR HENDERSON, 1863-1935, British Foreign Secretary.
1935 CARL VON OSSIETZKY, 1889-1938, German journalist.
1936 CARLOS DE SAAVEDRA LAMAS, 1878- , Argentine Minister for Foreign Affairs.
1937 EDGAR ALGERNON ROBERT GASCOYNE, VISCOUNT CECIL OF CHELWOOD, 1864- , English, Lord Privy Seal.
1938 OFFICE INTERNATIONAL NANSEN POUR LES RÉFUGIÉS, founded 1921 in Geneva by Fridtjof Nansen.
1944 COMITÉ INTERNATIONAL DE LA CROIX-ROUGE, founded 1863 in Geneva.
1945 CORDELL HULL, 1871-1955, American Secretary of State.
1946 EMILY GREENE BALCH, 1867- , American, honorary president of Women's International League for Peace and Freedom; and JOHN R. MOTT, 1865-1955, American, former president International Missionary Council.
1947 THE FRIENDS' SERVICE COUNCIL, founded in 1647 in London, and the AMERICAN FRIENDS' SERVICE COMMITTEE.
1949 JOHN BOYD ORR OF BRECHIN, 1880- , Scottish physician, President of National Peace Council.
1950 RALPH BUNCHE, 1904- , American, professor, director, Division of Trusteeship, United Nations.
1951 LÉON JOUHAUX, 1879-1954, French trade-union leader.
1952 ALBERT SCHWEITZER, 1875- , French missionary surgeon, founder Lambaréné Hospital.
1953 GEORGE CATLETT MARSHALL, 1880- , American general, president of American Red Cross, ex-Secretary of State and of Defense, originator of "Marshall Plan."
1954 OFFICE OF THE UNITED NATIONS HIGH COMMISSIONER FOR REFUGEES, Geneva.
1957 LESTER BOWLES PEARSON, 1897- , Canadian former Secretary of State for External Affairs.
1958 DOMINIQUE GEORGES HENRI PIRE, 1910- , Belgian: for his work with refugees.
1959 PHILIP NOEL-BAKER, 1889- , British: for his work to further brotherhood.

ABBREVIATIONS COMMONLY IN USE

The following alphabetical list of abbreviations with their meanings includes those that are commonly employed in reference works, textbooks, journals, and newspapers. Where two or more forms of an abbreviation are shown, as by capital letter or lower-case letter, or with a following period or without such period, each form is in recognized use. Where an abbreviation covers two or more meanings, each meaning is given in alphabetical sequence, and where, in such listing, a given abbreviation is sometimes used in a form which applies to one meaning only, that separate usage is shown in parentheses following the meaning to which it applies. In abbreviations of foreign words, as Latin, the meaning of the abbreviation is first given, followed by the word or phrase, in brackets, in the language of origin. Many of the abbreviations here given are not capitalized when used independently; but in combination with proper names they are properly capitalized, as *ft.* (fort) in Ft. Henry; *mt.* (mount) in Mt. Shasta; *coll.* (college) in Claverling Coll.; *dept.* (department) in Dept. of Biology.

A

a. about; accepted; acting; active; adjective; after; afternoon; aged; alto; amateur; ampere; anode; anonymous; answer; anterior; approved; are (measure); area; argent; at; before [L *ante*]; year [L *anno*].
a., a assists.
a., A acre.
ā., āā., Ā, ĀĀ ana.
A. Absolute (temperature); Academy; America.
A argon.
A, A angstrom (unit).
AA, A.A. achievement age; Airman Apprentice; Alcoholics Anonymous; anti-aircraft.
A.A.A. Agricultural Adjustment Administration (Agency) (AAA); Amateur Athletic Association; American Athletic Association; Anti-aircraft Artillery; Automobile Association of America.
AAAA, A.A.A.A. Amateur Athletic Association of America; Associated Actors and Artists of America.
A.A.A.L. American Academy of Arts and Letters.
AAAS, A.A.A.S. American Academy of Arts and Sciences; American Association for the Advancement of Science.
AACS, A.A.C.S. Army Air Communications System; Airways and Air Communications Service.
AAE, A.A.E. American Association of Engineers.
AAEE, A.A.E.E. American Association of Electrical Engineers.
AAF, A.A.F. Army Air Forces.
A. and M. Agricultural and Mechanical.
A.A.O.N.M.S. Ancient Arabic Order of Nobles of the Mystic Shrine.
A.A.P.S.S. American Academy of Political and Social Sciences.
Aar. Aaron.
A.A.S.R., A.&A.S.R. Ancient and Accepted Scottish Rite.
A.A.U. Amateur Athletic Union.
A.A.U.P. American Association of University Professors.
A.A.U.W. American Association of University Women.
a.b., ab (times) at bat.
a.b., A.B. able-bodied (seaman).
Ab alabamine.
A.B. Bachelor of Arts [L *Artium Baccalaureus*].
AB adapter booster; airborne.
AB, A.B. Aviation Boatswain's Mate.
aband. abandoned.
abbr., abbrev. abbreviation.
ABC the alphabet; (Alphabetical) Railway Guide (Britain). **(A.B.C.)**; American Broadcasting System; Argentina, Brazil, and Chile; Audit Bureau of Circulation.
ABCD America, Britain, China, Dutch East Indies.
Aber. Aberdeen.
ab ex. from without [L *ab extra*].
ab init. from the beginning [L *ab initio*].
abl. ablative.
abn airborne.
abo aboriginal.
abp., Abp. archbishop.
abr. abridged; abridgment.
abs. absent; absolute (temperature); absolutely; abstract.
abstr. abstract; abstracted.
a.c., a-c, ac, a c, A.C., A-C, AC, A C alternating current.
ABS, A.B.S. American Bureau of Shipping.
a/c, A/C account; account current.
Ac actinium; altocumulus; Acts of the Apostles.
AC, A.C. Air Controlman; Air Corps.
A.C. after Christ; Athletic Club; Armored Corps; Army Corps.
a.c. before meals [L *ante cibum*].
acad. academic; academy.
acc. acceptance; accompanied; according; account; accountant.
acc., accus. accusative.
Acc altocumulus castellatus.
ACC, A.C.C. Air Coordinating Committee.
accel. accelerando; accelerate.
acct. account; accountant.
A.C.E. Association for Childhood Education.
ack. acknowledge; acknowledgment.
A.C.L.S., ACLS American Council of Learned Societies.
A.C.P. American College of Physicians.
acpt. acceptance.
ACS, A.C.S. American Chemical Society.
A/cs Pay. accounts payable.
A/cs Rec. accounts receivable.
act. actinium; active.
A.C.T. Australian Capital Territory.
actg. acting.
ACTH adrenocorticotropic hormone.
A.C.W.A. Amalgamated Clothing Workers of America.
ad. adapted; adaptor; add; advertisement.
a.d. after date; before the day [L *ante diem*].
AD air depot.
A.D. active duty; year of our Lord [L *anno domini*].
A.D.A. American Dental Association; Americans for Democratic Action **(ADA)**.
A.D.C., a.d.c., ADC aide-de-camp.
add. addenda; addendum; addition; additional; address.
ad fin. at the end, to one end [L *ad finem*].
ad h. l. to this place, on this passage [L *ad hunc locum*].
ad inf. to infinity [L *ad infinitum*].
ad init. to, or at, the beginning [L *ad initium*].
ad int. in the meantime [L *ad interim*].
adj. adjacent; adjective; adjourned; adjunct; adjustment.
Adj., Adjt. Adjutant.
ad lib., ad libit. at pleasure; to the amount desired [L *ad libitum*].
ad loc. at the place [L *ad locum*].
adm. administrative; administrator; admitted.
Adm. Admiral; Admiralty.
admrx., admx. administratrix.
ads. advertisements.
ads, A.D.S. autograph document signed.
A.D.S. American Dialect Society.
ad us. according to custom [L *ad usum*].
adv. against [L *adversus*]; advance; adverb; adverbial; advertisement; advise; advocate; in proportion to the value [L *ad valorem*].
ad val. according to value [L *ad valorem*].
advt. advertisement.
ae., aet., aetat., of age [L *aetatis*].
AE, A.E. Agricultural Engineer; Aviation Electrician's Mate.
A.E.A. Actors' Equity Association.
A.E. and P. Ambassador Extraordinary and Plenipotentiary.
AEC, A.E.C. American Engineering Council; Atomic Energy Commission.
AEF, A.E.F. American Expeditionary Force.
aeron. aeronautics.
a.f., A.F., AF, a-f audio frequency.
Af. Africa; African.
AF, A.F., A.-Fr. Anglo-French.
AF, A.F. Air Force.
A.F.A.M., A.F.&A.M. Ancient Free and Accepted Masons.
AFB Air Force Base.
Afg., Afgh. Afghanistan.
AFL, A.F.L., A.F. of L. American Federation of Labor.
AFL-CIO, A.F.L.-C.I.O. American Federation of Labor—Congress of Industrial Organizations.
A.F.M. American Federation of Musicians.
Afr. Africa; African.
A.F.T.R.A. American Federation of Television and Radio Artists.
A.F.T. American Federation of Teachers.
Ag. August.
Ag silver [L *argentum*].
A.G. Adjutant General **(AG)**; Attorney General; joint-stock company [G *Aktiengesellschaft*].
AG, A.G. Aerographer's Mate.
AGA, A.G.A. American Gas Association.
agcy. agency.
AGF Army Ground Forces.
A.G.M.A. American Guild of Musical Artists.
agr., agri., agric. agricultural; agriculture; agriculturist.
agt., Agt. agent; agreement.
A.G.V.A. American Guild of Variety Artists.
a.h. ampere-hour.
A.H. in the year of the Hegira [L *anno Hegirae*].
AHQ, A.HQ. Army Headquarters.
A.I. in the year of the discovery [L *anno inventionis*].
AI aircraft interception.
A.I.A. American Institute of Architects.
A.I.C. American Institute of Chemists.
A.I.Ch.E. American Institute of Chemical Engineers.
A.I.E.E. American Institute of Electrical Engineers.
A Int Air Intelligence.
a.k.a. also known as.
A.K.C. American Kennel Club.
al. other things (persons) [L *alia, alii*].
AL, A.L. Aviation Electronicsman.
Al aluminum.
Ala. Alabama.
ALA, A.L.A. American Library Association; Automobile Legal Association.
Alas. Alaska.
Alb. Albania; Albanian; Albany; Alberta.
Alba. Alberta (Canada).
Ald., Aldm. Alderman.
Alex. Alexander.
Alf. Alfonso; Alfred **(Alfr.)**.
alg. algebra.
Alg. Algeria; Algerian.
A.L.P. American Labor Party.
A.L.R. American Law Reports.
als, A.L.S. autograph letter signed.
alt. alteration; alternate; alternating; alternations; alternative; altitude; alto.
Alta. Alberta (Canada).
alter. alteration.
alum. aluminum.
a.m., A.M. before noon [L *ante meridiem*].
Am. America.
Am americium.
AM, am. amplitude.
am. ammeter.
AM, A.M., a-m, a.m. amplitude modulation.
AM Aviation Structural Mechanic.
Am. alabamine; America; American.
A.M. in the year of the world [L *anno mundi*]; Master of Arts [L *Artium Magister*]; associate member; the wonderful year (1666) [L *annus mirabilis*].
A.M.A. Agricultural Marketing Administration **(AMA)**; American Management Association; American Medical Association; American Missionary Association; American Municipal Association; Automobile Manufacturer's Association.
Amb. Ambassador.
A.M.D.G., AMDG to the greater glory of God [L *ad majorem Dei gloriam*].
A.M.E. African Methodist Episcopal.
Amer. America; American.
AMG, A.M.G. Allied Military Government (of Occupied Territory).
Am. Ind. American Indian.
Amo Amos.
amp. ampere; amperage.
amph. amphibian; amphibious; amphoric.
amp—hr. ampere-hour.
AMS, A.M.S. Agricultural Marketing Service; Army Medical Staff.
Am. Sp. American Spanish.
amt. amount.
AMVETS American Veterans (of World War II and the Korean War).
AN Airman.
AN, A.N. A.-N. Anglo-Norman.
an. anonymous; before [L *ante*].
anal. analogous; analogy; analysis; analtyic(al).
anat. anatomical; anatomy.
anc. ancient; anciently.
ANC, A.N.C. Army Nurses Corps.
and. moderately slow [It. *andante*].
And. Andorra.
Ang. Anglesey; Anglican; Angola.
Angl. Anglian; Anglican; Anglicized.
Anglo-Ind. Anglo-Indian.
Anglo-Ir. Anglo-Irish.
Anglo-L. Anglo-Latin.
anim. animated [It. *animato*].
ann. annals; annual; annuities; annuity; years [L *anni*].
anon., Anon. anonymous.
ans. answer; answered.
ant. antenna; antiquarian; antiquity; antonym.
Ant. Antarctic; Antarctica; Anthony **(Anth.)**; Antigua; Antrim.
anthrop., anthropol. anthropological; anthropology.
antilog. antilogarithm.
antiq. antiquarian; antiquities.
A.N.Z.A.C., ANZAC Australian and New Zealand Army Corps.
ANZUS Australia, New Zealand, United States.
a/o, A/O account of.
AO, A.O. Aviation Ordnanceman.
A.O.H. Ancient Order of Hibernians.
AOL, A.O.L., a.o.l. absent over leave.
aor. aorist.
Ap. apostle; April.
ap. apothecary.
AP, A.P. a.p. antipersonnel; armor-piercing.
A.P., AP airplane; Associated Press.
aph. aphetic.
Apl. April.
apmt. appointment.
APO, A.P.O. Army Post Office.
Apoc. Apocalypse; Apocrypha; Apocryphal.
app. apparatus; apparent; apparently; appended; appendix; appoint; apprentice.
appar. apparatus; apparent; apparently.
approx. approximate; approximately.
Apr., Apr April.
APS, A.P.S. Army Postal Service.
apt(s). apartment(s).
aq., Aq. aqueous; water [L *aqua*].
AQ, A.Q. achievement quotient.
AR, A.R. Airman Recruit.
a.r. in the year of the reign [L *anno regni*].
Ar. argon; silver [L *argentum*].
ar. argent; aromatic; arrival, arrive.
Ar., Arab. Arabia; Arabian; Arabic.
A.R.A. Agricultural Research Administration **(ARA)**; Air Reserve Association; American Railway Association.
Aram. Aramaic.
A.R.C. American Red Cross **(ARC)**; Automobile Racing Club.
arch. archaic; archaism; archery; archipelago; architect; architectural; architecture.
Arch., Archbp. Archbishop.
archaeol., archeol. archaeology; archeology.
Archd. Archdeacon; Archduke.
archit. architecture.
archt. architect.
A.R.C.S. Associate of the Royal College of Science (Surgeons).
arg. argent; silver [L *argentum*].
Arg. Argentina; Argyll.
arith. arithmetic; arithmetical.
Ariz. Arizona.
Ark. Arkansas.
Arm. Armagh; Armenia; Armenian; Armoric.
Ar.M. Master of Architecture [L *Architecturae Magister*].
ARP, A.R.P. air-raid precautions.
arr. arrange; arranged; arrangements; arrival; arrive; arrived.
ARS, A.R.S. Agricultural Research Service.
art. article; artificial; artist.
Art., Arth. Arthur.
A.R.V. American Revised (Standard) Version (of the Bible).
AS anti-submarine.
AS, AS., A.S. Anglo-Saxon.
as., asym. asymmetric.
As altostratus; arsenic.
ASA, A.S.A. American Standards Association.
asb. asbestos.
ASCAP, A.S.C.A.P. American Society of Composers, Authors and Publishers.
ASCE, A.S.C.E. Am. Soc. C. E. American Society of Civil Engineers.
ASF, A.S.F. Army Service Forces.
asgd. assigned.
ASME, A.S.M.E., Am. Soc. M. E. American Society of Mechanical Engineers.
ASN, A.S.N. Army Service Number.
A.S.P.C.A. American Society for the Prevention of Cruelty to Animals.
ass. assistant, association.
Ass., Assyr. Assyrian.
assd. assigned.
assn. association.
assoc. associate; association.
A.S.S.R. Autonomous Soviet Socialist Republic.
asst., Asst. assistant.
Assyr. Assyria; Assyrian.
ASTM, A.S.T.M. American Society for Testing Materials; American Society of Tropical Medicine.
ASTP, A.S.T.P. Army Specialized Training Program.
astr., astron. astronomer; astronomical; astronomy.
astrol. astrologer; astrological; astrology.
AT, A.T. anti-tank; Aviation Electronics Technician.
at. atmosphere; atomic.
At astatine.
ATC, A.T.C. Air Transport Command.
athl. athlete; athletic; athletics.
Atl. Atlantic.
atm. atmosphere; atmospheric.
at. no. atomic number.
A.T.S. American Temperance Society; American Tract Society.
ATS, A.T.S. Army Transport Service; Auxiliary Territorial Service.
att., attn., atten. attention.
att., atty. attorney.
attrib. attribute; attributive; attributively.
Atty. Gen. Attorney General.
at. wt. atomic weight.
a.u., A.u., A.U. angstrom unit; astronomical unit.
Au gold [L *aurum*].
A.U.C. from (the year of) the building of the city (of Rome) [L *ab urbe condita*].
aud. audible; audit; auditor.
Aug. August **(Aug)**; Augustan; Augustus.
AUS, A.U.S. Army of the United States.
Aus., Aust. Austria.
Aus., Austl. Australia; Australian.
auth. authentic; author; authority; authorized.
Auth. Ver. Authorized Version (of the Bible).
auto. automatic; automotive.

aux., auxil. auxiliary.
A.V. Artillery Volunteers; Authorized Version (of the Bible).
a.v., a/v, A/V according to value [L *ad valorem*].
Av. avenue.
av., avdp. avoirdupois.
av., avg. average.
AVC, A.V.C. American Veterans Committee.
AVC, a.v.c. automatic volume control.
ave. avenue.
avg. average.
avn. aviation.
avoir. avoirdupois.
A/W actual weight; all water.
AWL, A.W.L., a.w.l. absent with leave.
AWOL, A.W.O.L., a.w.o.l., awol absent without leave.
AWS, A.W.S. American War Standards.
AWVS, A.W.V.S. American Women's Volunteer Services.
ax. axis; axiom.
AYH American Youth Hostels.
Ayr. Ayrshire.
az. azimuth; azure.

B

b., b base; base hit; baseman.
b., B. bachelor; balboa; base; bass; basso; bat; battery; bay; belga; bench; bicuspid; bolivar; boliviano; book; born; brass; breadth; brother.
B. bacillus; Baumé; Bible; Boston; British; Brotherhood.
B bishop (chess); boron.
B/- bag; bale.
B- bomber.
Ba barium.
B.A. Bachelor of Arts [L *Baccalaureus Artium*]; British Academy; British Association (for the Advancement of Science); Buenos Aires.
Bab. Babylonia.
bach. bachelor.
bact., bacteriol. bacteriological; bacteriologist; bacteriology.
B.A.E. Bachelor of Aeronautical Engineering; Bachelor of Arts in Education; Bureau of Agricultural Economics **(BAE)**.
B. Ag., B. Agr. Bachelor of Agriculture.
B. Ag. Sc. Bachelor of Agricultural Science.
BAL British anti-lewisite.
Bal. Baluchistan.
bal. balance; balancing.
Balt. Baltic.
B and B brandy and benedictine.
Bap., Bapt. Baptist.
bapt. baptized.
bar. barometer; barometric; barrel; barrister.
BAR, B.A.R. Browning automatic rifle.
Bar. Baruch.
B. Ar., B. Arch. Bachelor of Architecture.
Barb. Barbados; Barbara.
barit. baritone.
Bart. Baronet.
B.A.S. Bachelor of Agricultural Science; Bachelor of Applied Science.
bat., batt. battalion; battery.
Bav. Bavaria; Bavarian.
b.b., bb base(s) on balls.
B.B.A., B.Bus. Ad. Bachelor of Business Administration.
B.B.C., BBC British Broadcasting Company.
bbl, bbl. barrel; barrels.
B.C., b.c. bass clarinet; battery commander; bicycle club.
B.C. Bachelor of Chemistry; Bachelor of Commerce; before Christ; British Columbia.
B.C.E. Bachelor of Chemical Engineering; Bachelor of Civil Engineering.
BCF, B.C.F. British Caribbean Federation.
BCG bacillus Calmette-Guérin (TB vaccine).
B.C.L. Bachelor of Civil Law.
B.C.P. Book of Common Prayer.
B.C.S. Bachelor of Chemical Science.
bd. board; bond; bound; bundle.
B.D. Bachelor of Divinity.
b.d., B/D bank draft; bills discounted; brought down.
bd. ft. board feet.
bdg. binding.
B.D.S. Bachelor of Dental Surgery.
BDSA Business and Defense Services Administration.
Be beryllium.
Bé Baumé (scale).
B.E. Bachelor of Education; Bachelor of Engineering; Bank of England; Board of Education.
B.E., B/E, b/e bills of exchange.
B.E.A. British East Africa.
BEC Bureau of Employees' Compensation.
Beds, Beds. Bedfordshire.
B.E.E. Bachelor of Electrical Engineering.
bef. before.
BEF., B.E.F. British Expeditionary Force(s).
Bel., Belg. Belgian; Belgium.
B.E.M. British Empire Medal.
Ben., Benj. Benjamin.
Bened. Benedict.
Beng. Bengali.
Berks, Berks. Berkshire.
Bern. Bernard.
Berw. Berwick.
B. ès L. Bachelor of Letters [F *Bachelier ès Lettres*].
B. ès S. Bachelor of Sciences [F *Bachelier ès Sciences*].
bet., betw. between.
bev, BEV billion electron volts.
BEW, B.E.W. Board of Economic Warfare.
b.f., bf. bold face.
b.f., B/F brought forward.
B.F. Bachelor of Finance; Bachelor of Forestry.
B.F.A. Bachelor of Fine Arts.
BFDC Bureau of Foreign and Domestic Commerce.
bg. bag(s).
bhp brake horsepower.
Bi bismuth.
B.I. British India.
Bib. Bible; Biblical.
bibl., Bibl. biblical; bibliographical.
bibliog. bibliography.
bicarb. bicarbonate of soda.
b.i.d. twice a day [L *bis in die*].
biochem. biochemistry.
biog. biographer; biographical; biography.
biol. biological; biologist; biology.
BIS, B.I.S. Bank for International Settlements; British Information Services.
B.J. Bachelor of Journalism.
Bk berkelium.
bk. bank; block; book.
bkg. banking.
bkkpg. bookkeeping.
bklr. black letter.
bkpt. bankrupt.
bks., Bks. barracks; books.
b.l., B/L bill of lading; breechloading.
bl. bale; barrel; black; blue.
B.L., B.LL. Bachelor of Laws.
B.L.A. Bachelor of Liberal Arts.
bld bold face.
bldg., blg. building.
B.L.E. Brotherhood of Locomotive Engineers.
B.L.F.E. Brotherhood of Locomotive Firemen and Enginemen.
B. Lit., B. Litt. Bachelor of Letters (Literature) [L *Baccalaureus Litterarum*].
blk. black; block.
bln., bln balloon.
bls. bales; barrels.
B.L.S. Bachelor of Library Science; Bureau of Labor Statistics **(BLS)**.
blvd. boulevard.
b.m. board measure.
BM Boatswain's Mate.
B.M. Bachelor of Medicine [L *Baccalaureus Medicinae*]; Bachelor of Music [L *Baccalaureus Musicae*]; Bureau of Mines; benchmark **(BM)**.
B.M.E. Bachelor of Mechanical Engineering; Bachelor of Mining Engineering.
B. Mech. E. Bachelor of Mechanical Engineering.
BMI Broadcast Music, Inc.
BMR basal metabolic rate.
B. Mus. Bachelor of Music.
Bn., bn. baron; battalion.
B.N. bank note.
B.N.A. Basle Anatomical Nomenclature [L *Basle Nomina Anatomica*]; British North America.
b.o. back order; bad order; box office; branch office; broker's order; buyer's option.
B.O., BO Board of Ordnance.
B/O brought over.
Boh., Bohem. Bohemia; Bohemian.
Bol. Bolivia.
bor. borough.
bot. botanical; botanist; botany; bottle.
B.O.T. Board of Trade.
boul. boulevard.
bp, b.p. below proof; boiling point.
b.p. B/P bill of parcels; bills payable.
B.P. Bachelor of Pharmacy.
B.P., B.Ph., B.Phil. Bachelor of Philosophy [L *Baccalaureus Philosophiae*].
bp. birthplace; bishop.
B.Pd., B.Pe. Bachelor of Pedagogy.
B.P.D.P.A. Brotherhood of Painters, Decorators and Paperhangers of America.
B.P.E. Bachelor of Physical Education.
B.P.H. Bachelor of Public Health.
BPI, B.P.I. Bureau of Public Inquiries.
B.P.O.E. Benevolent and Protective Order of Elks.
BPPC Brussels Pact Permanent Commission.
br. branch; brand; brief; bridge; brig; bronze; brother.
Br. Breton, Britain, British.
Br bromine.
B.R., B/R Bill of Rights.
b.r., B/R bills receivable.
Braz. Brazil; Brazilian.
Brazil. Brazilian.
B.R.C.A. Brotherhood of Railway Carmen of America.
B.R.C.S. British Red Cross Society.
B.Rec., b.rec. bills receivable.
Brec. Brecknockshire.
brev. brevet; brevetted.
Br. Gu. British Guiana.
Br. Hond. British Honduras.
Br. I. British India.
Brig. Brigade; Brigadier.
Brig. Gen. Brigadier General.
Brit. Britain; Britannia; Britannica; British.
bro. brother.
bros. brothers.
B.R.T. Brotherhood of Railroad Trainmen.
b.s. balance sheet.
b.s., B/S bill of sale.
B.S. Bachelor of Surgery.
B/S bags; bales.
B.S., B.Sc. Bachelor of Science [L *Baccalaureus Scientiae*].
B.S. Bachelor of Surgery.
B.S.A. Bachelor of Scientific Agriculture; Bibliographical Society of America; Boy Scouts of America; British South Africa.
B.S.C.P. Brotherhood of Sleeping Car Porters.
B.S.Ed. Bachelor of Science in Education.
B.S.S., B.S.Sc. B.S. in S.S. Bachelor of (Science in) Social Sciences.
Bt. Baronet.
BT Boilerman.
B.T., B.Th. Bachelor of Theology.
B.T.U., BTU, B.th.u., Btu British thermal unit.
Btry Battery.
BU Builder.
bu. bureau.
bu., bu bushel; bushels.
buck. buckram.
Bucks, Bucks. Buckinghamshire.
bul., bull. bulletin.
Bulg. Bulgaria; Bulgarian.
B.V. Blessed Virgin [L *Beata Virgo*]; farewell [L *bene vale*].
B.V.M. Blessed Virgin Mary [L *Beata Virgo Maria*].
bvt. brevet; brevetted.
B.W.I. British West Indies.
bx. box.
Bz. benzene.

C

c carat; constant; coulomb.
c. about [L *circa, circiter, circum*]; calends; candle; capacity; cape; carbon; carton; case; catcher; cathode; cent; canter; centime; cents; century; chancery; chapter; chief; child; church; cirrus; city; cloudy; companion; consul; copper; copy; copyright; cost; cubic; current; gallon [L *congius*]; hundredweight.
c., c, C., C centigrade; centimeter.
C. Catholic; Celsius; Celtic; Chancellor; Congress; Conservative; Court.
C carbon; one hundred.
c.a., C.A. chartered accountant; chartered agent; chief accountant; claim agent; commercial agent; consular agent; controller of accounts; county alderman.
ca. about [L *circa*]; candle; cathode; centiare.
ca., ca centare(s).
Ca calcium.
C.A. Catholic Action; Central America; Coast Artillery; Confederate Army.
CA, C.A. chronological age; Corps Area.
C/A commercial account; credit account; current account.
CAA, C.A.A. Civil Aeronautics Administration (Authority).
CAB, C.A.B. Civil Aeronautics Board; Consumers Advisory Board; Cooperative Analysis of Broadcasting.
CAC, C.A.C. Coast Artillery Corps; Combat Aircrew **(C.A.)**.
Caern. Caernarvonshire.
C.A.F., c.a.f. cost and freight; cost, assurance, and freight.
Caith. Caithness.
cal. calendar; calends; caliber; calomel; small calorie **(cal)**.
cal calorie (small).
Cal., Calv. Calvin.
Cal., Cal caliber; large calorie.
Calif., Cal. California.
Cam., Camb. Cambridge.
Cam camouflage.
Cambs. Cambridgeshire.
can. canon; canto.
Can. Canada; Canadian.
Canad. Canadian.
canc. cancel; cancellation; canceled.
cant. canton; cantonment.
Cant. Canterbury; Canticles; Cantonese.
Cantab. Cambridge [L *Cantabrigia*].
cap. capital; capitalize; capitulum; chapter [L *caput*].
CAP, C.A.P. Civil Air Patrol.
caps capital letters.
Capt. Captain.
cap'y capacity **(cap.)**.
car. carat.
Car. Carlow; Charles [L *Carolus*].
CAR, C.A.R. Civil Air Regulations.
Card. Cardiganshire **(Cards.)**; Cardinal.
CARE Cooperative for American Remittances Everywhere.
Carm., Carmarths. Carmarthenshire.
carp. carpentry.
cat. catalog; catechism.
cath. cathedral.
Cath. Catherine; Catholic.
cav. cavalier; cavalry.
CAVU, C.A.V.U. ceiling and visibility unlimited.
c.b. center of buoyancy; confined to barracks.
Cb columbium; cumulonimbus.
CB Construction Battalion.
C.B. Bachelor of Surgery [L *Baccalaureus Chirurgiae*]; Companion of the Bath.
CBC, C.B.C. Canadian Broadcasting Company.
C.B.D., c.b.d. cash before delivery.
CBI China, Burma, India.
CBS, C.B.S. Columbia Broadcasting System.
c.c., C.C. carbon copy; cash credit; cashier's check; chief clerk; circuit court; city council; city councilor; civil court; common councilman; company clerk; company commander; consular clerk; contra credit; county clerk; county commissioner; county council; county court; current account [F *compte courant*].
cc. chapters.
cc, cc., c.c. cubic centimeter.
Cc cirrocumulus.
C.C. Cape Colony.
CC cyanogen chloride (poison gas).
C.C.A. Chief Clerk of the Admiralty; Circuit Court of Appeals; Commission for Conventional Armaments **(CCA)**.
CCC, C.C.C. Civilian Conservation Corps; Commodity Credit Corporation.
CCF, C.C.F. Cooperative Commonwealth Federation (of Canada).
CCS, C.C.S. Combined Chiefs of Staff.
C.D. Civilian Defense.
c.d. cash discount.
cd., cd cord; cords.
Cd. cadmium.
cd. ft. cord foot (feet).
CDR, Cdr. Commander.
CE Council of Europe; Construction Electrician's Mate; Corps of Engineers.
C.E. Chemical Engineer; Chief Engineer; Christian Endeavor; Church of England; Civil Engineer; Common Era.
Ce cerium.
CEA Council of Economic Advisers.
CED Committee for Economic Development.
C.E.F. Canadian Expeditionary Force.
Cel., Cels. Celsius.
Celt. Celtic.
cen., cent. central; century.
cent. centered; centigrade; centimeter; one hundred [L *centum*].
CERN European Council for Nuclear Research [F *Centre Européen des Recherches Nucléaires*].
cert. certificate; certify; certiorari.
certif. certificate; certificated.
cet. par. other things being equal [L *ceteris paribus*].
Cey. Ceylon.
c.f., cf center field(er).
c.f., C.F. cost and freight.
cf. compare [L *confer*].
Cf californium.
c/f, C/F carry forward.
c.f.i., C.F.I. cost, freight, and insurance.
cfm, c.f.m. cubic feet per minute.
CFR, C.F.R. Code of Federal Regulations.
cfs, c.f.s. cubic feet per second.
c.g., C.G. captain of the guard; center of gravity; consul general.
cg., cg, cgm. centigram(s).
C.G. Coast Guard; Commanding General **(CG)**.
C.G.H. Cape of Good Hope.
cgs, CGS, c.g.s. centimeter-gram-second.
C.G.T. General Confederation of Labor [F *Confédération Générale du Travail*], [Sp. *Confederación General de Trabajadres*].
c.h., C.H. clearing-house; courthouse; customhouse.
C.H. Companion of Honor.
Ch. Chaldean; Chaldee; China; Chinese.
Ch Chronicles.
ch. chain; champion; chargé d'affaires; check (chess); chestnut; chervonets; chief; child; children; chirurgeon; church; of surgery [L *chirurgiae*].
ch., chap. chaplain; chapter.
c.-h., c-hr. candle-hour.
Ch., Chas. Charles.
Chal., Chald. Chaldaic; Chaldean; Chaldee.
Chanc. Chancelor; Chancery.
char. character; charter.
Ch.B. Bachelor of Surgery [L *Chirurgiae Baccalaureus*].
Ch. Clk. Chief Clerk.
Ch.E., Che.E. Chemical Engineer.
chem. chemical; chemist; chemistry.
Ches., Chesh. Cheshire.
chg. charge.
chgd. charged.
Chi. Chicago.
Chin. Chinese.
Ch.J. Chief Justice.
Ch.M. Master of Surgery [L *Chirurgiae Magister*].
chm. checkmate.
chm., chmn. chairman.
Chr. Christ; Christian.
Chr., Chris. Christopher.
chron., chronol. chronological; chronology.
Chron. Chronicles.
chs. chapters.
Ci cirrus.
C.I. Channel Islands.
CIA Central Intelligence Agency.
Cia., cia. company [Sp. *Compania*].
CIAA Coordinator of Inter-American Affairs.
CIC, C.I.C. Counter-Intelligence Corps.
C.I.D. Criminal Investigation Department (Brit.).
C.I.E. Companion (of the Order) of the Indian Empire.
Cie., cie. company [F *compagnie*].
c.i.f., C.I.F. cost, insurance, and freight.
C. in C., C in C, CINC, Cinc Commander in Chief.
CIO, C.I.O. Congress of Industrial Organizations.
cir., circ. about [L *circa, circiter, circum*]; circular; circulation; circumference.
cit. citation; cited; citizen.
civ. civil; civilian.
C.J. body of law [L *corpus juris*]; Chief Judge; Chief Justice.
ck. cask; check; cook.
c.l. carload; carload lots; center line; civil law; craft loss (insurance).
cl. claim; class; classification; clause; clearance; clergyman;

clerk; cloth.
cl., cl centiliter.
Cl chlorine.
Cla. Clackmannan.
clar. clarinet.
Clar. Clarence; clarendon (type).
class. classic; classical; classification; classified; classify.
Clem. Clement.
cler. clerical.
climatol. climatological; climatology.
clin. clinic; clinical.
clk. clerk; clock.
clm. column.
C.L.U. Chartered Life Underwriter.
CM Mechanic.
c.m. church missionary; circular mil; common meter; corresponding member; court-martial.
C.M. Master of Surgery [L *Chirurgiae Magister*].
cm., cm centimeter(s).
Cm cumulonimbus mammatus; curium.
Cmdr. Commander.
C.M.G. Companion (of the Order) of St. Michael and St. George.
cml. commercial.
CMTC, C.M.T.C. Citizens' Military Training Camp.
c.n., C/N circular note; credit note.
CN Constructionman.
Cn cumulonimbus.
CNS, C.N.S. central nervous system.
C.O. commanding officer; conscientious objector.
co., Co. company; county.
c/o, C/O, c.o. care of; carried over; cash order.
Co cobalt; Corinthians.
coad. coadjutor.
coch., cochl. a spoon; spoonful [L *cochleare*].
c.o.d., C.O.D. cash on delivery; collect on delivery.
Cod., cod. codex.
Codd., codd. codices.
coef. coefficient.
C. of C. Chamber of Commerce.
C. of S. Chief of Staff.
cog., cogn. cognate.
col. collected; collector; college; colonial; colony; color; colored; column.
Col. Colombia; Colonel; Colorado; Columbia.
Col., Col, Coloss. Colossians.
coll. colleague; collect; collection; collective; collector; college; collegiate; colloquial.
collab. collaborated; collaboration; collaborator.
collat. collateral.
colloq. colloquial; colloquialism; colloquially.
Colo., Col. Colorado.
colog cologarithm.
com. comedy; comic; comma; commentary; commerce; commercial; common; commonly; commune; communication; community.
Com. Commission; Commissioner; Committee; Commodore; Communist.
Com., Comdr. Commander.
Com., Como., COMO, Commodore.
comb. combination.
comdg. commanding.
Comdt. Commandant.
Com. in Ch., Cominch Commander in Chief.
coml. commercial.
comm. commander; commentary; commerce; commercial; commissary; commission; committee; commonwealth; commutator.
comp. companion; comparative; compare; comparison; compilation; compiled; compiler; complete; compose; composer; composition; compositor; compound; compounded; comprising.
compar. comparative.
compt. compartment; comptometer.
Comr. Commissioner.
Com. Ver. Common Version (of the Bible).
con. against [L *contra*]; concerto; conclusion; condense; conduct; connection; consols; consolidate; continued; wife [L *conjunx*].
Con. Conformist; Consul.
conc. concentrate; concentrated; concentration; concerning.
conch. conchology.
cond. condition; conducted; conductivity; conductor.
conf. compare [L *confer*]; confection (medicine); conference; confessor.
Confed. Confederate; Confederation.
cong. gallon [L *congius*].
Cong. Congregational; Congress; Congressional.
conj. conjugation; conjunction; conjunctive.
Conn. Connecticut.
cons. consecrated; conserve; consigned; consignment; consolidated; consonant; construction; consulting.
cons., Cons. constable; constitution; constitutional; consul.
consol. consolidated.
const., Const. constable; constant; constitution.
constr. construction; construed.
cont. containing; contents; continent; continue; continued; contract; contraction; contrary; control.
Cont. Continental.
contd. continued.
contemp. contemporary.
contg. containing.
contin. continued; let it be continued [L *continuetur*].
contr. contract; contraction; contralto; control.
contrib. contribution; contributor.
coop., co-op. cooperative.
cop. copper; copyright; copyrighted.
Cop. Copernican; Coptic.
cor. corner; cornet; coroner; corpus; correct; corrected; correction; correlative; correspondence; correspondent; corresponding; corrupt.
Cor. Corinthians.
Corn. Cornish; Cornwall.
corol., coroll. corollary.
Corp. Corporal.
corp., corpn. corporation.
corr. correct; corrected; correspond; correspondence; correspondent; corresponding; corrupt; corrupted; corruption.
correl. correlative.
corresp. correspondence.
c.o.s., C.O.S. cash on shipment.
cos. companies; counties.
cos cosine.
cosec cosecant.
cosh hyperbolic cosine.
cot cotangent.
coth hyperbolic cotangent.
covers coversed sine.
c.p., C.P. chemically pure; common pleas; court of probate.
CP command post; Construction Apprentice.
cp. compare.
cp, c.p. candlepower.
Cp. Christoph.
C.P. Canadian Press; Cape Province; center of pressure; Chief Patriarch; Common Prayer; Communist Party.
C.P.A. Certified Public Accountant.
cpd. compound.
C.P.H. Certificate in Public Health.
Cpl. corporal.
cpm, c.p.m. cycles per minute.
CPO, C.P.O., c.p.o. Chief Petty Officer.
C.P.S. keeper of the privy seal [L *custos privati sigilli*].
cps coupons; cycles per second **(c.p.s.)**.
cpt. counterpoint.
CPTP, C.P.T.P. Civil Pilot Training Program.
CQ Change of Quarters.
cr. created; credit; creditor; crescendo; creek; crown(s).
Cr. Cranch's U.S. Supreme Court Reports.
Cr chromium.
CR Construction Recruit.
C.R. Costa Rica.
craniol. craniological; craniology.
craniom. craniometry.
cres., cresc. crescendo.
crim.con. criminal conversation.
crit. critic; critical; criticism; criticized.
CRM counter-radar measures.
crs. credits; creditors.
cryst. crystalline; crystallized; crystallography; crystals.
Cs. conscious.
Cs cesium; cirrostratus.
CS Commissaryman.
C.S. Christian Science; Christian Scientist; Confederate States.
C.S., c.s. capital stock; civil service.
C/S, cs. case; cases.
C.S.A. Civilian Supply Administration **(CSA)**; Confederate States of America.
csc cosecant.
CSC Civil Service Commission.
C.S.C. Conspicuous Service Cross.
csch hyperbolic cosecant.
csk. cask; countersink.
CSO, C.S.O. Chief Signal Officer **(CSigO)**; Chief Staff Officer.
CSS Commodity Stabilization Service.
C.S.T., CST, c.s.t. central standard time.
CT Communications Technician.
C.T., CT, c.t. central time.
ct. cent(s); certificate; county; court; one hundred [L *centum*].
Ct. Connecticut; Count.
CTC, C.T.C. Citizens' Training Camp.
ctg. cartridge.
ctn cotangent.
ctnh hyperbolic cotangent.
ctr. center.
cts. centimes; cents; certificates.
cu., cu cubic.
Cu copper [L *cuprum*]; cumulus.
cu. cm., cu cm cubic centimeter(s).
cu.ft. cubic foot; cubic feet.
cu.in. cubic inch(es).
Cumb, Cumb. Cumberland.
cur. currency; current.
cu.yd. cubic yard(s).
C.V. Common Version (of the Bible).
cv., cvt. convertible.
cw, CW continuous wave.
CWA, C.W.A. Civil Works Administration.
C.W.O., c.w.o. cash with order.
CWO, C.W.O. Chief Warrant Officer.
CWS Chemical Warfare Service.
cwt. hundredweight.
cy. capacity; currency; cycles.
cyc. cyclopedia; cyclopedic.
cyl cylinder.
CYO, C.Y.O. Catholic Youth Organization.
C.Z. Canal Zone.

D

d. dam (genealogy); date; daughter; day; days; dead; decree; degree; delete; democrat; democratic; density; deputy; deserter; dextro- (chemistry); diameter; died; dime; dinar; director; dividend; dollar; door; dorsal; dose; dowager; drachma; dyne; give [L *da*].
d penny; pence [L *denarius, denarii*].
D. December; Department; Doctor; Don; Duchess; Duke; Dutch; God [L *Deus*]; Lord [L *Dominus*].
D deuterium; five hundred.
da. daughter; day; days.
D.A. District Attorney; Delayed Action.
DA Dental Apprentice.
D.A.B., DAB Dictionary of American Biography.
D.A.E., DAE Dictionary of American English.
dal., dal decaliter.
Dak. Dakota.
Dall. Dallas's U.S. Supreme Court Reports.
Dan. Daniel **(Dan, Danl.)**; Danish; Danzig.
D.A.R., DAR Daughters of the American Revolution.
dat. dative.
dau. daughter.
DAV, D.A.V. Disabled American Veterans.
Dav. David.
D.B. Domesday Book.
d.b. daybook.
db, db. decibel; decibels.
d.b.a. doing business as.
D.B.E. Dame (Commander, Order) of the British Empire.
d.b.h. diameter at breast height.
D. Bib. Douai Bible.
dbl. double.
DC Damage Controlman; Dental Corps; Disarmament Commission.
d.c., d-c, dc, d c, D.C., D-C, DC, D C direct current.
D.C. District of Columbia; Doctor of Chiropractic; from the beginning [It. *da capo*].
D.C.L. Doctor of Canon Law; Doctor of Civil Law.
D.C.M. Distinguished Conduct Medal (Brit.).
D.C.S. Deputy Clerk of Sessions; Doctor of Christian Science; Doctor of Commercial Science.
d.d., D/D demand draft.
d.d., D/d, D/D days after date; days' date.
dd., d/d. delivered.
DD Department of Defense.
D.D. Doctor of Divinity [L *Divinitatis Doctor*].
D.D.S. Doctor of Dental Science (or Surgery).
D.E. Destroyer Escort **(DE)**; Doctor of Engineering; Doctor of Entomology.
deb., deben. debenture.
Deb. Deborah.
dec. deceased; declaration; declension; declination; decrease; decrescendo.
dec., decim. decimeter.
Dec., Dec December.
decd. deceased.
decl. declension.
decoct. decoction.
decresc. decrescendo.
ded., dedic. dedication.
def. defective; defendant; defense; deferred; defined; definite; definition.
deg. degree.
D.E.I. Dutch East Indies.
del. delegate; delete; deliver; he (she) drew it [L *delineavit*].
Del. Delaware.
deliq. deliquescent.
Dem. Democrat; Democratic.
demon. demonstrative.
Den. Denmark.
Denb., Denbh. Denbighshire.
denom. denomination.
dent. dental; dentist; dentistry.
dep. departs; departure; deposed; deposit; depot.
Dep. dependency.
dep., dept. department; deponent; deputy.
der., deriv. derivation; derivative; derive; derived.
Derby, Derby. Derbyshire.
dermatol. dermatological; dermatologist; dermatology.
Derry Londonderry (county in Ireland).
desc. descendant.
descr. descriptive.
D. ès L. Doctor of Letters [F *Docteur ès Lettres*].
D. ès S. Doctor of Sciences [F *Docteur ès Sciences*].
det. detach; detachment; detail; detector.
Deu, Deut. Deuteronomy.
dev. deviation.
Dev., Devon. Devonshire.
devel. development.
D.F. Dean of the Faculty; Defender of the Faith [L *Defensor Fidei*]; Federal District [Pg. *Districto Federal;* Sp. *Distrito Federal*].
DF, D/F, D.F. direction finding.
D.F.C., DFC Distinguished Flying Cross.
D.G. by the grace of God [L *Dei gratia*].
dg., dg decigram(s).
d.h. deadhead; that is to say [G *das heisst*].
di., dia., diam. diameter.
Di didymium.
diag., diagr. diagram.
dial. dialect; dialectal; dialectic; dialectical.
dict. dictation; dictator; dictionary.
diet. dietetics.
diff. difference; different; differential.
dil. dilute.
dim. dimension.
dim., dimin. diminuendo; diminutive.
din. dinar.
dioc. diocesan; diocese.
dipl. diplomat; diplomatic.
dir. director.
dis. distance; distant.
disc. discount; discover; discovered.
disch. discharged.
diss. dissertations.
dist. discount; distance; distant; distinguish; distinguished; district.
Dist. Atty. District Attorney.
distr. distribute; distributed; distribution; distributive; distributor.
div. divergence; diversion; divide; divided; dividend; divine; division; divisor; divorced.
Div. Divinity.
DK Disbursing Clerk.
dk. deck; dock.
dkg., dkg dekagram(s).
dkl., dkl dekaliter(s).
dkm., dkm dekameter(s).
dks., dks dekastere(s).
dl., dl deciliter(s).
D/L demand loan.
D.Lit., D.Litt. Doctor of Letters (Literature) [L *Doctor Lit(t)erarum*].
dlr. dealer
D.L.S. Doctor of Library Science.
dm., dm decameter(s); decimeter(s).
D.M. Deputy Master.
DM., Dm. Deutschemark.
DM Draftsman.
DMB Defense Mobilization Board.
D.M.D. Doctor of Dental Medicine [L *Dentariae Medicinae Doctor*].
D.Mus. Doctor of Music.
D.N. our Lord [L *Dominus noster*].
DN Dentalman.
DNA deoxyribonucleic acid.
DNB, D.N.B. Dictionary of National Biography (British).
do. ditto.
D.O. Doctor of Optometry; Doctor of Osteopathy.
D.O.A. dead on arrival.
doc. document.
dol. dolce; dollar; dollars **(dols.)**.
Dom. Dominica; Dominican.
dom. domestic; dominion.
Don. Donegal.
Dor. Dorian; Doric.
Dora, D.O.R.A. The Defence of the Realm Acts (England).
Dors., Dorset, Dorset. Dorsetshire.
Dow. Dowager.
doz., doz dozen(s).
DP, D.P. degree of polymerization; diametral pitch; displaced person.
D.P.H. Doctor of Public Health; Doctor of Public Hygiene **(D.P.Hy.)**.
D.Ph., D.Phil. Doctor of Philosophy.
dpt. department; deponent.
D.P.W. Department of Public Works.
D.R., D/R, d.r. dead reckoning; deposit receipt.
dr. debit; debtor; drachma; dram(s); drawer.
Dr. doctor; drive.
dram. pers. dramatis personae.
d.s. daylight saving; days after sight; document signed.
ds., ds decistere(s).
Ds dysprosium.
D.S., d.s. (repeat) from this sign [It. *dal segno*].
D.S., D.Sc. Doctor of Science [L *Doctor Scientiae*].
D.S.C. Department of Street Cleaning; Distinguished Service Cross.
D.S.M., DSM Distinguished Service Medal.
D.S.O. Distinguished Service Order; District Staff Officer.
d.s.p. died without issue [L *decessit sine prole*].
DSRD, D.S.R.D. Department of Scientific Research and Development.
DST, D.S.T. Daylight Saving Time.
D.S.T. Doctor of Sacred Theology.
DT Dental Technician.
d.t. delirium tremens **(d.t's)**; double time.
D.T., D.Th., D.Theol. Doctor of Theology.
Du. Duke; Dutch.
Dub. Dublin.
Dumb. Dumbarton.
Dumf. Dumfries.
dup., dupl. duplicate.
Dur. Durham.
D.V. Douay Version (of the Bible).
D.v., D.V. God willing [L *Deo volente*].
D.V.M. Doctor of Veterinary Medicine.
D.V.M.S. Doctor of Veterinary Medicine and Surgery.
D/W dock warrant.
D.W.S. Department of Water Supply.
d.w.t. dead weight tons.
dwt. pennyweight.
DX, D.X. distance; distant.
Dy dysprosium.
dyn., dynam. dynamics.
dz. dozen.

E

e., E., E English.
e., e error(s).
e., E., E east, eastern.
e, e. erg.
E., E English.
E. Earl; Earth.
ea. each.
EAM National Liberation Front [Gr. *Ethniko Apelefitherotiko Metopo*].
e. and o.e. errors and omissions excepted.

Eb erbium.
Eb. Ebenezer.
EbN east by north.
EbS east by south.
E.C. Engineering Corps; Established Church.
e.c. for the sake of example [L *exempli causa*].
ECA Economic Cooperation Administration.
ECAFE Economic Commission for Asia and the Far East.
eccl., eccles. ecclesiastical.
Ecc, Eccl., Eccles. Ecclesiastes.
Ecclus. Ecclesiasticus.
ECE Economic Commission for Europe.
ECLA Economic Commission for Latin America.
ECME Economic Commission for the Middle East.
ecol. ecological; ecology.
econ. economic; economics; economy.
ECOSOC Economic and Social Council (of the United Nations).
ECSC European Coal and Steel Community.
Ecua. Ecuador.
E.D. Eastern Department; ex-dividend.
ed. edited; edition; editor.
Ed., Edw., Edwd. Edward.
Ed.B. Bachelor of Education.
EDC European Defense Community.
Ed.D. Doctor of Education.
EDES National Democratic Greek Union [Gr. *Ethnikos Dimokratikos Ellenikos Syndesmos*].
Edg. Edgar.
edit. edited; edition; editor.
Edm. Edmund.
Ed.M. Master of Education.
E.D.T., e.d.t. eastern daylight time.
educ. education; educational.
E.E. Early English; Electrical Engineer; Electrical Engineering; Envoy Extraordinary.
e.e. errors excepted.
E.E. & M.P. Envoy Extraordinary & Minister Plenipotentiary.
EEDC Economic Employment and Development Commission.
EEG electroencephalogram.
EE.UU. United States [Sp. *Estados Unidos*].
eff. efficiency.
efflor. efflorescent.
e.g. for example [L *exempli gratia*].
Eg. Egypt; Egyptian; Egyptology.
Egb. Egbert.
Egyptol. Egyptology.
EHF, E.H.F., ehf, e.h.f. extremely high frequency.
EHFA, E.H.F.A. Electric Home and Farm Authority.
E.I., E.Ind. East Indian; East Indies.
EIB, E.I.B. Export-Import Bank.
EKKA National and Communal Liberation [Gr. *Ethniki kai Koinoniki Apeleftherosis*].
el., elev. elevation.
E.L. East Lothian.
ELAS National Popular Liberation Army [Gr. *Ethnikos Laikos Apeleftherotikos Stratos*].
elec., elect. electric; electrical; electrician; electricity **(electr.)**.
elem. elementary; elements.
elev. elevation.
Eliz. Elizabeth; Elizabethan.
ellipt. elliptical.
elong. elongation.
e. long. east longitude.
EM Electrician's Mate; enlisted man (men).
Em. Emily; Emma; Emmanuel.
Em., eman. emanation (chemistry).
emb., embryol. embryology.
e.m.f., emf, E.M.F., EMF electromotive force.
Emp. Empire; Emperor; Empress.
emph. emphasis; emphatic.
e.m.u., emu, E.M.U. electromagnetic units.
emul. emulsion.
EN Engineman.
enc., encl. enclosed; enclosure.
ency., encyc., encycl. encyclopedia.
ENE, E.N.E., ene, e.n.e. east-northeast.
eng. engine; engineer; engineering; engraved; engraver; engraving.
Eng. England; English.
Eng.D. Doctor of Engineering.
engin. engineering.
engr. engineer; engraved; engraver; engraving.
enl. enlarged; enlisted.
Ens. Ensign.
entom., entomol. entomological; entomology.
env. envelope.
e.o. from office [L *ex officio*].
EOKA National Organization of Cypriot Fighters [Gr. *Ethniki Organosis Kyprion Agoniston*].
e.p. en passant (chess).
EP extended play.
Ep., Epis., Epist. Epistle(s).
Eph. Ephraim.
Eph., Eph, Ephes. Ephesians.
epil. epilogue.
Epiph. Epiphany.
Epis., Episc. Episcopal.
epit. epitaph; epitome.
EPON United Panhellenic Organization of Youth [Gr. *Eniaia Panellinois Organosis Neon*].
EPU European Payments Union.
E.Q., EQ educational quotient.
eq. equal; equalizer; equation; equator; equivalent **(equiv.)**.
e.r., er earned runs.
E.R. East Riding; King Edward; [L *Eduardus Rex*]; Queen Elizabeth [L *Elizabeth Regina*].
Er erbium.
E.R.A. Educational Research Association; Emergency Relief Administration **(ERA)**.
ERP, E.R.P. European Recovery Program.
erron. erroneous; erroneously.
E.R.V. English Revised Version (Bible).
ES Specialist (emergency service).
ESA Economic Stabilization Administration.
ESC Economic and Social Council.
eschat. eschatology.
Esd. Esdras.
ESE, E.S.E., ese, e.s.e. east-southeast.
Esk. Eskimo.
esp., espec. especially.
ESP extrasensory perception.
Esq., Esqr. Esquire.
Ess. Essex.
est. established **(estab.)**; estate; estimated; estuary.
EST, E.S.T., e.s.t. eastern standard time.
Est. Estonia.
Esth. Esther **(Est)**; Esthonia.
e.s.u., esu electrostatic unit.
ET Electronics Technician.
e.t.a., ETA estimated time of arrival.
et al. and others [L *et alii*]; and elsewhere [L *et alibi*].
etc. and so forth; and others [L *et ceteri, ceterae,* or *cetera*].
etch. etched; etcher; etching.
eth. ether; ethical; ethics.
Eth. Ethiopia; Ethiopian; Ethiopic.
ethnog. ethnographical; ethnography.
ethnol. ethnological; ethnology.
ETO European Theater of Operations.
et seq. and the following; and what follows [L *et sequens, et sequentes* or *sequentia*].
ety., etym., etymol. etymological; etymology.
Eu europium.
E.U. United States [F *États-Unis*].
Eug. Eugene.
euphem. euphemism; euphemistic.
Eur. Europe; European.
E.V. English Version (of the Bible).
e.v., ev electron volt(s).
evac. evacuation.
evan., evang. evangelical; evangelist.
Evang. Evangelical.
evap. evaporation.
ex. examination; examine; examined; example; except; excepted; exception; exchange; excursion; executed; executive.
Ex., Ex, Exod. Exodus.
exam. examination; examined; examinee; examinor.
exc. excellent; except; excepted; exception; excursion.
Exc. Excellency.
exch. exchange; exchequer.
excl. exclusive.
excl., exclam. exclamation.
exec. executive; executor.
ex int. without interest.
ex lib. from the library (of) [L *ex libris*].
ex off. from office [L *ex officio*].
exp. expenses; expiration; expired; export; exportation; exported; exporter; express.
exper. experimental.
expt. experiment.
exptl. experimental.
exr. executor.
exsec exsecant.
ext. extension; external; externally; extinct; extra; extract.
Ez., Ez, Ezr. Ezra.
Eze., Eze, Ezek. Ezekiel.

F

f activity coefficient.
f. farthing; fathom; feet; female; feminine; fine; fluid (ounce); folio; following; foot; formed; forte; franc; frequency; from; let it be made [L *fiat*]; strong [L *forte*].
f., F function (of).
f., f, F., F farad.
f., f foul(s).
f, f/, f:, F, F/, F: F-number.
F. February; Fellow; France; Friday; son [L *filius*].
F, F. Fahrenheit; French.
FA Fireman Apprentice.
F.A., FA Field Artillery; Fine Arts; Food Administration.
f.a. fire alarm; freight agent.
f.a., F.A. free alongside.
F.A.A.A.S. Fellow of the American Association for the Advancement of Science.
fac. factor; factory.
fac., facsim. facsimile.
F.A.C.D. Fellow of the American College of Dentists.
F.A.C.P. Fellow of the American College of Physicians.
F.A.C.S. Fellow of the American College of Surgeons.
F.A.G.O. Fellow of the American Guild of Organists.
Fah., Fahr. Fahrenheit.
F.A.I.A. Fellow of the American Institute of Architects.
fam. familiar; family.
F.A.M., F. & A.M. Free and Accepted Masons.
FAO, F.A.O. Food and Agricultural Organization (of the United Nations).
F.A.P.S. Fellow of the American Physical Society.
f.a.s., F.A.S. free alongside ship.
FAS Foreign Agricultural Service.
F.A.S. Fellow of the Actuarial Society (Can.); Fellow of the Antiquarian Society (Brit.); Fellow of the Anthropological Society (Brit.).
F.A.S.A. Fellow of the Acoustical Society of America.
fasc. a bundle [L *fasciculus*].
fath. fathom.
f.b. freight bill.
f.b., fb fullback.
F.B.A. Fellow of the British Academy.
F.B.I. Federal Bureau of Investigation **(FBI)**; Federation of British Industries.
fbm feet board measure (board feet).
f.c. follow copy (printing).
Fc fractocumulus.
FC Fire Controlman.
FCA, F.C.A. Farm Credit Administration.
F.C.C. Federal Communications Commission **(FCC)**; Federal Council of Churches; First Class Certificate; Food Control Committee.
FCDA Federal Civil Defense Administration.
FCIC, F.C.I.C. Federal Crop Insurance Corporation.
fcp. foolscap.
F.D. Defender of the Faith [L *Fidei Defensor*]; Finance Department **(FD)**; Fire Department.
FDA, F.D.A. Food and Drug Administration; Food Distribution Administration.
FDIC, F.D.I.C. Federal Deposit Insurance Corporation.
Fe iron [L *ferrum*].
FEA Foreign Economic Administration.
FEB Fair Employment Board.
Feb., Feb February.
FEC Far Eastern Commission.
fec. he (or she) made it [L *fecit*].
fed. federal; federated; federation.
fem. female; feminine.
FEPC, F.E.P.C. Fair Employment Practices Committee.
Fer., Ferm. Fermanagh.
FERA, F.E.R.A. Federal Emergency Relief Administration.
feud. feudal; feudalism.
ff. folios; following; fortissimo **(ff)**.
f.f. fixed focus.
f.f.a., F.F.A. free foreign agent; free from alongside.
F.F.A. Future Farmers of America.
FFC Foreign Funds Control.
F.F.I. French Forces of the Interior.
F.F.V. First Families of Virginia.
F.G.S.A. Fellow of the Geological Society of America.
FHA Federal Housing Administration.
FHLBB Federal Home Loan Bank Board.
f.h.p. friction horsepower.
F.I. Falkland Islands.
fict. fiction.
fid. fidelity; fiduciary.
fifo, FIFO first in, first out.
fig. figuratively; figure(s).
filo, FILO first in, last out.
fin. finance; financial; finished.
Fin. Finland; Finnish.
fl. floor; florin(s); flourished; flower; fluid; flute.
fl fluid.
Fl. Flanders; Flemish.
Fl fluorine.
Fla. Florida.
FLB Federal Land Bank.
fl. dr. fluid dram(s).
Flem. Flemish.
flex. flexible.
Flint. Flintshire.
Flo. Florence.
flor. he (she) flourished [L *floruit*].
Flor. Florida.
fl. oz. fluid ounce(s).
fluores. fluorescent.
F.M. Field Manual **(FM)**; Field Marshal; Foreign Missions.
Fm fermium.
fm. fathom; from.
FM frequency modulation.
FMCS Federal Mediation and Conciliation Service.
FMk, FMK markka.
FNMA Federal National Mortgage Association.
F.O. field officer; Foreign Office.
f.o.b., F.O.B. free on board.
F.O.E. Fraternal Order of Eagles.
fol. folio; following.
foll. following.
f.o.r., F.O.R. free on rails.
for. foreign; forestry.
fort. fortification; fortified.
f.p. fireplug; fire policy; floating policy; fully paid.
f.p., fp forward pass.
f.p., fp, fp. freezing point.
f.p., F.P., fp foot-pound(s).
fp. foolscap.
FPA, F.P.A. Food Products Administration; Foreign Press Association.
FPC, F.P.C. Federal Power Commission.
FPHA, F.P.H.A. Federal Public Housing Administration.
fpm, f.p.m. feet per minute.
FPO fleet post office.
fps, f.p.s. feet per second; foot-pound-second (system).
f.r. right-hand page [L *folio recto*].
fr. fragment; franc; from.
Fr. Brother [L *Frater*]; Father; France; Francis; French; Friar; Friday; wife [G *Frau*].
Fr francium.
Frl. Miss [G *Fräulein*].
F.R.B., FRB Federal Reserve Bank; Federal Reserve Board.
FRC, F.R.C. Federal Radio Commission; Federal Relief Commission.
F.R.C.P. Fellow of the Royal College of Physicians.
F.R.C.S. Fellow of the Royal College of Surgeons.
Fred., Fredk. Frederick.
freq. frequency; frequently; frequentative.
F.R.G.S. Fellow of the Royal Geographical Society.
Fri. Friday.
Fris., Frs. Frisian.
F.R.S. Federal Reserve System **(FRS)**; Fellow of the Royal Society.
F.R.S.A. Fellow of the Royal Society of Arts.
frt. freight.
F.S. Field Service; Fleet Surgeon.
f.s. foot-second.
Fs fractostratus.
FSA, F.S.A. Farm Security Administration; Federal Security Agency.
FSCC, F.S.C.C. Federal Surplus Commodities Corporation.
FSH follicle-stimulating hormone.
FSR, F.S.R. Field Service Regulations.
ft. feet; foot; fort; fortification; fortified.
FT Fire Control Technician.
FTC, F.T.C. Federal Trade Commission.
ft.-c. foot-candle.
fth., fthm. fathom.
ft.-l. foot-lambert.
ft.-lb. foot-pound.
fur. furlong; furnish; furnished.
furl. furlough.
furn. furnished; furniture.
fut. future.
f.v. on the back of the page [L *folio verso*].
FWA, F.W.A. Federal Works Agency.
fwd. forward.
F.Z.S. Fellow of the Zoological Society.

G

g. gender; general; genitive; grand; guide.
g., G. conductance; gauge; gold; gourde **(gde.)**; grain(s); guilder(s); guinea(s); gulf.
g., g goal(s); goalie; goalkeeper.
g., g, G., G gram(s).
g general intelligence; gravity.
G gun.
G, G. German.
G. Germany; specific gravity.
g.a., G.A., G/A general average.
Ga. Gallic; Georgia.
Ga gallium.
G.A. General Agent; General Assembly.
Gael. Gaelic.
gal., gall. gallon(s).
Gal. Galatians **(Gal)**; Galen; Galway.
galv. galvanic; galvanism; galvanized.
GAO General Accounting Office.
G.A.R., GAR Grand Army of the Republic.
GARIOA Government and Relief in Occupied Areas.
GATT General Agreement on Tariffs and Trade.
G.A.W. guaranteed annual wage.
gaz. gazette; gazetteer.
G.B. Great Britain.
G.B.E. (Knight or Dame) Grand (Cross or Order) of the British Empire.
GCA, G.C.A. ground control approach (radar).
g-cal. gram calorie(s).
G.C.B. (Knight) Grand Cross of the Bath.
G.C.D. g.c.d., gcd, greatest common divisor.
g.c.f., gcf, G.C.F. greatest common factor.
GCI ground controlled interception (aircraft).
G.C.L.H. Grand Cross of the Legion of Honor.
g.c.m., gcm, G.C.M. greatest common measure.
GCT, G.C.T., G.c.t. Greenwich civil time.
G.C.V.O. (Knight) Grand Cross of the (Royal) Victorian Order.
G.D. Grand Duchess; Grand Duchy; Grand Duke.
Gd gadolinium.
gds. goods.
Ge germanium.
geb. born [G *geboren*].
gel. gelatinous.
gen. gender; genera; general; generally; generator; generic; genitive; genus.
Gen. General; Genesis **(Gen)**; Geneva; Genevan.
geneal. genealogical; genealogy.
genit. genitive.
genl. general.
gent. gentleman; gentlemen.
Geo. George.
geod. geodesy; geodetic.
Geof., Geoff. Geoffrey.
geog. geographer; geographic; geographical; geography.
geol. geologic; geological; geologist; geology.
geom. geometer; geometric; geometrical; geometry.
GEPC German External Property Commission.
ger. gerund.
Ger., Germ. German; Germany.
Gert. Gertrude.
gest. died [G *gestorben*].
G.F.T.U. General Federation of Trade Unions.
g.gr. great gross.
GHA Greenwich hour angle.
GHQ, G.H.Q. General Headquarters.
g.i., G.I. gastro-intestinal.
GI, G.I. general issue; government issue.
gi. gill(s).
Gib. Gibraltar.
Gil., Gilb. Gilbert.
Gk. Greek.

Gl glucinum.
gl. glass; gloss.
Glam., Glamorg. Glamorganshire.
gld. guilder(s).
Glos., Glos, Gloucs. Gloucestershire.
gloss. glossary.
glt. gilt (bookbinding).
gm. gram(s).
G.M. general manager; Grand Master.
GM Gunner's Mate.
G.m.a.t. Greenwich mean astronomical time.
Gmc. Germanic.
G.m.t. GMT, G.M.T. Greenwich mean time.
GNP, G.N.P. gross national product.
GO general orders.
G.O.P. Grand Old Party (Republican party).
Goth. gothic; Gothic.
gov., govt. government.
Gov. Governor.
G.P. general paresis; general practitioner; Graduate in Pharmacy.
g.p.m. gallons per minute.
GPO, G.P.O. General Post Office; Government Printing Office.
g.p.s. gallons per second.
G.P.U. State Political Department [Russian *Gosudarstvennoye politicheskoye upravleniye*].
GQ, G.Q., g.q. general quarters.
G.R. King George [L *Georgius Rex*].
gr. grade; grain(s); gram(s); grammar; great; gross; group.
Gr. Grecian; Greece; Greek.
grad. graduate; graduated.
gram. grammar; grammarian; grammatical.
Gr.Br., Gr.Brit. Great Britain.
Greg. Gregory.
gro. gross.
gr.wt. gross weight.
GS German silver.
G.S. General Secretary; General Staff; Girl Scouts.
GSA General Services Administration.
G.S.A. Girl Scouts of America.
GSC, G.S.C. General Staff Corps.
GSO, G.S.O. General Staff Officer.
gt. gilt; great.
g.t.c., G.T.C. good till canceled.
gtd. guaranteed.
gtt. a drop [L *gutta*].
g.u., g.-u. genito-urinary.
guar. guaranteed.
Guat. Guatemala.
Gui. Guiana.
Guin. Guinea.
gun. gunnery.
guttat. by drops [L *guttatim*].
g.v. gravimetric volume.
gym. gymnasium; gymnastics.
gyn., gynecol. gynecological; gynecology.

H

h. harbor; hard; hardness; heavy sea; height; hence; high; horns (music); hour(s); hundred; husband.
h., h hit(s).
h, H Henry.
h, hy henry *(electr.)*.
H hydrogen; intensity of magnetic field.
h.a. this year [L *hoc anno*].
HA Hospital Apprentice.
ha hectare.
Hab., Hab Habakkuk.
hab. corp. have the body [L *habeas corpus*].
Hag., Hag Haggai.
Hal halogen.
Hants, Hants. Hampshire.
Har. Harold.
hav. haversine.
Hb hemoglobin.
hb., hb halfback.
H.B.M. His (Her) Britannic Majesty.
H.C. House of Commons.
hcap., hcp. handicap.
h.c.f., hcf, H.C.F. highest common factor.
hd. hand; head.
hdbk. handbook.
hdkf., hkf. handkerchief.
hdqrs. headquarters.
He helium.
H.E. His Eminence; His Excellency.
HE, H.E. high explosive.
Heb., Heb, Hebr. Hebrew; Hebrews.
Hen. Henry.
her. heraldic; heraldry.
Herb. Herbert.
Heref., Herefs. Herefordshire.
herp., herpetol. herpetology.
Herts, Herts. Hertfordshire.
hex. hexachord; hexagon; hexagonal.
h.f., h-f, HF high frequency.
hf. half.
Hf hafnium.
hf.bd. half-bound (bookbinding).
hf.mor. half-morocco (bookbinding).
HG, H.G. High German.
H.G. His (Her) Grace; Home Guard.
hg hectogram(s); heliogram.
Hg mercury [L *hydrargyrum*].
hgt. height.
H.H. His (Her) Highness; His Holiness.
hhd hogshead.
HHFA Housing and Home Finance Agency.
H.I. Hawaiian Islands.
H.I.H. His (Her) Imperial Highness.
H.I.M. His (Her) Imperial Majesty.
Hind. Hindi; Hindu; Hindustan; Hindustani.
hist. histology; historian; historical; history.
H.J. here lies [L *hic jacet*].
H.J.S. here lies buried [L *hic jacet sepultus*].
hl hectoliter(s).
H.L. House of Lords.
HL mustard-lewisite (poison gas).
HLBB Home Loan Bank Board.
h.m. in this month [L *hoc mense*].
HM Hospital Corpsman.
hm hectometer(s).
H.M. His (Her) Majesty.
H.M.S. His (Her) Majesty's Service (Ship, Steamer).
HN Hospitalman.
HN nitrogen mustard gas.
H.O. head office; Home Office.
ho. house.
Ho holmium.
HOLC, H.O.L.C. Home Owners' Loan Corporation.
hon. honorably; honorary.
Hon. Honorable.
Hond. Honduras.
hor. horizon; horizontal; horology.
horol. horology.
hort., hortic. horticultural; horticulture.
Hos., Hos Hosea.
hosp. hospital.
How. Howard's U.S. Supreme Court Reports; howitzer.
HP, H.P. high power.
hp, HP, h.p., H.P. high pressure; horsepower.
hp.-hr. horsepower-hour.
h.q. look for this [L *hoc quaere*].
HQ, H.Q., hq, h.q. headquarters.
h.r., hr home run(s).
hr. hour(s).
Hr. Mister [G *Herr*].
H.R. Home Rule; House of Representatives.
H.R.E. Holy Roman Empire.
H.R.H. His (Her) Royal Highness.
H.R.I.P. here rests in peace [L *hic requiescat in pace*].
hrs. hours.
H.S. here is buried [L *hic sepultus*]; here lies [L *hic situ*]; High School; Home Secretary; in this sense [L *hoc sensu*].
H.S.H. His (Her) Serene Highness.
H.S.M. His (Her) Serene Majesty.
H.T. Hawaiian Territory.
h.t. at this time [L *hoc tempore*]; in or under this title [L *hoc titulo*].
ht. heat; height.
Hts. Heights.
Hu. Hugh; Hugo.
Hub. Hubert.
Hun., Hung. Hungarian; Hungary.
Hunts, Hunts. Huntingdonshire.
H.V., h.v., hv high voltage.
h.w. high water.
Hy. Henry.
hyd., hydros. hydrostatics.
hydraul. hydraulic; hydraulics.
hyg. hygiene; hygroscopic.
hyp., hypoth. hypotenuse; hypothesis; hypothetical.

I

i. incisor; interest; intransitive; island.
I. Island(s); Isle(s).
I Ides; Independent; iodine.
Ia. Iowa.
IADB Inter-American Defense Board.
IAS indicated air speed.
IATA International Air Transport Association.
ib., ibid. in the same place [L *ibidem*].
IBRD International Bank for Reconstruction and Development.
IBT International Brotherhood of Teamsters.
IC Interior Communications Electrician.
ICA International Cooperation Administration.
ICAO International Civil Aviation Organization.
ICC Indian Claims Commission; Interstate Commerce Commission **(I.C.C.)**.
Ice., Icel. Iceland; Icelandic.
ICES International Council for the Exploration of the Sea.
ICFTU International Confederation of Free Trade Unions.
ichth. ichthyology.
ICITO Interim Commission International Trade Organization.
ICJ International Court of Justice.
ICNAF International Commission for Northwest Atlantic Fisheries.
I.C.S. Indian Civil Service.
i.c.w. interrupted continuous wave.
I.C.Z. Isthmian Canal Zone.
id. the same [L *idem*].
Id. Idaho.
ID, I.D., i.d. inside diameter.
I.D. Infantry Division; Intelligence Department; Iraqi dinar.
Ida. Idaho.
i.e. that is [L *id est*].
I.E., IE Indo-European.
IF, I.F. i.f., i-f intermediate frequency.
IFCTU International Federation of Christian Trade Unions.
IFF Identification friend or foe (British radar device).
I.F.S. Irish Free State.
I.G. amalgamation [G *Interessengemeinschaft*]; Indo-Germanic; Inspector General.
ign. ignites; ignition; unknown [L *ignotus*].
IHP, I.H.P. ihp, i.h.p., i.-hp. indicated horsepower.
I.H.S., IHS Jesus: often, incorrectly, Jesus Savior of Men [L *Iesus hominum salvator*] or, in this sign [L *in hoc signo*].
Il illinium.
ILA International Longshoremen's Association.
ILC International Law Commission.
ill., illus., illust. illustrate; illustrated; illustration; illustrator.
Ill. Illinois.
illit. illiterate.
ILO, I.L.O. International Labor Organization.
I.L.P. Independent Labour Party.
ILS instrument landing system.
I.M. Isle of Man.
IM Instrumentman.
IMC International Materials Conference.
IMCO Intergovernmental Maritime Consultative Organization.
IMF International Monetary Fund.
imit. imitation; imitative.
immun. immunology.
IMO International Meteorological Organization.
imp. imperative; imperfect; imperial; impersonal; import; important; imported; importer; imprimatur; improper.
Imp. Imperator.
imper. imperative.
imperf. imperfect; imperforate.
impers. impersonal.
impf. imperfect.
imp. gal. imperial gallon.
impv. imperative.
in. inch(es).
In indium.
inbd. inboard.
inc. inclosure; including; inclusive; income; incorporated; increase.
inch., incho. inchoative.
incl. inclosure; including.
incog. incognito.
incor., incorp. incorporated.
incorr. incorrect.
incr. increased; increasing.
I.N.D. in the name of God [L *in nomine Dei*].
ind. independence; independent; index; indicated; indicative; indigo; indirect; industrial.
Ind. India; Indian; Indiana; Indies.
indecl. indeclinable.
indef. indefinite.
inden., indent. indention.
indic. indicating; indicative; indicator.
individ. individual.
induc. induction.
ined. not published.
in ex. at length [L *in extenso*].
in f. at the end [L *in fine*].
inf. below [L *infra*]; inferior; infinitive; information.
Inf. Infantry.
infin. infinitive.
infl. influence; influenced.
init. initial; in the beginning [L *initio*].
inj. injection.
in.-lb. inch-pound.
in lim. at the outset (on the threshold) [L *in limine*].
in loc. in its place [L *in loco*].
in loc. cit. in the place cited [L *in loco citato*].
inorg. inorganic.
I.N.R.I. Jesus of Nazareth, King of the Jews [L *Iesus Nazarenus Rex Iudaeorum*].
ins. inches; inspector; insular; insulated; insulation; insurance.
I.N.S., INS International News Service.
insc., inscr. inscribe; inscribed; inscription.
insep. inseparable.
insol. insoluble.
insp. inspected; inspector.
inst. instant; instantaneous; instrument; instrumental.
Inst. Institute; Institution.
instr. instruction; instructor; instrument; instrumental.
insur. insurance.
int. intelligence; interest; interior; interjection; internal; international; interval; intransitive.
intens. intensive.
inter. intermediate.
interj. interjection.
internat. international.
interp. interpreter.
Interpol. International Police Organization.
interrog. interrogative.
intr., intrans. intransitive.
in trans. on the way [L *in transitu*].
Int. Rev. Internal Revenue.
introd. introduction; introductory.
inv. invented; invention; inventor; invoice.
Inv. Inverness.
invert. invertebrate.
invt. inventory.
Io ionium.
I.O.B.B. Independent Order of B'nai B'rith.
I.O.F. Independent Order of Foresters.
I.O.O.F. Independent Order of Odd Fellows.
IOPH International Office of Public Health.
I.O.R.M. Improved Order of Red Men.
I.O.U., IOU I owe you.
i.p. in passing (chess).
i.p., ip innings pitched.
I.P.A., IPA International Phonetic Alphabet (Association).
IPI International Press Institute.
IPPC International Penal and Penitentiary Commission.
I.P.R. Institute of Pacific Relations.
i.q. the same as [L *idem quod*].
IQ, I.Q. intelligence quotient.
i.q.e.d. what was to be proved [L *id quod erat demonstrandum*].
Ir. Ireland; Irish.
Ir iridium.
Iran. Iranian; Iranic.
Ire. Ireland.
IRO International Refugee Organization.
irreg. irregularly.
IRS Internal Revenue Service.
is. island(s); isle(s).
Is., Is, Isa. Isaiah.
Isab. Isabella.
isl. island(s).
iso. isotropic.
isom. isometric.
isoth. isothermal.
Isr. Israel.
isth. isthmus.
it., ital. italic; italics.
It., Ital. Italian; Italy.
itin. itinerant; itinerary.
ITO International Trade Organization.
ITU International Telecommunication Union; International Typographical Union.
IU, I.U. international unit(s).
IWC International Wheat Council.
IWW, I.W.W. Industrial Workers of the World.

J

j, J joule.
J. Judge; Justice.
j.a., j/a, J.A., J/A joint account.
JA, J.A. Judge Advocate.
Ja., Jan., Jan January.
Jac. Jacob; James [L *Jacobus*].
J.A.G. Judge Advocate General.
Jam. Jamaica.
Jap. Japan; Japanese.
Jas., Jas James.
Jav. Javanese.
J.C.D. Doctor of Canon Law [L *Juris Canonici Doctor*]; Doctor of Civil Law [L *Juris Civilis Doctor*].
jct., jctn. junction.
J.D. Doctor of Laws [L *Jurum Doctor*].
Je., Je June.
Jer. Jeremiah **(Jer)**; Jerome; Jeremy.
j.g., jg junior grade.
Jl July.
Jn John.
Jno. John.
JO Journalist.
Joe Joel.
join. joinery.
Jon. Jonah **(Jon)**; Jonathan **(Jona.)**.
Jos. Joseph; Joshua **(Jos)**; Josiah.
Josh. Joshua.
jour. journal; journalist; journeyman.
J.P. Justice of the Peace.
jr., Jr. junior.
Ju, Judg. Judges.
Jud. Jude; Judith.
Jul. Jules; Julius; July **(Jul)**.
Jun., Jun June.
jun., Jun. junior.
junc., junct. junction.
Jur.D. Doctor of Law [L *Juris Doctor*].
jurisp. jurisprudence.
jus., just. justice.
juv. juvenile.
j.v. junior varsity.
jwlr. jeweler.
Jy. July.

K

k. capacity; carat; constant.
k., K. calends [L *kalendae*]; crown; karat; kilogram; king; king (chess); knight; kopeck; koruna; krone.
k kilo.
K Kelvin (temperature scale); potassium [L *kalium*].
ka. kathode (cathode).
kal. kalends (calends).
Kan., Kans., Kas. Kansas.
Kath. Katharine; Katherine.
K.B. King's Bench; Knight Bachelor.
KB king's bishop (chess).
KBP king's bishop's pawn (chess).
kc., kc kilocycle(s).
K.C. King's Counsel; Knight Commander; Knights of Columbus.
kcal kilocalorie.
K.C.B. Knight Commander (of the Order) of the Bath.
K.C.V.O. Knight Commander of the (Royal) Victorian Order.
K.D. knocked down.
Ken. Kentucky.
kg., kg keg(s); kilogram(s).
K.G. Knight (of the Order) of the Garter.
Ki Kings (book of the Bible).
Kild. Kildare.
Kilk. Kilkenny.
kilo. kilogram(s); kilometer(s).
kilog. kilogram(s).
kilol. kiloliter(s).
kilom. kilometer(s).
Kin. Kinross.
Kinc. Kincardine.
kingd. kingdom.
KKE Greek Communist Party [Gr. *Kommunistiko Komma Ellados*].
K.K.K., KKK Ku Klux Klan.
KKt king's knight (chess).
KKtP king's knight's pawn (chess).
kl., kl kiloliter(s).
km., km kilometer(s).
kn. kronen.
Knt knight (chess).
KO, K.O., k.o. knockout.
K. of C. Knight (Knights) of Columbus.
K. of P. Knight (Knights) of Pythias.
kop. kopeck(s).

K.P. Knight (of the Order of St.) Patrick; Knights of Pythias.
KP king's pawn (chess).
KP, K.P. kitchen police.
kr., Kr. kreuzer; krona; krone.
Kr krypton.
KR king's rook.
KRP king's rook's pawn (chess).
kt. karat (carat).
Kt, Kt. knight (chess).
K.T. Knight (Knights) Templar.
kv., kv kilovolt(s).
kva, kv.-a. kilovolt-ampere.
kvar reactive kilovolt-ampere.
kw., kw kilowatt(s).
K.W.H., kw-h, kw.-hr., kw-hr kilowatt-hour(s).
Ky. Kentucky.

L

l. book [L *liber*]; lake; land; lat; latitude; law; leaf; league; left; lempira; length; leu; lev; lex; line; link; lira; lire; lit; low; place [L *locus*].
l, l. liter(s); lumen(s).
l., l (games) lost.
l., l- levo.
L. licentiate; Linnaeus; Lodge (fraternal).
L, £, l. pound (sterling) [L *libra*].
L, L. lambert; Latin.
L length; lewisite; longitude.
L, l coefficient of inductance.
La. Louisiana.
La lanthanum.
L.A. Legislative Assembly; Library Association; Local Agent; Los Angeles.
lab. laboratory.
Lab. Labrador.
L.A.M. Master of Liberal Arts [L *Liberalium Artium Magister*].
lam. laminated.
Lam., Lam Lamentations.
Lancs, Lancs. Lancashire.
lang. language.
Lap. Lapland.
Lapp. Lappish.
laryngol. laryngological; laryngology.
lat. latitude.
Lat. Latin; Latvia.
Latv. Latvia.
Laur. Laurence.
Lawr. Lawrence.
Laz. Lazarus.
lb. pound(s) [L *libra*].
L.B. Bachelor of Letters [L *Litterarum Baccalaureus*]; Local Board.
lb.ap. pound, apothecary.
lb.av. pound, avoirdupois.
lbs. pounds.
lb.t. pound, troy.
l.c. left center; lower case; in the place cited [L *loco citato*].
l/c, L/C letter of credit; lower case.
L.C. Library of Congress.
LCC. Landing Craft, Control.
L.C.D., l.c.d., lcd, lowest common denominator.
LCI. Landing Craft, Infantry.
l.c.l., L.C.L. less than carload lot.
l.c.m., lcm, L.C.M. lowest, or least, common multiple.
LCM Landing Craft, Mechanized.
LCR Landing Craft, Rubber.
LCS Landing Craft, Support.
LCT Landing Craft, Tank; local civil time **(L.C.T.)**.
LCVP Landing Craft, Vehicle, Personnel.
ld. lead (printing).
Ld. Lord.
L.D., LD., LD Low Dutch.
ldg. landing; leading; loading.
L.Div. Licentiate in Divinity.
ld. lmt. load limit.
ldry laundry.
L.D.S. Licentiate in Dental Surgery.
l.e., le left end.
LE, £E Egyptian pound.
lea. league; leather; leave.
lect. lecture.
led. ledger.
leg. legal; legate; legato; legislation; legislative.
legis. legislation; legislative; legislature.
Leics. Leicestershire.
Leit. Leitrim.
Lem. Lemuel.
Leon. Leonard.
L. ès S. Licentiate in Sciences [F *Licencié ès Sciences*].
Lev., Lev, Levit. Leviticus.
lex. lexicon.
lexicog. lexicographer; lexicographical; lexicography.
l.f., lf. lightface (printing).
l.f., lf left field(er); left forward.

l.f., l-f, LF, L.F. low frequency.
l.g., lg left guard.
LG, LG., L.G. Low German.
lg., lge. large.
LGk. Late Greek.
lgth. length.
LH luteinizing hormone.
l.h., L.H., LH left hand.
l.h., lh, l.h.b., lhb left halfback.
LHA local hour angle.
L.H.D. Doctor of Humanities [L *Litterarum Humaniorum*, or *In Litteris Humanioribus, Doctor*].
MA Machine Accountant; Maritime Administration.
MA, M.A. mental age.
Ma masurium.
MAC mean aerodynamic chord.
Mac., Macc. Maccabees.
Maced. Macedonia; Macedonian.
mach., machin. machine; machinery; machinist.
Mad., Madm. Madam.
mag. magazine; magnet; magnetism; magnitude.
M. Agr., M. Agric. Master of Agriculture.
maj. majority.
Maj. Major.
Mal. Malachi **(Mal)**; Malay; Malayan; Malta.
malac. malacology.
Man. manila (paper); Manitoba.
Manch. Manchukuo; Manchuria.
manuf. manufacture; manufactured; manufacturer; manufacturing.
mar. marine; maritime; marginal.
Mar., Mar March.
March. Marchioness.
marg. margin; marginal.
Marg. Margaret.
Marq. Marquis.
mas., masc. masculine.
Mass. Massachusetts.
mat. matinee; matins; maturity.
MATS Military Air Transport Service.
math. mathematical; mathematician; mathematics.
matr., matric. matriculate; matriculation.
Matt. Matthew; Matthias.
max. maximum.
Max. Maximilian.
M.B. Bachelor of Medicine [L *Medicinae Baccalaureus*].
M.B.A. Master in (or of) Business Administration.
MBS Mutual Broadcasting Service.
mc, m.c., mc. megacycle; millicurie.
M.C. Maritime Commission; Master Commandant; Master of Ceremonies; Medical Corps; Member of Congress.
M.Ch. Master of Surgery [L *Magister Chirurgiae*].
M.C.L. Master of Civil Law.
Md. Maryland.
M.D. Doctor of Medicine [L *Medicinae Doctor*]; Medical Department; mentally deficient.
MD, M.D. Middle Dutch.
M/D, m/d memorandum of deposit; months' date.
Mddx., Mdx. Middlesex.
M.D.S. Master of Dental Surgery.
mdse. merchandise.
MDu. Middle Dutch.
m.e. marbled edges (bookbinding).
Me. Maine.
Me methyl.
M.E. Mechanical (Military, Mining) Engineer; Methodist Episcopal.
ME Metalsmith.
ME, ME., M.E. Middle English.
Mea. Meath.
meas. measure; measurable.
mech., mechan. mechanical; mechanics; mechanism.
med. medical; medicine; medieval; medium.
M.Ed. Master of Education.
Med.Gk. Medieval Greek.
Medit. Mediterranean.
Med.L, Med.L. Medieval Latin.
meg. megohm.
mem. member; memoir; memorandum; memorial.
memo. memorandum.
mensur. mensuration.
m.e.p. mean effective pressure.
mer. meridian; meridional.
merc. mercantile; mercurial; mercury.
Meri., Merions. Merionethshire.
Messrs., Messrs Messieurs.
met. metaphor; metaphysics; meteorological; metronome; metropolitan.
metal., metall. metallurgical; metallurgy.
metaph. metaphor; metaphorical; metaphysics.

metath. metathesis; metathetical.
meteor., meteorol. meteorological; meteorology.
meth. method; methylated.
Meth. Methodist.
meton. metonymy.
metrol. metrological; metrology.
metrop. metropolitan.
mev, Mev, MEV, m.e.v. million electron volts.
Mex. Mexican; Mexico.
MF, M.F., mf, m.f. medium frequency.
mf, mf. mezzoforte; millifarad.
mf., mfd microfarad.
MF., MF, Middle French.
mfg. manufacturing.
M.Flem. Middle Flemish.
mfr. manufacture; manufacturer.
MG Military Government.
mg., mg, mgm milligram(s).
Mg magnesium.
MGB Soviet Ministry of State Security [Russian *Ministerstvo Gosudarstvennoi Bezopasnosti*].
Mgr. Manager; Monseigneur; Monsignor.
mgt. management.
MH Medal of Honor.
mh, mh. millihenry.
MHG, MHG., M.H.G. Middle High German.
mi., mi mile(s).
Mic., Mic Micah.
Mich. Michael; Michigan.
micros. microscope; microscopic; microscopy.
mid. middle; midshipman.
Mid.Dan. Middle Danish.
Middix., Midx. Middlesex.
Mid.L. Midlothian.
Mid.Sw. Middle Swedish.
mil. mileage; military; militia; million.
milit. military.
mim., mimeo. mimeograph; mimeographed.
min. mineralogical; mineralogy; minim(s); minimum; mining; minor; minute(s).
mineral. mineralogy.
Minn. Minnesota.
m.i.p. marine insurance policy; mean indicated pressure.
misc. miscellaneous; miscellany.
Miss. Mississippi.
mixt. mixture.
mk. mark; markka.
Mk Mark.
mkd. marked.
m.k.s., mk., M.K.S. meter-kilogram-second (system).
mkt. market.
ml., ml milliliter(s).
M.L., ML., ML Medieval (or Middle) Latin.
ML Molder.
M.L.A. Modern Language Association.
MLD, M.L.D., m.l.d. minimum lethal dose.
MLG, M.L.G., MLG. Middle Low German.
Mlle., Mlle Mademoiselle.
Mlles., Mlles Mesdemoiselles.
m.m. with the necessary changes [L *mutatis mutandis*].
mm., mm millimeter(s); thousands [L *millia*].
MM. their Majesties; Messieurs.
MM Machinist's Mate.
Mme., Mme Madame.
Mmes., Mmes Mesdames.
m.m.f. magnetomotive force.
mmfd. micromicrofarad.
m.n. the name being changed [L *mutato nomine*].
MN Mineman.
Mn manganese.
mo. month(s); monthly.
Mo. Missouri; Monday.
Mo molybdenum.
M.O. mail order; medical officer; money order.
mob. mobile; mobilization; mobilized.
mod. moderate; moderato; modern.
Mod.Gr. Modern Greek.
Moham. Mohammedan.
M.O.I. Ministry of Information.
mol. molecular; molecule.
mol.wt. molecular weight.
mon. monastery; monetary.
Mon. Monaghan (county, Ireland); Monday; Monsignor.
Mon., Mons. Monmouthshire.
Mong. Mongolia; Mongolian.
monocl. monoclinic.
monog., monogr. monograph.
Mons. Monsieur.
Monsig. Monsignor.
Mont. Montana; Montgomeryshire **(Montgom.)**.
mor. morocco (bookbinding).
Mor. Morocco.
morn. morning.

morph., morphol. morphological; morphology.
mort. mortuary.
MOS military occupational specialty.
mos. months.
mot. motor, motorized.
mp, m.p., M.P. melting point.
M.P. Member of Parliament; Military Police **(MP)**.
mp mezzo piano.
M.Pd. Master of Pedagogy.
M.P.E. Master of Physical Education.
mph, m.p.h., MPH miles per hour.
M.P.P.D.A., MPPDA Motion Picture Producers and Distributors of America, Inc.
MR Machinery Repairman; motivational research.
Mr. Mister.
M.R.A. Moral Re-Armament.
MRC Metals Reserve Company.
MRP Popular Republican Movement [F *Mouvement Républicaine Populaire*].
Mrs. Mistress.
m.s., ms, MS, MS months after sight.
m.s., ms, MS., MS manuscript.
M.S. Master of Science [L *Magister Scientiae*]; sacred to the memory of [L *memoriae sacrum*]; mine sweeper.
MS. motorship.
MSA Mutual Security Agency.
M.Sc. Master of Science.
msg. message.
Msgr. Monsignor.
M.Sgt., M/Sgt Master Sergeant.
m.s.l., M.S.L. mean sea level.
mss., mss, MSS., MSS manuscripts.
MST, M.S.T., m.s.t. mountain standard time.
Ms-Th mesothorium.
mt. mount; mountain.
m.t. mean time; metric ton; motor transport; mountain time.
Mt Matthew.
mtg. meeting; mortgage.
mtge. mortgage.
mtl. material.
m.t.l., M.T.L. mean tidal level.
mtn. mountain.
MTO Mediterranean Theater of Operations.
Mt. Rev. Most Reverend.
mts. mountains.
MU Musician.
mun. municipal; municipality.
mus. museum; music; musician.
Mus.B., Mus.Bac. Bachelor of Music [L *Musicae Baccalaureus*].
Mus.D., Mus.Doc., Mus.Dr. Doctor of Music [L *Musicae Doctor*].
Mus.M. Master of Music [L *Musicae Magister*].
mut. mutilated; mutual.
mv., mv millivolt(s).
Mv mendelevium.
MVA, M.V.A. Missouri Valley Authority.
MVD Soviet Ministry of Internal Affairs [Russian *Ministerstvo Vnutrennikh Del*].
M.W. Most Worshipful; Most Worthy.
M.W.A. Modern Woodmen of America.
Mx. Middlesex.
MY May.
mya. myriare(s).
mycol. mycological; mycology.
myg. myriagram(s).
myl. myrialiter(s).
mym. myriameter(s).
myth., mythol. mythological; mythology.

N

n. born [L *natus*]; name; navigation; navy; nephew; net; neuter; new; night; nominative; noon; normal; note; number; our [L *noster*].
N. Nationalist; Norse; November.
N., N, n. north; northern.
N areal radiant intensity; Avogadro's number; knight (chess); nitrogen.
Na sodium [L *natrium*].
N.A. National Academician; National Army; North America.
N.A.A. National Aeronautic Association; National Automobile Association.
N.A.A.C.P., NAACP National Association for the Advancement of Colored People.
NAC North Atlantic Council.
NACA National Advisory Committee for Aeronautics.
N.A.D. National Academy of Design.

Nah., Nah Nahum.
N.Am. North American.
N.A.M. National Association of Manufacturers.
N.Am.Ind. North American Indian.
nat. national; native; natural; naturalist.
Nat., Nath. Nathan; Nathaniel.
nat.hist. natural history.
natl. national.
NATO North Atlantic Treaty Organization.
NATS National Air Transport Service.
naut. nautical.
nav. naval; navigable; navigation.
navig. navigation; navigator.
n.b., N.B. note well [L *nota bene*].
N.B. New Brunswick.
Nb niobium.
NBA, N.B.A. National Boxing Association.
NBC National Broadcasting Company.
NbE north by east.
NBS, N.B.S. National Bureau of Standards.
NbW north by west.
n.c., N.C. nitrocellulose.
N.C., N.Car. North Carolina.
NC Nurse Corps.
NCAA, N.C.A.A. National Collegiate Athletic Association.
NCO, N.C.O., n.c.o. non-commissioned officer.
N.D., n.d. no date.
Nd neodymium.
N.D., N.Dak. North Dakota.
NDRC, N.D.R.C. National Defense Research Committee.
NE, N.E., ne, n.e. northeast; northeastern.
Ne neon.
N.E. New England.
NEA, N.E.A. National Education Association.
Neb., Nebr. Nebraska.
N.E.C., NEC National Electric Code; National Emergency Council.
n.e.c. not elsewhere classified.
NED, N.E.D. New English Dictionary (Oxford English Dictionary).
neg. negative; negatively.
Neh., Neh Nehemiah.
n.e.i. not elsewhere indicated.
N.E.I., NEI Netherlands East Indies.
neol. neologism.
NEP, Nep., nep. New Economic Policy.
n.e.s. not elsewhere specified (or stated).
Neth. Netherlands.
Neth. Ind. Netherlands Indies.
neur., neurol. neurological; neurology.
neut. neuter; neutral.
Nev. Nevada.
Newf. Newfoundland.
New M. New Mexico.
New Test. New Testament.
n.f. noun feminine.
N.F. National Formulary; Newfoundland; Norman French.
N.F., n/f. no funds.
n.g., N.G. no good.
NG, N.G. National Guard.
NGr, N.Gr., NGr. New Greek.
N. Gui. Ter. New Guinea Territory.
N.H. New Hampshire.
NHA, N.H.A. National Housing Administration (Agency).
N.Heb. New Hebrides.
NHG, NHG., N.H.G. New High German.
NHI National Health Insurance (Brit.).
n.h.p. nominal horsepower.
Ni nickel.
N.I., N.Ire. Northern Ireland.
NIA National Intelligence Authority.
Nic., Nicar. Nicaragua.
Nig. Nigeria; Nigerian.
NIRA, N.I.R.A. National Industrial Recovery Act.
N.J. New Jersey.
NKVD People's Commissariat of Internal Affairs [Russ. *Narodnyi Komissariat Vnutrennikh Del*].
n.l. new line (printing); north latitude; not clear [L *non liquet*]; not far [L *non longe*]; not lawful [<L *non licet*].
NL, NL., N.L. New Latin.
N.lat. north latitude.
NLRB, N.L.R.B. National Labor Relations Board.
n.m. nautical mile; noun masculine.
N.M., N.Mex. New Mexico.
N.M.U., NMU National Maritime Union.
NNE, N.N.E., nne, n.n.e. north-northeast.
NNW, N.N.W., nnw, n.n.w. north-northwest.
no., No. north; northern; number.
nob. for (or on) our part [L *nobis*].
nol. pros. unwilling to prosecute [L *nolle prosequi*].
nom. nomenclature; nominal; nominative.
nomin. nominative.
noncom. non-commissioned (officer).
non cul. not guilty [L *non culpabilis*].
non dest. notwithstanding [L *non destante*].
non obs., non obst. notwithstanding [<L *non obstante*].
non pros. he (or she) does not prosecute [L *non prosequitur*].
non seq. it does not follow [L *non sequitur*].
nor., Nor. north; northern.
Nor. Norman; Norway; Norwegian.
Norf. Norfolk.
norm normal.
Northants., Northants Northamptonshire.
Northum., Northumb. Northumberland.
Norw. Norway; Norwegian.
nos., Nos. numbers.
Notts., Notts Nottinghamshire.
nov. novelist.
Nov., Nov November.
n.p. net proceeds; new paragraph; no paging; no place (of publication).
Np neptunium.
N.P. no protest; Notary Public; unless before [L *nisi prius*].
NP neuropsychiatric.
NPN non-protein nitrogen.
n.p. or d. no place or date.
n.p.t., N.P.T. normal pressure and temperature.
nr. near.
NRA, N.R.A. National Recovery Administration.
NRAB National Railroad Adjustment Board.
n.s. near side (shipping); new series; new style.
N.S., n.s. not specified.
Ns nimbostratus.
N.S. New Style; Nova Scotia.
NSA National Shipping Authority.
NSC National Security Council.
NSF National Science Foundation.
N.S.F., N/S/F not sufficient funds.
N.S.P.C.A. National Society for the Prevention of Cruelty to Animals.
N.S.P.C.C. National Society for the Prevention of Cruelty to Children.
NSRB National Security Resources Board.
N.S.W. New South Wales.
nt. net.
Nt niton.
NT., N.T. New Testament.
n.t.p., N.T.P. normal temperature and pressure.
nt.wt. net weight.
num. number(s); numeral(s).
Num., Num, Numb. Numbers.
numis., numism. numismatic; numismatics.
nv. non-voting (stock).
NW, N.W., nw, n.w. northwest; northwestern.
NWLB National War Labor Board.
N.W.T. Northwest Territories (Canada).
N.Y. New York.
NYA, N.Y.A. National Youth Administration.
N.Y.C. New York City.
N.Z., N.Zeal. New Zealand.

O

o. off; only; pint [L *octarius*].
o., O. octavo; old; order.
o ohm.
o- ortho-.
O. Ocean; October; Ohio; Ontario; Oregon.
O Old; oxygen.
OAPC, O.A.P.C. Office of Alien Property Custodian.
OAr., O.Ar. Old Arabic.
OAS Organization of American States.
ob. he (or she) died [L *obiit*]; in passing [L *obiter*]; oboe.
Ob., Ob, Obad. Obadiah.
obb. obbligato.
obdt., obt. obedient.
O.B.E. Officer (of the Order) of the British Empire.
obit. obituary.
obj. object; objection; objective.
obl. oblique; oblong.
obs. obscure; observation; observatory; obsolete; obsolescence.
obsol., obsoles. obsolescent.
ob.s.p. died without issue [L *obiit sine prole*].
obstet. obstetrical; obstetrics.
obv. obverse.
o.c. in the work cited [L *opere citato*].
oc. ocean.
o/c old charter; overcharge.
O.C., OC Office of Censorship; Officer Commanding; original cover.
OCAS Organization of Central American States.
occ. occasion; occasionally; occident; occidental.
occult. occultism.
OCD, O.C.D. Office of Civilian Defense.
oceanog. oceanography.
OCIAA, O.C.I.A.A. Office of the Coordinator of Inter-American Affairs.
OCS Office of Contract Settlement; Officer Candidate School.
oct. octavo.
Oct. Octavius; Octavus; October **(Oct).**
octupl. octuplicate.
o.d. olive drab; on demand; outside diameter.
O.D. Officer of the Day; Ordnance Department; overdraft; overdrawn.
OD., OD, ODu. Old Dutch.
ODM Office of Defense Mobilization.
ODT, O.D.T. Office of Defense Transportation.
o.e. omissions expected.
OE, OE., O.E. Old English.
O.E.D. Oxford English Dictionary.
OEEC Organization for European Economic Cooperation.
OEM, O.E.M. Office for Emergency Management.
O.E.S. Office of Economic Stabilization **(OES)**; Order of the Eastern Star.
OF, OF., O.F. Old French.
off. offered; office; official; officinal.
Off. Offaly.
O.F.S. Orange Free State.
O.G. Officer of the Guard; original gum (philately).
OHE Office of the Housing Expediter.
OHG, OHG., O.H.G. Old High German.
O.H.M.S. On His (Her) Majesty's Service.
OIAA Office of Inter-American Affairs.
OIC Office of Information and Culture (of the State Department).
O.Ir. Old Irish.
OIT Office of International Trade.
O.K., etc. See text.
Okla. Oklahoma.
Ol. Oliver.
Old Test. Old Testament.
oleo. oleomargarine.
O.M. Order of Merit.
OM Opticalman.
Om. Ostmark(s).
ON, ON., O.N. Old Norse.
ONI Office of Naval Intelligence.
onomat. onomatopoeia; onomatopoeic.
Ont. Ontario.
OOD, O.O.D. Officer of the Day; Officer of the Deck.
O.P., OP, o.p., op out of print; overprint; overproof.
OP observation post.
op. operation; opposite; work [L *opus*]; works [L *opera*].
OPA, O.P.A. Office of Price Administration.
op. cit. in the work cited [L *opere citato*].
ophthal. ophthalmology.
opp. oppose; opposed; opposite.
opt. optative; optical; optics.
o.r. owner's risk.
OR, O.R. Operating room.
orat. oratorical; oratory.
O.R.C. Officers' Reserve Corps **(ORC)**; Orange River Colony.
orch. orchestra; orchestral.
ord. ordained; order; ordinal; ordinance; ordinary; ordnance.
ordn. ordnance.
Ore., Oreg. Oregon.
org. organic; organism; organized.
orig. original; originally.
Ork. Orkney (Islands).
ornith., ornithol. ornithological; ornithologist; ornithology.
orth. orthopedic; orthopedics.
O.S. ordinary seaman.
O.S., o/s, O/S Old Style.
OS, OS., O.S Old Saxon.
Os osmium.
osc. oscillating; oscillator.
Osc. Oscar.
O.S.F. Order of Saint Francis.
O.Sp. Old Spanish.
OSRD, O.S.R.D. Office of Scientific Research and Development.
OSS, O.S.S. Office of Strategic Services.
OSSR, O.S.S.R. Office of Selective Service Records.
osteo. osteopath; osteopathy.
Osw. Oswald.
O.T. Old Testament; on truck; overtime.
OTC, O.T.C. Officer's Training Corps.
otol. otology.
OTS, O.T.S. Officer Training School.
ott. octave [It. *ottava*].
O.U.A.M. Order of United American Mechanics.
OWI, O.W.I. Office of War Information.
OWMR, O.W.M.R. Office of War Mobilization and Reconversion.
Ox., Oxf. Oxford; Oxfordshire.
Oxon. Oxford [L *Oxonia*]; Oxfordshire **(Oxon).**
oz, oz. ounce(s).
oz.ap. ounce, apothecary.
oz.av. ounce, avoirdupois.
ozs. ounces.
oz.t. ounce, troy.

P

p. after [L *post*]; by [L *per*]; by weight [L *pondere*]; first [L *primus*]; for [L *pro*]; in part [L *partim*]; page; part; participle; past; penny; perch (measure); period; perishable; peseta **(pta)**; peso; pint; pipe; pole (measure); population; post; power; pressure.
p, p. pitcher; softly [It. *piano*].
P. bishop [L *pontifex*]; father [F *père*; L *pater*]; pastor; pengö; people [L *populus*], peso; piaster; pope [L *papa*]; president; priest; prince; prompter (theater).
P parental; phosphorus; pressure; prisoner; pawn (chess).
p- para- (chemistry).
P- pursuit.
p.a. for the year [L *pro anno*]; participial adjective; particular average (insurance); public address; yearly [L *per annum*].
pa. paper.
Pa. Pennsylvania.
Pa protoactinium.
P.A. Passenger Agent; Petroleum Administration **(PA)**; Post Adjutant; Purchasing Agent.
P.A., P/A power of attorney; private account.
PA public-address (system).
PABA, paba para-aminobenzoic acid.
Pac., Pacif. Pacific.
P.A.C. Pan-American Congress; Political Action Committee **(PAC).**
p. ae. equal parts [L *partes aequales*].
paint. painting.
Pal. Palestine.
paleob. paleobotany.
paleog. paleography.
paleontol. paleontology.
palm. palmistry.
pam., pamph. pamphlet.
Pan. Panama.
P. and L. profit and loss.
PAO Panhellenic Guerrilla Organization [Gr. *Panelliniki Antartiki Organosis*].
pap. paper.
PAPA Philippine Alien Property Administration.
par. paragraph; parallel; parenthesis.
Par., Para. Paraguay.
paren. parenthesis.
parens. parentheses.
parl. parliamentary.
part. participle; particular.
pass. everywhere [L *passim*]; passage; passenger; passive.
pat. patent; patented; patrol; pattern.
Pat. Patrick.
patd. patented.
path., pathol. pathology.
Pat. Off. Patent Office.
P.A.U., PAU Pan American Union.
PAW, P.A.W. Petroleum Administration for War.
P.A.Y.E. pay as you earn; pay as you enter.
payt. payment.
Pb lead [L *plumbum*].
P.B. passbook; Plymouth Brethren; Pharmacopoeia Britannica; prayer book.
PBA Public Buildings Administration.
PBS Public Buildings Service.
PBX, P.B.X. private branch (telephone) exchange.
p.c. after meals [L *post cibos*]; postcard.
p.c., Pc., P.C., PC. percent, percentage.
pc. piece; price.
p/c, P/C petty cash; price current.
PC Preparatory Commission.
P.C. Past Commander; Philippine Constabulary; Police Constable; Post Commander; Privy Council.
PCA Progressive Citizens of America.
Pcs. preconscious.
pct. percent.
p.d., P.D. by the day [L *per diem*]; pitch diameter; potential difference.
pd. paid.
Pd palladium.
P.D. Police Department.
PD phenyl dichloride.
Pd.B. Bachelor of Pedagogy [L *Pedagogiae Doctor*].
Pd.D. Doctor of Pedagogy [L *Pedagogiae Baccalaureus*].
Pd.M. Master of Pedagogy [L *Pedagogiae Magister*].
P.E. Petroleum Engineer; Presiding Elder; probable error; Protestant Episcopal.
Pe Peter.
ped. pedal; pedestal; pedestrian.
PEEA Political Committee of National Liberation [Gr. *Politiki Epitropi Ethnikou Apeleftheroseos*].
Peeb. Peebles.
P.E.I. Prince Edward Island.
Pemb. Pembrokeshire.
pen. peninsula; penitent; penitentiary.
P.E.N. (International Association of) Poets, Playwrights, Editors, Essayists, and Novelists.
Penn., Penna. Pennsylvania.
penol. penology.
per. period; person.
per an., per ann. by the year [L *per annum*].
perd., perden. dying away [It. *perdendo*].
perf. perfect; perforated; performer.
perh. perhaps.
perm. permanent.
perp. perpendicular; perpetual.
pers. person; personal.
Pers. Persia; Persian.
persp. perspective.
pert. pertaining.
Peru., Peruv. Peruvian.
pet. petroleum.
Pet. Peter; Peters' U.S. Supreme Court Reports.
petn. petition.
petrog. petrography.
petrol. petrology.
p.f. louder [It. *più forte*]; power factor.
pf. perfect; pfennig; preferred.
Pfc, Pfc. Private, first class.
pfd. preferred.
pfg. pfennig.
Pg. Portugal; Portuguese.
P.G. Past Grand (Master); paying guest; postgraduate.
PGA, P.G.A. Professional Golfers Association.
PH Photographer's Mate.
P.H. Purple Heart.
ph. phrase.
Ph phenyl.
PHA Public Housing Administration (Authority).
phar., pharm. pharmaceutical; pharmacist; pharmacopoeia; pharmacy.
Phar. B. Bachelor of Pharmacy [L *Pharmaciae Baccalaureus*].
Phar. D. Doctor of Pharmacy [L *Pharmaciae Doctor*].
Phar. M. Master of Pharmacy [L *Pharmaciae Magister*].
pharmacol. pharmacology.
Ph.B. Bachelor of Philosophy [L *Philosophiae Baccalaureus*].
Ph.C. Pharmaceutical Chemist; Philosopher of Chiropractic.
Ph.D. Doctor of Philosophy [L

Philosophiae Doctor].

phil., philos. philosopher; philosophical; philosophy.

Phil. Philemon; Philip; Philippians; Philippines.

Phila. Philadelphia.

Philem. Philemon.

philol. philology.

philos. philosopher; philosophical; philosophy.

Phm Philemon.

phon. phonetic; phonetics; phonology.

phonet. phonetic; phonetics.

phonog. phonography.

phonol. phonology.

phot., photog. photograph; photographic; photography.

photom. photometrical; photometry.

Php Philippians.

phr. phrase.

phren., phrenol. phrenological; phrenology.

PHS Public Health Service.

phys. physical; physician; physicist; physics.

physiol. physiological; physiology.

phytogeog. phytogeography.

P.I. Philippine Islands.

PI Printer.

Pi., pias. piaster.

PICAO Provisional International Civil Aviation Organization.

pict. pictorial; picture.

pil. pill [L *pilula*].

pinx. he (or she) painted [L *pinxit*].

pizz. pizzicato.

pk. pack; park; peak; peck.

pkg. package(s).

pkt. packet.

pl. place; plate; plural.

plat. platform; platoon.

plen. plenipotentiary.

plf., plff. plaintiff.

plu. plural.

plup., plupf. pluperfect.

plur. plural; plurality.

p.m., P.M. after death [L *post mortem*]; afternoon [L *post meridiem*]; post-mortem.

P.M. Pacific Mail; Past Master; Paymaster; Police Magistrate; Postmaster; Prime Minister; Provost Marshal.

PM Patternmaker.

Pm promethium.

PMA Production and Marketing Administration.

pmk. postmark.

pmkd. postmarked.

PN Personnel Man (U.S. Navy).

p.n., P/N promissory note.

pneum. pneumatic; pneumatics.

p.n.g. a person who is not acceptable [L *persona non grata*].

pnxt. he (or she) painted [L *pinxit*].

p.o., po put-out(s).

p.o., P.O. personnel officer; petty officer; postal order; post office.

Po polonium.

P.O.D. pay on delivery **(p.o.d.)**; Post Office Department.

Pod.D. Doctor of Podiatry.

poet. poetic; poetical; poetry.

P. of H. Patrons of Husbandry.

pol., polit. political; politics.

Pol. Poland; Polish.

pol. econ., polit. econ. political economy.

pom., pomol. pomological; pomology.

pon., pont. pontoon.

pop. popular; popularly; population.

p.o.r. pay on return.

port. portrait.

Port. Portugal; Portuguese.

pos., posit. position; positive.

poss. possible; possibly; possession; possessive.

post. postal.

pot. potential; pottery.

P.O.W., POW Prisoner of War.

p.p., P.P. parcel post; parish priest; postpaid.

pp., p.p. past participle.

pp. pages; privately printed.

pp pianissimo.

PP pellagra preventive (factor).

p.p.c., P.P.C. to take leave [F *pour prendre congé*].

ppd. postpaid; prepaid.

pph. pamphlet.

p.p.i. policy proof of interest.

PPI plan position indicator.

ppl. participle.

ppm., p.p.m. parts per million.

ppp. pianissimo.

ppr., p.pr. present participle.

p.p.s., P.P.S. additional postscript [L *post postscriptum*].

p.q. previous question.

P.Q. Province of Quebec.

pr. pair; pairs; paper; power; preferred; preposition; present; price; priest; prince; printing; pronoun.

PR Parachute Rigger.

Pr., Prov. Provençal.

Pr praseodymium.

P.R. Puerto Rico; proportional representation.

PRA, P.R.A. Public Roads Administration.

preb. prebend; prebendary.

prec. preceding.

precanc. precanceled.

pred. predicate; predication; predicative; prediction.

pref. preface; prefatory; preference; preferred; prefix.

prehist. prehistoric; prehistory.

prelim. preliminary.

prem. premium.

prep. preparation; preparatory; prepare; preposition.

pres. present; presidency; president; presidential; presumptive.

Presb., Presbyt. Presbyter; Presbyterian.

pret. preterit.

prev. previous; previously.

prim. primary; primitive.

prin. principal; principally; principle.

print. printer; printing.

priv. private; privately; privative.

P.R.O., PRO public relations officer.

pro. professional.

prob. probable; probably; problem.

proc. proceedings; process; proclamation.

prod. produce; produced; product.

prof. professor.

prog. progress; progressive.

prom. promenade; promontory.

pron. pronominal; pronoun; pronounced; pronunciation.

pronom. pronominal.

prop. proper; properly; property; proposition; proprietor; proprietary.

propr. proprietor; proprietary.

pros. prosody.

Prot. Protectorate; Protestant.

pro tem. for the time being [L *pro tempore*].

prov. proverbial; providence; provident; province; provincial; provision; provost.

Prov. Provençal; Proverbs **(Pro)**.

Prov. Eng. Provincial English.

prox. next (month) [L *proximo*].

prs. pairs.

prtd. printed.

prtg. printing.

Prus., Pruss. Prussia; Prussian.

p.r.v., P.R.V. to return a call [F *pour rendre visite*].

p.s., P.S. passenger steamer; permanent secretary; postscript; prompt side (theater); public sale.

ps. pieces; pseudonym.

Ps., Ps, Psa. Psalms.

P.S. Police Sergeant; Privy Seal; Public School.

pseud. pseudonym.

p.s.f., psf pounds per square foot.

p.s.i., psi pounds per square inch.

p.ss., P.SS. postscripts.

PST, P.S.T., P.s.t. Pacific standard time.

psych., psychol. psychological; psychologist; psychology.

psychoanal. psychoanalysis.

p.t. for the time being [L *pro tempore*]; post town; postal telegraph.

pt. part; payment; pint(s); point(s); port; preterit.

Pt platinum.

P.T. Physical Training.

P.T.A., PTA Parent-Teachers' Association.

pta. peseta.

ptbl. portable.

ptg. printing.

p.t.o., P.T.O. please turn over.

pts. parts; pints.

pty. proprietary.

Pu plutonium.

pub. public; publication; published; publisher; publishing.

publ. publication; published; publisher.

pulv. pulverized.

punct. punctuation.

pur., purch. purchaser; purchasing.

p.v. par value; post village; priest vicar.

Pvt. Private.

P.W., PW Prisoner of War.

PWA, P.W.A. Public Works Administration.

PWD, P.W.D. Psychological Warfare Division: Public Works Department.

pwr. power.

pwt. pennyweight.

PX post exchange.

pxt. he (or she) painted it [L *pinxit*].

pymt. payment.

pyro., pyrotech. pyrotechnics.

Q

q. farthing [L *quadrans*]; quart; quarter; quarterly; quarto; quasi-; queen; query; question; quetzal; quintal; quire.

Q. Quebec.

Q queen (chess).

q., Q. quarto.

q.b., qb quarterback.

Q.B. Queen's Bench.

QB queen's bishop (chess).

QBP queen's bishop's pawn (chess).

Q.C. Quartermaster Corps; Queen's College; Queen's Counsel.

q.d. as if one should say [L *quasi dicat*]; as if said [L *quasi dictum*]; as if he had said [L *quasi dixisset*].

q.e. which is [L *quod est*].

Q.E.D. which was to be demonstrated [L *quod erat demonstrandum*].

Q.E.F. which was to be done [L *quod erat faciendum*].

Q.E.I., q.e.i. which was to be found out [L *quod erat inveniendum*].

Q.F. quick-firing.

QKt queen's knight (chess).

QKtP queen's knight's pawn (chess).

q.l. as much as you please [L *quantum libet*].

ql. quintal.

qlty. quality.

QM, Q.M. Quartermaster.

QMC, Q.M.C. Quartermaster Corps.

QMG, Q.M.G., Q.M.Gen. Quartermaster General.

qn. question.

QP queen's pawn (chess).

q.pl., Q.P. as much as you wish [L *quantum placeat*].

Qq. quartos.

qq.v. which see (plural) [L *quos vide*].

qr. farthing [L *quadrans*]; quarter; quarterly; quire.

QR queen's rook (chess).

QRP queen's rook's pawn (chess).

qrs. farthings [L *quadrantes*]; quarters; quires.

qrtly. quarterly.

q.s. as much as suffices [L *quantum sufficit*]; quarter section.

qt., qt quart.

qt. quantity.

q.t. quiet.

qto. quarto.

qts., qts quarts.

qu. quart; queen; query; question.

quad. quadrangle; quadrant; quadrat; quadrilateral; quadruple.

quar., quart. quarter; quarterly.

Que. Quebec.

ques. question.

quin., quint. quintuple; quintuplet.

quor. quorum.

quot. quotation; quoted.

q.v. as much as you will [L *quantum vis*]; which see [L *quod vide*].

qy. query.

R

r. range; rare; received; recipe; residence; resides; retired; right; right-hand page [L *recto*]; rises; rod; rook (chess); rubber; ruble.

r roentgen(s).

r., r run(s).

r., R. commonwealth [L *res publica*]; king [L *rex*]; queen [L *regina*]; rabbi; railroad; railway; rector; redactor; river; road; royal; ruble; rupee; take [L *recipe*].

r, R resistance (electrical); royal; ruble.

R gas constant (chemistry); radical (chemistry); radius; ratio; Réaumur; Republican; respond or response (ecclesiastical); ring (chemistry); rook (chess).

Ra radium.

R.A. Rear Admiral; right ascension; Royal Academy (Academician); Royal Artillery.

RA, R.A. Regular Army.

RAAF, R.A.A.F. Royal Australian Air Force.

rad. radical; radio; radius; root [L *radix*].

Rad. Radnorshire.

raddol. becoming softer [It. *raddolcendo*].

RAF, R.A.F. Royal Air Force.

ral., rall. gradually slower [It. *rallentando*].

R.A.M. Royal Academy of Music; Royal Arch Mason.

R.A.R. radio acoustic ranging.

RB Renegotiation Board.

Rb rubidium.

r.b.i., rbi run(s) batted in.

R.C. Red Cross; Reserve Corps; Roman Catholic.

RCAF, R.C.A.F. Royal Canadian Air Force.

R.C.Ch. Roman Catholic Church.

R.C.M. radar countermeasure.

R.C.M.P. Royal Canadian Mounted Police.

rcd. received.

RCP Royal College of Physicians.

rcpt. receipt.

RCS Royal College of Surgeons.

Rct. Recruit.

rd. rod; round.

rd., Rd. reduce; rix-dollar; road.

R.D. Rural Delivery.

Rd radium.

RD Radarman.

R/D refer to drawer (banking).

RDB Research and Development Board.

R.D.I. Royal Designer for Industry.

r.e., re right end.

Re. rupee.

Re rhenium.

R.E. real estate; Reformed Episcopal; Right Excellent; Royal Engineers.

REA, R.E.A. Rural Electrification Administration.

react. reactance (electricity).

Réaum. Réaumur.

rec. receipt; received; receptacle; recipe; record; recorded; recorder; recording.

recd. received.

recip. reciprocal; reciprocity.

recit. recitative.

rec. sec. recording secretary.

rect. receipt; rectified; rector; rectory.

rec't receipt.

red. reduced; reduction.

redisc. rediscount.

redup., redupl. reduplicated; reduplication.

ref. referee; reference; referred; refining; reformation; reformed; refunding.

Ref.Ch. Reformed Church.

refl. reflection; reflective; reflectively; reflex; reflexive.

refrig. refrigeration.

Ref. Sp. Reformed Spelling.

reg. regent; regiment; region; register; registered; registrar; registry; regular; regularly; regulation; regulator.

Reg. Queen [L *regina*]; Reginald.

regt. regent; regiment.

rel. relating; relative; relatively; released; religion; religious.

rel. pron. relative pronoun.

rem. remittance.

Renf. Renfrew.

rep. repair; repeat; report; reporter; representative; reprint; republic.

Rep. Republican.

repr. representing; reprinted.

Repub. Republic; Republican.

req. required; requisition.

res. research; reserve; residence; resides; residue; resigned; resistance; resistor; resolution.

resp. respective; respectively; respiration; respondent.

restr. restaurant.

Resurr. Resurrection.

ret. retain; retired; returned.

retd. retained; returned.

retrog. retrogressive.

rev. revenue; reverse; reversed; review; reviewed; revise; revised; revision; revolution; revolving.

Rev. Revelation **(Rev)**; Reverend.

Rev. Ver. Revised Version.

r.f., rf right field(er); right forward.

r.f., R.F., RF radio frequency; range finder; rapid fire; reducing flame; right field.

R.F. French Republic [F *République française*].

R.F.A. Royal Field Artillery.

R.F.C. Reconstruction Finance Corporation **(RFC)**; Royal Flying Corps.

RFD, R.F.D. Rural Free Delivery.

r.g., rg right guard.

r.h. relative humidity; right hand.

r.h., rh, r.h.b., rhb right halfback.

Rh Rhesus (blood factor); rhodium.

R.H. Royal Highness.

rhap. rhapsody.

rhbdr. rhombohedral.

rheo. rheostat(s).

rhet. rhetoric; rhetorical.

rhin., rhinol. rhinology.

rhomb. rhombic.

r.h.p. rated horsepower.

R.I. King and Emperor [L *Rex et Imperator*]; Queen and Empress [L *Regina et Imperatrix*]; Rhode Island.

Rich., Richd. Richard.

R.I.I.A. Royal Institute of International Affairs.

R.I.P. may he (she, or they) rest in peace [L *requiescat*, or *requiescant, in pace*].

rip. supplementary (music) [It. *ripieno*].

rit. retarded (music) [It. *ritardando*].

riv. river.

RJ road junction.

rkva reactive kilovolt-ampere.

rm. ream; room.

Rm., R.M., RM, r.m. Reichsmark(s).

RM Radioman.

R.M.A. Royal Military Academy (Woolwich).

R.M.C. Royal Military College (Sandhurst).

rms. reams; rooms.

rms, r.m.s. root mean square.

R.M.S. Railway Mail Service; Royal Mail Service (or Ship).

Rn radon.

R.N. registered nurse; Royal Navy.

R.N.A.S. Royal Naval Air Service.

R.N.R. Royal Naval Reserve.

R.N.V.R. Royal Naval Volunteer Reserve.

R.N.W.M.P. Royal Northwest Mounted Police.

RNZAF, R.N.Z.A.F. Royal New Zealand Air Force.

ro. recto; rood.

Rob., Robt. Robert.

Rog. Roger.

rom. roman (type).

Rom. Roman; Romance; Romania; Romanian; Romans **(Rom)**.

R.O.P. record of production; run of paper.

Ros. Roscommon.

rot. rotating; rotation.

ROTC, R.O.T.C. Reserve Officers' Training Corps (Camp).

roul. roulette (philately).

Roum. Roumania; Roumanian.

Rox. Roxburgh.

roy. royal.

R.P. Reformed Presbyterian; Regius Professor.

R.P.D. Doctor of Political Science [L *Rerum Politicarum Doctor*].

RPF Reunion of the French People [F *Rassemblement du Peuple Français*].

rpm, r.p.m. revolutions per minute.

R.P.O. Railway Post Office.

rps, r.p.s. revolutions per second.

rpt. report.

r.q., R.Q. respiratory quotient.

R.R. railroad **(RR.)**; Right Reverend.

rr. very rarely [L *rarissime*].

RRB, R.R.B. Railroad Retirement Board.

rs, RS. reis; rupees.

R.S. Recording Secretary; Reformed Spelling; Revised Statutes.

R.S.A. Royal Society of Arts.

R.S.F.S.R., RSFSR Russian Soviet Federated Socialist Republic.

R.S.S. Fellow of the Royal Society [L *Regiae Societatis Sodalis*].

R.S.V., RSV Revised Standard Version (Bible).

r.s.v.p., R.S.V.P. please reply [F *répondez s'il vous plaît*].

r.t., rt right tackle.

rt. right.

Rt. Hon. Right Honorable.

Rt. Rev. Right Reverend.

Rts. rights.

Ru Ruth; ruthenium.

rub. ruble.
Rud. Rudolf.
Rum. Rumania; Rumanian.
Rus., Russ. Russia; Russian.
russ. russia (leather).
Rut., Rutd., Rutl. Rutland; Rutlandshire.
R.V. Revised Version (Bible).
rva, RVA reactive volt-ampere.
R.W. Right Worshipful; Right Worthy.
Ry., ry. railway.

S

s. sacral; second; section; see; semi-; series; set; shilling [L *solidus*]; sign; signed; silver; singular; sire; solo; son; sou; steamer; stem; stock; substantive; sucre; sun; surplus.
s., S. buried [L *sepultus*]; fellow [L *socius* or *sodalis*]; lies [L *situs*]; saint; school; scribe; secondary; senate; singular; socialist; society; soprano; steel.
s stere(s).
S., S, s., s south; southern.
s- symmetrical (chemistry).
S. Sabbath; Saturday; Saxon; Senate; September; Signor; Sunday.
S., Sig. signature.
S Seaman; sulfur; knight (in chess) [G *Springer*].
s.a. according to art [L *secundum artem*]; without year [L *sine anno*]; semiannual; small arms.
S.A. Salvation Army; sex appeal; South Africa; South America; South Australia; Storm Division of Troops [G *Sturmabteilung*].
Sa samarium.
Sa Samuel.
SA arsine.
SA Seaman Apprentice.
Sab. Sabbath.
SACEUR Supreme Allied Commander Europe.
S.A.E., SAE Society of Automotive Engineers.
S.Afr. South Africa; South African.
Salop, Salop. Shropshire [Anglo-French *Salopescira* from Old English *Scrobbesbyrig(scīr)* Shrewsbury(shire)].
Salv. Salvador; Salvator.
Sam., Saml. Samuel.
S.Am., S.Amer. South America; South American.
S.Am.Ind. South American Indian.
Sans., Sansk. Sanskrit.
S.ap. apothecary's scruple.
Sar., Sard. Sardinia.
S.A.R. Sons of the American Revolution **(SAR)**; South African Republic.
S.A.S. Fellow of the Society of Antiquaries [L *Societatis Antiquarium Socius*].
Sask. Saskatchewan.
sat. saturated; saturation.
Sat. Saturday; Saturn.
sav. savings.
Sax. Saxon; Saxony.
s.b., sb stolen base(s).
S.B. Bachelor of Science [L *Scientiae Baccalaureus*]; Shipping Board.
sb. substantive.
sb antimony [L *stibium*].
SBA Small Business Administration.
SbE south by east.
SbW south by west.
s.c. salvage charges; sharp cash; single column; sized and calendered; small capitals (printing); supercalendered.
S.C. Sanitary Corps; Signal Corps; South Carolina; Supreme Court.
SC Security Council (of the United Nations).
sc. he (she) carved or engraved it [L *sculpsit*]; namely [L *scilicet*]; scale; scene; science; screw; scruple.
Sc. Scotch; Scots; Scottish.
Sc scandium; stratocumulus.
Scan., Scand. Scandinavia; Scandinavian.
SCAP Supreme Commander for the Allied Powers.
s.caps small capitals (printing).
Sc.B. Bachelor of Science [L *Scientiae Baccalaureus*].
Sc.D. Doctor of Science [L *Scientiae Doctor*].
sch. school; schooner.
sched. schedule.
schol. scholar; scholastic.
sci. science; scientific.
sci. fa. show cause [L *scire facias*].
scil. namely [L *scilicet*].
Sc.M. Master of Science [L *Scientiae Magister*].
Scot. Scotch; Scotland; Scots; Scottish.
scr. scrip; script; scruple.
Script. Scriptural; Scriptures.
SCS, S.C.S. Soil Conservation Service.
sculp., sculpt. he (or she) carved it [L *sculpsit*]; sculptor; sculptures.
S.C.V. Sons of Confederate Veterans.
SD Steward.
s.d. indefinitely (without date) [L *sine die*].
s.d., S.D. standard deviation.
S.D. Doctor of Science [L *Scientiae Doctor*]; sends greetings [L *salutem dicit*].
S.D., S.Dak. South Dakota.
Se selenium.
SE, S.E., se, s.e. southeast; southeastern.
SEATO Southeast Asia Treaty Organization.
sec. according to [L *secundum*]; secant **(sec)**; second(s); secondary; secretary; section(s); sector.
SEC, S.E.C. Securities and Exchange Commission.
sec.-ft. second-foot.
sech hyperbolic secant.
sec.leg. according to law [L *secundum legem*].
sec.reg. according to rule [L *secundum regulam*].
secs. seconds; sections.
sect. section; sectional.
secy. secretary.
seg. segment.
seismol. seismology.
sel. selected; selection(s).
Selk. Selkirk.
Sem. Seminary; Semitic.
sen., Sen. senate; senator; senior.
sent. sentence.
sep. sepal; separate.
Sep., Sept. September **(Sep)**; Septuagint.
seq. sequel.
seq., seqq. the following [L *sequens, sequentia*].
ser. serial; series; sermon.
Serb. Serbia; Serbian(s).
Serg., Sergt. Sergeant.
serv. servant; service.
Serv. Servia; Servian.
sess. session.
s.f. show cause [L *scire facias*].
sf., sforz., sfz. with emphasis (music) [It. *sforzando, sforzato*].
SFSR, S.F.S.R. Soviet Federated Socialist Republic.
s.g. specific gravity.
sg, s.g. senior grade.
sgd. signed.
Sgt. Sergeant.
SH Ship's Serviceman.
sh. share; sheep (bookbinding); sheet; shilling(s); shunt.
SHA sidereal hour angle.
SHAEF Supreme Headquarters, Allied Expeditionary Forces.
Shak. Shakespeare.
SHAPE Supreme Headquarters Allied Powers (Europe).
SHD, S.H.D. Subsistence Homesteads Division.
Shet. Shetland (Islands).
SHF, S.H.F., shf, s.h.f. superhigh frequency.
shipt., shpt. shipment.
shp, s.hp. shaft horsepower.
shr. share(s).
S.H.S. Fellow of the Historical Society [L *Societatis Historiae Socius*].
shtg. shortage.
sh.tn. short ton.
s.h.v. under this word [L *sub hac voce*].
Si silicon.
S.I. Sandwich Islands; Staten Island (N.Y.).
Sib. Siberia; Siberian.
Sic. Sicilian; Sicily.
sig., Sig. signal; signature; signor; signore; signori.
sigill. seal [L *sigillum*].
sim. simile.
Sim. Simeon; Simon.
sin sine.
sing. singular.
sinh hyperbolic sine.
sist. sister.
s.j. under consideration [L *sub judice*].
S.J. Society of Jesus [L *Societas Jesu*].
S.J.D. Doctor of Juridical Science [L *Scientiae Juridicae Doctor*].
SK Storekeeper, U. S. Navy.
sk. sack.
Skr., Skt. Sanskrit.
s.l. without place (of publication) [L *sine loco*].
s.l.a.n. without place, date, or name [L *sine loco, anno, vel nomine*].
S. lat. south latitude.
Slav. Slavic; Slavonian.
sld. sailed; sealed.
SLIC, S.L.I.C. Savings and Loan Insurance Corporation.
s.l.p. without lawful issue [L *sine legitima prole*].
sm. small.
Sm samarium.
S.M. Master of Science [L *Scientiae Magister*]; Sergeant Major; Soldier's Medal; State Militia.
sm.c., sm.caps small capitals.
smorz. dying away (music) [It. *smorzando*].
s.m.p. without male issue [L *sine mascula prole*].
SN Seaman.
s.n. without name [L *sine nomine*].
Sn. sanitary.
Sn tin [L *stannum*].
s.o., so struck out.
s.o. seller's option.
SO Sonarman.
So. south; southern.
soc. socialist; society.
sociol. sociological; sociology.
Soc. Is. Society Islands.
S. of Sol. Song of Solomon.
sol. solicitor; soluble; solution.
Sol. Solomon.
Sol Song of Solomon.
soln. solution
Som. Somaliland; Somerset; Somersetshire.
Somerset, Somerset. Somersetshire.
son. sonata.
SOP, S.O.P. standard operating procedure.
sop. soprano.
sos., sost., sosten. sustained (music) [It. *sostenuto*].
sov. sovereign.
Sov. Un. Soviet Union.
s.p. single phase; single pole; supraprotest; without issue [L *sine prole*].
SP shore patrol; shore police.
sp. special; species; specific; specimen; spelling; spirit(s).
Sp. Spain; Spaniard; Spanish.
Sp. Am. Spanish American.
SPAR always ready [L *Semper Paratus*] (See text).
S.P.A.S. Fellow of the American Philosophical Society [L *Societatis Philosophiae Americanae Socius*].
SPC South Pacific Commission.
S.P.C.A. Society for the Prevention of Cruelty to Animals.
S.P.C.C. Society for the Prevention of Cruelty to Children.
spec. special; specification; speculation.
specif. specifically.
spg. spring.
sp. gr. specific gravity.
sp. ht. specific heat.
sph. spherical.
spirit. spiritual; spiritualism.
spkr. speaker.
spp. species (plural).
S.P.R. Society for Psychical Research.
spt. seaport.
sq. square; sequence; the following [L *sequentia*].
Sq. Squadron; Square (street).
sq. ft. square foot; square feet.
sq. in. square inch(es).
sq. m., sq. mi. square mile(s).
sq. rd. square rod(s).
sq. yd. square yard(s).
S.P.Q.R. Senate and People of Rome [L *Senatus Populusque Romanus*].
SR Seaman Recruit.
Sr. Senior; Señor; Sir; Sister.
Sr strontium.
S.R. Sons of the Revolution.
Sra. Señora.
S.R.O. standing room only.
Srta. Señorita.
s.s., ss., ss shortstop.
ss namely (in law) [L *scilicet*].
SS. Saints.
S.S. Silver Star; Sunday School; written above [L *supra scriptum*].
SS, S.S., S/S steamship.
SS., SS Nazi guards [G *Schutzstaffeln*].
SSA, S.S.A. Social Security Act (Administration).
SSB, S.S.B. Social Security Board.
S.S.C. Society of the Holy Cross [L *Societas Sanctae Crucis*].
SSE, S.S.E., sse, s.s.e. south-southeast.
S.Sgt., S/Sgt Staff Sergeant.
SSR, S.S.R. Soviet Socialist Republic.
SSS Selective Service System.
SSW, S.S.W., ssw, s.s.w. south-southwest.
s.t. short ton.
st. stand; stanza; stere; stet; stitch; stone; street; strophe.
st., St. statute(s); street.
St. Saint; Straight; Strait.
St stratus (clouds).
sta. station; stationary; stator.
Sta. Santa; Station.
stac., stacc. staccato.
Staffs, Staffs. Staffordshire.
stan. stanchion.
Staph. staphylococcus.
stat. immediately [L *statim*]; static; stationary; statistics; statuary; statue; statute; statutes.
S.T.B. Bachelor of Sacred Theology [L *Sacrae Theologiae Baccalaureus*].
stbd. starboard.
std. standard.
S.T.D. Doctor of Sacred Theology [L *Sacrae Theologiae Doctor*].
Ste. Sainte.
sten. stencil; stenographer; stenography.
steno., stenog. stenographer; stenography.
ster., stg. sterling.
stereo. stereotype.
St.Ex. Stock Exchange.
stge. storage.
stip. stipend; stipendiary; stipulation.
Stir. Stirling; Stirlingshire.
stk. stock.
stor. storage.
S.T.P. Professor of Sacred Theology [L *Sacrae Theologiae Professor*].
stp. stamped.
STP standard temperature and pressure.
str. steamer; strait; string(s).
Strep. streptococcus.
string. in accelerated time (music) [It. *stringendo*].
stud. student.
Su. Sunday; Susan.
sub. subaltern; subject; submarine; subscription; substitute; suburb; suburban; understand (or supply) [L *subaudi*].
subd. subdivision.
subj. subject; subjective; subjectively; subjunctive.
subs. subscription; subsidiary.
subseq. subsequent; subsequently.
subst. substantive; substitute.
succ. successor.
suf., suff. suffix.
Suff. Suffolk.
Suff., Suffr. Suffragan.
sug., sugg. suggested; suggestion.
Sun., Sund. Sunday.
sup. above [L *supra*]; superfine; superior; superlative; supplement; supplementary; supply; supreme.
super. superfine; superintendent; superior; supernumerary.
superl. superlative.
supp., suppl. supplement; supplementary.
supr. supreme.
supt., Supt. superintendent.
sur. surcharged; surplus.
Sur. Surrey.
surg. surgeon; surgery; surgical.
surr. surrender; surrendered.
surv. survey; surveying; surveyor; surviving.
Sus., Suss. Sussex.
susp. suspended.
Suth. Sutherland; Sutherlandshire.
s.v. sailing vessel; under this word [L *sub verbo, sub voce*].
SV Surveyor.
SW, S.W., sw, s.w. southwest; southwestern.
Sw., Swe., Swed. Sweden; Swedish.
S.W.A., S.W.Afr. South-West Africa.
Swit., Switz., Swtz. Switzerland.
SWNCC The State-War-Navy Coordinating Committee.
SWPC, S.W.P.C. Smaller War Plants Corporation.
syl., syll. syllable.
sym. symbol; symmetrical; symphony.
sym- symmetrical.
syn. synchronize; synonym; synonymous; synonymy.
synd. syndicate.
synop. synopsis.
syr. syrup (pharmacy).
Syr. Syria; Syriac; Syrian.
syst. system; systematic.

T

t. in the time of [L *tempore*]; talari; tare; target; teaspoon(s); telephone; temperature; tempo; tenor; tense (grammar); terminal; territorial; territory; thaler; time; tome; ton(s); town; township; transit; transitive; troy (weight); volume [L *tomus*].
T. tablespoon(s); Testament; Tuesday; Trinity; Turkish.
T tantalum; Technician; temperature (absolute); tension (surface); time.
T- triple bond (chemistry).
TA Steward Apprentice.
Ta tantalum.
TAB Technical Assistance Board.
tab. table(s).
TAC Tactical Air Command; Technical Assistance Committee.
tal. qual. as they come (or average quality) [L *talis qualis*].
tan tangent.
tanh hyperbolic tangent.
TAP Technical Assistance Program.
tart. tartaric.
Tas., Tasm. Tasmania; Tasmanian.
taut. tautological; tautology.
t.b. trial balance.
TB, T.B., Tb., Tb, t.b., tb tubercle bacillus; tuberculosis.
Tb terbium.
tbs., tbsp. tablespoon(s); tablespoonful(s).
TC Trusteeship Council (of the United Nations).
Tc technetium.
tc. tierce(s).
tchr. teacher.
TCS traffic control station.
td., td, TD touchdown.
TD Tank Destroyer; Tradevman (training device man, U.S. Navy.).
T.D. Traffic Director; Treasury Department **(TD)**.
t.d.n. total digestible nutrients.
TE Teleman, U. S. Navy.
Te tellurium.
tech. technical; technology.
technol. technology.
Tech.Sgt. Technical Sergeant.
tel. telegram; telegraph; telegraphic; telephone.
telecom. telecommunication(s).
teleg. telegram; telegraph; telegraphic; telegraphy.
temp. in the time of [L *tempore*]; temperature; temporary.
ten. tenement; tenor; tenuto (music).
Tenn. Tennessee.
ter., terr. terrace; territorial; territory.
terat., teratol. teratology.
term. terminal; termination; terminology.
test. testamentary; testator.
Test. Testament.
tetr., tetrag. tetragonal.
Teut. Teuton; Teutonic.
Tex. Texas; Texan.
T.F. Territorial Force (British).
tfr. transfer.
t.g. type genus.
tgt. target.
Th. Theodore; Thomas.
Th., Thur., Thurs. Thursday.
Th Thessalonians; thorium.
T.H. Territory of Hawaii.
Thad. Thaddeus.
Th.B. Bachelor of Theology [L *Theologiae Baccalaureus*].
Th.D. Doctor of Theology [L *Theologiae Doctor*].
Th-Em thoron (thorium emanation).
theat. theater; theatrical.
Theo. Theodora; Theodore; Theodosia.
theol. theologian; theological; theology.
theor. theorem.
theos. theosophical; theosophist; theosophy.
therap. therapeutic; therapeutics.
therm. thermometer.
thermochem. thermochemical; thermochemistry.
thermodynam. thermodynamics.
Thess. Thessalonians; Thessaly.
Tho., Thos. Thomas.
Ti Timothy; titanium.
t.i.d. three times daily [L *ter in die*].
Tim. Timothy.
tinct. tincture.
Tip. Tipperary.
tit. title.

Tit., Tit Titus.
tk. truck.
TKO, t.k.o., T.K.O. technical knockout.
Tl thallium.
T/L time loan.
T.L. trade-last.
t.l.o. total loss only.
t.m. true mean.
TM Torpedoman's Mate.
Tm thulium.
TMort trench mortar.
TN Steward.
tn. ton; train.
Tn thoron.
tng. training.
TNT, T.N.T. trinitrotoluene; trinitrotoluol.
t.o. turnover; turn over.
Tob. Tobias; Tobit.
tonn. tonnage.
top., topog. topographical; topography.
tox., toxicol. toxicology.
t.p. title page.
tp. township; troop.
t.p.r. temperature, pulse, respiration.
tps. townships.
TR Steward Recruit.
Tr terbium.
Tr. Troop.
tr. tare; tincture; trace; train; transitive; translated; translation; translator; transpose; treasurer; trust.
trag. tragedy; tragic.
trans. transactions; transfer; transferred; transitive; translated; translation; translator; transportation; transpose; transverse.
transf. transfer; transference; transferred.
transl. translated; translation(s).
transp. transparent; transportation.
trav. traveler; travels.
treas. treasurer; treasury.
trf, t.r.f., t-r-f tuned radio frequency.
trfd. transferred.
tricl. triclinic (crystal).
trig., trigon. trigonometric; trigonometry.
trim. trimetric (crystal).
tripl. triplicate.
trit. triturate.
trop. tropic; tropical; tropics.
t.s. tensile strength.
T.Sgt., T/Sgt Technical Sergeant.
tsp. teaspoon(s); teaspoonful(s).
T.T. Tanganyika Territory.
Tu., Tues. Tuesday.
Tu., Tu thulium; tungsten.
T.U. Trade Union; Training Unit.
T.U.C., TUC Trades Union Congress.
Turk. Turkey; Turkish.
TV television; terminal velocity.
TVA, T.V.A. Tennessee Valley Authority (Administration).
twp. township.
Ty. Territory.
typ., typo., typog. typographer; typographic; typographical; typography.
typw. typewriter; typewritten.
Tyr. Tyrone.

U

u. and [G *und*].
u., U. uncle; university; upper.
U uranium.
UAW, U.A.W. United Automobile, Aircraft and Agricultural Implement Workers; United Automobile Workers of America (Union).
U.B. United Brethren (in Christ).
u.c. upper case (printing).
ucs. unconscious.
U.C.V. United Confederate Veterans.
U.D.C. Union of Democratic Control (Britain); United Daughters of the Confederacy.
UFO, ufo unidentified flying object(s).
UHF, U.H.F., uhf, u.h.f. ultra-high frequency.
U.J.D. Doctor of Civil and Canon Law [L *Utriusque Juris Doctor*].
U.K. United Kingdom.
Ukr. Ukraine.
ult. ultimate; ultimately.
ult., ulto. last month [L *ultimo*].
UM Underwater Mechanic.
UMT Universal Military Training.
UMTS Universal Military Training Service (System).
UMW, U.M.W. United Mine Workers of America.
UN, U.N. United Nations.
unabr. unabridged.
unb., unbd. unbound (bookbinding).
UNCCP United Nations Conciliation Commission for Palestine.
UNCIO United Nations Conference on International Organization.
UNCIP United Nations Commission for India and Pakistan.
UNCOK United Nations Commission on Korea.
UNCURK United Nations Commission for the Unification and Rehabilitation of Korea.
undsgd. undersigned.
undtkr. undertaker.
UNEDA United Nations Economic Development Administration.
UNESCO, Unesco United Nations Educational, Scientific, and Cultural Organization.
ung. ointment [L *unguentum*].
UNICEF United Nations Children's Fund.
Unit. Unitarian; Unitarianism.
univ. universal; universally; university.
Univ. Universalist.
UNKRA United Nations Korean Reconstruction Agency.
unl. unlimited.
unm. unmarried.
UNO United Nations Organization.
unof. unofficial.
unp. unpaged.
unpub. unpublished.
UNRPR United Nations Relief for Palestine Refugees.
UNRRA United Nations Relief and Rehabilitation Administration.
UNRWA United Nations Relief and Works Agency.
UNSCOB United Nations Special Committee on the Balkans.
U. of S. Afr. Union of South Africa.
U.P., UP Union Pacific (Railroad); United Press.
up. upper.
UPU Universal Postal Union.
U.P.W.A. United Packinghouse Workers of America; United Public Workers of America.
Ur uranium.
urol. urology.
Uru. Uruguay.
u.s. in the place mentioned above [L *ubi supra*]; as above. [L *ut supra*].
US, U.S. United States.
USA, U.S.A. Union of South Africa; United States Army; United States of America.
USAF, U.S.A.F. United States Air Force.
USAFI, U.S.A.F.I. United States Armed Forces Institute.
U.S.C. & G.S. United States Coast and Geodetic Survey.
U.S.C.A. United States Code Annotated.
USCC, U.S.C.C. United States Commercial Company.
USCG, U.S.C.G. United States Coast Guard.
USDA, U.S.D.A. United States Department of Agriculture.
USES, U.S.E.S. United States Employment Service.
USFET United States Forces in the European Theater.
USGS, U.S.G.S. United States Geological Survey.
USHA, U.S.H.A. United States Housing Authority (Administration).
USIA, U.S.I.A. United States Information Agency.
USIBA United States International Book Association.
USIS, U.S.I.S. United States Information Service.
USM, U.S.M. United States Mail; United States Marines; United States Mint.
USMA, U.S.M.A. United States Military Academy.
USMC, U.S.M.C. United States Marine Corps; United States Maritime Commission.
USN, U.S.N. United States Navy.
USNA, U.S.N.A. United States National Army; United States Naval Academy.
USNG, U.S.N.G. United States National Guard.
USNR, U.S.N.R. United States Naval Reserve.
USO, U.S.O. United Service Organizations.
USP, U.S.P. United States Patent; United States Pharmacopoeia **(U. S. Pharm.)**.
U.S.P.H.S., USPHS United States Public Health Service.
U.S.P.O. United States Post Office.
U.S.R. United States Reserves.
U.S.R.C. United States Reserve Corps.
U.S.S. United States Senate; United States Ship (Steamer) **(USS)**.
U.S.S.B., USSB United States Shipping Board.
U.S.S.Ct. United States Supreme Court.
U.S.S.R., USSR Union of Soviet Socialist Republics.
usu. usual; usually.
U.S.V. United States Volunteers.
u.s.w., usw and so forth [G *und so weiter*].
U.S.W.A. United Shoe Workers of America; United Steel Workers of America.
u.t. universal time.
ut. utility.
Ut. Utah.
ut dict. as directed [L *ut dictum*].
ut sup. as above [L *ut supra*].
U.T.W.A. United Textile Workers of America.
ux. wife [L *uxor*].

V

v. of [G *von*]; see [L *vide*]; valve; ventral; verb; verse; version; versus; vicar; vice-; village; vision; vocative; voice; voltage; volunteer(s); von.
v., V. volt; volume.
V. Venerable; Viscount.
V vanadium; vector; velocity; victory; volume.
v.a. verb active; verbal adjective.
va volt-ampere(s).
Va. Virginia.
V.A. Veterans Administration **(VA)**; Vicar Apostolic; Vice Admiral; (Order of) Victoria and Albert.
vac. vacuum.
val. valentine; valuation; value.
VAR visual-aural range.
var. variant; variation; variety; variometer; various.
var reactive volt-ampere.
Vat. Vatican.
v. aux. verb auxiliary.
vb. verb; verbal.
vb. n. verbal noun.
V.C. Veterinary Corps; Vice Chairman; Vice Chancellor; Vice Consul; Victoria Cross.
v.d. vapor density; various dates.
Vd vanadium.
Vd., Vds. you [Sp. *usted, ustedes*].
V.D., VD, v.d. venereal disease.
veg. vegetable; vegetation.
vel. vellum (bookbinding).
Ven. Venerable; Venice; Venus.
Venez. Venezuela.
vent. ventilating; ventilation; ventilator.
ver. verse(s); version.
verb. sap., verb. sat. a word to the wise is sufficient [L *verbum sapienti, verbum satis*].
vers versed sine, versine.
Ver. St. United States [G *Vereinigte Staaten*].
vert. vertebra; vertebrate; vertical.
ves. vessel; vestry; vesicle; vesicular.
vet. veteran; veterinarian; veterinary.
veter. veterinary.
VFW, V.F.W. Veterans of Foreign Wars.
V.G. Vicar General.
v.g. for example [L *verbi gratia*].
VHF, V.H.F., vhf, v.h.f. very high frequency.
v.i. see below [L *vide infra*]; verb intransitive.
V.I. Virgin Islands.
Vi virginium.
vib. vibration.
vic. vicar; vicarage.
Vic., Vict. Victoria; Victorian.
vid. for example (see) [L *vide*].
vil. village.
v.imp. verb impersonal.
VIP, V.I.P. very important person.
v.irr. verb irregular.
vis. visibility; visual.
Vis., Visc., Visct. Viscount; Viscountess.
viv. lively (music) [It. *vivace*].
viz., viz namely [L *videlicet*].
VL, V.L. Vulgar Latin.
VLF, V.L.F., vlf, v.l.f. very low frequency.
vm. voltmeter.
V.M.D. Doctor of Veterinary Medicine [L *Veterinariae Medicinae Doctor*].
v.n., v.neut. verb neuter.
voc. vocational; vocative.
vocab. vocabulary.
vol. volcano; volume; volunteer.
volc. volcanic; volcano.
vols. volumes.
vox pop. voice of the people [L *vox populi*].
voy. voyage.
v.p. various pagings; various places; verb passive; voting pool (stocks).
V.P. Vice President.
V.R. Queen Victoria [L *Victoria Regina*].
v.r. verb reflexive.
V. Rev. Very Reverend.
v.s. see above [L *vide supra*]; vibration seconds (sound); volumetric solution.
vs. versus.
V.S. Veterinary Surgeon.
VSS versions.
v.t. verb transitive.
VT variable timing.
Vt. Vermont.
Vul., Vulg. Vulgate.
vulg. vulgar; vulgarity.
v.v. vice versa.
vv. verses; violins.

W

w. wanting; warden; warehousing; week(s); weight; wide; width; wife; with; word; work.
W, W., w, w. watt; west; western.
w., w won.
W. Wales; Washington; Wednesday; Welsh.
W tungsten [G *wolfram*].
W.A. West Africa; Western Australia.
WAA, W.A.A. War Assets Administration.
W.A.A.C., WAAC Women's Army Auxiliary Corps.
WAAF, W.A.A.F. Women's Auxiliary Air Force (Brit.).
WAAS, W.A.A.S. Women's Auxiliary Army Service (Brit.).
WAC, W.A.C. Women's Army Corps.
WAD, W.A.D. Works Allotment Division.
w.a.e. when actually employed.
WAF, W.A.F. Women in the Air Force.
WAFS, W.A.F.S. Women's Auxiliary Ferrying Squadron.
Wal. Walloon.
Wal., Wallach. Wallachian.
Wall. Wallace's U.S. Supreme Court Reports.
war. warrant.
War, War. Warwickshire.
warrty. warranty.
Wash. Washington.
WASP, W.A.S.P. Women's Air Force Service Pilots.
Wat. Waterford.
watt-hr watt-hour(s).
WAVES, W.A.V.E.S. Women Appointed for Voluntary Emergency Service.
w.b. warehouse book; water ballast; westbound.
W.B., W/B, W/b, W.b. waybill.
WbN west by north.
WbS west by south.
w.c. water closet; without charge.
W.C.T.U. Women's Christian Temperance Union.
WD, W.D. War Department.
Wed. Wednesday.
West., Westm. Westminster.
Westm. Westmeath; Westmorland.
Westmd. Westmorland.
Wex. Wexford.
w.f., wf wrong font (printing).
WFA, W.F.A. War Food Administration.
W.Flem. West Flemish.
WFTU, W.F.T.U. World Federation of Trade Unions.
WGmc., W.Ger. West Germanic.
wh, wh., whr, whr., w.-hr. watt-hour(s).
Wheat. Wheaton's U.S. Supreme Court Reports.
whf. wharf.
WHO World Health Organization.
w.i. when issued (stocks); wrought iron.
W.I. West India; West Indian; West Indies.
Wick. Wicklow.
Wig. Wigtown.
Wilts, Wilts. Wiltshire.
W.Ind. West Indies.
Wis., Wisc. Wisconsin.
Wisd. (book of) Wisdom.
wk. weak; week; work.
wkly. weekly.
wks. weeks; works.
w.l., WL water line; wavelength.
W.L. West Lothian.
WLB, W.L.B. War Labor Board.
wldr. welder.
W. long. west longitude.
wm wattmeter.
Wm. William.
W.M. Worshipful Master.
WMC, W.M.C. War Manpower Commission.
wmk. watermark.
WMO World Meteorological Organization.
WNW, W.N.W., wnw, w.n.w. west-northwest.
WO, W.O. wait order; Warrant Officer.
Worcs, Worcs. Worcestershire.
W.O.W. Woodmen of the World.
WOWS, W.O.W.S. Women Ordnance Workers.
w.p., W.P. weather permitting; wire payment.
WPA, W.P.A. Work Projects Administration; Works Progress Administration.
WPB, W.P.B. War Production Board.
WRA, W.R.A. War Relocation Authority.
WRAC, W.R.A.C. Women's Royal Army Corps.
WRAF, W.R.A.F. Women's Royal Air Force.
WREN, W.R.N.S. Women's Royal Naval Service.
wrnt. warrant.
WSA, W.S.A. War Shipping Administration.
WSW, W.S.W., wsw, w.s.w. west-southwest.
wt., wt weight.
W.Va. West Virginia.
WVS, W.V.S. Women's Volunteer Service (Brit.).
Wyo. Wyoming.

X

x an abscissa; an unknown quantity.
X Christ; Christian; a ten-dollar bill; xenon.
x.c., X.C. ex-coupon.
x.d., X.D., x-div. ex-dividend.
Xe xenon.
xi., X.I., x-int. ex-interest.
Xmas Christmas.
Xn. Christian.
Xnty., Xty. Christianity.
XP an emblem for Christ (see text).
x-ref. cross-reference.
X-rts. ex-rights.
Xtian. Christian.
xyl. xylograph.

Y

y. yard(s); year(s); younger; youngest.
y an ordinate; an unknown quantity.
Y. Young Men's (Women's) Christian Association.
Y admittance in ohms; yen; yttrium.
Yb ytterbrium.
yd. yard(s).
yds. yards.
yeo. yeomanry.
Yid. Yiddish.
Y.M., Y.M.C.A., YMCA Young Men's Christian Association.
Y.M.H.A., YMHA Young Men's Hebrew Association.
YN Yeoman, U. S. Navy.
Yorks, Yorks. Yorkshire.
Y.P.S.C.E. Young People's Society of Christian Endeavor.
yr. year(s); younger; your.
yrs. years; yours.
Y.T. Yukon Territory.
Yt yttrium.
Yuc. Yucatan.
Y.W., Y.W.C.A., YWCA Young Women's Christian Association.
Y.W.H.A. Young Women's Hebrew Association.

Z

z., Z. zone.
z an unknown quantity.
Z atomic number; zenith distance.
z.B. for example [G *zum Beispiel*].
Zec, Zech. Zechariah.
Zep, Zeph. Zephaniah.
Z/F zone of fire.
Zn zinc.
zool. zoological; zoologist; zoology.
Zr zirconium.

CENSUS

PLACES WITHIN THE UNITED STATES AND CANADA

The following list contains the names and population statistics of all incorporated and unincorporated places in the United States and its possessions having a population of 5,000 or more according to the 1960 census, and of places in Canada having a population of 10,000 or more according to the 1956 census for that country. The approximate location of the places named is briefly indicated in abbreviated form.

Further details regarding places of historical or exceptional commercial importance are to be found under the alphabetical arrangement in the regular text pages of the dictionary.

A

Abbeville s. cen. La.	10,414
— nw. S.C.	5,436
Aberdeen ne. Md.	9,679
— ne. Miss.	6,450
— ne. S. Dak.	23,073
— w. Wash.	18,741
Abilene e. cen. Kans.	6,746
— n. cen. Tex.	90,368
Abington e. Mass.	10,607
— se. Pa.	55,831
Ada se. Okla.	14,347
Adams nw. Mass.	11,949
Addison ne. Ill.	6,741
Adrian se. Mich.	20,347
Aguadilla nw. Puerto Rico	15,952
Aiea Oahu, Hawaii	11,826
Aiken sw. S.C.	11,243
Ajo sw. Ariz.	7,049
Akron ne. Ohio	290,351
Alabama state, se. U.S.	3,266,740
Alameda w. Calif.	61,316
— se. Idaho	10,660
Alamogordo s. cen. N.M.	21,723
Alamo Heights s. cen. Tex.	7,552
Alamosa s. cen. Colo.	6,205
Alaska state, nw. N. America	226,167
Albany w. Calif.	14,804
— sw. Ga.	53,890
— e. N.Y.; cap.	129,726
— nw. Ore.	12,926
Albemarle w. N.C.	12,261
Alberta prov., w. Can.	1,123,116
Albert Lea s. Minn.	17,108
Albertville ne. Ala.	8,250
Albion s. Mich.	12,749
— w. N.Y.	5,182
Albuquerque cen. N.M.	201,189
Alcoa e. Tenn.	6,395
Alexander City e. cen. Ala.	13,140
Alexandria e. cen. Ind.	5,582
— cen. La.	40,279
— w. cen. Minn.	6,713
— ne. Va.	91,023
Algona n. Iowa	5,702
Alhambra sw. Calif.	54,807
Alice s. Tex.	20,861
Aliquippa w. Pa.	26,369
Alisal w. cen. Calif.	16,473
Allen Park se. Mich.	37,052
Allentown e. Pa.	108,347
Alliance nw. Nebr.	7,845
— ne. Ohio	28,362
Alma ne. Mich.	8,978
— s. Que., Can.	10,822
Alpena ne. Mich.	14,682
Altadena s. Calif.	40,568
Altamont sw. Ore.	10,811
Alton sw. Ill.	43,047
Altoona w. Pa.	69,407
Altus sw. Okla.	21,225
Alum Rock w. Calif.	18,942
Alva nw. Okla.	6,258
Alvin se. Tex.	5,643
Amarillo n. Tex.	137,969
Ambler se. Pa.	6,765
Ambridge w. Pa.	13,865
American Fork n. cen. Utah	6,373
American Samoa U.S. island group, S. Pacific Ocean	20,040
Americus sw. Ga.	13,472
Ames cen. Iowa	27,003
Amesbury ne. Mass.	9,625
Amherst w. cen. Mass.	10,306
— n. cen. N.S., Can.	10,301
— n. Ohio	6,750
Amityville se. N.Y.	8,318
Amory ne. Miss.	6,474
Amsterdam e. cen. N.Y.	28,772
Anaconda sw. Mont.	12,054
Anacortes nw. Wash.	8,414
Anadarko sw. Okla.	6,299
Anaheim sw. Calif.	104,184
Anchorage s. cen. Alaska	44,237
Andalusia s. Ala.	10,263
Anderson cen. Ind.	49,061
— nw. S.C.	41,316
Andover ne. Mass.	15,878
Andrews w. cen. Tex.	11,135
Angleton se. Tex.	7,312
Annapolis e. Md.; cap.	23,385
Ann Arbor se. Mich.	67,340
Anniston e. Ala.	33,657
Anoka se. Minn.	10,562
Ansonia sw. Conn.	19,819
Antigo ne. Wis.	9,691
Antioch w. cen. Calif.	17,305
Appleton e. Wis.	48,411
Aransas Pass se. Tex.	6,956
Arbutus–Halethorpe–Relay n. Md.	22,402
Arcadia sw. Calif.	41,005
— s. cen. Fla.	5,889
Arcata nw. Calif.	5,235
Archbald ne. Pa.	5,471
Arden–Arcade cen. Calif.	73,352
Ardmore s. cen. Okla.	20,184
Arecibo n. Puerto Rico	28,460
Arizona state, sw. U.S.	1,302,161
Arkadelphia s. Ark.	8,069
Arkansas state, s. cen. U.S.	1,786,272
Arkansas City s. cen. Kans.	14,262
Arlington e. cen. Mass.	49,953
— se. N.Y.	8,317
— e. Tex.	44,775
Arlington County ne. Va.	163,401
Arlington Heights ne. Ill.	27,878
Arnold w. Pa.	9,437
Artesia sw. Calif.	9,993
— se. N.M.	12,000
Arvada e. cen. Colo.	19,242
Arvida n. Que., Can.	12,919
Arvin s. Calif.	5,310
Asbury Park e. cen. N.J.	17,366
Asheboro cen. N.C.	9,449
Asheville w. N.C.	60,192
Ashland ne. Ky.	31,283
— n. cen. Ohio	17,419
— sw. Ore.	9,119
— e. cen. Pa.	5,237
— n. Wis.	10,132
Ashtabula ne. Ohio	24,559
Aston se. Pa.	10,595
Astoria nw. Ore.	11,239
Atascadero sw. Calif.	5,983
Atchison ne. Kans.	12,529
Athens n. cen. Ala.	9,330
— cen. Ga.	31,355
— se. Ohio	16,470
— se. Tenn.	12,103
— e. Tex.	7,086
Atherton n. Calif.	7,717
Athol n. cen. Mass.	10,161
Atlanta nw. Ga.; cap.	487,455
Atlantic sw. Iowa	6,890
Atlantic City se. N.J.	59,544
Atmore sw. Ala.	8,173
Attalla ne. Ala.	8,257
Attleboro se. Mass.	27,118
Atwater w. cen. Calif.	7,318
Auburn e. cen. Ala.	16,261
— cen. Calif.	5,586
— ne. Ind.	6,350
— cen. Mass.	14,047
— sw. Me.	24,449
— w. cen. N.Y.	35,249
— w. cen. Wash.	11,933
Auburndale cen. Fla.	5,595
Audubon sw. N.J.	10,440
Augusta e. Ga.	70,626
— s. cen. Kans.	6,434
— se. Me.	21,680
Aurora ne. Colo.	48,548
— ne. Ill.	63,715
Austin se. Minn.	27,908
— cen. Tex.; cap.	186,545
Avalon sw. Pa.	6,859
Avondale s. cen. Ariz.	6,151
Avon Lake n. Ohio	9,403
Avon Park s. Fla.	6,073
Azusa sw. Calif.	20,497

B

Babylon se. N.Y.	11,062
Baden w. Pa.	6,109
Bainbridge sw. Ga.	12,714
Baker ne. Ore.	9,986
Bakersfield s. cen. Calif.	56,848
Balch Springs ne. Tex.	6,821
Baldwin se. N.Y.	30,204
— sw. Pa.	24,489
Baldwin Park sw. Calif.	33,951
Baldwinsville cen. N.Y.	5,985
Ballinger w. cen. Tex.	5,043
Ballwin e. Mo.	5,710
Baltimore cen. Md.	939,024
Bangor s. cen. Me.	38,912
— e. Pa.	5,766
Banning s. cen. Calif.	10,250
Baraboo s. Wis.	6,672
Barberton n. cen. Ohio	33,805
Barnes nw. Ore.	5,076
Barnstable se. Mass.	13,465
Barre n. cen. Vt.	10,387
Barrie s. Ont., Can.	16,851
Barrington ne. Ill.	5,434
— sw. N.J.	7,943
— e. R.I.	13,826
Barstow s. cen. Calif.	11,644
Bartlesville ne. Okla.	27,893
Bartonville nw. Ill.	7,253
Bartow cen. Fla.	12,849
Bastrop ne. La.	15,193
Batavia ne. Ill.	7,496
— w. N.Y.	18,210
Batesville ne. Ark.	6,207
Bath s. Me.	10,717
— sw. N.Y.	6,166
Baton Rouge ne. La.; cap.	152,419
Battle Creek s. Mich.	44,169
Bay ne. Ohio	14,489
Bayamón n. Puerto Rico	15,267
Bay City e. Mich.	53,604
— se. Tex.	11,656
Bayonne ne. N.J.	74,215
Bay Saint Louis se. Miss.	5,073
Baytown se. Tex.	28,159
Beachwood n. Ohio	6,089
Beacon se. N.Y.	13,922
Beardstown w. cen. Ill.	6,294
Beatrice se. Nebr.	12,132
Beaufort s. cen. S.C.	6,298
Beaumont se. Tex.	119,175
Beaver w. Pa.	6,160
Beaver Dam se. Wis.	13,118
Beaver Falls w. Pa.	16,240
Beaverton nw. Ore.	5,937
Beckley sw. W. Va.	18,642
Bedford s. cen. Ind.	13,024
— ne. Ohio	15,223
— sw. Va.	5,921
Bedford Heights n. Ohio	5,275
Beech Grove s. cen. Ind.	10,973
Beeville s. Tex.	13,811
Belen w. cen. N.M.	5,031
Belfast s. Me.	6,140
Bell sw. Calif.	19,450
Bellaire e. Ohio	11,502
— se. Tex.	19,872
Bellefontaine w. Ohio	11,424
Bellefontaine Neighbors e. Mo.	13,650
Bellefonte cen. Pa.	6,088
Belle Glade se. Fla.	11,273
Belleville sw. Ill.	37,264
— ne. N.J.	35,005
— s. Ont., Can.	20,605
Bellevue n. cen. Ky.	9,336
— e. Nebr.	8,831

1479

Bellevue 1480 Charleston

Place	Pop.
Bellevue w. cen. Ohio	8,286
— sw. Pa.	11,412
— w. cen. Wash.	12,809
Bellflower sw. Calif.	44,846
Bell Gardens sw. Calif.	26,467
Bellingham nw. Wash.	34,688
Bellmawr se. N.J.	11,853
Bellmead e. cen. Tex.	5,127
Bellmore se. N.Y.	12,784
Bellwood ne. Ill.	20,729
Belmar e. cen. N.J.	5,190
Belmont w. cen. Calif.	15,996
— e. Mass.	28,715
— sw. N.C.	5,007
Beloit s. Wis.	32,846
Belpre se. Ohio	5,418
Belton nw. S.C.	5,106
— cen. Tex.	8,163
Belvidere ne. Ill.	11,223
Bemidji n. cen. Minn.	9,958
Bend cen. Ore.	11,936
Benicia w. cen. Calif.	6,070
Bennettsville ne. S.C.	6,963
Bennington sw. Vt.	8,023
Bensenville ne. Ill.	9,141
Benton s. cen. Ark.	10,399
— se. Ill.	7,023
Benton Harbor sw. Mich.	19,136
Benton Heights sw. Mich.	6,112
Berea ne. Ohio	16,592
Bergenfield ne. N.J.	27,203
Berkeley w. cen. Calif.	111,268
— ne. Ill.	5,792
— e. cen. Mo.	18,676
Berkley se. Mich.	23,275
Berlin n. N.H.	17,821
Bernardsville n. cen. N.J.	5,515
Berwick ne. Pa.	13,353
Berwyn ne. Ill.	54,224
Bessemer cen. Ala.	33,054
Bethany cen. Okla.	12,342
Bethel s. Conn.	5,624
— sw. Pa.	23,650
Bethesda cen. Md.	56,527
Bethlehem e. Pa.	75,408
Bethpage-Old Bethpage se. N.Y.	20,515
Bettendorf e. cen. Iowa	11,534
Beverly ne. Mass.	36,108
Beverly Hills sw. Calif.	30,817
— se. Mich.	8,633
Bexley sw. Ohio	14,319
Biddeford sw. Me.	19,255
Big Rapids w. cen. Mich.	8,686
Big Spring cen. Tex.	31,230
Billings cen. Mont.	52,851
Biloxi se. Miss.	44,053
Binghamton cen. N.Y.	75,941
Birmingham cen. Ala.	340,887
— se. Mich.	25,525
Bisbee se. Ariz.	9,914
Bismarck cen. N.Dak.; cap.	27,670
Blackfoot se. Idaho	7,378
Blacksburg w. Va.	7,070
Blackwell n. cen. Okla.	9,588
Blaine e. Minn.	7,570
Blakely ne. Pa.	6,374
Bloomfield ne. N.J.	51,867
Bloomingdale n. N.J.	5,293
Bloomington cen. Ill.	36,271
— s. cen. Ind.	31,357
— se. Minn.	50,498
Bloomsburg ne. Pa.	10,655
Blue Ash sw. Ohio	8,341
Bluefield sw. W. Va.	19,256
Blue Island ne. Ill.	19,618
Bluffton ne. Ind.	6,238
Blythe se. Calif.	6,023
Blytheville ne. Ark.	20,797
Boca Raton se. Fla.	6,961
Bogalusa se. La.	21,423
Bogota ne. N.J.	7,965
Boise sw. Idaho; cap.	34,481
Bonham ne. Tex.	7,357
Boone cen. Iowa	12,468
Boonton ne. N.J.	7,981
Boonville cen. Mo.	7,090
Borger n. Tex.	20,911
Bossier nw. La.	32,776
Boston e. Mass.; cap.	696,197
Boulder n. cen. Colo.	37,718
Bound Brook n. cen. N.J.	10,263
Bountiful n. cen. Utah	17,039
Bowling Green s. cen. Ky.	28,338
— nw. Ohio	13,574
Boynton Beach se. Fla.	10,467
Bozeman sw. Mont.	13,361
Brackenridge w. cen. Pa.	5,697
Braddock w. Pa.	12,337
Bradenton w. cen. Fla.	19,380
Bradford nw. Pa.	15,061
Bradley ne. Ill.	8,082
Brady cen. Tex.	5,338
Brainerd cen. Minn.	12,898
Braintree e. Mass.	31,069
Brampton s. Ont., Can.	12,587
Brandon sw. Man., Can.	24,796
Brantford se. Ont., Can.	51,869
Brattleboro se. Vt.	9,315
Brawley se. Calif.	12,703
Brazil w. cen. Ind.	8,853
Brea sw. Calif.	8,487
Breckenridge n. cen. Tex.	6,273
Breckenridge Hills e. Mo.	6,299
Brecksville n. Ohio	5,435
Bremerton nw. Wash.	28,922
Brenham cen. Tex.	7,740
Brentwood e. cen. Mo.	12,250
— se. N.Y.	15,387
— sw. Pa.	13,706
Brewer s. cen. Me.	9,009
Brewton s. cen. Ala.	6,309
Briarcliff Manor se. N.Y.	5,105
Bridgeport sw. Conn.	156,748
— se. Pa.	5,306
Bridgeton e. Mo.	7,820
— sw. N.J.	20,966
Bridge View ne. Ill.	7,334
Bridgeville sw. Pa.	7,112
Brigham n. Utah	11,728
Brighton e. cen. Colo.	7,055
Bristol cen. Conn.	45,499
— se. Pa.	12,364
— e. cen. Pa. (twp.)	59,298
— e. R.I.	14,570
— ne. Tenn.	17,582
— sw. Va.	17,144
British Columbia prov., sw. Can.	1,398,464
Broadview ne. Ill.	8,588
Broadview Heights n. Ohio	6,209
Brockport w. N.Y.	5,256
Brockton e. Mass.	72,813
Brockville s. Ont., Can.	13,885
Broken Arrow ne. Okla.	5,928
Bronxville se. N.Y.	6,744
Brookfield ne. Ill.	20,429
— n. cen. Mo.	5,694
— se. Wis.	19,812
Brookhaven sw. Miss.	9,885
— se. Pa.	5,280
Brookings e. S. Dak.	10,558
Brookline e. Mass.	54,044
Brooklyn ne. Ohio	10,733
Brooklyn Center se. Minn.	24,356
Brooklyn Park se. Minn.	10,197
Brook Park n. Ohio	12,856
Brown Deer se. Wis.	11,280
Brownfield w. cen. Tex.	10,286
Brownsville nw. Fla.	38,417
— sw. Pa.	6,055
— w. Tenn.	5,424
— s. Tex.	48,040
Brownwood cen. Tex.	16,974
Brunswick se. Ga.	21,703
— n. Ohio	6,453
— se. Me.	9,444
Bryan nw. Ohio	7,361
— e. Tex.	27,542
Buchanan sw. Mich.	5,341
Buckhannon n. cen. W.Va.	6,386
Bucyrus n. cen. Ohio	12,276
Buena Park sw. Calif.	46,401
Buena Vista w. cen. Va.	6,300
Buffalo nw. N.Y.	532,759
Bunkie e. cen. La.	5,188
Burbank sw. Calif.	90,155
Burkburnett n. Tex.	7,621
Burley s. Idaho	7,508
Burlingame w. cen. Calif.	24,036
Burlington se. Iowa	32,430
— w. N.J.	12,687
— cen. N.C.	33,199
— nw. Vt.	35,531
— se. Wis.	5,856
Burrillville n. R.I.	9,119
Butler nw. N.J.	5,414
— w. Pa.	20,975
Butte sw. Mont.	27,877

C

Place	Pop.
Cadillac nw. Mich.	10,112
Caguas e. cen. Puerto Rico	32,030
Cahokia sw. Ill.	15,829
Cairo sw. Ga.	7,427
Cairo s. Ill.	9,348
Caldwell sw. Idaho	12,230
— ne. N.J.	6,924
Calexico se. Calif.	7,992
Calgary s. Alta., Can.	181,780
California state, w. U.S.	15,717,204
— nw. Pa.	5,978
Calumet City ne. Ill.	25,000
Calumet Park ne. Ill.	8,448
Camas sw. Wash.	5,666
Cambridge e. Mass.	107,716
— cen. Md.	12,239
— e. cen. Ohio	14,562
Camden s. Ark.	15,823
— w. N.J.	117,159
— cen. S.C.	6,842
Cameron cen. Tex.	5,640
Campbell w. cen. Calif.	11,863
— ne. Ohio	13,406
Campbellsville se. Ky.	6,966
Camp Hill se. Pa.	8,559
Canada dom., N.America	16,080,791
Canandaigua w. N.Y.	9,370
Canon City s. cen. Colo.	8,973
Canonsburg sw. Pa.	11,877
Canton w. cen. Ill.	13,588
— e. Mass.	12,771
— w. Miss.	9,707
— w. N.C.	5,068
— n. N.Y.	5,046
— e. Ohio	113,631
Canyon n. Tex.	5,864
Cap-de-la-Madeleine s. Que., Can.	22,943
Cape Girardeau e. Mo.	24,947
Carbondale s. Ill.	14,670
— ne. Pa.	13,595
Caribou n. Me.	8,305
Carlinville sw. Ill.	5,440
Carlisle s. Pa.	16,623
Carlsbad s. Calif.	9,253
— se. N.M.	25,541
Carlstadt ne. N.J.	6,042
Carmi se. Ill.	6,152
Carmichael ne. Calif.	20,455
Carnegie sw. Pa.	11,887
Carol City s. Fla.	21,749
Carpentersville n. Ill.	17,424
Carrizo Springs sw. Tex.	5,699
Carroll w. Iowa	7,682
Carrollton w. Ga.	10,973
Carson s. Calif.	38,059
Carson City w. Nev.	5,163
Carteret ne. N.J.	20,502
Cartersville nw. Ga.	8,668
Carthage sw. Mo.	11,264
— e. Tex.	5,262
Caruthersville se. Mo.	8,643
Casa Grande s. Ariz.	8,311
Casper e. cen. Wyo.	38,930
Castle Shannon sw. Pa.	11,836
Castro Valley cen. Calif.	37,120
Cataño ne. Puerto Rico	25,218
Catasauqua e. Pa.	5,062
Catonsville n. cen. Md.	37,372
Catskill e. N. Y.	5,825
Cayce s. S.C.	8,517
Cayey se. Puerto Rico	19,755
Cedarburg e. Wis.	5,191
Cedar City sw. Utah	7,543
Cedar Falls ne. Iowa	21,195
Cedar Grove ne. N.J.	14,603
Cedarhurst se. N. Y.	6,954
Cedar Lake nw. Ind.	5,766
Cedar Rapids e. Iowa	92,035
Cedartown nw. Ga.	9,340
Celina w. cen. Ohio	7,659
Centereach se. N.Y.	8,524
Center Line se Mich.	10,164
Centerville s. Iowa	6,629
— sw. Pa.	5,088
Central Falls ne. R.I.	19,858
Centralia s. cen. Ill.	13,904
— sw. Wash.	8,586
Centreville sw. Ill.	12,769
Chadron nw. Nebr.	5,079
Chambersburg s. Pa.	17,670
Chamblee w. cen. Ga.	6,635
Champaign e. cen. Ill.	49,583
Chandler sw. Ariz.	9,531
Chanute se. Kans.	10,849
Chapel Hill n. cen. N.C.	12,573
Chariton s. Iowa	5,042
Charleroi sw. Pa.	8,148
Charles City n. Iowa	9,964
Charleston e. Ill.	10,505
— se. Mo.	5,911
— s. S.C.	65,925

Charleston 1481 Douglas

Place	Pop.
Charleston w. W. Va.; cap.	85,796
Charlestown se. Ind.	5,726
Charlotte s. Mich.	7,657
—sw. N.C.	201,564
Charlotte Amalie Virgin Islands; cap.	12,740
Charlottesville n. cen. Va.	29,427
Charlottetown cen. P.E.I., Can.; cap.	16,707
Chatham e. N.J.	9,517
—s. Ont., Can.	22,262
Chattahoochee nw. Fla.	9,699
Chattanooga se. Tenn.	130,009
Cheboygan n. Mich.	5,859
Cheektowaga–Northwest w. N.Y.	52,362
Cheektowaga–Southwest w. N.Y.	12,766
Chehalis sw. Wash.	5,199
Chelsea e. Mass.	33,749
Cheltenham se. Pa.	35,990
Cheraw ne. S.C.	5,171
Cherokee nw. Iowa	7,724
Chester se. Pa.	63,658
—n. cen. S.C.	6,906
Cheverly s. cen. Md.	5,223
Cheviot sw. Ohio	10,701
Cheyenne se. Wyo.; cap.	43,505
Chicago ne. Ill.	3,550,404
Chicago Heights ne. Ill.	34,331
Chicago Ridge ne. Ill.	5,748
Chickasaw sw. Ala.	10,002
Chickasha s. Okla.	14,866
Chico n. cen. Calif.	14,757
Chicopee s. cen. Mass.	61,553
Chicoutimi cen. Que., Can.	24,878
Childress n. Tex.	6,399
Chillicothe n. Mo.	9,236
—s. cen. Ohio	24,957
Chino sw. Calif.	10,305
Chippewa Falls w. cen. Wis.	11,708
Chisholm ne. Minn.	7,144
Christiansted Virgin Islands.	5,088
Chula Vista s. Calif.	42,034
Cicero ne. Ill.	69,130
Cincinnati sw. Ohio	502,550
Circleville s. cen. Ohio	11,059
Clairton sw. Pa.	18,389
Clanton cen. Ala.	5,683
Claremont sw. Calif.	12,633
—w. N.H.	13,563
Claremore ne. Okla.	6,639
Clarendon Hills ne. Ill.	5,885
Clark ne. N.J.	12,195
Clarksburg n. cen. W. Va.	28,112
Clarksdale nw. Miss.	21,105
Clarkston se. Wash.	6,209
Clarksville se. Ind.	8,088
—n. cen. Tenn.	22,021
Clawson se. Mich.	14,795
Clayton e. Mo.	15,245
Clearfield w. cen. Pa.	9,270
—n. Utah	8,833
Clear Lake ne. Iowa	6,158
Clearwater w. Fla.	34,653
Cleburne ne. Tex.	15,381
Cleveland nw. Miss.	10,172
—ne. Ohio	876,050
—se. Tenn.	16,196
—se. Tex.	5,838
Cleveland Heights ne. Ohio	61,813
Cliffside Park n. N.J.	17,642
Clifton ne. N.J.	82,084
Clifton Forge w. Va.	5,268
Clifton Heights se. Pa.	8,005
Clinton cen. Ill.	7,355
—w. Ind.	5,843
—e. Iowa	33,589
—n. cen. Mass.	12,848
—w. Mo.	6,925
—cen. N.C.	7,461
—w. Okla.	9,617
—nw. S.C.	7,937
Cloquet e. Minn.	9,013
Closter ne. N.J.	7,767
Clovis s. cen. Calif.	5,546
—e. N.M.	23,713
Coalinga cen. Calif.	5,965
Coamo s. cen. Puerto Rico	12,163
Coatesville se. Pa.	12,971
Cocoa e. cen. Fla.	12,294
Coeur d'Alene n. Idaho	14,291
Coffeyville se. Kans.	17,382
Cohoes e. N.Y.	20,129
Coldwater s. Mich.	8,880
Coleman cen. Tex.	6,371
College Park w. Ga.	23,469
—w. cen. Md.	18,482
College Station e. cen. Tex.	11,396
Collingdale se. Pa.	10,268
Collingswood w. N.J.	17,370
Collinsville w. Ill.	14,217
Collister sw. Idaho	5,436
Colonial Heights se. Va.	9,587
Colonie e. cen. N.Y.	6,992
Colorado state, w. U.S.	1,753,947
Colorado City cen. Tex.	6,457
Colorado Springs cen. Colo.	70,194
Colton s. Calif.	18,666
Columbia s. Miss.	7,117
—cen. Mo.	36,650
—se. Pa.	12,075
—cen. S.C.; cap.	97,433
—cen. Tenn.	17,624
Columbia Heights e. Minn.	17,533
Columbus w. Ga.	116,779
—se. Ind.	20,778
—e. Miss.	24,771
—e. Nebr.	12,476
—cen. Ohio; cap.	471,316
Commack se. N.Y.	9,613
Commerce sw. Calif.	9,555
—ne. Tex.	5,789
Commerce Town ne. Colo.	8,970
Compton s. Calif.	71,812
Concord w. cen. Calif.	36,208
—w. N.C.	17,799
—s. N.H.; cap.	28,991
Concordia n. Kans.	7,022
Conneaut ne. Ohio	10,557
Connecticut state, ne. U.S.	2,535,234
Connellsville sw. Pa.	12,814
Connersville e. Ind.	17,698
Conroe se. Tex.	9,192
Conshohocken se. Pa.	10,259
Conway cen. Ark.	9,791
—e. S.C.	8,563
Cookeville n. cen. Tenn.	7,805
Coon Rapids e. Minn.	14,931
Coos Bay sw. Ore.	7,084
Copiague se. N.Y.	14,081
Coral Gables s. Fla.	34,793
Coraopolis sw. Pa.	9,643
Corbin s. Ky.	7,119
Cordele sw. Ga.	10,609
Corinth n. Miss.	11,453
Corner Brook w. Newf., Can.	23,225
Corning w. N.Y.	17,085
Cornwall s. Ont., Can.	18,158
Corona s. Calif.	13,336
Coronado s. Calif.	18,039
Corpus Christi s. Tex.	167,690
Corry nw. Pa.	7,744
Corsicana ne. Tex.	20,344
Corte Madera w. Calif.	5,962
Cortez sw. Colo.	6,764
Cortland cen. N.Y.	19,181
Corvallis w. Ore.	20,669
Coshocton e. cen. Ohio	13,106
Costa Mesa sw. Calif.	37,550
Council Bluffs w. Iowa	54,361
Coventry w. cen. R.I.	15,432
Covina sw. Calif.	20,124
Covington n. cen. Ga.	8,167
—n. Ky.	60,376
—se. La.	6,754
—w. Tenn.	5,298
—nw. Va.	11,062
Crafton sw. Pa.	8,418
Cranford ne. N.J.	26,424
Cranston e. R.I.	66,766
Crawfordsville w. Ind.	14,231
Cresskill ne. N.J.	7,290
Crest Hill ne. Ill.	5,887
Crestline n. cen. Ohio	5,521
Creston s. Iowa	7,667
Crestview nw. Fla.	7,467
Crestwood e. Mo.	11,106
Creve Coeur cen. Ill.	6,684
—e. Mo.	5,122
Crockett e. Tex.	5,356
Crookston nw. Minn.	8,546
Crossett se. Ark.	5,370
Croton-on-Hudson se. N.Y.	6,812
Crowley sw. La.	15,617
Crown Point nw. Ind.	8,443
Crystal se. Minn.	24,283
Crystal City s. cen. Tex.	9,101
Crystal Lake n. Ill.	8,314
Cudahy se. Wis.	17,975
Cuero se. Tex.	7,338
Cullman n. Ala.	10,883
Culver City sw. Calif.	32,163
Cumberland w. Md.	33,415
—ne. R.I.	18,792
Cushing cen. Okla.	8,619
Cutler Ridge se. Fla.	7,005
Cuyahoga Falls ne. Ohio	47,922
Cynthiana n. Ky.	5,641

D

Place	Pop.
Daigleville se. La.	5,906
Dalhart n. Tex.	5,160
Dallas nw. Ore.	5,072
—ne. Tex.	679,684
Dalles, The n. Ore.	10,493
Dalton n. Ga.	17,868
Daly City w. Calif.	44,791
Danbury w. Conn.	22,928
Dania se. Fla.	7,065
Dansville w. cen. N.Y.	5,460
Danvers ne. Mass.	21,926
Danville e. Ill.	41,856
—cen. Ky.	9,010
—e. cen. Pa.	6,889
—s. cen. Va.	46,577
Darby se. Pa.	14,059
—se. Pa. (twp.)	12,598
Darlington ne. S.C.	6,710
Dartmouth se. Mass.	14,607
—s. cen. N.S., Can.	21,093
Davenport e. Iowa	88,981
Davis ne. Calif.	8,910
Dawson sw. Ga.	5,062
Dayton n. Ky.	9,050
—sw. Ohio	262,332
Daytona Beach e. Fla.	37,395
Dearborn se. Mich.	112,007
Decatur n. Ala.	29,217
—nw. Ga.	22,026
—cen. Ill.	78,004
—w. Ind.	8,327
Decorah ne. Iowa	6,435
Dedham e. Mass.	23,869
Deerfield ne. Ill.	11,786
Deerfield Beach se. Fla.	9,573
Deer Park se. N.Y.	16,726
—sw. Ohio	8,423
Defiance nw. Ohio	14,553
De Funiak Springs nw. Fla.	5,282
De Kalb n. Ill.	18,486
De Land e. Fla.	10,775
Delano s. cen. Calif.	11,913
Delaware state, e. U. S.	446,292
—sw. N.J.	31,522
—w. cen. Ohio	13,282
Del City cen. Okla.	12,934
Del Paso Heights–Robla ne. Calif.	11,495
Delphos nw. Ohio	6,961
Delray Beach se. Fla.	12,230
Del Rio sw. Tex.	18,612
Deming sw. N.M.	6,764
Demopolis w. cen. Ala.	7,377
Denham Springs e. cen. La.	5,991
Denison ne. Tex.	22,748
Denton ne. Tex.	26,844
Denver cen. Colo.; cap.	493,887
De Pere e. cen. Wis.	10,045
Depew w. N.Y.	13,580
Derby e. cen. Colo.	10,124
—sw. Conn.	12,132
—s. cen. Kans.	6,458
De Ridder sw. La.	7,188
Des Moines cen. Iowa; cap.	208,982
De Soto e. Mo.	5,804
Des Plaines ne. Ill.	34,886
Detroit se. Mich.	1,670,144
Detroit Lakes w. Minn.	5,633
Devils Lake ne. N.Dak.	6,299
Dexter se. Mo.	5,519
Dickinson sw. N.Dak.	9,971
Dickson nw. Tenn.	5,028
Dickson City e. Pa.	7,738
Dillon nw. S.C.	6,173
Dinuba e. cen. Calif.	6,103
District Heights sw. Md.	7,524
District of Columbia federal district, e. U. S.	763,956
Dixon n. Ill.	19,565
Dobbs Ferry se. N.Y.	9,260
Dock Junction se. Ga.	5,417
Dodge City s. Kans.	13,520
Dolton ne. Ill.	18,746
Donaldsonville e. cen. La.	6,082
Donelson n. cen. Tenn.	17,195
Donna s. Tex.	7,522
Donora sw. Pa.	11,131
Dormont sw. Pa.	13,098
Dorval sw. Que., Can.	14,055
Dothan se. Ala.	31,440
Douglas se. Ariz.	11,925
—se. Ga.	8,736

Place	Population
Dover cen. Del.; cap.	7,250
— se. N.H.	19,131
— n. N.J.	13,034
— e. cen. Ohio	11,300
Dowagiac sw. Mich.	7,208
Downers Grove ne. Ill.	21,154
Downey sw. Calif.	82,505
Downingtown se. Pa.	5,598
Doylestown se. Pa.	5,917
Dracut ne. Mass.	13,684
Drummondville s. Que., Can.	26,284
Duarte sw. Calif.	13,962
Dublin cen. Ga.	13,814
Du Bois w. cen. Pa.	10,667
Dubuque e. Iowa	56,606
Duluth e. Minn.	106,884
Dumas nw. Tex.	8,477
Dumont n. N.J.	18,882
Dunbar w. W.Va.	11,006
Duncan s. cen. Okla.	20,009
Dundalk cen. Md.	82,428
Dunedin w. cen. Fla.	8,444
Dunellen cen. N.J.	6,840
Dunkirk w. N.Y.	18,205
Dunmore ne. Pa.	18,917
Dunn e. cen. N.C.	7,566
Duquesne sw. Pa.	15,019
Du Quoin s. Ill.	6,558
Durango sw. Colo.	10,530
Durant s. cen. Okla.	10,467
Durham n. cen. N.C.	78,302
Duryea ne. Pa.	5,626
Dyersburg nw. Tenn.	12,499

E

Place	Population
Eagle Pass s. cen. Tex.	12,094
Eagleton Village e. Tenn.	5,068
Easley nw. S.C.	8,283
East Alton sw. Ill.	7,630
East Aurora w. N.Y.	6,791
East Bakersfield s. cen. Calif.	42,828
East Chicago nw. Ind.	57,669
East Cleveland ne. Ohio	37,991
East Detroit e. Mich.	45,756
East Gary nw. Ind.	9,309
East Grand Forks nw. Minn.	6,998
East Grand Rapids sw. Mich.	10,924
Easthampton w. Mass.	12,326
East Hartford n. cen. Conn.	43,977
East Haven s. cen. Conn.	21,388
Eastlake ne. Ohio	12,467
East Lansing s. cen. Mich.	30,198
East Liverpool e. Ohio	22,306
East Los Angeles sw. Calif.	104,270
Eastman s. cen. Ga.	5,118
East Moline nw. Ill.	16,732
Easton e. Pa.	31,955
East Orange e. N.J.	77,259
East Palestine e. Ohio	5,232
East Paterson ne. N.J.	19,344
East Peoria w. cen. Ill.	12,310
East Point e. Ga.	35,633
East Providence ne. R.I.	41,955
East Ridge se. Tenn.	19,570
East Rochester w. N.Y.	8,152
East Rockaway se. N.Y.	10,721
East Rutherford ne. N.J.	7,769
East Saint Louis sw. Ill.	81,712
East Stroudsburg e. Pa.	7,674
Eastview s. Ont., Can.	19,283
East Whittier sw. Calif.	19,884
East Wilmington se. N.C.	5,520
Eaton w. Ohio	5,034
Eau Claire w. Wis.	37,987
Eaugallie e. cen. Fla.	12,300
Economy w. Pa.	5,925
Ecorse se. Mich.	17,328
Edgewood se. Pa.	5,124
Edina se. Minn.	28,501
Edinburg s. Tex.	18,706
Edmond cen. Okla.	8,577
Edmonds nw. Wash.	8,016
Edmonton cen. Alta., Can.; cap.	226,002
Edmundston nw. N.B., Can.	11,997
Edna se. Tex.	5,038
Edwardsville sw. Ill.	9,996
— ne. Pa.	5,711
Effingham se. Ill.	8,172
Elberton ne. Ga.	7,107
El Cajon sw. Calif.	37,618
El Campo se. Tex.	7,700
El Centro s. Calif.	16,811
El Cerrito w. Calif.	25,437
El Dorado s. Ark.	25,292
— s. Kans.	12,523
Elgin n. Ill.	49,447

Place	Population
Elizabeth e. N.J.	107,698
Elizabeth City ne. N.C.	14,062
Elizabethton ne. Tenn.	10,896
Elizabethtown w. cen. Ky.	9,641
— se. Pa.	6,780
Elk City w. cen. Okla.	8,196
Elk Grove Village ne. Ill.	6,608
Elkhart n. Ind.	40,274
Elkins e. cen. W. Va.	8,307
Elko ne. Nev.	6,298
Elkton ne. Md.	5,989
Ellensburg cen. Wash.	8,625
Ellenville s. N.Y.	5,003
Ellwood City w. Pa.	12,413
Elmhurst ne. Ill.	36,991
Elmira s. N.Y.	46,517
Elmira Heights s. N.Y.	5,157
Elmira Southeast s. cen. N.Y.	6,689
Elmont se. N.Y.	30,138
El Monte sw. Calif.	13,163
Elmwood Park ne. Ill.	23,866
El Paso sw. Tex.	276,687
El Paso de Robles sw. Calif.	6,677
El Reno cen. Okla.	11,015
El Rio sw. Calif.	6,966
El Segundo sw. Calif.	14,219
Elsmere ne. Del.	7,319
Elwood cen. Ind.	11,793
Ely ne. Minn.	5,438
Elyria n. Ohio	43,782
Emmaus e. Pa.	10,262
Emporia cen. Kans.	18,190
— s. Va.	5,535
Endicott s. cen. N.Y.	18,775
Enfield n. cen. Conn.	31,464
Englewood cen. Colo.	33,398
— ne. N.J.	26,057
Enid n. cen. Okla.	38,859
Ennis ne. Tex.	9,347
Enterprise se. Ala.	11,410
Ephrata se. Pa.	7,688
— cen. Wash.	6,548
Erie nw. Pa.	138,440
Escanaba n. Mich.	15,391
Escondido sw. Calif.	16,377
Essex Junction nw. Vt.	5,340
Estherville n. Iowa	7,927
Etna sw. Pa.	5,519
Euclid ne. Ohio	62,998
Eufaula se. Ala.	8,357
Eugene w. Ore.	50,977
Eunice sw. La.	11,326
Eureka nw. Calif.	28,137
Eustis cen. Fla.	6,189
Evansdale n. Iowa	5,738
Evanston ne. Ill.	79,283
Evansville s. Ind.	141,543
Eveleth ne. Minn.	5,721
Everett e. Mass.	43,544
— nw. Wash.	40,304
Evergreen Park ne. Ill.	24,178
Excelsior Springs nw. Mo.	6,473
Exeter se. N.H.	5,896

F

Place	Population
Fairbanks e. cen. Alaska	13,311
Fairborn sw. Ohio	19,453
Fairbury se. Nebr.	5,572
Fairfax cen. Calif.	5,813
— ne. Va.	13,585
Fairfield cen. Ala.	15,816
— ne. Calif.	14,968
— sw. Conn.	46,183
— se. Ill.	6,362
— s. Iowa	8,054
— sw. Ohio	9,734
Fairhaven se. Mass.	14,339
Fair Haven e. cen. N.J.	5,678
Fair Lawn ne. N.J.	36,421
Fairmont s. Minn.	9,745
— nw. W. Va.	27,477
Fair Oaks sw. Ga.	7,969
Fair Plain sw. Mich.	7,998
Fairport w. N.Y.	5,507
Fairview ne. N.J.	9,399
— se. N.Y.	8,626
Fairview Park ne. Ohio	14,624
Fairway w. Kans.	5,398
Fajardo ne. Puerto Rico	12,424
Falcon Heights e. Minn.	5,927
Falfurrias s. Tex.	6,515
Fall River se. Mass.	99,942
Falls ne. Pa.	29,082
Falls Church ne. Va.	10,192
Falls City se. Nebr.	5,598
Fanwood ne. N.J.	7,963

Place	Population
Fargo se. N. Dak.	46,662
Faribault s. Minn.	16,926
Farmers Branch ne. Tex.	13,441
Farmingdale s. N.Y.	6,128
Farmington se. Mich.	6,881
— e. Mo.	5,618
— nw. N.M.	23,786
Farrel w. Pa.	13,793
Fayetteville nw. Ark.	20,274
— cen. N.C.	47,106
— s. cen. Tenn.	6,804
Fenton se. Mich.	6,142
Fergus Falls w. Minn.	13,733
Ferguson e. Mo.	22,149
Fernandina Beach ne. Fla.	7,276
Ferndale se. Mich.	31,347
Festus e. cen. Mo.	7,021
Findlay cen. Ohio	30,344
Fitchburg cen. Mass.	43,021
Fitzgerald s. cen. Ga.	8,781
Flagstaff n. Ariz.	18,214
Flin Flon nw. Man., Can.	10,234
Flint cen. Mich.	196,940
Flora s.cen. Ill.	5,331
Floral Park se. N.Y.	17,499
Florence nw. Ala.	31,649
— n. Ky.	5,837
— e. S.C.	24,722
Florence Graham sw. Calif.	38,164
Florham Park n. N.J.	7,222
Florida state, se. U.S.	4,951,560
Florissant e. Mo.	38,166
Folcroft se. Pa.	7,013
Fond du Lac e. Wis.	32,719
Fontana s. cen. Calif.	14,659
Ford City w. cen. Pa.	5,440
Forest City sw. N.C.	6,556
Forest Grove nw. Ore.	5,628
Forest Hill se. Ont., Can.	19,480
Forest Hills sw. Pa.	8,796
Forest Park sw. Ga.	14,201
— ne. Ill.	14,452
Forrest City e. Ark.	10,544
Fort Atkinson se. Wis.	7,908
Fort Collins n. Colo.	25,027
Fort Dodge cen. Iowa	28,399
Fort Lauderdale se. Fla.	83,648
Fort Lee ne. N.J.	21,815
Fort Madison se. Iowa	15,247
Fort Morgan ne. Colo.	7,379
Fort Myers se. Fla.	22,523
Fort Payne ne. Ala.	7,029
Fort Pierce e. Fla.	25,256
Fort Scott se. Kans.	9,410
Fort Smith w. Ark.	52,991
Fort Stockton w. cen. Tex.	6,373
Fort Thomas n. Ky.	14,896
Fort Valley w. cen. Ga.	8,310
Fort Walton Beach nw. Fla.	12,147
Fort Wayne ne. Ind.	161,776
Fort William w. cen. Ont., Can.	39,464
Fort Worth n. Tex.	356,268
Forty Fort ne. Pa.	6,431
Fostoria nw. Ohio	15,732
Fountain City ne. Tenn.	10,365
Fountain Hill e. Pa.	5,428
Fox Point e. Wis.	7,315
Frackville e. cen. Pa.	5,654
Framingham e. Mass.	44,526
Frankfort cen. Ind.	15,302
— n. cen. Ky.; cap.	18,365
Franklin s. cen. Ind.	15,302
— s. Ky.	5,319
— s. cen. La.	8,673
— cen. Mass.	6,391
— cen. N.H.	6,742
— sw. Ohio	7,917
— nw. Pa.	9,586
— w. cen. Tenn.	6,977
— s. Va.	7,264
— e. Wis.	10,000
Franklin Park ne. Ill.	18,322
Franklin Square se. N.Y.	32,483
Fraser se. Mich.	7,027
Frederick nw. Md.	21,744
— sw. Okla.	5,879
Fredericksburg ne. Va.	13,639
Fredericton s. N.B., Can.; cap.	18,303
Fredonia w. N.Y.	8,477
Freehold e. cen. N.J.	9,140
Freeland e. cen. Pa.	5,068
Freeport n. Ill.	26,628
— se. N.Y.	34,419
— se. Tex.	11,619
Fremont e. Nebr.	19,698
— n. Ohio	17,573
Fresno cen. Calif.	133,929

Fridley 1483 Homestead

Place	Population
Fridley e. Minn.	15,173
Front Royal n. cen. Va.	7,949
Frostburg nw. Md.	6,722
Fullerton s. Calif.	56,180
Fulton cen. Mo.	11,131
—n. cen. N.Y.	14,261

G

Place	Population
Gadsden ne. Ala.	58,088
Gaffney nw. S.C.	10,435
Gainesville n. cen. Fla.	29,701
—n. Ga.	16,523
—ne. Tex.	13,083
Galax sw. Va.	5,254
Galena Park se. Tex.	10,852
Galesburg nw. Ill.	37,243
Galion n. cen. Ohio	12,650
Gallatin n. cen. Tenn.	7,901
Gallipolis s. Ohio	8,775
Gallup nw. N.M.	14,089
Galt s. Ont., Can.	23,738
Galveston se. Tex.	67,175
Gardena sw. Calif.	35,943
Garden City se. Ga.	5,451
—sw. Kans.	11,811
—se. Mich.	38,017
—se. N.Y.	23,948
Garden City Park–Herricks se. N.Y.	15,364
Garden Grove sw. Calif.	84,238
Gardiner s. Me.	6,897
Gardner n. Mass.	19,038
Garfield ne. N.J.	29,253
Garfield Heights ne. Ohio	38,455
Garland ne. Tex.	38,501
Garwood ne. N.J.	5,426
Gary nw. Ind.	178,320
Gastonia sw. N.C.	32,276
Genesco nw. Ill.	5,169
Geneva ne. Ill.	7,646
—w. cen. N.Y.	17,286
—ne. Ohio	5,677
Georgetown ne. Ky.	6,986
—e. S.C.	12,261
—cen. Tex.	5,218
Georgia state, se. U.S.	3,943,116
Gettysburg s. Pa.	7,960
Gilroy cen. Calif.	7,348
Girard ne. Ohio	12,997
Glace Bay ne. N.S., Can.	24,416
Gladewater e. Tex.	5,742
Gladstone nw. Mich.	5,267
—nw. Mo.	14,502
Glasgow s. Ky.	10,069
—n. Mont.	6,398
Glassboro sw. N.J.	10,253
Glassport sw. Pa.	8,418
Glencoe n. Ill.	10,472
Glen Cove se. N.Y.	23,817
Glendale s. cen. Ariz.	15,696
—s. Calif.	119,442
—e. Wis.	9,537
Glendive e. cen. Mont.	7,058
Glendora sw. Calif.	20,752
Glen Ellyn ne. Ill.	15,972
Glenolden se. Pa.	7,249
Glen Ridge ne. N.J.	8,322
Glen Rock ne. N.J.	12,896
Glens Falls e. N.Y.	18,580
Glenview ne. Ill.	18,132
Globe e. cen. Ariz.	6,217
Gloucester n. Mass.	25,789
Gloucester City e. N.J.	15,511
Gloversville ne. N.Y.	21,741
Golden n. cen. Colo.	7,118
Golden Valley se. Minn.	14,559
Goldsboro cen. N.C.	28,873
Gonzales s. cen. Tex.	5,829
Goosport sw. La.	16,778
Goshen n. Ind.	13,718
Goulds s. Fla.	5,121
Grafton ne. N.Dak.	5,885
—n. W.Va.	5,791
Graham n. cen. N.C.	7,723
—n. Tex.	8,505
Granby s. Que., Can.	27,095
Grand Forks ne. N.Dak.	34,451
Grand Haven w. Mich.	11,066
Grand Island s. Nebr.	25,742
Grand Junction w. Colo.	18,694
Grand Ledge s. cen. Mich.	5,165
Grand'Mère s. Que., Can.	14,023
Grand Prairie ne. Tex.	30,386
Grand Rapids sw. Mich.	177,313
—n. cen. Minn.	7,265
Grandview w. cen. Mo.	6,027
Grandview Heights cen. Ohio	8,270
Grandville w. cen. Mich.	7,975
Granite City sw. Ill.	40,073
Grants cen. N.M.	10,274
Grants Pass sw. Ore.	10,118
Great Bend cen. Kans.	16,670
Great Falls cen. Mont.	55,357
Great Neck se. N.Y.	10,171
Greeley n. Colo.	26,314
Green Bay e. Wis.	62,888
Greenbelt w. cen. Md.	7,479
Greencastle w. cen. Ind.	8,506
Greendale e. Wis.	6,843
Greeneville ne. Tenn.	11,759
Greenfield se. Ind.	9,049
—nw. Mass.	14,389
—sw. Ohio	5,422
—e. Wis.	17,636
Greenhills sw. Ohio	5,407
Greenlawn se. N.Y.	5,422
Greensboro n. N.C.	119,574
Greensburg se. Ind.	6,605
—sw. Pa.	17,383
Greentree sw. Pa.	5,226
Greenville s. Ala.	6,894
—cen. Mich.	7,440
—w. Miss.	41,502
—e. N.C.	22,860
—w. Ohio	10,585
—nw. Pa.	8,765
—nw. S.C.	66,188
—ne. Tex.	19,087
Greenwich sw. Conn.	53,793
Greenwood se. Ind.	7,169
—w. Miss.	20,436
—w. S.C.	16,644
Greer nw. S.C.	8,967
Grenada cen. Miss.	7,914
Gretna se. La.	21,967
Griffin w. cen. Ga.	21,735
Griffith nw. Ind.	9,483
Grinnell cen. Iowa	7,367
Grosse Pointe se. Mich.	6,631
Grosse Pointe Farms se. Mich.	12,172
Grosse Pointe Park se. Mich.	15,457
Grosse Pointe Woods se. Mich.	18,580
Groton se. Conn.	10,111
Grove City s. Ohio	8,107
—w. Pa.	8,368
Grover sw. Calif.	5,210
Groves se. Tex.	17,304
Guam U.S. island, N. Pacific Ocean	66,910
Guayama se. Puerto Rico	19,152
Guelph s. Ont., Can.	33,860
Gulfport sw. Fla.	9,730
—s. Miss.	30,204
Guntersville ne. Ala.	6,592
Guthrie cen. Okla.	9,502
Guttenberg ne. N.J.	5,118
Guymon nw. Okla.	5,768

H

Place	Population
Hackensack ne. N.J.	30,521
Hackettstown nw. N.J.	5,276
Haddon sw. N.J.	17,099
Haddonfield w. N.J.	13,201
Haddon Heights w. N.J.	9,260
Hagerstown w. cen. Md.	36,660
Hagginwood cen. Calif.	11,469
Haines City cen. Fla.	9,135
Haledon ne. N.J.	6,161
Hales Corners se. Wis.	5,549
Halifax cen. N.S., Can.; cap.	93,301
Hallandale s. Fla.	10,483
Haltom City ne. Tex.	23,133
Hamburg w. N.Y.	9,145
Hamburg-Lakeshore nw. N.Y.	11,527
Hamden s. Conn.	41,056
Hamilton w. cen. N.J.	65,035
—sw. Ohio	72,354
—s. Ont., Can.	239,625
Hammond nw. Ind.	111,698
—se. La.	10,563
Hammonton s. cen. N.J.	9,854
Hampton e. Va.	89,258
Hamtramck se. Mich.	34,137
Hancock nw. Mich.	5,022
Hanford cen. Calif.	10,133
Hannibal ne. Mo.	20,028
Hanover w. N.H.	5,649
—s. Pa.	15,538
Hapeville w. Ga.	10,082
Harahan s. La.	9,275
Harlingen s. Tex.	41,207
Harmony w. Pa.	5,106
Harper Woods se. Mich.	19,995
Harriman e. Tenn.	5,931
Harrisburg s. Ill.	9,171
—s. cen. Pa.; cap.	79,697
Harrison nw. Ark.	6,580
—ne. N.J.	11,743
—sw. Pa.	15,710
Harrisonburg n. cen. Va.	11,916
Harrodsburg cen. Ky.	6,061
Hartford cen. Conn.; cap.	162,178
—e. Wis.	5,627
Hartford City e. Ind.	8,053
Hartselle n. Ala.	5,000
Hartsville ne. S.C.	6,392
Harvey ne. Ill.	29,071
Harwood Heights ne. Ill.	5,688
Hasbrouck Heights ne. N.J.	13,046
Hastings sw. Mich.	6,375
—se. Minn.	8,965
—s. Nebr.	21,412
Hastings-on-Hudson se. N.Y.	8,979
Hatboro se. Pa.	7,315
Hattiesburg s. Miss.	34,989
Haverford se. Pa.	54,019
Haverhill n. Mass.	46,346
Haverstraw se. N.Y.	5,771
Havre n. Mont.	10,740
Havre de Grace ne. Md.	8,510
Hawaii state, N. Pacific Ocean	632,772
Hawthorne sw. Calif.	33,035
—ne. N.J.	17,735
Hayden w. Fla.	5,471
Hays cen. Kans.	11,947
Haysville s. cen. Kans.	5,836
Hayward w. Calif.	72,700
Hazard se. Ky.	5,958
Hazel Crest ne. Ill.	6,205
Hazel Park se. Mich.	25,631
Hazelwood e. Mo.	6,045
Hazleton e. cen. Pa.	32,056
Hearne e. cen. Tex.	5,072
Helena e. Ark.	11,500
—w. Mont.; cap.	20,227
Hellertown e. Pa.	6,716
Hemet se. Calif.	5,416
Hempfield sw. Pa.	29,704
Hempstead se. N.Y.	34,641
Henderson n. Ky.	16,892
—n. cen. N.C.	12,740
—s. Nev.	12,525
—ne. Tex.	9,666
Hendersonville w. N.C.	5,911
Henrietta Northeast w. N.Y.	6,403
Henryetta e. Okla.	6,551
Hereford nw. Tex.	7,652
Herkimer e. cen. N.Y.	9,396
Hermosa Beach sw. Calif.	16,115
Herrin s. Ill.	9,474
Hershey n. Pa.	6,851
Hialeah se. Fla.	66,972
Hibbing ne. Minn.	17,731
Hickory w. N.C.	19,328
Hicksville se. N.Y.	50,405
Highland nw. Ind.	16,284
Highland Park n. Ill.	25,532
—se. Mich.	38,063
—ne. N.J.	11,049
—ne. Tex.	10,411
High Point e. cen. N.C.	62,063
Hillcrest Heights s. Md.	15,295
Hillgrove sw. Calif.	14,669
Hilliard cen. Ohio	5,633
Hillsboro s. Ohio	5,474
—nw. Ore.	8,232
—ne. Tex.	7,402
Hillsborough w. cen. Calif.	7,554
Hillsdale s. Mich.	7,629
—ne. N.J.	8,734
Hillside ne. Ill.	7,794
—ne. N.J.	22,304
Hilo ne. Hawaii, Hawaii	25,966
Hingham e. Mass.	15,378
Hinsdale n. Ill.	12,859
Hinton se. W.Va.	5,197
Hitchcock se. Tex.	5,216
Hobart n. Ind.	18,680
—sw. Okla.	5,132
Hobbs se. N.M.	26,275
Hoboken ne. N.J.	48,441
Holdenville cen. Okla.	5,712
Holdredge s. Nebr.	5,226
Holland sw. Mich.	24,777
Hollidaysburg sw. Pa.	6,475
Hollister w. cen. Calif.	6,071
Holly Springs n. Miss.	5,621
Hollywood se. Fla.	35,237
Holyoke sw. Mass.	52,689
Homestead s. Fla.	9,152

Homestead 1484 La Verne

Homestead sw. Pa.	7,502	**Jacksonville** e. Tex.	9,590	**Kirksville** ne. Mo.	13,123
Hometown ne. Ill.	7,479	**Jacksonville Beach** ne. Fla.	12,049	**Kirkwood** e. Mo.	29,421
Homewood cen. Ala.	20,289	**Jacques-Cartier** s. Que., Can.	33,132	**Kissimmee** e. Fla.	6,845
— ne. Ill.	13,371	**Jamestown** sw. N.Y.	41,418	**Kitchener** s. Ont., Can.	59,562
Honesdale ne. Pa.	5,569	— se. N.Dak.	15,163	**Kittanning** w. Pa.	6,793
Honolulu se. Oahu; cap. of Hawaii	294,197	**Janesville** se. Wis.	35,164	**Kittery** sw. Me.	8,051
Hoopeston e. Ill.	6,606	**Jasper** n. Ala.	10,799	**Klamath Falls** s. Ore.	16,949
Hope sw. Ark.	8,399	— s. Ind.	6,737	**Knoxville** s. Iowa	7,817
Hopewell se. Va.	17,895	**Jasper Place** sw. Alta., Can.	15,957	— e. Tenn.	111,827
Hopkins se. Minn.	11,370	**Jeanerette** s. La.	5,568	**Kokomo** cen. Ind.	47,197
Hopkinsville w. Ky.	19,465	**Jeannette** sw. Pa.	16,565	**Kosciusko** cen. Miss.	6,800
Hoquiam sw. Wash.	10,762	**Jefferson** s. Pa.	8,280		
Hornell sw. N.Y.	13,907	**Jefferson City** cen. Mo.; cap.	28,228	**L**	
Horseheads s. cen. N.Y.	7,207	**Jefferson Heights** se. La.	19,353		
Hot Springs w. cen. Ark.	28,337	**Jeffersonville** se. Ind.	19,522	**Labrador** nw. Newf., Can.	10,814
Houlton e. Me.	5,976	**Jenkintown** se. Pa.	5,017	**La Canada-Flintridge** sw. Calif.	18,338
Houma s. La.	22,561	**Jennings** sw. La.	11,887	**Lacey** w. cen. Wash.	6,630
Houston e. cen. Tex.	938,219	— e. cen. Mo.	19,965	**Lachine** s. Que., Can.	34,494
Hubbard ne. Ohio	7,137	**Jericho** se. N.Y.	10,795	**Lackawanna** w. N.Y.	29,564
Hudson n. cen. Mass.	7,897	**Jersey City** ne. N.J.	276,101	**Laconia** cen. N.H.	15,288
— e. N.Y.	11,075	**Jersey Shore** n. cen. Pa.	5,613	**La Crosse** sw. Wis.	47,575
Hudson Falls e. cen. N.Y.	7,752	**Jerseyville** sw. Ill.	7,420	**Ladue** e. cen. Mo.	9,466
Hueytown cen. Ala.	5,997	**Jesup** s. Ga.	7,304	**Lafayette** w. cen. Calif.	7,114
Hugo se. Okla.	6,287	**Jim Thorpe** e. Pa.	5,945	— nw. Ind.	42,330
Hull e. Mass.	7,055	**Johnson City** s. cen. N.Y.	19,118	— s. La.	40,400
— sw. Que., Can.	49,243	— ne. Tenn.	29,892	**La Fayette** nw. Ga.	5,588
Humacao e. Puerto Rico	8,009	**Johnston** ne. R.I.	17,160	**Lafayette Southwest** s. La.	6,682
Humboldt w. Tenn.	8,482	**Johnstown** e. cen. N.Y.	10,390	**La Follette** ne. Tenn.	6,204
Huntingdon cen. Pa.	7,234	— sw. Pa.	53,949	**La Grande** ne. Ore.	9,014
Huntington ne. Ind.	16,185	**Joliet** ne. Ill.	66,780	**La Grange** w. Ga.	23,632
— e. N.Y.	11,255	**Joliette** s. Que., Can.	16,940	— ne. Ill.	15,285
— w. W.Va.	83,627	**Jonesboro** ne. Ark.	21,418	**La Grange Park** ne. Ill.	13,793
Huntington Beach sw. Calif.	11,492	**Jonquière** s. Que., Can.	25,550	**Laguna Beach** sw. Calif.	9,288
Huntington Park sw. Calif.	29,920	**Joplin** sw. Mo.	38,958	**La Habra** sw. Calif.	25,136
Huntington Station s. N.Y.	23,438	**Juncos** e. Puerto Rico	6,252	**La Junta** s. Colo.	8,026
Huntington Woods se. Mich.	8,746	**Junction City** ne. Kans.	18,700	**Lake Charles** sw. La.	63,392
Huntsville e. Ala.	72,365	**Juneau** se. Alaska; cap.	6,797	**Lake City** n. Fla.	9,465
— e. Tex.	11,999			— e. cen. S.C.	6,059
Huron n. Ohio	5,197	**K**		**Lake Forest** n. Ill.	10,687
— e. cen. S.Dak.	14,180			**Lake Jackson** se. Tex.	9,651
Hurst n. Tex.	10,165	**Kailua-Lanikai** w. cen. Hawaii,		**Lakeland** w. cen. Fla.	41,350
Hutchinson cen. Kans.	37,574	Hawaii	25,622	**Lake Providence** ne. La.	5,781
— w. cen. Minn.	6,207	**Kalamazoo** sw. Mich.	82,089	**Lakeview** cen. Mich.	10,384
Hyannis se. Mass.	5,139	**Kalispell** nw. Mont.	10,151	**Lake Wales** cen. Fla.	8,346
Hyattsville w. cen. Md.	15,168	**Kane** n. Pa.	5,380	**Lakewood** sw. Calif.	67,126
		Kaneohe e. Oahu, Hawaii	14,414	— e. cen. Colo.	19,338
I		**Kankakee** e. Ill.	27,666	— e. cen. N.J.	13,004
		Kannapolis sw. N.C.	34,647	— ne. Ohio	66,154
Idaho state, nw. U.S.	667,191	**Kansas** state, cen. U.S.	2,178,611	**Lake Worth** se. Fla.	20,758
Idaho Falls e. Idaho	33,161	**Kansas City** e. Kans.	121,901	**Lamar** se. Colo.	7,369
Ilion e. cen. N.Y.	10,199	— w. Mo.	475,539	**Lamarque** se. Tex.	13,969
Illinois state, n. cen. U.S.	10,081,158	**Kaplan** s. La.	5,267	**La Mesa** sw. Calif.	30,441
Imperial Beach sw. Calif.	17,773	**Kaukauna** e. cen. Wis.	10,096	— nw. Tex.	12,438
Independence n. Iowa	7,069	**Keansburg** n. N.J.	6,854	**Lamont** sw. Calif.	6,177
— se. Kans.	11,222	**Kearney** s. Nebr.	14,210	**Lampasas** cen. Tex.	5,061
— nw. Mo.	62,328	**Kearns** n. Utah	17,172	**Lancaster** sw. Calif.	26,012
— n. Ohio	6,868	**Kearny** ne. N.J.	37,472	— s. N.B., Can.	12,371
Indiana state, n. cen. U.S.	4,662,498	**Keene** sw. N.H.	17,562	— w. N.Y.	12,254
— w. cen. Pa.	13,005	**Keizer** nw. Ore.	5,288	— cen. Ohio	29,916
Indianapolis cen. Ind.; cap.	476,258	**Kellogg** n. Idaho	5,061	— se. Pa.	61,055
Indianola s. cen. Iowa	7,062	**Kelso** sw. Wash.	8,379	— se. Pa. (twp.)	10,020
— w. Miss.	6,714	**Kendallville** ne. Ind.	6,765	— n. cen. S.C.	7,999
Indio sw. Calif.	9,745	**Kenilworth** ne. N.J.	8,379	— se. Wis.	7,501
Inglewood sw. Calif.	63,390	**Kenmore** w. N.Y.	21,261	**Lanett** e. Ala.	7,674
— n. cen. Tenn.	26,527	**Kenner** se. La.	17,037	**Langley Park** s. cen. Md.	11,510
Inkster se. Mich.	39,097	**Kennett** se. Mo.	9,098	**Lansdale** se. Pa.	12,612
International Falls n. Minn.	6,778	**Kennewick** se. Wash.	14,244	**Lansdowne** se. Pa.	12,601
Inwood se. N.Y.	10,362	**Kenogami** s. Que., Can.	11,309	**Lansdowne-Baltimore Highlands**	
Iola se. Kans.	6,885	**Kenora** w. cen. Ont., Can.	10,278	n. Md.	13,134
Ionia cen. Mich.	6,754	**Kenosha** se. Wis.	67,899	**Lansford** e. Pa.	5,958
Iowa state, cen. U.S.	2,757,537	**Kent** ne. Ohio	17,836	**Lansing** ne. Ill.	18,098
Iowa City e. Iowa	33,443	— nw. Wash.	9,017	— s. Mich.; cap.	107,807
Iowa Falls cen. Iowa	5,565	**Kenton** nw. Ohio	8,747	**Lantona** se. Fla.	5,021
Irondequoit w. N.Y.	55,337	**Kentucky** state, e. cen. U.S.	3,038,156	**Lapeer** e. Mich.	6,160
Iron Mountain w. Mich.	9,299	**Keokuk** se. Iowa	16,316	**La Porte** nw. Ind.	21,157
Ironton s. Ohio	15,745	**Kermit** w. cen. Tex.	10,465	**La Puente** sw. Calif.	24,723
Ironwood nw. Mich.	10,265	**Kerrville** cen. Tex.	8,901	**Laramie** se. Wyo.	17,520
Irving n. cen. Tex.	45,985	**Ketchikan** s. Alaska	6,483	**Larchmont** se. N.Y.	6,789
Irvington ne. N.J.	59,379	**Kettering** sw. Ohio	54,462	**Laredo** s. Tex.	60,678
— se. N.Y.	5,494	**Kewanee** nw. Ill.	16,324	**Largo** w. Fla.	5,302
Isabella nw. Puerto Rico	7,305	**Keyport** e. N.J.	6,440	**Larkspur** w. Calif.	5,710
Ishpeming n. Mich.	8,857	**Keyser** ne. W.Va.	6,192	**Larned** sw. Kans.	5,001
Ithaca s. cen. N.Y.	28,799	**Key West** s. Fla.	33,956	**La Salle** n. cen. Ill.	11,897
Ivywild e. cen. Colo.	11,065	**Kilgore** ne. Tex.	10,092	— s. Que., Can.	18,973
		Killeen cen. Tex.	23,377	**Las Cruces** s. N.M.	29,367
J		**Kimberly** e. Wis.	5,322	**Las Vegas** se. Nev.	64,405
		Kingsford nw. Mich.	5,084	— ne. N.M. (city)	7,790
Jacinto City se. Tex.	9,547	**Kings Mountain** w. N.C.	8,008	— ne. N.M. (town)	6,028
Jackson s. Mich.	50,720	**Kings Point** se. N.Y.	5,410	**Latrobe** sw. Pa.	11,932
— w. cen. Miss.; cap.	144,422	**Kingsport** ne. Tenn.	26,314	**La Tuque** s. Que., Can.	11,096
— s. Ohio	6,980	**Kingston** se. N.Y.	29,260	**Laurel** n. Md.	8,503
— w. Tenn.	33,849	— se. Ont., Can.	48,618	— se. Miss.	27,889
Jacksonville e. Ala.	5,678	— ne. Pa.	20,261	**Laurens** nw. S.C.	9,598
— n. cen. Ark.	14,488	**Kingsville** s. Tex.	25,297	**Laurinburg** s. cen. N.C.	8,242
— ne. Fla.	201,030	**Kinloch** e. cen. Mo.	6,501	**Lauzon** s. Que., Can.	12,877
— w. cen. Ill.	21,690	**Kinston** e. N.C.	24,819	**Laval-des-Rapides** s. Que., Can.	11,248
— e. N.C.	13,491	**Kirkland** nw. Wash.	6,025	**La Verne** sw. Calif.	6,516

Lawndale   1485   Mesquite

Place	Population
Lawndale sw. Calif.	21,740
Lawrence se. Ind.	10,103
—e. Kans.	32,858
—ne. Mass.	70,933
—se. N.Y.	5,907
Lawrenceberg se. Ind.	5,004
Lawrenceburg s. cen. Tenn.	8,042
Lawrenceville se. Ill.	5,492
Lawton sw. Okla.	61,697
Layton n. Utah	9,027
Lead w. S.Dak.	6,211
Leaksville n. N.C.	6,427
Leaside s. Ont., Can.	16,538
Leavenworth ne. Kans.	22,052
Leawood e. Kans.	7,466
Lebanon cen. Ind.	9,523
—cen. Mo.	8,220
—w. N.H.	9,299
—sw. Ohio	5,993
—w. Ore.	5,858
—w. Pa.	30,045
—cen. Tenn.	10,512
Leed cen. Ala.	6,162
Leesburg n. cen. Fla.	11,172
Lees Summit w. cen. Mo.	8,267
Lehighton e. Pa.	6,318
Leland sw. Miss.	6,295
Le Mars nw. Iowa	6,767
Lemon Grove sw. Calif.	19,348
Lennox sw. Calif.	31,224
Lenoir w. N.C.	10,257
Leominster n. Mass.	27,929
Leonia ne. N.J.	8,384
Lethbridge s. Alta., Can.	29,462
Leucadia sw. Calif.	5,665
Levelland nw. Tex.	10,153
Lévis s. Que., Can.	13,644
Levittown cen. N.J.	11,861
—se. N.Y.	65,276
Lewisberg cen. Pa.	5,523
Lewisburg s. cen. Tenn.	6,338
Lewiston nw. Idaho	12,691
—sw. Me.	40,804
Lewiston Orchards w. Idaho	9,680
Lewistown n. cen. Mont.	7,408
—cen. Pa.	12,640
Lexington cen. Ky.	62,810
—ne. Mass.	27,691
—w. cen. N.C.	16,093
—s. cen. Nebr.	5,572
—w. cen. Va.	7,537
Lexington Park n. Md.	7,039
Liberal s. Kans.	13,813
Liberty w. cen. Mo.	8,909
—e. Tex.	6,127
Libertyville ne. Ill.	8,560
Lima nw. Ohio	51,037
Lincoln cen. Ill.	16,890
—se. Nebr.; cap.	128,521
—ne. R.I.	13,551
Lincoln Heights sw. Ohio (Hamilton County)	7,798
—cen. Ohio (Richland County)	8,004
Lincoln Park se. Mich.	53,933
—n. N.J.	6,048
Lincolnton w. cen. N.C.	5,699
Lincolnwood ne. Ill.	11,744
Linda n. cen. Calif.	6,129
Linden ne. N.J.	39,931
Lindenhurst se. N.Y.	20,695
Lindenwold s. N.J.	7,335
Lindsay s. cen. Calif.	5,397
—s. Ont., Can.	10,110
Linton sw. Ind.	5,736
Litchfield cen. Ill.	7,330
—s. cen. Minn.	5,078
Lititz se. Pa.	5,987
Little Chute e. Wis.	5,099
Little Falls cen. Minn.	7,551
—n. N.J.	9,730
—e. cen. N.Y.	8,935
Little Ferry ne. N.J.	6,175
Littlefield nw. Tex.	7,236
Little Rock cen. Ark.; cap.	107,813
Little Silver e. N.J.	5,202
Littleton e. cen. Colo.	13,670
Liveoak n. Fla.	6,544
Livermore cen. Calif.	16,058
Livingston s. Mont.	8,229
—ne. N.J.	23,124
Livonia se. Mich.	66,702
Lockhart se. Tex.	6,084
Lock Haven cen. Pa.	11,748
Lockland sw. Ohio	5,292
Lockport n. cen. Ill.	7,560
—w. N.Y.	26,443
Lock Raven n. Md.	23,278
Locust Grove se. N.Y.	11,558
Lodi w. cen. Calif.	22,229
—ne. N.J.	23,502
Logan s. cen. Ohio	6,417
—n. Utah	18,731
Logansport cen. Ind.	21,106
Lombard ne. Ill.	22,561
Lomita sw. Calif.	14,893
Lompoc sw. Calif.	14,415
London w. cen. Ohio	6,379
—s. Ont., Can.	101,693
Long Beach s. Calif.	344,168
—se. N.Y.	26,473
Long Branch e. N.J.	26,228
—s. Ont., Can.	10,249
Longmeadow sw. Mass.	10,565
Longmont n. Colo.	11,489
Longueuil s. Que., Can.	14,332
Longview ne. Tex.	40,050
—sw. Wash.	23,349
Lorain n. Ohio	68,932
Los Alamos n. cen. N.M.	12,584
Los Altos w. Calif.	19,696
Los Angeles s. Calif.	2,479,015
Los Banos cen. Calif.	5,272
Los Gatos cen. Calif.	9,036
Louisiana state, s. U.S.	3,257,022
Louisville nw. Ky.	390,639
—e. cen. Miss.	5,066
—ne. Ohio	5,116
Loveland n. Colo.	9,734
—sw. Ohio	5,008
Loves Park n. cen. Ill.	9,086
Lovington se. N.M.	9,660
Lowell ne. Mass.	92,107
Lower Burrell sw. Pa.	11,952
Lower Merion se. Pa.	59,420
Lower Southampton e. Pa.	12,619
Lualualei-Maili Oahu, Hawaii	5,045
Lubbock nw. Tex.	128,691
Ludington w. Mich.	9,421
Ludlow n. Ky.	6,233
—sw. Mass.	13,805
Lufkin e. Tex.	17,641
Lumberton s. cen. N.C.	15,305
Luzerne ne. Pa.	5,118
Lynbrook se. N.Y.	19,881
Lynchburg cen. Va.	54,790
Lyndhurst ne. N.J.	21,867
—nw. Ohio	16,805
Lynn ne. Mass.	94,478
Lynn Gardens ne. Tenn.	5,261
Lynwood sw. Calif.	31,614
—nw. Wash.	7,207
Lyons ne. Ill.	9,936

M

Place	Population
Mableton nw. Ga.	7,127
Macomb w. Ill.	12,135
Macon cen. Ga.	69,764
Madeira sw. Ohio	6,744
Madera cen. Calif.	14,430
Madison sw. Ill.	6,861
—s. Ind.	10,097
—ne. N.J.	15,122
—se. S. Dak.	5,420
—n. Tenn.	13,583
—s. cen. Wis.; cap.	126,706
Madison Heights se. Mich.	33,343
Madisonville s. Ky.	13,110
Magna n. Utah	6,442
Magnolia sw. Ark.	10,651
Magog s. Que., Can.	12,720
Mahanoy City e. cen. Pa.	8,536
Maine state, ne. U.S.	969,265
Malden e. Mass.	57,676
—se. Mo.	5,007
Malone ne. N.Y.	8,737
Malvern cen. Ark.	9,566
Malverne se. N.Y.	9,968
Mamaroneck se. N.Y.	17,673
Manassas Park ne. Va.	5,342
Manati n. cen. Puerto Rico	9,671
Manchester ne. Conn.	42,102
—s. N.H.	88,282
Mandan s. cen. N.Dak.	10,525
Manhattan n. cen. Kans.	22,993
Manhattan Beach sw. Calif.	33,934
Manistee w. Mich.	8,324
Manitoba prov., w. Can.	850,040
Manitowoc e. Wis.	32,275
Mankato se. Minn.	23,797
Mansfield nw. La.	5,839
—n. cen. Ohio	47,325
Manteca cen. Calif.	8,242
Manville cen. N.J.	10,995
Maple Heights ne. Ohio	31,667
Maple Shade sw. N.J.	12,947
Maplewood e. Minn.	18,519
—e. Miss.	12,552
—ne. N.J.	23,977
Maquoketa e. Iowa	5,909
Marblehead ne. Mass.	18,521
Margate City se. N.J.	9,474
Marianna e. Ark.	5,134
—n. Fla.	7,152
Marietta n. cen. Ga.	25,565
—se. Ohio	16,847
Marinette ne. Wis.	13,329
Marion s. Ill.	11,274
—s. cen. Ind.	37,854
—e. cen. Iowa	10,882
—n. cen. Ohio	37,079
—ne. S.C.	7,174
—w. Va.	8,385
Markham ne. Ill.	11,704
Marlborough cen. Mass.	18,819
Marlin e. Tex.	6,918
Marple se. Pa.	19,722
Marquette n. Mich.	19,824
Marshall w. Mich.	6,736
—se. Minn.	6,681
—cen. Mo.	9,572
—nw. Tex.	23,846
Marshalltown cen. Iowa	22,521
Marshfield cen. Wis.	14,153
Martinez w. Calif.	9,604
Martinsburg ne. W.Va.	15,179
Martins Ferry e. Ohio	11,919
Martinsville s. cen. Ind.	7,525
—s. Va.	18,798
Maryland state, e. U.S.	3,100,689
Maryville n. cen. Calif.	9,553
—nw. Mo.	7,807
—e. Tenn.	10,348
Mason City n. Iowa	30,642
Massachusetts state, ne. U.S.	5,148,578
Massapequa se. N.Y.	32,900
Massapequa Park se. N.Y.	19,904
Massena ne. N.Y.	15,478
Massillon ne. Ohio	31,236
Matawan e. cen. N.J.	5,097
Mathis se. Tex.	6,075
Mattoon e. Ill.	19,088
Maumee nw. Ohio	12,063
Mayagüez w. cen. Puerto Rico	50,808
Mayfield s. Ky.	10,762
Mayfield Heights ne. Ohio	13,478
Maynard e. cen. Mass.	7,695
Maysville ne. Ky.	8,484
Maywood sw. Calif.	14,588
—ne. Ill.	27,330
—ne. N.J.	11,460
McAlester e. Okla.	17,419
McAllen s. Tex.	32,728
McComb s. Miss.	12,020
McCook s. Nebr.	8,301
McKeesport sw. Pa.	45,489
McKees Rocks sw. Pa.	13,185
McKinney ne. Tex.	13,763
McMinnville nw. Ore.	7,656
—se. Tenn.	9,013
McPherson cen. Kans.	9,996
Meadville nw. Pa.	16,671
Mechanicsburg s. cen. Pa.	8,123
Mechanicville e. N.Y.	6,831
Medford ne. Mass.	64,971
—sw. Ore.	24,425
Media se. Pa.	5,803
Medicine Hat se. Alta., Can.	20,826
Medina ne. N.Y.	6,681
—ne. Ohio	8,235
Melbourne e. cen. Fla.	11,982
Melrose ne. Mass.	29,619
Melrose Park ne. Ill.	22,291
Melvindale se. Mich.	13,089
Memphis sw. Tenn.	497,524
Menasha e. cen. Wis.	14,647
Mendota n. cen. Ill.	6,154
Mendota Heights se. Minn.	5,028
Menlo Park w. cen. Calif.	26,957
Menominee nw. Mich.	11,289
Menomonee Falls se. Wis.	18,276
Menomonie w. Wis.	8,624
Merced cen. Calif.	20,068
Mercedes s. Tex.	10,943
Meriden cen. Conn.	51,850
Meridian e. Miss.	49,374
Merriam ne. Kans.	5,084
Merrick se. N.Y.	18,789
Merrill n. cen. Wis.	9,451
Mesa s. cen. Ariz.	33,772
Mesquite ne. Tex.	27,526

Place	Population
Methuen ne. Mass.	28,114
Metropolis s. Ill.	7,339
Metuchen ne. N.J.	14,041
Mexia se. Tex.	6,121
Mexico e. Mo.	12,889
Miami se. Fla.	291,688
— ne. Okla.	12,869
Miami Beach se. Fla.	63,145
Miamisburg sw. Ohio	9,893
Miami Shores se. Fla.	8,865
Miami Springs se. Fla.	11,229
Michigan state, n. cen. U.S.	7,823,194
Michigan City n. Ind.	36,653
Middleborough e. Mass.	6,003
Middleburg Heights n. Ohio	7,282
Middle River ne. Md.	10,825
Middlesborough se. Ky.	12,607
Middlesex ne. N.J.	10,520
Middletown cen. Conn.	33,250
— e. cen. N.J.	39,675
— se. N.Y.	23,475
— sw. Ohio	42,115
— e. Pa. (Bucks County)	26,894
— s. cen. Pa. (Dauphin County)	11,182
— se. R.I.	12,675
Midland e. cen. Mich.	27,779
— w. Pa.	6,425
— nw. Tex.	62,625
Midland Park ne. N.J.	7,543
Midlothian ne. Ill.	6,605
Midvale n. Utah	5,802
Midway-Hardwick cen. Ga.	16,909
Midwest City cen. Okla.	36,058
Milan nw. Tenn.	5,208
Miles City e. Mont.	9,665
Milford w. cen. Conn.	41,662
— e. Del.	5,795
— cen. Mass.	13,722
Millbrae w. cen. Calif.	15,873
Millburn ne. N.J.	18,799
Millbury s. cen. Mass.	9,623
Millcreek nw. Pa.	28,441
Milledgeville cen. Ga.	11,117
Millington sw. Tenn.	6,059
Millinocket n. cen. Me.	7,318
Milltown cen. N.J.	5,435
Millvale sw. Pa.	6,624
Mill Valley w. cen. Calif.	10,411
Millville s. N.J.	19,096
Milpitas w. Calif.	6,572
Milton e. Mass.	26,375
— e. cen. Pa.	7,972
Milwaukee se. Wis.	741,324
Milwaukie nw. Ore.	9,099
Mimico s. Ont., Can.	13,687
Minden n. La.	12,785
Mineola se. N.Y.	20,519
Mineral Wells n. cen. Tex.	11,053
Minersville e. cen. Pa.	6,606
Minneapolis e. Minn.	482,872
Minnesota state, n. cen. U.S.	3,413,864
Minnetonka se. Minn.	25,037
Minot nw. N.Dak.	30,604
Mirada Hills sw. Calif.	22,444
Miramar se. Fla.	5,485
Mishawaka n. Ind.	33,361
Mission s. Tex.	14,081
Mississippi state, s. cen. U.S.	2,178,141
Missoula w. Mont.	27,090
Missouri state, cen. U.S.	4,319,813
Mitchell se. S.Dak.	12,555
Moberly n. cen. Mo.	13,170
Mobile sw. Ala.	202,779
Modesto w. cen. Calif.	36,585
Moline nw. Ill.	42,705
Monaca w. Pa.	8,394
Monahans w. cen. Tex.	8,567
Moncton se. N.B., Can.	36,003
Monessen sw. Pa.	18,424
Monett sw. Mo.	5,359
Monmouth w. Ill.	10,372
Monongahela City sw. Pa.	8,388
Monroe s. Ga.	6,826
— n. La.	52,219
— s. Mich.	22,968
— sw. N.C.	10,882
— s. Wis.	8,050
Monroeville sw. Pa.	22,446
Monrovia sw. Calif.	27,079
Montague nw. Mass.	7,836
Montana state, nw. U.S.	674,767
Montclair se. Calif.	13,546
— ne. N.J.	43,129
Montebello sw. Calif.	32,097
Monterey w. Calif.	22,618
Monterey Park sw. Calif.	37,821
Montevideo sw. Minn.	5,693
Montgomery e. cen. Ala.; cap.	134,393
Monticello se. N.Y.	5,222
Montoursville n. cen. Pa.	5,211
Montpelier n. cen. Vt.; cap.	8,782
Montreal s. Que., Can.	1,109,439
Montreal North s. Que., Can.	25,407
Montrose w. Colo.	5,044
Mont-Royal s. Que., Can.	16,990
Mooresville w. cen. N.C.	6,918
Moorhead w. Minn.	22,934
Moose Jaw s. Sask., Can.	29,603
Morehead City e. N.C.	5,583
Morgan City s. La.	13,540
Morganton w. N.C.	9,186
Morgantown n. W.Va.	22,487
Morrilton cen. Ark.	5,997
Morris e. Ill.	7,935
Morristown cen. N.J.	17,712
— ne. Tenn.	21,267
Morrisville se. Pa.	7,790
Morton cen. Ill.	5,325
Morton Grove ne. Ill.	20,533
Moscow nw. Idaho	11,183
Moses Lake cen. Wash.	11,299
Moss Point s. Miss.	6,631
Moultrie s. Ga.	15,764
Mound e. Minn.	5,440
Mounds View e. Minn.	6,416
Mountain Brook n. cen. Ala.	12,680
Mountain Home s. Idaho	9,344
Mountainside ne. N.J.	6,325
Mountain View w. cen. Calif.	30,889
Mount Airy nw. N.C.	7,055
Mount Carmel w. Ill.	8,594
— e. cen. Pa.	10,760
Mount Clemens se. Mich.	21,016
Mount Ephraim sw. N.J.	5,447
Mount Healthy sw. Ohio	6,553
Mount Holly w. cen. N.J.	13,271
Mount Kisco se. N.Y.	6,805
Mountlake Terrace s. Wash.	9,122
Mount Lebanon sw. Pa.	35,361
Mount Oliver sw. Pa.	5,980
Mount Pleasant se. Iowa	7,339
— cen. Mich.	14,875
— sw. Pa.	6,107
— se. S.C.	5,116
— ne. Tex.	8,027
Mount Prospect ne. Ill.	18,906
Mount Rainier w. cen. Md.	9,855
Mount Sterling ne. Ky.	5,370
Mount Vernon s. Ill.	15,566
— sw. Ind.	5,970
— se. N.Y.	76,010
— cen. Ohio	13,284
— nw. Wash.	7,921
Mullins e. S.C.	6,229
Muncie e. Ind.	68,603
Mundelein ne. Ill.	10,526
Munhall sw. Pa.	17,312
Munster nw. Ind.	10,313
Murfreesboro cen. Tenn.	18,991
Murphysboro s. Ill.	8,673
Murray sw. Ky.	9,303
— n. cen. Utah	16,806
Muscatine e. Iowa	20,997
Muskegon w. Mich.	46,485
Muskegon Heights w. Mich.	19,552
Muskogee e. cen. Okla.	38,059
Myrtle Beach e. S.C.	7,834

N

Place	Population
Nacogdoches e. Tex.	12,674
Nampa sw. Idaho	18,013
Nanaimo sw. B.C., Can.	12,705
Nanticoke ne. Pa.	15,601
Napa w. Calif.	22,170
Naperville n. Ill.	12,933
Napoleon nw. Ohio	6,739
Narberth se. Pa.	5,109
Nashua s. N.H.	39,096
Nashville ne. Tenn.; cap.	170,874
Natchez sw. Miss.	23,791
Natchitoches nw. La.	13,924
Natick e. Mass.	28,831
National City s. Calif.	32,771
Naugatuck s. cen. Conn.	19,511
Nazareth e. Pa.	6,209
Nebraska state, cen. U.S.	1,411,330
Nebraska City se. Nebr.	7,252
Nederland se. Tex.	12,036
Needham e. Mass.	25,793
Neenah e. cen. Wis.	18,057
Negaunee n. Mich.	6,126
Neosho sw. Mo.	7,452
Neptune e. N.J.	21,487
Nether Providence se. Pa.	10,380
Nevada state, w. U.S.	285,278
— w. Mo.	8,416
New Albany s. Ind.	37,812
— n. Miss.	5,151
Newark w. cen. Calif.	9,884
— ne. Del.	11,404
— ne. N.J.	405,220
— nw. N.Y.	12,868
— cen. Ohio	41,798
New Bedford s. Mass.	102,477
New Berlin se. Wis.	15,788
New Bern e. N.C.	15,717
Newberry nw. S.C.	8,208
New Braunfels s. cen. Tex.	15,631
New Brighton e. Minn.	6,448
— w. Pa.	8,397
New Britain cen. Conn.	82,201
New Brunswick prov., e. Can.	554,616
— e. cen. N.J.	40,139
Newburgh se. N.Y.	30,979
Newburyport e. Mass.	14,004
New Castle e. Ind.	20,349
— w. Pa.	44,790
New Cumberland s. Pa.	9,257
Newfoundland prov., e. Can.	415,074
— isl., e. Can.	404,260
New Hampshire state, ne. U.S.	606,921
New Hanover s. cen. N.J.	28,528
New Haven s. Conn.	152,048
New Hyde Park se. N.Y.	10,808
New Iberia w. La.	29,062
New Jersey state, e. U.S.	6,066,782
New Kensington w. Pa.	23,485
New London se. Conn.	34,182
— e. cen. Wis.	5,288
New Martinsville nw. W. Va.	5,607
New Mexico state, sw. U.S.	951,023
New Milford ne. N.J.	18,810
Newnan w. Ga.	12,169
New Orleans se. La.	627,525
New Philadelphia e. cen. Ohio	14,241
Newport ne. Ark.	7,007
— n. Ky.	30,070
— w. Ore.	5,344
— se. R.I.	47,049
— se. Tenn.	6,448
— n. Vt.	5,019
Newport Beach sw. Calif.	26,564
Newport News se. Va.	113,662
New Providence ne. N.J.	10,243
New Rochelle se. N.Y.	76,812
New Shrewsbury e. cen. N.J.	7,313
New Smyrna Beach ne. Fla.	8,781
Newton cen. Iowa	15,381
— cen. Kans.	14,877
— e. Mass.	92,384
— n. N.J.	6,563
— w. N.C.	6,658
Newton Falls ne. Ohio	5,038
New Toronto s. Ont., Can.	11,560
New Ulm s. Minn.	11,114
New Waterford ne. N.S., Can.	10,381
New Westminster sw. B.C., Can.	31,665
New York state, ne. U.S.	16,782,304
New York City se. N.Y.	7,781,984
Niagara Falls n. N.Y.	102,394
— s. Ont., Can.	23,563
Niles ne. Ill.	20,393
— sw. Mich.	13,842
— ne. Ohio	19,545
Nitro sw. W.Va.	6,894
Noblesville cen. Ind.	7,664
Nogales s. Ariz.	7,286
Noranda sw. Que., Can.	10,323
Norfolk ne. Nebr.	13,111
— se. Va.	305,872
Normal cen. Ill.	13,357
Norman cen. Okla.	33,412
Norridge ne. Ill.	14,087
Norristown se. Pa.	38,925
North Adams nw. Mass.	19,905
Northampton w. cen. Mass.	30,058
— e. Pa.	8,866
North Andover n. Mass.	10,908
North Arlington ne. N.J.	17,477
North Atlanta w. Ga.	12,661
North Attleborough cen. Mass.	14,777
North Augusta w. S.C.	10,348
North Bay se. Ont., Can.	21,020
North Bellmore se. N.Y.	19,639
North Belmont sw. N.C.	8,328
North Bend sw. Ore.	7,512
North Bergen ne. N.J.	42,387
North Braddock sw. Pa.	13,204
Northbrook ne. Ill.	11,635

Place	Population
North Canton ne. Ohio	7,727
North Carolina state, e. U.S.	4,556,155
North Chicago n. Ill.	20,517
North College Hill sw. Ohio	12,035
North Dakota state, n. cen. U.S.	632,446
Northfield se. Minn.	8,707
— se. N.J.	5,849
North Haledon n. N.J.	6,026
North Highlands cen. Calif.	21,271
North Kansas City nw. Mo.	5,657
North Kingston sw. R.I.	18,977
Northlake ne. Ill.	12,318
North Las Vegas se. Nev.	18,422
North Little Rock cen. Ark.	58,032
North Mankato s. Minn.	5,927
North Merrick se. N.Y.	12,976
North Miami se. Fla.	28,708
North Miami Beach se. Fla.	21,405
North New Hyde Park se. N.Y.	17,929
North Olmsted ne. Ohio	16,290
North Pelham se. N.Y.	5,326
North Plainfield ne. N.J.	16,993
North Platte cen. Nebr.	17,184
Northport w. cen. Ala.	5,245
— se. N.Y.	5,972
North Providence ne. R.I.	18,220
North Richland Hills n. Tex.	8,662
North Riverside ne. Ill.	7,989
North Royalton n. Ohio	9,290
North Sacramento cen. Calif.	12,922
North Saint Paul e. Minn.	8,520
North Shreveport nw. La.	7,701
North Syracuse cen. N.Y.	7,412
North Tarrytown se. N.Y.	8,818
North Tonawanda w. N.Y.	34,757
North Valley Stream se. N.Y.	17,239
North Vancouver se. B.C., Can.	19,951
North Versailles sw. Pa.	13,583
Northwest Territories terr., n. Can.	19,313
Norwalk sw. Calif.	88,739
— sw. Conn.	67,775
— n. Ohio	12,900
Norwich se. Conn.	38,506
— cen. N.Y.	9,175
Norwood e. Mass.	24,898
— sw. Ohio	34,580
— se. Pa.	6,729
Nova Scotia prov., e. Can.	694,717
Novato w. Calif.	17,881
Novi se. Mich.	6,390
Nutley ne. N.J.	29,513
Nyack se. N.Y.	6,062

O

Place	Population
Oak Creek se. Wis.	9,372
Oakdale sw. La.	6,618
Oakland w. Calif.	367,548
— ne. N.J.	9,446
Oakland Park se. Fla.	5,331
Oak Lawn ne. Ill.	27,471
Oakmont se. Pa.	7,504
Oak Park ne. Ill.	61,093
— se. Mich.	36,632
Oak Ridge se. Tenn.	27,169
Oakwood nw. Ohio	10,493
Oberlin n. cen. Ohio	8,198
Ocala n. cen. Fla.	13,598
Ocean City se. N.J.	7,618
Oceanside sw. Calif.	24,971
— se. N.Y.	30,448
Ocean Springs se. Miss.	5,025
Oconomowoc se. Wis.	6,682
Odessa w. Tex.	80,338
Oelwein ne. Iowa	8,282
Ogden n. Utah	70,197
Ogdensburg n. N.Y.	16,122
Ohio state, n. cen. U.S.	9,706,397
Oil City nw. Pa.	17,692
Oildale s. cen. Calif.	19,969
Oklahoma state, s. cen. U.S.	2,328,284
Oklahoma City cen. Okla.; cap.	324,253
Okmulgee e. Okla.	15,951
Olathe e. cen. Kans.	10,987
Old Forge ne. Pa.	8,928
Old Town s. cen. Me.	8,626
Olean sw. N.Y.	21,868
Olivette e. Mo.	8,257
Olney se. Ill.	8,780
Olympia w. Wash.; cap.	18,273
Olyphant ne. Pa.	5,864
Omaha e. Nebr.	301,598
Oneida cen. N.Y.	11,677
Oneida Rolling Mill Park sw. Ohio	6,504
Oneonta cen. N.Y.	13,412
Ontario prov., e. Can.	5,404,933
— s. Calif.	46,617
Ontario e. Ore.	5,101
Opalocka se. Fla.	9,810
Opelika e. Ala.	15,678
Opelousas s. cen. La.	17,417
Opp s. Ala.	5,535
Opportunity e. cen. Wash.	12,465
Oradell ne. N.J.	7,487
Orange s. Calif.	26,444
— ne. N.J.	35,789
— e. Tex.	25,605
Orangeburg cen. S.C.	13,852
Oregon state, nw. U.S.	1,768,687
— nw. Ohio	13,319
Oregon City nw. Ore.	7,996
Orem n. cen. Utah	18,394
Orillia s. Ont., Can.	13,857
Orinda Village w. Calif.	5,568
Orlando e. cen. Fla.	88,135
Ormond Beach e. cen. Fla.	8,658
Orono se. Minn.	5,643
Oroville n. cen. Calif.	6,115
Orrville ne. Ohio	6,511
Osceola ne. Ark.	6,189
Oshawa s. Ont., Can.	50,412
Oshkosh e. cen. Wis.	45,110
Oskaloosa s. Iowa	11,053
Ossining se. N.Y.	18,662
Oswego nw. N.Y.	22,155
— nw. Ore.	8,906
Ottawa n. Ill.	19,408
— e. Kans.	10,673
— se. Ont.; cap. of Canada	222,129
Ottumwa e. Iowa	33,871
Outremont s. Que., Can.	29,990
Overland e. cen. Mo.	22,763
Overland Park ne. Kans.	21,110
Overlea n. cen. Md.	10,795
Owatonna s. Minn.	13,409
Owego s. N.Y.	5,417
Owens s. N.C.	5,207
Owensboro w. Ky.	42,471
Owen Sound s. Ont., Can.	16,976
Owosso cen. Mich.	17,006
Oxford cen. Mass.	6,985
— n. cen. Miss.	5,283
— n. cen. N.C.	6,978
— sw. Ohio	7,828
Oxnard sw. Calif.	40,265
Ozark se. Ala.	9,534

P

Place	Population
Pacifica w. cen. Calif.	20,995
Pacific Grove w. Calif.	12,121
Paducah w. Ky.	34,479
Pagedale e. Mo.	5,106
Painesville ne. Ohio	16,116
Palatine ne. Ill.	11,504
Palatka ne. Fla.	11,028
Palestine e. Tex.	13,974
Palisades Park ne. N.J.	11,943
Palm Beach se. Fla.	6,055
Palmdale sw. Calif.	8,511
Palmerton e. Pa.	5,942
Palmetto sw. Fla.	5,556
Palm Springs sw. Calif.	13,468
Palmyra w. N.J.	7,036
— se. Pa.	6,999
Palo Alto w. Calif.	52,287
Palos Verdes Estates sw. Calif.	9,564
Pampa n. Tex.	24,664
Pana cen. Ill.	6,432
Panama City n. Fla.	33,275
Panama Canal Zone U.S. terr., Central America	41,684
Paradise e. cen. Calif. (Butte County)	8,268
— cen. Calif. (Stanislaus County)	5,616
Paragould ne. Ark.	9,947
Paramount sw. Calif.	27,249
Paramus ne. N.J.	23,238
Paris e. Ill.	9,823
— ne. Ky.	7,791
— nw. Tenn.	9,325
— ne. Tex.	20,977
Parkersburg nw. W. Va.	44,797
Park Forest ne. Ill.	29,993
Park Ridge n. Ill.	32,659
— ne. N.J.	6,389
Parkville Carney n. Md.	27,236
Parma ne. Ohio	82,845
Parsippany-Troy Hills n. N.J.	25,557
Parsons se. Kans.	13,929
Pasadena sw. Calif.	116,407
— se. Tex.	58,737
Pascagoula s. Miss.	17,139
Pasco se. Wash.	14,522
Passaic ne. N.J.	53,963
Patchogue se. N.Y.	8,838
Paterson ne. N.J.	143,663
Paulsboro w. N.J.	8,121
Pauls Valley s. cen. Okla.	6,856
Pawhuska n. Okla.	5,414
Pawtucket ne. R.I.	81,001
Peabody ne. Mass.	32,202
Pearl s. cen. Miss.	5,081
Pecos w. Tex.	12,728
Peekskill se. N.Y.	18,737
Pekin cen. Ill.	28,146
Pelham Manor se. N.Y.	6,114
Pella s. cen. Iowa	5,198
Pembroke s. Ont., Can.	15,434
Pendleton ne. Ore.	14,434
Penn Hills sw. Pa.	51,512
Pennsauken sw. N.J.	33,771
Penns Grove sw. N.J.	6,176
Pennsylvania state, e. U.S.	11,319,366
Penn Yan w. cen. N.Y.	5,770
Pensacola nw. Fla.	56,752
Penticton s. cen. B.C., Can.	11,894
Peoria cen. Ill.	103,162
Peoria Heights cen. Ill.	7,064
Perrine s. Fla.	6,424
Perry nw. Fla.	8,030
— cen. Ga.	6,032
— cen. Iowa	6,442
— n. Okla.	5,210
Perrysburg nw. Ohio	5,519
Perryton n. Tex.	7,903
Perryville e. Mo.	5,117
Perth Amboy e. N.J.	38,007
Peru n. cen. Ill.	10,460
— e. cen. Ind.	14,453
Petaluma w. Calif.	14,035
Peterborough s. Ont., Can.	42,698
Petersburg se. Va.	36,750
Petoskey n. Mich.	6,138
Pharr s. Tex.	14,106
Phenix City e. Ala.	27,630
Philadelphia e. cen. Miss.	5,017
— se. Pa.	2,002,512
Phillipsburg w. N.J.	18,502
Phoenix s. cen. Ariz.; cap.	437,170
Phoenixville se. Pa.	13,797
Picayune s. Miss.	7,834
Pico Rivera sw. Calif.	49,150
Piedmont w. Calif.	11,117
Pierre cen. S.D.; cap.	10,088
Pikesville n. cen. Md.	18,737
Pine Bluff s. cen. Ark.	44,037
Pine Lawn e. Mo.	5,943
Pinellas w. cen. Fla.	10,848
Pineville cen. La.	8,638
Pinole w. cen. Calif.	6,064
Pipestone sw. Minn.	5,324
Piqua w. Ohio	19,219
Pitcairn sw. Pa.	5,383
Pitman sw. N.J.	8,644
Pittsburg w. cen. Calif.	19,062
— e. Kans.	18,678
Pittsburgh sw. Pa.	604,332
Pittsburg West w. Calif.	5,188
Pittsfield w. Mass.	57,879
Pittston ne. Pa.	12,407
Placentia sw. Calif.	5,861
Plainedge se. N.Y.	21,973
Plainfield cen. Ind.	5,460
— n. cen. N.J.	45,330
Plains ne. Pa.	10,995
Plainview se. N.Y.	27,710
— n. Tex.	18,735
Plant City w. cen. Fla.	15,711
Plaquemine s. La.	7,689
Plattsburg ne. N.Y.	23,844
Plattsmouth e. Nebr.	6,244
Plattville sw. Wis.	6,957
Pleasant Hills w. cen. Calif.	23,844
— sw. Pa.	8,573
Pleasantville se. N.J.	15,172
— se. N.Y.	5,877
Pleasure Ridge Park n. cen. Ky.	10,612
Plum sw. Pa.	10,241
Plymouth n. Ind.	7,558
— e. Mass.	6,488
— se. Mich.	8,766
— se. Minn.	9,576
— ne. Pa.	10,401
— e. Wis.	5,128
Pocatello se. Idaho	28,534
Pointe-aux-Trembles sw. Que., Can.	11,981
Pointe-Claire sw. Que., Can.	15,208
Point Pleasant e. cen. N.J.	10,182
— w. W.Va.	5,785

Place	Population
Pomona s. Calif.	67,157
Pompano Beach se. Fla.	15,992
Pompton Lakes ne. N.J.	9,445
Ponca City n. cen. Okla.	24,411
Ponce s. Puerto Rico	115,192
Pontiac e. cen. Ill.	8,435
— se. Mich.	82,233
Poplar Bluff se. Mo.	15,926
Portage nw. Ind.	11,822
— s. cen. Wis.	7,822
Portage la Prairie s. Man., Can.	10,525
Port Alberni sw. B.C., Can.	10,373
Portales e. N.M.	9,695
Port Allen se. La.	5,026
Port Angeles nw. Wash.	12,653
Port Arthur w. cen. Ont., Can.	38,136
— e. Tex.	66,676
Port Chester se. N.Y.	24,960
Port Clinton n. cen. Ohio	6,870
Port Colborne s. Ont., Can.	14,028
Porterville cen. Calif.	7,991
Port Hueneme s. Calif.	11,067
Port Huron e. Mich.	36,084
Port Jervis se. N.Y.	9,268
Portland cen. Conn.	5,587
— e. Ind.	6,999
— sw. Me.	72,566
— nw. Ore.	372,676
Port Lavaca se. Tex.	8,864
Port Neches se. Tex.	8,696
Portsmouth se. N.H.	25,833
— s. Ohio	33,637
— se. R.I.	8,251
— se. Va.	114,773
Port Townsend nw. Wash.	5,074
Port Vue sw. Pa.	6,635
Port Washington se. N.Y.	15,657
— e. Wis.	5,984
Potsdam nw. N.Y.	7,765
Pottstown se. Pa.	26,144
Pottsville e. Pa.	21,659
Poughkeepsie se. N.Y.	38,330
Prairie du Chien sw. Wis.	5,649
Prairie Village ne. Kans.	25,356
Pratt s. Kans.	8,156
Prattville cen. Ala.	6,616
Prescott cen. Ariz.	12,861
Presque Isle n. Me.	12,886
Price e. cen. Utah	6,802
Prichard sw. Ala.	47,371
Prince Albert cen. Sask., Can.	20,366
Prince Edward Island prov., e. Can.	99,285
Prince George cen. B.C., Can.	10,563
Prince Rupert w. B.C., Can.	10,498
Princeton n. cen. Ill.	6,250
— s. Ind.	7,906
— w. Ky.	5,618
— cen. N.J.	11,890
— s. W.Va.	8,393
Prospect Park ne. N.J.	5,201
— se. Pa.	6,596
Providence ne. R.I.; cap.	207,498
Provo n. cen. Utah	36,047
Pryor Creek ne. Okla.	6,476
Pueblo se. Colo.	91,181
Puerto Rico West Indies, commonwealth within U.S.	2,353,297
Pulaski s. cen. Tenn.	6,616
— w. Va.	10,469
Pullman se. Wash.	12,957
Punxsutawney w. cen. Pa.	8,805
Putnam ne. Conn.	6,952
Puyallup w. cen. Wash.	12,063

Q

Place	Population
Quakertown se. Pa.	6,305
Quebec prov., e. Can.	4,628,378
— s. Que., Can.; cap.	170,703
Quincy nw. Fla.	8,874
— w. Ill.	43,793
— e. Mass.	87,409
Quitman s. Ga.	5,071

R

Place	Population
Racine se. Wis.	89,144
Radford sw. Va.	9,371
Radnor se. Pa.	21,697
Rahway ne. N.J.	27,699
Raleigh cen. N.C.	93,931
Ramsey ne. N.J.	9,527
Rancho-Cordova cen. Calif.	7,429
Randolph e. Mass.	18,900
Rankin sw. Pa.	5,164
Rantoul e. cen. Ill.	22,116
Rapid City sw. S.Dak.	24,399
Raritan n. cen. N.J. (borough)	6,137
— n. cen. N.J. (twp.)	15,334
Raton ne. N.M.	8,146
Ravenna ne. Ohio	10,918
Rawlins s. cen. Wyo.	8,968
Raymondville s. Tex.	9,385
Rayne sw. La.	8,634
Raytown w. Mo.	17,083
Reading ne. Mass.	19,259
— sw. Ohio	12,832
— se. Pa.	98,177
Red Bank e. N.J.	12,482
Red Bank-White Oak se. Tenn.	10,777
Red Bluff n. Calif.	7,202
Red Deer s. cen. Alta., Can.	12,338
Redding n. Calif.	12,773
Redlands s. Calif.	26,829
Red Lion s. Pa.	5,594
Red Oak sw. Iowa	6,421
Redondo Beach sw. Calif.	46,986
Red Wing se. Minn.	10,528
Redwood City w. Calif.	46,290
Reedley cen. Calif.	5,850
Regina s. Sask., Can.; cap.	89,755
Reidsville n. cen. N.C.	14,267
Reno w. Nev.	51,470
Rensselaer e. N.Y.	10,506
Renton w. cen. Wash.	18,453
Reserve se. La.	5,297
Revere e. Mass.	40,080
Reynoldsburg cen. Ohio	7,793
Rhinelander n. cen. Wis.	8,790
Rhode Island state, ne. U.S.	859,488
Rialto se. Calif.	18,567
Rice Lake nw. Wis.	7,303
Richardson se. Tex.	16,810
Richfield se. Minn.	42,523
Richland se. Wash.	23,548
Richland Hills n. Tex.	7,804
Richmond w. Calif.	71,854
— e. Ind.	44,149
— cen. Ky.	12,168
— e. Va.	219,958
Richmond Heights e. Mo.	14,622
— n. Ohio	5,068
Ridgecrest sw. Calif.	5,099
Ridgefield ne. N.J.	10,788
Ridgefield Park ne. N.J.	12,701
Ridgewood ne. N.J.	25,391
Ridgway nw. Pa.	6,387
Ridley se. Pa.	35,738
Ridley Park se. Pa.	7,387
Rimouski s. Que., Can.	14,630
Rio Grande City s. Tex.	5,835
Ripon se. Wis.	6,163
Rittman ne. Ohio	5,410
Riverdale ne. Ill.	12,008
River Edge ne. N.J.	13,264
River Forest ne. Ill.	12,695
River Grove ne. Ill.	8,464
Riverhead se. N.Y.	5,830
River Oaks ne. Tex.	8,444
River Rouge se. Mich.	18,147
Riverside sw. Calif.	84,332
— ne. Ill.	9,750
— w. cen. N.J.	8,474
— sw. Ont., Can.	13,335
Riverton nw. Wyo.	6,845
Riverview se. Mich.	7,237
Riviera Beach se. Fla.	13,046
Roanoke e. Ala.	5,288
— w. cen. Va.	97,110
Roanoke Rapids ne. N.C.	13,320
Robbins e. cen. Ill.	7,511
Robbinsdale e. Minn.	16,381
Robinson se. Ill.	7,226
Robstown s. Tex.	10,266
Rochelle n. cen. Ill.	7,008
Rochelle Park ne. N.J.	6,119
Rochester se. Mich.	5,431
— se. Minn.	40,663
— e. N.H.	15,927
— w. N.Y.	318,611
— w. Pa.	5,952
Rockaway ne. N.J.	5,413
Rock Falls nw. Ill.	10,261
Rockford n. Ill.	126,706
Rock Hill e. Mo.	6,523
— n. S.C.	29,404
Rockingham s. N.C.	5,512
Rock Island nw. Ill.	51,863
Rockland e. Mass.	13,119
— s. Me.	8,769
Rock Springs sw. Wyo.	10,371
Rockville n. Conn.	9,478
— w. cen. Md.	26,090
Rockville Centre se. N.Y.	26,355
Rockwood e. Tenn.	5,345
Rocky Mount ne. N.C.	32,147
Rocky River ne. Ohio	18,097
Roeland Park e. Kans.	8,949
Rogers nw. Ark.	5,700
Rolla cen. Mo.	11,132
Rolling Meadows ne. Ill.	10,879
Rome nw. Ga.	32,226
— cen. N.Y.	51,646
Roosevelt se. N.Y.	12,883
Rosedale n. cen. Ohio	8,204
Roselle ne. N.J.	21,032
Roselle Park ne. N.J.	12,546
Rosemead sw. Calif.	15,476
Rosenberg se. Tex.	9,698
Roseville cen. Calif.	13,421
— se. Mich.	50,195
— se. Minn.	23,997
Ross sw. Pa.	25,952
Roswell se. N.M.	39,593
Rotterdam e. N.Y.	16,871
Round Lake Beach ne. Ill.	5,011
Rouyn sw. Que., Can.	17,076
Roxboro n. N.C.	5,147
Roy n. Utah	9,239
Royal Oak se. Mich.	80,612
Rumford sw. Me.	7,233
Rumson e. cen. N.J.	6,405
Runnemede sw. N.J.	8,396
Rushville e. Ind.	7,264
Russell cen. Kans.	6,113
Russellville nw. Ala.	6,628
— ne. Ark.	8,921
— s. Ky.	5,861
Ruston n. La.	13,991
Rutherford ne. N.J.	20,473
Rutland s. cen. Vt.	18,325
Rye se. N.Y.	14,225

S

Place	Population
Saco sw. Me.	10,936
Sacramento cen. Calif.	191,667
Saddle Brook ne. N.J.	13,834
Saginaw e. cen. Mich.	98,265
Saint Albans nw. Vt.	8,806
— w. cen. W.Va.	15,103
Saint Ann e. Mo.	12,155
Saint Anthony se. Minn.	5,084
Saint Augustine ne. Fla.	14,734
Saint Bernard sw. Ohio	6,778
Saint Boniface se. Man., Can.	28,851
Saint Catharines s. Ont., Can.	39,708
Saint Charles n. Ill.	9,269
— e. Mo.	21,189
Saint Clair e. cen. Pa.	5,159
Saint Clair Shores e. Mich.	76,657
Saint Cloud cen. Minn.	33,815
Sainte Foy sw. Que., Can.	14,615
Saint Francis se. Wis.	10,065
Saint George sw. Utah	5,130
Saint Helens nw. Ore.	5,022
Saint Hyacinthe s. Que., Can.	20,439
Saint James w. Man., Can.	26,502
Saint Jean s. Que., Can.	24,367
Saint Jérôme s. Que., Can.	20,645
Saint John e. Mo.	7,342
— s. N.B., Can.	52,491
Saint Johns cen. Mich.	5,629
Saint John's se. Newf., Can.; cap.	57,078
Saint Johnsbury ne. Vt.	6,809
Saint Joseph sw. Mich.	11,755
— nw. Mo.	79,673
Saint Lambert s. Que., Can.	12,224
Saint Laurent s. Que., Can.	38,291
Saint Louis e. Mo.	750,026
Saint Louis Park e. Minn.	43,310
Saint Martinsville s. cen. La.	6,468
Saint Marys nw. Ohio	7,737
— nw. Pa.	8,065
Saint Matthews n. cen. Ky.	8,738
Saint Michel e. cen. Que., Can.	24,706
Saint Paul e. Minn.; cap.	313,411
Saint Peter s. Minn.	8,484
Saint Petersburg w. Fla.	181,298
Saint Petersburg Beach w. Fla.	6,268
Saint Thomas s. Ont., Can.	19,129
— isl., Virgin Islands	16,046
Salamanca sw. N.Y.	8,480
Salem s. Ill.	6,165
— ne. Mass.	39,211
— sw. N.J.	8,941
— ne. Ohio	13,854
— nw. Ore.	49,142
— w. cen. Va.	16,058
Salem Heights nw. Ore.	10,770

Salina — Swarthmore

Place	Population
Salina cen. Kans.	43,202
Salinas w. Calif.	28,957
Salisbury se. Md.	16,302
— w. cen. N.C.	21,297
Salt Lake City n. Utah; cap.	189,454
San Angelo w. cen. Tex.	58,815
San Anselmo w. Calif.	11,584
San Antonio s. Tex.	587,718
San Benito s. Tex.	16,422
San Bernardino s. Calif.	91,922
San Bruno w. Calif.	29,063
San Buenaventura s. Calif.	29,114
San Carlos w. cen. Calif.	21,370
San Clemente s. Calif.	8,527
Sandersville e. cen. Ga.	5,425
San Diego sw. Calif.	573,224
Sand Springs ne. Okla.	7,754
Sandusky n. cen. Ohio	31,989
San Fernando sw. Calif.	16,093
Sanford e. cen. Fla.	19,175
— sw. Me.	10,936
— cen. N.C.	12,253
San Francisco w. Calif.	742,855
San Gabriel sw. Calif.	22,561
Sanger cen. Calif.	8,072
San Germán sw. Puerto Rico	7,779
San Jose w. Calif.	204,196
San Juan n. Puerto Rico; cap.	432,508
San Leandro w. Calif.	65,962
San Lorenzo w. Calif.	23,773
San Luis Obispo w. Calif.	20,437
San Marcos s. cen. Tex.	12,713
San Marino s. Calif.	13,658
San Mateo w. Calif.	69,870
San Pablo w. cen. Calif.	19,687
San Pedro se. Tex.	7,634
San Rafael w. Calif.	20,460
San Remo se. N.Y.	11,996
Santa Ana sw. Calif.	100,350
Santa Barbara sw. Calif.	58,768
Santa Clara w. Calif.	58,880
Santa Cruz w. Calif.	25,596
Santa Fe n. cen. N.M.; cap.	34,676
Santa Fe Springs sw. Calif.	16,342
Santa Maria sw. Calif.	20,027
Santa Monica sw. Calif.	83,249
Santa Paula sw. Calif.	13,279
Santa Rosa w. Calif.	31,027
Sapulpa ne. Okla.	14,282
Saranac Lake n. N.Y.	6,421
Saranap w. cen. Calif.	6,450
Sarasota w. Fla.	34,083
Saratoga w. cen. Calif.	14,861
Saratoga Springs e. N.Y.	16,630
Sarnia s. Ont., Can.	43,447
Saskatchewan prov., w. Can.	880,665
Saskatoon s. Sask., Can.	72,858
Saugus ne. Mass.	20,666
Sault Sainte Marie n. Mich.	18,722
— s. cen. Ont., Can.	37,329
Sausalito w. Calif.	5,331
Savannah e. Ga.	149,245
Sayre ne. Pa.	7,917
Sayreville e. cen. N.J.	22,553
Scarsdale se. N.Y.	17,968
Schenectady e. cen. N.Y.	81,682
Schiller Park ne. Ill.	5,687
Schuylkill Haven e. cen. Pa.	6,470
Scotch Plains ne. N.J.	18,491
Scotia e. N.Y.	7,625
Scott sw. Pa.	19,094
Scottdale se. Ariz.	10,026
— sw. Pa.	6,244
Scottsbluff w. Nebr.	13,377
Scottsboro n. Ala.	6,449
Scranton ne. Pa.	111,443
Sea Cliff se. N.Y.	5,669
Seaford se. N.Y.	14,718
Seal Beach sw. Calif.	6,994
Searcy n. cen. Ark.	7,272
Seaside w. cen. Calif.	19,353
Seat Pleasant s. cen. Md.	5,365
Seattle w. Wash.	557,087
Sebring s. cen. Fla.	6,939
Secaucus ne. N.J.	12,154
Security e. cen. Colo.	9,017
Sedalia w. cen. Mo.	23,874
Seguin s. cen. Tex.	14,299
Selma e. cen. Ala.	28,385
— cen. Calif.	6,934
Seminole cen. Okla.	11,464
— nw. Tex.	5,737
Seneca nw. S.C.	5,227
Seneca Falls w. cen. N.Y.	7,439
Seven Hills n. Ohio	5,708
Sewickley sw. Pa.	6,157
Seymour sw. Conn.	10,100
Seymour s. Ind.	11,629
Shadyside e. Ohio	5,028
Shaker Heights ne. Ohio	36,460
Shakopee se. Minn.	5,201
Shaler sw. Pa.	24,939
Shamokin e. cen. Pa.	13,674
Shannontown e. cen. S.C.	7,064
Sharon e. Mass.	5,888
— nw. Pa.	25,267
Sharon Hill se. Pa.	7,123
Sharpsburg sw. Pa.	6,096
Sharpsville w. Pa.	6,061
Shawano e. cen. Wis.	6,103
Shawinigan Falls s. Que., Can.	28,597
Shawinigan South s. Que., Can.	10,947
Shawnee ne. Kans.	9,072
— e. cen. Okla.	24,326
Sheboygan e. Wis.	45,747
Sheffield nw. Ala.	13,491
Sheffield Lake n. Ohio	6,884
Shelby w. N.C.	17,698
— n. cen. Ohio	9,106
Shelbyville se. Ind.	14,317
— s. cen. Tenn.	10,466
Shelton sw. Conn.	18,190
— w. Wash.	5,651
Shenandoah sw. Iowa	6,567
— e. Pa.	11,073
Sherbrooke s. Que., Can.	58,668
Sheridan n. cen. Wyo.	11,651
Sherman ne. Tex.	24,988
Shillington se. Pa.	5,639
Shippensburg nw. Pa.	6,138
Shively n. cen. Ky.	15,155
Shoreview e. Minn.	7,157
Shorewood se. Wis.	15,990
Shreveport n. La.	164,372
Sidney w. cen. Nebr.	8,004
— w. Ohio	14,663
Sierra Madre sw. Calif.	9,732
Sikeston se. Mo.	13,765
Sillery s. Que., Can.	13,154
Silsbee e. cen. Tex.	6,277
Silver City sw. N.M.	6,972
Silver Springs w. cen. Md.	66,348
Silverton w. Ohio	6,682
Sinton se. Tex.	6,008
Sioux City w. Iowa	89,159
Sioux Falls se. S.Dak.	65,466
Skokie ne. Ill.	59,364
Skowhegan sw. Me.	6,667
Slaton nw. Tex.	6,568
Slidell se. La.	6,356
Sloan w. N.Y.	5,803
Smithfield e. cen. N.C.	6,117
Smyrna s. Ga.	10,157
Snyder nw. Tex.	13,850
Socorro w. cen. N.M.	5,271
Solon n. Ohio	6,333
Solvay cen. N.Y.	8,732
Somerset s. Ky.	7,112
— s. Mass.	12,196
— sw. Pa.	6,347
Somersworth e. N.H.	8,529
Somerville e. Mass.	94,697
— n. cen. N.J.	12,458
Sorel s. Que., Can.	16,476
Souderton se. Pa.	5,381
South Amboy e. N.J.	8,422
South Bakersfield s. cen. Calif.	30,249
South Bend n. Ind.	132,445
South Boston s. Va.	5,974
Southbridge s. Mass.	15,889
South Carolina state, se. U.S.	2,382,594
South Charleston w. cen. W.Va.	19,180
South Dakota state, n. cen. U.S.	680,514
Southern Pines cen. N.C.	5,198
South Euclid ne. Ohio	27,569
South Farmingdale se. N.Y.	16,318
Southfield se. Mich.	31,501
South Gate sw. Calif.	53,831
— se. Mich.	29,404
South Hadley sw. Mass.	14,956
South Haven sw. Mich.	6,149
South Holland se. Ill.	10,412
South Houston se. Tex.	7,523
South Huntington se. N.Y.	7,084
Southington cen. Conn.	9,952
South Kingston sw. R.I.	11,942
South Miami se. Fla.	9,846
South Milwaukee se. Wis.	20,307
South Modesto cen. Calif.	5,465
South Norfolk se. Va.	22,035
South Ogden n. cen. Utah	7,405
South Orange ne. N.J.	16,175
South Pasadena sw. Calif.	19,706
South Plainfield ne. N.J.	17,879
South Portland sw. Me.	22,788
South River e. cen. N.J.	13,397
South Sacramento–Fruitridge cen. Calif.	16,443
South Saint Paul e. Minn.	22,032
South Salt Lake n. cen. Utah	9,520
South San Francisco w. Calif.	39,418
South San Gabriel sw. Calif.	26,213
South Sioux City ne. Nebr.	7,200
South Tucson s. Ariz.	7,004
South Westbury se. N.Y.	11,977
South Williamsport cen. Pa.	6,972
Spanish Fork n. cen. Utah	6,472
Sparks w. Nev.	16,618
Sparrows Point–Fort Howard–Edgmere cen. Md.	11,775
Sparta sw. Wis.	6,080
Spartanburg nw. S.C.	44,352
Speedway s. cen. Ind.	9,624
Spenard s. cen. Alaska	9,074
Spencer w. Iowa	8,864
— cen. Mass.	5,593
Spokane e. Wash.	181,608
Spotswood e. N.J.	5,788
Springdale nw. Ark.	10,076
— w. cen. Pa.	5,602
Springfield cen. Ill.; cap.	83,271
— s. cen. Mass.	174,463
— sw. Mo.	95,865
— ne. N.J.	14,467
— w. Ohio	82,723
— w. cen. Ore.	19,616
— se. Pa. (Delaware County)	26,733
— se. Pa. (Montgomery County)	20,652
— nw. Tenn.	9,221
— ne. Va.	10,783
— se. Vt.	6,600
Springfield Place s. Mich.	5,136
Spring Garden s. Pa.	11,387
Spring Hill nw. La.	6,437
Spring Valley n. Ill.	5,371
— se. N.Y.	6,538
Springville n. cen. Utah	7,913
Stamford sw. Conn.	92,713
— w. cen. Tex.	5,259
Stanton sw. Calif.	11,163
Starkville ne. Miss.	9,041
State College cen. Pa.	22,409
Statesboro e. Ga.	8,356
Statesville w. cen. N.C.	19,844
Staunton n. cen. Va.	22,232
Steelton se. Pa.	11,266
Steger ne. Ill.	6,432
Stephenville cen. Tex.	7,359
Sterling ne. Colo.	10,751
— n. Ill.	15,688
Steubenville e. Ohio	32,495
Stevens Point cen. Wis.	17,837
Stickney ne. Ill.	6,239
Stillwater e. Minn.	8,310
— cen. Okla.	23,965
Stockton cen. Calif.	86,321
Stoneham ne. Mass.	17,821
Stoneleigh–Rodgers Forge n. Md.	15,645
Storm Lake nw. Iowa	7,728
Storrs n. cen. Conn.	6,054
Stoughton e. Mass.	16,328
— s. Wis.	5,555
Stow ne. Ohio	12,194
Stowe sw. Pa.	11,730
Stratford sw. Conn.	45,012
— s. Ont., Can.	19,972
Streator n. cen. Ill.	16,868
Strongsville n. Ohio	8,504
Stroudsburg e. Pa.	6,070
Struthers ne. Ohio	15,631
Sturgeon Bay e. Wis.	7,353
Sturgis s. Mich.	8,915
Stuttgart e. Ark.	9,661
Sudbury e. cen. Ont., Can.	46,482
Suffern se. N.Y.	5,094
Suffolk se. Va.	12,609
Suitland–Silver Hills w. cen. Md.	10,300
Sulphur s. La.	11,429
Sulphur Springs ne. Tex.	9,160
Summit ne. Ill.	10,374
— ne. N.J.	23,677
Sumner w. cen. Wash.	5,874
Sumter cen. S.C.	23,062
Sunbury e. cen. Pa.	13,687
Sunnyside s. Wash.	6,208
Sunnyvale w. cen. Calif.	52,898
Superior nw. Wis.	33,563
Susanville ne. Calif.	5,598
Swainsboro e. cen. Ga.	5,943
Swampscott ne. Mass.	13,294
Swarthmore se. Pa.	5,753

Place	Population
Sweetwater nw. Tex.	13,914
Swift Current sw. Sask., Can.	10,612
Swissvale sw. Pa.	15,089
Swoyersville ne. Pa.	6,751
Sycamore n. cen. Ill.	6,961
Sydney ne. N.S., Can.	32,162
Sylacauga e. cen. Ala.	12,857
Sylvania n. cen. Ohio	5,187
Syracuse cen. N.Y.	216,038

T

Place	Population
Tacoma w. Wash.	147,979
Tahlequah e. Okla.	5,840
Takoma Park w. Md.	16,799
Talladega e. cen. Ala.	17,742
Tallahassee n. Fla.; cap.	48,174
Tallmadge ne. Ohio	10,246
Tallulah n. La.	9,413
Tamaqua e. Pa.	10,173
Tampa w. Fla.	274,970
Tarboro ne. N.C.	8,411
Tarentum w. Pa.	8,232
Tarpon Springs nw. Fla.	6,768
Tarrant City cen. Ala.	7,810
Tarrytown se. N.Y.	11,109
Taunton se. Mass.	41,132
Taylor ne. Pa.	6,148
— e. cen. Tex.	9,434
Taylorville cen. Ill.	8,801
Teaneck ne. N.J.	42,085
Tecumseh s. Mich.	7,045
Tell City s. Ind.	6,609
Tempe s. cen. Ariz.	24,894
Temple e. cen. Tex.	30,419
Temple City sw. Calif.	31,838
Tenafly ne. N.J.	14,264
Tennessee state, e. cen. U.S.	3,567,089
Terre Haute w. Ind.	72,500
Terrell ne. Tex.	13,803
Terrell Hills s. cen. Tex.	5,572
Terryville w. cen. Conn.	5,231
Texarkana sw. Ark.	19,788
— ne. Tex.	30,218
Texas state, s. cen. U.S.	9,579,677
Texas City se. Tex.	30,065
Thetford Mines s. Que., Can.	19,511
The Village cen. Okla.	12,118
Thibodaux w. La.	13,403
Thief River Falls nw. Minn.	7,151
Thomaston w. Ga.	9,336
Thomasville s. Ga.	18,246
— w. cen. N.C.	15,190
Thornton cen. Colo.	11,353
Three Rivers s. Mich.	7,092
— s. Que., Can.	50,483
Tiffin n. cen. Ohio	21,478
Tifton s. Ga.	9,903
Timmins e. cen. Ont., Can.	27,551
Timonium–Lutherville cen. Md.	12,265
Tinley Park ne. Ill.	6,392
Tipton cen. Ind.	5,604
Titusville e. cen. Fla.	6,410
— nw. Pa.	8,356
Tiverton se. R.I.	9,461
Toccoa n. Ga.	7,303
Toledo nw. Ohio	318,003
Tomah w. cen. Wis.	5,321
Toms River se. N.J.	6,062
Tonawanda w. N.Y.	21,561
— w. N.Y. (unincorporated)	83,771
Tooele nw. Utah	9,133
Topeka ne. Kans.; cap.	119,484
Toppenish s. Wash.	5,667
Toronto e. Ohio	7,780
— s. Ont., Can.; cap.	667,706
Torrance s. Calif.	100,991
Torrington nw. Conn.	30,045
Totowa ne. N.J.	10,897
Towson n. cen. Md.	19,090
Tracy w. cen. Calif	11,289
Trail sw. B.C., Can.	11,395
Traverse City nw. Mich.	18,432
Trenton se. Mich.	18,439
— n. Mo.	6,262
— w. N.J.; cap.	114,167
— s. Ont., Can.	11,492
Trinidad s. Colo.	10,691
Trois-Rivières See THREE RIVERS, Can.	
Troy se. Ala.	10,234
— se. Mich.	19,058
— e. cen. N.Y.	67,492
— w. Ohio	13,685
Truro cen. N.S., Can.	12,250
Tuckahoe se. N.Y.	6,423
Tucson s. Ariz.	212,892
Tucumcari ne. N.M.	8,143
Tulare cen. Calif.	13,824
Tullahoma s. cen. Tenn.	12,242
Tulsa ne. Okla.	261,685
Tupelo n. Miss.	17,221
Tupper Lake n. N.Y.	5,200
Turlock cen. Calif.	9,116
Turtle Creek sw. Pa.	10,607
Tuscaloosa w. Ala.	63,370
Tuscumbia nw. Ala.	8,994
Tutuila isl., American Samoa	16,785
Twin Falls s. Idaho	20,126
Two Rivers e. Wis.	12,393
Tyler ne. Tex.	51,230
Tyrone w. cen. Pa.	7,792

U

Place	Population
Uhrichsville e. Ohio	6,201
Ukiah nw. Calif.	9,900
Union ne. N.J.	51,499
— nw. S.C.	10,191
Union Beach e. cen. N.J.	5,862
Union City w. Calif.	6,618
— ne. N.J.	52,180
— nw. Tenn.	8,837
Uniondale se. N.Y.	20,041
Uniontown sw. Pa.	17,942
United States republic, North America	179,323,175
University City e. Mo.	51,249
University Heights ne. Ohio	16,641
University Park ne. Tex.	23,302
Upland sw. Calif.	15,918
Upper Arlington cen. Ohio	28,486
Upper Darby se. Pa.	93,158
Upper Moreland se. Pa.	21,032
Urbana e. Ill.	27,294
— w. cen. Ohio	10,461
Urbandale s. cen. Iowa	5,821
Utah state, w. U.S.	890,627
Utica cen. N.Y.	100,410
Utuado w. cen. Puerto Rico	9,901
Uvalde s. cen. Tex.	10,293

V

Place	Population
Vacaville n. cen. Calif.	10,898
Valdosta s. Ga.	30,652
Vallejo w. cen. Calif.	60,877
Valley City se. N. Dak.	7,809
Valleyfield s. Que., Can.	23,584
Valley Station n. Ky.	10,553
Valley Stream se. N.Y.	38,629
Valparaiso nw. Fla.	5,975
— n. Ind.	15,227
Van Buren n. Ark.	6,787
Vancouver sw. B.C., Can.	365,844
— sw. Wash.	32,464
Vandalia s. cen. Ill.	5,537
— sw. Ohio	6,342
Vandergrift w. Pa.	8,742
Van Wert nw. Ohio	11,323
Venice sw. Ill.	5,380
Ventnor City s. N.J.	8,688
Verdun s. Que., Can.	78,262
Vermillion se. S. Dak.	6,102
Vermont state, ne. U.S.	389,881
Vernon n. Tex.	12,141
Vernon Valley se. N.Y.	5,998
Vero Beach se. Fla.	8,849
Verona ne. N.J.	13,782
Vicksburg w. Miss.	29,130
Victoria sw. B.C., Can.; cap.	54,584
— se. Tex.	33,047
Victoriaville s. Que., Can.	16,031
Vidalia se. Ga.	7,569
Vienna ne. Va.	11,440
— nw. W. Va.	9,381
Villa Park ne. Ill.	20,391
Ville Platte s. cen. La.	7,512
Vincennes s. Ind.	18,046
Vineland s. N.J.	37,685
Vinita ne. Okla.	6,027
Virginia state, e. U.S.	3,966,949
— ne. Minn.	10,034
Virginia Beach se. Va.	8,091
Virgin Islands U.S. terr., West Indies	31,904
Visalia cen. Calif.	15,791
Vista s. Calif.	14,795

W

Place	Population
Wabash ne. Ind.	12,621
Waco e. cen. Tex.	97,808
Wadsworth ne. Ohio	10,635
Wahiawa cen. Oahu, Hawaii	15,512
Wahpeton se. N. Dak.	5,876
Waianae–Makaha w. Oahu, Hawaii	6,844
Wailuku nw. Maui, Hawaii	6,969
Wakefield ne. Mass.	24,295
Wakefield-Peacedale s. R.I.	5,569
Waldwick ne. N.J.	10,495
Walla Walla se. Wash.	24,536
Wallingford s. cen. Conn.	29,920
Wallington ne. N.J.	9,261
Walnut Creek w. cen. Calif.	9,903
Walnut Heights w. Calif.	5,080
Walpole e. Mass.	14,068
Walsenburg s. Colo.	5,071
Walterboro s. S.C.	5,417
Waltham e. Mass.	55,413
Wanaque n. N.J.	7,126
Wantagh se. N.Y.	34,172
Wapakoneta w. Ohio	6,756
Ware sw. Mass.	6,650
Warminster se. Pa.	15,994
Warner Robins cen. Ga.	18,633
Warr Acres cen. Okla.	7,135
Warren s. Ark.	6,752
— se. Mich.	89,246
— ne. Ohio	59,648
— nw. Pa.	14,505
— e. R.I.	8,750
Warrensburg w. Mo.	9,689
Warrensville Heights n. Ohio	10,609
Warrington nw. Fla.	16,752
Warsaw n. Ind.	7,234
Warwick cen. R.I.	68,504
Wasco sw. Calif.	6,841
Waseca s. Minn.	5,898
Washington state, nw. U.S.	2,853,214
— D.C.; cap. of U.S.	763,956
— cen. Ill.	5,919
— s. Ind.	10,846
— se. Iowa	6,037
— e. Mo.	7,961
— ne. N.C.	9,939
— nw. N.J.	5,723
— e. cen. Ohio	12,388
— sw. Pa.	23,545
Washington Park sw. Ill.	6,601
Washington Terrace n. cen. Utah	6,441
Waterbury w. cen. Conn.	107,130
Waterloo ne. Iowa	71,755
— cen. N.Y.	5,098
— s. Ont., Can.	16,373
Watertown e. Mass.	39,092
— n. cen. N.Y.	33,306
— ne. S. Dak.	14,077
— se. Wis.	13,943
Waterville s. cen. Me.	18,695
Watervliet e. N.Y.	13,917
Watseka e. cen. Ill.	5,219
Watsonville w. Calif.	13,293
Waukegan n. Ill.	55,719
Waukesha se. Wis.	30,004
Waupun se. Wis.	7,935
Wausau cen. Wis.	31,943
Wauwatosa se. Wis.	56,923
Waverly ne. Iowa	6,357
— s. N.Y.	5,950
Waxahachie ne. Tex.	12,749
Waycross s. Ga.	20,944
Wayne se. Mich.	16,034
— n. N.J.	29,353
Waynesboro e. Ga.	5,359
— s. Pa.	10,427
— n. cen. Va.	15,694
Waynesburg s. Pa.	5,188
Waynesville N.C.	6,159
Weatherford ne. Tex.	9,759
Webb City sw. Mo.	6,740
Webster s. cen. Mass.	12,072
Webster City cen. Iowa	8,520
Webster Groves e. Mo.	28,990
Weehawken ne. N.J.	13,504
Weirton n. W. Va.	28,201
Welch s. W. Va.	5,313
Welland s. Ont., Can.	16,405
Wellesley ne. Mass.	26,071
Wellington n. Kans.	8,809
Wellsburg nw. W. Va.	5,514
Wellston e. cen. Mo.	7,979
— s. Ohio	5,728
Wellsville w. N.Y.	5,967
— e. Ohio	7,117
Wenatchee cen. Wash.	16,726
Weslaco s. Tex.	15,649
West Allis se. Wis.	68,157
West Bend se. Wis.	9,969
Westbrook sw. Me.	13,820
Westbury se. N.Y.	14,757
West Caldwell ne. N.J.	8,314

Location	Population
Westchester ne. Ill.	18,092
West Chester se. Pa.	15,705
West Chicago ne. Ill.	6,854
West Columbia n. S.C.	6,410
West Concord s. cen. N.C.	5,510
West Covina sw. Calif.	50,645
West Des Moines cen. Iowa	11,949
West Elmira s. N.Y.	5,763
West End Anniston ne. Ala.	5,485
Westerly sw. R.I.	9,698
Western Springs ne. Ill.	10,838
Westerville n. cen. Ohio	7,011
Westfield sw. Mass.	26,302
— e. N.J.	31,447
West Frankfort s. Ill.	9,027
West Hartford cen. Conn.	62,382
West Haven s. Conn.	43,002
West Haverstraw se. N.Y.	5,020
West Hazleton e. Pa.	6,278
West Helena e. cen. Ark.	8,385
West Hempstead-Lakeview se. N.Y.	24,783
West Hollywood sw. Calif.	28,870
West Lafayette w. Ind.	12,680
Westlake n. Ohio	12,906
West Long Branch e. cen. N.J.	5,337
West Memphis e. cen. Ark.	19,374
West Miami se. Fla.	5,296
West Mifflin sw. Pa.	27,289
West Milwaukee se. Wis.	5,043
Westminster sw. Calif.	25,750
— cen. Colo.	13,850
— n. cen. Md.	6,123
West Monroe n. La.	15,215
Westmont ne. Ill.	5,997
— sw. Pa.	6,573
Westmount s. Que., Can.	24,800
West New York ne. N.J.	35,547
West Norriton se. Pa.	8,342
Weston n. cen. W. Va.	8,754
West Orange ne. N.J.	39,895
West Palm Beach se. Fla.	56,208
West Paterson n. N.J.	7,602
West Pittston ne. Pa.	6,998
West Plains s. Mo.	5,836
West Point e. Miss.	8,550
West Saint Paul e. Minn.	13,101
West Seneca w. N.Y.	23,138
West Springfield sw. Mass.	24,924
West University Place se. Tex.	14,628
West View sw. Pa.	8,079
West Virginia state, e. U.S.	1,860,421
West Warwick cen. R.I.	21,414
Westwego se. La.	9,815
West Winter Haven cen. Fla.	5,050
Westwood ne. N.J.	9,046
Westwood Lakes n. Fla.	22,517
West York s. Pa.	5,526
Wethersfield n. cen. Conn.	20,561
Wewoka cen. Okla.	5,954
Weymouth e. Mass.	48,177
Wharton n. N.J.	5,006
— s. Tex.	5,734
Wheaton n. Ill.	24,312
— w. cen. Md.	54,635
Wheat Ridge cen. Colo.	21,619
Wheeling ne. Ill.	7,169
— n. W.Va.	53,400
White Bear Lake e. cen. Minn.	12,849
Whitefish Bay se. Wis.	18,390
Whitehall s. Ohio	20,818
— sw. Pa.	16,075
Whitehaven nw. Tenn.	13,894
White Oak sw. Pa.	9,047
White Plains se. N.Y.	50,485
White Settlement ne. Tex.	11,513
Whitewater se. Wis.	6,380
Whiting n. Ind.	8,137
Whitinsville s. Mass.	5,102
Whitman e. Mass.	10,485
Whitney sw. Idaho	13,603
Whittier sw. Calif.	33,663
Wichita s. Kans.	254,698
Wichita Falls cen. Tex.	101,724
Wickliffe ne. Ohio	15,760
Wilkes-Barre ne. Pa.	63,551
Wilkins sw. Pa.	8,272
Wilkinsburg sw. Pa.	30,066
Willard n. Ohio	5,457
Williamsburg se. Va.	6,832
Williamson sw. W.Va.	6,746
Williamsport n. Pa.	41,967
Williamston e. cen. N.C.	44,013
Williamstown nw. Mass.	5,428
Williamsville n. N.Y.	6,316
Willimantic ne. Conn.	13,881
Williston nw. N.Dak.	11,866
Williston Park se. N.Y.	8,255
Willmar w. Minn.	10,417
Willoughby ne. Ohio	15,058
Willowick ne. Ohio	18,749
Wilmette n. Ill.	28,268
Wilmington n. Del.	95,827
— se. N.C.	44,013
— sw. Ohio	8,915
Wilson e. cen. N.C.	28,753
— e. Pa.	8,465
Wilton Manor se. Fla.	8,257
Winchester e. Ind.	5,742
— cen. Ky.	10,187
— ne. Mass.	19,376
— n. Va.	15,110
Windber sw. Pa.	6,994
Winder ne. Ga.	5,555
Windsor s. Ont., Can.	121,980
Windsor Heights sw. Iowa	5,906
Winfield s. Kans.	11,117
Winnetka ne. Ill.	13,368
Winnfield n. cen. La.	7,022
Winnipeg s. Man., Can.; cap.	255,093
Winona se. Minn.	24,895
Winooski nw. Vt.	7,420
Winslow ne. Ariz.	8,862
Winsted nw. Conn.	8,136
Winston-Salem nw. N.C.	111,135
Winter Garden cen. Fla.	5,513
Winter Haven cen. Fla.	16,277
Winter Park e. cen. Fla.	17,162
Winthrop e. Mass.	20,303
Winton ne. Pa.	5,456
Wisconsin state, n. cen. U.S.	3,951,777
Wisconsin Rapids cen. Wis.	15,042
Woburn ne. Mass.	31,214
Woodbine-Radnor-Glencliff w. Tenn.	14,485
Woodbridge e. N.J.	78,846
Woodbury w. N.J.	12,453
Woodland w. Calif.	13,524
Woodlawn-Rockdale-Millford Mills n. Md.	19,254
Woodmere se. N.Y.	14,011
Woodmont-Green Hills-Glendale w. Tenn.	23,161
Wood-Ridge ne. N.J.	7,964
Wood River s. Ill.	11,694
Woodson Terrace e. Mo.	6,048
Woodstock n. Ill.	8,897
— s. Ont., Can.	18,347
Woodward nw. Okla.	7,747
Woonsocket n. R.I.	47,080
Wooster ne. Ohio	17,046
Worcester cen. Mass.	186,587
Worland n. cen. Wyo.	5,806
Worth ne. Ill.	8,196
Worthington sw. Minn.	9,015
— n. cen. Ohio	9,239
Wyandotte se. Mich.	43,519
Wyckoff ne. N.J.	11,205
Wyoming state, w. cen. U.S.	330,066
— w. Mich.	45,829
— sw. Ohio	7,736
Wyomissing se. Pa.	5,044
Wytheville sw. Va.	5,634

X

Xenia sw. Ohio	20,445

Y

Yakima s. cen. Wash.	43,284
Yankton se. S.Dak.	9,279
Yauco sw. Puerto Rico	9,024
Yazoo City w. Miss.	11,236
Yeadon se. Pa.	11,610
Yoakum s. Tex.	5,761
Yonkers se. N.Y.	190,634
York se. Nebr.	6,173
— s. cen. Pa.	11,610
Youngstown ne. Ohio	166,689
Ypsilanti se. Mich.	20,957
Yuba City n. cen. Calif.	11,507
Yukon terr., nw. Can.	12,190
Yuma sw. Ariz.	23,974

Z

Zanesville se. Ohio	39,077
Zion n. Ill.	11,941

FOREIGN CENSUS

PLACES WITHIN FOREIGN COUNTRIES OF THE WORLD

The following list contains the names of principal cities and foreign countries of the world and, for convenience, the approximate location, by familiar abbreviations, of the places named. Populated places of more than 25,000 persons are included, except in countries of dense population, as in China, India, Japan, the Philippine Islands, or in countries for which the information is not available. In the countries excepted, places of more than 50,000 are included. Also included are places of lesser population but of commercial or political importance, as islands and small states. Variant place names of historical importance are, in some instances, shown in parentheses following the entry.

Further details regarding countries, ancient and modern, and places of historical or exceptional commercial importance are to be found under the alphabetical arrangement in the regular text pages of the dictionary.

The populations given accord with latest official sources. The population, in each case, is shown in thousands.

Unless otherwise identified as shire, county, province, island, state, etc., the places listed are cities or towns. The asterisk before a population figure indicates that the census unit is a district or that suburbs are included.

The abbreviation, cap., indicates that the place described is the capital of the country or subdivision after which the abbreviation is placed. Thus the entry reading "Alma Ata, sw. Kazakh S.S.R.; cap." shows that Alma Ata, in the southwestern part of Kazakh Soviet Socialist Republic, is the capital of that republic, and the entry "Allahabad, s. Uttar Pradesh (cap.), India" shows that Allahabad, in the southern part of Uttar Pradesh, India, is the capital of that state.

A

Aachen sw. North Rhine-Westphalia, West Germany 129
Aalborg n. Denmark 79
Aalst See ALOST.
Aarhus cen. Denmark 116
Abadan sw. Iran 64
Abashiri Japan 39
Abeokuta sw. Nigeria 56
Aberdare Glamorganshire, Wales 41
Aberdeen county, ne. Scotland 308
— Aberdeen co., Scotland 182
Abertillery Monmouthshire, England 27
Abidjan sw. French West Africa 48
Åbo See TURKU.
Abruzzi e Molise region, e. cen. Italy 1,682
Abuyog Leyte, P.I. 39
Acapulco sw. Mexico 10
Accra s. Ghana; cap. 135
Accrington Lancashire, England 40
Aconcagua province, cen. Chile 118
Acre territory, w. Brazil 116
Acton Middlesex, England 67
Adalia See ANTALYA.
Adana s. cen. Turkey 117
Adapazari nw. Turkey 36
Addis Ababa cen. Ethiopia; cap. 250
Adelaide se. South Australia; cap. *382
Aden cap. of Colony 56
Aden Colony sw. Arabian peninsula 80
Aden Protectorate sw. Arabian peninsula 650
Adrano e. Sicily, Italy 28
Adrianople (Edirne) nw. Turkey 30
Aegean Islands division, Greece 523
Afghanistan kingdom, w. Asia 12,000
Agra Uttar Pradesh, n. India 375
Agrigento Sicily, Italy 34
— province 4,698
Aguascalientes state, cen. Mexico 188
— s. Aguascalientes; cap. 93
Agusan province, P.I. 126
Ahmedabad n. Bombay, w. India 788
Ahmednagar n. Bombay, India 70
Ahwaz sw. Iran 49
Airdrie Lanark co., Scotland 30
Aix-la-Chapelle See AACHEN.
Ajmer w. India 196
Akhmim e. cen. Egypt 32
Akita nw. Honshu, Japan 126
Akmolinsk n. cen. Kazakh S.S.R. 400
Akola cen. Madhya Pradesh, India 62
Aktyubinsk nw. European R.S.F.S.R. 45
Akyab w. Burma 48
Alagoas state, e. Brazil 1,106
Alagoinhas e. Bahia, Brazil 21
Albacete s. Spain 71
Albania republic, s. Europe 1,112
Alcoy se. Spain 40
Aldershot Hampshire, England 36
Alegrete w. Rio Grande do Sul, Brazil 20
Aleksandrovsk-Sakhalinskiy w. Asiatic R.S.F.S.R. 100
Aleppo nw. Syria 324
Alessandria province, n. Italy 477
— Piedmont; cap. 55
Alexandria n. cen. Egypt 925
Algeria French territory, n. Africa 8,618
Algiers n. Algeria; cap. 266
Alicante se. Spain 104
Aligarh w. Uttar Pradesh, n. India 141
Alkmaar nw. Netherlands 37

Allahabad s. Uttar Pradesh (cap.), India 332
Alleppey s. India 115
Alma Ata sw. Kazakh S.S.R.; cap. 330
Almelo ne. Netherlands 40
Almería se. Spain 76
Alor islands, Indonesia 68
Alost cen. Belgium 42
Altona nw. North Rhine-Westphalia, West Germany 237
Altrincham n. Cheshire, England 39
Alwar ne. Rajasthan, India 54
Amagasaki w. cen. Honshu, Japan 279
Amapá territory, n. Brazil 38
Amargosa e. Bahia, Brazil 39
Amazonas state, nw. Brazil 532
Amb state, W. Pakistan 48
Ambala nw. India 107
Ambato nw. Ecuador 33
Ambon (Amboina) island, Indonesia 66
Amersfoort cen. Netherlands 55
Amiens n. France 78
Amman (or 'Amman) n. cen. Jordan; cap. 90
Amoy se. Fukien, China 138
Amraoti w. Madhya Pradesh, w. India 61
Amritsar nw. India 325
Amsterdam nw. Netherlands; cap. 803
Anápolis s. Goyaz, Brazil 18
Ancona The Marches and Ancona (cap.), Italy 52
Andaluciá region, s. Spain 5,605
Andaman and Nicobar Islands state, e. Bay of Bengal, India 30
Andizhan e. Uzbek S.S.R. 115
Andorra state, sw. Europe 5
Andria s. Italy 61
Angeles Pampanga, P.I. *37
Angers nw. France 86
Anglesey county, nw. Wales 50
Anglo-Egyptian Sudan See SUDAN.
Angola Portuguese colony, w. Africa 4,111
Angora See ANKARA.
Angostura See CIUDAD BOLÍVAR.
Angus county, e. cen. Scotland 274
Anhwei province, e. China 21,705
Ankara nw. Turkey; cap. 286
Anking China 121
Annam province, North Vietnam 6,750
Anshan ne. China 165
Antâkya See ANTIOCH.
Antalya sw. Anatolia (cap.), Turkey 27
Antequera s. Spain 32
Antigua island, Leeward Islands 40
Antioch n. Turkey 72
Antioquia department, nw. Colombia 1,118
Antique province, P.I. 233
Antofagasta province, n. Chile 145
— sw. Antofagasta (cap.), Chile 49
Antrim county, ne. Northern Ireland 231
Antung ne. China 271
Antwerp n. Belgium 263
Anzhero-Sudzhensk sw. Asiatic R.S.F.S.R. 116
Aomori n. Honshu, Japan 106
Aparri ne. Luzon, P.I. 25
Apeldoorn cen. Netherlands 62
Apuania Tuscany, Italy 41
Apulia (Puglia) region, se. Italy 3,214
Aquila Abruzzi e Molise, Italy 20
Aracajú e. Sergipe, Brazil 68
Aracati n. Ceará, Brazil 35
Arad nw. Rumania 87
Aragon region, ne. Spain 1,094

Araguari w. Minas Gerais, Brazil 25
Arak (Sultanabad) e. Iran 55
Arauco province, cen. Chile 66
Archangel See ARKHANGELSK.
Ardebil nw. Iran 81
Arequipa department, s. Peru 263
— cap. 61
Arezzo province, Italy 329
— city 28
Argenteuil n. cen. France 53
Argentina republic, s. South America 16,108
Argyll county, w. cen. Scotland 63
Arkhangelsk nw. European R.S.F.S.R. 238
Armagh county, s. Northern Ireland 114
Armavir sw. European R.S.F.S.R. 102
Armenia w. cen. Colombia 29
Armenian Soviet Socialist Republic se. European U.S.S.R. 1,600
Arnhem n. Netherlands 97
Aroroy Masbate, P.I. 31
Arrah w. Bihar, India 53
Artibonite department, Haiti 568
Artigas department, Uruguay 57
Aruba island, Neth. Antilles 51
Asahigawa w. cen. Hokkaido, Japan 123
Asansol W. Bengal, India 55
Ascoli Piceno prov., The Marches, Italy 327
Ashikaga Honshu, Japan *52
Ashington Northumberland, England *29
Ashkhabad s. cen. Turkmen S.S.R.; cap. 142
Ashton-under-Lyne Lancashire, England 46
Asir region, Saudi Arabia 750
Asmara cen. Eritrea; cap. 64
Asnières n. cen. France 72
Assam state, ne. India 9,129
Asti province, Italy 224
— Piedmont, Italy 29
Astrakhan se. European R.S.F.S.R. 276
Asturias region, n. Spain 888
— Cebu, P.I. 25
Asunción sw. Paraguay; cap. 204
Asyut e. cen. Egypt 88
Atacama province, n. Chile 84
Athens se. Greece; cap. 487
Atlántico department, n. Colombia 268
Aubervilliers n. cen. France *52
Auckland North I., New Zealand *328
Augsburg s. Bavaria, West Germany 185
Aurangabad nw. Hyderabad, India 66
Aussig See USTI NAD LABEM.
Australia Commonwealth of Nations 7,579
Austria republic, cen. Europe 6,918
Avellaneda ne. Buenos Aires, Arg. 100
Avellino province, Italy 495
— Campania, Italy 29
Avignon se. France 47
Ayacucho department, Peru 358
Ayr county, sw. Scotland 321
— Ayr county 43
Aysen province, s. Chile 17
Ayutthaya s. Thailand 16
Azcapotzalco n. Federal District, Mex. 48
Azerbaijan Soviet Socialist Republic sw. U.S.S.R. 3,400
Azores islands, N. Atlantic; prov., Portugal 318
Azul s. Buenos Aires, Arg. 27

B

Babol (Babul or Barfrush) n. cen. Iran 37
Babushkin s. cen. European R.S.F.S.R. 103

Entry	Page
Bacău e. Rumania	34
Bacolod Negros, P.I.	42
Badajoz sw. Spain	79
Badalona ne. Spain	42
Baden-Württemberg sw. W. Germany	3,884
Bagé s. Rio Grande do Sul, Brazil	35
Baghdad (Bagdad) e. cen. Iraq; cap.	364
Baguio n. Luzon, P.I.; summer cap.	29
Bahama Islands (Bahamas) Brit. colony, The West Indies	66
Bahawalpur state, W. Pakistan	1,820
Bahia state, e. Brazil	4,900
— city See SALVADOR.	
Bahía Blanca Buenos Aires, Arg.	93
Bahrein Islands se. coast of Arabia	109
Baja California Norte state, nw. Mexico	226
Baku e. Azerbaijan S.S.R.; cap.	*901
Balearic Islands (Baleares) province, se. coast of Spain	422
Bali island, Indonesia	1,074
Balikpapan e. Borneo, Indonesia	29
Ballarat s. Victoria, Australia	40
Bally s. W. Bengal, India	50
Baluchistan province, W. Pakistan	1,178
Bamako cen. French West Africa	60
Bamberg n. Bavaria, West Germany	75
Banaras Uttar Pradesh, e. India	375
Bandar ne. Madras, India	59
Bandra n. Bombay, India	72
Bandung w. Java, Indonesia	765
Banff county, ne. Scotland	50
Bangalore e. Mysore (cap.), s. India	778
Banggai Archipelago Indonesia	50
Bangka island, Indonesia	230
Bangkok w. Thailand; cap.	688
Bangui n. Fr. Equatorial Africa	41
Banjermasin sw. Borneo, Indonesia	150
Bantayan island, cen. Philippine Is.	40
Barbados British colony, The West Indies	193
Barcelona province, ne. Spain	2,232
— city, ne. Spain	1,280
Bareilly w. Uttar Pradesh, n. India	208
Barfrush See BABOL.	
Bari Apulia, Italy; cap.	267
Barili Cebu, P.I.	*29
Barisal E. Pakistan	89
Barking Essex, England	78
Barnaul sw. Asiatic R.S.F.S.R.	255
Barnes Surrey, England	40
Barnsley West Riding, Yorkshire, England	75
Baroda n. Bombay, w. India	211
Barquisimeto Lara (cap.), Venezuela	*105
Barrackpore West Bengal, e. India	59
Barranquilla n. Atlántico (cap.), Colombia	150
Barrow-in-Furness Lancashire, England	67
Barry Glamorganshire, Wales	40
Basel (Basle, Bâle) Switzerland	183
Basey Samar, P.I.	*35
Basilan island, sw. Philippine Is.	95
Basilicata region, Italy	628
Basra s. Iraq	93
Bassein s. Burma	50
Bastia Corsica, France	37
Basutoland Brit. protectorate, s. Africa	564
Bataan province, s. Luzon, P.I.	93
Batangas province, sw. Luzon, P.I.	510
Batavia See JAKARTA.	
Bath Somerset, England	79
Batley West Riding, Yorkshire, England	40
Batum or **Batumi** sw. Georgian S.S.R.	70
Bauan sw. Luzon, P.I.	*37
Bavaria state, s. West Germany	9,118
Bayern See BAVARIA.	
Bayreuth ne. Bavaria, West Germany	58
Bebington Cheshire, England	48
Bechuana British territory, s. Africa	294
Bechuanaland Protectorate s. Africa	223
Beckenham Kent, England	74
Bedford Bedfordshire, England	53
Bedfordshire county, se. cen. England	312
Beeston and Stapleford Nottingham, England	50
Beira cen. Mozambique	43
Beirut (Beyrouth) w. cen. Lebanon; cap.	178
Békés se. Hungary	29
Békéscsaba se. Hungary	45
Belém ne. Pará (cap.), Brazil	230
Belfast s. Antrim, Northern Ireland; cap.	444
Belgaum s. Bombay, India	75
Belgian Congo colony, cen. Africa	11,121
Belgium kingdom, nw. Europe	8,512
Belgrade (Beograd) ne. Yugoslavia; cap.	388
Belitung island, Indonesia	80
Belize w. Brit. Honduras; cap.	22
Bellary cen. Madras, India	56
Belluno province, Veneto, Italy	236
Belo Horizonte s. Minas Gerais (cap.), Brazil	346
Belorussian S.S.R. w. European U.S.S.R.	7,220
Benares See BANARAS.	
Bendigo cen. Victoria, Australia	30
Benevento Campania, Italy	48
— province	332
Bengal, East province, E. Pakistan	42,119
Bengal, West state, e. India	24,786
Bengasi (Bengazi or **Benghazi)** ne. Libya; co-cap. (with Tripoli)	60
Benguela province, w. Angola	1,101
Benha n. Egypt	*36
Beni department, n. Bolivia, S. America	119
Beni Suef ne. Egypt	56
Benoni Transvaal, Union of South Africa	74
Beograd See BELGRADE.	
Beppu n. Kyushu, Japan	93
Berbera e. Brit. Somaliland	25
Berchem Belgium	45
Berdichev cen. Ukrainian S.S.R.	66
Berdyansk See OSIPENKO.	
Berezniki e. cen. European R.S.F.S.R.	63
Bergamo province, Lombardy, Italy	697
— city, Bergamo (cap.), n. Italy	103
Bergen sw. Norway	110
Bergen op Zoom s. Netherlands	27
Berkshire county, s. cen. England	403
Berlin n. East Germany; cap.	1,189
— See WESTERN BERLIN.	
Bermuda British colony, N. Atlantic	37
Bern or **Berne** nw. Switzerland; cap.	146
Berwick county, se. Scotland	26
Besançon e. cen. France	51
Beşiktaş nw. Turkey	72
Bettigiri See GADAG.	
Beuthen (Bytom) w. Poland	93
Bexley Kent, England	88
Beyoğlu See PERA.	
Beyrouth See BEIRUT.	
Bezhitsa w. European R.S.F.S.R.	82
Béziers s. France	59
Bezwada cen. Madras, India	160
Bhadgaon Nepal	93
Bhagalpur ne. Bihar, India	115
Bhatpara s. West Bengal, India	154
Bhaunagar se. Saurashtra, w. India	137
Bhiwani w. India	50
Bhopal state, cen. India	838
— cap.	102
Bhutan protectorate, s. cen. Asia	300
Białystok ne. Poland	56
Bié province, Angola	701
Biel nw. Switzerland	48
Bielefeld North Rhine-Westphalia, West Germany	153
Bihar state, ne. India	40,218
— cap.	54
Bikaner n. Rajasthan, India	117
Bilaspur state, nw. India	127
Bilbao n. Spain	229
Biliran island, e. cen. Philippine Is.	55
Billiton See BELITUNG.	
Bilston Staffordshire, England	33
Binhdinh cen. South Vietnam	75
Bío-Bío province, cen. Chile	129
Birkenhead Cheshire, England	143
Birmingham Warwickshire, England	1,112
Bisk (Biisk, Biysk) w. Asiatic R.S.F.S.R.	112
Bismarck Archipelago Territory of New Guinea	137
Bitolj (Bitola) s. Yugoslavia	31
Bizerte n. Tunisia	39
Blackburn Lancashire, England	111
Blackpool Lancashire, England	147
Blagoveshchensk se. Asiatic R.S.F.S.R.	59
Blaydon-on-Tyne Durham, England	30
Blida n. Algeria	30
Bloemfontein Orange Free State (cap.), Union of South Africa	*83
Blois n. cen. France	21
Blumenau e. Santa Catarina, Brazil	22
Blyth Northumberland, England	34
Bobruisk se. Belorussian S.S.R.	84
Bochum w. North Rhine-Westphalia, West Germany	290
Bogor n. Java, Indonesia	104
Bogotá cen. Cundinamarca; cap. of Colombia and of department	326
Bohemia province, w. Czechoslovakia	8,762
Bohol island, s. cen. Philippine Is.	553
Bokhara See BUKHARA.	
Boksburg Transvaal, Union of South Africa	53
Bolívar department, n. Colombia	855
Bolivia republic, w. cen. South America	3,028
Bologna province, Italy	763
— city, Emilia-Romagna region; cap.	339
Bolton Lancashire, England	167
Bolzano province, n. Italy	334
— city, Bolzano (cap.), Italy	68
Bombay state, w. India	35,943
— w. Bombay (cap.), India	2,839
Bône ne. Algeria	77
Bonn w. North Rhine-Westphalia, West Germany; cap.	111
Bootle Lancashire, England	74
Borås sw. Sweden	58
Bordeaux sw. France	238
Borisoglebsk s. cen. European R.S.F.S.R.	52
Borneo island, region, Indonesia	3,000
Bor Sa'id See PORT SAID.	
Bottrop cen. North Rhine-Westphalia, West Germany	92
Bougainville island, Solomon Islands, Terr. of New Guinea	37
Bougie n. Algeria	21
Boulogne-Billancourt n. France	79
Boulogne-sur-Mer n. France	34
Bourges cen. France	41
Bournemouth Hampshire, England	145
Boutilimit nw. Fr. W. Africa	35
Boyacá department, n. Colombia	756
Bozen See BOLZANO (city).	
Brabant province, Belgium	1,798
Bradford West Riding, Yorkshire, England	292
Braga nw. Portugal	33
Bragança e. Pará, Brazil	4
Brăila se. Rumania	96
Brakpan Transvaal, Union of South Africa	83
Brandenburg w. cen. East Germany	71
Braşov (Oraşul Stalin) e. cen. Rumania	83
Bratislava sw. Slovakia (cap.), Czechoslovakia	*184
Braunschweig See BRUNSWICK.	
Brazil republic, South America	52,645
Brazzaville se. Middle Congo; cap. of French Equatorial Africa	83
Brecknockshire county, se. Wales	56
Breda s. Netherlands	85
Bremen state, nw. West Germany	568
— city, n. Bremen (cap.), West Germany	456
Bremerhaven n. Bremen, West Germany	112
Brentford and Chiswick Middlesex, England	59
Brescia province, n. Italy	857
— city, Lombardy, n. Italy	142
Breslau See WROCŁAW.	
Brest nw. France	62
Brest-Litovsk See BRZEŚĆ NAD BUGIEM.	
Brighton Sussex, England	156
Brindisi province, Italy	311
— Apulia, Italy	49
Brisbane se. Queens. (cap.), Australia	402
Bristol Gloucestershire, England	442
British Bechuanaland See BECHUANALAND.	
British Guiana British colony, n. South America	378
British Honduras British colony, n. Central America	59
British North Borneo ne. portion of Borneo	334
British Solomon Islands protectorate, sw. Pacific	100
British Somaliland (Somaliland Protectorate) ne. Africa	600
British West Indies w. Atlantic; including Bahamas, Barbados, Jamaica, Leeward Islands, Trinidad, Tobago, and Windward Islands. See THE WEST INDIES.	
Brno s. Moravia, Czechoslovakia	273
Broach n. Bombay, India	56
Broken Hill N.S. Wales, Australia	*27
Bromberg See BYDGOSZCZ.	
Bromley Kent, England	64
Bromma e. Sweden	72
Bruges (Brugge) Belgium	52
Brunei sultanate, n. Borneo	41
Brünn See BRNO.	
Brunswick Lower Saxony, nw. West Germany	223
Brusa nw. Turkey	100
Brussels Belgium; cap.	*1,299

Brüx See Most.
Bryansk e. European R.S.F.S.R. 185
Brześć nad Bugiem Belorussian S.S.R. 50
Bucaramanga ne. Santander (cap.), Colombia 41
Bucharest s. Rumania; cap. *1,041
Buckinghamshire s. cen. England 386
Budapest n. cen. Hungary; cap. 1,058
Budaun cen. Uttar Pradesh, n. India 52
Budweis See České Budějovice.
Buenos Aires province, e. Argentina 4,406
— ne. Buenos Aires, Argentina; cap. 3,000
Buga cen. Valle del Cauca, Colombia 19
Buganda kingdom, s. Uganda 1,318
Buitenzorg See Bogor.
Bukhara se. Uzbek S.S.R. 50
Bulacan province, P.I. 411
Bulan se. Luzon, P.I. 14
Bulawayo sw. Southern Rhodesia 82
Bulgaria republic se. Europe 9,022
Bundaberg Queens., Australia 16
Burdwan sw. West Bengal, India 63
Burg n. East Germany 27
Burgas se. Bulgaria 43
Burgenland province, Austria 276
Burgos n. Spain 74
Burhanpur w. India 54
Burma republic, se. Asia 18,489
Burnley Lancashire, England 85
Bursa See Brusa.
Burton-upon-Trent Staffordshire, England 49
Burujird sw. Iran 46
Bury Lancashire, England 59
Bushire w. Iran 25
Bussum cen. Netherlands 33
Bute island county, sw. Scotland 19
Buzau se. cen. Rumania 44
Bydgoszcz nw. Poland 134
Bytom See Beuthen.

C

Caaguazú department, Paraguay 72
Caazapá department, s. Paraguay 72
Cabanatuan n. cen. Luzon, P.I. 15
Cabimas w. Venezuela 32
Cabo Verde See Cape Verde Is.
Cachoeira do Sul Rio Grande do Sul, Brazil 24
Cádiz s. Spain 100
Caen nw. France 47
Caernarvonshire county, nw. Wales 124
Caerphilly Glamorganshire, Wales *36
Cagayan province, n. Luzon, P.I. 311
Cagliari province, Sardinia, Italy 667
— city, cap. of Sardinia 137
Cairns e. Queensland, Australia 17
Cairo n. Egypt; cap. 2,101
Caithness county, ne. Scotland 23
Cajamarca department, Peru 494
Calabria region, sw. Italy 2,043
Calais n. France 42
Calamba cen. Luzon, P.I. 32
Calatrava Negros, P.I. 39
Calcutta s. West Bengal (cap.), India 2,550
Caldas department, w. Colombia 770
Cali s. Valle del Cauca, Colombia 88
Calicut s. Madras, India 158
Callao department, sw. Peru 82
— city, sw. Peru 69
Caloocan cen. Luzon, P.I. 39
Caltanissetta province, Sicily, Italy 298
— city, cen. Sicily 37
Camagüey province, cen. Cuba 488
— cen. Camagüey (cap.), Cuba 80
Cambodia kingdom, sw. Indochina 4,073
Cambridge Cambridgeshire, England 81
Cambridgeshire county, se. England 167
Cameroons, British U.N. Trust Territory, w. Africa 1,027
—, French (Cameroun) U.N. Trust Territory, w. Africa 3,073
Camiguin island s. Philippine Is. 60
Campania department, sw. Italy 4,389
Campeche state, s. Mexico 122
— city, nw. Campeche 31
Campina Grande ne. Brazil 73
Campinas São Paulo, Brazil 102
Campobasso province, Italy 406
— city, Abruzzi e Molise, Italy 28
Campo Grande s. Mato Grosso, Brazil 32
Campos ne. Rio de Janeiro, Brazil 63
Canary Islands province, Spain, off nw. coast of Africa 793
Canberra se. N.S. Wales; cap. of Australia *15
Candia n. cen. Crete 53
Canea nw. Crete; cap. 34
Cannes se. France 37
Cannock Staffordshire, England *41
Canterbury Kent, England 28
— district, New Zealand 280
Cantho s. South Vietnam 27
Canton cen. Kwangtung (cap.), China 1,413
Cape of Good Hope Province (Cape Province) s. Union of South Africa 4,053
Cape Town or **Capetown** Cape of Good Hope Prov. (cap.), Union of South Africa *471
Cape Verde Islands (Cabo Verde) Portuguese island colony, off nw. coast of Africa 147
Capital district, Paraguay 204
Capiz province, n. Panay, P.I. 442
Carabobo state, Venezuela 243
Caracas n. cen. Venezuela; cap. 488
Cárdenas n. Matanzas, Cuba 37
Cardiff Glamorganshire, Wales 244
Cardiganshire county, sw. Wales 53
Carlisle Cumberland, England 67
Carlow county, s. Leinster, Ireland 34
Carmarthenshire county, s. cen. Wales 172
Caroline Islands U.N. Trust Territory of the Pacific Islands 37
Carshalton Surrey, England 62
Cartagena n. Bolívar, Colombia 73
— se. Spain 113
Carúpano ne. Venezuela 31
Casablanca n. Fr. Morocco 551
Caserta province, sw. Italy 599
— city, cen. Campania 25
Cassel See Kassel.
Castellammare di Stabia s. cen. Italy 45
Castellón de la Plana e. Spain 53
Castrop-Rauxel w. North Rhine-Westphalia, West Germany 70
Cataiñgan Masbate, P.I. *53
Catamarca province, nw. Argentina 145
Catanduanes island, ne. P.I. 112
Catania province, Sicily, Italy 797
— city, e. Sicily 296
Catanzaro province, Calabria, Italy 717
— city se. Calabria 28
Catbalogan Samar (cap.), P.I. 11
Cauca department, sw. Colombia 356
Caulfield s. Victoria, Australia 80
Cautín province, cen. Chile 375
Cavan county, s. Ulster, Ireland 70
Cavite sw. Luzon, P.I. *35
Cawnpore cen. Uttar Pradesh, n. India 705
Caxias do Sul Rio Grande do Sul, s. Brazil 32
Cayenne e. French Guiana; cap. 11
Ceará state, ne. Brazil 2,735
— city See Fortaleza.
Cebu island, s. cen. Philippine Is. 1,123
— Cebu, P.I. 166
Cegléd n. cen. Hungary 40
Celaya cen. Guanajuato, Mexico 34
Celebes island, Indonesia 5,500
Celle Lower Saxony, n. West Germany 59
Cephalonia island, w. Greece 59
Ceram island, Indonesia 96
České Budějovice s. Bohemia, Czechoslovakia 38
Ceuta n. Morocco 60
Ceylon island, Commonwealth of Nations, Indian Ocean 6,657
Chacabuco n. Buenos Aires, Argentina 12
Chaco province, n. Argentina 444
Chad territory, French Equatorial Africa 2,241
Chadderton Lancashire, England 31
Chahar province, ne. China 3,500
Châlons-sur-Marne ne. France 28
Chalon-sur-Saône ne. France 30
Chamartín de la Rosa cen. Spain 65
Chandernagore West Bengal, w. India 48
Changan See Sian.
Changchun (Hsinking) ne. China 630
Changhua n. Taiwan 62
Changsha ne. Hunan (cap.), China 396
Changteh e. China 50
Chankiang sw. Kwantung, China 268
Channel Islands English Channel, w. of Normandy 103
Chaotung cen. China 50
Chapayevsk e. cen. European R.S.F.S.R. 58
Chapra nw. Bihar, India 55
Charleroi Belgium 27
Chartres n. cen. France 23
Chatham Kent, England 47
Chefoo See Yentai.
Cheju island, s. South Korea 276
— city, s. South Korea 58
Chekiang province, se. China 19,942
Chelmsford Essex, England 38
Cheltenham Gloucestershire, England 63
Chelyabinsk sw. Asiatic R.S.F.S.R. 612
Chemnitz East Germany 250
Chemulpo See Inchon.
Chengteh n. Jehol (cap.), ne. China 60
Chengtu e. Szechwan (cap.), China 749
Cherbourg nw. France 34
Cheremkhovo s. cen. Asiatic R.S.F.S.R. 124
Cheribon See Tjirebon.
Cherkassy cen. Ukrainian S.S.R. 51
Chernigov ne. Ukrainian S.S.R. 67
Chernovtsy w. Ukrainian S.S.R. 142
Cheshire county, nw. cen. England 1,258
Chester Cheshire, England 48
Chesterfield Derbyshire, England 67
Chiangmai nw. Thailand 26
Chiapas state, s. Mexico 904
Chiayi cen. Taiwan 119
Chiba se. Honshu, Japan 134
Chieti province, Abruzzi e Molise, Italy 400
— city 25
Chigasaki cen. Honshu, Japan 47
Chihuahua state, n. Mexico 845
— cen. Chihuahua; cap. 86
Chile republic, w. South America 5,916
Chillán cen. Nuble (cap.), Chile 42
Chiloé province, s. Chile 102
Chimkent se. Kazakh S.S.R. 130
China republic, se. Asia 461,006
Chinchow sw. Manchuria 142
Chinghai See Tsinghai.
Chinhsien n. China 148
Chinkiang sw. Kiangsu, China 207
Chinnampo w. North Korea 61
Chios (Khios) Aegean island, Greece 72
Chiquimula se. Guatemala 32
Chiquinquirá w. Boyacá, Colombia 8
Chita s. cen. Asiatic R.S.F.S.R. 162
Chitral state, West Pakistan 107
Chittagong se. East Bengal, W. Pakistan 269
Chkalov se. European R.S.F.S.R. 226
Chocó department, w. Colombia 111
Choisy-le-Roi n. cen. France 28
Cholon s. South Vietnam 481
Chongjin ne. North Korea 184
Chonju South Korea 101
Chorley Lancashire, England 32
Chorzów sw. Poland 111
Chosen See Korea.
Chowkiakow e. China 200
Christchurch South Island, New Zealand 174
Chubut territory, s. Argentina 540
Chunchon South Korea 55
Chungju South Korea 66
Chungking s. Szechwan, China 1,013
Chuquisaca department, s. Bolivia 283
Ciego de Ávila w. Camagüey, Cuba 67
Ciénaga n. Magdalena, Colombia 23
Cienfuegos cen. Cuba 53
Ciudad Bolívar se. Venezuela *31
Ciudad Juárez ne. Chihuahua, Mexico 123
Ciudad Madero se. Tamaulipas, Mexico 41
Ciudad Real province, s. cen. Spain 567
Ciudad Trujillo s. Dominican Republic; cap. 182
Clackmannan county, cen. Scotland 38
Clare county, n. Munster, Ireland 85
Cleethorpes Lincolnshire, England 29
Clermont-Ferrand s. cen. France 94
Clichy-la-Garenne near Paris, France 53
Cluj nw. Rumania 118
Clydebank Dumbarton co., Scotland 45
Coahuila state, ne. Mexico 720
Coatbridge Lanark co., Scotland 48
Coblenz s. Rhineland-Palatinate, West Germany 64
Coburg n. Bavaria, West Germany 44
— se. Victoria, Australia 60
Cocanada cen. Madras, India 100
Cochabamba department, cen. Bolivia 490
— cen. Cochabamba (cap.), Bolivia 81
Cochin China province, South Vietnam 5,625
Coimbatore s. Madras, India 195
Colchagua province, cen. Chile 138
Colchester Essex, England 57
Colima state, sw. Mexico 112
Colmar ne. France 44
Cologne (Köln) w. North Rhine-Westphalia, West Germany 590

Place	Description	Number
Colombes	n. cen. France	61
Colombia	republic, nw. South America	11,477
Colombo	sw. Ceylon; cap.	355
Colón	n. cen. Panama	52
Colonia	department, sw. Uruguay	133
Como	province, n. Italy	564
—	city, n. Italy	62
Comoro Islands	w. Indian Ocean	156
Conakry	w. French Guinea; cap.	38
Concepción	e. Entre Ríos, Arg.	15
—	province, cen. Chile	308
—	nw. Concepción (cap.), Chile	85
—	n. cen. Luzon, P.I.	33
—	department, e. Paraguay	63
Concordia	ne. Entre Ríos, Arg.	42
Congo	province, Angola	787
Conjeeveram	s. Madras, India	74
Connacht or Connaught	province, nw. Ireland	472
Constanta	se. Rumania	79
Constantine	ne. Algeria	80
Coorg	state, s. India	229
Copenhagen	e. Denmark; cap.	768
Copiapó	cen. Atacama (cap.), Chile	15
Coquilhatville	w. Belgian Congo	10
Coquimbo	province, n. Chile	246
Córdoba	province, cen. Argentina	1,455
—	ne. Córdoba (cap.), Arg.	351
—	province, cen. Spain	782
—	s. Spain	165
Corfu	island, nw. Greece	110
—	city	35
Cork	county, Munster, s. Ireland	344
—	s. Cork, Munster, Ireland	76
Cornwall	county, sw. England	346
Coro	n. Venezuela	28
Coronel	e. Concepción, Chile	15
Corrientes	province, e. Argentina	571
—	nw. Corrientes (cap.), Argentina	56
Corsica (Corse)	island department, France; nw. Mediterranean	268
Corum	n. Turkey	23
Corumbá	sw. Mato Grosso, Brazil	19
Coruña, La	nw. Spain	134
Coseley	Staffordshire, England	34
Cosenza	province, s. Italy	686
—	Calabria, Italy	50
Costa Rica	republic, Central America	801
Cotabato	province, P.I.	440
Cottbus	e. East Germany	49
Coulsdon	Surrey, England	37
Courbevoie	n. cen. France	54
Courtrai	Belgium	40
Coventry	Warwickshire, England	258
Coyoacán	Federal District, Mexico	46
Cracow (Kraków)	sw. Poland	299
Craiova	sw. Rumania	85
Cremona	province, Italy	383
—	Lombardy, Italy	55
Crete (Krete)	Greek island, e. Mediterranean	370
Crewe	Cheshire, England	52
Croatia	republic, Yugoslavia	3,749
Crosby	Lancashire, England	58
Croydon	Surrey, England	249
Cruz Alta	Rio Grande do Sul, s. Brazil	19
Cuba	republic, w. West Indies	5,854
Cúcuta	e. Norte de Santander (cap.), Colombia	41
Cuddalore	s. Madras, India	61
Cuenca	s. Ecuador	46
Cumaná	n. Venezuela	*46
Cumberland	county, nw. England	285
Cundinamarca	department, cen. Colombia	1,174
Cuneo	province, Italy	580
—	Piedmont, Italy	22
Curaçao	Netherlands Antilles; island, n. of Venezuela	95
Curepipe	cen. Mauritius	27
Curicó	province, cen. Chile	81
—	cen. Curicó (cap.), Chile	21
Curitiba	e. Paraná (cap.), Brazil	142
Cutch	state, w. India	568
Cuttack	e. Orissa, India	102
Cuyabá	cen. Mato Grosso, Brazil	54
Cuzco (Cusco)	department, s. Peru	487
—	city, s. Peru	41
Cyclades	islands, Aegean Sea, s. of Greece	121
Cyprus	British island colony; e. Mediterranean	450
Cyrenaica	division, ne. Libya	304
Czechoslovakia	republic, cen. Europe	12,090
Czestochowa	sw. Poland	101

D

Place	Description	Number
Dacca	East Bengal (cap.), East Pakistan	401
Dagenham	Essex, England	115
Dagupan	w. cen. Luzon, P.I.	*43
Dahomey	se. French West Africa	1,476
Dairen	Manchuria, n. China	544
Dakar	w. French West Africa; cap.	186
Damanhur	n. Egypt	85
Damão	Portuguese India	69
Damascus	sw. Syria; cap.	291
Damietta	ne. Egypt	53
Danzig (Gdańsk)	n. Poland	191
Daraga	s. Luzon, P.I.	29
Darbhanga	n. Bihar, India	69
Darlington	Durham, England	85
Darmstadt	s. Hesse, West Germany	94
Dartford	Kent, England	40
Darwen	Lancashire, England	31
Daugavpils	See DVINSK.	
Daulatabad	See MALAYER.	
Davao	se. Mindanao, P.I.	81
Debrecen	e. Hungary	120
Decin	nw. Czechoslovakia	*30
Dehra Dun	nw. Uttar Pradesh, India	78
Delft	w. cen. Netherlands	62
Delhi	state, n. India	1,742
—	e. Delhi; cap.	914
Delmenhorst	n. Lower Saxony, West Germany	57
Denbighshire	county, n. cen. Wales	171
Denmark	kingdom, nw. Europe	4,281
D'Entrecasteaux Islands	Papua	40
Dera Ghazi Khan	cen. West Pakistan	36
Dera Ismail Khan	nw. West Pakistan	42
Derby	Derbyshire, England	141
Derbyshire	county, cen. England	826
Dessau	cen. East Germany	88
Dessye	en. cen. Ethiopia	36
Deurne	Belgium	57
Deventer	ne. Netherlands	44
Devonshire	county, sw. England	798
Dewsbury	West Riding, Yorkshire, England	53
Dhahran	e. Saudi Arabia	10
Dhulia	ne. Bombay, India	*54
Diamantina	cen. Minas Gerais, Brazil	10
Dibai (Debai)	sheikdom, Trucial Oman	24
Diégo-Suarez	n. Madagascar	22
Dieppe	n. cen. France	21
Dijon	e. cen. France	92
Dili	e. Portuguese Timor; cap.	2
Dinajpur	n. East Bengal, East Pakistan	34
Dindigul	s. Madras, India	56
Dire Dawa	e. cen. Ethiopa	20
Distrito Federal	district, ne. Arg.	3,000
—	se. Brazil	2,377
—	cen. Mexico	3,012
—	n. Venezuela	700
Divinópolis	s. Brazil	21
Diyarbakir	se. Turkey	45
Djakarta	See JAKARTA.	
Djibouti	Fr. Somaliland; cap.	17
Djokjakarta	See JOGJAKARTA.	
Dneprodzerzhinsk	e. cen. Ukrainian S.S.R.	163
Dnepropetrovsk	e. cen. Ukrainian S.S.R.	576
Dodecanese	Greek islands, se. Aegean	121
Dominica	island, Windward Islands, The West Indies	49
Dominican Republic	e. Hispaniola, West Indies	2,121
Doncaster	West Riding, Yorkshire, England	82
Donegal	county, nw. Ulster, Ireland	136
Dordrecht	cen. Netherlands	65
Dorsetshire	county, s. England	291
Dortmund	w. North Rhine-Westphalia, West Germany	500
Douglas	Isle of Man; cap.	20
Dover	Kent, England	35
Down	county, se. Northern Ireland	241
Drama	ne. Greece	33
Drammen	se. Norway	26
Dresden	s. East Germany	467
Dublin	county, e. Leinster, Ireland	636
—	e. Dublin co., Leinster, Ireland; cap.	506
Dubrovnik	s. Yugoslavia	16
Dudley	Worcestershire, England	63
Duisburg	w. North Rhine-Westphalia, West Germany	409
Dumbarton	county, cen. Scotland	164
Dumfries	county, s. Scotland	86
Dundee	Angus co., Scotland	177
Dunedin	South Island, New Zealand	*95
Dunfermline	Fife co., Scotland	44
Dunkirk (Dunkerque)	n. France	10
Dun Laoghaire or Dunleary (Kingstown)	s. Dublin, Ireland	44
Durango	state, cen. Mexico	628
—	s. Durango (cap.), Mexico	60
Durazno	cen. Uruguay	27
Durban	Natal, Union of South Africa	372
Durham	county, ne. England	1,463
Durres (Durazzo)	w. Albania	14
Düsseldorf	North Rhine-Westphalia (cap.), West Germany	498
Dutch Guiana	See SURINAM.	
Dvinsk	Latvian S.S.R.	45
Dzaudzhikau	N. Ossetian A.S.S.R.; cap.	127
Dzerdzhinsk	e. cen. European R.S.F.S.R.	147
Dzhambul	se. Kazakh S.S.R.	63

E

Place	Description	Number
Ealing	Middlesex, England	187
East Bengal	See EAST PAKISTAN.	
Eastbourne	Sussex, England	58
East Germany (German Democratic Republic)		17,314
East Ham	Essex, England	121
East London	Cape of Good Hope Province, Union of South Africa	*79
East Lothian	county, se. Scotland	52
East Pakistan	prov., Pakistan	42,100
East Riding	div., Yorkshire, England	511
Ebbw Vale	Monmouthshire, England	29
Eccles	Lancashire, England	44
Ecuador	republic, nw. South America	3,077
Ede	cen. Netherlands	45
Edinburgh	Midlothian co., Scotland; cap.	467
Edirne	See ADRIANOPLE.	
Edmonton	Middlesex, England	104
Eger	n. Hungary	29
Egypt	province, United Arab Republic, ne. Africa	19,087
Eindhoven	s. Netherlands	135
Eire	See IRELAND.	
Elbing	w. Poland	21
Elche	se. Spain	31
El Faiyum	See FAIYUM.	
Elisabethville	Belgian Congo	62
El Ladhiqiya	See LATAKIA.	
Ellore	ne. Madras, India	65
El Mansura	See MANSURA.	
El Obeid	cen. Sudan	72
El Salvador	republic, Central America	1,856
Ely, Isle of	county, e. England	89
Emden	Lower Saxony, West Germany	37
Emilia-Romagna	region, n. Italy	3,538
Enfield	Middlesex, England	110
Engels	s. cen. European R.S.F.S.R.	73
England	se. division of Great Britain	42,611
Enna	Sicily, Italy	27
Enschede	ne. Netherlands	79
Entre Ríos	province, ne. Argentina	776
Enzeli	See PAHLEVI.	
Epsom and Ewell	Surrey, England	68
Erfurt	sw. East Germany	175
Erith	Kent, England	46
Eritrea	federated province of Ethiopia, ne. Africa	1,080
Erivan	sw. Armenian S.S.R.; cap.	385
Erzurum	ne. Turkey	54
Esbjerg	sw. Denmark	48
Esch-sur-Alzette	s. Luxemburg	27
Esher	Surrey, England	51
Eskilstuna	se. Sweden	53
Eskisehir (Eskishehir)	nw. Turkey	88
Espírito Santo	state, e. Brazil	870
Essen	w. North Rhine-Westphalia, West Germany	605
Essex	county, se. England	2,044
Eston	North Riding, Yorkshire, England	33
Estonian Soviet Socialist Republic or Estonia	ne. Europe	1,000
Etawah	cen. Uttar Pradesh, India	53
Ethiopia	kingdom, ne. Africa	10,000
Etterbeek	cen. Belgium	50
Euboea (Evvoia)	island, Greece	165
Exeter	Devonshire, England	75

F

Place	Description	Number
Faeroe Islands	Danish island group	32
Faiyum	province, n. Egypt	671

Faizabad e. Afghanistan 25
— n. Uttar Pradesh, India *57
Falkirk Stirling co., Scotland 38
Farnworth Lancashire, England 28
Farrukhabad cen. Uttar Pradesh, India *69
Fas See FEZ.
Fayal island, Azores 26
Feira de Santana Bahia, Brazil 27
Felling Durham, England 25
Fenyanghsien ne. China 65
Fermanagh county, sw. Northern Ireland 53
Fernando Po island, Spanish Guinea 34
Ferozepore nw. India 82
Ferrara province, Italy 421
— city, Emilia-Romagna, Italy 134
Ferrol del Caudillo, El nw. Spain 77
Fez n. cen. Morocco 201
Fife county, e. cen. Scotland 307
Fiji (Fiji Islands) British colony, S. Pacific 277
Finchley Middlesex, England 70
Finland republic, ne. Europe 4,029
Firenze province, Italy 918
— city See FLORENCE.
Fiume department, ne. Italy 107
— city See RIJEKA.
Flensburg Schleswig-Holstein, West Germany 163
Flintshire county, ne. Wales 145
Florence (Firenze) Tuscany, Italy 376
Flores island, Indonesia 495
Florianópolis e. Santa Catarina (cap.), Brazil 49
Florida department, s. Uruguay 108
Focşani e. Rumania 28
Foggia province, Italy 661
— city, Apulia 84
Folkestone Kent, England 45
Foochow e. Fukien (cap.), China 328
Forlì province, Italy 485
— city, Emilia-Romagna 39
Formosa See TAIWAN.
— territory, n. Argentina 112
Fortaleza n. Ceará (cap.), Brazil 214
Fort-de-France w. Martinique; cap. 49
Fort-Lamy cen. Fr. Equatorial Africa 24
France republic, w. Europe 42,000
Frankfort on the Main sw. Hesse, West Germany 524
Frankfort on the Oder cen. East Germany 52
Fredericksberg e. Denmark 119
Freetown w. Sierra Leone; cap. 65
Freiburg s. West Germany 110
Fremantle sw. Western Australia 28
French Equatorial Africa territory, cen. Africa 4,407
French Guiana overseas department, ne. South America 28
French Guinea overseas territory, Fr. W. Africa 2,125
French India five separated dependencies incorporated into India, 1954 317
French Oceania French possessions, S. Pacific 55
French Somaliland overseas territory, ne. Africa 58
French Sudan territory, Fr. W. Africa 3,137
French West Africa colonial federation, nw. Africa 16,377
Fresnillo cen. Zacatecas, Mexico 30
Fribourg w. Switzerland 29
Friendly Islands See TONGA ISLANDS.
Frosinone province, w. cen. Italy 468
Frunze n. cen. Kirghiz S.S.R.; cap. 190
Fuchow See FOOCHOW.
Fukien province, se. China 11,100
Fukui s. Honshu, Japan 101
Fukuoka n. Kyushu, Japan 393
Fukushima n. Honshu, Japan 93
Fukuyama sw. Honshu, Japan 67
Funchal Madeira 55
Fürth n. cen. Bavaria, West Germany 100
Fusan See PUSAN.
Fuse s. Honshu, Japan 150
Fushun ne. China 280
Fuyu ne. China 65
Fyn island, n. Denmark 356
Fyzabad See FAIZABAD, India.

G

Gabon territory, sw. French Equatorial Africa 410
Gadag s. Bombay, India *56
Galati se. Rumania 80
Galle s. Ceylon 49
Galway county, s. Connacht, Ireland 165
Gambia British colony and protectorate, w. Africa 253
Garden Reach se. West Bengal, India 110
Gateshead Durham, England 117
Gavle e. Sweden 46
Gaya cen. Bihar, India 134
Gaza sw. Palestine; in Gaza Strip, administered by U.N. since 1957 34
Gaziantep s. Turkey 73
Gdańsk See DANZIG.
Gdynia nw. Poland 78
Geelong s. Victoria, Australia *45
Gelligaer Glamorganshire, Wales 36
Gelsenkirchen w. North Rhine-Westphalia, West Germany 310
Geneva sw. Switzerland 146
Genoa (Genova) Liguria, Italy 681
Gensan See WONSAN.
Gent See GHENT.
Georgetown n. British Guiana; cap. 74
George Town Penang, Malayan Federation *189
Georgian Soviet Socialist Republic se. European U.S.S.R. 3,555
Gera e. cen. East Germany 89
Germany state, cen. Europe 68,200
See EAST GERMANY, WEST GERMANY.
Germiston Transvaal, Union of South Africa *131
Ghana state in Commonwealth of Nations, w. Africa 4,118
Ghazni ne. Afghanistan 27
Ghent nw. Belgium *228
Gibraltar British crown colony, w. Mediterranean 23
Giessen n. Hesse, West Germany 46
Gifu cen. Honshu, Japan 212
Gijón nw. Spain 111
Gilbert and Ellice Islands British colony, sw. Pacific 37
Gilgit subdivision, nw. Kashmir, India 77
Gillingham Kent, England 68
Girardot sw. Cundinamarca, Colombia ... 22
Giza n. Egypt 66
Gladbeck North Rhine-Westphalia, West Germany 72
Glamorganshire county, se. Wales 1,202
Glasgow Lanark co., Scotland 1,131
Gliwice (Gleiwitz) s. Poland 96
Gloucester Gloucestershire, England 67
Gloucestershire county, sw. England 939
Gôa district, Portuguese India 46
Godoy Cruz w. Argentina 46
Goiânia s. Goiás (cap.), Brazil 41
Goiás state, cen. Brazil 1,235
Gold Coast a province of Ghana 2,050
Gomel se. Belorussian S.S.R. 144
Gómez Palacio ne. Durango, Mexico 45
Gorizia Friuli-Venezia Giulia, Italy 31
Gorki (Gor'kiy) cen. European R.S.F.S.R. 876
Görlitz se. East Germany 86
Gorlovka se. Ukrainian S.S.R. 240
Gosport Hampshire, England 58
Göteborg sw. Sweden 354
Göttingen w. Lower Saxony, West Germany 78
Gouda w. cen. Netherlands 37
Granada w. cen. Nicaragua 22
— s. Spain 154
Gran Canaria island, Canary Is. 281
Gravesend Kent, England 43
Graz Styria (cap.), cen. Austria 226
Great Britain kingdom, nw. Europe; including England, Scotland, Wales, Isle of Man, and Channel Islands ... 48,840
Great Yarmouth s. Norfolk, England 51
Greece kingdom, se. Europe 7,603
Greenland Danish colony, ne. of Canada .. 21
Greenock Renfrew co., Scotland 76
Greenwich bor. of London, s. England 91
Grenada Windward Islands, The West Indies 72
Grenoble se. France 97
Grimsby Lincolnshire, England 95
Grodno w. Belorussian S.S.R. 50
Groningen n. Netherlands 132
Grosseto Tuscany, Italy 25
Grozny s. cen. European R.S.F.S.R. 226
Grudziadz nw. Poland 38
Guadalajara cen. Jalisco (cap.), Mexico .. 378
Guadeloupe island, French colony, W. Indies 279
Guairá department, Paraguay 90
Guajira commissary, Colombia 54
Gualeguaychú se. Entre Ríos, Argentina .. 37
Guanajuato state, cen. Mexico 1,325
Guantánamo e. Oriente, Cuba 42
Guaporé territory, n. Brazil 27
Guatemala republic, nw. Central America 2,787
— s. Guatemala; cap. 283
Guayaquil w. Ecuador 263
Guayas province, Ecuador 547
Guernsey Channel Islands, England 44
Guerrero state, sw. Mexico 915
Guiana, British See BRITISH GUIANA.
—, **Dutch** See SURINAM.
—, **French** See FRENCH GUIANA.
Guildford Surrey, England 47
Guimaras Island w. cen. Philippine Is. 40
Guinea See FRENCH GUINEA, PORTUGUESE GUINEA, SPANISH GUINEA.
Guinobatan se. Luzon, P.I. *32
Gujranwala West Pakistan 114
Gulbarga w. Hyderabad, India 77
Gunsan See KUNSAN.
Guntur cen. Madras, India 124
Gwalior and Lashkar cen. India 240
Györ nw. Hungary 55

H

Haarlem nw. Netherlands 156
Habana province, w. Cuba 1,235
— w. Habana; cap. of Cuba and prov. 660
Hachinohe n. Honshu, Japan 104
Hachioji s. cen. Honshu, Japan 83
Hadhramaut district, Aden Protectorate . 350
Haeju sw. North Korea 82
Hagen w. North Rhine-Westphalia, West Germany 146
Hagonoy s. cen. Luzon, P.I. *37
Hague, The ('s Gravenhage) w. cen. Netherlands; de facto cap. 533
Haifa w. Israel 129
Haiphong ne. North Vietnam *143
Haiti republic, West Indies 3,112
Hakodate s. Hokkaido, Japan 229
Halberstadt w. East Germany 45
Haleb See ALEPPO.
Halesowen Worcestershire, England 40
Halifax West Riding, Yorkshire 98
Halle sw. East Germany 223
Halmahera island, Indonesia 83
Halmstad sw. Sweden 35
Hälsingborg s. Sweden 71
Hama w. Syria 71
Hamada s. Honshu, Japan 41
Hamadan nw. Iran 119
Hamamatsu s. cen. Honshu, Japan 152
Hamburg nw. West Germany 1,605
Hamhung (Kanko) cen. North Korea ... 112
Hamilton Lanark co., Scotland 40
Hamm w. North Rhine-Westphalia, West Germany 60
Hampshire county, s. England 1,292
Hanamkonda See WARANGAL.
Hangchow n. Chekiang (cap.), China 437
Hankow e. Hupeh, China 749
Hanoi e. cen. North Vietnam 387
Hanover cen. Lower Saxony, West Germany 442
Hanyang e. Hupeh, China 101
Harar e. cen. Ethiopia 45
Harbin cen. Manchuria, China 760
Harburg-Wilhelmsburg port of Hamburg, West Germany 116
Hargeisa Br. Somaliland; cap. 15
Harrogate West Riding, Yorkshire, England 50
Harrow Middlesex, England 219
Hasselt ne. Belgium 29
Hastings Sussex, England 66
Havana See HABANA.
Havant and Waterloo s. Hampshire, England 32
Hayes and Harlington Middlesex, England *66
Hebron sw. Jordan 26
Heerlen se. Netherlands *56
Heidelberg n. Baden-Württemberg, West Germany 116
Heijo See PYONGYANG.
Heilbronn n. Baden-Württemberg, West Germany 64
Heilungkiang province, ne. China 2,563
Helder, Den nw. Netherlands 29
Helmond s. Netherlands 28
Helsinki (Helsingfors) s. Finland; cap. .. 363

Hendon Middlesex, England 156
Hengelo ne. Netherlands 46
Henzada cen. Burma 31
Herat w. Afghanistan 75
Herefordshire county, sw. England 127
Herford cen. Lower Saxony, West Germany 50
Herne w. North Rhine-Westphalia, West Germany 111
Hersfeld e. Hesse, West Germany 21
Herstal e. Belgium 27
Hertfordshire county, sw. England 610
Hesse (Hessen) state, West Germany 4,323
Heston and Isleworth Middlesex, England *107
Heywood Lancashire, England 25
Hidalgo state, cen. Mexico 850
High Wycombe Buckinghamshire, England 40
Hildesheim Lower Saxony, West Germany 72
Hilla cen. Iraq 51
Hilversum cen. Netherlands 85
Himachal Pradesh terr., n. India 989
Himeji s. Honshu, Japan 212
Hinckley Leicestershire, England 39
Hindenburg See ZABRZE.
Hirosaki n. Honshu, Japan 66
Hiroshima w. Honshu, Japan 285
Hobart se. Tasmania (cap.), Australia *76
Hoboken Belgium 32
Hodeida w. Yemen 26
Hodmezövasarhely se. Hungary 59
Hof n. Bavaria, West Germany 61
Hofei cen. China 70
Hokkaido island, Japan 4,296
Holguín Oriente, Cuba 36
Holland See NETHERLANDS.
Homs w. cen. Syria 100
Honan province, e. cen. China 28,473
Honduras republic, n. Central America 1,533
Hong Kong British colony, s. China 1,857
Honshu main Japanese island 62,587
Hopeh province, ne. China 28,529
Hornchurch Essex, England 104
Hornsey Middlesex, England 98
Horsens cen. Denmark 36
Hove Sussex, England 69
Howrah s. West Bengal, e. India 443
Hradec Králové nw. Czechoslovakia 51
Hsinking See CHANGCHUN.
Huancayo cen. Peru 27
Hubli s. Bombay, India 95
Huddersfield West Riding, Yorkshire, England 129
Hué n. South Vietnam *407
Huelva sw. Spain 57
Huila department, w. Colombia 225
Hull Yorkshire, England 299
Hunan province, se. China 27,186
Hungary republic, e. Europe 9,205
Hungnam (Konan) e. cen. North Korea 144
Huntingdon co., e. cen. England 69
Hupeh province, e. China 21,034
Hurstville N.S. Wales, Australia 34
Hwaining See ANKING.
Hyde Cheshire, England 32
Hyderabad state, cen. India 18,652
— s. Hyderabad (cap.), India *1,085
— s. West Pakistan 241

I

Iasy See JASSY.
Ibadan sw. Nigeria 336
Ibagué cen. Tolima (cap.), Colombia 27
Içel s. Turkey 38
Iceland republic; isl., N. Atlantic 144
Ichang sw. Hupeh, China 108
Ichinomiya s. Honshu, Japan 71
Ifni Sp. terr., nw. Africa 46
Ilford Essex, England 185
Ilhéus se. Bahia, Brazil 23
Ilkeston Derbyshire, England 34
Iloilo s. Panay, P.I. 110
Imabari s. Shikoku, Japan 60
Imperia Liguria, Italy 22
Imphal e. Manipur (cap.), India 99
Inchon nw. South Korea 108
India peninsula, republic 356,755
Indonesia republic, se. Asia 79,260
Indore cen. India 308
Ingolstadt w. Bavaria, West Germany 40
Innsbruck Tirol (cap.), w. Austria 95
Invercargill s. S. Island, New Zealand *31
Inverness county, nw. Scotland 85
Ionio province, Apulia, s. Italy 322
Ipoh cen. Malayan Federation 81
Ipswich Suffolk, England 105
— se. Queensland, Australia 26
Iquique w. Tarapacá (cap.), Chile 38
Iquitos n. Peru 32
Iran kingdom, sw. Asia 19,000
Irapuato w. Guanajuato, Mexico 49
Iraq (or Irak) republic, sw. Asia 4,800
Ireland republic, British Isles 2,961
Irkutsk s. cen. Asiatic R.S.F.S.R. 342
Isabela province, n. Luzon, P.I. 264
Isfahan w. cen. Iran 184
Isle of Man See MAN, ISLE OF.
Isle of Wight div. of Hampshire, England 95
Issy-les-Moulineaux n. cen. France 40
Istanbul nw. Turkey 1,000
Italian Somaliland See SOMALIA.
Italy republic, se. Europe 47,138
Itami s. Honshu, Japan 56
Ivanovo e. European R.S.F.S.R. 319
Ivry-sur-Seine n. France 40
Ixelles cen. Belgium 91
Izabal department, e. Guatemala 55
Izhevsk European R.S.F.S.R. 252
Izmir See SMYRNA.
Izmit nw. Turkey 36

J

Jaén s. Spain 62
Jaffa w. cen. Israel 101
Jaffna n. Ceylon 63
Jaipur ne. Rajasthan (cap.), India 291
Jakarta (Djarkarta) nw. Java, Indonesia; cap. 2,800
Jalapa s. Guatemala 6
— cen. Veracruz (cap.), Mexico 51
Jalisco state, cen. Mexico 1,747
Jamaica island, The West Indies 1,237
Jammu s. Jammu and Kashmir, India 50
Jammu and Kashmir state, n. India 4,370
Jamnagar w. Saurashtra, India 106
Jamshedpur se. Bihar, e. India 219
Japan archipelago empire, e. Asia 83,199
Jaro n. cen. Leyte, P.I. 26
Jarrow Durham, England 28
Jassy ne. Rumania 94
Java island, Indonesia 39,755
Jehol province, ne. China 6,109
Jelgava (Yelgava) cen. Latvian S.S.R. 34
Jena sw. East Germany 83
Jerez or **Jerez de la Frontera** s. Spain 108
Jersey Channel Islands 57
Jerusalem (New City) e. Israel; cap. 120
— (Old City) w. Jordan 50
Jesselton n. North Borneo; cap. 12
Jhang-Maghiana West Pakistan 73
Jhansi sw. Uttar Pradesh, India 128
João Pessoa e. Paraíba (cap.), Brazil 91
Jodhpur cen. Rajasthan, India 174
Jogjakarta s. Java, Indonesia 295
Johannesburg Transvaal, Union of South Africa *763
Johore state, Malayan Federation 738
Johore Bahru s. Johore (cap.), Malayan Federation 39
Joinvile ne. Santa Catarina, Brazil 41
Jolo Island sw. Philippine Is. 114
Jönköping s. Sweden 45
Jordan kingdom, w. Asia 1,250
Jubbulpore n. Madhya Pradesh, India 208
Juiz de Fora s. Minas Gerais, Brazil 87
Jujuy province, nw. Argentina 167
Jullundur cen. Punjab, India 169
Jumet Belgium 29
Jundiai se. São Paulo, Brazil 40
Junín n. Buenos Aires, Arg. 37
— department, Peru 339
Jutiapa department, se. Guatemala 136
Jutland n. Denmark 1,902
Jyväskylä cen. Finland 31

K

Kabul e. Afghanistan; cap. 206
Kadievka or **Kadiyevka** cen. Ukrainian S.S.R. 170
Kadikoy nw. Turkey 78
Kaesong sw. North Korea 89
Kagoshima s. Kyushu, Japan 230
Kai Islands Indonesia 51
Kaifeng Honan (cap.), e. China 303
Kaiserslautern Rhineland-Palatinate, West Germany 62
Kaishu See HAEJU.
Kaizuka s. Honshu, Japan 54
Kalamata s. Greece 35
Kalgan Chahar (cap.), n. China 151
Kalinin e. European R.S.F.S.R. 240
Kaliningrad w. European R.S.F.S.R. 372
Kalisz sw. Poland 48
Kaluga e. European R.S.F.S.R. 122
Kamensk-Ural'skiy se. European R.S.F.S.R. 122
Kanazawa n. cen. Honshu, Japan 252
Kandahar sw. Afghanistan 77
Kandy cen. Ceylon 51
Kanko See HAMHUNG.
Kano n. Nigeria 107
Kansu province, n. cen. China 6,897
Kaohsiung s. Taiwan 275
Kaolan See LANCHOW.
Kaposvár w. Hungary 33
Kapsan ne. North Korea 58
Karachi s. West Pakistan; cap., Pakistan 1,006
Karafuto Japanese name for SAKHALIN.
Karaganda ne. Kazakh S.S.R. 350
Karbala s. cen. Iraq 123
Karikal e. Madras, India 71
Karlsruhe w. Baden-Württemberg, West Germany 198
Kaschau See KOŠICE.
Kashan w. cen. Iran 45
Kashmir See JAMMU AND KASHMIR.
Kassa See KOŠICE.
Kassala e. Sudan 40
Kassel n. Hesse, West Germany 161
Kastrop See CASTROP-RAUXEL.
Kasur cen. West Pakistan 53
Katmandu e. cen. Nepal; cap. *109
Katowice (Stalinogród) sw. Poland 128
Kaunas s. Lithuanian S.S.R. 195
Kavalla ne. Greece 47
Kawagoe e. cen. Honshu, Japan 53
Kawaguchi se. Honshu, Japan 125
Kawasaki se. Honshu, Japan 319
Kayseri s. cen. Turkey 65
Kazakh Soviet Socialist Republic sw. U.S.S.R. 8,500
Kazan s. cen. European R.S.F.S.R. 565
Kazvin nw. Iran 77
Kecskemét s. Hungary 88
Kedah state, Malayan Federation 554
Kediri se. Java, Indonesia 48
Keelung (Chilung) n. Taiwan 100
Keighley West Riding, Yorkshire, England 57
Kei Islands See KAI ISLANDS.
Keijo See SEOUL.
Kelantan state, Malayan Federation 449
Kemerovo sw. Asiatic R.S.F.S.R. 240
Kénitra See PORT-LYAUTEY.
Kenjiko See KYOMIPO.
Kent county, se. England 1,563
Kenya British colony, protectorate, e. Africa 3,692
Kephallenia See CEPHALONIA.
Kerch se. Crimea, European R.S.F.S.R. 104
Kerkrade se. Netherlands 41
Kerkyra See CORFU.
Kerman se. Iran 53
Kermansha w. Iran 103
Kerry county, w. Munster, Ireland 134
Kettering Northamptonshire, England 37
Kew s. Victoria, Australia 31
Khabarovsk se. Asiatic R.S.F.S.R. 280
Kharagpur sw. West Bengal, India 130
Kharkov ne. Ukrainian S.S.R. 877
Khartoum cen. Sudan; cap. 75
Khaskovo cen. Bulgaria 27
Kherson s. Ukrainian S.S.R. 134
Khios See CHIOS.
Khonkaen ne. Thailand 10
Khorramshahr sw. Iran 30
Kiamusze ne. China 168
Kiangsi province, se. China 12,725
Kiangsu province, e. China 36,052
Kiangtu See YANGCHOW.
Kidderminster Worcestershire, England 37
Kiel n. Schleswig-Holstein (cap.), West Germany 254
Kielce sw. Poland 58
Kiev n. cen. Ukrainian S.S.R.; cap. 991
Kildare county, cen. Leinster, Ireland 65
Kilkenny county, s. Leinster, Ireland 67
Kilmarnock Ayr co., Scotland 42
Kimberley Cape of Good Hope Province, Union of South Africa *55
Kincardine county, e. Scotland 47
Kineshma cen. European R.S.F.S.R. 75
Kingston se. Jamaica; cap. 109
Kingston-upon-Hull See HULL.

Kingston-upon-Thames — Louth

Kingston-upon-Thames Surrey, England 40
Kingstown See DUN LAOGHAIRE.
Kinhwa ne. China ... 211
Kinross county, e. Scotland ... 7
Kirghiz Soviet Socialist Republic sw. U.S.S.R. ... 1,490
Kirin province, ne. China ... 6,981
— Kirin (cap.), China ... 247
Kirkcaldy Fife co., Scotland ... 49
Kirkcudbright county, s. Scotland ... 31
Kirkuk ne. Iraq ... 69
Kirov e. European R.S.F.S.R. ... 211
Kirovabad cen. Azerbaijan S.S.R. ... 111
Kirovgrad e. European R.S.F.S.R. ... 35
Kirovograd cen. Ukrainian S.S.R. ... 115
Kirovsk n. European R.S.F.S.R. ... 50
Kiryu e. cen. Honshu, Japan ... 95
Kishinev cen. Moldavian S.S.R. ... 190
Kishiwada cen. Honshu, Japan ... 99
Kiskunfélegyháza cen. Hungary ... 37
Kiskunhalas cen. Hungary ... 32
Kislovodsk s. European R.S.F.S.R. ... 51
Kitzingen nw. Bavaria, West Germany ... 16
Kiukiang n. Kiangsi, China ... 137
Kizel e. cen. R.S.F.S.R. ... 40
Kladno nw. Bohemia, Czechoslovakia ... 41
Klagenfurt cen. Austria ... 63
Klaipeda See MEMEL.
Klang s. Selangor, Malayan Federation ... 34
Kobe w. Honshu, Japan ... 765
København See COPENHAGEN.
Koblenz See COBLENZ.
Kochi s. Shikoku, Japan ... 162
Kofu s. cen. Honshu, Japan ... 121
Kogarah N.S. Wales, Australia ... 39
Kohat n. West Pakistan ... 41
Koil See ALIGARH.
Kokand e. Uzbek S.S.R. ... 85
Kokura n. Kyushu, Japan ... 199
Kolar Gold Fields se. Mysore, India ... 159
Kolarovgrad ne. Bulgaria ... 31
Kolding sw. Denmark ... 31
Kolhapur s. Bombay, India ... 137
Köln See COLOGNE.
Kolomna e. European R.S.F.S.R. ... 75
Kom Ombo se. Egypt ... 39
Komotine ne. Greece ... 30
Komsomolsk se. European R.S.F.S.R. ... 169
Konan See HUNGNAM.
Kongmoon See SUNWUI.
Königgrätz See HRADEC KRÁLOVÉ.
Königsberg See KALININGRAD.
Königshütte See CHORZÓW.
Konstantinovka Ukrainian S.S.R. ... 95
Konya sw. Turkey ... 65
Köpenick Berlin, East Germany ... 114
Korea, North republic, ne. Asia ... 9,102
—, South republic, ne. Asia ... 20,188
Koriyama ne. Honshu, Japan ... 70
Kortrijk See COURTRAI.
Košice ne. Czechoslovakia ... 52
Kostroma e. European R.S.F.S.R. ... 156
Köthen cen. East Germany ... 43
Kottbus See COTTBUS.
Kovno See KAUNAS.
Kovrov e. European R.S.F.S.R. ... 67
Kowloon Hong Kong, China ... 700
Kozhikode See CALICUT.
Kragujevac e. cen. Yugoslavia ... 33
Kraków See CRACOW.
Kramatorsk e. Ukrainian S.S.R. ... 117
Krasnodar sw. European R.S.F.S.R. ... 271
Krasnoyarsk cen. Asiatic R.S.F.S.R. ... 328
Krefeld w. North Rhine-Westphalia, West Germany ... 170
Kremenchug cen. Ukrainian S.S.R. ... 89
Krete See CRETE.
Krivoi Rog cen. Ukrainian S.S.R. ... 322
Kronstadt nw. European R.S.F.S.R. ... 50
Krugersdorp Transvaal, Union of South Africa ... 72
Kuala Lumpur s. Selangor; cap. of Malayan Federation ... 176
Kuching w. Sarawak, Borneo; cap. ... 38
Kudus cen. Java, Indonesia ... 54
Kuibyshev se. European R.S.F.S.R. ... 760
Kum See QUM.
Kumamoto w. Kyushu, Japan ... 267
Kumasi n. Ashanti, Ghana ... 78
Kumbakonam se. Madras, India ... 67
Kunming Yünnan (cap.), s. China ... 307
Kunsan (Gunsan) South Korea ... 74
Kuntsevo e. European R.S.F.S.R. ... 111
Kuopio se. Finland ... 33
Kure w. Honshu, Japan ... 188
Kurgan sw. Asiatic R.S.F.S.R. ... 106
Kursk sw. European R.S.F.S.R. ... 179

Kurume n. Kyushu, Japan ... 100
Kushiro se. Hokkaido, Japan ... 93
Kutaisi or Kutasis w. Georgian S.S.R. ... 114
Kutch See CUTCH.
Kuwait sheikdom, nw. coast, Persian Gulf ... 207
Kuybyshev See KUIBYSHEV.
Kwangchowan See CHANKIANG.
Kwanghwa nw. Hupeh, China ... 29
Kwangju sw. South Korea ... 139
Kwangsi province, s. China ... 14,603
Kwangtung province, s. China ... 27,825
Kweichow province, s. China ... 10,518
Kweihsui (Kweisui) n. China ... 108
Kweilin e. cen. China ... 88
Kweiyang cen. Kweichow (cap.), China ... 240
Kyomipo (Kenjiho) North Korea ... 53
Kyongsong See SEOUL.
Kyoto s. Honshu, Japan ... 1,101
Kyushu island, sw. Japan ... 11,399

L

Laaland island, Denmark ... 86
La Chaux-de-Fonds w. Switzerland ... 33
La Coruña See CORUÑA, LA.
Lagos sw. Nigeria; cap. ... 230
Lahore West Pakistan ... 849
Lahti s. Finland ... 44
Lampang nw. Thailand ... 23
Lanark county, s. cen. Scotland ... 1,614
Lancashire county, nw. England ... 5,116
Lancaster Lancashire, England ... 52
Lanchow (Kaolan) n. China ... 204
Langreo See SAMA DE LANGREO.
Laoag nw. Luzon, P.I. ... 44
Laohokow See KWANGHWA.
Laoighis county, e. Leinster, Ireland ... 48
Laos kingdom, nw. Indochina ... 1,200
La Pampa province, cen. Argentina ... 167
La Paz department, nw. Bolivia ... 948
— w. La Paz (cap.); de facto seat of govt. of Bolivia ... 321
La Plata ne. Buenos Aires (cap.), Argentina ... 218
Larache w. Morocco ... 44
La Rioja province, w. Argentina ... 109
— city ... 23
La Rochelle sw. France ... 46
La Serena nw. Coquimbo (cap.), Chile ... 21
Las Palmas de Gran Canaria Canary Islands ... *153
La Spezia province, Italy ... 231
— city, Liguria ... 110
Las Villas province, Cuba ... 939
Latakia w. Syria ... 87
Latina province, Italy ... 283
Latium See LAZIO.
Latvian Soviet Socialist Republic ne. Europe ... 1,800
Launceston n. Tasmania ... 40
Lausanne w. Switzerland ... 107
Lavalleja department, se. Uruguay ... 117
Lazio (Latium) department, w. cen. Italy ... 3,346
Leamington Warwickshire, England ... 36
Leatherhead Surrey, England ... 27
Lebanon republic, n. of Israel, w. Asia ... 1,165
Lecce province, Italy ... 623
— Apulia, Italy ... 60
Leeds West Riding, Yorkshire, England ... 505
Leeuwarden n. Netherlands ... 58
Leeward Islands The West Indies ... 109
Legaspi se. Luzon, P.I. ... *78
Leghorn (Livorno) Tuscany, Italy ... 140
Legnica See LIEGNITZ.
Le Havre nw. France ... 105
Leicester Leicestershire, England ... 285
Leicestershire county, cen. England ... 631
Leiden See LEYDEN.
Leigh Lancashire, England ... 49
Leinster province, Ireland ... 1,335
Leipzig s. East Germany ... 607
Leitrim county, ne. Connacht, Ireland ... 45
Leix See LAOIGHIS.
Le Mans nw. France ... 91
Lemberg See LVOV.
Lemnos island, Aegean Sea, Greece ... 24
Leninabad Tadzhik S.S.R. ... 50
Leninakan e. Armenian S.S.R. ... 103
Leningrad ne. European R.S.F.S.R. ... *3,182
Leninsk-Kuznetskiy sw. Asiatic R.S.F.S.R. ... 119
Lens n. France ... 34
León w. Guanajuato, Mexico ... 123
— nw. Nicaragua ... 32
— province, nw. Spain ... 545

León nw. Spain ... 59
Leopoldville province, Belgian Congo ... 2,445
— Belgian Congo; cap. ... 160
Lepaya See LIEPAJA.
Le Puy se. France ... 18
Lérida ne. Spain ... 53
Lesbos (Lesvos) island, Greece ... 136
Leuven See LOUVAIN.
Levallois-Perret n. cen. France ... 61
Levkas island, Greece ... 32
Leyden sw. Netherlands ... 87
Leyte island, se. Philippine Is. ... 1,007
Leyton Essex, England ... 105
Lhasa se. Tibet (cap.), w. China ... 20
Liaopeh province, China ... 3,798
Liaoyang ne. China ... 100
Libau See LIEPAJA.
Liberec nw. Czechoslovakia ... *53
Liberia republic, w. Africa ... 2,750
Libia See LIBYA.
Libya federal kingdom, N. Africa ... 1,166
Lichtenberg Berlin, East Germany ... 158
Lidcombe N.S. Wales, Australia ... 20
Liége Belgium ... 156
Liegnitz (Legnica) w. Poland ... 24
Liepaja w. Latvian S.S.R. ... 57
Lierre Belgium ... 29
Liguria region, nw. Italy ... 1,557
Lille n. France ... 180
Lima department, Peru ... 849
— sw. Peru; cap. ... 534
Limburg province, Belgium ... 460
— province, Netherlands ... 684
Limeira e. São Paulo, Brazil ... 27
Limerick county, s. Munster, Ireland ... 143
— n. Limerick, Munster, Ireland ... 50
Limoges w. cen. France ... 96
Linares province, cen. Chile ... 135
— cen. Linares (cap.), Chile ... 17
— s. Spain ... 31
Linchwan See FUCHOW.
Lincoln Lincolnshire, England ... 69
Lincolnshire county, ne. England ... 706
Línea, La s. Spain ... 34
Lingayen w. cen. Luzon, P.I. ... 31
Lingga Archipelago Indonesia ... 31
Linköping se. Sweden ... 55
Linz nw. Austria ... 185
Lipa sw. Luzon, P.I. ... 49
Lipetsk w. European R.S.F.S.R. ... 123
Lisbon (Lisboa) w. Portugal; cap. ... 790
Lithuanian Soviet Socialist Republic (Lithuania) ne. Europe ... 2,700
Liuchow (Maping) se. China ... 208
Liverpool Lancashire, England ... 786
Livorno province, Italy ... 281
— city See LEGHORN.
Livramento cen. Rio Grande do Sul, Brazil ... 29
Ljubljana nw. Yugoslavia ... 121
Llanelly Carmarthenshire, Wales ... 38
Llanquihue province, s. Chile ... 117
Llwchwr Glamorganshire, Wales ... 26
Loanda See LUANDA.
Lobito Angola ... 24
Łódź province, sw. Poland ... 1,518
— city, sw. Poland ... 622
Logroño n. Spain ... 52
Loja province, s. Ecuador ... 216
Lokeren Belgium ... 25
Lolland See LAALAND.
Lomas de Zamora ne. Buenos Aires, Argentina ... 128
Lombardy (Lombardia) region, n. Italy ... 6,836
Lomblem island, Indonesia ... 47
Lombok island, Indonesia ... 701
Lomé Togo, w. Africa ... 30
London city and administrative county, England; cap., United Kingdom ... *8,346
Londonderry county, nw. Northern Ireland ... 105
— w. Londonderry, Northern Ireland ... 50
Long Eaton Derbyshire, cen. England ... 28
Longford county, nw. Leinster, Ireland ... 36
Longxuyen South Vietnam ... 148
Loon Bohol, P.I. ... 29
Lorca se. Spain ... 24
Loreto department, Peru ... 169
Lorient nw. France ... 46
Los Ríos province, w. cen. Ecuador ... 137
Lota w. Concepción, Chile ... 37
Loughborough Leicestershire, England ... 35
Lourdes sw. France ... 12
Lourenço Marques s. Mozambique; cap. ... 93
Louth county, ne. Leinster, Ireland ... 66

Louvain (Leuven) cen. Belgium 37
Lower Austria (Niederösterreich) province, Austria 1,250
Lower Hutt North Island, New Zealand *44
Lower Saxony n. West Germany 6,795
Lowestoft Suffolk, England 45
Loyalty Islands sw. Pacific 12
Luanda Angola; cap. 61
Luang Prabang n. Laos 142
Lubang Islands P.I. 15
Lübeck nw. West Germany 237
Lublin e. Poland 99
Lucania See BASILICATA.
Lucca province, Tuscany, Italy 364
— city, n. Tuscany 39
Lucena s. Spain 23
— Philippine Is. *33
Lucerne cen. Switzerland 61
Lucknow Uttar Pradesh (cap.), India 498
Lüdenscheid s. cen. North Rhine–Westphalia, West Germany 51
Ludhiana cen. Punjab, India 149
Ludwigsburg cen. Baden–Württemberg, West Germany 50
Ludwigshafen Rhineland–Palatinate, West Germany 150
Lugansk See VOROSHILOVGRAD.
Lugo nw. Spain 54
Luimneach See LIMERICK.
Lund s. Sweden 34
Lungkiang See TSITSIHAR.
Luque s. Paraguay 25
Luton Bedfordshire, England 110
Luxemburg (Luxembourg) grand duchy, w. Europe 298
— s. Luxemburg; cap. 61
— province, e. Belgium 213
Luxor n. Egypt 24
Luzern See LUCERNE.
Luzon island, n. Philippine Is. 9,074
Lvov (Lwów) w. Ukrainian S.S.R. 387
Lyallpur Punjab, West Pakistan 179
Lyons se. France 440
Lysva e. European R.S.F.S.R. 51
Lytham St. Anne's Lancashire, England 30
Lyublin See LUBLIN.

M

Maastricht See MAESTRICHT.
Macao (Macáu) Portuguese colony, se. China 374
—Macao; cap. 312
Macassar (Makassar) sw. Celebes, Indonesia 250
Macclesfield Cheshire, England 36
Macedonia division, n. Greece 1,691
— republic, s. Yugoslavia 1,152
Maceió e. Alagoas (cap.), Brazil 102
Macerata province, The Marches, Italy .. 301
Madagascar French island colony, se. Africa 4,463
Madeira district of Portugal, e. Atlantic Ocean 269
Madhya Pradesh state, n. cen. India .. 21,327
Madiun e. cen. Java, Indonesia 45
Madras state, s. India 56,952
— e. Madras; cap. 1,429
Madrid cen. Spain; cap. 1,618
Madura island, Indonesia 2,440
Madurai s. Madras, India 362
Maestricht s. Netherlands 75
Magallanes province, s. Chile 48
— city See PUNTA ARENAS.
Magdalena department, n. Colombia 342
Magdeburg w. cen. East Germany 236
Magelang cen. Java, Indonesia 53
Magnitogorsk se. European R.S.F.S.R. .. 284
Mahalla el Kubra n. Egypt 140
Mährisch–Ostrau See MORAVSKÁ OSTRAVA.
Maidenhead Berkshire, England 27
Maidstone Kent, England 54
Maiduguri ne. Nigeria 24
Maikop s. cen. European R.S.F.S.R. 67
Mainz w. Rhineland–Palatinate (cap.), West Germany 88
Maiquetía n. Venezuela 37
Maisons–Alfort n. cen. France 36
Majorca (Mallorca) island, Spain 327
Makassar See MACASSAR.
Makati cen. Luzon, P.I. 34
Makeevka e. Ukrainian S.S.R. 311
Makhachkala s. European R.S.F.S.R. 106
Makó se. Hungary 34
Malabon cen. Luzon, P.I. *46
Malacca Settlement of Malacca; cap. 55
Malacca, Settlement of s. Malayan Federation 239
Málaga s. Spain 276
Malang e. central Java, Indonesia 500
Malatya se. Turkey 49
Malayan Federation (Federation of Malaya) Commonwealth of Nations, se. Asia 5,420
Malayer w. Iran 32
Maldive Islands (Maldiva) Brit. protected sultanate, nw. Indian Ocean 82
Maldonado department, s. Uruguay 68
Malines See MECHLIN.
Malita se. Mindanao, P.I. 31
Mallawi e. cen. Egypt 34
Malleco province, cen. Chile 154
Mallorca See MAJORCA.
Malmö s. Sweden 192
Malta British colony; island, cen. Mediterranean 278
Man, Isle of n. Irish Sea 55
Manabí province, Ecuador 382
Manado See MENADO.
Managua w. cen. Nicaragua; cap. 107
Manama Bahrein Islands; cap. 40
Manaus (Manáos) e. Amazonas (cap.), Brazil 111
Manchester Lancashire, England 703
Manchuria region, ne. China 36,903
Mandalay cen. Burma 163
Mangalore sw. Madras, India 117
Manila cen. Luzon, P.I. 984
Manipur state, ne. India 579
Manissa w. cen. Turkey 39
Manizales cen. Caldas (cap.), Colombia ... 51
Manly se. N.S. Wales, Australia 34
Mannheim nw. Baden–Württemberg, West Germany 244
Manresa ne. Spain 31
Mansfield Nottinghamshire, England 51
Mansura (El Mansura) ne. Egypt 102
Mantova province, n. Italy 424
Mantua (Mantova) Lombardy, Italy 51
Manzanillo sw. Oriente, Cuba 36
Maping See LIUCHOW.
Maracaibo nw. Venezuela *232
Maracay n. Venezuela *65
Maragheh nw. Iran 35
Maranhão state, n. Brazil 1,600
— city See SÃO LUÍS.
Maras cen. Turkey 35
Marches, The (Le Marche) region, cen. Italy 1,361
Mardan n. West Pakistan 48
Mar del Plata e. Buenos Aires, Arg. 117
Mardin se. Turkey 22
Margate Kent, England 42
Marianao n. Habana, Cuba 120
Marianas Islands w. Pacific 48
Maribor n. Yugoslavia 66
Marie Galante island, Fr. Lesser Antilles 30
Marília s. São Paulo, Brazil 36
Marinduque island, n. cen. Philippine Is. .. 85
Mariupol See ZHDANOV.
Marrakesh sw. Morocco 238
Marruecos See MOROCCO.
Marsala w. Sicily, Italy 71
Marseille (Marseilles) se. France 552
Marshall Islands archipelago, U. N. Trust Territory of Pacific Islands, w. Pacific ... 11
Martinique island, French West Indies ... 262
Masan South Korea 91
Masbate island, cen. Philippine Is. 162
Mascara n. Algeria 26
Mashonaland Southern Rhodesia 1,165
Masqat See MUSCAT.
Masulipatam e. Madras, India 59
Matanzas province, w. Cuba 361
— nw. Matanzas, Cuba 59
Mataró ne. Spain 25
Matera province, Italy 183
Mathura w. Uttar Pradesh, India 106
Mato Grosso state, w. Brazil 528
Matsue sw. Honshu, Japan 74
Matsumoto cen. Honshu, Japan 86
Matsuyama w. Shikoku, Japan 164
Matsuye See MATSUE.
Mattancheri w. Travancore–Cochin, India 53
Maule province, cen. Chile 70
Mauritania territory, Fr. West Africa 524
Mauritius island, British colony, w. Indian Ocean 475
Mayo county, w. Connacht, Ireland 148
Mazagan nw. Morocco 40
Mazar-i-Sharif n. Afghanistan 42
Mazatlán s. Sinaloa, Mexico 41
Meath county, ne. Leinster, Ireland 66
Mechlin (Malines) n. Belgium 60
Medan ne. Sumatra, Indonesia 500
Medellín cen. Antioquia, Colombia 144
Medina ne. Saudi Arabia 12
Meerut w. Uttar Pradesh, n. India 241
Meissen se. East Germany 48
Meknès Morocco 160
Melbourne s. Victoria (cap.), Australia *1,227
Melilla Spanish enclave, Morocco 81
Melitopol se. Ukrainian S.S.R. 75
Memel w. Lithuanian S.S.R. 41
Menado ne. Celebes, Indonesia 27
Mendoza province, w. Argentina 591
— n. Mendoza (cap.), Arg. 105
Mengtze s. Yünnan, China 9
Menorca See MINORCA.
Mercedes ne. Buenos Aires, Arg. 38
— sw. Uruguay 30
Mérida w. Yucatán (cap.), Mexico 145
— state, Venezuela 210
Merionethshire nw. Wales 41
Merksem Belgium 29
Mersin See IÇEL.
Merthyr Tydfil Glamorganshire, Wales ... 61
Merton and Morden Surrey, England 74
Meshed ne. Iran 191
Messina province, Sicily, Italy 666
— Sicily, Italy 202
Meta intendancy, cen. Colombia 52
Metz ne. France 65
Mexicali n. Bajo California, Mexico 65
Mexico republic, s. North America 25,706
— state, cen. Mexico 1,390
Mexico City cen. Federal District; cap. of dist. and republic 2,234
Mezotur e. Hungary 27
Miagao s. Panay, P.I. 30
Michoacán state, w. Mexico 1,415
Michurinsk w. European R.S.F.S.R. 70
Middelburg sw. Netherlands 21
Middle Congo Fr. Equatorial Africa 684
Middlesbrough North Riding, Yorkshire, England 147
Middlesex county, se. England 2,269
Middleton Lancashire, England 32
Midlothian county, se. Scotland 566
Mieres nw. Spain 9
Milan Lombardy (cap.), Italy 1,268
Milano province, n. Italy 2,500
Minas Gerais state, e. Brazil 7,839
Mindanao island, s. Philippine Is. 2,428
Mindoro island, nw. Philippine Is. 167
Minhow See FOOCHOW.
Minorca island, Spain, w. Mediterranean Sea 43
Minsk cen. Belorussian S.S.R.; cap. 412
Minya n. cen. Egypt 69
Mirzapur se. Uttar Pradesh, India 71
Misamis Occidental province, Philippine Islands 208
Misamis Oriental province, Philippine Is. 370
Misiones territory, ne. Argentina 244
— department, Paraguay 42
Miskolc ne. Hungary 104
Misurata n. Libya 5
Mitcham Surrey, England 67
Mito se. Honshu, Japan 67
Miyako island, Ryukyu Islands 56
Miyakonojo s. Kyushu, Japan 75
Miyazaki se. Kyushu, Japan 103
Modena province, Italy 497
— city, cen. Emilia–Romagna, Italy 111
Mogadishu se. Somalia, U.N. Trust Territory; cap. 85
Mogador sw. Morocco 24
Mogilev sw. Belorussian S.S.R. 106
Moji n. Kyushu, Japan 124
Mokpo South Korea 111
Moldavian Soviet Socialist Republic European U.S.S.R. 2,660
Molenbeek–Saint–Jean Belgium 64
Molotov See PERM.
Moluccas islands, Indonesia 600
Mombasa se. Kenya 85
Momostenango sw. Guatemala 8
Monaco principality; enclave, se. France ... 19
Monaghan county, s. Ulster, Ireland 57
Monghyr cen. Bihar, India 63
Mongolia republic, e. cen. Asia 2,078
Monmouthshire sw. England 425
Monrovia sw. Liberia; cap. 12
Mons sw. Belgium 26
Montenegro constituent republic of Yugoslavia 377

Monterrey

Monterrey w. Nuevo León, Mexico 332
Montevideo s. Uruguay; cap. 708
Montgomeryshire e. cen. Wales 46
Montluçon cen. France 45
Montpellier s. France 81
Montreuil-sous-Bois n. France 70
Monza Lombardy, Italy 70
Moppo See MOKPO.
Moradabad w. Uttar Pradesh, n. India ... 164
Moratuwa w. Ceylon 50
Moravia (Morava) region, cen.
  Czechoslovakia 8,762
Moravská Ostrava n. Moravia,
  Czechoslovakia 181
Moray county, ne. Scotland 48
Morecambe and Heysham Lancashire,
  England 37
Morelia n. Michoacán (cap.), Mexico 63
Morelos state, s. Mexico 276
Morioka ne. Honshu, Japan 118
Morley West Riding, Yorkshire, England .. 40
Morocco (Marruecos) kingdom, nw.
  Africa 10,000
— city See MARRAKESH.
Moscow w. cen. European R.S.F.S.R.;
  cap. 5,100
Most (Brüx) nw. Czechoslovakia 33
Mostaganem n. Algeria 50
Mosul n. Iraq 203
Motherwell and Wishaw Lanark co.,
  Scotland 68
Moulmein s. Burma 71
Mountain Ash ne. Glamorganshire, Wales 31
Mouscron (Moeskroen) Belgium 36
Mozambique Portuguese colony, se.
  Africa 5,732
Muharraq island, Bahrein archipelago ... 26
Mühlhausen sw. East Germany 48
Mukalla s. Aden Protectorate 19
Mukden See SHENYANG.
Mülheim w. North Rhine-Westphalia,
  West Germany 149
Mulhouse e. France 86
Multan sw. Punjab, West Pakistan 190
Muna island, Indonesia 109
München-Gladbach w. North Rhine-
  Westphalia, West Germany 125
Munich (München) se. West Germany ... 831
Munster province, Ireland 917
Münster w. North Rhine-Westphalia,
  West Germany 120
Murcia se. Spain 218
Murmansk nw. European R.S.F.S.R. 168
Muroran sw. Hokkaido, Japan 110
Muscat (Masqat) Muscat and Oman;
  cap. 5
Muscat and Oman sultanate, se.
  Arabia 550
Mutankiang ne. China 200
Muttra See MATHURA.
Muzaffarpur nw. Bihar, India 54
Myingyan nw. Burma 29
Mymensingh E. Bengal, E. Pakistan 52
Mysore state, s. India 9,071
— s. Mysore, India 244
Mytilene island See LESBOS.
— city, Lesbos, Greece 29
Mytishchi w. cen. European
  R.S.F.S.R. 60

N

Nablus nw. Jordan 23
Naga Cebu, P.I. 8
Nagahama s. Honshu, Japan 47
Nagano cen. Honshu, Japan 101
Nagaoka ne. Honshu, Japan 66
Nagasaki w. Kyushu, Japan 242
Nagercoil s. Travancore-Cochin,
  India 52
Nagoya s. Honshu, Japan 1,031
Nagpur ne. Bombay, India 449
Naha Ryukyu Islands; cap. 45
Nairn county, n. cen. Scotland 9
Nairobi s. cen. Kenya; cap. 119
Najin North Korea 34
Nakatsu n. Kyushu, Japan 51
Nakhon Ratchasima s. Thailand 22
Nakhon Sithammarat s. Thailand 15
Nalchik s. European R.S.F.S.R. 48
Namangan e. Uzbek S.S.R. 104
Namdinh (Nam-Dinh) North Vietnam ... 40
Namhoi s. Kwangtung, China 95
Nampo See CHINNAMPO.
Namur Belgium 31
Nanchang Kiangsi (cap.), se. China 192
Nancy ne. France 108

1500

Nanking s. Kiangsu, China 1,201
Nanning s. Kwangsi, China 203
Nanping e. Fukien, China 53
Nantes nw. France 187
Nantung n. Kiangsu, China 133
Nanyang s. Honan, China 50
Naokata n. Kyushu, Japan 54
Naples (Napoli) Campania, Italy *866
Napoli province, w. cen. Italy 2,079
Nara sw. Honshu, Japan 78
Narayanganj E. Bengal, E. Pakistan 68
Nariño department, sw. Colombia 465
Nasik cen. Bombay, India *55
Natal province, e. Union of South
  Africa 2,202
— e. Rio Grande do Norte (cap.), Brazil .. 98
Nawa See NAHA.
Nawabshah West Pakistan 34
Nayarit state, cen. Mexico 291
Neath Glamorganshire, Wales 32
Negapattinam se. Madras, India *53
Negombo w. Ceylon 32
Negri Sembilan state, Malayan
  Federation 296
Negros island, sw. Philippine Is. 1,218
Neiva n. Huila (cap.), Colombia 37
Nellore e. Madras, India 56
Nelson Lancashire, England 38
Nepal kingdom, s. Asia 5,600
Netherlands (Holland) kingdom, nw.
  Europe 9,625
Netherlands Antilles islands in the
  West Indies 155
Netherlands New Guinea western
  New Guinea, sw. Pacific 1,000
Neuchâtel w. Switzerland 28
Neuilly-sur-Seine n. cen. France 59
Neumünster s. Schleswig-Holstein, West
  Germany 73
Neuquén territory, w. Argentina 86
Neuss w. North Rhine-Westphalia, West
  Germany 63
Nevers cen. France 34
New Britain island, Bismarck
  Archipelago, Terr. of New Guinea 90
New Caledonia island, French colony, sw.
  Pacific 61
Newcastle e. N.S Wales, Australia 127
Newcastle-under-Lyme Staffordshire,
  England 76
Newcastle-upon-Tyne Northumberland,
  England 291
New Delhi s. Delhi, India; cap. 278
New Guinea island, sw. Pacific 2,094
New Guinea, Territory of U.N. Trust
  Territory, sw. Pacific 1,094
New Hebrides island group, British-French
  condominium, sw. Pacific 49
New Ireland island, Bismarck Arch. 34
Newport Monmouthshire, England 105
New Providence Island Bahamas 29
New South Wales state, se.
  Australia 2,985
New Zealand British Commonwealth of
  Nations, sw. Pacific 1,940
Nias island, Indonesia 187
Nicaragua republic, Central America .. 1,053
Nice se. France 207
Nicobar Islands See ANDAMAN AND NICOBAR
  ISLANDS.
Nicosia n. cen. Cyprus; cap. 34
Nictheroy See NITERÓI.
Niger Fr. West Africa 2,041
Nigeria, Federation of w. Africa 26,000
Niigata ne. Honshu, Japan 220
Nijmegen cen. Netherlands 107
Nikolaev s. cen. Ukrainian S.S.R. 206
Nikopol se. Ukrainian S.S.R. 57
Nîmes se. France 75
Ningpo e. Chekiang, China 250
Ningtu e. China 65
Nish (Nis) e. cen. Yugoslavia 51
Nishinomiya w. Honshu, Japan 126
Niterói s. Rio de Janeiro (cap.), Brazil .. 175
Nizamabad cen. Hyderabad, India 65
Nizhni Tagil w. Siberian R.S.F.S.R. 297
Nobeoka se. Kyushu, Japan 88
Norfolk county, e. England 547
Norrköping se. Sweden 85
Norte de Santander department, n.
  Colombia 433
Northampton Northamptonshire,
  England 104
Northamptonshire co., cen. England 360
Northern Ireland The six counties, Antrim,
  Armagh, Down, Fermanagh, London-
  derry, and Tyrone, of ne. Ireland ... 1,370

Oruro

Northern Rhodesia British protectorate,
  se. Africa 1,947
North Riding division, Yorkshire,
  England 525
Northumberland county, ne. England ... 798
North Vietnam se. Asia 9,800
Northwood and Ruislip See RUISLIP
  NORTHWOOD.
Norway kingdom, n. Europe 3,157
Norwich Norfolk, England 121
Nottingham Nottinghamshire, England .. 306
Nottinghamshire co., e. cen. England ... 841
Nouméa New Caledonia; cap. 10
Nova Gôa (Pangim) Portuguese India;
  cap. 32
Nova Iguaçu s. Rio de Janeiro, Brazil 59
Nova Lisboa cen. Angola 28
Novara province, Piedmont, Italy 423
— city, ne. Piedmont 58
Novi Sad ne. Yugoslavia 77
Novocherkassk s. cen. European
  R.S.F.S.R. 81
Novorossiisk sw. European R.S.F.S.R. ... 95
Novosibirsk sw. Asiatic R.S.F.S.R. 731
Nuble province, cen. Chile 243
Nueva Ecija province, Luzon, P.I. 468
Nueva Esparta state, Venezuela 76
Nueva Rosita e. Coahuila, Mexico 26
Nueva Vizcaya province, cen. Luzon,
  Philippine Islands 83
Nuevo Laredo n. Tamaulipas, Mexico ... 58
Nuevo León state, ne. Mexico 736
Nuhutjut (Nuhu Chut) island, Indonesia . 25
Numazu s. cen. Honshu, Japan 101
Nuneaton Warwickshire, England 54
Nuoro province, Sardinia, Italy 257
Nuremberg (Nürnberg) n. Bavaria,
  West Germany 360
Nyasaland Protectorate, Federation of
  Rhodesia and Nyasaland, se. Africa .. 2,349
Nyíregyháza ne. Hungary 56

O

Oaxaca state, s. Mexico 1,415
— cen. Oaxaca (cap.), Mexico 46
Oberhausen w. North Rhine-Westphalia,
  West Germany 202
Odense s. Denmark 100
Odessa s. cen. Ukrainian S.S.R. 607
Offaly county, cen. Leinster, Ireland 54
Offenbach Hesse, West Germany 88
Ogbomosho sw. Nigeria 85
O'Higgins province, cen. Chile 200
Oita ne. Kyushu, Japan 94
Okayama w. Honshu, Japan 163
Okazaki s. Honshu, Japan 96
Okinawa Ryukyu Islands, sw. of Kyushu,
  Japan 579
Oldbury Worcestershire, England 54
Oldenburg Lower Saxony, West
  Germany 122
Oldham Lancashire, England 121
Olomouc cen. Moravia, Czechoslovakia .. *58
Olsztyn (Allenstein) n. Poland 29
Oman See MUSCAT AND OMAN.
Omdurman cen. Sudan 127
Omsk w. Asiatic R.S.F.S.R. 505
Omuta (Omuda) nw. Kyushu, Japan 192
Onoda sw. Honshu, Japan 58
Onomichi sw. Honshu, Japan 61
Opava w. cen. Czechoslovakia *30
Oporto nw. Portugal 285
Oradea (Oradea-Mare) n. Rumania 82
Oran n. Algeria 245
Orange e. N.S Wales, Australia 14
Orange Free State province, e. Union of
  South Africa 879
Orasul Stalin cen. Rumania 83
Ordzhonikidze See DZAUDZHIKAU.
Ordzhonikidzegrad See BEZHITSA.
Örebro s. cen. Sweden 66
Orekhovo-Zuyevo w. European
  R.S.F.S.R. 109
Orel w. European R.S.F.S.R. 128
Orense w. Spain 55
Oriente province, e. Cuba 1,356
Orihuela se. Spain 38
Orissa state, e. India 14,644
Orizaba cen. Veracruz, Mexico 55
Orkney Islands island county,
  ne. Scotland 22
Orléans n. cen. France 65
Orpington Kent, England 63
Orsk se. European R.S.F.S.R. 157
Oruro department, sw. Bolivia 210
— ne. Oruro (cap.), Bolivia 63

Osaka w. cen. Honshu, Japan 1,956
Oshima islands, Ryukyu Islands 214
Osijek n. Yugoslavia 50
Osipenko se. Ukrainian S.S.R. 51
Oslo se. Norway; cap. 417
Osnabrück w. Lower Saxony, West Germany 109
Osorno province, cen. Chile 107
— n. Osorno (cap.), Chile 25
Ostend (Oostende) w. Belgium 50
Ostrava See MORAVSKÁ OSTRAVA.
Ostrów Wielkopolski w. cen. Poland 31
Otago district, New Zealand 237
Otaru w. Hokkaido, Japan 178
Otsu nw. Honshu, Japan 71
Oubangui-Chari See UBANGI-SHARI.
Oujda or Oudjda ne. Morocco 89
Oulu w. Finland 37
Ovalle w. Coquimbo, Chile 15
Oviedo nw. Spain 106
Oxford Oxfordshire, England 98
Oxfordshire county, s. cen. England 275
Oyo sw. Nigeria 79

P

Pabianice e. cen. Poland 37
Pachuca s. Hidalgo (cap.), Mexico 57
Pacific Islands, Trust Territory of the (Carolines, Marshalls, Marianas) 55
Padang w. Sumatra, Indonesia 109
Paderborn North Rhine-Westphalia, West Germany 29
Padua (Padova) province, n. Italy 714
— city, Veneto, n. Italy 166
Pagadian nw. Mindanao, P.I. 46
Pahang state, Malayan Federation 250
Pahlevi nw. Iran 46
Paisley Renfrew co., Scotland 94
Pakhoi s. Kwangtung, China 36
Pakistan republic, Commonwealth of Nations, s. Asia 75,687
See EAST PAKISTAN, WEST PAKISTAN.
Pakse s. Laos 130
Palawan island, w. cen. Philippine Is. 54
Palembang se. Sumatra, Indonesia 210
Palencia Spain 42
Palermo province, sw. Italy 1,014
— Sicily (cap.), Italy 484
Palghat sw. Madras, India 55
Palma (de Mallorca) Majorca; cap. 137
Palma, La island, Canary Islands 64
Palmerston North s. North Island, N.Z. *32,845
Palmira se. Valle del Cauca, Colombia 45
Pamplona ne. Spain 72
Panama (Panamá) republic, Central America 805
— cen. Panama; cap. 128
Panay island, w. cen. Philippine Is. 1,424
Pančevo n. Serbia, Yugoslavia 31
Panevežys cen. Lithuanian S.S.R. 26
Pangim See NOVA GÔA.
Pankow n. Berlin, East Germany 144
Pantin n. cen. France 35
Paoting Hopeh (cap.), e. China 130
Paotow n. cen. China 63
Papeete Tahiti; cap. 12
Papua Australian territory, se. New Guinea 376
Pará state, n. Brazil 1,143
— city See BELÉM.
Paraguarí department, w. Paraguay 151
Paraguay republic, s. cen. South America 1,406
Parahyba See JOÃO PESSOA.
Paraíba state, ne. Brazil 1,731
Paramaribo n. Surinam; cap. 75
Paraná Entre Ríos (cap.), Argentina 84
— state, s. Brazil 2,149
Paranaguá e. Paraná, Brazil 16
Pardubice e. Bohemia, Czechoslovakia 44
Paris n. cen. France; cap. 2,691
Parma province, n. Italy 391
— city, Emilia-Romagna, Italy 122
Parnaíba n. Piauí, Brazil 31
Passau e. Bavaria, W. Germany 34
Passo Fundo Rio Grande do Sul, s. Brazil 25
Pasto Nariño (cap.), Colombia 27
Patan e. cen. Nepal *105
Paternò Sicily, sw. Italy 33
Patna nw. Bihar (cap.), e. India 282
Patras (Patrai) s. Greece 79
Pavia province, nw. Italy 506
— Lombardy, Italy 51
Paysandú w. Uruguay 46
Pazardzhik s. Bulgaria 31
Pécs sw. Hungary 78
Peebles county, se. Scotland 15
Pegu s. Burma 26
Peiping See PEKING.
Pekalongan n. cen. Java, Indonesia 66
Peking n. Hopeh, China; cap. 1,797
Peleng island, Indonesia 38
Peloponnesus division, s. Greece 1,128
Pelotas s. Rio Grande do Sul, Brazil 79
Pemba island n. of Zanzibar, e. Africa 115
Pembrokeshire county, sw. Wales 90
Penang Island w. Malayan Federation 263
Penge Kent, England 25
— city (George Town), Penang 189
Penghu Islands islands e. of Taiwan 78
Penhsihu (Penki) ne. China 100
Penza cen. European R.S.F.S.R. 158
Pera (Beyoğlu) nw. Turkey 280
Perak state, w. Malayan Federation 954
Pereira s. Caldas, Colombia 31
Pergamino n. Buenos Aires, Arg. 31
Perlis state, nw. Malayan Federation 71
Perm (Molotov) w. Asiatic R.S.F.S.R. 450
Pernambuco state, e. Brazil 3,431
— city. See RECIFE.
Perovo w. cen. European R.S.F.S.R. 132
Perpignan se. France 64
Persia See IRAN.
Perth county, cen. Scotland 128
— Perth co., Scotland 40
— sw. Western Australia; cap. *273
Peru republic, nw. South America 8,405
Perugia province, cen. Italy 580
— Umbria (cap.), Italy 38
Pervoural'sk n. European R.S.F.S.R. 49
Pesaro The Marches, Italy 29
Pescadores See PENGHU ISLANDS.
Pescara province, Italy 239
— city, Abruzzi e Molise, Italy 1,682
Peshawar n. West Pakistan 151
Petah Tiqva w. Israel 32
Peterborough, Soke of county, Northamptonshire, England 63
— city, ne. Northamptonshire 53
Petropavlovsk n. Kazakh S.S.R. 118
Petrópolis s. Rio de Janeiro, Brazil 61
Petrozavodsk s. Karelo-Finnish S.S.R. 118
Pforzheim n. Baden-Württemberg, West Germany 53
Philippeville ne. Algeria 41
Philippines, Republic of the 19,234
Philippopolis See PLOVDIV.
Phnômpenh See PNOM-PENH.
Piacenza province, n. Italy 299
— city, Emilia-Romagna, Italy 52
Piauí state, ne. Brazil 1,064
Pichincha province, n. Ecuador 394
Pico Island Azores 27
Piedmont (Piemonte) region, nw. Italy 3,513
Pietermaritzburg Natal (cap.), Union of South Africa 64
Pilsen (Plzeň) w. Bohemia, Czechoslovakia 121
Pinar del Río province, w. Cuba 398
— cen. Pinar del Río (cap.), Cuba 26
Pinsk w. Belorussian S.S.R. 32
Piotrków e. cen. Poland 40
Pisa province, n. Italy 349
— city, Tuscany, Italy 55
Pistoia province, cen. Italy 219
— city, Tuscany, Italy 29
Plauen s. cen. East Germany 85
Pleven n. cen. Bulgaria 39
Plock nw. Poland 29
Ploești s. cen. Rumania 96
Plovdiv s. cen. Bulgaria 125
Plymouth Devonshire, England 210
Plzeň See PILSEN.
Pnom-Penh Cambodia; cap. 364
Podolsk w. European R.S.F.S.R. 113
Pola See PULA.
Poland republic, e. Europe 24,971
Poltava ne. Ukrainian S.S.R. 130
Pondicherry e. Madras, India 223
Ponta Delgada Azores 21
Ponta Grossa e. Paraná, Brazil 44
Pontevedra nw. Spain 45
Pontianak nw. Borneo, Indonesia 45
Pontypool Monmouthshire, England 43
Pontypridd Glamorganshire, Wales 39
Poole Dorset, England 84
Poona cen. Bombay, India 485
Pori (Björneborg) sw. Finland 43
Port Adelaide s. South Australia 33
Port Arthur s. Manchuria, China 25
Port-au-Prince s. Haiti; cap. 142
Port Elizabeth Cape of Good Hope Province, Union of South Africa *146
Port Louis nw. Mauritius; cap. 57
Port-Lyautey (Kénitra) nw. Morocco 57
Port Moresby Papua; cap. 18
Pôrto See OPORTO.
Pôrto Alegre e. Rio Grande do Sul (cap.), Brazil 384
Port-of-Spain Trinidad; cap. of Trinidad and Tobago 93
Porto-Novo Dahomey (cap.), se. French West Africa 30
Port Said (Bor Sa'id) ne. Egypt 179
Portsmouth Hampshire, England 234
Port Sudan e. Sudan 54
Port Talbot Glamorganshire, Wales 44
Portugal republic, sw. Europe 8,510
Portuguese Guinea colony, nw. Africa 517
Portuguese India colony, w. India 637
Portuguese Timor See TIMOR.
Posadas w. Misiones, Argentina 38
Posen See POZNAŃ.
Potenza province, s. Italy 445
— city, s. Italy 31
Potosí department, sw. Bolivia 534
— ne. Potosí (cap.), Bolivia 46
Potsdam East Germany 114
Poznań w. cen. Poland 272
Prague (Praha) nw. Bohemia, Czechoslovakia; cap. 922
Pressburg See BRATISLAVA.
Preston Lancashire, England 119
Prestwick Ayrshire, s. Scotland 11
Pretoria Transvaal, Union of South Africa; administrative cap. *244
Probolinggo e. Java, Indonesia 37
Prokopyevsk sw. Asiatic R.S.F.S.R. 260
Prostějov (Prossnitz) cen. Moravia, Czechoslovakia 32
Przemyśl s. Poland 37
Pskov nw. European R.S.F.S.R. 60
Puebla state, s. Mexico 1,620
— n. Puebla, Mexico 207
Puente de Vallecas se. Spain 56
Puerto Cabello n. Venezuela 34
Puerto La Cruz n. Venezuela 28
Puerto Montt cen. Llanquihue (cap.), Chile 44
Puglia See APULIA.
Pula (Pola) nw. Croatia, Yugoslavia 23
Punjab state, nw. India 12,638
Punta Arenas s. Chile 25
Purwokerto w. Java, Indonesia 33
Pusan (Fusan) s. South Korea 474
Pyatigorsk sw. European R.S.F.S.R. 63
Pyongyang (Heijo) North Korea; cap. 343

Q

Qena (Qina) e. cen. Egypt 40
Queensland state, ne. Australia 1,106
Quelpart See CHEJU, island.
Querétaro state, cen. Mexico 286
— w. Querétaro 49
Quetta West Pakistan 83
Quezaltenango sw. Guatemala 47
Quezon City s. Luzon, P.I.; cap. 108
Quilmes ne. Buenos Aires, Argentina 85
Quito n. Ecuador; cap. 213
Qum n. cen. Iran 82

R

Rabat n. Morocco; cap. 161
Rabaul Territory of New Guinea 7
Radnorshire county, e. cen. Wales 20
Radom cen. Poland 70
Ragusa province, Sicily, Italy 223
— city, s. Sicily 50
— Yugoslavia. See DUBROVNIK.
Raichur Hyderabad, se. India 54
Raipur e. Madhya Pradesh, India 63
Rajahmundry ne. Madras, India 105
Rajasthan state, nw. India 15,297
Rampur cen. Uttar Pradesh, India 134
Ramsgate Kent, England 36
Rancagua O'Higgins (cap.), Chile 31
Ranchi s. Bihar, India 107
Randers n. Denmark 40
Randwick e. N.S. Wales, Australia 101
Rangoon s. Burma; cap. 500
Rashid See ROSETTA.
Rashin See NAJIN.
Ratisbon See REGENSBURG.

Ravenna province, cen. Italy 294
— city, Emilia-Romagna, Italy 37
Ravensburg se. Baden-Württemberg, West Germany 26
Rawalpindi West Pakistan 237
Rawtenstall Lancashire, England 25
Reading Berkshire, England 114
Recife e. Pernambuco (cap.), Brazil 522
Recklinghausen w. North Rhine-Westphalia, West Germany 105
Redditch e. Worcester, England 29
Regensburg e. cen. Bavaria, West Germany 117
Reggio di Calabria province, s. Italy 640
— city, Calabria, Italy 141
Reggio nell' Emilia province, n. Italy 389
— city, n. cen. Italy 106
Reichenbach s. East Germany 35
Reichenberg See LIBEREC.
Reigate Surrey, England 42
Reims (Rheims) ne. France 106
Reinosa See REYNOSA.
Rembang cen. Java, Indonesia 13
Remscheid ne. North Rhine-Westphalia, West Germany 103
Renaix (Ronse) Belgium 25
Rendsburg n. cen. Schleswig-Holstein, West Germany 35
Renfrew county, sw. Scotland 330
Rennes nw. France 103
Resht nw. Iran .... 10[illegible]
Resistencia e. Chaco (cap.), Argentina 65
Réunion French colony, island, w. Indian Ocean 242
Reus ne. Spain 29
Reutlingen cen. Baden-Württemberg, West Germany 46
Reval or **Revel** See TALLINN.
Reykjavik sw. Iceland; cap. 56
Reynosa n. Tamaulipas, Mexico 34
Rhineland-Palatinate state, West Germany 2,994
Rhodes (Ródhos) island, Aegean Sea 55
Rhodesia and Nyasaland, Federation of se. Africa 7,260
Rhondda Glamorganshire, Wales 111
Riau Archipelago See RIOUW ARCHIPELAGO.
Ribeirão Prêto n. São Paulo, Brazil 65
Richmond Surrey, England 42
Rieti province, Italy 180
— city, cen. Italy 25
Riga cen. Latvian S.S.R.; cap. 565
Rijeka (Fiume) w. Yugoslavia 72
Riobamba cen. Ecuador 27
Río Cuarto cen. Córdoba, Argentina 40
Rio de Janeiro state, se. Brazil 2,326
— Federal District, se. Brazil; cap. 2,335
Rio Grande s. Rio Grande do Sul, Brazil .. 64
Rio Grande do Norte state, ne. Brazil 984
Rio Grande do Sul state, s. Brazil 4,213
Río Muni district, Spanish Guinea, West Africa 135
Río Negro territory, cen. Argentina 132
Riouw Archipelago Indonesia 77
Riyadh e. cen. Saudi Arabia; co-cap. with Mecca 80
Rizaiyeh (Urmia) nw. Iran 46
Rizal s. Luzon, P.I. 89
Rocha se. Uruguay 25
Rochdale Lancashire, England 88
Rochester Kent, England 44
Rockhampton e. Queens., Australia 33
Roeselare See ROULERS.
Roma province, w. cen. Italy 2,156
Rome (Roma) s. cen. Italy; cap. 1,657
Romford Essex, England 88
Ronda s. Spain 33
Ronse See RENAIX.
Roodepoort-Maraisburg Transvaal, Union of South Africa 72
Roosendaal s. Netherlands 24
Rosario cen. Santa Fe, Argentina 464
Roscommon county, e. Connacht, Ireland 72
Rosetta n. Egypt *29
Roskilde e. Denmark 26
Ross and Cromarty county, n. Scotland 60
Rostock n. East Germany 115
Rostov or **Rostov-on-Don** sw. European R.S.F.S.R. 552
Rotherham West Riding, Yorkshire, England 82
Roti island, Indonesia 59
Rotterdam w. cen. Netherlands 646
Roubaix n. France 99
Rouen n. cen. France 101
Roulers (Roeselare) Belgium 32
Rovigo province, Veneto, Italy 355
Rowley Regis Staffordshire, England 49
Roxburgh county, se. Scotland 46
Rufisque w. cen. Fr. West Africa 30
Ruislip Northwood Middlesex, England .. 68
Rumania republic, se. Europe 15,873
Ruse or **Russe** n. Bulgaria 53
Russian Soviet Federated Socialist Republic A republic, including 14 autonomous republics, of e. Europe (European R.S.F.S.R.) and n. Asia (Asiatic R.S.F.S.R.), comprising the northern portion of the Union of Soviet Socialist Republics 113,200
Rutherglen Lanark co., Scotland 25
Rutlandshire county, England 21
Ryazan w. European R.S.F.S.R. 136
Rybinsk See SHCHERBAKOV.
Ryde e. N.S. Wales, Australia 36
Ryukyu Islands sw. Pacific 917
Rzhev w. European R.S.F.S.R. 54

S

Saar, The state, sw. West Germany 852
Saarbrücken sw. The Saar; cap. 90
Sabadell ne. Spain 59
Sabzawar ne. Iran 33
Sado island, w. of Honshu, Japan 126
Safi w. Morocco 26
Saga nw. Kyushu, Japan 50
Sagay w. Negros, P.I. 54
Saharanpur nw. Uttar Pradesh, India 148
Saigon South Vietnam; cap. 698
Saint Albans Hertsfordshire, England 44
Saint Annes and Lytham See LYTHAM ST. ANNE'S.
Saint-Denis n. cen. France 69
— Réunion; cap. 25
Saint-Étienne se. France 156
Saint Gall ne. Switzerland 68
Saint Helens Lancashire, England 110
Saint Helier Jersey, Channel Islands *25
Saint-Josse-ten-Noode cen. Belgium 28
Saint Kilda s. Victoria, Australia 58
Saint Kitts (St. Christopher) Leeward Islands, West Indies 29
Saint-Louis w. French West Africa 63
Saint Lucia island, Windward Islands 70
Saint-Maur-des-Fossés n. cen. France. 55
Saint-Nicolas (Sint-Niklaas) Belgium ... 43
Saint-Ouen n. cen. France 45
Saint-Quentin n. France 47
Saint Vincent island, Windward Islands .. 57
Sakai s. Honshu, Japan 216
Sakhalin e. Asiatic R.S.F.S.R. 300
Salamanca w. Spain 80
Salayar or **Salajar** island, Indonesia 76
Sale Cheshire, England 43
Salé nw. Morocco 57
Salem s. Madras, India 202
Salerno province, s. Italy 834
— city, Campania, Italy 55
Salford Lancashire, England 178
Salisbury Wiltshire, England 32
— ne. Southern Rhodesia; cap. 90
Salonika (Thessalonike) n. Greece 225
Salop See SHROPSHIRE.
Salta province, n. Argentina 290
— cen. Salta (cap.), Arg. 67
Saltillo se. Coahuila (cap.), Mexico 70
Salto department, nw. Uruguay 104
— city, Salto (cap.), Uruguay 44
Salvador (Bahia) Bahia (cap.), Brazil 396
Salvador, El See EL SALVADOR.
Salzburg w. Austria 101
Salzgitter See WATENSTEDT-SALZGITTER.
Sama de Langreo nw. Spain 40
Samar island, e. cen. Philippine Is. 500
Samara See KUIBYSHEV.
Samarkand e. Uzbek S.S.R. 170
Sambhal w. Uttar Pradesh, India 54
Samoa island group, South Pacific 94
Samos island, Greece 57
Samsun n. cen. Turkey 44
San Antonio w. Santiago, Chile 27
San Cristóbal nw. Venezuela *56
Sancti-Spíritus cen. Cuba 92
San Fernando ne. Buenos Aires, Argentina 45
— n. Colchagua (cap.), Chile 14
— cen. Luzon, P.I. 35
— s. Spain 32
— w. Trinidad, The West Indies 29
Sangi or **Sangihe Islands** Indonesia 135
San Isidro Leyte, P.I. 12
San José cen. Costa Rica; cap. 87
— s. Uruguay 30
San Juan province, w. Argentina 260
— s. San Juan (cap.), Argentina 81
Sankt Gallen See SAINT GALL.
Sankt Pölten n. Austria 40
Sanlúcar de Barrameda sw. Spain 40
San Luis province, cen. Argentina 168
— n. San Luis (cap.), Argentina 26
San Luis Potosí state, cen. Mexico 856
— s. San Luis Potosí (cap.), Mexico 127
San Marcos department, sw. Guatemala 231
San Marino republic, state and city 24
San Martín ne. Buenos Aires, Argentina 10
— department, cen. Peru 142
San Pablo cen. Luzon, P.I. *50
San Pedro department, cen. Paraguay 65
San Pedro Sula nw. Honduras 21
San Rafael Mendoza, w. Argentina 33
San Remo Liguria, n. Italy 32
San Salvador s. El Salvador; cap. 162
San Sebastián n. Spain 114
San Severo Apulia, se. Italy 46
Santa Ana w. El Salvador 52
Santa Catarina state, se. Brazil 1,578
Santa Clara cen. Cuba 54
Santa Cruz department, e. Bolivia 286
— w. Santa Cruz (cap.), Bolivia 43
— se. Mindanao, P.I. *55
Santa Cruz de Tenerife Canary Islands, Spain 104
Santa Fe province, ne. Argentina 1,700
— e. Santa Fe (cap.), Arg. 168
Santai w. cen. China 70
Santa Lucía s. Uruguay 15
Santa Maria cen. Rio Grande do Sul, Brazil 46
Santa Marta n. Magdalena (cap.), Colombia 25
Santander department, n. Colombia 615
— province, n. Spain 405
— city, Santander (cap.), n. Spain 103
Santander Norte See NORTE DE SANTANDER.
Santarém n. Pará, Brazil 15
Santiago province, cen. Chile 1,262
— cen. Santiago; cap. of Chile and province 952
— province, n. Dominican Republic 56
— nw. Spain 26
Santiago de Cuba s. Oriente (cap.), Cuba 118
Santiago del Estero province, n. Argentina 538
— w. Santiago del Estro (cap.), Arg. 63
Santo Domingo See CIUDAD TRUJILLO.
Santos e. São Paulo, Brazil 208
São João del Rei s. Minas Gerais, Brazil 38
São Luís n. Maranhão (cap.), Brazil 81
São Paulo state, s. Brazil 9,243
— s. São Paulo (cap.), Brazil 2,042
São Salvador See SALVADOR, Brazil.
São Tomé e Príncipe Portuguese islands off w. cen. Africa 60
Sapporo w. Hokkaido (cap.), Japan 314
Saragossa (Zaragoza) ne. Spain 264
Sarajevo s. cen. Yugoslavia 118
Saratov s. cen. European R.S.F.S.R. 518
Sarawak Brit. crown colony, Borneo 626
Sardinia (Sardegna) region (island), Italy 1,274
Sargodha West Pakistan 78
Sari n. Iran 26
Sariwon (Shariin) North Korea 43
Sasebo nw. Kyushu, Japan 194
Sassari Sardinia, Italy 66
Saudi Arabia kingdom, sw. Asia 5,500
Saugor n. Madhya Pradesh, India 64
Savona province, n. Italy 239
— city, Liguria, Italy 60
Savu Islands Indonesia 33
Scarborough North Riding, Yorkshire, England 44
Schaerbeek cen. Belgium 124
Scheveningen nw. Netherlands 66
Schiedam w. cen. Netherlands 70
Schleswig-Holstein state, n. West Germany 2,594
Schöneberg in Berlin enclave, West Germany 189
Schouten Islands Netherlands New Guinea 25
Schwerin nw. East Germany 88
Scotland division, n. Great Britain 5,095
Scunthorpe Lincolnshire, England 54
Scutari (Shkodër) nw. Albania 34
Scutari (Üsküdar) nw. Turkey 70
Secunderabad cen. Hyderbad, India 225
Segovia cen. Spain 30

Seishin See CHONGJIN.
Selangor state, Malayan Federation 911
Selkirk county, se. Scotland 22
Semarang n. cen. Java, Indonesia 500
Semipalatinsk ne. Kazakh S.S.R. 136
Sendai n. Honshu, Japan 342
Senegal territory, Fr. West Africa 1,740
Senta n. Yugoslavia 25
Seoul w. cen. South Korea; cap. 1,446
Seraing Belgium 42
Serampore West Bengal, India 53
Serbia constituent repub., Yugoslavia .. 6,523
Seremban w. Negri Sembilan (cap.), Malayan Federation 35
Sergipe state, ne. Brazil 650
Serov w. Asiatic R.S.F.S.R. 64
Serowe e. Bechuanaland Protectorate 6
Serpukhov w. European R.S.F.S.R. 102
Serrai (Serres) ne. Greece 33
Sétif ne. Algeria 40
Setúbal s. cen. Portugal 44
Sevastopol s. Crimea, R.S.F.S.R. 133
Seville (Sevilla) sw. Spain 377
Seychelles British island group, w. Indian Ocean 35
Seyhan See ADANA.
Sfax se. Tunisia 55
s'Gravenhage See HAGUE, THE.
Shahjahanpur cen. Uttar Pradesh, India 105
Shahr-i-Tajam See SARI.
Shakhty sw. European R.S.F.S.R. 180
Shanghai s. Kiangsu, China 4,982
Shanhaikwan ne. Hopeh, China 70
Shansi province, ne. China 15,025
Shantung province, ne. China 36,671
Shariin See SARIWON.
Shasi s. Hupeh, China 113
Shcherbakov nw. European R.S.F.S.R. .. 162
Sheffield West Riding, Yorkshire, England 513
Shensi province, nw. cen. China 9,492
Shenyang (Mukden) ne. China 1,123
Sherbro Island Sierra Leone 108
's Hertogenbosch s. Netherlands 53
Shetland Islands n. of Scotland 19
Shibin el Kom n. Egypt 42
Shikarpur n. Sind, West Pakistan 62
Shikoku island, s. Japan 4,220
Shimizu s. cen. Honshu, Japan 88
Shimonoseki w. Honshu, Japan 194
Shinshu See SINCHANG.
Shipley West Riding, Yorkshire, England 33
Shiraz sw. Iran 110
Shizuoka s. cen. Honshu, Japan 239
Shkodër See SCUTARI.
Sholapur e. Bombay, India 266
Shrewsbury Shropshire, England 44
Shropshire (Salop) co., w. cen. England . 290
Shumen See KOLAROVGRAD.
Shunsen See CHUNCHON.
Shuya w. cen. European R.S.F.S.R. 58
Sialkot ne. West Pakistan 156
Siam See THAILAND.
Sian (Changan) cen. Shensi (cap.); cen. China 631
Siangtan Hunan, s. cen. China 83
Siargao island, Philippine Islands 30
Šiauliai n. Lithuanian S.S.R. 32
Sibenik w. Yugoslavia 16
Sibiu w. cen. Rumania 61
Sicily (Sicilia) island, sw. Italy 642
Sidi-bel-Abbès n. Algeria 52
Siena province, n. cen. Italy 277
— Tuscany, Italy 36
Sierra Leone British colony, nw. Africa 124
— British protectorate, nw. Africa 1,734
Sikkim protectorate of India 135
Silva Pôrto cen. Angola 33
Simferopol s. Crimea, R.S.F.S.R. 159
Sinaloa state, w. Mexico 622
Sinchang North Korea 77
Sind division, West Pakistan 4,619
Singapore British colony, Singapore island 940
— city (cap.) 680
Sining Tsinghai (cap.), nw. China 59
Sinkiang-Uigur Autonomous Region China 4,873
Sinuiju w. North Korea 118
Siquijor island sw. Philippine Is. 57
Siracusa province, Sicily, s. Italy 320
— city See SYRACUSE.
Sivas n. Turkey 52
Skoplje (Skopje) s. Yugoslavia 92
Slavyansk e. Ukrainian S.S.R. 75
Sligo county, n. Connacht, Ireland 62
Slivno (Sliven) e. cen. Bulgaria 35
Slough Buckinghamshire, England 66
Slovakia (Slovensko) region, e. Czechoslovakia 3,402
Slovenia constituent republic, Yugoslavia 1,389
Smethwick Staffordshire, England 76
Smolensk w. European R.S.F.S.R. 131
Smyrna (Izmir) w. cen. Turkey 231
Soerabaja See SURABAYA.
Soerakarta See SOLO.
Sofia (Sofiya) e. cen. Bulgaria; cap. 434
Sohag e. cen. Egypt 43
Solingen w. North Rhine-Westphalia, West Germany 148
Solo cen. Java, Indonesia 165
Solomon Islands e. British S.I. Protectorate and Territory of New Guinea 39
Somalia ne. Africa 940
Somaliland See BRITISH SOMALILAND, FRENCH SOMALILAND.
Somaliland Protectorate See BRITISH SOMALILAND.
Sombor n. Yugoslavia 34
Somersetshire county, sw. England 551
Sondrio province, Lombardy, nw. Italy .. 153
Songjin sw. North Korea 68
Sonora state, nw. Mexico 507
Soochow s. Kiangsu, China 336
Sorocaba s. São Paulo, Brazil 70
Sosnowiec sw. Poland 77
Sotteville-lès-Rouen n. cen. France 18
Sousse e. cen. Tunisia 37
South Africa, Union of s. Africa 11,000
Southall Middlesex, England 60
Southampton Hampshire, England 178
— city and county, Hampshire, England 1,196
South Australia state, s. Australia 646
Southend-on-Sea Essex, England 151
Southern Rhodesia British colony, se. Africa 1,764
Southgate Middlesex, England 73
South Melbourne s. Victoria, Australia ... 43
Southport Lancashire, England 83
South Shields Durham, England 107
South Suburban s. West Bengal, India ... 103
South-West Africa self-governing mandate, sw. Africa 350
Sovetsk (Tilsit) n. European R.S.F.S.R. 58
Spain state, sw. Europe 27,976
Spalato See SPLIT.
Spandau Berlin enclave, West Germany .. 166
Spenborough West Riding, Yorkshire, England 37
Spice Islands See MOLUCCAS.
Split (Spalato) w. cen. Yugoslavia 49
Springs Transvaal, Union of South Africa 111
Srinagar cen. Jammu and Kashmir (cap.), n. India 208
Stafford Staffordshire, England 40
Staffordshire county, w. cen. England . 1,431
Staines Middlesex, England 40
Stalin ne. Bulgaria See VARNA.
— cen. Rumania See BRAȘOV.
Stalinabad w. Tadzhik S.S.R.; cap. 191
Stalingrad sw. European R.S.F.S.R. 525
Stalino se. Ukrainian S.S.R. 625
Stalinogorsk w. European R.S.F.S.R. ... 129
Stalinogród See KATOWICE.
Stalinsk s. Asiatic R.S.F.S.R. 374
Stanislav Ukrainian S.S.R. 61
Stanisławow See STANISLAV.
Stanleyville cen. Belgian Congo 18
Stara Zagora cen. Bulgaria 37
Stavanger sw. Norway 50
Stavropol se. European F.S.F.R. 123
Steglitz Berlin enclave, West Germany ... 157
Stettin (Szczecin) nw. Poland 73
Stirling county, s. cen. Scotland 187
Stockholm se. Sweden; cap. 746
Stockport Cheshire, England 142
Stockton-on-Tees Durham, England 17
Stoke-on-Trent Staffordshire, England . 275
Stolp (Slupsk) n. Poland 34
Stourbridge Worcestershire, England 37
Stralsund n. East Germany 50
Strasbourg e. France 167
Stretford Lancashire, England 61
Stuttgart cen. Baden-Württemberg (cap.), West Germany 482
Subotica n. Yugoslavia 112
Suchitepéquez department, s. Guatemala 123
Süchow sw. Shantung, China 160
Sucre Chuquisaca; cap. of Bolivia and dept. 40
Sudan, Republic of the ne. Africa 8,053
Suez ne. Egypt 108
Suffolk county, se. England 440
Sukkur n. Sind, West Pakistan 77
Sulaimaniya ne. Iraq 41
Sulawesi See CELEBES.
Sultanabad See ARAK.
Sulu Archipelago Philippine Islands 241
Sumatra island, Indonesia 12,000
Sumba island, Indonesia 182
Sumbawa island, Indonesia 315
Sumy ne. Ukrainian S.S.R. 64
Sunderland Durham, England 182
Sunwui s. Kwangtung, China 50
Surabaya ne. Java, Indonesia 848
Surakarta See SOLO.
Surat n. Bombay, India 223
Surbiton Surrey, England 61
Surigao province, ne. Mindanao, P.I. 265
Surinam (Dutch Guiana) Netherlands colony, n. South America 207
Surrey county, se. England 1,602
Susa See SOUSSE.
Sussex county, se. England 936
Sutherland county, n. Scotland 15
Sutton and Cheam Surrey, England 81
Sutton Coldfield Warwickshire, England 48
Sutton-in-Ashfield Nottinghamshire, England 41
Suva Fiji; cap. 23
Sverdlovsk w. Asiatic R.S.F.S.R. 707
Swansea Glamorganshire, Wales 161
Swat n. West Pakistan 569
Swatow Kwangtung, se. China 168
Swaziland Brit. protectorate, s. Africa ... 160
Sweden kingdom, nw. Europe 7,047
Swindon Wiltshire, England 69
Switzerland republic, cen. Europe 4,715
Sydney e. N.S. Wales (cap.), Australia *1,484
Syracuse (Siracusa) Sicily, Italy 61
Syria province, United Arab Republic, sw. Asia 3,135
Syzran se. European R.S.F.S.R. 169
Szczecin See STETTIN.
Szechwan province, s. cen. China 47,107
Szeged se. Hungary 132

T

Tabasco state, se. Mexico 362
Tablas island, nw. Philippine Is. 50
Tabriz nw. Iran 259
Táchira state, w. Venezuela 308
Tacloban Leyte, P.I. 31
Tacuarembó department, n. Uruguay 108
Tadzhik Soviet Socialist Republic U.S.S.R., cen. Asia 1,455
Taegu se. South Korea 314
Taejon cen. South Korea 127
Taganrog sw. European R.S.F.S.R. 189
Tahiti island, Fr. Oceania 30
Taichung w. Taiwan 142
Taiku See TAEGU.
Tainan sw. Taiwan 173
Taipeh (Taihoku) Taiwan; cap. 541
Taiping cen. Perak (cap.), Malayan Federation 41
Taiwan (Formosa) island se. of China 6,126
Taiyüan Shansi (cap.), n. cen. China 316
Takamatsu ne. Shikoku, Japan 124
Takao See KAOHIUNG.
Takaoka n. cen. Honshu, Japan 142
Takasaki e. cen. Honshu, Japan 93
Talca province, cen. Chile 157
— s. Talca (cap.), Chile 50
Talcahuano w. Concepción, Chile 36
Talien See DAIREN.
Talisay w. Negros, P.I. 11
Tallinn (Tallin) nw. Estonian S.S.R.; cap. 257
Tamatave ne. Madagascar 30
Tamaulipas state, ne. Mexico 717
Tambov w. European R.S.F.S.R. 150
Tampere (Tammerfors) sw. Finland 103
Tampico se. Tamaulipas, Mexico 94
Tananarive cen. Madagascar; cap. 165
Tandil s. Buenos Aires, Argentina 39
Tanganyika U.N. Trust Territory (Brit.), e. Africa 7,477
Tangier nw. Africa 80
Tangshan Hopeh, e. China 137
Tanimbar Islands Indonesia 32

Tanjay e. Negros, P.I.	33
Tanjore se. Madras, India	100
Tanta n. Egypt	140
Taonan ne. China	56
Tapachula Chiapas, s. Mexico	30
Tapul Group islands, Philippine Islands	46
Tarabulus See TRIPOLI.	
Taranto province See IONIO.	
— Apulia, Italy	167
Tarapacá province, n. Chile	104
Tarbes sw. France	43
Targu-Mures nw. Rumania	47
Tarija department, s. Bolivia	136
— w. Tarija (cap.), Bolivia	17
Tárlac n. cen. Luzon, P.I.	21
Tarnopol See TERNOPOL.	
Tarnów sw. Poland	33
Tarragona ne. Spain	39
Tarrasa ne. Spain	44
Tarsus s. Turkey, Asia Minor	34
Tartu (Yurev) e. Estonian S.S.R.	71
Tashkent e. Uzbek S.S.R.; cap.	778
Tasmania state, se. Australia	257
Tatabánya n. Hungary	40
Tatung n. China	80
Taubaté São Paulo, s. Brazil	36
Taunton Somersetshire, England	33
Tavoy s. Burma	33
Tawitawi islands, Philippine Islands	48
Tbilisi See TIFLIS.	
Tchad See CHAD.	
Tegal n. cen. Java, Indonesia	43
Tegucigalpa w. Honduras; cap.	72
Teheran or Tehran n. Iran; cap.	544
Tel Aviv w. cen. Israel (including Jaffa)	335
Tembuland district, e. Cape of Good Hope Province, Union of South Africa	304
Temesvár See TIMIŞOARA.	
Tempelhof s. cen. Western Berlin, West Germany	119
Temuco n. Cautín (cap.), Chile	42
Tenerife island, Canary Is.	262
Teplice (Teplitz) nw. Bohemia, Czechoslovakia	45
Teramo province, s. cen. Italy	272
— city (cap.)	62
Teresina Piauí (cap.), ne. Brazil	53
Terni Umbria, Italy	58
Ternopol w. Ukrainian S.S.R.	36
Tetuán w. Morocco; cap.	94
Thailand (Siam) kingdom, se. Asia	17,316
The Hague See HAGUE, THE.	
Therezina See TERESINA.	
Thessalonike See SALONIKA.	
Thessaly (Thessalia) division, w. cen. Greece	626
Thorn See TORUŃ.	
Thule w. Greenland	5
Tibet w. China	3,000
Ticao island, w. cen. Philippine Is.	38
Tientsin ne. China	1,718
Tiflis se. Georgian S.S.R.; cap.	635
Tilburg s. Netherlands	114
Tilsit See SOVETSK.	
Timaru e. South Island, N.Z.	*23
Timisoara sw. Rumania	112
Timor island, s. Indonesia	1,657
Timor, Portuguese e. Timor	424
Tinnevelly s. Madras, India	61
Tipperary county, e. Munster, Ireland	136
Tipton Staffordshire, England	39
Tirana or Tiranë cen. Albania; cap.	60
Tiruchirappali e. cen. Madras, India	218
Titagarh s. West Bengal, India	57
Tjirebon nw. Java, Indonesia	54
Tlaxcala state, cen. Mexico	294
Tlemcen nw. Algeria	50
Tobata n. Kyushu, Japan	88
Togo French repub., w. Africa	944
Togoland region in Ghana	383
Tokuno island, Ryukyu Islands	53
Tokushima e. Shikoku, Japan	121
Tokyo s. cen. Honshu, Japan; cap.	5,385
Tolima department, w. Colombia	548
Tollygunge s. West Bengal, India	150
Toluca s. Mexico (cap.), Mexico	53
Tomelloso s. cen. Spain	29
Tomsk w. Asiatic R.S.F.S.R.	224
Tonga Islands (Friendly Islands) archipelago, S. Pacific	47
Tongyong s. South Korea	42
Tonkin See NORTH VIETNAM.	
Toowoomba Queensland, Australia	33
Torino province, n. Italy	1,427
— city See TURIN.	
Törökszentmiklos e. cen. Hungary	30
Torquay Devonshire, England	53
Torre Annunziata Campania, s. Italy	50
Torre del Greco Campania, s. Italy	45
Torreón sw. Coahuila, Mexico	129
Toruń (Thorn) nw. Poland	68
Toscana See TUSCANY.	
Totonicapán sw. Guatemala	9
Tottenham Middlesex, England	127
Toulon se. France	116
Toulouse s. France	226
Tourane South Vietnam	51
Tourcoing n. France	73
Tournai Belgium	32
Tours w. cen. France	76
Townsville ne. Queensland, Australia	34
Toyama n. cen. Honshu, Japan	154
Toyohashi s. Honshu, Japan	146
Trabzon See TREBIZOND.	
Transvaal province, n. Union of South Africa	4,236
Trapani province, w. Sicily	418
— city, Sicily	64
Trebizond ne. Turkey	34
Trengganu state, Malayan Federation	226
Trent (Trento) n. Italy	40
Trento province, ne. Italy	395
Tres Arroyos s. Buenos Aires, Argentina	29
Treves or Trèves See TRIER.	
Treviso province, ne. Italy	611
— Veneto, Italy	29
Trichinopoly See TIRUCHIRAPPALI.	
Trichur Kerala, sw. India	58
Trier sw. Rhineland-Palatinate, West Germany	76
Trieste e. Italy	263
Trinidad and Tobago British island colony, The West Indies	531
Tripoli (Tarabulus) nw. Lebanon	*86
— nw. Libya; co-cap. (with Bengasi)	123
Tripura territory, e. India	649
Trivandrum Kerala (cap.), sw. India	185
Trondheim s. cen. Norway	57
Troyes ne. France	53
Trujillo n. Peru	37
— state, Venezuela	285
Tsangwu See WUCHOW.	
Tsinan Shantung (cap.), China	575
Tsinghai province, nw. China	1,346
Tsingtao e. Shantung, China	846
Tsitsihar ne. China	174
Tsu sw. Honshu, Japan	76
Tucumán province, nw. Argentina	605
— cen. Tucumán (cap.), Argentina	153
Tuguegarao ne. Luzon, P.I.	12
Tukangbesi Islands Indonesia	56
Tula n. European R.S.F.S.R.	320
Tunbridge Wells Kent, England	38
Tunis n. Tunisia; cap.	365
Tunisia republic, n. Africa	3,232
Turin (Torino) Piedmont, Italy	712
Turkey republic, sw. Asia and se. Europe	20,935
Turkmen Soviet Socialist Republic U.S.S.R., cen. Asia	1,170
Turku (Åbo) sw. Finland	101
Turnhout Belgium	32
Tuscany region, nw. Italy	3,152
Tuticorin s. Madras, India	76
Twickenham Middlesex, England	106
Tynemouth se. Northumberland, England	67
Tyrone county, w. cen. Northern Ireland	132
Tyumen w. Asiatic R.S.F.S.R.	125

U

Ubangi-Shari colony, e. French Equatorial Africa	1,062
Ube w. Honshu, Japan	129
Uberaba w. Minas Gerais, Brazil	43
Uccle Belgium	56
Udaipur cen. Rajasthan, India	60
Udine province, ne. Italy	795
— Veneto, Italy	42
Ufa se. European R.S.F.S.R.	471
Uganda Protectorate e. cen. Africa	4,958
Uji-yamada sw. Honshu, Japan	32
Ujjain Madhya Pradesh, India	129
Ukrainian Soviet Socialist Republic e. cen. Europe	30,960
Ulan Bator n. cen. Mongolia; cap.	100
Ulan-Ude s. cen. Asiatic R.S.F.S.R.	158
Ulm sw. Baden-Württemberg, West Germany	69
Ulyanovsk se. European R.S.F.S.R.	183
Umbria region, cen. Italy	802
Union of South Africa See SOUTH AFRICA, UNION OF.	
Union of Soviet Socialist Republics e. Europe, w. and n. Asia	201,130
United Arab Republic See EGYPT, SYRIA.	
United Kingdom Gt. Britain and N. Ireland, nw. Europe	50,210
Upper Austria province, Austria	1,108
Upper Volta province, Fr. West Africa	3,044
Upsala se. Sweden	63
Uralsk nw. Kazakh S.S.R.	66
Urawa cen. Honshu, Japan	115
Urfa se. Turkey	32
Urga See ULAN BATOR.	
Urmia See RIZAIYEH.	
Uruguaiana Rio Grande do Sul, Brazil	33
Uruguay republic, se. South America	2,353
Urumchi nw. China	70
Usti nad Labem n. Bohemia, Czechoslovakia	*56
Utrecht cen. Netherlands	185
Utsunomiya se. Honshu, Japan	107
Uttar Pradesh state, n. India	63,254
Uwajima w. Shikoku, Japan	57
Uxbridge Middlesex, England	56
Uzbek Soviet Socialist Republic s. cen. Asia	6,282

V

Vaasa w. Finland	35
Valdepeñas s. cen. Spain	29
Valdivia province, s. Chile	192
— w. Valdivia (cap.), Chile	35
Valencia e. Spain	509
— n. Venezuela	*88
Valenciennes n. France	38
Valladolid n. Spain	124
Valle del Cauca department, w. Colombia	732
Valparaiso province, w. Chile	425
— Valparaiso (cap.), Chile	210
Varese Lombardy, Italy	30
Varna e. Bulgaria	78
Västerås cen. Sweden	60
Vejle cen. Denmark	29
Vélez-Málaga s. Spain	27
Veliki Bečkerek See ZRENJANIN.	
Vellore cen. Madras, India	107
Velsen nw. Netherlands	48
Veneto region, n. Italy	3,909
Venezia province, Veneto, n. Italy	740
— city See VENICE.	
Venezuela republic, n. South America	4,986
Venice n. Italy	316
Venlo se. Netherlands	27
Veracruz state, e. Mexico	2,030
— e. Veracruz, Mexico	104
Vercelli Piedmont, Italy	36
Verona province, ne. Italy	645
— city, Veneto, ne. Italy	178
Versailles n. cen. France	63
Verviers Belgium	41
Vicenza province, Veneto, n. Italy	608
— city	54
Victoria state, se. Australia	2,055
— Hong Kong colony; cap.	887
Vienna (Wien) ne. Austria; cap.	1,761
Vietnam See NORTH VIETNAM, SOUTH VIETNAM.	
Vigo nw. Spain	138
Viipuri See VYBORG.	
Vijayavada See BEZWADA.	
Villahermosa Tabasco (cap.), Mexico	25
Villa María cen. Córdoba, Arg.	35
Villarrica s. Paraguay	40
Villeurbanne se. France	81
Vilna or Vilnyus Lithuanian S.S.R.; cap.	200
Vilvoorde Belgium	26
Viña del Mar Valparaiso, Chile	70
Vincennes n. cen. France	49
Vinnitsa cen. Ukrainian S.S.R.	92
Vitebsk ne. Belorussian S.S.R.	128
Viterbo cen. Italy	36
Vitoria n. Spain	44
Vitória e. Espírito Santo (cap.), Brazil	46
Vittoria se. Sicily, Italy	38
Vizagapatam ne. Madras, India	70
Vizianagaram ne. Madras, India	52
Vlaardingen w. cen. Netherlands	30
Vladimir e. European R.S.F.S.R.	121
Vladivostok se. Asiatic R.S.F.S.R.	265
Vologda nw. European R.S.F.S.R.	127
Volos ne. Greece	41
Volsk s. cen. European R.S.F.S.R.	55
Voorburg S. Holland prov., Netherlands	36

Voronezh sw. European R.S.F.S.R. 400
Voroshilovgrad e. Ukrainian S.S.R. 251
Voroshilovsk e. Ukrainian S.S.R. 101
Vyatka See KIROV.
Vyborg (Viipuri) nw. European
 R.S.F.S.R. 57

W

Wad Medani e. Sudan 61
Wakamatsu n. Kyushu, Japan 89
Wakayama sw. Honshu, Japan 191
Wakefield West Riding, Yorkshire,
 England 60
Wałbrzych sw. Poland 73
Walcheren island, Netherlands 67
Wales principality, w. Great
 Britain 2,172
Wallasey Cheshire, England 101
Wallsend Northumberland, England 49
Walsall Staffordshire, England 118
Walthamstow Essex, England 121
Walton and Weybridge Surrey, England .. 38
Wanhsien e. Szechwan, China 110
Wanne-Eickel w. West Germany 86
Wǎnsan See WONSAN.
Wanstead and Woodford, Essex,
 England 61
Warangal cen. Hyderabad, India 130
Warrington Lancashire, England 81
Warsaw (Warszawa) e. cen. Poland;
 cap. 479
Warwickshire county, cen. England 1,860
Watenstedt-Salzgitter Lower Saxony,
 West Germany 101
Waterford county, e. Munster, Ireland 76
— e. Waterford, Ireland 28
Watford Hertfordshire, England 73
Waverly n. N.S. Wales, Australia 75
Wednesbury Staffordshire, England 34
Weihai (Weihaiwei) n. Shantung, China . 175
Weimar sw. East Germany 82
Weissensee e. East Germany 82
Wellington North Island, N.Z.; cap. *173
Wembley Middlesex, England 131
Wenchow se. Chekiang, China 159
West Bengal See BENGAL, WEST.
West Bromwich Staffordshire, England ... 88
Western Australia state, w. Australia ... 503
Western Berlin n. West Germany 2,204
Western Samoa, Territory of See SAMOA.
West Germany (Federal Republic of
 Germany) 49,728
West Ham Essex, England 158
West Hartlepool Durham, England 78

West Indies, The federation of British
 colonies, Caribbean 2,537
West Lothian county, se. Scotland 86
Westmeath county, nw. Leinster,
 Ireland 67
Westmorland county, nw. England 68
Weston-Super-Mare Somersetshire,
 England 40
West Pakistan province, Pakistan 34,047
West Riding division, Yorkshire,
 England 3,352
Wexford county, se. Leinster, Ireland 92
White Russian S.S.R. See BELORUSSIAN S.S.R.
Wicklow county, e. Leinster, Ireland 68
Widnes Lancashire, England 49
Wiener Neustadt e. Austria 36
Wiesbaden sw. Hesse (cap.), W. Germany 218
Wigan Lancashire, England 83
Wight, Isle of county, England 96
Wigtown county, sw. Scotland 32
Wilhelmshaven n. Lower Saxony, West
 Germany 101
Willemstad Neth. W. Indies; cap. 41
Willesden Middlesex, England 180
Wilmersdorf Western Berlin, West
 Germany 141
Wilno (Vilnyus) See VILNA.
Wiltshire county, s. cen. England 387
Wimbledon Surrey, England 58
Windward Islands s. The West Indies 252
Winterthur n. Switzerland 67
Witten w. North Rhine-Westphalia,
 West Germany 76
Włocławek nw. Poland 48
Woking Surrey, England 48
Wollongong-Port Kembla e. New
 South Wales, Australia 18
Wolverhampton Staffordshire, England . 163
Wonsan e. North Korea 113
Wood Green Middlesex, England 52
Worcester Worcestershire, England 60
Worcestershire county, sw. cen. England 522
Worksop Nottinghamshire, England 31
Worms s. Rhineland-Palatinate, West
 Germany 52
Worthing Sussex, England 69
Wrocław (Breslau) sw. Poland 171
Wuchang se. China 199
Wuchow e. Kwangsi, China 207
Wuhu se. Anhwei, China 203
Wuppertal n. North Rhine-Westphalia,
 West Germany 362
Würzburg nw. Bavaria, West Germany ... 78
Wusih Kiangsu, e. China 272

X

Xalapa See JALAPA.
Xanthe w. Thrace, Greece 32

Y

Yahata See YAWATA.
Yamagata nw. Honshu, Japan 105
Yangchow cen. Kiangsu (cap.), e. China . 127
Yarkand Sinkiang-Uigur, nw. China 57
Yarmouth See GREAT YARMOUTH.
Yaroslavl nw. European R.S.F.S.R. 374
Yawata n. Kyushu, Japan 210
Yecla se. Spain 20
Yegoryevsk w. European R.S.F.S.R. 56
Yelets sw. European R.S.F.S.R. 50
Yentai (Chefoo) Shantung, China 227
Yezd cen. Iran 54
Yokkaichi sw. Honshu, Japan 123
Yokohama se. Honshu, Japan 952
Yokosuka se. Honshu, Japan 251
Yonezawa nw. Honshu, Japan 55
York Yorkshire, England 105
Yorkshire county, ne. England 4,626
Yucatán state, se. Mexico 518
Yugoslavia republic, se. Europe 15,751
Yünnan province, s. China 9,171

Z

Zaandam nw. Netherlands 41
Zabrze sw. Poland 104
Zacatecas state, cen. Mexico 662
Zagazig ne. Egypt 82
Zagreb nw. Yugoslavia 290
Zamboanga sw. Mindanao, P.I. 17
Zanzibar British protectorate; island,
 w. Indian Ocean 149
— e. Zanzibar; cap. 45
Zaporozhe se. Ukrainian S.S.R. 374
Zaragoza See SARAGOSSA.
Zealand island, Denmark 1,610
Zeeland province, sw. Netherlands 260
Zeist cen. Netherlands 32
Zetland See SHETLAND.
Zhdanov se. Ukrainian S.S.R. 273
Zhitomir nw. Ukrainian S.S.R. 95
Zlatoust e. Asiatic R.S.F.S.R. 143
Zonguldak nw. Turkey 37
Zrenjanin n. Yugoslavia 41
Zurich n. Switzerland 390
Zwickau East Germany 122
Zwolle ne. Netherlands 42